W9-CSP-430

Table of Radioactive Isotopes

Table of Radioactive Isotopes

Edgardo Browne and **Richard B. Firestone**

Virginia S. Shirley, Editor

Lawrence Berkeley Laboratory
University of California

A Wiley-Interscience Publication
JOHN WILEY & SONS
New York • Chichester • Brisbane • Toronto • Singapore

Library of Congress Cataloging-in-Publication Data:

Browne, Edgardo.
Table of radioactive isotopes.

"A Wiley-Interscience publication."

1. Radioisotopes—Tables. I. Firestone, Richard B.
II. Shirley, Virginia S. III. Title.
QD601.2.B76 1986 541.3′884′0212 86-9069
ISBN 0-471-84909-X

Printed in the United States of America

10 9 8 7 6 5 4 3 2 1

Preface

The Isotopes Project at the University of California's Lawrence Berkeley Laboratory produced this *Table of Radioactive Isotopes* in response to user demand for a convenient source of adopted radiation data. We felt the need for such a book at the time of the publication of the 7th edition of the *Table of Isotopes*,[1] as several concurrent effects became apparent. Among the major of these were the increasing and more diverse applied use of radioisotopes, the continuing rapid growth in the volume of nuclear data, the restructuring of the data evaluation and dissemination efforts on an international level,[2] and the rapid development of computer capability for the managing, analysis, and distribution of data.[3] Our historic ties to the groups which produced the 7th and earlier[4] editions of the *Table of Isotopes* led us into the present effort. In the early days of the project, C. Michael Lederer and Janis M. Dairiki initiated the user surveys and pioneering development work necessary for the transition to the present book. We believe that the thoroughness, consistency, and accuracy, so evident from the very first Berkeley compilation, *A Table of Induced Radioactivities*,[5] have been maintained as we made the large changes dictated by present-day science and computer technology.

In considering the data needs of the applied users for whom the *Table of Radioactive Isotopes* was tailored, we tried to satisfy the ever increasing demand for *adopted* properties for *all* radiations emitted by nuclei. We have therefore included tables of adopted properties, which were derived from experimental data plus reliable calculations (e.g., continuous radiation spectra), along with those based on statistical analyses of existing experimental data alone (e.g., γ-ray spectra). We further calculated other derived adopted properties (e.g., average photon energies per disintegration) for which we sensed strong user demand. Our mutually complementary analysis/calculation handling of the nuclear properties have led to what we feel is a high degree of uniformity, completeness, and self-consistency in both the data and appendices sections of the book.

In a concession to the continuing value of direct human, rather than computer, effort, the mass-chain decay schemes were updated by hand from the corresponding schemes drawn for the 7th edition of the *Table of Isotopes*. Mirriam Schwartz was the laboratory illustrator for both books, and transferred her experience, dedication, and expertise to the present project, from its immediate predecessor. We are indebted to her for the 262 drawings in the *Table of Radioactive Isotopes*.

We want to thank the many Lawrence Berkeley Laboratory staff members who helped us with the book. We appreciate the support of the LBL Nuclear Science Division and its successive directors Joseph Cerny, Earl Hyde, and presently James Symons. The division staff member playing a major role in this final hectic year of production has been Traudel Prussin, our data analyst. She has cheerfully served as a computer technician and courier, and done numerous other tasks in-between. J. Michael Nitschke was very helpful in reviewing the data for $A \geqslant 256$, and division guest Peter Lemmertz assisted briefly with work on $A=243$.

We appreciate the help of the LBL Computing Division, and its director Leroy Kerth, in providing excellent resources for our computer needs. Herbert Albrecht assisted with preliminary programming work, Alexander Merola and Marvin Atchley facilitated our electronic data management, and Martin Gelbaum and Robert Rendler helped with our use of UNIX. Among others in the Computing Division to whom we are indebted are Rosemary Allen, Wayne Graves, Gilman Johnson, Raymond Partyka, Jerry Borges, William Jaquith, Margaret Morley, and Virginia Sventek.

There are many LBL staff members in Information Services whom we want to acknowledge for help with the technical aspects of the book. We are indebted to Theodore Kirksey, Deputy for Information Services, for his overall assistance, and for his special help, along with that of Gloria Haire, in connection with the excellent service provided by the LBL library. Richard Bailey, head of the laboratory's Technical Information Department, coordinated publication procedures within the laboratory, and formalities with the publisher. Mirriam Schwartz, in addition to serving as illustrator, was our production assistant, and did much of the work on designing and coordinating the page layouts and appendices. Ralph Dennis, supervisor of the Technical Illustration Group, has cooperated in timely fashion to provide us with help as needed. Others in Information Services we want to acknowledge are Steven Ow-Ling and Stanley Combs, for expediting phototypesetter output, Alice Ramirez, for typesetting of labels for drawings and appendices, and Charles Dees and Warren Lockhart for preparing the negatives from which the book was printed. We consulted Thomas Budinger of the LBL Biology and Medicine Division regarding the best formats for nuclear data useful in medicine, and want to thank him for his help in that connection.

We are indebted to our program manager in the Office of Basic Energy Sciences, Stanley Whetstone, and others in the U.S. Department of Energy for the funding and trust which made the project possible. Likewise, we want to thank the many officials and project leaders who showed confidence in this book and supported its production. Our special thanks go to Sol Pearlstein, Mulki Bhat, and Jagdish K. Tuli of the National Nuclear Data Center at Brookhaven, to Murray Martin of the Nuclear Data Project at Oak Ridge, to the members of the National Academy of Sciences Panel on Basic Nuclear Data Compilations, especially to present chairman Peter Parker and former chairman Thomas Tombrello, and to the members of the IAEA International Nuclear Data Committee (INDC), especially to chairman Alex Lorenz. Lastly, we want to thank the many nuclear-data evaluators worldwide for their contributions to the evaluation journals and computer files. Their massive efforts provided the starting material from which this book was produced.

Edgardo Browne
Richard B. Firestone

Berkeley, California
January, 1986

---

[1] *Table of Isotopes*, 7th edition, C.M. Lederer and V.S. Shirley, editors; E. Browne, J.M. Dairiki, and R.E. Doebler, principal authors; A.A. Shihab-Eldin, L.J. Jardine, J.K. Tuli, and A.B. Buyrn, authors; John Wiley and Sons, Inc., New York (1978).

[2] The Isotopes Project at the Lawrence Berkeley Laboratory is a member, along with projects at the University of Pennsylvania, the Idaho National Engineering Laboratory, the Oak Ridge National Laboratory, and the Brookhaven National Laboratory, of the United States Nuclear Data Network (USNDN). Centered at the National Nuclear Data Center (NNDC) at Brookhaven, the USNDN is in turn a member of The International Network for Nuclear Structure Data Evaluation, which functions under the auspices of the Nuclear Data Section of the International Atomic Energy Agency (IAEA). In addition to the data groups in the United States, evaluation centers in Belgium, Canada, China, England, France, West Germany, Japan, Kuwait, Sweden, and the Soviet Union also contribute to the international nuclear-data evaluation effort.

[3] Most of the data in the *Table of Radioactive Isotopes* have been derived from the *Evaluated Nuclear Structure Data File* (ENSDF), a computer file of evaluated nuclear-structure and radioactivity-decay data, which is maintained by the National Nuclear Data Center (NNDC), Brookhaven National Laboratory, on behalf of The International Network for Nuclear Structure Data Evaluation. The file, with a traditional 80-character record sequential structure, has been reorganized into numerical indexed files at the Lawrence Berkeley Laboratory, and stored in a random-access computer database (LBL/ENSDF). This latter is managed with DATATRIEVE, a Digital Equipment Corporation database-management system. The tables and text presented in this book have been created with data from the LBL/ENSDF database, using the text-formatting program TROFF for driving the Graphics Systems phototypesetter on the UNIX (Trademark of Bell Laboratories) operating system. Nuclear-structure data not explicitly shown in the *Table of Radioactive Isotopes* can be retrieved from the LBL/ENSDF database (stored in a dedicated disk, operated by a cluster of five VAX-11/8600 computers (VAX is a Trademark of Digital Equipment Corporation)). Remote access to the LBL/ENSDF database is possible through the Advanced Research Projects Agency (ARPA), High Energy Physics (HEP), and Tymshare's Inc. public data communications (TYMNET) networks, linking numerous computers in the United States and abroad.

[4] Previous editions: (1) G.T. Seaborg, *Rev. Mod. Phys.* **16**, 1 (1944); (2) G.T. Seaborg and I. Perlman, *Rev. Mod. Phys.* **20**, 585 (1948); (3) J.M. Hollander, I. Perlman, and G.T. Seaborg, *Rev. Mod. Phys.* **25**, 469 (1953); (4) D. Strominger, J.M. Hollander, and G.T. Seaborg, *Rev. Mod. Phys.* **30**, 585 (1958); (5) C.M. Lederer, J.M. Hollander, and I. Perlman, *Table of Isotopes*, John Wiley and Sons, Inc., New York (1967).

[5] J.J. Livingood and G.T. Seaborg, *Rev. Mod. Phys.* **12**, 30 (1940).

Description of Figures and Tables

1. Sources of data

The *Table of Radioactive Isotopes* has been produced from various sources of evaluated radioactivity data. The main source, the Evaluated Nuclear Structure Data File (ENSDF),[1] covers data for nuclei with $A \geqslant 45$, evaluated by members of the Nuclear Structure and Decay Data (NSDD) network, and published in the journal *Nuclear Data Sheets.* The principal source of data for $21 \leqslant A \leqslant 44$ was the evaluation of Endt and Van der Leun,[2] for $5 \leqslant A \leqslant 20$, the evaluations of Ajzenberg-Selove,[3-7] and for $1 \leqslant A \leqslant 4$, the compilations included in References 8-11. Most nuclei covered by the 13th edition of the *Chart of the Nuclides* (1983)[12] have entries in the *Table of Radioactive Isotopes;* and throughout, the *Table of Isotopes*[9] has been used for data checking and updating. Data from the original literature have been added for newly discovered isotopes and for isotopes for which the *Chart of the Nuclides* (1983)[12] showed a need for selective updating. Original sources have been used also for isotopes with $A \leqslant 44$, for which radioactivity data are presented differently in the published compilations. The following have been included throughout: isotopic abundances from Holden, et al.,[13] decay energies and atomic mass excesses from Wapstra, et al.,[14] isotope production methods from the *Table of Isotopes,*[9] and a few half-lives from Walker.[15]

2. General Presentation

For each mass number (A), there is a mass-chain decay scheme showing the isobars drawn to scale, along with some of their adopted properties. The tabular entries for each isotope, ordered by increasing atomic number (Z), are headed by a concise table of properties. This *header* is followed by relevant tables of radiation data. The latter list energies in *keV* and intensities either on an absolute scale (%), referring to 100 disintegrations of the parent isotope, or on a relative scale (*rel*). For isotopes with well known decay schemes, the tables include total average radiation energies per disintegration at the beginning. Tables of atomic electrons and continuous radiation are included only if absolute intensities are known. Values are rounded to reflect uncertainties of $\leqslant 25$ in the last two significant digits; statistical uncertainties, in italics, reflect the confidence in those digits. Quantities with uncertainties of between 50 and 100% are reported as approximate, and those with uncertainties of more than 100% are rewritten as upper limits equated to the values plus uncertainties. Half-life units have been standardized to include the following ranges: 1 to 1000 for *ms*, *µs*, *ns*, and *ps*; 1 to 60 for *s* and *min*; 1 to 24 for *h*; 1 to 365.25 for *d*; and $\geqslant 1$ for *yr*. Also, half-lives for very short lived isotopes (those with half-life <1 *ps*) are reported in small fractions of seconds.

3. Mass-Chain Decay Scheme

This schematic drawing, formatted and labeled in traditional decay-scheme fashion, precedes the tabular entries for each mass number (A). The accompanying literature citation, referring to the most recent published evaluation for that A, is usually the primary source of data for the drawing and tables. The ground-states for the isobars are represented as heavy lines whose vertical positions show their energies relative to the most beta-stable nucleus of the chain. The energy scale is shown at the left of the drawing, and the decay energies (Q-values) for the various decay modes are shown under the isobars. Q-values for nuclei far from stability are not given in *The 1983 Atomic Mass Table,*[14] and are, instead, theoretical values from calculations by Myers.[16] These latter were used for scaling purposes, but are not shown on the drawings. Isomeric states, represented by medium heavy lines, are drawn above the ground-states, though not necessarily to scale because of space constraints. Relative positions of levels have been maintained throughout, and standard graphics techniques have been used to show two or more states of unknown relative position. Except for a few isomers which decay only by spontaneous fission (*SF*), every ground-state or isomer shown on a drawing has a corresponding entry in the tabular listings. Short-lived (<100 *ms*) isomers which decay only through isomeric transitions (*IT*) are not included in the *Table of Radioactive Isotopes.* Delayed-particle emission from levels in precursor daughters is indicated by light lines drawn approximately to scale. Groups of three levels are used schematically if there are several daughter levels or if their energies are not known. α-decay parent isotopes are shown directly above their corresponding daughter nuclei; their vertical positions are not related to the energy scale.

Adopted values for level half-lives, energies, spins, and parities are shown on the drawings. Decay modes are indicated by easily recognized descriptive arrows; and percentage branchings are given for competing decay processes. ϵ is used to designate either electron-capture decay (*EC*), positron decay (β^+), or combined EC+β^+. However, Q_ϵ shows the actual energy or mass difference between two isobars, i.e., Q_{EC}, rather than $Q_{\beta+}$ ($= Q_{EC} - 2m_ec^2$). Fission and particle-breakup decays are represented by star patterns indicating the various resultant particles.

Figures shown on the mass-chain decay schemes have been rounded to an uncertainty of ≤5 in the last digit, except for the decay energies (Q-values) from Wapstra, et al.[14] These are quoted as given.

4. Header

This concise table heads the tabular data for isotopes and isomers shown on the mass-chain decay scheme. It contains the isotope's atomic number, mass number, element symbol, half-life, decay modes (and percentage branchings), mass excess, specific activity, principal means of production, and isotopic abundance. It sometimes also contains level widths for light nuclei and some prompt- and delayed-particle information. Conventional notation has been used throughout, and the various properties in the *header* are readily identified. "$t_{1/2}$ unknown" is shown for isotopes for which the half-life has not been measured. For isomers, a correction for excitation energy has been included in the mass excess values from *The 1983 Atomic Mass Table.*[14]

The specific activity per integral second is given in units of *Ci/g* for isotopes with a half-life $\leq 10^{10}$ *yr*. This quantity, defined as the number of disintegrations per gram of source during the initial second, is given by:

$$SpA(Ci/g) = \int_0^1 \lambda N dt = \frac{1.627 \times 10^{13}}{A} (1 - e^{-\lambda}),$$

where

N is the number of atoms at time t,
A is the mass number, and
$\lambda = \frac{\ln 2}{t_{1/2}}$ is the decay constant, in which $t_{1/2}$ is the half-life.

The conventional specific activity (in *Ci/g*) gives the instantaneous disintegration rate per gram of source. For isotopes with a half-life ≤1 *s*, a significant amount of the source has already decayed during the first second; and hence this definition of specific activity overestimates the emitted radiation. It should be noted that, for a half-life much longer than 1 *s*, the specific activity given as defined above approaches the instantaneous value

$$SpA(Ci/g) = \frac{1.127 \times 10^{13}}{At_{1/2}}.$$

5. Alpha Particles

This table gives alpha-particle energies and intensities. Except for the reporting of simple spectra, energy values have been adopted from the decay schemes, as described in Section 1 of *Methods of Analysis.* Intensities are experimental values taken from ENSDF[1] or from the original literature. These are given either on an absolute scale (%), referring to 100 disintegrations of the parent isotope, or on an arbitrary relative scale (*rel*). Intensities are given only for isotopes which have more than one measured α-particle group. The total average alpha energy per disintegration $\langle\alpha\rangle$ (described in Section 1.4 of *Methods of Analysis*), with its statistical uncertainty, is given in the table heading. Values for the individual α-particle energies and corresponding intensities, shown with their uncertainties, are given in the first and second columns of the table, respectively.

6. Delayed Particles

This table gives energies and intensities for particles emitted following the β^-, β^+, or *EC* decay of "precursor" isotopes. Energy values are shown as measured (laboratory frame of reference); but occasionally, energy values derived from decay schemes and particle separation energies from *The 1983 Atomic Mass Table*[14] are shown instead. Intensities are given either on an absolute scale (%), referring to 100 disintegrations of the precursor parent, or on an arbitrary relative scale (*rel*). Values for the delayed-particle energies and corresponding intensities, shown with their uncertainties, are given in the first and second columns of the table, respectively. In a few rare and clearly labeled cases, similar tables of prompt particles are also included.

7. Photons

This table gives γ-ray energies, intensities, and multipolarities. The total average photon energy per disintegration $\langle\gamma\rangle$ (described in Section 1.3 of *Methods of Analysis*), with its statistical uncertainty, is given in the table heading. The multipolarities are shown in the first column of the table, following the symbol γ. This latter is appended with the appropriate decay mode whenever competing modes exceed 1%. Multipolarities in parentheses are considered uncertain, and those between brackets have been inferred from level-scheme considerations. For γ rays of mixed multipolarity, only the predominant component is given if the mixing is $<0.1\%$; otherwise, the predominant component is given first, followed by the weaker component and its percentage admixture. Commas between multipolarities indicate multiple possibilities.

γ-ray and x-ray energies and intensities, adopted as described in Sections 1-3 of *Methods of Analysis*, are shown in the second and third columns of the table, respectively. γ-ray intensities are given either on an absolute scale (%), referring to 100 disintegrations of the parent isotope, or on a relative scale (*rel*), normalized to the value 100 for the most intense γ ray. X-ray intensities are always given on an absolute scale, and have been calculated from the number of vacancies in the K, L_1, L_2, and L_3 atomic shells, produced both by conversion (including $E0$) and by electron-capture processes. Except when they are the only source of electromagnetic radiation, x-rays are not listed if their intensities are $\lesssim 0.01\%$ of that for the most intense γ ray emitted by the isotope. X-ray components are grouped and identified according to the classical (Siegbahn) notation, and complex transitions can be resolved with the help of the x-ray tables in Appendix C5 and the breakdown shown below.

Classical designation	Individual transitions
$K_{\alpha 1}$	$K_{\alpha 1}$
$K_{\alpha 2}$	$K_{\alpha 2}+K_{\alpha 3}$
$K_{\beta 1}$	$K_{\beta 1}+K_{\beta 3}+K_{\beta 5}$
$K_{\beta 2}$	$K_{\beta 2}+K_{\beta 4}+\cdots$
L_{α}	$L_{\alpha 1}+L_{\alpha 2}$
L_{β}	$L_{\beta 1}+L_{\beta 2}+L_{\beta 3}+L_{\beta 4}+L_{\beta 5}+L_{\beta 6}$
L_{γ}	$L_{\gamma 1}+L_{\gamma 2}+L_{\gamma 3}+L_{\gamma 6}$
L_{η}	L_{η}
L_{ℓ}	L_{ℓ}

Individual energy and intensity values in the table are followed by their statistical uncertainties. The overall systematic uncertainties in the γ-ray intensities are indicated with a footnote in the intensity column, and shown at the end of the table. If more than one decay mode is involved, the systematic uncertainty for each mode is given. These have been calculated by adding, in quadrature, the uncertainties in the percentage branching to those of the γ-ray normalization factors. Systematic uncertainties of more than 50% are shown as "approximate".

γ rays from an isomeric state in a daughter nucleus have footnotes in the intensity column to show that the reported value is the measured "equilibrium intensity"

$$I_{\gamma} = \frac{t_{1/2}(p)-t_{1/2}(d)}{t_{1/2}(p)} I \ .$$

Here, $t_{1/2}(p)$ and $t_{1/2}(d)$ are the half-lives of the parent and daughter nuclei, respectively, and I is the decay-scheme value of the γ-ray intensity. If $t_{1/2}(d) \geqslant 0.25$ times $t_{1/2}(p)$, implying that equilibrium cannot be reached within a reasonable time, the γ rays deexciting the isomeric states are shown with the footnotes, but no values of I, in the intensity column. Another commonly used footnote indicates the total combined intensity for an unresolved doublet or multiplet. If determined, the energies alone are reported for successive members of the multiplet. Miscellaneous footnotes and cross references handle the interpretation of spectra for mixed sources, uncertain parent assignments, etc.

8. Atomic Electrons

This table gives internal conversion and Auger-electron energies and absolute intensities, analyzed as described in Section 4 of *Methods of Analysis*. Because there are so many possible electron lines from even a small number of electromagnetic and electron-capture transitions, the table has been reduced in size, but not usefulness, by grouping the lines into energy bins. These range in size from 1 *keV* (for intense lines) up to 100 *keV* (for groups of weaker lines whose combined intensities make minor contributions to the total intensity). The total average electron energy per disintegration $\langle e \rangle$, with its statistical uncertainty, is shown in the table heading. The individual bin intervals and the average energies per disintegration in those intervals are given in the first and second columns of the table, respectively. The corresponding absolute intensities, with their statistical uncertainties, are given in the third column. A table of atomic electrons is included only for those isotopes for which absolute transition intensities are known and for which the internal conversion and Auger processes are the predominant mechanisms for the production of atomic electrons. If the electrons are associated with only one decay mode (usually *IT*), this is clearly indicated with a footnote.

9. Continuous Radiation

This table gives average energies per disintegration and absolute intensities of β^-, β^+, and internal bremsstrahlung (*IB*) radiation, all of which have continuous energy spectra. As described in Section 5 of *Methods of Analysis*, these spectra have been analyzed and binned into the following standard energy intervals: 0-10, 10-20, 20-40, 40-100, 100-300, 300-600, 600-1300, 1300-2500, 2500-5000, 5000-7500, and >7500 *keV*. The highest bin energy is the energy for the highest transition. Often in the ϵ decay of light nuclei, however, the electron-capture branching is so weak that it is neglected in the calculations. The energy of the highest bin is then 1022 *keV* lower than the transition endpoint (pure β^+ decay). A table of continuous radiation is included only for those isotopes for which the decay scheme and beta or electron-capture absolute intensities are known. These latter have been renormalized whenever there was a discrepancy between the sum of individual transition intensities and the percentage decay branching. No intensities for bremsstrahlung radiation have been calculated for the 0-10-*keV* bin since these diverge at 0 *keV*.

The total average energies per disintegration ($\langle \beta^- \rangle$, $\langle \beta^+ \rangle$, $\langle IB \rangle$) are given separately for β^-, β^+, and internal bremsstrahlung in the table heading. The individual bin energy ranges, radiation types, and values for the corresponding average energies and absolute intensities are given in columns 1 through 4 of the table, respectively. Also, the total β^+ intensity is reported as the final entry in the fourth column. The calculated continuous radiation intensities presented in this table presume that the isotope's decay scheme, as presently known, is complete; however, for isotopes with decay energies exceeding 5 *MeV*, substantial photon intensity may have possibly remained undetected. This is because γ rays from high-density and/or weakly-populated level regions in daughter nuclei are difficult to measure. The resulting beta intensities are then systematically skewed to favor the populations to lower-energy levels in the daughter nuclei. Consequently, the continuous radiation intensities given in the table may have systematic shifts toward higher energy bins, increasing the total average energy per disintegration. One should be cautious when using the table in these instances.

10. Appendices

Following the main section of the book are these useful appendices: *Appendix A. Physical Constants* (commonly used fundamental constants and conversion factors); *Appendix B. Nuclear Spectroscopy Standards* (γ-ray and α-particle standards for detector calibration); *Appendix C. Atomic Data* (theoretical internal conversion coefficients, electron-capture subshell ratios, electron binding energies, atomic yields, x-ray energies and intensities, and Auger-electron intensities); *Appendix D. Absorption of Radiation in Matter* (half-thicknesses in several absorbers for γ rays with energies ranging to 100 *MeV*, range and stopping power for selected charged particles in several media, positron annihilation radiation corrected for annihilation-in-flight, and experimental average radiation energies per disintegration for isotopes from the *Table of Radioactive Isotopes*; this latter table includes annihilation radiation intensities and total positron branchings, none of which are shown in the tables of the main section); and *Appendix E. Properties of the Elements* (atomic weights, densities, melting and boiling points, first ionization potentials, and specific heats of the elements).

Methods of Analysis

1. γ-Ray and Alpha-Particle Energies

1.1. Formulation of the Problem

The statistical procedures described below provide a method for adopting γ-ray and alpha-particle energies from experimentally determined energies and level-scheme correlations. The transition energies are related to the level energies by the equation

$$E_{ji} = \sum_{k=1}^{m} g_{ki}\xi_k - \Delta_j \ . \tag{1}$$

Here, E_{ji} refers to the i-th transition measured in the j-th experiment. The transition energies are in the center of mass reference frame and are related to the laboratory energies by

$$E_{ji}(c.m.) = E_{ji}(lab) + E_R \ , \tag{2}$$

where the recoil energy (E_R) is

$$E_R = \begin{cases} \dfrac{E_\gamma^2}{2M_R c^2} & \text{for } \gamma \text{ emission} \\ E_\alpha \dfrac{M_\alpha}{M_R} + \dfrac{E_\alpha^2}{2M_R c^2} & \text{for } \alpha \text{ particle decay} \end{cases} \tag{3}$$

M_α (=4.0026 amu) is the alpha-particle mass and M_R, the mass (in amu) of the recoiling daughter nucleus ($\cong A$ for γ-ray emission and $\cong A$-4 for alpha decay). E_γ and E_α are the γ-ray and alpha-particle energies, respectively, and c is the speed of light. The recoil mass-energy (in MeV) is $M_R c^2 \cong 931.5 M_R$.

ξ_k in equation (1) is the energy of the k-th level, m is the number of levels in the daughter nucleus, and g_{ki}, the place of the γ ray in the level scheme. More specifically,

$$g_{ki} = \begin{cases} 1 & \text{if the } i\text{-th transition deexcites the } k\text{-th level} \\ 0 & \text{if the } i\text{-th transition neither populates nor deexcites the } k\text{-th level} \\ -1 & \text{if the } i\text{-th transition populates the } k\text{-th level} \end{cases} \tag{4}$$

The term Δ_j in equation (1) allows the zero energy in each set of γ rays from the j-th experiment to vary independently. This *shift*, which is a systematic uncertainty in the energy calibration, has often been found to be statistically significant. Equation (1) represents a linear regression in which the maximum likelihood estimates of the level energies (ξ_k) and *shifts* (Δ_j) can be obtained in closed form, as follows.

Equation (1) can be written in matrix notation as

$$\tilde{E} = G\tilde{\xi} \ . \tag{5}$$

The components of $\tilde{E}$ and $\tilde{\xi}$ are $E_1 = E_{11}$, $E_2 = E_{12}, \ldots, E_r = E_{ji}, \ldots, E_t = E_{pn}$, and $\xi_1, \xi_2, \ldots\ldots, \xi_m$, $\Delta_1, \Delta_2, \ldots, \Delta_p$, respectively, where t is the total number of γ rays measured in p experiments, and n is the number of γ rays from each individual experiment. Weighing the data by the inverse square of the experimental uncertainties, the solution to equation (5) is[17]

$$\hat{\xi} = (G^T W G)^{-1} G^T W \tilde{E} \ , \tag{6}$$

where G^T is the transpose of the placement matrix G, and W is the diagonal weight matrix

$$W_{rr} = \frac{1}{(\Delta E_{rr})^2 \sigma^2} \tag{7}$$

In equation (7),

$$\sigma^2 = \frac{1}{n-m}(\tilde{E}^T W \tilde{E} - \tilde{E}^T W G \hat{\xi}) \tag{8}$$

for n transitions deexciting m levels.

The variances in the fitted level energies and *shifts* are

$$\Delta\xi_k = (U_{kk})^{1/2} \quad k=1,2,...m+p \ , \tag{9}$$

where U_{kk}'s are the diagonal elements of the variance matrix

$$U = \sigma^2 (G^T W G)^{-1}. \tag{10}$$

The adopted alpha-particle and γ-ray energies are given by

$$\hat{E}_i = \sum_{k=1}^{m} g_{ki}\ \xi_k \tag{11}$$

and their corresponding variances from the fit become[18]

$$\Delta E_i = (V_{uu} + V_{ll} - 2V_{ul})^{1/2}\ \sigma \ , \tag{12}$$

with the subscripts u and l referring to the upper and lower levels, respectively. V is the covariance matrix

$$V = (G^T W G)^{-1} . \tag{13}$$

1.2. Method of Calculation

The adopted alpha-particle and γ-ray energies are calculated with the computer code GAMUT[19] from evaluated input data in the LBL/ENSDF[20] database. Decay data from radioactive parents populating a common daughter nucleus are generally used. If energy uncertainties are not in the file, default values of 1 *keV* for γ-ray and 10 *keV* for alpha-particle energies are used. Both of these default values are increased by a factor of 2.5 if the input energies are given as "approximate". "Systematic" energies are not used in the calculations, and identical experimental energies given for several parent isotopes decaying into a common daughter nucleus are used only once. The *shift* parameter Δ_j is considered in the calculations only if there are sufficient degrees of freedom, and its inclusion is optional. Also, both the selection of input data and the use of assigned uncertainties rather than default values are options that can be changed whenever appropriate.

The adopted energies are tested by a Chi-square analysis similar to that described in the *Review of Particle Properties.*[8] The deviations of the input (experimental) energies from the adopted (fitted) energies are assumed to have a Chi-square distribution in which the χ^2_{ji} for the i-th transition in the j-th experiment is

$$\chi^2_{ji} = \frac{(E_{ji} - E_i)^2}{\Delta E^2_{ji}} \ . \tag{14}$$

The analysis involves the removing of extreme outlying data (those with χ^2_{ji} values with less than 0.5% likelihood). For these extreme outliers, the input uncertainty (ΔE_{ji}) is increased so that $\chi^2_{ji} = 1$, and the calculation is then repeated. Normally, a single iteration to correct extreme outliers is sufficient, but if ΔE_{ji} becomes greater than E_{ji}, the energy value is rejected. Once the unacceptable outliers are removed, all of the γ rays from the same parent isotope (data set) with $\chi^2_j/f > 1.0$ are adjusted. Here

$$\frac{\chi^2_j}{f} = \frac{1}{n-1} \sum_{i=1}^{n} \frac{(E_{ji} - E_i)^2}{\Delta E^2_{ji}} \ . \tag{15}$$

The uncertainties of all input energies for these data sets are increased so that $\chi^2_j = 1$, and then the calculation is reiterated. If χ^2_j remains greater than 1, further increases in uncertainties and iterations are performed until either $\chi^2_j \leq 1$ or ten iterations have been completed. Data with unacceptable outliers or slow convergence in the Chi-square analysis are individually reviewed and adjusted. Finally, the adopted energies are stored in the LBL/ENSDF[20] database, and subsequently reported in the photon and alpha-particle tables.

1.3. Average Photon Energies per Disintegration

The average photon energy per disintegration is

$$\langle\gamma\rangle = \frac{1}{100} \sum_i E_i I_i \ , \tag{16}$$

where E_i and I_i are the i-th photon energies and absolute intensities, respectively. The statistical uncertainty in the average energy, to the first order in a Taylor series expansion, is

$$\Delta\langle\gamma\rangle=\frac{1}{100}\left\{\sum_i(E_i\Delta I_i)^2+(I_i\Delta E_i)^2\right\}^{1/2}, \tag{17}$$

where ΔE_i and ΔI_i are the uncertainties in the photon energies and intensities, respectively. The uncertainty in the average energy ($\Delta\langle\gamma\rangle$) does not include any component of systematic uncertainty, and may be over-estimated because of the inherent correlations between photon energies and intensities.

For fission products with decay Q-values $\geqslant 5$ *MeV*, the actual average energies per disintegration are known to be systematically higher than those calculated.[21] This may be a consequence of incomplete decay-scheme information. Statistical models predict that, for isotopes with large Q-values, substantial numbers of low-intensity γ rays, many with high energies, contribute significantly to the total average photon energy. For such isotopes, the average energies in these tables should be considered as lower limits and used with caution.

1.4. Average Alpha-Particle Energies per Disintegration

The average alpha-particle energy per disintegration is

$$\langle\alpha\rangle=\frac{1}{100}\sum_i E_iI_i\ . \tag{18}$$

Here, E_i is the energy of the i-th alpha particle group, and

$$I_i=B\frac{Y_i}{\sum_k Y_k}, \tag{19}$$

the corresponding absolute intensity per 100 disintegrations. In equation (19), Y_i is the relative intensity, and B, the percentage alpha branching. The summation is over all k alpha particle groups.

Alpha-particle energies are constrained by the following equation:

$$E_i=E_0-e_i\ , \tag{20}$$

where E_0 is the energy of the ground-state alpha particle group, and e_i, the energy of the level populated by the i-th alpha particle group. The statistical uncertainty in the average energy, to the first order in a Taylor series expansion, is

$$\Delta\langle\alpha\rangle=\frac{B}{100}\left[\Delta E_0^2+\frac{1}{(\sum_k Y_k)^2}\left\{\sum_i Y_i^2\Delta e_i^2+\frac{1}{(\sum_k Y_k)^2}\sum_i e_i^2\Delta Y_i^2\,(\sum_k Y_k(1-\delta_{ki})\,)^2\right\}\right]^{1/2}, \tag{21}$$

where ΔE_0 is the uncertainty in E_0, Δe_i, the uncertainty in e_i, ΔY_i, the uncertainty in Y_i, and

$$\delta_{ik}\ =\ \begin{cases}1 & \text{for } i=k\\ 0 & \text{for } i\neq k\end{cases} \tag{22}$$

For pure alpha emitters, the percentage alpha branching B is 100.

The uncertainties Δe_i are typically $\leqslant 1$ *keV* for level energies derived from γ-ray measurements. ΔE_0 is generally between 1 and 10 *keV*, and e_i is usually $\lesssim 1000$ *keV*. Thus the resulting uncertainties in the average energies from equation (21) are usually $\leqslant 1\%$, their small size due primarily to the narrow range of alpha-particle energies.

2. γ-ray Intensities

2.1. Formulation of the Problem

The following statistical procedures provide a method for analyzing γ-ray intensities so that one consistent set of branching ratios is adopted for the deexcitation of each nuclear level. γ-ray intensities depopulating a level are typically measured on unrelated scales, which are independent of the various parent isotopes or nuclear reactions populating the level. Following the procedures of Tepel[22] and Lederer,[23] the scales are assumed to be linearly related by factors α_j, such that the expression

$$Q=\sum_i\sum_j\frac{(I_{ij}\alpha_j-\overline{I}_i)^2}{(\alpha_j\Delta I_{ij})^2} \tag{1}$$

is minimized. Here I_{ij} is the intensity of the i-th γ-ray in the j-th measurement, ΔI_{ij} is the uncertainty in I_{ij}, $\overline{I}_i$ is the adopted branching ratio for the i-th γ ray, and the summations are over all γ rays. Equation (1) can be rewritten as

$$Q=\sum_i\sum_j\omega_{ij}(I_{ij}-\beta_j\overline{I}_i)^2\ , \tag{2}$$

where $\omega_{ij}=\Delta I_{ij}^{-2}$, $\beta_j=\alpha_j^{-1}$, and $\beta_1=1$. Equation (2) can be minimized, following the method of Tepel,[22] by iteratively solving the following system of equations:

$$\overline{I}_i=\frac{\sum_j\omega_{ij}I_{ij}\beta_j}{\sum_j\omega_{ij}\beta_j^2} \tag{3}$$

$$\beta_j=\frac{1}{\alpha_j}=\frac{\sum_i\omega_{ij}I_{ij}\overline{I}_i}{\sum_i\omega_{ij}\overline{I}_i^2} \tag{4}$$

Initially, $\beta_j=1$ is assumed for all values of j. $\overline{I}_i$'s are then calculated from equation (3) and substituted into equation (4) to recalculate β_j. This process is iterated until β_j converges. The intensities are then converted to the original relative scale

$$\overline{I}_{ij}=\overline{I}_i\beta_j\ , \tag{5}$$

where $\overline{I}_{ij}$ is the adopted intensity for the i-th γ-ray in the j-th measurement. The uncertainty in $\overline{I}_i$ is calculated from the final parameters and the covariance matrix C, as shown in the equation

$$\Delta\overline{I}_i=(\beta_j^2C_{ii}+\overline{I}_i^2C_{jj}+2\beta_j\overline{I}_iC_{ij})^{1/2}\ . \tag{6}$$

This procedure is equally valid for the analysis of γ-ray intensities, whether for different measurements with the same parent isotope or nuclear reaction, or for the decay of a single level, which has been populated by various parent isotopes or nuclear reactions.

2.2. Method of Calculation

Adopted γ-ray intensities are calculated with the computer code GAMUT.[19] Input data for each parent isotope populating levels in a common daughter nucleus are taken from the LBL/ENSDF[20] database. The minimization is necessary only for the cases in which several parent isotopes decay to levels deexcited by several γ-rays. The iteration is continued until β_j varies by less than 0.1%. "Systematic" intensities in the database are not used, and "calculated" or "approximate" intensities, are assumed to have 50% uncertainties. By default, 20% uncertainty is assumed for γ-ray intensities with no input uncertainties in the file, and 2% is used for those intensities which have <2% uncertainty in the file. These default uncertainties are occasionally overridden in special cases.

A Chi-square analysis similar to that described for γ-ray energies in Section 1.2 of *Methods of Analysis* is applied to the intensities. The uncertainties ΔI_{ij} of far outliers are increased to give $\chi_{ij}^2=1.0$, where

$$\chi_{ij}^2=\frac{(\overline{I}_i-I_{ij})^2}{\Delta I_{ij}^2}\ . \tag{7}$$

The calculation is then repeated until no far outliers remain. If $\chi_j^2/f\ >1.0$ for the j-th data set with N_j γ rays, where

$$\frac{\chi_j^2}{f}=\frac{1}{N_j-1}\sum_{i=1}^{N_j}\frac{(\overline{I}_i-I_{ij})^2}{\Delta I_{ij}^2}\ , \tag{8}$$

the uncertainties in that data set are then increased to give $\chi_j^2/f\ =1$. The calculation is repeated until either $\chi_j^2/f\ \leqslant 1$ or ten iterations are completed. Inconsistent data are reevaluated on a case by case basis. γ-ray intensities are given in the photon tables either on an absolute scale (%), referring to 100 disintegrations of the parent isotope, or on a relative scale (*rel*), normalized to the value 100 for the most intense γ ray.

3. Atomic X-rays

3.1. X-ray Energies

Photon tables include energies and absolute intensities for x-rays arising from vacancies created in the K, L_1, L_2, and L_3 atomic shells by internal-conversion and electron-capture processes. X-ray energies are determined from the differences between the corresponding atomic-shell binding energies. An unweighted average energy is adopted for complex transitions which have multiple final atomic-shell designations (e.g., $L_{\beta 2,15}$), and an average energy, weighted by theoretical x-ray emission rates, is adopted for combined transitions. Classical Siegbahn notation is used to designate the various single-line (e.g., $K_{\alpha 1}$) and combined (e.g., $K_{\beta 1}' = K_{\beta 1} + K_{\beta 3} + K_{\beta 5}$) transitions.

3.2. Atomic Vacancies from Internal Conversion

The number of primary vacancies per 100 disintegrations, created by internal conversion of the m-th γ ray in the j-th atomic shell (V_{mj}), is

$$V_{mj} = I_m \ \alpha_{jm} \ , \qquad (1)$$

where I_m is the absolute γ-ray intensity per 100 disintegrations and α_{jm}, its internal conversion coefficient for the j-th atomic shell.

Theoretical internal conversion coefficients are from calculations by Band, et al.[24] for $Z<30$, and by Rösel, et al.,[25] for other values of Z. Conversion coefficients for specific energies within each reported range are determined by interpolating with a cubic spline function, as recommended in Reference 25. For $Z \geqslant 84$, the published conversion coefficient values are adjusted to compensate for the more recent atomic binding energies of Porter and Freedman.[26] The K, L_1, and L_2 conversion coefficients for E0 transitions are calculated with the theory of Church and Weneser.[27] For transitions with energies greater than 1022 keV, the conversion component from pair production is calculated[28] with the method of Wilkinson.[29,30]

3.3. Atomic Vacancies from Electron Capture

The number of primary vacancies per 100 disintegrations, created by the ϵ-th electron-capture transition in the j-th atomic shell ($V_{\epsilon j}$), is

$$V_{\epsilon j} = I_\epsilon \frac{\lambda_j}{\lambda_{EC}} \ , \qquad (2)$$

where I_ϵ is the absolute intensity per 100 disintegrations for the ϵ-th electron-capture transition, λ_j is the electron-capture probability per unit time to the j-th atomic shell, and λ_{EC} ($= \sum_j \lambda_j$), the total electron-capture probability per unit time.

The total transition probability, in the event positron emission is possible (i.e., whenever the transition energy exceeds 2 $m_e c^2$), is given by

$$\lambda_{tot} = \lambda_{EC} + \lambda_{\beta+} \ , \qquad (3)$$

where EC and $\beta+$ refer to electron-capture and positron decay, respectively. Theoretical electron-capture-to-positron ratios ($\lambda_{EC} \ / \ \lambda_{\beta+}$), taken from Gove and Martin,[31] are used to determine λ_{EC} for each transition.

The electron-capture probability (λ_j) is

$$\lambda_j = \frac{G_\beta^2}{2\pi^3} n_j C_j F_j \ , \qquad (4)$$

where G_β is the fundamental weak coupling constant, n_j is the relative occupation number for the partially filled j-th atomic shell ($n_j = 1$ for closed shells), and C_j contains the transition nuclear matrix elements. The function F_j for the j-th atomic shell, which corresponds to the integrated Fermi function of β decay, is

$$F_j = \left(\frac{\pi}{2} \right) \beta_j^2 B_j \left(W_0 + 1 - |E_j| \right)^2 , \qquad (5)$$

where $W_0 (= Q_\epsilon - E_\ell)$ is the electron-capture transition energy in units of $m_e c^2$ (the electron rest-mass energy), Q_ϵ is the electron-capture decay energy (the difference between the atomic masses of the parent and daughter nuclei), E_ℓ is the energy of the level populated in the daughter nucleus, and $|E_j|$ is the

j-th atomic-shell binding energy, also in units of $m_e c^2$. β_j and B_j are the Coulomb amplitude of the bound-state electron radial wavefunction, and the exchange and overlap correction, respectively, obtained from Bambynek, et al.[32]

Secondary vacancies are produced by the redistribution of electrons in the atomic shells. The number of secondary L_j vacancies (V'_{Lj}) per K vacancy is

$$V'_{Lj} = \omega_K \, P_{K\alpha(4-j)} + (1-\omega_K) \sum_k (1+\delta_{Lj,Xk}) P_{K-LjXk} \, , \tag{6}$$

where $P_{K\alpha(4-j)}$ is the number of $K_{\alpha(4-j)}$ x-rays per total K x-rays, P_{K-LjXk} is the number of $K-L_jX_k$ Auger electrons per total K Auger electrons, ω_K is the fluorescence yield (the probability for emission of a K x-ray per K vacancy), and $\delta_{Lj,Xk}$ is the Kronecker delta function. It should be noted that, for $X_k = L_j$, two vacancies in the L_j subshell are produced per $K-L_jL_j$ Auger electron.

3.4. Absolute x-ray Intensities

The absolute K x-ray intensity per 100 disintegrations (I_{Ki}) is

$$I_{Ki} = P_{Ki} \omega_K \left(\sum_m V_{mK} + \sum_\epsilon V_{\epsilon K} \right) , \tag{7}$$

where i refers to the specific x-ray transition, P_{Ki} is the number of K_i x-rays per total K x-rays, ω_K is the fluorescence yield (defined as above), V_{mK} is the number of K vacancies per 100 disintegrations created by internal conversion of the m-th γ ray (see equation 1), and $V_{\epsilon K}$ is the number of K vacancies per 100 disintegrations created by the ϵ-th electron-capture transition (see equation 2).

Absolute L x-ray intensities per 100 disintegrations are given by:

$$I_{L1i} = V_{L1} P_{L1i} \omega_{L1} \tag{8}$$

$$I_{L2i} = (V_{L2} + V_{L1} f_{12}) P_{L2i} \omega_{L2} \tag{9}$$

$$I_{L3i} = [V_{L3} + V_{L2} f_{23} + V_{L1}(f_{13} + f_{12} f_{23} + f'_{13})] P_{L3i} \omega_{L3} \tag{10}$$

In these equations, P_{Lji} ($j=1,2,3$) is the number of specific L_{ji} x-rays per total L_j x-rays from the j-th atomic subshell, ω_{Lj} is the L_j fluorescence yield, $f_{jj'}$ is the Coster-Kronig yield for the $L_j \rightarrow L_{j'}$ transition, and f'_{13} is the intrashell radiative yield for the $L_1 \rightarrow L_3$ transition. The total number of L_j vacancies (V_{Lj}) per 100 disintegrations is

$$V_{Lj} = \sum_m V_{mLj} + \sum_\epsilon V_{\epsilon Lj} + V_K \sum_{Lj} V'_{Lj} \, , \tag{11}$$

where V_{mLj} is the number of primary L_j vacancies per 100 disintegrations created by internal conversion of the m-th γ ray (see equation 1), $V_{\epsilon Lj}$ is the number of primary L_j vacancies per 100 disintegrations created by the ϵ-th electron-capture transition (see equation 2), V'_{Lj} is the number of secondary L_j vacancies per K vacancy, and V_K is the total number of primary K vacancies (created by internal conversion and electron capture) per 100 disintegrations. This latter is given by

$$V_K = \sum_m V_{mK} + \sum_\epsilon V_{\epsilon K} \, . \tag{12}$$

3.5. Method of Calculation

The computer code ATOMS[33] is used to calculate the relevant numbers of atomic vacancies, and energies and absolute intensities for K x-rays ($K_{\alpha1,2,3}$ and $K_{\beta1-5}$) and L x-rays ($L_{\alpha1,2}$, $L_{\beta1-6}$, $L_{\gamma1-3,6}$, L_η, and L_ℓ). Input data, i.e., the energies, absolute intensities, and multipolarities for the γ rays, are adopted values from the photon tables. Highly converted and $E0$ transitions, though not shown in the photon tables, are used in the calculation of x-ray data. γ rays with unknown multipolarities are assumed to be $M1$, $E2$, or $E1$, with a median internal conversion coefficient for these multipolarities used. The code also includes use of a median internal conversion coefficient, along with an uncertainty which spans the maximum range, for γ rays of mixed multipolarity, but unknown mixing ratio. 2% uncertainty is assigned to the theoretical internal conversion coefficients for pure transitions. Most input electron-capture intensities are decay-scheme values from ENSDF.[1] The resulting uncertainties in the x-ray intensities contain contributions from the uncertainties in the γ-ray and the electron-capture data. Throughout, the x-ray intensities determined were compared with values given by Reus and Westmeier.[34] Tables 7a-7d in Appendix C5 list energies and intensities for all of the individual x-rays per 100 vacancies in the K, L_1, L_2, and L_3 atomic shells.

4. Atomic Electrons

4.1. Electron Energies and Absolute Intensities

The tables of atomic electrons include electrons emitted both from internal conversion in all of the occupied atomic shells and from the Auger process in the K- and L-atomic shells. γ rays converted in the j-th atomic shell emit electrons with energy

$$E_e(j)=E_\gamma-B(j)\,, \tag{1}$$

where E_γ is the γ-ray energy, and $B(j)$, the binding energy of the j-th atomic shell.[35] The electron's absolute intensity ($I_e(j)$) is

$$I_e(j)=\alpha_j(ML,EL\pm1)I_\gamma\,, \tag{2}$$

where $\alpha_j(ML,EL\pm1)$ is the conversion coefficient in the j-th atomic shell for the multipolarity(ies) $(ML,EL\pm1)$, and I_γ, the absolute γ-ray intensity. Theoretical conversion coefficients are from calculations by Band, et al.,[24] for $Z<30$, and by Rösel, et al.,[25] for $Z\geqslant30$. Conversion coefficients for specific energies within each reported range are determined by interpolating with a cubic spline function, as recommended in Reference 25.

Auger electrons are emitted during subsequent redistribution of the atomic electrons which fill vacancies created by electron capture or internal conversion. The energy of Auger electrons (E) for the $W-XY,\ {}^{2S+1}L_J$ transition is given by

$$E(W-XY,\ {}^{2S+1}L_J)=B(W)-B(X)-B(Y)-\Delta({}^{2S+1}L_J)\,, \tag{3}$$

where W is the atomic shell of the primary vacancy, X and Y are the inner atomic shells involved in the Auger process, and $\Delta({}^{2S+1}L_J)$ is a correction term for the ${}^{2S+1}L_J$ configuration. This term has been neglected in calculations for the *Table of Radioactive Isotopes*, and the following approximate energies are used for the L_i-XY $(i=1,2,3)$ Auger electrons:

$$E(L_i-XY)\simeq B(L_i)-2B(M_5)\,. \tag{4}$$

Precise calculated Auger-electron energies for vacancies in the K, L, M, and N atomic subshells are given by Larkins.[35]

The absolute intensity (I) of K-Auger electrons is given by

$$I_{K-XY}=P_{K-XY}V_K(1-\omega_K)\,, \tag{5}$$

where P_{K-XY} is the probability per K-shell vacancy for emission of a K-XY Auger electron, V_K is the number of K-shell vacancies per 100 disintegrations, and ω_K is the fluorescence yield (the probability for emission of a K x-ray per K vacancy). The probabilities P_{K-XY} are taken from Chen, et al.,[36] the ω_K values, from Krause,[37] and the vacancies V_K are calculated as explained in Section 3 of *Methods of Analysis*.

The combined L_i-XY-Auger electron intensities are given by

$$I_{L-XY}=V_{L_i}-I_x(L_i)\,, \tag{6}$$

where V_{L_i} is the number of L_i-shell vacancies (also described in Section 3), and $I_x(L_i)$, the combined L_i x-ray intensity per 100 disintegrations.

4.2. Binning of Electron Data

The calculated electron spectra are extremely complex, as a consequence of the large numbers of conversion-electron and Auger-electron transitions possible for a single γ ray. Because it is not realistic to present each transition separately, the electron lines are grouped into energy bins. This is done initially with 1-*keV* bins; and if there are fewer than 30 of these bins, all of them are shown. If this is not the case, the most intense 1-*keV* bins are presented; and the remaining ones are grouped into 50- or 100-*keV* bins, according to the following procedure. First, the 1-*keV* bins are sorted in order of decreasing value for the product $\overline{E}_i\times\overline{I}_i$, in which $\overline{E}_i$ is the centroid energy, and $\overline{I}_i$, the absolute intensity of the i-th bin. Second, all bins in which $\dfrac{F_{R+1}-F_R}{F_R}\geqslant0.04$ (4%), with $F_R=\dfrac{1}{100}\sum_{i=1}^{R}\overline{E}_i\overline{I}_i$ are presented as 1-*keV* bins. Finally, the intensity between discrete 1-*keV* bins is combined into wider bins, not exceeding 50 *keV* for energies below 1500 *keV* and up to 100 *keV* otherwise. It should be noted that these wider bins lie between existing electron lines, and may have differing energy ranges. The transitionless regions of the electron spectrum are presented implicitly.

The average electron energy for a bin ($\langle e \rangle$) is

$$\langle e \rangle = \frac{1}{100} \sum_{i=1}^{N} \overline{E}_i \overline{I}_i \ , \tag{7}$$

where $\overline{E}_i$ and $\overline{I}_i$ are defined as above, and N is the number of 1-keV intervals within the wider bin.

The statistical uncertainty shown for the intensity of each bin is given by

$$\Delta \overline{I}_l = \left\{ \sum_{i=1}^{N} \Delta \overline{I}_i^2 \right\}^{1/2} , \tag{8}$$

where l refers to the l-th bin, and $\Delta \overline{I}_i$ is the statistical uncertainty of the absolute intensity for the i-th 1-keV bin.

The total average electron energy per disintegration ($\langle e \rangle$) (shown in the headings of the tables of atomic electrons) is

$$\langle e \rangle = \frac{1}{100} \sum_{i=1}^{M} \overline{E}_i \overline{I}_i \ , \tag{9}$$

and its statistical uncertainty,

$$\Delta \langle e \rangle = \frac{1}{100} \left\{ \sum_{i=1}^{M} \overline{E}_i^2 \Delta \overline{I}_i^2 \right\}^{1/2} . \tag{10}$$

Here M is the total number of 1-*keV* bins in the electron spectrum. Equation (10) does not contain a contribution from the uncertainty in $\overline{E}_i$. This contribution is normally negligible (<1%), and to incorporate it would be complicated. A greater component of uncertainty in the total average electron energy results from incomplete decay-scheme knowledge. This is particularly true for the decays of isotopes with large decay energies.[21] In addition to having the obvious problems associated with decay-scheme error (e.g., mis-assigned γ rays), these isotopes often have large numbers of weak unobserved transitions, particularly high-energy ones. The atomic electron tables in the *Table of Radioactive Isotopes* are derived from measured photon data, and one must bear in mind the resulting consequence of incomplete decay-scheme knowledge.

5. Continuous Radiation

5.1. β^- and β^+ Spectra

A beta-emitting isotope has a continuous spectrum of beta particles, which results from the sharing (between the beta groups, the neutrinos, and the recoiling nucleus), of the available decay energy. Beta-particle spectra are seldom measured; however, beta-decay theory is sufficiently adequate to enable one to calculate them accurately if decay schemes are well known.

The probability, per unit time, that a nucleus will decay by beta emission is given by

$$\lambda_n = \frac{G_\beta^2}{2\pi^2} \int_1^{W_0} P(W) dW \ , \tag{1}$$

where

$$P(W) = pW(W_0 - W)^{1/2} S_n(Z,W) \ . \tag{2}$$

In these equations, G_β is the weak interaction coupling constant, W and W_0 are the beta-particle and maximum energies, respectively, p $(=(W^2-1)^{1/2})$ is the beta-particle linear momentum, $S_n(Z,W)$ is the shape factor for the n-th forbidden beta decay, and Z, the atomic number of the daughter isotope. $S_n(Z,W)$ contains the corrections for screening of the nuclear Coulomb field by atomic electrons, for finite nuclear size, and for the energy dependence of the degree of forbiddenness. Energies are given in units of $m_e c^2$ (electron rest-mass energy); i.e., $W = 1 + \frac{E}{m_e c^2}$, where E is the kinetic energy of the beta particle.

The intensity per 100 disintegrations, in the interval E, $E+\Delta E$, is

$$I_\beta(\%)=100\sum_i \frac{I_i \int_E^{E+\Delta E} P(E)dE}{\int_0^{E_{i0}} P(E)dE}=100\sum_i \frac{I_i \int_W^{W+\Delta W} P(W)dW}{\int_1^{W_{i0}} P(W)dW}, \tag{3}$$

and the corresponding average energy per disintegration,

$$\langle\beta\pm\rangle=\sum_i \frac{I_i \int_E^{E+\Delta E} P(E)EdE}{\int_0^{E_{i0}} P(E)dE}=\sum_i \frac{I_i \int_W^{W+\Delta W} P(W)WdW}{\int_1^{W_{i0}} P(W)dW}. \tag{4}$$

Here I_i and E_{i0} are the absolute intensity and maximum kinetic energy (endpoint) of the i-th beta particle group, respectively. For β^- decay, the endpoint energy is

$$E_{i0}=Q_{\beta^-}+E_p-E_i\ , \tag{5}$$

and for β^+ decay, the corresponding energy is

$$E_{i0}=Q_\epsilon+E_p-E_i-2m_ec^2\ . \tag{6}$$

In equations (5) and (6), Q_{β^-} and Q_ϵ are the available decay energies, E_p is the excitation energy in the parent isotope ($E_p=0$ for the ground state), and E_i, the level energy in the daughter isotope populated by the i-th beta-particle group.

5.2. Internal Bremsstrahlung

Internal bremsstrahlung (IB) is the continuous-energy electromagnetic radiation that accompanies β^-, β^+, and electron-capture decay. Internal bremsstrahlung usually has an intensity which is much weaker than that of the γ rays, but often comparable to that of the x-rays or atomic electrons, from the same isotope. Energy-wise internal bremsstrahlung has a spectrum ranging from 0 *keV* up to the beta-particle endpoint, and is often the only source of high-energy photon radiation.

5.3. Internal Bremsstrahlung from Beta Decay

The following expressions give the absolute intensity, per 100 disintegrations of the parent isotope, and the average energy per disintegration for a beta emitter's internal bremsstrahlung in the energy interval k to $k+\Delta k$:

$$IB(\%)=100\sum_i I_i \int_k^{k+\Delta k} S(k)dk \tag{7}$$

$$\langle IB_{\beta\pm}\rangle=\sum_i I_i \int_k^{k+\Delta k} kS(k)dk\ . \tag{8}$$

Here I_i is the absolute intensity of the i-th beta-particle group, and $S(k)dk$, the probability per beta-particle disintegration for emission of a photon with energy between k and $k+dk$. Knipp and Uhlenbeck[38] give $S(k)dk$ as

$$S(k)dk=\frac{\int_{1+k}^{W_0} P(W_e)dW_e\ \phi(W_e,k)dk}{\int_1^{W_0} P(W_e)dW_e}, \tag{9}$$

where W_e is the energy of the i-th beta particle after the emission of a photon of energy k. The function $P(W)$ is defined as in Section 5.1, and $\phi(W_e,k)dk$ is the probability that a beta particle emitted with energy W_e will produce a photon with energy between k and $k+dk$. $\phi(W_e,k)$ has the analytical form

$$\phi(W_e, k)=\alpha\frac{p}{\pi p_e k}\left(\frac{W_e^2+W^2}{W_e p}ln(W+p) - 2\right). \tag{10}$$

Here p_e is the linear momentum of the i-th beta particle after the emission of a photon of energy k, and α ($\simeq 1/137$) is the fine structure constant. Equations (9) and (10) apply to allowed beta-particle transitions; and although they do not include Coulomb effects in the wave functions, they have been shown to provide good approximations for allowed and also for forbidden transitions.[39] It should be noted that $\phi(W_e, k)$ diverges for $k=0$, but that the average energy does not.

5.4. Internal Bremsstrahlung from Electron-Capture Decay

Internal bremsstrahlung (IB) accompanying electron-capture decay is significant only for capture from the $1s$, $2s$, $2p$, and $3p$ atomic shells. The following expression gives the absolute IB intensity, per 100 disintegrations of the parent isotope, in the energy interval from k to $k+\Delta k$:

$$IB(\%)=\frac{100}{\omega_{tot}}\sum_i I_i \int_k^{k+\Delta k} d\omega_{nr}\,. \tag{11}$$

Here I_i is the absolute intensity of the i-th electron-capture transition, ω_{tot} is the total electron capture probability, and $d\omega_{nr}$, the probability for emission of a photon with energy between k and $k+dk$ for electron capture from the nr atomic shell. The corresponding internal-bremsstrahlung average energy per disintegration is

$$\langle IB_\epsilon\rangle=\frac{1}{\omega_{tot}}\sum_i I_i \int_k^{k+\Delta k} kd\omega_{nr}\,. \tag{12}$$

The following expressions from Bambynek, et al.,[32] based on the theory of Glauber and Martin,[40] give the photon emission probabilities for internal bremsstrahlung.

For s-shell capture, the photon emission probability per K-shell capture is

$$\frac{dw_{ns}}{\omega_K}=\frac{\alpha}{\pi}\frac{\left|\Phi_{ns}(0)\right|^2}{\left|\Phi_{1s}(0)\right|^2}\frac{k(q_{ns}-k)^2}{q_{1s}^2}R_{ns}\,dk\,, \tag{13}$$

where $q_{ns}(=Q_\epsilon - E_{ns})$ is the total available energy shared between the photon and the neutrino (Q_ϵ is the electron-capture decay energy and E_{ns}, the binding energy of the ns-shell), and α ($\simeq 1/137$) is the fine structure constant. The correction factor $R_{ns}(k)$, containing the most important relativistic and Coulomb effects, is

$$R_{ns}(k)=\frac{1}{2}(1+B_{ns}^2)\,, \tag{14}$$

where

$$B_{1s}(k)=1-\frac{4}{3}\frac{\eta_1}{1+\eta_1}\left[1+\frac{\eta_1}{(1-\eta_1)}\left(2\kappa(\lambda_1)-1\right)\right] \tag{15}$$

and

$$B_{2s}(k)=1-\frac{\eta_2}{(1-\frac{\eta_2^2}{4})}\left(\frac{4}{3}+\frac{5}{6}\eta_2\right)-\frac{\eta_2^2}{(1-\frac{\eta_2^2}{4})^2}\left(\frac{8}{3}(1-\eta_2^2)\kappa(\lambda_2)-3-\eta_2+\frac{5}{4}\eta_2^2\right). \tag{16}$$

In these latter,

$$\eta_1=(1+\frac{k}{E_{1s}})^{-1/2} \text{ and } \eta_2=(\frac{1}{4}+\frac{\mathrm{k}}{\mathrm{E}_{2s}})^{-1/2}\,, \tag{17}$$

$$\lambda_1=\frac{(1-\eta_1)}{(1+\eta_1)} \text{ and } \lambda_2=\frac{(2-\eta_2)}{(2+\eta_2)}\,, \tag{18}$$

and

$$\kappa(\lambda_n)=\ln(1+\lambda_n)-\eta\sum_{j=1}^{\infty}\frac{(-\lambda_n)^j}{j(j-\eta)}\,. \tag{19}$$

For a daughter nucleus with atomic number Z, the parameter η is

$$\eta = \frac{Z\alpha E_{ns}}{(1-E_{ns}^2)^{1/2}} . \qquad (20)$$

For p-shell capture, the photon emission probability per K-shell capture is

$$\frac{d\omega_{np}}{\omega_K} = \frac{4}{\pi Z^2 \alpha} \left(Q_{np}(k) \right)^2 k \frac{(q_{np}-k)^2}{q_{1s}^2} dk , \qquad (21)$$

where $Q_{2p}(k)$ and $Q_{3p}(k)$, also incorporating the most important relativistic and Coulomb effects, are

$$Q_{2p}(k) = \frac{\eta_2^2}{4(1-\frac{\eta_2^2}{4})^2} \left(1+\frac{2}{3}\eta_2 - \frac{7}{12}\eta_2^2 + \frac{4}{3}\eta_2^2 \kappa(\lambda_2) \right) , \qquad (22)$$

$$Q_{3p}(k) = \frac{4}{27} \frac{\eta_3^2}{(1-\frac{\eta_3^2}{9})^3} \left[(1-\frac{\eta_3}{3}) \left(1+\eta_3 - 2(\frac{\eta_3}{3})^2 - 8(\frac{\eta_3}{3})^3 \right) + \frac{4}{3}\eta_3^2 (1-\frac{\eta_3^2}{3}) \kappa(\lambda_3) \right] , \qquad (23)$$

and

$$\eta_3 = \left(\frac{1}{9} + \frac{k}{B_{1s}} \right)^{-1/2} . \qquad (24)$$

For first-forbidden unique transitions, the photon emission probability per K-shell capture (neglecting Coulomb effects) is

$$\frac{d\omega_{1s}}{\omega_K} = \frac{\alpha}{\pi} \frac{k(q_{1s}-k)^2}{q_{1s}^2} [(1-\frac{k}{q_{1s}})^2 + \frac{k^2}{q_{1s}^2}] dk . \qquad (25)$$

5.5. Method of Calculation

Average energies per disintegration, and the corresponding absolute intensities, for β^-, β^+, and *IB* radiation are calculated and reported in standard intervals (bins) of 0-10, 10-20, 20-40, 40-100, 100-300, 300-600, 600-1300, 1300-2500, 2500-5000, 5000-7500, and >7500 *keV*. For β^- and β^+ decay, the last interval extends up to the β^- or β^+ endpoint energy (defined in equations (5) and (6)) for the lowest populated daughter level. For electron-capture decay, the last interval extends up to $Q_\epsilon + E_p - E_\ell$ (Q_ϵ and E_p are defined in Section 5.1, and E_ℓ is the energy of the lowest populated daughter level). No absolute intensity is given for bremsstrahlung in the 0-10-*keV* interval because of the divergence at zero energy. The total average energies per disintegration are given in the headings of the continuous radiation tables.

The energies for levels in the daughter nuclei and the absolute beta-particle and electron-capture intensities used in the calculations are primarily from the *LBL/ENSDF*[20] database. Decay energies (Q-values) are from *The 1983 Atomic Mass Table*[14] for $A < 200$, and from *ENSDF*[1] for $A \geqslant 200$. Beta transitions are assumed to be allowed if the degree of forbiddeness is not specified in the database. Non-unique forbidden transitions, for which there are no exact solutions, are treated as allowed, and results accurate to within several percent are achieved. Beta-spectra shape factors $S_n(Z,W)$ are calculated with a computer code developed by Gove and Martin.[31]

Decay-scheme knowledge is presumed to be complete throughout, and one must be cautious in using the continuous radiation tables, especially for isotopes with decay energies exceeding 5 *MeV*. Errors in the assignment of γ rays are always possible; and above about 5 *MeV*, substantial photon intensity often remains undetected. This latter effect tends to skew systematically the beta spectra in favor of the lower-energy levels (i.e., those populated by the higher-energy beta particles) in the daughter nuclei, resulting in systematic shifts in the continuous-radiation intensities towards higher-energy bins. The total average energies per disintegration are then higher than they would be if the decay schemes were completely known.

References

1) *Evaluated Nuclear Structure Data File* (ENSDF), a computer file of evaluated nuclear structure and radioactivity decay data. The file is maintained by the National Nuclear Data Center (NNDC), Brookhaven National Laboratory, on behalf of The International Network for Nuclear Structure Data Evaluation.

2) *Energy Levels of A=21-44 Nuclei (VI),* P.M. Endt and C. van der Leun, *Nucl. Phys.* **A310**, 1-752 (1978).

3) *Energy Levels of Light Nuclei A=5-10,* F. Ajzenberg-Selove, *Nucl. Phys.* **A413**, 1-214 (1984).

4) *Energy Levels of Light Nuclei A=11-12,* F. Ajzenberg-Selove, *Nucl. Phys.* **A433**, 1-158 (1985).

5) *Energy Levels of Light Nuclei A=13-15,* F. Ajzenberg-Selove, *Nucl. Phys.* **A360**, 1-186 (1981).

6) *Energy Levels of Light Nuclei A=16-17,* F. Ajzenberg-Selove, *Nucl. Phys.* **A375**, 1-168 (1982); **A392**, 185(Errata) (1983).

7) *Energy Levels of Light Nuclei A=18-20,* F. Ajzenberg-Selove, *Nucl. Phys.* **A392**, 1-216 (1983); **A413**, 168(Errata) (1984).

8) *Review of Particle Properties,* Particle Data Group, *Rev. Mod. Phys.* **56**, No. 2, Part II, S1 (1984).

9) *Table of Isotopes*, 7th edition, C.M. Lederer and V.S. Shirley, editors; E. Browne, J.M. Dairiki, and R.E. Doebler, principal authors; A.A. Shihab-Eldin, L.J. Jardine, J.K. Tuli, and A.B. Buyrn, authors; John Wiley and Sons, Inc., New York (1978).

10) *Energy Levels of Light Nuclei A=3,* S. Fiarman and S.S. Hanna, *Nucl. Phys.* **A251**, 1 (1975).

11) *Energy Levels of Light Nuclei A=4,* S. Fiarman and W.E. Meyerhof, *Nucl. Phys.* **A206**, 1 (1973).

12) *Chart of the Nuclides*, 13th edition (1983); revised by F.W. Walker, D.G. Miller, and F. Feiner, Knolls Atomic Power Laboratory, General Electric Company, Schenectady, NY 12301.

13) *Isotopic Compositions of the Elements 1983,* N.E. Holden, R.L. Martin, and I.L. Barnes, *Pure Appl. Chem.* **56**, 675 (1984).

14) *The 1983 Atomic Mass Table,* A.H. Wapstra, G. Audi, K. Bos, and R. Hoekstra, *Nucl. Phys.* **A432**, 1-362 (1985).

15) F.W. Walker, private communications (1984-1985).

16) *Droplet Model of Atomic Nuclei,* W.D. Myers, IFI Plenum Data Co., Plenum Publ. Corp., New York (1977).

17) *Statistics for Physicists,* B.R. Martin, Academic Press Inc. (London) Ltd., London (1971).

18) R.G. Helmer, R.C. Greenwood, and R.J. Gehrke, *Nucl. Instrum. Methods* **155**, 189 (1978).

19) *GAMUT, a Computer Code for Statistical Analysis of γ-ray Data,* R.B. Firestone and E. Browne, Univ. of California Lawrence Berkeley Lab., Report **LBL-16870**, 295 (1984).

20) *Lawrence Berkeley Laboratory Evaluated Nuclear Structure Data File* (LBL/ENSDF), a numerical version of ENSDF (Reference 1), implemented into DATATRIEVE (a database management system of Digital Equipment Corporation) by the LBL Isotopes Project; Univ. of California Lawrence Berkeley Lab., Report **LBL-15207**, 99 (1983).

21) T. Yoshida and R. Nakasima, *J. Sci. Technol. (Tokyo)* **18**, 393 (1981).

22) J.W. Tepel, *International Atomic Energy Agency, Vienna*, Report **INDC(NDS)-115/NE**, 121 (1980); private communication (1983).

23) C.M. Lederer, private communication (1982).

24) I.M. Band, M.B. Trzhaskovskaya, and M.A. Listengarten, *At. Data and Nucl. Data Tables* **18**, 433 (1976).

25) F. Rösel, H.M. Fries, K. Alder, and H.C. Pauli, *At. Data and Nucl. Data Tables* **21**, 91 (1978) [for $Z=30\text{-}67$]; **21**, 291 (1978) [for $Z=68\text{-}104$].

26) F.T. Porter and M.S. Freedman, *J. Phys. Chem. Ref. Data* **7**, 1267 (1978).

27) E.L. Church and J. Weneser, *Phys. Rev.* **103**, 1035 (1956).

28) R.B. Firestone, Paper 84/26, submitted to the Consultants' Meeting organized by the IAEA and held at the Fachinformationszentrum in Kalsruhe, FRG, 3-6 April 1984, *International Atomic Energy Agency, Vienna*, Report **INDC(NDS)-157/GE**.

29) D.H Wilkinson, *Nucl. Phys.* **A133**, 1 (1969).

30) D.H. Wilkinson, *Nucl. Instrum. Methods* **82**, 122 (1970).

31) N.B. Gove and M.J. Martin, *Nucl. Data Tables* **A10**, 205 (1971).

32) W. Bambynek, H. Behrens, M.H. Chen, B. Crasemann, M.L. Fitzpatrick, K.W.D. Ledingham, H. Genz, M. Mutterer, and R.L. Intemann, *Rev. Mod. Phys.* **49**, 77 (1977).

33) *ATOMS, a computer code for calculating x-ray and Auger-electron intensities from* K, L_1, L_2, *and* L_3 *atomic-shell vacancies produced by nuclear decay*, R.B. Firestone, Univ. of California Lawrence Berkeley Lab., Report **LBL-15207**, 95 (1983).

34) U. Reus and W. Westmeier, *At. Data and Nucl. Data Tables* **29**, 193 (1983).

35) F.B. Larkins, *At. Data and Nucl. Data Tables* **20**, 313 (1977).

36) M.H. Chen, B. Crasemann, and H. Mark, *At. Data and Nucl. Data Tables* **24**, 13 (1979).

37) M.O. Krause, *J. Phys. Chem. Ref. Data* **8**, 307 (1979).

38) J.K. Knipp and G.E. Uhlenbeck, *Physica* **3**, 425 (1936).

39) C.S.W. Chang and D.L. Falkoff, *Phys. Rev.* **76**, 365 (1949).

40) R.J. Glauber and P.C. Martin, *Phys. Rev.* **104**, 158 (1956).

CONTENTS

A = 1

RMP **56**, S1 (1984)

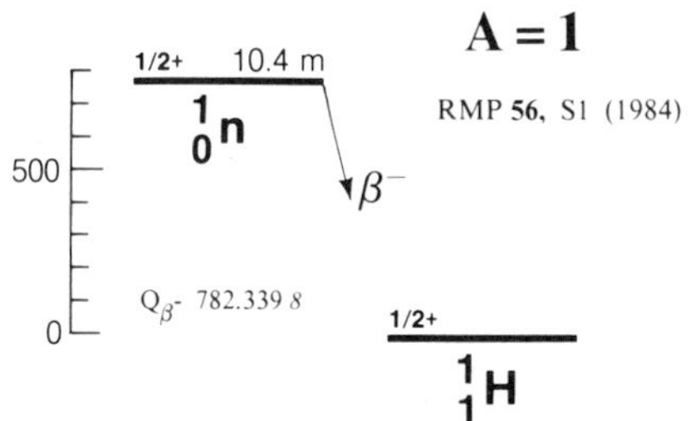

$^{1}_{0}$n (10.37 *18* min)

Mode: β-
Δ: 8071.369 *13* keV
SpA: 1.81×10^{10} Ci/g
Prod: spontaneous fission (^{252}Cf); induced fission; nuclear reactions using radioactive sources, e.g., ^{9}Be(α,n) (alphas from ^{210}Po, ^{225}Ac, ^{226}Ra, ^{238}Pu, ^{241}Am, etc.) and ^{9}Be(γ,n) (photons from ^{88}Y, ^{124}Sb, electron accelerators, etc.); nuclear reactions using accelerators, e.g., ^{2}H(d,^{3}He), ^{3}H(d,α), ^{3}H(p,t), ^{7}Li(d,n), ^{9}Be(d,n), ^{9}Be(^{3}He,n), and ^{12}C(^{3}He,n)

Continuous Radiation (^{1}n)

⟨β-⟩=301 keV;⟨IB⟩=0.26 keV

E_{bin}(keV)		⟨ ⟩(keV)	(%)
0 - 10	β-	0.0250	0.432
	IB	0.0147	
10 - 20	β-	0.110	0.72
	IB	0.0140	0.097
20 - 40	β-	0.60	1.97
	IB	0.026	0.090
40 - 100	β-	6.2	8.6
	IB	0.063	0.099
100 - 300	β-	82	40.4
	IB	0.110	0.066
300 - 600	β-	185	43.6
	IB	0.032	0.0089
600 - 782	β-	27.4	4.22
	IB	0.00059	9.3×10^{-5}

$^{1}_{1}$H ($>1\times10^{30}$ yr)

Δ: 7289.030 *11* keV
%: 99.985 *1*

A = 2

Table of Isotopes (1978)

$^{2}_{1}$H (stable)

Δ: 13135.824 *22* keV
%: 0.015 *1* (Lake Michigan water); 0.0044 to 0.0184 (other sources); lower for electrolytic hydrogen

A = 3

NP **A251**, 1 (1975)

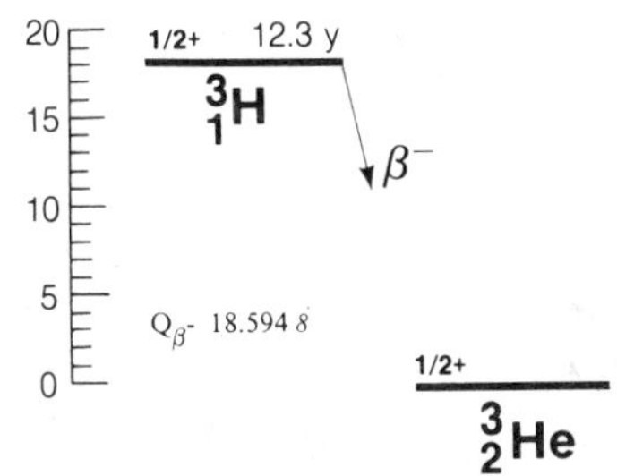

$^{3}_{1}$H (12.33 *6* yr)

Mode: β-
Δ: 14949.91 *3* keV
SpA: 9664 Ci/g
Prod: ^{6}Li(n,α)

Continuous Radiation (^{3}H)

⟨β-⟩=5.7 keV;⟨IB⟩=0.000112 keV

E_{bin}(keV)		⟨ ⟩(keV)	(%)
0 - 10	β-	3.83	85
	IB	0.000111	
10 - 19	β-	1.85	15.1
	IB	1.65×10^{-6}	1.48×10^{-5}

$^{3}_{2}$He(stable)

Δ: 14931.32 *3* keV
%: 1.38 *3* $\times10^{-4}$ (atmospheric sources); 6×10^{-8} to 0.0041 (other sources)

A = 4

NP **A206**, 1 (1973)

$^{4}_{2}$He(stable)

Δ: 2424.92 *5* keV
%: 99.999862 *3*

A = 5

NP **A413**, 1 (1984)

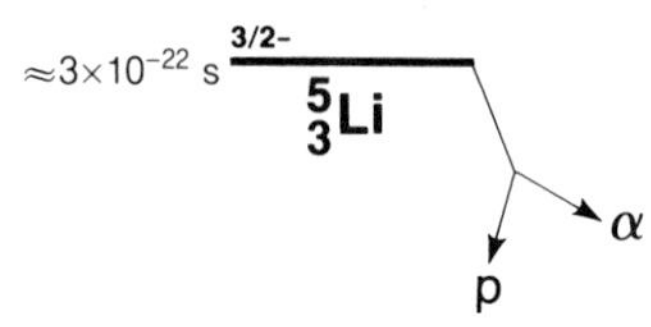

Q_ϵ 290 *70*

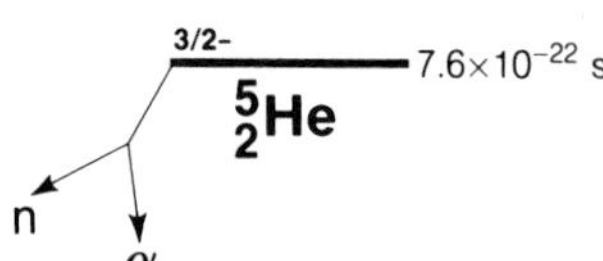

$^{5}_{2}$He(7.60 *25*×10^{-22} s)

Γ: 600 *20* keV
Mode: n+α
Δ: 11390 *50* keV
Prod: ^{3}H(d,γ); ^{3}H(t,n); ^{3}He(t,p); ^{4}He(p,π^{+}); ^{4}He(d,p); ^{6}Li(γ,p); ^{6}Li(e,ep); ^{6}Li(n,d); ^{6}Li(p,2p); ^{6}Li(d,^{3}He); ^{6}Li(α,αp); ^{7}Li(γ,d); ^{7}Li(π^{+},2p); ^{7}Li(π^{-},2n); ^{7}Li(n,t); ^{7}Li(p,^{3}He); ^{7}Li(p,pd); ^{7}Li(d,α); ^{9}Be(p,pα); ^{9}Be(p,d^{3}He); ^{9}Be(^{3}He,^{7}Be); ^{9}Be(α,2α); ^{9}Be(α,^{8}Be); ^{10}B(d,^{7}Be)

$^{5}_{3}$Li(~3×10^{-22} s)

Γ: ~1500 keV
Mode: p+α
Δ: 11680 *50* keV
Prod: ^{3}He(d,γ); ^{3}He(t,n); ^{3}He(^{3}He,p); ^{4}He(d,n); ^{4}He(^{3}He,d); ^{6}Li(p,d); ^{6}Li(π^{+},p); ^{6}Li(p,d); ^{6}Li(d,t); ^{6}Li(^{3}He,α); ^{7}Li(π^{+},d); ^{7}Li(p,t); ^{9}Be(^{3}He,^{7}Li); ^{9}Be(α,^{8}Li); ^{10}B(d,^{7}Li)

A = 6

NP **A413**, 23 (1984)

5000
0
0+ 807 ms
$^{6}_{2}$He
β^{-}
1+
$^{6}_{3}$Li
0+
5.0×10^{-21} s
$^{6}_{4}$Be
p
p
α

Q_{β^-} 3506.7 *7*

Q_ϵ 4288 *5*

$^{6}_{2}$He(806.7 *15* ms)

Mode: β-
Δ: 17592.3 *10* keV
SpA: 1.564×10^{12} Ci/g
Prod: ^{9}Be(n,α); ^{7}Li(γ,p)

Continuous Radiation (^{6}He)

⟨β-⟩=1567 keV;⟨IB⟩=4.6 keV

E_{bin}(keV)		⟨ ⟩(keV)	(%)
0 - 10	β-	0.00112	0.0201
	IB	0.050	
10 - 20	β-	0.00462	0.0303
	IB	0.050	0.35
20 - 40	β-	0.0250	0.082
	IB	0.098	0.34
40 - 100	β-	0.273	0.377
	IB	0.28	0.44
100 - 300	β-	5.2	2.44
	IB	0.84	0.47
300 - 600	β-	31.7	6.8
	IB	0.99	0.23
600 - 1300	β-	272	27.8
	IB	1.45	0.168
1300 - 2500	β-	950	51
	IB	0.82	0.048
2500 - 3507	β-	308	11.1
	IB	0.039	0.00146

$^{6}_{3}$Li(stable)

Δ: 14085.6 *7* keV
%: 7.5 *2*; commercial sources may be depleted in ^{6}Li

$^{6}_{4}$Be(5.0 *3* ×10^{-21} s)

Γ: 92 *6* keV
Mode: 2p+α
Δ: 18374 *5* keV
Prod: ^{3}He(^{3}He,γ); ^{4}He(^{3}He,n); ^{6}Li(p,n); ^{6}Li(^{3}He,t); ^{6}Li(^{6}Li,^{6}He); ^{9}Be(^{3}He,^{6}He)

A = 7

NP **A413**, 50 (1984)

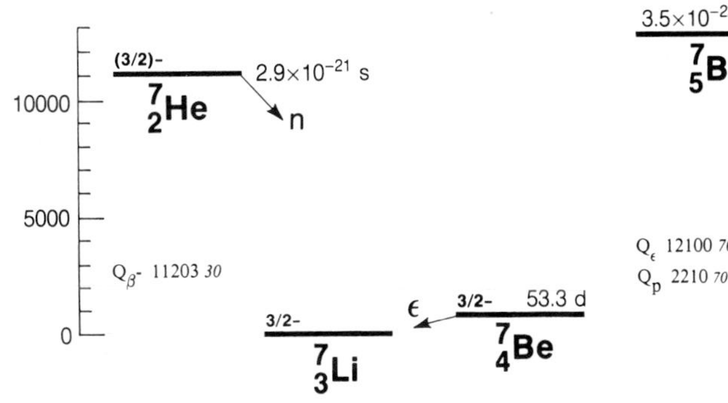

$Q_ε$ 861.90 *4*

$^{7}_{2}$He(2.9 *5* ×10^{-21} s)

Γ: 160 *30* keV
Mode: n
Δ: 26110 *30* keV
Prod: ^{7}Li(π^{-},γ); ^{7}Li(n,p); ^{7}Li(t,^{3}He); ^{9}Be(^{6}Li,^{8}B)

n: 380 *30*

$^{7}_{3}$Li(stable)

Δ: 14906.8 *8* keV
%: 92.5 *2*; commercial sources may be enriched in ^{7}Li

$^{7}_{4}$Be(53.29 *7* d)

Mode: ε
Δ: 15768.7 *8* keV
SpA: 3.500×10^{5} Ci/g
Prod: ^{6}Li(d,n); ^{10}B(p,α); ^{12}C(^{3}He,2α)

Photons (^{7}Be)

⟨γ⟩=49.6 *10* keV

$γ_{mode}$	γ(keV)	γ(%)
γ M1(+0.6%E2)	477.606 *2*	10.39 *21*

Continuous Radiation (^{7}Be)

⟨IB⟩=0.168 keV

E_{bin}(keV)		⟨ ⟩(keV)	(%)
10 - 20	IB	1.8×10^{-5}	0.000115
20 - 40	IB	0.000141	0.00045
40 - 100	IB	0.0021	0.0029
100 - 300	IB	0.038	0.018
300 - 600	IB	0.100	0.023
600 - 862	IB	0.028	0.0042

$^{7}_{5}$B (3.5 *5* ×10^{-22} s)

Γ: 1300 *200* keV
Mode: 2p, 3p
Δ: 27870 *70* keV
Prod: ^{7}Li(π^{+},π^{-}); ^{10}B(^{3}He,^{6}He)

A = 8

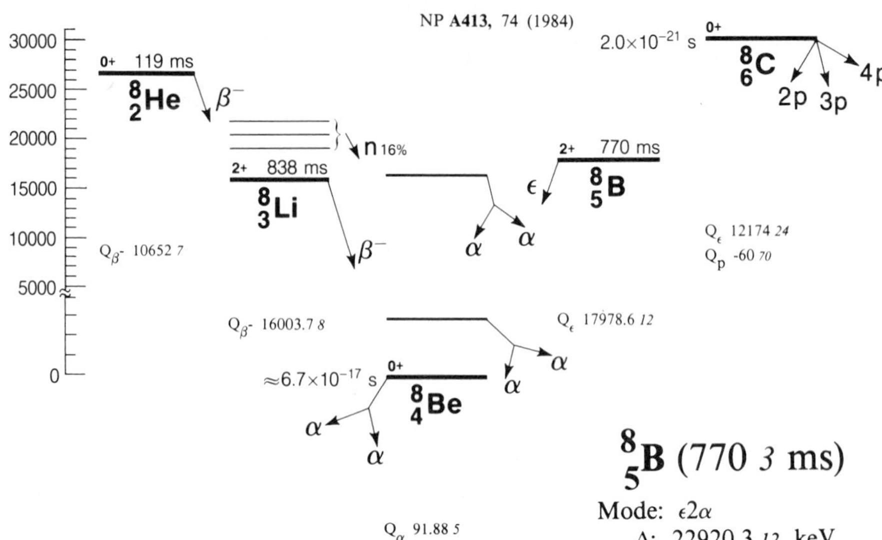

$^{8}_{2}$He(119.0 *15* ms)

Mode: β-, β-n(16 *1* %)
Δ: 31598 *7* keV
SpA: 2.03×10^{12} Ci/g
Prod: protons on C; protons on O; ^{11}B(γ,3p); fission

β-n: 610, 1030, 2500, 3000

Photons (^{8}He)

⟨γ⟩=848 *10* keV

γ_{mode}	γ(keV)	γ(%)
$\gamma_{\beta\text{-n}}$ M1(+0.6%E2)	477.606 *2*	5.1 *6*
$\gamma_{\beta\text{-}}$M1	980.74 *10*	84 *1*

$^{8}_{5}$B (770 *3* ms)

Mode: ε2α
Δ: 22920.3 *12* keV
SpA: 1.207×10^{12} Ci/g
Prod: ^{6}Li(^{3}He,n)

Continuous Radiation (^{8}He)

⟨β-⟩=4345 keV;⟨IB⟩=22 keV

E_{bin}(keV)		⟨ ⟩(keV)	(%)
0 - 10	β-	0.000109	0.00196
	IB	0.087	
10 - 20	β-	0.000451	0.00296
	IB	0.086	0.60
20 - 40	β-	0.00245	0.0080
	IB	0.17	0.60
40 - 100	β-	0.0272	0.0375
	IB	0.51	0.78
100 - 300	β-	0.54	0.254
	IB	1.62	0.90
300 - 600	β-	3.63	0.78
	IB	2.2	0.52
600 - 1300	β-	39.8	3.99
	IB	4.4	0.49
1300 - 2500	β-	277	14.1
	IB	5.6	0.31
2500 - 5000	β-	1644	43.6
	IB	6.0	0.18
5000 - 7500	β-	1911	31.4
	IB	1.53	0.026
7500 - 9671	β-	469	5.8
	IB	0.077	0.00097

Continuous Radiation (^{8}Li)

⟨β-⟩=6243 keV;⟨IB⟩=37 keV

E_{bin}(keV)		⟨ ⟩(keV)	(%)
0 - 10	β-	3.73×10^{-5}	0.00068
	IB	0.102	
10 - 20	β-	0.000149	0.00098
	IB	0.102	0.71
20 - 40	β-	0.00080	0.00261
	IB	0.20	0.70
40 - 100	β-	0.0087	0.0120
	IB	0.60	0.92
100 - 300	β-	0.174	0.082
	IB	1.9	1.08
300 - 600	β-	1.20	0.257
	IB	2.8	0.64
600 - 1300	β-	14.1	1.40
	IB	5.7	0.64
1300 - 2500	β-	113	5.7
	IB	8.0	0.44
2500 - 5000	β-	994	25.6
	IB	10.9	0.31
5000 - 7500	β-	2133	34.1
	IB	5.3	0.089
7500 - 12964	β-	2987	32.8
	IB	1.8	0.021

Delayed Alphas (^{8}B)

⟨α⟩~15767 keV

α(keV)*	α(%)
1570 *30*	~14
8359 *3*	~186

* broad peaks

Continuous Radiation (^{8}B)

⟨β+⟩=6548 keV;⟨IB⟩=41 keV

E_{bin}(keV)		⟨ ⟩(keV)	(%)
0 - 10	β+	0.00483	0.073
	IB	0.100	
10 - 20	β+	0.0270	0.176
	IB	0.100	0.69
20 - 40	β+	0.160	0.52
	IB	0.20	0.69
40 - 100	β+	1.51	2.13
	IB	0.59	0.90
100 - 300	β+	7.0	4.14
	IB	1.9	1.05
300 - 600	β+	0.84	0.185
	IB	2.7	0.63
600 - 1300	β+	9.5	0.95
	IB	5.7	0.64
1300 - 2500	β+	79	3.98
	IB	8.1	0.45
2500 - 5000	β+	736	18.9
	IB	11.8	0.34
5000 - 7500	β+	1757	28.0
	IB	6.5	0.109
7500 - 15550	β+	3957	41
	IB	3.3	0.038
	Σβ+		100

$^{8}_{3}$Li(838 *6* ms)

Mode: β-2α
Δ: 20945.4 *8* keV
SpA: 1.145×10^{12} Ci/g
Prod: ^{7}Li(n,γ); ^{7}Li(d,p)

Alpha Particles (^{8}Li)

α(keV)
1570 *30* *

* broad peak

$^{8}_{4}$Be(6.7 *17* $\times10^{-17}$ s)

Γ: 6.8 *17* eV
Mode: 2α
Δ: 4941.73 *11* keV
Prod: nuclear reactions using accelerators, e.g. ^{4}He(α,γ), ^{6}Li(d,γ), ^{7}Li(p,γ), ^{9}Be(γ,n), ^{10}B(n,t), ^{12}C(^{3}He,^{7}Be), ^{13}C(α,^{9}Be), ^{16}O(α,^{12}C), etc.

Alpha Particles (^{8}Be)

α(keV)
46.06 *3*

$^{8}_{6}$C (2.0 *4* $\times10^{-21}$ s)

Γ: 230 *50* keV
Mode: (2p, 3p, 4p)?
Δ: 35095 *24* keV
Prod: ^{14}N(^{3}He,^{9}Li); ^{12}C(α,^{8}He)

A = 9

NP **A413**, 105 (1984)

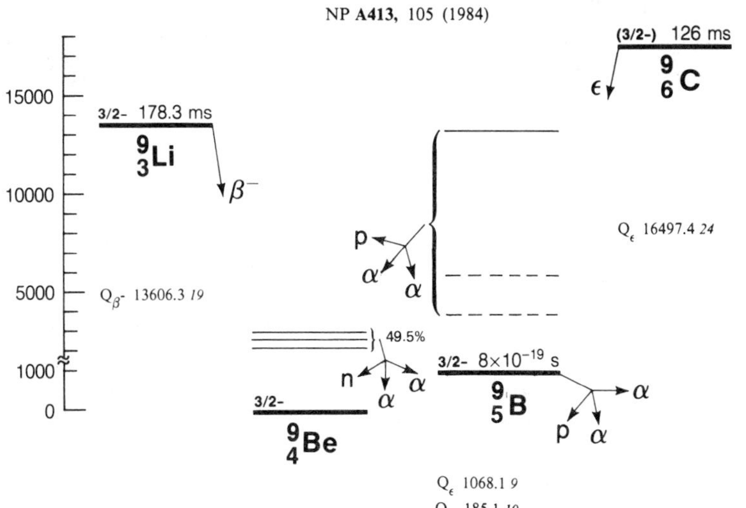

Q$_\epsilon$ 1068.1 *9*
Q$_p$ 185.1 *10*

$^{9}_{3}$Li(178.3 *4* ms)

Mode: β-, β-(n+2α)(49.5 *5* %)
Δ: 24953.9 *20* keV
SpA: 1.771×10^{12} Ci/g
Prod: ^{9}Be(n,p); ^{9}Be(d,2p); ^{10}Be(^{11}B,^{12}C); ^{3}H(^{7}Li,p); protons on Ir

Delayed Alphas (^{9}Li)

α(keV)	α(rel)
200	12 *4* •
300	
700	100

• 200α + 300α

Delayed Neutrons (^{9}Li)

n(keV)	n(rel)
300	100 *13* •
650	6.8 *13*
~1000	10.0 *16* •

• broad peaks

Continuous Radiation (^{9}Li)

⟨β-⟩=5732 keV;⟨IB⟩=34 keV

E_{bin}(keV)		⟨ ⟩(keV)	(%)
0 - 10	β-	0.000180	0.00330
	IB	0.097	
10 - 20	β-	0.00072	0.00472
	IB	0.097	0.67
20 - 40	β-	0.00379	0.0124
	IB	0.19	0.67
40 - 100	β-	0.0405	0.056
	IB	0.57	0.88
100 - 300	β-	0.74	0.352
	IB	1.8	1.02
300 - 600	β-	4.40	0.95
	IB	2.6	0.60
600 - 1300	β-	35.5	3.64
	IB	5.3	0.59
1300 - 2500	β-	151	7.9
	IB	7.3	0.40
2500 - 5000	β-	1062	27.5
	IB	9.7	0.28
5000 - 7500	β-	2044	32.8
	IB	4.5	0.075
7500 - 13606	β-	2434	26.8
	IB	1.46	0.017

$^{9}_{4}$Be(stable)

Δ: 11347.7 *4* keV
%: 100

$^{9}_{5}$B (8 *3* $\times10^{-19}$ s)

Γ: 0.54 *21* keV
Mode: p+2α
Δ: 12415.8 *10* keV
Prod: ^{6}Li(^{3}He,γ); ^{6}Li(α,n); ^{6}Li(^{6}Li,t); ^{7}Li(^{3}He,n); ^{9}Be(p,n); ^{9}Be(^{3}He,t); ^{9}Be(^{6}Li,^{6}He); ^{10}B(γ,n); ^{10}B(p,d); ^{10}B(d,t); ^{10}B(^{3}He,α); ^{11}B(p,t); ^{12}C(π^+,π^-t); ^{12}C(p,α); ^{12}C(^{3}H,^{6}He); ^{12}C(α,^{7}Li); ^{12}C(^{16}O,^{19}F)

$^{9}_{6}$C (126.5 *9* ms)

Mode: ε, ε(p+2α)
Δ: 28913.2 *22* keV
SpA: 1.801×10^{12} Ci/g
Prod: ^{10}B(p,2n); ^{11}B(p,3n)

Delayed Protons (^{9}C)

p(keV)	p(rel)
3450 *250*	
4200 *300 ?*	
5000 *200 ?*	
6100 *100*	
9280 *240*	100
12300 *100*	83 *14*

A = 10

NP **A413**, 131 (1984)

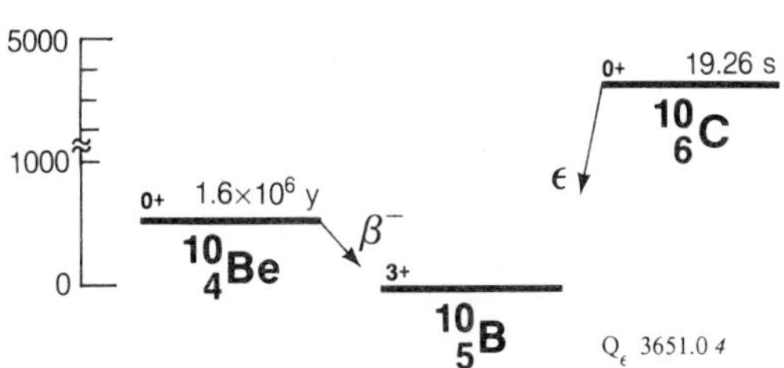

Q$_{\beta^-}$ 556.2 *5*

$^{10}_{4}$Be(1.6 *2* $\times 10^6$ yr)

Mode: β-
Δ: 12607.0 *4* keV
SpA: 0.022 Ci/g
Prod: ^{9}Be(n,γ); ^{9}Be(d,p)

Continuous Radiation (^{10}Be)

⟨β-⟩=203 keV;⟨IB⟩=0.126 keV

E_{bin}(keV)		⟨ ⟩(keV)	(%)
0 - 10	β-	0.066	1.25
	IB	0.0103	
10 - 20	β-	0.249	1.64
	IB	0.0096	0.067
20 - 40	β-	1.23	4.06
	IB	0.017	0.060
40 - 100	β-	11.2	15.7
	IB	0.038	0.060
100 - 300	β-	108	55
	IB	0.047	0.030
300 - 556	β-	82	22.1
	IB	0.0043	0.00122

$^{10}_{5}$B (stable)

Δ: 12050.8 *3* keV
%: 19.9 *2*

$^{10}_{6}$C (19.26 *5* s)

Mode: ε
Δ: 15701.7 *5* keV
SpA: 5.755$\times 10^{10}$ Ci/g
Prod: ^{10}B(p,n)

Photons (^{10}C)

⟨γ⟩=723 *14* keV

γ_{mode}	γ(keV)	γ(%)
γ E2	718.29 *9*	100
γ M1	1021.78 *14*	1.468 *14*

Continuous Radiation (^{10}C)

⟨β+⟩=809 keV;⟨IB⟩=1.52 keV

E_{bin}(keV)		⟨ ⟩(keV)	(%)
0 - 10	β+	0.00192	0.0286
	IB	0.032	
10 - 20	β+	0.0118	0.077
	IB	0.032	0.22
20 - 40	β+	0.079	0.255
	IB	0.062	0.21
40 - 100	β+	1.00	1.37
	IB	0.17	0.26
100 - 300	β+	19.3	9.2
	IB	0.45	0.26
300 - 600	β+	102	22.3
	IB	0.42	0.101
600 - 1300	β+	501	54
	IB	0.33	0.042
1300 - 2500	β+	186	12.6
	IB	0.0170	0.00122
	Σβ+		100

A = 11

NP **A433**, 4 (1985)

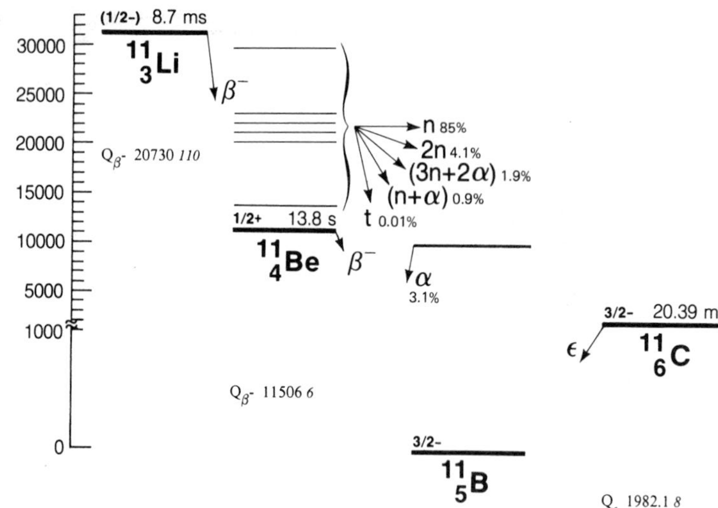

$^{11}_{3}$Li(8.7 *1* ms)

Mode: β-, β-n(85 *1* %), β-2n(4.1 *4* %), β-(3n+2α)(1.9 *2* %), β-(n+α)(0.9 *3* %), β-t(0.010 *4* %)
Δ: 40900 *110* keV
SpA: 1.480$\times 10^{12}$ Ci/g
Prod: protons on Ir; protons on U

Delayed Alphas (^{11}Li)

α(keV)†	α(%)
46.06 *3*	1.6 *2*
1570 *30*	0.30 *5*
~1600	0.9 *3*

† broad peaks

Photons (^{11}Li)

⟨γ⟩=1464 *120* keV

γ_{mode}	γ(keV)	γ(%)
$\gamma_{\beta\text{-n}}$ E1	218.8 *3*	1.6 *6*
$\gamma_{\beta\text{-}}$E1	320.04 *10*	9.2 *7*
$\gamma_{\beta\text{-n}}$	2589.4 *3*	5.8 *18*
$\gamma_{\beta\text{-n}}$ E1	2592.6 *4*	
$\gamma_{\beta\text{-n}}$ E2	2811.3 *3*	2.8 *12*
$\gamma_{\beta\text{-n}}$ E2	3366.94 *19*	35 *3*
$\gamma_{\beta\text{-n}}$ E1	5955.4 *3*	~0.39

Continuous Radiation (^{11}Li)

⟨β-⟩=8580 keV;⟨IB⟩=58 keV

E_{bin}(keV)		⟨ ⟩(keV)	(%)
0 - 10	β-	2.90$\times 10^{-5}$	0.00053
	IB	0.117	
10 - 20	β-	0.000116	0.00076
	IB	0.117	0.81
20 - 40	β-	0.00061	0.00202
	IB	0.23	0.81
40 - 100	β-	0.0066	0.0092
	IB	0.69	1.06
100 - 300	β-	0.128	0.060
	IB	2.3	1.24
300 - 600	β-	0.82	0.177
	IB	3.2	0.75
600 - 1300	β-	8.4	0.85
	IB	6.9	0.77
1300 - 2500	β-	60	3.04
	IB	10.1	0.56
2500 - 5000	β-	570	14.6
	IB	15.7	0.45
5000 - 7500	β-	1479	23.5
	IB	10.0	0.165
7500 - 20730	β-	6462	60
	IB	8.8	0.094

$^{11}_{4}$Be(13.81 *8* s)

Mode: β-, β-α(3.1 *4* %)
Δ: 20174 *6* keV
SpA: 7.24×10^{10} Ci/g
Prod: ^{11}B(n,p); ^{9}Be(t,p)

Delayed Alphas (^{11}Be)

α(keV)
466 *8*
771 *8*

Photons (^{11}Be)

⟨γ⟩=1455 *43* keV

γ_{mode}	γ(keV)	γ(%)†
$\gamma_{\beta\text{-}\alpha}$M1(+0.6%E2)	477.606 *2*	0.39 *4*
$\gamma_{\beta\text{-}}$[M1+E2]	692.33 *10*	0.0344 *11*
$\gamma_{\beta\text{-}}$E1	1771.37 *22*	0.263 *14*
$\gamma_{\beta\text{-}}$M1	2124.49 *3*	35.5
$\gamma_{\beta\text{-}}$M1+0.3%E2	2895.19 *20*	0.081 *3*
$\gamma_{\beta\text{-}}$M1	4443.9 *3*	0.054 *3*
$\gamma_{\beta\text{-}}$E1	4666.06 *22*	1.82 *5*
$\gamma_{\beta\text{-}}$M1	5019.08 *20*	0.467 *16*
$\gamma_{\beta\text{-}}$E1	5851.5 *2*	2.13 *9*
$\gamma_{\beta\text{-}}$E1	6789.59 *22*	4.48 *22*
$\gamma_{\beta\text{-}}$E1	7974.78 *20*	1.90 *14*

† uncert(syst): 5.1% for β-, 13% for β-α

Continuous Radiation (^{11}Be)

⟨β-⟩=4652 keV;⟨IB⟩=25 keV

E_{bin}(keV)		⟨ ⟩(keV)	(%)
0 - 10	β-	0.000408	0.0076
	IB	0.088	
10 - 20	β-	0.00157	0.0103
	IB	0.088	0.61
20 - 40	β-	0.0081	0.0267
	IB	0.17	0.60
40 - 100	β-	0.084	0.117
	IB	0.52	0.79
100 - 300	β-	1.48	0.70
	IB	1.64	0.91
300 - 600	β-	8.1	1.77
	IB	2.3	0.53
600 - 1300	β-	56	5.9
	IB	4.5	0.51
1300 - 2500	β-	252	13.0
	IB	5.9	0.33
2500 - 5000	β-	1297	34.3
	IB	7.1	0.20
5000 - 7500	β-	1914	31.1
	IB	2.5	0.042
7500 - 11506	β-	1123	13.1
	IB	0.40	0.0049

$^{11}_{5}$B (stable)

Δ: 8668.0 *4* keV
%: 80.1 *2*

$^{11}_{6}$C (20.39 *2* min)

Mode: ϵ
Δ: 10650.1 *9* keV
SpA: 8.381×10^{8} Ci/g
Prod: ^{11}B(p,n); ^{10}B(p,γ); ^{10}B(d,n); ^{14}N(p,α)

Continuous Radiation (^{11}C)

⟨β+⟩=385 keV;⟨IB⟩=0.41 keV

E_{bin}(keV)		⟨ ⟩(keV)	(%)
0 - 10	β+	0.0080	0.120
	IB	0.018	
10 - 20	β+	0.0492	0.319
	IB	0.017	0.121
20 - 40	β+	0.321	1.05
	IB	0.033	0.114
40 - 100	β+	3.89	5.4
	IB	0.084	0.131
100 - 300	β+	63	30.2
	IB	0.168	0.099
300 - 600	β+	204	46.5
	IB	0.080	0.020
600 - 1300	β+	114	16.2
	IB	0.0102	0.00143
1300 - 1982	IB	0.00119	8.0×10^{-5}
	Σβ+		99.759

A = 12

NP **A433,** 43 (1985)

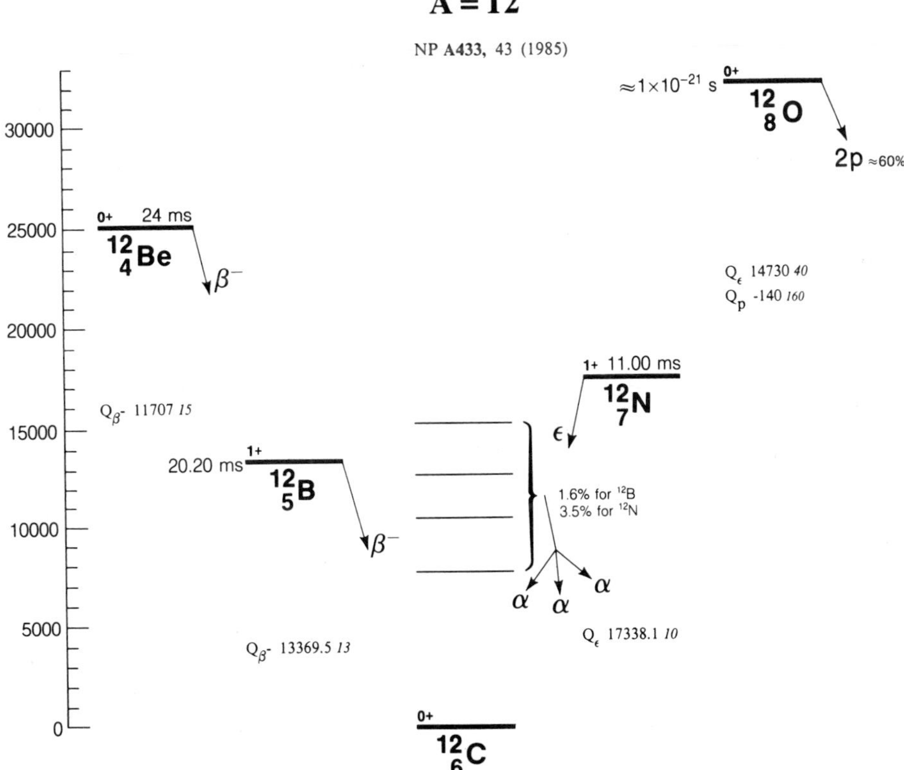

$^{12}_{4}$Be(24 *3* ms)

Mode: β-
Δ: 25077 *15* keV
SpA: 1.36×10^{12} Ci/g
Prod: protons on ^{18}O; protons on ^{15}N; protons on ^{19}F; protons on ^{23}Na; protons on ^{27}Al; protons on ^{16}O

Continuous Radiation (^{12}Be)

⟨β-⟩=5616 keV;⟨IB⟩=32 keV

E_{bin}(keV)		⟨ ⟩(keV)	(%)
0 - 10	β-	5.5×10^{-5}	0.00102
	IB	0.098	
10 - 20	β-	0.000212	0.00139
	IB	0.097	0.68
20 - 40	β-	0.00111	0.00363
	IB	0.19	0.67
40 - 100	β-	0.0119	0.0164
	IB	0.58	0.88
100 - 300	β-	0.234	0.110
	IB	1.9	1.02
300 - 600	β-	1.59	0.341
	IB	2.6	0.60
600 - 1300	β-	18.4	1.84
	IB	5.3	0.60
1300 - 2500	β-	145	7.3
	IB	7.2	0.40
2500 - 5000	β-	1199	31.0
	IB	9.4	0.27
5000 - 7500	β-	2277	36.6
	IB	3.9	0.066
7500 - 11707	β-	1975	22.7
	IB	0.83	0.0099

$^{12}_{5}$B (20.20 *2* ms)

Mode: β-, β-3α(1.58 *30* %)
Δ: 13369.5 *13* keV
SpA: 1.3563×10^{12} Ci/g
Prod: ^{11}B(d,p)

β-3α: 191.5; 0-3000 range (summed energies)

Photons (^{12}B)

⟨γ⟩=57.0 *11* keV

γ_{mode}	γ(keV)	γ(%)
$\gamma_{\beta\text{-}}$E2	3215.0 *11*	0.00065 *18*
$\gamma_{\beta\text{-}}$E2	4438.0 *3*	1.28 *3*

Continuous Radiation (^{12}B)

⟨β-⟩=6355 keV;⟨IB⟩=38 keV

E_{bin}(keV)		⟨ ⟩(keV)	(%)
0 - 10	β-	4.95×10^{-5}	0.00094
	IB	0.103	
10 - 20	β-	0.000186	0.00123
	IB	0.103	0.71
20 - 40	β-	0.00095	0.00313
	IB	0.20	0.71
40 - 100	β-	0.0101	0.0139
	IB	0.61	0.93
100 - 300	β-	0.195	0.092
	IB	2.0	1.08
300 - 600	β-	1.31	0.282
	IB	2.8	0.64
600 - 1300	β-	15.0	1.50
	IB	5.8	0.64
1300 - 2500	β-	116	5.9
	IB	8.0	0.44
2500 - 5000	β-	963	24.9
	IB	11.2	0.32
5000 - 7500	β-	2041	32.6
	IB	5.6	0.093
7500 - 13370	β-	3219	34.8
	IB	2.1	0.025

$^{12}_{6}$C (stable)

Δ: ≡0.0 for ^{12}C scale
%: 98.90 *3*

$^{12}_{7}$N (11.000 *16* ms)

Mode: ϵ, ϵ3α(3.5 *5* %)
Δ: 17338.1 *10* keV
SpA: 1.3563×10^{12} Ci/g
Prod: ^{12}C(p,n); ^{10}B(^{3}He,n)

ϵ3α: 191.5; 0-3500 range (summed energies)

Photons (^{12}N)

⟨γ⟩=96 *5* keV

γ_{mode}	γ(keV)	γ(%)
γ_{ϵ}M1	2399 *5*	~6×10^{-5}
γ_{ϵ}E2	3215.0 *11*	0.0012 *3*
γ_{ϵ}E2	4438.0 *3*	2.02 *9*
γ_{ϵ}[M1]	5055 *5*	~0.0008
γ_{ϵ}M1	10666 *3*	~0.00010
γ_{ϵ}M1	12703 *5*	~0.006
γ_{ϵ}M1	15099 *3*	0.039 *13*

Continuous Radiation (^{12}N)

⟨β+⟩=7729 keV;⟨IB⟩=51 keV

E_{bin}(keV)		⟨ ⟩(keV)	(%)
0 - 10	β+	8.1×10^{-6}	0.000119
	IB	0.111	
10 - 20	β+	5.3×10^{-5}	0.000343
	IB	0.111	0.77
20 - 40	β+	0.000368	0.00119
	IB	0.22	0.76
40 - 100	β+	0.00499	0.0068
	IB	0.66	1.00
100 - 300	β+	0.115	0.054
	IB	2.1	1.17
300 - 600	β+	0.82	0.176
	IB	3.0	0.71
600 - 1300	β+	9.7	0.96
	IB	6.5	0.72
1300 - 2500	β+	76	3.83
	IB	9.4	0.52
2500 - 5000	β+	666	17.1
	IB	14.1	0.40
5000 - 7500	β+	1609	25.6
	IB	8.5	0.141
7500 - 17338	β+	5368	52
	IB	6.1	0.066
	Σβ+		100

$^{12}_{8}$O (~1×10^{-21} s)

Γ: ~400 keV
Mode: 2p(60 *30* % est)
Δ: 32060 *40* keV
Prod: ^{16}O(α,^{8}He); ^{12}C(π^{+},π^{-})

A = 13

NP **A360,** 1 (1981)

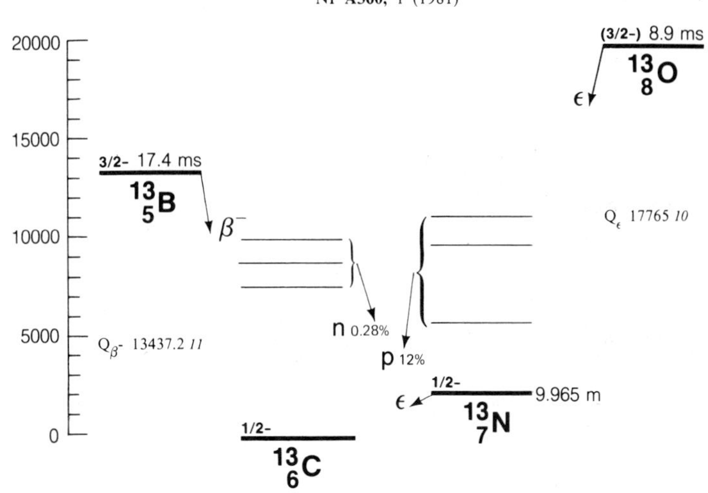

$^{13}_{5}$B (17.36 *16* ms)

Mode: β-, β-n(0.28 *4* %)
Δ: 16562.3 *11* keV
SpA: 1.252×10^{12} Ci/g
Prod: ^{11}B(t,p)

Delayed Neutrons (^{13}B)

n(keV)	n(%)
2401 *3*	0.094 *20*
3613 *20*	0.16 *3*
4570 *5*	0.022 *7*

Photons (^{13}B)

⟨γ⟩=280 *30* keV

γ_{mode}	γ(keV)	γ(%)
γM1	3680 *5*	7.6 *8*

Continuous Radiation (^{13}B)

⟨β-⟩=6321 keV;⟨IB⟩=38 keV

E_{bin}(keV)		⟨ ⟩(keV)	(%)
0 - 10	β-	4.68 ×10^{-5}	0.00089
	IB	0.103	
10 - 20	β-	0.000176	0.00116
	IB	0.102	0.71
20 - 40	β-	0.00090	0.00296
	IB	0.20	0.71
40 - 100	β-	0.0095	0.0132
	IB	0.61	0.92
100 - 300	β-	0.185	0.087
	IB	2.0	1.08
300 - 600	β-	1.26	0.269
	IB	2.8	0.64
600 - 1300	β-	14.5	1.45
	IB	5.8	0.64
1300 - 2500	β-	115	5.8
	IB	8.0	0.44
2500 - 5000	β-	981	25.3
	IB	11.1	0.32
5000 - 7500	β-	2067	33.1
	IB	5.5	0.092
7500 - 13437	β-	3142	34.0
	IB	2.1	0.024

$^{13}_{6}$C (stable)

Δ: 3125.025 *16* keV
%: 1.10 *3*

$^{13}_{7}$N (9.965 *4* min)

Mode: ε
Δ: 5345.5 *3* keV
SpA: 1.4506×10^{9} Ci/g
Prod: ^{10}B(α,n); ^{12}C(d,n); ^{13}C(p,n); ^{12}C(p,γ); ^{16}O(p,α)

Continuous Radiation (^{13}N)

⟨β+⟩=491 keV;⟨IB⟩=0.64 keV

E_{bin}(keV)		⟨ ⟩(keV)	(%)
0 - 10	β+	0.00447	0.065
	IB	0.022	
10 - 20	β+	0.0289	0.187
	IB	0.021	0.149
20 - 40	β+	0.196	0.64
	IB	0.041	0.143
40 - 100	β+	2.48	3.40
	IB	0.109	0.168
100 - 300	β+	44.1	21.1
	IB	0.24	0.141
300 - 600	β+	183	40.7
	IB	0.158	0.039
600 - 1300	β+	262	33.8
	IB	0.040	0.0056
1300 - 2220	IB	0.0020	0.000126
	Σβ+		99.805

$^{13}_{8}$O (8.9 *2* ms)

Mode: ε, εp(12 *3* %)
Δ: 23111 *10* keV
SpA: 1.25×10^{12} Ci/g
Prod: ^{14}N(p,2n)

Delayed Protons (^{13}O)

p(keV)	p(%)†
917 *8*	0.36 *15*
1447 *2*	10.7
2340 *10*	0.14 *3*
2856 *7*	0.046 *16*
3175 *50*	0.032 *11 ?*
3672 *50*	0.14 *7*
5015 *8*	0.035 *11*
5890 *50*	0.049 *11 ?*
6438 *11*	0.37 *3*
6953 *7*	0.9 *1*
7769 *50*	0.05 *3*

† 29% uncert(syst)

Photons (^{13}O)

⟨γ⟩=25 *7* keV

γ_{mode}	γ(keV)	γ(%)
$\gamma_{\epsilon p}$E2	4438.0 *3*	0.56

Continuous Radiation (^{13}O)

⟨β+⟩=7895 keV;⟨IB⟩=52 keV

E_{bin}(keV)		⟨ ⟩(keV)	(%)
0 - 10	β+	5.5 ×10^{-6}	7.9 ×10^{-5}
	IB	0.112	
10 - 20	β+	3.78 ×10^{-5}	0.000244
	IB	0.112	0.77
20 - 40	β+	0.000271	0.00088
	IB	0.22	0.77
40 - 100	β+	0.00378	0.0051
	IB	0.66	1.01
100 - 300	β+	0.089	0.0415
	IB	2.2	1.19
300 - 600	β+	0.66	0.140
	IB	3.1	0.71
600 - 1300	β+	8.0	0.80
	IB	6.6	0.73
1300 - 2500	β+	67	3.38
	IB	9.5	0.52
2500 - 5000	β+	637	16.3
	IB	14.4	0.41
5000 - 7500	β+	1597	25.4
	IB	8.8	0.146
7500 - 16743	β+	5586	54
	IB	6.6	0.071
	Σβ+		100

A = 14

NP **A360**, 56 (1981)

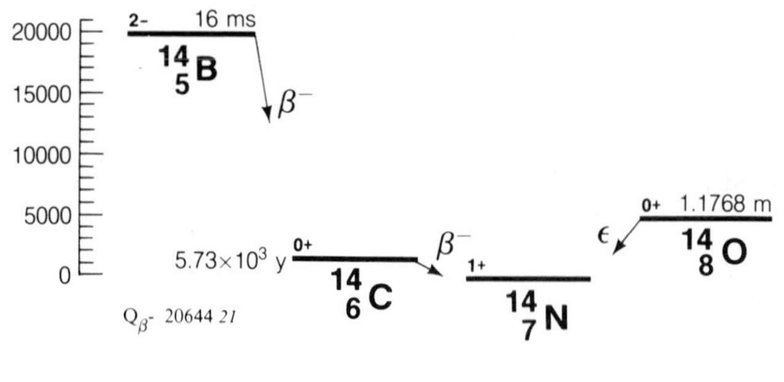

$^{14}_{5}$B (16.1 *12* ms)

Mode: β-
Δ: 23664 *21* keV
SpA: 1.16×10^{12} Ci/g
Prod: ^{10}Be(^{6}Li,2p)

Photons (^{14}B)

⟨γ⟩=6208 *1100* keV

γ_{mode}	γ(keV)	γ(%)†
γ E1+8.5%M2	613.6 *22*	<4
γ [E2]	634.2 *14*	0.7 *3*
γ E1+1.0%M2	1247.8 *22*	<5
γ E1	6094.3 *13*	90.0
γ E3	6728.3 *11*	9.0 *18*
γ [M2]	7341.5 *22*	<2

† 2.0% uncert(syst)

Continuous Radiation (^{14}B)

⟨β-⟩=7116 keV;⟨IB⟩=45 keV

E_{bin}(keV)		⟨ ⟩(keV)	(%)
0 - 10	β-	3.23×10^{-5}	0.00061
	IB	0.108	
10 - 20	β-	0.000121	0.00080
	IB	0.107	0.74
20 - 40	β-	0.00062	0.00204
	IB	0.21	0.74
40 - 100	β-	0.0066	0.0091
	IB	0.64	0.97
100 - 300	β-	0.129	0.060
	IB	2.1	1.14
300 - 600	β-	0.88	0.188
	IB	2.9	0.68
600 - 1300	β-	10.4	1.03
	IB	6.2	0.69
1300 - 2500	β-	85	4.29
	IB	8.8	0.49
2500 - 5000	β-	786	20.2
	IB	12.8	0.37
5000 - 7500	β-	1865	29.7
	IB	7.2	0.119
7500 - 20644	β-	4368	44.5
	IB	4.1	0.045

$^{14}_{6}$C (5730 *40* yr)

Mode: β-
Δ: 3019.910 *25* keV
SpA: 4.46 Ci/g
Prod: ^{14}N(n,p)

Continuous Radiation (^{14}C)

⟨β-⟩=49.5 keV;⟨IB⟩=0.0084 keV

E_{bin}(keV)		⟨ ⟩(keV)	(%)
0 - 10	β-	0.53	10.3
	IB	0.0025	
10 - 20	β-	1.72	11.4
	IB	0.0018	0.0129
20 - 40	β-	6.9	22.9
	IB	0.0023	0.0082
40 - 100	β-	30.7	46.8
	IB	0.0018	0.0032
100 - 156	β-	9.7	8.5
	IB	5.6×10^{-5}	5.1×10^{-5}

$^{14}_{7}$N (stable)

Δ: 2863.436 *24* keV
%: 99.634 *9*

$^{14}_{8}$O (1.1768 *3* min)

Mode: ε
Δ: 8006.56 *8* keV
SpA: 1.1357×10^{10} Ci/g
Prod: ^{14}N(p,n); ^{12}C(^{3}He,n)

Photons (^{14}O)

⟨γ⟩=2299.5 *3* keV

γ_{mode}	γ(keV)	γ(%)
γ [M1]	1634.8 *3*	0.0557 *4*
γ [M1]	2312.69 *7*	99.388 *12*
γ E2+12%M1	3947.2 *3*	0.00226 *2*

Continuous Radiation (^{14}O)

⟨β+⟩=776 keV;⟨IB⟩=1.42 keV

E_{bin}(keV)		⟨ ⟩(keV)	(%)
0 - 10	β+	0.00161	0.0233
	IB	0.031	
10 - 20	β+	0.0110	0.071
	IB	0.031	0.21
20 - 40	β+	0.077	0.251
	IB	0.060	0.21
40 - 100	β+	1.02	1.40
	IB	0.166	0.26
100 - 300	β+	20.4	9.6
	IB	0.43	0.24
300 - 600	β+	109	23.9
	IB	0.39	0.094
600 - 1300	β+	505	55
	IB	0.29	0.037
1300 - 2500	β+	136	9.4
	IB	0.021	0.00135
2500 - 4123	β+	4.49	0.151
	IB	0.00126	4.3×10^{-5}
	Σβ+		99.913

A = 15

NP **A360**, 107 (1981)

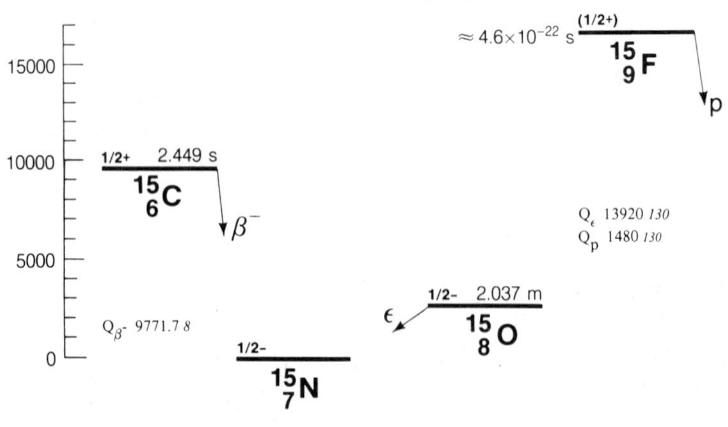

$^{15}_{6}$C (2.449 *5* s)

Mode: β-
Δ: 9873.2 *8* keV
SpA: 2.675×10^{11} Ci/g
Prod: ^{14}C(d,p)

Photons (^{15}C)

⟨γ⟩=3609 *110* keV

γ_{mode}	γ(keV)	γ(%)
γ E1	5297.79 *4*	68 *2*
γ E1+0.03%M2	7299.18 *17*	0.0074 *8*
γ E1	8310.32 *14*	0.032 *4*
γ E1	8568.8 *10*	0.0042 *7*
γ E1	9047.1 *7*	0.031 *3*

Continuous Radiation (^{15}C)

⟨β-⟩=2871 keV;⟨IB⟩=12.6 keV

E_{bin}(keV)		⟨ ⟩(keV)	(%)
0 - 10	β-	0.00064	0.0123
	IB	0.069	
10 - 20	β-	0.00235	0.0155
	IB	0.068	0.47
20 - 40	β-	0.0118	0.0388
	IB	0.136	0.47
40 - 100	β-	0.121	0.167
	IB	0.40	0.61
100 - 300	β-	2.20	1.04
	IB	1.23	0.68
300 - 600	β-	13.6	2.93
	IB	1.61	0.38
600 - 1300	β-	129	13.1
	IB	2.9	0.33
1300 - 2500	β-	647	33.8
	IB	3.1	0.17
2500 - 5000	β-	1181	35.1
	IB	2.4	0.072
5000 - 7500	β-	703	11.5
	IB	0.60	0.0104
7500 - 9772	β-	196	2.41
	IB	0.034	0.00043

$^{15}_{7}$N (stable)

Δ: 101.50 *4* keV
%: 0.366 *9*

$^{15}_{8}$O (2.037 *3* min)

Mode: ϵ
Δ: 2855.5 *5* keV
SpA: 6.135×10^{9} Ci/g
Prod: ^{14}N(d,n); ^{14}N(p,γ); ^{16}O(^{3}He,α); ^{12}C(α,n); ^{16}O(p,d)

Continuous Radiation (^{15}O)

⟨β+⟩=735 keV;⟨IB⟩=1.28 keV

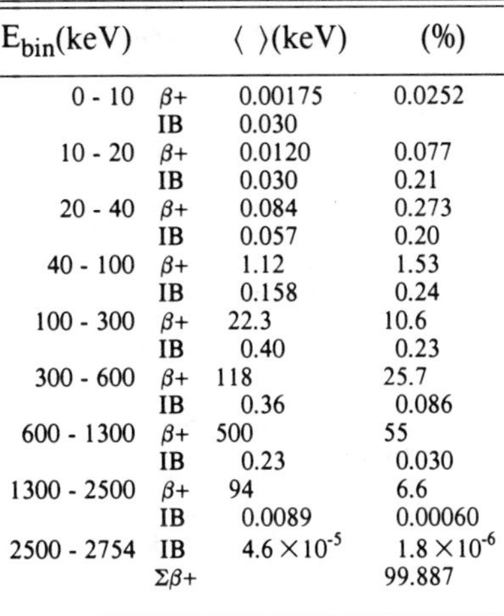

E_{bin}(keV)		⟨ ⟩(keV)	(%)
0 - 10	β+	0.00175	0.0252
	IB	0.030	
10 - 20	β+	0.0120	0.077
	IB	0.030	0.21
20 - 40	β+	0.084	0.273
	IB	0.057	0.20
40 - 100	β+	1.12	1.53
	IB	0.158	0.24
100 - 300	β+	22.3	10.6
	IB	0.40	0.23
300 - 600	β+	118	25.7
	IB	0.36	0.086
600 - 1300	β+	500	55
	IB	0.23	0.030
1300 - 2500	β+	94	6.6
	IB	0.0089	0.00060
2500 - 2754	IB	4.6×10^{-5}	1.8×10^{-6}
	Σβ+		99.887

$^{15}_{9}$F (4.6 *9* ×10^{-22} s)

Γ: 1000 *200* keV
Mode: p
Δ: 16770 *130* keV
Prod: ^{20}Ne(^{3}He,^{8}Li)

p: 1370 *120*

A = 16

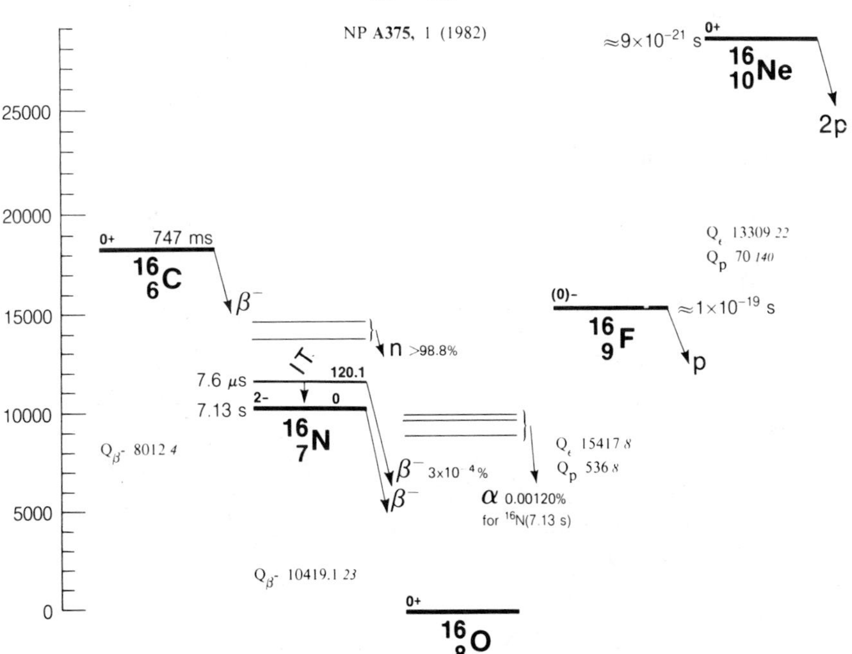

$^{16}_{6}$C (747 *8* ms)

Mode: β-, β-n(>98.8 %)
Δ: 13694 *4* keV
SpA: 6.15×10^{11} Ci/g
Prod: ^{14}C(t,p); ^{14}C(^{18}O,^{16}O); ^{11}B(^{7}Li,2p)

Delayed Neutrons (^{16}C)

n(keV)	n(%)
810 *5*	84.4 *17*
1714 *5*	15.6 *17*

Continuous Radiation (^{16}C)

⟨β-⟩=2049 keV;⟨IB⟩=7.1 keV

E_{bin}(keV)		⟨ ⟩(keV)	(%)
0 - 10	β-	0.00088	0.0167
	IB	0.059	
10 - 20	β-	0.00319	0.0211
	IB	0.059	0.41
20 - 40	β-	0.0160	0.053
	IB	0.116	0.40
40 - 100	β-	0.164	0.228
	IB	0.34	0.52
100 - 300	β-	3.01	1.42
	IB	1.02	0.57
300 - 600	β-	18.7	4.04
	IB	1.28	0.30
600 - 1300	β-	175	17.8
	IB	2.1	0.24
1300 - 2500	β-	835	43.8
	IB	1.7	0.099
2500 - 4657	β-	1017	32.6
	IB	0.40	0.0136

$^{16}_{7}$N (7.13 *2* s)

Mode: β-, β-α(0.00120 *5* %)
Δ: 5682.1 *23* keV
SpA: 9.42×10^{10} Ci/g
Prod: ^{15}N(d,p); ^{16}O(n,p); ^{19}F(n,α); ^{15}N(n,γ)

Delayed Alphas (^{16}N)

α(keV)	α(%)†
1282.4 *5*	4.6 *8* × 10^{-8}
1852 *21*	0.00120 *
2014 *3*	6.5 *11* × 10^{-7}

† 4.2% uncert(syst)
* broad peak

Photons (^{16}N)

⟨γ⟩=4591 *840* keV

$γ_{mode}$	γ(keV)	γ(%)†
γE2	986.39 *14*	0.0033 *8*
γ[M1]	1754.9 *5*	0.10 *3*
γE1	1954.6 *8*	0.034 *10*
γE2+~33%M1	2741.1 *5*	0.89 *18*
γM2	2822.2 *9*	0.0013 *5*
γE3	6129.22 *4*	68.8
γE2	6915.5 *6*	0.034 *7*
γE1	7115.20 *14*	4.7 *3*
γM2	8869.2 *5*	0.08 *3*

† 2.9% uncert(syst)

Continuous Radiation (^{16}N)

⟨β-⟩=2693 keV;⟨IB⟩=11.7 keV

E_{bin}(keV)		⟨ ⟩(keV)	(%)
0 - 10	β-	0.00096	0.0185
	IB	0.066	
10 - 20	β-	0.00340	0.0225
	IB	0.065	0.45
20 - 40	β-	0.0168	0.055
	IB	0.129	0.45
40 - 100	β-	0.169	0.234
	IB	0.38	0.58
100 - 300	β-	3.00	1.42
	IB	1.16	0.65
300 - 600	β-	18.0	3.88
	IB	1.51	0.35
600 - 1300	β-	157	16.0
	IB	2.7	0.30
1300 - 2500	β-	699	36.8
	IB	2.7	0.154
2500 - 5000	β-	933	28.6
	IB	2.2	0.065
5000 - 7500	β-	558	9.0
	IB	0.73	0.0124
7500 - 10419	β-	325	3.88
	IB	0.088	0.00110

$^{16}_{7}$N (7.58 *9* μs)

Mode: IT, β-(0.00033 *8* %)
Δ: 5802.2 *24* keV
SpA: 1.017×10^{12} Ci/g
Prod: ^{15}N(d,p)

Photons (^{16}N)

⟨γ⟩=120 keV

$γ_{mode}$	γ(keV)	γ(%)
γ	120.1 *5*	100

Continuous Radiation (^{16}N)

⟨β-⟩=0.0166 keV;⟨IB⟩=9.1 $\times10^{-5}$ keV

E_{bin}(keV)		⟨ ⟩(keV)	(%)
0 - 10	β-	3.11×10^{-10}	6.0×10^{-9}
	IB	3.1×10^{-7}	
10 - 20	β-	1.11×10^{-9}	7.3×10^{-9}
	IB	3.1×10^{-7}	2.1×10^{-6}
20 - 40	β-	5.5×10^{-9}	1.81×10^{-8}
	IB	6.1×10^{-7}	2.1×10^{-6}
40 - 100	β-	5.6×10^{-8}	7.8×10^{-8}
	IB	1.8×10^{-6}	2.8×10^{-6}
100 - 300	β-	1.06×10^{-6}	4.99×10^{-7}
	IB	5.8×10^{-6}	3.2×10^{-6}
300 - 600	β-	7.1×10^{-6}	1.52×10^{-6}
	IB	8.1×10^{-6}	1.9×10^{-6}
600 - 1300	β-	8.0×10^{-5}	8.0×10^{-6}
	IB	1.64×10^{-5}	1.8×10^{-6}
1300 - 2500	β-	0.00062	3.12×10^{-5}
	IB	2.2×10^{-5}	1.20×10^{-6}
2500 - 5000	β-	0.00470	0.000122
	IB	2.6×10^{-5}	7.6×10^{-7}
5000 - 7500	β-	0.0076	0.000123
	IB	8.7×10^{-6}	1.49×10^{-7}
7500 - 10539	β-	0.00362	4.34×10^{-5}
	IB	9.6×10^{-7}	1.19×10^{-8}

$^{16}_{8}$O (stable)

Δ: -4737.03 *5* keV
%: 99.762 *15*

$^{16}_{9}$F (~1×10^{-19} s)

Γ: 40 *20*
Mode: p
Δ: 10680 *8* keV
Prod: ^{14}N(^{3}He,n); ^{14}N(^{10}B,^{8}Li); ^{16}O(p,n); ^{16}O(^{3}He,t); ^{19}F(^{3}He,^{6}He)

p: 514 *13*

$^{16}_{10}$Ne(~9×10^{-21} s)

Γ: ~50 kev
Mode: 2p
Δ: 23989 *20* keV
Prod: ^{20}Ne(α,^{8}He); ^{16}O($π^{+}$,$π^{-}$)

2p: 1250 (summed energy)

A = 17

NP **A375**, 72 (1982)

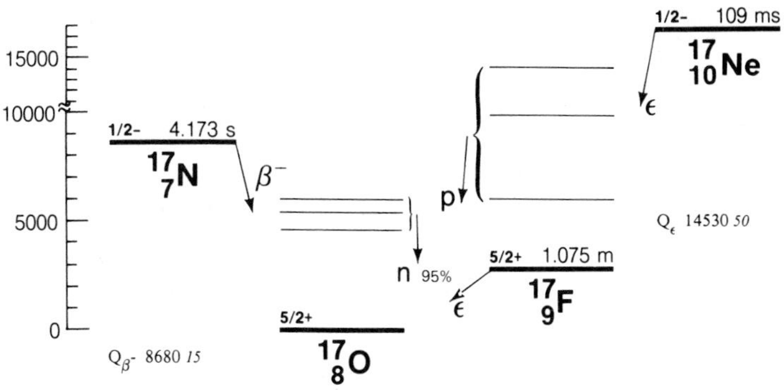

$^{17}_{7}$N (4.173 *4* s)

Mode: β-, β-n(95.1 *7* %)
Δ: 7871 *15* keV
SpA: 1.4652×10^{11} Ci/g
Prod: ^{15}N(t,p); ^{14}C(α,p); ^{17}O(n,p); ^{10}Be(^{11}B,α)

Delayed Neutrons (^{17}N)

n(keV)	n(%)
382.8 *9*	38.0 *13*
884 *21*	~0.6
1170.9 *8*	50.1 *13*
1700.3 *17*	6.9 *5*

Photons (^{17}N)

⟨γ⟩=36.3 *19* keV

γ_{mode}	γ(keV)	γ(%)†
γ_β-E2	870.725 *20*	3.30 *14*
γ_β-E1	2184.2 *2*	0.34

† 15% uncert(syst)

Continuous Radiation (^{17}N)

⟨β-⟩=1707 keV;⟨IB⟩=5.5 keV

E_{bin}(keV)		⟨ ⟩(keV)	(%)
0 - 10	β-	0.00166	0.0321
	IB	0.052	
10 - 20	β-	0.0059	0.039
	IB	0.052	0.36
20 - 40	β-	0.0291	0.096
	IB	0.102	0.36
40 - 100	β-	0.292	0.406
	IB	0.30	0.45
100 - 300	β-	5.2	2.45
	IB	0.88	0.49
300 - 600	β-	30.8	6.7
	IB	1.06	0.25
600 - 1300	β-	258	26.4
	IB	1.62	0.19
1300 - 2500	β-	876	47.3
	IB	1.09	0.063
2500 - 5000	β-	469	15.4
	IB	0.29	0.0092
5000 - 7500	β-	67	1.14
	IB	0.028	0.00050
7500 - 8680	β-	2.28	0.0292
	IB	0.000149	1.8×10^{-6}

$^{17}_{8}$O (stable)

Δ: -809.3 *4* keV
%: 0.038 *3*

$^{17}_{9}$F (1.075 *4* min)

Mode: ε
Δ: 1951.54 *24* keV
SpA: 1.023×10^{10} Ci/g
Prod: ^{16}O(d,n); ^{14}N(α,n)

Continuous Radiation (^{17}F)

⟨β+⟩=739 keV;⟨IB⟩=1.3 keV

E_{bin}(keV)		⟨ ⟩(keV)	(%)
0 - 10	β+	0.00152	0.0217
	IB	0.030	
10 - 20	β+	0.0110	0.071
	IB	0.030	0.21
20 - 40	β+	0.080	0.258
	IB	0.058	0.20
40 - 100	β+	1.09	1.48
	IB	0.159	0.25
100 - 300	β+	22.0	10.4
	IB	0.41	0.23
300 - 600	β+	117	25.5
	IB	0.36	0.087
600 - 1300	β+	502	55
	IB	0.24	0.031
1300 - 2500	β+	97	6.8
	IB	0.0103	0.00068
2500 - 2761	IB	6.4×10^{-5}	2.5×10^{-6}
	Σβ+		99.85

$^{17}_{10}$Ne(109 *1* ms)

Mode: ε, εp
Δ: 16480 *50* keV
SpA: 9.56×10^{11} Ci/g
Prod: ^{19}F(p,3n); ^{16}O(^{3}He,2n)

Delayed Protons (^{17}Ne)

p(keV)	p(%)
1341 *10*	0.034 *2*
1786 *10*	0.37 *4*
2047 *25*	0.14 *1*
2338 *30*	0.48 *7*
3271 *3*	0.31 *5*
3459 *3*	0.17 *5*
3773 *15*	16.2 *7*
4200 *3*	0.16 *3*
4593 *10*	54.0 *7*
5117 *10*	10.6 *2*
5464 *30*	0.35 *10*
6690 *30*	0.18 *5*
7035 *10*	6.80 *11*
1341 *10*	0.034 *2*
7375 *10*	6.14 *25*
7741 *25*	1.76 *6*
9969 *3*	0.07 *2*

Photons (^{17}Ne)

⟨γ⟩=45 *4* keV

γ_{mode}	γ(keV)	γ(%)†
$\gamma_{\epsilon p}$E3	6129.22 *4*	0.161 *15*
$\gamma_{\epsilon p}$E2	6915.5 *6*	0.18 *4*
$\gamma_{\epsilon p}$E1	7115.20 *14*	0.32 *3*

† 11% uncert(syst)

Continuous Radiation (^{17}Ne)

⟨β+⟩=3633 keV;⟨IB⟩=17 keV

E_{bin}(keV)		⟨ ⟩(keV)	(%)
0 - 10	β+	4.59×10^{-5}	0.00065
	IB	0.079	
10 - 20	β+	0.000351	0.00226
	IB	0.079	0.55
20 - 40	β+	0.00267	0.0086
	IB	0.157	0.55
40 - 100	β+	0.0386	0.052
	IB	0.46	0.71
100 - 300	β+	0.90	0.420
	IB	1.46	0.81
300 - 600	β+	6.2	1.33
	IB	2.0	0.46
600 - 1300	β+	64	6.4
	IB	3.8	0.43
1300 - 2500	β+	386	19.9
	IB	4.5	0.25
2500 - 5000	β+	1836	49.7
	IB	4.1	0.122
5000 - 7500	β+	1242	21.1
	IB	0.61	0.0107
7500 - 13508	β+	98	1.16
	IB	0.036	0.00043
	Σβ+		100

A = 18

NP A392, 1 (1983)

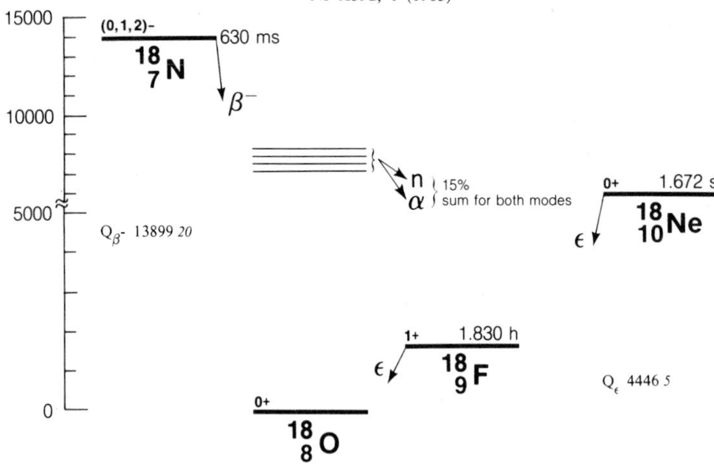

$^{18}_{7}$N (630 *30* ms)

Mode: β-, (β-n+β-α)(15 *6* %)
Δ: 13117 *20* keV
SpA: 6.0×10^{11} Ci/g
Prod: ^{18}O(n,p)

Photons (^{18}N)

⟨γ⟩=5333 *80* keV

γ_{mode}	γ(keV)	γ(%)
$\gamma_{\beta-}$E1	535.11 *5*	2.85 *14*
$\gamma_{\beta-}$E2	821.62 *8*	60.6 *18*
$\gamma_{\beta-}$M1	1074.7 *3*	0.80 *12*
$\gamma_{\beta-}$E1	1177.1 *4*	0.42 *13*
$\gamma_{\beta-}$E2	1571.9 *8*	0.64 *13*
$\gamma_{\beta-}$E1	1609.7 *3*	0.9 *3*
$\gamma_{\beta-}$E2	1651.47 *6*	60.5 *18*
$\gamma_{\beta-}$	1688.2 *9*	0.63 *11*
$\gamma_{\beta-}$	1893.9 *5*	0.37 *6*
$\gamma_{\beta-}$M1+1.6%E2	1937.96 *9*	4.49 *14*
$\gamma_{\beta-}$E2	1981.82 *9*	97.9 *20*
$\gamma_{\beta-}$M1	2424.66 *25*	17.5 *7*
$\gamma_{\beta-}$	2428.9 *5*	1.41 *14*
$\gamma_{\beta-}$	2437.2 *10*	0.61 *10*
$\gamma_{\beta-}$E1	2473.01 *9*	20.4 *10*
$\gamma_{\beta-}$[M1]	2673.2 *4*	1.63 *16*
$\gamma_{\beta-}$E1	3114.9 *4*	0.92 *14*
$\gamma_{\beta-}$[M1]	3315.1 *4*	0.63 *25*
$\gamma_{\beta-}$E1	3547.5 *3*	2.01 *14*
$\gamma_{\beta-}$E2	3919.55 *13*	0.65 *7*
$\gamma_{\beta-}$	4366.6 *5*	0.84 *21*
$\gamma_{\beta-}$[E1]	5787.6 *4*	3.6 *3*
$\gamma_{\beta-}$E1	6197.0 *4*	1.40 *14*
$\gamma_{\beta-}$	6258.3 *9*	0.95 *15* ?
$\gamma_{\beta-}$	7127.7 *9*	2.5 *5*

Continuous Radiation (^{18}N)

⟨β-⟩=3887 keV;⟨IB⟩=19 keV

E_{bin}(keV)		⟨ ⟩(keV)	(%)
0 - 10	β-	0.000259	0.00500
	IB	0.082	
10 - 20	β-	0.00092	0.0061
	IB	0.081	0.56
20 - 40	β-	0.00458	0.0151
	IB	0.162	0.56
40 - 100	β-	0.0465	0.064
	IB	0.48	0.73
100 - 300	β-	0.86	0.406
	IB	1.51	0.84
300 - 600	β-	5.6	1.19
	IB	2.1	0.48
600 - 1300	β-	58	5.8
	IB	4.0	0.45
1300 - 2500	β-	367	18.8
	IB	4.8	0.27
2500 - 5000	β-	1688	45.8
	IB	4.8	0.142
5000 - 7500	β-	1445	24.0
	IB	1.08	0.019
7500 - 11917	β-	323	3.94
	IB	0.070	0.00087

$^{18}_{8}$O (stable)

Δ: -782.2 *9* keV
%: 0.200 *12*

$^{18}_{9}$F (1.8295 *8* h)

Mode: ϵ
Δ: 873.2 *7* keV
SpA: 9.516×10^{7} Ci/g
Prod: ^{18}O(p,n); ^{16}O(t,n); ^{16}O(^{3}He,p); ^{19}F(n,2n); ^{19}F(d,t); ^{20}Ne(d,α)

Continuous Radiation (^{18}F)

⟨β+⟩=242 keV;⟨IB⟩=0.21 keV

E_{bin}(keV)		⟨ ⟩(keV)	(%)
0 - 10	β+	0.0124	0.178
	IB	0.0121	
10 - 20	β+	0.088	0.57
	IB	0.0114	0.079
20 - 40	β+	0.62	2.01
	IB	0.021	0.073
40 - 100	β+	7.6	10.5
	IB	0.049	0.076
100 - 300	β+	100	49.9
	IB	0.072	0.045
300 - 600	β+	133	33.7
	IB	0.019	0.0050
600 - 1300	β+	0.352	0.058
	IB	0.027	0.0031
1300 - 1655	IB	0.0028	0.00020
	Σβ+		96.9

$^{18}_{10}$Ne(1.672 *5* s)

Mode: ϵ
Δ: 5319 *5* keV
SpA: 3.069×10^{11} Ci/g
Prod: ^{16}O(^{3}He,n); ^{19}F(p,2n)

Photons (^{18}Ne)

⟨γ⟩=83.5 *23* keV

γ_{mode}	γ(keV)	γ(%)†
γ M1	658.3 *9*	0.164 *23*
γ M1	1040.2 *9*	7.83
γ E1	1079.44 *10*	0.0013 *3*
γ M1	1698.5 *12*	0.056 *13*

† 2.7% uncert(syst)

Continuous Radiation (^{18}Ne)

⟨β+⟩=1500 keV;⟨IB⟩=4.3 keV

E_{bin}(keV)		⟨ ⟩(keV)	(%)
0 - 10	β+	0.000295	0.00416
	IB	0.049	
10 - 20	β+	0.00226	0.0145
	IB	0.049	0.34
20 - 40	β+	0.0171	0.055
	IB	0.096	0.33
40 - 100	β+	0.244	0.332
	IB	0.28	0.42
100 - 300	β+	5.5	2.58
	IB	0.81	0.45
300 - 600	β+	35.4	7.6
	IB	0.94	0.22
600 - 1300	β+	297	30.4
	IB	1.35	0.156
1300 - 2500	β+	920	50.2
	IB	0.71	0.042
2500 - 3424	β+	242	8.8
	IB	0.026	0.00098
	Σβ+		100

A = 19

NP **A392**, 65 (1983)

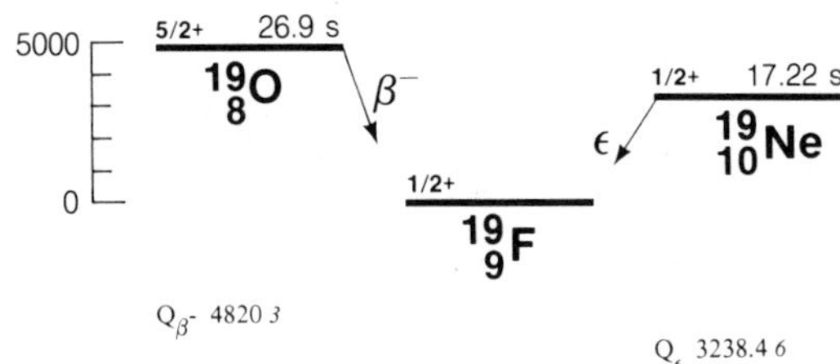

$^{19}_{8}$O (26.91 *8* s)

Mode: β-
Δ: 3332 *3* keV
SpA: 2.178×10^{10} Ci/g
Prod: ^{18}O(n,γ); ^{18}O(d,p)

Photons (^{19}O)

⟨γ⟩=939 *16* keV

γ_{mode}	γ(keV)	γ(%)
γ E1	109.894 *5*	2.54 *10*
γ E2	197.142 *4*	95.9 *21*
γ E1	1148.58 *20*	~0.00050
γ E2	1235.82 *20*	0.017 *2*
γ M1	1356.844 *6*	50.4 *11*
γ E1	1444.085 *6*	2.64 *6*
γ M1	1553.971 *5*	1.39 *3*
γ M1+2.9%E2	1597.780 *23*	0.0192 *5*
γ M1	2353.98 *10*	0.00180 *23*
γ E2	2582.52 *3*	0.0189 *5*
γ M1	3710.64 *10*	0.00110 *15*
γ E1	3797.87 *10*	0.00130 *14*
γ M1	3907.74 *10*	0.00380 *17*
γ M1+2.4%E2	4180.06 *3*	0.0792 *17*

Continuous Radiation (^{19}O)

⟨β-⟩=1742 keV;⟨IB⟩=5.6 keV

E_{bin}(keV)		⟨ ⟩(keV)	(%)
0 - 10	β-	0.00174	0.0338
	IB	0.053	
10 - 20	β-	0.0060	0.0400
	IB	0.053	0.37
20 - 40	β-	0.0292	0.096
	IB	0.104	0.36
40 - 100	β-	0.286	0.398
	IB	0.30	0.46
100 - 300	β-	4.86	2.31
	IB	0.90	0.50
300 - 600	β-	28.4	6.1
	IB	1.09	0.26
600 - 1300	β-	241	24.6
	IB	1.69	0.19
1300 - 2500	β-	874	46.9
	IB	1.17	0.068
2500 - 4710	β-	593	19.4
	IB	0.21	0.0071

$^{19}_{9}$F (stable)

Δ: -1487.40 *14* keV
%: 100

$^{19}_{10}$Ne(17.22 *2* s)

Mode: ϵ
Δ: 1751.0 *6* keV
SpA: 3.380×10^{10} Ci/g
Prod: ^{19}F(p,n)

Photons (^{19}Ne)

⟨γ⟩=0.045 *5* keV

γ_{mode}	γ(keV)	γ(%)
γ E1	109.894 *5*	0.0121 *20*
γ E2	197.142 *4*	0.0021 *3*
γ M1	1356.844 *6*	0.0019 *3*
γ E1	1444.085 *6*	0.000100 *16*
γ M1	1553.971 *5*	5.2 *8* $\times10^{-5}$

Continuous Radiation (^{19}Ne)

⟨β+⟩=963 keV;⟨IB⟩=2.1 keV

E_{bin}(keV)		⟨ ⟩(keV)	(%)
0 - 10	β+	0.00076	0.0107
	IB	0.037	
10 - 20	β+	0.0058	0.0373
	IB	0.036	0.25
20 - 40	β+	0.0437	0.141
	IB	0.071	0.25
40 - 100	β+	0.62	0.84
	IB	0.20	0.31
100 - 300	β+	13.3	6.3
	IB	0.54	0.31
300 - 600	β+	78	16.9
	IB	0.55	0.131
600 - 1300	β+	481	50.9
	IB	0.55	0.066
1300 - 2500	β+	390	25.0
	IB	0.071	0.0049
2500 - 3238	IB	0.00082	3.0×10^{-5}
	Σβ+		99.9

A = 20

NP **A392**, 120 (1983)

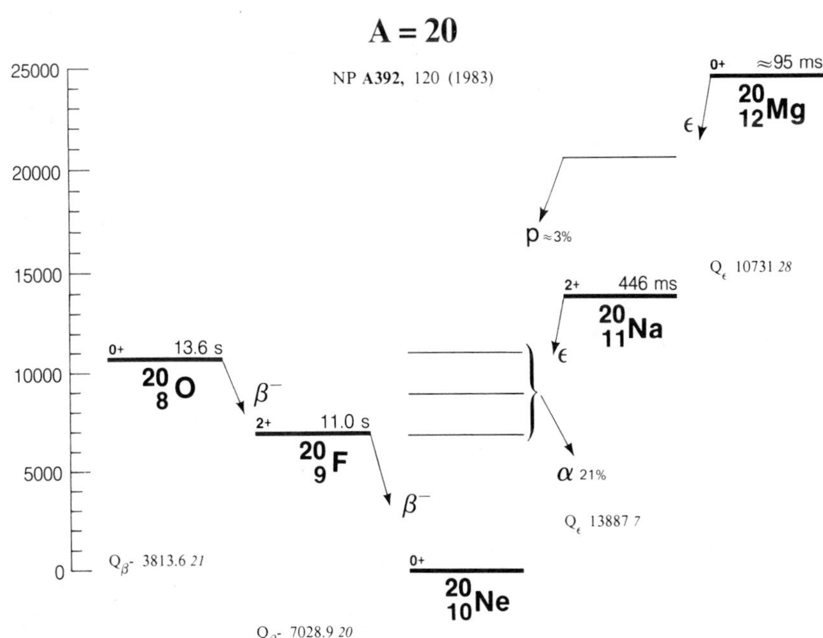

$^{20}_{8}$O (13.57 *10* s)

Mode: β-
Δ: 3796.3 *21* keV
SpA: 4.05×10^{10} Ci/g
Prod: ^{18}O(t,p); ^{18}O(^{18}O,^{16}O)

Photons (^{20}O)

⟨γ⟩=1057 keV

γ$_{mode}$	γ(keV)	γ(%)
γ M1	1056.9 *2*	100

Continuous Radiation (^{20}O)

⟨β-⟩=1198 keV;⟨IB⟩=3.0 keV

E$_{bin}$(keV)		⟨ ⟩(keV)	(%)
0 - 10	β-	0.00331	0.064
	IB	0.042	
10 - 20	β-	0.0115	0.076
	IB	0.042	0.29
20 - 40	β-	0.056	0.184
	IB	0.082	0.29
40 - 100	β-	0.55	0.76
	IB	0.23	0.36
100 - 300	β-	9.4	4.45
	IB	0.66	0.37
300 - 600	β-	53	11.5
	IB	0.73	0.17
600 - 1300	β-	385	39.9
	IB	0.91	0.107
1300 - 2500	β-	738	42.6
	IB	0.26	0.0163
2500 - 2757	β-	12.0	0.467
	IB	0.00019	7.4×10^{-6}

$^{20}_{9}$F (11.03 *6* s)

Mode: β-
Δ: -17.33 *20* keV
SpA: 4.96×10^{10} Ci/g
Prod: ^{19}F(n,γ); ^{19}F(d,p)

Photons (^{20}F)

⟨γ⟩=1634 keV

γ$_{mode}$	γ(keV)	γ(%)
γ E2	1633.602 *15*	100
γ E1+0.6%M2+0.2%E3	3332.54 *19*	0.0090 *4*

Continuous Radiation (^{20}F)

⟨β-⟩=2481 keV;⟨IB⟩=9.6 keV

E$_{bin}$(keV)		⟨ ⟩(keV)	(%)
0 - 10	β-	0.00066	0.0129
	IB	0.066	
10 - 20	β-	0.00226	0.0150
	IB	0.065	0.45
20 - 40	β-	0.0109	0.0358
	IB	0.130	0.45
40 - 100	β-	0.107	0.149
	IB	0.38	0.58
100 - 300	β-	1.92	0.91
	IB	1.17	0.65
300 - 600	β-	12.1	2.59
	IB	1.51	0.35
600 - 1300	β-	120	12.1
	IB	2.6	0.30
1300 - 2500	β-	678	35.1
	IB	2.5	0.145
2500 - 5000	β-	1655	48.8
	IB	1.03	0.034
5000 - 5395	β-	14.0	0.275
	IB	0.00025	4.8×10^{-6}

$^{20}_{10}$Ne(stable)

Δ: -7046.2 *20* keV
%: 90.51 *9*

$^{20}_{11}$Na(446 *3* ms)

Mode: ϵ, ϵα(20.5 *16* %)
Δ: 6841 *7* keV
SpA: 6.42×10^{11} Ci/g
Prod: ^{20}Ne(p,n); ^{12}C(^{10}B,2n); ^{12}C(^{11}B,3n); protons on Mg; protons on Na

Delayed Alphas (^{20}Na)

α(keV)	α(%)
2148 *5*	16.4 *13*
2477 *5*	0.67 *5*
3210 *70*	0.034 *7*
3801 *7*	0.25 *2*
4438 *5*	2.84 *22*
4683 *7*	0.087 *9*
4894 *7*	0.193 *15*
5272 *15*	0.036 *4*
5701 *20*	0.0016 *5*

Photons (^{20}Na)

⟨γ⟩=1339 keV

γ$_{mode}$	γ(keV)	γ(%)†
γ$_ϵ$E2	1633.602 *15*	79.4
γ$_ϵ$[M1+E2]	2452 *5*	~0.0008
γ$_ϵ$M1	2853 *7*	0.0078 *11*
γ$_ϵ$E1+0.6%M2+0.2%E3	3332.54 *19*	~0.008
γ$_ϵ$[E1]	4252 *4*	~0.0016
γ$_ϵ$E1	4653 *4*	0.0024 *4*
γ$_ϵ$E1	5308 *3*	0.00143 *24*
γ$_ϵ$[M1+E2]	5624 *4*	~0.0032
γ$_ϵ$M1	6635 *4*	~0.010
γ$_ϵ$[M1+E2]	8239 *4*	~0.021
γ$_ϵ$M1	8639 *3*	0.100 *13*
γ$_ϵ$M1	9249 *3*	0.03 *1*
γ$_ϵ$[M1+E2]	9626 *4*	0.034 *12*
γ$_ϵ$E2	10272 *3*	0.0073 *18*
γ$_ϵ$M1	11258 *4*	0.169 *22*

† 2.0% uncert(syst) for ϵ

Continuous Radiation (^{20}Na)

⟨β+⟩=4757 keV;⟨IB⟩=26 keV

E$_{bin}$(keV)		⟨ ⟩(keV)	(%)
0 - 10	β+	3.96×10^{-5}	0.00055
	IB	0.089	
10 - 20	β+	0.000319	0.00205
	IB	0.089	0.62
20 - 40	β+	0.00249	0.0080
	IB	0.18	0.61
40 - 100	β+	0.0367	0.0499
	IB	0.52	0.80
100 - 300	β+	0.85	0.396
	IB	1.67	0.92
300 - 600	β+	5.6	1.20
	IB	2.3	0.54
600 - 1300	β+	51	5.2
	IB	4.6	0.52
1300 - 2500	β+	264	13.7
	IB	6.0	0.33
2500 - 5000	β+	1311	35.0
	IB	7.2	0.21
5000 - 7500	β+	1839	29.7
	IB	2.7	0.046
7500 - 11231	β+	1285	15.0
	IB	0.45	0.0055
	Σβ+		100

$^{20}_{12}$Mg(~95 ms)

Mode: ϵ, ϵp(~3 %)
Δ: 17572 *27* keV
SpA: ~8×10^{11} Ci/g
Prod: ^{20}Ne(^{3}He,3n); ^{24}Mg(α,^{8}He)

Delayed Protons (^{20}Mg)

p(keV)	p(%)
3950 *60*	~1
4160 *50*	~2

A = 21

NP **A310**, 15 (1978)

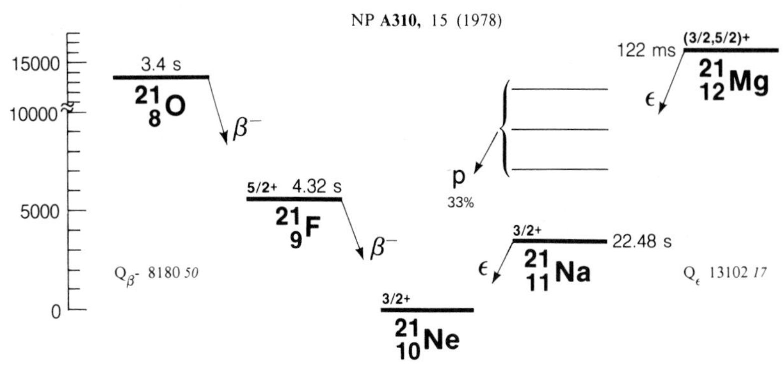

Q_{β^-} 5687 *8*

Q_ϵ 3546.9 *15*

$^{21}_{8}$O (3.42 *10* s)

Mode: β-
Δ: 8130 *50* keV
SpA: 1.42×10^{11} Ci/g
Prod: ^{9}Be(^{18}O,2pα)

Photons (^{21}O)

⟨γ⟩=2947 *53* keV

γ_{mode}	γ(keV)	γ(%)†
γ (E2)	280.05 *8*	14.8 *6*
γ [M1+E2]	933.41 *24*	5.8 *6*
γ [M1+E2]	1450.32 *10*	9.8 *6*
γ [M1+E2]	1729.24 *11*	4.1 *6*
γ [M1+E2]	1730.35 *7*	45.6 *6*
γ [E2]	1754.77 *8*	11.3 *6*
γ [M1+E2]	1787.20 *7*	14.2 *6*
γ [M1+E2]	1884.05 *9*	6.8 *6*
γ [M1+E2]	3179.43 *9*	5.2 *6*
γ [M1+E2]	3459.44 *9*	3.0 *6*
γ [M1+E2]	3517.39 *7*	15.4 *6*
γ [M1+E2]	4571.89 *25*	4.7 *6*
γ [M1+E2]	4583.5 *3*	5.3 *6*

† 10% uncert(syst)

Continuous Radiation (^{21}O)

⟨β-⟩=2317 keV;⟨IB⟩=8.8 keV

E_{bin}(keV)		⟨ ⟩(keV)	(%)
0 - 10	β-	0.00105	0.0204
	IB	0.063	
10 - 20	β-	0.00365	0.0242
	IB	0.062	0.43
20 - 40	β-	0.0178	0.058
	IB	0.123	0.43
40 - 100	β-	0.176	0.244
	IB	0.36	0.55
100 - 300	β-	3.09	1.46
	IB	1.10	0.61
300 - 600	β-	18.5	3.99
	IB	1.40	0.33
600 - 1300	β-	161	16.4
	IB	2.4	0.27
1300 - 2500	β-	701	36.8
	IB	2.2	0.127
2500 - 5000	β-	1304	38.5
	IB	1.03	0.033
5000 - 6450	β-	130	2.41
	IB	0.0165	0.00030

$^{21}_{9}$F (4.32 *3* s)

Mode: β-
Δ: -48 *8* keV
SpA: 1.149×10^{11} Ci/g
Prod: ^{18}O(α,p); ^{19}F(t,p)

Photons (^{21}F)

⟨γ⟩=547 *10* keV

γ_{mode}	γ(keV)	γ(%)†
γ (M1+0.5%E2)	350.725 *8*	90
γ (M1+2.0%E2)	1395.131 *17*	15.3 *3*
γ (E2+0.2%M3)	1745.831 *19*	0.774 *13*
γ	1890.4 *3*	0.0018 *3* ?
γ	2779.43 *20*	0.00159 *15*
γ (E1+0.6%M2)	3533.2 *4*	0.00292 *15*
γ (M1+2.2%E2)	3735.2 *5*	0.00249 *23*
γ (E1+0.2%M2)	3883.9 *4*	0.00096 *13*
γ	4174.36 *20*	0.0319 *6*
γ	4333.57 *18*	0.0475 *12*
γ	4525.01 *20*	0.0095 *3*
γ	4684.22 *18*	0.0280 *10*

† 3.3% uncert(syst)

Continuous Radiation (^{21}F)

⟨β-⟩=2354 keV;⟨IB⟩=8.9 keV

E_{bin}(keV)		⟨ ⟩(keV)	(%)
0 - 10	β-	0.00083	0.0162
	IB	0.064	
10 - 20	β-	0.00284	0.0188
	IB	0.063	0.44
20 - 40	β-	0.0136	0.0449
	IB	0.126	0.44
40 - 100	β-	0.134	0.186
	IB	0.37	0.56
100 - 300	β-	2.36	1.11
	IB	1.13	0.62
300 - 600	β-	14.4	3.11
	IB	1.44	0.34
600 - 1300	β-	137	13.9
	IB	2.5	0.28
1300 - 2500	β-	720	37.4
	IB	2.3	0.132
2500 - 5000	β-	1468	43.8
	IB	0.87	0.029
5000 - 5687	β-	12.5	0.243
	IB	0.00032	6.3×10^{-6}

$^{21}_{10}$Ne(stable)

Δ: -5735.4 *20* keV
%: 0.27 *2*

$^{21}_{11}$Na(22.48 *3* s)

Mode: ε
Δ: -2188.6 *21* keV
SpA: 2.353×10^{10} Ci/g
Prod: ^{24}Mg(p,α); ^{20}Ne(p,γ); ^{20}Ne(d,n)

Photons (^{21}Na)

⟨γ⟩=16.1 *18* keV

γ_{mode}	γ(keV)	γ(%)
γ M1+0.5%E2	350.725 *8*	5.0 *1*

Continuous Radiation (^{21}Na)

⟨β+⟩=1103 keV;⟨IB⟩=2.6 keV

E_{bin}(keV)		⟨ ⟩(keV)	(%)
0 - 10	β+	0.000498	0.0069
	IB	0.040	
10 - 20	β+	0.00401	0.0257
	IB	0.040	0.28
20 - 40	β+	0.0312	0.100
	IB	0.078	0.27
40 - 100	β+	0.454	0.62
	IB	0.22	0.34
100 - 300	β+	10.1	4.74
	IB	0.62	0.35
300 - 600	β+	62	13.3
	IB	0.66	0.156
600 - 1300	β+	433	45.1
	IB	0.76	0.090
1300 - 2500	β+	598	36.0
	IB	0.169	0.0111
2500 - 3547	β+	0.0150	0.00060
	IB	0.0019	6.8×10^{-5}
	Σβ+		99.9

$^{21}_{12}$Mg(122 *3* ms)

Mode: ε, εp(33 *11* %)
Δ: 10914 *16* keV
SpA: 7.72×10^{11} Ci/g
Prod: ^{23}Na(p,3n); ^{20}Ne(^{3}He,2n)

Delayed Protons (^{21}Mg)

p(keV)	p(%)†*
902 *19*	1.73 *6*
1060 *4*	0.45 *7*
1257 *10*	2.43 *21*
1498 *10*	0.66 *6*
1773 *2*	5.4 *3*
1939 *5*	10.4 *5*
2043 *14*	0.87 *10*
2157 *24*	0.49 *4*
2474 *19*	0.80 *21*
2589 *29*	0.31 *11*
3168 *3*	0.34 *3*
3271 *24*	0.62 *4*
3378 *14*	0.47 *6*
3488 *33*	0.14 *1*
3600 *480*	0.14 *2*
3743 *33*	0.38 *5*
3874 *19*	1.07 *6*
4044 *24*	0.26 *2*
4143 *19*	0.10 *2*
4515 *14*	0.26 *5*
4670 *4*	1.56 *8*
5586 *14*	0.05 *1*
5701 *14*	0.08 *1*
6083 *24*	0.06 *1*
6227 *4*	0.57 *4*

† 33% uncert(syst)
* 9% additional intensity unresolved

Photons (^{21}Mg)

⟨γ⟩=427 *33* keV

γ_{mode}	γ(keV)	γ(%)†
γ_ϵM1+0.7%E2	331.93 *10*	51 *5*
γ_ϵM1+2.0%E2	1384.0 *3*	10.1 *19*
$\gamma_{\epsilon p}$E2	1633.602 *15*	7.48 *21*
γ_ϵ[E2]	1715.9 *3*	0.8 *3*
$\gamma_{\epsilon p}$E1+0.6%M2+0.2%E3	3332.54 *19*	0.66 *6*
$\gamma_{\epsilon p}$[E1]	4252 *4*	0.87 *10*
$\gamma_{\epsilon p}$E1	4965.84 *20*	0.0039 *13*

† 34% uncert(syst)

Continuous Radiation (^{21}Mg)

⟨β+⟩=4724 keV;⟨IB⟩=26 keV

E_{bin}(keV)		⟨ ⟩(keV)	(%)
0 - 10	β+	3.25×10^{-5}	0.000449
	IB	0.089	
10 - 20	β+	0.000276	0.00177
	IB	0.089	0.61
20 - 40	β+	0.00223	0.0072
	IB	0.18	0.61
40 - 100	β+	0.0338	0.0459
	IB	0.52	0.80
100 - 300	β+	0.81	0.377
	IB	1.66	0.92
300 - 600	β+	5.5	1.18
	IB	2.3	0.53
600 - 1300	β+	54	5.4
	IB	4.6	0.51
1300 - 2500	β+	276	14.4
	IB	5.9	0.33
2500 - 5000	β+	1335	35.3
	IB	7.1	0.21
5000 - 7500	β+	1761	28.7
	IB	2.6	0.045
7500 - 12080	β+	1292	14.8
	IB	0.54	0.0066
	Σβ+		100

$^{22}_{8}$O (910 *350* ms)

Mode: β-
Δ: 9440 *90* keV
SpA: 3.9×10^{11} Ci/g
Prod: ^{40}Ar on ^{9}Be

$^{22}_{9}$F (4.23 *4* s)

Mode: β-
Δ: 2830 *30* keV
SpA: 1.118×10^{11} Ci/g
Prod: ^{22}Ne(n,p); ^{18}O(^{6}Li,2p); ^{22}Ne on ^{181}Ta

Photons (^{22}F)

⟨γ⟩=5391 *61* keV

γ_{mode}	γ(keV)	γ(%)
γ	823.4 *12*	~0.30
γ E2	1274.53 *2*	100.0
γ	1899.9 *6*	8.7 *4*
γ E2	2082.4 *4*	81.9 *20*
γ [M1+E2]	2165.9 *5*	61.6 *14*
γ (M1)	2283.8 *6*	5.1 *3*
γ [M1+E2]	2987.6 *9*	7.0 *3*
γ	3983.4 *10*	1.2 *2*
γ (E2)	4248.1 *6*	1.0 *2*
γ (M1+2.2%E2)	4366.0 *6*	11.3 *6*

Continuous Radiation (^{22}F)

⟨β-⟩=2358 keV;⟨IB⟩=8.9 keV

E_{bin}(keV)		⟨ ⟩(keV)	(%)
0 - 10	β-	0.00085	0.0166
	IB	0.064	
10 - 20	β-	0.00291	0.0193
	IB	0.063	0.44
20 - 40	β-	0.0140	0.0460
	IB	0.125	0.43
40 - 100	β-	0.137	0.190
	IB	0.37	0.56
100 - 300	β-	2.43	1.15
	IB	1.12	0.62
300 - 600	β-	15.0	3.23
	IB	1.44	0.34
600 - 1300	β-	142	14.4
	IB	2.5	0.28
1300 - 2500	β-	719	37.5
	IB	2.3	0.132
2500 - 5000	β-	1420	42.4
	IB	0.93	0.031
5000 - 7500	β-	55	0.97
	IB	0.022	0.00039
7500 - 9578	β-	3.52	0.0437
	IB	0.00054	6.8×10^{-6}

A = 22

NP **A310**, 38 (1978)

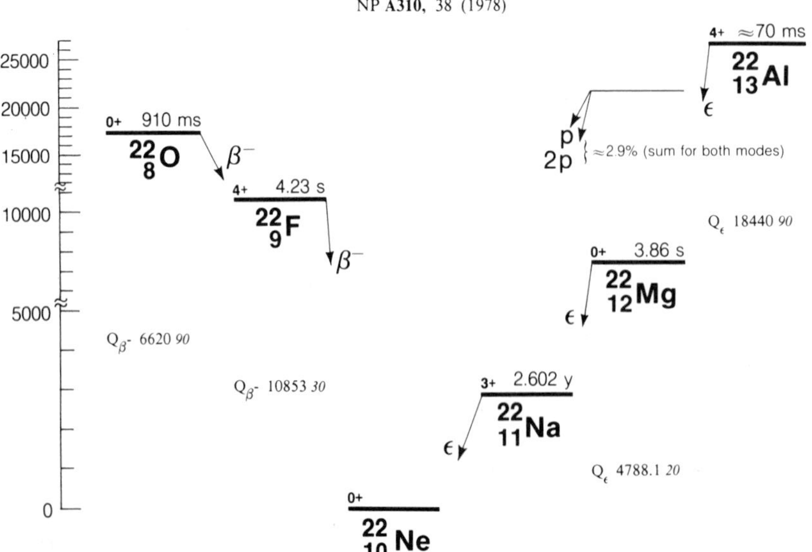

$^{22}_{10}$Ne(stable)

Δ: -8026.6 *16* keV
%: 9.22 *9* (for air only)

$^{22}_{11}$Na(2.602 *2* yr)

Mode: ϵ
Δ: -5184.6 *17* keV
SpA: 6245 Ci/g
Prod: ^{19}F(α,n); ^{24}Mg(d,α)

Photons (^{22}Na)

⟨γ⟩=1273.73 *21* keV

γ_{mode}	γ(keV)	γ(%)
Ne $K_{\alpha 2}$	0.848	0.053 *6*
Ne $K_{\alpha 1}$	0.849	0.106 *11*
γ E2	1274.53 *2*	99.937 *15*

Continuous Radiation (^{22}Na)

⟨β+⟩=195 keV;⟨IB⟩=0.23 keV

E_{bin}(keV)		⟨ ⟩(keV)	(%)
0 - 10	β+	0.0122	0.170
	IB	0.0099	
10 - 20	β+	0.096	0.61
	IB	0.0092	0.064
20 - 40	β+	0.71	2.30
	IB	0.0166	0.058
40 - 100	β+	8.9	12.3
	IB	0.037	0.059
100 - 300	β+	105	53
	IB	0.051	0.032
300 - 600	β+	80	21.6
	IB	0.028	0.0065
600 - 1300	β+	0.280	0.0299
	IB	0.071	0.0081
1300 - 2500	β+	0.123	0.0084
	IB	0.0040	0.00029
2500 - 2842	IB	1.13×10^{-7}	4.4×10^{-9}
	Σβ+		89.4

$^{22}_{12}$Mg(3.857 *9* s)

Mode: ϵ
Δ: -396.6 *15* keV
SpA: 1.217×10^{11} Ci/g
Prod: ^{20}Ne(^{3}He,n)

Photons (^{22}Mg)

⟨γ⟩=700.2 *19* keV

γ_{mode}	γ(keV)	γ(%)
γ (M1)	72.92 *10*	59.5 *13*
γ (E2)	582.0 *7*	100.00
γ (M1)	1278.82 *20*	5.71 *10*
γ [E2]	1933.7 *7*	0.09 *3*

Continuous Radiation (^{22}Mg)

⟨β+⟩=1369 keV;⟨IB⟩=3.7 keV

E_{bin}(keV)		⟨ ⟩(keV)	(%)
0 - 10	β+	0.000289	0.00399
	IB	0.046	
10 - 20	β+	0.00245	0.0157
	IB	0.046	0.32
20 - 40	β+	0.0197	0.063
	IB	0.090	0.31
40 - 100	β+	0.296	0.401
	IB	0.26	0.40
100 - 300	β+	6.8	3.18
	IB	0.75	0.42
300 - 600	β+	42.9	9.3
	IB	0.85	0.20
600 - 1300	β+	335	34.6
	IB	1.15	0.135
1300 - 2500	β+	870	48.2
	IB	0.49	0.031
2500 - 3183	β+	113	4.24
	IB	0.0072	0.00028
	Σβ+		100

$^{22}_{13}$Al(~70 ms)

Mode: ϵ, (ϵp + ϵ2p)(~2.9 %)
Δ: 18040 *90* keV
SpA: ~7×10^{11} Ci/g
Prod: ^{24}Mg(^{3}He,p4n)

ϵp: 7839 *15*, 8149 *21*
ϵ2p: 4139 *20*, 5636 *20* (summed energies)

$^{23}_{9}$F (2.23 *14* s)

Mode: β-
Δ: 3350 *170* keV
SpA: 1.89×10^{11} Ci/g
Prod: ^{10}Be(^{18}O,αp)

Photons (^{23}F)

γ_{mode}	γ(keV)	γ(rel)
γ (M1)	492.5 *5*	11 *3*
γ	613.3 *5*	~0.6
γ	814.60 *18*	25 *5*
γ (E2)	1016.7 *4*	20 *6*
γ (M1)	1116.6 *6*	~0.8
γ (E2)	1298.0 *7*	~2
γ (M1)	1609.1 *4*	~2
γ	1701.34 *14*	100 *5*
γ (M1)	1822.15 *21*	47 *3*
γ	1919.5 *4*	19.3 *25*
γ	2129.2 *4*	68 *11*
γ (M1)	2314.6 *5*	8 *3*
γ (M1)	2414.5 *4*	15 *3*
γ	2515.88 *12*	~6
γ	2734.0 *4*	11.9 *16*
γ (M1)	3431.1 *3*	25.4 *16*
γ	3830.4 *4*	6.8 *9*

A = 23

NP **A310**, 67 (1978)

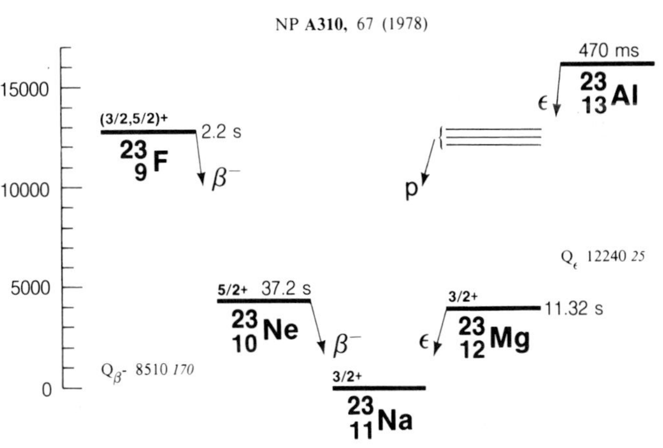

$^{23}_{10}$Ne(37.24 *12* s)

Mode: β-
Δ: -5156 *3* keV
SpA: 1.305×10^{10} Ci/g
Prod: ^{22}Ne(n,γ); ^{22}Ne(d,p); ^{23}Na(n,p)

Photons (^{23}Ne)

⟨γ⟩=165 *5* keV

γ_{mode}	γ(keV)	γ(%)†
γ (M1+0.3%E2)	439.85 *13*	32.9
γ (M1+3.5%E2)	1636.5 *3*	0.99 *3*
γ (E2+3.0%M3)	2076.3 *3*	0.102 *7*
γ (M1+0.9%E2)	2542.45 *19*	0.0270 *20*
γ (M1)	2982.25 *17*	0.0378 *20*

† 3.0% uncert(syst)

Continuous Radiation (^{23}Ne)

⟨β-⟩=1902 keV;⟨IB⟩=6.3 keV

E_{bin}(keV)		⟨ ⟩(keV)	(%)
0 - 10	β-	0.00139	0.0271
	IB	0.057	
10 - 20	β-	0.00466	0.0308
	IB	0.056	0.39
20 - 40	β-	0.0220	0.073
	IB	0.111	0.39
40 - 100	β-	0.213	0.296
	IB	0.32	0.49
100 - 300	β-	3.69	1.75
	IB	0.97	0.54
300 - 600	β-	22.2	4.80
	IB	1.20	0.28
600 - 1300	β-	200	20.3
	IB	1.9	0.22
1300 - 2500	β-	878	46.3
	IB	1.45	0.084
2500 - 4376	β-	798	26.5
	IB	0.25	0.0079

$^{23}_{11}$Na(stable)

Δ: -9531.4 *9* keV
%: 100

$^{23}_{12}$Mg(11.317 *11* s)

Mode: ε
Δ: -5473.1 *15* keV
SpA: 4.204×10^{10} Ci/g
Prod: ^{23}Na(p,n)

Photons (^{23}Mg)

⟨γ⟩=36.2 *13* keV

γ_{mode}	γ(keV)	γ(%)
γ (M1+0.3%E2)	439.85 *13*	8.2 *3*
γ (E2)	1951.04 *22*	0.0024 *3*
γ (M1)	2390.85 *22*	0.0045 *6*

Continuous Radiation (^{23}Mg)

⟨β+⟩=1337 keV;⟨IB⟩=3.5 keV

E_{bin}(keV)		⟨ ⟩(keV)	(%)
0 - 10	β+	0.000283	0.0039
	IB	0.046	
10 - 20	β+	0.00240	0.0154
	IB	0.045	0.31
20 - 40	β+	0.0193	0.062
	IB	0.089	0.31
40 - 100	β+	0.290	0.394
	IB	0.25	0.39
100 - 300	β+	6.7	3.14
	IB	0.73	0.41
300 - 600	β+	43.0	9.3
	IB	0.83	0.20
600 - 1300	β+	347	35.8
	IB	1.11	0.130
1300 - 2500	β+	866	48.5
	IB	0.43	0.027
2500 - 4058	β+	73	2.77
	IB	0.0087	0.00032
	Σβ+		99.9

$^{23}_{13}$Al(470 *30* ms)

Mode: ε, εp
Δ: 6767 *25* keV
SpA: 5.5×10^{11} Ci/g
Prod: ^{24}Mg(p,2n)

εp: 832 *29*

$^{24}_{10}$Ne(3.38 *2* min)

Mode: β-
Δ: -5950 *10* keV
SpA: 2.314×10^{9} Ci/g
Prod: ^{22}Ne(t,p)

Photons (^{24}Ne)

⟨γ⟩=541 keV

γ_{mode}	γ(keV)	γ(%)
γ (M3)	472.28 *9*	100 *
γ [M1+E2]	874.35 *14*	7.9 *2*

* with ^{24}Na(20.2 ms) in equilib

Continuous Radiation (^{24}Ne)

⟨β-⟩=803 keV;⟨IB⟩=1.54 keV

E_{bin}(keV)		⟨ ⟩(keV)	(%)
0 - 10	β-	0.0094	0.184
	IB	0.032	
10 - 20	β-	0.0313	0.208
	IB	0.031	0.22
20 - 40	β-	0.147	0.484
	IB	0.061	0.21
40 - 100	β-	1.37	1.91
	IB	0.170	0.26
100 - 300	β-	21.2	10.2
	IB	0.44	0.25
300 - 600	β-	102	22.3
	IB	0.42	0.100
600 - 1300	β-	466	50.5
	IB	0.35	0.043
1300 - 1998	β-	213	14.3
	IB	0.023	0.00166

A = 24

NP **A310**, 96 (1978)

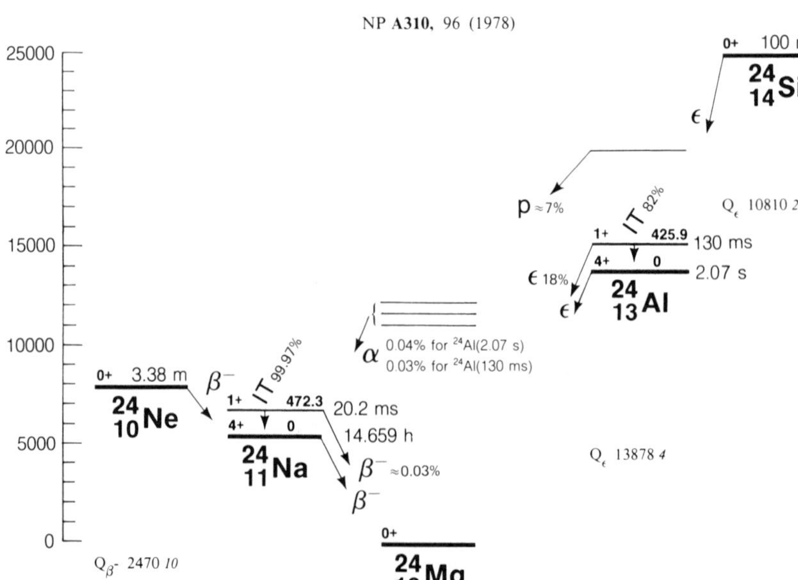

$^{24}_{11}$Na(14.659 *4* h)

Mode: β-
Δ: -8419.5 *9* keV
SpA: 8.9074×10^{6} Ci/g
Prod: ^{23}Na(n,γ)

Photons (^{24}Na)

⟨γ⟩=4123 keV

γ_{mode}	γ(keV)	γ(%)
γ[M1+E2]	996.78 *6*	0.00130 *12*
γ E2	1368.598 *6*	100
γ (E2)	2753.995 *14*	99.944 *5*
γ (E2+0.3%M1)	2869.48 *4*	0.00025 *4*
γ (E2+0.4%M1)	3866.13 *5*	0.051 *4*
γ E2	4237.90 *4*	0.00083 *12*

Continuous Radiation (^{24}Na)

⟨β-⟩=554 keV;⟨IB⟩=0.80 keV

E_{bin}(keV)		⟨ ⟩(keV)	(%)
0 - 10	β-	0.0183	0.360
	IB	0.024	
10 - 20	β-	0.060	0.399
	IB	0.024	0.164
20 - 40	β-	0.277	0.91
	IB	0.045	0.157
40 - 100	β-	2.52	3.52
	IB	0.122	0.19
100 - 300	β-	36.5	17.6
	IB	0.29	0.166
300 - 600	β-	154	34.1
	IB	0.21	0.052
600 - 1300	β-	359	42.9
	IB	0.082	0.0112
1300 - 2500	β-	1.74	0.131
	IB	5.0×10^{-5}	3.0×10^{-6}
2500 - 4145	β-	0.0221	0.00074
	IB	5.9×10^{-6}	2.0×10^{-7}

$^{24}_{11}$Na(20.18 *10* ms)

Mode: IT(99.97 %), β-(~0.03 %)
Δ: -7947.2 *9* keV
SpA: 6.78×10^{11} Ci/g
Prod: daughter ^{24}Ne; ^{23}Na(n,γ); ^{23}Na(d,p); ^{27}Al(n,α); ^{22}Ne(^{3}He,p)

Photons (^{24}Na)

⟨γ⟩=472.00 keV

γ_{mode}	γ(keV)	γ(%)
γ (M3)	472.28 *9*	99.97

Continuous Radiation (^{24}Na)

⟨β-⟩=0.83 keV;⟨IB⟩=0.0034 keV

E_{bin}(keV)		⟨ ⟩(keV)	(%)
0 - 10	β-	1.71×10^{-7}	3.35×10^{-6}
	IB	2.1×10^{-5}	
10 - 20	β-	5.7×10^{-7}	3.75×10^{-6}
	IB	2.1×10^{-5}	0.000144
20 - 40	β-	2.66×10^{-6}	8.8×10^{-6}
	IB	4.1×10^{-5}	0.000143
40 - 100	β-	2.55×10^{-5}	3.55×10^{-5}
	IB	0.000121	0.00019
100 - 300	β-	0.000450	0.000212
	IB	0.00038	0.00021
300 - 600	β-	0.00283	0.00061
	IB	0.00049	0.000115
600 - 1300	β-	0.0287	0.00290
	IB	0.00089	0.000101
1300 - 2500	β-	0.174	0.0090
	IB	0.00092	5.2×10^{-5}
2500 - 5000	β-	0.58	0.0164
	IB	0.00050	1.59×10^{-5}
5000 - 5986	β-	0.0437	0.00083
	IB	3.0×10^{-6}	5.7×10^{-8}

$^{24}_{12}$Mg(stable)

Δ: -13933.1 *7* keV
%: 78.99 *3*

$^{24}_{13}$Al(2.066 *10* s)

Mode: ϵ, ϵα(0.035 *6* %)
Δ: -55 *4* keV
SpA: 1.933×10^{11} Ci/g
Prod: ^{24}Mg(p,n)

Delayed Alphas (^{24}Al)

α(keV)	α(%)
1587 *5*	0.0072 *20*
1982 *5*	0.026 *6*
2280 *10*	0.00011 *3*
2337 *10*	0.0010 *3*
2369 *10*	0.00075 *20*
3040 *15*	0.00002 *1*

Photons (^{24}Al)

⟨γ⟩=8475 *150* keV

γ_{mode}	γ(keV)	γ(%)
γ	587.91 *13*	0.149 *8*
γ[M1+E2]	775.14 *6*	0.053 *8*
γ	822.83 *11*	0.021 *8*
γ	861.77 *8*	~0.022 ?
γ	909.05 *13*	0.122 *6*
γ[M1+E2]	996.78 *6*	0.136 *7*
γ	1059.71 *6*	0.285 *17*
γ	1076.83 *4*	14.8 *3*
γ	1090.62 *9*	0.140 *7*
γ	1274.77 *8*	<0.11
γE2	1368.598 *6*	96.0 *25*
γ	1434.1 *4*	0.063 *6*
γ	1704.0 *6*	0.016 *4*
γ(E2)	1771.88 *5*	0.40 *1*
γ[M1+E2]	1887.37 *5*	<0.06

Photons (^{24}Al) (continued)

γ_{mode}	γ(keV)	γ(%)
γ[E1]	1899.62 *11*	0.82 *2*
γ	1952.35 *11*	0.094 *6*
γ	2127.5 *4*	0.054 *7*
γ	2136.49 *7*	0.168 *9*
γ[E1]	2381.14 *11*	0.037 *10*
γ	2428.83 *6*	0.774 *18*
γ	2566.9 *4*	0.065 *7*
γ	2576.7 *6*	0.030 *12* ?
γ(E2)	2753.995 *14*	41.2 *9*
γ(E2+0.3%M1)	2869.48 *4*	1.10 *3*
γ	3203.88 *5*	3.09 *7*
γ[E1]	3377.81 *11*	0.043 *7*
γ[E1]	3493.29 *10*	0.04 *1*
γ[M1+E2]	3505.54 *5*	1.98 *6*
γ(E2+0.4%M1)	3866.13 *5*	5.31 *21*
γ	4200.52 *5*	4.02 *22*
γE2	4237.90 *4*	3.62 *20*
γ[M1+E2]	4280.56 *6*	0.66 *4*
γ	4315.99 *5*	14.2 *9*
γ(E2)	4641.13 *5*	3.42 *25*
γ	5062.12 *9*	0.036 *13*
γ	5177.59 *8*	0.98 *10*
γ	5340.06 *8*	0.115 *13*
γ[M1+E2]	5392.62 *5*	18.3 *18*
γ[M1+E2]	5979.13 *10*	0.093 *9*
γ[E1]	6246.86 *10*	0.54 *4*
γ	7069.46 *5*	43.0 *13*
γE2	7347.36 *10*	0.153 *16*
γE3	7615.07 *10*	0.224 *15*
γ	7930.95 *8*	1.34 *10*
γ[E2]	8145.95 *5*	0.028 *7*
γ	9450.1 *4*	0.11 *2*
γ	9943.4 *15*	0.027 *6*

Continuous Radiation (^{24}Al)

⟨β+⟩=2027 keV;⟨IB⟩=7.2 keV

E_{bin}(keV)		⟨ ⟩(keV)	(%)
0 - 10	β+	0.000126	0.00172
	IB	0.058	
10 - 20	β+	0.00112	0.0072
	IB	0.057	0.40
20 - 40	β+	0.0093	0.0300
	IB	0.114	0.39
40 - 100	β+	0.145	0.197
	IB	0.33	0.51
100 - 300	β+	3.49	1.63
	IB	1.00	0.55
300 - 600	β+	23.4	5.03
	IB	1.24	0.29
600 - 1300	β+	213	21.7
	IB	2.0	0.23
1300 - 2500	β+	844	45.0
	IB	1.64	0.094
2500 - 5000	β+	765	24.1
	IB	0.69	0.021
5000 - 7500	β+	166	2.79
	IB	0.083	0.00147
7500 - 9755	β+	11.8	0.151
	IB	0.00084	1.01×10^{-5}
	Σβ+		99.95

$^{24}_{13}$Al(130 *4* ms)

Mode: IT(82 *3* %), ϵ(18 *3* %), ϵα(0.028 *6* %)
Δ: 371 *4* keV
SpA: 6.75×10^{11} Ci/g
Prod: ^{24}Mg(p,n)

Delayed Alphas (^{24}Al)

α(keV)	α(%)
1140 *10*	0.0009 *3*
1339 *10*	0.00009 *4*
1419 *5*	0.016 *5*
1785 *10*	0.009 *3*
1837 *10*	0.0017 *6*
2572 *10*	0.00008 *3*
2620 *10*	0.00004 *2*

Photons (^{24}Al)

⟨γ⟩=687 *32* keV

γ_{mode}	γ(keV)	γ(%)†
γ_{IT}[M3]	425.90 *7*	82 *3*
γ_ϵE2	1368.598 *6*	5.3 *10*
γ_ϵ(E2+0.3%M1)	2869.48 *4*	0.09 *3*
γ_ϵE2	4237.90 *4*	0.30 *9*
γ_ϵ[M1+E2]	8595.0 *11*	0.6 *1*
γ_ϵ	8688.6 *25*	~0.2
γ_ϵ	9825.9 *20*	~0.2
γ_ϵM1	9963.1 *11*	1.6 *2*

† uncert(syst): 29% for ε, 2.2% for IT

Continuous Radiation (^{24}Al)

⟨β+⟩=956 keV;⟨IB⟩=5.6 keV

E_{bin}(keV)		⟨ ⟩(keV)	(%)
0 - 10	β+	5.8×10^{-6}	7.9×10^{-5}
	IB	0.0164	
10 - 20	β+	5.1×10^{-5}	0.000329
	IB	0.0163	0.113
20 - 40	β+	0.000428	0.00138
	IB	0.033	0.113
40 - 100	β+	0.0067	0.0090
	IB	0.096	0.147
100 - 300	β+	0.160	0.075
	IB	0.31	0.17
300 - 600	β+	1.08	0.232
	IB	0.43	0.100
600 - 1300	β+	9.9	1.01
	IB	0.88	0.098
1300 - 2500	β+	42.5	2.24
	IB	1.19	0.066
2500 - 5000	β+	159	4.16
	IB	1.60	0.046
5000 - 7500	β+	319	5.1
	IB	0.76	0.0128
7500 - 14304	β+	425	4.66
	IB	0.26	0.0030
	Σβ+		18

$^{24}_{14}$Si(100 *42* ms)

Mode: ε, εp(~7 %)
Δ: 10755 *19* keV
SpA: 7×10^{11} Ci/g
Prod: ^{24}Mg(^{3}He,3n)

εp: 3912.9 *37*

Continuous Radiation (^{25}Ne)

⟨β-⟩=3177 keV;⟨IB⟩=14.0 keV

E_{bin}(keV)		⟨ ⟩(keV)	(%)
0 - 10	β-	0.000419	0.0082
	IB	0.075	
10 - 20	β-	0.00141	0.0093
	IB	0.074	0.52
20 - 40	β-	0.0067	0.0221
	IB	0.148	0.51
40 - 100	β-	0.065	0.090
	IB	0.43	0.67
100 - 300	β-	1.16	0.55
	IB	1.36	0.75
300 - 600	β-	7.4	1.58
	IB	1.8	0.42
600 - 1300	β-	75	7.6
	IB	3.4	0.38
1300 - 2500	β-	469	24.1
	IB	3.8	0.21
2500 - 5000	β-	2008	55
	IB	2.8	0.085
5000 - 7110	β-	616	11.1
	IB	0.132	0.0024

A = 25

NP **A310,** 127 (1978)

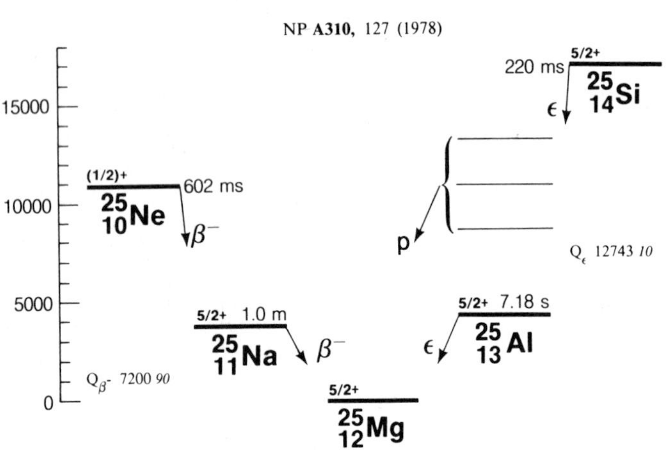

$^{25}_{10}$Ne(602 *8* ms)

Mode: β-
Δ: -2160 *100* keV
SpA: 4.45×10^{11} Ci/g
Prod: ^{22}Ne on ^{232}Th; ^{9}Be(^{18}O,2p)

Photons (^{25}Ne)

⟨γ⟩=391 *24* keV

γ_{mode}	γ(keV)	γ(%)
γ (M1)	89.53 *10*	95.6 *6*
γ (M1+E2)	979.77 *16*	18.2 *19*
γ (E2)	1069.30 *19*	2.3 *4*
γ (M1)	1132.8 *10*	~0.4
γ [M1+E2]	2112.5 *10*	0.62 *19*
γ (M1+E2)	2202 *1*	1.1 *3*
γ [M1]	3220 *3*	0.53 *15* ?
γ	3598 *3*	~0.22
γ	3688 *3*	0.96 *24*

$^{25}_{11}$Na(59.6 *7* s)

Mode: β-
Δ: -9359 *6* keV
SpA: 7.53×10^9 Ci/g
Prod: ^{25}Mg(n,p); ^{23}Na(t,p)

Photons (^{25}Na)

⟨γ⟩=435 *3* keV

γ_{mode}	γ(keV)	γ(%)†
γ (M1+1.7%E2)	389.698 *22*	12.65 *22*
γ (E2)	585.032 *21*	12.96 *18*
γ (M1+0.2%E2)	836.83 *3*	0.104 *3*
γ (M1+11%E2)	974.720 *23*	14.89 *22*
γ (M1+5.9%E2)	989.846 *22*	0.166 *3*
γ (E2)	1379.528 *21*	0.231 *5*
γ (M1+3.5%E2)	1611.703 *11*	9.45 *14*
γ (M1+26%E2)	1964.525 *23*	0.1462 *22*
γ (M1+E2)	2216.31 *3*	0.0932 *21*
γ (M1+29%E2)	2801.28 *3*	0.0492 *16*

† 5.3% uncert(syst)

Continuous Radiation (^{25}Na)

⟨β-⟩=1505 keV;⟨IB⟩=4.4 keV

E_{bin}(keV)		⟨ ⟩(keV)	(%)
0 - 10	β-	0.00286	0.056
	IB	0.049	
10 - 20	β-	0.0095	0.063
	IB	0.048	0.33
20 - 40	β-	0.0441	0.146
	IB	0.095	0.33
40 - 100	β-	0.416	0.58
	IB	0.27	0.42
100 - 300	β-	6.9	3.28
	IB	0.80	0.45
300 - 600	β-	38.8	8.4
	IB	0.94	0.22
600 - 1300	β-	294	30.3
	IB	1.36	0.157
1300 - 2500	β-	835	45.7
	IB	0.77	0.046
2500 - 3834	β-	330	11.5
	IB	0.064	0.0023

$^{25}_{12}$Mg(stable)

Δ: -13192.5 *7* keV
%: 10.00 *1*

$^{25}_{13}$Al(7.183 *12* s)

Mode: ε
Δ: -8915.4 *11* keV
SpA: 5.989×10^{10} Ci/g
Prod: ^{24}Mg(p,γ); ^{25}Mg(p,n); ^{24}Mg(d,n); ^{28}Si(p,α)

Photons (^{25}Al)

⟨γ⟩=13.1 *3* keV

γ_{mode}	γ(keV)	γ(%)†
γ (M1+1.7%E2)	389.698 *22*	0.0202 *21*
γ (E2)	585.032 *21*	0.023 *3*
γ (M1+11%E2)	974.720 *23*	0.0237 *24*
γ (M1+3.5%E2)	1611.703 *11*	0.788

† 2.7% uncert(syst)

Continuous Radiation (^{25}Al)

⟨β+⟩=1453 keV;⟨IB⟩=4.1 keV

E_{bin}(keV)		⟨ ⟩(keV)	(%)
0 - 10	β+	0.000207	0.00284
	IB	0.048	
10 - 20	β+	0.00185	0.0118
	IB	0.048	0.33
20 - 40	β+	0.0154	0.0495
	IB	0.094	0.33
40 - 100	β+	0.238	0.322
	IB	0.27	0.41
100 - 300	β+	5.6	2.63
	IB	0.79	0.44
300 - 600	β+	36.7	7.9
	IB	0.92	0.22
600 - 1300	β+	309	31.7
	IB	1.28	0.149
1300 - 2500	β+	931	51
	IB	0.62	0.038
2500 - 4277	β+	171	6.3
	IB	0.020	0.00072
	Σβ+		100

$^{25}_{14}$Si(220 *3* ms)

Mode: ε, εp
Δ: 3827 *10* keV
SpA: 6.23×10^{11} Ci/g
Prod: ^{27}Al(p,3n); ^{24}Mg(^{3}He,2n)

Delayed Protons (^{25}Si)

p(keV)	p(rel)
790 *50*	14
930 *50*	23
1730 *30*	11
1870 *30*	40
2090 *30*	24
2220 *30*	25
2900 *30*	10
3130 *40*	15
3330 *30*	39
3480 *40*	11
3700 *30*	8
4080 *20*	100
4360 *30*	7
4630 *30*	17
4750 *50*	2
5100 *30*	6
5350 *40*	20
5660 *50*	2

A = 26

NP **A310**, 156 (1978)

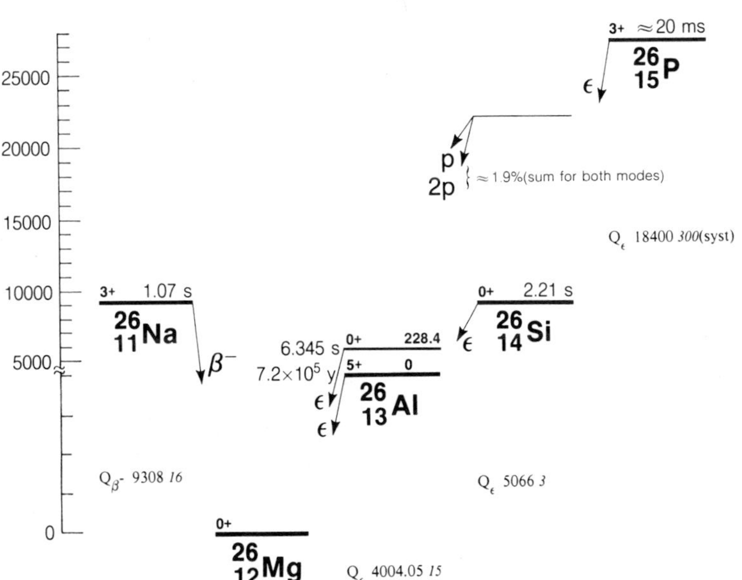

$^{26}_{11}$Na(1.072 *9* s)

Mode: β-
Δ: -6906 *16* keV
SpA: 2.981×10^{11} Ci/g
Prod: ^{26}Mg(n,p); protons on U; ^{10}B(^{18}O,2p); ^{18}O(^{13}C,pα)

Photons (^{26}Na)

⟨γ⟩=2170 *14* keV

γ_{mode}	γ(keV)	γ(%)†
γ (M1)	815.3 *16*	0.042 *15*
γ (M1+0.4%E2)	1003.0 *11*	0.8 *3*
γ (M1+1.5%E2)	1129.66 *10*	5.8 *3*
γ (M1+2.9%E2)	1365.4 *14*	0.5 *1*
γ [M1+E2]	1393.6 *12*	0.23 *5*
γ (M1+9.0%E2)	1411.7 *11*	3.2 *4*
γ	1774.1 *18*	1.6 *3*
γ E2	1808.65 *7*	98.9
γ (M1)	1896.5 *12*	1.98 *15*
γ (E2)	1961.8 *15*	~0.030
γ (M1)	2132.6 *11*	0.59 *10*
γ	2182.8 *18*	0.18 *4*
γ (E2)	2509.8 *13*	0.54 *14*
γ [M1+E2]	2523.2 *12*	1.32 *12*
γ (M1+1.1%E2)	2541.3 *11*	2.43 *16*
γ (E2)	2777.0 *21*	0.24 *5*
γ E2	2938.22 *12*	0.58 *7*
γ [M1+E2]	3026.1 *12*	0.13 *3*
γ (E2)	3091.4 *15*	0.32 *10*
γ E2	4331.7 *12*	0.19 *4*
γ E2	4834.5 *12*	0.35 *7*

† 0.30% uncert(syst)

Continuous Radiation (^{26}Na)

⟨β-⟩=3344 keV;⟨IB⟩=15.1 keV

E_{bin}(keV)		⟨ ⟩(keV)	(%)
0 - 10	β-	0.000432	0.0085
	IB	0.077	
10 - 20	β-	0.00143	0.0095
	IB	0.076	0.53
20 - 40	β-	0.0067	0.0222
	IB	0.151	0.52
40 - 100	β-	0.064	0.090
	IB	0.45	0.68
100 - 300	β-	1.13	0.54
	IB	1.40	0.77
300 - 600	β-	7.1	1.53
	IB	1.9	0.44
600 - 1300	β-	72	7.2
	IB	3.5	0.40
1300 - 2500	β-	437	22.5
	IB	4.1	0.23
2500 - 5000	β-	1923	52
	IB	3.3	0.099
5000 - 7499	β-	904	15.9
	IB	0.26	0.0047

$^{26}_{12}$Mg(stable)

Δ: -16214.0 *8* keV
%: 11.01 *2*

$^{26}_{13}$Al(7.2 *3* ×10^5 yr)

Mode: ε
Δ: -12209.9 *8* keV
SpA: 0.0191 Ci/g
Prod: ^{26}Mg(p,n); ^{25}Mg(d,n); ^{28}Si(d,α)

Photons (^{26}Al)

⟨γ⟩=1839.6 *24* keV

γ_{mode}	γ(keV)	γ(%)†
Mg $K_{\alpha2}$	1.254	0.040 *4*
Mg $K_{\alpha1}$	1.254	0.080 *8*
γ (M1+1.5%E2)	1129.66 *10*	2.47 *18*
γ E2	1808.65 *7*	99.76
γ E2	2938.22 *12*	0.25 *3*

†<0.1% uncert(syst)

Continuous Radiation (^{26}Al)

⟨β+⟩=400 keV;⟨IB⟩=0.96 keV

E_{bin}(keV)		⟨ ⟩(keV)	(%)
0 - 10	β+	0.00177	0.0243
	IB	0.018	
10 - 20	β+	0.0157	0.100
	IB	0.018	0.122
20 - 40	β+	0.128	0.413
	IB	0.034	0.117
40 - 100	β+	1.88	2.57
	IB	0.089	0.138
100 - 300	β+	36.8	17.5
	IB	0.21	0.119
300 - 600	β+	155	34.5
	IB	0.18	0.042
600 - 1300	β+	207	27.0
	IB	0.27	0.030
1300 - 2195	IB	0.150	0.0097
	Σβ+		82

$^{26}_{13}$Al(6.345 *3* s)

Mode: ε
Δ: -11981.5 *8* keV
SpA: 6.478×10^{10} Ci/g
Prod: ^{23}Na(α,n); ^{26}Mg(p,n)

Continuous Radiation (^{26}Al)

⟨β+⟩=1439 keV;⟨IB⟩=4.0 keV

E_{bin}(keV)		⟨ ⟩(keV)	(%)
0 - 10	β+	0.000207	0.00284
	IB	0.048	
10 - 20	β+	0.00185	0.0118
	IB	0.047	0.33
20 - 40	β+	0.0154	0.0495
	IB	0.093	0.32
40 - 100	β+	0.238	0.323
	IB	0.27	0.41
100 - 300	β+	5.7	2.64
	IB	0.78	0.44
300 - 600	β+	37.0	8.0
	IB	0.91	0.21
600 - 1300	β+	314	32.2
	IB	1.26	0.147
1300 - 2500	β+	932	51
	IB	0.59	0.035
2500 - 4232	β+	151	5.6
	IB	0.0169	0.00062
	Σβ+		99.9

$^{26}_{14}$Si(2.210 *21* s)

Mode: ε
Δ: -7144 *3* keV
SpA: 1.685×10^{11} Ci/g
Prod: ^{24}Mg(^{3}He,n)

Photons (^{26}Si)

⟨γ⟩=232.7 *14* keV

γ_{mode}	γ(keV)	γ(%)†
γ	416.8 *3*	<0.08
γ	829.4 *8*	21.89 *11*
γ	1622.3 *10*	2.73 *5*
γ	1654.7 *7*	0.032 *7*
γ	1844.2 *20*	0.258 *6*
γ	2510.8 *7*	0.0617 *22*

† 1.9% uncert(syst)

Continuous Radiation (^{26}Si)

⟨β+⟩=1619 keV;⟨IB⟩=4.9 keV

E_{bin}(keV)		⟨ ⟩(keV)	(%)
0 - 10	β+	0.000150	0.00204
	IB	0.051	
10 - 20	β+	0.00141	0.0090
	IB	0.051	0.35
20 - 40	β+	0.0121	0.0390
	IB	0.100	0.35
40 - 100	β+	0.193	0.261
	IB	0.29	0.44
100 - 300	β+	4.68	2.18
	IB	0.86	0.48
300 - 600	β+	30.9	6.7
	IB	1.02	0.24
600 - 1300	β+	267	27.4
	IB	1.52	0.18
1300 - 2500	β+	915	49.4
	IB	0.92	0.055
2500 - 4838	β+	402	14.0
	IB	0.082	0.0030
	Σβ+		99.9

$^{26}_{15}$P (~20 ms)

Mode: ε, (εp + ε2p)(~1.9 %)
Δ: 11260 *300* keV syst
SpA: ~6×10^{11} Ci/g
Prod: ^{28}Si(^{3}He,4np)

εp: 6827 *50*, 7269 *15*
ε2p: 4920 *20* (summed energy)

A = 27

NP **A310**, 183 (1978)

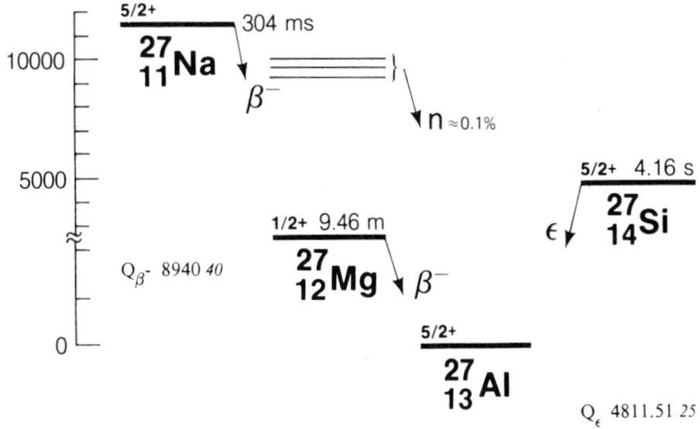

Q_{β^-} 2610.6 *6*

$^{27}_{11}$Na(304 *7* ms)

Mode: β-, β-n(~0.1 %)
Δ: -5650 *40* keV
SpA: 5.41×10^{11} Ci/g
Prod: protons on U; ^{11}B(^{18}O,2p); protons on In

Photons (^{27}Na)

⟨γ⟩=1134 *21* keV

γ_{mode}	γ(keV)	γ(%)†
$\gamma_{\beta-}$[M1+E2]	955.38 *9*	0.87 *9*
$\gamma_{\beta-}$[M1+4.6%E2]	984.70 *7*	87.4
$\gamma_{\beta-}$[M1+E2]	1169.44 *24*	0.59 *6*
$\gamma_{\beta-}$	1666.8 *6*	0.064 *13*
$\gamma_{\beta-}$(E2)	1698.04 *10*	11.9 *6*
$\gamma_{\beta-}$[M1+E2]	1729.0 *4*	0.32 *4*
$\gamma_{\beta-}$[M1+E2]	1792.8 *4*	0.15 *4*
$\gamma_{\beta-n}$ E2	1808.65 *7*	0.13 *4*
$\gamma_{\beta-}$(E2)	1940.04 *8*	0.46 *5*
$\gamma_{\beta-}$	2442.3 *4*	0.42 *6*
$\gamma_{\beta-}$[M1+E2]	2451.8 *4*	0.026 *9*
$\gamma_{\beta-}$[M1+E2]	2506.1 *4*	0.25 *4*
$\gamma_{\beta-}$[M1+E2]	2612.8 *6*	0.17 *5*
$\gamma_{\beta-}$	2836.1 *6*	0.10 *4*
$\gamma_{\beta-}$	3490.7 *4*	0.12 *4*
$\gamma_{\beta-}$[M1+E2]	4007.6 *9*	0.18 *4*

† 2.3% uncert(syst)

Continuous Radiation (^{27}Na)

⟨β-⟩=3626 keV;⟨IB⟩=17 keV

E_{bin}(keV)		⟨ ⟩(keV)	(%)
0 - 10	β-	0.000386	0.0076
	IB	0.080	
10 - 20	β-	0.00128	0.0085
	IB	0.079	0.55
20 - 40	β-	0.0060	0.0198
	IB	0.158	0.55
40 - 100	β-	0.057	0.080
	IB	0.47	0.71
100 - 300	β-	0.99	0.470
	IB	1.47	0.81
300 - 600	β-	6.1	1.31
	IB	2.0	0.46
600 - 1300	β-	59	5.9
	IB	3.8	0.43
1300 - 2500	β-	363	18.6
	IB	4.5	0.25
2500 - 5000	β-	1945	52
	IB	4.0	0.121
5000 - 7500	β-	1243	21.4
	IB	0.44	0.0080
7500 - 7955	β-	9.1	0.120
	IB	0.000139	1.8×10^{-6}

$^{27}_{12}$Mg(9.462 *11* min)

Mode: β-
Δ: -14586.2 *9* keV
SpA: 7.355×10^{8} Ci/g
Prod: ^{26}Mg(n,γ); ^{26}Mg(d,p); protons on In

Photons (^{27}Mg)

⟨γ⟩=913 *45* keV

γ_{mode}	γ(keV)	γ(%)†
γ (M1)	170.677 *14*	0.79 *4*
γ (E2)	843.757 *22*	73
γ (M1+11%E2)	1014.429 *22*	29.1 *8*

† 1.4% uncert(syst)

Continuous Radiation (^{27}Mg)

⟨β-⟩=702 keV;⟨IB⟩=1.21 keV

E_{bin}(keV)		⟨ ⟩(keV)	(%)
0 - 10	β-	0.0123	0.243
	IB	0.029	
10 - 20	β-	0.0401	0.266
	IB	0.028	0.20
20 - 40	β-	0.183	0.61
	IB	0.055	0.19
40 - 100	β-	1.67	2.34
	IB	0.151	0.23
100 - 300	β-	25.3	12.2
	IB	0.38	0.22
300 - 600	β-	119	26.2
	IB	0.34	0.081
600 - 1300	β-	471	52
	IB	0.22	0.028
1300 - 1767	β-	84	5.9
	IB	0.0049	0.00036

$^{27}_{13}$Al(stable)

Δ: -17196.8 *7* keV
%: 100

$^{27}_{14}$Si(4.16 *2* s)

Mode: ϵ
Δ: -12385.3 *7* keV
SpA: 9.25×10^{10} Ci/g
Prod: ^{27}Al(p,n)

Photons (^{27}Si)

⟨γ⟩=5.2 *8* keV

γ_{mode}	γ(keV)	γ(%)†
γ (M1)	170.677 *14*	0.00047 *7* ?
γ (E2)	843.757 *22*	~0.00050
γ (M1+11%E2)	1014.429 *22*	0.0172 *24*
γ (M1+1.3%E2)	1719.5 *6*	0.0122 *16*
γ (M1+18%E2)	2210.5 *6*	0.18
γ (M1+3.4%E2)	2733.9 *6*	0.0033 *9*
γ (M1)	2981.1 *9*	0.0259 *14*

† 7.2% uncert(syst)

Continuous Radiation (^{27}Si)

⟨β+⟩=1716 keV;⟨IB⟩=5.3 keV

E_{bin}(keV)		⟨ ⟩(keV)	(%)
0 - 10	β+	0.000122	0.00166
	IB	0.053	
10 - 20	β+	0.00115	0.0073
	IB	0.053	0.37
20 - 40	β+	0.0099	0.0317
	IB	0.104	0.36
40 - 100	β+	0.157	0.213
	IB	0.30	0.46
100 - 300	β+	3.85	1.79
	IB	0.90	0.50
300 - 600	β+	25.9	5.6
	IB	1.09	0.25
600 - 1300	β+	238	24.2
	IB	1.67	0.19
1300 - 2500	β+	951	50.7
	IB	1.08	0.064
2500 - 4812	β+	498	17.4
	IB	0.100	0.0036
	Σβ+		99.9

$^{28}_{11}$Na(30.5 *4* ms)

Mode: β-, β-n(0.58 *12* %)
Δ: -1140 *140* keV
SpA: 5.81×10^{11} Ci/g
Prod: protons on U; protons on In

Photons (^{28}Na)

⟨γ⟩=1135 *86* keV

γ_{mode}	γ(keV)	γ(%)†
$\gamma_{\beta-}$E2	1473.4 *6*	37
$\gamma_{\beta-}$E2	2389.3 *8*	18.6 *10*
$\gamma_{\beta-}$[M1+E2]	3083.6 *6*	1.30 *19*
$\gamma_{\beta-}$	3087.4 *9*	2.6 *3*
$\gamma_{\beta-}$[E1]	5271.7 *10*	0.48 *11*

† 14% uncert(syst)

Continuous Radiation (^{28}Na)

⟨β-⟩=6104 keV;⟨IB⟩=36 keV

E_{bin}(keV)		⟨ ⟩(keV)	(%)
0 - 10	β-	8.2×10^{-5}	0.00160
	IB	0.101	
10 - 20	β-	0.000271	0.00180
	IB	0.101	0.70
20 - 40	β-	0.00127	0.00420
	IB	0.20	0.69
40 - 100	β-	0.0123	0.0171
	IB	0.59	0.91
100 - 300	β-	0.223	0.105
	IB	1.9	1.06
300 - 600	β-	1.47	0.315
	IB	2.7	0.63
600 - 1300	β-	16.7	1.67
	IB	5.6	0.63
1300 - 2500	β-	130	6.6
	IB	7.7	0.43
2500 - 5000	β-	1068	27.6
	IB	10.5	0.30
5000 - 7500	β-	2070	33.3
	IB	5.0	0.084
7500 - 13880	β-	2818	30.4
	IB	1.9	0.022

$^{28}_{12}$Mg(20.90 *3* h)

Mode: β-
Δ: -15018.8 *21* keV
SpA: 5.355×10^{6} Ci/g
Prod: ^{26}Mg(t,p); ^{26}Mg(α,2p); ^{30}Si(p,3p)

Photons (^{28}Mg)

⟨γ⟩=1374 *79* keV

γ_{mode}	γ(keV)	γ(%)†
γ M1	30.6382 *7*	95 *1*
γ	400.57 *3*	36 *4*
γ	647.98 *9*	0.085 *9*
γ	941.77 *5*	36 *4*
γ M1+2.6%E2	1342.33 *6*	54 *5*
γ	1372.96 *6*	4.7 *5*
γ M1+3.8%E2	1589.72 *8*	4.7 *5*

† 1.0% uncert(syst)

A = 28

NP **A310**, 208 (1978)

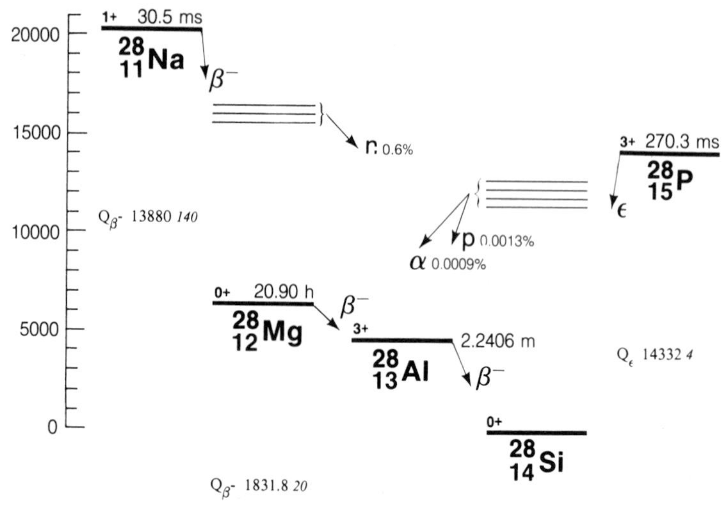

Continuous Radiation (^{28}Mg)

⟨β-⟩=152 keV;⟨IB⟩=0.076 keV

E_{bin}(keV)		⟨ ⟩(keV)	(%)
0 - 10	β-	0.157	3.12
	IB	0.0078	
10 - 20	β-	0.492	3.28
	IB	0.0071	0.049
20 - 40	β-	2.11	7.0
	IB	0.0122	0.043
40 - 100	β-	15.7	22.5
	IB	0.025	0.040
100 - 300	β-	102	55
	IB	0.023	0.0157
300 - 600	β-	30.6	8.9
	IB	0.00082	0.00025
600 - 860	β-	0.167	0.0249
	IB	6.0×10^{-6}	9.4×10^{-7}

$^{28}_{13}$Al(2.2406 *5* min)

Mode: β-
Δ: -16850.6 *7* keV
SpA: 2.9894×10^{9} Ci/g
Prod: ^{27}Al(n,γ); daughter ^{28}Mg; ^{27}Al(d,p)

Photons (^{28}Al)

⟨γ⟩=1779 *18* keV

γ_{mode}	γ(keV)	γ(%)
γ [E2]	1778.988 *9*	100

Continuous Radiation (^{28}Al)

⟨β-⟩=1242 keV;⟨IB⟩=3.2 keV

E_{bin}(keV)		⟨ ⟩(keV)	(%)
0 - 10	β-	0.00403	0.080
	IB	0.043	
10 - 20	β-	0.0130	0.087
	IB	0.043	0.30
20 - 40	β-	0.060	0.197
	IB	0.084	0.29
40 - 100	β-	0.55	0.77
	IB	0.24	0.37
100 - 300	β-	8.9	4.25
	IB	0.69	0.38
300 - 600	β-	49.8	10.8
	IB	0.76	0.18
600 - 1300	β-	367	38.0
	IB	0.98	0.114
1300 - 2500	β-	787	44.8
	IB	0.31	0.020
2500 - 2863	β-	29.1	1.12
	IB	0.00078	3.1×10^{-5}

$^{28}_{14}$Si(stable)

Δ: -21492.4 *6* keV
%: 92.23 *1*

$^{28}_{15}$P (270.3 *5* ms)

Mode: ε, εp(0.0013 *4* %), εα(0.00086 *25* %)
Δ: -7161 *4* keV
SpA: 5.3655×10^{11} Ci/g
Prod: ^{28}Si(p,n)

Delayed Protons (^{28}P)

p(keV)	p(%)
468 *10*	0.00008 *3*
680 *7*	0.00066 *19*
826 *7*	0.00007 *3*
956 *5*	0.00033 *10*
1087 *10*	0.00004 *2*
1269 *5*	0.00015 *4*

Delayed Alphas (^{28}P)

α(keV)	α(%)
1312 *5*	0.000074 *26*
1434 *5*	0.00023 *7*
1668 *5*	0.000021 *11*
1789 *5*	0.000047 *18*
1976 *5*	0.000046 *18*
2105 *5*	0.00031 *7*
2200 *5*	0.000063 *23*
2351 *5*	0.000018 *10*
2497 *5*	0.000036 *15*
2665 *7*	0.000014 *7*

Photons (^{28}P)

⟨γ⟩=3697 *48* keV

γ_{mode}	γ(keV)	γ(%)†
γ	1352.20 *15*	0.140 *8*
γ	1516.52 *16*	0.204 *14*
γ	1522.79 *9*	~0.9
γ	1657.1 *3*	0.40 *17*
γ	1658.82 *19*	~0.8
γ [E2]	1778.988 *9*	97.5
γ	2312.30 *14*	0.196 *6*
γ [E2]	2838.36 *14*	2.35 *5*
γ	2953.37 *24*	0.081 *3*
γ	3039.22 *15*	2.70 *7*
γ	3104.6 *3*	0.024 *4*
γ	3181.51 *19*	0.032 *3*
γ	3251.9 *5*	0.032 *4*
γ	3278.8 *4*	0.098 *4*
γ	3315.81 *24*	0.101 *6*
γ	3970.97 *22*	0.116 *24*
γ M1+2.0%E2	4497.00 *14*	11.0 *3*
γ	5108.8 *5*	0.050 *3*
γ M1+E2	6019.52 *15*	1.75 *8*
γ	6153.81 *22*	0.112 *10*
γ	6479.2 *4*	0.385 *8*
γ	6808.90 *18*	3.33 *11*
γ	7414.9 *5*	0.21 *6*
γ M1	7535.69 *18*	8.5 *3*
γ M1+1.0%E2	7601.0 *3*	0.55 *3*
γ [E2]	7932.38 *22*	2.15 *11*
γ	8015.3 *10*	0.040 *6*
γ	8257.8 *4*	0.052 *6*
γ	8428.4 *15*	0.029 *4*
γ	8887.50 *23*	0.086 *8*
γ	9379.5 *3*	0.0202 *25*
γ [E2]	9477.4 *9*	<0.11
γ	9793.8 *10*	0.013 *3* ?

† 1.9% uncert(syst)

Continuous Radiation (^{28}P)

⟨β+⟩=4591 keV;⟨IB⟩=25 keV

E_{bin}(keV)		⟨ ⟩(keV)	(%)
0 - 10	β+	2.17×10^{-5}	0.000292
	IB	0.087	
10 - 20	β+	0.000214	0.00136
	IB	0.087	0.60
20 - 40	β+	0.00191	0.0061
	IB	0.17	0.60
40 - 100	β+	0.0314	0.0425
	IB	0.51	0.78
100 - 300	β+	0.80	0.372
	IB	1.63	0.90
300 - 600	β+	5.7	1.21
	IB	2.2	0.52
600 - 1300	β+	58	5.9
	IB	4.5	0.50
1300 - 2500	β+	324	16.8
	IB	5.7	0.32
2500 - 5000	β+	1254	33.9
	IB	6.8	0.20
5000 - 7500	β+	1667	27.0
	IB	2.6	0.044
7500 - 11531	β+	1282	14.8
	IB	0.50	0.0061
	Σβ+		100

A = 29

NP **A310**, 243 (1978)

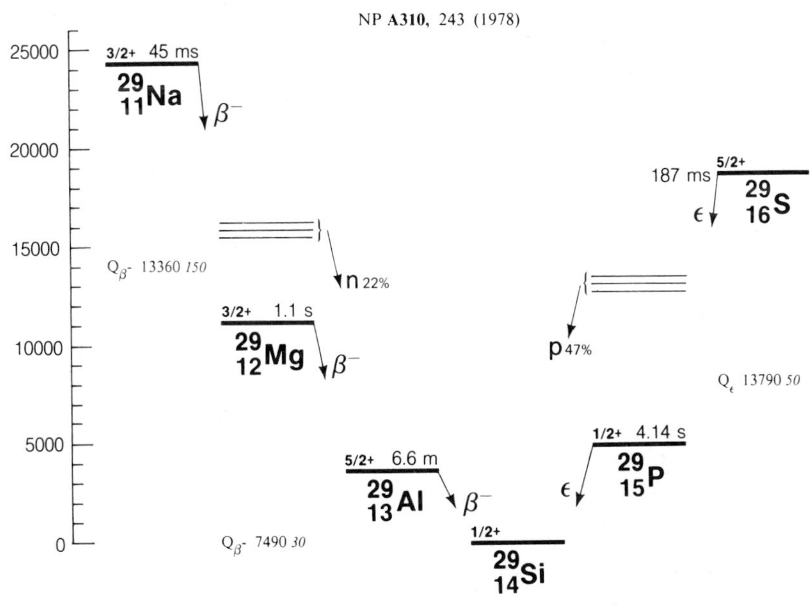

$^{29}_{11}$Na(44.9 *12* ms)

Mode: β-, β-n(22 *3* %)
Δ: 2640 *150* keV
SpA: 5.61×10^{11} Ci/g
Prod: protons on U; protons on In

Photons (^{29}Na)

⟨γ⟩=1823 *230* keV

γ_{mode}	γ(keV)	γ(%)†
$\gamma_{\beta-}$	1041.2 *8*	1.98 *18*
$\gamma_{\beta\text{-}n}$ E2	1473.4 *6*	8.7 *14*
$\gamma_{\beta-}$	1585.5 *6*	6.3 *5*
$\gamma_{\beta-}$	1638.0 *6*	6.3 *5*
$\gamma_{\beta-}$	2129.5 *10*	1.4 *3*
$\gamma_{\beta-}$	2146.3 *12*	0.7 *3*
$\gamma_{\beta-}$	2560.5 *6*	36
$\gamma_{\beta-}$	2614.6 *9*	2.41 *25*
$\gamma_{\beta-}$	2658.1 *15*	0.63 *23*
$\gamma_{\beta-}$	3170.6 *11*	3.6 *4*
$\gamma_{\beta-}$	3170.7 *15*	3.6 *4*
$\gamma_{\beta-}$	3227.5 *12*	4.1 *4*
$\gamma_{\beta-}$	3677.0 *15*	1.17 *23*
$\gamma_{\beta-}$	3992.3 *9*	0.54 *18*

† 25% uncert(syst)

Continuous Radiation (^{29}Na)

⟨β-⟩=4860 keV;⟨IB⟩=27 keV

E_{bin}(keV)		⟨ ⟩(keV)	(%)
0 - 10	β-	0.000197	0.00387
	IB	0.090	
10 - 20	β-	0.00066	0.00434
	IB	0.090	0.63
20 - 40	β-	0.00307	0.0101
	IB	0.18	0.62
40 - 100	β-	0.0296	0.0412
	IB	0.53	0.81
100 - 300	β-	0.53	0.249
	IB	1.70	0.94
300 - 600	β-	3.40	0.73
	IB	2.4	0.55

Continuous Radiation (^{29}Na) (continued)

E_{bin}(keV)		⟨ ⟩(keV)	(%)
600 - 1300	β-	36.3	3.65
	IB	4.7	0.53
1300 - 2500	β-	250	12.8
	IB	6.2	0.34
2500 - 5000	β-	1454	38.7
	IB	7.3	0.21
5000 - 7500	β-	1745	28.4
	IB	2.8	0.047
7500 - 13360	β-	1370	15.5
	IB	0.68	0.0080

$^{29}_{12}$Mg(1.09 *12* s)

Mode: β-

Δ: -10728 *30* keV

SpA: 2.6×10^{11} Ci/g

Prod: ^{18}O(^{13}C,2p); protons on In

Photons (^{29}Mg)

⟨γ⟩=1858 *120* keV

γ_{mode}	γ(keV)	γ(%)†
γ M1+0.1%E2	960.5 *3*	15.0 *10*
γ [M1+E2]	1307.5 *3*	4.6 *6*
γ E2	1398.09 *18*	16.4 *11*
γ M1+6.6%E2	1430.29 *24*	6.9 *6*
γ M1	1467.6 *3*	3.6 *4*
γ M1+2.5%E2	1754.29 *18*	9.9 *7*
γ E2	1786.49 *24*	2.84 *11*
γ M1	2035.0 *6*	2.5 *6*
γ [M1+E2]	2224.08 *25*	36
γ [M1+E2]	2865.6 *3*	4.1 *5*
γ [M1+E2]	3061.7 *3*	1.4 *4*
γ [M1+E2]	3184.48 *20*	1.1 *3*

† 14% uncert(syst)

Continuous Radiation (^{29}Mg)

⟨β-⟩=2537 keV;⟨IB⟩=10.1 keV

E_{bin}(keV)		⟨ ⟩(keV)	(%)
0 - 10	β-	0.00092	0.0181
	IB	0.066	
10 - 20	β-	0.00302	0.020
	IB	0.065	0.45
20 - 40	β-	0.0140	0.0462
	IB	0.130	0.45
40 - 100	β-	0.132	0.184
	IB	0.38	0.58
100 - 300	β-	2.27	1.08
	IB	1.17	0.65
300 - 600	β-	13.9	2.99
	IB	1.51	0.35
600 - 1300	β-	132	13.4
	IB	2.7	0.30
1300 - 2500	β-	680	35.4
	IB	2.6	0.148
2500 - 5000	β-	1422	41.8
	IB	1.43	0.044
5000 - 7490	β-	286	5.05
	IB	0.078	0.00143

$^{29}_{13}$Al(6.56 *6* min)

Mode: β-

Δ: -18215 *5* keV

SpA: 9.88×10^{8} Ci/g

Prod: ^{26}Mg(α,p); ^{27}Al(t,p)

Photons (^{29}Al)

⟨γ⟩=1376 *14* keV

γ_{mode}	γ(keV)	γ(%)
γ [M1+E2]	397.82 *4*	0.029 *11*
γ M1	754.77 *5*	0.310 *6*
γ M1+0.2%E2	1039.09 *14*	~0.007
γ M1+1.4%E2	1152.58 *4*	1.033 *25*
γ M1+3.7%E2	1273.359 *15*	91.3
γ M1+6.4%E2	1793.83 *14*	0.026 *4*
γ E2	2028.10 *5*	3.51 *7*
γ M1+10%E2	2425.89 *4*	5.2 *5*

Continuous Radiation (^{29}Al)

⟨β-⟩=977 keV;⟨IB⟩=2.1 keV

E_{bin}(keV)		⟨ ⟩(keV)	(%)
0 - 10	β-	0.0076	0.149
	IB	0.037	
10 - 20	β-	0.0244	0.162
	IB	0.036	0.25
20 - 40	β-	0.111	0.366
	IB	0.071	0.25
40 - 100	β-	1.00	1.40
	IB	0.20	0.31
100 - 300	β-	15.5	7.4
	IB	0.54	0.31
300 - 600	β-	78	16.9
	IB	0.56	0.133
600 - 1300	β-	431	45.7
	IB	0.60	0.071
1300 - 2407	β-	452	27.9
	IB	0.101	0.0069

$^{29}_{14}$Si(stable)

Δ: -21895.0 *6* keV

%: 4.67 *1*

$^{29}_{15}$P (4.142 *15* s)

Mode: ε

Δ: -16950.5 *18* keV

SpA: 8.65×10^{10} Ci/g

Prod: ^{28}Si(d,n); protons on Si

Photons (^{29}P)

⟨γ⟩=27.0 *4* keV

γ_{mode}	γ(keV)	γ(%)†
γ M1+1.4%E2	1152.58 *4*	0.065 *4*
γ M1+3.7%E2	1273.359 *15*	1.320
γ M1+10%E2	2425.89 *4*	0.389 *10*

† 1.7% uncert(syst)

Continuous Radiation (^{29}P)

⟨β+⟩=1771 keV;⟨IB⟩=5.6 keV

E_{bin}(keV)		⟨ ⟩(keV)	(%)
0 - 10	β+	0.000102	0.00137
	IB	0.054	
10 - 20	β+	0.00100	0.0064
	IB	0.054	0.37
20 - 40	β+	0.0089	0.0285
	IB	0.106	0.37
40 - 100	β+	0.146	0.197
	IB	0.31	0.47
100 - 300	β+	3.63	1.69
	IB	0.92	0.51
300 - 600	β+	24.6	5.3
	IB	1.12	0.26
600 - 1300	β+	226	23.0
	IB	1.7	0.20
1300 - 2500	β+	933	49.6
	IB	1.19	0.070
2500 - 4945	β+	583	20.1
	IB	0.133	0.0048
	Σβ+		99.9

$^{29}_{16}$S (187 *4* ms)

Mode: ε, εp(47 *5* %)

Δ: -3160 *50* keV

SpA: 5.47×10^{11} Ci/g

Prod: ^{31}P(p,3n); ^{28}Si(^{3}He,2n)

Delayed Protons (^{29}S)

p(keV)	p(%)†
739.5 *9*	3.4 *3*
1006 *25*	0.16 *6*
1257 *10*	3.8 *4*
1766 *15*	0.38 *5*
1910 *15*	0.30 *4*
2129.8 *9*	11.9 *4*
2457.0 *9*	0.53 *4*
2531 *10*	1.08 *7*
2883 *15*	0.082 *15*
2961 *15*	0.18 *2*
3101 *15*	0.18 *2*
3211 *15*	0.16 *2*
3296 *15*	0.34 *3?*
3456 *15*	0.38 *5*
3587 *15*	0.21 *4*
3720.6 *22*	2.34 *12*
3770 *15*	0.71 *6*
3870 *20*	0.27 *6*
4186 *20*	1.08 *7*
4338 *20*	0.33 *4*
4685 *20*	0.27 *4*
4835 *20*	0.23 *4*
5174 *15*	0.69 *7*
5304 *15*	0.92 *7*
5438.2 *22*	15.8 *4*
5585 *20*	0.87 *7*
5853 *30*	0.14 *3*
6446 *30*	0.16 *3*

† 10% uncert(syst)

Photons (^{29}S)

$\langle\gamma\rangle$=1199 *54* keV

γ_{mode}	γ(keV)	γ(%)†
γ_ϵ[M1+E2]	468.9 *3*	~0.8
γ_ϵM1	570.3 *1*	0.36 *10*
γ_ϵM1+3.9%E2	1039.1 *3*	2.5 *5*
γ_ϵ[M1+E2]	1152.0 *3*	0.22 *7*
γ_ϵM1+2.8%E2	1383.51 *7*	30.7 *22*
γ_ϵM1+5.9%E2	1722.3 *3*	0.68 *23*
$\gamma_{\epsilon p}$[E2]	1778.988 *9*	11.6 *6*
γ_ϵE2	1953.84 *17*	5.2 *6*
γ_ϵM1+4.6%E2	2422.7 *3*	17.4 *17*
$\gamma_{\epsilon p}$[E2]	2838.36 *14*	0.16 *6*

† 10% uncert(syst)

Continuous Radiation (^{29}S)

$\langle\beta+\rangle$=4057 keV;$\langle$IB$\rangle$=21 keV

E_{bin}(keV)		$\langle\ \rangle$(keV)	(%)
0 - 10	β+	2.74×10^{-5}	0.000372
	IB	0.082	
10 - 20	β+	0.000258	0.00164
	IB	0.082	0.57
20 - 40	β+	0.00222	0.0071
	IB	0.163	0.56
40 - 100	β+	0.0357	0.0483
	IB	0.48	0.73
100 - 300	β+	0.90	0.417
	IB	1.52	0.84
300 - 600	β+	6.4	1.36
	IB	2.1	0.48
600 - 1300	β+	66	6.7
	IB	4.0	0.45
1300 - 2500	β+	382	19.7
	IB	5.0	0.28
2500 - 5000	β+	1411	38.6
	IB	5.4	0.158
5000 - 7500	β+	1452	23.8
	IB	1.7	0.029
7500 - 11385	β+	739	8.7
	IB	0.24	0.0030
	Σβ+		99

A = 30

NP **A310**, 271 (1978)

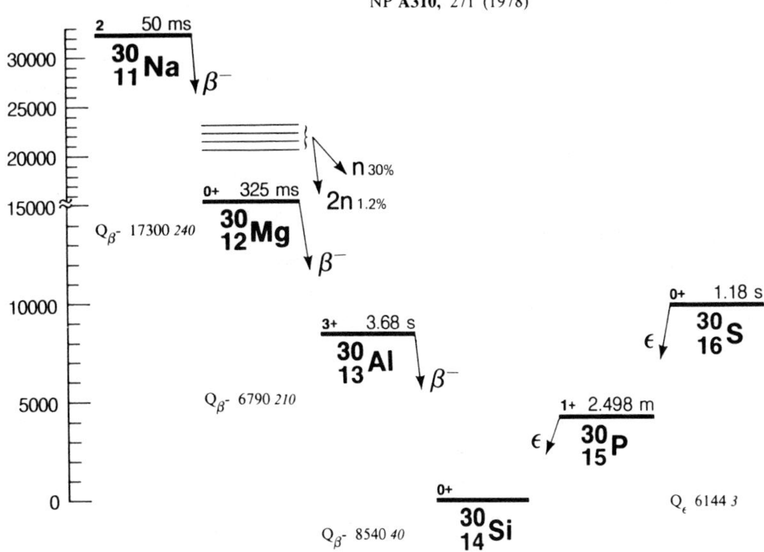

$^{30}_{11}$Na(50 *3* ms)

Mode: β-, β-n(30 *4* %), β-2n(1.15 *25* %)
Δ: 8200 *250* keV
SpA: 5.4×10^{11} Ci/g
Prod: protons on U

Photons (^{30}Na)

$\langle\gamma\rangle$=2824 *110* keV

γ_{mode}	γ(keV)	γ(%)†
$\gamma_{\beta-}$	305.6 *4*	5.8 *6*
$\gamma_{\beta-}$	337.3 *9*	2.9 *3*
$\gamma_{\beta-}$	985.7 *7*	6.5 *4*
$\gamma_{\beta-n}$	1041.2 *8*	11.5 *5*
$\gamma_{\beta-}$[E2]	1483.0 *4*	46.0
$\gamma_{\beta-}$	1506.1 *7*	3.8 *3*
$\gamma_{\beta-}$	1552.6 *11*	1.9 *3*
$\gamma_{\beta-}$	1560.4 *10*	1.5 *3*
$\gamma_{\beta-n}$	1638.0 *6*	1.5 *3*
$\gamma_{\beta-}$	1788.6 *6*	2.5 *3*
$\gamma_{\beta-}$	1820.9 *9*	2.9 *4*
$\gamma_{\beta-}$	1871.5 *10*	0.8 *3*
$\gamma_{\beta-}$	1978.7 *7*	10.9 *7*
$\gamma_{\beta-}$	2499.1 *7*	0.9 *3*
$\gamma_{\beta-}$	2685.8 *12*	1.0 *4*
$\gamma_{\beta-}$	3179.2 *6*	4.8 *7*

Photons (^{30}Na) (continued)

γ_{mode}	γ(keV)	γ(%)†
$\gamma_{\beta-}$	3430.4 *12*	1.1 *4*
$\gamma_{\beta-}$	3484.7 *5*	4.5 *7*
$\gamma_{\beta-}$	3542.2 *8*	3.1 *6*
$\gamma_{\beta-}$	3625.1 *8*	0.9 *3*
$\gamma_{\beta-}$	3930.7 *7*	2.9 *6*
$\gamma_{\beta-}$	4685.6 *18*	1.0 *5*
$\gamma_{\beta-}$	4967.6 *6*	5.2 *12*
$\gamma_{\beta-}$	5021.9 *8*	5.0 *11*
$\gamma_{\beta-}$	5094.6 *10*	2.7 *6*
$\gamma_{\beta-}$	5413.5 *7*	2.3 *6*

† 15% uncert(syst)

Continuous Radiation (^{30}Na)

$\langle\beta-\rangle$=5686 keV;$\langle$IB$\rangle$=33 keV

E_{bin}(keV)		$\langle\ \rangle$(keV)	(%)
0 - 10	β-	0.000122	0.00240
	IB	0.097	
10 - 20	β-	0.000406	0.00269
	IB	0.097	0.67
20 - 40	β-	0.00191	0.0063
	IB	0.19	0.67
40 - 100	β-	0.0184	0.0256
	IB	0.57	0.87
100 - 300	β-	0.331	0.156
	IB	1.8	1.01
300 - 600	β-	2.16	0.463
	IB	2.6	0.60
600 - 1300	β-	23.8	2.39
	IB	5.3	0.59
1300 - 2500	β-	176	8.9
	IB	7.2	0.40
2500 - 5000	β-	1246	32.6
	IB	9.4	0.27
5000 - 7500	β-	1866	30.3
	IB	4.3	0.071
7500 - 15817	β-	2372	25.0
	IB	1.9	0.021

$^{30}_{12}$Mg(325 *30* ms)

Mode: β-
Δ: -9100 *210* keV
SpA: 4.8×10^{11} Ci/g
Prod: protons on U

Photons (^{30}Mg)

$\langle\gamma\rangle$=583 *59* keV

γ_{mode}	γ(keV)	γ(%)†
γ	244.0 *5*	85 *14*
γ	443.7 *5*	71
γ[E2]	687.7 *6*	2.0 *6*
γ	2168.6 *12*	2.1 *7*

† 14% uncert(syst)

$^{30}_{13}$Al(3.68 *3* s)

Mode: β-
Δ: -15890 *40* keV
SpA: 9.30×10^{10} Ci/g
Prod: ^{30}Si(n,p); ^{18}O(^{18}O,αpn)

Photons (^{30}Al)

$\langle\gamma\rangle$=3495 *42* keV

γ_{mode}	γ(keV)	γ(%)†
γ[M1+E2]	1039.6 *23*	0.20 *9*
γ M1+3.3%E2	1262.9 *8*	40.6 *12*
γ M1+3.1%E2	1311.2 *8*	2.56 *12*
γ[M1+E2]	1332.0 *6*	0.93 *9*
γ M1+0.9%E2	1534.5 *9*	0.108 *25*
γ M1+1.7%E2	1732.8 *10*	1.87 *12*
γ E2	2235.0 *9*	65.1
γ M1+21%E2	2574.0 *14*	0.96 *7*

Photons (^{30}Al)
(continued)

γ_{mode}	γ(keV)	γ(%)†
γ M1+30%E2	2594.8 *11*	5.80 *14*
γ [M1+E2]	2995.6 *17*	0.52 *4*
γ E2	3497.8 *16*	32.8 *10*
γ M1	3769.4 *2*	0.095 *19*
γ E2	4808.8 *22*	2.13 *17*

† 1.1% uncert(syst)

Continuous Radiation (^{30}Al)
$\langle\beta-\rangle$=2296 keV;$\langle$IB$\rangle$=8.6 keV

E_{bin}(keV)		$\langle\ \rangle$(keV)	(%)
0 - 10	β-	0.00112	0.0221
	IB	0.063	
10 - 20	β-	0.00363	0.0241
	IB	0.062	0.43
20 - 40	β-	0.0166	0.055
	IB	0.123	0.43
40 - 100	β-	0.155	0.216
	IB	0.36	0.55
100 - 300	β-	2.63	1.25
	IB	1.10	0.61
300 - 600	β-	15.9	3.42
	IB	1.40	0.33
600 - 1300	β-	148	15.0
	IB	2.4	0.27
1300 - 2500	β-	733	38.3
	IB	2.2	0.126
2500 - 5000	β-	1348	40.5
	IB	0.84	0.028
5000 - 6305	β-	47.3	0.88
	IB	0.0051	9.2×10^{-5}

$^{30}_{14}$Si(stable)

Δ: -24433.2 *6* keV
%: 3.10 *1*

$^{30}_{15}$P (2.498 *4* min)

Mode: ϵ
Δ: -20207.4 *17* keV
SpA: 2.503×10^{9} Ci/g
Prod: ^{27}Al(α,n); ^{32}S(d,α); ^{29}Si(p,γ); ^{31}P(γ,n); ^{30}Si(p,n)

Photons (^{30}P)
$\langle\gamma\rangle$=1.42 *6* keV

γ_{mode}	γ(keV)	γ(%)†
γ M1+3.3%E2	1262.9 *8*	0.00087 *6*
γ M1+0.9%E2	1534.5 *9*	9.9 *17* $\times10^{-5}$
γ E2	1552.5 *2*	0.00339 *23*
γ E2	2235.0 *9*	0.059 *3*
γ E2	3497.8 *16*	0.00070 *5*
γ M1	3769.4 *2*	7.8 *14* $\times10^{-5}$

† 4.5% uncert(syst)

Continuous Radiation (^{30}P)
$\langle\beta+\rangle$=1435 keV;$\langle$IB$\rangle$=4.0 keV

E_{bin}(keV)		$\langle\ \rangle$(keV)	(%)
0 - 10	β+	0.000160	0.00216
	IB	0.048	
10 - 20	β+	0.00158	0.0101
	IB	0.047	0.33
20 - 40	β+	0.0140	0.0450
	IB	0.093	0.32
40 - 100	β+	0.229	0.310
	IB	0.27	0.41
100 - 300	β+	5.6	2.62
	IB	0.78	0.44
300 - 600	β+	37.1	8.0
	IB	0.90	0.21
600 - 1300	β+	314	32.3
	IB	1.26	0.146
1300 - 2500	β+	930	51
	IB	0.58	0.035
2500 - 4226	β+	148	5.5
	IB	0.020	0.00074
	$\Sigma\beta$+		99.86

$^{30}_{16}$S (1.18 *4* s)

Mode: ϵ
Δ: -14063 *3* keV
SpA: 2.41×10^{11} Ci/g
Prod: ^{28}Si(^{3}He,n)

Photons (^{30}S)
$\langle\gamma\rangle$=586 *4* keV

γ_{mode}	γ(keV)	γ(%)†
γ M1	677.2 *4*	78.4
γ [M1+E2]	709.00 *15*	0.30 *7*
γ M1	2341.3 *9*	2.27 *4*
γ [M1+E2]	3018.4 *10*	0.010 *5*

† 2.3% uncert(syst)

Continuous Radiation (^{30}S)
$\langle\beta+\rangle$=2082 keV;$\langle$IB$\rangle$=7.3 keV

E_{bin}(keV)		$\langle\ \rangle$(keV)	(%)
0 - 10	β+	6.4×10^{-5}	0.00085
	IB	0.060	
10 - 20	β+	0.00066	0.00420
	IB	0.059	0.41
20 - 40	β+	0.0061	0.0195
	IB	0.117	0.41
40 - 100	β+	0.103	0.138
	IB	0.34	0.52
100 - 300	β+	2.62	1.21
	IB	1.04	0.58
300 - 600	β+	18.1	3.88
	IB	1.3	0.30
600 - 1300	β+	173	17.6
	IB	2.2	0.25
1300 - 2500	β+	820	43.0
	IB	1.8	0.104
2500 - 5000	β+	1068	34.1
	IB	0.46	0.0152
5000 - 6144	β+	0.110	0.00219
	IB	0.00068	1.31×10^{-5}
	$\Sigma\beta$+		99.85

A = 31

NP **A310**, 296 (1978)

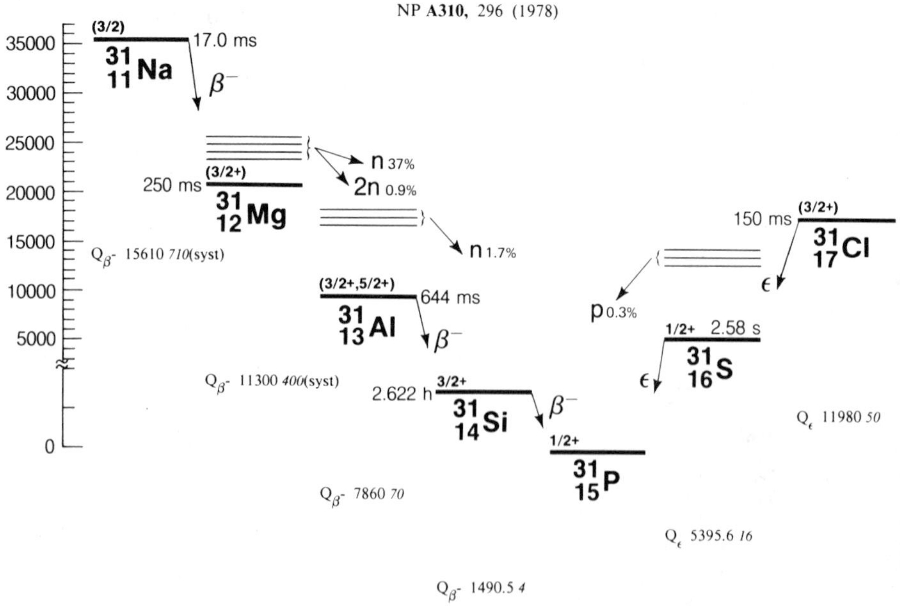

35000
30000
25000
20000
15000
10000
5000
0
(3/2) $^{31}_{11}$Na 17.0 ms β^-
n 37%
2n 0.9%
250 ms (3/2+) $^{31}_{12}$Mg
n 1.7%
Q_{β^-} 15610 *710*(syst)
(3/2+,5/2+) $^{31}_{13}$Al 644 ms β^-
Q_{β^-} 11300 *400*(syst)
2.622 h 3/2+ $^{31}_{14}$Si β^-
1/2+ $^{31}_{15}$P
Q_{β^-} 7860 *70*
Q_{β^-} 1490.5 *4*
150 ms (3/2+) $^{31}_{17}$Cl ϵ
p 0.3%
1/2+ 2.58 s $^{31}_{16}$S ϵ
Q_ϵ 11980 *50*
Q_ϵ 5395.6 *16*

$^{31}_{11}$Na(17.0 *4* ms)

Mode: β-, β-n(37 *5* %), β-2n(0.87 *24* %)
Δ: 11810 *580* keV
SpA: 5.25×10^{11} Ci/g
Prod: protons on U; protons on In

Photons (^{31}Na)

⟨γ⟩=843 *78* keV

γ_{mode}	γ(keV)	γ(%)†
$\gamma_{\beta-}$	50.6 *8*	>7
$\gamma_{\beta-}$	170.6 *6*	4.9 *3*
$\gamma_{\beta-}$	221.2 *6*	2.15 *21*
$\gamma_{\beta-n}$	305.6 *4*	1.86 *20*
$\gamma_{\beta-}$	622.9 *8*	3.3 *3*
$\gamma_{\beta-}$	673.2 *12*	1.43 *11*
$\gamma_{\beta-}$	808.2 *7*	1.6 *3*
$\gamma_{\beta-}$	894.1 *12*	0.80 *23*
$\gamma_{\beta-n}$	985.7 *7*	1.7 *3*
$\gamma_{\beta-}$	1213.8 *7*	1.23 *11*
$\gamma_{\beta-n}$ [E2]	1483.0 *4*	14
$\gamma_{\beta-n}$	1820.9 *9*	2.9 *3*
$\gamma_{\beta-n}$	1978.7 *7*	3.1 *4*
$\gamma_{\beta-}$	2022.0 *5*	3.9 *5*
$\gamma_{\beta-}$	2192.6 *6*	3.1 *4*
$\gamma_{\beta-}$	2243.2 *5*	10.6 *7*
$\gamma_{\beta-}$	3537.0 *12*	0.8 *3*

† 30% uncert(syst)

$^{31}_{12}$Mg(250 *30* ms)

Mode: β-, β-n(1.7 *3* %)
Δ: -3790 *410* keV syst
SpA: 4.9×10^{11} Ci/g
Prod: protons on In

Photons (^{31}Mg)

⟨γ⟩=1750 *130* keV

γ_{mode}	γ(keV)	γ(%)†
$\gamma_{\beta-}$	665.8 *6*	7.6 *5*
$\gamma_{\beta-}$	903.6 *8*	1.6 *3*
$\gamma_{\beta-}$	947.1 *6*	22.8 *13*
$\gamma_{\beta-}$	1612.9 *5*	26
$\gamma_{\beta-}$	1626.3 *6*	17.9 *5*
$\gamma_{\beta-}$	1820.5 *6*	2.4 *3*
$\gamma_{\beta-}$	2486.3 *7*	1.0 *3*
$\gamma_{\beta-}$	2529.8 *8*	2.3 *4*
$\gamma_{\beta-}$	2675.9 *7*	1.6 *4*
$\gamma_{\beta-}$	2949.9 *10*	2.9 *4*
$\gamma_{\beta-}$	3196.9 *9*	3.1 *5*
$\gamma_{\beta-}$	3433.3 *6*	3.9 *5*
$\gamma_{\beta-}$	3623.0 *7*	4.4 *6*
$\gamma_{\beta-}$	4203.3 *10*	2.0 *3*
$\gamma_{\beta-}$	4809.6 *10*	~0.6

† 31% uncert(syst)

$^{31}_{13}$Al(644 *25* ms)

Mode: β-
Δ: -15090 *70* keV
SpA: 3.46×10^{11} Ci/g
Prod: ^{18}O(^{18}O,αp); protons on U

Photons (^{31}Al)

γ_{mode}	γ(keV)	γ(rel)
γ [M1+E2]	621.93 *25*	13.6 *10*
γ [M1+E2]	752.3 *3*	25.4 *11*
γ [M1+E2]	1564.5 *3*	23.8 *22*
γ E2+5.3%M1	1694.85 *25*	80.9 *22*
γ [M1+E2]	2316.74 *25*	100.0 *25*
γ [M1+E2]	2787.7 *8*	4.9 *21* ?

$^{31}_{14}$Si(2.622 *5* h)

Mode: β-
Δ: -22950.2 *7* keV
SpA: 3.856×10^{7} Ci/g
Prod: ^{30}Si(n,γ)

Photons (^{31}Si)

⟨γ⟩~0.9 keV

γ_{mode}	γ(keV)	γ(%)
γ M1+8.3%E2	1266.2 *3*	~0.07

Continuous Radiation (^{31}Si)

⟨β-⟩=595 keV;⟨IB⟩=0.91 keV

E_{bin}(keV)		⟨ ⟩(keV)	(%)
0 - 10	β-	0.0184	0.364
	IB	0.026	
10 - 20	β-	0.059	0.389
	IB	0.025	0.17
20 - 40	β-	0.261	0.86
	IB	0.048	0.167
40 - 100	β-	2.30	3.22
	IB	0.13	0.20
100 - 300	β-	32.8	15.8
	IB	0.31	0.18
300 - 600	β-	142	31.4
	IB	0.25	0.060
600 - 1300	β-	405	47.0
	IB	0.114	0.0154
1300 - 1491	β-	12.3	0.92
	IB	0.00019	1.43×10^{-5}

$^{31}_{15}$P (stable)

Δ: -24440.7 *6* keV
%: 100

$^{31}_{16}$S (2.584 *18* s)

Mode: ε
Δ: -19045.2 *15* keV
SpA: 1.235×10^{11} Ci/g
Prod: ^{31}P(p,n); ^{28}Si(α,n)

Photons (^{31}S)

⟨γ⟩=15.3 *3* keV

γ_{mode}	γ(keV)	γ(%)†
γ M1+8.3%E2	1266.2 *3*	1.103 *22*
γ [M1+E2]	1869.0 *10*	0.00088 *11*
γ [M1+E2]	2239.7 *6*	0.0048 *8*
γ [M1]	3135.0 *10*	0.0318 *9*
γ M1+16%E2	3505.7 *6*	0.0073 *5*

† 3.0% uncert(syst)

Continuous Radiation (^{31}S)

⟨β+⟩=1996 keV;⟨IB⟩=6.8 keV

E_{bin}(keV)		⟨ ⟩(keV)	(%)
0 - 10	β+	6.5×10^{-5}	0.00087
	IB	0.058	
10 - 20	β+	0.00067	0.00427
	IB	0.058	0.40
20 - 40	β+	0.0062	0.0198
	IB	0.115	0.40
40 - 100	β+	0.104	0.141
	IB	0.33	0.51
100 - 300	β+	2.68	1.24
	IB	1.01	0.56
300 - 600	β+	18.7	4.0
	IB	1.26	0.29
600 - 1300	β+	183	18.5
	IB	2.0	0.23
1300 - 2500	β+	873	45.8
	IB	1.62	0.093
2500 - 5000	β+	919	30.2
	IB	0.32	0.0103
5000 - 5396	IB	0.000107	2.1×10^{-6}
	Σβ+		99.9

$^{31}_{17}$Cl(150 *25* ms)

Mode: ε, εp(0.3 %)
Δ: -7070 *50* keV
SpA: 5.2×10^{11} Ci/g
Prod: ^{32}S(p,2n)

εp: 989 *15*, 1528 *20*

A = 32

NP **A310**, 322 (1978)

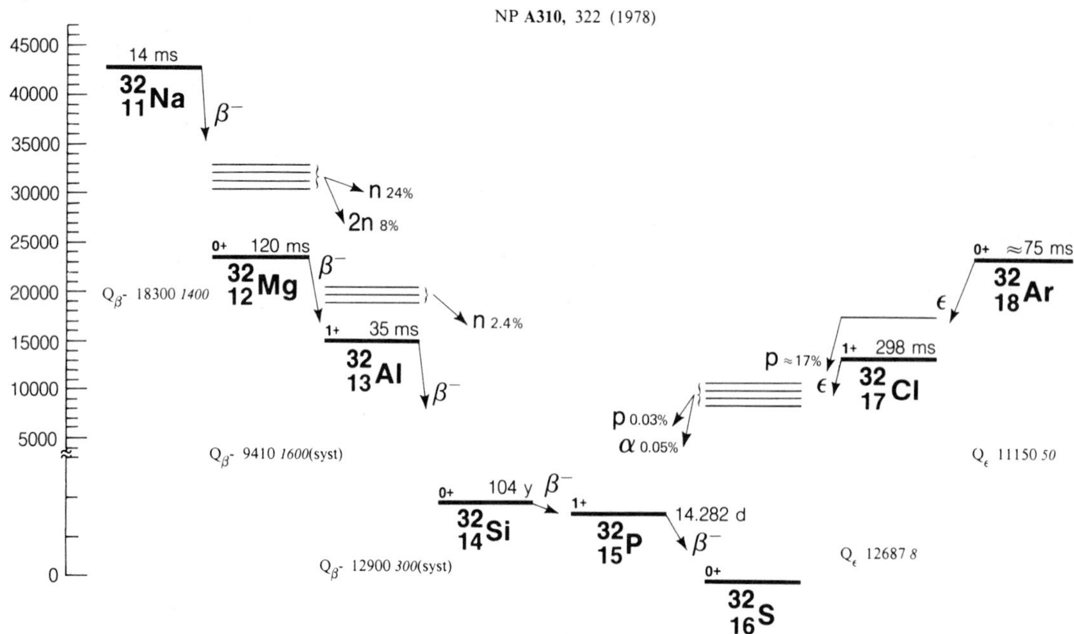

$^{32}_{11}$Na(14.0 *7* ms)

Mode: β-, β-n(24 *7* %), β-2n(8 *4* %)
Δ: 16530 *740* keV
SpA: 5.09×10^{11} Ci/g
Prod: protons on U; protons on In

Photons (^{32}Na)

⟨γ⟩=2140 *140* keV

γ_{mode}	γ(keV)	γ(%)†
$\gamma_{\beta\text{-n}}$	50.6 *8*	>4
$\gamma_{\beta\text{-n}}$	170.6 *6*	5.4 *15*
$\gamma_{\beta\text{-n}}$	221.2 *6*	2.6 *8*
$\gamma_{\beta\text{-}}$	239.5 *12*	16.6 *19*
$\gamma_{\beta\text{-}}$	694.4 *12*	2.3 *10*
$\gamma_{\beta\text{-}}$	885.5 *7*	60
$\gamma_{\beta\text{-n}}$	894.1 *12*	2.6 *6*
$\gamma_{\beta\text{-}}$	1232.2 *12*	2.9 *8*
$\gamma_{\beta\text{-}}$	1436.1 *10*	6.1 *12*
$\gamma_{\beta\text{-2n}}$[E2]	1483.0 *4*	~1.2
$\gamma_{\beta\text{-}}$	1782.6 *8*	5.0 *12*
$\gamma_{\beta\text{-}}$	1973.0 *12*	8.6 *15*
$\gamma_{\beta\text{-}}$	2151.6 *7*	32 *4*
$\gamma_{\beta\text{-}}$	2550.7 *10*	5.5 *12*
$\gamma_{\beta\text{-}}$	3934.1 *10*	8.0 *18*

† 15% uncert(syst)

$^{32}_{12}$Mg(120 *20* ms)

Mode: β-, β-n(2.4 *5* %)
Δ: -1770 *1600* keV
SpA: 5.1×10^{11} Ci/g
Prod: protons on In; protons on Ir

Photons (^{32}Mg)

⟨γ⟩=869 *59* keV

γ_{mode}	γ(keV)	γ(%)†
$\gamma_{\beta\text{-}}$	735.5 *12*	10.6 *22*
$\gamma_{\beta\text{-}}$	2466.9 *12*	4.1 *20*
$\gamma_{\beta\text{-}}$	2765.3 *9*	25

† 4.0% uncert(syst)

$^{32}_{13}$Al(35 *5* ms)

Mode: β-
Δ: -11180 *300* keV syst
SpA: 5.1×10^{11} Ci/g
Prod: protons on In

Photons (^{32}Al)

⟨γ⟩=503 *74* keV

γ_{mode}	γ(keV)	γ(%)†
γ [E2]	1941.4 *3*	12.0
γ [E2]	3042.6 *12*	8.9 *24*

† 4.2% uncert(syst)

Continuous Radiation (^{32}Al)

⟨β-⟩=5851 keV;⟨IB⟩=34 keV

E_{bin}(keV)		⟨ ⟩(keV)	(%)
0 - 10	β-	0.000107	0.00211
	IB	0.099	
10 - 20	β-	0.000347	0.00230
	IB	0.099	0.68
20 - 40	β-	0.00160	0.0053
	IB	0.20	0.68
40 - 100	β-	0.0151	0.0210
	IB	0.58	0.89
100 - 300	β-	0.266	0.126
	IB	1.9	1.04
300 - 600	β-	1.73	0.371
	IB	2.6	0.61
600 - 1300	β-	19.3	1.93
	IB	5.5	0.61
1300 - 2500	β-	147	7.4
	IB	7.5	0.41
2500 - 5000	β-	1139	29.6
	IB	9.9	0.29
5000 - 7500	β-	2045	33.0
	IB	4.5	0.076
7500 - 12900	β-	2499	27.5
	IB	1.46	0.017

$^{32}_{14}$Si(104 *13* yr)

Mode: β-
Δ: -24080.8 *24* keV
SpA: 107 Ci/g
Prod: ^{30}Si(t,p); protons on Cl

Continuous Radiation (^{32}Si)

$\langle\beta-\rangle$=69 keV;$\langle$IB$\rangle$=0.0166 keV

E_{bin}(keV)		$\langle\ \rangle$(keV)	(%)
0 - 10	β-	0.425	8.5
	IB	0.0036	
10 - 20	β-	1.26	8.4
	IB	0.0029	0.020
20 - 40	β-	4.93	16.5
	IB	0.0041	0.0148
40 - 100	β-	27.8	40.9
	IB	0.0052	0.0088
100 - 225	β-	34.4	25.7
	IB	0.00089	0.00075

$^{32}_{15}$P (14.282 *5* d)

Mode: β-

Δ: -24305.8 *6* keV

SpA: 2.8571×10^5 Ci/g

Prod: ^{31}P(n,γ); ^{34}S(d,α); ^{32}S(n,p)

Continuous Radiation (^{32}P)

$\langle\beta-\rangle$=695 keV;$\langle$IB$\rangle$=1.18 keV

E_{bin}(keV)		$\langle\ \rangle$(keV)	(%)
0 - 10	β-	0.0141	0.279
	IB	0.029	
10 - 20	β-	0.0447	0.297
	IB	0.028	0.20
20 - 40	β-	0.199	0.66
	IB	0.055	0.19
40 - 100	β-	1.75	2.45
	IB	0.150	0.23
100 - 300	β-	25.7	12.4
	IB	0.38	0.22
300 - 600	β-	120	26.3
	IB	0.33	0.080
600 - 1300	β-	470	52
	IB	0.21	0.028
1300 - 1710	β-	77	5.5
	IB	0.0039	0.00029

$^{32}_{16}$S (stable)

Δ: -26016.18 *24* keV

%: 95.02 *9*

$^{32}_{17}$Cl(298 *2* ms)

Mode: ϵ, ϵα(0.05 %), ϵp(0.03 %)

Δ: -13330 *8* keV

SpA: 4.59×10^{11} Ci/g

Prod: ^{32}S(p,n)

Delayed Protons (^{32}Cl)

p(keV)	p(%)
762 *5*	0.0052 *8*
991 *5*	0.0113 *17*
1051 *5*	0.0019 *4*
1324 *5*	0.0052 *8*
1381 *5*	0.00078 *20*
1856 *5*	0.0016 *3*

Delayed Alphas (^{32}Cl)

α(keV)	α(%)
1526 *5*	0.0011 *2*
1673 *5*	0.0146 *20*
1998 *5*	0.0002 *1*
2201 *5*	0.030 *4*
2417 *5*	0.0040 *7*
2656 *5*	0.00069 *20*
2927 *5*	0.0017 *3*
3072 *5*	0.00024 *10*
3135 *5*	0.00084 *20*
3364 *5*	0.00051 *10*
3601 *5*	0.00006 *3*

Photons (^{32}Cl)

$\langle\gamma\rangle$=3753 *130* keV

γ_{mode}	γ(keV)	γ(%)†
γ[E2]	1548.4 *16*	3.6 *6*
γE2	2051.1 *19*	0.42 *3*
γ[E2]	2231.1 *11*	92 *4*
γ[M1+E2]	2418 *4*	0.040 *17*
γM1+32%E2	2464.2 *9*	4.0 *4*
γ[M1+E2]	2831 *4*	0.014 *6*
γ[E2]	2885.4 *16*	1.0 *4*
γ[E2]	3318.5 *13*	2.5 *4*
γ[E2]	4282.0 *15*	2.58 *9*
γ[M1+E2]	4433.7 *16*	0.83 *18*
γ[M1]	4695.1 *13*	2.8 *6*
γ[M1+E2]	4770.5 *15*	20.5 *20*
γ[M1+E2]	4882 *4*	0.45 *19*
γ[M1+E2]	4964 *3*	0.63 *16*
γ[E2]	5549.3 *14*	1.6 *3*
γ[E2]	7112 *4*	0.014 *6*
γ[M1]	7195 *3*	0.41 *10*

† 4.3% uncert(syst)

Continuous Radiation (^{32}Cl)

$\langle\beta+\rangle$=3770 keV;$\langle$IB$\rangle$=18 keV

E_{bin}(keV)		$\langle\ \rangle$(keV)	(%)
0 - 10	β+	1.75×10^{-5}	0.000232
	IB	0.080	
10 - 20	β+	0.000190	0.00121
	IB	0.080	0.55
20 - 40	β+	0.00181	0.0058
	IB	0.159	0.55
40 - 100	β+	0.0316	0.0425
	IB	0.47	0.72
100 - 300	β+	0.84	0.390
	IB	1.48	0.82
300 - 600	β+	6.1	1.31
	IB	2.0	0.47
600 - 1300	β+	66	6.6
	IB	3.9	0.43
1300 - 2500	β+	401	20.6
	IB	4.6	0.26
2500 - 5000	β+	1603	43.8
	IB	4.6	0.134
5000 - 7500	β+	1411	23.3
	IB	1.02	0.018
7500 - 11665	β+	282	3.50
	IB	0.044	0.00056
	Σβ+		100

$^{32}_{18}$Ar(~75 ms)

Mode: ϵ, ϵp(~17 %)

Δ: -2180 *50* keV

SpA: ~5×10^{11} Ci/g

Prod: protons on V

ϵp: 3350.5 *50*

A = 33

NP A310, 350 (1978)

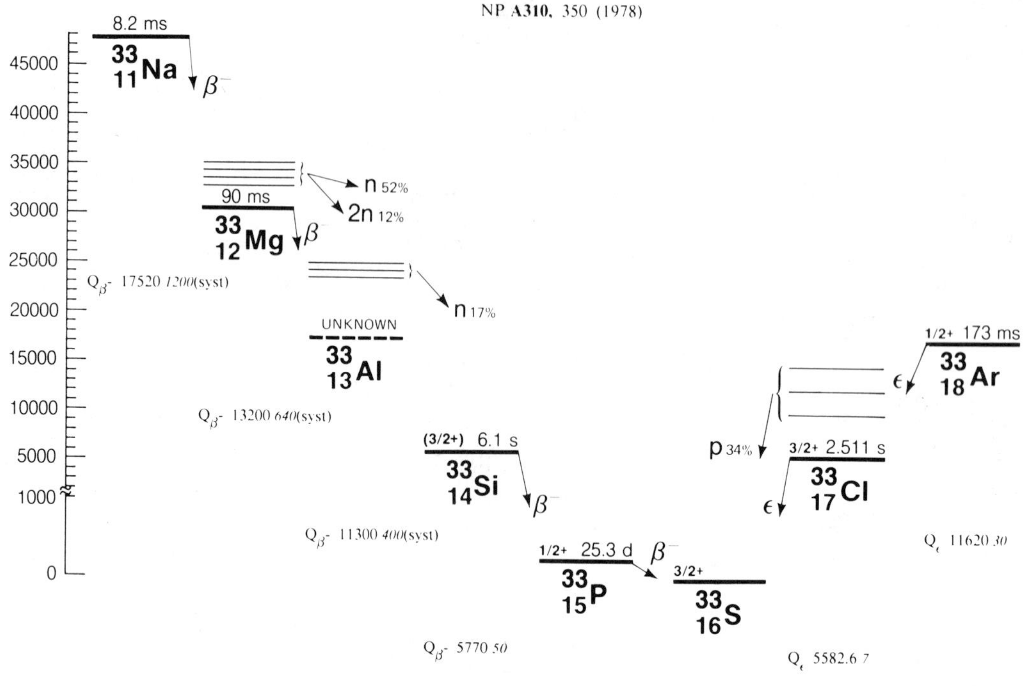

$^{33}_{11}$Na(8.2 *4* ms)

Mode: β-, β-n(52 *20* %), β-2n(12 *5* %)
Δ: 21450 *1100* keV
SpA: 4.93×10^{11} Ci/g
Prod: protons on U; protons on In

Photons (^{33}Na)

⟨γ⟩=330 *80* keV

γ_{mode}	γ(keV)	γ(%)†
$\gamma_{\beta-}$	484.9 *10*	~2
$\gamma_{\beta-}$	546.5 *10*	6.4 *21*
$\gamma_{\beta-}$	704.3 *10*	3.7 *18*
$\gamma_{\beta-n}$	885.5 *7*	16
$\gamma_{\beta-}$	1242.6 *18*	4.2 *16*
$\gamma_{\beta-n}$	2550.7 *10*	~3

† 44% uncert(syst)

$^{33}_{12}$Mg(90 *20* ms)

Mode: β-, β-n(17 *5* %)
Δ: 3930 *500* keV syst
SpA: 4.9×10^{11} Ci/g
Prod: protons on In; protons on Ir

$^{33}_{14}$Si(6.11 *21* s)

Mode: β-
Δ: -20570 *50* keV
SpA: 5.29×10^{10} Ci/g
Prod: ^{18}O(^{18}O,2pn); ^{40}Ar on Th

Photons (^{33}Si)

γ_{mode}	γ(keV)	γ(rel)
γ M1	416.1 *4*	6.7 *6*
γ [M1+E2]	690.9 *7*	0.77 *12*
γ [M1+E2]	1107.0 *8*	0.88 *13*
γ M1+26%E2	1431.5 *4*	13.1 *10*
γ [E2]	1847.63 *24*	100 *1*
γ M1+2.6%E2	2538.5 *7*	9.3 *8*

$^{33}_{15}$P (25.34 *12* d)

Mode: β-
Δ: -26338.0 *15* keV
SpA: 1.561×10^{5} Ci/g
Prod: ^{33}S(n,p); ^{37}Cl(γ,α)

Continuous Radiation (^{33}P)

⟨β-⟩=76 keV;⟨IB⟩=0.020 keV

E_{bin}(keV)		⟨ ⟩(keV)	(%)
0 - 10	β-	0.381	7.6
	IB	0.0040	
10 - 20	β-	1.13	7.6
	IB	0.0033	0.023
20 - 40	β-	4.46	14.9
	IB	0.0049	0.017
40 - 100	β-	26.4	38.6
	IB	0.0067	0.0113
100 - 249	β-	44.0	31.3
	IB	0.00157	0.00129

$^{33}_{16}$S (stable)

Δ: -26586.51 *21* keV
%: 0.75 *1*

$^{33}_{17}$Cl(2.511 *3* s)

Mode: ε
Δ: -21003.9 *7* keV
SpA: 1.1897×10^{11} Ci/g
Prod: ^{32}S(d,n); ^{33}S(p,n)

Photons (^{33}Cl)

$\langle\gamma\rangle$=26.6 *3* keV

γ_{mode}	γ(keV)	γ(%)†
γ M1+2.2%E2	840.91 *5*	0.517 *16*
γ [E2]	1125.32 *11*	0.0060 *15*
γ M1+11%E2	1471.53 *11*	0.0252 *14*
γ M1+24%E2	1966.2 *1*	0.454 *7*
γ [E2]	2025.45 *11*	0.0067 *8*
γ M1+0.1%E2	2312.4 *1*	0.0101 *6*
γ M1+1.3%E2	2866.3 *1*	0.436 *8*
γ [M1+E2]	4052.7 *5*	0.00047 *13*
γ	4745.6 *10*	0.00037 *10*

† 13% uncert(syst)

Continuous Radiation (^{33}Cl)

$\langle\beta+\rangle$=2080 keV;$\langle$IB$\rangle$=7.3 keV

E_{bin}(keV)		$\langle\ \rangle$(keV)	(%)
0 - 10	β+	5.2×10^{-5}	0.00069
	IB	0.060	
10 - 20	β+	0.00057	0.00360
	IB	0.059	0.41
20 - 40	β+	0.0054	0.0172
	IB	0.117	0.41
40 - 100	β+	0.093	0.126
	IB	0.34	0.52
100 - 300	β+	2.44	1.13
	IB	1.04	0.58
300 - 600	β+	17.2	3.68
	IB	1.3	0.30
600 - 1300	β+	169	17.1
	IB	2.2	0.25
1300 - 2500	β+	835	43.7
	IB	1.8	0.104
2500 - 5000	β+	1057	34.0
	IB	0.45	0.0147
5000 - 5583	IB	0.00073	1.43×10^{-5}
	$\Sigma\beta$+		99.74

$^{33}_{18}$Ar(173 *2* ms)

Mode: ϵ, ϵp(34 *6* %)

Δ: -9380 *30* keV

SpA: 4.84×10^{11} Ci/g

Prod: ^{35}Cl(p,3n); ^{32}S(^{3}He,2n); ^{24}Mg(^{12}C,3n)

Delayed Protons (^{33}Ar)

p(keV)	p(%)
1092 *34*	0.015 *6*
1226 *34*	0.019 *6*
1323 *29*	0.27 *10*
1473 *34*	0.020 *10*
1539 *39*	~0.04
1641 *19*	0.40 *2*
1781 *19*	0.43 *2*
1888 *29*	
1961 *29*	~0.02
2108 *34*	2.50 *4*
2233 *34*	
2365 *34*	0.047 *7*
2488 *15*	0.31 *2*
2606 *29*	0.08 *3*
2749 *24*	0.81 *10*
2885 *39*	0.060 *10*
2956 *29*	0.21 *3*
3069 *29*	0.57 *2*
3170 *4*	26.7 *3*
3300 *19*	0.54 *3*
3364 *29*	0.352 *14*
3483 *34*	
3637 *34*	0.007 *3*
3742 *34*	~0.004
3853 *19*	0.58 *2*
3982 *34*	0.016 *5*
4833 *39*	0.013 *6*
5032 *19*	0.230 *14*
5180 *29*	0.037 *3*
5320 *39*	0.010 *3*
5627 *19*	
5723 *24*	
5846 *29*	0.017 *3*
6119 *39*	0.011 *2*
6485 *29*	0.0034 *17*

Photons (^{33}Ar)

$\langle\gamma\rangle$=406 *29* keV

γ_{mode}	γ(keV)	γ(%)†
γ_{ϵ}[M1+E2]	810.51 *16*	48 *4*
$\gamma_{\epsilon p}$[E2]	1548.4 *16*	0.023 *10*
$\gamma_{\epsilon p}$[E2]	2231.1 *11*	0.72 *12*

† 16% uncert(syst)

Continuous Radiation (^{33}Ar)

$\langle\beta+\rangle$=3965 keV;$\langle$IB$\rangle$=20 keV

E_{bin}(keV)		$\langle\ \rangle$(keV)	(%)
0 - 10	β+	1.55×10^{-5}	0.000205
	IB	0.082	
10 - 20	β+	0.000177	0.00112
	IB	0.082	0.57
20 - 40	β+	0.00174	0.0056
	IB	0.162	0.56
40 - 100	β+	0.0312	0.0420
	IB	0.48	0.73
100 - 300	β+	0.84	0.390
	IB	1.52	0.84
300 - 600	β+	6.1	1.31
	IB	2.1	0.48
600 - 1300	β+	65	6.5
	IB	4.0	0.45
1300 - 2500	β+	384	19.8
	IB	4.9	0.27
2500 - 5000	β+	1509	41.4
	IB	5.1	0.149
5000 - 7500	β+	1491	24.4
	IB	1.41	0.024
7500 - 11620	β+	510	6.2
	IB	0.110	0.00137
	$\Sigma\beta$+		99.9

A = 34

NP **A310**, 371 (1978)

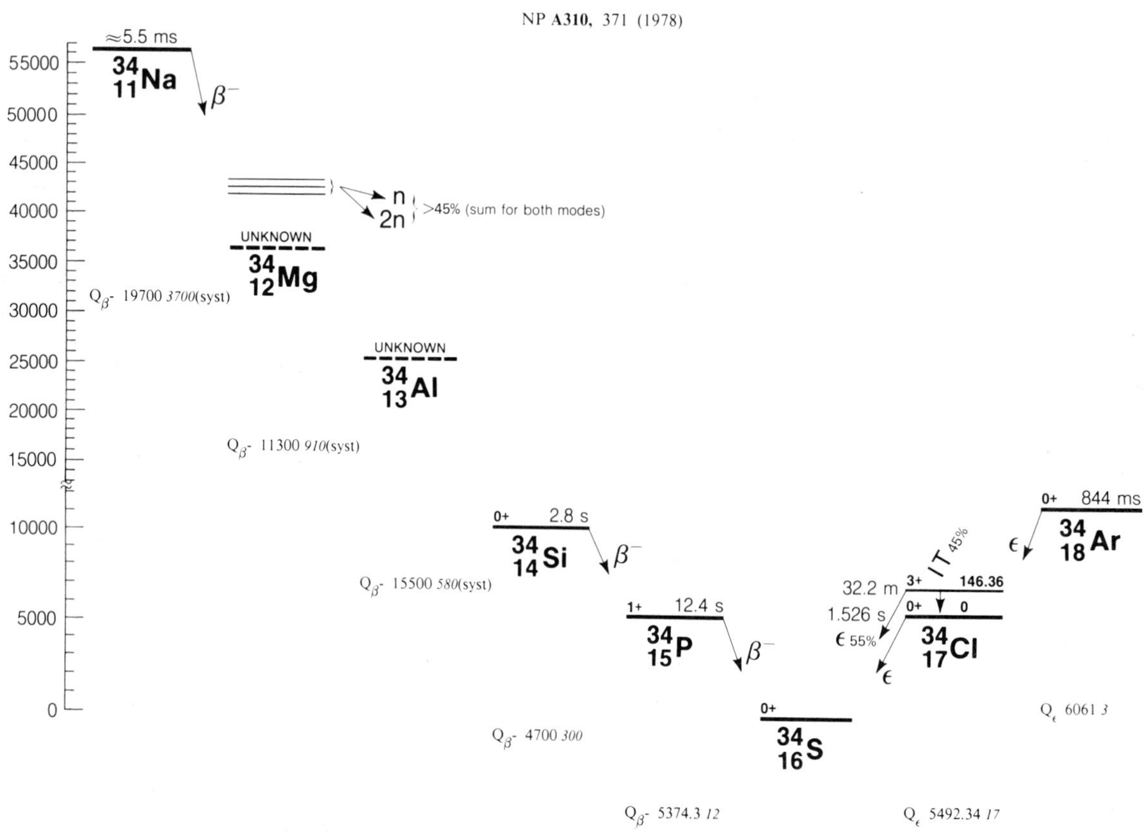

$^{34}_{11}$Na(5.5 *10* ms)

Mode: β-, β-n, β-2n
Δ: 26640 *3600* keV
SpA: 4.8×10^{11} Ci/g
Prod: protons on In

Photons (^{34}Na)

γ_{mode}	γ(keV)
$\gamma_{\beta\text{-}2n}$	885.5 *7*

$^{34}_{14}$Si(2.77 *20* s)

Mode: β-
Δ: -19860 *300* keV
SpA: 1.06×10^{11} Ci/g
Prod: ^{18}O(^{18}O,2p)

Photons (^{34}Si)

γ_{mode}	γ(keV)	γ(rel)
γ	429.06 *13*	94 *3*
γ	1178.51 *15*	100 *5*
γ	1607.55 *19*	56 *8*

$^{34}_{15}$P (12.43 *8* s)

Mode: β-
Δ: -24557.9 *12* keV
SpA: 2.596×10^{10} Ci/g
Prod: ^{37}Cl(n,α); ^{34}S(n,p); ^{18}O(^{18}O,pn)

Photons (^{34}P)

⟨γ⟩=334.2 *17* keV

γ_{mode}	γ(keV)	γ(%)†
γ [E2]	770.8 *11*	~0.0033
γ [E2]	1787.4 *10*	0.045 *15*
γ M1	1946.8 *11*	0.042 *15*
γ M1+18%E2	1987.3 *4*	0.131 *20*
γ [E2]	2127.7 *4*	15.00 *5*
γ [M1]	4074.4 *11*	0.069 *9*
γ [E2]	4114.9 *6*	0.19 *3*

† 13% uncert(syst)

Continuous Radiation (^{34}P)

⟨β-⟩=2306 keV;⟨IB⟩=8.6 keV

E_{bin}(keV)		⟨ ⟩(keV)	(%)
0 - 10	β-	0.00139	0.0276
	IB	0.063	
10 - 20	β-	0.00445	0.0295
	IB	0.062	0.43
20 - 40	β-	0.0200	0.066
	IB	0.124	0.43
40 - 100	β-	0.181	0.253
	IB	0.36	0.55
100 - 300	β-	2.96	1.41
	IB	1.10	0.61
300 - 600	β-	17.2	3.72
	IB	1.41	0.33
600 - 1300	β-	151	15.4
	IB	2.4	0.27
1300 - 2500	β-	715	37.5
	IB	2.2	0.127
2500 - 5000	β-	1409	41.8
	IB	0.85	0.028
5000 - 5374	β-	10.1	0.199
	IB	0.000163	3.2×10^{-6}

$^{34}_{16}$S (stable)

Δ: -29932.25 *21* keV
%: 4.21 *8*

$^{34}_{17}$Cl(1.5262 *25* s)

Mode: ϵ
Δ: -24439.9 *3* keV
SpA: 1.747×10^{11} Ci/g
Prod: daughter ^{34}Cl(32.2 min); ^{31}P(α,n); ^{34}S(p,n)

Continuous Radiation (^{34}Cl)

⟨β+⟩=2051 keV;⟨IB⟩=7.1 keV

E_{bin}(keV)		⟨ ⟩(keV)	(%)
0 - 10	β+	5.2×10^{-5}	0.00069
	IB	0.059	
10 - 20	β+	0.00057	0.00360
	IB	0.059	0.41
20 - 40	β+	0.0054	0.0173
	IB	0.116	0.40
40 - 100	β+	0.094	0.126
	IB	0.34	0.52
100 - 300	β+	2.46	1.14
	IB	1.03	0.57
300 - 600	β+	17.4	3.73
	IB	1.29	0.30
600 - 1300	β+	173	17.6
	IB	2.1	0.24
1300 - 2500	β+	856	44.8
	IB	1.7	0.099
2500 - 5000	β+	1002	32.5
	IB	0.38	0.0124
5000 - 5492	IB	0.00023	4.5×10^{-6}
	Σβ+		99.92

$^{34}_{17}$Cl(32.23 *14* min)

Mode: IT(44.5 *8* %), ϵ(55.5 *8* %)
Δ: -24293.6 *3* keV
SpA: 1.716×10^{8} Ci/g
Prod: ^{31}P(α,n); ^{35}Cl(n,2n); ^{24}Mg(^{12}C,pn)

Photons (^{34}Cl)

⟨γ⟩=1561 *11* keV

γ_{mode}	γ(keV)	γ(%)†
S $K_{\alpha2}$	2.307	0.026 *3*
S $K_{\alpha1}$	2.308	0.052 *5*
S $K_{\beta1}$'	2.464	0.0036 *4*
Cl $K_{\alpha2}$	2.621	0.108 *12*
Cl $K_{\alpha1}$	2.622	0.215 *24*
Cl $K_{\beta1}$'	2.816	0.0201 *22*
γ_{IT} (M3)	146.36 *3*	40.5 *5*
γ_ϵM1+2.5%E2	1176.1 *4*	14.11 *12*
γ_ϵM1+1.0%E2	1572.8 *5*	0.016 *5*
γ_ϵM1+18%E2	1987.3 *4*	0.186 *6*
γ_ϵ[E2]	2127.7 *4*	42.9 *5*
γ_ϵ[E2]	2560.4 *6*	0.034 *3*
γ_ϵM1+1.3%E2	2748.8 *5*	0.022 *3*
γ_ϵ[E2]	3303.6 *5*	12.31 *13*
γ_ϵ[E2]	4114.9 *6*	0.273 *6*

† uncert(syst): 1.7% for ϵ, 1.7% for IT

Atomic Electrons (^{34}Cl)

⟨e⟩=5.55 *22* keV

e_{bin}(keV)	⟨e⟩(keV)	e(%)
9	0.00077	0.008
144	5.08	3.54 *15*
146	0.457	0.313 *13*
1174	0.00387	0.000330 *7*
1176	0.000302	2.57 *6* ×10^{-5}
1570	3.5 ×10^{-6}	2.2 *7* ×10^{-7}
1573	2.7 ×10^{-7}	1.7 *6* ×10^{-8}
1985	3.51 ×10^{-5}	1.77 *7* ×10^{-6}
1987	2.74 ×10^{-6}	1.38 *5* ×10^{-7}
2125	0.0084	0.000396 *16*
2127	0.00066	3.08 *13* ×10^{-5}
2558	5.8 ×10^{-6}	2.26 *21* ×10^{-7}
2560	4.5 ×10^{-7}	1.76 *17* ×10^{-8}
2746	3.4 ×10^{-6}	1.25 *16* ×10^{-7}
2749	2.7 ×10^{-7}	9.8 *13* ×10^{-9}
3301	0.00183	5.55 *23* ×10^{-5}
3303	0.000143	4.32 *18* ×10^{-6}
4112	3.66 ×10^{-5}	8.9 *4* ×10^{-7}
4115	2.85 ×10^{-6}	6.9 *3* ×10^{-8}

Continuous Radiation (^{34}Cl)

⟨β+⟩=455 keV;⟨IB⟩=0.97 keV

E_{bin}(keV)		⟨ ⟩(keV)	(%)
0 - 10	β+	0.000342	0.00455
	IB	0.018	
10 - 20	β+	0.00369	0.0234
	IB	0.017	0.121
20 - 40	β+	0.0344	0.110
	IB	0.034	0.118
40 - 100	β+	0.57	0.77
	IB	0.094	0.145
100 - 300	β+	12.4	5.8
	IB	0.25	0.141
300 - 600	β+	61	13.5
	IB	0.24	0.058
600 - 1300	β+	214	24.0
	IB	0.25	0.030
1300 - 2500	β+	167	10.1
	IB	0.061	0.0040
2500 - 3511	IB	0.0023	8.3×10^{-5}
	Σβ+		54

$^{34}_{18}$Ar(844 *4* ms)

Mode: ϵ
Δ: -18379 *3* keV
SpA: 2.681×10^{11} Ci/g
Prod: ^{32}S(^{3}He,n)

Photons (^{34}Ar)

⟨γ⟩=839 *14* keV

γ_{mode}	γ(keV)	γ(%)†
γ[M1]	461.00 *4*	0.91 *9*
γ[M1]	666.54 *5*	2.5
γ[M1]	2579.9 *5*	0.863 *25*
γ[M1]	3128.7 *5*	1.30 *3*

† 4.0% uncert(syst)

Continuous Radiation (^{34}Ar)

⟨β+⟩=2289 keV;⟨IB⟩=8.5 keV

E_{bin}(keV)		⟨ ⟩(keV)	(%)
0 - 10	β+	3.96×10^{-5}	0.00052
	IB	0.063	
10 - 20	β+	0.000452	0.00286
	IB	0.063	0.43
20 - 40	β+	0.00443	0.0142
	IB	0.124	0.43
40 - 100	β+	0.079	0.106
	IB	0.36	0.55
100 - 300	β+	2.10	0.97
	IB	1.11	0.62
300 - 600	β+	14.7	3.16
	IB	1.41	0.33
600 - 1300	β+	145	14.6
	IB	2.4	0.27
1300 - 2500	β+	743	38.7
	IB	2.2	0.126
2500 - 5000	β+	1385	42.4
	IB	0.74	0.025
5000 - 6061	β+	0.0170	0.000339
	IB	0.00164	3.1×10^{-5}
	Σβ+		99.9

A = 35

NP **A310**, 397 (1978)

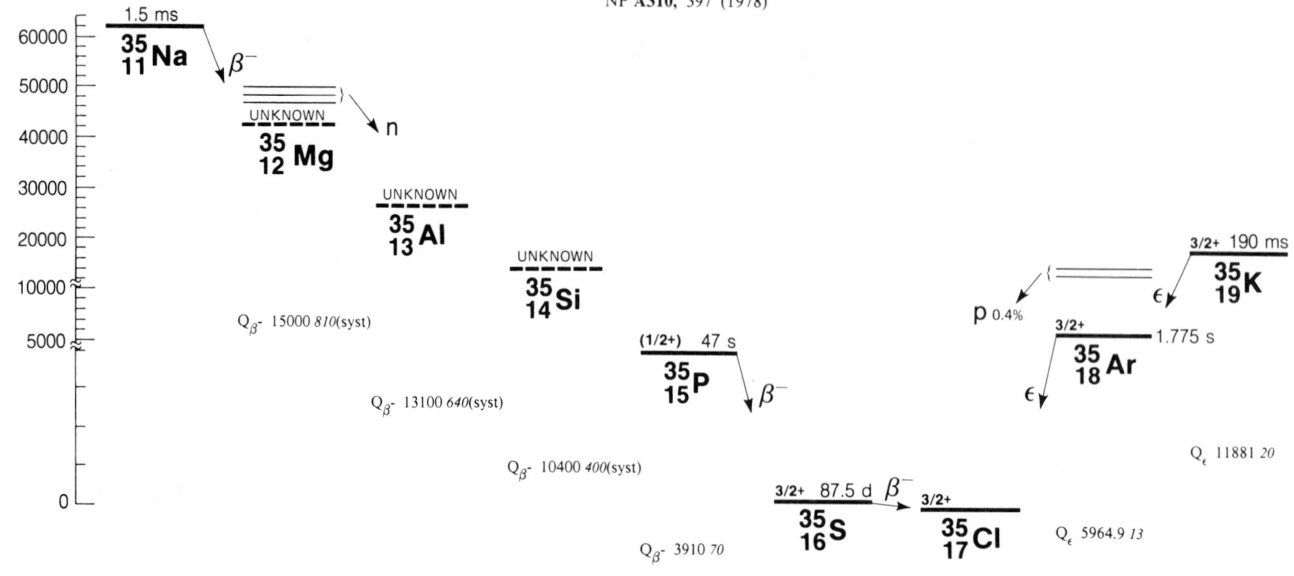

$^{35}_{11}$Na(1.5 *5* ms)

Mode: β-, β-n
SpA: 4.7×10^{11} Ci/g
Prod: protons on Ir

$^{35}_{15}$P (47.3 *7* s)

Mode: β-
Δ: -24940 *80* keV
SpA: 6.76×10^{9} Ci/g
Prod: ^{37}Cl(γ,2p); ^{37}Cl(t,αp); ^{36}S(t,α); ^{19}F(^{18}O,2p)

Photons (^{35}P)

⟨γ⟩=1572keV

γ_{mode}	γ(keV)	γ(%)
γ [M1+E2]	1572.24 *15*	100

Continuous Radiation (^{35}P)

⟨β-⟩=988 keV;⟨IB⟩=2.2 keV

E_{bin}(keV)		⟨ ⟩(keV)	(%)
0 - 10	β-	0.0071	0.140
	IB	0.037	
10 - 20	β-	0.0225	0.149
	IB	0.037	0.25
20 - 40	β-	0.100	0.332
	IB	0.072	0.25

Continuous Radiation (^{35}P) (continued)

E_{bin}(keV)		⟨ ⟩(keV)	(%)
40 - 100	β-	0.90	1.26
	IB	0.20	0.31
100 - 300	β-	13.9	6.7
	IB	0.55	0.31
300 - 600	β-	73	15.9
	IB	0.57	0.135
600 - 1300	β-	450	47.4
	IB	0.60	0.072
1300 - 2338	β-	450	28.1
	IB	0.092	0.0063

$^{35}_{16}$S (87.51 *12* d)

Mode: β-
Δ: -28846.89 *21* keV
SpA: 4.263×10^{4} Ci/g
Prod: ^{34}S(n,γ); ^{37}Cl(d,α)

Continuous Radiation (^{35}S)

⟨β-⟩=48.6 keV;⟨IB⟩=0.0086 keV

E_{bin}(keV)		⟨ ⟩(keV)	(%)
0 - 10	β-	0.64	13.0
	IB	0.0024	
10 - 20	β-	1.82	12.2
	IB	0.0018	0.0128
20 - 40	β-	6.6	22.1
	IB	0.0023	0.0082
40 - 100	β-	28.2	43.0
	IB	0.0020	0.0035
100 - 167	β-	11.4	9.7
	IB	8.9×10^{-5}	8.1×10^{-5}

$^{35}_{17}$Cl(stable)

Δ: -29013.72 *7* keV
%: 75.77 *5*

$^{35}_{18}$Ar(1.775 *4* s)

Mode: ε
Δ: -23048.8 *13* keV
SpA: 1.503×10^{11} Ci/g
Prod: ^{32}S(α,n); ^{35}Cl(p,n)

Photons (^{35}Ar)

⟨γ⟩=27.4 *3* keV

γ_{mode}	γ(keV)	γ(%)†
γ [M1+E2]	930.66 *16*	0.014 *3*
γ M1+1.1%E2	1219.10 *17*	1.244 *19*
γ [M1+E2]	1273.8 *4*	0.00015 *5*
γ [M1+E2]	1474.53 *19*	0.0103 *16*
γ E2+11%M1	1762.96 *16*	0.287 *5*
γ [M1+E2]	2155.0 *15*	0.00142 *10*
γ M1+6.4%E2	2693.58 *15*	0.1363 *22*
γ [M1]	2748.3 *4*	0.0060 *8*
γ M1+0.5%E2	3002.0 *4*	0.0901 *17*
γ [M1+E2]	3917.8 *15*	0.0065 *4*
γ [M1+E2]	3967.3 *4*	0.0013 *3*

† 2.7% uncert(syst)

Continuous Radiation (^{35}Ar)

$\langle\beta+\rangle$=2269 keV;$\langle$IB$\rangle$=8.3 keV

E_{bin}(keV)		$\langle\ \rangle$(keV)	(%)
0 - 10	β+	3.67 ×10^{-5}	0.000485
	IB	0.063	
10 - 20	β+	0.000419	0.00265
	IB	0.062	0.43
20 - 40	β+	0.00411	0.0131
	IB	0.124	0.43
40 - 100	β+	0.074	0.099
	IB	0.36	0.55
100 - 300	β+	1.97	0.91
	IB	1.10	0.61
300 - 600	β+	14.1	3.02
	IB	1.40	0.33
600 - 1300	β+	144	14.5
	IB	2.4	0.27
1300 - 2500	β+	767	39.9
	IB	2.1	0.123
2500 - 5000	β+	1342	41.4
	IB	0.68	0.022
5000 - 5965	IB	0.00138	2.6 ×10^{-5}
	Σβ+		99.93

$^{35}_{19}$K (190 *30* ms)

Mode: ε, εp(0.37 *15* %)
Δ: -11168 *20* keV
SpA: 4.5×10^{11} Ci/g
Prod: protons on Sc

Delayed Protons (^{35}K)

p(keV)	p(%)
1282 *20*	0.020 *4*
1425 *20*	0.124 *10*
1555 *20*	0.051 *4*
1705 *20*	0.052 *7*
1875 *20*	0.030 *5*
1980 *20*	0.021 *4*
2282 *20*	0.029 *5*
2425 *20*	0.024 *5*
2575 *20*	0.012 *4*
2807 *20*	0.007 *2*

Photons (^{35}K)

$\langle\gamma\rangle$=3274 *90* keV

γ_{mode}	γ(keV)	γ(%)†
γ_ϵ	887.3 *3*	0.46 *15*
γ_ϵ[M1+E2]	1044.6 *4*	0.66 *20*
γ_ϵ[M1+E2]	1184.10 *25*	7.3 *4*
γ_ϵ	1427.0 *3*	1.52 *25*
γ_ϵ	1507.7 *4*	0.96 *20*
γ_ϵ[M1+E2]	1750.78 *25*	14.2 *5*
γ_ϵ[E2]	1798.8 *3*	1.8 *3*
γ_ϵ[M1+E2]	2589.92 *10*	26.4 *10*
γ_ϵ	2638.0 *3*	2.8 *4*
γ_ϵ	2697.8 *6*	1.2 *3*
γ_ϵ	2934.7 *3*	1.8 *3*
γ_ϵ[M1+E2]	2982.80 *12*	50.8 *20*
γ_ϵ[M1]	3541.8 *5*	1.5 *3*
γ_ϵ[M1+E2]	3821.9 *3*	1.8 *4*
γ_ϵ[M1+E2]	4388.5 *3*	1.8 *4*
γ_ϵ	4528.1 *4*	1.3 *4*
γ_ϵ[M1+E2]	4725.8 *6*	0.61 *25*
γ_ϵ	4785.5 *11*	1.0 *4*
γ_ϵ[M1+E2]	5572.48 *15*	3.1 *8*

† 5.7% uncert(syst)

Continuous Radiation (^{35}K)

$\langle\beta+\rangle$=3569 keV;$\langle$IB$\rangle$=17 keV

E_{bin}(keV)		$\langle\ \rangle$(keV)	(%)
0 - 10	β+	1.40 ×10^{-5}	0.000184
	IB	0.078	
10 - 20	β+	0.000168	0.00106
	IB	0.078	0.54
20 - 40	β+	0.00171	0.0054
	IB	0.154	0.53
40 - 100	β+	0.0315	0.0423
	IB	0.45	0.70
100 - 300	β+	0.87	0.403
	IB	1.43	0.79
300 - 600	β+	6.4	1.38
	IB	1.9	0.45
600 - 1300	β+	70	7.0
	IB	3.7	0.41
1300 - 2500	β+	436	22.4
	IB	4.3	0.24
2500 - 5000	β+	1682	46.5
	IB	4.0	0.118
5000 - 7500	β+	1076	17.9
	IB	0.88	0.0151
7500 - 10859	β+	297	3.56
	IB	0.082	0.00100
	Σβ+		100

A = 36

NP **A310**, 420 (1978)

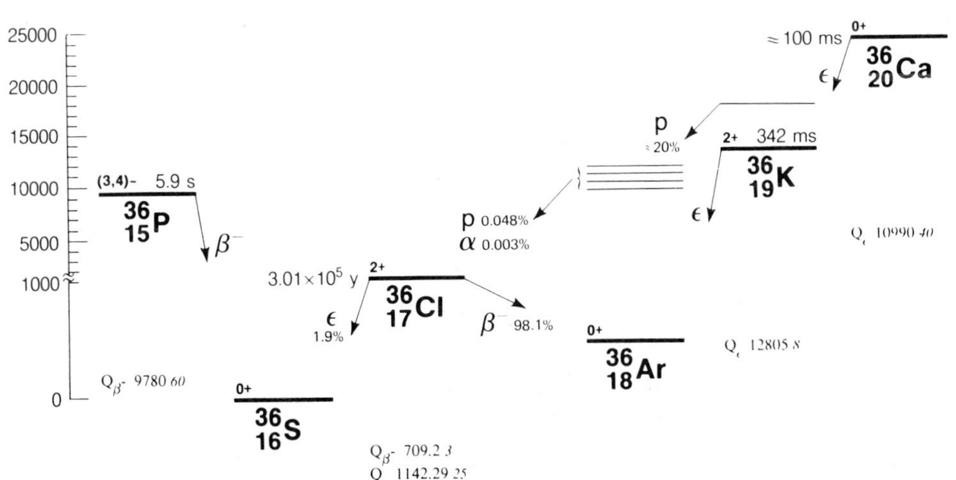

$^{36}_{15}$P (5.9 *4* s)

Mode: β-
Δ: -20890 *60* keV
SpA: 5.0×10^{10} Ci/g
Prod: ^{37}Cl(n,2p)

Photons (^{36}P)

⟨γ⟩=4608 *360* keV

γ_{mode}	γ(keV)	γ(%)†
γ[E1]	902.0 *5*	77 *11*
γ	1638.1 *8*	38 *6*
γ[E2]	3290.7 *3*	100

† 10% uncert(syst)

$^{36}_{16}$S (stable)

Δ: -30664.44 *25* keV
%: 0.02 *1*

$^{36}_{17}$Cl(3.01 *2* $\times 10^{5}$ yr)

Mode: β-(98.1 *1* %), ϵ(1.9 *1* %)
Δ: -29522.15 *7* keV
SpA: 0.03299 Ci/g
Prod: ^{35}Cl(n,γ)

Photons (^{36}Cl)

⟨γ⟩=0.00308 *23* keV

γ_{mode}	γ(keV)	γ(%)†
S $K_{\alpha 2}$	2.307	0.042 *4*
S $K_{\alpha 1}$	2.308	0.085 *9*
S $K_{\beta 1}$'	2.464	0.0059 *6*

† 5.3% uncert(syst)

Continuous Radiation (^{36}Cl)

⟨β-⟩=246 keV;⟨β+⟩=0.0075 keV;⟨IB⟩=0.20 keV

E_{bin}(keV)		⟨ ⟩(keV)	(%)
0 - 10	β-	0.081	1.60
	β+	1.94×10^{-5}	0.000259
	IB	0.0122	
10 - 20	β-	0.248	1.65
	β+	0.000182	0.00116
	IB	0.0115	0.080
20 - 40	β-	1.06	3.52
	β+	0.00125	0.00407
	IB	0.021	0.073
40 - 100	β-	8.4	11.9
	β+	0.0058	0.0091
	IB	0.050	0.078
100 - 300	β-	87	43.5
	β+	0.000281	0.000267
	IB	0.079	0.048
300 - 600	β-	143	34.8
	IB	0.020	0.0054
600 - 1142	β-	6.9	1.10
	IB	0.0035	0.00047
	Σβ+		0.0149

$^{36}_{18}$Ar(stable)

Δ: -30231.4 *3* keV
%: 0.337 *3*

$^{36}_{19}$K (342 *2* ms)

Mode: ϵ, ϵp(0.048 *14* %), ϵα(0.0034 *13* %)
Δ: -17426 *8* keV
SpA: 3.925×10^{11} Ci/g
Prod: ^{36}Ar(p,n); ^{45}Sc(p,7n3p)

Delayed Protons (^{36}K)

p(keV)	p(%)
501 *10*	0.00035 *12*
693 *5*	0.0075 *22*
849 *5*	0.0019 *6*
970 *5*	0.033 *9*
1333 *5*	0.0035 *10*
1530 *10*	0.00030 *12*
1874 *10*	0.00011 *5*
1992 *10*	0.00048 *19*
2048 *10*	0.00047 *18*
2458 *10*	0.00029 *12*
2640 *10*	0.00020 *9*

Delayed Alphas (^{36}K)

α(keV)	α(%)
2015 *5*	0.0015 *5*
2213 *10*	0.00044 *18*
2430 *15*	~0.00003
2553 *15*	~0.00002
2725 *5*	0.0010 *4*
3146 *15*	~0.00002
3271 *15*	~0.00003
3375 *15*	0.00006 *3*
3479 *15*	0.00011 *5*
3516 *15*	0.00008 *4*
3922 *15*	0.00004 *2*
4443 *20*	0.00004 *2*

Photons (^{36}K)

⟨γ⟩=4458 *85* keV

γ_{mode}	γ(keV)	γ(%)†
γ[E2]	1970.90 *21*	82.0
γ[M1+E2]	2170.93 *17*	3.0 *3* ?
γE1+0.3%M2	2208.07 *21*	29.9 *14*
γ[E1]	2433.75 *18*	31.8 *15*
γ	2444.46 *20*	0.17 *6*
γE2+<31%M1	2470.89 *21*	4.8 *19*
γ[M1+E2]	2699.6 *4*	0.39 *3*
γ[M1+E2]	2897.4 *3*	0.99 *7*
γM1	2923.8 *4*	0.17 *6*
γ[E1]	3160.2 *3*	0.21 *7*
γ[E3]	4178.85 *22*	2.05 *8*
γ[E2]	4441.65 *21*	8.0 *5*
γ[M1+E2]	4641.66 *21*	0.72 *8*
γ[M1+E2]	4760.3 *4*	0.44 *7*
γ	4896.4 *8*	0.34 *7*
γ	4950.5 *9*	0.16 *5* ?
γ	4977.6 *10*	0.14 *5* ?
γ[M1+E2]	5170.3 *4*	0.30 *7*
γ[M1+E2]	5207.2 *3*	0.31 *7*
γ[M1+E2]	5368.0 *3*	0.54 *7*
γ[M1+E2]	5738.5 *4*	0.36 *6*
γ	6162.5 *14*	0.031 *11*
γ[M1+E2]	6585.5 *5*	0.16 *6*
γ[E2]	6612.29 *21*	6.6 *5*

Photons (^{36}K)
(continued)

γ_{mode}	γ(keV)	γ(%)†
γ	6730.9 *4*	0.45 *6*
γ	6867.0 *7*	0.09 *3*
γ	7177.8 *4*	0.38 *5*
γ[M1+E2]	7532 *1*	0.127 *25*
γ[M1]	7709.0 *4*	0.172 *25*
γ	7971.0 *7*	0.127 *20*
γ[M1]	8133.1 *14*	0.053 *16*
γ[M1]	9218.8 *12*	0.049 *12*

† 2.4% uncert(syst)

Continuous Radiation (^{36}K)

⟨β+⟩=3517 keV;⟨IB⟩=16.6 keV

E_{bin}(keV)		⟨ ⟩(keV)	(%)
0 - 10	β+	1.76×10^{-5}	0.000231
	IB	0.078	
10 - 20	β+	0.000210	0.00133
	IB	0.078	0.54
20 - 40	β+	0.00213	0.0068
	IB	0.154	0.53
40 - 100	β+	0.0393	0.053
	IB	0.45	0.69
100 - 300	β+	1.08	0.499
	IB	1.42	0.79
300 - 600	β+	7.9	1.69
	IB	1.9	0.45
600 - 1300	β+	83	8.4
	IB	3.6	0.41
1300 - 2500	β+	488	25.2
	IB	4.2	0.23
2500 - 5000	β+	1597	44.8
	IB	3.8	0.112
5000 - 7500	β+	1059	17.5
	IB	0.87	0.0150
7500 - 10835	β+	281	3.45
	IB	0.050	0.00063
	Σβ+		99.96

$^{36}_{20}$Ca(~100 ms)

Mode: ϵ, ϵp(~20 %)
Δ: -6440 *40* keV
SpA: $\sim 5 \times 10^{11}$ Ci/g
Prod: ^{40}Ca(^{3}He,α3n)

ϵp: 2519 *21*

A = 37

NP **A310**, 450 (1978)

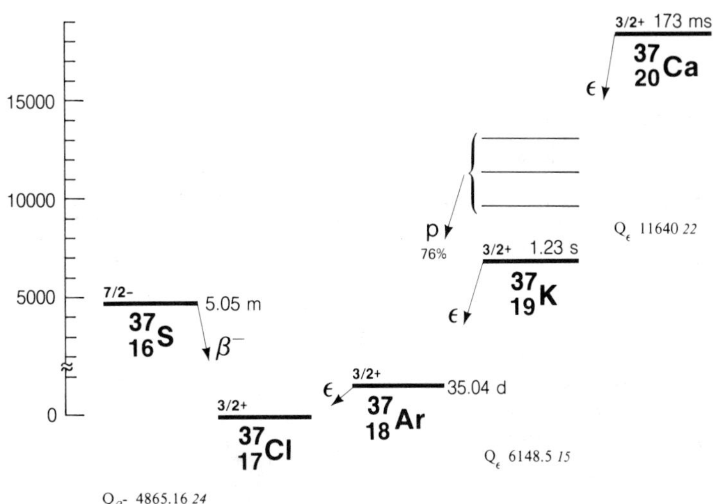

Q_{β^-} 4865.16 *24*

Q_ϵ 813.9 *6*

$^{37}_{16}$S (5.05 *2* min)

Mode: β-

Δ: -26896.6 *3* keV

SpA: 1.005×10^9 Ci/g

Prod: ^{37}Cl(n,p); ^{40}Ar(n,α); ^{36}S(n,γ)

Photons (^{37}S)

⟨γ⟩=2932 *18* keV

γ$_{mode}$	γ(keV)	γ(%)†
γ M1+29%E2	906.77 *9*	0.054 *7*
γ [M1+E2]	1169.55 *20*	0.034 *7*
γ E2+33%M1	3086.63 *20*	0.062 *21*
γ M2+3.1%E3	3103.98 *9*	94.0
γ E1+0.6%M2	3741.59 *10*	0.26 *3*
γ E3	4010.67 *12*	0.027 *10*
γ [E1]	4396.57 *20*	~0.0038

† 0.64% uncert(syst)

Continuous Radiation (^{37}S)

⟨β-⟩=800 keV;⟨IB⟩=1.64 keV

E$_{bin}$(keV)		⟨ ⟩(keV)	(%)
0 - 10	β-	0.0132	0.262
	IB	0.031	
10 - 20	β-	0.0416	0.277
	IB	0.031	0.21
20 - 40	β-	0.184	0.61
	IB	0.059	0.21
40 - 100	β-	1.60	2.25
	IB	0.165	0.25
100 - 300	β-	23.4	11.3
	IB	0.43	0.24
300 - 600	β-	110	24.1
	IB	0.41	0.097
600 - 1300	β-	457	50.3
	IB	0.35	0.043
1300 - 2500	β-	133	8.6
	IB	0.123	0.0071
2500 - 4865	β-	75	2.30
	IB	0.039	0.00131

$^{37}_{17}$Cl(stable)

Δ: -31761.75 *10* keV

%: 24.23 *5*

$^{37}_{18}$Ar(35.04 *4* d)

Mode: ϵ

Δ: -30947.9 *6* keV

SpA: 1.0071×10^5 Ci/g

Prod: ^{37}Cl(p,n); ^{37}Cl(d,2n); ^{34}S(α,n); ^{39}K(d,α); ^{40}Ca(n,α); ^{36}Ar(n,γ)

Photons (^{37}Ar)

⟨γ⟩=0.230 *17* keV

γ$_{mode}$	γ(keV)	γ(%)
Cl L$_\beta$	0.239	0.0017 *6*
Cl K$_{\alpha2}$	2.621	2.7 *3*
Cl K$_{\alpha1}$	2.622	5.5 *6*
Cl K$_{\beta1}$'	2.816	0.51 *5*

Continuous Radiation (^{37}Ar)

⟨IB⟩=0.139 keV

E$_{bin}$(keV)		⟨ ⟩(keV)	(%)
10 - 20	IB	2.0×10^{-5}	0.000131
20 - 40	IB	0.000117	0.00038
40 - 100	IB	0.0018	0.0024
100 - 300	IB	0.034	0.0163
300 - 600	IB	0.087	0.020
600 - 814	IB	0.0164	0.0025

$^{37}_{19}$K (1.226 *7* s)

Mode: ϵ

Δ: -24799.4 *14* keV

SpA: 1.900×10^{11} Ci/g

Prod: ^{40}Ca(p,α); ^{36}Ar(^{3}He,d)

Photons (^{37}K)

⟨γ⟩=40 *5* keV

γ$_{mode}$	γ(keV)	γ(%)
γ M1+2.5%E2	2796.0 *3*	1.42 *16*
γ M1+6.1%E2	3601.8 *7*	0.014 *2*

Continuous Radiation (^{37}K)

⟨β+⟩=2353 keV;⟨IB⟩=8.8 keV

E$_{bin}$(keV)		⟨ ⟩(keV)	(%)
0 - 10	β+	3.06×10^{-5}	0.000401
	IB	0.064	
10 - 20	β+	0.000366	0.00231
	IB	0.064	0.44
20 - 40	β+	0.00371	0.0118
	IB	0.126	0.44
40 - 100	β+	0.068	0.091
	IB	0.37	0.56
100 - 300	β+	1.86	0.86
	IB	1.13	0.63
300 - 600	β+	13.3	2.86
	IB	1.45	0.34
600 - 1300	β+	136	13.7
	IB	2.5	0.28
1300 - 2500	β+	727	37.8
	IB	2.3	0.132
2500 - 5000	β+	1474	44.6
	IB	0.83	0.028
5000 - 6149	β+	0.57	0.0112
	IB	0.0024	4.6×10^{-5}
	Σβ+		99.92

$^{37}_{20}$Ca(173 *4* ms)

Mode: ϵ, ϵp(76 *3* %)

Δ: -13160 *22* keV

SpA: 4.319×10^{11} Ci/g

Prod: ^{39}K(p,3n); ^{40}Ca(p,d2n); ^{36}Ar(^{3}He,2n)

Delayed Protons (^{37}Ca)

p(keV)	p(%)
870 *15*	6.4 *7*
1709 *10*	4.0 *4*
1925 *10*	4.5 *3*
2498 *20*	1.15 *14*
2580 *20*	2.08 *20*
2745 *20*	1.2 *5*
3063 *15*	1.9 *6*
3103 *3*	46.7
3173 *10*	5.9 *5*
3339 *30*	1.0 *3*
3487 *20*	0.51 *19*
4046 *20*	0.65 *9*

Photons (^{37}Ca)

$\langle\gamma\rangle$=115 keV

γ_{mode}	γ(keV)	$\gamma(\%)^{\dagger}$
γ_{ϵ}	1369 *2*	8.4

† 30% uncert(syst)

Continuous Radiation (^{37}Ca)

$\langle\beta+\rangle$=3266 keV;$\langle$IB$\rangle$=14.9 keV

E_{bin}(keV)		$\langle\ \rangle$(keV)	(%)
0 - 10	β+	1.44×10^{-5}	0.000188
	IB	0.075	
10 - 20	β+	0.000180	0.00114
	IB	0.075	0.52
20 - 40	β+	0.00189	0.0060
	IB	0.148	0.51
40 - 100	β+	0.0359	0.0482
	IB	0.44	0.67
100 - 300	β+	1.01	0.467
	IB	1.37	0.76
300 - 600	β+	7.5	1.61
	IB	1.8	0.42
600 - 1300	β+	82	8.2
	IB	3.4	0.38
1300 - 2500	β+	506	26.0
	IB	3.8	0.22
2500 - 5000	β+	1759	49.4
	IB	3.1	0.092
5000 - 7500	β+	699	11.7
	IB	0.59	0.0101
7500 - 11640	β+	212	2.55
	IB	0.053	0.00066
	$\Sigma\beta$+		99.9

A = 38

NP **A310**, 474 (1978)

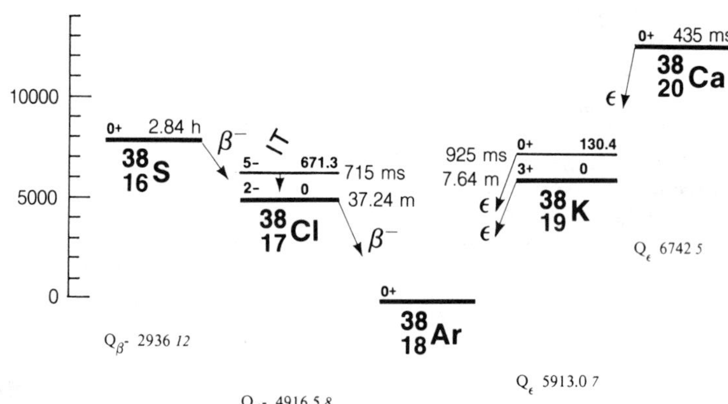

$^{38}_{16}$S (2.84 *1* h)

Mode: β-
Δ: -26862 *12* keV
SpA: 2.904×10^{7} Ci/g
Prod: ^{37}Cl(α,3p); ^{36}S(t,p); ^{40}Ar(γ,2p)

Photons (^{38}S)

$\langle\gamma\rangle$=1725 *40* keV

γ_{mode}	γ(keV)	$\gamma(\%)^{\dagger}$
γ	1746.2 *2*	2.9 *3*
γ[E1]	1941.7 *2*	84
γ[E1]	2751.6 *5*	1.60 *17*

† 2.4% uncert(syst)

Continuous Radiation (^{38}S)

$\langle\beta-\rangle$=480 keV;$\langle$IB$\rangle$=0.76 keV

E_{bin}(keV)		$\langle\ \rangle$(keV)	(%)
0 - 10	β-	0.0471	0.94
	IB	0.021	
10 - 20	β-	0.145	0.96
	IB	0.020	0.138
20 - 40	β-	0.61	2.04
	IB	0.038	0.132
40 - 100	β-	4.79	6.8
	IB	0.100	0.155
100 - 300	β-	54	26.7
	IB	0.23	0.131
300 - 600	β-	168	38.1
	IB	0.166	0.040
600 - 1300	β-	144	18.8
	IB	0.136	0.0162
1300 - 2500	β-	101	5.6
	IB	0.053	0.0033
2500 - 2936	β-	8.2	0.313
	IB	0.00030	1.17×10^{-5}

$^{38}_{17}$Cl(37.24 *5* min)

Mode: β-
Δ: -29798.23 *14* keV
SpA: 1.3285×10^{8} Ci/g
Prod: ^{37}Cl(n,γ); ^{37}Cl(d,p); daughter ^{38}S

Photons (^{38}Cl)

$\langle\gamma\rangle$=1421 *44* keV

γ_{mode}	γ(keV)	$\gamma(\%)^{\dagger}$
γ E1	1642.69 *15*	31.0 *4*
γ[E2]	2167.68 *11*	42
γ[E3]	3810.27 *19*	0.0256 *21*

† 4.8% uncert(syst)

Continuous Radiation (^{38}Cl)

$\langle\beta-\rangle$=1553 keV;$\langle$IB$\rangle$=5.2 keV

E_{bin}(keV)		$\langle\ \rangle$(keV)	(%)
0 - 10	β-	0.0132	0.262
	IB	0.046	
10 - 20	β-	0.0412	0.274
	IB	0.046	0.32
20 - 40	β-	0.179	0.59
	IB	0.090	0.31
40 - 100	β-	1.51	2.12
	IB	0.26	0.40
100 - 300	β-	19.9	9.7
	IB	0.75	0.42
300 - 600	β-	74	16.6
	IB	0.91	0.21
600 - 1300	β-	188	21.2
	IB	1.45	0.166
1300 - 2500	β-	473	25.2
	IB	1.26	0.072
2500 - 4917	β-	795	24.1
	IB	0.42	0.0142

$^{38}_{17}$Cl(715 *3* ms)

Mode: IT
Δ: -29126.90 *24* keV
SpA: 2.658×10^{11} Ci/g
Prod: ^{37}Cl(n,γ)

Photons (^{38}Cl)

⟨γ⟩=671.0 *3* keV

γ_{mode}	γ(keV)	γ(%)†
γ M3	671.33 *20*	99.95

† <0.1% uncert(syst)

$^{38}_{18}$Ar(stable)

Δ: -34714.7 *8* keV
%: 0.063 *1*

$^{38}_{19}$K (7.636 *18* min)

Mode: ε
Δ: -28801.7 *9* keV
SpA: 6.475×10^{8} Ci/g
Prod: ^{35}Cl(α,n); ^{40}Ca(d,α); ^{36}Ar(^{3}He,p)

Photons (^{38}K)

⟨γ⟩=2170.4 *6* keV

γ_{mode}	γ(keV)	γ(%)
γ [M1+E2]	1768.2 *5*	0.0094 *13*
γ [E2]	2167.68 *11*	99.858 *13*
γ [E2]	3935.8 *5*	0.142 *11*

Continuous Radiation (^{38}K)

⟨β+⟩=1204 keV;⟨IB⟩=3.1 keV

E_{bin}(keV)		⟨ ⟩(keV)	(%)
0 - 10	β+	0.000141	0.00185
	IB	0.043	
10 - 20	β+	0.00168	0.0106
	IB	0.042	0.29
20 - 40	β+	0.0170	0.054
	IB	0.083	0.29
40 - 100	β+	0.307	0.413
	IB	0.24	0.36
100 - 300	β+	7.9	3.69
	IB	0.67	0.38
300 - 600	β+	52	11.1
	IB	0.74	0.17
600 - 1300	β+	394	40.8
	IB	0.93	0.109
1300 - 2500	β+	742	43.0
	IB	0.30	0.019
2500 - 3745	β+	8.6	0.336
	IB	0.0158	0.00056
	Σβ+		99.5

$^{38}_{19}$K (924.6 *15* ms)

Mode: ε
Δ: -28671.3 *10* keV
SpA: 2.259×10^{11} Ci/g
Prod: ^{35}Cl(α,n); ^{39}K(γ,n); ^{40}Ca(d,α); ^{36}Ar(t,n)

Continuous Radiation (^{38}K)

⟨β+⟩=2320 keV;⟨IB⟩=8.6 keV

E_{bin}(keV)		⟨ ⟩(keV)	(%)
0 - 10	β+	2.95×10^{-5}	0.000387
	IB	0.064	
10 - 20	β+	0.000353	0.00223
	IB	0.063	0.44
20 - 40	β+	0.00358	0.0114
	IB	0.125	0.43
40 - 100	β+	0.066	0.089
	IB	0.37	0.56
100 - 300	β+	1.81	0.84
	IB	1.12	0.62
300 - 600	β+	13.1	2.81
	IB	1.43	0.33
600 - 1300	β+	137	13.8
	IB	2.5	0.28
1300 - 2500	β+	751	39.0
	IB	2.2	0.129
2500 - 5000	β+	1418	43.4
	IB	0.76	0.025
5000 - 6043	β+	0.00301	6.0×10^{-5}
	IB	0.0020	3.8×10^{-5}
	Σβ+		99.92

$^{38}_{20}$Ca(435 *9* ms)

Mode: ε
Δ: -22060 *5* keV
SpA: 3.41×10^{11} Ci/g
Prod: ^{40}Ca(γ,2n); ^{36}Ar(^{3}He,n); ^{24}Mg(^{16}O,2n)

Photons (^{38}Ca)

⟨γ⟩=347 *25* keV

γ_{mode}	γ(keV)	γ(%)
γ [M1]	328.3 *2*	2.6 *4*
γ [E2]	458.7 *2*	0.026 *4*
γ [M1]	1567.7 *5*	21.0 *16*
γ [M1]	3210 *2*	0.29 *4*

Continuous Radiation (^{38}Ca)

⟨β+⟩=2433 keV;⟨IB⟩=9.3 keV

E_{bin}(keV)		⟨ ⟩(keV)	(%)
0 - 10	β+	2.56×10^{-5}	0.000334
	IB	0.065	
10 - 20	β+	0.000321	0.00202
	IB	0.065	0.45
20 - 40	β+	0.00336	0.0107
	IB	0.128	0.44
40 - 100	β+	0.063	0.085
	IB	0.37	0.57
100 - 300	β+	1.77	0.82
	IB	1.15	0.64
300 - 600	β+	12.8	2.75
	IB	1.48	0.35
600 - 1300	β+	132	13.3
	IB	2.6	0.29
1300 - 2500	β+	708	36.8
	IB	2.4	0.139
2500 - 5000	β+	1547	45.6
	IB	1.01	0.032
5000 - 6612	β+	31.7	0.62
	IB	0.00100	2.0×10^{-5}
	Σβ+		99.9

A = 39

NP **A310**, 504 (1978)

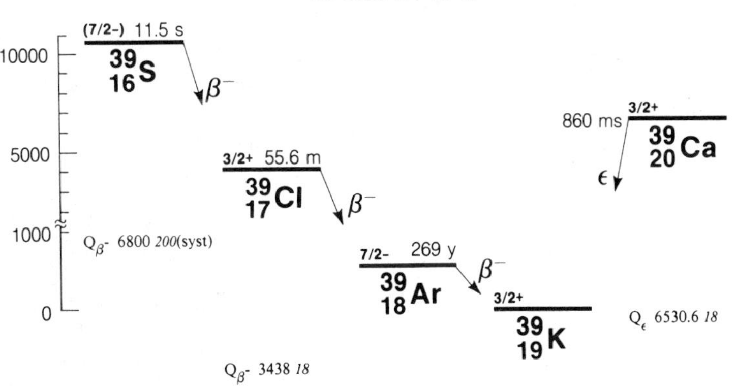

Qβ- 565 *5*

$^{39}_{16}$S (11.5 *5* s)

Mode: β-

Δ: -23000 *200* keV syst

SpA: 2.44×10^{10} Ci/g

Prod: ^{40}Ar(n,2p); ^{40}Ar on Th

Photons (^{39}S)

⟨γ⟩=1786 *90* keV

γ_{mode}	γ(keV)	γ(%)†
γ	396.11 *23*	37 *3*
γ	396.7 *6*	3.5 *11*
γ	484.85 *24*	10.8 *13*
γ	874.31 *18*	12.8 *15*
γ	903.8 *6*	3.5 *11*
γ	1300.52 *16*	52 *5*
γ	1696.62 *17*	44 *4*

† 6.1% uncert(syst)

Continuous Radiation (^{39}S)

⟨β-⟩=2282 keV;⟨IB⟩=8.4 keV

E_{bin}(keV)		⟨ ⟩(keV)	(%)
0 - 10	β-	0.00118	0.0233
	IB	0.063	
10 - 20	β-	0.00374	0.0248
	IB	0.062	0.43
20 - 40	β-	0.0167	0.055
	IB	0.123	0.43
40 - 100	β-	0.151	0.211
	IB	0.36	0.55
100 - 300	β-	2.50	1.19
	IB	1.10	0.61
300 - 600	β-	15.1	3.25
	IB	1.41	0.33
600 - 1300	β-	143	14.5
	IB	2.4	0.27
1300 - 2500	β-	746	38.8
	IB	2.2	0.124
2500 - 5000	β-	1373	41.9
	IB	0.72	0.024
5000 - 5500	β-	2.33	0.0455
	IB	5.3×10^{-5}	1.04×10^{-6}

$^{39}_{17}$Cl(55.6 *2* min)

Mode: β-

Δ: -29804 *19* keV

SpA: 8.67×10^{7} Ci/g

Prod: ^{40}Ar(α,αp); ^{40}Ar(γ,p); ^{37}Cl(t,p)

Photons (^{39}Cl)

⟨γ⟩=1452 *21* keV

γ_{mode}	γ(keV)	γ(%)†
γ [E1]	250.16 *6*	47.0 *16*
γ [M1+E2]	840.72 *12*	~0.05
γ	856.5 *13*	~0.022
γ	985.68 *14*	2.17 *11*
γ [E1]	1090.87 *11*	2.55 *11*
γ [E1]	1235.83 *15*	0.11 *4*
γ E2	1267.10 *5*	54.3
γ [M1+E2]	1311.7 *7*	0.29 *3*
γ M2+4.0%E3	1517.25 *6*	38.6 *8*
γ [E1]	1561.8 *7*	0.31 *3*
γ M1+4.5%E2	2092.9 *10*	0.081 *11*

† 2.4% uncert(syst)

Continuous Radiation (^{39}Cl)

⟨β-⟩=821 keV;⟨IB⟩=1.64 keV

E_{bin}(keV)		⟨ ⟩(keV)	(%)
0 - 10	β-	0.0132	0.262
	IB	0.032	
10 - 20	β-	0.0414	0.275
	IB	0.032	0.22
20 - 40	β-	0.181	0.60
	IB	0.061	0.21
40 - 100	β-	1.55	2.18
	IB	0.17	0.26
100 - 300	β-	22.0	10.6
	IB	0.45	0.26
300 - 600	β-	100	21.9
	IB	0.43	0.103
600 - 1300	β-	449	48.6
	IB	0.39	0.047
1300 - 2500	β-	230	14.7
	IB	0.068	0.0043
2500 - 3438	β-	18.1	0.66
	IB	0.0020	7.5×10^{-5}

$^{39}_{18}$Ar(269 *3* yr)

Mode: β-

Δ: -33242 *5* keV

SpA: 34.1 Ci/g

Prod: neutrons on KCl; ^{38}Ar(n,γ)

Continuous Radiation (^{39}Ar)

⟨β-⟩=218 keV;⟨IB⟩=0.147 keV

E_{bin}(keV)		⟨ ⟩(keV)	(%)
0 - 10	β-	0.100	1.99
	IB	0.0110	
10 - 20	β-	0.303	2.02
	IB	0.0103	0.072
20 - 40	β-	1.26	4.21
	IB	0.019	0.065
40 - 100	β-	9.7	13.8
	IB	0.042	0.067
100 - 300	β-	98	49.3
	IB	0.058	0.037
300 - 565	β-	109	28.7
	IB	0.0068	0.0020

$^{39}_{19}$K (stable)

Δ: -33806.6 *11* keV

%: 93.2581 *30*

$^{39}_{20}$Ca(859.6 *14* ms)

Mode: ε

Δ: -27276.0 *21* keV

SpA: 2.310×10^{11} Ci/g

Prod: ^{39}K(p,n); ^{40}Ca(γ,n)

Photons (^{39}Ca)

⟨γ⟩=0.057 *15* keV

γ_{mode}	γ(keV)	γ(%)
γ M1+32%E2	2522.0 *4*	0.0023 *6*

Continuous Radiation (^{39}Ca)

⟨β+⟩=2559 keV;⟨IB⟩=10.1 keV

E_{bin}(keV)		⟨ ⟩(keV)	(%)
0 - 10	β+	2.03×10^{-5}	0.000264
	IB	0.067	
10 - 20	β+	0.000254	0.00160
	IB	0.067	0.46
20 - 40	β+	0.00266	0.0085
	IB	0.132	0.46
40 - 100	β+	0.0504	0.068
	IB	0.39	0.59
100 - 300	β+	1.42	0.65
	IB	1.20	0.66
300 - 600	β+	10.4	2.23
	IB	1.55	0.36
600 - 1300	β+	112	11.3
	IB	2.7	0.31
1300 - 2500	β+	660	34.1
	IB	2.7	0.154
2500 - 5000	β+	1747	51
	IB	1.20	0.039
5000 - 6531	β+	28.2	0.55
	IB	0.0060	0.000112
	Σβ+		99.92

A = 40

NP **A310**, 529 (1978)

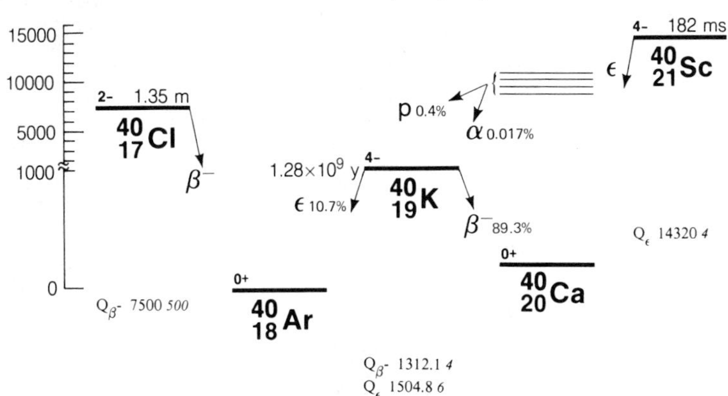

$^{40}_{17}$Cl(1.35 *2* min)

Mode: β-
Δ: -27540 *500* keV
SpA: 3.47×10^{9} Ci/g
Prod: ^{40}Ar(n,p)

Photons (^{40}Cl)

⟨γ⟩=4037 *94* keV

γ_{mode}	γ(keV)	γ(%)
γ	222.8 *4*	0.15 *5*
γ[E1]	237.88 *18*	0.15 *5*
γ	260.9 *3*	0.77 *8*
γ	302.9 *3*	~0.05
γ[E2]	315.3 *3*	0.023 *8*
γ	361.9 *4*	0.069 *15*
γ[E2]	368.9 *3*	~0.015
γ	380.6 *4*	0.08 *3*
γ[E1]	472.63 *21*	0.23 *8*
γ	479.20 *23*	0.77 *8*
γ	643.22 *18*	6.2 *5*
γ[E2]	659.90 *14*	2.39 *23*
γ[E1]	787.95 *24*	0.77 *8*
γ	881.09 *19*	2.54 *23*
γ	1042.4 *3*	0.46 *15*
γ	1050.8 *3*	0.46 *8*
γ[M1+E2]	1063.20 *12*	2.31 *15*
γ[E2]	1087.55 *21*	~0.08
γ	1092.9 *3*	0.23 *3*
γ[E1]	1156.87 *18*	0.46 *8*
γ	1187.0 *3*	0.69 *8*
γ	1318.6 *4*	0.39 *5*
γ	1333.5 *8*	0.31 *5*
γ	1353.72 *22*	0.19 *8*
γ[M1+E2]	1394.75 *15*	1.15 *15*
γ[E2]	1432.1 *3*	1.46 *3*
γ[E2]	1460.832 *10*	77.0
γ	1558.76 *21*	0.46 *5*
γ	1579.5 *4*	0.31 *8*
γ	1588.92 *25*	0.92 *15*
γ[M1+E2]	1747.43 *18*	2.70 *23*
γ	1777.09 *25*	0.0154 *23*
γ[M1+E2]	1798.04 *15*	2.5 *3*
γ	2050.36 *25*	1.00 *15*
γ	2061.5 *3*	0.39 *15*
γ[E1]	2220.04 *15*	7.3 *6*
γ[M1+E2]	2457.91 *11*	5.1 *5*
γ[E2]	2523.99 *12*	2.08 *23*
γ	2621.92 *19*	15.1 *12*
γ	2840.24 *23*	29.7 *23*
γ	3101.09 *17*	10.8 *15*
γ	3193.8 *10*	~0.08
γ[E2]	3208.20 *18*	0.46 *8*
γ	3356.5 *4*	0.31 *8*
γ	3511.11 *25*	0.15 *6*

Photons (^{40}Cl) (continued)

γ_{mode}	γ(keV)	γ(%)
γ	3704.8 *6*	0.77 *8*
γ	3759.8 *4*	0.077 *23*
γ	3785.0 *6*	0.62 *8*
γ[E2]	3918.65 *11*	3.9 *4*
γ	3941.78 *20*	0.15 *4*
γ	4082.64 *19*	0.23 *5*
γ	4147.8 *10*	0.85 *8*
γ	4178.8 *3*	0.23 *5*
γ	4324.3 *3*	0.15 *4*
γ	4357.7 *3*	0.39 *5*
γ	4480.8 *3*	0.23 *5*
γ	4580.5 *3*	0.08 *3*
γ	4737.6 *4*	0.039 *8*
γ	4768.8 *3*	0.46 *8*
γ	5165.4 *6*	~0.08
γ	5309.7 *10*	~0.15
γ	5400.2 *8*	0.15 *6*
γ	5629.1 *10*	~0.08
γ	5880.2 *4*	3.8 *3*
γ	5950.1 *10*	~0.039
γ	6053.2 *8*	0.31 *5*
γ	6208.1 *8*	~0.039
γ	6338.6 *8*	0.23 *4*
γ	6475.6 *8*	0.154 *23*

Continuous Radiation (^{40}Cl)

⟨β-⟩=1571 keV;⟨IB⟩=5.0 keV

E_{bin}(keV)		⟨ ⟩(keV)	(%)
0 - 10	β-	0.00478	0.095
	IB	0.049	
10 - 20	β-	0.0150	0.100
	IB	0.048	0.34
20 - 40	β-	0.066	0.219
	IB	0.095	0.33
40 - 100	β-	0.58	0.81
	IB	0.27	0.42
100 - 300	β-	8.7	4.17
	IB	0.80	0.45
300 - 600	β-	44.7	9.7
	IB	0.95	0.22
600 - 1300	β-	296	30.8
	IB	1.40	0.162
1300 - 2500	β-	761	41.8
	IB	0.93	0.054
2500 - 5000	β-	352	11.0
	IB	0.38	0.0116
5000 - 7500	β-	107	1.86
	IB	0.034	0.00061

$^{40}_{18}$Ar(stable)

Δ: -35039.6 *13* keV
%: 99.600 *3*

$^{40}_{19}$K (1.277 *8* ×10^{9} yr)

Mode: β-(89.33 *11* %), ϵ(10.67 *11* %)
Δ: -33534.8 *11* keV
SpA: 5.65×10^{-6} Ci/g
Prod: natural source
%: 0.0117 *1*

Photons (^{40}K)

⟨γ⟩=156.1 *16* keV

γ_{mode}	γ(keV)	γ(%)†
Ar L_{β}	0.276	0.016 *5*
Ar $K_{\alpha2}$	2.955	2.8 *3*
Ar $K_{\alpha1}$	2.957	5.7 *6*
Ar $K_{\beta1}$'	3.190	0.68 *7*
γ_{ϵ}[E2]	1460.832 *10*	10.67

† 1.0% uncert(syst)

Continuous Radiation (^{40}K)

⟨β-⟩=455 keV;⟨β+⟩=0.00089 keV;⟨IB⟩=0.62 keV

E_{bin}(keV)		⟨ ⟩(keV)	(%)
0 - 10	β-	0.0249	0.495
	β+	2.94×10^{-8}	3.87×10^{-7}
	IB	0.020	
10 - 20	β-	0.077	0.51
	β+	3.41×10^{-7}	2.16×10^{-6}
	IB	0.020	0.136
20 - 40	β-	0.332	1.10
	β+	3.25×10^{-6}	1.04×10^{-5}
	IB	0.037	0.13
40 - 100	β-	2.77	3.9
	β+	4.91×10^{-5}	6.7×10^{-5}
	IB	0.100	0.155
100 - 300	β-	36.8	17.9
	β+	0.00058	0.000295
	IB	0.23	0.133
300 - 600	β-	144	32.0
	β+	0.000260	7.4×10^{-5}
	IB	0.161	0.039

Continuous Radiation (^{40}K)
(continued)

E_{bin}(keV)		⟨ ⟩(keV)	(%)
600 - 1300	β-	270	33.4
	IB	0.052	0.0074
1300 - 1505	β-	0.00425	0.000326
	IB	1.06×10^{-7}	7.9×10^{-9}
	Σβ+		0.00045

$^{40}_{20}$Ca(stable)

Δ: -34846.9 *12* keV
%: 96.941 *13*

$^{40}_{21}$Sc(182.3 *7* ms)

Mode: ε, εp(0.44 *7* %), εα(0.017 *5* %)
Δ: -20527 *4* keV
SpA: 3.978×10^{11} Ci/g
Prod: ^{40}Ca(p,n); ^{40}Ca(^{3}He,2np)

Delayed Protons (^{40}Sc)

p(keV)	p(%)
1006 *3*	0.072 *11*
1060 *8*	0.044 *8*
1071 *6*	0.055 *10*
1095 *3*	0.110 *17*
1241 *3*	0.032 *5*
1445 *4*	0.0088 *15*
1463 *8*	0.0026 *7*
1552 *3*	0.0050 *9*
1609 *5*	~0.0009
1678 *4*	0.0042 *9*
1752 *4*	0.0013 *4*
1835 *4*	0.0139 *22*
1953 *4*	0.00046 *20*
1986 *8*	0.00030 *20*
2065 *4*	0.0028 *5*
2089 *4*	0.0094 *14*
2121 *4*	0.0125 *19*
2197 *5*	0.0017 *4*
2211 *10*	0.00035 *20*
2305 *5*	0.00076 *30*
2365 *8*	0.00092 *30*
2386 *5*	0.0128 *20*
2423 *9*	0.00081 *30*
2457 *5*	0.0038 *2*
2516 *5*	0.00035 *20*
2562 *8*	0.0020 *4*
2578 *7*	0.0020 *4*
2641 *7*	0.00069 *20*
2716 *6*	0.0011 *3*
2743 *6*	0.0023 *4*
2816 *5*	0.0068 *11*
2912 *5*	0.00051 *20*
3012 *7*	~0.0003
3045 *9*	0.00083 *20*
3205 *10*	0.00024 *10*
3308 *10*	0.00073 *30*
3376 *10*	~0.0003
3584 *10*	~0.0001
3613 *10*	0.00024 *10*
3649 *10*	~0.0001

Delayed Alphas (^{40}Sc)

α(keV)	α(%)
2089 *6*	0.00088 *20*
2620 *8*	~0.00016
2780 *8*	~0.0002
2802 *8*	0.00032 *10*
2837 *8*	0.00021 *10*
3082 *7*	0.00078 *20*
3132 *7*	0.00083 *20*
3203 *7*	0.00069 *20*
3316 *5*	0.0059 *12*
3401 *7*	0.00042 *20*
3552 *12*	~0.0001
3643 *12*	~0.0001
3748 *5*	0.0038 *8*
3839 *7*	0.00024 *10*
3988 *7*	0.00036 *10*
4058 *6*	0.00066 *20*
4160 *7*	0.00023 *10*
4218 *7*	~0.00009
4320 *6*	0.00028 *10*
4462 *7*	0.00050 *20*
4519 *9*	~0.00016

Photons (^{40}Sc)
⟨γ⟩=6085 *120* keV

γ_{mode}	γ(keV)	γ(%)†
γ[E2]	752.1 *6*	41 *4*
γ[M1+E2]	1122.2 *7*	11.9 *20*
γ[M1+E2]	1874.3 *6*	24.9 *15*
γ[M1+E2]	2042.3 *6*	25.4 *15*
γ[M1+E2]	2840 *3*	2.1 *10*
γ[M1+E2]	3164.4 *6*	11.9 *20*
γ[E3]	3732.1 *8*	99.5
γ[M1+E2]	3916.5 *6*	12.9 *20*

† 0.10% uncert(syst)

Continuous Radiation (^{40}Sc)
⟨β+⟩=3391 keV;⟨IB⟩=15.6 keV

E_{bin}(keV)		⟨ ⟩(keV)	(%)
0 - 10	β+	1.10×10^{-5}	0.000143
	IB	0.077	
10 - 20	β+	0.000145	0.00091
	IB	0.076	0.53
20 - 40	β+	0.00157	0.00498
	IB	0.152	0.53
40 - 100	β+	0.0306	0.0410
	IB	0.45	0.68
100 - 300	β+	0.88	0.406
	IB	1.40	0.78
300 - 600	β+	6.6	1.41
	IB	1.9	0.44
600 - 1300	β+	73	7.3
	IB	3.5	0.40
1300 - 2500	β+	463	23.8
	IB	4.1	0.23
2500 - 5000	β+	1813	50.2
	IB	3.4	0.102
5000 - 7500	β+	909	15.2
	IB	0.55	0.0096
7500 - 9561	β+	125	1.57
	IB	0.0168	0.00021
	Σβ+		100

A = 41

NP **A310**, 563 (1978)

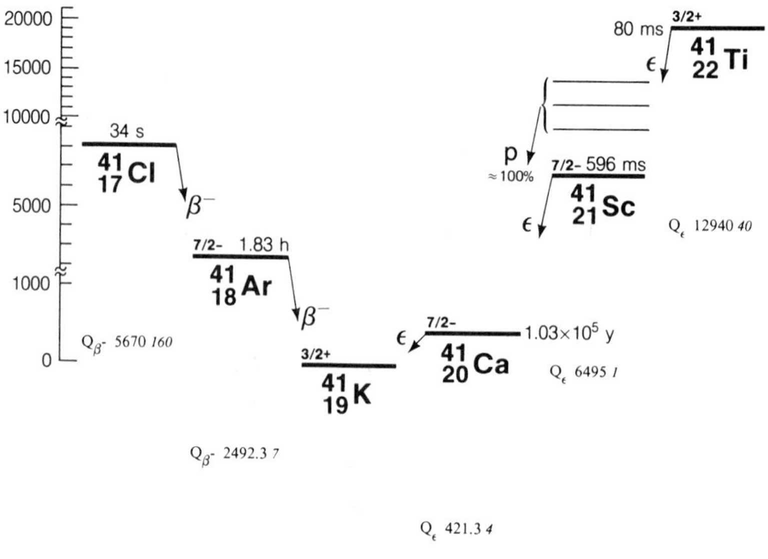

$^{41}_{17}$Cl(34 *3* s)

Mode: β-
Δ: -27400 *160* keV
SpA: 8.0×10^{9} Ci/g
Prod: ^{40}Ar on ^{181}Ta

Photons (^{41}Cl)

γ_{mode}	γ(keV)
γ	167 *1*
γ	349 *1*
γ	515 *1*
γ	516 *1*
γ	519 *1*
γ	838 *1*
γ	868 *1*
γ	1187 *1*
γ	1353 *1*
γ	1359 *1*

$^{41}_{18}$Ar(1.827 *7* h)

Mode: β-
Δ: -33067.4 *14* keV
SpA: 4.184×10^{7} Ci/g
Prod: ^{40}Ar(n,γ)

Photons (^{41}Ar)

⟨γ⟩=1283.6 *3* keV

γ_{mode}	γ(keV)	γ(%)†
γ (M2)	1293.64 *4*	99.16
γ (E2)	1677.0 *3*	0.052 *5*

†<0.1% uncert(syst)

Continuous Radiation (^{41}Ar)

⟨β-⟩=464 keV;⟨IB⟩=0.59 keV

E_{bin}(keV)		⟨ ⟩(keV)	(%)
0 - 10	β-	0.0322	0.64
	IB	0.021	
10 - 20	β-	0.100	0.67
	IB	0.020	0.141
20 - 40	β-	0.432	1.43
	IB	0.039	0.134
40 - 100	β-	3.61	5.08
	IB	0.102	0.158
100 - 300	β-	47.0	22.9
	IB	0.23	0.131
300 - 600	β-	172	38.5
	IB	0.146	0.036
600 - 1300	β-	236	30.5
	IB	0.038	0.0054
1300 - 2492	β-	4.78	0.284
	IB	0.00146	9.4×10^{-5}

$^{41}_{19}$K (stable)

Δ: -35559.7 *11* keV
%: 6.7302 *30*

$^{41}_{20}$Ca(1.03 *4* ×10^{5} yr)

Mode: ϵ
Δ: -35138.3 *12* keV
SpA: 0.085 Ci/g
Prod: ^{40}Ca(n,γ)

Photons (^{41}Ca)

⟨γ⟩=0.42 *3* keV

γ_{mode}	γ(keV)	γ(%)
K L_β	0.323	0.022 *7*
K $K_{\alpha2}$	3.311	3.8 *4*
K $K_{\alpha1}$	3.314	7.6 *8*
K $K_{\beta1}$'	3.590	1.07 *11*

Continuous Radiation (^{41}Ca)

⟨IB⟩=0.0113 keV

E_{bin}(keV)		⟨ ⟩(keV)	(%)
10 - 20	IB	2.2×10^{-5}	0.000151
20 - 40	IB	9.8×10^{-5}	0.00032
40 - 100	IB	0.00104	0.00145
100 - 300	IB	0.0079	0.0042
300 - 421	IB	0.0019	0.00056

$^{41}_{21}$Sc(596.3 *17* ms)

Mode: ϵ
Δ: -28643.4 *15* keV
SpA: 2.728×10^{11} Ci/g
Prod: ^{40}Ca(p,γ); ^{40}Ca(d,n)

Photons (^{41}Sc)

⟨γ⟩=1.00 *9* keV

γ_{mode}	γ(keV)	γ(%)
γ [M1+E2]	2575.1 *6*	0.023 *3*
γ M1+7.8%E2	2959.2 *4*	0.0139 *14*

Continuous Radiation (^{41}Sc)

⟨β+⟩=2539 keV;⟨IB⟩=10.0 keV

E_{bin}(keV)		⟨ ⟩(keV)	(%)
0 - 10	β+	1.81×10^{-5}	0.000235
	IB	0.067	
10 - 20	β+	0.000237	0.00149
	IB	0.066	0.46
20 - 40	β+	0.00256	0.0081
	IB	0.132	0.46
40 - 100	β+	0.0500	0.067
	IB	0.39	0.59

Continuous Radiation (^{41}Sc) (continued)

E_{bin}(keV)		⟨ ⟩(keV)	(%)
100 - 300	β+	1.43	0.66
	IB	1.19	0.66
300 - 600	β+	10.6	2.27
	IB	1.54	0.36
600 - 1300	β+	114	11.4
	IB	2.7	0.31
1300 - 2500	β+	666	34.4
	IB	2.7	0.152
2500 - 5000	β+	1724	50.5
	IB	1.21	0.039
5000 - 6495	β+	23.2	0.453
	IB	0.0116	0.00022
	Σβ+		99.83

$^{41}_{22}$Ti(80 *2* ms)

Mode: ϵp
Δ: -15700 *40* keV
SpA: 3.969×10^{11} Ci/g
Prod: ^{40}Ca(^{3}He,2n)

Delayed Protons (^{41}Ti)†

p(keV)	p(%)
1000 *15*	9.4 *6*
1248 *15*	0.95 *22*
1546 *15*	5.24 *19*
1983 *25*	0.75 *15*
2063 *30*	0.99 *12*
2271 *10*	6.33 *24*
2409 *20*	3.57 *9*
2662 *20*	1.96 *20*
2814 *15*	1.19 *12*
3077 *15*	14.6 *10*
3148 *20*	0.97 *27*
3339 *30*	0.56 *10*
3487 *20*	0.68 *10*
3605 *15*	2.35 *10*
3690 *15*	3.76 *20*
3749 *10*	7.5 *5*
3836 *25*	0.58 *5*
3904 *10*	0.36 *5*
4187 *15*	3.74 *14*
4379 *15*	1.75 *10*
4564 *20*	0.53 *7*
4638 *10*	5.36 *19*
4734 *4*	24.3 *5*
4832 *20*	0.73 *7*
4876 *20*	0.82 *10*
4925 *20*	0.70 *7*
5177 *30*	0.36 *7*

† 1.4% uncert(syst)

Photons (^{41}Ti)

⟨γ⟩=72 *9* keV

γ_{mode}	γ(keV)	γ(%)
γ[E3]	3732.1 *8*	1.94 *24*

Continuous Radiation (^{41}Ti)

⟨β+⟩=3428 keV;⟨IB⟩=15.7 keV

E_{bin}(keV)		⟨ ⟩(keV)	(%)
0 - 10	β+	8.6×10^{-6}	0.000111
	IB	0.077	
10 - 20	β+	0.000117	0.00074
	IB	0.077	0.53
20 - 40	β+	0.00131	0.00416
	IB	0.153	0.53
40 - 100	β+	0.0264	0.0353
	IB	0.45	0.69
100 - 300	β+	0.78	0.357
	IB	1.42	0.79
300 - 600	β+	5.9	1.26
	IB	1.9	0.44
600 - 1300	β+	66	6.6
	IB	3.6	0.41
1300 - 2500	β+	437	22.4
	IB	4.2	0.23
2500 - 5000	β+	1935	53
	IB	3.4	0.104
5000 - 7500	β+	896	15.3
	IB	0.44	0.0077
7500 - 9822	β+	88	1.09
	IB	0.0135	0.000170
	Σβ+		100

A = 42

NP **A310**, 599 (1978)

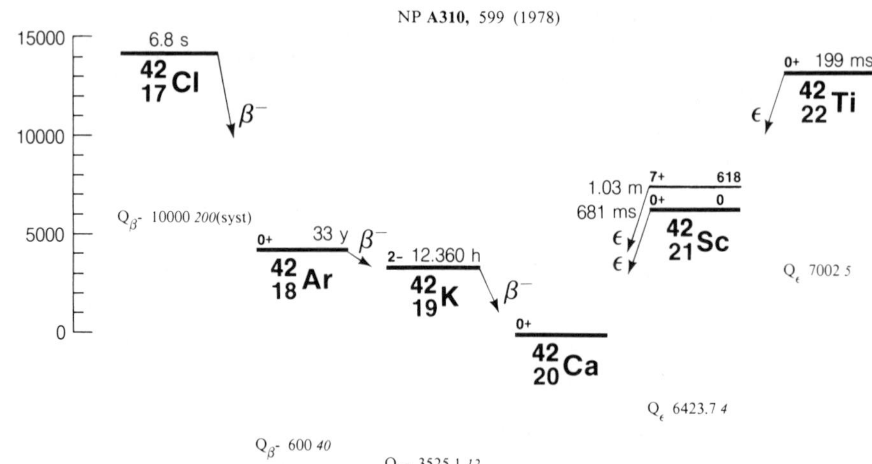

$^{42}_{17}$Cl(6.8 *3* s)

Mode: β-
Δ: -24420 *200* keV syst
SpA: 3.76×10^{10} Ci/g
Prod: protons on Nb; protons on U

$^{42}_{18}$Ar(32.9 *11* yr)

Mode: β-
Δ: -34420 *40* keV
SpA: 259 Ci/g
Prod: multiple n-capture from ^{40}Ar; ^{40}Ar(t,p)

Continuous Radiation (^{42}Ar)

⟨β-⟩=233 keV;⟨IB⟩=0.166 keV

E_{bin}(keV)		⟨ ⟩(keV)	(%)
0 - 10	β-	0.092	1.85
	IB	0.0116	
10 - 20	β-	0.282	1.88
	IB	0.0109	0.076
20 - 40	β-	1.18	3.92
	IB	0.020	0.069
40 - 100	β-	9.0	12.8
	IB	0.046	0.073
100 - 300	β-	93	46.6
	IB	0.067	0.042
300 - 600	β-	129	32.9
	IB	0.0100	0.0029

$^{42}_{19}$K (12.360 *3* h)

Mode: β-
Δ: -35022.7 *15* keV
SpA: 6.0367×10^{6} Ci/g
Prod: ^{41}K(n,γ)

Photons (^{42}K)

⟨γ⟩=290 *9* keV

γ_{mode}	γ(keV)	γ(%)†
γ E2	312.35 *25*	0.350 *21*
γ [E2]	328.41 *22*	$\sim2\times10^{-5}$
γ [E1]	693.9 *8*	0.0034 *8*
γ M1+3.2%E2	898.99 *20*	0.054 *3*
γ [E1]	1022.6 *8*	0.0209 *15*
γ [E2]	1227.65 *8*	0.0024 *11*
γ E2	1524.58 *8*	18.8
γ E1	1921.8 *8*	0.043 *4*
γ (E2)	2423.79 *22*	0.021 *3*

† 3.2% uncert(syst)

Continuous Radiation (^{42}K)

⟨β-⟩=1425 keV;⟨IB⟩=4.1 keV

E_{bin}(keV)		⟨ ⟩(keV)	(%)
0 - 10	β-	0.0068	0.135
	IB	0.047	
10 - 20	β-	0.0204	0.136
	IB	0.046	0.32
20 - 40	β-	0.084	0.278
	IB	0.091	0.31
40 - 100	β-	0.67	0.94
	IB	0.26	0.40
100 - 300	β-	9.7	4.65
	IB	0.76	0.42
300 - 600	β-	48.2	10.5
	IB	0.88	0.21
600 - 1300	β-	298	31.3
	IB	1.26	0.145
1300 - 2500	β-	753	40.8
	IB	0.73	0.043
2500 - 3525	β-	315	11.2
	IB	0.047	0.00169

$^{42}_{20}$Ca(stable)

Δ: -38547.8 *12* keV
%: 0.647 *3*

$^{42}_{21}$Sc(681.3 *7* ms)

Mode: ε
Δ: -32124.1 *13* keV
SpA: 2.4742×10^{11} Ci/g
Prod: ^{39}K(α,n); ^{40}Ca(t,n); ^{42}Ca(p,n)

Photons (^{42}Sc)

⟨γ⟩=0.19 *5* keV

γ_{mode}	γ(keV)	γ(%)
γ E2	312.35 *25*	0.010 *3*
γ E2	1524.58 *8*	0.010 *3*

Continuous Radiation (^{42}Sc)

⟨β+⟩=2507 keV;⟨IB⟩=9.8 keV

E_{bin}(keV)		⟨ ⟩(keV)	(%)
0 - 10	β+	1.87×10^{-5}	0.000243
	IB	0.066	
10 - 20	β+	0.000245	0.00154
	IB	0.066	0.46
20 - 40	β+	0.00265	0.0084
	IB	0.131	0.45
40 - 100	β+	0.052	0.069
	IB	0.38	0.59
100 - 300	β+	1.48	0.68
	IB	1.18	0.66
300 - 600	β+	10.9	2.34
	IB	1.53	0.36
600 - 1300	β+	117	11.8
	IB	2.7	0.30
1300 - 2500	β+	680	35.2
	IB	2.6	0.148
2500 - 5000	β+	1683	49.6
	IB	1.10	0.036
5000 - 6424	β+	15.0	0.293
	IB	0.0057	0.000107
	Σβ+		99.9

$^{42}_{21}$Sc(1.027 *7* min)

Mode: ϵ
Δ: -31506.6 *17* keV
SpA: 4.34×10^{9} Ci/g
Prod: ^{39}K(α,n); ^{42}Ca(p,n)

Photons (^{42}Sc)

⟨γ⟩=3189 *6* keV

γ_{mode}	γ(keV)	γ(%)
γ [E2]	328.41 *22*	1.0 *4*
γ [E2]	437.03 *8*	99.9
γ M1+3.2%E2	898.99 *20*	0.73 *23*
γ [E2]	1227.65 *8*	99.0 *4*
γ E2	1524.58 *8*	99.70 *12*
γ (E2)	2423.79 *22*	0.28 *9*

Continuous Radiation (^{42}Sc)

⟨β+⟩=1255 keV;⟨IB⟩=3.3 keV

E_{bin}(keV)		⟨ ⟩(keV)	(%)
0 - 10	β+	9.7×10^{-5}	0.00126
	IB	0.044	
10 - 20	β+	0.00127	0.0080
	IB	0.043	0.30
20 - 40	β+	0.0137	0.0435
	IB	0.085	0.30
40 - 100	β+	0.262	0.352
	IB	0.24	0.37
100 - 300	β+	7.1	3.29
	IB	0.70	0.39
300 - 600	β+	47.3	10.2
	IB	0.78	0.18
600 - 1300	β+	375	38.8
	IB	1.01	0.118
1300 - 2500	β+	802	45.8
	IB	0.37	0.023
2500 - 3852	β+	24.0	0.93
	IB	0.025	0.00089
	Σβ+		99.3

$^{42}_{22}$Ti(199 *6* ms)

Mode: ϵ
Δ: -25122 *6* keV
SpA: 3.76×10^{11} Ci/g
Prod: ^{40}Ca(^{3}He,n); ^{28}Si(^{16}O,2n)

Photons (^{42}Ti)

⟨γ⟩=368 *86* keV

γ_{mode}	γ(keV)	γ(%)†
γ [M1]	611.2 *2*	56
γ [M1+E2]	636.3 *5*	~0.7
γ [M1+E2]	975.2 *4*	~0.6
γ [E2]	1586.4 *3*	~0.06
γ [M1]	2222.6 *5*	0.67 *22*

† 25% uncert(syst)

Continuous Radiation (^{42}Ti)

⟨β+⟩=2610 keV;⟨IB⟩=10.3 keV

E_{bin}(keV)		⟨ ⟩(keV)	(%)
0 - 10	β+	1.50×10^{-5}	0.000194
	IB	0.068	
10 - 20	β+	0.000206	0.00129
	IB	0.067	0.47
20 - 40	β+	0.00230	0.0073
	IB	0.133	0.46
40 - 100	β+	0.0461	0.062
	IB	0.39	0.60
100 - 300	β+	1.35	0.62
	IB	1.21	0.67
300 - 600	β+	10.0	2.15
	IB	1.57	0.37
600 - 1300	β+	108	10.9
	IB	2.8	0.32
1300 - 2500	β+	641	33.1
	IB	2.8	0.158
2500 - 5000	β+	1779	52
	IB	1.30	0.042
5000 - 5980	β+	70	1.33
	IB	0.0044	8.4×10^{-5}
	Σβ+		100

A = 43

NP **A310**, 630 (1978)

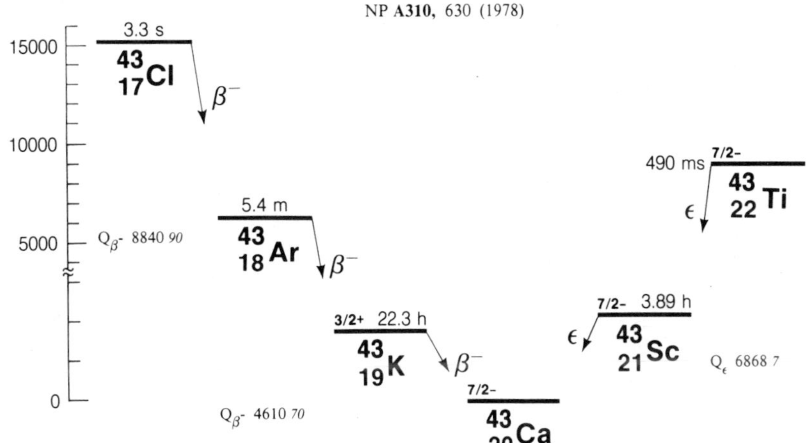

Q_{ϵ} 2220.6 *20*

Q_{β^-} 1817 *10*

$^{43}_{17}$Cl(3.3 *2* s)

Mode: β-
Δ: -23130 *60* keV
SpA: 7.2×10^{10} Ci/g
Prod: protons on Nb; protons on U

$^{43}_{18}$Ar(5.37 *6* min)

Mode: β-
Δ: -31980 *70* keV
SpA: 8.13×10^{8} Ci/g
Prod: protons on Ti; ^{48}Ca(γ,αn); protons on V

Photons (^{43}Ar)

⟨γ⟩=1538 *39* keV

γ_{mode}	γ(keV)	γ(%)†
γ	167.1 *1*	0.085 *17*
γ	231.4 *1*	0.095 *19*
γ	236.2 *1*	0.116 *23*
γ	302.9 *2*	0.065 *16*
γ	413.9 *1*	1.46 *22*
γ	439.3 *2*	0.095 *19*
γ	479.2 *1*	4.0 *4*
γ	548.5 *1*	0.46 *7*
γ	561.1 *1*	3.2 *5*
γ	587.0 *1*	0.31 *6*
γ	639.7 *3*	0.092 *18*
γ	667.5 *2*	0.065 *16*
γ	738.1 *1*	15.5 *15*
γ	755.0 *3*	0.061 *15*
γ	812.4 *4*	0.044 *11*
γ	878.2 *8*	0.031 *8*
γ	890.4 *1*	4.0 *4*
γ	910.5 *9*	0.034 *9*
γ	922.5 *5*	0.058 *14*

Photons (^{43}Ar) (continued)

γ_{mode}	γ(keV)	γ(%)†
γ	974.9 *1*	34.1
γ	1080.0 *2*	0.16 *3*
γ	1110.1 *1*	1.06 *16*
γ	1121.0 *2*	0.109 *22*
γ	1132.6 *1*	0.42 *6*
γ	1138.1 *1*	0.30 *6*
γ	1146.4 *2*	0.31 *6*
γ	1184.3 *3*	0.17 *4*
γ	1202.4 *3*	3.3 *5*
γ	1207.1 *3*	2.6 *4*
γ	1235.7 *2*	0.21 *4*
γ	1255.6 *3*	0.109 *22*
γ	1277.9 *5*	0.085 *17*
γ	1304.3 *7*	0.106 *21*
γ	1311.4 *1*	0.77 *12*
γ	1369 *1*	0.014 *3*
γ	1369.9 *1*	6.8 *7*
γ	1398.7 *1*	0.32 *7*
γ	1419.3 *1*	0.44 *7*
γ	1439.8 *1*	12.6 *13*
γ	1443 *1*	1.63 *25*

Photons (^{43}Ar)
(continued)

γ_{mode}	γ(keV)	γ(%)†
γ	1487.8 5	0.092 18
γ	1509.7 1	0.71 11
γ	1550.0 1	0.78 12
γ	1559.9 1	0.54 8
γ	1590.4 2	0.36 5
γ	1605.7 8	0.089 18
γ	1621.7 5	0.19 4
γ	1713.3 6	0.109 22
γ	1724.6 2	0.32 6
γ	1750.0 5	0.068 14
γ	1758.2 2	0.35 5
γ	1783.7 2	0.43 6
γ	1849.6 8	0.085 17
γ	1866.1 1	2.4 4
γ	1889.2 7	0.102 20
γ	1905.9 6	0.13 3
γ	1950.8 3	0.34 5
γ	2057.9 3	0.36 6
γ	2097.8 5	0.33 7
γ	2102.3 5	0.33 7
γ	2176.2 7	0.085 17
γ	2189.2 3	0.69 10
γ	2287.6 2	0.16 3
γ	2318.9 2	1.06 16
γ	2333.9 2	2.8 4
γ	2344.5 2	7.4 7
γ	2345 1	0.041 10
γ	2401.8 3	0.095 19
γ	2438.9 5	0.22 4
γ	2479.9 1	2.2 3
γ	2506.7 15	0.034 9
γ	2535.7 7	0.041 10
γ	2603.4 4	0.031 8
γ	2701.9 5	0.044 11
γ	2739.5 7	0.072 14
γ	2870.1 2	0.26 5
γ	2894.2 2	0.61 9
γ	2976.2 3	0.095 19
γ	3264.3 2	0.50 8
γ	3309.9 2	0.25 5
γ	3380.6 7	0.027 7
γ	3393.0 2	0.38 6
γ	3455.1 4	0.068 14
γ	3646.4 5	0.017 4
γ	3714.3 2	0.32 7

† 11% uncert(syst)

Continuous Radiation (^{43}Ar)
$\langle\beta-\rangle$=1356 keV;$\langle$IB$\rangle$=4.0 keV

E_{bin}(keV)		$\langle\ \rangle$(keV)	(%)
0 - 10	β-	0.0086	0.171
	IB	0.044	
10 - 20	β-	0.0269	0.179
	IB	0.044	0.30
20 - 40	β-	0.117	0.387
	IB	0.086	0.30
40 - 100	β-	1.00	1.40
	IB	0.25	0.38
100 - 300	β-	14.0	6.7
	IB	0.71	0.40
300 - 600	β-	63	13.8
	IB	0.81	0.19
600 - 1300	β-	298	31.8
	IB	1.16	0.134
1300 - 2500	β-	590	32.8
	IB	0.74	0.043
2500 - 4610	β-	391	12.8
	IB	0.136	0.0047

$^{43}_{19}$K (22.3 1 h)

Mode: β-
Δ: -36592 10 keV
SpA: 3.268$\times 10^{6}$ Ci/g
Prod: ^{40}Ar(α,p)

Photons (^{43}K)
$\langle\gamma\rangle$=964 12 keV

γ_{mode}	γ(keV)	γ(%)†
γ	184.0 2	0.26 5
γ M1+1.2%E2	220.58 6	4.08 21
γ M1+3.6%E2	372.81 5	86.7 17
γ [E1]	396.93 8	11.36 23
γ M1+9.4%E2	404.28 11	<0.23
γ (E2)	593.39 4	11.0 3
γ E1	617.51 9	80.0
γ [E1]	801.20 10	0.146 10
γ [M2]	990.31 7	0.33 6
γ	1015.1 10	0.16 7
γ [E1]	1021.78 10	1.87 7
γ [E1]	1394.58 9	0.102 12

† 2.5% uncert(syst)

Continuous Radiation (^{43}K)
$\langle\beta-\rangle$=314 keV;$\langle$IB$\rangle$=0.31 keV

E_{bin}(keV)		$\langle\ \rangle$(keV)	(%)
0 - 10	β-	0.067	1.33
	IB	0.0150	
10 - 20	β-	0.205	1.37
	IB	0.0143	0.099
20 - 40	β-	0.87	2.88
	IB	0.027	0.093
40 - 100	β-	6.9	9.7
	IB	0.066	0.103
100 - 300	β-	74	36.8
	IB	0.122	0.073
300 - 600	β-	168	39.2
	IB	0.051	0.0131
600 - 1300	β-	61	8.4
	IB	0.0098	0.0013
1300 - 1817	β-	2.69	0.186
	IB	0.00019	1.38$\times 10^{-5}$

$^{43}_{20}$Ca(stable)

Δ: -38409.4 12 keV
%: 0.135 3

$^{43}_{21}$Sc(3.891 12 h)

Mode: ϵ
Δ: -36188.7 23 keV
SpA: 1.873$\times 10^{7}$ Ci/g
Prod: ^{40}Ca(α,p); daughter ^{43}Ti

Photons (^{43}Sc)
$\langle\gamma\rangle$=83 4 keV

γ_{mode}	γ(keV)	γ(%)†
Ca L_{β}	0.377	0.0033 9
Ca $K_{\alpha2}$	3.688	0.53 3
Ca $K_{\alpha1}$	3.692	1.06 5
Ca $K_{\beta1}$'	4.013	0.166 8
γ M1+1.2%E2	220.58 6	0.0008 3
γ M1+3.6%E2	372.81 5	22.0
γ (E2)	593.39 4	0.0021 7
γ	1338.03 20	0.00176 22
γ [M1+E2]	1558.60 21	0.0082 5
γ [M1+E2]	1931.4 2	0.0148 7

† 4.6% uncert(syst)

Atomic Electrons (^{43}Sc)
$\langle$e$\rangle$=0.45 3 keV

e_{bin}(keV)	$\langle$e$\rangle$(keV)	e(%)
3	0.242	7.3 8
4	0.062	1.70 17
10	0.0978451	1.0
217	3.0 $\times 10^{-6}$	~1$\times 10^{-6}$
220	2.7 $\times 10^{-7}$	~1$\times 10^{-7}$
369	0.0436	0.0118 6
372	0.00383	0.00103 6
589	4.4 $\times 10^{-6}$	7.5 25 $\times 10^{-7}$
593	3.8 $\times 10^{-7}$	6.5 22 $\times 10^{-8}$
1334	7.6 $\times 10^{-7}$	5.7 20 $\times 10^{-8}$
1338	6.6 $\times 10^{-8}$	4.9 17 $\times 10^{-9}$
1555	3.8 $\times 10^{-6}$	2.4 3 $\times 10^{-7}$
1558	3.3 $\times 10^{-7}$	2.09 22 $\times 10^{-8}$
1927	5.7 $\times 10^{-6}$	2.9 3 $\times 10^{-7}$
1931	4.9 $\times 10^{-7}$	2.53 22 $\times 10^{-8}$

Continuous Radiation (^{43}Sc)
$\langle\beta+\rangle$=419 keV;$\langle$IB$\rangle$=0.80 keV

E_{bin}(keV)		$\langle\ \rangle$(keV)	(%)
0 - 10	β+	0.00080	0.0105
	IB	0.019	
10 - 20	β+	0.0104	0.066
	IB	0.018	0.128
20 - 40	β+	0.110	0.350
	IB	0.035	0.122
40 - 100	β+	1.98	2.68
	IB	0.093	0.144
100 - 300	β+	42.6	20.3
	IB	0.21	0.121
300 - 600	β+	170	38.1
	IB	0.159	0.039
600 - 1300	β+	204	26.6
	IB	0.19	0.021
1300 - 2221	IB	0.081	0.0053
	$\Sigma\beta$+		88

$^{43}_{22}$Ti(490 10 ms)

Mode: ϵ
Δ: -29321 7 keV
SpA: 2.87$\times 10^{11}$ Ci/g
Prod: ^{40}Ca(α,n)

Continuous Radiation (^{43}Ti)

⟨β+⟩=2725 keV;⟨IB⟩=11.1 keV

E_{bin}(keV)		⟨ ⟩(keV)	(%)
0 - 10	β+	1.33×10^{-5}	0.000172
	IB	0.069	
10 - 20	β+	0.000182	0.00114
	IB	0.069	0.48
20 - 40	β+	0.00204	0.0065
	IB	0.137	0.47
40 - 100	β+	0.0409	0.055
	IB	0.40	0.61
100 - 300	β+	1.20	0.55
	IB	1.24	0.69
300 - 600	β+	9.0	1.92
	IB	1.63	0.38
600 - 1300	β+	98	9.9
	IB	2.9	0.33
1300 - 2500	β+	603	31.0
	IB	3.0	0.17
2500 - 5000	β+	1911	55
	IB	1.58	0.051
5000 - 6868	β+	103	1.97
	IB	0.0162	0.00030
	Σβ+		99.9

A = 44

NP **A310**, 659 (1978)

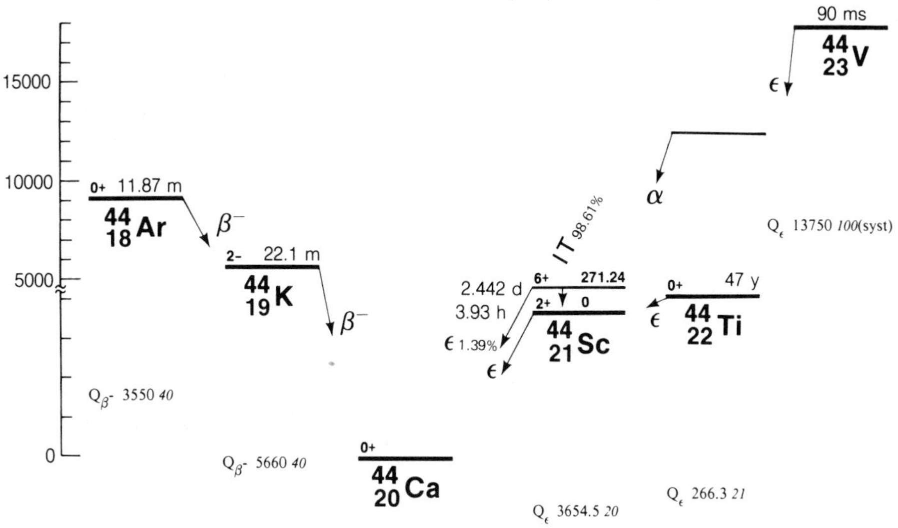

15000
10000
5000
0
90 ms
$^{44}_{23}$V
ε
α
Q$_ε$ 13750 *100*(syst)
0+ 11.87 m
$^{44}_{18}$Ar
β⁻
2- 22.1 m
$^{44}_{19}$K
β⁻
IT 98.61%
2.442 d 6+ 271.24
3.93 h 2+ 0
ε 1.39%
$^{44}_{21}$Sc
ε
0+ 47 y
ε
$^{44}_{22}$Ti
Q$_{β^-}$ 3550 *40*
Q$_{β^-}$ 5660 *40*
0+
$^{44}_{20}$Ca
Q$_ε$ 3654.5 *20*
Q$_ε$ 266.3 *21*

$^{44}_{18}$Ar(11.87 *5* min)

Mode: β-
Δ: -32262 *20* keV
SpA: 3.598×10^{8} Ci/g
Prod: protons on Ti; protons on V; ^{48}Ca(γ,α)

Photons (^{44}Ar)

⟨γ⟩=1829 *110* keV

$γ_{mode}$	γ(keV)	γ(%)†
γ	137.3 *3*	0.12 *3*
γ	182.6 *1*	62
γ	382.9 *1*	0.49 *12*
γ	408.1 *1*	3.9 *6*
γ	426.7 *1*	2.5 *4*
γ	519.4 *4*	0.044 *11*
γ	531.2 *3*	0.12 *3*
γ	693.8 *2*	0.22 *6*
γ	809.1 *1*	1.8 *3*
γ	866.1 *10*	1.54 *23*
γ	884.9 *1*	0.031 *8*
γ	894.2 *1*	0.63 *16*
γ	911.1 *2*	0.14 *4*
γ	975.0 *4*	0.21 *5*
γ	1051.3 *1*	3.7 *6*
γ	1076.6 *1*	0.92 *14*
γ	1114.7 *1*	2.1 *3*
γ	1276.6 *1*	1.45 *22*
γ	1460.0 *1*	2.1 *3*
γ	1585.7 *2*	0.49 *12*
γ	1639.7 *2*	0.64 *16*
γ	1703.4 *1*	53 *5*
γ	1765.4 *8*	0.11 *3*
γ	1886.1 *1*	30 *3*
γ	2143.5 *4*	0.74 *18*
γ	2279.9 *3*	0.16 *4*
γ	2325.8 *2*	0.77 *19*

† 18% uncert(syst)

Continuous Radiation (^{44}Ar)

⟨β-⟩=660 keV;⟨IB⟩=1.10 keV

E_{bin}(keV)		⟨ ⟩(keV)	(%)
0 - 10	β-	0.0177	0.351
	IB	0.028	
10 - 20	β-	0.055	0.366
	IB	0.027	0.19
20 - 40	β-	0.239	0.79
	IB	0.052	0.18
40 - 100	β-	2.04	2.86
	IB	0.143	0.22
100 - 300	β-	28.6	13.8
	IB	0.36	0.20
300 - 600	β-	127	27.9
	IB	0.30	0.073
600 - 1300	β-	443	49.7
	IB	0.18	0.024
1300 - 2500	β-	60	4.25
	IB	0.0035	0.00025
2500 - 3167	β-	0.174	0.0065
	IB	1.15×10^{-5}	4.4×10^{-7}

$^{44}_{19}$K (22.13 *19* min)

Mode: β-
Δ: -35810 *40* keV
SpA: 1.931×10^{8} Ci/g
Prod: ^{44}Ca(n,p)

Photons (^{44}K)

⟨γ⟩=2377 *14* keV

$γ_{mode}$	γ(keV)	γ(%)†
γ	174.36 *25*	0.058 *12*
γ	209.99 *25*	~0.017
γ[E1]	263.54 *5*	0.11 *3*
γ[E2]	353.655 *13*	~0.017
γ	368.217 *11*	2.25 *4*
γ	374.83 *4*	0.19 *5*

Photons (^{44}K) (continued)

$γ_{mode}$	γ(keV)	γ(%)†
γ	403.93 *15*	0.064 *17*
γ	463.3 *4*	<0.035
γ	646.63 *16*	0.09 *3*
γ[E1]	651.361 *8*	3.02 *12*
γ	682.34 *3*	~0.08
γ E2	726.492 *13*	3.77 *12*
γ	733.081 *16*	0.16 *7*
γ	747.643 *15*	2.09 *12*
γ M1+8.5%E2	761.21 *5*	0.12 *6*
γ	766.8 *5*	~0.05
γ	876.546 *22*	1.73 *4*
γ	891.108 *25*	0.09 *4*
γ	938.58 *15*	0.17 *6*
γ[E1]	1005.011 *14*	0.029 *6*
γ	1019.572 *13*	0.85 *4*
γ[E1]	1024.744 *11*	6.67 *12*
γ[M1+E2]	1050.56 *3*	0.56 *8*
γ	1074.2 *4*	~0.09
γ	1101.292 *12*	<0.023
γ	1106.27 *25*	0.10 *3*
γ	1107.90 *4*	0.67 *5*
γ	1119.7 *4*	~0.017
γ[E2]	1126.081 *9*	7.60 *12*
γ E2	1157.008 *3*	58.00
γ	1195.4 *10*	~0.05
γ	1222.48 *7*	0.48 *5*
γ	1244.755 *23*	0.83 *3*
γ	1272.8 *4*	~0.08
γ[E2]	1364.18 *20*	<0.023
γ	1377.6 *5*	0.12 *6*
γ	1427.5 *4*	0.11 *5*
γ	1428.66 *15*	~0.023
γ M1+1.5%E2	1499.459 *11*	7.87 *24*
γ	1576.13 *7*	0.17 *5*
γ	1625.1 *7*	~0.035
γ	1634.53 *11*	0.23 *5*
γ	1658.69 *13*	0.12 *6*
γ[E1]	1701.90 *3*	0.10 *4*
γ	1752.635 *10*	4.06 *6*
γ[E1]	1777.973 *14*	2.12 *5*
γ	1810.9 *6*	~0.07
γ	1884.9 *5*	~0.023
γ E2	1887.27 *5*	0.12 *6*
γ	1893.2 *4*	0.11 *5*
γ	1896.0 *9*	~0.11
γ	1916.1 *5*	~0.13
γ	1977.1 *4*	~0.05
γ	1992.6 *5*	~0.07

Photons (^{44}K)
(continued)

γ_{mode}	γ(keV)	γ(%)†
γ [M1+E2]	2144.19 *4*	0.75 *5*
γ [E1]	2150.797 *10*	22.7 *5*
γ	2167.8 *6*	~0.08
γ	2200.2 *4*	<0.023
γ	2269.468 *24*	~0.029
γ	2324.3 *10*	~0.035
γ	2338.3 *6*	~0.041
γ [E2]	2423.3 *6*	~0.041
γ	2497.3 *9*	~0.03
γ [E1]	2504.433 *10*	0.65 *5*
γ	2518.995 *12*	9.69 *17*
γ	2598.4 *6*	~0.04
γ (M1+28%E2)	2619.17 *12*	0.21 *4*
γ [E2]	2656.425 *11*	0.97 *4*
γ	2710.9 *7*	<0.035
γ	2740.4 *8*	~0.006
γ	2847.7 *7*	~0.029
γ	2847.7 *7*	~0.029
γ	2936.9 *6*	0.046 *17*
γ	2973 *1*	~0.017
γ	2982.48 *11*	0.128 *17*
γ	3067.0 *8*	<0.023 ?
γ	3103.1 *4*	0.064 *23*
γ	3158.09 *13*	0.151 *23*
γ [E1]	3201.30 *3*	0.70 *5*
γ	3217.3 *6*	~0.029
γ	3227.1 *8*	~0.017
γ [E1]	3242.0 *5*	~0.035
γ	3252.031 *14*	0.157 *23*
γ [M1+E2]	3279.5 *5*	~0.017
γ [E2]	3301.14 *4*	0.31 *4*
γ	3307.744 *10*	0.017 *6*
γ	3395.487 *23*	1.66 *5*
γ	3415.5 *5*	0.058 *23*
γ [E1]	3661.371 *10*	6.09 *12*
γ	3675.931 *12*	0.014 *6* ?
γ	3726.84 *7*	0.029 *6*
γ	3755.2 *9*	~0.008
γ [E1]	3868.53 *20*	0.064 *17*
γ	3967.8 *10*	~0.006
γ	4005.07 *10*	<0.0023
γ	4074 *1*	<0.009
γ	4162.6 *8*	~0.005 ?
γ	4167.8 *5*	~0.009
γ	4210.1 *10*	~0.005
γ	4210.2 *7*	~0.005
γ	4337.9 *8*	<0.010
γ	4403.8 *5*	~0.0035
γ	4408.947 *14*	0.053 *9*
γ [E2]	4436.5 *5*	~0.007
γ	4471.5 *6*	~0.006
γ	4648.9 *5*	~0.008
γ	4865.82 *11*	0.162 *6*
γ	4892.3 *8*	~0.003
γ	5025.42 *20*	~0.0017
γ	5161.96 *10*	0.064 *4*
γ	5560.6 *5*	~0.0029

† 16% uncert(syst)

Continuous Radiation (^{44}K)
$\langle\beta-\rangle$=1414 keV;$\langle$IB$\rangle$=4.6 keV

E_{bin}(keV)		$\langle\ \rangle$(keV)	(%)
0 - 10	β-	0.0106	0.211
	IB	0.044	
10 - 20	β-	0.0329	0.219
	IB	0.043	0.30
20 - 40	β-	0.141	0.469
	IB	0.085	0.29
40 - 100	β-	1.18	1.67
	IB	0.24	0.37
100 - 300	β-	15.9	7.7
	IB	0.70	0.39
300 - 600	β-	68	15.0
	IB	0.82	0.19
600 - 1300	β-	291	31.4
	IB	1.22	0.140

Continuous Radiation (^{44}K)
(continued)

E_{bin}(keV)		$\langle\ \rangle$(keV)	(%)
1300 - 2500	β-	406	23.0
	IB	0.99	0.057
2500 - 5000	β-	614	17.8
	IB	0.44	0.0142
5000 - 5660	β-	18.0	0.349
	IB	0.00067	1.32×10^{-5}

$^{44}_{20}$Ca(stable)

Δ: -41469.9 *13* keV
%: 2.086 *5*

$^{44}_{21}$Sc(3.927 *8* h)

Mode: ϵ
Δ: -37815.4 *22* keV
SpA: 1.814×10^{7} Ci/g
Prod: daughter ^{44}Ti; daughter ^{44}Sc(2.442 d); ^{41}K(α,n); ^{45}Sc(γ,n); alphas on Ca

Photons (^{44}Sc)
$\langle\gamma\rangle$=1172 *230* keV

γ_{mode}	γ(keV)	γ(%)†
Ca L_β	0.377	0.00015 *4*
Ca $K_{\alpha2}$	3.688	0.0251 *13*
Ca $K_{\alpha1}$	3.692	0.0501 *25*
Ca $K_{\beta1}$'	4.013	0.0078 *4*
γ [E1]	263.54 *5*	5.3 *20* $\times10^{-6}$
γ [E1]	651.361 *8*	0.00015 *4*
γ [E1]	1024.744 *11*	0.00032 *9*
γ E2	1157.008 *3*	99.9
γ M1+1.5%E2	1499.459 *11*	0.910 *18*
γ [M1+E2]	2144.19 *4*	0.00159 *18*
γ [E1]	2150.797 *10*	0.0011 *3*
γ [E2]	2656.425 *11*	0.112 *3*
γ [E2]	3301.14 *4*	0.00066 *7*
γ	3307.744 *10*	8 *4* $\times10^{-7}$

†<0.1% uncert(syst)

Atomic Electrons (^{44}Sc)
$\langle e\rangle$=0.94 *14* keV

e_{bin}(keV)	$\langle e\rangle$(keV)	e(%)
3	0.114	3.5 *4*
4	0.029	0.80 *8*
10	0.0463302	0.5
260	1.4×10^{-7}	$\sim6\times10^{-8}$
263	1.3×10^{-8}	$\sim5\times10^{-9}$
647	8.7×10^{-7}	$\sim1\times10^{-7}$
651	7.5×10^{-8}	$\sim1\times10^{-8}$
1021	1.2×10^{-6}	$\sim1\times10^{-7}$
1024	1.0×10^{-7}	$\sim1\times10^{-8}$
1153	0.69	0.060 *12*
1157	0.060	0.0051 *10*
1495	0.00401	0.000268 *8*
1499	0.000346	2.31 *6* $\times10^{-5}$
2140	5.7×10^{-6}	2.6 *4* $\times10^{-7}$
2144	4.9×10^{-7}	2.3 *3* $\times10^{-8}$
2147	2.5×10^{-6}	1.1 *3* $\times10^{-7}$

Atomic Electrons (^{44}Sc)
(continued)

e_{bin}(keV)	$\langle e\rangle$(keV)	e(%)
2150	2.1×10^{-7}	1.0 *3* $\times10^{-8}$
2652	0.000356	1.34 *7* $\times10^{-5}$
2656	3.07×10^{-5}	1.16 *6* $\times10^{-6}$
3297	1.84×10^{-6}	5.6 *6* $\times10^{-8}$
3301	1.59×10^{-7}	4.8 *5* $\times10^{-9}$
3304	2.0×10^{-9}	$\sim6\times10^{-11}$
3307	1.7×10^{-10}	$\sim5\times10^{-12}$

Continuous Radiation (^{44}Sc)
$\langle\beta+\rangle$=597 keV;$\langle$IB$\rangle$=1.11 keV

E_{bin}(keV)		$\langle\ \rangle$(keV)	(%)
0 - 10	β+	0.000434	0.0056
	IB	0.026	
10 - 20	β+	0.0056	0.0355
	IB	0.025	0.17
20 - 40	β+	0.060	0.191
	IB	0.048	0.168
40 - 100	β+	1.12	1.50
	IB	0.131	0.20
100 - 300	β+	26.8	12.6
	IB	0.32	0.19
300 - 600	β+	140	30.7
	IB	0.27	0.065
600 - 1300	β+	418	48.6
	IB	0.20	0.025
1300 - 2497	β+	10.8	0.80
	IB	0.090	0.0055
	$\Sigma\beta$+		94

$^{44}_{21}$Sc(2.442 *4* d)

Mode: IT(98.61 *2* %), ϵ(1.39 *2* %)
Δ: -37544.2 *22* keV
SpA: 1.2154×10^{6} Ci/g
Prod: ^{41}K(α,n); ^{45}Sc(γ,n); alphas on Ca

Photons (^{44}Sc)
$\langle\gamma\rangle$=280.6 *6* keV

γ_{mode}	γ(keV)	γ(%)†
Sc L_ℓ	0.348	0.0031 *8*
Sc L_η	0.353	0.0024 *6*
Sc L_α	0.396	0.0034 *9*
Sc L_β	0.427	0.0043 *12*
Sc $K_{\alpha2}$	4.086	0.62 *4*
Sc $K_{\alpha1}$	4.091	1.24 *7*
Sc $K_{\beta1}$'	4.461	0.201 *12*
γ_{IT} (E4)	271.241 *10*	86.6
γ_ϵ	1001.83 *3*	1.390 *9*
γ_ϵ[E2]	1126.081 *9*	1.390 *9*
γ_ϵE2	1157.008 *3*	1.390 *9*

† uncert(syst): 1.4% for ϵ, <0.1% for IT

Atomic Electrons (^{44}Sc)*

$\langle e\rangle$=32.6 *12* keV

e_{bin}(keV)	$\langle e\rangle$(keV)	e(%)
3	0.0210	0.60 *7*
4	0.31	8.3 *9*
9	0.0971972	1.0
267	29.2	10.9 *4*
271	2.98	1.10 *4*

* with IT

Continuous Radiation (^{44}Sc)

$\langle IB\rangle$=0.00091 keV

E_{bin}(keV)		$\langle\ \rangle$(keV)	(%)
10 - 20	IB	3.7×10^{-7}	2.5×10^{-6}
20 - 40	IB	1.64×10^{-6}	5.3×10^{-6}
40 - 100	IB	2.3×10^{-5}	3.1×10^{-5}
100 - 300	IB	0.00037	0.00018
300 - 600	IB	0.00051	0.000128
600 - 641	IB	1.65×10^{-6}	2.7×10^{-7}

$^{44}_{22}$Ti(47.3 *12* yr)

Mode: ϵ

Δ: -37549.1 *14* keV

SpA: 172 Ci/g

Prod: ^{45}Sc(p,2n); ^{45}Sc(d,3n)

Photons (^{44}Ti)

$\langle\gamma\rangle$=138.4 *14* keV

γ_{mode}	γ(keV)	γ(%)†
Sc L_ℓ	0.348	0.029 *8*
Sc L_η	0.353	0.021 *5*
Sc L_α	0.396	0.031 *8*
Sc L_β	0.428	0.040 *11*
Sc $K_{\alpha2}$	4.086	5.6 *3*
Sc $K_{\alpha1}$	4.091	11.1 *6*
Sc $K_{\beta1}$'	4.461	1.81 *9*
γ(E1)	67.85 *4*	91.0
γ(M1)	78.38 *4*	96.6 *7*
γ[M2]	146.23 *6*	0.10 *3*

† 2.2% uncert(syst)

Atomic Electrons (^{44}Ti)

$\langle e\rangle$=10.9 *4* keV

e_{bin}(keV)	$\langle e\rangle$(keV)	e(%)
3	0.186	5.3 *5*
4	2.8	75 *8*
10	0.859299	9
63	4.47	7.1 *3*
67	0.421	0.62 *3*
74	1.98	2.67 *11*
78	0.190	0.244 *10*
142	0.0057	0.0041 *12*
146	0.00056	0.00038 *12*

Continuous Radiation (^{44}Ti)

$\langle IB\rangle$=0.00101 keV

E_{bin}(keV)		$\langle\ \rangle$(keV)	(%)
10 - 20	IB	2.5×10^{-5}	0.00018
20 - 40	IB	7.2×10^{-5}	0.00024
40 - 100	IB	0.00027	0.00043
100 - 198	IB	2.2×10^{-5}	1.9×10^{-5}

$^{44}_{23}$V (90 *25* ms)

Mode: ϵ, $\epsilon\alpha$

Δ: -23800 *100* keV syst

SpA: 3.7×10^{11} Ci/g

Prod: ^{40}Ca(^{6}Li,2n)

$\epsilon\alpha$: 2770 *20*

A = 45

NDS **40**, 149 (1983)

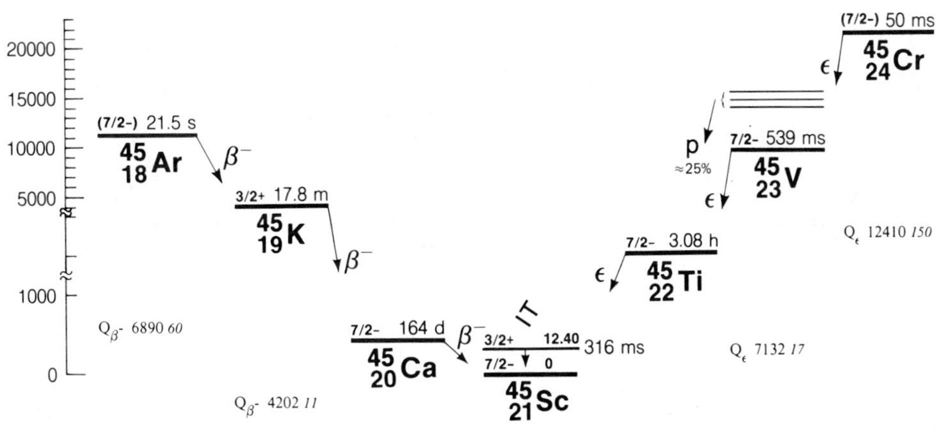

$^{45}_{18}$Ar(21.48 *15* s)

Mode: β-
Δ: -29720 *60* keV
SpA: 1.149×10^{10} Ci/g
Prod: protons on V; ^{48}Ca(n,2n2p)

Photons (^{45}Ar)

⟨γ⟩=2974 *79* keV

γ_{mode}	γ(keV)	γ(%)†
γ M1,E1	61.35 *4*	25.0 *6*
γ M1,E2	474.43 *15*	2.4 *5*
γ	549.06 *7*	2.77 *25*
γ	557.77 *7*	2.24 *21*
γ	598.4 *4*	0.34 *7* ?
γ	608.99 *13*	0.32 *6*
γ	619.12 *7*	3.04 *25*
γ	685.3 *2*	1.23 *18*
γ	846.1 *5*	0.30 *5*
γ	949.9 *3*	0.98 *2*
γ	1019.99 *5*	35
γ	1043.0 *3*	0.88 *14*
γ	1053.7 *7*	0.39 *8*
γ M2,E3	1081.34 *6*	1.3 *6*
γ	1106.82 *8*	10.85 *24*
γ	1122.91 *24*	1.36 *6*
γ	1138.6 *4*	0.66 *18*
γ	1142.5 *16*	0.37 *8*
γ	1168.5 *7*	0.36 *7*
γ	1172.7 *4*	0.84 *17*
γ	1209.9 *3*	0.51 *10* ?
γ	1323.5 *6*	0.39 *9*
γ	1424.3 *3*	0.640 *13*
γ	1435.3 *4*	0.95 *21*
γ	1473.7 *3*	1.4 *3*
γ	1485.9 *9*	0.34 *7*
γ	1548.7 *4*	0.88 *5*
γ	1639.10 *9*	9.6 *3*
γ	1671.95 *24*	0.36 *6*
γ	1722.6 *3*	1.18 *24*
γ	1808.20 *16*	13.4 *5*
γ	1840.3 *5*	0.76 *14*
γ M2,E3	2283.0 *4*	0.84 *25*
γ	2357.23 *16*	7.9 *16*
γ	2489.6 *12*	0.58 *12*
γ	2516.6 *4*	0.84 *14*
γ	2549.6 *12*	0.65 *13*
γ	2687.3 *3*	6.3 *6*
γ	2747.8 *7*	0.58 *9*
γ	2796.6 *7*	1.20 *24*
γ	2883.2 *5*	0.28 *11*
γ	3336.9 *5*	0.50 *18*
γ	3707.2 *3*	33.7 *10*
γ	3996.23 *18*	0.28 *7*
γ	4043.8 *11*	0.27 *6*
γ	4356.8 *5*	0.44 *9*
γ	4568.9 *11*	0.54 *11*

† 14% uncert(syst)

Continuous Radiation (^{45}Ar)

⟨β-⟩=1578 keV;⟨IB⟩=4.9 keV

E_{bin}(keV)		⟨ ⟩(keV)	(%)
0 - 10	β-	0.00370	0.073
	IB	0.050	
10 - 20	β-	0.0116	0.077
	IB	0.049	0.34
20 - 40	β-	0.0509	0.169
	IB	0.097	0.34
40 - 100	β-	0.446	0.63
	IB	0.28	0.43
100 - 300	β-	7.0	3.32
	IB	0.82	0.46
300 - 600	β-	38.4	8.3
	IB	0.97	0.23

Continuous Radiation (^{45}Ar)
(continued)

E_{bin}(keV)		⟨ ⟩(keV)	(%)
600 - 1300	β-	291	30.0
	IB	1.44	0.166
1300 - 2500	β-	800	44.0
	IB	0.92	0.054
2500 - 5000	β-	430	13.2
	IB	0.25	0.0082
5000 - 5809	β-	11.8	0.226
	IB	0.00060	1.14×10^{-5}

$^{45}_{19}$K (17.81 *25* min)

Mode: β-
Δ: -36611 *11* keV
SpA: 2.35×10^{8} Ci/g
Prod: ^{48}Ca(d,αn); ^{46}Ca(γ,p); ^{48}Ca(p,α)

Photons (^{45}K)

⟨γ⟩=1856 *50* keV

γ_{mode}	γ(keV)	γ(%)†
γ (M1)	174.24 *3*	75 *4*
γ [M1+E2]	349.0 *4*	0.0048 *11*
γ	417.33 *18*	0.378 *21*
γ	453.98 *17*	0.043 *5*
γ M1+E2	465.16 *14*	0.096 *5*
γ [E1]	492.38 *15*	1.28 *5*
γ	512.21 *21*	1.2 *4*
γ	522.4 *4*	~0.004
γ	623.3 *4*	~0.007 ?
γ	771.5 *4*	~0.007 ?
γ [M1+E2]	814.1 *4*	0.0059 *21*
γ	834.5 *12*	0.021 *5*
γ	847.5 *12*	0.021 *5*
γ	871.30 *18*	0.128 *16*
γ	891.13 *20*	~0.6
γ	919.13 *17*	0.87 *5*
γ [E1]	957.53 *15*	7.7 *4*
γ	1098.5 *4*	0.085 *16*
γ (M1+E2)	1260.25 *11*	8.6 *4*
γ	1336.45 *18*	2.8 *8*
γ (E2)	1434.49 *12*	4.26 *21*
γ	1705.57 *15*	53 *3*
γ (M1+10%E2)	1725.39 *13*	0.96 *5*
γ	1750.9 *6*	0.037 *5*
γ [E1]	1859.9 *3*	0.053 *5*
γ	1879.80 *15*	0.203 *16*
γ (E2)	1899.63 *13*	0.47 *3*
γ	1940.5 *8*	~0.011
γ (E2)	2074.3 *4*	~0.009
γ	2179.35 *16*	0.96 *5*
γ	2217.76 *15*	0.021 *5*
γ	2231.0 *8*	0.011 *4*
γ	2348.7 *4*	0.11 *4* ?
γ	2353.59 *16*	14.3 *8*
γ	2391.99 *15*	0.011 *4*
γ	2522.9 *4*	0.021 *5* ?
γ	2596.66 *15*	2.77 *16*
γ [M1+E2]	2666.6 *5*	0.043 *5*
γ	2770.89 *15*	0.022 *4*
γ	2802.4 *4*	0.362 *21*
γ (E2)	2840.8 *5*	0.037 *5*
γ	2849.3 *4*	0.085 *5*
γ	2976.6 *4*	0.107 *5*
γ [E1]	3120.1 *3*	0.330 *21*
γ [E1]	3294.3 *3*	0.224 *16*
γ	3367.4 *8*	0.0037 *5*
γ	3406.4 *8*	0.0037 *5*
γ	3479.7 *5*	0.037 *5* ?
γ	3490.4 *4*	0.069 *5*

Photons (^{45}K)
(continued)

γ_{mode}	γ(keV)	γ(%)†
γ	3653.9 *5*	0.0096 *21* ?
γ	3704.8 *6*	0.0085 *16*

† 5.1% uncert(syst)

Continuous Radiation (^{45}K)

⟨β-⟩=993 keV;⟨IB⟩=2.3 keV

E_{bin}(keV)		⟨ ⟩(keV)	(%)
0 - 10	β-	0.0106	0.210
	IB	0.037	
10 - 20	β-	0.0330	0.219
	IB	0.036	0.25
20 - 40	β-	0.143	0.473
	IB	0.070	0.24
40 - 100	β-	1.22	1.71
	IB	0.20	0.30
100 - 300	β-	17.5	8.4
	IB	0.54	0.31
300 - 600	β-	83	18.2
	IB	0.56	0.133
600 - 1300	β-	416	44.6
	IB	0.63	0.074
1300 - 2500	β-	386	23.3
	IB	0.22	0.0133
2500 - 4202	β-	89	3.00
	IB	0.023	0.00079

$^{45}_{20}$Ca(163.8 *18* d)

Mode: β-
Δ: -40813.4 *13* keV
SpA: 1.771×10^{4} Ci/g
Prod: ^{44}Ca(n,γ)

Photons (^{45}Ca)

γ_{mode}	γ(keV)	γ(%)
γ M2	12.40 *2*	3.3 *3* ×10^{-6} *

* with ^{45}Sc(316 ms) in equilib

Continuous Radiation (^{45}Ca)

⟨β-⟩=77 keV;⟨IB⟩=0.021 keV

E_{bin}(keV)		⟨ ⟩(keV)	(%)
0 - 10	β-	0.398	8.0
	IB	0.0040	
10 - 20	β-	1.16	7.7
	IB	0.0033	0.023
20 - 40	β-	4.45	14.9
	IB	0.0050	0.018
40 - 100	β-	25.7	37.6
	IB	0.0070	0.0117
100 - 257	β-	45.4	31.8
	IB	0.0018	0.00144

$^{45}_{21}$Sc(stable)

Δ: -41069.9 *13* keV
%: 100

$^{45}_{21}$Sc(316 *9* ms)

Mode: IT
Δ: -41057.5 *13* keV
SpA: 3.21×10^{11} Ci/g
Prod: ^{45}Sc(p,p'); daughter ^{45}Ca

Photons (^{45}Sc)

⟨γ⟩=0.318 *22* keV

γ_{mode}	γ(keV)	γ(%)†
Sc L_ℓ	0.348	0.012 *3*
Sc L_η	0.353	0.0084 *23*
Sc L_α	0.396	0.012 *3*
Sc L_β	0.428	0.016 *5*
Sc $K_{\alpha2}$	4.086	2.17 *24*
Sc $K_{\alpha1}$	4.091	4.3 *5*
Sc $K_{\beta1}$'	4.461	0.70 *8*
Sc $K_{\beta2}$'	4.540	7.2 *8* $\times 10^{-9}$
γ M2	12.40 *2*	0.176

† 11% uncert(syst)

Atomic Electrons (^{45}Sc)

⟨e⟩=5.2 *4* keV

e_{bin}(keV)	⟨e⟩(keV)	e(%)
3	0.072	2.1 *3*
4	1.08	29 *4*
7	0.346	5
8	3.0	38 *4*
12	0.70	5.8 *6*

$^{45}_{22}$Ti(3.080 *8* h)

Mode: ε
Δ: -39007 *3* keV
SpA: 2.261×10^{7} Ci/g
Prod: ^{45}Sc(p,n); ^{45}Sc(d,2n)

Photons (^{45}Ti)

⟨γ⟩=3.47 *17* keV

γ_{mode}	γ(keV)	γ(%)
Sc $K_{\alpha2}$	4.086	0.77 *4*
Sc $K_{\alpha1}$	4.091	1.53 *8*
Sc $K_{\beta1}$'	4.461	0.250 *12*
γ M2	12.40 *2*	2.8 *3* $\times 10^{-5}$ *
γ E1	363.6 *5*	0.0057 *13*
γ M1(+1.7%E2)	424.4 *5*	0.0137 *20*
γ M1+7.6%E2	431.4 *6*	~0.0014
γ M1+25%E2	529.7 *6*	0.0011 *4*
γ E1+0.4%M2	542.1 *6*	0.0009 *4*
γ M1+1.1%E2	719.1 *3*	0.154 *12*
γ E2	961.1 *4*	0.0030 *4*
γ E1	973.5 *4*	0.0058 *7*
γ M1,E2	1031.6 *5*	0.0048 *6*
γ E2+<5.9%M1	1235.9 *5*	0.0118 *13*
γ E2+14%M1	1407.5 *3*	0.085 *9*
γ M1+18%E2	1660.4 *3*	0.041 *4*

* with ^{45}Sc(316 ms) in equilib

Atomic Electrons (^{45}Ti)

⟨e⟩=0.53 *4* keV

e_{bin}(keV)	⟨e⟩(keV)	e(%)
3	0.026	0.73 *8*
4	0.38	10.3 *10*
8	0.00048	0.0061 *7*
10	0.118444	1.2
12	0.000111	0.00093 *10*
359	9.6×10^{-6}	2.7 *6* $\times 10^{-6}$
363	8.5×10^{-7}	2.4 *5* $\times 10^{-7}$
420	2.5×10^{-5}	6.0 *9* $\times 10^{-6}$
424	2.2×10^{-6}	5.3 *8* $\times 10^{-7}$
427	2.8×10^{-6}	$\sim 6 \times 10^{-7}$
431	2.5×10^{-7}	$\sim 6 \times 10^{-8}$
525	2.0×10^{-6}	3.8 *14* $\times 10^{-7}$
529	1.8×10^{-7}	3.3 *12* $\times 10^{-8}$
538	8.0×10^{-7}	1.5 *7* $\times 10^{-7}$
542	7.0×10^{-8}	1.3 *6* $\times 10^{-8}$
715	0.000155	2.17 *17* $\times 10^{-5}$
719	1.37×10^{-5}	1.91 *15* $\times 10^{-6}$
957	3.1×10^{-6}	3.2 *4* $\times 10^{-7}$
961	2.7×10^{-7}	2.8 *4* $\times 10^{-8}$
969	2.5×10^{-6}	2.6 *3* $\times 10^{-7}$
973	2.2×10^{-7}	2.3 *3* $\times 10^{-8}$
1027	3.9×10^{-6}	3.8 *7* $\times 10^{-7}$
1031	3.5×10^{-7}	3.3 *7* $\times 10^{-8}$
1231	8.5×10^{-6}	6.9 *8* $\times 10^{-7}$
1235	7.4×10^{-7}	6.0 *7* $\times 10^{-8}$
1236	7.2×10^{-9}	5.9 *7* $\times 10^{-10}$
1403	5.3×10^{-5}	3.7 *4* $\times 10^{-6}$
1407	4.6×10^{-6}	3.3 *4* $\times 10^{-7}$
1656	1.96×10^{-5}	1.19 *13* $\times 10^{-6}$
1660	1.73×10^{-6}	1.04 *11* $\times 10^{-7}$

Continuous Radiation (^{45}Ti)

⟨β+⟩=373 keV;⟨IB⟩=0.78 keV

E_{bin}(keV)		⟨ ⟩(keV)	(%)
0 - 10	β+	0.00075	0.0098
	IB	0.017	
10 - 20	β+	0.0102	0.064
	IB	0.0167	0.116
20 - 40	β+	0.111	0.353
	IB	0.032	0.110
40 - 100	β+	2.06	2.77
	IB	0.083	0.128
100 - 300	β+	45.1	21.4
	IB	0.18	0.106
300 - 600	β+	178	39.9
	IB	0.137	0.033
600 - 1300	β+	148	20.4
	IB	0.22	0.024
1300 - 2063	IB	0.093	0.0062
	Σβ+		85

$^{45}_{23}$V (539 *18* ms)

Mode: ε
Δ: -31875 *17* keV
SpA: 2.62×10^{11} Ci/g
Prod: ^{40}Ca(^{7}Li,2n)

$^{45}_{24}$Cr(50 *6* ms)

Mode: ε, εp(~25 %)
Δ: -19460 *150* keV
SpA: 3.6×10^{11} Ci/g
Prod: ^{32}S(^{16}O,3n)

εp: 2060 *50*

A = 46

NDS **24**, 1 (1978)

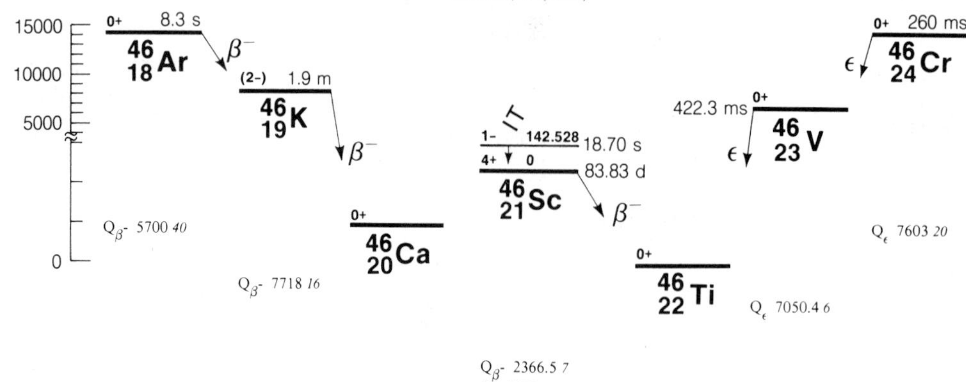

Q$_{\beta^-}$ 2366.5 *7*
Q$_\epsilon$ 1379 *4*

$^{46}_{18}$Ar(8.3 *5* s)

Mode: β-
Δ: -29720 *40* keV
SpA: 2.83×10^{10} Ci/g
Prod: protons on V; ^{48}Ca(n,n2p)

Photons (^{46}Ar)

γ_{mode}	γ(keV)	γ(rel)
γ	288.1 *7*	0.7 *2*
γ	584.7 *15*	0.4 *1*
γ	1020.3 *12*	0.8 *3*
γ	1944.3 *2*	100 *20*

$^{46}_{19}$K (1.92 *7* min)

Mode: β-
Δ: -35420 *16* keV
SpA: 2.12×10^{9} Ci/g
Prod: ^{48}Ca(d,α)

Photons (^{46}K)

⟨γ⟩=2991 *180* keV

γ_{mode}	γ(keV)	γ(%)†
γ E2	1346.9 *10*	91.0 *9*
γ	1439 *2*	~3
γ [M1+E2]	1669.7 *19*	4.6 *18* ?
γ	1780 *2*	8.2 *18*
γ [E1]	2274 *2*	~8
γ [E2]	3016.6 *20*	~9 ?
γ [M1+E2]	3700 *5*	28.2 *9* ?

† 4.4% uncert(syst)

Continuous Radiation (^{46}K)

⟨β-⟩=2322 keV;⟨IB⟩=9.1 keV

E_{bin}(keV)		⟨ ⟩(keV)	(%)
0 - 10	β-	0.00339	0.067
	IB	0.061	
10 - 20	β-	0.0106	0.070
	IB	0.061	0.42
20 - 40	β-	0.0458	0.152
	IB	0.120	0.42
40 - 100	β-	0.392	0.55
	IB	0.35	0.54
100 - 300	β-	5.7	2.74
	IB	1.07	0.60
300 - 600	β-	28.1	6.1
	IB	1.37	0.32
600 - 1300	β-	180	18.7
	IB	2.4	0.27
1300 - 2500	β-	563	30.3
	IB	2.3	0.132
2500 - 5000	β-	1352	37.9
	IB	1.36	0.043
5000 - 6373	β-	192	3.58
	IB	0.023	0.00041

$^{46}_{20}$Ca(stable)

Δ: -43138 *4* keV
%: 0.004 *3*

$^{46}_{21}$Sc(83.83 *2* d)

Mode: β-
Δ: -41759.2 *13* keV
SpA: 3.3861×10^{4} Ci/g
Prod: ^{45}Sc(n,γ)

Photons (^{46}Sc)

⟨γ⟩=2009.47 *7* keV

γ_{mode}	γ(keV)	γ(%)
γ E2	889.25 *3*	99.984 *1*
γ E2	1120.51 *5*	99.987 *1*

Continuous Radiation (^{46}Sc)

⟨β-⟩=112 keV;⟨IB⟩=0.043 keV

E_{bin}(keV)		⟨ ⟩(keV)	(%)
0 - 10	β-	0.259	5.2
	IB	0.0058	
10 - 20	β-	0.77	5.1
	IB	0.0051	0.036
20 - 40	β-	3.06	10.2
	IB	0.0084	0.029
40 - 100	β-	20.2	29.1
	IB	0.0149	0.024
100 - 300	β-	85	49.5
	IB	0.0085	0.0063
300 - 600	β-	2.75	0.87
	IB	2.0×10^{-5}	5.8×10^{-6}
600 - 1300	β-	0.0140	0.00164
	IB	3.8×10^{-6}	5.2×10^{-7}
1300 - 1477	β-	0.000358	2.66×10^{-5}
	IB	4.9×10^{-9}	3.7×10^{-10}

$^{46}_{21}$Sc(18.70 *5* s)

Mode: IT
Δ: -41616.7 *13* keV
SpA: 1.288×10^{10} Ci/g
Prod: ^{45}Sc(n,γ)

Photons (^{46}Sc)

$\langle\gamma\rangle$=88.6 *18* keV

γ_{mode}	γ(keV)	γ(%)†
Sc L_ℓ	0.348	0.010 *3*
Sc L_η	0.353	0.0072 *18*
Sc L_α	0.396	0.010 *3*
Sc L_β	0.427	0.013 *4*
Sc $K_{\alpha2}$	4.086	1.90 *10*
Sc $K_{\alpha1}$	4.091	3.78 *20*
Sc $K_{\beta1}$'	4.461	0.62 *3*
γ E3	142.528 *3*	62.0

† 3.2% uncert(syst)

Atomic Electrons (^{46}Sc)

$\langle e\rangle$=52.3 *13* keV

e_{bin}(keV)	$\langle e\rangle$(keV)	e(%)
3	0.063	1.81 *19*
4	0.95	25 *3*
10	0.297624	3
138	46.2	33.5 *10*
142	4.83	3.40 *10*

$^{46}_{22}$Ti(stable)

Δ: -44125.7 *13* keV
%: 8.0 *1*

$^{46}_{23}$V (422.33 *20* ms)

Mode: ε
Δ: -37075.3 *14* keV
SpA: 2.8528×10^{11} Ci/g
Prod: ^{46}Ti(p,n)

Continuous Radiation (^{46}V)

$\langle\beta+\rangle$=2814 keV;$\langle$IB$\rangle$=11.7 keV

E_{bin}(keV)		$\langle\ \rangle$(keV)	(%)
0 - 10	β+	1.08×10^{-5}	0.000138
	IB	0.071	
10 - 20	β+	0.000154	0.00097
	IB	0.070	0.49
20 - 40	β+	0.00178	0.0056
	IB	0.139	0.48
40 - 100	β+	0.0367	0.0490
	IB	0.41	0.63
100 - 300	β+	1.10	0.504
	IB	1.27	0.70
300 - 600	β+	8.3	1.77
	IB	1.67	0.39
600 - 1300	β+	92	9.2
	IB	3.0	0.34
1300 - 2500	β+	573	29.5
	IB	3.2	0.18
2500 - 5000	β+	1975	56
	IB	1.8	0.057
5000 - 7050	β+	165	3.12
	IB	0.027	0.00049
	Σβ+		99.9

$^{46}_{24}$Cr(260 *60* ms)

Mode: ε
Δ: -29472 *20* keV
SpA: 3.3×10^{11} Ci/g
Prod: ^{32}S(^{16}O,2n)

Continuous Radiation (^{46}Cr)

$\langle\beta+\rangle$=3088 keV;$\langle$IB$\rangle$=13.3 keV

E_{bin}(keV)		$\langle\ \rangle$(keV)	(%)
0 - 10	β+	7.5×10^{-6}	9.6×10^{-5}
	IB	0.074	
10 - 20	β+	0.000112	0.00070
	IB	0.074	0.51
20 - 40	β+	0.00133	0.00421
	IB	0.146	0.51
40 - 100	β+	0.0283	0.0377
	IB	0.43	0.66
100 - 300	β+	0.87	0.398
	IB	1.35	0.75
300 - 600	β+	6.6	1.42
	IB	1.8	0.42
600 - 1300	β+	75	7.5
	IB	3.3	0.37
1300 - 2500	β+	493	25.3
	IB	3.6	0.21
2500 - 5000	β+	2084	57
	IB	2.4	0.076
5000 - 6581	β+	429	7.9
	IB	0.062	0.00109
	Σβ+		100

A = 47

NDS **22**, 59 (1977)

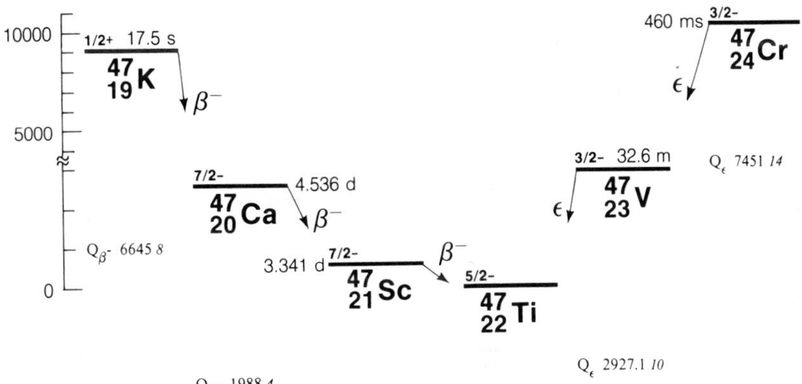

$^{47}_{19}$K (17.5 *3* s)

Mode: β-

Δ: -35698 *8* keV

SpA: 1.345×10^{10} Ci/g

Prod: ^{48}Ca(γ,p); ^{48}Ca(t,α)

Photons (^{47}K)

⟨γ⟩=2596 *51* keV

γ_{mode}	γ(keV)	γ(%)
γ [E1]	564.7 *3*	14.6 *15*
γ [E1]	585.8 *3*	85 *5*
γ [E2]	2013.1 *3*	100

Continuous Radiation (^{47}K)

⟨β-⟩=1805 keV;⟨IB⟩=5.8 keV

E_{bin}(keV)		⟨ ⟩(keV)	(%)
0 - 10	β-	0.00222	0.0439
	IB	0.055	
10 - 20	β-	0.0069	0.0461
	IB	0.054	0.38
20 - 40	β-	0.0303	0.100
	IB	0.107	0.37
40 - 100	β-	0.266	0.372
	IB	0.31	0.48
100 - 300	β-	4.22	2.01
	IB	0.93	0.52
300 - 600	β-	24.5	5.3
	IB	1.14	0.27
600 - 1300	β-	214	21.9
	IB	1.8	0.21
1300 - 2500	β-	905	47.9
	IB	1.27	0.074
2500 - 4067	β-	657	22.4
	IB	0.161	0.0056

$^{47}_{20}$Ca(4.536 *2* d)

Mode: β-

Δ: -42343 *4* keV

SpA: 6.125×10^{5} Ci/g

Prod: ^{46}Ca(n,γ)

Photons (^{47}Ca)

⟨γ⟩=1063 *25* keV

γ_{mode}	γ(keV)	γ(%)†
γ [E1]	41.03 *5*	0.0064 *7*
γ [M1+E2]	489.22 *8*	6.74 *22*
γ [E1]	530.25 *9*	0.105 *15*
γ [M2]	766.82 *9*	0.195 *15*
γ [E2]	807.85 *8*	6.89 *22*
γ M1+E2	1297.06 *8*	74.9
γ [M1+E2]	1878.0 *5*	0.028 *3*

† 2.5% uncert(syst)

Continuous Radiation (^{47}Ca)

⟨β-⟩=345 keV;⟨IB⟩=0.43 keV

E_{bin}(keV)		⟨ ⟩(keV)	(%)
0 - 10	β-	0.078	1.56
	IB	0.0157	
10 - 20	β-	0.239	1.59
	IB	0.0150	0.104
20 - 40	β-	1.00	3.32
	IB	0.028	0.098
40 - 100	β-	7.7	11.0
	IB	0.071	0.110
100 - 300	β-	77	38.8
	IB	0.142	0.084
300 - 600	β-	129	31.3
	IB	0.090	0.022
600 - 1300	β-	90	9.9
	IB	0.066	0.0080
1300 - 1988	β-	39.3	2.64
	IB	0.0043	0.00031

$^{47}_{21}$Sc(3.341 *3* d)

Mode: β-

Δ: -44331.3 *22* keV

SpA: 8.315×10^{5} Ci/g

Prod: daughter ^{47}Ca

Photons (^{47}Sc)

⟨γ⟩=108 *3* keV

γ_{mode}	γ(keV)	γ(%)
Ti $K_{\alpha2}$	4.505	0.0231 *14*
Ti $K_{\alpha1}$	4.511	0.046 *3*
Ti $K_{\beta1}$'	4.932	0.0077 *5*
γ M1+0.2%E2	159.381 *15*	68 *2*

Continuous Radiation (^{47}Sc)

⟨β-⟩=162 keV;⟨IB⟩=0.089 keV

E_{bin}(keV)		⟨ ⟩(keV)	(%)
0 - 10	β-	0.168	3.35
	IB	0.0083	
10 - 20	β-	0.505	3.36
	IB	0.0076	0.053
20 - 40	β-	2.05	6.8
	IB	0.0132	0.046
40 - 100	β-	14.6	20.9
	IB	0.028	0.044
100 - 300	β-	99	53
	IB	0.030	0.019
300 - 600	β-	45.4	12.4
	IB	0.0023	0.00066
600 - 601	β-	6.0×10^{-7}	1.01×10^{-7}
	IB	2.2×10^{-15}	3.6×10^{-16}

$^{47}_{22}$Ti(stable)

Δ: -44931.9 *11* keV

%: 7.3 *1*

$^{47}_{23}$V (32.6 *3* min)

Mode: ε

Δ: -42004.8 *13* keV

SpA: 1.227×10^{8} Ci/g

Prod: ^{45}Sc(α,2n); ^{47}Ti(p,n); ^{46}Ti(d,n); ^{47}Ti(d,2n)

Photons (^{47}V)

⟨γ⟩=8.40 *20* keV

γ_{mode}	γ(keV)	γ(%)†
Ti $K_{\alpha2}$	4.505	0.198 *10*
Ti $K_{\alpha1}$	4.511	0.394 *19*
Ti $K_{\beta1}$'	4.932	0.066 *3*
γ M1+0.2%E2	159.381 *15*	0.107 *5*
γ M1+8.5%E2	244.2 *4*	0.094 *4*
γ [M1+E2]	1243.5 *5*	0.0025 *6*
γ E2	1390.41 *25*	0.0797 *23*
γ M1+E2	1549.79 *25*	0.0669 *19*
γ M1+0.2%E2	1794.0 *3*	0.192
γ [E2]	2003.5 *5*	0.0036 *4*
γ M1+E2	2007.3 *10*	0.0100 *6*
γ M1	2162.9 *5*	0.0674 *15*
γ	2366.3 *4*	0.0085 *6*
γ	2525.6 *4*	0.0088 *4*
γ [M1+E2]	2548.7 *5*	0.0067 *4*
γ [E2]	2793.3 *5*	0.00065 *12*

† 4.4% uncert(syst)

Atomic Electrons (^{47}V)

⟨e⟩=0.128 *10* keV

e_{bin}(keV)	⟨e⟩(keV)	e(%)
4	0.098	2.39 *24*
5	0.00139	0.029 *3*
10	0.026848	0.27
154	0.00087	0.00056 *3*
159	8.2 ×10^{-5}	5.2 *3* ×10^{-5}
239	0.00057	0.00024 *3*
244	5.3 ×10^{-5}	2.16 *24* ×10^{-5}
1239	1.9 ×10^{-6}	1.5 *4* ×10^{-7}
1243	1.7 ×10^{-7}	1.4 *4* ×10^{-8}
1385	5.80 ×10^{-5}	4.19 *15* ×10^{-6}
1390	5.18 ×10^{-6}	3.73 *13* ×10^{-7}
1545	4.1 ×10^{-5}	2.64 *20* ×10^{-6}
1549	3.6 ×10^{-6}	2.35 *17* ×10^{-7}
1789	9.8 ×10^{-5}	5.5 *3* ×10^{-6}
1793	8.7 ×10^{-6}	4.9 *3* ×10^{-7}
1794	5.1 ×10^{-8}	2.87 *17* ×10^{-9}
1999	1.87 ×10^{-6}	9.4 *11* ×10^{-8}
2002	4.9 ×10^{-6}	2.46 *18* ×10^{-7}
2003	1.67 ×10^{-7}	8.4 *9* ×10^{-9}
2007	4.4 ×10^{-7}	2.19 *17* ×10^{-8}
2158	3.04 ×10^{-5}	1.41 *4* ×10^{-6}
2162	2.71 ×10^{-6}	1.26 *4* ×10^{-7}
2361	3.1 ×10^{-6}	1.3 *3* ×10^{-7}
2366	2.8 ×10^{-7}	1.2 *3* ×10^{-8}
2521	3.1 ×10^{-6}	1.2 *3* ×10^{-7}
2525	2.8 ×10^{-7}	1.10 *25* ×10^{-8}
2544	2.81 ×10^{-6}	1.10 *9* ×10^{-7}
2548	2.51 ×10^{-7}	9.8 *8* ×10^{-9}
2788	2.6 ×10^{-7}	9.4 *17* ×10^{-9}
2793	2.4 ×10^{-8}	8.4 *15* ×10^{-10}

Continuous Radiation (^{47}V)

⟨β+⟩=803 keV;⟨IB⟩=1.7 keV

E_{bin}(keV)		⟨ ⟩(keV)	(%)
0 - 10	β+	0.000190	0.00245
	IB	0.032	
10 - 20	β+	0.00271	0.0170
	IB	0.032	0.22
20 - 40	β+	0.0309	0.098
	IB	0.061	0.21
40 - 100	β+	0.62	0.82
	IB	0.17	0.26
100 - 300	β+	16.2	7.6
	IB	0.45	0.26
300 - 600	β+	97	21.0
	IB	0.43	0.103
600 - 1300	β+	504	54
	IB	0.40	0.049
1300 - 2500	β+	186	12.6
	IB	0.135	0.0080
2500 - 2927	IB	0.0050	0.00019
	Σβ+		97

$^{47}_{24}$Cr(460.0 *15* ms)

Mode: ε

Δ: -34554 *14* keV

SpA: 2.696×10^{11} Ci/g

Prod: ^{46}Ti(^{3}He,2n)

Continuous Radiation (^{47}Cr)

⟨β+⟩=3011 keV;⟨IB⟩=12.9 keV

E_{bin}(keV)		⟨ ⟩(keV)	(%)
0 - 10	β+	8.0×10^{-6}	0.000102
	IB	0.073	
10 - 20	β+	0.000119	0.00074
	IB	0.073	0.50
20 - 40	β+	0.00142	0.00448
	IB	0.144	0.50
40 - 100	β+	0.0301	0.0401
	IB	0.42	0.65
100 - 300	β+	0.92	0.422
	IB	1.33	0.73
300 - 600	β+	7.0	1.50
	IB	1.8	0.41
600 - 1300	β+	79	7.9
	IB	3.2	0.37
1300 - 2500	β+	514	26.3
	IB	3.5	0.20
2500 - 5000	β+	2065	57
	IB	2.3	0.072
5000 - 7451	β+	346	6.4
	IB	0.066	0.00119
	Σβ+		99.9

A = 48

NDS **23**, 1 (1978)

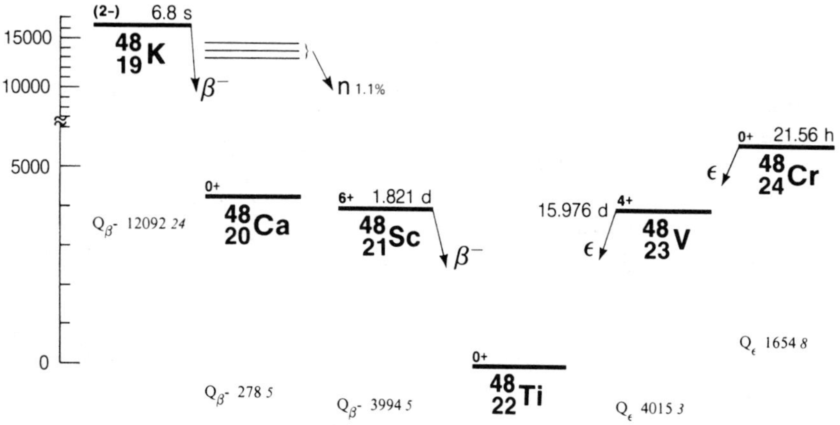

15000
10000
5000
0
(2-) 6.8 s
48 19 K
β⁻
n 1.1%
Q$_{β^-}$ 12092 *24*
0+
48 20 Ca
6+ 1.821 d
48 21 Sc
β⁻
15.976 d 4+
ε
48 23 V
0+ 21.56 h
ε
48 24 Cr
0+
48 22 Ti
Q$_{β^-}$ 278 *5*
Q$_{β^-}$ 3994 *5*
Q$_ε$ 4015 *3*
Q$_ε$ 1654 *8*

$^{48}_{19}$K (6.8 *2* s)

Mode: β-, β-n(1.14 *15* %)

Δ: -32124 *24* keV

SpA: 3.286×10^{10} Ci/g

Prod: ^{48}Ca(n,p); protons on U

Photons (^{48}K)

⟨γ⟩=6775 *160* keV

$γ_{mode}$	γ(keV)	γ(%)†
$γ_{β-}$[E2]	671.23 *12*	3.6 *4*
$γ_{β-}$E1	675.16 *5*	17.3 *6*
$γ_{β-}$	715.57 *10*	1.4 *4*
$γ_{β-}$	753.3 *8*	~0.9
$γ_{β-}$[M1+E2]	780.11 *3*	31.9 *16*
$γ_{β-}$	793.06 *8*	9.9 *6*
$γ_{β-}$[M1+E2]	862.66 *7*	4.4 *6*
$γ_{β-}$[E1]	866.59 *12*	3.4 *6*
$γ_{β-}$	1300.9 *3*	9 *3*
$γ_{β-}$	1315.62 *6*	13.1 *10*
$γ_{β-}$	1525.41 *9*	4.0 *6*
$γ_{β-}$[E1]	1537.81 *5*	15.1 *9*
$γ_{β-}$	1633.6 *3*	6.4 *16*
$γ_{β-}$	1783.1 *3*	8.9 *14*
$γ_{β-}$[M1+E2]	2031.18 *9*	3.0 *4*
$γ_{β-}$	2073.31 *8*	1.9 *6*
$γ_{β-}$	2178.25 *9*	2.4 *5*
$γ_{β-}$	2283.09 *8*	2.6 *5*
$γ_{β-}$	2388.04 *7*	11.0 *8*
$γ_{β-}$[E1]	2788.85 *7*	16.6 *10*
$γ_{β-}$	3063.17 *8*	3.8 *7*
$γ_{β-}$E2	3831.50 *8*	80
$γ_{β-}$E3	4506.60 *9*	3.8 *10*
$γ_{β-}$[E1]	6613.7 *5*	13.6 *8*
$γ_{β-}$[E1]	7300.9 *5*	2.5 *6*

† 1.2% uncert(syst)

Continuous Radiation (^{48}K)

⟨β-⟩=2634 keV;⟨IB⟩=10.7 keV

E_{bin}(keV)		⟨ ⟩(keV)	(%)
0 - 10	β-	0.00103	0.0204
	IB	0.067	
10 - 20	β-	0.00322	0.0214
	IB	0.067	0.46
20 - 40	β-	0.0141	0.0467
	IB	0.132	0.46
40 - 100	β-	0.124	0.174
	IB	0.39	0.59
100 - 300	β-	2.01	0.96
	IB	1.20	0.67
300 - 600	β-	12.1	2.60
	IB	1.56	0.36
600 - 1300	β-	116	11.7
	IB	2.8	0.32
1300 - 2500	β-	635	32.9
	IB	2.8	0.159
2500 - 5000	β-	1579	46.0
	IB	1.56	0.049
5000 - 7500	β-	287	4.98
	IB	0.096	0.0017
7500 - 8260	β-	3.41	0.0443
	IB	0.000113	1.45×10^{-6}

$^{48}_{20}$Ca(>2.0×10^{16} yr)

Δ: -44216 *4* keV

%: 0.187 *3*

$^{48}_{21}$Sc(1.821 *4* d)

Mode: β-

Δ: -44493 *5* keV

SpA: 1.494×10^{6} Ci/g

Prod: ^{51}V(n,α); ^{50}Ti(d,α); ^{48}Ca(p,n); ^{48}Ca(d,2n)

Photons (^{48}Sc)

$\langle\gamma\rangle$=3349 *6* keV

γ_{mode}	γ(keV)	γ(%)†
γ [M1+E2]	175.357 *4*	7.47 *7*
γ E2	983.501 *2*	100.0
γ E2	1037.4961 *19*	97.5 *3*
γ [E2]	1212.849 *4*	2.38 *2*
γ E2	1312.046 *3*	100.0

† 0.30% uncert(syst)

Continuous Radiation (^{48}Sc)

$\langle\beta-\rangle$=221 keV;$\langle$IB$\rangle$=0.155 keV

E_{bin}(keV)		$\langle\ \rangle$(keV)	(%)
0 - 10	β-	0.108	2.15
	IB	0.0111	
10 - 20	β-	0.327	2.18
	IB	0.0103	0.072
20 - 40	β-	1.36	4.51
	IB	0.019	0.065
40 - 100	β-	10.3	14.6
	IB	0.043	0.068
100 - 300	β-	93	47.6
	IB	0.062	0.039
300 - 600	β-	114	28.7
	IB	0.0104	0.0028
600 - 661	β-	1.32	0.214
	IB	5.0×10^{-6}	8.2×10^{-7}

$^{48}_{22}$Ti(stable)

Δ: -48487.1 *11* keV
%: 73.8 *1*

$^{48}_{23}$V (15.976 *3* d)

Mode: ϵ
Δ: -44472 *3* keV
SpA: 1.7027×10^{5} Ci/g
Prod: ^{45}Sc(α,n); ^{48}Ti(p,n); ^{48}Ti(d,2n); ^{50}Cr(d,α)

Photons (^{48}V)

$\langle\gamma\rangle$=2401 *4* keV

γ_{mode}	γ(keV)	γ(%)†
Ti L_ℓ	0.395	0.014 *4*
Ti L_η	0.401	0.010 *3*
Ti L_α	0.452	0.034 *9*
Ti L_β	0.476	0.033 *9*
Ti $K_{\alpha2}$	4.505	2.83 *14*
Ti $K_{\alpha1}$	4.511	5.6 *3*
Ti $K_{\beta1}$'	4.932	0.94 *5*
γ [M1+E2]	803.18 *8*	0.150 *20*
γ [M1+E2]	928.281 *17*	0.77 *5*
γ	944.060 *7*	7.76 *9*
γ E2	983.501 *2*	99.99
γ E2	1312.046 *3*	97.49 *20*
γ M1+3.8%E2	1437.14 *8*	0.120 *20*
γ M1+6.5%E2	2240.300 *16*	2.41 *4*
γ [E1]	2375.0 *5*	~0.010
γ [E2]	2420.61 *8*	~0.010

† 0.20% uncert(syst)

Atomic Electrons (^{48}V)

$\langle$e$\rangle$=2.01 *14* keV

e_{bin}(keV)	$\langle$e$\rangle$(keV)	e(%)
4	1.39	34 *4*
5	0.0199	0.41 *4*
10	0.38272	4
798	0.00019	2.4 *6* $\times10^{-5}$
803	1.8 $\times10^{-5}$	2.2 *5* $\times10^{-6}$
923	0.00082	8.9 *16* $\times10^{-5}$
928	7.4 $\times10^{-5}$	8.0 *14* $\times10^{-6}$
939	0.007	0.0007 *3*
943	0.00059	6.2 *25* $\times10^{-5}$
944	7.7 $\times10^{-6}$	$\sim8\times10^{-7}$
979	0.1131	0.01156 *23*
983	0.01019	0.001037 *21*
1307	0.0758	0.00580 *12*
1311	0.00671	0.000512 *10*
1312	7.00 $\times10^{-5}$	5.34 *11* $\times10^{-6}$
1432	7.3 $\times10^{-5}$	5.1 *9* $\times10^{-6}$
1437	6.6 $\times10^{-6}$	4.6 *8* $\times10^{-7}$
2235	0.00107	4.77 *12* $\times10^{-5}$
2240	9.52 $\times10^{-5}$	4.25 *11* $\times10^{-6}$
2370	2.8 $\times10^{-6}$	$\sim1\times10^{-7}$
2374	2.5 $\times10^{-7}$	$\sim1\times10^{-8}$
2375	1.3 $\times10^{-9}$	$\sim6\times10^{-11}$
2416	4.4 $\times10^{-6}$	$\sim2\times10^{-7}$
2420	4.0 $\times10^{-7}$	$\sim2\times10^{-8}$

Continuous Radiation (^{48}V)

$\langle\beta+\rangle$=144 keV;$\langle$IB$\rangle$=0.64 keV

E_{bin}(keV)		$\langle\ \rangle$(keV)	(%)
0 - 10	β+	0.00092	0.0119
	IB	0.0075	
10 - 20	β+	0.0129	0.081
	IB	0.0068	0.047
20 - 40	β+	0.143	0.456
	IB	0.0125	0.044
40 - 100	β+	2.60	3.52
	IB	0.031	0.049
100 - 300	β+	46.7	22.8
	IB	0.070	0.040
300 - 600	β+	91	22.1
	IB	0.115	0.026
600 - 1300	β+	4.05	0.63
	IB	0.34	0.039
1300 - 2500	β+	0.227	0.0152
	IB	0.049	0.0035
2500 - 3031	IB	7.1×10^{-6}	2.7×10^{-7}
	$\Sigma\beta$+		50

$^{48}_{24}$Cr(21.56 *3* h)

Mode: ϵ
Δ: -42818 *7* keV
SpA: 3.028×10^{6} Ci/g
Prod: ^{46}Ti(α,2n)

Photons (^{48}Cr)

$\langle\gamma\rangle$=420.9 *12* keV

γ_{mode}	γ(keV)	γ(%)†
V L_ℓ	0.446	0.037 *11*
V L_η	0.454	0.028 *8*
V L_α	0.511	0.15 *4*
V L_β	0.531	0.13 *4*
V $K_{\alpha2}$	4.945	7.3 *9*
V $K_{\alpha1}$	4.952	14.5 *18*
V $K_{\beta1}$'	5.427	2.5 *3*
γ M1+E2	112.44 *2*	99 *1*
γ E2	308.33 *2*	100.0

† <0.1% uncert(syst)

Atomic Electrons (^{48}Cr)

$\langle$e$\rangle$~18 keV

e_{bin}(keV)	$\langle$e$\rangle$(keV)	e(%)
4	2.7	61 *9*
5	0.72	14.7 *23*
107	12	~12
112	1.3	~1
303	1.42	0.469 *9*
308	0.135	0.0437 *9*

Continuous Radiation (^{48}Cr)

$\langle\beta+\rangle$=1.33 keV;$\langle$IB$\rangle$=0.47 keV

E_{bin}(keV)		$\langle\ \rangle$(keV)	(%)
0 - 10	β+	0.000306	0.00394
	IB	0.00102	
10 - 20	β+	0.00422	0.0265
	IB	0.000101	0.00071
20 - 40	β+	0.0426	0.136
	IB	0.00022	0.00075
40 - 100	β+	0.51	0.72
	IB	0.0019	0.0026
100 - 300	β+	0.77	0.58
	IB	0.041	0.019
300 - 600	IB	0.18	0.039
600 - 1233	IB	0.25	0.032
	$\Sigma\beta$+		1.47

A = 49

NDS **24**, 175 (1978)

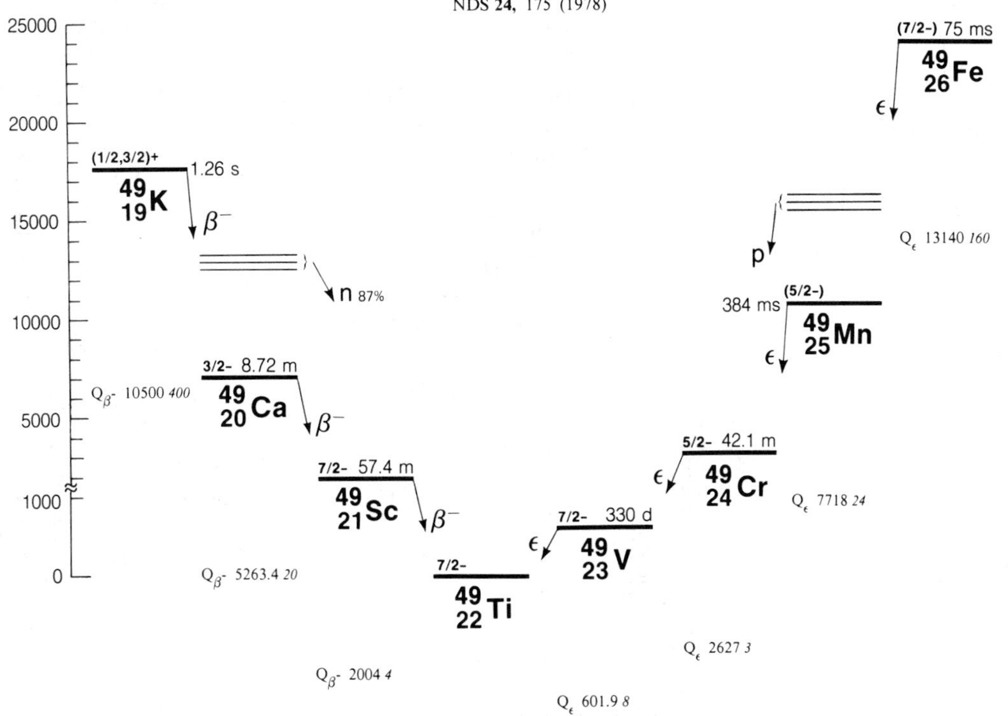

$^{49}_{19}$K (1.26 *5* s)

Mode: β-, β-n(87 *8* %)
Δ: -30790 *400* keV
SpA: 1.41×10^{11} Ci/g
Prod: protons on U

Photons (^{49}K)

γ_{mode}	γ(keV)	γ(rel)
$\gamma_{\beta-}$[M1+E2]	2026.1 *18*	100 *19*
$\gamma_{\beta-}$[E1]	2253.1 *18*	84 *19*
$\gamma_{\beta-}$[E1]	4279.1 *23*	97 *25*

$^{49}_{20}$Ca(8.716 *11* min)

Mode: β-
Δ: -41291 *4* keV
SpA: 4.400×10^{8} Ci/g
Prod: ^{48}Ca(n,γ)

Photons (^{49}Ca)

⟨γ⟩=3165 *42* keV

γ_{mode}	γ(keV)	γ(%)†
γ[M1+E2]	143.35 *18*	0.035 *9*
γ[E1]	856.0 *3*	0.13 *3*
γ[M1+E2]	987.55 *14*	0.08 *3*
γ[E1]	1144.9 *4*	0.11 *3*
γ[E1]	1288.2 *4*	0.07 *3*
γ	1409.02 *20*	0.63 *6*
γ	2228.6 *3*	0.19 *5*

Photons (^{49}Ca)
(continued)

γ_{mode}	γ(keV)	γ(%)†
γ	2372.0 *3*	0.49 *9*
γ[E2]	3084.54 *10*	92.1
γ[M1+E2]	4072.02 *10*	7.0 *7*
γ[M1+E2]	4738.32 *20*	0.21 *6*

† 1.1% uncert(syst)

Continuous Radiation (^{49}Ca)

⟨β-⟩=870 keV;⟨IB⟩=1.8 keV

E_{bin}(keV)		⟨ ⟩(keV)	(%)
0 - 10	β-	0.0123	0.245
	IB	0.034	
10 - 20	β-	0.0382	0.254
	IB	0.033	0.23
20 - 40	β-	0.164	0.54
	IB	0.065	0.22
40 - 100	β-	1.38	1.94
	IB	0.18	0.28
100 - 300	β-	19.4	9.4
	IB	0.48	0.27
300 - 600	β-	90	19.7
	IB	0.48	0.114
600 - 1300	β-	446	47.8
	IB	0.46	0.055
1300 - 2500	β-	313	20.2
	IB	0.049	0.0035
2500 - 5000	β-	0.234	0.0083
	IB	5.0×10^{-5}	1.69×10^{-6}
5000 - 5263	β-	0.000222	4.38×10^{-6}
	IB	2.1×10^{-9}	4.2×10^{-11}

$^{49}_{21}$Sc(57.4 *1* min)

Mode: β-
Δ: -46555 *4* keV
SpA: 6.685×10^{7} Ci/g
Prod: ^{48}Ca(d,n)

Photons (^{49}Sc)

⟨γ⟩=1.04 *18* keV

γ_{mode}	γ(keV)	γ(%)
γ	1622.6 *6*	0.010 *3*
γ[M1+E2]	1761.9 *3*	0.05 *1*

Continuous Radiation (^{49}Sc)

⟨β-⟩=822 keV;⟨IB⟩=1.60 keV

E_{bin}(keV)		⟨ ⟩(keV)	(%)
0 - 10	β-	0.0124	0.247
	IB	0.033	
10 - 20	β-	0.0385	0.256
	IB	0.032	0.22
20 - 40	β-	0.165	0.55
	IB	0.062	0.22
40 - 100	β-	1.38	1.94
	IB	0.17	0.27
100 - 300	β-	19.7	9.5
	IB	0.46	0.26
300 - 600	β-	95	20.8
	IB	0.44	0.104
600 - 1300	β-	479	52
	IB	0.37	0.046
1300 - 2004	β-	227	15.2
	IB	0.025	0.0018

$^{49}_{22}$Ti(stable)

Δ: -48558.1 *11* keV

%: 5.5 *1*

$^{49}_{23}$V (330 *15* d)

Mode: ε

Δ: -47956.3 *13* keV

SpA: 8075 Ci/g

Prod: ^{52}Cr(p,α); ^{48}Ti(d,n)

Photons (^{49}V)

⟨γ⟩=0.87 *3* keV

γ_{mode}	γ(keV)	γ(%)
Ti L_ℓ	0.395	0.028 *8*
Ti L_η	0.401	0.021 *5*
Ti L_α	0.452	0.068 *17*
Ti L_β	0.476	0.066 *18*
Ti $K_{\alpha2}$	4.505	5.7 *3*
Ti $K_{\alpha1}$	4.511	11.4 *6*
Ti $K_{\beta1}$'	4.932	1.91 *9*

Atomic Electrons (^{49}V)

⟨e⟩=3.6 *3* keV

e_{bin}(keV)	⟨e⟩(keV)	e(%)
4	2.8	69 *7*
5	0.040	0.83 *9*
10	0.7776	8

Continuous Radiation (^{49}V)

⟨IB⟩=0.053 keV

E_{bin}(keV)		⟨ ⟩(keV)	(%)
10 - 20	IB	3.6×10^{-5}	0.00025
20 - 40	IB	0.000123	0.00040
40 - 100	IB	0.00158	0.0022
100 - 300	IB	0.024	0.0119
300 - 600	IB	0.027	0.0069
600 - 602	IB	5.1×10^{-10}	8.5×10^{-11}

$^{49}_{24}$Cr(42.09 *15* min)

Mode: ε

Δ: -45329 *3* keV

SpA: 9.12×10^{7} Ci/g

Prod: ^{48}Ti(α,3n); ^{47}Ti(α,2n); ^{46}Ti(α,n); ^{51}V(p,3n); ^{50}Cr(n,2n)

Photons (^{49}Cr)

⟨γ⟩=109.0 *14* keV

γ_{mode}	γ(keV)	γ(%)†
γ [M1+E2]	62.289 *3*	16.7 *3*
γ M1+~20%E2	90.639 *3*	54.2 *11*
γ E2	152.928 *3*	30.9 *6*
γ [M1+E2]	1361.53 *9*	0.051 *4*
γ [M1+E2]	1423.82 *9*	0.0119 *16*
γ [M1+E2]	1508.3 *3*	0.0103 *16*
γ [M1+E2]	1514.45 *9*	0.029 *3*
γ [M1+E2]	1570.5 *3*	0.0195 *22*
γ [M1+E2]	2091.1 *13*	0.00041 *9*
γ [M1+E2]	2144.3 *9*	0.00094 *14*
γ [M1+E2]	2218.6 *18*	0.00020 *6*
γ [M1+E2]	2234.9 *9*	0.00023 *7*

† 1.6% uncert(syst)

Continuous Radiation (^{49}Cr)

⟨β+⟩=595 keV;⟨IB⟩=1.23 keV

E_{bin}(keV)		⟨ ⟩(keV)	(%)
0 - 10	β+	0.000284	0.00364
	IB	0.026	
10 - 20	β+	0.00420	0.0263
	IB	0.025	0.17
20 - 40	β+	0.0493	0.156
	IB	0.048	0.167
40 - 100	β+	1.00	1.33
	IB	0.131	0.20
100 - 300	β+	25.4	11.9
	IB	0.32	0.19
300 - 600	β+	135	29.7
	IB	0.28	0.067
600 - 1300	β+	418	48.3
	IB	0.25	0.030
1300 - 2500	β+	15.0	1.10
	IB	0.144	0.0088
2500 - 2627	IB	3.1×10^{-5}	1.22×10^{-6}
	Σβ+		93

$^{49}_{25}$Mn(384 *17* ms)

Mode: ε

Δ: -37611 *24* keV

SpA: 2.78×10^{11} Ci/g

Prod: ^{28}Si on Mg

Photons (^{49}Mn)

⟨γ⟩=17 *7* keV

γ_{mode}	γ(keV)	γ(%)
γ [M1+E2]	272.3 *4*	6 *3*

Continuous Radiation (^{49}Mn)

⟨β+⟩=3133 keV;⟨IB⟩=13.7 keV

E_{bin}(keV)		⟨ ⟩(keV)	(%)
0 - 10	β+	6.3×10^{-6}	8.1×10^{-5}
	IB	0.075	
10 - 20	β+	9.8×10^{-5}	0.00062
	IB	0.074	0.51
20 - 40	β+	0.00121	0.00383
	IB	0.147	0.51
40 - 100	β+	0.0265	0.0353
	IB	0.43	0.66
100 - 300	β+	0.83	0.379
	IB	1.36	0.75
300 - 600	β+	6.4	1.36
	IB	1.8	0.42
600 - 1300	β+	72	7.3
	IB	3.4	0.38
1300 - 2500	β+	480	24.6
	IB	3.7	0.21
2500 - 5000	β+	2088	57
	IB	2.6	0.081
5000 - 7500	β+	486	8.9
	IB	0.107	0.0019
7500 - 7718	IB	2.4×10^{-5}	3.2×10^{-7}
	Σβ+		99.9

$^{49}_{26}$Fe(75 *10* ms)

Mode: ε, εp

Δ: -24470 *160* keV

SpA: 3.3×10^{11} Ci/g

Prod: ^{40}Ca(^{12}C,3n)

εp: 1920 *50*

A = 50

NDS **42**, 369 (1984)

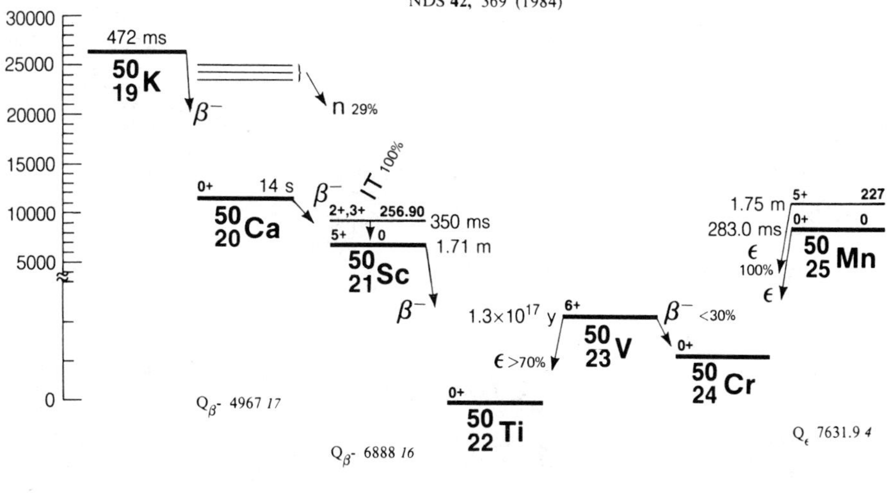

$^{50}_{19}$K (472 *4* ms)

Mode: β-, β-n(29 *3* %)
SpA: 2.506×10^{11} Ci/g
Prod: protons on U; protons on Ir

Delayed Neutrons (^{50}K)

n(keV)	n(%)†
150 *7*	4.5 *5*
510 *10*	3.7 *11*
700 *30*	~0.4
970 *50*	~0.6
1250 *40*	0.9 *4*
1300 *40*	0.22 *7*
1450 *40*	0.22 *7*
2030 *60*	1.6 *4*
2170 *60*	0.7 *3*
2260 *70*	0.7 *3*
2480 *60*	7.4 *9*
2830 *90*	5.2 *8*
3340 *130*	2.1 *4*
3980 *150*	0.15 *5*
4010 *160*	0.59 *15*
4600 *170*	0.15 *4*
5010 *170*	0.15 *4*

† 16.5% uncert(syst)

Photons (^{50}K)

γ_{mode}	γ(keV)
$\gamma_{\beta-}$[E2]	1025.7 *16*
$\gamma_{\beta-}$[M1+E2]	1973 *4*
$\gamma_{\beta-n}$[M1+E2]	2026.1 *18*
$\gamma_{\beta-}$	3007 *3*
$\gamma_{\beta-n}$	3354 *4*
$\gamma_{\beta-}$	4032 *3*
$\gamma_{\beta-n}$	4073 *4*
$\gamma_{\beta-}$	4880 *4*

$^{50}_{20}$Ca(13.9 *6* s)

Mode: β-
Δ: -39571 *8* keV
SpA: 1.58×10^{10} Ci/g
Prod: ^{48}Ca(t,p)

Photons (^{50}Ca)

⟨γ⟩=1782 *23* keV

γ_{mode}	γ(keV)	γ(%)†
Sc L_ℓ	0.348	~0.009
Sc L_η	0.353	~0.007
Sc L_α	0.396	~0.010
Sc L_β	0.427	~0.012
Sc $K_{\alpha2}$	4.086	~2
Sc $K_{\alpha1}$	4.091	~4
Sc $K_{\beta1}$'	4.461	~0.6
γ M1(+E2)	71.552 *5*	52 *5*
γ [M3]	256.894 *10*	98 *3* *
γ	1519.300 *17*	59.9 *12*
γ	1590.850 *17*	36.6 *7*

† 0.11% uncert(syst)
* with ^{50}Sc(350 ms) in equilib

Continuous Radiation (^{50}Ca)

⟨β-⟩=1354 keV;⟨IB⟩=3.7 keV

E_{bin}(keV)		⟨ ⟩(keV)	(%)
0 - 10	β-	0.00432	0.086
	IB	0.046	
10 - 20	β-	0.0134	0.089
	IB	0.045	0.31
20 - 40	β-	0.058	0.193
	IB	0.089	0.31
40 - 100	β-	0.504	0.71
	IB	0.26	0.39
100 - 300	β-	7.8	3.71
	IB	0.74	0.41
300 - 600	β-	42.7	9.3
	IB	0.84	0.20

Continuous Radiation (^{50}Ca) (continued)

E_{bin}(keV)		⟨ ⟩(keV)	(%)
600 - 1300	β-	327	33.7
	IB	1.15	0.133
1300 - 2500	β-	872	48.3
	IB	0.50	0.029
2500 - 3119	β-	104	3.92
	IB	0.0062	0.00024

$^{50}_{21}$Sc(1.710 *8* min)

Mode: β-
Δ: -44538 *16* keV
SpA: 2.192×10^{9} Ci/g
Prod: ^{50}Ti(n,p); ^{48}Ca(t,n); daughter ^{50}Ca

Photons (^{50}Sc)

⟨γ⟩=3192 *29* keV

γ_{mode}	γ(keV)	γ(%)
γ [E2]	523.812 *23*	88.7 *18*
γ [E2]	1121.141 *7*	99.5 *20*
γ [M1+E2]	1472.38 *10*	0.61 *4*
γ [E2]	1553.785 *10*	100
γ	1682.02 *7*	0.28 *3*
γ	2205.81 *7*	1.27 *3*
γ [M1+E2]	2705.05 *18*	0.105 *16*
γ	2765.7 *3*	0.145 *18*
γ	3132.2 *3*	0.251 *15*
γ [E2]	3826.13 *18*	0.044 *10*

Continuous Radiation (^{50}Sc)

$\langle\beta-\rangle$=1626 keV;$\langle$IB$\rangle$=4.9 keV

E_{bin}(keV)		$\langle\ \rangle$(keV)	(%)
0 - 10	β-	0.00323	0.064
	IB	0.051	
10 - 20	β-	0.0100	0.067
	IB	0.051	0.35
20 - 40	β-	0.0434	0.144
	IB	0.100	0.35
40 - 100	β-	0.372	0.52
	IB	0.29	0.44
100 - 300	β-	5.7	2.72
	IB	0.86	0.48
300 - 600	β-	31.5	6.8
	IB	1.02	0.24
600 - 1300	β-	253	25.9
	IB	1.54	0.18
1300 - 2500	β-	909	48.7
	IB	0.96	0.057
2500 - 4213	β-	428	15.0
	IB	0.077	0.0028

$^{50}_{21}$Sc(350 *30* ms)

Mode: IT

Δ: -44281 *16* keV

SpA: 2.81×10^{11} Ci/g

Prod: ^{50}Ti(n,p); ^{48}Ca(t,n)

Photons (^{50}Sc)

$\langle\gamma\rangle$=246 *5* keV

γ_{mode}	γ(keV)	$\gamma(\%)^{\dagger}$
Sc $K_{\alpha2}$	4.086	0.177 *11*
Sc $K_{\alpha1}$	4.091	0.352 *22*
Sc $K_{\beta1}$'	4.461	0.057 *4*
γ[M3]	256.894 *10*	95.9

† 0.11% uncert(syst)

Atomic Electrons (^{50}Sc)

$\langle$e$\rangle$=8.8 *4* keV

e_{bin}(keV)	$\langle$e$\rangle$(keV)	e(%)
3	0.0059	0.168 *18*
4	0.088	2.4 *3*
9	0.027377	0.3
252	7.9	3.12 *14*
256	0.77	0.299 *13*

$^{50}_{22}$Ti(stable)

Δ: -51426.2 *11* keV

%: 5.4 *1*

$^{50}_{23}$V (1.3 *5* $\times10^{17}$ yr)

Mode: ϵ(>70 %), β-(<30 %)

Δ: -49219.7 *15* keV

Prod: natural source

%: 0.250 *2*

Photons (^{50}V)

γ_{mode}	γ(keV)	γ(%)
γ_{ϵ}[E2]	1553.785 *10*	>70

$^{50}_{24}$Cr(stable)

Δ: -50257.9 *16* keV

%: 4.345 *9*

$^{50}_{25}$Mn(283.0 *4* ms)

Mode: ϵ

Δ: -42626.0 *16* keV

SpA: 2.974×10^{11} Ci/g

Prod: ^{50}Cr(p,n)

Continuous Radiation (^{50}Mn)

$\langle\beta+\rangle$=3099 keV;$\langle$IB$\rangle$=13.5 keV

E_{bin}(keV)		$\langle\ \rangle$(keV)	(%)
0 - 10	β+	6.5×10^{-6}	8.3×10^{-5}
	IB	0.074	
10 - 20	β+	0.000101	0.00063
	IB	0.074	0.51
20 - 40	β+	0.00124	0.00393
	IB	0.146	0.51
40 - 100	β+	0.0272	0.0362
	IB	0.43	0.66
100 - 300	β+	0.85	0.389
	IB	1.35	0.75
300 - 600	β+	6.5	1.40
	IB	1.8	0.42
600 - 1300	β+	74	7.4
	IB	3.3	0.38
1300 - 2500	β+	489	25.0
	IB	3.7	0.21
2500 - 5000	β+	2084	57
	IB	2.5	0.079
5000 - 7500	β+	445	8.2
	IB	0.094	0.00169
7500 - 7632	IB	5.5×10^{-6}	7.3×10^{-8}
	$\Sigma\beta$+		99.9

$^{50}_{25}$Mn(1.75 *3* min)

Mode: ϵ

Δ: -42399 *6* keV

SpA: 2.14×10^{9} Ci/g

Prod: ^{50}Cr(p,n)

Photons (^{50}Mn)

$\langle\gamma\rangle$=3351 *83* keV

γ_{mode}	γ(keV)	$\gamma(\%)^{\dagger}$
Cr $K_{\alpha2}$	5.405	0.062 *3*
Cr $K_{\alpha1}$	5.415	0.123 *6*
Cr $K_{\beta1}$'	5.947	0.0208 *10*
γ	661.6 *3*	22.8 *9*
γ	711.2 *5*	0.46 *18*
γ[E2]	783.29 *10*	91.3 *18*
γ[E2]	1097.96 *19*	94 *4*
γ[E2]	1282.5 *3*	30.1 *18*
γ[M1+E2]	1443.25 *19*	63 *5*
γ	1793.5 *6*	0.46 *9* ?
γ	1944.1 *3*	3.5 *5*
γ	1993.6 *5*	0.82 *18*
γ	2017.0 *5*	0.55 *18*
γ	2404.4 *8*	0.15 *5*
γ[E2]	2541.18 *25*	0.55 *14*
γ	2811.5 *6*	0.18 *5*
γ	3114.9 *5*	0.91 *18*

† 10% uncert(syst)

Atomic Electrons (^{50}Mn)

$\langle$e$\rangle$=0.498 *25* keV

e_{bin}(keV)	$\langle$e$\rangle$(keV)	e(%)
5	0.026	0.54 *6*
6	0.00035	0.0059 *6*
656	0.043	0.007 *3*
661 - 705	0.0048	0.0007 *3*
710 - 711	7.1×10^{-5}	$\sim1\times10^{-5}$
777	0.188	0.0242 *11*
783	0.0176	0.00225 *10*
1092	0.117	0.0107 *6*
1097	0.0108	0.00098 *6*
1276	0.0308	0.00241 *18*
1282	0.00284	0.000222 *16*
1437	0.053	0.0037 *4*
1788 - 1793	0.00028	1.6 *5* $\times10^{-5}$
1938 - 2016	0.0028	0.00014 *3*
2398 - 2404	7.4×10^{-5}	3.1 *11* $\times10^{-6}$
2535 - 2541	0.00033	1.3 *3* $\times10^{-5}$
2806 - 2811	8.5×10^{-5}	3.0 *9* $\times10^{-6}$
3109 - 3114	0.0004	1.3 *3* $\times10^{-5}$

Continuous Radiation (^{50}Mn)

$\langle\beta+\rangle$=1516 keV;$\langle$IB$\rangle$=4.6 keV

E_{bin}(keV)		$\langle\ \rangle$(keV)	(%)
0 - 10	β+	3.82×10^{-5}	0.000488
	IB	0.049	
10 - 20	β+	0.00059	0.00371
	IB	0.049	0.34
20 - 40	β+	0.0073	0.0229
	IB	0.096	0.33
40 - 100	β+	0.157	0.209
	IB	0.28	0.43
100 - 300	β+	4.64	2.14
	IB	0.81	0.46
300 - 600	β+	32.7	7.0
	IB	0.96	0.23
600 - 1300	β+	290	29.7
	IB	1.40	0.162
1300 - 2500	β+	937	51
	IB	0.80	0.048
2500 - 4695	β+	251	9.1
	IB	0.096	0.0033
	$\Sigma\beta$+		99

A = 51

NDS **23**, 163 (1978)

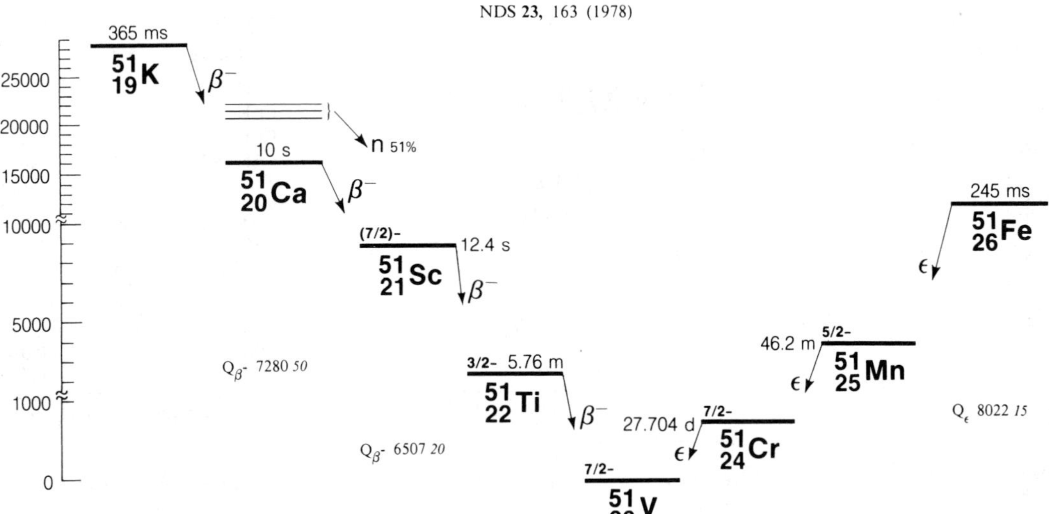

$^{51}_{19}$K (365 *5* ms)

Mode: β-, β-n(51 *9* %)
SpA: 2.71×10^{11} Ci/g
Prod: protons on U; protons on Ir

$^{51}_{20}$Ca(10.0 *8* s)

Mode: β-
Δ: -35940 *50* keV
SpA: 2.14×10^{10} Ci/g
Prod: daughter ^{51}K

Photons (^{51}Ca)

⟨γ⟩=2742 *95* keV

γ_{mode}	γ(keV)	γ(%)†
γ	352.4 *23*	0.17 *5*
γ	532.2 *4*	3.8 *8*
γ	547.7 *1*	19.7 *20*
γ	861.6 *1*	35
γ	1167.5 *3*	23.5 *23*
γ	1314.8 *6*	4.5 *9*
γ	1323.7 *2*	13.8 *14*
γ	1394.0 *2*	27 *3*
γ	1424.0 *7*	2.4 *7*
γ	1480.1 *3*	22.4 *22*
γ	1485.3 *4*	16.2 *16*
γ	1644.4 *4*	4.8 *10*
γ	1714.8 *3*	7.9 *8*
γ	1847.1 *3*	6.2 *12*
γ	1996.5 *5*	3.8 *8*
γ	2028 *3*	0.9 *3*
γ	2333.4 *18*	1.2 *4*
γ	2378.7 *14*	1.1 *3*
γ	2912 *1*	1.4 *4*
γ	3038.9 *7*	2.4 *7*
γ	3196.5 *11*	3.1 *9*
γ	3771.7 *11*	2.1 *6*
γ	4379.0 *15* ?	1.2 *4*

† 10% uncert(syst)

Continuous Radiation (^{51}Ca)

⟨β-⟩=2107 keV;⟨IB⟩=7.5 keV

E_{bin}(keV)		⟨ ⟩(keV)	(%)
0 - 10	β-	0.00180	0.0357
	IB	0.060	
10 - 20	β-	0.0056	0.0374
	IB	0.059	0.41
20 - 40	β-	0.0244	0.081
	IB	0.117	0.41
40 - 100	β-	0.213	0.298
	IB	0.34	0.52
100 - 300	β-	3.37	1.61
	IB	1.04	0.58
300 - 600	β-	19.6	4.24
	IB	1.3	0.30
600 - 1300	β-	176	17.9
	IB	2.2	0.25
1300 - 2500	β-	801	42.2
	IB	1.8	0.106
2500 - 5000	β-	1069	33.1
	IB	0.60	0.019
5000 - 6418	β-	37.8	0.71
	IB	0.0035	6.3×10^{-5}

$^{51}_{21}$Sc(12.4 *1* s)

Mode: β-
Δ: -43220 *20* keV
SpA: 1.735×10^{10} Ci/g
Prod: ^{48}Ca(α,p)

Photons (^{51}Sc)

⟨γ⟩=2350 *26* keV

γ_{mode}	γ(keV)	γ(%)†
γ	331.2 *3*	2.86 *16*
γ[M1+E2]	386.7 *3*	1.83 *9*
γ[M1+E2]	576.5 *3*	3.35 *12*
γ[M1+E2]	706.7 *3*	0.95 *7*
γ	717.9 *3*	7.1 *3*
γ	775.4 *4*	0.61 *7*
γ	887.4 *4*	0.84 *9*
γE2	907.3 *3*	9.3 *3*
γ[E2]	977.6 *6*	0.63 *8*
γ	1032.7 *14*	0.26 *7*
γ[M1+E2]	1123.9 *8*	1.40 *16*
γ[E2]	1163.7 *4*	0.33 *5*
γ[M1+E2]	1166.6 *6*	0.64 *7*
γ[M1+E2]	1254.1 *8*	0.40 *7*
γ[M1+E2]	1294.0 *3*	6.1 *3*
γ	1351.9 *4*	0.71 *7*
γ[E2]	1437.4 *3*	52.0
γ	1474.6 *3*	1.94 *10*
γ	1482.1 *3*	2.08 *10*
γ[M1+E2]	1567.6 *3*	14.9 *5*
γ	1625.2 *3*	3.38 *16*
γ	1750.6 *14*	0.14 *4*
γ	1800.1 *20*	0.30 *6*
γ	2051.1 *3*	8.3 *3*
γ[M1+E2]	2144.1 *3*	31.8 *10*
γ	2181.3 *3*	1.94 *8*
γ	2619.1 *20*	0.33 *4*
γ[E2]	2691.5 *8*	0.23 *3*
γ	2738.1 *20*	0.17 *4*
γ	2919.5 *4*	0.44 *4*

† 1.1% uncert(syst)

Continuous Radiation (^{51}Sc)

⟨β-⟩=1841 keV;⟨IB⟩=6.1 keV

E_{bin}(keV)		⟨ ⟩(keV)	(%)
0 - 10	β-	0.00260	0.052
	IB	0.055	
10 - 20	β-	0.0081	0.054
	IB	0.054	0.38
20 - 40	β-	0.0350	0.116
	IB	0.107	0.37
40 - 100	β-	0.301	0.422
	IB	0.31	0.48
100 - 300	β-	4.66	2.22
	IB	0.93	0.52
300 - 600	β-	26.2	5.7
	IB	1.15	0.27
600 - 1300	β-	217	22.2
	IB	1.8	0.21
1300 - 2500	β-	822	43.9
	IB	1.38	0.080
2500 - 5000	β-	770	24.6
	IB	0.31	0.0106
5000 - 5070	β-	0.0260	0.00052
	IB	3.4×10^{-8}	6.8×10^{-10}

$^{51}_{22}$Ti(5.76 *1* min)

Mode: β-

Δ: -49727.2 *14* keV

SpA: 6.394×10^{8} Ci/g

Prod: ^{50}Ti(n,γ); ^{48}Ca(α,n)

Photons (^{51}Ti)

⟨γ⟩=369 *5* keV

γ_{mode}	γ(keV)	γ(%)†
γ M1+17%E2	320.084 *6*	93.0
γ M1+E2	608.56 *4*	1.18 *9*
γ E2	928.64 *4*	6.9 *4*

† 0.43% uncert(syst)

Continuous Radiation (^{51}Ti)

⟨β-⟩=868 keV;⟨IB⟩=1.8 keV

E_{bin}(keV)		⟨ ⟩(keV)	(%)
0 - 10	β-	0.0116	0.231
	IB	0.034	
10 - 20	β-	0.0359	0.239
	IB	0.033	0.23
20 - 40	β-	0.154	0.509
	IB	0.065	0.22
40 - 100	β-	1.29	1.81
	IB	0.18	0.28
100 - 300	β-	18.4	8.8
	IB	0.48	0.27
300 - 600	β-	89	19.5
	IB	0.47	0.113
600 - 1300	β-	464	49.7
	IB	0.44	0.054
1300 - 2153	β-	295	19.2
	IB	0.044	0.0031

$^{51}_{23}$V (stable)

Δ: -52199.7 *15* keV

%: 99.750 *2*

$^{51}_{24}$Cr(27.704 *4* d)

Mode: ϵ

Δ: -51448.3 *16* keV

SpA: 9.2416×10^{4} Ci/g

Prod: ^{50}Cr(n,γ)

Photons (^{51}Cr)

⟨γ⟩=32.5 *3* keV

γ_{mode}	γ(keV)	γ(%)†
V L_{ℓ}	0.446	0.033 *9*
V L_{η}	0.454	0.025 *6*
V L_{α}	0.511	0.13 *3*
V L_{β}	0.531	0.11 *3*
V $K_{\alpha2}$	4.945	6.5 *3*
V $K_{\alpha1}$	4.952	12.9 *6*
V $K_{\beta1}$'	5.427	2.21 *11*
γ M1+17%E2	320.084 *6*	9.83

† 1.4% uncert(syst)

Atomic Electrons (^{51}Cr)

⟨e⟩=3.08 *25* keV

e_{bin}(keV)	⟨e⟩(keV)	e(%)
4	2.38	54 *6*
5	0.64	13.1 *13*
315	0.0507	0.0161 *6*
319	0.00463	0.00145 *6*
320	0.000144	4.5 *3* $\times10^{-5}$

Continuous Radiation (^{51}Cr)

⟨IB⟩=0.096 keV

E_{bin}(keV)		⟨ ⟩(keV)	(%)
10 - 20	IB	4.3×10^{-5}	0.00030
20 - 40	IB	0.000129	0.00043
40 - 100	IB	0.00164	0.0022
100 - 300	IB	0.028	0.0138
300 - 600	IB	0.059	0.0141
600 - 751	IB	0.0053	0.00083

$^{51}_{25}$Mn(46.2 *1* min)

Mode: ϵ

Δ: -48239.7 *15* keV

SpA: 7.979×10^{7} Ci/g

Prod: ^{50}Cr(d,n); ^{50}Cr(p,γ)

Photons (^{51}Mn)

⟨γ⟩=5.49 *11* keV

γ_{mode}	γ(keV)	γ(%)†
Cr $K_{\alpha2}$	5.405	0.215 *11*
Cr $K_{\alpha1}$	5.415	0.425 *21*
Cr $K_{\beta1}$'	5.947	0.072 *4*
γ M1+6.1%E2	604.1 *4*	0.015 *5*
γ E2	749.10 *9*	0.265
γ E2	808.29 *17*	0.0148 *24*
γ [M1+E2]	1148.05 *21*	0.078 *3*
γ M1+3.0%E2	1164.46 *23*	0.076 *3*
γ [M1+E2]	1353.2 *4*	0.0080 *13*
γ [M1+E2]	1557.37 *17*	0.0040 *5*
γ [E2]	1899.41 *25*	0.0064 *8*
γ [M1+E2]	2001.35 *12*	0.0371 *19*
γ [M1+E2]	2079.62 *16*	0.0069 *5*
γ [M1+E2]	2312.49 *20*	0.0130 *11*

† 2.6% uncert(syst)

Atomic Electrons (^{51}Mn)

⟨e⟩=0.093 *9* keV

e_{bin}(keV)	⟨e⟩(keV)	e(%)
5	0.091	1.86 *19*
6	0.00121	0.0205 *21*
598	3.0×10^{-5}	5.0 *16* $\times10^{-6}$
603	2.8×10^{-6}	4.6 *14* $\times10^{-7}$
604	4.1×10^{-8}	6.8 *21* $\times10^{-9}$
743	0.00059	7.9 *4* $\times10^{-5}$
748	5.41×10^{-5}	7.2 *3* $\times10^{-6}$
749	1.22×10^{-6}	1.63 *7* $\times10^{-7}$
802	2.9×10^{-5}	3.6 *6* $\times10^{-6}$
808	2.7×10^{-6}	3.4 *5* $\times10^{-7}$
1142	8.3×10^{-5}	7.3 *9* $\times10^{-6}$
1147	7.7×10^{-6}	6.7 *8* $\times10^{-7}$
1158	7.2×10^{-5}	6.2 *3* $\times10^{-6}$
1164	6.7×10^{-6}	5.75 *25* $\times10^{-7}$
1347	7.1×10^{-6}	5.3 *10* $\times10^{-7}$
1352	6.5×10^{-7}	4.8 *9* $\times10^{-8}$
1353	6.8×10^{-9}	5.0 *16* $\times10^{-10}$
1551	3.1×10^{-6}	2.0 *3* $\times10^{-7}$
1557	2.9×10^{-7}	1.8 *3* $\times10^{-8}$
1893	4.4×10^{-6}	2.3 *3* $\times10^{-7}$
1899	4.0×10^{-7}	2.1 *3* $\times10^{-8}$
1995	2.36×10^{-5}	1.18 *10* $\times10^{-6}$
2001	2.18×10^{-6}	1.09 *9* $\times10^{-7}$
2074	4.3×10^{-6}	2.05 *21* $\times10^{-7}$
2079	3.9×10^{-7}	1.89 *19* $\times10^{-8}$
2306	7.4×10^{-6}	3.2 *3* $\times10^{-7}$
2312	6.9×10^{-7}	3.0 *3* $\times10^{-8}$

Continuous Radiation (^{51}Mn)

⟨β+⟩=935 keV;⟨IB⟩=2.2 keV

E_{bin}(keV)		⟨ ⟩(keV)	(%)
0 - 10	β+	0.000107	0.00137
	IB	0.036	
10 - 20	β+	0.00166	0.0104
	IB	0.035	0.24
20 - 40	β+	0.0202	0.064
	IB	0.069	0.24
40 - 100	β+	0.429	0.57
	IB	0.19	0.30
100 - 300	β+	12.0	5.6
	IB	0.53	0.30
300 - 600	β+	76	16.5
	IB	0.54	0.128
600 - 1300	β+	479	50.7
	IB	0.59	0.070
1300 - 2500	β+	367	23.6
	IB	0.20	0.0122
2500 - 3209	IB	0.018	0.00068
	Σβ+		97

$^{51}_{26}$Fe(245 *7* ms)

Mode: ϵ

Δ: -40218 *15* keV

SpA: 3.00×10^{11} Ci/g

Prod: ^{50}Cr(^{3}He,2n)

Continuous Radiation (^{51}Fe)

⟨β+⟩=3291 keV;⟨IB⟩=14.8 keV

E_{bin}(keV)		⟨ ⟩(keV)	(%)
0 - 10	β+	4.90×10^{-6}	6.2×10^{-5}
	IB	0.076	
10 - 20	β+	7.9×10^{-5}	0.000496
	IB	0.076	0.53
20 - 40	β+	0.00101	0.00319
	IB	0.151	0.52
40 - 100	β+	0.0227	0.0302
	IB	0.45	0.68
100 - 300	β+	0.72	0.332
	IB	1.40	0.77
300 - 600	β+	5.6	1.21
	IB	1.9	0.44
600 - 1300	β+	65	6.5
	IB	3.5	0.40
1300 - 2500	β+	440	22.5
	IB	4.0	0.23
2500 - 5000	β+	2092	57
	IB	3.1	0.094
5000 - 7500	β+	688	12.4
	IB	0.18	0.0032
7500 - 8022	IB	0.00035	4.6×10^{-6}
	Σβ+		99.9

A = 52

NDS **25**, 235 (1978)

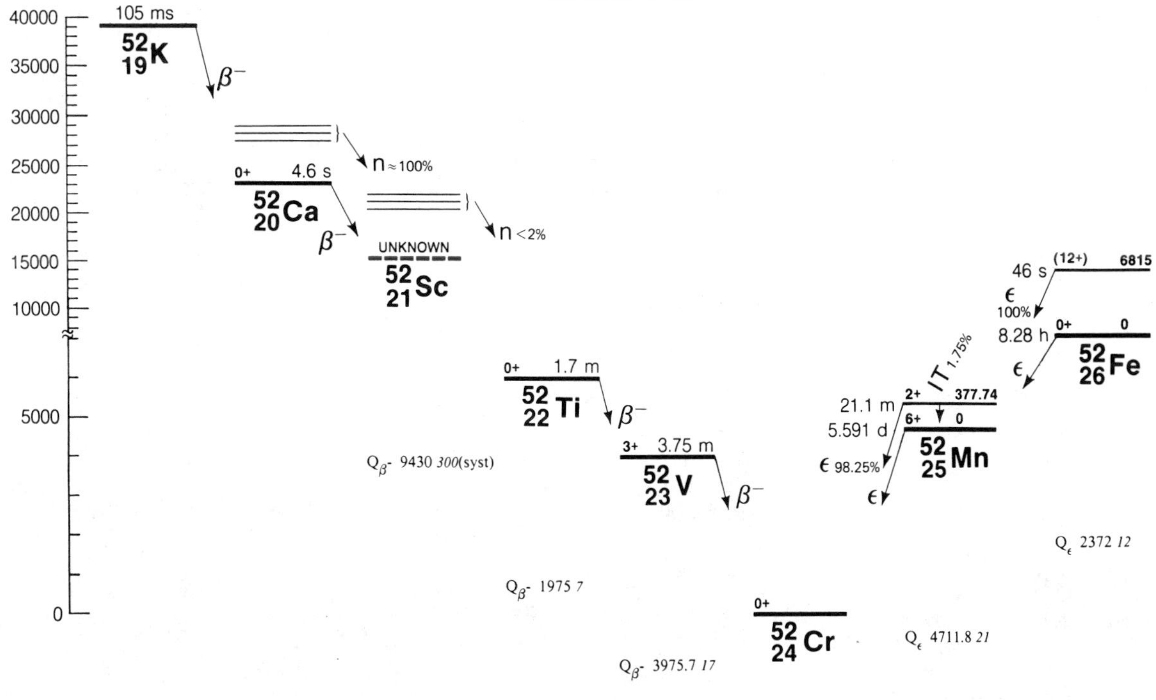

$^{52}_{19}$K (105 *5* ms)

Mode: β-, β-n(~100 %)
SpA: 3.13×10^{11} Ci/g
Prod: protons on Ir

$^{52}_{20}$Ca(4.6 s)

Mode: β-, β-n(<2 %)
SpA: 4×10^{10} Ci/g
Prod: daughter ^{53}K; daughter ^{52}K

$^{52}_{22}$Ti(1.7 *1* min)

Mode: β-
Δ: -49464 *7* keV
SpA: 2.12×10^{9} Ci/g
Prod: ^{50}Ti(t,p)

Photons (^{52}Ti)

⟨γ⟩=127 *6* keV

γ_{mode}	γ(keV)	γ(%)
V L_ℓ	0.446	0.030 *13*
V L_η	0.454	0.022 *9*
V L_α	0.511	0.12 *5*
V L_β	0.530	0.10 *4*
V $K_{\alpha2}$	4.945	5.6 *19*
V $K_{\alpha1}$	4.952	11 *4*
V $K_{\beta1}'$	5.427	1.9 *6*
γ M1+0.5%E2	17.0 *3*	18 *6*
γ	124.453 *3*	99

Atomic Electrons (^{52}Ti)

⟨e⟩=23 *9* keV

e_{bin}(keV)	⟨e⟩(keV)	e(%)
4	2.1	48 *16*
5	0.57	12 *4*
12	8.1	70 *25*
16	1.6	10 *4*
119	9.4	~8
124	1.0	~0.8

Continuous Radiation (^{52}Ti)

⟨β-⟩=742 keV;⟨IB⟩=1.34 keV

E_{bin}(keV)		⟨ ⟩(keV)	(%)
0 - 10	β-	0.0152	0.301
	IB	0.030	
10 - 20	β-	0.0469	0.312
	IB	0.030	0.21
20 - 40	β-	0.20	0.66
	IB	0.057	0.20
40 - 100	β-	1.67	2.35
	IB	0.159	0.24
100 - 300	β-	23.4	11.3
	IB	0.41	0.23
300 - 600	β-	109	23.9
	IB	0.37	0.089
600 - 1300	β-	476	52
	IB	0.27	0.034
1300 - 1833	β-	132	9.1
	IB	0.0100	0.00072

$^{52}_{23}$V (3.75 *1* min)

Mode: β-
Δ: -51439.6 *18* keV
SpA: 9.63×10^{8} Ci/g
Prod: ^{51}V(n,γ); alphas on Ti

Photons (^{52}V)

⟨γ⟩=1445 *14* keV

γ_{mode}	γ(keV)	γ(%)†
γ M1+E2	398.135 *16*	0.0100 *8*
γ (M1+4.6%E2)	647.487 *16*	0.0242 *20*
γ [M1+E2]	704.55 *16*	~0.0018
γ E2	935.527 *13*	0.061 *3*
γ [M1+E2]	1045.617 *19*	0.0043 *12*
γ E2	1333.654 *15*	0.587 *10*
γ E2	1434.082 *9*	100
γ E2+2.5%M1	1530.719 *14*	0.116 *2*
γ M1+3.5%E2	1727.61 *7*	0.0073 *10*
γ [E2]	1981.123 *18*	0.0026 *5*
γ [M1+E2]	2038.19 *16*	~0.0005
γ [M1+E2]	2337.61 *16*	0.0019 *3*
γ E2	2964.755 *16*	0.0005 *2*
γ E2	3161.64 *7*	0.00078 *12*
γ [E2]	3771.62 *16*	0.00058 *13*

† 1.0% uncert(syst)

Continuous Radiation (^{52}V)

⟨β-⟩=1068 keV;⟨IB⟩=2.5 keV

E_{bin}(keV)		⟨ ⟩(keV)	(%)
0 - 10	β-	0.0079	0.156
	IB	0.039	
10 - 20	β-	0.0243	0.162
	IB	0.039	0.27
20 - 40	β-	0.104	0.344
	IB	0.076	0.26
40 - 100	β-	0.87	1.22
	IB	0.21	0.33
100 - 300	β-	12.6	6.1
	IB	0.60	0.34
300 - 600	β-	65	14.1
	IB	0.63	0.150
600 - 1300	β-	415	43.4
	IB	0.73	0.086
1300 - 2500	β-	575	34.5
	IB	0.160	0.0106
2500 - 2542	β-	0.065	0.00259
	IB	6.7×10^{-8}	2.7×10^{-9}

$^{52}_{24}$Cr(stable)

Δ: -55415.2 *16* keV
%: 83.789 *12*

$^{52}_{25}$Mn(5.591 *3* d)

Mode: ε
Δ: -50703.4 *24* keV
SpA: 4.4912×10^{5} Ci/g
Prod: ^{52}Cr(p,n); ^{52}Cr(d,2n); ^{54}Fe(d,α); ^{52}Cr(^{3}He,p2n); ^{50}Cr(α,pn); ^{52}Cr(α,p3n)

Photons (^{52}Mn)

⟨γ⟩=3156 *28* keV

γ_{mode}	γ(keV)	γ(%)
Cr L_ℓ	0.500	0.023 *6*
Cr L_η	0.510	0.017 *4*
Cr L_α	0.572	0.14 *4*
Cr L_β	0.592	0.12 *3*
Cr $K_{\alpha2}$	5.405	5.16 *25*
Cr $K_{\alpha1}$	5.415	10.2 *5*
Cr $K_{\beta1}'$	5.947	1.73 *9*
γ [M1+E2]	200.655 *21*	0.076 *2*
γ E2	346.097 *19*	0.98 *2*
γ M1+E2	398.135 *16*	0.086 *6*
γ [M1+E2]	399.56 *3*	0.183 *8*
γ [M1+E2]	502.046 *18*	0.21 *2*
γ [M1+E2]	600.21 *3*	0.39 *2*
γ (M1+4.6%E2)	647.487 *16*	0.399 *19*
γ E2	744.229 *13*	90.0 *19*
γ [M1+E2]	848.139 *19*	3.32 *8*
γ [M1+E2]	901.60 *3*	0.044 *4*
γ E2	935.527 *13*	94.5 *20*
γ [M1+E2]	1045.617 *19*	0.071 *19*
γ [M1+E2]	1246.267 *14*	4.21 *10*
γ [M1+E2]	1247.69 *3*	0.38 *4*
γ E2	1333.654 *15*	5.09 *11*
γ E2	1434.082 *9*	100
γ	1441 *1*	~0.003
γ [M1+E2]	1645.82 *3*	0.047 *3*
γ	1839.11 *17*	0.005 *1*
γ [E2]	1981.123 *18*	0.042 *8*
γ [M1+E2]	2257.38 *25*	0.0027 *6*

Atomic Electrons (^{52}Mn)

⟨e⟩=2.67 *22* keV

e_{bin}(keV)	⟨e⟩(keV)	e(%)
5	2.18	45 *5*
195 - 200	0.0024	~0.0012
340 - 346	0.0126	0.00371 *10*
392 - 399	0.0017	0.00043 *14*
496 - 501	0.0008	0.00017 *6*
594 - 641	0.0018	0.00030 *5*
647	6.6×10^{-5}	1.01 *7* $\times10^{-5}$
738	0.202	0.0274 *8*
842 - 848	0.0056	0.00067 *11*
896 - 901	6.9×10^{-5}	7.7 *13* $\times10^{-6}$
930	0.147	0.0158 *5*
1040 - 1045	9.2×10^{-5}	8.8 *25* $\times10^{-6}$
1240 - 1247	0.0049	0.00039 *4*
1328 - 1333	0.00543	0.000409 *11*
1428	0.0902	0.00631 *14*
1433 - 1440	0.00830	0.000579 *13*
1640 - 1645	3.8×10^{-5}	2.34 *21* $\times10^{-6}$
1833 - 1839	3.0×10^{-6}	1.6 *5* $\times10^{-7}$
1975 - 1981	3.1×10^{-5}	1.5 *3* $\times10^{-6}$
2251 - 2257	1.7×10^{-6}	7.6 *16* $\times10^{-8}$

Continuous Radiation (^{52}Mn)

⟨β+⟩=71 keV;⟨IB⟩=0.71 keV

E_{bin}(keV)		⟨ ⟩(keV)	(%)
0 - 10	β+	0.00065	0.0083
	IB	0.0044	
10 - 20	β+	0.0099	0.062
	IB	0.0034	0.024
20 - 40	β+	0.115	0.365
	IB	0.0062	0.022
40 - 100	β+	2.14	2.89
	IB	0.0154	0.024
100 - 300	β+	34.1	16.9
	IB	0.050	0.026
300 - 600	β+	34.8	9.2

Continuous Radiation (^{52}Mn)
(continued)

E_{bin}(keV)		⟨ ⟩(keV)	(%)
	IB	0.157	0.035
600 - 1300	IB	0.45	0.051
1300 - 1598	IB	0.030	0.0022
	Σβ+		29

$^{52}_{25}$Mn(21.1 *2* min)

Mode: ϵ(98.25 *5* %), IT(1.75 *5* %)
Δ: -50325.7 *24* keV
SpA: 1.713×10^{8} Ci/g
Prod: daughter ^{52}Fe

Photons (^{52}Mn)
⟨γ⟩=1422 *28* keV

γ_{mode}	γ(keV)	γ(%)†
Cr $K_{\alpha 2}$	5.405	0.124 *6*
Cr $K_{\alpha 1}$	5.415	0.245 *12*
Cr $K_{\beta 1}$'	5.947	0.0415 *20*
γ_{IT} E4	377.738 *11*	1.68
γ_ϵM1+E2	398.135 *16*	0.00050 *8*
γ_ϵ[M1+E2]	704.55 *16*	0.028 *9*
γ_ϵE2	935.527 *13*	0.024 *9*
γ_ϵE2	1333.654 *15*	0.029 *4*
γ_ϵE2	1434.082 *9*	98.2 *20*
γ_ϵE2+2.5%M1	1530.719 *14*	0.046 *3*
γ_ϵM1+3.5%E2	1727.61 *7*	0.215 *10*
γ_ϵ[M1+E2]	2038.19 *16*	0.0079 *10*
γ_ϵ[M1+E2]	2337.61 *16*	0.0068 *10*
γ_ϵE2	2964.755 *16*	0.00020 *8*
γ_ϵE2	3161.64 *7*	0.023 *3*
γ_ϵ	3381.53 *12*	0.0025 *5*
γ_ϵ[E2]	3771.62 *16*	0.0021 *4*
γ_ϵ[M1]	3951.0 *14*	0.0007 *3*
γ_ϵ	4815.51 *12*	0.0025 *4*

†uncert(syst): <0.1% for ϵ, 2.9% for IT

Atomic Electrons (^{52}Mn)
⟨e⟩=0.401 *9* keV

e_{bin}(keV)	⟨e⟩(keV)	e(%)
5	0.054	1.10 *11*
6	0.00119	0.0203 *21*
371	0.224	0.0604 *18*
377	0.0247	0.00654 *19*
392 - 398	3.1 ×10^{-6}	~8×10^{-7}
699 - 704	6.3 ×10^{-5}	9 *3* ×10^{-6}
930 - 935	4.0 ×10^{-5}	4.3 *15* ×10^{-6}
1328 - 1333	3.1 ×10^{-5}	2.4 *3* ×10^{-6}
1428	0.0886	0.00620 *18*
1433	0.00809	0.000565 *16*
1434 - 1530	9.6 ×10^{-5}	6.55 *20* ×10^{-6}
1722 - 1727	0.000161	9.3 *4* ×10^{-6}
2032 - 2038	5.4 ×10^{-6}	2.6 *3* ×10^{-7}
2332 - 2337	4.2 ×10^{-6}	1.8 *3* ×10^{-7}
2959 - 2964	1.1 ×10^{-7}	3.7 *17* ×10^{-9}
3123 - 3161	1.20 ×10^{-5}	3.8 *4* ×10^{-7}
3376 - 3381	1.0 ×10^{-6}	3.1 *8* ×10^{-8}
3766 - 3771	9.8 ×10^{-7}	2.6 *5* ×10^{-8}
3945 - 3950	3.1 ×10^{-7}	8 *3* ×10^{-9}
4810 - 4815	9.2 ×10^{-7}	1.9 *4* ×10^{-8}

Continuous Radiation (^{52}Mn)
⟨β+⟩=1133 keV;⟨IB⟩=2.9 keV

E_{bin}(keV)		⟨ ⟩(keV)	(%)
0 - 10	β+	6.9×10^{-5}	0.00088
	IB	0.041	
10 - 20	β+	0.00107	0.0067
	IB	0.040	0.28
20 - 40	β+	0.0131	0.0414
	IB	0.079	0.27
40 - 100	β+	0.279	0.373
	IB	0.22	0.34
100 - 300	β+	8.0	3.71
	IB	0.63	0.36
300 - 600	β+	53	11.5
	IB	0.69	0.164
600 - 1300	β+	400	41.6
	IB	0.87	0.102
1300 - 2500	β+	670	39.4
	IB	0.32	0.020
2500 - 3655	β+	2.03	0.080
	IB	0.038	0.00135
	Σβ+		97

$^{52}_{26}$Fe(8.275 *8* h)

Mode: ϵ
Δ: -48331 *12* keV
SpA: 7.283×10^{6} Ci/g
Prod: ^{50}Cr(α,2n); ^{55}Mn(p,4n); ^{3}He on Cr

Photons (^{52}Fe)
⟨γ⟩=174 *4* keV

γ_{mode}	γ(keV)	γ(%)
Mn L_ℓ	0.556	0.017 *5*
Mn L_η	0.568	0.012 *3*
Mn L_α	0.637	0.13 *4*
Mn L_β	0.657	0.10 *3*
Mn $K_{\alpha 2}$	5.888	3.9 *3*
Mn $K_{\alpha 1}$	5.899	7.7 *7*
Mn $K_{\beta 1}$'	6.490	1.36 *12*
γ M1+E2	168.684 *11*	99.2
γ E4	377.738 *11*	1.68 *4* *

* with ^{52}Mn(21.1 min) in equilib

Atomic Electrons (^{52}Fe)
⟨e⟩~7.5 keV

e_{bin}(keV)	⟨e⟩(keV)	e(%)
1	0.25	39 *5*
5	1.21	23 *3*
6	0.35	5.9 *7*
162	5.0	~3
168	0.5	~0.3
371	0.224	0.060 *19*
377	0.0247	0.0065 *20*

Continuous Radiation (^{52}Fe)
⟨β+⟩=189 keV;⟨IB⟩=0.86 keV

E_{bin}(keV)		⟨ ⟩(keV)	(%)
0 - 10	β+	0.00053	0.0068
	IB	0.0098	
10 - 20	β+	0.0085	0.053
	IB	0.0088	0.061
20 - 40	β+	0.104	0.330
	IB	0.0165	0.058
40 - 100	β+	2.10	2.83
	IB	0.042	0.065
100 - 300	β+	43.6	21.0
	IB	0.094	0.054
300 - 600	β+	120	28.2
	IB	0.137	0.031
600 - 1300	β+	23.3	3.58
	IB	0.45	0.050
1300 - 1826	IB	0.101	0.0071
	Σβ+		56

$^{52}_{26}$Fe(45.9 *6* s)

Mode: ϵ
Δ: -41516 *130* keV
SpA: 4.69×10^{9} Ci/g
Prod: ^{40}Ca(^{14}N,np)

Photons (^{52}Fe)
⟨γ⟩=3982 *290* keV

γ_{mode}	γ(keV)	γ(%)†
γ	622 *1*	100
γ (M1+1.2%E2)	870.0 *8*	100
γ (E2)	929 *1*	100
γ (M1+8.8%E2)	1416.0 *8*	95 *2*
γ (E2)	2038 *1*	~5
γ (E2)	2286.0 *8*	5 *2*

† approximate

Continuous Radiation (^{52}Fe)
⟨β+⟩=1980 keV;⟨IB⟩=6.8 keV

E_{bin}(keV)		⟨ ⟩(keV)	(%)
0 - 10	β+	1.75×10^{-5}	0.000223
	IB	0.058	
10 - 20	β+	0.000283	0.00177
	IB	0.058	0.40
20 - 40	β+	0.00358	0.0113
	IB	0.114	0.40
40 - 100	β+	0.080	0.106
	IB	0.33	0.51
100 - 300	β+	2.48	1.14
	IB	1.00	0.56
300 - 600	β+	18.4	3.93
	IB	1.25	0.29
600 - 1300	β+	185	18.7
	IB	2.0	0.23
1300 - 2500	β+	881	46.3
	IB	1.62	0.094
2500 - 5000	β+	893	29.5
	IB	0.36	0.0116
5000 - 5350	IB	0.00034	6.6×10^{-6}
	Σβ+		99.7

A = 53

NDS **43**, 481 (1984)

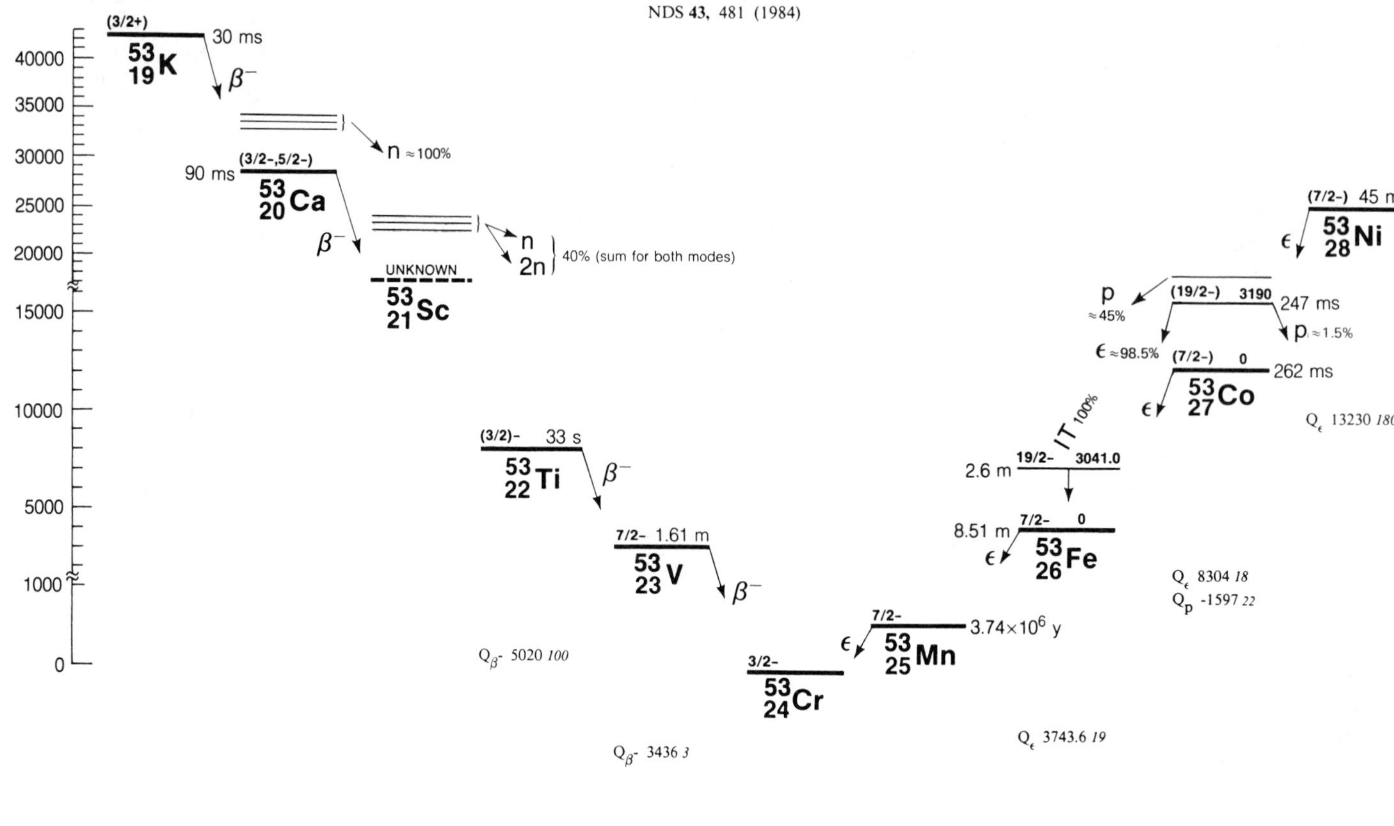

Q$_\epsilon$ 596.1 *9*

$^{53}_{19}$K (30 *5* ms)

Mode: β-, β-n(~100 %)
SpA: 3.1×10^{11} Ci/g
Prod: protons on Ir

$^{53}_{20}$Ca(90 *15* ms)

Mode: β-, (β-n + β-2n)(40 *10* %)
SpA: 3.1×10^{11} Ci/g
Prod: daughter ^{54}K; daughter ^{53}K

$^{53}_{22}$Ti(32.7 *9* s)

Mode: β-
Δ: -46830 *100* keV
SpA: 6.44×10^{9} Ci/g
Prod: ^{48}Ca(^{7}Li,pn)

Photons (^{53}Ti)

⟨γ⟩=1963 *72* keV

γ_{mode}	γ(keV)	γ(%)†
V K$_{\alpha2}$	4.945	0.107 *8*
V K$_{\alpha1}$	4.952	0.212 *16*
V K$_{\beta1}$'	5.427	0.036 *3*
γ [M1]	100.83 *8*	20.3 *12*
γ [M1]	127.62 *8*	46 *5*
γ [E2]	228.45 *8*	40
γ [M1+E2]	680.0 *4*	4.0 *8*
γ	1001.0 *11*	4.5 *13*
γ	1034.4 *8*	2.5 *9*
γ	1321.2 *7*	6.0 *12*
γ	1422.0 *7*	10.7 *16*
γ [M1+E2]	1675.59 *20*	25 *3*
γ [M1+E2]	1729.2 *6*	4.8 *7*
γ [M1+E2]	1776.41 *21*	4.1 *6*
γ	1855.6 *4*	3.2 *5*
γ [M1+E2]	1904.03 *20*	12.3 *8*
γ	1956.4 *4*	3.5 *4*
γ [M1+E2]	2355.6 *4*	3.1 *5*
γ [M1+E2]	2456.4 *4*	5.3 *5*
γ [M1+E2]	2601.0 *4*	6.2 *8*
γ [M1+E2]	2701.9 *4*	2.9 *6*
γ [M1+E2]	2702 *2*	1.0 *3*
γ [M1+E2]	2829.5 *4*	2.0 *4*

† 4.8% uncert(syst)

Atomic Electrons (^{53}Ti)

⟨e⟩=2.58 *15* keV

e_{bin}(keV)	⟨e⟩(keV)	e(%)
4 - 5	0.050	1.11 *11*
95	0.38	0.40 *3*
100	0.038	0.038 *3*
122	0.60	0.49 *6*
127	0.058	0.046 *5*
223	1.26	0.57 *6*
228	0.121	0.053 *6*
675 - 680	0.0083	0.0012 *4*
996 - 1034	0.0068	0.00067 *23*
1316 - 1321	0.0042	0.00032 *12*
1417 - 1422	0.0071	0.00050 *16*
1670 - 1729	0.0210	0.00125 *13*
1771 - 1855	0.0044	0.00025 *4*
1899 - 1956	0.0096	0.00050 *4*
2350 - 2355	0.0017	7.1 *12* ×10^{-5}
2451 - 2456	0.0028	0.000115 *12*
2596 - 2601	0.0032	0.000121 *16*
2696 - 2702	0.0019	7.1 *11* ×10^{-5}
2824 - 2829	0.00099	3.5 *7* ×10^{-5}

Continuous Radiation (^{53}Ti)

⟨β-⟩=1332 keV;⟨IB⟩=3.7 keV

E_{bin}(keV)		⟨ ⟩(keV)	(%)
0 - 10	β-	0.0056	0.111
	IB	0.045	
10 - 20	β-	0.0173	0.115
	IB	0.044	0.31
20 - 40	β-	0.074	0.246
	IB	0.087	0.30
40 - 100	β-	0.63	0.88
	IB	0.25	0.38
100 - 300	β-	9.4	4.49
	IB	0.72	0.40
300 - 600	β-	49.5	10.8
	IB	0.82	0.19
600 - 1300	β-	343	35.7
	IB	1.11	0.129
1300 - 2500	β-	728	41
	IB	0.54	0.032
2500 - 4792	β-	202	6.7
	IB	0.063	0.0022

$^{53}_{23}$V (1.61 *4* min)

Mode: β-

Δ: -51847 *3* keV

SpA: 2.20×10^9 Ci/g

Prod: ^{53}Cr(n,p); ^{54}Cr(γ,p)

Photons (^{53}V)

⟨γ⟩=1038 *180* keV

γ_{mode}	γ(keV)	γ(%)†
γ M1+23%E2	247.6 *7*	0.18 *4*
γ M1+0.5%E2	282.89 *23*	0.76 *15*
γ [E2]	442.67 *24*	0.39 *8*
γ M1+0.6%E2	530.5 *7*	0.18 *4*
γ [M1]	563.57 *24*	0.39 *8*
γ M1+11%E2	1006.24 *21*	90
γ E2	1289.12 *24*	10 *2*

† 2.2% uncert(syst)

Continuous Radiation (^{53}V)

⟨β-⟩=1005 keV;⟨IB⟩=2.2 keV

E_{bin}(keV)		⟨ ⟩(keV)	(%)
0 - 10	β-	0.0087	0.173
	IB	0.038	
10 - 20	β-	0.0269	0.179
	IB	0.037	0.26
20 - 40	β-	0.115	0.381
	IB	0.072	0.25
40 - 100	β-	0.96	1.35
	IB	0.20	0.31
100 - 300	β-	14.0	6.7
	IB	0.56	0.32
300 - 600	β-	71	15.5
	IB	0.58	0.139
600 - 1300	β-	437	46.0
	IB	0.63	0.076
1300 - 2430	β-	482	29.7
	IB	0.109	0.0074

$^{53}_{24}$Cr(stable)

Δ: -55283.4 *16* keV

%: 9.501 *11*

$^{53}_{25}$Mn(3.74 *4* ×10^6 yr)

Mode: ϵ

Δ: -54687.2 *16* keV

SpA: 0.001803 Ci/g

Prod: ^{53}Cr(p,n); ^{52}Cr(d,n); ^{50}Cr(α,p); protons on Fe

Photons (^{53}Mn)

⟨γ⟩=1.34 *4* keV

γ_{mode}	γ(keV)	γ(%)
Cr L_ℓ	0.500	0.033 *9*
Cr L_η	0.510	0.024 *6*
Cr L_α	0.572	0.20 *5*
Cr L_β	0.592	0.16 *4*
Cr $K_{\alpha2}$	5.405	7.4 *4*
Cr $K_{\alpha1}$	5.415	14.6 *7*
Cr $K_{\beta1}$'	5.947	2.47 *12*

Atomic Electrons (^{53}Mn)

⟨e⟩=3.2 *3* keV

e_{bin}(keV)	⟨e⟩(keV)	e(%)
5	3.1	64 *7*
6	0.041	0.70 *7*

Continuous Radiation (^{53}Mn)

⟨IB⟩=0.051 keV

E_{bin}(keV)		⟨ ⟩(keV)	(%)
10 - 20	IB	5.2×10^{-5}	0.00037
20 - 40	IB	0.000133	0.00044
40 - 100	IB	0.00156	0.0021
100 - 300	IB	0.023	0.0115
300 - 596	IB	0.025	0.0065

$^{53}_{26}$Fe(8.51 *2* min)

Mode: ϵ

Δ: -50943.6 *22* keV

SpA: 4.1660×10^8 Ci/g

Prod: ^{50}Cr(α,n); ^{52}Cr(α,3n); ^{54}Fe(d,t); ^{54}Fe(n,2n); ^{54}Fe(p,pn); ^{55}Mn(p,3n)

Photons (^{53}Fe)

⟨γ⟩=185 *16* keV

γ_{mode}	γ(keV)	γ(%)†
Mn $K_{\alpha2}$	5.888	0.237 *12*
Mn $K_{\alpha1}$	5.899	0.468 *23*
Mn $K_{\beta1}$'	6.490	0.083 *4*
γ M1+23%E2	377.92 *10*	42
γ E2	1288.02 *10*	<0.08
γ [E2]	1397.6 *3*	~0.008
γ E2+12%M1	1619.92 *10*	0.50 *8*
γ M1+3.0%E2	2273.5 *3*	0.38 *4*
γ [M1+E2]	2307.7 *3*	0.013 *4*
γ [M1+E2]	2685.6 *3*	0.080 *21*
γ [M1+E2]	2748.8 *4*	0.14 *3*
γ E2+39%M1	2946.6 *4*	0.050 *17*
γ [M1+E2]	3248.8 *8*	~0.038

† 19% uncert(syst)

Atomic Electrons (^{53}Fe)

⟨e⟩=0.354 *24* keV

e_{bin}(keV)	⟨e⟩(keV)	e(%)
1	0.0152	2.34 *24*
5	0.073	1.41 *14*
6	0.0210	0.36 *4*
371	0.222	0.060 *6*
377	0.0216	0.0057 *6*
1281	4.8 ×10^{-5}	<7×10^{-6}
1287	4.5 ×10^{-6}	<7×10^{-7}
1391	8.7 ×10^{-6}	~6×10^{-7}
1397	8.2 ×10^{-7}	~6×10^{-8}
1613	0.00045	2.8 *5* ×10^{-5}
1619	4.2 ×10^{-5}	2.6 *4* ×10^{-6}
2267	0.00024	1.06 *12* ×10^{-5}
2273	2.25 ×10^{-5}	9.9 *11* ×10^{-7}
2301	8.1 ×10^{-6}	3.5 *12* ×10^{-7}
2307	7.6 ×10^{-7}	3.3 *11* ×10^{-8}
2679	4.7 ×10^{-5}	1.7 *5* ×10^{-6}
2685	4.4 ×10^{-6}	1.6 *4* ×10^{-7}
2742	8.0 ×10^{-5}	2.9 *7* ×10^{-6}
2748	7.5 ×10^{-6}	2.7 *7* ×10^{-7}
2940	2.8 ×10^{-5}	1.0 *3* ×10^{-6}
2946	2.6 ×10^{-6}	9 *3* ×10^{-8}
3242	2.0 ×10^{-5}	~6×10^{-7}
3248	1.9 ×10^{-6}	~6×10^{-8}

Continuous Radiation (^{53}Fe)

⟨β+⟩=1105 keV;⟨IB⟩=2.9 keV

E_{bin}(keV)		⟨ ⟩(keV)	(%)
0 - 10	β+	6.9×10^{-5}	0.00088
	IB	0.040	
10 - 20	β+	0.00111	0.0070
	IB	0.040	0.27
20 - 40	β+	0.0140	0.0443
	IB	0.077	0.27
40 - 100	β+	0.307	0.408
	IB	0.22	0.34
100 - 300	β+	8.8	4.08
	IB	0.62	0.35
300 - 600	β+	58	12.5
	IB	0.67	0.159
600 - 1300	β+	415	43.3
	IB	0.84	0.098
1300 - 2500	β+	618	36.8
	IB	0.33	0.020
2500 - 3744	β+	4.73	0.185
	IB	0.040	0.00143
	Σβ+		97

$^{53}_{26}$Fe(2.58 *6* min)

Mode: IT

Δ: -47902.9 *22* keV

SpA: 1.37×10^{9} Ci/g

Prod: ^{55}Mn(p,3n); ^{54}Fe(p,pn); ^{54}Fe(d,t); ^{54}Fe(n,2n); ^{52}Cr(α,3n)

Photons (^{53}Fe)

⟨γ⟩=3012 *150* keV

γ_{mode}	γ(keV)	γ(%)†
γ [E4]	701.17 *9*	98.60
γ M1+1.2%E2	1011.59 *8*	85 *9*
γ M1+1.2%E2	1328.20 *8*	86 *8*
γ [M5]	1712.74 *12*	1.28 *10*
γ [E2]	2339.76 *8*	12.8 *20*
γ [E6]	3040.89 *12*	0.059 *10*

† 0.30% uncert(syst)

$^{53}_{27}$Co(262 *25* ms)

Mode: ε

Δ: -42640 *18* keV

SpA: 2.9×10^{11} Ci/g

Prod: ^{39}K(^{16}O,2n)

Continuous Radiation (^{53}Co)

⟨β+⟩=3430 keV;⟨IB⟩=15.8 keV

E_{bin}(keV)		⟨ ⟩(keV)	(%)
0 - 10	β+	3.88×10^{-6}	4.92×10^{-5}
	IB	0.078	
10 - 20	β+	6.5×10^{-5}	0.000408
	IB	0.078	0.54
20 - 40	β+	0.00086	0.00270
	IB	0.154	0.54
40 - 100	β+	0.0199	0.0264
	IB	0.45	0.70
100 - 300	β+	0.65	0.296
	IB	1.43	0.79
300 - 600	β+	5.09	1.09
	IB	1.9	0.45
600 - 1300	β+	59	5.9
	IB	3.7	0.41
1300 - 2500	β+	408	20.8
	IB	4.2	0.24
2500 - 5000	β+	2075	56
	IB	3.5	0.105
5000 - 7500	β+	882	15.6
	IB	0.27	0.0049
7500 - 8304	IB	0.00122	1.58×10^{-5}
	Σβ+		99.9

$^{53}_{27}$Co(247 *12* ms)

Mode: ε(~98.5 %), p(~1.5 %)

Δ: -39450 *18* keV

SpA: 2.89×10^{11} Ci/g

Prod: ^{40}Ca(^{16}O,p2n); ^{54}Fe(p,2n)

Protons (^{53}Co)

p(keV)	p(%)
1550 *30*	~1.5

Continuous Radiation (^{53}Co)

⟨β+⟩=3450 keV;⟨IB⟩=16.0 keV

E_{bin}(keV)		⟨ ⟩(keV)	(%)
0 - 10	β+	3.62×10^{-6}	4.59×10^{-5}
	IB	0.078	
10 - 20	β+	6.1×10^{-5}	0.000381
	IB	0.077	0.54
20 - 40	β+	0.00080	0.00252
	IB	0.154	0.53
40 - 100	β+	0.0186	0.0246
	IB	0.45	0.69
100 - 300	β+	0.60	0.277
	IB	1.43	0.79
300 - 600	β+	4.76	1.02
	IB	1.9	0.45
600 - 1300	β+	55	5.6
	IB	3.7	0.41
1300 - 2500	β+	386	19.7
	IB	4.3	0.24
2500 - 5000	β+	2028	55
	IB	3.6	0.110
5000 - 7500	β+	975	17.2
	IB	0.32	0.0058
7500 - 8453	IB	0.0020	2.5×10^{-5}
	Σβ+		98.4

$^{53}_{28}$Ni(45 *15* ms)

Mode: ε, εp(~45 %)

Δ: -29410 *180* keV

SpA: 3.1×10^{11} Ci/g

Prod: ^{40}Ca(^{16}O,3n)

Delayed Protons (^{53}Ni)

p(keV)	p(%)
1903 *50*	~45

A = 54

NDS **23**, 455 (1978)

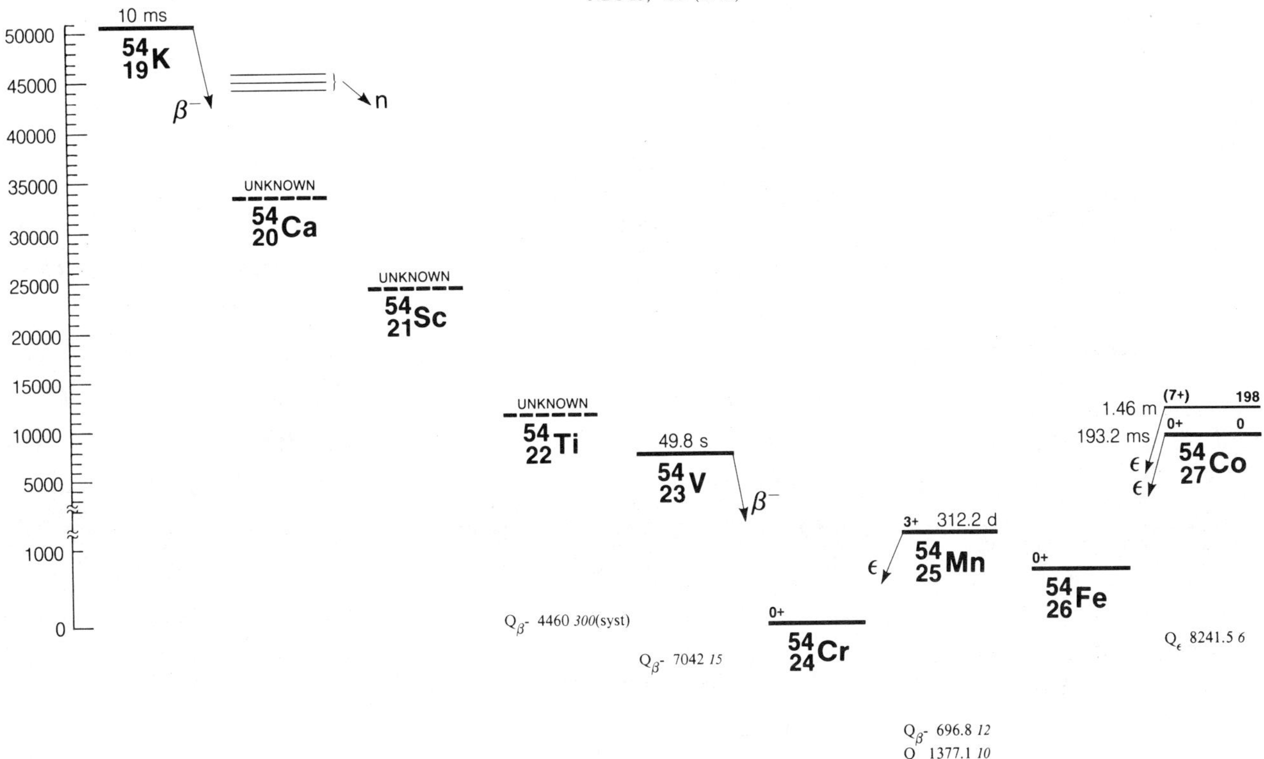

$^{54}_{19}$K (10 *5* ms)

Mode: β-, β-n
SpA: 3.0×10^{11} Ci/g
Prod: protons on Ir

$^{54}_{23}$V (49.8 *5* s)

Mode: β-
Δ: -49889 *15* keV
SpA: 4.17×10^{9} Ci/g
Prod: ^{54}Cr(n,p)

Photons (^{54}V)

⟨γ⟩=4107 *85* keV

γ_{mode}	γ(keV)	γ(%)†
γ[E2]	563.19 *23*	4.23 *21*
γ	625.9 *3*	0.7 *3*
γ	638.7 *3*	3.6 *4*
γ	646.0 *4*	2.2 *4*
γE2	834.826 *21*	97.1 *7*
γ	923.39 *25*	8.1 *8*
γE2	988.69 *20*	80.1 *7*
γ	1008.92 *25*	1.4 *6*
γ	1335.60 *24*	2.5 *5*
γ[E2]	1398.27 *21*	4.6 *7*
γ	1463.33 *15*	8.6 *7*
γM1+5.4%E2	1784.33 *18*	8.2 *7*
γ[M1+E2]	1831.0 *4*	4.8 *10*
γ[M1+E2]	1961.45 *17*	10.0 *10*

Photons (^{54}V) (continued)

γ_{mode}	γ(keV)	γ(%)†
γ[M1+E2]	1974.23 *20*	4.5 *10*
γM1+1.0%E2	2238.7 *3*	1.3 *4*
γ	2258.97 *17*	45.6 *15*
γ	2324.3 *3*	2.2 *6*
γ	2394.2 *3*	3.0 *15*
γ[M1+E2]	2601.7 *4*	2.7 *5*
γ[E2]	2619.13 *18*	2.9 *14*
γ	2626.8 *8*	1.6 *6*
γ[E2]	2962.9 *3*	3.5 *9*
γ	3382.8 *3*	4.0 *8*

† 1.5% uncert(syst)

Continuous Radiation (^{54}V)

⟨β-⟩=1310 keV;⟨IB⟩=3.5 keV

E_{bin}(keV)		⟨ ⟩(keV)	(%)
0 - 10	β-	0.00507	0.101
	IB	0.045	
10 - 20	β-	0.0157	0.104
	IB	0.044	0.31
20 - 40	β-	0.067	0.223
	IB	0.087	0.30
40 - 100	β-	0.57	0.80
	IB	0.25	0.38
100 - 300	β-	8.5	4.07
	IB	0.72	0.40
300 - 600	β-	45.8	9.9
	IB	0.81	0.19
600 - 1300	β-	340	35.1
	IB	1.08	0.126
1300 - 2500	β-	832	46.5

Continuous Radiation (^{54}V) (continued)

E_{bin}(keV)		⟨ ⟩(keV)	(%)
	IB	0.44	0.027
2500 - 3387	β-	83	3.11
	IB	0.0054	0.00021

$^{54}_{24}$Cr(stable)

Δ: -56931.0 *16* keV
%: 2.365 *5*

$^{54}_{25}$Mn(312.20 *7* d)

Mode: ε
Δ: -55554.0 *18* keV
SpA: 7745.2 Ci/g
Prod: ^{56}Fe(d,α); ^{51}V(α,n); ^{53}Cr(d,n); ^{54}Fe(n,p)

Photons (^{54}Mn)

⟨γ⟩=835.95 *6* keV

γ_{mode}	γ(keV)	γ(%)
Cr L_ℓ	0.500	0.033 *9*
Cr L_η	0.510	0.024 *6*
Cr L_α	0.572	0.20 *5*
Cr L_β	0.592	0.16 *4*
Cr $K_{\alpha2}$	5.405	7.4 *4*
Cr $K_{\alpha1}$	5.415	14.6 *7*
Cr $K_{\beta1}$'	5.947	2.47 *12*
γE2	834.826 *21*	99.975 *2*

Atomic Electrons (^{54}Mn)

⟨e⟩=3.4 *3* keV

e_{bin}(keV)	⟨e⟩(keV)	e(%)
5	3.1	64 *7*
6	0.041	0.70 *7*
829	0.185	0.0224 *5*
834	0.0173	0.00208 *4*

Continuous Radiation (^{54}Mn)

⟨IB⟩=0.038 keV

E_{bin}(keV)		⟨ ⟩(keV)	(%)
10 - 20	IB	5.2×10^{-5}	0.00037
20 - 40	IB	0.000131	0.00044
40 - 100	IB	0.00151	0.0021
100 - 300	IB	0.021	0.0103
300 - 542	IB	0.0149	0.0041

$^{54}_{26}$Fe(stable)

Δ: -56250.8 *14* keV
%: 5.8 *1*

$^{54}_{27}$Co(193.23 *14* ms)

Mode: ϵ
Δ: -48009.3 *15* keV
SpA: 2.9307×10^{11} Ci/g
Prod: ^{54}Fe(p,n)

Continuous Radiation (^{54}Co)

⟨β+⟩=3399 keV;⟨IB⟩=15.6 keV

E_{bin}(keV)		⟨ ⟩(keV)	(%)
0 - 10	β+	3.97×10^{-6}	5.04×10^{-5}
	IB	0.078	
10 - 20	β+	6.7×10^{-5}	0.000418
	IB	0.077	0.54
20 - 40	β+	0.00088	0.00277
	IB	0.154	0.53
40 - 100	β+	0.0204	0.0270
	IB	0.45	0.69
100 - 300	β+	0.66	0.303
	IB	1.42	0.79
300 - 600	β+	5.2	1.11
	IB	1.9	0.45
600 - 1300	β+	60	6.0
	IB	3.6	0.41
1300 - 2500	β+	415	21.2
	IB	4.2	0.24
2500 - 5000	β+	2080	56
	IB	3.4	0.103
5000 - 7500	β+	838	14.9
	IB	0.25	0.0044
7500 - 8242	IB	0.00096	1.25×10^{-5}
	Σβ+		99.9

$^{54}_{27}$Co(1.46 *3* min)

Mode: ϵ
Δ: -47811 *4* keV
SpA: 2.38×10^{9} Ci/g
Prod: ^{54}Fe(p,n); ^{54}Fe(d,2n)

Photons (^{54}Co)

⟨γ⟩=2948 *20* keV

γ_{mode}	γ(keV)	γ(%)†
γ E2	411 *1*	100
γ E2	1130 *1*	100
γ E2	1407 *1*	100

† 5% uncert(syst)

Continuous Radiation (^{54}Co)

⟨β+⟩=2045 keV;⟨IB⟩=7.3 keV

E_{bin}(keV)		⟨ ⟩(keV)	(%)
0 - 10	β+	1.41×10^{-5}	0.000179
	IB	0.059	
10 - 20	β+	0.000238	0.00149
	IB	0.059	0.41
20 - 40	β+	0.00311	0.0098
	IB	0.116	0.40
40 - 100	β+	0.072	0.095
	IB	0.34	0.52
100 - 300	β+	2.26	1.04
	IB	1.02	0.57
300 - 600	β+	17.0	3.63
	IB	1.29	0.30
600 - 1300	β+	173	17.5
	IB	2.1	0.24
1300 - 2500	β+	855	44.8
	IB	1.8	0.102
2500 - 5000	β+	998	32.4
	IB	0.49	0.0156
5000 - 5492	IB	0.00140	2.7×10^{-5}
	Σβ+		99.5

A = 55

NDS **44**, 463 (1985)

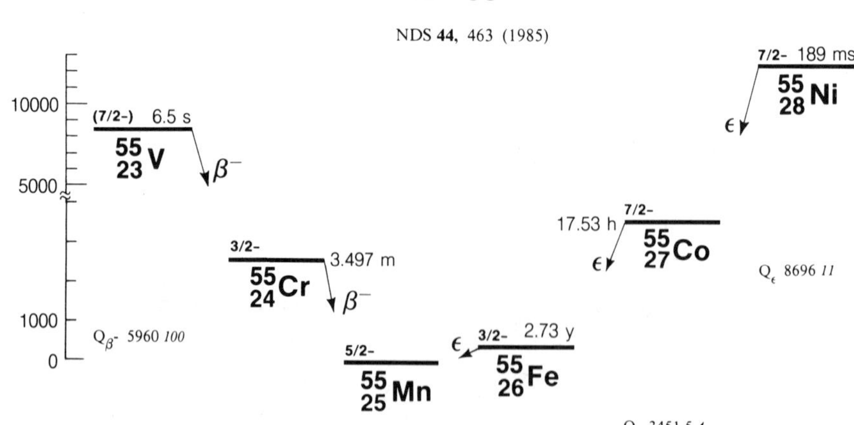

$^{55}_{23}$V (6.54 *15* s)

Mode: β-
Δ: -49150 *100* keV
SpA: 2.98×10^{10} Ci/g
Prod: ^{48}Ca(^{9}Be,np)

Photons (^{55}V)

⟨γ⟩=689 *13* keV

γ_{mode}	γ(keV)	γ(%)†
γ [M1+E2]	224.04 *8*	1.02 *15*
γ [M1+E2]	242.02 *11*	0.80 *15*
γ [M1+E2]	314.78 *9*	1.02 *7*
γ [M1+E2]	334.01 *7*	1.09 *15*
γ [M1+E2]	362.95 *8*	1.45 *7*
γ M1(+E2)	517.65 *9*	72.6
γ [M1+E2]	565.82 *10*	4.50 *15*
γ [M1+E2]	880.60 *8*	18.1 *3*
γ [E2]	921.00 *11*	4.6 *5*
γ [M1+E2]	961.33 *20*	2.18 *7*
γ [E2]	1214.61 *7*	4.0 *3*

† 3.2% uncert(syst)

Continuous Radiation (^{55}V)

⟨β-⟩=2393 keV;⟨IB⟩=9.1 keV

E_{bin}(keV)		⟨ ⟩(keV)	(%)
0 - 10	β-	0.00131	0.0260
	IB	0.064	
10 - 20	β-	0.00407	0.0271
	IB	0.064	0.44
20 - 40	β-	0.0175	0.058
	IB	0.127	0.44
40 - 100	β-	0.150	0.210
	IB	0.37	0.57
100 - 300	β-	2.35	1.12
	IB	1.14	0.63
300 - 600	β-	13.8	2.98
	IB	1.46	0.34
600 - 1300	β-	131	13.2
	IB	2.5	0.29
1300 - 2500	β-	702	36.4
	IB	2.4	0.136
2500 - 5000	β-	1531	45.7
	IB	0.91	0.030
5000 - 5718	β-	12.8	0.2[illegible]0
	IB	0.00027	5.4×10^{-6}

$^{55}_{24}$Cr(3.497 *3* min)

Mode: β-
Δ: -55105.9 *16* keV
SpA: 9.760×10^{8} Ci/g
Prod: ^{54}Cr(n,γ)

Photons (^{55}Cr)

⟨γ⟩=0.67 *5* keV

γ_{mode}	γ(keV)	γ(%)†
γ M1+0.5%E2	126.3 *5*	0.00174 *22*
γ [E2]	1402.1 *5*	0.00133 *19*
γ M1+4.1%E2	1528.3 *3*	0.037
γ M1+E2	2242.0 *8*	0.00041 *11*
γ M1+E2	2252.8 *6*	0.0031 *4*
γ M1+2.2%E2	2268.4 *9*	0.00011 *4*
γ E2	2368.3 *7*	0.00019 *4*

† 8.1% uncert(syst)

Continuous Radiation (^{55}Cr)

⟨β-⟩=1101 keV;⟨IB⟩=2.6 keV

E_{bin}(keV)		⟨ ⟩(keV)	(%)
0 - 10	β-	0.0074	0.146
	IB	0.040	
10 - 20	β-	0.0228	0.152
	IB	0.039	0.27
20 - 40	β-	0.097	0.322
	IB	0.077	0.27
40 - 100	β-	0.81	1.14
	IB	0.22	0.34
100 - 300	β-	11.8	5.7
	IB	0.61	0.34
300 - 600	β-	61	13.3
	IB	0.66	0.156
600 - 1300	β-	406	42.4
	IB	0.77	0.091
1300 - 2500	β-	620	36.8
	IB	0.18	0.0121
2500 - 2603	β-	0.91	0.0361
	IB	3.6×10^{-6}	1.45×10^{-7}

$^{55}_{25}$Mn(stable)

Δ: -57709.2 *15* keV
%: 100

$^{55}_{26}$Fe(2.73 *3* yr)

Mode: ϵ
Δ: -57477.4 *14* keV
SpA: 2381 Ci/g
Prod: ^{54}Fe(n,γ)

Photons (^{55}Fe)

⟨γ⟩=1.63 *5* keV

γ_{mode}	γ(keV)	γ(%)†
Mn L_ℓ	0.556	0.037 *10*
Mn L_η	0.568	0.025 *6*
Mn L_α	0.637	0.28 *7*
Mn L_β	0.657	0.22 *6*
Mn $K_{\alpha2}$	5.888	8.2 *4*
Mn $K_{\alpha1}$	5.899	16.2 *8*
Mn $K_{\beta1}$'	6.490	2.86 *14*

Atomic Electrons (^{55}Fe)

⟨e⟩=3.8 *3* keV

e_{bin}(keV)	⟨e⟩(keV)	e(%)
1	0.53	81 *8*
5	2.5	49 *5*
6	0.73	12.4 *13*

Continuous Radiation (^{55}Fe)

⟨IB⟩=0.0041 keV

E_{bin}(keV)		⟨ ⟩(keV)	(%)
10 - 20	IB	6.1×10^{-5}	0.00044
20 - 40	IB	0.000117	0.00039
40 - 100	IB	0.00089	0.00127
100 - 232	IB	0.00157	0.00115

$^{55}_{27}$Co(17.53 *3* h)

Mode: ϵ
Δ: -54026.0 *15* keV
SpA: 3.250×10^{6} Ci/g
Prod: ^{54}Fe(d,n); ^{54}Fe(p,γ); ^{56}Fe(p,2n)

Photons (^{55}Co)

⟨γ⟩=1221 *36* keV

γ_{mode}	γ(keV)	γ(%)†
Fe L_ℓ	0.615	0.010 *3*
Fe L_η	0.628	0.0065 *16*
Fe L_α	0.705	0.088 *22*
Fe L_β	0.726	0.068 *18*
Fe $K_{\alpha2}$	6.391	2.23 *13*
Fe $K_{\alpha1}$	6.404	4.41 *25*
Fe $K_{\beta1}$'	7.058	0.79 *5*
γ [M1+E2]	91.85 *15*	1.16 *7*
γ M1+0.6%E2	385.34 *15*	0.54 *4*
γ M1+1.6%E2	411.37 *23*	1.07 *7*
γ M1+0.6%E2	477.19 *15*	20.2 *14*
γ E2	519.87 *23*	0.82 *8*
γ M1+4.3%E2	803.66 *20*	1.87 *13*
γ [M1+E2]	827.30 *24*	0.21 *6*
γ M1+11%E2	931.24 *16*	75
γ (M1)	984.59 *22*	0.52 *10*
γ [M1+E2]	1212.64 *23*	0.26 *3*
γ E2	1316.57 *17*	7.09 *14*
γ E2	1369.93 *22*	2.92 *22*
γ E2	1408.42 *18*	16.9 *3*
γ	1555.6 *3*	0.046 *10*
γ	1622.3 *3*	0.045 *5*
γ	1792.03 *25*	0.082 *13*
γ	1941.0 *3*	0.014 *6*
γ [M1+E2]	2143.9 *3*	0.090 *8*
γ	2177.4 *3*	0.29 *4*
γ [M1+E2]	2578.7 *6*	0.043 *5*
γ	2872.2 *3*	0.118 *6*
γ	2938.8 *3*	0.057 *10*
γ	3108.6 *3*	0.0053 *22*

† 4.7% uncert(syst)

Atomic Electrons (^{55}Co)

⟨e⟩=1.6 *3* keV

e_{bin}(keV)	⟨e⟩(keV)	e(%)
1	0.144	20.0 *21*
5	0.052	0.97 *10*
6	0.77	13.3 *14*
7	0.0113	0.162 *17*
85	0.3	~0.4
378 - 411	0.0071	0.00179 *9*
470	0.065	0.0138 *10*
476 - 519	0.0114	0.00231 *13*
797 - 827	0.0041	0.00052 *3*
924	0.117	0.0127 *7*
930 - 977	0.0120	0.00129 *7*
1206 - 1212	0.00037	3.0 *4* $\times 10^{-5}$
1309 - 1316	0.00962	0.000734 *19*
1363 - 1408	0.0250	0.00179 *4*
1549 - 1622	7.7 $\times 10^{-5}$	4.9 *11* $\times 10^{-6}$
1785 - 1791	6.3 $\times 10^{-5}$	3.5 *11* $\times 10^{-6}$
1934 - 1940	1.0 $\times 10^{-5}$	5.4 *25* $\times 10^{-7}$
2137 - 2177	0.00027	1.26 *24* $\times 10^{-5}$
2572 - 2578	3.1 $\times 10^{-5}$	1.22 *15* $\times 10^{-6}$
2865 - 2938	0.000101	3.5 *6* $\times 10^{-6}$
3101 - 3108	2.9 $\times 10^{-6}$	9 *4* $\times 10^{-8}$

Continuous Radiation (^{55}Co)

⟨β+⟩=430 keV;⟨IB⟩=1.14 keV

E_{bin}(keV)		⟨ ⟩(keV)	(%)
0 - 10	β+	0.000246	0.00313
	IB	0.019	
10 - 20	β+	0.00410	0.0256
	IB	0.018	0.128
20 - 40	β+	0.052	0.165
	IB	0.035	0.123
40 - 100	β+	1.12	1.50
	IB	0.095	0.147
100 - 300	β+	28.0	13.2
	IB	0.23	0.134
300 - 600	β+	130	28.9
	IB	0.22	0.053
600 - 1300	β+	263	31.6
	IB	0.34	0.038
1300 - 2500	β+	7.3	0.54
	IB	0.18	0.0115
2500 - 2520	IB	3.7 $\times 10^{-7}$	1.49 $\times 10^{-8}$
	Σβ+		76

$^{55}_{28}$Ni(189 *5* ms)

Mode: ϵ

Δ: -45330 *11* keV

SpA: 2.88 $\times 10^{11}$ Ci/g

Prod: ^{54}Fe(^{3}He,2n)

Continuous Radiation (^{55}Ni)

⟨β+⟩=3623 keV;⟨IB⟩=17 keV

E_{bin}(keV)		⟨ ⟩(keV)	(%)
0 - 10	β+	2.97 $\times 10^{-6}$	3.75 $\times 10^{-5}$
	IB	0.080	
10 - 20	β+	5.2 $\times 10^{-5}$	0.000324
	IB	0.080	0.55
20 - 40	β+	0.00070	0.00222
	IB	0.159	0.55
40 - 100	β+	0.0168	0.0223
	IB	0.47	0.72
100 - 300	β+	0.56	0.255
	IB	1.48	0.82
300 - 600	β+	4.44	0.95
	IB	2.0	0.47
600 - 1300	β+	52	5.2
	IB	3.8	0.43
1300 - 2500	β+	368	18.8
	IB	4.6	0.25
2500 - 5000	β+	2027	54
	IB	4.0	0.121
5000 - 7500	β+	1170	20.3
	IB	0.43	0.0077
7500 - 8696	β+	0.71	0.0095
	IB	0.0038	4.8 $\times 10^{-5}$
	Σβ+		99.9

A = 56

NDS **20**, 253 (1977)

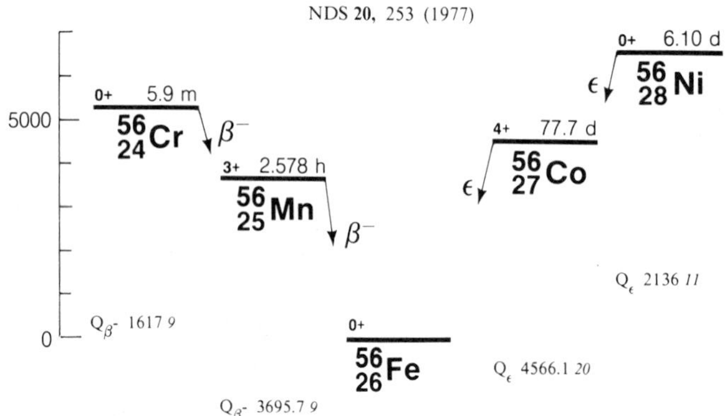

$^{56}_{24}$Cr(5.94 *10* min)

Mode: β-

Δ: -55291 *10* keV

SpA: 5.647 $\times 10^{8}$ Ci/g

Prod: ^{54}Cr(t,p)

Photons (^{56}Cr)

⟨γ⟩=70 *12* keV

γ_{mode}	γ(keV)	γ(%)
Mn L_ℓ	0.556	0.034 *12*
Mn L_η	0.568	0.023 *8*
Mn L_α	0.637	0.26 *9*
Mn L_β	0.657	0.21 *7*
Mn $K_{\alpha 2}$	5.888	7.6 *17*
Mn $K_{\alpha 1}$	5.899	15 *3*
Mn $K_{\beta 1}$'	6.490	2.7 *6*
γ M1+1.1%E2	26.6043 *4*	33 *7*
γ M1+41%E2	83.8992 *14*	71 *14*
γ (E2)	110.5035 *15*	0.24 *5*

Atomic Electrons (^{56}Cr)

⟨e⟩=39 *10* keV

e_{bin}(keV)	⟨e⟩(keV)	e(%)
1	0.51	78 *19*
5	2.4	46 *11*
6	0.68	12 *3*
20	11.3	56 *13*
26	1.9	7.4 *19*
77	19.8	26 *13*
83	2.3	~3
104	0.069	0.067 *14*
110	0.0076	0.0069 *14*

Continuous Radiation (^{56}Cr)

⟨β-⟩=591 keV;⟨IB⟩=0.90 keV

E_{bin}(keV)		⟨ ⟩(keV)	(%)
0 - 10	β-	0.0239	0.474
	IB	0.025	
10 - 20	β-	0.073	0.488
	IB	0.025	0.17
20 - 40	β-	0.310	1.03
	IB	0.048	0.166
40 - 100	β-	2.52	3.55
	IB	0.129	0.20
100 - 300	β-	33.3	16.2
	IB	0.31	0.18
300 - 600	β-	140	30.9
	IB	0.25	0.060
600 - 1300	β-	400	46.3
	IB	0.116	0.0156
1300 - 1508	β-	14.8	1.09
	IB	0.00026	2.0 $\times 10^{-5}$

$^{56}_{25}$Mn(2.5785 *6* h)

Mode: β-
Δ: -56908.4 *15* keV
SpA: 2.1702×10^{7} Ci/g
Prod: ^{55}Mn(n,γ)

Photons (^{56}Mn)

⟨γ⟩=1692 *23* keV

γ_{mode}	γ(keV)	γ(%)†
γ [M1+E2]	787.80 *4*	0.00034 *4*
γ E2	846.812 *20*	98.9 *3*
γ [M1+E2]	1037.879 *19*	0.040 *5*
γ E2	1238.317 *23*	0.099 *10*
γ [M1+E2]	1360.29 *3*	0.0048 *5*
γ M1+3.5%E2	1810.80 *4*	27.2 *7*
γ M1+4.0%E2	2113.19 *6*	14.3 *4*
γ [E2]	2276.17 *3*	0.00034 *7*
γ [M1+E2]	2522.95 *9*	0.99 *3*
γ [M1+E2]	2598.57 *3*	0.0188 *20*
γ [E2]	2657.58 *4*	0.653 *20*
γ [E2]	2959.96 *6*	0.306 *10*
γ [E2]	3369.72 *9*	0.170 *9*

† <0.1% uncert(syst)

Continuous Radiation (^{56}Mn)

⟨β-⟩=830 keV;⟨IB⟩=1.9 keV

E_{bin}(keV)		⟨ ⟩(keV)	(%)
0 - 10	β-	0.0344	0.69
	IB	0.031	
10 - 20	β-	0.104	0.70
	IB	0.030	0.21
20 - 40	β-	0.431	1.43
	IB	0.059	0.20
40 - 100	β-	3.29	4.66
	IB	0.164	0.25
100 - 300	β-	35.0	17.4
	IB	0.44	0.25
300 - 600	β-	101	23.0
	IB	0.45	0.106
600 - 1300	β-	247	27.0
	IB	0.54	0.063
1300 - 2500	β-	429	24.5
	IB	0.17	0.0108
2500 - 2849	β-	14.3	0.55
	IB	0.00036	1.42×10^{-5}

$^{56}_{26}$Fe(stable)

Δ: -60604.1 *15* keV
%: 91.72 *30*

$^{56}_{27}$Co(77.7 *5* d)

Mode: ϵ
Δ: -56038.0 *25* keV
SpA: 3.001×10^{4} Ci/g
Prod: ^{56}Fe(p,n); ^{55}Mn(α,3n); daughter ^{56}Ni; ^{56}Fe(d,2n); ^{58}Ni(d,α)

Photons (^{56}Co)

⟨γ⟩=3378 *32* keV

γ_{mode}	γ(keV)	γ(%)†
Fe L_{ℓ}	0.615	0.031 *8*
Fe L_{η}	0.628	0.021 *5*
Fe L_{α}	0.705	0.29 *7*
Fe L_{β}	0.726	0.22 *6*
Fe $K_{\alpha2}$	6.391	7.3 *4*
Fe $K_{\alpha1}$	6.404	14.4 *7*
Fe $K_{\beta1}$'	7.058	2.58 *13*
γ [M1+E2]	263.46 *5*	0.021 *4*
γ [M1+E2]	411.23 *5*	0.025 *5*
γ [M1+E2]	486.86 *10*	0.050 *10*
γ [M1+E2]	674.69 *4*	0.030 *10*
γ [M1+E2]	733.64 *4*	0.192 *22*
γ [M1+E2]	787.80 *4*	0.307 *7*
γ E2	846.812 *20*	99.9
γ [M1+E2]	896.63 *6*	0.075 *10*
γ [M1+E2]	977.48 *4*	1.40 *3*
γ [M1+E2]	997.10 *4*	0.14 *3*
γ [M1+E2]	1037.879 *19*	14.1 *3*
γ [M1+E2]	1089.11 *8*	0.050 *20*
γ [M1+E2]	1140.46 *6*	0.126 *15*
γ [E2]	1160.09 *6*	0.091 *10*
γ [M1+E2]	1175.15 *3*	2.26 *6*
γ [M1+E2]	1199.03 *5*	0.043 *11*
γ E2	1238.317 *23*	67.0 *13*
γ [M1+E2]	1272.08 *10*	0.0200 *10*
γ	1335.59 *6*	0.1209 *24*
γ [M1+E2]	1360.29 *3*	4.29 *9*
γ [M1+E2]	1442.86 *5*	0.174 *4*
γ [E2]	1462.48 *5*	0.072 *6*
γ [E2]	1640.53 *4*	0.063 *3*
γ [M1+E2]	1771.51 *4*	15.5 *3*
γ M1+3.5%E2	1810.80 *4*	0.650 *20*
γ [M1+E2]	1963.98 *7*	0.702 *14*
γ [M1+E2]	2015.34 *4*	3.03 *6*
γ [M1+E2]	2034.96 *4*	7.77 *16*
γ M1+4.0%E2	2113.19 *6*	0.376 *8*
γ [M1+E2]	2213.00 *3*	0.376 *8*
γ [E2]	2276.17 *3*	0.120 *18*
γ	2373.44 *6*	0.061 *8*
γ [M1+E2]	2522.95 *9*	0.054 *6*
γ [M1+E2]	2598.57 *3*	16.7 *3*
γ [E2]	2657.58 *4*	0.0156 *7*
γ [E2]	2959.96 *6*	0.0080 *3*
γ [M1+E2]	3009.78 *4*	1.03 *9*
γ [M1+E2]	3202.25 *6*	3.02 *21*
γ [M1+E2]	3253.60 *4*	7.4 *5*
γ [E2]	3273.23 *4*	1.73 *12*
γ [E2]	3369.72 *9*	0.0094 *5*
γ [E2]	3451.27 *4*	0.89 *9*
γ [M1+E2]	3548.2 *1*	0.173 *17*
γ	3600.86 *23*	0.0150 *20*
γ	3611.70 *7*	0.0078 *10*

† <0.1% uncert(syst)

Atomic Electrons (^{56}Co)

⟨e⟩=3.6 *3* keV

e_{bin}(keV)	⟨e⟩(keV)	e(%)
1	0.47	65 *7*
5	0.170	3.1 *3*
6	2.5	43 *4*
256 - 263	0.00041	~0.00016
404 - 411	0.00018	4.5 *19* ×10^{-5}
480 - 486	0.00026	5.5 *20* ×10^{-5}
668 - 674	9.0 ×10^{-5}	1.3 *5* ×10^{-5}
727 - 733	0.00051	7.0 *15* ×10^{-5}
781 - 787	0.00073	9.3 *16* ×10^{-5}
840	0.225	0.0268 *8*
846 - 890	0.0219	0.00259 *7*
970 - 996	0.0027	0.00028 *3*
1031 - 1037	0.0234	0.00227 *25*
1082 - 1088	7.8 ×10^{-5}	7 *3* ×10^{-6}
1133 - 1174	0.0036	0.00031 *3*
1192 - 1198	6.1 ×10^{-5}	5.1 *13* ×10^{-6}
1231	0.0891	0.00724 *20*

Atomic Electrons (^{56}Co) (continued)

e_{bin}(keV)	⟨e⟩(keV)	e(%)
1237 - 1271	0.00855	0.000691 *19*
1328 - 1360	0.0054	0.00040 *3*
1436 - 1462	0.000290	2.01 *12* ×10^{-5}
1633 - 1640	6.8 ×10^{-5}	4.15 *24* ×10^{-6}
1764 - 1810	0.0156	0.00089 *6*
1957 - 2034	0.0100	0.000496 *22*
2106 - 2112	0.000309	1.47 *4* ×10^{-5}
2206 - 2275	0.000405	1.82 *10* ×10^{-5}
2366 - 2373	3.9 ×10^{-5}	1.7 *4* ×10^{-6}
2516 - 2598	0.0123	0.00047 *3*
2650 - 2657	1.15 ×10^{-5}	4.34 *25* ×10^{-7}
2953 - 3009	0.00069	2.31 *22* ×10^{-5}
3195 - 3273	0.0078	0.000241 *13*
3363 - 3451	0.00057	1.65 *16* ×10^{-5}
3541 - 3611	0.000117	3.3 *3* ×10^{-6}

Continuous Radiation (^{56}Co)

⟨β+⟩=120 keV;⟨IB⟩=0.44 keV

E_{bin}(keV)		⟨ ⟩(keV)	(%)
0 - 10	β+	7.9×10^{-5}	0.00100
	IB	0.0066	
10 - 20	β+	0.00130	0.0081
	IB	0.0051	0.036
20 - 40	β+	0.0163	0.051
	IB	0.0099	0.034
40 - 100	β+	0.331	0.444
	IB	0.028	0.043
100 - 300	β+	6.8	3.24
	IB	0.086	0.047
300 - 600	β+	28.6	6.3
	IB	0.120	0.028
600 - 1300	β+	83	9.7
	IB	0.143	0.017
1300 - 2481	β+	1.70	0.127
	IB	0.046	0.0028
	Σβ+		20

$^{56}_{28}$Ni(6.10 *2* d)

Mode: ϵ
Δ: -53902 *11* keV
SpA: 3.822×10^{5} Ci/g
Prod: ^{54}Fe(α,2n); ^{56}Fe(^{3}He,3n)

Photons (^{56}Ni)

⟨γ⟩=1721 *20* keV

γ_{mode}	γ(keV)	γ(%)†
Co L_{ℓ}	0.678	0.042 *11*
Co L_{η}	0.693	0.027 *7*
Co L_{α}	0.776	0.44 *11*
Co L_{β}	0.799	0.34 *9*
Co $K_{\alpha2}$	6.915	10.1 *5*
Co $K_{\alpha1}$	6.930	19.8 *10*
Co $K_{\beta1}$'	7.649	3.60 *18*
γ M1+0.03%E2	158.39 *3*	98.8
γ M1	269.512 *17*	36.5 *8*
γ E2	480.452 *17*	36.5 *8*
γ M1	749.962 *19*	49.5 *12*
γ M1(+0.08%E2)	811.86 *3*	86.0 *17*
γ E2	1561.81 *3*	14.0 *6*

† 2.0% uncert(syst)

Atomic Electrons (^{56}Ni)

⟨e⟩=6.9 *3* keV

e_{bin}(keV)	⟨e⟩(keV)	e(%)
1	0.65	82 *8*
6	2.7	45 *5*
7	0.74	10.9 *11*
8	0.048	0.63 *6*
151	1.65	1.09 *3*
157	0.165	0.105 *3*
158	0.00635	0.00403 *12*
262	0.281	0.107 *3*
269	0.0284	0.0106 *3*
473	0.267	0.0564 *17*
480	0.0266	0.00554 *16*
742	0.106	0.0143 *5*
749	0.0104	0.00139 *4*
804	0.169	0.0210 *9*
811	0.0165	0.00204 *9*
1554	0.0161	0.00104 *5*
1561	0.00155	0.000100 *5*

Continuous Radiation (^{56}Ni)

⟨IB⟩=0.018 keV

E_{bin}(keV)		⟨ ⟩(keV)	(%)
10 - 20	IB	0.000104	0.00078
20 - 40	IB	0.000151	0.00051
40 - 100	IB	0.00135	0.0019
100 - 300	IB	0.0122	0.0065
300 - 416	IB	0.0020	0.00060

A = 57

NDS **20**, 327 (1977)

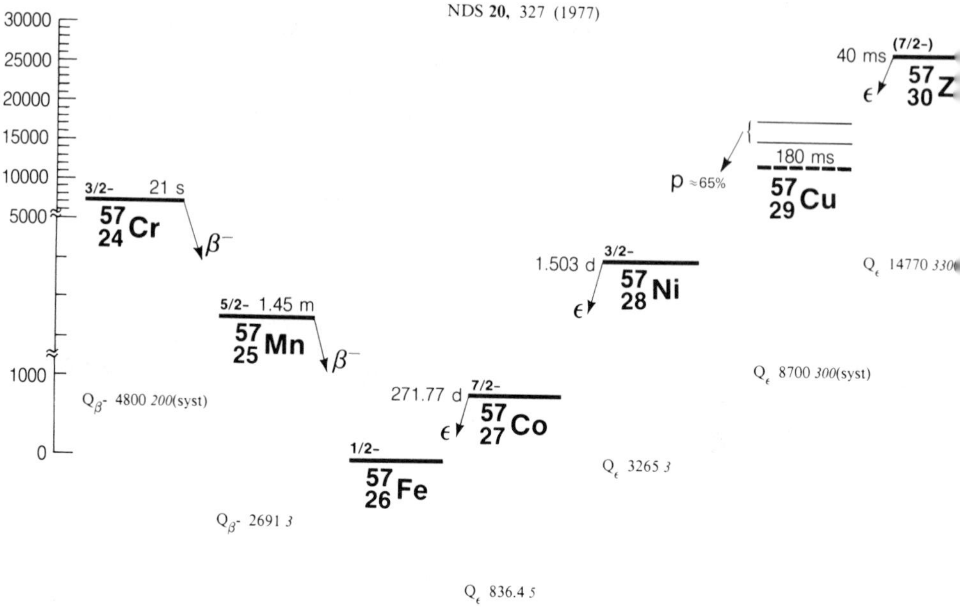

30000
25000
20000
15000
10000
5000
1000
0

3/2- 21 s
$^{57}_{24}$Cr
β⁻
Q_{β^-} 4800 *200*(syst)

5/2- 1.45 m
$^{57}_{25}$Mn
β⁻
Q_{β^-} 2691 *3*

1/2-
$^{57}_{26}$Fe

271.77 d 7/2-
ε
$^{57}_{27}$Co
Q_ϵ 836.4 *5*

1.503 d 3/2-
ε
$^{57}_{28}$Ni
Q_ϵ 3265 *3*

p ≈65%
180 ms
$^{57}_{29}$Cu
Q_ϵ 8700 *300*(syst)

40 ms (7/2-)
ε
$^{57}_{30}$Z
Q_ϵ 14770 *330*

$^{57}_{24}$Cr(21.1 *10* s)

Mode: β-
Δ: -52690 *200* keV syst
SpA: 9.2×10^{9} Ci/g
Prod: ^{48}Ca(^{11}B,pn)

Photons (^{57}Cr)

⟨γ⟩=438 *20* keV

γ_{mode}	γ(keV)	γ(%)†
γ (M1+31%E2)	83.4 *2*	7.8 *16*
γ	105.8 *3*	2.73 *17*
γ	342.9 *5*	0.46 *12*
γ	479.1 *5*	0.18 *7*
γ	684.1 *5*	0.51 *20*
γ	766.5 *5*	0.33 *8*
γ	850.2 *6*	7.6 *16*
γ	1055.8 *3*	0.79 *9*
γ	1129.9 *3*	1.32 *12*
γ	1209.2 *3*	0.55 *10*
γ	1292.2 *5*	0.71 *13*
γ	1327.3 *3*	0.30 *11*
γ	1335.7 *5*	1.1 *5*
γ	1409.3 *3*	0.88 *14*
γ	1492.7 *3*	0.90 *12*
γ	1535.0 *3*	4.6 *3*
γ	1642.2 *5*	0.20 *7*
γ	1752.1 *5*	4.9
γ	1835.2 *6*	1.30 *12*
γ	1852.0 *4*	1.12 *14*
γ	2063.5 *8*	0.21 *7*
γ	2232.9 *6*	0.41 *7*
γ	2257.2 *9*	~0.23
γ	2410.0 *7*	0.82 *9*
γ	2493.1 *7*	0.76 *7*
γ	2618.3 *8*	0.42 *13*

† 11% uncert(syst)

Continuous Radiation (^{57}Cr)

⟨β-⟩=1940 keV;⟨IB⟩=6.6 keV

E_{bin}(keV)		⟨ ⟩(keV)	(%)
0 - 10	β-	0.00276	0.055
	IB	0.057	
10 - 20	β-	0.0085	0.057
	IB	0.056	0.39
20 - 40	β-	0.0365	0.121
	IB	0.111	0.39
40 - 100	β-	0.307	0.431
	IB	0.32	0.49
100 - 300	β-	4.62	2.21
	IB	0.97	0.54
300 - 600	β-	25.5	5.5
	IB	1.20	0.28
600 - 1300	β-	205	21.0
	IB	2.0	0.22
1300 - 2500	β-	777	41.3
	IB	1.57	0.091
2500 - 4800	β-	928	29.3
	IB	0.40	0.0138

$^{57}_{25}$Mn(1.45 *3* min)

Mode: β-
Δ: -57488 *3* keV
SpA: 2.27×10^{9} Ci/g
Prod: ^{54}Cr(α,p); ^{57}Fe(n,p)

Photons (^{57}Mn)

⟨γ⟩=75.5 *10* keV

γ_{mode}	γ(keV)	γ(%)†
Fe L_ℓ	0.615	0.034 *10*
Fe L_η	0.628	0.023 *6*
Fe L_α	0.705	0.31 *8*
Fe L_β	0.726	0.24 *7*
Fe $K_{\alpha2}$	6.391	7.9 *7*
Fe $K_{\alpha1}$	6.404	15.6 *15*
Fe $K_{\beta1}$'	7.058	2.8 *3*
γ M1	14.4119 *4*	10.7 *8*
γ M1+1.4%E2	122.0612 *15*	10.8 *4*
γ E2	136.4730 *15*	1.35 *5*
γ [M1+E2]	230.264 *19*	0.164 *8*
γ [M1+E2]	339.66 *3*	0.120 *10*
γ [M1+E2]	352.324 *19*	1.56 *4*
γ [M1+E2]	366.736 *19*	0.296 *17*
γ [M1+E2]	569.922 *22*	0.383 *14*
γ [M1+E2]	691.982 *22*	4.15 *11*
γ [E2]	706.393 *22*	0.175 *8*
γ [M1+E2]	870.66 *4*	0.193 *8*
γ [M1+E2]	920.90 *5*	0.070 *8*
γ [E2]	992.72 *4*	0.107 *8*
γ [M1+E2]	1018.86 *10*	0.033 *4*
γ [M1+E2]	1260.55 *5*	0.242 *12*
γ [M1+E2]	1612.87 *5*	0.545 *20*
γ [M1+E2]	1725.24 *10*	0.123 *8*

† 7.3% uncert(syst)

Atomic Electrons (^{57}Mn)

⟨e⟩=10.8 *6* keV

e_{bin}(keV)	⟨e⟩(keV)	e(%)
1	0.51	71 *9*
5	0.185	3.4 *4*
6	2.7	47 *6*
7	5.7	78 *7*
14	1.11	8.1 *7*
115	0.267	0.233 *9*
121 - 136	0.288	0.224 *8*
223 - 230	0.004	~0.0020
333 - 366	0.020	0.0057 *21*
563 - 569	0.0015	0.00027 *7*
685 - 706	0.0125	0.0018 *4*
864 - 870	0.00040	4.6 *7* $\times 10^{-5}$
914 - 920	0.000134	1.47 *25* $\times 10^{-5}$
986 - 1018	0.000263	2.65 *19* $\times 10^{-5}$
1253 - 1260	0.00032	2.58 *24* $\times 10^{-5}$
1606 - 1612	0.00057	3.6 *3* $\times 10^{-5}$
1718 - 1725	0.000122	7.1 *6* $\times 10^{-6}$

Continuous Radiation (^{57}Mn)

⟨β-⟩=1100 keV;⟨IB⟩=2.6 keV

E_{bin}(keV)		⟨ ⟩(keV)	(%)
0 - 10	β-	0.0080	0.158
	IB	0.040	
10 - 20	β-	0.0246	0.164
	IB	0.039	0.27
20 - 40	β-	0.104	0.346
	IB	0.077	0.27
40 - 100	β-	0.86	1.21
	IB	0.22	0.34
100 - 300	β-	12.3	5.9
	IB	0.61	0.34
300 - 600	β-	62	13.6
	IB	0.66	0.155
600 - 1300	β-	399	41.8
	IB	0.78	0.092
1300 - 2500	β-	622	36.6
	IB	0.20	0.0129
2500 - 2677	β-	3.34	0.131
	IB	3.0×10^{-5}	1.18×10^{-6}

$^{57}_{26}$Fe(stable)

Δ: -60178.9 *15* keV

%: 2.2 *1*

$^{57}_{27}$Co(271.77 *5* d)

Mode: ϵ

Δ: -59342.5 *15* keV

SpA: 8429.1 Ci/g

Prod: γ rays on ^{58}Ni; ^{56}Fe(d,n); ^{56}Fe(p,γ); ^{55}Mn(α,2n)

Photons (^{57}Co)

⟨γ⟩=125.2 *6* keV

γ_{mode}	γ(keV)	γ(%)†
Fe L_ℓ	0.615	0.071 *19*
Fe L_η	0.628	0.048 *12*
Fe L_α	0.705	0.65 *16*
Fe L_β	0.726	0.50 *13*
Fe $K_{\alpha 2}$	6.391	16.4 *9*
Fe $K_{\alpha 1}$	6.404	32.5 *17*
Fe $K_{\beta 1}$'	7.058	5.8 *3*
γ M1	14.4119 *4*	9.54 *12*
γ M1+1.4%E2	122.0612 *15*	85.5 *4*
γ E2	136.4730 *15*	10.69 *17*
γ [M1+E2]	230.264 *19*	0.000339 *25*
γ [M1+E2]	339.66 *3*	0.0046 *4*
γ [M1+E2]	352.324 *19*	0.00322 *21*
γ [M1+E2]	366.736 *19*	0.00061 *4*
γ [M1+E2]	569.922 *22*	0.0146 *6*
γ [M1+E2]	691.982 *22*	0.158 *4*
γ [E2]	706.393 *22*	0.0067 *4*

† 0.22% uncert(syst)

Atomic Electrons (^{57}Co)

⟨e⟩=17.6 *6* keV

e_{bin}(keV)	⟨e⟩(keV)	e(%)
1	1.06	147 *15*
5	0.39	7.1 *7*
6	5.7	98 *10*
7	5.11	70 *3*
14	0.99	7.3 *3*
115	2.11	1.84 *3*
121	0.222	0.183 *3*
129	1.86	1.43 *4*
136	0.202	0.149 *4*
223	8.4 $\times 10^{-6}$	$\sim 4 \times 10^{-6}$
229	8.0 $\times 10^{-7}$	$\sim 3 \times 10^{-7}$
230	6.4 $\times 10^{-8}$	$\sim 3 \times 10^{-8}$
333	4.6 $\times 10^{-5}$	$\sim 1 \times 10^{-5}$
339	4.6 $\times 10^{-6}$	$\sim 1 \times 10^{-6}$
345	3.0 $\times 10^{-5}$	9 *4* $\times 10^{-6}$
351	2.8 $\times 10^{-6}$	8 *4* $\times 10^{-7}$
352	1.4 $\times 10^{-7}$	$\sim 4 \times 10^{-8}$
360	5.2 $\times 10^{-6}$	1.4 *7* $\times 10^{-6}$
366	5.1 $\times 10^{-7}$	1.4 *7* $\times 10^{-7}$
563	5.2 $\times 10^{-5}$	9 *3* $\times 10^{-6}$
569	5.1 $\times 10^{-6}$	9 *3* $\times 10^{-7}$
685	0.00042	6.1 *13* $\times 10^{-5}$
691	4.0 $\times 10^{-5}$	5.8 *13* $\times 10^{-6}$
699	2.05 $\times 10^{-5}$	2.93 *19* $\times 10^{-6}$
706	1.99 $\times 10^{-6}$	2.82 *19* $\times 10^{-7}$

Continuous Radiation (^{57}Co)

⟨IB⟩=0.082 keV

E_{bin}(keV)		⟨ ⟩(keV)	(%)
10 - 20	IB	8.1×10^{-5}	0.00060
20 - 40	IB	0.000150	0.00050
40 - 100	IB	0.00162	0.0022
100 - 300	IB	0.027	0.0133
300 - 600	IB	0.049	0.0121
600 - 700	IB	0.00165	0.00026

$^{57}_{28}$Ni(1.503 *4* d)

Mode: ϵ

Δ: -56077 *3* keV

SpA: 1.524×10^{6} Ci/g

Prod: ^{59}Co(p,3n); ^{54}Fe(α,n); ^{58}Ni(n,2n); ^{58}Ni(d,dn)

Photons (^{57}Ni)

⟨γ⟩=1509 *30* keV

γ_{mode}	γ(keV)	γ(%)†
Co L_ℓ	0.678	0.025 *7*
Co L_η	0.693	0.016 *4*
Co L_α	0.776	0.27 *7*
Co L_β	0.799	0.20 *5*
Co $K_{\alpha 2}$	6.915	6.1 *4*
Co $K_{\alpha 1}$	6.930	12.0 *7*
Co $K_{\beta 1}$'	7.649	2.17 *13*
γ [M1+E2]	127.226 *25*	12.9 *8*
γ [M1+E2]	161.93 *10*	0.017 *3*
γ [M1+E2]	252.69 *8*	<0.031
γ [M1+E2]	379.92 *8*	0.075 *15*
γ [M1+E2]	673.42 *18*	0.045 *9*
γ	907.02 *22*	0.086 *17*
γ	1046.39 *14*	0.125 *25*
γ [M1+E2]	1223.49 *25*	0.086 *17*
γ [E2]	1377.62 *4*	77.9 *16*
γ	1730.6 *3*	0.056 *11*
γ [E2]	1757.53 *7*	7.1 *6*
γ [M1+E2]	1896.90 *22*	0.022 *4*
γ [M1+E2]	1919.46 *8*	14.7 *9*
γ [M1+E2]	2132.9 *3*	0.037 *7*
γ	2730.63 *20*	0.023 *5*
γ	2803.89 *14*	0.13 *3*
γ [M1+E2]	3176.9 *3*	0.019 *4*

† 2.0% uncert(syst)

Atomic Electrons (^{57}Ni)

⟨e⟩=4.4 *13* keV

e_{bin}(keV)	⟨e⟩(keV)	e(%)
1	0.39	50 *5*
6	1.64	27 *3*
7	0.45	6.6 *7*
8	0.029	0.38 *4*
120	1.6	~1
126	0.18	~0.14
154 - 161	0.0013	~0.0008
245 - 252	0.0004	~0.00017
372 - 379	0.0007	0.00019 *9*
666 - 673	0.00015	2.3 *6* $\times 10^{-5}$
899 - 906	0.00015	1.7 *7* $\times 10^{-5}$
1039 - 1046	0.00018	1.7 *7* $\times 10^{-5}$
1216 - 1223	0.00013	1.09 *22* $\times 10^{-5}$
1370	0.102	0.0075 *3*
1723 - 1757	0.0080	0.00046 *4*
1889 - 1919	0.0150	0.00078 *6*
2125 - 2132	3.4 $\times 10^{-5}$	1.6 *3* $\times 10^{-6}$
2723 - 2803	0.000102	3.6 *9* $\times 10^{-6}$
3169 - 3176	1.3 $\times 10^{-5}$	4.3 *8* $\times 10^{-7}$

Continuous Radiation (^{57}Ni)

⟨β+⟩=143 keV;⟨IB⟩=0.91 keV

E_{bin}(keV)		⟨ ⟩(keV)	(%)
0 - 10	β+	0.000304	0.00386
	IB	0.0081	
10 - 20	β+	0.0052	0.0327
	IB	0.0067	0.046
20 - 40	β+	0.068	0.215
	IB	0.0124	0.043
40 - 100	β+	1.43	1.92
	IB	0.032	0.050
100 - 300	β+	29.7	14.3
	IB	0.084	0.047
300 - 600	β+	86	20.0
	IB	0.162	0.036
600 - 1300	β+	25.5	3.83
	IB	0.50	0.056
1300 - 1888	IB	0.108	0.0075
	Σβ+		40

$^{57}_{29}$Cu(180 ms)

decay not observed

Δ: -47380 *300* keV syst
SpA: 3×10^{11} Ci/g
Prod: protons on Ni

$^{57}_{30}$Zn(40 *10* ms)

Mode: ε, εp(~65 %)
Δ: -32610 *130* keV
SpA: 2.9×10^{11} Ci/g
Prod: ^{40}Ca(^{20}Ne,3n)

Delayed Protons (^{57}Zn)

p(keV)	p(%)
1920 *50*	~30
2530 *50*	~15
4570 *50*	~20

Photons (^{57}Zn)

γ_{mode}	γ(keV)	γ(%)
$\gamma_{\epsilon p}$E2	2701 *3*	~30

$^{58}_{25}$Mn(1.088 *12* min)

Mode: β-
Δ: -55830 *30* keV
SpA: 2.96×10^{9} Ci/g
Prod: ^{58}Fe(n,p)

Photons (^{58}Mn)

⟨γ⟩=2364 *27* keV

γ_{mode}	γ(keV)	γ(%)†
γ [M1+E2]	459.185 *20*	21.4 *6*
γ [M1+E2]	466.525 *24*	1.27 *4*
γ E2+2.5%M1	523.90 *3*	3.70 *11*
γ [E2]	632.87 *6*	0.56 *5*
γ E2	810.791 *6*	88.2
γ M1+31%E2	863.974 *6*	14.9 *4*
γ [E2]	925.706 *25*	1.68 *6*
γ [E2]	1156.76 *5*	1.05 *4*
γ E2	1265.78 *3*	9.1 *3*
γ [M1+E2]	1301.33 *7*	0.69 *4*
γ M1+20%E2	1323.152 *20*	59.4 *18*
γ	1446.45 *22*	0.097 *18*
γ [M1+E2]	1488.08 *16*	0.16 *4*
γ [M1+E2]	1558.56 *6*	0.512 *18*
γ E2	1674.753 *6*	11.6 *3*
γ [E2]	1712.50 *9*	0.07 *3*
γ [M1+E2]	1767.84 *7*	3.18 *13*
γ [E2]	1789.666 *25*	2.87 *9*
γ [M1+E2]	2011.97 *17*	<0.0016
γ M1+10%E2	2065.63 *14*	0.159 *18*
γ [M1+E2]	2179.02 *9*	0.48 *4*
γ [M1+E2]	2227.01 *7*	0.30 *7*
γ [E2]	2236.39 *9*	0.344 *18*
γ M1(+0.3%E2)	2272.94 *20*	0.079 *18*
γ [M1+E2]	2422.51 *6*	1.12 *3*
γ [E2]	2433.09 *25*	0.123 *18*
γ	2513.7 *4*	0.053 *18* ?
γ [M1+E2]	2638.18 *9*	1.33 *4*
γ	2699.98 *25*	0.106 *9*
γ [M1+E2]	2818.76 *23*	0.85 *3*
γ	2855.41 *23*	0.062 *9*
γ [M1+E2]	3090.95 *7*	0.097 *18*
γ [M1+E2]	3502.11 *9*	<0.035
γ [E2]	3629.51 *23*	0.071 *18*
γ	3681.7 *5*	0.071 *18*
γ	3779.5 *4*	0.150 *18*
γ	4530.07 *23*	~0.035

† 3.1% uncert(syst)

A = 58

NDS **42**, 457 (1984)

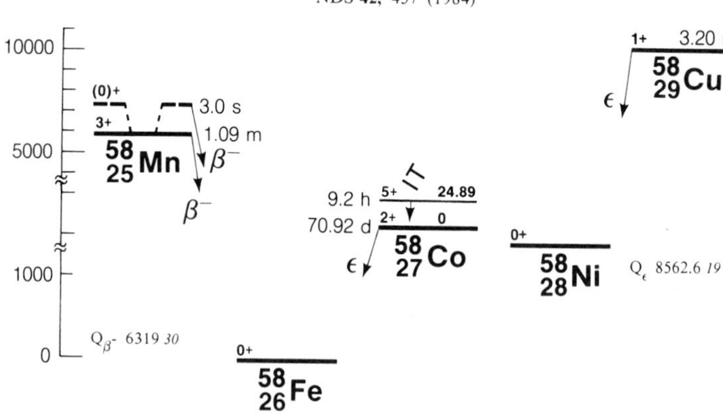

Continuous Radiation (^{58}Mn)

⟨β-⟩=1746 keV;⟨IB⟩=5.6 keV

E_{bin}(keV)		⟨ ⟩(keV)	(%)
0 - 10	β-	0.00336	0.067
	IB	0.053	
10 - 20	β-	0.0104	0.069
	IB	0.053	0.37
20 - 40	β-	0.0442	0.147
	IB	0.104	0.36
40 - 100	β-	0.369	0.52
	IB	0.30	0.46
100 - 300	β-	5.5	2.62
	IB	0.90	0.50
300 - 600	β-	29.7	6.4
	IB	1.09	0.26
600 - 1300	β-	232	23.9
	IB	1.7	0.20
1300 - 2500	β-	841	44.8
	IB	1.20	0.071
2500 - 4242	β-	638	21.4
	IB	0.18	0.0060

$^{58}_{25}$Mn(3.0 *1* s)

Mode: β-
Δ: -55830 *30* keV
SpA: 5.79×10^{10} Ci/g
Prod: ^{58}Fe(n,p)

$^{58}_{26}$Fe(stable)

Δ: -62152.2 *15* keV
%: 0.28 *1*

$^{58}_{27}$Co(70.916 *15* d)

Mode: ε
Δ: -59844.3 *18* keV
SpA: 3.1746×10^{4} Ci/g
Prod: ^{55}Mn(α,n); ^{59}Co(n,2n); ^{58}Ni(n,p)

Photons (^{58}Co)

$\langle\gamma\rangle$=824 *3* keV

γ_{mode}	γ(keV)	γ(%)†
Fe L_ℓ	0.615	0.033 *9*
Fe L_η	0.628	0.022 *6*
Fe L_α	0.705	0.30 *8*
Fe L_β	0.726	0.23 *6*
Fe $K_{\alpha2}$	6.391	7.7 *4*
Fe $K_{\alpha1}$	6.404	15.3 *8*
Fe $K_{\beta1}'$	7.058	2.75 *13*
γ E2	810.791 *6*	99.5 *3*
γ M1+31%E2	863.974 *6*	0.72 *3*
γ E2	1674.753 *6*	0.56 *3*

† <0.10% uncert(syst)

Atomic Electrons (^{58}Co)

$\langle e\rangle$=3.6 *3* keV

e_{bin}(keV)	$\langle e\rangle$(keV)	e(%)
1	0.50	69 *7*
5	0.181	3.3 *3*
6	2.7	46 *5*
7	0.039	0.56 *6*
804	0.241	0.0299 *9*
810	0.0233	0.00288 *8*
857	0.00130	0.000151 *7*
863	0.000125	1.45 *7* $\times10^{-5}$
1668	0.00054	3.23 *17* $\times10^{-5}$
1674	5.1 $\times10^{-5}$	3.06 *16* $\times10^{-6}$

Continuous Radiation (^{58}Co)

$\langle\beta+\rangle$=30.0 keV;$\langle$IB$\rangle$=0.73 keV

E_{bin}(keV)		$\langle\ \rangle$(keV)	(%)
0 - 10	β+	0.000397	0.00503
	IB	0.0030	
10 - 20	β+	0.0065	0.0404
	IB	0.00150	0.0105
20 - 40	β+	0.079	0.251
	IB	0.0027	0.0093
40 - 100	β+	1.49	2.02
	IB	0.0070	0.0108
100 - 300	β+	19.8	10.1
	IB	0.043	0.021
300 - 600	β+	8.6	2.48
	IB	0.18	0.040
600 - 1300	β+	0.00216	0.000268
	IB	0.47	0.056
1300 - 2308	IB	0.0125	0.00093
	$\Sigma\beta$+		14.9

$^{58}_{27}$Co(9.15 *10* h)

Mode: IT
Δ: -59819.4 *18* keV
SpA: 5.90$\times10^{6}$ Ci/g
Prod: ^{55}Mn(α,n); ^{59}Co(n,2n)

Photons (^{58}Co)

$\langle\gamma\rangle$=1.83 *6* keV

γ_{mode}	γ(keV)	γ(%)
Co L_ℓ	0.678	0.040 *11*
Co L_η	0.693	0.023 *6*
Co L_α	0.776	0.43 *11*
Co L_β	0.800	0.29 *8*
Co $K_{\alpha2}$	6.915	7.8 *4*
Co $K_{\alpha1}$	6.930	15.4 *8*
Co $K_{\beta1}'$	7.649	2.79 *15*
γ M3	24.889 *21*	0.0369 *7*

Atomic Electrons (^{58}Co)

$\langle e\rangle$=20.8 *4* keV

e_{bin}(keV)	$\langle e\rangle$(keV)	e(%)
1	0.60	75 *8*
6	2.11	35 *4*
7	0.58	8.4 *9*
8	0.037	0.49 *5*
17	12.0	69.6 *20*
24	5.52	23.0 *7*

$^{58}_{28}$Ni(stable)

Δ: -60225.1 *15* keV
%: 68.27 *1*

$^{58}_{29}$Cu(3.204 *7* s)

Mode: ϵ
Δ: -51662.4 *25* keV
SpA: 5.459$\times10^{10}$ Ci/g
Prod: ^{58}Ni(p,n)

Photons (^{58}Cu)

$\langle\gamma\rangle$=523 *31* keV

γ_{mode}	γ(keV)	γ(%)†
Ni $K_{\alpha2}$	7.461	0.34 *6*
Ni $K_{\alpha1}$	7.478	0.66 *11*
Ni $K_{\beta1}'$	8.265	0.120 *20*
γ [M1]	40.4 *3*	4.8 *8*
γ [E2]	167.1 *3*	0.91 *10*
γ [M1+E2]	818.9 *3*	~0.11
γ	854.9 *3*	0.66 *6*
γ E2+47%M1	1321.5 *3*	1.17 *5*
γ [M1+E2]	1448.25 *17*	11.5 *3*
γ E2	1454.52 *15*	16.0
γ [E2]	1488.6 *3*	1.06 *6*
γ	1547.0 *4*	0.072 *18*
γ M1(+4.3%E2)	1584.1 *4*	0.202 *22*
γ	1673.8 *4*	0.082 *11*
γ M1+30%E2	1810.03 *16*	0.400 *22*
γ [E2]	2077.6 *7*	0.069 *14*
γ [M1+E2]	2140.4 *3*	0.067 *18*
γ [M1+E2]	2444.9 *8*	0.077 *18*
γ E2	2776.0 *3*	0.083 *18*
γ M1	2902.74 *21*	0.52 *3*
γ E2	3038.6 *4*	0.104 *13*
γ	3083.7 *6*	0.099 *16*
γ [E2]	3264.50 *16*	0.70 *5*
γ [M1]	3594.82 *25*	0.43 *3*
γ [E2]	3899.3 *8*	0.053 *19*

† 13% uncert(syst)

Atomic Electrons (^{58}Cu)

$\langle e\rangle$=1.19 *14* keV

e_{bin}(keV)	$\langle e\rangle$(keV)	e(%)
1	0.022	2.5 *5*
6	0.029	0.45 *9*
7	0.079	1.16 *22*
8	0.0015	0.019 *4*
32	0.80	2.5 *4*
39	0.102	0.26 *4*
40	0.0020	0.0051 *9*
159	0.106	0.067 *8*
811 - 854	0.0017	0.00020 *7*
1313 - 1321	0.00186	0.000142 *6*
1440 - 1539	0.042	0.00293 *22*
1546 - 1583	0.00027	1.69 *18* $\times10^{-5}$
1665 - 1673	8.2 $\times10^{-5}$	4.9 *15* $\times10^{-6}$
1802 - 1809	0.000470	2.61 *14* $\times10^{-5}$
2069 - 2140	0.000145	6.9 *11* $\times10^{-6}$
2437 - 2444	7.3 $\times10^{-5}$	3.0 *7* $\times10^{-6}$
2768 - 2775	7.4 $\times10^{-5}$	2.7 *5* $\times10^{-6}$
2894 - 2902	0.00043	1.48 *9* $\times10^{-5}$
3030 - 3083	0.000156	5.1 *6* $\times10^{-6}$
3256 - 3264	0.00057	1.74 *13* $\times10^{-5}$
3586 - 3594	0.000318	8.9 *6* $\times10^{-6}$
3891 - 3898	3.9 $\times10^{-5}$	1.0 *3* $\times10^{-6}$

Continuous Radiation (^{58}Cu)

$\langle\beta+\rangle$=3298 keV;$\langle$IB$\rangle$=15.1 keV

E_{bin}(keV)		$\langle\ \rangle$(keV)	(%)
0 - 10	β+	4.20$\times10^{-6}$	5.3$\times10^{-5}$
	IB	0.076	
10 - 20	β+	7.7$\times10^{-5}$	0.000476
	IB	0.076	0.52
20 - 40	β+	0.00107	0.00335
	IB	0.150	0.52
40 - 100	β+	0.0260	0.0345
	IB	0.44	0.68
100 - 300	β+	0.87	0.397
	IB	1.39	0.77
300 - 600	β+	6.8	1.45
	IB	1.9	0.43
600 - 1300	β+	75	7.5
	IB	3.5	0.39
1300 - 2500	β+	459	23.6
	IB	4.0	0.22
2500 - 5000	β+	1876	51
	IB	3.3	0.099
5000 - 7500	β+	881	15.4
	IB	0.32	0.0057
7500 - 8563	β+	0.0080	0.000106
	IB	0.0027	3.5$\times10^{-5}$
	$\Sigma\beta$+		99.8

A = 59

NDS 39, 641 (1983)

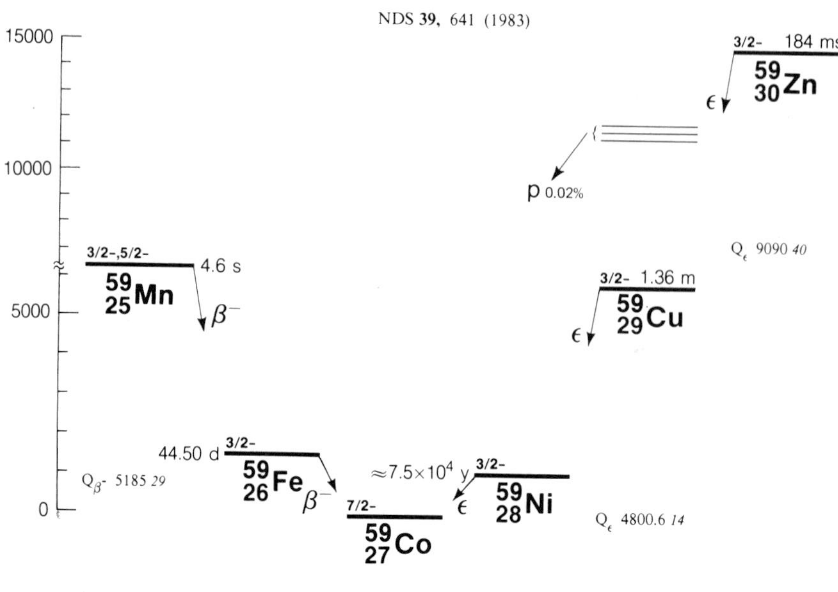

$^{59}_{25}$Mn(4.6 *1* s)

Mode: β-

Δ: -55477 *29* keV

SpA: 3.86×10^{10} Ci/g

Prod: ^{48}Ca(^{13}C,pn)

Photons (^{59}Mn)

γ_{mode}	γ(keV)	γ(rel)
γ [M1+E2]	286.6 *3*	12.9 *10*
γ [M1+E2]	439.4 *4*	4.5 *13*
γ [M1+E2]	472.5 *3*	69 *5*
γ [M1+E2]	570.4 *3*	59 *4*
γ [M1+E2]	590.8 *3*	22.5 *18*
γ [M1+E2]	688.8 *4*	~0.6 ?
γ [M1+E2]	726.0 *3*	100
γ [M1+E2]	874.6 *4*	6.7 *8*
γ [M1+E2]	1161.2 *3*	1.7 *7*

$^{59}_{26}$Fe(44.496 *7* d)

Mode: β-

Δ: -60661.9 *15* keV

SpA: 4.9738×10^{4} Ci/g

Prod: ^{58}Fe(n,γ); ^{59}Co(n,p)

Photons (^{59}Fe)

⟨γ⟩=1188 *22* keV

γ_{mode}	γ(keV)	γ(%)
γ M1+0.04%E2	142.652 *4*	1.02 *4*
γ [M1+E2]	190.0 *3*	0.013 *3*
γ M1+4.6%E2	192.349 *7*	3.08 *10*
γ M1+1.9%E2	335.000 *8*	0.27 *1*
γ [M1+E2]	382.3 *3*	0.018 *3*
γ E2	1099.251 *6*	56.5 *15*
γ E2	1291.596 *8*	43.2 *11*
γ M1+3.8%E2	1481.6 *3*	0.059 *6*

Continuous Radiation (^{59}Fe)

⟨β-⟩=118 keV;⟨IB⟩=0.052 keV

E_{bin}(keV)		⟨ ⟩(keV)	(%)
0 - 10	β-	0.291	5.8
	IB	0.0061	
10 - 20	β-	0.85	5.7
	IB	0.0053	0.037
20 - 40	β-	3.25	10.9
	IB	0.0089	0.031
40 - 100	β-	19.7	28.7
	IB	0.0168	0.027
100 - 300	β-	76	44.1
	IB	0.0137	0.0094
300 - 600	β-	16.5	4.79
	IB	0.00086	0.00023
600 - 1300	β-	0.76	0.086
	IB	0.00025	3.3×10^{-5}
1300 - 1565	β-	0.0482	0.00352
	IB	1.21×10^{-6}	9.0×10^{-8}

$^{59}_{27}$Co(stable)

Δ: -62226.5 *15* keV

%: 100

$^{59}_{28}$Ni(7.5 *13* $\times10^{4}$ yr)

Mode: ε

Δ: -61153.6 *15* keV

SpA: 0.081 Ci/g

Prod: ^{58}Ni(n,γ); ^{59}Co(d,2n)

Photons (^{59}Ni)

⟨γ⟩=2.32 *8* keV

γ_{mode}	γ(keV)	γ(%)
Co L_ℓ	0.678	0.042 *11*
Co L_η	0.693	0.027 *7*
Co L_α	0.776	0.44 *11*
Co L_β	0.799	0.33 *9*
Co $K_{\alpha2}$	6.915	9.9 *5*
Co $K_{\alpha1}$	6.930	19.6 *10*
Co $K_{\beta1}$'	7.649	3.56 *17*

Atomic Electrons (^{59}Ni)

⟨e⟩=4.1 *3* keV

e_{bin}(keV)	⟨e⟩(keV)	e(%)
1	0.64	81 *8*
6	2.7	44 *5*
7	0.73	10.8 *11*
8	0.047	0.62 *6*

Continuous Radiation (^{59}Ni)

⟨β+⟩=3.731×10^{-6} keV;⟨IB⟩=0.30 keV

E_{bin}(keV)		⟨ ⟩(keV)	(%)
0 - 10	β+	5.7×10^{-8}	7.3×10^{-7}
	IB	0.0021	
10 - 20	β+	6.5×10^{-7}	4.15×10^{-6}
	IB	0.000106	0.00079
20 - 40	β+	2.66×10^{-6}	9.3×10^{-6}
	IB	0.000165	0.00055
40 - 100	β+	3.69×10^{-7}	8.6×10^{-7}
	IB	0.0018	0.0024
100 - 300	IB	0.037	0.018
300 - 600	IB	0.142	0.032
600 - 1073	IB	0.118	0.0162
	Σβ+		1.50×10^{-5}

$^{59}_{29}$Cu(1.358 *8* min)

Mode: ε
Δ: -56353.0 *21* keV
SpA: 2.336×10^{9} Ci/g
Prod: ^{58}Ni(p,γ); ^{58}Ni(d,n)

Photons (^{59}Cu)

⟨γ⟩=440 *5* keV

γ_{mode}	γ(keV)	γ(%)
Ni K$_{\alpha2}$	7.461	0.197 *10*
Ni K$_{\alpha1}$	7.478	0.387 *19*
Ni K$_{\beta1}$'	8.265	0.070 *4*
γ M1+1.3%E2	339.37 *7*	8.02 *16*
γ [M1+E2]	423.45 *8*	2.52 *9*
γ [M1+E2]	464.98 *7*	5.76 *12*
γ [M1+E2]	538.68 *9*	0.15 *1*
γ [M1+E2]	545.67 *10*	0.22 *2*
γ [M1]	836.51 *7*	2.19 *4*
γ [M1+E2]	849.72 *11*	~0.03 ?
γ M1+1.0%E2	878.04 *7*	11.75 *24*
γ [E2]	962.13 *9*	0.05 *1*
γ E2+2.1%M1	998.0 *6*	0.052 *16*
γ M1+16%E2	1189.08 *9*	0.370 *14*
γ [E2]	1214.78 *10*	0.029 *10*
γ [M1+E2]	1225.77 *20*	0.035 *12*
γ [M1+E2]	1269.77 *7*	0.210 *12*
γ [M1+E2]	1301.49 *7*	14.6 *3*
γ [M1+E2]	1340.39 *8*	1.40 *3*
γ [M1+E2]	1395.38 *7*	0.320 *13*
γ [M1+E2]	1536.81 *18*	0.052 *9*
γ E2+45%M1	1679.75 *8*	0.230 *12*
γ [M1+E2]	1734.74 *6*	1.17 *3*
γ [M1+E2]	1803.0 *14*	0.022 *5*
γ [M1+E2]	1949.86 *17*	0.118 *8*
γ [M1+E2]	2414.82 *18*	0.044 *6*
γ [M1+E2]	2681.0 *14*	0.018 *4*

Atomic Electrons (^{59}Cu)

⟨e⟩=0.254 *16* keV

e_{bin}(keV)	⟨e⟩(keV)	e(%)
1	0.0127	1.46 *15*
6	0.0172	0.27 *3*
7	0.046	0.68 *7*
8	0.00089	0.0109 *11*
331	0.0522	0.0158 *5*
338	0.00527	0.00156 *5*
339	5.1 ×10^{-5}	1.50 *11* ×10^{-5}
415	0.019	0.0047 *18*
422 - 423	0.0020	0.00047 *17*
457	0.037	0.008 *3*
530 - 545	0.0020	0.00037 *7*
828 - 849	0.00525	0.00063 *3*
870	0.0240	0.00276 *8*
954 - 997	0.00025	2.6 *4* ×10^{-5}
1181 - 1225	0.00074	6.2 *3* ×10^{-5}
1261 - 1269	0.00035	2.75 *25* ×10^{-5}
1293	0.0215	0.00166 *13*
1300 - 1340	0.00430	0.000326 *17*
1387 - 1395	0.00048	3.5 *3* ×10^{-5}
1528 - 1536	7.1 ×10^{-5}	4.7 *8* ×10^{-6}
1671 - 1734	0.00173	0.000101 *5*
1795 - 1802	2.6 ×10^{-5}	1.5 *3* ×10^{-6}
1942 - 1949	0.000132	6.8 *6* ×10^{-6}
2406 - 2414	4.2 ×10^{-5}	1.74 *24* ×10^{-6}
2673 - 2680	1.6 ×10^{-5}	6.0 *13* ×10^{-7}

Continuous Radiation (^{59}Cu)

⟨β+⟩=1490 keV;⟨IB⟩=4.6 keV

E_{bin}(keV)		⟨ ⟩(keV)	(%)
0 - 10	β+	2.73×10^{-5}	0.000344
	IB	0.049	
10 - 20	β+	0.000496	0.00308
	IB	0.048	0.33
20 - 40	β+	0.0069	0.0216
	IB	0.094	0.33
40 - 100	β+	0.165	0.219
	IB	0.27	0.42
100 - 300	β+	5.2	2.39
	IB	0.80	0.45
300 - 600	β+	36.3	7.8
	IB	0.94	0.22
600 - 1300	β+	300	30.9
	IB	1.39	0.160
1300 - 2500	β+	838	46.0
	IB	0.90	0.053
2500 - 4801	β+	310	10.9
	IB	0.160	0.0054
	Σβ+		98

$^{59}_{30}$Zn(183.7 *23* ms)

Mode: ε, εp(0.023 *8* %)
Δ: -47260 *40* keV
SpA: 2.70×10^{11} Ci/g
Prod: ^{3}He on Ni

Delayed Protons (^{59}Zn)

p(keV)	p(%)
913 *10*	0.0007 *3*
1063 *5*	0.0014 *5*
1264 *10*	~0.0004
1331 *10*	~0.0004
1376 *5*	0.0023 *7*
1778 *5*	0.0045 *13*
1817 *5*	0.0026 *8*
1857 *5*	0.0017 *6*
2025 *5*	0.0016 *5*
2089 *5*	0.0028 *9*
2182 *10*	0.0011 *4*
2197 *10*	0.0010 *4*
2250 *10*	0.0008 *3*
2410 *15*	0.0016 *6*
2455 *15*	0.0005 *2*

Photons (^{59}Zn)

⟨γ⟩=39 *12* keV

γ_{mode}	γ(keV)	γ(%)
γ [M1+E2]	491.4 *2*	5 *2*
γ M1+3.6%E2	913.8 *4*	~2

Continuous Radiation (^{59}Zn)

⟨β+⟩=3790 keV;⟨IB⟩=18 keV

E_{bin}(keV)		⟨ ⟩(keV)	(%)
0 - 10	β+	2.15×10^{-6}	2.70×10^{-5}
	IB	0.082	
10 - 20	β+	4.07×10^{-5}	0.000253
	IB	0.082	0.57
20 - 40	β+	0.00058	0.00184
	IB	0.162	0.56
40 - 100	β+	0.0148	0.0195
	IB	0.48	0.73
100 - 300	β+	0.509	0.232
	IB	1.51	0.84
300 - 600	β+	4.08	0.87
	IB	2.1	0.48
600 - 1300	β+	47.9	4.80
	IB	4.0	0.45
1300 - 2500	β+	338	17.3
	IB	4.8	0.27
2500 - 5000	β+	1961	52
	IB	4.5	0.136
5000 - 7500	β+	1419	24.2
	IB	0.63	0.0112
7500 - 9090	β+	18.9	0.247
	IB	0.0095	0.000121
	Σβ+		99.88

A = 60

NDS **28**, 103 (1979)

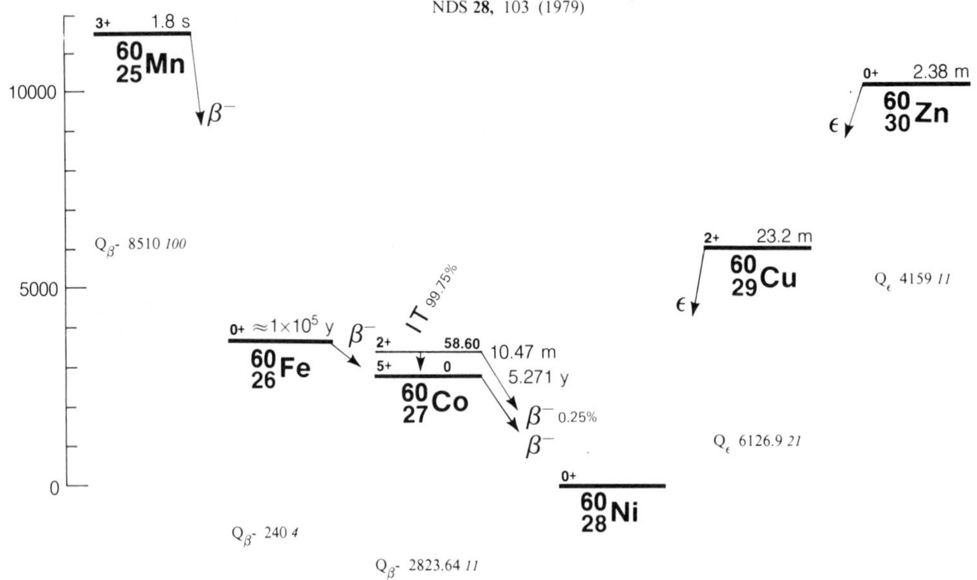

$^{60}_{25}$Mn(1.79 *10* s)

Mode: β-
Δ: -52900 *100* keV
SpA: 8.7×10^{10} Ci/g
Prod: ^{48}Ca(^{18}O,αpn)

Photons (^{60}Mn)

⟨γ⟩=2690 *78* keV

γ_{mode}	γ(keV)	γ(%)†
γ [M1+E2]	493.4 *3*	20.7 *11*
γ [M1+E2]	678.6 *5*	2.5 *6*
γ E2	824.1 *3*	85
γ E2	1290.6 *5*	11.8 *11*
γ [M1+E2]	1475.9 *4*	12.4 *12*
γ [M1+E2]	1969.3 *4*	60 *3*
γ [E2]	2299.9 *5*	15.0 *17*

† 2.3% uncert(syst)

Continuous Radiation (^{60}Mn)

⟨β-⟩=2653 keV;⟨IB⟩=10.6 keV

E_{bin}(keV)		⟨ ⟩(keV)	(%)
0 - 10	β-	0.00107	0.0212
	IB	0.068	
10 - 20	β-	0.00331	0.0220
	IB	0.068	0.47
20 - 40	β-	0.0141	0.0469
	IB	0.134	0.47
40 - 100	β-	0.120	0.168
	IB	0.39	0.60
100 - 300	β-	1.86	0.89
	IB	1.22	0.68
300 - 600	β-	11.0	2.37
	IB	1.59	0.37
600 - 1300	β-	106	10.7
	IB	2.8	0.32
1300 - 2500	β-	612	31.6

Continuous Radiation (^{60}Mn)
(continued)

E_{bin}(keV)		⟨ ⟩(keV)	(%)
	IB	2.9	0.163
2500 - 5000	β-	1824	52
	IB	1.43	0.046
5000 - 6396	β-	98	1.87
	IB	0.0073	0.000136

$^{60}_{26}$Fe(~1×10^5 yr)

Mode: β-
Δ: -61407 *4* keV
SpA: ~0.06 Ci/g
Prod: protons on Cu

Photons (^{60}Fe)

⟨γ⟩=1.17 *6* keV

γ_{mode}	γ(keV)	γ(%)
γ M3(+E4)	58.603 *7*	2.0 *1* ? *

* with ^{60}Co(10.47 min) in equilib

Continuous Radiation (^{60}Fe)

⟨β-⟩=51 keV;⟨IB⟩=0.0096 keV

E_{bin}(keV)		⟨ ⟩(keV)	(%)
0 - 10	β-	0.64	12.9
	IB	0.0026	
10 - 20	β-	1.79	12.0
	IB	0.0019	0.0137
20 - 40	β-	6.3	21.2
	IB	0.0025	0.0091
40 - 100	β-	27.4	41.5
	IB	0.0024	0.0043
100 - 181	β-	15.0	12.4
	IB	0.000169	0.000149

$^{60}_{27}$Co(5.271 *1* yr)

Mode: β-
Δ: -61647.1 *15* keV
SpA: 1130.36 Ci/g
Prod: ^{59}Co(n,γ)

Photons (^{60}Co)

⟨γ⟩=2504.4 *2* keV

γ_{mode}	γ(keV)	γ(%)
γ E2	346.95 *4*	0.0076 *5*
γ M1+40%E2	826.33 *4*	0.0075 *7*
γ E2	1173.237 *4*	99.90 *2*
γ E2	1332.501 *5*	99.9824 *5*
γ E2	2158.86 *4*	0.00114 *11*
γ E4	2505.738 *6*	2.0 *4* ×10^{-6} ?

Continuous Radiation (^{60}Co)

⟨β-⟩=96 keV;⟨IB⟩=0.033 keV

E_{bin}(keV)		⟨ ⟩(keV)	(%)
0 - 10	β-	0.324	6.5
	IB	0.0050	
10 - 20	β-	0.95	6.3
	IB	0.0043	0.030
20 - 40	β-	3.65	12.2
	IB	0.0068	0.024
40 - 100	β-	22.3	32.3
	IB	0.0113	0.019
100 - 300	β-	68	42.6
	IB	0.0052	0.0039
300 - 600	β-	0.220	0.061
	IB	0.00023	5.5×10^{-5}
600 - 1300	β-	0.355	0.0396
	IB	0.000128	1.64×10^{-5}
1300 - 1491	β-	0.0206	0.00152
	IB	3.5×10^{-7}	2.6×10^{-8}

$^{60}_{27}$Co(10.47 *4* min)

Mode: IT(99.75 *3* %), β-(0.25 *3* %)
Δ: -61588.5 *15* keV
SpA: 2.991×10^{8} Ci/g
Prod: ^{59}Co(n,γ)

Photons (^{60}Co)

⟨γ⟩=6.6 *4* keV

γ_{mode}	γ(keV)	γ(%)†
Co L_ℓ	0.678	0.039 *11*
Co L_η	0.693	0.025 *6*
Co L_α	0.776	0.42 *11*
Co L_β	0.799	0.31 *8*
Co $K_{\alpha2}$	6.915	8.8 *5*
Co $K_{\alpha1}$	6.930	17.4 *9*
Co $K_{\beta1}$'	7.649	3.16 *17*
γ_{IT} M3(+E4)	58.603 *7*	2.01
$\gamma_{\beta-}$M1+40%E2	826.33 *4*	0.0058 *7*
$\gamma_{\beta-}$E2	1332.501 *5*	0.25
$\gamma_{\beta-}$E2	2158.86 *4*	0.00088 *11*

† uncert(syst): 3.0% for IT, 12% for β-

Atomic Electrons (^{60}Co)*

⟨e⟩=52.1 *12* keV

e_{bin}(keV)	⟨e⟩(keV)	e(%)
1	0.61	77 *8*
6	2.40	39 *4*
7	0.65	9.6 *10*
8	0.042	0.56 *6*
51	40.2	79.0 *22*
58	8.14	14.1 *4*

* with IT

Continuous Radiation (^{60}Co)

⟨β-⟩=1.48 keV;⟨IB⟩=0.0023 keV

E_{bin}(keV)		⟨ ⟩(keV)	(%)
0 - 10	β-	6.5×10^{-5}	0.00130
	IB	6.3×10^{-5}	
10 - 20	β-	0.000200	0.00133
	IB	6.2×10^{-5}	0.00043
20 - 40	β-	0.00084	0.00278
	IB	0.000118	0.00041
40 - 100	β-	0.0066	0.0094
	IB	0.00032	0.00050
100 - 300	β-	0.084	0.0410
	IB	0.00078	0.00045
300 - 600	β-	0.337	0.075
	IB	0.00063	0.000152
600 - 1300	β-	0.99	0.114
	IB	0.00031	4.1×10^{-5}
1300 - 1550	β-	0.055	0.00406
	IB	1.28×10^{-6}	9.5×10^{-8}

$^{60}_{28}$Ni(stable)

Δ: -64470.7 *15* keV
%: 26.10 *1*

$^{60}_{29}$Cu(23.2 *3* min)

Mode: ϵ
Δ: -58344 *3* keV
SpA: 1.350×10^{8} Ci/g
Prod: ^{60}Ni(p,n); ^{60}Ni(d,2n)

Photons (^{60}Cu)

⟨γ⟩=2962 *46* keV

γ_{mode}	γ(keV)	γ(%)†
Ni L_ℓ	0.743	0.0037 *10*
Ni L_η	0.760	0.0021 *5*
Ni L_α	0.851	0.046 *12*
Ni L_β	0.876	0.032 *8*
Ni $K_{\alpha2}$	7.461	0.91 *5*
Ni $K_{\alpha1}$	7.478	1.80 *9*
Ni $K_{\beta1}$'	8.265	0.326 *16*
γ [M1+E2]	120.38 *11*	0.194 *18*
γ E2	346.95 *4*	2.0 *7* $\times10^{-5}$
γ M1+E2	467.33 *11*	3.52 *18*
γ [M1+E2]	497.87 *14*	1.67 *9*
γ M1(+0.3%E2)	643.20 *17*	0.97 *5*
γ M1+0.8%E2	680.5 *4*	~0.035 ?
γ M1+11%E2	739.5 *5*	0.08 *3*
γ [M1+E2]	747.9 *7*	0.057 *25*
γ M1+40%E2	826.33 *4*	21.9 *10*
γ [E2]	839.16 *17*	0.46 *7*
γ M1,E2	896.35 *23*	0.13 *5*
γ [M1]	909.16 *15*	2.02 *9*
γ [E2]	952.36 *14*	2.73 *18*
γ [M1+E2]	965.19 *13*	0.30 *6*
γ E2	984.49 *20*	~0.08
γ [M1+E2]	994.0 *10*	~0.04 ?
γ M1+15%E2	1027.5 *4*	~0.09 ?
γ [M1+E2]	1035.20 *13*	3.70 *18*
γ M1,E2	1110.52 *16*	1.06 *18*
γ E2	1173.237 *4*	0.26 *9*
γ M1,E2	1234.6 *3*	0.11 *4*
γ E2+19%M1	1293.65 *11*	1.85 *18*
γ [M1+E2]	1307.1 *4*	0.11 *3*
γ E2	1332.501 *5*	88.0
γ M1,E2	1420.0 *3*	0.114 *18*
γ	1424.7 *3*	0.070 *18*
γ M1	1451.1 *4*	0.17 *3*

Photons (^{60}Cu)
(continued)

γ_{mode}	γ(keV)	γ(%)†
γ	1579.6 *6*	0.09 *4*
γ [M1+E2]	1713.1 *7*	<0.04 ?
γ M1	1735.49 *24*	0.062 *18*
γ M1,E2	1767.0 *3*	0.10 *4*
γ [M1+E2]	1791.51 *13*	45.4 *23*
γ [M1+E2]	1861.51 *13*	4.8 *3*
γ M1,E2	1919.66 *21*	0.70 *7*
γ M1+32%E2	1936.84 *16*	2.20 *9*
γ M1,E2	2060.9 *3*	0.79 *4*
γ E2	2158.86 *4*	3.33 *17*
γ	2175.8 *10*	0.052 *15*
γ	2263.8 *4*	0.11 *4*
γ [M1+E2]	2334.6 *3*	~0.035
γ	2376.8 *7*	0.062 *13*
γ	2389.9 *3*	0.12 *4*
γ M1,E2	2403.5 *4*	0.77 *8*
γ E4	2505.738 *6*	5.2 *10* $\times10^{-7}$?
γ [M1+E2]	2539.4 *7*	0.026 *12*
γ [M1+E2]	2675.3 *7*	0.13 *3*
γ M1,E2	2687.83 *21*	0.44 *7*
γ M1+29%E2	2745.96 *21*	1.06 *9*
γ	2889.5 *7*	0.020 *8*
γ [M1+E2]	2986.4 *5*	0.123 *18*
γ [E2]	3124.01 *13*	4.8 *3*
γ [M1+E2]	3160.9 *3*	0.58 *3*
γ [M1]	3194.01 *13*	2.02 *9*
γ	3203.1 *7*	0.033 *11* ?
γ	3216.2 *3*	~0.035
γ [M1+E2]	3246.4 *7*	0.026 *9*
γ E2	3269.33 *16*	0.77 *4*
γ E2	3393.4 *3*	0.053 *18*
γ	3427.8 *7*	0.026 *9*
γ	3511.5 *12*	~0.018
γ	3516.4 *6*	~0.018
γ M1	3736.0 *4*	0.026 *9*
γ [E2]	3871.9 *7*	0.011 *4*
γ [E2]	4007.7 *7*	0.08 *3*
γ M1	4020.30 *21*	0.77 *8*
γ E2	4078.44 *21*	0.062 *18*
γ [E2]	4318.9 *5*	0.044 *9*
γ	4334.6 *10*	0.0123 *18*
γ [E2]	4493.3 *3*	0.040 *8*
γ	4548.6 *3*	0.040 *8*
γ [E2]	4578.9 *7*	0.022 *6*
γ	4760.2 *7*	~0.008
γ	4844.0 *12*	0.009 *3*
γ	5048.3 *7*	~0.0018

† 1.1% uncert(syst)

Atomic Electrons (^{60}Cu)

⟨e⟩=0.73 *4* keV

e_{bin}(keV)	⟨e⟩(keV)	e(%)
1	0.059	6.8 *7*
6	0.080	1.24 *13*
7	0.215	3.2 *3*
8 - 47	0.0042	0.051 *5*
112	0.030	~0.027
119 - 120	0.0035	~0.0030
339 - 346	3.8×10^{-7}	1.1 *5* $\times10^{-7}$
459	0.022	0.0049 *17*
466 - 497	0.013	0.0026 *7*
635 - 680	0.00316	0.00050 *3*
731 - 747	0.00041	5.5 *14* $\times10^{-5}$
818	0.054	0.0066 *4*
825 - 838	0.0067	0.00082 *4*
888 - 908	0.0047	0.00052 *3*
944 - 993	0.0079	0.00083 *6*
1019 - 1034	0.0079	0.00077 *8*
1102 - 1110	0.0020	0.00018 *3*
1165 - 1172	0.00051	4.4 *13* $\times10^{-5}$
1226 - 1234	0.00018	1.5 *6* $\times10^{-5}$
1285 - 1306	0.0033	0.000254 *22*
1324	0.133	0.01003 *23*
1412 - 1450	0.00048	3.4 *4* $\times10^{-5}$
1571 - 1579	9.3×10^{-5}	6 *3* $\times10^{-6}$

Atomic Electrons (^{60}Cu) (continued)

e_{bin}(keV)	⟨e⟩(keV)	e(%)
1705 - 1766	0.00022	1.3 *3* ×10^{-5}
1783	0.050	0.00278 *22*
1790 - 1861	0.0103	0.00057 *3*
1911 - 1936	0.00324	0.000168 *7*
2053 - 2151	0.00405	0.000190 *9*
2158 - 2255	0.00044	2.02 *18* ×10^{-5}
2263 - 2334	4.3 ×10^{-5}	1.8 *7* ×10^{-6}
2368 - 2403	0.00089	3.7 *4* ×10^{-5}
2497 - 2539	2.4 ×10^{-5}	9 *4* ×10^{-7}
2667 - 2745	0.00143	5.2 *4* ×10^{-5}
2881 - 2978	0.000108	3.6 *5* ×10^{-6}
2985 - 2986	9.1 ×10^{-6}	3.0 *5* ×10^{-7}
3116 - 3215	0.0060	0.000192 *8*
3238 - 3268	0.00064	1.97 *11* ×10^{-5}
3385 - 3427	5.9 ×10^{-5}	1.7 *4* ×10^{-6}
3503 - 3516	2.3 ×10^{-5}	6.5 *23* ×10^{-7}
3728 - 3735	1.9 ×10^{-5}	5.1 *16* ×10^{-7}
3864 - 3871	8.4 ×10^{-6}	2.2 *8* ×10^{-7}
3999 - 4078	0.00063	1.57 *14* ×10^{-5}
4311 - 4334	3.8 ×10^{-5}	8.8 *14* ×10^{-7}
4485 - 4578	6.5 ×10^{-5}	1.44 *19* ×10^{-6}
4752 - 4843	9.6 ×10^{-6}	2.0 *7* ×10^{-7}

Continuous Radiation (^{60}Cu)

⟨β+⟩=894 keV;⟨IB⟩=2.5 keV

E_{bin}(keV)		⟨ ⟩(keV)	(%)
0 - 10	β+	7.9 ×10^{-5}	0.00100
	IB	0.034	
10 - 20	β+	0.00144	0.0089
	IB	0.033	0.23
20 - 40	β+	0.0197	0.062
	IB	0.065	0.22
40 - 100	β+	0.461	0.61
	IB	0.18	0.28
100 - 300	β+	13.2	6.2
	IB	0.50	0.28
300 - 600	β+	79	17.2
	IB	0.53	0.125
600 - 1300	β+	427	45.7
	IB	0.68	0.079
1300 - 2500	β+	342	20.8
	IB	0.42	0.025
2500 - 4794	β+	32.9	1.18
	IB	0.098	0.0032
	Σβ+		92

$^{60}_{30}$Zn(2.38 *5* min)

Mode: ε

Δ: -54185 *11* keV

SpA: 1.31×10^{9} Ci/g

Prod: ^{58}Ni(α,2n); ^{58}Ni(^{3}He,n)

Photons (^{60}Zn)

⟨γ⟩=398 *21* keV

γ_{mode}	γ(keV)	γ(%)†
Cu L_ℓ	0.811	0.016 *5*
Cu L_η	0.831	0.009 *3*
Cu L_α	0.930	0.22 *6*
Cu L_β	0.957	0.14 *4*
Cu $K_{\alpha2}$	8.028	3.9 *5*
Cu $K_{\alpha1}$	8.048	7.6 *10*
Cu $K_{\beta1}$'	8.905	1.37 *18*
γ (M1+41%E2)	62.0 *9*	18.8 *23*
γ	273.0 *9*	8.0 *9*
γ	334.0 *9*	6.6 *9*
γ [M1+E2]	365 *1*	2.3 *5*
γ	572.4 *3*	1.9 *5*
γ [M1+E2]	669.0 *9*	47
γ	947 *1*	0.75 *19* ?

† 6.4% uncert(syst)

Atomic Electrons (^{60}Zn)

⟨e⟩=18.6 *19* keV

e_{bin}(keV)	⟨e⟩(keV)	e(%)
1	0.26	28 *4*
7	0.91	12.9 *20*
8	0.25	3.2 *5*
9	0.017	0.19 *3*
53	14.4	27 *4*
61	2.2	3.7 *5*
264	0.16	~0.06
272	0.017	~0.006
325	0.08	~0.025
333	0.009	~0.0026
356	0.027	0.008 *4*
364	0.0028	0.0008 *4*
563	0.008	~0.0014
571	0.0008	~0.00014
660	0.18	0.027 *6*
668	0.018	0.0027 *6*
938	0.0014	0.00015 *7*
946	0.00014	1.5 *7* ×10^{-5}

Continuous Radiation (^{60}Zn)

⟨β+⟩=1202 keV;⟨IB⟩=3.2 keV

E_{bin}(keV)		⟨ ⟩(keV)	(%)
0 - 10	β+	3.47 ×10^{-5}	0.000436
	IB	0.043	
10 - 20	β+	0.00065	0.00406
	IB	0.042	0.29
20 - 40	β+	0.0093	0.0293
	IB	0.082	0.29
40 - 100	β+	0.230	0.305
	IB	0.23	0.36
100 - 300	β+	7.3	3.36
	IB	0.67	0.37
300 - 600	β+	50.6	10.9
	IB	0.74	0.18
600 - 1300	β+	389	40.3
	IB	0.97	0.114
1300 - 2500	β+	715	41.4
	IB	0.42	0.026
2500 - 4097	β+	40.4	1.53
	IB	0.038	0.00137
	Σβ+		98

A = 61

NDS 38, 463 (1983)

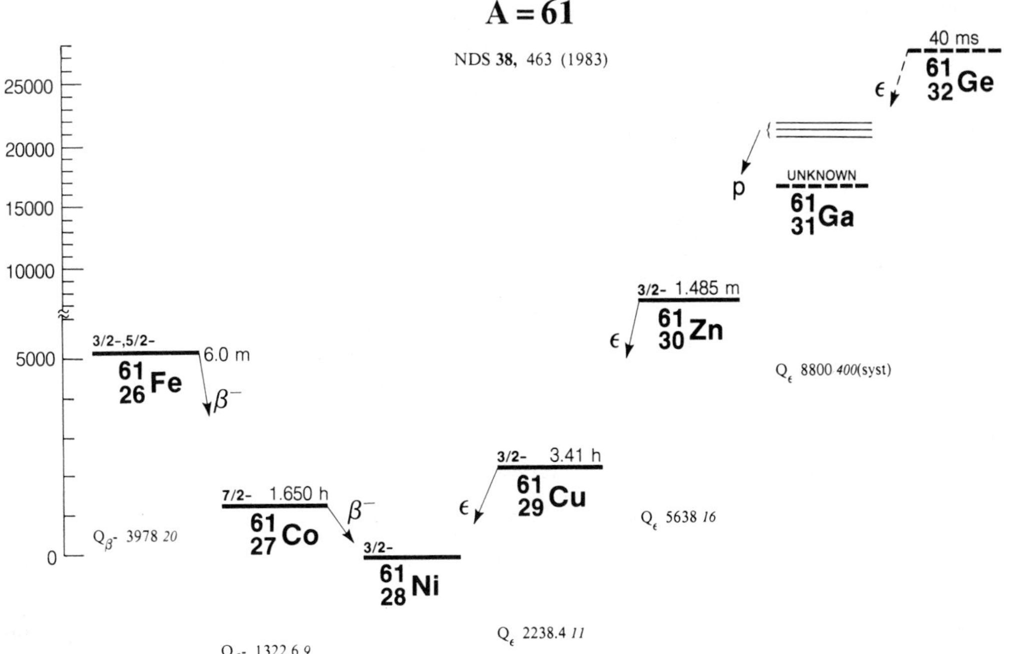

$^{61}_{26}$Fe(5.98 *6* min)

Mode: β-

Δ: -58919 *20* keV

SpA: 5.15×10^{8} Ci/g

Prod: ^{64}Ni(n,α); ^{64}Ni(d,αp)

Photons (^{61}Fe)

⟨γ⟩=1391 *58* keV

γ_{mode}	γ(keV)	γ(%)†
γ	120.34 *8*	5.3 *4*
γ	177.64 *7*	2.01 *17*
γ	297.98 *6*	22.2 *17*
γ	333.20 *24*	0.22 *4*
γ	349.82 *18*	0.16 *4*
γ	440.81 *12*	0.22 *4*
γ	542.6 *4*	0.07 *3* ?
γ	561.39 *20*	0.06 *3* ?
γ	603.3 *3*	0.07 *3*
γ	618.45 *11*	0.93 *7*
γ	657.15 *17*	0.22 *10*
γ	686.08 *15*	0.40 *8*
γ	697.31 *23*	0.11 *3*
γ	748.14 *14*	0.81 *7*
γ	769.4 *3*	0.16 *4*
γ	806.42 *15*	0.19 *5*
γ	925.78 *15*	0.34 *4*
γ	945.3 *3*	0.11 *3*
γ	977.62 *16*	~0.07 ?
γ	984.06 *15*	0.61 *13*
γ	988.84 *22*	0.61 *13*
γ [E2]	1027.51 *8*	42.7 *22*
γ	1097.96 *14*	0.70 *6*
γ	1205.15 *8*	44
γ	1275.59 *15*	0.61 *13*
γ	1285.74 *21*	0.37 *6*
γ	1381.47 *21*	0.40 *5*
γ	1404.1 *4*	0.12 *5*
γ	1539.00 *16*	0.27 *5*
γ [M1+E2]	1618.93 *17*	0.37 *4*
γ	1645.95 *11*	7.0 *4*
γ	1659.34 *16*	0.78 *9*
γ	1836.97 *16*	0.14 *3*

Photons (^{61}Fe) (continued)

γ_{mode}	γ(keV)	γ(%)†
γ	1879.3 *3*	0.26 *4*
γ	1889.0 *3*	0.18 *4*
γ	1899.4 *5*	0.074 *22*
γ	1972.88 *20*	0.061 *17*
γ	1999.6 *3*	0.13 *4*
γ	2011.56 *14*	4.4 *3*
γ	2177.3 *3*	0.21 *4*
γ	2231.0 *3*	0.109 *17*
γ	2484.5 *4*	0.122 *13*
γ	2754.5 *4*	0.77 *9*
γ	2920.1 *5*	0.070 *13*
γ	3191.1 *6*	0.083 *13*
γ	3204.7 *3*	0.044 *9*
γ	3239.2 *6*	0.057 *13*
γ	3365.0 *7*	0.044 *9*

† 10% uncert(syst)

Continuous Radiation (^{61}Fe)

⟨β-⟩=1093 keV;⟨IB⟩=2.6 keV

E_{bin}(keV)		⟨ ⟩(keV)	(%)
0 - 10	β-	0.0098	0.194
	IB	0.039	
10 - 20	β-	0.0301	0.20
	IB	0.039	0.27
20 - 40	β-	0.127	0.421
	IB	0.076	0.26
40 - 100	β-	1.03	1.45
	IB	0.22	0.33
100 - 300	β-	14.0	6.8
	IB	0.60	0.34
300 - 600	β-	66	14.4
	IB	0.65	0.154
600 - 1300	β-	382	40.1
	IB	0.78	0.092
1300 - 2500	β-	617	35.9
	IB	0.22	0.0137
2500 - 2951	β-	13.3	0.51
	IB	0.00037	1.45×10^{-5}

$^{61}_{27}$Co(1.650 *5* h)

Mode: β-

Δ: -62897.1 *17* keV

SpA: 3.113×10^{7} Ci/g

Prod: ^{64}Ni(p,α); ^{64}Ni(d,αn); ^{61}Ni(n,p); ^{59}Co(t,p); ^{58}Fe(α,p)

Photons (^{61}Co)

⟨γ⟩=98.4 *21* keV

γ_{mode}	γ(keV)	γ(%)†
Ni L_ℓ	0.743	0.0051 *13*
Ni L_η	0.760	0.0029 *7*
Ni L_α	0.851	0.063 *16*
Ni L_β	0.876	0.044 *11*
Ni $K_{\alpha2}$	7.461	1.27 *6*
Ni $K_{\alpha1}$	7.478	2.50 *12*
Ni $K_{\beta1}$'	8.265	0.455 *22*
γ M1	67.415 *3*	84.7
γ [E2]	625.670 *12*	0.153 *15*
γ E2+23%M1	841.213 *12*	0.76 *5*
γ M1+3.3%E2	908.626 *12*	3.70 *23*

† 0.45% uncert(syst)

Atomic Electrons (^{61}Co)

⟨e⟩=7.39 *13* keV

e_{bin}(keV)	⟨e⟩(keV)	e(%)
1	0.082	9.5 *10*
6	0.111	1.73 *18*
7	0.30	4.4 *5*
8	0.0058	0.070 *7*
59	6.16	10.42 *21*
66	0.681	1.026 *21*
67	0.0410	0.0617 *13*
617	0.0007	~0.00012

Atomic Electrons (^{61}Co) (continued)

e_{bin}(keV)	⟨e⟩(keV)	e(%)
625	7.2 $\times10^{-5}$	~1 $\times10^{-5}$
833	0.00202	0.000243 *16*
840	0.000201	2.39 *16* $\times10^{-5}$
900	0.0073	0.00082 *5*
908	0.00073	8.0 *5* $\times10^{-5}$

Continuous Radiation (^{61}Co)

⟨β-⟩=459 keV;⟨IB⟩=0.59 keV

E_{bin}(keV)		⟨ ⟩(keV)	(%)
0 - 10	β-	0.0447	0.89
	IB	0.021	
10 - 20	β-	0.136	0.90
	IB	0.020	0.139
20 - 40	β-	0.56	1.86
	IB	0.038	0.132
40 - 100	β-	4.28	6.1
	IB	0.101	0.156
100 - 300	β-	47.3	23.4
	IB	0.22	0.131
300 - 600	β-	159	35.4
	IB	0.149	0.037
600 - 1255	β-	249	31.5
	IB	0.040	0.0057

$^{61}_{28}$Ni(stable)

Δ: -64219.6 *15* keV

%: 1.13 *1*

$^{61}_{29}$Cu(3.408 *10* h)

Mode: ε

Δ: -61981.2 *18* keV

SpA: 1.507$\times10^{7}$ Ci/g

Prod: ^{60}Ni(d,n); ^{63}Cu(γ,2n); ^{59}Co(α,2n)

Photons (^{61}Cu)

⟨γ⟩=199 *7* keV

γ_{mode}	γ(keV)	γ(%)†
Ni L_ℓ	0.743	0.017 *5*
Ni L_η	0.760	0.0099 *25*
Ni L_α	0.851	0.22 *6*
Ni L_β	0.876	0.15 *4*
Ni $K_{\alpha2}$	7.461	4.31 *21*
Ni $K_{\alpha1}$	7.478	8.5 *4*
Ni $K_{\beta1}$'	8.265	1.54 *8*
γ M1	67.415 *3*	3.94 *13*
γ [M1+E2]	190.956 *24*	0.005 *1*
γ [E2]	215.545 *3*	0.0045 *11*
γ [M1+E2]	276.619 *16*	0.0262 *25*
γ M1+0.2%E2	282.9593 *19*	12.5
γ [M1]	373.055 *3*	2.15 *6*
γ [M1+E2]	443.572 *22*	0.0038 *12*
γ [M1+E2]	529.235 *11*	0.41 *8*
γ [M1+E2]	544.226 *16*	
γ [E2]	588.598 *4*	1.200 *24*
γ [E2]	625.670 *12*	0.050 *4*
γ [M1+E2]	629.888 *25*	0.0063 *25*
γ E2+40%M1	656.012 *3*	10.66 *25*
γ [M1+E2]	701.006 *24*	0.0113 *25*

Photons (^{61}Cu) (continued)

γ_{mode}	γ(keV)	γ(%)†
γ M1+5.4%E2	816.624 *22*	0.361 *8*
γ [M1+E2]	820.842 *17*	0.0225 *25*
γ E2+23%M1	841.213 *12*	0.249 *10*
γ [M1+E2]	902.286 *11*	0.0900 *25*
γ M1+3.3%E2	908.626 *12*	1.21 *4*
γ E2+14%M1	947.4 *4*	~0.0025 ?
γ E2	1014.8 *4*	<0.0012 ?
γ [M1+E2]	1032.166 *21*	0.066 *11*
γ M1+3.2%E2	1064.912 *17*	0.060 *5*
γ [M1+E2]	1073.456 *12*	0.054 *4*
γ [M1+E2]	1089.1 *7*	<0.0006 ?
γ [M1+E2]	1099.579 *22*	0.284 *10*
γ [M1+E2]	1117.827 *11*	0.049 *4*
γ E2+0.01%M1	1132.325 *17*	0.099 *4*
γ M1+2.6%E2	1185.241 *11*	3.69 *8*
γ [M1+E2]	1446.503 *12*	0.0500 *25*
γ M1+0.7%E2	1542.208 *21*	0.0337 *13*
γ M1+11%E2	1609.621 *21*	0.0275 *25*
γ [M1+E2]	1662.043 *12*	0.0525 *25*
γ [M1+E2]	1729.455 *12*	0.080 *5*
γ [M1]	1840.49 *4*	0.0018 *5*
γ M1+7.0%E2	1997.7 *7*	0.0038 *12*
γ [M1+E2]	2123.44 *4*	0.0400 *12*

† 18% uncert(syst)

Atomic Electrons (^{61}Cu)

⟨e⟩=2.20 *12* keV

e_{bin}(keV)	⟨e⟩(keV)	e(%)
1	0.28	32 *3*
6	0.38	5.8 *6*
7	1.01	14.9 *15*
8	0.0195	0.238 *24*
59	0.286	0.485 *18*
66 - 67	0.0336	0.0506 *18*
183 - 215	0.00055	0.00028 *10*
268	0.0005	~0.00019
275	0.102	0.037 *7*
276 - 282	0.0106	0.0038 *7*
365 - 372	0.0133	0.00364 *16*
435 - 443	2.9 $\times10^{-5}$	7 *3* $\times10^{-6}$
521 - 528	0.0023	0.00044 *14*
580 - 629	0.0072	0.00124 *5*
648	0.039	0.0061 *7*
655 - 700	0.0040	0.00061 *7*
808 - 840	0.00168	0.000205 *5*
894 - 939	0.00286	0.000318 *10*
946 - 947	5.6 $\times10^{-7}$	6 *3* $\times10^{-8}$
1006 - 1031	0.000140	1.36 *25* $\times10^{-5}$
1057 - 1099	0.00077	7.1 *5* $\times10^{-5}$
1109 - 1131	0.000292	2.60 *12* $\times10^{-5}$
1177 - 1184	0.00615	0.000523 *13*
1438 - 1534	0.000113	7.7 *4* $\times10^{-6}$
1541 - 1609	3.9 $\times10^{-5}$	2.45 *19* $\times10^{-6}$
1654 - 1729	0.000166	9.8 *6* $\times10^{-6}$
1832 - 1840	2.0 $\times10^{-6}$	1.1 *3* $\times10^{-7}$
1989 - 1997	4.0 $\times10^{-6}$	2.0 *6* $\times10^{-7}$
2115 - 2123	4.2 $\times10^{-5}$	1.98 *12* $\times10^{-6}$

Continuous Radiation (^{61}Cu)

⟨β+⟩=306 keV;⟨IB⟩=1.09 keV

E_{bin}(keV)		⟨ ⟩(keV)	(%)
0 - 10	β+	0.000209	0.00264
	IB	0.0148	
10 - 20	β+	0.00375	0.0234
	IB	0.0135	0.093
20 - 40	β+	0.0507	0.160
	IB	0.026	0.089
40 - 100	β+	1.14	1.52
	IB	0.069	0.106

Continuous Radiation (^{61}Cu) (continued)

E_{bin}(keV)		⟨ ⟩(keV)	(%)
100 - 300	β+	27.6	13.1
	IB	0.169	0.096
300 - 600	β+	115	25.7
	IB	0.19	0.045
600 - 1300	β+	162	20.9
	IB	0.41	0.046
1300 - 2238	IB	0.19	0.0124
	Σβ+		61

$^{61}_{30}$Zn(1.485 *3* min)

Mode: ε

Δ: -56343 *16* keV

SpA: 2.068$\times10^{9}$ Ci/g

Prod: ^{58}Ni(α,n)

Photons (^{61}Zn)

⟨γ⟩=524 *6* keV

γ_{mode}	γ(keV)	γ(%)†
Cu L_ℓ	0.811	0.00075 *20*
Cu L_η	0.831	0.00044 *11*
Cu L_α	0.930	0.010 *3*
Cu L_β	0.957	0.0068 *18*
Cu $K_{\alpha2}$	8.028	0.190 *10*
Cu $K_{\alpha1}$	8.048	0.371 *20*
Cu $K_{\beta1}$'	8.905	0.067 *4*
γ	148.9 *5*	0.17 *7*
γ [M1+E2]	265.78 *8*	0.55 *3*
γ M1+1.1%E2	421.4 *13*	<0.09
γ [M1+E2]	425.37 *13*	0.148 *8*
γ M1+E2	474.83 *8*	16.8 *3*
γ M1+0.3%E2	593.4 *3*	~0.06
γ	604.4 *18*	<0.09
γ	638.1 *9*	<0.08
γ M1+3.7%E2	690.21 *10*	1.87 *14*
γ [M1+E2]	697.51 *10*	0.43 *4*
γ	751.46 *11*	0.31 *4*
γ E2	919.62 *10*	0.094 *16*
γ M1+2.1%E2	934.2 *3*	0.09 *4*
γ M1+11%E2	970.02 *11*	2.57 *12*
γ [M1+E2]	1132.52 *16*	~0.18
γ	1147.2 *5*	0.156 *23*
γ M1+E2	1185.40 *10*	1.72 *11*
γ E2	1310.82 *10*	0.936 *19*
γ E2+7.4%M1	1394.44 *8*	1.22 *3*
γ M1+4.6%E2	1457.54 *8*	0.31 *9*
γ [M1+E2]	1481.92 *10*	0.788 *23*
γ [M1+E2]	1502.22 *10*	0.140 *16*
γ [M1+E2]	1538.1 *3*	0.085 *8*
γ	1565.5 *25*	<0.11
γ [M1]	1613.92 *13*	0.296 *23*
γ M1+16%E2	1660.22 *8*	7.80 *18*
γ E2	1732.2 *13*	<0.14
γ [M1+E2]	1882.90 *13*	0.480 *10*
γ M1+32%E2	1904.3 *3*	0.091 *12*
γ M1+7.9%E2	1932.35 *8*	0.66 *4*
γ M1+0.3%E2	1997.40 *14*	1.18 *3*
γ [M1+E2]	2088.74 *11*	0.628 *13*
γ [M1+E2]	2208.8 *2*	0.84 *5*
γ [M1+E2]	2357.71 *12*	0.328 *23*
γ	2382.0 *3*	0.109 *16*
γ [M1+E2]	2457.7 *3*	0.66 *4*
γ [M1+E2]	2472.21 *13*	0.08 *3*
γ M1+27%E2	2544.3 *11*	0.076 *14*
γ [M1+E2]	2683.62 *22*	0.68 *5*
γ [M1+E2]	2792.71 *14*	0.803 *23*
γ	2840.17 *15*	0.248 *5*
γ	2856.8 *3*	0.43 *6*

Photons (^{61}Zn)
(continued)

γ_{mode}	γ(keV)	γ(%)†
γ [M1+E2]	2932.5 *3*	0.094 *23*
γ [M1+E2]	3019.1 *11*	0.187 *23*
γ [M1+E2]	3092.2 *14*	0.117 *23*
γ	3521.0 *16*	0.14 *3*

† 5.1% uncert(syst)

Atomic Electrons (^{61}Zn)
⟨e⟩=0.27 *4* keV

e_{bin}(keV)	⟨e⟩(keV)	e(%)
1	0.0123	1.29 *13*
7	0.044	0.63 *7*
8	0.0125	0.158 *16*
9	0.00082	0.0092 *10*
140	0.016	~0.012
148	0.0019	~0.0013
257	0.013	~0.005
412 - 424	0.0017	0.00040 *13*
466	0.11	0.025 *8*
474	0.012	0.0025 *8*
584 - 629	0.0006	9 *4* $\times10^{-5}$

Atomic Electrons (^{61}Zn)
(continued)

e_{bin}(keV)	⟨e⟩(keV)	e(%)
637	1.4 $\times10^{-5}$	<4 $\times10^{-6}$
681	0.0056	0.00082 *7*
689 - 697	0.0022	0.00033 *5*
742 - 751	0.0009	0.00012 *5*
911 - 933	0.00048	5.2 *10* $\times10^{-5}$
961 - 969	0.0060	0.00062 *3*
1124 - 1146	0.00062	5.5 *20* $\times10^{-5}$
1176 - 1184	0.0034	0.000291 *24*
1302 - 1310	0.00175	0.000134 *4*
1385 - 1394	0.00211	0.000152 *4*
1449 - 1537	0.00208	0.000141 *11*
1556 - 1613	0.00048	3.0 *4* $\times10^{-5}$
1651	0.0099	0.000598 *17*
1659 - 1731	0.00107	6.4 *6* $\times10^{-5}$
1874 - 1931	0.00154	8.1 *3* $\times10^{-5}$
1988 - 2088	0.00214	0.000106 *3*
2200 - 2208	0.00095	4.3 *3* $\times10^{-5}$
2349 - 2381	0.00045	1.90 *17* $\times10^{-5}$
2449 - 2543	0.00084	3.42 *25* $\times10^{-5}$
2675 - 2683	0.00067	2.50 *21* $\times10^{-5}$
2784 - 2856	0.00131	4.6 *4* $\times10^{-5}$
2924 - 3018	0.00026	8.7 *10* $\times10^{-6}$
3083 - 3091	0.000105	3.4 *7* $\times10^{-6}$
3512 - 3520	0.00010	2.9 *8* $\times10^{-6}$

Continuous Radiation (^{61}Zn)
⟨β+⟩=1857 keV;⟨IB⟩=6.5 keV

E_{bin}(keV)		⟨ ⟩(keV)	(%)
0 - 10	β+	1.78 $\times10^{-5}$	0.000224
	IB	0.055	
10 - 20	β+	0.000336	0.00208
	IB	0.055	0.38
20 - 40	β+	0.00479	0.0150
	IB	0.108	0.37
40 - 100	β+	0.118	0.157
	IB	0.31	0.48
100 - 300	β+	3.77	1.74
	IB	0.94	0.52
300 - 600	β+	26.4	5.7
	IB	1.16	0.27
600 - 1300	β+	219	22.5
	IB	1.9	0.22
1300 - 2500	β+	784	41.8
	IB	1.55	0.090
2500 - 5000	β+	823	26.5
	IB	0.46	0.0148
5000 - 5638	IB	0.0023	4.5 $\times10^{-5}$
	Σβ+		98

$^{61}_{32}$Ge(40 *15* ms) ?

Mode: ϵ, ϵp
SpA: 2.7$\times10^{11}$ Ci/g
Prod: ^{24}Mg on Ca

ϵp: 3098 *69*

A = 62

NDS **26**, 5 (1979)

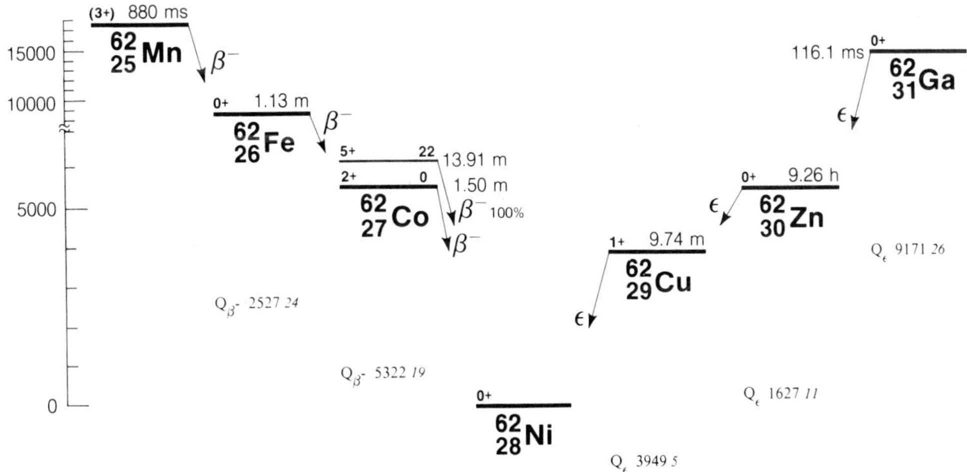

$^{62}_{25}$Mn(880 *150* ms)

Mode: β-
SpA: 1.43×10^{11} Ci/g
Prod: ^{76}Ge on W

Photons (^{62}Mn)

γ$_{mode}$	γ(keV)	γ(rel)
γ	876.8 *3*	100 *10*
γ	942.1 *4*	29 *8*
γ	1299.0 *4*	28 *9*
γ	1457.4 *5*	16 *7*
γ	1815.0 *6*	23 *8*
γ	2016.0 *8*	~11

$^{62}_{26}$Fe(1.13 *3* min)

Mode: β-
Δ: -58896 *15* keV
SpA: 2.66×10^{9} Ci/g
Prod: ^{64}Ni(n,p2n)

Photons (^{62}Fe)

γ$_{mode}$	γ(keV)
γ[M1+E2]	506.1 *1*

$^{62}_{27}$Co(1.50 *4* min)

Mode: β-
Δ: -61424 *19* keV
SpA: 2.01×10^{9} Ci/g
Prod: ^{64}Ni(d,α); ^{62}Ni(n,p); ^{65}Cu(n,α)

Photons (^{62}Co)

⟨γ⟩=1599 *24* keV

γ$_{mode}$	γ(keV)	γ(%)†
γ	1067.9 *3*	~0.02
γ E2+<9.0%M1	1129.00 *7*	11.3 *7*
γ E2	1173.04 *7*	83.8
γ M1+28%E2	1886.5 *12*	~0.4
γ M1+2.2%E2	1985.2 *4*	1.5 *4*
γ E2	2084.4 *4*	~0.3
γ	2097.88 *23*	0.88 *18*
γ E2	2302.01 *6*	14.7 *5*
γ M1+16%E2	2346.1 *8*	1.3 *3*
γ [E2]	3158.2 *4*	0.83 *17*
γ [E2]	3257.4 *4*	~0.008
γ	3270.88 *24*	0.21 *5* *
γ	3369.8 *3*	~0.3
γ [E2]	3519.1 *8*	~0.08
γ	4063.3 *10*	0.3 *1*

† 1.0% uncert(syst)
* with ^{62}Co(1.50 or 13.91 min)

Continuous Radiation (^{62}Co)

⟨β-⟩=1639 keV;⟨IB⟩=5.1 keV

E$_{bin}$(keV)		⟨ ⟩(keV)	(%)
0 - 10	β-	0.00412	0.082
	IB	0.051	
10 - 20	β-	0.0127	0.085
	IB	0.051	0.35
20 - 40	β-	0.054	0.178
	IB	0.100	0.35
40 - 100	β-	0.444	0.62
	IB	0.29	0.44
100 - 300	β-	6.4	3.09
	IB	0.85	0.48
300 - 600	β-	34.3	7.4
	IB	1.02	0.24
600 - 1300	β-	257	26.5
	IB	1.55	0.18
1300 - 2500	β-	831	44.8
	IB	1.02	0.060
2500 - 5000	β-	510	17.2
	IB	0.135	0.0047
5000 - 5322	β-	0.0377	0.00074
	IB	4.9×10^{-7}	9.6×10^{-9}

$^{62}_{27}$Co(13.91 *5* min)

Mode: β-
Δ: -61402 *20* keV
SpA: 2.179×10^{8} Ci/g
Prod: ^{64}Ni(d,α); ^{62}Ni(n,p); ^{65}Cu(n,α); ^{64}Ni(p,^{3}He); ^{64}Ni(p,2pn)

Photons (^{62}Co)*

⟨γ⟩=2690 *25* keV

γ$_{mode}$	γ(keV)	γ(%)†
γ	777.65 *24*	1.8 *2*
γ [E2]	874.94 *24*	1.3 *2*
γ E2+<9.0%M1	1129.00 *7*	1.32 *12*
γ E2	1163.62 *19*	68.1 *14*
γ E2	1173.04 *7*	97.9
γ	1718.8 *3*	6.8 *4*
γ	1753.41 *25*	0.6 *2*
γ E2	2003.91 *24*	18.6 *5*
γ E2	2104.75 *24*	6.5 *3*
γ E2	2302.01 *6*	1.73 *14*
γ	2882.37 *24*	1.1 *1*

† 1.0% uncert(syst)
* see also ^{62}Co(1.50 min)

Continuous Radiation (^{62}Co)

⟨β-⟩=1058 keV;⟨IB⟩=2.6 keV

E$_{bin}$(keV)		⟨ ⟩(keV)	(%)
0 - 10	β-	0.0100	0.198
	IB	0.038	
10 - 20	β-	0.0307	0.204
	IB	0.037	0.26
20 - 40	β-	0.129	0.429
	IB	0.073	0.25
40 - 100	β-	1.05	1.48
	IB	0.21	0.32
100 - 300	β-	14.2	6.9
	IB	0.58	0.33
300 - 600	β-	67	14.6
	IB	0.63	0.148

Continuous Radiation (^{62}Co)

(continued)

E$_{bin}$(keV)		⟨ ⟩(keV)	(%)
600 - 1300	β-	354	37.6
	IB	0.76	0.090
1300 - 2500	β-	582	33.3
	IB	0.26	0.0159
2500 - 3008	β-	39.2	1.49
	IB	0.0017	6.7×10^{-5}

$^{62}_{28}$Ni(stable)

Δ: -66745.7 *15* keV
%: 3.59 *1*

$^{62}_{29}$Cu(9.74 *2* min)

Mode: ϵ
Δ: -62797 *5* keV
SpA: 3.112×10^{8} Ci/g
Prod: daughter ^{62}Zn; ^{59}Co(α,n); ^{62}Ni(p,n)

Photons (^{62}Cu)

⟨γ⟩=7.14 *18* keV

γ$_{mode}$	γ(keV)	γ(%)†
Ni L$_{\ell}$	0.743	0.00093 *25*
Ni L$_{\eta}$	0.760	0.00053 *13*
Ni L$_{\alpha}$	0.851	0.012 *3*
Ni L$_{\beta}$	0.876	0.0080 *21*
Ni K$_{\alpha2}$	7.461	0.231 *11*
Ni K$_{\alpha1}$	7.478	0.455 *22*
Ni K$_{\beta1}$'	8.265	0.083 *4*
γ	479.2 *5*	0.0004 ?
γ [M1+E2]	856.2 *4*	0.0004 ?
γ E2	875.74 *7*	0.147 *7*
γ	1067.9 *3*	~0.0006
γ E2+<9.0%M1	1129.00 *7*	0.0315 *15*
γ E2	1173.04 *7*	0.335
γ E2+6.6%M1	1717.6 *4*	0.0026 *4*
γ M1+2.2%E2	1985.2 *4*	0.0011 *3*
γ E2	2084.4 *4*	0.005 *1*
γ	2097.88 *23*	0.0029 *4*
γ E2	2302.01 *6*	0.0410 *16*
γ [E2]	3158.2 *4*	0.00059 *12*
γ [E2]	3257.4 *4*	0.00013 *6*
γ	3270.88 *24*	0.00070 *10*
γ	3369.8 *3*	0.0078 *5*
γ	3861.7 *11*	0.00027 *7*

† 4.0% uncert(syst)

Atomic Electrons (^{62}Cu)

⟨e⟩=0.094 *6* keV

e$_{bin}$(keV)	⟨e⟩(keV)	e(%)
1	0.0152	1.75 *18*
6	0.0206	0.32 *3*
7	0.056	0.81 *8*
8	0.00107	0.0131 *13*
867 - 875	0.000433	4.99 *24* ×10^{-5}
1060 - 1067	9.6 ×10^{-7}	~9×10^{-8}
1121 - 1165	0.00066	5.72 *24* ×10^{-5}

Atomic Electrons (^{62}Cu)
(continued)

e_{bin}(keV)	⟨e⟩(keV)	e(%)
1709 - 1717	3.4 $\times10^{-6}$	2.0 *3* $\times10^{-7}$
1977 - 2076	6.2 $\times10^{-6}$	3.0 *5* $\times10^{-7}$
2083 - 2097	3.0 $\times10^{-6}$	1.5 *3* $\times10^{-7}$
2294 - 2301	4.22 $\times10^{-5}$	1.84 *8* $\times10^{-6}$
3150 - 3249	6.0 $\times10^{-7}$	1.9 *3* $\times10^{-8}$
3256 - 3270	4.9 $\times10^{-7}$	1.5 *3* $\times10^{-8}$
3362 - 3369	5.3 $\times10^{-6}$	1.6 *3* $\times10^{-7}$
3853 - 3861	1.7 $\times10^{-7}$	4.5 *13* $\times10^{-9}$

Continuous Radiation (^{62}Cu)
⟨β+⟩=1280 keV;⟨IB⟩=3.7 keV

E_{bin}(keV)		⟨ ⟩(keV)	(%)
0 - 10	β+	3.24 $\times10^{-5}$	0.000408
	IB	0.044	
10 - 20	β+	0.00059	0.00366
	IB	0.044	0.30
20 - 40	β+	0.0081	0.0256
	IB	0.086	0.30
40 - 100	β+	0.196	0.259
	IB	0.25	0.38
100 - 300	β+	6.1	2.83
	IB	0.71	0.40
300 - 600	β+	42.9	9.2
	IB	0.80	0.19
600 - 1300	β+	353	36.3
	IB	1.10	0.128
1300 - 2500	β+	833	47.0
	IB	0.53	0.033
2500 - 3949	β+	45.2	1.73
	IB	0.089	0.0031
	Σβ+		97

$^{62}_{30}$Zn(9.26 *2* h)

Mode: ϵ
Δ: -61170 *10* keV
SpA: 5.458$\times10^{6}$ Ci/g
Prod: ^{63}Cu(p,2n); ^{63}Cu(d,3n); ^{60}Ni(^{3}He,n); ^{60}Ni(α,2n)

Photons (^{62}Zn)
⟨γ⟩=353 *13* keV

γ_{mode}	γ(keV)	γ(%)†
Cu L_ℓ	0.811	0.051 *13*
Cu L_η	0.831	0.029 *7*
Cu L_α	0.930	0.69 *18*
Cu L_β	0.957	0.46 *12*
Cu $K_{\alpha2}$	8.028	12.8 *7*
Cu $K_{\alpha1}$	8.048	24.9 *13*
Cu $K_{\beta1}$'	8.905	4.50 *23*
γ M1	40.83 *4*	25.2 *13*
γ [M1+E2]	202.65 *4*	0.0108 *13*
γ M1+0.2%E2	243.48 *4*	2.49 *13*
γ M1+10%E2	247.01 *5*	1.88 *10*
γ M1+E2	260.54 *5*	1.34 *8*
γ [M1+E2]	304.90 *5*	0.285 *15*
γ	349.69 *6*	0.44 *3*
γ M1+1.5%E2	385.38 *7*	0.0172 *15*
γ [M1+E2]	394.05 *4*	2.21 *4*
γ	489.23 *6*	0.0157 *15*
γ M1+E2	507.55 *5*	14.6 *8*
γ M1+E2	548.38 *5*	15.2 *8*
γ M1	596.70 *5*	25.7
γ	627.61 *7*	0.0008 *3*
γ [M1+E2]	637.53 *5*	0.252 *15*
γ [M1+E2]	644.87 *6*	0.0141 *8*
γ	657.56 *15*	0.0013 *3*
γ	671.97 *5*	0.0044 *5*
γ	731.29 *14*	0.0023 *3*
γ [M1+E2]	792.15 *6*	0.0087 *8*
γ	827.63 *13*	0.0030 *4*
γ [M1+E2]	881.30 *7*	0.0144 *10*
γ	915.44 *5*	0.0152 *10*
γ	1141.83 *7*	0.0342 *21*
γ [M1+E2]	1186.20 *7*	0.0039 *13*
γ [M1+E2]	1388.84 *7*	0.0116 *8*
γ [M1+E2]	1429.67 *7*	0.027 *3*
γ [M1+E2]	1485.18 *18*	~0.0005
γ [M1+E2]	1526.01 *19*	0.0057 *13*

† 7.4% uncert(syst)

Atomic Electrons (^{62}Zn)
⟨e⟩=10.4 *4* keV

e_{bin}(keV)	⟨e⟩(keV)	e(%)
1	0.82	87 *9*
7	3.0	43 *4*
8	0.84	10.6 *11*
9	0.055	0.62 *6*
32	4.7	14.7 *8*
40	0.62	1.56 *9*
194 - 243	0.059	0.0250 *11*
246 - 260	0.039	0.015 *8*
296 - 341	0.010	0.0033 *12*
349 - 393	0.025	0.0064 *23*
480 - 507	0.10	0.020 *5*
539 - 547	0.088	0.016 *4*
588	0.091	0.0154 *12*
596 - 644	0.0103	0.00172 *12*
649 - 671	1.9 $\times10^{-5}$	2.9 *10* $\times10^{-6}$
722 - 730	6.6 $\times10^{-6}$	9 *4* $\times10^{-7}$
783 - 827	3.6 $\times10^{-5}$	4.5 *7* $\times10^{-6}$
872 - 915	7.3 $\times10^{-5}$	8.2 *15* $\times10^{-6}$
1133 - 1177	6.2 $\times10^{-5}$	5.4 *17* $\times10^{-6}$
1380 - 1429	6.4 $\times10^{-5}$	4.5 *4* $\times10^{-6}$
1476 - 1525	9.5 $\times10^{-6}$	6.3 *13* $\times10^{-7}$

Continuous Radiation (^{62}Zn)
⟨β+⟩=21.7 keV;⟨IB⟩=0.52 keV

E_{bin}(keV)		⟨ ⟩(keV)	(%)
0 - 10	β+	9.4 $\times10^{-5}$	0.00118
	IB	0.0037	
10 - 20	β+	0.00173	0.0108
	IB	0.00122	0.0086
20 - 40	β+	0.0237	0.075
	IB	0.0021	0.0072
40 - 100	β+	0.51	0.68
	IB	0.0060	0.0092
100 - 300	β+	9.3	4.58
	IB	0.042	0.021
300 - 600	β+	11.8	3.05
	IB	0.153	0.034
600 - 1300	β+	0.000147	2.44 $\times10^{-5}$
	IB	0.29	0.035
1300 - 1627	IB	0.020	0.00142
	Σβ+		8.4

$^{62}_{31}$Ga(116.1 *3* ms)

Mode: ϵ
Δ: -51999 *28* keV
SpA: 2.618$\times10^{11}$ Ci/g
Prod: ^{64}Zn(p,3n)

Continuous Radiation (^{62}Ga)
⟨β+⟩=3855 keV;⟨IB⟩=19 keV

E_{bin}(keV)		⟨ ⟩(keV)	(%)
0 - 10	β+	1.74 $\times10^{-6}$	2.18 $\times10^{-5}$
	IB	0.083	
10 - 20	β+	3.43 $\times10^{-5}$	0.000212
	IB	0.082	0.57
20 - 40	β+	0.000507	0.00159
	IB	0.163	0.57
40 - 100	β+	0.0132	0.0174
	IB	0.48	0.74
100 - 300	β+	0.465	0.212
	IB	1.53	0.85
300 - 600	β+	3.78	0.81
	IB	2.1	0.48
600 - 1300	β+	45.2	4.52
	IB	4.0	0.45
1300 - 2500	β+	326	16.6
	IB	4.9	0.28
2500 - 5000	β+	1944	52
	IB	4.7	0.141
5000 - 7500	β+	1506	25.6
	IB	0.71	0.0125
7500 - 9171	β+	29.1	0.379
	IB	0.0131	0.000166
	Σβ+		99.87

A = 63

NDS **28**, 559 (1979)

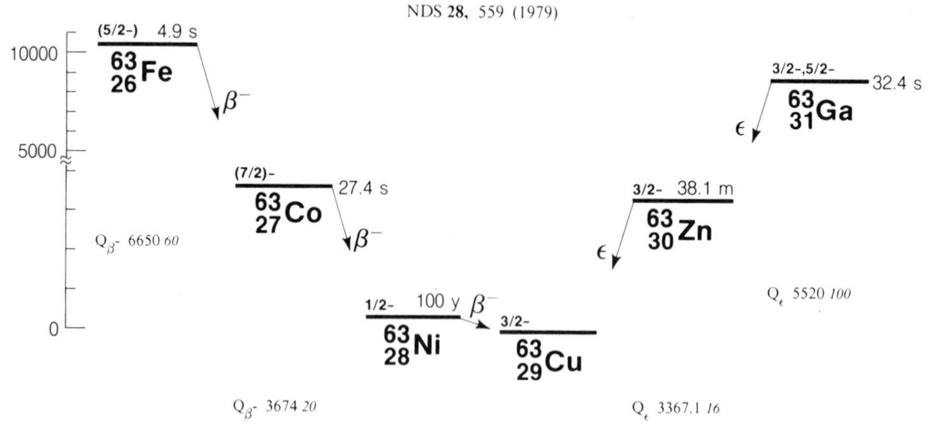

$^{63}_{26}$Fe(4.9 *5* s)

Mode: β-
Δ: -55190 *60* keV
SpA: 3.4×10^{10} Ci/g
Prod: ^{76}Ge on W

Photons (^{63}Fe)

γ_{mode}	γ(keV)	γ(rel)
γ	994.8 *10*	100 *20*
γ	1364.7 *10*	37 *7*
γ	1427.5 *10*	46 *9*

$^{63}_{27}$Co(27.4 *5* s)

Mode: β-
Δ: -61839 *20* keV
SpA: 6.45×10^{9} Ci/g
Prod: ^{64}Ni(n,np); ^{64}Ni(γ,p)

Photons (^{63}Co)

⟨γ⟩=119 *9* keV

γ_{mode}	γ(keV)	γ(%)†
Ni L_ℓ	0.743	0.021 *6*
Ni L_η	0.760	0.012 *3*
Ni L_α	0.851	0.26 *7*
Ni L_β	0.876	0.17 *5*
Ni $K_{\alpha2}$	7.461	5.1 *3*
Ni $K_{\alpha1}$	7.478	9.9 *7*
Ni $K_{\beta1}'$	8.265	1.81 *12*
γ (E2)	87.32 *11*	48.4
γ (M1)	155.74 *19*	1.74 *19*
γ [M1+E2]	913.5 *3*	0.45 *6*
γ [M1+E2]	981.96 *22*	2.6 *3*
γ [E2]	1069.28 *23*	1.60 *19*
γ	2174.7 *5*	1.2 *4* ?

† 3.0% uncert(syst)

Atomic Electrons (^{63}Co)

⟨e⟩=39.0 *17* keV

e_{bin}(keV)	⟨e⟩(keV)	e(%)
1	0.34	39 *4*
6	0.44	6.9 *8*
7	1.19	17.4 *20*
8	0.023	0.28 *3*
79	32.7	41.4 *21*
86	4.23	4.90 *25*
147	0.034	0.023 *3*
155	0.0036	0.0023 *3*
905	0.00101	0.000111 *20*
913	0.000100	1.09 *20* $\times10^{-5}$
974	0.0052	0.00053 *9*
981	0.00051	5.2 *9* $\times10^{-5}$
1061	0.0032	0.00030 *4*
1068	0.00031	2.9 *4* $\times10^{-5}$
2166	0.0009	4.2 *18* $\times10^{-5}$
2174	9.0 $\times10^{-5}$	4.1 *17* $\times10^{-6}$

Continuous Radiation (^{63}Co)

⟨β-⟩=1531 keV;⟨IB⟩=4.5 keV

E_{bin}(keV)		⟨ ⟩(keV)	(%)
0 - 10	β-	0.00421	0.084
	IB	0.049	
10 - 20	β-	0.0130	0.086
	IB	0.049	0.34
20 - 40	β-	0.055	0.182
	IB	0.096	0.33
40 - 100	β-	0.454	0.64
	IB	0.28	0.43
100 - 300	β-	6.6	3.18
	IB	0.82	0.46
300 - 600	β-	35.5	7.7
	IB	0.96	0.23
600 - 1300	β-	273	28.1
	IB	1.41	0.162
1300 - 2500	β-	898	48.5
	IB	0.80	0.048
2500 - 3587	β-	318	11.4
	IB	0.045	0.00168

$^{63}_{28}$Ni(100.1 *20* yr)

Mode: β-
Δ: -65512.8 *15* keV
SpA: 56.7 Ci/g
Prod: ^{62}Ni(n,γ)

Continuous Radiation (^{63}Ni)

⟨β-⟩=17.1 keV;⟨IB⟩=0.00112 keV

E_{bin}(keV)		⟨ ⟩(keV)	(%)
0 - 10	β-	1.77	37.1
	IB	0.00069	
10 - 20	β-	3.96	27.0
	IB	0.00029	0.0021
20 - 40	β-	8.2	29.0
	IB	0.000140	0.00055
40 - 66	β-	3.21	6.9
	IB	7.2×10^{-6}	1.63×10^{-5}

$^{63}_{29}$Cu(stable)

Δ: -65578.7 *15* keV
%: 69.17 *2*

$^{63}_{30}$Zn(38.1 *3* min)

Mode: ε
Δ: -62211.6 *22* keV
SpA: 7.83×10^{7} Ci/g
Prod: ^{60}Ni(α,n); ^{63}Cu(p,n); ^{63}Cu(d,2n)

Photons (^{63}Zn)

⟨γ⟩=152 *4* keV

γ_{mode}	γ(keV)	γ(%)†
Cu L_ℓ	0.811	0.0034 *9*
Cu L_η	0.831	0.0020 *5*
Cu L_α	0.930	0.047 *12*
Cu L_β	0.957	0.031 *8*
Cu $K_{\alpha2}$	8.028	0.86 *4*
Cu $K_{\alpha1}$	8.048	1.67 *8*
Cu $K_{\beta1}'$	8.905	0.302 *15*
γ [M1+E2]	244.00 *17*	0.005 *1* ?
γ M1+1.1%E2	365.00 *8*	0.0118 *25*
γ [M1+E2]	443.27 *13*	0.017 *4*
γ M1(+6.0%E2)	450.03 *4*	0.242 *17*
γ [M1+E2]	475.47 *24*	~0.006
γ [M1+E2]	515.27 *10*	0.022 *8*
γ [M1+E2]	534.4 *3*	0.0050 *17*
γ M1(+2.2%E2)	584.99 *6*	0.034 *4*
γ	624.18 *15*	0.014 *3*
γ M1+1.1%E2	669.76 *4*	8.4
γ [M1+E2]	674.74 *22*	0.015 *3*
γ	685.5 *3*	0.0042 *17*
γ [E2]	742.45 *5*	0.069 *8*
γ [M1+E2]	754.4 *3*	0.0067 *25*
γ [M1+E2]	765.62 *14*	0.0067 *25*
γ [M1+E2]	877.40 *7*	~0.0034
γ M1(+0.5%E2)	899.14 *22*	0.0126 *25*
γ [M1+E2]	924.58 *12*	0.0101 *20*
γ M1+21%E2	962.17 *3*	6.6 *3*
γ [M1+E2]	988.90 *8*	0.0039 *11*
γ M1+7.3%E2	1049.1 *3*	0.0045 *12*
γ [M1+E2]	1123.85 *6*	0.113 *12*
γ [M1+E2]	1130.61 *13*	0.0134 *25*
γ	1149.61 *13*	0.0193 *25*
γ	1169.81 *10*	0.0079 *17*
γ [M1+E2]	1208.88 *9*	0.0126 *25*
γ	1233.47 *24*	0.0025 *8*
γ E2	1327.16 *7*	0.071 *4*
γ [M1+E2]	1341.6 *3*	0.0025 *8*
γ [M1+E2]	1374.61 *11*	0.0353 *25*
γ	1389.79 *7*	0.044 *6*
γ [M1]	1392.65 *8*	0.099 *15*
γ M1+27%E2	1412.20 *4*	0.76 *3*
γ	1445.8 *3*	0.0025 *8*
γ	1479.4 *4*	~0.0017
γ M1(+2.0%E2)	1547.15 *5*	0.125 *5*
γ [M1+E2]	1573.87 *7*	0.0168 *17*
γ [E2]	1667.02 *12*	0.0014 *6*
γ	1696.7 *10*	~0.002
γ	1754.78 *10*	0.0045 *10*
γ [M1+E2]	1827.5 *3*	0.0043 *11*
γ [E2]	1861.30 *22*	0.0143 *20*
γ [E2]	1866.28 *8*	0.0202 *22*
γ	1927.4 *4*	0.0059 *12*
γ [M1+E2]	2011.3 *3*	0.0109 *17*
γ	2026.99 *13*	0.057 *6*
γ	2047.19 *10*	0.0038 *11*
γ [M1+E2]	2062.40 *9*	0.035 *3*
γ [M1+E2]	2081.52 *25*	0.0151 *17*
γ [E2]	2092.77 *13*	0.0025 *8*
γ	2110.85 *24*	0.0063 *13*
γ	2181.9 *7*	~0.0013
γ	2188.2 *3*	~0.0017
γ	2219.8 *4*	0.0030 *8*
γ [M1+E2]	2336.76 *11*	0.076 *5*
γ [M1+E2]	2497.3 *3*	0.0218 *25*
γ	2512.1 *5*	0.0101 *17*
γ [M1+E2]	2536.02 *7*	0.068 *7*
γ	2696.72 *13*	0.041 *4*
γ	2716.92 *9*	0.0134 *17*
γ	2780.58 *24*	0.0160 *17*
γ	2806.5 *4*	0.0042 *8*
γ	2858.0 *3*	0.0034 *8*
γ	2889.6 *4*	0.0025 *8*
γ	3044.7 *8*	0.0050 *8*
γ	3100.8 *8*	0.00059 *17*

† 4.8% uncert(syst)

Atomic Electrons (^{63}Zn)

⟨e⟩=0.366 *22* keV

e_{bin}(keV)	⟨e⟩(keV)	e(%)
1	0.055	5.8 *6*
7	0.201	2.9 *3*
8	0.056	0.71 *7*
9	0.0037	0.042 *4*
235 - 244	0.00018	~8 ×10^{-5}
356 - 364	8.6 ×10^{-5}	2.4 *5* ×10^{-5}
434 - 475	0.00157	0.00035 *3*
506 - 533	0.00017	3.4 *12* ×10^{-5}
576 - 623	0.00019	3.2 *5* ×10^{-5}
661	0.0260	0.00393 *20*
666 - 685	0.00269	0.000402 *20*
733 - 765	0.00033	4.5 *5* ×10^{-5}
868 - 916	6.5 ×10^{-5}	7.2 *10* ×10^{-6}
923 - 924	2.4 ×10^{-6}	2.6 *6* ×10^{-7}
953	0.0144	0.00151 *8*
961 - 988	0.00145	0.000150 *8*
1040 - 1048	9.6 ×10^{-6}	9.2 *22* ×10^{-7}
1115 - 1161	0.00031	2.8 *3* ×10^{-5}
1169 - 1208	2.6 ×10^{-5}	2.1 *4* ×10^{-6}
1224 - 1233	3.7 ×10^{-6}	3.0 *14* ×10^{-7}
1318 - 1366	0.000189	1.42 *7* ×10^{-5}
1374 - 1411	0.00146	0.000104 *4*
1437 - 1478	5.3 ×10^{-6}	3.6 *12* ×10^{-7}
1538 - 1573	0.000209	1.35 *6* ×10^{-5}
1658 - 1754	9.1 ×10^{-6}	5.3 *13* ×10^{-7}
1819 - 1865	5.1 ×10^{-5}	2.75 *22* ×10^{-6}
1918 - 2010	1.9 ×10^{-5}	9.7 *14* ×10^{-7}
2018 - 2110	0.000128	6.2 *8* ×10^{-6}
2173 - 2219	5.5 ×10^{-6}	2.5 *7* ×10^{-7}
2328 - 2336	8.3 ×10^{-5}	3.6 *3* ×10^{-6}
2488 - 2535	0.000101	4.0 *3* ×10^{-6}
2688 - 2780	5.7 ×10^{-5}	2.1 *3* ×10^{-6}
2798 - 2889	8.0 ×10^{-6}	2.8 *5* ×10^{-7}
3036 - 3100	4.3 ×10^{-6}	1.4 *3* ×10^{-7}

Continuous Radiation (^{63}Zn)

⟨β+⟩=919 keV;⟨IB⟩=2.5 keV

E_{bin}(keV)		⟨ ⟩(keV)	(%)
0 - 10	β+	5.9 ×10^{-5}	0.00074
	IB	0.035	
10 - 20	β+	0.00111	0.0069
	IB	0.034	0.24
20 - 40	β+	0.0158	0.0496
	IB	0.067	0.23
40 - 100	β+	0.384	0.509
	IB	0.19	0.29
100 - 300	β+	11.6	5.4
	IB	0.52	0.29
300 - 600	β+	73	15.8
	IB	0.55	0.130
600 - 1300	β+	434	46.1
	IB	0.68	0.080
1300 - 2500	β+	400	25.0
	IB	0.35	0.021
2500 - 3367	IB	0.041	0.00152
	Σβ+		93

$^{63}_{31}$Ga(32.4 *5* s)

Mode: ϵ

Δ: -56690 *100* keV

SpA: 5.47×10^9 Ci/g

Prod: ^{63}Cu(α,4n); ^{60}Ni(^{6}Li,3n); ^{58}Ni(^{6}Li,n); ^{64}Zn(p,2n)

Photons (^{63}Ga)

⟨γ⟩=354 *21* keV

γ_{mode}	γ(keV)	γ(%)†
Zn $K_{\alpha2}$	8.616	0.131 *8*
Zn $K_{\alpha1}$	8.639	0.257 *15*
Zn $K_{\beta1}'$	9.572	0.047 *3*
γ M1(+0.1%E2)	192.68 *18*	5.7 *4*
γ [M1+E2]	247.70 *19*	3.40 *22*
γ M1+0.4%E2	389.0 *3*	~0.38
γ [E2]	415.0 *4*	~0.29
γ M1+0.3%E2	457.18 *25*	0.60 *16*
γ [M1+E2]	626.81 *17*	10.2 *6*
γ M1+0.2%E2	636.75 *19*	11.1
γ M1+E2	649.86 *19*	4.9 *3*
γ	768.21 *17*	2.1 *3*
γ [M1+E2]	1054.4 *4*	~0.26
γ [M1+E2]	1064.9 *4*	2.2 *4*
γ	1147.3 *3*	0.34 *8*
γ	1202.3 *3*	~0.27
γ	1395.02 *20*	4.1 *8*
γ [M1+E2]	1498.5 *4*	~0.32
γ [M1+E2]	1691.2 *4*	3.0 *6*

† 19% uncert(syst)

Atomic Electrons (^{63}Ga)

⟨e⟩=0.43 *6* keV

e_{bin}(keV)	⟨e⟩(keV)	e(%)
1 - 7	0.0186	0.96 *9*
8	0.0210	0.27 *3*
9 - 10	0.0064	0.075 *8*
183	0.103	0.056 *5*
191 - 193	0.0129	0.0067 *5*
238	0.10	~0.042
247	0.011	~0.004
379 - 415	0.0071	0.0018 *6*
448 - 457	0.0037	0.00083 *19*
617	0.048	0.0077 *17*
626	0.0049	0.00078 *17*
627	0.042	0.0066 *13*
636 - 637	0.0048	0.00076 *13*
640	0.022	0.0034 *7*
649 - 650	0.0026	0.00040 *7*
759 - 768	0.006	0.0008 *4*
1045 - 1065	0.0063	0.00059 *11*
1138 - 1147	0.00062	5.4 *21* ×10^{-5}
1193 - 1202	0.00045	~4 ×10^{-5}
1385 - 1395	0.0060	0.00043 *15*
1489 - 1498	0.0006	3.8 *18* ×10^{-5}
1682 - 1691	0.0047	0.00028 *5*

Continuous Radiation (^{63}Ga)

⟨β+⟩=1887 keV;⟨IB⟩=6.5 keV

E_{bin}(keV)		⟨ ⟩(keV)	(%)
0 - 10	β+	1.14 ×10^{-5}	0.000143
	IB	0.056	
10 - 20	β+	0.000224	0.00139
	IB	0.056	0.39
20 - 40	β+	0.00330	0.0104
	IB	0.110	0.38
40 - 100	β+	0.085	0.112
	IB	0.32	0.49
100 - 300	β+	2.85	1.31
	IB	0.96	0.54
300 - 600	β+	21.5	4.60
	IB	1.19	0.28
600 - 1300	β+	208	21.1
	IB	1.9	0.22
1300 - 2500	β+	882	46.8
	IB	1.52	0.088
2500 - 5000	β+	773	25.5
	IB	0.40	0.0129
5000 - 5520	IB	0.00126	2.5 ×10^{-5}
	Σβ+		99

A = 64

NDS **28**, 179 (1979)

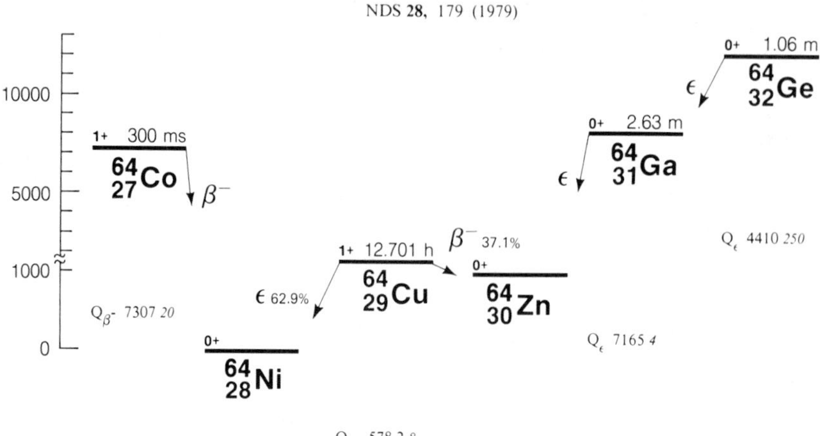

$^{64}_{27}$Co(300 *30* ms)

Mode: β-
Δ: -59791 *20* keV
SpA: 2.29×10^{11} Ci/g
Prod: ^{64}Ni(n,p)

Photons (^{64}Co)

⟨γ⟩=181 *28* keV

γ_{mode}	γ(keV)	γ(%)†
γ	931.1 *3*	5 *1*
γ [E2]	1345.78 *6*	10

† approximate

Continuous Radiation (^{64}Co)

⟨β-⟩=3288 keV;⟨IB⟩=14.7 keV

E_{bin}(keV)		⟨ ⟩(keV)	(%)
0 - 10	β-	0.00069	0.0137
	IB	0.076	
10 - 20	β-	0.00213	0.0142
	IB	0.076	0.52
20 - 40	β-	0.0091	0.0301
	IB	0.150	0.52
40 - 100	β-	0.076	0.107
	IB	0.44	0.68
100 - 300	β-	1.17	0.56
	IB	1.39	0.77
300 - 600	β-	7.0	1.51
	IB	1.9	0.43
600 - 1300	β-	70	7.0
	IB	3.5	0.39
1300 - 2500	β-	441	22.6
	IB	4.0	0.22
2500 - 5000	β-	1993	54
	IB	3.1	0.094
5000 - 7307	β-	776	13.8
	IB	0.19	0.0036

$^{64}_{28}$Ni(stable)

Δ: -67098.0 *16* keV
%: 0.91 *1*

$^{64}_{29}$Cu(12.701 *2* h)

Mode: ε(62.9 *4* %), β-(37.1 *4* %)
Δ: -65423.5 *15* keV
SpA: 3.8552×10^{6} Ci/g
Prod: ^{63}Cu(n,γ)

Photons (^{64}Cu)

⟨γ⟩=7.7 *20* keV

γ_{mode}	γ(keV)	γ(%)†
Ni L_ℓ	0.743	0.020 *5*
Ni L_η	0.760	0.011 *3*
Ni L_α	0.851	0.24 *6*
Ni L_β	0.876	0.17 *4*
Ni $K_{\alpha2}$	7.461	4.86 *24*
Ni $K_{\alpha1}$	7.478	9.6 *5*
Ni $K_{\beta1}$'	8.265	1.74 *9*
γ_ϵ[E2]	1345.78 *6*	0.48

† uncert(syst): 30% for ε

Atomic Electrons (^{64}Cu)

⟨e⟩=1.90 *13* keV

e_{bin}(keV)	⟨e⟩(keV)	e(%)
1	0.31	36 *4*
6	0.42	6.6 *7*
7	1.14	16.8 *17*
8	0.0220	0.27 *3*
1337	0.00072	5.4 *16* ×10^{-5}
1345	7.1 ×10^{-5}	5.3 *16* ×10^{-6}

Continuous Radiation (^{64}Cu)

⟨β-⟩=71 keV;⟨β+⟩=49.8 keV;⟨IB⟩=0.61 keV

E_{bin}(keV)		⟨ ⟩(keV)	(%)
0 - 10	β-	0.052	1.05
	β+	0.000190	0.00239
	IB	0.0072	
10 - 20	β-	0.157	1.05
	β+	0.00338	0.0211
	IB	0.0057	0.040
20 - 40	β-	0.63	2.11
	β+	0.0450	0.142
	IB	0.0103	0.036
40 - 100	β-	4.52	6.5
	β+	0.96	1.28
	IB	0.024	0.038
100 - 300	β-	36.1	18.7
	β+	18.3	8.9
	IB	0.052	0.029
300 - 600	β-	29.1	7.7
	β+	30.2	7.5
	IB	0.109	0.024
600 - 1300	β+	0.273	0.0445
	IB	0.36	0.040
1300 - 1675	IB	0.040	0.0029
	Σβ+		18

$^{64}_{30}$Zn(stable)

Δ: -66001.7 *17* keV
%: 48.6 *3*

$^{64}_{31}$Ga(2.630 *11* min)

Mode: ε
Δ: -58837 *4* keV
SpA: 1.115×10^{9} Ci/g
Prod: ^{63}Cu(α,3n); ^{64}Zn(p,n); ^{64}Zn(d,2n)

Photons (^{64}Ga)

$\langle\gamma\rangle$=2400 33 keV

γ_{mode}	γ(keV)	γ(%)†
Zn L_ℓ	0.884	0.00106 23
Zn L_η	0.907	0.00064 16
Zn L_α	1.012	0.016 3
Zn L_β	1.043	0.011 3
Zn $K_{\alpha2}$	8.616	0.291 12
Zn $K_{\alpha1}$	8.639	0.570 24
Zn $K_{\beta1}$'	9.572	0.105 4
γ [M1]	756.70 14	1.36 7
γ E2+39%M1	807.93 7	14.8 3
γ E2	918.84 7	8.76 19
γ E2	991.57 7	46.8
γ [M1]	1186.05 22	0.12 3
γ [M1]	1276.46 8	6.04 14
γ	1351.67 12	<0.14
γ [M1+E2]	1387.37 7	12.8 3
γ	1411.5 12	0.15 4
γ [M1]	1455.59 9	2.03 10
γ	1462.58 12	0.33 5
γ [M1]	1514.85 14	0.20 4
γ [M1+E2]	1566.50 8	2.26 11
γ E2	1617.72 14	1.78 9
γ [M1+E2]	1625.76 12	1.13 8
γ E2	1799.48 8	3.89 14
γ [M1+E2]	1995.83 18	1.73 10
γ	2103.0 14	0.16 4
γ [M1+E2]	2195.28 7	10.02 23
γ	2270.49 10	2.27 11
γ [M1+E2]	2374.41 7	7.58 19
γ [M1+E2]	2433.67 12	0.64 6
γ [M1]	2543.9 5	0.19 4
γ [M1+E2]	2654.8 5	0.30 5
γ [M1+E2]	2803.74 17	0.70 7
γ	2912.8 14	<0.19
γ [M1]	3186.81 9	0.16 5
γ	3262.02 12	0.13 4
γ	3321.8 12	0.084 23
γ [M1]	3365.93 7	14.2 6
γ [M1]	3425.20 12	4.40 23
γ [M1+E2]	3462.7 5	0.047 9
γ	3616.8 9	0.11 4
γ [M1]	3795.26 18	1.33 8
γ [M1]	4454.2 5	0.81 5
γ	4608.3 9	~0.033
γ	4712.2 14	~0.019

† 4.9% uncert(syst)

Atomic Electrons (^{64}Ga)

$\langle e\rangle$=0.434 9 keV

e_{bin}(keV)	$\langle e\rangle$(keV)	e(%)
1	0.0191	1.84 19
7	0.0221	0.30 3
8	0.047	0.61 6
9	0.0142	0.164 17
747 - 757	0.0046	0.00061 4
798	0.0489	0.00613 22
807 - 808	0.00577	0.000715 24
909	0.0260	0.00286 8
918 - 919	0.00306	0.000333 9
982	0.125	0.0127 7
990	0.0123	0.00124 7
991 - 992	0.00234	0.000236 12
1176 - 1186	0.00026	2.2 5 $\times10^{-5}$
1267 - 1276	0.0119	0.00094 4
1378	0.0218	0.00158 11
1386 - 1411	0.00273	0.000197 13
1446 - 1515	0.0043	0.000299 18
1557 - 1626	0.0087	0.000546 22
1790 - 1799	0.00592	0.000331 12
1986 - 1996	0.00237	0.000119 9
2093 - 2186	0.0116	0.00053 3
2194 - 2270	0.0036	0.000160 24
2365 - 2434	0.0099	0.000416 22
2534 - 2544	0.00021	8.4 15 $\times10^{-6}$
2645 - 2655	0.00034	1.28 19 $\times10^{-5}$
2794 - 2804	0.00075	2.7 3 $\times10^{-5}$
3177 - 3262	0.00026	8.2 17 $\times10^{-6}$
3312 - 3366	0.0136	0.000404 20
3416 - 3463	0.00417	0.000122 7
3607 - 3617	8.8 $\times10^{-5}$	2.4 9 $\times10^{-6}$
3786 - 3795	0.00117	3.1 2 $\times10^{-5}$
4445 - 4454	0.00066	1.48 9 $\times10^{-5}$
4599 - 4608	2.3 $\times10^{-5}$	5.1 24 $\times10^{-7}$
4702 - 4712	1.3 $\times10^{-5}$	2.8 13 $\times10^{-7}$

Continuous Radiation (^{64}Ga)

$\langle\beta+\rangle$=1772 keV;$\langle$IB$\rangle$=6.4 keV

E_{bin}(keV)		$\langle\ \rangle$(keV)	(%)
0 - 10	β+	2.04 $\times10^{-5}$	0.000255
	IB	0.052	
10 - 20	β+	0.000399	0.00247
	IB	0.052	0.36
20 - 40	β+	0.0059	0.0184
	IB	0.102	0.35
40 - 100	β+	0.149	0.197
	IB	0.30	0.45
100 - 300	β+	4.85	2.23
	IB	0.88	0.49
300 - 600	β+	34.1	7.3
	IB	1.08	0.25
600 - 1300	β+	275	28.4
	IB	1.7	0.20
1300 - 2500	β+	691	38.4
	IB	1.48	0.085
2500 - 5000	β+	696	19.8
	IB	0.76	0.024
5000 - 7165	β+	71	1.33
	IB	0.023	0.00042
	Σβ+		98

$^{64}_{32}$Ge(1.06 4 min)

Mode: ε

Δ: -54430 250 keV

SpA: 2.75$\times10^{9}$ Ci/g

Prod: ^{64}Zn(^{3}He,3n); ^{54}Fe(^{12}C,2n)

Photons (^{64}Ge)

$\langle\gamma\rangle$=365 10 keV

γ_{mode}	γ(keV)	γ(%)†
γ E2	42.9 4	
γ [M1+E2]	85.3 5	~8
γ [M1]	128.2 2	10.7 7
γ [M1+E2]	384.1 3	4.7 5
γ [M1]	427.0 3	37.4 10
γ [M1]	667.1 3	16.9 10
γ [M1]	774.5 3	7.0 6

† 10% uncert(syst)

A = 65

NDS **16**, 351 (1975)

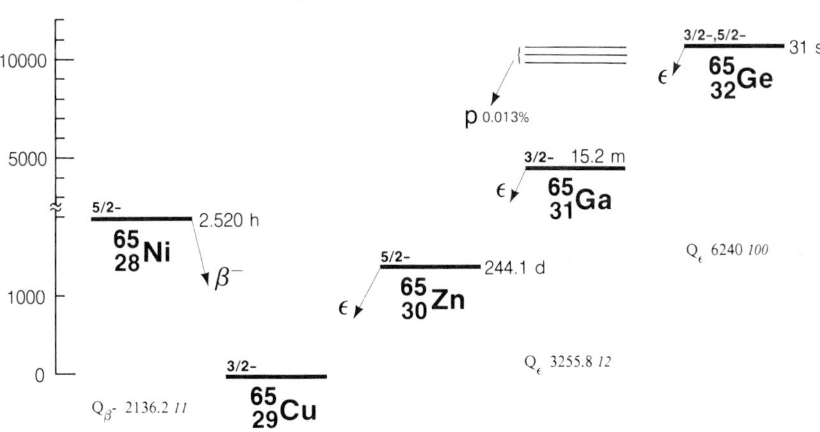

$^{65}_{28}$Ni(2.520 *2* h)

Mode: β-
Δ: -65124.8 *16* keV
SpA: 1.9131×10^7 Ci/g
Prod: ^{64}Ni(n,γ)

Photons (^{65}Ni)

⟨γ⟩=549 *12* keV

γ_{mode}	γ(keV)	γ(%)†
γ [E2]	344.93 *13*	0.00088 *7* ?
γ M1+0.2%E2	366.32 *3*	4.61 *12*
γ [M1+E2]	507.89 *6*	0.287 *16*
γ [M1+E2]	609.41 *6*	0.141 *9*
γ M1+0.8%E2	770.66 *13*	0.085 *9*
γ [E2]	852.82 *13*	0.075 *9*
γ [M1+E2]	954.33 *14*	0.0094 *9*
γ M1+16%E2	1115.584 *22*	14.8 *3*
γ [E2]	1481.90 *3*	23.5
γ [M1+E2]	1623.47 *6*	0.475 *16*
γ [M1+E2]	1724.98 *6*	0.388 *16*

† 3.4% uncert(syst)

Continuous Radiation (^{65}Ni)

⟨β-⟩=632 keV;⟨IB⟩=1.17 keV

E_{bin}(keV)		⟨ ⟩(keV)	(%)
0 - 10	β-	0.0479	0.95
	IB	0.026	
10 - 20	β-	0.145	0.97
	IB	0.025	0.17
20 - 40	β-	0.59	1.97
	IB	0.048	0.167
40 - 100	β-	4.42	6.3
	IB	0.131	0.20
100 - 300	β-	44.0	22.1
	IB	0.33	0.19
300 - 600	β-	103	23.8
	IB	0.30	0.072
600 - 1300	β-	295	31.9
	IB	0.28	0.034
1300 - 2136	β-	184	12.0
	IB	0.027	0.0019

$^{65}_{29}$Cu(stable)

Δ: -67261.0 *18* keV
%: 30.83 *2*

$^{65}_{30}$Zn(244.1 *2* d)

Mode: ϵ
Δ: -65910.2 *18* keV
SpA: 8230 Ci/g
Prod: ^{64}Zn(n,γ)

Photons (^{65}Zn)

⟨γ⟩=569.3 *11* keV

γ_{mode}	γ(keV)	γ(%)
Cu L_ℓ	0.811	0.046 *12*
Cu L_η	0.831	0.027 *7*
Cu L_α	0.930	0.63 *16*
Cu L_β	0.958	0.42 *11*
Cu $K_{\alpha2}$	8.028	11.6 *6*
Cu $K_{\alpha1}$	8.048	22.6 *11*
Cu $K_{\beta1}$'	8.905	4.08 *20*
γ [E2]	344.93 *13*	0.00303 *22*
γ M1+0.8%E2	770.66 *13*	0.0030 *3*
γ M1+16%E2	1115.518 *22*	50.75 *10*

Atomic Electrons (^{65}Zn)

⟨e⟩=4.4 *3* keV

e_{bin}(keV)	⟨e⟩(keV)	e(%)
1	0.75	79 *8*
7	2.7	39 *4*
8	0.76	9.6 *10*
9	0.050	0.56 *6*
336	5.7 ×10^{-5}	1.69 *14* ×10^{-5}
344	6.0 ×10^{-6}	1.74 *15* ×10^{-6}
762	7.9 ×10^{-6}	1.04 *11* ×10^{-6}
770	7.9 ×10^{-7}	1.03 *10* ×10^{-7}
1107	0.0934	0.00844 *14*
1114	0.00912	0.000818 *14*
1115	0.000137	1.233 *24* ×10^{-5}

Continuous Radiation (^{65}Zn)

⟨β+⟩=2.08 keV;⟨IB⟩=0.29 keV

E_{bin}(keV)		⟨ ⟩(keV)	(%)
0 - 10	β+	6.3×10^{-5}	0.00079
	IB	0.0030	
10 - 20	β+	0.00113	0.0070
	IB	0.00030	0.0022
20 - 40	β+	0.0146	0.0462
	IB	0.00034	0.00117
40 - 100	β+	0.267	0.363
	IB	0.00164	0.0023
100 - 300	β+	1.78	1.04
	IB	0.021	0.0099
300 - 600	β+	0.0142	0.00461
	IB	0.091	0.020
600 - 1300	IB	0.18	0.022
1300 - 1351	IB	8.9×10^{-5}	6.8×10^{-6}
	Σβ+		1.46

$^{65}_{31}$Ga(15.2 *2* min)

Mode: ϵ
Δ: -62654.4 *20* keV
SpA: 1.902×10^8 Ci/g
Prod: ^{63}Cu(α,2n); ^{64}Zn(d,n); ^{64}Zn(p,γ); ^{56}Fe(^{11}B,2n)

Photons (^{65}Ga)

⟨γ⟩=257 *11* keV

γ_{mode}	γ(keV)	γ(%)†
Zn L_ℓ	0.884	0.023 *5*
Zn L_η	0.907	0.014 *4*
Zn L_α	1.012	0.35 *8*
Zn L_β	1.042	0.23 *6*
Zn $K_{\alpha2}$	8.616	6.2 *5*
Zn $K_{\alpha1}$	8.639	12.2 *10*
Zn $K_{\beta1}$'	9.572	2.24 *18*
γ E2	53.93 *10*	4.9 *5*
γ M1+<5.4%E2	61.22 *10*	11.6 *11*
γ M1+>8.8%E2	91.85 *11*	<0.4 ?
γ M1+7.4%E2	115.15 *9*	55 *8*
γ M1+3.9%E2	153.07 *10*	9.0 *9*
γ M1+38%E2	207.0 *1*	2.58 *25*
γ [M1+E2]	422.44 *14*	0.057 *14*
γ	479.6 *6*	~0.020
γ [M1+E2]	560.14 *11*	0.11 *3*
γ [M1+E2]	575.13 *16*	<0.049
γ [M1+E2]	602.91 *14*	0.064 *10*
γ M1+E2	653.80 *13*	0.76 *7*
γ [M1+E2]	660.05 *14*	0.12 *4*
γ [M1+E2]	702.81 *11*	0.106 *17*
γ E2	715.02 *13*	0.164 *17*
γ [M1+E2]	751.89 *13*	8.2 *5*
γ M1+4.1%E2	768.95 *13*	1.29 *11*
γ [M1+E2]	794.66 *11*	0.271 *25*
γ [M1]	813.11 *13*	0.130 *18*
γ [M1+E2]	855.88 *11*	0.177 *15*
γ E2+20%M1	864.8 *4*	0.044 *14*
γ [E2]	867.04 *13*	0.118 *25*
γ [M1+E2]	909.81 *11*	0.52 *5*
γ [M1+E2]	932.36 *13*	1.82 *14*
γ [E2]	993.58 *15*	0.050 *7*
γ [M1+E2]	1047.51 *13*	0.91 *6*
γ [M1+E2]	1137.07 *13*	0.148 *12*
γ	1214.8 *8*	<0.012
γ [M1+E2]	1228.92 *13*	0.73 *5*
γ	1247.9 *8*	0.0049 *12*
γ [M1+E2]	1262.94 *11*	0.077 *7*
γ [E2]	1290.14 *13*	0.012 *4*
γ [M1+E2]	1344.07 *12*	0.22 *3*
γ [M1+E2]	1354.79 *10*	0.80 *9*
γ	1368.1 *10*	<0.007
γ [M1+E2]	1416.01 *10*	0.228 *25*
γ [M1+E2]	1469.94 *10*	0.074 *12*
γ	1502.6 *8*	0.0098 *25*
γ	1524.1 *8*	~0.005
γ	1576.6 *4*	~0.010
γ	1587.9 *8*	~0.0049
γ	1685.5 *4*	0.018 *5*
γ	1740.21 *20*	0.014 *6*
γ	1826.91 *20*	0.037 *4*
γ	1874.71 *20*	0.075 *9*
γ	1887.91 *20*	0.0086 *25*
γ	1966.5 *4*	0.065 *7*
γ	2009.8 *6*	~0.006
γ	2081.5 *8*	~0.007
γ	2087.6 *8*	~0.006
γ	2102.21 *20*	0.0135 *25*
γ	2163.8 *8*	0.015 *4*
γ [M1+E2]	2212.27 *21*	0.135 *11*
γ	2251.91 *20*	<0.010
γ	2289.9 *4*	0.016 *4*
γ [M1+E2]	2304.12 *22*	~0.010
γ	2343.2 *6*	0.0123 *25*
γ [M1]	2365.34 *21*	0.039 *5*
γ	2404.4 *4*	0.025 *4*
γ [E2]	2419.27 *22*	~0.0033
γ	2433.0 *8*	0.0061 *25*
γ	2458.5 *6*	0.0086 *25*
γ	2468.7 *8*	<0.0020
γ	2526.1 *11*	~0.003
γ	2549.7 *4*	0.031 *5*
γ	2570.9 *3*	0.0042 *16*
γ	2584.0 *3*	~0.0020
γ	2627.7 *8*	~0.0023
γ	2635.8 *3*	0.0049 *12*
γ	2648.8 *13*	~0.0016
γ	2679.5 *8*	0.0049 *12*
γ	2685.7 *13*	~0.0020
γ	2722.1 *8*	0.0014 *6*

Photons (^{65}Ga)
(continued)

γ_{mode}	γ(keV)	γ(%)†
γ	2792.5 3	0.0049 12
γ	2801.7 3	0.0023 6
γ	2835.6 3	0.0135 25
γ	2853.3 4	0.0014 4
γ	2891.3 17	0.0010 4
γ	2901.5 18	0.0010 4
γ	2908.2 18	~0.0006
γ	2941.2 18	<0.0006
γ	2963.4 19	~0.0006
γ	2996.1 13	0.0020 6
γ	3004.4 13	0.0020 6
γ	3014.9 13	~0.0006
γ	3025.8 20	<0.0012
γ	3056.1 13	0.0016 6
γ	3114.4 13	<0.0006

† 12% uncert(syst)

Atomic Electrons (^{65}Ga)
⟨e⟩=23.2 14 keV

e_{bin}(keV)	⟨e⟩(keV)	e(%)
1	0.43	41 5
7	0.47	6.4 8
8	1.00	13.0 16
9 - 10	0.31	3.5 4
44	12.3	28 3
52	1.27	2.47 24
53	2.33	4.4 5
54 - 92	0.57	1.00 8
105	3.5	3.3 6
114 - 153	0.80	0.64 7
197 - 207	0.113	0.057 10
413 - 422	0.00060	0.00014 6
470 - 480	0.00014	~3 $\times 10^{-5}$
550 - 593	0.0011	0.00020 5
602 - 650	0.0039	0.00061 11
653 - 702	0.00092	0.000136 19
703 - 752	0.034	0.0045 7
759 - 803	0.0057	0.00074 5
812 - 857	0.00115	0.000135 15
864 - 910	0.00164	0.000183 23
923 - 932	0.0054	0.00059 7
984 - 994	0.000149	1.52 21 $\times 10^{-5}$
1038 - 1048	0.00234	0.000226 23
1127 - 1137	0.00035	3.1 3 $\times 10^{-5}$
1205 - 1253	0.00173	0.000141 12
1262 - 1290	4.3 $\times 10^{-5}$	3.4 6 $\times 10^{-6}$
1334 - 1368	0.00199	0.000148 14
1406 - 1416	0.00042	3.0 3 $\times 10^{-5}$
1460 - 1524	0.000153	1.04 15 $\times 10^{-5}$
1567 - 1588	1.9 $\times 10^{-5}$	1.2 5 $\times 10^{-6}$
1676 - 1740	3.9 $\times 10^{-5}$	2.3 7 $\times 10^{-6}$
1817 - 1888	0.00014	7.5 15 $\times 10^{-6}$
1957 - 2010	7.9 $\times 10^{-5}$	4 1 $\times 10^{-6}$
2072 - 2164	4.4 $\times 10^{-5}$	2.1 4 $\times 10^{-6}$
2203 - 2294	0.000203	9.2 8 $\times 10^{-6}$
2303 - 2395	8.2 $\times 10^{-5}$	3.5 4 $\times 10^{-6}$
2403 - 2469	2.2 $\times 10^{-5}$	8.9 19 $\times 10^{-7}$
2516 - 2584	3.8 $\times 10^{-5}$	1.5 3 $\times 10^{-6}$
2618 - 2712	1.5 $\times 10^{-5}$	5.8 10 $\times 10^{-7}$
2721 - 2802	6.6 $\times 10^{-6}$	2.4 5 $\times 10^{-7}$
2826 - 2908	1.5 $\times 10^{-5}$	5.4 11 $\times 10^{-7}$
2932 - 3026	5.2 $\times 10^{-6}$	1.7 4 $\times 10^{-7}$
3046 - 3114	1.6 $\times 10^{-6}$	5.3 19 $\times 10^{-8}$

Continuous Radiation (^{65}Ga)
⟨β+⟩=791 keV;⟨IB⟩=2.2 keV

E_{bin}(keV)		⟨ ⟩(keV)	(%)
0 - 10	β+	6.6 $\times 10^{-5}$	0.00083
	IB	0.031	
10 - 20	β+	0.00129	0.0080
	IB	0.030	0.21
20 - 40	β+	0.0188	0.059
	IB	0.059	0.21
40 - 100	β+	0.467	0.62
	IB	0.166	0.26
100 - 300	β+	14.0	6.5
	IB	0.45	0.25
300 - 600	β+	84	18.3
	IB	0.46	0.110
600 - 1300	β+	434	46.7
	IB	0.60	0.070
1300 - 2500	β+	258	16.9
	IB	0.37	0.021
2500 - 3202	IB	0.028	0.00105
	Σβ+		89

$^{65}_{32}$Ge(30.9 7 s)

Mode: ε, εp(0.013 5 %)
Δ: -56410 100 keV
SpA: 5.55 $\times 10^{9}$ Ci/g
Prod: ^{64}Zn(^{3}He,2n); ^{40}Ca(^{28}Si,2pn)

εp: 1100-2200 range

Photons (^{65}Ge)
⟨γ⟩=816 26 keV

γ_{mode}	γ(keV)	γ(%)†
Ga $K_{\alpha2}$	9.225	1.19 19
Ga $K_{\alpha1}$	9.252	2.3 4
Ga $K_{\beta1}$'	10.263	0.44 7
γ	62.01 16	27 5
γ[M1+E2]	190.74 15	10.3 6
γ	458.94 20	2.0 3
γ	587.67 16	2.6 4
γ[M1+E2]	618.39 20	1.52 23
γ	649.68 14	33
γ	753.0 3	1.29 20
γ[M1+E2]	809.13 16	21.5 13
γ	825.9 5	0.36 13 ?
γ[M1+E2]	884.96 22	0.33 13
γ	970.5 4	0.23 10 ?
γ	1070.20 21	0.92 10
γ[E2]	1075.70 22	0.82 10
γ	1150.8 7	~0.13 ?
γ	1183.5 3	0.46 10 ?
γ	1205.8 4	1.22 13 ?
γ	1229.65 21	2.2 3
γ	1237.1 3	1.25 10 ?
γ	1512.0 4	0.33 7 ?
γ	1600.7 5	0.69 10
γ	1616.5 5	0.73 10 ?
γ	1688.59 23	2.21 20
γ	1817.32 23	0.40 10
γ	1879.32 20	0.96 23
γ	1901.6 5	0.40 10 ?
γ	2099.6 4	1.48 10 ?
γ	2120.3 5	0.3 1 ?
γ	2161.7 4	0.53 10 ?
γ	2219.0 20	~0.23
γ	2279.8 5	0.33 10 ?
γ	2387.8 7	0.36 17 ?
γ	2448.0 4	1.39 17

Photons (^{65}Ge)
(continued)

γ_{mode}	γ(keV)	γ(%)†
γ	2470.1 6	0.33 7 ?
γ	2703.5 15	~0.20
γ	2717.2 15	0.33 10
γ	2968.5 12	0.50 7
γ	3085.1 4	0.23 10 ?
γ	3279.2 6	0.30 7 ?

† 9.1% uncert(syst)

Atomic Electrons (^{65}Ge)
⟨e⟩=5.1 7 keV

e_{bin}(keV)	⟨e⟩(keV)	e(%)
1	0.080	7.1 13
8	0.24	3.0 6
9 - 10	0.073	0.80 14
52	3.2	6.3 12
61	0.41	0.68 13
62	0.068	0.111 21
180	0.6	~0.34
189 - 191	0.08	0.044 22
449 - 459	0.016	~0.0036
577 - 618	0.023	0.0038 12
639	0.13	0.020 10
648 - 650	0.015	0.0024 10
743 - 753	0.0044	0.00059 25
799 - 826	0.086	0.0108 14
875 - 885	0.0012	0.00013 5
960 - 971	0.0006	~6 $\times 10^{-5}$
1060 - 1076	0.0044	0.00041 7
1140 - 1184	0.0011	1.0 3 $\times 10^{-4}$
1195 - 1237	0.0085	0.00070 14
1502 - 1601	0.0015	9.5 22 $\times 10^{-5}$
1606 - 1688	0.0040	0.00024 6
1807 - 1901	0.0022	0.000118 25
2089 - 2162	0.0027	0.000129 25
2209 - 2280	0.00062	2.8 9 $\times 10^{-5}$
2377 - 2470	0.0022	9.0 17 $\times 10^{-5}$
2693 - 2717	0.00052	1.9 6 $\times 10^{-5}$
2958 - 2968	0.00047	1.6 4 $\times 10^{-5}$
3075 - 3085	0.00021	7 3 $\times 10^{-6}$
3269 - 3279	0.00027	8.2 22 $\times 10^{-6}$

Continuous Radiation (^{65}Ge)
⟨β+⟩=2052 keV;⟨IB⟩=7.6 keV

E_{bin}(keV)		⟨ ⟩(keV)	(%)
0 - 10	β+	9.4 $\times 10^{-6}$	0.000117
	IB	0.059	
10 - 20	β+	0.000191	0.00118
	IB	0.058	0.40
20 - 40	β+	0.00290	0.0091
	IB	0.115	0.40
40 - 100	β+	0.076	0.101
	IB	0.34	0.51
100 - 300	β+	2.59	1.19
	IB	1.02	0.57
300 - 600	β+	19.3	4.13
	IB	1.28	0.30
600 - 1300	β+	180	18.4
	IB	2.1	0.24
1300 - 2500	β+	783	41.3
	IB	1.9	0.109
2500 - 5000	β+	1066	33.5
	IB	0.69	0.022
5000 - 6178	β+	0.293	0.0058
	IB	0.0076	0.000144
	Σβ+		99

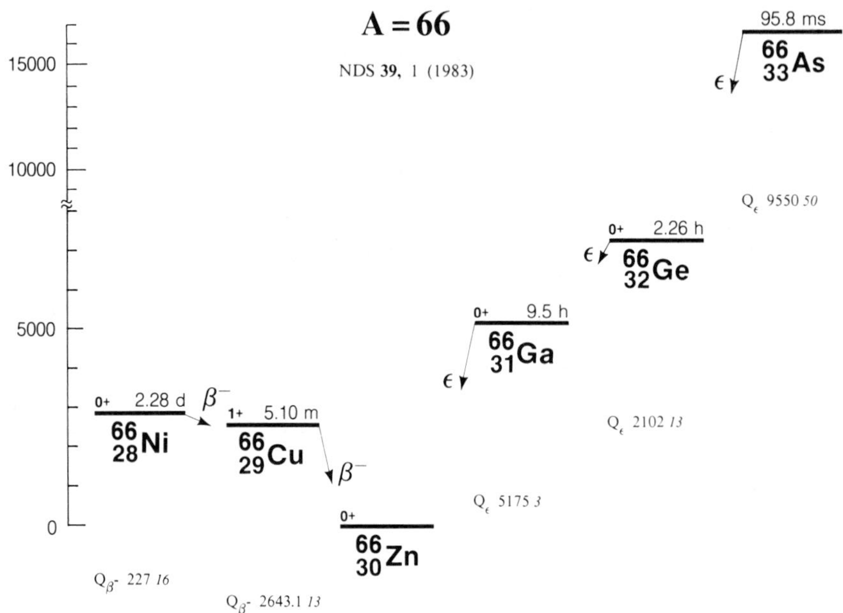

$^{66}_{28}$Ni(2.275 *17* d)

Mode: β-
Δ: -66028 *16* keV
SpA: 8.70×10^5 Ci/g
Prod: fission; multiple n-capture from ^{64}Ni

Continuous Radiation (^{66}Ni)

⟨β-⟩=65 keV;⟨IB⟩=0.0155 keV

E_{bin}(keV)		⟨ ⟩(keV)	(%)
0 - 10	β-	0.497	10.0
	IB	0.0034	
10 - 20	β-	1.42	9.5
	IB	0.0027	0.019
20 - 40	β-	5.2	17.5
	IB	0.0039	0.0137
40 - 100	β-	26.7	39.5
	IB	0.0048	0.0081
100 - 227	β-	31.4	23.5
	IB	0.00082	0.00069

$^{66}_{29}$Cu(5.10 *2* min)

Mode: β-
Δ: -66255.6 *19* keV
SpA: 5.580×10^8 Ci/g
Prod: ^{65}Cu(n,γ); daughter ^{66}Ni

Photons (^{66}Cu)

⟨γ⟩=78 *15* keV

γ_{mode}	γ(keV)	γ(%)†
γ E2+36%M1	833.58 *4*	0.17 *4*
γ E2	1039.35 *4*	7.4
γ E2	1333.21 *10*	0.0027 *4*

† 24% uncert(syst)

Continuous Radiation (^{66}Cu)

⟨β-⟩=1076 keV;⟨IB⟩=2.5 keV

E_{bin}(keV)		⟨ ⟩(keV)	(%)
0 - 10	β-	0.0092	0.182
	IB	0.039	
10 - 20	β-	0.0282	0.187
	IB	0.039	0.27
20 - 40	β-	0.118	0.393
	IB	0.076	0.26
40 - 100	β-	0.96	1.35
	IB	0.21	0.33
100 - 300	β-	13.3	6.4
	IB	0.60	0.34
300 - 600	β-	65	14.3
	IB	0.64	0.151
600 - 1300	β-	401	42.1
	IB	0.74	0.088
1300 - 2500	β-	593	35.0
	IB	0.18	0.0119
2500 - 2643	β-	2.09	0.082
	IB	1.36×10^{-5}	5.4×10^{-7}

$^{66}_{30}$Zn(stable)

Δ: -68898.7 *16* keV
%: 27.9 *2*

$^{66}_{31}$Ga(9.49 *8* h)

Mode: ε
Δ: -63724 *3* keV
SpA: 5.00×10^6 Ci/g
Prod: ^{63}Cu(α,n); ^{69}Ga(γ,3n); daughter ^{66}Ge

Photons (^{66}Ga)

⟨γ⟩=1879 *46* keV

γ_{mode}	γ(keV)	γ(%)†
Zn L_ℓ	0.884	0.020 *4*
Zn L_η	0.907	0.012 *3*
Zn L_α	1.012	0.30 *6*
Zn L_β	1.043	0.20 *5*
Zn $K_{\alpha2}$	8.616	5.52 *23*
Zn $K_{\alpha1}$	8.639	10.8 *5*
Zn $K_{\beta1}$'	9.572	1.98 *8*
γ	171.97 *23*	0.0106 *4*
γ [M1]	290.71 *7*	0.053 *4* ?
γ M1,E1	410.26 *7*	0.091 *23*
γ [M1]	448.94 *12*	0.110 *4*
γ	459.87 *13*	0.091 *4* ?
γ [M1]	578.74 *13*	0.061 *4*
γ [M1]	686.27 *6*	0.262 *8*
γ E2+36%M1	833.58 *4*	6.03 *12*
γ [M1]	853.12 *7*	0.0758 *19*
γ [M1]	856.73 *10*	0.119 *4*
γ [M1]	907.43 *12*	0.114 *8* ?
γ M1,E1	914.41 *8*	0.072 *4* ?
γ M1,E1	981.15 *8*	0.0493 *19*
γ M1,E1	1008.87 *12*	0.061 *9*
γ E2	1039.35 *4*	37.9
γ [M1]	1060.37 *21*	0.013 *4*
γ M1,E1	1148.00 *9*	0.080 *6*
γ [M1]	1190.41 *7*	0.144 *15*
γ E2	1232.50 *5*	0.523 *15*
γ [M1]	1232.97 *7*	0.053 *15*
γ E2	1333.21 *10*	1.232 *25*
γ [M1]	1356.36 *4*	0.30 *11*
γ [M1]	1356.83 *7*	0.11 *4*
γ [M1]	1357.26 *7*	0.27 *8*
γ [M1]	1419.13 *11*	0.644 *13*
γ [M1]	1459.05 *11*	0.1016 *23*
γ	1508.50 *7*	0.576 *12*

Photons (^{66}Ga) (continued)

γ_{mode}	γ(keV)	γ(%)†
γ [M1+E2]	1740.99 12	0.053 23
γ [M1+E2]	1899.21 7	0.436 11
γ [M1+E2]	1918.75 5	2.14 4
γ E2	2066.06 6	0.0326 15
γ [M1+E2]	2173.59 14	0.12 3 ?
γ	2173.9 3 ?	
γ [M1+E2]	2189.92 5	5.71 11
γ	2213.63 8	0.138 5
γ [M1+E2]	2292.61 11	0.042 4
γ	2393.55 19	0.248 6
γ [M1+E2]	2422.89 6	1.96 4
γ E2	2492.57 20	0.0239 23 ?
γ [M1+E2]	2589.31 7	0.028 3
γ M1+E2	2752.30 5	23.2 8
γ E2	2780.31 12	0.128 5
γ [M1+E2]	2934.15 13	0.216 11
γ	2977.7 5	0.023 3 ?
γ	2993.4 5	0.032 3 ?
γ	3047.18 8	0.057 4
γ [M1]	3229.23 5	1.48 11
γ [M1+E2]	3256.43 6	0.106 15
γ	3381.37 6	1.40 11
γ [M1+E2]	3422.85 7	0.83 8
γ	3432.86 19	0.28 3
γ	3724.9 13	0.0023 4 ?
γ	3736.9 8	0.0121 19 ?
γ [M1+E2]	3767.68 13	0.136 19
γ M1	3791.61 6	1.02 11
γ	3806.4 13	0.0023 4 ?
γ	3811.2 5	0.0080 15 ?
γ	3827.0 5	0.0064 11 ?
γ	4086.48 8	1.14 19
γ [M1]	4295.73 7	3.5 7
γ M1	4462.14 7	0.72 15
γ M1	4806.97 14	1.5 4

† 3.2% uncert(syst)

Atomic Electrons (^{66}Ga)

⟨e⟩=2.13 11 keV

e_{bin}(keV)	⟨e⟩(keV)	e(%)
1	0.36	35 4
7	0.42	5.7 6
8	0.89	11.5 12
9	0.27	3.1 3
162 - 172	0.0008	~0.0005
281 - 291	0.00060	0.000212 16
401 - 449	0.0016	0.00037 12
450 - 460	0.0007	~0.00015
569 - 579	0.000279	4.9 3 $\times10^{-5}$
677 - 686	0.00099	0.000146 7
824 - 857	0.0221	0.00268 15
898 - 914	0.00049	5.4 19 $\times10^{-5}$
971 - 1009	0.00023	2.4 6 $\times10^{-5}$
1030	0.095	0.0092 4
1038 - 1060	0.0111	0.00107 4
1138 - 1181	0.00042	3.6 5 $\times10^{-5}$
1189 - 1233	0.00132	0.000108 5
1324 - 1357	0.0038	0.000286 22
1409 - 1459	0.00133	9.4 3 $\times10^{-5}$
1499 - 1508	0.00079	5.2 15 $\times10^{-5}$
1731 - 1741	8.2 $\times10^{-5}$	4.7 18 $\times10^{-6}$
1890 - 1919	0.00365	0.000192 9
2056 - 2066	4.42 $\times10^{-5}$	2.15 10 $\times10^{-6}$
2164 - 2214	0.0076	0.000347 19
2283 - 2293	5.1 $\times10^{-5}$	2.25 22 $\times10^{-6}$
2384 - 2483	0.00258	0.000107 6
2491 - 2589	3.4 $\times10^{-5}$	1.33 12 $\times10^{-6}$
2743	0.0226	0.00082 4
2751 - 2780	0.00273	0.000099 4
2924 - 2993	0.000273	9.3 6 $\times10^{-6}$
3038 - 3047	4.8 $\times10^{-5}$	1.6 3 $\times10^{-6}$
3220 - 3256	0.00154	4.8 4 $\times10^{-5}$
3372 - 3433	0.00215	6.3 7 $\times10^{-5}$
3715 - 3811	0.00105	2.78 25 $\times10^{-5}$
3817 - 3827	4.9 $\times10^{-6}$	1.3 3 $\times10^{-7}$

Atomic Electrons (^{66}Ga) (continued)

e_{bin}(keV)	⟨e⟩(keV)	e(%)
4077 - 4086	0.00085	2.1 5 $\times10^{-5}$
4286 - 4296	0.0029	6.8 12 $\times10^{-5}$
4452 - 4462	0.00059	1.32 25 $\times10^{-5}$
4797 - 4807	0.0012	2.4 6 $\times10^{-5}$

Continuous Radiation (^{66}Ga)

⟨β+⟩=986 keV;⟨IB⟩=3.7 keV

E_{bin}(keV)		⟨ ⟩(keV)	(%)
0 - 10	β+	5.6 $\times10^{-5}$	0.00070
	IB	0.031	
10 - 20	β+	0.00107	0.0066
	IB	0.030	0.21
20 - 40	β+	0.0149	0.0469
	IB	0.059	0.20
40 - 100	β+	0.322	0.432
	IB	0.170	0.26
100 - 300	β+	5.9	2.87
	IB	0.52	0.29
300 - 600	β+	21.4	4.76
	IB	0.69	0.161
600 - 1300	β+	109	11.3
	IB	1.16	0.133
1300 - 2500	β+	463	24.5
	IB	0.80	0.046
2500 - 5000	β+	386	13.0
	IB	0.20	0.0065
5000 - 5175	IB	5.7 $\times10^{-5}$	1.13 $\times10^{-6}$
	Σβ+		57

$^{66}_{32}$Ge(2.26 5 h)

Mode: ε

Δ: -61622 13 keV

SpA: 2.10$\times10^{7}$ Ci/g

Prod: ^{64}Zn(α,2n); ^{56}Fe(^{12}C,2n); ^{64}Zn(^{3}He,n)

Photons (^{66}Ge)

⟨γ⟩=440 5 keV

γ_{mode}	γ(keV)	γ(%)†
Ga L_ℓ	0.957	0.054 12
Ga L_η	0.984	0.033 9
Ga L_α	1.098	0.87 18
Ga L_β	1.132	0.58 15
Ga L_γ	1.297	0.0020 6
Ga $K_{\alpha2}$	9.225	15.4 9
Ga $K_{\alpha1}$	9.252	30.2 18
Ga $K_{\beta1}$'	10.263	5.7 3
Ga $K_{\beta2}$'	10.444	0.0333 20
γ M1+0.7%E2	22.25 4	1.58 11
γ	38.61 19	<0.17
γ	40.00 19	~0.37
γ M1,E1	41.91 10	~0.14
γ [M1]	42.81 4	1.1 3 ?
γ M1	43.83 3	29.0 9
γ M1,E1	53.55 5	<0.31
γ	55.4 12	~0.08
γ (M1+<0.6%E2)	65.05 3	7.2 3
γ M1,E1	71.66 4	0.17 3
γ [M1]	90.97 3	0.39 11
γ M1,7.4%E1	96.36 5	0.20 6
γ (M1)	108.880 25	10.5 3
γ [M1]	125.207 22	0.31 3
γ [M1]	147.812 20	1.30 25

Photons (^{66}Ge) (continued)

γ_{mode}	γ(keV)	γ(%)†
γ [M1]	154.769 22	0.31 3
γ [M1]	169.36 3	~0.17
γ M1+5.7%E2	182.05 3	5.7 3
γ M1+7.6%E2	190.258 22	5.7 4
γ	196.2 3	<0.11
γ	201.69 14	0.08 3
γ [M1+E2]	224.86 5	0.20 6 ?
γ M1,E2	234.09 3	0.08 3
γ [M1]	245.74 3	5.41 20
γ (M1+6.1%E2)	273.019 22	10.5 4
γ (M1)	290.93 3	~0.25
γ [M1]	302.581 22	2.51 20
γ [M1]	315.82 4	0.82 3
γ [M1]	324.13 3	0.14 3
γ M1(+1.0%E2)	338.070 23	8.7 6
γ	370.6 4	<0.11
γ M1	381.90 3	28.2
γ [M1]	415.09 4	0.42 6
γ [M1]	427.787 24	0.54 11
γ [M1]	470.59 4	7.4 4
γ [M1]	471.94 3	3.27 23
γ [M1]	483.90 11	0.08 3
γ [M1]	492.84 3	0.62 3
γ	529.9 4	0.08 3
γ M1	536.67 3	6.18 20
γ	555.16 11	0.113 23 ?
γ [M1]	597.14 3	0.25 3
γ	619.64 11	0.085 17 ?
γ [M1]	639.95 5	0.59 3
γ [M1]	662.19 3	0.11 3
γ	664.11 10	0.08 3
γ M1	706.02 3	4.29 20
γ	723.67 19	0.08 3
γ	740.08 8	~0.06
γ [M1]	756.92 11	0.65 3
γ	782.37 24	~0.06
γ [M1]	799.72 11	~0.023
γ	811.74 9	~0.023
γ [M1]	821.97 11	<0.06
γ (M1)	865.80 11	0.25 6 ?
γ	892.66 11	~0.06
γ	908.09 8	~0.037
γ [M1+E2]	919.39 7	0.14 6
γ	930.34 8	0.028 11
γ	935.46 12	0.042 9
γ	974.17 9	~0.025
γ	995.9 3	0.062 17
γ	1010.5 4	~0.017
γ [M1+E2]	1020.03 4	~0.014
γ	1059.6 8	~0.014
γ	1094.8 6	~0.05
γ	1101.31 4	0.149 20
γ	1120.44 24	~0.028
γ	1144.12 5	~0.034
γ [M1+E2]	1165.12 7	0.23 6
γ [M1+E2]	1174.79 4	0.121 20
γ	1221.96 7	0.44 3
γ	1250.3 10	~0.014
γ [M1+E2]	1265.76 4	0.079 14
γ	1322.60 3	0.45 3
γ	1329.5 8	~0.08
γ	1339.7 3	~0.014
γ [M1+E2]	1347.17 7	0.065 23
γ [M1+E2]	1387.49 22	~0.014
γ [M1+E2]	1412.22 7	0.37 3
γ M1	1456.05 7	0.08 3
γ [M1+E2]	1478.45 22	<0.039
γ [M1+E2]	1490.61 5	0.11 3
γ [M1+E2]	1507.7 3	~0.17
γ [M1+E2]	1512.86 3	0.68 9
γ	1536.2 8	~0.011
γ	1548.4 8	<0.017
γ [M1+E2]	1660.50 22	~0.05
γ M1	1769.38 22	0.10 3

† 1.4% uncert(syst)

Atomic Electrons (^{66}Ge)

⟨e⟩=18.6 *12* keV

e_{bin}(keV)	⟨e⟩(keV)	e(%)
1	1.04	92 *11*
8	3.2	40 *4*
9	0.9	9.9 *11*
10	0.051	0.50 *6*
12	0.98	8.2 *8*
21 - 22	0.32	1.5 *3*
30	1.0	<7
32	0.25	0.76 *19*
33	6.12	18.3 *7*
39 - 42	0.4	~0.9
43	1.1	2.6 *7*
44 - 54	0.34	0.7 *3*
55	0.88	1.61 *9*
61 - 96	0.32	0.49 *17*
99	0.52	0.53 *3*
108 - 155	0.139	0.116 *7*
159 - 202	0.36	0.205 *21*
214 - 263	0.247	0.099 *6*
272 - 316	0.063	0.0217 *12*
323 - 372	0.312	0.0871 *24*
381 - 428	0.0343	0.0089 *3*
460 - 493	0.076	0.0164 *7*
520 - 555	0.0362	0.0069 *3*

Atomic Electrons (^{66}Ge)
(continued)

e_{bin}(keV)	⟨e⟩(keV)	e(%)
587 - 630	0.0041	0.00067 *6*
639 - 664	0.00114	0.00018 *3*
696 - 740	0.0180	0.00258 *12*
747 - 789	0.00270	0.000360 *23*
798 - 822	0.00017	$\sim 2\times10^{-5}$
855 - 898	0.0006	$\sim 7\times10^{-5}$
907 - 935	0.00067	7.3 *20* $\times10^{-5}$
964 - 1011	0.00028	2.8 *9* $\times10^{-5}$
1019 - 1060	3.5×10^{-5}	$\sim 3\times10^{-6}$
1084 - 1134	0.00054	4.9 *13* $\times10^{-5}$
1143 - 1175	0.00088	7.6 *13* $\times10^{-5}$
1212 - 1255	0.0010	8.2 *22* $\times10^{-5}$
1264 - 1312	0.00070	5.3 *18* $\times10^{-5}$
1319 - 1347	0.00039	2.9 *10* $\times10^{-5}$
1377 - 1412	0.00079	5.6 *5* $\times10^{-5}$
1446 - 1491	0.00042	2.9 *5* $\times10^{-5}$
1497 - 1548	0.0017	0.000111 *17*
1650 - 1660	8.0×10^{-5}	$\sim 5\times10^{-6}$
1759 - 1769	0.00016	9 *3* $\times10^{-6}$

Continuous Radiation (^{66}Ge)

⟨β+⟩=82 keV;⟨IB⟩=0.94 keV

E_{bin}(keV)		⟨ ⟩(keV)	(%)
0 - 10	β+	0.000141	0.00177
	IB	0.0063	
10 - 20	β+	0.00281	0.0174
	IB	0.0047	0.035
20 - 40	β+	0.0409	0.129
	IB	0.0072	0.025
40 - 100	β+	0.95	1.27
	IB	0.019	0.029
100 - 300	β+	19.7	9.5
	IB	0.065	0.035
300 - 600	β+	45.1	10.7
	IB	0.18	0.040
600 - 1300	β+	16.2	2.27
	IB	0.56	0.064
1300 - 2058	IB	0.091	0.0063
	Σβ+		24

$^{66}_{33}$As(95.8 *4* ms)

Mode: ϵ

Δ: -52070 *50* keV

SpA: 2.464×10^{11} Ci/g

Prod: ^{58}Ni(^{10}B,2n); ^{40}Ca(^{32}S,npα); ^{50}Cr(^{19}F,3n)

A = 67

NDS **39**, 741 (1983)

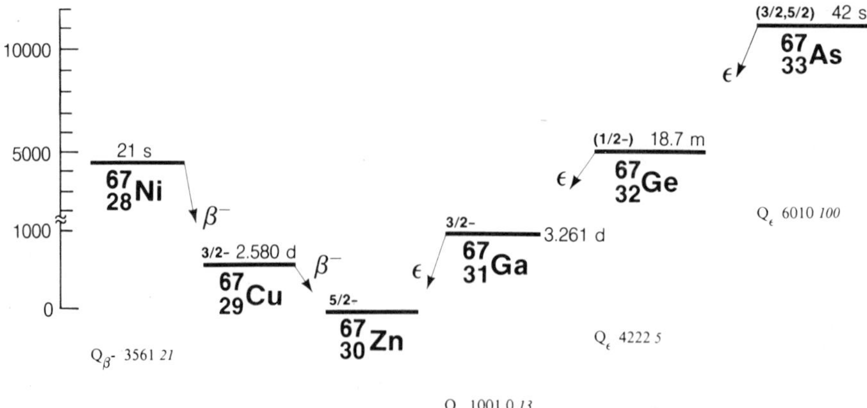

$^{67}_{28}$Ni(21 *1* s)

Mode: β-
Δ: -63742 *19* keV
SpA: 7.9×10^{9} Ci/g
Prod: ^{70}Zn(n,α)

Photons (^{67}Ni)

γ_{mode}	γ(keV)	γ(rel)
γ	100.0 *4*	~37
γ	140.6 *8*	29 *4*
γ	208.0 *6*	50 *4*
γ	553.1 *4*	32 *7*
γ	708.5 *3*	68 *11*
γ	751.5 *2*	17 *4*
γ	779.1 *4*	20 *7*
γ	874.1 *4*	65 *26* *
γ	1072.2 *3*	100 *37* *
γ	1100.4 *5*	23 *5*
γ	1226.5 *8*	15 *5*
γ	1653.9 *4*	74
γ	1760.2 *7*	29 *7*
γ	1809 *1*	22 *9*
γ	1938 *1*	18 *4*
γ	1975 *1*	48 *8*

* may include impurity

$^{67}_{29}$Cu(2.580 *4* d)

Mode: β-
Δ: -67303 *8* keV
SpA: 7.554×10^{5} Ci/g
Prod: ^{64}Ni(α,p); ^{67}Zn(n,p); ^{65}Cu(t,p); multiple n-capture from ^{65}Cu

Photons (^{67}Cu)

⟨γ⟩=115.2 *16* keV

γ_{mode}	γ(keV)	γ(%)
Zn L_ℓ	0.884	0.0069 *15*
Zn L_η	0.907	0.0042 *11*
Zn L_α	1.012	0.105 *22*
Zn L_β	1.042	0.069 *18*
Zn $K_{\alpha2}$	8.616	1.88 *8*
Zn $K_{\alpha1}$	8.639	3.68 *16*
Zn $K_{\beta1}$'	9.572	0.68 *3*
γ M1+0.6%E2	91.267 *4*	7.02 *12*
γ E2	93.312 *4*	16.1 *3*
γ M1+0.8%E2	184.578 *5*	48.6 *8*
γ M1+0.2%E2	208.951 *6*	0.112 *4*
γ M1+0.7%E2	300.218 *6*	0.797 *16*
γ M1+3.1%E2	393.529 *6*	0.223 *7*

Atomic Electrons (^{67}Cu)

⟨e⟩=14.0 *3* keV

e_{bin}(keV)	⟨e⟩(keV)	e(%)
1	0.127	12.3 *13*
7	0.143	1.93 *20*
8	0.30	3.9 *4*
9	0.091	1.06 *11*
10	0.00074	0.0077 *8*
82	0.43	0.53 *3*
84	10.1	12.1 *3*
90	0.051	0.056 *4*
91	0.0081	0.0089 *7*

Atomic Electrons (^{67}Cu) (continued)

e_{bin}(keV)	⟨e⟩(keV)	e(%)
92	1.39	1.51 *4*
93	0.216	0.232 *7*
175	0.97	0.555 *20*
183	0.101	0.0550 *19*
184	0.0205	0.0111 *6*
185	0.00088	0.000479 *17*
199	0.00181	0.00091 *4*
208	0.000195	9.4 *4* ×10^{-5}
209	3.09 ×10^{-5}	1.48 *6* ×10^{-5}
291	0.00773	0.00266 *9*
299	0.00081	0.000272 *9*
300	0.000129	4.29 *14* ×10^{-5}
384	0.00148	0.000387 *21*
392	0.000153	3.89 *21* ×10^{-5}
393	2.43 ×10^{-5}	6.2 *4* ×10^{-6}
394	1.31 ×10^{-6}	3.33 *17* ×10^{-7}

Continuous Radiation (^{67}Cu)

⟨β-⟩=142 keV;⟨IB⟩=0.070 keV

E_{bin}(keV)		⟨ ⟩(keV)	(%)
0 - 10	β-	0.218	4.35
	IB	0.0073	
10 - 20	β-	0.64	4.30
	IB	0.0066	0.046
20 - 40	β-	2.53	8.5
	IB	0.0113	0.039
40 - 100	β-	16.7	24.0
	IB	0.023	0.036
100 - 300	β-	93	51.0
	IB	0.021	0.0142
300 - 576	β-	28.5	8.0
	IB	0.00118	0.00034

$^{67}_{30}$Zn(stable)

Δ: -67879.3 *16* keV
%: 4.1 *1*

$^{67}_{31}$Ga(3.261 *1* d)

Mode: ϵ
Δ: -66878.4 *17* keV
SpA: 5.9763×10^{5} Ci/g
Prod: ^{66}Zn(d,n); ^{65}Cu(α,2n)

Photons (^{67}Ga)

⟨γ⟩=154.9 *13* keV

γ_{mode}	γ(keV)	γ(%)†
Zn L_ℓ	0.884	0.061 *13*
Zn L_η	0.907	0.037 *9*
Zn L_α	1.012	0.92 *19*
Zn L_β	1.043	0.61 *16*
Zn $K_{\alpha2}$	8.616	16.7 *7*
Zn $K_{\alpha1}$	8.639	32.7 *14*
Zn $K_{\beta1}$'	9.572	5.99 *25*
γ M1+0.6%E2	91.267 *4*	2.95 *5*
γ E2	93.312 *4*	37.0 *7*
γ M1+0.8%E2	184.578 *5*	20.4 *4*
γ M1+0.2%E2	208.951 *6*	2.33 *6*
γ M1+0.7%E2	300.218 *6*	16.6 *3*
γ M1+3.1%E2	393.529 *6*	4.64 *9*

Photons (^{67}Ga) (continued)

γ_{mode}	γ(keV)	γ(%)†
γ M1+2.0%E2	494.166 *9*	0.068 *2*
γ [M1+E2]	703.115 *8*	0.0108 *8*
γ E2	794.381 *8*	0.0509 *15*
γ E2+4.7%M1	814.7 *5*	~2×10^{-5}
γ M1+43%E2	887.692 *8*	0.144 *3*

† 15% uncert(syst)

Atomic Electrons (^{67}Ga)

⟨e⟩=33.2 *7* keV

e_{bin}(keV)	⟨e⟩(keV)	e(%)
1	1.10	106 *11*
7	1.27	17.1 *18*
8	2.7	35 *4*
9 - 10	0.82	9.5 *10*
82	0.181	0.222 *14*
84	23.2	27.8 *8*
90 - 91	0.0248	0.0274 *18*
92	3.20	3.47 *10*
175 - 209	0.502	0.283 *9*
291 - 300	0.181	0.0620 *17*
384 - 394	0.0346	0.0090 *4*
485 - 494	0.000386	8.0 *3* ×10^{-5}
693 - 703	4.7 ×10^{-5}	6.8 *12* ×10^{-6}
785 - 815	0.000225	2.86 *19* ×10^{-5}
878 - 888	0.000449	5.11 *19* ×10^{-5}

Continuous Radiation (^{67}Ga)

⟨IB⟩=0.132 keV

E_{bin}(keV)		⟨ ⟩(keV)	(%)
10 - 20	IB	0.00034	0.0028
20 - 40	IB	0.00021	0.00071
40 - 100	IB	0.00167	0.0023
100 - 300	IB	0.029	0.0142
300 - 600	IB	0.075	0.018
600 - 908	IB	0.022	0.0032

$^{67}_{32}$Ge(18.7 *5* min)

Mode: ϵ
Δ: -62656 *5* keV
SpA: 1.50×10^{8} Ci/g
Prod: ^{64}Zn(α,n)

Photons (^{67}Ge)

⟨γ⟩=475 *9* keV

γ_{mode}	γ(keV)	γ(%)†
Ga L_ℓ	0.957	0.0038 *8*
Ga L_η	0.984	0.0024 *6*
Ga L_α	1.098	0.061 *13*
Ga L_β	1.132	0.041 *10*
Ga L_γ	1.297	0.00015 *5*
Ga $K_{\alpha2}$	9.225	1.10 *5*
Ga $K_{\alpha1}$	9.252	2.16 *10*
Ga $K_{\beta1}$'	10.263	0.406 *18*
Ga $K_{\beta2}$'	10.444	0.00238 *11*
γ [M1+E2]	167.05 *5*	84 *3*
γ [M1+E2]	253.34 *20*	0.32 *3*
γ M1+0.4%E2	359.43 *15*	1.47 *15*
γ [M1+E2]	468.93 *19*	0.127 *15*
γ M1+33%E2	551.85 *21*	0.083 *10*
γ [M1+E2]	558.30 *23*	0.059 *5*
γ M1+12%E2	661.31 *15*	0.30 *3*

Photons (^{67}Ge)
(continued)

γ_{mode}	γ(keV)	γ(%)†
γ [M1+E2]	728.22 *22*	0.44 *10*
γ [M1+E2]	728.73 *21*	2.35 *25*
γ E2	744.22 *18*	0.069 *10*
γ [M1+E2]	811.64 *19*	0.78 *8*
γ M1+2.1%E2	828.36 *15*	2.99 *25*
γ M1+E2	898.64 *20*	0.96 *10*
γ M1+10%E2	911.27 *18*	3.1 *3*
γ M1+E2	914.65 *17*	3.0 *3*
γ M1+46%E2	981.56 *19*	1.13 *10*
γ M1+E2	1081.70 *17*	1.03 *10*
γ M1+30%E2	1196.5 *10*	0.028 *3*
γ E2	1203 *1*	0.0127 *25*
γ M1+E2	1280.57 *19*	0.38 *4*
γ	1317.6 *8*	0.078 *10*
γ M1+E2	1450.48 *19*	0.66 *6*
γ M1+2.5%E2	1472.94 *17*	4.9
γ [M1+E2]	1639.99 *17*	0.64 *10*
γ [M1+E2]	1642.85 *17*	0.88 *10*
γ	1668.8 *10*	0.0127 *25*
γ	1679.3 *7*	0.0113 *24*
γ	1708.4 *5*	0.044 *5*
γ	1791.3 *5*	0.064 *10*
γ M1+11%E2	1809.90 *17*	1.32 *15*
γ	1845.5 *7*	0.025 *5*
γ	1976.1 *5*	0.137 *15*
γ	2080.4 *7*	0.0216 *25*
γ	2143.6 *4*	0.059 *20*
γ	2396.9 *4*	0.069 *10*
γ	2452.6 *5*	0.0157 *15*
γ	2526.6 *5*	0.206 *20*
γ	2563.6 *4*	0.225 *25*
γ	2668.8 *8*	0.035 *3*
γ	2730.7 *4*	0.093 *10*
γ	2865.8 *4*	0.103 *10*
γ	3042.2 *4*	0.034 *5*
γ	3058.2 *3*	0.191 *20*
γ	3123.1 *7*	0.069 *10*
γ	3144.2 *10*	0.0132 *15*
γ	3162.0 *7*	0.113 *10*
γ	3225.3 *3*	0.029 *3*
γ	3307.8 *10*	0.0176 *20*
γ	3361.9 *10*	0.0201 *20*
γ	3368.6 *9*	0.0176 *20*
γ	3401.6 *4*	0.152 *15*
γ	3465.1 *14*	0.0083 *15*
γ	3612.4 *20*	0.0059 *20*
γ	3632.1 *14*	0.0059 *20*
γ	3655.4 *7*	0.0044 *15*
γ	3728.0 *9*	0.0127 *25*

† 4.1% uncert(syst)

Atomic Electrons (^{67}Ge)
⟨e⟩=2.84 *12* keV

e_{bin}(keV)	⟨e⟩(keV)	e(%)
1	0.074	6.5 *7*
8	0.227	2.8 *3*
9 - 10	0.068	0.74 *7*
157	2.14	1.36 *7*
166	0.237	0.143 *8*
243 - 253	0.0050	0.00205 *18*
349 - 359	0.0140	0.0040 *4*
459 - 469	0.00085	0.000186 *20*
541 - 558	0.00085	0.000155 *19*
651 - 661	0.00140	0.000215 *19*
718 - 744	0.0113	0.00158 *18*
801 - 828	0.0130	0.00159 *10*
888 - 915	0.0220	0.00244 *13*
971 - 982	0.0034	0.00035 *4*
1071 - 1082	0.00266	0.000248 *22*
1186 - 1203	0.000101	8.5 *8* $\times10^{-6}$
1270 - 1318	0.00096	7.5 *7* $\times10^{-5}$
1440 - 1473	0.0107	0.00074 *3*
1630 - 1708	0.00275	0.000168 *14*
1781 - 1845	0.00225	0.000125 *12*
1966 - 1976	0.00017	8.4 *23* $\times10^{-6}$
2070 - 2143	9.2 $\times10^{-5}$	4.4 *13* $\times10^{-6}$
2387 - 2453	9.0 $\times10^{-5}$	3.7 *8* $\times10^{-6}$

Atomic Electrons (^{67}Ge)
(continued)

e_{bin}(keV)	⟨e⟩(keV)	e(%)
2516 - 2564	0.00044	1.7 *3* $\times10^{-5}$
2658 - 2731	0.000127	4.7 *8* $\times10^{-6}$
2855 - 2866	0.000099	3.5 *8* $\times10^{-6}$
3032 - 3123	0.00027	8.9 *14* $\times10^{-6}$
3134 - 3225	0.000142	4.5 *7* $\times10^{-6}$
3297 - 3391	0.00017	5.0 *8* $\times10^{-6}$
3400 - 3465	2.1 $\times10^{-5}$	6.2 *10* $\times10^{-7}$
3602 - 3655	1.4 $\times10^{-5}$	3.8 *8* $\times10^{-7}$
3718 - 3728	1.1 $\times10^{-5}$	2.9 *7* $\times10^{-7}$

Continuous Radiation (^{67}Ge)
⟨β+⟩=1178 keV;⟨IB⟩=3.5 keV

E_{bin}(keV)		⟨ ⟩(keV)	(%)
0 - 10	β+	3.22 $\times10^{-5}$	0.000402
	IB	0.041	
10 - 20	β+	0.00065	0.00404
	IB	0.041	0.28
20 - 40	β+	0.0098	0.0308
	IB	0.079	0.28
40 - 100	β+	0.253	0.334
	IB	0.23	0.35
100 - 300	β+	8.0	3.68
	IB	0.65	0.36
300 - 600	β+	52	11.2
	IB	0.74	0.17
600 - 1300	β+	340	35.7
	IB	1.05	0.122
1300 - 2500	β+	713	40.0
	IB	0.60	0.036
2500 - 4222	β+	65	2.44
	IB	0.108	0.0037
	$\Sigma\beta$+		93

$^{67}_{33}$As(42.5 *12* s)

Mode: ϵ

Δ: -56650 *100* keV

SpA: 3.93$\times10^{9}$ Ci/g

Prod: ^{58}Ni(^{14}N,αn)

Photons (^{67}As)
⟨γ⟩=451 *41* keV

γ_{mode}	γ(keV)	γ(%)†
Ge L_ℓ	1.037	0.032 *9*
Ge L_η	1.068	0.018 *6*
Ge L_α	1.188	0.56 *15*
Ge L_β	1.224	0.31 *10*
Ge L_γ	1.412	0.0024 *8*
Ge $K_{\alpha2}$	9.855	6.0 *10*
Ge $K_{\alpha1}$	9.886	11.7 *19*
Ge $K_{\beta1}'$	10.980	2.3 *4*
Ge $K_{\beta2}'$	11.184	0.037 *6*
γ E2	18.26 *21*	0.16 *3*
γ E2+>5.9%M1	104.40 *18*	4.55 *12*
γ E2+>4.0%M1	120.95 *18*	9.52 *19*
γ M1+E2	122.66 *18*	19.2
γ [M1+E2]	225.35 *20*	1.48 *15*
γ (M1+E2)	243.61 *20*	7.7 *4*
γ	247.9 *3*	1.5 *6*
γ	500.3 *3*	
γ (E2+44%M1)	588.9 *3*	1.98 *21*
γ	632.9 *3*	2.5 *6*
γ	685.47 *20*	2.3 *6*
γ [M1+E2]	693.3 *3*	4.8 *6*
γ [M1+E2]	776.38 *22*	~1.0
γ	789.87 *21*	4.8 *15*
γ	808.13 *21*	6.1 *12*
γ [M1+E2]	897.33 *22*	3.1 *10*

Photons (^{67}As)
(continued)

γ_{mode}	γ(keV)	γ(%)†
γ	1050.3 *4*	0.29 *13*
γ	1151.4 *5*	0.58 *21*
γ	1171.2 *4*	1.5 *4*
γ	1274.1 *5*	1.3 *6*
γ	1293.9 *4*	~1
γ	1385.3 *7*	~0.33
γ	1576.0 *10*	0.7 *4*
γ	1657.2 *4*	~0.38
γ	1912.8 *7*	
γ	2128.4 *10*	~0.38
γ	2156.4 *7*	
γ	2218.2 *10*	3.8 *13*
γ	2280.0 *10*	~0.29
γ	2474.5 *7*	

† 13% uncert(syst)

Atomic Electrons (^{67}As)
⟨e⟩=20 *3* keV

e_{bin}(keV)	⟨e⟩(keV)	e(%)
1	0.66	54 *10*
7	2.0	28 *5*
8	0.41	4.9 *9*
9	0.76	8.8 *16*
10 - 11	0.35	3.6 *6*
17	4.0	24 *4*
18	0.66	3.6 *7*
93	2.23	2.39 *8*
103 - 104	0.371	0.359 *10*
110	3.23	2.95 *8*
112	3.7	~3
120 - 123	1.1	0.9 *3*
214 - 248	0.45	0.19 *7*
578 - 622	0.025	0.0042 *10*
632 - 674	0.010	0.0015 *7*
682 - 693	0.027	0.0040 *7*
765 - 808	0.042	0.0053 *16*
886 - 897	0.012	0.0013 *4*
1039 - 1050	0.0007	~7$\times10^{-5}$
1140 - 1171	0.0044	0.00038 *13*
1263 - 1294	0.0051	0.00040 *19*
1374 - 1385	0.0006	~4$\times10^{-5}$
1565 - 1657	0.0017	0.00011 *4*
2117 - 2217	0.0051	0.00023 *9*
2218 - 2280	0.00041	1.8 *9* $\times10^{-5}$

Continuous Radiation (^{67}As)
⟨β+⟩=2084 keV;⟨IB⟩=7.3 keV

E_{bin}(keV)		⟨ ⟩(keV)	(%)
0 - 10	β+	7.5$\times10^{-6}$	9.3$\times10^{-5}$
	IB	0.060	
10 - 20	β+	0.000158	0.00098
	IB	0.059	0.41
20 - 40	β+	0.00247	0.0077
	IB	0.117	0.41
40 - 100	β+	0.067	0.089
	IB	0.34	0.52
100 - 300	β+	2.36	1.08
	IB	1.04	0.58
300 - 600	β+	18.1	3.88
	IB	1.30	0.30
600 - 1300	β+	180	18.2
	IB	2.1	0.24
1300 - 2500	β+	823	43.3
	IB	1.8	0.102
2500 - 4970	β+	1061	33.5
	IB	0.46	0.0152
	$\Sigma\beta$+		100

A = 68

NDS 33, 481 (1981)

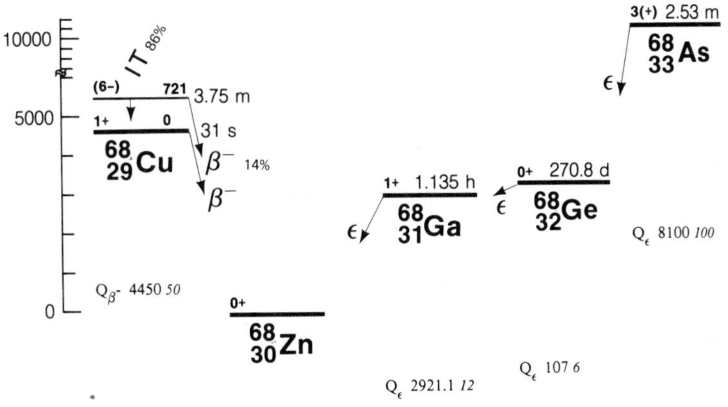

$^{68}_{29}$Cu(31 1 s)

Mode: β-
Δ: -65560 50 keV
SpA: 5.29×10^{9} Ci/g
Prod: ^{71}Ga(n,α); ^{68}Zn(n,p); ^{70}Zn(γ,np)

Photons (^{68}Cu)

⟨γ⟩=986 45 keV

γ_{mode}	γ(keV)	γ(%)†
γ	570.9 10	
γ E2	578.3 3	2.1 6
γ	736.6 15	
γ E2+32%M1	805.80 8	0.36 12
γ E2	1077.29 8	60 4
γ M1+3.3%E2	1261.18 10	14.9 10
γ	1293.3 11	
γ	1432.8 12	
γ	1530.3 11	0.87 22
γ [E1]	1675.8 12	
γ M1+6.8%E2	1744.67 19	2.2 4
γ E2	1883.08 8	0.52 18
γ	2108.6 11	1.6 4
γ E2	2338.45 12	1.6 7
γ E2	2821.93 21	0.11 3

† 1.6% uncert(syst)

Continuous Radiation (^{68}Cu)

⟨β-⟩=1487 keV;⟨IB⟩=4.5 keV

E_{bin}(keV)		⟨ ⟩(keV)	(%)
0 - 10	β-	0.0070	0.138
	IB	0.048	
10 - 20	β-	0.0214	0.142
	IB	0.047	0.33
20 - 40	β-	0.090	0.298
	IB	0.093	0.32
40 - 100	β-	0.72	1.02
	IB	0.27	0.41
100 - 300	β-	9.8	4.74
	IB	0.78	0.43
300 - 600	β-	47.0	10.3
	IB	0.91	0.21
600 - 1300	β-	285	29.9
	IB	1.34	0.154
1300 - 2500	β-	720	39.3
	IB	0.86	0.050
2500 - 4450	β-	424	14.1
	IB	0.128	0.0044

$^{68}_{29}$Cu(3.75 5 min)

Mode: IT(86.0 11 %), β-(14.0 11 %)
Δ: -64839 50 keV
SpA: 7.362×10^{8} Ci/g
Prod: ^{68}Zn(n,p); ^{71}Ga(n,α)

Photons (^{68}Cu)

⟨γ⟩=1007 85 keV

γ_{mode}	γ(keV)	γ(%)†
Cu L_{ℓ}	0.811	0.036 10
Cu L_{η}	0.831	0.021 5
Cu L_{α}	0.930	0.49 13
Cu L_{β}	0.957	0.32 9
Cu $K_{\alpha2}$	8.028	8.7 7
Cu $K_{\alpha1}$	8.048	17.0 15
Cu $K_{\beta1}$'	8.905	3.1 3
γ_{IT} [M1+8.8%E2]	84.4 4	72 3
γ_{IT} [M3]	111.0 5	17.4 8
$\gamma_{\beta-}$(M1)	151.3 10	4.9 5
$\gamma_{\beta-}$	497.7 9	
γ_{IT} [E1]	525.7 4	75 15
$\gamma_{\beta-}$	570.9 10	
$\gamma_{\beta-}$E2	578.3 3	
$\gamma_{\beta-}$	586.0 11	0.49 13
γ_{IT} [M2]	610.1 5	1.13 23
γ_{IT} [M4]	636.7 5	9.7 14
$\gamma_{\beta-}$	670.1 12	<0.56 ?
$\gamma_{\beta-}$	736.6 15	
$\gamma_{\beta-}$E2+32%M1	805.80 8	0.9 3
$\gamma_{\beta-}$	1014.3 12	
$\gamma_{\beta-}$(E1)	1040.7 10	8.5 9
$\gamma_{\beta-}$[E2]	1073.5 9	1.5 4 ?
$\gamma_{\beta-}$E2	1077.29 8	12.9 15
$\gamma_{\beta-}$	1148.9 13	
$\gamma_{\beta-}$	1221.9 10	
$\gamma_{\beta-}$M1+3.3%E2	1261.18 10	
$\gamma_{\beta-}$	1293.3 11	<0.8
$\gamma_{\beta-}$E2	1339.5 10	10.7 11
$\gamma_{\beta-}$	1384.8 20	
$\gamma_{\beta-}$	1432.8 12	
$\gamma_{\beta-}$	1530.3 11	
$\gamma_{\beta-}$	1540.0 12	0.76 22
$\gamma_{\beta-}$[E1]	1675.8 12	<1
$\gamma_{\beta-}$M1+6.8%E2	1744.67 19	
$\gamma_{\beta-}$E2	1883.08 8	1.3 5
$\gamma_{\beta-}$	2108.6 11	
$\gamma_{\beta-}$E2	2338.45 12	

† uncert(syst): 12% for β-, 2.0% for IT

Atomic Electrons (^{68}Cu)*

⟨e⟩=78 4 keV

e_{bin}(keV)	⟨e⟩(keV)	e(%)
1	0.58	61 8
7	2.04	29 4
8	0.57	7.2 9
9	0.037	0.42 5
75	8.8	12 4
83	1.1	1.3 4
102	54.9	54 3
110	8.6	7.9 5
517	0.17	0.032 7
525	0.017	0.0032 7
601	0.0111	0.0018 4
609	0.00114	0.00019 4
628	0.59	0.094 14
636	0.065	0.0102 15

* with IT

Continuous Radiation (^{68}Cu)

⟨β-⟩=110 keV;⟨IB⟩=0.21 keV

E_{bin}(keV)		⟨ ⟩(keV)	(%)
0 - 10	β-	0.00255	0.0507
	IB	0.0044	
10 - 20	β-	0.0078	0.052
	IB	0.0043	0.030
20 - 40	β-	0.0327	0.109
	IB	0.0083	0.029
40 - 100	β-	0.262	0.369
	IB	0.023	0.035
100 - 300	β-	3.44	1.67
	IB	0.060	0.034
300 - 600	β-	15.1	3.33
	IB	0.057	0.0135
600 - 1300	β-	59	6.6
	IB	0.050	0.0061
1300 - 2500	β-	31.5	1.91
	IB	0.0088	0.00057
2500 - 2801	β-	0.302	0.0118
	IB	4.7×10^{-6}	1.8×10^{-7}

$^{68}_{30}$Zn(stable)

Δ: -70006.1 *17* keV
%: 18.8 *4*

$^{68}_{31}$Ga(1.135 *5* h)

Mode: ϵ
Δ: -67085.0 *21* keV
SpA: 4.060×10^{7} Ci/g
Prod: daughter ^{68}Ge; ^{65}Cu(α,n); ^{68}Zn(p,n); ^{67}Zn(d,n); ^{69}Ga(n,2n); ^{70}Ge(γ,np)

Photons (^{68}Ga)

⟨γ⟩=38 *3* keV

γ_{mode}	γ(keV)	γ(%)†
Zn L_ℓ	0.884	0.0050 *11*
Zn L_η	0.907	0.0030 *8*
Zn L_α	1.012	0.076 *16*
Zn L_β	1.043	0.050 *13*
Zn $K_{\alpha2}$	8.616	1.39 *6*
Zn $K_{\alpha1}$	8.639	2.72 *11*
Zn $K_{\beta1}$'	9.572	0.498 *21*
γ E2	578.3 *3*	0.030 *7*
γ E2+32%M1	805.80 *8*	0.089 *7*
γ E2	1077.29 *8*	3.0
γ M1+3.3%E2	1261.18 *10*	0.090 *4*
γ M1+6.8%E2	1744.67 *19*	0.0090 *22*
γ E2	1883.08 *8*	0.130 *7*
γ E2	2338.45 *12*	0.009 *5*
γ E2	2821.93 *21*	0.00045 *11*

† 10% uncert(syst)

Atomic Electrons (^{68}Ga)

⟨e⟩=0.50 *3* keV

e_{bin}(keV)	⟨e⟩(keV)	e(%)
1	0.091	8.8 *9*
7	0.105	1.42 *15*
8	0.222	2.9 *3*
9	0.067	0.78 *8*
569 - 578	0.00022	3.9 *8* $\times10^{-5}$
796 - 806	0.000334	4.2 *3* $\times10^{-5}$
1068	0.0071	0.00067 *7*
1076 - 1077	0.00084	7.8 *7* $\times10^{-5}$
1252 - 1261	0.000180	1.44 *6* $\times10^{-5}$
1646 - 1745	3.0×10^{-5}	1.8 *3* $\times10^{-6}$
1873 - 1883	0.000190	1.01 *5* $\times10^{-5}$
2329 - 2338	1.2×10^{-5}	5.0 *21* $\times10^{-7}$
2812 - 2822	4.9×10^{-7}	1.7 *4* $\times10^{-8}$

Continuous Radiation (^{68}Ga)

⟨β+⟩=740 keV;⟨IB⟩=2.0 keV

E_{bin}(keV)		⟨ ⟩(keV)	(%)
0 - 10	β+	7.0×10^{-5}	0.00088
	IB	0.030	
10 - 20	β+	0.00137	0.0085
	IB	0.029	0.20
20 - 40	β+	0.0199	0.063
	IB	0.057	0.20
40 - 100	β+	0.496	0.66
	IB	0.158	0.24
100 - 300	β+	14.9	6.9
	IB	0.42	0.24
300 - 600	β+	91	19.8
	IB	0.42	0.100
600 - 1300	β+	465	50.2
	IB	0.51	0.060
1300 - 2500	β+	168	11.5
	IB	0.35	0.021
2500 - 2921	IB	0.0134	0.00051
	Σβ+		89

$^{68}_{32}$Ge(270.8 *3* d)

Mode: ϵ
Δ: -66978 *6* keV
SpA: 7090 Ci/g
Prod: ^{66}Zn(α,2n)

Photons (^{68}Ge)

⟨γ⟩=4.12 *11* keV

γ_{mode}	γ(keV)	γ(%)
Ga L_ℓ	0.957	0.046 *10*
Ga L_η	0.984	0.029 *7*
Ga L_α	1.098	0.75 *15*
Ga L_β	1.133	0.50 *13*
Ga L_γ	1.297	0.0019 *6*
Ga $K_{\alpha2}$	9.225	13.2 *5*
Ga $K_{\alpha1}$	9.252	25.8 *11*
Ga $K_{\beta1}$'	10.263	4.85 *20*
Ga $K_{\beta2}$'	10.444	0.0284 *12*

Atomic Electrons (^{68}Ge)

⟨e⟩=4.4 *3* keV

e_{bin}(keV)	⟨e⟩(keV)	e(%)
1	0.89	78 *8*
8	2.7	34 *3*
9	0.77	8.4 *9*
10	0.044	0.43 *4*

Continuous Radiation (^{68}Ge)

⟨IB⟩=0.0046 keV

E_{bin}(keV)		⟨ ⟩(keV)	(%)
10 - 20	IB	0.00119	0.0110
20 - 40	IB	0.000137	0.00049
40 - 100	IB	0.000159	0.00029
100 - 107	IB	7.1×10^{-8}	6.8×10^{-8}

$^{68}_{33}$As(2.527 *13* min)

Mode: ϵ
Δ: -58880 *100* keV
SpA: 1.092×10^{9} Ci/g
Prod: ^{70}Ge(p,3n)

Photons (^{68}As)

⟨γ⟩=2679 *74* keV

γ_{mode}	γ(keV)	γ(%)†
Ge $K_{\alpha2}$	9.855	0.148 *6*
Ge $K_{\alpha1}$	9.886	0.288 *12*
Ge $K_{\beta1}$'	10.980	0.0556 *23*
γ (M1+5.6%E2)	612.27 *24*	9.6 *5*
γ (M1+8.4%E2)	651.50 *22*	31.7 *16*
γ [E2]	703.0 *7*	0.69 *8*
γ [E2]	738.8 *4*	3.00 *23*
γ M1+6.6%E2	762.08 *20*	33.3 *18*
γ	988.7 *4*	1.00 *8*
γ E2	1016.42 *22*	77
γ [E2]	1053.0 *4*	1.77 *15*
γ [M1+E2]	1170.0 *5*	0.92 *8*
γ [M1+E2]	1245.6 *5*	0.46 *8*
γ E2	1252.7 *3*	4.5 *3*
γ [E2]	1263.76 *23*	5.1 *3*
γ [M1+E2]	1309.9 *5*	0.54 *8*
γ [E2]	1333.2 *4*	1.15 *8*
γ (M1+3.0%E2)	1413.6 *3*	15.1 *8*
γ [M1+E2]	1622.9 *4*	4.16 *23*
γ E1	1633.5 *10*	1.00 *8*
γ	1640.1 *4*	1.69 *8*
γ [E2]	1646.2 *5*	1.00 *8*
γ E2	1778.49 *23*	19.9 *12*
γ E2	1815.1 *4*	1.39 *15*
γ [M1+E2]	2007.7 *5*	3.5 *3*
γ [E2]	2025.8 *3*	0.54 *8*
γ [M1+E2]	2072.0 *4*	0.31 *8*
γ	2229.3 *10*	~0.15
γ	2272.0 *7*	0.77 *8*
γ [M1+E2]	2385.0 *4*	0.92 *8*
γ [E2]	2458.1 *7*	~1
γ	2459.1 *10*	5.0 *8*
γ [M1+E2]	2506.5 *10*	1.39 *8*
γ [M1+E2]	2793.7 *10*	1.23 *8*
γ	3058 *3*	0.85 *15*
γ [E2]	3088.4 *5*	1.00 *8*
γ [M1+E2]	3222.9 *10*	~0.15
γ	3288.4 *7*	0.31 *8*
γ [E2]	3401.3 *4*	0.23 *8*
γ [M1+E2]	3551.9 *10*	0.39 *8*

† 7.8% uncert(syst)

Atomic Electrons (^{68}As)

⟨e⟩=0.841 *24* keV

e_{bin}(keV)	⟨e⟩(keV)	e(%)
1 - 11	0.048	1.25 *9*
601	0.047	0.0079 *5*
611 - 612	0.0058	0.00095 *6*
640	0.143	0.0224 *15*
650 - 692	0.0212	0.00322 *17*
702 - 739	0.0172	0.00237 *18*
751	0.125	0.0167 *10*
761 - 762	0.0152	0.00200 *10*
978 - 989	0.0026	0.00026 *10*
1005	0.237	0.0236 *19*
1015	0.0246	0.00242 *19*
1016 - 1053	0.0100	0.00096 *6*
1159 - 1170	0.00256	0.000220 *22*
1235 - 1264	0.0264	0.00212 *10*
1299 - 1333	0.0042	0.000318 *21*
1402	0.0299	0.00213 *14*
1412 - 1414	0.00356	0.000252 *14*
1612 - 1646	0.0139	0.00086 *6*
1767	0.033	0.00187 *11*
1777 - 1815	0.0064	0.000360 *19*
1997 - 2072	0.0073	0.00036 *3*
2218 - 2272	0.00112	5.0 *11* $\times 10^{-5}$
2374 - 2459	0.0085	0.00035 *7*
2495 - 2506	0.00194	7.8 *6* $\times 10^{-5}$
2783 - 2794	0.00160	5.8 *4* $\times 10^{-5}$
3047 - 3088	0.00210	6.8 *8* $\times 10^{-5}$
3212 - 3288	0.00049	1.5 *4* $\times 10^{-5}$
3390 - 3401	0.00027	7.9 *24* $\times 10^{-6}$
3541 - 3552	0.00043	1.22 *23* $\times 10^{-5}$

Continuous Radiation (^{68}As)

⟨β+⟩=2011 keV;⟨IB⟩=7.2 keV

E_{bin}(keV)		⟨ ⟩(keV)	(%)
0 - 10	β+	7.6×10^{-6}	9.5×10^{-5}
	IB	0.058	
10 - 20	β+	0.000160	0.00099
	IB	0.058	0.40
20 - 40	β+	0.00251	0.0079
	IB	0.114	0.40
40 - 100	β+	0.068	0.090
	IB	0.33	0.51
100 - 300	β+	2.41	1.10
	IB	1.01	0.56
300 - 600	β+	18.5	3.96
	IB	1.26	0.29
600 - 1300	β+	185	18.7
	IB	2.1	0.24
1300 - 2500	β+	847	44.7
	IB	1.8	0.101
2500 - 5000	β+	950	30.3
	IB	0.53	0.018
5000 - 7084	β+	8.4	0.160
	IB	0.00062	1.18×10^{-5}
	Σβ+		99

A = 69

NDS **35**, 101 (1982)

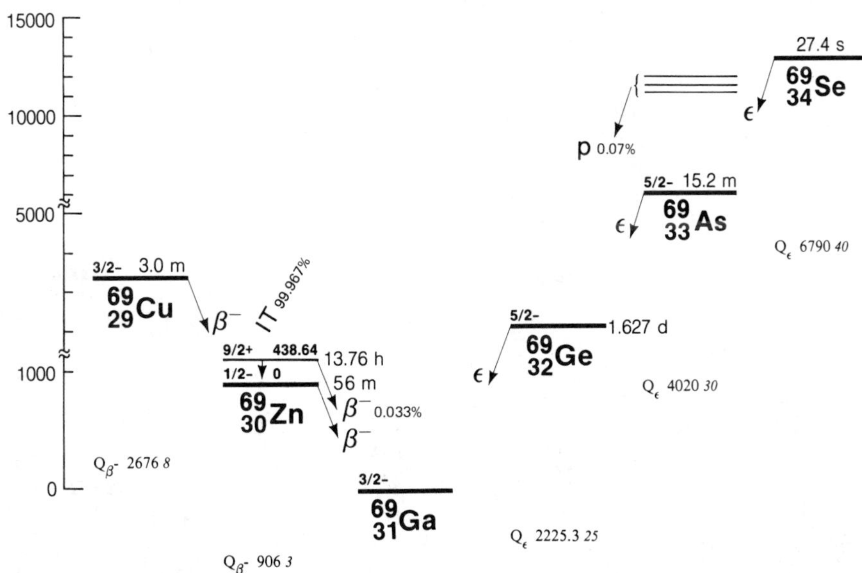

$^{69}_{29}$Cu(3.0 *1* min)

Mode: β-
Δ: -65741 *8* keV
SpA: 9.1×10^{8} Ci/g
Prod: ^{70}Zn(γ,p)

Photons (^{69}Cu)

⟨γ⟩=222 *24* keV

γ_{mode}	γ(keV)	γ(%)†
γ	84 *1*	0.4 *1* ?
γ	109.5 *20*	<0.2
γ [M1+E2]	172.7 *8*	<0.30
γ [E2]	530.2 *3*	3.0 *6*
γ	594.5 *4*	1.0 *2*
γ [M1+E2]	648.5 *5*	~1
γ [M1+E2]	833.3 *4*	6.2 *12*
γ	897.5 *5*	0.30 *6*
γ [M1+E2]	991.4 *10*	0.60 *12*
γ [M1+E2]	1005.9 *8*	10
γ [E2]	1178.7 *5*	1.0 *2*
γ	1427.7 *5*	0.90 *18*
γ [M1+E2]	1496 *3*	0.10 *2*
γ [M1+E2]	1824.7 *11*	0.10 *2*
γ [E2]	2027 *3*	<0.050 ?
γ	2170 *20*	<0.030 ?
γ [M2]	2400 *100*	<0.030 ?

† 20% uncert(syst)

Continuous Radiation (^{69}Cu)

⟨β-⟩=1026 keV;⟨IB⟩=2.4 keV

E_{bin}(keV)		⟨ ⟩(keV)	(%)
0 - 10	β-	0.0113	0.224
	IB	0.038	
10 - 20	β-	0.0346	0.230
	IB	0.037	0.26
20 - 40	β-	0.145	0.482
	IB	0.073	0.25
40 - 100	β-	1.17	1.64
	IB	0.21	0.32
100 - 300	β-	15.7	7.6
	IB	0.57	0.32
300 - 600	β-	73	16.0
	IB	0.60	0.142
600 - 1300	β-	398	42.2
	IB	0.68	0.081
1300 - 2500	β-	535	31.5
	IB	0.169	0.0111
2500 - 2676	β-	3.16	0.124
	IB	2.8×10^{-5}	1.12×10^{-6}

$^{69}_{30}$Zn(55.6 *16* min)

Mode: β-
Δ: -68417.0 *18* keV
SpA: 4.90×10^{7} Ci/g
Prod: daughter ^{69}Zn(13.76 h); ^{68}Zn(n,γ); ^{71}Ga(d,α)

Photons (^{69}Zn)

⟨γ⟩=0.0061 *10* keV

γ_{mode}	γ(keV)	γ(%)
γ [M1+E2]	297.88 *7*	5.3 *23* $\times10^{-7}$
γ M1+<26%E2	318.62 *6*	0.0012 *2*
γ M1+E2	553.35 *7*	1.4 *5* $\times10^{-5}$
γ M1+0.8%E2	871.97 *6*	0.00025 *8*

Continuous Radiation (^{69}Zn)

⟨β-⟩=321 keV;⟨IB⟩=0.31 keV

E_{bin}(keV)		⟨ ⟩(keV)	(%)
0 - 10	β-	0.069	1.37
	IB	0.0154	
10 - 20	β-	0.209	1.39
	IB	0.0147	0.102
20 - 40	β-	0.86	2.85
	IB	0.027	0.095
40 - 100	β-	6.5	9.2
	IB	0.068	0.107
100 - 300	β-	69	34.5
	IB	0.128	0.077
300 - 600	β-	177	40.8
	IB	0.053	0.0136
600 - 906	β-	68	9.9
	IB	0.0032	0.00049

$^{69}_{30}$Zn(13.76 *2* h)

Mode: IT(99.967 *3* %), β-(0.033 *3* %)
Δ: -67978.4 *18* keV
SpA: 3.301×10^{6} Ci/g
Prod: ^{68}Zn(n,γ); ^{71}Ga(d,α); ^{69}Ga(n,p)

Photons (^{69}Zn)

⟨γ⟩=416.1 *13* keV

γ_{mode}	γ(keV)	γ(%)†
Zn $K_{\alpha2}$	8.616	0.65 *3*
Zn $K_{\alpha1}$	8.639	1.27 *5*
Zn $K_{\beta1}$'	9.572	0.234 *10*
$\gamma_{\beta-}$[E2]	255.48 *7*	6.2 *22* $\times10^{-5}$
γ_{IT} M4	438.634 *18*	94.8
$\gamma_{\beta-}$M1+<0.4%E2	574.10 *5*	0.033

† uncert(syst): 0.30% for IT, 9.1% for β-

Atomic Electrons (^{69}Zn)*

⟨e⟩=22.5 *4* keV

e_{bin}(keV)	⟨e⟩(keV)	e(%)
1	0.043	4.2 *4*
7	0.049	0.67 *7*
8	0.104	1.36 *14*
9	0.032	0.37 *4*
10	0.00026	0.0027 *3*
429	19.5	4.55 *9*
437	2.09	0.478 *10*
438	0.593	0.135 *3*
439	0.0593	0.0135 *3*

* with IT

Continuous Radiation (^{69}Zn)

⟨β-⟩=0.096 keV;⟨IB⟩=8.4×10^{-5} keV

E_{bin}(keV)		⟨ ⟩(keV)	(%)
0 - 10	β-	2.84×10^{-5}	0.00057
	IB	4.7×10^{-6}	
10 - 20	β-	8.5×10^{-5}	0.00057
	IB	4.4×10^{-6}	3.1×10^{-5}
20 - 40	β-	0.000342	0.00114
	IB	8.2×10^{-6}	2.9×10^{-5}
40 - 100	β-	0.00246	0.00351
	IB	2.0×10^{-5}	3.1×10^{-5}
100 - 300	β-	0.0241	0.0120
	IB	3.5×10^{-5}	2.1×10^{-5}
300 - 600	β-	0.059	0.0136
	IB	1.11×10^{-5}	3.1×10^{-6}
600 - 770	β-	0.0102	0.00158
	IB	2.0×10^{-7}	3.2×10^{-8}

$^{69}_{31}$Ga(stable)

Δ: -69323 *3* keV
%: 60.1 *2*

$^{69}_{32}$Ge(1.627 *4* d)

Mode: ε
Δ: -67097 *4* keV
SpA: 1.163×10^{6} Ci/g
Prod: ^{69}Ga(d,2n); ^{67}Zn(α,2n); ^{70}Ge(n,2n); ^{12}C on ^{59}Co; protons on Ga; ^{70}Ge(γ,n)

Photons (^{69}Ge)

⟨γ⟩=710 *45* keV

γ_{mode}	γ(keV)	γ(%)†
Ga L_{ℓ}	0.957	0.036 *8*
Ga L_{η}	0.984	0.022 *6*
Ga L_{α}	1.098	0.57 *12*
Ga L_{β}	1.133	0.38 *10*
Ga L_{γ}	1.297	0.0014 *4*
Ga $K_{\alpha2}$	9.225	10.3 *4*
Ga $K_{\alpha1}$	9.252	20.2 *8*
Ga $K_{\beta1}$'	10.263	3.79 *16*
Ga $K_{\beta2}$'	10.444	0.0222 *9*
γ [M1]	197.50 *22*	0.025 *4* ?
γ [M1+E2]	234.80 *6*	0.37 *3*
γ [E2]	255.48 *7*	0.025 *7*
γ [M1+E2]	297.88 *7*	0.025 *7*
γ M1+<26%E2	318.62 *6*	1.55 *11*
γ [M1+E2]	381.19 *11*	~0.025
γ [M1+E2]	419.07 *6*	0.072 *7*
γ [M1+E2]	532.68 *6*	0.270 *22*
γ M1+E2	553.35 *7*	0.69 *4*
γ M1+<0.4%E2	574.10 *5*	13.3 *11*
γ	587.34 *9*	0.30 *3*
γ [M1+E2]	762.54 *7*	0.230 *22*
γ [E2]	788.15 *6*	0.34 *3*
γ	817.20 *8*	0.036 *4*
γ M1+0.8%E2	871.97 *6*	11.9 *9*
γ E2+13%M1	913.86 *11*	0.061 *14*
γ [M1+E2]	951.75 *6*	0.029 *4*
γ (E2)	1052.00 *7*	0.43 *3*
γ M1+14%E2	1106.77 *6*	36
γ [M1+E2]	1151.67 *7*	0.043 *7* ?
γ M1+4.6%E2	1207.22 *7*	0.39 *3*
γ [M1+E2]	1317.39 *9*	~0.0029

Photons (^{69}Ge)
(continued)

γ$_{mode}$	γ(keV)	γ(%)†
γ E2	1336.63 *7*	4.5 *4*
γ (E2+12%M1)	1349.87 *7*	0.32 *4*
γ [E2]	1404.72 *21*	0.018 *4*
γ [M1+E2]	1449.55 *3*	0.047 *4*
γ	1469.8 *8*	0.011 *4*
γ E2	1487.95 *10*	0.097 *7*
γ [M1+E2]	1525.83 *6*	0.266 *22*
γ [M1+E2]	1572.86 *8*	0.230 *18*
γ	1615.1 *10*	0.011 *4*
γ (M1+~36%E2)	1723.33 *21*	~0.040
γ	1725.3 *8*	~0.011 ?
γ M1+2.4%E2	1891.48 *8*	0.48 *4*
γ	1923.96 *7*	0.151 *11*
γ (M1+2.4%E2)	2023.63 *6*	0.54 *4*
γ	2043.9 *8*	0.036 *4*

† 11% uncert(syst)

Atomic Electrons (^{69}Ge)
⟨e⟩=3.69 *24* keV

e$_{bin}$(keV)	⟨e⟩(keV)	e(%)
1	0.69	61 *6*
8	2.12	26 *3*
9	0.60	6.6 *7*
187 - 235	0.016	~0.007
245 - 288	0.0019	0.00072 *17*
297 - 319	0.0251	0.0081 *5*
371 - 419	0.0012	0.00030 *9*
522 - 564	0.069	0.0123 *9*
573 - 587	0.0090	0.00157 *15*
752 - 788	0.00257	0.00033 *3*
807 - 817	0.00011	1.3 *5* $\times 10^{-5}$
862 - 903	0.039	0.0045 *3*
913 - 952	0.000116	1.23 *15* $\times 10^{-5}$
1042 - 1052	0.00129	0.000124 *10*
1096	0.083	0.0075 *8*
1105 - 1152	0.0099	0.00089 *8*
1197 - 1207	0.00091	7.6 *5* $\times 10^{-5}$
1307 - 1350	0.0109	0.00082 *6*
1394 - 1439	0.000123	8.6 *8* $\times 10^{-6}$
1448 - 1526	0.00096	6.4 *14* $\times 10^{-5}$
1562 - 1615	0.00044	2.84 *25* $\times 10^{-5}$
1713 - 1725	8.2 $\times 10^{-5}$	4.8 *18* $\times 10^{-6}$
1881 - 1924	0.00093	4.9 *4* $\times 10^{-5}$
2013 - 2044	0.00084	4.2 *3* $\times 10^{-5}$

Continuous Radiation (^{69}Ge)
⟨β+⟩=116 keV;⟨IB⟩=0.74 keV

E$_{bin}$(keV)		⟨ ⟩(keV)	(%)
0 - 10	β+	6.7 $\times 10^{-5}$	0.00084
	IB	0.0076	
10 - 20	β+	0.00134	0.0083
	IB	0.0060	0.044
20 - 40	β+	0.0197	0.062
	IB	0.0098	0.034
40 - 100	β+	0.467	0.62
	IB	0.027	0.042
100 - 300	β+	11.0	5.3
	IB	0.086	0.047
300 - 600	β+	43.4	9.8
	IB	0.169	0.038
600 - 1300	β+	61	7.8
	IB	0.33	0.038
1300 - 2225	IB	0.104	0.0068
	Σβ+		24

$^{69}_{33}$As(15.2 *2* min)

Mode: ϵ
Δ: -63080 *30* keV
SpA: 1.792$\times 10^{8}$ Ci/g
Prod: ^{70}Ge(p,2n); ^{58}Ni(^{16}O,αp); ^{59}Co(^{12}C,2n)

Photons (^{69}As)
⟨γ⟩=182.3 *24* keV

γ$_{mode}$	γ(keV)	γ(%)†
Ge L$_{\ell}$	1.037	0.0056 *12*
Ge L$_{\eta}$	1.068	0.0033 *8*
Ge L$_{\alpha}$	1.188	0.097 *20*
Ge L$_{\beta}$	1.227	0.059 *15*
Ge L$_{\gamma}$	1.412	0.00068 *21*
Ge K$_{\alpha2}$	9.855	1.56 *7*
Ge K$_{\alpha1}$	9.886	3.04 *13*
Ge K$_{\beta1}$'	10.980	0.587 *25*
Ge K$_{\beta2}$'	11.184	0.0097 *4*
γ	68.2 *5 ?*	
γ E2	86.781 *24*	3.43 *7*
γ M1	141.36 *5*	0.065 *11*
γ M1+3.9%E2	145.950 *24*	4.94 *10*
γ M1+35%E2	232.731 *24*	10.88
γ M1+1.2%E2	287.31 *5*	1.41 *3*
γ M1+3.5%E2	374.09 *5*	0.446 *11*
γ M2	398.02 *15*	1.02 *7*
γ E2	414.47 *15*	0.152 *22*
γ [E1]	438.40 *12*	0.27 *4*
γ M1+E2	559.16 *17*	0.14 *3*
γ [M1+E2]	621.03 *24*	0.054 *11*
γ	656.7 *6*	0.054 *11*
γ	681.0 *3*	0.09 *3*
γ M1+34%E2	762.38 *23*	0.054 *11*
γ [E1]	812.48 *12*	0.81 *4*
γ M1+31%E2	821.97 *14*	0.35 *3*
γ (E2)	846.47 *17*	0.11 *3*
γ E2+15%M1	862.20 *20*	0.57 *4*
γ M1+27%E2	927.41 *16*	0.076 *11*
γ M1+E2	933.25 *17*	0.446 *22*
γ M1+E2	963.32 *14*	0.054 *11*
γ	1040.90 *16*	0.076 *11*
γ M1+9.0%E2	1073.36 *16*	0.326 *11*
γ	1105.11 *17*	
γ [M1+E2]	1160.14 *16*	0.065 *11*
γ [M1+E2]	1165.72 *23*	<0.044
γ (M1+20%E2)	1182.26 *16*	0.044 *11*
γ M1+E2	1196.05 *14*	0.196 *11*
γ	1216.10 *10*	0.098 *11*
γ [M1+E2]	1220.52 *16*	0.261 *11*
γ	1236.95 *18*	0.054 *11*
γ	1241.3 *4*	0.11 *4*
γ	1287.48 *24*	0.065 *11*
γ	1303.1 *3*	0.033 *11*
γ [M1+E2]	1307.30 *16*	0.239 *11*
γ	1320.46 *22*	0.054 *11*
γ	1378.30 *19*	0.207 *11*
γ (M1+1.1%E2)	1381.53 *24*	0.272 *11*
γ (M1+30%E2)	1414.98 *17*	0.196 *22*
γ (E2)	1431.0 *4*	~0.07
γ	1434.31 *18*	0.098 *11*
γ [M1+E2]	1453.02 *23*	0.033 *11*
γ	1479.19 *17*	0.218 *11*
γ	1516.4 *3*	0.044 *11*
γ	1525.10 *20*	0.141 *22*
γ	1534.38 *13*	0.370 *22*
γ [M1+E2]	1539.80 *23*	0.082 *3*
γ	1585.90 *20*	0.141 *22*
γ	1611.03 *19*	0.109 *11*
γ	1614.26 *24*	0.27 *3*
γ	1639.5 *4*	<0.05
γ [M1+E2]	1658.27 *15*	0.103 *8*
γ	1680.33 *13*	0.294 *22*
γ	1767.11 *13*	0.35 *4*
γ [M1+E2]	1804.22 *15*	0.120 *11*
γ	1822.0 *5*	~0.033
γ	1872.69 *17*	0.044 *11*
γ [M1+E2]	1891.00 *15*	0.078 *5*
γ	2000.7 *3*	0.403 *22*

Photons (^{69}As)
(continued)

γ$_{mode}$	γ(keV)	γ(%)†
γ	2014.05 *16*	0.283 *22*
γ	2149.6 *3*	0.283 *11*
γ [E1]	2168.4 *5*	0.087 *11*
γ	2246.77 *16*	0.070 *5*
γ	2274.5 *10*	0.037 *3*
γ	2333.8 *3*	0.044 *11*
γ	2362.03 *25*	0.065 *11*
γ	2398.84 *24*	0.120 *11*
γ	2499.8 *5*	0.044 *11*
γ [M1+E2]	2502.69 *22*	0.030 *3*
γ	2507.98 *25*	0.034 *5*
γ [E1]	2582.9 *5*	0.087 *11*
γ	2686.14 *24*	~0.07
γ [M1+E2]	2735.41 *22*	0.054 *11*
γ	2946.7 *4*	0.042 *3*
γ	2987.0 *10*	0.009 *3*
γ	3058.9 *5*	0.0228 *22*

† 2.0% uncert(syst)

Atomic Electrons (^{69}As)
⟨e⟩=4.61 *10* keV

e$_{bin}$(keV)	⟨e⟩(keV)	e(%)
1 - 8	0.217	10.1 *9*
9	0.198	2.29 *23*
10 - 11	0.092	0.94 *9*
76	2.85	3.76 *11*
85	0.288	0.338 *10*
86	0.155	0.181 *5*
87 - 130	0.0755	0.0862 *24*
135	0.207	0.154 *17*
140 - 146	0.029	0.0198 *20*
222	0.36	0.164 *24*
231 - 276	0.067	0.0276 *25*
286 - 287	0.00236	0.000824 *20*
363 - 403	0.0421	0.0109 *6*
413 - 438	0.00145	0.00034 *4*
548 - 559	0.0011	0.00020 *6*
610 - 657	0.00060	9.6 *23* $\times 10^{-5}$
670 - 681	0.00038	~6 $\times 10^{-5}$
751 - 762	0.00025	3.3 *6* $\times 10^{-5}$
801 - 846	0.00343	0.00042 *3*
851 - 862	0.00241	0.000283 *20*
916 - 963	0.00207	0.000224 *16*
1030 - 1073	0.00114	0.000107 *7*
1149 - 1196	0.00089	7.6 *7* $\times 10^{-5}$
1205 - 1241	0.00121	0.000100 *11*
1276 - 1320	0.00088	6.8 *6* $\times 10^{-5}$
1367 - 1415	0.00142	0.000103 *9*
1420 - 1468	0.0009	6.3 *20* $\times 10^{-5}$
1478 - 1575	0.00131	8.6 *13* $\times 10^{-5}$
1584 - 1680	0.00130	8.0 *12* $\times 10^{-5}$
1756 - 1822	0.00076	4.3 *9* $\times 10^{-5}$
1862 - 1891	0.000196	1.05 *13* $\times 10^{-5}$
1990 - 2014	0.00090	4.5 *8* $\times 10^{-5}$
2138 - 2236	0.00051	2.4 *4* $\times 10^{-5}$
2245 - 2334	0.000105	4.6 *9* $\times 10^{-6}$
2351 - 2399	0.00022	9.1 *17* $\times 10^{-6}$
2489 - 2583	0.000204	8.1 *8* $\times 10^{-6}$
2675 - 2735	0.00014	5.3 *14* $\times 10^{-6}$
2936 - 2987	5.3 $\times 10^{-5}$	1.8 *3* $\times 10^{-6}$
3048 - 3059	2.3 $\times 10^{-5}$	7.6 *16* $\times 10^{-7}$

Continuous Radiation (^{69}As)

⟨β+⟩=1215 keV;⟨IB⟩=3.7 keV

E_{bin}(keV)		⟨ ⟩(keV)	(%)
0 - 10	β+	2.34×10^{-5}	0.000292
	IB	0.042	
10 - 20	β+	0.000493	0.00305
	IB	0.042	0.29
20 - 40	β+	0.0076	0.0239
	IB	0.082	0.28
40 - 100	β+	0.204	0.269
	IB	0.23	0.36
100 - 300	β+	6.7	3.07
	IB	0.67	0.38
300 - 600	β+	45.1	9.7
	IB	0.77	0.18
600 - 1300	β+	340	35.3
	IB	1.10	0.128
1300 - 2500	β+	772	43.5
	IB	0.65	0.039
2500 - 4020	β+	51	1.95
	IB	0.144	0.0050
	Σβ+		94

$^{69}_{34}$Se(27.4 *2* s)

Mode: ϵ, ϵp(0.07 *1* %)
Δ: -56290 *40* keV
SpA: 5.89×10^{9} Ci/g
Prod: ^{40}Ca(^{32}S,2pn)

ϵp: 1000-3200 range

Photons (^{69}Se)

⟨γ⟩=583 *24* keV

γ_{mode}	γ(keV)	γ(%)†
γ	66.45 *5*	21.8 *20*
γ	97.97 *7*	66
γ	332.80 *13*	1.3 *3*
γ	399.25 *13*	3.50 *20*
γ	497.22 *14*	1.12 *20*
γ	624.87 *13*	1.8 *7*
γ	637.03 *20*	2.1 *3*
γ	691.31 *13*	17.2 *20*
γ	789.28 *13*	3.6 *4*
γ	835.5 *3*	2.3 *5*
γ	911.8 *3*	0.53 *20*
γ	978.3 *3*	1.85 *20*
γ	1329.4 *20*	0.59 *20*
γ	1360.1 *3*	1.06 *13*
γ	1394.4 *7*	0.92 *13*
γ	1456.6 *4*	1.0 *5*
γ	1475.5 *3*	1.6 *7*
γ	1526.2 *3*	0.79 *20*
γ	1556.9 *5*	0.92 *13*
γ	1562.8 *10*	0.33 *13*
γ	1592.7 *3*	0.92 *13*
γ	1646.0 *7*	0.73 *13*
γ	1652.2 *3*	0.79 *13*
γ	1690.6 *3*	0.99 *13*
γ	1699.9 *3*	1.39 *13*
γ	1744.0 *7*	0.40 *7*
γ	1766.3 *3*	1.3 *3*
γ	1849.0 *5*	0.40 *13*
γ	1864.3 *3*	0.33 *13*
γ	1911.7 *4*	0.99 *13*
γ	1956.5 *10*	0.33 *13*
γ	2051.4 *3*	0.59 *20*

† 11% uncert(syst)

Continuous Radiation (^{69}Se)

⟨β+⟩=2437 keV;⟨IB⟩=9.3 keV

E_{bin}(keV)		⟨ ⟩(keV)	(%)
0 - 10	β+	4.33×10^{-6}	5.4×10^{-5}
	IB	0.065	
10 - 20	β+	9.5×10^{-5}	0.00058
	IB	0.065	0.45
20 - 40	β+	0.00153	0.00478
	IB	0.128	0.45
40 - 100	β+	0.0429	0.056
	IB	0.38	0.58
100 - 300	β+	1.56	0.71
	IB	1.15	0.64
300 - 600	β+	12.3	2.64
	IB	1.48	0.35
600 - 1300	β+	131	13.2
	IB	2.6	0.29
1300 - 2500	β+	708	36.8
	IB	2.5	0.140
2500 - 5000	β+	1547	45.7
	IB	1.02	0.033
5000 - 6790	β+	36.6	0.71
	IB	0.00154	3.0×10^{-5}
	Σβ+		100

A = 70

NDS 25, 1 (1978)

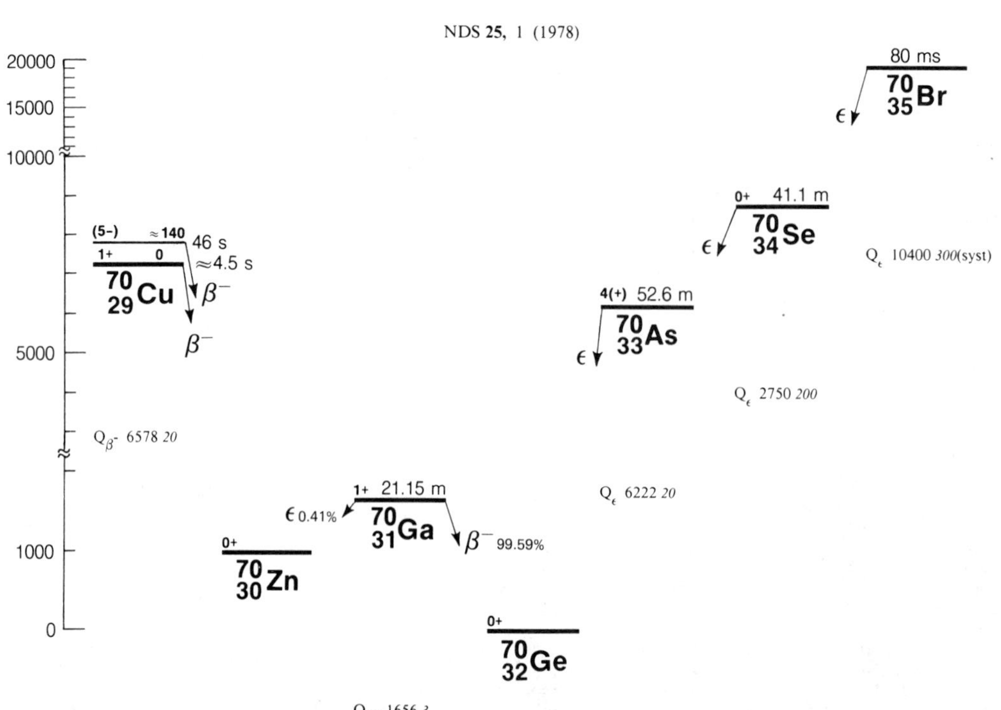

$^{70}_{29}$Cu(4.5 *10* s)

Mode: β-
Δ: -62982 *20* keV
SpA: 3.3×10^{10} Ci/g
Prod: ^{70}Zn(n,p)

Photons (^{70}Cu)

γ_{mode}	γ(keV)	γ(%)
γ[E2]	884.8 *2*	54 *10*

Continuous Radiation (^{70}Cu)

⟨β-⟩=2782 keV;⟨IB⟩=11.4 keV

E_{bin}(keV)		⟨ ⟩(keV)	(%)
0 - 10	β-	0.00106	0.0210
	IB	0.070	
10 - 20	β-	0.00327	0.0218
	IB	0.069	0.48
20 - 40	β-	0.0139	0.0461
	IB	0.138	0.48
40 - 100	β-	0.115	0.162
	IB	0.40	0.62
100 - 300	β-	1.74	0.83
	IB	1.25	0.70
300 - 600	β-	10.2	2.19
	IB	1.65	0.38
600 - 1300	β-	98	9.9
	IB	3.0	0.34
1300 - 2500	β-	573	29.5
	IB	3.1	0.18
2500 - 5000	β-	1884	53
	IB	1.8	0.056
5000 - 6578	β-	215	4.00
	IB	0.028	0.00051

$^{70}_{29}$Cu(46 *5* s)

Mode: β-
Δ: ~-62842 keV
SpA: 3.5×10^{9} Ci/g
Prod: ^{70}Zn(n,p)

Photons (^{70}Cu)

⟨γ⟩=2838 *78* keV

γ_{mode}	γ(keV)	γ(%)
γ	386.5 *4*	8 *3* ?
γ[E2]	884.8 *2*	100
γ[E2]	901.7 *2*	87 *4*
γ	1108.7 *4*	8 *2* ?
γ[E1]	1251.7 *2*	57 *4*
γ	1271 *1*	1.0 *4* ?
γ	1428 *2*	3 *1* ?
γ	1476.5 *20*	~1 ?
γ	1520 *2*	1.4 *4* ?
γ	1690.6 *5*	4.7 *9* ?
γ	1953.5 *10*	2.1 *4* ?
γ	2061.4 *5*	3.7 *9* ?
γ	3062 *2*	1.4 *5* ?

$^{70}_{30}$Zn(stable)

Δ: -69560 *3* keV
%: 0.6 *1*

$^{70}_{31}$Ga(21.15 *5* min)

Mode: β-(99.59 *5* %), ϵ(0.41 *5* %)
Δ: -68905 *3* keV
SpA: 1.270×10^{8} Ci/g
Prod: ^{69}Ga(n,γ)

Photons (^{70}Ga)

⟨γ⟩=7.6 *14* keV

γ_{mode}	γ(keV)	γ(%)
Zn L_{ℓ}	0.884	0.00019 *4*
Zn L_{η}	0.907	0.00011 *3*
Zn L_{α}	1.012	0.0029 *6*
Zn L_{β}	1.043	0.0019 *5*
Zn $K_{\alpha2}$	8.616	0.0518 *21*
Zn $K_{\alpha1}$	8.639	0.101 *4*
Zn $K_{\beta1}$'	9.572	0.0186 *8*
$\gamma_{\beta-}$[E2]	176.283 *20*	0.30 *6*
$\gamma_{\beta-}$E2	1039.33 *8*	0.68 *14*

Atomic Electrons (^{70}Ga)*

⟨e⟩=0.0182 *10* keV

e_{bin}(keV)	⟨e⟩(keV)	e(%)
1	0.0034	0.33 *3*
7	0.0039	0.053 *5*
8	0.0083	0.108 *11*
9	0.0025	0.029 *3*
10	2.03×10^{-5}	0.000213 *22*

* with ϵ

Continuous Radiation (^{70}Ga)

⟨β-⟩=644 keV;⟨IB⟩=1.06 keV

E_{bin}(keV)		⟨ ⟩(keV)	(%)
0 - 10	β-	0.0233	0.464
	IB	0.027	
10 - 20	β-	0.071	0.475
	IB	0.026	0.18
20 - 40	β-	0.297	0.99
	IB	0.051	0.18
40 - 100	β-	2.33	3.30
	IB	0.139	0.22
100 - 300	β-	29.5	14.4
	IB	0.35	0.20
300 - 600	β-	124	27.4
	IB	0.29	0.071
600 - 1300	β-	435	48.8
	IB	0.17	0.022
1300 - 1656	β-	52	3.75
	IB	0.0021	0.000152

$^{70}_{32}$Ge(stable)

Δ: -70561.5 *15* keV
%: 20.5 *5*

$^{70}_{33}$As(52.6 *3* min)

Mode: ϵ
Δ: -64340 *20* keV
SpA: 5.11×10^{7} Ci/g
Prod: ^{70}Ge(p,n); ^{70}Ge(d,2n)

Photons (^{70}As)

⟨γ⟩=3174 *55* keV

γ_{mode}	γ(keV)	γ(%)†
Ge L_{ℓ}	1.037	0.0053 *12*
Ge L_{η}	1.068	0.0031 *8*
Ge L_{α}	1.188	0.091 *19*
Ge L_{β}	1.227	0.056 *14*
Ge L_{γ}	1.412	0.00067 *20*
Ge $K_{\alpha2}$	9.855	1.49 *6*
Ge $K_{\alpha1}$	9.886	2.90 *12*
Ge $K_{\beta1}$'	10.980	0.560 *24*
Ge $K_{\beta2}$'	11.184	0.0093 *4*
γ[E2]	176.283 *20*	2.61 *25*
γ	240.1 *3*	0.21 *4*
γ[M1+E2]	252.4 *3*	2.9 *3*
γ[M1+E2]	294.3 *3*	0.19 *4*
γ[M1+E2]	298.2 *3*	0.38 *8*
γ	372.7 *20*	~1
γ[E2]	446.03 *25*	0.16 *3*
γ[M1+E2]	449.94 *25*	0.11 *3*
γ[E2]	491.84 *20*	0.98 *8*
γ[E1]	496.8 *5*	2.5 *5* ?
γ	594.8 *3*	16.3 *16*
γ[M1+E2]	607.2 *3*	3.9 *4*
γ	614.7 *10*	4.0 *6*
γ[M1+E2]	653.0 *3*	0.60 *12*
γ M1+49%E2	668.12 *20*	21.2 *4*
γ	685.7 *10*	2.0 *8* ?
γ	695.7 *10*	~2 ?
γ M1	744.23 *25*	20.8 *4*
γ	759.9 *4*	~0.25
γ	827.8 *3*	0.35 *7*
γ	889.1 *3*	3.1 *3*
γ	893.0 *3*	1.96 *16*
γ[E2]	901.5 *3*	1.39 *16*
γ[M1+E2]	905.4 *3*	12.2 *12*
γ[E2]	941.77 *22*	1.39 *16*
γ	953.7 *3*	0.47 *10*
γ E2	1039.33 *8*	82
γ E2	1099.0 *3*	4.4 *4*
γ E2	1114.14 *23*	21.2 *4*
γ[M1+E2]	1118.05 *22*	3.2 *3*
γ	1218.2 *5*	0.18 *4*
γ	1250 *20*	3.8 *12*
γ	1295.8 *5*	0.17 *8*
γ	1331.6 *3*	0.61 *12*
γ	1335.5 *3*	0.61 *12*
γ	1339.07 *25*	8.9 *9*
γ[E2]	1351.4 *3*	0.59 *12*
γ[M1+E2]	1412.34 *24*	8.58 *17*
γ	1418.0 *5*	0.50 *16*
γ	1495.9 *3*	1.55 *16*
γ	1506.8 *5*	~0.39
γ	1511.8 *5*	~0.28
γ	1523.0 *5*	5.1 *5*
γ	1566.3 *5*	~0.28
γ	1587.7 *4*	0.44 *9*
γ E2	1707.43 *21*	17.9 *4*
γ	1781.5 *3*	3.9 *4*
γ	1883.0 *5*	0.51 *11* ?
γ	1944.8 *4*	<0.16
γ	1948.7 *4*	0.16 *3*
γ	2007.17 *24*	2.9 *3*
γ E2	2019.51 *25*	16.7 *16*

Photons (^{70}As)
(continued)

γ_{mode}	γ(keV)	γ(%)†
γ	2064 *3*	<0.12 ?
γ	2095 *3*	~0.18
γ[E2]	2157.36 *23*	0.37 *8*
γ	2219.3 *9*	~0.11
γ	2255.8 *4*	~0.13
γ	2325.4 *5*	~0.11 ?
γ	2332.3 *5*	~0.08
γ	2420.9 *7*	<0.16
γ	2424.8 *7*	<0.16 ?
γ	2425.0 *10*	<0.16 ?
γ	2449.6 *3*	0.35 *7*
γ	2517.5 *9*	<0.15
γ	2637.1 *5*	0.30 *7*
γ	2780 *3*	0.074 *15* ?
γ	2853.2 *21*	<0.08 ?
γ	2963.5 *9*	<0.15
γ	3125.3 *10*	0.065 *13*
γ	3290 *40*	0.18 *4*
γ	3470 *40*	0.15 *3*
γ	3920 *30*	0.16 *3*
γ	4089.2 *21*	<0.033 ?
γ	4332.5 *5*	<0.033 ?
γ	4432.1 *10*	<0.016 ?
γ	4700 *50*	~0.016

† 3.3% uncert(syst)

Atomic Electrons (^{70}As)
$\langle e\rangle$=1.99 *8* keV

e_{bin}(keV)	$\langle e\rangle$(keV)	e(%)
1	0.102	8.3 *9*
8	0.103	1.23 *13*
9	0.189	2.19 *22*
10	0.083	0.86 *9*
11	0.0044	0.041 *4*
165	0.34	0.206 *21*
175 - 176	0.049	0.0277 *25*
229 - 240	0.008	~0.0036
241	0.10	~0.041
251 - 298	0.028	0.010 *3*
362 - 373	0.017	~0.005
435 - 481	0.0141	0.00300 *25*
486 - 497	0.0102	0.0021 *3*
584	0.08	~0.014
593 - 642	0.060	0.0100 *19*
652 - 653	0.00039	6.0 *14* $\times10^{-5}$
657	0.108	0.0165 *7*
667 - 696	0.029	0.0043 *10*
733	0.0797	0.0109 *3*
743 - 760	0.0106	0.00142 *8*
817 - 828	0.0011	0.00014 *6*
878 - 905	0.066	0.0074 *8*
931 - 954	0.0066	0.00071 *8*
1028	0.244	0.0237 *9*
1038 - 1039	0.0295	0.00284 *10*
1088 - 1099	0.0137	0.00126 *11*
1103	0.0578	0.00524 *15*
1107 - 1118	0.0162	0.00146 *10*
1207 - 1250	0.008	0.0006 *3*
1285 - 1334	0.017	0.0013 *4*
1335 - 1351	0.0032	0.00024 *5*
1401 - 1418	0.0204	0.00145 *9*
1485 - 1577	0.013	0.00086 *18*
1586 - 1588	7.3 $\times10^{-5}$	4.6 *15* $\times10^{-6}$
1696 - 1781	0.0401	0.00235 *10*
1872 - 1949	0.0010	5.5 *14* $\times10^{-5}$
1996 - 2095	0.032	0.00159 *14*
2146 - 2245	0.00085	3.9 *7* $\times10^{-5}$
2254 - 2332	0.00025	1.1 *4* $\times10^{-5}$
2410 - 2506	0.00077	3.2 *10* $\times10^{-5}$
2516 - 2517	8.9 $\times10^{-6}$	~4 $\times10^{-7}$
2626 - 2637	0.00033	1.3 *4* $\times10^{-5}$
2769 - 2853	0.00012	4.4 *16* $\times10^{-6}$
2952 - 2963	7.7 $\times10^{-5}$	~3 $\times10^{-6}$
3114 - 3125	6.6 $\times10^{-5}$	2.1 *6* $\times10^{-6}$
3279 - 3290	0.00018	5.4 *15* $\times10^{-6}$

Atomic Electrons (^{70}As)
(continued)

e_{bin}(keV)	$\langle e\rangle$(keV)	e(%)
3459 - 3470	0.00014	4.1 *11* $\times10^{-6}$
3909 - 3920	0.00014	3.6 *9* $\times10^{-6}$
4078 - 4089	1.5 $\times10^{-5}$	~4 $\times10^{-7}$
4321 - 4332	1.4 $\times10^{-5}$	~3 $\times10^{-7}$
4421 - 4432	7.1 $\times10^{-6}$	~2 $\times10^{-7}$
4689 - 4700	1.4 $\times10^{-5}$	3.0 *14* $\times10^{-7}$

Continuous Radiation (^{70}As)
$\langle\beta+\rangle$=863 keV;$\langle$IB$\rangle$=2.5 keV

E_{bin}(keV)		$\langle\ \rangle$(keV)	(%)
0 - 10	β+	4.11 $\times10^{-5}$	0.00051
	IB	0.033	
10 - 20	β+	0.00086	0.0053
	IB	0.033	0.23
20 - 40	β+	0.0134	0.0419
	IB	0.063	0.22
40 - 100	β+	0.355	0.468
	IB	0.18	0.27
100 - 300	β+	11.4	5.3
	IB	0.49	0.28
300 - 600	β+	74	16.0
	IB	0.52	0.122
600 - 1300	β+	441	46.9
	IB	0.68	0.079
1300 - 2500	β+	336	21.2
	IB	0.46	0.027
2500 - 3770	β+	1.14	0.0445
	IB	0.048	0.0018
	$\Sigma\beta$+		90

$^{70}_{34}$Se(41.1 *3* min)

Mode: ϵ

Δ: -61590 *200* keV

SpA: 6.53×10^{7} Ci/g

Prod: ^{75}As(d,7n); ^{59}Co(^{14}N,3n); ^{58}Ni(^{14}N,pn); ^{3}He on Ge; ^{70}Ge(^{3}He,3n)

Photons (^{70}Se)
$\langle\gamma\rangle$=276.7 *24* keV

γ_{mode}	γ(keV)	γ(%)†
As L_ℓ	1.120	0.078 *17*
As L_η	1.155	0.046 *12*
As L_α	1.282	1.4 *3*
As L_β	1.324	0.85 *22*
As L_γ	1.524	0.010 *3*
As $K_{\alpha2}$	10.508	20.2 *12*
As $K_{\alpha1}$	10.544	39.4 *23*
As $K_{\beta1}$'	11.724	7.8 *5*
As $K_{\beta2}$'	11.948	0.241 *14*
γ E2	32.05 *3*	1.90 *9*
γ	39.59 *5*	0.40 *6*
γ[M1]	49.50 *3*	35.4 *7*
γ[M1]	86.18 *4*	0.78 *6*
γ[M1]	113.51 *5*	1.56 *6*
γ[M1]	129.56 *9*	0.29 *6*
γ[M1]	132.53 *4*	3.48 *7*
γ[M1]	135.67 *4*	2.59 *6*
γ[M1+E2]	153.23 *5*	0.43 *6*
γ	160.93 *8*	0.78 *6*
γ	198.7 *3*	0.17 *6*
γ[M1+E2]	202.73 *5*	4.84 *10*
γ[M1+E2]	223.43 *6*	0.63 *6*
γ[M1+E2]	244.13 *5*	2.82 *9*
γ	247.11 *9*	0.43 *6*
γ[M1+E2]	255.90 *6*	1.81 *6*

Photons (^{70}Se)
(continued)

γ_{mode}	γ(keV)	γ(%)†
γ	263.15 *6*	2.74 *6*
γ[M1+E2]	290.49 *5*	0.43 *12*
γ[M1+E2]	293.63 *4*	2.79 *9*
γ	296.60 *8*	0.43 *9*
γ	301.8 *4*	0.35 *9*
γ	312.65 *6*	0.23 *6*
γ	343.85 *15*	0.66 *6*
γ[M1+E2]	376.66 *4*	9.33 *19*
γ[M1+E2]	413.86 *6*	2.07 *6*
γ[M1+E2]	426.16 *4*	28.8
γ	458.42 *25*	0.26 *6*
γ[M1+E2]	500.03 *6*	1.27 *17*
γ	545.81 *12*	0.32 *3*
γ[M1+E2]	549.53 *6*	0.29 *3*
γ	561.86 *13*	0.17 *3*
γ	564.83 *12*	0.17 *3*
γ	858.46 *12*	0.12 *3*
γ	1325.2 *4*	0.35 *9*
γ	1569.4 *4*	0.29 *3*
γ	1618.9 *4*	0.17 *3*

† 1.0% uncert(syst)

Atomic Electrons (^{70}Se)
$\langle e\rangle$=41.4 *15* keV

e_{bin}(keV)	$\langle e\rangle$(keV)	e(%)
1	1.56	117 *13*
2	0.093	6.1 *7*
9	3.8	41 *5*
10 - 12	1.17	11.3 *11*
20	14.0	69 *4*
28	1.1	~4
31	7.0	22.9 *11*
32	1.21	3.80 *19*
38	8.1	21.4 *18*
39 - 86	1.42	2.93 *18*
102 - 149	0.56	0.45 *5*
152 - 201	0.36	0.19 *9*
203 - 251	0.33	0.14 *4*
254 - 302	0.14	0.049 *13*
311 - 344	0.012	~0.0037
365 - 414	0.54	0.14 *3*
425 - 458	0.047	0.011 *3*
488 - 538	0.017	0.0034 *7*
544 - 565	0.0028	0.00051 *14*
847 - 858	0.00039	4.6 *20* $\times10^{-5}$
1313 - 1325	0.0007	5.3 *20* $\times10^{-5}$
1557 - 1619	0.00079	5.0 *11* $\times10^{-5}$

Continuous Radiation (^{70}Se)
$\langle\beta+\rangle$=447 keV;$\langle$IB$\rangle$=1.68 keV

E_{bin}(keV)		$\langle\ \rangle$(keV)	(%)
0 - 10	β+	7.3 $\times10^{-5}$	0.00091
	IB	0.020	
10 - 20	β+	0.00158	0.0098
	IB	0.020	0.141
20 - 40	β+	0.0250	0.078
	IB	0.036	0.126
40 - 100	β+	0.67	0.88
	IB	0.099	0.153
100 - 300	β+	20.1	9.4
	IB	0.25	0.144
300 - 600	β+	108	23.9
	IB	0.27	0.064
600 - 1300	β+	297	34.8
	IB	0.55	0.061
1300 - 2500	β+	20.9	1.50
	IB	0.43	0.027
2500 - 2668	IB	0.00075	2.9 $\times10^{-5}$
	$\Sigma\beta$+		70

$^{70}_{35}$Br(80.2 *8* ms)

Mode: ϵ
Δ: -51190 *360* keV syst
SpA: 2.325×10^{11} Ci/g
Prod: protons on U; protons on Nb; ^{58}Ni(^{14}N,2n)

A = 71

NDS 27, 517 (1979)

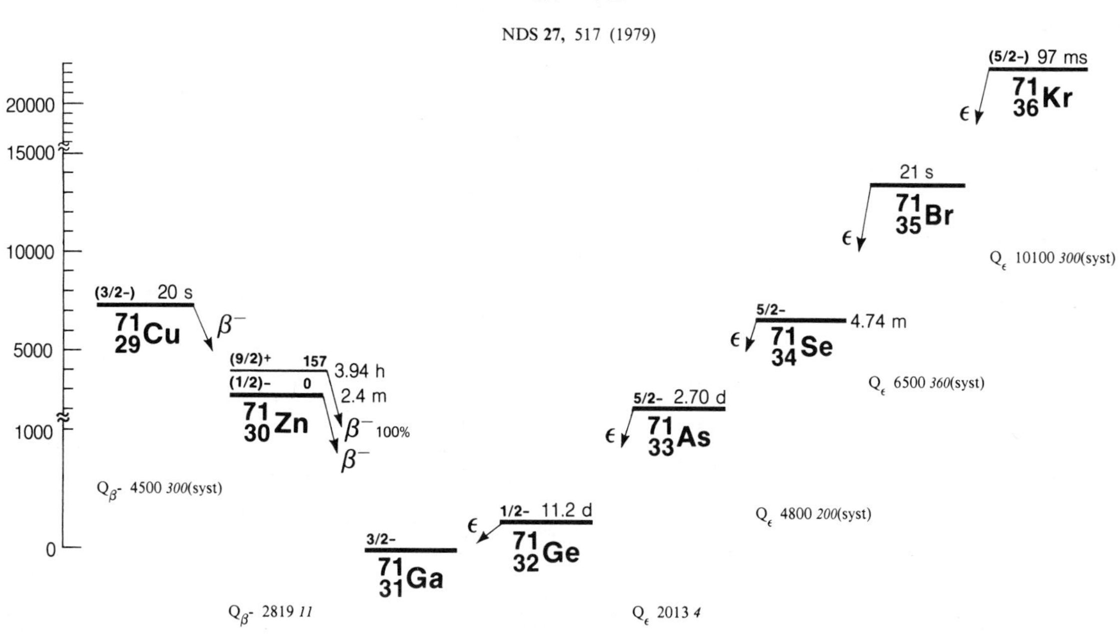

$^{71}_{29}$Cu(19.5 *16* s)

Mode: β-
Δ: -62820 *300* keV syst
SpA: 8.0×10^{9} Ci/g
Prod: ^{76}Ge on W

Photons (^{71}Cu)

γ_{mode}	γ(keV)	γ(rel)
γ [E2]	128.6 *2*	24.0 *17*
γ [M1+E2]	184.8 *3*	5.3 *14*
γ [E2]	197.5 *2*	12.7 *14*
γ [M1+E2]	489.7 *4*	100 *8*
γ [M1+E2]	520.4 *3*	5.3 *22*
γ [E1]	586.5 *4*	30.3 *22*
γ [M1+E2]	595.2 *5*	30.5 *19*
γ	668.4 *10*	5.0 *19*
γ [M1+E2]	674.8 *3*	25.5 *24*
γ	1233.6 *5*	13 *5*
γ [E1]	1504.8 *5*	10 *3*
γ	1791.3 *8*	13 *4*
γ [E2]	2021.7 *12*	~4

$^{71}_{30}$Zn(2.45 *10* min)

Mode: β-
Δ: -67322 *11* keV
SpA: 1.08×10^{9} Ci/g
Prod: ^{70}Zn(n,γ)

Photons (^{71}Zn)

⟨γ⟩=306 *17* keV

γ_{mode}	γ(keV)	γ(%)†
γ M1+E2	121.62 *3*	3.0 *3*
γ [M1+E2]	390.03 *4*	3.8 *3*
γ [M1+E2]	398.70 *7*	0.61 *6*
γ [M1+E2]	422.99 *8*	0.038 *3*
γ [M1+E2]	453.25 *5*	0.188 *16*
γ M1+0.07%E2	487.36 *5*	0.118 *13*
γ M1+12%E2	511.65 *3*	30
γ [M1+E2]	520.32 *8*	0.081 *7*
γ [E2]	574.87 *5*	0.026 *4*
γ	666.83 *8*	0.90 *10*
γ	721.38 *10*	0.54 *6*
γ [M1+E2]	910.35 *7*	7.8 *6*
γ E2+6.0%M1	964.90 *5*	0.76 *6*
γ	1109.4 *5*	0.16 *3*
γ	1120.07 *8*	2.18 *22*

Photons (^{71}Zn) (continued)

γ_{mode}	γ(keV)	γ(%)†
γ	1144.36 *9*	0.080 *10*
γ	1241.70 *8*	0.032 *3*
γ	1267.1 *14*	0.0090 *10*
γ	1384.18 *24*	0.035 *3*
γ	1553.04 *19*	0.026 *3*
γ	1631.72 *8*	0.38 *3*
γ	1904.50 *23*	0.170 *16*
γ	2064.68 *19*	0.045 *6*
γ	2294.51 *23*	0.026 *3*

† approximate

Continuous Radiation (^{71}Zn)

⟨β-⟩=1048 keV;⟨IB⟩=2.4 keV

E_{bin}(keV)		⟨ ⟩(keV)	(%)
0 - 10	β-	0.0109	0.217
	IB	0.038	
10 - 20	β-	0.0334	0.223
	IB	0.038	0.26
20 - 40	β-	0.140	0.465
	IB	0.074	0.26

Continuous Radiation (^{71}Zn)
(continued)

E_{bin}(keV)		⟨ ⟩(keV)	(%)
40 - 100	β-	1.12	1.58
	IB	0.21	0.32
100 - 300	β-	14.9	7.2
	IB	0.58	0.33
300 - 600	β-	70	15.3
	IB	0.61	0.146
600 - 1300	β-	397	42.0
	IB	0.71	0.084
1300 - 2500	β-	553	32.6
	IB	0.18	0.0118
2500 - 2819	β-	10.9	0.421
	IB	0.00024	9.4×10^{-6}

$^{71}_{30}$Zn(3.94 *5* h)

Mode: β-
Δ: -67165 *12* keV
SpA: 1.120×10^{7} Ci/g
Prod: ^{70}Zn(n,γ)

Photons (^{71}Zn)
⟨γ⟩=1575 *32* keV

γ_{mode}	γ(keV)	γ(%)†
γ [E1]	98.63 *10*	0.062 *7*
γ M1+E2	121.62 *3*	2.85 *23*
γ M1+0.4%E2	142.75 *4*	5.6 *6*
γ [E1]	386.39 *4*	93
γ [M1+E2]	390.03 *4*	2.6 *3*
γ [M1+E2]	398.70 *7*	0.024 *4*
γ [M1+E2]	422.99 *8*	0.00151 *23*
γ [M1+E2]	453.25 *5*	1.08 *8*
γ M1+0.07%E2	487.36 *5*	62 *3*
γ M1+12%E2	511.65 *3*	28.5 *18*
γ [M1+E2]	520.32 *8*	0.0032 *4*
γ [M2]	529.14 *6*	0.046 *19*
γ	565.84 *10*	0.195 *19*
γ [E2]	574.87 *5*	0.151 *24*
γ	588.57 *13*	0.050 *5*
γ [E2]	596.00 *5*	27.9 *19*
γ M1+35%E2	620.29 *4*	57 *3*
γ M1+0.7%E2	753.54 *6*	3.3 *3*
γ	771.39 *6*	2.05 *19*
γ [M1+E2]	910.35 *7*	0.307 *20*
γ	952.04 *16*	0.0102 *9*
γ	956.92 *12*	0.195 *19*
γ	964.53 *8*	~0.5
γ E2+6.0%M1	964.90 *5*	4.4 *4*
γ	974.77 *13*	0.35 *4*
γ	988.83 *8*	1.21 *9*
γ	994.60 *18*	0.030 *4*
γ [M2]	1006.68 *6*	0.74 *19*
γ	1011.56 *12*	0.68 *7*
γ	1086.16 *9*	0.041 *7* ?
γ	1107.5 *4*	~0.7
γ [E2]	1107.65 *5*	2.0 *3*
γ	1139.93 *7*	0.20 *3*
γ	1190.7 *8*	0.011 *3*
γ	1208.20 *16*	0.021 *4*
γ	1226.62 *6*	0.019 *3*
γ	1232.50 *17*	0.028 *4*
γ	1244.48 *10*	0.061 *8*
γ	1282.67 *8*	0.27 *3*
γ	1306.65 *15*	0.112 *9*
γ [M1+E2]	1311.53 *14*	0.102 *9*
γ	1322.19 *16*	0.23 *3*
γ	1340.04 *17*	0.0102 *19*
γ	1343.30 *12*	0.046 *6*
γ	1380.99 *18*	0.36 *4*
γ [E2]	1395.41 *11*	0.084 *9*
γ [E1]	1410.16 *17*	0.0065 *19*
γ	1476.18 *8*	0.60 *6*
γ	1486.05 *13*	0.046 *5*
γ	1493.9 *4*	0.050 *6* ?
γ	1503.9 *5*	0.012 *3*

Photons (^{71}Zn)
(continued)

γ_{mode}	γ(keV)	γ(%)†
γ	1613.01 *7*	0.012 *3*
γ [E1]	1697.92 *14*	0.0047 *9*
γ	1708.58 *17*	0.084 *9*
γ	1719.85 *16*	0.037 *9*
γ	1760.21 *7*	0.93 *9*
γ [E1]	1840.66 *14*	0.046 *5*
γ	1905.87 *18*	0.0045 *6*
γ	1963.58 *13*	0.0056 *9*
γ	2001.26 *18*	0.0037 *9*
γ [E1]	2318.19 *14*	0.65 *9*
γ	2488.61 *19*	0.0047 *9*

† 4.3% uncert(syst)

Continuous Radiation (^{71}Zn)
⟨β-⟩=541 keV;⟨IB⟩=0.80 keV

E_{bin}(keV)		⟨ ⟩(keV)	(%)
0 - 10	β-	0.0405	0.81
	IB	0.023	
10 - 20	β-	0.122	0.81
	IB	0.023	0.158
20 - 40	β-	0.492	1.64
	IB	0.044	0.152
40 - 100	β-	3.59	5.1
	IB	0.118	0.18
100 - 300	β-	39.2	19.3
	IB	0.28	0.161
300 - 600	β-	139	31.0
	IB	0.21	0.052
600 - 1300	β-	348	40.6
	IB	0.097	0.0131
1300 - 2011	β-	10.6	0.78
	IB	0.00029	2.1×10^{-5}

$^{71}_{31}$Ga(stable)

Δ: -70141.5 *23* keV
%: 39.9 *2*

$^{71}_{32}$Ge(11.15 *15* d)

Mode: ϵ
Δ: -69905.9 *18* keV
SpA: 1.649×10^{5} Ci/g
Prod: ^{70}Ge(n,γ); ^{71}Ga(d,2n); ^{72}Ge(n,2n)

Photons (^{71}Ge)
⟨γ⟩=4.16 *11* keV

γ_{mode}	γ(keV)	γ(%)
Ga L_ℓ	0.957	0.046 *10*
Ga L_η	0.984	0.029 *7*
Ga L_α	1.098	0.75 *15*
Ga L_β	1.133	0.50 *13*
Ga L_γ	1.297	0.0018 *6*
Ga $K_{\alpha2}$	9.225	13.3 *6*
Ga $K_{\alpha1}$	9.252	26.1 *11*
Ga $K_{\beta1}$'	10.263	4.91 *20*
Ga $K_{\beta2}$'	10.444	0.0287 *12*

Atomic Electrons (^{71}Ge)
⟨e⟩=4.5 *3* keV

e_{bin}(keV)	⟨e⟩(keV)	e(%)
1	0.90	79 *8*
8	2.7	34 *4*
9	0.78	8.5 *9*
10	0.044	0.44 *4*

Continuous Radiation (^{71}Ge)
⟨IB⟩=0.0070 keV

E_{bin}(keV)		⟨ ⟩(keV)	(%)
10 - 20	IB	0.00121	0.0111
20 - 40	IB	0.00019	0.00068
40 - 100	IB	0.00092	0.00133
100 - 236	IB	0.00153	0.00112

$^{71}_{33}$As(2.70 *3* d)

Mode: ϵ
Δ: -67893 *4* keV
SpA: 6.81×10^{5} Ci/g
Prod: ^{69}Ga(α,2n); ^{70}Ge(d,n)

Photons (^{71}As)
⟨γ⟩=276 *5* keV

γ_{mode}	γ(keV)	γ(%)†
Ge L_ℓ	1.037	0.042 *9*
Ge L_η	1.068	0.024 *6*
Ge L_α	1.188	0.72 *15*
Ge L_β	1.227	0.44 *11*
Ge L_γ	1.412	0.0052 *16*
Ge $K_{\alpha2}$	9.855	11.7 *5*
Ge $K_{\alpha1}$	9.886	22.7 *10*
Ge $K_{\beta1}$'	10.980	4.38 *20*
Ge $K_{\beta2}$'	11.184	0.073 *3*
γ M2	23.48 *8*	0.019 *4*
γ E2	174.94 *4*	83.1
γ M1	247.41 *5*	0.133 *17*
γ [M1+E2]	279.27 *10*	0.175 *17*
γ [E1]	306.26 *6*	0.017 *4*
γ	308.35 *5*	0.0199 *17*
γ E2	326.60 *9*	2.66 *25*
γ [M1+E2]	348.28 *12*	0.033 *8*
γ E1,M1	350.08 *5*	0.216 *25*
γ [E1]	373.75 *14*	0.075 *8*
γ [M1+E2]	387.31 *13*	
γ M1+11%E2	391.41 *9*	0.50 *5*
γ	392.13 *11*	0.058 *8*
γ	431.16 *11*	0.025 *6*
γ	444.91 *14*	~0.0017
γ	448.47 *10*	~0.0033 ?
γ [M1+E2]	465.22 *6*	0.083 *8*
γ	480.09 *16*	0.0158 *17*
γ	486.8 *10*	
γ M1+E2	499.88 *5*	2.83 *25*
γ [M1+E2]	504.26 *7*	0.150 *17*
γ [M1+E2]	526.68 *10*	0.66 *7*
γ [M1+E2]	533.32 *3*	0.0116 *17*
γ [E1]	570.55 *12*	0.025 *6*
γ M1+32%E2	572.35 *5*	0.166 *17*
γ [M1+E2]	595.68 *13*	0.083 *8*
γ	609.7 *7*	0.0208 *17*
γ	615.21 *14*	0.42 *4*
γ	622.69 *4*	0.0133 *8*
γ	633.30 *4*	0.048 *6*
γ	639.53 *11*	<0.07
γ	659.37 *13*	0.058 *8*
γ [M1+E2]	680.01 *14*	0.066 *17*
γ	698.41 *13*	0.025 *6*
γ	702.5 *3*	~0.033 ?
γ M1+E2	708.25 *5*	0.216 *25*

Photons (^{71}As)
(continued)

γ_{mode}	γ(keV)	γ(%)†
γ	712.00 *10*	0.0066 *17* ?
γ[M1+E2]	712.63 *7*	0.283 *25*
γ	727.503 *20*	0.0183 *8*
γ[E2]	747.29 *3*	0.133 *17*
γ	754.12 *10*	~0.0015
γ	759.36 *14*	0.0116 *8*
γ	788.90 *5*	0.0091 *17*
γ	798.6 *3*	0.0258 *25*
γ	808.234 *23*	0.0299 *25*
γ	814.0 *5*	0.038 *3*
γ[M1+E2]	831.28 *6*	0.050 *8*
γ	839.88 *13*	0.0012 *5* ?
γ[M1+E2]	851.62 *10*	0.166 *17*
γ	881.64 *13*	0.0216 *17*
γ	881.85 *3*	0.0307 *25*
γ	886.93 *10*	0.0042 *8* ?
γ	906.78 *13*	0.033 *4*
γ	911.4 *3*	0.0083 *8*
γ[M1+E2]	920.63 *12*	0.249 *25*
γ	935.149 *18*	0.028 *3*
γ	964.48 *11*	0.066 *8*
γ E2+49%M1	993.87 *10*	0.0020 *8* ?
γ	996.09 *3*	0.0100 *8*
γ	1006.76 *14*	0.050 *8*
γ[E2]	1026.56 *10*	0.33 *3*
γ	1033.76 *6*	0.183 *17*
γ[M1+E2]	1037.57 *7*	0.175 *17*
γ	1039.32 *5*	0.0091 *8*
γ	1044.80 *6*	0.0052 *6*
γ	1058.89 *6*	0.025 *6*
γ	1083.83 *3*	0.0116 *8*
γ[M1+E2]	1095.56 *12*	4.2 *4*
γ	1098.7 *3*	0.091 *8*
γ	1104.13 *3*	0.0116 *8*
γ	1129.26 *5*	0.0058 *8*
γ	1139.41 *11*	0.789 *16*
γ	1190.7 *3*	0.0024 *3*
γ[E2]	1212.51 *6*	0.33 *3*
γ	1218.37 *4*	0.0025 *5*
γ	1231.72 *13*	0.058 *8*
γ	1238.3 *10*	0.0052 *6*
γ	1243.50 *5*	0.0017 *3*
γ E2	1246.97 *10*	0.00075 *25* ?
γ	1255.66 *5*	0.0042 *4*
γ	1267.07 *6*	0.0208 *17*
γ	1280.79 *6*	0.0017 *7*
γ	1298.47 *14*	0.199 *17*
γ	1308.23 *16*	0.0059 *7*
γ	1331.71 *14*	0.066 *8*
γ	1378.71 *7*	0.0009 *4*
γ	1383.84 *4*	0.0035 *3*
γ	1423.7 *3*	0.0141 *17*
γ	1454.20 *4*	0.00075 *25*
γ	1460.8 *6*	0.077 *7*
γ	1568.44 *4*	~0.0017
γ	1582.3 *10*	0.0030 *4*
γ	1593.67 *9*	0.0047 *6*
γ	1598.6 *3*	0.0241 *25*
γ	1605.74 *3*	0.0141 *8*
γ	1617.14 *4*	0.0108 *8*
γ	1629.134 *18*	0.0249 *8*
γ	1730.4 *10*	0.00175 *25*
γ	1743.371 *24*	0.0307 *8*
γ	1763.0 *3*	0.0208 *17*
γ	1848.0 *10*	0.0024 *4*
γ	1938.0 *3*	0.0081 *10*
γ	1965.00 *7*	0.00116 *25* ?

† 0.60% uncert(syst)

Atomic Electrons (^{71}As)
⟨e⟩=16.9 *3* keV

e_{bin}(keV)	⟨e⟩(keV)	e(%)
1	0.80	65 *7*
8	0.81	9.6 *10*
9	1.48	17.1 *18*
10	0.65	6.7 *7*
11	0.035	0.32 *3*
12	0.40	3.2 *7*
22 - 23	0.17	0.75 *14*
164	11.05	6.74 *14*
174	1.35	0.779 *16*
236 - 279	0.0076	0.0030 *9*
295 - 339	0.084	0.0264 *22*
347 - 392	0.0067	0.00176 *17*
420 - 469	0.0013	0.00029 *6*
479 - 527	0.033	0.0067 *13*
532 - 572	0.00103	0.00018 *3*
585 - 633	0.0033	0.00055 *17*
638 - 687	0.00075	0.000113 *25*
691 - 736	0.0038	0.00054 *5*
743 - 789	0.00024	3.1 *6* $\times10^{-5}$
797 - 841	0.00105	0.000127 *14*
850 - 896	0.00033	3.7 *8* $\times10^{-5}$
900 - 935	0.00103	0.000113 *13*
953 - 996	0.00032	3.3 *9* $\times10^{-5}$
1005 - 1048	0.00223	0.000218 *20*
1057 - 1104	0.0126	0.00116 *13*
1118 - 1139	0.0017	0.00015 *5*
1180 - 1227	0.00104	8.6 *8* $\times10^{-5}$
1230 - 1280	7.1 $\times10^{-5}$	5.6 *11* $\times10^{-6}$
1281 - 1321	0.00050	3.8 *10* $\times10^{-5}$
1330 - 1379	2.0 $\times10^{-5}$	1.5 *4* $\times10^{-6}$
1382 - 1424	2.5 $\times10^{-5}$	1.8 *6* $\times10^{-6}$
1443 - 1461	0.00013	9 *3* $\times10^{-6}$
1557 - 1629	0.000129	8.1 *11* $\times10^{-6}$
1719 - 1763	7.7 $\times10^{-5}$	4.4 *9* $\times10^{-6}$
1837 - 1927	1.3 $\times10^{-5}$	6.9 *17* $\times10^{-7}$
1937 - 1965	2.7 $\times10^{-6}$	1.4 *3* $\times10^{-7}$

Continuous Radiation (^{71}As)
⟨β+⟩=104 keV;⟨IB⟩=1.02 keV

E_{bin}(keV)		⟨ ⟩(keV)	(%)
0 - 10	β+	0.000123	0.00153
	IB	0.0063	
10 - 20	β+	0.00255	0.0158
	IB	0.0070	0.054
20 - 40	β+	0.0386	0.121
	IB	0.0091	0.032
40 - 100	β+	0.95	1.26
	IB	0.024	0.037
100 - 300	β+	22.5	10.7
	IB	0.071	0.039
300 - 600	β+	66	15.3
	IB	0.17	0.038
600 - 1300	β+	14.4	2.19
	IB	0.60	0.066
1300 - 1838	IB	0.135	0.0094
	Σβ+		30

$^{71}_{34}$Se(4.74 *5* min)

Mode: ε
Δ: -63090 *200* keV syst
SpA: 5.58$\times10^{8}$ Ci/g
Prod: ^{70}Ge(α,3n); ^{14}N on Cu; ^{58}Ni(^{16}O,2pn); ^{60}Ni(^{16}O,αn)

Photons (^{71}Se)
⟨γ⟩=627 *5* keV

γ_{mode}	γ(keV)	γ(%)†
As $K_{\alpha2}$	10.508	0.90 *4*
As $K_{\alpha1}$	10.544	1.75 *8*
As $K_{\beta1}$'	11.724	0.346 *15*
As $K_{\beta2}$'	11.948	0.0107 *5*
γ E2	143.51 *8*	2.3 *3*
γ M1	147.53 *5*	47.5 *10*
γ[M1+E2]	358.77 *9*	1.76 *10*
γ[M1+E2]	362.79 *11*	~0.47
γ	484.56 *10*	0.57 *10*
γ[M1+E2]	681.42 *14*	0.66 *10*
γ[M1+E2]	685.44 *14*	0.062 *5*
γ	723.02 *8*	2.61 *10*
γ	727.04 *10*	0.12 *5*
γ	773.91 *8*	0.104 *14* ?
γ[E2]	777.30 *9*	0.052 *14* ?
γ	830.59 *6*	9.74 *19*
γ	834.61 *9*	0.38 *10*
γ	843.32 *7*	1.14 *5*
γ	847.34 *8*	0.62 *5*
γ	870.54 *7*	6.60 *13*
γ[M1+E2]	924.83 *8*	0.907 *24*
γ	937.16 *7*	0.893 *18*
γ	957.21 *14*	0.157 *10*
γ	978.11 *5*	1.216 *24*
γ	981.85 *22*	0.10 *3*
γ	990.85 *7*	0.209 *19*
γ	1000.45 *22*	0.104 *10*
γ	1003.77 *6*	0.266 *10*
γ	1057.3 *4*	0.081 *10*
γ	1095.42 *6*	9.83 *20*
γ	1099.44 *9*	1.38 *14*
γ	1137.59 *16*	0.147 *10*
γ	1186.78 *9*	0.152 *10*
γ	1242.95 *6*	7.20 *14*
γ[M1+E2]	1265.52 *6*	0.47 *6*
γ[M1+E2]	1269.55 *9*	0.114 *14*
γ	1295.92 *6*	0.475 *10*
γ	1299.94 *9*	0.071 *14*
γ	1315.97 *14*	0.204 *10*
γ	1319.99 *14*	0.100 *10*
γ	1324.08 *11*	0.195 *10*
γ	1395.0 *4*	0.076 *19*
γ	1403.2 *4*	0.052 *19*
γ	1443.44 *7*	0.209 *10*
γ	1445.59 *18*	0.124 *14*
γ	1456.5 *4*	0.052 *14*
γ	1463.50 *14*	0.143 *19*
γ	1468.49 *9*	0.318 *14*
γ	1499.87 *17*	0.043 *10*
γ	1504.90 *9*	0.104 *10*
γ	1529.13 *12*	0.057 *19* ?
γ	1559.18 *11*	~0.033 ?
γ	1582.2 *4*	0.081 *10*
γ	1604.48 *7*	0.465 *24*
γ	1608.51 *10*	0.095 *24*
γ	1636.69 *13*	0.019 *5* ?
γ	1683.6 *4*	0.047 *10*
γ	1701.64 *13*	0.047 *10* ?
γ	1729.80 *9*	0.124 *10*
γ	1752.01 *8*	0.081 *10*
γ	1760.19 *9*	0.185 *10*
γ	1769.5 *5*	0.047 *10*
γ	1834.34 *5*	0.765 *15*
γ	1926.7 *5*	0.090 *10*
γ	1930.29 *10*	0.057 *10* ?
γ	1981.87 *6*	0.252 *10*
γ	2282.18 *9*	0.085 *10*
γ	2286.20 *11*	~0.029
γ	2359.69 *11*	0.124 *10*
γ	2381.0 *4*	0.038 *10*
γ	2412.0 *7*	0.062 *10*
γ	2418.4 *4*	0.024 *5*
γ	2429.71 *9*	0.047 *10* ?
γ	2507.22 *11*	0.067 *10*
γ	2520.4 *3*	0.062 *10*
γ	2609.4 *9*	0.057 *19*
γ	2854.3 *7*	0.029 *5*
γ	2909.8 *4*	0.047 *10*
γ	2926.9 *3*	0.062 *10*
γ	3002.4 *7*	~0.009
γ	3025.68 *9*	0.014 *5* ?

Photons (^{71}Se)
(continued)

γ_{mode}	γ(keV)	γ(%)†
γ	3078.7 *10*	0.014 *5*
γ	3095.3 *7*	0.029 *5*
γ	3173.20 *10*	0.047 *5*
γ	3189.9 *3*	0.067 *10*
γ	3246.3 *9*	0.014 *5*
γ	3359.3 *5*	~0.009
γ	3458.1 *5*	0.014 *5*
γ	3590.5 *7*	0.019 *5*

† 7.2% uncert(syst)

Atomic Electrons (^{71}Se)
$\langle e \rangle$=3.19 *9* keV

e_{bin}(keV)	$\langle e \rangle$(keV)	e(%)
1 - 2	0.064	4.7 *5*
9	0.168	1.84 *19*
10 - 12	0.052	0.50 *5*
132	0.54	0.41 *5*
136	1.84	1.35 *4*
142	0.071	0.050 *6*
143 - 144	0.0123	0.0086 *11*
146	0.213	0.146 *4*
147 - 148	0.0371	0.0252 *7*
347 - 363	0.042	0.012 *4*
473 - 485	0.0049	~0.0010
670 - 715	0.015	0.0021 *7*
721 - 765	0.0019	0.00026 *7*
772 - 819	0.031	0.0037 *16*
823 - 871	0.033	0.0039 *10*
913 - 957	0.0068	0.00074 *11*
966 - 1004	0.0053	0.00055 *13*
1045 - 1094	0.027	0.0025 *8*
1095 - 1138	0.00111	0.000100 *21*
1175 - 1187	0.00034	2.9 *10* $\times 10^{-5}$
1231 - 1270	0.017	0.0014 *4*
1284 - 1324	0.0021	0.00017 *3*
1383 - 1432	0.00060	4.2 *10* $\times 10^{-5}$
1434 - 1468	0.00122	8.4 *15* $\times 10^{-5}$
1488 - 1582	0.00056	3.7 *6* $\times 10^{-5}$
1593 - 1690	0.00113	7.0 *15* $\times 10^{-5}$
1700 - 1769	0.00070	4.0 *6* $\times 10^{-5}$

Atomic Electrons (^{71}Se)
(continued)

e_{bin}(keV)	$\langle e \rangle$(keV)	e(%)
1822 - 1918	0.0014	7.4 *17* $\times 10^{-5}$
1925 - 1982	0.00039	2.0 *5* $\times 10^{-5}$
2270 - 2369	0.00035	1.52 *23* $\times 10^{-5}$
2379 - 2430	0.00017	7.2 *12* $\times 10^{-6}$
2495 - 2520	0.00016	6.3 *12* $\times 10^{-6}$
2597 - 2609	6.9 $\times 10^{-5}$	2.7 *10* $\times 10^{-6}$
2842 - 2927	0.000157	5.4 *8* $\times 10^{-6}$
2990 - 3083	7.1 $\times 10^{-5}$	2.3 *4* $\times 10^{-6}$
3094 - 3190	0.000128	4.0 *6* $\times 10^{-6}$
3234 - 3246	1.5 $\times 10^{-5}$	4.8 *17* $\times 10^{-7}$
3347 - 3446	2.3 $\times 10^{-5}$	6.9 *21* $\times 10^{-7}$
3457 - 3458	1.6 $\times 10^{-6}$	4.6 *16* $\times 10^{-8}$
3579 - 3590	2.0 $\times 10^{-5}$	5.5 *16* $\times 10^{-7}$

Continuous Radiation (^{71}Se)
$\langle \beta + \rangle$=1394 keV;$\langle$IB$\rangle$=4.5 keV

E_{bin}(keV)		$\langle\ \rangle$(keV)	(%)
0 - 10	β+	1.64 $\times 10^{-5}$	0.000204
	IB	0.046	
10 - 20	β+	0.000357	0.00220
	IB	0.046	0.32
20 - 40	β+	0.0057	0.0179
	IB	0.090	0.31
40 - 100	β+	0.158	0.208
	IB	0.26	0.40
100 - 300	β+	5.4	2.48
	IB	0.75	0.42
300 - 600	β+	38.4	8.3
	IB	0.88	0.21
600 - 1300	β+	312	32.2
	IB	1.32	0.152
1300 - 2500	β+	804	44.5
	IB	0.91	0.053
2500 - 4800	β+	235	8.3
	IB	0.22	0.0074
	$\Sigma\beta$+		96

$^{71}_{35}$Br(21.4 *6* s)

Mode: ϵ

Δ: -56590 *300* keV syst

SpA: 7.31$\times 10^{9}$ Ci/g

Prod: ^{58}Ni(^{16}O,p2n)

Photons (^{71}Br)
$\langle \gamma \rangle$=156 *6* keV

γ_{mode}	γ(keV)	γ(%)†
γ	48.78 *5*	1.40 *25*
γ	122.72 *5*	5.3 *4*
γ	171.6 *1*	6.4 *5*
γ	233.7 *1*	6.7 *5*
γ	260.5 *1*	8.3 *4*
γ	282.4 *1*	2.6 *5*
γ	387.4 *2*	1.74 *25*
γ	474.6 *2*	2.1 *3*
γ	647.6 *3*	1.24 *25*
γ	756.9 *2*	4.1 *4*
γ	796.4 *4*	4.6 *5*

† 30% uncert(syst)

$^{71}_{36}$Kr(97 *9* ms)

Mode: ϵ

Δ: -46490 *420* keV syst

SpA: 2.29$\times 10^{11}$ Ci/g

Prod: protons on ^{93}Nb

A = 72

NDS **31**, 103 (1980)

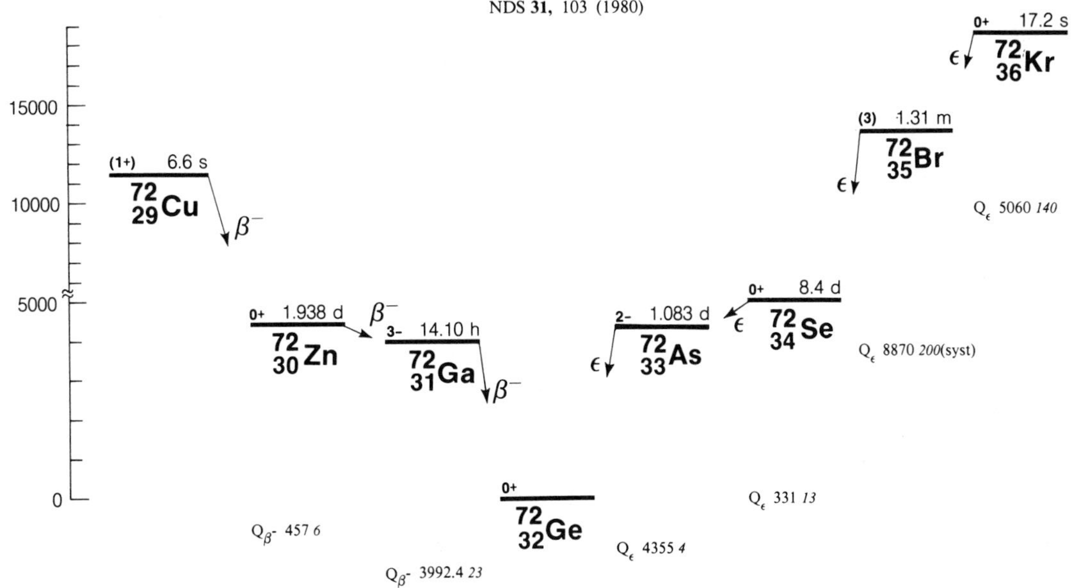

$^{72}_{29}$Cu(6.6 *1* s)

Mode: β-
SpA: 2.25×10^{10} Ci/g
Prod: ^{76}Ge on W

Photons (^{72}Cu)

γ_{mode}	γ(keV)	γ(rel)
γ	534.6 *4*	0.8 *2*
γ	612.2 *3*	1.4 *2*
γ [E2]	652.4 *3*	100 *3*
γ	797.5 *5*	2.0 *3*
γ [E2]	846.5 *3*	11.4 *8*
γ	858.2 *5*	2.5 *3*
γ	988.0 *5*	1.7 *3*
γ [M1+E2]	1004.6 *3*	17.7 *8*
γ	1015.7 *4*	4.2 *5*
γ	1146.4 *6*	1.3 *3*
γ	1251.4 *5*	7.8 *7*
γ	1516.8 *5*	1.7 *3*
γ	1540.1 *5*	8.8 *7*
γ [E2]	1657.7 *5*	14.9 *9*
γ	1918.5 *8*	1.5 *4*
γ	1993.8 *8*	6.2 *5*
γ	2006 *1*	3.2 *4*
γ	2255 *1*	6.4 *8*
γ	2408 *2*	4.8 *9*
γ	3008 *2*	3.2 *3*
γ	3054 *2*	1.7 *4*
γ	3099 *2*	2.7 *4*
γ	3478 *3*	3.2 *4*
γ	3708 *4*	1.7 *3*
γ	3866 *3*	7.0 *8*

$^{72}_{30}$Zn(1.938 *4* d)

Mode: β-
Δ: -68134 *6* keV
SpA: 9.360×10^{5} Ci/g
Prod: fission

Photons (^{72}Zn)

⟨γ⟩=152 *3* keV

γ_{mode}	γ(keV)	γ(%)†
Ga L_ℓ	0.957	0.045 *13*
Ga L_η	0.984	0.028 *9*
Ga L_α	1.098	0.73 *21*
Ga L_β	1.132	0.49 *16*
Ga L_γ	1.297	0.0017 *6*
Ga $K_{\alpha2}$	9.225	13 *3*
Ga $K_{\alpha1}$	9.252	26 *5*
Ga $K_{\beta1}$'	10.263	4.9 *10*
Ga $K_{\beta2}$'	10.444	0.029 *6*
γ (M1)	16.4 *3*	7.4 *15*
γ (M1)	41.9 *1*	0.83 *8*
γ M1	46.80 *11*	0.58 *8*
γ (M1)	79.40 *12*	1.74 *8*
γ (M1)	88.70 *8*	2.16 *8*
γ (M2)	102.80 *8*	2.32 *8*
γ (M1)	112.10 *9*	2.08 *8*
γ (E1)	144.70 *9*	83
γ (E1)	191.50 *9*	9.38 *19*

† 2.4% uncert(syst)

Atomic Electrons (^{72}Zn)

⟨e⟩=15.8 *12* keV

e_{bin}(keV)	⟨e⟩(keV)	e(%)
1	0.90	79 *17*
6	5.0	83 *17*
8	2.7	34 *7*
9	0.77	8.5 *18*
10	0.044	0.43 *9*
15	1.4	9.1 *19*
16 - 47	0.59	2.5 *3*
69 - 89	0.330	0.443 *17*
92	1.26	1.36 *7*
102	0.272	0.267 *15*
103 - 112	0.0420	0.0399 *16*
134	2.12	1.58 *7*
143 - 191	0.426	0.274 *8*

Continuous Radiation (^{72}Zn)

⟨β-⟩=85 keV;⟨IB⟩=0.026 keV

E_{bin}(keV)		⟨ ⟩(keV)	(%)
0 - 10	β-	0.376	7.5
	IB	0.0044	
10 - 20	β-	1.09	7.3
	IB	0.0037	0.026
20 - 40	β-	4.12	13.8
	IB	0.0058	0.020
40 - 100	β-	23.8	34.8
	IB	0.0089	0.0147
100 - 296	β-	56	36.6
	IB	0.0031	0.0024

$^{72}_{31}$Ga(14.10 *2* h)

Mode: β-

Δ: -68591.2 *25* keV

SpA: 3.087×10^{6} Ci/g

Prod: ^{71}Ga(n,γ)

Photons (^{72}Ga)

⟨γ⟩=2709 *14* keV

γ_{mode}	γ(keV)	γ(%)†
Ge $K_{\alpha2}$	9.855	0.086 *10*
Ge $K_{\alpha1}$	9.886	0.167 *19*
Ge $K_{\beta1}$'	10.980	0.032 *4*
Ge $K_{\beta2}$'	11.184	0.00053 *6*
γ [E1]	50.94 *3*	0.0100 *14*
γ (E1)	112.578 *25*	0.187 *23* ?
γ	113.62 *5*	0.0057 *10*
γ E2	142.64 *3*	0.0111 *9*
γ [E1]	230.85 *11*	0.023 *7*
γ [M1+E2]	289.41 *4*	0.191 *13*
γ [M1+E2]	306.14 *5*	0.0210 *19*
γ	317.9 *3*	0.0220 *19*
γ [M1+E2]	336.69 *3*	0.107 *3*
γ [M1+E2]	381.59 *4*	0.268 *10*
γ [M1+E2]	401.88 *6*	0.0325 *19*
γ [M1+E2]	428.68 *4*	0.261 *23*
γ [E1]	449.86 *3*	0.094 *18*
γ [E1]	479.61 *4*	0.088 *5*
γ [E1]	495.99 *11*	0.056 *4*
γ [M1+E2]	520.86 *4*	0.055 *5*
γ	587.57 *12*	0.121 *7*
γ M1+6.6%E2	601.019 *20*	5.56 *10*
γ E2+1.0%M1	630.017 *19*	24.9 *4*
γ M1	735.61 *3*	0.367 *7*
γ	738.17 *9*	0.055 *4*
γ [E2]	772.65 *4*	0.0341 *25*
γ [E1]	786.55 *3*	3.20 *6*
γ [M1+E2]	810.26 *3*	2.01 *4*
γ E2	834.088 *19*	95.63
γ [E1]	861.20 *3*	0.913 *20*
γ [E1]	878.53 *4*	0.072 *5*
γ E2	894.340 *22*	9.88 *20*
γ [E1]	924.66 *11*	0.142 *4*
γ [M1+E2]	938.30 *3*	0.077 *3* ?
γ [E1]	939.58 *4*	0.259 *7*
γ	940.62 *5*	0.0070 *9*
γ [E1]	970.71 *4*	1.102 *22*
γ [M1+E2]	975.60 *11*	0.033 *10*
γ E2	999.94 *3*	0.799 *16*
γ [M1+E2]	1029.27 *11*	0.0039 *5*
γ	1032.4 *4*	0.065 *6*
γ [E2]	1037.23 *11*	0.0210 *19*
γ [E1]	1050.875 *23*	6.96 *12*
γ [E2]	1155.4 *3*	0.0105 *19* ?
γ	1163.29 *7*	0.077 *6*
γ [E1]	1193.91 *18*	0.035 *8* ?
γ [E1]	1215.22 *3*	0.792 *15*
γ [M1+E2]	1231.030 *23*	1.45 *3*
γ [E1]	1260.12 *3*	1.13 *3*
γ [E1]	1276.85 *4*	1.57 *3*
γ	1291.4 *4*	0.056 *5*
γ	1390.48 *4*	0.082 *3*
γ [E2]	1464.097 *24*	3.53 *6*
γ [E1]	1475.99 *5*	0.0179 *17*
γ	1501.0 *3*	0.0191 *10*
γ	1519.5 *5*	0.032 *6*
γ	1541.3 *6*	0.0163 *10*
γ [M1+E2]	1568.31 *3*	0.200 *6*
γ [E1]	1571.72 *4*	0.822 *20*
γ [E1]	1596.80 *3*	4.26 *13*
γ	1613.14 *8*	0.039 *6*
γ [E2]	1629.94 *3*	0.033 *6*
γ [E1]	1680.882 *24*	0.843 *25*
γ [E2]	1710.94 *4*	0.386 *10*
γ [M1+E2]	1711.20 *11*	0.045 *11*
γ	1837.7 *3*	0.209 *11*
γ [E1]	1861.12 *3*	5.25 *10*
γ [E1]	1877.86 *4*	0.231 *6*
γ [E2]	1920.31 *11*	0.158 *5*
γ	1991.48 *5*	0.113 *3*
γ	2029.0 *3*	0.123 *6*

Photons (^{72}Ga) (continued)

γ_{mode}	γ(keV)	γ(%)†
γ [E1]	2105.99 *5*	0.0225 *22*
γ [E1]	2109.55 *4*	1.039 *20*
γ [M1+E2]	2116.4 *3*	0.022 *6*
γ E1	2201.73 *4*	25.9 *5*
γ	2214.14 *7*	0.181 *11*
γ [E1]	2248.62 *5*	0.0112 *11*
γ [E2]	2402.37 *3*	0.0242 *12* ?
γ [E2]	2402.91 *10*	0.0157 *17*
γ E1+2.3%M2	2491.12 *3*	7.67 *22*
γ E1+1.0%M2	2507.86 *4*	12.78 *22*
γ	2514.95 *3*	0.249 *8*
γ	2583.5 *4*	0.014 *3* ?
γ [E2]	2605.52 *11*	0.018 *4*
γ E1	2621.48 *5*	0.131 *3*
γ	2633.75 *4*	0.0145 *15*
γ [M1+E2]	2785.3 *3*	0.0297 *18* ?
γ E1	2844.14 *7*	0.42 *3*
γ	2897.2 *8*	0.0048 *10*
γ [E1]	2940.06 *5*	0.0105 *10* ?
γ [E2]	2950.5 *3*	0.0038 *10* ?
γ	2981.34 *22*	0.055 *6*
γ M2	3035.79 *4*	0.0046 *9*
γ	3067.1 *6*	0.0029 *10*
γ [E2]	3094.34 *11*	0.0166 *18*
γ	3325.18 *3*	0.0031 *9*
γ	3338.0 *3*	0.0033 *9* ?

†<0.1% uncert(syst)

Atomic Electrons (^{72}Ga)

⟨e⟩=4.1 *4* keV

e_{bin}(keV)	⟨e⟩(keV)	e(%)
1 - 50	0.030	0.73 *8*
101 - 143	0.0138	0.0128 *15*
220 - 231	0.00029	0.00013 *4*
278 - 326	0.0082	0.0028 *9*
335 - 382	0.0044	0.0012 *4*
391 - 439	0.0040	0.00096 *23*
448 - 496	0.00056	0.000119 *6*
510 - 521	0.00047	9.2 *21* ×10^{-5}
576 - 601	0.0335	0.00568 *16*
619	0.164	0.0264 *7*
629 - 630	0.0203	0.00322 *7*
680	2.9	0.42 *6*
690	0.39	0.056 *13*
725 - 773	0.00198	0.000272 *13*
775 - 810	0.0150	0.00189 *12*
823	0.392	0.0476 *10*
833	0.0410	0.00492 *10*
834 - 883	0.045	0.00515 *12*
893 - 941	0.00538	0.000599 *13*
960 - 1000	0.00466	0.000476 *11*
1018 - 1051	0.0104	0.00100 *4*
1144 - 1194	0.00024	2.1 *5* ×10^{-5}
1204 - 1249	0.00609	0.000498 *20*
1259 - 1291	0.00221	0.000174 *7*
1379 - 1390	0.00014	1.0 *3* ×10^{-5}
1453 - 1541	0.0080	0.000553 *21*
1557 - 1630	0.00599	0.000379 *13*
1670 - 1711	0.00171	0.000101 *4*
1827 - 1920	0.00595	0.000321 *12*
1980 - 2029	0.00031	1.5 *3* ×10^{-5}
2095 - 2191	0.0220	0.00101 *3*
2200 - 2249	0.00269	0.000122 *4*
2391 - 2490	0.00681	0.000275 *10*
2491 - 2583	0.0118	0.000472 *12*
2594 - 2634	0.000166	6.36 *25* ×10^{-6}
2774 - 2844	0.000382	1.35 *7* ×10^{-5}
2886 - 2981	7.5 ×10^{-5}	2.5 *4* ×10^{-6}
3025 - 3094	3.2 ×10^{-5}	1.05 *9* ×10^{-6}
3314 - 3338	8.4 ×10^{-6}	2.5 *5* ×10^{-7}

Continuous Radiation (^{72}Ga)

⟨β-⟩=498 keV;⟨IB⟩=0.89 keV

E_{bin}(keV)		⟨ ⟩(keV)	(%)
0 - 10	β-	0.072	1.43
	IB	0.021	
10 - 20	β-	0.216	1.44
	IB	0.020	0.138
20 - 40	β-	0.87	2.91
	IB	0.038	0.132
40 - 100	β-	6.4	9.1
	IB	0.100	0.155
100 - 300	β-	60	30.3
	IB	0.24	0.136
300 - 600	β-	123	28.8
	IB	0.20	0.048
600 - 1300	β-	150	17.2
	IB	0.20	0.024
1300 - 2500	β-	146	8.4
	IB	0.066	0.0041
2500 - 3158	β-	11.9	0.445
	IB	0.00078	3.0×10^{-5}

$^{72}_{32}$Ge(stable)

Δ: -72583.6 *15* keV

%: 27.4 *6*

$^{72}_{33}$As(1.083 *4* d)

Mode: ϵ

Δ: -68228 *4* keV

SpA: 1.674×10^{6} Ci/g

Prod: daughter ^{72}Se; ^{69}Ga(α,n); ^{72}Ge(p,n)

Photons (^{72}As)

⟨γ⟩=881 *14* keV

γ_{mode}	γ(keV)	γ(%)†
Ge L_{ℓ}	1.037	0.0071 *16*
Ge L_{η}	1.068	0.0041 *10*
Ge L_{α}	1.188	0.122 *25*
Ge L_{β}	1.227	0.074 *19*
Ge L_{γ}	1.412	0.0009 *3*
Ge $K_{\alpha2}$	9.855	1.98 *11*
Ge $K_{\alpha1}$	9.886	3.86 *21*
Ge $K_{\beta1}$'	10.980	0.74 *4*
Ge $K_{\beta2}$'	11.184	0.0123 *7*
γ [E1]	50.94 *3*	0.0014 *3*
γ (E1)	112.578 *25*	0.026 *3*
γ	113.62 *5*	0.017 *3*
γ E2	142.64 *3*	0.0105 *25*
γ [E1]	230.85 *11*	9 *3* ×10^{-5}
γ [M1+E2]	289.41 *4*	0.00073 *7*
γ [M1+E2]	306.14 *5*	0.00053 *5*
γ [M1+E2]	336.69 *3*	0.00604 *22*
γ [M1+E2]	381.59 *4*	0.00102 *7*
γ [M1+E2]	401.88 *6*	0.00082 *6*
γ [M1+E2]	428.68 *4*	0.068 *6*
γ [E1]	449.86 *3*	0.013 *3*
γ [E1]	479.61 *4*	0.0227 *13*
γ [M1+E2]	520.86 *4*	0.00102 *10*
γ [M2]	587.57 *12*	0.00307 *20*
γ M1+6.6%E2	601.019 *20*	0.313 *8*
γ E2+1.0%M1	630.017 *19*	7.86 *13*
γ M1	735.61 *3*	0.0113 *15*
γ	738.17 *9*	0.0090 *11*
γ [E2]	772.65 *4*	0.0108 *8*
γ [E1]	786.55 *3*	0.449 *11*
γ [M1+E2]	810.26 *3*	0.0076 *5*

Photons (^{72}As) (continued)

γ_{mode}	γ(keV)	γ(%)†
γ E2	834.088 19	79.5
γ [E1]	861.20 3	0.00348 23
γ [E1]	878.53 4	0.0187 13
γ E2	894.340 22	0.775 16
γ [E1]	905.18 24	0.029 6
γ [M1+E2]	938.30 3	0.0501 25
γ [E1]	939.58 4	0.00654 23
γ	940.62 5	0.0207 24
γ	950.26 16	~0.006
γ [E1]	970.71 4	0.0206 8
γ E2	999.94 3	0.025 3
γ [M1+E2]	1029.27 11	0.0249 15
γ [E1]	1050.875 23	0.977 17
γ [E1]	1148.41 17	0.012 3
γ [E2]	1155.4 3	0.0051 11
γ	1163.29 7	0.0128 12
γ [E1]	1193.91 18	0.084 3
γ [E1]	1215.22 3	0.206 6
γ [M1+E2]	1231.030 23	0.082 3
γ [E1]	1260.12 3	0.0043 3
γ [E1]	1276.85 4	0.0395 12
γ	1390.48 4	0.241 7
γ [E2]	1464.097 24	1.113 19
γ [E1]	1475.99 5	0.506 10
γ	1522.05 16	0.013 3
γ [M1+E2]	1568.31 3	0.130 5
γ [E1]	1571.72 4	0.0153 6
γ	1581.51 16	0.025 6
γ [E1]	1596.80 3	0.0162 10
γ	1613.14 8	0.0065 10
γ [E2]	1629.94 3	0.00100 22
γ [E1]	1680.882 24	0.118 3
γ [E2]	1710.94 4	0.252 8
γ [E1]	1861.12 3	0.0200 11
γ [E1]	1877.86 4	0.00584 20
γ [E2]	1920.31 11	~0.06
γ	1939.1 3	0.0080 8 ?
γ	1991.48 5	0.333 10
γ [E1]	2086.69 17	0.0143 8
γ [E1]	2105.99 5	0.636 24
γ [E1]	2109.55 4	0.270 8
γ [M1+E2]	2116.4 3	0.0278 16 ?
γ E1	2201.73 4	0.484 15
γ	2214.14 7	0.0299 16
γ [E1]	2248.62 5	0.316 12
γ [E1]	2339.51 6	0.0365 21
γ [E2]	2402.37 3	0.0158 7
γ [E2]	2402.91 10	0.101 4
γ E1+2.3%M2	2491.12 3	0.0292 16
γ E1+1.0%M2	2507.86 4	0.323 8
γ	2514.95 3	0.0349 14
γ	2521.96 15	0.0461 24
γ	2577.0 4	0.0183 24
γ [M1+E2]	2586.04 25	0.076 17
γ E1	2621.48 5	0.387 8
γ	2633.75 4	5.5 7 $\times10^{-5}$
γ [E1]	2716.69 17	0.0119 24
γ [M1+E2]	2785.3 3	0.0143 16 ?
γ	2833.4 3	0.0087 16 ?
γ E1	2844.14 7	0.069 7
γ [E1]	2859.32 17	0.019 4
γ [E1]	2940.06 5	0.297 18
γ [E2]	2950.5 3	0.0048 8 ?
γ	2981.34 22	~0.2
γ M2	3035.79 4	8.6 18 $\times10^{-5}$
γ [E2]	3094.34 11	0.107 13
γ [E1]	3112.13 5	0.064 10
γ	3149.78 16	0.066 10
γ	3161.16 22	0.0080 8
γ	3207.0 4	0.0064 8
γ	3256.4 5	0.0024 4
γ	3303.79 22	0.0072 16
γ	3325.18 3	1.2 3 $\times10^{-5}$
γ	3338.0 3	0.016 3 ?
γ	3399.0 5	0.0032 5
γ [E1]	3550.75 17	0.0205 17
γ [E1]	3803.56 6	0.100 6
γ [E2]	3872.2 4	0.0052 10
γ	3995.21 22	0.0493 10

† 2.1% uncert(syst)

Atomic Electrons (^{72}As)

⟨e⟩=11.5 21 keV

e_{bin}(keV)	⟨e⟩(keV)	e(%)
1 - 50	0.64	16.7 13
101 - 143	0.009	0.008 4
220 - 231	1.1 $\times10^{-6}$	5.1 23 $\times10^{-7}$
278 - 326	0.00014	4.4 22 $\times10^{-5}$
335 - 382	2.9 $\times10^{-5}$	8 3 $\times10^{-6}$
391 - 439	0.00089	0.00021 6
448 - 480	9.0 $\times10^{-5}$	1.92 13 $\times10^{-5}$
510 - 521	8.8 $\times10^{-6}$	1.7 9 $\times10^{-6}$
576 - 619	0.0534	0.00864 22
629 - 630	0.00639	0.001016 23
680	9.2	1.4 3
690	1.2	0.18 5
725 - 773	0.00014	1.9 4 $\times10^{-5}$
775 - 810	0.00091	0.000117 5
823	0.325	0.0396 11
833 - 879	0.0398	0.00478 12
883 - 930	0.00347	0.000392 10
937 - 971	7.4 $\times10^{-5}$	7.8 13 $\times10^{-6}$
989 - 1029	0.000167	1.66 13 $\times10^{-5}$
1040 - 1051	0.00143	0.000137 5
1137 - 1183	0.000157	1.34 11 $\times10^{-5}$
1192 - 1231	0.000492	4.06 15 $\times10^{-5}$
1249 - 1277	5.5 $\times10^{-5}$	4.3 3 $\times10^{-6}$
1379 - 1390	0.00043	3.1 9 $\times10^{-5}$
1453 - 1522	0.00309	0.000212 7
1557 - 1630	0.000352	2.25 14 $\times10^{-5}$
1670 - 1711	0.000608	3.59 14 $\times10^{-5}$
1850 - 1939	3.6 $\times10^{-5}$	1.92 21 $\times10^{-6}$
1980 - 2076	0.00045	2.3 6 $\times10^{-5}$
2085 - 2116	0.00088	4.21 15 $\times10^{-5}$
2191 - 2249	0.000758	3.43 10 $\times10^{-5}$
2328 - 2403	0.000202	8.5 3 $\times10^{-6}$
2480 - 2577	0.00055	2.18 11 $\times10^{-5}$
2585 - 2634	0.000338	1.29 3 $\times10^{-5}$
2706 - 2785	2.9 $\times10^{-5}$	1.04 10 $\times10^{-6}$
2822 - 2859	8.2 $\times10^{-5}$	2.88 23 $\times10^{-6}$
2929 - 3025	0.000247	8.4 6 $\times10^{-6}$
3034 - 3112	0.000182	5.9 5 $\times10^{-6}$
3139 - 3207	8.1 $\times10^{-5}$	2.6 5 $\times10^{-6}$
3245 - 3338	2.5 $\times10^{-5}$	7.5 14 $\times10^{-7}$
3388 - 3399	3.1 $\times10^{-6}$	9.1 21 $\times10^{-8}$
3540 - 3551	1.57 $\times10^{-5}$	4.4 4 $\times10^{-7}$
3792 - 3872	8.1 $\times10^{-5}$	2.12 12 $\times10^{-6}$
3984 - 3995	4.5 $\times10^{-5}$	1.12 19 $\times10^{-6}$

Continuous Radiation (^{72}As)

⟨β+⟩=1024 keV;⟨IB⟩=3.1 keV

E_{bin}(keV)		⟨ ⟩(keV)	(%)
0 - 10	β+	3.03 $\times10^{-5}$	0.000378
	IB	0.037	
10 - 20	β+	0.00063	0.00392
	IB	0.037	0.25
20 - 40	β+	0.0098	0.0307
	IB	0.071	0.25
40 - 100	β+	0.257	0.340
	IB	0.20	0.31
100 - 300	β+	8.1	3.74
	IB	0.57	0.32
300 - 600	β+	53	11.4
	IB	0.65	0.152
600 - 1300	β+	364	38.0
	IB	0.90	0.104
1300 - 2500	β+	554	32.5
	IB	0.56	0.033
2500 - 4355	β+	45.4	1.66
	IB	0.117	0.0041
	Σβ+		88

$^{72}_{34}$Se(8.40 8 d)

Mode: ε

Δ: -67897 12 keV

SpA: 2.159$\times10^{5}$ Ci/g

Prod: ^{75}As(d,5n); ^{70}Ge(α,2n)

Photons (^{72}Se)

⟨γ⟩=34.2 11 keV

γ_{mode}	γ(keV)	γ(%)†
As L_ℓ	1.120	0.073 16
As L_η	1.155	0.043 11
As L_α	1.282	1.3 3
As L_β	1.325	0.81 21
As L_γ	1.524	0.011 3
As $K_{\alpha2}$	10.508	21.1 9
As $K_{\alpha1}$	10.544	41.1 18
As $K_{\beta1}$'	11.724	8.1 4
As $K_{\beta2}$'	11.948	0.251 11
γ E1	46.0 3	58

† 3.5% uncert(syst)

Atomic Electrons (^{72}Se)

⟨e⟩=22.0 7 keV

e_{bin}(keV)	⟨e⟩(keV)	e(%)
1	1.40	105 11
2	0.104	6.8 7
9	3.9	43 5
10	1.12	10.8 11
11	0.048	0.45 5
12	0.058	0.50 5
34	13.1	38.4 15
44	1.51	3.40 14
45	0.346	0.77 3
46	0.323	0.71 3

Continuous Radiation (^{72}Se)

⟨IB⟩=0.0105 keV

E_{bin}(keV)		⟨ ⟩(keV)	(%)
10 - 20	IB	0.0042	0.036
20 - 40	IB	0.00026	0.00092
40 - 100	IB	0.00111	0.00160
100 - 285	IB	0.0033	0.0022

$^{72}_{35}$Br(1.31 4 min)

Mode: ε

Δ: -59030 200 keV syst

SpA: 1.98$\times10^{9}$ Ci/g

Prod: ^{58}Ni(^{16}O,pn)

Photons (^{72}Br)

$\langle\gamma\rangle$=1875 *42* keV

γ_{mode}	γ(keV)	γ(%)†
Se L_ℓ	1.204	0.0070 *22*
Se L_η	1.245	0.0042 *14*
Se L_α	1.379	0.14 *4*
Se L_β	1.427	0.08 *3*
Se L_γ	1.648	0.0011 *4*
Se $K_{\alpha2}$	11.182	1.9 *4*
Se $K_{\alpha1}$	11.222	3.8 *8*
Se $K_{\beta1}'$	12.494	0.76 *17*
Se $K_{\beta2}'$	12.741	0.037 *8*
γ E2	75.0 *3*	3.05 *20*
γ [E2]	379.78 *23*	3.6 *6*
γ [M1+E2]	454.77 *18*	13.1 *8*
γ [E2]	513.3 *9*	2.0 *8* ?
γ [M1+E2]	537.78 *23*	1.3 *4*
γ	559.37 *21*	2.6 *3*
γ	709.94 *23*	1.6 *4*
γ	752.4 *3*	2.9 *4*
γ (E2)	774.8 *3*	7.1 *4*
γ [M1+E2]	833.3 *8*	2.0 *8*
γ [E2]	862.00 *17*	70.2
γ	1014.14 *25*	~0.7
γ	1054.68 *21*	3.7 *6*
γ [E2]	1061.64 *24*	5.5 *5*
γ	1089.2 *3*	3.2 *4*
γ [E2]	1125.22 *23*	5.3 *6*
γ [M1+E2]	1136.63 *24*	7.0 *7*
γ	1227.4 *3*	1.1 *4*
γ [M1+E2]	1269.30 *22*	~0.8
γ [E2]	1316.77 *20*	17.3 *11*
γ	1349.88 *25*	2.2 *4*
γ	1434.5 *3*	1.0 *4*
γ	1509.45 *23*	3.3 *6*
γ [M1+E2]	1571.2 *4*	3.8 *3*
γ	1648.82 *25*	1.5 *4*
γ [M1+E2]	1724.07 *24*	3.4 *3*
γ [E2]	1807.08 *23*	1.8 *4*
γ	1909.2 *3*	1.3 *4*
γ [E2]	2150.1 *8*	1.0 *3*
γ	2371.43 *25*	7.5 *8*
γ [E2]	2433.2 *4*	1.3 *3*
γ	2465.0 *8*	1.0 *3*

† 1.4% uncert(syst)

Atomic Electrons (^{72}Br)

$\langle e\rangle$=55 *22* keV

e_{bin}(keV)	$\langle e\rangle$(keV)	e(%)
1 - 12	0.60	14.6 *24*
62	3.9	6.2 *4*
73	0.41	0.55 *4*
74 - 75	0.463	0.63 *3*
367 - 380	0.04	~0.012
442 - 455	0.17	0.039 *9*
501 - 547	0.055	0.0106 *25*
558 - 559	0.0021	0.00038 *18*
697 - 740	0.019	0.0027 *9*
751 - 775	0.045	0.0059 *4*
821 - 862	0.373	0.0439 *17*
924	42.9	~5
935	5.8	~0.6
1001 - 1049	0.030	0.0029 *4*
1053 - 1089	0.008	0.0007 *4*
1113 - 1137	0.043	0.0039 *3*
1215 - 1257	0.0048	0.00039 *14*
1268 - 1317	0.052	0.0040 *3*
1337 - 1350	0.0049	0.00036 *13*
1422 - 1434	0.0020	0.00014 *6*
1497 - 1571	0.0158	0.00103 *15*
1636 - 1724	0.0106	0.00063 *8*
1794 - 1807	0.0038	0.00021 *4*
1897 - 1909	0.0022	0.00011 *4*
2137 - 2150	0.0019	8.7 *23* $\times10^{-5}$
2359 - 2452	0.014	0.00058 *11*
2463 - 2465	0.00014	5.8 *19* $\times10^{-6}$

Continuous Radiation (^{72}Br)

$\langle\beta+\rangle$=2762 keV;$\langle$IB$\rangle$=11.3 keV

E_{bin}(keV)		$\langle\ \rangle$(keV)	(%)
0 - 10	β+	2.84×10^{-6}	3.52×10^{-5}
	IB	0.070	
10 - 20	β+	6.4×10^{-5}	0.000396
	IB	0.069	0.48
20 - 40	β+	0.00107	0.00333
	IB	0.137	0.48
40 - 100	β+	0.0309	0.0406
	IB	0.40	0.62
100 - 300	β+	1.15	0.53
	IB	1.25	0.69
300 - 600	β+	9.3	1.99
	IB	1.64	0.38
600 - 1300	β+	103	10.3
	IB	3.0	0.33
1300 - 2500	β+	607	31.4
	IB	3.0	0.17
2500 - 5000	β+	1780	50.9
	IB	1.7	0.055
5000 - 7500	β+	262	4.79
	IB	0.045	0.00080
	$\Sigma\beta$+		99.9

$^{72}_{36}$Kr(17.2 *3* s)

Mode: ϵ

Δ: -53970 *240* keV syst

SpA: 8.93×10^{9} Ci/g

Prod: ^{58}Ni(^{16}O,2n)

Photons (^{72}Kr)

$\langle\gamma\rangle$=359 *17* keV

γ_{mode}	γ(keV)	γ(%)†
Br $K_{\alpha2}$	11.878	0.41 *18*
Br $K_{\alpha1}$	11.924	0.8 *4*
Br $K_{\beta1}'$	13.290	0.16 *7*
Br $K_{\beta2}'$	13.562	0.011 *5*
γ [M1]	124.4 *3*	8.0 *11* ?
γ [M1]	147.49 *25*	~4 ?
γ [E2]	162.73 *16*	14 *2*
γ [M1]	252.41 *16*	6.2 *7*
γ [E2]	310.21 *20*	29.5 *11*
γ [E2]	415.14 *16*	36.4
γ [E2]	576.7 *3*	12.4 *7*

† 6.0% uncert(syst)

Atomic Electrons (^{72}Kr)

$\langle e\rangle$=5.0 *15* keV

e_{bin}(keV)	$\langle e\rangle$(keV)	e(%)
2	0.031	2.0 *9*
10	0.07	0.6 *3*
11	0.0020	0.017 *8*
12	0.021	0.18 *8*
13	0.0018	0.013 *6*
111	0.50	0.45 *7*
123	0.062	0.050 *7*
124	0.0113	0.0091 *13*
134	0.21	~0.16
146	0.025	~0.017
147	0.0046	~0.0031
149	1.5	<2
161	0.2	<0.25
162	0.025	<0.030
163	0.012	<0.014
239	0.133	0.056 *7*
251	0.0151	0.0060 *8*
252	0.0028	0.00110 *14*
297	1.10	0.371 *20*
308	0.113	0.0367 *20*
309	0.0171	0.0055 *3*
310	0.0236	0.0076 *4*
402	0.69	0.173 *12*
413	0.072	0.0173 *12*
414	0.0078	0.00189 *14*
415	0.0144	0.00347 *25*
563	0.121	0.0214 *15*
575	0.0134	0.00233 *17*
576	0.00203	0.000352 *25*
577	0.00041	7.0 *5* $\times10^{-5}$

A = 73

NDS **29**, 1 (1980)

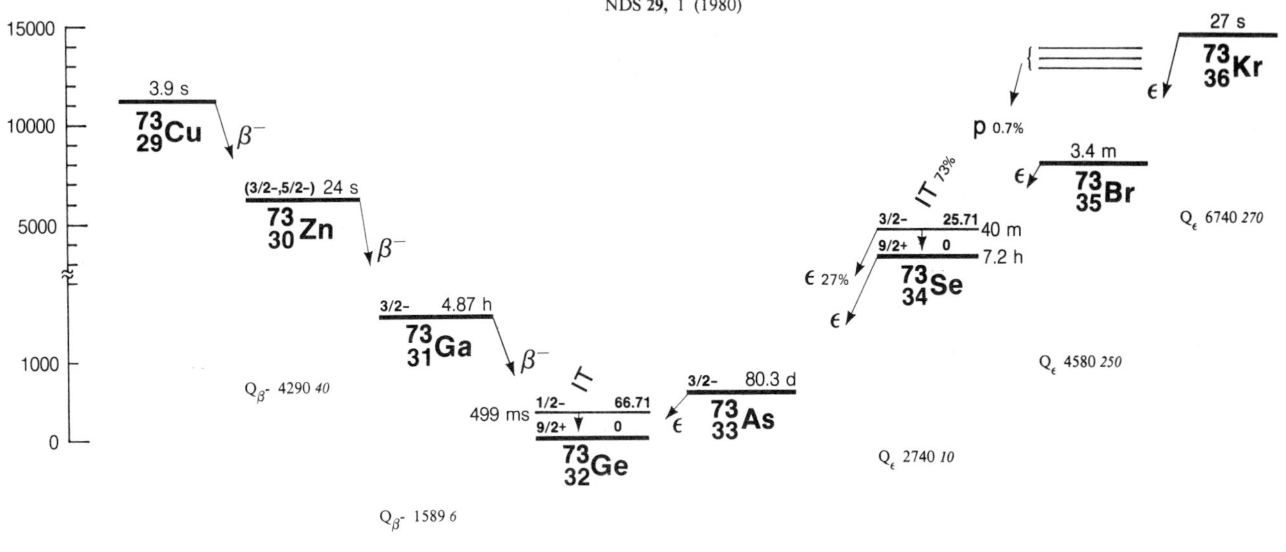

$^{73}_{29}$Cu(3.9 *3* s)

Mode: β-
SpA: 3.6×10^{10} Ci/g
Prod: ^{76}Ge on W

Photons (^{73}Cu)

γ_{mode}	γ(keV)	γ(rel)
γ	199.2 *3*	17 *2*
γ	306.8 *3*	10 *2*
γ	449.7 *3*	100 *5*
γ	502.0 *3*	12 *2*
γ	674.4 *3*	9 *3*

$^{73}_{30}$Zn(23.5 *10* s)

Mode: β-
Δ: -65410 *40* keV
SpA: 6.5×10^{9} Ci/g
Prod: protons on Ge; ^{76}Ge on W

Photons (^{73}Zn)

γ_{mode}	γ(keV)	γ(rel)
γ [M1+E2]	218.1 *2*	100 *3*
γ [M1+E2]	278.4 *4*	1.6 *2*
γ [M1+E2]	415.2 *4*	1.7 *2*
γ [M1+E2]	495.6 *3*	24.7 *13*
γ [M1+E2]	693.1 *3*	6.3 *4*
γ [M1+E2]	910.5 *4*	31.9 *13*
γ	1113.0 *4*	4.9 *5*
γ [M1+E2]	1197.3 *4*	12.9 *9*
γ	1428.3 *5*	1.5 *4*
γ	1474.3 *5*	2.2 *5*

Photons (^{73}Zn) (continued)

γ_{mode}	γ(keV)	γ(rel)
γ [M1+E2]	1612.9 *4*	15.4 *9*
γ	1692.5 *6*	3.6 *5*
γ	1924.7 *8*	9.3 *7*
γ	1979.9 *8*	6.2 *6*
γ [M1+E2]	2109 *1*	12.1 *9*
γ	2344 *2*	2.3 *4*
γ	2772 *3*	1.7 *3*
γ	2989 *3*	2.1 *3*

$^{73}_{31}$Ga(4.87 *3* h)

Mode: β-
Δ: -69705 *6* keV
SpA: 8.81×10^{6} Ci/g
Prod: ^{73}Ge(n,p); ^{76}Ge(d,αn); fission

Photons (^{73}Ga)

⟨γ⟩=352 *5* keV

γ_{mode}	γ(keV)	γ(%)†
Ge L_ℓ	1.037	0.092 *20*
Ge L_η	1.068	0.052 *13*
Ge L_α	1.188	1.6 *3*
Ge L_β	1.224	0.89 *23*
Ge L_γ	1.412	0.0074 *23*
Ge $K_{\alpha2}$	9.855	16.4 *7*
Ge $K_{\alpha1}$	9.886	31.9 *14*
Ge $K_{\beta1}'$	10.980	6.1 *3*
Ge $K_{\beta2}'$	11.184	0.102 *5*
γ E2	13.273 *17*	0.0910 *24* *
γ M2	53.439 *9*	10.29 *24* *
γ M1+0.5%E2	68.69 *19*	0.40 *8*
γ	216.4 *3*	0.096 *24*
γ [E1]	284.89 *19*	0.32 *5*
γ [M1+E2]	297.30 *5*	79.8 *16*

Photons (^{73}Ga) (continued)

γ_{mode}	γ(keV)	γ(%)†
γ [M1+E2]	325.69 *5*	11.17 *24*
γ [E1]	350.74 *5*	0.21 *3*
γ [E1]	379.13 *5*	0.487 *24*
γ [E1]	488.16 *9*	0.359 *24*
γ	501.6 *4*	0.15 *5*
γ [E2]	541.60 *9*	0.30 *3*
γ	561.9 *3*	0.18 *4*
γ [E1]	576.92 *10*	0.15 *4*
γ [M1+E2]	739.39 *4*	4.23 *24*
γ [M1+E2]	767.79 *6*	1.44 *8*
γ	993.6 *3*	0.14 *3*
γ [M1]	1065.08 *5*	1.28 *8*

† 2.1% uncert(syst)
* with ^{73}Ge(499 ms) in equilib

Atomic Electrons (^{73}Ga)

⟨e⟩=56.5 *15* keV

e_{bin}(keV)	⟨e⟩(keV)	e(%)
1	1.87	152 *16*
2 - 8	1.70	39.5 *17*
9	2.08	24.0 *25*
10 - 11	0.96	9.9 *10*
12	7.6	63.5 *22*
13	1.28	9.7 *3*
42	31.7	75.0 *25*
52	5.82	11.2 *4*
53 - 69	1.07	1.99 *7*
205 - 216	0.005	~0.0023
274 - 315	2.3	0.8 *3*
324 - 368	0.031	0.009 *3*
378 - 379	0.000284	7.5 *4* $\times10^{-5}$
477 - 502	0.0024	0.00050 *13*
530 - 577	0.0045	0.00083 *11*
728 - 768	0.028	0.0038 *4*
982 - 994	0.00037	3.7 *15* $\times10^{-5}$
1054 - 1065	0.00372	0.000352 *23*

Continuous Radiation (^{73}Ga)

⟨β-⟩=442 keV;⟨IB⟩=0.56 keV

E_{bin}(keV)		⟨ ⟩(keV)	(%)
0 - 10	β-	0.052	1.04
	IB	0.020	
10 - 20	β-	0.158	1.05
	IB	0.019	0.134
20 - 40	β-	0.64	2.15
	IB	0.037	0.127
40 - 100	β-	4.80	6.8
	IB	0.096	0.149
100 - 300	β-	50.1	24.9
	IB	0.21	0.124
300 - 600	β-	155	34.8
	IB	0.139	0.034
600 - 1300	β-	230	29.2
	IB	0.038	0.0054
1300 - 1522	β-	1.17	0.087
	IB	2.3×10^{-5}	1.7×10^{-6}

$^{73}_{32}$Ge(stable)

Δ: -71294.7 *15* keV
%: 7.8 *2*

$^{73}_{32}$Ge(499 *11* ms)

Mode: IT
Δ: -71228.0 *15* keV
SpA: 1.67×10^{11} Ci/g
Prod: daughter ^{73}As

Photons (^{73}Ge)

⟨γ⟩=11.14 *22* keV

γ_{mode}	γ(keV)	γ(%)
Ge L_ℓ	1.037	0.093 *20*
Ge L_η	1.068	0.052 *13*
Ge L_α	1.188	1.6 *3*
Ge L_β	1.224	0.90 *23*
Ge L_γ	1.412	0.0075 *23*
Ge $K_{\alpha2}$	9.855	16.5 *7*
Ge $K_{\alpha1}$	9.886	32.0 *14*
Ge $K_{\beta1}$'	10.980	6.2 *3*
Ge $K_{\beta2}$'	11.184	0.103 *5*
γ E2	13.273 *17*	0.092 *3*
γ M2	53.439 *9*	10.5 *3*

Atomic Electrons (^{73}Ge)

⟨e⟩=54.8 *12* keV

e_{bin}(keV)	⟨e⟩(keV)	e(%)
1	1.88	152 *16*
2	0.568	26.2 *9*
8	1.14	13.6 *14*
9	2.09	24.1 *25*
10	0.92	9.5 *10*
11	0.049	0.45 *5*
12	7.7	63.9 *22*
13	1.29	9.8 *3*
42	32.3	76 *3*
52	5.91	11.4 *4*
53	1.03	1.93 *6*

$^{73}_{33}$As(80.30 *6* d)

Mode: ϵ
Δ: -70955 *4* keV
SpA: 2.2275×10^{4} Ci/g
Prod: ^{72}Ge(d,n); ^{74}Ge(p,2n)

Photons (^{73}As)

⟨γ⟩=15.8 *3* keV

γ_{mode}	γ(keV)	γ(%)
Ge L_ℓ	1.037	0.14 *3*
Ge L_η	1.068	0.081 *20*
Ge L_α	1.188	2.5 *5*
Ge L_β	1.225	1.4 *4*
Ge L_γ	1.412	0.014 *4*
Ge $K_{\alpha2}$	9.855	30.6 *13*
Ge $K_{\alpha1}$	9.886	59 *3*
Ge $K_{\beta1}$'	10.980	11.5 *5*
Ge $K_{\beta2}$'	11.184	0.190 *8*
γ E2	13.273 *17*	0.092 *3* *
γ M2	53.439 *9*	10.5 *3* *

* with ^{73}Ge(499 ms) in equilib

Atomic Electrons (^{73}As)

⟨e⟩=59.4 *13* keV

e_{bin}(keV)	⟨e⟩(keV)	e(%)
1	2.9	231 *24*
2	0.568	26.2 *9*
8	2.11	25 *3*
9	3.9	45 *5*
10	1.71	17.6 *18*
11	0.091	0.84 *9*
12	7.7	63.9 *22*
13	1.29	9.8 *3*
42	32.3	76 *3*
52	5.91	11.4 *4*
53	1.03	1.93 *6*

Continuous Radiation (^{73}As)

⟨IB⟩=0.0091 keV

E_{bin}(keV)		⟨ ⟩(keV)	(%)
10 - 20	IB	0.0031	0.029
20 - 40	IB	0.00022	0.00079
40 - 100	IB	0.00106	0.00152
100 - 272	IB	0.0028	0.0019

$^{73}_{34}$Se(7.15 *8* h)

Mode: ϵ
Δ: -68215 *11* keV
SpA: 6.00×10^{6} Ci/g
Prod: ^{70}Ge(α,n); ^{75}As(d,4n); ^{74}Se(γ,n); ^{75}As(p,3n); ^{74}Se(n,2n)

Photons (^{73}Se)

⟨γ⟩=422 *35* keV

γ_{mode}	γ(keV)	γ(%)†
As L_ℓ	1.120	0.028 *6*
As L_η	1.155	0.016 *4*
As L_α	1.282	0.51 *11*
As L_β	1.325	0.31 *8*
As L_γ	1.524	0.0041 *13*
As $K_{\alpha2}$	10.508	8.1 *5*
As $K_{\alpha1}$	10.544	15.7 *10*
As $K_{\beta1}$'	11.724	3.1 *2*
As $K_{\beta2}$'	11.948	0.096 *6*
γ M1+0.06%E2	67.17 *6*	70 *8*
γ M2+0.03%E3	360.60 *11*	97
γ [E3]	427.77 *11*	0.078 *15*
γ	442.72 *11*	0.0504 *25*
γ	509.89 *11*	0.266 *8*
γ	557.57 *9*	0.0533 *19*
γ	575.51 *8*	0.146 *7*
γ	600.5 *3*	0.020 *3*
γ [E2]	609.39 *14*	0.049 *4*
γ	682.36 *12*	0.0194 *19*
γ	700.30 *12*	0.0446 *19*
γ	765.16 *8*	0.1271 *25*
γ	783.10 *9*	0.0582 *19*
γ	793.1 *5*	0.0640 *19*
γ	813.40 *15*	0.0087 *10*
γ	818.77 *8*	0.0369 *19*
γ	847.28 *8*	0.081 *6*
γ	857.1 *3*	0.023 *6*
γ	865.22 *8*	0.524 *16*
γ	872.97 *16*	0.038 *7*
γ	887.73 *12*	~0.011
γ	900.89 *8*	0.135 *3*
γ	926.43 *9*	0.0039 *10*
γ	930.22 *15*	0.0049 *10*
γ	968.23 *10*	
γ	982.91 *9*	0.0340 *10*
γ	993.60 *9*	0.0049 *10*
γ	1002.79 *16*	0.0039 *10*
γ	1018.57 *9*	0.0533 *19*
γ	1036.51 *7*	0.0145 *10*
γ	1110.77 *6*	0.201 *4*
γ	1154.1 *3*	0.0049 *10*
γ	1158.3 *4*	0.0029 *10*
γ	1207.88 *12*	0.0039 *10*
γ	1215.5 *8*	
γ	1225.82 *13*	~0.0029
γ	1250.03 *20*	0.0039 *10*
γ	1275.05 *12*	0.0068 *10*
γ	1309.00 *12*	0.0039 *10*
γ	1317.96 *13*	0.0058 *10*
γ	1323.94 *20*	0.0068 *10*
γ	1340.67 *6*	0.0689 *19*
γ	1406.11 *15*	~0.0019
γ	1422.79 *6*	0.136 *5*
γ	1439.23 *17*	~0.0019
γ	1451.73 *20*	0.0058 *19*
γ	1482.36 *11*	0.0223 *10*
γ	1547.57 *10*	0.0310 *10*
γ	1670.83 *11*	0.0049 *10*
γ	1738.5 *5*	~0.0019
γ	1752.94 *11*	0.0107 *10*
γ	1801.66 *8*	0.019 *5*
γ	1847.9 *3*	0.0078 *10*
γ	1883.78 *7*	0.0301 *19*
γ	1889.70 *20*	0.0029 *10*
γ	1972.8 *3*	<0.0019
γ	2006.3 *4*	~0.0019
γ	2024.0 *3*	~0.0019
γ	2048.2 *8*	<0.0019
γ	2055.0 *3*	0.0029 *10*
γ	2156.26 *11*	0.0049 *10*
γ	2170.6 *3*	~0.0019
γ	2516.86 *14*	0.0049 *10*

† 10% uncert(syst)

Atomic Electrons (^{73}Se)

⟨e⟩=18.0 *12* keV

e_{bin}(keV)	⟨e⟩(keV)	e(%)
1	0.53	40 *5*
2	0.040	2.6 *3*
9	1.51	16.5 *19*
10 - 12	0.47	4.5 *5*
55	9.4	17.1 *19*
66	1.24	1.89 *21*
67	0.218	0.33 *4*
349	4.0	1.14 *12*
359 - 361	0.53	0.149 *13*
416 - 443	0.0050	0.00118 *19*
498 - 546	0.0024	0.00048 *20*
556 - 600	0.0015	0.00025 *8*
608 - 609	4.7 $\times10^{-5}$	7.8 *6* $\times10^{-6}$
670 - 700	0.00029	4.3 *14* $\times10^{-5}$
753 - 802	0.00099	0.00013 *3*
807 - 856	0.0020	0.00024 *8*
857 - 901	0.00078	8.8 *21* $\times10^{-5}$
915 - 930	2.6 $\times10^{-5}$	2.9 *9* $\times10^{-6}$
971 - 1019	0.00026	2.7 *6* $\times10^{-5}$
1025 - 1037	3.8 $\times10^{-5}$	3.7 *13* $\times10^{-6}$
1099 - 1146	0.00051	4.6 *15* $\times10^{-5}$
1153 - 1196	9.7 $\times10^{-6}$	8 *3* $\times10^{-7}$
1206 - 1250	1.6 $\times10^{-5}$	1.3 *4* $\times10^{-6}$
1263 - 1312	4.5 $\times10^{-5}$	3.5 *7* $\times10^{-6}$
1316 - 1341	0.00014	1.1 *3* $\times10^{-5}$
1394 - 1440	0.00027	1.9 *6* $\times10^{-5}$
1450 - 1547	9.6 $\times10^{-5}$	6.4 *13* $\times10^{-6}$
1659 - 1753	2.8 $\times10^{-5}$	1.6 *3* $\times10^{-6}$
1790 - 1888	9.1 $\times10^{-5}$	5.0 *9* $\times10^{-6}$
1889 - 1973	1.5 $\times10^{-6}$	~8 $\times10^{-8}$
1994 - 2055	1.1 $\times10^{-5}$	5.5 *14* $\times10^{-7}$
2144 - 2170	9.2 $\times10^{-6}$	4.3 *11* $\times10^{-7}$
2505 - 2517	6.0 $\times10^{-6}$	2.4 *7* $\times10^{-7}$

Continuous Radiation (^{73}Se)

⟨β+⟩=368 keV;⟨IB⟩=1.52 keV

E_{bin}(keV)		⟨ ⟩(keV)	(%)
0 - 10	β+	8.3 $\times10^{-5}$	0.00104
	IB	0.0168	
10 - 20	β+	0.00180	0.0112
	IB	0.017	0.123
20 - 40	β+	0.0285	0.090
	IB	0.031	0.106
40 - 100	β+	0.76	1.00
	IB	0.082	0.128
100 - 300	β+	22.4	10.5
	IB	0.21	0.118
300 - 600	β+	115	25.5
	IB	0.23	0.054
600 - 1300	β+	230	28.5
	IB	0.55	0.060
1300 - 2500	β+	0.54	0.0390
	IB	0.39	0.025
2500 - 2673	IB	3.9 $\times10^{-5}$	1.55 $\times10^{-6}$
	Σβ+		66

$^{73}_{34}$Se(39.8 *13* min)

Mode: IT(73 *3* %), ε(27 *3* %)

Δ: -68189 *11* keV

SpA: 6.47 $\times10^{7}$ Ci/g

Prod: ^{70}Ge(α,n); ^{72}Ge(α,3n); ^{74}Se(γ,n); ^{75}As(p,3n); ^{74}Se(n,2n)

Photons (^{73}Se)

⟨γ⟩=38.5 *3* keV

γ_{mode}	γ(keV)	γ(%)†
Se L_ℓ	1.204	0.036 *8*
Se L_η	1.245	0.021 *6*
Se L_α	1.379	0.70 *15*
Se L_β	1.421	0.38 *10*
Se L_γ	1.648	0.0012 *4*
As $K_{\alpha2}$	10.508	1.04 *5*
As $K_{\alpha1}$	10.544	2.02 *9*
Se $K_{\alpha2}$	11.182	2.43 *15*
Se $K_{\alpha1}$	11.222	4.7 *3*
As $K_{\beta1}$'	11.724	0.398 *17*
As $K_{\beta2}$'	11.948	0.0123 *5*
Se $K_{\beta1}$'	12.494	0.95 *6*
Se $K_{\beta2}$'	12.741	0.046 *3*
γ_{IT} E3	25.71 *4*	0.0134 *6*
γ_ϵM1+0.06%E2	67.17 *6*	1.81 *15*
γ_ϵ[M1]	84.25 *6*	1.42 *7*
γ_ϵM1+E2	139.53 *5*	0.0794 *23*
γ_ϵ	169.68 *6*	0.047 *4*
γ_ϵM1+E2	181.10 *5*	0.320 *7*
γ_ϵM1+E2	253.93 *5*	1.65 *3*
γ_ϵM1+E2	262.02 *5*	0.117 *3*
γ_ϵ	309.20 *7*	0.1153 *23*
γ_ϵM1+E2	320.62 *5*	0.570 *11*
γ_ϵM1+E2	393.46 *5*	1.136 *23*
γ_ϵM1+E2	401.54 *5*	0.873 *17*
γ_ϵ	442.72 *11*	0.0183 *11*
γ_ϵ	490.30 *7*	0.0926 *23*
γ_ϵ	509.89 *11*	0.096 *7*
γ_ϵ	571.22 *7*	0.169 *5*
γ_ϵ[M1+E2]	577.7 *3*	0.693 *14*
γ_ϵ[M1+E2]	588.31 *7*	0.061 *3*
γ_ϵ	644.03 *10*	0.0082 *7*
γ_ϵ	646.64 *8*	~0.0007
γ_ϵ[M1+E2]	655.48 *6*	0.0830 *17*
γ_ϵ	693.34 *7*	0.0152 *7*
γ_ϵ[M1+E2]	702.51 *10*	0.0641 *13*
γ_ϵ	724.95 *9*	0.0329 *10*
γ_ϵ[M1+E2]	769.67 *9*	0.1052 *23*
γ_ϵ	792.22 *12*	0.0120 *10*
γ_ϵ	823.75 *5*	0.076 *4*
γ_ϵ	832.86 *7*	0.038 *4*
γ_ϵ	850.23 *18*	0.189 *7*
γ_ϵ	860.22 *24*	0.0008 *3*
γ_ϵ	908.65 *7*	0.0145 *5*
γ_ϵ	934.88 *10*	0.0156 *5*
γ_ϵ	993.42 *6*	0.0530 *11*
γ_ϵ	1002.54 *7*	0.0460 *10*
γ_ϵ	1010.51 *7*	~0.005
γ_ϵ	1019.63 *7*	0.0387 *8*
γ_ϵ	1045.57 *9*	~0.007
γ_ϵ	1077.68 *4*	0.412 *8*
γ_ϵ	1086.79 *6*	0.024 *5*
γ_ϵ	1124.87 *19*	0.0023 *8*
γ_ϵ	1133.99 *19*	0.0015 *7*
γ_ϵ	1188.81 *10*	0.0049 *16*
γ_ϵ	1210.02 *24*	0.0054 *8*
γ_ϵ	1215.25 *10*	0.0130 *12*
γ_ϵ	1232.33 *10*	0.0175 *8*
γ_ϵ	1302.11 *6*	0.0453 *9*
γ_ϵ	1326.83 *16*	0.0018 *5*
γ_ϵ	1395.32 *19*	0.0015 *3*
γ_ϵ	1407.12 *10*	0.0010 *3*
γ_ϵ	1538.75 *17*	0.0021 *3*
γ_ϵ	1588.84 *16*	0.0023 *3*
γ_ϵ	1606.39 *20*	0.0010 *3*
γ_ϵ	1711.2 *4*	0.0016 *3*
γ_ϵ	1719.09 *11*	0.0026 *3*
γ_ϵ	1888.77 *11*	0.0015 *3*
γ_ϵ	1898.04 *17*	~0.0010
γ_ϵ	1905.85 *11*	0.0013 *3*
γ_ϵ	1910.24 *10*	0.0012 *3*
γ_ϵ	1974.90 *12*	0.0030 *3*
γ_ϵ	1982.29 *16*	0.00362 *16*
γ_ϵ	2091.33 *10*	0.0013 *3*
γ_ϵ	2127.40 *19*	~0.00033
γ_ϵ	2135.7 *3*	~0.0005

Photons (^{73}Se) (continued)

γ_{mode}	γ(keV)	γ(%)†
γ_ϵ	2144.48 *18*	~0.0005
γ_ϵ	2400.52 *11*	~0.0005
γ_ϵ	2425.8 *5*	0.0010 *3*
γ_ϵ	2484.78 *10*	0.0008 *3*

† uncert(syst): 14% for ε, 4.1% for IT

Atomic Electrons (^{73}Se)

⟨e⟩=18.6 *6* keV

e_{bin}(keV)	⟨e⟩(keV)	e(%)
1	0.92	64 *7*
2 - 9	0.246	3.5 *3*
10	0.44	4.5 *5*
11 - 12	0.145	1.31 *14*
13	1.80	13.8 *7*
24	12.2	50.3 *25*
25	0.0363	0.142 *7*
26	2.23	8.7 *4*
55 - 84	0.435	0.71 *4*
128 - 170	0.043	0.028 *12*
180 - 181	0.0036	~0.0020
242 - 262	0.07	0.029 *13*
297 - 321	0.016	0.0053 *18*
382 - 431	0.031	0.0081 *17*
441 - 490	0.0008	0.00017 *8*
498 - 510	0.0008	0.00015 *7*
559 - 588	0.0071	0.00125 *20*
632 - 681	0.00065	0.000100 *13*
691 - 725	0.00052	7.5 *11* $\times10^{-5}$
758 - 792	0.00059	7.7 *8* $\times10^{-5}$
812 - 860	0.0011	0.00013 *3*
897 - 935	9.2 $\times10^{-5}$	1.0 *3* $\times10^{-5}$
982 - 1020	0.00039	3.9 *8* $\times10^{-5}$
1034 - 1078	0.0011	0.00010 *3*
1085 - 1134	1.6 $\times10^{-5}$	1.4 *4* $\times10^{-6}$
1177 - 1220	8.6 $\times10^{-5}$	7.1 *14* $\times10^{-6}$
1231 - 1232	4.1 $\times10^{-6}$	3.3 *11* $\times10^{-7}$
1290 - 1327	9.7 $\times10^{-5}$	7.5 *23* $\times10^{-6}$
1383 - 1407	4.7 $\times10^{-6}$	3.4 *10* $\times10^{-7}$
1527 - 1606	9.4 $\times10^{-6}$	6.0 *12* $\times10^{-7}$
1699 - 1719	6.9 $\times10^{-6}$	4.0 *9* $\times10^{-7}$
1877 - 1975	1.63 $\times10^{-5}$	8.5 *12* $\times10^{-7}$
1981 - 2079	2.2 $\times10^{-6}$	1.1 *3* $\times10^{-7}$
2090 - 2144	2.0 $\times10^{-6}$	9 *3* $\times10^{-8}$
2389 - 2485	2.9 $\times10^{-6}$	1.2 *3* $\times10^{-7}$

Continuous Radiation (^{73}Se)

⟨β+⟩=155 keV;⟨IB⟩=0.51 keV

E_{bin}(keV)		⟨ ⟩(keV)	(%)
0 - 10	β+	1.66 $\times10^{-5}$	0.000207
	IB	0.0065	
10 - 20	β+	0.000361	0.00223
	IB	0.0065	0.046
20 - 40	β+	0.0057	0.0179
	IB	0.0122	0.042
40 - 100	β+	0.153	0.202
	IB	0.034	0.052
100 - 300	β+	4.70	2.18
	IB	0.088	0.050
300 - 600	β+	26.8	5.9
	IB	0.092	0.022
600 - 1300	β+	105	11.6
	IB	0.150	0.017
1300 - 2500	β+	18.7	1.32
	IB	0.123	0.0074
2500 - 2766	IB	0.00125	4.9 $\times10^{-5}$
	Σβ+		21

$^{73}_{35}$Br(3.4 *3* min)

Mode: ϵ
Δ: -63640 *250* keV
SpA: 7.6×10^{8} Ci/g
Prod: ^{12}C on Cu; ^{59}Co(^{16}O,2n)

Photons (^{73}Br)

⟨γ⟩=507 *15* keV

γ_{mode}	γ(keV)	γ(%)†
γ E3	25.71 *4*	*
γ [M1]	64.85 *8*	34
γ [M1]	125.48 *8*	7.8 *7*
γ	275.05 *13*	3.4 *3*
γ	335.68 *13*	11.6 *7*
γ	374.14 *15*	2.7 *3*
γ	400.53 *12*	6.8 *3*
γ	489.45 *13*	1.4 *3*
γ	539.44 *15*	2.7 *7*
γ	550.08 *13*	1.0 *3*
γ	614.93 *12*	2.7 *3*
γ	638.74 *18*	1.4 *3*
γ	699.37 *17*	13.6 *10*
γ	788.10 *14*	1.0 *3*
γ	848.72 *13*	6.8 *3*
γ	861.7 *3*	1.0 *3*
γ	870.00 *14*	1.7 *3*
γ	913.58 *12*	6.5 *7*
γ	930.62 *14*	7.5 *7*
γ	995.47 *13*	2.4 *3*

† 15% uncert(syst)
* with ^{73}Se(40 min)

$^{73}_{36}$Kr(27.0 *12* s)

Mode: ϵ, ϵp(0.68 *12* %)
Δ: -56890 *140* keV
SpA: 5.65×10^{9} Ci/g
Prod: protons on Zr; ^{58}Ni(^{16}O,n)

ϵp: 1300-3300 range

Photons (^{73}Kr)

⟨γ⟩=561 *37* keV

γ_{mode}	γ(keV)	γ(%)†
γ_ϵ	63.1 *10*	19.1 *13* ?
γ_ϵ	151.2 *3*	12.5 *13*
γ_ϵ	178.20 *24*	65.8 *13*
γ_ϵ	213.8 *3*	8.6 *13*
γ_ϵ	220.1 *8*	2.2 *9* ?
γ_ϵ	241.4 *4*	7.2 *7*
γ_ϵ	303.7 *4*	3.9 *7*
γ_ϵ	329.4 *3*	4.6 *13*
γ_ϵ	392.0 *3*	7.2 *13*
γ_ϵ	396.4 *5*	~5 ?
γ_ϵ	424.1 *10*	4.1 *16* ?
γ_ϵ	454.9 *5*	15 *4* ?
γ_ϵ	459.1 *10*	6 *3* ?
γ_ϵ	473.7 *4*	10.5 *13*
γ_ϵ	503.5 *10*	9 *3* ?
γ_ϵ	636.1 *5*	8.6 *20* ?
γ_ϵ	960.2 *10*	3.6 *14* ?
$\gamma_{\epsilon p}$	862.00 *17*	0.23

† 4.0% uncert(syst)

A = 74

NDS **17**, 519 (1976)

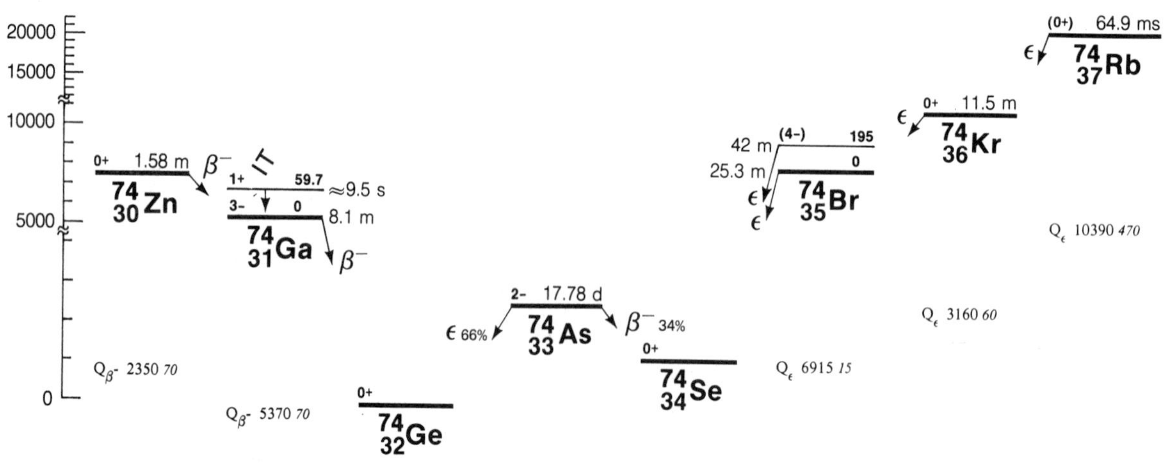

$^{74}_{30}$Zn(1.583 *17* min)

Mode: β-
Δ: -65707 *19* keV
SpA: 1.599×10^{9} Ci/g
Prod: protons on Ge; fission

Photons (^{74}Zn)

⟨γ⟩=198 *6* keV

γ_{mode}	γ(keV)	γ(%)†
Ga $K_{\alpha2}$	9.225	6.3 *5*
Ga $K_{\alpha1}$	9.252	12.4 *10*
Ga $K_{\beta1}$'	10.263	2.34 *18*
Ga $K_{\beta2}$'	10.444	0.0137 *11*
γ E1,M1	50.2 *3*	19 *1*
γ E1,M1	53.3 *3*	10.3 *10*
γ (M1)	56.5 *1*	83 *
γ	86.1 *5*	3.73 *25*
γ	116.7 *5*	3.90 *25*
γ E1,M1	140.1 *4*	38.1 *24*
γ M1	190.3 *4*	26.8 *17*
γ	347.3 *5*	6.4 *5*

† 7.0% uncert(syst)
* with ^{74}Ga(9.5 s) in equilib

Atomic Electrons (^{74}Zn)

⟨e⟩=26.2 *19* keV

e_{bin}(keV)	⟨e⟩(keV)	e(%)
1	0.43	38 *5*
8	1.31	16.3 *20*
9	0.37	4.1 *5*
10	0.021	0.207 *25*
40	3.3	8.4 *7*
43	1.63	3.8 *4*
46	11.7	25.4 *20*
49	0.44	0.89 *11*
50	0.072	0.144 *18*
52	0.21	0.40 *6*
53	0.035	0.065 *10*
55	1.51	2.74 *22*
56	0.252	0.45 *4*
76	1.6	~2
85	0.24	~0.28
86	0.04	~0.04
106	0.8	~0.7
115	0.08	~0.07
116	0.024	~0.021
117	0.017	~0.014
130	1.15	0.89 *12*
139	0.128	0.092 *15*
140	0.021	0.0150 *25*
180	0.56	0.311 *21*
189	0.061	0.0325 *22*
190	0.0101	0.0053 *4*
337	0.09	~0.025
346	0.009	~0.0027
347	0.0015	~0.0004

Continuous Radiation (^{74}Zn)

⟨β-⟩=892 keV;⟨IB⟩=1.8 keV

E_{bin}(keV)		⟨ ⟩(keV)	(%)
0 - 10	β-	0.0125	0.248
	IB	0.034	
10 - 20	β-	0.0383	0.255
	IB	0.034	0.24
20 - 40	β-	0.160	0.53
	IB	0.066	0.23
40 - 100	β-	1.29	1.82
	IB	0.19	0.28
100 - 300	β-	17.6	8.5
	IB	0.50	0.28
300 - 600	β-	85	18.5
	IB	0.49	0.118
600 - 1300	β-	460	49.0
	IB	0.48	0.058
1300 - 2290	β-	329	21.2
	IB	0.053	0.0037

$^{74}_{31}$Ga(8.1 *1* min)

Mode: β-
Δ: -68060 *70* keV
SpA: 3.13×10^{8} Ci/g
Prod: ^{76}Ge(d,α); ^{74}Ge(n,p)

Photons (^{74}Ga)

⟨γ⟩=3066 *23* keV

γ_{mode}	γ(keV)	γ(%)†
γ	233.44 *7*	0.16 *3* ?
γ	259.06 *18*	0.11 *3* ?
γ	302.52 *14*	0.11 *4* ?
γ	365.1 *15*	0.09 *3* ?
γ	443.97 *11*	~0.06 ?
γ	471.72 *16*	0.39 *5* ?
γ	485.26 *13*	1.07 *6* ?
γ E2+40%M1	492.89 *6*	5.1 *5*
γ	497.77 *9*	1.01 *9*
γ	504.75 *23*	0.10 *3* ?
γ	521.15 *14*	0.12 *3* ?
γ	541.2 *5*	0.16 *3* ?
γ	545.26 *14*	0.064 *18* ?
γ	551.88 *14*	0.11 *3* ?
γ E2	595.93 *4*	91.88
γ	603.99 *9*	2.94 *18*
γ E2+9.0%M1	608.48 *4*	14.7 *4*
γ	639.22 *10*	0.83 *5*
γ	652.56 *15*	0.06 *3* ?
γ	701.59 *7*	0.85 *5*
γ [E2]	715.27 *13*	0.24 *3*
γ [E2]	734.19 *7*	0.112 *16*
γ	784.15 *9*	0.71 *4*
γ	809.47 *14*	0.25 *4*
γ [E2]	867.93 *5*	8.91 *18*
γ [E2]	886.85 *12*	0.39 *3*
γ	942.56 *7*	1.29 *3*
γ	961.03 *6*	1.64 *4*
γ	975.16 *9*	0.28 *6*
γ [M1+E2]	993.64 *6*	0.65 *3*
γ	999.52 *12*	0.31 *3* ?
γ	1023.9 *6*	0.13 *3* ?
γ M1+10%E2	1101.37 *6*	5.44 *11*
γ	1131.67 *14*	0.85 *5* ?
γ	1134.53 *13*	0.39 *4* ?
γ	1160.37 *11*	0.64 *5*
γ	1177.69 *9*	0.24 *3*
γ	1184.47 *11*	0.28 *3*
γ [E2]	1204.41 *4*	7.63 *15*
γ	1293.6 *5*	0.27 *3*
γ	1312.90 *9*	0.67 *5*
γ [E1]	1332.20 *8*	1.75 *9*
γ	1337.08 *8*	1.61 *4* ?

Photons (^{74}Ga) (continued)

γ_{mode}	γ(keV)	γ(%)†
γ	1357.97 *24*	0.32 *3* ?
γ	1443.30 *8*	3.72 *7*
γ	1470.7 *5*	0.193 *18* ?
γ	1478.53 *12*	0.30 *3*
γ	1489.44 *7*	2.89 *6*
γ	1510.3 *6*	0.24 *3*
γ	1570.52 *8*	0.96 *4*
γ [M1+E2]	1602.11 *6*	0.275 *18*
γ	1630.34 *13*	~0.09 ?
γ	1676.74 *8*	0.74 *4*
γ [E1]	1745.18 *9*	4.83 *10*
γ	1806.5 *5*	0.29 *3*
γ	1829.96 *8*	1.89 *5*
γ [E1]	1940.67 *7*	5.49 *11*
γ	1971.41 *11*	0.22 *5*
γ	1999.67 *13*	0.40 *4*
γ	2004.75 *15*	0.51 *5* ?
γ	2014.48 *9*	1.32 *6*
γ	2023.78 *11*	0.48 *4*
γ	2036.28 *21*	0.17 *4*
γ	2074.6 *5*	0.27 *4* ?
γ	2097.91 *7*	0.89 *5*
γ	2131.08 *13*	0.202 *9*
γ	2138.71 *8*	0.84 *4*
γ	2197.70 *13*	0.33 *3*
γ [E2]	2198.03 *7*	0.53 *3*
γ	2233.11 *13*	~0.17 ?
γ	2257.21 *10*	1.80 *5*
γ	2279.08 *11*	2.36 *6*
γ [E1]	2353.64 *9*	45.2 *9*
γ	2362.51 *9*	0.175 *18*
γ	2438.43 *7*	0.28 *3*
γ	2486.19 *14*	~0.09
γ	2504.41 *21*	0.66 *9*
γ	2579.87 *11*	1.30 *3*
γ	2616.3 *5*	0.239 *18* ?
γ	2690.58 *12*	1.01 *3*
γ	2737.84 *21*	0.165 *9*
γ	2747.17 *8*	0.85 *5*
γ	2771.95 *12*	0.119 *9*
γ	2786.00 *21*	0.63 *5*
γ	2790.85 *12*	0.52 *4* ?
γ	2970.98 *9*	1.09 *3*
γ	2997.28 *21*	0.101 *9*
γ	3018.4 *3*	0.064 *9* ?
γ	3031.00 *15*	0.17 *5* ?
γ	3031.7 *3*	0.221 *18*
γ	3110.0 *15*	~0.028 ?
γ	3211.17 *13*	0.744 *18* ?
γ	3232.41 *11*	0.662 *18*
γ	3240.1 *6*	0.074 *9*
γ	3273.40 *9*	0.037 *9* ?
γ	3299.03 *11*	0.377 *9*
γ	3354.09 *13*	0.735 *18*
γ	3605.74 *21*	0.331 *18*
γ	3626.8 *3*	~0.018 ?
γ	3639.51 *16*	0.074 *18* ?
γ	3717.0 *5*	0.028 *9* ?
γ	3762.3 *3*	0.083 *9*
γ	3818.1 *10*	0.046 *9*
γ	3995.21 *13*	0.028 *9* ?

† 0.15% uncert(syst)

Continuous Radiation (^{74}Ga)

⟨β-⟩=979 keV;⟨IB⟩=2.3 keV

E_{bin}(keV)		⟨ ⟩(keV)	(%)
0 - 10	β-	0.0176	0.350
	IB	0.036	
10 - 20	β-	0.054	0.358
	IB	0.035	0.24
20 - 40	β-	0.223	0.74
	IB	0.069	0.24
40 - 100	β-	1.74	2.46
	IB	0.19	0.30
100 - 300	β-	21.3	10.4

Continuous Radiation (^{74}Ga) (continued)

E_{bin}(keV)		⟨ ⟩(keV)	(%)
	IB	0.53	0.30
300 - 600	β-	84	18.6
	IB	0.55	0.131
600 - 1300	β-	369	39.6
	IB	0.64	0.076
1300 - 2500	β-	411	24.6
	IB	0.23	0.0143
2500 - 4774	β-	92	2.95
	IB	0.036	0.00124

$^{74}_{31}$Ga(9.5 *10* s)

Mode: IT
Δ: -68000 *70* keV
SpA: 1.55×10^{10} Ci/g
Prod: ^{74}Ge(n,p)

Photons (^{74}Ga)

⟨γ⟩=43.1 *4* keV

γ_{mode}	γ(keV)	γ(%)
Ga $K_{\alpha 2}$	9.225	3.45 *19*
Ga $K_{\alpha 1}$	9.252	6.8 *4*
Ga $K_{\beta 1}$'	10.263	1.27 *7*
Ga $K_{\beta 2}$'	10.444	0.0075 *4*
γ (M1)	56.5 *1*	74.4

Atomic Electrons (^{74}Ga)

⟨e⟩=13.2 *5* keV

e_{bin}(keV)	⟨e⟩(keV)	e(%)
1	0.234	20.6 *22*
8	0.71	8.9 *10*
9	0.202	2.22 *24*
10	0.0115	0.113 *12*
46	10.5	22.7 *10*
55	1.36	2.46 *10*
56	0.226	0.401 *17*

$^{74}_{32}$Ge(stable)

Δ: -73423.4 *14* keV
%: 36.5 *7*

$^{74}_{33}$As(17.78 *3* d)

Mode: ε(65.8 *16* %), β-(34.2 *16* %)
Δ: -70861.1 *22* keV
SpA: 9.924×10^{4} Ci/g
Prod: ^{71}Ga(α,n); ^{75}As(n,2n); ^{75}As(γ,n); ^{74}Ge(p,n) ^{74}Se(n,p)

Photons (^{74}As)

⟨γ⟩=467 *10* keV

γ_{mode}	γ(keV)	γ(%)†
Ge L_ℓ	1.037	0.019 *4*
Ge L_η	1.068	0.011 *3*
Ge L_α	1.188	0.32 *7*
Ge L_β	1.227	0.19 *5*
Ge L_γ	1.412	0.0023 *7*
Ge $K_{\alpha 2}$	9.855	5.23 *22*
Ge $K_{\alpha 1}$	9.886	10.2 *4*
Ge $K_{\beta 1}$'	10.980	1.96 *8*
Ge $K_{\beta 2}$'	11.184	0.0325 *14*
γ_ϵ	233.44 *7*	0.00021 *6*
γ_ϵE2+40%M1	492.89 *6*	0.0067 *18*
γ_ϵE2	595.93 *4*	60.2
γ_ϵE2+9.0%M1	608.48 *4*	0.57 *3*
$\gamma_{\beta-}$E2+4.1%M1	634.18 *6*	0.021 *4* ?
$\gamma_{\beta-}$E2	634.79 *5*	15.4
γ_ϵ[E2]	715.27 *13*	0.0070 *10*
γ_ϵ[E2]	734.19 *7*	0.0033 *5*
γ_ϵ[E2]	867.93 *5*	0.00464 *18*
γ_ϵ[E2]	886.85 *12*	0.0257 *9*
γ_ϵ[M1+E2]	993.64 *6*	0.0194 *7*
γ_ϵM1+10%E2	1101.37 *6*	0.0072 *18*
γ_ϵ[E2]	1204.41 *4*	0.295 *6*
$\gamma_{\beta-}$[E2]	1268.96 *6*	0.0079 *4* ?
γ_ϵ[M1+E2]	1602.11 *6*	0.0082 *5*
γ_ϵ[E2]	2198.03 *7*	0.0157 *11*

† uncert(syst): 2.4% for ε, 4.7% for β-

Atomic Electrons (^{74}As)

⟨e⟩=2.17 *9* keV

e_{bin}(keV)	⟨e⟩(keV)	e(%)
1	0.36	29 *3*
8	0.36	4.3 *4*
9	0.66	7.7 *8*
10	0.29	3.0 *3*
222 - 233	8.7 ×10^{-6}	~4×10^{-6}
482 - 493	6.8 ×10^{-5}	1.4 *6* ×10^{-5}
585	0.439	0.0751 *23*
595	0.0467	0.00786 *25*
596 - 608	0.0122	0.00204 *6*
704 - 734	6.0 ×10^{-5}	8.5 *8* ×10^{-6}
857 - 887	0.000127	1.46 *6* ×10^{-5}
983 - 994	6.5 ×10^{-5}	6.6 *6* ×10^{-6}
1090 - 1101	2.1 ×10^{-5}	1.9 *4* ×10^{-6}
1193 - 1204	0.00082	6.9 *3* ×10^{-5}
1591 - 1602	1.65 ×10^{-5}	1.03 *8* ×10^{-6}
2187 - 2198	2.44 ×10^{-5}	1.12 *8* ×10^{-6}

Continuous Radiation (^{74}As)

⟨β-⟩=137 keV;⟨β+⟩=128 keV;⟨IB⟩=1.02 keV

E_{bin}(keV)		⟨ ⟩(keV)	(%)
0 - 10	β-	0.0240	0.479
	β+	7.9×10^{-5}	0.00098
	IB	0.0128	
10 - 20	β-	0.072	0.481
	β+	0.00164	0.0102
	IB	0.0129	0.092
20 - 40	β-	0.291	0.97
	β+	0.0249	0.078
	IB	0.022	0.078
40 - 100	β-	2.11	3.01
	β+	0.62	0.83
	IB	0.058	0.090
100 - 300	β-	20.6	10.4
	β+	16.2	7.7
	IB	0.139	0.080
300 - 600	β-	47.5	11.0
	β+	62	14.0
	IB	0.167	0.039

Continuous Radiation (^{74}As) (continued)

E_{bin}(keV)		⟨ ⟩(keV)	(%)
600 - 1300	β-	67	7.9
	β+	48.2	6.4
	IB	0.44	0.048
1300 - 2500	β-	0.119	0.0090
	β+	1.36	0.100
	IB	0.17	0.0114
2500 - 2562	IB	9.6×10^{-6}	3.8×10^{-7}
	Σβ+		29

$^{74}_{34}$Se(stable)

Δ: -72215.0 *15* keV
%: 0.9 *1*

$^{74}_{35}$Br(25.3 *3* min)

Mode: ε
Δ: -65300 *15* keV
SpA: 1.004×10^{8} Ci/g
Prod: daughter ^{74}Kr

Photons (^{74}Br)

⟨γ⟩=3650 *43* keV

γ_{mode}	γ(keV)	γ(%)†
Se L_ℓ	1.204	0.0065 *14*
Se L_η	1.245	0.0039 *10*
Se L_α	1.379	0.13 *3*
Se L_β	1.427	0.076 *19*
Se L_γ	1.648	0.0011 *3*
Se $K_{\alpha 2}$	11.182	1.85 *8*
Se $K_{\alpha 1}$	11.222	3.59 *15*
Se $K_{\beta 1}$'	12.494	0.73 *3*
Se $K_{\beta 2}$'	12.741	0.0349 *15*
γ E2	218.99 *6*	17.6 *11*
γ	521.06 *9*	0.019 *7*
γ	615.20 *7*	0.22 *8* ?
γ E2+4.1%M1	634.18 *6*	18 *3*
γ E2	634.79 *5*	63.4
γ	871.45 *14*	~0.25
γ	935.90 *13*	0.76 *13*
γ	984.95 *7*	4.08 *13*
γ	1022.79 *9*	5.20 *19*
γ	1045.18 *9*	0.46 *7*
γ	1109.62 *11*	0.57 *6*
γ	1161.19 *13*	0.19 *6*
γ	1203.94 *7*	1.50 *17*
γ	1225.64 *9*	1.27 *13*
γ	1249.37 *8*	0.21 *7* ?
γ [E2]	1268.96 *6*	6.88 *19*
γ	1310.41 *19*	0.51 *6*
γ	1409.72 *13*	0.63 *13*
γ	1460.36 *10*	1.01 *10*
γ	1474.16 *13*	1.08 *13*
γ	1512.85 *23*	1.90 *13*
γ	1524.81 *11*	0.32 *6*
γ	1679.35 *9*	0.86 *10*
γ	1701.05 *11*	0.82 *13*
γ	1715.80 *18*	1.27 *13*
γ	1743.80 *10*	1.33 *13*
γ	1842.85 *23*	2.28 *13*
γ	1882.19 *12*	1.52 *13*
γ	1949.57 *11*	1.52 *13*
γ	1981.06 *11*	1.33 *6*
γ	2087.4 *17*	1.33 *25*
γ	2088.73 *18*	<0.32 ?
γ	2130.71 *12*	2.85 *13*
γ	2157.92 *24*	0.32 *13*

Photons (^{74}Br) (continued)

γ_{mode}	γ(keV)	γ(%)†
γ	2270.80 *11*	1.78 *13*
γ [M1+E2]	2355.57 *19*	0.76 *13*
γ	2378.57 *11*	0.32 *13*
γ	2386.7 *3*	0.44 *13*
γ	2396.24 *11*	2.79 *13*
γ	2436.79 *21*	0.70 *13*
γ	2464.58 *17*	0.95 *13*
γ [E1]	2519.32 *12*	0.57 *13*
γ	2541.2 *3*	0.32 *13*
γ	2615.22 *11*	7.29 *19*
γ	2661.61 *15*	5.20 *19*
γ	2685.97 *11*	~0.25
γ	2703.90 *17*	1.52 *13*
γ [E2]	2770.75 *19*	2.03 *13*
γ	2879.76 *16*	0.44 *13*
γ	2904.96 *10*	1.65 *13*
γ [E1]	2934.49 *11*	0.76 *13*
γ	2951 *5*	0.63 *19*
γ	2975.6 *3*	1.52 *19*
γ [M1+E2]	2989.73 *18*	0.32 *13*
γ	3098.74 *16*	0.44 *13*
γ [E1]	3110.9 *3*	0.32 *13*
γ	3119.08 *18*	0.95 *13*
γ	3190.51 *25*	0.95 *13*
γ	3240.57 *21*	0.63 *13*
γ	3249.99 *12*	6.09 *25*
γ [E1]	3267.51 *24*	0.51 *13*
γ	3295.77 *16*	2.73 *13*
γ	3338.06 *17*	0.44 *13*
γ	3409.49 *25*	0.38 *13*
γ	3412.0 *17*	0.70 *13*
γ	3459.55 *20*	1.20 *13*
γ	3488.7 *4*	0.38 *13*
γ [E1]	3526.1 *3*	0.63 *13*
γ	3539.72 *11*	0.63 *13*
γ [E2]	3624.49 *18*	5.52 *13*
γ [E1]	3631.9 *4*	2.54 *19*
γ	3733.50 *17*	1.78 *13*
γ [E1]	3745.1 *3*	0.63 *13*
γ [E1]	3788.23 *12*	3.99 *19*
γ	3852.4 *3*	1.27 *13*
γ [E1]	3901.66 *23*	1.39 *13*
γ	3972.82 *16*	2.28 *13*
γ	4044.25 *25*	0.82 *13*
γ	4094.30 *20*	0.51 *13*
γ	4222.0 *11*	0.44 *13*
γ [E1]	4266.6 *4*	1.08 *19*
γ	4342.4 *4*	1.33 *19*
γ [E1]	4379.9 *3*	4.12 *25*
γ	4487.2 *3*	~0.19
γ [E1]	4536.41 *24*	~0.13
γ	4649.5 *7*	0.38 *13*

† 2.5% uncert(syst)

Atomic Electrons (^{74}Br)

$\langle e\rangle$=3.14 *11* keV

e_{bin}(keV)	$\langle e\rangle$(keV)	e(%)
1	0.125	8.6 *9*
2 - 9	0.038	0.87 *7*
10	0.30	3.1 *3*
11	0.10	0.91 *9*
12	0.0063	0.051 *5*
206	1.47	0.71 *5*
217	0.150	0.069 *4*
218 - 219	0.063	0.0288 *13*
508 - 521	0.00016	~3 $\times10^{-5}$
603 - 615	0.0013	~0.00022
622	0.62	0.100 *6*
633 - 635	0.079	0.0125 *7*
859 - 871	0.0009	~0.00011
923 - 972	0.014	0.0014 *5*
983 - 1033	0.018	0.0018 *5*
1044 - 1045	0.00014	1.4 *5* $\times10^{-5}$
1097 - 1110	0.0015	0.00014 *5*
1149 - 1191	0.0037	0.00032 *11*
1202 - 1249	0.0039	0.00032 *9*

Atomic Electrons (^{74}Br) (continued)

e_{bin}(keV)	$\langle e\rangle$(keV)	e(%)
1256 - 1298	0.0225	0.00179 *8*
1309 - 1310	0.00012	9 *3* $\times10^{-6}$
1397 - 1410	0.0013	9 *3* $\times10^{-5}$
1448 - 1525	0.0085	0.00057 *10*
1667 - 1744	0.0075	0.00044 *7*
1830 - 1882	0.0063	0.00034 *7*
1937 - 1981	0.0045	0.00023 *4*
2075 - 2158	0.0070	0.00033 *6*
2258 - 2355	0.0039	0.00017 *3*
2366 - 2464	0.0072	0.00030 *4*
2507 - 2603	0.0096	0.00037 *8*
2614 - 2704	0.0101	0.00038 *6*
2758 - 2771	0.00319	0.000116 *8*
2867 - 2963	0.0058	0.000198 *24*
2974 - 2990	0.00067	2.3 *6* $\times10^{-5}$
3086 - 3178	0.0030	9.5 *13* $\times10^{-5}$
3189 - 3283	0.0113	0.00035 *5*
3294 - 3338	0.00086	2.6 *5* $\times10^{-5}$
3397 - 3489	0.0030	8.9 *12* $\times10^{-5}$
3513 - 3612	0.0079	0.000219 *10*
3619 - 3632	0.00306	8.5 *5* $\times10^{-5}$
3721 - 3788	0.0060	0.000161 *12*
3840 - 3901	0.0026	6.7 *8* $\times10^{-5}$
3960 - 4044	0.0033	8.4 *12* $\times10^{-5}$
4082 - 4094	0.00054	1.3 *4* $\times10^{-5}$
4209 - 4266	0.00139	3.3 *5* $\times10^{-5}$
4330 - 4380	0.0049	0.000112 *8*
4475 - 4536	0.00030	7 *3* $\times10^{-6}$
4637 - 4649	0.00038	8 *3* $\times10^{-6}$

Continuous Radiation (^{74}Br)

$\langle\beta+\rangle$=1010 keV;$\langle$IB$\rangle$=3.4 keV

E_{bin}(keV)		$\langle\ \rangle$(keV)	(%)
0 - 10	β+	3.01 $\times10^{-5}$	0.000373
	IB	0.036	
10 - 20	β+	0.00068	0.00418
	IB	0.036	0.25
20 - 40	β+	0.0111	0.0347
	IB	0.069	0.24
40 - 100	β+	0.312	0.410
	IB	0.20	0.30
100 - 300	β+	10.4	4.78
	IB	0.55	0.31
300 - 600	β+	67	14.5
	IB	0.62	0.146
600 - 1300	β+	383	40.9
	IB	0.93	0.107
1300 - 2500	β+	396	23.4
	IB	0.76	0.044
2500 - 5000	β+	154	4.89
	IB	0.159	0.0053
5000 - 6061	β+	0.00058	1.16 $\times10^{-5}$
	IB	0.00117	2.2 $\times10^{-5}$
	$\Sigma\beta$+		89

$^{74}_{35}$Br(41.5 *15* min)

Mode: ϵ

Δ: -65105 *62* keV

SpA: 6.12$\times10^{7}$ Ci/g

Prod: ^{65}Cu(^{12}C,3n); ^{63}Cu(^{14}N,2np); deuterons on Se

Photons (^{74}Br)

$\langle\gamma\rangle$=3160 *53* keV

γ_{mode}	γ(keV)	γ(%)†
Se L_ℓ	1.204	0.0049 *11*
Se L_η	1.245	0.0029 *7*
Se L_α	1.379	0.095 *20*
Se L_β	1.427	0.058 *15*
Se L_γ	1.648	0.00081 *25*
Se $K_{\alpha2}$	11.182	1.41 *6*
Se $K_{\alpha1}$	11.222	2.74 *11*
Se $K_{\beta1}$'	12.494	0.553 *23*
Se $K_{\beta2}$'	12.741	0.0266 *11*
γ E2	218.99 *6*	5.0 *5*
γ	368.5 *3*	0.16 *5*
γ	521.06 *9*	0.64 *9*
γ	615.20 *7*	7.5 *6*
γ E2+4.1%M1	634.18 *6*	20 *3*
γ E2	634.79 *5*	91.9
γ	679.25 *11*	0.64 *9*
γ	724.69 *12*	<0.9
γ E2	728.32 *7*	35 *3*
γ	744.64 *8*	2.8 *3*
γ	763.90 *17*	0.23 *5*
γ	778.48 *25*	0.64 *9*
γ	838.77 *9*	5.5 *6*
γ	850.46 *20*	0.92 *9*
γ E2	867.96 *20*	0.74 *9*
γ	979.75 *20*	0.46 *9*
γ	984.95 *7*	1.75 *11*
γ [E1]	986.4 *3*	~0.7
γ	1022.53 *17*	0.40 *5*
γ	1045.18 *9*	0.38 *4*
γ	1079.89 *17*	0.46 *9*
γ [E1]	1080.5 *3*	0.46 *9*
γ	1145.8 *8*	0.51 *9* ?
γ	1193.87 *16*	0.092 *18*
γ	1198.1 *3*	0.37 *9*
γ	1200.31 *11*	5.3 *6*
γ	1203.94 *7*	0.65 *8*
γ	1249.37 *8*	7.0 *6*
γ	1261.7 *8*	~0.6 ?
γ [E2]	1268.96 *6*	7.7 *4*
γ	1289.3 *8*	0.46 *18* ?
γ	1294.44 *11*	1.9 *3*
γ	1299.54 *24*	0.21 *4*
γ	1366.61 *18*	2.3 *3*
γ	1421.6 *3*	0.64 *9*
γ	1455.36 *20*	1.8 *3*
γ	1460.36 *10*	0.83 *8*
γ	1468.56 *20*	0.55 *9*
γ	1472.95 *9*	1.47 *18*
γ	1494.8 *6*	0.40 *6*
γ	1507.98 *21*	0.22 *5*
γ	1515.7 *3*	~0.27
γ	1555.51 *17*	0.32 *6*
γ	1567.28 *22*	0.33 *6*
γ	1649.64 *16*	0.36 *6*
γ	1679.35 *9*	0.71 *8*
γ	1714.93 *16*	6.0 *6*
γ	1745.72 *20*	0.13 *5*
γ	1769.9 *4*	0.16 *5*
γ	1837.6 *3*	0.83 *9* ?
γ [M1+E2]	1843.3 *3*	0.92 *9*
γ	1853.72 *24*	0.46 *9*
γ [M1+E2]	1890.0 *3*	0.74 *9*
γ	1928.61 *11*	0.64 *9*
γ	1932.94 *18*	0.46 *9*
γ	1952.8 *3*	0.32 *6*
γ	1994.98 *19*	0.46 *9*
γ	2027.85 *25*	0.46 *9*
γ	2098.4 *4*	0.46 *9*
γ	2115.2 *8*	0.36 *15* ?
γ	2131.4 *8*	0.76 *11* ?
γ	2150.7 *8*	0.74 *18* ?
γ	2158.2 *8*	0.46 *18* ?
γ	2167.4 *8*	0.64 *18* ?
γ	2183.4 *3*	1.01 *9*
γ	2207.4 *8*	0.55 *18* ?
γ	2217.09 *24*	0.46 *9*
γ	2228.6 *8*	~0.27 ?
γ	2276.2 *8*	0.64 *18* ?
γ	2283.81 *16*	2.8 *4*
γ	2311.90 *21*	3.4 *5*
γ	2333.7 *3*	0.83 *9*

Photons (^{74}Br) (continued)

γ_{mode}	γ(keV)	γ(%)†
γ	2370.7 8	0.69 18 ?
γ	2388.51 20	1.10 18
γ	2396.97 19	0.14 5
γ	2408.55 17	0.9 4
γ	2443.23 16	0.37 9
γ	2471.98 25	0.74 9
γ[M1+E2]	2477.4 3	0.55 11
γ	2484.79 23	0.37 9
γ	2502.68 18	0.17 5
γ	2558.0 4	0.37 9
γ	2615.96 18	<0.09 ?
γ	2661.61 15	0.64 18
γ	2679.8 3	0.74 9
γ	2695.55 24	0.74 9
γ	2701.83 19	1.19 18
γ	2708.36 21	0.55 9
γ	2744.1 6	0.37 9
γ	2754.5 5	0.28 9
γ	2823.1 4	
γ	2945.99 19	0.3 1 ?
γ	3032.3 6	<0.09 ?
γ	3040.19 21	1.01 18 ?
γ	3136.84 17	0.64 9
γ	3153.23 21	1.01 18
γ	3172.5 3	1.10 18
γ	3227.25 20	1.10 18
γ	3247.36 22	<0.046 ?
γ	3295.77 16	0.34 10
γ	3298.7 4	0.9 3
γ	3323.54 22	0.55 9
γ	3336.2 3	1.19 18
γ	3393.9 4	0.55 9
γ	3431.04 20	1.29 18
γ	3625.0 15	0.14 5
γ	3633.2 15	~0.09
γ	3684.6 6	0.17 6
γ	3786.7 15	0.23 6
γ	3806.6 3	1.10 18
γ	3861.41 20	1.38 18
γ	3881.51 22	0.74 9
γ	3902.4 14	0.08 4
γ	3910.0 14	~0.028
γ	3951.16 20	1.10 18
γ	3957.69 22	3.5 5
γ	3972.8 15	~0.07
γ	4027.0 4	1.10 18
γ	4064.5 3	0.19 6
γ	4122.2 4	0.11 5
γ	4200.4 15	0.12 5
γ	4344.0 20	~0.09
γ	4380.4 12	0.20 6

† 0.76% uncert(syst)

Atomic Electrons (^{74}Br)

⟨e⟩=2.38 7 keV

e_{bin}(keV)	⟨e⟩(keV)	e(%)
1	0.095	6.6 7
2 - 9	0.0289	0.67 5
10	0.226	2.33 24
11	0.077	0.70 7
12	0.0048	0.039 4
206	0.41	0.201 19
217 - 219	0.060	0.0275 19
356 - 368	0.0026	~0.0007
508 - 521	0.005	0.0010 5
603 - 615	0.045	0.008 3
622	0.85	0.137 6
633	0.093	0.0147 7
634 - 679	0.0197	0.00308 24
712	0.0020	<0.0006
716	0.209	0.0292 24
723 - 766	0.043	0.0058 8
777 - 826	0.019	0.0023 10
837 - 868	0.0095	0.00112 20
967 - 1010	0.0091	0.00093 23
1021 - 1068	0.0030	0.00029 6

Atomic Electrons (^{74}Br) (continued)

e_{bin}(keV)	⟨e⟩(keV)	e(%)
1078 - 1081	0.00022	2.1 5 $\times10^{-5}$
1133 - 1181	0.0015	0.00013 4
1185 - 1204	0.015	0.0013 4
1237 - 1282	0.046	0.0037 5
1287 - 1300	0.00107	8.3 18 $\times10^{-5}$
1354 - 1367	0.0049	0.00036 12
1409 - 1456	0.0075	0.00052 10
1459 - 1555	0.0062	0.00041 7
1566 - 1649	0.00072	4.4 13 $\times10^{-5}$
1667 - 1757	0.012	0.00072 18
1768 - 1854	0.0041	0.00023 3
1877 - 1953	0.0038	0.000200 24
1982 - 2028	0.0014	7.2 16 $\times10^{-5}$
2086 - 2183	0.0066	0.00031 4
2195 - 2284	0.0067	0.00030 5
2299 - 2397	0.0098	0.00042 7
2407 - 2503	0.0036	0.000145 17
2545 - 2616	0.00055	2.2 6 $\times10^{-5}$
2649 - 2744	0.0058	0.000215 24
2753 - 2754	3.8 $\times10^{-5}$	1.4 5 $\times10^{-6}$
2933 - 3032	0.0025	8.2 16 $\times10^{-5}$
3038 - 3137	0.00103	3.3 6 $\times10^{-5}$
3141 - 3235	0.0038	0.000121 17
3246 - 3336	0.0035	0.000106 16
3381 - 3431	0.0021	6.2 11 $\times10^{-5}$
3612 - 3684	0.00045	1.2 3 $\times10^{-5}$
3774 - 3869	0.0037	9.6 12 $\times10^{-5}$
3880 - 3973	0.0052	0.000132 21
4014 - 4110	0.0015	3.7 7 $\times10^{-5}$
4121 - 4200	0.00014	3.3 11 $\times10^{-6}$
4331 - 4380	0.00030	7.0 17 $\times10^{-6}$

Continuous Radiation (^{74}Br)

⟨β+⟩=1408 keV;⟨IB⟩=5.1 keV

E_{bin}(keV)		⟨ ⟩(keV)	(%)
0 - 10	β+	2.24 $\times10^{-5}$	0.000278
	IB	0.044	
10 - 20	β+	0.000503	0.00310
	IB	0.044	0.31
20 - 40	β+	0.0083	0.0258
	IB	0.086	0.30
40 - 100	β+	0.231	0.304
	IB	0.25	0.38
100 - 300	β+	7.6	3.50
	IB	0.73	0.41
300 - 600	β+	48.3	10.5
	IB	0.88	0.21
600 - 1300	β+	275	29.3
	IB	1.43	0.163
1300 - 2500	β+	603	32.8
	IB	1.20	0.069
2500 - 5000	β+	471	14.9
	IB	0.42	0.0134
5000 - 6475	β+	3.14	0.061
	IB	0.0082	0.000154
	Σβ+		91

$^{74}_{36}$Kr(11.50 *11* min)

Mode: ε

Δ: -62140 *60* keV

SpA: 2.208$\times10^{8}$ Ci/g

Prod: protons on Br; ^{60}Ni(^{16}O,2n); ^{63}Cu(^{14}N,3n); protons on Zr

Photons (^{74}Kr)

⟨γ⟩=286 7 keV

γ_{mode}	γ(keV)	γ(%)†
γ	9.78 4	4.7 16
γ	26.37 11	0.16 6
γ	62.75 7	10.6 9
γ	67.28 11	1.37 12
γ	72.53 7	0.25 6
γ	79.75 7	0.28 6
γ	83.40 23	~0.12
γ	89.53 7	31
γ	89.7 3	0.94 19
γ[M1+E2]	93.66 6	3.49 19
γ[M1+E2]	123.22 7	9.4 4
γ	132.52 20	0.66 6 ?
γ	140.22 7	9.1 4
γ	149.59 11	2.25 12
γ	166.60 12	0.34 6
γ	179.3 3	0.31 6
γ	202.97 6	19.5 8
γ	210.5 4	~0.09
γ	212.75 6	0.94 9
γ[M1+E2]	216.87 8	10.2 5
γ	225.0 5	0.28 6
γ	229.34 11	0.25 6
γ	233.88 7	5.3 5
γ	239.12 11	0.16 6
γ	296.62 6	11.3 5
γ	300.27 22	0.97 12
γ	306.41 6	10.5 5
γ	311.4 3	0.87 16
γ	369.80 19	0.31 6
γ	373.5 7	0.31 9
γ[M1+E2]	396.18 16	0.84 9
γ	444.7 6	0.50 9
γ[M1+E2]	488.46 19	0.25 6
γ[M1+E2]	519.39 17	0.50 9
γ	524.5 6	~0.08
γ	530.4 8	0.19 6
γ	534.3 6	0.31 9
γ	536.40 17	0.31 9
γ	606.4 8	0.25 6
γ	608.93 16	1.06 12
γ[M1+E2]	611.68 19	0.19 6
γ	618.8 6	0.25 6
γ	628.68 19	0.25 9
γ	691.43 18	0.28 6
γ	701.21 18	1.72 16
γ	738.9 8	0.19 6
γ[M1+E2]	757.1 4	0.59 9
γ	765.3 8	0.19 6
γ	797.5 13	0.22 6
γ	831.6 6	0.16 6
γ	861.9 15	~0.12
γ[M1+E2]	880.3 4	0.16 6
γ	899.9 10	0.22 6
γ	969.8 4	0.25 6
γ	978.1 8	0.16 6
γ	1013.7 15	0.22 6
γ	1060.8 15	0.22 6

† approximate

Continuous Radiation (^{74}Kr)

⟨β+⟩=713 keV;⟨IB⟩=2.1 keV

E_{bin}(keV)		⟨ ⟩(keV)	(%)
0 - 10	β+	3.83 $\times10^{-5}$	0.000474
	IB	0.029	
10 - 20	β+	0.00089	0.0055
	IB	0.029	0.20
20 - 40	β+	0.0151	0.0470
	IB	0.055	0.19
40 - 100	β+	0.432	0.57
	IB	0.152	0.23
100 - 300	β+	14.5	6.7
	IB	0.41	0.23
300 - 600	β+	91	19.8
	IB	0.41	0.098
600 - 1300	β+	453	49.2

Continuous Radiation (^{74}Kr)
(continued)

E_{bin}(keV)		⟨ ⟩(keV)	(%)
	IB	0.57	0.065
1300 - 2500	β+	154	10.5
	IB	0.47	0.027
2500 - 3070	IB	0.0165	0.00063
	Σβ+		87

$^{74}_{37}$Rb(64.9 *5* ms)

Mode: ϵ
Δ: -51750 *470* keV
SpA: 2.199×10^{11} Ci/g
Prod: protons on Nb

Continuous Radiation (^{74}Rb)
⟨β+⟩=4457 keV;⟨IB⟩=23 keV

E_{bin}(keV)		⟨ ⟩(keV)	(%)
0 - 10	β+	5.9×10^{-7}	7.2×10^{-6}
	IB	0.088	
10 - 20	β+	1.42×10^{-5}	8.7×10^{-5}
	IB	0.088	0.61
20 - 40	β+	0.000250	0.00078
	IB	0.18	0.61
40 - 100	β+	0.0077	0.0101
	IB	0.52	0.79
100 - 300	β+	0.307	0.139
	IB	1.65	0.91
300 - 600	β+	2.62	0.56
	IB	2.3	0.53
600 - 1300	β+	32.3	3.22
	IB	4.5	0.51
1300 - 2500	β+	244	12.4
	IB	5.8	0.32
2500 - 5000	β+	1684	44.2
	IB	6.3	0.19
5000 - 7500	β+	2093	34.5
	IB	1.51	0.026
7500 - 9368	β+	401	5.01
	IB	0.053	0.00065
	Σβ+		100

A = 75

NDS **32**, 211 (1981)

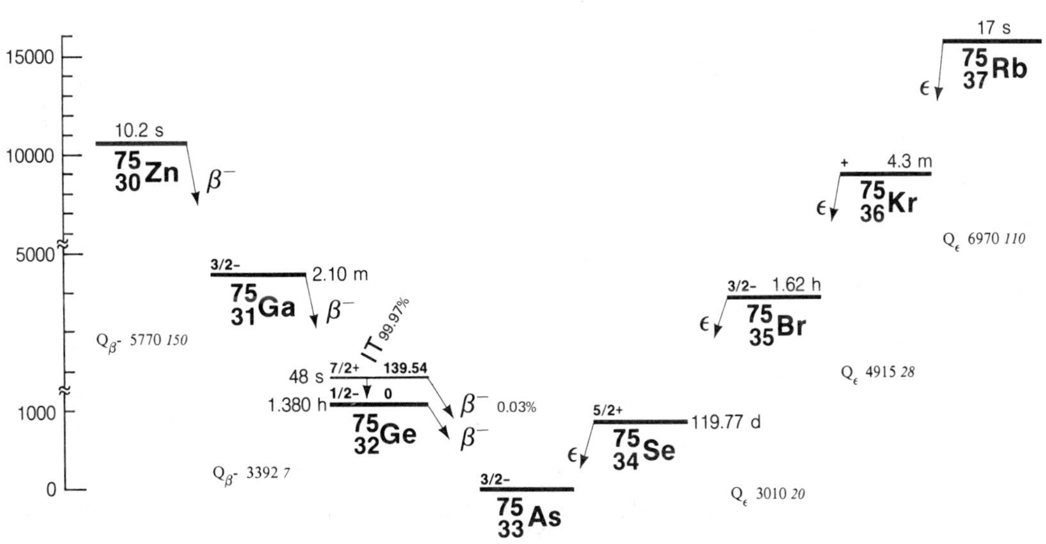

$^{75}_{30}$Zn(10.2 *3* s)

Mode: β-

Δ: -62700 *150* keV

SpA: 1.43×10^{10} Ci/g

Prod: fission

$^{75}_{31}$Ga(2.10 *3* min)

Mode: β-

Δ: -68466 *7* keV

SpA: 1.191×10^{9} Ci/g

Prod: ^{76}Ge(n,pn); ^{76}Ge(γ,p); fission

Photons (^{75}Ga)

γ_{mode}	γ(keV)	γ(rel)
γ	124.24 *23*	0.8 *3*
γ E3	139.54 *3*	*
γ [E1]	177.26 *19*	13.5 *10*
γ [E2]	204.14 *19*	5.3 *8*
γ [M1]	252.94 *15*	100
γ	279.39 *20*	2.9 *6*
γ	310.75 *19*	6.6 *10*
γ [E2]	316.81 *19*	3.3 *7*
γ	321.68 *18*	2.0 *4*
γ [E2]	428.29 *20*	1.1 *3*
γ	444.3 *3*	0.9 *1* ?
γ [E2]	444.77 *22*	1.6 *2*
γ [E2]	457.08 *19*	4.5 *4*
γ [E2]	568.56 *21*	2.4 *4*
γ	574.63 *16*	31.6 *20*
γ [E1]	584.31 *22*	1.5 *1*
γ [M1]	632.43 *17*	5.5 *3*
γ	647.78 *24*	1.8 *2* ?
γ	761.13 *23*	1.2 *3*
γ [M1+E2]	783.33 *24*	2.1 *2*
γ [M1]	885.37 *16*	11.1 *7*
γ	926.90 *20*	6.6 *5*
γ [E2]	987.47 *24*	1.4 *2*
γ	1044.44 *25*	2.6 *2*
γ	1174.42 *23*	1.5 *2*
γ	1182.39 *20*	1.4 *2*
γ	1221.9 *3*	1.3 *2*
γ [E2]	1240.4 *3*	0.3 *1*
γ	1248.57 *20*	5.5 *3*
γ	1358.89 *20*	0.4 *1*
γ	1427.36 *23*	0.9 *2*
γ	1501.51 *20*	4.5 *5*
γ	1543.6 *3*	1.1 *2*
γ	1745.69 *20*	<0.10
γ	1796.52 *25*	0.6 *1*
γ	2089.8 *3*	~0.2 ?
γ [E1]	2103.8 *3*	~0.2 ?

* with ^{75}Ge(48 s)

$^{75}_{32}$Ge(1.3797 *7* h)

Mode: β-

Δ: -71857.5 *21* keV

SpA: 3.0283×10^{7} Ci/g

Prod: ^{74}Ge(n,γ); ^{75}As(n,p)

Photons (^{75}Ge)

⟨γ⟩=35 *3* keV

γ_{mode}	γ(keV)	γ(%)†
γ M1+2.3%E2	66.060 *5*	0.113 *3*
γ [E2]	80.943 *5*	$1.6_{6} \times 10^{-6}$
γ [E2]	96.7337 *17*	$2.33_{8} \times 10^{-5}$
γ [E1]	121.1192 *22*	0.000116 *4*
γ [E1]	136.0017 *24*	<0.0008 ?
γ M1+18%E2	198.596 *5*	1.18 *3*
γ M1+0.3%E2	264.6556 *25*	11.3
γ	270.17 *16*	0.0034 *11*
γ M1+16%E2	279.5381 *22*	0.0056 *11*
γ	338.14 *12*	0.0045 *11*
γ	353.03 *12*	0.0203 *23*
γ [E1]	400.6568 *15*	$7.75_{25} \times 10^{-5}$
γ	419.08 *12*	0.184 *7*
γ	468.77 *16*	0.221 *9*
γ	617.68 *12*	0.112 *5*

† 9.7% uncert(syst)

Continuous Radiation (^{75}Ge)

⟨β-⟩=421 keV;⟨IB⟩=0.50 keV

E_{bin}(keV)		⟨ ⟩(keV)	(%)
0 - 10	β-	0.0477	0.95
	IB	0.019	
10 - 20	β-	0.145	0.97
	IB	0.019	0.129
20 - 40	β-	0.60	1.99
	IB	0.035	0.123
40 - 100	β-	4.58	6.5
	IB	0.092	0.143
100 - 300	β-	53	25.9
	IB	0.20	0.115
300 - 600	β-	171	38.5
	IB	0.118	0.029
600 - 1178	β-	192	25.2
	IB	0.024	0.0035

$^{75}_{32}$Ge(47.7 *7* s)

Mode: IT(99.970 *6* %), β-(0.030 *6* %)

Δ: -71718.0 *21* keV

SpA: 3.13×10^{9} Ci/g

Prod: ^{74}Ge(n,γ); ^{75}As(n,p); ^{76}Ge(n,2n)

Photons (^{75}Ge)

⟨γ⟩=56.8 *17* keV

γ_{mode}	γ(keV)	γ(%)†
Ge L_ℓ	1.037	0.030 *7*
Ge L_η	1.068	0.018 *5*
Ge L_α	1.188	0.52 *11*
Ge L_β	1.226	0.32 *8*
Ge L_γ	1.412	0.0033 *10*
Ge $K_{\alpha2}$	9.855	8.1 *4*
Ge $K_{\alpha1}$	9.886	15.7 *8*
Ge $K_{\beta1}'$	10.980	3.02 *16*
Ge $K_{\beta2}'$	11.184	0.050 *3*
γ_{IT}	61.80 *8*	0.0116 *23* ?
γ_{IT}	77.75 *9*	~0.0031 ?
$\gamma_{\beta-}$[E2]	80.943 *5*	$1.2_{5} \times 10^{-8}$
$\gamma_{\beta-}$[E2]	96.7337 *17*	$1.19_{20} \times 10^{-5}$
$\gamma_{\beta-}$[E1]	121.1192 *22*	$5.9_{10} \times 10^{-5}$
$\gamma_{\beta-}$[E1]	136.0017 *24*	0.00020 *3*
γ_{IT} E3	139.54 *3*	38.8

Photons (^{75}Ge) (continued)

γ_{mode}	γ(keV)	γ(%)†
$\gamma_{\beta-}$M1+16%E2	279.5381 *22*	$4.4_{20} \times 10^{-5}$
$\gamma_{\beta-}$[E1]	400.6568 *15*	3.9×10^{-5}

† uncert(syst): 3.1% for IT, 20% for β-

Atomic Electrons (^{75}Ge)*

⟨e⟩=80.7 *24* keV

e_{bin}(keV)	⟨e⟩(keV)	e(%)
1	0.60	48 *5*
8	0.56	6.6 *7*
9	1.02	11.8 *13*
10	0.45	4.6 *5*
11	0.024	0.221 *24*
51	0.012	~0.024
60	0.0012	~0.0021
61	0.001	~0.0017
62	0.0004	~0.0006
67	0.0018	~0.003
76	0.00019	~0.00025
77	0.00012	<0.00030
78	5.0×10^{-5}	$\sim 6 \times 10^{-5}$
128	64.3	50.0 *18*
138	11.9	8.6 *3*
139	1.83	1.32 *5*
140	0.107	0.077 *3*

* with IT

Continuous Radiation (^{75}Ge)

⟨β-⟩=0.097 keV;⟨IB⟩=9.5×10^{-5} keV

E_{bin}(keV)		⟨ ⟩(keV)	(%)
0 - 10	β-	2.07×10^{-5}	0.000412
	IB	4.7×10^{-6}	
10 - 20	β-	6.3×10^{-5}	0.000419
	IB	4.4×10^{-6}	3.1×10^{-5}
20 - 40	β-	0.000257	0.00086
	IB	8.3×10^{-6}	2.9×10^{-5}
40 - 100	β-	0.00194	0.00275
	IB	2.1×10^{-5}	3.2×10^{-5}
100 - 300	β-	0.0205	0.0102
	IB	3.9×10^{-5}	2.3×10^{-5}
300 - 600	β-	0.053	0.0122
	IB	1.65×10^{-5}	4.2×10^{-6}
600 - 917	β-	0.0216	0.00315
	IB	1.10×10^{-6}	1.7×10^{-7}

$^{75}_{33}$As(stable)

Δ: -73035.1 *16* keV

%: 100

$^{75}_{34}$Se(119.77 *1* d)

Mode: ϵ

Δ: -72171.3 *15* keV

SpA: 1.4536×10^{4} Ci/g

Prod: ^{74}Se(n,γ); ^{75}As(d,2n); ^{75}As(p,n)

Photons (^{75}Se)

$\langle\gamma\rangle$=392 *4* keV

γ_{mode}	γ(keV)	γ(%)†
As L_ℓ	1.120	0.058 *13*
As L_η	1.155	0.034 *9*
As L_α	1.282	1.06 *22*
As L_β	1.325	0.64 *17*
As L_γ	1.524	0.009 *3*
As $K_{\alpha2}$	10.508	16.6 *8*
As $K_{\alpha1}$	10.544	32.4 *15*
As $K_{\beta1}$'	11.724	6.4 *3*
As $K_{\beta2}$'	11.948	0.198 *9*
γ M2+0.1%E3	24.386 *3*	0.033 *6*
γ M1+2.3%E2	66.060 *5*	1.140 *23*
γ [E2]	80.943 *5*	0.0071 *24*
γ [E2]	96.7337 *17*	3.48 *8*
γ [E1]	121.1192 *22*	17.3 *4*
γ [E1]	136.0017 *24*	59.0 *12*
γ M1+18%E2	198.596 *5*	1.47 *3*
γ [M1+E2]	249.3 *3*	9.5 *24* $\times10^{-5}$
γ M1+0.3%E2	264.6556 *25*	59.2 *12*
γ	270.17 *16*	5.4 *20* $\times10^{-6}$
γ M1+16%E2	279.5381 *22*	25.2 *5*
γ E3	303.9235 *20*	1.34 *3*
γ	338.14 *12*	0.00015 *4*
γ	353.03 *12*	0.00069 *9*
γ [E2]	373.62 *24*	0.00248 *24*
γ [E1]	400.6568 *15*	11.56 *23*
γ	419.08 *12*	0.0062 *4*
γ	468.77 *16*	0.00035 *6*
γ M1+2.1%E2	542.02 *17*	0.000130 *24*
γ [E2]	556.90 *17*	3.5 *12* $\times10^{-5}$
γ M1+13%E2	572.22 *24*	0.0295 *24*
γ	617.68 *12*	0.00380 *22*
γ E2	821.55 *17*	0.000165 *12*

† 1.0% uncert(syst)

Atomic Electrons (^{75}Se)

$\langle e\rangle$=14.2 *4* keV

e_{bin}(keV)	$\langle e\rangle$(keV)	e(%)
1	1.10	83 *9*
2	0.084	5.5 *6*
9	3.1	34 *4*
10	0.88	8.5 *9*
11 - 12	0.083	0.75 *6*
13	0.68	5.4 *10*
23 - 69	0.58	1.9 *3*
79 - 81	0.0015	0.0018 *4*
85	2.29	2.69 *12*
95 - 97	0.413	0.433 *17*
109	0.71	0.65 *3*
120 - 121	0.095	0.079 *3*
124	1.94	1.56 *7*
134 - 136	0.258	0.191 *7*
187 - 199	0.064	0.0339 *20*
237 - 249	4.0 $\times10^{-6}$	~2 $\times10^{-6}$
253	0.96	0.381 *11*
258 - 265	0.125	0.0473 *12*
268	0.484	0.181 *7*
269 - 304	0.277	0.0955 *19*
326 - 374	7.4 $\times10^{-5}$	2.1 *3* $\times10^{-5}$
389 - 419	0.0612	0.0157 *6*
457 - 469	3.3 $\times10^{-6}$	~7 $\times10^{-7}$
530 - 572	0.000206	3.7 *3* $\times10^{-5}$
606 - 618	2.1 $\times10^{-5}$	3.5 *16* $\times10^{-6}$
810 - 822	8.5 $\times10^{-7}$	1.05 *7* $\times10^{-7}$

Continuous Radiation (^{75}Se)

$\langle IB\rangle$=0.029 keV

E_{bin}(keV)		$\langle\ \rangle$(keV)	(%)
10 - 20	IB	0.0042	0.036
20 - 40	IB	0.00028	0.00098
40 - 100	IB	0.00147	0.0021
100 - 300	IB	0.0149	0.0077
300 - 600	IB	0.0060	0.00169
600 - 864	IB	0.00028	4.3 $\times10^{-5}$

$^{75}_{35}$Br(1.62 *3* h)

Mode: ε

Δ: -69161 *20* keV

SpA: 2.58 $\times10^{7}$ Ci/g

Prod: ^{74}Se(d,n); ^{74}Se(p,γ); ^{65}Cu(^{12}C,2n); protons on Br; protons on Y

Photons (^{75}Br)

$\langle\gamma\rangle$=473 *6* keV

γ_{mode}	γ(keV)	γ(%)†
Se L_ℓ	1.204	~0.04
Se L_η	1.245	~0.024
Se L_α	1.379	~0.8
Se L_β	1.423	~0.4
Se L_γ	1.648	0.0026 *8*
Se $K_{\alpha2}$	11.182	4.56 *19*
Se $K_{\alpha1}$	11.222	8.8 *4*
Se $K_{\beta1}$'	12.494	1.78 *8*
Se $K_{\beta2}$'	12.741	0.086 *4*
γ M1+7.0%E2	112.22 *8*	1.75 *18*
γ M1+4.3%E2	141.30 *6*	6.9 *6*
γ [M1+E2]	195.57 *9*	0.09 *3*
γ [M1+E2]	236.15 *7*	0.83 *6*
γ E1	286.56 *8*	92
γ M1+2.4%E2	292.97 *8*	2.79 *14*
γ M1(+20%E2)	299.50 *9*	0.25 *4*
γ	309.46 *13*	0.092 *18*
γ [E1]	315.64 *8*	0.63 *6*
γ M1+E2	319.75 *24*	0.101 *18*
γ	324.77 *17*	0.25 *4*
γ [M1+E2]	349.49 *12*	0.18 *5*
γ M1+31%E2	377.44 *7*	4.1 *4*
γ E1	427.86 *8*	4.5 *5*
γ M1+13%E2	431.72 *8*	4.0 *5*
γ E2	461.04 *24*	0.12 *3*
γ	467.66 *12*	0.13 *3*
γ [E2]	484.26 *13*	0.29 *4*
γ [M1+E2]	488.01 *15*	0.18 *4*
γ [M1+E2]	490.79 *12*	0.34 *4*
γ	514.1 *5*	~0.09
γ	534.73 *12*	0.138 *18*
γ [E1]	551.79 *9*	0.31 *4*
γ [M1+E2]	566.49 *9*	0.47 *5*
γ E2(+M1)	573.01 *7*	2.08 *23*
γ	579.9 *3*	0.092 *18*
γ [E1]	586.06 *10*	0.19 *3*
γ	598.28 *16*	0.34 *5*
γ	608.96 *11*	1.76 *18*
γ [M1+E2]	646.21 *15*	0.16 *3*
γ [M1+E2]	659.30 *11*	0.37 *5*
γ [E1]	664.00 *9*	0.120 *18*
γ [M1+E2]	676.43 *21*	0.120 *18*
γ	701.62 *13*	0.19 *3*
γ	734.02 *12*	1.61 *18*
γ	770.88 *10*	0.49 *6*
γ [E2]	780.98 *15*	0.110 *18*
γ	788.78 *20*	0.35 *4*
γ [E1]	859.57 *9*	0.25 *3*
γ	890.8 *3*	0.26 *3*
γ	897.78 *16*	0.52 *6*
γ	912.17 *10*	1.06 *11*
γ	946.3 *3*	0.15 *3*
γ [M1+E2]	952.27 *11*	1.74 *18*
γ [M1+E2]	958.80 *12*	0.28 *5*
γ [E1]	961.85 *15*	0.46 *6*
γ	975.13 *15*	0.092 *18*
γ [E1]	1074.06 *15*	0.110 *18*
γ	1144.58 *20*	0.19 *3*
γ [E1]	1245.36 *12*	0.50 *6*
γ	1380.6 *3*	0.120 *18*
γ	1448.96 *13*	0.34 *4*
γ	1515.9 *3*	0.120 *18*
γ	1561.18 *13*	0.129 *18*

† 1.1% uncert(syst)

Atomic Electrons (^{75}Br)

$\langle e\rangle$=6.8 *25* keV

e_{bin}(keV)	$\langle e\rangle$(keV)	e(%)
1	0.9	~65
2	0.0234	1.41 *15*
5	2.3	~46
6	0.5	~7
7 - 9	0.10	1.3 *5*
10	0.73	7.5 *8*
11	0.247	2.25 *23*
12	0.0156	0.127 *13*
100	0.16	0.17 *3*
111 - 112	0.026	0.023 *4*
129	0.38	0.30 *5*
140 - 183	0.061	0.043 *7*
194 - 236	0.043	0.019 *9*
274	0.818	0.299 *7*
280 - 325	0.177	0.0619 *17*
337 - 377	0.072	0.020 *4*
415 - 461	0.072	0.0172 *15*
466 - 514	0.0112	0.0023 *3*
522 - 572	0.026	0.0047 *7*
573 - 609	0.014	0.0024 *8*
634 - 676	0.0049	0.00075 *8*
689 - 734	0.008	0.0012 *4*
758 - 789	0.0042	0.00054 *14*
847 - 896	0.0032	0.00036 *9*
898 - 946	0.0117	0.00126 *18*
949 - 975	0.00209	0.000220 *19*
1061 - 1074	0.00019	1.8 *3* $\times10^{-5}$
1132 - 1145	0.00050	4.4 *16* $\times10^{-5}$
1233 - 1245	0.00075	6.1 *6* $\times10^{-5}$
1368 - 1381	0.00025	1.9 *6* $\times10^{-5}$
1436 - 1516	0.00092	6.4 *16* $\times10^{-5}$
1549 - 1561	0.00024	1.6 *5* $\times10^{-5}$

Continuous Radiation (^{75}Br)

⟨β+⟩=499 keV;⟨IB⟩=1.8 keV

E_{bin}(keV)		⟨ ⟩(keV)	(%)
0 - 10	β+	5.7×10^{-5}	0.00071
	IB	0.021	
10 - 20	β+	0.00128	0.0079
	IB	0.022	0.155
20 - 40	β+	0.0208	0.065
	IB	0.040	0.137
40 - 100	β+	0.57	0.75
	IB	0.109	0.168
100 - 300	β+	17.5	8.1
	IB	0.29	0.162
300 - 600	β+	96	21.0
	IB	0.31	0.073
600 - 1300	β+	338	38.0
	IB	0.58	0.065
1300 - 2500	β+	46.6	3.32
	IB	0.47	0.028
2500 - 2898	IB	0.0027	0.000106
	Σβ+		71

$^{75}_{36}$Kr(4.3 *1* min)

Mode: ε
Δ: -64246 *20* keV
SpA: 5.82×10^{8} Ci/g
Prod: protons on Zr; protons on Br; ^{79}Br(p,5n)

Photons (^{75}Kr)

⟨γ⟩=307 *9* keV

γ_{mode}	γ(keV)	γ(%)†
Br L_ℓ	1.293	0.018 *4*
Br L_η	1.339	0.011 *3*
Br L_α	1.481	0.37 *8*
Br L_β	1.533	0.22 *6*
Br L_γ	1.777	0.0030 *9*
Br $K_{\alpha2}$	11.878	5.1 *3*
Br $K_{\alpha1}$	11.924	9.8 *6*
Br $K_{\beta1}$'	13.290	2.02 *12*
Br $K_{\beta2}$'	13.562	0.137 *8*
γ[M1]	22.07 *14*	1.79 *10*
γE2	88.40 *15*	3.49 *14*
γ[E1]	119.48 *19*	1.98 *7*
γE1	132.36 *12*	68
γ	153.21 *18*	8.2 *5*
γE1	154.43 *15*	21.2 *11*
γ	179.33 *20*	0.26 *3*
γ	216.33 *20*	0.116 *20*
γ	219.53 *17*	0.27 *7* ?
γ	220.75 *17*	0.18 *8* ?
γ	228.5 *3*	0.075 *20*
γ	241.61 *14*	1.22 *14*
γ	273.03 *20*	0.54 *7*
γ	295.9 *3*	0.24 *5*
γ	352.43 *20*	0.75 *7*
γ	622.9 *3*	0.18 *6*
γ	627.58 *25*	0.23 *5*
γ	670.36 *23*	~0.07
γ	673.57 *21*	0.34 *3*
γ	692.43 *21*	~0.07
γ	698.40 *25*	0.25 *3*
γ	700.48 *20*	0.59 *5*
γ	746.93 *17*	0.39 *4*
γ	769.00 *14*	0.71 *5*
γ	781.88 *22*	0.136 *20*
γ	787.6 *3*	0.38 *4*
γ	793.05 *17*	1.73 *12*
γ	824.78 *21*	1.02 *6*
γ	901.36 *15*	0.23 *3*
γ	916.2 *3*	0.29 *3*
γ	940.00 *25*	0.31 *3*
γ	952.9 *3*	0.20 *3*
γ	1227.87 *21*	0.41 *5*
γ	1238.42 *25*	0.54 *4*
γ	1249.4 *3*	0.44 *5*
γ	1346.3 *3*	1.41 *9*
γ	1356.8 *3*	0.37 *5*
γ	1469.47 *19*	1.14 *10*
γ	1601.82 *20*	1.84 *8*
γ	1612.37 *25*	0.28 *4*
γ	1668.5 *3*	0.32 *4*

† 8.8% uncert(syst)

Atomic Electrons (^{75}Kr)

⟨e⟩=12.1 *9* keV

e_{bin}(keV)	⟨e⟩(keV)	e(%)
2	0.38	24 *3*
9	1.21	14.1 *10*
10	0.81	7.9 *9*
11 - 22	0.69	4.5 *3*
75	3.08	4.12 *18*
87	0.552	0.64 *3*
88 - 118	0.205	0.211 *7*
119	2.66	2.24 *20*
131 - 132	0.37	0.282 *22*
140	1.0	~0.7
141	0.62	0.441 *24*
151 - 179	0.27	0.18 *6*
203 - 242	0.09	0.043 *17*
260 - 296	0.026	0.010 *5*
339 - 352	0.015	~0.0043
609 - 657	0.0029	0.00048 *16*
660 - 700	0.0068	0.00101 *25*
733 - 782	0.015	0.0019 *5*
786 - 825	0.0053	0.00066 *21*
888 - 927	0.0028	0.00031 *7*
938 - 953	0.0008	9 *3* $\times10^{-5}$
1214 - 1249	0.0035	0.00029 *6*
1333 - 1357	0.0042	0.00032 *8*
1456 - 1469	0.0025	0.00017 *5*
1588 - 1668	0.0049	0.00031 *7*

Continuous Radiation (^{75}Kr)

⟨β+⟩=1552 keV;⟨IB⟩=5.8 keV

E_{bin}(keV)		⟨ ⟩(keV)	(%)
0 - 10	β+	8.2×10^{-6}	0.000101
	IB	0.049	
10 - 20	β+	0.000191	0.00118
	IB	0.049	0.34
20 - 40	β+	0.00325	0.0101
	IB	0.096	0.33
40 - 100	β+	0.096	0.126
	IB	0.28	0.43
100 - 300	β+	3.49	1.60
	IB	0.82	0.46
300 - 600	β+	26.3	5.6
	IB	1.00	0.23
600 - 1300	β+	242	24.7
	IB	1.63	0.19
1300 - 2500	β+	882	47.4
	IB	1.37	0.079
2500 - 4795	β+	398	14.0
	IB	0.51	0.0166
	Σβ+		93

$^{75}_{37}$Rb(17.2 *8* s)

Mode: ε
Δ: -57280 *110* keV
SpA: 8.6×10^{9} Ci/g
Prod: protons on Y; ^{32}S on ^{50}Cr

Photons (^{75}Rb)

γ_{mode}	γ(keV)
γ	178.8 *5*

A = 76

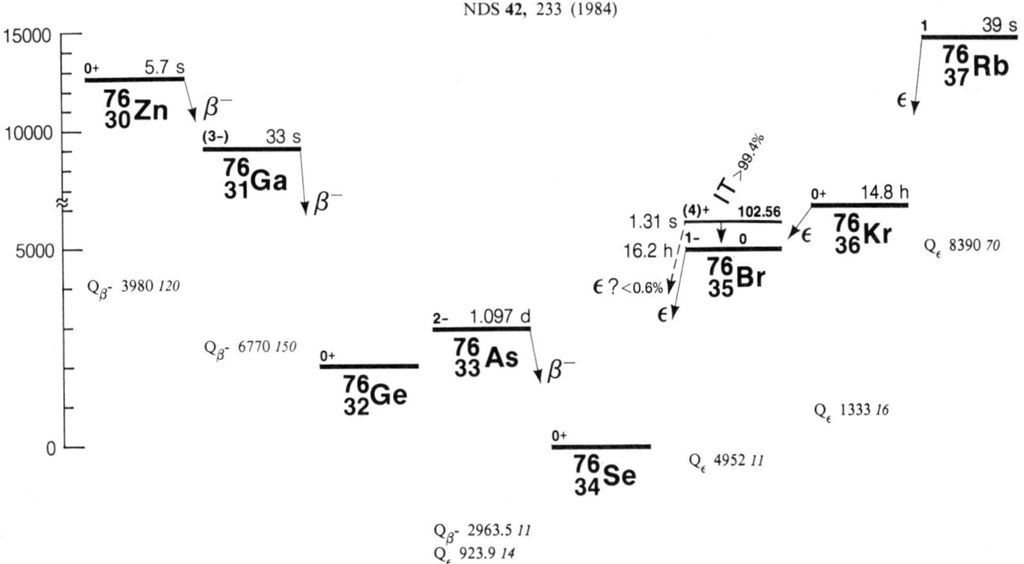

$^{76}_{30}$Zn(5.7 *3* s)

Mode: β-
Δ: -62460 *190* keV
SpA: 2.45×10^{10} Ci/g
Prod: fission

Photons (^{76}Zn)

$γ_{mode}$	γ(keV)
γ	75.9 *6*
γ	172.6 *10*
γ	199.6 *7*
γ	275.5 *7*
γ	281.8 *10*
γ	290.2 *8*
γ	366.1 *8*
γ	748.9 *10*
γ	755.3 *7*
γ	831.1 *7*
γ	1030.8 *7*
γ	2091.1 *10*

$^{76}_{31}$Ga(32.6 *6* s)

Mode: β-
Δ: -66440 *150* keV
SpA: 4.51×10^{9} Ci/g
Prod: ^{76}Ge(n,p), fission

Photons (^{76}Ga)

⟨γ⟩=2801 *30* keV

$γ_{mode}$	γ(keV)	γ(%)†
γ	336.15 *18*	5.3 *13* ?
γ [M1+E2]	431.00 *5*	9.2 *7*
γ E2+12%M1	545.517 *25*	26.0 *13*
γ E2	562.93 *3*	66
γ [M1+E2]	661.45 *18*	0.74 *7* ?
γ	843.77 *19*	1.14 *11* ?
γ [E2]	847.16 *4*	3.53 *20*
γ	885.85 *10*	1.32 *10* ?
γ [E2]	911.43 *10*	1.00 *7*
γ	927.07 *10*	0.92 *5* ?
γ [M1+E2]	976.51 *5*	4.62 *13*
γ	1014.15 *15*	0.36 *5* ?
γ	1043.6 *4*	0.30 *3*
γ	1051.64 *15*	0.47 *7*
γ E2	1108.44 *4*	15.8 *3*
γ [E1]	1175.81 *24*	0.47 *12*
γ	1182.0 *3*	0.51 *7* ?
γ [M1+E2]	1208.31 *10*	1.53 *11*
γ	1249.11 *20*	0.64 *7*
γ	1259.44 *10*	0.30 *7*
γ	1273.06 *10*	1.20 *7* ?
γ [E1]	1282.31 *8*	0.28 *7*
γ	1310.70 *22*	0.28 *5* ?
γ [E2]	1348.15 *11*	0.75 *5*
γ [E2]	1358.67 *14*	0.18 *6*
γ	1443.65 *20*	0.26 *7* ?
γ	1461.23 *25*	0.33 *7* ?
γ	1482.63 *16*	0.50 *7*
γ	1489.70 *16*	0.23 *7*
γ	1502.5 *3*	0.49 *7* ?
γ	1546.08 *20*	0.43 *9*
γ [E1]	1583.95 *8*	0.20 *7*
γ [E2]	1612.89 *13*	0.45 *6*
γ	1634.07 *15*	1.14 *5* ?
γ [E2]	1639.30 *9*	5.54 *13*
γ [M1+E2]	1642.76 *7*	0.93 *7*
γ [M1+E2]	1660.31 *13*	0.77 *5*
γ [E2]	1721.68 *15*	0.15 *5*
γ [M1+E2]	1733.19 *14*	0.73 *7*
γ	1811.39 *8*	0.84 *5*
γ	1878.26 *17*	0.36 *4*
γ	1892.71 *17*	0.40 *3* ?
γ	1902.23 *11*	0.42 *3*
γ	1912.74 *7*	0.59 *3*

Photons (^{76}Ga) (continued)

$γ_{mode}$	γ(keV)	γ(%)†
γ	1924.6 *3*	0.20 *3* ?
γ	1940.25 *13*	0.69 *5*
γ	1980.96 *15*	0.22 *4*
γ	2040.75 *13*	0.33 *5*
γ [M1+E2]	2073.75 *5*	4.24 *11*
γ	2091.59 *20*	0.18 *4*
γ [E1]	2129.46 *7*	2.20 *7*
γ [E2]	2184.81 *9*	0.50 *4*
γ	2203.87 *11*	1.37 *10*
γ	2214.38 *6*	2.24 *7*
γ [M1+E2]	2278.69 *13*	0.44 *3*
γ	2347.5 *2*	0.44 *5*
γ	2356.89 *7*	2.47 *11*
γ	2369.18 *17*	0.28 *9* ?
γ	2435.63 *23*	0.37 *5*
γ	2476.85 *20*	0.22 *5*
γ [M1+E2]	2481.13 *15*	0.20 *4*
γ	2489.6 *4*	0.20 *4*
γ [M1+E2]	2524.30 *10*	0.80 *5*
γ [M1+E2]	2578.55 *7*	2.24 *7*
γ	2591.06 *16*	0.27 *5*
γ [M1+E2]	2619.25 *6*	2.25 *7*
γ [E2]	2668.8 *4*	0.16 *3* ?
γ	2680.93 *21*	0.32 *3*
γ	2692.37 *8*	0.15 *4*
γ	2700.71 *25*	0.20 *3* ?
γ	2759.88 *6*	1.10 *5*
γ	2778.48 *20*	0.80 *8*
γ	2782.77 *12*	1.01 *8* ?
γ	2843.37 *7*	1.60 *7*
γ	2868.03 *17*	0.35 *5*
γ	2882.9 *4*	0.14 *5* ?
γ	2914.68 *17*	0.74 *6* ?
γ	2919.80 *7*	9.1 *3*
γ [E1]	2970.91 *13*	0.40 *5*
γ	2981.1 *3*	0.20 *4* ?
γ	3034.61 *20*	0.52 *5*
γ [M1+E2]	3069.79 *10*	0.92 *5*
γ	3130.91 *14*	0.21 *4* ?
γ [E2]	3141.46 *7*	4.24 *21*
γ	3145.3 *4*	0.30 *6* ?
γ	3190.50 *24*	0.21 *3* ?
γ	3275.91 *20*	0.58 *5*
γ	3283.6 *5*	0.17 *4*
γ	3323.98 *20*	~0.11
γ	3328.7 *3*	0.20 *6* ?

Photons (^{76}Ga)
(continued)

γ_{mode}	γ(keV)	γ(%)†
γ	3334.6 3	0.19 4 ?
γ	3366.55 25	0.145 20 ?
γ	3388.86 7	2.83 17
γ	3402.41 18	0.132 20 ?
γ	3465.6 3	0.14 3 ?
γ	3496.7 6	0.11 3
γ	3559.3 3	0.59 5 ?
γ	3675.56 25	0.45 5 ?
γ	3736.8 3	0.16 4 ?
γ [E2]	3752.20 18	0.17 3
γ	3842.3 4	0.092 20
γ	3913.53 21	0.13 3 ?
γ	3925.21 20	0.34 3
γ	3951.77 7	4.2 3
γ	3994.3 10	0.22 3
γ	4122.2 3	0.25 3 ?
γ [E2]	4253.17 15	0.22 3

† 4.6% uncert(syst)

Continuous Radiation (^{76}Ga)
$\langle\beta$-$\rangle$=1795 keV;$\langle$IB$\rangle$=6.1 keV

E_{bin}(keV)		$\langle\ \rangle$(keV)	(%)
0 - 10	β-	0.0065	0.128
	IB	0.053	
10 - 20	β-	0.0198	0.132
	IB	0.052	0.36
20 - 40	β-	0.082	0.274
	IB	0.103	0.36
40 - 100	β-	0.65	0.92
	IB	0.30	0.46
100 - 300	β-	8.4	4.09
	IB	0.89	0.50
300 - 600	β-	37.9	8.3
	IB	1.09	0.25
600 - 1300	β-	234	24.4
	IB	1.7	0.20
1300 - 2500	β-	691	37.4
	IB	1.39	0.080
2500 - 5000	β-	785	23.6
	IB	0.51	0.0167
5000 - 6207	β-	36.6	0.69
	IB	0.0033	5.9×10^{-5}

$^{76}_{32}$Ge(stable)

Δ: -73214.5 16 keV
%: 7.8 2

$^{76}_{33}$As(1.097 3 d)

Mode: β-
Δ: -72290.7 18 keV
SpA: 1.567×10^{6} Ci/g
Prod: ^{75}As(n,γ)

Photons (^{76}As)
$\langle\gamma\rangle$=419 11 keV

γ_{mode}	γ(keV)	γ(%)†
γ [E1]	301.92 11	0.0090 13
γ [M1+E2]	358.15 5	0.011 4 ?
γ [E1]	403.12 7	0.0234 13
γ [M1+E2]	438.24 11	0.0020 4 ?
γ [E2]	456.90 5	0.0356 22
γ	463.7 7	~0.0009 ?
γ	466.6 10	~0.004 ?
γ M1+E2	472.87 3	0.050 4
γ	484.70 17	0.0058 14
γ	499.24 17	0.022 9
γ E2	559.08 3	45
γ [E2]	563.27 3	1.20 6
γ [M1+E2]	571.61 3	0.141 9
γ	575.37 5	0.067 5
γ	602.6 4	~0.0009 ?
γ [E1]	641.41 4	0.09 3
γ E2+3.6%M1+E0	657.062 23	6.2 3
γ	665.1 10	<0.07 ?
γ [E2]	665.41 3	0.377 20
γ E2+22%M1	695.20 7	0.0090 9
γ [M1+E2]	727.28 10	0.0186 13
γ [E1]	740.16 3	0.116 9
γ	755.1 5	<0.0009
γ E2	771.78 4	0.122 9
γ	776.6 5	~0.0009 ?
γ [E2]	796.39 11	~0.004
γ E2	809.91 7	0.0171 9
γ	852.9 10	~0.0022 ?
γ	857.1 8	<0.0018 ?
γ	863.9 4	0.0113 9
γ	867.69 4	0.132 8
γ [E1]	882.23 3	0.056 3
γ	907.6 4	~0.0018 ?
γ [M1+E2]	911.10 11	~0.00040
γ	921.7 4	~0.0009 ?
γ [E2]	954.60 17	~0.0018 ?
γ	957.7 5	~0.0018 ?
γ [E1]	980.98 4	0.0423 20
γ	1032.27 7	~0.0009
γ	1060.7 3	0.0018 5 ?
γ [E1]	1098.31 5	0.0036 5
γ [M1+E2]	1129.92 3	0.125 9
γ	1130.1 10	~0.018 ?
γ (E1)	1213.02 3	1.43 9
γ E2	1216.14 3	3.42 15
γ M1+19%E2	1228.67 3	1.15 6
γ	1306.82 4	0.113 16
γ	1324.59 5	~0.020
γ [E2]	1392.68 11	
γ	1439.30 4	0.267 11
γ [E1]	1453.84 4	0.107 6
γ [E2]	1466.97 7	~0.0009
γ	1533.09 4	0.0243 13
γ	1563.1 10	0.0018 5
γ [M1+E2]	1568.16 11	0.0076 4
γ [E2]	1611.66 17	0.0077 5
γ [E2]	1787.74 3	0.297 16
γ	1804.04 6	~0.0013 ?
γ [E1]	1870.08 3	0.056 5
γ	1881.4 4	~0.0009 ?
γ [M1+E2]	1955.94 10	0.0089 8
γ	2096.35 3	0.576 25
γ [E1]	2110.89 3	0.332 16
γ [E2]	2127.22 11	0.0015 3
γ	2429.14 3	0.033 3
γ	2655.41 4	0.0440 22
γ	2669.95 4	0.00027 5 ?

† 4.4% uncert(syst)

Continuous Radiation (^{76}As)
$\langle\beta$-$\rangle$=1068 keV;$\langle$IB$\rangle$=2.6 keV

E_{bin}(keV)		$\langle\ \rangle$(keV)	(%)
0 - 10	β-	0.0195	0.389
	IB	0.038	
10 - 20	β-	0.059	0.392
	IB	0.038	0.26
20 - 40	β-	0.237	0.79
	IB	0.074	0.26
40 - 100	β-	1.71	2.43
	IB	0.21	0.32
100 - 300	β-	17.1	8.5
	IB	0.58	0.33
300 - 600	β-	66	14.6
	IB	0.63	0.149
600 - 1300	β-	358	37.9
	IB	0.77	0.090
1300 - 2500	β-	587	33.6
	IB	0.26	0.0163
2500 - 2964	β-	37.5	1.43
	IB	0.00157	6.1×10^{-5}

$^{76}_{34}$Se(stable)

Δ: -75254.1 15 keV
%: 9.0 2

$^{76}_{35}$Br(16.2 2 h)

Mode: ϵ
Δ: -70302 11 keV
SpA: 2.55×10^{6} Ci/g
Prod: ^{75}As(α,3n)

Photons (^{76}Br)
$\langle\gamma\rangle$=2226 29 keV

γ_{mode}	γ(keV)	γ(%)†
Se L_ℓ	1.204	0.025 5
Se L_η	1.245	0.015 4
Se L_α	1.379	0.48 10
Se L_β	1.427	0.29 7
Se L_γ	1.648	0.0041 13
Se $K_{\alpha2}$	11.182	7.1 3
Se $K_{\alpha1}$	11.222	13.7 6
Se $K_{\beta1}$'	12.494	2.78 11
Se $K_{\beta2}$'	12.741	0.133 6
γ	209.72 20	0.059 12 ?
γ	281.42 20	0.16 3 ?
γ [E1]	301.92 11	0.015 5
γ	309.22 20	0.14 3 ?
γ	318.42 20	0.13 3 ?
γ [M1+E2]	358.15 5	0.42 13
γ	399.84 6	0.34 4
γ [E1]	403.12 7	0.038 12
γ [M1+E2]	438.24 11	0.25 5 ?
γ [E2]	456.90 5	0.066 5 ?
γ M1+E2	472.87 3	1.85 8
γ	484.70 17	0.013 3
γ	490.12 7	0.36 4
γ	499.24 17	0.16 7 ?
γ	504.66 7	0.229 15 ?
γ	546.5 5	0.163 22 ?
γ E2	559.08 3	74
γ [E2]	563.27 3	3.6 6
γ [M1+E2]	571.61 3	0.260 20
γ	598.92 20	0.41 16
γ	604.5 3	0.22 7
γ	635.7 6	0.074 22 ?
γ [E1]	641.41 4	0.14 4 ?
γ E2+3.6%M1+E0	657.062 23	15.9 6
γ [E2]	665.41 3	0.69 3
γ	681.54 10	0.422 22

Photons (^{76}Br) (continued)

γ_{mode}	γ(keV)	γ(%)†
γ	696.07 10	0.49 3
γ[M1+E2]	727.28 10	0.63 9
γ	730.94 7	0.58 7
γ[E1]	740.16 3	0.19 4
γ E2	771.78 4	0.414 22
γ	789.20 16	0.47 3
γ[E2]	796.39 11	∼0.6 ?
γ	803.73 17	0.53 4
γ	812.5 5	0.14 4 ?
γ	836.49 12	0.38 7
γ	867.69 4	0.298 18
γ[E1]	882.23 3	0.417 19
γ	886.44 8	0.333 22
γ	899.08 18	0.170 22 ?
γ	900.98 8	0.155 15
γ[M1+E2]	911.10 11	∼0.05 ?
γ	922.35 10	
γ	934.09 9	0.074 15
γ	942.0 4	0.19 2
γ[E2]	954.60 17	∼0.07
γ[E1]	980.98 4	0.316 17
γ	1030.01 17	0.57 6
γ	1032.86 12	0.58 6
γ	1038.9 4	∼0.07
γ	1060 2	∼0.044 ?
γ[E1]	1098.31 5	0.0059 16
γ[M1+E2]	1129.92 3	4.60 21
γ	1147.1 6	0.059 15 ?
γ	1158.2 5	0.148 15
γ	1162.83 7	0.163 22 ?
γ	1180.77 19	0.09 4
γ	1193 2	0.10 4 ?
γ(E1)	1213.02 3	2.3 5
γ E2	1216.14 3	8.8 4
γ	1224.27 14	0.28 10
γ M1+19%E2	1228.67 3	2.12 8
γ	1254.8 6	0.08 3 ?
γ	1271 2	0.059 22 ?
γ	1282.07 6	0.07 3 ?
γ	1288.43 23	0.052 22 ?
γ	1298 2	0.089 15 ?
γ	1300.6 3	0.155 15
γ	1306.82 4	0.185 22 ?
γ	1315.1 3	0.052 15 ?
γ	1324.59 5	∼0.044 ?
γ	1372.35 7	0.55 4
γ	1380.82 5	2.52 13
γ[E2]	1392.68 11	
γ	1429.01 17	0.27 2
γ	1439.30 4	0.607 25
γ[E1]	1453.84 4	0.80 4
γ	1461 2	0.13 3 ?
γ	1471.10 6	2.31 12
γ	1504.5 4	0.09 4 ?
γ	1533.09 4	0.055 4 ?
γ	1536.4 6	0.17 7 ?
γ	1560.14 19	0.459 22 ?
γ[M1+E2]	1568.16 11	0.98 7
γ[E2]	1611.66 17	0.28 6
γ	1642 3	0.13 4 ?
γ	1655.29 18	0.118 22 ?
γ	1662.51 10	0.14 5 ?
γ	1671.42 16	0.24 7
γ	1741.9 10	0.118 15
γ	1769.8 4	0.42 2
γ[E2]	1787.74 3	0.55 3
γ	1802.0 6	∼0.030 ?
γ	1815 2	0.148 15 ?
γ	1833.8 8	∼0.19
γ	1853.67 5	14.7 7
γ[E1]	1870.08 3	0.092 20
γ	1883 2	0.13 4 ?
γ	1901 2	0.12 4 ?
γ	1943.95 6	0.47 7
γ[M1+E2]	1955.94 10	0.30 4
γ	1975.3 7	∼0.10
γ	1991 2	0.08 3 ?
γ	2045.9 9	0.178 15 ?
γ	2071.9 4	∼0.27
γ	2082 2	0.12 4 ?
γ	2096.35 3	1.31 6
γ[E1]	2110.89 3	2.48 10
γ[E2]	2127.22 11	0.19 4
γ	2135.36 10	0.94 7

Photons (^{76}Br) (continued)

γ_{mode}	γ(keV)	γ(%)†
γ	2170 2	0.10 4 ?
γ	2182.8 3	0.13 4
γ	2229.15 10	∼0.10
γ	2235 2	0.13 6 ?
γ	2299 2	0.14 4 ?
γ	2309.6 10	0.10 3
γ	2340.26 8	0.09 4 ?
γ	2391.48 6	4.7 3
γ	2411.2 4	∼0.06 ?
γ	2429.14 3	0.054 13
γ	2481.70 9	0.133 22
γ	2510.72 5	1.95 11
γ	2546.7 20	∼0.006 ?
γ	2601.00 6	0.70 4
γ	2631.0 4	0.13 4 ?
γ	2655.41 4	0.100 6
γ	2669.95 4	0.0020 4
γ	2688.6 4	0.36 4 ?
γ	2713.1 5	0.074 22 ?
γ	2754.4 3	0.074 22 ?
γ	2792.40 10	5.6 3
γ	2837 3	0.11 4 ?
γ	2844 3	0.15 4 ?
γ	2900.06 16	0.27 10
γ	2950.54 6	7.4 4
γ	2982.8 4	0.09 3 ?
γ	2997.30 8	0.96 7
γ	3044.95 8	0.022 7 ?
γ	3064 2	0.074 22 ?
γ	3069.78 5	0.044 15 ?
γ	3093.20 19	0.163 15 ?
γ	3160.06 7	0.148 15
γ	3351.46 10	0.252 22
γ	3370.2 5	0.089 15
γ	3411.5 3	0.289 15
γ	3508 3	0.059 22 ?
γ	3525.32 18	0.178 15
γ	3604.01 8	1.55 11
γ	3639.8 4	0.148 15
γ	3877.7 10	∼0.015 ?
γ	3892 2	∼0.030
γ	3913.5 10	∼0.015 ?
γ	3929.2 5	0.089 15
γ	3963.5 10	0.022 7
γ	3970.5 3	0.010 4
γ	4019.4 4	0.059 15 ?
γ	4047.1 6	0.052 15
γ	4065 3	0.022 7 ?
γ	4084 3	∼0.015 ?
γ	4173.0 9	0.022 7 ?
γ	4436.8 10	0.052 15 ?
γ	4455 3	0.0067 22 ?
γ	4492 3	0.0059 22 ?
γ	4606.2 6	0.022 7

† 2.7% uncert(syst)

Atomic Electrons (^{76}Br)

⟨e⟩=3.37 14 keV

e_{bin}(keV)	⟨e⟩(keV)	e(%)
1	0.48	33 3
2 - 9	0.145	3.36 25
10	1.14	11.7 12
11	0.38	3.5 4
197 - 210	0.0037	∼0.0019
269 - 318	0.012	0.0040 14
345 - 390	0.013	0.0036 12
398 - 444	0.0051	0.0012 3
455 - 505	0.030	0.0064 12
534 - 545	0.0012	0.00022 11
546	0.712	0.130 4
547 - 592	0.136	0.0244 12
597 - 641	0.0013	0.00021 5
644	0.113	0.0175 8
653 - 696	0.0244	0.00371 22
715 - 759	0.0092	0.00127 18
770 - 813	0.0082	0.0010 3
824 - 870	0.0033	0.00039 9

Atomic Electrons (^{76}Br) (continued)

e_{bin}(keV)	⟨e⟩(keV)	e(%)
874 - 921	0.0028	0.00032 6
929 - 968	0.0015	0.00015 5
979 - 1029	0.0034	0.00033 8
1030 - 1060	0.00036	3.5 9 $\times 10^{-5}$
1086 - 1134	0.0161	0.00144 10
1145 - 1193	0.0013	0.000110 22
1200 - 1242	0.0399	0.00331 12
1253 - 1302	0.00146	0.000114 18
1305 - 1325	0.00016	1.2 4 $\times 10^{-5}$
1360 - 1381	0.0065	0.00048 12
1416 - 1461	0.0073	0.0005 1
1469 - 1568	0.0044	0.00029 3
1599 - 1671	0.0018	0.000111 16
1729 - 1821	0.0027	0.00015 3
1832 - 1931	0.025	0.0014 4
1942 - 2033	0.00124	6.3 9 $\times 10^{-5}$
2044 - 2135	0.0071	0.00034 3
2157 - 2235	0.00066	3.0 7 $\times 10^{-5}$
2286 - 2379	0.0063	0.00027 7
2390 - 2482	0.00104	4.3 8 $\times 10^{-5}$
2498 - 2588	0.0035	0.000137 25
2599 - 2688	0.00086	3.2 5 $\times 10^{-5}$
2700 - 2792	0.0073	0.00026 6
2824 - 2900	0.00067	2.3 5 $\times 10^{-5}$
2938 - 3032	0.0104	0.00035 7
3043 - 3093	0.00034	1.11 18 $\times 10^{-5}$
3147 - 3160	0.00018	5.6 12 $\times 10^{-6}$
3339 - 3411	0.00072	2.2 3 $\times 10^{-5}$
3495 - 3591	0.0018	5.1 9 $\times 10^{-5}$
3602 - 3640	0.00035	9.7 13 $\times 10^{-6}$
3865 - 3963	0.00019	4.9 8 $\times 10^{-6}$
3969 - 4065	0.00014	3.6 6 $\times 10^{-6}$
4071 - 4160	3.7 $\times 10^{-5}$	9 3 $\times 10^{-7}$
4171 - 4173	2.5 $\times 10^{-6}$	6.0 20 $\times 10^{-8}$
4424 - 4492	6.6 $\times 10^{-5}$	1.5 4 $\times 10^{-6}$
4593 - 4606	2.2 $\times 10^{-5}$	4.9 16 $\times 10^{-7}$

Continuous Radiation (^{76}Br)

⟨β+⟩=642 keV;⟨IB⟩=2.6 keV

E_{bin}(keV)		⟨ ⟩(keV)	(%)
0 - 10	β+	5.1 $\times 10^{-5}$	0.00064
	IB	0.022	
10 - 20	β+	0.00114	0.0071
	IB	0.024	0.17
20 - 40	β+	0.0183	0.057
	IB	0.042	0.147
40 - 100	β+	0.479	0.64
	IB	0.121	0.19
100 - 300	β+	12.3	5.8
	IB	0.36	0.20
300 - 600	β+	47.3	10.7
	IB	0.48	0.111
600 - 1300	β+	134	14.5
	IB	0.92	0.105
1300 - 2500	β+	350	19.1
	IB	0.49	0.030
2500 - 4952	β+	98	3.50
	IB	0.109	0.0036
	Σβ+		54

$^{76}_{35}$Br(1.31 2 s)

Mode: IT(>99.4 %), ϵ?(<0.6 %)

Δ: -70199 11 keV

SpA: 8.80×10^{10} Ci/g

Prod: alphas on ^{75}As

Photons (^{76}Br)

⟨γ⟩=40 *3* keV

γ_{mode}	γ(keV)	γ(%)†
Br L_ℓ	1.293	0.082 *18*
Br L_η	1.339	0.048 *12*
Br L_α	1.481	1.7 *3*
Br L_β	1.534	0.98 *25*
Br L_γ	1.777	0.014 *4*
Br $K_{\alpha2}$	11.878	22.4 *9*
Br $K_{\alpha1}$	11.924	43.4 *18*
Br $K_{\beta1}'$	13.290	9.0 *4*
Br $K_{\beta2}'$	13.562	0.608 *25*
γ_{IT} M1	45.466 *20*	48.5
γ_{IT} M2	57.097 *20*	9.1 *4*
γ_{IT} [E3]	102.56 *3*	<0.10
γ_ϵE2	559.08 *3*	<0.6 ?
γ_ϵE2	771.78 *4*	<0.6 ?

† uncert(syst): 0.30% for IT

Atomic Electrons (^{76}Br)*

⟨e⟩=65.2 *12* keV

e_{bin}(keV)	⟨e⟩(keV)	e(%)
2	1.71	108 *11*
10	3.6	35 *4*
11	0.107	0.94 *10*
12	1.14	9.8 *10*
13	0.097	0.74 *8*
32	14.6	45.7 *11*
44	35.6	81.5 *23*
45	0.425	0.941 *23*
55	5.39	9.7 *3*
56	1.31	2.36 *7*
57	1.28	2.24 *7*

* with IT

$^{76}_{36}$Kr(14.8 *1* h)

Mode: ε

Δ: -68969 *12* keV

SpA: 2.786×10^6 Ci/g

Prod: ^{79}Br(p,4n); ^{74}Se(α,2n)

Photons (^{76}Kr)

⟨γ⟩=426 *16* keV

γ_{mode}	γ(keV)	γ(%)†
Br L_ℓ	1.293	0.072 *16*
Br L_η	1.339	0.043 *11*
Br L_α	1.481	1.5 *3*
Br L_β	1.534	0.86 *22*
Br L_γ	1.777	0.013 *4*
Br $K_{\alpha2}$	11.878	20.1 *10*
Br $K_{\alpha1}$	11.924	39.0 *20*
Br $K_{\beta1}'$	13.290	8.0 *4*
Br $K_{\beta2}'$	13.562	0.55 *3*
γ [M1]	35.72 *17*	0.086 *19*
γ [M1]	38.11 *18*	0.14 *4* ?
γ [M1]	39.62 *12*	0.055 *12*
γ M1	45.466 *20*	19.5 *20* *
γ M2	57.097 *20*	0.055 *12* ? *
γ M1(+E2)	63.60 *11*	0.066 *16*
γ	76.3 *10*	0.14 *4* ?
γ M1(+E2)	90.98 *15*	0.21 *4*
γ M1(+E2)	96.67 *11*	0.109 *23*
γ [E3]	102.56 *3*	<0.0006 *
γ M1(+E2)	103.22 *12*	3.3 *4*
γ	104.94 *12*	0.16 *4*
γ	113.4 *3*	0.14 *4* ?
γ	121.3 *3*	0.20 *4* ?
γ (M1)	134.94 *11*	2.5 *3*
γ (M1)	136.29 *11*	1.05 *12*
γ	141.9 *3*	0.20 *6* ?
γ [E2]	149.50 *11*	<0.19 ?
γ [M1]	166.60 *13*	0.16 *4*
γ M1+50%E2	170.67 *18*	0.14 *4* ?
γ	179.9 *10*	0.17 *4* ?
γ	192.0 *4*	0.043 *12*
γ M1+26%E2	199.89 *11*	1.17 *12*
γ M1+50%E2	214.45 *10*	0.30 *3*
γ M1+50%E2	232.66 *23*	0.055 *12*
γ	234.7 *3*	0.13 *3* ?
γ	239.0 *3*	0.27 *4* ?
γ E1	252.06 *10*	6.2 *8*
γ (E1)	270.19 *9*	21.1 *23*
γ (E1)	271.54 *10*	4.3 *4*
γ	295.85 *19*	0.19 *4*
γ (M1,E2)	298.97 *13*	0.86 *12*
γ (M1,E2)	300.33 *13*	0.43 *12*
γ [E1]	309.82 *10*	2.6 *3*
γ E1	315.66 *9*	39 *4*
γ [E1]	317.01 *10*	0.47 *16*
γ [E1]	355.28 *10*	4.7 *8*
γ M1+50%E2	363.92 *14*	0.59 *8*
γ E1	406.49 *9*	12.1 *12*
γ	428.5 *5*	0.17 *4* ?
γ [M1+E2]	431.75 *22*	0.078 *16*
γ	438.6 *4*	0.14 *4* ?
γ [M1+E2]	446.39 *15*	0.39 *6*
γ [E1]	451.95 *9*	9.8 *12*
γ	459.79 *23*	0.043 *12* ?
γ	473.1 *3*	0.34 *5* ?
γ [M1+E2]	484.50 *17*	0.066 *16*
γ	490.19 *20*	0.18 *4*
γ	499.41 *22*	0.47 *8*
γ	520.9 *3*	0.20 *4* ?
γ [M1+E2]	543.06 *15*	0.29 *3*
γ	548.5 *4*	0.19 *4* ?
γ [M1+E2]	552.46 *24*	1.68 *20*
γ [E1]	570.52 *13*	0.14 *5* ?
γ	576.0 *3*	0.12 *3*
γ [M1+E2]	581.17 *17*	
γ	581.34 *15*	0.51 *8*
γ [M1+E2]	582.69 *15*	1.13 *12*
γ	599.2 *4*	0.20 *7* ?
γ	619.44 *17*	0.39 *4*
γ	640.9 *4*	0.20 *7* ?
γ	666.0 *4*	0.14 *6* ?
γ [M1+E2]	684.39 *17*	0.15 *3*
γ	730.72 *21*	0.19 *4*
γ [M1+E2]	795.67 *21*	0.29 *3*
γ [E1]	822.65 *24*	0.25 *3*
γ [E1]	852.88 *14*	0.14 *3*
γ [E1]	868.12 *24*	0.25 *3*
γ [E1]	890.99 *16*	0.12 *3* ?
γ [E1]	898.34 *14*	0.16 *4*
γ	911.0 *10*	0.12 *3* ?
γ [E1]	936.45 *16*	0.109 *23* ?
γ [E1]	1002.26 *21*	0.12 *3* ?
γ	1030.3 *5*	0.27 *10* ?
γ	1070.3 *5*	0.3 *1* ?

† 5.1% uncert(syst)

* with ^{76}Br(1.31 s) in equilib

Atomic Electrons (^{76}Kr)

⟨e⟩=16.7 *12* keV

e_{bin}(keV)	⟨e⟩(keV)	e(%)
2	1.49	95 *10*
10	3.2	31 *3*
11	0.096	0.84 *9*
12	1.03	8.8 *9*
13 - 26	0.197	1.12 *10*
32	5.9	18.5 *19*
34 - 40	0.022	0.061 *8*
44	1.11	2.5 *3*
45 - 89	0.58	1.01 *22*
90	1.1	~1
91 - 140	0.68	0.62 *13*
142 - 190	0.111	0.064 *11*
192 - 239	0.127	0.056 *6*
250 - 299	0.38	0.145 *11*
300	0.00026	8.7 *24* ×10^{-5}
302	0.32	0.104 *11*
304 - 350	0.090	0.0276 *20*
354 - 393	0.070	0.0180 *16*
405 - 452	0.069	0.0159 *17*
458 - 507	0.012	0.0026 *7*
519 - 568	0.026	0.0049 *7*
569 - 618	0.0150	0.0026 *4*
619 - 666	0.0021	0.00032 *12*
671 - 717	0.0020	0.00029 *7*
729 - 731	0.00011	1.5 *7* ×10^{-5}
782 - 823	0.00231	0.00029 *3*
839 - 885	0.00143	0.000166 *14*
889 - 936	0.00074	8.2 *21* ×10^{-5}
989 - 1030	0.0011	0.00011 *4*
1057 - 1070	0.0009	9 *4* ×10^{-5}

Continuous Radiation (^{76}Kr)

⟨IB⟩=0.20 keV

E_{bin}(keV)		⟨ ⟩(keV)	(%)
10 - 20	IB	0.0073	0.060
20 - 40	IB	0.00038	0.00138
40 - 100	IB	0.0018	0.0025
100 - 300	IB	0.031	0.0151
300 - 600	IB	0.103	0.024
600 - 1017	IB	0.057	0.0080

$^{76}_{37}$Rb(39.1 *6* s)

Mode: ε

Δ: -60580 *70* keV

SpA: 3.76×10^9 Ci/g

Prod: protons on Y; ^{32}S on ^{50}Cr; protons on Ta

Photons (^{76}Rb)

γ_{mode}	γ(keV)	γ(rel)
γ	63.9 *20*	2.3 *5* ?
γ	243.9 *20*	~2 ?
γ	253.9 *20*	~2 ?
γ [E2]	345.94 *9*	12.7 *12*
γ	355.50 *9*	17.3 *20*
γ	376.44 *16*	0.5 *2*
γ [M1+E2]	403.87 *16*	0.3 *1*
γ [E2]	423.98 *9*	92 *5*
γ	431.22 *20*	0.6 *2*
γ	453.35 *17*	5.6 *6*
γ [M1+E2]	465.89 *15*	0.5 *2*
γ	479.73 *15*	1.7 *3*
γ [E2]	610.67 *22*	3.2 *7*
γ [E2]	652.87 *22*	0.5 *2*
γ	766.80 *22*	4.5 *9*
γ (M1+4.6%E2)	797.66 *12*	9.5 *10*
γ	822.27 *15*	4.4 *5*
γ	868.9 *20*	~2 ?
γ	883.60 *12*	12.3 *12*
γ [E2]	917.60 *12*	11.3 *11*
γ	936.9 *20*	2.5 *12* ?
γ	973.05 *14*	5.6 *6*
γ	1005.56 *19*	3.0 *4*
γ [E1]	1036.04 *22*	0.5 *2*
γ	1120.7 *3*	0.7 *2*
γ	1174.09 *14*	6.3 *6*
γ [E2]	1221.63 *13*	5.3 *8*
γ [M1+E2]	1263.54 *11*	2.0 *3*
γ (M1+13%E2)	1308.8 *3*	3.0 *3*
γ [E2]	1321.47 *16*	1.5 *4*

Photons (^{76}Rb)
(continued)

γ_{mode}	γ(keV)	γ(rel)
γ	1334.43 *16*	1.5 *4*
γ	1349.48 *14*	1.9 *4*
γ	1553.20 *24*	1.7 *4*
γ [M1+E2]	1667.40 *15*	0.8 *2*
γ	1680.36 *14*	15.0 *15*

Photons (^{76}Rb)
(continued)

γ_{mode}	γ(keV)	γ(rel)
γ [E2]	1687.51 *13*	3.1 *6*
γ	1718.2 *3*	3.0 *6*
γ	1803.20 *17*	16.1 *20*
γ [E1]	1833.68 *20*	4.3 *6*
γ	2104.33 *15*	3.6 *6*

Photons (^{76}Rb)
(continued)

γ_{mode}	γ(keV)	γ(rel)
γ	2147.12 *11*	1.2 *4*
γ	2350.84 *25*	4.2 *6*
γ	2392.6 *3*	4.6 *6*
γ	2571.09 *12*	100
γ	2600.46 *18*	6.5 *8*
γ	2816.6 *3*	7.8 *9*

A = 77

NDS **29**, 75 (1980)

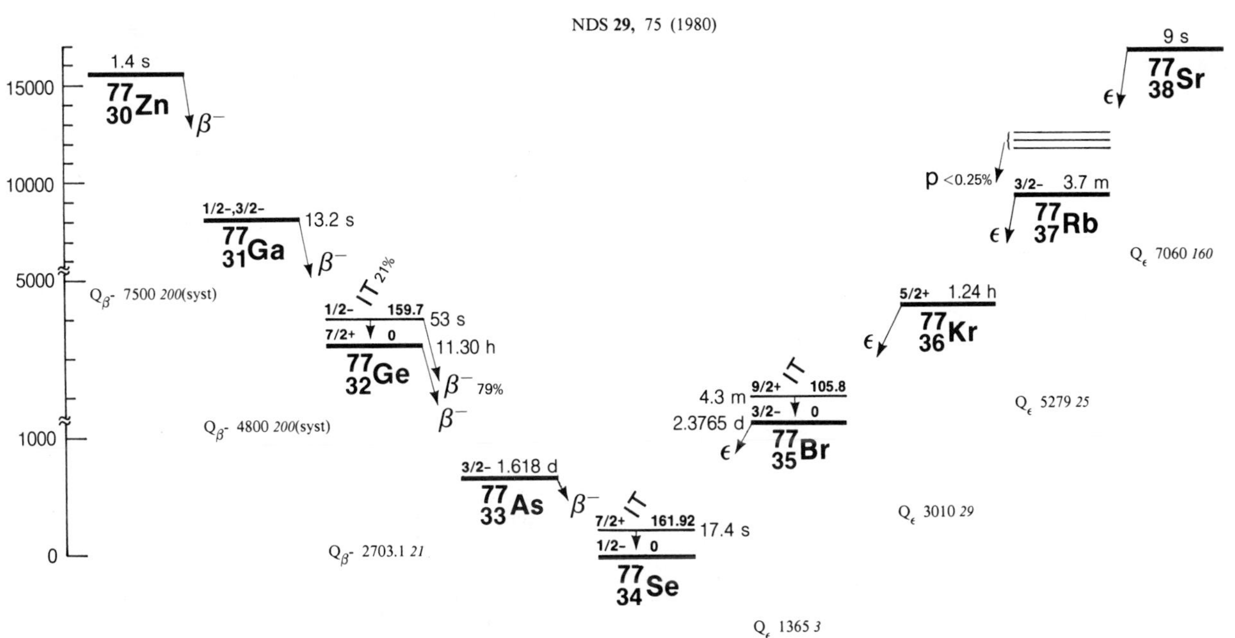

$^{77}_{30}$Zn(1.4 *3* s)

Mode: β-
Δ: -58910 *280* keV syst
SpA: 8.3×10^{10} Ci/g
Prod: fission

Photons (^{77}Zn)

γ_{mode}	γ(keV)	γ(rel)
γ	151.9 *7*	
γ	160.9 *8*	
γ	188.8 *8*	100 *20*
γ	284.2 *7*	
γ	306.5 *9*	
γ	312.8 *8*	
γ	399.2 *7*	
γ	473.0 *7*	75 *15*
γ	551.1 *8*	

Photons (^{77}Zn)
(continued)

γ_{mode}	γ(keV)	γ(rel)
γ	683.4 *7*	
γ	853.6 *9*	
γ	926.4 *9*	
γ	1096.6 *9*	
γ	1666.1 *9*	
γ	1832.5 *9*	
γ	1836 *1*	

$^{77}_{31}$Ga(13.2 *2* s)

Mode: β-

Δ: -66410 *200* keV syst

SpA: 1.081×10^{10} Ci/g

Prod: fission

Photons (^{77}Ga)

γ$_{mode}$	γ(keV)	γ(rel)
γ (E3)	159.71 *10*	*
γ	197.3 *8*	
γ	243.3 *7*	
γ	402.7 *8*	
γ	421.5 *8*	
γ	459.4 *8*	48 *10*
γ [M1+E2]	470.0 *10*	100 *20*
γ	618.8 *7*	
γ	642.4 *8*	
γ	740.5 *10*	
γ	862.1 *6*	
γ	888.5 *10*	
γ	1242.9 *10*	
γ	1504.5 *8*	
γ	2187.9 *10*	

* with ^{77}Ge(53 s)

$^{77}_{32}$Ge(11.30 *1* h)

Mode: β-

Δ: -71215.4 *19* keV

SpA: 3.602×10^{6} Ci/g

Prod: ^{76}Ge(n,γ)

Photons (^{77}Ge)

⟨γ⟩=1022 *6* keV

γ$_{mode}$	γ(keV)	γ(%)†
As L$_\ell$	1.120	0.0019 *5*
As L$_\eta$	1.155	0.0011 *3*
As L$_\alpha$	1.282	0.034 *8*
As L$_\beta$	1.325	0.020 *6*
As L$_\gamma$	1.524	0.00027 *9*
As K$_{\alpha2}$	10.508	0.54 *6*
As K$_{\alpha1}$	10.544	1.04 *12*
As K$_{\beta1}$'	11.724	0.206 *23*
As K$_{\beta2}$'	11.948	0.0064 *7*
γ	150.20 *5*	0.036 *10*
γ E2	156.357 *16*	0.76 *3*
γ	159.03 *4*	0.218 *8*
γ	177.32 *3*	0.169 *5*
γ M1,E2	194.75 *3*	1.68 *5*
γ	208.712 *24*	0.892 *22*
γ M2+1.0%E3	211.020 *17*	29.2 *7*
γ M1+2.6%E2	215.483 *16*	27.1 *7*
γ	219.10 *3*	<0.28
γ	254.65 *12*	0.199 *4*
γ M1+E2	264.415 *14*	51.0
γ	267.879 *21*	<0.6
γ	313.49 *13*	0.0194 *5*
γ	326.06 *6*	0.026 *8*
γ	337.54 *6*	0.219 *5*
γ M1,E2	338.496 *20*	0.633 *15*
γ	338.6 *5*	
γ	350.06 *2*	0.0158 *5*
γ E1	367.377 *14*	13.3 *3*
γ	398.992 *21*	0.0908 *18*
γ E1	416.308 *16*	20.6 *4*
γ	419.73 *3*	1.164 *23*
γ	430.54 *12*	0.0097 *5*
γ	439.59 *7*	0.191 *4*
γ	444.50 *3*	0.0163 *5*
γ	461.345 *18*	1.198 *24*
γ	469.94 *5*	0.0077 *5*
γ E3	475.435 *16*	0.937 *19*
γ	503.92 *11*	0.0663 *13*
γ	520.25 *3*	0.28 *3*
γ	531.51 *16*	0.0439 *9*
γ	534.99 *4*	~0.0010
γ	557.75 *11*	~0.041
γ M1+1.9%E2	557.979 *14*	15.2 *3*
γ	569.43 *6*	0.0745 *15*
γ	582.519 *19*	0.738 *15*
γ	610.739 *23*	0.0582 *12*
γ	613.82 *3*	0.084 *9*
γ	614.474 *21*	0.50 *4*
γ	624.69 *12*	0.172 *3*
γ E1	631.791 *13*	6.59 *13*
γ	634.46 *4*	1.97 *4*
γ	639.16 *12*	0.0383 *8*
γ	655.00 *5*	0.0117 *5*
γ	659.670 *25*	0.0291 *6*
γ	665.46 *3*	0.0046 *5*
γ	672.99 *3*	0.50 *5*
γ	673.63 *12*	0.126 *13*
γ	680.41 *3*	0.0367 *7*
γ	685.19 *5*	0.063 *7*
γ	685.36 *7*	0.024 *3*
γ	698.519 *25*	0.216 *4*
γ	705.18 *3*	0.101 *2*
γ	712.30 *3*	0.783 *16*
γ M1,E2	714.334 *14*	6.77 *14*
γ	730.62 *5*	0.0194 *5*
γ	743.60 *3*	0.168 *3*
γ	745.730 *16*	0.914 *18*
γ	749.841 *17*	0.836 *17*
γ	766.69 *2*	0.743 *15*
γ	775.85 *8*	0.0143 *5*
γ	781.237 *16*	0.959 *19*
γ	784.66 *3*	1.244 *25*
γ	789.01 *3*	0.0918 *18*
γ M1,E2	794.30 *3*	0.262 *5*
γ	798.88 *6*	0.0454 *9*
γ	802.859 *24*	0.0265 *5*
γ	810.326 *16*	2.15 *4*
γ	813.346 *23*	0.1245 *25*
γ	823.18 *4*	0.568 *11*
γ	825.856 *19*	0.041 *10*
γ	843.172 *20*	0.197 *4*
γ	857.57 *15*	0.0291 *6*
γ	875.152 *21*	0.740 *15*
γ	883.963 *24*	0.0148 *5*
γ	889.11 *12*	0.0133 *5*
γ [M1+E2]	896.473 *17*	0.1158 *23*
γ	900.66 *11*	0.1143 *23*
γ	906.959 *17*	0.900 *18*
γ	913.789 *21*	0.346 *7*
γ	921.013 *23*	0.0673 *13*
γ	923.045 *24*	0.653 *13*
γ	924.275 *25*	
γ [E1]	925.353 *17*	0.68 *7*
γ	926.05 *8*	0.061 *6*
γ	928.72 *7*	0.988 *20*
γ	939.21 *4*	0.269 *5*
γ	945.70 *11*	0.0291 *6* ?
γ	946.03 *7*	0.0291 *6* ?
γ	959.195 *25*	0.0663 *13*
γ	966.63 *13*	0.036 *10*
γ	970.33 *12*	0.026 *8*
γ	974.284 *19*	
γ	985.73 *6*	0.0934 *19*
γ	996.52 *4*	0.0995 *20*
γ	1006.47 *6*	0.0128 *5* ?
γ	1021.96 *3*	0.0056 *5*
γ	1030.3 *10*	
γ [E2]	1052.828 *20*	0.0296 *6*
γ	1055.23 *3*	
γ	1055.59 *5*	
γ	1061.67 *3*	0.142 *3*
γ	1080.68 *5*	0.228 *5*
γ	1085.07 *7*	5.72 *11*
γ	1104.16 *3*	0.0321 *6*
γ	1114.76 *5*	0.0974 *19*
γ	1124.939 *25*	0.1112 *22*
γ	1129.61 *5*	
γ	1134.71 *15*	0.0260 *5*
γ	1151.839 *22*	0.185 *4*
γ	1155.45 *15*	0.0184 *10*
γ [E1]	1165.07 *11*	0.0464 *9*
γ	1186.34 *3*	0.0352 *7*
γ	1193.228 *16*	2.43 *5*
γ	1201.21 *6*	0.0770 *15*
γ	1215.43 *3*	0.1204 *24*
γ	1234.50 *13*	0.0265 *5*
γ	1242.159 *18*	0.377 *8*
γ [E1]	1263.845 *16*	0.803 *16*
γ	1279.93 *3*	0.164 *3*
γ	1295.38 *5*	0.057 *6*
γ	1296.09 *7*	0.085 *9*
γ	1309.251 *20*	0.460 *9*
γ [E1]	1312.776 *17*	0.339 *7*
γ	1319.64 *3*	0.285 *6*
γ	1323.262 *21*	0.0153 *5*
γ	1326.12 *3*	0.0362 *7*
γ	1339.209 *19*	0.0648 *13*
γ	1354.05 *11*	0.0138 *5*
γ	1358.182 *23*	0.0276 *6*
γ	1365.55 *13*	
γ	1368.299 *17*	3.2 *3*
γ	1385.615 *25*	0.0071 *5*
γ	1397.3 *5*	0.0071 *5* ?
γ	1410.7 *10*	
γ	1452.50 *12*	0.1143 *23*
γ	1454.52 *6*	0.0352 *7*
γ	1464.99 *12*	0.056 *6*
γ	1466.45 *3*	0.057 *6*
γ	1476.31 *4*	0.229 *5*
γ	1478.94 *3*	0.120 *13*
γ	1478.985 *21*	0.081 *7*
γ	1495.563 *16*	0.471 *9*
γ [E1]	1528.255 *18*	0.0439 *9*
γ	1538.742 *18*	0.135 *3*
γ	1556.94 *3*	0.0117 *5*
γ	1569.43 *4*	0.0541 *11*
γ	1573.662 *20*	0.622 *12*
γ	1580.9 *10*	
γ	1624.12 *13*	0.0051 *10*
γ	1639.60 *13*	0.0061 *15*
γ	1642.93 *11*	0.0199 *15*
γ	1709.81 *2*	0.290 *6*
γ	1719.63 *3*	0.377 *8*
γ	1722.30 *3*	0.0485 *10*
γ	1727.125 *24*	0.140 *3*
γ	1735.668 *20*	0.0362 *7*
γ	1759.15 *12*	~0.006
γ	1784.599 *21*	0.0071 *5*
γ	1792.47 *13*	0.0454 *10*
γ	1809.79 *13*	0.036 *5*
γ	1828.69 *21*	0.009 *3*
γ	1831.36 *21*	0.0260 *5*
γ	1846.354 *21*	0.162 *3*
γ	1878.65 *4*	0.0347 *7*
γ	1881.44 *5*	0.0122 *15*
γ	1912.02 *11*	0.0224 *5*
γ	1929.34 *11*	0.0250 *10*
γ	1948.83 *13*	0.0077 *10*
γ	2000.077 *18*	0.530 *11*
γ	2037.79 *5*	0.0582 *12*
γ	2077.176 *21*	0.220 *4*
γ	2089.67 *3*	0.328 *7*
γ	2126.107 *22*	0.193 *4*
γ	2248.80 *5*	0.0168 *20*
γ	2279.39 *11*	0.0071 *10*
γ	2328.32 *11*	0.0209 *10*
γ	2341.584 *20*	0.446 *9*
γ	2354.07 *3*	0.0046 *5*

† 0.98% uncert(syst)

Atomic Electrons (^{77}Ge)

$\langle e \rangle$=8.1 *9* keV

e_{bin}(keV)	$\langle e \rangle$(keV)	e(%)
1 - 12	0.169	4.2 *4*
138 - 183	0.322	0.204 *14*
193 - 197	0.06	~0.032
199	3.79	1.90 *6*
204	0.633	0.311 *10*
207	0.013	~0.006
209	0.428	0.204 *7*
210 - 243	0.218	0.103 *3*
253	1.7	~0.7
254 - 256	0.009	~0.003
263	0.19	~0.07
264 - 313	0.034	0.013 *7*
314 - 356	0.090	0.0258 *9*
366 - 415	0.123	0.0306 *18*
416 - 464	0.052	0.0114 *13*
468 - 508	0.0073	0.00151 *21*
519 - 568	0.104	0.0191 *5*
569 - 614	0.0090	0.0015 *4*
620 - 669	0.032	0.0051 *8*
671 - 719	0.0444	0.0063 *3*
729 - 777	0.019	0.0025 *5*
780 - 826	0.013	0.0016 *4*
831 - 877	0.0033	0.00038 *11*
882 - 929	0.0120	0.00132 *20*
934 - 974	0.00086	9.0 *16* $\times10^{-5}$
984 - 1022	0.00035	3.6 *10* $\times10^{-5}$
1041 - 1085	0.015	0.0014 *5*
1092 - 1140	0.00103	9.2 *17* $\times10^{-5}$
1144 - 1193	0.0059	0.00050 *15*
1200 - 1242	0.0012	9.4 *23* $\times10^{-5}$
1252 - 1301	0.0030	0.00023 *3*
1308 - 1357	0.0066	0.00049 *15*
1358 - 1397	0.00070	5.1 *15* $\times10^{-5}$
1441 - 1484	0.0020	0.000139 *21*
1494 - 1574	0.0016	0.000101 *20*
1612 - 1710	0.00113	6.7 *13* $\times10^{-5}$
1715 - 1810	0.00050	2.9 *4* $\times10^{-5}$
1817 - 1912	0.00040	2.2 *4* $\times10^{-5}$
1917 - 2000	0.00081	4.1 *10* $\times10^{-5}$
2026 - 2125	0.00111	5.3 *7* $\times10^{-5}$
2237 - 2330	0.00057	2.5 *6* $\times10^{-5}$
2340 - 2354	6.7 $\times10^{-5}$	2.9 *6* $\times10^{-6}$

Continuous Radiation (^{77}Ge)

$\langle\beta-\rangle$=664 keV;$\langle$IB$\rangle$=1.24 keV

E_{bin}(keV)		$\langle\ \rangle$(keV)	(%)
0 - 10	β-	0.0444	0.88
	IB	0.027	
10 - 20	β-	0.133	0.89
	IB	0.026	0.18
20 - 40	β-	0.54	1.79
	IB	0.050	0.18
40 - 100	β-	3.84	5.5
	IB	0.138	0.21
100 - 300	β-	36.5	18.3
	IB	0.35	0.20
300 - 600	β-	110	24.6
	IB	0.32	0.077
600 - 1300	β-	326	36.2
	IB	0.28	0.035
1300 - 2500	β-	188	11.8
	IB	0.041	0.0027
2500 - 2703	β-	0.478	0.0187
	IB	5.5 $\times10^{-6}$	2.2 $\times10^{-7}$

$^{77}_{32}$Ge(52.9 *6* s)

Mode: β-(79 *2* %), IT(21 *2* %)
Δ: -71055.7 *19* keV
SpA: 2.75$\times10^{9}$ Ci/g
Prod: ^{76}Ge(n,γ)

Photons (^{77}Ge)

$\langle\gamma\rangle$=65 *7* keV

γ_{mode}	γ(keV)	γ(%)†
Ge L_ℓ	1.037	0.0048 *11*
Ge L_η	1.068	0.0028 *8*
Ge L_α	1.188	0.082 *19*
Ge L_β	1.226	0.050 *14*
Ge L_γ	1.412	0.00054 *17*
Ge $K_{\alpha2}$	9.855	1.30 *14*
Ge $K_{\alpha1}$	9.886	2.5 *3*
As $K_{\alpha2}$	10.508	0.043 *6*
As $K_{\alpha1}$	10.544	0.084 *12*
Ge $K_{\beta1}$'	10.980	0.49 *5*
Ge $K_{\beta2}$'	11.184	0.0081 *9*
As $K_{\beta1}$'	11.724	0.0166 *23*
As $K_{\beta2}$'	11.948	0.00052 *7*
γ_{IT} (E3)	159.71 *10*	11.3
$\gamma_{\beta-}$M1,E2	194.75 *3*	0.48 *4*
$\gamma_{\beta-}$M1+2.6%E2	215.483 *16*	21
$\gamma_{\beta-}$	350.06 *2*	0.00132 *13*
$\gamma_{\beta-}$	398.992 *21*	0.0076 *7*
$\gamma_{\beta-}$	419.73 *3*	0.097 *9*
$\gamma_{\beta-}$	614.474 *21*	0.041 *5*

† uncert(syst): 9.5% for IT, 14% for β-

Atomic Electrons (^{77}Ge)

$\langle e \rangle$=15.4 *12* keV

e_{bin}(keV)	$\langle e \rangle$(keV)	e(%)
1	0.098	7.9 *11*
2	0.00021	0.0136 *23*
8	0.090	1.07 *15*
9	0.172	2.0 *3*
10	0.075	0.77 *11*
11	0.0040	0.037 *5*
12	0.000119	0.00103 *17*
149	12.0	8.1 *8*
158	2.02	1.28 *13*
160	0.33	0.208 *21*
183	0.032	0.0174 *16*
193	0.0038	0.00199 *19*
195	0.00066	0.00034 *3*
204	0.49	0.24 *3*
214	0.055	0.026 *4*
215	0.0096	0.0044 *6*
338	2.0 $\times10^{-5}$	~6$\times10^{-6}$
349	2.2 $\times10^{-6}$	~6$\times10^{-7}$
350	3.9 $\times10^{-7}$	~1$\times10^{-7}$
387	8.7 $\times10^{-5}$	~2$\times10^{-5}$
397	8.8 $\times10^{-6}$	~2$\times10^{-6}$
398	7.5 $\times10^{-7}$	~2$\times10^{-7}$
399	1.6 $\times10^{-6}$	~4$\times10^{-7}$
408	0.0010	~0.00025
418	0.00011	~3$\times10^{-5}$
420	1.9 $\times10^{-5}$	~5$\times10^{-6}$
603	0.00021	~3$\times10^{-5}$
613	2.2 $\times10^{-5}$	~4$\times10^{-6}$
614	3.8 $\times10^{-6}$	~6$\times10^{-7}$

Continuous Radiation (^{77}Ge)

$\langle\beta-\rangle$=933 keV;$\langle$IB$\rangle$=2.3 keV

E_{bin}(keV)		$\langle\ \rangle$(keV)	(%)
0 - 10	β-	0.0058	0.115
	IB	0.033	
10 - 20	β-	0.0178	0.119
	IB	0.033	0.23
20 - 40	β-	0.075	0.248
	IB	0.064	0.22
40 - 100	β-	0.60	0.85
	IB	0.18	0.28
100 - 300	β-	8.4	4.06
	IB	0.52	0.29
300 - 600	β-	43.1	9.4
	IB	0.57	0.134
600 - 1300	β-	295	30.7
	IB	0.71	0.083
1300 - 2500	β-	569	32.7
	IB	0.22	0.0138
2500 - 2863	β-	16.1	0.62
	IB	0.00042	1.65 $\times10^{-5}$

$^{77}_{33}$As(1.6179 *21* d)

Mode: β-
Δ: -73918.5 *22* keV
SpA: 1.0481$\times10^{6}$ Ci/g
Prod: daughter ^{77}Ge

Photons (^{77}As)

$\langle\gamma\rangle$=7.9 *10* keV

γ_{mode}	γ(keV)	γ(%)†
γ [E1]	62.03 *6*	0.0040 *16*
γ M1+50%E2	81.18 *4*	0.00045 *12*
γ E1+1.0%M2	87.86 *5*	0.194 *13*
γ [E2]	125.62 *7*	0.0007 *3*
γ M1+50%E2	139.08 *6*	0.0100 *19*
γ [M1+E2]	141.61 *6*	<1.4$\times10^{-5}$
γ E3	161.92 *5*	0.161 *19* *
γ M1+50%E2	189.65 *5*	~1$\times10^{-6}$
γ M1+0.9%E2	200.46 *4*	0.00067 *17*
γ M1+3.3%E2	238.97 *4*	1.6
γ E2	249.78 *4*	0.410 *23*
γ M1+8.7%E2	270.83 *4*	0.0068 *5*
γ [E1]	277.51 *5*	1.8 *4* $\times10^{-5}$
γ M1+1.6%E2	281.64 *4*	0.048 *4*
γ (M1+E2)	331.26 *6*	<0.00035 ?
γ (E2)	342.08 *6*	<3$\times10^{-5}$
γ [E1]	405.67 *10*	<4$\times10^{-5}$
γ [E1]	419.12 *7*	<9$\times10^{-5}$
γ E2	439.43 *4*	0.00086 *20*
γ M1+3.3%E2	520.61 *4*	0.47 *3*

† 25% uncert(syst)
* with ^{77}Se(17.4 s) in equilib

Continuous Radiation (^{77}As)

$\langle\beta-\rangle$=226 keV;$\langle$IB$\rangle$=0.164 keV

E_{bin}(keV)		$\langle\ \rangle$(keV)	(%)
0 - 10	β-	0.119	2.37
	IB	0.0113	
10 - 20	β-	0.357	2.38
	IB	0.0105	0.073
20 - 40	β-	1.43	4.78
	IB	0.019	0.067
40 - 100	β-	10.2	14.6
	IB	0.044	0.070

Continuous Radiation (^{77}As) (continued)

E_{bin}(keV)		⟨ ⟩(keV)	(%)
100 - 300	β-	89	45.3
	IB	0.066	0.041
300 - 600	β-	122	30.1
	IB	0.0126	0.0035
600 - 683	β-	3.02	0.486
	IB	1.9×10^{-5}	3.0×10^{-6}

$^{77}_{34}$Se(stable)

Δ: -74601.6 *15* keV
%: 7.6 *2*

$^{77}_{34}$Se(17.45 *10* s)

Mode: IT
Δ: -74439.7 *15* keV
SpA: 8.23×10^{9} Ci/g
Prod: ^{76}Se(n,γ); daughter ^{77}Br

Photons (^{77}Se)

⟨γ⟩=87.4 *19* keV

γ_{mode}	γ(keV)	γ(%)†
Se L_ℓ	1.204	0.025 *6*
Se L_η	1.245	0.015 *4*
Se L_α	1.379	0.48 *10*
Se L_β	1.426	0.29 *8*
Se L_γ	1.648	0.0036 *11*
Se $K_{\alpha2}$	11.182	6.8 *3*
Se $K_{\alpha1}$	11.222	13.2 *6*
Se $K_{\beta1}$'	12.494	2.66 *12*
Se $K_{\beta2}$'	12.741	0.128 *6*
γ E3	161.92 *5*	52.4

† 2.2% uncert(syst)

Atomic Electrons (^{77}Se)

⟨e⟩=72.4 *17* keV

e_{bin}(keV)	⟨e⟩(keV)	e(%)
1	0.50	35 *4*
2	0.032	1.94 *20*
9	0.104	1.12 *12*
10	1.09	11.2 *12*
11	0.37	3.4 *4*
12	0.0232	0.189 *20*
149	57.6	38.6 *11*
160	10.8	6.72 *20*
162	1.90	1.17 *4*

$^{77}_{35}$Br(2.37650 *25* d)

Mode: ε
Δ: -73237 *3* keV
SpA: 7.1356×10^{5} Ci/g
Prod: ^{75}As(α,2n)

Photons (^{77}Br)

⟨γ⟩=323 *3* keV

γ_{mode}	γ(keV)	γ(%)†
Se L_ℓ	1.204	0.055 *12*
Se L_η	1.245	0.033 *8*
Se L_α	1.379	1.06 *22*
Se L_β	1.427	0.64 *16*
Se L_γ	1.648	0.009 *3*
Se $K_{\alpha2}$	11.182	15.6 *6*
Se $K_{\alpha1}$	11.222	30.2 *12*
Se $K_{\beta1}$'	12.494	6.10 *25*
Se $K_{\beta2}$'	12.741	0.293 *12*
γ [E1]	62.03 *6*	~0.053
γ M1,E2	81.18 *4*	0.022 *6*
γ E1+1.0%M2	87.86 *5*	1.45 *4*
γ [E2]	125.62 *7*	0.0096 *12*
γ M1,E2	139.08 *6*	0.134 *5*
γ [M1+E2]	141.61 *6*	0.0026 *7*
γ [E1]	144.50 *7*	0.0060 *12*
γ E3	161.92 *5*	1.140 *23* *
γ M1,E2	180.71 *4*	0.294 *7*
γ M1,E2	187.29 *4*	0.0597 *24*
γ M1,E2	189.65 *5*	~0.0024
γ M1+0.9%E2	200.46 *4*	1.26 *5*
γ [M1+E2]	231.55 *8*	0.065 *5*
γ M1+3.3%E2	238.97 *4*	23.9
γ [E1]	243.39 *5*	0.038 *5*
γ E2	249.78 *4*	3.08 *7*
γ M1+8.7%E2	270.83 *4*	0.336 *12*
γ [E1]	277.51 *5*	0.0334 *24*
γ M1+1.6%E2	281.64 *4*	2.38 *5*
γ M1+2.9%E2	297.25 *4*	4.30 *19*
γ M1,E2	303.82 *4*	1.218 *24*
γ [E1]	325.22 *7*	0.024 *5*
γ (M1+E2)	331.26 *6*	0.069 *7*
γ (E2)	342.08 *6*	0.0065 *12*
γ [E2]	378.43 *4*	0.062 *5*
γ M1,E2	378.93 *8*	~0.010
γ M1,E2	385.00 *4*	0.865 *24*
γ [E1]	390.87 *6*	0.023 *3*
γ [E1]	405.67 *10*	0.0076 *14*
γ [E1]	419.12 *7*	0.0170 *22*
γ [E2]	424.10 *6*	0.0227 *24*
γ E2	439.43 *4*	1.62 *4*
γ [E1]	472.05 *6*	0.0081 *22*
γ M1+6.6%E2	484.54 *4*	1.034 *24*
γ [E2]	504.56 *10*	0.0093 *19*
γ M1,E2	518.01 *8*	0.17 *5*
γ M1+3.3%E2	520.61 *4*	23.1 *5*
γ [E1]	523.43 *6*	0.041 *7*
γ M1,E2	565.72 *5*	0.442 *14*
γ [E2]	568.07 *5*	0.886 *19*
γ	574.65 *4*	1.228 *25*
γ M1+2.8%E2	578.89 *4*	3.06 *7*
γ	585.47 *4*	1.62 *3*
γ [M1+E2]	610.48 *6*	0.0222 *22*
γ	662.45 *9*	0.0836 *24*
γ [E1]	704.15 *6*	0.0165 *19*
γ [E2]	749.55 *6*	0.031 *3*
γ M1+8.9%E2	755.37 *4*	1.72 *3*
γ M1,E2	766.18 *4*	0.0430 *24*
γ [M1+E2]	791.23 *8*	0.0096 *22*
γ M1,E2	817.85 *4*	2.15 *5*
γ	824.43 *4*	0.0136 *14*
γ	885.73 *9*	0.0086 *10*
γ [E1]	911.47 *6*	0.0026 *5*
γ [E1]	929.66 *10*	0.0029 *10*
γ	947.76 *11*	0.00072 *24*
γ [M1+E2]	980.88 *9*	0.0038 *7*
γ [M1+E2]	991.69 *9*	0.0229 *12*
γ M1,E2	1005.14 *4*	0.956 *19*
γ	1186.73 *11*	0.0017 *5*
γ [E2]	1230.65 *9*	0.00096 *24*

† 2.1% uncert(syst)
* with ^{77}Se(17.4 s) in equilib

Atomic Electrons (^{77}Br)

⟨e⟩=7.8 *3* keV

e_{bin}(keV)	⟨e⟩(keV)	e(%)
1	1.05	73 *7*
2 - 9	0.319	7.4 *6*
10	2.50	26 *3*
11	0.85	7.7 *8*
69 - 113	0.168	0.220 *19*
124 - 144	0.0258	0.0202 *9*
149	1.25	0.840 *24*
160 - 200	0.349	0.212 *4*
219	0.0030	~0.0014
226	0.536	0.237 *8*
230 - 232	0.00088	0.00038 *10*
237	0.244	0.103 *3*
238 - 285	0.160	0.0596 *14*
291 - 341	0.0468	0.0159 *5*
342 - 391	0.0174	0.00467 *16*
393 - 439	0.0293	0.00686 *18*
459 - 508	0.181	0.0357 *10*
511 - 555	0.0336	0.00633 *13*
562 - 610	0.043	0.0075 *11*
650 - 691	0.00049	7 *3* $\times10^{-5}$
702 - 750	0.00833	0.00112 *3*
754 - 791	0.00131	0.000174 *5*
805 - 824	0.0112	0.00139 *5*
873 - 917	4.0×10^{-5}	4.6 *14* $\times10^{-6}$
928 - 968	1.7×10^{-5}	1.7 *3* $\times10^{-6}$
979 - 1005	0.00390	0.000392 *12*
1174 - 1218	6.9×10^{-6}	5.8 *16* $\times10^{-7}$
1229 - 1231	3.4×10^{-7}	2.8 *6* $\times10^{-8}$

Continuous Radiation (^{77}Br)

⟨β+⟩=1.12 keV;⟨IB⟩=0.35 keV

E_{bin}(keV)		⟨ ⟩(keV)	(%)
0 - 10	β+	1.73×10^{-5}	0.000214
	IB	0.00113	
10 - 20	β+	0.000372	0.00230
	IB	0.0057	0.048
20 - 40	β+	0.0056	0.0176
	IB	0.00042	0.00149
40 - 100	β+	0.120	0.161
	IB	0.0019	0.0027
100 - 300	β+	0.98	0.55
	IB	0.032	0.0154
300 - 600	β+	0.0216	0.0069
	IB	0.122	0.027
600 - 1300	IB	0.18	0.023
1300 - 1365	IB	0.000148	1.12×10^{-5}
	Σβ+		0.74

$^{77}_{35}$Br(4.28 *10* min)

Mode: IT
Δ: -73131 *3* keV
SpA: 5.70×10^{8} Ci/g
Prod: ^{76}Se(p,γ)

Photons (^{77}Br)

⟨γ⟩=19.3 *4* keV

γ_{mode}	γ(keV)	γ(%)†
Br L_ℓ	1.293	0.049 *11*
Br L_η	1.339	0.030 *8*
Br L_α	1.481	0.99 *20*
Br L_β	1.532	0.59 *15*
Br L_γ	1.777	0.0068 *21*
Br $K_{\alpha2}$	11.878	12.1 *6*

Photons (^{77}Br)
(continued)

γ_{mode}	γ(keV)	γ(%)†
Br $K_{\alpha 1}$	11.924	23.4 11
Br $K_{\beta 1}$'	13.290	4.84 23
Br $K_{\beta 2}$'	13.562	0.328 15
γ E3	105.81 8	13.6

† 2.2% uncert(syst)

Atomic Electrons (^{77}Br)
$\langle e \rangle$=85.8 19 keV

e_{bin}(keV)	$\langle e \rangle$(keV)	e(%)
2	1.06	67 7
10	1.94	18.9 20
11	0.058	0.51 5
12	0.62	5.3 6
13	0.052	0.40 4
92	60.8	65.9 20
104	18.0	17.3 5
106	3.28	3.10 9

$^{77}_{36}$Kr(1.24 1 h)

Mode: ϵ
Δ: -70227 29 keV
SpA: 3.28×10^7 Ci/g
Prod: ^{79}Br(p,3n); ^{74}Se(α,n)

Photons (^{77}Kr)
$\langle \gamma \rangle$=214 4 keV

γ_{mode}	γ(keV)	γ(%)†
Br L_ℓ	1.293	0.019 5
Br L_η	1.339	0.011 3
Br L_α	1.481	0.38 9
Br L_β	1.533	0.22 6
Br L_γ	1.777	0.0029 9
Br $K_{\alpha 2}$	11.878	4.9 4
Br $K_{\alpha 1}$	11.924	9.4 8
Br $K_{\beta 1}$'	13.290	1.95 16
Br $K_{\beta 2}$'	13.562	0.132 11
γ (E2)	23.94 10	<0.04
γ	77.0 20	11.0 6 ?
γ E3	105.81 8	1.28 8 *
γ E1	129.75 8	80
γ M1,E2	146.40 9	37.6 16
γ (M1+50%E2)	161.9 3	0.22 3
γ	166.6 3	0.160 24
γ E1	276.15 10	2.88 16
γ M1,E2	288.12 15	0.184 16
γ M1,E2	312.06 15	3.4 8
γ	588.43 17	0.11 4
γ M1,E2	606.60 21	0.37 5
γ	698.0 3	0.048 8
γ	734.83 17	0.42 10
γ M1,E2	748.33 22	~0.032
γ	837.4 3	0.15 4
γ	861.4 3	0.20 4
γ	864.58 17	0.032 8
γ M1,E2	894.72 21	0.11 3
γ	968.1 3	0.024 8
γ	992.0 3	0.072 16
γ	1001.0 10	0.026 5 ?
γ	1031.4 4	0.014 6
γ	1158.33 21	0.048 8
γ	1300.05 20	0.38 7
γ	1446.44 19	0.11 3
γ	1479.8 5	0.0136 24

Photons (^{77}Kr)
(continued)

γ_{mode}	γ(keV)	γ(%)†
γ	1576.19 20	0.032 8
γ	1702.0 20	0.067 6 ?
γ	1999.4 4	0.027 8
γ	2031.9 7	0.0056 24
γ	2064.0 7	0.013 3
γ	2068.2 5	0.015 4
γ	2129.2 4	0.018 4
γ	2335 15	0.015 6 ?

† 1.3% uncert(syst)
* with ^{77}Br(4.3 min) in equilib

Atomic Electrons (^{77}Kr)
$\langle e \rangle$=20.1 6 keV

e_{bin}(keV)	$\langle e \rangle$(keV)	e(%)
2	0.39	25 4
10	1.0	9 3
11 - 24	0.61	3.9 12
92	5.7	6.2 4
104	1.70	1.63 11
106	0.310	0.293 19
116	3.24	2.79 7
128 - 130	0.451	0.352 7
133	5.6	4.20 23
145	0.76	0.53 3
146 - 167	0.187	0.127 10
263 - 312	0.143	0.049 7
575 - 607	0.0040	0.00068 10
685 - 733	0.0023	0.00032 13
735 - 748	0.00024	3.3 15 $\times10^{-5}$
824 - 865	0.0015	0.00018 5
881 - 895	0.00056	6.4 16 $\times10^{-5}$
955 - 1001	0.00040	4.1 11 $\times10^{-5}$
1018 - 1031	4.3 $\times10^{-5}$	~4 $\times10^{-6}$
1145 - 1158	0.00013	1.2 4 $\times10^{-5}$
1287 - 1300	0.0009	7 3 $\times10^{-5}$
1433 - 1480	0.00028	1.9 7 $\times10^{-5}$
1563 - 1576	6.5 $\times10^{-5}$	4.2 15 $\times10^{-6}$
1689 - 1702	0.00013	7.6 22 $\times10^{-6}$
1986 - 2068	0.000102	5.1 10 $\times10^{-6}$
2116 - 2129	2.9 $\times10^{-5}$	1.4 4 $\times10^{-6}$
2322 - 2335	2.3 $\times10^{-5}$	1.0 4 $\times10^{-6}$

Continuous Radiation (^{77}Kr)
$\langle\beta+\rangle$=681 keV;$\langle$IB$\rangle$=1.9 keV

E_{bin}(keV)		$\langle\ \rangle$(keV)	(%)
0 - 10	β+	4.36 $\times10^{-5}$	0.00054
	IB	0.028	
10 - 20	β+	0.00101	0.0062
	IB	0.028	0.20
20 - 40	β+	0.0170	0.053
	IB	0.053	0.18
40 - 100	β+	0.486	0.64
	IB	0.147	0.23
100 - 300	β+	15.9	7.3
	IB	0.39	0.22
300 - 600	β+	98	21.3
	IB	0.38	0.091
600 - 1300	β+	455	49.8
	IB	0.49	0.057
1300 - 2500	β+	113	7.8
	IB	0.38	0.022
2500 - 3010	IB	0.0071	0.00028
	$\Sigma\beta$+		87

$^{77}_{37}$Rb(3.70 15 min)

Mode: ϵ
Δ: -64950 40 keV
SpA: 6.6×10^8 Ci/g
Prod: protons on Y-Nb targets; ^{63}Cu(^{16}O,2n); ^{59}Co(^{20}Ne,2n); ^{79}Br(^{3}He,5n)

Photons (^{77}Rb)
$\langle \gamma \rangle$=807 36 keV

γ_{mode}	γ(keV)	γ(%)†
Kr L_ℓ	1.383	0.016 4
Kr L_η	1.435	0.0094 25
Kr L_α	1.581	0.34 8
Kr L_β	1.639	0.20 5
Kr L_γ	1.907	0.0028 9
Kr $K_{\alpha 2}$	12.598	4.3 4
Kr $K_{\alpha 1}$	12.651	8.4 8
Kr $K_{\beta 1}$'	14.109	1.78 16
Kr $K_{\beta 2}$'	14.408	0.160 14
γ M1+50%E2	39.68 15	0.29 12 ?
γ E1	66.50 4	59
γ	77.7 6	0.12 4
γ	106 1	0.09 4
γ	129.4 7	~0.06 ?
γ M1,E2	149.94 8	4.7 5
γ M1,E2	178.82 5	24.2 6
γ [E2]	214.55 8	~0.06 ?
γ	236.66 13	0.41 12
γ [E1]	245.32 6	1.3 6
γ	254.23 14	1.89 18
γ	254.50 9	0.71 12
γ	289.9 9	0.18 6
γ	306.39 24	0.41 18
γ	354.59 20	0.35 12
γ	362 3	5.9 12 ?
γ [M1+E2]	393.37 7	11.4 9
γ	433.05 14	1.00 12
γ	469.05 7	1.24 12
γ	510.7 6	~0.8
γ	524.95 24	~0.12
γ	545.21 13	1.18 12
γ	553.2 5	0.35 12
γ	568.2 3	0.18 6
γ	577.0 5	1.24 18 ?
γ	597.2 3	0.44 9
γ	608.39 23	2.7 6
γ	609.12 17	0.65 18
γ	617.68 13	~0.18
γ	626.69 6	3.3 3
γ	635.06 11	0.24 6
γ	647.87 7	2.8 3
γ	666.9 6	1.36 24 ?
γ	674.89 23	0.41 12
γ	691 2	~0.6 ?
γ	712.52 8	0.28 7
γ	713.32 19	0.28 7
γ	724.03 14	0.53 12
γ	729.7 7	~0.24
γ	744.81 11	0.189 24
γ	747.1 3	~0.12
γ	756.9 8	~0.12 ?
γ	776.1 7	0.47 12
γ	779.5 6	1.12 18
γ	783.20 17	0.71 18
γ	792.11 7	0.32 9
γ	800.94 7	0.55 12
γ	805.51 6	2.1 6
γ	805.6 7	0.59 18
γ	826.0 15	~0.06
γ	834.6 7	~0.24
γ	852.56 18	0.83 6
γ	860.0 7	0.65 12
γ	872.01 7	0.29 12
γ	910.23 9	0.65 6
γ	946.6 5	0.41 12
γ	958.73 9	1.8 3
γ	966.36 7	0.68 6
γ	970.93 6	2.8 4

Photons (^{77}Rb) (continued)

γ_{mode}	γ(keV)	γ(%)†
γ	988.32 7	2.0 3
γ	991.71 14	0.307 24
γ	1013.1 5	1.12 24
γ	1024.8 6	0.47 12
γ	1037.43 6	1.30 18
γ	1067.91 8	0.75 12
γ	1099 3	3.0 6 ?
γ	1108.66 12	0.35 12
γ	1124.00 9	0.201 18
γ	1153.74 7	1.10 15
γ	1176.57 16	0.57 12
γ	1199.17 12	0.27 9
γ	1235.22 14	0.083 18
γ	1243.07 16	0.165 24
γ	1267 3	2.4 6 ?
γ	1311.37 7	0.336 18
γ	1328 3	9.4 24 ?
γ	1359.6 7	~0.18
γ	1377.6 7	0.41 18
γ	1408.0 12	~0.06
γ	1427.16 10	0.20 7
γ	1447.7 12	~0.12
γ	1509.5 7	~0.18
γ	1541.5 15	0.29 12
γ	1605.98 9	0.425 18
γ	1631.5 15	~0.24
γ	1662.3 12	~0.18
γ	1668.21 10	0.407 24
γ	1672.48 10	0.29 6
γ	1715.6 8	~0.06 ?
γ	1801.89 24	0.136 24
γ	1838.36 8	0.71 6
γ	2339.87 18	0.124 18
γ	2508.15 22	0.130 18
γ	2577.31 21	0.29 3
γ	2594.37 17	0.195 18
γ	2762.37 22	0.59 6
γ	2822.62 21	0.142 18

† 15% uncert(syst)

Atomic Electrons (^{77}Rb)

⟨e⟩=16.6 8 keV

e_{bin}(keV)	⟨e⟩(keV)	e(%)
2 - 10	0.41	21 3
11	0.59	5.4 7
12 - 14	0.27	2.19 23
25	0.8	3.1 12
38	0.36	1.0 4
39 - 40	0.069	0.17 6
52	8.2	15.8 14
63	0.08	~0.13
65	1.13	1.75 15
66 - 115	0.28	0.39 5
127 - 129	0.0021	~0.0016
136	0.69	0.51 5
148 - 150	0.115	0.077 7
164	2.37	1.44 6
177 - 222	0.397	0.222 10
231 - 276	0.14	0.06 3
288 - 306	0.012	~0.004
340 - 379	0.33	0.089 25
391 - 433	0.041	0.0104 23
455 - 496	0.021	0.0046 19
509 - 554	0.017	0.0031 10
563 - 609	0.039	0.0066 18
612 - 661	0.052	0.0083 21
665 - 713	0.010	0.0015 4
715 - 762	0.0059	0.00079 20
765 - 812	0.026	0.0033 8
820 - 860	0.0085	0.00101 24
870 - 910	0.0027	0.00030 11
932 - 977	0.035	0.0036 8
986 - 1036	0.0116	0.00115 24
1037 - 1085	0.011	0.0010 3
1094 - 1139	0.0057	0.00051 12
1152 - 1199	0.0028	0.00024 6
1221 - 1267	0.0071	0.00057 20
1297 - 1345	0.026	0.0020 7
1358 - 1406	0.0012	9 4 $\times10^{-5}$
1408 - 1448	0.0008	5.4 18 $\times10^{-5}$
1495 - 1592	0.0019	0.00012 3
1604 - 1701	0.0026	0.00016 3
1714 - 1802	0.00028	1.6 5 $\times10^{-5}$
1824 - 1838	0.0014	7.6 21 $\times10^{-5}$
2326 - 2340	0.00021	8.8 24 $\times10^{-6}$
2494 - 2593	0.00096	3.7 6 $\times10^{-5}$
2748 - 2822	0.00110	4.0 8 $\times10^{-5}$

Continuous Radiation (^{77}Rb)

⟨β+⟩=1653 keV;⟨IB⟩=5.2 keV

E_{bin}(keV)		⟨ ⟩(keV)	(%)
0 - 10	β+	8.6 $\times10^{-6}$	0.000106
	IB	0.052	
10 - 20	β+	0.000208	0.00128
	IB	0.051	0.36
20 - 40	β+	0.00363	0.0113
	IB	0.101	0.35
40 - 100	β+	0.109	0.143
	IB	0.29	0.45
100 - 300	β+	4.02	1.84
	IB	0.87	0.48
300 - 600	β+	29.7	6.4
	IB	1.05	0.25
600 - 1300	β+	257	26.3
	IB	1.61	0.19
1300 - 2500	β+	866	46.7
	IB	1.09	0.064
2500 - 5000	β+	496	16.9
	IB	0.132	0.0046
	$\Sigma\beta$+		98

$^{77}_{38}$Sr(9 1 s)

Mode: ϵ, ϵp(<0.25 %)
Δ: -57890 150 keV
SpA: 1.57$\times10^{10}$ Ci/g
Prod: ^{40}Ca(^{40}Ca,2pn)
ϵp: 1000-3500 range

Photons (^{77}Sr)

γ_{mode}	γ(keV)	γ(rel)
γ	144.75 10	11 2
γ	146.87 10	100
γ	159.9 1	7 2
γ	184.8 7	
γ	306.77 14	
γ	331.6 7	
γ	368.2 10	
γ	471 1	
γ	501.8 10	
γ	741.9 10	

A = 78

NDS 33, 189 (1981)

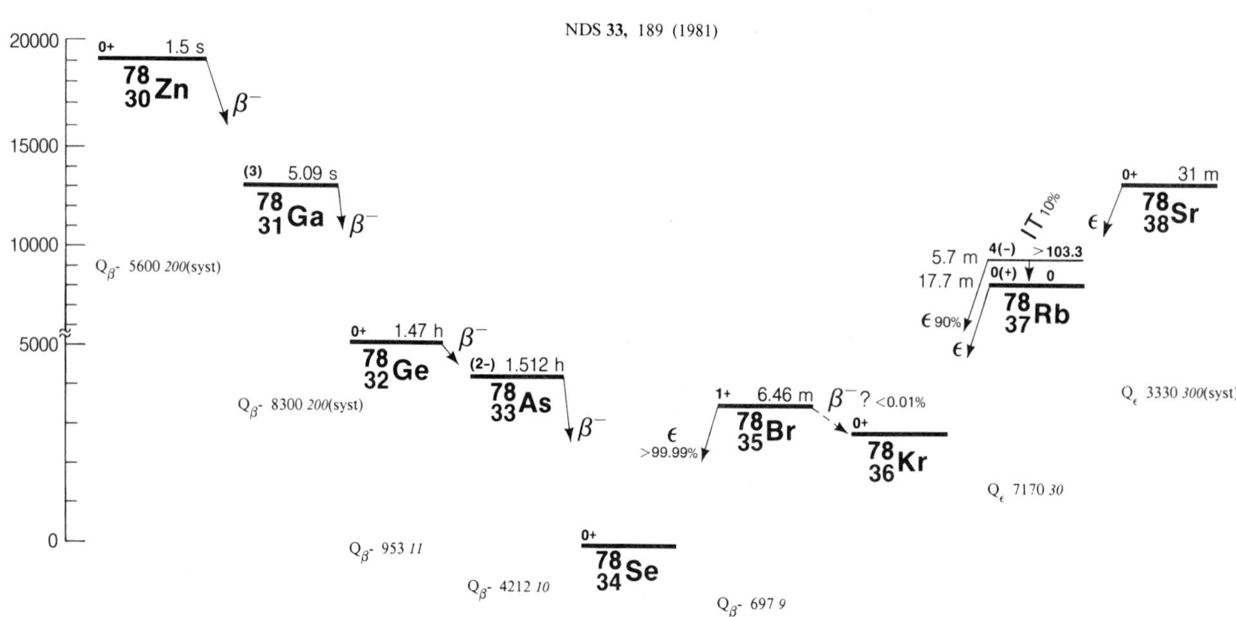

$^{78}_{30}$Zn(1.47 *15* s)

Mode: β-

Δ: -57960 *280* keV syst

SpA: 7.8×10^{10} Ci/g

Prod: fission

Photons (^{78}Zn)

⟨γ⟩=1529 *39* keV

γ_{mode}	γ(keV)	γ(%)†
γ	60.24 *3*	2.59 *18*
γ	92.13 *6*	0.59 *11*
γ	112.29 *3*	3.4 *3*
γ [M1+E2]	119.36 *3*	5.6 *3*
γ	132.63 *7*	0.7 *3*
γ [M1+E2]	170.72 *7*	1.7 *4*
γ	172.53 *4*	2.6 *4*
γ	181.68 *5*	28.1 *13*
γ	187.83 *7*	2.3 *3*
γ	224.76 *5*	43.9 *13*
γ	262.20 *17*	2.9 *3*
γ	275.26 *23*	0.8 *4*
γ	281.39 *8*	16.5 *10*
γ	303.35 *9*	3.2 *3*
γ	321.99 *14*	3.3 *4*
γ	341.62 *8*	2.7 *10*
γ	344.12 *6*	3.9 *12*
γ	354.21 *5*	7.3 *11*
γ	386.10 *7*	1.8 *3*
γ	395.48 *8*	0.79 *18*
γ	412.59 *8*	0.83 *18*
γ	440.23 *20*	~1.0
γ	446.34 *7*	1.0 *3*
γ	453.91 *8*	19.7 *9*
γ	537.46 *15*	1.01 *13*
γ	635.59 *8*	20.9 *11*
γ [M1+E2]	722.8 *3*	0.75 *22*
γ	727.72 *9*	0.8 *4*
γ	744.03 *21*	0.9 *4*
γ	749.69 *9*	3.9 *5*
γ	762.22 *16*	~0.5
γ [M1+E2]	788.28 *25*	1.2 *4*
γ	797.7 *6*	~0.6
γ	807.85 *17*	3.3 *5*
γ	818.40 *13*	1.8 *3*
γ	860.35 *8*	24.5 *11*
γ	909.00 *16*	2.2 *4*
γ	957.58 *15*	1.27 *22*
γ	979.71 *8*	6.6 *6*
γ [M1+E2]	1006.23 *12*	8.7 *6*
γ	1157.48 *14*	0.8 *3*
γ [M1+E2]	1174.59 *14*	~3
γ [M1+E2]	1345.30 *13*	5.0 *4*
γ [M1+E2]	1558.0 *4*	~0.31
γ	1570.06 *13*	1.1 *3*
γ [M1+E2]	1623.78 *25*	~0.4
γ [M1+E2]	1675.14 *24*	1.3 *3*
γ	1927.12 *24*	0.7 *3*
γ [M1+E2]	1970.7 *3*	0.7 *3*
γ	2026.7 *3*	1.2 *3*
γ	2103.3 *3*	~0.8
γ	2205.64 *14*	4.8 *5*
γ	2489.4 *3*	0.8 *3*
γ [M1+E2]	2564.2 *4*	1.0 *4*
γ	2626.8 *5*	1.0 *3*
γ [M1+E2]	2693.5 *4*	1.1 *3*
γ	3553.8 *4*	1.3 *4*

† 2.7% uncert(syst)

Continuous Radiation (^{78}Zn)

⟨β-⟩=1819 keV;⟨IB⟩=6.0 keV

E_{bin}(keV)		⟨ ⟩(keV)	(%)
0 - 10	β-	0.00354	0.070
	IB	0.054	
10 - 20	β-	0.0109	0.073
	IB	0.054	0.37
20 - 40	β-	0.0460	0.153
	IB	0.107	0.37
40 - 100	β-	0.375	0.53
	IB	0.31	0.47
100 - 300	β-	5.4	2.59
	IB	0.92	0.52
300 - 600	β-	28.9	6.3
	IB	1.13	0.27
600 - 1300	β-	224	23.0
	IB	1.8	0.21
1300 - 2500	β-	803	42.9
	IB	1.35	0.079
2500 - 4740	β-	758	24.5
	IB	0.29	0.0100

$^{78}_{31}$Ga(5.09 *5* s)

Mode: β-

Δ: -63560 *200* keV syst

SpA: 2.66×10^{10} Ci/g

Prod: fission

Photons (^{78}Ga)

⟨γ⟩=2539 *52* keV

γ_{mode}	γ(keV)	γ(%)†
γ	346.09 *17*	5.2 *8*
γ	458.17 *10*	5.9 *3*
γ	532.88 *22*	0.24 *7*
γ [M1+E2]	567.24 *9*	18.2 *9*
γ [E2]	619.53 *12*	77 *4*
γ	675.14 *15*	6.3 *3*
γ [M1+E2]	863.3 *3*	0.8 *4*
γ [E2]	892.2 *3*	~0.34
γ [E2]	927.5 *3*	1.00 *23*
γ [E2]	951.06 *15*	7.5 *4*
γ	963.2 *5*	~0.8
γ	1021.22 *14*	1.2 *3*
γ	1025.41 *10*	12.4 *7*
γ	1061.49 *20*	0.68 *20*
γ [E2]	1186.77 *12*	20.1 *9*
γ	1212.69 *16*	2.0 *3*
γ [M1+E2]	1223.59 *18*	4.6 *4*
γ [M1+E2]	1252.38 *15*	1.85 *23*
γ	1308.46 *24*	~0.35
γ [M1+E2]	1382.8 *3*	~0.7
γ	1476.13 *17*	1.4 *5*
γ	1479.39 *13*	8.2 *5*
γ [M1+E2]	1519.65 *19*	1.62 *23*
γ	1564.36 *21*	1.20 *18*
γ	1573.6 *3*	1.04 *19*
γ	1604.62 *19*	1.60 *23*
γ	1670.86 *15*	1.64 *20*
γ	1675.4 *3*	0.85 *18*
γ	1745.57 *20*	0.72 *17*
γ [M1+E2]	1819.61 *15*	2.8 *5*
γ	1819.92 *18*	0.8 *3*
γ	1934.29 *16*	9.3 *6*
γ	2043.43 *25*	1.32 *25*
γ	2046.62 *14*	5.5 *6*
γ	2238.09 *16*	1.2 *3*
γ	2241.1 *5*	0.8 *3* ?
γ [E2]	2333.85 *24*	1.8 *3*
γ	2358.5 *5*	1.03 *23*
γ	2501.52 *17*	2.5 *5* ?
γ [E2]	2501.59 *24*	~0.6
γ [E2]	2706.4 *2*	3.4 *4*

Photons (^{78}Ga)

(continued)

γ_{mode}	γ(keV)	γ(%)†
γ	2770.95 *19*	1.3 *3*
γ	3083.98 *21*	~0.39
γ	3093.2 *3*	1.8 *3*
γ	3464.7 *5*	4.4 *5*
γ	3508.1 *10*	~0.09 ?
γ	4459.1 *10*	1.31 *23*

† 5.2% uncert(syst)

Continuous Radiation (^{78}Ga)

⟨β-⟩=2587 keV;⟨IB⟩=10.5 keV

E_{bin}(keV)		⟨ ⟩(keV)	(%)
0 - 10	β-	0.00138	0.0273
	IB	0.066	
10 - 20	β-	0.00424	0.0282
	IB	0.065	0.45
20 - 40	β-	0.0179	0.060
	IB	0.130	0.45
40 - 100	β-	0.147	0.207
	IB	0.38	0.58
100 - 300	β-	2.18	1.04
	IB	1.17	0.65
300 - 600	β-	12.4	2.68
	IB	1.53	0.36
600 - 1300	β-	114	11.6
	IB	2.7	0.31
1300 - 2500	β-	602	31.3
	IB	2.8	0.158
2500 - 5000	β-	1593	45.7
	IB	1.58	0.050
5000 - 7500	β-	264	4.70
	IB	0.064	0.00117
7500 - 7681	β-	0.088	0.00117
	IB	3.4×10^{-7}	4.5×10^{-9}

$^{78}_{32}$Ge(1.467 *17* h)

Mode: β-

Δ: -71863 *4* keV

SpA: 2.74×10^{7} Ci/g

Prod: fission; ^{82}Se(n,αn)

Photons (^{78}Ge)

⟨γ⟩=277 *4* keV

γ_{mode}	γ(keV)	γ(%)†
As $K_{\alpha2}$	10.508	0.055 *3*
As $K_{\alpha1}$	10.544	0.107 *6*
As $K_{\beta1}'$	11.724	0.0211 *11*
γ [E1]	277.3 *3*	95.6
γ [E1]	293.9 *5*	4.0 *8*

† 1.0% uncert(syst)

Atomic Electrons (^{78}Ge)

⟨e⟩=1.00 *4* keV

e_{bin}(keV)	⟨e⟩(keV)	e(%)
1	0.0037	0.27 *3*
2	0.00026	0.0171 *18*
9	0.0103	0.113 *12*
10	0.0029	0.028 *3*
11	0.000126	0.00118 *13*
12	0.000151	0.00130 *14*
265	0.84	0.317 *13*
276	0.091	0.0329 *14*
277	0.0157	0.00565 *23*
282	0.032	0.0113 *22*
292	0.0033	0.00111 *22*
293	0.00017	6.0 *12* $\times 10^{-5}$
294	0.00059	0.00020 *4*

Continuous Radiation (^{78}Ge)

⟨β-⟩=226 keV;⟨IB⟩=0.164 keV

E_{bin}(keV)		⟨ ⟩(keV)	(%)
0 - 10	β-	0.114	2.28
	IB	0.0113	
10 - 20	β-	0.345	2.30
	IB	0.0106	0.074
20 - 40	β-	1.39	4.65
	IB	0.019	0.067
40 - 100	β-	10.1	14.4
	IB	0.045	0.070
100 - 300	β-	90	45.7
	IB	0.066	0.041
300 - 600	β-	122	30.3
	IB	0.0123	0.0034
600 - 676	β-	2.39	0.387
	IB	1.25×10^{-5}	2.0×10^{-6}

$^{78}_{33}$As(1.512 *3* h)

Mode: β-

Δ: -72816 *10* keV

SpA: 2.658×10^{7} Ci/g

Prod: ^{81}Br(n,α); fission; ^{78}Se(n,p)

Photons (^{78}As)

⟨γ⟩=1321 *40* keV

γ_{mode}	γ(keV)	γ(%)†
γ	156.50 *10*	0.092 *22*
γ	174.44 *12*	0.18 *4*
γ	351.50 *10*	0.162 *16*
γ	354.77 *16*	1.89 *22*
γ	391.0 *4*	0.124 *16*
γ	450.08 *12*	0.081 *22*
γ	462.44 *12*	0.54 *5*
γ	468.8 *4*	0.097 *16*
γ [E2]	497.26 *12*	0.184 *22*
γ	503.88 *14*	0.42 *4*
γ	545.41 *7*	2.92 *16*
γ	551.8 *4*	0.17 *3*
γ E2	613.85 *5*	54
γ	636.88 *13*	0.205 *22*
γ	657.89 *13*	0.27 *3*
γ	686.17 *10*	0.92 *11*
γ [E2]	687.28 *9*	0.65 *11*
γ E2+16%M1	694.87 *5*	17.4 *9*
γ	722.03 *15*	0.146 *16*
γ	757.11 *14*	0.086 *22*
γ	828.04 *8*	8.6 *5*

Photons (^{78}As) (continued)

γ_{mode}	γ(keV)	γ(%)†
γ	842.68 *8*	0.97 *16*
γ	882.0 *3*	0.19 *3*
γ [E2]	884.89 *10*	0.46 *4*
γ (E2)	888.78 *8*	2.11 *16*
γ	903.8 *3*	0.08 *3*
γ	959.00 *13*	0.46 *4*
γ	967.28 *19*	0.16 *5*
γ	988.83 *16*	0.092 *22*
γ [E1]	1005.10 *12*	0.32 *4*
γ [M1+E2]	1018.68 *15*	0.140 *22*
γ	1079.88 *12*	1.78 *11*
γ	1144.94 *11*	1.78 *11*
γ	1169.16 *18*	0.12 *3*
γ [E1]	1199.01 *10*	0.70 *5*
γ	1228.85 *8*	0.19 *3*
γ	1240.28 *6*	6.2 *5*
γ	1290.48 *13*	0.10 *3*
γ [E2]	1308.71 *5*	12.4 *5*
γ	1339.06 *10*	0.39 *5*
γ	1373.45 *8*	5.0 *3*
γ [E2]	1382.15 *9*	0.70 *5*
γ	1440.54 *13*	0.39 *4*
γ	1529.95 *7*	2.59 *16*
γ	1641.97 *14*	0.16 *4*
γ [M1+E2]	1713.54 *15*	1.89 *11*
γ	1720.93 *14*	0.35 *6*
γ	1737.76 *16*	0.11 *3*
γ	1792.04 *13*	0.97 *11*
γ	1835.88 *12*	1.51 *11*
γ [E1]	1893.87 *10*	0.32 *8*
γ	1921.03 *13*	0.81 *11*
γ	1923.70 *9*	0.71 *9*
γ [E2]	1995.98 *9*	1.35 *10*
γ	2064.0 *3*	0.11 *4*
γ	2068.31 *8*	0.81 *11*
γ	2187.83 *13*	0.36 *4*
γ	2224.81 *7*	0.86 *11*
γ [E2]	2327.37 *15*	~0.11
γ	2537.54 *9*	0.26 *4*
γ	2615.88 *13*	0.70 *11*
γ	2628.4 *6*	0.081 *22*
γ	2680.80 *12*	1.73 *16*
γ	2758.9 *3*	0.113 *22*
γ	2797.6 *3*	0.21 *3*
γ	2838.64 *8*	~0.05
γ	3097.6 *5*	0.059 *16*

† 11% uncert(syst)

Continuous Radiation (^{78}As)

⟨β-⟩=1234 keV;⟨IB⟩=3.5 keV

E_{bin}(keV)		⟨ ⟩(keV)	(%)
0 - 10	β-	0.0169	0.337
	IB	0.041	
10 - 20	β-	0.052	0.344
	IB	0.040	0.28
20 - 40	β-	0.213	0.71
	IB	0.079	0.27
40 - 100	β-	1.65	2.33
	IB	0.23	0.35
100 - 300	β-	19.7	9.6
	IB	0.64	0.36
300 - 600	β-	75	16.6
	IB	0.73	0.17
600 - 1300	β-	276	30.0
	IB	1.02	0.118
1300 - 2500	β-	534	29.0
	IB	0.65	0.038
2500 - 4212	β-	328	11.0
	IB	0.093	0.0033

$^{78}_{34}$Se(stable)

Δ: -77028.1 *15* keV

%: 23.6 *6*

$^{78}_{35}$Br(6.46 *4* min)

Mode: ϵ(>99.99 %), β-?(<0.01 %)

Δ: -73454 *4* keV

SpA: 3.728×10^{8} Ci/g

Prod: ^{75}As(α,n); ^{78}Se(p,n); ^{77}Se(p,γ); ^{77}Se(d,n)

Photons (^{78}Br)

⟨γ⟩=89.2 *24* keV

γ_{mode}	γ(keV)	γ(%)†
Se L_ℓ	1.204	0.0041 *9*
Se L_η	1.245	0.0024 *6*
Se L_α	1.379	0.079 *16*
Se L_β	1.427	0.048 *12*
Se L_γ	1.648	0.00067 *21*
Se $K_{\alpha 2}$	11.182	1.17 *5*
Se $K_{\alpha 1}$	11.222	2.27 *9*
Se $K_{\beta 1}$'	12.494	0.459 *19*
Se $K_{\beta 2}$'	12.741	0.0221 *9*
γ_ϵ	450.08 *12*	0.00080 *22*
$\gamma_{\beta-}$E2	455.028 *23*	0.0075 ?
γ_ϵ[E2]	497.26 *12*	0.00069 *14*
γ_ϵE2	613.85 *5*	13.6
γ_ϵ[E2]	687.28 *9*	0.0024 *6*
γ_ϵE2+16%M1	694.87 *5*	0.062 *3*
γ_ϵ[E2]	884.89 *10*	0.069 *3*
γ_ϵ[M1+E2]	1018.68 *15*	0.00012 *5*
γ_ϵ	1144.94 *11*	0.0177 *14*
γ_ϵ	1228.85 *8*	0.0135 *13*
γ_ϵ[E2]	1308.71 *5*	0.0440 *13*
γ_ϵ	1339.06 *10*	0.0117 *11*
γ_ϵ[E2]	1382.15 *9*	0.0026 *4*
γ_ϵ[M1+E2]	1713.54 *15*	0.0016 *5* ?
γ_ϵ	1720.93 *14*	0.045 *3*
γ_ϵ	1923.70 *9*	0.049 *3*
γ_ϵ[E2]	1995.98 *9*	0.0051 *8*
γ_ϵ	2284.63 *21*	0.0050 *11*
γ_ϵ[E2]	2327.37 *15*	~9 $\times 10^{-5}$
γ_ϵ	2392.3 *3*	0.0035 *4*
γ_ϵ	2476.7 *4*	0.0177 *14*
γ_ϵ	2537.54 *9*	0.0178 *19* ?
γ_ϵ	2641.5 *4*	0.0027 *3*
γ_ϵ	2770.0 *11*	0.00061 *12*
γ_ϵ	2898.46 *22*	0.00034 *10*
γ_ϵ	3006.1 *3*	0.00045 *8*

† uncert(syst): 2.9% for ϵ

Atomic Electrons (^{78}Br)

⟨e⟩=0.480 *22* keV

e_{bin}(keV)	⟨e⟩(keV)	e(%)
1	0.079	5.5 *6*
2	0.0060	0.36 *4*
9	0.0180	0.193 *20*
10	0.188	1.94 *20*
11	0.064	0.58 *6*
437 - 485	1.7×10^{-5}	3.7 *16* $\times 10^{-6}$
496 - 497	1.1×10^{-6}	2.2 *10* $\times 10^{-7}$
601	0.110	0.0183 *6*
612 - 614	0.0141	0.00230 *7*
675 - 695	0.000458	6.7 *4* $\times 10^{-5}$
872 - 885	0.000345	3.95 *20* $\times 10^{-5}$
1006 - 1019	4.7×10^{-7}	4.7 *21* $\times 10^{-8}$
1132 - 1145	4.5×10^{-5}	4.0 *14* $\times 10^{-6}$
1216 - 1229	3.2×10^{-5}	2.6 *9* $\times 10^{-6}$
1296 - 1339	0.000158	1.22 *8* $\times 10^{-5}$
1369 - 1382	7.5×10^{-6}	5.5 *8* $\times 10^{-7}$
1701 - 1721	8.2×10^{-5}	4.8 *13* $\times 10^{-6}$
1911 - 1996	9.0×10^{-5}	4.7 *11* $\times 10^{-6}$
2272 - 2327	7.4×10^{-6}	3.2 *10* $\times 10^{-7}$
2380 - 2477	2.9×10^{-5}	1.18 *24* $\times 10^{-6}$
2525 - 2537	2.4×10^{-5}	9.5 *23* $\times 10^{-7}$
2629 - 2641	3.6×10^{-6}	1.4 *3* $\times 10^{-7}$
2757 - 2770	7.8×10^{-7}	2.8 *8* $\times 10^{-8}$
2886 - 2898	4.2×10^{-7}	1.5 *5* $\times 10^{-8}$
2993 - 3006	5.5×10^{-7}	1.8 *5* $\times 10^{-8}$

Continuous Radiation (^{78}Br)

⟨β+⟩=1022 keV;⟨IB⟩=3.1 keV

E_{bin}(keV)		⟨ ⟩(keV)	(%)
0 - 10	β+	2.30×10^{-5}	0.000285
	IB	0.038	
10 - 20	β+	0.00052	0.00320
	IB	0.037	0.26
20 - 40	β+	0.0085	0.0267
	IB	0.072	0.25
40 - 100	β+	0.242	0.318
	IB	0.21	0.32
100 - 300	β+	8.3	3.83
	IB	0.58	0.32
300 - 600	β+	58	12.5
	IB	0.64	0.150
600 - 1300	β+	411	43.0
	IB	0.87	0.101
1300 - 2500	β+	545	32.8
	IB	0.59	0.034
2500 - 3574	β+	0.114	0.00452
	IB	0.109	0.0039
	Σβ+		92

$^{78}_{36}$Kr(stable)

Δ: -74151 *8* keV
%: 0.35 *2*

$^{78}_{37}$Rb(17.66 *8* min)

Mode: ϵ
Δ: -66980 *30* keV
SpA: 1.365×10^{8} Ci/g
Prod: ^{65}Cu(^{16}O,3n); ^{69}Ga(^{12}C,3n); protons on Zr; protons on Nb; ^{71}Ga(^{12}C,5n); ^{64}Zn(^{16}O,pn)

Photons (^{78}Rb)

⟨γ⟩=3227 *37* keV

γ_{mode}	γ(keV)	γ(%)†
Kr L_ℓ	1.383	0.0079 *23*
Kr L_η	1.435	0.0046 *15*
Kr L_α	1.581	0.17 *5*
Kr L_β	1.640	0.10 *3*
Kr L_γ	1.907	0.0014 *5*
Kr $K_{\alpha2}$	12.598	2.1 *4*
Kr $K_{\alpha1}$	12.651	4.1 *7*
Kr $K_{\beta1}$'	14.109	0.86 *15*
Kr $K_{\beta2}$'	14.408	0.078 *13*
γ	47.11 *10*	0.31 *6* ?
γ [E2,M1]	416.89 *3*	0.203 *13*
γ [E2,M1]	445.30 *3*	0.062 *5*
γ E2	455.028 *23*	62.5
γ [E2]	562.147 *25*	11.4 *6*
γ [M1+E2]	607.96 *3*	0.15 *3*
γ [E2]	636.37 *4*	0.380 *24*
γ E2	664.47 *3*	1.16 *6*
γ	675.92 *5*	0.350 *24*
γ	687.51 *5*	0.53 *3*
γ M1+20%E2	692.870 *22*	12.5 *5*
γ	701.2 *5*	0.8 *3* ?
γ [E2]	738.68 *3*	1.67 *8*
γ [E1]	859.53 *5*	2.81 *19*
γ E2,M1	1109.76 *3*	1.18 *6*
γ E2	1147.893 *24*	7.9 *4*
γ	1203.29 *5*	0.087 *13*
γ	1295.47 *5*	0.47 *3*
γ [M1+E2]	1300.83 *3*	3.17 *18*
γ	1317.89 *4*	0.212 *24*
γ	1350.12 *8*	0.106 *19*
γ	1425.40 *7*	0.26 *2*
γ	1428.00 *6*	0.15 *3*
γ	1467.06 *8*	0.063 *13*
γ	1508.21 *4*	0.344 *19*
γ	1595.18 *8*	0.419 *25*
γ	1652.54 *6*	0.57 *3*
γ	1734.93 *6*	0.66 *4*
γ [E2]	1755.84 *3*	0.81 *4*
γ	1767.09 *5*	0.43 *5*
γ	1772.91 *4*	0.050 *10*
γ	1779.13 *4*	2.94 *19*
γ	1785.67 *5*	0.85 *4*
γ	1806.19 *6*	0.381 *25*
γ	1821.92 *9*	1.13 *12*
γ	1832.9 *7*	2.8 *3* ?
γ	1844.63 *6*	0.29 *3*
γ	1855.12 *5*	0.60 *3*
γ	1885.82 *6*	0.169 *25*
γ	1906.30 *5*	0.244 *25*
γ	1930.14 *5*	0.237 *25*
γ [E1]	1943.98 *4*	0.369 *25*
γ	1988.33 *5*	0.76 *4*
γ	2000.37 *6*	0.14 *3*
γ	2052.97 *8*	0.56 *4*
γ	2082.56 *4*	0.55 *3*
γ	2118.26 *6*	0.73 *4*
γ	2137.38 *5*	1.74 *9*
γ	2201.07 *4*	0.68 *4*
γ	2213.28 *4*	0.319 *19*
γ	2240.68 *4*	1.36 *7*
γ	2284.49 *5*	0.26 *3*
γ	2289.54 *4*	1.51 *8*
γ	2333.41 *5*	0.91 *5*
γ	2391.13 *4*	0.31 *3*
γ	2420.26 *4*	4.56 *25*
γ [E1]	2427.01 *8*	0.61 *3*
γ	2443.34 *5*	0.66 *4*
γ	2514.24 *5*	3.31 *19*
γ	2521.85 *4*	0.188 *25*
γ	2537.49 *6*	0.188 *22*
γ	2557.86 *6*	0.29 *3*
γ	2681.42 *8*	0.74 *4*
γ	2745.33 *5*	0.58 *4*
γ	2789.64 *3*	0.369 *19*
γ	2882.79 *6*	0.30 *3*
γ	2892.43 *5*	2.13 *12*
γ	2920.364 *18*	0.66 *4*
γ	2941.35 *4*	1.16 *6*
γ	2982.39 *4*	5.2 *3*
γ	2990.59 *5*	0.73 *4*
γ	3023.15 *5*	0.57 *4*

Photons (^{78}Rb)
(continued)

γ_{mode}	γ(keV)	γ(%)†
γ	3053.72 *7*	0.162 *25*
γ	3083.98 *4*	4.69 *25*
γ	3173.27 *8*	0.231 *25*
γ	3215.10 *11*	0.156 *19*
γ	3230.42 *4*	0.77 *4*
γ	3272.90 *8*	0.188 *19*
γ	3288.47 *9*	0.17 *3*
γ	3295.01 *9*	0.194 *25*
γ	3309.42 *15*	0.22 *3*
γ	3325.55 *11*	0.19 *3*
γ	3351.82 *15*	0.206 *25*
γ	3362.07 *15*	0.21 *3*
γ	3374.27 *8*	0.54 *6*
γ	3437.40 *4*	4.50 *25*
γ	3438.17 *4*	10.8 *6*
γ	3482.49 *3*	0.96 *5*
γ	3521.74 *9*	0.094 *25*
γ	3538.99 *4*	1.81 *9*
γ	3552.72 *5*	0.59 *3*
γ	3575.00 *5*	2.44 *12*
γ	3634.20 *5*	0.138 *25*
γ	3662.10 *5*	0.181 *19*
γ	3746.56 *7*	0.28 *3*
γ	3773.30 *9*	0.219 *25*
γ	3829.28 *8*	0.144 *19*
γ	3863.52 *6*	0.331 *25*
γ	3893.18 *4*	3.8 *3*
γ	3913.68 *16*	0.47 *3*
γ	3937.49 *3*	1.11 *6*
γ	3994.24 *6*	0.27 *3*
γ	4007.72 *5*	0.374 *19*
γ	4040.28 *5*	0.369 *19*
γ	4044.49 *10*	0.72 *3*
γ	4074.59 *11*	0.26 *4*
γ	4089.20 *5*	0.188 *25*
γ	4095.63 *18*	0.13 *5*
γ	4175.21 *11*	0.175 *19*
γ	4201.56 *7*	0.169 *19*
γ	4420.74 *9*	0.55 *3*
γ	4556.36 *6*	0.144 *19*
γ	4725.59 *8*	0.28 *3*
γ	4737.32 *10*	0.206 *25*
γ	4877.87 *11*	0.069 *13*
γ	5112.59 *16*	0.144 *19*
γ	5180.59 *8*	0.356 *19*
γ	5243.46 *19*	0.481 *25*
γ	5332.87 *11*	0.313 *25*
γ	5369.39 *15*	0.106 *13*
γ	5529.06 *9*	0.181 *19*
γ	5567.59 *16*	0.26 *3*

† 3.2% uncert(syst)

Atomic Electrons (^{78}Rb)

⟨e⟩=3.1 *6* keV

e_{bin}(keV)	⟨e⟩(keV)	e(%)
2	0.17	9.8 *21*
10	0.028	0.27 *5*
11	0.29	2.6 *5*
12	0.103	0.83 *16*
13 - 14	0.030	0.23 *3*
33	0.6	~2
45	0.21	~0.5
403 - 431	0.0046	0.00112 *23*
441	1.05	0.237 *9*
443 - 445	0.000119	2.7 *5* $\times 10^{-5}$
453	0.120	0.0266 *10*
455	0.0224	0.00492 *19*
548	0.125	0.0228 *15*
560 - 608	0.0181	0.00322 *17*
622 - 664	0.0164	0.00253 *16*
673 - 701	0.100	0.0147 *7*
724 - 739	0.0128	0.00177 *10*
845 - 860	0.0069	0.00082 *6*
1095 - 1134	0.0340	0.00301 *13*

Atomic Electrons (^{78}Rb) (continued)

e_{bin}(keV)	⟨e⟩(keV)	e(%)
1146 - 1189	0.00392	0.000341 *16*
1201 - 1203	2.8×10^{-5}	2.3 *8* $\times10^{-6}$
1281 - 1318	0.0131	0.00102 *7*
1336 - 1350	0.00027	2.0 *7* $\times10^{-5}$
1411 - 1453	0.00113	7.9 *17* $\times10^{-5}$
1465 - 1508	0.00081	5.4 *16* $\times10^{-5}$
1581 - 1652	0.0021	0.00013 *3*
1721 - 1820	0.0200	0.00112 *14*
1822 - 1916	0.0036	0.000192 *24*
1928 - 2000	0.0022	0.000113 *20*
2039 - 2137	0.0063	0.00030 *4*
2187 - 2284	0.0068	0.00030 *4*
2288 - 2377	0.0023	9.7 *17* $\times10^{-5}$
2389 - 2481	0.0127	0.00052 *9*
2493 - 2558	0.0067	0.00027 *5*
2667 - 2745	0.0020	7.5 *12* $\times10^{-5}$
2775 - 2870	0.0017	5.9 *10* $\times10^{-5}$
2878 - 2976	0.0134	0.00046 *6*
2980 - 3070	0.0079	0.00026 *5*
3082 - 3173	0.00105	3.4 *5* $\times10^{-5}$
3201 - 3295	0.0023	7.1 *8* $\times10^{-5}$
3307 - 3374	0.00159	4.8 *6* $\times10^{-5}$
3423 - 3522	0.022	0.00064 *9*
3525 - 3620	0.0066	0.000185 *22*
3632 - 3662	0.00026	7.0 *13* $\times10^{-6}$
3732 - 3829	0.00082	2.2 *3* $\times10^{-5}$
3849 - 3937	0.0073	0.000187 *25*
3980 - 4075	0.00270	6.7 *6* $\times10^{-5}$
4081 - 4175	0.00040	9.6 *18* $\times10^{-6}$
4187 - 4201	0.00021	5.0 *10* $\times10^{-6}$
4406 - 4421	0.00066	1.50 *25* $\times10^{-5}$
4542 - 4556	0.00017	3.8 *7* $\times10^{-6}$
4711 - 4737	0.00056	1.20 *16* $\times10^{-5}$
4864 - 4878	8.0×10^{-5}	1.6 *4* $\times10^{-6}$

Continuous Radiation (^{78}Rb)

⟨β+⟩=1300 keV;⟨IB⟩=4.6 keV

E_{bin}(keV)		⟨ ⟩(keV)	(%)
0 - 10	β+	1.81×10^{-5}	0.000223
	IB	0.042	
10 - 20	β+	0.000434	0.00267
	IB	0.042	0.29
20 - 40	β+	0.0075	0.0234
	IB	0.081	0.28
40 - 100	β+	0.221	0.290
	IB	0.23	0.36
100 - 300	β+	7.4	3.43
	IB	0.68	0.38
300 - 600	β+	47.6	10.3
	IB	0.81	0.19
600 - 1300	β+	317	33.2
	IB	1.28	0.147
1300 - 2500	β+	546	31.3
	IB	1.05	0.061
2500 - 5000	β+	363	10.9
	IB	0.33	0.0109
5000 - 7170	β+	17.9	0.339
	IB	0.00148	2.8×10^{-5}
	Σβ+		90

$^{78}_{37}$Rb(5.74 *6* min)

Mode: ε(90 *2* %), IT(10 *2* %)
Δ: >-66877 keV
SpA: 4.20×10^{8} Ci/g
Prod: ^{79}Br(^{3}He,4n); ^{65}Cu(^{16}O,3n); ^{64}Zn(^{16}O,pn); ^{69}Ga(^{12}C,3n)

Photons (^{78}Rb)

⟨γ⟩=2328 *25* keV

γ_{mode}	γ(keV)	γ(%)†
Kr $K_{\alpha2}$	12.598	0.357 *15*
Kr $K_{\alpha1}$	12.651	0.69 *3*
Kr $K_{\beta1}$'	14.109	0.146 *6*
Kr $K_{\beta2}$'	14.408	0.0132 *6*
γ_{IT} E1,M1	103.3 *1*	8
γ_ϵ	341.26 *6*	1.54 *8*
γ_ϵ[E2,M1]	416.89 *3*	2.25 *14*
γ_ϵ[E2,M1]	445.30 *3*	0.69 *4*
γ_ϵE2	455.028 *23*	81
γ_ϵ[E2]	562.147 *25*	0.28 *8*
γ_ϵ[M1+E2]	607.96 *3*	0.036 *7*
γ_ϵ	611.39 *8*	1.02 *6*
γ_ϵ[E2]	636.37 *4*	0.093 *9*
γ_ϵE2	664.47 *3*	38.3 *20*
γ_ϵ	675.92 *5*	0.064 *8*
γ_ϵ	687.51 *5*	0.034 *8*
γ_ϵM1+20%E2	692.870 *22*	12.3 *5*
γ_ϵE2	725.02 *4*	6.2 *3*
γ_ϵE2	735.02 *4*	1.48 *8*
γ_ϵ[E2]	738.68 *3*	0.41 *3*
γ_ϵE2+19%M1	753.43 *4*	3.89 *24*
γ_ϵ[E1]	771.97 *6*	0.44 *4*
γ_ϵ	848.68 *11*	0.66 *5*
γ_ϵE2	858.34 *7*	2.00 *11*
γ_ϵ	1086.81 *7*	0.70 *4*
γ_ϵ	1095.92 *8*	0.87 *5*
γ_ϵE2,M1	1109.76 *3*	13.2 *6*
γ_ϵE2	1147.893 *24*	7.8 *4*
γ_ϵE2+30%M1	1180.32 *5*	0.70 *4*
γ_ϵ	1199.34 *4*	7.2 *4*
γ_ϵ	1232.47 *6*	1.33 *7*
γ_ϵ	1265.57 *11*	1.25 *7*
γ_ϵ	1270.20 *5*	1.35 *7*
γ_ϵ	1288.26 *6*	0.58 *6*
γ_ϵ	1295.47 *5*	0.030 *7*
γ_ϵ[M1+E2]	1300.83 *3*	0.78 *7*
γ_ϵ	1317.89 *4*	0.72 *4*
γ_ϵ	1326.47 *5*	0.40 *5*
γ_ϵ	1350.12 *8*	0.62 *6*
γ_ϵ	1360.65 *5*	0.62 *4*
γ_ϵ	1375.59 *6*	1.22 *7*
γ_ϵE2	1417.89 *4*	0.95 *6*
γ_ϵ	1425.40 *7*	0.60 *5*
γ_ϵ	1529.74 *9*	1.77 *9*
γ_ϵE1+0.2%M2	1630.30 *5*	4.94 *24*
γ_ϵ	1644.63 *4*	7.1 *4*
γ_ϵ	1668.77 *5*	0.94 *6*
γ_ϵ[E2]	1755.84 *3*	0.199 *18*
γ_ϵ	1767.09 *5*	0.032 *5*
γ_ϵ	1772.91 *4*	0.17 *3*
γ_ϵ	1779.13 *4*	0.94 *5*
γ_ϵ	1785.67 *5*	0.156 *19*
γ_ϵ	1796.29 *6*	1.34 *10*
γ_ϵ	1844.63 *6*	0.122 *19*
γ_ϵ	1852.56 *5*	3.08 *24*
γ_ϵ	1879.89 *7*	1.39 *7*
γ_ϵ	1901.68 *5*	0.60 *4*
γ_ϵ[E1]	1943.98 *4*	6.7 *4*
γ_ϵ	1952.93 *6*	1.72 *9*
γ_ϵ	1988.33 *5*	0.048 *11*
γ_ϵ	2000.37 *6*	0.011 *3*
γ_ϵ	2013.27 *5*	3.16 *16*
γ_ϵ	2041.67 *6*	0.89 *5*
γ_ϵ	2052.97 *8*	0.097 *24*
γ_ϵ	2114.06 *5*	0.83 *5*
γ_ϵ	2118.26 *6*	1.68 *8*
γ_ϵ	2209.81 *5*	0.79 *6*
γ_ϵ	2222.60 *9*	0.93 *6*
γ_ϵ	2240.68 *4*	0.25 *3*
γ_ϵ[E1]	2427.01 *8*	0.83 *5*

Photons (^{78}Rb) (continued)

γ_{mode}	γ(keV)	γ(%)†
γ_ϵ	2443.34 *5*	0.042 *10*
γ_ϵ	2537.49 *6*	0.080 *13*
γ_ϵ	2626.68 *5*	2.84 *15*
γ_ϵ	2655.09 *5*	1.17 *7*
γ_ϵ	2990.59 *5*	0.055 *7*
γ_ϵ	3270.41 *5*	0.66 *4*
γ_ϵ	3319.53 *4*	2.89 *15*
γ_ϵ	3552.72 *5*	0.044 *6*
γ_ϵ	4007.72 *5*	0.028 *4*

† uncert(syst): 2.5% for ε, 25% for IT

Atomic Electrons (^{78}Rb)*

⟨e⟩=2.50 *5* keV

e_{bin}(keV)	⟨e⟩(keV)	e(%)
2 - 14	0.103	2.28 *17*
327 - 341	0.034	~0.010
403 - 431	0.051	0.0124 *25*
441	1.36	0.308 *10*
443 - 445	0.00132	0.00030 *6*
453	0.156	0.0344 *11*
548 - 597	0.010	0.0018 *6*
606 - 636	0.0018	0.00029 *6*
650	0.311	0.048 *3*
662 - 676	0.0417	0.0063 *3*
679	0.080	0.0118 *6*
686 - 735	0.074	0.0104 *4*
737 - 772	0.0299	0.00404 *23*
834 - 858	0.0151	0.00178 *15*
1072 - 1094	0.0050	0.00046 *11*
1095	0.049	0.0045 *3*
1096 - 1134	0.0348	0.00308 *13*
1146 - 1185	0.025	0.0021 *6*
1197 - 1232	0.0060	0.00050 *12*
1251 - 1301	0.0114	0.00090 *14*
1304 - 1350	0.0059	0.00044 *8*
1359 - 1404	0.0060	0.00043 *7*
1411 - 1425	0.0018	0.00013 *3*
1515 - 1530	0.0040	0.00027 *8*
1616 - 1669	0.025	0.0015 *3*
1742 - 1838	0.0114	0.00063 *10*
1843 - 1942	0.0162	0.00085 *7*
1944 - 2041	0.0082	0.00041 *8*
2051 - 2118	0.0045	0.00021 *4*
2195 - 2240	0.0034	0.00015 *3*
2413 - 2443	0.00107	4.4 *3* $\times10^{-5}$
2523 - 2612	0.0040	0.00015 *4*
2625 - 2655	0.0023	8.6 *16* $\times10^{-5}$
2976 - 2990	7.9×10^{-5}	2.7 *6* $\times10^{-6}$
3256 - 3319	0.0049	0.000147 *24*
3538 - 3553	5.9×10^{-5}	1.7 *4* $\times10^{-6}$
3993 - 4008	3.5×10^{-5}	8.8 *18* $\times10^{-7}$

* with ε

Continuous Radiation (^{78}Rb)

⟨β+⟩=1510 keV;⟨IB⟩=5.0 keV

E_{bin}(keV)		⟨ ⟩(keV)	(%)
0 - 10	β+	7.8×10^{-6}	9.6×10^{-5}
	IB	0.047	
10 - 20	β+	0.000188	0.00116
	IB	0.046	0.32
20 - 40	β+	0.00329	0.0103
	IB	0.091	0.32
40 - 100	β+	0.099	0.130
	IB	0.26	0.40
100 - 300	β+	3.68	1.68
	IB	0.78	0.44
300 - 600	β+	27.5	5.9
	IB	0.95	0.22
600 - 1300	β+	239	24.5

Continuous Radiation (^{78}Rb)
(continued)

E_{bin}(keV)		⟨ ⟩(keV)	(%)
	IB	1.47	0.169
1300 - 2500	β+	733	39.9
	IB	1.09	0.063
2500 - 5000	β+	505	16.2
	IB	0.26	0.0087
5000 - 6818	β+	1.82	0.0349
	IB	9.7×10^{-5}	1.8×10^{-6}
	Σβ+		88

$^{78}_{38}$Sr(30.6 *23* min)

$t_{1/2}$ possibly shorter

Mode: ε
Δ: -63650 *300* keV syst
SpA: 7.9×10^{7} Ci/g
Prod: ^{64}Zn(^{16}O,2n); ^{66}Zn(^{16}O,4n)

A = 79

NDS **37**, 393 (1982)

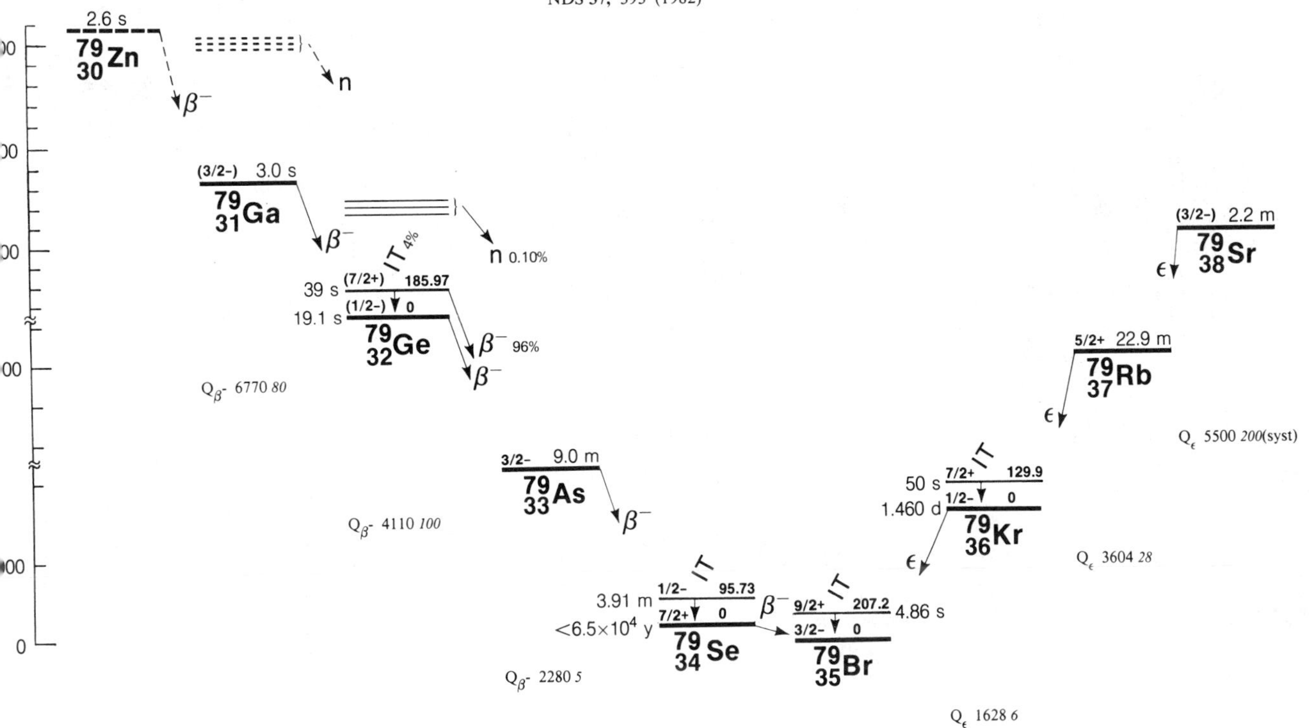

$^{79}_{30}$Zn(2.63 *9* s)

attributed to ^{79}Zn + ^{79}Ga

Mode: β-, β-n
SpA: 4.77×10^{10} Ci/g
Prod: fission

Photons (^{79}Zn)†

γ_{mode}	γ(keV)	γ(rel)
γ	702.0 *2*	75 *15*
γ	866.3 *2*	100 *20*
γ	874.3 *2*	31 *6*

† with ^{79}Zn + ^{79}Ga

$^{79}_{31}$Ga(3.00 *9* s)

Mode: β-, β-n(0.098 *10* %)
Δ: -62760 *130* keV
SpA: 4.25×10^{10} Ci/g
Prod: fission

Photons (^{79}Ga)&

⟨γ⟩=2078 21 keV

γ_{mode}	γ(keV)	γ(%)†
Ge $K_{\alpha2}$	9.855	0.30 7
Ge $K_{\alpha1}$	9.886	0.59 13
Ge $K_{\beta1}$'	10.980	0.114 25
Ge $K_{\beta2}$'	11.184	0.0019 4
γ M1+38%E2	90.72 3	3.2 3
γ M1+38%E2	142.402 25	4.3 3
γ [E3]	185.97 4	*
γ M1,E2	204.58 4	0.62 5
γ [E1]	278.73 4	0.307 24
γ [E1]	421.13 4	1.84 10
γ	437.90 4	1.38 7
γ [E2]	464.70 3	24.2 10
γ	516.38 3	21.5 15
γ	571.31 4	0.368 24
γ	580.12 3	0.242 24
γ	580.30 4	0.121 24
γ [E2]	607.10 3	3.53 17
γ	670.84 3	1.48 10
γ	697.14 5	0.344 24
γ	707.39 12	0.194 19
γ	722.52 3	0.69 4
γ	797.19 7	0.140 12
γ	802.42 8	0.179 15
γ	813.15 13	0.150 15
γ	824.75 5	0.244 15
γ	882.80 4	0.160 12
γ	901.71 4	0.60 4
γ	910.04 10	0.184 15
γ	918.33 4	1.06 7
γ	949.16 3	1.04 7
γ	952.96 3	1.19 7
γ	957.15 3	1.06 7
γ	1008.83 4	0.92 7
γ	1025.01 4	0.465 24
γ	1039.88 4	0.48 12
γ	1039.88 3	2.18 24
γ	1052.49 4	0.319 19
γ	1091.56 3	4.65 19
γ	1104.30 4	1.28 5
γ	1167.92 20	0.198 17
γ	1187.21 3	12.8 5
γ	1211.99 12	0.184 17
γ	1244.37 7	0.58 4
γ	1300.11 4	0.271 19
γ	1326.19 4	0.92 5
γ	1382.90 8	0.42 3
γ	1462.90 3	5.5 3
γ	1533.07 3	0.167 17
γ	1553.63 4	0.213 22
γ	1585.12 7	0.44 3
γ	1605.30 3	3.70 17
γ	1623.79 3	4.31 19
γ	1639.87 10	0.48 4
γ	1733.7 3	0.227 24
γ	1741.52 20	0.29 3
γ	1816.10 9	0.182 17
γ	1864.60 5	0.61 3
γ	1884.01 4	2.93 15
γ	1906.30 4	0.174 15
γ	1923.07 9	0.223 19
γ	1953.63 7	0.44 3
γ	1970.82 8	0.29 3
γ	1997.02 4	3.90 19
γ	2048.70 3	1.77 10
γ	2069.99 4	0.97 7
γ	2140.15 4	7.0 4
γ	2164.93 12	1.06 7
γ	2167.35 9	0.12 5
γ	2253.05 4	1.89 10
γ	2258.07 9	1.45 7
γ	2287.64 5	1.23 7
γ	2310.69 10	0.31 7
γ	2314.52 20	~0.8
γ	2387.76 9	0.34 3
γ	2426.42 20	0.27 4
γ	2430.04 6	1.14 7
γ	2483.1 3	0.29 5
γ	2500.27 13	0.39 4
γ	2533.92 8	0.39 4
γ	2560.77 12	0.223 22
γ	2598.15 15	0.77 5

Photons (^{79}Ga)& (continued)

γ_{mode}	γ(keV)	γ(%)†
γ	2682.51 9	0.73 5
γ	2703.16 12	0.87 5
γ	2731.8 3	0.32 4
γ	2740.55 15	0.53 4
γ	2774.43 9	1.26 7
γ	2802.42 20	0.38 4
γ	2824.90 9	1.94 10
γ	2835.75 12	0.73 5
γ	2869.31 13	1.33 7
γ	2975.55 4	0.237 17
γ	2997.4 4	0.172 17
γ	3011.52 13	0.201 17
γ	3086.4 6	0.12 5
γ	3090.2 5	0.189 24
γ	3167.85 12	2.06 12
γ	3205.23 15	0.63 4
γ	3244.8 4	0.201 22
γ	3276.0 6	0.15 4
γ	3286.6 3	0.177 17
γ	3290.9 4	0.126 12
γ	3304.2 3	0.162 17
γ [M1+E2]	3318.92 21	0.50 4
γ	3352.11 12	1.50 10
γ	3367.0 5	0.203 24
γ	3376.00 21	0.433 24
γ	3427.68 21	0.319 24
γ	3446.6 3	0.249 19
γ	3514.3 4	0.247 24
γ	3539.6 4	0.131 12
γ [E1]	3597.63 21	0.44 4
γ [M1+E2]	3618.9 3	0.32 3
γ	3667.7 4	0.56 4
γ	3803.0 3	0.090 12
γ	3892.36 21	0.356 24
γ [E1]	3897.6 3	0.319 24
γ	4184.1 4	0.145 12

† 4.1% uncert(syst)
* with ^{79}Ge(39 s)
& see also ^{79}Zn

Atomic Electrons (^{79}Ga)

⟨e⟩=2.7 4 keV

e_{bin}(keV)	⟨e⟩(keV)	e(%)
1 - 11	0.099	2.6 4
80	1.1	1.3 3
89	0.13	0.14 4
90 - 91	0.054	0.059 12
131	0.5	~0.36
141	0.06	~0.04
142	0.010	~0.007
175	0.092	0.053 9
185 - 205	0.055	0.029 11
268 - 279	0.00278	0.00104 8
410 - 438	0.022	0.0052 16
454	0.291	0.064 4
463 - 465	0.0368	0.0079 4
505	0.14	~0.027
515 - 560	0.019	0.0036 16
569 - 607	0.0303	0.0051 3
660 - 707	0.009	0.0013 4
711 - 723	0.0027	0.00039 16
786 - 825	0.0024	0.00030 6
872 - 918	0.0057	0.00064 15
938 - 957	0.0088	0.00094 20
998 - 1041	0.011	0.00103 25
1051 - 1093	0.013	0.0012 4
1103 - 1104	0.00031	2.8 9 $\times10^{-5}$
1157 - 1201	0.027	0.0023 8
1211 - 1244	0.0012	10 3 $\times10^{-5}$
1289 - 1326	0.0027	0.00021 5
1372 - 1383	0.00075	5.5 17 $\times10^{-5}$
1452 - 1543	0.010	0.00068 19
1552 - 1640	0.0139	0.00086 16

Atomic Electrons (^{79}Ga) (continued)

e_{bin}(keV)	⟨e⟩(keV)	e(%)
1723 - 1816	0.00101	5.8 10 $\times10^{-5}$
1854 - 1952	0.0060	0.00032 6
1953 - 2049	0.0078	0.00039 7
2059 - 2156	0.0113	0.00053 10
2164 - 2258	0.0042	0.00019 3
2277 - 2314	0.0028	0.00012 3
2377 - 2472	0.0023	9.6 14 $\times10^{-5}$
2482 - 2561	0.00117	4.6 6 $\times10^{-5}$
2587 - 2682	0.0017	6.3 10 $\times10^{-5}$
2692 - 2791	0.0036	0.000131 14
2801 - 2869	0.0043	0.000151 20
2964 - 3011	0.00063	2.1 3 $\times10^{-5}$
3075 - 3168	0.0024	7.5 13 $\times10^{-5}$
3194 - 3293	0.00141	4.4 5 $\times10^{-5}$
3303 - 3376	0.0027	8.0 9 $\times10^{-5}$
3417 - 3514	0.00078	2.3 3 $\times10^{-5}$
3529 - 3619	0.00081	2.26 14 $\times10^{-5}$
3657 - 3668	0.00053	1.4 3 $\times10^{-5}$
3792 - 3891	0.00062	1.59 16 $\times10^{-5}$
3892 - 3897	2.94 $\times10^{-5}$	7.5 5 $\times10^{-7}$
4173 - 4184	0.000129	3.1 5 $\times10^{-6}$

Continuous Radiation (^{79}Ga)

⟨β-⟩=2146 keV;⟨IB⟩=7.9 keV

E_{bin}(keV)		⟨ ⟩(keV)	(%)
0 - 10	β-	0.00262	0.052
	IB	0.060	
10 - 20	β-	0.0081	0.054
	IB	0.059	0.41
20 - 40	β-	0.0340	0.113
	IB	0.117	0.41
40 - 100	β-	0.277	0.39
	IB	0.34	0.52
100 - 300	β-	4.02	1.93
	IB	1.04	0.58
300 - 600	β-	22.0	4.77
	IB	1.30	0.31
600 - 1300	β-	183	18.7
	IB	2.2	0.25
1300 - 2500	β-	743	39.4
	IB	1.9	0.110
2500 - 5000	β-	1093	32.7
	IB	0.82	0.026
5000 - 6770	β-	101	1.86
	IB	0.0158	0.00029

$^{79}_{32}$Ge(19.1 3 s)

Mode: β-

Δ: -69530 100 keV

SpA: 7.34×10^9 Ci/g

Prod: fission

Photons (^{79}Ge)

⟨γ⟩=309 7 keV

γ_{mode}	γ(keV)	γ(%)†
As $K_{\alpha2}$	10.508	0.46 8
As $K_{\alpha1}$	10.544	0.89 15
As $K_{\beta1}$'	11.724	0.18 3
As $K_{\beta2}$'	11.948	0.0054 9
γ (M1)	100.43 8	2.70 15

Photons (^{79}Ge) (continued)

γ_{mode}	γ(keV)	γ(%)†
γ M1+12%E2	109.63 5	21.4
γ M1+50%E2	230.54 3	1.7 3
γ	287.77 23	0.13 3
γ	306.95 18	0.113 21
γ	325.47 9	0.246 21
γ	503.21 8	2.16 11
γ	524.46 6	0.084 14
γ	551.18 11	0.124 11
γ	603.64 6	1.09 6
γ	634.09 4	0.26 4
γ	724.10 9	0.47 4
γ	749.07 9	0.43 4
γ	774.71 11	0.291 17
γ	808.26 9	0.347 21
γ	825.00 11	0.37 3
γ	871.75 6	0.347 21
γ	1031.56 9	0.250 17
γ	1181.37 8	0.94 4
γ	1259.75 7	0.332 15
γ	1265.70 7	0.257 15
γ	1275.30 6	0.411 21
γ	1396.20 7	2.10 9
γ	1417.99 7	0.56 4
γ	1505.83 6	9.2 4
γ	1538.41 9	0.244 15
γ	1557.23 9	0.304 17
γ	1571.30 17	0.33 3
γ	1845.34 8	0.218 13
γ	1869.33 8	0.66 4
γ	2594.2 3	0.261 19

† 4.7% uncert(syst)

Atomic Electrons (^{79}Ge)

$\langle e \rangle$=3.3 4 keV

e_{bin}(keV)	$\langle e \rangle$(keV)	e(%)
1 - 2	0.033	2.5 4
9	0.085	0.94 18
10 - 12	0.027	0.26 4
89	0.191	0.215 15
98	2.4	2.5 4
99 - 100	0.0273	0.0275 16
108	0.32	0.30 6
109 - 110	0.056	0.051 9
219 - 231	0.09	~0.039
276 - 325	0.011	0.0037 13
491 - 539	0.019	0.0038 17
550 - 592	0.006	0.0010 5
602 - 634	0.0021	0.00034 12
712 - 749	0.0038	0.00052 16
763 - 808	0.0024	0.00031 9
813 - 860	0.0023	0.00028 8
870 - 872	0.00013	1.4 5 $\times 10^{-5}$
1020 - 1032	0.00066	6.5 23 $\times 10^{-5}$
1169 - 1181	0.0021	0.00018 6
1248 - 1275	0.0021	0.00017 3
1384 - 1418	0.0051	0.00037 9
1494 - 1571	0.018	0.0012 3
1833 - 1869	0.0013	7.2 15 $\times 10^{-5}$
2582 - 2594	0.00032	1.2 3 $\times 10^{-5}$

Continuous Radiation (^{79}Ge)

$\langle \beta- \rangle$=1684 keV;$\langle$IB$\rangle$=5.3 keV

E_{bin}(keV)		$\langle\ \rangle$(keV)	(%)
0 - 10	β-	0.0039	0.077
	IB	0.052	
10 - 20	β-	0.0120	0.080
	IB	0.052	0.36
20 - 40	β-	0.0505	0.168
	IB	0.102	0.35
40 - 100	β-	0.410	0.58
	IB	0.29	0.45
100 - 300	β-	5.8	2.81
	IB	0.88	0.49
300 - 600	β-	31.3	6.8
	IB	1.06	0.25
600 - 1300	β-	245	25.2
	IB	1.61	0.19
1300 - 2500	β-	849	45.6
	IB	1.09	0.064
2500 - 4110	β-	552	18.7
	IB	0.138	0.0048

$^{79}_{32}$Ge(39 1 s)

Mode: β-(96 1 %), IT(4 1 %)
Δ: -69344 100 keV
SpA: 3.63$\times 10^9$ Ci/g
Prod: ^{82}Se(n,α); fission

Photons (^{79}Ge)

$\langle \gamma \rangle$=1777 75 keV

γ_{mode}	γ(keV)	γ(%)†
Ge $K_{\alpha 2}$	9.855	0.16 4
Ge $K_{\alpha 1}$	9.886	0.32 8
As $K_{\alpha 2}$	10.508	0.51 14
As $K_{\alpha 1}$	10.544	1.0 3
Ge $K_{\beta 1}$'	10.980	0.061 16
Ge $K_{\beta 2}$'	11.184	0.0010 3
As $K_{\beta 1}$'	11.724	0.20 5
As $K_{\beta 2}$'	11.948	0.0061 16
$\gamma_{\beta-}$M1+12%E2	109.63 5	11.6 23
γ_{IT}[E3]	185.97 4	2.8
$\gamma_{\beta-}$	216.5 14	11.7 18
$\gamma_{\beta-}$M1+50%E2	230.54 3	61
$\gamma_{\beta-}$	287.9 3	1.1 3
$\gamma_{\beta-}$	307.08 23	0.86 18
$\gamma_{\beta-}$	325.60 11	2.09 18
$\gamma_{\beta-}$	447.01 5	5.6 3 ?
$\gamma_{\beta-}$	484.96 6	6.3 4
$\gamma_{\beta-}$	524.46 6	4.3 7
$\gamma_{\beta-}$[M2]	542.34 5	32.6 18
$\gamma_{\beta-}$	551.31 14	1.06 10
$\gamma_{\beta-}$	634.09 4	13.4 23
$\gamma_{\beta-}$	644.73 4	3.1 4
$\gamma_{\beta-}$	724.23 11	4.0 4
$\gamma_{\beta-}$	745.14 5	7.8 4 ?
$\gamma_{\beta-}$	749.20 11	3.6 4
$\gamma_{\beta-}$	755.1 14	~18 ?
$\gamma_{\beta-}$	765.63 6	4.67 25
$\gamma_{\beta-}$	774.84 14	2.46 18
$\gamma_{\beta-}$[E1]	781.58 7	12.3 9
$\gamma_{\beta-}$	825.13 14	3.13 25
$\gamma_{\beta-}$	875.26 4	3.07 18
$\gamma_{\beta-}$[E1]	902.48 8	3.7 3
$\gamma_{\beta-}$	1015.25 5	9.8 5
$\gamma_{\beta-}$	1031.69 11	2.15 12
$\gamma_{\beta-}$	1117.64 5	9.9 5
$\gamma_{\beta-}$[M1+E2]	1192.15 6	4.7 3 ?
$\gamma_{\beta-}$	1256.42 5	9.9 6
$\gamma_{\beta-}$	1259.75 7	2.83 12
$\gamma_{\beta-}$	1418.12 9	4.7 4
$\gamma_{\beta-}$	1538.54 11	2.09 12

Photons (^{79}Ge) (continued)

γ_{mode}	γ(keV)	γ(%)†
$\gamma_{\beta-}$	1557.36 11	2.58 12
$\gamma_{\beta-}$	1571.43 22	2.76 25
$\gamma_{\beta-}$	1845.47 10	1.84 12
$\gamma_{\beta-}$	2594.3 4	2.21 18

† uncert(syst): 25% for IT, 6.3% for β-

Atomic Electrons (^{79}Ge)

$\langle e \rangle$=8.9 17 keV

e_{bin}(keV)	$\langle e \rangle$(keV)	e(%)
1 - 12	0.22	5.5 11
98	1.3	1.3 4
108 - 110	0.20	0.19 5
175	1.8	1.0 3
185	0.28	0.15 4
186	0.045	0.024 6
205	0.6	~0.27
215 - 216	0.08	~0.036
219	2.7	~1
229	0.32	~0.14
230 - 276	0.08	0.033 14
286 - 326	0.07	0.021 9
435 - 484	0.11	0.024 9
485 - 524	0.033	0.006 3
530	0.62	0.117 8
539 - 551	0.087	0.0161 11
622 - 645	0.09	0.014 6
712 - 755	0.16	0.021 7
763 - 782	0.038	0.0049 6
813 - 825	0.011	0.0014 5
863 - 902	0.017	0.0019 5
1003 - 1032	0.032	0.0032 10
1106 - 1118	0.024	0.0022 7
1180 - 1192	0.0141	0.00119 10
1245 - 1260	0.027	0.0022 6
1406 - 1418	0.009	0.00064 20
1527 - 1571	0.0130	0.00084 14
1834 - 1845	0.0028	0.00015 4
2582 - 2594	0.0027	0.000104 24

Continuous Radiation (^{79}Ge)

$\langle \beta- \rangle$=1250 keV;$\langle$IB$\rangle$=3.4 keV

E_{bin}(keV)		$\langle\ \rangle$(keV)	(%)
0 - 10	β-	0.0068	0.136
	IB	0.042	
10 - 20	β-	0.0210	0.140
	IB	0.042	0.29
20 - 40	β-	0.088	0.292
	IB	0.082	0.29
40 - 100	β-	0.71	1.00
	IB	0.23	0.36
100 - 300	β-	9.8	4.74
	IB	0.67	0.38
300 - 600	β-	49.8	10.8
	IB	0.76	0.18
600 - 1300	β-	331	34.6
	IB	1.02	0.119
1300 - 2500	β-	667	37.7
	IB	0.49	0.030
2500 - 4296	β-	191	6.6
	IB	0.040	0.00142

$^{79}_{33}$As(9.01 *15* min)

Mode: β-
Δ: -73639 *6* keV
SpA: 2.64×10^{8} Ci/g
Prod: daughter ^{79}Ge; ^{80}Se(n,pn); ^{80}Se(γ,p); fission

Photons (^{79}As)

⟨γ⟩=57.9 *17* keV

γ_{mode}	γ(keV)	γ(%)†
Se L_ℓ	1.204	0.084 *20*
Se L_η	1.245	0.051 *14*
Se L_α	1.379	1.6 *4*
Se L_β	1.426	1.0 *3*
Se L_γ	1.648	0.011 *4*
Se $K_{\alpha2}$	11.182	21.1 *21*
Se $K_{\alpha1}$	11.222	41 *4*
Se $K_{\beta1}$'	12.494	8.3 *8*
Se $K_{\beta2}$'	12.741	0.40 *4*
γ E3	95.73 *3*	16.5 *15* *
γ [E1]	364.8 *4*	1.88 *6*
γ	402.6 *5*	0.097 *23*
γ [M1+E2]	432.0 *4*	1.50 *3*
γ [M1+E2]	446.8 *4*	0.26 *3*
γ	476.2 *4*	0.36 *4*
γ	552.3 *5*	0.135 *24*
γ	715.3 *4*	0.30 *3*
γ	724.2 *7*	0.114 *12*
γ [M1+E2]	878.8 *4*	1.41 *6*
γ	993.3 *7*	0.13 *4*

† 13% uncert(syst)
* with ^{79}Se(3.91 min)

Atomic Electrons (^{79}As)

⟨e⟩=142 *10* keV

e_{bin}(keV)	⟨e⟩(keV)	e(%)
1	1.7	119 *16*
2 - 12	5.0	55 *5*
83	99.5	120 *11*
94	30.4	32 *3*
95	1.56	1.64 *15*
96	3.8	4.0 *4*
352 - 401	0.0138	0.00386 *25*
402 - 447	0.025	0.0060 *13*
464 - 476	0.0035	0.0008 *4*
540 - 552	0.0010	0.00018 *9*
703 - 724	0.0020	0.00028 *9*
866 - 879	0.0066	0.00077 *7*
981 - 993	0.00040	4.1 *18* ×10^{-5}

Continuous Radiation (^{79}As)

⟨β-⟩=876 keV;⟨IB⟩=1.8 keV

E_{bin}(keV)		⟨ ⟩(keV)	(%)
0 - 10	β-	0.0138	0.273
	IB	0.034	
10 - 20	β-	0.0422	0.281
	IB	0.033	0.23
20 - 40	β-	0.176	0.58
	IB	0.065	0.23
40 - 100	β-	1.40	1.97
	IB	0.18	0.28
100 - 300	β-	18.6	9.0

Continuous Radiation (^{79}As) (continued)

E_{bin}(keV)		⟨ ⟩(keV)	(%)
	IB	0.49	0.28
300 - 600	β-	87	19.1
	IB	0.48	0.115
600 - 1300	β-	454	48.5
	IB	0.46	0.056
1300 - 2185	β-	315	20.3
	IB	0.049	0.0034

$^{79}_{34}$Se(<6.5×10^{4} yr)

Mode: β-
Δ: -75919.1 *16* keV
SpA: >0.07 Ci/g
Prod: fission

Continuous Radiation (^{79}Se)

⟨β-⟩=53 keV;⟨IB⟩=0.0094 keV

E_{bin}(keV)		⟨ ⟩(keV)	(%)
0 - 10	β-	0.481	9.7
	IB	0.0027	
10 - 20	β-	1.49	9.9
	IB	0.0020	0.0142
20 - 40	β-	6.2	20.5
	IB	0.0026	0.0092
40 - 100	β-	33.2	49.7
	IB	0.0021	0.0038
100 - 151	β-	11.5	10.1
	IB	5.8×10^{-5}	5.4×10^{-5}

$^{79}_{34}$Se(3.91 *5* min)

Mode: IT
Δ: -75823.4 *16* keV
SpA: 6.08×10^{8} Ci/g
Prod: ^{78}Se(n,γ)

Photons (^{79}Se)

⟨γ⟩=13.7 *3* keV

γ_{mode}	γ(keV)	γ(%)†
Se L_ℓ	1.204	0.048 *11*
Se L_η	1.245	0.030 *8*
Se L_α	1.379	0.93 *19*
Se L_β	1.426	0.56 *15*
Se L_γ	1.648	0.0063 *19*
Se $K_{\alpha2}$	11.182	12.1 *7*
Se $K_{\alpha1}$	11.222	23.4 *13*
Se $K_{\beta1}$'	12.494	4.7 *3*
Se $K_{\beta2}$'	12.741	0.228 *13*
γ E3	95.73 *3*	9.5

† 3.2% uncert(syst)

Atomic Electrons (^{79}Se)

⟨e⟩=81.5 *25* keV

e_{bin}(keV)	⟨e⟩(keV)	e(%)
1	0.99	68 *7*
2	0.056	3.4 *4*
9	0.186	1.99 *22*
10	1.94	20.0 *22*
11	0.66	6.0 *7*
12	0.041	0.34 *4*
83	57.1	69 *3*
94	17.5	18.5 *8*
95	0.90	0.94 *4*
96	2.19	2.29 *10*

$^{79}_{35}$Br(stable)

Δ: -76070.0 *24* keV
%: 50.69 *5*

$^{79}_{35}$Br(4.86 *4* s)

Mode: IT
Δ: -75862.8 *24* keV
SpA: 2.738×10^{10} Ci/g
Prod: ^{78}Se(p,γ); ^{79}Br(n,n')

Photons (^{79}Br)

⟨γ⟩=158.6 *15* keV

γ_{mode}	γ(keV)	γ(%)†
Br L_ℓ	1.293	0.014 *3*
Br L_η	1.339	0.0083 *21*
Br L_α	1.481	0.28 *6*
Br L_β	1.533	0.17 *4*
Br L_γ	1.777	0.0021 *7*
Br $K_{\alpha2}$	11.878	3.72 *19*
Br $K_{\alpha1}$	11.924	7.2 *4*
Br $K_{\beta1}$'	13.290	1.49 *8*
Br $K_{\beta2}$'	13.562	0.101 *5*
γ E3	207.2 *4*	75.8

† 0.79% uncert(syst)

Atomic Electrons (^{79}Br)

⟨e⟩=48.0 *14* keV

e_{bin}(keV)	⟨e⟩(keV)	e(%)
2	0.29	18.6 *20*
10	0.60	5.8 *6*
11	0.0178	0.156 *17*
12	0.190	1.62 *17*
13	0.0161	0.122 *13*
194	39.2	20.2 *7*
205	3.94	1.92 *7*
206	2.65	1.29 *5*
207	1.19	0.577 *21*

$^{79}_{36}$Kr(1.460 4 d)

Mode: ϵ

Δ: -74442 6 keV

SpA: 1.132×10[6] Ci/g

Prod: ^{79}Br(p,n); ^{79}Br(d,2n); ^{78}Kr(n,γ)

Photons (^{79}Kr)

⟨γ⟩=184.4 24 keV

γ_{mode}	γ(keV)	γ(%)†
Br L_ℓ	1.293	0.054 12
Br L_η	1.339	0.032 8
Br L_α	1.481	1.09 22
Br L_β	1.534	0.65 17
Br L_γ	1.777	0.009 3
Br $K_{\alpha2}$	11.878	15.1 6
Br $K_{\alpha1}$	11.924	29.2 12
Br $K_{\beta1}$'	13.290	6.03 25
Br $K_{\beta2}$'	13.562	0.409 17
γ M1	44.24 7	0.216 25
γ M1+5.9%E2	136.16 6	0.85 11
γ M1,E2	180.40 7	~0.10
γ M1(+<3.8%E2)	208.58 6	0.77 5
γ M1+0.9%E2	217.11 6	2.37 10
γ M1(+0.4%E2)	261.34 6	12.7
γ	280.54 11	0.0103 14
γ M1+18%E2	299.55 7	1.54 8
γ M1+1.7%E2	306.54 6	2.6 1
γ (M1)	344.74 6	0.236 19
γ M1+3.5%E2	388.98 6	1.51 8
γ M1+11%E2	397.50 5	9.3 3
γ	434.51 6	0.039 5
γ	500.31 8	0.017 5
γ	506.47 11	0.089 18
γ M1+4.4%E2	523.14 10	0.23 6
γ	525.48 8	0.39 8
γ	538.5 10	0.0066 19
γ	570.67 7	0.0055 8
γ M1+8.3%E2	606.08 5	8.12 20
γ	614.91 7	0.10 3
γ	650.93 15	0.0114 25
γ	715.05 11	0.0061 11
γ [M1+E2]	726.24 7	0.019 4
γ	734.17 14	0.013 4
γ [M1+E2]	809.18 9	0.097 6
γ M1(+<1.4%E2)	832.01 6	1.26 6
γ	851.21 11	0.038 4
γ	870.33 14	0.0076 25
γ	895.45 12	0.0112 20
γ	914.57 14	0.007 3
γ M1,E2	934.82 7	0.121 8
γ M1,E2	1025.79 7	0.151 10
γ M1,E2	1070.98 7	0.061 10
γ	1076.1 4	0.0066 25
γ	1112.55 11	0.07 3
γ M1+43%E2	1115.22 8	0.37 4
γ	1131.67 14	0.042 5
γ M1,E2	1332.32 6	0.43 3

† 3.2% uncert(syst)

Atomic Electrons (^{79}Kr)

⟨e⟩=5.3 3 keV

e_{bin}(keV)	⟨e⟩(keV)	e(%)
2	1.12	71 7
10	2.42	23.6 24
11	0.072	0.63 6
12	0.77	6.6 7
13 - 44	0.146	0.75 6
123 - 167	0.077	0.060 8
179 - 217	0.103	0.0508 19

Atomic Electrons (^{79}Kr) (continued)

e_{bin}(keV)	⟨e⟩(keV)	e(%)
248	0.261	0.105 5
260 - 307	0.121	0.0431 19
331 - 376	0.0220	0.0060 3
384	0.116	0.0303 13
387 - 435	0.0182	0.00459 16
487 - 525	0.0066	0.0013 4
537 - 571	4.7 ×10[-5]	8 4 ×10[-6]
593 - 637	0.0632	0.0106 3
649 - 651	7.8 ×10[-6]	1.2 6 ×10[-6]
702 - 734	0.00022	3.1 6 ×10[-5]
796 - 838	0.0072	0.00088 5
849 - 895	8.9 ×10[-5]	1.0 3 ×10[-5]
901 - 935	0.00060	6.5 6 ×10[-5]
1012 - 1058	0.00086	8.4 7 ×10[-5]
1063 - 1111	0.00154	0.000140 16
1112 - 1132	0.00028	2.5 4 ×10[-5]
1182 - 1227	2.4 ×10[-5]	2.0 9 ×10[-6]
1239 - 1240	2.2 ×10[-6]	~2×10[-7]
1319 - 1332	0.00137	0.000104 9

Continuous Radiation (^{79}Kr)

⟨β+⟩=18.5 keV;⟨IB⟩=0.74 keV

E_{bin}(keV)		⟨ ⟩(keV)	(%)
0 - 10	β+	4.48×10[-5]	0.00055
	IB	0.00136	
10 - 20	β+	0.00102	0.0063
	IB	0.0077	0.061
20 - 40	β+	0.0164	0.051
	IB	0.0020	0.0069
40 - 100	β+	0.411	0.55
	IB	0.0055	0.0084
100 - 300	β+	7.9	3.89
	IB	0.042	0.021
300 - 600	β+	10.2	2.61
	IB	0.18	0.039
600 - 1300	β+	0.000218	3.62×10[-5]
	IB	0.47	0.054
1300 - 1628	IB	0.032	0.0023
	Σβ+		7.1

$^{79}_{36}$Kr(50 3 s)

Mode: IT

Δ: -74312 6 keV

SpA: 2.84×10[9] Ci/g

Prod: ^{79}Br(p,n); protons on Zr

Photons (^{79}Kr)

⟨γ⟩=40.0 8 keV

γ_{mode}	γ(keV)	γ(%)†
Kr L_ℓ	1.383	0.044 10
Kr L_η	1.435	0.026 7
Kr L_α	1.581	0.92 19
Kr L_β	1.639	0.54 14
Kr L_γ	1.907	0.0068 21
Kr $K_{\alpha2}$	12.598	10.8 5
Kr $K_{\alpha1}$	12.651	21.0 10
Kr $K_{\beta1}$'	14.109	4.43 22
Kr $K_{\beta2}$'	14.408	0.399 20
γ E3	129.86 6	27.2

† 2.2% uncert(syst)

Atomic Electrons (^{79}Kr)

⟨e⟩=89.1 23 keV

e_{bin}(keV)	⟨e⟩(keV)	e(%)
2	0.95	56 6
10	0.143	1.37 14
11	1.47	13.5 14
12	0.53	4.3 5
13	0.096	0.77 8
14	0.059	0.43 5
116	65.8	56.9 19
128	16.9	13.2 4
130	3.16	2.43 8

$^{79}_{37}$Rb(22.9 5 min)

Mode: ϵ

Δ: -70837 27 keV

SpA: 1.039×10[8] Ci/g

Prod: ^{65}Cu(^{16}O,2n); ^{79}Br(^{3}He,3n); ^{71}Ga(^{12}C,4n); protons on Ta, U, Th

Photons (^{79}Rb)

⟨γ⟩=595 9 keV

γ_{mode}	γ(keV)	γ(%)†
Kr L_ℓ	1.383	0.034 8
Kr L_η	1.435	0.020 5
Kr L_α	1.581	0.72 15
Kr L_β	1.639	0.42 11
Kr L_γ	1.907	0.0058 18
Kr $K_{\alpha2}$	12.598	9.0 4
Kr $K_{\alpha1}$	12.651	17.4 7
Kr $K_{\beta1}$'	14.109	3.67 16
Kr $K_{\beta2}$'	14.408	0.330 14
γ [E1]	17.31 6	0.231 23
γ [M1]	19.09 6	0.162 14
γ (M1)	52.40 6	0.291 9
γ (M1)	63.88 5	0.0832 23
γ [E1]	107.80 5	0.069 5
γ [E1]	111.35 7	0.039 8
γ [M1,E2]	116.28 6	0.208 7
γ E3	129.86 6	10.74 23 *
γ [M1,E2]	141.67 5	0.670 23
γ (E1)	143.45 6	13.9 3
γ E2	147.18 6	10.49 23
γ [E1]	149.31 6	0.670 23
γ M1(+<20%E2)	154.81 6	7.90 18
γ M1+32%E2	160.76 4	8.59 18
γ M1+0.6%E2	182.83 5	19.2 4
γ [M1,E2]	201.31 6	0.178 18
γ M1,E2	218.69 7	0.51 7
γ M1+0.2%E2	219.15 7	0.62 7
γ	244 5	0.092 23
γ M1(+E2)	275.01 13	0.106 23
γ [E1]	286.29 7	0.901 23
γ [E1]	302.15 21	0.16 7
γ M1+E2	302.82 21	~0.09
γ [E1]	304.13 6	0.104 18
γ	312.65 7	0.065 14
γ E1+3.8%M2	320.14 20	0.076 23
γ	337 5	0.62 12
γ E1	350.63 6	7.23 14
γ	382.58 11	0.23 5
γ M1	384.14 6	1.18 7
γ	388 5	0.21 7
γ M1	397.64 5	6.05 14
γ (E2)	401.98 8	0.808 23
γ	417 5	0.23 7
γ	428.93 8	0.139 23
γ [M1+E2]	461.52 5	1.224 24
γ	476.32 13	0.055 14
γ [E2]	486.91 7	0.224 18
γ E1	505.44 5	10.6 5
γ M1,E2	506.00 7	2.1 5
γ	524.26 10	0.217 12

Photons (^{79}Rb)
(continued)

γ_{mode}	γ(keV)	γ(%)†
γ [E1]	533.45 6	1.29 3
γ [E1]	541.09 6	0.739 23
γ	543.34 11	0.139 14
γ	549 5	0.65 7
γ [E2]	558.40 5	0.254 14
γ [E1]	569.32 5	0.924 23
γ [E2]	603.19 6	0.670 23
γ [E1]	604.97 7	0.092 16
γ M1,E2	622.28 5	8.78 18
γ	644 5	0.231 12
γ	655 5	0.231 12
γ	662.81 9	0.069 23
γ E1	688.26 5	23.1
γ	706.0 10	0.058 7
γ	724.48 14	0.062 7
γ	728 5	0.12 5
γ	774.17 6	0.601 18
γ [M1+E2]	786.45 6	0.162 9
γ	788.48 12	0.185 9
γ	792.45 9	0.023 5
γ	815.57 14	0.021 7
γ	835.4 19	0.058 9
γ	841.71 8	0.058 7
γ	857.87 7	0.035 12
γ	884 5	0.092 23
γ	891.94 14	~0.018
γ	907.30 14	0.083 9
γ	911.7 19	0.028 5
γ	915.84 6	0.601 12
γ	921.74 6	0.444 12
γ	930.15 13	0.046 5
γ	934.93 6	0.328 7
γ [M1+E2]	941.26 6	0.173 14
γ	949.24 13	0.069 16
γ	950.86 14	0.0139 23
γ	955.41 9	0.028 5
γ	964.1 24	0.0069 23
γ	968.9 14	0.0162 23
γ	974.14 8	0.035 5 ?
γ	976.22 12	0.0115 23
γ	1009.41 17	0.037 9
γ	1017.8 19	0.028 7
γ [E1]	1072.74 8	0.030 7 ?
γ	1076.55 7	0.051 7
γ	1084.9 24	0.0069 23
γ [E1]	1090.57 7	0.0115 23
γ	1117.20 17	0.044 7
γ	1124.29 11	~0.009
γ	1132.33 7	0.055 7
γ	1137.69 7	0.120 9
γ	1139.35 15	0.012 5
γ	1148 3	0.023 9
γ	1151.08 17	~0.014
γ	1157.92 13	0.021 7
γ	1160.01 21	~0.014
γ	1170.17 16	0.042 7
γ [M1+E2]	1184.09 6	0.249 5
γ	1197 3	0.014 5
γ	1199.3 19	0.023 5
γ	1208.03 7	0.074 7
γ	1225.86 7	0.083 9
γ	1245.48 8	0.037 5
γ	1252.7 19	0.012 5
γ	1257.55 21	0.021 5
γ	1267.9 19	0.018 5
γ	1273.9 24	~0.014
γ	1279.11 11	0.016 7 ?
γ	1286.20 20	0.0046 9
γ [E1]	1291.88 6	0.067 7
γ	1298.44 7	0.129 7
γ	1308.5 14	0.025 5
γ	1316.7 14	0.028 5
γ	1338 3	0.0069 23
γ	1357.0 19	0.014 5
γ	1366.8 14	0.021 5
γ	1373 5	0.018 5
γ	1379 5	0.0115 23
γ	1390.2 19	0.0092 23
γ	1395.1 10	0.0254 23
γ	1404 3	0.0069 23
γ	1408.5 14	0.0162 23
γ	1412.4 14	0.0162 23
γ	1416.92 9	0.042 5
γ	1427.17 6	0.074 5

Photons (^{79}Rb)
(continued)

γ_{mode}	γ(keV)	γ(%)†
γ	1453.89 13	0.0115 23
γ [E1]	1474.71 6	0.074 5
γ	1480.14 6	0.166 5
γ	1485.2 14	0.018 5
γ	1491.2 24	~0.009
γ	1498 5	~0.07
γ	1504.72 9	0.095 5
γ	1509 9	0.014 5
γ	1517.6 14	0.0162 23
γ	1521.93 11	0.035 7
γ	1530 3	~0.007
γ	1536 3	0.012 5
γ	1548 4	~0.007
γ	1553 4	0.014 5
γ	1558.59 10	0.021 5
γ	1563 3	0.014 5
γ	1577.68 9	0.016 5
γ	1592.8 19	0.016 5
γ	1598 4	~0.009
γ	1606.5 10	0.0208 23
γ	1613.1 5	0.0254 23
γ	1629.73 11	0.0162 23
γ	1633.9 5	0.0231 23
γ	1636 3	~0.0046
γ	1663.60 11	0.014 5
γ	1665.38 11	0.021 5
γ	1673.9 24	~0.009
γ	1678.38 14	0.076 5
γ	1682.69 11	0.044 5
γ	1708.8 24	0.012 5
γ	1716.5 19	0.016 5
γ	1723.4 19	0.018 5
γ	1730 3	~0.007
γ	1743 5	0.069 23
γ	1748 5	~0.009
γ	1753 5	~0.009
γ	1769.2 19	0.016 5
γ	1776.0 14	0.018 5
γ	1784.6 14	0.016 5
γ	1792.5 19	0.014 5
γ	1802.0 24	0.012 5
γ	1820.1 19	0.012 5
γ [M1+E2]	1834.08 12	~0.009 ?
γ	1854 6	0.035 12
γ	1857 6	0.012 5
γ	1885 6	~0.0046
γ [M1+E2]	1897.95 12	~0.0046
γ	1908.9 10	0.0092 23
γ	1922.55 18	0.028 7
γ	1924.67 18	0.012 5
γ	1929.3 24	0.0115 23
γ	1943.9 10	0.025 5
γ [M1+E2]	1947.95 14	0.0115 23
γ	1951.4 19	0.0115 23
γ	1956.43 19	0.0092 23
γ	1961.9 10	0.0139 23
γ	1975.51 19	~0.018
γ	1982.50 14	~0.028
γ	1994.7 14	0.012 5
γ	2000.4 24	0.0069 23
γ	2007.6 14	0.014 5
γ	2021 7	<0.014
γ [M1+E2]	2052.76 12	0.025 5
γ	2060 5	0.069 23
γ	2067.1 19	0.0069 23
γ	2084.9 19	0.0069 23
γ	2102.9 19	0.018 5
γ	2113 5	~0.007
γ	2148 3	~0.014
γ	2157 3	0.025 9
γ	2169.7 19	~0.014
γ	2175 3	~0.012
γ [E1]	2184.23 13	0.039 5 ?
γ [E1]	2202.07 12	0.023 7
γ	2214.6 24	0.016 5
γ	2234 3	0.012 5
γ	2238 5	~0.009
γ [M1+E2]	2293.19 13	0.018 7
γ [M1+E2]	2295.58 12	0.037 9
γ	2339.7 10	0.0115 23
γ	2357.6 14	0.0092 23
γ [E1]	2400.98 13	0.023 7
γ [E1]	2403.38 12	0.0046 9
γ	2415.2 24	0.018 5

Photons (^{79}Rb)
(continued)

γ_{mode}	γ(keV)	γ(%)†
γ	2420 4	~0.009
γ	2432 3	0.035 7
γ [E2]	2434.86 13	0.032 7
γ	2441 3	~0.0046
γ	2446.9 19	0.0115 23
γ	2452 6	~0.009
γ [M1+E2]	2453.94 13	~0.007
γ [E2]	2456.34 12	0.014 5
γ	2487 3	0.012 5
γ	2490 4	~0.009
γ	2539.4 14	0.0092 23
γ	2542.6 10	0.0162 23
γ	2582.6 24	0.0092 23
γ [E1]	2586.20 11	0.0254 23
γ	2591 3	0.0069 23
γ	2595.5 14	0.0092 23
γ	2608.8 19	~0.0046
γ	2620 3	~0.0046
γ	2632 3	~0.009
γ	2637 3	~0.014
γ	2657 6	0.012 5
γ	2684.6 10	0.0231 23
γ	2693.8 14	0.0115 23
γ	2700.9 24	0.0115 23
γ	2703 5	0.0046 9
γ	2742.1 14	0.0139 23
γ	2745.0 14	0.0092 23
γ	2750.3 14	0.0069 23
γ	2794 3	0.014 5
γ	2813 7	~0.0046
γ	2817 5	~0.0046
γ	2861.4 14	0.0069 23
γ	2872 5	~0.0046
γ	2875.6 24	0.0092 23
γ	2983.3 14	0.0069 23
γ	2996 3	0.0069 23
γ	3007 3	0.0069 23
γ	3021.5 14	0.0092 23

† 5.2% uncert(syst)

* with ^{79}Kr(50 s) in equilib

Atomic Electrons (^{79}Rb)

$\langle e \rangle$=43.9 8 keV

e_{bin}(keV)	$\langle e \rangle$(keV)	e(%)
2	0.75	43.4 44
3 - 51	2.3	22.6 10
52 - 97	0.0115	0.0157 7
102 - 115	0.07	~0.06
116	25.8	22.3 7
127	0.11	~0.09
128	6.62	5.17 15
129	0.496	0.384 17
130	1.24	0.95 3
133	2.68	2.02 6
135 - 183	3.19	2.11 12
187 - 229	0.076	0.037 9
242 - 290	0.021	0.0076 11
298 - 337	0.070	0.021 3
349 - 398	0.136	0.0358 11
400 - 447	0.024	0.0055 10
460 - 506	0.083	0.0169 15
510 - 558	0.024	0.0045 5
567 - 608	0.079	0.0131 16
620 - 663	0.0128	0.00204 22
674 - 723	0.074	0.0109 5
724 - 772	0.0039	0.00051 16
774 - 821	0.0015	0.00019 5
827 - 870	0.00077	9.0 24 $\times 10^{-5}$
878 - 927	0.0067	0.00074 13
928 - 976	0.00118	0.000125 19
995 - 1018	0.00022	2.2 7 $\times 10^{-5}$
1057 - 1103	0.00049	4.5 8 $\times 10^{-5}$
1110 - 1158	0.00094	8.3 14 $\times 10^{-5}$
1160 - 1208	0.00131	0.000112 8
1212 - 1260	0.00050	4.1 8 $\times 10^{-5}$
1265 - 1309	0.00070	5.4 10 $\times 10^{-5}$
1315 - 1358	0.00015	1.11 23 $\times 10^{-5}$

Atomic Electrons (^{79}Rb)
(continued)

e_{bin}(keV)	⟨e⟩(keV)	e(%)
1365 - 1413	0.00047	3.4 *5* $\times10^{-5}$
1415 - 1460	0.000162	1.12 *10* $\times10^{-5}$
1466 - 1563	0.00119	8.0 *12* $\times10^{-5}$
1576 - 1674	0.00057	3.5 *4* $\times10^{-5}$
1676 - 1776	0.00042	2.4 *4* $\times10^{-5}$
1778 - 1870	0.00022	1.19 *21* $\times10^{-5}$
1883 - 1982	0.00037	1.93 *23* $\times10^{-5}$
1986 - 2085	0.00026	1.3 *3* $\times10^{-5}$
2089 - 2188	0.00030	1.40 *18* $\times10^{-5}$
2200 - 2295	0.00018	8.2 *12* $\times10^{-6}$
2325 - 2421	0.00022	9.2 *12* $\times10^{-6}$
2426 - 2525	0.000143	5.8 *8* $\times10^{-6}$
2528 - 2623	0.000143	5.5 *7* $\times10^{-6}$
2630 - 2728	0.000118	4.4 *5* $\times10^{-6}$
2731 - 2817	6.1 $\times10^{-5}$	2.2 *4* $\times10^{-6}$
2847 - 2875	3.1 $\times10^{-5}$	1.07 *23* $\times10^{-6}$
2969 - 3021	4.3 $\times10^{-5}$	1.4 *3* $\times10^{-6}$

Continuous Radiation (^{79}Rb)
⟨β+⟩=743 keV;⟨IB⟩=2.6 keV

E_{bin}(keV)		⟨ ⟩(keV)	(%)
0 - 10	β+	2.75 $\times10^{-5}$	0.000339
	IB	0.029	
10 - 20	β+	0.00066	0.00406
	IB	0.030	0.21
20 - 40	β+	0.0115	0.0358
	IB	0.056	0.19
40 - 100	β+	0.340	0.446
	IB	0.156	0.24
100 - 300	β+	11.7	5.4
	IB	0.42	0.24
300 - 600	β+	76	16.4
	IB	0.46	0.108
600 - 1300	β+	417	44.7
	IB	0.71	0.081
1300 - 2500	β+	238	15.4
	IB	0.66	0.038
2500 - 3604	β+	0.0098	0.000390
	IB	0.051	0.0019
	Σβ+		82

$^{79}_{38}$Sr(2.25 *10* min)

Mode: ϵ
Δ: -65340 *200* keV syst
SpA: 1.06$\times10^{9}$ Ci/g
Prod: ^{64}Zn(^{16}O,n)

Photons (^{79}Sr)
⟨γ⟩=176 *8* keV

γ_{mode}	γ(keV)	γ(%)†
γ M1,E1	39.46 *9*	28
γ M1,E1	105.05 *6*	21.8 *6*
γ	134.95 *11*	3.1 *6*
γ	140.95 *9*	4.5 *6*
γ	144.51 *8*	3.1 *14*
γ	167.57 *10*	~0.6
γ	219.06 *9*	5.9 *8*
γ	246.00 *9*	3.1 *14*
γ	308.52 *13*	1.3 *3*
γ	317.84 *11*	5.0 *8*
γ	324.11 *9*	5.3 *8*
γ	366.47 *20*	2.2 *3*
γ	413.57 *12*	7.6 *6*
γ	612.48 *21*	4.8 *6*

† 18% uncert(syst)

A = 80

NDS **36**, 127 (1982)

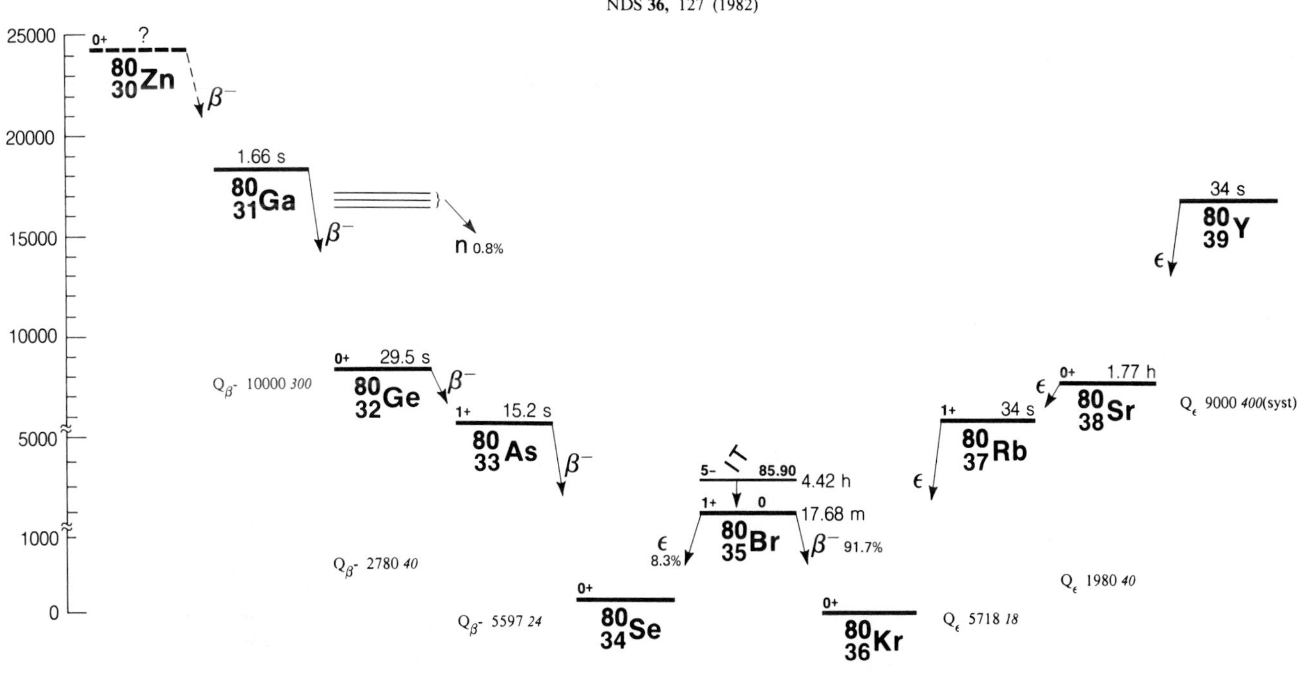

$^{80}_{30}$Zn($t_{1/2}$ unknown)

Mode: β-

Prod: fission

Photons (^{80}Zn)

γ_{mode}	γ(keV)	γ(rel)
γ	642 *1*	46 *9*
γ	712 *1*	100 *20*
γ	715 *1*	70 *14*

$^{80}_{31}$Ga(1.66 *2* s)

Mode: β-, β-n(0.84 *6* %)

Δ: -59380 *300* keV

SpA: 6.94×10^{10} Ci/g

Prod: fission

Photons (^{80}Ga)

⟨γ⟩=2773 *29* keV

γ_{mode}	γ(keV)	γ(%)†
γ	398.54 *5*	0.42 *8*
γ	466.69 *5*	1.14 *4*
γ	519.91 *14*	1.05 *9*
γ	523.15 *3*	10.8 *3*
γ	570.99 *5*	4.87 *17*
γ	586.11 *3*	5.5 *3*
γ[E2]	659.08 *4*	84.0 *25*
γ	692.14 *4*	0.487 *25*
γ	707.47 *8*	0.252 *25*
γ	771.11 *5*	0.395 *17*
γ	808.37 *4*	0.61 *3*
γ	833.89 *4*	4.70 *25*
γ[M1+E2]	914.40 *4*	4.45 *17*
γ	989.44 *5*	0.95 *4*
γ	1004.74 *4*	0.76 *3*
γ	1040.54 *4*	1.44 *6*
γ	1047.49 *10*	0.25 *8*
γ	1064.71 *5*	0.75 *4*
γ[E2]	1083.39 *4*	52.1 *17*
γ	1109.25 *3*	20.0 *7*
γ	1130.63 *7*	0.98 *4*
γ	1135.88 *4*	3.53 *17*
γ	1154.78 *11*	0.65 *4*
γ	1157.89 *6*	0.286 *25*
γ	1235.67 *6*	5.2 *3*
γ	1244.77 *8*	0.66 *3*
γ	1249.61 *5*	0.244 *25*
γ	1294.25 *5*	0.58 *4*
γ	1306.73 *5*	1.90 *8*
γ	1312.93 *4*	7.14 *25*
γ	1471.87 *4*	0.56 *3*
γ	1547.32 *19*	0.31 *4*
γ	1561.22 *8*	0.25 *5*
γ[E2]	1573.47 *4*	3.70 *17*
γ	1585.22 *5*	0.529 *25*
γ	1680.56 *5*	4.54 *17*
γ	1772.75 *6*	1.37 *10*
γ	1850.02 *5*	0.56 *3*
γ	1867.39 *10*	0.260 *17*
γ	1882.10 *14*	0.17 *3*
γ	1941.47 *11*	0.496 *25*
γ	1999.14 *9*	0.52 *3*
γ	2008.71 *9*	0.437 *25*
γ	2016.46 *21*	0.235 *17*
γ	2114.51 *7*	0.96 *4*
γ	2140.61 *5*	0.74 *5*
γ	2160.46 *21*	0.40 *4*
γ	2283.15 *7*	1.08 *5*

Photons (^{80}Ga)
(continued)

γ_{mode}	γ(keV)	γ(%)†
γ	2351.56 *6*	0.328 *25*
γ	2396.47 *10*	0.344 *25*
γ	2554.88 *12*	0.202 *17*
γ	2581.10 *5*	0.94 *4*
γ	2599.21 *19*	0.72 *4*
γ	2637.9 *5*	0.18 *3*
γ	2665.11 *15*	0.48 *4*
γ	2750.09 *6*	0.54 *3*
γ	2764.40 *6*	0.91 *5*
γ	2821.96 *9*	0.34 *3*
γ	2948.38 *7*	0.82 *4*
γ	3043.95 *10*	0.412 *25*
γ	3090.50 *14*	0.46 *3*
γ	3108.36 *9*	1.13 *6*
γ	3335.5 *4*	0.29 *5*
γ	3664.45 *5*	3.11 *17*
γ	3764.45 *16*	0.40 *3*
γ	3818.6 *3*	0.49 *4*
γ	3919.3 *4*	0.25 *4*
γ	3971.5 *3*	0.35 *5*
γ	4238.53 *20*	0.45 *3*
γ	4412.53 *24*	0.60 *4*
γ	4443.3 *3*	0.84 *7*
γ	4678.80 *16*	0.55 *3*
γ	4729.8 *3*	0.35 *3*
γ	5354.83 *20*	0.210 *17*
γ	5387.73 *24*	1.18 *5*

† 5.9% uncert(syst)

Continuous Radiation (^{80}Ga)

⟨β-⟩=3536 keV;⟨IB⟩=16.7 keV

E_{bin}(keV)		⟨ ⟩(keV)	(%)
0 - 10	β-	0.00083	0.0164
	IB	0.078	
10 - 20	β-	0.00255	0.0170
	IB	0.077	0.54
20 - 40	β-	0.0108	0.0358
	IB	0.154	0.53
40 - 100	β-	0.089	0.125
	IB	0.45	0.69
100 - 300	β-	1.33	0.63
	IB	1.43	0.79
300 - 600	β-	7.7	1.65
	IB	1.9	0.45
600 - 1300	β-	73	7.4
	IB	3.7	0.41
1300 - 2500	β-	427	22.0
	IB	4.3	0.24
2500 - 5000	β-	1709	46.8
	IB	3.9	0.116
5000 - 7500	β-	1137	19.1
	IB	0.72	0.0125
7500 - 10000	β-	181	2.23
	IB	0.030	0.00038

$^{80}_{32}$Ge(29.5 *4* s)

Mode: β-

Δ: -69380 *30* keV

SpA: 4.72×10^{9} Ci/g

Prod: fission

Photons (^{80}Ge)

⟨γ⟩=600 *22* keV

γ_{mode}	γ(keV)	γ(%)†
γ	110.2 *4*	11.5 *5* ?
γ	198.9 *10*	
γ[M1+E2]	265.20 *6*	48.0 *24*
γ	310.52 *5*	2.26 *14*
γ	319.29 *6*	0.80 *8*
γ	360.70 *5*	1.44 *10*
γ	369.47 *6*	1.06 *7*
γ[M1+E2]	414.78 *7*	0.61 *6*
γ	576.10 *6*	0.65 *5*
γ	626.28 *7*	1.49 *19*
γ[M1+E2]	679.99 *5*	3.94 *24*
γ	782.1 *4*	1.4 *5* ?
γ[M1+E2]	936.80 *6*	7.2 *4*
γ	1013.8 *4*	4.8 *10* ?
γ	1115.8 *4*	5.0 *7* ?
γ	1135.8 *10*	
γ	1255.9 *4*	6.0 *8* ?
γ	1305.3 *10*	
γ	1564.1 *4*	8.6 *8* ?
γ[M1+E2]	1872.93 *20*	0.250 *19*

† 4.2% uncert(syst)

Continuous Radiation (^{80}Ge)

⟨β-⟩=1049 keV;⟨IB⟩=2.4 keV

E_{bin}(keV)		⟨ ⟩(keV)	(%)
0 - 10	β-	0.0099	0.197
	IB	0.038	
10 - 20	β-	0.0305	0.203
	IB	0.038	0.26
20 - 40	β-	0.127	0.423
	IB	0.074	0.26
40 - 100	β-	1.02	1.44
	IB	0.21	0.32
100 - 300	β-	13.9	6.7
	IB	0.58	0.33
300 - 600	β-	68	14.9
	IB	0.62	0.146
600 - 1300	β-	410	43.1
	IB	0.71	0.084
1300 - 2500	β-	552	32.8
	IB	0.17	0.0108
2500 - 2780	β-	5.2	0.201
	IB	9.2×10^{-5}	3.6×10^{-6}

$^{80}_{33}$As(15.2 *2* s)

Mode: β-

Δ: -72165 *24* keV

SpA: 9.07×10^{9} Ci/g

Prod: ^{80}Se(n,p); fission

Photons (^{80}As)

⟨γ⟩=637 *30* keV

γ_{mode}	γ(keV)	γ(%)†
γ	321.1 *7*	5.5 *13* *
γ E2	665.94 *15*	42
γ E2+M1	783.0 *3*	0.97 *13*
γ (E2)	811.4 *5*	0.34 *13*
γ	861.9 *3*	0.80 *8*
γ	908.6 *7*	0.71 *13* *
γ[M1+E2]	1064.6 *4*	0.13 *4*
γ	1206.97 *20*	4.2 *4*
γ (E2,M1)	1293.9 *3*	0.97 *17*

Photons (^{80}As)
(continued)

γ_{mode}	γ(keV)	γ(%)†
γ	1415.8 5	~0.08
γ	1422.6 7	<0.08 *
γ E2	1448.9 3	1.13 17
γ	1633.7 7	1.18 21 *
γ	1644.8 3	7.6 4
γ [M1+E2]	1847.5 3	1.13 17
γ (E2)	1959.8 3	0.46 8
γ	1968.7 7	~0.13 *
γ	2156.9 5	~0.08
γ	2357.7 5	0.88 17
γ	2461.2 7	~0.21 *
γ [E2]	2513.5 3	~0.17
γ	2598.0 7	~0.13 *
γ	2774.1 10	0.29 13
γ	2836.1 10	0.25 8
γ	2939.8 5	~0.08
γ	3023.6 5	~0.08
γ	3060.6 6	<0.08

† 9.5% uncert(syst)
* with ^{80}As or ^{82}As(14.0 s) or ^{82}As(19.1 s)

Continuous Radiation (^{80}As)
⟨β-⟩=2249 keV;⟨IB⟩=8.3 keV

E_{bin}(keV)		⟨ ⟩(keV)	(%)
0 - 10	β-	0.00222	0.0442
	IB	0.062	
10 - 20	β-	0.0069	0.0456
	IB	0.061	0.42
20 - 40	β-	0.0288	0.096
	IB	0.121	0.42
40 - 100	β-	0.234	0.330
	IB	0.35	0.54
100 - 300	β-	3.38	1.62
	IB	1.08	0.60
300 - 600	β-	18.6	4.02
	IB	1.37	0.32
600 - 1300	β-	159	16.2
	IB	2.3	0.27
1300 - 2500	β-	717	37.7
	IB	2.1	0.122
2500 - 5000	β-	1329	39.6
	IB	0.82	0.027
5000 - 5597	β-	21.8	0.422
	IB	0.00070	1.37×10^{-5}

$^{80}_{34}$Se(stable)

Δ: -77762.1 18 keV
%: 49.7 7

$^{80}_{35}$Br(17.68 2 min)

Mode: β-(91.7 2 %), ε(8.3 2 %)
Δ: -75891.0 24 keV
SpA: 1.3289×10^{8} Ci/g
Prod: ^{79}Br(n,γ); daughter ^{80}Br(4.42 h); ^{81}Br(γ,n); ^{79}Br(d,p)

Photons (^{80}Br)
⟨γ⟩=53 4 keV

γ_{mode}	γ(keV)	γ(%)†
Se L_ℓ	1.204	0.0033 7
Se L_η	1.245	0.0020 5
Se L_α	1.379	0.064 13
Se L_β	1.427	0.038 10
Se L_γ	1.648	0.00054 17
Se $K_{\alpha2}$	11.182	0.94 4
Se $K_{\alpha1}$	11.222	1.82 8
Se $K_{\beta1}$'	12.494	0.368 15
Se $K_{\beta2}$'	12.741	0.0177 7
$\gamma_{\beta-}$E2	616.9 3	6.7
$\gamma_{\beta-}$E2+2.9%M1	639.93 18	0.252 17
γ_ϵE2	665.94 15	1.08
γ	677.6 10	0.008 3 ?
γ	688.0 10	0.012 3 ?
$\gamma_{\beta-}$(E2)	704.37 19	0.19 3
γ_ϵE2+M1	783.0 3	<0.013 ?
γ_ϵ(E2)	811.4 5	0.040 13
$\gamma_{\beta-}$[E2]	1256.8 3	0.079 6
γ	1339.1 8 ?	
γ_ϵE2	1448.9 3	<0.016

† uncert(syst): 9.0% for ε, 9.6% for β-

Atomic Electrons (^{80}Br)*
⟨e⟩=0.296 18 keV

e_{bin}(keV)	⟨e⟩(keV)	e(%)
1	0.063	4.4 5
2	0.0048	0.29 3
9	0.0144	0.154 16
10	0.150	1.55 16
11	0.051	0.46 5
12	0.0032	0.026 3
653	0.0076	0.00116 11
664	0.00080	0.000120 11
665	2.26×10^{-5}	3.4 3 $\times10^{-6}$
666	0.000145	2.18 20 $\times10^{-5}$
770	3.3×10^{-5}	$<9\times10^{-6}$
781	3.4×10^{-6}	$<9\times10^{-7}$
782	1.3×10^{-7}	$<3\times10^{-8}$
783	6.2×10^{-7}	$<1.6\times10^{-7}$
799	0.00020	2.5 9 $\times10^{-5}$
810	2.2×10^{-5}	2.7 9 $\times10^{-6}$
811	3.9×10^{-6}	4.8 16 $\times10^{-7}$
1436	1.9×10^{-5}	$\sim1\times10^{-6}$
1447	2.0×10^{-6}	$\sim1\times10^{-7}$
1449	3.5×10^{-7}	$\sim2\times10^{-8}$

* with ε

Continuous Radiation (^{80}Br)
⟨β-⟩=716 keV;⟨β+⟩=8.0 keV;⟨IB⟩=1.45 keV

E_{bin}(keV)		⟨ ⟩(keV)	(%)
0 - 10	β-	0.0163	0.324
	β+	6.4×10^{-6}	8.0×10^{-5}
	IB	0.029	
10 - 20	β-	0.0498	0.332
	β+	0.000143	0.00088
	IB	0.029	0.20
20 - 40	β-	0.207	0.69
	β+	0.00230	0.0072
	IB	0.055	0.19
40 - 100	β-	1.63	2.30
	β+	0.060	0.080
	IB	0.153	0.24
100 - 300	β-	20.9	10.1
	β+	1.54	0.73
	IB	0.40	0.23
300 - 600	β-	93	20.3
	β+	4.95	1.14

Continuous Radiation (^{80}Br)
(continued)

E_{bin}(keV)		⟨ ⟩(keV)	(%)
	IB	0.39	0.092
600 - 1300	β-	418	45.3
	β+	1.44	0.216
	IB	0.36	0.044
1300 - 2001	β-	183	12.2
	IB	0.034	0.0024
	Σβ+		2.2

$^{80}_{35}$Br(4.42 1 h)

Mode: IT
Δ: -75805.1 24 keV
SpA: 8.862×10^{6} Ci/g
Prod: ^{79}Br(n,γ)

Photons (^{80}Br)
⟨γ⟩=24.1 5 keV

γ_{mode}	γ(keV)	γ(%)†
Br L_ℓ	1.293	0.094 21
Br L_η	1.339	0.052 13
Br L_α	1.481	1.9 4
Br L_β	1.534	1.1 3
Br L_γ	1.777	0.016 5
Br $K_{\alpha2}$	11.878	23.2 10
Br $K_{\alpha1}$	11.924	44.8 19
Br $K_{\beta1}$'	13.290	9.3 4
Br $K_{\beta2}$'	13.562	0.63 3
γ E1	37.052 2	39.1
γ M3	48.85 3	0.324 9

† 2.9% uncert(syst)

Atomic Electrons (^{80}Br)
⟨e⟩=60.8 12 keV

e_{bin}(keV)	⟨e⟩(keV)	e(%)
2	1.91	121 12
10	3.7	36 4
11	0.111	0.97 10
12	1.18	10.1 10
13	0.100	0.76 8
24	12.7	53.7 17
35	27.4	77 3
36	0.329	0.93 3
37	0.400	1.09 3
47	10.9	23.2 8
49	2.14	4.40 15

$^{80}_{36}$Kr(stable)

Δ: -77892 8 keV
%: 2.25 2

$^{80}_{37}$Rb(34 *4* s)

Mode: ϵ

Δ: -72173 *19* keV

SpA: 4.1×10^{9} Ci/g

Prod: daughter ^{80}Sr; ^{71}Ga(^{12}C,3n)

Photons (^{80}Rb)

⟨γ⟩=184 *19* keV

γ_{mode}	γ(keV)	γ(%)†
Kr $K_{\alpha2}$	12.598	0.336 *14*
Kr $K_{\alpha1}$	12.651	0.65 *3*
Kr $K_{\beta1}$'	14.109	0.138 *6*
Kr $K_{\beta2}$'	14.408	0.0124 *5*
γ E2	616.9 *3*	25
γ E2+2.9%M1	639.93 *18*	1.60 *15*
γ (E2)	704.37 *19*	1.88 *20*
γ [E2]	1256.8 *3*	0.50 *5*

† 12% uncert(syst)

Atomic Electrons (^{80}Rb)

⟨e⟩=0.39 *3* keV

e_{bin}(keV)	⟨e⟩(keV)	e(%)
2	0.026	1.52 *16*
10	0.0045	0.042 *4*
11	0.046	0.42 *4*
12	0.0164	0.133 *14*
13	0.0030	0.0238 *24*
14	0.00184	0.0132 *14*
603	0.23	0.038 *5*
615	0.026	0.0042 *5*
617	0.0048	0.00078 *10*
626	0.0138	0.00221 *21*
638	0.00154	0.000242 *23*
640	0.00029	4.5 *4* ×10^{-5}
690	0.0138	0.00200 *23*
702	0.00144	0.000205 *23*
703	9.2 ×10^{-5}	1.31 *15* ×10^{-5}
704	0.00028	4.0 *5* ×10^{-5}
1242	0.00166	0.000134 *15*
1255	0.000178	1.42 *16* ×10^{-5}
1257	2.9 ×10^{-5}	2.27 *25* ×10^{-6}
1257	4.6 ×10^{-6}	3.6 *4* ×10^{-7}

Continuous Radiation (^{80}Rb)

⟨β+⟩=2039 keV;⟨IB⟩=8.0 keV

E_{bin}(keV)		⟨ ⟩(keV)	(%)
0 - 10	β+	4.19×10^{-6}	5.2×10^{-5}
	IB	0.059	
10 - 20	β+	0.000101	0.00062
	IB	0.058	0.41
20 - 40	β+	0.00178	0.0055
	IB	0.115	0.40
40 - 100	β+	0.054	0.071
	IB	0.34	0.51
100 - 300	β+	2.08	0.95
	IB	1.02	0.57
300 - 600	β+	16.5	3.53
	IB	1.28	0.30
600 - 1300	β+	170	17.3
	IB	2.2	0.25
1300 - 2500	β+	832	43.6
	IB	2.0	0.114
2500 - 5000	β+	1018	32.7
	IB	0.91	0.028
5000 - 5718	IB	0.0109	0.00021
	Σβ+		98

$^{80}_{38}$Sr(1.772 *25* h)

Mode: ϵ

Δ: -70190 *30* keV

SpA: 2.21×10^{7} Ci/g

Prod: ^{14}N on Ga; daughter ^{80}Y

Photons (^{80}Sr)

⟨γ⟩=337 *24* keV

γ_{mode}	γ(keV)	γ(%)†
Rb L_{ℓ}	1.482	0.060 *13*
Rb L_{η}	1.542	0.035 *9*
Rb L_{α}	1.694	1.3 *3*
Rb L_{β}	1.760	0.76 *20*
Rb L_{γ}	2.051	0.012 *4*
Rb $K_{\alpha2}$	13.336	16.2 *7*
Rb $K_{\alpha1}$	13.395	31.2 *14*
Rb $K_{\beta1}$'	14.959	6.7 *3*
Rb $K_{\beta2}$'	15.286	0.72 *3*
γ	174.9 *4*	10.1 *12*
γ	236.1 *7*	4.3 *4*
γ	317.4 *8*	1.05 *12*
γ	378.6 *4*	4.3 *4*
γ	414.1 *4*	3.2 *4*
γ [M1+E2]	553.5 *4*	7.0 *8*
γ [M1+E2]	589.0 *4*	39 *4*

† 7.7% uncert(syst)

Atomic Electrons (^{80}Sr)

⟨e⟩=6.6 *8* keV

e_{bin}(keV)	⟨e⟩(keV)	e(%)
2	1.28	70 *7*
11	0.87	7.7 *8*
12	1.42	12.3 *13*
13	0.89	6.8 *7*
15	0.081	0.55 *6*
160	1.0	~0.6
173	0.14	~0.08
175	0.027	~0.015
221	0.21	~0.09
234	0.026	~0.011
236	0.005	~0.0022
302	0.026	~0.009
315	0.0027	~0.0009
316	0.0004	~0.00013
317	0.0006	~0.00019
363	0.07	~0.020
377	0.008	~0.0022
378	0.0014	~0.00038
379	0.00020	~5×10^{-5}
399	0.04	~0.011
412	0.005	~0.0013
414	0.0010	~0.00024
538	0.074	0.014 *3*
551	0.0079	0.0014 *3*
552	0.0005	~9×10^{-5}
553	0.0016	0.00029 *6*
574	0.37	0.065 *11*
587	0.042	0.0072 *13*
589	0.0081	0.0014 *3*

Continuous Radiation (^{80}Sr)

⟨β+⟩=30.1 keV;⟨IB⟩=1.06 keV

E_{bin}(keV)		⟨ ⟩(keV)	(%)
0 - 10	β+	1.16×10^{-5}	0.000143
	IB	0.00160	
10 - 20	β+	0.000286	0.00175
	IB	0.0106	0.078
20 - 40	β+	0.00501	0.0156
	IB	0.0031	0.0108
40 - 100	β+	0.145	0.190
	IB	0.0086	0.013
100 - 300	β+	4.15	1.94
	IB	0.052	0.026
300 - 600	β+	16.3	3.69
	IB	0.20	0.045
600 - 1300	β+	9.5	1.36
	IB	0.64	0.072
1300 - 1980	IB	0.144	0.0098
	Σβ+		7.2

$^{80}_{39}$Y (33.8 *6* s)

Mode: ϵ

Δ: -61190 *400* keV syst

SpA: 4.13×10^{9} Ci/g

Prod: ^{24}Mg on Ni; ^{28}Si on Ni

Photons (^{80}Y)

γ_{mode}	γ(keV)	γ(rel)
γ [E2]	385.87 *10*	100 *20*
γ [E2]	595.06 *14*	42 *8*
γ	690.49 *18*	3.0 *6*
γ	756.48 *12*	11.0 *22*
γ [E2]	782.81 *16*	6.0 *12*
γ	851.90 *14*	9.0 *18*
γ	1185.22 *15*	15 *3*
γ	1277.5 *3*	3.0 *6*
γ	1394.6 *3*	1.0 *2*

A = 81

NDS **15**, 137 (1975)

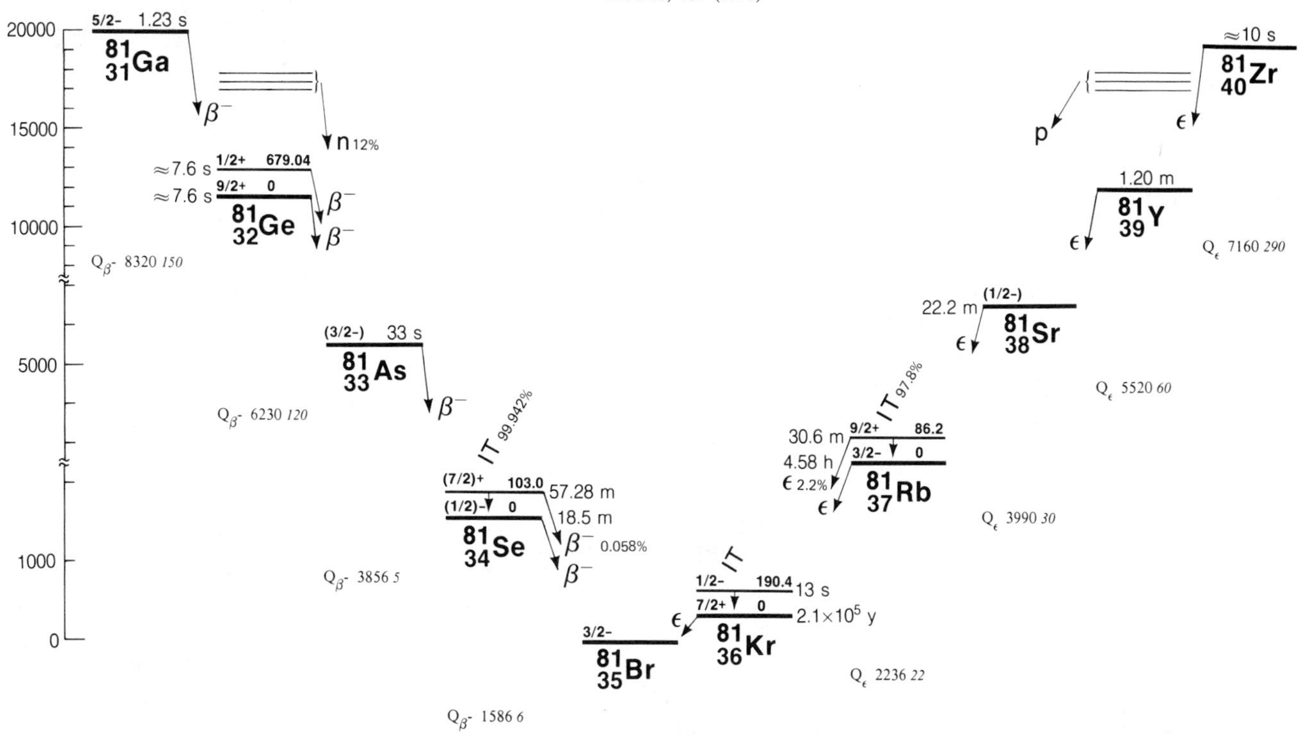

$^{81}_{31}$Ga(1.23 *1* s)

Mode: β-, β-n(12.0 *9* %)
Δ: -57990 *190* keV
SpA: 8.66×10^{10} Ci/g
Prod: fission

Photons (^{81}Ga)

⟨γ⟩=2082 *40* keV

γ$_{mode}$	γ(keV)	γ(%)†
γβ-[E1]	216.5 *1*	35 *4*
γβ-	256.6 *3*	0.16 *3*
γβ-	262.0 *1*	0.27 *5*
γβ-	437.4 *1*	0.15 *3*
γβ-	482.5 *1*	3.1 *3*
γβ-	501.9 *2*	0.33 *7*
γβ-	530.2 *1*	4.2 *4*
γβ-	562.4 *1*	0.69 *14*
γβ-	574.8 *1*	0.81 *16*
γβ-	613.9 *1*	0.23 *5*
γβ-	626.4 *1*	0.21 *4*
γβ-n[E2]	659.08 *4*	1.8 *4*
γβ-	659.1 *1*	5.2 *5*
γβ-	698.7 *1*	1.57 *16*
γβ-[E2]	711.2 *1*	16.5 *16*
γβ-	728.3 *1*	0.56 *11*
γβ-	730.8 *1*	0.89 *18*
γβ-	776.2 *1*	1.22 *12*
γβ-	805.3 *1*	0.70 *14*
γβ-	828.3 *1*	20.6 *21*
γβ-	865.8 *1*	0.70 *14*
γβ-n[M1+E2]	914.40 *4*	~0.026
γβ-	920.7 *3*	0.10 *2*

Photons (^{81}Ga) (continued)

γ$_{mode}$	γ(keV)	γ(%)†
γβ-	936.6 *1*	9.0 *9*
γβ-	962.6 *1*	0.38 *8*
γβ-	991.1 *1*	0.63 *13*
γβ-	1016.4 *2*	0.88 *18*
γβ-	1019.8 *1*	2.30 *23*
γβ-	1083.2 *1*	0.78 *16*
γβ-	1104.9 *1*	0.40 *8*
γβ-	1116.6 *1*	1.12 *11*
γβ-	1137.1 *1*	0.64 *13*
γβ-	1164.5 *1*	1.22 *12*
γβ-	1189.2 *1*	0.56 *11*
γβ-	1203.2 *1*	0.18 *4*
γβ-	1272.7 *1*	6.3 *6*
γβ-	1286.4 *1*	5.7 *6*
γβ-[M1+E2]	1303.2 *1*	1.68 *17*
γβ-	1339.8 *1*	0.35 *7*
γβ-	1352.9 *1*	1.19 *12*
γβ-	1405.1 *1*	0:90 *18*
γβ-	1448.2 *1*	0.48 *10*
γβ-	1483.5 *1*	0.27 *5*
γβ-	1548.5 *1*	4.6 *5*
γβ-n[E2]	1573.47 *4*	~0.022
γβ-	1604.9 *1*	0.52 *10*
γβ-	1633.5 *1*	0.82 *16*
γβ-	1667.6 *1*	0.91 *18*
γβ-	1671.4 *4*	0.120 *24*
γβ-	1710.2 *2*	0.70 *14*
γβ-	1713.3 *1*	3.6 *4*
γβ-	1730.9 *1*	0.51 *10*
γβ-	1770.5 *1*	0.15 *3*
γβ-	1779.0 *1*	1.78 *18*
γβ-	1805.6 *1*	0.45 *9*
γβ-	1818.2 *1*	0.18 *4*
γβ-	1852.4 *2*	0.52 *10*
γβ-	1874.4 *1*	1.02 *10*
γβ-	1941.0 *1*	1.03 *10*
γβ-	1955.6 *1*	0.86 *17*

Photons (^{81}Ga) (continued)

γ$_{mode}$	γ(keV)	γ(%)†
γβ-	1982.4 *2*	0.25 *5*
γβ-	2041.6 *1*	0.52 *10*
γβ-	2049.6 *1*	0.5 *1*
γβ-	2125.7 *1*	0.98 *20*
γβ-	2138.4 *1*	1.33 *10*
γβ-	2180.7 *1*	2.35 *24*
γβ-	2216.2 *2*	0.52 *10*
γβ-	2271.7 *2*	0.39 *8*
γβ-	2281.7 *1*	0.89 *18*
γβ-	2288.6 *2*	0.69 *14*
γβ-	2311.1 *2*	0.65 *13*
γβ-	2335.8 *2*	0.45 *9*
γβ-	2362.9 *2*	0.5 *1*
γβ-	2379.0 *4*	0.52 *10*
γβ-	2419.9 *2*	0.70 *14*
γβ-	2436.5 *2*	0.45 *9*
γβ-	2444.2 *1*	5.3 *5*
γβ-	2464.7 *1*	0.67 *13*
γβ-	2541.4 *1*	0.69 *14*
γβ-	2549.9 *1*	0.70 *14*
γβ-	2650.4 *2*	0.38 *8*
γβ-	2726.1 *1*	0.98 *20*
γβ-	2754.8 *3*	0.52 *10*
γβ-	2792.0 *2*	0.63 *13*
γβ-	2955.0 *2*	0.44 *9*
γβ-	2986.4 *2*	0.49 *10*
γβ-	3448.6 *2*	0.40 *8*
γβ-	3489.0 *1*	0.41 *8*
γβ-	3503.2 *1*	3.1 *3*
γβ-	3665.5 *1*	1.04 *10*
γβ-	3698.0 *1*	0.59 *12*
γβ-	3773.1 *2*	0.68 *14*
γβ-	4035.2 *1*	1.96 *20*
γβ-	4470.4 *2*	0.76 *15*

† 7.5% uncert(syst)

$^{81}_{32}$Ge(7.6 *6* s)

2 isomers reported with similar $t_{1/2}$

Mode: β-

Δ: -65631 *120* keV

Δ: -64952 *120* keV

Prod: fission

Photons (^{81}Ge(both isomers))

γ_{mode}	γ(keV)	γ(%)†
γ[M1+E2]	93.1 *1*	22 *4*
γ	133.7 *1*	2.6 *3*
γ[M1+E2]	197.3 *1*	12.3 *12*
γ[E2]	242.8 *1*	1.7 *3*
γ[M1+E2]	290.3 *1*	6.5 *7*
γ[M1+E2]	336.0 *1*	41 *4*
γ	391.3 *1*	1.4 *3*
γ	401.8 *1*	4.8 *5*
γ	456.3 *2*	1.0 *2*
γ	463.1 *3*	0.80 *16*
γ	468.0 *1*	2.9 *6*
γ	482.4 *1*	4.6 *5*
γ	507.4 *3*	0.80 *16*
γ	609.1 *3*	0.79 *16*
γ	616.3 *2*	1.10 *22*
γ	637.3 *3*	1.0 *2*
γ	665.9 *3*	4.5 *9*
γ	706.1 *1*	4.0 *8*
γ	709.1 *3*	2.6 *5*
γ	737.7 *1*	12.8 *13*
γ	747.4 *1*	4.3 *4*
γ	751.5 *1*	4.4 *4*
γ	758.5 *1*	6.4 *6*
γ	771.3 *2*	2.5 *5*
γ	792.9 *1*	22.8 *23*
γ	859.1 *1*	4.7 *5*
γ	875.8 *1*	6.6 *7*
γ	990.3 *4*	0.80 *16*
γ	1005.7 *3*	0.66 *13*
γ	1013.0 *2*	5.6 *6*
γ	1038.5 *4*	0.70 *14*
γ	1056.5 *2*	1.7 *3*
γ	1058.6 *2*	2.1 *4*
γ	1083.2 *3*	1.6 *3*
γ	1095.5 *3*	1.20 *24*
γ	1100.3 *2*	1.3 *3*
γ	1144.8 *2*	4.8 *5*
γ	1156.4 *2*	4.9 *5*
γ	1225.8 *2*	1.7 *3*
γ	1238.9 *3*	1.0 *2*
γ	1256.1 *2*	0.47 *9*
γ	1297.4 *3*	1.20 *24*
γ	1429.5 *1*	4.7 *5*
γ	1435.7 *2*	1.9 *4*
γ	1495.5 *1*	11.6 *12*
γ	1582.3 *2*	3.7 *7*
γ	1629.5 *2*	2.7 *5*
γ	1686.5 *3*	0.45 *9*
γ	1869.8 *2*	2.1 *4*
γ	1882.5 *1*	7.6 *8*
γ	1886.8 *2*	0.66 *13*
γ	2103.9 *3*	1.10 *22*
γ	2174.3 *2*	3.6 *7*
γ	2207.5 *3*	1.20 *24*
γ	2228.2 *5*	0.83 *17*
γ	2331.3 *2*	1.9 *4*
γ	2337.4 *4*	1.4 *3*
γ	2436.0 *1*	0.40 *8*
γ	2526.5 *2*	1.6 *3*
γ	2629.9 *2*	2.4 *5*
γ	2754.8 *3*	1.9 *4*
γ	2800.2 *2*	2.4 *5*
γ	2845.8 *2*	1.20 *24*
γ	2859.0 *1*	0.40 *8*
γ	2904.7 *3*	0.76 *15*
γ	3136.6 *3*	0.69 *14*

Photons (^{81}Ge(both isomers)) (continued)

γ_{mode}	γ(keV)	γ(%)†
γ	3195.1 *2*	2.8 *6*
γ	3469.5 *2*	1.20 *24*
γ	3562.7 *2*	2.2 *4*

† per 100 β- of both isomers

$^{81}_{33}$As(33 *1* s)

Mode: β-

Δ: -72535 *6* keV

SpA: 4.18×10^{9} Ci/g

Prod: ^{82}Se(n,pn); ^{82}Se(γ,p); fission

Photons (^{81}As)

⟨γ⟩=138 *11* keV

γ_{mode}	γ(keV)	γ(%)†
γ E3+3%M4	102.98 *9*	*
γ	156.24 *23*	0.170 *24*
γ [M1+E2]	467.60 *19*	12.0
γ	490.72 *20*	5.1 *1*
γ [E1]	520.85 *25*	0.88 *8*
γ	756.0 *3*	0.103 *12*
γ [E1]	835.8 *3*	0.197 *24*
γ	874.83 *20*	0.083 *12*
γ	915.29 *24*	0.061 *12*
γ [E1]	938.41 *24*	0.20 *4*
γ [M1+E2]	949.9 *3*	0.21 *4*
γ [M1+E2]	1406.01 *23*	0.599 *12*
γ	1561.9 *3*	0.226 *24*
γ	1661.5 *3*	0.079 *12*
γ	1688.4 *3*	0.067 *7*
γ	1841.62 *24*	0.042 *6*
γ	1864.75 *24*	0.068 *7*
γ	1882.2 *3*	<0.0048
γ	2029.5 *3*	0.134 *12*
γ	2078.98 *25*	0.041 *6*
γ	2102.10 *24*	0.220 *24*
γ	2145.6 *3*	0.034 *6*
γ	2301.8 *3*	0.164 *17*
γ	2332.34 *23*	0.040 *6*
γ	2341.0 *3*	0.047 *7*
γ	2569.69 *23*	0.103 *12*
γ	2659.5 *4*	0.092 *11*
γ	2731.8 *15*	0.058 *8*
γ	2832.0 *3*	0.170 *18*
γ	3119.5 *15*	0.022 *6*

† approximate
* with ^{81}Se(57.28 min)

Continuous Radiation (^{81}As)

⟨β-⟩=1617 keV;⟨IB⟩=4.9 keV

E_{bin}(keV)		⟨ ⟩(keV)	(%)
0 - 10	β-	0.00457	0.091
	IB	0.051	
10 - 20	β-	0.0140	0.093
	IB	0.051	0.35
20 - 40	β-	0.059	0.195
	IB	0.100	0.35
40 - 100	β-	0.470	0.66
	IB	0.29	0.44
100 - 300	β-	6.5	3.12
	IB	0.85	0.47
300 - 600	β-	33.0	7.2
	IB	1.02	0.24
600 - 1300	β-	250	25.7

Continuous Radiation (^{81}As) (continued)

E_{bin}(keV)		⟨ ⟩(keV)	(%)
	IB	1.53	0.18
1300 - 2500	β-	889	47.7
	IB	0.96	0.057
2500 - 3856	β-	439	15.3
	IB	0.085	0.0030

$^{81}_{34}$Se(18.5 *1* min)

Mode: β-

Δ: -76391.8 *19* keV

SpA: 1.254×10^{8} Ci/g

Prod: ^{80}Se(n,γ); daughter ^{81}Se(57.28 min); ^{80}Se(d,p); ^{81}Br(n,p); ^{82}Se(γ,n)

Photons (^{81}Se)

⟨γ⟩=9.9 *5* keV

γ_{mode}	γ(keV)	γ(%)†
γ	178.42 *7*	0.0086 *9*
γ M1+E2	275.988 *11*	0.87
γ	290.12 *7*	0.017 *4*
γ M1+E2	290.12 *4*	0.75 *8*
γ	538.22 *7*	0.059 *6*
γ	552.36 *4*	0.110 *10*
γ [M1+E2]	566.11 *4*	0.26 *3*
γ	649.92 *7*	0.032 *4*
γ	828.34 *4*	0.32 *3*

† 13% uncert(syst)

Continuous Radiation (^{81}Se)

⟨β-⟩=611 keV;⟨IB⟩=0.96 keV

E_{bin}(keV)		⟨ ⟩(keV)	(%)
0 - 10	β-	0.0262	0.52
	IB	0.026	
10 - 20	β-	0.080	0.53
	IB	0.025	0.18
20 - 40	β-	0.332	1.10
	IB	0.049	0.170
40 - 100	β-	2.59	3.66
	IB	0.133	0.20
100 - 300	β-	32.3	15.8
	IB	0.32	0.19
300 - 600	β-	132	29.3
	IB	0.27	0.064
600 - 1300	β-	412	46.9
	IB	0.139	0.018
1300 - 1586	β-	30.6	2.23
	IB	0.00087	6.5×10^{-5}

$^{81}_{34}$Se(57.28 *5* min)

Mode: IT(99.942 %), β-(0.058 %)

Δ: -76288.8 *19* keV

SpA: 4.052×10^{7} Ci/g

Prod: ^{80}Se(n,γ); ^{80}Se(d,p); ^{81}Br(n,p); ^{82}Se(γ,n)

Photons (^{81}Se)

$\langle\gamma\rangle$=15.0 7 keV

γ_{mode}	γ(keV)	γ(%)†
Se L_ℓ	1.204	0.048 11
Se L_η	1.245	0.029 7
Se L_α	1.379	0.93 19
Se L_β	1.426	0.56 14
Se L_γ	1.648	0.0068 21
Se $K_{\alpha2}$	11.182	12.2 6
Se $K_{\alpha1}$	11.222	23.7 11
Se $K_{\beta1}'$	12.494	4.80 23
Se $K_{\beta2}'$	12.741	0.231 11
γ_{IT} E3+3%M4	102.98 9	9.7
$\gamma_{\beta-}$M2	260.29 20	0.057 6
$\gamma_{\beta-}$M1+E2	275.988 11	0.057 6
$\gamma_{\beta-}$	491.9 7	9.7 19 $\times10^{-5}$
$\gamma_{\beta-}$	767.9 7	0.00047 5

† 7.2% uncert(syst)

Atomic Electrons (^{81}Se)*

$\langle e\rangle$=87.8 20 keV

e_{bin}(keV)	$\langle e\rangle$(keV)	e(%)
1	0.97	67 7
2	0.061	3.7 4
9	0.188	2.01 21
10	1.96	20.2 21
11	0.66	6.0 6
12	0.042	0.34 4
90	62.8	69.6 22
101	7.42	7.32 22
102	10.5	10.3 3
103	3.21	3.12 10

* with IT

Continuous Radiation (^{81}Se)

$\langle\beta-\rangle$=0.247 keV;$\langle$IB$\rangle$=0.00030 keV

E_{bin}(keV)		$\langle\ \rangle$(keV)	(%)
0 - 10	β-	2.76×10^{-5}	0.00055
	IB	1.13×10^{-5}	
10 - 20	β-	8.4×10^{-5}	0.00056
	IB	1.09×10^{-5}	7.6×10^{-5}
20 - 40	β-	0.000346	0.00115
	IB	2.1×10^{-5}	7.2×10^{-5}
40 - 100	β-	0.00264	0.00374
	IB	5.4×10^{-5}	8.4×10^{-5}
100 - 300	β-	0.0303	0.0149
	IB	0.000116	6.8×10^{-5}
300 - 600	β-	0.101	0.0226
	IB	7.0×10^{-5}	1.7×10^{-5}
600 - 1153	β-	0.113	0.0149
	IB	1.37×10^{-5}	2.0×10^{-6}

$^{81}_{35}$Br(stable)

Δ: -77977 5 keV
%: 49.31 5

$^{81}_{36}$Kr(2.10 20 $\times10^5$ yr)

Mode: ε
Δ: -77697 6 keV
SpA: 0.021 Ci/g
Prod: ^{80}Kr(n,γ)

Photons (^{81}Kr)

$\langle\gamma\rangle$=16.5 14 keV

γ_{mode}	γ(keV)	γ(%)†
Br L_ℓ	1.293	0.058 13
Br L_η	1.339	0.034 9
Br L_α	1.481	1.16 24
Br L_β	1.534	0.69 18
Br L_γ	1.777	0.010 3
Br $K_{\alpha2}$	11.878	16.0 7
Br $K_{\alpha1}$	11.924	30.9 13
Br $K_{\beta1}'$	13.290	6.4 3
Br $K_{\beta2}'$	13.562	0.432 18
γ M1+E2	275.988 11	3.6

† 14% uncert(syst)

Atomic Electrons (^{81}Kr)

$\langle e\rangle$=4.8 3 keV

e_{bin}(keV)	$\langle e\rangle$(keV)	e(%)
2	1.19	75 8
10	2.6	25.0 25
11	0.076	0.67 7
12	0.81	7.0 7
13	0.069	0.53 5
263	0.068	0.026 3
274	0.0077	0.0028 3
276	0.00140	0.00051 6

Continuous Radiation (^{81}Kr)

$\langle$IB$\rangle$=0.0118 keV

E_{bin}(keV)		$\langle\ \rangle$(keV)	(%)
10 - 20	IB	0.0070	0.057
20 - 40	IB	0.00033	0.00120
40 - 100	IB	0.00109	0.00159
100 - 281	IB	0.0029	0.0019

$^{81}_{36}$Kr(13 1 s)

Mode: IT
Δ: -77507 6 keV
SpA: 1.04×10^{10} Ci/g
Prod: daughter ^{81}Rb; protons on Zr

Photons (^{81}Kr)

$\langle\gamma\rangle$=129.8 20 keV

γ_{mode}	γ(keV)	γ(%)†
Kr L_ℓ	1.383	0.020 4
Kr L_η	1.435	0.012 3
Kr L_α	1.581	0.41 9
Kr L_β	1.639	0.24 6
Kr L_γ	1.907	0.0033 10

Photons (^{81}Kr)
(continued)

γ_{mode}	γ(keV)	γ(%)†
Kr $K_{\alpha2}$	12.598	5.12 23
Kr $K_{\alpha1}$	12.651	9.9 4
Kr $K_{\beta1}'$	14.109	2.10 9
Kr $K_{\beta2}'$	14.408	0.189 9
γ E3	190.4 2	67

† 1.5% uncert(syst)

Atomic Electrons (^{81}Kr)

$\langle e\rangle$=59.5 12 keV

e_{bin}(keV)	$\langle e\rangle$(keV)	e(%)
2	0.43	25 3
10	0.068	0.65 7
11	0.70	6.4 7
12	0.25	2.02 21
13	0.046	0.36 4
14	0.028	0.201 21
176	47.5	27.0 7
188	4.81	2.55 7
189	4.06	2.15 6
190	1.65	0.867 22

$^{81}_{37}$Rb(4.58 1 h)

Mode: ε
Δ: -75461 23 keV
SpA: 8.447×10^6 Ci/g
Prod: ^{79}Br(α,2n); ^{79}Br(^{3}He,n); protons on Rb

Photons (^{81}Rb)

$\langle\gamma\rangle$=325 6 keV

γ_{mode}	γ(keV)	γ(%)
Kr L_ℓ	1.383	0.062 14
Kr L_η	1.435	0.036 9
Kr L_α	1.581	1.3 3
Kr L_β	1.640	0.75 19
Kr L_γ	1.907	0.011 4
Kr $K_{\alpha2}$	12.598	16.6 7
Kr $K_{\alpha1}$	12.651	32.1 14
Kr $K_{\beta1}'$	14.109	6.8 3
Kr $K_{\beta2}'$	14.408	0.61 3
γ M1	49.5 1	0.068 16
γ M1	64.5 4	0.056 12
γ E3	190.4 2	64.3 14 *
γ	218.8 6	0.019 5
γ M1+E2	244.3 2	0.310 9
γ [E2]	266.2 5	0.037 5
γ	283.1 5	0.044 9
γ	319.5 4	0.044 5
γ	339.4 4	0.058 7
γ M1	357.7 2	0.760 23
γ	386.0 3	0.084 9
γ M1	389.0 2	0.457 23
γ	399.7 5	0.026 7
γ M1	446.3 1	23.3 7
γ E1	456.9 1	3.03 9
γ	476.8 1	0.524 14
γ [E2]	499.4 2	0.119 5
γ M1,E2	510.4 3	5.4 9
γ E2,M1	537.6 10	0.19 7
γ M1,E2	548.9 1	0.473 14
γ M1,E2	568.9 1	0.585 16
γ	602.3 3	0.0513 23
γ	608.5 2	0.259 12
γ	689.9 3	0.0303 23
γ	701.5 5	0.054 19

Photons (^{81}Rb)
(continued)

γ_{mode}	γ(keV)	γ(%)
γ	729.1 1	0.296 9
γ	758.3 2	0.0513 23
γ	782.5 5	0.0140 23
γ M1,E2	803.5 2	0.836 23
γ M1,E2	834.8 2	0.816 23
γ	903.2 6	~0.0047
γ	912.5 6	~0.0047
γ	977.1 2	0.566 19
γ	1041.1 2	0.54 3
γ	1048.4 3	0.047 5
γ	1069.3 3	0.0606 23
γ	1087.7 5	0.0070 23
γ	1090.4 5	0.0116 23
γ	1108.0 2	0.0513 23
γ	1136.4 4	0.0116 23
γ	1363.8 6	~0.0047
γ	1381.5 5	0.0093 23
γ	1427.8 2	0.0326 23
γ	1487.4 5	~0.009
γ	1536.0 8	~0.0047
γ	1554.9 3	0.042 5
γ	1874.0 4	0.0140 23

* with ^{81}Kr(13 s) in equilib

Atomic Electrons (^{81}Rb)
⟨e⟩=60.6 14 keV

e_{bin}(keV)	⟨e⟩(keV)	e(%)
2 - 10	1.52	79 8
11	2.25	20.7 21
12 - 50	1.08	8.5 7
63 - 64	0.0019	0.0030 5
176	45.3	25.7 8
188	4.58	2.43 7
189	3.87	2.05 6
190	1.57	0.826 24
204 - 252	0.020	0.009 3
264 - 305	0.0028	0.0010 4
318 - 358	0.0140	0.0041 3
372 - 400	0.0086	0.00228 23
432 - 477	0.308	0.0709 22
485 - 535	0.076	0.015 3
536 - 569	0.0071	0.00127 16
588 - 608	0.0022	0.00038 14
676 - 715	0.0019	0.00027 9
727 - 768	0.00051	6.8 18 $\times10^{-5}$
781 - 820	0.0096	0.00120 7
833 - 835	0.00056	6.7 5 $\times10^{-5}$
889 - 912	3.7 $\times10^{-5}$	4.1 17 $\times10^{-6}$
963 - 977	0.0021	0.00021 8
1027 - 1076	0.0022	0.00021 6
1086 - 1135	0.00020	1.8 5 $\times10^{-5}$
1349 - 1381	3.5 $\times10^{-5}$	2.6 8 $\times10^{-6}$
1413 - 1428	7.9 $\times10^{-5}$	5.6 17 $\times10^{-6}$
1473 - 1555	0.00013	8.3 21 $\times10^{-6}$
1860 - 1874	2.7 $\times10^{-5}$	1.5 5 $\times10^{-6}$

Continuous Radiation (^{81}Rb)
⟨β+⟩=132 keV;⟨IB⟩=1.23 keV

E_{bin}(keV)		⟨ ⟩(keV)	(%)
0 - 10	β+	5.6 $\times10^{-5}$	0.00069
	IB	0.0063	
10 - 20	β+	0.00132	0.0081
	IB	0.0119	0.088
20 - 40	β+	0.0225	0.070
	IB	0.0116	0.040
40 - 100	β+	0.63	0.83
	IB	0.031	0.047
100 - 300	β+	17.1	8.0
	IB	0.091	0.050
300 - 600	β+	65	14.7
	IB	0.19	0.043

Continuous Radiation (^{81}Rb)
(continued)

E_{bin}(keV)		⟨ ⟩(keV)	(%)
600 - 1300	β+	49.6	6.9
	IB	0.66	0.073
1300 - 2046	IB	0.23	0.0153
	Σβ+		31

$^{81}_{37}$Rb(30.6 3 min)

Mode: IT(97.8 3 %), ε(2.2 3 %)
Δ: -75375 23 keV
SpA: 7.59$\times10^{7}$ Ci/g
Prod: ^{79}Br(α,2n)

Photons (^{81}Rb)
⟨γ⟩=12.21 21 keV

γ_{mode}	γ(keV)	γ(%)
γ_{ϵ}M1	49.5 1	0.67 4
γ_{IT}[E3]	86.2 2	4.71 20
γ_{ϵ}E3	190.4 2	0.0123 12 *
γ_{ϵ}[E2]	266.2 5	~0.00010
γ_{ϵ}	368.3 3	0.0094 5
γ_{ϵ}M1	446.3 1	0.0183 20
γ_{ϵ}E1	456.9 1	0.0075 14
γ_{ϵ}	463.3 3	0.0183 20
γ_{ϵ}	465.5 3	0.0183 20
γ_{ϵ}[E2]	499.4 2	0.0254 15
γ_{ϵ}M1,E2	548.9 1	0.092 6
γ_{ϵ}	551.5 15	0.0051 20
γ_{ϵ}	643.6 1	0.100 4
γ_{ϵ}	657.5 2	0.0119 5
γ_{ϵ}	682.3 1	0.0427 18
γ_{ϵ}	729.2 8	0.0285 20
γ_{ϵ}	732.1 2	0.0183 10
γ_{ϵ}	761.9 2	0.0081 6
γ_{ϵ}	824.2 5	0.0132 10
γ_{ϵ}	873.8 3	0.0078 6
γ_{ϵ}	885.0 2	0.0333 14
γ_{ϵ}	932.4 2	0.0333 14
γ_{ϵ}	981.6 2	0.0253 14
γ_{ϵ}	1011 1	<0.0031
γ_{ϵ}	1014.4 4	0.0105 7
γ_{ϵ}	1087 1	0.0104 15
γ_{ϵ}	1099.9 2	0.065 3
γ_{ϵ}	1136 1	0.0037 7
γ_{ϵ}	1157.0 4	0.0061 6
γ_{ϵ}	1194.6 2	0.096 4
γ_{ϵ}	1206.0 15	<0.0010
γ_{ϵ}	1286.9 4	0.0060 4
γ_{ϵ}	1297.0 4	0.0064 4
γ_{ϵ}	1633.2 5	0.0061 4
γ_{ϵ}	1638.4 4	0.0110 7
γ_{ϵ}	1682.7 4	0.0133 8
γ_{ϵ}	1687.9 4	0.0093 7
γ_{ϵ}	1694.4 4	0.0159 8
γ_{ϵ}	1732 1	~0.0010
γ_{ϵ}	1743.5 3	0.0492 21
γ_{ϵ}	1781.8 5	0.0048 4
γ_{ϵ}	1853 1	0.00092 20
γ_{ϵ}	1902.6 7	0.0042 4

* with ^{81}Kr(13 s) in equilib

$^{81}_{38}$Sr(22.15 22 min)

Mode: ε
Δ: -71470 40 keV
SpA: 1.048$\times10^{8}$ Ci/g
Prod: ^{85}Rb(p,5n); ^{55}Mn(^{32}S,αpn); ^{52}Cr(^{32}S,2pn)

Photons (^{81}Sr)
⟨γ⟩=509 4 keV

γ_{mode}	γ(keV)	γ(%)
Rb L_ℓ	1.482	0.0106 24
Rb L_η	1.542	0.0061 16
Rb L_α	1.694	0.23 5
Rb L_β	1.760	0.13 4
Rb L_γ	2.051	0.0022 7
Rb $K_{\alpha2}$	13.336	2.86 17
Rb $K_{\alpha1}$	13.395	5.5 3
Rb $K_{\beta1}$'	14.959	1.19 7
Rb $K_{\beta2}$'	15.286	0.127 8
γ E1	55.95 3	0.096 7
γ M1+E2	113.02 3	0.178 15
γ	131.56 14	~0.024
γ M1+E2	142.15 3	3.14 10
γ M1+E2	147.76 3	31.3 10
γ M1+E2	153.54 3	35.1 11
γ E2	158.96 6	0.127 20
γ M1+E2	188.27 3	16.0 5
γ	197.32 8	0.093 25
γ	206.98 7	0.31 3
γ	217.73 4	0.34 4
γ E1	245.24 4	0.73 5
γ M1+E2	255.16 3	1.61 7
γ E2	289.95 5	0.10 5
γ M1+E2	301.30 3	1.52 8
γ	347.8 3	~0.05
γ M1+E2	386.55 4	1.72 9
γ	410.83 11	0.46 7
γ E2	421.29 6	0.98 8
γ	422.47 15	0.39 5
γ M1+E2	443.34 4	18.2 6
γ	463.08 17	0.30 6
γ	465.80 5	1.25 8
γ	477.15 16	0.67 7
γ	486.69 6	1.32 9
γ	523.71 4	1.39 5
γ	541.51 14	0.12 5
γ	548.65 5	0.58 3
γ	558.8 4	~0.038
γ M1+E2	574.67 3	7.01 22
γ	607.88 3	1.42 5
γ	630.57 6	0.302 24
γ	644.56 17	0.213 22
γ	663.6 3	0.087 18
γ	670.4 3	0.109 18
γ	702.14 9	1.05 4
γ	711.90 6	1.44 5
γ M1+E2	720.81 3	3.67 11
γ	769.5 4	~0.022
γ	807.04 18	0.087 15
γ	811.01 12	0.167 18
γ	841.34 15	0.222 20
γ	851.39 6	0.62 3
γ	895.10 25	0.151 18
γ M1+E2	909.03 3	2.83 9
γ	923.05 8	0.32 2
γ M1+E2	938.45 3	3.07 10
γ	953.8 4	0.053 15
γ	978.66 7	0.388 22
γ	998.6 4	0.035 13
γ	1066.96 8	0.399 22
γ	1080.72 11	0.510 25
γ	1090.75 17	0.164 16
γ	1110.26 10	0.297 20
γ	1136.63 11	0.207 15
γ	1193.76 5	0.63 3
γ	1211.2 4	0.069 13
γ	1243.0 4	0.036 13
γ	1252.82 10	0.329 18
γ	1273.45 19	0.169 18
γ	1317.54 24	0.053 13
γ	1324.7 4	0.116 16
γ	1360.6 3	0.104 15
γ	1365.68 22	0.136 15
γ	1382.44 8	0.422 22
γ	1400.33 6	0.435 22
γ	1468.8 4	0.093 13
γ	1502.5 7	0.045 13
γ	1533.6 7	~0.022
γ	1544.6 5	0.031 13
γ	1554.15 11	0.284 18

Photons (^{81}Sr) (continued)

γ_{mode}	γ(keV)	γ(%)
γ	1602.9 *7*	0.069 *15*
γ	1616.31 *17*	0.109 *15*
γ	1627.5 *7*	~0.020
γ	1631.9 *7*	0.038 *13*
γ	1654.4 *3*	0.089 *15*
γ	1659.6 *5*	0.024 *9*
γ	1679.6 *7*	0.084 *13*
γ	1698.35 *18*	0.040 *11*
γ	1722.2 *7*	0.086 *13*

Atomic Electrons (^{81}Sr)

$\langle e \rangle$=8.1 *8* keV

e_{bin}(keV)	$\langle e \rangle$(keV)	e(%)
2 - 11	0.38	13.7 *14*
12	0.25	2.17 *24*
13 - 56	0.194	1.35 *13*
98 - 132	0.27	0.23 *4*
133	1.86	1.41 *5*
138	1.97	1.42 *5*
140 - 144	0.057	0.040 *3*
146	0.230	0.158 *6*
147 - 159	0.339	0.224 *6*
173	1.5	~0.9
182 - 230	0.30	0.16 *7*
240 - 288	0.14	0.056 *16*
290 - 333	0.008	0.0028 *9*
346 - 387	0.039	0.011 *3*
396 - 422	0.037	0.0091 *13*
428	0.28	0.066 *15*
441 - 487	0.081	0.018 *3*
509 - 557	0.021	0.0040 *14*
558 - 608	0.090	0.0160 *19*

Atomic Electrons (^{81}Sr) (continued)

e_{bin}(keV)	$\langle e \rangle$(keV)	e(%)
615 - 664	0.0049	0.00078 *21*
668 - 712	0.041	0.0059 *7*
719 - 768	0.0036	0.00049 *4*
769 - 811	0.0013	0.00016 *5*
826 - 851	0.0039	0.00047 *14*
880 - 923	0.0339	0.00374 *14*
936 - 983	0.0038	0.00040 *6*
997 - 999	1.5×10^{-5}	1.5 *7* $\times 10^{-6}$
1052 - 1095	0.0046	0.00043 *8*
1108 - 1137	0.00080	7.1 *21* $\times 10^{-5}$
1179 - 1228	0.0023	0.00019 *6*
1238 - 1273	0.0015	0.00012 *3*
1302 - 1350	0.00106	8.0 *15* $\times 10^{-5}$
1359 - 1400	0.0024	0.00017 *4*
1454 - 1552	0.00117	7.7 *15* $\times 10^{-5}$
1554 - 1653	0.00083	5.1 *8* $\times 10^{-5}$
1654 - 1722	0.00049	2.9 *6* $\times 10^{-5}$

Continuous Radiation (^{81}Sr)

$\langle \beta + \rangle$=967 keV;$\langle$IB$\rangle$=3.4 keV

E_{bin}(keV)		$\langle\ \rangle$(keV)	(%)
0 - 10	$\beta+$	1.60×10^{-5}	0.000198
	IB	0.035	
10 - 20	$\beta+$	0.000399	0.00245
	IB	0.036	0.25
20 - 40	$\beta+$	0.0071	0.0222
	IB	0.068	0.24
40 - 100	$\beta+$	0.219	0.287
	IB	0.19	0.30
100 - 300	$\beta+$	7.9	3.62
	IB	0.54	0.31
300 - 600	$\beta+$	54	11.7
	IB	0.62	0.145
600 - 1300	$\beta+$	365	38.4
	IB	0.94	0.108
1300 - 2500	$\beta+$	530	31.2
	IB	0.78	0.045
2500 - 3990	$\beta+$	9.1	0.348
	IB	0.168	0.0060
	$\Sigma\beta+$		86

$^{81}_{39}$Y (1.200 *25* min)

Mode: ϵ
Δ: -65950 *70* keV
SpA: 1.93×10^{9} Ci/g
Prod: daughter ^{81}Zr

Photons (^{81}Y)

$\langle \gamma \rangle$=3.7 *3* keV

γ_{mode}	γ(keV)	γ(%)†
γ[E2]	79.17 *4*	0.902 *11*
γ	124.17 *4*	1.1
γ	408.18 *6*	0.39 *7*

† 9.1% uncert(syst)

$^{81}_{40}$Zr(~10 s)

Mode: ϵ, ϵp
Δ: -58790 *300* keV
SpA: $\sim 1.3 \times 10^{10}$ Ci/g
Prod: ^{52}Cr(^{32}S,3n); ^{54}Fe(^{32}S,αn)

A=82

NDS **15**, 315 (1975)

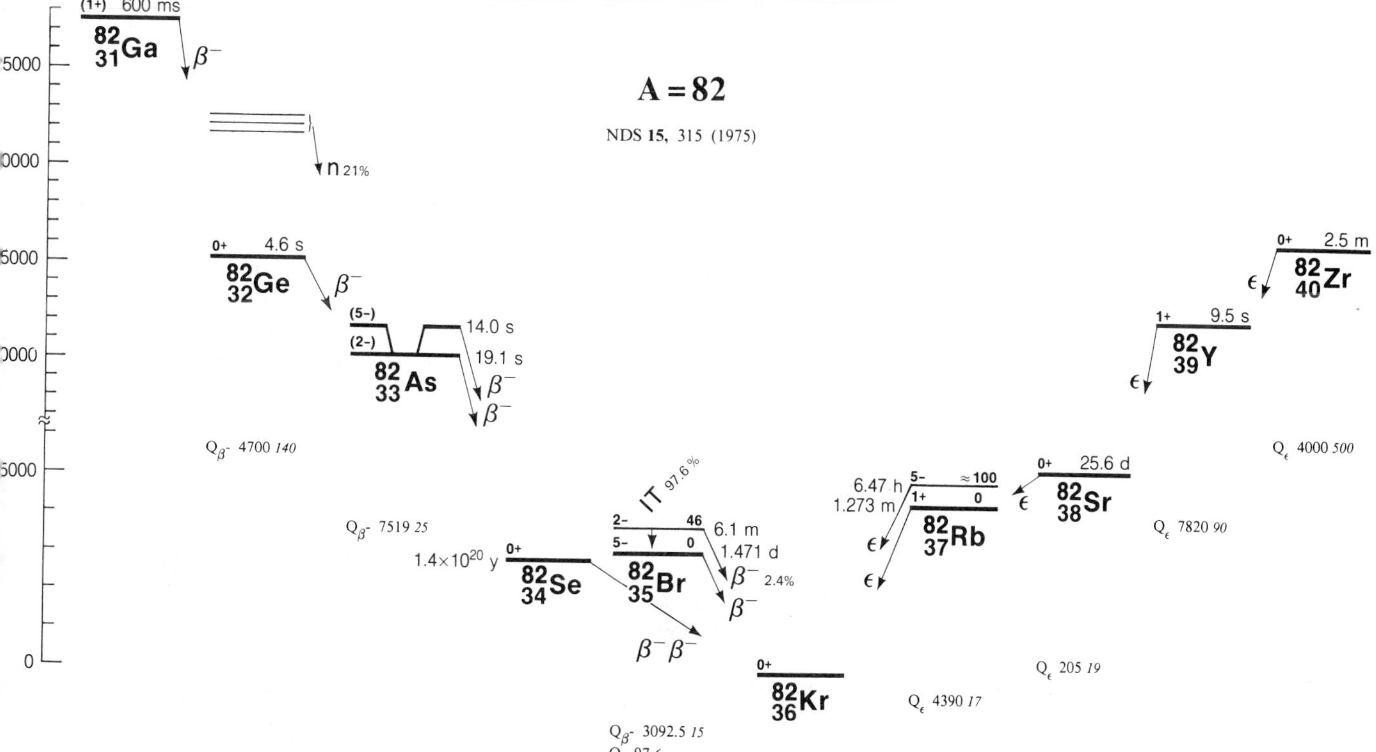

$^{82}_{31}$Ga(600 *10* ms)

Mode: β-, β-n(21.4 *22* %)
SpA: 1.360×10^{11} Ci/g
Prod: fission

Photons (^{82}Ga)

γ_{mode}	γ(keV)	γ(rel)
$\gamma_{\beta-}$	216.5 *1*	6.5 *13*
$\gamma_{\beta-}$	415.7 *3*	2.1 *4*
$\gamma_{\beta-}$	530.2 *1*	2.3 *5*
$\gamma_{\beta-}$	711.2 *1*	16 *3*
$\gamma_{\beta-}$[M1+E2]	867.5 *1*	13.4 *13*
$\gamma_{\beta-}$	938.3 *2*	5.7 *11*
$\gamma_{\beta-}$	985.1 *2*	5.3 *11*
$\gamma_{\beta-}$[E2]	1348.1 *1*	100 *10*
$\gamma_{\beta-}$	1354.1 *5*	4.0 *8*
$\gamma_{\beta-}$	1909.3 *1*	10.6 *11*
$\gamma_{\beta-}$[E2]	2215.0 *2*	22.0 *22*

$^{82}_{32}$Ge(4.6 *4* s)

Mode: β-
Δ: -65380 *140* keV
SpA: 2.78×10^{10} Ci/g
Prod: fission

Photons (^{82}Ge)

⟨γ⟩=1078 *200* keV

γ_{mode}	γ(keV)	γ(%)†
γ	139.7 *1*	1.4 *3*
γ [M1+E2]	248.8 *1*	3.6 *7*
γ [E1]	834.2 *1*	8.4 *17*
γ	951.8 *2*	1.5 *3*
γ [E1]	1091.9 *1*	90

† approximate

$^{82}_{33}$As(19.1 *5* s)

Mode: β-
Δ: -70078 *25* keV
SpA: 7.07×10^{9} Ci/g
Prod: ^{82}Se(n,p); daughter ^{82}Ge; fission

Photons (^{82}As)&

⟨γ⟩=609 *60* keV

γ_{mode}	γ(keV)	γ(%)†
γ	186.1 *5*	~0.6 *
γ [E2]	654.4 *1*	15
γ [E2]	755.2 *2*	1.80 *15*
γ [M1+E2]	1076.4 *4*	1.18 *19*
γ [E2]	1079.9 *4*	1.68 *13*
γ	1549.8 *2*	*
γ [E2]	1731.3 *1*	4.1 *3*
γ	1970.9 *3*	1.58 *13*
γ	2346.2 *10*	1.6 *8*
γ	2353.4 *10*	2.0 *9*
γ	2441.2 *5*	~2
γ	2513.5 *5*	~1
γ	2603.8 *5*	~2
γ	2722.7 *5*	1.4 *6*
γ	2834.8 *10*	1.6 *8*
γ	3668.8 *10*	~1
γ	3773 *1*	0.9 *3*

† 30% uncert(syst)
* with ^{82}As(19.1 or 14.0 s)
& see also ^{80}As

Continuous Radiation (^{82}As)

⟨β-⟩=3275 keV;⟨IB⟩=14.8 keV

E_{bin}(keV)		⟨ ⟩(keV)	(%)
0 - 10	β-	0.00096	0.0191
	IB	0.075	
10 - 20	β-	0.00296	0.0197
	IB	0.075	0.52
20 - 40	β-	0.0125	0.0414
	IB	0.149	0.52
40 - 100	β-	0.102	0.143
	IB	0.44	0.67
100 - 300	β-	1.49	0.71
	IB	1.38	0.76
300 - 600	β-	8.5	1.83
	IB	1.8	0.43
600 - 1300	β-	78	7.9
	IB	3.5	0.39
1300 - 2500	β-	437	22.6
	IB	3.9	0.22
2500 - 5000	β-	1902	52
	IB	3.2	0.096
5000 - 7500	β-	848	14.9
	IB	0.24	0.0043
7500 - 7519	β-	0.00073	9.7×10^{-6}
	IB	9.6×10^{-11}	1.28×10^{-12}

$^{82}_{33}$As(14.0 *5* s)

Mode: β-
Δ: -70078 *25* keV
SpA: 9.6×10^{9} Ci/g
Prod: ^{82}Se(n,p); fission

Photons (^{82}As)&

⟨γ⟩=2728 *130* keV

γ_{mode}	γ(keV)	γ(%)†
γ [E1]	343.5 *1*	24 *3*
γ [M1+E2]	560.5 *1*	13.3 *22*
γ [E2]	654.4 *1*	74
γ [M1+E2]	815.1 *4*	7.2 *13*
γ [M1+E2]	818.6 *4*	28 *3*
γ	903.0 *5*	~3
γ [M1+E2]	1076.4 *4*	8.9 *15*
γ [E2]	1079.9 *4*	24 *3*
γ	1540.9 *2*	8.1 *15*
γ	1718.0 *7*	~3
γ [E2]	1731.3 *1*	28 *4*
γ [M1+E2]	1895.4 *1*	40 *4*

† 7% uncert(syst)
& see also ^{80}As and ^{82}As(19.1 s)

Continuous Radiation (^{82}As)

⟨β-⟩=1965 keV;⟨IB⟩=6.7 keV

E_{bin}(keV)		⟨ ⟩(keV)	(%)
0 - 10	β-	0.00272	0.054
	IB	0.057	
10 - 20	β-	0.0084	0.056
	IB	0.057	0.39
20 - 40	β-	0.0353	0.117
	IB	0.112	0.39
40 - 100	β-	0.287	0.404
	IB	0.33	0.50
100 - 300	β-	4.13	1.98
	IB	0.98	0.55
300 - 600	β-	22.7	4.92
	IB	1.22	0.29
600 - 1300	β-	193	19.7
	IB	2.0	0.23
1300 - 2500	β-	824	43.5
	IB	1.59	0.092
2500 - 5000	β-	914	29.1
	IB	0.39	0.0134
5000 - 5784	β-	7.8	0.150
	IB	0.00038	7.3×10^{-6}

$^{82}_{34}$Se(1.4 *3* ×10^{20} yr)

Mode: β-β-
Δ: -77596.1 *21* keV
Prod: natural source
%: 9.2 *5*

$^{82}_{35}$Br(1.4708 *8* d)

Mode: β-
Δ: -77499 *5* keV
SpA: 1.0826×10^{6} Ci/g
Prod: ^{81}Br(n,γ)

Photons (^{82}Br)

⟨γ⟩=2642 *18* keV

γ_{mode}	γ(keV)	γ(%)†
Kr $K_{\alpha2}$	12.598	0.14 *7*
Kr $K_{\alpha1}$	12.651	0.27 *13*
Kr $K_{\beta1}$'	14.109	0.06 *3*
Kr $K_{\beta2}$'	14.408	0.0052 *24*
γ [M1+E2]	92.190 *16*	0.752 *25*
γ [E1]	100.89 *8*	0.070 *7*
γ [E1]	129.29 *4*	0.030 *6*
γ [M1+E2]	137.40 *5*	0.152 *3*
γ [M1+E2]	179.8 *2*	~0.010
γ [E1]	221.459 *25*	2.27 *7*
γ [M1+E2]	273.480 *8*	0.84 *4*
γ	332.90 *3*	0.090 *4*
γ [E1]	401.16 *6*	0.091 *8*
γ [E1]	554.35 *1*	70.9 *14*
γ [E1]	559.5 *3*	~0.013
γ [M1+E2]	619.11 *1*	43.1 *13*
γ [M1+E2]	698.368 *11*	28.2 *8*
γ [E1]	735.64 *7*	0.075 *8*
γ [E2]	776.516 *14*	83.6
γ [E1]	827.831 *6*	24.0 *8*
γ	932.1 *2*	0.010 *4*
γ	952.02 *15*	0.368 *17*
γ [E1]	1007.59 *3*	1.31 *3*
γ [E2]	1044.077 *6*	27.4 *6*
γ [M1+E2]	1072.9 *1*	0.079 *13*
γ [E1]	1081.29 *6*	0.618 *17*

Photons (^{82}Br)
(continued)

γ_{mode}	γ(keV)	γ(%)†
γ[E2]	1099.9 2	0.0058 25
γ	1174.0 4	0.018 8
γ[M1+E2]	1180.1 1	0.086 8
γ[M1+E2]	1317.473 10	26.9 5
γ[E2]	1474.88 2	16.63 25
γ	1650.37 4	0.791 16
γ[E1]	1779.66 5	0.115 13
γ	1871.6 2	0.025 8
γ[E2]	1956.8 1	~0.033

† 1.0% uncert(syst)

Atomic Electrons (^{82}Br)
$\langle e\rangle$=2.4 3 keV

e_{bin}(keV)	$\langle e\rangle$(keV)	e(%)
2 - 14	0.041	0.9 3
78	0.4	~0.5
87 - 136	0.11	0.11 5
137 - 180	0.0019	~0.0012
207 - 221	0.0415	0.0199 9
259 - 273	0.036	0.014 5
319 - 333	0.0021	~0.0007
387 - 401	0.00060	0.000155 13
540	0.259	0.0480 21
545 - 559	0.0334	0.00604 22
605	0.35	0.058 8
617 - 619	0.046	0.0075 10
684	0.193	0.028 3
696 - 736	0.0253	0.0036 3
762	0.526	0.069 3
775	0.0579	0.0075 3
776 - 814	0.065	0.0081 3
826 - 828	0.0069	0.00083 4
918 - 952	0.0014	0.00015 5
993 - 1008	0.00277	0.000278 12
1030	0.113	0.0110 5
1042 - 1086	0.0161	0.00154 5
1098 - 1100	2.9×10^{-6}	2.6 10 $\times10^{-7}$
1160 - 1180	0.00039	3.4 4 $\times10^{-5}$
1303	0.084	0.0064 4
1316 - 1317	0.0106	0.00081 5
1461 - 1475	0.0522	0.00357 14
1636 - 1650	0.0017	0.00010 3
1765 - 1857	0.000205	1.15 14 $\times10^{-5}$
1870 - 1957	8.6×10^{-5}	4.4 19 $\times10^{-6}$

Continuous Radiation (^{82}Br)
$\langle\beta$-$\rangle$=137 keV;$\langle$IB$\rangle$=0.064 keV

E_{bin}(keV)		$\langle\ \rangle$(keV)	(%)
0 - 10	β-	0.221	4.41
	IB	0.0071	
10 - 20	β-	0.65	4.36
	IB	0.0063	0.044
20 - 40	β-	2.57	8.6
	IB	0.0108	0.038
40 - 100	β-	16.9	24.3
	IB	0.021	0.034
100 - 300	β-	95	52
	IB	0.018	0.0124
300 - 444	β-	21.6	6.4
	IB	0.00041	0.000129

$^{82}_{35}$Br(6.13 8 min)

Mode: IT(97.6 %), β-(2.4 %)
Δ: -77453 5 keV
SpA: 3.74×10^{8} Ci/g
Prod: ^{81}Br(n,γ)

Photons (^{82}Br)
$\langle\gamma\rangle$=7.3 4 keV

γ_{mode}	γ(keV)	γ(%)
Br L_ℓ	1.293	0.055 12
Br L_η	1.339	0.029 7
Br L_α	1.481	1.12 23
Br L_β	1.534	0.60 15
Br L_γ	1.777	0.010 3
Br $K_{\alpha2}$	11.878	12.4 6
Br $K_{\alpha1}$	11.924	24.1 11
Br $K_{\beta1}$'	13.290	4.97 23
Br $K_{\beta2}$'	13.562	0.337 16
γ_{IT} M3	46 2	0.24 5
$\gamma_{\beta-}$[M1+E2]	619.09 4	<0.0036
$\gamma_{\beta-}$[M1+E2]	698.37 1	0.0261 6
$\gamma_{\beta-}$[E2]	711.2 1	0.00135 6
$\gamma_{\beta-}$[E2]	776.52 1	0.20
$\gamma_{\beta-}$[M1+E2]	1072.99 7	0.00157 8
$\gamma_{\beta-}$[E1]	1081.29 6	0.00034 7
$\gamma_{\beta-}$[E2]	1168.50 11	0.000279 20
$\gamma_{\beta-}$	1173.4 10	4.0 8 $\times10^{-6}$
$\gamma_{\beta-}$[M1+E2]	1180.27 2	0.00303 14
$\gamma_{\beta-}$[M1+E2]	1180.95 1	~0.00020
$\gamma_{\beta-}$[M1+E2]	1317.44 3	0.00086 4
$\gamma_{\beta-}$[E2]	1395.1 2	$\sim4\times10^{-5}$
$\gamma_{\beta-}$[E2]	1474.88 2	0.0152 3
$\gamma_{\beta-}$[M1+E2]	1703.19 4	0.000219 20
$\gamma_{\beta-}$[M1+E2]	1879.5 1	0.000319 20
$\gamma_{\beta-}$[E2]	1956.75 4	0.00127 6
$\gamma_{\beta-}$[M1]	2479.6 1	0.000185 10
$\gamma_{\beta-}$[E2]	2656.0 1	5.6 10 $\times10^{-5}$

Atomic Electrons (^{82}Br)*
$\langle e\rangle$=38.0 7 keV

e_{bin}(keV)	$\langle e\rangle$(keV)	e(%)
2	1.10	70 7
10	1.99	19.5 20
11	0.060	0.52 5
12	0.63	5.4 6
13	0.054	0.41 4
33	22.0	67.6 19
44	10.2	23.0 7
46	2.01	4.38 13

* with IT

Continuous Radiation (^{82}Br)
$\langle\beta$-$\rangle$=31.8 keV;$\langle$IB$\rangle$=0.086 keV

E_{bin}(keV)		$\langle\ \rangle$(keV)	(%)
0 - 10	β-	0.000219	0.00435
	IB	0.00108	
10 - 20	β-	0.00067	0.00445
	IB	0.00106	0.0074
20 - 40	β-	0.00278	0.0092
	IB	0.0021	0.0073
40 - 100	β-	0.0218	0.0307
	IB	0.0060	0.0092
100 - 300	β-	0.280	0.136
	IB	0.017	0.0096
300 - 600	β-	1.28	0.281
	IB	0.020	0.0046
600 - 1300	β-	7.7	0.80
	IB	0.027	0.0031
1300 - 2500	β-	19.3	1.06
	IB	0.0123	0.00076

Continuous Radiation (^{82}Br)
(continued)

E_{bin}(keV)		$\langle\ \rangle$(keV)	(%)
2500 - 3139	β-	3.22	0.120
	IB	0.00023	8.4×10^{-6}

$^{82}_{36}$Kr(stable)

Δ: -80591 5 keV
%: 11.6 1

$^{82}_{37}$Rb(1.273 2 min)

Mode: ϵ
Δ: -76202 17 keV
SpA: 1.793×10^{9} Ci/g
Prod: daughter ^{82}Sr

Photons (^{82}Rb)
$\langle\gamma\rangle$=118 8 keV

γ_{mode}	γ(keV)	γ(%)†
Kr $K_{\alpha2}$	12.598	0.76 3
Kr $K_{\alpha1}$	12.651	1.47 6
Kr $K_{\beta1}$'	14.109	0.311 13
Kr $K_{\beta2}$'	14.408	0.0280 12
γ	466.9 4	~0.013
γ	522.8 5	0.0040 13
γ	696.86 15	0.068 3
γ[M1+E2]	698.37 1	0.133 4
γ[E2]	711.2 1	0.051 3
γ[E2]	776.52 1	13.4
γ	869.3 4	~0.0012
γ	975.2 1	0.0075 9
γ	992.2 1	0.0016 7
γ	1021.4 5	~0.0013
γ[E2]	1044.1 5	~0.0008
γ[E1]	1081.4 7	~0.00027
γ	1086.8 5	~0.0012
γ	1123.6 7	~0.0007
γ[E2]	1168.2 2	0.0012 5
γ[M1+E2]	1180.27 2	0.0174 13
γ[E2]	1395.14 3	0.470 9
γ[E2]	1474.88 2	0.079 3
γ	1607.7 3	0.0020 3
γ	1673.5 2	0.0063 4
γ	1698.7 3	~0.0013
γ[M1+E2]	1703.19 4	0.0449 9
γ	1711.9 4	0.0015 3
γ	1785.13 7	0.0027 5
γ[M1+E2]	1879.18 15	0.0090 5
γ[E2]	1956.75 4	0.0060 5
γ	2167.59 4	0.0375 13
γ	2241.98 17	~0.0008
γ	2410.26 5	0.0209 11
γ[M1]	2479.65 4	0.0362 13
γ	2508.9 2	0.0025 7
γ	2578.7 2	0.00094 9
γ[E2]	2655.85 15	0.0023 5
γ	2788.4 5	0.00102 7
γ	2940.0 3	0.0028 5
γ	2966.3 7	0.00054 4
γ	2994.0 2	0.0067 13
γ	3059.2 5	0.00060 4
γ	3104.5 5	0.00013 4
γ	3355.6 5	0.00025 3
γ	3457.4 7	0.000099 20
γ	3815 1	0.00040 3
γ	3836 1	0.000196 20
γ	3881 1	7.8 19 $\times10^{-5}$
γ	3911 1	9.4 13 $\times10^{-5}$
γ	3956 1	8.0 13 $\times10^{-5}$

† 7.3% uncert(syst)

Atomic Electrons (^{82}Rb)

⟨e⟩=0.319 *15* keV

e_{bin}(keV)	⟨e⟩(keV)	e(%)
2	0.058	3.4 *4*
10	0.0101	0.096 *10*
11	0.103	0.95 *10*
12	0.037	0.30 *3*
13 - 14	0.0109	0.084 *6*
453 - 467	0.00016	$\sim3\times10^{-5}$
508 - 523	3.8×10^{-5}	$\sim7\times10^{-6}$
683 - 711	0.00184	0.00027 *3*
762	0.084	0.0111 *9*
775 - 777	0.0110	0.00142 *10*
855 - 869	5.1×10^{-6}	$\sim6\times10^{-7}$
961 - 1007	3.7×10^{-5}	3.8 *11* $\times10^{-6}$
1019 - 1067	4.7×10^{-6}	4.6 *22* $\times10^{-7}$
1072 - 1122	5.9×10^{-6}	5 *3* $\times10^{-7}$
1123 - 1168	6.6×10^{-5}	5.7 *6* $\times10^{-6}$
1178 - 1180	7.8×10^{-6}	6.6 *5* $\times10^{-7}$
1381 - 1395	0.00156	0.000113 *5*
1461 - 1475	0.000248	1.70 *8* $\times10^{-5}$
1593 - 1689	0.000129	7.7 *5* $\times10^{-6}$
1697 - 1785	2.23×10^{-5}	1.30 *12* $\times10^{-6}$
1865 - 1957	3.70×10^{-5}	1.95 *11* $\times10^{-6}$
2153 - 2242	6.7×10^{-5}	3.1 *8* $\times10^{-6}$
2396 - 2495	0.000110	4.5 *4* $\times10^{-6}$
2507 - 2578	1.9×10^{-6}	7.5 *15* $\times10^{-8}$
2642 - 2656	4.4×10^{-6}	1.6 *4* $\times10^{-7}$
2774 - 2788	1.5×10^{-6}	5.5 *12* $\times10^{-8}$
2926 - 2994	1.5×10^{-5}	4.9 *10* $\times10^{-7}$
3045 - 3104	1.05×10^{-6}	3.4 *6* $\times10^{-8}$
3341 - 3355	3.5×10^{-7}	1.03 *22* $\times10^{-8}$
3443 - 3457	1.3×10^{-7}	3.9 *10* $\times10^{-9}$
3801 - 3897	9.7×10^{-7}	2.5 *3* $\times10^{-8}$
3909 - 3956	1.14×10^{-7}	2.9 *6* $\times10^{-9}$

Continuous Radiation (^{82}Rb)

⟨β+⟩=1409 keV;⟨IB⟩=4.8 keV

E_{bin}(keV)		⟨ ⟩(keV)	(%)
0 - 10	β+	9.7×10^{-6}	0.000120
	IB	0.047	
10 - 20	β+	0.000234	0.00144
	IB	0.046	0.32
20 - 40	β+	0.00409	0.0128
	IB	0.091	0.32
40 - 100	β+	0.124	0.162
	IB	0.26	0.40
100 - 300	β+	4.56	2.08
	IB	0.76	0.43
300 - 600	β+	33.9	7.3
	IB	0.90	0.21
600 - 1300	β+	295	30.3
	IB	1.37	0.157
1300 - 2500	β+	882	48.4
	IB	1.00	0.058
2500 - 4390	β+	194	7.1
	IB	0.29	0.0099
	Σβ+		95

$^{82}_{37}$Rb(6.472 *6* h)

Mode: ϵ

Δ: ~-76102 keV

SpA: 5.905×10^{6} Ci/g

Prod: ^{79}Br(α,n); ^{82}Kr(d,2n); protons on Rb

Photons (^{82}Rb)

⟨γ⟩=2530 *17* keV

γ_{mode}	γ(keV)	γ(%)†
Kr L_ℓ	1.383	0.048 *10*
Kr L_η	1.435	0.027 *7*
Kr L_α	1.581	1.00 *21*
Kr L_β	1.640	0.58 *15*
Kr L_γ	1.907	0.009 *3*
Kr $K_{\alpha2}$	12.598	13.0 *6*
Kr $K_{\alpha1}$	12.651	25.1 *11*
Kr $K_{\beta1}$'	14.109	5.31 *23*
Kr $K_{\beta2}$'	14.408	0.478 *21*
γ [M1+E2]	92.190 *16*	0.65 *6*
γ [E1]	100.89 *8*	0.110 *9*
γ [E1]	129.29 *4*	0.127 *9*
γ [M1+E2]	137.15 *3*	0.118 *9*
γ	183.27 *2*	2.13 *4*
γ [E1]	221.46 *2*	2.07 *4*
γ [M1+E2]	273.480 *8*	1.04 *6*
γ	389.4 *1*	~0.08
γ [E1]	401.16 *6*	0.51 *9*
γ	455.28 *3*	1.27 *9*
γ	499.31 *5*	0.21 *6*
γ [E1]	554.35 *1*	62.5 *8*
γ	583.80 *5*	1.34 *4*
γ	606.37 *4*	2.01 *6*
γ [M1+E2]	619.11 *1*	38.0 *8*
γ [M1+E2]	698.37 *1*	26.4 *7*
γ	703.56 *4*	0.144 *25*
γ	755.76 *7*	0.313 *17*
γ [E2]	776.52 *1*	84.5
γ	836.0 *7*	~0.17
γ	952.02 *15*	0.64 *4*
γ	963.7 *3*	0.093 *25*
γ	976.9 *2*	0.068 *8*
γ	987.1 *5*	~0.034
γ [E1]	1007.59 *3*	7.18 *9*
γ [E2]	1044.08 *2*	32.1 *9*
γ [M1+E2]	1072.99 *7*	0.73 *5*
γ [E1]	1081.29 *6*	1.31 *4*
γ	1086 *1*	0.025 *8*
γ [E2]	1099.81 *5*	0.75 *3*
γ [M1+E2]	1180.1 *1*	0.118 *17*
γ	1190.81 *6*	0.287 *17*
γ	1218 *1*	~0.04 ?
γ	1228.9 *4*	0.059 *25*
γ [M1+E2]	1317.43 *2*	23.7 *6*
γ	1330.8 *5*	0.524 *17*
γ	1374.8 *2*	0.093 *25*
γ [E2]	1395.4 *5*	~0.025
γ	1441.7 *1*	0.161 *25*
γ [E2]	1474.88 *2*	15.5 *3*
γ	1506.8 *5*	~0.017
γ	1543.0 *4*	0.042 *17*
γ	1555.3 *4*	0.034 *8*
γ	1641.3 *4*	0.025 *8*
γ	1650.37 *4*	1.183 *24*
γ	1707.8 *3*	0.034 *8*
γ [E1]	1779.66 *5*	0.262 *17*
γ	1835.20 *9*	0.118 *9*
γ	1871.5 *3*	0.027 *8*
γ [E2]	1956.6 *1*	0.059 *9*
γ	1961.3 *5*	~0.017
γ	1974.0 *1*	0.110 *8*
γ	1996.5 *2*	0.047 *8*
γ	2002.0 *3*	0.027 *8*
γ	2073.0 *3*	0.024 *7*
γ	2130.8 *4*	0.019 *6*
γ	2242.95 *10*	0.101 *8*
γ	2247.47 *13*	0.079 *7*
γ	2305 *1*	~0.008
γ	2315.0 *5*	0.30 *6*

† 1.1% uncert(syst)

Atomic Electrons (^{82}Rb)

⟨e⟩=6.2 *4* keV

e_{bin}(keV)	⟨e⟩(keV)	e(%)
2	1.00	59 *6*
10	0.172	1.64 *17*
11	1.76	16.2 *17*
12	0.63	5.1 *5*
13 - 14	0.187	1.43 *11*
78	0.30	~0.4
87 - 135	0.10	0.11 *4*
137	0.0006	~0.0004
169	0.18	~0.11
181 - 221	0.066	0.034 *8*
259 - 273	0.045	0.017 *7*
375 - 401	0.0048	0.0012 *4*
441 - 485	0.017	0.0039 *18*
497 - 499	0.00025	5.1 *25* $\times10^{-5}$
540	0.229	0.0423 *18*
552 - 592	0.053	0.0093 *14*
604	0.0013	0.00022 *11*
605	0.31	0.051 *7*
606 - 619	0.041	0.0067 *9*
684 - 721	0.205	0.030 *3*
734 - 756	0.0017	0.00023 *9*
762	0.532	0.070 *3*
775 - 822	0.0700	0.0090 *3*
834 - 836	8.6×10^{-5}	$\sim1\times10^{-5}$
938 - 987	0.0031	0.00033 *10*
993 - 1042	0.162	0.0158 *6*
1044 - 1086	0.0114	0.00107 *3*
1098 - 1100	0.000372	3.38 *17* $\times10^{-5}$
1166 - 1215	0.0016	0.00013 *3*
1216 - 1229	3.2×10^{-5}	2.6 *11* $\times10^{-6}$
1303 - 1331	0.085	0.0065 *4*
1360 - 1395	0.00032	2.3 *8* $\times10^{-5}$
1427 - 1475	0.0492	0.00336 *13*
1492 - 1555	0.00021	1.4 *4* $\times10^{-5}$
1627 - 1708	0.0027	0.00016 *4*
1757 - 1835	0.00067	3.7 *5* $\times10^{-5}$
1857 - 1956	0.00022	1.16 *17* $\times10^{-5}$
1959 - 2059	0.00038	1.9 *3* $\times10^{-5}$
2071 - 2131	3.9×10^{-5}	1.8 *6* $\times10^{-6}$
2229 - 2315	0.00083	3.6 *7* $\times10^{-5}$

Continuous Radiation (^{82}Rb)

⟨β+⟩=91 keV;⟨IB⟩=1.24 keV

E_{bin}(keV)		⟨ ⟩(keV)	(%)
0 - 10	β+	5.5×10^{-5}	0.00068
	IB	0.0045	
10 - 20	β+	0.00132	0.0081
	IB	0.0107	0.080
20 - 40	β+	0.0223	0.070
	IB	0.0080	0.028
40 - 100	β+	0.61	0.81
	IB	0.021	0.033
100 - 300	β+	15.9	7.5
	IB	0.072	0.039
300 - 600	β+	49.2	11.4
	IB	0.20	0.045
600 - 1300	β+	23.5	3.05
	IB	0.74	0.082
1300 - 2500	β+	1.39	0.100
	IB	0.19	0.0128
2500 - 2670	IB	6.8×10^{-5}	2.7×10^{-6}
	Σβ+		23

$^{82}_{38}$Sr(25.55 *15* d)

Mode: ϵ

Δ: -75997 *9* keV

SpA: 6.23×10^{4} Ci/g

Prod: ^{85}Rb(p,4n); ^{22}Ne on Cu; protons on Mo

Photons (^{82}Sr)

$\langle\gamma\rangle$=7.85 *21* keV

γ_{mode}	γ(keV)	γ(%)
Rb L_ℓ	1.482	0.064 *14*
Rb L_η	1.542	0.037 *9*
Rb L_α	1.694	1.4 *3*
Rb L_β	1.760	0.81 *21*
Rb L_γ	2.051	0.014 *4*
Rb $K_{\alpha2}$	13.336	17.0 *7*
Rb $K_{\alpha1}$	13.395	32.7 *13*
Rb $K_{\beta1}$'	14.959	7.0 *3*
Rb $K_{\beta2}$'	15.286	0.75 *3*

Atomic Electrons (^{82}Sr)

$\langle$e$\rangle$=4.76 *24* keV

e_{bin}(keV)	$\langle$e$\rangle$(keV)	e(%)
2	1.35	73 *8*
11	0.91	8.1 *8*
12	1.49	12.9 *13*
13	0.93	7.1 *7*
15	0.085	0.58 *6*

Continuous Radiation (^{82}Sr)

$\langle$IB$\rangle$=0.0122 keV

E_{bin}(keV)		$\langle\ \rangle$(keV)	(%)
10 - 20	IB	0.0099	0.074
20 - 40	IB	0.00046	0.0017
40 - 100	IB	0.00087	0.00132
100 - 205	IB	0.00070	0.00055

$^{82}_{39}$Y (9.5 *4* s)

Mode: ϵ

Δ: -68180 *90* keV

SpA: 1.40×10^{10} Ci/g

Prod: ^{60}Ni(^{24}Mg,pn)

Photons (^{82}Y)

$\langle\gamma\rangle$=185 *35* keV

γ_{mode}	γ(keV)	γ(%)†
γ[E2]	573.6 *1*	25
γ[M1+E2]	601.7 *1*	3.25 *25*
γ[E2]	737.5 *1*	2.25 *25*
γ[E2]	1175.0 *2*	~0.50

† 24% uncert(syst)

$^{82}_{40}$Zr(2.5 *1* min)

Mode: ϵ

Δ: -64180 *510* keV

SpA: 9.2×10^{8} Ci/g

Prod: ^{32}S on ^{54}Fe

A = 83

NDS **15**, 169 (1975)

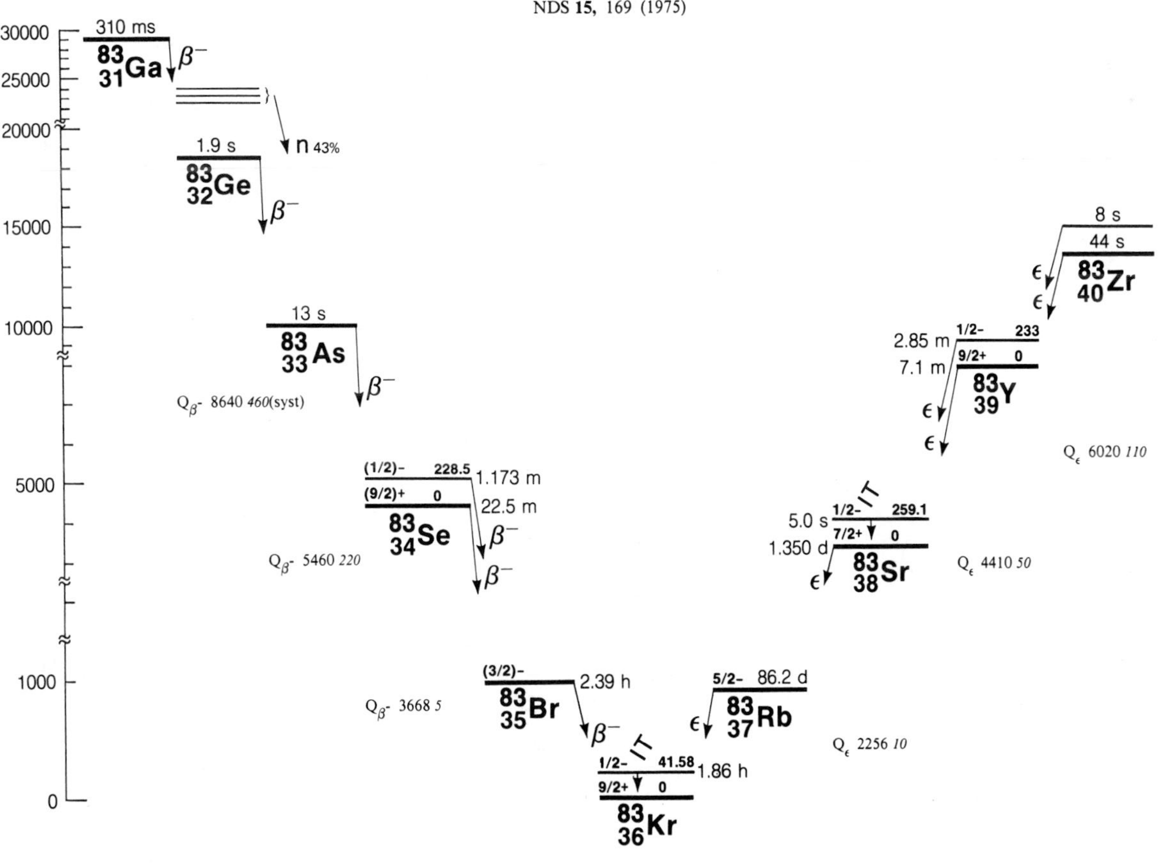

$^{83}_{31}$Ga(310 *10* ms)

Mode: β-, β-n(43 *7* %)
SpA: 1.75×10^{11} Ci/g
Prod: fission

$^{83}_{32}$Ge(1.9 *4* s)

Mode: β-
Δ: -61240 *400* keV syst
SpA: 6.0×10^{10} Ci/g
Prod: fission

$^{83}_{33}$As(13.3 *6* s)

Mode: β-
Δ: -69880 *220* keV
SpA: 1.0×10^{10} Ci/g
Prod: fission

Photons (^{83}As)

γ_{mode}	γ(keV)	γ(rel)
γ	310.9 *3*	2.7 *3*
γ	582.0 *1*	9.9 *4*
γ	734.5 *1*	100
γ	780.5 *2*	2.8 *3*
γ	803.4 *2*	9.2 *4*
γ	833.8 *1*	19.5 *17*
γ	1013.7 *2*	6.1 *5*
γ	1057.6 *2*	4.7 *10*
γ	1113.1 *1*	34 *3*
γ	1158.4 *2*	4.6 *5*
γ	1331.0 *2*	15.8 *11*
γ	1549.6 *2*	3.0 *2*
γ	1822.4 *1*	7.9 *7*
γ	1895.4 *1*	17.6 *15*
γ	1917.2 *1*	14.8 *17*
γ	2076.7 *1*	28 *3*
γ	2202.9 *2*	22.2 *19*
γ	2316.5 *10*	~2
γ	2660.1 *10*	2.6 *4*
γ	2857.9 *5*	16.3 *15*
γ	3241.9 *10*	4.1 *4*
γ	3865 *2*	1.1 *5*

$^{83}_{34}$Se(22.5 *2* min)

Mode: β-
Δ: -75343 *4* keV
SpA: 1.007×10^{8} Ci/g
Prod: ^{82}Se(n,γ)

Photons (^{83}Se)

⟨γ⟩=2562 *24* keV

γ_{mode}	γ(keV)	γ(%)†
γ[M1+E2]	208.43 *8*	1.85 *14*
γ	225.24 *5*	31.9 *14*
γ	296.20 *11*	0.27 *7*
γ	322.3 *3*	~0.07

Photons (^{83}Se) (continued)

γ_{mode}	γ(keV)	γ(%)†
γ	340.27 *11*	0.41 *7*
γ[M1+E2]	356.71 *4*	68.6
γ	371.44 *9*	0.55 *7*
γ	389.29 *7*	0.62 *7*
γ	442.39 *5*	1.09 *7*
γ	451.62 *9*	0.82 *7*
γ	457.57 *6*	3.50 *21*
γ	472.80 *16*	0.14 *3*
γ	485.85 *6*	2.26 *14*
γ	510.03 *5*	44.3 *17*
γ	553.50 *8*	4.3 *10*
γ	571.98 *8*	4.46 *21*
γ	581.70 *20*	0.34 *7*
γ	593.51 *9*	0.75 *7*
γ	609.29 *9*	3.09 *21*
γ	621.77 *8*	0.41 *7*
γ	664.96 *9*	3.22 *21*
γ	679.48 *7*	1.10 *7*
γ	705.91 *10*	0.21 *7*
γ	712.22 *8*	2.68 *14*
γ	718.17 *6*	16.3 *7*
γ	735.28 *6*	0.82 *7*
γ	799.10 *5*	16.1 *6*
γ	836.63 *7*	15.9 *7*
γ	866.75 *5*	8.8 *3*
γ	883.76 *7*	7.8 *3*
γ	887.90 *8*	4.73 *21*
γ	933.77 *10*	0.69 *7*
γ	943.42 *5*	0.89 *7*
γ	988.0 *1*	0.62 *7*
γ	995.88 *7*	1.30 *14*
γ	1036.70 *12*	0.21 *7*
γ	1036.83 *12*	0.21 *7*
γ	1042.2 *2*	1.03 *7*
γ	1064.15 *7*	5.9 *3*
γ	1082.01 *9*	2.68 *21*
γ	1110.43 *17*	0.41 *7*
γ	1191.76 *9*	4.18 *21*
γ	1207.0 *2*	0.89 *7*
γ	1225.92 *9*	1.23 *14*
γ	1245.12 *12*	0.69 *7*
γ	1259.4 *1*	0.89 *7*
γ	1294.20 *9*	1.6 *3*
γ	1299.25 *10*	5.8 *3*
γ	1305.73 *10*	0.55 *7*
γ	1317.11 *10*	4.1 *4*
γ	1341.33 *8*	5.7 *3*
γ	1352.59 *7*	4.8 *3*
γ	1420.86 *7*	1.10 *7*
γ	1435.9 *3*	0.82 *7*
γ	1447.49 *9*	0.48 *7*
γ	1475.3 *3*	0.69 *7*
γ	1554.80 *8*	2.54 *14*
γ	1664.55 *17*	0.55 *7*
γ	1684.3 *10*	0.21 *7*
γ	1716.30 *12*	0.62 *21*
γ	1780.04 *8*	1.9 *3*
γ	1827.16 *8*	1.37 *14*
γ	1847.68 *9*	0.69 *7*
γ	1854.41 *10*	1.51 *14*
γ	1871.22 *8*	1.44 *14*
γ	1894.81 *9*	7.8 *3*
γ	1973.3 *4*	0.62 *7*
γ	2045.3 *5*	0.69 *7*
γ	2072.5 *7*	~0.27
γ	2085.5 *4*	0.55 *7*
γ	2167.4 *4*	0.34 *7*
γ[E1]	2290.06 *8*	9.3 *3*
γ[E1]	2337.18 *8*	3.43 *21*
γ	2420.0 *4*	0.41 *7*

† 0.87% uncert(syst)

Continuous Radiation (^{83}Se)

⟨β-⟩=478 keV;⟨IB⟩=0.75 keV

E_{bin}(keV)		⟨ ⟩(keV)	(%)
0 - 10	β-	0.056	1.11
	IB	0.020	
10 - 20	β-	0.169	1.13
	IB	0.020	0.138
20 - 40	β-	0.69	2.30
	IB	0.038	0.131
40 - 100	β-	5.2	7.4
	IB	0.099	0.154
100 - 300	β-	55	27.5
	IB	0.23	0.131
300 - 600	β-	153	35.0
	IB	0.17	0.042
600 - 1300	β-	156	19.2
	IB	0.142	0.0170
1300 - 2500	β-	106	6.2
	IB	0.035	0.0022
2500 - 2869	β-	1.63	0.063
	IB	3.8×10^{-5}	1.49×10^{-6}

$^{83}_{34}$Se(1.173 *5* min)

Mode: β-
Δ: -75115 *4* keV
SpA: 1.921×10^{9} Ci/g
Prod: ^{82}Se(n,γ)

Photons (^{83}Se)

⟨γ⟩=955 *16* keV

γ_{mode}	γ(keV)	γ(%)†
γ	188.87 *7*	0.173 *17*
γ	231.52 *6*	0.329 *17*
γ	322.2 *3*	0.052 *17*
γ[M1+E2]	356.71 *4*	17.3
γ	371.64 *10*	0.087 *17*
γ	391.34 *10*	0.104 *17*
γ	442.39 *5*	0.082 *7*
γ	510.03 *5*	<0.36 ?
γ	631.25 *6*	0.47 *4*
γ	673.91 *5*	15.1 *5*
γ	698.7 *4*	0.087 *17*
γ	799.10 *5*	1.21 *7*
γ	866.75 *5*	<0.07
γ	987.96 *6*	15.3 *5*
γ	997.65 *8*	1.28 *7*
γ	1020.69 *7*	1.97 *10*
γ	1030.62 *5*	20.9 *10*
γ	1053.65 *8*	1.49 *7*
γ	1063.34 *8*	3.39 *17*
γ	1116.04 *10*	0.54 *4*
γ	1303.25 *9*	0.93 *5*
γ	1548.9 *4*	0.052 *17*
γ	1558.5 *3*	1.19 *7*
γ	1659.96 *8*	1.78 *9*
γ	1694.59 *7*	0.74 *5*
γ	1778.97 *18*	0.71 *5*
γ	2022.1 *5*	0.069 *17*
γ	2051.29 *7*	11.0 *4*
γ	2452.86 *19*	0.040 *10*
γ	2734.3 *5*	0.045 *10*
γ	2809.57 *19*	0.022 *7*
γ	2945.1 *8*	0.022 *7*
γ	3091.0 *5*	0.038 *10*

† 2.3% uncert(syst)

Continuous Radiation (^{83}Se)

⟨β-⟩=1263 keV;⟨IB⟩=3.4 keV

E_{bin}(keV)		⟨ ⟩(keV)	(%)
0 - 10	β-	0.0087	0.173
	IB	0.043	
10 - 20	β-	0.0267	0.178
	IB	0.042	0.29
20 - 40	β-	0.111	0.370
	IB	0.083	0.29
40 - 100	β-	0.88	1.25
	IB	0.24	0.37
100 - 300	β-	11.9	5.7
	IB	0.68	0.38
300 - 600	β-	57	12.4
	IB	0.77	0.18
600 - 1300	β-	342	36.0
	IB	1.02	0.119
1300 - 2500	β-	673	37.6
	IB	0.49	0.030
2500 - 3897	β-	178	6.2
	IB	0.035	0.00124

$^{83}_{35}$Br(2.39 *2* h)

Mode: β-

Δ: -79011 *5* keV

SpA: 1.580×10^{7} Ci/g

Prod: daughter ^{83}Se

Photons (^{83}Br)

⟨γ⟩=7.44 *14* keV

γ_{mode}	γ(keV)	γ(%)†
γ M1+0.02%E2	9.400 *8*	*
γ E3	32.178 *17*	*
γ [M1+E2]	119.331 *18*	0.00159 *13*
γ [M1+E2]	128.52 *3*	0.00019 *3* ?
γ [E2]	520.461 *23*	0.0622 *18*
γ	529.651 *10*	1.30 *3*
γ [E1]	552.639 *22*	0.0219 *9*
γ	562.039 *23*	1.17 *16* ×10^{-5}
γ [E2]	648.981 *20*	0.0125 *7*
γ [E1]	681.16 *3*	0.0043 *3*

† 31% uncert(syst)
* with ^{83}Kr(1.86 h)

Continuous Radiation (^{83}Br)

⟨β-⟩=326 keV;⟨IB⟩=0.32 keV

E_{bin}(keV)		⟨ ⟩(keV)	(%)
0 - 10	β-	0.072	1.43
	IB	0.0155	
10 - 20	β-	0.218	1.45
	IB	0.0148	0.103
20 - 40	β-	0.89	2.96
	IB	0.028	0.096
40 - 100	β-	6.6	9.4
	IB	0.069	0.108
100 - 300	β-	68	33.8
	IB	0.132	0.079
300 - 600	β-	174	39.9
	IB	0.057	0.0146
600 - 931	β-	76	11.1
	IB	0.0041	0.00063

$^{83}_{36}$Kr(stable)

Δ: -79983 *3* keV

%: 11.5 *1*

$^{83}_{36}$Kr(1.86 *1* h)

Mode: IT

Δ: -79941 *3* keV

SpA: 2.030×10^{7} Ci/g

Prod: daughter ^{83}Rb; daughter ^{83}Br

Photons (^{83}Kr)

⟨γ⟩=2.57 *7* keV

γ_{mode}	γ(keV)	γ(%)†
Kr L_{ℓ}	1.383	0.098 *23*
Kr L_{η}	1.435	0.059 *15*
Kr L_{α}	1.581	2.1 *5*
Kr L_{β}	1.645	1.3 *4*
Kr L_{γ}	1.907	0.034 *11*
γ M1+0.02%E2	9.400 *8*	5.0
Kr $K_{\alpha2}$	12.598	4.67 *24*
Kr $K_{\alpha1}$	12.651	9.0 *5*
Kr $K_{\beta1}$'	14.109	1.91 *10*
Kr $K_{\beta2}$'	14.408	0.172 *9*
γ E3	32.178 *17*	0.0517 *15*

† 6.1% uncert(syst)

Atomic Electrons (^{83}Kr)

⟨e⟩=38.2 *7* keV

e_{bin}(keV)	⟨e⟩(keV)	e(%)
2	1.83	106 *11*
7	4.5	60 *4*
8	1.07	13.9 *14*
9	1.25	13.6 *10*
10	0.062	0.59 *6*
11	0.63	5.8 *6*
12	0.228	1.85 *20*
13	0.042	0.33 *4*
14	0.026	0.184 *19*
18	4.39	24.6 *9*
30	8.2	26.9 *10*
31	12.0	39.4 *14*
32	3.97	12.4 *4*

$^{83}_{37}$Rb(86.2 *1* d)

Mode: ϵ

Δ: -79044 *21* keV

SpA: 1.8250×10^{4} Ci/g

Prod: ^{81}Br(α,2n); daughter ^{83}Sr

Photons (^{83}Rb)

⟨γ⟩=507 *13* keV

γ_{mode}	γ(keV)	γ(%)†
Kr L_{ℓ}	1.383	0.15 *4*
Kr L_{η}	1.435	0.089 *25*
Kr L_{α}	1.581	3.2 *8*
Kr L_{β}	1.645	2.0 *6*
Kr L_{γ}	1.907	0.051 *18*

Photons (^{83}Rb)
(continued)

γ_{mode}	γ(keV)	γ(%)†
γ M1+0.02%E2	9.400 *8*	6.0 *14* *
Kr $K_{\alpha2}$	12.598	19.9 *10*
Kr $K_{\alpha1}$	12.651	38.6 *20*
Kr $K_{\beta1}$'	14.109	8.1 *4*
Kr $K_{\beta2}$'	14.408	0.73 *4*
γ E3	32.178 *17*	0.037 *5* *
γ [M1+E2]	119.331 *18*	0.0114 *11*
γ [M1+E2]	128.52 *3*	0.00138 *23*
γ	228.23 *5*	0.0129 *4*
γ [E2]	520.461 *23*	46
γ	529.651 *10*	30.2 *14*
γ [E1]	552.639 *22*	16.3 *6*
γ	562.039 *23*	0.0087 *9*
γ [E2]	648.981 *20*	0.090 *5*
γ [E1]	681.16 *3*	0.031 *3*
γ [M1+E2]	790.06 *5*	0.676 *18*
γ [E1]	799.46 *5*	0.244 *9*

† 4.3% uncert(syst)
* with ^{83}Kr(1.86 h) in equilib

Atomic Electrons (^{83}Rb)

⟨e⟩=36.5 *20* keV

e_{bin}(keV)	⟨e⟩(keV)	e(%)
2	2.8	163 *22*
7	5.4	72 *17*
8	1.3	17 *4*
9	1.5	16 *4*
10	0.26	2.5 *3*
11	2.7	25 *3*
12 - 14	1.26	10.1 *9*
18	3.1	17.5 *22*
30	5.8	19.2 *24*
31	8.5	28 *4*
32	2.8	8.8 *11*
105 - 129	0.0038	~0.0035
214 - 228	0.0007	~0.00034
506 - 553	1.01	0.20 *3*
560 - 562	8.4 ×10^{-6}	1.5 *7* ×10^{-6}
635 - 681	0.00096	0.000151 *8*
776 - 799	0.0050	0.00065 *5*

Continuous Radiation (^{83}Rb)

⟨IB⟩=0.030 keV

E_{bin}(keV)		⟨ ⟩(keV)	(%)
10 - 20	IB	0.0085	0.066
20 - 40	IB	0.00042	0.00154
40 - 100	IB	0.00141	0.0020
100 - 300	IB	0.0099	0.0054
300 - 600	IB	0.0069	0.00162
600 - 931	IB	0.0027	0.00040

$^{83}_{38}$Sr(1.3504 *12* d)

Mode: ϵ

Δ: -76788 *21* keV

SpA: 1.1650×10^{6} Ci/g

Prod: ^{85}Rb(p,3n); ^{75}As(^{12}C,p3n); daughter ^{83}Y(7.1 min)

Photons (^{83}Sr)

⟨γ⟩=540 22 keV

γmode	γ(keV)	γ(%)†
γ M1	5.28 8	0.065 4
Rb $K_{\alpha2}$	13.336	23.0 15
Rb $K_{\alpha1}$	13.395	44 3
Rb $K_{\beta1}$'	14.959	9.6 6
Rb $K_{\beta2}$'	15.286	1.02 7
γ M2	42.33 8	1.57 15
γ M1+0.8%E2	94.33 15	0.41 3
γ	157.5 3	~0.039
γ	160.05 19	0.151 24
γ E2+36%M1	290.10 19	0.53 7
γ (E2)	381.09 17	7.7 9
γ (E2)	381.57 12	11.9 9
γ E2(+6.8%M1)	389.71 16	1.55 13
γ E1	418.63 12	5.0 4
γ E1	423.91 11	1.56 11
γ M1(+1.9%E2)	438.41 16	0.90 6
γ	559.66 19	0.19 3
γ	565.2 3	0.12 3
γ	630.90 24	0.036 12
γ	637.62 22	0.053 15 ?
γ	645.27 23	0.080 18
γ	651.5 3	0.098 24
γ	659.46 18	0.36 4
γ	673.4 3	0.068 9
γ	678.47 20	0.053 6
γ	683.0 5	0.021 9
γ	712.1 3	0.027 6
γ	715.50 21	0.098 9
γ	722.1 3	0.015 6
γ	731.86 17	0.080 9
γ	737.14 17	0.211 15
γ	753.65 20	0.09 3
γ	758.93 20	0.41 7
γ E2(+24%M1)	762.67 15	30
γ M1,E2	778.65 16	1.94 15
γ [E1]	805.00 15	0.071 12
γ	808.6 9	~0.024
γ M1,E2	819.51 14	0.83 5
γ	830.3 3	<0.015 ?
γ	839.1 3	<0.009 ?
γ	848.75 18	0.214 21
γ [E1]	853.70 18	0.134 15 *
γ	868.97 21	0.027 6
γ	880.1 3	0.036 6
γ	890.78 25	0.166 18
γ	904.08 21	0.062 6
γ	908.05 18	0.327 24
γ	916.97 21	0.128 12
γ	934.6 3	0.045 9
γ	944.93 20	0.146 12
γ	994.42 22	0.58 4
γ	1003.7 3	0.024 9
γ	1011.70 24	0.033 9
γ	1019.45 23	0.068 9
γ	1035.4 3	0.045 12
γ	1039.00 16	0.059 21
γ	1044.28 16	0.36 5
γ	1051.12 23	0.104 15
γ	1054.81 23	0.214 21
γ	1079.7 3	~0.006
γ	1086.53 19	0.039 6
γ	1098.30 16	0.264 18
γ	1103.58 17	0.027 6
γ	1130.84 25	0.039 9
γ [M1+E2]	1147.33 18	1.23 7
γ	1160.22 15	1.53 8
γ	1174.16 25	0.056 6
γ	1202.55 15	0.169 12
γ	1208.4 3	~0.006
γ	1215.18 18	0.226 12
γ	1233.92 25	0.027 6
γ [E1]	1238.13 13	0.223 15
γ [E1]	1243.41 13	0.077 6
γ	1251.21 25	0.015 6
γ	1271.87 23	0.080 6
γ	1277.9 5	0.045 6
γ	1297.3 3	0.137 12
γ	1324.94 21	0.25 3
γ	1331.3 3	~0.015
γ	1374.95 20	0.062 9
γ	1383.3 7	0.048 12

Photons (^{83}Sr)
(continued)

γmode	γ(keV)	γ(%)†
γ	1384.7 3	0.104 21
γ	1440.9 3	0.024 6
γ [M1+E2]	1528.42 16	0.095 9
γ [E1]	1562.62 19	1.96 10
γ	1592.78 24	0.024 6
γ	1598.06 24	0.036 9
γ	1606.1 8	~0.009
γ	1613.1 3	0.015 6
γ	1624.8 9	~0.009
γ	1649.23 23	0.021 3
γ	1653.44 22	0.080 9
γ	1666.49 23	0.080 9
γ	1706.51 20	0.030 6
γ	1710.57 21	0.039 6
γ	1748.84 20	0.027 6
γ	1756.52 21	0.024 6
γ	1766.3 3	0.024 9
γ	1798.85 20	
γ	1874.5 3	0.036 6
γ	1911.5 3	0.042 6
γ [E1]	1947.04 14	0.056 6
γ [E1]	1952.32 14	0.86 5
γ	2015.25 20	0.045 6
γ	2047.86 21	0.107 9
γ	2090.19 21	0.140 9
γ	2134.47 21	0.030 6
γ	2147.47 19	0.184 12

† 9.4% uncert(syst)
* doublet

Atomic Electrons (^{83}Sr)

⟨e⟩=26.1 14 keV

e_{bin}(keV)	⟨e⟩(keV)	e(%)
2	1.90	103 12
3 - 5	0.233	6.6 3
11	1.24	11.0 12
12	2.02	17.5 20
13	1.27	9.7 11
15	0.116	0.79 9
27	13.7	50 5
40	3.2	7.9 8
41	0.58	1.44 14
42	0.78	1.85 18
79 - 94	0.059	0.074 5
142 - 160	0.028	~0.019
275 - 290	0.024	0.0087 20
366 - 409	0.67	0.180 14
417 - 438	0.0177	0.00419 22
544 - 565	0.0028	0.00051 23
616 - 663	0.0039	0.00061 20
668 - 717	0.0015	0.00021 5
720 - 763	0.244	0.033 3
777 - 824	0.0076	0.00095 6
828 - 876	0.0024	0.00029 8
878 - 919	0.0025	0.00028 7
929 - 945	0.0009	~0.00010
979 - 1024	0.0031	~0.00032
1029 - 1078	0.0026	0.00025 8
1079 - 1129	0.0011	0.00010 3
1131 - 1174	0.0106	0.00093 15
1187 - 1236	0.0019	0.00016 4
1238 - 1282	0.00074	5.8 12 $\times10^{-5}$
1295 - 1331	0.00080	6.1 18 $\times10^{-5}$
1360 - 1385	0.00058	4.2 9 $\times10^{-5}$
1426 - 1513	0.00034	2.3 3 $\times10^{-5}$
1526 - 1625	0.00346	0.000223 13
1634 - 1710	0.00058	3.5 11 $\times10^{-5}$
1734 - 1766	0.00016	9.4 21 $\times10^{-6}$
1859 - 1952	0.00148	7.7 4 $\times10^{-5}$
2000 - 2090	0.00058	2.8 10 $\times10^{-5}$
2119 - 2147	0.00041	1.9 4 $\times10^{-5}$

Continuous Radiation (^{83}Sr)

⟨β+⟩=123 keV;⟨IB⟩=1.11 keV

E_{bin}(keV)		⟨ ⟩(keV)	(%)
0 - 10	β+	3.29 $\times10^{-5}$	0.000407
	IB	0.0057	
10 - 20	β+	0.00081	0.00498
	IB	0.0130	0.094
20 - 40	β+	0.0141	0.0440
	IB	0.0107	0.037
40 - 100	β+	0.396	0.52
	IB	0.029	0.045
100 - 300	β+	10.6	5.02
	IB	0.090	0.049
300 - 600	β+	45.1	10.1
	IB	0.19	0.042
600 - 1300	β+	67	8.6
	IB	0.55	0.061
1300 - 2256	IB	0.23	0.0146
	Σβ+		24

$^{83}_{38}$Sr(4.95 12 s)

Mode: IT
Δ: -76529 21 keV
SpA: 2.56$\times10^{10}$ Ci/g
Prod: daughter ^{83}Y

Photons (^{83}Sr)

⟨γ⟩=227.8 24 keV

γmode	γ(keV)	γ(%)†
Sr L_{ℓ}	1.582	0.0083 18
Sr L_{η}	1.649	0.0049 12
Sr L_{α}	1.806	0.19 4
Sr L_{β}	1.879	0.11 3
Sr L_{γ}	2.196	0.0016 5
Sr $K_{\alpha2}$	14.098	2.12 9
Sr $K_{\alpha1}$	14.165	4.09 17
Sr $K_{\beta1}$'	15.832	0.90 4
Sr $K_{\beta2}$'	16.192	0.123 5
γ E3	259.1 1	87.5

† 1.0% uncert(syst)

Atomic Electrons (^{83}Sr)

⟨e⟩=31.2 6 keV

e_{bin}(keV)	⟨e⟩(keV)	e(%)
2	0.184	9.3 10
12	0.29	2.37 24
14	0.112	0.81 8
15	0.0030	0.0195 20
16	0.0070	0.045 5
243	25.4	10.46 24
257	4.32	1.68 4
259	0.858	0.332 8

$^{83}_{39}$Y (7.06 *8* min)

Mode: ϵ

Δ: -72380 *60* keV

SpA: 3.21×10^8 Ci/g

Prod: ^{75}As(^{12}C,4n); ^{84}Sr(p,2n); protons on Mo

Photons (^{83}Y)

⟨γ⟩=407 *7* keV

γ_{mode}	γ(keV)	γ(%)†
γ (M1)	35.5 *1*	19.6 *13*
γ	138.8 *4*	0.102 *13*
γ	195.4 *4*	0.160 *19*
γ	227.5 *2*	0.115 *13*
γ	234.4 *3*	0.064 *13*
γ	245.3 *4*	0.064 *19*
γ E3	259.1 *1*	2.05 *13* *
γ	270.5 *3*	0.077 *10*
γ	391.6 *2*	1.47 *10*
γ	409.3 *4*	0.070 *13*
γ	420.3 *3*	1.87 *13*
γ	421.8 *3*	0.077 *15*
γ	434.6 *5*	0.077 *19*
γ	454.4 *2*	2.43 *6*
γ	489.9 *2*	5.89 *13*
γ	494.5 *2*	0.74 *10*
γ	525.6 *4*	0.075 *19*
γ	545.4 *3*	0.34 *6*
γ	547.1 *3*	0.29 *6*
γ	581.1 *3*	0.147 *19*
γ	603.0 *3*	0.064 *13*
γ	618.2 *2*	0.62 *5*
γ	654.5 *2*	0.198 *19*
γ	682.1 *1*	1.26 *6*
γ	717.6 *2*	0.461 *19*
γ	721.2 *2*	1.08 *5*
γ	743.5 *1*	1.25 *6*
γ	764.7 *6*	0.41 *5*
γ	781.5 *10*	
γ	790.8 *2*	0.45 *5*
γ	800.4 *1*	0.51 *3*
γ	827 *1*	
γ	858.7 *1*	3.26 *13*
γ	875.8 *3*	0.19 *3*
γ	882.1 *1*	6.4
γ	893.9 *4*	0.12 *3*
γ	916.5 *5*	0.083 *19*
γ	927.3 *1*	0.89 *5*
γ	931.5 *5*	0.10 *3*
γ	943.6 *5*	0.052 *16*
γ	951.8 *1*	1.98 *10*
γ	962.8 *1*	0.54 *3*
γ	1001.2 *6*	0.083 *19*
γ	1062.6 *10*	0.048 *16*
γ	1092.4 *6*	0.064 *19*
γ	1097.2 *8*	~0.038
γ	1115.0 *5*	0.064 *19*
γ	1154.4 *10*	
γ	1165 *1*	
γ	1197.9 *2*	0.37 *3*
γ	1203.6 *5*	0.070 *13*
γ	1233.3 *6*	0.064 *13*
γ	1239.2 *2*	0.58 *3*
γ	1264.3 *4*	0.096 *16*
γ	1336.5 *1*	3.14 *13*
γ	1366.4 *6*	0.064 *19*
γ	1371.9 *1*	1.02 *5*
γ	1378.8 *8*	0.032 *12*
γ	1392.4 *3*	0.141 *13*
γ	1401.0 *15*	~0.013
γ	1407.1 *10*	0.033 *10*
γ	1434.2 *2*	0.205 *19*
γ	1463.4 *3*	0.47 *6*
γ	1473.8 *6*	0.070 *13*
γ	1498.8 *2*	0.44 *4*
γ	1508.2 *6*	0.070 *10*
γ	1527.2 *10*	0.033 *10*
γ	1532.2 *10*	0.020 *10*
γ	1540.1 *7*	0.030 *10*
γ	1584.3 *10*	0.027 *13*
γ	1604.6 *8*	0.046 *13*
γ	1611.7 *10*	
γ	1640.1 *10*	0.062 *13*
γ	1710.0 *15*	~0.026
γ	1717.0 *8*	0.043 *15*
γ	1745.3 *9*	0.046 *13*
γ	1752.6 *4*	0.20 *3*
γ	1811 *1*	
γ	1819.9 *10*	
γ	1826.3 *10*	
γ	1846.8 *5*	0.077 *20*
γ	1855.0 *4*	0.106 *15*
γ	1872 *1*	
γ	1879.8 *7*	0.13 *3*
γ	1882.2 *8*	~0.042
γ	1909.6 *10*	
γ	1915.7 *4*	0.27 *3*
γ	1928 *2*	~0.026
γ	1942.3 *10*	0.028 *10*
γ	1964.5 *6*	0.081 *13*
γ	1973.0 *9*	0.036 *12*
γ	2011.1 *5*	0.118 *18*
γ	2016.9 *6*	0.098 *15*
γ	2049.0 *15*	0.027 *13*
γ	2054.1 *12*	0.040 *13*
γ	2068.0 *6*	0.060 *11*
γ	2073.6 *12*	0.038 *15*
γ	2089.8 *10*	~0.026
γ	2096.3 *10*	0.035 *14*
γ	2104.9 *8*	0.062 *12*
γ	2112.2 *7*	0.068 *12*
γ	2126.1 *10*	0.040 *12*
γ	2147.0 *25*	0.022 *10*
γ	2153.9 *10*	
γ	2178.7 *8*	0.077 *14*
γ	2187 *1*	0.032 *13*
γ	2224.5 *15*	0.032 *13*
γ	2250 *1*	~0.026
γ	2260.4 *9*	0.035 *13*
γ	2317.4 *6*	0.06 *1*
γ	2368 *1*	~0.013
γ	2393.8 *9*	0.042 *20*
γ	2400.0 *16*	~0.023
γ	2694.0 *12*	0.025 *12*
γ	2729 *2*	0.029 *13*
γ	2734 *2*	~0.013
γ	2829.8 *17*	0.032 *13*
γ	2869.6 *15*	0.043 *13*
γ	2879.2 *15*	0.077 *15*
γ	2905.3 *9*	0.21 *3*
γ	2909 *2*	~0.026
γ	2944 *1*	0.157 *19*
γ	2973 *2*	~0.017
γ	2981 *2*	0.035 *13*
γ	3060 *2*	0.016 *6*
γ	3220.0 *25*	0.022 *10*
γ	3251 *3*	0.024 *12*
γ	3297 *3*	0.035 *12*
γ	3420 *4*	~0.013

† 9.4% uncert(syst)

* with ^{83}Sr(5.0 s) in equilib

Continuous Radiation (^{83}Y)

⟨β+⟩=1375 keV;⟨IB⟩=4.8 keV

E_{bin}(keV)		⟨ ⟩(keV)	(%)
0 - 10	β+	8.4×10^{-6}	0.000103
	IB	0.046	
10 - 20	β+	0.000217	0.00133
	IB	0.046	0.32
20 - 40	β+	0.0040	0.0124
	IB	0.089	0.31
40 - 100	β+	0.127	0.167
	IB	0.26	0.39
100 - 300	β+	4.87	2.22
	IB	0.75	0.42
300 - 600	β+	36.3	7.8
	IB	0.88	0.21
600 - 1300	β+	304	31.3
	IB	1.35	0.154
1300 - 2500	β+	842	46.3
	IB	1.04	0.060
2500 - 4410	β+	187	6.8
	IB	0.31	0.0105
	Σβ+		95

$^{83}_{39}$Y (2.85 *2* min)

Mode: ϵ

Δ: -72147 *120* keV

SpA: 7.93×10^8 Ci/g

Prod: ^{54}Fe(^{32}S,3p); ^{56}Fe(^{32}S,αp); ^{55}Mn(^{32}S,2p2n); ^{84}Sr(p,2n); protons on Mo

Photons (^{83}Y)

⟨γ⟩=439 *15* keV

γ_{mode}	γ(keV)	γ(%)†
Sr L_ℓ	1.582	0.013 *3*
Sr L_η	1.649	0.0073 *19*
Sr L_α	1.806	0.28 *6*
Sr L_β	1.879	0.17 *4*
Sr L_γ	2.196	0.0026 *8*
Sr $K_{\alpha2}$	14.098	3.25 *20*
Sr $K_{\alpha1}$	14.165	6.3 *4*
Sr $K_{\beta1}'$	15.832	1.37 *8*
Sr $K_{\beta2}'$	16.192	0.187 *12*
γ E3	259.1 *1*	90.2 *
γ	421.8 *3*	32 *3*
γ	494.5 *2*	13.5 *18*

† 0.67% uncert(syst)

* with ^{83}Sr(5.0 s) in equilib

Atomic Electrons (^{83}Y)

⟨e⟩=33.1 *17* keV

e_{bin}(keV)	⟨e⟩(keV)	e(%)
2	0.28	14.0 *16*
12	0.44	3.6 *4*
14	0.172	1.24 *14*
15	0.0046	0.030 *3*
16	0.0107	0.068 *8*
243	26.2	10.8 *7*
257	4.5	1.73 *11*
259	0.88	0.342 *22*
406	0.5	~0.11
420	0.05	~0.013
421	0.008	~0.0020
422	0.0025	~0.0006
478	0.14	~0.030
492	0.016	~0.0032
493	0.0007	~0.00014
494	0.0032	~0.0007
495	3.2 ×10^{-5}	~7×10^{-6}

Continuous Radiation (^{83}Y)

⟨β+⟩=1390 keV;⟨IB⟩=5.0 keV

E_{bin}(keV)		⟨ ⟩(keV)	(%)
0 - 10	β+	7.3×10^{-6}	8.9×10^{-5}
	IB	0.046	
10 - 20	β+	0.000188	0.00115
	IB	0.046	0.32
20 - 40	β+	0.00347	0.0108
	IB	0.090	0.31
40 - 100	β+	0.111	0.145
	IB	0.26	0.40
100 - 300	β+	4.30	1.96
	IB	0.76	0.42
300 - 600	β+	32.8	7.0
	IB	0.90	0.21
600 - 1300	β+	294	30.1
	IB	1.39	0.159
1300 - 2500	β+	886	48.6
	IB	1.14	0.065
2500 - 4384	β+	173	6.4
	IB	0.40	0.0136
	Σβ+		94

$^{83}_{40}$Zr(44 *2* s)

Mode: ε
Δ: -66360 *100* keV
SpA: 3.06×10^{9} Ci/g
Prod: ^{54}Fe(^{32}S,2pn)

Photons (^{83}Zr)

⟨γ⟩=141 *5* keV

γ_{mode}	γ(keV)	γ(%)†
γ M1	55.55 *4*	15
γ E2	105.01 *4*	11.6 *5*
γ	254.76 *4*	9.6 *5*
γ	304.21 *4*	8.3 *5*
γ	359.80 *5*	5.5 *3*
γ	474.38 *6*	10.4 *8*
γ	1132.8 *3*	~0.15

† 29% uncert(syst)

$^{83}_{40}$Zr(8 s)

Mode: ε
Δ: >-66360 keV
SpA: 1.6×10^{10} Ci/g
Prod: ^{54}Fe(^{32}S,2pn)

A=84

NDS **27**, 339 (1979)

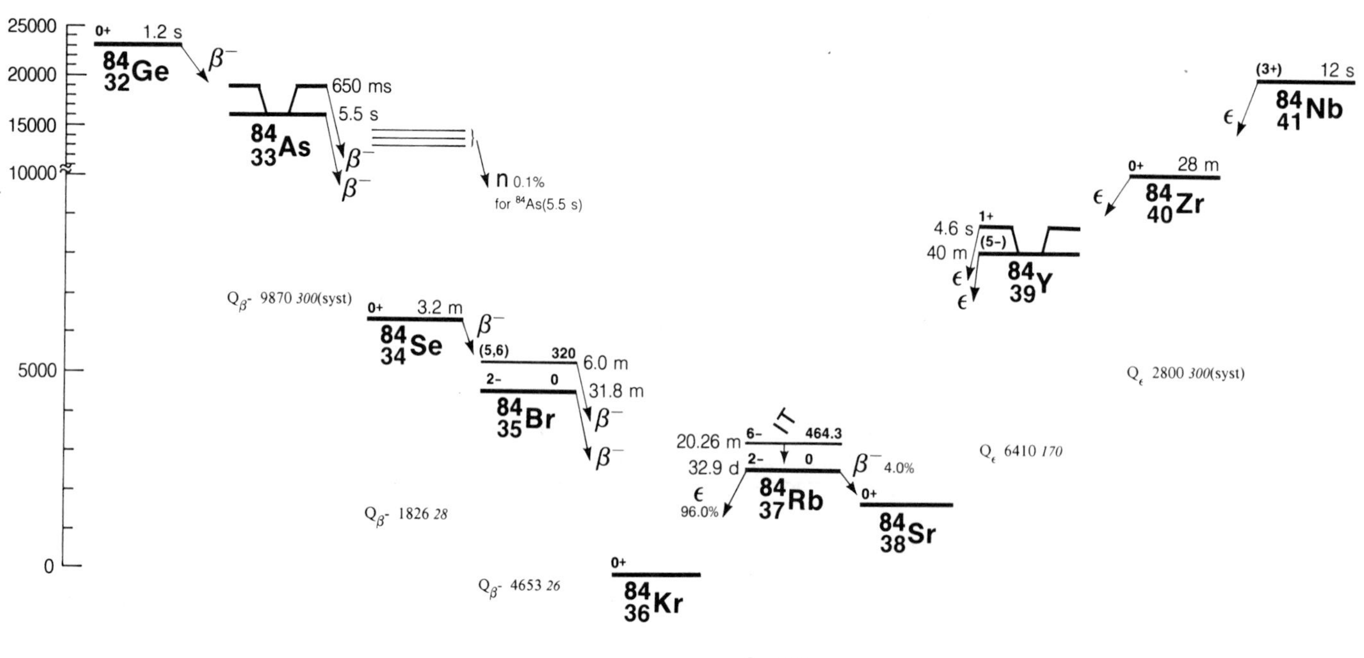

$^{84}_{32}$Ge(1.2 *3* s)

Mode: β-
SpA: 8.5×10^{10} Ci/g
Prod: fission

$^{84}_{33}$As(5.5 *3* s)

Mode: β-, β-n(0.13 *6* %)
Δ: -66080 *300* keV syst
SpA: 2.29×10^{10} Ci/g
Prod: fission

Photons (^{84}As)

⟨γ⟩=1574 *150* keV

γ_{mode}	γ(keV)	γ(%)†
γ	577.36 *17*	3.9 *5*
γ	667.03 *19*	20.7 *16*
γ	1244.38 *19*	3.0 *4*
γ	1248.6 *4*	1.32 *10*
γ	1317.0 *5*	1.71 *25*
γ	1443.8 *5*	2.4 *9*
γ[E2]	1454.96 *20*	49
γ	1843.56 *20*	3.4 *5*
γ	2086.5 *4*	4.7 *4*
γ	2460.9 *3*	4.0 *3*
γ	2722.9 *5*	0.88 *10*
γ	3038.2 *3*	1.52 *15*
γ	3474.9 *10*	0.8 *3*
γ	4435.8 *20*	1.0 *4*
γ	4885.9 *12*	~0.44
γ	4945.8 *7*	1.42 *24*
γ	5087.6 *10*	~0.9
γ	5150.9 *10*	1.0 *4*

† 22% uncert(syst)

Continuous Radiation (^{84}As)

⟨β-⟩=4039 keV;⟨IB⟩=20 keV

E_{bin}(keV)		⟨ ⟩(keV)	(%)
0 - 10	β-	0.00055	0.0109
	IB	0.083	
10 - 20	β-	0.00170	0.0113
	IB	0.083	0.58
20 - 40	β-	0.0072	0.0237
	IB	0.165	0.57
40 - 100	β-	0.059	0.083
	IB	0.49	0.75
100 - 300	β-	0.87	0.418
	IB	1.55	0.86
300 - 600	β-	5.1	1.10
	IB	2.1	0.49
600 - 1300	β-	50.5	5.10
	IB	4.1	0.46
1300 - 2500	β-	324	16.6
	IB	5.1	0.28
2500 - 5000	β-	1707	45.7
	IB	5.2	0.154
5000 - 7500	β-	1621	26.9
	IB	1.16	0.020
7500 - 9870	β-	331	4.08
	IB	0.059	0.00074

$^{84}_{33}$As(650 *150* ms)

Mode: β-
Δ: -66080 *300* keV syst
SpA: 1.3×10^{11} Ci/g
Prod: fission

$^{84}_{34}$Se(3.2 *2* min)

Mode: β-
Δ: -75952 *15* keV
SpA: 7.0×10^{8} Ci/g
Prod: fission

Photons (^{84}Se)

⟨γ⟩=420 *82* keV

γ_{mode}	γ(keV)	γ(%)
γ[E1]	408.2 *4*	100
γ	498.5 *6*	2.4 *8*

Continuous Radiation (^{84}Se)

⟨β-⟩=539 keV;⟨IB⟩=0.78 keV

E_{bin}(keV)		⟨ ⟩(keV)	(%)
0 - 10	β-	0.0318	0.63
	IB	0.024	
10 - 20	β-	0.097	0.65
	IB	0.023	0.159
20 - 40	β-	0.402	1.34
	IB	0.044	0.153
40 - 100	β-	3.12	4.42
	IB	0.118	0.18
100 - 300	β-	38.1	18.6
	IB	0.28	0.160
300 - 600	β-	148	32.9
	IB	0.21	0.051
600 - 1300	β-	346	41.2
	IB	0.082	0.0109
1300 - 1418	β-	3.14	0.236
	IB	2.3×10^{-5}	1.7×10^{-6}

$^{84}_{35}$Br(31.80 *8* min)

Mode: β-
Δ: -77778 *26* keV
SpA: 7.038×10^{7} Ci/g
Prod: ^{87}Rb(n,α); fission

Photons (^{84}Br)

⟨γ⟩=1720 *47* keV

γ_{mode}	γ(keV)	γ(%)†
γ	230.21 *18*	0.29 *4*
γ	340.03 *24*	0.070 *16*
γ[E1]	354.74 *16*	0.29 *4*
γ	382.11 *15*	0.57 *8*
γ	394.3 *3*	
γ[E2]	447.66 *20*	0.050 *8*
γ	561.5 *3*	0.082 *20*
γ[E1]	605.03 *18*	1.72 *25*

Photons (^{84}Br)
(continued)

γ_{mode}	γ(keV)	γ(%)†
γ	688.8 *10*	0.090 *25*
γ	736.84 *19*	1.27 *20*
γ[E1]	802.39 *15*	5.9 *6*
γ(E2)	881.69 *8*	41.0
γ	946.9 *3*	0.37 *8*
γ[E2]	955.8 *20*	0.06 *3*
γ	987.13 *20*	0.78 *12*
γ	1005.86 *23*	0.45 *12*
γ E2(+0.08%M1)	1016.05 *12*	6.6 *5*
γ	1082.9 *3*	0.139 *25*
γ	1119.4 *3*	0.139 *25*
γ	1142.8 *14*	0.033 *12*
γ	1184.49 *18*	0.107 *21*
γ(E2)	1213.41 *16*	2.5 *3*
γ	1255.4 *4*	0.045 *8*
γ	1438.1 *4*	0.061 *16*
γ(E2)	1463.7 *2*	1.71 *23*
γ	1533.1 *4*	0.098 *20*
γ[E1]	1578.04 *22*	0.66 *12*
γ	1607.7 *3*	0.39 *6*
γ[M1+E2]	1741.38 *25*	1.60 *25*
γ	1779.7 *10*	0.061 *16*
γ	1808.25 *21*	0.041 *12*
γ[E1]	1818.43 *15*	0.24 *4*
γ[M1+E2]	1877.4 *3*	1.11 *16*
γ[E2]	1897.72 *13*	13.4 *9*
γ[E1]	2029.70 *24*	2.0 *4*
γ	2094.3 *5*	0.21 *4*
γ	2200.52 *18*	1.15 *16*
γ[E1]	2219.2 *5*	~0.07
γ	2484.25 *20*	6.6 *7*
γ[E1]	2594.07 *21*	0.14 *3*
γ[E2]	2623.05 *25*	0.30 *6*
γ[E2]	2759.1 *3*	0.49 *8*
γ	2824.27 *19*	1.11 *16*
γ	2988.5 *5*	0.16 *4*
γ[E1]	3045.72 *22*	2.5 *4*
γ	3202.7 *5*	0.21 *4*
γ[E1]	3235.2 *4*	2.0 *3*
γ	3365.91 *21*	2.8 *4*
γ[E1]	3927.37 *22*	6.7 *7*
γ	4084.3 *5*	0.29 *4*
γ[E1]	4116.8 *5*	0.0037 *8*

† 4.4% uncert(syst)

Continuous Radiation (^{84}Br)

⟨β-⟩=1245 keV;⟨IB⟩=3.8 keV

E_{bin}(keV)		⟨ ⟩(keV)	(%)
0 - 10	β-	0.0276	0.55
	IB	0.040	
10 - 20	β-	0.084	0.56
	IB	0.039	0.27
20 - 40	β-	0.340	1.13
	IB	0.077	0.27
40 - 100	β-	2.52	3.58
	IB	0.22	0.34
100 - 300	β-	25.8	12.9
	IB	0.63	0.35
300 - 600	β-	72	16.4
	IB	0.72	0.170
600 - 1300	β-	233	25.2
	IB	1.07	0.123
1300 - 2500	β-	470	25.6
	IB	0.78	0.045
2500 - 4653	β-	441	14.0
	IB	0.18	0.0062

$^{84}_{35}$Br(6.0 *2* min)

Mode: β-
Δ: -77458 *100* keV
SpA: 3.73×10^{8} Ci/g
Prod: ^{87}Rb(n,α); fission

Photons (^{84}Br)

⟨γ⟩=2823 *170* keV

γ_{mode}	γ(keV)	γ(%)
γ E1,M1	424.3 *8*	100 *10*
γ [E2]	447.66 *20*	2.9 *3*
γ (E2)	881.69 *8*	98 *10*
γ E2(+0.08%M1)	1016.05 *12*	1.00 *8*
γ (E2)	1463.7 *2*	101 *10*
γ [E2]	1897.72 *13*	2.03 *15*

Continuous Radiation (^{84}Br)

⟨β-⟩=897 keV;⟨IB⟩=1.9 keV

E_{bin}(keV)		⟨ ⟩(keV)	(%)
0 - 10	β-	0.0131	0.260
	IB	0.035	
10 - 20	β-	0.0401	0.267
	IB	0.034	0.24
20 - 40	β-	0.167	0.56
	IB	0.066	0.23
40 - 100	β-	1.32	1.87
	IB	0.19	0.29
100 - 300	β-	17.6	8.5
	IB	0.50	0.28
300 - 600	β-	84	18.3
	IB	0.50	0.119
600 - 1300	β-	455	48.5
	IB	0.49	0.059
1300 - 2204	β-	339	21.8
	IB	0.055	0.0038

$^{84}_{36}$Kr(stable)

Δ: -82431 *3* keV
%: 57.0 *3*

$^{84}_{37}$Rb(32.87 *11* d)

Mode: ε(96.0 *5* %), β-(4.0 *5* %)
Δ: -79746 *4* keV
SpA: 4.729×10^{4} Ci/g
Prod: ^{81}Br(α,n); ^{85}Rb(γ,n)

Photons (^{84}Rb)

⟨γ⟩=622 *10* keV

γ_{mode}	γ(keV)	γ(%)†
Kr L_{ℓ}	1.383	0.043 *9*
Kr L_{η}	1.435	0.025 *6*
Kr L_{α}	1.581	0.90 *18*
Kr L_{β}	1.640	0.52 *13*
Kr L_{γ}	1.907	0.0081 *25*
Kr $K_{\alpha2}$	12.598	11.6 *5*
Kr $K_{\alpha1}$	12.651	22.6 *9*
Kr $K_{\beta1}$'	14.109	4.77 *20*

Photons (^{84}Rb) (continued)

γ_{mode}	γ(keV)	γ(%)†
Kr $K_{\beta2}$'	14.408	0.429 *18*
γ_{ϵ}(E2)	881.69 *8*	67.9
γ_{ϵ}E2(+0.08%M1)	1016.05 *12*	0.38 *3*
γ_{ϵ}[E2]	1897.72 *13*	0.77 *5*

† uncert(syst): 1.8% for ε

Atomic Electrons (^{84}Rb)

⟨e⟩=3.77 *20* keV

e_{bin}(keV)	⟨e⟩(keV)	e(%)
2	0.90	52 *5*
10	0.154	1.47 *15*
11	1.58	14.5 *15*
12	0.57	4.6 *5*
13	0.104	0.83 *8*
14	0.064	0.46 *5*
867	0.353	0.0407 *18*
880	0.0385	0.00438 *19*
881	0.0064	0.00072 *3*
882	0.00078	8.8 *4* ×10^{-5}
1002	0.00161	0.000161 *14*
1014	0.000174	1.72 *15* ×10^{-5}
1016	3.2 ×10^{-5}	3.2 *3* ×10^{-6}
1883	0.00169	9.0 *6* ×10^{-5}
1896	0.000180	9.5 *7* ×10^{-6}
1897	2.93 ×10^{-5}	1.55 *11* ×10^{-6}
1898	4.2 ×10^{-7}	2.22 *16* ×10^{-8}

Continuous Radiation (^{84}Rb)

⟨β-⟩=13.3 keV;⟨β+⟩=144 keV;⟨IB⟩=1.42 keV

E_{bin}(keV)		⟨ ⟩(keV)	(%)
0 - 10	β-	0.00309	0.062
	β+	4.37×10^{-5}	0.00054
	IB	0.0071	
10 - 20	β-	0.0093	0.062
	β+	0.00104	0.0064
	IB	0.0126	0.093
20 - 40	β-	0.0372	0.124
	β+	0.0176	0.055
	IB	0.0131	0.046
40 - 100	β-	0.264	0.377
	β+	0.486	0.64
	IB	0.035	0.055
100 - 300	β-	2.54	1.27
	β+	12.7	6.0
	IB	0.108	0.059
300 - 600	β-	7.1	1.61
	β+	43.0	9.8
	IB	0.22	0.049
600 - 1300	β-	3.40	0.497
	β+	75	8.4
	IB	0.70	0.077
1300 - 2500	β+	13.1	0.94
	IB	0.32	0.020
2500 - 2685	IB	0.00152	6.0×10^{-5}
	Σβ+		26

$^{84}_{37}$Rb(20.26 *4* min)

Mode: IT
Δ: -79282 *4* keV
SpA: 1.1045×10^{8} Ci/g
Prod: ^{81}Br(α,n); ^{85}Rb(n,2n); ^{85}Rb(γ,n)

Photons (^{84}Rb)

⟨γ⟩=376 *26* keV

γ_{mode}	γ(keV)	γ(%)†
γ M3+E4	216.1 *3*	34 *3*
γ [M1+E2]	248.20 *19*	63
γ E4(+M5)	464.3 *3*	32 *3*

† 13% uncert(syst)

$^{84}_{38}$Sr(stable)

Δ: -80640 *4* keV
%: 0.56 *1*

$^{84}_{39}$Y (40 *1* min)

Mode: ε
Δ: -74230 *170* keV
SpA: 5.60×10^{7} Ci/g
Prod: ^{75}As(^{12}C,3n); ^{84}Sr(d,2n); ^{88}Sr(p,5n); ^{55}Mn(^{32}S,2pn); ^{56}Fe(^{32}S,3pn); ^{16}O on Ga; protons on Nb

Photons (^{84}Y)

⟨γ⟩=2832 *160* keV

γ_{mode}	γ(keV)	γ(%)†
Sr L_{ℓ}	1.582	0.0049 *11*
Sr L_{η}	1.649	0.0028 *7*
Sr L_{α}	1.806	0.111 *23*
Sr L_{β}	1.880	0.064 *16*
Sr L_{γ}	2.196	0.0011 *3*
Sr $K_{\alpha2}$	14.098	1.31 *6*
Sr $K_{\alpha1}$	14.165	2.51 *11*
Sr $K_{\beta1}$'	15.832	0.551 *24*
Sr $K_{\beta2}$'	16.192	0.075 *3*
γ [M1+E2]	288.37 *15*	0.78 *20*
γ	462.85 *15*	9.4 *11*
γ [M1+E2]	602.14 *9*	8.8 *13*
γ [M1+E2]	660.59 *9*	14.6 *12*
γ	680.6 *3*	4.1 *4*
γ	704.50 *21*	4.3 *5*
γ (E2)	793.02 *9*	98
γ (E2)	974.35 *9*	74 *5*
γ [E1]	994.09 *14*	4.1 *4*
γ (E2)	1039.81 *9*	45.2 *9*
γ	1092.25 *20*	4.4 *4*
γ	1110.25 *20*	3.2 *3*
γ	1144.25 *20*	3.0 *5*
γ [E2]	1255.03 *15*	5.88 *20*
γ [M1+E2]	1262.72 *12*	2.45 *20*
γ [E2]	1453.61 *12*	2.0 *7*
γ	1502.65 *15*	6.4 *8*
γ [E1]	1614.51 *16*	1.76 *20*
γ [E1]	1654.67 *13*	2.55 *20*
γ	1744.30 *19*	2.25 *20*
γ	1763.55 *20*	1.96 *20*
γ	1810.8 *3*	0.9 *3*

Photons (^{84}Y) (continued)

γ_{mode}	γ(keV)	γ(%)†
γ [M1+E2]	2006.46 *18*	~0.29
γ [M1+E2]	2294.83 *15*	2.2 *3*

† approximate

Atomic Electrons (^{84}Y)

$\langle e \rangle$=2.43 *16* keV

e_{bin}(keV)	$\langle e \rangle$(keV)	e(%)
2	0.107	5.4 *6*
12	0.175	1.45 *15*
14 - 16	0.075	0.54 *5*
272 - 288	0.035	0.013 *5*
447	0.11	~0.025
461 - 463	0.016	0.0034 *17*
586	0.088	0.015 *3*
600 - 602	0.0120	0.0020 *4*
644	0.127	0.0198 *25*
658 - 704	0.074	0.0110 *25*
777	0.69	0.089 *18*
791	0.078	0.0098 *20*
793	0.015	0.0019 *4*
958	0.39	0.041 *3*
972 - 994	0.062	0.0064 *4*
1024	0.218	0.0213 *10*
1038 - 1076	0.043	0.0041 *5*
1090 - 1128	0.023	0.0021 *5*
1142 - 1144	0.0012	0.00011 *4*
1239 - 1263	0.0363	0.00292 *12*
1438 - 1502	0.025	0.0017 *4*
1598 - 1654	0.0074	0.00045 *3*
1728 - 1810	0.0122	0.00070 *13*
1990 - 2006	0.0008	~4×10^{-5}
2279 - 2295	0.0054	0.00024 *3*

Continuous Radiation (^{84}Y)

$\langle\beta+\rangle$=1143 keV;$\langle$IB$\rangle$=3.6 keV

E_{bin}(keV)		$\langle\ \rangle$(keV)	(%)
0 - 10	β+	1.69×10^{-5}	0.000208
	IB	0.040	
10 - 20	β+	0.000435	0.00267
	IB	0.040	0.28
20 - 40	β+	0.0080	0.0248
	IB	0.077	0.27
40 - 100	β+	0.250	0.327
	IB	0.22	0.34
100 - 300	β+	8.9	4.11
	IB	0.62	0.35
300 - 600	β+	59	12.8
	IB	0.71	0.167
600 - 1300	β+	356	37.7
	IB	1.02	0.118
1300 - 2500	β+	581	33.1
	IB	0.70	0.041
2500 - 5000	β+	137	4.77
	IB	0.151	0.0051
5000 - 5617	IB	0.00026	5.0×10^{-6}
	$\Sigma\beta$+		93

$^{84}_{39}$Y (4.6 *2* s)

Mode: ϵ

Δ: -74230 *170* keV

SpA: 2.71×10^{10} Ci/g

Prod: ^{84}Sr(p,n)

Photons (^{84}Y)

$\langle\gamma\rangle$=278 *81* keV

γ_{mode}	γ(keV)	γ(%)†
Sr K$_{\alpha2}$	14.098	0.188 *9*
Sr K$_{\alpha1}$	14.165	0.361 *17*
Sr K$_{\beta1}$'	15.832	0.079 *4*
Sr K$_{\beta2}$'	16.192	0.0108 *5*
γ (E2)	793.02 *9*	35

† approximate

Atomic Electrons (^{84}Y)

$\langle e \rangle$=0.33 *7* keV

e_{bin}(keV)	$\langle e \rangle$(keV)	e(%)
2	0.0153	0.78 *8*
12	0.025	0.209 *22*
14	0.0099	0.072 *8*
15	0.00027	0.00172 *18*
16	0.00062	0.0039 *4*
777	0.25	0.032 *9*
791	0.028	0.0035 *10*
793	0.0054	0.00068 *20*

Continuous Radiation (^{84}Y)

$\langle\beta+\rangle$=2341 keV;$\langle$IB$\rangle$=9.4 keV

E_{bin}(keV)		$\langle\ \rangle$(keV)	(%)
0 - 10	β+	2.43×10^{-6}	2.98×10^{-5}
	IB	0.064	
10 - 20	β+	6.3×10^{-5}	0.000384
	IB	0.063	0.44
20 - 40	β+	0.00116	0.00361
	IB	0.125	0.43
40 - 100	β+	0.0376	0.0491
	IB	0.37	0.56
100 - 300	β+	1.51	0.69
	IB	1.12	0.62
300 - 600	β+	12.3	2.64
	IB	1.44	0.34
600 - 1300	β+	134	13.5
	IB	2.5	0.29
1300 - 2500	β+	734	38.1
	IB	2.4	0.139
2500 - 5000	β+	1451	43.8
	IB	1.18	0.037
5000 - 6410	β+	8.6	0.169
	IB	0.031	0.00058
	$\Sigma\beta$+		99

$^{84}_{40}$Zr(27.6 *9* min)

Mode: ϵ

Δ: -71430 *350* keV syst

SpA: 8.1×10^{7} Ci/g

Prod: protons on ^{89}Y; protons on ^{90}Zr; ^{58}Ni(^{28}Si,2p); ^{58}Ni(^{29}Si,2pn)

Photons (^{84}Zr)

γ_{mode}	γ(keV)	γ(rel)
γ	41.1 *2*	37 *4*
γ	44.9 *2*	48 *5*
γ	112.5 *1*	100
γ	131.6 *2*	28 *4*
γ	320.0 *1*	30 *4*
γ	372.9 *1*	41 *4*
γ	557.0 *3*	20 *3*
γ	600.0 *2*	20 *4*
γ	666.7 *3*	39 *4*

$^{84}_{41}$Nb(12 *3* s)

Mode: ϵ

SpA: 1.1×10^{10} Ci/g

Prod: ^{58}Ni(^{32}S,npα)

Photons (^{84}Nb)

$\langle\gamma\rangle$=706 *120* keV

γ_{mode}	γ(keV)	γ(%)
γ [E2]	540.0 *5*	100
γ [E2]	722.8 *5*	23 *8*

A = 85

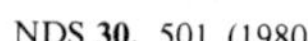

NDS **30**, 501 (1980)

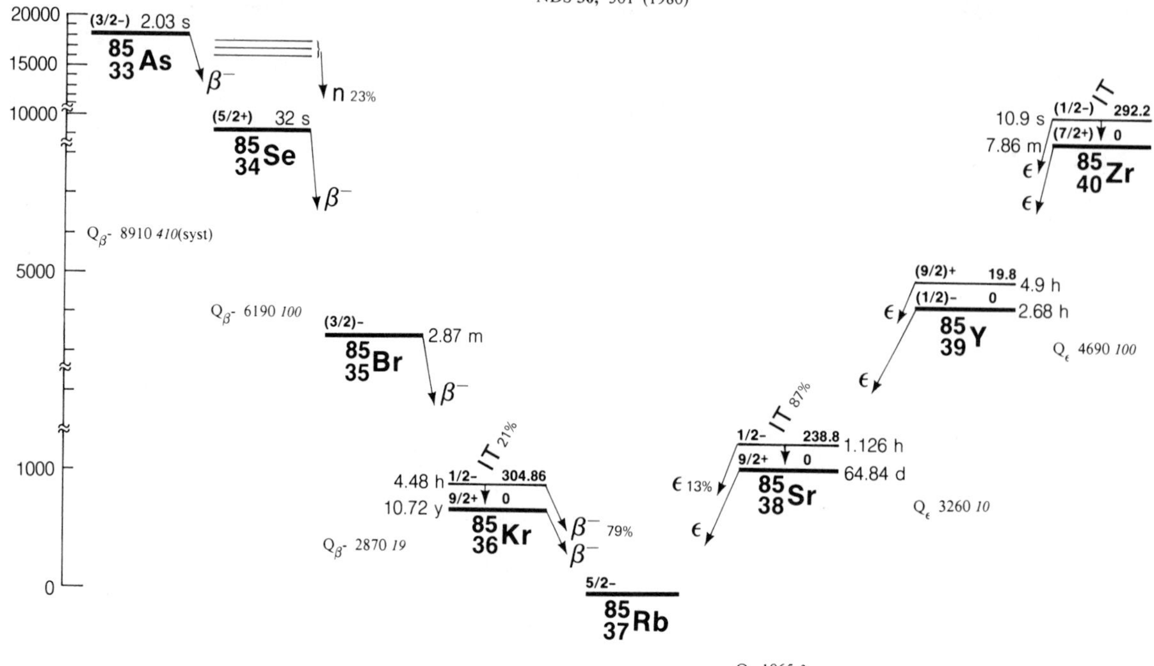

$^{85}_{33}$As(2.028 *12* s)

Mode: β-, β-n(23 *3* %)
Δ: -63510 *400* keV syst
SpA: 5.54×10^{10} Ci/g
Prod: fission

Delayed Neutrons (^{85}As)

n(keV)	n(%)
56 *1*	0.7
140 *3*	1.8
245 *6*	0.22
271 *2*	0.7
495 *3*	4.4
516 *3*	3.9
708 *3*	2.3
925 *4*	3.2
1154 *7*	1.6
1187 *8*	1.5
1420 *7*	1.8
1506 *11*	0.8

Photons (^{85}As)

⟨γ⟩=358 *48* keV

γ_{mode}	γ(keV)	γ(%)
$\gamma_{\beta-}$	461.5 *3*	0.39 *8*
$\gamma_{\beta\text{-n}}$	577.36 *17*	0.98 *15*
$\gamma_{\beta\text{-n}}$	667.03 *19*	6.9 *6*
$\gamma_{\beta-}$	694.3 *5*	0.21 *7*
$\gamma_{\beta-}$	1111.5 *5*	2.0 *6*

Photons (^{85}As)
(continued)

γ_{mode}	γ(keV)	γ(%)
$\gamma_{\beta\text{-n}}$	1244.38 *19*	0.65 *11*
$\gamma_{\beta\text{-n}}$ [E2]	1454.96 *20*	16
$\gamma_{\beta\text{-n}}$	1843.56 *20*	0.51 *7*
$\gamma_{\beta-}$	3345.0 *5*	0.28 *3*
$\gamma_{\beta-}$	3749.4 *7*	0.44 *7*
$\gamma_{\beta-}$	4756 *1*	
$\gamma_{\beta-}$	4913 *1*	
$\gamma_{\beta-}$	5054 *1*	

$^{85}_{34}$Se(31.7 *9* s)

Mode: β-
Δ: -72420 *100* keV
SpA: 4.14×10^{9} Ci/g
Prod: fission

Photons (^{85}Se)

⟨γ⟩=2215 *47* keV

γ_{mode}	γ(keV)	γ(%)
γ [M1+E2]	345.11 *22*	45.7 *23*
γ	432.8 *4*	2.7 *3*
γ	597.4 *3*	1.33 *14*
γ	610.36 *25*	3.5 *4*
γ	839.5 *3*	2.7 *3*
γ	940.3 *4*	2.7 *3*
γ	955.46 *25*	5.5 *6*
γ	988.0 *3*	1.74 *18*

Photons (^{85}Se)
(continued)

γ_{mode}	γ(keV)	γ(%)
γ	1081.6 *3*	2.9 *3*
γ [M1+E2]	1191.1 *4*	1.92 *18*
γ	1207.78 *23*	5.6 *6*
γ	1227.0 *5*	
γ	1246.9 *3*	1.55 *14*
γ	1373.2 *3*	1.97 *18*
γ	1426.7 *3*	7.5 *8*
γ	1449.8 *3*	2.42 *23*
γ	1552.9 *3*	0.82 *9*
γ	1598.3 *3*	1.19 *14*
γ	1700.7 *4*	2.06 *23*
γ	1724.1 *3*	2.38 *23*
γ	1794.9 *3*	1.37 *14*
γ	1805.5 *5*	
γ	1943.4 *3*	2.06 *23*
γ	2029.4 *4*	1.01 *9*
γ	2091.2 *4*	0.69 *14*
γ	2233.1 *3*	2.24 *23*
γ	2245.6 *4*	0.27 *5*
γ	2303.9 *4*	0.50 *9*
γ	2416.8 *3*	2.9 *3*
γ	2447.0 *4*	0.46 *9*
γ	2454.7 *3*	3.6 *4*
γ	2543.0 *3*	0.73 *14*
γ	2550.4 *4*	1.19 *14*
γ	2565.6 *4*	0.96 *18*
γ	2583.9 *4*	1.74 *18*
γ	2601.3 *3*	1.7 *4*
γ	2723.9 *5*	1.14 *14*
γ	2872.0 *5*	0.64 *14*
γ	3007.4 *5*	3.2 *3*
γ	3396.4 *3*	8.4 *8*
γ	3479.2 *4*	0.69 *14*
γ	3539.3 *4*	1.28 *14*
γ	3555.1 *5*	0.50 *9*
γ	3624.6 *3*	1.55 *14*
γ	3654.7 *4*	3.9 *4*
γ	3682.9 *3*	2.19 *23*
γ	3741.5 *3*	0.69 *14*

Photons (^{85}Se)
(continued)

γ_{mode}	γ(keV)	γ(%)
γ	3773.3 *4*	3.0 *3*
γ	3826.4 *5*	1.14 *14*

Continuous Radiation (^{85}Se)
$\langle\beta-\rangle$=1755 keV;$\langle$IB$\rangle$=6.1 keV

E_{bin}(keV)		$\langle\ \rangle$(keV)	(%)
0 - 10	β-	0.0063	0.125
	IB	0.051	
10 - 20	β-	0.0193	0.129
	IB	0.051	0.35
20 - 40	β-	0.081	0.269
	IB	0.101	0.35
40 - 100	β-	0.65	0.91
	IB	0.29	0.45
100 - 300	β-	8.8	4.24
	IB	0.86	0.48
300 - 600	β-	43.4	9.5
	IB	1.05	0.24
600 - 1300	β-	275	28.8
	IB	1.65	0.19
1300 - 2500	β-	574	31.9
	IB	1.37	0.078
2500 - 5000	β-	789	22.9
	IB	0.63	0.020
5000 - 6190	β-	64	1.21
	IB	0.0059	0.000108

$^{85}_{35}$Br(2.87 *3* min)

Mode: β-
Δ: -78607 *19* keV
SpA: 7.69×10^{8} Ci/g
Prod: fission

Photons (^{85}Br)
$\langle\gamma\rangle$=66.0 *10* keV

γ_{mode}	γ(keV)	γ(%)†
γ	96.84 *7*	0.0377 *23*
γ	147.60 *17*	0.039 *9*
γ	175.91 *5*	0.058 *3*
γ	201.84 *7*	0.0236 *18*
γ	235.30 *7*	0.008 *3*
γ	249.85 *8*	0.0205 *11*
γ	263.77 *15*	0.011 *3*
γ	272.02 *10*	0.072 *4*
γ M4	304.86 *3*	*
γ	421.6 *3*	0.092 *7*
γ	433.70 *16*	0.0147 *18*
γ	455.59 *15*	0.021 *3*
γ	541.64 *18*	0.021 *3*
γ	546.6 *3*	0.016 *3*
γ	600.88 *21*	0.020 *4*
γ	689.31 *5*	0.040 *7*
γ	766.24 *16*	0.012 *3*
γ	772.07 *9*	0.012 *3*
γ	794.71 *7*	0.104 *7*
γ	798.00 *10*	0.047 *6*
γ	802.41 *6*	2.56 *5*
γ	809.97 *18*	0.015 *3*
γ	824.06 *21*	0.012 *3*
γ	831.45 *7*	0.051 *5*
γ	861.79 *5*	0.228 *13*
γ	865.22 *5*	0.177 *10*
γ	913.34 *7*	0.134 *8*
γ	919.08 *7*	0.65 *3*
γ	924.61 *6*	1.63 *8*

Photons (^{85}Br)
(continued)

γ_{mode}	γ(keV)	γ(%)†
γ	946.2 *3*	0.018 *4*
γ	1030.01 *8*	0.022 *4*
γ	1031.84 *12*	0.054 *5*
γ	1037.70 *5*	0.102 *8*
γ	1047.29 *16*	0.021 *3* ?
γ	1072.17 *16*	0.030 *3*
γ	1131.59 *15*	0.031 *4*
γ[E2]	1140.72 *7*	0.096 *5*
γ	1260.42 *13*	0.047 *4*
γ	1416.50 *9*	0.067 *5*
γ	1727.01 *5*	0.381 *16*
γ	1808.20 *24*	0.016 *4*
γ	1832.40 *6*	0.150 *8*
γ	2438.5 *4*	0.018 *4*
γ	2463.76 *17*	0.017 *4* ?

† 6.1% uncert(syst)
* with ^{85}Kr(4.48 h)

Continuous Radiation (^{85}Br)
$\langle\beta-\rangle$=1038 keV;$\langle$IB$\rangle$=2.4 keV

E_{bin}(keV)		$\langle\ \rangle$(keV)	(%)
0 - 10	β-	0.0118	0.234
	IB	0.038	
10 - 20	β-	0.0360	0.240
	IB	0.038	0.26
20 - 40	β-	0.150	0.497
	IB	0.073	0.26
40 - 100	β-	1.17	1.65
	IB	0.21	0.32
100 - 300	β-	15.0	7.3
	IB	0.58	0.32
300 - 600	β-	68	14.9
	IB	0.61	0.145
600 - 1300	β-	400	42.0
	IB	0.70	0.083
1300 - 2500	β-	554	33.1
	IB	0.155	0.0104
2500 - 2565	β-	0.221	0.0088
	IB	4.4×10^{-7}	1.8×10^{-8}

$^{85}_{36}$Kr(10.72 *2* yr)

Mode: β-
Δ: -81477 *3* keV
SpA: 392.3 Ci/g
Prod: ^{84}Kr(n,γ); fission

Photons (^{85}Kr)
$\langle\gamma\rangle$=2.2 keV

γ_{mode}	γ(keV)	γ(%)†
γ M2	513.996 *16*	0.434

† 2.3% uncert(syst)

Continuous Radiation (^{85}Kr)
$\langle\beta-\rangle$=251 keV;$\langle$IB$\rangle$=0.20 keV

E_{bin}(keV)		$\langle\ \rangle$(keV)	(%)
0 - 10	β-	0.110	2.20
	IB	0.0124	
10 - 20	β-	0.328	2.18
	IB	0.0117	0.081
20 - 40	β-	1.30	4.33
	IB	0.021	0.075
40 - 100	β-	9.0	12.8
	IB	0.051	0.080
100 - 300	β-	81	40.7
	IB	0.082	0.050
300 - 600	β-	153	36.7
	IB	0.019	0.0051
600 - 687	β-	6.3	1.01
	IB	4.4×10^{-5}	7.2×10^{-6}

$^{85}_{36}$Kr(4.480 *8* h)

Mode: β-(79.0 *6* %), IT(21.0 *6* %)
Δ: -81172 *3* keV
SpA: 8.229×10^{6} Ci/g
Prod: ^{84}Kr(n,γ); fission; ^{82}Se(α,n)

Photons (^{85}Kr)
$\langle\gamma\rangle$=156 *3* keV

γ_{mode}	γ(keV)	γ(%)
Kr L_{ℓ}	1.383	0.0043 *10*
Kr L_{η}	1.435	0.0025 *6*
Rb L_{ℓ}	1.482	0.0023 *5*
Rb L_{η}	1.542	0.0013 *3*
Kr L_{α}	1.581	0.091 *19*
Kr L_{β}	1.640	0.053 *14*
Rb L_{α}	1.694	0.051 *10*
Rb L_{β}	1.760	0.029 *8*
Kr L_{γ}	1.907	0.00081 *25*
Rb L_{γ}	2.051	0.00046 *14*
Kr $K_{\alpha2}$	12.598	1.14 *6*
Kr $K_{\alpha1}$	12.651	2.21 *11*
Rb $K_{\alpha2}$	13.336	0.62 *3*
Rb $K_{\alpha1}$	13.395	1.21 *6*
Kr $K_{\beta1}$'	14.109	0.468 *23*
Kr $K_{\beta2}$'	14.408	0.0421 *21*
Rb $K_{\beta1}$'	14.959	0.260 *12*
Rb $K_{\beta2}$'	15.286	0.0277 *13*
$\gamma_{\beta-}$M1	151.18 *3*	75.1
γ_{IT} M4	304.86 *3*	13.7

Atomic Electrons (^{85}Kr)
$\langle$e$\rangle$=26.3 *6* keV

e_{bin}(keV)	$\langle$e$\rangle$(keV)	e(%)
2	0.141	8.0 *9*
10	0.0151	0.144 *15*
11	0.189	1.73 *18*
12	0.111	0.93 *10*
13	0.045	0.34 *4*
14	0.0063	0.045 *5*
15	0.0031	0.0213 *22*
136	4.31	3.17 *10*
149	0.531	0.356 *11*
151	0.103	0.0685 *21*
291	17.5	6.02 *21*
303	2.81	0.93 *3*
305	0.533	0.175 *6*

Continuous Radiation (^{85}Kr)

⟨β-⟩=229 keV;⟨IB⟩=0.20 keV

E_{bin}(keV)		⟨ ⟩(keV)	(%)
0 - 10	β-	0.066	1.31
	IB	0.0111	
10 - 20	β-	0.198	1.32
	IB	0.0106	0.073
20 - 40	β-	0.81	2.69
	IB	0.020	0.068
40 - 100	β-	5.9	8.5
	IB	0.048	0.075
100 - 300	β-	59	29.7
	IB	0.085	0.051
300 - 600	β-	131	30.7
	IB	0.029	0.0077
600 - 841	β-	31.7	4.77
	IB	0.00104	0.000167

$^{85}_{37}$Rb(stable)

Δ: -82164 3 keV

%: 72.165 13

$^{85}_{38}$Sr(64.84 2 d)

Mode: ε

Δ: -81099 4 keV

SpA: 2.3692×10^{4} Ci/g

Prod: ^{84}Sr(n,γ); ^{85}Rb(p,n); ^{85}Rb(d,2n)

Photons (^{85}Sr)

⟨γ⟩=518.3 3 keV

γ_{mode}	γ(keV)	γ(%)†
Rb L_ℓ	1.482	0.065 14
Rb L_η	1.542	0.037 9
Rb L_α	1.694	1.4 3
Rb L_β	1.760	0.82 21
Rb L_γ	2.051	0.013 4
Rb $K_{\alpha2}$	13.336	17.3 7
Rb $K_{\alpha1}$	13.395	33.3 14
Rb $K_{\beta1}$'	14.959	7.2 3
Rb $K_{\beta2}$'	15.286	0.77 3
γ M2	513.996 16	99.27
γ M1+E2	868.6 4	0.0120 5

†<0.1% uncert(syst)

Atomic Electrons (^{85}Sr)

⟨e⟩=8.4 3 keV

e_{bin}(keV)	⟨e⟩(keV)	e(%)
2	1.37	74 8
11	0.93	8.2 8
12	1.51	13.1 13
13	0.95	7.3 7
15	0.087	0.59 6
499	3.14	0.631 13
512	0.367	0.0717 14
514	0.0709	0.0138 3
853	6.6 ×10^{-5}	7.8 5 ×10^{-6}
867	7.3 ×10^{-6}	8.4 6 ×10^{-7}
868	1.22 ×10^{-6}	1.41 10 ×10^{-7}
869	1.73 ×10^{-7}	1.99 14 ×10^{-8}

Continuous Radiation (^{85}Sr)

⟨IB⟩=0.046 keV

E_{bin}(keV)		⟨ ⟩(keV)	(%)
10 - 20	IB	0.0100	0.074
20 - 40	IB	0.00052	0.0019
40 - 100	IB	0.00168	0.0024
100 - 300	IB	0.019	0.0095
300 - 600	IB	0.0143	0.0039
600 - 1065	IB	0.00066	9.0×10^{-5}

$^{85}_{38}$Sr(1.1258 12 h)

Mode: IT(87 1 %), ε(13 1 %)

Δ: -80860 4 keV

SpA: 3.274×10^{7} Ci/g

Prod: ^{84}Sr(n,γ); ^{85}Rb(p,n); ^{85}Rb(d,2n)

Photons (^{85}Sr)

⟨γ⟩=216 5 keV

γ_{mode}	γ(keV)	γ(%)†
Rb L_ℓ	1.482	0.0087 19
Rb L_η	1.542	0.0050 13
Sr L_ℓ	1.582	0.06 3
Sr L_η	1.649	~0.034
Rb L_α	1.694	0.19 4
Rb L_β	1.760	0.11 3
Sr L_α	1.806	1.4 7
Sr L_β	1.872	~0.7
Rb L_γ	2.051	0.0018 6
Sr L_γ	2.196	0.00059 22
Rb $K_{\alpha2}$	13.336	2.34 11
Rb $K_{\alpha1}$	13.395	4.51 21
Sr $K_{\alpha2}$	14.098	0.45 3
Sr $K_{\alpha1}$	14.165	0.87 6
Rb $K_{\beta1}$'	14.959	0.97 4
Rb $K_{\beta2}$'	15.286	0.104 5
Sr $K_{\beta1}$'	15.832	0.190 13
Sr $K_{\beta2}$'	16.192	0.0260 19
γ_ϵM1	151.18 3	12.4 25
γ_{IT} M1+17%E2	231.78 4	84.1 17
γ_{IT} M4	238.75 8	0.34 6

† uncert(syst): 7.7% for ε, 1.1% for IT

Atomic Electrons (^{85}Sr)

⟨e⟩=14.0 20 keV

e_{bin}(keV)	⟨e⟩(keV)	e(%)
2	1.7	87 36
5	4.0	80 36
7	1.2	18 8
11	0.126	1.11 12
12	0.27	2.28 24
13	0.129	0.98 10
14	0.024	0.172 20
15	0.0124	0.084 9
16	0.00148	0.0095 11
136	0.71	0.52 11
149	0.088	0.059 12
151	0.017	0.0113 23
216	3.60	1.67 9
223	1.23	0.55 10
230	0.45	0.195 12
231	0.068	0.0295 15
232	0.0207	0.0090 9
237	0.24	0.101 18
238	0.042	0.018 3
239	0.0064	0.0027 5

Continuous Radiation (^{85}Sr)

⟨IB⟩=0.046 keV

E_{bin}(keV)		⟨ ⟩(keV)	(%)
10 - 20	IB	0.00130	0.0096
20 - 40	IB	7.0×10^{-5}	0.00026
40 - 100	IB	0.00026	0.00036
100 - 300	IB	0.0047	0.0022
300 - 600	IB	0.019	0.0043
600 - 1153	IB	0.021	0.0028

$^{85}_{39}$Y (2.68 5 h)

Mode: ε

Δ: -77839 11 keV

SpA: 1.38×10^{7} Ci/g

Prod: ^{84}Sr(d,n); ^{84}Sr(p,γ); protons on Mo; protons on Zr

Photons (^{85}Y)

⟨γ⟩=400 61 keV

γ_{mode}	γ(keV)	γ(%)
γ [M1+E2]	151.31 14	0.0028 13
γ [M1+E2]	193.62 13	0.048 10
γ	215.91 15	0.193 18
γ M1+17%E2	231.78 4	*
γ M4	238.75 8	*
γ	409.52 14	0.84 6
γ [M1+E2]	504.51 10	60
γ [E2]	698.13 11	0.18 4
γ	914.03 12	9.0 5
γ	1278.2 4	0.241 24
γ	1320.7 4	0.37 9

* with ^{85}Sr(1.126 h)

Continuous Radiation (^{85}Y)

$\langle\beta+\rangle$=483 keV;$\langle$IB$\rangle$=2.2 keV

E_{bin}(keV)		$\langle\ \rangle$(keV)	(%)
0 - 10	$\beta+$	2.98×10^{-5}	0.000367
	IB	0.020	
10 - 20	$\beta+$	0.00076	0.00469
	IB	0.023	0.163
20 - 40	$\beta+$	0.0139	0.0434
	IB	0.038	0.132
40 - 100	$\beta+$	0.429	0.56
	IB	0.105	0.162
100 - 300	$\beta+$	14.6	6.8
	IB	0.28	0.159
300 - 600	$\beta+$	86	18.8
	IB	0.33	0.077
600 - 1300	$\beta+$	314	35.4
	IB	0.72	0.079
1300 - 2500	$\beta+$	68	4.62
	IB	0.67	0.040
2500 - 3021	IB	0.0143	0.00054
	$\Sigma\beta+$		66

$^{85}_{39}$Y (4.86 13 h)

Mode: ϵ

Δ: -77819 11 keV

SpA: 7.59×10^{6} Ci/g

Prod: ^{84}Sr(d,n); ^{75}As(^{12}C,2n); ^{85}Rb(^{3}He,3n); ^{85}Rb(α,4n); ^{84}Sr(p,γ); protons on Mo; protons on Zr

Photons (^{85}Y)

$\langle\gamma\rangle$=760 9 keV

γ_{mode}	γ(keV)	γ(%)†
Sr L_ℓ	1.582	0.029 6
Sr L_η	1.649	0.017 4
Sr L_α	1.806	0.65 13
Sr L_β	1.880	0.38 10
Sr L_γ	2.196	0.0065 20
Sr $K_{\alpha2}$	14.098	7.7 3
Sr $K_{\alpha1}$	14.165	14.8 6
Sr $K_{\beta1}$'	15.832	3.24 13
Sr $K_{\beta2}$'	16.192	0.443 19
γ	129.8 5	0.072 6
γ [M1+E2]	151.31 14	0.020 8
γ	180.0 5	0.091 4
γ [M1+E2]	193.62 13	0.355 17
γ M1+17%E2	231.78 4	23.4 14
γ M4	238.75 8	0.014 3 *
γ	438.41 22	0.172 11
γ	468.43 18	0.122 11
γ [M1+E2]	504.51 10	1.55 6
γ [M1+E2]	535.68 9	3.55 14
γ [E2]	546.82 12	1.22 6
γ	558.33 21	0.271 17
γ	568.51 11	1.72 7
γ	577.16 20	0.233 17
γ [M1+E2]	587.76 12	0.122 11
γ [E1]	611.96 13	1.11 6
γ	616.76 11	0.89 6
γ	637.85 18	0.100 17
γ	657.75 22	0.061 11
γ	662.05 19	0.083 17
γ	667.58 25	0.150 17 ?
γ [E2]	698.13 11	1.33 6
γ	718.68 17	0.083 11
γ	724.66 14	0.454 22
γ	735.32 25	~0.020
γ	747.46 22	0.037 7
γ	763.00 19	0.177 22
γ [E2]	767.46 9	3.7 4

Photons (^{85}Y) (continued)

γ_{mode}	γ(keV)	γ(%)†
γ	768.76 12	1.33 14
γ [E1]	769.90 15	0.30 6
γ [M1+E2]	788.02 11	1.61 6
γ	796.48 12	0.244 17
γ	800.0 3	0.040 13
γ	810.89 13	0.194 17
γ	817.01 12	0.80 4
γ	821.50 19	0.222 17
γ	843.2 8	0.017 6
γ	861.89 12	1.00 4
γ	865.56 17	0.122 11
γ	898.77 22	0.089 6
γ	910.14 13	0.205 17
γ [E1]	915.37 24	0.13 4
γ	941.46 24	0.089 9
γ	945.35 17	0.172 13
γ	959.2 6	0.020 7
γ	965.0 3	~0.010
γ	988.98 14	0.061 6
γ	996.73 11	0.55 3
γ	1026.17 22	0.111 11
γ	1030.31 10	2.07 8
γ	1055.6 3	0.030 8
γ	1067.4 3	0.04 1
γ	1089.86 12	0.061 11
γ	1109.5 3	0.111 11
γ	1115.9 3	0.030 7
γ [M1+E2]	1123.44 10	1.83 8
γ	1131.19 16	0.035 7
γ	1170.2 6	0.04 1
γ	1173.52 17	0.051 11
γ	1187.10 11	0.283 17
γ	1214.73 25	0.045 9
γ M1+E2	1220.76 14	2.03 9
γ	1235.35 13	0.040 8
γ	1239.6 8	0.017 6
γ	1254.9 3	0.100 11
γ	1262.09 10	0.66 3
γ [M1+E2]	1323.69 10	0.70 3
γ	1338.41 12	0.172 11
γ	1356.52 11	0.68 4
γ	1376.88 20	0.030 7
γ [M1+E2]	1395.51 13	0.45 3
γ	1398.65 14	0.100 22
γ	1404.77 11	3.16 13
γ	1415.07 11	0.416 22
γ E2	1425.9 4	0.040 8
γ [M1+E2]	1469.15 22	<0.022
γ	1520.10 16	0.036 7
γ [E2]	1555.47 10	0.222 11
γ	1561.84 21	0.040 6
γ	1566.38 13	0.233 11
γ	1570.09 25	~0.014
γ	1575.1 3	~0.010
γ	1584.49 11	1.22 7
γ	1588.95 18	0.343 22
γ [M1+E2]	1627.29 14	0.277 17
γ M1+E2	1657.7 4	0.133 22
γ	1688.05 20	0.150 11
γ [E2]	1700.93 21	0.083 11
γ	1705.50 16	0.61 3
γ	1747.2 6	0.040 8
γ	1750.40 24	0.083 11
γ [M1+E2]	1815.08 21	0.072 11
γ	1826.9 3	0.040 7
γ	1854.53 13	0.416 22
γ	1892.19 9	1.83 12
γ	1919.83 20	0.144 11
γ	1934.32 12	0.222 17
γ	1940.44 9	0.59 3
γ	2038.51 20	0.061 6
γ	2042.9 3	0.017 6
γ [M1+E2]	2046.86 21	0.066 6
γ	2070.2 8	0.017 6
γ	2086.31 13	0.150 11
γ	2094.7 4	0.017 4
γ	2120.16 9	0.81 6
γ	2123.97 9	5.1 3
γ	2166.10 12	0.45 3
γ	2172.21 10	2.33 12
γ	2189.81 21	0.026 4
γ	2205.3 3	0.030 3
γ	2239.4 3	0.050 4
γ	2256.2 10	~0.008

Photons (^{85}Y) (continued)

γ_{mode}	γ(keV)	γ(%)†
γ	2289.9 3	0.026 4
γ	2318.0 5	0.0100 17
γ	2351.93 9	0.58 4
γ	2486.0 3	0.026 3
γ	2536.5 3	0.0172 22
γ	2550.39 14	0.233 17
γ	2578.6 3	0.078 11
γ	2582.8 3	0.028 6
γ	2642.36 17	0.133 11
γ	2717.7 3	0.0122 22
γ	2743.59 20	0.019 3
γ	2748.59 23	0.111 6
γ	2759.1 3	0.0028 11
γ	2768.3 3	0.029 3
γ	2778.0 5	0.035 4
γ	2782.16 14	0.343 17
γ	2786.4 5	~0.022
γ	2799.8 4	0.0205 22
γ	2810.4 3	0.024 4
γ	2814.6 3	0.122 9
γ	2843.7 3	0.0050 11
γ	2857.1 3	0.0100 17
γ	2897.4 5	0.0100 17
γ	2936.4 8	0.0050 11
γ	2975.36 20	0.0161 22
γ	2980.36 23	0.0161 22
γ	2990.9 3	0.0161 22
γ	3018.2 5	0.0144 22
γ	3031.5 4	0.027 3
γ	3063.3 4	0.025 3
γ	3075.4 3	0.0078 17
γ	3088.9 3	0.0089 17
γ	3129.2 5	~0.007

† 16% uncert(syst)

* with ^{85}Sr(1.126 h) in equilib

Atomic Electrons (^{85}Y)

$\langle$e$\rangle$=3.70 16 keV

e_{bin}(keV)	$\langle$e$\rangle$(keV)	e(%)
2	0.63	32 3
12	1.03	8.6 9
14	0.41	2.9 3
15 - 16	0.036	0.232 18
114 - 151	0.022	~0.019
164 - 194	0.048	0.027 11
216	1.00	0.46 4
223	0.050	0.022 5
230	0.124	0.054 5
231 - 239	0.036	0.0156 12
422 - 468	0.0042	0.0010 4
488 - 536	0.088	0.0172 15
542 - 588	0.027	0.0048 14
596 - 642	0.013	0.0022 5
646 - 682	0.0142	0.00211 15
696 - 745	0.0053	0.00075 17
746 - 795	0.056	0.0074 6
796 - 843	0.0057	0.00071 21
846 - 894	0.0071	0.00083 23
897 - 945	0.0018	0.00019 4
949 - 997	0.0026	0.00026 8
1010 - 1056	0.009	0.0009 3
1065 - 1114	0.0103	0.00094 6
1115 - 1157	0.00146	0.000130 11
1168 - 1215	0.0092	0.00077 5
1219 - 1262	0.0037	0.00030 6
1308 - 1357	0.0059	0.00044 5
1361 - 1410	0.013	0.00090 21
1413 - 1453	0.00019	$1.4\ 4\times10^{-5}$
1467 - 1566	0.00165	0.000107 13
1568 - 1657	0.0055	0.00034 6
1672 - 1750	0.0024	0.00014 3
1799 - 1892	0.0054	0.00029 6
1904 - 1940	0.0021	0.000111 21
2022 - 2122	0.013	0.00061 13
2124 - 2205	0.0060	0.00028 6
2223 - 2318	0.00019	$8.4\ 14\times10^{-6}$

Atomic Electrons (^{85}Y) (continued)

e_{bin}(keV)	⟨e⟩(keV)	e(%)
2336 - 2352	0.0011	4.8 *12* $\times10^{-5}$
2470 - 2567	0.00068	2.7 *4* $\times10^{-5}$
2576 - 2642	0.00026	10.0 *22* $\times10^{-6}$
2702 - 2799	0.00127	4.6 *6* $\times10^{-5}$
2808 - 2897	7.1 $\times10^{-5}$	2.5 *3* $\times10^{-6}$
2920 - 3018	0.000154	5.2 *6* $\times10^{-6}$
3029 - 3127	8.6 $\times10^{-5}$	2.8 *4* $\times10^{-6}$
3129	1.9 $\times10^{-7}$	~6 $\times10^{-9}$

Continuous Radiation (^{85}Y)

⟨β+⟩=564 keV;⟨IB⟩=2.1 keV

E_{bin}(keV)		⟨ ⟩(keV)	(%)
0 - 10	β+	1.27 $\times10^{-5}$	0.000156
	IB	0.022	
10 - 20	β+	0.000325	0.00199
	IB	0.026	0.18
20 - 40	β+	0.0060	0.0185
	IB	0.042	0.144
40 - 100	β+	0.186	0.244
	IB	0.117	0.18
100 - 300	β+	6.7	3.06
	IB	0.33	0.19
300 - 600	β+	44.4	9.6
	IB	0.40	0.093
600 - 1300	β+	278	29.4
	IB	0.60	0.070
1300 - 2500	β+	235	15.0
	IB	0.53	0.030
2500 - 3280	IB	0.066	0.0025
	Σβ+		57

$^{85}_{40}$Zr(7.86 *4* min)

Mode: ϵ

Δ: -73150 *100* keV

SpA: 2.812$\times10^{8}$ Ci/g

Prod: ^{89}Y(p,5n); ^{59}Co(^{32}S,αpn); ^{56}Fe(^{32}S,2pn); ^{55}Mn(^{32}S,pn); ^{90}Zr(p,p5n)

Photons (^{85}Zr)

⟨γ⟩=538 *17* keV

γ_{mode}	γ(keV)	γ(%)†
Y L_ℓ	1.686	0.0061 *13*
Y L_η	1.762	0.0034 *9*
Y L_α	1.922	0.14 *3*
Y L_β	2.005	0.082 *21*
Y L_γ	2.269	0.0025 *7*
Y $K_{\alpha2}$	14.883	1.60 *7*
Y $K_{\alpha1}$	14.958	3.07 *14*
Y $K_{\beta1}$'	16.735	0.68 *3*
Y $K_{\beta2}$'	17.125	0.101 *5*
γ[E2]	266.30 *19*	2.37 *17*
γ	357.90 *22*	1.12 *12*
γ[E2]	416.26 *24*	25.0 *18*
γ	416.31 *20*	0.25 *4*
γ	454.26 *17*	42
γ	480.1 *3*	0.21 *8*
γ	622.8 *4*	0.25 *4*
γ[M1+E2]	636.7 *4*	0.75 *8*
γ	744.13 *21*	0.29 *4*
γ	782.14 *23*	1.58 *12*
γ	800.0 *3*	0.67 *8*

Photons (^{85}Zr) (continued)

γ_{mode}	γ(keV)	γ(%)†
γ	810.9 *3*	0.42 *4*
γ	836.7 *3*	0.62 *12*
γ[M1+E2]	838.0 *3*	~0.25
γ	874.7 *3*	0.37 *4*
γ	957.3 *3*	0.71 *8*
γ[M1+E2]	986.8 *5*	0.25 *4*
γ	990.8 *4*	0.46 *4*
γ	1118.6 *4*	0.33 *4*
γ	1170.3 *5*	0.29 *4*
γ	1191.5 *5*	0.25 *4*
γ	1198.39 *17*	4.5 *3*
γ	1291.0 *3*	0.37 *4*
γ	1339.6 *4*	0.12 *4*
γ	1410.25 *22*	1.04 *8*
γ	1419.1 *5*	0.21 *4*
γ	1518.0 *5*	0.17 *4*
γ[E1]	1567.4 *4*	0.42 *4*
γ	1730.14 *23*	0.71 *8*
γ[M1+E2]	1768.15 *20*	1.79 *12*
γ	1876.1 *3*	0.42 *4*
γ	1934.1 *5*	0.46 *4*
γ[E1]	1937.9 *4*	0.62 *4*
γ	1955.3 *4*	0.42 *4*

† 7.3% uncert(syst)

Atomic Electrons (^{85}Zr)

⟨e⟩=2.1 *3* keV

e_{bin}(keV)	⟨e⟩(keV)	e(%)
2	0.135	6.4 *7*
12	0.0190	0.154 *16*
13	0.186	1.46 *15*
14 - 17	0.089	0.60 *5*
249	0.160	0.064 *5*
264 - 266	0.0254	0.0096 *7*
341 - 358	0.027	~0.008
399	0.61	0.154 *13*
414	0.075	0.0181 *15*
416	0.0152	0.0036 *3*
437	0.5	~0.12
452	0.06	~0.014
454 - 480	0.016	0.0035 *17*
606 - 637	0.0106	0.00172 *25*
727 - 765	0.010	0.0014 *5*
780 - 821	0.0122	0.0015 *3*
834 - 875	0.0026	0.00030 *9*
940 - 989	0.0069	0.00072 *15*
990 - 991	4.1 $\times10^{-5}$	4.1 *15* $\times10^{-6}$
1102 - 1119	0.0013	0.00012 *4*
1153 - 1198	0.018	0.0015 *5*
1274 - 1323	0.0016	0.00013 *4*
1337 - 1340	0.00014	1.1 *4* $\times10^{-5}$
1393 - 1419	0.0039	0.00028 *8*
1501 - 1567	0.00128	8.3 *13* $\times10^{-5}$
1713 - 1768	0.0079	0.00045 *4*
1859 - 1955	0.0042	0.00022 *3*

Continuous Radiation (^{85}Zr)

⟨β+⟩=1328 keV;⟨IB⟩=4.9 keV

E_{bin}(keV)		⟨ ⟩(keV)	(%)
0 - 10	β+	8.1 $\times10^{-6}$	0.000099
	IB	0.044	
10 - 20	β+	0.000215	0.00132
	IB	0.045	0.31
20 - 40	β+	0.00407	0.0127
	IB	0.086	0.30
40 - 100	β+	0.133	0.173
	IB	0.25	0.38
100 - 300	β+	5.07	2.32
	IB	0.72	0.40
300 - 600	β+	36.8	7.9
	IB	0.86	0.20
600 - 1300	β+	291	30.1
	IB	1.36	0.155
1300 - 2500	β+	807	44.3
	IB	1.11	0.065
2500 - 4670	β+	188	6.8
	IB	0.38	0.0127
	Σβ+		92

$^{85}_{40}$Zr(10.9 *3* s)

Mode: IT, ϵ

Δ: -72858 *100* keV

SpA: 1.18$\times10^{10}$ Ci/g

Prod: ^{89}Y(p,5n)

Photons (^{85}Zr)

γ_{mode}	γ(keV)	γ(rel)
γ_{IT}[E3]	292.2 *3*	~100
γ_ϵ	416.5 *3*	9

A = 86

NDS 25, 553 (1978)

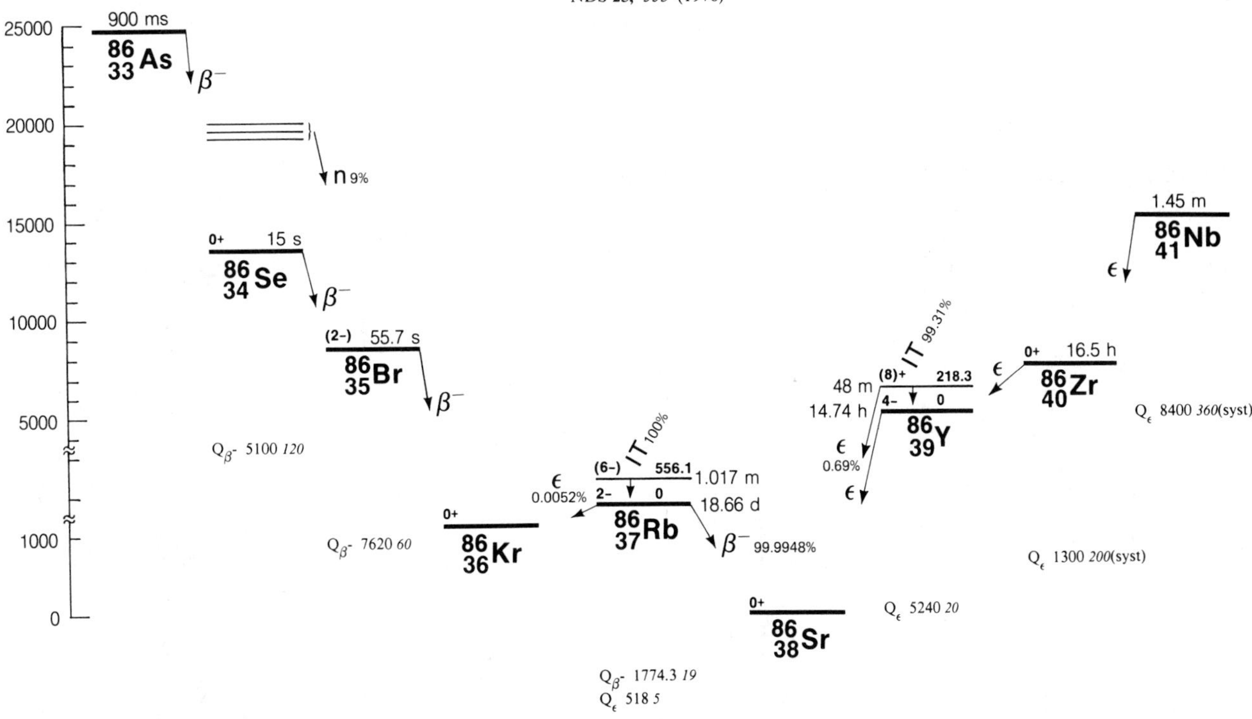

$^{86}_{33}$As(900 *200* ms)

Mode: β-, β-n(9.4 *18* %)
SpA: 1.02×10^{11} Ci/g
Prod: fission

Photons (^{86}As)

γ_{mode}	γ(keV)
$\gamma_{\beta-}$[E2]	704.1 *10*

$^{86}_{34}$Se(15.3 *9* s)

Mode: β-
Δ: -70540 *130* keV
SpA: 8.4×10^{9} Ci/g
Prod: fission

Photons (^{86}Se)

γ_{mode}	γ(keV)	γ(rel)
γ	47.6 *3*	9.0 *18*
γ	156.5 *3*	3.5 *7*
γ	207.5 *3*	6.0 *12*
γ	788.1 *11*	4.0 *4* ?
γ	1183.3 *8*	5.0 *5* ?
γ	1400.0 *6*	14.0 *14*
γ	2012.3 *5*	24.0 *24*
γ	2241.4 *5*	17 *3*
γ	2443.2 *5*	100 *10*
γ	2661.7 *5*	49 *5*

$^{86}_{35}$Br(55.7 *5* s)

Mode: β-
Δ: -75640 *60* keV
SpA: 2.341×10^{9} Ci/g
Prod: ^{86}Kr(n,p); fission

Photons (^{86}Br)

⟨γ⟩=3252 *87* keV

γ_{mode}	γ(keV)	γ(%)†
γ	500.9 *3*	~0.5
γ[E1]	749.30 *19*	0.71 *13*
γ[M1+E2]	784.74 *13*	3.8 *6*
γ	803.1 *3*	2.8 *6*
γ	1216.77 *12*	6.0 *7*
γ	1285.60 *24*	7.7 *6*
γ	1361.27 *10*	10.5 *11*
γ	1389.54 *15*	9.9 *6*
γ	1465.2 *3*	5.6 *3*
γ[E1]	1534.04 *16*	9.3 *14*
γ[E2]	1564.51 *8*	64 *4*
γ[E2]	1769.2 *12*	1.2 *3* ?
γ	1966.06 *18*	7.2 *5*
γ[E2]	2349.23 *14*	9.6 *6*
γ	2750.78 *15*	21.2 *14*
γ	2925.75 *13*	2.7 *3*
γ	5406.5 *6*	4.6 *6*
γ	5518.8 *9*	2.8 *3*
γ	6161.4 *20*	~0.07
γ[E1]	6209.6 *6*	0.58 *13*
γ	6722 *3*	~0.045
γ	6768.5 *15*	0.10 *3*

† 6.0% uncert(syst)

Continuous Radiation (^{86}Br)

⟨β-⟩=1930 keV;⟨IB⟩=7.0 keV

E_{bin}(keV)		⟨ ⟩(keV)	(%)
0 - 10	β-	0.0052	0.103
	IB	0.055	
10 - 20	β-	0.0158	0.106
	IB	0.054	0.38
20 - 40	β-	0.066	0.220
	IB	0.107	0.37
40 - 100	β-	0.53	0.74
	IB	0.31	0.48
100 - 300	β-	7.1	3.42
	IB	0.93	0.52
300 - 600	β-	34.8	7.6
	IB	1.15	0.27
600 - 1300	β-	232	24.1
	IB	1.9	0.21
1300 - 2500	β-	703	38.0
	IB	1.59	0.090
2500 - 5000	β-	753	22.2
	IB	0.85	0.026
5000 - 7500	β-	199	3.46
	IB	0.062	0.00112
7500 - 7620	β-	0.056	0.00074
	IB	1.18×10^{-7}	1.57×10^{-9}

$^{86}_{36}$Kr(stable)

Δ: -83262 *5* keV
%: 17.3 *2*

$^{86}_{37}$Rb(18.66 *2* d)

Mode: β-(99.9948 *5* %), ϵ(0.0052 *5* %)
Δ: -82743 *3* keV
SpA: 8.137×10^{4} Ci/g
Prod: ^{85}Rb(n,γ)

Photons (^{86}Rb)

⟨γ⟩=94.5 *9* keV

γ_{mode}	γ(keV)	γ(%)†
$\gamma_{\beta-}$(E2)	1076.69 *6*	8.78

† 0.91% uncert(syst)

Continuous Radiation (^{86}Rb)

⟨β-⟩=667 keV;⟨IB⟩=1.18 keV

E_{bin}(keV)		⟨ ⟩(keV)	(%)
0 - 10	β-	0.0339	0.67
	IB	0.027	
10 - 20	β-	0.103	0.68
	IB	0.027	0.19
20 - 40	β-	0.418	1.39
	IB	0.052	0.18
40 - 100	β-	3.10	4.41
	IB	0.142	0.22
100 - 300	β-	33.2	16.5
	IB	0.36	0.20
300 - 600	β-	109	24.3
	IB	0.32	0.077
600 - 1300	β-	403	43.8
	IB	0.24	0.030
1300 - 1774	β-	119	8.3
	IB	0.0077	0.00057

$^{86}_{37}$Rb(1.017 *3* min)

Mode: IT

Δ: -82187 *3* keV

SpA: 2.138×10^{9} Ci/g

Prod: ^{85}Rb(n,γ); ^{87}Rb(n,2n)

Photons (^{86}Rb)

⟨γ⟩=546.1 *7* keV

γ_{mode}	γ(keV)	γ(%)†
Rb L_{ℓ}	1.482	0.0012 *3*
Rb L_{η}	1.542	0.00069 *17*
Rb L_{α}	1.694	0.025 *5*
Rb L_{β}	1.759	0.015 *4*
Rb L_{γ}	2.051	0.00022 *7*
Rb $K_{\alpha2}$	13.336	0.307 *17*
Rb $K_{\alpha1}$	13.395	0.59 *3*
Rb $K_{\beta1}$'	14.959	0.128 *7*
Rb $K_{\beta2}$'	15.286	0.0136 *7*
γ (E4)	556.07 *18*	98.19

† 0.10% uncert(syst)

Atomic Electrons (^{86}Rb)

⟨e⟩= 9.9 *3* keV

e_{bin}(keV)	⟨e⟩(keV)	e(%)
2	0.025	1.36 *15*
11	0.0165	0.146 *16*
12	0.027	0.233 *25*
13	0.0169	0.129 *14*
15	0.00154	0.0105 *11*
541	8.4	1.56 *6*
554	1.20	0.216 *9*
556	0.231	0.0415 *17*

$^{86}_{38}$Sr(stable)

Δ: -84518 *3* keV

%: 9.86 *1*

$^{86}_{39}$Y (14.74 *2* h)

Mode: ϵ

Δ: -79278 *20* keV

SpA: 2.472×10^{6} Ci/g

Prod: ^{85}Rb(α,3n); ^{86}Sr(p,n); ^{88}Sr(p,3n); ^{85}Rb(^{3}He,2n)

Photons (^{86}Y)

⟨γ⟩=3252 *24* keV

γ_{mode}	γ(keV)	γ(%)†
Sr L_{ℓ}	1.582	0.045 *10*
Sr L_{η}	1.649	0.026 *6*
Sr L_{α}	1.806	1.01 *21*
Sr L_{β}	1.880	0.59 *15*
Sr L_{γ}	2.196	0.010 *3*
Sr $K_{\alpha2}$	14.098	11.9 *5*
Sr $K_{\alpha1}$	14.165	23.0 *10*
Sr $K_{\beta1}$'	15.832	5.03 *21*
Sr $K_{\beta2}$'	16.192	0.69 *3*
γ [E2]	132.41 *6*	0.165 *8*
γ	145.03 *11*	0.031 *3*
γ	182.31 *9*	0.11 *3*
γ M1	187.92 *8*	1.26 *4*
γ E2	190.96 *6*	1.01 *3*
γ	209.98 *10*	0.396 *17*
γ	235.42 *11*	0.396 *17*
γ	238.14 *10*	0.132 *25*
γ E1	252.19 *6*	0.371 *16*
γ	255.82 *25*	0.074 *25*
γ	264.59 *9*	0.536 *25*
γ E1	307.00 *7*	3.47 *8*
γ M1	331.15 *12*	0.833 *25*
γ [E1]	355.30 *21*	0.099 *25*
γ	370.25 *9*	0.82 *4*
γ	380.44 *22*	0.45 *3*
γ [M1+E2]	382.94 *8*	3.63 *12*
γ	426.08 *9*	0.305 *16*
γ [E1]	439.41 *8*	0.20 *7*
γ E1	443.15 *6*	16.9 *5*
γ	444.18 *10*	0.64 *16*
γ	448.13 *7*	0.074 *25*
γ	468.95 *14*	0.297 *25*
γ	503.0 *3*	0.09 *3*
γ [E2]	512.39 *6*	
γ M1+E2	515.43 *8*	4.89 *14*
γ (M1)	580.53 *6*	4.79 *14*
γ E1	608.39 *7*	2.01 *15*
γ	618.36 *9*	0.21 *3*
γ E1	627.77 *6*	32.6 *10*
γ	634.84 *8*	0.091 *25* ?
γ [M1+E2]	644.80 *7*	2.2 *3*
γ	645.86 *11*	9.2 *11*
γ [M1+E2]	648.54 *7*	
γ [E1]	689.23 *11*	0.17 *3*
γ	702.3 *8*	0.25 *8*
γ [M1]	703.35 *6*	15.4 *4*
γ E2	709.99 *7*	2.62 *7*
γ	719.19 *17*	0.22 *3*
γ	740.80 *7*	1.36 *5*
γ [E1]	767.62 *8*	2.4 *3*
γ	768.46 *9*	0.32 *11*
γ M1+E2	777.60 *7*	22.4 *6*
γ	783.67 *12*	0.26 *3*
γ E1	826.08 *8*	3.30 *8*
γ	833.79 *12*	1.5 *3*
γ [E2]	835.76 *7*	4.4 *6*
γ	882.94 *9*	0.25 *8*
γ	887.53 *8*	0.44 *4*
γ [E1]	955.54 *6*	1.04 *4*
γ	971.59 *12*	0.27 *3*
γ	1018.07 *9*	~0.18
γ [M1]	1024.12 *7*	3.79 *17*
γ (E2)	1076.69 *6*	82.5

Photons (^{86}Y)
(continued)

γ_{mode}	γ(keV)	γ(%)†
γ [E1]	1087.95 *7*	0.041 *8*
γ	1092.92 *7*	0.69 *4*
γ	1102.12 *17*	0.198 *25*
γ [M1+E2]	1132.37 *11*	0.297 *25*
γ [E1]	1143.19 *9*	0.10 *3* ?
γ	1150.31 *11*	
γ (E2)	1153.17 *6*	30.5 *9*
γ	1154.1 *15*	
γ (M1)	1163.10 *8*	1.18 *4*
γ E1	1253.19 *6*	1.53 *5*
γ	1270.25 *8*	0.65 *10*
γ	1283.88 *7*	0.29 *11*
γ	1295.0 *4*	0.29 *8*
γ	1296.06 *13*	0.54 *3*
γ	1327.86 *10*	0.09 *4*
γ [M1]	1349.21 *9*	2.95 *9*
γ [E1]	1405.36 *7*	0.18 *5*
γ [E1]	1415.29 *9*	0.33 *9*
γ [E2]	1507.95 *9*	0.35 *4*
γ	1533.25 *9*	0.22 *3*
γ	1536.07 *7*	0.12 *3*
γ [M1+E2]	1565.49 *22*	0.18 *5* ?
γ	1696.33 *7*	0.635 *16*
γ [M1+E2]	1711.7 *4*	0.17 *3*
γ	1724.20 *8*	0.55 *4*
γ [E1]	1790.86 *8*	1.00 *4*
γ [M1]	1801.71 *7*	1.65 *5*
γ E2	1854.28 *8*	17.2 *5*
γ E1	1920.78 *8*	20.8 *7*
γ	1969.2 *10*	0.049 *8*
γ	2017.3 *3*	0.132 *16*
γ	2088.32 *19*	0.248 *25*
γ [E1]	2108.70 *7*	0.049 *8*
γ	2180.8 *4*	0.033 *8*
γ	2292.07 *22*	0.124 *8*
γ	2482.03 *8*	0.116 *8*
γ	2556.4 *4*	0.027 *8*
γ [E1]	2568.44 *9*	2.25 *11*
γ	2610.09 *19*	1.24 *7*
γ [E2]	2642.16 *22*	0.17 *4*
γ [E2]	2788.4 *4*	~0.011
γ	2794.9 *3*	0.206 *16*
γ	2827.8 *11*	0.058 *16*
γ	2863.7 *17*	0.009 *4*
γ	2865.90 *19*	0.38 *7*
γ	2997.44 *9*	~0.008
γ	3069.65 *22*	0.116 *17*
γ	3334.0 *4*	0.124 *16*
γ	3641.3 *17*	0.041 *8*
γ	3877 *6*	~0.05

† 0.48% uncert(syst)

Atomic Electrons (^{86}Y)

⟨e⟩=5.50 *21* keV

e_{bin}(keV)	⟨e⟩(keV)	e(%)
2	0.97	49 *5*
12	1.60	13.3 *14*
14	0.63	4.6 *5*
15 - 16	0.056	0.36 *3*
116 - 145	0.077	0.065 *5*
166 - 172	0.068	0.040 *5*
175	0.147	0.084 *3*
180 - 222	0.093	0.047 *12*
233 - 265	0.038	0.016 *6*
291 - 339	0.0604	0.0201 *4*
353 - 383	0.118	0.032 *7*
410 - 453	0.131	0.0305 *16*
467 - 515	0.073	0.0146 *20*
564 - 612	0.178	0.0298 *8*
616 - 646	0.11	0.017 *5*
673 - 719	0.158	0.0229 *8*
725 - 752	0.017	0.0023 *5*
761	0.155	0.0204 *13*
765 - 810	0.0320	0.00408 *16*
818 - 867	0.042	0.0051 *6*
871 - 888	0.0022	0.00025 *9*
939 - 972	0.0038	0.00041 *5*
1002 - 1024	0.0214	0.00212 *12*

Atomic Electrons (^{86}Y) (continued)

e_{bin}(keV)	⟨e⟩(keV)	e(%)
1061	0.381	0.0359 *14*
1072 - 1116	0.0549	0.00510 *18*
1127 - 1163	0.153	0.0134 *6*
1237 - 1284	0.0084	0.00067 *9*
1293 - 1333	0.0112	0.00084 *4*
1347 - 1389	0.00168	0.000124 *8*
1399 - 1415	0.00066	4.7 *11* $\times10^{-5}$
1492 - 1565	0.0028	0.000186 *21*
1680 - 1775	0.0049	0.00029 *4*
1786 - 1854	0.0561	0.00306 *9*
1905 - 2001	0.0328	0.00172 *6*
2015 - 2108	0.00063	3.0 *7* $\times10^{-5}$
2165 - 2181	6.8 $\times10^{-5}$	3.1 *11* $\times10^{-6}$
2276 - 2292	0.00024	1.1 *3* $\times10^{-5}$
2466 - 2556	0.00300	0.000118 *7*
2566 - 2642	0.0030	0.000114 *21*
2772 - 2866	0.00115	4.1 *7* $\times10^{-5}$
2981 - 3069	0.00021	6.7 *15* $\times10^{-6}$
3318 - 3334	0.00020	5.9 *14* $\times10^{-6}$
3625 - 3641	6.3 $\times10^{-5}$	1.7 *5* $\times10^{-6}$
3861 - 3877	7.3 $\times10^{-5}$	~2$\times10^{-6}$

Continuous Radiation (^{86}Y)

⟨β+⟩=219 keV;⟨IB⟩=1.42 keV

E_{bin}(keV)		⟨ ⟩(keV)	(%)
0 - 10	β+	3.19$\times10^{-5}$	0.000393
	IB	0.0092	
10 - 20	β+	0.00081	0.00497
	IB	0.0165	0.115
20 - 40	β+	0.0144	0.0450
	IB	0.018	0.061
40 - 100	β+	0.416	0.55
	IB	0.048	0.074
100 - 300	β+	11.3	5.3
	IB	0.143	0.079
300 - 600	β+	51	11.3
	IB	0.24	0.055
600 - 1300	β+	117	13.8
	IB	0.61	0.069
1300 - 2500	β+	36.3	2.19
	IB	0.32	0.020
2500 - 4163	β+	3.30	0.123
	IB	0.0155	0.00054
	Σβ+		33

$^{86}_{39}$Y (48 *1* min)

Mode: IT(99.31 *4* %), ε(0.69 *4* %)
Δ: -79060 *20* keV
SpA: 4.55$\times10^{7}$ Ci/g
Prod: ^{85}Rb(α,3n); ^{87}Sr(p,2n)

Photons (^{86}Y)

⟨γ⟩=215.5 *24* keV

γ_{mode}	γ(keV)	γ(%)†
Y L_ℓ	1.686	0.063 *19*
Y L_η	1.762	0.035 *11*
Y L_α	1.922	1.5 *4*
Y L_β	1.998	0.76 *24*
Y L_γ	2.171	0.011 *4*
γ_{IT} E3	10.2 *1*	4.0 *8* $\times10^{-5}$
Y $K_{\alpha2}$	14.883	1.00 *5*
Y $K_{\alpha1}$	14.958	1.92 *10*
Y $K_{\beta1}'$	16.735	0.430 *22*
Y $K_{\beta2}'$	17.125	0.063 *3*
γ_ϵ(E2)	98.6 *1*	0.331 *19*
γ_{IT} E2(+31%M1)	208.1 *2*	93.6

Photons (^{86}Y) (continued)

γ_{mode}	γ(keV)	γ(%)†
γ_ϵ(E2)	627.20 *20*	0.69 *7*
γ_ϵ(E2)	1076.69 *6*	0.69 *7*
γ_ϵ(E2)	1153.17 *6*	0.69 *7*

† uncert(syst): 8.2% for ε, 1.0% for IT

Atomic Electrons (^{86}Y)*

⟨e⟩=20.4 *14* keV

e_{bin}(keV)	⟨e⟩(keV)	e(%)
2	1.6	75 *16*
8	6.2	77 *15*
10	1.7	17 *4*
12	0.0119	0.097 *10*
13	0.117	0.92 *10*
14	0.0128	0.089 *10*
15	0.039	0.26 *3*
16	0.0033	0.0199 *21*
17	0.00124	0.0074 *8*
191	9.2	4.80 *17*
206	1.27	0.618 *23*
208	0.259	0.125 *5*

* with IT

Continuous Radiation (^{86}Y)

⟨β+⟩=2.88 keV;⟨IB⟩=0.0139 keV

E_{bin}(keV)		⟨ ⟩(keV)	(%)
0 - 10	β+	2.30$\times10^{-7}$	2.84$\times10^{-6}$
	IB	0.000124	
10 - 20	β+	5.9$\times10^{-6}$	3.62$\times10^{-5}$
	IB	0.000148	0.00104
20 - 40	β+	0.000108	0.000336
	IB	0.00023	0.00082
40 - 100	β+	0.00332	0.00436
	IB	0.00064	0.00099
100 - 300	β+	0.113	0.052
	IB	0.00167	0.00095
300 - 600	β+	0.66	0.144
	IB	0.0019	0.00046
600 - 1300	β+	2.05	0.239
	IB	0.0047	0.00052
1300 - 2500	β+	0.057	0.00427
	IB	0.0044	0.00027
2500 - 2503	IB	6.7$\times10^{-14}$	2.5$\times10^{-15}$
	Σβ+		0.44

$^{86}_{40}$Zr(16.5 *1* h)

Mode: ε
Δ: -77980 *200* keV syst
SpA: 2.208$\times10^{6}$ Ci/g
Prod: ^{89}Y(p,4n)

Photons (^{86}Zr)

⟨γ⟩=294.3 *20* keV

γ_{mode}	γ(keV)	γ(%)†
Y L_ℓ	1.686	0.13 *3*
Y L_η	1.762	0.072 *18*
Y L_α	1.922	3.0 *6*
Y L_β	2.005	1.7 *4*
Y L_γ	2.266	0.050 *14*
Y $K_{\alpha2}$	14.883	33.5 *19*
Y $K_{\alpha1}$	14.958	64 *4*
Y $K_{\beta1}'$	16.735	14.4 *8*
Y $K_{\beta2}'$	17.125	2.12 *12*

Photons (^{86}Zr) (continued)

γ_{mode}	γ(keV)	γ(%)†
γ E1	29.10 *8*	21.6 *15*
γ M1+E2	94.2 *1*	0.046 *3*
γ M1+E2	127.7 *1*	0.070 *9*
γ M1	135.6 *1*	0.47 *5*
γ M1	160.7 *1*	0.096 *9*
γ M1	173.7 *1*	0.070 *9*
γ [E2]	207.9 *2*	0.082 *8*
γ M1,E1	214.9 *1*	0.082 *8*
γ E2	242.8 *1*	95.80
γ M1	612.00 *8*	5.7 *3*
γ	620.6 *2*	0.27 *3*
γ [E1]	641.10 *8*	

† 0.10% uncert(syst)

Atomic Electrons (^{86}Zr)

⟨e⟩=29.9 *9* keV

e_{bin}(keV)	⟨e⟩(keV)	e(%)
2	2.9	136 *15*
12	8.8	73 *5*
13	3.9	31 *3*
14	0.43	3.0 *3*
15	1.29	8.8 *10*
16 - 17	0.150	0.91 *8*
27	2.37	8.8 *6*
29 - 77	0.52	1.77 *13*
92 - 136	0.072	0.062 *12*
144 - 191	0.0217	0.0130 *8*
198 - 215	0.0046	0.0023 *5*
226	8.05	3.57 *7*
240	0.855	0.356 *7*
241 - 243	0.457	0.1890 *25*
595 - 621	0.065	0.0108 *5*

Continuous Radiation (^{86}Zr)

⟨IB⟩=0.24 keV

E_{bin}(keV)		⟨ ⟩(keV)	(%)
10 - 20	IB	0.0129	0.086
20 - 40	IB	0.00079	0.0030
40 - 100	IB	0.0020	0.0029
100 - 300	IB	0.032	0.0154
300 - 600	IB	0.114	0.026
600 - 1028	IB	0.077	0.0108

$^{86}_{41}$Nb(1.45 *5* min)

Mode: ε
Δ: -69580 *300* keV syst
SpA: 1.50$\times10^{9}$ Ci/g
Prod: protons on Ag

Photons (^{86}Nb)

γ_{mode}	γ(keV)	γ(rel)
γ E2	628.5 *5*	6 *2*
γ E2	751.9 *5*	100
γ E2	914.7 *5*	66 *5*
γ E2	1003.4 *5*	32 *3*

NDS **27**, 389 (1979)

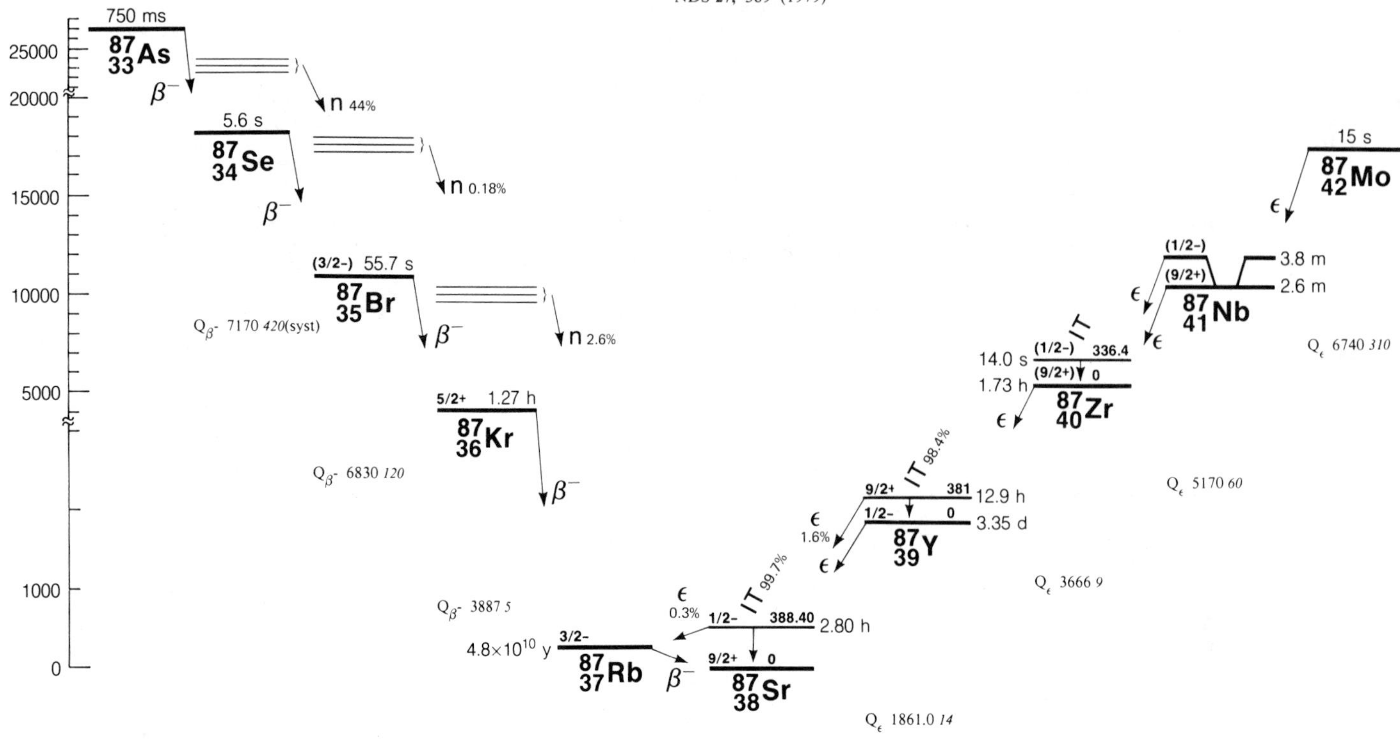

$^{87}_{33}$As(750 *60* ms)

Mode: β-, β-n(44 *14* %)
SpA: 1.13×10^{11} Ci/g
Prod: fission

$^{87}_{34}$Se(5.55 *20* s)

Mode: β-, β-n(0.18 *3* %)
Δ: -66710 *400* keV syst
SpA: 2.20×10^{10} Ci/g
Prod: fission

Photons (^{87}Se)

γ$_{mode}$	γ(keV)	γ(rel)
γ	242.5 *3*	100 *20*
γ	334.0 *3*	94 *19*
γ	468.0 *3*	48 *10*
γ	573.2 *3*	52 *10*
γ	701.5 *3*	11.0 *22*
γ	710.5 *3*	20 *4*
γ	1167.6 *3*	18 *4*
γ	1305.0 *3*	18 *4*
γ	1878.1 *3*	24 *5*
γ	3683.8 *5*	4.4 *9*
γ	3744.5 *5*	6.1 *12*
γ	3926.3 *5*	7.7 *15*

$^{87}_{35}$Br(55.69 *13* s)

Mode: β-, β-n(2.57 *15* %)
Δ: -73880 *120* keV
SpA: 2.314×10^{9} Ci/g
Prod: fission
β-n(avg): 250 *60*

Delayed Neutrons (^{87}Br)

n(keV)	n(%)
18.0 *15*	0.47 *12*
40.3 *15*	0.077 *19*
52.2 *23*	0.35 *9*
70.8 *17*	0.17 *4*
80 *3*	0.11 *3*
121 *3*	0.15 *4*
135.8 *19*	0.15 *4*
147.5 *18*	0.13 *3*
169 *3*	0.067 *17*
182 *3*	0.18 *5*
211.1 *24*	0.056 *14*
248 *4*	0.21 *5*
256.2 *25*	0.073 *18*
312.4 *24*	0.056 *14*
339 *3*	0.027 *7*
386.1 *22*	0.035 *9*
401 *3*	0.031 *8*
407 *3*	0.047 *11*
437.7 *24*	0.071 *18*
457 *3*	0.035 *9*
638 *4*	0.017 *4*
668 *4*	0.089 *22*

Photons (^{87}Br)

⟨γ⟩=4069 *40* keV

γ$_{mode}$	γ(keV)	γ(%)†
γ	158.01 *6*	0.20 *3* ?
γ	230.08 *7*	0.861 *17*
γ	263.84 *7*	0.37 *3*
γ	380.45 *7*	0.18 *3* ?
γ	421.85 *5*	5.00 *14*
γ	461.63 *6*	0.662 *22*
γ	493.33 *23*	0.176 *19* ?
γ	529.42 *6*	1.20 *5*
γ[E2]	531.94 *6*	8.03 *24*
γ	585.81 *7*	0.84 *5*
γ	610.53 *6*	1.14 *3*
γ	651.92 *7*	1.70 *5*
γ	681.23 *15*	0.51 *3* ?
γ	698.54 *14*	0.15 *4* ?
γ	714.10 *14*	0.20 *3* ?
γ	831.20 *8*	1.20 *6*
γ	874.37 *8*	0.60 *4* ?
γ	894.0 *4*	0.45 *5* ?
γ	920.94 *25*	0.77 *6* ?
γ	944.25 *7*	1.88 *9*
γ	952.85 *10*	0.97 *6*
γ	955.0 *4*	0.42 *5* ?
γ	1021.60 *8*	1.88 *6*
γ	1069.2 *3*	0.31 *3*
γ	1095.4 *14*	6.08 *12*
γ	1146.49 *10*	0.35 *3*
γ	1185.56 *8*	0.19 *4* ?
γ	1278.1 *8*	0.28 *7* ?
γ	1338.25 *10*	0.85 *5*
γ	1349.48 *8*	0.42 *6*
γ	1360.61 *10*	4.9 *3*
γ	1412.6 *4*	0.87 *13*
γ	1419.79 *5*	32.0 *10*

Photons (^{87}Br) (continued)

γ_{mode}	γ(keV)	γ(%)†
γ	1449.40 *7*	1.72 *6*
γ	1476.18 *6*	11.7 *4*
γ	1577.80 *6*	8.6 *3*
γ	1607.40 *7*	1.85 *6*
γ	1640.40 *11*	0.45 *4*
γ	1659.68 *14*	0.31 *4*
γ	1768.5 *3*	1.03 *7*
γ	1798.40 *10*	0.96 *11*
γ	1836.8 *3*	1.40 *11*
γ	1868.77 *13*	0.28 *3* ?
γ	1881.41 *6*	4.2 *3*
γ	1934.55 *10*	0.40 *4* ?
γ	2005.59 *6*	7.7 *6*
γ	2058.70 *15*	0.21 *5* ?
γ [E2]	2071.70 *8*	3.48 *11*
γ	2122.69 *11*	1.78 *7*
γ	2169.7 *3*	0.67 *5*
γ	2232.9 *5*	0.33 *5* ?
γ	2254.4 *7*	0.34 *4* ?
γ	2259.0 *6*	0.36 *4* ?
γ	2299.38 *13*	0.60 *3* ?
γ	2339.76 *13*	0.46 *5* ?
γ	2372.63 *10*	1.39 *13*
γ	2378.3 *6*	0.30 *5* ?
γ	2398.17 *12*	0.77 *4* ?
γ	2452.15 *8*	1.01 *6*
γ	2454.56 *13*	0.64 *13* ?
γ	2498.8 *3*	0.83 *11* ?
γ	2509.58 *12*	0.36 *4* ?
γ	2517.81 *18*	0.9 *3*
γ	2546.0 *8*	0.15 *5* ?
γ	2575.7 *3*	0.90 *5*
γ	2638.4 *5*	0.83 *8* ?
γ	2641.99 *17*	0.82 *8*
γ	2663.8 *3*	0.32 *3*
γ	2694.8 *5*	0.36 *3*
γ	2705.19 *13*	2.56 *10*
γ	2713.33 *19*	0.35 *5*
γ	2747.97 *15*	1.22 *12* ?
γ	2810.6 *11*	0.12 *4* ?
γ	2821.22 *12*	2.53 *10* ?
γ	2836.77 *9*	2.17 *9*
γ	2869.15 *14*	0.47 *5*
γ	2997.5 *3*	3.29 *13* ?
γ	3027.17 *7*	1.81 *8*
γ	3063.1 *8*	0.30 *5* ?
γ	3080.31 *12*	0.24 *3* ?
γ	3091.82 *20*	0.47 *3* ?
γ	3120.08 *11*	0.44 *4* ?
γ	3132.98 *14*	0.42 *4*
γ	3142.81 *18*	0.36 *6*
γ	3167.9 *5*	0.52 *5* ?
γ	3176.4 *6*	1.87 *7*
γ	3182.6 *11*	0.44 *6* ?
γ	3201.17 *20*	0.11 *4*
γ	3207.23 *18*	0.16 *3*
γ	3218.16 *10*	0.38 *4*
γ	3248.7 *4*	0.6 *3* ?
γ	3256.9 *4*	0.44 *4*
γ	3271.6 *6*	0.46 *8* ?
γ	3460.90 *12*	0.43 *4*
γ	3496.6 *8*	0.31 *4*
γ	3523.51 *24*	0.17 *6* ?
γ	3541.91 *11*	0.65 *4*
γ	3580.4 *5*	0.17 *3*
γ	3611.0 *11*	0.10 *3* ?
γ	3794.70 *17*	0.96 *19*
γ	3809.5 *5*	0.56 *6*
γ	3895.6 *7*	0.20 *3*
γ	3903.3 *9*	0.19 *4*
γ	3917.3 *4*	2.94 *11* ?
γ	3953.0 *3*	0.32 *4* ?
γ	3953.7 *5*	0.32 *4* ?
γ	4088.99 *20*	0.30 *5*
γ	4181.34 *11*	6.6 *5*
γ	4297.6 *5*	0.65 *6* ?
γ	4465.0 *13*	0.11 *5*
γ	4523.34 *17*	0.35 *4*
γ	4572.7 *4*	1.47 *10*
γ	4620.91 *19*	0.52 *5*
γ	4645.4 *4*	1.49 *8*
γ	4663.36 *21*	0.43 *7*
γ	4710.72 *13*	0.70 *5*
γ	4752.9 *6*	0.33 *4*

Photons (^{87}Br) (continued)

γ_{mode}	γ(keV)	γ(%)†
γ	4784.96 *17*	2.67 *13*
γ	4824.4 *6*	0.23 *8*
γ	4829.6 *6*	0.07 *3* ?
γ	4836.2 *6*	0.10 *4*
γ	4872.5 *4*	0.50 *6*
γ	4961.65 *11*	2.83 *16*
γ	4999.5 *5*	0.19 *4* ?
γ	5022.3 *5*	0.26 *6*
γ	5044.9 *5*	0.60 *6* ?
γ	5089.1 *7*	0.28 *3*
γ	5104.11 *24*	0.71 *5* ?
γ	5118.1 *12*	0.32 *10*
γ	5120.52 *13*	0.88 *8*
γ	5195.27 *21*	0.75 *6*
γ	5201.3 *6*	0.96 *7* ?
γ	5214.43 *18*	0.33 *3*
γ	5262.6 *3*	0.09 *3* ?
γ	5474.2 *5*	0.61 *5* ?
γ	5543.6 *8*	0.048 *10*
γ	5588.6 *10*	0.022 *6*
γ	5595.12 *24*	0.096 *16*
γ	5636.02 *24*	0.10 *3*
γ	5659.3 *8*	0.064 *16*
γ	5672.1 *8*	0.096 *19*
γ	5687.0 *4*	0.16 *6*
γ	5687.4 *7*	0.16 *6*
γ	5720 *3*	~0.026
γ	5794.5 *3*	0.080 *22*
γ	5821 *5*	~0.013
γ	6095.2 *19*	

† 7.8% uncert(syst)

Continuous Radiation (^{87}Br)

⟨β-⟩=1871 keV;⟨IB⟩=6.6 keV

E_{bin}(keV)		⟨ ⟩(keV)	(%)
0 - 10	β-	0.0065	0.129
	IB	0.054	
10 - 20	β-	0.0198	0.132
	IB	0.053	0.37
20 - 40	β-	0.083	0.274
	IB	0.106	0.37
40 - 100	β-	0.65	0.92
	IB	0.31	0.47
100 - 300	β-	8.4	4.10
	IB	0.92	0.51
300 - 600	β-	38.8	8.5
	IB	1.13	0.26
600 - 1300	β-	223	23.5
	IB	1.8	0.21
1300 - 2500	β-	626	33.5
	IB	1.57	0.090
2500 - 5000	β-	937	28.3
	IB	0.59	0.019
5000 - 6830	β-	37.4	0.69
	IB	0.0053	0.000099

$^{87}_{36}$Kr(1.272 *10* h)

Mode: β-

Δ: -80706 *5* keV

SpA: 2.832×10^{7} Ci/g

Prod: ^{86}Kr(n,γ); fission; daughter ^{87}Br

Photons (^{87}Kr)

⟨γ⟩=792 *16* keV

γ_{mode}	γ(keV)	γ(%)†
γ	129.5 *3*	0.045 *10*
γ M1+6.4%E2	402.637 *19*	49.6 *20*
γ	510.84 *14*	0.069 *10*
γ	552.2 *4*	0.029 *4*
γ	582.32 *17*	0.035 *10*
γ	673.91 *6*	1.89 *5*
γ	814.30 *6*	0.164 *5*
γ	836.43 *5*	0.769 *20*
γ [M1+E2]	845.50 *4*	7.34 *20*
γ	894.07 *11*	0.045 *3*
γ	901.6 *3*	0.026 *5*
γ	946.81 *10*	0.129 *5*
γ	976.82 *8*	0.057 *4*
γ	1063.16 *10*	0.027 *6*
γ	1175.49 *5*	1.11 *4*
γ	1338.01 *5*	0.63 *4*
γ	1382.62 *6*	0.288 *10*
γ	1389.93 *12*	0.119 *5*
γ	1461.87 *17*	0.050 *5*
γ	1532.91 *18*	0.36 *6* ?
γ	1578.12 *5*	0.129 *10*
γ	1611.29 *10*	0.104 *15*
γ	1740.64 *5*	2.04 *5*
γ	1842.67 *24*	0.139 *10*
γ	2011.91 *5*	2.88 *11*
γ	2378.40 *17*	0.094 *5*
γ	2408.67 *14*	0.228 *20*
γ	2554.92 *7*	9.2 *5*
γ	2558.08 *7*	3.92 *25*
γ	2652.55 *24*	0.023 *4*
γ	2811.30 *14*	0.322 *15*
γ	2960.71 *7*	0.069 *20*
γ [E1]	3055.17 *24*	0.084 *5*
γ	3308.58 *12*	0.446 *25*

† 4.8% uncert(syst)

Continuous Radiation (^{87}Kr)

⟨β-⟩=1330 keV;⟨IB⟩=3.8 keV

E_{bin}(keV)		⟨ ⟩(keV)	(%)
0 - 10	β-	0.0132	0.263
	IB	0.044	
10 - 20	β-	0.0403	0.268
	IB	0.043	0.30
20 - 40	β-	0.166	0.55
	IB	0.085	0.30
40 - 100	β-	1.28	1.81
	IB	0.24	0.37
100 - 300	β-	15.3	7.5
	IB	0.70	0.39
300 - 600	β-	60	13.3
	IB	0.81	0.19
600 - 1300	β-	266	28.4
	IB	1.15	0.133
1300 - 2500	β-	711	38.4
	IB	0.67	0.040
2500 - 3887	β-	276	9.8
	IB	0.046	0.00168

$^{87}_{37}$Rb(4.80 *13* $\times 10^{10}$ yr)

Mode: β-

Δ: -84592 *3* keV

Prod: natural source

%: 27.835 *13*

Continuous Radiation (^{87}Rb)

$\langle\beta-\rangle$=82 keV;$\langle$IB$\rangle$=0.024 keV

E_{bin}(keV)		$\langle\ \rangle$(keV)	(%)
0 - 10	β-	0.397	8.0
	IB	0.0042	
10 - 20	β-	1.15	7.7
	IB	0.0035	0.025
20 - 40	β-	4.30	14.4
	IB	0.0054	0.019
40 - 100	β-	24.1	35.4
	IB	0.0081	0.0135
100 - 282	β-	52	34.5
	IB	0.0026	0.0021

$^{87}_{38}$Sr(stable)

Δ: -84875 *3* keV

%: 7.00 *1* (variations for sources containing Rb, due to ^{87}Rb decay)

$^{87}_{38}$Sr(2.795 *13* h)

Mode: IT(99.70 *8* %), ε(0.30 *8* %)

Δ: -84486 *3* keV

SpA: 1.289×10^{7} Ci/g

Prod: daughter ^{87}Y; ^{86}Sr(n,γ); ^{87}Rb(p,n)

Photons (^{87}Sr)

$\langle\gamma\rangle$=321 *3* keV

γ_{mode}	γ(keV)	γ(%)†
Sr L_ℓ	1.582	0.012 *3*
Sr L_η	1.649	0.0068 *17*
Sr L_α	1.806	0.27 *6*
Sr L_β	1.879	0.16 *4*
Sr L_γ	2.196	0.0026 *8*
Rb $K_{\alpha2}$	13.336	0.0498 *21*
Rb $K_{\alpha1}$	13.395	0.096 *4*
Sr $K_{\alpha2}$	14.098	3.06 *13*
Sr $K_{\alpha1}$	14.165	5.89 *25*
Rb $K_{\beta1}$'	14.959	0.0207 *9*
Rb $K_{\beta2}$'	15.286	0.00221 *9*
Sr $K_{\beta1}$'	15.832	1.29 *6*
Sr $K_{\beta2}$'	16.192	0.176 *8*
γ_{IT} M4	388.40 *5*	82.3

† 1.0% uncert(syst)

Atomic Electrons (^{87}Sr)

$\langle$e$\rangle$=67.4 *13* keV

e_{bin}(keV)	$\langle$e$\rangle$(keV)	e(%)
2	0.26	13.3 *14*
11	0.0027	0.0237 *24*
12	0.42	3.4 *4*
13	0.0027	0.0209 *21*
14	0.162	1.17 *12*
15	0.0046	0.030 *3*
16	0.010	0.064 *7*
372	56.1	15.1 *3*
386	8.66	2.24 *5*
388	1.75	0.451 *10*

$^{87}_{39}$Y (3.346 *13* d)

Mode: ε

Δ: -83014 *3* keV

SpA: 4.486×10^{5} Ci/g

Prod: ^{86}Sr(d,n); ^{87}Sr(p,n); ^{87}Sr(d,2n)

Photons (^{87}Y)

$\langle\gamma\rangle$=787 *6* keV

γ_{mode}	γ(keV)	γ(%)†
Sr L_ℓ	1.582	0.079 *17*
Sr L_η	1.649	0.045 *11*
Sr L_α	1.806	1.8 *4*
Sr L_β	1.880	1.0 *3*
Sr L_γ	2.196	0.018 *5*
Sr $K_{\alpha2}$	14.098	20.9 *9*
Sr $K_{\alpha1}$	14.165	40.2 *17*
Sr $K_{\beta1}$'	15.832	8.8 *4*
Sr $K_{\beta2}$'	16.192	1.20 *5*
γ M4	388.40 *5*	84.8 *
γ M1	484.90 *5*	92.2

† 1.0% uncert(syst)
* with ^{87}Sr(2.80 h) in equilib

Atomic Electrons (^{87}Y)

$\langle$e$\rangle$=75.5 *14* keV

e_{bin}(keV)	$\langle$e$\rangle$(keV)	e(%)
2	1.72	87 *9*
12	2.8	23.3 *24*
14	1.10	8.0 *8*
15	0.030	0.192 *20*
16	0.069	0.44 *5*
372	57.9	15.5 *4*
386	8.92	2.31 *5*
388	1.80	0.465 *10*
469	1.091	0.233 *5*
483	0.123	0.0255 *6*
485	0.0244	0.00505 *11*

Continuous Radiation (^{87}Y)

$\langle\beta+\rangle$=0.420 keV;$\langle$IB$\rangle$=0.26 keV

E_{bin}(keV)		$\langle\ \rangle$(keV)	(%)
0 - 10	β+	1.73×10^{-6}	2.13×10^{-5}
	IB	0.00019	
10 - 20	β+	4.32×10^{-5}	0.000265
	IB	0.0114	0.080
20 - 40	β+	0.00075	0.00233
	IB	0.00068	0.0025
40 - 100	β+	0.0194	0.0258
	IB	0.0020	0.0029
100 - 300	β+	0.297	0.151
	IB	0.033	0.0159
300 - 600	β+	0.103	0.0303
	IB	0.119	0.027
600 - 1300	IB	0.094	0.0126
1300 - 1473	IB	0.00060	4.5×10^{-5}
	Σβ+		0.21

$^{87}_{39}$Y (12.9 *4* h)

Mode: IT(98.43 *16* %), ε(1.57 *16* %)

Δ: -82632 *3* keV

SpA: 2.79×10^{6} Ci/g

Prod: ^{86}Sr(d,n); daughter ^{87}Zr; ^{87}Sr(p,n)

Photons (^{87}Y)

$\langle\gamma\rangle$=299.6 *9* keV

γ_{mode}	γ(keV)	γ(%)†
Y L_ℓ	1.686	0.014 *3*
Y L_η	1.762	0.0078 *20*
Y L_α	1.922	0.32 *7*
Y L_β	2.005	0.19 *5*
Y L_γ	2.268	0.0057 *16*
Sr $K_{\alpha2}$	14.098	0.145 *16*
Sr $K_{\alpha1}$	14.165	0.28 *3*
Y $K_{\alpha2}$	14.883	3.55 *15*
Y $K_{\alpha1}$	14.958	6.8 *3*
Sr $K_{\beta1}$'	15.832	0.061 *7*
Sr $K_{\beta2}$'	16.192	0.0084 *9*
Y $K_{\beta1}$'	16.735	1.52 *6*
Y $K_{\beta2}$'	17.125	0.224 *10*
γ_{IT} M4	381.4 *6*	78.05

† 0.16% uncert(syst)

Atomic Electrons (^{87}Y)

$\langle$e$\rangle$=74.8 *13* keV

e_{bin}(keV)	$\langle$e$\rangle$(keV)	e(%)
2	0.32	15.2 *16*
12	0.062	0.51 *6*
13	0.41	3.2 *3*
14	0.053	0.37 *4*
15	0.137	0.93 *10*
16	0.0120	0.074 *8*
17	0.0044	0.026 *3*
364	61.9	17.0 *3*
379	9.88	2.61 *5*
381	2.05	0.538 *11*

Continuous Radiation (^{87}Y)

$\langle\beta+\rangle$=4.00 keV;$\langle$IB$\rangle$=0.028 keV

E_{bin}(keV)		$\langle\ \rangle$(keV)	(%)
0 - 10	β+	6.1×10^{-7}	7.6×10^{-6}
	IB	0.00018	
10 - 20	β+	1.57×10^{-5}	9.6×10^{-5}
	IB	0.00027	0.0019
20 - 40	β+	0.000284	0.00089
	IB	0.00034	0.00118
40 - 100	β+	0.0086	0.0114
	IB	0.00091	0.00141
100 - 300	β+	0.277	0.129
	IB	0.0024	0.00136
300 - 600	β+	1.41	0.314
	IB	0.0036	0.00081
600 - 1300	β+	2.30	0.295
	IB	0.0121	0.00131
1300 - 2242	IB	0.0079	0.00051
	Σβ+		0.75

$^{87}_{40}$Zr(1.733 *8* h)

Mode: ϵ

Δ: -79348 *8* keV

SpA: 2.0780×10^{7} Ci/g

Prod: ^{89}Y(p,3n); ^{86}Sr(^{3}He,2n); ^{86}Sr(α,3n); protons on Y; protons on Nb

Photons (^{87}Zr)

γ_{mode}	γ(keV)	γ(rel)
γ	202.3 *6*	<2 ?
γ	252.0 *8* ?	
γ [E1]	359.4 *12* ?	
γ M4	381.4 *6*	*
γ	611.9 *7*	4 *1*
γ	634.7 *7*	~2
γ [E2]	773.2 *7*	8 *1*
γ [E2]	795.2 *9*	10 *2*
γ	797.6 *7*	6 *2*
γ	837.0 *7*	~2
γ	922.4 *9*	
γ	974.1 *8*	4 *1*
γ	1025.2 *5*	28 *2*
γ	1049.6 *8*	~2
γ	1148.4 *8*	
γ	1159.8 *7*	8 *1*
γ	1204.6 *10*	5 *1*
γ	1210.9 *7*	33 *2*
γ	1217.1 *10*	3 *1*
γ [M1+E2]	1227.5 *6*	100 *2*
γ	1388.7 *9*	3 *1*
γ	1400.4 *10*	4 *1*
γ	1406.9 *11*	~1
γ	1590.9 *9*	
γ	1656.7 *11*	3 *1*
γ	1693.8 *10*	~1
γ	1809.2 *12*	3 *1*
γ	1822.8 *7*	3 *1*
γ	1858.9 *11*	~2
γ	1862.2 *8*	~1
γ	2173.6 *9*	~2
γ	2185.0 *8*	3 *1*
γ	2223.0 *12*	9 *2*
γ	2616.2 *8*	~3
γ	2884.2 *11*	~2

* with ^{87}Y(12.9 h)

$^{87}_{40}$Zr(14.0 *2* s)

Mode: IT

Δ: -79012 *8* keV

SpA: 9.04×10^{9} Ci/g

Prod: ^{89}Y(p,3n); ^{90}Zr(p,p3n)

Photons (^{87}Zr)

⟨γ⟩=238 *3* keV

γ_{mode}	γ(keV)	γ(%)†
Zr L_{ℓ}	1.792	0.056 *10*
Zr L_{η}	1.876	0.031 *5*
Zr L_{α}	2.042	1.32 *21*
Zr L_{β}	2.133	0.76 *13*
Zr L_{γ}	2.370	0.034 *7*
Zr $K_{\alpha2}$	15.691	12.1 *6*
Zr $K_{\alpha1}$	15.775	23.3 *12*
Zr $K_{\beta1}'$	17.663	5.3 *3*
Zr $K_{\beta2}'$	18.086	0.86 *5*
γ (E3)	135.1 *2*	27.4 *6*
γ [M1]	201.31 *19*	96.3

† 1.5% uncert(syst)

Atomic Electrons (^{87}Zr)

⟨e⟩=98 *3* keV

e_{bin}(keV)	⟨e⟩(keV)	e(%)
2	1.22	54 *6*
3	0.062	2.5 *3*
13	1.17	8.8 *10*
14	0.32	2.34 *25*
15	0.51	3.3 *4*
16	0.095	0.60 *7*
17	0.042	0.24 *3*
18	0.0113	0.064 *7*
117	63.3	54.1 *24*
133	21.2	16.0 *7*
135	4.49	3.33 *15*
183	4.82	2.63 *11*
199	0.60	0.302 *13*
201	0.126	0.062 *3*

$^{87}_{41}$Nb(2.60 *7* min)

Mode: ϵ

Δ: -74180 *60* keV

SpA: 8.29×10^{8} Ci/g

Prod: ^{90}Zr(p,4n); protons on Ag; ^{89}Y(α,6n); ^{58}Ni(^{32}S,3p); ^{59}Co(^{32}S,2p2n)

Photons (^{87}Nb)

γ_{mode}	γ(keV)	γ(rel)
γ [M1]	201.31 *19*	100 *4*
γ	269.1 *10*	6.0 *12*
γ	470.71 *20*	73.6 *17*
γ	600.3 *3*	8.7 *11*
γ	616.69 *19*	29 *3*
γ	801.6 *4*	5.3 *22*
γ	887.1 *6*	7.2 *22*
γ	983.1 *15*	4.8 *14*
γ	1066.8 *4*	26 *6*
γ	1083.2 *4*	3 *1*
γ	1168.1 *15*	3.6 *11*
γ	1285.1 *14*	4.0 *12*
γ	1559.1 *15*	3.0 *9*
γ	1683.5 *3*	11 *5*
γ	1884.8 *4*	32 *4*
γ	2153.4 *7*	3.8 *11*

$^{87}_{41}$Nb(3.82 *9* min)

Mode: ϵ

Δ: -74180 *60* keV

SpA: 5.65×10^{8} Ci/g

Prod: ^{90}Zr(p,4n); protons on Ag; ^{89}Y(α,6n); ^{58}Ni(^{32}S,3p); ^{59}Co(^{32}S,2p2n)

Photons (^{87}Nb)

⟨γ⟩=253.6 *23* keV

γ_{mode}	γ(keV)	γ(%)†
Zr L_{ℓ}	1.792	0.062 *11*
Zr L_{η}	1.876	0.035 *6*
Zr L_{α}	2.042	1.49 *24*
Zr L_{β}	2.133	0.85 *14*
Zr L_{γ}	2.371	0.038 *7*

Photons (^{87}Nb) (continued)

γ_{mode}	γ(keV)	γ(%)†
Zr $K_{\alpha2}$	15.691	13.7 *7*
Zr $K_{\alpha1}$	15.775	26.3 *13*
Zr $K_{\beta1}'$	17.663	6.0 *3*
Zr $K_{\beta2}'$	18.086	0.97 *5*
γ (E3)	135.1 *2*	29.2 *
γ [M1]	201.31 *19*	102.6 *

† 1.0% uncert(syst)

* with ^{87}Zr(14.0 s) in equilib

Atomic Electrons (^{87}Nb)

⟨e⟩=104 *3* keV

e_{bin}(keV)	⟨e⟩(keV)	e(%)
2	1.36	60 *7*
3	0.071	2.8 *3*
13	1.32	9.9 *11*
14	0.36	2.6 *3*
15	0.57	3.8 *4*
16	0.107	0.68 *7*
17	0.047	0.27 *3*
18	0.0127	0.072 *8*
117	67.4	57.5 *24*
133	22.6	17.0 *7*
135	4.77	3.54 *15*
183	5.14	2.80 *12*
199	0.64	0.321 *13*
201	0.134	0.067 *3*

Continuous Radiation (^{87}Nb)

⟨β+⟩=1664 keV;⟨IB⟩=6.4 keV

E_{bin}(keV)		⟨ ⟩(keV)	(%)
0 - 10	β+	3.89×10^{-6}	4.76×10^{-5}
	IB	0.052	
10 - 20	β+	0.000107	0.00065
	IB	0.052	0.36
20 - 40	β+	0.00208	0.0065
	IB	0.101	0.35
40 - 100	β+	0.071	0.092
	IB	0.29	0.45
100 - 300	β+	2.90	1.32
	IB	0.87	0.49
300 - 600	β+	23.1	4.94
	IB	1.07	0.25
600 - 1300	β+	227	23.0
	IB	1.8	0.20
1300 - 2500	β+	919	49.0
	IB	1.57	0.089
2500 - 4834	β+	492	17.2
	IB	0.69	0.022
	Σβ+		96

$^{87}_{42}$Mo(15 *2* s)

Mode: ϵ

Δ: -67440 *310* keV

SpA: 8.4×10^{9} Ci/g

Prod: ^{58}Ni(^{32}S,2pn)

Photons (^{87}Mo)

γ_{mode}	γ(keV)	γ(rel)
γ	262.5 *4*	100 *9*
γ	397.0 *6*	33 *6*

A = 88

NDS **18**, 87 (1976)

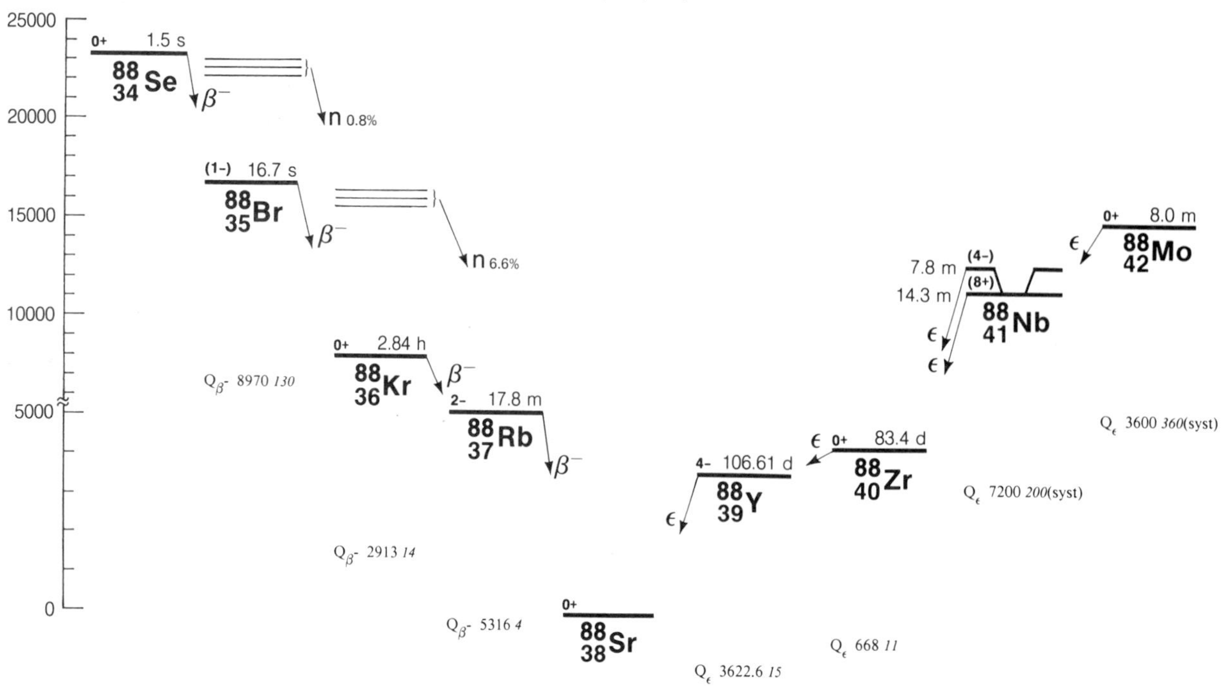

$^{88}_{34}$Se(1.53 *6* s)

Mode: β-, β-n(0.75 *6* %)
SpA: 6.7×10^{10} Ci/g
Prod: fission

Photons (^{88}Se)

γ_{mode}	γ(keV)	γ(rel)
γ	113.5 *3*	8.8 *18*
γ	159.2 *3*	100 *20*
γ	249.5 *3*	29 *6*
γ	259.2 *3*	82 *16*
γ	272.7 *3*	41 *8*
γ	293.3 *3*	17 *3*
γ	408.7 *3*	34 *7*
γ	566.0 *3*	16 *3*
γ	1495.0 *3*	33 *7*
γ	1644.5 *3*	58 *12*
γ	1744.5 *3*	62 *12*
γ	1903.7 *3*	64 *13*
γ	2894.8 *3*	37 *8*

$^{88}_{35}$Br(16.7 *2* s)

Mode: β-, β-n(6.6 *4* %)
Δ: -70720 *130* keV
SpA: 7.52×10^{9} Ci/g
Prod: fission
β-n: 127, 160, 205, 235, 260?
390, 455?, 540, 615?, 670?
β-n(avg): 330 *30*

Photons (^{88}Br)

⟨γ⟩=3053 *81* keV

γ_{mode}	γ(keV)	γ(%)†
$\gamma_{\beta-}$	120.75 *22*	~0.15 ?
$\gamma_{\beta-}$	288.83 *12*	~0.15
$\gamma_{\beta-}$	459.95 *12*	0.69 *8*
$\gamma_{\beta-n}$ [E2]	531.94 *6*	0.92
$\gamma_{\beta-}$	682.25 *24*	~0.23
$\gamma_{\beta-}$	697.98 *11*	~0.23
$\gamma_{\beta-}$	764.82 *10*	~0.8
$\gamma_{\beta-}$[E2]	775.19 *8*	77.0 *15*
$\gamma_{\beta-}$	793.6 *4*	1.2 *4* ?
$\gamma_{\beta-}$[M1+E2]	802.09 *9*	15.8 *8*
$\gamma_{\beta-}$	868.92 *9*	4.1 *4*
$\gamma_{\beta-}$	1053.64 *9*	1.77 *15*
$\gamma_{\beta-}$[M1+E2]	1073.76 *10*	1.31 *15*
$\gamma_{\beta-}$	1285.26 *20*	0.54 *23*
$\gamma_{\beta-}$[M1+E2]	1352.09 *19*	0.69 *15*
$\gamma_{\beta-}$	1368.42 *23*	~1
$\gamma_{\beta-}$	1428.58 *15*	~0.31
$\gamma_{\beta-}$[M1+E2]	1440.64 *8*	5.0 *4*
$\gamma_{\beta-}$	1467.87 *19*	~0.31 ?
$\gamma_{\beta-}$	1480.6 *7*	~0.31
$\gamma_{\beta-}$	1566.90 *8*	2.3 *4*
$\gamma_{\beta-}$[E2]	1577.27 *12*	3.5 *4*
$\gamma_{\beta-}$	1644.11 *8*	3.1 *4*
$\gamma_{\beta-}$	1691.9 *4*	0.46 *15*
$\gamma_{\beta-}$	1855.72 *11*	0.85 *8*
$\gamma_{\beta-}$[M1+E2]	1875.84 *13*	0.39 *15*
$\gamma_{\beta-}$	2052.72 *12*	0.54 *15*
$\gamma_{\beta-}$	2061.1 *5*	~0.31
$\gamma_{\beta-}$[M1+E2]	2154.17 *18*	0.46 *15*
$\gamma_{\beta-}$[E2]	2215.82 *11*	0.54 *15*
$\gamma_{\beta-}$	2269.94 *18*	0.85 *15* ?
$\gamma_{\beta-}$	2287.97 *19*	0.54 *15* ?
$\gamma_{\beta-}$	2428.2 *4*	0.77 *15* ?
$\gamma_{\beta-}$	2491.53 *12*	~0.31
$\gamma_{\beta-}$	2503.60 *12*	0.39 *15*
$\gamma_{\beta-}$	2523.0 *4*	0.39 *15*
$\gamma_{\beta-}$	2624.44 *9*	1.77 *23*
$\gamma_{\beta-}$	2828.3 *3*	0.54 *15*
$\gamma_{\beta-}$	2875.23 *10*	2.00 *23* ?
$\gamma_{\beta-}$	2933.95 *24*	1.9 *4*
$\gamma_{\beta-}$	2945.67 *20*	2.39 *23*

Photons (^{88}Br) (continued)

γ_{mode}	γ(keV)	γ(%)†
$\gamma_{\beta-}$	3019.39 *11*	1.15 *15* ?
$\gamma_{\beta-}$	3187.15 *12*	2.16 *23*
$\gamma_{\beta-}$	3278.77 *10*	2.85 *23*
$\gamma_{\beta-}$	3400.0 *3*	0.92 *23*
$\gamma_{\beta-}$	3493.33 *11*	1.46 *23*
$\gamma_{\beta-}$	3932.13 *10*	5.0 *4*
$\gamma_{\beta-}$	4021.60 *12*	2.62 *23*
$\gamma_{\beta-}$	4147.86 *9*	4.2 *12*
$\gamma_{\beta-}$	4254.9 *3*	0.92 *23* ?
$\gamma_{\beta-}$	4310.9 *5*	0.77 *23* ?
$\gamma_{\beta-}$	4492.35 *24*	0.85 *23*
$\gamma_{\beta-}$	4562.82 *19*	3.2 *4* ?
$\gamma_{\beta-}$	4663.44 *9*	0.85 *23* ?
$\gamma_{\beta-}$	4713.5 *5*	1.08 *23* ?
$\gamma_{\beta-}$	4721.1 *6*	1.7 *4*
$\gamma_{\beta-}$	4849.8 *12*	0.69 *23*
$\gamma_{\beta-}$	4985.2 *6*	1.77 *23*
$\gamma_{\beta-}$	5019.7 *3*	2.08 *23* ?
$\gamma_{\beta-}$	5195.7 *6*	0.77 *23*
$\gamma_{\beta-}$	5212.9 *12*	0.54 *23*
$\gamma_{\beta-}$	5297.6 *5*	0.54 *23* ?
$\gamma_{\beta-}$	5478.2 *5*	~0.39 ?
$\gamma_{\beta-}$	5762.4 *12*	~0.31
$\gamma_{\beta-}$	5981.6 *12*	~0.46
$\gamma_{\beta-}$	6996.9 *24*	<0.39

† uncert(syst): 2.6% for β-, 8.8% for β-n

Continuous Radiation (^{88}Br)

⟨β-⟩=2784 keV;⟨IB⟩=12.2 keV

E_{bin}(keV)		⟨ ⟩(keV)	(%)
0 - 10	β-	0.00507	0.101
	IB	0.067	
10 - 20	β-	0.0155	0.103
	IB	0.066	0.46
20 - 40	β-	0.064	0.213
	IB	0.131	0.46

Continuous Radiation (^{88}Br)
(continued)

E_{bin}(keV)		⟨ ⟩(keV)	(%)
40 - 100	β-	0.496	0.70
	IB	0.38	0.59
100 - 300	β-	6.1	2.98
	IB	1.19	0.66
300 - 600	β-	25.2	5.6
	IB	1.56	0.36
600 - 1300	β-	136	14.2
	IB	2.8	0.32
1300 - 2500	β-	476	25.3
	IB	3.1	0.17
2500 - 5000	β-	1366	37.8
	IB	2.5	0.076
5000 - 7500	β-	751	12.8
	IB	0.31	0.0055
7500 - 8970	β-	23.1	0.297
	IB	0.00140	1.8×10^{-5}

$^{88}_{36}$Kr(2.84 *3* h)

Mode: β-
Δ: -79687 *14* keV
SpA: 1.254×10^{7} Ci/g
Prod: fission

Photons (^{88}Kr)
⟨γ⟩=1955 *19* keV

γ_{mode}	γ(keV)	γ(%)†
Rb L_ℓ	1.482	0.0094 *25*
Rb L_η	1.542	0.0055 *16*
Rb L_α	1.694	0.21 *5*
Rb L_β	1.759	0.12 *4*
Rb L_γ	2.051	0.0018 *6*
Rb $K_{\alpha2}$	13.336	2.5 *3*
Rb $K_{\alpha1}$	13.395	4.8 *5*
Rb $K_{\beta1}'$	14.959	1.03 *11*
Rb $K_{\beta2}'$	15.286	0.109 *11*
γ M1	27.530 *11*	1.94 *14*
γ [M1+E2]	28.310 *15*	0.028 *10*
γ [E2]	122.30 *4*	0.197 *7*
γ M1+E2	165.924 *19*	3.10 *6*
γ [E2]	168.809 *19*	<0.007
γ	176.74 *22*	0.024 *7*
γ [M1+E2]	196.339 *17*	26.0 *5*
γ [M1+E2]	240.74 *3*	0.253 *7*
γ	311.72 *3*	0.107 *7*
γ	334.733 *14*	0.145 *7*
γ	350.23 *3*	0.017 *7*
γ	362.263 *11*	2.25 *6*
γ [M1+E2]	363.043 *14*	~0.05
γ [M1+E2]	390.572 *10*	0.64 *4*
γ	391.23 *4*	~0.08
γ	421.69 *4*	0.128 *24*
γ	471.818 *14*	0.727 *15*
γ	500.128 *14*	0.097 *7*
γ	517.05 *3*	0.035 *10*
γ	570.63 *5*	0.062 *7*
γ	573.36 *3*	0.073 *7*
γ	579.18 *4*	0.024 *10*
γ	603.24 *17*	0.042 *10*
γ	666.051 *19*	0.087 *14*
γ	677.40 *3*	0.235 *14*
γ	730.99 *5*	0.035 *10*
γ	741.30 *15*	0.035 *10*
γ	774.16 *4*	0.097 *14*
γ	779.14 *3*	0.097 *21*
γ	788.32 *3*	0.533 *14*
γ	790.24 *4*	0.125 *10*
γ	798.79 *5*	0.028 *10*
γ	822.04 *3*	0.090 *10*
γ	834.859 *3*	13.0 *3*

Photons (^{88}Kr)
(continued)

γ_{mode}	γ(keV)	γ(%)†
γ	850.353 *25*	0.173 *10*
γ	862.388 *11*	0.671 *17*
γ	879.29 *3*	0.024 *7*
γ	883.08 *12*	0.042 *7*
γ	945.07 *3*	0.294 *14*
γ	950.59 *4*	0.038 *10*
γ	961.95 *3*	0.083 *10*
γ	985.813 *16*	1.31 *3*
γ	990.26 *3*	0.142 *17*
γ	1039.636 *23*	0.484 *17*
γ	1049.656 *23*	0.142 *10*
γ	1054.6 *3*	0.031 *10*
γ	1090.40 *3*	0.062 *14*
γ	1141.40 *3*	1.28 *3*
γ	1179.543 *24*	0.996 *21*
γ	1185.08 *3*	0.69 *3*
γ	1210.006 *25*	0.142 *24*
γ	1212.612 *25*	0.14 *5*
γ	1245.23 *4*	0.363 *17*
γ	1250.75 *3*	1.121 *22*
γ	1298.91 *5*	0.093 *21*
γ	1303.11 *12*	0.066 *24*
γ	1324.989 *24*	0.16 *4*
γ	1335.84 *3*	0.066 *10*
γ	1352.519 *25*	0.159 *21*
γ	1369.416 *19*	1.48 *6*
γ	1407.05 *4*	0.218 *17*
γ	1464.83 *5*	0.114 *14*
γ	1518.417 *24*	2.15 *6*
γ	1529.765 *16*	10.93 *22*
γ	1603.84 *3*	0.46 *3*
γ	1608.0 *3*	0.069 *17*
γ	1661.17 *5*	0.090 *21*
γ	1686.06 *3*	0.66 *7*
γ	1789.2 *3*	0.045 *17*
γ	1793.3 *3*	0.035 *14*
γ	1801.3 *4*	0.038 *14*
γ	1892.79 *13*	0.138 *24*
γ	1908.71 *5*	0.100 *14*
γ	2029.883 *16*	4.53 *9*
γ [E1]	2035.456 *18*	3.74 *10*
γ	2186.18 *3*	0.29 *6*
γ [E1]	2195.804 *20*	13.2 *3*
γ [E1]	2231.790 *17*	3.39 *7*
γ [M1+E2]	2259.71 *16*	0.031 *14*
γ [E1]	2352.097 *25*	0.730 *21*
γ	2364.608 *16*	0.031 *14*
γ [E1]	2392.137 *15*	34.6 *7*
γ	2408.82 *5*	0.104 *10*
γ	2535.55 *14*	0.042 *4*
γ [E1]	2548.430 *24*	0.623 *12*
γ [E1]	2771.08 *5*	0.149 *7*

† 4.6% uncert(syst)

Atomic Electrons (^{88}Kr)
⟨e⟩=6.0 *12* keV

e_{bin}(keV)	⟨e⟩(keV)	e(%)
2	0.20	11.1 *19*
11	0.132	1.17 *17*
12	1.47	12 *1*
13	0.24	1.8 *9*
15	0.0124	0.084 *12*
25	0.275	1.08 *8*
26 - 28	0.24	0.9 *4*
107 - 122	0.100	0.092 *4*
151	0.38	~0.25
154 - 177	0.07	~0.041
181	2.2	~1
194	0.26	~0.13
195 - 241	0.10	0.050 *22*
297 - 335	0.007	0.0023 *9*
347 - 391	0.06	0.018 *7*
406 - 422	0.0020	~0.00048
457 - 502	0.010	0.0022 *9*
515 - 564	0.0012	0.00022 *7*
569 - 603	0.00049	8 *3* $\times 10^{-5}$

Atomic Electrons (^{88}Kr)
(continued)

e_{bin}(keV)	⟨e⟩(keV)	e(%)
651 - 677	0.0021	0.00032 *11*
716 - 764	0.0013	0.00017 *5*
772 - 820	0.059	0.007 *3*
822 - 868	0.011	0.0014 *4*
877 - 883	3.4×10^{-5}	3.9 *11* $\times 10^{-6}$
930 - 975	0.0067	0.00069 *19*
984 - 1024	0.0022	0.00022 *6*
1034 - 1075	0.00101	9.7 *21* $\times 10^{-5}$
1088 - 1126	0.0038	0.00033 *12*
1139 - 1185	0.0058	0.00050 *11*
1195 - 1243	0.0049	0.00040 *9*
1245 - 1288	0.00081	6.4 *14* $\times 10^{-5}$
1297 - 1337	0.00108	8.1 *17* $\times 10^{-5}$
1350 - 1392	0.0046	0.00034 *9*
1405 - 1450	0.00033	2.3 *7* $\times 10^{-5}$
1463 - 1530	0.033	0.0021 *6*
1589 - 1686	0.0030	0.00018 *4*
1774 - 1801	0.00026	1.4 *4* $\times 10^{-5}$
1878 - 1908	0.00050	2.6 *6* $\times 10^{-5}$
2015 - 2035	0.0143	0.00071 *12*
2171 - 2259	0.0231	0.00105 *4*
2337 - 2409	0.0464	0.00195 *8*
2520 - 2548	0.00086	3.40 *15* $\times 10^{-5}$
2756 - 2771	0.000183	6.6 *4* $\times 10^{-6}$

Continuous Radiation (^{88}Kr)
⟨β-⟩=359 keV;⟨IB⟩=0.63 keV

E_{bin}(keV)		⟨ ⟩(keV)	(%)
0 - 10	β-	0.141	2.82
	IB	0.0149	
10 - 20	β-	0.421	2.81
	IB	0.0142	0.099
20 - 40	β-	1.67	5.6
	IB	0.027	0.092
40 - 100	β-	11.3	16.2
	IB	0.067	0.105
100 - 300	β-	80	42.0
	IB	0.149	0.086
300 - 600	β-	61	15.7
	IB	0.134	0.032
600 - 1300	β-	70	7.6
	IB	0.163	0.019
1300 - 2500	β-	126	7.1
	IB	0.060	0.0038
2500 - 2913	β-	7.9	0.301
	IB	0.00026	1.02×10^{-5}

$^{88}_{37}$Rb(17.8 *1* min)

Mode: β-
Δ: -82600 *5* keV
SpA: 1.200×10^{8} Ci/g
Prod: ^{87}Rb(n,γ); fission; daughter ^{88}Kr

Photons (^{88}Rb)
⟨γ⟩=629 *8* keV

γ_{mode}	γ(keV)	γ(%)†
γ [M1+E2]	338.93 *9*	0.060 *6*
γ [M1+E2]	416.18 *13*	0.0036 *13*
γ [E1]	439.08 *10*	0.014 *3*
γ [E1]	484.41 *4*	0.028 *6*
γ [E1]	898.065 *17*	14.1 *3*
γ [E1]	1027.43 *8*	0.011 *4*

Photons (^{88}Rb)
(continued)

γ_{mode}	γ(keV)	γ(%)†
γ [E1]	1218.22 13	0.051 6
γ [E1]	1366.36 9	0.103 13
γ [M1+E2]	1382.47 4	0.742 25
γ [E1]	1679.72 6	0.045 9
γ [M1+E2]	1779.87 4	0.216 13
γ [M1+E2]	1798.65 13	0.062 13
γ [E2]	1836.077 18	21.4 4
γ [M1+E2]	2111.40 6	0.118 11
γ [M1+E2]	2118.80 9	0.051 6
γ	2388.1 6	0.028 9
γ [M1+E2]	2577.77 6	0.180 9
γ [E1]	2677.92 4	1.96 4
γ	2734.121 23	0.102 4
γ [E1]	3009.44 6	0.244 9
γ [E1]	3016.84 9	~0.0043
γ [E2]	3218.52 4	0.214 6
γ [M1]	3486.55 8	0.131 4
γ	3524.03 6	~0.006
γ [E2]	4035.6 4	0.0107 21
γ [E1]	4742.75 11	0.143 6
γ	4852.85 9	0.0090 13

† 5.6% uncert(syst)

Continuous Radiation (^{88}Rb)
$\langle\beta-\rangle$=2068 keV;$\langle$IB$\rangle$=7.6 keV

E_{bin}(keV)		$\langle\ \rangle$(keV)	(%)
0 - 10	β-	0.0082	0.162
	IB	0.057	
10 - 20	β-	0.0247	0.165
	IB	0.057	0.39
20 - 40	β-	0.101	0.335
	IB	0.112	0.39
40 - 100	β-	0.75	1.06
	IB	0.33	0.50
100 - 300	β-	8.1	3.99
	IB	0.99	0.55
300 - 600	β-	29.8	6.6
	IB	1.25	0.29
600 - 1300	β-	184	19.1
	IB	2.1	0.24
1300 - 2500	β-	610	32.6
	IB	1.9	0.109
2500 - 5000	β-	1227	35.8
	IB	0.80	0.026
5000 - 5316	β-	8.8	0.174
	IB	0.000112	2.3×10^{-6}

$^{88}_{38}$Sr(stable)

Δ: -87916 3 keV
%: 82.58 1

$^{88}_{39}$Y (106.61 2 d)

Mode: ϵ
Δ: -84294 3 keV
SpA: 1.3918×10^4 Ci/g
Prod: ^{88}Sr(p,n); ^{88}Sr(d,2n)

Photons (^{88}Y)
$\langle\gamma\rangle$=2684 16 keV

γ_{mode}	γ(keV)	γ(%)†
Sr L_ℓ	1.582	0.067 15
Sr L_η	1.649	0.038 10
Sr L_α	1.806	1.5 3
Sr L_β	1.880	0.87 22
Sr L_γ	2.196	0.015 5
Sr $K_{\alpha2}$	14.098	17.7 7
Sr $K_{\alpha1}$	14.165	34.0 14
Sr $K_{\beta1}'$	15.832	7.4 3
Sr $K_{\beta2}'$	16.192	1.02 4
γ [E1]	484.41 4	0.0010 3
γ	850.6 8	0.066 13
γ [E1]	898.065 17	92.7 17
γ [M1+E2]	1382.47 4	0.025 4
γ [E2]	1836.077 18	99.35
γ	2734.121 23	0.666 25
γ [E2]	3218.52 4	0.0073 12

†<0.1% uncert(syst)

Atomic Electrons (^{88}Y)
$\langle$e$\rangle$=5.4 3 keV

e_{bin}(keV)	$\langle$e$\rangle$(keV)	e(%)
2	1.45	73 8
12	2.37	19.7 20
14	0.93	6.8 7
15	0.025	0.162 17
16	0.058	0.37 4
468	1.1×10^{-6}	$\sim 2 \times 10^{-7}$
482	1.2×10^{-7}	$\sim 3 \times 10^{-8}$
484	2.4×10^{-8}	$\sim 5 \times 10^{-9}$
835	0.00029	3.5 16 $\times 10^{-5}$
848	3.1×10^{-5}	3.6 17 $\times 10^{-6}$
849	1.6×10^{-6}	$\sim 2 \times 10^{-7}$
850	5.5×10^{-6}	7 3 $\times 10^{-7}$
851	8.0×10^{-7}	9 5 $\times 10^{-8}$
882	0.224	0.0254 11
896	0.0244	0.00272 12
898	0.00477	0.000531 23
1366	8.8×10^{-5}	6.4 12 $\times 10^{-6}$
1380	9.4×10^{-6}	6.8 12 $\times 10^{-7}$
1381	1.1×10^{-7}	8 3 $\times 10^{-9}$
1382	1.9×10^{-6}	1.35 25 $\times 10^{-7}$
1820	0.264	0.0145 6
1834	0.0286	0.00156 6
1836	0.00488	0.000266 11
2718	0.0010	3.8 10 $\times 10^{-5}$
2732	0.00011	4.1 11 $\times 10^{-6}$
2734	1.9×10^{-5}	7.0 19 $\times 10^{-7}$
3202	1.27×10^{-5}	4.0 7 $\times 10^{-7}$
3216	1.34×10^{-6}	4.2 7 $\times 10^{-8}$
3217	4.1×10^{-8}	1.29 23 $\times 10^{-9}$
3218	2.3×10^{-7}	7.3 13 $\times 10^{-9}$

Continuous Radiation (^{88}Y)
$\langle\beta+\rangle$=0.79 keV;$\langle$IB$\rangle$=0.20 keV

E_{bin}(keV)		$\langle\ \rangle$(keV)	(%)
0 - 10	β+	4.89×10^{-7}	6.0×10^{-6}
	IB	0.00020	
10 - 20	β+	1.23×10^{-5}	7.5×10^{-5}
	IB	0.0114	0.080
20 - 40	β+	0.000216	0.00067
	IB	0.00071	0.0026
40 - 100	β+	0.0060	0.0080
	IB	0.0021	0.0030
100 - 300	β+	0.156	0.074
	IB	0.031	0.0149
300 - 600	β+	0.53	0.121
	IB	0.094	0.022

Continuous Radiation (^{88}Y)
(continued)

E_{bin}(keV)		$\langle\ \rangle$(keV)	(%)
600 - 1300	β+	0.095	0.0148
	IB	0.057	0.0073
1300 - 1787	IB	0.0071	0.00050
	$\Sigma\beta$+		0.22

$^{88}_{40}$Zr(83.4 3 d)

Mode: ϵ
Δ: -83626 10 keV
SpA: 1.779×10^4 Ci/g
Prod: protons on Nb

Photons (^{88}Zr)
$\langle\gamma\rangle$=391.9 5 keV

γ_{mode}	γ(keV)	γ(%)†
Y L_ℓ	1.686	0.071 16
Y L_η	1.762	0.040 10
Y L_α	1.922	1.7 3
Y L_β	2.005	0.96 25
Y L_γ	2.270	0.030 9
Y $K_{\alpha2}$	14.883	18.5 8
Y $K_{\alpha1}$	14.958	35.5 15
Y $K_{\beta1}'$	16.735	7.9 3
Y $K_{\beta2}'$	17.125	1.17 5
γ [E3]	392.9 1	97.30

† 0.10% uncert(syst)

Atomic Electrons (^{88}Zr)
$\langle$e$\rangle$=15.3 5 keV

e_{bin}(keV)	$\langle$e$\rangle$(keV)	e(%)
2	1.57	74 8
12	0.220	1.79 18
13	2.15	16.9 17
14	0.236	1.64 17
15	0.71	4.8 5
16	0.060	0.37 4
17	0.0228	0.137 14
376	8.8	2.33 9
391	1.28	0.327 13
393	0.260	0.066 3

Continuous Radiation (^{88}Zr)
$\langle$IB$\rangle$=0.018 keV

E_{bin}(keV)		$\langle\ \rangle$(keV)	(%)
10 - 20	IB	0.0129	0.086
20 - 40	IB	0.00073	0.0028
40 - 100	IB	0.00125	0.0019
100 - 275	IB	0.0026	0.0018

$^{88}_{41}$Nb(14.3 *3* min)

Mode: ε
Δ: -76430 *200* keV syst
SpA: 1.49×10^{8} Ci/g
Prod: ^{79}Br(^{12}C,3n); deuterons on Zr; alphas on Y; ^{32}S on ^{60}Ni; ^{32}S on ^{59}Co; ^{90}Zr(p,3n)

Photons (^{88}Nb)

⟨γ⟩=3195 *65* keV

γ_{mode}	γ(keV)	γ(%)†
Zr L_ℓ	1.792	0.058 *11*
Zr L_η	1.876	0.031 *5*
Zr L_α	2.042	1.39 *23*
Zr L_β	2.134	0.77 *13*
Zr L_γ	2.375	0.035 *7*
Zr $K_{\alpha2}$	15.691	13.5 *10*
Zr $K_{\alpha1}$	15.775	25.9 *18*
Zr $K_{\beta1}$'	17.663	5.9 *4*
Zr $K_{\beta2}$'	18.086	0.95 *7*
γ (E2)	76.8 *10*	23.5 *14*
γ	271.8 *7*	29.4 *15*
γ	399.61 *24*	31 *3*
γ	502.9 *10*	80 *6*
γ [E2]	671.4 *7*	66.3 *21*
γ [E2]	1057.0 *3*	99.94
γ [E2]	1082.6 *3*	98 *5*

†<0.1% uncert(syst)

Atomic Electrons (^{88}Nb)

⟨e⟩=51.1 *25* keV

e_{bin}(keV)	⟨e⟩(keV)	e(%)
2 - 18	3.7	75 *7*
59	31.8	54 *4*
74	5.8	7.8 *6*
75	3.04	4.1 *3*
76	1.61	2.10 *15*
254 - 272	1.4	~0.5
382 - 400	0.6	~0.17
485 - 503	1.1	0.22 *10*
653 - 671	0.79	0.121 *6*
1039 - 1083	1.21	0.115 *4*

Continuous Radiation (^{88}Nb)

⟨β+⟩=1181 keV;⟨IB⟩=4.3 keV

E_{bin}(keV)		⟨ ⟩(keV)	(%)
0 - 10	β+	7.7×10^{-6}	9.4×10^{-5}
	IB	0.041	
10 - 20	β+	0.000210	0.00128
	IB	0.042	0.29
20 - 40	β+	0.00409	0.0127
	IB	0.080	0.28
40 - 100	β+	0.138	0.180
	IB	0.23	0.35
100 - 300	β+	5.5	2.49
	IB	0.66	0.37
300 - 600	β+	41.2	8.8
	IB	0.76	0.18
600 - 1300	β+	341	35.2
	IB	1.18	0.135
1300 - 2500	β+	750	42.6
	IB	1.02	0.059
2500 - 4313	β+	43.5	1.63
	IB	0.32	0.0113
	Σβ+		91

$^{88}_{41}$Nb(7.8 *2* min)

Mode: ε
Δ: -76430 *200* keV syst
SpA: 2.74×10^{8} Ci/g
Prod: ^{90}Zr(p,3n); ^{32}S on ^{60}Ni; ^{32}S on ^{59}Co

Photons (^{88}Nb)

⟨γ⟩=2825 *78* keV

γ_{mode}	γ(keV)	γ(%)†
Zr L_ℓ	1.792	0.0049 *9*
Zr L_η	1.876	0.0026 *4*
Zr L_α	2.042	0.116 *19*
Zr L_β	2.135	0.065 *11*
Zr L_γ	2.385	0.0031 *6*
Zr $K_{\alpha2}$	15.691	1.22 *7*
Zr $K_{\alpha1}$	15.775	2.33 *13*
Zr $K_{\beta1}$'	17.663	0.53 *3*
Zr $K_{\beta2}$'	18.086	0.086 *5*
γ	262.4 *4*	10.2 *5*
γ	399.61 *24*	44 *3*
γ	451.1 *3*	25 *1*
γ [M1+E2]	534.2 *3*	13.7 *7*
γ [E1]	638.3 *3*	25.9 *11*
γ [E1]	662.0 *4*	4.3 *6*
γ [M1+E2]	760.83 *24*	16.9 *7*
γ [E1]	918.5 *5*	11.3 *9* ?
γ [E2]	1057.0 *3*	93 *5*
γ [E2]	1082.6 *3*	59 *5*
γ [E1]	1399.1 *3*	5.9 *9*
γ [E2]	1817.8 *3*	10.2 *7*
γ	1975.5 *8*	6.0 *7* ?

† 3.2% uncert(syst)

Atomic Electrons (^{88}Nb)

⟨e⟩=3.7 *6* keV

e_{bin}(keV)	⟨e⟩(keV)	e(%)
2 - 3	0.107	4.7 *5*
13	0.117	0.88 *10*
14 - 18	0.098	0.66 *5*
244	0.5	~0.19
260 - 262	0.07	~0.028
382	0.8	~0.21
397 - 400	0.12	~0.029
433	0.35	~0.08
449 - 451	0.051	0.011 *6*
516	0.19	0.038 *5*
532 - 534	0.028	0.0052 *7*
620	0.104	0.0168 *10*
636 - 662	0.0329	0.0051 *4*
743	0.142	0.0191 *15*
758 - 761	0.0195	0.00257 *16*
900 - 918	0.035	0.0039 *3*
1039	0.51	0.049 *3*
1054 - 1057	0.069	0.0065 *3*
1065	0.31	0.029 *3*
1080 - 1083	0.043	0.0039 *3*
1381 - 1399	0.0128	0.00093 *14*
1800 - 1817	0.036	0.00199 *15*
1957 - 1975	0.015	0.00078 *23*

Continuous Radiation (^{88}Nb)

⟨β+⟩=1434 keV;⟨IB⟩=5.3 keV

E_{bin}(keV)		⟨ ⟩(keV)	(%)
0 - 10	β+	5.4×10^{-6}	6.7×10^{-5}
	IB	0.047	
10 - 20	β+	0.000149	0.00091
	IB	0.047	0.33
20 - 40	β+	0.00291	0.0090
	IB	0.092	0.32
40 - 100	β+	0.098	0.128
	IB	0.26	0.40
100 - 300	β+	3.98	1.81
	IB	0.78	0.43
300 - 600	β+	30.9	6.6
	IB	0.93	0.22
600 - 1300	β+	282	28.8
	IB	1.46	0.166
1300 - 2500	β+	897	48.9
	IB	1.22	0.070
2500 - 4744	β+	221	8.0
	IB	0.46	0.0155
	Σβ+		94

$^{88}_{42}$Mo(8.0 *2* min)

Mode: ε
Δ: -72830 *300* keV syst
SpA: 2.67×10^{8} Ci/g
Prod: ^{32}S on ^{59}Co; ^{32}S on ^{58}Ni

Photons (^{88}Mo)

γ_{mode}	γ(keV)	γ(rel)
γ	80.0 *5*	80 *16*
γ	130.9 *5*	60 *12*
γ	170.7 *5*	100 *20*

A = 89

NDS **16**, 445 (1975)

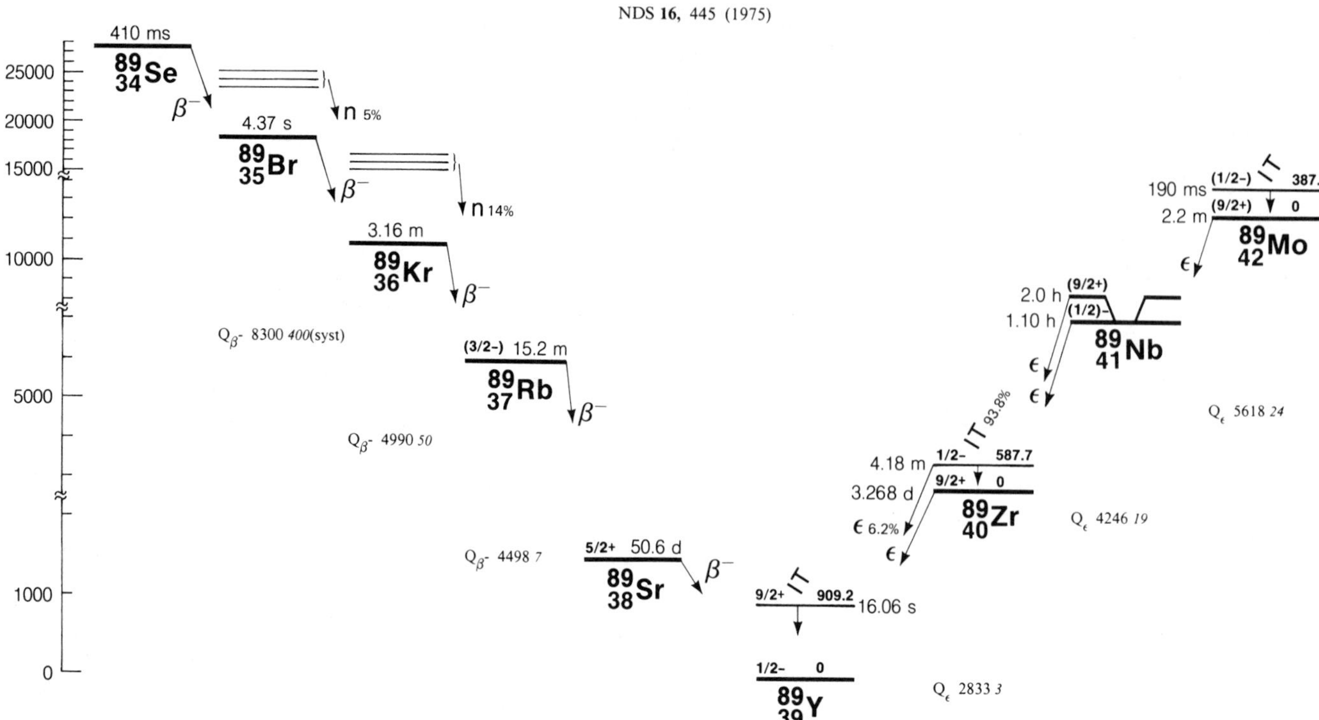

$^{89}_{34}$Se(410 *40* ms)

Mode: β-, β-n(5.0 *15* %)
SpA: 1.49×10^{11} Ci/g
Prod: fission

$^{89}_{35}$Br(4.37 *3* s)

Mode: β-, β-n(14.2 *8* %)
Δ: -68420 *400* keV syst
SpA: 2.682×10^{10} Ci/g
Prod: fission
β-n: 270, 400, 610, 680, 740, 800, 900
β-n(avg): 430 *60*

Photons (^{89}Br)

⟨γ⟩=1669 *22* keV

γmode	γ(keV)	γ(%)†
γβ-	28.51 *10*	
γβ-	282.1 *2*	0.064 *19*
γβ-	353.08 *20*	0.12 *3*
γβ-	356.27 *19*	0.14 *3*
γβ-	382.87 *5*	0.384 *19*
γβ-	385.11 *5*	0.307 *19*
γβ-	397.94 *5*	0.099 *7*
γβ-	411.49 *4*	2.11 *13*
γβ-	456.3 *2*	0.064 *19*
γβ-	498.04 *20*	0.30 *4*
γβ-	554.37 *5*	0.141 *6*

Photons (^{89}Br) (continued)

γmode	γ(keV)	γ(%)†
γβ-	558.50 *6*	0.103 *6*
γβ-	580.03 *4*	0.44 *3*
γβ-	595.99 *6*	0.288 *19*
γβ-	609.38 *20*	0.15 *3*
γβ-	621.56 *5*	0.117 *5*
γβ-	738.72 *5*	0.169 *7*
γβ-n [E2]	775.19 *8*	5.6
γβ-	789.76 *5*	0.339 *19*
γβ-n [M1+E2]	802.09 *9*	0.202 *13*
γβ-	807.33 *8*	0.75 *6*
γβ-	896.37 *10*	0.116 *10*
γβ-	953.53 *4*	4.54 *13*
γβ-	962.70 *4*	1.54 *6*
γβ-	991.5 *3*	0.34 *6*
γβ-	997.93 *4*	4.54 *13*
γβ-	1012.25 *4*	1.02 *3*
γβ-	1026.46 *4*	1.92 *6*
γβ-	1036.13 *7*	0.250 *13*
γβ-	1064 *1*	~0.13
γβ-	1069.24 *7*	0.230 *13*
γβ-	1080.61 *8*	0.205 *13*
γβ-	1097.82 *3*	6.4
γβ-	1122.08 *10*	0.31 *3*
γβ-	1190.11 *8*	0.282 *19*
γβ-	1351.31 *5*	2.24 *13*
γβ-	1355.54 *20*	0.28 *3*
γβ-	1379.80 *4*	1.41 *10*
γβ-	1391.64 *7*	0.47 *3*
γβ-	1399.77 *13*	0.26 *3*
γβ-	1454.22 *7*	1.15 *6*
γβ-	1483.00 *4*	2.05 *13*
γβ-	1487.25 *20*	0.51 *3*
γβ-	1507.70 *4*	2.62 *13*
γβ-	1529.79 *17*	0.29 *4*
γβ-	1533.62 *20*	0.24 *3*
γβ-n [E2]	1577.27 *12*	0.052 *10*
γβ-	1606.93 *10*	0.371 *19*
γβ-	1693.71 *12*	0.166 *19*
γβ-	1697.00 *17*	0.119 *19*

Photons (^{89}Br) (continued)

γmode	γ(keV)	γ(%)†
γβ-	1741.34 *12*	0.218 *19*
γβ-	1763.80 *15*	0.160 *19*
γβ-	1782 *1*	~0.38
γβ-	1858.73 *5*	1.14 *4*
γβ-	1865.98 *6*	0.86 *3*
γβ-	1915.23 *8*	0.390 *19*
γβ-	1919.96 *12*	0.218 *19*
γβ-	1928.44 *13*	0.339 *19*
γβ-	1957.22 *15*	0.29 *3*
γβ-	1966.51 *12*	0.43 *3*
γβ-	1994.4 *1*	0.73 *3*
γβ-	2006.72 *13*	0.53 *3*
γβ-	2034.49 *13*	0.38 *3*
γβ-	2075.61 *13*	0.28 *3*
γβ-	2108.16 *6*	1.52 *10*
γβ-	2118.40 *6*	2.05 *13*
γβ-	2165.04 *12*	0.38 *3*
γβ-	2179.69 *10*	0.32 *3*
γβ-	2208.59 *12*	0.36 *3*
γβ-	2216.60 *12*	0.72 *6*
γβ-	2228.42 *20*	0.198 *19*
γβ-	2234.81 *15*	0.26 *3*
γβ-	2298.47 *8*	0.90 *6*
γβ-	2318.79 *13*	0.28 *3*
γβ-	2336.18 *15*	0.29 *3*
γβ-	2345.33 *10*	0.78 *6*
γβ-	2503.16 *15*	0.36 *3*
γβ-	2578.03 *16*	0.88 *10*
γβ-	2733.97 *15*	0.40 *4*
γβ-	2785.8 *2*	0.23 *3*
γβ-	2815.3 *2*	0.24 *3*
γβ-	2894.20 *15*	1.10 *8*
γβ-	3001.60 *13*	0.48 *4*
γβ-	3017.0 *7*	0.19 *6*
γβ-	3132.30 *15*	0.30 *3*
γβ-	3139.5 *2*	0.40 *4*
γβ-	3174.6 *2*	0.48 *3*
γβ-	3187.48 *15*	0.50 *4*
γβ-	3289.8 *3*	0.39 *5*

Photons (^{89}Br)
(continued)

γ_{mode}	γ(keV)	γ(%)†
$\gamma_{\beta-}$	3326.4 2	0.70 6
$\gamma_{\beta-}$	3400.0 8	0.19 6
$\gamma_{\beta-}$	3489.1 2	0.36 4
$\gamma_{\beta-}$	3495.1 3	0.22 3
$\gamma_{\beta-}$	3531.5 2	0.61 5
$\gamma_{\beta-}$	3588.0 3	0.35 3
$\gamma_{\beta-}$	3837.3 3	0.49 5
$\gamma_{\beta-}$	3877.0 8	0.19 6
$\gamma_{\beta-}$	4086.3 3	1.92 19
$\gamma_{\beta-}$	4166.3 3	4.1 3
$\gamma_{\beta-}$	4185.6 3	0.66 6
$\gamma_{\beta-}$	4209.9 3	0.35 3
$\gamma_{\beta-}$	4232.5 4	0.28 3
$\gamma_{\beta-}$	4238.2 3	0.46 5
$\gamma_{\beta-}$	4260.5 4	0.42 5
$\gamma_{\beta-}$	4288.6 3	0.53 5
$\gamma_{\beta-}$	4319.0 3	0.96 9
$\gamma_{\beta-}$	4353.6 3	1.25 11
$\gamma_{\beta-}$	4402.9 4	0.39 4
$\gamma_{\beta-}$	4501.6 3	0.94 9
$\gamma_{\beta-}$	4510.4 4	0.33 4
$\gamma_{\beta-}$	4516.9 4	0.49 5
$\gamma_{\beta-}$	4610.5 4	0.36 4
$\gamma_{\beta-}$	4672.9 4	0.27 3
$\gamma_{\beta-}$	4678.6 4	0.49 5

† uncert(syst): 7.8% for β-, 8.6% for β-n

$^{89}_{36}$Kr(3.16 *4* min)

Mode: β-
Δ: -76720 *50* keV
SpA: 6.67×10^{8} Ci/g
Prod: fission

Photons (^{89}Kr)
⟨γ⟩=1825 *17* keV

γ_{mode}	γ(keV)	γ(%)†
γ	197.28 6	1.81 14
γ[E1]	197.90 8	~0.21
γ	205.42 11	0.123 24
γ[M1+E2]	220.97 4	19.9 11
γ	240.92 6	~0.012
γ[E1]	264.33 7	0.66 4
γ[M1+E2]	267.98 7	0.084 18
γ	286.45 20	0.026 8
γ	295.25 10	~0.016
γ	304.82 6	~0.022
γ	318.75 17	0.044 14
γ[M1+E2]	338.38 6	0.34 3
γ	345.16 6	1.17 8
γ	356.19 5	4.12 22
γ	365.02 4	0.90 6
γ	369.46 4	1.37 8
γ	381.10 8	0.046 12
γ	402.37 7	0.32 4
γ	411.59 5	2.55 14
γ[E1]	419.41 12	0.038 10
γ	438.20 5	0.96 6
γ[E1]	466.27 5	0.80 6
γ	487.6 3	0.08 3
γ[M1+E2]	490.96 11	0.32 4
γ	497.59 5	6.6 5
γ[E2]	498.48 7	1.13 20
γ	523.20 18	0.034 12
γ	542.4 4	0.030 12
γ	548.27 23	0.030 12
γ[M1+E2]	557.38 11	0.159 16
γ	577.17 5	5.6 3
γ	586.00 4	16.4 9
γ	599.65 6	0.088 12
γ	609.49 7	~0.018

Photons (^{89}Kr)
(continued)

γ_{mode}	γ(keV)	γ(%)†
γ[E1]	626.36 7	0.60 4
γ[E2]	629.86 8	0.34 3
γ	651.89 23	0.038 14
γ	660.7 6	0.048 16
γ	663.13 9	0.078 18
γ	665.88 18	0.113 16
γ	669.49 11	0.042 14
γ	671.59 16	0.105 20
γ	674.28 6	0.231 22
γ	687.25 16	0.070 18
γ[E1]	696.38 5	1.77 12
γ	707.19 6	0.50 3
γ[M1+E2]	710.19 6	0.78 6
γ	729.82 8	0.29 3
γ	738.51 4	4.18 22
γ	747.34 5	0.11 3
γ	754.04 21	0.092 24
γ[E1]	762.81 6	0.92 12
γ	762.87 21	0.40 8
γ[M1+E2]	776.61 6	1.11 18
γ	783.47 9	~0.022
γ	826.91 5	0.76 6
γ	835.72 5	1.09 8
γ[M1+E2]	857.61 11	0.285 24
γ	867.25 6	5.9 3
γ	870.78 8	0.159 18
γ[M1+E2]	904.47 4	7.1 4
γ	917.27 11	0.074 12
γ[E2]	931.16 6	0.62 4
γ	933.82 10	0.038 12
γ	939.8 3	0.066 14
γ	944.38 8	0.163 16
γ	953.21 8	0.105 16
γ	960.61 8	0.32 3
γ[E1]	964.74 7	0.058 14
γ	970.00 10	0.094 14
γ	974.51 6	0.98 6
γ[E2]	997.58 5	0.66 4
γ	1010.98 19	0.107 14
γ	1038.5 5	0.030 12
γ[E1]	1044.56 7	0.41 3
γ	1048.14 23	0.062 12
γ	1058.4 3	~0.030
γ	1063.67 17	0.070 16
γ	1067.62 7	0.068 16
γ	1076.64 6	0.23 3
γ	1088.28 7	0.36 3
γ	1098.68 24	0.064 24
γ	1103.53 5	0.90 6
γ	1107.97 4	2.91 18
γ	1116.80 4	1.65 10
γ	1131.53 8	0.159 22
γ	1152.85 14	0.064 16
γ[M1+E2]	1162.64 6	0.213 20
γ	1168.02 8	0.034 14
γ[M1+E2]	1172.44 6	0.98 8
γ	1182.60 15	0.165 22
γ	1186.77 14	0.183 18
γ	1195.49 6	0.084 14
γ	1201.52 9	~0.018
γ	1210.92 10	~0.022
γ[M1+E2]	1229.07 7	0.143 18
γ	1235.85 6	0.59 5
γ	1241.7 4	0.088 16
γ	1251.87 10	0.038 16
γ[E2]	1267.57 11	~0.024
γ	1273.92 5	1.35 8
γ	1278.98 10	~0.032
γ	1299.06 19	0.044 14
γ	1303.48 22	0.099 14
γ[M1+E2]	1309.40 8	0.068 14
γ	1324.50 4	3.04 18
γ[E1]	1336.03 9	0.13 3
γ	1340.03 21	0.193 24
γ	1367.38 10	0.147 18
γ	1372.36 10	0.125 16
γ	1381.4 3	0.058 16
γ	1412.78 6	0.263 22
γ	1421.61 6	0.22 4
γ	1441.0 4	~0.020
γ	1455.5 7	0.052 22
γ	1457.73 7	0.074 24
γ	1460.69 25	0.121 24
γ	1464.54 25	0.177 24

Photons (^{89}Kr)
(continued)

γ_{mode}	γ(keV)	γ(%)†
γ	1469.9 3	0.19 3
γ[E1]	1472.99 4	6.8 4
γ	1482.88 18	0.044 20
γ	1488.54 11	0.094 20
γ	1501.18 6	1.31 10
γ	1506.39 9	0.111 20
γ[M1+E2]	1530.37 7	3.3 2
γ	1533.90 7	5.1 3
γ	1555.48 14	0.151 18
γ	1574.23 5	0.189 18
γ	1583.06 5	0.090 14
γ[M1+E2]	1600.87 6	0.072 14
γ	1634.28 7	0.82 6
γ	1644.00 9	0.34 3
γ	1657.8 5	0.040 12
γ	1667.78 9	0.127 14
γ	1677.17 9	0.139 22
γ	1680.16 18	0.084 20
γ	1683.8 3	0.131 24
γ	1693.32 12	0.25 10
γ[E1]	1693.96 4	4.4 3
γ	1707.7 4	0.024 10
γ	1711.19 15	0.034 12
γ	1721.47 15	0.223 18
γ	1730.1 6	0.030 12
γ	1735.7 4	0.056 12
γ	1765.31 22	0.048 12
γ	1777.80 6	0.76 6
γ	1788.38 17	0.105 16
γ	1791.45 25	0.046 14
γ	1804.6 6	0.030 12
γ	1811.00 16	0.139 16
γ	1823.97 6	0.066 14
γ	1827.5 4	0.064 12
γ	1831.37 10	0.086 12
γ	1837.6 4	0.12 3
γ	1839.70 12	0.35 3
γ	1851.08 9	0.050 12
γ	1864.97 9	0.080 14
γ[E1]	1868.81 7	0.195 18
γ	1880.09 22	0.157 16
γ	1886.7 6	0.034 12
γ	1898.60 25	0.030 12
γ	1903.54 6	1.03 10
γ	1925.80 23	~0.016
γ[E1]	1935.24 8	0.034 12
γ[M1+E2]	1939.24 5	0.64 4
γ	1966.60 13	0.131 14
γ	1977.0 4	0.038 12
γ	1998.77 6	0.117 22
γ	2002.66 16	0.036 16
γ	2012.42 5	1.55 10
γ	2021.25 5	0.243 20
γ	2039.8 4	~0.018
γ	2046.63 9	0.261 20
γ	2079.5 9	0.030 12
γ	2082.52 18	0.058 14
γ	2100.82 5	0.94 6
γ	2141.47 15	0.062 12
γ	2143.99 21	0.064 12
γ	2150.57 13	~0.020
γ[E2]	2160.21 4	0.53 4
γ	2168.1 6	0.042 14
γ	2190.1 3	~0.026
γ	2196.23 7	0.13 5
γ	2207.51 13	0.046 14
γ	2232.3 4	0.024 10
γ	2280.40 6	0.20 4
γ	2285.4 4	0.046 20
γ	2322.28 19	0.052 14
γ	2330.65 8	0.036 14
γ[E1]	2377.44 5	0.80 6
γ	2401.13 5	0.72 6
γ	2440.65 12	0.046 16
γ	2468.0 4	~0.016
γ	2490.23 25	0.024 10
γ	2503.88 11	0.050 12
γ	2522.18 14	0.050 12
γ	2535.44 15	0.094 14
γ	2545.88 13	0.050 14
γ	2550.91 19	0.030 12
γ	2555.5 8	0.034 12
γ[E1]	2598.40 5	0.107 16
γ	2622.7 3	~0.022

Photons (^{89}Kr)
(continued)

γ_{mode}	γ(keV)	γ(%)†
γ[E1]	2645.41 *6*	0.42 *3*
γ	2659.3 *5*	0.086 *16*
γ	2703.4 *9*	0.034 *14*
γ	2722.85 *25*	0.036 *14*
γ	2742.23 *8*	0.028 *12*
γ	2751.05 *8*	0.123 *14*
γ	2756.78 *18*	0.066 *14*
γ	2760.5 *7*	0.046 *16*
γ	2775.73 *9*	~0.030
γ	2782.21 *7*	0.76 *6*
γ	2789.02 *25*	0.052 *18*
γ	2793.95 *9*	0.68 *4*
γ	2804.3 *8*	0.040 *16*
γ	2819.61 *18*	0.131 *16*
γ	2853.18 *25*	0.24 *3*
γ[E1]	2866.38 *6*	1.73 *10*
γ	2873.52 *10*	0.096 *18*
γ	2879.5 *4*	0.32 *3*
γ	2916.8 *4*	0.030 *10*
γ	2947.02 *15*	0.078 *14*
γ	2999.1 *3*	0.044 *12*
γ[M1+E2]	3017.80 *12*	0.25 *3*
γ	3029.31 *22*	0.269 *24*
γ	3049.9 *7*	0.040 *12*
γ	3098.57 *21*	0.038 *12*
γ	3107.24 *8*	0.193 *18*
γ	3140.48 *13*	1.03 *8*
γ	3153.98 *23*	~0.026
γ	3160.0 *6*	0.062 *12*
γ	3172.23 *14*	0.099 *14*
γ	3212.95 *19*	0.032 *12*
γ	3220.05 *13*	0.43 *3*
γ	3257.2 *5*	0.052 *12*
γ	3271.5 *5*	0.054 *12*
γ	3299.9 *4*	0.038 *12*
γ	3318.1 *6*	0.082 *18*
γ	3321.8 *3*	0.070 *16*
γ	3341.59 *21*	0.036 *14*
γ	3347.6 *6*	0.068 *16*
γ	3352.1 *9*	0.042 *14*
γ	3361.71 *8*	1.03 *8*
γ	3371.10 *9*	0.62 *6*
γ	3400.29 *23*	0.135 *14*
γ	3440.1 *4*	0.044 *12*
γ	3462.96 *15*	~0.042
γ	3504.09 *18*	~0.020
γ	3532.99 *15*	1.33 *8*
γ	3568.1 *7*	0.056 *18*
γ	3583.66 *18*	0.257 *20*
γ	3629.4 *5*	0.080 *14*
γ	3634.09 *21*	0.038 *12*
γ	3639.8 *4*	0.038 *12*
γ	3652.5 *5*	0.058 *12*
γ	3665.6 *4*	0.084 *12*
γ	3678.0 *3*	0.066 *12*
γ	3717.63 *13*	0.84 *6*
γ	3721.4 *4*	0.048 *20*
γ	3733.4 *4*	0.14 *5*
γ	3756.47 *23*	~0.016
γ	3781.70 *13*	0.131 *12*
γ	3809.8 *4*	0.020 *8*
γ	3827.97 *15*	0.137 *16*
γ	3837.8 *5*	0.082 *10*
γ	3843.2 *3*	0.109 *12*
γ	3882.7 *6*	0.040 *8*
γ	3899.0 *3*	~0.034
γ	3901.6 *3*	0.133 *20*
γ	3923.10 *18*	0.41 *3*
γ	3965.78 *21*	0.207 *16*
γ	3977.43 *23*	0.27 *5*
γ	3977.7 *4*	0.070 *12*
γ	3996.0 *4*	0.141 *12*
γ	4005.1 *7*	0.028 *8*
γ	4045.66 *21*	0.020 *8*
γ	4048.93 *15*	0.115 *12*
γ	4081.24 *18*	0.074 *10*
γ	4118.17 *21*	0.014 *6*
γ	4144.07 *18*	0.026 *8*
γ	4146.71 *13*	~0.016
γ	4162.8 *6*	0.028 *6*
γ	4176.4 *11*	~0.012
γ	4184.3 *3*	0.050 *8*
γ	4253.5 *10*	0.014 *6*
γ	4266.6 *3*	0.028 *6*
γ	4279.6 *7*	0.020 *6*
γ	4307.3 *4*	~0.010
γ	4321.4 *11*	0.010 *4*
γ	4340.7 *3*	0.103 *10*
γ	4367.67 *13*	0.042 *6*
γ	4405.2 *3*	~0.008
γ	4448.3 *12*	0.010 *4*
γ	4478.44 *23*	0.014 *4*
γ	4487.6 *3*	0.133 *12*
γ	4631.63 *21*	0.028 *6*
γ	4655.8 *7*	0.010 *4*
γ	4686.5 *4*	~0.008
γ	4701.7 *9*	0.010 *4*

† 5.5% uncert(syst)

Continuous Radiation (^{89}Kr)

⟨β-⟩=1385 keV;⟨IB⟩=4.2 keV

E_{bin}(keV)		⟨ ⟩(keV)	(%)
0 - 10	β-	0.0156	0.310
	IB	0.044	
10 - 20	β-	0.0471	0.314
	IB	0.044	0.30
20 - 40	β-	0.192	0.64
	IB	0.086	0.30
40 - 100	β-	1.42	2.02
	IB	0.25	0.38
100 - 300	β-	15.3	7.6
	IB	0.71	0.40
300 - 600	β-	57	12.6
	IB	0.83	0.19
600 - 1300	β-	284	30.1
	IB	1.21	0.139
1300 - 2500	β-	590	32.6
	IB	0.81	0.047
2500 - 4990	β-	438	13.9
	IB	0.18	0.0063

$^{89}_{37}$Rb(15.2 *1* min)

Mode: β-

Δ: -81713 *8* keV

SpA: 1.389×10^{8} Ci/g

Prod: fission; ^{86}Kr(α,p)

Photons (^{89}Rb)

⟨γ⟩=2071 *50* keV

γ_{mode}	γ(keV)	γ(%)†
γ[E1]	118.46 *9*	~0.012
γ	205.60 *24*	~0.012
γ	272.65 *6*	1.42 *7*
γ	289.99 *5*	0.54 *9*
γ[M1+E2]	466.90 *12*	0.070 *17*
γ[E1]	562.63 *7*	0.046 *6*
γ	596.01 *23*	0.023 *6*
γ	657.90 *5*	10.0 *5*
γ	699.71 *10*	0.023 *6*
γ	767.04 *12*	0.162 *17*
γ	776.36 *9*	0.070 *17*
γ	801.60 *13*	~0.017
γ	822.2 *3*	0.029 *12*
γ	947.88 *5*	9.2 *5*
γ[M1+E2]	975.64 *6*	0.058 *12*
γ	1025.6 *5*	0.23 *8*
γ[E2]	1032.08 *5*	58 *3*
γ	1057.14 *13*	~0.023
γ	1081.67 *19*	0.023 *6*
γ	1138.6 *4*	~0.012
γ	1160.67 *25*	0.035 *6*
γ	1211.9 *5*	~0.012
γ	1220.53 *6*	0.220 *17*
γ	1228.66 *10*	0.122 *17*
γ	1233.94 *14*	~0.029
γ	1248.28 *5*	42.6 *23*
γ[M1+E2]	1419.80 *8*	0.093 *12*
γ	1429.8 *5*	~0.012
γ[M1+E2]	1473.47 *14*	0.35 *3*
γ	1501.30 *11*	0.197 *17*
γ[E1]	1538.26 *6*	2.55 *17*
γ	1596.4 *5*	0.017 *6*
γ	1644.30 *19*	0.023 *6*
γ	1770.1 *3*	~0.012
γ[M1+E2]	1940.37 *13*	0.331 *23*
γ	1980.0 *4*	0.023 *6*
γ[M1+E2]	2007.70 *6*	2.38 *17*
γ	2057.7 *5*	0.23 *9*
γ	2110.1 *3*	0.017 *6*
γ	2196.15 *6*	13.3 *9*
γ	2231.5 *4*	0.023 *6*
γ	2280.34 *5*	0.180 *17*
γ	2372.5 *4*	~0.012
γ[M1+E2]	2451.86 *9*	0.052 *6*
γ[E1]	2570.32 *6*	9.9 *5*
γ	2668.2 *5*	~0.012
γ	2685.7 *4*	0.029 *6*
γ	2707.39 *9*	2.03 *12*
γ	2818.3 *5*	~0.012
γ	2948.1 *4*	0.017 *6*
γ	2955.7 *4*	~0.006
γ	3037.7 *4*	~0.012
γ	3141.9 *3*	0.052 *6*
γ	3228.20 *6*	0.075 *6*
γ	3263.8 *3*	0.017 *6*
γ	3303.38 *24*	~0.006
γ	3508.97 *11*	1.15 *7*
γ	3651.96 *19*	0.035 *12*
γ	3782.0 *5*	~0.012
γ	3846.0 *4*	0.029 *6*
γ	3987.7 *4*	0.017 *6*
γ	4093.9 *6*	0.075 *12*

† 6.9% uncert(syst)

Continuous Radiation (^{89}Rb)

⟨β-⟩=1018 keV;⟨IB⟩=2.7 keV

E_{bin}(keV)		⟨ ⟩(keV)	(%)
0 - 10	β-	0.0211	0.419
	IB	0.036	
10 - 20	β-	0.064	0.428
	IB	0.035	0.24
20 - 40	β-	0.265	0.88
	IB	0.068	0.24
40 - 100	β-	2.04	2.88
	IB	0.19	0.30
100 - 300	β-	24.4	11.9
	IB	0.53	0.30
300 - 600	β-	95	21.1
	IB	0.56	0.133
600 - 1300	β-	315	35.0
	IB	0.72	0.084
1300 - 2500	β-	346	19.7
	IB	0.43	0.025
2500 - 4498	β-	236	7.7
	IB	0.084	0.0029

$^{89}_{38}$Sr(50.55 *9* d)

Mode: β-
Δ: -86210 *4* keV
SpA: 2.902×10^{4} Ci/g
Prod: ^{88}Sr(d,p); ^{88}Sr(n,γ); ^{86}Kr(α,n)

Photons (^{89}Sr)

⟨γ⟩=0.086 *7* keV

γ_{mode}	γ(keV)	γ(%)†
γ (M4)	909.15 *7*	0.0095 *

† 8.6% uncert(syst)
* with ^{89}Y(16.06 s) in equilib

Continuous Radiation (^{89}Sr)

⟨β-⟩=583 keV;⟨IB⟩=0.90 keV

E_{bin}(keV)		⟨ ⟩(keV)	(%)
0 - 10	β-	0.0359	0.72
	IB	0.025	
10 - 20	β-	0.109	0.73
	IB	0.024	0.169
20 - 40	β-	0.443	1.48
	IB	0.047	0.162
40 - 100	β-	3.28	4.67
	IB	0.126	0.20
100 - 300	β-	35.7	17.6
	IB	0.31	0.18
300 - 600	β-	127	28.3
	IB	0.25	0.060
600 - 1300	β-	399	45.2
	IB	0.125	0.0168
1300 - 1492	β-	17.9	1.32
	IB	0.00029	2.2×10^{-5}

$^{89}_{39}$Y (stable)

Δ: -87702 *3* keV
%: 100

$^{89}_{39}$Y (16.06 *4* s)

Mode: IT
Δ: -86793 *3* keV
SpA: 7.725×10^{9} Ci/g
Prod: daughter ^{89}Zr; ^{89}Y(n,n')

Photons (^{89}Y)

⟨γ⟩=901.4 *4* keV

γ_{mode}	γ(keV)	γ(%)†
Y $K_{\alpha2}$	14.883	0.156 *18*
Y $K_{\alpha1}$	14.958	0.30 *3*
Y $K_{\beta1}$'	16.735	0.067 *8*
Y $K_{\beta2}$'	17.125	0.0098 *11*
γ (M4)	909.15 *7*	99.14

†<0.1% uncert(syst)

Atomic Electrons (^{89}Y)

⟨e⟩=7.7 *7* keV

e_{bin}(keV)	⟨e⟩(keV)	e(%)
2	0.0132	0.62 *9*
12	0.0018	0.0150 *22*
13	0.018	0.142 *21*
14	0.0020	0.0138 *20*
15	0.0060	0.041 *6*
16	0.00051	0.0031 *5*
17	0.00019	0.00115 *17*
892	6.6	0.74 *8*
907	0.83	0.091 *10*
909	0.169	0.0186 *20*

$^{89}_{40}$Zr(3.268 *3* d)

Mode: ε
Δ: -84869 *4* keV
SpA: 4.489×10^{5} Ci/g
Prod: ^{89}Y(p,n); ^{89}Y(d,2n)

Photons (^{89}Zr)

⟨γ⟩=925.7 *13* keV

γ_{mode}	γ(keV)	γ(%)†
Y L_{ℓ}	1.686	0.055 *12*
Y L_{η}	1.762	0.030 *8*
Y L_{α}	1.922	1.3 *3*
Y L_{β}	2.005	0.73 *19*
Y L_{γ}	2.269	0.022 *6*
Y $K_{\alpha2}$	14.883	14.3 *6*
Y $K_{\alpha1}$	14.958	27.4 *11*
Y $K_{\beta1}$'	16.735	6.11 *25*
Y $K_{\beta2}$'	17.125	0.90 *4*
γ (M4)	909.15 *7*	99.01 *
γ [M1+E2]	1620.8 *2*	0.070 *7*
γ [M1+E2]	1657.3 *2*	0.099 *10*
γ [M1+E2]	1712.9 *8*	0.76 *7*
γ [E2]	1744.5 *2*	0.129 *10*

†<0.1% uncert(syst)
* with ^{89}Y(16.06 s) in equilib

Atomic Electrons (^{89}Zr)

⟨e⟩=11.5 *3* keV

e_{bin}(keV)	⟨e⟩(keV)	e(%)
2	1.20	57 *6*
12	0.169	1.38 *14*
13	1.66	13.0 *13*
14	0.182	1.27 *13*
15	0.55	3.7 *4*
16	0.046	0.28 *3*
17	0.0176	0.106 *11*
892	6.6	0.74 *3*
907	0.83	0.091 *4*
909	0.169	0.0186 *7*
1604	0.00023	1.43 *16* ×10^{-5}
1618	2.4 ×10^{-5}	1.50 *17* ×10^{-6}
1619	6.4 ×10^{-7}	3.9 *10* ×10^{-8}
1620	4.3 ×10^{-6}	2.7 *3* ×10^{-7}
1640	0.00032	1.92 *22* ×10^{-5}
1655	3.4 ×10^{-5}	2.08 *25* ×10^{-6}
1657	6.0 ×10^{-6}	3.6 *4* ×10^{-7}
1696	0.00235	0.000139 *15*
1711	0.00026	1.50 *17* ×10^{-5}
1713	4.5 ×10^{-5}	2.6 *3* ×10^{-6}
1727	0.00039	2.23 *19* ×10^{-5}

Atomic Electrons (^{89}Zr) (continued)

e_{bin}(keV)	⟨e⟩(keV)	e(%)
1742	4.2 ×10^{-5}	2.42 *21* ×10^{-6}
1744	7.3 ×10^{-6}	4.2 *4* ×10^{-7}

Continuous Radiation (^{89}Zr)

⟨β+⟩=90 keV;⟨IB⟩=1.38 keV

E_{bin}(keV)		⟨ ⟩(keV)	(%)
0 - 10	β+	3.36×10^{-5}	0.000413
	IB	0.0044	
10 - 20	β+	0.00088	0.0054
	IB	0.0141	0.095
20 - 40	β+	0.0163	0.0507
	IB	0.0083	0.029
40 - 100	β+	0.495	0.65
	IB	0.021	0.033
100 - 300	β+	14.3	6.7
	IB	0.072	0.039
300 - 600	β+	52	11.9
	IB	0.20	0.045
600 - 1300	β+	22.3	3.27
	IB	0.81	0.089
1300 - 1924	IB	0.25	0.017
	Σβ+		23

$^{89}_{40}$Zr(4.18 *1* min)

Mode: IT(93.76 *6* %), ε(6.24 *6* %)
Δ: -84281 *4* keV
SpA: 5.047×10^{8} Ci/g
Prod: ^{89}Y(p,n); daughter Nb

Photons (^{89}Zr)

⟨γ⟩=618 *4* keV

γ_{mode}	γ(keV)	γ(%)†
Y $K_{\alpha2}$	14.883	0.86 *4*
Y $K_{\alpha1}$	14.958	1.65 *7*
Zr $K_{\alpha2}$	15.691	0.78 *3*
Zr $K_{\alpha1}$	15.775	1.49 *5*
Y $K_{\beta1}$'	16.735	0.369 *15*
Y $K_{\beta2}$'	17.125	0.0545 *23*
Zr $K_{\beta1}$'	17.663	0.340 *12*
Zr $K_{\beta2}$'	18.086	0.0550 *21*
γ_{IT} M4	587.70 *18*	89.5
γ_{ϵ}M1+2.4%E2	1507.4 *5*	6.04

† uncert(syst): 0.96% for ε, 0.63% for IT

Atomic Electrons (^{89}Zr)

⟨e⟩=24.7 *4* keV

e_{bin}(keV)	⟨e⟩(keV)	e(%)
2	0.138	6.3 *6*
3	0.0045	0.178 *18*
12	0.0102	0.083 *9*
13	0.175	1.35 *14*
14	0.031	0.227 *23*
15	0.066	0.44 *5*
16	0.0089	0.056 *6*
17	0.0037	0.0218 *22*
18	0.00072	0.0041 *4*
570	20.7	3.64 *8*

Atomic Electrons (^{89}Zr) (continued)

e_{bin}(keV)	⟨e⟩(keV)	e(%)
585	2.92	0.499 *10*
587	0.524	0.0892 *19*
588	0.0904	0.0154 *3*
1490	0.0214	0.00143 *3*
1505	0.00233	0.000155 *3*
1507	0.000404	2.68 *6* $\times10^{-5}$

Continuous Radiation (^{89}Zr)

⟨β+⟩=7.1 keV;⟨IB⟩=0.090 keV

E_{bin}(keV)		⟨ ⟩(keV)	(%)
0 - 10	β+	2.07 $\times10^{-6}$	2.54 $\times10^{-5}$
	IB	0.00032	
10 - 20	β+	5.4 $\times10^{-5}$	0.000333
	IB	0.00092	0.0062
20 - 40	β+	0.00100	0.00312
	IB	0.00062	0.0022
40 - 100	β+	0.0304	0.0400
	IB	0.00163	0.0025
100 - 300	β+	0.88	0.413
	IB	0.0053	0.0029
300 - 600	β+	3.22	0.73
	IB	0.0135	0.0030
600 - 1300	β+	2.03	0.265
	IB	0.051	0.0056
1300 - 2500	β+	0.91	0.056
	IB	0.0165	0.00111
2500 - 3421	IB	0.00031	1.13 $\times10^{-5}$
	Σβ+		1.51

$^{89}_{41}$Nb(1.10 *3* h)

Mode: ε

Δ: -80622 *19* keV

SpA: 3.201 $\times10^{7}$ Ci/g

Prod: ^{12}C on Br; protons on Zr

Photons (^{89}Nb)

⟨γ⟩=1048 *36* keV

γ_{mode}	γ(keV)	γ(%)†
Zr L_ℓ	1.792	0.019 *3*
Zr L_η	1.876	0.0098 *15*
Zr L_α	2.042	0.44 *7*
Zr L_β	2.135	0.25 *4*
Zr L_γ	2.385	0.0119 *23*
Zr $K_{\alpha2}$	15.691	4.63 *16*
Zr $K_{\alpha1}$	15.775	8.9 *3*
Zr $K_{\beta1}'$	17.663	2.02 *7*
Zr $K_{\beta2}'$	18.086	0.327 *12*
γ (M1)	507.2 *3*	81 *7*
γ M4	587.70 *18*	95.5 *
γ [M1+E2]	650.0 *8*	<2 ?
γ [M1+E2]	769.4 *5*	6.2 *6*
γ [M1+E2]	1276.6 *5*	1.5 *5*

† 10% uncert(syst)
* with ^{89}Zr(4.18 min) in equilib

Atomic Electrons (^{89}Nb)

⟨e⟩=28.4 *5* keV

e_{bin}(keV)	⟨e⟩(keV)	e(%)
2	0.38	16.9 *17*
3	0.027	1.06 *11*
13	0.45	3.3 *3*
14	0.121	0.89 *9*
15	0.194	1.27 *13*
16	0.036	0.231 *24*
17	0.0159	0.092 *9*
18	0.0043	0.0244 *25*
489	1.10	0.224 *21*
505	0.126	0.0250 *23*
507	0.0262	0.0052 *5*
570	22.1	3.88 *9*
585	3.11	0.532 *12*
587	0.559	0.0952 *21*
588	0.0964	0.0164 *4*
632	0.009	<0.0027
647	0.0009	<0.00029
648	6.2 $\times10^{-5}$	<1.9 $\times10^{-5}$
650	0.0002	<6 $\times10^{-5}$
751	0.051	0.0068 *8*
767	0.0058	0.00076 *10*
769	0.00119	0.000155 *20*
1259	0.0068	0.00054 *17*
1274	0.00076	5.9 *19* $\times10^{-5}$
1276	0.00013	1.0 *3* $\times10^{-5}$
1276	2.1 $\times10^{-5}$	1.7 *5* $\times10^{-6}$

Continuous Radiation (^{89}Nb)

⟨β+⟩=789 keV;⟨IB⟩=3.0 keV

E_{bin}(keV)		⟨ ⟩(keV)	(%)
0 - 10	β+	1.50 $\times10^{-5}$	0.000184
	IB	0.030	
10 - 20	β+	0.000410	0.00251
	IB	0.032	0.22
20 - 40	β+	0.0079	0.0247
	IB	0.058	0.20
40 - 100	β+	0.263	0.344
	IB	0.164	0.25
100 - 300	β+	9.9	4.54
	IB	0.45	0.25
300 - 600	β+	67	14.6
	IB	0.50	0.119
600 - 1300	β+	402	42.8
	IB	0.84	0.095
1300 - 2500	β+	309	19.6
	IB	0.85	0.049
2500 - 3658	β+	0.267	0.0105
	IB	0.095	0.0035
	Σβ+		82

$^{89}_{41}$Nb(2.03 *7* h)

Mode: ε

Δ: -80622 *19* keV

SpA: 1.73 $\times10^{7}$ Ci/g

Prod: ^{12}C on Br; ^{89}Y(α,4n); protons on Mo

Photons (^{89}Nb)

⟨γ⟩=620 *15* keV

γ_{mode}	γ(keV)	γ(%)†
Zr L_ℓ	1.792	0.018 *3*
Zr L_η	1.876	0.0097 *15*
Zr L_α	2.042	0.44 *7*
Zr L_β	2.135	0.24 *4*
Zr L_γ	2.385	0.0118 *23*
Zr $K_{\alpha2}$	15.691	4.60 *16*
Zr $K_{\alpha1}$	15.775	8.8 *3*
Zr $K_{\beta1}'$	17.663	2.01 *7*
Zr $K_{\beta2}'$	18.086	0.325 *12*
γ	173.4 *4*	0.076 *18* ?
γ [M1+E2]	206.12 *20*	0.07 *3*
γ	229.7 *5*	0.14 *5* ?
γ	348.1 *5*	~0.047 ?
γ [M1+E2]	356.3 *3*	0.246 *25*
γ	361.4 *10*	0.091 *18*
γ E2	480.8 *4*	0.14 *4*
γ (M1)	507.2 *3*	0.80 *7*
γ	520.4 *3*	~0.09 ?
γ [E1]	532.8 *3*	0.54 *7*
γ M4	587.70 *18*	1.2 *1* *
γ	617.0 *4*	0.043 *18*
γ	625.7 *4*	0.094 *18*
γ	658.2 *10*	~0.07
γ [E1]	739.0 *3*	0.228 *22*
γ	758.2 *6*	0.091 *14* ?
γ	785.8 *7*	0.036 *14*
γ	795.1 *6*	0.043 *14* ?
γ	846.5 *7*	0.080 *22*
γ [E2]	863.5 *3*	0.47 *7*
γ [M1+E2]	920.80 *24*	1.48 *14*
γ	964.3 *4*	0.11 *4*
γ	991.7 *3*	0.094 *18*
γ	1004.9 *10*	0.098 *18*
γ	1060.9 *3*	0.27 *4*
γ [M1+E2]	1126.9 *3*	2.2 *3*
γ [M1+E2]	1242.7 *3*	0.25 *3*
γ M1,E2	1259.38 *18*	1.27 *11*
γ [E1]	1303.4 *4*	0.33 *4*
γ	1332.75 *22*	1.27 *11*
γ	1378.1 *5*	0.07 *3*
γ	1412.4 *20*	~0.018
γ	1448.6 *3*	0.40 *4*
γ [M1+E2]	1465.49 *20*	0.90 *7*
γ E1	1511.85 *21*	1.99 *14*
γ [M1+E2]	1581.32 *23*	0.54 *5*
γ [E2]	1627.68 *16*	3.62
γ [E1]	1642.0 *3*	0.203 *18*
γ M1,E2	1833.79 *15*	3.37 *25*
γ	1948.6 *5*	0.07 *3* ?
γ	2101.5 *3*	0.62 *7*
γ	2128.9 *3*	0.58 *7*
γ	2132.4 *15*	0.13 *5* ?
γ	2221.5 *5*	0.188 *22*
γ	2279.4 *15*	~0.036
γ	2297.6 *7*	0.116 *18*
γ	2389 *1*	0.054 *14* ?
γ	2572.8 *3*	2.75 *22*
γ	2612.3 *4*	0.31 *3*
γ	2714.4 *20*	~0.036
γ	2730.7 *5*	0.07 *3* ?
γ	2740.4 *15*	~0.029
γ [M1+E2]	2754.6 *3*	0.47 *7*
γ	2890.0 *5*	0.206 *25*
γ	2928.0 *5*	0.188 *22*
γ	2960.40 *21*	1.81 *18*
γ	2981.4 *8*	0.047 *14*
γ [E1]	3016.6 *4*	0.217 *22*
γ [M1+E2]	3093.14 *15*	3.1 *3*
γ	3141.6 *9*	0.022 *7*

Photons (^{89}Nb) (continued)

γ_{mode}	γ(keV)	γ(%)†
γ	3281.3 *7*	0.036 *7*
γ	3467.3 *6*	0.043 *7*
γ	3512.8 *6*	0.065 *11*
γ	3531.4 *15*	~0.033 ?
γ	3534.4 *15*	~0.018 ?
γ	3557.6 *7*	0.054 *11*
γ	3576.2 *5*	0.203 *22*
γ	3837.4 *9*	0.065 *14*
γ	3907.4 *15*	0.009 *4* ?
γ	3931.4 *15*	~0.0036 ?
γ	3948.4 *15*	0.0054 *22* ?
γ	3965.9 *12*	0.011 *4*

† 5.5% uncert(syst)
* with ^{89}Zr(4.18 min) in equilib

Atomic Electrons (^{89}Nb)

⟨e⟩=1.71 *7* keV

e_{bin}(keV)	⟨e⟩(keV)	e(%)
2	0.38	16.7 *17*
3	0.027	1.06 *11*
13	0.44	3.3 *3*
14	0.120	0.89 *9*
15	0.193	1.26 *13*
16 - 18	0.056	0.345 *25*
155 - 204	0.017	0.010 *5*
206 - 230	0.010	~0.005
330 - 361	0.0118	0.0035 *7*
463 - 507	0.0164	0.0034 *3*
515 - 533	0.0033	0.00063 *8*
570	0.278	0.049 *4*
585	0.039	0.0067 *6*
587 - 626	0.0095	0.00161 *13*
640 - 658	0.0006	~9 ×10^{-5}
721 - 768	0.0017	0.00023 *4*
777 - 795	0.00030	3.9 *17* ×10^{-5}
828 - 863	0.0043	0.00051 *7*
903 - 946	0.0115	0.00126 *12*
962 - 1005	0.0010	0.00010 *3*
1043 - 1061	0.0012	0.00012 *4*
1109 - 1127	0.0126	0.00114 *16*
1225 - 1259	0.0078	0.00063 *5*
1285 - 1333	0.0053	0.00041 *12*

Atomic Electrons (^{89}Nb) (continued)

e_{bin}(keV)	⟨e⟩(keV)	e(%)
1360 - 1394	0.00031	2.2 *9* ×10^{-5}
1410 - 1449	0.0048	0.00034 *4*
1463 - 1512	0.0046	0.000308 *19*
1563 - 1642	0.0166	0.00104 *6*
1816 - 1833	0.0119	0.00066 *6*
1931 - 1948	0.00018	9 *4* ×10^{-6}
2083 - 2132	0.0032	0.00015 *3*
2204 - 2297	0.00080	3.6 *7* ×10^{-5}
2371 - 2389	0.00012	5.1 *18* ×10^{-6}
2555 - 2612	0.0065	0.00025 *6*
2696 - 2754	0.00148	5.4 *7* ×10^{-5}
2872 - 2963	0.0044	0.00015 *3*
2979 - 3075	0.0069	0.000225 *23*
3091 - 3141	0.00089	2.9 *3* ×10^{-5}
3263 - 3281	6.7 ×10^{-5}	2.1 *6* ×10^{-6}
3449 - 3540	0.00037	1.06 *15* ×10^{-5}
3555 - 3576	0.00037	1.04 *22* ×10^{-5}
3819 - 3913	0.00013	3.5 *8* ×10^{-6}
3929 - 3966	2.8 ×10^{-5}	7.2 *19* ×10^{-7}

Continuous Radiation (^{89}Nb)

⟨β+⟩=1089 keV;⟨IB⟩=4.0 keV

E_{bin}(keV)		⟨ ⟩(keV)	(%)
0 - 10	β+	9.1 ×10^{-6}	0.000112
	IB	0.036	
10 - 20	β+	0.000244	0.00149
	IB	0.039	0.27
20 - 40	β+	0.00455	0.0141
	IB	0.071	0.25
40 - 100	β+	0.136	0.179
	IB	0.20	0.31
100 - 300	β+	4.38	2.03
	IB	0.60	0.34
300 - 600	β+	28.3	6.1
	IB	0.74	0.17
600 - 1300	β+	236	24.3
	IB	1.16	0.133
1300 - 2500	β+	702	38.6
	IB	0.87	0.051
2500 - 4246	β+	117	4.36
	IB	0.31	0.0104
	Σβ+		76

$^{89}_{42}$Mo(2.15 *20* min)

Mode: ϵ
Δ: -75004 *16* keV
SpA: 9.8×10^{8} Ci/g
Prod: ^{92}Mo(p,p3n)

Photons (^{89}Mo)

γ_{mode}	γ(keV)	γ(rel)
γ	658.6 *10*	100 *20*
γ	803 *1*	11.0 *22*
γ	1155.1 *10*	29 *6*
γ	1272 *1*	47 *9*

$^{89}_{42}$Mo(190 *15* ms)

Mode: IT
Δ: -74617 *16* keV
SpA: 1.78×10^{11} Ci/g
Prod: ^{92}Mo(p,p3n)

Photons (^{89}Mo)

γ_{mode}	γ(keV)
γ [M1+E2]	118.2 *2*
γ [M4]	268.5 *2*

A = 90

NDS **16**, 55 (1975)

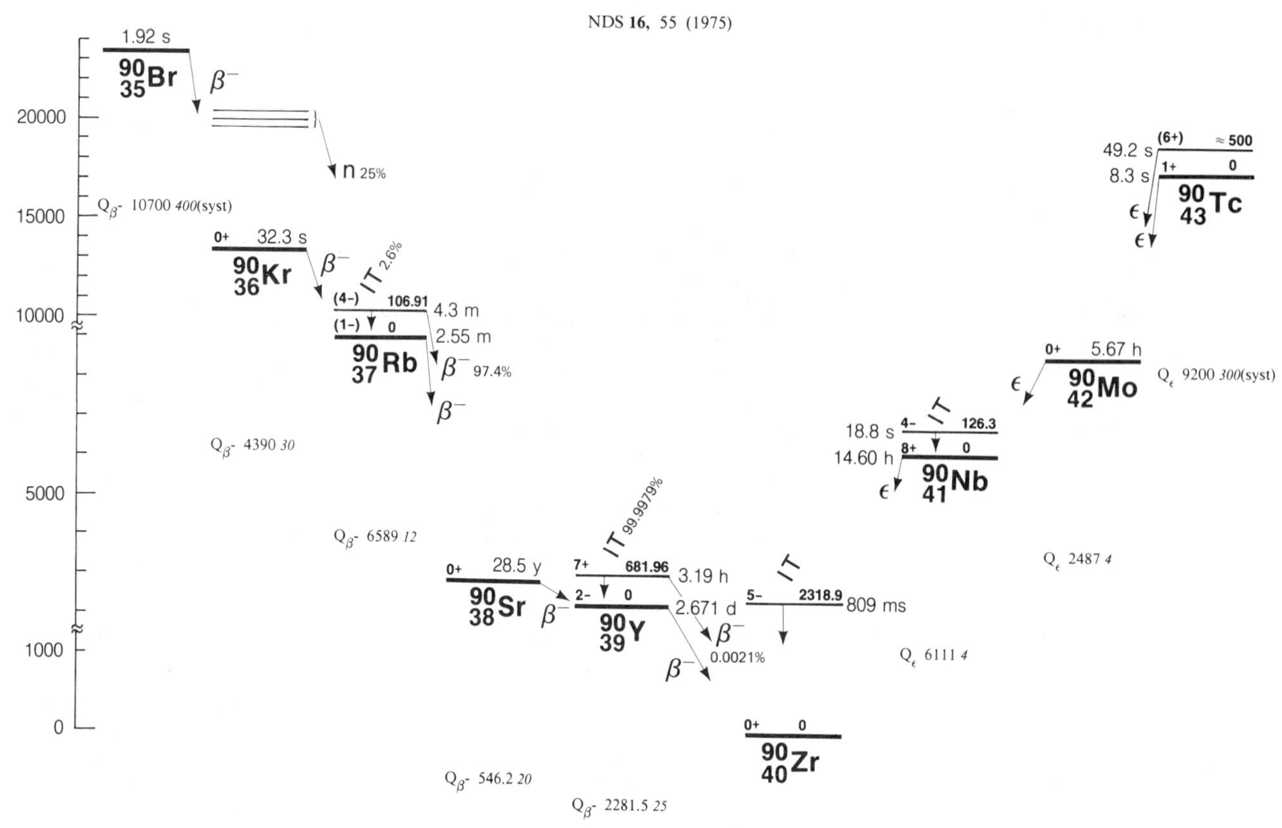

$^{90}_{35}$Br(1.92 *2* s)

Mode: β-, β-n(24.6 *17* %)
Δ: -64260 *400* keV syst
SpA: 5.48×10^{10} Ci/g
Prod: fission

Photons (^{90}Br)

⟨γ⟩=1543 *24* keV

γ_{mode}	γ(keV)	γ(%)†
$\gamma_{\beta\text{-n}}$	28.51 *10*	
$\gamma_{\beta\text{-n}}$	382.87 *5*	0.70 *4*
$\gamma_{\beta\text{-n}}$	411.48 *8*	3.87 *19*
$\gamma_{\beta\text{-}}$	539.50 *8*	0.66 *8*
$\gamma_{\beta\text{-}}$	578.06 *18*	0.60 *7*
$\gamma_{\beta\text{-n}}$	580.03 *4*	0.357 *21*
$\gamma_{\beta\text{-}}$[M1+E2]	655.17 *6*	7.7 *5*
$\gamma_{\beta\text{-}}$[E2]	707.05 *6*	38.0 *15*
$\gamma_{\beta\text{-}}$	740.82 *8*	2.85 *15*
$\gamma_{\beta\text{-}}$	786.02 *12*	1.60 *11*
$\gamma_{\beta\text{-}}$	840.29 *10*	0.95 *8*
$\gamma_{\beta\text{-}}$	913.38 *20*	0.26 *3*
$\gamma_{\beta\text{-}}$	939.9 *2*	0.43 *5*
$\gamma_{\beta\text{-}}$	955.84 *8*	2.20 *11*
$\gamma_{\beta\text{-n}}$	962.70 *4*	1.25 *8*
$\gamma_{\beta\text{-n}}$	991.5 *3*	0.27 *3*
$\gamma_{\beta\text{-n}}$	991.5 *3*	0.28 *5*
$\gamma_{\beta\text{-n}}$	997.93 *4*	0.33 *3*
$\gamma_{\beta\text{-n}}$	1026.46 *4*	0.137 *5*
$\gamma_{\beta\text{-n}}$	1097.82 *3*	0.91 *8*
$\gamma_{\beta\text{-}}$	1123.37 *15*	0.50 *4*
$\gamma_{\beta\text{-}}$	1133.5 *4*	0.19 *4*
$\gamma_{\beta\text{-}}$	1233.21 *8*	4.18 *15*
$\gamma_{\beta\text{-}}$	1323.57 *12*	0.80 *8*
$\gamma_{\beta\text{-}}$	1344.76 *15*	0.51 *5*
$\gamma_{\beta\text{-}}$[E2]	1362.32 *10*	11.2 *9*
$\gamma_{\beta\text{-}}$	1396.16 *13*	0.95 *8*
$\gamma_{\beta\text{-}}$	1418.34 *11*	0.80 *8*
$\gamma_{\beta\text{-}}$	1440.21 *12*	0.32 *3*
$\gamma_{\beta\text{-}}$	1465.9 *3*	0.36 *4*
$\gamma_{\beta\text{-}}$	1486.0 *4*	0.56 *8*
$\gamma_{\beta\text{-}}$	1542.20 *13*	1.33 *11*
$\gamma_{\beta\text{-}}$	1610.80 *11*	2.24 *11*
$\gamma_{\beta\text{-}}$	1625.11 *13*	1.67 *11*
$\gamma_{\beta\text{-}}$	1828.9 *2*	0.41 *4*
$\gamma_{\beta\text{-}}$	1836.49 *10*	1.44 *8*
$\gamma_{\beta\text{-}}$	1888.79 *13*	0.56 *5*
$\gamma_{\beta\text{-}}$	2002.9 *5*	0.25 *5*
$\gamma_{\beta\text{-}}$	2024.1 *3*	0.44 *6*
$\gamma_{\beta\text{-}}$	2040.79 *14*	0.80 *8*
$\gamma_{\beta\text{-}}$	2073.0 *5*	0.15 *4*
$\gamma_{\beta\text{-}}$	2105.7 *2*	0.57 *5*
$\gamma_{\beta\text{-}}$	2121.2 *3*	0.37 *4*
$\gamma_{\beta\text{-}}$	2172.9 *4*	0.13 *5*
$\gamma_{\beta\text{-}}$	2297.1 *3*	0.28 *5*
$\gamma_{\beta\text{-}}$	2318.6 *2*	1.06 *8*
$\gamma_{\beta\text{-}}$	2411.8 *3*	0.43 *7*
$\gamma_{\beta\text{-}}$	2433.09 *15*	0.99 *4*
$\gamma_{\beta\text{-}}$	2543.93 *20*	0.29 *4*
$\gamma_{\beta\text{-}}$	2612.2 *2*	0.55 *5*
$\gamma_{\beta\text{-}}$	2665.4 *3*	0.34 *5*
$\gamma_{\beta\text{-}}$	2682.6 *2*	0.46 *5*
$\gamma_{\beta\text{-}}$	2730.5 *2*	1.67 *11*
$\gamma_{\beta\text{-}}$	2763.9 *4*	0.190 *15*
$\gamma_{\beta\text{-}}$	2801.0 *3*	0.35 *7*
$\gamma_{\beta\text{-}}$	2864.1 *2*	1.18 *11*
$\gamma_{\beta\text{-}}$	2987.1 *2*	1.10 *8*
$\gamma_{\beta\text{-}}$	3022.3 *4*	0.24 *6*
$\gamma_{\beta\text{-}}$	3116.7 *4*	0.38 *5*
$\gamma_{\beta\text{-}}$	3231.3 *3*	3.4 *3*
$\gamma_{\beta\text{-}}$	3318.5 *3*	1.18 *11*
$\gamma_{\beta\text{-}}$	3505.8 *4*	0.25 *5*
$\gamma_{\beta\text{-}}$	3669.1 *4*	0.28 *4*
$\gamma_{\beta\text{-}}$	3949.0 *3*	0.36 *4*
$\gamma_{\beta\text{-}}$	3955.0 *3*	0.32 *3*
$\gamma_{\beta\text{-}}$	4023.1 *6*	0.25 *7*
$\gamma_{\beta\text{-}}$	4290.9 *3*	0.65 *6*
$\gamma_{\beta\text{-}}$	4297.2 *3*	0.63 *6*
$\gamma_{\beta\text{-}}$	4320.9 *5*	0.35 *5*
$\gamma_{\beta\text{-}}$	4367.3 *5*	0.61 *8*
$\gamma_{\beta\text{-}}$	4372.8 *5*	0.54 *8*
$\gamma_{\beta\text{-}}$	4495.1 *4*	0.33 *5*
$\gamma_{\beta\text{-}}$	4762.3 *5*	0.26 *4*
$\gamma_{\beta\text{-}}$	4787.4 *4*	0.22 *4*
$\gamma_{\beta\text{-}}$	5154.0 *5*	0.11 *4*
$\gamma_{\beta\text{-}}$	5221.6 *5*	0.30 *5*

† uncert(syst): 11% for β-, 13% for β-n

$^{90}_{36}$Kr(32.32 *9* s)

Mode: β-
Δ: -74960 *30* keV
SpA: 3.837×10^{9} Ci/g
Prod: fission

Photons (^{90}Kr)

⟨γ⟩=1237 *11* keV

γ_{mode}	γ(keV)	γ(%)†
Rb $K_{\alpha2}$	13.336	1.16 *6*
Rb $K_{\alpha1}$	13.395	2.25 *12*
Rb $K_{\beta1}$'	14.959	0.48 *3*
Rb $K_{\beta2}$'	15.286	0.052 *3*
γ M1+21%E2	106.036 *21*	0.416 *25*
γ (M3)	106.91 *3*	*
γ M1+20%E2	120.927 *22*	3.26 *22*
γ M1+20%E2	121.797 *22*	32.9 *11*
γ	180.67 *6*	~0.036
γ	220.72 *12*	~0.036
γ	227.83 *3*	0.116 *11*
γ	234.432 *23*	2.46 *11*
γ	242.178 *24*	9.2 *3*
γ	249.323 *23*	1.27 *11*
γ	305.05 *3*	0.051 *11*
γ	309.06 *6*	0.127 *11*
γ	356.23 *3*	0.10 *4*
γ	386.58 *4*	0.119 *11*
γ	392.6 *4*	~0.022
γ	396.32 *12*	0.047 *11*
γ	419.11 *3*	0.297 *11*
γ	429.99 *6*	0.14 *3*
γ	433.450 *25*	1.21 *4*
γ	433.93 *5*	0.09 *3*
γ [M1+E2]	465.33 *17*	0.065 *11*
γ	470.32 *8*	0.221 *14*
γ	476.10 *11*	0.123 *11*
γ	492.62 *3*	1.12 *3*
γ	498.69 *4*	0.141 *11*
γ	507.51 *4*	0.058 *18*
γ	539.485 *23*	28.6 *7*
γ	554.38 *3*	4.71 *11*
γ	565.21 *6*	0.192 *14*
γ	569.20 *4*	0.561 *18*
γ	576.99 *11*	0.051 *14*
γ	586.00 *17*	0.047 *7*
γ	614.41 *4*	0.195 *14*
γ	619.07 *4*	1.01 *3*
γ [M1+E2]	620.96 *21*	~0.036
γ	626.61 *7*	0.264 *18*
γ	658.1 *5*	0.029 *11*
γ	661.282 *25*	0.308 *11*
γ	677.90 *7*	0.355 *18*
γ	691.04 *11*	0.369 *14*
γ	705.38 *11*	0.116 *11*
γ	731.29 *3*	1.38 *4*
γ	739.3 *4*	0.022 *7* ?
γ	745.88 *7*	0.058 *18*
γ	925.56 *7*	0.206 *14*
γ	941.80 *4*	1.24 *3*
γ	947.24 *18*	0.054 *18*
γ	967.33 *11*	0.199 *18*
γ	980.31 *6*	0.174 *14*
γ	1031.60 *7*	0.058 *14*
γ	1039.14 *4*	0.387 *18*
γ	1103.90 *4*	0.319 *18*
γ	1118.719 *25*	36.2 *8*
γ	1165.59 *4*	0.77 *3*
γ	1240.34 *11*	0.326 *22*
γ	1293.7 *4*	0.054 *14* ?
γ [M1+E2]	1303.03 *7*	0.087 *14*
γ	1309.68 *10*	0.257 *14*
γ	1341.16 *16*	0.145 *18* ?
γ	1386.67 *6*	0.181 *18*
γ	1423.77 *3*	2.73 *6*
γ	1460.27 *17*	0.062 *18*
γ	1466.25 *5*	0.228 *18*
γ	1530.46 *9*	~0.036
γ	1537.82 *3*	8.98 *18*
γ	1552.163 *25*	2.04 *5*
γ	1620.07 *16*	0.141 *14* ?
γ	1658.198 *23*	1.23 *3*
γ	1692.70 *18*	0.072 *18*
γ	1695.6 *4*	~0.012
γ	1751.0 *3*	0.054 *11* ?
γ [E1]	1779.993 *25*	6.23 *14*
γ	1819.15 *17*	0.069 *11*
γ	1885.35 *5*	0.210 *14*
γ	1899.70 *5*	0.177 *14*
γ	1980.92 *8*	0.159 *11*
γ	2005.73 *5*	0.109 *18*

Photons (^{90}Kr)
(continued)

γ_{mode}	γ(keV)	γ(%)†
γ [E1]	2127.53 *5*	1.28 *4*
γ	2149.52 *9*	0.257 *11*
γ	2160.39 *17*	0.029 *9*
γ	2191.38 *17*	0.105 *11*
γ	2205.72 *17*	0.036 *11*
γ	2352.74 *14*	0.083 *14*
γ	2417.50 *13*	0.177 *14*
γ	2421.73 *7*	0.047 *14*
γ	2432.32 *13*	0.141 *14*
γ	2468.60 *7*	0.43 *4*
γ	2479.19 *14*	~0.036
γ	2497.78 *14*	~0.014
γ	2726.78 *7*	0.81 *3*
γ	2770.75 *21*	0.054 *11*
γ	2855.17 *7*	0.30 *6*
γ	2865.76 *13*	0.174 *14*
γ	2949.0 *3*	~0.036 ?
γ	3010.80 *13*	0.029 *9*
γ	3205.1 *3*	0.032 *8* ?
γ	3217.2 *3*	~0.010 ?
γ	3256.2 *12*	0.019 *8*
γ	3268.9 *3*	0.062 *11* ?
γ	3344.3 *3*	0.105 *14* ?
γ	3465.1 *9*	0.033 *11*
γ	3855.3 *4*	0.112 *11*
γ	4166.5 *10*	0.029 *11*

† 8.0% uncert(syst)

* with ^{90}Rb(4.3 min)

Atomic Electrons (^{90}Kr)

⟨e⟩=8.5 *4* keV

e_{bin}(keV)	⟨e⟩(keV)	e(%)
2 - 15	0.33	7.1 *6*
91	0.086	0.095 *6*
92	0.32	0.35 *10*
104 - 105	0.088	0.084 *19*
106	0.50	0.47 *3*
107	4.96	4.66 *17*
119	0.074	0.062 *4*
120	0.73	0.611 *22*
121 - 165	0.159	0.130 *4*
179 - 226	0.13	~0.06
227	0.4	~0.19
228 - 249	0.14	0.061 *24*
290 - 309	0.0053	0.0018 *8*
341 - 387	0.0054	0.0015 *5*
391 - 434	0.026	0.0061 *22*
450 - 499	0.020	0.0042 *13*
505 - 508	7.4 $\times 10^{-5}$	1.5 *7* $\times 10^{-5}$
524	0.24	~0.046
537 - 586	0.08	0.015 *5*
599 - 646	0.013	0.0022 *6*
656 - 705	0.0056	0.00084 *22*
716 - 746	0.008	0.0012 *5*
910 - 952	0.0069	0.00074 *21*
965 - 980	0.0008	8 *3* $\times 10^{-5}$
1016 - 1039	0.0016	0.00016 *5*
1089 - 1119	0.12	0.011 *4*
1150 - 1166	0.0025	0.00021 *7*
1225 - 1240	0.0010	8 *3* $\times 10^{-5}$
1278 - 1326	0.0016	0.000122 *22*
1339 - 1387	0.00054	3.9 *12* $\times 10^{-5}$
1409 - 1451	0.0079	0.00056 *16*
1458 - 1552	0.027	0.0018 *5*
1605 - 1695	0.0034	0.00021 *5*
1736 - 1819	0.0097	0.000548 *22*
1870 - 1966	0.00110	5.8 *10* $\times 10^{-5}$
1979 - 2005	0.00025	1.3 *3* $\times 10^{-5}$
2112 - 2205	0.00257	0.000121 *8*
2338 - 2432	0.00079	3.3 *5* $\times 10^{-5}$
2453 - 2498	0.00084	3.4 *8* $\times 10^{-5}$
2712 - 2771	0.0014	5.2 *11* $\times 10^{-5}$
2840 - 2934	0.00081	2.8 *5* $\times 10^{-5}$
2947 - 3011	5.1 $\times 10^{-5}$	1.7 *5* $\times 10^{-6}$

Atomic Electrons (^{90}Kr)
(continued)

e_{bin}(keV)	⟨e⟩(keV)	e(%)
3190 - 3269	0.00018	5.7 *10* $\times 10^{-6}$
3329 - 3344	0.00015	4.6 *11* $\times 10^{-6}$
3450 - 3465	4.7 $\times 10^{-5}$	1.4 *5* $\times 10^{-6}$
3840 - 3855	0.00015	4.0 *8* $\times 10^{-6}$
4151 - 4166	3.9 $\times 10^{-5}$	9 *4* $\times 10^{-7}$

Continuous Radiation (^{90}Kr)

⟨β-⟩=1299 keV;⟨IB⟩=3.6 keV

E_{bin}(keV)		⟨ ⟩(keV)	(%)
0 - 10	β-	0.0089	0.177
	IB	0.044	
10 - 20	β-	0.0273	0.182
	IB	0.043	0.30
20 - 40	β-	0.114	0.378
	IB	0.085	0.29
40 - 100	β-	0.89	1.26
	IB	0.24	0.37
100 - 300	β-	11.6	5.6
	IB	0.69	0.39
300 - 600	β-	55	12.0
	IB	0.79	0.19
600 - 1300	β-	339	35.5
	IB	1.07	0.124
1300 - 2500	β-	649	37.0
	IB	0.55	0.033
2500 - 4390	β-	243	8.0
	IB	0.079	0.0027

$^{90}_{37}$Rb(2.55 *5* min)

Mode: β-

Δ: -79353 *13* keV

SpA: 8.17 $\times 10^{8}$ Ci/g

Prod: fission

Photons (^{90}Rb)

⟨γ⟩=1846 *19* keV

γ_{mode}	γ(keV)	γ(%)
γ [M1+E2]	314.66 *4*	0.0147 *5*
γ	543.8 *3*	0.06 *3*
γ [E2]	551.10 *6*	0.0151 *9*
γ	720.65 *6*	0.0223 *10*
γ	739.00 *24*	0.047 *8*
γ	752.23 *11*	0.067 *8*
γ	765.44 *8*	0.0015 *4*
γ [E2]	824.21 *6*	0.64 *3*
γ [E2]	831.66 *4*	27.8 *8*
γ	886.06 *8*	0.061 *11*
γ	892.1 *3*	0.028 *11*
γ	921.29 *14*	0.026 *3*
γ	985.27 *13*	0.025 *5*
γ	997.78 *6*	0.431 *14*
γ	1004.0 *3*	~0.0010
γ	1026.44 *15*	0.0154 *22*
γ	1038.63 *6*	0.295 *11*
γ M1+20%E2	1060.65 *3*	6.63 *12*
γ	1109.29 *10*	0.058 *22*
γ	1140.45 *6*	0.112 *4*
γ [M1+E2]	1146.81 *7*	0.040 *5*
γ	1176.36 *7*	0.036 *17*
γ	1271.75 *6*	0.0647 *20*
γ	1302.1 *3*	0.100 *14*
γ	1326.82 *11*	0.125 *14*
γ M1+18%E2	1375.30 *3*	0.297 *12*
γ	1430.3 *3*	0.047 *8*

Photons (^{90}Rb)
(continued)

γ_{mode}	γ(keV)	γ(%)
γ	1438.32 *11*	0.031 *11*
γ	1456.79 *14*	0.029 *3*
γ	1488.9 *3*	0.030 *5*
γ	1522.1 *4*	0.036 *8*
γ	1547.9 *3*	0.061 *14*
γ	1590.46 *20*	0.133 *17*
γ	1631.59 *12*	0.161 *14*
γ M1(+E2)	1665.59 *5*	0.310 *8*
γ	1669.0 *3*	0.14 *5*
γ	1686.2 *5*	0.042 *11*
γ	1738.85 *8*	0.0247 *8*
γ	1747.08 *14*	0.029 *3*
γ	1804.05 *6*	0.573 *19*
γ	1829.93 *8*	0.147 *14*
γ	1842.3 *5*	0.078 *16*
γ	1870.88 *13*	0.072 *17*
γ [E2]	1892.30 *4*	0.380 *14*
γ	1941.78 *6*	0.0116 *8*
γ	1973.3 *5*	0.036 *14*
γ	1996.1 *4*	0.036 *14*
γ	2119.7 *8*	0.069 *14*
γ [E2]	2139.34 *10*	0.309 *22*
γ	2148.25 *8*	0.21 *3*
γ M1+5.0%E2	2207.44 *7*	0.431 *17*
γ	2216.36 *11*	0.50 *3*
γ	2239.7 *3*	~0.15
γ	2246.0 *4*	~0.06
γ	2256.43 *7*	0.0123 *6*
γ	2298.17 *15*	0.042 *16*
γ	2381.02 *9*	0.072 *19*
γ	2473.61 *10*	0.58 *5*
γ	2476.51 *20*	~0.10
γ [E2]	2497.24 *6*	0.047 *3*
γ	2688.4 *3*	0.117 *22*
γ (E2+9.0%M1)	2724.10 *11*	0.136 *15*
γ	2788.9 *5*	~0.09
γ	2924.8 *3*	0.069 *22*
γ	2980.38 *6*	0.089 *19*
γ	3032.72 *7*	0.061 *7*
γ [M1]	3039.08 *7*	0.70 *3*
γ	3081.44 *20*	0.147 *25*
γ	3205.20 *8*	0.473 *25*
γ	3295.03 *6*	0.81 *4*
γ	3303.84 *10*	0.84 *3*
γ	3317.05 *6*	0.267 *8*
γ	3361.85 *11*	0.92 *4*
γ	3383.27 *7*	6.37 *19*
γ	3534.23 *9*	3.81 *11*
γ	3538.7 *3*	0.15 *3*
γ	3626.88 *23*	0.12 *5*
γ	3664.0 *5*	0.078 *16*
γ	3814.46 *12*	0.55 *4*
γ	3929.3 *5*	~0.042
γ	3958.6 *5*	0.075 *22*
γ	4019.3 *4*	~0.033
γ	4061.7 *3*	0.23 *5*
γ	4087.28 *18*	0.242 *17*
γ	4135.47 *10*	6.37 *22*
γ	4192.72 *19*	0.047 *4*
γ	4278.4 *8*	0.047 *10*
γ	4332.06 *20*	0.37 *7*
γ	4355.64 *6*	0.420 *22*
γ	4365.86 *9*	7.54 *23*
γ	4501.32 *20*	0.033 *14*
γ	4599.3 *3*	0.142 *11*
γ	4635.1 *14*	0.022 *4*
γ	4646.09 *12*	2.11 *8*
γ	4685.4 *3*	0.019 *8*
γ	4790.2 *5*	0.058 *14*
γ	4918.90 *18*	0.072 *8*
γ	4934.8 *7*	0.033 *7*
γ	4973.67 *20*	0.195 *14*
γ	5007.7 *9*	0.022 *4*
γ	5070.2 *3*	0.14 *3*
γ	5187.26 *6*	1.10 *4*
γ	5254.08 *12*	0.220 *14*
γ	5299.5 *9*	0.016 *3*
γ	5332.94 *20*	0.409 *19*
γ	5600.1 *5*	0.031 *5*

Continuous Radiation (^{90}Rb)
⟨β-⟩=1981 keV;⟨IB⟩=7.4 keV

E_{bin}(keV)		⟨ ⟩(keV)	(%)
0 - 10	β-	0.0080	0.159
	IB	0.054	
10 - 20	β-	0.0245	0.163
	IB	0.054	0.37
20 - 40	β-	0.102	0.338
	IB	0.106	0.37
40 - 100	β-	0.79	1.12
	IB	0.31	0.47
100 - 300	β-	10.1	4.91
	IB	0.93	0.52
300 - 600	β-	44.8	9.8
	IB	1.15	0.27
600 - 1300	β-	229	24.3
	IB	1.9	0.22
1300 - 2500	β-	485	26.4
	IB	1.8	0.104
2500 - 5000	β-	1057	29.9
	IB	1.05	0.033
5000 - 6589	β-	156	2.87
	IB	0.022	0.00041

$^{90}_{37}$Rb(4.30 *8* min)

Mode: β-(97.4 *5* %), IT(2.6 *5* %)
Δ: -79246 *13* keV
SpA: 4.85×10^8 Ci/g
Prod: fission

Photons (^{90}Rb)
⟨γ⟩=3300 *38* keV

γ_{mode}	γ(keV)	γ(%)†
Rb $K_{\alpha2}$	13.336	0.397 *24*
Rb $K_{\alpha1}$	13.395	0.76 *5*
Rb $K_{\beta1}$'	14.959	0.165 *10*
Rb $K_{\beta2}$'	15.286	0.0176 *11*
γ_{IT} (M3)	106.91 *3*	0.229 *6*
$\gamma_{\beta-}$	196.9 *4*	0.30 *6*
$\gamma_{\beta-}$[M1+E2]	314.66 *4*	0.85 *3*
$\gamma_{\beta-}$	395.00 *18*	~0.08
$\gamma_{\beta-}$	442.43 *23*	0.12 *3*
$\gamma_{\beta-}$	522.10 *6*	0.41 *3*
$\gamma_{\beta-}$[E2]	551.10 *6*	0.87 *5*
$\gamma_{\beta-}$	720.65 *6*	0.57 *3*
$\gamma_{\beta-}$	739.00 *24*	0.00048 *9*
$\gamma_{\beta-}$	765.44 *8*	0.082 *21*
$\gamma_{\beta-}$	779.45 *12*	0.28 *6*
$\gamma_{\beta-}$[E2]	824.21 *6*	7.5 *4*
$\gamma_{\beta-}$[E2]	831.66 *4*	94 *3*
$\gamma_{\beta-}$	871.91 *8*	0.55 *4*
$\gamma_{\beta-}$	886.06 *8*	0.0042 *8*
$\gamma_{\beta-}$	921.29 *14*	0.23 *3*
$\gamma_{\beta-}$	952.45 *5*	1.75 *6*
$\gamma_{\beta-}$	985.27 *13*	0.094 *20*
$\gamma_{\beta-}$	1004.0 *3*	~0.06
$\gamma_{\beta-}$	1013.8 *1*	0.26 *3*
$\gamma_{\beta-}$	1021.20 *19*	0.08 *3*
$\gamma_{\beta-}$	1026.44 *15*	0.140 *20*
$\gamma_{\beta-}$M1+20%E2	1060.65 *3*	9.59 *20*
$\gamma_{\beta-}$[M1+E2]	1087.06 *9*	0.072 *15*
$\gamma_{\beta-}$	1140.45 *6*	0.83 *4*
$\gamma_{\beta-}$[M1+E2]	1146.81 *7*	0.0157 *20*
$\gamma_{\beta-}$	1176.36 *7*	0.0025 *11*
$\gamma_{\beta-}$	1242.74 *4*	3.15 *18*
$\gamma_{\beta-}$	1271.75 *6*	1.65 *7*
$\gamma_{\beta-}$	1298.3 *5*	0.21 *4*
$\gamma_{\beta-}$	1326.82 *11*	0.00126 *18*
$\gamma_{\beta-}$M1+18%E2	1375.30 *3*	17.1 *5*
$\gamma_{\beta-}$[M1+E2]	1377.35 *8*	2.4 *8*
$\gamma_{\beta-}$	1391.59 *24*	0.45 *8*
$\gamma_{\beta-}$	1425.05 *23*	0.28 *3*

Photons (^{90}Rb)
(continued)

γ_{mode}	γ(keV)	γ(%)†
$\gamma_{\beta-}$	1438.32 *11*	0.00031 *12*
$\gamma_{\beta-}$	1456.79 *14*	0.26 *4*
$\gamma_{\beta-}$	1459.69 *25*	0.20 *5*
$\gamma_{\beta-}$	1485.40 *17*	0.22 *7*
$\gamma_{\beta-}$	1488.9 *3*	0.36 *5*
$\gamma_{\beta-}$	1576.9 *4*	0.12 *4*
$\gamma_{\beta-}$	1603.48 *19*	0.48 *5*
$\gamma_{\beta-}$	1658.82 *23*	0.45 *6*
$\gamma_{\beta-}$M1(+E2)	1665.59 *5*	4.64 *12*
$\gamma_{\beta-}$[M1+E2]	1692.00 *8*	0.28 *5*
$\gamma_{\beta-}$	1696.11 *6*	1.71 *6*
$\gamma_{\beta-}$	1738.85 *8*	1.98 *8*
$\gamma_{\beta-}$	1747.08 *14*	0.26 *3*
$\gamma_{\beta-}$	1764.71 *10*	~0.09
$\gamma_{\beta-}$	1793.84 *6*	0.87 *5*
$\gamma_{\beta-}$	1837.97 *7*	0.86 *6*
$\gamma_{\beta-}$	1877.34 *20*	0.46 *5*
$\gamma_{\beta-}$[E2]	1892.30 *4*	0.550 *24*
$\gamma_{\beta-}$	1902.92 *25*	0.14 *6*
$\gamma_{\beta-}$	1941.78 *6*	0.64 *4*
$\gamma_{\beta-}$	2128.26 *6*	5.42 *15*
$\gamma_{\beta-}$[E2]	2139.34 *10*	0.12 *6*
$\gamma_{\beta-}$	2200.8 *3*	0.50 *6*
$\gamma_{\beta-}$M1+5.0%E2	2207.44 *7*	0.168 *9*
$\gamma_{\beta-}$	2246.0 *4*	~0.30
$\gamma_{\beta-}$	2256.43 *7*	0.68 *4*
$\gamma_{\beta-}$	2298.17 *15*	0.38 *14*
$\gamma_{\beta-}$	2311.09 *23*	0.30 *9*
$\gamma_{\beta-}$	2335.2 *7*	0.22 *9*
$\gamma_{\beta-}$	2442.91 *7*	0.27 *7*
$\gamma_{\beta-}$	2473.61 *10*	0.0058 *5*
$\gamma_{\beta-}$[E2]	2497.24 *6*	0.70 *5*
$\gamma_{\beta-}$	2538.37 *24*	0.18 *7*
$\gamma_{\beta-}$	2543.90 *14*	0.57 *7*
$\gamma_{\beta-}$	2591.97 *19*	0.67 *7*
$\gamma_{\beta-}$	2618.03 *4*	0.64 *9*
$\gamma_{\beta-}$(E2+9.0%M1)	2724.10 *11*	0.52 *5*
$\gamma_{\beta-}$	2740.3 *4*	~0.15
$\gamma_{\beta-}$(E2+40%M1)	2752.63 *8*	11.9 *4*
$\gamma_{\beta-}$	2834.19 *14*	1.92 *11*
$\gamma_{\beta-}$	2900.2 *3*	~0.11
$\gamma_{\beta-}$	2911.5 *5*	~0.13
$\gamma_{\beta-}$	3032.72 *7*	0.45 *5*
$\gamma_{\beta-}$[M1]	3039.08 *7*	0.274 *15*
$\gamma_{\beta-}$	3148.84 *14*	2.58 *10*
$\gamma_{\beta-}$	3198.2 *5*	0.15 *6*
$\gamma_{\beta-}$	3214.4 *7*	0.14 *6*
$\gamma_{\beta-}$	3317.05 *6*	14.9 *4*
$\gamma_{\beta-}$	3370.7 *4*	0.41 *6*
$\gamma_{\beta-}$	3383.27 *7*	0.433 *14*
$\gamma_{\beta-}$	3503.53 *6*	2.45 *10*
$\gamma_{\beta-}$	3534.23 *9*	0.0384 *18*
$\gamma_{\beta-}$	3572.80 *17*	1.60 *9*
$\gamma_{\beta-}$	3620.8 *3*	0.60 *23*
$\gamma_{\beta-}$	3626.88 *23*	0.9 *4*
$\gamma_{\beta-}$	3814.46 *12*	0.0152 *14*
$\gamma_{\beta-}$	3972.2 *5*	0.38 *7*
$\gamma_{\beta-}$	4115.6 *4*	0.37 *6*
$\gamma_{\beta-}$	4192.72 *19*	0.88 *7*
$\gamma_{\beta-}$	4209.45 *14*	0.94 *9*
$\gamma_{\beta-}$	4257.51 *19*	0.76 *6*
$\gamma_{\beta-}$	4365.86 *9*	0.076 *3*
$\gamma_{\beta-}$	4454.02 *19*	1.22 *8*
$\gamma_{\beta-}$	4646.09 *12*	0.058 *4*
$\gamma_{\beta-}$	4725.77 *23*	0.11 *3*
$\gamma_{\beta-}$	4996.0 *3*	0.07 *3*

Atomic Electrons (^{90}Rb)*

⟨e⟩=2.47 *9* keV

e_{bin}(keV)	⟨e⟩(keV)	e(%)
2	0.034	1.83 *20*
11	0.0213	0.189 *21*
12	0.035	0.30 *3*
13	0.0218	0.167 *19*
15	0.00199	0.0135 *15*
92	1.85	2.01 *10*
105	0.425	0.405 *20*
107	0.086	0.080 *4*

* with IT

Continuous Radiation (^{90}Rb)

⟨β-⟩=1399 keV;⟨IB⟩=4.3 keV

E_{bin}(keV)		⟨ ⟩(keV)	(%)
0 - 10	β-	0.0099	0.197
	IB	0.044	
10 - 20	β-	0.0303	0.202
	IB	0.044	0.30
20 - 40	β-	0.126	0.418
	IB	0.086	0.30
40 - 100	β-	0.98	1.39
	IB	0.25	0.38
100 - 300	β-	12.5	6.1
	IB	0.71	0.40
300 - 600	β-	56	12.2
	IB	0.83	0.20
600 - 1300	β-	299	31.7
	IB	1.21	0.140
1300 - 2500	β-	555	31.1
	IB	0.85	0.049
2500 - 5000	β-	454	13.5
	IB	0.31	0.0100
5000 - 5864	β-	21.6	0.413
	IB	0.00138	2.7×10^{-5}

$^{90}_{38}$Sr(28.5 *2* yr)

Mode: β-

Δ: -85942 *3* keV

SpA: 139.4 Ci/g

Prod: fission

Continuous Radiation (^{90}Sr)

⟨β-⟩=196 keV;⟨IB⟩=0.124 keV

E_{bin}(keV)		⟨ ⟩(keV)	(%)
0 - 10	β-	0.143	2.88
	IB	0.0099	
10 - 20	β-	0.426	2.84
	IB	0.0092	0.064
20 - 40	β-	1.68	5.6
	IB	0.0164	0.057
40 - 100	β-	11.5	16.5
	IB	0.037	0.058
100 - 300	β-	96	49.2
	IB	0.047	0.030
300 - 546	β-	86	23.0
	IB	0.0046	0.00134

$^{90}_{39}$Y (2.671 *4* d)

Mode: β-

Δ: -86488 *3* keV

SpA: 5.432×10^{5} Ci/g

Prod: ^{89}Y(n,γ); daughter ^{90}Sr

Photons (^{90}Y)

⟨γ⟩=0.00031 keV

γ_{mode}	γ(keV)	γ(%)†
γ E2	2186.187 *5*	1.4 *3*$\times10^{-6}$

Continuous Radiation (^{90}Y)

⟨β-⟩=934 keV;⟨IB⟩=2.0 keV

E_{bin}(keV)		⟨ ⟩(keV)	(%)
0 - 10	β-	0.0169	0.337
	IB	0.035	
10 - 20	β-	0.052	0.344
	IB	0.035	0.24
20 - 40	β-	0.213	0.71
	IB	0.068	0.24
40 - 100	β-	1.62	2.29
	IB	0.19	0.29
100 - 300	β-	19.3	9.4
	IB	0.52	0.29
300 - 600	β-	79	17.5
	IB	0.53	0.126
600 - 1300	β-	401	42.5
	IB	0.57	0.068
1300 - 2282	β-	432	26.9
	IB	0.096	0.0063

$^{90}_{39}$Y (3.19 *1* h)

Mode: IT(99.9979 *2* %), β-(0.0021 *2* %)

Δ: -85806 *3* keV

SpA: 1.092×10^{7} Ci/g

Prod: ^{87}Rb(α,n); ^{89}Y(n,γ); ^{89}Y(d,p); ^{93}Nb(n,α); ^{88}Sr(α,d)

Photons (^{90}Y)

⟨γ⟩=634 *18* keV

γ_{mode}	γ(keV)	γ(%)†
Y L_ℓ	1.686	0.0083 *18*
Y L_η	1.762	0.0047 *12*
Y L_α	1.922	0.19 *4*
Y L_β	2.005	0.11 *3*
Y L_γ	2.267	0.0033 *10*
Y $K_{\alpha2}$	14.883	2.14 *11*
Y $K_{\alpha1}$	14.958	4.11 *21*
Y $K_{\beta1}$'	16.735	0.92 *5*
Y $K_{\beta2}$'	17.125	0.135 *7*
$\gamma_{\beta-}$E3	132.59 *3*	8.7 *9*$\times10^{-5}$ *
γ_{IT} M1+7.5%E2	202.47 *3*	96.6 *19*
$\gamma_{\beta-}$[E2]	425.46 *20*	0.00034 *5* *
γ_{IT} M4	479.49 *4*	91
γ_{IT} (E5)	681.96 *5*	0.36 *7*
$\gamma_{\beta-}$E2	2186.187 *5*	8.7 *17*$\times10^{-8}$ *
$\gamma_{\beta-}$E5	2318.900 *5*	0.00173 *17* *

† uncert(syst): 4.0% for IT, 9.5% for β-

* with ^{90}Zr(809 ms) in equilib

Atomic Electrons (^{90}Y)*

⟨e⟩=47.5 *16* keV

e_{bin}(keV)	⟨e⟩(keV)	e(%)
2	0.184	8.7 *9*
12	0.025	0.207 *22*
13	0.25	1.95 *21*
14	0.027	0.19 *2*
15	0.082	0.56 *6*
16	0.0070	0.043 *5*
17	0.0026	0.0159 *17*
185	5.0	2.68 *15*
200	0.63	0.314 *20*
202	0.128	0.063 *4*
462	35.0	7.6 *3*
477	5.13	1.08 *5*
479	1.06	0.221 *10*
665	0.051	0.0076 *16*
680	0.0067	0.00099 *20*
682	0.0014	0.00020 *4*

* with IT

$^{90}_{40}$Zr(stable)

Δ: -88769.7 *24* keV

%: 51.45 *2*

$^{90}_{40}$Zr(809.2 *20* ms)

Mode: IT

Δ: -86450.8 *24* keV

SpA: 1.041×10^{11} Ci/g

Prod: ^{93}Nb(p,α); ^{90}Zr(n,n')

Photons (^{90}Zr)

⟨γ⟩=2016 *120* keV

γ_{mode}	γ(keV)	γ(%)†
Zr L_ℓ	1.792	0.0090 *18*
Zr L_η	1.876	0.0051 *9*
Zr L_α	2.042	0.21 *4*
Zr L_β	2.133	0.124 *24*
Zr L_γ	2.369	0.0055 *12*
Zr $K_{\alpha2}$	15.691	1.95 *20*
Zr $K_{\alpha1}$	15.775	3.7 *4*
Zr $K_{\beta1}$'	17.663	0.85 *9*
Zr $K_{\beta2}$'	18.086	0.138 *15*
γ E3	132.59 *3*	4.2 *4*
γ [E2]	425.46 *20*	16.6 *25*
γ E2	2186.187 *5*	0.0042 *8*
γ E5	2318.900 *5*	84

† 0.60% uncert(syst)

Atomic Electrons (^{90}Zr)

⟨e⟩=16.7 *11* keV

e_{bin}(keV)	⟨e⟩(keV)	e(%)
2	0.20	8.8 *12*
3	0.0099	0.39 *6*
13	0.19	1.41 *20*
14	0.051	0.38 *5*
15	0.082	0.53 *8*
16	0.0152	0.097 *14*
17	0.0067	0.039 *6*
18	0.0018	0.0103 *15*
115	10.3	9.0 *9*

Atomic Electrons (^{90}Zr) (continued)

e_{bin}(keV)	⟨e⟩(keV)	e(%)
130	3.5	2.7 3
132	0.65	0.49 5
133	0.097	0.073 8
407	0.41	0.100 16
423	0.051	0.0119 19
425	0.0105	0.0025 4
2168	1.11×10^{-5}	5.1 10 $\times10^{-7}$
2184	1.23×10^{-6}	5.6 11 $\times10^{-8}$
2186	2.2×10^{-7}	9.9 20 $\times10^{-9}$
2301	0.96	0.042 3
2316	0.107	0.0046 3
2317	0.0056	0.000243 15
2318	0.0188	0.00081 5
2319	0.00104	4.5 3 $\times10^{-5}$

$^{90}_{41}$Nb(14.60 5 h)

Mode: ε

Δ: -82659 5 keV

SpA: 2.385×10^{6} Ci/g

Prod: ^{90}Zr(p,n); ^{90}Zr(d,2n); descendant ^{90}Mo; ^{93}Nb(γ,3n); ^{92}Mo(γ,np)

Photons (^{90}Nb)

⟨γ⟩=3657 44 keV

γ_{mode}	γ(keV)	γ(%)†
Zr L_ℓ	1.792	0.059 10
Zr L_η	1.876	0.031 5
Zr L_α	2.042	1.40 22
Zr L_β	2.135	0.79 13
Zr L_γ	2.381	0.037 7
Zr $K_{\alpha2}$	15.691	14.3 5
Zr $K_{\alpha1}$	15.775	27.4 10
Zr $K_{\beta1}$'	17.663	6.25 22
Zr $K_{\beta2}$'	18.086	1.01 4
γ E3	132.59 3	4.13 8 *
γ E2	141.174 15	66.7 13
γ	329.050 16	0.12 4
γ [E1]	337.49 15	0.025 8
γ	371.298 3	1.80 7
γ [M1+E2]	420.27 5	0.0271 25
γ [E2]	425.46 20	0.0049 8 *
γ	501.27 23	0.013 5
γ M1	518.58 6	0.69 5
γ [E1]	561.590 11	0.120 3
γ	757.93 4	0.040 4
γ	792.03 19	0.011 3
γ E1	827.72 4	1.107 22
γ (E2)	890.62 4	1.72 8
γ	1051.50 3	0.213 8
γ	1057.8 10	0.017 5
γ	1093.12 8	0.094 6
γ E1	1129.195 6	92.7 19
γ	1192.7 10	0.164 16
γ E3	1270.364 13	1.30 3
γ	1470.491 24	0.459 16
γ	1492.9 10	
γ M1,E2	1574.995 23	0.517 16
γ M1,E2	1611.72 3	2.38 7
γ	1658.06 3	0.328 16
γ M1,E2	1716.223 21	0.500 16
γ M1,E2	1843.295 22	0.689 16
γ M1,E2	1913.146 25	1.271 25
γ	1984.49 3	0.681 25
γ	2000.1 3	0.066 8
γ E2	2186.187 5	18.0 4 *
γ (E1)	2222.287 18	0.623 25
γ E5	2318.900 5	82.0 16 *
γ	2321.88 24	0.74 16

Photons (^{90}Nb) (continued)

γ_{mode}	γ(keV)	γ(%)†
γ	2360 1	~0.007
γ	2740.9 3	0.0074 25
γ	2747.8 3	0.0049 16
γ	3316.0 10	

† 0.61% uncert(syst)

* with ^{90}Zr(809 ms) in equilib

Atomic Electrons (^{90}Nb)

⟨e⟩=45.5 7 keV

e_{bin}(keV)	⟨e⟩(keV)	e(%)
2 - 3	1.31	58 6
13	1.37	10.3 11
14 - 18	1.15	7.7 5
115	10.1	8.81 25
123	22.1	17.9 5
130	3.46	2.66 8
132 - 133	0.732	0.554 14
139	3.78	2.73 8
311 - 353	0.042	~0.012
369 - 418	0.006	0.0017 8
420 - 425	3.3×10^{-5}	7.7 10 $\times10^{-6}$
483 - 519	0.0105	0.00208 14
544 - 562	0.00065	0.000118 5
740 - 774	0.00033	4.5 16 $\times10^{-5}$
789 - 828	0.00380	0.000468 12
873 - 891	0.0134	0.00153 8
1034 - 1075	0.0014	0.00013 4
1091 - 1129	0.238	0.0214 5
1175 - 1193	0.00066	5.6 20 $\times10^{-5}$
1252 - 1270	0.0124	0.000988 25
1452 - 1470	0.0015	0.00010 3
1557 - 1656	0.0126	0.00079 4
1658 - 1743	0.00191	0.000113 7
1758 - 1843	0.00243	0.000133 8
1895 - 1984	0.0062	0.00033 3
1998 - 2000	1.9×10^{-5}	10 3 $\times10^{-7}$
2168 - 2222	0.0551	0.00254 6
2301 - 2360	1.07	0.0465 12
2723 - 2747	2.5×10^{-5}	9.2 25 $\times10^{-7}$

Continuous Radiation (^{90}Nb)

⟨β+⟩=350 keV;⟨IB⟩=2.0 keV

E_{bin}(keV)		⟨ ⟩(keV)	(%)
0 - 10	β+	2.14×10^{-5}	0.000263
	IB	0.0150	
10 - 20	β+	0.00058	0.00358
	IB	0.021	0.144
20 - 40	β+	0.0112	0.0350
	IB	0.029	0.100
40 - 100	β+	0.366	0.481
	IB	0.078	0.121
100 - 300	β+	13.0	6.0
	IB	0.21	0.118
300 - 600	β+	77	16.9
	IB	0.27	0.062
600 - 1300	β+	251	29.0
	IB	0.70	0.077
1300 - 2500	β+	8.8	0.66
	IB	0.68	0.042
2500 - 2522	IB	6.9×10^{-7}	2.8×10^{-8}
	Σβ+		53

$^{90}_{41}$Nb(18.82 9 s)

Mode: IT

Δ: -82533 5 keV

SpA: 6.54×10^{9} Ci/g

Prod: daughter ^{90}Mo; ^{90}Zr(p,n); ^{90}Zr(d,2n)

Photons (^{90}Nb)

⟨γ⟩=82.4 16 keV

γ_{mode}	γ(keV)	γ(%)†
Nb L_ℓ	1.902	0.028 5
Nb L_η	1.996	0.0136 21
Nb L_α	2.166	0.68 11
Nb L_β	2.278	0.38 6
Nb L_γ	2.537	0.022 4
Nb $K_{\alpha2}$	16.521	6.4 3
Nb $K_{\alpha1}$	16.615	12.3 5
Nb $K_{\beta1}$'	18.618	2.83 11
Nb $K_{\beta2}$'	19.074	0.477 20
γ E2	122.902 22	64.0

† 3.1% uncert(syst)

Atomic Electrons (^{90}Nb)

⟨e⟩=39.4 9 keV

e_{bin}(keV)	⟨e⟩(keV)	e(%)
2	0.59	24.5 25
3	0.044	1.62 17
14	0.75	5.3 6
16	0.29	1.82 19
17	0.0182	0.110 11
18	0.0235	0.129 13
19	0.0044	0.0235 24
104	30.5	29.4 8
120	4.48	3.72 11
121	1.45	1.20 3
122	0.591	0.483 14
123	0.670	0.547 15

$^{90}_{42}$Mo(5.67 5 h)

Mode: ε

Δ: -80172 6 keV

SpA: 6.14×10^{6} Ci/g

Prod: ^{93}Nb(p,4n); ^{90}Zr(α,4n)

Photons (^{90}Mo)

⟨γ⟩=571 15 keV

γ_{mode}	γ(keV)	γ(%)†
Nb L_ℓ	1.902	0.103 18
Nb L_η	1.996	0.049 8
Nb L_α	2.166	2.5 4
Nb L_β	2.279	1.38 23
Nb L_γ	2.543	0.081 16
Nb $K_{\alpha2}$	16.521	24.2 9
Nb $K_{\alpha1}$	16.615	46.1 17
Nb $K_{\beta1}$'	18.618	10.7 4
Nb $K_{\beta2}$'	19.074	1.79 7
γ (M1)	43.23 4	2.17 23
γ E2	122.902 22	64.1 23 *
γ M1	163.46 9	6.0 5
γ M1	203.66 10	6.4 5
γ E3	257.87 4	78 3

Photons (^{90}Mo)
(continued)

γ_{mode}	γ(keV)	γ(%)†
γ M1	323.73 *18*	6.3 *5*
γ	421.5 *3*	0.25 *8*
γ [M1+E2]	424.9 *3*	0.36 *8*
γ	441.0 *5*	0.93 *23*
γ M1,E2	445.91 *19*	6.0 *6*
γ M1	472.79 *24*	1.42 *15*
γ M1,E2	490.3 *3*	0.70 *8*
γ	518.2 *7*	~0.16
γ M1,E2	942.1 *3*	5.5 *6*
γ	946.9 *5*	0.70 *23*
γ	987.8 *10*	0.14 *5*
γ M1,E2	990.9 *5*	1.01 *8*
γ M1,E2	1271.8 *6*	4.1 *4*
γ M1,E2	1388.0 *3*	1.86 *23*
γ	1446.5 *20*	~0.047
γ M1,E2	1455.2 *6*	1.9 *5*
γ [M1+E2]	1463.7 *5*	0.70 *23*
γ	1482.0 *7*	<0.47 ?

† 3.6% uncert(syst)
* with ^{90}Nb(18.8 s) in equilib

Atomic Electrons (^{90}Mo)
⟨e⟩=79.0 *17* keV

e_{bin}(keV)	⟨e⟩(keV)	e(%)
2	2.11	88 *9*
3	0.179	6.6 *7*
14	2.8	20.1 *21*
16 - 43	2.77	13.5 *9*
104	30.5	29.4 *12*
120	4.49	3.73 *15*
121 - 163	3.26	2.60 *6*
185 - 204	0.41	0.220 *15*
239	25.8	10.8 *4*
255	3.98	1.56 *6*
256 - 305	2.30	0.885 *20*
321 - 324	0.0259	0.0081 *5*
403 - 446	0.175	0.041 *6*
454 - 499	0.043	0.0094 *8*
516 - 518	0.00026	~5×10^{-5}
923 - 972	0.055	0.0059 *6*
985 - 991	0.00100	0.000101 *9*
1253 - 1272	0.0231	0.00184 *19*
1369 - 1388	0.0095	0.00070 *9*
1428 - 1464	0.0134	0.00093 *17*
1479 - 1482	0.00010	~7×10^{-6}

Continuous Radiation (^{90}Mo)
⟨β+⟩=122 keV;⟨IB⟩=1.42 keV

E_{bin}(keV)		⟨ ⟩(keV)	(%)
0 - 10	β+	2.00×10^{-5}	0.000244
	IB	0.0057	
10 - 20	β+	0.00056	0.00341
	IB	0.017	0.111
20 - 40	β+	0.0110	0.0340
	IB	0.0113	0.040
40 - 100	β+	0.358	0.468
	IB	0.029	0.044
100 - 300	β+	11.7	5.4
	IB	0.087	0.048
300 - 600	β+	54	12.1
	IB	0.18	0.041
600 - 1300	β+	56	7.5
	IB	0.72	0.079
1300 - 2105	IB	0.36	0.024
	Σβ+		26

$^{90}_{43}$Tc(8.3 *4* s)

Mode: ϵ
Δ: -70970 *300* keV syst
SpA: 1.45×10^{10} Ci/g
Prod: ^{92}Mo(p,3n)

Photons (^{90}Tc)
⟨γ⟩=740 *96* keV

γ_{mode}	γ(keV)	γ(%)†
Mo $K_{\alpha2}$	17.374	0.130 *6*
Mo $K_{\alpha1}$	17.479	0.248 *11*
Mo $K_{\beta1}$'	19.602	0.0586 *25*
Mo $K_{\beta2}$'	20.091	0.0102 *5*
γ [E2]	948.1 *2*	78

† 13% uncert(syst)

Atomic Electrons (^{90}Tc)
⟨e⟩=0.67 *8* keV

e_{bin}(keV)	⟨e⟩(keV)	e(%)
3	0.0124	0.48 *5*
14	0.00139	0.0097 *10*
15	0.0130	0.088 *9*
17	0.0061	0.036 *4*
19	0.00047	0.0024 *3*
20	8.9 ×10^{-5}	0.00046 *5*
928	0.56	0.060 *8*
945	0.063	0.0067 *9*
946	0.00182	0.00019 *3*
948	0.0138	0.00145 *20*

$^{90}_{43}$Tc(49.2 *4* s)

Mode: ϵ
Δ: ~-70470 keV syst
SpA: 2.530×10^{9} Ci/g
Prod: ^{92}Mo(p,3n)

Photons (^{90}Tc)
⟨γ⟩=3103 *120* keV

γ_{mode}	γ(keV)	γ(%)
γ	134.6 *5*	5.2 *8*
γ	231.2 *8*	2.68 *5*
γ [E1]	262.5 *8*	1.3 *4*
γ	310.3 *6*	2.6 *6*
γ	481.7 *3*	13.6 *9*
γ	543.4 *5*	4.2 *4*
γ [E1]	546.8 *8*	9.3 *6*
γ	592.9 *8*	1.2 *4*
γ	801.2 *5*	2.8 *4*
γ [E2]	809.8 *3*	34.3 *15*
γ	944.7 *4*	36.6 *20*
γ	948.1 *2*	100
γ	983.5 *5*	5.4 *5*
γ [E2]	1035.5 *10*	1.1 *4*
γ	1054.3 *3*	100 *6*
γ	1147.9 *5*	1.7 *3*
γ	1291.4 *5*	7.3 *6*
γ	1363.7 *8*	3.6 *5*
γ	2091.7 *9*	1.9 *4*
γ	2355.3 *10*	0.7 *3*

A = 91

NDS **31**, 181 (1980)

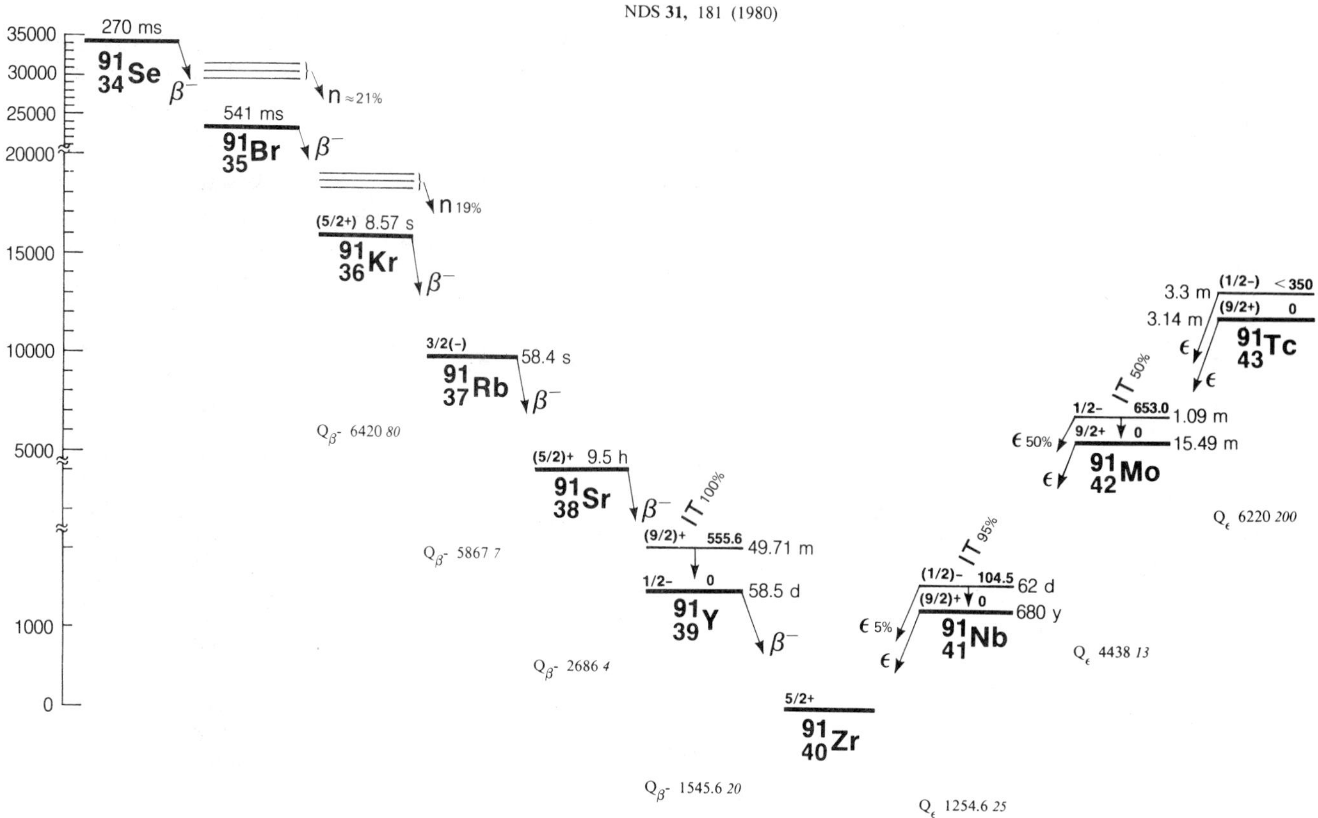

$^{91}_{34}$Se(270 *50* ms)

Mode: β-, β-n(21 *10* %)
SpA: 1.7×10^{11} Ci/g
Prod: fission

$^{91}_{35}$Br(541 *5* ms)

Mode: β-, β-n(19.2 *13* %)
SpA: 1.292×10^{11} Ci/g
Prod: fission

$^{91}_{36}$Kr(8.57 *4* s)

Mode: β-
Δ: -71370 *80* keV
SpA: 1.390×10^{10} Ci/g
Prod: fission

Photons (^{91}Kr)

⟨γ⟩=1746 *19* keV

γ_{mode}	γ(keV)	γ(%)†
Rb $K_{\alpha2}$	13.336	0.96 *9*
Rb $K_{\alpha1}$	13.395	1.86 *18*
Rb $K_{\beta1}$'	14.959	0.40 *4*
Rb $K_{\beta2}$'	15.286	0.043 *4*
γ M1(+1.4%E2)	108.81 *3*	43.5 *25*
γ	215.11 *5*	0.18 *4*
γ	384.45 *21*	0.07 *3*
γ	397.79 *4*	1.57 *11*
γ	400.32 *9*	0.21 *5*
γ	412.12 *6*	2.35 *13*
γ	446.78 *4*	1.65 *9*
γ	451.23 *14*	0.052 *17* ?
γ	469.61 *15*	0.065 *22*
γ	474.63 *7*	0.91 *6*
γ	481.44 *6*	1.24 *8*
γ	489.46 *9*	0.43 *6*
γ	502.06 *9*	1.6 *3*
γ	506.60 *4*	19.1 *13*
γ	542.21 *13*	~0.06
γ	546.01 *6*	0.41 *4*
γ	555.59 *4*	1.94 *10*
γ	569.03 *18*	0.20 *3*
γ	588.23 *6*	0.90 *5*
γ	612.90 *4*	7.7 *4*
γ	630.17 *5*	2.22 *13*
γ	662.44 *7*	1.28 *8*
γ	671.47 *6*	0.70 *5*
γ	680.17 *12*	0.12 *3*
γ	712.12 *7*	0.22 *3*
γ	721.71 *5*	0.66 *4*
γ	748.70 *6*	0.57 *4*
γ	761.12 *6*	1.04 *7*
γ	765.65 *11*	~0.05
γ	771.75 *8*	0.35 *4*
γ	779.95 *7*	0.10 *4*

Photons (^{91}Kr) (continued)

γ_{mode}	γ(keV)	γ(%)†
γ	785.59 *8*	0.47 *5*
γ	797.70 *7*	0.24 *3*
γ	802.24 *9*	0.122 *22*
γ	807.25 *8*	0.57 *4*
γ	814.19 *15*	0.13 *3*
γ	817.72 *9*	0.46 *4*
γ	822.17 *8*	0.40 *4*
γ	825.98 *12*	0.41 *4*
γ	846.28 *12*	0.11 *4*
γ	858.79 *13*	0.26 *4*
γ	874.95 *7*	1.27 *7*
γ	880.01 *18*	0.126 *22*
γ	893.56 *7*	0.17 *4*
γ	895.28 *12*	0.29 *7*
γ	900.69 *11*	0.16 *4*
γ	953.12 *7*	0.33 *4*
γ	956.07 *7*	0.32 *4*
γ	992.09 *12*	0.13 *5*
γ	995.06 *6*	0.80 *6*
γ	1008.86 *19*	0.19 *3*
γ	1025.01 *6*	2.87 *22*
γ	1027.96 *6*	0.65 *13*
γ	1041.80 *15*	0.22 *3* ?
γ	1059.07 *12*	0.27 *3*
γ	1069.26 *6*	0.087 *22*
γ	1085.9 *3*	0.117 *22* ?
γ	1091.60 *9*	0.34 *3*
γ	1102.35 *9*	0.76 *8*
γ	1108.66 *6*	7.2 *4*
γ	1129.6 *3*	0.11 *4*
γ	1136.77 *6*	1.04 *8*
γ	1158.90 *6*	0.10 *5*
γ	1178.07 *6*	1.29 *7*
γ	1195.49 *6*	0.26 *3*
γ	1199.24 *18*	0.10 *3*
γ	1202.3 *3*	0.12 *3*
γ	1215.51 *8*	0.66 *5*

Photons (^{91}Kr) (continued)

γ_{mode}	γ(keV)	γ(%)†
γ	1227.52 22	0.122 22 ?
γ	1230.98 18	0.052 22
γ	1247.6 3	0.174 4
γ	1267.71 6	0.67 5
γ	1277.00 16	0.21 4
γ	1281.14 12	0.62 5
γ	1293.06 12	0.49 5
γ	1304.29 6	1.25 8
γ	1311.66 16	0.44 5
γ	1315.75 10	0.59 6
γ	1324.32 8	0.55 5
γ	1327.07 18	0.13 4
γ	1338.10 16	0.17 4 ?
γ	1353.43 9	0.60 9
γ	1356.38 8	0.75 9
γ	1359.98 9	0.22 5
γ	1365.25 17	0.23 6
γ	1368.18 7	0.33 6
γ	1387.10 11	0.55 6
γ	1392.84 6	0.55 5
γ	1401.87 12	0.23 5
γ	1419.66 11	0.84 6
γ	1425.67 15	0.10 4
γ	1438.86 12	0.36 4
γ	1456.48 10	0.35 10
γ	1459.43 10	0.28 8
γ	1468.66 12	0.16 4
γ	1474.13 13	0.09 3
γ	1479.86 19	0.54 6 ?
γ	1500.71 12	0.70 9
γ	1501.65 6	4.8 3
γ	1506.45 6	0.83 17
γ	1518.03 20	0.09 3
γ	1525.1 3	0.16 4
γ	1528.30 13	0.91 6
γ	1537.31 10	0.33 4
γ	1547.67 12	0.37 5
γ	1555.32 17	0.61 17
γ	1557.34 9	0.48 17
γ	1563.6 3	0.17 4
γ	1577.90 22	0.10 3
γ	1583.29 6	0.38 4
γ	1589.06 14	0.11 3
γ	1614.10 14	1.04 7 ?
γ	1626.7 3	0.33 10
γ	1633.53 13	~0.14
γ	1650.47 11	0.17 4
γ	1659.88 14	0.10 3
γ	1666.63 9	0.27 7
γ	1666.73 12	0.79 7 ?
γ	1675.47 19	0.38 4
γ	1681.27 24	0.17 3
γ	1698.06 12	0.15 5
γ	1709.85 25	0.24 8
γ	1724.86 9	0.19 4
γ	1727.81 8	0.50 4
γ	1741.93 11	0.80 6
γ	1753.01 14	0.19 3
γ	1779.06 14	0.82 7
γ	1783.23 12	0.38 5
γ	1789.32 13	0.41 4
γ	1823.00 13	0.30 4
γ	1827.41 13	0.20 4
γ	1834.5 3	0.13 3
γ	1843.0 2	0.13 4
γ	1856.5 3	0.08 4
γ	1866.44 12	0.17 4
γ	1871.54 9	0.19 6
γ	1875.06 17	0.47 6
γ	1879.55 15	0.24 4
γ	1884.5 3	0.104 17
γ	1913.96 24	0.07 3
γ	1965.14 18	0.67 6
γ	1983.60 9	0.17 4
γ	1995.07 18	~0.043
γ	2003.89 15	<0.08
γ	2039.38 21	0.39 4
γ	2057.42 14	0.42 4
γ	2072.41 17	0.31 4
γ	2087.01 12	0.17 4
γ	2139.92 8	0.71 8
γ	2195.82 12	0.35 4
γ	2242.47 13	0.17 3
γ	2251.4 3	0.14 4 ?

Photons (^{91}Kr) (continued)

γ_{mode}	γ(keV)	γ(%)†
γ	2268.49 18	0.23 5
γ	2280.64 11	0.15 5
γ	2322.92 19	0.11 4 ?
γ	2377.30 18	0.36 4
γ	2381.64 17	0.22 3
γ	2391.92 13	0.11 4
γ	2394.52 17	0.13 4
γ	2413.39 11	0.34 5
γ	2424.9 3	0.15 4
γ	2446.75 12	0.25 7
γ	2450.67 14	0.68 8
γ	2457.57 13	0.35 5
γ	2473.16 19	0.41 9
γ	2479.9 5	0.21 6
γ	2484.42 8	2.78 17
γ	2495.75 11	0.69 7
γ	2539.69 19	0.170 22
γ	2550.37 16	0.183 22
γ	2555.80 25	0.10 4
γ	2558.03 13	0.17 7
γ	2559.47 14	0.35 6
γ	2585.80 25	0.10 3
γ	2593.23 8	0.54 5
γ	2607.03 12	0.24 5
γ	2620.36 19	0.67 6
γ	2627.60 21	0.174 22
γ	2642.40 25	0.20 4
γ	2662.65 24	0.09 3
γ	2686.83 19	0.09 4
γ	2732.05 25	0.24 7
γ	2735.87 9	1.48 13
γ	2752.80 8	0.74 6
γ	2769.50 25	0.21 5
γ	2809.9 3	~0.12
γ	2811.17 12	0.27 3
γ	2844.67 9	0.33 5
γ	2855.34 13	0.40 5
γ	2870.93 19	0.85 7
γ	2893.52 11	0.40 5
γ	2904.5 3	~0.07
γ	2919.98 11	0.27 4
γ	2926.9 3	0.09 3
γ	2930.7 4	0.20 4 ?
γ	2965.9 4	0.13 4
γ	2981.88 13	1.30 8
γ	3002.33 11	0.26 9
γ	3004.80 12	~0.11
γ	3041.7 3	0.13 6
γ	3044.61 19	~0.05 ?
γ	3052.6 11	~0.16
γ	3056.95 16	0.87 9
γ	3097.39 16	0.37 4
γ	3109.42 24	0.36 8
γ	3113.61 12	2.13 13
γ	3181.0 4	0.11 4
γ	3265.5 5	0.061 22
γ	3325.07 25	0.22 3
γ	3393.78 24	0.28 4 ?
γ	3403.50 24	0.16 4
γ	3435.9 5	0.10 4
γ	3444.4 5	0.21 4
γ	3490.0 3	0.07 4
γ	3578.8 4	0.11 3
γ	3705.1 3	0.07 3
γ	3910.08 24	0.048 17
γ	3974.3 3	0.048 17
γ	4129.20 21	0.052 17
γ	4199.5 3	0.074 22 ?
γ	4437.1 4	0.070 13 ?

† 5.1% uncert(syst)

Atomic Electrons (^{91}Kr)

⟨e⟩=6.2 4 keV

e_{bin}(keV)	⟨e⟩(keV)	e(%)
2 - 15	0.27	5.8 6
94	4.5	4.8 4
107	0.59	0.56 7
108 - 109	0.116	0.107 9
200 - 215	0.012	~0.006
369 - 412	0.069	0.018 6
432 - 480	0.054	0.012 3
481 - 489	0.016	~0.0033
491	0.18	~0.037
500 - 546	0.046	0.009 3
554 - 598	0.06	0.011 4
611 - 656	0.035	0.0056 14
660 - 707	0.0070	0.0010 3
710 - 759	0.011	0.0015 4
761 - 807	0.0124	0.0016 3
811 - 860	0.0093	0.0011 3
865 - 901	0.0039	0.00045 9
938 - 980	0.0058	0.00060 14
990 - 1028	0.015	0.0015 4
1040 - 1087	0.0051	0.00047 10
1090 - 1137	0.028	0.0026 8
1144 - 1187	0.0057	0.00049 12
1193 - 1232	0.0032	0.00026 6
1246 - 1293	0.0090	0.00071 12
1296 - 1345	0.0096	0.00073 10
1350 - 1393	0.0055	0.00041 6
1400 - 1444	0.0050	0.00035 6
1453 - 1548	0.026	0.0017 3
1553 - 1652	0.0081	0.00050 7
1658 - 1753	0.0062	0.00037 5
1764 - 1860	0.0069	0.00038 4
1864 - 1963	0.0024	0.000125 22
1965 - 2057	0.0027	0.000131 19
2070 - 2140	0.0018	8.4 18 $\times10^{-5}$
2181 - 2279	0.0019	8.7 12 $\times10^{-5}$
2280 - 2380	0.00160	6.8 10 $\times10^{-5}$
2381 - 2481	0.0096	0.00039 5
2482 - 2578	0.0033	0.000130 14
2584 - 2672	0.0026	9.8 13 $\times10^{-5}$
2685 - 2769	0.0044	0.000160 24
2795 - 2893	0.0039	0.000137 15
2902 - 3001	0.0037	0.000124 17
3002 - 3098	0.0058	0.00019 3
3107 - 3181	0.00058	1.9 3 $\times10^{-5}$
3250 - 3325	0.00041	1.25 25 $\times10^{-5}$
3379 - 3475	0.00119	3.5 5 $\times10^{-5}$
3488 - 3579	0.00017	4.7 12 $\times10^{-6}$
3690 - 3705	9.1 $\times10^{-5}$	2.5 10 $\times10^{-6}$
3895 - 3974	0.00013	3.3 9 $\times10^{-6}$
4114 - 4199	0.00017	4.0 9 $\times10^{-6}$
4422 - 4437	9.0 $\times10^{-5}$	2.0 5 $\times10^{-6}$

Continuous Radiation (^{91}Kr)

⟨β-⟩=2073 keV;⟨IB⟩=7.5 keV

E_{bin}(keV)		⟨ ⟩(keV)	(%)
0 - 10	β-	0.00346	0.069
	IB	0.058	
10 - 20	β-	0.0106	0.071
	IB	0.058	0.40
20 - 40	β-	0.0446	0.148
	IB	0.114	0.40
40 - 100	β-	0.357	0.503
	IB	0.33	0.51
100 - 300	β-	4.95	2.38
	IB	1.00	0.56
300 - 600	β-	25.8	5.6
	IB	1.26	0.29
600 - 1300	β-	196	20.2
	IB	2.1	0.24
1300 - 2500	β-	724	38.5
	IB	1.8	0.104
2500 - 5000	β-	1045	31.2
	IB	0.76	0.024
5000 - 6420	β-	77	1.44
	IB	0.0085	0.000155

$^{91}_{37}$Rb(58.4 *4* s)

Mode: β-
Δ: -77794 *8* keV
SpA: 2.110×10^9 Ci/g
Prod: fission

Photons (^{91}Rb)

⟨γ⟩=2335 *33* keV

γmode	γ(keV)	γ(%)†
Sr L$_\ell$	1.582	0.027 *6*
Sr L$_\eta$	1.649	0.016 *4*
Sr L$_\alpha$	1.806	0.61 *13*
Sr L$_\beta$	1.878	0.35 *9*
Sr L$_\gamma$	2.196	0.0052 *16*
Sr K$_{\alpha2}$	14.098	6.9 *5*
Sr K$_{\alpha1}$	14.165	13.2 *10*
Sr K$_{\beta1}$'	15.832	2.90 *21*
Sr K$_{\beta2}$'	16.192	0.40 *3*
γ E2(+<8.4%M1)	93.670 *18*	33.7 *17*
γ	345.50 *3*	8.3 *4*
γ	439.172 *23*	2.09 *10*
γ	509.39 *9*	0.17 *3*
γ	593.26 *3*	1.29 *7*
γ	602.890 *25*	2.83 *13*
γ	702.70 *25*	0.104 *17*
γ	749.81 *23*	0.111 *20* ?
γ	816.5 *5*	0.101 *24*
γ	875.06 *12*	0.108 *17*
γ	911.98 *23*	0.071 *24*
γ	917.68 *9*	0.19 *3*
γ	948.39 *3*	1.17 *6*
γ	993.73 *12*	0.30 *3*
γ	1006.05 *12*	0.094 *24*
γ	1022.66 *5*	0.44 *4*
γ [M1]	1035.32 *11*	0.13 *5*
γ	1042.06 *3*	2.19 *10*
γ	1137.25 *5*	3.88 *20*
γ	1149.4 *3*	0.07 *3* ?
γ	1173.09 *14*	0.10 *3* ?
γ	1205.66 *22*	0.115 *24*
γ	1230.92 *5*	0.29 *3*
γ	1238.7 *6*	0.067 *20*
γ	1250.7 *8*	0.047 *20*
γ	1274.13 *7*	0.25 *3*
γ	1300.08 *12*	0.162 *24*
γ	1367.79 *7*	0.76 *5*
γ	1388.50 *10*	0.22 *3*
γ	1482.17 *10*	1.45 *10*
γ	1503.60 *18*	0.09 *3*
γ	1594.10 *13*	0.41 *4* ?
γ	1615.92 *5*	2.46 *13*
γ	1624.49 *14*	0.50 *3*
γ	1625.55 *5*	0.71 *5*
γ	1628.58 *11*	0.90 *6*
γ	1646.63 *8*	0.26 *3*
γ	1711.96 *17*	0.21 *4*
γ	1719.96 *22*	0.31 *5*
γ	1740.30 *8*	1.42 *10*
γ	1766.34 *14*	0.17 *4* ?
γ	1795.0 *3*	0.12 *4* ?
γ	1823.44 *12*	0.36 *6*
γ	1841.37 *14*	0.13 *4*
γ	1849.28 *8*	3.30 *17*
γ	1859.56 *18*	0.15 *3*
γ	1873.7 *3*	0.11 *5* ?
γ	1917.11 *12*	0.76 *6*
γ	1942.95 *8*	0.40 *4*
γ	1952.99 *13*	0.07 *3*
γ	1971.04 *5*	6.7 *3*
γ	2013.62 *11*	0.27 *4*
γ	2036.37 *18*	0.37 *5*
γ	2064.71 *5*	0.79 *6*
γ	2143.28 *11*	0.67 *5*
γ	2161.65 *22*	0.12 *4* ?
γ	2196.0 *4*	0.19 *4*
γ	2208.5 *7*	0.10 *4*
γ	2218.80 *5*	0.28 *4*
γ	2236.95 *11*	0.14 *3*
γ	2254.70 *16*	0.12 *3*
γ	2263.05 *19*	0.15 *3*

Photons (^{91}Rb) (continued)

γmode	γ(keV)	γ(%)†
γ	2322.73 *12*	0.45 *5*
γ	2448.1 *3*	0.15 *5* ?
γ	2505.95 *13*	1.42 *8*
γ [E1]	2564.29 *5*	12.5 *6*
γ	2606.7 *5*	0.14 *4*
γ	2724.2 *7*	0.17 *5*
γ	2783.33 *17*	0.33 *5*
γ	2789.75 *19*	0.49 *5*
γ	2847.40 *11*	0.65 *7*
γ	2872.3 *4*	0.19 *6* ?
γ	2897.71 *16*	0.21 *4*
γ	2912.0 *4*	0.330 *24*
γ	2925.60 *11*	1.52 *10*
γ	2958.47 *16*	0.13 *4*
γ	2990.70 *17*	0.20 *5*
γ	3007.35 *17*	0.27 *5*
γ	3107.9 *9*	0.16 *5*
γ	3147.33 *15*	0.65 *6*
γ	3224.4 *3*	0.33 *4* ?
γ	3271.09 *11*	0.44 *6*
γ	3285.68 *18*	0.16 *5*
γ	3301.8 *4*	0.14 *5* ?
γ	3337.47 *17*	0.22 *6*
γ	3346.2 *3*	0.18 *8* ?
γ	3352.84 *17*	0.20 *7*
γ	3376.5 *3*	~0.27
γ	3395.5 *4*	0.32 *6* ?
γ	3410.9 *3*	~0.08 ?
γ	3446.51 *17*	1.48 *10*
γ [E1]	3599.58 *11*	10.4 *5*
γ	3604.16 *15*	0.37 *17*
γ	3639.12 *10*	1.21 *10*
γ	3643.78 *21*	0.79 *8* ?
γ	3682.96 *17*	~0.08
γ	3736.83 *13*	0.57 *13*
γ	3745.7 *3*	0.20 *5* ?
γ	3800.9 *4*	0.15 *3* ?
γ	3839.4 *3*	0.61 *6* ?
γ	3844.7 *3*	1.02 *8* ?
γ	3888.55 *18*	0.28 *4*
γ	3906.2 *9*	0.09 *3*
γ	3938.4 *3*	0.18 *4* ?
γ	3949.64 *15*	0.64 *5*
γ [E1]	3984.61 *10*	0.41 *4*
γ	4043.31 *15*	0.74 *5*
γ	4061.3 *5*	0.11 *4*
γ	4063.83 *19*	~0.04
γ [E1]	4078.28 *10*	4.08 *20*
γ	4095.68 *15*	0.24 *3*
γ	4157.50 *19*	0.70 *5*
γ [E1]	4171.79 *15*	0.28 *3*
γ	4189.35 *15*	0.23 *3*
γ	4224.8 *6*	0.101 *20*
γ	4234.04 *18*	0.219 *24*
γ	4249.0 *3*	0.34 *3* ?
γ	4253.7 *3*	0.377 *8* ?
γ [E1]	4265.45 *15*	1.43 *8*
γ	4297.3 *3*	0.115 *17* ?
γ	4358.45 *16*	0.054 *17*
γ	4391.0 *3*	0.054 *17* ?
γ	4452.9 *3*	0.145 *17* ?
γ	4544.1 *5*	0.054 *13*
γ	4699.42 *25*	~0.017 ?

† 5.9% uncert(syst)

Atomic Electrons (^{91}Rb)

⟨e⟩=34.9 *17* keV

e$_{bin}$(keV)	⟨e⟩(keV)	e(%)
2	0.61	31 *4*
12	0.92	7.7 *9*
14 - 16	0.40	2.8 *3*
78	26.2	33.8 *22*
91	2.83	3.09 *20*
92	2.45	2.67 *17*
93 - 94	1.05	1.13 *6*
329 - 346	0.20	~0.06
423 - 439	0.031	0.007 *4*
493 - 509	0.0019	0.00039 *19*
577 - 603	0.035	0.0060 *20*
687 - 734	0.0013	0.00018 *6*
748 - 750	7.9 ×10^{-5}	1.1 *5* ×10^{-5}
800 - 817	0.00054	7 *3* ×10^{-5}
859 - 902	0.0016	0.00018 *5*
910 - 948	0.0053	0.00056 *20*
978 - 1026	0.012	0.0011 *3*
1033 - 1042	0.0011	0.00010 *3*
1121 - 1157	0.014	0.0013 *4*
1171 - 1215	0.0013	0.00011 *3*
1223 - 1272	0.0013	0.000101 *24*
1274 - 1300	0.00052	4.0 *14* ×10^{-5}
1352 - 1388	0.0029	0.00021 *6*
1466 - 1503	0.0043	0.00029 *9*
1578 - 1646	0.0135	0.00084 *14*
1696 - 1795	0.0054	0.00031 *6*
1807 - 1901	0.0109	0.00059 *12*
1915 - 2013	0.017	0.00085 *21*
2020 - 2064	0.0025	0.000120 *25*
2127 - 2221	0.0030	0.000140 *21*
2235 - 2322	0.0015	6.4 *12* ×10^{-5}
2432 - 2506	0.0029	0.00012 *3*
2548 - 2606	0.0174	0.00068 *4*
2708 - 2789	0.0017	6.3 *10* ×10^{-5}
2831 - 2925	0.0049	0.000171 *23*
2942 - 3007	0.00101	3.4 *6* ×10^{-5}
3092 - 3147	0.0013	4.2 *8* ×10^{-5}
3208 - 3302	0.00172	5.3 *7* ×10^{-5}
3321 - 3411	0.0020	6.0 *10* ×10^{-5}
3430 - 3446	0.0023	6.8 *14* ×10^{-5}
3583 - 3683	0.0164	0.000455 *25*
3721 - 3801	0.0014	3.7 *7* ×10^{-5}
3823 - 3922	0.0032	8.4 *10* ×10^{-5}
3934 - 4027	0.0024	6.1 *7* ×10^{-5}
4041 - 4095	0.0055	0.000134 *7*
4141 - 4238	0.00303	7.2 *6* ×10^{-5}
4247 - 4342	0.00200	4.7 *3* ×10^{-5}
4356 - 4453	0.00029	6.5 *11* ×10^{-6}
4528 - 4544	7.5 ×10^{-5}	1.6 *5* ×10^{-6}
4683 - 4699	2.3 ×10^{-5}	~5×10^{-7}

Continuous Radiation (^{91}Rb)

⟨β-⟩=1542 keV;⟨IB⟩=5.0 keV

E$_{bin}$(keV)		⟨ ⟩(keV)	(%)
0 - 10	β-	0.0097	0.192
	IB	0.047	
10 - 20	β-	0.0296	0.197
	IB	0.047	0.32
20 - 40	β-	0.123	0.408
	IB	0.092	0.32
40 - 100	β-	0.96	1.35
	IB	0.26	0.41
100 - 300	β-	12.1	5.9
	IB	0.77	0.43
300 - 600	β-	54	11.9
	IB	0.92	0.21
600 - 1300	β-	294	31.2
	IB	1.38	0.159
1300 - 2500	β-	545	30.0
	IB	1.07	0.061
2500 - 5000	β-	616	18.4
	IB	0.40	0.0130
5000 - 5867	β-	19.5	0.375
	IB	0.00093	1.8×10^{-5}

$^{91}_{38}$Sr(9.52 *6* h)

Mode: β-
Δ: -83661 *5* keV
SpA: 3.617×10^{6} Ci/g
Prod: fission; ^{94}Zr(n,α)

Photons (^{91}Sr)

⟨γ⟩=1047 *69* keV

γ_{mode}	γ(keV)	γ(%)†
Y $K_{\alpha 2}$	14.883	0.62 *3*
Y $K_{\alpha 1}$	14.958	1.18 *5*
Y $K_{\beta 1}'$	16.735	0.264 *12*
Y $K_{\beta 2}'$	17.125	0.0390 *18*
γ [E1]	118.45 *7*	0.073 *3*
γ [M1+E2]	261.07 *6*	0.448 *9*
γ [M1+E2]	272.77 *7*	0.26 *4*
γ (M1+E2)	274.56 *7*	1.03 *3*
γ [M1+E2]	359.00 *6*	0.050 *3*
γ [E1]	379.52 *7*	0.147 *3*
γ	393.01 *6*	0.050 *3*
γ	486.66 *8*	0.080 *3*
γ [M1+E2]	506.65 *7*	0.043 *3*
γ [E1]	520.67 *7*	0.033 *3*
γ [E2]	533.84 *6*	0.077 *3*
γ M4	555.56 *7*	61.5 *12* *
γ [E1]	592.80 *9*	0.094 *3*
γ M1+0.2%E2	620.07 *6*	1.77 *4*
γ	626.78 *7*	0.043 *3*
γ [E1]	631.28 *6*	0.554 *11*
γ (M2+E1)	652.29 *8*	2.97 *17*
γ [M1+E2]	653.00 *7*	8.0 *3*
γ	654.08 *7*	0.37 *13*
γ [M1+E2]	660.79 *7*	0.100 *3*
γ [E2]	749.72 *7*	23.6 *5*
γ [M1+E2]	761.22 *7*	0.574 *11*
γ [E2]	793.52 *6*	0.063 *3*
γ [M1+E2]	820.71 *8*	0.160 *3*
γ	823.65 *10*	0.067 *3*
γ [E1]	879.67 *6*	0.187 *4*
γ [M1+E2]	892.84 *6*	0.070 *3*
γ [E1]	901.34 *8*	0.094 *3*
γ M1,E2	925.77 *7*	3.84 *8*
γ	973.89 *9*	0.040 *3*
γ	992.14 *10*	0.043 *3*
γ M1+E2	1024.28 *7*	33
γ [M1+E2]	1054.58 *7*	0.224 *5*
γ [E1]	1140.74 *7*	0.127 *3*
γ [M1+E2]	1280.85 *7*	0.932 *19*
γ	1305.28 *7*	0.017 *3*
γ [M1+E2]	1327.35 *7*	0.040 *3*
γ	1353.41 *10*	0.023 *3*
γ [E1]	1413.51 *7*	0.979 *20*
γ [M1+E2]	1473.70 *7*	0.167 *3*
γ	1486.38 *9*	0.013 *3*
γ [E2]	1545.83 *6*	0.067 *3*
γ [M1+E2]	1553.62 *8*	0.017 *3*
γ	1626.17 *11*	0.013 *3*
γ	1646.21 *12*	0.0030 *3*
γ	1651.06 *8*	0.291 *6*
γ	1723.61 *10*	0.160 *3*
γ	2016.41 *12*	0.004 *1*
γ	2412.13 *10*	0.0043 *10*

† 6.9% uncert(syst)
* with ^{91}Y(49.71 min) in equilib

Atomic Electrons (^{91}Sr)

⟨e⟩=18.8 *4* keV

e_{bin}(keV)	⟨e⟩(keV)	e(%)
2 - 17	0.166	3.4 *3*
101 - 118	0.0050	0.0048 *3*
244 - 275	0.094	0.037 *8*
342 - 391	0.0038	0.00106 *18*
470 - 519	0.0031	0.00062 *11*
520 - 534	0.000165	3.11 *14* ×10^{-5}
539	15.3	2.84 *8*
553	2.14	0.387 *11*
555	0.379	0.0683 *19*
556 - 603	0.0795	0.0141 *3*
610 - 659	0.15	0.023 *5*
660 - 661	2.19 ×10^{-5}	3.3 *3* ×10^{-6}
733 - 776	0.227	0.0309 *12*
791 - 824	0.00171	0.000213 *20*
863 - 909	0.0243	0.00268 *15*
923 - 972	0.00326	0.000352 *18*
973 - 1022	0.20	0.019 *4*
1024 - 1055	0.0052	0.00051 *8*
1124 - 1141	0.000302	2.68 *11* ×10^{-5}
1264 - 1310	0.00466	0.000368 *19*
1325 - 1353	9.7 ×10^{-5}	7.3 *20* ×10^{-6}
1396 - 1413	0.00198	0.000142 *6*
1457 - 1553	0.00103	7.0 *3* ×10^{-5}
1609 - 1707	0.00253	0.000153 *9*
1721 - 1723	4.8 ×10^{-5}	2.8 *8* ×10^{-6}
1999 - 2016	9.4 ×10^{-6}	4.7 *16* ×10^{-7}
2395 - 2412	9.0 ×10^{-6}	3.7 *12* ×10^{-7}

Continuous Radiation (^{91}Sr)

⟨β-⟩=648 keV;⟨IB⟩=1.23 keV

E_{bin}(keV)		⟨ ⟩(keV)	(%)
0 - 10	β-	0.0402	0.80
	IB	0.026	
10 - 20	β-	0.122	0.81
	IB	0.025	0.18
20 - 40	β-	0.496	1.65
	IB	0.049	0.17
40 - 100	β-	3.68	5.2
	IB	0.134	0.21
100 - 300	β-	39.7	19.7
	IB	0.33	0.19
300 - 600	β-	128	28.8
	IB	0.30	0.072
600 - 1300	β-	265	31.0
	IB	0.28	0.034
1300 - 2500	β-	208	12.0
	IB	0.078	0.0050
2500 - 2686	β-	2.14	0.084
	IB	2.2×10^{-5}	8.5×10^{-7}

$^{91}_{39}$Y (58.51 *6* d)

Mode: β-
Δ: -86347 *3* keV
SpA: 2.4524×10^{4} Ci/g
Prod: fission

Photons (^{91}Y)

⟨γ⟩=3.6 *8* keV

γ_{mode}	γ(keV)	γ(%)†
γ E2	1205.0 *5*	0.30

† 23% uncert(syst)

Continuous Radiation (^{91}Y)

⟨β-⟩=603 keV;⟨IB⟩=0.96 keV

E_{bin}(keV)		⟨ ⟩(keV)	(%)
0 - 10	β-	0.0352	0.70
	IB	0.026	
10 - 20	β-	0.106	0.71
	IB	0.025	0.17
20 - 40	β-	0.433	1.44
	IB	0.048	0.167
40 - 100	β-	3.19	4.54
	IB	0.13	0.20
100 - 300	β-	34.3	17.0
	IB	0.32	0.18
300 - 600	β-	123	27.3
	IB	0.27	0.064
600 - 1300	β-	411	46.0
	IB	0.143	0.019
1300 - 1546	β-	31.8	2.33
	IB	0.00077	5.8×10^{-5}

$^{91}_{39}$Y (49.71 *4* min)

Mode: IT
Δ: -85791 *3* keV
SpA: 4.156×10^{7} Ci/g
Prod: fission; daughter ^{91}Sr

Photons (^{91}Y)

⟨γ⟩=527.7 *11* keV

γ_{mode}	γ(keV)	γ(%)†
Y L_{ℓ}	1.686	0.0036 *8*
Y L_{η}	1.762	0.0020 *5*
Y L_{α}	1.922	0.082 *17*
Y L_{β}	2.005	0.048 *12*
Y L_{γ}	2.268	0.0014 *4*
Y $K_{\alpha 2}$	14.883	0.92 *4*
Y $K_{\alpha 1}$	14.958	1.76 *7*
Y $K_{\beta 1}'$	16.735	0.393 *16*
Y $K_{\beta 2}'$	17.125	0.0580 *25*
γ M4	555.56 *7*	94.90

† 0.21% uncert(syst)

Atomic Electrons (^{91}Y)

⟨e⟩=27.9 *5* keV

e_{bin}(keV)	⟨e⟩(keV)	e(%)
2	0.079	3.7 *4*
12	0.0109	0.089 *9*
13	0.107	0.84 *9*
14	0.0117	0.082 *8*
15	0.035	0.240 *25*
16	0.0030	0.0182 *19*
17	0.00113	0.0068 *7*
539	23.6	4.39 *9*
553	3.31	0.598 *12*
555	0.586	0.1055 *21*
556	0.0961	0.0173 *4*

$^{91}_{40}$Zr(stable)

Δ: -87892.8 *24* keV
%: 11.22 *2*

$^{91}_{41}$Nb(680 *130* yr)

Mode: ϵ
Δ: -86638 *3* keV
SpA: 5.8 Ci/g
Prod: ^{90}Zr(d,n)

Photons (^{91}Nb)

⟨γ⟩=10.27 *23* keV

γ_{mode}	γ(keV)	γ(%)
Zr L_ℓ	1.792	0.075 *13*
Zr L_η	1.876	0.039 *6*
Zr L_α	2.042	1.8 *3*
Zr L_β	2.135	0.99 *16*
Zr L_γ	2.386	0.048 *9*
Zr $K_{\alpha2}$	15.691	18.6 *6*
Zr $K_{\alpha1}$	15.775	35.6 *12*
Zr $K_{\beta1}'$	17.663	8.1 *3*
Zr $K_{\beta2}'$	18.086	1.31 *5*

Atomic Electrons (^{91}Nb)

⟨e⟩=4.9 *3* keV

e_{bin}(keV)	⟨e⟩(keV)	e(%)
2	1.52	67 *7*
3	0.108	4.3 *4*
13	1.79	13.4 *14*
14	0.49	3.6 *4*
15	0.78	5.1 *5*
16	0.145	0.93 *9*
17	0.064	0.37 *4*
18	0.0173	0.098 *10*

Continuous Radiation (^{91}Nb)

⟨β+⟩=0.178 keV;⟨IB⟩=0.45 keV

E_{bin}(keV)		⟨ ⟩(keV)	(%)
0 - 10	β+	5.4×10^{-6}	6.7×10^{-5}
	IB	0.000149	
10 - 20	β+	0.000139	0.00085
	IB	0.0145	0.092
20 - 40	β+	0.00234	0.0073
	IB	0.00103	0.0040
40 - 100	β+	0.0479	0.065
	IB	0.0022	0.0031
100 - 300	β+	0.127	0.091
	IB	0.037	0.017
300 - 600	IB	0.162	0.036
600 - 1255	IB	0.24	0.031
	Σβ+		0.164

$^{91}_{41}$Nb(62 d)

Mode: IT(95 *2* %), ϵ(5 *2* %)
Δ: -86534 *3* keV
SpA: 2.3×10^4 Ci/g
Prod: ^{89}Y(α,2n); ^{90}Zr(d,n); ^{93}Nb(γ,2n)

Photons (^{91}Nb)

⟨γ⟩=51 *25* keV

γ_{mode}	γ(keV)	γ(%)†
Nb L_ℓ	1.902	0.076 *14*
Nb L_η	1.996	0.031 *5*
Nb L_α	2.166	1.9 *3*
Nb L_β	2.282	0.91 *16*
Nb L_γ	2.554	0.057 *12*
Zr $K_{\alpha2}$	15.691	0.91 *3*
Zr $K_{\alpha1}$	15.775	1.75 *6*
Nb $K_{\alpha2}$	16.521	14.1 *9*
Nb $K_{\alpha1}$	16.615	26.8 *18*
Zr $K_{\beta1}'$	17.663	0.400 *14*
Zr $K_{\beta2}'$	18.086	0.0646 *24*
Nb $K_{\beta1}'$	18.618	6.2 *4*
Nb $K_{\beta2}'$	19.074	1.04 *7*
γ_{IT} (M4)	104.50 *10*	0.556
γ_ϵE2	1205.0 *5*	~4

† uncert(syst): approximate for ϵ, 2.1% for IT

Atomic Electrons (^{91}Nb)

⟨e⟩=90 *4* keV

e_{bin}(keV)	⟨e⟩(keV)	e(%)
2	1.54	65 *8*
3	0.146	5.4 *6*
13	0.088	0.66 *7*
14	1.66	11.9 *14*
15	0.038	0.25 *3*
16	0.65	4.0 *5*
17	0.043	0.26 *3*
18	0.052	0.29 *3*
19	0.0095	0.051 *6*
86	54.9	64 *4*
102	25.2	24.7 *15*
104	5.8	5.6 *3*
1187	0.016	~0.0014
1202	0.0017	~0.00015
1203	7.7×10^{-5}	$\sim6\times10^{-6}$
1205	0.00037	$\sim3\times10^{-5}$

Continuous Radiation (^{91}Nb)

⟨IB⟩=0.0078 keV

E_{bin}(keV)		⟨ ⟩(keV)	(%)
10 - 20	IB	0.00072	0.0046
20 - 40	IB	4.6×10^{-5}	0.00018
40 - 100	IB	5.9×10^{-5}	8.9×10^{-5}
100 - 300	IB	0.00058	0.00028
300 - 600	IB	0.0021	0.00048
600 - 1300	IB	0.0042	0.00051
1300 - 1359	IB	3.5×10^{-6}	2.7×10^{-7}

$^{91}_{42}$Mo(15.49 *1* min)

Mode: ϵ
Δ: -82200 *13* keV
SpA: 1.3334×10^8 Ci/g
Prod: ^{92}Mo(p,d); ^{92}Mo(n,2n)

Photons (^{91}Mo)

⟨γ⟩=19.4 *3* keV

γ_{mode}	γ(keV)	γ(%)†
Nb L_ℓ	1.902	0.0050 *9*
Nb L_η	1.996	0.0023 *4*
Nb L_α	2.166	0.122 *19*
Nb L_β	2.280	0.066 *11*
Nb L_γ	2.547	0.0040 *8*
Nb $K_{\alpha2}$	16.521	1.19 *4*
Nb $K_{\alpha1}$	16.615	2.28 *8*
Nb $K_{\beta1}'$	18.618	0.526 *18*
Nb $K_{\beta2}'$	19.074	0.089 *3*
γ (M4)	104.50 *10*	*
γ [E2]	603.94 *21*	0.00107 *11*
γ	1050.53 *23*	0.0053 *7*
γ [E2]	1082.07 *8*	0.0207 *13*
γ [M1+E2]	1155.02 *21*	0.0036 *10* ?
γ	1446.97 *18*	0.0125 *10*
γ [M1+E2]	1581.24 *10*	0.226 *8*
γ [E1]	1605.47 *18*	0.0102 *10*
γ [M1+E2]	1637.02 *11*	0.329 *12*
γ [E2]	1740.14 *22*	0.0135 *7*
γ [E1]	1790.50 *19*	0.0299 *16*
γ [E1]	2530.5 *4*	0.0043 *13*
γ	2631.75 *23*	0.118 *3*
γ [M1+E2]	2792.02 *19*	0.0099 *10*
γ	3028.18 *20*	0.085 *3*
γ	3149.0 *4*	0.055 *3*
γ	3187.5 *6*	0.0053 *7*
γ	3837.3 *7*	0.0030 *7*
γ	3886.4 *7*	0.0023 *7*
γ	3916.4 *7*	0.0016 *7*
γ	4180.3 *12*	0.0020 *7*

† 5.5% uncert(syst)
* with ^{91}Nb(62 d)

Atomic Electrons (^{91}Mo)

⟨e⟩=0.316 *18* keV

e_{bin}(keV)	⟨e⟩(keV)	e(%)
2	0.101	4.2 *4*
3	0.0091	0.34 *3*
14	0.139	0.99 *10*
16	0.055	0.34 *3*
17 - 19	0.0086	0.049 *3*
585 - 604	1.7×10^{-5}	2.8 *12* $\times10^{-6}$
1032 - 1080	0.000159	1.50 *13* $\times10^{-5}$
1136 - 1155	2.3×10^{-5}	2.0 *5* $\times10^{-6}$
1428 - 1447	4.6×10^{-5}	3.2 *11* $\times10^{-6}$
1562 - 1637	0.00246	0.000154 *7*
1721 - 1790	0.000115	6.6 *3* $\times10^{-6}$
2512 - 2530	7.4×10^{-6}	2.9 *8* $\times10^{-7}$
2613 - 2631	0.00027	1.04 *24* $\times10^{-5}$
2773 - 2792	2.7×10^{-5}	9.8 *10* $\times10^{-7}$
3009 - 3028	0.00018	6.0 *13* $\times10^{-6}$
3130 - 3187	0.000124	4.0 *8* $\times10^{-6}$
3818 - 3916	1.29×10^{-5}	3.3 *6* $\times10^{-7}$
4161 - 4180	3.6×10^{-6}	9 *3* $\times10^{-8}$

Continuous Radiation (^{91}Mo)

⟨β+⟩=1453 keV;⟨IB⟩=5.5 keV

E_{bin}(keV)		⟨ ⟩(keV)	(%)
0 - 10	β+	4.63×10^{-6}	5.7×10^{-5}
	IB	0.047	
10 - 20	β+	0.000131	0.00080
	IB	0.048	0.33
20 - 40	β+	0.00262	0.0081
	IB	0.092	0.32
40 - 100	β+	0.091	0.118
	IB	0.27	0.41
100 - 300	β+	3.74	1.70
	IB	0.78	0.44
300 - 600	β+	29.3	6.3
	IB	0.94	0.22
600 - 1300	β+	270	27.6
	IB	1.50	0.17

Continuous Radiation (^{91}Mo)
(continued)

E_{bin}(keV)		⟨ ⟩(keV)	(%)
1300 - 2500	$\beta+$	907	49.2
	IB	1.30	0.074
2500 - 4438	$\beta+$	243	8.8
	IB	0.51	0.017
	$\Sigma\beta+$		94

$^{91}_{42}$Mo(1.087 *13* min)

Mode: IT(50.1 *13* %), ε(49.9 *13* %)
Δ: -81547 *13* keV
SpA: 1.891×10^9 Ci/g
Prod: ^{90}Zr(^{3}He,2n); ^{92}Mo(γ,n)

Photons (^{91}Mo)
⟨γ⟩=943 *15* keV

γ_{mode}	γ(keV)	γ(%)†
Nb L_ℓ	1.902	0.0050 *9*
Nb L_η	1.996	0.0023 *4*
Mo L_ℓ	2.016	0.0015 *3*
Mo L_η	2.120	0.00072 *11*
Nb L_α	2.166	0.121 *19*
Nb L_β	2.280	0.066 *11*
Mo L_α	2.293	0.038 *6*
Mo L_β	2.425	0.022 *4*
Nb L_γ	2.547	0.0040 *8*
Mo L_γ	2.697	0.0014 *3*
Nb $K_{\alpha2}$	16.521	1.19 *4*
Nb $K_{\alpha1}$	16.615	2.26 *8*
Mo $K_{\alpha2}$	17.374	0.351 *19*
Mo $K_{\alpha1}$	17.479	0.67 *4*
Nb $K_{\beta1}$'	18.618	0.523 *18*
Nb $K_{\beta2}$'	19.074	0.088 *3*
Mo $K_{\beta1}$'	19.602	0.157 *9*
Mo $K_{\beta2}$'	20.091	0.0274 *16*
γ_ϵ(M4)	104.50 *10*	*
γ_ϵ[M1+E2]	425.84 *10*	0.169 *24*
γ_{IT}[M4]	652.98 *9*	48.2
γ_ϵ[M1+E2]	732.59 *13*	0.169 *24*
γ_ϵ[M1+E2]	1032.52 *13*	0.532 *19*
γ_ϵ[E2]	1082.07 *8*	0.50 *3*
γ_ϵ[M1+E2]	1158.42 *12*	0.281 *19*
γ_ϵ[M1+E2]	1207.98 *9*	18.8 *6*
γ_ϵ[M1+E2]	1507.91 *9*	24.4 *7*
γ_ϵ[M1+E2]	2240.48 *12*	0.73 *3*

† uncert(syst): 4.9% for ε, 2.6% for IT
* with ^{91}Nb(62 d)

Atomic Electrons (^{91}Mo)
⟨e⟩=12.2 *5* keV

e_{bin}(keV)	⟨e⟩(keV)	e(%)
2 - 20	0.401	7.6 *5*
407 - 426	0.0044	0.00108 *20*
633	9.9	1.57 *8*
650	1.41	0.217 *10*
652	0.223	0.0342 *16*
714 - 733	0.0019	0.00026 *4*
1014 - 1032	0.00379	0.000373 *24*
1063 - 1082	0.00329	0.000309 *21*
1139 - 1158	0.00175	0.000153 *13*
1189 - 1208	0.112	0.0094 *6*
1489 - 1508	0.115	0.0077 *5*
2221 - 2240	0.00240	0.000108 *7*

Continuous Radiation (^{91}Mo)
⟨β+⟩=541 keV;⟨IB⟩=2.1 keV

E_{bin}(keV)		⟨ ⟩(keV)	(%)
0 - 10	$\beta+$	4.11×10^{-6}	5.02×10^{-5}
	IB	0.019	
10 - 20	$\beta+$	0.000116	0.00071
	IB	0.020	0.136
20 - 40	$\beta+$	0.00232	0.0072
	IB	0.037	0.128
40 - 100	$\beta+$	0.080	0.104
	IB	0.105	0.162
100 - 300	$\beta+$	3.18	1.45
	IB	0.30	0.169
300 - 600	$\beta+$	23.4	5.03
	IB	0.35	0.082
600 - 1300	$\beta+$	177	18.5
	IB	0.56	0.064
1300 - 2500	$\beta+$	301	17.5
	IB	0.54	0.031
2500 - 4986	$\beta+$	36.3	1.26
	IB	0.149	0.0052
	$\Sigma\beta+$		44

$^{91}_{43}$Tc(3.14 *2* min)

Mode: ε
Δ: -75980 *200* keV
SpA: 6.57×10^8 Ci/g
Prod: ^{92}Mo(p,2n)

Photons (^{91}Tc)
⟨γ⟩=1546 *18* keV

γ_{mode}	γ(keV)	γ(%)†
Mo L_ℓ	2.016	0.0076 *13*
Mo L_η	2.120	0.0035 *5*
Mo L_α	2.293	0.19 *3*
Mo L_β	2.425	0.107 *18*
Mo L_γ	2.699	0.0071 *14*
Mo $K_{\alpha2}$	17.374	1.77 *6*
Mo $K_{\alpha1}$	17.479	3.37 *12*
Mo $K_{\beta1}$'	19.602	0.80 *3*
Mo $K_{\beta2}$'	20.091	0.139 *5*
γ[E1]	206.15 *13*	0.049 *8*
γ	217.15 *12*	0.144 *16*
γ[E1]	277.98 *10*	0.62 *6*
γ	297.08 *11*	0.238 *24*
γ	337.56 *12*	1.15 *19*
γ[E2]	375.83 *17*	0.51 *6*
γ	482.70 *11*	1.09 *8*
γ[M1+E2]	502.99 *10*	0.83 *8*
γ	548.48 *12*	1.68 *10*
γ	562.0 *5*	0.182 *19*
γ	628.41 *11*	0.75 *14*
γ[M4]	652.98 *9*	0.80 *16* *
γ	668.89 *13*	~0.32
γ	690.8 *12*	0.069 *14*
γ	810.84 *11*	5.1 *3*
γ	814.03 *11*	1.8 *3*
γ	845.56 *11*	1.22 *13*
γ	852.19 *14*	0.86 *11*
γ	878.65 *16*	1.10 *8* ?
γ	902.82 *20*	1.62 *8*
γ	935.92 *20*	0.213 *19*
γ	984.97 *17*	0.222 *14*
γ	992.7 *5*	0.222 *14*
γ	1076.39 *9*	0.94 *6*
γ	1088.82 *11*	0.59 *4*
γ	1111.11 *8*	3.18 *11*
γ	1146.7 *4*	0.146 *18*
γ	1244.53 *18*	0.107 *14*
γ	1255.6 *3*	0.107 *16*
γ	1285.88 *15*	0.29 *3*
γ	1322.53 *15*	0.70 *3*
γ	1354.37 *9*	0.73 *3*
γ[E2]	1362.11 *8*	4.4 *8*
γ	1379.1 *3*	0.088 *16*
γ(E2)	1414.28 *18*	0.78 *5*
γ	1491.4 *4*	0.187 *16*
γ	1549.9 *4*	0.085 *11* &
γ	1564.90 *9*	7.01 *21*
γ[E1]	1605.37 *7*	7.89 *24*
γ[E1]	1640.09 *8*	9.2 *3*
γ	1650.42 *20*	0.57 *3*
γ	1671.1 *3*	0.146 *13*
γ[E2]	1731.46 *17*	0.139 *13*
γ	1752.0 *3*	0.101 *13* &
γ[E1]	1762.66 *17*	0.114 *13*
γ	1795.42 *20*	0.11 *3* &
γ	1890.02 *20*	0.096 *14* &
γ	1902.45 *10*	6.08 *21*
γ	2173.08 *15*	0.309 *16*
γ	2233.77 *9*	1.33 *6*
γ	2296.32 *20*	0.482 *24*
γ	2397.28 *15*	0.083 *8*
γ	2450.91 *9*	13.8 *4*
γ	2492.26 *13*	0.618 *24*
γ	2516.82 *20*	0.078 *8*
γ	2527.4 *3*	0.666 *24*
γ	2540.7 *7*	0.070 *10*
γ[E2]	2580.83 *18*	0.125 *10*
γ	2664.0 *4*	0.054 *10*
γ	2716.46 *7*	1.87 *6*
γ	2724.1 *3*	0.131 *10*
γ	2781.08 *16*	3.22 *13* ?
γ	2793.72 *20*	0.256 *19*
γ	2820.4 *3*	0.096 *10*
γ	2859.62 *20*	0.138 *10*
γ[E1]	2887.40 *15*	1.40 *5*
γ[E1]	3009.97 *20*	0.254 *21*
γ[E1]	3118.26 *21*	0.245 *11*
γ	3167.6 *4*	0.070 *8*
γ	3197.4 *4*	0.184 *22*
γ[E2]	3235.78 *23*	0.032 *5* &
γ[E1]	3249.44 *16*	0.011 *3* &
γ	3279.8 *3*	0.042 *8*
γ	3307.8 *6*	0.021 *6* &
γ	3374 *1*	0.018 *6*
γ	3381.2 *5*	0.054 *6*
γ	3403.8 *3*	0.080 *8*
γ	3419.4 *4*	0.205 *14*
γ	3453.6 *5*	0.117 *10*
γ[E2]	3475.25 *23*	0.016 *3* &
γ	3541.8 *3*	0.096 *8*
γ	3593.17 *20*	0.168 *13*
γ	3627.90 *21*	0.014 *5*
γ	3645.7 *15*	0.042 *6*
γ	3651.6 *15*	0.042 *6*
γ	3668.37 *22*	0.022 *7*
γ	3701.5 *3*	0.062 *3*
γ[E1]	3736.76 *16*	0.118 *8*
γ	3776.3 *20*	0.024 *5* &
γ	3833.4 *9*	0.021 *3*
γ	3886.5 *3*	0.029 *5*
γ	3907.8 *8*	0.040 *5* &
γ	3937.2 *5*	0.038 *5*
γ	4046.7 *4*	0.051 *6*
γ	4056.3 *4*	0.096 *6*
γ	4075.6 *4*	0.085 *10* &
γ	4086.2 *4*	0.069 *6* &
γ	4118.9 *4*	0.059 *6*
γ	4199.1 *8*	0.030 *5*
γ	4217.8 *3*	0.053 *5*
γ	4229.8 *9*	0.030 *5*
γ	4401.1 *10*	0.018 *3*
γ	4445.8 *15*	0.027 *5*
γ	4592.9 *9*	0.018 *6*

† 11% uncert(syst)
* with ^{91}Mo(1.09 min)
& with ^{91}Tc(3.14 or 3.3 min)

Atomic Electrons (^{91}Tc)

⟨e⟩=1.11 *6* keV

e_{bin}(keV)	⟨e⟩(keV)	e(%)
3	0.169	6.5 *7*
14	0.0189	0.132 *14*
15	0.177	1.20 *12*
17	0.083	0.49 *5*
19 - 20	0.0076	0.039 *4*
186 - 217	0.014	~0.007
258 - 297	0.021	0.0079 *21*
318	0.032	~0.010
335 - 376	0.026	0.0075 *11*
463 - 503	0.035	0.0073 *18*
528 - 562	0.024	0.0045 *18*
608 - 628	0.008	0.0013 *6*
633	0.16	0.026 *5*
649	0.0026	~0.0004
650	0.023	0.0036 *7*
652 - 691	0.0061	0.00093 *13*
791	0.032	0.0041 *18*
794 - 843	0.030	0.0037 *9*
845 - 883	0.017	0.0020 *6*
900 - 936	0.0026	0.00028 *8*
965 - 993	0.0025	0.00026 *7*
1056 - 1091	0.022	0.0020 *6*
1108 - 1147	0.0027	0.00024 *7*
1225 - 1266	0.0021	0.00017 *4*
1283 - 1322	0.0031	0.00024 *8*
1334 - 1379	0.027	0.0020 *3*
1394 - 1414	0.0040	0.000289 *19*
1471 - 1565	0.027	0.0017 *5*
1585 - 1671	0.0414	0.00257 *9*
1711 - 1795	0.00154	8.8 *10* $\times 10^{-5}$
1870 - 1902	0.019	0.0010 *3*
2153 - 2233	0.0045	0.00020 *5*
2276 - 2296	0.0013	5.7 *15* $\times 10^{-5}$
2377 - 2397	0.00022	9.1 *25* $\times 10^{-6}$
2431	0.031	0.0013 *4*
2448 - 2540	0.0077	0.00031 *5*
2561 - 2644	0.00050	1.94 *18* $\times 10^{-5}$
2661 - 2724	0.0048	0.00018 *4*
2761 - 2859	0.0087	0.00031 *7*
2867 - 2887	0.00240	8.4 *4* $\times 10^{-5}$
2990 - 3010	0.00043	1.43 *12* $\times 10^{-5}$
3098 - 3197	0.00096	3.0 *3* $\times 10^{-5}$
3216 - 3307	0.00024	7.3 *10* $\times 10^{-6}$
3354 - 3453	0.00099	2.9 *3* $\times 10^{-5}$
3455 - 3541	0.00024	6.8 *12* $\times 10^{-6}$
3573 - 3669	0.00058	1.63 *24* $\times 10^{-5}$
3681 - 3776	0.00036	9.7 *8* $\times 10^{-6}$
3813 - 3907	0.000176	4.5 *6* $\times 10^{-6}$
3917 - 3937	7.5 $\times 10^{-5}$	1.9 *4* $\times 10^{-6}$
4027 - 4119	0.00069	1.70 *16* $\times 10^{-5}$
4179 - 4229	0.00021	5.1 *6* $\times 10^{-6}$
4381 - 4445	8.3 $\times 10^{-5}$	1.9 *3* $\times 10^{-6}$
4573 - 4593	3.2 $\times 10^{-5}$	7 *3* $\times 10^{-7}$

Continuous Radiation (^{91}Tc)

⟨β+⟩=1635 keV;⟨IB⟩=6.8 keV

E_{bin}(keV)		⟨ ⟩(keV)	(%)
0 - 10	β+	4.30 $\times 10^{-6}$	5.2 $\times 10^{-5}$
	IB	0.049	
10 - 20	β+	0.000125	0.00076
	IB	0.050	0.35
20 - 40	β+	0.00257	0.0080
	IB	0.096	0.33
40 - 100	β+	0.091	0.119
	IB	0.28	0.43
100 - 300	β+	3.74	1.70
	IB	0.83	0.46
300 - 600	β+	28.5	6.1
	IB	1.03	0.24
600 - 1300	β+	241	24.8
	IB	1.8	0.20
1300 - 2500	β+	683	37.4
	IB	1.8	0.100
2500 - 5000	β+	678	20.7
	IB	0.88	0.028
5000 - 6220	β+	0.82	0.0163
	IB	0.027	0.00050
	Σβ+		91

$^{91}_{43}$Tc(3.3 *1* min)

Mode: ϵ

Δ: <-75630 keV

SpA: 6.25×10^{8} Ci/g

Prod: ^{92}Mo(p,2n)

Photons (^{91}Tc)$^{\&}$

⟨γ⟩=948 *25* keV

γ_{mode}	γ(keV)	γ(%)†
Mo L_{ℓ}	2.016	0.0061 *11*
Mo L_{η}	2.120	0.0028 *4*
Mo L_{α}	2.293	0.152 *24*
Mo L_{β}	2.425	0.086 *15*
Mo L_{γ}	2.698	0.0057 *11*
Mo $K_{\alpha2}$	17.374	1.42 *6*
Mo $K_{\alpha1}$	17.479	2.69 *12*
Mo $K_{\beta1}$'	19.602	0.64 *3*
Mo $K_{\beta2}$'	20.091	0.111 *5*
γ [E1]	206.15 *13*	0.027 *5*
γ [E1]	277.98 *10*	0.022 *3*
γ [M1+E2]	502.99 *10*	50.4 *21*
γ	606.88 *13*	1.45 *11*
γ [M4]	652.98 *9*	70 *3* *
γ	927.73 *9*	3.72 *21*
γ	1328.46 *12*	2.51 *11*
γ [E2]	1362.11 *8*	2.5 *4*
γ	1430.72 *12*	1.96 *11*
γ	1534.60 *11*	2.45 *11*
γ [E1]	1605.37 *7*	0.117 *23*
γ [E1]	1640.09 *8*	0.33 *4*
γ	2037.59 *9*	0.52 *5*
γ [E2]	3046.7 *3*	0.37 *3*
γ [E2]	3081.4 *3*	0.117 *21*
γ [M1+E2]	3530.8 *3*	0.71 *4*
γ [M1+E2]	4033.8 *3*	0.085 *21*

† 4.7% uncert(syst)

* with ^{91}Mo(1.09 min)

& see also ^{91}Tc(3.14 min)

Atomic Electrons (^{91}Tc)

⟨e⟩=18.2 *9* keV

e_{bin}(keV)	⟨e⟩(keV)	e(%)
3 - 20	0.36	6.7 *6*
186 - 206	0.00080	0.00042 *7*
258 - 278	0.00040	0.000152 *19*
483	0.91	0.189 *22*
500 - 503	0.135	0.027 *3*
587 - 607	0.016	0.0027 *12*
633	14.3	2.27 *14*
650	2.04	0.314 *20*
652	0.323	0.049 *3*
908 - 928	0.023	0.0025 *10*
1308 - 1342	0.022	0.0017 *3*
1359 - 1362	0.00159	0.000117 *17*
1411 - 1431	0.008	0.00055 *18*
1515 - 1605	0.009	0.00062 *20*
1620 - 1640	0.00075	4.6 *6* $\times 10^{-5}$
2018 - 2037	0.0015	7.6 *22* $\times 10^{-5}$
3027 - 3081	0.00134	4.4 *3* $\times 10^{-5}$
3511 - 3530	0.00176	5.0 *4* $\times 10^{-5}$
4014 - 4033	0.00020	4.9 *11* $\times 10^{-6}$

Continuous Radiation (^{91}Tc)

⟨β+⟩=1966 keV;⟨IB⟩=8.1 keV

E_{bin}(keV)		⟨ ⟩(keV)	(%)
0 - 10	β+	2.51 $\times 10^{-6}$	3.06 $\times 10^{-5}$
	IB	0.057	
10 - 20	β+	7.3 $\times 10^{-5}$	0.000446
	IB	0.057	0.40
20 - 40	β+	0.00150	0.00465
	IB	0.111	0.39
40 - 100	β+	0.053	0.070
	IB	0.32	0.50
100 - 300	β+	2.23	1.01
	IB	0.98	0.55
300 - 600	β+	17.6	3.76
	IB	1.24	0.29
600 - 1300	β+	174	17.6
	IB	2.1	0.24
1300 - 2500	β+	797	42.0
	IB	2.1	0.117
2500 - 5000	β+	976	31.1
	IB	1.09	0.034
5000 - 5917	IB	0.017	0.00033
	Σβ+		96

A = 92

NDS **30**, 573 (1980)

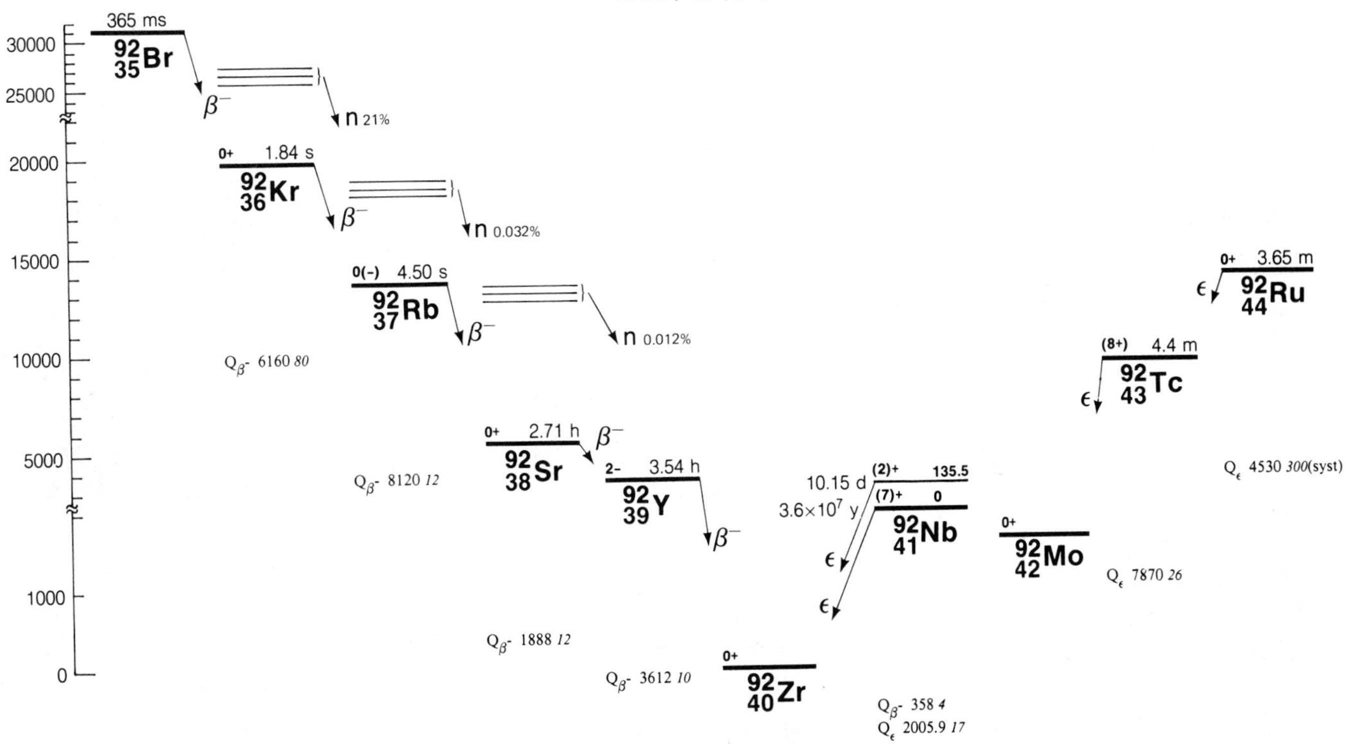

$^{92}_{35}$Br(365 *7* ms)

Mode: β-, β-n(21 *8* %)
SpA: 1.50×10^{11} Ci/g
Prod: fission

$^{92}_{36}$Kr(1.840 *8* s)

Mode: β-, β-n(0.032 *3* %)
Δ: -68680 *80* keV
SpA: 5.553×10^{10} Ci/g
Prod: fission

Photons (^{92}Kr)

⟨γ⟩=1452 *40* keV

γ$_{mode}$	γ(keV)	γ(%)†
γ (M1+E2)	142.35 *5*	66 *3*
γ	159.15 *9*	0.11 *3*
γ	167.90 *8*	0.132 *20*
γ	185.56 *20*	0.112 *20*
γ	191.07 *7*	0.86 *7*
γ	214.88 *7*	0.37 *3*
γ	281.93 *15*	0.28 *3*
γ	316.72 *7*	6.0 *3*
γ	333.43 *7*	0.046 *20*
γ	342.26 *6*	2.18 *13*
γ	350.22 *7*	0.28 *3*
γ	372.56 *10*	0.12 *3*
γ	394.81 *14*	0.12 *3*
γ	436.25 *9*	0.22 *4*
γ	440.01 *8*	0.62 *5*
γ	481.25 *9*	0.18 *3*
γ	484.61 *6*	3.3 *2*
γ	492.58 *7*	0.58 *5*
γ	534.92 *8*	0.42 *4*
γ	548.30 *6*	14.4 *8*
γ	585.88 *13*	0.23 *4*
γ	623.60 *8*	1.39 *7*
γ	632.63 *13*	0.19 *5*
γ	678.09 *7*	0.38 *3*
γ	683.66 *20*	0.17 *3*
γ	728.24 *13*	0.12 *3*
γ	737.27 *9*	0.53 *7*
γ	785.69 *9*	0.46 *7*
γ	812.57 *6*	15.0 *8*
γ	826.0 *6*	0.13 *6*
γ	868.35 *9*	~0.09
γ	876.26 *6*	4.4 *3*
γ	920.86 *9*	0.28 *4*
γ	928.04 *10*	0.14 *4*
γ	1044.15 *7*	4.9 *3*
γ	1115.09 *20*	0.106 *20*
γ	1178.78 *20*	0.11 *3*
γ	1218.51 *5*	62 *3*
γ	1232.5 *6*	~0.05
γ	1240.46 *20*	0.16 *3*
γ	1250.02 *13*	0.099 *20*
γ	1259.0 *3*	0.079 *20* ?
γ	1285.0 *5*	0.053 *20*
γ	1291.4 *3*	~0.09
γ	1310.72 *14*	0.119 *20*
γ	1346.67 *20*	0.46 *20*
γ [E1]	1360.87 *5*	3.56 *20*
γ	1393.6 *3*	0.053 *20*
γ	1415.36 *10*	0.172 *20*
γ	1474.15 *24*	0.099 *20*
γ	1525.8 *3*	0.086 *20*
γ	1540.59 *15*	~0.05
γ	1554.34 *7*	0.40 *3*
γ	1594.76 *24*	0.053 *20*
γ	1620.7 *4*	~0.040 ?
γ	1659.31 *17*	0.046 *20*
γ	1663.38 *20*	0.092 *20*
γ	1675.0 *3*	0.059 *20* ?
γ	1762.65 *24*	0.073 *20*
γ	1896.60 *7*	0.86 *13*
γ	1932.9 *5*	0.059 *20*
γ	1973.41 *16*	0.15 *3*
γ	1980.59 *16*	0.086 *20*
γ	1987.28 *14*	0.25 *3*
γ	2004.16 *17*	0.053 *20*
γ	2039.04 *14*	0.42 *4*
γ	2075.4 *4*	0.092 *20*
γ	2079.36 *24*	0.092 *20*
γ	2095.14 *16*	0.13 *3*
γ	2128.99 *24*	0.09 *3*
γ	2270.62 *15*	0.053 *20*
γ	2277.45 *14*	0.119 *20*
γ	2402.02 *16*	0.046 *20*
γ	2413.81 *19*	~0.05
γ	2416.83 *15*	0.09 *3*
γ	2433.42 *24*	0.106 *20*
γ	2444.98 *14*	0.18 *3*
γ	2468.51 *13*	0.18 *3*
γ	2584.72 *16*	0.13 *6*
γ	2587.33 *14*	0.25 *6*
γ	2610.86 *13*	0.30 *3*
γ	2713.2 *6*	0.059 *20*
γ	2718.73 *15*	0.28 *3*
γ	2759.08 *14*	0.22 *3*
γ	2793.54 *18*	0.099 *20*
γ	2832.66 *18*	0.31 *3*
γ	2853.8 *3*	0.07 *3* ?
γ	3057.00 *23*	0.13 *3*
γ	3099.8 *5*	0.09 *3*
γ	3149.37 *19*	0.13 *3*
γ	3199.47 *17*	0.45 *3*
γ	3272.00 *25*	0.066 *20*
γ	3324.5 *3*	0.046 *13*

Photons (^{92}Kr)
(continued)

γ_{mode}	γ(keV)	γ(%)†
γ	3342.9 *4*	0.040 *13* ?
γ	3659.6 *4*	0.086 *20* ?
γ	3727.3 *8*	~0.040

† approximate

Continuous Radiation (^{92}Kr)
$\langle\beta-\rangle$=2091 keV;$\langle$IB$\rangle$=7.4 keV

E_{bin}(keV)		$\langle\ \rangle$(keV)	(%)
0 - 10	β-	0.00241	0.0479
	IB	0.060	
10 - 20	β-	0.0074	0.0494
	IB	0.059	0.41
20 - 40	β-	0.0312	0.103
	IB	0.117	0.41
40 - 100	β-	0.251	0.354
	IB	0.34	0.52
100 - 300	β-	3.59	1.72
	IB	1.03	0.57
300 - 600	β-	19.7	4.26
	IB	1.30	0.30
600 - 1300	β-	170	17.3
	IB	2.2	0.25
1300 - 2500	β-	785	41.2
	IB	1.8	0.106
2500 - 5000	β-	1113	35.0
	IB	0.49	0.0167
5000 - 5667	β-	0.487	0.0094
	IB	1.60×10^{-5}	3.1×10^{-7}

$^{92}_{37}$Rb(4.50 *2* s)

Mode: β-, β-n(0.0115 *7* %)
Δ: -74836 *14* keV
SpA: 2.526×10^{10} Ci/g
Prod: fission

β-n(avg): 180 *40*

Photons (^{92}Rb)
$\langle\gamma\rangle$=520 *20* keV

γ_{mode}	γ(keV)	γ(%)†
γ	96.7 *4*	0.039 *18* ?
γ	386.38 *19*	0.061 *10*
γ	393.65 *9*	0.304 *16*
γ	569.93 *8*	1.36 *8*
γ	703.64 *18*	0.112 *24* ?
γ	756.09 *14*	0.272 *24*
γ[E2]	814.80 *9*	8.0 *4*
γ	963.57 *11*	0.37 *3*
γ	1239.2 *4*	0.09 *3* ?
γ	1273.57 *17*	0.24 *3* ?
γ	1326.01 *13*	0.34 *4*
γ	1384.72 *12*	0.88 *16*
γ	1398.9 *4*	0.13 *4*
γ	1464.3 *5*	0.10 *4* ?
γ	1712.38 *17*	1.05 *8*
γ	1778.37 *14*	~0.09
γ	1789.3 *9*	~0.10
γ	1817.0 *4*	0.14 *3*
γ	1895.4 *6*	0.14 *4* ?
γ	1968.8 *4*	0.17 *5*
γ	2006.12 *19*	0.18 *5*
γ	2233.1 *5*	0.19 *5*
γ	2820.90 *18*	1.50 *11*
γ	2859.5 *5*	~0.07
γ	2913.2 *4*	0.18 *5*

Photons (^{92}Rb)
(continued)

γ_{mode}	γ(keV)	γ(%)†
γ	3110.0 *6*	0.24 *7*
γ	3502.8 *8*	~0.09
γ	3669.3 *4*	0.10 *5*
γ	3823.0 *4*	~0.09
γ	4239.2 *4*	~0.08
γ	4428.0 *9*	0.31 *6*
γ	4508.8 *7*	0.15 *4*
γ	4637.8 *4*	0.54 *7*
γ	4808.5 *7*	~0.26 ?
γ	4836.8 *6*	0.25 *6*
γ	4923.5 *9*	0.26 *5*
γ	5084.9 *7*	0.21 *10*
γ	5188.5 *7*	0.6 *1*
γ	5215.1 *7*	0.26 *10*
γ	5248.8 *12*	0.26 *6*
γ	5301.1 *10*	0.20 *6*
γ	5376.7 *15*	0.14 *6*
γ	5497.8 *13*	0.20 *6*
γ	5573.8 *17*	~0.20
γ	5585.5 *6*	0.40 *8*
γ	5632.3 *10*	0.48 *8*
γ	5738.3 *9*	0.17 *6*
γ	5879.5 *15*	0.17 *6*
γ	5899.6 *7*	0.22 *6*
γ	6003.3 *7*	0.14 *5*
γ	6029.8 *7*	0.19 *6*
γ	6115.9 *10*	0.20 *6*

† approximate

Continuous Radiation (^{92}Rb)
$\langle\beta-\rangle$=3525 keV;$\langle$IB$\rangle$=16.6 keV

E_{bin}(keV)		$\langle\ \rangle$(keV)	(%)
0 - 10	β-	0.00199	0.0396
	IB	0.078	
10 - 20	β-	0.0061	0.0405
	IB	0.077	0.54
20 - 40	β-	0.0251	0.083
	IB	0.154	0.53
40 - 100	β-	0.193	0.274
	IB	0.45	0.69
100 - 300	β-	2.36	1.15
	IB	1.43	0.79
300 - 600	β-	10.1	2.21
	IB	1.9	0.45
600 - 1300	β-	70	7.2
	IB	3.7	0.41
1300 - 2500	β-	362	18.7
	IB	4.4	0.24
2500 - 5000	β-	1816	48.7
	IB	3.9	0.118
5000 - 7500	β-	1244	21.2
	IB	0.48	0.0086
7500 - 8120	β-	20.1	0.263
	IB	0.00049	6.4×10^{-6}

$^{92}_{38}$Sr(2.71 *1* h)

Mode: β-
Δ: -82956 *12* keV
SpA: 1.257×10^{7} Ci/g
Prod: fission

Photons (^{92}Sr)
$\langle\gamma\rangle$=1339 *50* keV

γ_{mode}	γ(keV)	γ(%)†
γ	241.63 *3*	2.97 *9*
γ	352.6 *20*	0.054 *9*
γ	430.66 *5*	3.33 *18* ?
γ	463.5 *20*	0.036 *9*
γ	491.51 *10*	0.26 *4*
γ	650.92 *9*	0.37 *3*
γ	892.55 *9*	0.10 *3*
γ	953.40 *6*	3.60 *18* ?
γ	1142.42 *5*	2.88 *18*
γ[E1]	1384.06 *5*	90 *4*

† 11% uncert(syst)

Continuous Radiation (^{92}Sr)
$\langle\beta-\rangle$=177 keV;$\langle$IB$\rangle$=0.126 keV

E_{bin}(keV)		$\langle\ \rangle$(keV)	(%)
0 - 10	β-	0.180	3.60
	IB	0.0088	
10 - 20	β-	0.54	3.58
	IB	0.0081	0.056
20 - 40	β-	2.13	7.1
	IB	0.0143	0.050
40 - 100	β-	14.4	20.6
	IB	0.031	0.049
100 - 300	β-	96	50.8
	IB	0.040	0.025
300 - 600	β-	45.1	12.5
	IB	0.0134	0.0033
600 - 1300	β-	13.5	1.46
	IB	0.0099	0.00123
1300 - 1888	β-	5.8	0.394
	IB	0.00054	3.8×10^{-5}

$^{92}_{39}$Y (3.54 *1* h)

Mode: β-
Δ: -84844 *10* keV
SpA: 9.62×10^{6} Ci/g
Prod: ^{94}Zr(d,α); daughter ^{92}Sr; ^{92}Zr(n,p)

Photons (^{92}Y)
$\langle\gamma\rangle$=253 *9* keV

γ_{mode}	γ(keV)	γ(%)†
γ[E2]	448.52 *10*	2.34 *14*
γ[E1]	492.62 *7*	0.49 *3*
γE2	561.11 *8*	2.40 *14*
γ[E1]	844.31 *8*	1.25 *8*
γM1+0.2%E2	912.81 *7*	0.67 *4*
γE2	934.53 *5*	13.9 *9*
γ[M1+E2]	972.37 *15*	0.068 *7*
γE2+14%M1	1132.43 *10*	0.243 *15*
γM1+E2	1405.42 *6*	4.8 *3*
γ[E2]	1847.33 *7*	0.347 *20*
γ[M1+E2]	1885.17 *15*	0.028 *5*
γ[E1]	1988.4 *5*	0.0061 *21*
γ[E2]	2066.94 *10*	0.042 *14* ?
γ[M1+E2]	2105.6 *3*	0.019 *3*
γ	2339.93 *6*	0.014 *4*
γ[E1]	2436.9 *5*	0.0031 *14*
γ	2473.4 *5*	0.0051 *14* ?
γ[E2]	2819.67 *15*	0.0042 *13*
γ[E2]	3263.9 *9*	0.0011 *4*
γ[E1]	3371.4 *5*	0.0031 *4*

† 9.4% uncert(syst)

Continuous Radiation (^{92}Y)

$\langle\beta-\rangle$=1435 keV;$\langle$IB$\rangle$=4.2 keV

E_{bin}(keV)		$\langle\ \rangle$(keV)	(%)
0 - 10	β-	0.0096	0.191
	IB	0.046	
10 - 20	β-	0.0293	0.195
	IB	0.046	0.32
20 - 40	β-	0.121	0.403
	IB	0.090	0.31
40 - 100	β-	0.94	1.32
	IB	0.26	0.40
100 - 300	β-	11.6	5.6
	IB	0.76	0.42
300 - 600	β-	50.3	11.1
	IB	0.88	0.21
600 - 1300	β-	267	28.2
	IB	1.29	0.149
1300 - 2500	β-	751	40.5
	IB	0.79	0.047
2500 - 3612	β-	354	12.5
	IB	0.059	0.0022

$^{92}_{40}$Zr(stable)

Δ: -88456.6 *24* keV

%: 17.15 *1*

$^{92}_{41}$Nb(3.6 *3* $\times 10^7$ yr)

Mode: ϵ

Δ: -86451 *3* keV

SpA: 0.000108 Ci/g

Prod: ^{92}Mo(n,p); ^{93}Nb(γ,n)

Photons (^{92}Nb)

$\langle\gamma\rangle$=1505.9 *23* keV

γ_{mode}	γ(keV)	γ(%)
Zr L_ℓ	1.792	0.075 *13*
Zr L_η	1.876	0.039 *6*
Zr L_α	2.042	1.8 *3*
Zr L_β	2.136	1.00 *16*
Zr L_γ	2.386	0.048 *10*
Zr $K_{\alpha2}$	15.691	18.6 *6*
Zr $K_{\alpha1}$	15.775	35.6 *12*
Zr $K_{\beta1}$'	17.663	8.1 *3*
Zr $K_{\beta2}$'	18.086	1.31 *5*
γ E2	561.11 *8*	100.0
γ E2	934.53 *5*	100.0

Atomic Electrons (^{92}Nb)

$\langle e\rangle$=7.3 *3* keV

e_{bin}(keV)	$\langle e\rangle$(keV)	e(%)
2	1.52	67 *7*
3	0.111	4.4 *5*
13	1.79	13.4 *14*
14	0.49	3.6 *4*
15	0.78	5.1 *5*
16	0.145	0.93 *9*
17	0.064	0.37 *4*
18	0.0173	0.098 *10*
543	1.43	0.264 *5*
559	0.171	0.0306 *6*
561	0.0354	0.00632 *13*
917	0.639	0.0697 *14*
932	0.0727	0.00780 *16*
934	0.0149	0.00159 *3*

Continuous Radiation (^{92}Nb)

$\langle$IB$\rangle$=0.042 keV

E_{bin}(keV)		$\langle\ \rangle$(keV)	(%)
10 - 20	IB	0.0145	0.092
20 - 40	IB	0.00099	0.0038
40 - 100	IB	0.0018	0.0026
100 - 300	IB	0.0163	0.0084
300 - 510	IB	0.0081	0.0023

$^{92}_{41}$Nb(10.15 *2* d)

Mode: ϵ

Δ: -86315 *3* keV

SpA: 1.398$\times 10^5$ Ci/g

Prod: ^{89}Y(α,n); ^{93}Nbγ,n); ^{93}Nb(n,2n)

Photons (^{92}Nb)

$\langle\gamma\rangle$=967.9 *22* keV

γ_{mode}	γ(keV)	γ(%)†
Zr L_ℓ	1.792	0.075 *13*
Zr L_η	1.876	0.039 *6*
Zr L_α	2.042	1.8 *3*
Zr L_β	2.135	0.99 *16*
Zr L_γ	2.386	0.048 *9*
Zr $K_{\alpha2}$	15.691	18.6 *7*
Zr $K_{\alpha1}$	15.775	35.7 *12*
Zr $K_{\beta1}$'	17.663	8.1 *3*
Zr $K_{\beta2}$'	18.086	1.32 *5*
γ M1+0.2%E2	912.81 *7*	1.73 *9*
γ E2	934.53 *5*	99.0
γ [E2]	1847.33 *7*	0.90 *4*

† 0.20% uncert(syst)

Atomic Electrons (^{92}Nb)

$\langle e\rangle$=5.6 *3* keV

e_{bin}(keV)	$\langle e\rangle$(keV)	e(%)
2	1.52	68 *7*
3	0.109	4.3 *4*
13	1.79	13.4 *14*
14	0.49	3.6 *4*
15	0.78	5.1 *5*
16	0.145	0.93 *10*
17	0.064	0.37 *4*
18	0.0173	0.098 *10*
895	0.0114	0.00127 *7*
910	0.00124	0.000136 *7*
911	3.30 $\times 10^{-5}$	3.62 *19* $\times 10^{-6}$
912	0.000224	2.46 *13* $\times 10^{-5}$
913	3.62 $\times 10^{-5}$	3.97 *21* $\times 10^{-6}$
917	0.633	0.0690 *14*
932	0.0720	0.00772 *16*
934	0.0147	0.00158 *3*
1829	0.00275	0.000151 *9*
1845	0.000304	1.65 *9* $\times 10^{-5}$
1847	5.3 $\times 10^{-5}$	2.89 *17* $\times 10^{-6}$

Continuous Radiation (^{92}Nb)

$\langle\beta+\rangle$=0.0508 keV;$\langle$IB$\rangle$=0.39 keV

E_{bin}(keV)		$\langle\ \rangle$(keV)	(%)
0 - 10	β+	3.41 $\times 10^{-6}$	4.18 $\times 10^{-5}$
	IB	0.000143	
10 - 20	β+	8.5 $\times 10^{-5}$	0.00052
	IB	0.0145	0.092
20 - 40	β+	0.00137	0.00430
	IB	0.00102	0.0039
40 - 100	β+	0.0232	0.0321
	IB	0.0021	0.0031
100 - 300	β+	0.0262	0.0210
	IB	0.035	0.0168
300 - 600	IB	0.150	0.033
600 - 1207	IB	0.19	0.025
	Σβ+		0.058

$^{92}_{42}$Mo(stable)

Δ: -86808 *4* keV

%: 14.84 *4*

$^{92}_{43}$Tc(4.4 *3* min)

Mode: ϵ

Δ: -78938 *26* keV

SpA: 4.6$\times 10^8$ Ci/g

Prod: ^{92}Mo(d,2n); ^{92}Mo(p,n)

Photons (^{92}Tc)

$\langle\gamma\rangle$=2965 *310* keV

γ_{mode}	γ(keV)	γ(%)
Mo L_ℓ	2.016	0.027 *5*
Mo L_η	2.120	0.0132 *20*
Mo L_α	2.293	0.69 *11*
Mo L_β	2.424	0.39 *7*
Mo L_γ	2.692	0.025 *5*
Mo $K_{\alpha2}$	17.374	6.3 *3*
Mo $K_{\alpha1}$	17.479	12.0 *5*
Mo $K_{\beta1}$'	19.602	2.83 *13*
Mo $K_{\beta2}$'	20.091	0.492 *23*
γ E1	85.0 *4*	12.1 *8*
γ E2	147.9 *5*	76 *3*
γ E1	244.0 *4*	13.8 *5*
γ E2	329.0 *4*	80 *3*
γ E2	773.1 *5*	100 *5*
γ	1336.0 *10*	~1
γ E2	1509.6 *5*	100
γ	1568.9 *10*	0.7 *3*
γ	1595.6 *9*	<0.50
γ [E1]	1702.1 *9*	0.9 *4*
γ [M1+E2]	1787.1 *9*	~0.3
γ	2159.6 *10*	1.8 *1*
γ	2307.5 *10*	1.4 *1*
γ	2511.5 *13*	0.31 *15*
γ	2705.2 *10*	0.66 *10*
γ	2853.1 *10*	0.5 *1*
γ	2873.4 *15*	~0.5
γ	2904.2 *15*	~0.30
γ	2977.5 *14*	~0.03 ?
γ	3026.1 *15*	0.53 *15*
γ	3134.4 *11*	0.10 *2*
γ	3219.4 *11*	
γ	3911.9 *15*	0.19 *6*
γ	4038.8 *13*	0.09 *4*
γ	4085.5 *15*	0.15 *7*
γ	4135.9 *15*	0.15 *3*
γ	4367.8 *12*	~0.35
γ	4573.1 *13*	<0.7

Atomic Electrons (^{92}Tc)

⟨e⟩=38.9 *11* keV

e_{bin}(keV)	⟨e⟩(keV)	e(%)
3 - 20	1.64	31 *3*
65	1.40	2.15 *15*
82 - 85	0.253	0.306 *18*
128	24.0	18.7 *8*
145	4.34	2.99 *13*
147 - 148	0.94	0.639 *24*
224 - 244	0.305	0.135 *5*
309	3.83	1.24 *5*
326 - 329	0.627	0.192 *7*
753	0.96	0.127 *7*
770 - 773	0.137	0.0178 *8*
1316 - 1336	0.0042	~0.00032
1490 - 1576	0.48	0.032 *6*
1593 - 1682	0.0019	0.00011 *5*
1699 - 1787	0.0015	~9 $\times10^{-5}$
2140 - 2159	0.0050	0.00024 *7*
2287 - 2307	0.0037	0.00016 *4*
2491 - 2511	0.0008	3.1 *16* $\times10^{-5}$
2685 - 2705	0.0016	5.9 *16* $\times10^{-5}$
2833 - 2904	0.0030	0.00011 *3*
2957 - 3026	0.0012	4.2 *13* $\times10^{-5}$
3114 - 3134	0.00022	7.0 *19* $\times10^{-6}$
3892 - 3911	0.00037	1.0 *3* $\times10^{-5}$
4019 - 4116	0.00071	1.8 *4* $\times10^{-5}$
4133 - 4135	3.3 $\times10^{-5}$	8 *2* $\times10^{-7}$
4348 - 4367	0.0007	~1 $\times10^{-5}$
4553 - 4573	0.0006	~1 $\times10^{-5}$

Continuous Radiation (^{92}Tc)

⟨β+⟩=1702 keV;⟨IB⟩=6.8 keV

E_{bin}(keV)		⟨ ⟩(keV)	(%)
0 - 10	β+	3.24 $\times10^{-6}$	3.95 $\times10^{-5}$
	IB	0.052	
10 - 20	β+	9.4 $\times10^{-5}$	0.00057
	IB	0.052	0.36
20 - 40	β+	0.00194	0.0060
	IB	0.101	0.35
40 - 100	β+	0.069	0.090
	IB	0.29	0.45
100 - 300	β+	2.89	1.31
	IB	0.88	0.49
300 - 600	β+	22.6	4.83
	IB	1.09	0.26
600 - 1300	β+	208	21.3
	IB	1.8	0.21
1300 - 2500	β+	840	44.6
	IB	1.7	0.097
2500 - 5000	β+	628	21.3
	IB	0.80	0.025
5000 - 5110	IB	7.1 $\times10^{-5}$	1.41 $\times10^{-6}$
	Σβ+		93

$^{92}_{44}$Ru(3.65 *5* min)

Mode: ε

Δ: -74410 *300* keV syst

SpA: 5.59$\times10^{8}$ Ci/g

Prod: protons on Ag; ^{92}Mo(^{3}He,3n)

Photons (^{92}Ru)

⟨γ⟩=1424 *16* keV

γ_{mode}	γ(keV)	γ(%)†
Tc L_ℓ	2.133	0.19 *3*
Tc L_η	2.249	0.090 *15*
Tc L_α	2.424	4.7 *8*
Tc L_β	2.575	2.8 *5*
Tc L_γ	2.853	0.20 *4*
Tc $K_{\alpha2}$	18.251	39.5 *24*
Tc $K_{\alpha1}$	18.367	75 *5*
Tc $K_{\beta1}'$	20.613	17.9 *11*
Tc $K_{\beta2}'$	21.136	3.22 *20*
γ M1	47.37 *3*	27 *3*
γ E2	56.252 *20*	8.7 *9*
γ M1	134.39 *7*	64.8 *19*
γ E2	213.72 *12*	95 *3*
γ [M1+E2]	259.24 *12*	91 *3*
γ [E2]	306.61 *12*	0.31 *3*
γ	410.34 *9*	1.78 *4*
γ	436.41 *20*	0.44 *5*
γ	450.53 *7*	6.74 *19*
γ	570.01 *10*	0.62 *7*
γ	584.92 *9*	0.58 *7*
γ [M1+E2]	594.28 *10*	0.58 *7*
γ	618.2 *4*	0.34 *4*
γ	634.65 *11*	0.28 *4*
γ [M1+E2]	657.19 *25*	~0.19
γ	663.51 *20*	0.48 *7*
γ	827.9 *5*	0.47 *10*
γ	839.0 *5*	0.32 *7*
γ [M1+E2]	866.89 *10*	11.3 *6*
γ	903.63 *8*	0.78 *5*
γ	910.23 *8*	3.21 *12*
γ [M1+E2]	938.1 *4*	0.24 *10* ?
γ	944.9 *3*	2.7 *3*
γ [M1+E2]	946.96 *12*	2.7 *3*
γ	958.71 *20*	0.37 *6*
γ	967.91 *20*	0.38 *6*
γ [M1+E2]	974.36 *14*	0.31 *5*
γ	1024.11 *20*	0.54 *5*
γ	1064.0 *3*	0.40 *5*
γ	1118.6 *3*	0.24 *6*
γ [M1+E2]	1219.57 *9*	6.0 *3*
γ	1228.93 *8*	3.33 *19*
γ	1269.12 *10*	0.34 *5*
γ	1394.8 *4*	0.48 *6*
γ	1403.51 *10*	1.61 *10*
γ	1460.0 *6*	0.63 *5*
γ	1517.4 *3*	1.90 *10*
γ	1560.81 *25*	0.74 *8*
γ [M1+E2]	1604.59 *10*	3.61 *13*
γ [M1+E2]	1679.45 *7*	9.1 *4*
γ [M1+E2]	1738.98 *12*	0.32 *7*
γ [M1+E2]	1813.84 *7*	0.19 *5*
γ	1882.4 *5*	0.19 *5*
γ	1900.51 *20*	0.37 *3*
γ	1928.4 *3*	0.18 *4*
γ [M1+E2]	2059.53 *14*	3.42 *19*
γ [M1+E2]	2193.91 *14*	0.80 *8*
γ [E2]	2241.29 *14*	0.26 *9*
γ	2302.2 *10*	1.05 *10*
γ [M1+E2]	2427.6 *3*	0.67 *7*
γ [M1+E2]	2471.02 *24*	0.20 *4*
γ [E2]	2518.40 *24*	0.09 *4*
γ [M1+E2]	2997.6 *4*	0.09 *4* ?
γ [M1+E2]	3132.0 *4*	~0.09 ?

† 8.4% uncert(syst)

Atomic Electrons (^{92}Ru)

⟨e⟩=97 *4* keV

e_{bin}(keV)	⟨e⟩(keV)	e(%)
3	4.3	157 *18*
15 - 21	6.2	38.1 *24*
26	15.1	57 *6*
35	20.8	59 *6*
44	2.8	6.4 *7*
45 - 47	0.97	2.10 *17*
53	6.5	12.3 *12*
54	5.5	10.3 *10*
56	2.8	4.9 *5*
113	8.0	7.08 *25*
131 - 134	1.36	1.03 *3*
193	12.9	6.68 *24*
211 - 214	2.48	1.17 *4*
238	6.0	2.5 *8*
256 - 304	1.0	0.41 *11*
306 - 307	0.00057	0.000185 *16*
389 - 436	0.16	0.039 *15*
447 - 451	0.017	0.0039 *16*
549 - 597	0.028	0.0050 *9*
614 - 663	0.011	0.0018 *4*
807 - 846	0.106	0.0125 *10*
864 - 910	0.041	0.0047 *10*
917 - 965	0.051	0.0055 *8*
967 - 1003	0.0032	0.00032 *12*
1021 - 1064	0.0027	0.00026 *9*
1098 - 1119	0.0013	0.00012 *5*
1199 - 1229	0.057	0.0047 *6*
1248 - 1269	0.0016	0.00013 *5*
1374 - 1403	0.0092	0.00066 *18*
1439 - 1460	0.0026	0.00018 *6*
1496 - 1584	0.027	0.00172 *20*
1602 - 1679	0.046	0.00279 *19*
1718 - 1813	0.0024	0.000136 *20*
1861 - 1928	0.0025	0.00013 *3*
2038 - 2059	0.0138	0.00067 *5*
2173 - 2241	0.0040	0.000182 *19*
2281 - 2302	0.0030	0.00013 *4*
2407 - 2497	0.0033	0.000137 *12*
2515 - 2518	3.4 $\times10^{-5}$	1.3 *5* $\times10^{-6}$
2977 - 2997	0.00026	9 *3* $\times10^{-6}$
3111 - 3132	0.00028	9 *4* $\times10^{-6}$

Continuous Radiation (^{92}Ru)

⟨β+⟩=618 keV;⟨IB⟩=3.7 keV

E_{bin}(keV)		⟨ ⟩(keV)	(%)
0 - 10	β+	5.6 $\times10^{-6}$	6.8 $\times10^{-5}$
	IB	0.022	
10 - 20	β+	0.000167	0.00102
	IB	0.030	0.20
20 - 40	β+	0.00351	0.0109
	IB	0.045	0.160
40 - 100	β+	0.126	0.164
	IB	0.123	0.19
100 - 300	β+	4.96	2.26
	IB	0.36	0.20
300 - 600	β+	33.7	7.3
	IB	0.49	0.115
600 - 1300	β+	215	22.6
	IB	1.15	0.128
1300 - 2500	β+	356	20.6
	IB	1.20	0.070
2500 - 3819	β+	7.7	0.298
	IB	0.29	0.0101
	Σβ+		53

A = 93

NDS **B8**, 527 (1972)

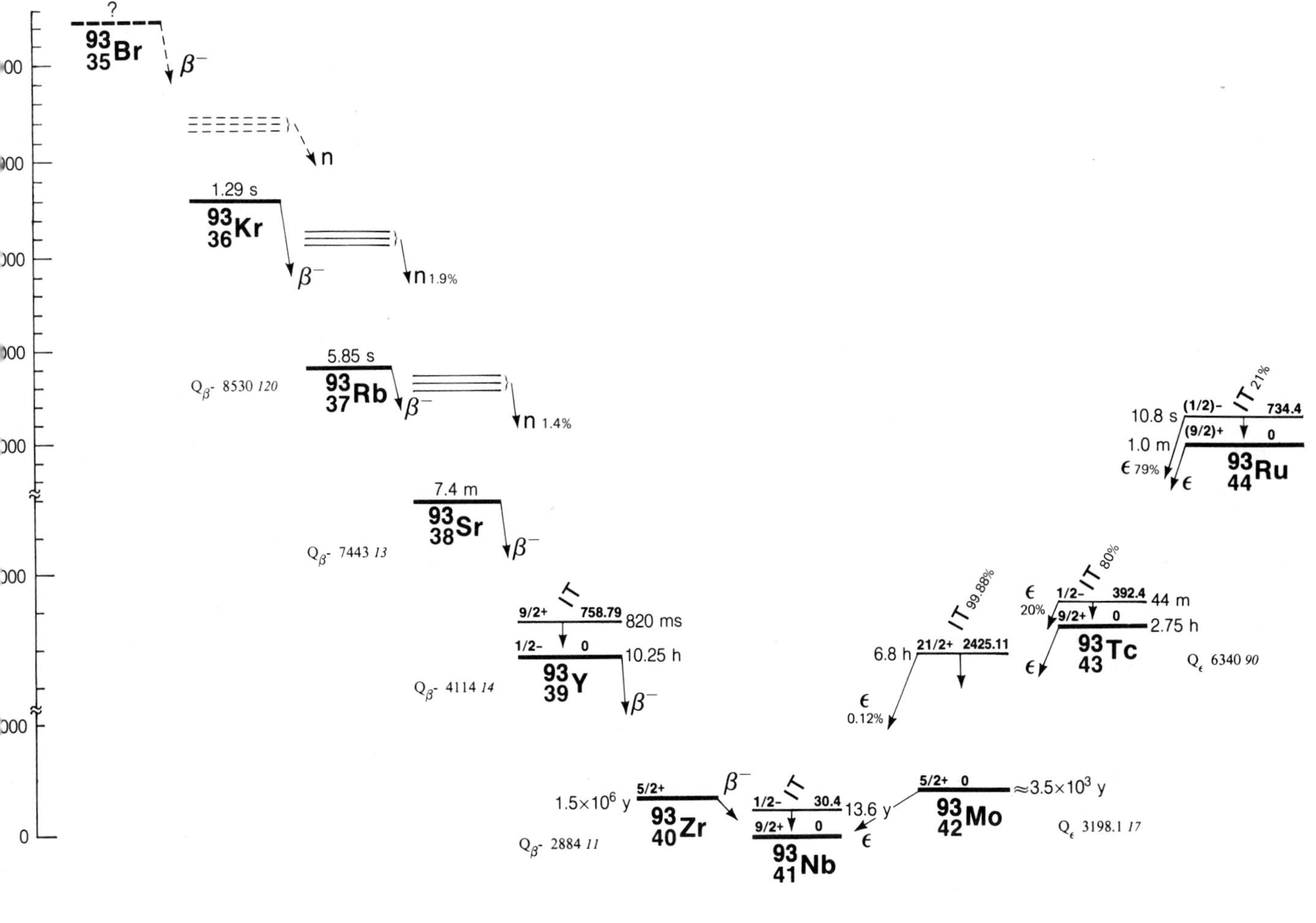

$^{93}_{35}$Br($t_{1/2}$ unknown)

Mode: β-, β-n

Prod: protons on U; protons on Nb

$^{93}_{36}$Kr(1.289 *12* s)

Mode: β-, β-n(1.92 *14* %)

Δ: -64150 *120* keV

SpA: 7.28×10^{10} Ci/g

Prod: fission

Photons (^{93}Kr)

⟨γ⟩=2277 *23* keV

γ_{mode}	γ(keV)	γ(%)†
γ	57.12 *3*	0.263 *12*
γ	70.56 *3*	1.57 *7*
γ	182.08 *3*	5.5 *3*
γ	191.13 *5*	0.079 *5*
γ	239.36 *11*	0.16 *3*
γ	252.64 *4*	20.0 *10*
γM1(+E2)	253.43 *3*	42.0 *22*
γ	254.88 *5*	0.71 *7*
γE2(+M1)	266.87 *3*	21.0 *10*
γ	292.77 *13*	0.092 *5*
γ	316.76 *9*	0.246 *20*
γM1	323.99 *3*	24.6 *12*
γ	398.96 *7*	0.121 *10*
γ	401.69 *6*	0.047 *7*
γ	480.43 *12*	0.089 *12*
γ	491.55 *14*	0.081 *12*
γ	496.60 *3*	1.85 *10*
γ	519.96 *14*	0.098 *12*
γ	529.67 *3*	0.50 *3*
γ	553.72 *4*	0.079 *12*
γ	555.99 *12*	0.106 *12*
γ	567.16 *4*	0.170 *12*

Photons (^{93}Kr)
(continued)

γ_{mode}	γ(keV)	γ(%)†
γ	570.23 *4*	1.22 *6*
γ	578.79 *15*	0.086 *10*
γ	616.59 *10*	0.103 *7*
γ	623.76 *12*	0.053 *6*
γ	643.48 *10*	0.093 *22*
γ	644.87 *6*	0.276 *25*
γ	686.86 *12*	0.138 *10*
γ	713.4 *5*	0.057 *10*
γ	717.0 *7*	0.052 *12*
γ	722.79 *5*	0.278 *17*
γ	733.76 *5*	0.90 *5*
γ	736.79 *14*	0.054 *7*
γ	770.57 *18*	0.140 *25*
γ	777.62 *6*	0.204 *15*
γ	820.59 *4*	3.79 *20*
γ	844.18 *3*	0.57 *3*
γ	852.80 *7*	0.096 *7*
γ	891.32 *5*	0.032 *10*
γ	895.16 *11*	0.177 *15*
γ	898.1 *3*	0.044 *10*
γ	921.23 *7*	0.231 *17*
γ	964.94 *9*	0.221 *17*
γ	976.16 *5*	0.72 *4*
γ	1000.34 *16*	0.047 *10*

Photons (^{93}Kr)
(continued)

γ_{mode}	γ(keV)	γ(%)†
γ	1005.91 9	0.167 12
γ	1026.26 3	2.21 12
γ	1046.56 7	0.123 12
γ	1051.31 14	0.076 12
γ	1054.62 13	0.108 12
γ	1058.79 12	0.31 4
γ	1060.59 10	0.39 4
γ	1080.52 25	0.042 15
γ	1083.38 3	0.83 4
γ	1096.82 4	0.130 25
γ	1126.31 9	0.069 12
γ	1136.28 10	0.079 15
γ	1139.22 14	0.197 17
γ	1157.11 11	0.322 25
γ	1191.53 8	0.236 15
γ	1214.89 6	1.80 10
γ	1235.74 11	0.135 22
γ	1239.04 5	1.13 6
γ	1290.51 13	0.24 3
γ	1296.17 4	1.92 10
γ	1309.60 5	0.106 12
γ	1313.40 7	0.300 25
γ	1318.35 10	0.93 7
γ	1350.25 3	0.76 4
γ	1361.01 8	0.231 17
γ	1364.77 4	0.70 5
γ	1374.67 6	0.43 3
γ	1382.48 13	0.19 4
γ	1387.73 6	1.38 10
γ	1421.90 4	0.98 5
γ	1435.34 4	1.03 7
γ	1445.61 15	0.207 22
γ	1458.65 5	0.40 3
γ	1471.09 10	0.39 4
γ	1505.79 4	2.29 12
γ	1508.66 14	0.22 3
γ	1526.14 9	0.219 25
γ	1528.99 13	0.148 22
γ	1543.20 15	0.349 25
γ	1556.37 9	0.253 20
γ	1563.03 5	0.96 5
γ	1576.48 13	0.091 25
γ	1586.94 6	0.86 5
γ	1596.71 9	1.40 7
γ	1613.64 10	0.35 6
γ	1616.57 13	0.069 25
γ	1627.08 10	2.02 10
γ	1637.88 14	0.51 5
γ	1641.15 5	1.48 7
γ	1651.92 6	0.71 4
γ	1663.12 12	0.42 3
γ	1666.06 17	0.084 22
γ	1682.10 10	0.098 25
γ	1685.40 7	0.56 5
γ	1687.76 16	0.15 5
γ	1697.85 4	1.43 7
γ	1704.56 7	0.258 25
γ	1711.29 5	0.51 5
γ	1713.18 6	0.31 5
γ	1742.52 7	1.30 7
γ	1745.37 15	0.42 4
γ	1755.96 7	0.32 3
γ	1779.75 4	0.59 3
γ	1785.88 17	0.125 25
γ	1789.00 7	0.32 3
γ	1794.85 8	0.89 5
γ	1798.27 12	0.185 25
γ	1803.89 12	0.226 20
γ	1822.23 21	~0.17
γ	1823.41 19	0.34 15
γ	1840.16 14	0.27 7
γ	1850.13 9	0.098 15
γ	1862.90 7	0.271 20
γ	1886.63 6	0.71 4
γ	1929.97 14	0.32 5
γ	1943.76 6	0.48 3
γ	1957.19 7	0.36 3
γ	1961.82 3	1.82 10
γ	1989.3 3	0.29 3
γ	1994.48 9	0.27 3
γ	2011.47 12	0.234 22
γ	2018.94 3	1.43 7
γ	2035.45 4	1.85 10
γ	2082.69 11	0.303 22

Photons (^{93}Kr)
(continued)

γ_{mode}	γ(keV)	γ(%)†
γ	2088.50 8	0.278 25
γ	2159.95 10	0.069 15
γ	2179.6 4	~0.10
γ	2181.58 6	1.18 10
γ	2235.61 14	0.074 22
γ	2239.21 12	0.182 25
γ	2308.29 13	0.076 17
γ	2342.71 6	0.18 6
γ	2349.96 4	7.5 4
γ	2366.25 17	0.13 5
γ	2368.81 12	0.14 5
γ	2411.48 6	0.315 22
γ	2424.54 9	0.177 20
γ	2491.06 9	0.47 7
γ	2496.09 6	2.34 12
γ	2517.63 9	0.079 17
γ	2521.53 16	0.48 3
γ	2532.04 3	0.133 15
γ	2548.19 9	0.63 5
γ	2557.35 6	0.60 4
γ	2561.62 9	1.02 6
γ	2589.16 4	0.52 3
γ	2602.60 4	4.28 22
γ	2606.73 16	0.73 6
γ	2663.42 11	0.52 5
γ	2678.16 6	0.27 5
γ	2700.52 13	0.22 3
γ	2720.39 19	0.204 25
γ	2739.42 5	0.52 3
γ	2755.7 3	0.219 22
γ	2773.0 4	0.209 22
γ	2782.3 3	0.56 4
γ	2796.54 6	0.369 25
γ	2809.98 6	0.450 25
γ	2826.88 16	0.199 20
γ	2838.37 13	0.185 22
γ	2846.1 7	0.7 3
γ	2852.70 14	0.19 4
γ	2856.02 4	2.21 12
γ	2913.42 15	0.212 25
γ	2944.7 3	0.18 3
γ	2948.37 17	0.62 4
γ	2956.66 8	0.61 4
γ	2972.3 3	0.45 4
γ	2998.35 13	0.64 15
γ	3000.67 11	0.34 15
γ	3014.5 3	0.32 10
γ	3026.76 14	0.177 25
γ	3097.8 7	0.079 20
γ	3105.34 13	0.300 25
γ	3150.45 19	0.21 5
γ	3196.8 4	0.15 5
γ	3214.53 19	0.219 22
γ	3220.52 12	0.182 20
γ	3226.81 13	1.01 7
γ	3230.17 13	0.15 5
γ	3250.35 13	0.160 17
γ	3261.06 24	0.089 15
γ	3281.15 17	0.081 20
γ	3285.73 17	0.180 20
γ	3295.02 10	0.21 4
γ	3298.20 7	0.65 5
γ	3304.30 17	0.11 3
γ	3307.6 3	0.10 3
γ	3355.92 18	0.22 7
γ	3358.76 14	~0.12
γ	3379.75 11	0.172 25
γ	3407.78 16	0.46 3
γ	3412.36 18	0.143 25
γ	3445.1 4	0.066 12
γ	3453.24 7	0.207 25
γ	3460.38 22	0.71 12
γ	3464.53 18	0.32 10
γ	3467.12 15	0.27 12
γ	3471.4 7	0.15 3
γ	3482.38 17	0.121 20
γ	3582.8 4	0.155 15
γ	3634.63 11	0.194 22
γ	3646.22 19	0.24 5
γ	3649.30 13	0.31 5
γ	3655.54 15	0.140 22
γ	3705.92 21	0.303 20
γ	3775.92 12	0.150 17
γ	3796.31 21	0.039 12

Photons (^{93}Kr)
(continued)

γ_{mode}	γ(keV)	γ(%)†
γ	3887.49 13	0.13 2
γ	4014.2 15	0.061 25
γ	4032.9 3	0.216 17

† 7.3% uncert(syst)

Continuous Radiation (^{93}Kr)

⟨β-⟩=2818 keV;⟨IB⟩=12.1 keV

E_{bin}(keV)		⟨ ⟩(keV)	(%)
0 - 10	β-	0.0052	0.104
	IB	0.068	
10 - 20	β-	0.0158	0.105
	IB	0.068	0.47
20 - 40	β-	0.064	0.214
	IB	0.135	0.47
40 - 100	β-	0.469	0.67
	IB	0.39	0.60
100 - 300	β-	4.73	2.36
	IB	1.22	0.68
300 - 600	β-	16.1	3.55
	IB	1.61	0.38
600 - 1300	β-	118	12.1
	IB	2.9	0.33
1300 - 2500	β-	509	26.6
	IB	3.2	0.18
2500 - 5000	β-	1580	44.1
	IB	2.3	0.070
5000 - 7500	β-	579	10.0
	IB	0.21	0.0039
7500 - 8277	β-	10.3	0.134
	IB	0.00030	3.9×10^{-6}

$^{93}_{37}$Rb(5.85 *3* s)

Mode: β-, β-n(1.39 *18* %)
Δ: -72679 *15* keV
SpA: 1.956×10^{10} Ci/g
Prod: fission

β-n(avg): 405

Photons (^{93}Rb)

⟨γ⟩=1405 *13* keV

γ_{mode}	γ(keV)	γ(%)†
$\gamma_{\beta-}$	163.55 18	0.083 19
$\gamma_{\beta-}$	205.36 17	~0.08
$\gamma_{\beta-}$[M1+E2]	213.40 4	4.8 3
$\gamma_{\beta-}$	219.15 4	1.98 11
$\gamma_{\beta-}$	351.78 10	0.047 5
$\gamma_{\beta-n}$	393.65 9	~0.019
$\gamma_{\beta-}$	405.01 18	0.039 6
$\gamma_{\beta-}$	432.55 3	12.5 6
$\gamma_{\beta-}$	473.85 21	0.020 9
$\gamma_{\beta-n}$	569.93 8	~0.042
$\gamma_{\beta-}$	596.01 16	0.159 21
$\gamma_{\beta-}$	602.59 20	0.036 11
$\gamma_{\beta-}$	610.46 16	0.126 23
$\gamma_{\beta-}$	661.66 10	0.20 4
$\gamma_{\beta-}$	709.97 4	3.8 3
$\gamma_{\beta-}$	722.02 13	0.036 5
$\gamma_{\beta-}$	768.34 10	0.083 14
$\gamma_{\beta-}$	776.57 21	0.038 12
$\gamma_{\beta-}$	793.68 6	0.77 4
$\gamma_{\beta-n}$ [E2]	814.80 9	1.36
$\gamma_{\beta-}$	822.40 8	0.121 21

Photons (^{93}Rb) (continued)

γ_{mode}	γ(keV)	γ(%)†
$\gamma_{\beta-}$	830.98 *18*	0.045 *10*
$\gamma_{\beta-}$	859.08 *15*	0.060 *8*
$\gamma_{\beta-}$	867.67 *11*	0.052 *6*
$\gamma_{\beta-}$	901.05 *13*	0.079 *10*
$\gamma_{\beta-}$	905.85 *10*	0.047 *10*
$\gamma_{\beta-}$	910.94 *12*	0.103 *11*
$\gamma_{\beta-}$	929.11 *5*	0.305 *21*
$\gamma_{\beta-}$	934.81 *6*	0.230 *18*
$\gamma_{\beta-n}$	963.57 *11*	~0.023
$\gamma_{\beta-}$	981.10 *24*	0.094 *21*
$\gamma_{\beta-}$	986.10 *5*	4.89 *25*
$\gamma_{\beta-}$	990.96 *14*	0.081 *16*
$\gamma_{\beta-}$	1034.90 *18*	0.047 *15*
$\gamma_{\beta-}$	1054.61 *10*	0.042 *9*
$\gamma_{\beta-}$	1059.40 *9*	0.046 *9*
$\gamma_{\beta-}$	1068.53 *9*	0.44 *4*
$\gamma_{\beta-}$	1077.61 *13*	0.033 *4*
$\gamma_{\beta-}$	1096.73 *9*	0.287 *18*
$\gamma_{\beta-}$	1100.67 *11*	0.130 *11*
$\gamma_{\beta-}$	1115.66 *15*	0.067 *10*
$\gamma_{\beta-}$	1119.89 *15*	0.052 *15*
$\gamma_{\beta-}$	1130.67 *17*	0.138 *15*
$\gamma_{\beta-}$	1138.01 *24*	0.145 *22*
$\gamma_{\beta-}$	1142.51 *4*	0.226 *19*
$\gamma_{\beta-}$	1148.21 *6*	1.10 *6*
$\gamma_{\beta-}$	1150.38 *7*	0.33 *3*
$\gamma_{\beta-}$	1164.17 *23*	0.065 *10*
$\gamma_{\beta-}$	1167.12 *24*	0.033 *9*
$\gamma_{\beta-}$	1202.14 *19*	0.034 *15*
$\gamma_{\beta-}$	1204.76 *19*	0.036 *15*
$\gamma_{\beta-}$	1208.53 *14*	0.111 *14*
$\gamma_{\beta-}$	1223.01 *18*	0.050 *11*
$\gamma_{\beta-}$	1238.28 *7*	1.06 *6*
$\gamma_{\beta-}$	1284.15 *13*	0.108 *25*
$\gamma_{\beta-}$	1287.08 *17*	0.080 *25*
$\gamma_{\beta-}$	1306.78 *8*	0.083 *10*
$\gamma_{\beta-}$	1315.60 *9*	0.271 *19*
$\gamma_{\beta-}$	1332.98 *7*	0.76 *8*
$\gamma_{\beta-}$	1349.55 *9*	0.101 *13*
$\gamma_{\beta-}$	1359.87 *12*	0.147 *14*
$\gamma_{\beta-}$	1365.39 *10*	0.234 *18*
$\gamma_{\beta-n}$	1384.72 *12*	~0.027
$\gamma_{\beta-}$	1385.29 *6*	4.1 *2*
$\gamma_{\beta-}$	1388.67 *24*	0.16 *4*
$\gamma_{\beta-}$	1396.85 *13*	0.041 *11*
$\gamma_{\beta-}$	1405.68 *11*	0.071 *9*
$\gamma_{\beta-}$	1437.14 *8*	0.30 *3*
$\gamma_{\beta-}$	1439.4 *3*	0.065 *21*
$\gamma_{\beta-}$	1453.21 *15*	0.036 *14*
$\gamma_{\beta-}$	1470.07 *14*	0.136 *15*
$\gamma_{\beta-}$	1473.28 *19*	0.039 *12*
$\gamma_{\beta-}$	1478.87 *14*	0.044 *9*
$\gamma_{\beta-}$	1484.09 *12*	0.063 *9*
$\gamma_{\beta-}$	1491.27 *16*	0.087 *13*
$\gamma_{\beta-}$	1495.63 *20*	0.165 *14*
$\gamma_{\beta-}$	1501.20 *9*	0.250 *16*
$\gamma_{\beta-}$	1507.89 *12*	0.170 *14*
$\gamma_{\beta-}$	1515.89 *13*	0.067 *13*
$\gamma_{\beta-}$	1530.46 *19*	0.046 *14*
$\gamma_{\beta-}$	1533.80 *19*	0.101 *15*
$\gamma_{\beta-}$	1547.74 *12*	0.204 *16*
$\gamma_{\beta-}$	1562.95 *9*	0.73 *5*
$\gamma_{\beta-}$	1566.38 *8*	0.042 *20*
$\gamma_{\beta-}$	1574.5 *1*	0.089 *10*
$\gamma_{\beta-}$	1578.05 *19*	0.110 *15*
$\gamma_{\beta-}$	1594.60 *9*	0.42 *3*
$\gamma_{\beta-}$	1612.95 *8*	1.20 *8*
$\gamma_{\beta-}$	1635.28 *13*	0.269 *22*
$\gamma_{\beta-}$	1662.13 *9*	0.262 *21*
$\gamma_{\beta-}$	1684.79 *11*	0.39 *3*
$\gamma_{\beta-}$	1690.4 *3*	0.044 *15*
$\gamma_{\beta-}$	1726.67 *12*	0.056 *11*
$\gamma_{\beta-}$	1735.69 *15*	~0.08
$\gamma_{\beta-}$	1738.28 *10*	~0.08
$\gamma_{\beta-}$	1743.65 *15*	0.080 *23*
$\gamma_{\beta-}$	1745.91 *15*	0.086 *23*
$\gamma_{\beta-}$	1750.04 *14*	0.181 *16*
$\gamma_{\beta-}$	1753.67 *25*	0.067 *14*
$\gamma_{\beta-n}$	1778.37 *14*	~0.007
$\gamma_{\beta-}$	1793.54 *13*	0.193 *18*
$\gamma_{\beta-}$	1803.6 *3*	0.171 *25*
$\gamma_{\beta-}$	1808.50 *7*	2.01 *10*
$\gamma_{\beta-}$	1812.78 *19*	0.179 *20*
$\gamma_{\beta-}$	1821.88 *9*	0.41 *3*

Photons (^{93}Rb) (continued)

γ_{mode}	γ(keV)	γ(%)†
$\gamma_{\beta-}$	1831.20 *12*	0.149 *18*
$\gamma_{\beta-}$	1836.57 *13*	~0.20
$\gamma_{\beta-}$	1837.99 *18*	0.34 *12*
$\gamma_{\beta-}$	1841.34 *21*	0.059 *17*
$\gamma_{\beta-}$	1869.69 *7*	1.36 *8*
$\gamma_{\beta-}$	1882.9 *3*	0.074 *15*
$\gamma_{\beta-}$	1886.53 *8*	0.104 *16*
$\gamma_{\beta-}$	1892.78 *12*	0.125 *15*
$\gamma_{\beta-}$	1900.96 *9*	0.331 *21*
$\gamma_{\beta-}$	1908.1 *3*	0.070 *23*
$\gamma_{\beta-}$	1910.84 *9*	0.82 *5*
$\gamma_{\beta-}$	1918.94 *11*	0.077 *15*
$\gamma_{\beta-}$	1927.61 *11*	0.54 *4*
$\gamma_{\beta-}$	1933.94 *11*	0.19 *3*
$\gamma_{\beta-}$	1956.53 *12*	0.125 *16*
$\gamma_{\beta-}$	1977.86 *11*	0.57 *4*
$\gamma_{\beta-}$	1982.93 *21*	0.050 *23*
$\gamma_{\beta-}$	1991.64 *25*	0.120 *16*
$\gamma_{\beta-}$	1997.69 *11*	0.042 *14*
$\gamma_{\beta-}$	2023.88 *18*	0.087 *19*
$\gamma_{\beta-}$	2026.83 *19*	0.166 *21*
$\gamma_{\beta-}$	2037.0 *8*	0.049 *23*
$\gamma_{\beta-}$	2043.84 *12*	0.219 *17*
$\gamma_{\beta-}$	2054.04 *9*	0.96 *5*
$\gamma_{\beta-}$	2058.87 *10*	0.251 *21*
$\gamma_{\beta-}$	2068.20 *11*	0.103 *11*
$\gamma_{\beta-}$	2087.60 *10*	0.125 *17*
$\gamma_{\beta-}$	2147.84 *16*	0.21 *3*
$\gamma_{\beta-}$	2168.27 *10*	0.315 *23*
$\gamma_{\beta-}$	2169.2 *4*	<0.08
$\gamma_{\beta-}$	2206.1 *3*	0.130 *19*
$\gamma_{\beta-}$	2229.46 *9*	0.68 *4*
$\gamma_{\beta-}$	2256.08 *20*	~0.05
$\gamma_{\beta-}$	2258.5 *3*	0.19 *4*
$\gamma_{\beta-}$	2262.1 *3*	0.101 *14*
$\gamma_{\beta-}$	2270.25 *11*	0.391 *23*
$\gamma_{\beta-}$	2292.87 *7*	0.384 *24*
$\gamma_{\beta-}$	2327.66 *12*	0.083 *13*
$\gamma_{\beta-}$	2334.2 *4*	0.046 *10*
$\gamma_{\beta-}$	2349.58 *11*	0.44 *4*
$\gamma_{\beta-}$	2359.14 *15*	0.234 *16*
$\gamma_{\beta-}$	2376.91 *20*	0.097 *15*
$\gamma_{\beta-}$	2386.79 *20*	0.161 *18*
$\gamma_{\beta-}$	2398.20 *19*	0.087 *13*
$\gamma_{\beta-}$	2403.85 *15*	0.046 *11*
$\gamma_{\beta-}$	2418.18 *19*	0.239 *24*
$\gamma_{\beta-}$	2451.7 *8*	0.12 *3*
$\gamma_{\beta-}$	2455.02 *12*	0.35 *4*
$\gamma_{\beta-}$	2462.25 *10*	0.35 *3*
$\gamma_{\beta-}$	2491.40 *11*	0.28 *3*
$\gamma_{\beta-}$	2505.32 *10*	0.59 *4*
$\gamma_{\beta-}$	2523.8 *3*	0.17 *6*
$\gamma_{\beta-}$	2550.26 *15*	0.193 *19*
$\gamma_{\beta-}$	2557.32 *13*	0.089 *15*
$\gamma_{\beta-}$	2568.72 *11*	0.274 *24*
$\gamma_{\beta-}$	2602.44 *14*	0.251 *24*
$\gamma_{\beta-}$	2613.84 *16*	0.093 *14*
$\gamma_{\beta-}$	2620.13 *14*	0.060 *14*
$\gamma_{\beta-}$	2625.2 *3*	0.066 *14*
$\gamma_{\beta-}$	2638.41 *12*	0.20 *3*
$\gamma_{\beta-}$	2646.63 *14*	0.13 *4*
$\gamma_{\beta-}$	2652.60 *15*	0.224 *23*
$\gamma_{\beta-}$	2661.11 *9*	0.222 *21*
$\gamma_{\beta-}$	2674.17 *22*	0.076 *15*
$\gamma_{\beta-}$	2705.02 *9*	0.74 *5*
$\gamma_{\beta-}$	2724.29 *13*	0.40 *6*
$\gamma_{\beta-}$	2734.18 *10*	0.042 *16*
$\gamma_{\beta-}$	2766.47 *10*	0.286 *21*
$\gamma_{\beta-}$	2773.1 *3*	0.087 *15*
$\gamma_{\beta-}$	2799.67 *11*	0.109 *19*
$\gamma_{\beta-}$	2812.48 *9*	0.077 *18*
$\gamma_{\beta-}$	2861.43 *9*	0.80 *5*
$\gamma_{\beta-}$	2869.16 *12*	0.315 *24*
$\gamma_{\beta-}$	2875.01 *16*	0.075 *17*
$\gamma_{\beta-}$	2880.70 *12*	0.274 *22*
$\gamma_{\beta-}$	2886.46 *9*	0.237 *25*
$\gamma_{\beta-}$	2890.58 *11*	0.29 *3*
$\gamma_{\beta-}$	2903.65 *22*	0.161 *19*
$\gamma_{\beta-}$	2954.87 *12*	0.32 *4*
$\gamma_{\beta-}$	2958.09 *16*	0.12 *3*
$\gamma_{\beta-}$	3027.68 *24*	0.035 *15*
$\gamma_{\beta-}$	3104.50 *15*	0.051 *17*
$\gamma_{\beta-}$	3113.96 *16*	0.268 *25*
$\gamma_{\beta-}$	3129.2 *8*	0.063 *19*

Photons (^{93}Rb) (continued)

γ_{mode}	γ(keV)	γ(%)†
$\gamma_{\beta-}$	3133.22 *15*	0.064 *19*
$\gamma_{\beta-}$	3171.79 *24*	0.139 *19*
$\gamma_{\beta-}$	3211.47 *16*	0.080 *16*
$\gamma_{\beta-}$	3226.48 *23*	0.217 *23*
$\gamma_{\beta-}$	3296.45 *18*	0.050 *21*
$\gamma_{\beta-}$	3338.28 *18*	0.097 *16*
$\gamma_{\beta-}$	3366.66 *12*	0.162 *20*
$\gamma_{\beta-}$	3371.05 *9*	0.81 *5*
$\gamma_{\beta-}$	3389.80 *11*	0.044 *14*
$\gamma_{\beta-}$	3403.67 *13*	0.329 *21*
$\gamma_{\beta-}$	3458.03 *10*	2.67 *14*
$\gamma_{\beta-}$	3477.58 *15*	0.194 *16*
$\gamma_{\beta-}$	3486.7 *3*	0.049 *12*
$\gamma_{\beta-}$	3501.97 *12*	0.41 *9*
$\gamma_{\beta-}$	3544.09 *16*	0.11 *4*
$\gamma_{\beta-}$	3547.1 *3*	0.10 *4*
$\gamma_{\beta-}$	3572.01 *13*	0.214 *19*
$\gamma_{\beta-}$	3585.66 *19*	0.076 *14*
$\gamma_{\beta-}$	3642.4 *4*	0.070 *15*
$\gamma_{\beta-}$	3664.81 *11*	0.39 *3*
$\gamma_{\beta-}$	3706.77 *22*	0.052 *13*
$\gamma_{\beta-}$	3721.11 *11*	0.109 *18*
$\gamma_{\beta-}$	3770.47 *13*	0.127 *15*
$\gamma_{\beta-}$	3789.10 *14*	0.109 *14*
$\gamma_{\beta-}$	3803.58 *9*	1.13 *6*
$\gamma_{\beta-}$	3821.59 *15*	0.07 *1*
$\gamma_{\beta-}$	3848.72 *15*	0.076 *18*
$\gamma_{\beta-}$	3867.33 *9*	1.85 *10*
$\gamma_{\beta-}$	3876.65 *10*	0.151 *16*
$\gamma_{\beta-}$	3883.95 *12*	0.324 *24*
$\gamma_{\beta-}$	3890.57 *10*	0.150 *16*
$\gamma_{\beta-}$	3934.51 *11*	0.70 *4*
$\gamma_{\beta-}$	3941.57 *21*	0.081 *16*
$\gamma_{\beta-}$	3954.96 *8*	0.028 *11*
$\gamma_{\beta-}$	4004.2 *3*	0.056 *14*
$\gamma_{\beta-}$	4010.6 *3*	0.038 *14*
$\gamma_{\beta-}$	4017.48 *15*	0.295 *21*
$\gamma_{\beta-}$	4157.6 *3*	0.069 *14*
$\gamma_{\beta-}$	4242.02 *13*	0.055 *9*
$\gamma_{\beta-}$	4250.8 *3*	0.035 *9*
$\gamma_{\beta-}$	4271.24 *15*	0.241 *16*
$\gamma_{\beta-}$	4281.95 *13*	0.121 *10*
$\gamma_{\beta-}$	4388.15 *23*	0.084 *10*
$\gamma_{\beta-}$	4461.04 *15*	0.055 *8*
$\gamma_{\beta-}$	4480.84 *21*	0.042 *8*
$\gamma_{\beta-}$	4615.10 *16*	0.031 *10*
$\gamma_{\beta-}$	4627.2 *3*	0.074 *10*
$\gamma_{\beta-}$	4645.02 *18*	0.031 *10*
$\gamma_{\beta-}$	4875.32 *20*	0.126 *10*
$\gamma_{\beta-}$	4890.0 *8*	0.019 *4*
$\gamma_{\beta-}$	4898.71 *23*	0.035 *5*
$\gamma_{\beta-}$	4948.4 *3*	0.050 *9*
$\gamma_{\beta-}$	4954.1 *3*	0.026 *6*
$\gamma_{\beta-}$	4971.8 *6*	0.024 *5*
$\gamma_{\beta-}$	4996.8 *5*	0.036 *6*
$\gamma_{\beta-}$	5138.9 *10*	0.016 *5*
$\gamma_{\beta-}$	5154.6 *10*	0.016 *5*
$\gamma_{\beta-}$	5164.8 *11*	0.014 *5*
$\gamma_{\beta-}$	5395.8 *3*	0.021 *5*
$\gamma_{\beta-}$	5409.0 *7*	0.029 *5*

† 13% uncert(syst)

Continuous Radiation (^{93}Rb)

$\langle\beta-\rangle$=2713 keV;$\langle$IB$\rangle$=11.6 keV

E_{bin}(keV)		$\langle\ \rangle$(keV)	(%)
0 - 10	β-	0.0064	0.128
	IB	0.066	
10 - 20	β-	0.0194	0.130
	IB	0.066	0.46
20 - 40	β-	0.079	0.262
	IB	0.131	0.45
40 - 100	β-	0.57	0.81
	IB	0.38	0.59
100 - 300	β-	5.6	2.82
	IB	1.18	0.66
300 - 600	β-	20.4	4.47

Continuous Radiation (^{93}Rb)
(continued)

E_{bin}(keV)		⟨ ⟩(keV)	(%)
	IB	1.55	0.36
600 - 1300	β-	132	13.7
	IB	2.8	0.32
1300 - 2500	β-	514	27.1
	IB	3.0	0.169
2500 - 5000	β-	1462	40.4
	IB	2.2	0.068
5000 - 7443	β-	578	10.2
	IB	0.154	0.0028

$^{93}_{38}$Sr(7.4 *2* min)

Mode: β-
Δ: -80121 *12* keV
SpA: 2.73×10^{8} Ci/g
Prod: ^{96}Zr(n,α); fission

Photons (^{93}Sr)
⟨γ⟩=2214 *30* keV

γ_{mode}	γ(keV)	γ(%)
Y L_ℓ	1.686	0.012 *3*
Y L_η	1.762	0.0071 *18*
Y L_α	1.922	0.28 *6*
Y L_β	2.004	0.17 *4*
Y L_γ	2.257	0.0045 *13*
Y $K_{\alpha2}$	14.883	2.93 *20*
Y $K_{\alpha1}$	14.958	5.6 *4*
Y $K_{\beta1}'$	16.735	1.26 *9*
Y $K_{\beta2}'$	17.125	0.185 *13*
γ	166.71 *13*	0.61 *16*
γ E3	168.59 *4*	18 *1* *
γ	260.08 *3*	7.2 *4*
γ [M1+E2]	285.66 *3*	0.266 *20*
γ	331.96 *5*	0.35 *3*
γ	342.59 *10*	0.07 *3*
γ	346.48 *4*	3.21 *17*
γ	377.15 *4*	1.45 *9*
γ	406.67 *9*	0.42 *4*
γ	424.63 *4*	0.25 *3*
γ	427.93 *7*	0.15 *3*
γ	432.65 *4*	1.45 *9*
γ	440.82 *13*	0.19 *4*
γ	446.20 *6*	2.31 *13*
γ	481.73 *7*	1.11 *10*
γ	483.61 *5*	1.63 *12*
γ	487.31 *11*	0.12 *5*
γ	518.37 *9*	0.126 *20*
γ	541.70 *4*	0.71 *4*
γ	545.74 *4*	0.39 *3*
γ	559.84 *8*	0.200 *20*
γ	571.98 *14*	0.21 *3*
γ	586.45 *15*	0.44 *15*
γ M1(+E2)	590.20 *4*	67 *4* *
γ	594.33 *7*	1.09 *14*
γ	596.21 *5*	1.30 *15*
γ	610.89 *4*	1.06 *7*
γ	630.98 *9*	0.19 *3*
γ	633.51 *7*	0.106 *20*
γ	650.40 *5*	0.186 *20*
γ	658.49 *5*	0.41 *4*
γ	663.51 *4*	1.61 *9*
γ	687.67 *6*	0.65 *6*
γ	690.30 *8*	0.99 *8*
γ	692.18 *7*	0.22 *6*
γ	710.29 *3*	21.3 *11*
γ	716.74 *5*	~0.29
γ	718.31 *4*	1.46 *20*
γ	764.03 *11*	0.029 *11*
γ	771.10 *4*	1.14 *7*
γ	775.44 *4*	0.26 *3*

Photons (^{93}Sr)
(continued)

γ_{mode}	γ(keV)	γ(%)
γ	782.77 *6*	0.21 *3*
γ	784.65 *7*	0.073 *20*
γ	788.55 *5*	0.75 *5*
γ	790.79 *5*	0.25 *3*
γ	795.14 *9*	0.226 *20*
γ	831.2 *7*	0.047 *20*
γ	834.81 *4*	1.64 *9*
γ	837.78 *25*	0.115 *16*
γ	858.45 *5*	0.71 *5*
γ [E2]	875.86 *5*	23.9 *13*
γ	888.18 *5*	21.6 *11*
γ	901.15 *6*	0.68 *4*
γ	910.48 *5*	0.80 *5*
γ	922.91 *5*	0.33 *3*
γ	927.92 *5*	0.63 *5*
γ	930.78 *6*	0.40 *3*
γ	952.41 *9*	0.106 *20*
γ	991.53 *19*	0.120 *20*
γ	1032.27 *10*	0.10 *3*
γ	1035.52 *4*	0.20 *3*
γ	1040.52 *4*	3.13 *20*
γ	1046.4 *4*	0.09 *3*
γ	1050.61 *22*	0.033 *14*
γ	1055.04 *5*	0.34 *3*
γ	1064.25 *7*	0.37 *3*
γ	1077.66 *7*	0.23 *3*
γ	1093.88 *5*	1.72 *10*
γ	1105.58 *9*	0.15 *3*
γ	1116.9 *3*	0.07 *3*
γ	1122.86 *6*	3.92 *20*
γ	1136.49 *7*	0.193 *20*
γ	1180.65 *9*	0.24 *3*
γ	1196.13 *5*	0.96 *5*
γ	1200.4 *10*	0.025 *11*
γ	1215.41 *5*	2.44 *13*
γ	1239.45 *9*	0.12 *3*
γ	1243.33 *6*	0.78 *5*
γ	1249.4 *4*	0.07 *3*
γ	1261.36 *6*	0.08 *3*
γ	1266.37 *5*	1.09 *8*
γ	1269.38 *5*	7.0 *3*
γ	1277.87 *6*	0.85 *6*
γ	1308.51 *5*	0.39 *3*
γ	1321.17 *4*	2.55 *13*
γ	1324.9 *3*	0.053 *20*
γ	1329.46 *15*	0.067 *13*
γ	1332.48 *6*	~0.5
γ	1334.37 *6*	0.66 *5*
γ	1378.97 *5*	0.35 *3*
γ	1386.99 *4*	3.39 *20*
γ	1433.93 *5*	0.88 *5*
γ	1438.94 *4*	0.49 *3*
γ	1466.31 *9*	0.100 *20*
γ	1469.33 *6*	0.51 *3*
γ	1482.96 *6*	0.100 *20*
γ	1491.95 *7*	0.54 *3*
γ	1506.39 *22*	0.047 *15*
γ	1512.02 *8*	0.054 *13*
γ	1520.04 *8*	0.31 *7*
γ	1538.85 *13*	0.100 *20*
γ	1542.60 *10*	0.040 *15*
γ	1551.54 *4*	1.00 *6*
γ	1609.78 *20*	0.193 *20*
γ	1633.88 *6*	1.42 *8*
γ	1641.92 *19*	0.043 *14*
γ	1647.51 *6*	0.87 *5*
γ	1652.1 *9*	0.035 *13*
γ	1667.96 *7*	~0.16
γ	1684.59 *8*	0.70 *5*
γ	1694.00 *5*	2.53 *14*
γ	1699.01 *4*	3.26 *20*
γ	1706.47 *9*	1.08 *7*
γ	1742.08 *23*	0.085 *15*
γ	1765.32 *5*	1.04 *5*
γ	1774.54 *7*	0.160 *20*
γ	1786.5 *4*	0.077 *12*
γ	1811.62 *4*	1.38 *8*
γ	1816.09 *5*	0.23 *3*
γ	1893.96 *6*	0.120 *20*
γ	1898.86 *21*	0.035 *13*
γ	1907.59 *6*	0.173 *20*
γ	1928.69 *4*	1.14 *7*
γ	1935.5 *9*	0.033 *11*

Photons (^{93}Sr)
(continued)

γ_{mode}	γ(keV)	γ(%)
γ	1944.67 *8*	0.55 *4*
γ	1952.3 *4*	0.097 *17*
γ	1971.6 *4*	0.033 *12*
γ	1978.1 *12*	0.026 *12*
γ	1979.85 *15*	0.036 *13*
γ	1984.67 *4*	0.080 *13*
γ	2010.54 *9*	0.119 *17*
γ	2054.6 *3*	0.133 *20*
γ	2063.61 *11*	0.61 *4*
γ	2076.5 *9*	0.059 *16*
γ	2094.0 *8*	0.073 *20*
γ	2104.71 *20*	0.31 *3*
γ	2108.4 *3*	0.086 *15*
γ	2129.04 *13*	0.10 *3*
γ	2172.32 *20*	0.070 *13*
γ	2179.61 *6*	0.29 *4*
γ	2203.4 *9*	0.086 *20*
γ	2221.9 *11*	~0.040
γ	2230.32 *8*	1.52 *9*
γ	2296.19 *9*	0.72 *5*
γ	2364.73 *7*	1.54 *9*
γ	2416.74 *9*	0.106 *20*
γ	2472.58 *23*	0.074 *10*
γ	2543.80 *7*	2.96 *16*
γ	2574.86 *5*	0.126 *20*
γ	2586.2 *3*	0.027 *8*
γ	2614.6 *4*	0.088 *11*
γ	2687.46 *5*	2.08 *12*
γ	2765.2 *8*	0.041 *13*
γ	2781.5 *5*	0.051 *7*
γ	2811.02 *23*	0.021 *5*
γ	2828.5 *3*	0.168 *17*
γ	2983.59 *23*	0.044 *14*
γ	2985.71 *20*	0.19 *3*
γ	2995.25 *22*	0.019 *5*
γ	3006.93 *10*	0.115 *11*
γ	3115.76 *15*	0.068 *9*

* with ^{93}Y(820 ms) in equilib

Atomic Electrons (^{93}Sr)
⟨e⟩=29.7 *13* keV

e_{bin}(keV)	⟨e⟩(keV)	e(%)
2 - 17	0.82	17.3 *16*
150	0.07	~0.05
152	20.6	13.6 *8*
164 - 165	0.011	~0.007
166	3.57	2.14 *13*
167	1.49	0.89 *5*
168	0.90	0.54 *3*
243 - 286	0.37	~0.15
315 - 360	0.12	0.037 *14*
375 - 424	0.042	0.010 *3*
425 - 470	0.073	0.016 *5*
479 - 529	0.017	0.0033 *8*
539 - 588	0.86	0.150 *17*
590 - 634	0.033	0.0056 *9*
641 - 690	0.029	0.0045 *10*
692 - 718	0.16	0.023 *9*
747 - 795	0.018	0.0023 *5*
814 - 859	0.170	0.0199 *13*
871 - 921	0.15	0.017 *5*
923 - 952	0.0011	0.00012 *3*
974 - 1023	0.013	0.0013 *5*
1029 - 1078	0.012	0.00115 *25*
1089 - 1136	0.018	0.0016 *5*
1164 - 1213	0.013	0.0011 *3*
1215 - 1264	0.032	0.0025 *7*
1266 - 1315	0.014	0.00107 *24*
1317 - 1362	0.0044	0.00033 *7*
1370 - 1417	0.013	0.0010 *3*
1422 - 1469	0.0040	0.00028 *5*
1475 - 1551	0.0062	0.00041 *7*
1593 - 1692	0.026	0.00159 *23*
1694 - 1786	0.0050	0.00029 *5*
1795 - 1894	0.0048	0.00027 *6*
1896 - 1994	0.0051	0.00027 *5*
2008 - 2106	0.0029	0.000143 *22*

Atomic Electrons (^{93}Sr)
(continued)

e_{bin}(keV)	⟨e⟩(keV)	e(%)
2108 - 2205	0.00128	5.9 *10* $\times 10^{-5}$
2213 - 2296	0.0049	0.00022 *4*
2348 - 2416	0.0035	0.00015 *4*
2456 - 2544	0.0061	0.00024 *6*
2558 - 2614	0.00047	1.8 *3* $\times 10^{-5}$
2670 - 2765	0.0042	0.00016 *4*
2779 - 2828	0.00036	1.3 *3* $\times 10^{-5}$
2967 - 3007	0.00067	2.3 *3* $\times 10^{-5}$
3099 - 3115	0.00012	3.9 *9* $\times 10^{-6}$

Continuous Radiation (^{93}Sr)
⟨β-⟩=757 keV;⟨IB⟩=1.52 keV

E_{bin}(keV)		⟨ ⟩(keV)	(%)
0 - 10	β-	0.0269	0.54
	IB	0.030	
10 - 20	β-	0.082	0.54
	IB	0.029	0.20
20 - 40	β-	0.334	1.11
	IB	0.056	0.20
40 - 100	β-	2.50	3.54
	IB	0.156	0.24
100 - 300	β-	28.7	14.1
	IB	0.40	0.23
300 - 600	β-	114	25.2
	IB	0.38	0.091
600 - 1300	β-	361	40.7
	IB	0.36	0.043
1300 - 2500	β-	232	13.6
	IB	0.099	0.0062
2500 - 3355	β-	18.4	0.68
	IB	0.00155	5.9 $\times 10^{-5}$

$^{93}_{39}$Y (10.25 *1* h)

Mode: β-
Δ: -84235 *11* keV
SpA: 3.287$\times 10^{6}$ Ci/g
Prod: fission

Photons (^{93}Y)
⟨γ⟩=88.9 *20* keV

γ_{mode}	γ(keV)	γ(%)
γ E2+M1	266.91 *4*	6.8 *4*
γ [M1+E2]	680.28 *6*	0.61 *3*
γ	714.47 *8*	0.0163 *20*
γ [E2]	947.18 *6*	1.94 *10*
γ	962.45 *10*	0.0109 *14*
γ [M1]	971.89 *23*	0.0163 *20*
γ	987.45 *11*	0.0095 *20*
γ [M1+E2]	1158.61 *10*	0.028 *3*
γ	1168.73 *20*	0.018 *3*
γ	1183.58 *7*	0.042 *3*
γ	1203.36 *7*	0.103 *6*
γ	1237.56 *7*	0.0272 *20*
γ [M1+E2]	1425.51 *9*	0.238 *14*
γ	1450.48 *7*	0.333 *20*
γ	1470.26 *7*	0.067 *4*
γ	1642.72 *9*	0.047 *3*
γ [M1+E2]	1652.16 *23*	0.022 *3*
γ [M1+E2]	1827.93 *20*	0.022 *3*
γ	1917.82 *6*	1.40 *7*
γ	2184.72 *6*	0.155 *9*
γ	2190.80 *9*	0.171 *10*
γ	2457.70 *10*	0.0061 *14*
γ	2473.63 *20*	0.0109 *14*

Continuous Radiation (^{93}Y)
⟨β-⟩=1171 keV;⟨IB⟩=3.0 keV

E_{bin}(keV)		⟨ ⟩(keV)	(%)
0 - 10	β-	0.0131	0.261
	IB	0.041	
10 - 20	β-	0.0399	0.265
	IB	0.040	0.28
20 - 40	β-	0.164	0.55
	IB	0.079	0.28
40 - 100	β-	1.25	1.77
	IB	0.23	0.35
100 - 300	β-	14.7	7.2
	IB	0.64	0.36
300 - 600	β-	60	13.3
	IB	0.71	0.167
600 - 1300	β-	330	34.7
	IB	0.91	0.107
1300 - 2500	β-	723	40.4
	IB	0.34	0.022
2500 - 2884	β-	41.5	1.59
	IB	0.00125	4.9 $\times 10^{-5}$

$^{93}_{39}$Y (820 *40* ms)

Mode: IT
Δ: -83476 *11* keV
SpA: 1.0$\times 10^{11}$ Ci/g
Prod: daughter ^{93}Sr; ^{94}Zr(t,α)

Photons (^{93}Y)
⟨γ⟩=680 *24* keV

γ_{mode}	γ(keV)	γ(%)
Y L_ℓ	1.686	0.034 *8*
Y L_η	1.762	0.020 *5*
Y L_α	1.922	0.78 *16*
Y L_β	2.003	0.46 *12*
Y L_γ	2.257	0.013 *4*
Y $K_{\alpha2}$	14.883	8.1 *5*
Y $K_{\alpha1}$	14.958	15.5 *9*
Y $K_{\beta1}$'	16.735	3.46 *20*
Y $K_{\beta2}$'	17.125	0.51 *3*
γ E3	168.59 *4*	51.0
γ M1(+E2)	590.20 *4*	100

Atomic Electrons (^{93}Y)
⟨e⟩=79 *3* keV

e_{bin}(keV)	⟨e⟩(keV)	e(%)
2	0.77	36 *4*
12	0.096	0.78 *9*
13	0.94	7.4 *8*
14	0.103	0.72 *8*
15	0.31	2.12 *23*
16	0.026	0.161 *18*
17	0.0100	0.060 *7*
152	58.3	38.4 *17*
166	10.1	6.1 *3*
167	4.21	2.53 *11*
168	2.56	1.52 *7*
169	0.374	0.222 *10*
573	1.12	0.196 *23*
588	0.129	0.022 *3*
590	0.026	0.0044 *6*

$^{93}_{40}$Zr(1.53 *10* $\times 10^{6}$ yr)

Mode: β-
Δ: -87119.3 *24* keV
SpA: 0.00251 Ci/g
Prod: fission

Photons (^{93}Zr)
⟨γ⟩=1.84 *5* keV

γ_{mode}	γ(keV)	γ(%)†
Nb L_ℓ	1.902	0.090 *16*
Nb L_η	1.996	0.0111 *17*
Nb L_α	2.166	2.2 *4*
Nb L_β	2.310	0.45 *8*
Nb L_γ	2.596	0.033 *7*
Nb $K_{\alpha2}$	16.521	3.08 *13*
Nb $K_{\alpha1}$	16.615	5.87 *25*
Nb $K_{\beta1}$'	18.618	1.36 *6*
Nb $K_{\beta2}$'	19.074	0.228 *10*
γ M4	30.4 *3*	0.000518 *

† 2.6% uncert(syst)
* with ^{93}Nb(13.6 yr) in equilib

Atomic Electrons (^{93}Zr)
⟨e⟩=28.1 *7* keV

e_{bin}(keV)	⟨e⟩(keV)	e(%)
2	1.49	63 *7*
3	0.119	4.4 *5*
11	1.60	14.1 *5*
14	0.36	2.6 *3*
16	0.141	0.87 *9*
17	0.0087	0.053 *6*
18	0.0113	0.062 *7*
19	0.00209	0.0113 *12*
28	19.1	68.5 *22*
30	5.23	17.4 *6*

Continuous Radiation (^{93}Zr)
⟨β-⟩=19.0 keV;⟨IB⟩=0.00128 keV

E_{bin}(keV)		⟨ ⟩(keV)	(%)
0 - 10	β-	1.46	30.4
	IB	0.00078	
10 - 20	β-	3.97	26.7
	IB	0.00034	0.0025
20 - 40	β-	10.1	35.5
	IB	0.000153	0.00060
40 - 60	β-	3.39	7.5
	IB	4.8 $\times 10^{-6}$	1.12 $\times 10^{-5}$

$^{93}_{41}$Nb(stable)

Δ: -87209.8 *25* keV
%: 100

$^{93}_{41}$Nb(13.6 *3* yr)

Mode: IT
Δ: -87179.4 *25* keV
SpA: 283 Ci/g
Prod: daughter ^{93}Zr; ^{93}Nb(n,n')

Photons (^{93}Nb)

$\langle\gamma\rangle$=1.84 *6* keV

γ_{mode}	γ(keV)	γ(%)
Nb L_ℓ	1.902	0.090 *16*
Nb L_η	1.996	0.0111 *17*
Nb L_α	2.166	2.2 *4*
Nb L_β	2.310	0.45 *8*
Nb L_γ	2.596	0.033 *8*
Nb $K_{\alpha 2}$	16.521	3.08 *16*
Nb $K_{\alpha 1}$	16.615	5.9 *3*
Nb $K_{\beta 1}$'	18.618	1.36 *7*
Nb $K_{\beta 2}$'	19.074	0.228 *13*
γ M4	30.4 *3*	0.000518

Atomic Electrons (^{93}Nb)

$\langle e\rangle$=28.1 *9* keV

e_{bin}(keV)	$\langle e\rangle$(keV)	e(%)
2	1.49	63 *7*
3	0.119	4.4 *5*
11	1.60	14.1 *6*
14	0.36	2.6 *3*
16	0.141	0.87 *10*
17	0.0087	0.053 *6*
18	0.0113	0.062 *7*
19	0.00209	0.0113 *12*
28	19.1	68 *3*
30	5.23	17.4 *8*

$^{93}_{42}$Mo(3500 *700* yr)

Mode: ϵ

Δ: -86804 *4* keV

SpA: 1.10 Ci/g

Prod: ^{92}Mo(n,γ); ^{93}Nb(p,n)

Photons (^{93}Mo)

$\langle\gamma\rangle$=12.5 *3* keV

γ_{mode}	γ(keV)	γ(%)†
Nb L_ℓ	1.902	0.17 *3*
Nb L_η	1.996	0.047 *7*
Nb L_α	2.166	4.0 *7*
Nb L_β	2.289	1.5 *3*
Nb L_γ	2.566	0.096 *20*
Nb $K_{\alpha 2}$	16.521	21.3 *8*
Nb $K_{\alpha 1}$	16.615	40.7 *15*
Nb $K_{\beta 1}$'	18.618	9.4 *4*
Nb $K_{\beta 2}$'	19.074	1.58 *6*
γ M4	30.4 *3*	0.00049 *

* with ^{93}Nb(13.6 yr) in equilib

† 6.1% uncert(syst)

Atomic Electrons (^{93}Mo)

$\langle e\rangle$=31.6 *11* keV

e_{bin}(keV)	$\langle e\rangle$(keV)	e(%)
2	3.0	125 *13*
3	0.27	9.9 *11*
11	1.52	13.3 *8*
14	2.5	17.7 *18*
16	0.97	6.0 *6*

Atomic Electrons (^{93}Mo) (continued)

e_{bin}(keV)	$\langle e\rangle$(keV)	e(%)
17	0.060	0.36 *4*
18	0.078	0.43 *4*
19	0.0145	0.078 *8*
28	18.2	65 *4*
30	5.0	16.5 *9*

Continuous Radiation (^{93}Mo)

$\langle$IB$\rangle$=0.024 keV

E_{bin}(keV)		$\langle\ \rangle$(keV)	(%)
10 - 20	IB	0.0161	0.098
20 - 40	IB	0.00129	0.0052
40 - 100	IB	0.00133	0.0020
100 - 300	IB	0.0050	0.0028
300 - 376	IB	0.00028	8.9 $\times 10^{-5}$

$^{93}_{42}$Mo(6.85 *7* h)

Mode: IT(99.88 *1* %), ϵ(0.12 *1* %)

Δ: -84379 *4* keV

SpA: 4.92$\times 10^{6}$ Ci/g

Prod: ^{93}Nb(d,2n); ^{90}Zr(α,n); ^{93}Nb(p,n); ^{92}Zr(α,3n)

Photons (^{93}Mo)

$\langle\gamma\rangle$=2313 *31* keV

γ_{mode}	γ(keV)	γ(%)†
Mo L_ℓ	2.016	0.032 *6*
Mo L_η	2.120	0.018 *3*
Mo L_α	2.293	0.80 *13*
Mo L_β	2.421	0.50 *8*
Mo L_γ	2.681	0.031 *6*
Mo $K_{\alpha 2}$	17.374	6.7 *3*
Mo $K_{\alpha 1}$	17.479	12.7 *5*
Mo $K_{\beta 1}$'	19.602	3.00 *12*
Mo $K_{\beta 2}$'	20.091	0.522 *22*
γ_{IT} [M1+E2]	114.106 *9*	0.679 *13*
γ_ϵ	155.89 *6*	0.010 *3*
γ_{IT} E4	263.143 *5*	56.7 *11*
γ_ϵ[E2]	385.42 *7*	0.06 *1*
γ_ϵ	541.31 *6*	0.06 *1*
γ_ϵ	572.84 *6*	0.05 *1*
γ_{IT} (E2)	684.753 *9*	99.7
γ_ϵ	689.14 *5*	0.07 *1*
γ_ϵ	845.03 *7*	~0.02
γ_ϵ[E2]	949.89 *3*	0.12 *1*
γ_{IT} E2+M1	1363.090 *18*	0.787 *16*
γ_ϵ	1417.86 *9*	0.03 *1*
γ_{IT} (E2)	1477.194 *17*	99.0 *19*

† uncert(syst): <0.1% for IT, 8.3% for ϵ

Atomic Electrons (^{93}Mo)*

$\langle e\rangle$=103.0 *21* keV

e_{bin}(keV)	$\langle e\rangle$(keV)	e(%)
2 - 20	1.84	37 *3*
94 - 137	0.31	~0.32
153 - 156	0.00031	~0.00020
243	71.5	29.4 *8*
260	7.52	2.89 *8*
261	15.1	5.78 *16*
263	4.97	1.89 *5*
366 - 385	0.0022	0.00061 *9*
522 - 570	0.0013	0.00024 *8*

Atomic Electrons (^{93}Mo)* (continued)

e_{bin}(keV)	$\langle e\rangle$(keV)	e(%)
572 - 573	1.2 $\times 10^{-5}$	2.2 *10* $\times 10^{-6}$
665	1.15	0.173 *8*
670 - 689	0.168	0.0246 *9*
826 - 845	0.00013	~2$\times 10^{-5}$
931 - 950	0.00092	0.000099 *8*
1343 - 1363	0.00436	0.000324 *14*
1399 - 1418	0.00011	8 *4* $\times 10^{-6}$
1457 - 1477	0.489	0.0335 *13*

* with IT

$^{93}_{43}$Tc(2.75 *5* h)

Mode: ϵ

Δ: -83606 *4* keV

SpA: 1.225$\times 10^{7}$ Ci/g

Prod: ^{92}Mo(d,n); ^{92}Mo(p,γ); protons on Cd

Photons (^{93}Tc)

$\langle\gamma\rangle$=1461 *180* keV

γ_{mode}	γ(keV)	γ(%)†
Mo L_ℓ	2.016	0.074 *13*
Mo L_η	2.120	0.034 *5*
Mo L_α	2.293	1.8 *3*
Mo L_β	2.425	1.04 *17*
Mo L_γ	2.699	0.069 *13*
Mo $K_{\alpha 2}$	17.374	17.2 *6*
Mo $K_{\alpha 1}$	17.479	32.7 *11*
Mo $K_{\beta 1}$'	19.602	7.7 *3*
Mo $K_{\beta 2}$'	20.091	1.35 *5*
γ [M1+E2]	114.106 *9*	0.066 *4*
γ E2+M1	1363.090 *18*	66
γ	1381.7 *15*	0.53 *7*
γ	1424.7 *14*	0.30 *3*
γ (E2)	1477.194 *17*	9.6 *5*
γ E2+M1	1520.3 *10*	23.9 *12*
γ	1538.8 *14*	0.66 *7*
γ [E2]	2409.0 *23*	0.32 *3*
γ	2479 *3*	0.046 *7*
γ	2730.6 *23*	0.32 *3*
γ	2739 *11*	0.073 *13*
γ	2822 *6*	0.033 *7*
γ	2901.9 *14*	0.092 *13*
γ	3026 *6*	0.013 *4*

† 3.0% uncert(syst)

Atomic Electrons (^{93}Tc)

$\langle e\rangle$=5.0 *3* keV

e_{bin}(keV)	$\langle e\rangle$(keV)	e(%)
3	1.63	63 *7*
14	0.183	1.28 *13*
15	1.72	11.6 *12*
17	0.81	4.8 *5*
19 - 20	0.074	0.38 *3*
94 - 114	0.030	~0.031
1343	0.32	0.024 *5*
1360 - 1405	0.047	0.0035 *6*
1422 - 1457	0.042	0.00288 *18*
1474 - 1538	0.126	0.0084 *5*
2389 - 2479	0.00114	4.8 *5* $\times 10^{-5}$
2711 - 2802	0.00101	3.7 *7* $\times 10^{-5}$
2819 - 2902	0.00022	7.6 *19* $\times 10^{-6}$
3006 - 3025	2.9 $\times 10^{-5}$	1.0 *3* $\times 10^{-6}$

Continuous Radiation (^{93}Tc)

⟨β+⟩=39.6 keV;⟨IB⟩=1.18 keV

E_{bin}(keV)		⟨ ⟩(keV)	(%)
0 - 10	β+	1.75×10^{-5}	0.000213
	IB	0.0020	
10 - 20	β+	0.000501	0.00306
	IB	0.017	0.103
20 - 40	β+	0.0100	0.0309
	IB	0.0053	0.019
40 - 100	β+	0.321	0.420
	IB	0.0107	0.0164
100 - 300	β+	9.0	4.25
	IB	0.051	0.026
300 - 600	β+	25.5	6.0
	IB	0.20	0.044
600 - 1300	β+	4.74	0.72
	IB	0.75	0.083
1300 - 1835	IB	0.145	0.0102
	Σβ+		11.4

$^{93}_{43}$Tc(43.5 *10* min)

Mode: IT(80 *2* %), ε(20 *2* %)
Δ: -83214 *4* keV
SpA: 4.65×10^{7} Ci/g
Prod: ^{92}Mo(d,n); ^{92}Mo(p,γ); ^{93}Nb(α,4n)

Photons (^{93}Tc)

⟨γ⟩=774 *26* keV

γ_{mode}	γ(keV)	γ(%)†
Mo L_ℓ	2.016	0.018 *3*
Mo L_η	2.120	0.0085 *13*
Tc L_ℓ	2.133	0.017 *3*
Tc L_η	2.249	0.0080 *13*
Mo L_α	2.293	0.46 *7*
Tc L_α	2.424	0.43 *7*
Mo L_β	2.426	0.26 *4*
Tc L_β	2.576	0.26 *5*
Mo L_γ	2.699	0.017 *3*
Tc L_γ	2.862	0.018 *4*
Mo $K_{\alpha2}$	17.374	4.27 *15*
Mo $K_{\alpha1}$	17.479	8.1 *3*
Tc $K_{\alpha2}$	18.251	3.75 *22*
Tc $K_{\alpha1}$	18.367	7.1 *4*
Mo $K_{\beta1}$'	19.602	1.91 *7*
Mo $K_{\beta2}$'	20.091	0.333 *12*
Tc $K_{\beta1}$'	20.613	1.7 *1*
Tc $K_{\beta2}$'	21.136	0.306 *19*
γ_{IT} (M4)	392.45 *10*	60
γ_ϵ	2644.5 *3*	15.6 *9*
γ_ϵ[M1+E2]	3129.0 *5*	2.22 *12*
γ_ϵ	3220.5 *7*	1.14 *12*
γ_ϵ	3298.2 *8*	0.45 *5*

† uncert(syst): 11% for ε, 4.2% for IT

Atomic Electrons (^{93}Tc)

⟨e⟩=76 *3* keV

e_{bin}(keV)	⟨e⟩(keV)	e(%)
3	0.79	30 *3*
14	0.045	0.32 *3*
15	0.58	3.9 *4*
16	0.24	1.53 *17*
17	0.215	1.26 *13*
18	0.158	0.88 *10*
19	0.0153	0.080 *8*
20	0.0163	0.081 *9*
21	0.0026	0.0126 *14*
371	61.1	16.5 *9*
389	8.3	2.13 *11*

Atomic Electrons (^{93}Tc)
(continued)

e_{bin}(keV)	⟨e⟩(keV)	e(%)
390	2.46	0.63 *3*
392	2.43	0.62 *3*
2625	0.033	0.0013 *4*
2642	0.0037	0.00014 *4*
2644	0.00068	2.6 *7* $\times10^{-5}$
3109	0.0053	0.000171 *13*
3126	0.00059	1.89 *16* $\times10^{-5}$
3129	0.000103	3.3 *3* $\times10^{-6}$
3129	3.4×10^{-6}	1.1 *3* $\times10^{-7}$
3201	0.0022	6.8 *17* $\times10^{-5}$
3218	0.00024	7.5 *20* $\times10^{-6}$
3220	4.4×10^{-5}	1.4 *4* $\times10^{-6}$
3278	0.00084	2.6 *7* $\times10^{-5}$
3295	9.1×10^{-5}	2.8 *7* $\times10^{-6}$
3296	3.0×10^{-6}	9 *4* $\times10^{-8}$
3298	1.7×10^{-5}	5.1 *14* $\times10^{-7}$

Continuous Radiation (^{93}Tc)

⟨IB⟩=0.033 keV

E_{bin}(keV)		⟨ ⟩(keV)	(%)
10 - 20	IB	0.0035	0.020
20 - 40	IB	0.00041	0.00168
40 - 100	IB	0.00043	0.00063
100 - 300	IB	0.0054	0.0026
300 - 600	IB	0.0162	0.0037
600 - 946	IB	0.0071	0.00103

$^{93}_{44}$Ru(59.7 *6* s)

Mode: ε
Δ: -77270 *90* keV
SpA: 2.020×10^{9} Ci/g
Prod: ^{92}Mo(α,3n); ^{92}Mo(^{3}He,2n)

Photons (^{93}Ru)

⟨γ⟩=159 *4* keV

γ_{mode}	γ(keV)	γ(%)†
γ (M4)	392.45 *10*	*
γ	560.96 *17*	0.110 *11*
γ	680.49 *9*	5.4
γ	711.1 *3*	0.050 *12*
γ	725.71 *20*	0.077 *5*
γ	1015.72 *9*	0.39 *4*
γ	1193.94 *10*	0.338 *22*
γ	1288.63 *20*	0.071 *12*
γ	1434.69 *10*	0.67 *3*
γ	1575.76 *20*	0.048 *7*
γ	1658.19 *15*	0.029 *7*
γ	1801.21 *10*	0.346 *16*
γ	1809.4 *4*	0.018 *4*
γ	1950.31 *20*	0.071 *7*
γ	1968.81 *23*	0.060 *6*
γ	2257.51 *10*	0.194 *19*
γ	2313.2 *8*	0.125 *18*
γ	2329.54 *16*	0.044 *8*
γ	2338.67 *14*	0.077 *16*
γ	2720.49 *17*	0.118 *6*
γ	2901.9 *3*	0.060 *11*
γ	2929.4 *6*	0.054 *7*
γ	3151.6 *3*	0.105 *11*
γ	3233.52 *18*	0.279 *14*
γ	3331.61 *23*	0.200 *20*
γ	3418.3 *3*	0.138 *14*
γ	3434.2 *5*	0.056 *7*
γ	3477.7 *3*	0.019 *3*
γ	3509.1 *3*	0.059 *10*
γ	3914.39 *16*	0.285 *16*
γ	3928.1 *3*	0.027 *5*

Photons (^{93}Ru)
(continued)

γ_{mode}	γ(keV)	γ(%)†
γ	3987.69 *16*	0.108 *14*
γ	4079.7 *5*	0.019 *5*
γ	4094.2 *5*	0.032 *5*
γ	4103.9 *8*	0.016 *5*
γ	4158.2 *3*	0.082 *5*
γ	4187.1 *3*	0.079 *5*
γ	4257.0 *5*	0.022 *5*
γ	4344.1 *5*	0.029 *6*
γ	4389.3 *3*	0.116 *12*
γ	4608.5 *3*	0.055 *9*
γ	4618.7 *4*	0.076 *9*
γ	4668.15 *16*	0.085 *9*
γ	4760.2 *5*	~0.0032
γ	4774.6 *5*	~0.0032
γ	4937.4 *5*	0.010 *3*
γ	4954.7 *8*	0.035 *5*
γ	5297.8 *8*	0.010 *3*

† 9.3% uncert(syst)
* with ^{93}Tc(44 min)

$^{93}_{44}$Ru(10.8 *3* s)

Mode: ε(79 *3* %), IT(21 *3* %)
Δ: -76536 *90* keV
SpA: 1.09×10^{10} Ci/g
Prod: ^{92}Mo(^{3}He,2n)

Photons (^{93}Ru)

⟨γ⟩=1130 *71* keV

γ_{mode}	γ(keV)	γ(%)†
γ_ϵ(M4)	392.45 *10*	*
γ_ϵ	642.75 *19*	0.43 *11*
γ_{IT}[M4]	734.4 *1*	20.4
γ_ϵ	927.78 *17*	1.55 *14*
γ_ϵ	1015.72 *9*	0.65 *25*
γ_ϵ	1022.79 *16*	0.65 *11*
γ_ϵ	1110.73 *12*	24.5 *10*
γ_ϵ	1395.76 *13*	36
γ_ϵ	2038.50 *15*	8.6 *4*

† uncert(syst): 14% for ε, 14% for IT
* with ^{93}Tc(44 min)

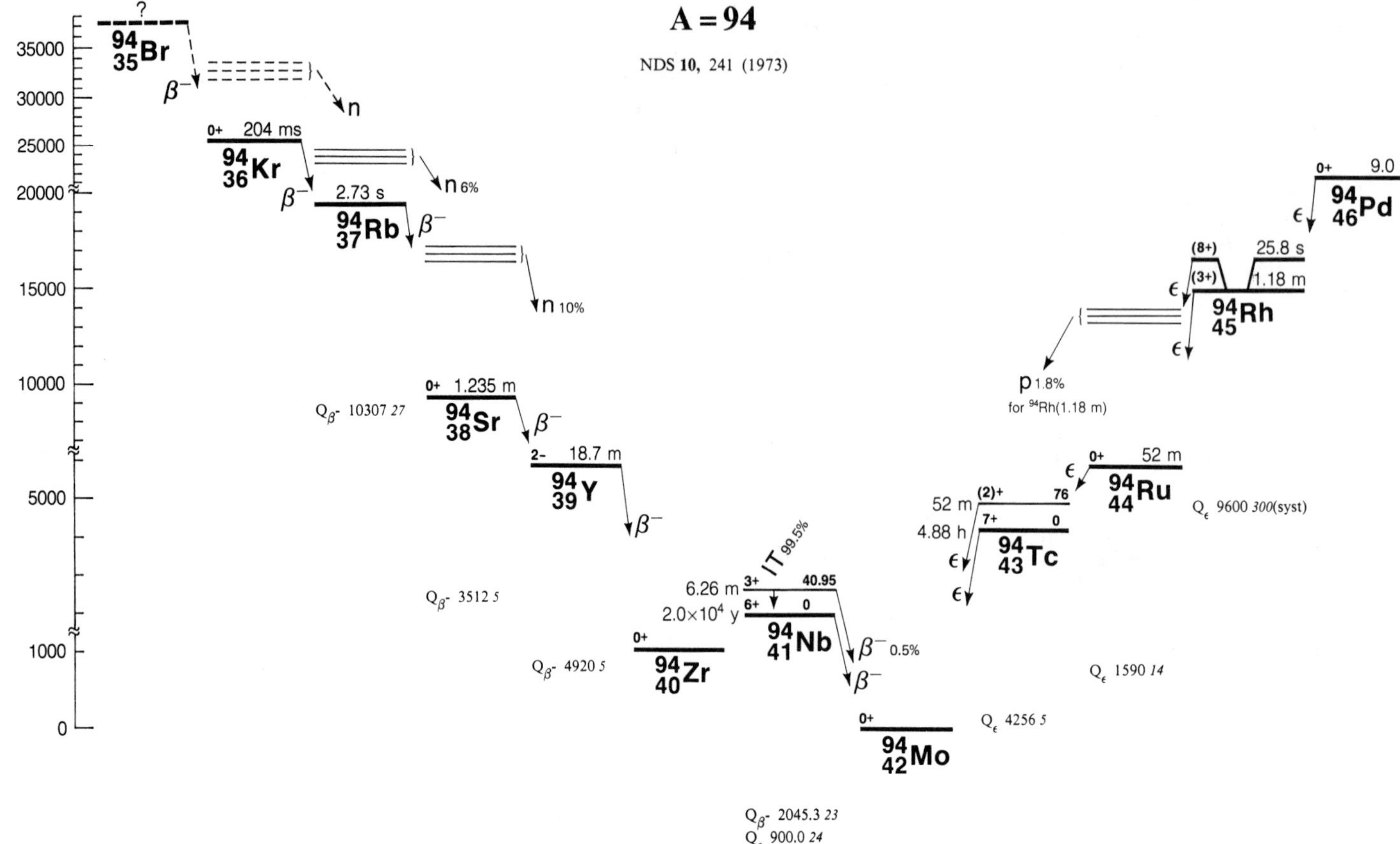

$^{94}_{35}$Br

Mode: β-, β-n

Prod: protons on U; protons on Nb

$^{94}_{36}$Kr(204 *8* ms)

Mode: β-, β-n(5.7 *22* %)

SpA: 1.67×10^{11} Ci/g

Prod: fission

Photons (^{94}Kr)

γ_{mode}	γ(keV)	γ(rel)
$\gamma_{\beta-}$	98 *2*	6.3 *5*
$\gamma_{\beta-}$	121 *2*	4.9 *5*
$\gamma_{\beta-}$	135.196 *6*	12.7 *10*
$\gamma_{\beta-}$	186.320 *7*	36.7 *22*
$\gamma_{\beta-}$	191.5 *10*	5.1 *5*
$\gamma_{\beta-}$	203.4 *10*	3.6 *4*
$\gamma_{\beta-}$	219.47 *5*	67 *6*
$\gamma_{\beta-}$	288.175 *16*	33.3 *20*
$\gamma_{\beta-}$	320.84 *4*	25.7 *16*
$\gamma_{\beta-}$	354.51 *6*	27.8 *19*
$\gamma_{\beta-}$	359.0 *1*	39.4 *24*
$\gamma_{\beta-}$	394.9 *8*	21.0 *14*
$\gamma_{\beta-}$	402.2 *8*	20.2 *15*
$\gamma_{\beta-}$	471.6 *8*	11.2 *8*
$\gamma_{\beta-}$	593.3 *10*	29 *2*
$\gamma_{\beta-}$	629.3 *1*	100
$\gamma_{\beta-}$	695.7 *10*	25.2 *19*
$\gamma_{\beta-}$	985 *1*	1.8 *3*

$^{94}_{37}$Rb(2.73 *1* s)

Mode: β-, β-n(10.4 *11* %)

Δ: -68529 *26* keV

SpA: 3.883×10^{10} Ci/g

Prod: fission

β-n(avg): 480

Photons (^{94}Rb)

⟨γ⟩=2725 *33* keV

γ_{mode}	γ(keV)	γ(%)†
$\gamma_{\beta-}$	117.7 *2*	0.087 *9*
$\gamma_{\beta-}$	207.1 *1*	0.192 *17* ?
$\gamma_{\beta-n}$ [M1+E2]	213.40 *4*	0.86 *17*
$\gamma_{\beta-n}$	219.15 *4*	0.92 *18*
$\gamma_{\beta-}$	253.0 *1*	0.59 *4*
$\gamma_{\beta-}$	332.6 *2*	0.044 *9*
$\gamma_{\beta-n}$	432.55 *3*	1.20 *24*
$\gamma_{\beta-}$	453.6 *1*	0.105 *17*
$\gamma_{\beta-}$	458.0 *1*	0.56 *4*
$\gamma_{\beta-}$(M1+11%E2)	503.8 *1*	1.74 *9*
$\gamma_{\beta-}$	558.0 *1*	0.165 *17*
$\gamma_{\beta-}$	601.7 *2*	0.279 *17*
$\gamma_{\beta-}$	633.7 *2*	0.131 *17*
$\gamma_{\beta-}$	658.5 *2*	0.113 *17*
$\gamma_{\beta-}$	660.9 *2*	0.218 *17*
$\gamma_{\beta-}$(M1+23%E2)	677.7 *1*	3.66 *17*
$\gamma_{\beta-n}$	709.97 *4*	0.35 *7*
$\gamma_{\beta-}$	710.7 *2*	0.57 *9*
$\gamma_{\beta-}$	723.7 *2*	0.47 *9*
$\gamma_{\beta-}$	734.5 *1*	0.18 *3*
$\gamma_{\beta-}$	783.8 *1*	0.55 *4*
$\gamma_{\beta-}$	806.5 *1*	0.12 *4*
$\gamma_{\beta-}$	812.9 *1*	2.35 *17*
$\gamma_{\beta-}$	826.1 *1*	0.65 *9*

Photons (^{94}Rb) (continued)

γ_{mode}	γ(keV)	γ(%)†
$\gamma_{\beta-}$(E2)	836.9 *1*	87.1
$\gamma_{\beta-}$	871.0 *2*	0.087 *9*
$\gamma_{\beta-}$	888.3 *2*	0.09 *3*
$\gamma_{\beta-}$	925.0 *1*	0.226 *17*
$\gamma_{\beta-}$	931.6 *1*	0.279 *17*
$\gamma_{\beta-}$	976.4 *2*	0.122 *9* ?
$\gamma_{\beta-n}$	986.10 *5*	0.58 *12*
$\gamma_{\beta-}$	1019.0 *2*	0.148 *17*
$\gamma_{\beta-}$	1045.7 *2*	0.51 *4*
$\gamma_{\beta-}$(E1)	1089.4 *2*	17.1 *9*
$\gamma_{\beta-n}$	1096.73 *9*	0.0120 *24*
$\gamma_{\beta-}$	1120.8 *2*	0.183 *17*
$\gamma_{\beta-n}$	1142.51 *4*	0.024 *5*
$\gamma_{\beta-n}$	1148.21 *6*	0.072 *14*
$\gamma_{\beta-}$	1151.7 *2*	0.56 *5*
$\gamma_{\beta-}$	1208.5 *2*	0.209 *17*
$\gamma_{\beta-}$	1244.9 *2*	0.27 *3*
$\gamma_{\beta-}$	1292.6 *2*	3.40 *17*
$\gamma_{\beta-}$M1,E2	1309.1 *2*	14.2 *8*
$\gamma_{\beta-}$	1324.0 *3*	0.087 *9*
$\gamma_{\beta-}$	1336.0 *3*	0.165 *17*
$\gamma_{\beta-}$	1339.4 *2*	0.37 *3* ?
$\gamma_{\beta-}$	1345 *1*	0.17 *4*
$\gamma_{\beta-}$	1384.4 *3*	0.51 *4*
$\gamma_{\beta-n}$	1385.29 *6*	0.096 *19*
$\gamma_{\beta-}$	1434.4 *2*	0.44 *4*
$\gamma_{\beta-}$	1453.5 *2*	0.22 *3* ?
$\gamma_{\beta-}$	1460.2 *5*	0.044 *17*
$\gamma_{\beta-}$	1485.6 *3*	0.078 *17*
$\gamma_{\beta-}$	1522.2 *3*	0.22 *3*
$\gamma_{\beta-}$	1534.3 *2*	0.57 *4*
$\gamma_{\beta-}$(E1)	1577.5 *2*	31.8 *16*
$\gamma_{\beta-}$	1594.5 *2*	0.30 *3*
$\gamma_{\beta-}$	1632.0 *2*	0.30 *3* ?
$\gamma_{\beta-}$	1703.3 *4*	0.139 *17* ?
$\gamma_{\beta-}$	1742.7 *3*	0.087 *9*
$\gamma_{\beta-}$	1755.8 *8*	0.35 *9*
$\gamma_{\beta-}$	1757.0 *4*	0.52 *17*
$\gamma_{\beta-}$	1766.8 *4*	0.131 *17*

Photons (^{94}Rb)
(continued)

γ_{mode}	γ(keV)	γ(%)†
$\gamma_{\beta-}$	1777.2 3	0.87 9
$\gamma_{\beta-}$(M1+43%E2)	1812.7 3	1.65 17
$\gamma_{\beta-}$	1866.9 3	2.9 3
$\gamma_{\beta-}$	1873.7 10	0.87 9
$\gamma_{\beta-}$	1902.2 3	0.20 3
$\gamma_{\beta-}$	1934.5 4	0.10 3
$\gamma_{\beta-}$	1964.6 4	0.052 9
$\gamma_{\beta-}$	1976.0 4	0.078 17
$\gamma_{\beta-}$	2014.0 4	0.37 4
$\gamma_{\beta-}$	2022.3 4	1.15 13
$\gamma_{\beta-}$	2084.7 4	0.71 7
$\gamma_{\beta-}$	2093.0 4	2.00 17
$\gamma_{\beta-}$	2098.9 4	0.27 3
$\gamma_{\beta-}$	2128.1 4	1.83 17
$\gamma_{\beta-}$	2144.2 4	0.70 7
$\gamma_{\beta-}$	2189.0 4	0.30 3
$\gamma_{\beta-}$	2209.9 4	1.74 17
$\gamma_{\beta-}$	2241.5 4	0.21 3
$\gamma_{\beta-}$	2271.4 5	2.1 3
$\gamma_{\beta-}$	2272.2 5	1.3 3
$\gamma_{\beta-}$	2317.1 5	0.24 3
$\gamma_{\beta-}$	2338.8 5	0.183 17
$\gamma_{\beta-}$	2354.4 5	0.183 17 ?
$\gamma_{\beta-}$	2373.1 5	0.087 17
$\gamma_{\beta-}$	2424.9 5	0.55 5
$\gamma_{\beta-}$	2433.9 5	0.157 17
$\gamma_{\beta-}$	2474.2 5	0.131 17 ?
$\gamma_{\beta-}$	2484.3 5	0.131 17
$\gamma_{\beta-}$	2501.0 5	0.33 4
$\gamma_{\beta-}$	2507.5 5	0.23 3 ?
$\gamma_{\beta-}$	2554.8 6	0.38 4
$\gamma_{\beta-}$	2574.9 6	0.27 3
$\gamma_{\beta-}$	2606.2 6	0.139 17
$\gamma_{\beta-}$	2633.8 6	0.139 17
$\gamma_{\beta-}$	2659.8 6	0.096 17
$\gamma_{\beta-}$	2663.5 6	0.087 9
$\gamma_{\beta-}$	2684.9 6	0.192 17
$\gamma_{\beta-}$	2692.1 6	0.39 4
$\gamma_{\beta-}$	2733.1 7	0.192 17
$\gamma_{\beta-}$	2748.5 7	0.08 3
$\gamma_{\beta-}$	2753.9 7	0.183 17
$\gamma_{\beta-}$	2759.0 7	0.078 17
$\gamma_{\beta-}$	2771.1 7	0.113 17
$\gamma_{\beta-}$	2798.4 7	0.052 9 ?
$\gamma_{\beta-}$	2821.1 7	0.26 3
$\gamma_{\beta-}$	2922.3 7	0.174 17
$\gamma_{\beta-}$	2931.9 7	0.28 4
$\gamma_{\beta-}$	2978.7 8	0.28 4 ?
$\gamma_{\beta-}$	3009.1 8	0.087 17
$\gamma_{\beta-}$	3016.6 8	0.096 17
$\gamma_{\beta-}$	3064.3 9	0.24 3
$\gamma_{\beta-}$	3076.6 9	0.23 3 ?
$\gamma_{\beta-}$	3116.3 10	0.157 17 ?
$\gamma_{\beta-}$	3131.9 10	0.34 4
$\gamma_{\beta-}$	3145.5 10	0.33 4
$\gamma_{\beta-}$	3168.6 10	0.131 17
$\gamma_{\beta-}$	3187.2 10	0.122 17 ?
$\gamma_{\beta-}$	3224.9 15	~0.17 ?
$\gamma_{\beta-}$	3229.4 10	0.33 4 ?
$\gamma_{\beta-}$	3250.1 10	0.19 3 ?
$\gamma_{\beta-}$	3265 1	0.113 17
$\gamma_{\beta-}$	3286.7 10	0.33 4 ?
$\gamma_{\beta-}$	3296.9 10	0.26 4 ?
$\gamma_{\beta-}$	3305.5 10	0.174 17 ?
$\gamma_{\beta-}$	3320.6 10	0.192 17
$\gamma_{\beta-}$	3341 1	0.30 4 ?
$\gamma_{\beta-}$	3362.2 10	0.20 3
$\gamma_{\beta-}$	3374 1	0.21 3 ?
$\gamma_{\beta-}$	3386.6 10	0.22 3 ?
$\gamma_{\beta-}$	3416.6 10	0.19 3
$\gamma_{\beta-}$	3431.4 10	0.27 3 ?
$\gamma_{\beta-}$	3471.4 10	0.16 3 ?
$\gamma_{\beta-}$	3483.8 10	0.078 17
$\gamma_{\beta-}$	3506 1	0.26 3
$\gamma_{\beta-}$	3529.8 10	0.27 4 ?
$\gamma_{\beta-}$	3575.7 10	0.17 3
$\gamma_{\beta-}$	3638.6 10	0.24 4
$\gamma_{\beta-}$	3681.8 10	0.20 3 ?
$\gamma_{\beta-}$	3809 1	0.25 3 ?
$\gamma_{\beta-}$	3836.4 10	0.68 7

Photons (^{94}Rb)
(continued)

γ_{mode}	γ(keV)	γ(%)†
$\gamma_{\beta-}$	3917.6 10	0.17 3 ?
$\gamma_{\beta-}$	3993.7 10	0.10 3
$\gamma_{\beta-}$	4008.2 10	0.026 9
$\gamma_{\beta-}$	4385.0 6	0.14 3
$\gamma_{\beta-}$	4661.1 5	0.19 4
$\gamma_{\beta-}$	4692.9 7	0.087 17
$\gamma_{\beta-}$	4811.4 5	0.18 3
$\gamma_{\beta-}$	4843.1 5	0.17 3
$\gamma_{\beta-}$	4994.0 5	0.26 4 ?
$\gamma_{\beta-}$	5086.2 7	0.09 3
$\gamma_{\beta-}$	5229.4 5	0.14 4
$\gamma_{\beta-}$	5452.1 7	0.07 3
$\gamma_{\beta-}$	5684.7 5	0.12 4
$\gamma_{\beta-}$	5807.9 10	~0.05
$\gamma_{\beta-}$	6346.9 15	~0.017

† uncert(syst): 4.6% for β-

Continuous Radiation (^{94}Rb)
$\langle\beta-\rangle$=3237 keV;$\langle$IB$\rangle$=14.6 keV

E_{bin}(keV)		$\langle\ \rangle$(keV)	(%)
0 - 10	β-	0.00119	0.0236
	IB	0.074	
10 - 20	β-	0.00366	0.0244
	IB	0.074	0.51
20 - 40	β-	0.0154	0.051
	IB	0.147	0.51
40 - 100	β-	0.124	0.175
	IB	0.43	0.66
100 - 300	β-	1.77	0.85
	IB	1.36	0.75
300 - 600	β-	9.9	2.13
	IB	1.8	0.42
600 - 1300	β-	89	9.0
	IB	3.4	0.38
1300 - 2500	β-	466	24.2
	IB	3.8	0.22
2500 - 5000	β-	1747	47.7
	IB	3.2	0.096
5000 - 7500	β-	901	15.5
	IB	0.33	0.0060
7500 - 9470	β-	23.8	0.301
	IB	0.0027	3.4×10^{-5}

$^{94}_{38}$Sr(1.235 5 min)

Mode: β-
Δ: -78836 7 keV
SpA: 1.612×10^{9} Ci/g
Prod: fission

Photons (^{94}Sr)
$\langle\gamma\rangle$=1427 9 keV

γ_{mode}	γ(keV)	γ(%)†
γ	102.1 10	
γ	252.9 2	0.037 9
γ	432.6 5	0.047 19
γ	520.6 1	0.30 5
γ	530.3 1	0.27 5
γ	621.6 1	1.95 20
γ[E1]	703.8 1	2.17 20
γ[M1+E2]	723.7 1	2.45 19
γ[M1+E2]	754.7 10	
γ	806.0 1	1.80 17
γ	906.8 1	0.41 6
γ[E1]	1427.6 1	94.2
γ	1560.7 3	0.055 15
γ	1648.0 2	0.082 17 ?
γ	1751.0 3	0.047 10 ?
γ	2063.5 4	0.037 10 ?
γ[E1]	2182.3 2	0.55 9
γ	2247.0 12	

† 0.59% uncert(syst)

Continuous Radiation (^{94}Sr)
$\langle\beta-\rangle$=835 keV;$\langle$IB$\rangle$=1.66 keV

E_{bin}(keV)		$\langle\ \rangle$(keV)	(%)
0 - 10	β-	0.0155	0.308
	IB	0.033	
10 - 20	β-	0.0475	0.316
	IB	0.032	0.22
20 - 40	β-	0.197	0.66
	IB	0.063	0.22
40 - 100	β-	1.55	2.19
	IB	0.17	0.27
100 - 300	β-	20	9.7
	IB	0.46	0.26
300 - 600	β-	92	20.1
	IB	0.45	0.107
600 - 1300	β-	460	49.3
	IB	0.41	0.050
1300 - 2500	β-	262	17.2
	IB	0.035	0.0025
2500 - 2890	β-	0.126	0.00487
	IB	3.4×10^{-6}	1.34×10^{-7}

$^{94}_{39}$Y (18.7 1 min)

Mode: β-
Δ: -82348 6 keV
SpA: 1.069×10^{8} Ci/g
Prod: fission; ^{96}Zr(d,α); ^{96}Zr(γ,np); ^{94}Zr(n,p)

Photons (^{94}Y)
$\langle\gamma\rangle$=772 28 keV

γ_{mode}	γ(keV)	γ(%)†
γ[E1]	308.2 3	0.056 11
γ[E2]	381.6 2	2.02 17
γ[E2]	550.9 1	4.9 3
γ[E1]	588 1	0.17 6
γ[M1+E2]	694.7 3	0.19 3
γ[M1+E2]	752.6 1	1.40 11
γ	887.5 4	0.078 17
γE2	918.74 5	56
γ	1001.8 3	0.062 17
γ[E2]	1066.5 3	0.062 17
γ[E1]	1138.9 1	6.0 4

Photons (^{94}Y)
(continued)

γ_{mode}	γ(keV)	γ(%)†
γ	1161.8 1	0.69 8
γ[M1+E2]	1232.6 2	0.33 3
γ	1236.6 2	0.13 3 ?
γ	1303.8 6	0.045 11
γ	1384.4 6	0.028 11
γ[E2]	1411.9 7	0.078 17
γ[M1+E2]	1447.4 2	0.25 4
γ	1587.9 6	0.034 11
γ	1630.0 5	0.034 11
γ[E2]	1671.4 1	2.46 22
γ	1891.6 2	0.39 5
γ	1904.6 8	0.034 11 ?
γ	1927.5 6	0.039 11
γ	1940.6 6	0.039 11
γ	1989.3 7	0.039 11 ?
γ	2140.6 2	0.95 11
γ	2255.3 7	0.034 11 ?
γ	2300.5 3	0.18 3
γ	2348.7 10	0.034 11
γ	2442.1 3	0.14 3
γ	2492.0 3	0.21 3 ?
γ	2527.3 4	0.20 3 ?
γ	2566.2 5	0.062 17 ?
γ	2662.4 10	0.028 11 ?
γ	2805.9 10	0.034 11 ?
γ	2846.3 3	0.37 4
γ	2898.7 6	0.101 22 ?
γ	2908.4 8	0.045 17 ?
γ	2966.6 10	~0.011 ?
γ	2998.4 10	~0.017 ?
γ	3190.3 10	0.018 8
γ	3264.4 7	0.062 17
γ	3318.7 7	0.050 17 ?
γ	3477.3 10	~0.022
γ	3541.5 10	~0.011
γ	3599.8 10	~0.017
γ	3666.5 15	~0.011
γ	3718.8 15	~0.008 ?
γ	3750.9 15	~0.011 ?
γ	3795.2 15	~0.011
γ	4002.1 15	~0.011
γ	4052.3 15	~0.006
γ	4098.4 15	~0.022

† 5.4% uncert(syst)

Continuous Radiation (^{94}Y)
⟨β-⟩=1813 keV;⟨IB⟩=6.1 keV

E_{bin}(keV)		⟨ ⟩(keV)	(%)
0 - 10	β-	0.0055	0.108
	IB	0.054	
10 - 20	β-	0.0167	0.111
	IB	0.053	0.37
20 - 40	β-	0.069	0.229
	IB	0.106	0.37
40 - 100	β-	0.53	0.76
	IB	0.31	0.47
100 - 300	β-	6.8	3.29
	IB	0.91	0.51
300 - 600	β-	32.0	7.0
	IB	1.12	0.26
600 - 1300	β-	220	22.8
	IB	1.8	0.20
1300 - 2500	β-	760	40.5
	IB	1.38	0.080
2500 - 4920	β-	794	25.2
	IB	0.34	0.0115

$^{94}_{40}$Zr(stable)

Δ: -87268 3 keV
%: 17.38 2

$^{94}_{41}$Nb(2.03 16 $\times 10^4$ yr)

Mode: β-
Δ: -86367.9 25 keV
SpA: 0.187 Ci/g
Prod: ^{93}Nb(n,γ)

Photons (^{94}Nb)
⟨γ⟩=1571.5 1 keV

γ_{mode}	γ(keV)	γ(%)
γ[E2]	702.630 18	99.814 6
γ[E2]	871.097 17	99.892 3

Continuous Radiation (^{94}Nb)
⟨β-⟩=146 keV;⟨IB⟩=0.073 keV

E_{bin}(keV)		⟨ ⟩(keV)	(%)
0 - 10	β-	0.206	4.13
	IB	0.0075	
10 - 20	β-	0.61	4.09
	IB	0.0068	0.047
20 - 40	β-	2.41	8.1
	IB	0.0117	0.041
40 - 100	β-	15.9	22.9
	IB	0.024	0.038
100 - 300	β-	96	52
	IB	0.022	0.0150
300 - 472	β-	30.3	8.8
	IB	0.00077	0.00024

$^{94}_{41}$Nb(6.26 1 min)

Mode: IT(99.53 9 %), β-(0.47 9 %)
Δ: -86326.9 25 keV
SpA: 3.192$\times 10^8$ Ci/g
Prod: ^{93}Nb(n,γ)

Photons (^{94}Nb)
⟨γ⟩=11.7 9 keV

γ_{mode}	γ(keV)	γ(%)†
Nb L_ℓ	1.902	0.082 14
Nb L_η	1.996	0.028 4
Nb L_α	2.166	2.0 3
Nb L_β	2.286	0.87 15
Nb L_γ	2.565	0.057 12
Nb $K_{\alpha2}$	16.521	12.4 5
Nb $K_{\alpha1}$	16.615	23.7 10
Nb $K_{\beta1}$'	18.618	5.47 23
Nb $K_{\beta2}$'	19.074	0.92 4
γ_{IT} M3	40.95 3	0.073
$\gamma_{\beta-}$[E2]	702.630 18	0.00315 20
$\gamma_{\beta-}$[E2]	871.097 17	0.50
$\gamma_{\beta-}$[M1+E2]	992.75 10	0.00075 10
$\gamma_{\beta-}$[E2]	1863.84 10	6.7 8 $\times 10^{-5}$

† uncert(syst): 3.0% for IT, 19% for β-

Atomic Electrons (^{94}Nb)*
⟨e⟩=32.8 6 keV

e_{bin}(keV)	⟨e⟩(keV)	e(%)
2	1.49	62 7
3	0.160	5.9 6
14	1.45	10.3 11
16	0.57	3.5 4
17	0.035	0.212 22
18	0.045	0.25 3
19	0.0084	0.045 5
22	12.4	56.7 20
38	6.80	17.8 6
39	6.57	17.0 6
40	1.23	3.03 10
41	2.01	4.95 17

* with IT

Continuous Radiation (^{94}Nb)
⟨β-⟩=2.14 keV;⟨IB⟩=0.0027 keV

E_{bin}(keV)		⟨ ⟩(keV)	(%)
0 - 10	β-	0.000229	0.00456
	IB	9.7$\times 10^{-5}$	
10 - 20	β-	0.00070	0.00464
	IB	9.4$\times 10^{-5}$	0.00065
20 - 40	β-	0.00285	0.0095
	IB	0.00018	0.00062
40 - 100	β-	0.0214	0.0303
	IB	0.00047	0.00072
100 - 300	β-	0.237	0.117
	IB	0.00103	0.00060
300 - 600	β-	0.80	0.179
	IB	0.00065	0.000161
600 - 1215	β-	1.08	0.139
	IB	0.000154	2.2$\times 10^{-5}$

$^{94}_{42}$Mo(stable)

Δ: -88413.2 24 keV
%: 9.25 2

$^{94}_{43}$Tc(4.883 17 h)

Mode: ε
Δ: -84157 5 keV
SpA: 6.827$\times 10^6$ Ci/g
Prod: ^{93}Nb(α,3n); ^{94}Mo(d,2n); ^{94}Mo(p,n); ^{93}Nb(^{3}He,2n); protons on Cd

Photons (^{94}Tc)
⟨γ⟩=2555 24 keV

γ_{mode}	γ(keV)	γ(%)
Mo L_ℓ	2.016	0.075 13
Mo L_η	2.120	0.035 5
Mo L_α	2.293	1.9 3
Mo L_β	2.425	1.06 18
Mo L_γ	2.699	0.070 14
Mo $K_{\alpha2}$	17.374	17.5 6
Mo $K_{\alpha1}$	17.479	33.4 12
Mo $K_{\beta1}$'	19.602	7.9 3
Mo $K_{\beta2}$'	20.091	1.37 5
γE2	82.9 4	0.21 4
γ[M1+E2]	449.2 3	3.3 3
γ[E2]	532.1 3	2.4 3
γ[E2]	702.630 18	99.6 20
γ[M1+E2]	742.27 17	1.2 2
γ[E2]	849.70 7	95.8 19
γ[E2]	871.097 17	99.9

Photons (^{94}Tc)
(continued)

γ_{mode}	γ(keV)	γ(%)
γ [M1+E2]	916.10 10	7.6 4
γ	1509.3 4	0.68 7
γ [E2]	1591.96 17	2.3 2
γ [E2]	1765.78 12	0.29 5

Atomic Electrons (^{94}Tc)
$\langle e \rangle$=8.2 3 keV

e_{bin}(keV)	$\langle e \rangle$(keV)	e(%)
3	1.67	65 7
14	0.186	1.31 13
15	1.75	11.9 12
17	0.82	4.8 5
19 - 20	0.075	0.39 3
63	0.25	0.39 8
80 - 83	0.088	0.109 18
429 - 449	0.083	0.019 3
512 - 532	0.049	0.0096 11
683	1.10	0.162 7
700 - 742	0.174	0.0247 9
830	0.80	0.096 4
847 - 850	0.113	0.0134 5
851	0.80	0.094 4
868 - 916	0.180	0.0205 7
1489 - 1572	0.0118	0.00076 8
1589 - 1592	0.00124	7.8 6 $\times 10^{-5}$
1746 - 1765	0.00121	6.9 11 $\times 10^{-5}$

Continuous Radiation (^{94}Tc)
$\langle \beta+ \rangle$=39.5 keV;$\langle$IB$\rangle$=1.13 keV

E_{bin}(keV)		$\langle\ \rangle$(keV)	(%)
0 - 10	β+	1.56 $\times 10^{-5}$	0.000190
	IB	0.0020	
10 - 20	β+	0.000447	0.00273
	IB	0.018	0.104
20 - 40	β+	0.0089	0.0276
	IB	0.0053	0.019
40 - 100	β+	0.288	0.377
	IB	0.0107	0.0165
100 - 300	β+	8.2	3.88
	IB	0.052	0.026
300 - 600	β+	25.5	5.9
	IB	0.19	0.043
600 - 1300	β+	5.4	0.83
	IB	0.70	0.077
1300 - 1833	IB	0.151	0.0105
	$\Sigma\beta$+		11.0

$^{94}_{43}$Tc(52 *1* min)

Mode: ϵ
Δ: -84082 5 keV
SpA: 3.85$\times 10^{7}$ Ci/g
Prod: ^{93}Nb(α,3n); ^{94}Mo(d,2n); ^{94}Mo(p,n); daughter ^{94}Ru; protons on Cd

Photons (^{94}Tc)
$\langle \gamma \rangle$=1226 16 keV

γ_{mode}	γ(keV)	γ(%)†
Mo L_ℓ	2.016	0.025 4
Mo L_η	2.120	0.0116 18
Mo L_α	2.293	0.62 10
Mo L_β	2.425	0.35 6
Mo L_γ	2.699	0.023 5

Photons (^{94}Tc)
(continued)

γ_{mode}	γ(keV)	γ(%)†
Mo $K_{\alpha2}$	17.374	5.82 20
Mo $K_{\alpha1}$	17.479	11.1 4
Mo $K_{\beta1}$'	19.602	2.61 9
Mo $K_{\beta2}$'	20.091	0.455 17
γ [E2]	871.097 17	94.2
γ [M1+E2]	992.75 10	2.3 3
γ	997.7 10	0.24 8
γ [M1+E2]	1195.8 3	0.75 9
γ	1264.5 3	0.22 8
γ	1357.0 15	0.19 8
γ [M1+E2]	1521.56 24	4.5 3
γ	1757.5 10	0.15 4
γ [E2]	1863.84 10	
γ	1868.34 8	5.7 3
γ	1928.8 14	~0.08
γ [E2]	2066.9 3	0.08 3
γ [E2]	2392.64 24	0.47 19
γ [M1+E2]	2529.4 3	0.31 8
γ	2576.8 20	0.12 5
γ	2641.2 11	
γ	2663.7 20	~0.07
γ	2739.42 8	3.9 4
γ	3021.3 9	~0.08
γ	3128.3 3	1.41 19
γ	3512.2 11	0.056 19
γ	3792.6 14	0.047 19
γ	3892.3 9	~0.015

† 0.47% uncert(syst)

Atomic Electrons (^{94}Tc)
$\langle e \rangle$=2.44 9 keV

e_{bin}(keV)	$\langle e \rangle$(keV)	e(%)
3	0.55	21.5 22
14	0.062	0.43 4
15	0.58	3.9 4
17	0.27	1.61 16
19 - 20	0.0249	0.129 11
851	0.76	0.089 4
868	0.086	0.0099 4
869 - 871	0.0215	0.00247 9
973 - 998	0.0192	0.00197 24
1176 - 1196	0.0048	0.00041 5
1244 - 1264	0.0010	8 4 $\times 10^{-5}$
1337 - 1357	0.0008	6 3 $\times 10^{-5}$
1502 - 1521	0.0223	0.00148 12
1738 - 1757	0.00050	2.9 11 $\times 10^{-5}$
1848 - 1928	0.018	0.0010 3
2047 - 2067	0.00031	1.5 5 $\times 10^{-5}$
2373 - 2392	0.0015	6.4 23 $\times 10^{-5}$
2509 - 2576	0.00128	5.1 10 $\times 10^{-5}$
2644 - 2739	0.0093	0.00034 8
3001 - 3021	0.00017	~6 $\times 10^{-6}$
3108 - 3128	0.0031	0.000099 24
3492 - 3512	0.00012	3.3 12 $\times 10^{-6}$
3773 - 3792	9.3 $\times 10^{-5}$	2.5 10 $\times 10^{-6}$
3872 - 3892	3.0 $\times 10^{-5}$	~8 $\times 10^{-7}$

Continuous Radiation (^{94}Tc)
$\langle \beta+ \rangle$=753 keV;$\langle$IB$\rangle$=3.2 keV

E_{bin}(keV)		$\langle\ \rangle$(keV)	(%)
0 - 10	β+	9.6 $\times 10^{-6}$	0.000117
	IB	0.028	
10 - 20	β+	0.000278	0.00170
	IB	0.033	0.22
20 - 40	β+	0.0056	0.0175
	IB	0.054	0.19
40 - 100	β+	0.194	0.253
	IB	0.153	0.24
100 - 300	β+	7.2	3.30
	IB	0.44	0.24
300 - 600	β+	47.1	10.2
	IB	0.52	0.122

Continuous Radiation (^{94}Tc)
(continued)

E_{bin}(keV)		$\langle\ \rangle$(keV)	(%)
600 - 1300	β+	316	33.1
	IB	0.92	0.104
1300 - 2500	β+	383	23.4
	IB	0.90	0.051
2500 - 3460	IB	0.17	0.0063
	$\Sigma\beta$+		70

$^{94}_{44}$Ru(51.8 *6* min)

Mode: ϵ
Δ: -82567 13 keV
SpA: 3.86$\times 10^{7}$ Ci/g
Prod: ^{92}Mo(α,2n); ^{94}Mo(α,4n)

Photons (^{94}Ru)
$\langle \gamma \rangle$=520 49 keV

γ_{mode}	γ(keV)	γ(%)†
Tc L_ℓ	2.133	0.087 15
Tc L_η	2.249	0.040 6
Tc L_α	2.424	2.2 4
Tc L_β	2.577	1.31 22
Tc L_γ	2.863	0.094 18
Tc $K_{\alpha2}$	18.251	19.9 7
Tc $K_{\alpha1}$	18.367	37.8 13
Tc $K_{\beta1}$'	20.613	9.0 3
Tc $K_{\beta2}$'	21.136	1.62 6
γ [M1+E2]	367.1 5	75
γ [M1+E2]	524.9 6	1.80 23
γ [M1+E2]	892.0 5	25 5

† 5.3% uncert(syst)

Atomic Electrons (^{94}Ru)
$\langle e \rangle$=8.3 6 keV

e_{bin}(keV)	$\langle e \rangle$(keV)	e(%)
3	1.96	72 7
15	0.84	5.5 6
16	1.27	8.1 8
17	0.078	0.45 5
18	0.84	4.7 5
20	0.071	0.35 4
21	0.0138	0.067 7
346	2.5	0.73 14
364	0.33	0.090 22
367	0.072	0.020 5
504	0.033	0.0065 10
522	0.0040	0.00077 14
524	0.00074	0.00014 3
525	0.000141	2.7 5 $\times 10^{-5}$
871	0.21	0.024 5
889	0.025	0.0028 7
891	0.0043	0.00048 11
892	0.0011	0.00012 4

Continuous Radiation (^{94}Ru)
$\langle$IB$\rangle$=0.27 keV

E_{bin}(keV)		$\langle\ \rangle$(keV)	(%)
10 - 20	IB	0.0168	0.095
20 - 40	IB	0.0053	0.024
40 - 100	IB	0.0023	0.0034
100 - 300	IB	0.031	0.0149
300 - 600	IB	0.106	0.024
600 - 1150	IB	0.105	0.0140

$^{94}_{45}$Rh(1.18 *1* min)

Mode: ϵ, ϵp(1.8 *5* %)
Δ: -72970 *300* keV syst
SpA: 1.687×10⁹ Ci/g
Prod: ^{96}Ru(p,3n)

Photons (^{94}Rh)&

⟨γ⟩=2705 *83* keV

γ_{mode}	γ(keV)	γ(%)
γ_ϵ(E1)	126.4 *2*	4.42 *20*
γ_ϵ(E2)	311.7 *1*	12 *3*
γ_ϵ(E1)	438.1 *2*	7.0 *3*
γ_ϵ	492.6 *3*	4.1 *3*
γ_ϵ	552.9 *3*	1.87 *20*
γ_ϵ(E2)	756.2 *1*	50 *5*
γ_ϵ	1068.1 *3*	4.6 *3*
γ_ϵ	1072.5 *2*	30.1 *10*
γ_ϵ	1100.7 *2*	2.85 *20*
γ_ϵ(E2)	1430.7 *1*	98
γ_ϵ	1539.7 *3*	4.25 *15*
γ_ϵ	1804.3 *10*	2.16 *20*
γ_ϵ	1902.5 *10*	2.16 *20*
γ_ϵ	2124.5 *10*	1.47 *20* ? *
γ_ϵ	2631.6 *10*	1.4 *3* ? *
γ_ϵ	2677.8 *10*	1.37 *20*
γ_ϵ	2778.6 *10*	1.1 *3* ? *
γ_ϵ	3007.7 *10*	0.98 *20* ? *
γ_ϵ	3210.3 *10*	1.28 *20* ? *
γ_ϵ	3256 *1*	1.96 *20* ? *

* with ^{94}Rh(1.18 min or 25.8 s)
& see also ^{94}Rh(25.8 s)

$^{94}_{45}$Rh(25.8 *2* s)

Mode: ϵ
Δ: -72970 *300* keV syst
SpA: 4.59×10⁹ Ci/g
Prod: ^{96}Ru(p,3n)

Photons (^{94}Rh)&

⟨γ⟩=2866 *77* keV

γ_{mode}	γ(keV)	γ(%)
γ(E1)	126.4 *2*	0.85 *10*
γ(E2)	146.1 *1*	75 *5*
γ(E2)	311.7 *1*	97 *4*
γ(E1)	438.1 *2*	2.8 *3*
γ(E2)	756.2 *1*	100 *3*
γ(E2)	1033.4 *3*	1.8 *2*
γ(E2)	1430.7 *1*	100
γ	2099.5 *10*	1.9 *2*
γ	2124.5 *10*	1.1 *2* ? *
γ	2631.6 *10*	1.0 *2* ? *
γ	2778.6 *10*	0.8 *2* ? *
γ	2966 *1*	0.9 *2*
γ	3007.7 *10*	0.7 *1* ? *
γ	3210.3 *10*	0.9 *1* ? *
γ	3256 *1*	1.4 *1* ? *

* with ^{94}Rh(25.8 s or 1.18 min)
& see also ^{94}Rh(1.18 min)

$^{94}_{46}$Pd(9.0 *5* s)

Mode: ϵ
SpA: 1.28×10¹⁰ Ci/g
Prod: ^{40}Ca on ^{58}Ni

Photons (^{94}Pd)

γ_{mode}	γ(keV)	γ(rel)
γ	54.6 *2*	11 *1*
γ	558.2 *2*	100 *3*
γ	723.9 *2*	12.1 *13*
γ	797.8 *2*	7.1 *12*

A = 95

NDS **38**, 1 (1983)

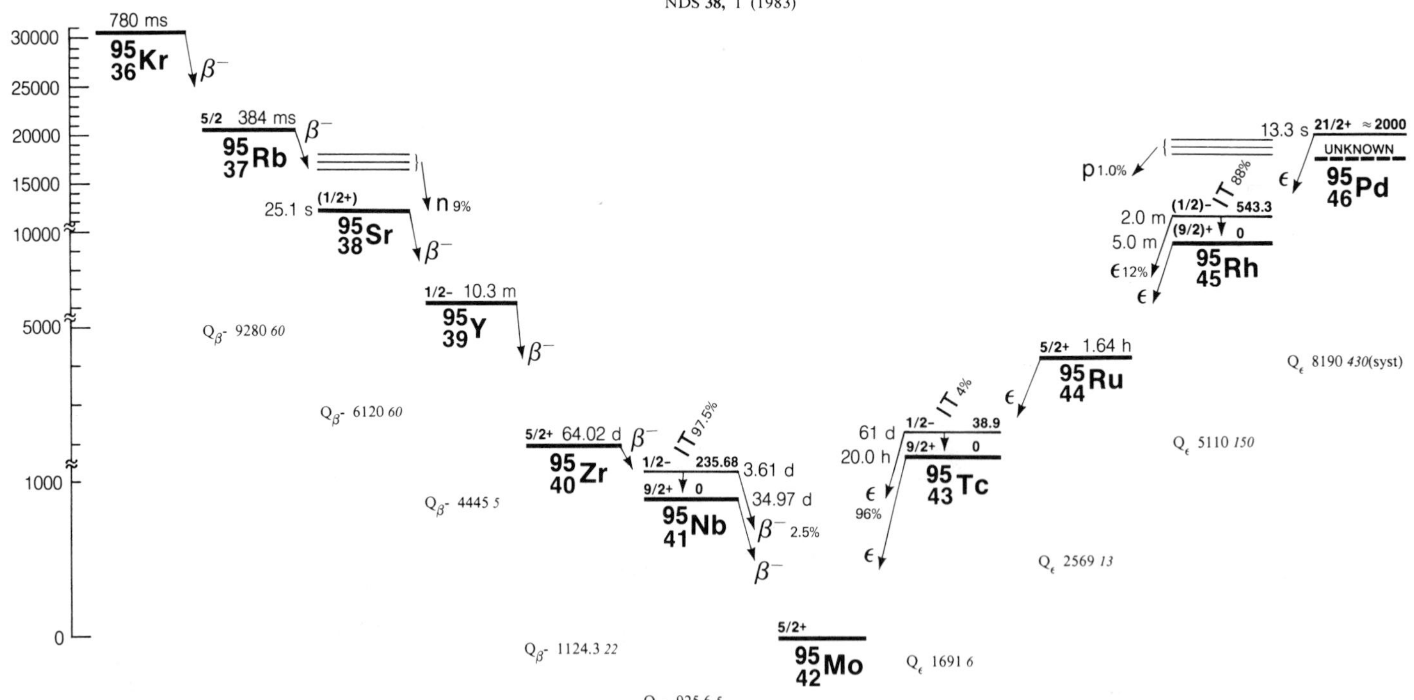

$^{95}_{36}$Kr(780 *30* ms)

Mode: β-
SpA: 1.01×10^{11} Ci/g
Prod: fission

$^{95}_{37}$Rb(384 *6* ms)

Mode: β-, β-n(8.8 *6* %)
Δ: -65808 *28* keV
SpA: 1.431×10^{11} Ci/g
Prod: fission

β-n(avg): 530 *10*

Photons (^{95}Rb)

⟨γ⟩=612 *15* keV

γ_{mode}	γ(keV)	γ(%)†
$\gamma_{\beta-}$[E2]	203.78 *24*	16.9 *9*
$\gamma_{\beta-}$	328.55 *24*	15.6 *8*
$\gamma_{\beta-}$[M1]	351.85 *24*	65 *3*
$\gamma_{\beta-n}$	458.0 *1*	0.0030 *6*
$\gamma_{\beta-n}$ (M1+11%E2)	503.8 *1*	0.090 *18*
$\gamma_{\beta-}$	564.88 *24*	2.60 *13*
$\gamma_{\beta-}$	578.3 *3*	3.90 *19*
$\gamma_{\beta-}$	659.9 *3*	7.2 *4*
$\gamma_{\beta-n}$ (M1+23%E2)	677.7 *1*	0.024 *5*
$\gamma_{\beta-}$	680.40 *24*	21.5 *11*
$\gamma_{\beta-}$	768.66 *24*	4.55 *23*
$\gamma_{\beta-n}$ (E2)	836.9 *1*	3.0 *6*
$\gamma_{\beta-n}$ (E1)	1089.4 *2*	0.14 *3*
$\gamma_{\beta-n}$ M1,E2	1309.1 *2*	0.14 *3*
$\gamma_{\beta-n}$	1434.4 *2*	0.015 *3*
$\gamma_{\beta-n}$ (E1)	1577.5 *2*	0.060 *12*
$\gamma_{\beta-n}$ (M1+43%E2)	1812.7 *3*	0.0090 *18*
$\gamma_{\beta-n}$	2271.4 *5*	0.072 *14*

† uncert(syst): 6.2% for β-

$^{95}_{38}$Sr(25.1 *2* s)

Mode: β-
Δ: -75090 *60* keV
SpA: 4.67×10^{9} Ci/g
Prod: fission

Photons (^{95}Sr)

⟨γ⟩=1023 *33* keV

γ_{mode}	γ(keV)	γ(%)†
γ	576.5 *5*	0.79 *9*
γ	685.9 *3*	24
γ	777.3 *5*	0.63 *8*
γ[E2]	826.9 *4*	3.0 *3*
γ	945.0 *4*	2.4 *3*
γ	982.9 *5*	1.26 *15*
γ	1277.6 *5*	2.2 *3*
γ	1335.2 *5*	0.72 *9*
γ	1360.6 *4*	0.57 *8*
γ	1722.0 *4*	0.54 *7*
γ	2031.3 *4*	0.50 *7*
γ	2046.5 *4*	0.37 *6*
γ	2095.2 *4*	0.13 *6*

Photons (^{95}Sr) (continued)

γ_{mode}	γ(keV)	γ(%)†
γ	2247.2 *4*	4.0 *5*
γ	2683.4 *5*	1.26 *17* ?
γ	2706.4 *5*	0.24 *4* ?
γ	2717.1 *4*	4.9 *6*
γ	2748.8 *5*	0.34 *6* ?
γ	2781.1 *4*	1.12 *15*
γ	2890.7 *5*	0.97 *13* ?
γ	2933.1 *4*	4.3 *5*
γ	3116.8 *5*	0.59 *9* ?
γ	3352.9 *4*	1.02 *14*
γ	3500.2 *5*	0.82 *12* ?
γ	3615.8 *5*	1.82 *24* ?
γ	3743.1 *5*	0.88 *13* ?
γ	4075.2 *5*	1.30 *19* ?
γ	4267.6 *5*	0.49 *9* ?

† 8.3% uncert(syst)

Continuous Radiation (^{95}Sr)

⟨β-⟩=2287 keV;⟨IB⟩=8.7 keV

E_{bin}(keV)		⟨ ⟩(keV)	(%)
0 - 10	β-	0.00296	0.059
	IB	0.062	
10 - 20	β-	0.0091	0.061
	IB	0.061	0.42
20 - 40	β-	0.0380	0.126
	IB	0.121	0.42
40 - 100	β-	0.303	0.428
	IB	0.35	0.54
100 - 300	β-	4.19	2.02
	IB	1.08	0.60
300 - 600	β-	21.9	4.76
	IB	1.37	0.32
600 - 1300	β-	170	17.5
	IB	2.4	0.27
1300 - 2500	β-	652	34.6
	IB	2.2	0.127
2500 - 5000	β-	1341	38.6
	IB	1.09	0.035
5000 - 6120	β-	98	1.85
	IB	0.0084	0.000153

$^{95}_{39}$Y (10.3 *2* min)

Mode: β-
Δ: -81214 *6* keV
SpA: 1.92×10^{8} Ci/g
Prod: fission; ^{96}Zr(γ,p)

Photons (^{95}Y)

⟨γ⟩=1287 *78* keV

γ_{mode}	γ(keV)	γ(%)†
γ	396.77 *25*	0.36 *7*
γ	432.36 *18*	2.0 *4*
γ	568.9 *3*	0.17 *3*
γ	580.4 *4*	0.15 *3*
γ	632.15 *22*	0.32 *7*
γ[E2]	954.13 *17*	19.0
γ	1002.15 *24*	0.38 *8*
γ	1048.6 *3*	1.03 *21*
γ	1173.8 *3*	0.70 *14*
γ	1213.7 *4*	~0.09
γ	1293.9 *3*	0.21 *4*
γ	1309.9 *3*	~0.06
γ	1324.13 *19*	5.3 *11*

Photons (^{95}Y) (continued)

γ_{mode}	γ(keV)	γ(%)†
γ	1357.1 *3*	0.72 *14*
γ	1409.2 *3*	~0.08
γ[M1+E2]	1418.6 *3*	0.51 *10*
γ	1511.7 *4*	0.68 *14*
γ	1618.4 *4*	1.7 *4*
γ	1631.8 *4*	~0.11
γ	1683.7 *4*	0.51 *10*
γ	1720.9 *3*	0.36 *7*
γ	1806.0 *3*	1.5 *3*
γ	1813.4 *8*	0.21 *4*
γ	1855.9 *4*	0.21 *4*
γ	1893.1 *3*	~0.7
γ	1904.5 *4*	0.19 *4*
γ	1926.0 *3*	0.65 *13*
γ	1940.3 *3*	2.8 *6*
γ	1956.27 *23*	0.38 *8*
γ[E1]	2176.0 *3*	8.2 *16*
γ	2252.6 *4*	0.46 *9*
γ[E1]	2296.0 *3*	1.4 *3*
γ[M1+E2]	2372.67 *25*	0.82 *16*
γ	2497.9 *8*	0.57 *11*
γ	2632.2 *5*	5.1 *10*
γ	2759.9 *10*	0.29 *6*
γ[E1]	3130.1 *3*	0.68 *14*
γ[E1]	3250.1 *3*	1.27 *25*
γ	3452.0 *8*	0.72 *14*
γ[E1]	3576.7 *4*	7.6 *15*
γ	3683.9 *15*	0.38 *8*
γ	3886.9 *20*	0.29 *6*
γ	3923.9 *20*	~0.19
γ	4068 *3*	~0.19

† 11% uncert(syst)

Continuous Radiation (^{95}Y)

⟨β-⟩=1352 keV;⟨IB⟩=4.3 keV

E_{bin}(keV)		⟨ ⟩(keV)	(%)
0 - 10	β-	0.0241	0.480
	IB	0.042	
10 - 20	β-	0.073	0.487
	IB	0.042	0.29
20 - 40	β-	0.299	0.99
	IB	0.082	0.28
40 - 100	β-	2.23	3.16
	IB	0.23	0.36
100 - 300	β-	24.0	11.9
	IB	0.67	0.38
300 - 600	β-	75	16.9
	IB	0.79	0.19
600 - 1300	β-	201	22.1
	IB	1.22	0.139
1300 - 2500	β-	483	25.8
	IB	0.96	0.055
2500 - 4445	β-	568	18.2
	IB	0.22	0.0075

$^{95}_{40}$Zr(64.02 *4* d)

Mode: β-
Δ: -85659 *3* keV
SpA: 2.1469×10^{4} Ci/g
Prod: ^{94}Zr(n,γ); fission

Photons (^{95}Zr)

⟨γ⟩=733 *7* keV

γ_{mode}	γ(keV)	γ(%)†
Nb $K_{\alpha2}$	16.521	0.177 *10*
Nb $K_{\alpha1}$	16.615	0.338 *19*
Nb $K_{\beta1}$'	18.618	0.078 *4*
Nb $K_{\beta2}$'	19.074	0.0131 *8*
γ M4	235.68 *2*	0.294 *16* *
γ M1+1.2%E2	724.199 *5*	44.1 *9*
γ M1+2.1%E2	756.729 *12*	54.5

† 0.37% uncert(syst)
* with ^{95}Nb(3.61 d) in equilib

Atomic Electrons (^{95}Zr)

⟨e⟩=2.98 *9* keV

e_{bin}(keV)	⟨e⟩(keV)	e(%)
2	0.0159	0.66 *7*
3	0.00144	0.054 *6*
14	0.0207	0.147 *16*
16	0.0081	0.050 *6*
17	0.00050	0.0030 *3*
18	0.00065	0.0036 *4*
19	0.000120	0.00065 *7*
217	1.47	0.68 *4*
233	0.321	0.138 *8*
235	0.060	0.0257 *15*
236	0.0102	0.00432 *25*
705	0.432	0.0612 *17*
722	0.0492	0.00682 *19*
724	0.0104	0.00143 *4*
738	0.505	0.0684 *14*
754	0.0575	0.00762 *15*
756	0.01025	0.00136 *3*
757	0.00183	0.000242 *5*

Continuous Radiation (^{95}Zr)

⟨β-⟩=117 keV;⟨IB⟩=0.050 keV

E_{bin}(keV)		⟨ ⟩(keV)	(%)
0 - 10	β-	0.271	5.4
	IB	0.0061	
10 - 20	β-	0.80	5.3
	IB	0.0054	0.037
20 - 40	β-	3.09	10.3
	IB	0.0089	0.031
40 - 100	β-	19.3	28.0
	IB	0.0166	0.027
100 - 300	β-	84	48.4
	IB	0.0119	0.0083
300 - 600	β-	8.4	2.41
	IB	0.00082	0.00021
600 - 1124	β-	1.07	0.154
	IB	6.2×10^{-5}	9.5×10^{-6}

$^{95}_{41}$Nb(34.97 *3* d)

Mode: β-
Δ: -86783.6 *21* keV
SpA: 3.930×10^{4} Ci/g
Prod: daughter ^{95}Zr; ^{96}Zr(p,2n)

Photons (^{95}Nb)

⟨γ⟩=764.31 *16* keV

γ_{mode}	γ(keV)	γ(%)†
γ M1+28%E2	204.114 *10*	0.028 *8*
γ [E2]	561.677 *13*	0.012 *3*
γ M1+2.6%E2	765.789 *9*	99.79

†<0.1% uncert(syst)

Continuous Radiation (^{95}Nb)

⟨β-⟩=43.5 keV;⟨IB⟩=0.0072 keV

E_{bin}(keV)		⟨ ⟩(keV)	(%)
0 - 10	β-	0.77	15.5
	IB	0.0022	
10 - 20	β-	2.10	14.1
	IB	0.00154	0.0109
20 - 40	β-	7.0	23.8
	IB	0.0019	0.0068
40 - 100	β-	25.7	39.8
	IB	0.00146	0.0026
100 - 300	β-	7.8	6.7
	IB	0.000100	7.6×10^{-5}
300 - 600	β-	0.076	0.0179
	IB	2.0×10^{-5}	5.1×10^{-6}
600 - 926	β-	0.0233	0.00340
	IB	1.17×10^{-6}	1.9×10^{-7}

$^{95}_{41}$Nb(3.61 *3* d)

Mode: IT(97.5 *1* %), β-(2.5 *1* %)
Δ: -86547.9 *21* keV
SpA: 3.81×10^{5} Ci/g
Prod: daughter ^{95}Zr; ^{97}Mo(d,α); ^{94}Zr(d,n); ^{96}Zr(p,2n)

Photons (^{95}Nb)

⟨γ⟩=71.2 *12* keV

γ_{mode}	γ(keV)	γ(%)†
Nb L_{ℓ}	1.902	0.057 *10*
Nb L_{η}	1.996	0.026 *4*
Nb L_{α}	2.166	1.39 *22*
Nb L_{β}	2.280	0.75 *13*
Nb L_{γ}	2.548	0.045 *9*
Nb $K_{\alpha2}$	16.521	12.6 *5*
Nb $K_{\alpha1}$	16.615	24.0 *10*
Nb $K_{\beta1}$'	18.618	5.55 *22*
Nb $K_{\beta2}$'	19.074	0.93 *4*
γ_{β}M1+28%E2	204.114 *10*	2.34 *7*
γ_{β}M1+37%E2	218.635 *18*	$<4\times10^{-8}$
γ_{IT} M4	235.68 *2*	24.9
γ_{β}M1	253.066 *16*	$<5\times10^{-7}$
γ_{β}M1+7.2%E2	582.062 *11*	0.055 *5*
γ_{β}E2+20%M1	616.492 *15*	0.00011 *5*
γ_{β}E2	786.174 *12*	0.0160 *14*
γ_{β}M1+0.5%E2	820.604 *12*	0.00040 *19*
γ_{β}M1+0.2%E2	835.126 *12*	$<2\times10^{-5}$?
γ_{β}E2	1039.237 *15*	$<2\times10^{-6}$

† uncert(syst): 4.3% for IT, 5.8% for β-

Atomic Electrons (^{95}Nb)*

⟨e⟩=161 *4* keV

e_{bin}(keV)	⟨e⟩(keV)	e(%)
2	1.14	48 *5*
3	0.105	3.9 *4*
14	1.47	10.5 *11*
16	0.57	3.6 *4*
17	0.036	0.215 *22*
18	0.046	0.25 *3*
19	0.0085	0.046 *5*
217	124.5	57.5 *16*
233	27.2	11.7 *3*
235	5.12	2.17 *6*
236	0.861	0.365 *10*

* with IT

Continuous Radiation (^{95}Nb)

⟨β-⟩=9.8 keV;⟨IB⟩=0.0114 keV

E_{bin}(keV)		⟨ ⟩(keV)	(%)
0 - 10	β-	0.00161	0.0320
	IB	0.00045	
10 - 20	β-	0.00485	0.0323
	IB	0.00043	0.0030
20 - 40	β-	0.0196	0.065
	IB	0.00082	0.0029
40 - 100	β-	0.142	0.202
	IB	0.0021	0.0033
100 - 300	β-	1.41	0.70
	IB	0.0045	0.0026
300 - 600	β-	4.04	0.92
	IB	0.0026	0.00065
600 - 1161	β-	4.18	0.55
	IB	0.00051	7.5×10^{-5}

$^{95}_{42}$Mo(stable)

Δ: -87709.2 *21* keV
%: 15.92 *4*

$^{95}_{43}$Tc(20.0 *1* h)

Mode: ϵ
Δ: -86018 *6* keV
SpA: 1.649×10^{6} Ci/g
Prod: ^{95}Mo(p,n); ^{94}Mo(d,n); ^{95}Mo(d,2n)

Photons (^{95}Tc)

⟨γ⟩=798 *14* keV

γ_{mode}	γ(keV)	γ(%)†
Mo L_{ℓ}	2.016	0.083 *15*
Mo L_{η}	2.120	0.039 *6*
Mo L_{α}	2.293	2.1 *3*
Mo L_{β}	2.426	1.18 *20*
Mo L_{γ}	2.699	0.078 *15*
Mo $K_{\alpha2}$	17.374	19.5 *7*
Mo $K_{\alpha1}$	17.479	37.0 *13*
Mo $K_{\beta1}$'	19.602	8.7 *3*
Mo $K_{\beta2}$'	20.091	1.52 *6*
γ	125.68 *10*	0.0103 *9*
γ M1	181.881 *16*	0.0025 *8*
γ M1+28%E2	204.114 *10*	0.31 *4*

Photons (^{95}Tc) (continued)

γ_{mode}	γ(keV)	γ(%)†
γ M1	307.926 14	0.0348 9
γ [E2]	467.067 17	<9×10^{-5} ?
γ	478.001 20	0.013 5
γ	494.98 3	<0.0014 ?
γ [E2]	561.677 13	0.012 3
γ [M1+E2]	593.111 18	0.022 7
γ	604.045 16	0.305 9
γ M1+2.6%E2	765.789 9	94.0 19
γ [E2]	774.991 10	0.017 5
γ	785.925 16	0.146 9
γ M1+0.7%E2	852.619 25	~0.0035
γ [E2]	869.601 16	0.318 8
γ E2	947.669 15	1.96 4
γ M1+22%E2	1056.730 25	~0.0015
γ M1+34%E2	1073.713 13	3.75 8
γ	1221.92 10	0.009 4
γ	1426.03 10	~3×10^{-5}
γ [E2]	1440.9 6	~0.0007
γ	1551.708 16	0.0221 9
γ [M1+E2]	1645.1 6	~0.0006

† 1.1% uncert(syst)

Atomic Electrons (^{95}Tc)

⟨e⟩=6.1 3 keV

e_{bin}(keV)	⟨e⟩(keV)	e(%)
3	1.85	72 7
14	0.207	1.45 15
15	1.95	13.2 13
17	0.91	5.4 6
19 - 20	0.083	0.43 4
106 - 126	0.0034	~0.0031
162 - 204	0.031	0.0165 18
288 - 308	0.00129	0.000446 13
447 - 495	0.00023	~5×10^{-5}
542 - 591	0.0035	0.00060 24
593 - 604	0.00043	7 3 ×10^{-5}
746	0.906	0.122 3
755 - 786	0.128	0.0167 4
833 - 870	0.00296	0.000347 14
928 - 948	0.0160	0.00172 4
1037 - 1074	0.0272	0.00258 6
1202 - 1222	4.3 ×10^{-5}	3.6 18 ×10^{-6}
1406 - 1441	3.4 ×10^{-6}	2.4 12 ×10^{-7}
1532 - 1625	8.3 ×10^{-5}	5.4 17 ×10^{-6}
1642 - 1645	3.0 ×10^{-7}	1.8 8 ×10^{-8}

Continuous Radiation (^{95}Tc)

⟨IB⟩=0.18 keV

E_{bin}(keV)		⟨ ⟩(keV)	(%)
10 - 20	IB	0.018	0.103
20 - 40	IB	0.0020	0.0085
40 - 100	IB	0.0022	0.0032
100 - 300	IB	0.030	0.0145
300 - 600	IB	0.092	0.021
600 - 925	IB	0.035	0.0052

$^{95}_{43}$Tc(61 2 d)

Mode: ϵ(96 1 %), IT(4 1 %)
Δ: -85979 6 keV
SpA: 2.25×10^{4} Ci/g
Prod: ^{95}Mo(p,n); ^{94}Mo(d,n); ^{95}Mo(d,2n)

Photons (^{95}Tc)

⟨γ⟩=716 7 keV

γ_{mode}	γ(keV)	γ(%)†
Mo L_ℓ	2.016	0.082 14
Mo L_η	2.120	0.038 6
Mo L_α	2.293	2.1 3
Mo L_β	2.425	1.16 19
Mo L_γ	2.699	0.078 15
Mo $K_{\alpha2}$	17.374	19.3 7
Mo $K_{\alpha1}$	17.479	36.6 13
Tc $K_{\alpha2}$	18.251	0.20 5
Tc $K_{\alpha1}$	18.367	0.37 9
Mo $K_{\beta1}$'	19.602	8.6 3
Mo $K_{\beta2}$'	20.091	1.51 6
Tc $K_{\beta1}$'	20.613	0.089 22
Tc $K_{\beta2}$'	21.136	0.016 4
γ_ϵM1+28%E2	204.114 10	66.2
γ_ϵM1+37%E2	218.635 18	0.045 2
γ_ϵ	245.45 25	0.0019 5
γ_ϵM1	253.066 16	0.640 13
γ_ϵ[M1]	262.95 25	<1.3×10^{-4} ?
γ_ϵ	291.66 4	0.0058 5
γ_ϵ[M1+E2]	318.0 3	0.0011 4
γ_ϵ[M1]	516.01 25	<0.0007
γ_ϵ	563.47 5	0.0099 13
γ_ϵM1+7.2%E2	582.062 11	31.4 6
γ_ϵ	589.28 25	0.0011 3
γ_ϵE2+20%M1	616.492 15	1.34 3
γ_ϵ	623.28 15	0.0060 20
γ_ϵE2	786.174 12	9.07 18
γ_ϵ[M1+E2]	799.60 5	0.0015 5
γ_ϵM1+0.5%E2	820.604 12	4.93 10
γ_ϵM1+0.2%E2	835.126 12	27.9 6
γ_ϵ	844.1 7	0.012 3
γ_ϵM1+0.7%E2	852.619 25	0.0219 7
γ_ϵE2	1039.237 15	2.91 6
γ_ϵM1+22%E2	1056.730 25	0.0093 3
γ_ϵ[M1+E2]	1098.07 25	<2.0×10^{-5} ?
γ_ϵ	1165.63 24	<1.0×10^{-4} ?
γ_ϵ	1221.92 10	0.00874 20
γ_ϵ[E2]	1302.18 25	<2.0×10^{-5} ?
γ_ϵ	1369.74 24	0.00014 3
γ_ϵ[M1+E2]	1416.09 5	0.00192 7
γ_ϵ	1426.03 10	~3×10^{-5}
γ_ϵ[M1+E2]	1620.20 5	0.0391 20

† uncert(syst): 1.8% for ϵ

Atomic Electrons (^{95}Tc)

⟨e⟩=13.9 4 keV

e_{bin}(keV)	⟨e⟩(keV)	e(%)
3	1.90	74 8
14	0.205	1.44 15
15	1.93	13.1 13
16	0.012	0.080 22
17	0.91	5.3 5
18 - 21	0.24	1.34 23
36	0.90	2.5 6
38 - 39	0.25	0.65 14
184	5.55	3.02 11
199	0.00358	0.00180 10
201	0.71	0.355 13
202 - 251	0.274	0.133 5
253 - 298	0.00105	0.00040 5
315 - 318	6.9 ×10^{-6}	2.2 8 ×10^{-6}
496 - 543	0.00011	2.1 10 ×10^{-5}
561	1.3 ×10^{-5}	~2×10^{-6}
562	0.429	0.0763 21
563 - 603	0.0792	0.01358 25
614 - 623	0.00266	0.000433 10
766 - 815	0.383	0.0478 9
818 - 853	0.0399	0.00481 10
1019 - 1057	0.0212	0.00208 5
1078 - 1098	7.0 ×10^{-8}	~6×10^{-9}
1146 - 1166	2.5 ×10^{-7}	~2×10^{-8}
1202 - 1222	4.0 ×10^{-5}	3.3 12 ×10^{-6}
1282 - 1302	5.6 ×10^{-8}	~4×10^{-9}

Atomic Electrons (^{95}Tc) (continued)

e_{bin}(keV)	⟨e⟩(keV)	e(%)
1350 - 1396	9.6 ×10^{-6}	6.9 5 ×10^{-7}
1406 - 1426	1.33 ×10^{-6}	9.4 8 ×10^{-8}
1600 - 1620	0.000181	1.13 8 ×10^{-5}

Continuous Radiation (^{95}Tc)

⟨β+⟩=0.89 keV;⟨IB⟩=0.25 keV

E_{bin}(keV)		⟨ ⟩(keV)	(%)
0 - 10	β+	8.9×10^{-7}	1.08×10^{-5}
	IB	0.00018	
10 - 20	β+	2.51×10^{-5}	0.000153
	IB	0.0169	0.098
20 - 40	β+	0.000488	0.00151
	IB	0.0020	0.0084
40 - 100	β+	0.0147	0.0194
	IB	0.0023	0.0034
100 - 300	β+	0.322	0.157
	IB	0.029	0.0138
300 - 600	β+	0.52	0.128
	IB	0.087	0.020
600 - 1300	β+	0.0300	0.00477
	IB	0.104	0.0125
1300 - 1730	IB	0.0094	0.00067
	Σβ+		0.31

$^{95}_{44}$Ru(1.64 1 h)

Mode: ϵ
Δ: -83449 12 keV
SpA: 2.011×10^{7} Ci/g
Prod: ^{92}Mo(α,n); ^{96}Ru(n,2n); ^{96}Ru(γ,n)

Photons (^{95}Ru)

⟨γ⟩=1102 14 keV

γ_{mode}	γ(keV)	γ(%)†
Tc L_ℓ	2.133	0.075 13
Tc L_η	2.249	0.035 5
Tc L_α	2.424	1.9 3
Tc L_β	2.577	1.14 19
Tc L_γ	2.863	0.082 15
Tc $K_{\alpha2}$	18.251	17.2 6
Tc $K_{\alpha1}$	18.367	32.8 11
Tc $K_{\beta1}$'	20.613	7.8 3
Tc $K_{\beta2}$'	21.136	1.41 5
γ [M1+E2]	157.12 7	0.05 2 ?
γ	254.69 7	0.22 1
γ [M1+E2]	290.47 5	3.73 10
γ M1+4.3%E2	301.05 5	2.13 6
γ	313.77 7	~0.22 ?
γ M1+11%E2	336.47 5	70.8
γ	348.30 7	0.21 1
γ	403.9 3	0.03 1 ?
γ [M1+E2]	421.41 11	0.07 1
γ	446.70 16	~0.08
γ	477.08 13	0.05 1
γ	505.42 6	0.12 5
γ M1+5.4%E2	551.77 6	1.72 6
γ	560.60 15	~0.05
γ [M1+E2]	572.31 25	0.13 6 ?
γ	576.24 11	~0.04
γ [M1+E2]	581.78 15	<0.12
γ E2	591.52 5	1.20 5
γ	606.38 10	0.11 5
γ [M1+E2]	607.6 4	0.17 7

Photons (^{95}Ru) (continued)

γ_{mode}	γ(keV)	γ(%)†
γ [E2]	626.93 5	18.0 5
γ E2	628.75 22	0.13 4
γ	652.96 7	1.03 4
γ	662.33 21	0.04 1 ?
γ	689.28 18	0.08 1
γ	711.68 15	0.18 2
γ	735.17 11	0.47 2
γ [M1+E2]	748.64 7	1.63 6
γ	755.98 7	0.22 2
γ	786.9 4	0.05 1 ?
γ	806.46 6	4.09 17
γ [M1+E2]	819.18 6	0.65 3
γ	834.33 8	0.07 1
γ M1+10%E2	842.24 6	1.28 5
γ [M1+E2]	876.69 24	0.22 2
γ E1	889.06 7	1.93 10
γ	891.25 12	~0.18
γ	893.61 9	0.16 7
γ	960.36 10	0.22 2
γ	976.10 21	0.06 1 ?
γ [M1+E2]	989.86 11	0.71 3
γ	1010.67 7	0.73 3
γ	1050.73 8	2.63 11
γ	1064.55 8	0.75 6
γ	1089.02 9	0.22 5
γ	1096.93 5	21.2 10
γ [E1]	1100.6 4	0.12 5 ?
γ	1104.27 8	0.21 6
γ [M1+E2]	1120.23 6	0.95 5
γ	1158.38 6	0.85 20
γ [M1+E2]	1158.53 8	0.71 18
γ M1+17%E2	1178.70 7	5.20 25
γ	1182.62 9	0.22 3
γ	1220.67 21	0.10 2
γ [E2]	1240.59 11	~0.08
γ [M1+E2]	1243.83 15	~0.04
γ	1261.40 9	0.33 2
γ	1297.5 3	~0.02
γ [M1+E2]	1324.10 14	<0.02
γ	1339.74 7	0.25 2
γ	1351.77 8	0.86 6
γ	1355.02 9	0.79 5
γ [M1+E2]	1400.95 14	0.030 6
γ [M1+E2]	1410.69 6	2.51 16
γ	1418.69 21	0.033 6
γ	1433.39 6	0.65 4
γ [M1+E2]	1448.99 9	0.13 1
γ	1459.42 6	2.12 13
γ [M1+E2]	1541.63 10	0.28 2
γ	1562.44 7	0.16 1
γ [M1+E2]	1625.15 14	0.09 1
γ	1642.24 8	0.10 1
γ	1691.48 9	0.09 1
γ	1697.70 12	0.12 1
γ [M1+E2]	1702.00 14	0.03 1
γ E2	1747.16 7	0.04 1
γ	1755.7 3	<0.02
γ [M1+E2]	1785.46 9	0.60 5
γ [M1+E2]	1832.09 10	0.24 2
γ	1852.90 7	0.12 1
γ	1931.25 8	0.29 3
γ	1978.70 9	0.014 5
γ	1988.16 12	0.68 6
γ	2047.43 8	0.35 3
γ [M1+E2]	2168.55 11	0.046 6
γ	2189.36 8	0.041 6
γ [E2]	2252.07 14	0.36 3
γ	2267.71 8	0.09 1
γ [E1]	2290.00 15	0.024 4
γ	2324.62 12	1.42 14
γ	2382.6 3	0.010 2
γ	2409.47 21	0.010 2

† 1.4% uncert(syst)

Atomic Electrons (^{95}Ru)

$\langle e \rangle$=8.19 25 keV

e_{bin}(keV)	$\langle e \rangle$(keV)	e(%)
3	1.70	62 6
15	0.73	4.8 5
16	1.10	7.1 7
17	0.068	0.39 4
18	0.73	4.1 4
20 - 21	0.073	0.36 3
136 - 157	0.011	~0.008
234 - 280	0.29	0.107 20
287 - 314	0.054	0.018 3
315	2.34	0.742 18
327	0.006	~0.0018
333	0.271	0.0813 20
334 - 383	0.087	0.0260 8
400 - 447	0.0038	0.0009 3
456 - 505	0.0028	0.00058 25
531 - 579	0.0548	0.0100 4
581 - 629	0.302	0.0497 21
632 - 668	0.012	0.0018 7
686 - 735	0.0250	0.0035 3
746 - 787	0.031	0.0040 16
798 - 842	0.0253	0.00311 21
868 - 894	0.0102	0.00117 12
939 - 987	0.0077	0.00080 8
989 - 1030	0.018	0.0018 6
1044 - 1089	0.11	0.010 4
1094 - 1137	0.030	0.0027 6
1155 - 1200	0.0401	0.00346 16
1218 - 1261	0.0020	0.00016 5
1276 - 1324	0.0012	9 3 $\times10^{-5}$
1331 - 1380	0.0042	0.00032 10
1390 - 1438	0.026	0.00184 24
1446 - 1541	0.0032	0.00021 3
1559 - 1642	0.00086	5.3 9 $\times10^{-5}$
1670 - 1764	0.0055	0.00043 3
1782 - 1852	0.00086	4.7 7 $\times10^{-5}$
1910 - 1988	0.0032	0.00016 4
2026 - 2047	0.0061	0.00041 10
2148 - 2247	0.00169	7.6 6 $\times10^{-5}$
2249 - 2324	0.0043	0.00019 5
2362 - 2409	5.6 $\times10^{-5}$	2.3 5 $\times10^{-6}$

Continuous Radiation (^{95}Ru)

$\langle\beta+\rangle$=72 keV;$\langle$IB$\rangle$=0.98 keV

E_{bin}(keV)		$\langle\ \rangle$(keV)	(%)
0 - 10	β+	7.6 $\times10^{-6}$	9.2 $\times10^{-5}$
	IB	0.0033	
10 - 20	β+	0.000225	0.00137
	IB	0.018	0.104
20 - 40	β+	0.00465	0.0144
	IB	0.0105	0.041
40 - 100	β+	0.160	0.209
	IB	0.018	0.028
100 - 300	β+	5.4	2.51
	IB	0.065	0.035
300 - 600	β+	26.8	5.9
	IB	0.155	0.035
600 - 1300	β+	39.4	5.07
	IB	0.47	0.052
1300 - 2233	IB	0.24	0.0158
	$\Sigma\beta$+		13.7

$^{95}_{45}$Rh(5.02 *10* min)

Mode: ϵ

Δ: -78340 *150* keV

SpA: 3.94$\times10^{8}$ Ci/g

Prod: ^{96}Ru(p,2n)

Photons (^{95}Rh)

$\langle\gamma\rangle$=1731 29 keV

γ_{mode}	γ(keV)	γ(%)†
Ru L_ℓ	2.253	0.026 5
Ru L_η	2.382	0.0118 18
Ru L_α	2.558	0.66 10
Ru L_β	2.731	0.41 7
Ru L_γ	3.033	0.031 6
Ru $K_{\alpha2}$	19.150	5.76 21
Ru $K_{\alpha1}$	19.279	10.9 4
Ru $K_{\beta1}$'	21.650	2.64 10
Ru $K_{\beta2}$'	22.210	0.492 19
γ	229.05 20	2.24 14
γ	401.30 21	0.45 8
γ	410.26 15	1.05 10
γ	622.2 4	2.59 21
γ	661.19 22	1.51 11
γ	666.0 3	0.39 5 ?
γ [E2]	677.73 18	5.67 21
γ	764.9 4	2.1 4
γ	894.86 23	1.68 14
γ	906.78 19	0.60 6
γ	941.63 18	70
γ	1079.03 20	1.30 10
γ	1175.2 4	1.5 3
γ	1292.33 22	0.39 5
γ	1305.11 23	~2
γ	1317.04 18	2.94 21
γ	1326.4 3	0.77 7 ?
γ	1338.92 23	0.70 14
γ [E2]	1351.88 20	20.4 6
γ	1378.4 3	0.54 6
γ	1388.1 8	0.20 8 ?
γ	1489.28 20	3.36 21
γ	1494.6 3	4.9 3
γ	1524.4 5	2.3 3
γ	1549.8 3	0.42 7
γ	1588.2 12	0.10 4 ?
γ	1714.2 12	0.35 14 ?
γ	1716.8 12	0.56 14 ?
γ	1749.17 22	0.63 7
γ	1925.2 3	0.76 7
γ	2121.1 3	1.58 11
γ	2155.6 3	0.77 17 ?
γ	2221.4 8	0.91 14 ?
γ	2268.0 3	0.41 7 ?
γ	2334.6 8	0.32 7 ?
γ [M1+E2]	2427.1 3	0.34 7
γ	2609.35 22	0.66 7
γ	2695.6 4	0.27 6 ?
γ	2714.4 9	0.24 7 ?
γ	2791.7 3	2.38 14
γ	3033.8 12	0.39 8 ?
γ	3041.7 5	0.46 4 ?
γ	3062.7 3	0.98 10
γ	3163.0 8	0.21 4 ?
γ	3235.9 12	0.22 4 ?
γ	3473.8 15	~0.08
γ	3551.0 3	0.35 6
γ	3686.5 8	0.13 3 ?
γ	3733.3 3	0.35 6
γ [E2]	3779.0 4	0.63 11
γ	3983.3 5	0.48 11 ?
γ	4177.5 12	0.070 21 ?

† 2.9% uncert(syst)

Atomic Electrons (^{95}Rh)

$\langle e \rangle$=2.62 24 keV

e_{bin}(keV)	$\langle e \rangle$(keV)	e(%)
3	0.59	20.2 21
16	0.58	3.6 4
18	0.055	0.30 3
19	0.203	1.08 11
21 - 22	0.0242	0.114 10
207	0.16	~0.08
226 - 229	0.030	~0.013
379 - 410	0.037	0.010 4
600 - 644	0.048	0.0078 24

Atomic Electrons (^{95}Rh) (continued)

e_{bin}(keV)	⟨e⟩(keV)	e(%)
656	0.075	0.0114 *6*
658 - 678	0.0140	0.00208 *16*
743 - 765	0.019	0.0025 *11*
873 - 907	0.017	0.0019 *6*
920	0.43	0.047 *21*
938 - 942	0.063	0.0067 *24*
1057 - 1079	0.008	0.0007 *3*
1153 - 1175	0.009	0.0007 *3*
1270 - 1317	0.032	0.0025 *6*
1323 - 1326	0.00047	3.6 *12* $\times10^{-5}$
1330	0.111	0.0083 *4*
1336 - 1385	0.0195	0.00144 *9*
1387 - 1388	2.1 $\times10^{-5}$	1.5 *7* $\times10^{-6}$
1467 - 1566	0.049	0.0033 *7*
1585 - 1588	5.2 $\times10^{-5}$	3.3 *16* $\times10^{-6}$
1692 - 1749	0.0090	0.00053 *9*
1903 - 1925	0.0027	0.00014 *4*
2099 - 2155	0.0077	0.00036 *8*
2199 - 2268	0.0042	0.00019 *4*
2313 - 2405	0.0021	8.9 *17* $\times10^{-5}$
2424 - 2427	0.00015	6.3 *12* $\times10^{-6}$
2587 - 2673	0.0025	9.7 *21* $\times10^{-5}$
2692 - 2791	0.0072	0.00026 *6*
3012 - 3062	0.0046	0.000153 *24*
3141 - 3235	0.00107	3.4 *7* $\times10^{-5}$
3452 - 3550	0.00101	2.9 *6* $\times10^{-5}$
3664 - 3757	0.0026	7.0 *9* $\times10^{-5}$
3776 - 3779	0.00021	5.5 *8* $\times10^{-6}$
3961 - 3983	0.0011	2.7 *7* $\times10^{-5}$
4155 - 4177	0.00015	3.6 *12* $\times10^{-6}$

Continuous Radiation (^{95}Rh)

⟨β+⟩=899 keV;⟨IB⟩=4.0 keV

E_{bin}(keV)		⟨ ⟩(keV)	(%)
0 - 10	β+	6.2 $\times10^{-6}$	7.5 $\times10^{-5}$
	IB	0.031	
10 - 20	β+	0.000189	0.00115
	IB	0.034	0.23
20 - 40	β+	0.00406	0.0126
	IB	0.064	0.23
40 - 100	β+	0.148	0.193
	IB	0.17	0.27
100 - 300	β+	5.8	2.66
	IB	0.51	0.28
300 - 600	β+	40.3	8.7
	IB	0.62	0.145
600 - 1300	β+	270	28.4
	IB	1.12	0.126
1300 - 2500	β+	524	29.5
	IB	1.11	0.064
2500 - 4168	β+	58	2.17
	IB	0.34	0.0116
	Σβ+		72

$^{95}_{45}$Rh(2.0 *4* min)

Mode: IT(88 *5* %), ϵ(12 *5* %)

Δ: -77797 *150* keV

SpA: 1.01×10^{9} Ci/g

Prod: ^{96}Ru(p,2n)

Photons (^{95}Rh)

⟨γ⟩=791 *30* keV

γ_{mode}	γ(keV)	γ(%)†
Rh L_ℓ	2.377	0.0074 *14*
Rh L_η	2.519	0.0034 *6*
Rh L_α	2.696	0.20 *3*
Rh L_β	2.889	0.124 *22*
Rh L_γ	3.203	0.0110 *21*
Rh $K_{\alpha2}$	20.074	1.60 *11*
Rh $K_{\alpha1}$	20.216	3.04 *21*
Rh $K_{\beta1}$'	22.717	0.74 *5*
Rh $K_{\beta2}$'	23.312	0.139 *10*
γ_{IT} (M4)	543.3 *3*	80
γ_ϵ[E2]	787.7 *4*	7.8 *6*
γ_ϵ	2821 *1*	0.80 *8*
γ_ϵ	3186.2 *8*	0.88 *8*
γ_ϵ	3407.1 *5*	2.01 *16*
γ_ϵ	3757.4 *20*	0.80 *16*
γ_ϵ	3824.4 *7*	1.29 *24* ?
γ_ϵ	4207.8 *20*	0.57 *12*
γ_ϵ	4242 *2*	0.68 *12*
γ_ϵ	4336.5 *20*	0.96 *16*

† uncert(syst): 42% for ϵ, 5.7% for IT

Atomic Electrons (^{95}Rh)*

⟨e⟩=42.9 *23* keV

e_{bin}(keV)	⟨e⟩(keV)	e(%)
3	0.174	5.7 *7*
16	0.0156	0.095 *11*
17	0.140	0.82 *10*
19	0.0183	0.095 *11*
20	0.052	0.26 *3*
22	0.0051	0.023 *3*
23	0.00158	0.0070 *8*
520	35.4	6.8 *4*
540	5.7	1.06 *7*
543	1.32	0.243 *16*

* with IT

Continuous Radiation (^{95}Rh)

⟨β+⟩=141 keV;⟨IB⟩=0.62 keV

E_{bin}(keV)		⟨ ⟩(keV)	(%)
0 - 10	β+	7.4 $\times10^{-7}$	9.0 $\times10^{-6}$
	IB	0.0044	
10 - 20	β+	2.28 $\times10^{-5}$	0.000139
	IB	0.0048	0.033
20 - 40	β+	0.000487	0.00150
	IB	0.0091	0.032
40 - 100	β+	0.0176	0.0229
	IB	0.025	0.038
100 - 300	β+	0.66	0.303
	IB	0.075	0.042
300 - 600	β+	3.98	0.87
	IB	0.096	0.022
600 - 1300	β+	21.8	2.30
	IB	0.18	0.020
1300 - 2500	β+	73	3.9
	IB	0.157	0.0091
2500 - 4866	β+	40.9	1.42
	IB	0.067	0.0022
	Σβ+		8.8

$^{95}_{46}$Pd(13.3 *3* s)

Mode: ϵ, ϵp(1.05 *15* %)

Δ: ~-68150 keV syst

SpA: 8.70×10^{9} Ci/g

Prod: ^{40}Ca on ^{58}Ni

Photons (^{95}Pd)

γ_{mode}	γ(keV)	γ(rel)
γ	168.8 *1*	12.4 *4*
γ	185.0 *1*	6.2 *2*
γ	347.6 *1*	1.10 *4*
γ	381.8 *1*	48.3 *14*
γ	524.0 *1*	10.6 *3*
γ	640.6 *3*	0.47 *11*
γ	680.3 *1*	8.1 *2*
γ	716.6 *1*	67.1 *13*
γ	731.9 *3*	0.59 *12*
γ	788.2 *1*	1.26 *13*
γ	794.7 *3*	0.32 *12*
γ	839.8 *3*	0.31 *2*
γ	859.9 *3*	0.44 *13*
γ	913.2 *1*	12.9 *4*
γ	935.7 *3*	0.52 *13*
γ	957.1 *2*	1.21 *15*
γ	1005.1 *2*	3.81 *20*
γ	1028.2 *2*	2.72 *16*
γ	1111.0 *2*	1.37 *18*
γ	1144.3 *3*	1.00 *18*
γ	1213.0 *5*	0.44 *16*
γ	1240.8 *5*	0.77 *16*
γ	1275.1 *3*	2.45 *18*
γ	1351.1 *1*	100 *3*
γ	1363.5 *5*	0.52 *15*
γ	1408.0 *5*	0.64 *15*
γ	1424.8 *3*	2.5 *2*
γ	1443.8 *5*	0.60 *15*
γ	1600.6 *3*	5.18 *22*
γ	1745.9 *5*	0.40 *15*
γ	1764.5 *5*	0.64 *15*
γ	1770.5 *5*	0.88 *16*
γ	1793.2 *5*	1.22 *16*
γ	1898.8 *5*	2.5 *2*
γ	1927.0 *5*	0.40 *14*
γ	1962.7 *5*	0.45 *14*

A = 96

NDS **35**, 281 (1982)

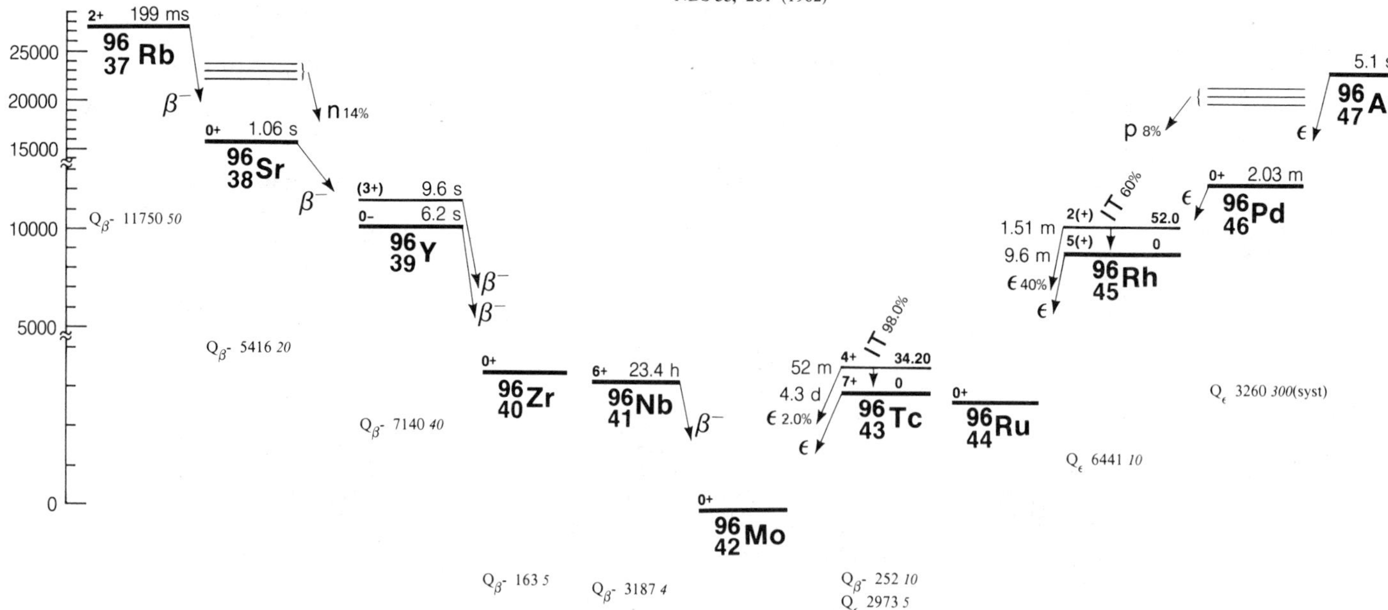

$^{96}_{37}$Rb(199 *3* ms)

Mode: β-, β-n(14.2 *14* %)
Δ: -61140 *30* keV
SpA: 1.643×10^{11} Ci/g
Prod: fission

β-n(avg): 445

Photons (^{96}Rb)

$\langle\gamma\rangle$=2004 *29* keV

γ_{mode}	γ(keV)	γ(%)†
$\gamma_{\beta\text{-n}}$ [E2]	203.78 *24*	2.3
$\gamma_{\beta\text{-}}$	320.59 *12*	0.190 *23*
$\gamma_{\beta\text{-n}}$	328.55 *24*	0.8
$\gamma_{\beta\text{-n}}$	331.6 *10*	0.16 *3*
$\gamma_{\beta\text{-}}$	345.31 *9*	0.16 *3*
$\gamma_{\beta\text{-}}$	347.49 *11*	0.28 *4*
$\gamma_{\beta\text{-n}}$ [M1]	351.85 *24*	8
$\gamma_{\beta\text{-}}$	366.75 *13*	0.084 *15*
$\gamma_{\beta\text{-}}$	374.43 *20*	0.144 *23*
$\gamma_{\beta\text{-}}$[E2]	398.88 *8*	0.36 *3*
$\gamma_{\beta\text{-}}$E2	414.31 *8*	2.74 *23*
$\gamma_{\beta\text{-n}}$	427.3 *10*	0.016 *3*
$\gamma_{\beta\text{-n}}$	435.5 *10*	0.041 *8*
$\gamma_{\beta\text{-}}$	455.38 *9*	0.43 *4*
$\gamma_{\beta\text{-}}$	468.82 *9*	1.19 *11*
$\gamma_{\beta\text{-}}$	485.17 *12*	0.68 *8*
$\gamma_{\beta\text{-}}$	522.63 *20*	0.190 *23*
$\gamma_{\beta\text{-}}$	555.22 *17*	0.64 *5*
$\gamma_{\beta\text{-n}}$	564.88 *24*	0.081 *16*
$\gamma_{\beta\text{-}}$	577.10 *13*	0.167 *23*
$\gamma_{\beta\text{-n}}$	578.3 *3*	0.16 *3*
$\gamma_{\beta\text{-n}}$	604.7 *10*	0.024 *5*
$\gamma_{\beta\text{-}}$	606.50 *11*	1.12 *11*
$\gamma_{\beta\text{-n}}$	622.3 *10*	0.016 *3*
$\gamma_{\beta\text{-}}$	643.94 *9*	0.99 *8*
$\gamma_{\beta\text{-}}$E2	649.6 *5*	0.61 *15*
$\gamma_{\beta\text{-n}}$	651.6 *10*	0.16 *3*
$\gamma_{\beta\text{-n}}$	659.9 *3*	0.41 *8*
$\gamma_{\beta\text{-}}$	673.29 *18*	0.15 *4*
$\gamma_{\beta\text{-}}$	677.10 *15*	0.19 *4*

Photons (^{96}Rb) (continued)

γ_{mode}	γ(keV)	γ(%)†
$\gamma_{\beta\text{-n}}$	680.40 *24*	1.4
$\gamma_{\beta\text{-}}$(E2+32%M1)	691.85 *7*	7.8 *5*
$\gamma_{\beta\text{-n}}$	703.5 *10*	0.024 *5*
$\gamma_{\beta\text{-}}$	732.70 *17*	0.32 *3*
$\gamma_{\beta\text{-}}$	765.63 *13*	0.160 *23*
$\gamma_{\beta\text{-n}}$	768.66 *24*	0.24 *5*
$\gamma_{\beta\text{-}}$(M1+27%E2)	813.18 *8*	6.8 *8*
$\gamma_{\beta\text{-}}$E2	814.92 *8*	76
$\gamma_{\beta\text{-}}$	854.64 *13*	0.55 *4*
$\gamma_{\beta\text{-}}$	867.50 *15*	0.205 *23*
$\gamma_{\beta\text{-n}}$	907.6 *10*	0.049 *10*
$\gamma_{\beta\text{-}}$	936.73 *10*	0.61 *8*
$\gamma_{\beta\text{-}}$	968.03 *20*	0.167 *23*
$\gamma_{\beta\text{-}}$	977.76 *9*	4.9 *4*
$\gamma_{\beta\text{-}}$	987.92 *15*	0.175 *23*
$\gamma_{\beta\text{-n}}$	1003.7 *10*	0.57 *11*
$\gamma_{\beta\text{-}}$	1027.8 *3*	0.198 *23*
$\gamma_{\beta\text{-}}$	1037.16 *7*	6.4 *5*
$\gamma_{\beta\text{-n}}$	1062.8 *10*	0.016 *3*
$\gamma_{\beta\text{-}}$	1075.58 *16*	0.220 *23*
$\gamma_{\beta\text{-n}}$	1087.3 *10*	0.024 *5*
$\gamma_{\beta\text{-}}$	1130.9 *3*	0.114 *15*
$\gamma_{\beta\text{-}}$	1160.67 *10*	0.84 *8*
$\gamma_{\beta\text{-n}}$	1163.2 *10*	0.016 *3*
$\gamma_{\beta\text{-}}$(M1+21%E2)	1179.94 *12*	3.19 *23*
$\gamma_{\beta\text{-}}$	1196.91 *15*	0.30 *3*
$\gamma_{\beta\text{-}}$	1212.60 *15*	0.40 *4*
$\gamma_{\beta\text{-}}$	1220.1 *3*	0.160 *23*
$\gamma_{\beta\text{-}}$	1252.7 *3*	0.175 *23*
$\gamma_{\beta\text{-n}}$	1259.7 *10*	0.24 *5*
$\gamma_{\beta\text{-}}$	1268.95 *12*	0.29 *3*
$\gamma_{\beta\text{-}}$	1298.35 *11*	1.24 *13*
$\gamma_{\beta\text{-}}$	1305.03 *20*	1.78 *19*
$\gamma_{\beta\text{-}}$	1335.78 *10*	2.46 *23*
$\gamma_{\beta\text{-}}$[M1+35%E2]	1402.23 *15*	2.10 *19*
$\gamma_{\beta\text{-}}$	1439.1 *3*	0.175 *23*
$\gamma_{\beta\text{-n}}$	1439.2 *10*	0.41 *8*
$\gamma_{\beta\text{-}}$	1454.53 *20*	0.205 *23* ?
$\gamma_{\beta\text{-}}$	1492.54 *10*	0.69 *7*
$\gamma_{\beta\text{-}}$[E2]	1506.76 *9*	4.3 *4*
$\gamma_{\beta\text{-}}$	1592.37 *17*	0.45 *5*
$\gamma_{\beta\text{-}}$	1596.98 *19*	0.129 *23*
$\gamma_{\beta\text{-}}$[E2]	1628.09 *10*	0.84 *8*
$\gamma_{\beta\text{-}}$	1650.0 *3*	0.26 *3*
$\gamma_{\beta\text{-}}$	1678.03 *20*	0.84 *8*
$\gamma_{\beta\text{-}}$	1714.25 *15*	0.76 *8*

Photons (^{96}Rb) (continued)

γ_{mode}	γ(keV)	γ(%)†
$\gamma_{\beta\text{-}}$	1756.4 *3*	0.32 *3*
$\gamma_{\beta\text{-}}$	1761.23 *20*	1.98 *23*
$\gamma_{\beta\text{-}}$	1770.7 *3*	0.175 *23*
$\gamma_{\beta\text{-}}$	1888.76 *15*	0.76 *8*
$\gamma_{\beta\text{-}}$	1904.65 *15*	0.48 *5*
$\gamma_{\beta\text{-}}$	1964.3 *4*	0.167 *23*
$\gamma_{\beta\text{-}}$	1994.84 *14*	0.33 *4*
$\gamma_{\beta\text{-}}$	1999.8 *3*	0.205 *23*
$\gamma_{\beta\text{-}}$	2034.8 *7*	0.068 *23*
$\gamma_{\beta\text{-}}$	2065.4 *3*	1.19 *12*
$\gamma_{\beta\text{-}}$	2083.85 *13*	1.07 *11*
$\gamma_{\beta\text{-}}$	2146.6 *5*	0.44 *5*
$\gamma_{\beta\text{-}}$	2196.4 *4*	0.205 *23* ?
$\gamma_{\beta\text{-}}$	2249.75 *15*	1.09 *11*
$\gamma_{\beta\text{-}}$	2307.45 *12*	1.34 *14*
$\gamma_{\beta\text{-}}$	2323.8 *3*	0.144 *23*
$\gamma_{\beta\text{-}}$	2380.73 *20*	1.09 *11*
$\gamma_{\beta\text{-}}$	2411.89 *18*	1.18 *12*
$\gamma_{\beta\text{-}}$	2429.8 *3*	0.47 *5* ?
$\gamma_{\beta\text{-}}$	2476.5 *3*	0.55 *6*
$\gamma_{\beta\text{-}}$	2493.5 *4*	0.35 *5*
$\gamma_{\beta\text{-}}$	2511.93 *20*	0.34 *5*
$\gamma_{\beta\text{-}}$	2541.3 *5*	0.20 *3*
$\gamma_{\beta\text{-}}$	2631.2 *4*	1.37 *15* ?
$\gamma_{\beta\text{-}}$	2751.33 *20*	0.84 *15*
$\gamma_{\beta\text{-}}$	2815.5 *4*	0.40 *5*
$\gamma_{\beta\text{-}}$	2940.2 *3*	0.49 *5*
$\gamma_{\beta\text{-}}$	3021.7 *7*	0.19 *3*
$\gamma_{\beta\text{-}}$	3047.1 *4*	0.30 *5*
$\gamma_{\beta\text{-}}$	3050.2 *4*	0.39 *5*
$\gamma_{\beta\text{-}}$	3365.9 *4*	0.55 *6*
$\gamma_{\beta\text{-}}$	3375.2 *4*	0.30 *5*
$\gamma_{\beta\text{-}}$	3507.5 *5*	0.26 *4* ?
$\gamma_{\beta\text{-}}$	3513.6 *5*	0.24 *4*
$\gamma_{\beta\text{-}}$	3527.0 *7*	0.32 *5*
$\gamma_{\beta\text{-}}$	3755.0 *3*	0.73 *11*
$\gamma_{\beta\text{-}}$	3842.3 *10*	0.144 *23* ?
$\gamma_{\beta\text{-}}$	3903.3 *5*	0.26 *4* ?
$\gamma_{\beta\text{-}}$	3906.8 *5*	0.23 *4*
$\gamma_{\beta\text{-}}$	3933.6 *5*	0.23 *4*
$\gamma_{\beta\text{-}}$	3984.3 *4*	0.47 *6* ?
$\gamma_{\beta\text{-}}$	4105.5 *7*	0.21 *5*
$\gamma_{\beta\text{-}}$	4227.9 *10*	0.33 *8*
$\gamma_{\beta\text{-}}$	4234.3 *10*	0.44 *8* ?
$\gamma_{\beta\text{-}}$	4275.8 *5*	0.30 *5* ?
$\gamma_{\beta\text{-}}$	4343.6 *4*	0.67 *14*

Photons (^{96}Rb)
(continued)

γ_{mode}	γ(keV)	γ(%)†
$\gamma_{\beta-}$	4352.9 *4*	0.45 *9*
$\gamma_{\beta-}$	4446.7 *13*	0.38 *11*
$\gamma_{\beta-}$	4604.7 *13*	0.21 *9*
$\gamma_{\beta-}$	5020.2 *20*	0.10 *5*
$\gamma_{\beta-}$	5167.8 *4*	0.13 *5*
$\gamma_{\beta-}$	5232.6 *10*	~0.12
$\gamma_{\beta-}$	5357.5 *10*	<0.30
$\gamma_{\beta-}$	5419.9 *15*	~0.15

† uncert(syst): 2.6% for β-

Continuous Radiation (^{96}Rb)
$\langle\beta-\rangle$=4083 keV;$\langle$IB$\rangle$=21 keV

E_{bin}(keV)		$\langle\ \rangle$(keV)	(%)
0 - 10	β-	0.000449	0.0089
	IB	0.081	
10 - 20	β-	0.00138	0.0092
	IB	0.081	0.56
20 - 40	β-	0.0058	0.0193
	IB	0.161	0.56
40 - 100	β-	0.0473	0.067
	IB	0.48	0.73
100 - 300	β-	0.69	0.332
	IB	1.51	0.84
300 - 600	β-	4.02	0.87
	IB	2.1	0.48
600 - 1300	β-	40.1	4.04
	IB	4.1	0.46
1300 - 2500	β-	264	13.5
	IB	5.2	0.29
2500 - 5000	β-	1512	40.2
	IB	5.7	0.167
5000 - 7500	β-	1697	27.9
	IB	1.57	0.027
7500 - 10935	β-	565	6.8
	IB	0.136	0.00168

$^{96}_{38}$Sr(1.06 *3* s)

Mode: β-
Δ: -72890 *40* keV
SpA: 8.14×10^{10} Ci/g
Prod: fission

Photons (^{96}Sr)
$\langle\gamma\rangle$=925 *21* keV

γ_{mode}	γ(keV)	γ(%)†
Y L_ℓ	1.686	0.0058 *13*
Y L_η	1.762	0.0032 *8*
Y L_α	1.922	0.13 *3*
Y L_β	2.005	0.077 *20*
Y L_γ	2.267	0.0023 *7*
Y $K_{\alpha2}$	14.883	1.51 *7*
Y $K_{\alpha1}$	14.958	2.90 *13*
Y $K_{\beta1}$'	16.735	0.65 *3*
Y $K_{\beta2}$'	17.125	0.096 *4*
γ M1	122.319 *3*	76.5
γ (M1+E2)	213.01 *7*	0.76 *15*
γ E1+0.3%M2	279.42 *6*	8.3 *4*
γ	356.20 *17*	0.54 *8*
γ M1+1.2%E2	530.00 *6*	9.0 *4*
γ	596.41 *7*	0.92 *15*
γ [E2]	652.32 *6*	0.46 *15*
γ	695.68 *20*	0.34 *4*
γ E1	809.42 *3*	71.9 *23*
γ [E1]	931.74 *3*	11.8 *8*

Photons (^{96}Sr)
(continued)

γ_{mode}	γ(keV)	γ(%)†
γ [M1+E2]	1051.87 *18*	0.38 *15*
γ	1165.62 *17*	~0.08
γ [E1]	1331.29 *18*	0.61 *15*
γ [E1]	1861.29 *18*	~0.15
γ [E1]	1983.60 *18*	1.91 *23*

† 1.6% uncert(syst)

Atomic Electrons (^{96}Sr)
$\langle$e$\rangle$=9.49 *20* keV

e_{bin}(keV)	$\langle$e$\rangle$(keV)	e(%)
2 - 17	0.41	8.1 *6*
105	7.48	7.10 *18*
120	0.983	0.819 *21*
196 - 213	0.07	0.035 *16*
262 - 279	0.124	0.0471 *22*
339 - 356	0.013	~0.0039
513 - 530	0.119	0.0230 *10*
579 - 596	0.008	0.0015 *7*
635 - 679	0.0076	0.0012 *3*
693 - 696	0.00030	4.3 *16* ×10^{-5}
792	0.207	0.0262 *10*
807 - 809	0.0275	0.00341 *11*
915 - 932	0.0334	0.00365 *25*
1035 - 1052	0.0022	0.00021 *8*
1149 - 1166	0.00029	~3×10^{-5}
1314 - 1331	0.0013	9.8 *22* ×10^{-5}
1844 - 1861	0.00026	1.4 *6* ×10^{-5}
1967 - 1983	0.0032	0.000160 *18*

Continuous Radiation (^{96}Sr)
$\langle\beta-\rangle$=1980 keV;$\langle$IB$\rangle$=6.8 keV

E_{bin}(keV)		$\langle\ \rangle$(keV)	(%)
0 - 10	β-	0.00271	0.054
	IB	0.058	
10 - 20	β-	0.0083	0.055
	IB	0.057	0.40
20 - 40	β-	0.0349	0.116
	IB	0.113	0.39
40 - 100	β-	0.280	0.395
	IB	0.33	0.50
100 - 300	β-	3.96	1.90
	IB	0.99	0.55
300 - 600	β-	21.6	4.68
	IB	1.24	0.29
600 - 1300	β-	185	18.8
	IB	2.0	0.23
1300 - 2500	β-	828	43.6
	IB	1.61	0.094
2500 - 5000	β-	941	30.4
	IB	0.35	0.0120
5000 - 5416	β-	0.51	0.0100
	IB	8.6×10^{-6}	1.70×10^{-7}

$^{96}_{39}$Y (6.2 *2* s)

Mode: β-
Δ: >-78300 keV
SpA: 1.79×10^{10} Ci/g
Prod: fission

Photons (^{96}Y)*
$\langle\gamma\rangle$=0.98 *14* keV

γ_{mode}	γ(keV)	γ(%)
Zr L_ℓ	1.792	0.0072 *21*
Zr L_η	1.876	0.0038 *10*
Zr L_α	2.042	0.17 *5*
Zr L_β	2.136	0.10 *3*
Zr L_γ	2.388	0.0047 *15*
Zr $K_{\alpha2}$	15.691	1.8 *4*
Zr $K_{\alpha1}$	15.775	3.4 *8*
Zr $K_{\beta1}$'	17.663	0.78 *17*
Zr $K_{\beta2}$'	18.086	0.13 *3*

* from E0 transition

Atomic Electrons (^{96}Y)
$\langle$e$\rangle$=149 *30* keV

e_{bin}(keV)	$\langle$e$\rangle$(keV)	e(%)
2	0.14	6.4 *16*
3	0.011	0.44 *12*
13	0.17	1.3 *3*
14	0.046	0.34 *8*
15	0.074	0.48 *12*
16	0.014	0.088 *22*
17	0.0061	0.035 *9*
18	0.0016	0.0094 *23*
1576	130.7	8.3 *19*
1591	18.0	1.1 *3*
1592	0.19	0.012 *4*

$^{96}_{39}$Y (9.6 *2* s)

Mode: β-
Δ: -78300 *40* keV
SpA: 1.181×10^{10} Ci/g
Prod: fission; ^{96}Zr(n,p)

Photons (^{96}Y)
$\langle\gamma\rangle$=3975 *42* keV

γ_{mode}	γ(keV)	γ(%)†
Zr $K_{\alpha2}$	15.691	0.38 *3*
Zr $K_{\alpha1}$	15.775	0.73 *6*
Zr $K_{\beta1}$'	17.663	0.167 *13*
Zr $K_{\beta2}$'	18.086	0.0270 *22*
γ [E1]	146.622 *10*	35.1 *15*
γ	174.48 *16*	2.0 *3* ?
γ	226.88 *16*	1.9 *4* ?
γ [M1+E2]	288.94 *11*	1.0 *3*
γ	328.82 *8*	0.44 *18*
γ [E2]	363.58 *9*	21.4 *10*
γ	475.44 *8*	3.3 *3*
γ [E2]	617.40 *9*	53 *3*
γ	631.43 *4*	8.4 *9*
γ [E2]	652.52 *11*	1.5 *4*
γ	690.29 *14*	1.0 *3*
γ [E2]	906.34 *12*	17.8 *10*
γ [E1]	914.95 *13*	57.3 *14*
γ [M1+E2]	960.24 *7*	3.6 *6*
γ	979.17 *20*	3.6 *3*
γ [E1]	1106.86 *7*	47.3 *16*
γ [E2]	1184.90 *14*	3.4 *5*
γ [E1]	1222.67 *9*	26.0 *9*
γ [E2]	1750.44 *15*	88.9
γ	1897.05 *15*	5.3 *5*
γ	2225.87 *17*	5.8 *8*

† 1.0% uncert(syst)

Atomic Electrons (^{96}Y)

⟨e⟩=4.92 22 keV

e_{bin}(keV)	⟨e⟩(keV)	e(%)
2 - 18	0.101	1.97 18
129	1.49	1.15 7
144	0.187	0.130 8
146 - 147	0.0389	0.0266 14
156	0.22	~0.14
172 - 209	0.16	~0.08
224 - 271	0.062	0.024 8
286 - 329	0.020	0.007 3
346	0.74	0.214 13
361 - 364	0.113	0.0314 16
457 - 475	0.048	0.010 5
599	0.64	0.106 7
613 - 653	0.18	0.030 5
672 - 690	0.008	0.0011 5
888	0.119	0.0133 9
897	0.157	0.0175 8
904 - 942	0.059	0.0064 5
958 - 979	0.020	0.0021 7
1089 - 1107	0.124	0.0113 5
1167 - 1205	0.074	0.0062 3
1220 - 1223	0.0073	0.00060 3
1732	0.286	0.0165 7
1748 - 1750	0.0371	0.00212 8
1879 - 1897	0.014	0.00075 22
2208 - 2226	0.014	0.00062 18

$^{96}_{40}$Zr(>4×10^{17} yr)

Δ: -85442 3 keV
%: 2.80 1

$^{96}_{41}$Nb(23.35 5 h)

Mode: β-
Δ: -85605 4 keV
SpA: 1.398×10^{6} Ci/g
Prod: ^{96}Zr(p,n); ^{98}Mo(d,α); ^{96}Zr(d,2n); ^{97}Mo(γ,p)

Photons (^{96}Nb)

⟨γ⟩=2462 22 keV

γ_{mode}	γ(keV)	γ(%)†
Mo K$_{\alpha2}$	17.374	0.206 20
Mo K$_{\alpha1}$	17.479	0.39 4
Mo K$_{\beta1}$'	19.602	0.093 9
Mo K$_{\beta2}$'	20.091	0.0161 16
γ [M1+E2]	108.85 3	~0.018
γ (M1)	120.44 6	0.025 6
γ [M1+E2]	128.10 6	~0.016
γ M1,E2	219.08 5	3.76 19
γ M1,E2	241.43 3	3.96 19
γ M1+11%E2	314.29 4	0.067 9
γ M1+<1.7%E2	316.54 4	0.036 6
γ (M1,E2)	349.78 5	0.67 13
γ (M1,E2)	350.28 3	1.08 6
γ M1,E2	352.57 6	0.87 5
γ E2	371.82 3	3.4 5
γ	409.8 4	~0.039
γ	413.1 4	~0.029
γ M1+8.9%E2	434.73 4	0.48 4
γ E2	460.01 3	28.5 8
γ M1,E2	480.67 3	6.20 21
γ (E2)	535.62 5	0.0111 17
γ (M1+E2)	568.86 3	56.8 15
γ (M1+E2)	591.21 5	1.06 10
γ [E2]	593.50 7	0.48 19
γ M1+16%E2	719.50 3	7.3 3
γ (E2)	721.60 5	0.73 7
γ E2	778.196 25	96.88

Photons (^{96}Nb)
(continued)

γ_{mode}	γ(keV)	γ(%)†
γ M1,E2	810.29 3	10.0 5
γ E2	812.54 3	2.9 3
γ (E2+48%M1)	847.60 6	1.32 14
γ E2	849.89 3	20.6 8
γ [E2]	885.40 4	0.0027 11
γ E2	1091.316 24	48.5 15
γ E2	1126.83 4	0.41 5
γ M1+14%E2	1200.165 25	19.8 6
γ [M1+E2]	1346.9 3	0.024 10
γ	1402.9 3	0.033 9
γ [E2]	1441.09 4	0.39 3
γ [E2]	1497.69 3	2.90 14
γ [E2]	1625.79 6	0.100 14

† 0.20% uncert(syst)

Atomic Electrons (^{96}Nb)

⟨e⟩=4.93 19 keV

e_{bin}(keV)	⟨e⟩(keV)	e(%)
3 - 20	0.053	0.99 11
89 - 128	0.019	0.019 8
199	0.33	0.17 7
216 - 219	0.058	0.027 9
221	0.28	0.13 5
239 - 241	0.048	0.020 7
294 - 333	0.093	0.028 5
347 - 350	0.0131	0.0038 7
352	0.124	0.035 5
353 - 393	0.021	0.0057 7
407 - 435	0.0116	0.00278 20
440	0.670	0.152 5
457 - 481	0.242	0.052 3
516 - 536	0.00022	4.3 19 ×10^{-5}
549	0.85	0.154 13
566 - 593	0.148	0.0261 22
699 - 722	0.095	0.0136 5
758	0.917	0.1209 24
775 - 813	0.263	0.0336 10
828	0.0112	0.00135 15
830	0.172	0.0207 9
845 - 885	0.0261	0.00307 11
1071	0.292	0.0272 10
1088 - 1127	0.0433	0.00397 11
1180 - 1200	0.129	0.0109 3
1327 - 1347	0.00014	1.0 4 ×10^{-5}
1383 - 1421	0.00189	0.000133 11
1438 - 1478	0.0127	0.00086 5
1495 - 1498	0.00170	0.000114 6
1606 - 1625	0.00045	2.8 4 ×10^{-5}

Continuous Radiation (^{96}Nb)

⟨β-⟩=249 keV;⟨IB⟩=0.20 keV

E_{bin}(keV)		⟨ ⟩(keV)	(%)
0 - 10	β-	0.106	2.12
	IB	0.0123	
10 - 20	β-	0.320	2.13
	IB	0.0116	0.081
20 - 40	β-	1.29	4.30
	IB	0.021	0.074
40 - 100	β-	9.2	13.1
	IB	0.050	0.079
100 - 300	β-	83	42.1
	IB	0.081	0.050
300 - 600	β-	142	34.2
	IB	0.020	0.0055
600 - 749	β-	12.8	2.0
	IB	0.00020	3.2×10^{-5}

$^{96}_{42}$Mo(stable)

Δ: -88792.1 20 keV
%: 16.68 4

$^{96}_{43}$Tc(4.28 7 d)

Mode: ε
Δ: -85819 6 keV
SpA: 3.18×10^{5} Ci/g
Prod: ^{93}Nb(α,n); ^{96}Mo(p,n); ^{96}Mo(d,2n); ^{95}Mo(d,n); ^{95}Mo(p,γ)

Photons (^{96}Tc)

⟨γ⟩=2506 45 keV

γ_{mode}	γ(keV)	γ(%)†
Mo L$_{\ell}$	2.016	0.083 15
Mo L$_{\eta}$	2.120	0.039 6
Mo L$_{\alpha}$	2.293	2.1 3
Mo L$_{\beta}$	2.426	1.18 20
Mo L$_{\gamma}$	2.701	0.079 15
Mo K$_{\alpha2}$	17.374	19.4 7
Mo K$_{\alpha1}$	17.479	36.8 13
Mo K$_{\beta1}$'	19.602	8.7 3
Mo K$_{\beta2}$'	20.091	1.51 6
γ [M1+E2]	108.85 3	~0.00034
γ (M1)	120.44 6	0.039 8
γ [M1+E2]	128.10 6	
γ M1,E2	219.08 5	0.059 4
γ M1,E2	241.43 3	0.090 7
γ M1+11%E2	314.29 4	2.49 23
γ M1+<1.7%E2	316.54 4	1.35 18
γ (M1,E2)	349.78 5	0.088 19
γ (M1,E2)	350.28 3	0.0196 16
γ M1,E2	352.57 6	0.0158 12
γ E2	371.82 3	0.077 12
γ M1+8.9%E2	434.73 4	0.75 5
γ E2	460.01 3	0.444 23
γ M1,E2	480.67 3	0.113 7
γ (E2)	535.62 5	0.41 4
γ (M1+E2)	568.86 3	0.89 4
γ (M1+E2)	591.21 5	0.14 3
γ [E2]	593.50 7	0.06 3
γ M1+16%E2	719.50 3	0.229 20
γ (E2)	721.60 5	0.095 19
γ E2	778.196 25	99.76
γ M1,E2	810.29 3	0.16 4
γ E2	812.54 3	82 4
γ (E2+48%M1)	847.60 6	0.019 4
γ E2	849.89 3	98 4
γ [E2]	885.40 4	0.10 4
γ E2	1091.316 24	1.10 7
γ E2	1126.83 4	15.2 11
γ M1+14%E2	1200.165 25	0.361 24
γ [E2]	1441.09 4	0.052 4
γ [E2]	1497.69 3	0.091 4
γ [E2]	1625.79 6	0.0010 2

†<0.1% uncert(syst)

Atomic Electrons (^{96}Tc)

⟨e⟩=8.2 *3* keV

e_{bin}(keV)	⟨e⟩(keV)	e(%)
3	1.84	72 *7*
14	0.206	1.44 *15*
15	1.93	13.1 *13*
17	0.91	5.3 *6*
19 - 20	0.083	0.43 *4*
89 - 120	0.0063	0.0062 *12*
199 - 241	0.014	0.0064 *15*
294 - 333	0.149	0.050 *4*
347 - 372	0.0039	0.00111 *14*
415 - 461	0.0319	0.0074 *3*
478 - 516	0.0075	0.00146 *15*
533 - 573	0.0191	0.0034 *3*
588 - 593	0.00042	7.1 *14* $\times 10^{-5}$
699 - 722	0.0039	0.00055 *4*
758	0.944	0.1245 *25*
775 - 790	0.1366	0.0176 *3*
793	0.73	0.092 *4*
807 - 813	0.104	0.0128 *5*
830	0.81	0.098 *4*
847 - 885	0.117	0.0138 *5*
1071 - 1107	0.095	0.0086 *7*
1124 - 1127	0.0122	0.00108 *8*
1180 - 1200	0.00234	0.000198 *12*
1421 - 1441	0.000261	1.84 *13* $\times 10^{-5}$
1478 - 1498	0.000445	3.01 *16* $\times 10^{-5}$

Continuous Radiation (^{96}Tc)

⟨IB⟩=0.044 keV

E_{bin}(keV)		⟨ ⟩(keV)	(%)
10 - 20	IB	0.018	0.102
20 - 40	IB	0.0020	0.0083
40 - 100	IB	0.0018	0.0027
100 - 300	IB	0.0139	0.0071
300 - 535	IB	0.0082	0.0023

$^{96}_{43}$Tc(51.5 *10* min)

Mode: IT(98.0 *5* %), ε(2.0 *5* %)

Δ: -85785 *6* keV

SpA: 3.80×10^7 Ci/g

Prod: ^{93}Nb(α,n); ^{96}Mo(p,n); ^{96}Mo(d,2n); ^{95}Mo(d,n); ^{95}Mo(p,γ)

Photons (^{96}Tc)

⟨γ⟩=48 *4* keV

γ_{mode}	γ(keV)	γ(%)†
Tc L_ℓ	2.133	0.087 *15*
Tc L_η	2.249	0.024 *4*
Tc L_α	2.424	2.2 *4*
Tc L_β	2.594	0.93 *16*
Tc L_γ	2.893	0.072 *15*
Mo $K_{\alpha2}$	17.374	0.423 *15*
Mo $K_{\alpha1}$	17.479	0.80 *3*
Tc $K_{\alpha2}$	18.251	9.5 *3*
Tc $K_{\alpha1}$	18.367	18.1 *6*
Mo $K_{\beta1}$'	19.602	0.190 *7*
Mo $K_{\beta2}$'	20.091	0.0331 *12*
Tc $K_{\beta1}$'	20.613	4.33 *15*
Tc $K_{\beta2}$'	21.136	0.78 *3*
γ_{IT} M3	34.20 *5*	0.02489
γ_ϵ[M1+E2]	108.85 *3*	~0.0010
γ_ϵ[M1+E2]	128.10 *6*	~0.0015

Photons (^{96}Tc) (continued)

γ_{mode}	γ(keV)	γ(%)†
γ_ϵM1,E2	219.08 *5*	0.00088 *11*
γ_ϵM1,E2	241.43 *3*	0.0053 *8*
γ_ϵ(M1,E2)	350.28 *3*	0.059 *3*
γ_ϵM1,E2	352.57 *6*	0.048 *3*
γ_ϵE2	371.82 *3*	0.0045 *9*
γ_ϵ[M1+E2]	374.97 *6*	0.0060 *15*
γ_ϵE2	460.01 *3*	0.0067 *8*
γ_ϵM1,E2	480.67 *3*	0.340 *10*
γ_ϵ(M1+E2)	568.86 *3*	0.0133 *17*
γ_ϵ[M1+E2]	615.90 *5*	0.059 *4*
γ_ϵM1+16%E2	719.50 *3*	0.294 *13*
γ_ϵE2	778.196 *25*	1.9
γ_ϵM1,E2	810.29 *3*	0.0024 *4*
γ_ϵ(E2+48%M1)	847.60 *6*	0.121 *9*
γ_ϵE2	849.89 *3*	0.283 *15*
γ_ϵ	852.7 *5*	0.0047 *19*
γ_ϵ[M1+E2]	966.18 *6*	0.051 *8*
γ_ϵ[M1+E2]	968.46 *8*	0.066 *8*
γ_ϵE2	1091.316 *24*	0.065 *10*
γ_ϵ[M1+E2]	1096.57 *5*	0.072 *5*
γ_ϵ	1107.43 *14*	0.0028 *9*
γ_ϵ	1109.71 *15*	~0.0022
γ_ϵ	1124.0 *5*	~0.0011
γ_ϵ	1173.3 *5*	0.0019 *8*
γ_ϵM1+14%E2	1200.165 *25*	1.09 *4*
γ_ϵ	1237.81 *14*	0.0051 *8*
γ_ϵ(E2+6.5%M1)	1317.4 *4*	0.0017 *8*
γ_ϵ[E2]	1441.09 *4*	
γ_ϵ[E2]	1497.69 *3*	0.118 *6*
γ_ϵ[E2]	1625.79 *6*	0.0091 *10*
γ_ϵ	1702.6 *5*	0.0017 *6*
γ_ϵ[M1+E2]	1816.05 *5*	0.043 *4*
γ_ϵ	1846.2 *8*	0.0030 *6*
γ_ϵ	1957.30 *14*	0.0097 *9*

† uncert(syst): 25% for ε, 0.51% for IT

Atomic Electrons (^{96}Tc)

⟨e⟩=26.2 *3* keV

e_{bin}(keV)	⟨e⟩(keV)	e(%)
3	1.77	65 *7*
13	5.50	41.8 *9*
14 - 15	0.45	3.0 *3*
16	0.61	3.9 *4*
17 - 21	0.50	2.78 *23*
31	5.92	19.0 *4*
32	7.84	24.9 *5*
34	3.61	10.69 *22*
89 - 128	0.0010	0.0010 *5*
199 - 241	0.00053	0.00024 *7*
330 - 375	0.0046	0.00138 *18*
440 - 481	0.0078	0.00169 *18*
549 - 596	0.00100	0.000171 *14*
613 - 616	0.000112	1.83 *19* $\times 10^{-5}$
699 - 719	0.00350	0.000499 *21*
758 - 807	0.020	0.0027 *6*
808 - 853	0.00389	0.000468 *19*
946 - 968	0.00095	0.000100 *9*
1071 - 1110	0.00098	9.1 *7* $\times 10^{-5}$
1121 - 1170	9.4 $\times 10^{-6}$	8 *4* $\times 10^{-7}$
1171 - 1218	0.00708	0.000599 *21*
1235 - 1238	2.8 $\times 10^{-6}$	2.3 *8* $\times 10^{-7}$
1297 - 1317	9.4 $\times 10^{-6}$	7 *3* $\times 10^{-7}$
1478 - 1498	0.00057	3.86 *22* $\times 10^{-5}$
1606 - 1702	4.7 $\times 10^{-5}$	2.9 *3* $\times 10^{-6}$
1796 - 1846	0.000188	1.04 *11* $\times 10^{-5}$
1937 - 1957	2.9 $\times 10^{-5}$	1.5 *5* $\times 10^{-6}$

Continuous Radiation (^{96}Tc)

⟨IB⟩=0.0050 keV

E_{bin}(keV)		⟨ ⟩(keV)	(%)
10 - 20	IB	0.00036	0.0021
20 - 40	IB	4.2 $\times 10^{-5}$	0.00017
40 - 100	IB	4.5 $\times 10^{-5}$	6.6 $\times 10^{-5}$
100 - 300	IB	0.00061	0.00029
300 - 600	IB	0.0021	0.00048
600 - 1300	IB	0.0018	0.00024
1300 - 1379	IB	7.6 $\times 10^{-7}$	5.8 $\times 10^{-8}$

$^{96}_{44}$Ru(stable)

Δ: -86071 *8* keV

%: 5.52 *5*

$^{96}_{45}$Rh(9.6 *3* min)

Mode: ε

Δ: -79630 *13* keV

SpA: 2.04×10^8 Ci/g

Prod: ^{96}Ru(p,n)

Photons (^{96}Rh)

⟨γ⟩=3211 *46* keV

γ_{mode}	γ(keV)	γ(%)
Ru L_ℓ	2.253	0.023 *4*
Ru L_η	2.382	0.0106 *16*
Ru L_α	2.558	0.60 *9*
Ru L_β	2.731	0.37 *6*
Ru L_γ	3.032	0.028 *5*
Ru $K_{\alpha2}$	19.150	5.18 *18*
Ru $K_{\alpha1}$	19.279	9.8 *3*
Ru $K_{\beta1}$'	21.650	2.38 *8*
Ru $K_{\beta2}$'	22.210	0.443 *17*
γ	237.92 *17*	0.16 *3*
γ	415.06 *18*	0.50 *12*
γ	421.8 *2*	0.5 *1*
γ[E2]	429.97 *17*	2.3 *2*
γ	471.49 *15*	0.18 *4*
γ	586.59 *16*	1.4 *2*
γ[E2]	594.06 *17*	0.74 *10*
γE2	631.67 *12*	80.0 *25*
γ	644.18 *11*	4.8 *3*
γE2	685.38 *10*	98 *5*
γ	741.90 *14*	31.5 *10*
γ	766.79 *19*	0.21 *4*
γE2	800.81 *14*	3.30 *15*
γ	808.81 *17*	0.65 *19*
γE2	832.70 *11*	100
γ	889.82 *18*	0.38 *5*
γ	915.56 *17*	1.3 *1*
γ	944.16 *17*	2.4 *3*
γ	995.76 *15*	0.51 *6*
γ	1006.80 *14*	0.227 *22*
γ(E1)	1070.35 *20*	1.7 *3*
γE2(+12%M1)	1098.58 *15*	
γ	1156.95 *15*	0.44 *15*
γ[E2]	1162.32 *21*	0.40 *15*
γ	1188.5 *3*	0.44 *15*
γ	1228.05 *18*	8.7 *10*
γ[M1+E2]	1230.77 *19*	7.5 *10*
γ	1242.20 *22*	1.3 *3*
γ	1275.85 *16*	3.1 *2*
γ	1364.56 *22*	0.25 *5*
γ[M1+E2]	1394.87 *20*	0.25 *5*
γ	1478.28 *15*	0.15 *3*
γ	1556.8 *5*	1.95 *20*

Photons (^{96}Rh)
(continued)

γ_{mode}	γ(keV)	γ(%)
γ	1558.6 4	1.0 4
γ	1593.0 3	0.30 5
γ	1605.60 22	2.7 1
γ	1643.06 22	0.42 7
γ	1648.80 22	2.0 1
γ	1656.1 3	0.49 20
γ	1692.17 14	1.88 17
γ	1700.5 4	0.25 5
γ	1737.65 18	5.7 3
γ	1773.58 16	0.20 4
γ	1788.61 17	1.9 1
γ	1800.8 3	0.5 1
γ	1859.71 18	1.63 14
γ	1885.57 22	0.30 4
γ	1907.38 19	0.27 8
γ [M1+E2]	1963.12 20	1.18 10
γ	1991.1 5	0.3 1
γ	1996.22 20	0.93 5
γ	2058.11 24	0.19 5
γ	2120.7 5	0.20 5
γ	2163.66 16	0.85 15
γ	2196.8 5	0.27 5
γ	2224.6 5	0.25 5
γ	2252.5 5	0.20 4
γ	2361.3 3	0.25 5
γ	2401.6 5	0.20 3
γ	2458.95 19	0.39 7
γ	2526.0 5	0.25 5
γ	2535.5 5	0.17 3
γ	2539.6 5	0.44 7
γ	2799.99 24	0.10 3
γ	2962.0 5	0.16 3
γ	3073.4 4	0.21 5
γ	3220.5 5	0.15 5
γ	3261.5 5	0.20 5
γ	3431.64 25	0.20 5

Atomic Electrons (^{96}Rh)
⟨e⟩=6.08 17 keV

e_{bin}(keV)	⟨e⟩(keV)	e(%)
3	0.53	18.1 19
16	0.53	3.3 3
18	0.050	0.27 3
19	0.183	0.97 10
21 - 22	0.0218	0.103 9
216 - 238	0.012	~0.006
393 - 430	0.103	0.025 3
449 - 471	0.0034	0.0008 4
564 - 594	0.032	0.0056 15
610	1.18	0.194 7
622 - 644	0.234	0.037 4
663	1.27	0.192 11
682 - 685	0.191	0.0280 12
720	0.26	0.036 17
739 - 787	0.079	0.0104 22
798 - 809	0.0057	0.00071 5
811	0.971	0.120 3
829 - 868	0.145	0.0175 3
887 - 922	0.025	0.0027 9
941 - 985	0.0064	0.00067 18
993 - 1007	0.00062	6.2 18 $\times10^{-5}$
1048 - 1070	0.0060	0.00057 9
1135 - 1166	0.0076	0.00066 17
1185 - 1231	0.107	0.0088 16
1239 - 1276	0.017	0.0013 5
1342 - 1392	0.0027	0.00020 4
1394 - 1395	3.5 $\times10^{-5}$	2.5 5 $\times10^{-6}$
1456 - 1555	0.013	0.00084 23
1556 - 1656	0.024	0.0015 3
1670 - 1766	0.037	0.0022 5
1770 - 1863	0.0097	0.00053 11
1882 - 1974	0.0101	0.00052 6
1988 - 2058	0.0011	5.7 13 $\times10^{-5}$
2099 - 2196	0.0043	0.00020 4
2202 - 2252	0.0014	6.4 15 $\times10^{-5}$
2339 - 2437	0.0024	0.000099 19

Atomic Electrons (^{96}Rh)
(continued)

e_{bin}(keV)	⟨e⟩(keV)	e(%)
2456 - 2539	0.0026	0.000104 18
2778 - 2800	0.00027	10 4 $\times10^{-6}$
2940 - 2962	0.00042	1.4 4 $\times10^{-5}$
3051 - 3073	0.00053	1.7 5 $\times10^{-5}$
3198 - 3261	0.00086	2.7 6 $\times10^{-5}$
3410 - 3431	0.00047	1.4 4 $\times10^{-5}$

Continuous Radiation (^{96}Rh)
⟨β+⟩=838 keV;⟨IB⟩=3.9 keV

E_{bin}(keV)		⟨ ⟩(keV)	(%)
0 - 10	β+	7.1 $\times10^{-6}$	8.7 $\times10^{-5}$
	IB	0.030	
10 - 20	β+	0.000219	0.00134
	IB	0.033	0.22
20 - 40	β+	0.00472	0.0146
	IB	0.062	0.22
40 - 100	β+	0.174	0.227
	IB	0.167	0.26
100 - 300	β+	7.1	3.25
	IB	0.48	0.27
300 - 600	β+	50.4	10.9
	IB	0.57	0.133
600 - 1300	β+	329	34.7
	IB	1.08	0.120
1300 - 2500	β+	420	24.8
	IB	1.24	0.071
2500 - 4923	β+	31.7	1.15
	IB	0.26	0.0093
	Σβ+		75

$^{96}_{45}$Rh(1.51 2 min)

Mode: IT(60 5 %), ε(40 5 %)
Δ: -79578 13 keV
SpA: 1.292$\times10^{9}$ Ci/g
Prod: ^{96}Ru(p,n)

Photons (^{96}Rh)
⟨γ⟩=839 47 keV

γ_{mode}	γ(keV)	γ(%)†
Ru L_ℓ	2.253	0.0051 9
Rh L_ℓ	2.377	0.057 10
Ru L_η	2.382	0.0023 4
Rh L_η	2.519	0.019 3
Ru L_α	2.558	0.131 21
Rh L_α	2.696	1.49 24
Ru L_β	2.731	0.081 13
Rh L_β	2.901	0.79 14
Ru L_γ	3.032	0.0062 12
Rh L_γ	3.222	0.071 14
Ru $K_{\alpha2}$	19.150	1.14 4
Ru $K_{\alpha1}$	19.279	2.17 8
Rh $K_{\alpha2}$	20.074	8.1 4
Rh $K_{\alpha1}$	20.216	15.3 7
Ru $K_{\beta1}'$	21.650	0.524 19
Ru $K_{\beta2}'$	22.210	0.098 4
Rh $K_{\beta1}'$	22.717	3.76 17
Rh $K_{\beta2}'$	23.312	0.70 3
γ_{IT} M3	52.0 1	0.092
γ_ϵ	471.49 15	0.49 7 ?
γ_ϵE2	685.38 10	3.7 16
γ_ϵ	766.79 19	0.39 6 ?
γ_ϵ	808.81 17	2.54 12

Photons (^{96}Rh)
(continued)

γ_{mode}	γ(keV)	γ(%)†
γ_ϵE2	832.70 11	39
γ_ϵ	944.16 17	0.69 12 ?
γ_ϵ	1006.80 14	0.79 6
γ_ϵE2(+12%M1)	1098.58 15	8.1 5
γ_ϵ	1242.20 22	0.63 12 ?
γ_ϵ[M1+E2]	1330.32 19	0.36 4
γ_ϵ(M1+1.1%E2)	1451.28 19	1.26 6
γ_ϵ	1478.28 15	0.40 7
γ_ϵ	1558.6 4	0.51 20 ?
γ_ϵ	1692.17 14	6.6 6
γ_ϵ[E2]	1743.51 15	0.39 4
γ_ϵ(M1+1.6%E2)	1743.51 17	0.82 8
γ_ϵ	1773.58 16	0.37 4
γ_ϵ	1907.38 19	1.04 4
γ_ϵ	2163.66 16	2.3 4
γ_ϵ	2257.6 3	1.68 12
γ_ϵ[E2]	2283.97 21	0.086 16
γ_ϵ[M1+E2]	2428.88 15	0.45 4
γ_ϵ	2458.95 19	0.73 6
γ_ϵ[E2]	2576.20 18	0.55 8
γ_ϵ	2841.4 5	0.14 3
γ_ϵ	3090.3 3	0.078 20
γ_ϵ	3119.6 5	0.17 4
γ_ϵ[E2]	3261.56 20	0.12 3

† uncert(syst): 13% for ε, 8.3% for IT

Atomic Electrons (^{96}Rh)
⟨e⟩=25.8 5 keV

e_{bin}(keV)	⟨e⟩(keV)	e(%)
3	1.33	44 5
16	0.195	1.20 12
17	0.71	4.2 4
18 - 23	0.44	2.26 16
29	9.9	34.4 12
49	10.0	20.5 7
51	1.15	2.23 8
52	1.46	2.83 10
449 - 471	0.009	0.0021 10
663 - 685	0.056	0.008 3
745 - 787	0.022	0.0028 11
806 - 833	0.44	0.054 6
922 - 944	0.0048	0.00052 22
985 - 1007	0.0052	0.00053 21
1076 - 1099	0.063	0.0059 4
1220 - 1242	0.0033	0.00027 11
1308 - 1330	0.0024	0.000182 21
1429 - 1478	0.0097	0.00067 5
1537 - 1558	0.0022	0.00014 7
1670 - 1751	0.033	0.0020 5
1770 - 1773	0.00017	9 3 $\times10^{-6}$
1885 - 1907	0.0037	0.00020 6
2142 - 2235	0.012	0.00056 13
2254 - 2284	0.00096	4.3 8 $\times10^{-5}$
2407 - 2458	0.0039	0.000160 25
2554 - 2576	0.0019	7.6 10 $\times10^{-5}$
2819 - 2841	0.00037	1.3 4 $\times10^{-5}$
3068 - 3119	0.00063	2.0 5 $\times10^{-5}$
3239 - 3261	0.00035	1.09 23 $\times10^{-5}$

Continuous Radiation (^{96}Rh)
⟨β+⟩=572 keV;⟨IB⟩=2.4 keV

E_{bin}(keV)		⟨ ⟩(keV)	(%)
0 - 10	β+	1.57 $\times10^{-6}$	1.90 $\times10^{-5}$
	IB	0.018	
10 - 20	β+	4.82 $\times10^{-5}$	0.000294
	IB	0.018	0.127
20 - 40	β+	0.00105	0.00323
	IB	0.036	0.126
40 - 100	β+	0.0392	0.0509

Continuous Radiation (^{96}Rh)
(continued)

E_{bin}(keV)		⟨ ⟩(keV)	(%)
	IB	0.102	0.157
100 - 300	β+	1.68	0.76
	IB	0.30	0.169
300 - 600	β+	13.0	2.79
	IB	0.37	0.086
600 - 1300	β+	110	11.3
	IB	0.62	0.071
1300 - 2500	β+	289	16.0
	IB	0.64	0.036
2500 - 5000	β+	158	5.2
	IB	0.28	0.0090
5000 - 5660	IB	0.0023	4.4×10^{-5}
	Σβ+		36

$^{96}_{46}$Pd(2.03 *3* min)

Mode: ε
Δ: -76370 *300* keV syst
SpA: 9.62×10^8 Ci/g
Prod: ^{40}Ca on ^{58}Ni

Photons (^{96}Pd)

γ_{mode}	γ(keV)	γ(rel)
γ	125.2 *2*	100 *3*
γ	336.4 *7*	3.0 *9*
γ	499.8 *3*	24.4 *18*
γ	599.1 *5*	10.8 *16*
γ	723.7 *3*	22.8 *20*
γ	762.6 *2*	71 *4*
γ	887.4 *7*	4.2 *10*
γ	1098.5 *3*	28 *2*
γ	1223.0 *7*	5.4 *12*

$^{96}_{47}$Ag(5.1 *4* s)

Mode: ε, εp(8.0 *23* %)
SpA: 2.15×10^{10} Ci/g
Prod: ^{40}Ca on ^{58}Ni; ^{40}Ca on ^{60}Ni

Photons (^{96}Ag)

γ_{mode}	γ(keV)	γ(rel)
γ_ϵ[E2]	107.1 *5*	40 *4*
γ_ϵ[E2]	325.5 *5*	88 *8*
γ_ϵ[E2]	684.4 *7*	96 *12*
γ_ϵ[E2]	1415.5 *5*	100 *10*

A = 97

NDS **10**, 1 (1973)

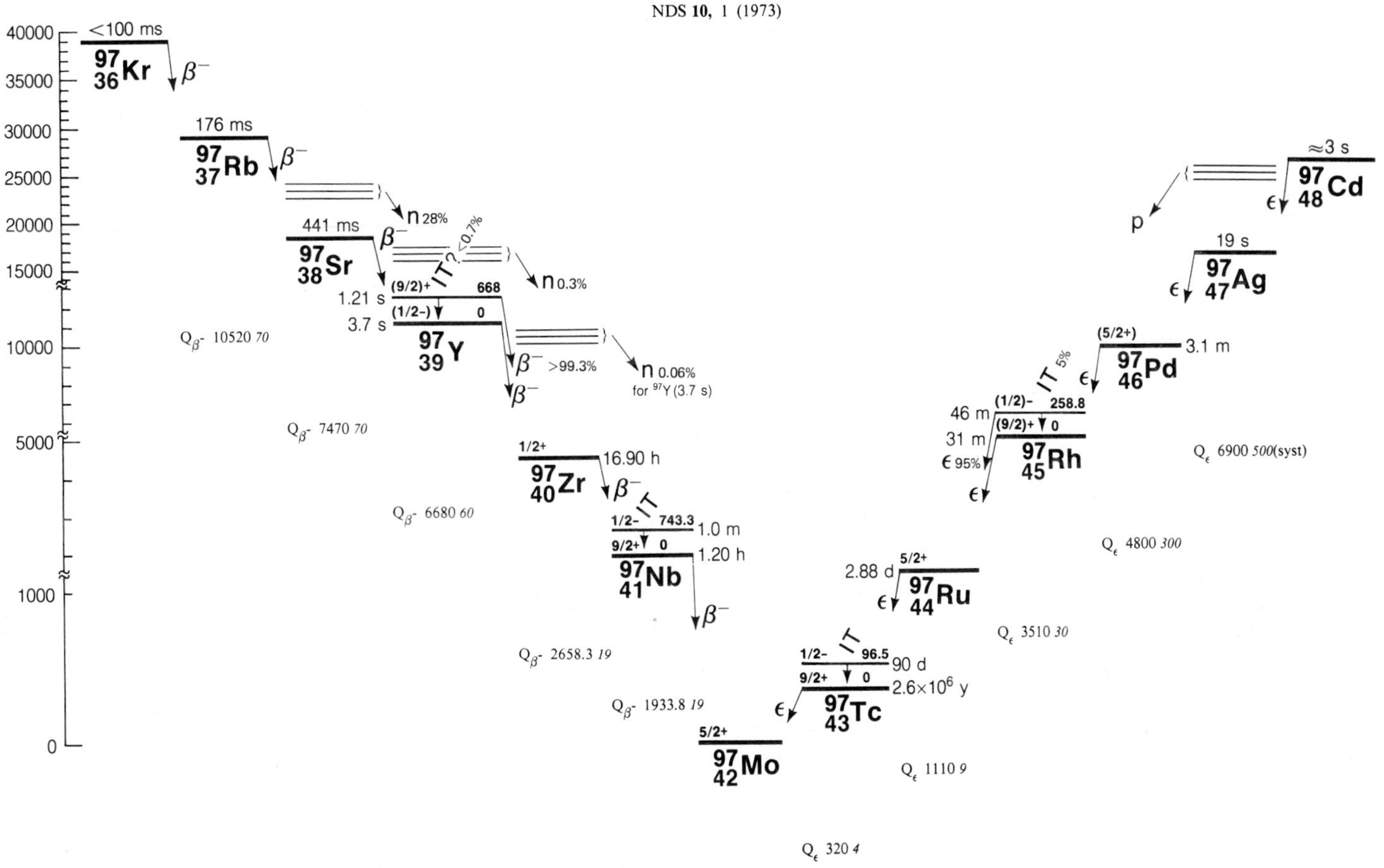

$^{97}_{36}$Kr(<100 ms)

Mode: β-
SpA: >1.7×10^{11} Ci/g
Prod: fission

$^{97}_{37}$Rb(176 *6* ms)

Mode: β-, β-n(28 *6* %)
Δ: -58280 *50* keV
SpA: 1.65×10^{11} Ci/g
Prod: fission

β-n(avg): 540

Photons (^{97}Rb)

γ_{mode}	γ(keV)	γ(rel)†
$\gamma_{\beta-}$	141 *1*	9.0 *5* ?
$\gamma_{\beta-}$	167 *1*	100 *5*
$\gamma_{\beta-}$	355 *1*	10.0 *5* ?
$\gamma_{\beta-}$	367 *1*	14.0 *7* ?
$\gamma_{\beta-n}$ E2	414.31 *8*	1.6 *3*
$\gamma_{\beta-}$	418 *1*	18.0 *9*
$\gamma_{\beta-n}$	455.38 *9*	0.021 *4*
$\gamma_{\beta-}$	519 *1*	38.0 *19*
$\gamma_{\beta-n}$	555.22 *17*	0.085 *17*
$\gamma_{\beta-}$	585 *1*	79 *4*
$\gamma_{\beta-}$	599 *1*	56 *3*
$\gamma_{\beta-n}$	606.50 *11*	0.0106 *21*
$\gamma_{\beta-n}$	643.94 *9*	0.127 *25*
$\gamma_{\beta-}$	652 *1*	29.0 *15* ?
$\gamma_{\beta-n}$ (E2+32%M1)	691.85 *7*	1.7 *4*
$\gamma_{\beta-}$	697 *1*	21 *1* ?
$\gamma_{\beta-n}$	765.63 *13*	0.0106 *21*
$\gamma_{\beta-n}$ (M1+27%E2)	813.18 *8*	1.19 *24*
$\gamma_{\beta-n}$ E2	814.92 *8*	10.6 *21*
$\gamma_{\beta-n}$	854.64 *13*	0.053 *11*
$\gamma_{\beta-n}$	977.76 *9*	0.085 *17*
$\gamma_{\beta-n}$	1037.16 *7*	0.34 *7*
$\gamma_{\beta-}$	1097 *1*	34.0 *17* ?
$\gamma_{\beta-n}$	1160.67 *10*	0.064 *13*
$\gamma_{\beta-n}$ (M1+21%E2)	1179.94 *12*	0.15 *3*
$\gamma_{\beta-}$	1258 *1*	52 *3* ?
$\gamma_{\beta-n}$	1268.95 *12*	0.032 *6*
$\gamma_{\beta-n}$	1298.35 *11*	0.0106 *21*
$\gamma_{\beta-n}$	1305.03 *20*	0.127 *25*
$\gamma_{\beta-n}$ [M1+35%E2]	1402.23 *15*	0.21 *4*
$\gamma_{\beta-n}$	1492.54 *10*	0.053 *11*
$\gamma_{\beta-n}$ [E2]	1506.76 *9*	0.73 *15*
$\gamma_{\beta-n}$	1592.37 *17*	
$\gamma_{\beta-n}$	2083.85 *13*	0.117 *23*

† γ(%) for β-n

$^{97}_{38}$Sr(441 *15* ms)

Mode: β-, β-n(0.27 *9* %)
Δ: -68800 *70* keV
SpA: 1.33×10^{11} Ci/g
Prod: fission

Photons (^{97}Sr)

⟨γ⟩=1844 *200* keV

γ_{mode}	γ(keV)	γ(%)†
γ	216.3 *3*	1.26 *25*
γ [M1+E2]	307.03 *17*	14 *3*
γ	310.6 *5*	1.7 *3*
γ	366.0 *3*	5.9 *12*
γ	412.35 *24*	2.9 *6*
γ	474.1 *3*	2.8 *6*
γ	479.75 *24*	3.2 *6*
γ	528.2 *4*	1.7 *3*
γ	652.1 *8*	14 *3*
γ	697.18 *25*	5.7 *11*
γ	767.0 *5*	2.0 *4*
γ	801.38 *24*	6.6 *13*
γ	892.1 *2*	5.2 *10*
γ	953.94 *22*	27 *5*
γ	1258.07 *22*	12.0 *24*
γ	1301.7 *5*	2.0 *4*
γ	1514.8 *3*	2.8 *6*
γ	1524.6 *5*	2.8 *6*
γ	1904.97 *21*	28
γ	2121.3 *3*	3.1 *6*
γ	2212.00 *19*	11.8 *24*
γ	2287.5 *4*	4.2 *8*

† 33% uncert(syst)

Continuous Radiation (^{97}Sr)

⟨β-⟩=2657 keV;⟨IB⟩=10.8 keV

E_{bin}(keV)		⟨ ⟩(keV)	(%)
0 - 10	β-	0.00147	0.0292
	IB	0.068	
10 - 20	β-	0.00452	0.0301
	IB	0.067	0.47
20 - 40	β-	0.0190	0.063
	IB	0.134	0.46
40 - 100	β-	0.153	0.216
	IB	0.39	0.60
100 - 300	β-	2.20	1.05
	IB	1.21	0.67
300 - 600	β-	12.3	2.66
	IB	1.58	0.37
600 - 1300	β-	113	11.5
	IB	2.8	0.32
1300 - 2500	β-	616	31.9
	IB	2.9	0.162
2500 - 5000	β-	1678	48.4
	IB	1.56	0.049
5000 - 7470	β-	235	4.19
	IB	0.057	0.00105

$^{97}_{39}$Y (3.7 *1* s)

Mode: β-, β-n(0.06 *1* %)
Δ: -76270 *60* keV
SpA: 2.87×10^{10} Ci/g
Prod: fission

Photons (^{97}Y)

⟨γ⟩=1821 *150* keV

γ_{mode}	γ(keV)	γ(%)†
γ [E2]	161.33 *16*	0.36 *7*
γ	296.98 *19*	1.29 *22*
γ	544.7 *3*	0.90 *18*
γ	594.63 *21*	0.34 *6*
γ	755.96 *20*	1.17 *21*
γ	1103.00 *15*	5.1 *10*
γ	1291.13 *25*	5.7 *11*

Photons (^{97}Y) (continued)

γ_{mode}	γ(keV)	γ(%)†
γ	1344.0 *3*	0.90 *18*
γ	1399.98 *19*	4.4 *8*
γ	1428.68 *25*	0.72 *14*
γ	1639.9 *3*	0.83 *17*
γ	1887.66 *24*	1.9 *4*
γ	1996.5 *3*	7.5 *15*
γ	2057.3 *4*	0.94 *19*
γ	2742.9 *3*	6.6 *13*
γ	3287.61 *22*	18
γ	3401.3 *3*	14 *3*
γ	3549.5 *4*	3.1 *6*

† 17% uncert(syst)

Continuous Radiation (^{97}Y)

⟨β-⟩=2158 keV;⟨IB⟩=8.1 keV

E_{bin}(keV)		⟨ ⟩(keV)	(%)
0 - 10	β-	0.00329	0.065
	IB	0.059	
10 - 20	β-	0.0101	0.067
	IB	0.059	0.41
20 - 40	β-	0.0423	0.141
	IB	0.116	0.40
40 - 100	β-	0.337	0.476
	IB	0.34	0.52
100 - 300	β-	4.67	2.25
	IB	1.03	0.57
300 - 600	β-	24.7	5.4
	IB	1.29	0.30
600 - 1300	β-	195	20
	IB	2.2	0.25
1300 - 2500	β-	717	38.3
	IB	2.0	0.112
2500 - 5000	β-	1040	30.0
	IB	1.03	0.032
5000 - 6680	β-	176	3.24
	IB	0.028	0.00052

$^{97}_{39}$Y (1.21 *3* s)

Mode: β-(>99.3 %), IT?(<0.7 %)
Δ: -75603 *60* keV
SpA: 7.32×10^{10} Ci/g
Prod: fission

Photons (^{97}Y)

⟨γ⟩=1790 *210* keV

γ_{mode}	γ(keV)	γ(%)
$\gamma_{\beta-}$[E2]	161.33 *16*	71 *14*
$\gamma_{\beta-}$	296.98 *19*	1.32 *23*
$\gamma_{\beta-}$	375.35 *22*	4.3 *9*
$\gamma_{\beta-}$	407.02 *23*	5.5 *11*
$\gamma_{\beta-}$[M1+E2]	427.31 *22*	0.89 *18*
$\gamma_{\beta-}$	456.73 *25*	1.7 *3*
$\gamma_{\beta-}$	542.66 *21*	0.89 *18*
$\gamma_{\beta-}$	594.63 *21*	0.97 *16*
γ_{IT} [M4]	667.8 *8*	<0.7 ?
$\gamma_{\beta-}$	755.96 *20*	3.3 *6*
$\gamma_{\beta-}$	969.98 *20*	39 *8*
$\gamma_{\beta-}$	999.4 *3*	2.5 *5*
$\gamma_{\beta-}$	1103.00 *15*	89 *18*
$\gamma_{\beta-}$	1244.1 *3*	7.8 *16*
$\gamma_{\beta-}$	1264.33 *19*	~3
$\gamma_{\beta-}$	1399.98 *19*	4.6 *8*

Continuous Radiation (^{97}Y)

⟨β-⟩=2407 keV;⟨IB⟩=9.3 keV

E_{bin}(keV)		⟨ ⟩(keV)	(%)
0 - 10	β-	0.00170	0.0338
	IB	0.064	
10 - 20	β-	0.0052	0.0349
	IB	0.064	0.44
20 - 40	β-	0.0220	0.073
	IB	0.126	0.44
40 - 100	β-	0.177	0.249
	IB	0.37	0.56
100 - 300	β-	2.52	1.21
	IB	1.13	0.63
300 - 600	β-	14.0	3.02
	IB	1.46	0.34
600 - 1300	β-	127	12.9
	IB	2.5	0.29
1300 - 2500	β-	665	34.6
	IB	2.4	0.139
2500 - 5000	β-	1537	45.3
	IB	1.05	0.034
5000 - 6083	β-	62	1.17
	IB	0.0051	9.2×10^{-5}

$^{97}_{40}$Zr(16.90 *5* h)

Mode: β-

Δ: -82950 *3* keV

SpA: 1.912×10^6 Ci/g

Prod: ^{96}Zr(n,γ); fission

Photons (^{97}Zr)

⟨γ⟩=869 *6* keV

γ_{mode}	γ(keV)	γ(%)†
Nb $K_{\alpha2}$	16.521	0.406 *15*
Nb $K_{\alpha1}$	16.615	0.77 *3*
Nb $K_{\beta1}$'	18.618	0.179 *7*
Nb $K_{\beta2}$'	19.074	0.0301 *12*
γ	111.59 *22*	0.056 *9*
γ	183.00 *16*	0.037 *9*
γ	218.90 *18*	0.176 *19*
γ [M1+E2]	254.27 *10*	1.25 *14*
γ	272.27 *14*	0.25 *4*
γ	294.9 *5*	0.08 *4*
γ	297.4 *5*	~0.037
γ	330.49 *15*	0.11 *3*
γ [M1+E2]	355.46 *8*	2.27 *23*
γ	400.45 *14*	0.32 *5*
γ	507.70 *8*	5.1 *5*
γ	513.49 *15*	0.56 *9*
γ	602.57 *10*	1.39 *14*
γ	690.70 *16*	0.25 *4*
γ	699.18 *15*	0.121 *19*
γ	703.76 *8*	0.93 *9*
γ M4	743.32 *8*	92.8 *
γ	804.57 *10*	0.65 *7*
γ [M1+E2]	829.84 *8*	0.223 *19*
γ	854.96 *8*	0.33 *4*
γ	971.44 *10*	0.29 *3*
γ	1021.19 *15*	1.34 *14*
γ	1110.49 *20*	0.111 *19*
γ	1147.94 *7*	2.6 *3*
γ [E2]	1276.13 *8*	0.97 *9*
γ [E1]	1362.65 *7*	1.34 *14*
γ [E2]	1750.50 *8*	1.34 *14*
γ [E2]	1851.70 *8*	0.35 *4*

† 0.33% uncert(syst)

* with ^{97}Nb(1.0 min) in equilib

Atomic Electrons (^{97}Zr)

⟨e⟩=15.2 *3* keV

e_{bin}(keV)	⟨e⟩(keV)	e(%)
2 - 19	0.107	2.03 *15*
93 - 112	0.023	~0.024
164 - 200	0.017	~0.009
216 - 254	0.10	0.043 *13*
270 - 312	0.010	0.0034 *12*
328 - 355	0.083	0.025 *5*
381 - 400	0.007	~0.0018
489 - 513	0.08	0.016 *7*
584 - 603	0.014	0.0025 *11*
672 - 704	0.011	0.0016 *5*
724	12.6	1.74 *4*
741	1.70	0.229 *5*
743	0.362	0.0487 *10*
786 - 830	0.0065	0.00082 *23*
836 - 855	0.0021	0.00025 *10*
952 - 971	0.0016	0.00017 *6*
1002 - 1021	0.007	0.0007 *3*
1092 - 1129	0.011	0.0010 *4*
1145 - 1148	0.0015	0.00013 *5*
1257 - 1276	0.0053	0.00042 *4*
1344 - 1363	0.0033	0.000242 *24*
1732 - 1750	0.0053	0.00031 *3*
1833 - 1851	0.00133	7.2 *7* $\times 10^{-5}$

Continuous Radiation (^{97}Zr)

⟨β-⟩=696 keV;⟨IB⟩=1.26 keV

E_{bin}(keV)		⟨ ⟩(keV)	(%)
0 - 10	β-	0.0303	0.60
	IB	0.028	
10 - 20	β-	0.092	0.61
	IB	0.028	0.19
20 - 40	β-	0.375	1.25
	IB	0.054	0.19
40 - 100	β-	2.78	3.95
	IB	0.148	0.23
100 - 300	β-	30.0	14.9
	IB	0.38	0.22
300 - 600	β-	106	23.5
	IB	0.34	0.082
600 - 1300	β-	416	45.5
	IB	0.27	0.033
1300 - 1915	β-	141	9.6
	IB	0.0129	0.00093

$^{97}_{41}$Nb(1.202 *12* h)

Mode: β-

Δ: -85608 *3* keV

SpA: 2.69×10^7 Ci/g

Prod: descendant ^{97}Zr; ^{98}Mo(γ,p)

Photons (^{97}Nb)

⟨γ⟩=666.8 *15* keV

γ_{mode}	γ(keV)	γ(%)†
γ [E2]	177.8 *3*	0.049 *10*
γ	239.00 *25*	0.049 *10*
γ [M1+E2]	480.4 *3*	0.148 *20*
γ	549.42 *22*	0.049 *10*
γ [M1+E2]	658.22 *15*	98.34
γ	719.44 *19*	0.089 *10*
γ	857.7 *3*	0.049 *10*
γ	909.81 *16*	0.039 *10*
γ [M1+E2]	1024.7 *5*	1.08 *11*

Photons (^{97}Nb)
(continued)

γ_{mode}	γ(keV)	γ(%)†
γ [E2]	1117.2 *4*	0.089 *10*
γ	1148.8 *3*	0.049 *10*
γ [M1+E2]	1268.86 *16*	0.157 *20*
γ	1515.9 *3*	0.118 *20*
γ	1629.24 *21*	0.030 *10*

† 0.11% uncert(syst)

Continuous Radiation (^{97}Nb)

⟨β-⟩=467 keV;⟨IB⟩=0.61 keV

E_{bin}(keV)		⟨ ⟩(keV)	(%)
0 - 10	β-	0.0432	0.86
	IB	0.021	
10 - 20	β-	0.131	0.88
	IB	0.020	0.141
20 - 40	β-	0.54	1.80
	IB	0.039	0.135
40 - 100	β-	4.07	5.8
	IB	0.102	0.158
100 - 300	β-	46.1	22.7
	IB	0.23	0.133
300 - 600	β-	160	35.9
	IB	0.153	0.038
600 - 1276	β-	256	32.1
	IB	0.043	0.0061

$^{97}_{41}$Nb(1.00 *13* min)

Mode: IT

Δ: -84865 *3* keV

SpA: 1.9×10^9 Ci/g

Prod: daughter ^{97}Zr

Photons (^{97}Nb)

⟨γ⟩=728.3 *8* keV

γ_{mode}	γ(keV)	γ(%)†
Nb $K_{\alpha2}$	16.521	0.403 *14*
Nb $K_{\alpha1}$	16.615	0.77 *3*
Nb $K_{\beta1}$'	18.618	0.178 *6*
Nb $K_{\beta2}$'	19.074	0.0299 *11*
γ M4	743.32 *8*	97.95

† 0.10% uncert(syst)

Atomic Electrons (^{97}Nb)

⟨e⟩=15.6 *3* keV

e_{bin}(keV)	⟨e⟩(keV)	e(%)
2	0.035	1.44 *15*
3	0.0030	0.113 *12*
14	0.047	0.33 *3*
16	0.0184	0.114 *12*
17	0.00114	0.0069 *7*
18	0.00147	0.0081 *8*
19	0.00027	0.00147 *15*
724	13.3	1.84 *4*
741	1.79	0.242 *5*
743	0.382	0.0514 *10*

$^{97}_{42}$Mo(stable)

Δ: -87542.1 *20* keV
%: 9.55 *2*

$^{97}_{43}$Tc(2.6 *4* $\times 10^6$ yr)

Mode: ϵ
Δ: -87222 *5* keV
SpA: 0.00142 Ci/g
Prod: daughter ^{97}Ru; ^{97}Mo(d,2n)

Photons (^{97}Tc)

⟨γ⟩=11.8 *3* keV

γ_{mode}	γ(keV)	γ(%)
Mo L_ℓ	2.016	0.083 *15*
Mo L_η	2.120	0.038 *6*
Mo L_α	2.293	2.1 *3*
Mo L_β	2.426	1.17 *20*
Mo L_γ	2.701	0.079 *15*
Mo $K_{\alpha 2}$	17.374	19.2 *7*
Mo $K_{\alpha 1}$	17.479	36.5 *13*
Mo $K_{\beta 1}$'	19.602	8.6 *3*
Mo $K_{\beta 2}$'	20.091	1.50 *6*

Atomic Electrons (^{97}Tc)

⟨e⟩=4.9 *3* keV

e_{bin}(keV)	⟨e⟩(keV)	e(%)
3	1.83	71 *7*
14	0.204	1.43 *15*
15	1.92	13.0 *13*
17	0.90	5.3 *5*
19	0.069	0.36 *4*
20	0.0131	0.067 *7*

Continuous Radiation (^{97}Tc)

⟨IB⟩=0.026 keV

E_{bin}(keV)		⟨ ⟩(keV)	(%)
10 - 20	IB	0.018	0.102
20 - 40	IB	0.0020	0.0082
40 - 100	IB	0.00156	0.0024
100 - 300	IB	0.0046	0.0029
300 - 320	IB	1.30×10^{-6}	4.3×10^{-7}

$^{97}_{43}$Tc(90.5 *10* d)

Mode: IT
Δ: -87126 *5* keV
SpA: 1.487×10^4 Ci/g
Prod: ^{96}Mo(d,n); ^{97}Mo(p,n); ^{97}Mo(d,2n); daughter ^{97}Ru

Photons (^{97}Tc)

⟨γ⟩=9.4 *4* keV

γ_{mode}	γ(keV)	γ(%)†
Tc L_ℓ	2.133	0.086 *16*
Tc L_η	2.249	0.033 *6*
Tc L_α	2.424	2.2 *4*
Tc L_β	2.583	1.13 *20*
Tc L_γ	2.873	0.083 *17*
Tc $K_{\alpha 2}$	18.251	14 *1*
Tc $K_{\alpha 1}$	18.367	26.6 *19*
Tc $K_{\beta 1}$'	20.613	6.4 *5*
Tc $K_{\beta 2}$'	21.136	1.14 *9*
γ M4	96.51 *9*	0.31

† 6.4% uncert(syst)

Atomic Electrons (^{97}Tc)

⟨e⟩=85 *3* keV

e_{bin}(keV)	⟨e⟩(keV)	e(%)
3	1.84	67 *8*
15	0.59	3.9 *5*
16	0.89	5.7 *7*
17	0.055	0.32 *4*
18	0.59	3.3 *4*
20	0.050	0.25 *3*
21	0.0097	0.047 *6*
75	46.3	61 *4*
93	12.3	13.1 *9*
94	15.4	16.5 *11*
96	6.8	7.0 *5*
97	0.048	0.050 *3*

$^{97}_{44}$Ru(2.88 *4* d)

Mode: ϵ
Δ: -86111 *8* keV
SpA: 4.67×10^5 Ci/g
Prod: ^{96}Ru(n,γ); ^{94}Mo(α,n)

Photons (^{97}Ru)

⟨γ⟩=240 *4* keV

γ_{mode}	γ(keV)	γ(%)†
Tc L_ℓ	2.133	0.090 *16*
Tc L_η	2.249	0.042 *6*
Tc L_α	2.424	2.3 *4*
Tc L_β	2.577	1.35 *23*
Tc L_γ	2.863	0.097 *18*
Tc $K_{\alpha 2}$	18.251	20.5 *7*
Tc $K_{\alpha 1}$	18.367	38.9 *14*
Tc $K_{\beta 1}$'	20.613	9.3 *3*
Tc $K_{\beta 2}$'	21.136	1.67 *6*
γ	108.78 *3*	0.108 *12*
γ	114.25 *4*	0.0017 *4*
γ	184.66 *4*	0.0046 *22*
γ M1+10%E2	215.70 *3*	86.0
γ E2	324.49 *3*	10.2 *4*
γ [M1+E2]	460.53 *3*	0.117 *5*
γ [M1+E2]	483.75 *10*	0.0022 *3*
γ	530.93 *3*	0.0026 *2*
γ [E2]	560.34 *5*	0.032 *2*
γ M1+0.01%E2	569.31 *3*	0.89 *4*
γ	639.716 *19*	0.0088 *11*
γ M1,E2	645.18 *4*	0.065 *4*
γ	670.202 *20*	0.0085 *4*
γ M1,E2	753.96 *3*	0.086 *6*
γ [E2]	785.01 *4*	0.082 *5*
γ M1,E2	855.42 *3*	0.049 *4*

Photons (^{97}Ru) (continued)

γ_{mode}	γ(keV)	γ(%)†
γ	898.17 *10*	0.00018 *5*
γ	969.67 *4*	0.00080 *9*

† 2.0% uncert(syst)

Atomic Electrons (^{97}Ru)

⟨e⟩=12.6 *4* keV

e_{bin}(keV)	⟨e⟩(keV)	e(%)
3	2.02	74 *8*
15	0.86	5.7 *6*
16	1.31	8.4 *9*
17	0.080	0.46 *5*
18	0.86	4.8 *5*
20 - 21	0.087	0.43 *4*
88 - 114	0.05	~0.06
164 - 185	0.0006	~0.0004
195	5.78	2.97 *11*
213	0.78	0.365 *17*
215	0.144	0.067 *3*
216	0.0274	0.0127 *6*
303	0.527	0.174 *8*
321 - 324	0.089	0.0276 *9*
439 - 484	0.0031	0.00070 *7*
510 - 558	0.0142	0.00260 *12*
560 - 569	0.00197	0.000348 *14*
619 - 668	0.00116	0.000185 *15*
733 - 782	0.00196	0.000262 *14*
784 - 785	2.13×10^{-5}	2.72 *16* $\times10^{-6}$
834 - 877	0.00050	6.0 *5* $\times10^{-5}$
895 - 898	1.6×10^{-7}	1.7 *8* $\times10^{-8}$
949 - 970	5.1×10^{-6}	5.3 *21* $\times10^{-7}$

Continuous Radiation (^{97}Ru)

⟨IB⟩=0.164 keV

E_{bin}(keV)		⟨ ⟩(keV)	(%)
10 - 20	IB	0.0168	0.095
20 - 40	IB	0.0053	0.024
40 - 100	IB	0.0023	0.0034
100 - 300	IB	0.029	0.0142
300 - 600	IB	0.085	0.020
600 - 1014	IB	0.026	0.0038

$^{97}_{45}$Rh(30.7 *6* min)

Mode: ϵ
Δ: -82600 *30* keV
SpA: 6.31×10^7 Ci/g
Prod: ^{96}Ru(d,n); ^{96}Ru(p,γ)

Photons (^{97}Rh)&

⟨γ⟩=744 *19* keV

γ_{mode}	γ(keV)	γ(%)†
γ (M1+E2)	188.6 *2*	1.0 *3*
γ	319.7 *5*	0.38 *8*
γ	350.7 *5*	0.38 *8*
γ	389.2 *4*	0.98 *15*
γ [M1+E2]	421.5 *3*	75
γ [M1+E2]	457.3 *4*	1.05 *15*
γ [E2]	651.2 *4*	1.05 *15*
γ	664.1 *8*	~0.08

Photons (^{97}Rh)[&]
(continued)

γ_{mode}	γ(keV)	γ(%)[†]
γ	703.0 *4*	1.05 *15*
γ	740.8 *5*	0.30 *8*
γ	765.5 *7*	~0.15
γ[E2]	777.2 *4*	1.58 *23*
γ	807.3 *4*	1.42 *15*
γ	839.8 *3*	12.4 *10*
γ[E2]	878.8 *3*	9.3 *8*
γ	1053.7 *4*	1.73 *22*
γ	1091.9 *6*	0.75 *15*
γ	1111.3 *4*	0.68 *15*
γ	1157.8 *5*	0.75 *15*
γ	1197.1 *6*	0.30 *8*
γ	1228.0 *5*	1.1 *3*
γ	1271.4 *8*	0.30 *8*
γ	1310.1 *6*	1.13 *15*
γ	1391.9 *7*	0.30 *8*
γ	1434.9 *7*	0.45 *15*
γ	1511.1 *6*	0.52 *15*
γ	1577.2 *6*	0.60 *15*
γ	1712.4 *7*	~0.30
γ	1750.5 *7*	0.45 *15*
γ	1764.3 *9*	0.45 *15*
γ	1876.0 *7*	0.30 *8*
γ	1887.0 *9*	~0.15
γ	1924.8 *8*	0.45 *15*
γ	1931.0 *9*	0.75 *22*
γ	1988.9 *9*	~0.15
γ	2185.7 *8*	~0.30
γ	2492.9 *10*	~0.22
γ	2753.8 *8*	0.22 *8*
γ	2947.4 *7*	~0.22

† 2.7% uncert(syst)
& see also ^{97}Rh(46 min)

$^{97}_{45}$Rh(46.2 *16* min)

Mode: ϵ(94.7 *10* %), IT(5.3 *10* %)
Δ: -82341 *30* keV
SpA: 4.20×10^{7} Ci/g
Prod: protons on Cd; ^{96}Ru(p,γ); ^{96}Ru(d,n)

Photons (^{97}Rh)
$\langle\gamma\rangle$=1896 *37* keV

γ_{mode}	γ(keV)	γ(%)[†]
γ_ϵ(M1+E2)	188.6 *2*	50.9
γ_ϵ	251.8 *4*	1.17 *15*
γ_{IT}	258.6 *4*	1.5
γ_ϵ	297.1 *5*	~0.31
γ_ϵ	311.7 *5*	0.31 *10* *
γ_ϵ	338.5 *5*	1.02 *10*
γ_ϵ	367.3 *4*	0.71 *10*
γ_ϵ	412.7 *6*	~0.25
γ_ϵ[M1+E2]	421.5 *3*	13.2 *15*
γ_ϵ	527.8 *2*	8.5 *7*
γ_ϵ	551.5 *6*	0.20 *5*
γ_ϵ	562.3 *4*	1.02 *15*
γ_ϵ	567.3 *4*	0.92 *15*
γ_ϵ	579.0 *8*	~0.41 *
γ_ϵ	582.4 *3*	2.59 *25*
γ_ϵ	605.7 *4*	0.46 *10*
γ_ϵ	610.9 *4*	0.61 *10*
γ_ϵ	683.8 *5*	0.31 *10* *
γ_ϵ	707.5 *6*	0.41 *15* *
γ_ϵ	719.0 *3*	3.1 *3*
γ_ϵ	731.7 *5*	0.46 *10* *
γ_ϵ	748.2 *5*	0.56 *10* *
γ_ϵ	771.1 *3*	4.8 *4*
γ_ϵ	820.5 *5*	0.76 *10*
γ_ϵ	845.8 *8*	~0.15 *
γ_ϵ	869.2 *6*	0.41 *10*

Photons (^{97}Rh)
(continued)

γ_{mode}	γ(keV)	γ(%)[†]
γ_ϵ	908.5 *5*	2.09 *20*
γ_ϵ	968.5 *9*	0.31 *10* *
γ_ϵ	995.5 *3*	3.6 *3*
γ_ϵ	1013.4 *3*	5.5 *5*
γ_ϵ	1060.9 *10*	~0.31 *
γ_ϵ	1117.2 *10*	0.31 *10*
γ_ϵ	1183.8 *6*	4.4 *5*
γ_ϵ	1187.0 *6*	4.4 *5*
γ_ϵ	1287.2 *6*	0.41 *10* *
γ_ϵ	1301.7 *9*	~0.10 *
γ_ϵ	1322.5 *8*	0.31 *10* *
γ_ϵ	1337.1 *5*	0.92 *10*
γ_ϵ	1345.1 *10*	0.15 *5* *
γ_ϵ	1376.6 *6*	0.51 *10*
γ_ϵ	1415.1 *9*	0.20 *5* *
γ_ϵ	1421.1 *8*	0.25 *5*
γ_ϵ	1426.4 *4*	2.29 *20*
γ_ϵ	1451.2 *7*	0.41 *10* *
γ_ϵ	1463.2 *5*	1.17 *15*
γ_ϵ	1474.5 *5*	1.53 *15*
γ_ϵ	1586.6 *4*	8.3 *6*
γ_ϵ	1656.9 *4*	1.88 *20*
γ_ϵ	1708.4 *8*	~0.15 *
γ_ϵ	1718.4 *6*	2.4 *4*
γ_ϵ	1739.7 *7*	1.27 *25*
γ_ϵ	1784.9 *6*	1.27 *15*
γ_ϵ	1813.3 *6*	0.41 *10* *
γ_ϵ	1876.0 *7*	0.15 *5*
γ_ϵ	1964.5 *7*	1.02 *10*
γ_ϵ	1978.2 *9*	0.20 *5* *
γ_ϵ	2007.4 *5*	3.9 *3*
γ_ϵ	2036.1 *5*	3.9 *3*
γ_ϵ	2110.3 *7*	0.61 *10*
γ_ϵ	2122.2 *6*	2.95 *25*
γ_ϵ	2151.4 *6*	1.22 *15*
γ_ϵ	2196.8 *6*	0.61 *10*
γ_ϵ	2236.5 *9*	0.97 *20*
γ_ϵ	2245.2 *5*	12.7 *9*
γ_ϵ	2318.5 *8*	~0.25
γ_ϵ	2338.8 *7*	0.20 *5* *
γ_ϵ	2375.7 *6*	0.92 *10*
γ_ϵ	2458.7 *9*	~0.2 *
γ_ϵ	2492.9 *10*	~0.05
γ_ϵ	2564.0 *9*	~0.15
γ_ϵ	2576.7 *9*	0.31 *10*
γ_ϵ	2608.0 *6*	2.29 *25*
γ_ϵ	2647.8 *6*	3.3 *4*
γ_ϵ	2737.6 *9*	~0.10
γ_ϵ	2767.1 *8*	~0.31 *
γ_ϵ	2777 *1*	~0.10 *
γ_ϵ	2788.9 *10*	~0.10 *
γ_ϵ	2801.0 *9*	0.20 *5* *
γ_ϵ	2843.8 *9*	~0.10 *
γ_ϵ	2852.9 *7*	0.25 *10* *
γ_ϵ	2899.3 *7*	0.31 *10* *
γ_ϵ	2930.9 *7*	0.61 *20*
γ_ϵ	3000.1 *9*	~0.051 *
γ_ϵ	3034.1 *7*	0.25 *5*
γ_ϵ	3076.2 *10*	~0.10
γ_ϵ	3101.8 *10*	~0.15 *
γ_ϵ	3108.6 *10*	0.25 *10* *
γ_ϵ	3185.7 *8*	0.20 *5*
γ_ϵ	3227.3 *8*	0.20 *5* *
γ_ϵ	3264 *1*	~0.05
γ_ϵ	3270.7 *10*	~0.05 *
γ_ϵ	3303.6 *10*	~0.05 *
γ_ϵ	3374.1 *6*	1.32 *20*
γ_ϵ	3400.8 *10*	~0.05 *
γ_ϵ	3441.5 *10*	~0.05 *
γ_ϵ	3458.3 *7*	0.25 *10* *
γ_ϵ	3494.5 *10*	~0.05 *

† uncert(syst): 3.9% for ϵ, 22% for IT
* with ^{97}Rh(46 or 31 min)

$^{97}_{46}$Pd(3.1 *1* min)

Mode: ϵ
Δ: -77800 *300* keV
SpA: 6.24×10^{8} Ci/g
Prod: ^{96}Ru(^{3}He,2n); ^{96}Ru(α,3n)

Photons (^{97}Pd)
$\langle\gamma\rangle$=1548 *26* keV

γ_{mode}	γ(keV)	γ(%)[†]
γ[M1+E2]	209.3 *5*	2.00 *12*
γ[M1+E2]	265.3 *1*	51
γ	354.4 *5*	1.08 *8*
γ[E2]	475.2 *1*	24.1 *13*
γ	556.4 *5*	0.283 *15* ?
γ	583.0 *5*	1.23 *11*
γ	590.7 *5*	1.80 *18*
γ	614.4 *5*	0.429 *20*
γ	658.5 *5*	1.35 *7*
γ	685.5 *5*	1.58 *12* ?
γ	745.7 *5*	1.05 *10*
γ	792.7 *1*	12.5 *7*
γ	862.7 *5*	0.92 *14*
γ	896.6 *5*	0.57 *7*
γ	933.7 *5*	2.20 *19*
γ	940.3 *3*	3.39 *18*
γ	947.4 *5*	0.45 *8* ?
γ	976.7 *5*	0.48 *10*
γ	1034.4 *5*	0.96 *17*
γ	1053.6 *5*	3.2 *6*
γ	1055.4 *5*	5.1 *3*
γ	1058.5 *5*	2.4 *2*
γ	1150.3 *5*	0.94 *13*
γ	1171.8 *3*	3.3 *4*
γ	1199.2 *5*	0.80 *9*
γ	1237.8 *5*	1.83 *10*
γ	1285.0 *5*	1.26 *11*
γ	1354.1 *5*	1.19 *12*
γ	1377.4 *5*	0.38 *6*
γ	1480.5 *5*	1.12 *10*
γ	1494.2 *2*	5.5 *3*
γ	1519.8 *5*	2.04 *15*
γ	1592.9 *5*	1.45 *9*
γ	1638.7 *3*	3.39 *22*
γ	1641.1 *3*	3.0 *3*
γ	1729.1 *5*	0.38 *3*
γ	1759.6 *1*	6.1 *4*
γ	1788.6 *5*	0.78 *4* ?
γ	1797.2 *5*	1.02 *5* ?
γ	1846.8 *3*	2.59 *19*
γ	1952.2 *5*	0.65 *17* ?
γ	1979.1 *5*	0.60 *3* ?
γ	1993.9 *5*	1.07 *9*
γ	2029.5 *5*	2.53 *20*
γ	2111.4 *10*	0.36 *3*
γ	2132.2 *10*	0.55 *6*
γ	2231.3 *10*	0.510 *25*
γ	2408 *1*	0.67 *8*
γ	2428.4 *10*	1.02 *13*
γ	2497.2 *10*	0.79 *8*
γ	2637.1 *10*	0.47 *9*
γ	2684.4 *10*	1.05 *8*
γ	2777.7 *10*	0.29 *7*
γ	2826.3 *10*	0.62 *9*
γ	2903.3 *10*	0.63 *13*
γ	2950.3 *10*	0.308 *15*
γ	2974.6 *10*	0.98 *17*
γ	3017 *1*	0.177 *15*
γ	3100.9 *10*	0.131 *25*
γ	3155.1 *10*	0.348 *25* ?
γ	3239.8 *10*	0.40 *5*
γ	3342.1 *10*	0.67 *12*
γ	3397.8 *10*	0.258 *20*

† 12% uncert(syst)

$^{97}_{47}$Ag(19 *2* s)

Mode: ϵ

Δ: -70900 *400* keV syst

SpA: 6.0×10^{9} Ci/g

Prod: ^{40}Ca on ^{62}Ni; ^{14}N on ^{92}Mo

Photons (^{97}Ag)

γ_{mode}	γ(keV)	γ(rel)
γ	686.2 *2*	100
γ	1294.6 *4*	45 *5*

$^{97}_{48}$Cd(~3 s)

Mode: ϵ, ϵp

SpA: ~3×10^{10} Ci/g

Prod: ^{40}Ca on ^{60}Ni; protons on Sn

A = 98

NDS **39**, 467 (1983)

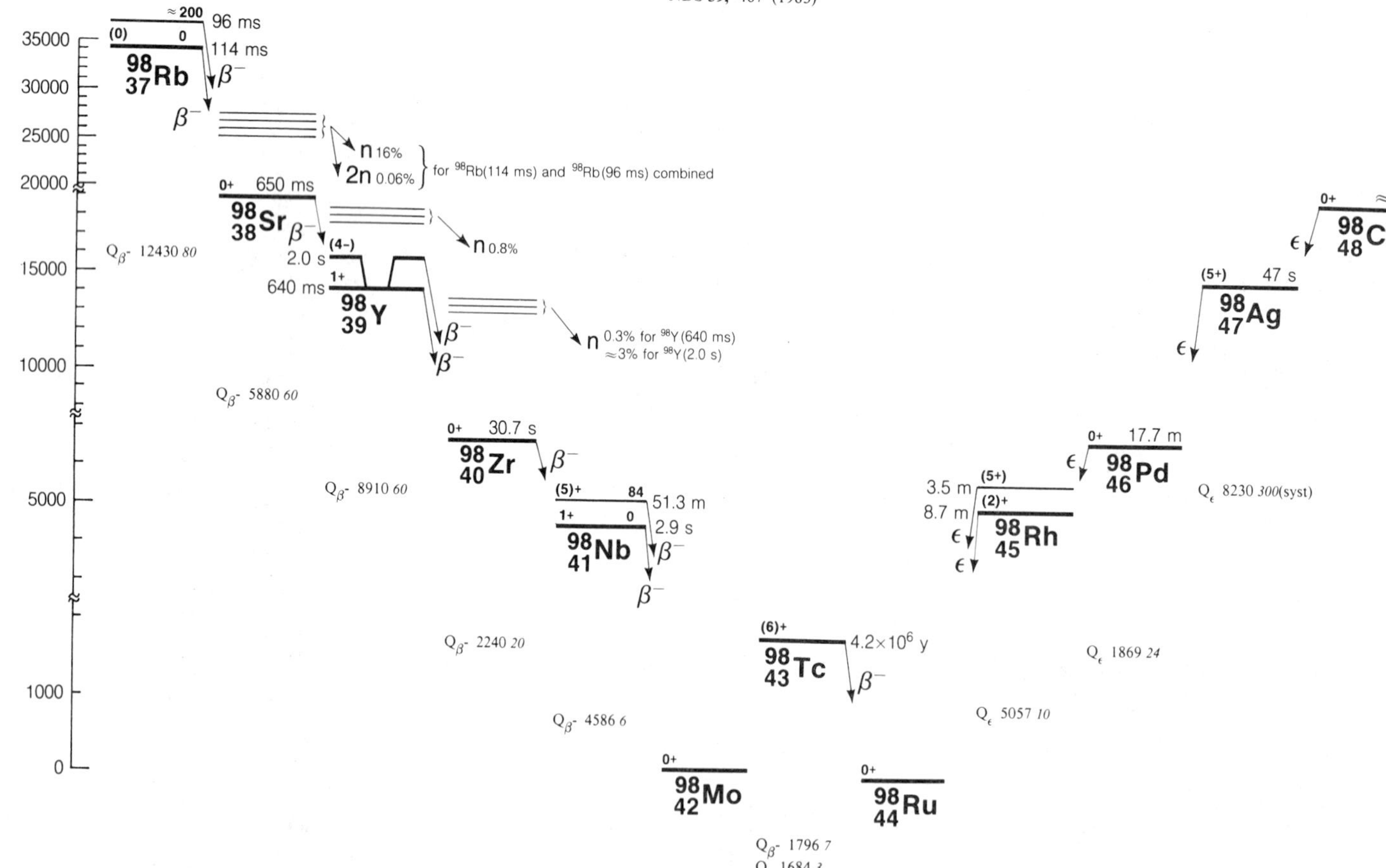

$^{98}_{37}$Rb(114 *5* ms)

Mode: β-, β-n(15.9 *13* %), β-2n(0.060 *9* %)
(includes ^{98}Rb(96 ms))
Δ: -54060 *60* keV
SpA: 1.66×10^{11} Ci/g
Prod: fission
β-n(avg): 490

Photons (^{98}Rb(114 + 96 ms))

γ_{mode}	γ(keV)	γ(rel)†
$\gamma_{\beta-}$E2	71.18 *12*	2.5 *5*
$\gamma_{\beta-}$	107.35 *10*	1.9 *2*
$\gamma_{\beta-}$	140.7 *4*	3 *1*
$\gamma_{\beta-n}$	141 *1*	1.8 *4*
$\gamma_{\beta-}$E2	144.479 *6*	100 *2*
$\gamma_{\beta-n}$	167 *1*	7.0 *14*
$\gamma_{\beta-}$	175.45 *20*	1.7 *2*
$\gamma_{\beta-}$E2	289.64 *10*	32.6 *16*
$\gamma_{\beta-}$	301.7 *5*	~0.2
$\gamma_{\beta-n}$	355 *1*	0.63 *13*
$\gamma_{\beta-n}$	378.9 *10*	0.035 *7*
$\gamma_{\beta-n}$	418 *1*	0.63 *13*
$\gamma_{\beta-n}$	433.3 *10*	0.42 *8*
$\gamma_{\beta-}$[E2]	433.85 *10*	2.7 *4*
$\gamma_{\beta-n}$	519 *1*	0.21 *4*
$\gamma_{\beta-n}$	585 *1*	2.2 *5*
$\gamma_{\beta-n}$	599 *1*	1.7 *3*
$\gamma_{\beta-}$	605.72 *18*	1.8 *1*
$\gamma_{\beta-}$	631.08 *17*	4.6 *3*
$\gamma_{\beta-n}$	644.8 *10*	0.70 *14*
$\gamma_{\beta-}$	656.16 *9*	11.5 *6*
$\gamma_{\beta-}$	668.4 *3*	1.0 *2*
$\gamma_{\beta-n}$	687.4 *10*	0.035 *7*
$\gamma_{\beta-}$	692.5 *3*	1.2 *2*
$\gamma_{\beta-}$	714.1 *3*	1.1 *2*
$\gamma_{\beta-}$	727.34 *11*	2.79 *19*
$\gamma_{\beta-n}$	749.7 *10*	0.035 *7*
$\gamma_{\beta-}$	776.9 *3*	~0.3
$\gamma_{\beta-}$	810.65 *20*	1.4 *2*
$\gamma_{\beta-}$	871.82 *11*	1.8 *3*
$\gamma_{\beta-}$	883.1 *4*	~0.3
$\gamma_{\beta-n}$	985.7 *10*	0.21 *4*
$\gamma_{\beta-}$	1080.1 *3*	0.9 *2* ?
$\gamma_{\beta-}$	1093.4 *3*	1.0 *2*
$\gamma_{\beta-}$	1105.2 *4*	0.8 *2*
$\gamma_{\beta-}$	1120.0 *5*	~0.5
$\gamma_{\beta-}$	1150.0 *5*	~0.6
$\gamma_{\beta-}$	1252.4 *8*	0.5 *2*
$\gamma_{\beta-}$	1324.3 *3*	2.0 *3*
$\gamma_{\beta-}$	1359.85 *20*	3.2 *3*
$\gamma_{\beta-}$	1386.5 *3*	3.3 *3*
$\gamma_{\beta-}$	1456.12 *16*	6.3 *5*
$\gamma_{\beta-}$	1600.59 *16*	4.0 *3*
$\gamma_{\beta-}$	1611.4 *5*	~0.5
$\gamma_{\beta-}$	1693.6 *3*	13.1 *7*
$\gamma_{\beta-}$	1772.20 *21*	2.8 *3*
$\gamma_{\beta-}$	1820.3 *3*	3.8 *4*
$\gamma_{\beta-}$	1866.1 *7*	1.1 *3*
$\gamma_{\beta-}$	1964.6 *3*	2.0 *3*
$\gamma_{\beta-}$	1979.8 *5*	1.5 *3*
$\gamma_{\beta-}$	2008.5 *8*	~0.5
$\gamma_{\beta-}$	2035.4 *10*	~0.5
$\gamma_{\beta-}$	2087.18 *19*	1.3 *3*
$\gamma_{\beta-}$	2092.7 *5*	1.4 *3*
$\gamma_{\beta-}$	2101.8 *6*	~0.5
$\gamma_{\beta-}$	2145.2 *4*	3.3 *4*
$\gamma_{\beta-}$	2172.14 *24*	17.2 *20*
$\gamma_{\beta-}$	2316.61 *24*	10.4 *10*
$\gamma_{\beta-}$	2487.8 *7*	1.5 *5*
$\gamma_{\beta-}$	2804.7 *6*	2.5 *7*
$\gamma_{\beta-}$	3010.6 *4*	9.4 *13*
$\gamma_{\beta-}$	3030.6 *5*	2.4 *4*
$\gamma_{\beta-}$	3146.8 *4*	2.8 *5*
$\gamma_{\beta-}$	3291.3 *4*	4.7 *7*

Photons (^{98}Rb(114 + 96 ms))
(continued)

γ_{mode}	γ(keV)	γ(rel)†
$\gamma_{\beta-}$	3481.2 *5*	1.6 *4*
$\gamma_{\beta-}$	3625.7 *5*	2.2 *5*
$\gamma_{\beta-}$	3680.8 *7*	3.1 *6*

† γ(%) for β-n

$^{98}_{37}$Rb(96 *3* ms)

Mode: β-, β-n(15.9 *13* %), β-2n(0.060 *9* %)
(includes ^{98}Rb(114 ms))
Δ: ~-53860 keV
SpA: 1.66×10^{11} Ci/g
Prod: fission
β-n(avg): 490

see ^{98}Rb(114 ms) for γ rays

$^{98}_{38}$Sr(650 *30* ms)

Mode: β-, β-n(0.8 *2* %)
Δ: -66490 *80* keV
SpA: 1.09×10^{11} Ci/g
Prod: daughter ^{98}Rb; fission

Photons (^{98}Sr)

⟨γ⟩=134 *4* keV

γ_{mode}	γ(keV)	γ(%)†
γ(E1)	36.30 *11*	6.44 *23*
γ	50.96 *14*	<0.23
γM1+E2	52.25 *15*	1.8 *5* ?
γE1,E2	119.022 *3*	23.0 *11*
γ	169.98 *14*	0.92 *23*
γ	203.67 *20*	<0.23 ?
γ	428.35 *15*	7.4 *5* ?
γ	444.298 *20*	8.1 *5*
γ	480.60 *11*	2.53 *23*
γ	563.32 *2*	2.99 *23*
γ[M1+E2]	599.62 *11*	0.92 *23*

† uncert(syst): 35% for β-

Continuous Radiation (^{98}Sr)

⟨β-⟩=2558 keV;⟨IB⟩=10.1 keV

E_{bin}(keV)		⟨ ⟩(keV)	(%)
0 - 10	β-	0.00150	0.0297
	IB	0.067	
10 - 20	β-	0.00461	0.0307
	IB	0.066	0.46
20 - 40	β-	0.0194	0.064
	IB	0.131	0.46
40 - 100	β-	0.156	0.220
	IB	0.38	0.59
100 - 300	β-	2.24	1.08
	IB	1.19	0.66
300 - 600	β-	12.6	2.72
	IB	1.54	0.36
600 - 1300	β-	116	11.8
	IB	2.7	0.31
1300 - 2500	β-	636	32.9
	IB	2.7	0.153

Continuous Radiation (^{98}Sr)
(continued)

E_{bin}(keV)		⟨ ⟩(keV)	(%)
2500 - 5000	β-	1721	49.8
	IB	1.26	0.041
5000 - 5880	β-	70	1.34
	IB	0.0040	7.4×10^{-5}

$^{98}_{39}$Y (640 *30* ms)

Mode: β-, β-n(0.3 *1* %)
Δ: -72370 *60* keV
SpA: 1.10×10^{11} Ci/g
Prod: fission

Photons (^{98}Y)

⟨γ⟩=889 *25* keV

γ_{mode}	γ(keV)	γ(%)†
Zr $K_{\alpha2}$	15.691	0.93 *18*
Zr $K_{\alpha1}$	15.775	1.8 *3*
Zr $K_{\beta1}$'	17.663	0.40 *8*
Zr $K_{\beta2}$'	18.086	0.065 *13*
γ E2	213.768 *22*	1.44 *14*
γ E2	268.44 *15*	2.52 *22*
γ[M1+E2]	367.57 *11*	0.24 *6*
γ[E2]	368.72 *13*	0.18 *4*
γ	385.92 *20*	0.30 *12*
γ[M1+E2]	521.37 *15*	0.72 *6*
γ	547.32 *20*	0.30 *12*
γ	599.82 *20*	0.30 *12*
γ E2	636.00 *15*	0.48 *8*
γ E2	736.28 *14*	0.42 *6*
γ	840.12 *20*	0.30 *12*
γ[E2]	890.08 *15*	0.36 *12*
γ E2	1222.65 *14*	12.0 *8*
γ E2	1590.22 *16*	4.86 *19*
γ[E2]	1744.01 *18*	1.38 *10*
γ	2305.33 *25*	0.66 *18*
γ	2419.96 *25*	1.62 *20*
γ	2573.76 *24*	0.84 *18*
γ	2941.32 *23*	5.8 *3*
γ	3064.2 *5*	0.48 *18*
γ	3203.5 *5*	0.78 *14*
γ	3227.7 *4*	1.38 *17*
γ	3310.03 *24*	2.4 *2*
γ	3375.5 *5*	0.66 *24*
γ	3468.4 *5*	0.66 *24*
γ	4450.3 *4*	3.42 *25*

† uncert(syst): 33% for β-

Atomic Electrons (^{98}Y)

⟨e⟩=40 *7* keV

e_{bin}(keV)	⟨e⟩(keV)	e(%)
2 - 18	0.24	4.8 *7*
196 - 214	0.203	0.102 *9*
250	0.175	0.070 *6*
266 - 268	0.0285	0.0107 *8*
350 - 386	0.021	0.0059 *14*
503 - 547	0.0155	0.0030 *5*
582 - 618	0.0084	0.0014 *3*
633 - 636	0.00078	0.000124 *17*
718 - 736	0.0043	0.00060 *8*
822	0.0016	~0.00019
836	34.7	4.2 *8*
838 - 840	0.00021	$\sim3\times10^{-5}$
851	4.9	0.57 *15*
852 - 890	0.041	0.0048 *16*
1205 - 1223	0.063	0.0052 *3*
1572 - 1590	0.0193	0.00122 *5*
1726 - 1744	0.0050	0.000291 *21*

Atomic Electrons (^{98}Y)
(continued)

e_{bin}(keV)	⟨e⟩(keV)	e(%)
2287 - 2305	0.0015	6.6 *23* $\times 10^{-5}$
2402 - 2420	0.0036	0.00015 *4*
2556 - 2573	0.0018	7.0 *21* $\times 10^{-5}$
2923 - 2941	0.0114	0.00039 *9*
3046 - 3064	0.0009	3.0 *12* $\times 10^{-5}$
3186 - 3227	0.0040	0.000126 *22*
3292 - 3375	0.0056	0.00017 *3*
3450 - 3468	0.0012	3.4 *13* $\times 10^{-5}$
4432 - 4450	0.0055	0.000124 *22*

Continuous Radiation (^{98}Y)
⟨β-⟩=3701 keV;⟨IB⟩=18 keV

E_{bin}(keV)		⟨ ⟩(keV)	(%)
0 - 10	β-	0.00083	0.0165
	IB	0.080	
10 - 20	β-	0.00256	0.0171
	IB	0.079	0.55
20 - 40	β-	0.0108	0.0358
	IB	0.158	0.55
40 - 100	β-	0.087	0.122
	IB	0.47	0.71
100 - 300	β-	1.24	0.60
	IB	1.47	0.81
300 - 600	β-	7.0	1.52
	IB	2.0	0.46
600 - 1300	β-	66	6.7
	IB	3.8	0.43
1300 - 2500	β-	386	19.9
	IB	4.6	0.26
2500 - 5000	β-	1685	45.7
	IB	4.4	0.13
5000 - 7500	β-	1418	23.7
	IB	0.79	0.0138
7500 - 8910	β-	137	1.75
	IB	0.0119	0.000141

$^{98}_{39}$Y (2.0 *2* s)

Mode: β-, β-n(3.4 *10* %)
Δ: -72370 *60* keV
SpA: 4.9×10^{10} Ci/g
Prod: fission

Photons (^{98}Y)
⟨γ⟩=3031 *76* keV

γ_{mode}	γ(keV)	γ(%)†
Zr $K_{\alpha2}$	15.691	0.30 *12*
Zr $K_{\alpha1}$	15.775	0.57 *24*
Zr $K_{\beta1}$'	17.663	0.13 *5*
Zr $K_{\beta2}$'	18.086	0.021 *9*
$\gamma_{\beta-}$[E1]	241.36 *20*	6.7 *11*
$\gamma_{\beta-}$[E1]	252.80 *11*	~4
$\gamma_{\beta-}$[M1+E2]	367.57 *11*	~0.13
$\gamma_{\beta-}$[E2]	368.72 *13*	1.4 *4*
$\gamma_{\beta-}$[E1]	583.12 *3*	17.4 *13*
$\gamma_{\beta-}$[E1]	620.368 *19*	72 *4*
$\gamma_{\beta-}$	647.44 *3*	54 *3*
$\gamma_{\beta-}$E2	736.28 *14*	~0.23
$\gamma_{\beta-}$[E1]	752.46 *20*	6.7 *19*
$\gamma_{\beta-}$E2	1222.65 *14*	93.1
$\gamma_{\beta-}$E2	1590.22 *16*	~3 ?
$\gamma_{\beta-}$	1787.2 *5*	4.0 *8*
$\gamma_{\beta-}$	1801.5 *5*	44 *3*

† uncert(syst): 1.9% for β-

Atomic Electrons (^{98}Y)
⟨e⟩=13 *5* keV

e_{bin}(keV)	⟨e⟩(keV)	e(%)
2 - 18	0.079	1.5 *5*
223 - 253	0.21	0.091 *21*
350 - 369	0.057	0.016 *4*
565 - 602	0.392	0.066 *4*
618 - 647	0.5	0.08 *3*
718 - 752	0.028	0.0038 *9*
836	9.5	~1
851	1.3	~0.16
852	0.010	~0.0012
1205	0.428	0.0355 *10*
1220 - 1223	0.0575	0.00471 *11*
1572 - 1590	0.011	0.0007 *3*
1769 - 1801	0.13	0.0075 *21*

Continuous Radiation (^{98}Y)
⟨β-⟩=2618 keV;⟨IB⟩=10.6 keV

E_{bin}(keV)		⟨ ⟩(keV)	(%)
0 - 10	β-	0.00170	0.0338
	IB	0.067	
10 - 20	β-	0.0052	0.0349
	IB	0.066	0.46
20 - 40	β-	0.0220	0.073
	IB	0.132	0.46
40 - 100	β-	0.176	0.249
	IB	0.39	0.59
100 - 300	β-	2.50	1.20
	IB	1.19	0.66
300 - 600	β-	13.8	2.99
	IB	1.55	0.36
600 - 1300	β-	124	12.6
	IB	2.8	0.31
1300 - 2500	β-	633	33.0
	IB	2.8	0.158
2500 - 5000	β-	1528	44.2
	IB	1.61	0.050
5000 - 7500	β-	317	5.7
	IB	0.070	0.0013
7500 - 7687	β-	0.0384	0.000509
	IB	1.56 $\times 10^{-7}$	2.1 $\times 10^{-9}$

$^{98}_{40}$Zr(30.7 *4* s)

Mode: β-
Δ: -81288 *20* keV
SpA: 3.71×10^{9} Ci/g
Prod: fission

Continuous Radiation (^{98}Zr)
⟨β-⟩=907 keV;⟨IB⟩=1.9 keV

E_{bin}(keV)		⟨ ⟩(keV)	(%)
0 - 10	β-	0.0135	0.269
	IB	0.035	
10 - 20	β-	0.0414	0.276
	IB	0.034	0.24
20 - 40	β-	0.172	0.57
	IB	0.067	0.23
40 - 100	β-	1.35	1.91
	IB	0.19	0.29
100 - 300	β-	17.6	8.5
	IB	0.51	0.29
300 - 600	β-	82	18.0
	IB	0.51	0.121
600 - 1300	β-	449	47.8
	IB	0.50	0.061
1300 - 2240	β-	356	22.7
	IB	0.061	0.0042

$^{98}_{41}$Nb(2.86 *6* s)

Mode: β-
Δ: -83528 *6* keV
SpA: 3.57×10^{10} Ci/g
Prod: daughter ^{98}Zr; ^{98}Mo(n,p)

Photons (^{98}Nb)
⟨γ⟩=84 *6* keV

γ_{mode}	γ(keV)	γ(%)†
γ[M1+E2]	326.27 *11*	0.077 *13*
γ[M1+E2]	447.52 *18*	0.47 *12*
γ[M1+25%E2]	644.98 *6*	0.85 *8*
γ[E2]	697.68 *11*	0.045 *9*
γ[M1+E2]	773.79 *16*	0.24 *7*
γE2	787.32 *8*	3.2
γE2+18%M1	971.25 *10*	0.80 *8*
γ[E2]	1023.95 *11*	1.60 *15*
γ	1250.0 *6*	0.22 *4*
γ[M1+E2]	1418.77 *16*	0.41 *6*
γ[E2]	1432.30 *9*	0.80 *9*
γ[E2]	1758.57 *11*	0.16 *3*
γ[E2]	1820.8 *6*	0.08 *3*

† 16% uncert(syst)

Continuous Radiation (^{98}Nb)
⟨β-⟩=1959 keV;⟨IB⟩=6.7 keV

E_{bin}(keV)		⟨ ⟩(keV)	(%)
0 - 10	β-	0.00300	0.060
	IB	0.057	
10 - 20	β-	0.0092	0.061
	IB	0.057	0.39
20 - 40	β-	0.0386	0.128
	IB	0.112	0.39
40 - 100	β-	0.308	0.434
	IB	0.33	0.50
100 - 300	β-	4.27	2.05
	IB	0.98	0.55
300 - 600	β-	22.9	4.95
	IB	1.22	0.29
600 - 1300	β-	190	19.4
	IB	2.0	0.23
1300 - 2500	β-	815	43.0
	IB	1.59	0.092
2500 - 4586	β-	926	29.9
	IB	0.35	0.0119

$^{98}_{41}$Nb(51.3 *4* min)

Mode: β-
Δ: -83444 *7* keV
SpA: 3.74×10^{7} Ci/g
Prod: ^{98}Mo(n,p)

Photons (^{98}Nb)
⟨γ⟩=2711 *61* keV

γ_{mode}	γ(keV)	γ(%)†
Mo $K_{\alpha2}$	17.374	0.20 *5*
Mo $K_{\alpha1}$	17.479	0.37 *10*
Mo $K_{\beta1}$'	19.602	0.088 *24*
Mo $K_{\beta2}$'	20.091	0.015 *4*
γ[M1+E2]	86.71 *13*	0.33 *7*
γ[E1]	146.91 *16*	0.13 *5*
γ	158.05 *18*	0.50 *9*
γ[E1]	172.90 *8*	3.1 *3*
γ[E1]	259.14 *9*	0.53 *9*
γ[M1+E2]	326.27 *11*	0.049 *10*
γ	335.37 *8*	10.7 *7*
γ	348.13 *18*	0.24 *5*

Photons (^{98}Nb) (continued)

γmode	γ(keV)	γ(%)†
γ	351.24 20	0.38 9
γ	399.6 3	0.23 5
γ [E2]	434.37 7	1.21 19
γ [M1+E2]	447.52 18	0.34 7
γ	454.98 9	0.74 11
γ [M1+E2]	543.93 8	0.60 9
γ	569.2 3	0.25 5
γ [M1+E2]	574.95 11	0.26 5
γ [E2]	603.26 17	0.45 9
γ [M1+25%E2]	644.98 6	5.6 3
γ [M1+E2]	672.63 13	0.61 9
γ	688.4 3	0.25 5
γ [E2]	697.68 11	0.30 5
γ [M1+E2]	713.94 7	8.6 6
γ E2	722.70 6	71 5
γ [M1+E2]	773.79 16	0.18 5
γ E2	787.32 8	93 6
γ [E2]	791.66 7	7.6 6
γ [M1+E2]	797.82 15	0.59 9
γ [E1]	820.19 14	0.22 5
γ [E2]	823.50 7	2.4 3
γ E2	833.56 8	10.2 7
γ [E2]	878.25 10	0.73 8
γ	886.1 3	0.19 5
γ	906.63 22	0.52 7
γ [E1]	909.69 12	1.02 9
γ [E2]	946.12 21	0.17 5
γ E2+18%M1	971.25 10	0.51 6
γ [E1]	987.41 12	0.48 7
γ [E2]	993.10 12	1.12 9
γ [E1]	996.40 10	2.05 19
γ [E1]	1004.06 18	0.24 5
γ [E2]	1023.95 11	1.02 8
γ	1048.7 3	0.22 5
γ	1064.67 20	0.34 6
γ [M1+E2]	1102.65 13	0.49 7
γ (M1+E2)	1110.96 15	0.83 11
γ [E2]	1122.03 16	0.14 5
γ	1168.92 8	18.0 13
γ [E2]	1189.4 3	0.17 5
γ [E1]	1194.05 13	0.55 7
γ	1199.3 4	0.13 5
γ [E2]	1204.11 18	0.17 5
γ [E2]	1214.16 17	0.15 5
γ [E2]	1221.68 16	0.29 6
γ [E1]	1230.39 10	1.49 19
γ [M1+E2]	1257.87 7	1.12 9
γ [E1]	1308.90 16	0.20 5
γ [M1+E2]	1317.61 13	0.82 8
γ [M1+E2]	1323.72 17	0.28 6
γ [E2]	1335.59 8	1.30 19
γ [E2]	1341.59 19	0.49 7
γ [E1]	1377.76 25	0.09 3
γ	1387.4 3	0.27 6
γ	1393.5 3	0.19 5
γ [M1+E2]	1399.70 21	0.13 5
γ	1407.69 21	0.26 5
γ	1415.4 4	0.12 5
γ [M1+E2]	1418.77 16	0.30 6
γ [E2]	1432.30 9	5.4 4
γ [E2]	1436.64 7	2.5 3
γ [M1+E2]	1467.24 20	0.76 7
γ [M1+E2]	1511.75 15	5.4 5
γ [E2]	1518.73 23	0.13 5
γ [M1+E2]	1541.03 18	2.14 19
γ [M1+E2]	1546.19 7	4.2 4
γ [E2]	1589.47 15	0.20 5
γ	1618.0 8	0.07 3
γ [E1]	1632.38 12	0.88 9
γ	1656.24 21	0.35 6
γ	1689.6 5	0.09 3
γ [M1+E2]	1701.74 8	9.3 7
γ [E2]	1758.57 11	0.102 18
γ [E2]	1779.46 10	0.29 6
γ [E2]	1786.1 4	0.14 5
γ	1792.1 4	0.18 5
γ [M1+E2]	1816.59 12	0.45 7
γ [M1+E2]	1885.45 23	2.8 3
γ [M1+E2]	1945.52 15	1.39 19
γ [E2]	1980.56 7	3.3 3
γ	2006.8 4	0.09 3
γ	2017.70 11	0.27 6
γ [E2]	2023.24 16	0.10 3
γ [M1+E2]	2037.65 17	0.08 3

Photons (^{98}Nb) (continued)

γmode	γ(keV)	γ(%)†
γ [M1+E2]	2113.64 21	0.21 7
γ	2127.7 5	0.07 3
γ	2133.6 9	~0.028
γ	2190.6 7	~0.06
γ	2209.2 7	~0.037
γ [E2]	2234.44 15	0.12 4
γ	2296.4 6	~0.06
γ	2378.93 21	0.12 4
γ [E2]	2424.43 9	0.14 5
γ	2433.1 8	~0.07
γ	2471.6 5	0.08 4
γ [E2]	2539.27 13	0.07 3
γ [E2]	2608.14 24	0.13 5
γ [E2]	2668.20 16	0.11 4
γ [E2]	2760.33 17	0.07 3
γ [E2]	2836.32 21	0.046 19

† 6.4% uncert(syst)

Atomic Electrons (^{98}Nb)

⟨e⟩=3.7 3 keV

e_{bin}(keV)	⟨e⟩(keV)	e(%)
3 - 20	0.051	0.96 24
67	0.22	~0.32
84 - 127	0.07	~0.08
138 - 147	0.08	~0.06
153	0.107	0.070 7
155 - 173	0.032	0.020 6
239 - 259	0.0106	0.0044 7
306	0.0019	~0.0006
315	0.31	~0.10
323 - 351	0.07	0.020 8
380 - 428	0.045	0.0109 15
432 - 455	0.019	0.0045 15
524 - 572	0.0184	0.0034 4
574 - 603	0.0074	0.00126 23
625 - 673	0.088	0.0140 8
678 - 688	0.0036	0.00053 8
694	0.092	0.0133 12
695 - 698	0.00048	6.9 9 $\times10^{-5}$
703	0.75	0.106 7
711 - 754	0.123	0.0171 9
767	0.86	0.113 7
771 - 778	0.076	0.0098 8
784	0.094	0.0120 8
785 - 834	0.165	0.0205 8
858 - 907	0.0146	0.00167 18
909 - 951	0.0050	0.00053 6
967 - 1004	0.0249	0.00253 12
1021 - 1065	0.0040	0.00038 8
1083 - 1122	0.0101	0.00092 9
1149 - 1197	0.09	0.008 3
1199 - 1238	0.0120	0.00098 7
1255 - 1304	0.0068	0.00053 5
1306 - 1342	0.0106	0.00080 8
1358 - 1407	0.0056	0.00041 6
1408 - 1447	0.044	0.00308 18
1464 - 1546	0.059	0.00388 23
1569 - 1656	0.0044	0.00027 3
1670 - 1766	0.043	0.00258 21
1772 - 1865	0.0126	0.00068 7
1883 - 1980	0.0194	0.00100 7
1987 - 2037	0.0018	8.8 16 $\times10^{-5}$
2094 - 2190	0.0013	6.2 12 $\times10^{-5}$
2206 - 2296	0.00058	2.6 6 $\times10^{-5}$
2359 - 2452	0.00114	4.8 9 $\times10^{-5}$
2469 - 2539	0.00026	1.0 3 $\times10^{-5}$
2588 - 2668	0.00073	2.8 6 $\times10^{-5}$
2740 - 2836	0.00033	1.2 3 $\times10^{-5}$

Continuous Radiation (^{98}Nb)

⟨β-⟩=785 keV;⟨IB⟩=1.55 keV

E_{bin}(keV)		⟨ ⟩(keV)	(%)
0 - 10	β-	0.0216	0.430
	IB	0.031	
10 - 20	β-	0.066	0.439
	IB	0.030	0.21
20 - 40	β-	0.273	0.91
	IB	0.059	0.20
40 - 100	β-	2.09	2.97
	IB	0.163	0.25
100 - 300	β-	25.5	12.4
	IB	0.43	0.24
300 - 600	β-	105	23.1
	IB	0.41	0.097
600 - 1300	β-	403	44.3
	IB	0.37	0.045
1300 - 2500	β-	242	15.1
	IB	0.064	0.0041
2500 - 3160	β-	6.8	0.253
	IB	0.00044	1.69 $\times10^{-5}$

$^{98}_{42}$Mo(stable)

Δ: -88113.2 20 keV
%: 24.13 6

$^{98}_{43}$Tc(4.2 3 $\times10^{6}$ yr)

Mode: β-
Δ: -86429 4 keV
SpA: 0.00087 Ci/g
Prod: ^{98}Mo(p,n); multiple n-capture from ^{96}Ru

Photons (^{98}Tc)

⟨γ⟩=1394.3 11 keV

γmode	γ(keV)	γ(%)
Ru $K_{\alpha2}$	19.150	0.088 3
Ru $K_{\alpha1}$	19.279	0.167 5
Ru $K_{\beta1}$'	21.650	0.0405 13
Ru $K_{\beta2}$'	22.210	0.0075 3
γ E2	652.41 5	99.7
γ E2	745.35 5	99.8

Atomic Electrons (^{98}Tc)

⟨e⟩=2.95 4 keV

e_{bin}(keV)	⟨e⟩(keV)	e(%)
3	0.0090	0.31 3
16	0.0090	0.055 6
18	0.00085	0.0046 5
19	0.0031	0.0166 17
21	0.00031	0.00146 15
22	6.2 $\times10^{-5}$	0.00029 3
630	1.40	0.223 5
649	0.166	0.0255 5
650	0.00814	0.001253 25
652	0.0385	0.00590 12
723	1.141	0.158 3
742	0.133	0.0180 4
743	0.00556	0.000748 15
745	0.0304	0.00408 8

Continuous Radiation (^{98}Tc)

⟨β-⟩=120 keV;⟨IB⟩=0.050 keV

E_{bin}(keV)		⟨ ⟩(keV)	(%)
0 - 10	β-	0.261	5.2
	IB	0.0062	
10 - 20	β-	0.77	5.1
	IB	0.0055	0.038
20 - 40	β-	2.99	10.0
	IB	0.0091	0.032
40 - 100	β-	18.8	27.2
	IB	0.017	0.028
100 - 300	β-	88	49.6
	IB	0.0117	0.0084
300 - 398	β-	9.0	2.77
	IB	8.7×10^{-5}	2.8×10^{-5}

$^{98}_{44}$Ru(stable)

Δ: -88225 *6* keV

%: 1.88 *5*

$^{98}_{45}$Rh(8.7 *2* min)

Mode: ε

Δ: >-83168 keV

SpA: 2.20×10^{8} Ci/g

Prod: daughter ^{98}Pd; ^{98}Ru(p,n); protons on Cd; ^{89}Y(^{12}C,3n)

Photons (^{98}Rh)

⟨γ⟩=832 *12* keV

γ_{mode}	γ(keV)	γ(%)†
Ru L_ℓ	2.253	0.0093 *16*
Ru L_η	2.382	0.0043 *7*
Ru L_α	2.558	0.24 *4*
Ru L_β	2.731	0.149 *25*
Ru L_γ	3.032	0.0114 *21*
Ru $K_{\alpha2}$	19.150	2.10 *7*
Ru $K_{\alpha1}$	19.279	3.99 *14*
Ru $K_{\beta1}'$	21.650	0.96 *3*
Ru $K_{\beta2}'$	22.210	0.180 *7*
γ E2+17%M1	597.7 *4*	0.75 *9* ?
γ E2,M1	614.5 *5*	0.82 *13*
γ E2	652.41 *5*	94.2
γ E2	745.35 *5*	5.3 *6*
γ E2+0.6%M1	762.2 *3*	1.7 *4*
γ	1164.4 *3*	4.5 *5*
γ E2	1414.6 *3*	1.14 *15*
γ	1816.8 *3*	4.7 *5*

† 0.64% uncert(syst)

Atomic Electrons (^{98}Rh)

⟨e⟩=2.22 *5* keV

e_{bin}(keV)	⟨e⟩(keV)	e(%)
3	0.214	7.4 *8*
16	0.214	1.32 *13*
18	0.0202	0.110 *11*
19	0.074	0.39 *4*
21 - 22	0.0088	0.042 *4*
576 - 615	0.028	0.0048 *11*
630	1.32	0.210 *4*
649	0.156	0.0241 *5*
650 - 652	0.0439	0.00674 *12*
723 - 762	0.090	0.0124 *11*
1142 - 1164	0.026	0.0022 *9*
1393 - 1415	0.0068	0.00048 *6*
1795 - 1816	0.018	0.0010 *3*

Continuous Radiation (^{98}Rh)

⟨β+⟩=1318 keV;⟨IB⟩=5.4 keV

E_{bin}(keV)		⟨ ⟩(keV)	(%)
0 - 10	β+	3.9×10^{-6}	4.74×10^{-5}
	IB	0.044	
10 - 20	β+	0.000120	0.00073
	IB	0.044	0.31
20 - 40	β+	0.00261	0.0080
	IB	0.086	0.30
40 - 100	β+	0.098	0.127
	IB	0.24	0.38
100 - 300	β+	4.23	1.92
	IB	0.72	0.40
300 - 600	β+	33.0	7.1
	IB	0.86	0.20
600 - 1300	β+	287	29.5
	IB	1.42	0.162
1300 - 2500	β+	811	44.6
	IB	1.42	0.080
2500 - 4404	β+	183	6.7
	IB	0.59	0.020
	Σβ+		90

$^{98}_{45}$Rh(3.5 *3* min)

Mode: ε

Δ: -83168 *12* keV

SpA: 5.5×10^{8} Ci/g

Prod: alphas on Ru; protons on Ru; deuterons on Ru

Photons (^{98}Rh)

⟨γ⟩=1469 *140* keV

γ_{mode}	γ(keV)	γ(%)
γ (M1+41%E2)	382.1 *11*	~3
γ E2+17%M1	597.7 *4*	4.6 *5*
γ E2,M1	614.5 *5*	5.0 *5*
γ E2	652.41 *5*	96
γ E2	745.35 *5*	78 *8*
γ E2+0.6%M1	762.2 *3*	<8
γ E2	823.7 *17*	<2 ?
γ E2,M1	1144.4 *11*	8.5 *9*
γ E2	1414.6 *3*	3.8 *4*

Continuous Radiation (^{98}Rh)

⟨β+⟩=928 keV;⟨IB⟩=4.1 keV

E_{bin}(keV)		⟨ ⟩(keV)	(%)
0 - 10	β+	6.3×10^{-6}	7.6×10^{-5}
	IB	0.034	
10 - 20	β+	0.000194	0.00118
	IB	0.035	0.24
20 - 40	β+	0.00419	0.0129
	IB	0.068	0.24
40 - 100	β+	0.156	0.203
	IB	0.19	0.29
100 - 300	β+	6.6	2.99
	IB	0.53	0.30
300 - 600	β+	48.8	10.5
	IB	0.62	0.144
600 - 1300	β+	359	37.5
	IB	1.07	0.120
1300 - 2500	β+	512	30.5
	IB	1.24	0.070
2500 - 3659	β+	1.44	0.057
	IB	0.29	0.0104
	Σβ+		82

$^{98}_{46}$Pd(17.7 *3* min)

Mode: ε

Δ: -81299 *21* keV

SpA: 1.084×10^{8} Ci/g

Prod: ^{96}Ru(α,2n); protons on Cd; ^{89}Y(^{14}N,5n)

Photons (^{98}Pd)

⟨γ⟩=410 *40* keV

γ_{mode}	γ(keV)	γ(%)†
γ (M1)	62.19 *21*	0.09 *3*
γ (M1)	67.74 *16*	8.5 *19*
γ M1	106.84 *16*	12.3 *24*
γ M1	112.39 *18*	47.4
γ (M1+E2)	174.58 *15*	10.0 *19*
γ (M1,E2)	663.28 *25*	25 *5*
γ [M1+E2]	725.5 *3*	4.7 *10*
γ [M1+E2]	837.86 *25*	14 *3*

† 4.6% uncert(syst)

Continuous Radiation (^{98}Pd)

⟨β+⟩=12.9 keV;⟨IB⟩=0.77 keV

E_{bin}(keV)		⟨ ⟩(keV)	(%)
0 - 10	β+	5.1×10^{-6}	6.2×10^{-5}
	IB	0.00078	
10 - 20	β+	0.000160	0.00098
	IB	0.0109	0.061
20 - 40	β+	0.00344	0.0106
	IB	0.017	0.076
40 - 100	β+	0.119	0.155
	IB	0.0053	0.0082
100 - 300	β+	3.43	1.62
	IB	0.041	0.020
300 - 600	β+	8.7	2.06
	IB	0.167	0.037
600 - 1300	β+	0.71	0.112
	IB	0.46	0.053
1300 - 1757	IB	0.068	0.0048
	Σβ+		4.0

$^{98}_{47}$Ag(46.7 *10* s)

Mode: ε

Δ: -73070 *300* keV syst

SpA: 2.45×10^{9} Ci/g

Prod: ^{14}N on ^{92}Mo; ^{40}Ca on ^{62}Ni

Photons (^{98}Ag)

⟨γ⟩=2377 *52* keV

γ_{mode}	γ(keV)	γ(%)†
γ [E2]	352.3 *7*	3.4 *5*
γ	452.0 *4*	11.0 *6*
γ [E2]	571.1 *3*	53.0 *11*
γ	612.0 *3*	6.5 *7*
γ [E2]	660.4 *10*	~6
γ [E2]	678.5 *3*	85 *4*
γ [E2]	863.1 *3*	100 *4*
γ	964.0 *7*	2.6 *3*
γ	1013.0 *7*	3.0 *4*
γ	1023.5 *10*	1.2 *3*
γ	1037.8 *7*	1.8 *3*
γ	1076.6 *10*	3.0 *7*
γ	1105.0 *7*	1.0 *3*
γ	1183.0 *5*	3.4 *3*
γ	1292 *1*	1.3 *3*
γ [M1+E2]	1328.9 *10*	2.4 *3*
γ	1427.2 *5*	4.5 *3*
γ	1468.1 *7*	2.6 *3*

Photons (^{98}Ag)
(continued)

γ_{mode}	γ(keV)	γ(%)†
γ	1523.4 *10*	1.1 *3*
γ	1561.3 *10*	1.3 *3*
γ	1584.6 *7*	3.5 *3*
γ	1651.8 *10*	1.5 *3*
γ	1702.2 *10*	1.1 *2*
γ [E2]	1901.9 *10*	1.9 *3*

† 0.10% uncert(syst)

$^{98}_{48}$Cd($\sim$8 s)

Mode: ϵ
SpA: $\sim 1.4\times10^{10}$ Ci/g
Prod: protons on Sn

A = 99

NDS **12**, 431 (1974)

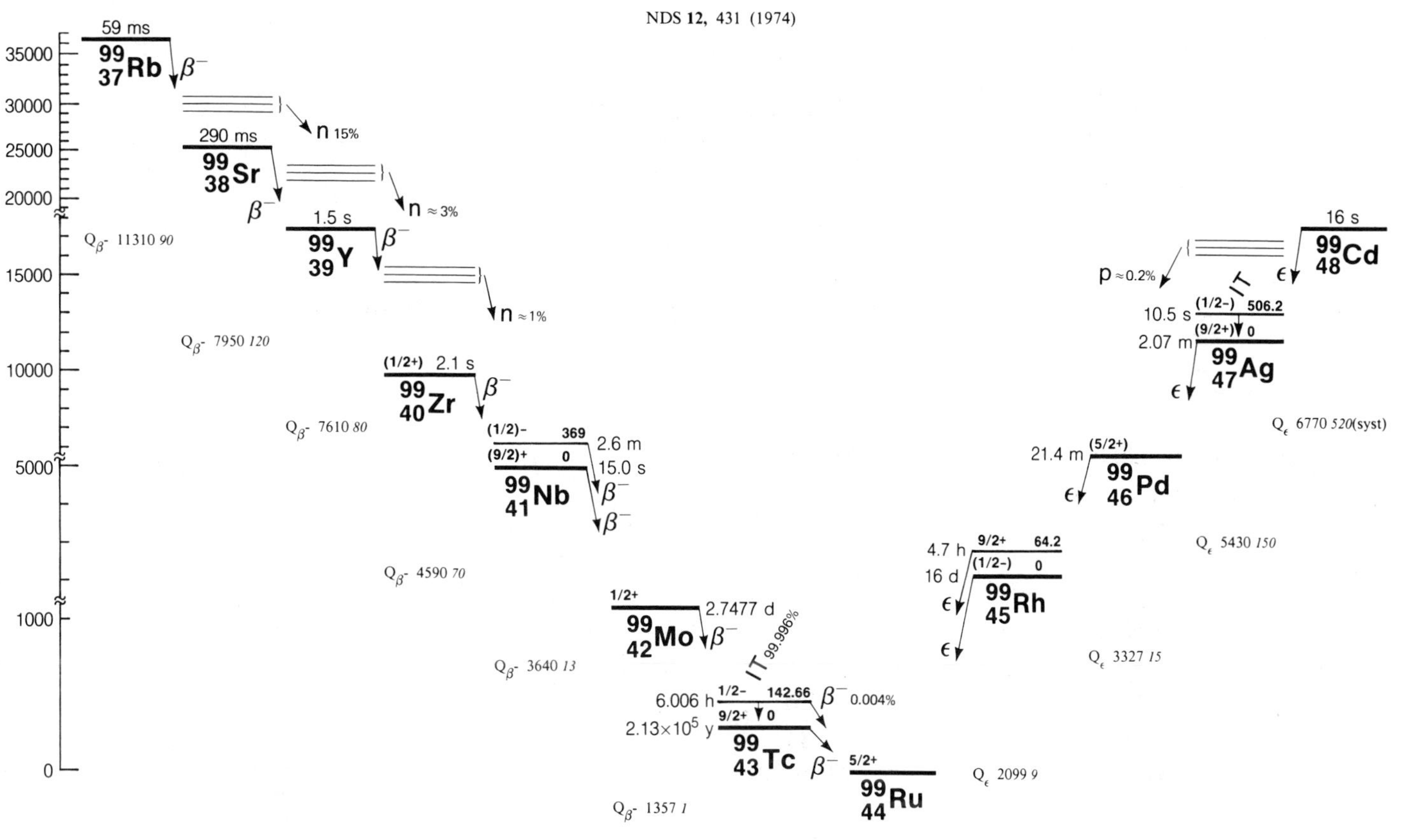

$^{99}_{37}$Rb(59 *4* ms)

Mode: β-, β-n(15 *3* %)
Δ: -50860 *120* keV
SpA: 1.64×10^{11} Ci/g
Prod: fission

Photons (^{99}Rb)
⟨γ⟩=35 *4* keV

γ_{mode}	γ(keV)	γ(%)†
$\gamma_{\beta\text{-n}}$E2	71.18 *12*	0.49 *10*
$\gamma_{\beta\text{-n}}$E2	144.479 *6*	8.1
$\gamma_{\beta\text{-n}}$E2	289.64 *10*	2.4 *5*
$\gamma_{\beta\text{-n}}$	656.16 *9*	0.70 *9*
$\gamma_{\beta\text{-n}}$	727.34 *11*	0.169 *23*
$\gamma_{\beta\text{-n}}$	871.82 *11*	0.11 *3*
$\gamma_{\beta\text{-n}}$	1080.1 *3*	0.81 *16*

† 22% uncert(syst)

$^{99}_{38}$Sr(290 *40* ms)

Mode: β-, β-n(3.4 *24* %)
Δ: -62180 *130* keV
SpA: 1.49×10^{11} Ci/g
Prod: fission

$^{99}_{39}$Y (1.5 *1* s)

Mode: β-, β-n(1.2 *8* %)
Δ: -70130 *100* keV
SpA: 6.1×10^{10} Ci/g
Prod: fission

Photons (^{99}Y)

⟨γ⟩=804 *50* keV

γ_{mode}	γ(keV)	γ(%)†
$\gamma_{\beta-}$	53.3 *3*	
$\gamma_{\beta-}$[M1]	121.76 *16*	54
$\gamma_{\beta-}$[E2]	130.17 *19*	7.6 *11*
$\gamma_{\beta-}$	193.85 *22*	2.7 *8*
$\gamma_{\beta-}$	276.52 *21*	3.2 *11*
$\gamma_{\beta-}$	406.15 *24*	2.2 *8*
$\gamma_{\beta-}$	415.3 *3*	0.8 *3*
$\gamma_{\beta-}$	453.65 *19*	5.9 *11*
$\gamma_{\beta-}$	472.36 *22*	~1
$\gamma_{\beta-}$	536.32 *22*	10.8 *16*
$\gamma_{\beta-}$	575.41 *20*	13.5 *16*
$\gamma_{\beta-}$	600.00 *22*	7.6 *22*
$\gamma_{\beta-}$	602.53 *22*	~5
$\gamma_{\beta-}$	614.0 *3*	6.5 *16*
$\gamma_{\beta-}$	639.9 *4*	4.3 *11*
$\gamma_{\beta-}$	724.29 *21*	24 *3*
$\gamma_{\beta-}$	730.17 *19*	2.4 *8*
$\gamma_{\beta-}$	782.2 *5*	5.9 *16*
$\gamma_{\beta-}$	930.7 *3*	5.4 *16*
$\gamma_{\beta-}$	1013.3 *3*	10 *3*

† uncert(syst): 19% for β-

Continuous Radiation (^{99}Y)

⟨β-⟩=3118 keV;⟨IB⟩=13.7 keV

E_{bin}(keV)		⟨ ⟩(keV)	(%)
0 - 10	β-	0.00094	0.0187
	IB	0.074	
10 - 20	β-	0.00291	0.0194
	IB	0.073	0.51
20 - 40	β-	0.0122	0.0406
	IB	0.146	0.50
40 - 100	β-	0.099	0.139
	IB	0.43	0.66
100 - 300	β-	1.42	0.68
	IB	1.34	0.74
300 - 600	β-	8.1	1.74
	IB	1.8	0.42
600 - 1300	β-	78	7.8
	IB	3.3	0.37
1300 - 2500	β-	471	24.2
	IB	3.7	0.21
2500 - 5000	β-	1973	54
	IB	2.7	0.082
5000 - 7500	β-	587	10.5
	IB	0.131	0.0024
7500 - 7610	β-	0.0081	0.000108
	IB	1.49×10^{-8}	2.0×10^{-10}

$^{99}_{40}$Zr(2.1 *1* s)

Mode: β-
Δ: -77740 *70* keV
SpA: 4.62×10^{10} Ci/g
Prod: fission

Photons (^{99}Zr)

⟨γ⟩=855 *46* keV

γ_{mode}	γ(keV)	γ(%)
γ[M1]	28.21 *17*	~2
γ[M1]	55.93 *16*	2.8 *6*
γ[E1]	81.91 *15*	4.6 *11*
γ[E1]	179.21 *16*	5.1 *9*
γ	387.26 *16*	9.6 *21*
γ	414.81 *16*	5.1 *11*
γ[M1+E2]	461.71 *17*	12.0 *22*
γ[M1+E2]	469.17 *16*	56
γ	489.93 *17*	0.8 *3*
γ[E2]	545.86 *15*	48 *3*
γ	581.0 *4*	~0.6
γ	594.02 *16*	27.6 *22*
γ	627.77 *18*	1.8 *6*
γ	650.0 *3*	2.2 *6*
γ	961.0 *4*	~0.6

Continuous Radiation (^{99}Zr)

⟨β-⟩=1548 keV;⟨IB⟩=4.6 keV

E_{bin}(keV)		⟨ ⟩(keV)	(%)
0 - 10	β-	0.00470	0.093
	IB	0.050	
10 - 20	β-	0.0144	0.096
	IB	0.049	0.34
20 - 40	β-	0.060	0.201
	IB	0.097	0.34
40 - 100	β-	0.480	0.68
	IB	0.28	0.43
100 - 300	β-	6.6	3.18
	IB	0.82	0.46
300 - 600	β-	34.6	7.5
	IB	0.97	0.23
600 - 1300	β-	269	27.6
	IB	1.43	0.165
1300 - 2500	β-	902	48.7
	IB	0.83	0.050
2500 - 4225	β-	335	11.9
	IB	0.050	0.0018

$^{99}_{41}$Nb(15.0 *2* s)

Mode: β-
Δ: -82327 *13* keV
SpA: 7.424×10^{9} Ci/g
Prod: fission; ^{100}Mo(n,np)

Photons (^{99}Nb)

⟨γ⟩=168 *62* keV

γ_{mode}	γ(keV)	γ(%)
γ[E2]	97.7 *1*	45.0
γ[M1+E2]	137.8 *2*	~90

$^{99}_{41}$Nb(2.6 *2* min)

Mode: β-
Δ: -81958 *14* keV
SpA: 7.3×10^{8} Ci/g
Prod: fission; ^{100}Mo(γ,p)

Photons (^{99}Nb)

⟨γ⟩=751 *12* keV

γ_{mode}	γ(keV)	γ(%)†
Mo L_{ℓ}	2.016	0.0077 *16*
Mo L_{η}	2.120	0.0038 *7*
Mo L_{α}	2.293	0.19 *4*
Mo L_{β}	2.423	0.110 *22*
Mo L_{γ}	2.687	0.0070 *15*
Mo $K_{\alpha2}$	17.374	1.69 *19*
Mo $K_{\alpha1}$	17.479	3.2 *4*
Mo $K_{\beta1}'$	19.602	0.76 *9*
Mo $K_{\beta2}'$	20.091	0.132 *15*
γ[E2]	97.7 *1*	6.7
γ[M1+E2]	137.8 *2*	1.32 *7*
γ[M1+E2]	174.4 *2*	0.273 *20*
γ[M1+E2]	197.5 *2*	0.246 *20*
γ[M1+E2]	253.5 *1*	3.70 *20*
γ[M1+E2]	263.8 *1*	0.76 *6*
γ	271.6 *3*	0.153 *13*
γ	280.5 *2*	0.259 *20*
γ[M1+E2]	351.2 *1*	2.77 *15*
γ	356.8 *3*	0.126 *13*
γ	365.2 *3*	0.293 *20*
γ[M1+E2]	379.6 *3*	0.180 *20*
γ	393.9 *3*	0.153 *20*
γ[E2]	427.6 *1*	0.63 *5*
γ	441.7 *2*	0.193 *13*
γ[M1+E2]	450.9 *1*	1.72 *9*
γ	500.2 *3*	0.31 *4*
γ	517.0 *3*	0.41 *7*
γ[M1]	525.4 *2*	1.22 *7*
γ	534.4 *4*	1.55 *9*
γ	535.5 *6*	~0.11
γ	539.2 *4*	0.146 *20*
γ[M1+E2]	548.9 *2*	0.50 *3*
γ	554.3 *2*	0.51 *3*
γ	593.6 *3*	0.36 *3*
γ	600.2 *3*	0.55 *4*
γ	631.8 *2*	1.62 *9*
γ	656.3 *4*	0.76 *6*
γ	668.0 *4*	0.120 *13*
γ	672.3 *5*	0.12 *3*
γ	674.5 *3*	0.80 *4*
γ	694.8 *3*	0.90 *5*
γ	713.6 *4*	0.153 *20*
γ	767.8 *5*	0.21 *3*
γ	780.3 *5*	0.13 *3*
γ	793.0 *2*	1.66 *9*
γ	847.0 *2*	1.00 *6*
γ	867.3 *5*	0.11 *3*
γ	890.2 *4*	0.44 *3*
γ	905.5 *3*	0.66 *5*
γ	927.8 *3*	1.01 *6*
γ	944.8 *4*	0.48 *3*
γ	948.4 *5*	0.293 *20*
γ	988.0 *4*	0.22 *3*
γ	1002.8 *4*	0.19 *3*
γ	1025.4 *3*	0.44 *3*
γ	1047.0 *8*	0.07 *3*
γ	1069.5 *3*	0.42 *3*
γ	1080.6 *3*	0.279 *20*
γ	1090.9 *5*	0.067 *20*
γ	1100.0 *3*	0.153 *20*
γ	1108.5 *3*	0.219 *13*
γ	1111.9 *5*	0.06 *3*
γ	1126.1 *3*	0.206 *20*
γ	1140.9 *4*	0.07 *3*
γ	1146.9 *4*	0.10 *3*
γ	1157.7 *5*	0.10 *3*
γ	1197.6 *5*	~0.027
γ	1220.1 *4*	0.060 *20*
γ	1228.9 *3*	0.200 *20*
γ	1244.5 *4*	0.11 *3*
γ	1253.6 *5*	0.213 *20*
γ	1258.1 *3*	0.58 *6*
γ	1303.8 *4*	0.133 *20*
γ	1314.6 *3*	0.58 *5*
γ	1345.1 *5*	~0.05
γ	1367.8 *4*	0.173 *20*
γ	1375.1 *4*	0.126 *20*
γ	1382.3 *4*	0.126 *20*
γ	1395.5 *4*	0.067 *20*
γ	1403.3 *4*	0.07 *3*
γ	1412.3 *5*	0.033 *13*

Photons (^{99}Nb) (continued)

γ_{mode}	γ(keV)	γ(%)†
γ	1446.7 4	0.12 3
γ	1473.6 3	0.47 3
γ	1531.5 4	0.29 3
γ	1542.2 3	0.36 3
γ	1569.0 4	0.140 20
γ	1587.9 4	0.17 3
γ	1647.9 3	0.226 20
γ	1660.9 6	~0.06
γ	1696.4 3	0.78 7
γ	1708.2 4	0.18 6
γ	1735.8 4	0.57 5
γ	1750.3 6	0.12 3
γ	1780.5 9	0.13 4
γ	1783.6 9	0.22 4
γ	1796.6 5	0.27 4
γ	1848.1 4	0.133 20
γ	1893.9 5	0.31 3
γ	1931.0 4	0.133 13
γ	1937.7 4	0.32 3
γ	1950.4 6	0.22 4
γ	1961.3 4	0.11 5
γ	1992.7 4	0.60 5
γ	2009.6 4	0.51 4
γ	2026.5 5	0.27 3
γ	2055.5 5	0.173 20
γ	2092.7 5	0.206 20
γ	2098.2 4	0.39 3
γ	2111.8 8	0.11 3
γ	2115.0 8	0.10 3
γ	2134.7 4	1.12 7
γ	2183.6 5	0.37 3
γ	2189.9 6	0.17 3
γ	2207.8 5	0.10 3
γ	2237.1 4	0.66 7
γ	2290.2 6	0.13 4
γ	2302.6 6	0.07 3
γ	2326.2 5	0.22 3
γ	2336.1 9	0.15 3
γ	2340.9 7	0.44 11
γ	2375.0 9	0.11 3
γ	2377.9 9	0.20 4
γ	2434.8 6	0.067 20
γ	2462.3 5	0.17 4
γ	2500.8 6	0.047 20
γ	2518.2 6	0.047 20
γ	2543.7 5	0.78 6
γ	2589.8 9	0.11 4
γ	2593.0 8	0.19 4
γ	2614.5 6	0.047 13
γ	2632.0 6	0.14 4
γ	2641.3 5	3.69 22
γ	2660.9 6	0.060 20
γ	2681.7 6	0.15 6
γ	2687.0 5	0.64 5
γ	2729.9 5	0.80 12
γ	2753.6 9	0.047 20
γ	2785.6 5	0.50 7
γ	2851.5 5	3.09 20
γ	2869.7 6	0.093 13
γ	2923.7 5	0.19 3
γ	2970.9 6	~0.027
γ	3001.7 5	0.15 3
γ	3028.5 5	0.067 20
γ	3090.6 9	~0.020
γ	3095.1 9	~0.020
γ	3141.0 6	0.080 20
γ	3177.2 6	0.047 20
γ	3263.3 7	~0.040

† 10% uncert(syst)

Atomic Electrons (^{99}Nb)

⟨e⟩=8.8 7 keV

e_{bin}(keV)	⟨e⟩(keV)	e(%)
3 - 20	0.45	8.7 10
78	5.5	7.0 8
95	1.35	1.42 17
97	0.25	0.26 3
98	0.042	0.043 5
118	0.32	~0.27
135 - 178	0.14	0.10 3
195 - 244	0.29	0.12 4
251 - 281	0.068	0.027 7
331 - 380	0.130	0.039 7
391 - 439	0.061	0.0145 14
441 - 480	0.0097	0.0021 5
497 - 546	0.067	0.0129 21
548 - 598	0.011	0.0019 6
600 - 648	0.024	0.0039 12
652 - 695	0.019	0.0028 7
711 - 748	0.0016	0.00021 9
760 - 793	0.013	0.0017 6
827 - 870	0.010	0.0012 3
886 - 928	0.015	0.0016 4
942 - 988	0.0027	0.00028 7
1000 - 1047	0.0029	0.00029 9
1050 - 1097	0.0060	0.00056 11
1099 - 1147	0.0025	0.00022 5
1155 - 1200	0.00043	3.6 13 $\times 10^{-5}$
1209 - 1258	0.0050	0.00040 9
1284 - 1325	0.0032	0.00025 7
1342 - 1383	0.0023	0.00017 3
1392 - 1427	0.00061	4.3 14 $\times 10^{-5}$
1444 - 1542	0.0043	0.00029 6
1549 - 1648	0.0021	0.000132 25
1658 - 1750	0.0056	0.00033 6
1761 - 1848	0.0025	0.000138 25
1874 - 1973	0.0049	0.00025 4
1990 - 2078	0.0045	0.00022 4
2090 - 2188	0.0057	0.00027 5
2189 - 2288	0.0024	0.000106 24
2290 - 2378	0.0030	0.000128 21
2415 - 2500	0.00083	3.4 7 $\times 10^{-5}$
2515 - 2614	0.0031	0.000123 22
2621 - 2710	0.0127	0.00048 9
2727 - 2785	0.0015	5.5 12 $\times 10^{-5}$
2832 - 2923	0.0078	0.00028 6
2951 - 3028	0.00054	1.8 4 $\times 10^{-5}$
3071 - 3157	0.00035	1.12 25 $\times 10^{-5}$
3174 - 3263	9.7 $\times 10^{-5}$	3.0 13 $\times 10^{-6}$

Continuous Radiation (^{99}Nb)

⟨β-⟩=1465 keV;⟨IB⟩=4.4 keV

E_{bin}(keV)		⟨ ⟩(keV)	(%)
0 - 10	β-	0.0102	0.203
	IB	0.047	
10 - 20	β-	0.0312	0.208
	IB	0.046	0.32
20 - 40	β-	0.129	0.430
	IB	0.091	0.32
40 - 100	β-	1.00	1.41
	IB	0.26	0.40
100 - 300	β-	12.3	6.0
	IB	0.76	0.43
300 - 600	β-	53	11.6
	IB	0.90	0.21
600 - 1300	β-	260	27.7
	IB	1.32	0.153
1300 - 2500	β-	732	39.4
	IB	0.85	0.050
2500 - 4005	β-	407	14.0
	IB	0.092	0.0032

$^{99}_{42}$Mo(2.7477 2 d)

Mode: β-

Δ: -85967.4 21 keV

SpA: 4.8001$\times 10^5$ Ci/g

Prod: ^{98}Mo(n,γ); fission

Photons (^{99}Mo)

⟨γ⟩=272 3 keV

γ_{mode}	γ(keV)	γ(%)†
Tc L_ℓ	2.133	0.0142 25
Tc L_η	2.249	0.0067 10
Tc L_α	2.424	0.36 6
Tc L_β	2.576	0.21 4
Tc L_γ	2.860	0.015 3
Tc $K_{\alpha 2}$	18.251	3.23 12
Tc $K_{\alpha 1}$	18.367	6.13 23
Tc $K_{\beta 1}$'	20.613	1.46 5
Tc $K_{\beta 2}$'	21.136	0.263 10
γ M1+0.01%E2	40.595 10	1.05 3
γ M1+1.4%E2	140.474 11	90.7 18 *
γ M4	142.658 11	0.029 6 *
γ	158.799 14	0.0169 9
γ E2	181.069 11	6.07 12
γ	242.274 22	0.00146 24
γ	249.111 17	0.0029 4
γ	366.424 12	1.16 4
γ	380.056 21	0.0091 4
γ	410.28 10	0.0019 5
γ	411.497 12	0.0146 6
γ	455.85 13	~0.0013
γ	457.631 22	0.0068 5
γ	469.64 7	0.0027 5
γ	528.809 14	0.0543 11
γ	537.80 15	0.0016 6
γ	580.710 18	0.0044 5
γ	581.31 12	~0.0010
γ	620.016 23	0.0023 9
γ	621.305 18	0.0112 11
γ	621.781 24	0.0259 9
γ	739.508 11	12.14
γ	761.778 19	0.00112 11
γ	777.919 11	4.35 11
γ[E1]	822.983 15	0.132 6
γ	960.763 14	0.102 5
γ	986.437 22	0.00134 12
γ	1001.358 14	0.00404 11
γ	1056.21 5	0.00097 24

† 1.8% uncert(syst)
* with ^{99}Tc(6.006 h) in equilib

Atomic Electrons (^{99}Mo)

⟨e⟩=17.6 4 keV

e_{bin}(keV)	⟨e⟩(keV)	e(%)
3	0.32	11.9 12
15 - 18	0.49	3.06 18
20	0.69	3.55 13
21 - 41	0.200	0.531 17
119	10.8	9.08 25
122	1.04	0.86 18
137	1.39	1.01 3
138	0.134	0.097 5
140	0.72	0.52 6
142 - 159	0.092	0.064 12
160	1.22	0.765 22
178 - 221	0.253	0.142 3
228 - 249	0.00021	~9 $\times 10^{-5}$
345 - 390	0.034	~0.010
407 - 456	0.00024	5.6 16 $\times 10^{-5}$
457 - 470	9.5 $\times 10^{-6}$	2.0 8 $\times 10^{-6}$
508 - 538	0.0008	0.00016 7
560 - 601	0.00045	7.6 23 $\times 10^{-5}$
617 - 622	5.7 $\times 10^{-5}$	9 3 $\times 10^{-6}$
718 - 762	0.14	0.019 6

Atomic Electrons (^{99}Mo) (continued)

e_{bin}(keV)	⟨e⟩(keV)	e(%)
775 - 823	0.0051	0.00065 *23*
940 - 986	0.00069	7 *3* $\times10^{-5}$
998 - 1035	8.0 $\times10^{-6}$	8 *3* $\times10^{-7}$
1053 - 1056	7.0 $\times10^{-7}$	7 *3* $\times10^{-8}$

Continuous Radiation (^{99}Mo)

⟨β-⟩=390 keV;⟨IB⟩=0.47 keV

E_{bin}(keV)		⟨ ⟩(keV)	(%)
0 - 10	β-	0.078	1.55
	IB	0.018	
10 - 20	β-	0.233	1.56
	IB	0.017	0.119
20 - 40	β-	0.94	3.12
	IB	0.032	0.113
40 - 100	β-	6.6	9.4
	IB	0.084	0.13
100 - 300	β-	57	28.8
	IB	0.18	0.105
300 - 600	β-	141	31.9
	IB	0.112	0.028
600 - 1214	β-	184	23.8
	IB	0.027	0.0038

$^{99}_{43}$Tc(2.13 *5* $\times10^{5}$ yr)

Mode: β-

Δ: -87324.4 *21* keV

SpA: 0.0170 Ci/g

Prod: fission; daughter ^{99}Mo

Photons (^{99}Tc)

γ_{mode}	γ(keV)	γ(%)
γ E2+29%M1	89.65 *18*	4.9 *17* $\times10^{-6}$

Continuous Radiation (^{99}Tc)

⟨β-⟩=85 keV;⟨IB⟩=0.026 keV

E_{bin}(keV)		⟨ ⟩(keV)	(%)
0 - 10	β-	0.385	7.7
	IB	0.0044	
10 - 20	β-	1.11	7.5
	IB	0.0037	0.026
20 - 40	β-	4.19	14.0
	IB	0.0057	0.020
40 - 100	β-	23.6	34.6
	IB	0.0088	0.0146
100 - 294	β-	55	36.2
	IB	0.0031	0.0024

$^{99}_{43}$Tc(6.006 *2* h)

~0.3% $t_{1/2}$ variation with chemical environment

Mode: IT(99.9963 *6* %), β-(0.0037 *6* %)

Δ: -87181.8 *21* keV

SpA: 5.2704$\times10^{6}$ Ci/g

Prod: daughter ^{99}Mo

Photons (^{99}Tc)

⟨γ⟩=123.9 *7* keV

γ_{mode}	γ(keV)	γ(%)†
Tc L_ℓ	2.133	0.0096 *17*
Tc L_η	2.249	0.0045 *7*
Tc L_α	2.424	0.24 *4*
Tc L_β	2.576	0.145 *24*
Tc L_γ	2.860	0.0103 *19*
Tc $K_{\alpha2}$	18.251	2.17 *8*
Tc $K_{\alpha1}$	18.367	4.12 *16*
Tc $K_{\beta1}$'	20.613	0.98 *4*
Tc $K_{\beta2}$'	21.136	0.177 *7*
γ_{IT} M1+1.4%E2	140.474 *11*	87.2
γ_{IT} M4	142.658 *11*	0.028 *6*
γ_{β}-E2+M1	232.71 *23*	8.8 *13* $\times10^{-6}$
γ_{β}-M1+E2	322.36 *21*	9.8 *7* $\times10^{-5}$

† uncert(syst): 0.57% for IT, 16% for β-

Atomic Electrons (^{99}Tc)*

⟨e⟩=14.2 *3* keV

e_{bin}(keV)	⟨e⟩(keV)	e(%)
3	0.218	7.9 *8*
15	0.091	0.60 *6*
16	0.138	0.89 *9*
17	0.0085	0.049 *5*
18	0.091	0.51 *5*
20	0.0077	0.038 *4*
21	0.00150	0.0073 *8*
119	10.38	8.69 *18*
122	1.00	0.82 *18*
137	1.33	0.965 *20*
138	0.126	0.091 *3*
140	0.69	0.49 *6*
142	0.074	0.052 *11*
143	0.013	0.0092 *20*

* with IT

$^{99}_{44}$Ru(stable)

Δ: -87618.0 *22* keV

%: 12.7 *1*

$^{99}_{45}$Rh(16 *1* d)

Mode: ϵ

Δ: -85519 *10* keV

SpA: 8.2$\times10^{4}$ Ci/g

Prod: ^{99}Ru(p,n); protons on Cd

Photons (^{99}Rh)

⟨γ⟩=528 *26* keV

γ_{mode}	γ(keV)	γ(%)†
Ru L_ℓ	2.253	0.125 *24*
Ru L_η	2.382	0.059 *10*
Ru L_α	2.558	3.2 *6*
Ru L_β	2.731	2.0 *4*
Ru L_γ	3.029	0.15 *3*
Ru $K_{\alpha2}$	19.150	27.4 *22*
Ru $K_{\alpha1}$	19.279	52 *4*
Ru $K_{\beta1}$'	21.650	12.6 *10*
Ru $K_{\beta2}$'	22.210	2.34 *19*
γ E2+29%M1	89.65 *18*	30 *6*

Photons (^{99}Rh) (continued)

γ_{mode}	γ(keV)	γ(%)†
γ	175.4 *3*	2.3 *5*
γ E2+M1	232.71 *23*	0.59 *15*
γ	253.2 *3*	0.39 *6*
γ (E2)	295.92 *25*	0.84 *17*
γ M1+E2	322.36 *21*	6.6 *13*
γ M1+E2	353.26 *25*	33 *7*
γ	442.91 *23*	1.7 *3*
γ	485.89 *25*	0.51 *6*
γ	528.63 *25*	40
γ	575.54 *23*	0.27 *4*
γ	618.28 *22*	4.0 *8*
γ	734.7 *4*	0.32 *6*
γ	764.4 *3*	0.40 *8*
γ	807.9 *3*	1.4 *3*
γ	850.8 *3*	0.24 *5*
γ	897.5 *3*	0.68 *14*
γ	940.2 *4*	1.5 *3*
γ	1000.7 *4*	0.68 *14*
γ	1062.0 *5*	0.16 *3*
γ	1089.7 *4*	0.32 *6*
γ	1208.8 *4*	0.16 *3*
γ	1292.4 *7*	0.36 *7*
γ	1324.9 *4*	0.16 *3*
γ	1383.1 *4*	0.16 *3*
γ	1442.9 *4*	0.080 *16*
γ	1484.0 *4*	0.16 *3*
γ	1504.9 *5*	0.080 *16*
γ	1532.6 *4*	0.56 *11*
γ	1572.3 *4*	0.23 *5*
γ	1616.6 *4*	0.24 *5*
γ	1662.0 *4*	0.060 *12*
γ	1749.4 *7*	0.060 *12*
γ	1969.9 *4*	0.16 *3*
γ	2059.5 *3*	0.032 *6*

† 2.5% uncert(syst)

Atomic Electrons (^{99}Rh)

⟨e⟩=43 *5* keV

e_{bin}(keV)	⟨e⟩(keV)	e(%)
3	2.9	100 *13*
16	2.8	17.2 *22*
18 - 22	1.35	7.1 *7*
68	24.0	36 *7*
86	2.8	3.3 *7*
87	4.2	4.9 *10*
89	1.4	1.5 *3*
153 - 175	0.37	~0.23
210 - 253	0.13	0.060 *21*
274 - 322	0.42	0.14 *3*
331 - 353	1.5	0.44 *10*
421 - 464	0.044	0.010 *4*
483 - 529	0.6	0.12 *5*
553 - 596	0.045	0.008 *4*
615 - 618	0.006	0.0010 *5*
713 - 762	0.0066	0.0009 *3*
764 - 808	0.012	0.0015 *7*
829 - 875	0.0064	0.0007 *3*
894 - 940	0.011	0.0012 *5*
979 - 1001	0.0045	0.00046 *19*
1040 - 1089	0.0029	0.00028 *9*
1187 - 1209	0.0009	7 *3* $\times10^{-5}$
1270 - 1303	0.0025	0.00020 *6*
1322 - 1361	0.0008	5.7 *22* $\times10^{-5}$
1380 - 1421	0.00042	2.9 *10* $\times10^{-5}$
1440 - 1532	0.0035	0.00024 *6*
1550 - 1640	0.0022	0.00014 *3*
1659 - 1749	0.00026	1.5 *5* $\times10^{-5}$
1948 - 2037	0.00065	3.3 *10* $\times10^{-5}$
2056 - 2059	1.3 $\times10^{-5}$	6.2 *20* $\times10^{-7}$

Continuous Radiation (^{99}Rh)

$\langle\beta+\rangle$=14.5 keV;$\langle$IB$\rangle$=0.87 keV

E_{bin}(keV)		$\langle\ \rangle$(keV)	(%)
0 - 10	β+	7.2×10^{-6}	8.8×10^{-5}
	IB	0.00084	
10 - 20	β+	0.000218	0.00133
	IB	0.0116	0.067
20 - 40	β+	0.00451	0.0139
	IB	0.0139	0.063
40 - 100	β+	0.147	0.193
	IB	0.0055	0.0084
100 - 300	β+	3.61	1.74
	IB	0.042	0.021
300 - 600	β+	7.7	1.83
	IB	0.18	0.040
600 - 1300	β+	3.01	0.426
	IB	0.55	0.063
1300 - 2010	IB	0.065	0.0045
	Σβ+		4.2

$^{99}_{45}$Rh(4.7 *1* h)

Mode: ε

Δ: -85454 *10* keV

SpA: 6.73×10^{6} Ci/g

Prod: ^{99}Ru(p,n); ^{98}Ru(d,n)

Photons (^{99}Rh)

$\langle\gamma\rangle$=543 *19* keV

γ_{mode}	γ(keV)	γ(%)†
Ru L_ℓ	2.253	0.085 *15*
Ru L_η	2.382	0.039 *6*
Ru L_α	2.558	2.2 *4*
Ru L_β	2.731	1.36 *23*
Ru L_γ	3.032	0.105 *20*
Ru $K_{\alpha2}$	19.150	19.1 *7*
Ru $K_{\alpha1}$	19.279	36.3 *13*
Ru $K_{\beta1}$'	21.650	8.8 *3*
Ru $K_{\beta2}$'	22.210	1.64 *6*
γ [E2+29%M1]	89.65 *18*	1.68 *16*
γ E2+M1	232.71 *23*	0.079 *14*
γ	253.2 *3*	0.50 *4* ?
γ [M1+E2]	276.9 *3*	1.60 *10*
γ M1+E2	322.36 *21*	0.88 *10*
γ [M1+E2]	340.8 *3*	69 *3*
γ [M1+E2]	378.8 *4*	<0.022
γ	485.89 *25*	0.66 *6*
γ	501.9 *5*	0.028 *6*
γ [M1+E2]	528.1 *3*	1.37 *22*
γ	558.2 *6*	0.20 *4*
γ	575.54 *23*	0.36 *4*
γ [M1+E2]	617.7 *3*	11.8 *10*
γ	644.0 *6*	0.11 *3*
γ	685.5 *3*	0.65 *22*
γ	702.0 *6*	0.16 *4*
γ	707.6 *6*	0.076 *22*
γ [E2]	719.6 *4*	1.20 *4*
γ	779.1 *6*	0.20 *6*
γ	808 *1*	0.065 *22* ?
γ	850.6 *6*	0.52 *11* ?
γ	899.9 *10*	0.17 *5* ?
γ	920.2 *3*	0.74 *4*
γ	936.7 *4*	2.16 *5*
γ	965.7 *6*	0.14 *3*
γ	984.8 *6*	0.17 *3*
γ	1002 *1*	0.087 *22* ?
γ	1119.1 *10*	0.20 *4* ?
γ	1158.1 *10*	0.21 *4*

Photons (^{99}Rh)
(continued)

γ_{mode}	γ(keV)	γ(%)†
γ	1172.2 *10*	0.10 *3*
γ	1243.4 *10*	0.09 *3* ?
γ	1261.0 *3*	10.9
γ	1277.7 *10*	0.11 *3*
γ	1306.2 *10*	0.087 *22*
γ	1499.5 *10*	0.076 *22*
γ	1762.9 *5*	0.017 *4*

† 10% uncert(syst)

Atomic Electrons (^{99}Rh)

$\langle e\rangle$=10.5 *7* keV

e_{bin}(keV)	$\langle e\rangle$(keV)	e(%)
3	1.96	67 *7*
16	1.94	12.0 *12*
18	0.184	1.0 *1*
19	0.68	3.6 *4*
21 - 22	0.081	0.38 *3*
68	1.32	1.96 *20*
86	0.157	0.182 *19*
87	0.234	0.27 *3*
89 - 90	0.089	0.100 *9*
211 - 255	0.14	0.057 *13*
274 - 300	0.057	0.020 *4*
319	2.8	0.89 *17*
320 - 322	0.0016	0.00050 *15*
338	0.38	0.11 *3*
340 - 379	0.084	0.025 *5*
464 - 506	0.039	0.0079 *15*
525 - 573	0.0114	0.0021 *5*
575 - 622	0.206	0.035 *3*
641 - 686	0.011	~0.0017
697 - 720	0.0169	0.00241 *11*
757 - 805	0.0023	0.00030 *11*
807 - 851	0.0042	0.00050 *22*
878 - 920	0.034	0.0037 *10*
934 - 982	0.0065	0.00069 *13*
984 - 1002	0.00010	1.0 *3* $\times10^{-5}$
1097 - 1136	0.0022	0.00020 *6*
1150 - 1172	0.0007	6.0 *23* $\times10^{-5}$
1221 - 1261	0.058	0.0047 *17*
1275 - 1306	0.00051	4.0 *14* $\times10^{-5}$
1477 - 1499	0.00034	2.3 *10* $\times10^{-5}$
1741 - 1762	6.7 $\times10^{-5}$	3.8 *14* $\times10^{-6}$

Continuous Radiation (^{99}Rh)

$\langle\beta+\rangle$=28.4 keV;$\langle$IB$\rangle$=0.96 keV

E_{bin}(keV)		$\langle\ \rangle$(keV)	(%)
0 - 10	β+	1.05×10^{-5}	0.000128
	IB	0.00151	
10 - 20	β+	0.000319	0.00194
	IB	0.0118	0.069
20 - 40	β+	0.0067	0.0206
	IB	0.0146	0.065
40 - 100	β+	0.225	0.294
	IB	0.0085	0.0131
100 - 300	β+	6.4	3.04
	IB	0.047	0.023
300 - 600	β+	18.2	4.26
	IB	0.18	0.040
600 - 1300	β+	3.45	0.53
	IB	0.59	0.066
1300 - 1823	IB	0.107	0.0075
	Σβ+		8.1

$^{99}_{46}$Pd(21.4 *2* min)

Mode: ε

Δ: -82192 *16* keV

SpA: 8.87×10^{7} Ci/g

Prod: ^{96}Ru(α,n)

Photons (^{99}Pd)

$\langle\gamma\rangle$=832 *12* keV

γ_{mode}	γ(keV)	γ(%)†
Rh L_ℓ	2.377	0.058 *10*
Rh L_η	2.519	0.026 *4*
Rh L_α	2.696	1.52 *24*
Rh L_β	2.889	0.95 *16*
Rh L_γ	3.203	0.085 *16*
Rh $K_{\alpha2}$	20.074	12.7 *5*
Rh $K_{\alpha1}$	20.216	24.1 *9*
Rh $K_{\beta1}$'	22.717	5.91 *21*
Rh $K_{\beta2}$'	23.312	1.11 *4*
γ M1(+E2)	136.23 *7*	73
γ	236.80 *24*	0.17 *3*
γ M1(+E2)	263.79 *7*	15.3 *7*
γ E1	293.55 *17*	1.31 *22*
γ E2	368.38 *18*	0.146 *22*
γ M1,E2	386.93 *9*	2.8 *3*
γ E2	400.02 *7*	3.58 *22*
γ M1	410.10 *13*	1.31 *7*
γ M1,E2	427.57 *10*	1.97 *11*
γ	524.29 *18*	0.34 *5*
γ	627.5 *3*	0.28 *7*
γ	646.91 *22*	0.59 *4*
γ M1,E2	650.71 *11*	1.31 *15*
γ	653.5 *2*	2.55 *15*
γ	653.8 *3*	1.46 *15*
γ	663.06 *25*	0.39 *7*
γ M1,E2	673.89 *13*	6.9 *5*
γ	685.10 *17*	0.131 *22* ?
γ	690.8 *6*	~0.07
γ	703.65 *15*	0.24 *3*
γ	714.5 *3*	~0.19
γ	718.85 *19*	0.43 *4*
γ	740.3 *4*	0.17 *3*
γ	758.6 *3*	0.16 *3*
γ	767.7 *4*	0.22 *4*
γ	774.3 *6*	0.12 *4*
γ M1,E2	786.94 *10*	3.36 *15*
γ	800.7 *3*	0.29 *7*
γ	810.12 *13*	2.04 *15*
γ	817.86 *17*	1.10 *7*
γ	852.60 *17*	0.31 *6*
γ	878.9 *3*	0.24 *4*
γ	886.8 *3*	0.34 *6* ?
γ	910.69 *22*	0.28 *7*
γ	954.09 *18*	0.18 *7*
γ	967.44 *15*	0.95 *7*
γ	1013.77 *24*	1.46 *15*
γ	1022.16 *24*	0.28 *10*
γ	1046.92 *22*	0.44 *4*
γ	1063.6 *2*	0.07 *3*
γ	1072.03 *15*	0.89 *7*
γ	1095.0 *3*	0.24 *7*
γ	1100.2 *3*	1.02 *15*
γ	1125.4 *5*	0.23 *6*
γ	1157.4 *4*	0.66 *11*
γ	1166.2 *3*	0.75 *11*
γ	1200.96 *24*	0.11 *4*
γ	1220.0 *4*	0.30 *7*
γ	1231.5 *3*	0.73 *7*
γ	1256.7 *4*	1.02 *15*
γ	1273.9 *5*	0.14 *4*
γ	1297.10 *24*	0.08 *4*
γ	1302.5 *3*	0.23 *7*
γ	1324.99 *25*	0.84 *11*
γ	1335.81 *16*	4.67 *22*
γ	1350.7 *5*	0.18 *4*
γ	1392.6 *4*	0.13 *6* ?
γ	1421.7 *3*	0.16 *7*
γ	1430.1 *4*	0.20 *7*
γ	1472.04 *16*	0.23 *7*
γ	1484.2 *8*	0.12 *5*
γ	1505.1 *3*	0.54 *7*

Photons (^{99}Pd)
(continued)

γ_{mode}	γ(keV)	γ(%)†
γ	1514.53 22	0.20 7
γ	1527.8 3	0.28 5
γ	1542.1 5	0.15 4
γ	1560.89 24	0.16 4
γ	1587.88 23	0.12 3
γ	1614.5 5	0.48 6
γ	1679.4 5	0.62 11
γ	1685.4 15	0.15 6
γ	1697.11 23	0.55 7
γ	1717.42 25	0.64 7
γ	1735.08 24	0.09 4
γ	1754.0 4	0.13 6
γ	1774.6 10	~0.11
γ	1790.1 4	0.21 4 ?
γ	1804.3 4	0.18 6
γ	1822.5 9	~0.13
γ	1840.0 3	0.19 5
γ	1851.67 23	0.45 7
γ	1862.7 4	0.18 4
γ	1879.8 7	0.17 4 ?
γ	1905.1 3	0.66 11
γ	1924.62 20	0.47 7
γ	1943.2 5	0.27 7
γ	1962.4 7	0.08 3
γ	1981.20 25	0.058 22
γ	1998.87 24	0.77 7
γ	2037.8 8	0.20 4
γ	2143.0 5	0.42 6
γ	2154.4 7	0.17 4
γ	2181.6 4	0.30 7
γ	2188.41 20	0.20 6
γ	2207.6 10	0.088 22
γ	2245.7 5	0.83 9
γ	2272.8 4	0.35 11
γ	2279.2 5	0.19 5
γ	2304.3 11	0.19 4
γ	2324.63 20	0.55 7
γ	2344.7 15	0.044 15
γ	2372.5 7	0.080 22
γ	2382.8 15	0.029 7
γ	2396.4 9	0.044 15
γ	2418.2 7	0.026 11
γ	2422.1 15	0.026 11
γ	2440.1 5	0.088 22
γ	2451.8 20	0.029 7
γ	2493.7 3	0.022 7
γ	2509.5 5	0.12 3
γ	2527.5 10	0.19 5
γ	2536.5 4	0.58 11
γ	2554.4 7	~0.029
γ	2558.8 3	0.19 6
γ	2576.3 5	0.029 7
γ	2616.1 10	0.026 7
γ	2636.3 7	0.095 22
γ	2660.2 9	0.041 7
γ	2672.8 4	~0.010
γ	2695.1 3	0.14 3
γ	2726.9 20	0.015 4
γ	2796.4 9	0.088 22
γ	2847.2 7	0.15 4
γ	2893.7 3	0.0051 22
γ	2983.4 7	0.010 4

† 2.7% uncert(syst)

Atomic Electrons (^{99}Pd)
⟨e⟩=17.5 5 keV

e_{bin}(keV)	⟨e⟩(keV)	e(%)
3	1.34	44 5
16	0.124	0.76 8
17	1.11	6.5 7
19	0.146	0.75 8
20	0.41	2.09 21
22 - 23	0.053	0.238 20
113	10.6	9.4 3
133	1.52	1.14 4
214 - 237	0.014	~0.006

Atomic Electrons (^{99}Pd)
(continued)

e_{bin}(keV)	⟨e⟩(keV)	e(%)
241	1.1	0.45 12
260 - 294	0.22	0.084 16
345 - 387	0.289	0.078 5
397 - 428	0.094	0.0233 17
501 - 524	0.006	0.0011 5
604 - 654	0.185	0.029 3
660 - 704	0.026	0.0039 4
711 - 759	0.0071	0.00096 21
764 - 810	0.075	0.0097 12
814 - 864	0.0081	0.00096 23
875 - 911	0.0029	0.00032 12
931 - 967	0.008	0.0009 3
991 - 1024	0.015	0.0015 4
1040 - 1077	0.014	0.0013 3
1092 - 1134	0.0060	0.00054 16
1143 - 1178	0.0057	0.00049 16
1197 - 1233	0.011	0.00091 23
1251 - 1299	0.0032	0.00025 6
1302 - 1351	0.030	0.0023 7
1369 - 1419	0.0030	0.00021 6
1421 - 1469	0.0017	0.00012 4
1471 - 1565	0.0068	0.00045 8
1584 - 1682	0.0076	0.00046 9
1685 - 1781	0.0058	0.00033 7
1787 - 1882	0.0068	0.00037 6
1901 - 1998	0.0065	0.00034 6
2015 - 2037	0.00072	$3.6\ 13 \times 10^{-5}$
2120 - 2207	0.0041	0.00019 3
2223 - 2322	0.0071	0.00032 5
2324 - 2422	0.00095	$4.0\ 7 \times 10^{-5}$
2429 - 2527	0.0027	0.000109 24
2531 - 2616	0.0013	$5.1\ 10 \times 10^{-5}$
2633 - 2726	0.00064	$2.4\ 5 \times 10^{-5}$
2773 - 2870	0.00070	$2.5\ 6 \times 10^{-5}$
2890 - 2983	3.0×10^{-5}	$1.0\ 4 \times 10^{-6}$

Continuous Radiation (^{99}Pd)
⟨β+⟩=430 keV;⟨IB⟩=2.3 keV

E_{bin}(keV)		⟨ ⟩(keV)	(%)
0 - 10	β+	7.3×10^{-6}	8.8×10^{-5}
	IB	0.0170	
10 - 20	β+	0.000230	0.00140
	IB	0.022	0.146
20 - 40	β+	0.00505	0.0156
	IB	0.041	0.151
40 - 100	β+	0.189	0.246
	IB	0.092	0.141
100 - 300	β+	7.5	3.42
	IB	0.26	0.146
300 - 600	β+	47.9	10.4
	IB	0.33	0.078
600 - 1300	β+	245	26.5
	IB	0.73	0.081
1300 - 2500	β+	129	8.5
	IB	0.80	0.046
2500 - 3126	IB	0.055	0.0021
	Σβ+		49

$^{99}_{47}$Ag(2.07 5 min)

Mode: ϵ

Δ: -76760 150 keV

SpA: 9.15×10^{8} Ci/g

Prod: ^{14}N on ^{92}Mo

Photons (^{99}Ag)
⟨γ⟩=1401 28 keV

γ_{mode}	γ(keV)	γ(%)†
γ	219.90 4	4.00 13
γ	243.7 2	0.38 6
γ	264.46 3	64
γ	287.65 7	0.51 13
γ	326.0 2	0.38 6
γ	352.4 1	0.63 13
γ	365.8 3	0.38 6
γ	371.3 3	0.19 6
γ	385.6 2	0.51 13
γ	391.7 3	0.32 6
γ	398.6 1	0.83 6
γ	416.6 5	0.25 6
γ	422.4 5	0.25 6
γ	438.3 2	0.32 6
γ	443.1 3	0.25 6
γ	463.73 7	1.08 13
γ	467.3 1	0.57 13
γ	488.1 3	0.25 6
γ	551.1 1	0.38 6
γ	568.20 4	3.75 13
γ	596.2 1	1.5 5
γ	602.9 1	0.70 19
γ	610.6 3	0.76 13
γ	636.0 1	1.40 13
γ	649.26 6	0.32 6
γ	653.2 1	~0.8
γ	686.99 5	3.17 19
γ	708.0 1	0.61 10
γ	725.4 2	0.76 6
γ	805.6 5	12.3 4
γ	815.6 10	6.7 6
γ	816.1 10	<1
γ	817.6 10	0.89 18
γ	832.29 4	13.1 5
γ	838.47 8	2.03 13
γ	853.73 9	0.51 4
γ	864.0 1	3.9 4
γ	881.1 3	0.22 5
γ	908.4 4	0.32 6
γ	911.4 1	0.19 6
γ	954.7 3	0.09 3
γ	963.2 3	1.02 13
γ	1010.1 2	0.51 6
γ	1034.8 2	0.38 6
γ	1068.8 3	0.51 6
γ	1076.3 1	0.51 6
γ	1102.60 7	3.17 13
γ	1158.9 2	0.51 13
γ	1175.0 2	1.21 6
γ	1182.2 4	0.51 19
γ	1203.98 8	1.97 13
γ	1233.1 2	0.32 6
γ	1261.9 5	0.51 19
γ	1275.8 1	2.4 6
γ	1281.5 5	0.63 19
γ	1304.9 2	0.63 25
γ	1339.3 2	0.86 10
γ	1356.1 2	0.63 6
γ	1368.6 3	0.32 10
γ	1402.8 2	0.15 3
γ	1416.5 2	1.33 19
γ	1423.9 4	~0.25
γ	1432.3 2	1.52 13
γ	1448.3 2	0.83 13
γ	1452.3 3	0.44 13
γ	1476.3 3	0.63 19
γ	1498.7 5	0.51 13
γ	1531.9 1	4.5 3
γ	1540.4 1	1.40 19
γ	1550.7 3	0.38 13
γ	1576.4 3	0.63 10
γ	1585.3 10	1.1 4
γ	1594.6 6	0.25 6
γ	1613.7 3	0.70 6
γ	1642.5 6	0.57 19
γ	1682.9 2	0.38 13
γ	1695.9 3	0.38 10
γ	1725.9 5	0.25 6
γ	1739.3 4	0.32 6
γ	1796.7 4	0.51 13
γ	1849.7 2	0.38 10
γ	1873.4 1	2.35 13

Photons (^{99}Ag) (continued)

γ_{mode}	γ(keV)	γ(%)†
γ	1881.0 2	1.27 13
γ	1907.1 4	1.21 6
γ	1975.5 5	0.38 10
γ	2010.5 7	0.13 4
γ	2028.8 4	0.11 4
γ	2068.1 4	0.47 5
γ	2204.6 3	0.57 6
γ	2206.0 4	0.19 6
γ	2264.0 6	~0.07
γ	2305.2 3	0.63 6
γ	2322.2 4	0.32 6
γ	2340.8 6	0.25 6
γ	2454.0 9	0.15 3
γ	2537.2 5	0.25 6
γ	2629.7 9	0.32 13
γ	2708.5 4	0.83 6
γ	2742.1 5	0.32 6
γ	2820.8 9	0.25 10
γ	2945.1 4	0.51 6
γ	3181.8 4	1.33 19
γ	3330 1	0.38 13
γ	3542.8 7	0.51 13

† 10% uncert(syst)

$^{99}_{47}$Ag(10.5 5 s)

Mode: IT
Δ: -76254 150 keV
SpA: 1.05×10^{10} Ci/g
Prod: ^{14}N on ^{92}Mo; ^{40}Ca on ^{63}Cu

Photons (^{99}Ag)

γ_{mode}	γ(keV)	γ(rel)
γ[E3]	163.6 5	30 2
γ[M1+E2]	342.6 5	100

$^{99}_{48}$Cd(16 3 s)

Mode: ε, εp(0.17 8 %)
Δ: -69990 500 keV syst
SpA: 7.0×10^{9} Ci/g
Prod: protons on Sn

Photons (^{99}Cd)

γ_{mode}	γ(keV)
γ[E3]	163.6 10 *
γ[M1+E2]	342.6 10 *
γ	671.8 10
γ	975.4 10
γ	1014.3 10
γ	1316.3 10

* with ^{99}Ag(10.5 s)

A = 100

NDS **11**, 279 (1974)

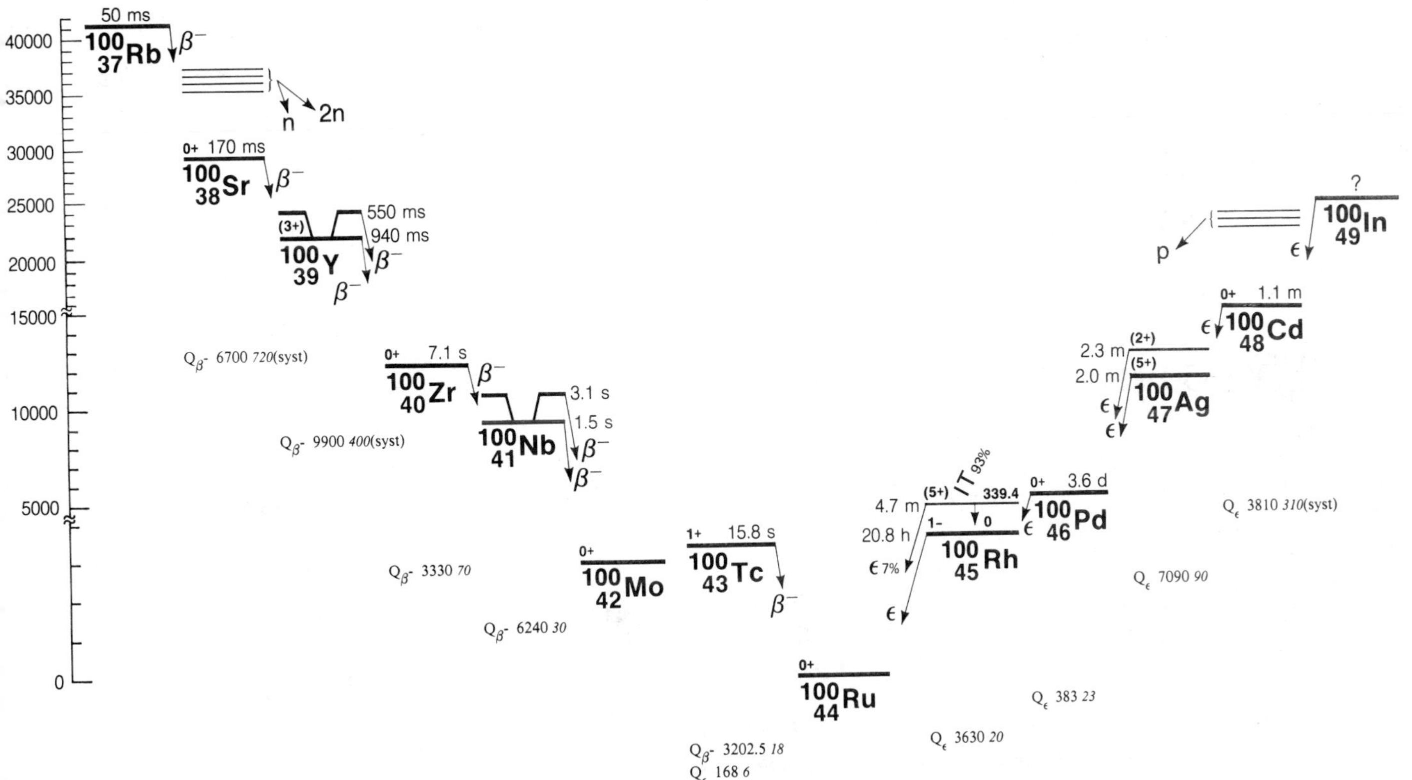

$^{100}_{37}$Rb(50 *10* ms)

Mode: β-, β-n, β-2n
SpA: 1.6×10^{11} Ci/g
Prod: fission

Photons (^{100}Rb)

γ_{mode}	γ(keV)	γ(rel)
$\gamma_{\beta\text{-n}}$	90.5 *10*	28 *6*
$\gamma_{\beta\text{-n}}$	120.8 *10*	15 *3*
$\gamma_{\beta\text{-}}$[E2]	129.2 *10*	100
$\gamma_{\beta\text{-}}$[E2]	288.4 *10*	36 *7*
$\gamma_{\beta\text{-2n}}$E2	144.479 *6* ?	

$^{100}_{38}$Sr(170 *80* ms)

Mode: β-
Δ: -60020 *610* keV syst
SpA: 1.6×10^{11} Ci/g
Prod: fission

$^{100}_{39}$Y (940 *30* ms)

Mode: β-
Δ: -66720 *410* keV syst
SpA: 8.5×10^{10} Ci/g
Prod: fission

Photons (^{100}Y)

⟨γ⟩=607 *51* keV

γ_{mode}	γ(keV)	γ(%)
γ [E2]	118.6 *2*	3.8 *15*
γ [E2]	212.7 *2*	75 *15*
γ [E2]	351.9 *2*	25 *3*
γ [E2]	498.0 *2*	6.0 *15*
γ	614.0 *2*	10.5 *23*
γ	665.8 *2*	9.8 *23*
γ [E2]	878.1 *2*	13.5 *23*
γ	1096.7 *2*	~3
γ	1195.5 *2*	3.8 *15*

$^{100}_{39}$Y (550 *150* ms)

Mode: β-
Δ: -66720 *410* keV syst
SpA: 1.2×10^{11} Ci/g
Prod: fission

Photons (^{100}Y)

γ_{mode}	γ(keV)	γ(rel)
γ [E2]	118.6 *2*	50 *10*
γ [E2]	212.7 *2*	100 *20*

$^{100}_{40}$Zr(7.1 *4* s)

Mode: β-
Δ: -76620 *80* keV
SpA: 1.51×10^{10} Ci/g
Prod: fission

Photons (^{100}Zr)

γ_{mode}	γ(keV)	γ(rel)
γ	103.8 *3*	2.0 *4*
γ	195.7 *3*	1.0 *2*
γ	336.0 *3*	2.0 *4*
γ	400.6 *3*	60 *12*
γ	440.9 *3*	4.0 *8*
γ	498.3 *3*	10 *2*
γ	504.3 *3*	100 *20*
γ	695.0 *3*	4.0 *8*
γ	749.4 *3*	2.0 *4*
γ	1257.4 *5*	5 *1*
γ	1654.4 *5*	7.0 *14*
γ	2436.0 *5*	7.0 *14*

$^{100}_{41}$Nb(1.5 *3* s)

Mode: β-
Δ: -79950 *30* keV
SpA: 6.0×10^{10} Ci/g
Prod: fission; daughter ^{100}Zr; ^{100}Mo(n,p)

Photons (^{100}Nb(1.5 + 3.1 s))

γ_{mode}	γ(keV)	γ(rel)
γ E2	159.43 *21*	8.0 *9*
γ	461.79 *20*	4.6 *6*
γ [M1+E2]	528.08 *18*	21.2 *25*
γ [E2]	535.47 *23*	100
γ	543.2 *3*	3.2 *5*
γ	549.5 *3*	1.5 *3*
γ [E2]	600.15 *19*	46 *5*
γ	622.7 *3*	1.1 *2*
γ	635.30 *21*	2.6 *5*
γ	638.26 *20*	2.8 *5*
γ	702.68 *24*	3.3 *5*
γ	707.36 *21*	5.7 *8*
γ	711.0 *3*	4.7 *7*
γ	768.46 *22*	6.2 *9*
γ	792.50 *20*	4.2 *6*
γ	927.89 *19*	5.4 *8*
γ	952.59 *24*	3.7 *6*
γ	966.00 *19*	14.7 *20*
γ [E2]	1063.55 *23*	7.2 *11*
γ	1071.33 *24*	3.1 *5*
γ	1245.2 *3*	3.1 *6*
γ [E1]	1280.33 *24*	20 *3*
γ	1427.79 *19*	3.4 *6*
γ	1499.85 *20*	3.7 *6*
γ	1515.2 *3*	3.0 *6*
γ	1566.14 *19*	3.8 *7*

$^{100}_{41}$Nb(3.1 *3* s)

Mode: β-
Δ: -79950 *30* keV
SpA: 3.3×10^{10} Ci/g
Prod: fission; ^{100}Mo(n,p)
see ^{100}Nb(1.5 s) for γ rays

$^{100}_{42}$Mo(stable)

Δ: -86186 *6* keV
%: 9.63 *2*

$^{100}_{43}$Tc(15.8 *1* s)

Mode: β-
Δ: -86017.4 *23* keV
SpA: 6.99×10^{9} Ci/g
Prod: ^{99}Tc(n,γ); ^{100}Mo(p,n); ^{103}Rh(n,α)

Photons (^{100}Tc)

⟨γ⟩=83 *5* keV

γ_{mode}	γ(keV)	γ(%)†
γ [E2]	378.7 *4*	0.029 *7*
γ E2	539.53 *4*	7.0
γ E2	590.76 *4*	5.7 *4*
γ [E2]	689.43 *19*	0.034 *7*
γ	734.80 *17*	0.009 *3*
γ M1,E2	822.53 *8*	0.076 *11*
γ [E2]	1025.1 *3*	0.034 *7*
γ [E2]	1201.3 *4*	0.043 *7*
γ	1325.56 *16*	0.0108 *23*
γ E2	1362.06 *7*	0.0574 *20*
γ [E2]	1511.95 *19*	0.44 *4*
γ [E1]	1559.30 *25*	~0.007
γ	1700.79 *24*	~0.0014
γ [E2]	1847.7 *3*	0.041 *7*
γ	1865.08 *17*	0.013 *3*
γ	1874.9 *10*	~0.0014
γ	2120.7 *6*	0.0035 *7*
γ	2127.6 *10*	~0.0014
γ [M1+E2]	2298.5 *5*	0.014 *4*
γ	2660.3 *6*	~0.0014

† 10% uncert(syst)

Continuous Radiation (^{100}Tc)

⟨β-⟩=1315 keV;⟨IB⟩=3.5 keV

E_{bin}(keV)		⟨ ⟩(keV)	(%)
0 - 10	β-	0.0073	0.146
	IB	0.045	
10 - 20	β-	0.0225	0.150
	IB	0.044	0.31
20 - 40	β-	0.094	0.312
	IB	0.087	0.30
40 - 100	β-	0.74	1.04
	IB	0.25	0.38
100 - 300	β-	9.7	4.69
	IB	0.71	0.40
300 - 600	β-	47.7	10.4
	IB	0.81	0.19
600 - 1300	β-	326	33.9
	IB	1.10	0.128
1300 - 2500	β-	812	45.0
	IB	0.49	0.029
2500 - 3203	β-	119	4.43
	IB	0.0085	0.00033

$^{100}_{44}$Ru(stable)

Δ: -89219.9 *22* keV
%: 12.6 *1*

$^{100}_{45}$Rh(20.8 *1* h)

Mode: ϵ

Δ: -85590 *20* keV

SpA: 1.507×10^{6} Ci/g

Prod: daughter ^{100}Pd; ^{100}Ru(p,n); ^{99}Ru(d,n); ^{99}Ru(p,γ); protons on Cd

Photons (^{100}Rh)

⟨γ⟩=2714 *90* keV

γ_{mode}	γ(keV)	γ(%)†
Ru L_ℓ	2.253	0.085 *15*
Ru L_η	2.382	0.039 *6*
Ru L_α	2.558	2.2 *4*
Ru L_β	2.732	1.36 *23*
Ru L_γ	3.033	0.105 *20*
Ru $K_{\alpha2}$	19.150	19.2 *7*
Ru $K_{\alpha1}$	19.279	36.4 *13*
Ru $K_{\beta1}$'	21.650	8.8 *3*
Ru $K_{\beta2}$'	22.210	1.64 *6*
γ	228.95 *24*	0.103 *21*
γ	301.87 *19*	~0.08
γ	302.31 *14*	0.66 *8*
γ M1,E2	370.44 *24*	~0.41
γ [E2]	378.7 *4*	~0.04
γ	398.83 *16*	0.14 *4*
γ	403.5 *3*	0.19 *4*
γ M1,E2	446.16 *8*	11.2 *4*
γ	465.12 *23*	0.10 *4*
γ	499.5 *4*	0.14 *4*
γ	519.05 *12*	0.72 *14*
γ E2	539.53 *4*	78.4 *21*
γ	553.30 *19*	~0.08
γ E1	588.17 *12*	4.12 *21*
γ E2	590.76 *4*	1.44 *21*
γ	600.62 *18*	0.27 *4*
γ	604.18 *18*	0.37 *6*
γ	651.51 *21*	0.51 *10*
γ	654.7 *3*	0.45 *10*
γ E2	686.9 *3*	0.66 *6*
γ [E2]	689.43 *19*	~0.010 ?
γ	734.80 *17*	0.27 *8*
γ [E1]	736.78 *25*	~0.10
γ	748.47 *14*	0.82 *4*
γ	775.8 *4*	~0.08
γ [M1+E2]	816.59 *25*	~0.6
γ M1,E2	822.53 *8*	20 *1*
γ	902.94 *21*	0.12 *4*
γ E1	1034.32 *13*	1.44 *6*
γ E1	1107.21 *10*	13.3 *4*
γ	1154.54 *16*	0.14 *6*
γ [E2]	1201.3 *4*	~0.06
γ	1204.80 *22*	~0.041
γ	1325.56 *16*	0.31 *4*
γ M1,E2	1341.57 *12*	4.84 *21*
γ E2	1362.06 *7*	15.1 *4*
γ	1386.30 *15*	0.39 *4*
γ [E2]	1511.95 *19*	0.13 *6*
γ E1	1553.36 *10*	21 *4*
γ [E1]	1559.30 *25*	0.62 *21*
γ	1627.42 *14*	1.58 *10*
γ	1700.79 *24*	0.226 *21*
γ	1707.83 *17*	0.226 *21*
γ [E2]	1847.7 *3*	~0.041
γ	1865.08 *17*	0.39 *8*
γ E1	1929.73 *9*	12.1 *4*
γ	1977.06 *15*	0.31 *4*
γ	1996.4 *4*	~0.041
γ	2166.95 *15*	0.19 *6*
γ	2193.7 *4*	~0.12
γ E1	2375.88 *10*	35.0 *25*
γ	2394.9 *5*	~0.10
γ	2469.25 *10*	~0.19
γ	2516.58 *15*	~0.06
γ	2530.34 *16*	2.7 *6*
γ	2545.0 *10*	~0.021
γ	2613.3 *3*	0.10 *4*
γ	2784.4 *4*	0.25 *4*
γ	2915.40 *10*	<0.08 ?
γ	2934.4 *5*	~0.012

Photons (^{100}Rh) (continued)

γ_{mode}	γ(keV)	γ(%)†
γ	3060.5 *8*	0.082 *21*
γ	3069.86 *16*	~0.041
γ	3323.9 *4*	0.012 *4*
γ	3419.0 *15*	~0.008
γ	3464.0 *9*	~0.0041

† 2.2% uncert(syst)

Atomic Electrons (^{100}Rh)

⟨e⟩=7.7 *3* keV

e_{bin}(keV)	⟨e⟩(keV)	e(%)
3	1.95	67 *7*
16	1.95	12.0 *12*
18	0.184	1.0 *1*
19	0.68	3.6 *4*
21 - 22	0.081	0.38 *3*
207 - 229	0.009	~0.0042
280 - 302	0.034	~0.012
348 - 396	0.026	0.0073 *24*
398 - 403	0.0007	0.00018 *8*
424	0.29	0.067 *8*
443 - 477	0.048	0.0109 *16*
496 - 516	0.012	0.0023 *11*
517	1.52	0.294 *10*
518 - 566	0.262	0.0486 *11*
569 - 604	0.039	0.0068 *8*
629 - 667	0.019	0.0030 *6*
684 - 732	0.011	0.0015 *5*
734 - 776	0.0018	0.00025 *8*
794	0.006	~0.0008
800	0.204	0.0255 *21*
813 - 823	0.0304	0.0037 *3*
881 - 903	0.0009	0.00010 *5*
1003 - 1034	0.00527	0.000519 *22*
1085 - 1132	0.0461	0.00424 *14*
1151 - 1201	0.0007	6.2 *25* ×10^{-5}
1202 - 1205	2.8 ×10^{-5}	~2×10^{-6}
1303 - 1342	0.115	0.0086 *3*
1359 - 1386	0.0133	0.00098 *5*
1490 - 1559	0.057	0.0037 *6*
1605 - 1705	0.0082	0.00051 *14*
1826 - 1908	0.0267	0.00140 *6*
1927 - 1996	0.0046	0.000235 *19*
2145 - 2193	0.0010	4.6 *15* ×10^{-5}
2354 - 2447	0.075	0.00317 *21*
2466 - 2544	0.0080	0.00032 *10*
2591 - 2613	0.00029	1.1 *5* ×10^{-5}
2762 - 2784	0.00067	2.4 *7* ×10^{-5}
2893 - 2934	0.00014	~5×10^{-6}
3038 - 3069	0.00031	1.0 *3* ×10^{-5}
3302 - 3397	4.7 ×10^{-5}	1.4 *4* ×10^{-6}
3416 - 3464	1.2 ×10^{-5}	3.5 *14* ×10^{-7}

Continuous Radiation (^{100}Rh)

⟨β+⟩=54 keV;⟨IB⟩=0.39 keV

E_{bin}(keV)		⟨ ⟩(keV)	(%)
0 - 10	β+	4.69×10^{-7}	5.7×10^{-6}
	IB	0.0021	
10 - 20	β+	1.44×10^{-5}	8.8×10^{-5}
	IB	0.0128	0.076
20 - 40	β+	0.000308	0.00095
	IB	0.0163	0.071
40 - 100	β+	0.0112	0.0145
	IB	0.0129	0.020
100 - 300	β+	0.435	0.199
	IB	0.056	0.029
300 - 600	β+	2.95	0.64
	IB	0.098	0.023
600 - 1300	β+	20.1	2.11
	IB	0.103	0.0123
1300 - 2500	β+	30.1	1.78
	IB	0.075	0.0042

Continuous Radiation (^{100}Rh) (continued)

E_{bin}(keV)		⟨ ⟩(keV)	(%)
2500 - 3630	β+	0.0508	0.00201
	IB	0.018	0.00065
	Σβ+		4.7

$^{100}_{45}$Rh(4.7 *3* min)

Mode: IT(93 *3* %), ϵ(7 *3* %)

Δ: -85251 *20* keV

SpA: 4.0×10^{8} Ci/g

Prod: alphas on Ru

Photons (^{100}Rh)

⟨γ⟩=221 *23* keV

γ_{mode}	γ(keV)	γ(%)†
γ_{IT}	42.05 *4*	1.05 *21*
γ_{IT} E1	74.73 *6*	70
γ_{IT} [M3]	264.7 *10*	31 *4*
γ_ϵE2	539.53 *4*	7 *2*
γ_ϵE2	686.9 *3*	7.0 *23*

† uncert(syst): 43% for ϵ, 5.4% for IT

$^{100}_{46}$Pd(3.63 *9* d)

Mode: ϵ

Δ: -85207 *12* keV

SpA: 3.60×10^{5} Ci/g

Prod: ^{103}Rh(p,4n); ^{103}Rh(d,5n); protons on Cd

Photons (^{100}Pd)

γ_{mode}	γ(keV)	γ(rel)
γ M1+2.2%E2	32.67 *5*	1.5 *5*
γ	42.05 *4*	1.5 *6*
γ	55.76 *13*	<2
γ E1	74.73 *6*	98 *8*
γ [M1+E2]	83.96 *8*	100
γ E1	126.01 *9*	11 *1*
γ	139.72 *14*	<1
γ	151.5 *3*	<1 ?
γ [E1]	158.68 *9*	2.0 *5*

$^{100}_{47}$Ag(2.0 *1* min)

Mode: ϵ

Δ: -78120 *90* keV

SpA: 9.4×10^{8} Ci/g

Prod: ^{102}Pd(p,3n)

Photons (^{100}Ag)

⟨γ⟩=2496 *55* keV

γ_{mode}	γ(keV)	γ(%)
γ	190.5 *5*	0.44 *4*
γ	222.5 *2*	1.2 *3*
γ	280.6 *2*	8.2 *5*
γ	352.9 *6*	1.02 *7*
γ	450.2 *1*	17.4 *10*
γ	528.9 *1*	1.98 *20*

Photons (^{100}Ag)
(continued)

γ_{mode}	γ(keV)	γ(%)
γ	569.5 *3*	0.74 *5*
γ	609.6 *1*	3.1 *3*
γ	639.9 *1*	8.2 *6*
γ E2	665.7 *1*	99
γ	730.9 *1*	7.5 *6*
γ E2	750.8 *1*	79 *5*
γ E2	773.3 *1*	24.3 *18*
γ	862.5 *1*	3.6 *3*
γ	890.3 *2*	2.57 *20*
γ [M1+E2]	922.3 *1*	1.3 *5*
γ	953.3 *5*	0.10 *3*
γ	960.7 *1*	1.68 *10*
γ	1053.9 *1*	13.8 *11*
γ	1115.8 *2*	3.4 *6*
γ [E2]	1260.5 *1*	5.1 *4*
γ	1278.0 *2*	4.7 *5*
γ	1405.3 *2*	3.3 *4*
γ	1503.7 *2*	11.3 *15*
γ [E2]	1587.9 *2*	0.9 *3*
γ	1639.0 *2*	~0.40
γ	1685.6 *3*	7.0 *7*
γ	1767.6 *3*	0.69 *20*
γ	1819.7 *3*	0.50 *20*
γ	1956.0 *4*	0.50 *20*
γ	2013.7 *10*	0.40 *10*
γ	2214.1 *3*	1.19 *20*

$^{100}_{47}$Ag(2.30 *15* min)

Mode: ϵ
Δ: >-78120 keV
SpA: 8.2×10^{8} Ci/g
Prod: ^{102}Pd(p,3n); ^{92}Mo(^{12}C,p3n); protons on Cd; daughter ^{100}Cd

Photons (^{100}Ag)
⟨γ⟩=1772 *140* keV

γ_{mode}	γ(keV)	γ(%)
γ	614.1 *4*	1.8 *9*
γ E2	665.7 *1*	86
γ E2	750.8 *1*	<26
γ [M1+E2]	922.3 *1*	7.2 *22*
γ	1115.8 *2*	10 *3*
γ	1205.5 *3*	4.7 *17*
γ [E2]	1523.6 *3*	8.6 *22*
γ [E2]	1587.9 *2*	5.3 *17*
γ	1639.0 *2*	~2
γ	1693.9 *3*	14.7 *17*
γ	1819.7 *3*	~4
γ	1956.0 *4*	~2
γ	2118.1 *4*	11 *3*

$^{100}_{48}$Cd(1.1 *3* min)

Mode: ϵ
Δ: -74310 *300* keV syst
SpA: 1.7×10^{9} Ci/g
Prod: protons on Sn; protons on Ag

Photons (^{100}Cd)

γ_{mode}	γ(keV)
γ	97.3 *10*
γ	123.8 *10*
γ	138.8 *10*
γ	178.1 *10*
γ	219.8 *10*
γ	367.6 *10*
γ	427.5 *10*
γ	567 *1*
γ	935.3 *10*

$^{100}_{49}$In($t_{1/2}$ unknown)

Mode: ϵ, ϵp
Prod: ^{40}Ca on ^{63}Cu

A = 101

NDS **28**, 343 (1979)

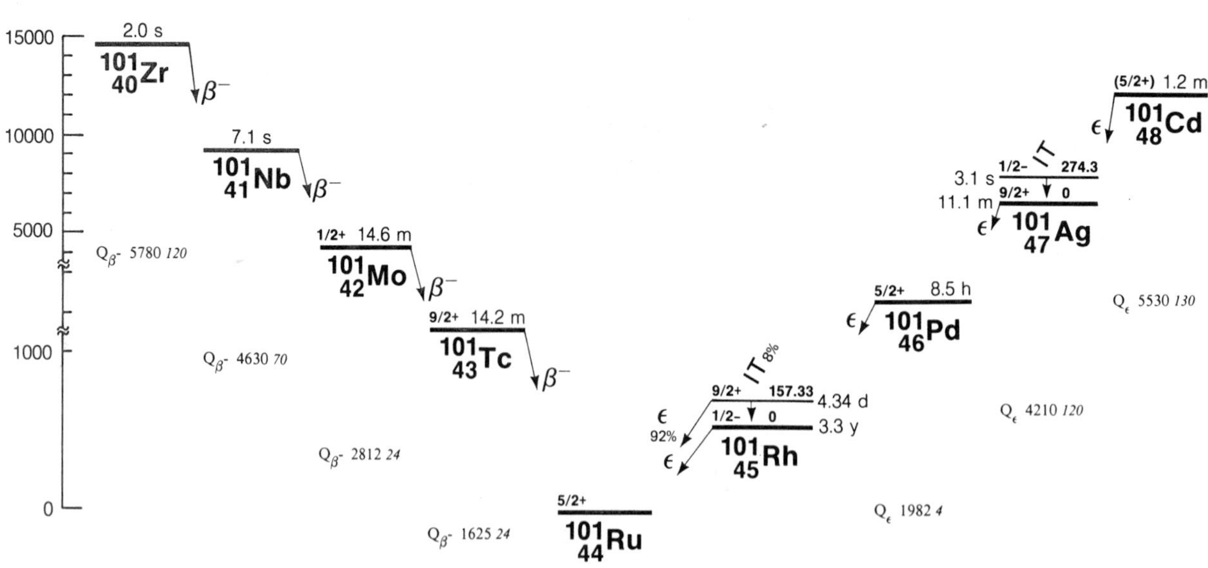

$^{101}_{40}$Zr(2.0 *3* s)

Mode: β-
Δ: -73100 *140* keV
SpA: 4.7×10^{10} Ci/g
Prod: fission

$^{101}_{41}$Nb(7.1 *3* s)

Mode: β-
Δ: -78880 *70* keV
SpA: 1.50×10^{10} Ci/g
Prod: fission

Photons (^{101}Nb)

γ_{mode}	γ(keV)
γ (M1)	13.47 *19*
γ (E2)	43.5 *10*
γ	118.51 *23*
γ	157.71 *19*
γ	180.6 *5*
γ	276.22 *16*
γ	289.69 *17*
γ	440.9 *2*
γ	465.92 *17*
γ	479.39 *17*
γ	796.7 *4*
γ	810.2 *4*

$^{101}_{42}$Mo(14.6 *1* min)

Mode: β-
Δ: -83513 *6* keV
SpA: 1.275×10^{8} Ci/g
Prod: ^{100}Mo(n,γ)

Photons (^{101}Mo)

⟨γ⟩=1514 *20* keV

γ_{mode}	γ(keV)	γ(%)†
Tc $K_{\alpha2}$	18.251	1.61 *8*
Tc $K_{\alpha1}$	18.367	3.05 *16*
Tc $K_{\beta1}$'	20.613	0.73 *4*
Tc $K_{\beta2}$'	21.136	0.131 *7*
γ M1(+E2)	80.96 *3*	3.84 *15*
γ	104.73 *8*	0.167 *15*
γ	105.98 *5*	0.248 *21*
γ	115.77 *5*	0.032 *6*
γ	169.0 *3*	0.024 *9*
γ [E2]	187.37 *12*	0.47 *6*
γ M2	191.943 *21*	18.8 *10*
γ	196.00 *3*	2.86 *15*
γ	212.03 *4*	0.51 *6*
γ	221.73 *14*	0.100 *11*
γ [M1+E2]	274.99 *13*	0.090 *11*
γ	318.07 *5*	0.235 *17*
γ	327.80 *4*	0.212 *15*
γ	333.69 *6*	0.78 *4*
γ	347.55 *5*	0.098 *9*
γ	353.12 *7*	0.141 *13*
γ	358.35 *10*	0.047 *13*
γ	367.65 *5*	0.11 *3*
γ	369.81 *10*	0.16 *4*
γ	371.46 *3*	0.18 *5*
γ	378.01 *24*	0.16 *4*

Photons (^{101}Mo) (continued)

γ_{mode}	γ(keV)	γ(%)†
γ [E1]	379.32 *12*	0.32 *5*
γ [M1+E2]	381.40 *7*	0.305 *23*
γ	384.3 *5*	0.049 *13*
γ	398.99 *3*	0.90 *5*
γ	408.76 *3*	1.60 *8*
γ	421.61 *3*	0.56 *6*
γ	422.4 *5*	0.09 *3*
γ	432.64 *15*	0.117 *23*
γ	441.97 *7*	0.051 *9*
γ	448.70 *5*	0.69 *4*
γ	452.18 *11*	0.073 *9*
γ	469.02 *15*	0.117 *11*
γ	482.55 *12*	0.081 *17*
γ	491.22 *5*	0.070 *8*
γ	497.24 *10*	0.13 *5*
γ [M1+E2]	499.680 *17*	1.47 *11*
γ	505.06 *13*	1.2 *6*
γ [M1+E2]	505.966 *18*	11.8 *8*
γ	510.19 *11*	0.98 *8*
γ [M1+E2]	512.86 *3*	1.75 *11*
γ	514.03 *20*	0.81 *8*
γ	515.56 *9*	0.51 *8*
γ	524.30 *7*	0.173 *13*
γ	533.63 *7*	0.400 *24*
γ	540.47 *21*	0.09 *3*
γ	560.02 *13*	0.070 *9*
γ	566.66 *4*	0.73 *11*
γ	571.85 *8*	0.186 *13*
γ	583.10 *16*	0.090 *24*
γ [M1+E2]	589.65 *19*	5.6 *13*
γ	590.928 *22*	16.4 *15*
γ	603.23 *10*	0.102 *13*
γ	606.99 *3*	0.21 *4*
γ	608.36 *3*	1.07 *6*
γ	611.15 *25*	0.14 *3*
γ	625.72 *16*	0.10 *3*
γ [M1+E2]	642.81 *3*	1.24 *6*
γ	649.85 *10*	0.028 *9*
γ	653.36 *11*	0.028 *9*
γ [E1]	660.59 *7*	0.224 *13*
γ	675.63 *25*	0.047 *9*
γ	686.54 *15*	0.068 *9*
γ	695.68 *3*	7.2 *6*
γ	701.96 *3*	0.38 *4*
γ	708.06 *5*	0.064 *17*
γ	713.11 *3*	3.4 *3*
γ [M1+E2]	728.16 *3*	0.092 *15*
γ	733.07 *8*	0.26 *4*
γ [M1+E2]	737.23 *5*	0.047 *13*
γ [E1]	739.63 *4*	0.301 *19*
γ	774.24 *7*	0.33 *3*
γ	775.87 *10*	0.107 *19*
γ [M1+E2]	778.37 *4*	0.96 *6*
γ	790.13 *20*	0.126 *11*
γ	798.14 *25*	0.071 *11*
γ [M1+E2]	804.36 *3*	1.00 *6*
γ	815.26 *10*	0.179 *15*
γ	847.57 *8*	0.085 *9*
γ	853.2 *1*	0.233 *13*
γ	858.68 *10*	0.111 *9*
γ [M1+E2]	869.83 *6*	0.34 *6*
γ	871.14 *3*	1.80 *19*
γ	877.42 *3*	3.4 *3*
γ	883.52 *5*	0.70 *8*
γ	886.75 *3*	0.23 *3*
γ	888.71 *7*	0.23 *3*
γ	894.88 *24*	0.058 *24*
γ	896.22 *10*	0.21 *3*
γ	903.62 *5*	0.201 *15*
γ	933.10 *6*	0.75 *23*
γ [M1+E2]	934.31 *3*	3.4 *3*
γ	943.74 *16*	0.16 *4*
γ	980.60 *8*	0.265 *15*
γ	988.27 *5*	0.19 *3*
γ	1007.81 *10*	0.177 *17*
γ	1011.26 *19*	2.26 *21*
γ [M1+E2]	1012.534 *25*	12.8 *8*
γ [E2]	1018.819 *25*	0.64 *9*
γ [M1+E2]	1019.66 *3*	0.47 *8*
γ	1030.31 *20*	0.071 *11*
γ	1049.86 *7*	0.346 *19*
γ	1064.17 *7*	0.21 *3*

Photons (^{101}Mo) (continued)

γ_{mode}	γ(keV)	γ(%)†
γ	1065.48 *8*	0.16 *3*
γ	1161.06 *3*	3.97 *21*
γ	1168.92 *7*	0.235 *17*
γ	1184.32 *7*	0.196 *17*
γ	1186.68 *6*	1.03 *6*
γ	1199.98 *4*	1.75 *9*
γ	1209.87 *10*	0.132 *11*
γ	1218.27 *24*	0.056 *9*
γ	1249.46 *13*	0.26 *6*
γ	1251.16 *3*	4.61 *24*
γ	1260.17 *7*	0.152 *17*
γ	1286.29 *17*	0.143 *9*
γ [M1+E2]	1291.23 *4*	0.115 *11*
γ	1293.34 *12*	0.209 *13*
γ [M1+E2]	1304.035 *25*	2.78 *15*
γ	1308.09 *12*	0.09 *4*
γ	1310.59 *6*	~0.06
γ	1314.48 *9*	0.231 *17*
γ	1325.74 *13*	0.26 *5*
γ	1336.52 *3*	0.177 *23*
γ	1339.46 *10*	0.175 *13*
γ	1346.14 *4*	1.03 *9*
γ	1351.27 *10*	0.041 *8*
γ	1355.91 *3*	1.67 *11*
γ [E1]	1377.89 *7*	0.244 *17*
γ	1380.53 *14*	0.107 *17*
γ [M1+E2]	1382.68 *5*	1.15 *6*
γ	1387.51 *9*	0.079 *8*
γ	1394.84 *6*	0.61 *4*
γ	1414.23 *8*	0.50 *3*
γ	1418.62 *6*	0.88 *5*
γ	1426.9 *9*	0.036 *8*
γ	1430.0 *6*	0.137 *24*
γ	1431.49 *4*	0.36 *3*
γ	1434.67 *8*	0.088 *9*
γ	1441.27 *3*	0.158 *9*
γ	1451.50 *25*	0.066 *8*
γ	1486.09 *7*	0.103 *8*
γ	1507.18 *19*	~0.039
γ	1514.18 *7*	0.186 *13*
γ [E1]	1518.00 *5*	0.220 *23*
γ	1520.12 *12*	0.24 *4*
γ	1523.37 *6*	0.290 *24*
γ	1526.72 *10*	0.113 *21*
γ	1529.85 *12*	0.27 *4*
γ [M1+E2]	1532.513 *25*	6.0 *3*
γ	1548.55 *19*	0.150 *13*
γ	1583.49 *5*	0.085 *8*
γ	1589.77 *5*	0.282 *15*
γ	1594.80 *5*	0.030 *13*
γ	1599.29 *5*	1.75 *9*
γ	1605.58 *5*	0.045 *8*
γ	1609.24 *13*	0.098 *8*
γ	1614.90 *5*	0.058 *6*
γ	1629.36 *7*	0.055 *9*
γ	1646.36 *23*	0.077 *9*
γ [E1]	1652.88 *12*	0.062 *8*
γ	1662.38 *10*	0.68 *4*
γ [E1]	1673.93 *3*	1.69 *9*
γ	1712.86 *7*	0.201 *13*
γ	1722.72 *10*	0.032 *13*
γ [E1]	1754.89 *3*	0.46 *3*
γ	1759.84 *7*	0.98 *6*
γ [E1]	1768.36 *5*	0.149 *9*
γ [E1]	1840.25 *3*	1.37 *8*
γ	1876.3 *9*	0.024 *8*
γ [M1+E2]	1882.35 *6*	0.090 *9*
γ [E2]	1888.64 *6*	0.041 *11*
γ	1921.70 *15*	0.039 *13*
γ	1941.68 *12*	0.055 *8*
γ [E2]	1946.832 *24*	0.094 *8*
γ	2024.36 *13*	0.075 *13*
γ	2028.1 *9*	0.103 *17*
γ [M1+E2]	2032.185 *24*	6.9 *4*
γ [E2]	2038.470 *24*	0.21 *3*
γ [M1+E2]	2041.25 *5*	2.11 *11*
γ [E2]	2047.54 *5*	0.088 *11*
γ	2088.81 *6*	0.79 *8*
γ	2112.80 *25*	0.15 *3*
γ	2114.29 *5*	0.45 *3*
γ	2131.4 *4*	0.036 *8*
γ	2223.27 *15*	0.165 *9*

Photons (^{101}Mo) (continued)

γ_{mode}	γ(keV)	γ(%)†
γ	2337.8 8	0.017 8
γ	2404.7 8	~0.015

† 10% uncert(syst)

Atomic Electrons (^{101}Mo)

⟨e⟩=13.4 5 keV

e_{bin}(keV)	⟨e⟩(keV)	e(%)
3 - 21	0.41	7.4 7
60	1.03	1.72 8
78 - 116	0.42	0.51 11
148 - 169	0.091	0.055 7
171	8.0	4.7 3
175	0.28	~0.16
184 - 187	0.0177	0.0096 8
189	1.24	0.66 4
191 - 222	0.39	0.20 3
254 - 297	0.014	0.0051 18
307 - 356	0.056	0.017 5
357 - 406	0.093	0.024 6
408 - 452	0.023	0.0054 16
462 - 510	0.39	0.080 6
511 - 560	0.022	0.0042 10
562 - 611	0.32	0.055 16
622 - 665	0.0219	0.00349 21
673 - 719	0.11	0.016 5
725 - 774	0.0149	0.00198 20
775 - 815	0.0146	0.00186 13
827 - 875	0.052	0.0060 13
877 - 923	0.036	0.0039 4
930 - 978	0.0073	0.00078 10
980 - 1029	0.133	0.0133 10
1030 - 1065	0.0024	0.00023 6
1140 - 1189	0.037	0.0032 7
1197 - 1246	0.023	0.0019 7
1247 - 1293	0.0221	0.00173 14
1301 - 1349	0.017	0.00125 24
1351 - 1398	0.0170	0.00124 13
1406 - 1451	0.0043	0.00031 5
1465 - 1562	0.0379	0.00251 17
1569 - 1662	0.0155	0.00096 15
1671 - 1768	0.0062	0.00036 6
1819 - 1919	0.00388	0.000212 11
1921 - 2017	0.0267	0.00133 10
2020 - 2114	0.0168	0.00082 5
2128 - 2223	0.00050	2.3 6 $\times10^{-5}$
2317 - 2404	9.0 $\times10^{-5}$	3.8 13 $\times10^{-6}$

Continuous Radiation (^{101}Mo)

⟨β-⟩=499 keV;⟨IB⟩=0.82 keV

E_{bin}(keV)		⟨ ⟩(keV)	(%)
0 - 10	β-	0.069	1.38
	IB	0.021	
10 - 20	β-	0.208	1.39
	IB	0.020	0.141
20 - 40	β-	0.83	2.78
	IB	0.039	0.135
40 - 100	β-	5.9	8.4
	IB	0.103	0.160
100 - 300	β-	53	27.0
	IB	0.24	0.140
300 - 600	β-	130	30.0
	IB	0.20	0.048
600 - 1300	β-	190	21.9
	IB	0.169	0.020
1300 - 2500	β-	118	7.2
	IB	0.030	0.0020
2500 - 2604	β-	0.114	0.00453
	IB	4.7 $\times10^{-7}$	1.9 $\times10^{-8}$

$^{101}_{43}$Tc(14.2 1 min)

Mode: β-
Δ: -86325 24 keV
SpA: 1.310$\times10^{8}$ Ci/g
Prod: daughter ^{101}Mo

Photons (^{101}Tc)

⟨γ⟩=337 11 keV

γ_{mode}	γ(keV)	γ(%)†
Ru $K_{\alpha2}$	19.150	0.451 24
Ru $K_{\alpha1}$	19.279	0.85 5
Ru $K_{\beta1}$'	21.650	0.207 11
Ru $K_{\beta2}$'	22.210	0.0385 21
γ [M1+E2]	97.16 17	~0.006
γ [M1+E2]	111.05 11	~0.006
γ [E2]	115.53 11	~0.006
γ M1+2.9%E2	127.220 19	2.86 14
γ E2	179.611 19	0.62 4
γ M1+E2	184.095 21	1.68 9
γ M1(+E2)	233.732 23	0.272 14
γ M1(+E2)	238.215 21	0.307 16
γ [M1+E2]	281.14 16	0.027 8
γ [M1+E2]	295.14 10	0.053 19
γ M1+1.2%E2	306.831 16	88 4
γ (M1)	311.315 24	0.28 4
γ [M1+E2]	322.08 11	0.053 18
γ	383.62 5	0.040 11
γ [M1+E2]	393.37 5	0.10 3
γ [E2]	417.826 24	0.029 13
γ [M1+E2]	422.36 11	0.032 5
γ [E2]	489.12 11	0.053 18
γ [E2]	516.06 11	0.108 13
γ [M1+E2]	531.44 4	1.02 5
γ M1+49%E2	545.046 23	5.99 22
γ [M1+E2]	616.34 11	~0.016
γ	617.35 5	0.051 5
γ	621.83 5	0.089 7
γ [M1+E2]	627.10 5	0.39 3
γ [M1+E2]	631.58 5	0.041 5
γ	673.4 6	0.038 8
γ (E2)	694.31 15	0.070 18
γ [E2]	715.54 4	0.69 4
γ (E2)	720.00 6	0.23 3
γ [E2]	811.19 5	0.060 15
γ [M1+E2]	842.76 4	0.230 13
γ	911.56 12	0.09 3
γ	928.66 5	0.127 11
γ [M1+E2]	938.41 5	0.087 9

† 4.5% uncert(syst)

Atomic Electrons (^{101}Tc)

⟨e⟩=5.36 19 keV

e_{bin}(keV)	⟨e⟩(keV)	e(%)
3	0.046	1.59 18
4 - 22	0.068	0.40 3
75 - 97	0.010	0.012 4
105	0.45	0.427 25
108 - 127	0.084	0.068 4
157	0.131	0.083 6
162	0.24	0.15 6
176 - 216	0.125	0.065 13
231 - 278	0.014	0.0055 11
281	5.0 $\times10^{-5}$	1.8 8 $\times10^{-5}$
285	3.46	1.21 6
289 - 300	0.0137	0.0047 7
304	0.435	0.143 7
306 - 322	0.099	0.0323 13
362 - 400	0.0064	0.0017 3
415 - 422	0.00029	6.9 16 $\times10^{-5}$
467 - 516	0.0236	0.0047 4
523 - 545	0.130	0.0246 9
594 - 632	0.0092	0.00152 12
651 - 698	0.0125	0.00181 10
712 - 720	0.00165	0.000232 10

Atomic Electrons (^{101}Tc) (continued)

e_{bin}(keV)	⟨e⟩(keV)	e(%)
789 - 821	0.00296	0.00036 3
840 - 843	0.00033	3.9 4 $\times10^{-5}$
889 - 938	0.0024	0.00027 5

Continuous Radiation (^{101}Tc)

⟨β-⟩=472 keV;⟨IB⟩=0.62 keV

E_{bin}(keV)		⟨ ⟩(keV)	(%)
0 - 10	β-	0.0434	0.86
	IB	0.021	
10 - 20	β-	0.132	0.88
	IB	0.020	0.142
20 - 40	β-	0.54	1.80
	IB	0.039	0.135
40 - 100	β-	4.08	5.8
	IB	0.103	0.160
100 - 300	β-	45.9	22.6
	IB	0.23	0.135
300 - 600	β-	157	35.2
	IB	0.158	0.039
600 - 1300	β-	264	32.8
	IB	0.050	0.0070
1300 - 1318	β-	0.0125	0.00096
	IB	5.5 $\times10^{-9}$	4.2 $\times10^{-10}$

$^{101}_{44}$Ru(stable)

Δ: -87950.6 22 keV
%: 17.0 1

$^{101}_{45}$Rh(3.3 3 yr)

Mode: ϵ
Δ: -87413 17 keV
SpA: 1073 Ci/g
Prod: ^{101}Ru(p,n); ^{99}Tc(α,2n); daughter ^{101}Pd; deuterons on Ru

Photons (^{101}Rh)

⟨γ⟩=300 9 keV

γ_{mode}	γ(keV)	γ(%)†
Ru L_{ℓ}	2.253	0.104 18
Ru L_{η}	2.382	0.047 7
Ru L_{α}	2.558	2.7 4
Ru L_{β}	2.732	1.7 3
Ru L_{γ}	3.036	0.129 24
Ru $K_{\alpha2}$	19.150	23.0 9
Ru $K_{\alpha1}$	19.279	43.5 17
Ru $K_{\beta1}$'	21.650	10.5 4
Ru $K_{\beta2}$'	22.210	1.96 8
γ [M1+E2]	97.16 17	~0.07
γ [M1+E2]	111.05 11	~0.07
γ [E2]	115.53 11	~0.07
γ M1+2.9%E2	127.220 19	73
γ	137.3 3	0.22 7
γ E2	179.611 19	0.00041 4
γ M1+E2	184.095 21	0.111 8
γ M1+~0.2%E2	197.98 14	70.8 15
γ	216.95 24	0.66 12
γ [M1+E2]	295.14 10	0.66 17
γ M1+1.2%E2	306.831 16	0.06 3

Photons (^{101}Rh) (continued)

γ_{mode}	γ(keV)	γ(%)†
γ (M1)	311.315 24	0.019 8
γ E2	325.20 14	13.4 11
γ	335.3 3	~0.07
γ	344.17 24	0.22 7
γ [M1+E2]	422.36 11	0.40 6
γ	462.5 3	0.073 22

† 8.2% uncert(syst)

Atomic Electrons (^{101}Rh)

$\langle e\rangle$=26.7 11 keV

e_{bin}(keV)	$\langle e\rangle$(keV)	e(%)
3	2.36	81 8
16	2.33	14.4 15
18	0.221	1.20 12
19	0.81	4.3 5
21 - 22	0.097	0.45 4
75 - 97	0.12	0.15 5
105	11.4	10.9 10
108 - 116	0.07	~0.06
124	1.71	1.38 14
127 - 162	0.41	0.32 3
176	5.32	3.03 9
177 - 217	0.93	0.47 3
273 - 322	0.87	0.288 21
325 - 344	0.0241	0.0074 6
400 - 440	0.0140	0.0035 6
459 - 462	0.00019	4.2 19 $\times10^{-5}$

Continuous Radiation (^{101}Rh)

$\langle$IB$\rangle$=0.028 keV

E_{bin}(keV)		$\langle\ \rangle$(keV)	(%)
10 - 20	IB	0.0115	0.066
20 - 40	IB	0.0130	0.060
40 - 100	IB	0.00132	0.0021
100 - 300	IB	0.00166	0.00109
300 - 411	IB	0.000107	3.3 $\times10^{-5}$

$^{101}_{45}$Rh(4.34 *1* d)

Mode: ϵ(92.3 6 %), IT(7.7 6 %)
Δ: -87256 17 keV
SpA: 2.979$\times10^5$ Ci/g
Prod: ^{101}Ru(p,n); ^{100}Ru(d,n); deuterons on Pd; daughter ^{101}Pd; ^{99}Tc(α,2n)

Photons (^{101}Rh)

$\langle\gamma\rangle$=304 11 keV

γ_{mode}	γ(keV)	γ(%)†
Ru L_ℓ	2.253	0.084 15
Rh L_ℓ	2.377	0.0068 13
Ru L_η	2.382	0.038 6
Rh L_η	2.519	0.0029 5
Ru L_α	2.558	2.2 3
Rh L_α	2.696	0.18 3
Ru L_β	2.732	1.34 22
Rh L_β	2.893	0.108 20
Ru L_γ	3.035	0.104 20
Rh L_γ	3.209	0.0096 20
Ru $K_{\alpha2}$	19.150	18.7 7
Ru $K_{\alpha1}$	19.279	35.4 12
Rh $K_{\alpha2}$	20.074	1.25 11
Rh $K_{\alpha1}$	20.216	2.37 21
Ru $K_{\beta1}$'	21.650	8.6 3
Ru $K_{\beta2}$'	22.210	1.60 6
Rh $K_{\beta1}$'	22.717	0.58 5
Rh $K_{\beta2}$'	23.312	0.109 10
γ_ϵM1+2.9%E2	127.220 19	0.64 14
γ_{IT} M4	157.33 3	0.25
γ_ϵE2	179.611 19	0.61 5
γ_ϵM1+E2	184.095 21	0.22 4
γ_ϵM1(+E2)	233.732 23	0.180 12
γ_ϵM1(+E2)	238.215 21	0.203 13
γ_ϵM1+1.2%E2	306.831 16	86
γ_ϵ(M1)	311.315 24	0.036 7
γ_ϵ	332.48 16	0.028 5
γ_ϵ	336.96 16	0.034 6
γ_ϵ[E2]	417.826 24	0.019 9
γ_ϵ[E2]	489.12 11	~0.009
γ_ϵ[E2]	496.2 4	0.013 3
γ_ϵM1+49%E2	545.046 23	3.96 16
γ_ϵ[M1+E2]	616.34 11	~0.0026
γ_ϵ[M1+E2]	623.4 4	0.018 3
γ_ϵ	643.79 16	~0.0035

† uncert(syst): 6.4% for ϵ, 11% for IT

Atomic Electrons (^{101}Rh)

$\langle e\rangle$=19.8 7 keV

e_{bin}(keV)	$\langle e\rangle$(keV)	e(%)
3	2.07	71 7
16	1.91	11.8 12
17 - 18	0.289	1.62 13
19	0.67	3.6 4
20 - 23	0.124	0.60 4
105 - 127	0.118	0.110 21
134	7.1	5.3 4
154	2.63	1.71 14
157	0.77	0.49 4
162 - 184	0.064	0.038 9
212 - 238	0.038	0.018 4
285	3.39	1.19 6
289 - 337	0.525	0.173 7
396 - 418	0.0007	0.00018 7
467 - 496	0.00057	0.000121 25
523 - 545	0.084	0.0159 6
594 - 644	0.00040	6.5 9 $\times10^{-5}$

Continuous Radiation (^{101}Rh)

$\langle$IB$\rangle$=0.033 keV

E_{bin}(keV)		$\langle\ \rangle$(keV)	(%)
10 - 20	IB	0.0106	0.061
20 - 40	IB	0.0121	0.056
40 - 100	IB	0.0017	0.0026
100 - 300	IB	0.0076	0.0043
300 - 388	IB	0.00045	0.000141

$^{101}_{46}$Pd(8.47 *6* h)

Mode: ϵ
Δ: -85431 18 keV
SpA: 3.66$\times10^6$ Ci/g
Prod: ^{103}Rh(p,3n); ^{99}Ru(α,2n)

Photons (^{101}Pd)

$\langle\gamma\rangle$=301 5 keV

γ_{mode}	γ(keV)	γ(%)†
Rh $K_{\alpha2}$	20.074	35.9 15
Rh $K_{\alpha1}$	20.216	68 3
Rh $K_{\beta1}$'	22.717	16.7 7
Rh $K_{\beta2}$'	23.312	3.13 13
γ M1+0.04%E2	24.471 10	3.90 15
γ (M1)	111.45 4	0.012 4
γ	130.02 6	~0.015
γ (E1)	133.4 3	0.021 8 ?
γ M4	157.33 3	*
γ	157.93 5	0.023 10
γ	171.0 5	~0.017 ?
γ (E1)	173.1 5	0.023 10 ?
γ	185.3 3	0.010 4 ?
γ M1	269.71 4	6.43 13
γ M1+7.8%E2	296.284 21	19.2
γ [M1+E2]	305.4 3	>0.04
γ E2	320.755 21	0.56 3
γ E2	355.32 9	0.223 13
γ	374.51 19	~0.006 ?
γ M1,E2	380.97 13	0.038 8
γ M1,E2	427.63 4	0.098 6
γ E2	435.13 6	0.063 8
γ M1(+E2)	453.70 4	0.605 23
γ	492.42 13	0.010 4
γ M1,E2	496.09 11	0.033 10
γ	565.15 5	0.21 8
γ M1,E2	565.99 3	3.44 8
γ M1(+E2)	590.46 3	12.06 24
γ M1,E2	611.62 4	0.094 10
γ [E1]	619.46 10	0.040 6
γ	702.64 25	0.017 6
γ M1,E2	723.08 4	0.29 6
γ M1(+E2)	723.91 3	2.05 17
γ M1,E2	748.39 3	0.501 19
γ	787.4 3	0.0048 23 ?
γ M1,E2	790.65 7	0.023 4
γ M1,E2	796.64 12	0.027 4
γ	821.11 12	0.019 8
γ M1,E2	853.93 6	0.088 8
γ	856.5 3	~0.008 ?
γ	870.71 20	0.021 6 ?
γ M1,E2	881.33 4	0.108 10
γ	905.72 4	~0.008 ?
γ	911.8 4	0.021 6
γ M1,E2	914.90 9	0.075 8
γ [M1+E2]	948.57 7	~0.008
γ	964.8 3	~0.019
γ M1,E2	992.78 4	0.94 6
γ	1014.68 19	0.023 8
γ M1,E2	1041.78 12	0.056 8
γ	1072.82 9	0.029 8
γ	1163.66 20	~0.010
γ [E1]	1165.5 3	~0.010
γ M1(+E2)	1177.61 3	0.353 19
γ M1(+E2)	1202.08 3	1.52 6
γ M1(+E2)	1218.28 6	0.520 19
γ M1	1289.06 3	2.28 6
γ M1,E2	1311.49 12	0.19 4
γ M1,E2	1313.53 3	0.088 17
γ	1342.53 9	0.025 4
γ [E1]	1390.9 3	0.0058 19
γ	1433.37 20	0.029 6
γ	1447.0 3	~0.0038 ?
γ	1512.4 3	0.025 6
γ [M1+E2]	1514.56 6	0.027 6
γ	1607.77 12	0.027 4
γ	1632.24 12	0.019 4

Photons (^{101}Pd)
(continued)

γ_{mode}	γ(keV)	γ(%)†
γ M1,E2	1638.81 9	0.100 10
γ	1646.5 10	~0.0017 ?
γ	1663.28 9	~0.0021
γ	1729.65 20	0.009 3

† 7.8% uncert(syst)
* with ^{101}Rh(4.34 d)

Atomic Electrons (^{101}Pd)
$\langle e \rangle$=14.1 5 keV

e_{bin}(keV)	$\langle e \rangle$(keV)	e(%)
1	0.88	70 3
3	3.8	123 13
16	0.35	2.14 22
17	3.1	18.5 19
19	0.41	2.13 22
20	1.16	5.9 6
21	1.88	8.9 4
22 - 24	0.545	2.33 9
88 - 135	0.013	0.012 4
148 - 185	0.006	0.0041 17
246 - 270	0.385	0.154 4
273	0.90	0.33 3
282 - 321	0.195	0.066 5
332 - 381	0.0139	0.00412 22
404 - 454	0.0241	0.0056 4
469 - 496	0.00104	0.00022 5
542 - 590	0.320	0.0568 24
596 - 619	0.00050	8.3 7 $\times 10^{-5}$
679 - 725	0.041	0.0059 4
745 - 794	0.00162	0.000214 15
796 - 833	0.00116	0.000140 18
847 - 892	0.00247	0.000284 23
902 - 949	0.00034	3.7 10 $\times 10^{-5}$
961 - 993	0.0095	0.00098 9
1011 - 1050	0.00071	6.9 11 $\times 10^{-5}$
1069 - 1073	2.4 $\times 10^{-5}$	2.3 9 $\times 10^{-6}$
1140 - 1179	0.0135	0.00115 8
1195 - 1218	0.0056	0.00047 3
1266 - 1313	0.0198	0.00155 5
1319 - 1368	0.00015	1.1 4 $\times 10^{-5}$
1388 - 1433	0.00016	1.1 4 $\times 10^{-5}$
1444 - 1514	0.00029	1.9 4 $\times 10^{-5}$
1585 - 1663	0.00079	4.9 5 $\times 10^{-5}$
1706 - 1729	3.6 $\times 10^{-5}$	2.1 9 $\times 10^{-6}$

Continuous Radiation (^{101}Pd)
$\langle\beta+\rangle$=17.4 keV;$\langle$IB$\rangle$=0.90 keV

E_{bin}(keV)		$\langle\ \rangle$(keV)	(%)
0 - 10	$\beta+$	6.2 $\times 10^{-6}$	7.5 $\times 10^{-5}$
	IB	0.00099	
10 - 20	$\beta+$	0.000192	0.00117
	IB	0.0110	0.062
20 - 40	$\beta+$	0.00411	0.0127
	IB	0.017	0.077
40 - 100	$\beta+$	0.143	0.186
	IB	0.0063	0.0097
100 - 300	$\beta+$	4.12	1.95
	IB	0.042	0.021
300 - 600	$\beta+$	11.4	2.67
	IB	0.18	0.039
600 - 1300	$\beta+$	1.73	0.267
	IB	0.56	0.063
1300 - 1800	IB	0.090	0.0063
	$\Sigma\beta+$		5.1

$^{101}_{47}$Ag(11.1 3 min)

Mode: ϵ
Δ: -81220 120 keV
SpA: 1.68$\times 10^{8}$ Ci/g
Prod: protons on Ag; ^{102}Pd(p,2n)

Photons (^{101}Ag)
$\langle\gamma\rangle$=823 23 keV

γ_{mode}	γ(keV)	γ(%)†
Pd L_ℓ	2.503	0.034 6
Pd L_η	2.660	0.0156 24
Pd L_α	2.838	0.91 14
Pd L_β	3.052	0.59 10
Pd L_γ	3.383	0.058 11
Pd $K_{\alpha2}$	21.020	7.4 3
Pd $K_{\alpha1}$	21.177	13.9 5
Pd $K_{\beta1}'$	23.811	3.46 13
Pd $K_{\beta2}'$	24.445	0.67 3
γ (M1+E2)	80.30 8	4.7 3
γ [E2]	180.74 10	0.26 10
γ M1	261.05 8	52
γ	274.70 11	1.66 16
γ [M1+E2]	326.92 9	1.87 16
γ	386.62 13	0.23 4
γ M1(+0.5%E2)	406.30 9	0.62 10
γ	420.1 4	0.08 4 ?
γ	439.24 9	2.82 9
γ	460.02 13	0.36 5
γ	470.65 11	0.31 5
γ	493.9 4	0.26 5
γ [M1+E2]	506.61 10	0.57 10
γ [E2]	507.66 11	2.0 5
γ	532.01 13	0.31 5
γ [M1+E2]	538.02 10	1.37 8
γ	543.38 10	2.29 16
γ	550.28 10	0.38 5
γ	575.68 13	0.35 5
γ	577.9 5	~0.05 ?
γ	581.68 12	0.21 5
γ [M1+E2]	586.00 11	1.0 3
γ [M1+E2]	587.96 9	9.9 5
γ	598.28 13	0.59 6
γ	611.40 12	0.65 7
γ [M1+E2]	617.40 11	0.16 5
γ	623.68 9	0.76 5
γ	654.41 9	1.56 10
γ E2	667.35 8	9.7 3
γ	677.67 14	1.3 3
γ E2	678.0 3	2.3 3
γ	734.72 9	3.2 3
γ E2	736.5 3	0.94 21
γ	799.77 17	0.29 3
γ	807.1 4	0.10 3
γ	825.86 14	0.24 5
γ [M1+E2]	867.14 17	0.28 6
γ	893.23 12	1.11 10
γ	899.26 12	0.47 6
γ	910.81 18	0.06 3
γ [M1+E2]	912.91 10	0.57 5
γ	930.66 13	0.31 5
γ	938.31 12	1.35 10
γ [M1+E2]	944.32 10	1.14 10
γ [E2]	1093.65 9	2.6 1
γ [E2]	1125.06 11	0.41 8
γ [M1+E2]	1173.96 8	8.84 21
γ [M1+E2]	1205.36 9	2.6 1
γ	1299.53 11	0.93 8
γ	1307.05 17	0.73 10
γ	1326.4 3	0.68 10
γ	1353.61 22	0.29 5
γ	1399.3 5	~0.10 ?
γ	1418.08 16	0.42 8
γ [E2]	1454.18 17	0.44 8
γ	1487.5 4	0.17 3 ?
γ	1632.41 25	0.47 16
γ	1632.8 4	0.42 16
γ	1641.7 3	0.16 3
γ	1671.9 4	0.14 4
γ	1815.5 7	0.25 9
γ	1901.2 5	0.24 8

Photons (^{101}Ag)
(continued)

γ_{mode}	γ(keV)	γ(%)†
γ	1959.32 24	0.24 8
γ	2041.76 16	0.61 5
γ	2053.1 3	0.81 7
γ	2131.7 4	0.31 10
γ	2307.9 3	0.62 10
γ	2444.4 7	~0.10
γ	2519.4 6	~0.08
γ	2634.8 3	0.29 4
γ	2664.3 8	0.18 5
γ	2699.1 3	0.36 10
γ	2854.0 8	0.14 4
γ	2888.1 6	0.14 4
γ	3143.3 8	0.17 4
γ	3197.4 9	0.07 3

† 15% uncert(syst)

Atomic Electrons (^{101}Ag)
$\langle e \rangle$=8.7 5 keV

e_{bin}(keV)	$\langle e \rangle$(keV)	e(%)
3	0.74	23.0 24
4 - 17	0.215	2.63 20
18	0.54	3.0 3
20 - 24	0.346	1.67 11
56	1.63	2.92 17
77	0.278	0.363 21
156 - 181	0.071	0.044 15
237	3.1	1.3 2
250	0.09	~0.035
257	0.38	0.148 22
258 - 303	0.225	0.081 8
323 - 362	0.022	0.0067 12
382 - 420	0.08	0.021 8
436 - 483	0.091	0.019 3
490 - 538	0.088	0.017 3
540 - 562	0.034	0.0061 11
564	0.186	0.0330 25
572 - 621	0.064	0.0109 13
623 - 667	0.239	0.0370 21
674 - 712	0.051	0.0072 22
731 - 775	0.0092	0.0012 3
783 - 826	0.0038	0.00047 15
843 - 890	0.023	0.0026 5
893 - 941	0.028	0.0031 5
1069 - 1101	0.0262	0.00244 12
1121 - 1171	0.076	0.0066 5
1173 - 1205	0.0239	0.00202 16
1275 - 1323	0.014	0.00105 24
1326 - 1375	0.0022	0.00016 5
1394 - 1430	0.0048	0.00034 7
1451 - 1487	0.0012	8.3 23 $\times 10^{-5}$
1608 - 1671	0.0056	0.00034 12
1791 - 1877	0.0019	0.00011 4
1898 - 1959	0.0011	5.7 21 $\times 10^{-5}$
2017 - 2107	0.0066	0.00032 6
2128 - 2131	0.00014	7 3 $\times 10^{-6}$
2284 - 2307	0.0022	1.0 3 $\times 10^{-4}$
2420 - 2519	0.00063	2.6 10 $\times 10^{-5}$
2610 - 2699	0.0026	0.00010 2
2830 - 2888	0.00082	2.9 7 $\times 10^{-5}$
3119 - 3197	0.00068	2.2 6 $\times 10^{-5}$

Continuous Radiation (^{101}Ag)

⟨β+⟩=804 keV;⟨IB⟩=4.3 keV

E_{bin}(keV)		⟨ ⟩(keV)	(%)
0 - 10	β+	4.71×10^{-6}	5.7×10^{-5}
	IB	0.029	
10 - 20	β+	0.000153	0.00093
	IB	0.031	0.21
20 - 40	β+	0.00348	0.0107
	IB	0.062	0.22
40 - 100	β+	0.136	0.177
	IB	0.160	0.25
100 - 300	β+	5.9	2.66
	IB	0.46	0.26
300 - 600	β+	42.6	9.2
	IB	0.57	0.133
600 - 1300	β+	295	30.9
	IB	1.16	0.129
1300 - 2500	β+	444	25.9
	IB	1.48	0.084
2500 - 3949	β+	16.8	0.64
	IB	0.37	0.0130
	Σβ+		69

Atomic Electrons (^{101}Ag)

⟨e⟩=116 *4* keV

e_{bin}(keV)	⟨e⟩(keV)	e(%)
3	1.30	39 *4*
4	0.87	24 *3*
18	0.60	3.3 *4*
19	0.86	4.6 *5*
21	0.33	1.56 *17*
22	0.35	1.61 *18*
24	0.046	0.188 *21*
25	0.0206	0.083 *9*
73	22.3	30.7 *19*
94	3.12	3.31 *21*
95	1.82	1.92 *12*
97	0.79	0.81 *5*
98	0.391	0.400 *25*
151	54.8	36.4 *23*
172	5.9	3.44 *22*
173	16.7	9.7 *6*
175	1.13	0.64 *4*
176	4.4	2.48 *15*

$^{101}_{47}$Ag(3.1 *1* s)

Mode: IT

Δ: -80946 *120* keV

SpA: 3.23×10^{10} Ci/g

Prod: ^{102}Pd(p,2n)

Photons (^{101}Ag)

⟨γ⟩=156 *6* keV

γ_{mode}	γ(keV)	γ(%)
Ag L_ℓ	2.634	0.085 *15*
Ag L_η	2.806	0.045 *7*
Ag L_α	2.984	2.3 *4*
Ag L_β	3.214	1.7 *3*
Ag L_γ	3.564	0.16 *3*
Ag $K_{\alpha2}$	21.990	16.2 *9*
Ag $K_{\alpha1}$	22.163	30.6 *16*
Ag $K_{\beta1}'$	24.934	7.7 *4*
Ag $K_{\beta2}'$	25.603	1.53 *8*
γ M1+14%E2	98.1 *2*	62 *4*
γ E3	176.2 *2*	47 *3*

$^{101}_{48}$Cd(1.2 *2* min)

Mode: ε

Δ: -75690 *180* keV

SpA: 1.5×10^{9} Ci/g

Prod: protons on Ag; protons on Sn

Photons (^{101}Cd)

⟨γ⟩=1697 *19* keV

γ_{mode}	γ(keV)	γ(%)
γ	46.1 *2*	0.400 *24*
γ M1+14%E2	98.1 *2*	47.5 *24*
γ	163.3 *2*	0.100 *6*
γ E3	176.2 *2*	5.7 *3* *
γ	210.4 *2*	0.100 *6*
γ	234.8 *2*	0.100 *6*
γ	278.2 *2*	0.400 *24*
γ	308.9 *2*	2.00 *12*
γ	334.7 *2*	0.62 *4*
γ	366.3 *2*	0.100 *6*
γ	394.6 *2*	0.400 *24*
γ	403.5 *2*	0.200 *12*
γ	405.2 *2*	0.62 *4*
γ	446.3 *2*	0.300 *18*
γ	476.5 *2*	1.80 *11*
γ	487.7 *2*	0.100 *6*
γ	493.8 *2*	0.400 *24*
γ M1,E2	523.4 *2*	5.1 *3*
γ	535.0 *2*	0.200 *12*
γ	570.7 *2*	0.200 *12*
γ M1,E2	609.1 *2*	1.90 *11*
γ	635.0 *2*	0.50 *3*
γ E1	637.9 *2*	2.40 *14*
γ	652.8 *2*	0.300 *18*
γ	677.4 *2*	0.76 *5*
γ	682.0 *2*	0.400 *24*
γ M1,E2	686.9 *2*	4.3 *3*
γ	690.0 *2*	0.50 *3*
γ	700.5 *2*	0.62 *4*
γ	705.8 *2*	2.50 *15*
γ M1,E2	728.8 *2*	0.80 *5*
γ	757.7 *2*	0.300 *18*
γ	772.4 *2*	0.85 *5*
γ	796.0 *2*	0.200 *12*
γ	798.0 *2*	0.200 *12*
γ M1,E2	924.7 *2*	6.9 *4*
γ	950.3 *2*	0.400 *24*
γ	969.6 *2*	0.400 *24*
γ	974.1 *2*	1.20 *7*
γ	1011.4 *2*	0.400 *24*
γ	1022.6 *2*	2.20 *13*
γ	1072.0 *2*	0.300 *18*
γ	1098.0 *2*	0.85 *5*
γ	1108.6 *2*	0.400 *24*
γ	1130.0 *2*	0.50 *3*
γ	1153.8 *2*	0.50 *3*
γ M1,E2	1187.3 *2*	2.80 *17*
γ	1196.4 *2*	0.71 *4*
γ	1203.0 *2*	4.9 *3*
γ	1227.0 *2*	0.72 *4*
γ M1,E2	1258.9 *2*	8.3 *5*
γ	1285.4 *2*	0.66 *4*
γ	1301.1 *2*	0.300 *18*
γ	1308.7 *2*	0.400 *24*
γ	1319.5 *2*	1.90 *11*
γ	1324.6 *2*	1.80 *11*
γ	1331.8 *2*	2.50 *15*
γ	1408.4 *2*	0.400 *24*
γ M1,E2	1417.1 *2*	5.8 *4*
γ	1430.7 *2*	0.52 *3*
γ	1491.6 *2*	0.57 *3*
γ	1586.8 *2*	1.10 *7*
γ	1592.6 *2*	1.90 *11*
γ	1631.4 *2*	2.50 *15*
γ	1642.4 *2*	0.76 *5*
γ	1656.5 *2*	1.30 *8*
γ	1685.7 *2*	0.71 *4*
γ	1690.9 *2*	3.4 *2*
γ	1696.7 *2*	4.3 *3*
γ	1722.5 *2*	11.5 *7*
γ	1761.7 *2*	1.20 *7*
γ	1783.1 *2*	0.300 *18*
γ	1796.6 *2*	1.90 *11*
γ	1801.6 *2*	0.57 *3*
γ	1820.6 *2*	0.300 *18*
γ	1842.0 *2*	0.50 *3*
γ	1853.0 *2*	0.400 *24*
γ	1859.7 *2*	4.3 *3*
γ	1892.0 *2*	0.66 *4*
γ	1960.9 *2*	2.40 *14*
γ	1983.5 *2*	0.400 *24*
γ	1990.2 *2*	1.40 *8*
γ	2012.9 *2*	0.400 *24*
γ	2032.5 *2*	0.400 *24*
γ	2055.4 *2*	0.400 *24*
γ	2059.0 *2*	0.100 *6*
γ	2130.0 *2*	1.40 *8*
γ	2139.2 *2*	0.400 *24*
γ	2146.0 *2*	0.100 *6*
γ	2190.7 *2*	0.300 *18*
γ	2243.5 *2*	0.50 *3*
γ	2293.3 *2*	0.200 *12*
γ	2391.2 *2*	0.200 *12*
γ	2431.5 *2*	0.200 *12*
γ	2776.1 *2*	0.57 *3*
γ	2841.9 *2*	1.80 *11*
γ	2875.7 *2*	0.200 *12*
γ	2940.0 *2*	0.200 *12*
γ	3064.9 *2*	0.200 *12*

* with ^{101}Ag(3.1 s) in equilib

A = 102

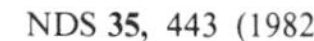

NDS **35**, 443 (1982)

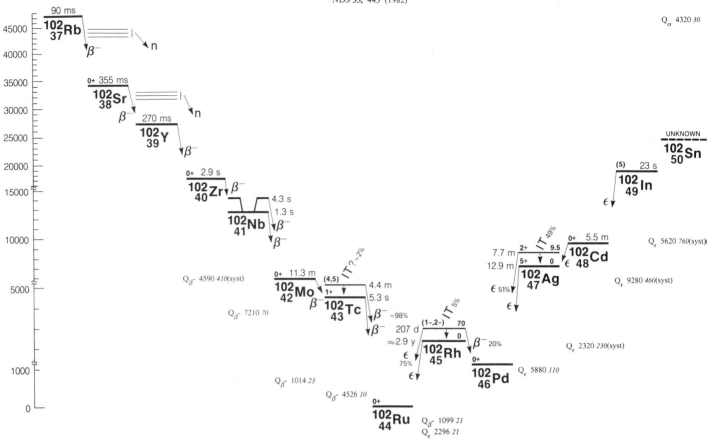

$^{102}_{37}$Rb(90 *20* ms)

Mode: β-, β-n
SpA: 1.6×10^{11} Ci/g
Prod: fission

$^{102}_{38}$Sr(355 *50* ms)

Mode: β-, β-n
SpA: 1.37×10^{11} Ci/g
Prod: fission

$^{102}_{39}$Y (270 *70* ms)

Mode: β-
SpA: 1.5×10^{11} Ci/g
Prod: fission

Photons (^{102}Y)

γ_{mode}	γ(keV)
γ (E2)	151.90 *10*
γ [E2]	326.51 *19*
γ	578.8 *2*
γ	1090.2 *2*

$^{102}_{40}$Zr(2.9 *2* s)

Mode: β-
Δ: -71760 *400* keV syst
SpA: 3.39×10^{10} Ci/g
Prod: fission

$^{102}_{41}$Nb(1.3 *2* s)

Mode: β-
Δ: -76350 *70* keV
SpA: 6.6×10^{10} Ci/g
Prod: daughter ^{102}Zr; fission

Photons (^{102}Nb)

γ_{mode}	γ(keV)
γ [E2]	296.0 4
γ [M1+E2]	397.6 3
γ [E2]	400.6 5
γ [M1+E2]	551.4 3
γ [E2]	847.4 4
γ [M1+E2]	949.0 3

$^{102}_{41}$Nb(4.3 4 s)

Mode: β-
Δ: -76350 70 keV
SpA: 2.38×10^{10} Ci/g
Prod: fission

Photons (^{102}Nb)

γ_{mode}	γ(keV)
γ [E2]	296.0 4 *
γ [M1+E2]	397.6 3
γ [E2]	400.6 5 *
γ [E2]	446.9 4
γ [M1+E2]	551.4 3 *
γ [E2]	847.4 4 *
γ [M1+E2]	949.0 3 *
γ [E2]	1235.4 3
γ [E2]	1633.0 3
γ [M1+E2]	1737.5 4
γ [E2]	2184.4 3

* with ^{102}Nb(4.3 + 1.3 s)

$^{102}_{42}$Mo(11.3 2 min)

Mode: β-
Δ: -83559 21 keV
SpA: 1.63×10^{8} Ci/g
Prod: fission

Photons (^{102}Mo)

⟨γ⟩=18.5 11 keV

γ_{mode}	γ(keV)	γ(%)†
Tc K$_{\alpha2}$	18.251	0.139 8
Tc K$_{\alpha1}$	18.367	0.264 16
Tc K$_{\beta1}$'	20.613	0.063 4
Tc K$_{\beta2}$'	21.136	0.0113 7
γ [M1]	42.79 5	0.025 6
γ [M1]	93.18 4	0.103 11
γ [M1]	135.97 4	0.232 11
γ [M1]	148.13 3	3.76 19
γ [M1]	211.60 3	3.8
γ [M1]	223.77 3	1.44 8
γ	266.56 5	0.084 8
γ [M1+E2]	359.74 3	0.27 8

† 13% uncert(syst)

Atomic Electrons (^{102}Mo)

⟨e⟩=0.96 4 keV

e$_{bin}$(keV)	⟨e⟩(keV)	e(%)
3 - 43	0.055	0.71 6
72 - 115	0.055	0.060 4
127	0.40	0.315 20
133 - 136	0.00476	0.00357 19
145	0.055	0.0376 24
148	0.0121	0.0082 5
191	0.23	0.123 17
203	0.082	0.040 3
209	0.030	0.0145 20
211 - 246	0.024	0.0106 12
264 - 267	0.0007	~0.00027
339 - 360	0.011	0.0031 10

Continuous Radiation (^{102}Mo)

⟨β-⟩=350 keV;⟨IB⟩=0.37 keV

E$_{bin}$(keV)		⟨ ⟩(keV)	(%)
0 - 10	β-	0.067	1.34
	IB	0.0165	
10 - 20	β-	0.203	1.36
	IB	0.0158	0.110
20 - 40	β-	0.83	2.76
	IB	0.030	0.103
40 - 100	β-	6.1	8.7
	IB	0.075	0.117
100 - 300	β-	63	31.5
	IB	0.149	0.088
300 - 600	β-	171	39.2
	IB	0.073	0.018
600 - 1014	β-	108	15.1
	IB	0.0085	0.00130

$^{102}_{43}$Tc(5.28 15 s)

Mode: β-
Δ: -84573 10 keV
SpA: 1.96×10^{10} Ci/g
Prod: daughter ^{102}Mo

Photons (^{102}Tc)

⟨γ⟩=81 4 keV

γ_{mode}	γ(keV)	γ(%)†
γ [E2]	256.58 9	0.019 10
γ E2	468.65 4	0.88 7
γ E2	475.12 3	6.7
γ E2(+0.04%M1)	628.10 4	0.77 5
γ [E2]	636.89 6	0.39 4
γ [E2]	734.00 7	0.10 3
γ [E2]	865.57 20	0.87 20
γ [E2]	1103.22 4	0.401 23
γ M1+5.9%E2	1105.53 6	0.69 7
γ E2	1362.10 8	0.37 8
γ [E2]	1580.64 6	0.079 15
γ	2201.1 10	0.23 8
γ	2434.1 10	0.23 8

† 7.5% uncert(syst)

Continuous Radiation (^{102}Tc)

⟨β-⟩=1945 keV;⟨IB⟩=6.6 keV

E$_{bin}$(keV)		⟨ ⟩(keV)	(%)
0 - 10	β-	0.00314	0.062
	IB	0.057	
10 - 20	β-	0.0096	0.064
	IB	0.057	0.39
20 - 40	β-	0.0403	0.134
	IB	0.112	0.39
40 - 100	β-	0.320	0.452
	IB	0.32	0.50
100 - 300	β-	4.40	2.12
	IB	0.98	0.55
300 - 600	β-	23.3	5.05
	IB	1.22	0.28
600 - 1300	β-	191	19.6
	IB	2.0	0.23
1300 - 2500	β-	820	43.3
	IB	1.56	0.091
2500 - 4526	β-	905	29.4
	IB	0.33	0.0112

$^{102}_{43}$Tc(4.35 7 min)

Mode: β-(98 2 %), IT?(2 2 %)
Δ: >-84573 keV
SpA: 4.23×10^{8} Ci/g
Prod: ^{102}Ru(n,p); fission

Photons (^{102}Tc)

⟨γ⟩=2487 27 keV

γ_{mode}	γ(keV)	γ(%)†
$\gamma_{\beta-}$[M1+E2]	345.97 7	0.109 13
$\gamma_{\beta-}$[M1+E2]	415.31 5	0.82 7
$\gamma_{\beta-}$E2+2.0%M1	418.53 5	4.3 3
$\gamma_{\beta-}$[M1+E2]	420.50 12	0.40 4
$\gamma_{\beta-}$E2	475.12 3	85.3
$\gamma_{\beta-}$	496.5 3	6.0 9
$\gamma_{\beta-}$E2(+0.04%M1)	628.10 4	25.0 11
$\gamma_{\beta-}$E2	631.33 4	15.8 9
$\gamma_{\beta-}$[M1+E2]	692.37 12	1.9 4
$\gamma_{\beta-}$[E2]	695.59 12	2.9 6
$\gamma_{\beta-}$E2	697.56 6	5.7 3
$\gamma_{\beta-}$	920.1 3	0.82 13
$\gamma_{\beta-}$E2+2.0%M1	1046.64 5	13.3 6
$\gamma_{\beta-}$	1074.7 5	1.3 3
$\gamma_{\beta-}$[E2]	1103.22 4	13.1 6
$\gamma_{\beta-}$M1+39%E2	1112.87 5	2.36 9
$\gamma_{\beta-}$	1127.8 3	1.17 7
$\gamma_{\beta-}$	1179.5 5	0.61 9
$\gamma_{\beta-}$	1197.4 4	7.5 9
$\gamma_{\beta-}$	1292.5 3	4.3 4
$\gamma_{\beta-}$	1317.8 17	0.91 7
$\gamma_{\beta-}$[E2]	1323.69 12	0.50 10 ?
$\gamma_{\beta-}$	1338.6 3	4.09 17
$\gamma_{\beta-}$	1488.5 7	0.67 3
$\gamma_{\beta-}$	1511.1 14	0.92 9
$\gamma_{\beta-}$	1594.8 5	2.77 17
$\gamma_{\beta-}$	1615.9 4	15.4 6
$\gamma_{\beta-}$	1711.1 3	2.81 17
$\gamma_{\beta-}$	1810.5 7	5.8 3
$\gamma_{\beta-}$	1907.0 7	1.64 13
$\gamma_{\beta-}$	1945.9 17	1.53 9
$\gamma_{\beta-}$	1966.7 3	1.45 9
$\gamma_{\beta-}$	2139.2 14	1.78 9
$\gamma_{\beta-}$	2226.1 5	5.7 3
$\gamma_{\beta-}$	2244.0 4	11.9 4
$\gamma_{\beta-}$	2339.1 3	0.55 7
$\gamma_{\beta-}$	2438.6 7	4.60 17
$\gamma_{\beta-}$	2535.1 7	0.49 3

† uncert(syst): 2.3% for β-

Continuous Radiation (^{102}Tc)

⟨β-⟩=780 keV;⟨IB⟩=1.53 keV

E_{bin}(keV)		⟨ ⟩(keV)	(%)
0 - 10	β-	0.0193	0.383
	IB	0.031	
10 - 20	β-	0.059	0.392
	IB	0.030	0.21
20 - 40	β-	0.244	0.81
	IB	0.059	0.20
40 - 100	β-	1.88	2.67
	IB	0.163	0.25
100 - 300	β-	23.4	11.4
	IB	0.43	0.24
300 - 600	β-	101	22.2
	IB	0.41	0.097
600 - 1300	β-	426	46.5
	IB	0.36	0.044
1300 - 2500	β-	221	13.8
	IB	0.061	0.0038
2500 - 3420	β-	7.0	0.261
	IB	0.00051	1.9×10^{-5}

$^{102}_{44}$Ru(stable)

Δ: -89099.5 *23* keV
%: 31.6 *2*

$^{102}_{45}$Rh(~2.9 yr)

Mode: ϵ
Δ: -86803 *21* keV
SpA: ~1209 Ci/g
Prod: ^{102}Ru(p,n); deuterons on Ru; ^{103}Rh(γ,n); ^{104}Pd(d,α); ^{103}Rh(n,2n); ^{103}Rh(d,t)

Photons (^{102}Rh)

⟨γ⟩=2162 *35* keV

γ_{mode}	γ(keV)	γ(%)
Ru L_ℓ	2.253	0.090 *16*
Ru L_η	2.382	0.041 *6*
Ru L_α	2.558	2.3 *4*
Ru L_β	2.733	1.44 *24*
Ru L_γ	3.040	0.114 *22*
Ru $K_{\alpha2}$	19.150	19.6 *7*
Ru $K_{\alpha1}$	19.279	37.1 *13*
Ru $K_{\beta1}$'	21.650	9.0 *3*
Ru $K_{\beta2}$'	22.210	1.67 *6*
γ [E2]	74.54 *13*	0.21 *9* ?
γ [M1+E2]	345.97 *7*	0.87 *10*
γ [M1+E2]	415.31 *5*	2.05 *19*
γ E2+2.0%M1	418.53 *5*	10.6 *7*
γ [M1+E2]	420.50 *12*	3.2 *3*
γ E2	475.12 *3*	95 *4*
γ E2(+0.04%M1)	628.10 *4*	8.5 *3*
γ E2	631.33 *4*	56 *2*
γ [M1+E2]	692.37 *12*	1.75 *20*
γ [E2]	695.59 *12*	2.7 *3*
γ E2	697.56 *6*	45.7 *18*
γ E2	766.90 *6*	34 *2*
γ E2+2.0%M1	1046.64 *5*	33.0 *15*
γ [E2]	1103.22 *4*	4.43 *18*
γ M1+39%E2	1112.87 *5*	18.9 *4*
γ [E2]	1323.69 *12*	0.46 *8*

Atomic Electrons (^{102}Rh)

⟨e⟩=11.5 *4* keV

e_{bin}(keV)	⟨e⟩(keV)	e(%)
3	2.03	70 *7*
16	1.99	12.3 *13*
18	0.188	1.02 *10*
19	0.69	3.7 *4*
21 - 52	0.39	0.97 *25*
71 - 75	0.15	0.21 *6*
324 - 346	0.040	0.0124 *24*
393	0.059	0.0150 *24*
396	0.333	0.084 *6*
398 - 421	0.168	0.041 *4*
453	2.33	0.515 *24*
472 - 475	0.371	0.079 *3*
606	0.127	0.0209 *12*
609	0.83	0.136 *6*
625 - 673	0.203	0.0316 *11*
675	0.58	0.086 *4*
689 - 698	0.095	0.0137 *4*
745	0.372	0.050 *3*
764 - 767	0.055	0.0072 *4*
1025 - 1047	0.272	0.0265 *12*
1081 - 1113	0.187	0.0172 *6*
1302 - 1324	0.0029	0.00022 *4*

Continuous Radiation (^{102}Rh)

⟨IB⟩=0.029 keV

E_{bin}(keV)		⟨ ⟩(keV)	(%)
10 - 20	IB	0.0114	0.065
20 - 40	IB	0.0124	0.058
40 - 100	IB	0.00074	0.00120
100 - 300	IB	0.0036	0.0019
300 - 423	IB	0.00050	0.000153

$^{102}_{45}$Rh(207 *3* d)

Mode: ϵ(75 *5* %), β-(20 *5* %), IT(5 *2* %)
Δ: -86733 *21* keV
SpA: 6184 Ci/g
Prod: ^{102}Ru(p,n); ^{101}Ru(d,n); ^{102}Ru(d,2n); ^{103}Rh(n,2n); ^{103}Rh(γ,n); ^{104}Pd(d,α); ^{103}Rh(d,t)

Photons (^{102}Rh)

⟨γ⟩=355 *15* keV

γ_{mode}	γ(keV)	γ(%)†
Ru L_ℓ	2.253	0.054 *10*
Ru L_η	2.382	0.025 *4*
Ru L_α	2.558	1.41 *22*
Ru L_β	2.732	0.87 *14*
Ru L_γ	3.033	0.067 *12*
Ru $K_{\alpha2}$	19.150	12.2 *4*
Ru $K_{\alpha1}$	19.279	23.2 *8*
Ru $K_{\beta1}$'	21.650	5.61 *20*
Ru $K_{\beta2}$'	22.210	1.04 *4*
γ_ϵ[E1]	217.11 *22*	<0.02
γ_ϵ[M1+E2]	224.31 *12*	~0.05 ?
γ_ϵ[E2]	256.58 *9*	0.020 *9*
γ_ϵ[M1+E2]	415.31 *5*	0.026 *3*
γ_ϵE2+2.0%M1	418.53 *5*	0.136 *10*
γ_ϵ[M1+E2]	456.42 *11*	0.08 *2*
γ_ϵE2	468.65 *4*	2.9 *2*
γ_ϵE2	475.12 *3*	46
$\gamma_{\beta-}$E2	556.61 *4*	1.9

Photons (^{102}Rh)
(continued)

γ_{mode}	γ(keV)	γ(%)†
γ_ϵE2(+0.04%M1)	628.10 *4*	5.5 *3*
γ_ϵE2	631.33 *4*	0.10 *3*
γ_ϵ[E2]	636.89 *6*	0.228 *23*
γ_ϵM1+6.6%E2	680.73 *5*	0.58 *4*
γ_ϵ[E2]	734.00 *7*	0.104 *20*
γ_ϵM1+1.9%E2	739.62 *5*	0.53 *8*
γ_ϵ[E2]	930.62 *12*	~0.03
γ_ϵ[M1+E2]	933.84 *12*	~0.02 ?
γ_ϵE2+2.0%M1	1046.64 *5*	0.422 *24*
γ_ϵ[E2]	1103.22 *4*	2.89 *8*
γ_ϵM1+5.9%E2	1105.53 *6*	0.40 *3*
γ_ϵM1+6.1%E2	1158.15 *5*	0.58 *4*
γ_ϵE2	1362.10 *8*	0.39 *5*
γ_ϵE2+23%M1	1561.94 *12*	0.11 *3*
γ_ϵ[E1]	1569.14 *22*	<0.02
γ_ϵ[E2]	1580.64 *6*	0.046 *8*
γ_ϵ[M1+E2]	1786.25 *5*	<0.02
γ_ϵ[E2]	2037.05 *12*	~0.03
γ_ϵ[E3]	2044.25 *23*	<0.002 ?
γ_ϵ[E2]	2261.36 *5*	<0.04

† uncert(syst): 6.7% for ϵ, 25% for β-

Atomic Electrons (^{102}Rh)

⟨e⟩=4.62 *20* keV

e_{bin}(keV)	⟨e⟩(keV)	e(%)
3	1.25	43 *4*
16	1.24	7.7 *8*
18	0.118	0.64 *7*
19	0.43	2.29 *23*
21 - 22	0.051	0.242 *21*
195 - 234	0.008	0.0036 *17*
253 - 257	0.00034	0.00014 *4*
393 - 434	0.0078	0.00191 *15*
447	0.073	0.0164 *12*
453	1.13	0.249 *17*
454 - 469	0.0117	0.00252 *13*
472	0.147	0.0312 *22*
606 - 637	0.101	0.0165 *9*
659 - 681	0.0088	0.00133 *8*
712 - 740	0.0086	0.00120 *14*
909 - 934	0.00049	5.3 *21* $\times 10^{-5}$
1025 - 1047	0.00348	0.000339 *18*
1081 - 1105	0.0257	0.00238 *9*
1136 - 1158	0.0046	0.00041 *3*
1340 - 1362	0.0024	0.000177 *19*
1540 - 1580	0.00087	5.6 *10* $\times 10^{-5}$
1764 - 1786	4.9×10^{-5}	$\sim 3 \times 10^{-6}$
2015 - 2044	0.00013	$\sim 7 \times 10^{-6}$
2239 - 2261	7.8×10^{-5}	$\sim 3 \times 10^{-6}$

Continuous Radiation (^{102}Rh)

⟨β-⟩=80 keV;⟨β+⟩=77 keV;⟨IB⟩=1.28 keV

E_{bin}(keV)		⟨ ⟩(keV)	(%)
0 - 10	β-	0.0119	0.237
	β+	7.3×10^{-6}	8.9×10^{-5}
	IB	0.0072	
10 - 20	β-	0.0361	0.241
	β+	0.000224	0.00136
	IB	0.0138	0.087
20 - 40	β-	0.147	0.490
	β+	0.00474	0.0147
	IB	0.021	0.083
40 - 100	β-	1.08	1.54
	β+	0.167	0.218
	IB	0.036	0.055
100 - 300	β-	11.2	5.6
	β+	5.7	2.65
	IB	0.100	0.056
300 - 600	β-	32.1	7.3
	β+	26.9	6.0

Continuous Radiation (^{102}Rh)
(continued)

E_{bin}(keV)		⟨ ⟩(keV)	(%)
	IB	0.18	0.041
600 - 1300	β-	35.4	4.65
	β+	43.9	5.4
	IB	0.61	0.067
1300 - 2366	β+	0.0280	0.00213
	IB	0.31	0.020
	Σβ+		14.3

$^{102}_{46}$Pd(stable)

Δ: -87902 *4* keV
%: 1.020 *12*

$^{102}_{47}$Ag(12.9 *3* min)

Mode: ε
Δ: -82020 *110* keV
SpA: 1.43×10^{8} Ci/g
Prod: ^{102}Pd(p,n); protons on Cd; ^{11}B on Mo; ^{14}N on Ag; daughter ^{102}Ag(7.7 min)

Photons (^{102}Ag)
⟨γ⟩=2607 *92* keV

γ_{mode}	γ(keV)	γ(%)†
Pd L_ℓ	2.503	0.021 *4*
Pd L_η	2.660	0.0098 *15*
Pd L_α	2.838	0.57 *9*
Pd L_β	3.052	0.37 *6*
Pd L_γ	3.383	0.037 *7*
Pd $K_{\alpha2}$	21.020	4.64 *16*
Pd $K_{\alpha1}$	21.177	8.8 *3*
Pd $K_{\beta1}$'	23.811	2.18 *8*
Pd $K_{\beta2}$'	24.445	0.421 *16*
γ E2	556.61 *4*	97.7
γ [E2]	603.47 *18*	1.7 *6*
γ E2	719.53 *19*	58 *6*
γ E2	835.1 *3*	13.7 *13*
γ	865.5 *3*	3.7 *4*
γ	891.3 *3*	4.1 *4*
γ	937.4 *3*	1.1 *3*
γ	964.2 *3*	1.4 *3*
γ E2+1.5%M1	977.98 *20*	2.1 *3*
γ [M1+E2]	1024.85 *24*	4.2 *4*
γ	1055.1 *3*	~0.8
γ	1067.1 *4*	~1.0
γ [M1+E2]	1256.9 *3*	12.7 *13*
γ	1305.5 *4*	2.3 *7*
γ	1394.2 *4*	~1
γ	1474.1 *4*	2.7 *7*
γ [M1+E2]	1522.8 *3*	2.7 *7*
γ [E2]	1534.58 *20*	2.2 *5*
γ E2+<0.4%M1	1555.6 *4*	2.5 *6*
γ E2	1581.44 *21*	13.7 *13*
γ [E2]	1744.38 *24*	17.3 *16*
γ	1799.3 *3*	2.8 *4*
γ	1890.2 *4*	0.7 *3*
γ	1924.7 *4*	~1 *
γ	2110.5 *5*	~0.6 *
γ [E2]	2242.3 *3*	1.0 *4*
γ	2310.0 *5*	~1 *
γ	2493.7 *5*	~0.8 *
γ	2612.8 *4*	3.5 *13* *
γ	2690.7 *5*	~1.0 *
γ	2726.7 *5*	1.4 *6* *
γ	2804.8 *5*	0.8 *3* *
γ	3397.8 *6*	~1 *
γ	3406.3 *6*	~2 *

† 0.56% uncert(syst)
* with ^{102}Ag(12.9 or 7.7 min)

Atomic Electrons (^{102}Ag)
⟨e⟩=5.18 *11* keV

e_{bin}(keV)	⟨e⟩(keV)	e(%)
3	0.47	14.5 *15*
4 - 17	0.135	1.66 *13*
18	0.34	1.89 *19*
20 - 24	0.218	1.05 *7*
532	2.00	0.375 *8*
553	0.261	0.0472 *10*
556 - 603	0.094	0.0166 *18*
695	0.78	0.113 *11*
716 - 720	0.121	0.0168 *14*
811	0.150	0.0185 *18*
832 - 867	0.088	0.0103 *24*
888 - 937	0.014	0.0016 *5*
940 - 978	0.033	0.0035 *6*
1001 - 1043	0.055	0.0055 *8*
1052 - 1067	0.0017	0.00016 *7*
1233 - 1281	0.115	0.0093 *11*
1302 - 1305	0.0017	0.00013 *6*
1370 - 1394	0.006	~0.0005
1450 - 1534	0.059	0.0040 *6*
1552 - 1581	0.085	0.0055 *5*
1720 - 1799	0.108	0.0062 *5*
1866 - 1924	0.008	0.00040 *15*
2086 - 2110	0.0022	~0.00010
2218 - 2309	0.009	0.00042 *17*
2469 - 2493	0.0028	~0.00011
2588 - 2688	0.015	0.00056 *19*
2690 - 2780	0.0068	0.00025 *8*
2801 - 2804	0.00032	1.1 *4* ×10^{-5}
3373 - 3406	0.008	0.00025 *10*

Continuous Radiation (^{102}Ag)
⟨β+⟩=957 keV;⟨IB⟩=4.6 keV

E_{bin}(keV)		⟨ ⟩(keV)	(%)
0 - 10	β+	4.73×10^{-6}	5.7×10^{-5}
	IB	0.034	
10 - 20	β+	0.000154	0.00094
	IB	0.035	0.24
20 - 40	β+	0.00350	0.0108
	IB	0.070	0.25
40 - 100	β+	0.138	0.178
	IB	0.19	0.29
100 - 300	β+	6.0	2.72
	IB	0.54	0.30
300 - 600	β+	44.4	9.6
	IB	0.65	0.151
600 - 1300	β+	318	33.3
	IB	1.21	0.135
1300 - 2500	β+	515	29.7
	IB	1.46	0.083
2500 - 4604	β+	74	2.64
	IB	0.41	0.0141
	Σβ+		78

$^{102}_{47}$Ag(7.7 *5* min)

Mode: ε(51 *5* %), IT(49 *5* %)
Δ: -82011 *110* keV
SpA: 2.39×10^{8} Ci/g
Prod: ^{102}Pd(p,n); ^{11}B on Mo; ^{14}N on Ag; protons on Cd; daughter ^{102}Cd

Photons (^{102}Ag)*
⟨γ⟩=1245 *72* keV

γ_{mode}	γ(keV)	γ(%)†
Pd $K_{\alpha2}$	21.020	2.80 *10*
Pd $K_{\alpha1}$	21.177	5.28 *19*
Pd $K_{\beta1}$'	23.811	1.31 *5*
Pd $K_{\beta2}$'	24.445	0.254 *10*
γ_ϵ[E2]	351.46 *22*	
γ_ϵE2	556.61 *4*	43 *4*
γ_ϵE2	719.53 *19*	4.5 *5*
γ_ϵE2+1.5%M1	977.98 *20*	2.65 *25*
γ_ϵ	1017.76 *22*	
γ_ϵ[E2]	1101.8 *6*	
γ_ϵE2+1.8%M1	1388.0 *3*	~2 ?
γ_ϵ	1461.2 *5*	4.5 *5*
γ_ϵ[E2]	1534.58 *20*	2.7 *6*
γ_ϵ	1588.9 *4*	1.2 *4*
γ_ϵ	1645.5 *4*	
γ_ϵ	1692.4 *5*	2.3 *11*
γ_ϵ	1834.8 *4*	9.8 *13*
γ_ϵ	2017.9 *5*	2.8 *11*
γ_ϵ	2054.3 *3*	6.7 *11*
γ_ϵ	2159.7 *3*	5.1 *9*
γ_ϵ	2566.9 *4*	~0.8
γ_ϵ	2682.0 *3*	1.7 *7*
γ_ϵ	2716.3 *3*	1.9 *7*
γ_ϵ	3238.6 *3*	4.9 *11*

† uncert(syst): 14% for ε
* see also ^{102}Ag(12.9 min)

Atomic Electrons (^{102}Ag)*
⟨e⟩=2.00 *10* keV

e_{bin}(keV)	⟨e⟩(keV)	e(%)
3	0.28	8.7 *9*
4 - 17	0.082	1.00 *8*
18	0.203	1.14 *12*
20	0.051	0.25 *3*
21	0.069	0.33 *3*
23 - 24	0.0115	0.049 *4*
532	0.87	0.163 *16*
553	0.114	0.0206 *20*
556 - 557	0.0257	0.0046 *4*
695	0.061	0.0088 *10*
716 - 720	0.0094	0.00131 *13*
954 - 978	0.0270	0.00282 *24*
1364 - 1388	0.011	~0.0008
1437 - 1534	0.041	0.0028 *6*
1565 - 1588	0.006	0.00037 *17*
1668 - 1692	0.010	~0.0006
1810 - 1834	0.042	0.0023 *8*
1994 - 2054	0.037	0.0018 *5*
2135 - 2159	0.019	0.0009 *3*
2543 - 2566	0.0026	~0.00010
2658 - 2716	0.011	0.00042 *12*
3214 - 3238	0.014	0.00043 *12*

* with ε

Continuous Radiation (^{102}Ag)

⟨β+⟩=462 keV;⟨IB⟩=2.3 keV

E_{bin}(keV)		⟨ ⟩(keV)	(%)
0 - 10	β+	2.78×10^{-6}	3.36×10^{-5}
	IB	0.0163	
10 - 20	β+	9.0×10^{-5}	0.00055
	IB	0.017	0.118
20 - 40	β+	0.00205	0.0063
	IB	0.034	0.121
40 - 100	β+	0.080	0.104
	IB	0.090	0.138
100 - 300	β+	3.44	1.56
	IB	0.26	0.145
300 - 600	β+	24.9	5.4
	IB	0.31	0.074
600 - 1300	β+	163	17.2
	IB	0.62	0.069
1300 - 2500	β+	210	12.3
	IB	0.76	0.043
2500 - 5000	β+	60	2.03
	IB	0.18	0.0062
5000 - 5333	IB	0.00021	4.0×10^{-6}
	Σβ+		39

$^{102}_{48}$Cd(5.5 *5* min)

Mode: ϵ
Δ: -79700 *200* keV syst
SpA: 3.3×10^{8} Ci/g
Prod: protons on Sn; protons on Ag

Photons (^{102}Cd)

⟨γ⟩=677 *14* keV

γ_{mode}	γ(keV)	γ(%)†
Ag L_ℓ	2.634	0.079 *14*
Ag L_η	2.806	0.037 *6*
Ag L_α	2.984	2.2 *3*
Ag L_β	3.219	1.46 *24*
Ag L_γ	3.578	0.14 *3*
Ag $K_{\alpha2}$	21.990	16.9 *6*
Ag $K_{\alpha1}$	22.163	31.8 *12*
Ag $K_{\beta1}$'	24.934	8.0 *3*
Ag $K_{\beta2}$'	25.603	1.59 *6*
γ M1	59.10 *18*	1.50 *19*
γ M1	97.60 *18*	3.13 *19*
γ M1	116.25 *15*	6.1 *4*
γ M1	120.63 *15*	2.4 *3*
γ [M1]	147.20 *18*	0.56 *19*
γ E2	213.47 *16*	4.6 *4*
γ	244.43 *17*	~0.31
γ [M1+E2]	360.67 *12*	3.7 *4*
γ M1+E2	415.16 *16*	7.6 *6*
γ (M1+E2)	481.31 *15*	62.5
γ	505.39 *16*	9.6 *13*
γ [M1+E2]	531.27 *20*	1.3 *4*
γ [M1+E2]	621.64 *17*	1.3 *6*
γ [M1+E2]	676.12 *14*	3.7 *5*
γ	920.55 *16*	~0.6
γ [M1+E2]	1036.79 *14*	12.8 *6*
γ [M1+E2]	1360.07 *20*	4.9 *6*

† 1.4% uncert(syst)

Atomic Electrons (^{102}Cd)

⟨e⟩=12.5 *3* keV

e_{bin}(keV)	⟨e⟩(keV)	e(%)
3	1.16	35 *4*
4	0.75	20.9 *21*
18	0.62	3.4 *4*
19	0.89	4.8 *5*
21	0.34	1.62 *17*
22	0.36	1.68 *17*
24 - 25	0.069	0.282 *22*
34	0.84	2.5 3
55 - 59	0.217	0.39 *4*
72	0.89	1.24 *8*
91	1.35	1.49 *10*
94	0.145	0.155 *10*
95	0.49	0.52 *7*
97 - 144	0.48	0.42 *3*
146 - 147	0.0030	0.0020 *6*
188	0.72	0.38 *3*
210 - 244	0.186	0.088 *11*
335 - 361	0.19	0.057 *8*
390 - 415	0.31	0.079 *7*
456	1.72	0.377 *23*
478 - 506	0.50	0.103 *19*
527 - 531	0.0049	0.00093 *24*
596 - 622	0.028	0.0047 *18*
651 - 676	0.070	0.0107 *15*
895 - 921	0.005	~0.0006
1011 - 1037	0.139	0.0137 *13*
1335 - 1360	0.040	0.0030 *4*

Continuous Radiation (^{102}Cd)

⟨β+⟩=92 keV;⟨IB⟩=0.84 keV

E_{bin}(keV)		⟨ ⟩(keV)	(%)
0 - 10	β+	5.2×10^{-5}	0.00063
	IB	0.0046	
10 - 20	β+	0.00167	0.0102
	IB	0.0092	0.056
20 - 40	β+	0.0357	0.110
	IB	0.028	0.113
40 - 100	β+	1.10	1.45
	IB	0.022	0.034
100 - 300	β+	20.5	10.0
	IB	0.065	0.035
300 - 600	β+	58	13.5
	IB	0.155	0.035
600 - 1300	β+	12.4	1.89
	IB	0.47	0.053
1300 - 1950	IB	0.090	0.0063
	Σβ+		27

$^{102}_{49}$In(23 *4* s)

Mode: ϵ
Δ: -70420 *420* keV syst
SpA: 4.7×10^{9} Ci/g
Prod: ^{40}Ca on ^{65}Cu; ^{16}O on ^{92}Mo; ^{14}N on ^{92}Mo

Photons (^{102}In)

γ_{mode}	γ(keV)	γ(rel)
γ	156.6 *10*	10 *2* ?
γ	396.5 *10*	12.0 *24*
γ	593 *1*	30 *6*
γ	776.8 *10*	100 *20*
γ	861.4 *10*	96 *19*
γ	923.7 *10*	10 *2* ?

A = 103

NDS **28**, 403 (1979)

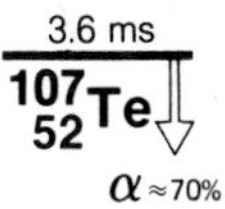

Q_{α} 3980 *50*

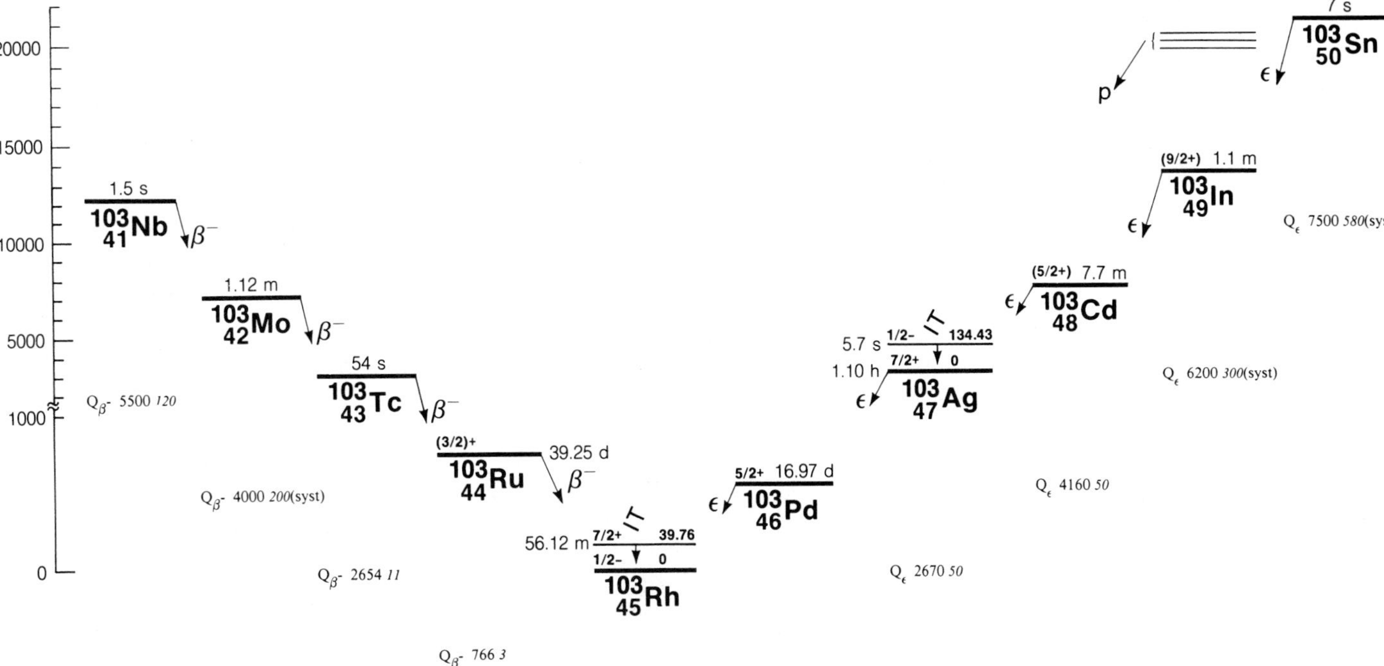

$^{103}_{41}$Nb(1.5 *2* s)

Mode: β-
Δ: -75110 *230* keV syst
SpA: 5.8×10^{10} Ci/g
Prod: fission

Photons (^{103}Nb)

γ_{mode}	γ(keV)
γ	102.80 *19*
γ	126.5 *10*
γ	241.5 *10*
γ	247.7 *10*
γ	538.1 *7*
γ	640.9 *7*

$^{103}_{42}$Mo(1.125 *25* min)

Mode: β-
Δ: -80610 *200* keV syst
SpA: 1.61×10^{9} Ci/g
Prod: fission

Photons (^{103}Mo)

γ_{mode}	γ(keV)	γ(rel)
γ	45.8 *2*	9.4 *19*
γ	83.4 *5*	
γ	218.3 *5*	
γ	424.0 *2*	100 *20*
γ	519.3 *5*	
γ	687.6 *2*	16 *3*

$^{103}_{43}$Tc(54.2 *8* s)

Mode: β-
Δ: -84606 *11* keV
SpA: 2.01×10^{9} Ci/g
Prod: fission; ^{104}Ru(n,np); ^{104}Ru(γ,p)

Photons (^{103}Tc)

⟨γ⟩=236 *5* keV

γ_{mode}	γ(keV)	γ(%)
γ[M1+E2]	133.2 *5*	0.34 *5*
γ[M1+E2]	136.07 *5*	15.4 *8*
γ	160.9 *1*	0.057 *16*
γ	172 *1*	0.062 *5*
γ[M1+E2]	174.31 *5*	2.6 *3*
γ	190.30 *5*	0.052 *10*
γ	210.40 *5*	9.2 *5* *

Photons (^{103}Tc)
(continued)

γ_{mode}	γ(keV)	γ(%)
γ[E2]	213.0 *2*	0.141 *8*
γ	232.0 *2*	0.094 *18*
γ	239.0 *2*	0.057 *18*
γ	245.3 *3*	0.041 *10*
γ	268.6 *1*	0.13 *5*
γ	270.1 *1*	0.55 *5*
γ	287.6 *1*	0.60 *5*
γ[E1]	294.8 *1*	~0.08
γ	315.25 *10*	0.143 *23*
γ[M1+E2]	343.55 *15*	3.74 *18*
γ[M1+E2]	346.37 *5*	16.2 *8*
γ	365.1 *2*	0.39 *5*
γ	368.7 *1*	0.18 *3*
γ	370.6 *1*	0.133 *24*
γ	378.0 *1*	0.096 *24*
γ	388.6 *1*	2.06 *13*
γ	401.6 *5*	0.63 *11*
γ	403.2 *5*	1.94 *11*
γ	418 *1*	~0.032
γ	428.3 *1*	0.18 *3*
γ[M1+E2]	432.1 *1*	0.050 *11*
γ	456.1 *1*	0.096 *24*
γ	487.3 *1*	0.152 *21*
γ	501.2 *1*	2.25 *13*
γ	525.5 *3*	0.063 *16*
γ	533.1 *5*	0.050 *24*
γ[M1+E2]	555.0 *1*	0.060 *11*
γ[M1+E2]	559.7 *3*	0.19 *5*
γ[M1+E2]	562.9 *1*	6.5 *5*
γ	583.6 *5*	0.021 *7*
γ	589.1 *1*	0.094 *19*
γ	592.1 *1*	0.097 *21*
γ	600.7 *2*	0.039 *11*
γ	607.6 *3*	0.029 *10*
γ	625.0 *5*	0.076 *24*
γ	638.6 *3*	0.26 *3*
γ	652.3 *3*	0.026 *8*

Photons (^{103}Tc) (continued)

γ_{mode}	γ(keV)	γ(%)
γ	658.5 *5*	0.26 *7*
γ	661.2 *1*	0.65 *5*
γ	741.4 *3*	0.042 *13*
γ	769.2 *2*	0.49 *3*
γ	772.0 *3*	~0.08
γ	774.8 *2*	0.49 *3*
γ	785.0 *5*	0.042 *13*
γ	804.6 *2*	0.21 *3*
γ	830.7 *5*	0.037 *11*
γ	851.9 *2*	0.20 *3*
γ	902.4 *3*	0.57 *5*
γ	905.3 *3*	0.32 *5*
γ	929.4 *2*	0.125 *16*
γ	937.8 *2*	0.47 *5*
γ	940.4 *3*	0.042 *15*
γ	952.0 *3*	0.026 *8*
γ	955.2 *5*	0.094 *24*
γ	1042.7 *5*	0.066 *19*
γ	1062.7 *2*	0.134 *21*
γ	1065.6 *2*	0.122 *18*
γ	1181 *1*	0.029 *10*
γ	1233.3 *10*	0.053 *16*
γ	1312 *1*	0.021 *7*
γ	1315 *1*	0.015 *5*

* doublet

Continuous Radiation (^{103}Tc)

⟨β-⟩=983 keV;⟨IB⟩=2.2 keV

E_{bin}(keV)		⟨ ⟩(keV)	(%)
0 - 10	β-	0.0124	0.245
	IB	0.037	
10 - 20	β-	0.0378	0.252
	IB	0.036	0.25
20 - 40	β-	0.157	0.52
	IB	0.071	0.25
40 - 100	β-	1.23	1.73
	IB	0.20	0.31
100 - 300	β-	15.9	7.7
	IB	0.55	0.31
300 - 600	β-	74	16.3
	IB	0.57	0.135
600 - 1300	β-	421	44.6
	IB	0.62	0.074
1300 - 2500	β-	469	28.5
	IB	0.122	0.0082
2500 - 2654	β-	1.07	0.0420
	IB	7.8×10^{-6}	3.1×10^{-7}

$^{103}_{44}$Ru(39.254 *8* d)

Mode: β-
Δ: -87260.6 *24* keV
SpA: 3.2295×10^{4} Ci/g
Prod: ^{102}Ru(n,γ); fission

Photons (^{103}Ru)

⟨γ⟩=485 *12* keV

γ_{mode}	γ(keV)	γ(%)†
Rh L_ℓ	2.377	0.084 *15*
Rh L_η	2.519	0.050 *8*
Rh L_α	2.696	2.2 *4*
Rh L_β	2.877	1.55 *24*
Rh L_γ	3.149	0.122 *19*
Rh $K_{\alpha2}$	20.074	2.48 *10*
Rh $K_{\alpha1}$	20.216	4.71 *19*
Rh $K_{\beta1}$'	22.717	1.15 *5*
Rh $K_{\beta2}$'	23.312	0.216 *9*
γ E3	39.755 *12*	0.0707 *18* *
γ (M1)	42.58 *4*	0.00107 *18*
γ M1	53.277 *9*	0.376 *18*
γ M1	62.51 *4*	0.00042 *13*
γ [M1+E2]	113.245 *17*	0.0036 *7*
γ M1(+E2)	114.942 *17*	0.0080 *7*
γ E1	241.860 *23*	0.0147 *15*
γ [E1]	292.60 *4*	<0.005
γ M1+2.8%E2	294.950 *19*	0.251 *8*
γ	317.74 *22*	0.0054 *9*
γ E2	357.46 *4*	0.009 *3*
γ E2	443.777 *12*	0.325 *8*
γ M1+4.0%E2	497.054 *10*	88.7 *23*
γ [M1+E2]	514.44 *4*	0.0048 *13*
γ (E2)	557.022 *14*	0.83 *17*
γ [M1+E2]	567.72 *4*	0.0016 *7*
γ M1+0.4%E2	610.298 *14*	5.64 *18*
γ [E2]	611.995 *18*	0.080 *9*
γ [E1]	651.750 *21*	0.00017 *7*

† 1.1% uncert(syst)
* with ^{103}Rh(56.12 min) in equilib

Atomic Electrons (^{103}Ru)

⟨e⟩=41.4 *9* keV

e_{bin}(keV)	⟨e⟩(keV)	e(%)
3	2.29	75 *8*
16	0.0242	0.148 *15*
17	1.77	10.7 *4*
19 - 36	0.530	1.85 *6*
37	27.3	74.5 *24*
39	6.06	15.4 *5*
40	1.00	2.51 *8*
42 - 90	0.0544	0.107 *5*
92 - 115	0.0044	0.0046 *19*
219 - 242	0.00039	0.000175 *16*
269 - 318	0.0136	0.00496 *17*
334 - 357	0.00048	0.00014 *4*
421 - 444	0.0112	0.00264 *7*
474	1.94	0.409 *13*
491 - 534	0.306	0.0617 *17*
544 - 589	0.098	0.0166 *6*
607 - 652	0.0140	0.00230 *7*

Continuous Radiation (^{103}Ru)

⟨β-⟩=69 keV;⟨IB⟩=0.021 keV

E_{bin}(keV)		⟨ ⟩(keV)	(%)
0 - 10	β-	0.54	10.8
	IB	0.0035	
10 - 20	β-	1.51	10.1
	IB	0.0028	0.020
20 - 40	β-	5.4	18.1
	IB	0.0042	0.0150
40 - 100	β-	25.2	37.7
	IB	0.0061	0.0101
100 - 300	β-	30.5	22.0
	IB	0.0038	0.0025
300 - 600	β-	5.2	1.25
	IB	0.00080	0.00022
600 - 766	β-	0.58	0.090
	IB	1.10×10^{-5}	1.7×10^{-6}

$^{103}_{45}$Rh(stable)

Δ: -88027 *3* keV
%: 100

$^{103}_{45}$Rh(56.12 *1* min)

Mode: IT
Δ: -87987 *3* keV
SpA: 3.2525×10^{7} Ci/g
Prod: daughter ^{103}Ru; daughter ^{103}Pd

Photons (^{103}Rh)

⟨γ⟩=1.65 *6* keV

γ_{mode}	γ(keV)	γ(%)†
Rh L_ℓ	2.377	0.081 *15*
Rh L_η	2.519	0.048 *8*
Rh L_α	2.696	2.1 *4*
Rh L_β	2.877	1.48 *24*
Rh L_γ	3.149	0.116 *19*
Rh $K_{\alpha2}$	20.074	2.14 *13*
Rh $K_{\alpha1}$	20.216	4.06 *25*
Rh $K_{\beta1}$'	22.717	1.00 *6*
Rh $K_{\beta2}$'	23.312	0.187 *12*
γ E3	39.755 *12*	0.068

† 5.0% uncert(syst)

Atomic Electrons (^{103}Rh)

⟨e⟩=37.5 *15* keV

e_{bin}(keV)	⟨e⟩(keV)	e(%)
3	2.19	72 *8*
16	0.0209	0.127 *15*
17	1.69	10.2 *6*
19	0.025	0.127 *15*
20	0.069	0.35 *4*
22	0.0068	0.031 *4*
23	0.00211	0.0093 *11*
36	0.196	0.54 *3*
37	26.4	72 *4*
39	5.9	14.9 *8*
40	0.97	2.43 *13*

$^{103}_{46}$Pd(16.97 *2* d)

Mode: ε
Δ: -87455 *4* keV
SpA: 7.470×10^{4} Ci/g
Prod: ^{102}Pd(n,γ); ^{103}Rh(d,2n); ^{103}Rh(p,n)

Photons (^{103}Pd)

⟨γ⟩=16.2 *4* keV

γ_{mode}	γ(keV)	γ(%)†
Rh L_ℓ	2.377	0.17 *3*
Rh L_η	2.519	0.090 *14*
Rh L_α	2.696	4.6 *7*
Rh L_β	2.883	3.0 *5*
Rh L_γ	3.179	0.25 *4*
Rh $K_{\alpha2}$	20.074	22.3 *8*

Photons (^{103}Pd)
(continued)

γ_{mode}	γ(keV)	γ(%)†
Rh $K_{\alpha1}$	20.216	42.4 *15*
Rh $K_{\beta1}$'	22.717	10.4 *4*
Rh $K_{\beta2}$'	23.312	1.94 *7*
γ E3	39.755 *12*	0.0683 *
γ M1	53.277 *9*	~3×10^{-5}
γ M1	62.51 *4*	0.00104 *3*
γ E1	241.860 *23*	6.7 *7*×10^{-7}
γ M1+2.8%E2	294.950 *19*	0.00280 *7*
γ [E1]	317.70 *4*	1.50 *7*×10^{-5}
γ E2	357.46 *4*	0.0221 *7*
γ E2	443.777 *12*	1.47 *5*×10^{-5}
γ M1+4.0%E2	497.054 *10*	0.00401 *12*

† 1.0% uncert(syst)
* with ^{103}Rh(56.12 min) in equilib

Atomic Electrons (^{103}Pd)
⟨e⟩=42.5 *8* keV

e_{bin}(keV)	⟨e⟩(keV)	e(%)
3	4.3	141 *14*
16	0.218	1.33 *14*
17	3.46	20.6 *14*
19 - 36	1.27	6.0 *4*
37	26.4	71.9 *16*
39	5.85	14.9 *3*
40	0.964	2.43 *5*
50 - 63	0.000112	0.000189 *8*
219 - 242	1.75 ×10^{-8}	7.9 *8* ×10^{-9}
272 - 318	0.000149	5.44 *15* ×10^{-5}
334 - 357	0.00119	0.000352 *11*
421 - 444	5.04 ×10^{-7}	1.19 *4* ×10^{-7}
474 - 497	0.000101	2.12 *7* ×10^{-5}

Continuous Radiation (^{103}Pd)
⟨IB⟩=0.057 keV

E_{bin}(keV)		⟨ ⟩(keV)	(%)
10 - 20	IB	0.0107	0.060
20 - 40	IB	0.0167	0.075
40 - 100	IB	0.0023	0.0035
100 - 300	IB	0.0170	0.0087
300 - 572	IB	0.0098	0.0027

$^{103}_{47}$Ag(1.095 *12* h)

Mode: ε
Δ: -84780 *50* keV
SpA: 2.78×10^{7} Ci/g
Prod: ^{103}Rh(α,4n); ^{104}Pd(p,2n); ^{102}Pd(d,n); ^{102}Pd(p,γ); protons on Cd; ^{11}B on Mo; ^{14}N on ^{107}Ag

Photons (^{103}Ag)
⟨γ⟩=563 *7* keV

γ_{mode}	γ(keV)	γ(%)†
Pd L_{ℓ}	2.503	0.080 *14*
Pd L_{η}	2.660	0.037 *6*
Pd L_{α}	2.838	2.2 *3*
Pd L_{β}	3.052	1.39 *23*
Pd L_{γ}	3.383	0.138 *25*
Pd $K_{\alpha2}$	21.020	17.4 *6*
Pd $K_{\alpha1}$	21.177	32.9 *11*
Pd $K_{\beta1}$'	23.811	8.2 *3*
Pd $K_{\beta2}$'	24.445	1.58 *6*
γ M1+1.3%E2	118.64 *3*	31.2 *7*
γ [E2]	125.23 *4*	~0.13
γ M1	148.11 *3*	28.3 *6*
γ [M1+E2]	166.84 *6*	0.051 *9*
γ M1+1.7%E2	186.09 *7*	0.059 *9*
γ M1	237.27 *11*	0.127 *17*
γ M1+0.7%E2	243.87 *4*	8.5
γ M1	265.06 *5*	~0.9
γ M1+2.1%E2	266.75 *3*	13.3 *4*
γ M1+3.6%E2	287.94 *4*	0.70 *3*
γ	298.30 *6*	0.153 *9*
γ	351.0 *8*	~0.05
γ	359.0 *2*	0.077 *25*
γ (M1)	368.10 *8*	0.068 *17*
γ (M1)	380.33 *19*	~0.15
γ	380.36 *13*	0.077 *15*
γ M1(+E2)	385.38 *11*	0.54 *4*
γ	389.40 *8*	0.10 *3*
γ (M1+E2)	431.91 *5*	0.170 *17*
γ E2	451.15 *7*	~0.024
γ [M1+E2]	454.79 *6*	~0.07
γ [E2]	455.55 *18*	~0.042
γ E2+35%M1	474.03 *7*	0.042 *17*
γ [M1+E2]	484.02 *7*	0.196 *25*
γ [E2]	498.96 *19*	~0.017
γ M1(+E2)	504.02 *11*	0.25 *9*
γ M1+32%E2	531.81 *4*	8.75 *18*
γ [E2]	546.65 *18*	0.042 *9*
γ [M1+E2]	575.12 *6*	0.76 *3*
γ M1(+E2)	580.01 *5*	0.94 *4*
γ M1,E2	608.49 *17*	0.119 *17*
γ M1,E2	625.74 *19*	0.17 *3*
γ E2	633.16 *8*	0.042 *17*
γ [M1+E2]	650.87 *7*	0.11 *3*
γ E2+24%M1	656.04 *8*	0.10 *3*
γ [M1+E2]	678.66 *12*	0.051 *9*
γ [E2]	683.71 *19*	0.08 *3*
γ [M1+E2]	698.65 *5*	0.229 *17*
γ E2	717.89 *6*	0.34 *3*
γ [M1+E2]	741.96 *5*	2.54 *7*
γ M1+E2	765.74 *8*	0.106 *13*
γ [E2]	774.81 *20*	0.077 *25*
γ M1+E2	801.99 *20*	0.119 *17*
γ	829.08 *14*	0.068 *17*
γ	874.25 *8*	0.28 *3*
γ M1,E2	884.37 *8*	0.22 *3*
γ M1,E2	888.66 *21*	0.085 *17*
γ E2	899.90 *7*	0.221 *25*
γ	911.53 *21*	0.068 *17*
γ [M1+E2]	938.81 *6*	0.63 *3*
γ [M1+E2]	1007.02 *5*	3.24 *10*
γ [M1+E2]	1029.90 *5*	1.30 *4*
γ	1042.96 *16*	0.13 *3*
γ [M1+E2]	1064.03 *6*	0.714 *25*
γ	1072.73 *13*	0.204 *17*
γ	1076.99 *20*	0.042 *17*
γ	1119.26 *10*	0.110 *17*
γ	1142.14 *10*	0.153 *13*
γ [M1+E2]	1155.13 *5*	3.05 *7*
γ M1+E2	1155.40 *21*	~0.026
γ	1157.91 *24*	0.038 *13*
γ M1+E2	1182.67 *6*	1.52 *3*
γ	1185.70 *25*	~0.05
γ	1267.37 *10*	0.17 *4*
γ	1271.63 *22*	~0.34
γ [M1+E2]	1273.77 *4*	9.4 *3*
γ	1280.23 *13*	0.20 *5*
γ	1292.5 *3*	0.081 *7*
γ	1303.10 *13*	0.067 *6*
γ	1325.39 *8*	0.408 *17*
γ	1337.09 *20*	0.078 *9*
γ	1386.00 *10*	0.519 *25*

Photons (^{103}Ag)
(continued)

γ_{mode}	γ(keV)	γ(%)†
γ	1416.6 *6*	0.034 *9*
γ	1422.97 *23*	0.065 *9*
γ	1428.33 *13*	0.051 *9*
γ	1445.85 *23*	0.042 *9*
γ	1485.90 *13*	0.213 *17*
γ	1514.2 *6*	0.030 *5*
γ	1537.1 *6*	0.059 *25*
γ	1546.97 *13*	0.172 *13*
γ	1592.13 *8*	0.070 *7*
γ	1604.53 *13*	0.119 *13*
γ	1689.71 *23*	0.043 *4*
γ	1701.74 *24*	0.080 *7*
γ	1709.47 *20*	0.064 *7*
γ	1743.58 *21*	0.042 *17*
γ	1747.7 *4*	0.042 *17*
γ	1775.64 *19*	0.167 *13*
γ	1811.02 *24*	0.054 *7*
γ	1838.81 *23*	0.105 *10*
γ	1845.7 *4*	0.047 *7*
γ	1953.4 *3*	0.077 *9* ?
γ	2098.96 *24*	0.042 *9*
γ	2141.36 *20*	0.098 *9*
γ	2156.74 *21*	0.050 *7*
γ	2164.24 *20*	0.042 *9*
γ	2175.1 *6*	0.034 *9*
γ	2179.6 *4*	0.026 *9*
γ	2233.54 *24*	0.055 *9*
γ	2242.4 *8*	0.026 *9*
γ	2267.4 *8*	0.026 *9*
γ	2275.38 *21*	0.084 *10*
γ	2298.7 *3*	0.059 *9*
γ	2342.82 *24*	0.042 *17*
γ	2345.8 *10*	~0.034
γ	2408.1 *2*	0.038 *5*
γ	2417.4 *3*	0.013 *4*
γ	2446.3 *4*	0.030 *5*

† 5.9% uncert(syst)

Atomic Electrons (^{103}Ag)
⟨e⟩=18.8 *3* keV

e_{bin}(keV)	⟨e⟩(keV)	e(%)
3	1.75	54 *6*
4 - 17	0.51	6.2 *5*
18	1.27	7.1 *7*
20 - 24	0.82	3.9 *3*
94	6.25	6.63 *21*
101	0.06	~0.06
115	0.97	0.84 *4*
118 - 122	0.239	0.202 *10*
124	3.95	3.19 *9*
125 - 142	0.013	0.010 *3*
145	0.566	0.392 *11*
147 - 186	0.139	0.0936 *21*
213 - 241	0.69	0.31 *4*
242	0.77	0.319 *13*
243 - 288	0.198	0.076 *3*
295 - 344	0.0076	0.0023 *7*
347 - 389	0.036	0.0099 *14*
408 - 456	0.0117	0.0028 *4*
460 - 507	0.205	0.0405 *11*
522 - 572	0.0651	0.0120 *3*
574 - 623	0.0096	0.00163 *11*
625 - 674	0.0095	0.00146 *12*
675 - 718	0.040	0.0057 *4*
738 - 778	0.0092	0.00124 *8*
798 - 829	0.0009	0.00011 *4*
850 - 899	0.0093	0.00107 *14*
900 - 939	0.0073	0.00079 *7*
983 - 1030	0.048	0.0048 *3*
1039 - 1077	0.0089	0.00085 *8*
1095 - 1142	0.0260	0.0023 *2*
1152 - 1186	0.0171	0.00147 *8*
1243 - 1292	0.080	0.0064 *5*
1300 - 1337	0.0029	0.00022 *7*
1362 - 1404	0.0036	0.00026 *8*
1413 - 1462	0.0013	9 *3* ×10^{-5}

Atomic Electrons (^{103}Ag) (continued)

e_{bin}(keV)	⟨e⟩(keV)	e(%)
1482 - 1580	0.0022	0.00015 3
1589 - 1687	0.00088	5.3 10 $\times 10^{-5}$
1689 - 1787	0.0014	8.0 16 $\times 10^{-5}$
1807 - 1845	0.00067	3.7 9 $\times 10^{-5}$
1929 - 1953	0.00031	1.6 5 $\times 10^{-5}$
2075 - 2172	0.00108	5.1 8 $\times 10^{-5}$
2174 - 2272	0.00069	3.1 6 $\times 10^{-5}$
2274 - 2345	0.00048	2.1 5 $\times 10^{-5}$
2384 - 2446	0.00028	1.15 22 $\times 10^{-5}$

Continuous Radiation (^{103}Ag)

$\langle\beta+\rangle$=181 keV;⟨IB⟩=1.9 keV

E_{bin}(keV)		⟨ ⟩(keV)	(%)
0 - 10	β+	7.6 $\times 10^{-6}$	9.2 $\times 10^{-5}$
	IB	0.0079	
10 - 20	β+	0.000246	0.00150
	IB	0.0144	0.090
20 - 40	β+	0.0055	0.0170
	IB	0.030	0.117
40 - 100	β+	0.206	0.267
	IB	0.042	0.064
100 - 300	β+	7.6	3.49
	IB	0.127	0.070
300 - 600	β+	43.5	9.5
	IB	0.24	0.054
600 - 1300	β+	124	14.5
	IB	0.79	0.086
1300 - 2500	β+	5.7	0.409
	IB	0.62	0.038
2500 - 2670	IB	0.00058	2.3 $\times 10^{-5}$
	$\Sigma\beta$+		28

$^{103}_{47}$Ag(5.7 3 s)

Mode: IT

Δ: -84646 50 keV

SpA: 1.809$\times 10^{10}$ Ci/g

Prod: ^{104}Pd(p,2n); daughter ^{103}Cd

Photons (^{103}Ag)

$\langle\gamma\rangle$=37.6 15 keV

γ_{mode}	γ(keV)	γ(%)†
Ag L_ℓ	2.634	0.072 13
Ag L_η	2.806	0.042 7
Ag L_α	2.984	2.0 3
Ag L_β	3.209	1.48 25
Ag L_γ	3.554	0.141 25
Ag $K_{\alpha2}$	21.990	11.6 7
Ag $K_{\alpha1}$	22.163	21.8 13
Ag $K_{\beta1}$'	24.934	5.5 3
Ag $K_{\beta2}$'	25.603	1.09 7
γ E3	134.43 5	21.2

† 5.0% uncert(syst)

Atomic Electrons (^{103}Ag)

⟨e⟩=96 3 keV

e_{bin}(keV)	⟨e⟩(keV)	e(%)
3	1.13	34 4
4	0.78	21.9 25
18	0.43	2.4 3
19	0.61	3.3 4
21	0.24	1.11 13
22	0.25	1.15 13
24	0.033	0.134 15
25	0.0147	0.059 7
109	52.1	48 3
131	32.5	24.8 13
134	8.0	5.9 3

$^{103}_{48}$Cd(7.70 17 min)

Mode: ϵ

Δ: -80620 18 keV

SpA: 2.37$\times 10^{8}$ Ci/g

Prod: ^{16}O on Mo; protons on Sn

Photons (^{103}Cd)

$\langle\gamma\rangle$=1766 34 keV

γ_{mode}	γ(keV)	γ(%)†
Ag L_ℓ	2.634	0.100 18
Ag L_η	2.806	0.048 8
Ag L_α	2.984	2.7 5
Ag L_β	3.218	1.9 3
Ag L_γ	3.576	0.18 4
Ag $K_{\alpha2}$	21.990	20.6 13
Ag $K_{\alpha1}$	22.163	38.8 25
Ag $K_{\beta1}$'	24.934	9.8 6
Ag $K_{\beta2}$'	25.603	1.94 13
γ (M1)	27.50 3	1.4 3
γ [M1]	69.34 5	0.082 14
γ E3	134.43 5	3.25 10 *
γ	187.98 19	~0.037
γ	243.35 11	0.59 7
γ	265.02 15	0.14 4
γ	298.32 17	0.10 4
γ	319.40 14	~0.025
γ	370.8 3	0.074 25
γ	376.93 9	0.16 7
γ	386.8 3	0.098 20 ?
γ [M1+E2]	386.94 5	3.71 10
γ	442.98 17	~0.17
γ [E2]	456.28 5	3.00 7
γ	463.53 22	0.16 5
γ	477.06 11	0.25 4
γ [M1+E2]	494.23 14	0.54 25
γ	496.29 15	0.18 6
γ	521.09 14	~0.025
γ	526.7 3	0.123 25
γ	530.82 21	0.54 9
γ	532.07 17	0.47 6
γ	544.38 18	0.37 10
γ	546.49 12	0.37 10
γ [M1+E2]	552.52 8	0.30 4
γ	562.15 15	0.16 5
γ	562.9 4	0.8 4
γ [M1+E2]	563.0 3	0.9 4
γ	598.44 22	0.12 5
γ [M1+E2]	620.00 15	0.37 4
γ	625.24 10	1.09 20
γ	626.17 9	1.83 6
γ	627.24 16	0.74 14
γ	642.41 15	0.32 9
γ	643.73 19	~0.14 ?
γ	647.35 16	0.17 6
γ	649.99 21	0.086 17 ?
γ	656.77 18	0.20 5
γ	663.44 14	0.22 4

Photons (^{103}Cd) (continued)

γ_{mode}	γ(keV)	γ(%)†
γ	666.99 17	0.25 7
γ	667.1 3	0.15 6
γ	676.65 10	0.25 6
γ	681.6 5	0.14 4
γ [E2]	689.34 16	0.14 5
γ	696.04 10	~0.10
γ [E2]	721.1 3	0.9 3
γ	721.73 19	0.18 7
γ	723.16 13	0.9 4
γ	734.24 14	0.21 10
γ	736.4 3	0.061 25
γ	737.58 12	0.11 4
γ	740.07 20	0.221 25
γ	749.75 16	0.32 4
γ	782.04 20	0.12 4
γ	789.67 21	0.15 4
γ	799.79 16	0.34 5
γ	807.67 19	0.28 4
γ [M1+E2]	815.91 11	0.58 5
γ	835.1 3	0.23 4
γ	840.03 21	0.34 14
γ	858.80 15	0.25 5
γ	866.27 16	0.34 11
γ	868.34 22	0.21 6
γ	871.06 9	0.15 4
γ	878.18 14	0.31 5
γ	881.91 24	0.32 12
γ	882.42 15	0.37 14
γ	883.1 5	0.098 25
γ	887.36 16	0.25 5
γ	906.4 9	0.14 6
γ	912.74 11	~0.07
γ [M1+E2]	920.29 13	0.22 4
γ	924.86 12	~0.12
γ	931.90 18	0.52 5
γ	939.28 12	0.25 5
γ	940.40 9	0.37 7 ?
γ	949.09 14	0.46 5
γ	962.09 17	0.31 11
γ	963.17 13	1.7 7
γ	982.0 4	0.22 5
γ	987.77 16	0.22 6
γ	998.9 3	0.074 15 ?
γ	1005.48 19	0.27 4
γ	1009.13 15	0.28 4
γ	1023.32 17	0.12 5
γ	1034.83 22	0.26 5
γ	1045.72 19	0.26 5
γ	1052.38 6	0.87 14
γ	1068.34 24	~0.22
γ	1071.76 7	0.58 5 ?
γ	1079.88 6	5.72 15
γ	1087.72 24	0.25 9
γ	1089.41 12	0.71 15
γ	1099.26 6	1.76 6
γ [E1]	1114.41 10	0.55 6
γ	1158.57 19	~0.12
γ [E1]	1183.74 10	0.41 6
γ	1207.9 3	0.23 9
γ	1246.04 17	~0.20
γ [M1+E2]	1284.14 7	~0.25
γ	1287.59 9	1.76 9
γ	1301.78 12	0.39 17
γ	1307.2 3	~0.11
γ [M1+E2]	1311.64 7	1.91 7
γ	1359.47 13	0.26 9
γ	1360.0 3	0.31 11
γ [E1]	1377.80 9	0.15 4
γ	1412.61 12	0.36 5
γ [E2]	1421.5 3	~0.09
γ	1428.53 19	0.39 4
γ	1434.27 7	0.31 4
γ	1441.14 9	0.53 5
γ [E1]	1447.13 9	0.74 15
γ	1448.69 7	5.83 22
γ	1461.77 6	12.3
γ	1476.19 7	2.25 10
γ	1499.09 20	0.27 4
γ	1517.84 18	0.17 4
γ	1529.25 17	0.60 6
γ	1551.99 12	2.83 11
γ	1556.93 10	2.64 11
γ	1567.63 15	0.37 14
γ [E1]	1570.68 10	~0.9

Photons (^{103}Cd)
(continued)

γ_{mode}	γ(keV)	γ(%)†
γ	1574.3 *3*	
γ	1627.93 *16*	0.15 *4*
γ	1636.7 *3*	0.23 *9*
γ	1637.6 *4*	0.21 *4* ?
γ	1646.4 *4*	0.15 *4*
γ	1668.89 *16*	0.23 *4*
γ[E1]	1677.97 *12*	0.16 *5*
γ	1685.2 *4*	0.17 *5*
γ	1693.18 *19*	0.64 *6*
γ	1694.1 *3*	0.16 *7*
γ[E2]	1705.11 *9*	0.53 *5*
γ	1718.61 *15*	0.44 *4*
γ	1748.41 *8*	1.53 *9*
γ	1756.24 *22*	0.17 *4*
γ	1766.59 *12*	0.66 *5*
γ	1775.91 *9*	0.28 *4*
γ	1808.91 *12*	0.36 *4*
γ	1821.95 *10*	1.11 *6*
γ[E1]	1834.07 *8*	1.02 *6*
γ	1856.65 *16*	0.53 *5*
γ	1879.94 *9*	3.49 *11*
γ	1906.92 *20*	0.17 *7*
γ[E1]	1918.96 *18*	0.43 *5*
γ	1930.55 *14*	2.00 *9*
γ	1954.57 *15*	0.25 *5* ?
γ	1955.9 *5*	0.209 *25*
γ	1958.05 *14*	0.308 *25*
γ[M1+E2]	1984.50 *9*	0.60 *5*
γ[E1]	1999.65 *10*	0.18 *6*
γ[M1+E2]	2012.00 *9*	1.34 *6*
γ	2022.52 *12*	1.07 *6*
γ[E1]	2064.91 *11*	0.37 *5*
γ	2067.11 *24*	0.11 *4*
γ	2097.42 *21*	0.33 *5*
γ	2117.85 *20*	0.12 *4*
γ	2124.92 *21*	0.27 *4*
γ	2132.94 *17*	2.05 *11*
γ	2136.44 *24*	
γ	2167.59 *24*	0.31 *4*
γ[E2]	2199.34 *11*	1.54 *7*
γ	2245.06 *16*	1.17 *6*
γ	2256.6 *4*	0.12 *4*
γ	2273.71 *16*	0.80 *5*
γ	2287.8 *3*	0.20 *4*
γ[E1]	2298.07 *12*	0.061 *25*
γ[E1]	2300.4 *3*	0.41 *9*
γ[E1]	2305.89 *18*	0.160 *25*
γ	2328.52 *16*	0.25 *4*
γ	2356.02 *16*	0.37 *4*
γ	2365.7 *8*	0.21 *4*
γ[E1]	2367.41 *12*	0.27 *9*
γ	2373.53 *11*	1.60 *6*
γ[E1]	2386.58 *9*	0.68 *5*
γ	2401.03 *10*	1.25 *6*
γ	2411.84 *12*	0.22 *4*
γ	2439.34 *12*	0.55 *4*
γ	2457.57 *16*	0.20 *4*
γ	2485.07 *16*	0.69 *5*
γ[E2]	2521.01 *9*	0.18 *4*
γ	2570.20 *18*	0.47 *4*
γ	2597.70 *17*	0.22 *4*
γ	2630.31 *23*	0.074 *25*
γ	2657.81 *23*	0.17 *4*
γ	2661.97 *19*	0.42 *4*
γ	2681.05 *20*	1.54 *6*
γ[E1]	2687.3 *3*	0.23 *5*
γ	2707.79 *16*	0.98 *6*
γ[E1]	2754.34 *12*	0.086 *25*
γ	2768.5 *3*	0.41 *5*
γ	2778.0 *4*	0.086 *25*
γ	2796.0 *3*	0.074 *25*
γ	2811.1 *3*	0.20 *4*
γ	2829.5 *3*	0.76 *6*
γ	2855.49 *22*	0.34 *5*

Photons (^{103}Cd)
(continued)

γ_{mode}	γ(keV)	γ(%)†
γ	2912.8 *5*	0.061 *25*
γ	2953.00 *16*	0.17 *4*
γ	2980.50 *16*	0.23 *4*
γ	3043.4 *4*	0.12 *4*
γ	3056.6 *4*	0.12 *4*
γ	3066.0 *4*	0.12 *4*
γ	3161.2 *3*	0.17 *4*
γ	3188.7 *3*	0.18 *4*
γ	3245.0 *5*	0.10 *4*

† 16% uncert(syst)
* with ^{103}Ag(5.7 s) in equilib

Atomic Electrons (^{103}Cd)
$\langle e \rangle$=21.8 *5* keV

e_{bin}(keV)	$\langle e \rangle$(keV)	e(%)
2	0.38	19 *4*
3	1.47	44 *5*
4	0.96	27 *3*
18	0.76	4.2 *5*
19	1.09	5.8 *7*
21 - 22	0.87	4.0 *3*
24	0.70	2.9 *6*
25 - 69	0.24	0.83 *12*
109	8.0	7.3 *3*
131	4.98	3.80 *14*
134	1.22	0.91 *3*
162 - 188	0.006	~0.003
218 - 265	0.06	~0.026
273 - 319	0.006	~0.0023
345 - 387	0.183	0.050 *5*
417 - 464	0.121	0.0278 *14*
469 - 518	0.039	0.0081 *19*
519 - 563	0.062	0.0115 *25*
573 - 622	0.069	0.0114 *25*
623 - 671	0.024	0.0037 *6*
673 - 721	0.034	0.0049 *13*
722 - 764	0.0076	0.0010 *3*
774 - 816	0.021	0.0026 *4*
831 - 880	0.023	0.0027 *5*
881 - 929	0.019	0.0021 *4*
931 - 980	0.025	0.0027 *9*
981 - 1027	0.014	0.0013 *3*
1031 - 1080	0.070	0.0066 *18*
1084 - 1133	0.0057	0.00052 *8*
1155 - 1204	0.0033	0.00028 *7*
1205 - 1246	0.0013	~0.00011
1259 - 1308	0.032	0.0025 *4*
1311 - 1360	0.0043	0.00032 *12*
1374 - 1423	0.040	0.0028 *8*
1425 - 1474	0.09	0.0061 *19*
1475 - 1570	0.038	0.0025 *5*
1602 - 1702	0.0152	0.00091 *10*
1704 - 1796	0.019	0.00107 *18*
1805 - 1904	0.024	0.0013 *3*
1905 - 1999	0.026	0.00136 *17*
2008 - 2107	0.0129	0.00062 *12*
2114 - 2199	0.0099	0.00046 *3*
2220 - 2305	0.0112	0.00050 *8*
2325 - 2414	0.0178	0.00075 *10*
2432 - 2520	0.0042	0.00017 *3*
2545 - 2636	0.0044	0.00017 *3*
2654 - 2752	0.0110	0.00041 *6*
2754 - 2852	0.0047	0.00017 *3*
2855 - 2952	0.00076	2.6 *6* $\times10^{-5}$
2955 - 3053	0.0018	6.1 *11* $\times10^{-5}$
3056 - 3136	0.00051	1.6 *5* $\times10^{-5}$
3157 - 3244	0.00091	2.9 *6* $\times10^{-5}$

Continuous Radiation (^{103}Cd)
$\langle\beta+\rangle$=348 keV;$\langle$IB$\rangle$=2.6 keV

E_{bin}(keV)		$\langle\ \rangle$(keV)	(%)
0 - 10	β+	2.94×10^{-6}	3.56×10^{-5}
	IB	0.0128	
10 - 20	β+	9.8×10^{-5}	0.00060
	IB	0.017	0.111
20 - 40	β+	0.00228	0.0070
	IB	0.043	0.164
40 - 100	β+	0.091	0.117
	IB	0.071	0.109
100 - 300	β+	3.82	1.74
	IB	0.22	0.121
300 - 600	β+	25.7	5.6
	IB	0.36	0.082
600 - 1300	β+	134	14.5
	IB	0.97	0.108
1300 - 2500	β+	166	9.4
	IB	0.77	0.046
2500 - 4160	β+	18.0	0.68
	IB	0.127	0.0044
	$\Sigma\beta$+		32

$^{103}_{49}$In(1.08 *12* min)

Mode: ϵ
Δ: -74420 *300* keV syst
SpA: 1.68×10^{9} Ci/g
Prod: ^{92}Mo(^{14}N,3n); ^{92}Mo(^{16}O,4np)

Photons (^{103}In)

γ_{mode}	γ(keV)	γ(rel)
γM1	187.93 *9*	100
γE2(+M1)	201.97 *12*	21 *3*

$^{103}_{50}$Sn(7 *3* s)

Mode: ϵ, ϵp
Δ: -66920 *500* keV syst
SpA: 1.5×10^{10} Ci/g
Prod: ^{50}Cr(^{58}Ni,2p3n); ^{54}Fe(^{58}Ni,4p5n)

A = 104

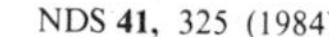

NDS **41**, 325 (1984)

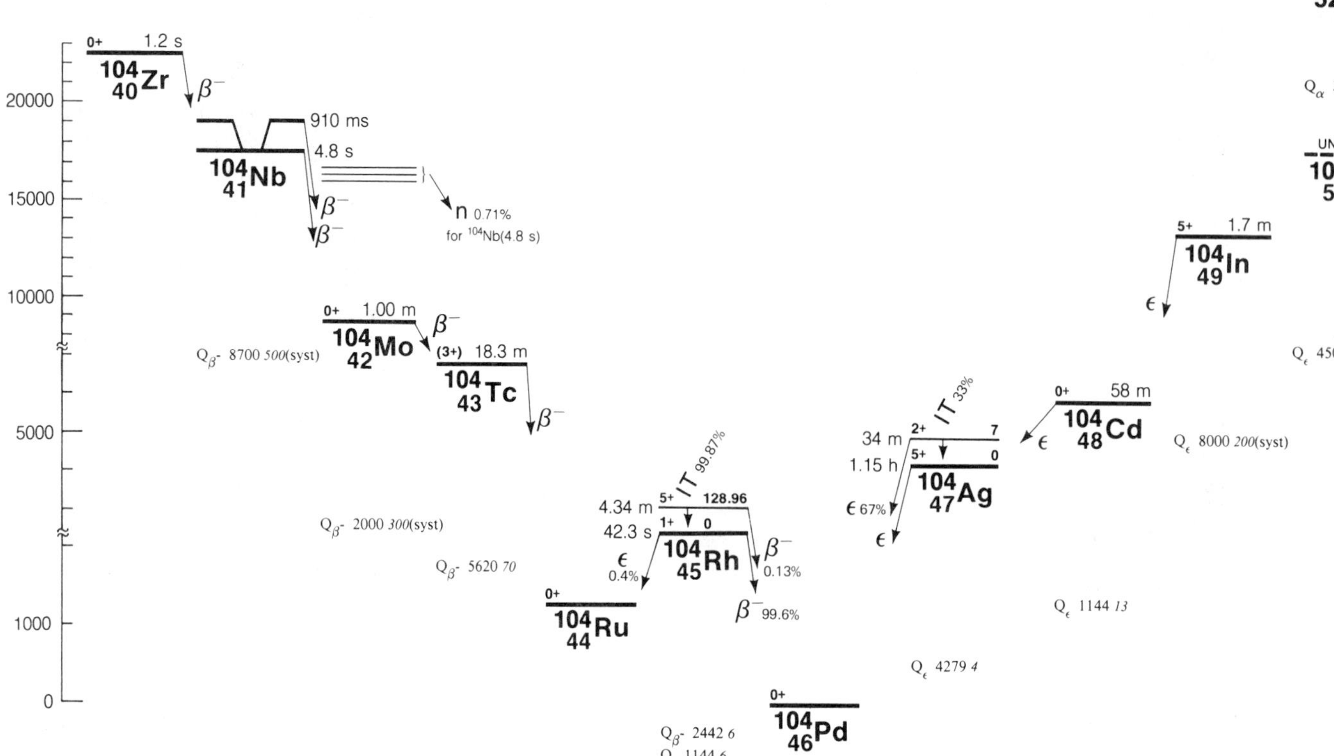

$^{104}_{40}$Zr(1.2 *3* s)

Mode: β-
SpA: 6.9×10^{10} Ci/g
Prod: fission

Photons (^{104}Zr)

γ_{mode}	γ(keV)	γ(rel)
γ	100.6 *4*	100 *20*
γ	140.9 *4*	33 *7*
γ	203.5 *4*	43 *9*
γ	212.9 *6*	60 *12*
γ	241.4 *3*	50 *10*
γ	263.0 *4*	45
γ	445.0 *4*	90 *18*
γ	504.7 *4*	<90

$^{104}_{41}$Nb(4.8 *4* s)

Mode: β-, β-n(0.71 %)
Δ: -71780 *410* keV syst
SpA: 2.10×10^{10} Ci/g
Prod: fission

Photons (^{104}Nb(4.8 s + 910 ms))

γ_{mode}	γ(keV)	γ(rel)
γ (E2)	73.69 *11*	0.80 *16*
γ E2	192.18 *9*	100 *20*
γ	216.12 *9*	0.40 *8*
γ	353.97 *18*	0.80 *16*
γ E2	368.37 *8*	20
γ	369.12 *13*	
γ	402.28 *13*	1.4 *3*
γ	477.52 *10*	17 *3*
γ [E2]	519.20 *10*	2.0 *4*
γ	555.28 *8*	8.4 *17*
γ	595.5 *3*	1.0 *2*
γ	609.60 *21*	1.0 *2*
γ [M1+E2]	620.20 *7*	19 *4*
γ	654.11 *9*	3.4 *7*
γ [E2]	693.89 *9*	6.0 *12*
γ	771.40 *8*	12.4 *25*
γ	802.48 *10*	1.0 *2*
γ [E2]	812.38 *10*	17 *3*
γ	836.32 *8*	18 *4*
γ	914.97 *23*	2.8 *6*
γ	1022.48 *12*	3.6 *7*
γ	1046.34 *17*	1.8 *4*
γ	1049.99 *20*	1.8 *4*
γ	1063.5 *3*	1.4 *3*
γ	1072.69 *16*	2.2 *4*
γ	1082.99 *20*	1.8 *4*
γ	1140.0 *4*	0.40 *8*
γ	1195.37 *23*	0.80 *16*
γ	1229.59 *20*	2.6 *5*
γ	1247.44 *23*	0.80 *16*
γ	1276.41 *17*	3.4 *7*
γ	1321.4 *4*	1.6 *3*
γ	1352.30 *18*	3.0 *6*
γ	1391.60 *9*	2.6 *5*
γ	1414.71 *18*	3.4 *7*
γ	1419.6 *3*	0.80 *16*

Photons (^{104}Nb(4.8 s + 910 ms)) (continued)

γ_{mode}	γ(keV)	γ(rel)
γ	1441.81 *17*	1.8 *4*
γ	1468.58 *18*	7.4 *15*
γ	1756.37 *21*	0.40 *8*
γ	1905.85 *23*	5.4 *11*
γ	2095.91 *16*	1.0 *2*
γ	2110.33 *19*	1.4 *3*
γ	2124.73 *21*	0.80 *16*
γ	2492.3 *4*	0.80 *16*
γ	2599.72 *23*	1.8 *4*
γ	2696.0 *3*	2.4 *5*

$^{104}_{41}$Nb(910 *100* ms)

Mode: β-
Δ: -71780 *410* keV syst
SpA: 8.3×10^{10} Ci/g
Prod: fission
see ^{104}Nb(4.8 s) for γ rays

$^{104}_{42}$Mo(1.00 *3* min)

Mode: β-
Δ: -80480 *310* keV syst
SpA: 1.80×10^9 Ci/g
Prod: fission

Photons (^{104}Mo)

⟨γ⟩=146 *28* keV

γ_{mode}	γ(keV)	γ(%)†
γ [M1]	36.45 *12*	14 *3*
γ [M1]	46.01 *11*	8.0 *16*
γ [M1]	50.07 *13*	3.9 *8*
γ [M1]	55.07 *13*	8.6 *17*
γ [M1]	68.70 *12*	55 *11*
γ E2	69.79 *20*	18 *4*
γ	86.10 *13*	~2
γ	87.10 *17*	~5
γ	91.07 *11*	4.9 *10*
γ	92.05 *19*	0.090 *18*
γ	98.20 *15*	<0.10
γ	101.09 *13*	0.26 *5*
γ [M1]	105.15 *11*	0.88 *18*
γ	114.71 *12*	0.070 *14*
γ	115.07 *17*	1.0 *2* ?
γ	133.10 *24*	<0.06
γ	151.16 *12*	0.73 *15*
γ	159.77 *12*	0.44 *9*
γ	178.15 *22*	0.37 *7*
γ	189.27 *15*	0.75 *15*
γ	196.21 *12*	0.37 *7*
γ	199.27 *15*	0.22 *4*
γ	221.87 *15*	0.30 *6*
γ	335.04 *14*	0.60 *12*
γ	376.09 *13*	4.7 *9*
γ	393.17 *17*	1.3 *3*
γ	421.14 *10*	2.6 *5*
γ	444.89 *20*	0.28 *6*
γ [M1+E2]	467.16 *12*	0.70 *14*
γ	535.86 *14*	<0.10
γ	604.19 *20*	0.39 *8*
γ	659.09 *20*	0.30 *6*
γ	710.59 *16*	0.33 *7*
γ	733.79 *20*	0.70 *14*
γ	768.69 *20*	0.70 *14*
γ	796.69 *16*	0.70 *14*

† 20% uncert(syst)

Continuous Radiation (^{104}Mo)

⟨β-⟩=685 keV;⟨IB⟩=1.19 keV

E_{bin}(keV)		⟨ ⟩(keV)	(%)
0 - 10	β-	0.0236	0.470
	IB	0.028	
10 - 20	β-	0.072	0.480
	IB	0.028	0.19
20 - 40	β-	0.298	0.99
	IB	0.053	0.19
40 - 100	β-	2.29	3.25
	IB	0.147	0.23
100 - 300	β-	28.1	13.7
	IB	0.37	0.21
300 - 600	β-	117	25.7
	IB	0.33	0.079
600 - 1300	β-	434	48.1
	IB	0.23	0.029
1300 - 1825	β-	104	7.2
	IB	0.0076	0.00055

$^{104}_{43}$Tc(18.3 *3* min)

Mode: β-
Δ: -82480 *70* keV
SpA: 9.88×10^7 Ci/g
Prod: fission; ^{104}Ru(n,p)

Photons (^{104}Tc)

⟨γ⟩=1999 *27* keV

γ_{mode}	γ(keV)	γ(%)†
Ru L_ℓ	2.253	0.0018 *4*
Ru L_η	2.382	0.00086 *16*
Ru L_α	2.558	0.047 *9*
Ru L_β	2.730	0.029 *6*
Ru L_γ	3.027	0.0022 *5*
Ru $K_{\alpha2}$	19.150	0.40 *4*
Ru $K_{\alpha1}$	19.279	0.76 *8*
Ru $K_{\beta1}$'	21.650	0.185 *19*
Ru $K_{\beta2}$'	22.210	0.034 *4*
γ	135.3 *10*	0.16 *8*
γ	150.8 *9*	0.40 *8*
γ	153.4 *10*	0.24 *8*
γ	160.4 *4*	1.7 *3*
γ	163.2 *10*	0.32 *8*
γ	170.0 *9*	0.24 *8*
γ	176.8 *5*	0.55 *16*
γ	179.1 *9*	0.40 *16*
γ	219.0 *8*	~0.32
γ	245.5 *8*	0.40 *16*
γ	272.0 *13*	0.16 *8*
γ	277.1 *13*	0.24 *8*
γ	280.8 *13*	0.16 *8*
γ	285.5 *7*	~0.32
γ	294.9 *7*	~0.6
γ	298.58 *8*	0.095 *24*
γ	314.62 *11*	0.17 *4*
γ	333.87 *17*	0.56 *8*
γ	349.00 *11*	~0.08
γ [M1+E2]	349.26 *5*	2.22 *24*
γ [M1+E2]	353.87 *6*	0.87 *16*
γ E2	358.00 *7*	79
γ	407.1 *9*	0.24 *8*
γ	413.2 *3*	0.10 *4*
γ	421.8 *10*	0.24 *8*
γ	459.38 *10*	0.095 *24*
γ	474.85 *14*	0.22 *6*
γ [M1+E2]	511.44 *10*	0.13 *3*
γ	519.38 *7*	0.79 *8*
γ	527.14 *10*	0.35 *6*
γ E2	530.44 *5*	13.9 *10*
γ E2+1.4%M1	535.05 *5*	13.1 *10*
γ	542.7 *8*	<0.24
γ	553.79 *7*	0.27 *6*
γ	565.37 *9*	0.079 *16*
γ	581.2 *5*	0.24 *8*
γ	584.09 *16*	0.55 *8*
γ	585.12 *11*	0.17 *5*
γ	605.2 *8*	0.63 *16*
γ	609.46 *7*	1.74 *24*
γ	614.07 *7*	1.03 *8*
γ	626.94 *8*	0.20 *4*
γ	630.03 *16*	0.40 *16*
γ E2	630.24 *9*	0.8 *4*
γ	648.49 *16*	0.21 *4*
γ	659.52 *11*	0.079 *16*
γ	667.95 *13*	0.31 *4*
γ	792.45 *6*	2.22 *24*
γ	795.45 *20*	0.15 *4*
γ	838.44 *7*	0.70 *7*
γ E2+9.2%M1	884.31 *5*	9.7 *10*
γ E2	893.05 *7*	9.1 *10*
γ	919.05 *13*	0.11 *4*
γ	977.2 *3*	0.119 *24*
γ	980.8 *3*	0.45 *6*
γ	983.90 *15*	0.135 *24*
γ	986.54 *15*	0.19 *3*
γ	1021.65 *7*	0.41 *4*
γ	1092.85 *10*	0.40 *4*
γ	1119.35 *13*	0.54 *6*
γ	1127.78 *11*	0.28 *8*
γ	1133.51 *12*	0.20 *8*

Photons (^{104}Tc)
(continued)

γ_{mode}	γ(keV)	γ(%)†
γ	1142.3 *3*	0.29 *4*
γ	1144.51 *7*	0.36 *4*
γ	1157.38 *6*	2.54 *24*
γ	1187.69 *7*	0.30 *3*
γ	1201.6 *3*	0.39 *5*
γ	1210.10 *15*	0.26 *3*
γ [E1]	1239.50 *9*	0.159 *24*
γ	1247.49 *7*	0.50 *6*
γ	1269.0 *3*	0.39 *5*
γ	1281.85 *7*	1.82 *16*
γ	1343.85 *13*	0.59 *6*
γ	1363.33 *21*	0.21 *4*
γ	1375.89 *8*	0.32 *5*
γ	1380.50 *7*	1.51 *16*
γ	1396.54 *9*	2.14 *24*
γ	1436.08 *17*	0.32 *8*
γ	1466.70 *8*	0.79 *8*
γ	1472.42 *9*	0.62 *6*
γ	1515.37 *8*	0.71 *8*
γ	1517.52 *14*	0.66 *8*
γ	1531.09 *10*	0.36 *7*
γ	1536.82 *11*	0.15 *4*
γ	1541.30 *9*	0.95 *8*
γ	1581.02 *15*	0.26 *4*
γ [E1]	1593.37 *10*	0.30 *4*
γ	1596.74 *7*	3.7 *3*
γ	1601.36 *8*	0.17 *4*
γ	1608.99 *15*	0.10 *3*
γ E1	1612.36 *6*	5.2 *5*
γ	1633.68 *17*	0.10 *3*
γ	1635.72 *7*	0.56 *6*
γ	1676.75 *6*	7.0 *6*
γ	1708.9 *3*	0.32 *8*
γ	1722.74 *7*	0.62 *6*
γ	1736.83 *9*	1.66 *16*
γ	1840.51 *16*	0.16 *8*
γ	1871.39 *14*	0.20 *8*
γ	1910.93 *7*	1.74 *16*
γ	1927.82 *13*	0.37 *5*
γ	1931.2 *4*	0.32 *5*
γ	1934.89 *15*	0.20 *3*
γ	1937.3 *4*	0.17 *3*
γ	1971.13 *17*	1.43 *16*
γ	1986.07 *9*	0.16 *8*
γ	1997.1 *3*	0.49 *6*
γ	2015.67 *9*	1.59 *16*
γ	2061.8 *3*	0.28 *4*
γ	2089.21 *15*	0.36 *4*
γ	2095.3 *3*	0.48 *5*
γ [E1]	2123.80 *8*	1.98 *16*
γ	2151.1 *3*	0.190 *24*
γ	2181.86 *9*	0.40 *4*
γ	2190.45 *13*	1.59 *16*
γ	2239.22 *14*	0.29 *4*
γ	2258.1 *3*	0.58 *6*
γ	2332.2 *3*	0.87 *8*
γ	2340.33 *12*	0.20 *5*
γ	2375.73 *13*	0.17 *4*
γ	2395.23 *17*	0.31 *4*
γ	2465.32 *15*	1.03 *8*
γ	2513.8 *3*	0.45 *5*
γ	2525.84 *18*	0.087 *16*
γ	2532.9 *3*	0.77 *8*
γ	2544.3 *3*	0.62 *6*
γ	2550.14 *12*	0.78 *7*
γ	2608.38 *9*	1.43 *16*
γ	2632.95 *16*	0.09 *3*
γ	2653.9 *4*	0.206 *24*
γ	2658.8 *4*	0.222 *24*
γ	2677.00 *17*	0.29 *4*
γ	2690.71 *13*	0.16 *3*
γ	2705.9 *3*	0.230 *24*
γ	2716.90 *9*	0.52 *6*
γ	2724.98 *13*	0.32 *4*
γ	2788.17 *17*	0.48 *5*
γ	2813.2 *4*	0.182 *24*
γ	2816.8 *4*	0.127 *24*
γ	2830.2 *4*	0.190 *24*
γ	2838.3 *4*	0.33 *4*
γ	2927.64 *15*	0.12 *4*
γ	2975.67 *21*	0.198 *24*
γ	2982.20 *16*	0.095 *16*
γ	3007.0 *4*	0.32 *4*
γ	3026.24 *17*	0.198 *24*

Photons (^{104}Tc) (continued)

γ_{mode}	γ(keV)	γ(%)†
γ	3056.27 18	0.28 3
γ	3085.18 12	0.135 24
γ	3143.42 8	0.71 8
γ	3149.15 9	1.03 8
γ	3187.3 4	0.36 4
γ	3225.75 13	0.28 3
γ	3260.02 13	0.151 24
γ	3276.88 15	0.119 24
γ	3318.60 17	0.26 3
γ	3370.48 18	0.26 3
γ	3374.49 20	0.21 3
γ	3418.2 4	0.32 6
γ	3517.24 16	0.143 24
γ	3637.7 5	0.25 3
γ	3704.3 5	0.087 16
γ	3714.3 5	0.42 5
γ	3811.92 15	0.11 3

† 10% uncert(syst)

Atomic Electrons (^{104}Tc)

⟨e⟩=6.4 3 keV

e_{bin}(keV)	⟨e⟩(keV)	e(%)
3 - 22	0.103	1.81 21
113 - 135	0.16	0.13 6
138	0.27	~0.19
141 - 179	0.33	0.21 7
197 - 245	0.06	0.028 14
250 - 299	0.08	0.032 10
311 - 315	0.019	~0.006
327	0.089	0.027 6
331 - 334	0.036	0.011 3
336	3.46	1.03 4
346 - 354	0.020	0.0057 11
355	0.478	0.135 5
357 - 407	0.119	0.0329 15
410 - 459	0.0065	0.0015 5
472 - 505	0.019	0.0038 13
508	0.278	0.055 4
509 - 511	8.3 $\times10^{-5}$	1.6 5 $\times10^{-5}$
513	0.257	0.050 4
516 - 565	0.104	0.0194 12
578 - 627	0.065	0.0109 24
629 - 668	0.0049	0.00076 23
770 - 816	0.025	0.0033 11
835 - 884	0.182	0.0209 15
890 - 919	0.0125	0.00141 12
955 - 1000	0.0084	0.00086 18
1018 - 1022	0.00034	3.3 12 $\times10^{-5}$
1071 - 1120	0.0095	0.00087 17
1122 - 1166	0.019	0.0016 5
1179 - 1225	0.0065	0.00054 11
1236 - 1282	0.012	0.0009 3
1322 - 1363	0.012	0.00086 21
1373 - 1414	0.012	0.0009 3
1433 - 1472	0.0065	0.00045 11
1493 - 1591	0.041	0.0026 4
1593 - 1687	0.035	0.0021 6
1701 - 1736	0.0090	0.00052 13
1818 - 1915	0.0109	0.00057 11
1925 - 2015	0.0131	0.00066 12
2040 - 2129	0.0087	0.00042 4
2148 - 2239	0.0089	0.00041 8
2255 - 2354	0.0039	0.00017 3
2373 - 2465	0.0040	0.00017 4
2492 - 2586	0.0113	0.00045 6
2605 - 2703	0.0058	0.000219 23
2705 - 2795	0.0023	8.3 13 $\times10^{-5}$
2808 - 2906	0.0018	6.3 10 $\times10^{-5}$
2924 - 3023	0.0021	7.1 10 $\times10^{-5}$
3026 - 3121	0.0026	8.5 15 $\times10^{-5}$

Atomic Electrons (^{104}Tc) (continued)

e_{bin}(keV)	⟨e⟩(keV)	e(%)
3127 - 3225	0.0044	0.000139 21
3238 - 3318	0.00129	3.9 6 $\times10^{-5}$
3348 - 3418	0.0019	5.6 8 $\times10^{-5}$
3495 - 3517	0.00033	9.6 24 $\times10^{-6}$
3616 - 3714	0.00174	4.7 7 $\times10^{-5}$
3790 - 3811	0.00025	6.6 21 $\times10^{-6}$

Continuous Radiation (^{104}Tc)

⟨β-⟩=1606 keV;⟨IB⟩=5.1 keV

E_{bin}(keV)		⟨ ⟩(keV)	(%)
0 - 10	β-	0.0066	0.131
	IB	0.050	
10 - 20	β-	0.0202	0.134
	IB	0.049	0.34
20 - 40	β-	0.084	0.279
	IB	0.097	0.34
40 - 100	β-	0.66	0.93
	IB	0.28	0.43
100 - 300	β-	8.6	4.17
	IB	0.83	0.46
300 - 600	β-	41.6	9.1
	IB	0.99	0.23
600 - 1300	β-	269	28.0
	IB	1.49	0.17
1300 - 2500	β-	723	39.3
	IB	1.05	0.061
2500 - 5000	β-	563	17.9
	IB	0.24	0.0081
5000 - 5262	β-	0.64	0.0125
	IB	6.0 $\times10^{-6}$	1.19 $\times10^{-7}$

$^{104}_{44}$Ru(stable)

Δ: -88098 5 keV
%: 18.7 2

$^{104}_{45}$Rh(42.3 4 s)

Mode: β-(99.55 10 %), ϵ(0.45 10 %)
Δ: -86954 3 keV
SpA: 2.544$\times10^{9}$ Ci/g
Prod: daughter ^{104}Rh(4.34 min); ^{103}Rh(n,γ); ^{103}Rh(d,p)

Photons (^{104}Rh)

⟨γ⟩=13 3 keV

γ_{mode}	γ(keV)	γ(%)†
Ru $K_{\alpha2}$	19.150	0.090 3
Ru $K_{\alpha1}$	19.279	0.170 6
Ru $K_{\beta1}$'	21.650	0.0411 14
Ru $K_{\beta2}$'	22.210	0.0077 3
γ_{ϵ}E2	358.00 7	0.0160 12
$\gamma_{\beta-}$[E2]	451.16 6	0.0062 6
$\gamma_{\beta-}$	452.14 12	0.0058 10
$\gamma_{\beta-}$[E0+E2]	459.32 10	~0.0036 ?
$\gamma_{\beta-}$	460.29 15	0.0016 4
$\gamma_{\beta-}$M1,E2	479.34 10	0.0029 4 ?
$\gamma_{\beta-}$	487.49 14	0.0032 11 ?
$\gamma_{\beta-}$[M1+E2]	497.45 10	0.0026 8 ?
$\gamma_{\beta-}$E2	555.83 5	2.0

Photons (^{104}Rh) (continued)

γ_{mode}	γ(keV)	γ(%)†
γ_{ϵ}E2	630.24 9	0.0010 4 ?
$\gamma_{\beta-}$E2	767.78 4	0.0111 22 ?
$\gamma_{\beta-}$[E2]	777.74 9	0.0062 6
$\gamma_{\beta-}$E2+37%M1	785.89 3	0.0036 3
$\gamma_{\beta-}$[E2]	1237.05 5	0.066 6
$\gamma_{\beta-}$	1238.03 12	0.0097 12 ?
$\gamma_{\beta-}$M1+5.4%E2	1265.23 10	0.0121 21
$\gamma_{\beta-}$E2	1341.72 5	0.00278 22
$\gamma_{\beta-}$[M1+E2]	1689.4 3	0.00044 12 ?
$\gamma_{\beta-}$	1793.85 12	0.00102 11

† uncert(syst): 25% for ϵ, 25% for β-

Atomic Electrons (^{104}Rh)*

⟨e⟩=0.0235 14 keV

e_{bin}(keV)	⟨e⟩(keV)	e(%)
3	0.0091	0.31 3
16	0.0091	0.056 6
18	0.00086	0.0047 5
19	0.0032	0.0168 17
21	0.00031	0.00149 15
22	6.3 $\times10^{-5}$	0.00029 3
336	0.00070	0.000208 16
355	9.7 $\times10^{-5}$	2.73 21 $\times10^{-5}$
357	1.43 $\times10^{-5}$	4.0 3 $\times10^{-6}$
358	7.2 $\times10^{-6}$	2.01 15 $\times10^{-6}$
608	1.5 $\times10^{-5}$	2.4 10 $\times10^{-6}$
627	1.8 $\times10^{-6}$	2.9 12 $\times10^{-7}$
630	4.0 $\times10^{-7}$	6 3 $\times10^{-8}$

* with ϵ

Continuous Radiation (^{104}Rh)

⟨β-⟩=983 keV;⟨IB⟩=2.2 keV

E_{bin}(keV)		⟨ ⟩(keV)	(%)
0 - 10	β-	0.0121	0.241
	IB	0.037	
10 - 20	β-	0.0371	0.247
	IB	0.036	0.25
20 - 40	β-	0.154	0.51
	IB	0.071	0.25
40 - 100	β-	1.20	1.70
	IB	0.20	0.31
100 - 300	β-	15.5	7.5
	IB	0.55	0.31
300 - 600	β-	73	15.9
	IB	0.57	0.135
600 - 1300	β-	422	44.6
	IB	0.62	0.074
1300 - 2442	β-	471	28.9
	IB	0.110	0.0074

$^{104}_{45}$Rh(4.34 5 min)

Mode: IT(99.87 1 %), β-(0.13 1 %)
Δ: -86825 3 keV
SpA: 4.16$\times10^{8}$ Ci/g
Prod: ^{103}Rh(n,γ); ^{103}Rh(d,p)

Photons (^{104}Rh)*

$\langle\gamma\rangle$=44.0 *4* keV

γ_{mode}	γ(keV)	γ(%)†
Rh L_ℓ	2.377	0.126 *22*
Rh L_η	2.519	0.067 *10*
Rh L_α	2.696	3.3 *5*
Rh L_β	2.883	2.2 *4*
Rh L_γ	3.179	0.19 *3*
Rh $K_{\alpha2}$	20.074	19.1 *7*
Rh $K_{\alpha1}$	20.216	36.3 *12*
Rh $K_{\beta1}$'	22.717	8.9 *3*
Rh $K_{\beta2}$'	23.312	1.67 *6*
γ_{IT} M3	31.842 *10*	0.00058 *5*
γ_{IT} E1	51.4225 *15*	48.3
γ_{IT} E3	77.533 *10*	2.08 *5*
γ_{IT} M1	97.114 *3*	2.99 *14*
$\gamma_{\beta-}$	179.32 *13*	~3×10^{-5}
$\gamma_{\beta-}$[M1+E2]	182.93 *6*	1.2 *3*×10^{-5}
$\gamma_{\beta-}$E2	263.05 *12*	~3×10^{-5}
$\gamma_{\beta-}$	332.6 *2*	0.00014 *3*
$\gamma_{\beta-}$M1,E2	362.25 *12*	~4×10^{-5}
$\gamma_{\beta-}$M1,E2	444.28 *11*	5.1 *9*×10^{-5}
$\gamma_{\beta-}$[E2]	451.16 *6*	0.00039 *5*
$\gamma_{\beta-}$	452.14 *12*	0.00038 *4*
$\gamma_{\beta-}$[E0+E2]	459.32 *10*	~0.00023
$\gamma_{\beta-}$	460.29 *15*	0.000104 *21*
$\gamma_{\beta-}$M1,E2	479.34 *10*	0.00013 *6* ?
$\gamma_{\beta-}$	487.49 *14*	0.00014 *3*
$\gamma_{\beta-}$[M1+E2]	497.45 *10*	0.000113 *20*
$\gamma_{\beta-}$E2	555.83 *5*	0.130
$\gamma_{\beta-}$	618.1 *5*	<8×10^{-5}
$\gamma_{\beta-}$	623.60 *14*	~8×10^{-5}
$\gamma_{\beta-}$E2	740.68 *4*	0.00087 *5*
$\gamma_{\beta-}$M1+40%E2	758.80 *4*	0.00092 *5*
$\gamma_{\beta-}$E2	767.78 *4*	0.0065 *3*
$\gamma_{\beta-}$[E2]	777.74 *9*	0.00038 *4*
$\gamma_{\beta-}$E2+37%M1	785.89 *3*	0.00085 *4*
$\gamma_{\beta-}$[E2]	839.88 *5*	8.7 *16*×10^{-5}
$\gamma_{\beta-}$M1+19%E2	857.99 *4*	0.00072 *5*
$\gamma_{\beta-}$[E2]	923.61 *6*	0.000184 *20*
$\gamma_{\beta-}$M1+29%E2	941.72 *5*	0.00067 *4*
$\gamma_{\beta-}$	1102.93 *12*	~9×10^{-6}
$\gamma_{\beta-}$	1121.04 *12*	~3×10^{-5}
$\gamma_{\beta-}$[E2]	1237.05 *5*	0.0042 *4*
$\gamma_{\beta-}$	1238.03 *12*	0.00063 *7*
$\gamma_{\beta-}$M1+5.4%E2	1265.23 *10*	~0.0005
$\gamma_{\beta-}$E2	1341.72 *5*	0.00066 *3*
$\gamma_{\beta-}$E2	1526.57 *4*	0.00082 *5*
$\gamma_{\beta-}$E2	1625.77 *4*	0.000384 *24*
$\gamma_{\beta-}$[E2]	1709.50 *6*	2.5 *5*×10^{-5}
$\gamma_{\beta-}$	1793.85 *12*	6.7 *7*×10^{-5}
$\gamma_{\beta-}$	1888.82 *12*	~2×10^{-5}

† uncert(syst): 1.0% for IT, 13% for β-

* with ^{104}Rh(42.3 s) in equilib

Atomic Electrons (^{104}Rh)*

$\langle e\rangle$=83.4 *11* keV

e_{bin}(keV)	$\langle e\rangle$(keV)	e(%)
3	3.2	104 *11*
9	0.111	1.28 *11*
16	0.187	1.14 *12*
17	1.67	9.8 *10*
19	0.219	1.14 *12*
20	0.62	3.1 *3*
22	0.061	0.27 *3*
23	0.0189	0.083 *9*
28	11.7	41.4 *10*
29	0.39	1.37 *12*
31	0.139	0.45 *4*
32	0.0271	0.085 *7*
48	2.51	5.21 *12*
51	0.583	1.14 *3*
54	20.9	38.4 *12*
74	16.5	22.2 *7*
75	16.7	22.4 *7*

Atomic Electrons (^{104}Rh)* (continued)

e_{bin}(keV)	$\langle e\rangle$(keV)	e(%)
77	7.85	10.2 *3*
78	0.0130	0.0167 *5*
94	0.113	0.121 *6*
96	0.0202	0.0209 *11*
97	0.0060	0.0062 *3*

* with IT

Continuous Radiation (^{104}Rh)

$\langle\beta$-$\rangle$=0.374 keV;$\langle$IB$\rangle$=0.00040 keV

E_{bin}(keV)		$\langle\ \rangle$(keV)	(%)
0 - 10	β-	0.000204	0.00409
	IB	1.8×10^{-5}	
10 - 20	β-	0.00061	0.00404
	IB	1.66×10^{-5}	0.000115
20 - 40	β-	0.00236	0.0079
	IB	3.1×10^{-5}	0.000107
40 - 100	β-	0.0153	0.0220
	IB	7.6×10^{-5}	0.000118
100 - 300	β-	0.091	0.0490
	IB	0.000149	8.8×10^{-5}
300 - 600	β-	0.116	0.0271
	IB	9.0×10^{-5}	2.2×10^{-5}
600 - 1247	β-	0.149	0.0189
	IB	2.4×10^{-5}	3.3×10^{-6}

$^{104}_{46}$Pd(stable)

Δ: -89397 *5* keV

%: 11.14 *8*

$^{104}_{47}$Ag(1.153 *17* h)

Mode: ϵ

Δ: -85118 *7* keV

SpA: 2.61×10^7 Ci/g

Prod: ^{103}Rh(α,3n); protons on Cd; ^{104}Pd(p,n); ^{103}Rh(^{3}He,2n); protons on Ag; descendant ^{104}Cd

Photons (^{104}Ag)

$\langle\gamma\rangle$=2563 *42* keV

γ_{mode}	γ(keV)	γ(%)†
Pd L_ℓ	2.503	0.081 *14*
Pd L_η	2.660	0.037 *6*
Pd L_α	2.838	2.2 *3*
Pd L_β	3.053	1.40 *23*
Pd L_γ	3.384	0.139 *25*
Pd $K_{\alpha2}$	21.020	17.6 *6*
Pd $K_{\alpha1}$	21.177	33.1 *12*
Pd $K_{\beta1}$'	23.811	8.2 *3*
Pd $K_{\beta2}$'	24.445	1.59 *6*
γ	179.32 *13*	0.93 *19* ?
γ[M1+E2]	182.93 *6*	0.46 *9* ?
γ	204.1 *10*	~0.6 ?
γ E2	263.05 *12*	1.0 *5*
γ	289.76 *20*	1.21 *19*
γ M1,E2	362.25 *12*	1.3 *3*
γ M1,E2	444.28 *11*	1.9 *3*
γ M1,E2	479.34 *10*	1.02 *14*
γ	487.49 *14*	1.1 *4*
γ[M1+E2]	497.45 *10*	0.9 *3* ?
γ E2	555.83 *5*	92.8 *19*

Photons (^{104}Ag) (continued)

γ_{mode}	γ(keV)	γ(%)†
γ	618.1 *5*	0.56 *19* ?
γ	623.60 *14*	2.5 *5*
γ	659.2 *3*	0.46 *9*
γ E2	740.68 *4*	6.9 *6*
γ M1+40%E2	758.80 *4*	7.3 *6*
γ E2	767.78 *4*	65.9 *19*
γ E2+37%M1	785.89 *3*	9.5 *8*
γ	806.0 *5*	0.28 *9* ?
γ[E2]	839.88 *5*	1.25 *21*
γ M1+19%E2	857.99 *4*	10.3 *10*
γ	862.9 *3*	6.9 *9*
γ	871.84 *21*	~0.28 ?
γ	883.1 *20*	~0.28 ?
γ	892.52 *19*	0.46 *9*
γ	908.24 *25*	4.5 *6*
γ[E2]	923.61 *6*	6.9 *7*
γ E2	926.0 *3*	12.5 *15*
γ M1+29%E2	941.72 *5*	25.2 *18*
γ	955.57 *21*	0.56 *9*
γ E1+28%M2	974.39 *19*	0.019 *3*
γ[M1+E2]	1022.75 *21*	0.56 *9*
γ	1075.45 *19*	2.1 *4*
γ	1102.93 *12*	0.28 *9* ?
γ	1121.04 *12*	0.84 *9*
γ	1133.2 *3*	0.18 *3*
γ	1192.1 *10*	~0.28 ?
γ	1247.25 *24*	0.55 *18*
γ M1+5.4%E2	1265.23 *10*	4.3 *6*
γ[M1+E2]	1284.09 *22*	0.74 *9*
γ	1316.11 *22*	0.28 *9* ?
γ	1323.1 *20*	~0.37 ?
γ E2	1341.72 *5*	7.4 *6*
γ[M1+E2]	1354.4 *3*	0.046 *9*
γ	1374.0 *3*	<0.19
γ	1418.6 *3*	0.121 *24*
γ[E2]	1425.1 *5*	0.22 *5*
γ	1451.0 *3*	1.11 *19*
γ	1456.1 *20*	~0.28 ?
γ	1478.8 *5*	0.28 *9* ?
γ E2	1526.57 *4*	6.5 *5*
γ	1544.8 *5*	0.46 *19* ?
γ	1551.7 *5*	~0.19
γ	1600.9 *3*	1.02 *19*
γ E2	1625.77 *4*	5.5 *6*
γ	1637.7 *4*	~0.19
γ	1687.1 *10*	~0.19 ?
γ[E2]	1709.50 *6*	0.92 *18*
γ	1723.1 *10*	~0.19 ?
γ	1742.16 *19*	0.32 *6*
γ	1761.0 *5*	
γ[E2]	1763.42 *20*	0.65 *19* ?
γ[M1+E2]	1781.53 *20*	3.2 *6*
γ	1788.9 *3*	<0.19 ?
γ	1792.1 *5*	~0.19 ?
γ	1813.55 *21*	0.93 *19*
γ	1834.24 *19*	<0.19
γ	1869.8 *5*	0.019 *4*
γ	1888.82 *12*	0.7 *3*
γ	1900.9 *3*	0.19 *3*
γ	1922.9 *4*	<0.07
γ	1957.0 *5*	0.037 *7*
γ	1986.1 *4*	0.65 *9*
γ	1992.1 *4*	0.17 *3*
γ	2015.02 *24*	0.24 *11*
γ	2115.1 *5*	<0.09 ?
γ	2157.1 *20*	<0.7 ?
γ	2159.9 *3*	<0.10
γ	2218.8 *3*	<0.19
γ	2244.7 *5*	0.019 *4*
γ	2266.7 *5*	~0.19
γ	2478.7 *4*	<0.11
γ[E2]	2549.3 *2*	<0.09
γ	2581.32 *21*	<0.09
γ	2613.5 *5*	0.056 *11*
γ	2778.0 *4*	0.13 *4*
γ	3097.9 *5*	0.019 *4*

† 1.9% uncert(syst)

Atomic Electrons (^{104}Ag)

⟨e⟩= 9.7 *3* keV

e_{bin}(keV)	⟨e⟩(keV)	e(%)
3	1.77	55 *6*
4	0.152	4.2 *4*
17	0.36	2.09 *21*
18	1.28	7.2 *7*
20	0.32	1.58 *16*
21	0.43	2.08 *21*
23 - 24	0.072	0.310 *25*
155 - 204	0.32	0.19 *7*
239 - 287	0.18	0.071 *23*
289 - 338	0.056	0.017 *4*
359 - 362	0.0091	0.0025 *7*
420 - 463	0.110	0.025 *4*
473 - 497	0.033	0.0068 *16*
531	1.90	0.358 *10*
552 - 599	0.341	0.061 *3*
614 - 659	0.012	0.0018 *6*
716 - 741	0.198	0.0273 *15*
743	0.81	0.109 *4*
755 - 803	0.272	0.0356 *16*
805 - 855	0.21	0.025 *4*
857 - 905	0.240	0.027 *3*
908	0.0009	~0.00010
917	0.256	0.0280 *21*
920 - 956	0.070	0.0075 *4*
971 - 1020	0.0056	0.00056 *10*
1022 - 1051	0.013	0.0013 *6*
1072 - 1121	0.011	0.00098 *24*
1130 - 1168	0.0017	~0.00015
1188 - 1223	0.0032	0.00026 *13*
1241 - 1284	0.043	0.0034 *4*
1292 - 1341	0.056	0.0043 *3*
1342 - 1374	0.0008	$\sim 6\times10^{-5}$
1394 - 1432	0.0086	0.00060 *17*
1447 - 1544	0.046	0.00305 *22*
1548 - 1637	0.038	0.00240 *22*
1663 - 1760	0.028	0.00159 *20*
1763 - 1845	0.0079	0.00045 *9*
1864 - 1962	0.0065	0.00034 *9*
1968 - 2014	0.0020	1.0 *3* $\times10^{-4}$
2091 - 2159	0.0018	$\sim 8\times10^{-5}$
2194 - 2266	0.0011	4.9 *21* $\times10^{-5}$
2454 - 2549	0.00037	$\sim 1\times10^{-5}$
2557 - 2613	0.00033	1.3 *6* $\times10^{-5}$
2754 - 2777	0.00042	1.5 *5* $\times10^{-5}$
3074 - 3097	5.3 $\times10^{-5}$	1.7 *5* $\times10^{-6}$

Continuous Radiation (^{104}Ag)

⟨β+⟩=86 keV;⟨IB⟩=1.42 keV

E_{bin}(keV)		⟨ ⟩(keV)	(%)
0 - 10	β+	6.7 $\times10^{-6}$	8.2 $\times10^{-5}$
	IB	0.0039	
10 - 20	β+	0.000218	0.00132
	IB	0.0116	0.069
20 - 40	β+	0.00486	0.0150
	IB	0.025	0.102
40 - 100	β+	0.181	0.235
	IB	0.021	0.033
100 - 300	β+	6.4	2.98
	IB	0.077	0.041
300 - 600	β+	30.4	6.8
	IB	0.21	0.047
600 - 1300	β+	42.4	5.3
	IB	0.75	0.083
1300 - 2500	β+	6.5	0.440
	IB	0.32	0.021
	Σβ+		15.7

$^{104}_{47}$Ag(33.5 *20* min)

Mode: ε(67 *5* %), IT(33 *5* %)

Δ: -85111 *7* keV

SpA: 5.4×10^7 Ci/g

Prod: ^{103}Rh(α,3n); daughter ^{104}Cd; protons on Cd; ^{104}Pd(p,n); ^{103}Rh(^{3}He,2n); protons on Ag

Photons (^{104}Ag)

⟨γ⟩=811 *29* keV

γ_{mode}	γ(keV)	γ(%)†
Pd L_ℓ	2.503	0.023 *4*
Ag L_ℓ	2.634	~0.041
Pd L_η	2.660	0.0105 *16*
Ag L_η	2.806	~0.00048
Pd L_α	2.838	0.62 *10*
Ag L_α	2.984	~1
Pd L_β	3.053	0.40 *7*
Pd L_γ	3.384	0.040 *7*
Ag L_β	3.412	~0.18
Ag L_γ	3.704	~0.007
Pd $K_{\alpha2}$	21.020	4.99 *17*
Pd $K_{\alpha1}$	21.177	9.4 *3*
Pd $K_{\beta1}'$	23.811	2.34 *8*
Pd $K_{\beta2}'$	24.445	0.453 *17*
γ_ϵ[E2]	451.16 *6*	
γ_ϵ	452.14 *12*	1.54 *18*
γ_ϵ	460.29 *15*	0.43 *9*
γ_ϵM1,E2	479.34 *10*	0.073 *11*
γ_ϵ	487.49 *14*	0.080 *19*
γ_ϵ[M1+E2]	497.45 *10*	0.065 *15*
γ_ϵE2	555.83 *5*	61
γ_ϵE2	767.78 *4*	0.61 *18* ?
γ_ϵ[E2]	777.74 *9*	0.43 *6*
γ_ϵE2+37%M1	785.89 *3*	1.36 *12*
γ_ϵ	934.81 *16*	0.30 *3*
γ_ϵE1+28%M2	974.39 *19*	0.0122 *21*
γ_ϵ	996.37 *20*	0.34 *3*
γ_ϵ	1133.2 *3*	0.116 *20*
γ_ϵ	1191.7 *3*	0.12 *3* ?
γ_ϵ	1238.03 *12*	2.59 *15*
γ_ϵ	1247.25 *24*	~1
γ_ϵM1+5.4%E2	1265.23 *10*	0.31 *5*
γ_ϵ	1297.9 *3*	0.55 *11*
γ_ϵE2	1341.72 *5*	1.05 *10*
γ_ϵ[M1+E2]	1354.4 *3*	0.030 *6*
γ_ϵ	1374.0 *3*	
γ_ϵ	1382.4 *3*	0.27 *5*
γ_ϵ	1418.6 *3*	0.20 *4*
γ_ϵ	1636.2 *5*	0.122 *24*
γ_ϵ	1637.7 *4*	0.18 *6*
γ_ϵ	1651.9 *4*	0.110 *22*
γ_ϵ[M1+E2]	1689.4 *3*	0.61 *6*
γ_ϵ	1720.70 *17*	1.16 *12*
γ_ϵ	1742.16 *19*	0.21 *4*
γ_ϵ	1782.26 *20*	1.4 *3*
γ_ϵ	1793.85 *12*	0.272 *23*
γ_ϵ	1869.8 *5*	0.0085 *17*
γ_ϵ	1890.3 *3*	0.09 *3* ?
γ_ϵ	1900.9 *3*	0.128 *22*
γ_ϵ	1922.9 *4*	
γ_ϵ	1936.2 *5*	0.043 *9*
γ_ϵ	1965.7 *5*	0.13 *3*
γ_ϵ	1977.5 *3*	0.58 *6*
γ_ϵ	1992.1 *4*	0.046 *8*
γ_ϵ	1999.2 *5*	0.067 *13*
γ_ϵ	2015.02 *24*	0.46 *9*
γ_ϵ	2066.3 *3*	0.15 *3*
γ_ϵ	2086.9 *5*	0.043 *9*
γ_ϵ	2139.3 *5*	1.07 *12*
γ_ϵ	2159.9 *3*	0.061 *12*
γ_ϵ	2215.7 *5*	0.061 *12*
γ_ϵ	2244.7 *5*	0.043 *9*
γ_ϵ	2254.3 *5*	<0.12 ?
γ_ϵ	2276.52 *17*	1.65 *12*
γ_ϵ	2338.08 *20*	0.61 *6*
γ_ϵ	2362.5 *4*	0.30 *3*
γ_ϵ	2419.7 *4*	0.49 *6*

Photons (^{104}Ag)
(continued)

γ_{mode}	γ(keV)	γ(%)†
γ_ϵ	2437.7 *4*	0.037 *7*
γ_ϵ	2478.7 *4*	0.067 *13*
γ_ϵ	2522.8 *4*	0.58 *6*
γ_ϵ	2557.5 *5*	0.030 *6*
γ_ϵ	2626.9 *3*	0.61 *6*
γ_ϵ	2658.1 *3*	0.21 *3*
γ_ϵ	2705.4 *5*	0.09 *3*
γ_ϵ	2729.6 *5*	0.79 *9*
γ_ϵ	2778.0 *4*	0.036 *7*
γ_ϵ	2852.2 *3*	0.21 *3*
γ_ϵ	2918.5 *4*	0.122 *24*
γ_ϵ	3008.4 *5*	0.055 *11*
γ_ϵ	3034.1 *5*	0.15 *3*
γ_ϵ	3097.9 *5*	0.0122 *24*
γ_ϵ	3116.6 *5*	0.015 *3*
γ_ϵ	3213.9 *3*	0.98 *12*
γ_ϵ	3408.0 *3*	0.98 *12*
γ_ϵ	3474.4 *4*	0.067 *6*
γ_ϵ	4009.3 *4*	0.0098 *20*

† uncert(syst): 7.5% for ε

Atomic Electrons (^{104}Ag)

⟨e⟩=5.0 *7* keV

e_{bin}(keV)	⟨e⟩(keV)	e(%)
3	1.3	40 *14*
4	0.9	25 *12*
6	0.39	~6
7 - 17	0.19	1.9 *6*
18	0.36	2.03 *21*
20 - 24	0.234	1.13 *8*
428 - 476	0.049	0.011 *5*
479 - 497	0.00052	0.00011 *4*
531	1.25	0.235 *18*
552	0.141	0.0255 *20*
553 - 556	0.060	0.0107 *6*
743 - 786	0.033	0.0044 *4*
910 - 950	0.0027	0.00029 *12*
971 - 996	0.0026	0.00027 *11*
1109 - 1133	0.0008	7 *3* $\times10^{-5}$
1167 - 1214	0.015	0.0012 *5*
1223 - 1265	0.011	0.0009 *3*
1274 - 1317	0.0098	0.00075 *11*
1330 - 1379	0.0026	0.00019 *4*
1382 - 1419	0.0011	8 *3* $\times10^{-5}$
1612 - 1696	0.0102	0.00061 *11*
1717 - 1793	0.0089	0.00051 *13*
1845 - 1941	0.0016	8.4 *17* $\times10^{-5}$
1953 - 2042	0.0052	0.00026 *5*
2063 - 2159	0.0045	0.00021 *6*
2191 - 2276	0.0065	0.00029 *8*
2314 - 2395	0.0047	0.00020 *4*
2413 - 2498	0.0022	9.0 *22* $\times10^{-5}$
2519 - 2603	0.0021	7.9 *21* $\times10^{-5}$
2623 - 2705	0.0034	0.00013 *3*
2726 - 2777	0.00041	1.5 *3* $\times10^{-5}$
2828 - 2918	0.00101	3.6 *7* $\times10^{-5}$
2984 - 3074	0.00064	2.1 *5* $\times10^{-5}$
3092 - 3190	0.0025	7.7 *21* $\times10^{-5}$
3210 - 3213	0.00033	1.03 *25* $\times10^{-5}$
3384 - 3474	0.0028	8.3 *19* $\times10^{-5}$
3985 - 4009	2.4 $\times10^{-5}$	6.1 *16* $\times10^{-7}$

Continuous Radiation (^{104}Ag)

⟨β+⟩=508 keV;⟨IB⟩=2.4 keV

E_{bin}(keV)		⟨ ⟩(keV)	(%)
0 - 10	β+	2.98×10^{-6}	3.61×10^{-5}
	IB	0.018	
10 - 20	β+	9.7×10^{-5}	0.00059
	IB	0.020	0.137
20 - 40	β+	0.00219	0.0068
	IB	0.040	0.144
40 - 100	β+	0.085	0.110
	IB	0.101	0.155
100 - 300	β+	3.58	1.63
	IB	0.29	0.164
300 - 600	β+	25.2	5.4
	IB	0.36	0.085
600 - 1300	β+	174	18.2
	IB	0.68	0.077
1300 - 2500	β+	302	17.6
	IB	0.70	0.040
2500 - 3730	β+	2.87	0.112
	IB	0.19	0.0069
	Σβ+		43

$^{104}_{48}$Cd(57.7 *10* min)

Mode: ε

Δ: -83974 *11* keV

SpA: 3.13×10^{7} Ci/g

Prod: ^{107}Ag(p,4n); ^{16}O on Mo; ^{106}Cd(p,p2n); ^{104}Pd(^{3}He,3n); protons on Sn

Photons (^{104}Cd)

⟨γ⟩=251 *10* keV

γ_{mode}	γ(keV)	γ(%)†
Ag $K_{\alpha 2}$	21.990	28.5 *14*
Ag $K_{\alpha 1}$	22.163	54 *3*
Ag $K_{\beta 1}$'	24.934	13.5 *7*
Ag $K_{\beta 2}$'	25.603	2.69 *14*
γ [M1+E2]	26.51 *24*	
γ M1	66.70 *14*	2.40 *19*
γ M1	83.60 *14*	47
γ M1	123.79 *20*	0.35 *4* ?
γ M1	150.30 *14*	0.113 *14*
γ [M1+E2]	559.10 *14*	6.3 *5*
γ [M1+E2]	625.79 *14*	2.16 *19*
γ M1,E2	709.39 *14*	19.5 *12*

† 2% uncert(syst)

Atomic Electrons (^{104}Cd)

⟨e⟩=29.6 *19* keV

e_{bin}(keV)	⟨e⟩(keV)	e(%)
3	1.97	59 *6*
4	1.27	36 *4*
18	1.06	5.8 *6*
19	1.51	8.1 *9*
21 - 25	1.32	6.1 *4*
41	1.16	2.82 *23*
58	16.8	29 *3*

Atomic Electrons (^{104}Cd) (continued)

e_{bin}(keV)	⟨e⟩(keV)	e(%)
63 - 67	0.277	0.44 *3*
80	2.9	3.6 *4*
83 - 125	0.79	0.93 *8*
146 - 150	0.0030	0.00205 *20*
534 - 559	0.160	0.0298 *25*
600 - 626	0.046	0.0076 *7*
684 - 709	0.35	0.050 *4*

Continuous Radiation (^{104}Cd)

⟨β+⟩=0.134 keV;⟨IB⟩=0.21 keV

E_{bin}(keV)		⟨ ⟩(keV)	(%)
0 - 10	β+	0.00474	0.059
	IB	0.00018	
10 - 20	β+	0.066	0.423
	IB	0.0068	0.036
20 - 40	β+	0.063	0.269
	IB	0.027	0.116
40 - 100	IB	0.0028	0.0043
100 - 300	IB	0.027	0.013
300 - 600	IB	0.086	0.019
600 - 1053	IB	0.062	0.0086
	Σβ+		0.75

$^{104}_{49}$In(1.7 *2* min)

Mode: ε

Δ: -75970 *200* keV syst

SpA: 1.06×10^{9} Ci/g

Prod: ^{92}Mo(^{16}O,p3n); ^{96}Ru(^{12}C,p3n)

Photons (^{104}In)

⟨γ⟩=2262 *95* keV

γ_{mode}	γ(keV)	γ(%)
Cd L_{ℓ}	2.767	0.0087 *15*
Cd L_{η}	2.957	0.0041 *6*
Cd L_{α}	3.133	0.24 *4*
Cd L_{β}	3.392	0.17 *3*
Cd L_{γ}	3.778	0.017 *3*
Cd $K_{\alpha 2}$	22.984	1.77 *7*
Cd $K_{\alpha 1}$	23.174	3.33 *12*
Cd $K_{\beta 1}$'	26.085	0.85 *3*
Cd $K_{\beta 2}$'	26.801	0.176 *7*
γ	321.28 *19*	3.5 *3*
γ	378.0 *4*	0.84 *7*
γ	403.5 *3*	0.57 *6*
γ	473.99 *20*	4.8 *4*
γ	502.7 *3*	4.3 *7*
γ	533.1 *3*	3.3 *3*
γ	622.28 *18*	12.1 *7*
γ E2	658.09 *20*	100
γ	817.3 *4*	1.15 *10*
γ E2	834.2 *3*	96 *10*
γ	840.3 *3*	0.81 *8*
γ E2	878.21 *19*	27.7 *19*
γ	884.9 *10*	0.48 *10*
γ [E2]	943.55 *25*	17.1 *21*
γ	1000.3 *4*	10.4 *12*
γ	1125.0 *3*	2.34 *22*
γ	1281.66 *25*	2.54 *23*
γ	1702.4 *3*	1.21 *10*
γ	2006.3 *3*	1.97 *17*

Atomic Electrons (^{104}In)

⟨e⟩=5.22 *22* keV

e_{bin}(keV)	⟨e⟩(keV)	e(%)
4	0.208	5.7 *6*
19 - 26	0.231	1.13 *7*
295	0.14	~0.05
317 - 351	0.052	0.016 *5*
374 - 403	0.022	0.0058 *24*
447 - 476	0.19	0.041 *14*
499 - 533	0.08	0.015 *6*
596	0.17	~0.028
618 - 622	0.027	0.0044 *18*
631	1.72	0.273 *15*
654	0.214	0.0327 *18*
655 - 658	0.0652	0.0099 *4*
791	0.011	0.0014 *7*
807	1.17	0.146 *15*
813 - 840	0.194	0.0233 *19*
851	0.317	0.037 *3*
858 - 885	0.054	0.0062 *4*
917	0.178	0.0195 *25*
940 - 974	0.11	0.011 *4*
996 - 1000	0.012	0.0012 *5*
1098 - 1125	0.019	0.0017 *7*
1255 - 1282	0.018	0.0014 *6*
1676 - 1702	0.0063	0.00038 *13*
1980 - 2006	0.009	0.00045 *15*

Continuous Radiation (^{104}In)

⟨β+⟩=1905 keV;⟨IB⟩=9.7 keV

E_{bin}(keV)		⟨ ⟩(keV)	(%)
0 - 10	β+	9.4×10^{-7}	1.14×10^{-5}
	IB	0.054	
10 - 20	β+	3.25×10^{-5}	0.000197
	IB	0.054	0.37
20 - 40	β+	0.00078	0.00240
	IB	0.108	0.38
40 - 100	β+	0.0330	0.0426
	IB	0.31	0.47
100 - 300	β+	1.60	0.72
	IB	0.94	0.52
300 - 600	β+	13.7	2.93
	IB	1.21	0.28
600 - 1300	β+	146	14.8
	IB	2.2	0.25
1300 - 2500	β+	726	38.0
	IB	2.7	0.149
2500 - 5000	β+	1013	31.8
	IB	2.1	0.066
5000 - 6508	β+	5.08	0.099
	IB	0.018	0.00035
	Σβ+		88

A = 105

NDS **27**, 1 (1979)

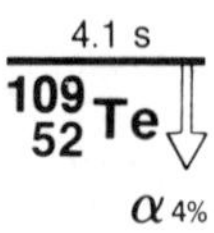

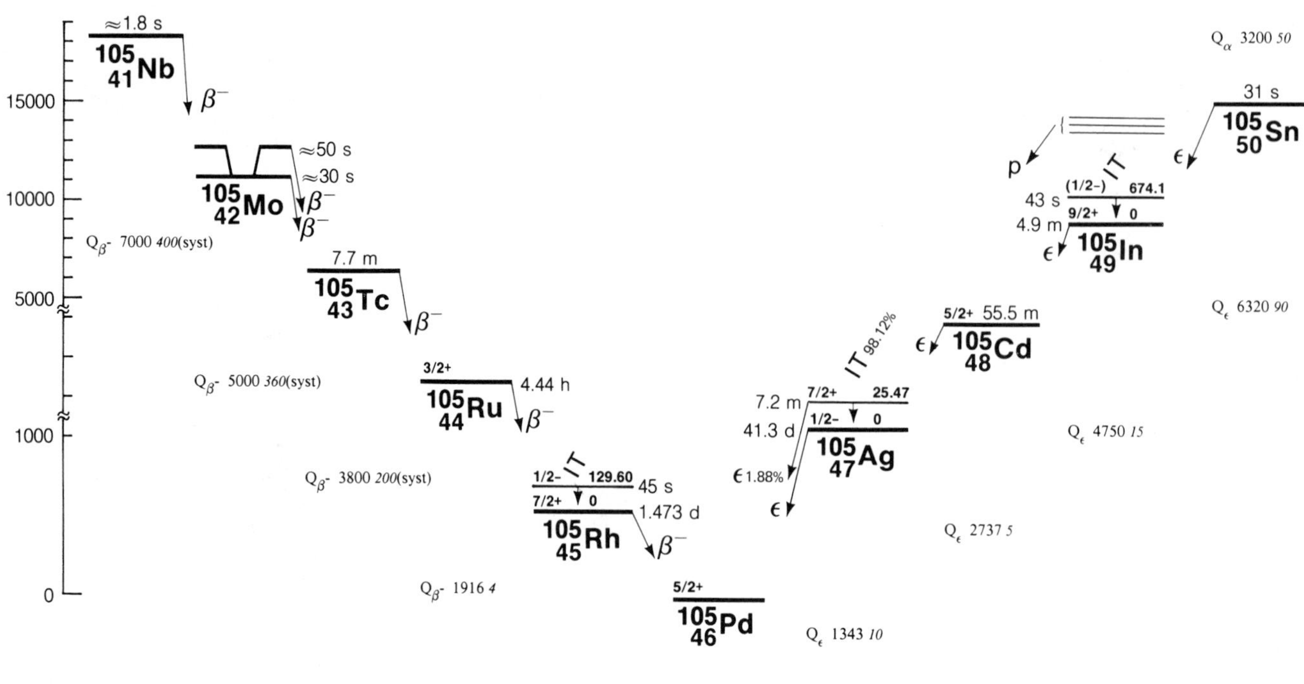

$^{105}_{41}$Nb(1.8 *8* s)

Mode: β-

Δ: -70140 *500* keV syst

SpA: 5.0×10^{10} Ci/g

Prod: fission

Photons (^{105}Nb)

γ_{mode}	γ(keV)
γ	189.45 *23* ?

$^{105}_{42}$Mo(~30 s)

Mode: β-

Δ: -77140 *300* keV syst

SpA: ~4×10^{9} Ci/g

Prod: fission

Photons (^{105}Mo(~30 + ~50 s))

γ_{mode}	γ(keV)	γ(rel)
γ	64.2 *5* ?	
γ	69.7 *5*	100 *20*
γ	77.8 *5*	8.0 *16*
γ	86.0 *5*	28 *6*
γ	89.2 *5* ?	
γ	123.5 *5*	2.1 *4*
γ	129.0 *5* ?	
γ	147.9 *5*	23 *5*
γ	161.0 *5*	7.1 *14*
γ	174.0 *5*	11.5 *23*
γ	187.5 *5* ?	
γ	198.4 *5*	2.1 *4*
γ	218.0 *5*	2.9 *6*
γ	237.5 *5*	1.7 *3*
γ	250.6 *5*	8.3 *17*
γ	286.4 *5*	1.6 *3*
γ	322.2 *5*	2.2 *4*
γ	376.0 *5*	6.1 *12*
γ	468.9 *5*	3.7 *7*
γ	620.4 *5*	2.4 *5*
γ	750.7 *5*	3.1 *6*
γ	1029.3 *5*	1.7 *3*
γ	1040.7 *5*	6.3 *13*
γ	1084.6 *5*	2.1 *4*

$^{105}_{42}$Mo(~50 s)

Mode: β-

Δ: -77140 *300* keV syst

SpA: ~2.1×10^{9} Ci/g

Prod: fission

see ^{105}Mo(~30 s) for γ rays

$^{105}_{43}$Tc(7.7 *2* min)

Mode: β-

Δ: -82140 *200* keV syst

SpA: 2.32×10^{8} Ci/g

Prod: fission

Photons (^{105}Tc)

⟨γ⟩=535 *17* keV

γ_{mode}	γ(keV)	γ(%)†
Ru L_{ℓ}	2.253	0.028 *10*
Ru L_{η}	2.382	0.014 *5*
Ru L_{α}	2.558	0.7 *3*
Ru L_{β}	2.729	0.46 *17*
Ru L_{γ}	3.029	0.035 *14*
Ru $K_{\alpha2}$	19.150	2.40 *23*
Ru $K_{\alpha1}$	19.279	4.5 *4*
γ [M1+3.5%E2]	20.56 *4*	1.1 *5*
Ru $K_{\beta1}'$	21.650	1.10 *10*
Ru $K_{\beta2}'$	22.210	0.205 *19*
γ [E1]	48.95 *7*	~0.027
γ	51.2 *4*	0.053 *21*
γ	55.4 *4*	0.11 *4* ?
γ [M1]	55.79 *3*	0.60 *10*
γ	72.5 *3*	0.16 *8*
γ [M1+44%E2]	75.30 *4*	1.5 *3*
γ	80.64 *5*	0.30 *10*
γ [E1]	82.53 *3*	2.9 *5*
γ [M1+E2]	87.40 *3*	<0.11
γ (M1+0.9%E2)	107.95 *4*	9.6 *10*
γ	112.5 *5*	~0.21
γ M1+7.5%E2	113.34 *4*	0.60 *10*
γ [M1+E2]	121.45 *5*	0.10 *4*
γ	121.9 *2*	~0.21 ?
γ	131.7 *2*	0.32 *11* ?
γ [E1]	138.31 *4*	2.9 *3*

Photons (^{105}Tc)
(continued)

γ_{mode}	γ(keV)	γ(%)†
γ (M1+17%E2)	143.18 4	10.7
γ [E1]	157.82 4	1.80 20
γ [M1]	159.27 4	7.0 7
γ [E1]	162.29 6	0.40 10
γ M1+5.2%E2	164.67 6	0.20 4
γ	169.28 22	0.41 10
γ	193.3 3	0.20 4
γ [M1+E2]	208.85 6	0.90 10
γ	212.4 2	0.30 10 ?
γ [E1]	213.61 4	0.90 10
γ [E1]	218.1 4	0.16 4
γ	224.6 7	~0.14 ?
γ [E1]	225.71 4	2.0 2
γ [E2]	229.41 6	0.20 10 ?
γ	242.6 3	0.063 14
γ [E1]	246.27 5	0.41 10
γ M1+14%E2	252.06 6	4.0 4
γ M1+5.2%E2	272.62 5	2.4 3
γ	280.2 5	0.70 11
γ	282.62 22	0.20 5
γ	300.3 5	0.33 13 ?
γ [E1]	301.01 5	0.80 11
γ	303.2 2	0.82 15 ?
γ	304.5 7	0.72 16 ?
γ	305.5 8	~0.7 ?
γ [M1]	307.03 23	0.70 19
γ	309.7 6	0.28 4
γ	314.8 3	0.50 21
γ [E1]	321.56 5	7.6 11
γ	322.31 22	1.0 3
γ [E1]	331.4 4	0.32 15
γ	333.5 4	~0.40
γ	333.94 22	~0.5
γ	352.9 6	~0.43 ?
γ [M1]	358.35 23	1.7 3
γ [M1+E2]	358.35 24	0.40 16
γ	397.61 22	~0.19
γ	407.4 5	~0.19
γ	415 1	~0.11
γ	418.6 3	0.30 11
γ	438.8 4	~0.17
γ	441.90 22	1.20 21
γ [M1]	445.74 23	1.10 21
γ [M1+E2]	462.84 9	3.0 3
γ	464.5 4	0.54 21 ?
γ [M1]	466.30 23	1.02 11
γ	470.0 3	0.30 10
γ [M1]	471.69 23	~0.21
γ	478.7 9	0.54 21 ?
γ	480.14 22	1.02 11
γ	484.1 3	0.30 5
γ [E1]	490.7 4	1.61 19
γ	535.92 22	<0.11
γ	538.14 10	~0.21
γ	540.9 3	~0.19
γ	543.0 5	~0.21 ?
γ [M1+E2]	565.0 3	0.90 11
γ	577.9 3	1.61 19
γ	581 1	0.32 11 ?
γ	604.6 4	0.7 3 ?
γ	608.5 5	1.6 5 ?
γ [M1+E2]	630.96 23	0.16 5
γ	640.3 3	1.80 16
γ	643.88 22	0.50 10
γ [M1]	646.3 5	0.80 10
γ	648.8 3	0.50 10
γ	656.9 4	<0.11
γ	664.95 22	0.29 11
γ	698 1	1.1 3 ?
γ	713.8 5	~0.39
γ	716.27 22	~0.39
γ	722.8 3	<0.11
γ	739.4 4	1.10 11
γ	756.7 3	0.80 11
γ	801 1	~0.21
γ	803.66 22	~0.17
γ	824.22 22	0.80 11
γ	883.4 3	<0.11
γ	896.0 5	1.0 5
γ	914.5 4	2.6 11 ?
γ	1003.7 3	0.48 21
γ	1008.3 3	1.20 11
γ	1048.11 22	~0.40
γ	1058.7 3	0.60 21

Photons (^{105}Tc)
(continued)

γ_{mode}	γ(keV)	γ(%)†
γ	1201.5 3	0.80 10
γ	1366.2 3	2.20 21
γ	1370.9 3	0.48 11
γ	1510.94 20	1.61 16
γ	1559.89 20	1.39 16
γ	1570.4 3	<0.11
γ	1673.23 20	<0.11
γ	1683.8 3	<0.11
γ	1882.7 3	0.107 21
γ	2053.9 4	1.02 21
γ	2081.9 4	1.02 21
γ	2155.3 3	1.50 16
γ	2167.2 4	0.21 4
γ	2174.1 4	0.30 4

† 14% uncert(syst)

Atomic Electrons (^{105}Tc)
⟨e⟩=16.0 14 keV

e_{bin}(keV)	⟨e⟩(keV)	e(%)
3	0.65	22 8
16	0.24	1.51 20
17	0.6	3.5 17
18	2.4	14 7
19	0.085	0.45 6
20	0.7	3.4 16
21 - 52	0.87	2.6 6
53	1.3	2.5 6
55 - 59	0.24	~0.4
60	0.37	0.62 11
65 - 70	0.09	~0.14
72	0.43	0.59 13
73 - 85	0.27	0.34 7
86	1.86	2.17 23
87 - 119	0.93	0.90 13
121	1.7	1.4 4
122 - 136	0.123	0.091 11
137	0.73	0.53 6
138 - 187	0.72	0.48 9
190 - 240	0.40	0.185 20
242 - 288	0.36	0.137 18
293 - 336	0.32	0.104 14
350 - 398	0.030	0.0080 18
404 - 450	0.172	0.040 4
457 - 491	0.064	0.0139 24
514 - 562	0.050	0.0091 19
564 - 609	0.035	0.0060 20
618 - 665	0.049	0.0078 15
676 - 723	0.028	0.0040 11
735 - 782	0.012	0.0015 5
798 - 824	0.007	0.0009 4
861 - 896	0.024	0.0028 12
911 - 915	0.0024	0.00026 13
982 - 1026	0.013	0.0013 4
1037 - 1059	0.0041	0.00039 18
1179 - 1202	0.0044	0.00037 14
1344 - 1371	0.013	0.0010 3
1489 - 1570	0.013	0.00087 21
1651 - 1683	0.0004	$\sim 3\times10^{-5}$
1861 - 1882	0.00039	$2.1\ 7\times10^{-5}$
2032 - 2081	0.0068	0.00033 8
2133 - 2174	0.0065	0.00030 7

Continuous Radiation (^{105}Tc)
⟨β-⟩=1415 keV;⟨IB⟩=4.1 keV

E_{bin}(keV)		⟨ ⟩(keV)	(%)
0 - 10	β-	0.0074	0.148
	IB	0.046	
10 - 20	β-	0.0228	0.152
	IB	0.046	0.32
20 - 40	β-	0.095	0.315
	IB	0.090	0.31
40 - 100	β-	0.74	1.05
	IB	0.26	0.40
100 - 300	β-	9.7	4.69
	IB	0.75	0.42
300 - 600	β-	46.5	10.1
	IB	0.88	0.21
600 - 1300	β-	296	31.0
	IB	1.24	0.144
1300 - 2500	β-	782	42.7
	IB	0.70	0.041
2500 - 3800	β-	279	9.9
	IB	0.046	0.00166

$^{105}_{44}$Ru(4.44 2 h)

Mode: β-

Δ: -85937 5 keV

SpA: 6.72×10^{6} Ci/g

Prod: ^{104}Ru(n,γ)

Photons (^{105}Ru)
⟨γ⟩=738 8 keV

γ_{mode}	γ(keV)	γ(%)†
Rh L_ℓ	2.377	0.021 4
Rh L_η	2.519	0.0112 17
Rh L_α	2.696	0.55 9
Rh L_β	2.882	0.37 6
Rh L_γ	3.182	0.032 6
Rh $K_{\alpha2}$	20.074	3.60 15
Rh $K_{\alpha1}$	20.216	6.8 3
Rh $K_{\beta1}'$	22.717	1.68 7
Rh $K_{\beta2}'$	23.312	0.314 14
γ	63.27 9	0.065 9
γ [M1+E2]	81.67 6	0.051 9
γ E3	129.60 4	5.60 14 •
γ [M1+E2]	139.40 6	0.047 9
γ M1+15%E2	149.17 5	1.74 14
γ (M1+E2)	163.53 6	0.154 19
γ [E1]	183.62 6	0.098 9
γ [M1+E2]	225.01 6	0.121 9
γ [M1+E2]	245.20 6	0.025 5
γ [M1+E2]	254.91 6	0.065 9
γ M1+6.4%E2	262.85 6	6.49 14
γ [E1]	286.59 7	0.028 5
γ [M1+E2]	306.68 6	0.079 9
γ (E1)	316.49 6	11.0 4
γ	326.11 8	1.05 12
γ (M1)	330.81 6	0.66 8
γ [E1]	339.76 6	0.014 5
γ	343.33 20	0.028 5
γ [E2]	350.08 6	0.285 14
γ (E1)	350.22 7	1.00 12
γ [E1]	369.51 8	0.047 9
γ (M1+8.3%E2)	393.40 6	3.73 8
γ	407.57 6	0.089 9
γ E1	413.49 7	2.22 19
γ [E2]	469.35 5	17.3 5
γ [M1+E2]	470.21 6	0.182 23
γ	479.63 20	0.0276 9
γ [M1+E2]	489.48 6	0.54 6
γ M1,E2	499.25 5	2.03 23
γ [M1+E2]	500.11 6	0.55 8
γ	513.75 7	0.20 5
γ [M1+E2]	539.28 6	0.112 9

Photons (^{105}Ru)
(continued)

γ_{mode}	γ(keV)	γ(%)†
γ [E1]	559.37 6	0.107 9
γ	571.11 7	~0.009
γ (E2)	575.09 6	0.84 9
γ [E1]	577.02 7	0.019 5
γ	591.20 7	0.079 9
γ [M1+E2]	597.07 7	0.029 7
γ [M1+E2]	620.94 7	0.070 9
γ [E1]	632.35 8	0.149 14
γ	635.42 14	0.014 5
γ [M1+E2]	638.66 5	0.219 23
γ	652.77 6	0.30 3
γ [E2]	656.25 6	2.03 23
γ E1	676.34 5	15.5 5
γ	700.99 9	0.019 5
γ [E2]	706.55 7	~0.009
γ M1+1.6%E2	724.27 5	46.7
γ	738.38 6	0.075 9
γ [E2]	805.93 5	0.045 9
γ [E2]	820.29 6	0.014 5
γ [M1+E2]	822.08 7	0.21 4
γ [M1+E2]	845.96 6	0.62 7
γ	846.93 20	0.028 5
γ [M1+E2]	851.98 6	0.154 19
γ (E2+40%M1)	875.86 6	2.47 9
γ	877.78 6	0.47 5
γ	907.68 6	0.52 6
γ [E1]	952.77 6	0.0149 14
γ [M1+E2]	969.46 5	2.08 7
γ	977.93 20	0.0019 5
γ	984.60 7	0.0103 19
γ	987.58 10	0.0070 14
γ	1017.48 8	0.32 3
γ	1059.59 14	0.027 7
γ [M1+E2]	1082.67 9	0.0079 19
γ	1085.40 19	0.0047 14
γ	1094.39 10	0.0033 9
γ [E2]	1172.16 7	0.0075 19
γ	1209.03 14	0.0061 19
γ [E1]	1215.61 6	0.070 9
γ [M1+E2]	1222.08 9	0.0182 23
γ	1228.89 15	0.0056 14
γ	1238.93 14	0.0019 5
γ [M1+E2]	1251.97 9	0.0191 23
γ [M1+E2]	1321.33 6	0.201 23
γ	1340.31 20	0.000467 9
γ	1357.24 9	0.0023 5
γ	1377.03 5	0.056 9
γ	1441.35 14	0.0061 19
γ	1448.33 20	0.0051 14
γ [E2]	1572.15 10	~0.0009
γ	1698.24 14	0.075 14
γ	1708.28 14	~0.00047
γ [M1+E2]	1721.32 9	0.033 9
γ	1765.4 3	~0.00019
γ	1809.65 20	~0.00023
γ	1829.6 3	~0.0007

† 1.1% uncert(syst)
* with ^{105}Rh(45 s) in equilib

Atomic Electrons (^{105}Ru)
$\langle e\rangle$=29.7 6 keV

e_{bin}(keV)	$\langle e\rangle$(keV)	e(%)
3 - 40	1.11	20.1 18
58 - 82	0.10	0.16 7
106	15.2	14.3 5
116	0.012	~0.011
126	5.18	4.10 17
127	3.47	2.74 9
129	1.69	1.31 4
130 - 164	0.384	0.288 15
180 - 225	0.017	0.0082 20
232 - 263	0.429	0.177 5
283 - 331	0.290	0.096 8
336 - 384	0.121	0.0328 13
390 - 413	0.0429	0.0109 6
446 - 491	0.615	0.136 5

Atomic Electrons (^{105}Ru)
(continued)

e_{bin}(keV)	$\langle e\rangle$(keV)	e(%)
496 - 539	0.0128	0.00254 23
548 - 597	0.0199	0.0036 3
598 - 639	0.039	0.0063 6
649 - 698	0.097	0.0148 5
700	4.1 $\times10^{-6}$	~6$\times10^{-7}$
701	0.624	0.0890 25
703 - 738	0.0916	0.0127 3
783 - 829	0.0118	0.00144 12
843 - 884	0.037	0.0044 3
904 - 953	0.0191	0.00202 17
955 - 994	0.0048	0.00049 9
1014 - 1062	0.00055	5.4 14 $\times10^{-5}$
1071 - 1094	3.4 $\times10^{-5}$	3.2 9 $\times10^{-6}$
1149 - 1192	0.00030	2.5 3 $\times10^{-5}$
1199 - 1239	0.00035	2.9 3 $\times10^{-5}$
1249 - 1252	1.81 $\times10^{-5}$	1.45 19 $\times10^{-6}$
1298 - 1340	0.00145	0.000112 14
1354 - 1377	0.00029	2.1 8 $\times10^{-5}$
1418 - 1448	5.5 $\times10^{-5}$	3.9 12 $\times10^{-6}$
1549 - 1572	5.3 $\times10^{-6}$	3.4 15 $\times10^{-7}$
1675 - 1765	0.00050	3.0 8 $\times10^{-5}$
1786 - 1829	3.9 $\times10^{-6}$	~2$\times10^{-7}$

Continuous Radiation (^{105}Ru)
$\langle\beta-\rangle$=411 keV;$\langle$IB$\rangle$=0.50 keV

E_{bin}(keV)		$\langle\ \rangle$(keV)	(%)
0 - 10	β-	0.059	1.17
	IB	0.019	
10 - 20	β-	0.177	1.18
	IB	0.018	0.126
20 - 40	β-	0.72	2.40
	IB	0.034	0.119
40 - 100	β-	5.3	7.5
	IB	0.089	0.139
100 - 300	β-	54	27.0
	IB	0.19	0.112
300 - 600	β-	160	36.2
	IB	0.118	0.029
600 - 1300	β-	188	24.4
	IB	0.029	0.0042
1300 - 1786	β-	2.74	0.193
	IB	0.000167	1.21 $\times10^{-5}$

$^{105}_{45}$Rh(1.4733 25 d)

Mode: β-
Δ: -87853 5 keV
SpA: 8.441$\times10^{5}$ Ci/g
Prod: daughter ^{105}Ru

Photons (^{105}Rh)
$\langle\gamma\rangle$=77 3 keV

γ_{mode}	γ(keV)	γ(%)†
Pd $K_{\alpha2}$	21.020	0.118 7
Pd $K_{\alpha1}$	21.177	0.223 12
Pd $K_{\beta1}$'	23.811	0.056 3
Pd $K_{\beta2}$'	24.445	0.0107 6
γ M1	38.70 4	0.023 3
γ M1+1.9%E2	280.52 3	0.167 10
γ M1+0.3%E2	306.28 5	5.13 19
γ M1+1.0%E2	319.22 3	19.0
γ M1+35%E2	442.33 7	0.042 6

† 2.1% uncert(syst)

Atomic Electrons (^{105}Rh)
$\langle e\rangle$=1.32 5 keV

e_{bin}(keV)	$\langle e\rangle$(keV)	e(%)
3	0.0120	0.37 4
4	0.00097	0.027 3
14	0.0173	0.120 17
17	0.0024	0.0141 16
18	0.0086	0.048 5
20	0.00216	0.0106 12
21	0.0029	0.0140 16
23	0.00037	0.00159 18
24	0.000118	0.00050 6
35	0.0052	0.0148 21
36	0.000113	0.00032 5
38	0.00109	0.0029 4
39	0.00021	0.00053 8
256	0.0090	0.00350 22
277	0.00117	0.00042 3
280	0.000266	9.5 6 $\times10^{-5}$
281	4.1 $\times10^{-9}$	1.48 24 $\times10^{-9}$
282	0.241	0.085 4
295	0.84	0.286 15
303	0.0310	0.0102 4
306	0.0070	0.00230 10
316	0.108	0.0343 19
319	0.0245	0.0077 4
418	0.00123	0.00029 4
439	0.00016	3.6 7 $\times10^{-5}$
442	3.6 $\times10^{-5}$	8.1 15 $\times10^{-6}$

Continuous Radiation (^{105}Rh)
$\langle\beta-\rangle$=152 keV;$\langle$IB$\rangle$=0.085 keV

E_{bin}(keV)		$\langle\ \rangle$(keV)	(%)
0 - 10	β-	0.239	4.79
	IB	0.0077	
10 - 20	β-	0.70	4.67
	IB	0.0070	0.049
20 - 40	β-	2.67	8.9
	IB	0.0122	0.043
40 - 100	β-	16.3	23.6
	IB	0.026	0.041
100 - 300	β-	81	44.4
	IB	0.030	0.019
300 - 566	β-	50.8	13.6
	IB	0.0027	0.00079

$^{105}_{45}$Rh(45 s)

Mode: IT
Δ: -87723 5 keV
SpA: 2.4$\times10^{9}$ Ci/g
Prod: daughter ^{105}Ru

Photons (^{105}Rh)

$\langle\gamma\rangle$=34.5 *6* keV

γ_{mode}	γ(keV)	γ(%)†
Rh L_ℓ	2.377	0.070 *12*
Rh L_η	2.519	0.038 *6*
Rh L_α	2.696	1.8 *3*
Rh L_β	2.882	1.27 *21*
Rh L_γ	3.181	0.108 *19*
Rh $K_{\alpha2}$	20.074	12.0 *5*
Rh $K_{\alpha1}$	20.216	22.8 *9*
Rh $K_{\beta1}$'	22.717	5.59 *22*
Rh $K_{\beta2}$'	23.312	1.05 *4*
γ E3	129.60 *4*	20.0

† 2.0% uncert(syst)

Atomic Electrons (^{105}Rh)

$\langle$e$\rangle$=94.8 *17* keV

e_{bin}(keV)	$\langle$e$\rangle$(keV)	e(%)
3	1.77	58 *6*
16	0.117	0.72 *7*
17	1.05	6.2 *6*
19	0.138	0.71 *7*
20	0.39	1.97 *21*
22	0.038	0.172 *18*
23	0.0119	0.052 *5*
106	54.4	51.1 *14*
126	17.5	13.9 *4*
127	12.4	9.8 *3*
129	6.03	4.67 *13*
130	1.02	0.791 *22*

$^{105}_{46}$Pd(stable)

Δ: -88419 *5* keV

%: 22.33 *8*

$^{105}_{47}$Ag(41.29 *7* d)

Mode: ϵ

Δ: -87076 *9* keV

SpA: 3.012×10^4 Ci/g

Prod: ^{103}Rh(α,2n); protons on Pd; deuterons on Pd

Photons (^{105}Ag)

$\langle\gamma\rangle$=509 *29* keV

γ_{mode}	γ(keV)	γ(%)†
Pd L_ℓ	2.503	0.111 *19*
Pd L_η	2.660	0.051 *8*
Pd L_α	2.838	3.0 *5*
Pd L_β	3.053	1.9 *3*
Pd L_γ	3.384	0.19 *4*
Pd $K_{\alpha2}$	21.020	24.1 *9*
Pd $K_{\alpha1}$	21.177	45.5 *16*
Pd $K_{\beta1}$'	23.811	11.3 *4*
Pd $K_{\beta2}$'	24.445	2.19 *8*
γ M1	38.70 *4*	0.0054 *7*
γ M1	64.05 *3*	11.15 *22*
γ	73.6 *22*	0.015 *3*
γ M1(+E2)	89.99 *3*	0.033 *4*
γ M1	112.47 *3*	0.0349 *25*
γ E2	155.43 *5*	0.408 *8*

Photons (^{105}Ag) (continued)

γ_{mode}	γ(keV)	γ(%)†
γ E1	158.97 *7*	0.031 *4*
γ	166.82 *8*	0.015 *8* ?
γ M2	182.88 *6*	0.358 *8*
γ [E1]	202.26 *7*	0.015 *5*
γ E2	216.21 *4*	0.014 *3*
γ	270.44 *8*	0.012 *4*
γ M1+1.9%E2	280.52 *3*	31.0 *6*
γ M1	284.94 *7*	0.096 *21*
γ M1	289.23 *4*	0.121 *8*
γ M1+0.3%E2	306.28 *5*	0.761 *15*
γ M1(+E2)	311.72 *3*	0.079 *4*
γ M1+1.0%E2	319.22 *3*	4.41 *9*
γ E1	325.37 *4*	0.200 *4*
γ (M1)	328.68 *4*	0.204 *8*
γ M1+1.2%E2	331.55 *4*	4.10 *8*
γ [E2]	344.50 *6*	0.129 *21* ?
γ E2	344.57 *4*	42 *8*
γ [M1+E2]	354.04 *4*	~0.008
γ E1	360.76 *5*	0.470 *9*
γ M1	370.25 *3*	0.732 *15*
γ [E2]	382.71 *5*	0.0046 *21*
γ M1+32%E2	392.73 *3*	1.99 *4*
γ M1(+E2)	401.71 *3*	0.191 *4*
γ M1+E2	408.06 *5*	0.042 *8*
γ (E1)	414.78 *4*	0.300 *8*
γ M1	421.00 *6*	0.121 *4*
γ E1	437.26 *4*	0.287 *8*
γ M1+35%E2	442.33 *7*	0.46 *5*
γ E2	443.44 *4*	10.77 *22*
γ (M1)	446.75 *5*	0.100 *8*
γ [M1+E2]	486.73 *10*	0.007 *3*
γ E1	527.25 *4*	0.108 *8*
γ M1,E2	560.78 *3*	0.562 *11*
γ	564.64 *11*	~0.007
γ	576.64 *8*	0.025 *4*
γ [E2]	580.13 *14*	0.008 *3*
γ	582.99 *15*	0.018 *4*
γ	598.7 *5*	0.015 *3*
γ [M1+E2]	609.85 *8*	0.0045 *14*
γ M1,E2	617.92 *4*	1.190 *24*
γ	640.69 *8*	0.029 *8*
γ E1	644.59 *4*	10.07 *20*
γ	645.70 *7*	0.058 *12*
γ [M1+E2]	648.54 *8*	~0.004
γ M1,E2	650.77 *3*	2.50 *5*
γ M1	673.25 *3*	0.969 *19*
γ (M1)	681.96 *3*	0.075 *17*
γ	709.9 *7*	~0.007
γ M1	727.28 *4*	0.146 *4*
γ E1	743.46 *4*	0.528 *12*
γ [E1]	768.81 *4*	0.010 *3*
γ [M1]	796.34 *14*	0.0029 *12*
γ E1	807.51 *4*	1.144 *25*
γ	844.89 *11*	0.025 *4*
γ [M1+E2]	860.39 *14*	0.0029 *12*
γ	921.21 *8*	0.017 *4*
γ [M1+E2]	929.06 *8*	0.014 *3*
γ M1	962.49 *3*	0.112 *4*
γ E1	1088.03 *3*	3.58 *7*
γ	1125.41 *11*	0.0112 *21*

† 0.96% uncert(syst)

Atomic Electrons (^{105}Ag)

$\langle$e$\rangle$=18.7 *6* keV

e_{bin}(keV)	$\langle$e$\rangle$(keV)	e(%)
3	2.43	75 *8*
4 - 14	0.216	5.9 *6*
17	0.50	2.9 *3*
18	1.75	9.8 *10*
20	0.44	2.17 *22*
21	0.59	2.9 *3*
23 - 39	0.101	0.43 *4*
40	5.29	13.3 *4*
49	0.012	~0.024
60	0.92	1.52 *4*
61 - 109	0.364	0.569 *23*

Atomic Electrons (^{105}Ag) (continued)

e_{bin}(keV)	$\langle$e$\rangle$(keV)	e(%)
112 - 159	0.384	0.260 *5*
163 - 213	0.0494	0.0274 *7*
216 - 246	0.0008	~0.00031
256	1.66	0.648 *19*
261 - 309	0.705	0.243 *4*
311 - 319	0.0309	0.00978 *23*
320	2.2	0.67 *14*
322 - 370	0.52	0.152 *19*
377 - 422	0.368	0.0883 *23*
434 - 483	0.0571	0.0130 *3*
484 - 527	0.00093	0.000183 *13*
536 - 585	0.0143	0.00264 *14*
594 - 641	0.130	0.0210 *6*
644 - 686	0.0269	0.00413 *11*
703 - 744	0.00546	0.000764 *16*
765 - 808	0.00598	0.000761 *20*
821 - 860	0.00027	3.3 *13* $\times10^{-5}$
897 - 938	0.00146	0.000157 *10*
959 - 962	0.000167	1.74 *6* $\times10^{-5}$
1064 - 1101	0.0141	0.00133 *3*
1122 - 1125	9.7 $\times10^{-6}$	9 *4* $\times10^{-7}$

Continuous Radiation (^{105}Ag)

$\langle\beta+\rangle$=0.00131 keV;$\langle$IB$\rangle$=0.19 keV

E_{bin}(keV)		$\langle\ \rangle$(keV)	(%)
0 - 10	β+	4.37×10^{-9}	5.3×10^{-8}
	IB	0.000169	
10 - 20	β+	1.38×10^{-7}	8.4×10^{-7}
	IB	0.0095	0.052
20 - 40	β+	2.96×10^{-6}	9.2×10^{-6}
	IB	0.021	0.092
40 - 100	β+	9.7×10^{-5}	0.000127
	IB	0.0025	0.0039
100 - 300	β+	0.00120	0.00066
	IB	0.025	0.0123
300 - 600	β+	7.3×10^{-6}	2.38×10^{-6}
	IB	0.080	0.018
600 - 1300	IB	0.047	0.0066
1300 - 1343	IB	5.9×10^{-7}	4.5×10^{-8}
	Σβ+		0.00080

$^{105}_{47}$Ag(7.23 *16* min)

Mode: IT(98.12 *3* %), ϵ(1.88 *3* %)

Δ: -87051 *9* keV

SpA: 2.47×10^8 Ci/g

Prod: daughter ^{105}Cd

Photons (^{105}Ag)

$\langle\gamma\rangle$=5.7 *4* keV

γ_{mode}	γ(keV)	γ(%)†
Ag L_ℓ	2.634	0.085 *16*
Ag L_η	2.806	0.051 *8*
Ag L_α	2.984	2.3 *4*
Ag L_β	3.206	1.7 *3*
Ag L_γ	3.521	0.148 *24*
Pd $K_{\alpha2}$	21.020	0.392 *14*
Pd $K_{\alpha1}$	21.177	0.74 *3*
Pd $K_{\beta1}$'	23.811	0.184 *7*
Pd $K_{\beta2}$'	24.445	0.0356 *14*
γ_{IT} E3	25.470 *11*	0.00412
γ_ϵM1	38.70 *4*	0.00110 *17*
γ_ϵM1	64.05 *3*	~0.00002
γ_ϵE2	216.21 *4*	0.00010 *4*

Photons (^{105}Ag) (continued)

γ_{mode}	γ(keV)	γ(%)†
γ_ϵM1+1.9%E2	280.52 3	0.037 12
γ_ϵM1+0.3%E2	306.28 5	0.185 20
γ_ϵM1+1.0%E2	319.22 3	0.90 10
γ_ϵ[M1+E2]	339.1 4	0.0031 6
γ_ϵE2	344.57 4	~0.00007
γ_ϵM1+35%E2	442.33 7	0.085 11
γ_ϵ[M1+E2]	475.1 4	~0.0004
γ_ϵ[M1+E2]	486.73 10	0.026 7
γ_ϵM1,E2	560.78 3	0.0042 12
γ_ϵ[M1+E2]	609.85 8	0.0161 19
γ_ϵ	629.9 4	0.0022 10
γ_ϵ[M1+E2]	648.54 8	~0.016
γ_ϵ	656.12 19	0.0036 14
γ_ϵ	781.4 4	0.0027 9
γ_ϵ	797 1	0.00090 18 ?
γ_ϵ[M1+E2]	929.06 8	0.052 15
γ_ϵ	1072.2 4	0.0042 13
γ_ϵ	1098.45 18	0.033 7

† uncert(syst): 1.2% for ϵ, 0.31% for IT

Atomic Electrons (^{105}Ag)

⟨e⟩=24.7 10 keV

e_{bin}(keV)	⟨e⟩(keV)	e(%)
3	1.52	45 5
4	0.88	25 3
14 - 21	0.054	0.294 18
22	17.2	78 4
23 - 24	0.00161	0.0069 6
25	5.0	19.9 11
35 - 39	0.00031	0.0009 4
192 - 216	1.8×10^{-5}	9 4 $\times 10^{-6}$
256 - 303	0.052	0.0179 16
306 - 339	0.0067	0.00213 19
418 - 462	0.0035	0.00082 9
483 - 487	0.000102	2.1 5 $\times 10^{-5}$
536 - 585	0.00039	6.7 8 $\times 10^{-5}$
606 - 653	0.00042	7 3 $\times 10^{-5}$
655 - 656	1.1×10^{-6}	$\sim 2 \times 10^{-7}$
757 - 797	3.7×10^{-5}	4.9 20 $\times 10^{-6}$
905 - 929	0.00060	6.6 17 $\times 10^{-5}$
1048 - 1095	0.00026	2.4 10 $\times 10^{-5}$
1098	5.5×10^{-6}	$\sim 5 \times 10^{-7}$

Continuous Radiation (^{105}Ag)

⟨β+⟩=0.00369 keV;⟨IB⟩=0.0064 keV

E_{bin}(keV)		⟨ ⟩(keV)	(%)
0 - 10	β+	1.65×10^{-8}	2.00×10^{-7}
	IB	3.3×10^{-6}	
10 - 20	β+	5.1×10^{-7}	3.13×10^{-6}
	IB	0.00018	0.00097
20 - 40	β+	1.07×10^{-5}	3.31×10^{-5}
	IB	0.00040	0.0017
40 - 100	β+	0.000317	0.000419
	IB	5.2×10^{-5}	7.9×10^{-5}
100 - 300	β+	0.00327	0.00181
	IB	0.00060	0.00029
300 - 600	β+	9.5×10^{-5}	3.05×10^{-5}
	IB	0.0023	0.00052
600 - 1300	IB	0.0028	0.00037
1300 - 1368	IB	1.27×10^{-6}	9.7×10^{-8}
	Σβ+		0.0023

$^{105}_{48}$Cd(55.5 4 min)

Mode: ϵ

Δ: -84339 10 keV

SpA: 3.226×10^{7} Ci/g

Prod: ^{106}Cd(n,2n); ^{106}Cd(γ,n); ^{102}Pd(α,n); ^{107}Ag(p,3n); ^{106}Cd(p,pn); protons on Sn

Photons (^{105}Cd)

⟨γ⟩=968 6 keV

γ_{mode}	γ(keV)	γ(%)†
Ag L_ℓ	2.634	0.16 3
Ag L_η	2.806	0.087 14
Ag L_α	2.984	4.4 8
Ag L_β	3.212	3.1 5
Ag L_γ	3.548	0.29 5
Ag $K_{\alpha2}$	21.990	15.7 6
Ag $K_{\alpha1}$	22.163	29.6 11
Ag $K_{\beta1}'$	24.934	7.4 3
γ E3	25.470 11	0.0040 3 •
Ag $K_{\beta2}'$	25.603	1.48 6
γ M1+21%E2	27.659 9	0.211 23
γ [M1+E2]	51.72 7	~0.019 ?
γ [M1+E2]	86.355 25	0.098 9
γ	107.6 3	~0.006
γ [M1+E2]	128.83 7	0.008 3 ?
γ	132.90 4	~0.019
γ	171.61 3	0.056 19
γ [M1+E2]	172.78 4	0.075 14
γ [M1+E2]	192.29 3	0.089 9
γ	221.75 12	0.033 5
γ [M1+E2]	229.929 25	0.070 9
γ	232.17 4	0.084 9
γ	249.32 6	~0.023
γ	249.50 3	0.047 19
γ	253.407 25	0.145 9
γ [M1+E2]	262.972 22	0.183 9
γ	283.27 3	0.155 9
γ [M1]	291.94 3	0.220 9
γ	295.53 5	~0.014
γ [E2]	307.596 25	0.056 23
γ [M1+E2]	307.829 20	0.84 6
γ	316.77 4	0.249 9
γ	324.98 20	0.023 9
γ [M1+E2]	340.639 18	0.39 4
γ [M1+E2]	340.87 3	0.07 3
γ	343.44 6	<0.028 ?
γ M1+1.1%E2	346.848 16	4.20 8
γ	353.90 12	0.042 9
γ	362.83 4	~0.019 ?
γ	371.27 10	0.042 9
γ	398.96 3	0.056 5
γ [M1+E2]	403.09 4	~0.014
γ [E1]	417.10 5	~0.014
γ [M1+E2]	422.22 3	0.089 5
γ E2	433.203 22	2.81 6
γ	444.23 8	<0.019 ?
γ [M1+E2]	454.15 5	0.108 9
γ	458.3 11	0.052 9
γ [M1+E2]	462.40 9	0.047 9
γ [M1+E2]	466.17 7	0.052 9
γ	486.84 4	0.066 9 ?
γ [M1+E2]	499.40 4	<0.06
γ	499.42 7	~0.038
γ	520.50 5	0.113 5
γ [M1+E2]	530.93 5	0.075 5
γ [E2]	538.61 5	0.708 23
γ [M1+E2]	544.85 6	0.014 5
γ	550.21 6	0.038 5 ?
γ [M1+E2]	558.07 6	0.038 5
γ [M1+E2]	570.567 24	0.108 9
γ [M1+E2]	575.87 4	~0.014 ?
γ	577.42 16	0.042 9
γ	580.03 4	0.056 5
γ [M1+E2]	583.15 4	0.084 5
γ M1+8.1%E2	590.37 4	0.117 5
γ [M1+E2]	598.50 4	0.117 9
γ	607.206 19	3.74 8
γ	609.42 4	0.108 9

Photons (^{105}Cd) (continued)

γ_{mode}	γ(keV)	γ(%)†
γ	613.5 4	0.24 7
γ [M1+E2]	613.80 3	~0.10
γ	617.46 6	0.08 3
γ	623.84 7	0.066 5
γ	630.84 6	0.038 14
γ	635.26 4	0.478 10
γ [E1]	640.434 22	0.061 9
γ	642.54 4	0.094 9
γ [M1+E2]	648.467 17	1.57 3
γ	656.50 3	0.070 5
γ [M1+E2]	658.36 4	0.060 5
γ [M1+E2]	662.73 4	0.080 9
γ [E2]	676.72 3	0.047 5
γ	682.23 3	0.033 9
γ [M1+E2]	691.70 3	0.042 9 ?
γ	695.78 5	0.075 9
γ [M1+E2]	697.54 3	0.089 5
γ	700.36 7	0.042 5
γ	703.47 4	0.300 9
γ	709.84 6	0.127 9
γ [M1+E2]	714.92 6	~0.019 ?
γ [M1+E2]	721.39 6	~0.019 ?
γ	727.53 13	0.056 9
γ E2	733.02 9	0.113 9
γ [E2]	739.19 4	~0.019
γ [M1+E2]	746.47 3	0.535 11
γ	749.65 7	0.038 9
γ	756.42 10	0.028 9
γ [E1]	757.97 5	0.084 9
γ [M1+E2]	762.86 3	~0.019
γ	770.11 5	0.056 9
γ [M1+E2]	775.44 3	0.192 9
γ	782.18 7	~0.019
γ	788.54 8	0.028 9
γ	800.37 8	0.047 9
γ	810.18 6	0.122 5
γ [M1+E2]	813.77 7	0.038 9 ?
γ [M1+E2]	825.80 4	0.122 14
γ [M1+E2]	828.43 4	<0.019 ?
γ	836.22 9	0.023 9
γ [M1+E2]	842.35 9	~0.033
γ [M1+E2]	842.74 7	0.066 19
γ	858.92 7	0.12 3
γ [M1+E2]	866.77 7	0.047 5
γ [M1+E2]	871.32 8	0.047 14
γ [M1+E2]	877.78 5	0.216 9
γ [M1+E2]	884.57 5	0.591 12
γ [E2]	889.16 4	0.253 9
γ	892.19 8	0.18 3
γ	896.53 6	0.084 9
γ [M1+E2]	921.62 3	0.478 10
γ [M1+E2]	928.57 5	~0.033
γ [E2]	934.151 17	1.271 25
γ	941.53 5	0.061 9
γ [E1]	948.028 22	0.854 17
γ [M1+E2]	954.67 3	0.028 9
γ [M1+E2]	961.810 16	4.69 9
γ	967.22 6	0.122 9
γ	972.43 5	0.056 9
γ	977.73 8	0.056 9
γ	981.6 1	0.136 14
γ [M1+E2]	984.48 7	0.070 19
γ	986.67 6	0.164 14
γ	992.66 8	0.042 9
γ [M1+E2]	998.09 3	0.15 4
γ [E1]	998.10 3	~0.14 ?
γ [M1+E2]	999.00 4	0.09 3
γ	1006.16 6	0.075 14
γ [M1+E2]	1013.33 6	0.103 14
γ	1021.48 20	0.033 14
γ [E2]	1031.13 3	~0.05
γ	1031.49 7	~0.07
γ	1032.92 7	0.070 19
γ [M1+E2]	1038.41 3	0.586 12
γ	1039.39 3	<0.028
γ	1042.62 5	0.21 8
γ [M1+E2]	1043.99 3	0.28 7
γ	1060.99 16	~0.019
γ [M1+E2]	1071.65 3	1.28 3
γ	1082.63 6	0.061 14
γ	1091.24 7	0.028 9
γ	1095.12 9	~0.014
γ [M1+E2]	1105.71 6	0.028 5
γ	1108.66 16	<0.028 ?

Photons (^{105}Cd)
(continued)

γ_{mode}	γ(keV)	γ(%)†
γ	1109.07 8	~0.038
γ	1119.32 16	0.061 9
γ [E1]	1124.64 3	0.094 9
γ	1137.19 20	0.038 9
γ	1145.01 6	0.014 5
γ	1148.37 10	~0.009
γ	1159.74 16	~0.033
γ [M1+E2]	1169.07 4	0.113 9
γ	1196.3 3	0.023 5
γ	1205.28 17	~0.019
γ [E1]	1210.998 23	0.197 14
γ	1217.47 7	0.019 5 ?
γ [E2]	1228.88 4	0.244 19
γ [E1]	1232.92 6	0.066 9
γ [E1]	1239.97 3	0.286 9
γ	1256.5 10	~0.08
γ [M1+E2]	1262.26 3	0.038 9
γ [E2]	1274.787 16	0.821 16
γ	1283.38 5	0.023 9
γ [E2]	1289.34 15	0.028 9
γ [E1]	1294.873 21	0.310 9
γ [E1]	1302.437 19	~0.09
γ [M1+E2]	1302.446 15	3.98 8
γ	1317.40 11	0.075 9
γ	1322.04 8	~0.06
γ	1322.66 3	0.11 3
γ	1327.31 6	0.047 9
γ [M1+E2]	1338.72 3	0.652 19
γ	1340.49 5	0.310 19
γ	1343.90 4	0.089 9
γ	1350.11 16	~0.019
γ	1360.77 3	0.56 7
γ [E1]	1360.89 7	0.12 5
γ	1375.80 6	0.113 9
γ [E2]	1388.30 5	2.70 5
γ [E1]	1403.29 3	0.389 14
γ [M1+E2]	1413.31 6	0.14 3
γ [M1+E2]	1415.96 5	1.45 12
γ	1416.01 10	0.19 7
γ	1422.57 7	0.028 9
γ	1431.88 7	0.042 9
γ	1459.95 16	0.084 14
γ	1465.1 4	~0.014
γ	1469.1 6	<0.019
γ	1485.64 6	0.169 9
γ [E1]	1489.71 3	0.446 9
γ	1507.88 9	0.023 5 ?
γ [E1]	1522.80 8	~0.019
γ [E2]	1532.372 22	0.066 9
γ [E1]	1553.08 4	0.033 9
γ [E1]	1557.842 21	2.05 4
γ [E2]	1582.60 5	0.633 13
γ [E1]	1586.81 3	0.206 9
γ [E2]	1610.271 20	0.263 14
γ [E1]	1635.741 22	1.046 21
γ	1644.036 24	0.872 17
γ	1665.27 4	0.95 8
γ	1665.67 3	0.35 7
γ	1693.33 3	3.54 7
γ [E2]	1697.00 3	0.117 14
γ [M1+E2]	1724.66 3	0.694 14
γ [M1+E2]	1740.95 7	0.033 9
γ	1797.5 4	0.023 9
γ [E1]	1809.50 4	0.023 5
γ [E1]	1823.28 6	0.15 3
γ	1831.66 14	0.155 9 ?
γ	1853.6 8	<0.009
γ	1860.18 7	0.038 9
γ	1867.14 6	0.047 9
γ [M1+E2]	1869.78 3	0.638 14
γ [E1]	1875.00 6	0.089 9
γ	1881.38 6	0.131 9
γ [E1]	1892.80 3	0.708 14
γ	1894.56 5	~0.05
γ [M1+E2]	1897.44 3	1.44 3
γ [E1]	1900.08 3	0.117 9
γ [E1]	1902.69 3	0.131 9
γ [E1]	1909.63 5	0.610 14
γ [E1]	1929.09 20	0.047 9
γ [E2]	1933.14 4	1.59 3
γ	1938.29 17	0.295 9
γ	1953.49 5	0.061 5
γ [M1+E2]	1960.80 4	0.89 5
γ [E1]	1961.35 6	~0.05

Photons (^{105}Cd)
(continued)

γ_{mode}	γ(keV)	γ(%)†
γ	1975.65 10	0.24 3
γ [E1]	1986.43 3	0.755 15
γ [E1]	1995.95 8	0.136 14
γ	2014.02 16	0.033 9
γ [M1+E2]	2028.46 6	0.633 19
γ [E1]	2053.73 6	0.159 9
γ [M1+E2]	2056.12 6	0.244 9
γ	2061.69 16	0.023 5
γ	2076.18 9	0.038 5
γ	2095.2 6	~0.009
γ	2117.29 8	0.089 9
γ [E1]	2156.34 4	0.375 14
γ [E2]	2203.35 5	<0.038 ?
γ	2203.65 8	0.09 4
γ [M1+E2]	2231.01 5	0.197 9
γ [E1]	2249.53 3	0.492 19
γ [E2]	2272.86 3	1.03 6
γ	2274.87 5	0.84 6 ?
γ	2289.10 6	0.030 3
γ	2300.34 5	0.516 19 ?
γ [E1]	2308.19 6	0.103 5
γ	2318.36 17	0.049 4
γ [E1]	2333.27 3	1.98 7
γ	2346.02 17	~0.0038 ?
γ	2364.6 13	0.049 3
γ [E2]	2375.10 6	0.0075 23
γ	2382.64 12	0.048 3
γ	2393.68 7	0.178 5 ?
γ [E1]	2400.57 6	0.044 3
γ	2423.02 9	0.352 14
γ [E1]	2429.14 8	0.06 3
γ	2447.44 6	<0.00047 ?
γ	2469.41 16	0.0061 14
γ	2512.1 5	0.0033 14
γ	2525.02 8	0.078 3
γ [E2]	2531.07 15	0.0061 14
γ	2554.3 4	0.0056 14
γ [M1+E2]	2558.73 15	0.0150 14
γ	2568.5 8	~0.0023
γ	2573.79 20	0.0150 14
γ	2594.5 5	0.0033 9
γ	2660.4 6	~0.0019

† 6.6% uncert(syst)
* with ^{105}Ag(7.2 min) in equilib

Atomic Electrons (^{105}Cd)
$\langle e \rangle$=31.0 15 keV

e_{bin}(keV)	$\langle e \rangle$(keV)	e(%)
2	0.073	3.4 4
3	2.6	78 9
4	1.61	45 5
18	0.58	3.2 3
19	0.83	4.4 5
21	0.32	1.51 16
22	17.1	78 6
24	1.3	5.3 12
25	4.9	19.6 16
26 - 61	0.46	1.5 3
82 - 130	0.043	~0.049
132 - 173	0.041	0.026 7
188 - 237	0.052	0.024 4
246 - 295	0.094	0.034 5
296 - 343	0.245	0.0761 24
346 - 395	0.0095	0.0027 4
396 - 445	0.124	0.0300 7
450 - 499	0.0060	0.0013 4
505 - 555	0.0270	0.00519 21
557 - 606	0.074	0.013 4
607 - 656	0.046	0.0074 7
657 - 706	0.0112	0.0016 3
708 - 757	0.0161	0.00221 12
758 - 807	0.0048	0.00061 9
809 - 858	0.0067	0.00080 10
859 - 903	0.0197	0.00226 14
909 - 958	0.078	0.0083 5
959 - 1007	0.0106	0.00109 15

Atomic Electrons (^{105}Cd)
(continued)

e_{bin}(keV)	$\langle e \rangle$(keV)	e(%)
1010 - 1058	0.0239	0.00231 17
1060 - 1109	0.0035	0.00033 3
1112 - 1160	0.0019	0.00016 3
1165 - 1214	0.00424	0.000353 18
1217 - 1264	0.0075	0.00060 4
1269 - 1318	0.044	0.00345 24
1319 - 1363	0.0230	0.00169 13
1372 - 1419	0.0177	0.00127 12
1422 - 1469	0.0028	0.00019 3
1482 - 1579	0.0122	0.000791 22
1582 - 1671	0.032	0.0019 4
1690 - 1784	0.0070	0.00041 5
1794 - 1892	0.0178	0.00096 4
1894 - 1993	0.0204	0.00106 4
1995 - 2095	0.0058	0.000288 14
2113 - 2205	0.0024	0.000109 12
2224 - 2321	0.0190	0.00084 6
2329 - 2429	0.0032	0.000135 17
2444 - 2543	0.00042	1.7 3 $\times 10^{-5}$
2548 - 2635	8.1 $\times 10^{-5}$	3.2 6 $\times 10^{-6}$
2657 - 2660	7.8 $\times 10^{-7}$	~3 $\times 10^{-8}$

Continuous Radiation (^{105}Cd)
$\langle \beta + \rangle$=204 keV;$\langle$IB$\rangle$=1.68 keV

E_{bin}(keV)		$\langle\ \rangle$(keV)	(%)
0 - 10	β+	4.14 $\times 10^{-6}$	5.01 $\times 10^{-5}$
	IB	0.0085	
10 - 20	β+	0.000138	0.00084
	IB	0.0132	0.084
20 - 40	β+	0.00318	0.0098
	IB	0.036	0.140
40 - 100	β+	0.125	0.162
	IB	0.046	0.071
100 - 300	β+	5.09	2.33
	IB	0.135	0.075
300 - 600	β+	33.0	7.2
	IB	0.20	0.046
600 - 1300	β+	143	15.9
	IB	0.58	0.063
1300 - 2500	β+	22.0	1.57
	IB	0.66	0.039
2500 - 2712	IB	0.0039	0.000154
	$\Sigma\beta$+		27

$^{105}_{49}$In(4.9 2 min)

Mode: ϵ
Δ: -79589 18 keV
SpA: 3.65×10^{8} Ci/g
Prod: ^{106}Cd(p,2n); ^{106}Cd(^{3}He,p3n); ^{107}Ag(^{3}He,5n); ^{90}Zr(^{19}F,4n)

Photons (^{105}In)

γ_{mode}	γ(keV)	γ(rel)
γ M1+0.3%E2	131.40 21	100 15
γ M1+1.4%E2	166.4 3	~1 ?
γ M1(+E2)	196.2 3	15.0 23
γ E2+12%M1	228.2 3	3.5 5
γ (M1+2.9%E2)	260.47 24	37 6
γ [M1+E2]	473.3 3	1.9 3
γ E2	510.7 4	~5
γ	570.3 5	2.5 4
γ M1+43%E2	604.71 24	20 3
γ E2+5.2%M1	639.7 3	12.0 18
γ E2	668.9 3	18 3

Photons (^{105}In) (continued)

γ_{mode}	γ(keV)	γ(rel)
γ E2+21%M1	701.5 3	2.4 4
γ E2	771.1 3	3.6 5
γ	799.9 5	0.50 8
γ	808.5 5	2.0 3
γ E2	832.9 3	15.0 23
γ	836.0 6	
γ	854.4 4	2.6 4
γ	879.2 4	1.00 15 ?
γ	943.4 4	1.60 24
γ	967.4 10	<0.9 ?
γ	1098.2 10	1.40 21 ?
γ	1114.9 4	2.0 3
γ	1126.8 4	4.0 6
γ	1139.7 3	1.8 3
γ	1191.1 4	3.0 5
γ	1255.9 4	5.0 8
γ	1309.3 6	<2
γ	1348.3 5	2.3 4
γ	1387.3 3	<15
γ	1608.8 5	0.7 1
γ	1762.1 10	0.90 13
γ	1877.4 15	2.4 4

$^{105}_{49}$In(43 4 s)

Mode: IT
Δ: -78915 18 keV
SpA: 2.48×10^9 Ci/g
Prod: ^{90}Zr(^{19}F,4n); ^{106}Cd(p,2n)

Photons (^{105}In)

$\langle\gamma\rangle$=636 64 keV

γ_{mode}	γ(keV)	γ(%)
In L_ℓ	2.905	0.0063 13
In L_η	3.112	0.0030 6
In L_α	3.286	0.17 3
In L_β	3.569	0.127 25
In L_γ	3.983	0.013 3
In $K_{\alpha2}$	24.002	1.20 13
In $K_{\alpha1}$	24.210	2.25 25
In $K_{\beta1}$'	27.265	0.58 7
In $K_{\beta2}$'	28.022	0.121 14
γ [M4]	674.1 3	94

Atomic Electrons (^{105}In)

$\langle e\rangle$=38 3 keV

e_{bin}(keV)	$\langle e\rangle$(keV)	e(%)
4	0.150	3.9 6
19	0.0105	0.054 8
20	0.088	0.44 6
23	0.023	0.100 15
24	0.025	0.104 15
26	0.00143	0.0054 8
27	0.0033	0.0124 18
646	31.3	4.8 5
670	5.1	0.76 8
673	1.02	0.152 16
674	0.221	0.033 4

$^{105}_{50}$Sn(31 6 s)

Mode: ε, εp
Δ: -73270 80 keV
SpA: 3.4×10^9 Ci/g
Prod: ^{50}Cr(^{58}Ni,2pn); ^{54}Fe(^{58}Ni,4p3n)

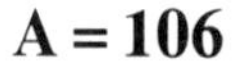

NDS 30, 305 (1980)

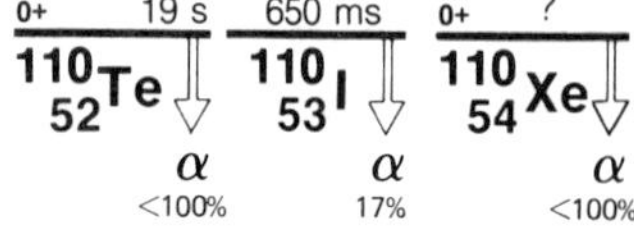

Q$_\alpha$ 2723 16 Q$_\alpha$ 3570 50 Q$_\alpha$ 3880 30

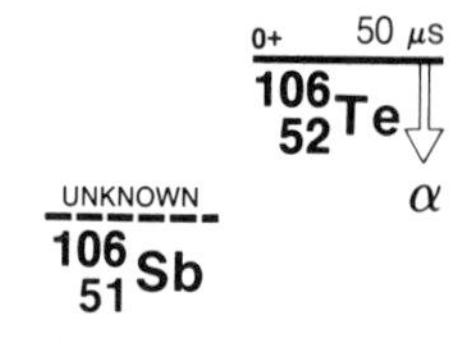

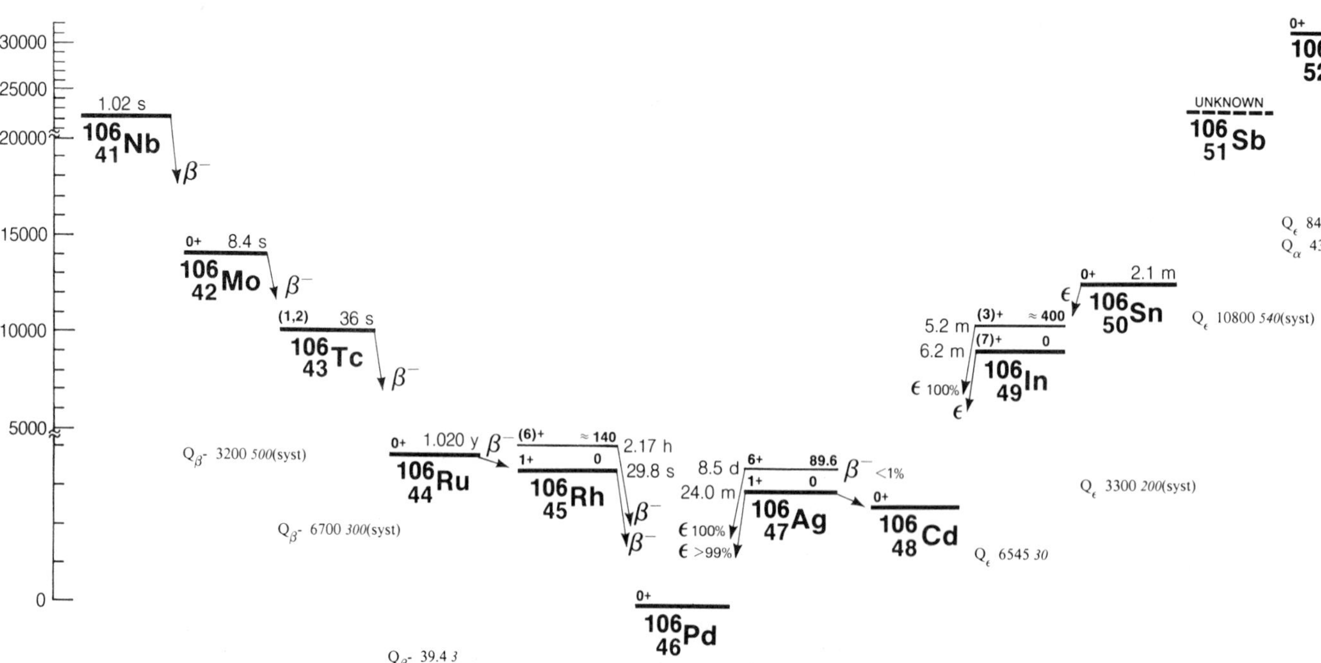

$^{106}_{41}$Nb(1.02 *5* s)

Mode: β-
SpA: 7.6×10^{10} Ci/g
Prod: fission

Photons (^{106}Nb)

⟨γ⟩=1001 *91* keV

γ_{mode}	γ(keV)	γ(%)†
γ [E2]	171.77 *12*	90.0 *18*
γ [E2]	351.0 *4*	35.1 *18*
γ [E2]	511.5 *5*	~5 ?
γ	539.2 *3*	14 *3*
γ	546.0 *4*	~5
γ	550.2 *4*	8 *3*
γ [E2]	711.0 *3*	10 *4*
γ [M1+E2]	714.3 *4*	27 *5*
γ	725.2 *4*	15 *5*
γ [E2]	785.2 *5*	~5
γ	897.0 *4*	~8
γ	1108.5 *5*	~5

† 4.4% uncert(syst)

$^{106}_{42}$Mo(8.4 *5* s)

Mode: β-
Δ: -76430 *400* keV syst
SpA: 1.22×10^{10} Ci/g
Prod: fission

Photons (^{106}Mo)

γ_{mode}	γ(keV)
γ	54.9 *5* ?
γ	85.9 *5*
γ	189.1 *5* ?
γ	249.6 *5*
γ	309.4 *5*
γ	315.3 *5*
γ	326.5 *5*
γ	378.9 *5*
γ	429.6 *2*
γ	465.8 *2* ?
γ	504.3 *2*
γ	595.3 *2*
γ	618.5 *2*
γ	634.4 *2*
γ	672.5 *5*

$^{106}_{43}$Tc(36 *1* s)

Mode: β-
Δ: -79630 *300* keV syst
SpA: 2.93×10^{9} Ci/g
Prod: fission

Photons (^{106}Tc)

γ_{mode}	γ(keV)	γ(rel)
γ [E2]	270.07 *3*	100
γ [M1+E2]	299.25 *6*	0.3 *1*
γ	353.80 *9*	0.7 *2*
γ [M1+E2]	376.87 *10*	~0.2
γ [E2]	401.59 *7*	~0.2
γ [E2]	444.63 *9*	1.2 *3*
γ [E2+M1]	522.25 *4*	13.4 *9*
γ [E2]	677.52 *8*	0.7 *1*
γ	682.82 *8*	0.7 *1*
γ [E2]	720.56 *4*	6.9 *7*
γ [E2]	792.31 *4*	9.3 *7*
γ [M1+E2]	821.49 *6*	1.8 *2*
γ	896.11 *20*	1.1 *2*
γ [M1+E2]	1122.15 *6*	3.0 *3*
γ	1240.61 *10*	0.15 *3*
γ	1248.78 *6*	1.0 *2*
γ [E2]	1392.21 *6*	0.7 *1*
γ	1478.52 *8*	0.8 *1*
γ	1504.31 *7*	2.1 *2*
γ	1589.74 *10*	0.5 *2*
γ	1615.54 *8*	3.0 *3*
γ	1643.11 *20*	1.3 *2*
γ	1667.31 *10*	0.5 *2*
γ	1710.80 *7*	1.0 *1*
γ	1840.50 *8*	1.7 *2*
γ	1969.33 *6*	15.9 *17*
γ	2068.90 *10*	0.4 *1*
γ	2153.61 *14*	0.6 *1*
γ	2239.40 *6*	24.4 *21*
γ	2267.21 *9*	2.5 *5*
γ	2362.74 *9*	0.4 *1*
γ	2431.35 *7*	4.5 *5*
γ	2571.80 *9*	2.0 *2*
γ	2701.41 *7*	10 *1*
γ	2758.64 *15*	1.7 *2*
γ	2777.03 *14*	4.3 *3*
γ	2789.44 *9*	14.1 *11*
γ	2916.33 *14*	5.8 *5*
γ	2945.91 *14*	6.1 *5*
γ	2989.33 *14*	1.1 *1*
γ	3031.51 *20*	0.9 *1*
γ	3047.09 *14*	5.0 *5*
γ	3059.50 *9*	1.1 *1*
γ	3094.03 *9*	1.1 *1*
γ	3186.39 *14*	9.0 *9*
γ	3259.39 *14*	4.9 *5*
γ	3280.87 *16*	0.9 *1*
γ	3364.09 *9*	1.8 *2*
γ	3550.93 *16*	2.2 *2*
γ	3660.3 *3*	0.7 *1*
γ	3930.3 *3*	0.6 *1*

$^{106}_{44}$Ru(1.020 *3* yr)

Mode: β-
Δ: -86330 *10* keV
SpA: 3306 Ci/g
Prod: fission

Continuous Radiation (^{106}Ru)

⟨β-⟩=10.0 keV;⟨IB⟩=0.00039 keV

E_{bin}(keV)		⟨ ⟩(keV)	(%)
0 - 10	β-	2.61	57
	IB	0.00032	
10 - 20	β-	4.31	30.0
	IB	6.4×10^{-5}	0.00049
20 - 39	β-	3.12	12.5
	IB	7.0×10^{-6}	3.1×10^{-5}

$^{106}_{45}$Rh(29.80 *8* s)

Mode: β-
Δ: -86370 *10* keV
SpA: 3.530×10^{9} Ci/g
Prod: daughter ^{106}Ru

Photons (^{106}Rh)

⟨γ⟩=206 *5* keV

γ_{mode}	γ(keV)	γ(%)†
γ E2(+M1)	328.49 *5*	0.00020 *5*
γ [E2]	333.00 *7*	0.0050 *21*
γ [E2]	428.45 *6*	0.072 *6*
γ E2+1.8%M1	429.668 *20*	0.0023 *6*
γ [M1+E2]	434.18 *5*	0.019 *4*
γ [E2]	439.29 *7*	0.010 *3*
γ E2	511.865 *3*	20.7
γ	552.4 *6*	~0.00033
γ	569.4 *6*	~0.00037
γ E2	578.38 *6*	0.0094 *6*
γ E2+1.3%M1	616.181 *21*	0.738 *23*
γ E2	621.92 *4*	9.8 *5*
γ [M1+E2]	680.31 *8*	0.0092 *12*
γ [M1+E2]	684.82 *7*	0.0050 *8*
γ [E2]	715.74 *9*	0.0099 *16*
γ E2	717.36 *5*	0.0068 *14*
γ [M1+E2]	751.09 *6*	0.0009 *3*
γ E2	873.47 *6*	0.43 *3*
γ [E1]	938.25 *10*	0.00054 *21*
γ E2+4.7%M1	1045.85 *3*	0.0052 *13*
γ M1+5.5%E2	1050.36 *5*	1.53 *6*
γ E2	1062.25 *8*	0.0296 *8*
γ [E2]	1108.75 *7*	0.0077 *10*
γ E2+30%M1	1114.48 *7*	0.0113 *11*
γ E2	1128.042 *21*	0.398 *8*
γ E2	1149.92 *8*	0.00273 *17*
γ M1+1.7%E2	1180.76 *6*	0.0147 *8*
γ E2	1194.56 *6*	0.0549 *20*
γ [E2]	1209.93 *10*	0.00031 *8* ?
γ	1258.95 *14*	0.00075 *14*
γ [E2]	1266.16 *15*	0.00120 *17*
γ [E2]	1305.37 *10*	0.00106 *12*
γ [M1+E2]	1311.11 *10*	0.00077 *14*
γ [E2]	1315.71 *16*	0.00335 *25*
γ [M1+E2]	1355.73 *13*	0.00035 *12*
γ [M1+E2]	1360.24 *13*	0.00159 *12*
γ [E1]	1372.42 *10*	0.00188 *14*
γ	1397.60 *8*	0.00252 *20*
γ [E2]	1489.64 *6*	0.0031 *7*
γ E2	1496.43 *9*	0.0269 *17*
γ [M1+E2]	1498.77 *18*	
γ [E2]	1562.22 *5*	0.158 *6*
γ [M1]	1571.46 *9*	0.00173 *18*
γ [M1+E2]	1577.19 *9*	0.00087 *17*
γ	1687.39 *14*	~0.00021 ?
γ	1693.12 *14*	0.00072 *21*
γ [M1+E2]	1730.66 *7*	0.00199 *17*
γ E2	1766.09 *8*	0.0277 *21*
γ [M1+E2]	1774.46 *18*	0.00093 *8*
γ [E2]	1784.17 *13*	0.00011 *3* ?
γ M1+3.3%E2	1796.93 *5*	0.0247 *12*
γ	1909.2 *3*	~0.00025
γ	1909.45 *8*	0.00101 *16*
γ M1+1.7%E2	1927.28 *10*	0.0136 *8*
γ	1955.0 *3*	0.00019 *4* ?
γ [E1]	1973.0 *3*	0.00017 *4* ?
γ [E1+0.5%M2]	1988.59 *9*	0.0244 *14*
γ [E2]	2093.28 *24*	0.00027 *6*
γ E2	2112.60 *9*	0.0350 *19*
γ [M1+E2]	2114.94 *18*	
γ M1+3.1%E2	2193.37 *8*	0.0048 *3*
γ [E2]	2242.51 *7*	0.00173 *11*
γ	2271.8 *3*	0.00108 *8*
γ [E2]	2308.79 *5*	0.0023 *3*
γ	2309.29 *14*	0.0025 *5*
γ E2	2316.51 *15*	0.0059 *3*
γ E2	2366.06 *15*	0.0224 *12*
γ (M1+0.7%E2)	2390.63 *18*	0.0062 *4*
γ M1+0.7%E2	2406.07 *13*	0.0141 *8*
γ [E2]	2439.14 *10*	0.00449 *23*

Photons (^{106}Rh) (continued)

γ_{mode}	γ(keV)	γ(%)†
γ	2455.9 15	0.00021 4
γ [E1]	2484.8 3	0.00075 8
γ	2500.45 9	5.4 14 ×10⁻⁵
γ	2525.4 3	~6 ×10⁻⁵
γ (M1+0.5%E2)	2542.93 21	0.00288 17
γ	2555.0 11	~4 ×10⁻⁵ ?
γ	2571.2 3	0.00137 8
γ [E2]	2626.80 18	8.5 21 ×10⁻⁵
γ	2651.7 3	0.00066 8
γ [M1]	2705.22 8	0.00219 24
γ E2	2709.45 24	0.0038 4
γ	2740.5 4	0.00021 4
γ	2788.5 5	8.1 21 ×10⁻⁵
γ E2	2809.2 3	0.00060 4
γ	2821.14 14	0.00118 8
γ	2865 1	~1 ×10⁻⁵
γ [E2]	2902.49 18	5.8 21 ×10⁻⁵
γ [E2]	2917.92 13	0.00095 10
γ	3037.3 3	0.00110 12
γ [M1]	3054.78 21	0.00029 4
γ	3165.4 13	6 3 ×10⁻⁶
γ	3249.8 5	8.1 17 ×10⁻⁵
γ	3273.3 5	4.1 14 ×10⁻⁵
γ	3300.5 11	8.7 21 ×10⁻⁶
γ	3375.9 14	1.14 21 ×10⁻⁵
γ	3401.8 9	1.26 19 ×10⁻⁵

† 2.9% uncert(syst)

Continuous Radiation (^{106}Rh)

⟨β-⟩=1411 keV;⟨IB⟩=4.0 keV

E_{bin}(keV)		⟨ ⟩(keV)	(%)
0 - 10	β-	0.0067	0.134
	IB	0.047	
10 - 20	β-	0.0207	0.138
	IB	0.046	0.32
20 - 40	β-	0.086	0.286
	IB	0.091	0.32
40 - 100	β-	0.67	0.95
	IB	0.26	0.40
100 - 300	β-	8.8	4.27
	IB	0.76	0.42
300 - 600	β-	43.4	9.5
	IB	0.88	0.21
600 - 1300	β-	301	31.2
	IB	1.24	0.143
1300 - 2500	β-	822	45.1
	IB	0.65	0.039
2500 - 3540	β-	235	8.4
	IB	0.029	0.00110

$^{106}_{45}$Rh(2.17 *3* h)

Mode: β-
Δ: ~-86230 keV
SpA: 1.364×10⁷ Ci/g
Prod: ^{108}Pd(d,α); ^{109}Ag(n,α); ^{106}Pd(n,p); protons on Cd; ^{106}Pd(d,2p); ^{107}Ag(n,2p)

Photons (^{106}Rh)

⟨γ⟩=2882 59 keV

γ_{mode}	γ(keV)	γ(%)†
Pd $K_{\alpha2}$	21.020	0.333 16
Pd $K_{\alpha1}$	21.177	0.63 3
Pd $K_{\beta1}$'	23.811	0.156 8
Pd $K_{\beta2}$'	24.445	0.0302 15
γ [E1]	178.3 4	0.048 16
γ M1(+E2)	195.01 4	0.60 6
γ M1+1.7%E2	221.721 15	6.50 12
γ E1	228.652 21	2.07 7
γ	319.5 4	0.87 17
γ E2(+M1)	328.49 5	1.18 6
γ M1(+E2)	374.62 5	0.28 4
γ E2+~0.4%M1	391.046 25	3.52 8
γ E2+9.0%M1	406.193 23	11.97 22
γ [M1+E2]	418.57 5	0.53 12
γ E2+1.8%M1	429.668 20	13.53 24
γ [M1+E2]	433.72 5	0.17 5
γ E1	451.004 20	24.9 5
γ	473.1 4	~0.9
γ [E1]	474.08 3	0.84 10
γ E2	511.865 3	87
γ [E1]	522.15 6	0.090 24
γ M1,E2	586.06 4	0.86 8
γ E2+11%M1	601.20 4	3.05 8
γ E2+1.3%M1	616.181 21	20.6 5
γ E1	646.01 4	2.78 8
γ M1,E2	679.66 3	0.58 7
γ M1,E2	680.439 10	1.85 20
γ E2+16%M1	703.11 5	4.68 20
γ E2	717.36 5	29.4 6
γ E1	748.38 4	19.8 4
γ E2+2.5%M1	793.19 4	5.77 11
γ E2	804.29 5	13.2 3
γ M1+40%E2	808.34 4	7.57 15
γ E2+3.0%M1	824.76 5	13.86 24
γ E2	847.43 5	3.7 4 *
γ E2	848.22 6	
γ [M1+E2]	875.45 4	0.63 9
γ [E2]	949.40 4	0.24 11
γ [E1]	956.33 4	0.49 9
γ M1,E2	1019.77 6	2.00 8
γ E2+4.7%M1	1045.85 3	30.8 6
γ [E1]	1076.87 5	0.051 19
γ [M1+E2]	1121.68 5	0.63 7
γ E2	1128.042 21	11.1 4
γ [M1+E2]	1136.82 5	0.43 6
γ	1178.05 4	0.19 4
γ E2	1199.38 4	10.8 5
γ E2	1222.86 4	8.0 4
γ [E2]	1394.39 5	2.85 21
γ [E2]	1420.46 5	~0.037
γ E2+16%M1	1527.87 5	17.4 11
γ [E2]	1565.57 5	0.61 25
γ E1	1572.50 4	6.8 4
γ E2+26%M1	1722.87 6	2.50 25
γ	1794.22 4	0.037 16
γ E2	1839.03 4	2.12 24
γ	2084.36 4	0.024 11

† 11% uncert(syst)
* combined intensity for doublet

Atomic Electrons (^{106}Rh)

⟨e⟩=7.4 3 keV

e_{bin}(keV)	⟨e⟩(keV)	e(%)
3 - 24	0.083	1.38 12
154 - 195	0.10	0.059 17
197	0.498	0.252 7
204 - 229	0.145	0.0680 13
295 - 328	0.108	0.036 8
350 - 375	0.153	0.0418 13
382	0.435	0.114 3
387 - 403	0.102	0.0255 12
405	0.445	0.110 3
406 - 426	0.0828	0.0197 6
427	0.228	0.0534 15
429 - 474	0.075	0.017 3

Atomic Electrons (^{106}Rh) (continued)

e_{bin}(keV)	⟨e⟩(keV)	e(%)
488	2.06	0.42 5
498	0.0007	~0.00014
508	0.23	0.046 6
509 - 522	0.104	0.0203 16
562 - 586	0.0736	0.0128 4
592	0.355	0.0600 19
598 - 646	0.0829	0.0135 3
655 - 680	0.109	0.0162 10
693	0.399	0.0575 16
700 - 748	0.183	0.0253 4
769 - 793	0.321	0.0412 8
800	0.154	0.0193 5
801 - 848	0.105	0.0128 16
851 - 875	0.008	0.0009 4
925 - 956	0.0047	0.00051 13
995 - 1020	0.0206	0.00207 17
1021	0.253	0.0248 7
1042 - 1077	0.0375	0.00359 7
1097 - 1137	0.105	0.0095 3
1154 - 1199	0.143	0.0121 4
1219 - 1223	0.0080	0.000656 25
1370 - 1417	0.0198	0.00144 11
1420 - 1504	0.097	0.0065 4
1524 - 1572	0.0376	0.00244 12
1699 - 1794	0.0144	0.00085 8
1815 - 1838	0.0112	0.00062 6
2060 - 2084	9.0 ×10⁻⁵	~4 ×10⁻⁶

Continuous Radiation (^{106}Rh)

⟨β-⟩=307 keV;⟨IB⟩=0.29 keV

E_{bin}(keV)		⟨ ⟩(keV)	(%)
0 - 10	β-	0.081	1.62
	IB	0.0147	
10 - 20	β-	0.246	1.64
	IB	0.0140	0.097
20 - 40	β-	1.00	3.33
	IB	0.026	0.091
40 - 100	β-	7.3	10.4
	IB	0.065	0.101
100 - 300	β-	71	35.7
	IB	0.119	0.071
300 - 600	β-	166	38.5
	IB	0.048	0.0124
600 - 923	β-	61	8.9
	IB	0.0031	0.00047

$^{106}_{46}$Pd(stable)

Δ: -89910 5 keV
%: 27.33 5

$^{106}_{47}$Ag(24.0 *1* min)

Mode: ϵ(>99 %), β-(<1 %)
Δ: -86944 6 keV
SpA: 7.39×10⁷ Ci/g
Prod: ^{103}Rh(α,n); ^{107}Ag(n,2n); ^{107}Ag(d,t)

Photons (^{106}Ag)

⟨γ⟩=100 9 keV

γ_{mode}	γ(keV)	γ(%)†
Pd L_ℓ	2.503	0.038 7
Pd L_η	2.660	0.018 3
Pd L_α	2.838	1.04 16
Pd L_β	3.053	0.67 11
Pd L_γ	3.383	0.066 12
Pd $K_{\alpha2}$	21.020	8.4 3
Pd $K_{\alpha1}$	21.177	15.8 6
Pd $K_{\beta1}$'	23.811	3.93 14
Pd $K_{\beta2}$'	24.445	0.76 3
γ[E2]	333.00 7	0.00053 22
γ[E2]	428.45 6	0.0077 6
γ[M1+E2]	434.18 5	0.0020 4
γ[E2]	439.29 7	0.0044 14
γE2	511.865 3	16.7
γE2	578.38 6	0.0066 5
γE2+1.3%M1	616.181 21	0.135 6
γE2	621.92 4	0.309 7
γ[M1+E2]	680.31 8	0.0049 7
γ[M1+E2]	684.82 7	0.0026 5
γ[E2]	715.74 9	0.00087 16
γE2	717.36 5	0.00114 15
γ[M1+E2]	751.09 6	0.00030 11
γE2	873.47 6	0.195 5
γ[E1]	938.25 10	~6 $\times10^{-6}$
γM1+5.5%E2	1050.36 5	0.163 5
γ[E2]	1108.75 7	0.0041 5
γE2+30%M1	1114.48 7	0.0060 4
γE2	1128.042 21	0.073 3
γE2	1149.92 8	0.00024 3
γM1+1.7%E2	1180.76 6	0.0049 3
γE2	1194.56 6	0.0385 21
γ[E2]	1209.93 10	1.9 5 $\times10^{-5}$
γ[E2]	1266.16 15	4.9 16 $\times10^{-5}$
γ[E2]	1305.37 10	6.6 9 $\times10^{-5}$
γ[M1+E2]	1311.11 10	4.8 10 $\times10^{-5}$
γ[E2]	1315.71 16	5.1 15 $\times10^{-5}$
γ[E1]	1372.42 10	2.3 8 $\times10^{-5}$
γ	1397.60 8	0.00292 22
γ[E2]	1489.64 6	0.0014 3
γ[M1+E2]	1498.77 18	0.00068 15
γ[E2]	1562.22 5	0.0168 5
γ[M1]	1571.46 9	0.00073 10
γ[M1+E2]	1577.19 9	0.00037 8
γ[M1+E2]	1730.66 7	0.00105 10
γE2	1766.09 8	0.00244 24
γM1+3.3%E2	1796.93 5	0.0082 4
γ	1909.45 8	0.00117 14
γM1+1.7%E2	1927.28 10	0.00085 5
γ[E1+0.5%M2]	1988.59 9	0.00029 10
γ[M1+E2]	2114.94 18	0.00044 19
γM1+3.1%E2	2193.37 8	0.00202 18
γ[E2]	2242.51 7	0.00092 8
γ[E2]	2308.79 5	0.00075 9
γE2	2316.51 15	0.00024 7
γE2	2366.06 15	0.00034 10
γ[E2]	2439.14 10	0.00028 4
γ	2500.45 9	6 3 $\times10^{-7}$
γ[E2]	2626.80 18	
γ[M1]	2705.22 8	0.00092 9

† 10% uncert(syst)

Atomic Electrons (^{106}Ag)

⟨e⟩=2.56 12 keV

e_{bin}(keV)	⟨e⟩(keV)	e(%)
3	0.84	26 3
4	0.072	2.01 20
17	0.173	1.00 10
18	0.61	3.4 4
20	0.153	0.75 8
21	0.206	0.99 10
23 - 24	0.034	0.148 12
309 - 333	3.5 $\times10^{-5}$	1.1 5 $\times10^{-5}$
404 - 439	0.00053	0.000129 13

Atomic Electrons (^{106}Ag) (continued)

e_{bin}(keV)	⟨e⟩(keV)	e(%)
488	0.39	0.081 8
508 - 554	0.064	0.0127 9
575 - 622	0.00878	0.00147 3
656 - 693	0.000158	2.4 4 $\times10^{-5}$
712 - 751	8.8 $\times10^{-6}$	1.2 3 $\times10^{-6}$
849 - 873	0.00230	0.000270 8
914 - 938	2.9 $\times10^{-8}$	3.1 14 $\times10^{-9}$
1026 - 1050	0.00171	0.000167 5
1084 - 1128	0.000716	6.49 23 $\times10^{-5}$
1146 - 1195	0.000356	3.04 14 $\times10^{-5}$
1206 - 1242	3.5 $\times10^{-7}$	2.8 13 $\times10^{-8}$
1263 - 1311	1.2 $\times10^{-6}$	9.5 25 $\times10^{-8}$
1312 - 1348	1.1 $\times10^{-7}$	8 3 $\times10^{-9}$
1369 - 1398	1.6 $\times10^{-5}$	1.2 4 $\times10^{-6}$
1465 - 1562	0.000123	8.0 3 $\times10^{-6}$
1568 - 1577	8.8 $\times10^{-7}$	5.6 9 $\times10^{-8}$
1706 - 1796	6.8 $\times10^{-5}$	3.85 16 $\times10^{-6}$
1885 - 1964	1.02 $\times10^{-5}$	5.3 9 $\times10^{-7}$
1985 - 1988	9.2 $\times10^{-8}$	4.6 13 $\times10^{-9}$
2091 - 2190	1.18 $\times10^{-5}$	5.5 6 $\times10^{-7}$
2193 - 2292	8.2 $\times10^{-6}$	3.62 25 $\times10^{-7}$
2305 - 2366	2.0 $\times10^{-6}$	8.6 16 $\times10^{-8}$
2415 - 2500	1.18 $\times10^{-6}$	4.9 6 $\times10^{-8}$
2681 - 2705	3.7 $\times10^{-6}$	1.38 13 $\times10^{-7}$

Continuous Radiation (^{106}Ag)

⟨β+⟩=492 keV;⟨IB⟩=3.1 keV

E_{bin}(keV)		⟨ ⟩(keV)	(%)
0 - 10	β+	7.4 $\times10^{-6}$	9.0 $\times10^{-5}$
	IB	0.020	
10 - 20	β+	0.000241	0.00147
	IB	0.023	0.155
20 - 40	β+	0.0054	0.0168
	IB	0.046	0.168
40 - 100	β+	0.211	0.275
	IB	0.106	0.163
100 - 300	β+	8.8	4.02
	IB	0.29	0.166
300 - 600	β+	60	12.9
	IB	0.38	0.087
600 - 1300	β+	307	33.4
	IB	0.96	0.104
1300 - 2500	β+	116	7.9
	IB	1.26	0.073
2500 - 2966	IB	0.052	0.0020
	Σβ+		59

$^{106}_{47}$Ag(8.46 *10* d)

Mode: ϵ

Δ: -86854 6 keV

SpA: 1.456×10^{5} Ci/g

Prod: ^{103}Rh(α,n); ^{107}Ag(γ,n); ^{107}Ag(n,2n); ^{107}Ag(d,t); protons on Cd

Photons (^{106}Ag)

⟨γ⟩=2867 30 keV

γ_{mode}	γ(keV)	γ(%)†
Pd L_ℓ	2.503	0.096 17
Pd L_η	2.660	0.044 7
Pd L_α	2.838	2.6 4
Pd L_β	3.053	1.7 3
Pd L_γ	3.387	0.17 3
Pd $K_{\alpha2}$	21.020	20.5 7
Pd $K_{\alpha1}$	21.177	38.7 13
Pd $K_{\beta1}$'	23.811	9.6 3
Pd $K_{\beta2}$'	24.445	1.86 7
γ[E1]	178.3 4	0.053 18 ?
γM1(+E2)	195.01 4	0.32 3
γM1+1.7%E2	221.721 15	6.70 17
γE1	228.652 21	2.13 7
γE2(+M1)	328.49 5	1.15 5
γ[E2]	333.00 7	0.0009 4
γM1(+E2)	374.62 5	0.26 3
γE2+~0.4%M1	391.046 25	3.89 10
γE2+9.0%M1	406.193 23	13.2 3
γ[M1+E2]	418.57 5	0.48 11
γ[E2]	428.45 6	0.0125 11
γE2+1.8%M1	429.668 20	13.1 3
γ[M1+E2]	433.72 5	0.09 3
γ[M1+E2]	434.18 5	0.0032 7
γE1	451.004 20	27.6 6
γ[E1]	474.08 3	0.93 5
γE2	511.865 3	88 3
γ[E1]	522.15 6	0.088 18
γM1,E2	586.06 4	0.46 4
γE2+11%M1	601.20 4	1.61 6
γE2+1.3%M1	616.181 21	21.7 6
γE1	646.01 4	1.47 6
γM1,E2	679.66 3	0.64 4
γM1,E2	680.439 10	2.05 22
γE2+16%M1	703.11 5	4.45 16
γE2	717.36 5	29.0 8
γE1	748.38 4	20.4 5
γE2+2.5%M1	793.19 4	5.3 3
γE2	804.29 5	12.5 4
γM1+40%E2	808.34 4	4.0 3
γE2+3.0%M1	824.76 5	15.3 3
γE2	847.43 5	2.8 6
γE2	848.22 6	1.6 5
γ[M1+E2]	875.45 4	0.33 4
γ[E2]	949.40 4	0.19 4
γ[E1]	956.33 4	0.48 8
γ	986.8 4	<0.0035
γM1,E2	1019.77 6	1.06 6
γE2+4.7%M1	1045.85 3	29.9 7
γM1+5.5%E2	1050.36 5	0.264 12
γ[M1+E2]	1053.79 6	0.97 14
γ[E1]	1076.87 5	0.053 18
γ[M1+E2]	1121.68 5	0.57 6
γE2	1128.042 21	11.7 3
γ[M1+E2]	1136.82 5	0.23 3
γE1(+0.2%M2)	1168.27 25	0.10 3
γ	1178.05 4	0.19 3
γE2	1199.38 4	11.9 5
γE2	1222.86 4	7.3 4
γ[E1]	1349.6 4	0.12 4 ?
γ[E2]	1394.39 5	1.51 11
γ[E2]	1420.46 5	~0.035
γE2+16%M1	1527.87 5	19.2 12
γ[E2]	1562.22 5	0.0273 14
γ[E2]	1565.57 5	0.48 4
γE1	1572.50 4	6.6 5
γ	1690.2 4	0.036 6
γE2+26%M1	1722.87 6	1.32 13
γ[E2]	1771.14 5	0.040 7
γ	1794.22 4	0.038 15
γE2	1839.03 4	1.93 21
γ	1909.1 6	0.013 4
γ	2084.36 4	0.023 4

† 5.7% uncert(syst)

Atomic Electrons (^{106}Ag)

⟨e⟩=12.4 *3* keV

e_{bin}(keV)	⟨e⟩(keV)	e(%)
3	2.06	64 *7*
4	0.195	5.4 *6*
17	0.42	2.45 *25*
18	1.49	8.4 *9*
20	0.37	1.84 *19*
21	0.50	2.43 *25*
23 - 24	0.084	0.36 *3*
154 - 195	0.055	0.032 *9*
197	0.514	0.260 *8*
204 - 229	0.150	0.0701 *15*
304 - 350	0.077	0.025 *3*
367 - 375	0.157	0.0427 *14*
382	0.482	0.126 *3*
387 - 404	0.109	0.0275 *11*
405	0.432	0.106 *3*
406 - 426	0.0799	0.0190 *5*
427	0.252	0.0591 *18*
428 - 474	0.0607	0.0137 *3*
488	2.09	0.428 *15*
498 - 522	0.340	0.0669 *17*
562 - 586	0.0390	0.0068 *3*
592	0.374	0.0632 *20*
598 - 646	0.0732	0.0119 *3*
655 - 680	0.110	0.0164 *7*
693	0.394	0.0568 *19*
700 - 748	0.185	0.0256 *5*
769 - 808	0.463	0.0588 *12*
821 - 851	0.083	0.0101 *10*
872 - 875	0.00054	6.1 *9* $\times10^{-5}$
925 - 962	0.0041	0.00045 *5*
983 - 1029	0.267	0.0262 *8*
1042 - 1077	0.0379	0.00364 *8*
1097 - 1144	0.108	0.0098 *3*
1154 - 1199	0.147	0.0124 *4*
1219 - 1223	0.0073	0.00060 *3*
1325 - 1370	0.0095	0.00069 *6*
1391 - 1420	0.00154	0.000111 *10*
1504 - 1572	0.146	0.0096 *5*
1666 - 1747	0.0079	0.00046 *4*
1768 - 1838	0.0104	0.00057 *6*
1885 - 1909	5.4 $\times10^{-5}$	2.9 *12* $\times10^{-6}$
2060 - 2084	8.8 $\times10^{-5}$	4.3 *15* $\times10^{-6}$

Continuous Radiation (^{106}Ag)

⟨IB⟩=0.035 keV

E_{bin}(keV)		⟨ ⟩(keV)	(%)
10 - 20	IB	0.0095	0.052
20 - 40	IB	0.021	0.091
40 - 100	IB	0.0018	0.0028
100 - 298	IB	0.0031	0.0021

$^{106}_{48}$Cd(stable)

Δ: -87132 *6* keV
%: 1.25 *3*

$^{106}_{49}$In(6.2 *1* min)

Mode: ϵ
Δ: -80590 *30* keV
SpA: 2.86$\times10^{8}$ Ci/g
Prod: ^{92}Mo(^{16}O,pn); ^{106}Cd(p,n)

Photons (^{106}In)

⟨γ⟩=2669 *180* keV

γ_{mode}	γ(keV)	γ(%)
γ [M1+E2]	161.22 *14*	~0.50
γ M1+7.4%E2	225.86 *15*	7.0 *4*
γ E2	541.02 *19*	12.7 *7*
γ E2	552.51 *17*	25.8 *13*
γ [M1+E2]	580.7 *3*	0.7 *1*
γ M1+13%E2	592.2 *3*	3.0 *3*
γ M1+10%E2	610.80 *11*	4.8 *12*
γ E2	632.65 *9*	100
γ E2	753.4 *3*	2.5 *2*
γ (M1+3.3%E2)	836.66 *14*	2.8 *2*
γ E2	861.17 *9*	92 *14*
γ	974.7 *4*	1.6 *1*
γ E2	997.88 *9*	48.2 *24*
γ E2	1009.38 *10*	30.4 *15*
γ E2	1140.1 *10*	2.7 *7*
γ	1145.1 *10*	3.9 *10*
γ E2	1471.97 *9*	3.0 *10*

$^{106}_{49}$In(5.2 *1* min)

Mode: ϵ
Δ: ~-80190 keV
SpA: 3.41$\times10^{8}$ Ci/g
Prod: ^{106}Cd(p,n)

Photons (^{106}In)

⟨γ⟩=2048 *66* keV

γ_{mode}	γ(keV)	γ(%)†
Cd L_ℓ	2.767	0.015 *3*
Cd L_η	2.957	0.0069 *11*
Cd L_α	3.133	0.40 *6*
Cd L_β	3.392	0.28 *5*
Cd L_γ	3.779	0.028 *5*
Cd $K_{\alpha2}$	22.984	3.00 *11*
Cd $K_{\alpha1}$	23.174	5.64 *20*
Cd $K_{\beta1}$'	26.085	1.44 *5*
Cd $K_{\beta2}$'	26.801	0.298 *11*
γ	575.3 *3*	0.37 *9*
γ M1+10%E2	610.80 *11*	2.0 *6*
γ E2	632.65 *9*	92
γ	811.2 *4*	1.10 *18*
γ E2	861.17 *9*	10.7 *17*
γ	980.6 *3*	0.46 *9*
γ M1+41%E2	1083.82 *11*	3.13 *18*
γ	1162.66 *10*	1.56 *9*
γ E2	1471.97 *9*	1.3 *4*
γ	1620.96 *10*	5.0 *3*
γ	1716.45 *10*	17.1 *9*
γ E2	1716.47 *9*	~2
γ [E1]	1737.9 *3*	1.47 *9*
γ	1745.8 *3*	1.38 *9*
γ	1933.66 *10*	8.4 *5*
γ	1997.6 *3*	2.12 *9*
γ	2087.36 *20*	2.21 *9*
γ	2256.96 *20*	5.0 *3*
γ	2285.6 *3*	0.83 *9*
γ	2304.7 *6*	0.46 *9*
γ	2487.7 *4*	1.01 *9*
γ	2697.1 *3*	0.92 *9*
γ	2794.8 *5*	0.83 *9*
γ	2862.0 *4*	1.47 *9*
γ	2918.2 *3*	3.86 *18*
γ	3120.3 *4*	0.64 *9*
γ	3223.1 *5*	0.74 *9*
γ	3394.6 *4*	1.01 *9*
γ	3494.7 *4*	2.12 *9*
γ	3889.3 *5*	0.83 *9*
γ	3912.1 *8*	0.46 *18*

† 11% uncert(syst)

Atomic Electrons (^{106}In)

⟨e⟩=3.25 *18* keV

e_{bin}(keV)	⟨e⟩(keV)	e(%)
4	0.35	9.7 *10*
19	0.205	1.07 *11*
20 - 26	0.185	0.85 *5*
549 - 584	0.050	0.0086 *23*
606	1.69	0.28 *3*
607 - 611	0.0066	0.0011 *3*
629	0.223	0.035 *4*
632 - 633	0.052	0.0083 *7*
784 - 811	0.013	0.0016 *7*
834	0.125	0.0150 *23*
857 - 861	0.0194	0.0023 *3*
954 - 981	0.0042	0.00044 *20*
1057 - 1084	0.0352	0.00332 *19*
1136 - 1163	0.012	0.0011 *4*
1445 - 1472	0.009	0.00065 *18*
1594 - 1620	0.027	0.0017 *6*
1690	0.09	0.0051 *20*
1711 - 1745	0.024	0.0014 *3*
1907 - 1997	0.049	0.0025 *7*
2061 - 2087	0.010	0.00047 *14*
2230 - 2304	0.026	0.0011 *3*
2461 - 2487	0.0039	0.00016 *5*
2670 - 2768	0.0059	0.00022 *4*
2791 - 2861	0.0054	0.00019 *5*
2892 - 2918	0.013	0.00046 *11*
3094 - 3120	0.0021	6.8 *18* $\times10^{-5}$
3196 - 3222	0.0023	7.3 *19* $\times10^{-5}$
3368 - 3394	0.0031	9.3 *22* $\times10^{-5}$
3468 - 3494	0.0064	0.00018 *4*
3863 - 3911	0.0037	9.5 *20* $\times10^{-5}$

Continuous Radiation (^{106}In)

⟨β+⟩=1607 keV;⟨IB⟩=8.6 keV

E_{bin}(keV)		⟨ ⟩(keV)	(%)
0 - 10	β+	1.55$\times10^{-6}$	1.88$\times10^{-5}$
	IB	0.048	
10 - 20	β+	5.4$\times10^{-5}$	0.000324
	IB	0.048	0.33
20 - 40	β+	0.00128	0.00395
	IB	0.099	0.35
40 - 100	β+	0.054	0.070
	IB	0.27	0.42
100 - 300	β+	2.56	1.15
	IB	0.82	0.46
300 - 600	β+	21.0	4.50
	IB	1.05	0.24
600 - 1300	β+	200	20.4
	IB	2.0	0.22
1300 - 2500	β+	736	39.5
	IB	2.5	0.140
2500 - 5000	β+	645	20.3
	IB	1.8	0.055
5000 - 6312	β+	1.55	0.0306
	IB	0.049	0.00092
	Σβ+		86

$^{106}_{50}$Sn(2.10 *15* min)

Mode: ϵ
Δ: -77290 *200* keV syst
SpA: 8.4$\times10^{8}$ Ci/g
Prod: ^{106}Cd(^{3}He,3n)

Photons (^{106}Sn)

$\langle\gamma\rangle$=980 *75* keV

γ_{mode}	γ(keV)	γ(%)†
In L_ℓ	2.905	0.113 *20*
In L_η	3.112	0.053 *8*
In L_α	3.286	3.1 *5*
In L_β	3.569	2.3 *4*
In L_γ	3.983	0.23 *4*
In $K_{\alpha2}$	24.002	22.3 *10*
In $K_{\alpha1}$	24.210	41.7 *18*
In $K_{\beta1}$'	27.265	10.8 *5*
In $K_{\beta2}$'	28.022	2.25 *10*
γ [M1]	49.30 *15*	2.1 *3* ?
γ [M1]	52.50 *15*	2.4 *4* ?
γ [M1]	101.81 *12*	1.07 *23*
γ [M1]	122.26 *9*	14.8 *15*
γ [M1]	224.06 *11*	14.5 *17*
γ [M1]	253.16 *9*	29
γ [M1]	325.95 *20*	6.7 *20*
γ [M1]	386.45 *16*	51 *7*
γ [M1]	477.22 *12*	32 *4*
γ [M1]	712.40 *19*	17 *5*
γ [E2]	863.67 *17*	11 *6*
γ	1097.0 *10*	1.5 *4*
γ [E2]	1189.62 *19*	16.8 *20*
γ	1421.0 *10*	2.1 *5*

† 17% uncert(syst)

Atomic Electrons (^{106}Sn)

$\langle e\rangle$=23.8 *9* keV

e_{bin}(keV)	$\langle e\rangle$(keV)	e(%)
4	2.7	71 *7*
19	0.194	1.00 *10*
20	1.63	8.1 *9*
21	1.58	7.4 *10*
23 - 24	0.88	3.8 *3*
25	1.72	7.0 *10*
26 - 74	1.51	3.07 *22*
94	3.6	3.8 *4*

Atomic Electrons (^{106}Sn) (continued)

e_{bin}(keV)	$\langle e\rangle$(keV)	e(%)
98 - 102	0.071	0.072 *13*
118	0.56	0.47 *5*
119 - 122	0.147	0.121 *10*
196	1.40	0.71 *9*
220 - 224	0.243	0.110 *11*
225	2.3	1.04 *21*
249 - 298	0.77	0.28 *5*
322 - 326	0.061	0.019 *5*
359	2.2	0.62 *9*
382 - 386	0.36	0.093 *11*
449	1.02	0.23 *3*
473 - 477	0.162	0.034 *4*
684 - 712	0.37	0.054 *15*
836 - 864	0.16	0.019 *8*
1069 - 1097	0.013	0.0012 *6*
1162 - 1190	0.164	0.0141 *16*
1393 - 1421	0.014	0.0010 *4*

Continuous Radiation (^{106}Sn)

$\langle\beta+\rangle$=84 keV;$\langle IB\rangle$=1.29 keV

E_{bin}(keV)		$\langle\ \rangle$(keV)	(%)
0 - 10	$\beta+$	1.10×10^{-5}	0.000133
	IB	0.0041	
10 - 20	$\beta+$	0.000383	0.00232
	IB	0.0053	0.035
20 - 40	$\beta+$	0.0091	0.0281
	IB	0.039	0.153
40 - 100	$\beta+$	0.359	0.465
	IB	0.021	0.033
100 - 300	$\beta+$	12.1	5.7
	IB	0.070	0.038
300 - 600	$\beta+$	44.3	10.1
	IB	0.19	0.043
600 - 1300	$\beta+$	26.8	3.74
	IB	0.73	0.080
1300 - 2036	IB	0.22	0.0152
	$\Sigma\beta+$		20

$^{106}_{52}$Te(50 *20* μs)

Mode: α

Δ: -58050 *630* keV syst

SpA: 1.5×10^{11} Ci/g

Prod: ^{58}Ni on ^{58}Ni; daughter ^{110}Xe

Alpha Particles (^{106}Te)

α(keV)
4160 *30*

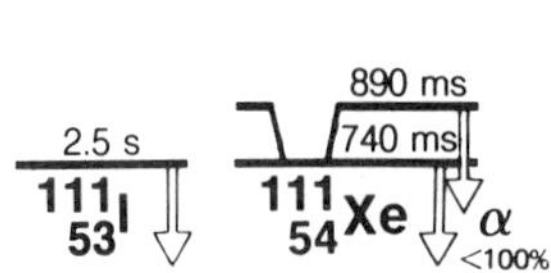

A = 107

NDS **34**, 643 (1981)

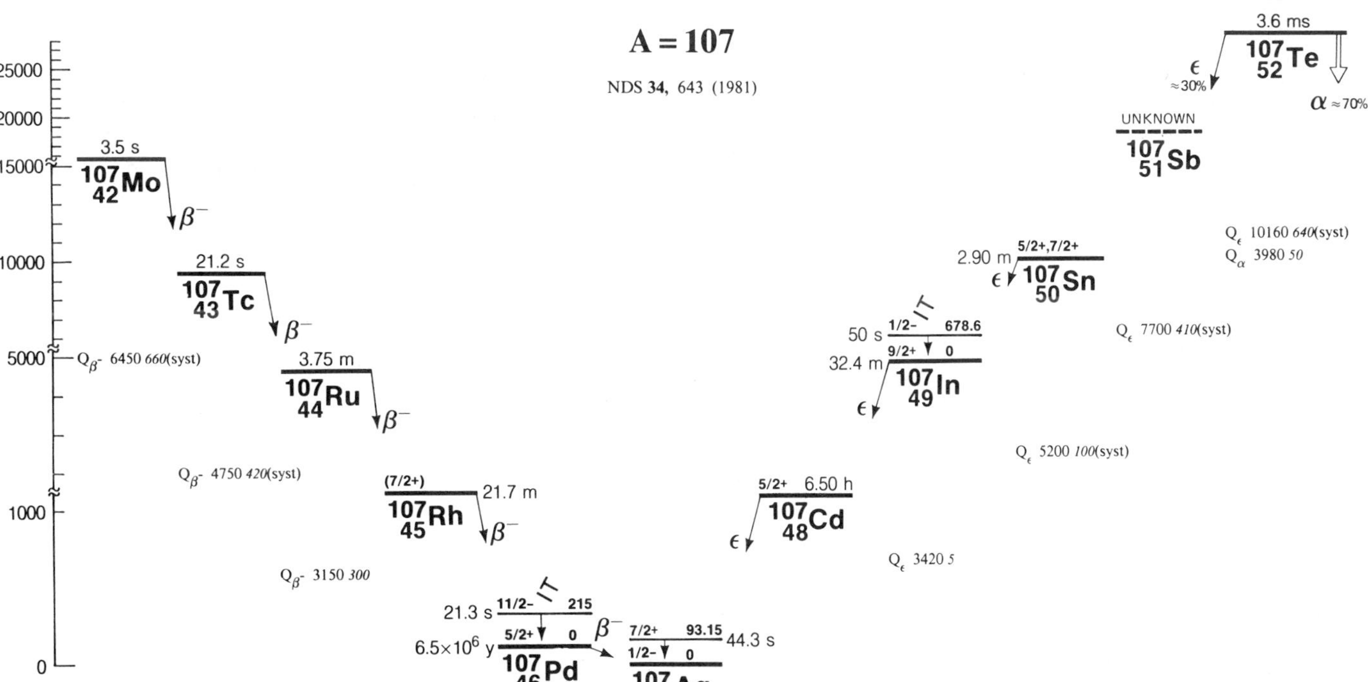

$^{107}_{42}$Mo(3.5 *5* s)

Mode: β-
Δ: -72510 *580* keV syst
SpA: 2.7×10^{10} Ci/g
Prod: fission

Photons (^{107}Mo)

γ_{mode}	γ(keV)
γ	64.2 *2*
γ	400.1 *2*
γ	483.3 *2*
γ	784.3 *2*

$^{107}_{43}$Tc(21.2 *2* s)

Mode: β-
Δ: -78960 *300* keV syst
SpA: 4.89×10^{9} Ci/g
Prod: fission

Photons (^{107}Tc)

⟨γ⟩=515 *8* keV

γ_{mode}	γ(keV)	γ(%)†
Ru L_ℓ	2.253	0.031 *6*
Ru L_η	2.382	0.015 *3*
Ru L_α	2.558	0.81 *15*
Ru L_β	2.728	0.50 *10*
Ru L_γ	3.017	0.037 *8*
Ru $K_{\alpha2}$	19.150	6.0 *6*
Ru $K_{\alpha1}$	19.279	11.5 *12*
Ru $K_{\beta1}$'	21.650	2.8 *3*
Ru $K_{\beta2}$'	22.210	0.52 *5*
γ E2+29%M1	39.401 *25*	0.35 *7*
γ (M1)	70.689 *24*	1.27 *19*
γ M1+49%E2	102.661 *24*	21.0 *22*
γ E2+28%M1	106.287 *24*	7.6 *6*
γ (E2)	108.45 *6*	0.70 *12*
γ [M1]	114.42 *4*	0.56 *7*
γ [M1]	138.35 *5*	0.30 *7*
γ [M1]	142.063 *25*	3.0 *3*
γ [M1]	145.52 *3*	2.3 *2*
γ	147.85 *6*	0.12 *4*
γ	169.46 *6*	0.21 *4*
γ [M1]	176.976 *23*	9.2 *7*
γ	180.03 *7*	0.3 *1*
γ	183.28 *5*	0.11 *2*
γ	185.11 *4*	0.65 *7*
γ	188.73 *5*	0.40 *4*
γ	195.72 *8*	0.11 *4*
γ	199.68 *7*	2.1 *2*
γ	216.21 *4*	0.40 *4*
γ	228.73 *11*	0.11 *1*
γ	257.60 *5*	0.08 *2*
γ	271.13 *8*	0.23 *7*
γ	291.03 *7*	0.07 *2*
γ	291.21 *8*	0.12 *4*
γ	291.40 *4*	4.1 *4*
γ	313.92 *7*	0.30 *2*
γ	315.38 *7*	0.63 *7*
γ	322.50 *4*	1.20 *12*
γ	335.36 *7*	1.30 *13*
γ	337.24 *7*	0.65 *7*
γ	346.3 *19*	
γ	354.57 *5*	2.50 *25*
γ	360.26 *5*	3.0 *4*
γ	361.03 *11*	0.09 *4*
γ	377.95 *7*	0.46 *5*

Photons (^{107}Tc) (continued)

γ_{mode}	γ(keV)	γ(%)†
γ	382.75 *8*	0.30 *4*
γ	386.07 *7*	0.98 *9*
γ	419.28 *5*	0.44 *11*
γ	423.08 *8*	0.3 *1*
γ	428.41 *11*	0.11 *4*
γ	443.81 *6*	0.63 *14*
γ	458.68 *5*	5.6 *6*
γ	460.85 *5*	0.54 *7*
γ	465.60 *8*	0.25 *6*
γ	470.81 *7*	0.33 *8*
γ	479.77 *6*	0.07 *2*
γ	483.21 *6*	0.07 *2*
γ	489.82 *8*	1.4 *3*
γ	490.89 *7*	0.9 *2*
γ	514.79 *6*	0.46 *4*
γ	521.19 *5*	0.11 *4*
γ	529.22 *8*	0.70 *7*
γ	530.52 *9*	0.25 *7*
γ	533.63 *8*	0.17 *7*
γ	534.46 *7*	0.32 *8*
γ	536.69 *23*	0.46 *4*
γ	539.88 *7*	0.17 *2*
γ	556.11 *5*	~0.04
γ	556.65 *8*	0.11 *4*
γ	562.85 *7*	0.11 *4*
γ	576.53 *9*	0.44 *14*
γ	577.64 *8*	0.09 *4*
γ	582.43 *6*	0.65 *7*
γ	585.87 *6*	0.11 *1*
γ	595.51 *5*	4.6 *2*
γ	603.37 *7*	0.17 *4*
γ	635.08 *7*	0.86 *9*
γ	638.71 *7*	0.56 *7*
γ	648.88 *7*	0.23 *2*
γ	652.25 *11*	0.40 *4*
γ	665.10 *7*	0.38 *11*
γ	665.62 *8*	0.12 *4*
γ	681.11 *6*	0.32 *4*
γ	704.50 *7*	0.14 *2*
γ	713.66 *23*	0.11 *2*
γ	722.79 *10*	0.17 *4*
γ	723.16 *7*	0.32 *11*
γ	723.20 *7*	0.35 *11*
γ	746.13 *6*	0.32 *3*
γ	751.83 *7*	0.30 *3*
γ	788.11 *8*	0.14 *4*
γ	807.16 *7*	0.21 *2*
γ	820.44 *6*	0.14 *2*
γ	822.0 *19*	
γ	825.86 *6*	0.07 *1*
γ	834.13 *7*	0.14 *2*
γ	850.73 *8*	0.32 *9*
γ	856.91 *7*	1.07 *7*
γ	863.34 *8*	0.09 *1*
γ	900.14 *7*	0.09 *1*
γ	923.10 *6*	0.04 *1*
γ	924.94 *7*	0.33 *4*
γ	969.83 *6*	0.56 *4*
γ	973.46 *6*	0.17 *4*
γ	981.32 *6*	0.56 *4*
γ	993.74 *7*	0.30 *3*
γ	1025.42 *7*	0.19 *2*
γ	1062.73 *8*	0.19 *2*
γ	1070.9 *19*	
γ	1076.12 *6*	0.09 *2*
γ	1094.23 *6*	0.23 *2*
γ	1118.15 *6*	1.10 *17*
γ	1126.02 *7*	0.09 *1*
γ	1218.63 *6*	1.10 *11*
γ	1263.68 *7*	0.53 *4*
γ	1264.88 *7*	0.23 *2*
γ	1284.13 *10*	0.14 *2*
γ	1296.38 *6*	0.51 *4*
γ	1361.3 *19*	
γ	1378.10 *6*	0.75 *7*
γ	1379.99 *6*	0.54 *9*
γ	1388.09 *6*	0.44 *5*
γ	1396.97 *6*	0.38 *7*
γ	1434.73 *5*	0.35 *4*
γ	1452.42 *6*	0.25 *3*
γ	1465.83 *6*	0.42 *4*
γ	1502.51 *5*	0.65 *7*

Photons (^{107}Tc) (continued)

γ_{mode}	γ(keV)	γ(%)†
γ	1555.08 *6*	0.14 *2*
γ	1573.19 *5*	1.50 *17*
γ	1650.94 *5*	0.61 *6*
γ	1654.57 *5*	0.21 *2*
γ	1667.72 *11*	0.33 *9*
γ	1742.03 *11*	0.54 *7*
γ	1993.8 *19*	
γ	2094.99 *24*	0.14 *2*
γ	2118.25 *13*	0.17 *2*
γ	2357.08 *13*	0.11 *2*
γ	2378.86 *24*	0.49 *5*
γ	2388.18 *13*	0.11 *3*
γ	2429.07 *14*	0.17 *2*
γ	2502.60 *13*	0.56 *6*
γ	2537.51 *13*	0.56 *6*
γ	2576.91 *13*	0.25 *2*

† 10% uncert(syst)

Atomic Electrons (^{107}Tc)

⟨e⟩=27.5 *19* keV

e_{bin}(keV)	⟨e⟩(keV)	e(%)
3	0.74	26 *4*
16	0.61	3.8 *5*
17	0.87	5.0 *11*
18 - 68	2.51	7.2 *7*
70 - 71	0.0177	0.025 *3*
81	10.0	12.5 *20*
84	4.3	5.1 *11*
86 - 92	0.57	0.65 *10*
99	1.20	1.21 *18*
100	1.2	1.2 *3*
102	0.47	0.46 *9*
103	1.1	1.1 *3*
105 - 148	1.27	1.08 *8*
155	0.82	0.53 *5*
158 - 207	0.64	0.37 *9*
213 - 258	0.024	0.010 *4*
268 - 315	0.36	0.13 *5*
319 - 364	0.24	0.070 *20*
375 - 423	0.038	0.0093 *24*
425 - 471	0.17	0.039 *13*
477 - 526	0.043	0.0086 *17*
527 - 576	0.073	0.013 *5*
577 - 617	0.025	0.0042 *10*
627 - 666	0.017	0.0027 *5*
678 - 724	0.014	0.0019 *6*
730 - 766	0.0042	0.00057 *18*
785 - 834	0.0070	0.00087 *20*
835 - 878	0.010	0.0012 *4*
897 - 925	0.0028	0.00031 *11*
948 - 994	0.0108	0.00113 *24*
1003 - 1041	0.0023	0.00022 *7*
1054 - 1096	0.008	0.00071 *24*
1104 - 1126	0.0013	0.00012 *3*
1197 - 1243	0.0094	0.00077 *21*
1260 - 1296	0.0038	0.00030 *8*
1356 - 1397	0.010	0.00074 *14*
1413 - 1452	0.0044	0.00031 *7*
1463 - 1555	0.0093	0.00061 *16*
1570 - 1667	0.0054	0.00033 *7*
1720 - 1742	0.0021	0.00012 *4*
2073 - 2118	0.00102	4.9 *11* ×10^{-5}
2335 - 2429	0.0026	0.000112 *20*
2480 - 2576	0.0040	0.00016 *3*

Continuous Radiation (^{107}Tc)

⟨β-⟩=1831 keV;⟨IB⟩=6.1 keV

E_{bin}(keV)		⟨ ⟩(keV)	(%)
0 - 10	β-	0.00389	0.077
	IB	0.055	
10 - 20	β-	0.0120	0.080
	IB	0.054	0.38
20 - 40	β-	0.0500	0.166
	IB	0.107	0.37
40 - 100	β-	0.395	0.56
	IB	0.31	0.48
100 - 300	β-	5.4	2.59
	IB	0.93	0.52
300 - 600	β-	27.9	6.1
	IB	1.14	0.27
600 - 1300	β-	217	22.3
	IB	1.8	0.21
1300 - 2500	β-	812	43.3
	IB	1.37	0.080
2500 - 4750	β-	768	24.9
	IB	0.28	0.0096

$^{107}_{44}$Ru(3.75 *5* min)

Mode: β-

Δ: -83710 *300* keV

SpA: 4.68×10^{8} Ci/g

Prod: ^{110}Pd(n,α); fission

Photons (^{107}Ru)

⟨γ⟩=206 *6* keV

γ_{mode}	γ(keV)	γ(%)†
γ	56.8 *3*	
γ	105.9 *3*	
γ	114.9 *5*	
γ (M1+E2)	194.11 *21*	10.6
γ	204.2 *10*	
γ	214.8 *10*	
γ	217.4 *10*	
γ	220.8 *5*	
γ	230.3 *3*	
γ	268.5 *3*	
γ	275.6 *10*	
γ	286.4 *10*	
γ	334.0 *10*	
γ	358.5 *5*	
γ	360.2 *6*	
γ	365.9 *10*	
γ	374.4 *3*	3.9 *3*
γ	394.3 *7*	
γ	405.9 *3*	2.33 *21*
γ	451.9 *10*	
γ	462.64 *21*	4.5 *3*
γ	485.8 *10*	
γ	489.3 *5*	1.4 *3*
γ	568.5 *3*	
γ	579.31 *22*	2.76 *21*
γ	636.1 *3*	
γ	683.4 *5*	
γ	703.7 *4*	
γ	737.8 *6*	
γ	809.6 *3*	
γ	821.2 *10*	
γ	843.6 *6*	
γ [M1+E2]	847.84 *22*	5.7 *4*
γ	897.9 *10*	
γ	912.0 *6*	
γ	931.7 *10*	
γ [M1+E2]	1041.95 *20*	3.18 *21*

Photons (^{107}Ru) (continued)

γ_{mode}	γ(keV)	γ(%)†
γ	1078.1 *3*	
γ	1112.4 *10*	
γ	1272.2 *3*	2.86 *21*
γ	1306.3 *5*	
γ	1444.0 *10*	

† 9.4% uncert(syst)

Continuous Radiation (^{107}Ru)

⟨β-⟩=1227 keV;⟨IB⟩=3.2 keV

E_{bin}(keV)		⟨ ⟩(keV)	(%)
0 - 10	β-	0.0085	0.169
	IB	0.043	
10 - 20	β-	0.0261	0.174
	IB	0.042	0.29
20 - 40	β-	0.109	0.362
	IB	0.083	0.29
40 - 100	β-	0.85	1.20
	IB	0.24	0.36
100 - 300	β-	11.2	5.4
	IB	0.67	0.38
300 - 600	β-	54	11.9
	IB	0.75	0.18
600 - 1300	β-	351	36.7
	IB	0.97	0.114
1300 - 2500	β-	728	41.0
	IB	0.39	0.024
2500 - 3150	β-	81	3.05
	IB	0.0052	0.00020

$^{107}_{45}$Rh(21.7 *4* min)

Mode: β-

Δ: -86862 *18* keV

SpA: 8.10×10^{7} Ci/g

Prod: ^{104}Ru(α,p); fission

Photons (^{107}Rh)

⟨γ⟩=313 *16* keV

γ_{mode}	γ(keV)	γ(%)†
γ	80.23 *16*	0.05 *1*
γ [M1+E2]	96.50 *25*	~0.009 ?
γ [M1+E2]	102.43 *15*	~0.02
γ E2	115.76 *12*	0.52 *4*
γ	175.25 *18*	0.07 *1* ?
γ [M1+E2]	198.93 *24*	0.04 *1*
γ	219.50 *17*	0.10 *2*
γ	232.51 *15*	0.21 *3*
γ [M1+E2]	266.13 *14*	0.28 *2*
γ	277.68 *13*	1.70 *12*
γ [M1+E2]	288.32 *13*	0.73 *5*
γ M1+E2	302.84 *15*	66 *5*
γ M1	312.29 *13*	4.8 *4*
γ	321.93 *14*	2.26 *16*
γ	348.28 *13*	2.27 *16*
γ [M1+E2]	357.92 *13*	0.41 *3*
γ [M1+E2]	367.37 *15*	1.91 *14*
γ [M1+E2]	381.89 *13*	0.65 *5*
γ	392.52 *13*	8.8 *7*
γ [E2]	431.96 *24*	0.026 *6* ?
γ [E2]	452.01 *14*	0.51 *4*
γ [M1+E2]	471.28 *23*	0.12 *1*
γ	511.8 *5*	~0.030
γ	521.3 *6*	~0.02
γ [E2]	554.44 *13*	0.078 *6*
γ [M1+E2]	567.77 *13*	1.15 *8*

Photons (^{107}Rh) (continued)

γ_{mode}	γ(keV)	γ(%)†
γ	644.0 *8*	0.032 *4*
γ M1+2.1%E2	670.21 *10*	2.22 *16*
γ	696.8 *8*	0.012 *3*
γ	709.65 *25*	0.075 *6*
γ	720.28 *25*	0.013 *2* ?
γ	753.9 *3*	0.013 *3* ?
γ E2+42%M1	789.88 *24*	0.089 *7*
γ	836.3 *6*	0.015 *3* ?
γ	845.8 *6*	0.025 *3* ?
γ	863.5 *10*	0.037 *3*
γ [E2]	1102.17 *24*	0.011 *3* ?
γ	1120.1 *10*	0.024 *4*
γ	1148.6 *6*	0.039 *4* ?

† 10% uncert(syst)

Continuous Radiation (^{107}Rh)

⟨β-⟩=436 keV;⟨IB⟩=0.55 keV

E_{bin}(keV)		⟨ ⟩(keV)	(%)
0 - 10	β-	0.0509	1.02
	IB	0.020	
10 - 20	β-	0.155	1.03
	IB	0.019	0.132
20 - 40	β-	0.63	2.11
	IB	0.036	0.126
40 - 100	β-	4.71	6.7
	IB	0.095	0.147
100 - 300	β-	50.9	25.2
	IB	0.21	0.121
300 - 600	β-	162	36.4
	IB	0.133	0.033
600 - 1300	β-	216	27.4
	IB	0.037	0.0052
1300 - 1512	β-	1.84	0.136
	IB	3.3×10^{-5}	2.5×10^{-6}

$^{107}_{46}$Pd(6.5 *3* ×10^{6} yr)

Mode: β-

Δ: -88374 *6* keV

SpA: 0.000513 Ci/g

Prod: fission

Continuous Radiation (^{107}Pd)

⟨β-⟩=9.3 keV;⟨IB⟩=0.00033 keV

E_{bin}(keV)		⟨ ⟩(keV)	(%)
0 - 10	β-	2.66	60
	IB	0.00028	
10 - 20	β-	4.49	31.3
	IB	4.4×10^{-5}	0.00034
20 - 33	β-	2.15	9.2
	IB	2.3×10^{-6}	1.06×10^{-5}

$^{107}_{46}$Pd(21.3 *5* s)

Mode: IT

Δ: -88159 *6* keV

SpA: 4.87×10^{9} Ci/g

Prod: ^{106}Pd(n,γ); ^{108}Pd(n,2n)

Photons (^{107}Pd)

$\langle\gamma\rangle$=152 5 keV

γ_{mode}	γ(keV)	γ(%)†
Pd L_ℓ	2.503	0.029 5
Pd L_η	2.660	0.0150 23
Pd L_α	2.838	0.77 12
Pd L_β	3.046	0.53 9
Pd L_γ	3.369	0.052 9
Pd $K_{\alpha2}$	21.020	5.6 3
Pd $K_{\alpha1}$	21.177	10.6 5
Pd $K_{\beta1}$'	23.811	2.63 12
Pd $K_{\beta2}$'	24.445	0.508 25
γ E3	214.9 10	69.0

† 2.9% uncert(syst)

Atomic Electrons (^{107}Pd)

$\langle e\rangle$=62.8 17 keV

e_{bin}(keV)	$\langle e\rangle$(keV)	e(%)
3	0.68	21.0 22
4	0.042	1.17 12
17	0.115	0.67 7
18	0.41	2.28 24
20	0.102	0.50 5
21	0.138	0.66 7
23	0.0175	0.075 8
24	0.0056	0.0235 25
191	44.6	23.4 9
211	4.86	2.30 8
212	8.7	4.10 15
214	2.69	1.26 5
215	0.476	0.222 8

$^{107}_{47}$Ag(stable)

Δ: -88407 5 keV

%: 51.839 5

$^{107}_{47}$Ag(44.3 2 s)

Mode: IT

Δ: -88314 5 keV

SpA: 2.362×10^9 Ci/g

Prod: daughter ^{107}Cd

Photons (^{107}Ag)

$\langle\gamma\rangle$=12.6 4 keV

γ_{mode}	γ(keV)	γ(%)†
Ag L_ℓ	2.634	0.086 16
Ag L_η	2.806	0.052 8
Ag L_α	2.984	2.3 4
Ag L_β	3.207	1.8 3
Ag L_γ	3.544	0.17 3
Ag $K_{\alpha2}$	21.990	10.4 6
Ag $K_{\alpha1}$	22.163	19.6 11
Ag $K_{\beta1}$'	24.934	4.9 3
Ag $K_{\beta2}$'	25.603	0.98 6
γ E3	93.146 19	4.67

† 4.3% uncert(syst)

Atomic Electrons (^{107}Ag)

$\langle e\rangle$=80.0 23 keV

e_{bin}(keV)	$\langle e\rangle$(keV)	e(%)
3	1.37	41 5
4	0.94	27 3
18	0.39	2.12 24
19	0.55	2.9 3
21	0.213	1.00 11
22	0.225	1.04 12
24	0.029	0.121 14
25	0.0132	0.053 6
68	29.1	43.1 22
89	3.11	3.48 17
90	34.6	38.6 19
92	0.60	0.64 3
93	8.8	9.5 5

$^{107}_{48}$Cd(6.50 2 h)

Mode: ε

Δ: -86990 7 keV

SpA: 4.506×10^6 Ci/g

Prod: ^{106}Cd(n,γ); ^{107}Ag(d,2n); ^{107}Ag(p,n)

Photons (^{107}Cd)

$\langle\gamma\rangle$=31.5 8 keV

γ_{mode}	γ(keV)	γ(%)†
Ag L_ℓ	2.634	0.18 3
Ag L_η	2.806	0.097 16
Ag L_α	2.984	5.0 8
Ag L_β	3.213	3.6 6
Ag L_γ	3.562	0.35 6
Ag $K_{\alpha2}$	21.990	31.2 15
Ag $K_{\alpha1}$	22.163	59 3
Ag $K_{\beta1}$'	24.934	14.8 7
Ag $K_{\beta2}$'	25.603	2.94 14
γ M1+0.6%E2	32.480 18	0.0047 5
γ E3	93.146 19	4.7 3 *
γ M1+0.4%E2	98.34 3	~0.0014
γ M1(+E2)	300.94 5	0.00261 16
γ M1+4.1%E2	324.833 25	0.0306 15
γ	356.59 14	0.00034 5
γ [M1+E2]	363.34 13	<0.00024 ?
γ E2	423.171 24	0.0293 15
γ [E1]	436.54 13	0.00026 3
γ M1+0.6%E2	461.68 13	0.00065 10
γ M1+5.5%E2	526.57 7	0.0045 3
γ [M1+E2]	550.0 3	0.00090 10
γ E1	597.28 3	0.0076 4
γ M1+7.3%E2	624.90 7	0.00350 21
γ [E2]	648.3 3	0.00018 3
γ E2+46%M1	719.93 7	0.00359 24
γ M1+0.3%E2	786.51 13	0.00163 16
γ (E2)	796.483 20	0.0649 20
γ [E1]	799.88 5	0.00196 16
γ (M1+E2)	818.27 7	0.00204 24
γ M1	828.963 21	0.163
γ [E1]	856.59 7	9.0 16 ×10^{-5}
γ E1	898.21 5	0.0097 4
γ [E1]	934.08 23	7.3 16 ×10^{-5}
γ E2	949.74 7	0.00130 13
γ	1049.95 7	0.00020 3
γ [E2]	1097.42 5	0.00196 16
γ M1,E2	1129.90 5	0.0072 3
γ [E2]	1165.77 23	0.00041 7
γ [E2]	1232.6 3	0.00047 10

Photons (^{107}Cd)
(continued)

γ_{mode}	γ(keV)	γ(%)†
γ [E1]	1258.91 23	4.9 10 ×10^{-5}
γ	1264.5 8	0.00014 3
γ	1294.0 8	2.0 5 ×10^{-5} ?
γ	1297.0 4	5.7 8 ×10^{-5} ?
γ [E1]	1325.8 3	8.2 13 ×10^{-5}

† 6.8% uncert(syst)

* with ^{107}Ag(44.3 s) in equilib

Atomic Electrons (^{107}Cd)

$\langle e\rangle$=85 3 keV

e_{bin}(keV)	$\langle e\rangle$(keV)	e(%)
3	2.8	84 9
4 - 32	6.1	74 6
68	29.2	43 3
73	0.00038	~0.0005
89	3.11	3.48 25
90	34.7	39 3
92	0.60	0.64 5
93	8.8	9.5 7
95 - 98	7.8 ×10^{-5}	8 3 ×10^{-5}
275 - 325	0.00189	0.00063 3
331 - 363	1.9 ×10^{-5}	6 3 ×10^{-6}
398 - 437	0.00125	0.000311 14
458 - 501	0.000110	2.20 13 ×10^{-5}
523 - 572	9.0 ×10^{-5}	1.64 7 ×10^{-5}
593 - 625	8.7 ×10^{-5}	1.45 7 ×10^{-5}
645 - 648	4.8 ×10^{-7}	7.4 12 ×10^{-8}
694 - 720	6.2 ×10^{-5}	8.9 6 ×10^{-6}
761 - 803	0.00316	0.000399 20
814 - 857	0.000326	3.95 23 ×10^{-5}
873 - 909	4.84 ×10^{-5}	5.53 23 ×10^{-6}
924 - 950	1.47 ×10^{-5}	1.59 14 ×10^{-6}
1024 - 1072	1.76 ×10^{-5}	1.65 15 ×10^{-6}
1094 - 1140	7.6 ×10^{-5}	6.9 6 ×10^{-6}
1162 - 1207	3.9 ×10^{-6}	3.2 6 ×10^{-7}
1229 - 1272	2.0 ×10^{-6}	1.6 3 ×10^{-7}
1290 - 1326	3.5 ×10^{-7}	2.7 4 ×10^{-8}

Continuous Radiation (^{107}Cd)

$\langle\beta+\rangle$=0.284 keV;$\langle$IB$\rangle$=0.53 keV

E_{bin}(keV)		$\langle\ \rangle$(keV)	(%)
0 - 10	β+	1.85×10^{-6}	2.24×10^{-5}
	IB	0.00019	
10 - 20	β+	5.9×10^{-5}	0.000358
	IB	0.0068	0.036
20 - 40	β+	0.00124	0.00383
	IB	0.027	0.116
40 - 100	β+	0.0354	0.0470
	IB	0.0030	0.0047
100 - 300	β+	0.247	0.149
	IB	0.037	0.017
300 - 600	β+	9.4×10^{-7}	3.12×10^{-7}
	IB	0.164	0.036
600 - 1300	IB	0.29	0.036
1300 - 1324	IB	2.1×10^{-6}	1.59×10^{-7}
	Σβ+		0.20

$^{107}_{49}$In(32.4 *3* min)

Mode: ε

Δ: -83570 *8* keV

SpA: 5.42×10^7 Ci/g

Prod: ^{106}Cd(d,n); ^{106}Cd(p,γ); ^{108}Cd(p,2n); ^{110}Cd(p,4n); ^{107}Ag(^{3}He,3n)

Photons (^{107}In)

⟨γ⟩=1189 *13* keV

γ_{mode}	γ(keV)	γ(%)†
Cd L_ℓ	2.767	0.072 *13*
Cd L_η	2.957	0.034 *5*
Cd L_α	3.133	2.0 *3*
Cd L_β	3.392	1.38 *23*
Cd L_γ	3.779	0.136 *25*
Cd $K_{\alpha2}$	22.984	14.5 *5*
Cd $K_{\alpha1}$	23.174	27.4 *10*
Cd $K_{\beta1}'$	26.085	7.0 *3*
Cd $K_{\beta2}'$	26.801	1.45 *6*
γ E1	36.53 *11*	0.062 *5*
γ M1+5.9%E2	204.98 *3*	48
γ M1(+E2)	300.53 *5*	0.162 *24*
γ M1(+E2)	303.55 *5*	0.64 *4*
γ M1(+E2)	320.95 *3*	10.3 *3*
γ [M1+E2]	348.90 *13*	0.17 *5*
γ M1(+E2)	365.31 *5*	3.86 *14*
γ M1(+E2)	381.42 *11*	0.067 *19*
γ	395.4 *3*	0.052 *19*
γ M1(+E2)	413.95 *12*	0.81 *10*
γ M1(+E2)	416.06 *16*	1.72 *14*
γ M1,E2	455.84 *16*	0.55 *5*
γ [M1+E2]	459.30 *7*	0.181 *19*
γ [M1+E2]	474.92 *11*	0.091 *19*
γ [E2]	488.11 *6*	0.081 *19*
γ [M1+E2]	499.03 *17*	0.23 *3*
γ M1+7.7%E2	505.51 *5*	12.0 *4*
γ M1,E2	519.28 *10*	0.26 *3*
γ M1,E2	554.15 *12*	0.76 *14*
γ [M1+E2]	598.51 *12*	0.24 *4*
γ [E2]	600.62 *16*	0.28 *5*
γ E2+16%M1	604.08 *5*	1.35 *7*
γ [M1+E2]	611.54 *8*	0.57 *5*
γ	617.32 *15*	0.28 *3*
γ M2	640.6 *1*	0.572 *24*
γ	669.82 *20*	0.076 *14*
γ [M1+E2]	677.79 *10*	0.181 *24*
γ [M1+E2]	687.03 *16*	0.095 *14*
γ M1,E2	702.37 *11*	0.45 *3*
γ	708.7 *6*	0.200 *24*
γ [M1+E2]	716.59 *15*	0.33 *14*
γ	723.5 *7*	0.21 *4*
γ E2	728.06 *6*	3.34 *14*
γ [M1+E2]	762.85 *7*	0.74 *6*
γ [M1+E2]	777.69 *12*	0.076 *14*
γ [M1+E2]	793.75 *9*	0.176 *24*
γ E2	809.06 *5*	3.29 *14*
γ [M1+E2]	840.23 *10*	0.181 *24*
γ	848.5 *3*	0.110 *19*
γ [M1+E2]	871.89 *8*	0.56 *3*
γ [E2]	903.05 *7*	0.48 *6*
γ	905.8 *3*	0.19 *5*
γ [E2]	915.09 *8*	1.14 *6*
γ [M1+E2]	919.46 *12*	0.33 *10*
γ (E2)	921.57 *15*	1.24 *14*
γ [M1+E2]	947.41 *6*	0.86 *6*
γ	953.7 *3*	0.35 *3*
γ	957.28 *24*	0.076 *14*
γ [M1+E2]	967.36 *9*	0.162 *19*
γ M1,E2	998.73 *9*	1.04 *6*
γ	998.76 *12*	0.19 *6*
γ [M1+E2]	1002.75 *16*	0.27 *3*
γ [E2]	1012.09 *9*	0.072 *24*
γ	1021.8 *7*	0.138 *24*
γ [M1+E2]	1036.12 *12*	0.20 *3*
γ [M1+E2]	1056.45 *8*	0.75 *4*
γ [M1+E2]	1063.37 *6*	1.25 *10*
γ	1067.68 *22*	0.48 *10*
γ [E2]	1081.98 *15*	0.167 *24*
γ [M1+E2]	1086.89 *16*	0.086 *19*
γ [M1+E2]	1113.15 *13*	0.105 *24*
γ	1139.9 *5*	0.21 *3*
γ [M1+E2]	1144.97 *17*	0.81 *10*
γ	1158.7 *3*	0.091 *24*
γ	1160.8 *7*	0.12 *3*
γ [M1+E2]	1172.42 *8*	0.16 *3*
γ [M1+E2]	1197.29 *12*	0.24 *5*
γ	1205.2 *8*	0.12 *3*
γ	1207.3 *4*	0.057 *24*
γ	1211.5 *10*	0.10 *3*
γ	1236.13 *24*	0.26 *7*
γ	1244.8 *3*	0.124 *24*
γ [M1+E2]	1268.35 *6*	5.49 *19*
γ	1325.52 *20*	0.51 *4*
γ	1343.69 *12*	1.41 *8*
γ	1366 *1*	~0.19
γ [M1+E2]	1377.40 *7*	1.50 *7*
γ	1403.56 *24*	0.24 *4*
γ [E2]	1411.10 *9*	1.16 *7*
γ [M1+E2]	1455.47 *9*	0.73 *7*
γ [M1+E2]	1500.83 *12*	1.04 *6*
γ	1507.9 *4*	~0.19
γ	1550.5 *10*	0.14 *3*
γ	1555.79 *22*	0.49 *5*
γ [M1+E2]	1571.43 *9*	1.43 *10*
γ	1588.12 *24*	0.10 *3*
γ	1591.4 *6*	0.12 *3*
γ [M1+E2]	1601.26 *12*	0.59 *7*
γ [M1+E2]	1607.13 *21*	0.119 *24*
γ [E2]	1641.03 *13*	0.12 *3*
γ	1653.5 *4*	0.32 *3*
γ	1663.77 *22*	0.28 *3*
γ [M1+E2]	1685.40 *12*	0.191 *24*
γ	1704.08 *24*	0.076 *19*
γ	1712.9 *4*	0.17 *4*
γ [M1+E2]	1717.22 *12*	0.31 *3*
γ	1733.0 *3*	0.13 *3*
γ [M1+E2]	1743.48 *15*	0.45 *3*
γ	1767.8 *3*	0.26 *4*
γ [M1+E2]	1779.28 *13*	1.39 *7*
γ [M1+E2]	1798.96 *13*	0.10 *3*
γ [M1+E2]	1801.36 *12*	0.14 *5*
γ	1831.24 *14*	0.73 *4*
γ	1842.71 *19*	0.124 *19*
γ [M1+E2]	1859.44 *15*	0.358 *24*
γ	1878.8 *3*	0.53 *3*
γ	1899.3 *4*	0.119 *19*
γ	1909.06 *24*	0.153 *19*
γ [M1+E2]	1922.20 *12*	1.81 *10*
γ	1935.6 *3*	0.119 *19*
γ	1955.22 *13*	0.64 *5*
γ [M1+E2]	1963.84 *13*	0.46 *3*
γ	1978.4 *3*	0.48 *3*
γ [M1+E2]	1983.52 *13*	1.21 *8*
γ [M1+E2]	2006.34 *12*	1.60 *7*
γ	2045.18 *22*	0.42 *3*
γ [M1+E2]	2064.42 *15*	1.72 *7*
γ	2078.6 *3*	0.15 *3*
γ [M1+E2]	2099.48 *13*	0.87 *7*
γ	2169.0 *6*	0.12 *3*
γ	2183.4 *3*	0.42 *5*
γ	2183.53 *24*	0.24 *7*
γ	2200.6 *5*	0.11 *3*
γ	2227.0 *3*	0.091 *19*
γ	2258.76 *13*	0.119 *19*
γ	2263.1 *3*	0.105 *24*
γ [M1+E2]	2284.78 *12*	1.14 *10*
γ [M1+E2]	2304.46 *13*	1.24 *14*
γ	2331.5 *3*	0.40 *3*
γ	2342.9 *3*	0.095 *14*
γ	2371.7 *5*	0.086 *19*
γ	2405.6 *5*	0.200 *19*
γ	2433.0 *3*	0.157 *14*
γ	2461.7 *4*	0.081 *10*
γ	2469.2 *4*	0.105 *14*
γ	2482.8 *5*	0.105 *24*
γ	2496.3 *3*	0.153 *19*
γ	2506.9 *4*	0.176 *14*
γ	2547.9 *3*	0.072 *10*
γ [M1+E2]	2554.53 *21*	0.114 *14*
γ	2584.1 *3*	0.210 *14*
γ	2637.9 *3*	0.095 *24*
γ	2652.4 *3*	0.210 *14*
γ	2665.1 *5*	0.44 *5*
γ [M1+E2]	2670.49 *21*	0.31 *5*
γ	2680.8 *3*	0.277 *24*
γ	2700.9 *3*	0.081 *10*
γ	2717.2 *4*	0.50 *3*
γ	2783.0 *5*	0.114 *14*
γ	2796.8 *3*	0.029 *10*
γ	2818.7 *4*	0.057 *10*
γ	2865.7 *5*	0.129 *19*
γ [M1+E2]	2875.47 *20*	0.85 *5*
γ	2986.0 *5*	0.100 *19*
γ	3005 *1*	0.048 *19*

† 8.4% uncert(syst)

Atomic Electrons (^{107}In)

⟨e⟩=11.8 *5* keV

e_{bin}(keV)	⟨e⟩(keV)	e(%)
4	1.71	47 *5*
10	0.014	~0.15
19	0.99	5.2 *5*
20 - 37	0.91	4.16 *23*
178	5.0	2.83 *24*
201	0.73	0.36 *3*
204 - 205	0.173	0.085 *6*
274 - 277	0.052	0.0189 *25*
294	0.60	0.21 *3*
297 - 345	0.31	0.094 *8*
348 - 395	0.131	0.0344 *23*
410 - 459	0.046	0.0107 *6*
461 - 475	0.0091	0.00194 *22*
479	0.330	0.0690 *25*
484 - 527	0.079	0.0156 *8*
550 - 599	0.058	0.0100 *5*
600 - 643	0.0463	0.0075 *3*
651 - 699	0.024	0.0036 *5*
701 - 736	0.070	0.0099 *4*
751 - 794	0.0479	0.0061 *3*
805 - 849	0.0175	0.00213 *14*
868 - 916	0.0410	0.0046 *3*
918 - 967	0.0196	0.00211 *21*
972 - 1021	0.0217	0.00222 *24*
1022 - 1068	0.0294	0.00283 *24*
1078 - 1118	0.0103	0.00093 *12*
1132 - 1181	0.0077	0.00067 *8*
1185 - 1233	0.0038	0.00032 *8*
1235 - 1268	0.051	0.0041 *4*
1299 - 1344	0.014	0.0010 *3*
1351 - 1400	0.0222	0.00163 *11*
1403 - 1452	0.0070	0.00049 *5*
1455 - 1552	0.0221	0.00146 *12*
1555 - 1653	0.0116	0.00073 *7*
1659 - 1753	0.0178	0.00103 *6*
1764 - 1859	0.0115	0.00064 *9*
1873 - 1963	0.0272	0.00141 *9*
1974 - 2073	0.0271	0.00134 *6*
2075 - 2174	0.0040	0.00019 *5*
2179 - 2278	0.0126	0.00056 *4*
2281 - 2379	0.0045	0.000196 *24*
2402 - 2496	0.0030	0.000122 *15*
2503 - 2583	0.00168	6.6 *9* ×10^{-5}
2611 - 2700	0.0070	0.00026 *3*
2713 - 2796	0.00090	3.3 *5* ×10^{-5}
2815 - 2875	0.0041	0.000145 *10*
2959 - 3004	0.00050	1.7 *4* ×10^{-5}

Continuous Radiation (^{107}In)

⟨β+⟩=316 keV;⟨IB⟩=2.2 keV

E_{bin}(keV)		⟨ ⟩(keV)	(%)
0 - 10	β+	3.64×10^{-6}	4.40×10^{-5}
	IB	0.0124	
10 - 20	β+	0.000125	0.00076
	IB	0.0146	0.098
20 - 40	β+	0.00296	0.0091
	IB	0.046	0.17
40 - 100	β+	0.120	0.156
	IB	0.068	0.105
100 - 300	β+	5.09	2.32
	IB	0.20	0.112
300 - 600	β+	33.6	7.3
	IB	0.30	0.069
600 - 1300	β+	169	18.3
	IB	0.75	0.083
1300 - 2500	β+	108	7.0
	IB	0.79	0.046
2500 - 3215	IB	0.070	0.0026
	Σβ+		35

$^{107}_{49}$In(50.4 6 s)

Mode: IT

Δ: -82891 8 keV

SpA: 2.078×10^{9} Ci/g

Prod: ^{106}Cd(d,n); ^{3}He on Ag

Photons (^{107}In)

⟨γ⟩=641.0 15 keV

γ_{mode}	γ(keV)	γ(%)†
In L_ℓ	2.905	0.0061 11
In L_η	3.112	0.0029 4
In L_α	3.286	0.17 3
In L_β	3.569	0.124 21
In L_γ	3.983	0.0127 23
In $K_{\alpha2}$	24.002	1.17 4
In $K_{\alpha1}$	24.210	2.19 8
In $K_{\beta1}'$	27.265	0.567 20
In $K_{\beta2}'$	28.022	0.119 4
γ M4	678.64 23	94.30

† 0.21% uncert(syst)

Atomic Electrons (^{107}In)

⟨e⟩=37.3 6 keV

e_{bin}(keV)	⟨e⟩(keV)	e(%)
4	0.146	3.8 4
19	0.0102	0.052 5
20	0.086	0.43 4
23	0.0226	0.097 10
24	0.0240	0.101 10
26	0.00140	0.0053 5
27	0.0032	0.0121 12
651	30.8	4.73 10
674	4.16	0.617 12
675	0.833	0.1235 25
678	1.005	0.148 3
679	0.216	0.0318 6

$^{107}_{50}$Sn(2.90 5 min)

Mode: ε

Δ: -78370 100 keV syst

SpA: 6.05×10^{8} Ci/g

Prod: ^{106}Cd(α,3n); ^{106}Cd(^{3}He,2n)

Photons (^{107}Sn)

γ_{mode}	γ(keV)	γ(rel)
γ M1(+E2)	362.4 8	
γ	377.4 10	
γ M1(+E2)	422.4 7	~5
γ M1	429.5 6	
γ E1	488.8 10	
γ	571.3 10	2.7 5
γ	596.3 10	1.8 4
γ	610.8 10	2.7 5
γ	625.3 10	
γ M4	678.64 23	100 17 *
γ	696.8 10	
γ	736.3 10	3.9 8
γ (M1,E2)	758.5 8	3.0 6
γ E2	803.3 8	6.3 13
γ	836.8 10	
γ	888.3 10	
γ	919.8 9	2.7 5
γ	981.8 10	3.0 6
γ M1+E2	1001.7 3	29 5
γ	1048.3 25	2.3 5
γ	1070.3 8	3.2 6
γ	1085.3 10	1.20 24
γ	1110.3 10	1.9 4
γ E2(+M1)	1130.2 6	100
γ	1167.3 10	1.8 4
γ	1174.3 10	4.4 9
γ	1185.8 7	12.5 25
γ	1217.3 10	2.2 4
γ	1310.3 10	1.6 3
γ	1336.2 8	1.7 3
γ	1358.5 8	6.5 13
γ	1383.3 10	1.3 3
γ (M1,E2)	1396.2 8	21 4
γ [M1+E2]	1424.1 7	9.6 19
γ	1445.3 10	2.6 5
γ	1473.3 10	4.8 10
γ (E2)	1542.3 10	30 6
γ	1581.3 25	1.5 3
γ	1704.3 10	6.1 12
γ	1732.4 8	2.9 6
γ	1808.3 10	25 5
γ [E1]	1911.4 8	4.9 10
γ	1936.3 10	1.5 3
γ	1944.3 25	0.90 18
γ	1963.3 10	2.8 6
γ	2004.3 10	8.0 16
γ	2041.3 10	0.80 16
γ	2063.3 10	7.5 15
γ	2094.8 6	8.8 18
γ	2116.9 6	9.9 20
γ	2186.3 10	1.9 4
γ	2216.3 10	7.6 15
γ	2302.3 10	4.1 8
γ	2316.0 7	5.7 11
γ	2379.3 10	0.90 18
γ	2448.3 10	1.0 2
γ	2466.4 8	1.0 2
γ	2483.3 10	0.90 18
γ	2546.4 5	10 2
γ	2570.3 10	1.4 3
γ	2644.3 10	0.90 18
γ	2650.3 10	0.90 18
γ	2659.3 10	0.90 18
γ	2673.3 10	1.7 3
γ	2716.3 10	1.9 4
γ	2825.3 10	13 3

Photons (^{107}Sn) (continued)

γ_{mode}	γ(keV)	γ(rel)
γ	2858.3 10	1.10 22
γ	3060.3 10	6.7 13
γ	3112.3 10	2.1 4
γ	3130.3 10	0.70 14
γ	3136.3 10	0.5 1
γ	3202.3 10	1.10 22
γ	3206.3 10	1.20 24
γ	3218.3 10	0.60 12
γ	3225.0 5	0.40 8
γ	3325.3 10	2.9 6
γ	3361.3 25	0.60 12
γ	3375.3 25	0.60 12
γ	3431.3 10	0.20 4
γ	3441.3 10	0.20 4
γ	3450.3 10	0.90 18
γ	3494.3 10	0.40 8
γ	3512.3 10	0.20 4
γ	3592.3 10	0.80 16

* with ^{107}In(50 s)

$^{107}_{52}$Te(3.6 5 ms)

Mode: α(~70 %), ε(~30 %)

Δ: -60510 500 keV syst

SpA: 1.52×10^{11} Ci/g

Prod: ^{96}Ru(^{16}O,5n)

Alpha Particles (^{107}Te)

α(keV)
3833 15

A = 108

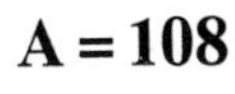

NDS 37, 289 (1982)

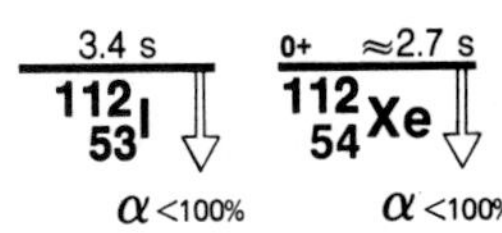

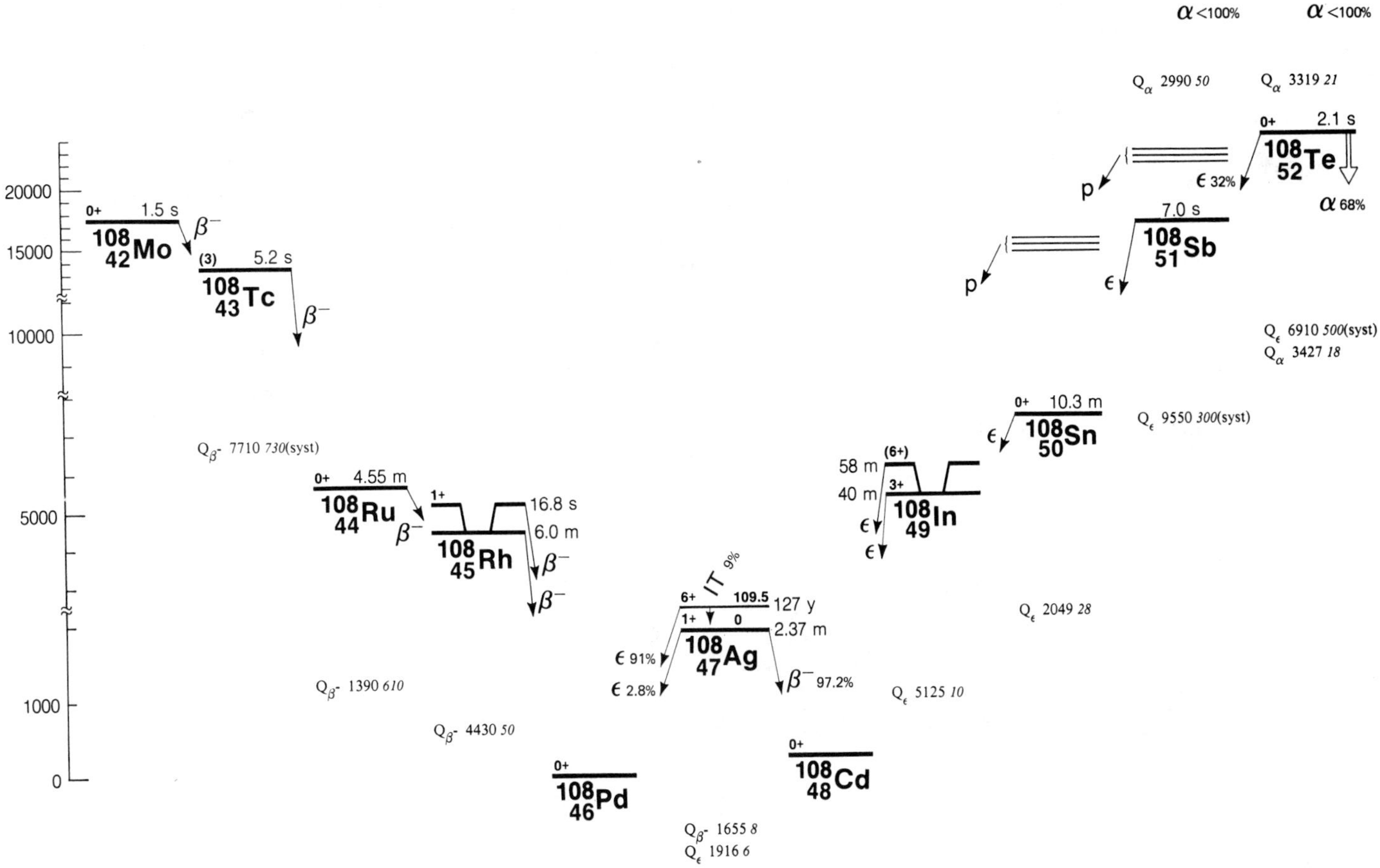

20000
15000
10000
5000
1000
0
0+ 1.5 s
108 42 Mo
β⁻
(3) 5.2 s
108 43 Tc
β⁻
Qβ- 7710 730(syst)
0+ 4.55 m
108 44 Ru
β⁻
1+
16.8 s
6.0 m
108 45 Rh
β⁻
β⁻
Qβ- 1390 610
Qβ- 4430 50
0+
108 46 Pd
IT 9%
6+ 109.5 127 y
1+ 0 2.37 m
108 47 Ag
ϵ 91%
ϵ 2.8%
β⁻ 97.2%
Qβ- 1655 8
Qϵ 1916 6
0+
108 48 Cd
Qϵ 5125 10
58 m (6+)
40 m 3+
108 49 In
ϵ
ϵ
Qϵ 2049 28
0+ 10.3 m
108 50 Sn
ϵ
Qϵ 9550 300(syst)
p
7.0 s
108 51 Sb
ϵ
Qϵ 6910 500(syst)
Qα 3427 18
p
ϵ 32%
0+ 2.1 s
108 52 Te
α 68%
Qα 2990 50
Qα 3319 21
3.4 s
112 53 I
α <100%
0+ ≈2.7 s
112 54 Xe
α <100%

$^{108}_{42}$Mo(1.5 *4* s)

Mode: β-
SpA: 5.6×10^{10} Ci/g
Prod: fission

Photons (^{108}Mo)

γ_{mode}	γ(keV)
γ	125.5 *2*
γ	258.53 *15*
γ	268.21 *6*

$^{108}_{43}$Tc(5.17 *7* s)

Mode: β-
Δ: -75990 *400* keV syst
SpA: 1.89×10^{10} Ci/g
Prod: fission

Photons (^{108}Tc)

⟨γ⟩=887 *21* keV

γ_{mode}	γ(keV)	γ(%)
Ru L_ℓ	2.253	0.0044 *8*
Ru L_η	2.382	0.0021 *3*
Ru L_α	2.558	0.114 *18*
Ru L_β	2.730	0.070 *12*
Ru L_γ	3.026	0.0053 *10*
Ru $K_{\alpha2}$	19.150	0.97 *5*
Ru $K_{\alpha1}$	19.279	1.84 *9*
Ru $K_{\beta1}$'	21.650	0.446 *22*
Ru $K_{\beta2}$'	22.210	0.083 *4*
γ	181.87 *11*	0.91 *16*
γ [E2]	242.22 *4*	82.4
γ [M1+E2]	267.00 *8*	1.07 *16*
γ [M1+E2]	309.66 *10*	0.62 *8*
γ [E2]	422.91 *8*	3.3 *8*
γ [M1+E2]	465.57 *5*	14.4 *12*
γ	576.58 *8*	1.8 *4*
γ [E2]	584.01 *8*	0.62 *21*
γ	669.06 *9*	1.24 *16*
γ [E2]	707.79 *4*	11.5 *8*
γ [M1+E2]	732.57 *7*	9.9 *8*
γ	850.93 *9*	3.3 *3*
γ [M1+E2]	1006.92 *8*	2.2 *3*
γ	1117.93 *8*	4.8 *6*
γ [E2]	1249.14 *9*	1.48 *25*
γ	1272.08 *20*	1.6 *4*
γ	1401.63 *10*	0.91 *16*
γ	1417.02 *9*	4.0 *3*
γ	1583.50 *7*	9.9 *8*
γ	1760.38 *10*	2.31 *16*
γ	1882.59 *10*	0.70 *12*

Atomic Electrons (^{108}Tc)

⟨e⟩=11.8 *4* keV

e_{bin}(keV)	⟨e⟩(keV)	e(%)
3 - 22	0.249	4.4 *4*
160 - 182	0.13	~0.08
220	8.6	3.92 *16*
239	1.34	0.560 *23*
242 - 288	0.41	0.168 *11*
306 - 310	0.0051	0.0017 *3*
401 - 423	0.12	0.029 *7*
443	0.34	0.077 *10*
462 - 466	0.053	0.0114 *12*
554 - 584	0.036	0.0065 *22*
647 - 686	0.156	0.0228 *19*
705 - 733	0.156	0.0219 *17*
829 - 851	0.026	0.0032 *13*
985 - 1007	0.020	0.0020 *3*
1096 - 1118	0.028	0.0026 *10*
1227 - 1272	0.019	0.0015 *3*
1380 - 1417	0.023	0.0016 *5*
1561 - 1583	0.041	0.0026 *9*
1738 - 1760	0.009	0.00051 *16*
1860 - 1882	0.0025	0.00014 *5*

Continuous Radiation (^{108}Tc)

⟨β-⟩=3103 keV;⟨IB⟩=13.6 keV

E_{bin}(keV)		⟨ ⟩(keV)	(%)
0 - 10	β-	0.00106	0.0211
	IB	0.074	
10 - 20	β-	0.00327	0.0218
	IB	0.073	0.51
20 - 40	β-	0.0137	0.0456
	IB	0.145	0.50
40 - 100	β-	0.110	0.155
	IB	0.43	0.65
100 - 300	β-	1.56	0.75
	IB	1.34	0.74
300 - 600	β-	8.7	1.88
	IB	1.8	0.41
600 - 1300	β-	82	8.3
	IB	3.3	0.37
1300 - 2500	β-	485	25.0
	IB	3.7	0.21
2500 - 5000	β-	1933	53
	IB	2.6	0.081
5000 - 7468	β-	592	10.5
	IB	0.145	0.0027

$^{108}_{44}$Ru(4.55 *5* min)

Mode: β-

Δ: -83700 *610* keV

SpA: 3.82×10^8 Ci/g

Prod: fission

Photons (^{108}Ru)

⟨γ⟩=61.0 *15* keV

γ_{mode}	γ(keV)	γ(%)†
γ	14.50 *17*	
γ	73.62 *14*	1.15 *6* ?
γ	91.28 *7*	2.38 *6* ?
γ	150.40 *14*	7.8 *3*
γ M1,E2	164.90 *14*	28.0 *8*

† 21% uncert(syst)

$^{108}_{45}$Rh(6.0 *3* min)

Mode: β-

Δ: -85090 *50* keV

SpA: 2.90×10^8 Ci/g

Prod: ^{108}Pd(n,p)

Photons (^{108}Rh)

⟨γ⟩=2264 *160* keV

γ_{mode}	γ(keV)	γ(%)†
Pd $K_{\alpha2}$	21.020	0.34 *4*
Pd $K_{\alpha1}$	21.177	0.64 *8*
Pd $K_{\beta1}$'	23.811	0.160 *20*
Pd $K_{\beta2}$'	24.445	0.031 *4*
γ [M1+E2]	404.35 *14*	27 *6*
γ E2	433.937 *5*	91
γ E2+10%M1	497.09 *9*	26 *5*
γ	581.02 *17*	59 *12*
γ E2	614.281 *6*	28 *6*
γ E2	722.938 *8*	7.3 *15*
γ [M1+E2]	901.44 *13*	30 *6*
γ E2	931.03 *9*	7.5 *15*

Photons (^{108}Rh)

(continued)

γ_{mode}	γ(keV)	γ(%)†
γ	947.01 *16*	50 *10*
γ	1092.24 *19*	2.9 *6*
γ	1234.16 *18*	7.9 *16*
γ	1528.02 *19*	0.91 *18*
γ	1815.17 *19*	5.9 *12*

† approximate

Atomic Electrons (^{108}Rh)

⟨e⟩=7.9 *8* keV

e_{bin}(keV)	⟨e⟩(keV)	e(%)
3	0.035	1.08 *17*
4 - 24	0.050	0.06 *2*
380	0.94	0.25 *6*
401	0.13	0.031 *9*
404	0.028	0.007 *2*
410	2.9	0.72 *14*
430	0.33	0.077 *15*
431 - 434	0.165	0.038 *5*
473	0.64	0.14 *3*
493 - 497	0.105	0.021 *3*
557	0.8	~0.14
577 - 581	0.12	0.021 *9*
590	0.48	0.081 *16*
611 - 614	0.076	0.0124 *21*
699 - 723	0.113	0.016 *3*
877	0.32	0.036 *8*
898 - 907	0.118	0.0130 *18*
923	0.36	~0.039
927 - 947	0.063	0.0067 *23*
1068 - 1092	0.021	0.0019 *8*
1210 - 1234	0.049	0.0040 *17*
1504 - 1527	0.0046	0.00030 *12*
1791 - 1815	0.025	0.0014 *5*

$^{108}_{45}$Rh(16.8 *5* s)

Mode: β-

Δ: -85090 *50* keV

SpA: 6.09×10^9 Ci/g

Prod: fission; daughter ^{108}Ru

Photons (^{108}Rh)

⟨γ⟩=319 *49* keV

γ_{mode}	γ(keV)	γ(%)†
Pd $K_{\alpha2}$	21.020	0.098 *21*
Pd $K_{\alpha1}$	21.177	0.19 *4*
Pd $K_{\beta1}$'	23.811	0.046 *10*
Pd $K_{\beta2}$'	24.445	0.0089 *19*
γ E2	433.937 *5*	43
γ E2+10%M1	497.09 *9*	5.1 *4*
γ	609.3 *10*	
γ E2	618.77 *5*	15.1 *13*
γ	891.7 *10*	
γ E2	931.03 *9*	1.49 *15*
γ [E2+3.1%M1]	1007.13 *5*	
γ	1105.92 *7*	
γ [E2]	1441.07 *5*	
γ	1500.6 *10*	

† 26% uncert(syst)

Atomic Electrons (^{108}Rh)

⟨e⟩=2.1 *4* keV

e_{bin}(keV)	⟨e⟩(keV)	e(%)
3	0.0101	0.31 *8*
4	0.00078	0.022 *5*
17	0.0020	0.012 *3*
18	0.0071	0.040 *10*
20	0.0018	0.0088 *21*
21	0.0024	0.012 *3*
23	0.00031	0.0013 *3*
24	9.8×10^{-5}	0.00041 *10*
410	1.4	0.34 *9*
430	0.16	0.036 *10*
431	0.035	0.0081 *21*
433	0.037	0.0084 *22*
434	0.0067	0.0015 *4*
473	0.128	0.0271 *22*
493	0.0145	0.00294 *24*
494	0.00250	0.00051 *4*
496	0.00273	0.00055 *5*
497	0.00111	0.000223 *18*
594	0.257	0.043 *4*
615	0.031	0.0051 *5*
616	0.00179	0.00029 *3*
618	0.0063	0.00102 *9*
619	0.00116	0.000188 *17*
907	0.0141	0.00155 *16*
927	0.00158	0.000171 *17*
928	0.000132	1.42 *14* $\times 10^{-5}$
930	0.00033	3.5 *4* $\times 10^{-5}$
931	6.0×10^{-5}	6.5 *7* $\times 10^{-6}$

Continuous Radiation (^{108}Rh)

⟨β-⟩=1779 keV;⟨IB⟩=5.8 keV

E_{bin}(keV)		⟨ ⟩(keV)	(%)
0 - 10	β-	0.00383	0.076
	IB	0.054	
10 - 20	β-	0.0118	0.078
	IB	0.054	0.37
20 - 40	β-	0.0492	0.164
	IB	0.106	0.37
40 - 100	β-	0.389	0.55
	IB	0.31	0.47
100 - 300	β-	5.3	2.55
	IB	0.91	0.51
300 - 600	β-	27.7	6.0
	IB	1.12	0.26
600 - 1300	β-	222	22.7
	IB	1.7	0.20
1300 - 2500	β-	859	45.7
	IB	1.25	0.073
2500 - 4430	β-	665	22.1
	IB	0.20	0.0069

$^{108}_{46}$Pd(stable)

Δ: -89522 *4* keV

%: 26.46 *9*

$^{108}_{47}$Ag(2.37 *1* min)

Mode: β-(97.15 *20* %), ε(2.85 *20* %)

Δ: -87605 *5* keV

SpA: 7.33×10^8 Ci/g

Prod: daughter ^{108}Ag(127 yr); ^{107}Ag(n,γ)

Photons (^{108}Ag)

$\langle\gamma\rangle$=15.6 *3* keV

γ_{mode}	γ(keV)	γ(%)†
Pd $K_{\alpha2}$	21.020	0.531 *18*
Pd $K_{\alpha1}$	21.177	1.00 *4*
Pd $K_{\beta1}'$	23.811	0.249 *9*
Pd $K_{\beta2}'$	24.445	0.0481 *18*
γ_ϵ[E2]	383.09 *13*	0.0009 *3*
γ_ϵ[E2]	388.36 *7*	0.0018 *6*
γ_ϵE2	433.937 *5*	0.50
γ_ϵE2+10%M1	497.09 *9*	0.00195 *21*
γ_ϵ[M1+E2]	510.05 *9*	<0.0035
γ_ϵE2	618.77 *5*	0.261 *13*
$\gamma_{\beta-}$E2	632.99 *4*	1.76
γ_ϵ[E2]	880.18 *10*	0.00319 *25*
γ_ϵE2	931.03 *9*	0.00057 *5*
γ_ϵ[E2+3.1%M1]	1007.13 *5*	0.0139 *7*
γ_ϵ	1105.92 *7*	0.00165 *15*
γ_ϵ[E2]	1441.07 *5*	0.00304 *20*
γ_ϵ	1539.86 *7*	0.00105 *10*

† uncert(syst): 7.0% for ϵ, 2.3% for β-

Atomic Electrons (^{108}Ag)*

$\langle e\rangle$=0.157 *7* keV

e_{bin}(keV)	$\langle e\rangle$(keV)	e(%)
3	0.053	1.65 *17*
4	0.0046	0.128 *13*
17	0.0109	0.063 *7*
18	0.039	0.216 *22*
20	0.0097	0.048 *5*
21	0.0130	0.063 *6*
23 - 24	0.00218	0.0094 *8*
359 - 388	0.00013	3.6 *8* $\times10^{-5}$
410	0.0161	0.0039 *3*
430 - 473	0.00277	0.00064 *3*
486 - 510	5.5 $\times10^{-5}$	~1 $\times10^{-5}$
594 - 619	0.00517	0.00087 *4*
856 - 880	3.7 $\times10^{-5}$	4.3 *3* $\times10^{-6}$
907 - 931	6.1 $\times10^{-6}$	6.7 *5* $\times10^{-7}$
983 - 1007	0.000137	1.39 *8* $\times10^{-5}$
1082 - 1106	1.1 $\times10^{-5}$	1.1 *4* $\times10^{-6}$
1417 - 1516	2.48 $\times10^{-5}$	1.73 *15* $\times10^{-6}$
1536 - 1539	6.4 $\times10^{-7}$	4.1 *14* $\times10^{-8}$

* with ϵ

Continuous Radiation (^{108}Ag)

$\langle\beta-\rangle$=608 keV;$\langle\beta+\rangle$=1.12 keV;$\langle$IB$\rangle$=1.02 keV

E_{bin}(keV)		$\langle\ \rangle$(keV)	(%)
0 - 10	β-	0.0269	0.53
	β+	2.13 $\times10^{-7}$	2.58 $\times10^{-6}$
	IB	0.026	
10 - 20	β-	0.082	0.55
	β+	6.8 $\times10^{-6}$	4.16 $\times10^{-5}$
	IB	0.025	0.18
20 - 40	β-	0.338	1.13
	β+	0.000152	0.000468
	IB	0.049	0.17
40 - 100	β-	2.58	3.66
	β+	0.0055	0.0072
	IB	0.132	0.20
100 - 300	β-	30.8	15.1
	β+	0.179	0.084
	IB	0.33	0.19
300 - 600	β-	123	27.3
	β+	0.66	0.151
	IB	0.28	0.067
600 - 1300	β-	405	45.6
	β+	0.271	0.0398
	IB	0.18	0.023

Continuous Radiation (^{108}Ag)
(continued)

E_{bin}(keV)		$\langle\ \rangle$(keV)	(%)
1300 - 1916	β-	46.6	3.34
	IB	0.0080	0.00056
	$\Sigma\beta$+		0.28

$^{108}_{47}$Ag(127 *21* yr)

Mode: ϵ(91.3 *6* %), IT(8.7 *6* %)
Δ: -87495 *5* keV
SpA: 26 Ci/g
Prod: ^{107}Ag(n,γ)

Photons (^{108}Ag)

$\langle\gamma\rangle$=1680 *18* keV

γ_{mode}	γ(keV)	γ(%)†
Pd L_ℓ	2.503	0.087 *15*
Ag L_ℓ	2.634	0.14 *3*
Pd L_η	2.660	0.040 *6*
Ag L_η	2.806	0.018 *3*
Pd L_α	2.838	2.4 *4*
Ag L_α	2.984	3.8 *6*
Pd L_β	3.053	1.52 *25*
Ag L_β	3.294	1.16 *22*
Pd L_γ	3.387	0.15 *3*
Ag L_γ	3.623	0.095 *20*
Pd $K_{\alpha2}$	21.020	18.7 *7*
Pd $K_{\alpha1}$	21.177	35.3 *12*
Ag $K_{\alpha2}$	21.990	5.6 *4*
Ag $K_{\alpha1}$	22.163	10.5 *8*
Pd $K_{\beta1}'$	23.811	8.8 *3*
Pd $K_{\beta2}'$	24.445	1.70 *6*
Ag $K_{\beta1}'$	24.934	2.65 *19*
Ag $K_{\beta2}'$	25.603	0.53 *4*
γ_{IT} M4	30.38 *6*	0.000223
γ_{IT} E1	79.131 *3*	76
γ_ϵE2	433.937 *5*	90.5
γ_ϵE2	614.281 *6*	89.8 *18*
γ_ϵE2	722.938 *8*	90.8 *18*

† uncert(syst): 0.66% for ϵ, 6.9% for IT

Atomic Electrons (^{108}Ag)

$\langle e\rangle$=15.1 *3* keV

e_{bin}(keV)	$\langle e\rangle$(keV)	e(%)
3	2.08	64 *7*
4 - 17	0.61	8.4 *7*
18	1.38	7.7 *8*
19 - 20	0.37	1.82 *17*
21	0.47	2.27 *23*
22 - 25	0.089	0.39 *3*
27	1.76	6.5 *5*
30	0.59	1.97 *15*
54	0.96	1.80 *13*
75 - 79	0.21	0.277 *13*
410	2.93	0.714 *15*
430 - 434	0.493	0.1145 *17*
590	1.56	0.264 *8*
611 - 614	0.245	0.0401 *10*
699	1.22	0.174 *5*
719 - 723	0.187	0.0260 *6*

Continuous Radiation (^{108}Ag)

$\langle$IB$\rangle$=0.031 keV

E_{bin}(keV)		$\langle\ \rangle$(keV)	(%)
10 - 20	IB	0.0087	0.047
20 - 40	IB	0.019	0.083
40 - 100	IB	0.00155	0.0025
100 - 254	IB	0.00160	0.00117

$^{108}_{48}$Cd(stable)

Δ: -89260 *6* keV
%: 0.89 *1*

$^{108}_{49}$In(39.6 *7* min)

Mode: ϵ
Δ: -84135 *11* keV
SpA: 4.40$\times10^{7}$ Ci/g
Prod: ^{107}Ag(α,3n); ^{108}Cd(p,n); ^{107}Ag(^{3}He,2n); ^{94}Mo(^{16}O,np)

Photons (^{108}In)

$\langle\gamma\rangle$=2217 *36* keV

γ_{mode}	γ(keV)	γ(%)†
Cd L_ℓ	2.767	0.049 *9*
Cd L_η	2.957	0.023 *4*
Cd L_α	3.133	1.35 *21*
Cd L_β	3.392	0.95 *16*
Cd L_γ	3.779	0.094 *17*
Cd $K_{\alpha2}$	22.984	10.1 *4*
Cd $K_{\alpha1}$	23.174	18.9 *7*
Cd $K_{\beta1}'$	26.085	4.82 *17*
Cd $K_{\beta2}'$	26.801	1.00 *4*
γ	156.28 *20*	0.160 *23*
γ	171.48 *20*	0.79 *5*
γ[M1+E2]	311.90 *23*	1.01 *7*
γ	391.38 *10*	0.36 *5*
γ	536.18 *10*	0.81 *7*
γE2	632.99 *4*	76.4
γ[M1+E2]	771.05 *18*	0.32 *5*
γE2	875.48 *8*	2.44 *15*
γ	884.2 *3*	0.28 *6*
γ	936.1 *3*	0.19 *6*
γM1,E2	968.24 *17*	4.38 *23*
γ	1017.9 *3*	0.15 *5*
γ[M1+E2]	1087.5 *3*	1.53 *11*
γ[M1+E2]	1280.14 *22*	0.50 *7*
γ	1293.8 *3*	0.72 *8*
γ	1408.6 *3*	~0.15
γ[M1+E2]	1445.5 *3*	0.24 *6*
γ	1475.1 *3*	0.59 *10*
γ	1513.18 *20*	0.91 *10*
γE1	1529.6 *3*	7.3 *4*
γ[E1]	1569.4 *3*	0.98 *10*
γE2	1601.22 *17*	4.05 *23*
γM1,E2	1732.1 *3*	3.82 *23*
γ[M1+E2]	1851.4 *3*	3.06 *23*
γ	1864.1 *4*	0.53 *15*
γ[E2]	1913.12 *22*	0.099 *23*
γM1,E2	1986.2 *3*	12.4 *7*
γ[M1+E2]	2048.54 *23*	3.1 *3*
γ	2112.5 *3*	0.25 *3*
γ[M1+E2]	2210.6 *3*	0.53 *8*
γ[M1+E2]	2224.3 *5*	1.38 *15*
γ	2278.4 *5*	0.63 *9*
γ	2317.0 *6*	0.31 *8*
γ[E2]	2365.1 *3*	0.53 *8*
γ[M1+E2]	2413.77 *24*	1.45 *8*
γ[E2]	2681.52 *23*	0.76 *8*

Photons (^{108}In) (continued)

γ_{mode}	γ(keV)	γ(%)†
γ	2816.1 *10*	0.70 *11*
γ [E2]	3046.74 *24*	2.44 *23*
γ [M1+E2]	3178.9 *3*	1.60 *23*
γ [E2]	3452.55 *23*	9.2 *5*
γ	3689.5 *22*	1.07 *15*
γ [E2]	3811.8 *3*	4.2 *3*
γ [E2]	3825.5 *5*	2.37 *23*
γ	4052.1 *25*	0.61 *23*
γ	4342.9 *23*	0.76 *23*

† 0.79% uncert(syst)

Atomic Electrons (^{108}In)

⟨e⟩=4.86 *17* keV

e_{bin}(keV)	⟨e⟩(keV)	e(%)
4	1.18	33 *3*
19	0.69	3.6 *4*
20	0.173	0.88 *9*
22	0.199	0.90 *9*
23	0.211	0.93 *10*
25 - 26	0.040	0.158 *12*
130	0.030	~0.023
145	0.12	~0.08
152 - 171	0.038	0.023 *9*
285 - 312	0.074	0.026 *3*
365 - 391	0.012	0.0032 *16*
509 - 536	0.016	0.0031 *14*
606	1.40	0.231 *5*
629	0.185	0.0294 *6*
632 - 633	0.0433	0.00685 *12*
744 - 771	0.0055	0.00073 *12*
849 - 884	0.0353	0.0041 *3*
909 - 942	0.050	0.0053 *6*
964 - 991	0.0084	0.00087 *10*
1014 - 1061	0.0149	0.00140 *17*
1083 - 1087	0.00219	0.000202 *21*
1253 - 1294	0.0095	0.00075 *16*
1382 - 1419	0.0026	0.00019 *5*
1442 - 1529	0.034	0.00229 *18*
1543 - 1601	0.0309	0.00196 *10*
1705 - 1732	0.0257	0.00151 *13*
1825 - 1913	0.0225	0.00123 *12*
1959 - 2048	0.091	0.0046 *3*
2086 - 2184	0.0036	0.000168 *25*
2198 - 2290	0.0114	0.00051 *6*
2313 - 2410	0.0099	0.00042 *3*
2655 - 2681	0.0034	0.000129 *12*
2789 - 2815	0.0024	8.7 *25* $\times 10^{-5}$
3020 - 3046	0.0100	0.00033 *3*
3152 - 3178	0.0064	0.00020 *3*
3426 - 3452	0.0346	0.00101 *6*
3663 - 3689	0.0031	8.6 *21* $\times 10^{-5}$
3785 - 3825	0.0233	0.00061 *4*
4025 - 4051	0.0017	4.3 *16* $\times 10^{-5}$
4316 - 4342	0.0021	4.8 *16* $\times 10^{-5}$

Continuous Radiation (^{108}In)

⟨β+⟩=694 keV;⟨IB⟩=3.8 keV

E_{bin}(keV)		⟨ ⟩(keV)	(%)
0 - 10	β+	3.78×10^{-6}	4.57×10^{-5}
	IB	0.024	
10 - 20	β+	0.000129	0.00078
	IB	0.025	0.17
20 - 40	β+	0.00305	0.0094
	IB	0.062	0.23
40 - 100	β+	0.122	0.158
	IB	0.132	0.20
100 - 300	β+	4.96	2.27
	IB	0.39	0.22
300 - 600	β+	31.3	6.8
	IB	0.53	0.124

Continuous Radiation (^{108}In) (continued)

E_{bin}(keV)		⟨ ⟩(keV)	(%)
600 - 1300	β+	181	19.2
	IB	1.12	0.126
1300 - 2500	β+	376	20.7
	IB	1.13	0.064
2500 - 4492	β+	100	3.63
	IB	0.39	0.0131
	Σβ+		53

$^{108}_{49}$In(58.0 *12* min)

Mode: ε

Δ: -84135 *11* keV

SpA: 3.00×10^{7} Ci/g

Prod: ^{107}Ag(α,3n); ^{108}Cd(p,n); ^{107}Ag(^{3}He,2n); ^{94}Mo(^{16}O,np)

Photons (^{108}In)

⟨γ⟩=2938 *110* keV

γ_{mode}	γ(keV)	γ(%)
Cd L_{ℓ}	2.767	0.078 *14*
Cd L_{η}	2.957	0.037 *6*
Cd L_{α}	3.133	2.1 *3*
Cd L_{β}	3.392	1.51 *25*
Cd L_{γ}	3.779	0.15 *3*
Cd $K_{\alpha2}$	22.984	15.9 *6*
Cd $K_{\alpha1}$	23.174	30.0 *11*
Cd $K_{\beta1}$'	26.085	7.6 *3*
Cd $K_{\beta2}$'	26.801	1.58 *6*
γ	206.13 *10*	0.49 *10*
γ M1	242.73 *7*	38 *3*
γ	266.52 *7*	3.0 *4*
γ [M1+E2]	268.32 *22*	0.25 *6*
γ (E2)	302.59 *12*	0.15 *3*
γ M1,E2	325.71 *7*	13.0 *10*
γ E2	350.53 *15*	0.40 *8*
γ M1+24%E2	373.79 *19*	0.58 *12*
γ [E1]	434.18 *22*	0.43 *9*
γ	448.88 *13*	0.45 *9*
γ E2	456.01 *15*	0.42 *9*
γ E1	516.39 *17*	0.34 *7*
γ E2	569.11 *10*	5.1 *7*
γ E2	632.99 *4*	99.7
γ	648.41 *10*	4.2 *6*
γ M1,E2	730.96 *6*	8.1 *9*
γ	754.88 *8*	3.8 *5*
γ	826.26 *13*	1.00 *20*
γ E2	875.48 *8*	95 *9*
γ	1008.62 *11*	0.62 *13*
γ E2	1032.88 *7*	25.7 *20*
γ M1+1.2%E2	1056.67 *6*	31 *3*
γ E1+0.2%M2	1093.27 *9*	4.7 *6*
γ E1+0.3%M2	1198.74 *9*	3.8 *6*
γ	1251.34 *11*	0.48 *10*
γ	1275.13 *11*	0.74 *15*
γ E2	1299.39 *6*	16 *4*
γ E2	1485.83 *8*	4.3 *6*
γ M1,E2	1606.43 *8*	7.5 *19*
γ	1805.41 *15*	0.77 *16*
γ	1859.13 *13*	0.56 *11*
γ	1910.88 *16*	0.18 *4*
γ	1947.48 *15*	0.59 *12*
γ	2308.00 *11*	0.79 *16*

Atomic Electrons (^{108}In)

⟨e⟩=13.5 *4* keV

e_{bin}(keV)	⟨e⟩(keV)	e(%)
4	1.88	52 *5*
19	1.09	5.7 *6*
20 - 22	0.59	2.82 *20*
23	0.34	1.48 *15*
25 - 26	0.064	0.251 *19*
179 - 206	0.06	~0.034
216	2.99	1.38 *11*
239	0.41	0.170 *14*
240 - 276	0.35	0.14 *5*
299	0.74	0.25 *4*
302 - 351	0.18	0.056 *8*
370 - 407	0.0088	0.0023 *3*
422 - 456	0.028	0.0065 *15*
490 - 516	0.0033	0.00068 *12*
542 - 569	0.130	0.024 *3*
606	1.83	0.302 *7*
622 - 648	0.36	0.058 *5*
704 - 751	0.19	0.027 *4*
754 - 800	0.011	0.0014 *7*
822 - 826	0.0015	0.00018 *8*
849	1.09	0.128 *12*
871 - 875	0.169	0.0193 *14*
982 - 1030	0.61	0.060 *4*
1032 - 1067	0.074	0.0071 *5*
1089 - 1093	0.0027	0.000243 *25*
1172 - 1199	0.0158	0.00134 *19*
1225 - 1273	0.13	0.0098 *23*
1274 - 1299	0.017	0.0013 *3*
1459 - 1486	0.032	0.0022 *3*
1580 - 1606	0.054	0.0034 *8*
1779 - 1859	0.0065	0.00036 *10*
1884 - 1947	0.0036	0.00019 *6*
2281 - 2307	0.0032	0.00014 *5*

Continuous Radiation (^{108}In)

⟨β+⟩=161 keV;⟨IB⟩=2.2 keV

E_{bin}(keV)		⟨ ⟩(keV)	(%)
0 - 10	β+	6.4×10^{-6}	7.8×10^{-5}
	IB	0.0072	
10 - 20	β+	0.000220	0.00134
	IB	0.0097	0.063
20 - 40	β+	0.0052	0.0160
	IB	0.038	0.149
40 - 100	β+	0.208	0.269
	IB	0.038	0.059
100 - 300	β+	8.2	3.77
	IB	0.116	0.064
300 - 600	β+	46.3	10.2
	IB	0.24	0.055
600 - 1300	β+	105	12.6
	IB	0.99	0.106
1300 - 2500	β+	2.22	0.162
	IB	0.74	0.047
2500 - 2584	IB	3.5×10^{-5}	1.37×10^{-6}
	Σβ+		27

$^{108}_{50}$Sn(10.30 *8* min)

Mode: ε

Δ: -82090 *30* keV

SpA: 1.689×10^{8} Ci/g

Prod: ^{106}Cd(α,2n)

Photons (^{108}Sn)

$\langle\gamma\rangle$=641 *8* keV

γ_{mode}	γ(keV)	γ(%)†
In L_ℓ	2.905	0.136 *25*
In L_η	3.112	0.065 *11*
In L_α	3.286	3.7 *6*
In L_β	3.569	2.7 *5*
In L_γ	3.981	0.28 *5*
In $K_{\alpha2}$	24.002	25.8 *15*
In $K_{\alpha1}$	24.210	48 *3*
In $K_{\beta1}$'	27.265	12.5 *7*
In $K_{\beta2}$'	28.022	2.62 *16*
γ M1(+<17%E2)	36.52 *9*	0.41 *10*
γ E2+47%M1	104.41 *8*	13.2 *4*
γ M1	168.36 *8*	19.0 *4*
γ M1(+E2)	236.24 *8*	5.94 *24*
γ E2	272.77 *7*	43.4 *9*
γ	363.1 *3*	<0.12
γ M1+37%E2	396.42 *7*	61.2 *12*
γ (M1+22%E2)	492.75 *10*	0.98 *18*
γ [M1+E2]	500.82 *9*	1.04 *24*
γ	565.08 *15*	<0.12
γ (E2)	669.18 *8*	21.5 *4*
γ	829.4 *5*	<0.12
γ	847.7 *4*	<0.12
γ	858.8 *6*	<0.12
γ (E2+45%M1)	889.17 *10*	3.1 *3*
γ	903.6 *6*	<0.12
γ (E2)	1161.93 *9*	0.86 *24*
γ	1231.1 *5*	<0.12
γ	1654.5 *5*	<0.12
γ	1684.9 *6*	<0.12
γ	1957.3 *6*	<0.12

† 0.93% uncert(syst)

Atomic Electrons (^{108}Sn)

$\langle e\rangle$=31 *3* keV

e_{bin}(keV)	$\langle e\rangle$(keV)	e(%)
4	3.3	85 *10*
9 - 19	0.57	5.2 *10*
20	1.89	9.4 *11*
23 - 32	1.27	5.2 *4*
33	0.75	2.3 *6*
36 - 37	0.24	0.67 *17*
76	7.2	9 *4*
100	1.5	1.5 *8*
101 - 104	1.3	~1
140	2.83	2.01 *6*
164 - 208	1.17	0.63 *6*
232 - 236	0.13	0.058 *12*
245	4.16	1.70 *5*
269 - 273	0.886	0.329 *8*
335 - 363	0.0026	~0.0008
368	2.60	0.706 *23*
392 - 396	0.45	0.114 *7*
465 - 501	0.071	0.0150 *20*
537 - 565	0.0013	~0.00024
641 - 669	0.444	0.069 *3*
801 - 848	0.0022	~0.00027
855 - 904	0.047	0.0055 *6*
1134 - 1162	0.0086	0.00076 *19*
1203 - 1231	0.0005	~4×10^{-5}
1627 - 1684	0.0007	~4×10^{-5}
1929 - 1957	0.0003	~2×10^{-5}

Continuous Radiation (^{108}Sn)

$\langle$IB$\rangle$=0.65 keV

E_{bin}(keV)		$\langle\ \rangle$(keV)	(%)
10 - 20	IB	0.0019	0.0108
20 - 40	IB	0.040	0.161
40 - 100	IB	0.0035	0.0054
100 - 300	IB	0.037	0.018
300 - 600	IB	0.17	0.038
600 - 1300	IB	0.38	0.046
1300 - 1776	IB	0.0166	0.00117

$^{108}_{51}$Sb(7.0 *5* s)

Mode: ϵ, ϵp

Δ: -72530 *300* keV syst

SpA: 1.42×10^{10} Ci/g

Prod: ^{112}Sn(p,5n); ^{58}Ni on ^{58}Ni

Photons (^{108}Sb)

γ_{mode}	γ(keV)	γ(rel)
γ [E2]	905.0 *5*	25 *8*
γ [E2]	1206.0 *5*	100

$^{108}_{52}$Te(2.1 *1* s)

Mode: α(68 *12* %), ϵ(32 *12* %), ϵp

Δ: -65620 *400* keV syst

SpA: 4.24×10^{10} Ci/g

Prod: ^{96}Ru(^{16}O,4n); ^{58}Ni on ^{58}Ni

ϵp: 2300-4240 range

Alpha Particles (^{108}Te)

α(keV)
3320 *20*

A = 109

NP **41**, 111 (1984)

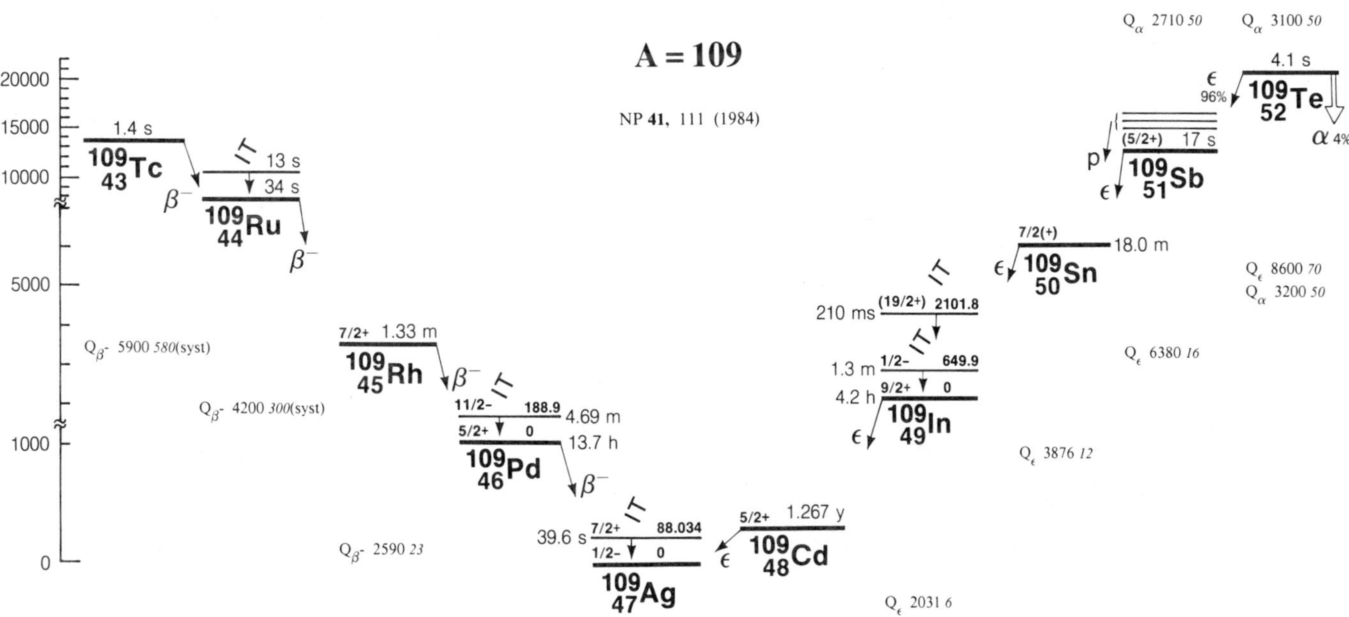

$^{109}_{43}$Tc(1.4 *4* s)

Mode: β-

Δ: -74910 *500* keV syst

SpA: 5.8×10^{10} Ci/g

Prod: fission

$^{109}_{44}$Ru(34.5 *10* s)

Mode: β-

Δ: -80810 *300* keV syst

SpA: 2.97×10^{9} Ci/g

Prod: fission

Photons (^{109}Ru)

γ_{mode}	γ(keV)	γ(rel)
γ	67.7 *2*	16.2 *15*
γ	172.5 *5*	1.8 *5*
γ	183.5 *5*	6.6 *4*
γ	193.9 *5*	5.1 *6*
γ	200.1 *5*	1.8 *5*
γ [M1+E2]	206.00 *19*	100
γ	218.1 *5*	1.3 *4*
γ [E2]	220.30 *19*	10.4 *15*
γ	225.7 *2*	78.1 *16*
γ	232.5 *5*	1.6 *2*
γ	239.5 *5*	3.7 *4*
γ	244.8 *5*	7.6 *7*
γ	248.9 *5*	3.5 *10*
γ	252.1 *5*	5.5 *5*
γ	312.8 *10*	1.0 *2*
γ	366.5 *5*	7.0 *12*
γ	382.2 *5*	2.4 *7*
γ [M1+E2]	426.30 *25*	46 *5*
γ	465.4 *2*	11.6 *15*
γ	626.2 *5*	~2
γ	671.6 *5*	5 *1*
γ	819.9 *5*	20.0 *16*
γ	839.8 *5*	6.0 *14*
γ	846.6 *5*	5.9 *10*
γ	890.0 *5*	11.6 *7*
γ	951.8 *5*	3.7 *10*
γ	1011.6 *5*	10.1 *14*
γ	1026.0 *5*	6.7 *6*
γ	1052.8 *5*	2.8 *5*
γ	1112.7 *5*	5.6 *10*
γ	1237.1 *5*	4.1 *5*
γ	1256.6 *5*	3.0 *5*
γ	1291.2 *5*	2.0 *4*
γ	1304.8 *5*	21.6 *20*
γ	1501.8 *5*	18.4 *18*
γ	1510.9 *5*	6.8 *10*
γ	1536.3 *5*	15.8 *16*
γ	1666.7 *5*	3.7 *7*
γ	1719.6 *5*	7.5 *10*
γ	1722.5 *5*	7.5 *10*
γ	1734.8 *5*	7 *1*
γ	1756.8 *5*	12.7 *15*
γ	1835.5 *5*	5.7 *12*
γ	1963.0 *5*	6.9 *10*
γ	1978.7 *5*	57 *6*

$^{109}_{44}$Ru(12.9 *10* s)

Mode: IT

Δ: >-80810 keV syst

SpA: 7.8×10^{9} Ci/g

Prod: fission

$^{109}_{45}$Rh(1.33 *3* min)

Mode: β-

Δ: -85014 *22* keV

SpA: 1.29×10^{9} Ci/g

Prod: fission

Photons (^{109}Rh)

⟨γ⟩=312 *20* keV

γ_{mode}	γ(keV)	γ(%)†
Pd L_{ℓ}	2.503	~0.033
Pd L_{η}	2.660	~0.017
Pd L_{α}	2.838	~0.9
Pd L_{β}	3.047	~0.6
Pd L_{γ}	3.362	~0.06
Pd $K_{\alpha2}$	21.020	4.8 *12*
Pd $K_{\alpha1}$	21.177	9.1 *23*
Pd $K_{\beta1}$'	23.811	2.3 *6*
Pd $K_{\beta2}$'	24.445	0.44 *11*
γ [M1+E2]	25.09 *6*	~0.03
γ M1(+22%E2)	35.42 *5*	1.30 *16*
γ (M1)	50.56 *6*	0.032 *11*
γ (M1)	58.99 *7*	0.032 *11*
γ M1(+43%E2)	81.80 *4*	0.70 *5*
γ E2	113.40 *4*	5.7 *3*
γ [M1+E2]	149.87 *6*	0.59 *5*
γ M1	152.94 *5*	0.65 *5*
γ [M1+E2]	166.22 *8*	~0.05
γ M1	178.04 *4*	7.6 *4*
γ M1+3.1%E2	200.12 *5*	0.49 *5*
γ M1	211.93 *6*	0.65 *11*
γ [M1+E2]	213.79 *5*	0.54 *11*
γ M1+8.8%E2	215.31 *6*	1.73 *11*
γ E1	245.05 *5*	1.30 *11*
γ M1+11%E2	249.20 *4*	5.8 *3*
γ [M1+E2]	264.35 *6*	0.38 *11*
γ E2	266.34 *6*	0.27 *5*
γ [E2]	274.30 *6*	~0.11
γ M1+21%E2	276.30 *5*	2.16 *16*
γ M1	291.44 *4*	7.5 *4*
γ	295.59 *5*	0.32 *5*
γ	320.8 *5*	~0.05
γ M1+E2	325.33 *6*	1.5 *3*
γ M1+23%E2	326.85 *4*	54
γ E2+M1	378.16 *5*	1.24 *11*
γ	391 *1*	~0.05 ?
γ M1+E2	426.16 *6*	7.7 *7*
γ [E2]	427.24 *4*	<0.27 ?
γ [M1+E2]	491.55 *6*	0.38 *5*
γ [M1+E2]	540.64 *4*	0.49 *5*
γ	597.1 *5*	~0.11
γ	617.9 *10*	~0.11
γ	692 *2*	~0.11 ?
γ	1041.7 *5*	0.11 *3*
γ	1072 *1*	~0.05
γ	1318.0 *5*	~0.11

† 11% uncert(syst)

Atomic Electrons (^{109}Rh)

⟨e⟩=19 *3* keV

e_{bin}(keV)	⟨e⟩(keV)	e(%)
1	0.0023	~0.3
3	0.8	~25
4	0.038	1.04 *24*
11	1.4	13 *5*
17 - 21	0.68	3.6 *6*
22	0.6	<6
23 - 26	0.21	~0.9
32	3.1	~10
35	0.8	~2
47 - 56	0.0090	0.018 *4*
57	0.5	~0.9
58 - 82	0.21	0.27 *12*
89	3.55	3.99 *24*
110	0.97	0.88 *5*
113 - 153	0.51	0.42 *5*
154	0.80	0.52 *3*
163 - 212	0.49	0.265 *15*
213 - 261	0.70	0.302 *12*
263 - 301	0.56	0.205 *11*
303	2.5	0.83 *9*
317 - 367	0.47	0.144 *11*
375 - 424	0.29	0.071 *7*
425 - 467	0.0168	0.0037 *4*
488 - 537	0.0132	0.0026 *3*
540 - 573	0.0016	~0.00029
594 - 618	0.0017	~0.00029
668 - 692	0.0013	~0.00019
1017 - 1048	0.0011	0.00011 *4*
1068 - 1072	5.0 ×10^{-5}	~5×10^{-6}
1294 - 1318	0.0006	~5×10^{-5}

Continuous Radiation (^{109}Rh)

⟨β-⟩=912 keV;⟨IB⟩=1.9 keV

E_{bin}(keV)		⟨ ⟩(keV)	(%)
0 - 10	β-	0.0141	0.281
	IB	0.035	
10 - 20	β-	0.0433	0.288
	IB	0.034	0.24
20 - 40	β-	0.180	0.60
	IB	0.067	0.23
40 - 100	β-	1.40	1.97
	IB	0.19	0.29
100 - 300	β-	17.8	8.6
	IB	0.51	0.29
300 - 600	β-	82	17.9
	IB	0.51	0.122
600 - 1300	β-	441	47.0
	IB	0.51	0.062
1300 - 2500	β-	369	23.3
	IB	0.071	0.0049
2500 - 2590	β-	0.067	0.00266
	IB	2.2×10^{-7}	8.7×10^{-9}

$^{109}_{46}$Pd(13.7 *1* h)

Mode: β-

Δ: -87604 *4* keV

SpA: 2.099×10^{6} Ci/g

Prod: ^{108}Pd(n,γ)

Photons (^{109}Pd)

⟨γ⟩=11.7 *6* keV

γ_{mode}	γ(keV)	γ(%)†
Ag L_ℓ	2.634	0.086 *17*
Ag L_η	2.806	0.053 *10*
Ag L_α	2.984	2.4 *4*
Ag L_β	3.206	1.8 *3*
Ag L_γ	3.543	0.17 *3*
Ag $K_{\alpha 2}$	21.990	9.9 *10*
Ag $K_{\alpha 1}$	22.163	18.7 *19*
Ag $K_{\beta 1}'$	24.934	4.7 *5*
Ag $K_{\beta 2}'$	25.603	0.93 *10*
γ M1+9.1%E2	44.73 *10*	0.00110 *12*
γ E3	88.0341 *11*	3.6 *3* *
γ M1+0.2%E2	103.84 *16*	0.00093 *20*
γ M1+E2	134.17 *10*	0.0013 *3*
γ M1+E2	145.11 *11*	0.00112 *20*
γ [M1+E2]	286.69 *20*	0.00014 *4*
γ E1	309.13 *17*	0.0049 *15*
γ M1+3.9%E2	311.37 *8*	0.032 *3*
γ M1+5.5%E2	390.52 *14*	0.00093 *20*
γ	395.61 *17*	0.00017 *7*
γ E1+5.9%M2	412.97 *11*	0.0066 *10*
γ E2	415.21 *15*	0.0107 *10*
γ E1+6.9%M2	423.91 *10*	0.00095 *20*
γ M1+2.6%E2	447.53 *17*	0.00083 *20*
γ [E1]	454.24 *16*	0.00054 *22*
γ [M1+E2]	496.89 *20*	7.6 *15* $\times10^{-5}$
γ M1+7.4%E2	551.37 *20*	0.00061 *15*
γ E1+6.4%M2	558.08 *11*	0.0024 *3*
γ	602.52 *9*	0.0080 *5*
γ [E2]	636.30 *9*	0.0100 *5*
γ [M1+E2]	647.25 *8*	0.0244
γ M1+0.09%E2	701.89 *14*	0.0031 *3*
γ	706.98 *17*	0.00159 *20*
γ [E1]	724.34 *9*	0.00019 *5*
γ	736.69 *11*	0.00168 *20*
γ	778.2 *3*	0.0015 *5*
γ [M1+E2]	781.42 *10*	0.0112 *12*
γ [M1+E2]	822.9 *3*	0.00019 *3*
γ E2	862.74 *19*	0.00013 *3*
γ	965.7 *3*	9 *3* $\times10^{-5}$
γ	1010.5 *3*	6.1 *20* $\times10^{-5}$

† 2.9% uncert(syst)
* with ^{109}Ag(39.6 s) in equilib

Atomic Electrons (^{109}Pd)

⟨e⟩=77 *4* keV

e_{bin}(keV)	⟨e⟩(keV)	e(%)
3 - 45	3.7	76 *7*
63	25.6	41 *4*
78	0.00024	0.00031 *7*
84	2.7	3.3 *3*
85	35.2	42 *4*
87	8.0	9.2 *9*
88 - 134	1.50	1.70 *16*
141 - 145	7.6 $\times10^{-5}$	5.3 *20* $\times10^{-5}$
261 - 309	0.00194	0.00067 *5*
365 - 414	0.00065	0.000168 *11*
415 - 454	4.2 $\times10^{-5}$	9.8 *15* $\times10^{-6}$
471 - 497	2.3 $\times10^{-6}$	4.8 *9* $\times10^{-7}$
526 - 558	4.7 $\times10^{-5}$	8.9 *10* $\times10^{-6}$
577 - 622	0.00072	0.000118 *11*
632 - 681	0.000162	2.47 *17* $\times10^{-5}$
698 - 737	3.2 $\times10^{-5}$	4.5 *13* $\times10^{-6}$
753 - 797	0.000194	2.6 *3* $\times10^{-5}$
819 - 863	2.0 $\times10^{-6}$	2.4 *4* $\times10^{-7}$
940 - 985	1.2 $\times10^{-6}$	1.3 *5* $\times10^{-7}$
1007 - 1010	6.5 $\times10^{-8}$	7 *3* $\times10^{-9}$

Continuous Radiation (^{109}Pd)

⟨β-⟩=361 keV;⟨IB⟩=0.39 keV

E_{bin}(keV)		⟨ ⟩(keV)	(%)
0 - 10	β-	0.064	1.28
	IB	0.0169	
10 - 20	β-	0.194	1.30
	IB	0.0162	0.113
20 - 40	β-	0.79	2.64
	IB	0.031	0.106
40 - 100	β-	5.9	8.3
	IB	0.078	0.121
100 - 300	β-	61	30.4
	IB	0.156	0.092
300 - 600	β-	173	39.5
	IB	0.079	0.020
600 - 1028	β-	119	16.6
	IB	0.0096	0.00145

$^{109}_{46}$Pd(4.69 *1* min)

Mode: IT
Δ: -87415 *4* keV
SpA: 3.674×10^{8} Ci/g
Prod: ^{108}Pd(n,γ)

Photons (^{109}Pd)

⟨γ⟩=111.1 *14* keV

γ_{mode}	γ(keV)	γ(%)†
Pd L_ℓ	2.503	0.040 *7*
Pd L_η	2.660	0.021 *3*
Pd L_α	2.838	1.08 *17*
Pd L_β	3.046	0.76 *12*
Pd L_γ	3.367	0.074 *13*
Pd $K_{\alpha 2}$	21.020	7.6 *3*
Pd $K_{\alpha 1}$	21.177	14.4 *5*
Pd $K_{\beta 1}'$	23.811	3.58 *13*
Pd $K_{\beta 2}'$	24.445	0.69 *3*
γ E3	188.9 *1*	55.8

† 1.3% uncert(syst)

Atomic Electrons (^{109}Pd)

⟨e⟩=77.4 *13* keV

e_{bin}(keV)	⟨e⟩(keV)	e(%)
3	0.96	30 *3*
4	0.057	1.57 *16*
17	0.157	0.91 *9*
18	0.55	3.1 *3*
20	0.139	0.69 *7*
21	0.187	0.90 *9*
23	0.0238	0.103 *11*
24	0.0076	0.032 *3*
165	52.5	31.9 *8*
185	5.67	3.06 *7*
186	12.8	6.88 *17*
188	3.69	1.96 *5*
189	0.652	0.345 *8*

$^{109}_{47}$Ag(stable)

Δ: -88720 *3* keV
%: 48.161 *5*

$^{109}_{47}$Ag(39.6 *2* s)

Mode: IT
Δ: -88632 *3* keV
SpA: 2.591×10^{9} Ci/g
Prod: daughter ^{109}Pd; daughter ^{109}Cd

Photons (^{109}Ag)

⟨γ⟩=11.0 *3* keV

γ_{mode}	γ(keV)	γ(%)†
Ag L_ℓ	2.634	0.086 *16*
Ag L_η	2.806	0.052 *8*
Ag L_α	2.984	2.3 *4*
Ag L_β	3.206	1.8 *3*
Ag L_γ	3.543	0.17 *3*
Ag $K_{\alpha 2}$	21.990	9.9 *5*
Ag $K_{\alpha 1}$	22.163	18.6 *9*
Ag $K_{\beta 1}'$	24.934	4.69 *23*
Ag $K_{\beta 2}'$	25.603	0.93 *5*
γ E3	88.0341 *11*	3.6

† 2.8% uncert(syst)

Atomic Electrons (^{109}Ag)

⟨e⟩=76.5 *18* keV

e_{bin}(keV)	⟨e⟩(keV)	e(%)
3	1.39	41 *5*
4	0.95	27 *3*
18	0.37	2.01 *22*
19	0.52	2.8 *3*
21	0.202	0.95 *10*
22	0.213	0.98 *11*
24	0.028	0.115 *12*
25	0.0126	0.051 *6*
63	25.5	40.8 *17*
84	2.73	3.24 *13*
85	35.1	41.5 *17*
87	8.0	9.1 *4*
88	1.49	1.69 *7*

$^{109}_{48}$Cd(1.2665 *11* yr)

Mode: ϵ
Δ: -88536 *4* keV
SpA: 2589.5 Ci/g
Prod: ^{108}Cd(n,γ); ^{109}Ag(d,2n)

Photons (^{109}Cd)

⟨γ⟩=26.0 *6* keV

γ_{mode}	γ(keV)	γ(%)†
Ag L_ℓ	2.634	0.18 *3*
Ag L_η	2.806	0.096 *15*
Ag L_α	2.984	5.0 *8*
Ag L_β	3.214	3.6 *6*
Ag L_γ	3.570	0.35 *6*
Ag $K_{\alpha2}$	21.990	28.9 *11*
Ag $K_{\alpha1}$	22.163	54.5 *21*
Ag $K_{\beta1}'$	24.934	13.7 *5*
Ag $K_{\beta2}'$	25.603	2.72 *11*
γ E3	88.0341 *11*	3.6 *

† 2.8% uncert(syst)

* with ^{109}Ag(39.6 s) in equilib

Atomic Electrons (^{109}Cd)

⟨e⟩=81.3 *18* keV

e_{bin}(keV)	⟨e⟩(keV)	e(%)
3	2.7	81 *8*
4	1.88	53 *6*
18	1.07	5.9 *6*
19	1.53	8.2 *9*
21	0.59	2.8 *3*
22	0.62	2.9 *3*
24	0.081	0.34 *4*
25	0.037	0.148 *15*
63	25.5	40.8 *16*
84	2.73	3.24 *13*
85	35.1	41.5 *16*
87	8.0	9.1 *4*
88	1.49	1.69 *7*

Continuous Radiation (^{109}Cd)

⟨IB⟩=0.033 keV

E_{bin}(keV)		⟨ ⟩(keV)	(%)
10 - 20	IB	0.0068	0.037
20 - 40	IB	0.026	0.110
40 - 96	IB	0.00041	0.00083

$^{109}_{49}$In(4.2 *1* h)

Mode: ϵ

Δ: -86505 *7* keV

SpA: 6.85×10^6 Ci/g

Prod: ^{107}Ag(α,2n); daughter ^{109}Sn

Photons (^{109}In)

⟨γ⟩=590 *9* keV

γ_{mode}	γ(keV)	γ(%)†
Cd L_ℓ	2.767	0.103 *18*
Cd L_η	2.957	0.048 *7*
Cd L_α	3.133	2.8 *4*
Cd L_β	3.392	2.0 *3*
Cd L_γ	3.779	0.20 *4*
Cd $K_{\alpha2}$	22.984	20.7 *7*

Photons (^{109}In) (continued)

γ_{mode}	γ(keV)	γ(%)†
Cd $K_{\alpha1}$	23.174	39.0 *14*
Cd $K_{\beta1}'$	26.085	9.9 *4*
Cd $K_{\beta2}'$	26.801	2.06 *8*
γ E2	59.6 *3*	0.18 *4*
γ	74.7 *10*	
γ	84 *1*	
γ	174.6 *10*	
γ M1	203.27 *14*	73.5
γ	222.93 *15*	<0.22
γ M2	259.3 *3*	0.059 *7*
γ M1,E2	287.8 *3*	1.76 *15*
γ [E2]	324.0 *3*	0.37 *7*
γ [M1+E2]	325.7 *3*	0.54 *3*
γ M1,E2	347.39 *22*	2.21 *15*
γ	420.37 *24*	0.99 *7*
γ M1,E2	426.20 *17*	4.23 *15*
γ	460.8 *3*	0.103 *15*
γ	482.2 *4*	0.125 *22*
γ	518.3 *3*	
γ	529.3 *3*	0.59 *3*
γ	542.1 *9*	0.073 *7*
γ	570.9 *3*	0.07 *3*
γ	580.8 *5*	0.081 *22*
γ	584.5 *4*	0.294 *22*
γ	599.2 *5*	<0.22
γ [E2]	613.6 *3*	2.50 *22*
γ	619.23 *24*	1.76 *22*
γ	619.7 *4*	0.191 *22*
γ M1,E2	623.63 *23*	6.0 *3*
γ	630.2 *3*	0.073 *22*
γ [E2]	649.7 *3*	3.01 *22*
γ	653.0 *3*	1.91 *15*
γ	678.7 *3*	0.99 *7*
γ	703.9 *6*	<0.07
γ	721.6 *3*	0.90 *7*
γ	728.2 *3*	0.26 *4*
γ	731.2 *3*	0.39 *4*
γ	753.9 *3*	0.44 *7*
γ [M1+E2]	793.83 *25*	0.59 *15*
γ	799.7 *3*	<0.7
γ	822.50 *24*	1.40 *7*
γ	831.5 *4*	0.26 *4*
γ	851.9 *3*	0.26 *4*
γ	862.5 *5*	0.17 *5*
γ	891.0 *3*	0.28 *5*
γ	900.6 *3*	0.47 *7*
γ	925.63 *23*	0.88 *15*
γ	930.2 *5*	0.29 *7*
γ	949.1 *3*	1.54 *22*
γ	962.5 *7*	0.096 *22*
γ	998.6 *3*	0.71 *4*
γ	1004.4 *3*	0.221 *22*
γ	1049.3 *3*	1.16 *6*
γ	1105.8 *4*	0.32 *3*
γ	1148.56 *22*	4.3 *4*
γ	1186.5 *5*	<0.29
γ	1196.0 *3*	1.76 *22*
γ	1272.28 *25*	0.59 *4*
γ	1346.2 *5*	0.59 *4*
γ	1351.83 *22*	0.74 *7*
γ	1386.0 *14*	0.037 *15*
γ	1418.93 *25*	1.32 *15*
γ	1429.7 *10*	0.088 *22*
γ	1475.54 *25*	0.44 *7*
γ	1539.2 *9*	0.140 *22*
γ	1569.2 *5*	0.132 *22*
γ	1622.20 *24*	2.06 *22*
γ	1772.4 *5*	0.44 *7*
γ	1848.6 *14*	<0.015
γ	1859.6 *5*	~0.029

† 0.68% uncert(syst)

Atomic Electrons (^{109}In)

⟨e⟩=15.9 *4* keV

e_{bin}(keV)	⟨e⟩(keV)	e(%)
4	2.45	67 *7*
19	1.42	7.4 *8*
20 - 22	0.76	3.7 *3*
23	0.43	1.91 *20*
25 - 60	0.83	2.1 *3*
177	7.55	4.28 *9*
196	0.010	<0.010
199	0.986	0.495 *10*
200 - 233	0.340	0.167 *3*
255 - 299	0.208	0.076 *9*
320 - 347	0.142	0.044 *4*
394 - 434	0.207	0.051 *4*
455 - 503	0.013	0.0027 *11*
515 - 558	0.010	0.0018 *5*
567 - 616	0.210	0.035 *3*
618 - 653	0.123	0.0195 *25*
675 - 724	0.022	0.0032 *9*
725 - 773	0.019	0.0025 *7*
790 - 836	0.025	0.0031 *10*
848 - 897	0.0082	0.0009 *3*
899 - 948	0.027	0.0030 *9*
949 - 998	0.008	0.0009 *3*
999 - 1046	0.010	0.0010 *4*
1049 - 1079	0.0025	0.00023 *10*
1102 - 1149	0.034	0.0030 *12*
1160 - 1196	0.014	0.0012 *5*
1246 - 1272	0.0041	0.00033 *13*
1320 - 1359	0.0089	0.00067 *18*
1382 - 1430	0.009	0.00063 *23*
1449 - 1542	0.0041	0.00028 *8*
1565 - 1622	0.011	0.00071 *25*
1746 - 1845	0.0024	0.00014 *5*
1848 - 1859	1.8 ×10^{-5}	~1.0×10^{-6}

Continuous Radiation (^{109}In)

⟨β+⟩=28.5 keV;⟨IB⟩=0.90 keV

E_{bin}(keV)		⟨ ⟩(keV)	(%)
0 - 10	β+	6.1×10^{-6}	7.4×10^{-5}
	IB	0.00155	
10 - 20	β+	0.000207	0.00126
	IB	0.0049	0.029
20 - 40	β+	0.00481	0.0148
	IB	0.034	0.139
40 - 100	β+	0.183	0.237
	IB	0.0091	0.0142
100 - 300	β+	5.9	2.75
	IB	0.043	0.022
300 - 600	β+	18.6	4.32
	IB	0.152	0.034
600 - 1300	β+	3.82	0.58
	IB	0.54	0.061
1300 - 1828	IB	0.112	0.0078
	Σβ+		7.9

$^{109}_{49}$In(1.34 *7* min)

Mode: IT
Δ: -85855 *7* keV
SpA: 1.28×10^{9} Ci/g
Prod: daughter ^{109}Sn; ^{107}Ag(α,2n)

Photons (^{109}In)

⟨γ⟩=610.1 *13* keV

γ_{mode}	γ(keV)	γ(%)†
In L_ℓ	2.905	0.0071 *13*
In L_η	3.112	0.0034 *5*
In L_α	3.286	0.20 *3*
In L_β	3.569	0.145 *24*
In L_γ	3.983	0.015 *3*
In $K_{\alpha2}$	24.002	1.37 *5*
In $K_{\alpha1}$	24.210	2.55 *9*
In $K_{\beta1}$'	27.265	0.660 *23*
In $K_{\beta2}$'	28.022	0.138 *5*
γ M4	649.85 *11*	93.70

† 0.21% uncert(syst)

Atomic Electrons (^{109}In)

⟨e⟩=41.6 *7* keV

e_{bin}(keV)	⟨e⟩(keV)	e(%)
4	0.171	4.5 *5*
19	0.0119	0.061 *6*
20	0.10	0.50 *5*
23	0.026	0.114 *12*
24	0.028	0.118 *12*
26	0.00163	0.0062 *6*
27	0.0038	0.0141 *14*
622	34.2	5.51 *11*
646	5.65	0.875 *18*
649	1.138	0.175 *4*
650	0.244	0.0376 *8*

$^{109}_{49}$In(210 *10* ms)

Mode: IT
Δ: -84403 *7* keV
SpA: 1.44×10^{11} Ci/g
Prod: ^{107}Ag(α,2n); ^{103}Rh(^{12}C,α2n)

Photons (^{109}In)

⟨γ⟩=2086 *140* keV

γ_{mode}	γ(keV)	γ(%)†
γ	~170	11.7 *23* ?
γ	~210	11.7 *23* ?
γ M1+0.2%E2	401.9 *5*	20.5 *19*
γ M3	673.5 *5*	97
γ M1+15%E2	1026.41 *15*	18.5 *19*
γ E2	1428 *5*	78 *3*

† 2.0% uncert(syst)

$^{109}_{50}$Sn(18.0 *2* min)

Mode: ε
Δ: -82630 *10* keV
SpA: 9.58×10^{7} Ci/g
Prod: ^{106}Cd(α,n)

Photons (^{109}Sn)

⟨γ⟩=2218 *76* keV

γ_{mode}	γ(keV)	γ(%)†
In L_ℓ	2.905	0.101 *18*
In L_η	3.112	0.047 *7*
In L_α	3.286	2.8 *4*
In L_β	3.570	2.0 *3*
In L_γ	3.984	0.21 *4*
In $K_{\alpha2}$	24.002	19.8 *7*
In $K_{\alpha1}$	24.210	37.0 *14*
In $K_{\beta1}$'	27.265	9.6 *4*
In $K_{\beta2}$'	28.022	2.00 *8*
γ [E1]	118.41 *13*	0.060 *12*
γ [M1+E2]	158.48 *16*	0.090 *18*
γ	181.37 *22*	0.060 *12*
γ	216.65 *22*	0.060 *12*
γ	220.80 *22*	0.120 *24*
γ	222.03 *12*	0.060 *12*
γ	229.27 *15*	0.120 *24*
γ	250.9 *3*	0.120 *24*
γ	261.39 *20*	0.33 *7*
γ [M1+E2]	310.06 *18*	0.15 *3*
γ	312.39 *20*	0.60 *12*
γ M1	331.04 *13*	9.7 *19*
γ	340.43 *16*	0.24 *5*
γ	353.6 *3*	1.02 *20*
γ	362.81 *21*	0.24 *5*
γ	373.21 *21*	0.24 *5*
γ	376.11 *17*	0.090 *18*
γ M1(+E2)	382.9 *3*	0.90 *18*
γ M1(+E2)	384.84 *16*	2.4 *5*
γ	401.73 *22*	0.090 *18*
γ M1(+E2)	437.19 *18*	1.5 *3*
γ	437.93 *18*	1.08 *22*
γ	453.23 *22*	0.15 *3*
γ	454.36 *23*	0.18 *4*
γ M1,E2	459.73 *19*	0.21 *4*
γ	465.78 *15*	0.15 *3*
γ	472.92 *18*	0.18 *4*
γ	478.6 *3*	0.120 *24*
γ	482.2 *3*	0.120 *24*
γ	495.41 *20*	0.39 *8*
γ (M1,E2)	501.3 *3*	0.27 *5*
γ E1	521.90 *15*	2.4 *5*
γ	522.20 *22*	<0.6
γ	539.9 *3*	0.21 *4*
γ	560.39 *24*	0.090 *18*
γ	594.70 *25*	0.090 *18*
γ	597.02 *23*	0.15 *3*
γ M1,E2	614.11 *12*	1.8 *4*
γ M1,E2	623.02 *17*	2.2 *4*
γ M4	649.85 *11*	28 *6* *
γ	659.95 *15*	1.4 *3*
γ	710.7 *3*	0.42 *8*
γ	732.79 *14*	0.63 *13*
γ	745.39 *15*	0.33 *7*
γ	785.31 *15*	0.27 *5*
γ M1,E2	790.76 *19*	1.6 *3*
γ	816.13 *22*	0.54 *11*
γ	817.11 *22*	0.54 *11*
γ	828.66 *16*	0.87 *17*
γ M1,E2	835.84 *18*	0.87 *17*
γ	848.5 *3*	0.15 *3*
γ	857.8 *3*	0.54 *11*
γ	869.34 *18*	0.69 *14*
γ	879.4 *4*	0.60 *12*
γ	888.57 *19*	0.63 *13*
γ	897.14 *21*	0.96 *19*
γ	967.4 *3*	0.48 *10*
γ	976.17 *13*	1.08 *22*
γ M1,E2	985.2 *3*	0.39 *8*
γ M1+15%E2	1026.41 *15*	5.1 *10*

Photons (^{109}Sn)
(continued)

γ_{mode}	γ(keV)	γ(%)†
γ [M1+E2]	1039.13 *19*	4.5 *9*
γ	1053.86 *24*	0.63 *13*
γ [M1+E2]	1072.18 *18*	0.24 *5*
γ	1078.47 *18*	0.090 *18*
γ (M1,E2)	1083.4 *4*	0.60 *12*
γ	1092.68 *24*	0.54 *11*
γ (E2)	1099.29 *9*	30
γ	1107.52 *17*	0.48 *10*
γ	1107.7 *3*	0.48 *10*
γ	1119.16 *19*	3.2 *6*
γ	1128.07 *17*	1.4 *3*
γ	1129.61 *24*	0.090 *18*
γ [E2]	1157.87 *15*	0.51 *10*
γ	1166.87 *18*	0.54 *11*
γ	1170.43 *25*	0.090 *18*
γ	1187.14 *23*	0.24 *5*
γ	1206.4 *3*	0.39 *8*
γ	1211.27 *18*	0.99 *20*
γ	1220.75 *17*	0.15 *3*
γ	1227.7 *3*	0.33 *7*
γ	1239.67 *24*	0.39 *8*
γ	1250.10 *22*	0.24 *5*
γ	1271.31 *20*	0.48 *10*
γ	1290.3 *4*	0.15 *3*
γ	1301.1 *4*	0.51 *10*
γ E2(+M1)	1321.32 *12*	11.9 *24*
γ	1350.1 *3*	0.90 *18*
γ	1375.38 *23*	0.39 *8*
γ	1388.34 *20*	0.24 *5*
γ	1409.16 *21*	0.69 *14*
γ	1430.53 *23*	0.30 *6*
γ	1442.77 *15*	0.72 *14*
γ	1455.43 *17*	0.66 *13*
γ [M1+E2]	1461.92 *16*	<2
γ	1463.59 *15*	3.0 *6*
γ	1464.27 *17*	3.6 *7*
γ [E1]	1488.57 *20*	4.1 *8*
γ	1492.36 *14*	1.4 *3*
γ	1501.46 *25*	0.15 *3*
γ	1524.31 *21*	0.51 *10*
γ	1546.3 *3*	0.120 *24*
γ	1565.25 *18*	0.24 *5*
γ	1574.32 *14*	5.5 *11*
γ [E1]	1580.32 *19*	1.14 *23*
γ	1603.1 *3*	0.39 *8*
γ	1621.4 *4*	0.45 *9*
γ	1655.9 *3*	0.48 *10*
γ [M1+E2]	1686.29 *16*	0.75 *15*
γ	1699.52 *19*	0.120 *24*
γ	1709.65 *22*	0.090 *18*
γ [M1+E2]	1713.40 *11*	0.99 *20*
γ	1722.31 *15*	1.05 *21*
γ	1759.37 *13*	1.11 *22*
γ	1818.8 *3*	0.18 *4*
γ	1825.1 *3*	0.63 *13*
γ	1825.52 *22*	0.63 *13*
γ	1843.7 *6*	0.72 *14*
γ	1858.60 *21*	0.39 *8*
γ [E1]	1890.38 *16*	1.5 *3*
γ [E1]	1911.35 *18*	5.8 *12*
γ [M1+E2]	1930.45 *13*	0.51 *10*
γ	1943.5 *3*	1.7 *3*
γ	1951.36 *15*	0.090 *18*
γ	1957.05 *12*	0.39 *8*
γ [E1]	1962.11 *22*	0.15 *3*
γ [E1]	2048.86 *15*	0.30 *6*
γ	2055.13 *22*	1.8 *4*
γ	2074.6 *4*	0.090 *18*
γ	2125.87 *11*	1.4 *3*
γ	2159.09 *23*	0.72 *14*
γ [E1]	2195.76 *18*	1.4 *3*
γ	2218.44 *19*	0.22 *4*
γ	2235.60 *22*	0.108 *22*
γ	2276.50 *22*	0.24 *5*
γ	2437.6 *4*	0.14 *3*
γ	2542.05 *15*	2.6 *5*
γ	2564.2 *7*	0.102 *20*
γ	2574.8 *3*	0.111 *22*
γ	2591.64 *13*	0.60 *12*
γ	2601.7 *4*	0.078 *16*
γ	2617.01 *25*	0.060 *12*
γ [M1+E2]	2785.56 *15*	1.7 *4*
γ	2813.09 *22*	0.39 *8*
γ	2851.90 *19*	0.060 *12*

Photons (^{109}Sn)
(continued)

γ_{mode}	γ(keV)	γ(%)†
γ	2858.64 13	1.02 20
γ [E2]	2871.25 13	0.081 16
γ	2919.8 7	0.015 3
γ [E2]	2942.99 22	0.111 22
γ	3013.4 3	0.126 25
γ [E2]	3029.73 13	0.045 9
γ	3034.8 4	0.123 25
γ	3050.64 15	0.36 7
γ	3065.6 2	0.078 16
γ	3140.0 3	0.042 8
γ	3316.7 3	0.090 18
γ	3360.9 6	0.021 4
γ	3395.8 4	0.111 22
γ	3418.4 4	0.0120 24
γ	3426.9 4	0.018 4

† 10% uncert(syst)
* with ^{109}In(1.3 min) in equilib

Atomic Electrons (^{109}Sn)
⟨e⟩=19.7 21 keV

e_{bin}(keV)	⟨e⟩(keV)	e(%)
4	2.41	63 7
19	0.172	0.89 9
20	1.45	7.2 7
23 - 27	0.86	3.65 25
90 - 131	0.028	0.023 7
153 - 201	0.049	0.027 7
212 - 261	0.043	0.019 7
282 - 284	0.037	0.013 6
303	0.53	0.17 4
306 - 355	0.21	0.062 9
357 - 402	0.142	0.039 7
409 - 456	0.129	0.031 5
459 - 501	0.048	0.0100 23
512 - 560	0.011	0.0020 5
567 - 614	0.096	0.0162 23
619	0.0059	0.00096 24
622	10.3	1.6 3
623 - 632	0.020	~0.0032
646	1.7	0.26 5
649 - 683	0.42	0.065 11
705 - 745	0.014	0.0020 7
757 - 801	0.052	0.0067 11
808 - 857	0.038	0.0046 8
858 - 897	0.020	0.0023 7
939 - 985	0.021	0.0022 6
998 - 1044	0.145	0.0144 18
1050 - 1099	0.37	0.034 3
1100 - 1142	0.026	0.0024 5
1154 - 1203	0.016	0.0014 3
1206 - 1250	0.0096	0.00078 18
1262 - 1301	0.104	0.0080 17
1317 - 1360	0.025	0.0019 3
1371 - 1415	0.011	0.00080 20
1426 - 1474	0.079	0.0055 14
1484 - 1580	0.048	0.0031 9
1593 - 1686	0.0187	0.00113 14
1694 - 1791	0.014	0.00080 19
1797 - 1890	0.034	0.00182 23
1903 - 1961	0.017	0.00088 17
2021 - 2098	0.015	0.00075 18
2122 - 2218	0.0096	0.00045 7

Atomic Electrons (^{109}Sn)
(continued)

e_{bin}(keV)	⟨e⟩(keV)	e(%)
2231 - 2276	0.0011	5.0 16 $\times 10^{-5}$
2410 - 2437	0.00059	2.4 8 $\times 10^{-5}$
2514 - 2613	0.014	0.00057 14
2758 - 2855	0.0140	0.00050 7
2858 - 2942	0.00067	2.3 3 $\times 10^{-5}$
2985 - 3065	0.0026	8.7 14 $\times 10^{-5}$
3112 - 3139	0.00015	4.7 14 $\times 10^{-6}$
3289 - 3368	0.00069	2.1 4 $\times 10^{-5}$
3390 - 3426	0.000144	4.2 7 $\times 10^{-6}$

Continuous Radiation (^{109}Sn)
⟨β+⟩=69 keV;⟨IB⟩=1.14 keV

E_{bin}(keV)		⟨ ⟩(keV)	(%)
0 - 10	β+	1.52×10^{-6}	1.84×10^{-5}
	IB	0.0030	
10 - 20	β+	5.4×10^{-5}	0.000325
	IB	0.0045	0.029
20 - 40	β+	0.00130	0.00399
	IB	0.041	0.165
40 - 100	β+	0.053	0.069
	IB	0.018	0.028
100 - 300	β+	2.20	1.00
	IB	0.071	0.038
300 - 600	β+	13.1	2.86
	IB	0.19	0.042
600 - 1300	β+	42.3	4.82
	IB	0.51	0.058
1300 - 2500	β+	11.6	0.72
	IB	0.29	0.018
2500 - 3876	β+	0.259	0.0100
	IB	0.0084	0.00030
	Σβ+		9.5

$^{109}_{51}$Sb(17.0 7 s)

Mode: ε
Δ: -76250 19 keV
SpA: 5.97×10^{9} Ci/g
Prod: ^{112}Sn(p,4n)

Photons (^{109}Sb)
⟨γ⟩=2881 110 keV

γ_{mode}	γ(keV)	γ(%)
Sn L_ℓ	3.045	0.022 3
Sn L_η	3.272	0.0116 12
Sn L_α	3.443	0.62 7
Sn L_β	3.742	0.55 7
Sn L_γ	4.220	0.064 9
Sn $K_{\alpha 2}$	25.044	4.61 15
Sn $K_{\alpha 1}$	25.271	8.6 3
Sn $K_{\beta 1}$'	28.474	2.25 7
Sn $K_{\beta 2}$'	29.275	0.474 16
γ	246.87 24	1.9 2
γ [M1+E2]	261.07 24	2.0 2
γ M1,E2	545.04 24	10.6 5
γ E2(+M1)	664.57 24	63 4
γ	678.77 24	19 1
γ [E2]	925.64 21	100
γ	950.84 24	2.0 1
γ [M1+E2]	1061.9 3	75 5
γ	1078.2 3	3.9 2
γ E2(+M1)	1343.7 3	2.2 2
γ	1495.88 24	30 2

$^{109}_{52}$Te(4.1 2 s)

Mode: ε(96.1 13 %), α(3.9 13 %), εp
Δ: -67650 70 keV
SpA: 2.32×10^{10} Ci/g
Prod: ^{92}Mo(^{20}Ne,3n); ^{96}Ru(^{16}O,3n)

εp: 2000-5500 range

Alpha Particles (^{109}Te)

α(keV)
3080 15

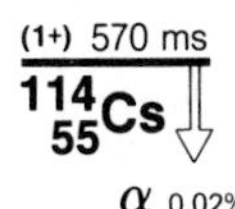

A = 110

NDS **38**, 545 (1983)

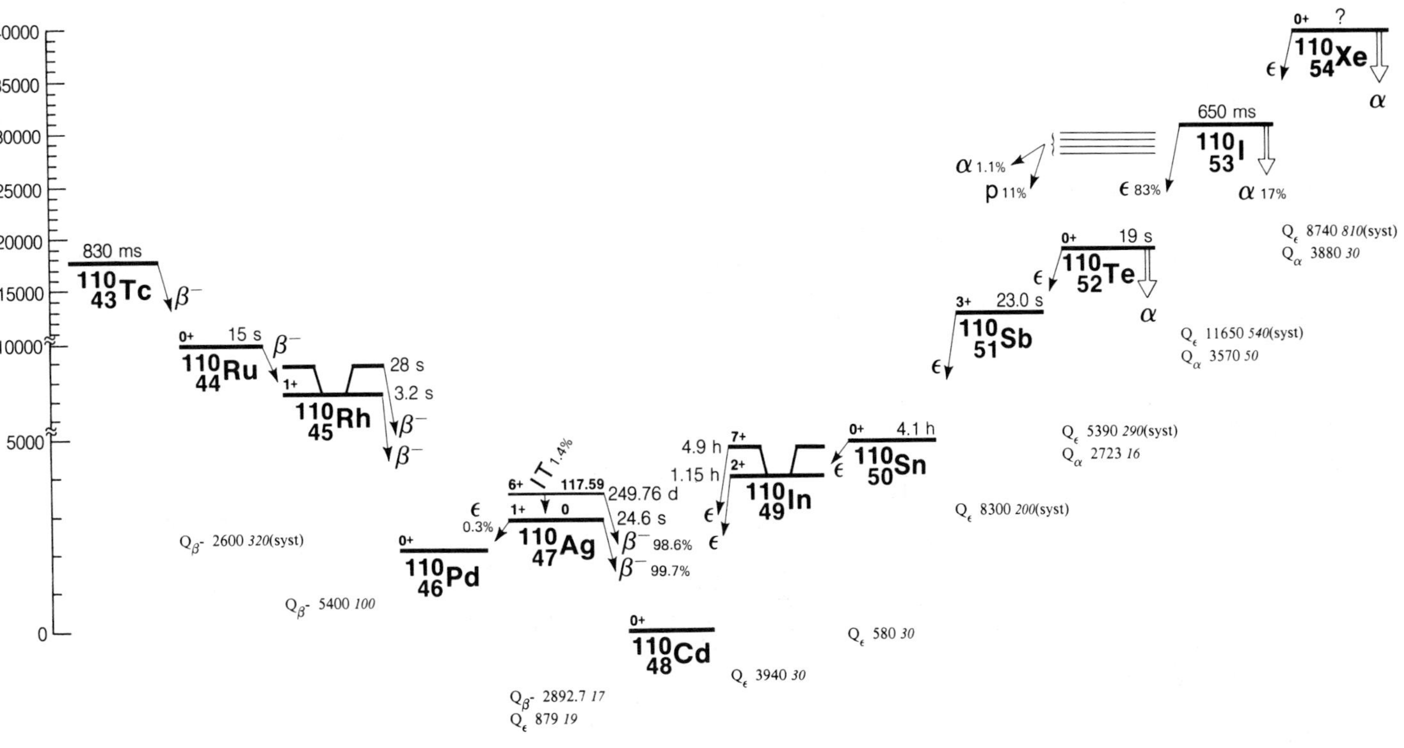

$^{110}_{43}$Tc(830 *40* ms)

Mode: β-
SpA: 8.4×10^{10} Ci/g
Prod: fission

Photons (^{110}Tc)

γ_{mode}	γ(keV)
γ (E2)	240.68 *11*

$^{110}_{44}$Ru(14.6 *10* s)

Mode: β-
Δ: -80340 *300* keV syst
SpA: 6.9×10^{9} Ci/g
Prod: fission

Photons (^{110}Ru)

γ_{mode}	γ(keV)	γ(rel)
γ	95.8 *3*	6.0 *5*
γ	112.1 *3*	100

$^{110}_{45}$Rh(3.2 *2* s)

Mode: β-
Δ: -82940 *100* keV
SpA: 2.88×10^{10} Ci/g
Prod: ^{110}Pd(n,p)

Photons (^{110}Rh)

⟨γ⟩=227 *39* keV

γ_{mode}	γ(keV)	γ(%)†
γ[E2]	373.83 *14*	51 *10*
γ E2+5.1%M1	439.87 *12*	5.1 *10*
γ[E2]	813.70 *15*	1.8 *4*

† 16% uncert(syst)

$^{110}_{45}$Rh(28.5 *15* s)

Mode: β-

Δ: -82940 *100* keV

SpA: 3.56×10^{9} Ci/g

Prod: ^{110}Pd(n,p)

Photons (^{110}Rh)

⟨γ⟩=2588 *72* keV

γ_{mode}	γ(keV)	γ(%)†
Pd $K_{\alpha2}$	21.020	0.448 *20*
Pd $K_{\alpha1}$	21.177	0.85 *4*
Pd $K_{\beta1}$'	23.811	0.210 *10*
Pd $K_{\beta2}$'	24.445	0.0407 *19*
γ [E2]	373.83 *14*	91
γ [M1+E2]	398.45 *14*	14.9 *14*
γ E2+5.1%M1	439.87 *12*	26 *3*
γ	478.46 *22*	4.0 *5*
γ	531.2 *5*	1.8 *5*
γ [E2]	546.32 *17*	36 *3*
γ	546.9 *4* ?	
γ	584.91 *17*	17.4 *14*
γ [E2]	653.53 *18*	17.0 *14*
γ	687.91 *16*	28 *4*
γ [E2]	813.70 *15*	9.1 *9*
γ [M1+E2]	838.32 *14*	21.7 *23*
γ	890.58 *18*	12.8 *14*
γ	904.73 *17*	27 *3*
γ	979.91 *20*	4.6 *5*
γ	1048.4 *4*	7.9 *14*
γ	1086.36 *18*	3.1 *9*
γ	1216.9 *3*	7.4 *14*
γ	1231.10 *25*	12.7 *18*
γ	1392.02 *25*	4.6 *9*
γ	1406.17 *24*	6.6 *12*
γ	1526.8 *4*	~2
γ	1578.48 *22*	~2
γ	1592.63 *21*	6.3 *14*
γ	1870.48 *23*	1.4 *5*
γ	1884.63 *22*	4.0 *9*

† 1.1% uncert(syst)

Atomic Electrons (^{110}Rh)

⟨e⟩=9.4 *3* keV

e_{bin}(keV)	⟨e⟩(keV)	e(%)
3 - 24	0.113	1.87 *15*
349	3.97	1.14 *5*
370	0.447	0.121 *5*
371 - 373	0.228	0.0614 *18*
374	0.55	0.146 *20*
395 - 398	0.086	0.022 *4*
416	0.83	0.199 *21*
436 - 478	0.22	0.050 *9*
507	0.027	~0.005
522	0.77	0.147 *13*
528 - 561	0.35	0.063 *20*
581 - 585	0.035	0.0061 *24*
629	0.266	0.042 *4*
650 - 654	0.042	0.0064 *5*
664	0.29	~0.044
684 - 688	0.045	0.007 *3*
789 - 838	0.40	0.050 *4*
866 - 905	0.35	0.040 *12*
956 - 980	0.037	0.0038 *16*
1024 - 1062	0.08	0.007 *3*
1083 - 1086	0.0028	0.00026 *12*
1193 - 1231	0.13	0.010 *3*
1368 - 1406	0.061	0.0044 *13*
1502 - 1592	0.047	0.0030 *9*
1846 - 1884	0.022	0.0012 *4*

Continuous Radiation (^{110}Rh)

⟨β-⟩=1083 keV;⟨IB⟩=2.6 keV

E_{bin}(keV)		⟨ ⟩(keV)	(%)
0 - 10	β-	0.0101	0.202
	IB	0.039	
10 - 20	β-	0.0311	0.207
	IB	0.039	0.27
20 - 40	β-	0.130	0.430
	IB	0.076	0.26
40 - 100	β-	1.01	1.43
	IB	0.22	0.33
100 - 300	β-	13.2	6.4
	IB	0.60	0.34
300 - 600	β-	64	13.9
	IB	0.64	0.153
600 - 1300	β-	398	41.7
	IB	0.76	0.089
1300 - 2500	β-	602	35.6
	IB	0.19	0.0123
2500 - 2953	β-	5.3	0.204
	IB	0.00018	6.8×10^{-6}

$^{110}_{46}$Pd(stable)

Δ: -88337 *18* keV

%: 11.72 *9*

$^{110}_{47}$Ag(24.6 *2* s)

Mode: β-(99.70 *6* %), ϵ(0.30 *6* %)

Δ: -87458 *3* keV

SpA: 4.11×10^{9} Ci/g

Prod: daughter ^{110}Ag(249.76 d); ^{109}Ag(n,γ)

Photons (^{110}Ag)

⟨γ⟩=31 *6* keV

γ_{mode}	γ(keV)	γ(%)†
$\gamma_{\beta-}$[E2]	295.4 *4*	0.0076 *13*
γ_{ϵ}[E2]	373.83 *14*	<0.02 ?
$\gamma_{\beta-}$[E1]	603.04 *4*	0.00038 *9*
$\gamma_{\beta-}$E2	657.7617 *23*	4.5
$\gamma_{\beta-}$[E2]	815.36 *7*	0.0382 *9*
$\gamma_{\beta-}$E2+35%M1	818.030 *3*	0.0089 *4*
$\gamma_{\beta-}$(E2)	1073.7 *4*	~0.0009
$\gamma_{\beta-}$M1+1.0%E2	1125.718 *15*	0.0153 *4*
$\gamma_{\beta-}$[E2]	1186.70 *12*	0.0027 *6*
$\gamma_{\beta-}$(E1+M2)	1421.06 *4*	0.00225 *7*
$\gamma_{\beta-}$E2	1475.786 *3*	0.00489 *19*
$\gamma_{\beta-}$(E2+M1)	1629.69 *8*	0.0022 *5*
$\gamma_{\beta-}$	1674.4 *7*	0.0072 *3*
$\gamma_{\beta-}$(E2)	1783.472 *15*	0.0044 *3*
$\gamma_{\beta-}$[E2]	2004.72 *12*	0.0036 *4*

† 5.1% uncert(syst)

Continuous Radiation (^{110}Ag)

⟨β-⟩=1185 keV;⟨IB⟩=3.0 keV

E_{bin}(keV)		⟨ ⟩(keV)	(%)
0 - 10	β-	0.0088	0.175
	IB	0.042	
10 - 20	β-	0.0269	0.179
	IB	0.041	0.29
20 - 40	β-	0.112	0.373
	IB	0.081	0.28
40 - 100	β-	0.87	1.24
	IB	0.23	0.35
100 - 300	β-	11.4	5.5
	IB	0.65	0.37
300 - 600	β-	55	12.1
	IB	0.72	0.17
600 - 1300	β-	365	38.1
	IB	0.91	0.106
1300 - 2500	β-	722	41.2
	IB	0.31	0.018
2500 - 2893	β-	29.8	1.15
	IB	0.00090	3.5×10^{-5}

$^{110}_{47}$Ag(249.76 *4* d)

Mode: β-(98.64 *6* %), IT(1.36 *6* %)

Δ: -87340 *3* keV

SpA: 4752.7 Ci/g

Prod: ^{109}Ag(n,γ)

Photons (^{110}Ag)

⟨γ⟩=2739 *21* keV

γ_{mode}	γ(keV)	γ(%)†
Ag $K_{\alpha2}$	21.990	0.202 *11*
Ag $K_{\alpha1}$	22.163	0.381 *20*
Cd $K_{\alpha2}$	22.984	0.157 *6*
Cd $K_{\alpha1}$	23.174	0.295 *11*
Ag $K_{\beta1}$'	24.934	0.096 *5*
Ag $K_{\beta2}$'	25.603	0.0191 *10*
Cd $K_{\beta1}$'	26.085	0.075 *3*
Cd $K_{\beta2}$'	26.801	0.0156 *6*
γ_{IT} M4	116.48 *5*	0.0080 *3*
$\gamma_{\beta-}$[M1+E2]	120.163 *15*	0.0180 *10*
$\gamma_{\beta-}$	133.331 *8*	0.073 *3*
$\gamma_{\beta-}$	219.348 *10*	0.0662 *19*
$\gamma_{\beta-}$	221.079 *12*	0.0681 *14*
$\gamma_{\beta-}$	229.412 *14*	0.0121 *8*
$\gamma_{\beta-}$[E1]	266.910 *11*	0.0407 *10*
$\gamma_{\beta-}$[M1+E2]	341.232 *12*	0.0021 *4*
$\gamma_{\beta-}$	360.22 *4*	0.0033 *7*
$\gamma_{\beta-}$[M1+E2]	365.450 *11*	0.086 *18*
$\gamma_{\beta-}$[E1]	387.073 *10*	~0.08
$\gamma_{\beta-}$	396.851 *14*	~0.06
$\gamma_{\beta-}$E1	409.312 *18*	0.0063 *6*
$\gamma_{\beta-}$	409.32 *6*	0.0064 *7*
$\gamma_{\beta-}$M1+13%E2	446.810 *3*	3.75 *8*
$\gamma_{\beta-}$	467.046 *20*	0.024 *4*
$\gamma_{\beta-}$	493.55 *4*	0.0104 *10*
$\gamma_{\beta-}$	544.596 *10*	0.0208 *10*
$\gamma_{\beta-}$	573.351 *9*	0.012 *3*
$\gamma_{\beta-}$[E1]	603.04 *4*	0.0062 *14*
$\gamma_{\beta-}$M1+33%E2	620.3594 *24*	2.81 *6*
$\gamma_{\beta-}$M1,E2	626.262 *11*	0.215 *11*
$\gamma_{\beta-}$	630.613 *9*	0.0379 *19*
$\gamma_{\beta-}$	649.72 *4* ?	
$\gamma_{\beta-}$E2	657.7617 *23*	94.6 *19*
$\gamma_{\beta-}$M1+12%E2	677.6218 *22*	10.36 *20*
$\gamma_{\beta-}$E2+24%M1	687.013 *3*	6.44 *13*
$\gamma_{\beta-}$E2+33%M1	706.6808 *25*	16.4 *3*
$\gamma_{\beta-}$	708.081 *14*	0.275 *18*
$\gamma_{\beta-}$	714.59 *4*	0.0085 *19*
$\gamma_{\beta-}$E2	744.276 *3*	4.71 *9*
$\gamma_{\beta-}$E2	763.943 *3*	22.3 *4*
$\gamma_{\beta-}$	774.735 *14*	~0.0019

Photons (^{110}Ag)
(continued)

γ_{mode}	γ(keV)	γ(%)†
γ_{β}-E2+35%M1	818.030 3	7.31 13
γ_{β}-E2	884.684 3	72.7 15
γ_{β}-E2	937.492 3	34.4 7
γ_{β}-	957.41 4	0.0076 10
γ_{β}-E1+0.3%M2	997.228 10	0.134 5
γ_{β}-[M1+E2]	1018.852 12	0.0142 9
γ_{β}-M1,E2	1085.505 12	0.066 5
γ_{β}-E1	1117.391 11	0.0410 24
γ_{β}-M1+1.0%E2	1125.718 15	0.0348 25
γ_{β}-	1163.222 12	0.045 8
γ_{β}-	1164.952 10	0.029 5
γ_{β}-[E2]	1186.70 12	0.00078 22
γ_{β}-	1250.969 9	0.023 7
γ_{β}-	1300.07 6	0.024 8
γ_{β}-	1334.339 14	0.141 5
γ_{β}-M1+16%E2	1384.2979 24	24.3 5
γ_{β}-(E1+M2)	1421.06 4	0.036 3
γ_{β}-E2	1475.786 3	4.02 7
γ_{β}-E2+M1	1505.038 3	13.0 3
γ_{β}-E2	1562.300 3	1.033 20
γ_{β}-	1592.759 14	0.0209 12
γ_{β}-(E2+M1)	1629.69 8	0.0058 10 ?
γ_{β}-(E2+M1)	1698.58 14	0.00180 19 ?
γ_{β}-	1775.43 4	0.0063 10
γ_{β}-(E2)	1783.472 15	0.0099 8
γ_{β}-[E2]	1903.526 12	0.0147 12
γ_{β}-[E2]	2004.72 12	0.00104 19 ?

† uncert(syst): 4.4% for IT, <0.1% for β-

Atomic Electrons (^{110}Ag)
$\langle e \rangle$=6.55 7 keV

e_{bin}(keV)	$\langle e \rangle$(keV)	e(%)
1 - 26	0.107	2.94 14
91	0.76	0.84 4
93 - 107	0.026	~0.025
113	0.464	0.411 18
116 - 133	0.130	0.111 6
193 - 240	0.017	0.008 3
263 - 267	0.000147	5.57 21 $\times10^{-5}$
315 - 362	0.0058	0.0017 3
365 - 409	0.0024	~0.0006
420 - 467	0.143	0.0338 8
490 - 518	0.00038	7 3 $\times10^{-5}$
541 - 576	0.00032	5.8 20 $\times10^{-5}$
594 - 630	0.0713	0.0119 3
631	1.63	0.259 7
651	0.191	0.0293 8
654	0.214	0.0327 9
657 - 678	0.187	0.0282 5
680	0.267	0.0392 10
681 - 718	0.130	0.0184 4
737	0.308	0.0418 12
740 - 775	0.0592	0.00783 15
791 - 818	0.1124	0.0141 3
858	0.823	0.096 3
881 - 885	0.127	0.0144 3
911	0.361	0.0397 11
931 - 971	0.0560	0.00600 13
992 - 1019	0.000251	2.53 18 $\times10^{-5}$
1059 - 1099	0.00123	0.000115 8
1113 - 1162	0.00064	5.7 15 $\times10^{-5}$
1163 - 1187	8.1 $\times10^{-6}$	7.0 22 $\times10^{-7}$
1224 - 1273	0.00030	2.4 9 $\times10^{-5}$
1296 - 1334	0.0010	7 3 $\times10^{-5}$
1358 - 1394	0.216	0.0159 4
1417 - 1449	0.0259	0.00179 5
1472 - 1566	0.112	0.0075 5
1589 - 1672	6.7 $\times10^{-5}$	4.1 6 $\times10^{-6}$
1695 - 1783	9.5 $\times10^{-5}$	5.4 7 $\times10^{-6}$
1877 - 1903	8.6 $\times10^{-5}$	4.6 4 $\times10^{-6}$
1978 - 2004	5.8 $\times10^{-6}$	2.9 5 $\times10^{-7}$

Continuous Radiation (^{110}Ag)
$\langle\beta-\rangle$=69 keV;$\langle$IB$\rangle$=0.035 keV

E_{bin}(keV)		$\langle\ \rangle$(keV)	(%)
0 - 10	β-	1.03	21.3
	IB	0.0034	
10 - 20	β-	2.52	17.0
	IB	0.0028	0.020
20 - 40	β-	6.6	22.8
	IB	0.0047	0.0164
40 - 100	β-	9.8	16.9
	IB	0.0096	0.0153
100 - 300	β-	29.8	15.8
	IB	0.0116	0.0074
300 - 600	β-	16.9	4.62
	IB	0.0020	0.00052
600 - 1300	β-	2.17	0.256
	IB	0.00053	7.4 $\times10^{-5}$
1300 - 1468	β-	0.0454	0.00338
	IB	5.8 $\times10^{-7}$	4.4 $\times10^{-8}$

$^{110}_{48}$Cd(stable)

Δ: -90351 3 keV
%: 12.49 9

$^{110}_{49}$In(1.152 8 h)

Mode: ϵ
Δ: -86410 30 keV
SpA: 2.474$\times10^{7}$ Ci/g
Prod: daughter ^{110}Sn; ^{107}Ag(α,n); ^{109}Ag(α,3n)

Photons (^{110}In)
$\langle\gamma\rangle$=929 32 keV

γ_{mode}	γ(keV)	γ(%)†
Cd L_{ℓ}	2.767	0.040 7
Cd L_{η}	2.957	0.019 3
Cd L_{α}	3.133	1.10 17
Cd L_{β}	3.392	0.78 13
Cd L_{γ}	3.779	0.077 14
Cd $K_{\alpha2}$	22.984	8.2 3
Cd $K_{\alpha1}$	23.174	15.4 5
Cd $K_{\beta1}$'	26.085	3.93 14
Cd $K_{\beta2}$'	26.801	0.82 3
γ [E1]	603.04 4	0.072 16
γ M1+33%E2	620.3594 24	0.0198 19
γ E2	657.7617 23	98
γ E2+24%M1	687.013 3	0.045 4
γ [E2]	815.36 7	0.284 20
γ E2+35%M1	818.030 3	0.83 3
γ E2	884.684 3	0.105 8
γ	958.1 3	0.088 20
γ	999.9 5	0.040 8 ?
γ	1023.13 25	0.049 10
γ (E2)	1073.7 4	0.098 9
γ	1085.72 14	0.039 7
γ M1+1.0%E2	1125.718 15	1.01 4
γ	1151.73 17	0.044 8
γ	1235.6 3	0.264 10
γ	1345.03 6	~0.011
γ	1410.10 15	0.036 8
γ (E1+M2)	1421.06 4	0.423 19
γ E2	1475.786 3	0.458 17
γ E2+M1	1505.038 3	0.092 9
γ	1603.0 5	0.127 10
γ	1628.8 3	0.078 8 ?
γ	1651.13 15	0.022 8 ?
γ	1666.0 4	0.041 8

Photons (^{110}In)
(continued)

γ_{mode}	γ(keV)	γ(%)†
γ (E2+M1)	1698.58 14	0.27 3
γ	1744.0 4	0.043 8
γ (E2)	1783.472 15	0.289 17
γ	1973.7 4	~0.07
γ	1976.4 4	~0.07
γ	2002.29 16	0.127 20
γ	2123.3 4	0.017 5
γ	2129.52 17	2.13 9
γ	2211.43 14	1.76 7
γ	2317.52 13	1.31 5
γ	2421.0 5	0.538 20
γ	2444.18 25	0.303 20
γ	2535.80 15	0.235 20
γ	2656.6 3	0.0421 20
γ [E1]	2745.67 19	0.085 7
γ	2787.27 17	0.087 9
γ	2808.29 19	0.548 20
γ	2817.63 15	0.065 8
γ	2869.18 14	0.042 6
γ	2975.26 13	0.136 9
γ	3043.58 14	0.131 9
γ	3078.7 5	0.303 20
γ	3101.93 25	~0.006
γ	3314.4 3	~0.005
γ [E1]	3403.41 19	0.025 4
γ	3466.03 19	0.0068 20
γ	3475.38 15	0.65 5
γ	3596.4 4	0.125 9
γ	3771.8 6	0.067 6

†<0.1% uncert(syst)

Atomic Electrons (^{110}In)
$\langle e \rangle$=4.08 15 keV

e_{bin}(keV)	$\langle e \rangle$(keV)	e(%)
4	0.96	27 3
19	0.56	2.9 3
20	0.141	0.72 7
22	0.162	0.73 8
23	0.172	0.76 8
25 - 26	0.0329	0.129 10
576 - 620	0.00104	0.00018 4
631	1.69	0.268 14
654	0.221	0.0338 18
657 - 687	0.0527	0.0080 4
789 - 818	0.0169	0.00214 6
858 - 885	0.00137	0.000159 11
931 - 973	0.0011	0.00012 4
996 - 1023	0.00048	4.8 21 $\times10^{-5}$
1047 - 1086	0.00133	0.000126 16
1099 - 1148	0.0120	0.00109 4
1151 - 1152	8.3 $\times10^{-6}$	7 3 $\times10^{-7}$
1209 - 1236	0.0019	0.00016 6
1318 - 1345	7.1 $\times10^{-5}$	~5 $\times10^{-6}$
1383 - 1421	0.005	~0.00036
1449 - 1504	0.00409	0.000281 10
1576 - 1672	0.0031	0.000190 24
1695 - 1783	0.00224	0.000128 8
1947 - 2002	0.0013	6.4 17 $\times10^{-5}$
2097 - 2185	0.016	0.00074 17
2207 - 2291	0.0056	0.00024 7
2314 - 2394	0.0025	0.00011 3
2417 - 2509	0.0022	9.0 19 $\times10^{-5}$
2532 - 2630	0.00024	9.4 20 $\times10^{-6}$
2653 - 2745	0.000234	8.6 7 $\times10^{-6}$
2761 - 2842	0.0026	9.3 18 $\times10^{-5}$
2865 - 2949	0.00042	1.4 4 $\times10^{-5}$
2971 - 3052	0.0014	4.5 9 $\times10^{-5}$
3075 - 3101	0.00014	4.6 15 $\times10^{-6}$
3288 - 3377	6.7 $\times10^{-5}$	2.0 4 $\times10^{-6}$
3399 - 3475	0.0020	5.8 13 $\times10^{-5}$
3570 - 3596	0.00037	1.04 23 $\times10^{-5}$
3745 - 3771	0.00019	5.2 11 $\times10^{-6}$

Continuous Radiation (^{110}In)

⟨β+⟩=623 keV;⟨IB⟩=3.7 keV

E_{bin}(keV)		⟨ ⟩(keV)	(%)
0 - 10	β+	4.06×10^{-6}	4.91×10^{-5}
	IB	0.024	
10 - 20	β+	0.000140	0.00085
	IB	0.025	0.169
20 - 40	β+	0.00333	0.0102
	IB	0.058	0.21
40 - 100	β+	0.138	0.178
	IB	0.129	0.20
100 - 300	β+	6.2	2.79
	IB	0.37	0.21
300 - 600	β+	45.3	9.7
	IB	0.45	0.106
600 - 1300	β+	303	32.0
	IB	0.99	0.109
1300 - 2500	β+	268	17.0
	IB	1.42	0.080
2500 - 3282	IB	0.20	0.0074
	Σβ+		62

$^{110}_{49}$In(4.9 *1* h)

Mode: ε

Δ: -86410 *30* keV

SpA: 5.81×10^{6} Ci/g

Prod: ^{109}Ag(α,3n); ^{107}Ag(α,n)

Photons (^{110}In)

⟨γ⟩=3101 *27* keV

γ_{mode}	γ(keV)	γ(%)†
Cd L_ℓ	2.767	0.103 *18*
Cd L_η	2.957	0.048 *7*
Cd L_α	3.133	2.8 *4*
Cd L_β	3.392	2.0 *3*
Cd L_γ	3.780	0.20 *4*
Cd $K_{\alpha2}$	22.984	20.8 *8*
Cd $K_{\alpha1}$	23.174	39.1 *15*
Cd $K_{\beta1}$'	26.085	10.0 *4*
Cd $K_{\beta2}$'	26.801	2.07 *8*
γ [M1+E2]	120.163 *15*	1.85 *23*
γ	121.2 *3*	~0.39
γ	133.331 *8*	0.00056 *4*
γ	219.348 *10*	0.00051 *3*
γ	221.079 *12*	0.00052 *3*
γ	229.412 *14*	0.0241 *16*
γ [E1]	266.910 *11*	0.000313 *18*
γ [M1+E2]	341.232 *12*	0.043 *9*
γ [M1+E2]	365.450 *11*	0.00066 *14*
γ [E1]	387.073 *10*	~0.0006
γ	396.851 *14*	~0.39
γ E1	409.312 *18*	0.65 *7*
γ M1+13%E2	446.810 *3*	0.0288 *16*
γ [M1+E2]	460.49 *7*	2.3 *5*
γ [E1]	461.79 *3*	4.7 *9*
γ	467.046 *20*	0.140 *22*
γ	493.55 *4*	$8.0\,_{8}\times10^{-5}$
γ M1,E2	560.33 *3*	1.87 *10*
γ [E1]	581.95 *3*	8.6 *3*
γ M1	584.81 *5*	6.49 *20*
γ [E1]	603.04 *4*	0.078 *18*
γ M1+33%E2	620.3594 *24*	0.055 *5*
γ M1,E2	626.262 *11*	1.48 *9*
γ M1+4.4%E2	641.69 *3*	25.9 *6*
γ M1(+E2)	648.45 *6*	0.39 *10*
γ E2	657.7617 *23*	98.3 *20*
γ M1+12%E2	677.6218 *22*	4.44 *20*
γ E2+24%M1	687.013 *3*	0.126 *12*
γ E2+33%M1	706.6808 *25*	0.126 *7*
γ M1+22%E2	707.42 *2*	29.5 *10*
γ	708.081 *14*	1.60 *22*
γ E2	744.276 *3*	2.02 *8*
γ M1	759.630 *11*	3.15 *10*
γ E2	763.943 *3*	0.171 *9*
γ	774.735 *14*	~0.011
γ	795.44 *6*	0.32 *3*
γ E2	795.44 *3*	0.32 *3*
γ E2+35%M1	818.030 *3*	2.27 *7*
γ E2	844.68 *5*	3.24 *10*
γ	871.10 *3*	0.31 *4*
γ E2	884.684 *3*	92.9 *19*
γ [E2]	901.56 *3*	1.97 *10*
γ E2	937.492 *3*	68.4 *14*
γ E1+0.3%M2	997.228 *10*	10.52 *21*
γ [M1+E2]	1018.852 *12*	0.295 *25*
γ	1019.501 *10*	0.69 *14*
γ E2(+M1)	1045.29 *6*	0.81 *4*
γ M1,E2	1085.505 *12*	1.36 *5*
γ E1	1117.391 *11*	4.22 *10*
γ M1+1.0%E2	1125.718 *15*	0.233 *8*
γ	1163.30 *13*	0.29 *10*
γ E2	1305.16 *6*	0.34 *3*
γ	1334.339 *14*	0.97 *6*
γ M1+16%E2	1384.2979 *24*	0.187 *9*
γ (E1+M2)	1421.06 *4*	0.462 *16*
γ E2	1475.786 *3*	1.24 *4*
γ E2+M1	1505.038 *3*	0.26 *4*
γ [E2]	1522.29 *5*	0.17 *3*
γ E2	1562.300 *3*	0.442 *20*
γ [E2]	1579.17 *3*	0.26 *6*
γ	1592.759 *14*	0.122 *18*
γ (E2+M1)	1629.69 *8*	0.46 *9* ?
γ	1697.116 *10*	0.256 *10*
γ (E2)	1783.472 *15*	0.07 *3*
γ	1802.41 *13*	0.62 *9*
γ [E2]	1903.526 *12*	0.31 *3*
γ [E2]	1982.78 *6*	0.383 *10*

† 0.11% uncert(syst)

Atomic Electrons (^{110}In)

⟨e⟩=12.5 *6* keV

e_{bin}(keV)	⟨e⟩(keV)	e(%)
4	2.44	67 *7*
19	1.42	7.4 *8*
20	0.36	1.81 *19*
22	0.41	1.85 *19*
23	0.44	1.92 *20*
25 - 26	0.083	0.326 *25*
93	0.7	~0.8
94 - 133	0.40	0.37 *16*
193 - 240	0.0025	~0.0012
263 - 267	1.1×10^{-6}	$4.3\,_{18}\times10^{-7}$
315 - 362	0.0027	0.0009 *4*
365 - 409	0.021	0.0056 *22*
420 - 467	0.141	0.032 *4*
490 - 534	0.043	0.0081 *7*
555 - 603	0.273	0.0484 *17*
615	0.518	0.0843 *25*
616 - 626	0.0121	0.0019 *4*
631	1.70	0.269 *8*
638 - 680	0.453	0.0695 *12*
681	0.53	0.077 *4*
683 - 718	0.110	0.0156 *5*
733 - 775	0.073	0.0099 *4*
791 - 841	0.0800	0.00989 *24*
844 - 845	0.0040	0.00048 *18*
858	1.05	0.123 *4*
867 - 902	0.188	0.0213 *5*
911	0.720	0.0790 *22*
933 - 971	0.154	0.0163 *3*
992 - 1042	0.025	0.0025 *3*
1045 - 1091	0.0182	0.00167 *5*
1099 - 1137	0.0201	0.00187 *17*
1159 - 1163	0.00029	$2.5\,_{11}\times10^{-5}$
1278 - 1308	0.0085	0.00065 *19*
1330 - 1358	0.0023	0.000169 *23*
1380 - 1421	0.005	~0.00039
1449 - 1536	0.0151	0.00102 *3*
1552 - 1629	0.0061	0.00039 *5*
1670 - 1757	0.0017	0.00010 *3*
1776 - 1802	0.0031	0.00018 *6*
1877 - 1956	0.00368	0.000192 *9*
1979 - 1982	0.000269	$1.36\,_{6}\times10^{-5}$

Continuous Radiation (^{110}In)

⟨β+⟩=0.078 keV;⟨IB⟩=0.156 keV

E_{bin}(keV)		⟨ ⟩(keV)	(%)
0 - 10	β+	8.3×10^{-8}	1.0×10^{-5}
	IB	0.00019	
10 - 20	β+	2.76×10^{-6}	1.67×10^{-5}
	IB	0.0038	0.021
20 - 40	β+	6.2×10^{-5}	0.000191
	IB	0.033	0.137
40 - 100	β+	0.00213	0.00279
	IB	0.0029	0.0046
100 - 300	β+	0.0428	0.0213
	IB	0.026	0.0127
300 - 600	β+	0.0331	0.0087
	IB	0.063	0.0147
600 - 1300	β+	5.6×10^{-5}	9.2×10^{-6}
	IB	0.027	0.0035
1300 - 1653	IB	0.00041	3.0×10^{-5}
	Σβ+		0.033

$^{110}_{50}$Sn(4.11 *10* h)

Mode: ε

Δ: -85830 *16* keV

SpA: 6.93×10^{6} Ci/g

Prod: ^{115}In(p,6n); ^{108}Cd(α,2n)

Photons (^{110}Sn)

⟨γ⟩=293 *4* keV

γ_{mode}	γ(keV)	γ(%)
In L_ℓ	2.905	0.111 *20*
In L_η	3.112	0.052 *8*
In L_α	3.286	3.1 *5*
In L_β	3.570	2.2 *4*
In L_γ	3.988	0.23 *4*
In $K_{\alpha2}$	24.002	21.5 *8*
In $K_{\alpha1}$	24.210	40.2 *14*
In $K_{\beta1}$'	27.265	10.4 *4*
In $K_{\beta2}$'	28.022	2.18 *8*
γ (M1)	283 *1*	97.0 *10*

Atomic Electrons (^{110}Sn)

⟨e⟩=13.1 *4* keV

e_{bin}(keV)	⟨e⟩(keV)	e(%)
4	2.6	69 *7*
19	0.187	0.96 *10*
20	1.58	7.8 *8*
23	0.41	1.79 *18*
24	0.44	1.86 *19*
26	0.026	0.097 *10*
27	0.060	0.222 *23*
255	6.6	2.60 *11*

Atomic Electrons (^{110}Sn) (continued)

e_{bin}(keV)	⟨e⟩(keV)	e(%)
279	0.89	0.321 13
282	0.177	0.063 3
283	0.0386	0.0136 6

Continuous Radiation (^{110}Sn)

⟨IB⟩=0.047 keV

E_{bin}(keV)		⟨ ⟩(keV)	(%)
10 - 20	IB	0.0020	0.0110
20 - 40	IB	0.039	0.160
40 - 100	IB	0.0024	0.0039
100 - 297	IB	0.0033	0.0022

$^{110}_{51}$Sb(23.0 *4* s)

Mode: ϵ

Δ: -77530 *200* keV syst

SpA: 4.39×10^9 Ci/g

Prod: ^{112}Sn(p,3n); ^{103}Rh(^{12}C,5n)

Photons (^{110}Sb)

⟨γ⟩=2797 *34* keV

γ_{mode}	γ(keV)	γ(%)†
Sn $K_{\alpha2}$	25.044	1.54 5
Sn $K_{\alpha1}$	25.271	2.87 9
Sn $K_{\beta1}$'	28.474	0.751 23
Sn $K_{\beta2}$'	29.275	0.158 5
γ	624.7 4	1.01 9
γ [E2]	636.8 3	4.3 4
γ	751.54 25	4.1 4
γ	767.2 4	0.8 3
γ	796.3 4	0.41 9
γ	827.32 24	9.4 6
γ	909.15 21	7.8 6
γ E2	984.92 9	31.2 19
γ	996.9 5	1.28 18
γ	1025.8 3	2.3 3
γ	1101.6 3	0.76 18
γ E2	1211.94 10	91.6
γ E2	1243.6 3	13.4 8
γ	1325.9 5	0.50 9
γ [M1+E2]	1333.8 4	1.56 18
γ	1339.5 6	0.48 9
γ	1352.0 6	1.28 18
γ	1376.5 6	0.22 9
γ	1419.6 6	0.31 9
γ	1433.1 4	1.28 18
γ [E2]	1482.7 4	4.2 3
γ	1609.6 4	2.29 18
γ [M1+E2]	1621.7 3	1.6 3
γ [M1+E2]	1703.0 9	0.26 9
γ	1736.46 24	7.1 4
γ	1765.4 5	4.0 3
γ	1780.4 12	0.33 9
γ	1971.1 6	4.9 4
γ	2010.7 3	0.41 14
γ	2029.1 5	3.7 3
γ	2121.07 23	7.3 4
γ	2172.8 6	0.11 5
γ	2235.0 5	2.3 3
γ	2328.8 6	1.10 18
γ	2382.6 10	0.82 18
γ	2418.0 4	0.71 18
γ [E2]	2545.8 5	0.54 18
γ	2673.3 7	2.1 4
γ [E2]	2833.6 3	~0.36
γ [E2]	2914.9 9	0.22 9

† 0.55% uncert(syst)

Atomic Electrons (^{110}Sb)

⟨e⟩=2.61 *9* keV

e_{bin}(keV)	⟨e⟩(keV)	e(%)
4	0.205	5.1 5
20	0.0131	0.065 7
21	0.109	0.52 5
24 - 28	0.066	0.268 18
596	0.016	~0.0027
608	0.087	0.0143 13
620 - 637	0.0173	0.00274 24
722	0.05	~0.007
738 - 767	0.025	0.0034 11
792 - 796	0.0008	10 5 ×10^{-5}
798	0.11	~0.013
823 - 827	0.017	0.0021 9
880	0.08	~0.009
905 - 909	0.012	0.0014 6
956	0.347	0.0363 23
968 - 997	0.088	0.0090 12
1021 - 1026	0.0031	0.00030 12
1072 - 1102	0.007	0.0007 3
1183	0.808	0.0683 14
1207	0.0931	0.00771 16
1208 - 1212	0.0304	0.00251 4
1214	0.115	0.0095 6
1239 - 1244	0.0175	0.00141 7
1297 - 1339	0.032	0.0024 4
1347 - 1390	0.0049	0.00036 10
1404 - 1433	0.009	0.0007 3
1454 - 1483	0.0351	0.00241 16
1580 - 1674	0.030	0.0019 4
1699 - 1780	0.067	0.0039 10
1942 - 2028	0.047	0.0024 5
2092 - 2172	0.037	0.0018 6
2206 - 2300	0.015	0.00068 17
2324 - 2417	0.0075	0.00032 7
2517 - 2545	0.0028	0.00011 3
2644 - 2673	0.009	0.00033 10
2804 - 2886	0.0027	9 3 ×10^{-5}
2910 - 2914	0.00013	4.5 15 ×10^{-6}

Continuous Radiation (^{110}Sb)

⟨β+⟩=1970 keV;⟨IB⟩=9.4 keV

E_{bin}(keV)		⟨ ⟩(keV)	(%)
0 - 10	β+	9.6×10^{-7}	1.16×10^{-5}
	IB	0.056	
10 - 20	β+	3.50×10^{-5}	0.000212
	IB	0.056	0.39
20 - 40	β+	0.00088	0.00271
	IB	0.114	0.39
40 - 100	β+	0.0392	0.0504
	IB	0.32	0.49
100 - 300	β+	1.95	0.88
	IB	0.98	0.54
300 - 600	β+	16.6	3.55
	IB	1.25	0.29
600 - 1300	β+	169	17.1
	IB	2.2	0.25
1300 - 2500	β+	765	40.3
	IB	2.5	0.140
2500 - 5000	β+	1007	31.6
	IB	1.8	0.056
5000 - 7088	β+	11.4	0.216
	IB	0.059	0.00110
	Σβ+		94

$^{110}_{52}$Te(18.6 *8* s)

Mode: ϵ, α

Δ: -72140 *200* keV syst

SpA: 5.41×10^9 Ci/g

Prod: ^{58}Ni on ^{58}Ni

Alpha Particles (^{110}Te)

α(keV)
2624 15

Photons (^{110}Te)

γ_{mode}	γ(keV)
γ	107.5 6
γ	219.1 6
γ	605.9 6
γ	894.8 6

$^{110}_{53}$I (650 *20* ms)

Mode: ϵ(83 *4* %), α(17 *4* %), ϵp(11 *4* %), ϵα(1.1 *4* %)

Δ: -60490 *500* keV syst

SpA: 9.7×10^{10} Ci/g

Prod: ^{58}Ni on ^{58}Ni

ϵp: 2500-6000 range

Alpha Particles (^{110}I)

α(keV)
3444 10

$^{110}_{54}$Xe($t_{1/2}$ unknown)

Mode: ϵ, α

Δ: -51750 *630* keV syst

Prod: ^{58}Ni on ^{58}Ni

Alpha Particles (^{110}Xe)

α(keV)
3737 30

A = 111

NDS 27, 453 (1979)

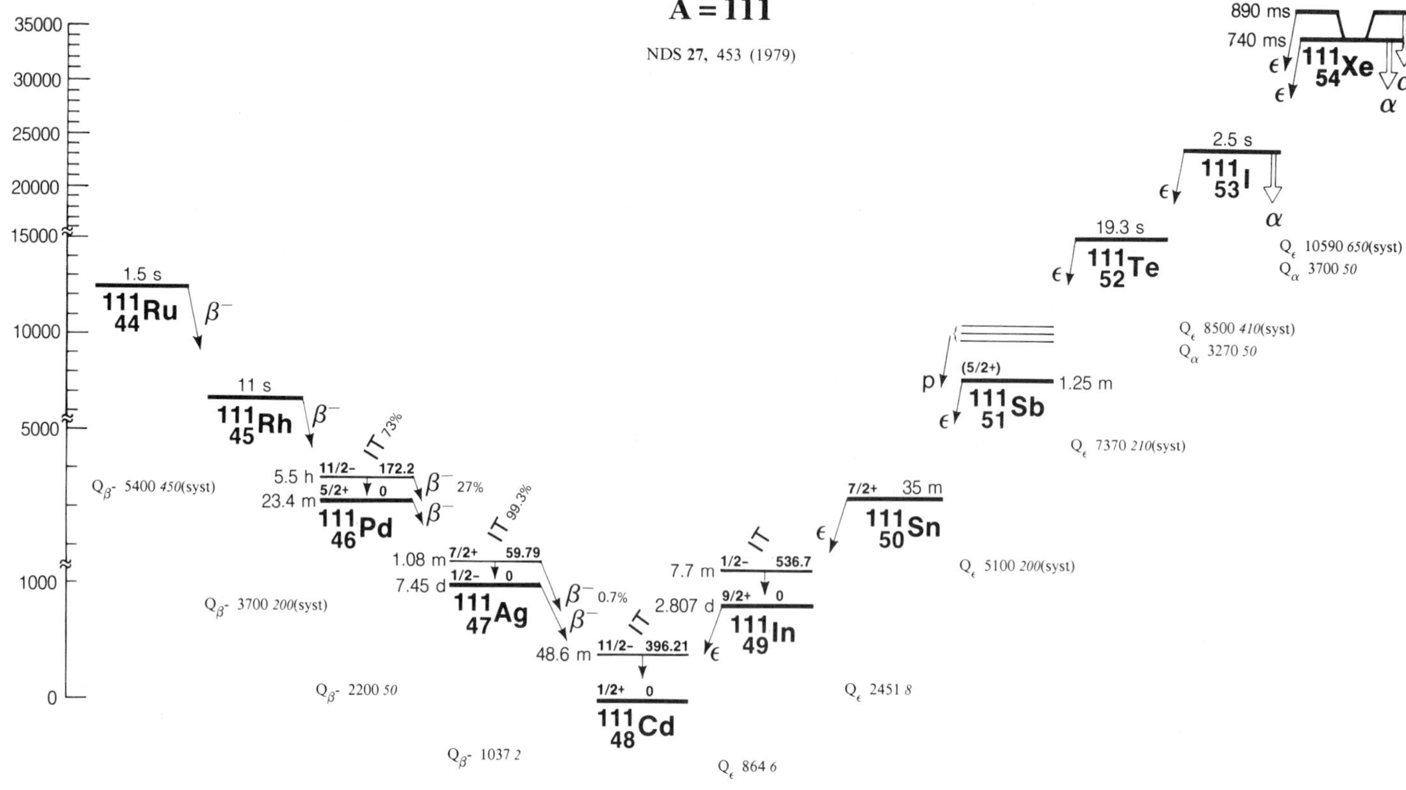

$^{111}_{44}$Ru(1.5 *3* s)

Mode: β-
Δ: -76920 *400* keV syst
SpA: 5.4×10^{10} Ci/g
Prod: fission

$^{111}_{45}$Rh(11 *1* s)

Mode: β-
Δ: -82320 *210* keV syst
SpA: 9.0×10^{9} Ci/g
Prod: fission

Photons (^{111}Rh)

γ_{mode}	γ(keV)
γ	275.3 *3*

$^{111}_{46}$Pd(23.4 *2* min)

Mode: β-
Δ: -86020 *50* keV
SpA: 7.24×10^{7} Ci/g
Prod: ^{110}Pd(n,γ); daughter ^{111}Pd(5.5 h)

Photons (^{111}Pd)
⟨γ⟩=45.0 *15* keV

γ_{mode}	γ(keV)	γ(%)
γ E3	59.79 *3*	0.55 *4* *
γ M1+1.4%E2	70.40 *4*	0.78 *11*
γ [E1]	86.99 *9*	0.0150 *16*
γ [M1]	101.57 *8*	0.0017 *4*
γ	141.52 *10*	0.00042 *17*
γ	166.0 *3*	~0.013 ?
γ [E2]	169.00 *10*	0.027 *8*
γ [E2]	202.10 *16*	0.0100 *25*
γ [M1+E2]	230.20 *8*	0.0234 *25*
γ [E1]	279.03 *10*	0.010 *3*
γ [M1]	289.62 *7*	0.102 *8*
γ	308.36 *15*	0.0092 *8*
γ [E2]	316.81 *6*	0.0184 *16*
γ [M1+E2]	352.2 *6*	0.0059 *25*
γ [E1]	376.61 *5*	0.444 *24*
γ [E2]	391.18 *6*	0.0260 *8*
γ [E1]	404.71 *15*	0.084 *5*
γ	414.0 *10*	0.025 *8*
γ [M1]	415.41 *9*	0.084 *23*
γ [M1+E2]	417.59 *16*	0.0040 *18*
γ [E2]	438.45 *8*	0.049 *5*
γ [E2]	476.61 *6*	0.058 *5*
γ	478.6 *4*	~0.019
γ [M1]	485.81 *8*	0.027 *5*
γ	493.41 *17*	~0.008
γ [M1+E2]	508.85 *9*	0.209 *25*
γ	516.44 *17*	~0.013
γ	516.8 *4*	~0.013
γ [M1+E2]	519.16 *15*	0.011 *4*
γ	540.75 *20*	0.026 *5*
γ [M1+E2]	547.02 *6*	0.368 *25*
γ	552.73 *9*	0.0159 *24*
γ	579.96 *6*	0.84
γ	603.48 *16*	0.023 *11*
γ	611.53 *16*	0.018 *4*
γ	623.13 *8*	0.276 *17*
γ	624.5 *5*	0.017 *4*
γ	634.56 *16*	~0.0033
γ	641.64 *15*	0.059 *4*
γ	650.37 *7*	0.552 *25*
γ	657.34 *21*	0.0234 *25*
γ	685.44 *15*	0.050 *6*
γ	709.75 *19*	0.126 *13*
γ	743.63 *24*	0.017 *4*
γ	745.9 *3*	0.013 *4*
γ	772.43 *17*	0.0042 *17*
γ	775.46 *18*	0.042 *4*
γ	793.5 *5*	0.018 *3*
γ	803.56 *14*	0.034 *6*
γ	808.49 *13*	~0.024
γ [E2]	808.77 *16*	0.0040 *16*
γ	816.3 *3*	0.009 *4*
γ	833.68 *16*	~0.008
γ	835.72 *13*	0.268 *17*
γ	890.55 *15*	0.0042 *17*
γ	920.67 *16*	0.024 *5*
γ	937.3 *10*	~0.007
γ	950.00 *14*	0.008 *3*
γ	955.2 *4*	0.038 *3*
γ	1002.25 *15*	0.061 *11*
γ	1014.9 *7*	~0.008
γ	1021.7 *4*	~0.007
γ	1026.56 *19*	~0.008
γ	1053.0 *7*	~0.0042
γ	1060.44 *24*	0.013 *4*
γ	1067.3 *4*	0.013 *4*
γ	1097.8 *4*	~0.007
γ	1120.37 *14*	0.14 *2*
γ	1246.0 *10*	~0.0033
γ	1269.4 *4*	0.0084 *25*
γ	1312.9 *4*	~0.008
γ	1348.0 *20*	~0.017
γ	1388.45 *12*	0.54 *4*
γ	1395.0 *10*	~0.0042
γ	1458.85 *12*	0.56 *4*
γ	1506.0 *10*	<0.0017

Photons (^{111}Pd)
(continued)

γ_{mode}	γ(keV)	γ(%)
γ	1542.8 4	0.0218 25
γ	1549.0 10	~0.0050
γ	1574.4 4	0.026 4
γ	1644.8 4	0.019 4
γ	1863.2 10	~0.0025

* with ^{111}Ag(1.08 min) in equilib

Continuous Radiation (^{111}Pd)
$\langle\beta-\rangle$=832 keV;$\langle$IB$\rangle$=1.67 keV

E_{bin}(keV)		$\langle\ \rangle$(keV)	(%)
0 - 10	β-	0.0183	0.364
	IB	0.033	
10 - 20	β-	0.056	0.372
	IB	0.032	0.22
20 - 40	β-	0.231	0.77
	IB	0.062	0.22
40 - 100	β-	1.77	2.50
	IB	0.17	0.27
100 - 300	β-	21.4	10.5
	IB	0.46	0.26
300 - 600	β-	91	20.1
	IB	0.45	0.107
600 - 1300	β-	443	47.5
	IB	0.42	0.051
1300 - 2140	β-	275	17.9
	IB	0.040	0.0028

$^{111}_{46}$Pd(5.5 *1* h)

Mode: IT(73 *3* %), β-(27 *3* %)
Δ: -85848 *50* keV
SpA: 5.13×10^6 Ci/g
Prod: ^{110}Pd(n,γ); ^{110}Pd(d,p)

Photons (^{111}Pd)
$\langle\gamma\rangle$=359 *7* keV

γ_{mode}	γ(keV)	γ(%)†
Pd L_ℓ	2.503	0.036 7
Ag L_ℓ	2.634	0.027 10
Pd L_η	2.660	0.019 3
Ag L_η	2.806	0.016 6
Pd L_α	2.838	0.97 16
Ag L_α	2.984	0.7 3
Pd L_β	3.045	0.68 11
Ag L_β	3.209	0.55 21
Pd L_γ	3.366	0.066 12
Ag L_γ	3.548	0.052 20
Pd $K_{\alpha2}$	21.020	6.7 4
Pd $K_{\alpha1}$	21.177	12.6 7
Ag $K_{\alpha2}$	21.990	3.2 6
Ag $K_{\alpha1}$	22.163	6.1 12
Pd $K_{\beta1}'$	23.811	3.12 17
Pd $K_{\beta2}'$	24.445	0.60 3
Ag $K_{\beta1}'$	24.934	1.5 3
Ag $K_{\beta2}'$	25.603	0.31 6
$\gamma_{\beta-}$E3	59.79 3	~0.10 *
$\gamma_{\beta-}$M1+1.4%E2	70.40 4	8.3 12
$\gamma_{\beta-}$[E1]	86.99 9	0.030 5
$\gamma_{\beta-}$[M1]	101.57 8	0.34 7
$\gamma_{\beta-}$	119.16 11	~0.10
$\gamma_{\beta-}$	166.0 3	~0.12
$\gamma_{\beta-}$	166.44 19	~0.10
$\gamma_{\beta-}$[E2]	169.00 10	0.51 14

Photons (^{111}Pd)
(continued)

γ_{mode}	γ(keV)	γ(%)†
γ_{IT} E3	172.18 8	33.6
$\gamma_{\beta-}$	271.92 12	0.166 16
$\gamma_{\beta-}$[M1]	289.62 7	1.04 10
$\gamma_{\beta-}$[M1+E2]	307.65 13	0.14 6
$\gamma_{\beta-}$[E2]	316.81 6	0.036 6
$\gamma_{\beta-}$	357.83 9	0.42 4
$\gamma_{\beta-}$[E1]	376.61 5	0.88 11
$\gamma_{\beta-}$[E2]	391.18 6	5.4 4
$\gamma_{\beta-}$[E2]	413.22 24	1.8 4
$\gamma_{\beta-}$[M1]	415.41 9	1.6 3
$\gamma_{\beta-}$[M1+E2]	417.59 16	0.056 22
$\gamma_{\beta-}$	439.59 14	0.24 4
$\gamma_{\beta-}$	444.11 20	0.126 14
$\gamma_{\beta-}$	454.1 5	0.26 10
$\gamma_{\beta-}$[M1+E2]	454.43 10	1.12 18
$\gamma_{\beta-}$	476.6 3	<0.040
$\gamma_{\beta-}$[M1]	485.81 8	0.51 7
$\gamma_{\beta-}$[M1+E2]	519.16 15	0.149 16
$\gamma_{\beta-}$	525.50 9	1.30 14
$\gamma_{\beta-}$	552.32 24	0.28 6
$\gamma_{\beta-}$	552.73 9	0.023 7
$\gamma_{\beta-}$[E2]	555.99 8	0.28 4
$\gamma_{\beta-}$[M1]	574.97 8	3.2 3
$\gamma_{\beta-}$[M1+E2]	583.5 5	0.26 6
$\gamma_{\beta-}$	595.5 3	0.126 12
$\gamma_{\beta-}$	617.67 13	~0.07
$\gamma_{\beta-}$	623.13 8	0.40 11
$\gamma_{\beta-}$[E2]	632.51 12	3.6 3
$\gamma_{\beta-}$[E2]	645.38 9	~0.10
$\gamma_{\beta-}$	654.50 15	0.150 16
$\gamma_{\beta-}$	667.85 14	0.98 10
$\gamma_{\beta-}$	694.13 9	2.0
$\gamma_{\beta-}$	697.1 3	0.05 3
$\gamma_{\beta-}$	703.58 12	0.66 8
$\gamma_{\beta-}$	715.9 10	~0.040
$\gamma_{\beta-}$	718.86 19	0.18 6
$\gamma_{\beta-}$	724.86 12	0.28 3
$\gamma_{\beta-}$	745.9 3	0.128 21
$\gamma_{\beta-}$	752.20 21	0.12 4
$\gamma_{\beta-}$[M1+E2]	762.08 9	1.26 10
$\gamma_{\beta-}$	797.42 12	1.02 6
$\gamma_{\beta-}$[E2]	808.77 16	0.056 21
$\gamma_{\beta-}$	816.3 3	0.09 4
$\gamma_{\beta-}$[M1+E2]	828.63 25	0.17 3
$\gamma_{\beta-}$[E2]	863.64 11	0.15 4
$\gamma_{\beta-}$	881.77 19	0.214 20
$\gamma_{\beta-}$[E1]	893.49 12	~0.06
$\gamma_{\beta-}$	915.8 5	~0.10
$\gamma_{\beta-}$	945.0 3	0.13 4
$\gamma_{\beta-}$	975.49 12	0.16 4
$\gamma_{\beta-}$	996.77 13	0.26 6
$\gamma_{\beta-}$	1000.41 25	0.10 4
$\gamma_{\beta-}$[E1]	1023.06 11	~0.14
$\gamma_{\beta-}$[E1]	1028.7 3	<0.06 ?
$\gamma_{\beta-}$	1045.4 5	0.10 4
$\gamma_{\beta-}$[E1]	1063.0 6	~0.14 ?
$\gamma_{\beta-}$	1075.9 10	~0.040
$\gamma_{\beta-}$	1087.9 4	0.20 6
$\gamma_{\beta-}$	1098.58 16	~0.12
$\gamma_{\beta-}$	1115.93 12	1.10 10
$\gamma_{\beta-}$	1139.7 4	~0.06
$\gamma_{\beta-}$[E1]	1142.1 3	~0.12
$\gamma_{\beta-}$	1163.22 18	0.34 4
$\gamma_{\beta-}$	1200.29 19	0.32 4
$\gamma_{\beta-}$	1222.53 20	~0.09
$\gamma_{\beta-}$	1269.9 10	<0.040
$\gamma_{\beta-}$[E1]	1282.37 17	1.04 10
$\gamma_{\beta-}$	1308.9 10	<0.06
$\gamma_{\beta-}$[E2]	1381.5 6	~0.040 ?
$\gamma_{\beta-}$	1418.0 3	0.060 20
$\gamma_{\beta-}$	1651.30 14	0.72 8
$\gamma_{\beta-}$	1690.9 1	1.28 10
$\gamma_{\beta-}$	1721.70 14	0.34 4
$\gamma_{\beta-}$	1775.25 18	0.46 6
$\gamma_{\beta-}$	1904.3 4	~0.08
$\gamma_{\beta-}$	1938.9 5	~0.10
$\gamma_{\beta-}$[E1]	1970.7 3	0.62 6
$\gamma_{\beta-}$	2064 1	~0.028
$\gamma_{\beta-}$	2085.9 10	<0.040

† uncert(syst): 12% for IT, 15% for β-
* with ^{111}Ag(1.08 min) in equilib

Atomic Electrons (^{111}Pd)
$\langle$e$\rangle$=81 *4* keV

e_{bin}(keV)	$\langle$e$\rangle$(keV)	e(%)
3 - 34	4.5	61 9
45	3.9	8.6 13
56	6.8	~12
59	1.5	~3
60 - 102	1.39	2.1 3
115 - 143	0.18	0.13 3
148	41.1	27.8 12
162 - 168	0.042	0.026 5
169	16.3	9.6 4
172	3.85	2.24 10
246 - 291	0.089	0.033 4
304 - 351	0.026	0.0078 24
354 - 392	0.39	0.104 7
409 - 454	0.075	0.0177 24
460 - 500	0.040	0.0083 22
515 - 558	0.089	0.0163 15
570 - 619	0.082	0.0136 11
620 - 669	0.050	0.0076 20
672 - 721	0.020	0.0029 7
722 - 762	0.0220	0.0030 3
772 - 816	0.015	0.0019 6
825 - 868	0.0043	0.00051 12
878 - 919	0.0022	0.00025 10
941 - 975	0.0042	0.00043 13
993 - 1042	0.0027	0.00027 7
1045 - 1090	0.010	0.0009 3
1095 - 1142	0.0042	0.00037 10
1159 - 1200	0.0030	0.00026 8
1219 - 1267	0.0036	0.00028 3
1269 - 1309	0.00068	5.3 14 $\times10^{-5}$
1356 - 1392	0.00060	4.4 15 $\times10^{-5}$
1414 - 1418	4.4 $\times10^{-5}$	3.1 14 $\times10^{-6}$
1626 - 1721	0.012	0.00070 16
1750 - 1775	0.0022	0.00012 4
1879 - 1970	0.0025	0.000130 21
2038 - 2085	0.00020	~10×10^{-6}

Continuous Radiation (^{111}Pd)
$\langle\beta-\rangle$=123 keV;$\langle$IB$\rangle$=0.20 keV

E_{bin}(keV)		$\langle\ \rangle$(keV)	(%)
0 - 10	β-	0.0298	0.60
	IB	0.0052	
10 - 20	β-	0.088	0.59
	IB	0.0050	0.035
20 - 40	β-	0.348	1.16
	IB	0.0095	0.033
40 - 100	β-	2.30	3.31
	IB	0.025	0.039
100 - 300	β-	15.6	8.2
	IB	0.060	0.034
300 - 600	β-	22.9	5.4
	IB	0.052	0.0124
600 - 1300	β-	54	5.9
	IB	0.044	0.0054
1300 - 2242	β-	28.4	1.82
	IB	0.0047	0.00033

$^{111}_{47}$Ag(7.45 *1* d)

Mode: β-
Δ: -88218 *4* keV
SpA: 1.5790×10^5 Ci/g
Prod: daughter ^{111}Pd; ^{110}Pd(d,n)

Photons (^{111}Ag)

⟨γ⟩=26.6 *11* keV

γ_{mode}	γ(keV)	γ(%)†
Cd $K_{\alpha2}$	22.984	0.065 *7*
Cd $K_{\alpha1}$	23.174	0.122 *13*
Cd $K_{\beta1}$'	26.085	0.031 *3*
Cd $K_{\beta2}$'	26.801	0.0065 *7*
γ M1+~1.0%E2	96.697 *8*	0.20 *6*
γ E2	245.384 *6*	1.24 *7*
γ [M1+E2]	278.03 *14*	0.00061 *16*
γ M1+36%E2	342.081 *6*	6.7
γ [M1+E2]	374.73 *14*	0.0029 *3*
γ	509.3 *3*	0.0013 *3*
γ	522.64 *23*	0.0010 *3*
γ [M1+E2]	524.74 *23*	0.0023 *4*
γ M1+E2	619.33 *23*	0.0042 *10*
γ E2	620.11 *14*	0.0181 *17*
γ M1+E2	621.43 *23*	0.0059 *15*
γ	754.7 *3*	0.0027 *5*
γ M1+E2	864.72 *23*	0.0033 *7*
γ M1+E2	866.82 *23*	0.00514 *13*

† 4.9% uncert(syst)

Atomic Electrons (^{111}Ag)

⟨e⟩=0.67 *3* keV

e_{bin}(keV)	⟨e⟩(keV)	e(%)
4 - 26	0.0162	0.26 *3*
70	0.063	0.09 *3*
93 - 97	0.014	0.014 *4*
219	0.145	0.066 *4*
241	0.0165	0.0068 *4*
242 - 278	0.0148	0.00609 *25*
315	0.342	0.109 *6*
338	0.0450	0.0133 *7*
339 - 375	0.0143	0.00419 *20*
483 - 525	0.000118	2.4 *4* $\times10^{-5}$
593 - 621	0.00063	0.000106 *11*
728 - 755	3.4 $\times10^{-5}$	4.6 *22* $\times10^{-6}$
838 - 867	0.000124	1.47 *14* $\times10^{-5}$

Continuous Radiation (^{111}Ag)

⟨β-⟩=354 keV;⟨IB⟩=0.38 keV

E_{bin}(keV)		⟨ ⟩(keV)	(%)
0 - 10	β-	0.068	1.36
	IB	0.0166	
10 - 20	β-	0.206	1.37
	IB	0.0160	0.111
20 - 40	β-	0.84	2.79
	IB	0.030	0.104
40 - 100	β-	6.1	8.7
	IB	0.076	0.118
100 - 300	β-	63	31.2
	IB	0.152	0.090
300 - 600	β-	170	38.8
	IB	0.076	0.019
600 - 1037	β-	114	15.9
	IB	0.0098	0.00148

$^{111}_{47}$Ag(1.080 *13* min)

Mode: IT(99.3 *2* %), β-(0.7 *2* %)
Δ: -88158 *4* keV
SpA: 1.560×10^{9} Ci/g
Prod: daughter ^{111}Pd

Photons (^{111}Ag)

⟨γ⟩=7.48 *25* keV

γ_{mode}	γ(keV)	γ(%)†
Ag L_{ℓ}	2.634	0.087 *16*
Ag L_{η}	2.806	0.055 *9*
Ag L_{α}	2.984	2.4 *4*
Ag L_{β}	3.204	1.8 *3*
Ag L_{γ}	3.532	0.17 *3*
Ag $K_{\alpha2}$	21.990	5.4 *3*
Ag $K_{\alpha1}$	22.163	10.1 *6*
Ag $K_{\beta1}$'	24.934	2.55 *15*
Ag $K_{\beta2}$'	25.603	0.51 *3*
γ_{IT} E3	59.79 *3*	0.531
$\gamma_{\beta-}$M1+~1.0%E2	96.697 *8*	0.00018 *7*
$\gamma_{\beta-}$M1+2.0%E2	171.28 *3*	~0.12
$\gamma_{\beta-}$E2	245.384 *6*	0.50 *3*
$\gamma_{\beta-}$[M1+E2]	278.03 *14*	0.0041 *11*
$\gamma_{\beta-}$M1+36%E2	342.081 *6*	0.0061 *21*
$\gamma_{\beta-}$[M1+E2]	374.73 *14*	0.020 *3*
$\gamma_{\beta-}$[M1+E2]	507.25 *20*	~0.04
$\gamma_{\beta-}$E2	620.11 *14*	0.122 *8*
$\gamma_{\beta-}$[E2]	752.63 *20*	0.043 *5*

† uncert(syst): 4.7% for IT, 33% for β-

Atomic Electrons (^{111}Ag)*

⟨e⟩=54.8 *19* keV

e_{bin}(keV)	⟨e⟩(keV)	e(%)
3	1.45	43 *5*
4	0.97	27 *3*
18	0.199	1.09 *12*
19	0.28	1.52 *17*
20	0.00017	0.00084 *15*
21	0.110	0.52 *6*
22	0.116	0.54 *6*
23	0.00020	0.00089 *16*
24	0.0151	0.062 *7*
25	0.0069	0.028 *3*
26	1.22 $\times10^{-5}$	4.7 *9* $\times10^{-5}$
34	7.6	22.2 *11*
56	35.0	62 *3*
59	7.7	13.0 *7*
60	1.33	2.22 *11*

* with IT

Continuous Radiation (^{111}Ag)

⟨β-⟩=1.56 keV;⟨IB⟩=0.00121 keV

E_{bin}(keV)		⟨ ⟩(keV)	(%)
0 - 10	β-	0.00107	0.0214
	IB	7.7×10^{-5}	
10 - 20	β-	0.00319	0.0213
	IB	7.2×10^{-5}	0.00050
20 - 40	β-	0.0126	0.0420
	IB	0.000131	0.00046
40 - 100	β-	0.084	0.121
	IB	0.00030	0.00048
100 - 300	β-	0.58	0.305
	IB	0.00048	0.00029
300 - 600	β-	0.74	0.178
	IB	0.000141	3.7×10^{-5}
600 - 851	β-	0.141	0.0212
	IB	4.8×10^{-6}	7.7×10^{-7}

$^{111}_{48}$Cd(stable)

Δ: -89255 *3* keV
%: 12.80 *6*

$^{111}_{48}$Cd(48.6 *3* min)

Mode: IT
Δ: -88859 *3* keV
SpA: 3.485×10^{7} Ci/g
Prod: ^{110}Cd(n,γ); daughter ^{111}In

Photons (^{111}Cd)

⟨γ⟩=284.5 *15* keV

γ_{mode}	γ(keV)	γ(%)†
Cd L_{ℓ}	2.767	0.074 *13*
Cd L_{η}	2.957	0.042 *7*
Cd L_{α}	3.133	2.0 *3*
Cd L_{β}	3.382	1.6 *3*
Cd L_{γ}	3.755	0.15 *3*
Cd $K_{\alpha2}$	22.984	12.0 *4*
Cd $K_{\alpha1}$	23.174	22.5 *8*
Cd $K_{\beta1}$'	26.085	5.73 *21*
Cd $K_{\beta2}$'	26.801	1.19 *5*
γ E3	150.822 *13*	29.1 *9*
γ E2	245.384 *6*	94.00

† 0.20% uncert(syst)

Atomic Electrons (^{111}Cd)

⟨e⟩=109.6 *17* keV

e_{bin}(keV)	⟨e⟩(keV)	e(%)
4	1.92	53 *5*
19	0.82	4.3 *4*
20	0.205	1.04 *11*
22	0.236	1.06 *11*
23	0.25	1.11 *11*
25	0.033	0.130 *13*
26	0.0151	0.058 *6*
124	54.3	43.8 *12*
147	30.8	20.9 *6*
150	6.45	4.29 *11*
151	1.18	0.784 *21*
219	11.00	5.03 *11*
241	1.25	0.520 *12*
242	0.666	0.275 *6*
245	0.456	0.186 *4*

$^{111}_{49}$In(2.807 *2* d)

Mode: ε
Δ: -88391 *6* keV
SpA: 4.191×10^{5} Ci/g
Prod: ^{109}Ag(α,2n)

Photons (^{111}In)

⟨γ⟩=405.2 *7* keV

γ_{mode}	γ(keV)	γ(%)†
Cd L_ℓ	2.767	0.120 *21*
Cd L_η	2.957	0.056 *9*
Cd L_α	3.133	3.3 *5*
Cd L_β	3.392	2.3 *4*
Cd L_γ	3.781	0.23 *4*
Cd $K_{\alpha2}$	22.984	24.1 *8*
Cd $K_{\alpha1}$	23.174	45.4 *16*
Cd $K_{\beta1}'$	26.085	11.6 *4*
Cd $K_{\beta2}'$	26.801	2.40 *9*
γ E3	150.822 *13*	0.0028 *6* *
γ M1+2.0%E2	171.28 *3*	90.24 *18*
γ E2	245.384 *6*	94.00

† 0.20% uncert(syst)
* with ^{111}Cd(48.6 min) in equilib

Atomic Electrons (^{111}In)

⟨e⟩=33.9 *5* keV

e_{bin}(keV)	⟨e⟩(keV)	e(%)
4	2.9	79 *8*
19	1.65	8.6 *9*
20	0.41	2.11 *21*
22	0.48	2.15 *22*
23	0.51	2.23 *23*
25	0.066	0.26 *3*
26	0.031	0.118 *12*
124	0.0051	0.0041 *8*
145	12.2	8.47 *19*
147	0.0029	0.0019 *4*
150	0.00060	0.00040 *8*
151	0.000110	7.3 *15* $\times 10^{-5}$
167	1.63	0.977 *21*
168	0.161	0.0963 *22*
171	0.428	0.251 *6*
219	11.00	5.03 *11*
241	1.25	0.520 *12*
242	0.666	0.275 *6*
245	0.456	0.186 *4*

Continuous Radiation (^{111}In)

⟨IB⟩=0.055 keV

E_{bin}(keV)		⟨ ⟩(keV)	(%)
10 - 20	IB	0.0039	0.021
20 - 40	IB	0.034	0.140
40 - 100	IB	0.0026	0.0042
100 - 300	IB	0.0121	0.0065
300 - 468	IB	0.0024	0.00073

$^{111}_{49}$In(7.7 *2* min)

Mode: IT
Δ: -87854 *6* keV
SpA: 2.20×10^8 Ci/g
Prod: ^{109}Ag(α,2n); ^{111}Cd(p,n); ^{112}Sn(γ,p)

Photons (^{111}In)

⟨γ⟩=469 *3* keV

γ_{mode}	γ(keV)	γ(%)†
In L_ℓ	2.905	0.0140 *25*
In L_η	3.112	0.0067 *10*
In L_α	3.286	0.39 *6*
In L_β	3.569	0.28 *5*
In L_γ	3.984	0.029 *5*
In $K_{\alpha2}$	24.002	2.66 *10*
In $K_{\alpha1}$	24.210	4.98 *18*
In $K_{\beta1}'$	27.265	1.29 *5*
In $K_{\beta2}'$	28.022	0.269 *10*
γ M4	536.7 *4*	87.0

† 0.50% uncert(syst)

Atomic Electrons (^{111}In)

⟨e⟩=67.4 *12* keV

e_{bin}(keV)	⟨e⟩(keV)	e(%)
4	0.34	8.8 *9*
19	0.0232	0.119 *12*
20	0.195	0.97 *10*
23	0.051	0.221 *23*
24	0.054	0.230 *24*
26	0.0032	0.0120 *12*
27	0.0074	0.028 *3*
509	54.6	10.73 *24*
532	7.73	1.45 *3*
533	2.02	0.380 *9*
536	1.98	0.370 *8*
537	0.423	0.0789 *18*

$^{111}_{50}$Sn(35.3 *8* min)

Mode: ϵ
Δ: -85939 *7* keV
SpA: 4.80×10^7 Ci/g
Prod: ^{110}Cd(α,3n); ^{112}Sn(γ,n)

Photons (^{111}Sn)

⟨γ⟩=186 *6* keV

γ_{mode}	γ(keV)	γ(%)†
In L_ℓ	2.905	0.075 *13*
In L_η	3.112	0.035 *5*
In L_α	3.286	2.1 *3*
In L_β	3.570	1.52 *25*
In L_γ	3.984	0.16 *3*
In $K_{\alpha2}$	24.002	14.8 *5*
In $K_{\alpha1}$	24.210	27.7 *10*
In $K_{\beta1}'$	27.265	7.15 *25*
In $K_{\beta2}'$	28.022	1.50 *6*
γ M1	265.1 *5*	0.066 *13*
γ E1	299.3 *3*	~0.021
γ	372.25 *18*	0.371 *16*
γ	384.0 *6*	0.017 *3*
γ	389.72 *22*	0.022 *4*
γ E2,M1	398.8 *5*	~0.005
γ E1	415.6 *7*	~0.019
γ M1(+E2)	457.4 *3*	0.34 *3*
γ M4	536.7 *4*	0.231 *24* *
γ	552.5 *6*	0.111 *13*
γ	564.49 *18*	0.233 *24*
γ	601.7 *5*	<0.013 ?
γ	607.5 *3*	0.074 *11* ?
γ	636.8 *3*	0.103 *21*
γ	649.1 *6*	0.17 *4*
γ	729.4 *6*	0.034 *7*

Photons (^{111}Sn)

(continued)

γ_{mode}	γ(keV)	γ(%)†
γ	746.6 *7*	0.034 *7*
γ	762.0 *3*	1.44 *3*
γ	814.1 *4*	0.117 *16*
γ	846.9 *6*	0.233 *24*
γ	954.20 *18*	0.490 *24*
γ	1006.4 *4*	0.017 *3*
γ	1026.5 *4*	0.225 *16*
γ E2	1101.0 *4*	0.66 *3*
γ	1111.2 *5*	0.069 *13*
γ M1+14%E2	1153.2 *3*	2.7
γ	1169.9 *6*	0.069 *21*
γ	1188.4 *8*	<0.008 ?
γ E2	1217.4 *6*	0.175 *16*
γ	1293.1 *6*	0.154 *21*
γ	1313.2 *10*	0.053 *11*
γ	1368.2 *4*	0.154 *21*
γ E2	1401.2 *10*	0.079 *16* ?
γ	1499.9 *4*	0.154 *24*
γ	1542.9 *3*	0.71 *5*
γ [M1+E2]	1610.5 *3*	1.25 *11*
γ	1915.1 *3*	1.83 *11*
γ	2107.4 *3*	0.42 *4*
γ	2163.1 *6*	0.024 *8*
γ	2179.7 *4*	0.252 *24*
γ	2212.2 *4*	0.19 *3*
γ	2289.4 *7*	0.048 *11*
γ	2323.1 *7*	0.30 *3*

† 15% uncert(syst)
* with ^{111}In(7.7 min) in equilib

Atomic Electrons (^{111}Sn)

⟨e⟩=3.99 *22* keV

e_{bin}(keV)	⟨e⟩(keV)	e(%)
4	1.80	47 *5*
19	0.129	0.66 *7*
20	1.08	5.4 *6*
23	0.28	1.23 *13*
24	0.30	1.28 *13*
26 - 27	0.059	0.220 *17*
237 - 271	0.0062	0.0026 *4*
295 - 344	0.012	~0.0035
356 - 390	0.0035	0.0010 *3*
429 - 457	0.0135	0.0031 *3*
509	0.145	0.028 *3*
525 - 574	0.039	0.0073 *6*
580 - 621	0.0053	0.0009 *3*
633 - 649	0.00065	0.00010 *4*
701 - 747	0.018	0.0024 *12*
758 - 786	0.0039	0.00051 *18*
810 - 847	0.0029	0.00036 *15*
926 - 954	0.0050	0.00054 *23*
978 - 1026	0.0023	0.00023 *9*
1073 - 1111	0.0076	0.00071 *4*
1125 - 1170	0.032	0.0028 *4*
1184 - 1217	0.00167	0.000140 *11*
1265 - 1313	0.0015	0.00012 *4*
1340 - 1373	0.0016	0.00012 *3*
1397 - 1401	8.5 $\times 10^{-5}$	6.1 *10* $\times 10^{-6}$
1472 - 1542	0.0054	0.00036 *11*
1583 - 1610	0.0097	0.00061 *7*
1887 - 1914	0.009	0.00049 *16*
2079 - 2179	0.0032	0.00015 *4*
2184 - 2262	0.0010	4.8 *13* $\times 10^{-5}$
2285 - 2322	0.0013	5.8 *17* $\times 10^{-5}$

Continuous Radiation (^{111}Sn)

$\langle\beta+\rangle$=195 keV;$\langle$IB$\rangle$=2.4 keV

E_{bin}(keV)		$\langle\ \rangle$(keV)	(%)
0 - 10	β+	5.3×10^{-6}	6.5×10^{-5}
	IB	0.0085	
10 - 20	β+	0.000188	0.00114
	IB	0.0095	0.064
20 - 40	β+	0.00456	0.0140
	IB	0.043	0.167
40 - 100	β+	0.189	0.244
	IB	0.046	0.070
100 - 300	β+	7.9	3.59
	IB	0.133	0.074
300 - 600	β+	47.9	10.5
	IB	0.24	0.055
600 - 1300	β+	137	16.2
	IB	0.97	0.104
1300 - 2451	β+	1.68	0.126
	IB	0.90	0.056
	Σβ+		31

$^{111}_{51}$Sb(1.250 *17* min)

Mode: ϵ

Δ: -80840 *200* keV syst

SpA: 1.349×10^{9} Ci/g

Prod: ^{112}Sn(p,2n); ^{103}Rh(^{12}C,4n)

Photons (^{111}Sb)

$\langle\gamma\rangle$=684 *38* keV

γ_{mode}	γ(keV)	γ(%)†
Sn L_ℓ	3.045	0.033 *5*
Sn L_η	3.272	0.0176 *19*
Sn L_α	3.443	0.93 *11*
Sn L_β	3.742	0.81 *10*
Sn L_γ	4.215	0.094 *14*
Sn $K_{\alpha2}$	25.044	6.7 *3*
Sn $K_{\alpha1}$	25.271	12.5 *5*
Sn $K_{\beta1}$'	28.474	3.27 *13*
Sn $K_{\beta2}$'	29.275	0.69 *3*
γ (E2)	100.02 *3*	3.5 *3*
γ (M1+E2)	154.26 *3*	71 *3*
γ M1,E2	388.86 *10*	3.7 *7*
γ	395.8 *5*	0.15 *3*
γ [M1]	488.88 *10*	42
γ	546.4 *7*	0.14 *3*
γ M1,E2	600.91 *10*	0.76 *15*
γ [E2]	643.14 *10*	1.9 *4*
γ E2	755.17 *10*	5.1 *3*
γ	778.09 *11*	2.8 *6*
γ	878.11 *10*	0.54 *11*
γ	896.7 *5*	2.0 *4*
γ	996.7 *5*	1.6 *3*
γ	1032.37 *10*	10.0 *9*
γ	1067.3 *7*	0.14 *3*
γ E2,M1	1122.1 *7*	0.28 *6*
γ	1147.3 *7*	4.2 *8*
γ	1150.9 *5*	~4 ?
γ	1179.3 *7*	0.88 *18*
γ	1276.3 *7*	0.15 *3*
γ M1	1322.8 *7*	0.19 *4*
γ	1477.0 *7*	0.16 *3*
γ	1538.3 *10*	0.20 *4*
γ	1841.1 *10*	0.49 *10*

† 12% uncert(syst)

Atomic Electrons (^{111}Sb)

$\langle$e$\rangle$=23.9 *8* keV

e_{bin}(keV)	$\langle$e$\rangle$(keV)	e(%)
4	0.92	22.7 *24*
20	0.057	0.28 *3*
21	0.47	2.25 *24*
24 - 28	0.288	1.16 *8*
71	2.8	4.0 *4*
96	1.19	1.25 *12*
99 - 100	0.305	0.307 *24*
125	13.1	10.4 *5*
150	1.99	1.33 *7*
153 - 154	0.491	0.320 *13*
360 - 396	0.21	0.057 *10*
460	1.4	0.31 *6*
484 - 517	0.23	0.047 *8*
542 - 572	0.019	0.0032 *7*
596 - 643	0.047	0.0076 *13*
726 - 774	0.131	0.018 *3*
777 - 778	0.0010	0.00013 *7*
849 - 897	0.031	0.0035 *13*
967 - 1003	0.10	0.010 *4*
1028 - 1067	0.015	0.0014 *6*
1093 - 1122	0.036	0.0032 *15*
1143 - 1179	0.013	0.0011 *4*
1247 - 1294	0.0030	0.00024 *5*
1318 - 1323	0.00027	2.1 *3* $\times10^{-5}$
1448 - 1538	0.0025	0.00017 *5*
1812 - 1840	0.0028	0.00015 *6*

Continuous Radiation (^{111}Sb)

$\langle\beta+\rangle$=1358 keV;$\langle$IB$\rangle$=6.7 keV

E_{bin}(keV)		$\langle\ \rangle$(keV)	(%)
0 - 10	β+	1.66×10^{-6}	2.01×10^{-5}
	IB	0.044	
10 - 20	β+	6.0×10^{-5}	0.000366
	IB	0.044	0.30
20 - 40	β+	0.00152	0.00467
	IB	0.092	0.32
40 - 100	β+	0.067	0.087
	IB	0.25	0.38
100 - 300	β+	3.30	1.49
	IB	0.73	0.41
300 - 600	β+	27.2	5.8
	IB	0.90	0.21
600 - 1300	β+	252	25.8
	IB	1.61	0.18
1300 - 2500	β+	815	44.3
	IB	2.0	0.110
2500 - 4946	β+	261	9.2
	IB	1.09	0.036
	Σβ+		87

$^{111}_{52}$Te(19.3 *4* s)

Mode: ϵ, ϵp

Δ: -73470 *70* keV

SpA: 5.17×10^{9} Ci/g

Prod: ^{94}Mo(^{20}Ne,3n); ^{102}Pd(^{12}C,3n)

ϵp: 5100

$^{111}_{53}$I (2.5 *2* s)

Mode: α, ϵ

Δ: -64970 *400* keV syst

SpA: 3.6×10^{10} Ci/g

Prod: ^{58}Ni on ^{58}Ni

Alpha Particles (^{111}I)

α(keV)
3152 *10*

$^{111}_{54}$Xe(740 *200* ms)

Mode: α, ϵ

Δ: -54380 *510* keV syst

SpA: 8.9×10^{10} Ci/g

Prod: ^{58}Ni on ^{58}Ni

Alpha Particles (^{111}Xe)

α(keV)
3480 *30*

$^{111}_{54}$Xe(890 *200* ms)

Mode: α, ϵ

Δ: -54380 *510* keV syst

SpA: 7.9×10^{10} Ci/g

Prod: ^{58}Ni on ^{58}Ni

Alpha Particles (^{111}Xe)

α(keV)
3580 *30*

A = 112

NDS **29**, 587 (1980)

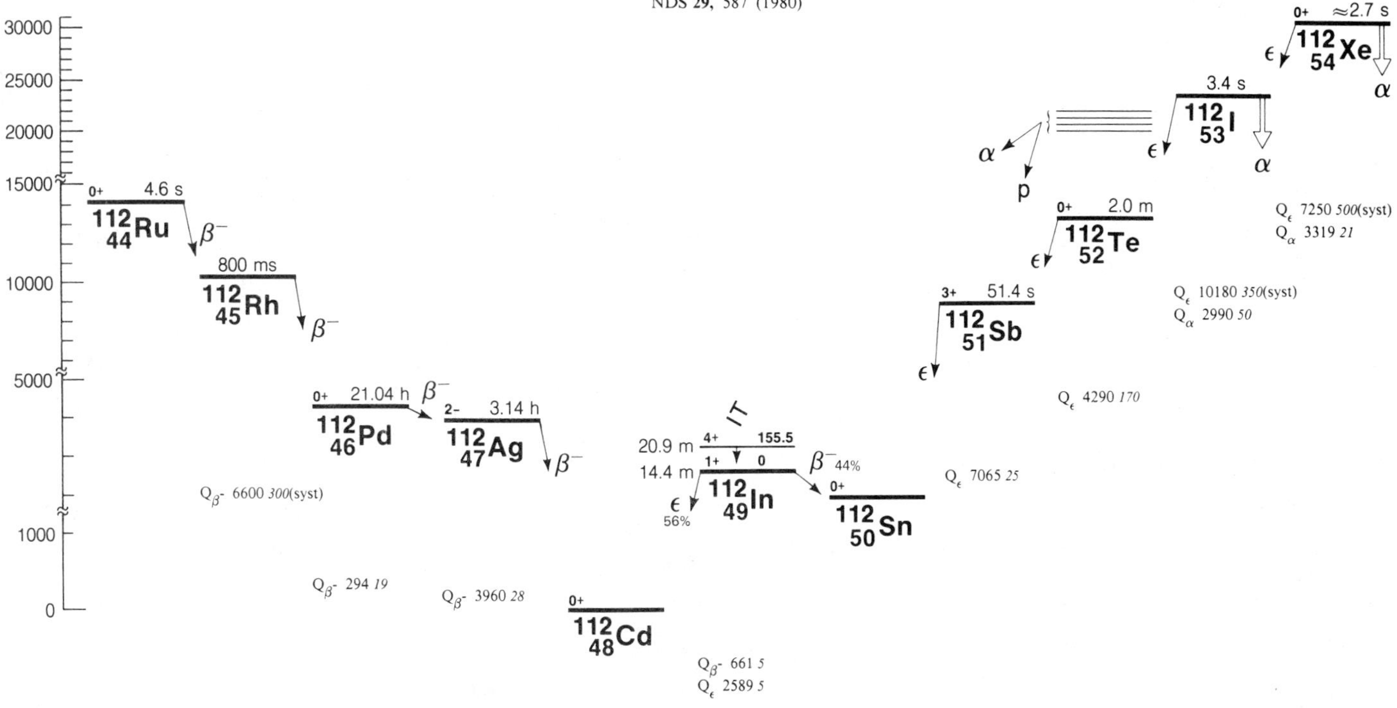

$^{112}_{44}$Ru(4.65 *14* s)

Mode: β-
SpA: 2.01×10^{10} Ci/g
Prod: fission

$^{112}_{45}$Rh(800 *100* ms)

Mode: β-
Δ: -79730 *300* keV syst
SpA: 8.4×10^{10} Ci/g
Prod: fission

Photons (^{112}Rh)

γ_{mode}	γ(keV)
γ [E2]	348.88 *17*

$^{112}_{46}$Pd(21.04 *5* h)

Mode: β-
Δ: -86329 *25* keV
SpA: 1.330×10^{6} Ci/g
Prod: fission

Photons (^{112}Pd)

γ_{mode}	γ(keV)
γ E1	18.5 *5*

$^{112}_{47}$Ag(3.14 *2* h)

Mode: β-
Δ: -86623 *28* keV
SpA: 8.91×10^{6} Ci/g
Prod: daughter ^{112}Pd; ^{115}In(n,α); ^{114}Cd(d,α); ^{112}Cd(n,p)

Photons (^{112}Ag)
⟨γ⟩=691 *27* keV

γ_{mode}	γ(keV)	γ(%)†
Cd $K_{\alpha2}$	22.984	0.054 *4*
Cd $K_{\alpha1}$	23.174	0.102 *7*
Cd $K_{\beta1}'$	26.085	0.0260 *19*
Cd $K_{\beta2}'$	26.801	0.0054 *4*
γ [E2]	121.04 *14*	0.077 *9*
γ	148.0 *2*	0.013 *3*
γ	159.6 *3*	0.013 *3*
γ [E1]	225.99 *14*	0.017 *4*
γ [E2]	244.71 *12*	0.013 *3*
γ	342.4 *3*	0.017 *4*
γ	356.0 *3*	0.030 *9*
γ [E2]	402.31 *13*	0.043 *9*
γ [E1]	410.82 *12*	0.116 *13*
γ	450.87 *16*	0.034 *9*

Photons (^{112}Ag)
(continued)

γ_{mode}	γ(keV)	γ(%)†
γ	529 *1*	0.017 *4*
γ [E1]	536.35 *12*	0.052 *9*
γ [E2]	558.84 *12*	0.013 *3*
γ	569.9 *3*	0.043 *11*
γ	585.5 *3*	0.039 *9*
γ [E2]	606.71 *8*	3.1 *3*
γ E2	617.48 *7*	43 *4*
γ	629.3 *4*	0.017 *4*
γ [M1+E2]	648.7 *3*	0.022 *4*
γ [M1+E2]	663.78 *16*	0.030 *9*
γ [M1+E2]	687.23 *18*	0.043 *9*
γ [E1]	692.88 *11*	1.08 *13*
γ M1+37%E2	694.89 *9*	3.0 *3*
γ	714.9 *3*	0.052 *9*
γ [E1]	718.52 *14*	0.168 *17*
γ M1+0.5%E2	751.89 *24*	0.039 *9*
γ [M1+E2]	762.34 *14*	0.052 *9*
γ	784.7 *6*	0.017 *4*
γ [E2]	798.06 *17*	0.52 *4*
γ	802.67 *25*	0.030 *9*
γ [E2]	815.93 *14*	0.133 *13*
γ [M1+E2]	843.76 *18*	0.013 *3*
γ M1+1.4%E2	851.42 *9*	1.07 *11*
γ	861.69 *15*	0.219 *22*
γ [M1+E2]	918.87 *14*	0.069 *9*
γ [M1+E2]	947.17 *13*	0.082 *9*
γ [E1]	956.9 *7*	0.0086 *22*
γ [E2]	982.66 *17*	0.052 *9*
γ [E2]	1007.05 *14*	0.103 *13*
γ [E1]	1037.95 *16*	0.073 *9*
γ	1063.8 *3*	0.077 *19*
γ [M1+E2]	1103.70 *12*	0.42 *4*
γ [E1]	1125.95 *19*	0.176 *17*
γ [E1]	1194.48 *16*	0.030 *9*
γ E2	1253.72 *11*	0.37 *4*
γ [E1]	1282.66 *17*	0.039 *9*
γ [E2]	1312.37 *10*	1.20 *13*
γ [E1]	1356.66 *15*	0.47 *4*
γ E1(+0.4%M2)	1387.77 *10*	5.4 *6*

Photons (^{112}Ag)
(continued)

γ_{mode}	γ(keV)	γ(%)†
γ	1398.03 17	0.030 9
γ [M1+E2]	1411.40 15	0.034 9
γ M1+8.8%E2	1446.78 25	0.043 9
γ	1451.39 19	0.150 17
γ [E2]	1468.89 10	0.62 7
γ [E2]	1499.57 15	0.017 4
γ [M1+E2]	1504.11 23	0.163 17
γ [E1]	1517.02 16	0.82 9 ?
γ (M1+1.9%E2)	1538.65 16	0.52 4
γ M1+1.3%E2	1613.76 12	2.8 3
γ	1652.5 8	0.030 9
γ [E2]	1683.8 3	0.043 9
γ [E2]	1715.7 3	0.043 9
γ [M1+E2]	1798.58 11	0.90 9
γ [E1]	1889.36 15	0.31 3
γ	1909.3 6	0.043 11
γ [E1]	1945.12 25	0.086 9
γ [E1]	2051.54 16	0.120 13
γ [M1+E2]	2056.61 20	0.60 4
γ	2066.1 16	0.013 3
γ E2(+0.3%M1)	2106.28 13	2.4 3
γ [M1+E2]	2148.2 3	0.099 9
γ [E2]	2156.12 17	0.069 9
γ [E1]	2211.89 14	0.43 4
γ	2362.5 6	0.039 9
γ [E1]	2506.83 15	1.08 9
γ [E1]	2551.82 24	0.108 13
γ	2685.9 4	0.25 3
γ [E2]	2723.74 14	0.099 9
γ [E1]	2752.9 3	0.095 9
γ [E1]	2829.36 14	0.43 4
γ [E2]	2962.1 7	0.017 4
γ	3393.1 12	0.0043 13

† 9.3% uncert(syst)

Atomic Electrons (^{112}Ag)
$\langle e\rangle$=1.33 8 keV

e_{bin}(keV)	$\langle e\rangle$(keV)	e(%)
4 - 26	0.0136	0.216 22
94	0.043	0.046 5
117 - 160	0.022	0.0181 25
199 - 245	0.0024	0.00112 20
316 - 356	0.0019	0.00059 23
376 - 424	0.0044	0.00113 15
447 - 451	0.00013	~3 $\times10^{-5}$
502 - 543	0.0018	0.00034 8
555 - 570	0.0007	0.00013 6
580	0.061	0.0105 11
581 - 586	9.5 $\times10^{-5}$	1.6 7 $\times10^{-5}$
591	0.82	0.139 13
603 - 607	0.0102	0.00169 17
613	0.094	0.0153 15
614	0.0148	0.00241 23
617	0.0255	0.0041 4
622 - 666	0.0083	0.00126 13
668	0.051	0.0077 8
683 - 725	0.0114	0.00164 11
736 - 785	0.0082	0.00107 9
789 - 835	0.0202	0.00246 23
840 - 862	0.00257	0.00030 3
892 - 930	0.00190	0.000209 21
943 - 983	0.00173	0.000178 16
1003 - 1038	0.0011	0.00010 3
1060 - 1104	0.0053	0.00049 5
1122 - 1168	0.00020	1.8 3 $\times10^{-5}$
1190 - 1227	0.0028	0.000232 24
1250 - 1286	0.0093	0.00072 8
1308 - 1357	0.00306	0.000231 15
1361 - 1408	0.0206	0.00151 15
1411 - 1451	0.0053	0.00037 4
1465 - 1538	0.0090	0.00060 3
1587 - 1683	0.0222	0.00139 13
1689 - 1772	0.0054	0.00031 4
1795 - 1889	0.00185	0.000101 8
1905 - 1945	0.00028	1.47 14 $\times10^{-5}$
2025 - 2121	0.0173	0.00083 7

Atomic Electrons (^{112}Ag)
(continued)

e_{bin}(keV)	$\langle e\rangle$(keV)	e(%)
2129 - 2211	0.00163	7.5 6 $\times10^{-5}$
2336 - 2362	0.00015	6.6 23 $\times10^{-6}$
2480 - 2551	0.00310	0.000125 9
2659 - 2752	0.0016	5.9 10 $\times10^{-5}$
2803 - 2829	0.00107	3.8 4 $\times10^{-5}$
2935 - 2961	7.2 $\times10^{-5}$	2.4 6 $\times10^{-6}$
3366 - 3392	1.3 $\times10^{-5}$	3.9 14 $\times10^{-7}$

Continuous Radiation (^{112}Ag)
$\langle\beta-\rangle$=1380 keV;$\langle$IB$\rangle$=4.0 keV

E_{bin}(keV)		$\langle\ \rangle$(keV)	(%)
0 - 10	β-	0.0123	0.244
	IB	0.045	
10 - 20	β-	0.0375	0.250
	IB	0.044	0.31
20 - 40	β-	0.154	0.51
	IB	0.087	0.30
40 - 100	β-	1.17	1.67
	IB	0.25	0.38
100 - 300	β-	13.9	6.8
	IB	0.72	0.40
300 - 600	β-	57	12.6
	IB	0.84	0.20
600 - 1300	β-	278	29.5
	IB	1.21	0.139
1300 - 2500	β-	659	35.8
	IB	0.76	0.045
2500 - 3960	β-	371	12.7
	IB	0.087	0.0031

$^{112}_{48}$Cd(stable)

Δ: -90582 3 keV
%: 24.13 11

$^{112}_{49}$In(14.4 2 min)

Mode: ϵ(56 3 %), β-(44 3 %)
Δ: -87994 5 keV
SpA: 1.165×10^8 Ci/g
Prod: ^{109}Ag(α,n); ^{112}Cd(p,n); ^{115}In(p,p3n); ^{113}In(γ,n)

Photons (^{112}In)
$\langle\gamma\rangle$=30.9 19 keV

γ_{mode}	γ(keV)	γ(%)†
Cd L_ℓ	2.767	0.035 6
Cd L_η	2.957	0.0166 25
Cd L_α	3.133	0.97 15
Cd L_β	3.392	0.68 11
Cd L_γ	3.779	0.068 12
Cd $K_{\alpha2}$	22.984	7.22 25
Cd $K_{\alpha1}$	23.174	13.6 5
Cd $K_{\beta1}'$	26.085	3.46 12
Cd $K_{\beta2}'$	26.801	0.72 3
γ_ϵ[E2]	121.04 14	
γ_ϵ[E2]	244.71 12	0.0010 3
γ_ϵ[E2]	402.31 13	0.015 3
γ_ϵ[E2]	558.84 12	0.0045 12 ?
γ_ϵ[E2]	606.71 8	0.669 13

Photons (^{112}In)
(continued)

γ_{mode}	γ(keV)	γ(%)†
γ_ϵE2	617.48 7	2.8
γ_ϵ[M1+E2]	687.23 18	0.0006 3
γ_ϵ[E1]	692.88 11	
γ_ϵM1+37%E2	694.89 9	~0.007
γ_ϵ[E2]	798.06 17	
γ_ϵ[E2]	815.93 14	
γ_ϵ[M1+E2]	843.76 18	~0.00019
γ_ϵM1+1.4%E2	851.42 9	0.085 7
γ_ϵE2	1253.72 11	0.132 8
γ_ϵ[E2]	1312.37 10	~0.0028
γ_ϵE1(+0.4%M2)	1387.77 10	
γ_ϵ[E2]	1468.89 10	0.049 5
γ_ϵ[M1+E2]	1504.11 23	
γ_ϵ(M1+1.9%E2)	1538.65 16	0.008 3
γ_ϵM1+1.3%E2	1613.76 12	
γ_ϵ[E2]	1683.8 3	0.010 4
γ_ϵ[E2]	2156.12 17	0.0010 5

† uncert(syst): 11% for ϵ

Atomic Electrons (^{112}In)
$\langle e\rangle$=1.87 10 keV

e_{bin}(keV)	$\langle e\rangle$(keV)	e(%)
4	0.85	23.3 24
19	0.49	2.6 3
20	0.124	0.63 6
22	0.143	0.64 7
23	0.151	0.67 7
25 - 26	0.0290	0.113 9
218 - 245	0.00015	7 3 $\times10^{-5}$
376 - 402	0.0007	0.00019 8
532 - 580	0.0133	0.00229 10
591	0.053	0.009 1
603 - 617	0.0109	0.00178 12
661 - 695	0.00015	2.2 9 $\times10^{-5}$
817 - 851	0.00137	0.000165 13
1227 - 1254	0.00114	9.3 6 $\times10^{-5}$
1286 - 1312	2.3 $\times10^{-5}$	1.8 8 $\times10^{-6}$
1442 - 1538	0.00043	2.9 3 $\times10^{-5}$
1657 - 1683	6.7 $\times10^{-5}$	4.1 14 $\times10^{-6}$
2129 - 2156	5.3 $\times10^{-6}$	2.5 11 $\times10^{-7}$

Continuous Radiation (^{112}In)
$\langle\beta-\rangle$=94 keV;$\langle\beta+\rangle$=151 keV;$\langle$IB$\rangle$=1.56 keV

E_{bin}(keV)		$\langle\ \rangle$(keV)	(%)
0 - 10	β-	0.058	1.15
	β+	3.53 $\times10^{-6}$	4.27 $\times10^{-5}$
	IB	0.0111	
10 - 20	β-	0.174	1.16
	β+	0.000121	0.00073
	IB	0.0119	0.081
20 - 40	β-	0.70	2.32
	β+	0.00287	0.0088
	IB	0.032	0.118
40 - 100	β-	4.85	6.9
	β+	0.116	0.150
	IB	0.052	0.081
100 - 300	β-	39.6	20.3
	β+	4.80	2.19
	IB	0.124	0.071
300 - 600	β-	48.0	12.0
	β+	30.2	6.6
	IB	0.163	0.037
600 - 1300	β-	0.55	0.089
	β+	109	12.3
	IB	0.57	0.061
1300 - 2500	β+	7.4	0.54
	IB	0.60	0.036
2500 - 2589	IB	0.00019	7.6 $\times10^{-6}$
	$\Sigma\beta$+		22

$^{112}_{49}$In(20.9 *2* min)

Mode: IT
Δ: -87839 *5* keV
SpA: 8.03×10^7 Ci/g
Prod: ^{109}Ag(α,n)

Photons (^{112}In)

⟨γ⟩=34.2 *19* keV

γ_{mode}	γ(keV)	γ(%)†
In L_ℓ	2.905	0.094 *25*
In L_η	3.112	0.043 *11*
In L_α	3.286	2.6 *7*
In L_β	3.571	1.9 *5*
In L_γ	3.989	0.19 *5*
In $K_{\alpha2}$	24.002	17 *3*
In $K_{\alpha1}$	24.210	31 *6*
In $K_{\beta1}$'	27.265	8.0 *16*
In $K_{\beta2}$'	28.022	1.7 *3*
γ M3	155.5 *2*	12.8

† 3.0% uncert(syst)

Atomic Electrons (^{112}In)

⟨e⟩=121 *18* keV

e_{bin}(keV)	⟨e⟩(keV)	e(%)
4	2.2	58 *13*
19	0.14	0.74 *17*
20	1.2	6.0 *14*
23	0.32	1.4 *3*
24	0.34	1.4 *3*
26	0.020	0.075 *17*
27	0.046	0.17 *4*
128	84.9	67 *13*
151	16.7	11.0 *22*
152	8.3	5.5 *11*
155	6.4	4.1 *8*

$^{112}_{50}$Sn(stable)

Δ: -88654 *5* keV
%: 0.97 *1*

$^{112}_{51}$Sb(51.4 *5* s)

Mode: ε
Δ: -81589 *25* keV
SpA: 1.947×10^9 Ci/g
Prod: ^{112}Sn(p,n)

Photons (^{112}Sb)

⟨γ⟩=1902 *63* keV

γ_{mode}	γ(keV)	γ(%)†
Sn L_ℓ	3.045	0.0106 *14*
Sn L_η	3.272	0.0055 *6*
Sn L_α	3.443	0.29 *3*
Sn L_β	3.742	0.26 *3*
Sn L_γ	4.220	0.030 *4*
Sn $K_{\alpha2}$	25.044	2.18 *7*
Sn $K_{\alpha1}$	25.271	4.07 *12*
Sn $K_{\beta1}$'	28.474	1.06 *3*
Sn $K_{\beta2}$'	29.275	0.224 *7*
γ	234.75 *19*	0.082 *8*
γ	279.86 *18*	0.055 *5*
γ	283.42 *15*	0.041 *4*
γ	301.6 *3*	0.157 *10*
γ	376.7 *8*	0.056 *11*
γ [M1+E2]	392.0 *3*	0.066 *8*
γ	401.34 *16*	0.036 *8*
γ	431.53 *23*	0.032 *5*
γ	445.9 *9*	0.041 *7*
γ	466.70 *21*	0.124 *12*
γ	558.1 *3*	0.084 *6*
γ	604.74 *14*	0.291 *17*
γ [E1]	612.25 *15*	0.38 *3*
γ	669.92 *19*	3.7 *6*
γ	700.28 *22*	0.064 *16*
γ	766.33 *16*	0.44 *3*
γ [M1+E2]	772.42 *17*	0.246 *18*
γ	797.5 *4*	0.256 *18*
γ	831.02 *16*	0.063 *13*
γ	868.4 *5*	0.048 *8*
γ [E1]	893.89 *22*	~0.26
γ [M1+E2]	894.22 *11*	2.65 *20*
γ	900.5 *3*	0.28 *5*
γ [M1+E2]	901.70 *21*	0.15 *5*
γ	921.2 *4*	0.070 *9*
γ	927.43 *15*	0.701 *19*
γ	962.8 *4*	0.16 *3*
γ [E2]	990.63 *9*	14.3 *4*
γ	1009.14 *21*	0.12 *3*
γ	1029.49 *22*	0.68 *5*
γ [E1]	1097.61 *12*	1.92 *19*
γ	1154.1 *16*	0.20 *6*
γ	1170.3 *17*	0.19 *5*
γ [M1+E2]	1219.09 *14*	0.202 *19*
γ [E2]	1256.39 *12*	96
γ [E2]	1264.21 *18*	1.13 *12*
γ	1276.92 *19*	0.08 *3*
γ	1282.73 *16*	0.096 *19*
γ	1292.56 *20*	~0.10
γ	1360.5 *3*	0.091 *5*
γ	1368.8 *7*	0.026 *5*
γ	1379.13 *15*	0.149 *7*
γ	1420.9 *5*	0.028 *6*
γ	1425.7 *4*	0.036 *6*
γ	1459.16 *9*	0.432 *19*
γ [M1+E2]	1464.37 *10*	0.442 *19*
γ	1477.5 *3*	0.259 *19*
γ	1498.95 *13*	1.37 *4*
γ [E2]	1527.28 *19*	0.74 *3*
γ	1534.4 *7*	0.030 *8*
γ	1550.8 *3*	0.117 *7*
γ	1566.15 *13*	1.59 *4*
γ	1582.1 *9*	0.019 *6*
γ	1620.4 *8*	0.112 *8*
γ	1630.7 *12*	0.024 *7*
γ [E2]	1656.3 *3*	0.54 *3*
γ [E2]	1688.66 *20*	0.259 *10*
γ [M1+E2]	1709.86 *14*	1.35 *5*
γ	1804.0 *4*	0.041 *4*
γ	1821.63 *14*	0.72 *3*
γ [M1+E2]	1836.02 *16*	0.99 *3*
γ	1878.9 *5*	0.023 *4*
γ [E2]	1892.32 *20*	0.643 *19*
γ	1925.8 *13*	0.034 *12*
γ	1985.7 *4*	0.038 *4*
γ	1991.9 *3*	0.218 *12*
γ	2016.3 *4*	0.084 *8*
γ [M1+E2]	2029.27 *18*	0.61 *4*
γ	2082.0 *3*	0.036 *3*
γ	2092.6 *3*	0.091 *7*
γ [E2]	2127.23 *17*	0.77 *5*
γ [E2]	2150.59 *13*	0.44 *3*

Photons (^{112}Sb) (continued)

γ_{mode}	γ(keV)	γ(%)†
γ	2160.56 *20*	1.02 *6*
γ	2199.23 *19*	0.298 *19*
γ	2247.1 *4*	0.064 *6*
γ	2267.54 *18*	0.35 *3*
γ [E2]	2296.76 *20*	1.11 *7*
γ	2397.9 *3*	0.180 *11*
γ	2448.7 *8*	0.027 *6*
γ	2454.5 *8*	0.025 *4*
γ [E2]	2475.46 *14*	0.73 *5*
γ	2556.3 *3*	0.211 *19*
γ	2610.4 *4*	0.090 *7*
γ	2669.8 *5*	0.0115 *19*
γ [E2]	2720.73 *15*	0.070 *6*
γ	2737.0 *9*	0.038 *3*
γ	2754.9 *12*	0.015 *3*
γ	2774.8 *12*	0.023 *4*
γ	2781.1 *14*	0.014 *4*
γ	2830.5 *7*	0.020 *3*
γ	2887.0 *10*	0.030 *3*
γ [E2]	2966.22 *16*	0.72 *5*
γ	2976.6 *7*	0.021 *3*
γ [E2]	3092.38 *18*	0.259 *19*
γ	3130.1 *8*	0.015 *3*
γ	3146.5 *9*	0.012 *3*
γ [E2]	3247.86 *20*	0.95 *6*
γ [E2]	3285.63 *20*	0.73 *5*
γ	3351.6 *9*	0.052 *4*
γ	3408.1 *7*	0.0182 *19*
γ	3431.4 *7*	0.0125 *19*
γ	3455.1 *7*	0.0125 *19*
γ	3653.6 *7*	0.187 *12*
γ	3700.3 *19*	0.013 *3*
γ	3723.5 *21*	0.012 *3*
γ	3826.8 *10*	0.0173 *19*
γ	3879.1 *8*	0.028 *3*
γ	3923.5 *8*	0.022 *3*
γ	4042.3 *9*	0.0058 *19*
γ	4077.0 *12*	0.011 *3*
γ	4085.8 *9*	0.027 *3*
γ	4152.1 *21*	0.0106 *19*
γ	4211.7 *12*	0.0125 *19*
γ	4389.8 *16*	0.0115 *19*
γ	4540.9 *18*	0.0077 *19*
γ	4566.6 *19*	0.0115 *19*
γ	4614.2 *21*	0.0086 *19*
γ	4665.1 *22*	0.0115 *19*
γ	4745 *4*	0.0048 *19*
γ	4910 *3*	0.0058 *19*
γ	5133 *4*	0.0067 *19*

† 5.2% uncert(syst)

Atomic Electrons (^{112}Sb)

⟨e⟩=2.03 *7* keV

e_{bin}(keV)	⟨e⟩(keV)	e(%)
4	0.29	7.2 *7*
20	0.0185	0.091 *9*
21	0.154	0.73 *8*
24 - 28	0.094	0.379 *25*
206 - 254	0.014	0.0062 *24*
272 - 302	0.011	0.0038 *18*
347 - 397	0.0070	0.0019 *4*
400 - 446	0.0052	0.0012 *4*
462 - 467	0.00053	0.00011 *5*
529 - 576	0.007	0.0012 *5*
583 - 612	0.0042	0.00072 *7*
641	0.05	~0.009
665 - 700	0.010	0.0015 *6*
737 - 772	0.014	0.0019 *5*
793 - 839	0.0018	0.00022 *7*
864 - 902	0.058	0.0066 *8*
917 - 959	0.0029	0.00031 *10*
961	0.158	0.0164 *8*
962 - 1009	0.032	0.0032 *3*
1025 - 1068	0.0092	0.00086 *9*
1093 - 1141	0.0043	0.00038 *10*
1150 - 1190	0.0024	0.00020 *3*

Atomic Electrons (^{112}Sb) (continued)

e_{bin}(keV)	⟨e⟩(keV)	e(%)
1215 - 1219	0.00030	2.4 *3* $\times 10^{-5}$
1227	0.82	0.067 *4*
1235 - 1248	0.0101	0.00082 *9*
1252	0.100	0.0080 *5*
1254 - 1293	0.0268	0.00213 *13*
1331 - 1379	0.0020	0.00015 *4*
1392 - 1435	0.0066	0.00046 *8*
1448 - 1495	0.012	0.00081 *24*
1498 - 1591	0.018	0.00120 *25*
1601 - 1688	0.0154	0.00093 *6*
1705 - 1803	0.0051	0.00029 *8*
1807 - 1897	0.0123	0.00067 *4*
1921 - 2016	0.0054	0.00027 *3*
2025 - 2123	0.0080	0.000382 *19*
2126 - 2218	0.0071	0.00033 *8*
2238 - 2296	0.0080	0.00035 *3*
2369 - 2454	0.0045	0.000184 *15*
2471 - 2556	0.0014	5.6 *10* $\times 10^{-5}$
2581 - 2669	0.00043	1.6 *4* $\times 10^{-5}$
2692 - 2780	0.00072	2.65 *24* $\times 10^{-5}$
2801 - 2886	0.00020	6.9 *14* $\times 10^{-6}$
2937 - 2976	0.00348	0.000118 *8*
3063 - 3146	0.00129	4.2 *3* $\times 10^{-5}$
3219 - 3285	0.0074	0.000230 *11*
3322 - 3407	0.00028	8.5 *14* $\times 10^{-6}$
3426 - 3454	4.8 $\times 10^{-5}$	1.4 *3* $\times 10^{-6}$
3624 - 3723	0.00071	1.9 *4* $\times 10^{-5}$
3798 - 3894	0.00021	5.4 *8* $\times 10^{-6}$
3919 - 4013	2.5 $\times 10^{-5}$	6.2 *17* $\times 10^{-7}$
4038 - 4123	0.000149	3.7 *6* $\times 10^{-6}$
4148 - 4211	4.3 $\times 10^{-5}$	1.02 *22* $\times 10^{-6}$
4361 - 4389	3.5 $\times 10^{-5}$	8.0 *19* $\times 10^{-7}$
4512 - 4610	8.2 $\times 10^{-5}$	1.8 *3* $\times 10^{-6}$
4613 - 4664	3.5 $\times 10^{-5}$	7.4 *17* $\times 10^{-7}$
4716 - 4744	1.4 $\times 10^{-5}$	3.0 *12* $\times 10^{-7}$
4881 - 4910	1.7 $\times 10^{-5}$	3.4 *12* $\times 10^{-7}$

Continuous Radiation (^{112}Sb)

⟨β+⟩=1768 keV;⟨IB⟩=8.7 keV

E_{bin}(keV)		⟨ ⟩(keV)	(%)
0 - 10	β+	1.18 $\times 10^{-6}$	1.42 $\times 10^{-5}$
	IB	0.052	
10 - 20	β+	4.29 $\times 10^{-5}$	0.000259
	IB	0.051	0.36
20 - 40	β+	0.00108	0.00332
	IB	0.106	0.37
40 - 100	β+	0.0478	0.062
	IB	0.30	0.45
100 - 300	β+	2.36	1.06
	IB	0.89	0.50
300 - 600	β+	19.7	4.21
	IB	1.14	0.27
600 - 1300	β+	189	19.3
	IB	2.0	0.23
1300 - 2500	β+	742	39.5
	IB	2.4	0.134
2500 - 5000	β+	814	25.9
	IB	1.70	0.052
5000 - 5808	IB	0.035	0.00067
	Σβ+		90

$^{112}_{52}$Te(2.0 *2* min)

Mode: ϵ
Δ: -77300 *160* keV
SpA: 8.4×10⁸ Ci/g
Prod: ^{112}Sn(^{3}He,3n)

Photons (^{112}Te)

γ_{mode}	γ(keV)	γ(rel)
γ	37.9 *9*	16 *5*
γ	51.4 *14*	5 *2*
γ	58.9 *4*	4.5 *12*
γ	69.3 *23*	5 *2*
γ	103.9 *6*	27 *5*
γ	132.5 *5*	23 *4*
γ	166.7 *6*	7 *2*
γ	236.3 *5*	9 *2*
γ	273.5 *11*	5 *2*
γ	279.7 *14*	~3
γ	295.2 *5*	86 *8*
γ	350.2 *6*	36 *3*
γ	357.0 *9*	6 *2*
γ[E2]	372.0 *6*	100
γ	417.8 *4*	57 *5*
γ	476.7 *4*	14 *3*
γ	493.3 *9*	30 *10*
γ	583.7 *14*	8 *3*
γ	597.8 *20*	~6
γ	609.1 *6*	4.3 *14*
γ	689.7 *11*	10 *3*
γ	697.8 *9*	12 *3*
γ[E2]	713.0 *6*	2.4 *9*
γ	742.3 *6*	11 *3*
γ	796.6 *6*	24 *7*
γ	806.7 *11*	9 *4*
γ	819.4 *6*	17 *4*
γ	881.2 *9*	10 *4*
γ	924.1 *17*	11 *5*
γ	927.7 *11*	11 *5*
γ	970.6 *6*	23 *6*
γ	1282 *3*	17 *7*
γ	1286.5 *23*	10 *4*
γ	1501.9 *17*	15 *4*
γ	1656.9 *9*	14 *5*
γ	1963.0 *11*	17 *5*

$^{112}_{53}$I (3.42 *11* s)

Mode: ϵ, α, ϵp, ϵα
Δ: -67120 *300* keV syst
SpA: 2.67×10¹⁰ Ci/g
Prod: ^{58}Ni on ^{58}Ni; ^{58}Ni on ^{63}Cu

ϵp: 2000-5000 range

Alpha Particles (^{112}I)

α(keV)
2880 *30*

Photons (^{112}I)

γ_{mode}	γ(keV)
γ_ϵ	688.9 *6*
γ_ϵ	786.9 *6*

$^{112}_{54}$Xe(2.7 *8* s)

Mode: α, ϵ
Δ: -59880 *400* keV syst
SpA: 3.29×10¹⁰ Ci/g
Prod: ^{58}Ni on ^{58}Ni

Alpha Particles (^{112}Xe)

α(keV)
3210 *20*

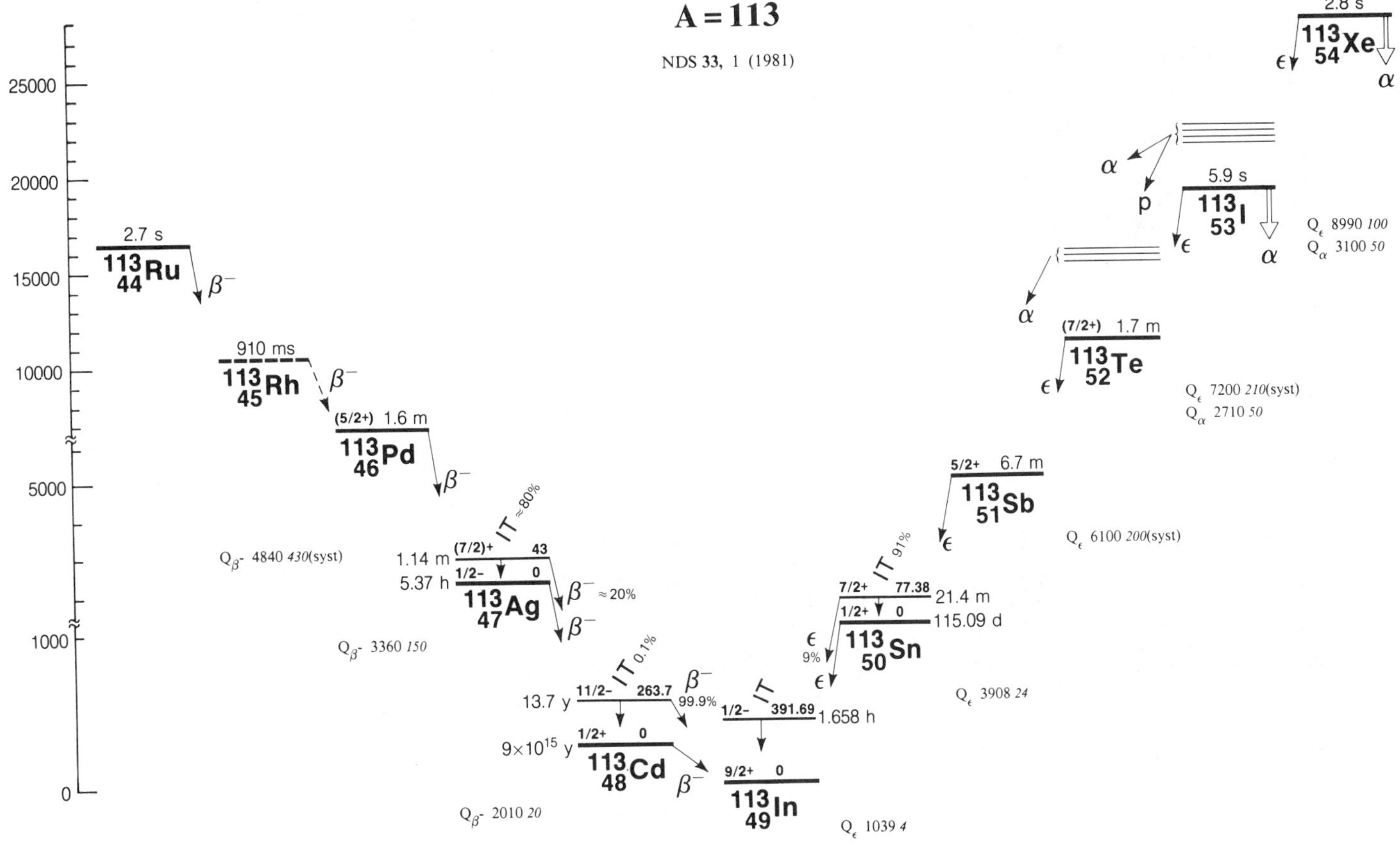

$^{113}_{44}$Ru(2.69 *10* s)

Mode: β-
SpA: 3.27×10^{10} Ci/g
Prod: fission

$^{113}_{45}$Rh(910 *80* ms)

attributed to ^{113}Rh or ^{111}Rh

Mode: β-
Δ: -78840 *400* keV syst
SpA: 7.7×10^{10} Ci/g
Prod: fission

Photons (^{113}Rh)

γ_{mode}	γ(keV)
γ	128.51 *10*

$^{113}_{46}$Pd(1.55 *8* min)

Mode: β-
Δ: -83680 *150* keV
SpA: 1.07×10^{9} Ci/g
Prod: fission

Photons (^{113}Pd)

γ_{mode}	γ(keV)	γ(rel)
γ[E3]	43.2 *10 ?*	*
γ	95.8 *15*	100
γ	171 *5*	1.5 *5*
γ	222.0 *2*	37 *3*
γ	398 *5*	~2
γ	472.2 *4*	3 *1*
γ	482.4 *4*	31.0 *7*
γ	567.9 *3*	22 *3*
γ	643.6 *2*	81 *4*
γ	739.4 *2*	76 *3*
γ	869.5 *6*	4.2 *8*

* with ^{113}Ag(1.14 min)

$^{113}_{47}$Ag(5.37 *5* h)

Mode: β-
Δ: -87041 *20* keV
SpA: 5.16×10^{6} Ci/g
Prod: fission; ^{114}Cd(γ,p)

Photons (^{113}Ag)

⟨γ⟩=72 *6* keV

γ_{mode}	γ(keV)	γ(%)†
γ M1	17.73 *5*	0.045 *5*
Cd $K_{\alpha2}$	22.984	0.105 *14*
Cd $K_{\alpha1}$	23.174	0.20 *3*
Cd $K_{\beta1}'$	26.085	0.050 *7*
Cd $K_{\beta2}'$	26.801	0.0104 *14*
γ [M1+E2]	96.52 *11*	0.037 *2*
γ [M1+E2]	125.52 *17*	<0.02
γ	133.35 *20*	0.066 *2*
γ [E1]	206.17 *8*	0.020 *2*
γ	217.11 *9*	0.028 *2*
γ [M1+E2]	250.04 *15*	0.0020 *5*
γ M1(+>45%E2)	258.60 *9*	1.64 *3*
γ [M1+E2]	267.74 *10*	<0.02
γ [M1+E2]	285.47 *10*	0.0077 *19*
γ [E2+<45%M1]	298.41 *5*	10
γ [E2]	316.14 *4*	1.34 *3*
γ (M1,E2)	332.89 *7*	0.598 *12*
γ M1,E2	339.25 *7*	0.638 *13*
γ [M1+E2]	364.26 *6*	0.140 *3*
γ	369.36 *12*	~0.010

Photons (^{113}Ag) (continued)

γ_{mode}	γ(keV)	γ(%)†
γ	374.38 12	0.025 2
γ [M1+E2]	381.98 6	0.145 3
γ [M1+E2]	392.26 8	0.020 2
γ [M1+E2]	409.98 10	0.0012 3
γ	410.65 10	0.012 2
γ	539.07 8	0.008 3
γ E2	583.88 9	0.21 3
γ [M1+E2]	585.04 11	~0.010
γ [E1]	610.57 10	0.045 10
γ	623.86 8	0.019 1
γ [E2]	672.14 7	0.03 1
γ [E2]	672.14 7	0.87 3
γ [M1+E2]	680.40 6	0.695 16
γ [E2]	708.40 9	0.0063 14
γ	733.62 11	~0.010
γ	809.78 7	0.015 2
γ	815.95 10	0.011 2
γ	827.51 8	~0.010
γ [E1]	878.31 6	0.052 2
γ [M1]	883.45 10	0.282 7
γ [E1]	896.04 6	0.058 10
γ [M1]	988.27 7	0.423 9
γ	1049.75 10	0.045 3
γ	1084.35 10	0.016 3
γ	1125.92 7	0.061 3
γ [M1+E2]	1180.65 7	0.037 3
γ [E1]	1194.45 6	0.378 10
γ [M1+E2]	1479.06 7	0.068 4

† approximate

Atomic Electrons (^{113}Ag)

⟨e⟩=1.39 14 keV

e_{bin}(keV)	⟨e⟩(keV)	e(%)
4	0.018	0.48 8
14	0.051	0.37 4
17 - 26	0.0284	0.153 10
70 - 107	0.052	0.061 20
122 - 133	0.006	0.0050 22
179 - 223	0.0041	0.0021 9
232	0.141	0.0608 21
241 - 268	0.0285	0.0112 4
272	0.68	0.25 5
281 - 285	0.00010	3.7 14 $\times10^{-5}$
289	0.090	0.0312 14
294	0.080	0.027 6
295 - 312	0.091	0.0301 25
313	0.036	0.0114 12
315 - 364	0.0302	0.0089 5
365 - 411	0.0026	0.00070 7
512 - 558	0.0048	0.00086 12
580 - 624	0.00142	0.00024 3
645 - 682	0.0316	0.00484 22
704 - 734	0.00015	~2 $\times10^{-5}$
783 - 827	0.00041	5.2 15 $\times10^{-5}$
852 - 896	0.00492	0.000572 22
962 - 988	0.00568	0.000588 23
1023 - 1058	0.00050	4.8 16 $\times10^{-5}$
1080 - 1126	0.00050	4.5 18 $\times10^{-5}$
1154 - 1194	0.00193	0.000165 7
1452 - 1479	0.00054	3.7 4 $\times10^{-5}$

Continuous Radiation (^{113}Ag)

⟨β-⟩=760 keV;⟨IB⟩=1.43 keV

E_{bin}(keV)		⟨ ⟩(keV)	(%)
0 - 10	β-	0.0217	0.432
	IB	0.031	
10 - 20	β-	0.066	0.441
	IB	0.030	0.21
20 - 40	β-	0.274	0.91
	IB	0.058	0.20
40 - 100	β-	2.09	2.96
	IB	0.161	0.25
100 - 300	β-	25.0	12.2
	IB	0.42	0.24
300 - 600	β-	103	22.8
	IB	0.39	0.093
600 - 1300	β-	443	48.1
	IB	0.32	0.040
1300 - 2010	β-	187	12.5
	IB	0.021	0.00147

$^{113}_{47}$Ag(1.14 3 min)

Mode: IT(~80 %), β-(~20 %)
Δ: -86998 20 keV
SpA: 1.45×10^{9} Ci/g
Prod: fission; ^{114}Cd(γ,p)

Photons (^{113}Ag)

⟨γ⟩=123 10 keV

γ_{mode}	γ(keV)	γ(%)†
Ag L_{ℓ}	2.634	~0.07
Ag L_{η}	2.806	~0.043
Ag L_{α}	2.984	~2
Ag L_{β}	3.204	~1
Ag L_{γ}	3.525	~0.13
γ_{β}-M1	17.73 5	0.37 7
Ag $K_{\alpha2}$	21.990	~2
Ag $K_{\alpha1}$	22.163	~3
Cd $K_{\alpha2}$	22.984	0.199 22
Cd $K_{\alpha1}$	23.174	0.37 4
Ag $K_{\beta1}$'	24.934	~0.8
Ag $K_{\beta2}$'	25.603	~0.15
Cd $K_{\beta1}$'	26.085	0.095 10
Cd $K_{\beta2}$'	26.801	0.0197 22
γ_{IT} [E3]	43.2 10	~0.07
γ_{β}-[M1+E2]	125.52 17	<0.20
γ_{β}-[M1]	142.22 14	1.6 3
γ_{β}-	188.2 5	<0.20
γ_{β}-[M1+E2]	250.04 15	0.62 12
γ_{β}-[M1+E2]	267.74 10	<0.20
γ_{β}-[M1+E2]	285.47 10	0.073 15
γ_{β}-[E2+<45%M1]	298.41 5	5.6 11
γ_{β}-[E2]	316.14 4	10
γ_{β}-[M1+E2]	392.26 8	6.2 12
γ_{β}-[M1+E2]	409.98 10	0.37 7
γ_{β}-	423.3 3	<0.10
γ_{β}-	487.0 5	0.22 4
γ_{β}-	548.8 3	0.20 4
γ_{β}-E2	583.88 9	2.0 4
γ_{β}-	589.1 4	0.098 20
γ_{β}-	691.0 3	0.40 8
γ_{β}-[E2]	708.40 9	2.0 4
γ_{β}-	708.7 3	0.49 10
γ_{β}-	731.3 4	0.34 7
γ_{β}-	737.0 5	<0.20
γ_{β}-	896.9 5	<0.20

† approximate

Atomic Electrons (^{113}Ag)

⟨e⟩=35 12 keV

e_{bin}(keV)	⟨e⟩(keV)	e(%)
3	1.2	~35
4 - 17	1.3	26 11
18	1.3	7 4
19 - 39	0.34	1.3 3
40	22.8	~57
42	0.034	~0.08
43	6.1	~14
99 - 142	0.39	0.33 6
161 - 188	0.017	~0.010
223 - 272	0.47	0.18 3
281 - 316	0.96	0.33 5
366 - 410	0.32	0.085 15
419 - 460	0.0043	~0.0009
483 - 522	0.0039	0.0008 4
545 - 589	0.051	0.0091 15
664 - 710	0.052	0.0076 15
727 - 737	0.0008	0.00011 4
870 - 897	0.0011	~0.00012

Continuous Radiation (^{113}Ag)

⟨β-⟩=106 keV;⟨IB⟩=0.157 keV

E_{bin}(keV)		⟨ ⟩(keV)	(%)
0 - 10	β-	0.0073	0.146
	IB	0.0046	
10 - 20	β-	0.0222	0.148
	IB	0.0045	0.031
20 - 40	β-	0.091	0.304
	IB	0.0086	0.030
40 - 100	β-	0.69	0.98
	IB	0.023	0.036
100 - 300	β-	7.8	3.85
	IB	0.055	0.031
300 - 600	β-	28.1	6.3
	IB	0.042	0.0102
600 - 1300	β-	65	7.6
	IB	0.020	0.0027
1300 - 1737	β-	4.92	0.349
	IB	0.00026	1.9 $\times10^{-5}$

$^{113}_{48}$Cd(9.3 19 $\times10^{15}$ yr)

Mode: β-
Δ: -89051 3 keV
Prod: natural source
%: 12.22 6

Continuous Radiation (^{113}Cd)

⟨β-⟩=91 keV;⟨IB⟩=0.030 keV

E_{bin}(keV)		⟨ ⟩(keV)	(%)
0 - 10	β-	0.355	7.1
	IB	0.0048	
10 - 20	β-	1.03	6.9
	IB	0.0040	0.028
20 - 40	β-	3.91	13.1
	IB	0.0064	0.023
40 - 100	β-	22.6	33.0
	IB	0.0103	0.0170
100 - 300	β-	63	39.8
	IB	0.0044	0.0033
300 - 316	β-	0.073	0.0239
	IB	4.1 $\times10^{-8}$	1.34 $\times10^{-8}$

$^{113}_{48}$Cd(13.7 *4* yr)

Mode: β-(99.9 %), IT(0.1 %)
Δ: -88787 *3* keV
SpA: 231 Ci/g
Prod: ^{112}Cd(n,γ); fission

Photons (^{113}Cd)

⟨γ⟩=0.071 *12* keV

γ_{mode}	γ(keV)	γ(%)
Cd L_ℓ	2.767	8.7 *23* ×10^{-5}
Cd L_η	2.957	6.3 *16* ×10^{-5}
Cd L_α	3.133	0.0024 *6*
Cd L_β	3.371	0.0022 *6*
Cd L_γ	3.744	0.00021 *6*
Cd $K_{\alpha2}$	22.984	0.0123 *25*
Cd $K_{\alpha1}$	23.174	0.023 *5*
Cd $K_{\beta1}$'	26.085	0.0059 *12*
Cd $K_{\beta2}$'	26.801	0.00123 *25*
γ_{IT} E5	263.71 *11*	0.023

Atomic Electrons (^{113}Cd)

⟨e⟩=0.28 *3* keV

e_{bin}(keV)	⟨e⟩(keV)	e(%)
4	0.0024	0.067 *15*
19	0.00084	0.0044 *10*
20	0.00021	0.00108 *24*
22	0.00024	0.00110 *25*
23	0.00026	0.00114 *25*
25	3.4 ×10^{-5}	0.00013 *3*
26	1.6 ×10^{-5}	6.0 *13* ×10^{-5}
237	0.119	0.05 *1*
260	0.097	0.037 *8*
263	0.057	0.022 *4*

Continuous Radiation (^{113}Cd)

⟨β-⟩=183 keV;⟨IB⟩=0.112 keV

E_{bin}(keV)		⟨ ⟩(keV)	(%)
0 - 10	β-	0.158	3.16
	IB	0.0093	
10 - 20	β-	0.474	3.16
	IB	0.0086	0.060
20 - 40	β-	1.89	6.3
	IB	0.0152	0.053
40 - 100	β-	12.9	18.4
	IB	0.033	0.053
100 - 300	β-	95	49.5
	IB	0.041	0.026
300 - 580	β-	73	19.3
	IB	0.0042	0.00122

$^{113}_{49}$In(stable)

Δ: -89367 *4* keV
%: 4.3 *2*

$^{113}_{49}$In(1.658 *1* h)

Mode: IT
Δ: -88975 *4* keV
SpA: 1.6726×10^{7} Ci/g
Prod: daughter ^{113}Sn

Photons (^{113}In)

⟨γ⟩=255 *5* keV

γ_{mode}	γ(keV)	γ(%)†
In L_ℓ	2.905	0.025 *4*
In L_η	3.112	0.0117 *18*
In L_α	3.286	0.68 *11*
In L_β	3.569	0.50 *8*
In L_γ	3.985	0.052 *9*
In $K_{\alpha2}$	24.002	4.58 *17*
In $K_{\alpha1}$	24.210	8.6 *3*
In $K_{\beta1}$'	27.265	2.21 *8*
In $K_{\beta2}$'	28.022	0.464 *18*
γ M4	391.690 *7*	64.2

† 1.1% uncert(syst)

Atomic Electrons (^{113}In)

⟨e⟩=134.0 *25* keV

e_{bin}(keV)	⟨e⟩(keV)	e(%)
4	0.92	24.0 *25*
19	0.062	0.32 *3*
20	0.52	2.6 *3*
23	0.137	0.59 *6*
24	0.146	0.62 *6*
26	0.0085	0.032 *3*
27	0.0198	0.074 *8*
364	104.7	28.8 *7*
387	16.0	4.14 *10*
388	5.95	1.54 *4*
391	4.53	1.16 *3*
392	0.960	0.245 *6*

$^{113}_{50}$Sn(115.09 *4* d)

Mode: ε
Δ: -88328 *4* keV
SpA: 1.0040×10^{4} Ci/g
Prod: ^{112}Sn(n,γ); ^{113}In(p,n); ^{113}In(d,2n)

Photons (^{113}Sn)

⟨γ⟩=280 *8* keV

γ_{mode}	γ(keV)	γ(%)†
In L_ℓ	2.905	0.15 *3*
In L_η	3.112	0.069 *11*
In L_α	3.286	4.1 *6*
In L_β	3.570	3.0 *5*
In L_γ	3.985	0.30 *6*
In $K_{\alpha2}$	24.002	28.3 *11*
In $K_{\alpha1}$	24.210	52.9 *20*
In $K_{\beta1}$'	27.265	13.7 *5*
In $K_{\beta2}$'	28.022	2.86 *11*

Photons (^{113}Sn) (continued)

γ_{mode}	γ(keV)	γ(%)†
γ M1+32%E2	255.07 *5*	1.82 *6*
γ	382.96 *9*	<6 ×10^{-5} ?
γ M4	391.690 *7*	64 *
γ E1	638.03 *8*	0.00095 *4*
γ [E3]	646.76 *5*	~6 ×10^{-6}

† 3.1% uncert(syst)
* with ^{113}In(1.658 h) in equilib

Atomic Electrons (^{113}Sn)

⟨e⟩=139 *4* keV

e_{bin}(keV)	⟨e⟩(keV)	e(%)
4	3.5	91 *9*
19	0.246	1.27 *13*
20	2.07	10.3 *11*
23	0.54	2.35 *24*
24	0.58	2.45 *25*
26	0.034	0.128 *13*
27	0.078	0.29 *3*
227	0.164	0.072 *8*
251	0.025	0.0100 *20*
254	0.0050	0.0020 *4*
255	0.00106	0.00042 *8*
355	1.0 ×10^{-6}	<6 ×10^{-7}
364	104.4	28.7 *11*
379	1.5 ×10^{-7}	<8 ×10^{-8}
382	3.1 ×10^{-8}	<16 ×10^{-9}
383	6.6 ×10^{-9}	<3 ×10^{-9}
387	16.0	4.13 *15*
388	5.94	1.53 *6*
391	4.53	1.16 *4*
392	0.96	0.245 *9*
610	6.6 ×10^{-6}	1.08 *5* ×10^{-6}
619	2.9 ×10^{-7}	~5 ×10^{-8}
634	8.1 ×10^{-7}	1.28 *6* ×10^{-7}
637	1.58 ×10^{-7}	2.48 *11* ×10^{-8}
638	3.44 ×10^{-8}	5.39 *24* ×10^{-9}
643	4.6 ×10^{-8}	~7 ×10^{-9}
646	9.1 ×10^{-9}	~1 ×10^{-9}
647	1.9 ×10^{-9}	~3 ×10^{-10}

Continuous Radiation (^{113}Sn)

⟨IB⟩=0.093 keV

E_{bin}(keV)		⟨ ⟩(keV)	(%)
10 - 20	IB	0.0019	0.0108
20 - 40	IB	0.039	0.161
40 - 100	IB	0.0031	0.0050
100 - 300	IB	0.022	0.0108
300 - 600	IB	0.027	0.0068
600 - 647	IB	3.0 ×10^{-5}	4.9 ×10^{-6}

$^{113}_{50}$Sn(21.4 *4* min)

Mode: IT(91.1 *23* %), ε(8.9 *23* %)
Δ: -88251 *4* keV
SpA: 7.77×10^{7} Ci/g
Prod: ^{112}Sn(n,γ); daughter ^{113}Sb

Photons (^{113}Sn)

$\langle\gamma\rangle$=13.5 *3* keV

γ_{mode}	γ(keV)	γ(%)†
In L_ℓ	2.905	0.0097 *17*
Sn L_ℓ	3.045	0.093 *14*
In L_η	3.112	0.0045 *7*
Sn L_η	3.272	0.044 *6*
In L_α	3.286	0.27 *4*
Sn L_α	3.443	2.6 *3*
In L_β	3.570	0.19 *3*
Sn L_β	3.747	2.2 *3*
In L_γ	3.984	0.020 *4*
Sn L_γ	4.230	0.26 *4*
In $K_{\alpha2}$	24.002	1.89 *7*
In $K_{\alpha1}$	24.210	3.54 *12*
Sn $K_{\alpha2}$	25.044	12.7 *5*
Sn $K_{\alpha1}$	25.271	23.7 *9*
In $K_{\beta1}$'	27.265	0.91 *3*
In $K_{\beta2}$'	28.022	0.192 *7*
Sn $K_{\beta1}$'	28.474	6.21 *24*
Sn $K_{\beta2}$'	29.275	1.31 *5*
γ_{IT} M3+1.7%E4	77.383 *18*	0.500

† uncert(syst): 2.5% for IT

Atomic Electrons (^{113}Sn)

$\langle e\rangle$=58.2 *20* keV

e_{bin}(keV)	$\langle e\rangle$(keV)	e(%)
4	2.7	68 *8*
19	0.0165	0.085 *9*
20	0.247	1.22 *13*
21	0.90	4.3 *5*
23	0.036	0.157 *16*
24	0.28	1.16 *12*
25	0.26	1.04 *11*
26	0.00226	0.0086 *9*
27	0.0096	0.036 *4*
28	0.046	0.163 *17*
48	24.5	50.8 *16*
73	22.8	31.2 *25*
76	2.02	2.64 *9*
77	4.4	5.7 *6*

Continuous Radiation (^{113}Sn)

$\langle$IB$\rangle$=0.029 keV

E_{bin}(keV)		$\langle\ \rangle$(keV)	(%)
10 - 20	IB	0.00017	0.00096
20 - 40	IB	0.0035	0.0143
40 - 100	IB	0.00030	0.00048
100 - 300	IB	0.0030	0.00144
300 - 600	IB	0.0115	0.0026
600 - 1116	IB	0.0106	0.00144

$^{113}_{51}$Sb(6.67 *7* min)

Mode: ϵ

Δ: -84421 *24* keV

SpA: 2.49×10^{8} Ci/g

Prod: ^{112}Sn(d,n); ^{114}Sn(p,2n)

Photons (^{113}Sb)

$\langle\gamma\rangle$=590 *11* keV

γ_{mode}	γ(keV)	γ(%)†
Sn L_ℓ	3.045	0.061 *9*
Sn L_η	3.272	0.031 *4*
Sn L_α	3.443	1.69 *22*
Sn L_β	3.744	1.46 *21*
Sn L_γ	4.222	0.17 *3*
Sn $K_{\alpha2}$	25.044	11.0 *6*
Sn $K_{\alpha1}$	25.271	20.6 *12*
Sn $K_{\beta1}$'	28.474	5.4 *3*
Sn $K_{\beta2}$'	29.275	1.13 *7*
γ M3+1.7%E4	77.383 *18*	*
γ M1,E2	88.238 *19*	2.7 *3*
γ	242.32 *15*	0.023 *5*
γ	273.19 *13*	0.038 *4*
γ	332.14 *18*	0.024 *11*
γ M1,E2	332.38 *4*	14.8 *6*
γ	409.76 *4*	0.128 *16*
γ	420.62 *4*	~0.24
γ	448.73 *20*	0.022 *9*
γ [M1]	498.00 *4*	80
γ	538.27 *9*	0.058 *4*
γ	603.45 *6*	0.0112 *24*
γ	608.32 *5*	0.400 *24*
γ	718.66 *17*	~0.032
γ	725.84 *14*	~0.012
γ	785.15 *12*	0.015 *3*
γ	800.99 *20*	0.027 *3*
γ	816.03 *14*	0.026 *3*
γ	886.49 *20*	0.080 *16*
γ	935.83 *4*	1.71 *9*
γ	940.70 *4*	2.62 *13*
γ	1013.21 *4*	0.91 *6*
γ	1018.08 *4*	0.480 *24*
γ	1058.34 *9*	0.054 *5*
γ	1128.79 *20*	0.027 *3*
γ	1146.58 *9*	0.45 *3*
γ	1148.17 *13*	0.11 *3*
γ	1205.77 *12*	0.022 *3*
γ	1233.87 *17*	0.45 *6*
γ	1236.40 *13*	0.17 *6*
γ	1241.97 *20*	0.11 *4*
γ	1245.92 *14*	0.22 *4*
γ M1,E2	1283.15 *12*	0.168 *16*
γ	1314.02 *14*	0.144 *16*
γ	1334.15 *14*	0.168 *16*
γ	1355.9 *3*	0.029 *3*
γ	1390.69 *20*	0.046 *4*
γ	1458.99 *16*	0.048 *5*
γ	1478.96 *9*	0.120 *16*
γ	1547.23 *16*	~0.06
γ	1547.36 *23*	~0.048
γ	1556.34 *9*	1.05 *8*
γ	1568.78 *13*	0.044 *5*
γ	1574.35 *19*	0.056 *6*
γ	1635.60 *23*	0.030 *4*
γ	1646.16 *13*	0.128 *16*
γ	1654.49 *17*	0.058 *6*
γ	1666.53 *14*	0.096 *16*
γ	1718.29 *20*	0.224 *24*
γ	1743.91 *14*	0.021 *3*
γ	1806.1 *3*	0.028 *3*
γ	1879.60 *16*	0.0192 *24*
γ	1889.4 *3*	0.062 *6*
γ	1918.7 *5*	0.008 *3*
γ	1956.98 *16*	0.057 *6*
γ	1967.97 *23*	0.0168 *24*
γ	2006.7 *6*	0.026 *3*
γ	2014.7 *6*	0.035 *5*
γ	2042.2 *4*	0.045 *6*
γ	2130.5 *4*	0.038 *5*
γ	2304.8 *7*	0.0128 *24*
γ	2337.2 *7*	0.0120 *24*
γ	2433.9 *5*	0.022 *4*
γ	2540.2 *4*	0.049 *6*

Photons (^{113}Sb)
(continued)

γ_{mode}	γ(keV)	γ(%)†
γ	2624.6 *6*	0.0120 *24*
γ	2791.5 *13*	0.0088 *24*
γ	2854.5 *5*	0.0104 *24*
γ	3143.7 *12*	0.0128 *24*
γ	3192.5 *12*	0.0112 *24*
γ	3605.6 *13*	0.017 *4*

† 2.5% uncert(syst)

* with ^{113}Sn(21.4 min)

Atomic Electrons (^{113}Sb)

$\langle e\rangle$=21.4 *13* keV

e_{bin}(keV)	$\langle e\rangle$(keV)	e(%)
4	1.66	41 *5*
20	0.094	0.46 *5*
21	0.78	3.7 *4*
24 - 28	0.47	1.92 *14*
48	5.1	10.6 *8*
59	2.0	3.3 *15*
73	4.7	6.5 *7*
76	0.42	0.55 *4*
77	0.91	1.19 *15*
84	0.8	~1.0
87 - 88	0.21	~0.24
213 - 244	0.0045	0.0020 *8*
269 - 273	0.0005	~0.00018
303	0.92	0.30 *3*
328 - 332	0.17	0.052 *9*
381 - 421	0.013	0.0032 *16*
444 - 449	9.8×10^{-5}	2.2 *11* $\times10^{-5}$
469	2.66	0.57 *3*
494 - 538	0.426	0.086 *3*
574 - 608	0.008	0.0014 *6*
689 - 726	0.0007	10 *5* $\times10^{-5}$
756 - 801	0.00088	0.00011 *3*
812 - 857	0.0009	~0.00010
882 - 931	0.044	0.0049 *17*
932 - 941	0.0046	0.00049 *17*
984 - 1029	0.015	0.0015 *5*
1054 - 1100	0.00029	2.6 *10* $\times10^{-5}$
1117 - 1148	0.0051	0.00045 *16*
1177 - 1217	0.0070	0.00058 *16*
1229 - 1279	0.0028	0.00022 *3*
1282 - 1330	0.0026	0.00020 *5*
1333 - 1361	0.00036	2.6 *10* $\times10^{-5}$
1386 - 1430	0.00034	2.4 *9* $\times10^{-5}$
1450 - 1547	0.008	0.00053 *17*
1552 - 1651	0.0028	0.00018 *3*
1654 - 1743	0.0015	9 *3* $\times10^{-5}$
1777 - 1876	0.00056	3.1 *7* $\times10^{-5}$
1879 - 1977	0.00060	3.1 *6* $\times10^{-5}$
1985 - 2042	0.00043	2.1 *5* $\times10^{-5}$
2101 - 2130	0.00019	9 *3* $\times10^{-6}$
2276 - 2336	0.00011	5.0 *12* $\times10^{-6}$
2405 - 2433	9.6×10^{-5}	4.0 *13* $\times10^{-6}$
2511 - 2595	0.00025	1.0 *3* $\times10^{-5}$
2620 - 2624	6.3×10^{-6}	2.4 *8* $\times10^{-7}$
2762 - 2854	7.6×10^{-5}	2.7 *7* $\times10^{-6}$
3114 - 3192	8.8×10^{-5}	2.8 *6* $\times10^{-6}$
3576 - 3605	5.6×10^{-5}	1.6 *5* $\times10^{-6}$

Continuous Radiation (^{113}Sb)

⟨β+⟩=701 keV;⟨IB⟩=4.5 keV

E_{bin}(keV)		⟨ ⟩(keV)	(%)
0 - 10	β+	3.19×10^{-6}	3.85×10^{-5}
	IB	0.026	
10 - 20	β+	0.000116	0.00070
	IB	0.026	0.18
20 - 40	β+	0.00290	0.0089
	IB	0.065	0.23
40 - 100	β+	0.126	0.162
	IB	0.144	0.22
100 - 300	β+	5.9	2.66
	IB	0.41	0.23
300 - 600	β+	44.3	9.5
	IB	0.51	0.120
600 - 1300	β+	309	32.4
	IB	1.17	0.128
1300 - 2500	β+	342	21.1
	IB	1.8	0.099
2500 - 3831	β+	0.301	0.0117
	IB	0.33	0.0120
	Σβ+		66

$^{113}_{52}$Te(1.7 *2* min)

Mode: ϵ

Δ: -78320 *200* keV syst

SpA: 9.8×10^{8} Ci/g

Prod: ^{112}Sn(α,3n)

Photons (^{113}Te)

⟨γ⟩=1308 *68* keV

γ_{mode}	γ(keV)	γ(%)†
Sb L_ℓ	3.189	0.0160 *21*
Sb L_η	3.437	0.0080 *8*
Sb L_α	3.604	0.44 *5*
Sb L_β	3.931	0.39 *5*
Sb L_γ	4.438	0.046 *7*
Sb $K_{\alpha2}$	26.111	3.09 *10*
Sb $K_{\alpha1}$	26.359	5.77 *18*
Sb $K_{\beta1}$'	29.712	1.52 *5*
Sb $K_{\beta2}$'	30.561	0.323 *11*
γ [E2]	238.5 *4*	0.46 *15*
γ	269.8 *6*	0.59 *15*
γ	392.0 *5*	~3
γ	437.7 *5*	~0.24
γ M1+0.8%E2	443.0 *3*	0.66 *20*
γ	473.1 *9*	0.55 *13*
γ	583.0 *6*	~0.7
γ	609.3 *6*	1.1 *4*
γ [E2]	644.74 *25*	6.4 *7*
γ M1+0.7%E2	647.1 *6*	0.79 *22*
γ [M1+E2]	736.8 *4*	1.0 *3*
γ M1+5.7%E2	814.1 *3*	22
γ	915.0 *5*	1.4 *4*
γ [M1+E2]	1018.5 *4*	13.0 *13*
γ	1039.2 *5*	1.5 *4*
γ	1071.7 *5*	1.5 *4* ?
γ	1181.0 *5*	12.3 *13*
γ	1206.6 *7*	1.3 *4*
γ	1245.4 *6*	~0.9
γ [E2]	1257.0 *4*	5.5 *7*
γ	1301.3 *7*	1.8 *7*
γ	1317.9 *8*	1.3 *4* ?
γ	1357.9 *5*	1.2 *3*
γ	1449.4 *7*	1.8 *4*
γ E2	1461.2 *6*	~2
γ	1516.0 *5*	1.5 *4*
γ [M1+E2]	1550.8 *5*	2.2 *7*
γ	1567.2 *9*	1.3 *4*
γ	1720.5 *5*	0.9 *3*
γ	1803.6 *8*	1.8 *7*
γ	1868.1 *10*	2.4 *7*

Photons (^{113}Te) (continued)

γ_{mode}	γ(keV)	γ(%)†
γ	1944.3 *12*	~0.9
γ	2047.8 *11*	1.5 *4*
γ	2094.1 *7*	2.9 *7*
γ	2115.3 *7*	2.0 *4*
γ	2221.2 *10*	2.0 *7*
γ	2534.5 *4*	2.6 *7*
γ	2552.4 *10*	1.5 *4*
γ	2606.5 *6*	1.8 *7*

† 23% uncert(syst)

Atomic Electrons (^{113}Te)

⟨e⟩=2.44 *15* keV

e_{bin}(keV)	⟨e⟩(keV)	e(%)
4	0.37	8.8 *9*
5 - 21	0.111	1.43 *12*
22	0.180	0.82 *8*
25	0.057	0.228 *23*
26	0.062	0.239 *24*
29 - 30	0.0121	0.042 *4*
208	0.063	0.03 *1*
234 - 270	0.06	0.026 *11*
362	0.09	~0.024
387 - 437	0.053	0.013 *3*
438 - 473	0.021	0.0047 *19*
552 - 583	0.034	~0.006
605 - 609	0.0033	~0.0006
614	0.132	0.0216 *24*
617 - 647	0.046	0.0073 *10*
706 - 737	0.022	0.0030 *9*
784	0.41	0.052 *12*
809 - 814	0.064	0.0079 *14*
884 - 915	0.017	0.0020 *10*
988	0.17	0.017 *3*
1009 - 1041	0.056	0.0055 *12*
1067 - 1072	0.0022	0.00020 *9*
1150	0.10	0.009 *4*
1176 - 1215	0.034	0.0029 *11*
1227 - 1271	0.071	0.0057 *8*
1287 - 1327	0.021	0.0016 *5*
1353 - 1358	0.0012	9 *4* $\times10^{-5}$
1419 - 1461	0.029	0.0020 *6*
1486 - 1566	0.041	0.0027 *6*
1690 - 1773	0.015	0.0009 *3*
1799 - 1867	0.016	0.0009 *3*
1914 - 1943	0.005	~0.00026
2017 - 2115	0.034	0.0017 *4*
2191 - 2220	0.010	0.00046 *20*
2504 - 2602	0.027	0.00106 *24*
2606	0.00016	$\sim6\times10^{-6}$

Continuous Radiation (^{113}Te)

⟨β+⟩=1651 keV;⟨IB⟩=9.3 keV

E_{bin}(keV)		⟨ ⟩(keV)	(%)
0 - 10	β+	1.01×10^{-6}	1.21×10^{-5}
	IB	0.049	
10 - 20	β+	3.76×10^{-5}	0.000227
	IB	0.049	0.34
20 - 40	β+	0.00097	0.00298
	IB	0.103	0.36
40 - 100	β+	0.0442	0.057
	IB	0.28	0.43
100 - 300	β+	2.23	1.00
	IB	0.84	0.47
300 - 600	β+	18.9	4.03
	IB	1.08	0.25
600 - 1300	β+	185	18.8
	IB	2.0	0.23
1300 - 2500	β+	739	39.4
	IB	2.7	0.150

Continuous Radiation (^{113}Te) (continued)

E_{bin}(keV)		⟨ ⟩(keV)	(%)
2500 - 5000	β+	706	22.6
	IB	2.1	0.066
5000 - 6100	β+	0.0358	0.00071
	IB	0.044	0.00083
	Σβ+		86

$^{113}_{53}$I (5.9 *5* s)

Mode: ϵ, α, ϵα

Δ: -71120 *50* keV

SpA: 1.60×10^{10} Ci/g

Prod: ^{58}Ni on ^{58}Ni

Alpha Particles (^{113}I)

α(keV)
2610 *40*

Photons (^{113}I)

γ_{mode}	γ(keV)	γ(rel)
γ	55.0 *2*	32 *2*
γ	160.0 *2*	14 *2*
γ	216.5 *2*	7 *2*
γ	320.4 *2*	33 *2*
γ	351.5 *2*	43 *2*
γ	406.1 *2*	8 *2*
γ	462.5 *2*	100
γ	523.0 *5*	7 *1*
γ	567.4 *2*	36 *3*
γ	608.6 *5*	6.2 *10*
γ	622.4 *2*	74 *3*
γ	628.0 *2*	13 *2*
γ	651.9 *5*	3.4 *10*
γ	690.2 *5*	8 *1*
γ	696.2 *5*	3.1 *10*
γ	774.0 *5*	8 *1*
γ	798.2 *2*	12 *2*
γ	802.1 *5*	8 *2*
γ	896.0 *5*	9.7 *10*
γ	929.1 *3*	8 *1*
γ	1161.0 *5*	8.7 *10*
γ	1422.4 *3*	11 *2*

$^{113}_{54}$Xe(2.8 *2* s)

Mode: ϵ, α, ϵp, ϵα

Δ: -62130 *80* keV

SpA: 3.16×10^{10} Ci/g

Prod: protons on Ce; protons on La; ^{58}Ni on ^{58}Ni

ϵp/α: 830 *50*

Alpha Particles (^{113}Xe)

α(keV)
2985 *15*

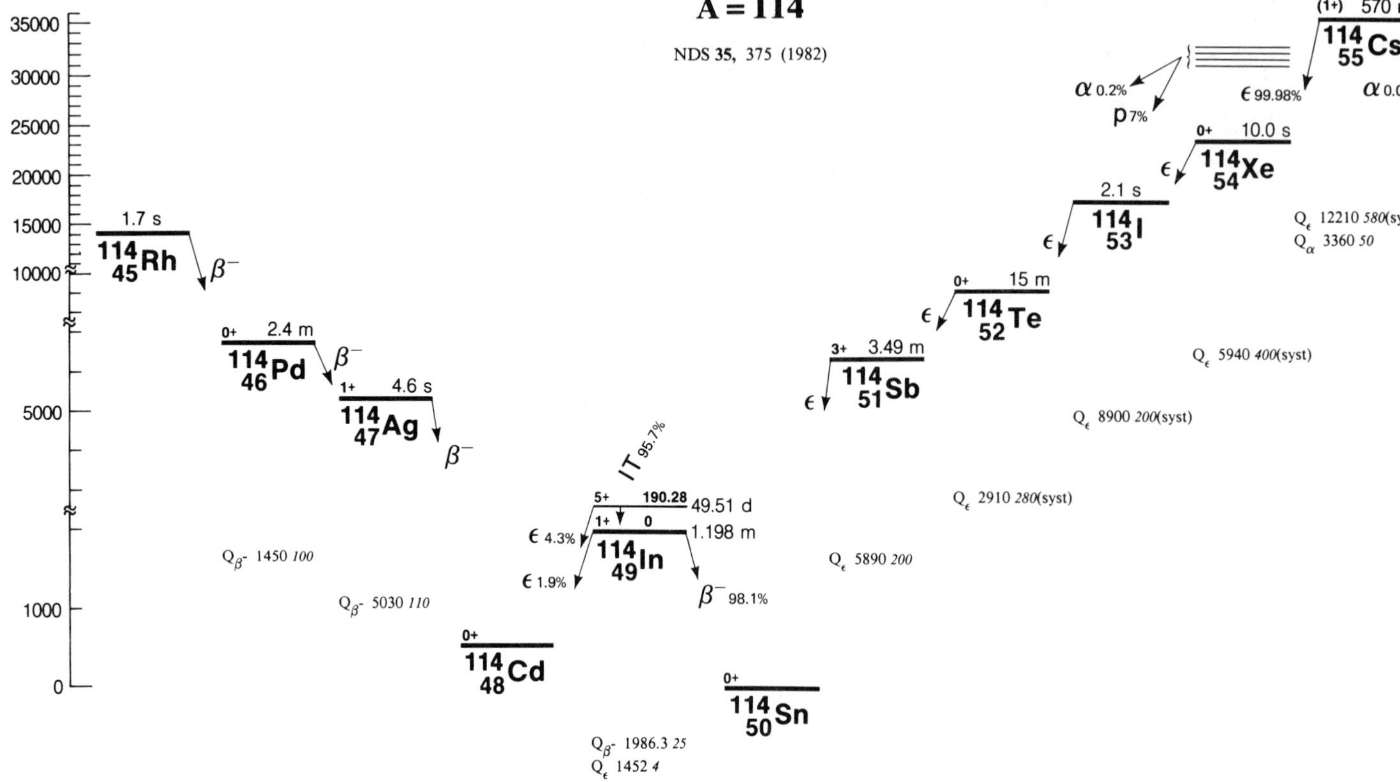

$^{114}_{45}$Rh(1.68 *7* s)

Mode: β-
SpA: 4.83×10^{10} Ci/g
Prod: fission

Photons (^{114}Rh)

γ_{mode}	γ(keV)
γ	332.86 *19*

$^{114}_{46}$Pd(2.4 *1* min)

Mode: β-
Δ: -83540 *150* keV
SpA: 6.9×10^{8} Ci/g
Prod: fission

Photons (^{114}Pd)

⟨γ⟩=14.1 *12* keV

γ_{mode}	γ(keV)	γ(%)†
γ[E1]	126.40 *13*	2.70 *14*
γ[E1]	136.30 *9*	0.50 *3*
γ	147.8 *2*	0.14 *2*
γ[E1]	231.70 *9*	3.00 *15*
γ[M1+E2]	358.10 *12*	0.8 *3*

† 40% uncert(syst)

Continuous Radiation (^{114}Pd)

⟨β-⟩=531 keV;⟨IB⟩=0.76 keV

E_{bin}(keV)		⟨ ⟩(keV)	(%)
0 - 10	β-	0.0363	0.72
	IB	0.023	
10 - 20	β-	0.111	0.74
	IB	0.023	0.157
20 - 40	β-	0.455	1.52
	IB	0.043	0.150
40 - 100	β-	3.44	4.89
	IB	0.116	0.18
100 - 300	β-	39.8	19.5
	IB	0.27	0.157
300 - 600	β-	147	32.7
	IB	0.20	0.049
600 - 1300	β-	334	39.7
	IB	0.077	0.0109
1300 - 1450	β-	5.2	0.390
	IB	5.6×10^{-5}	4.2×10^{-6}

$^{114}_{47}$Ag(4.6 *2* s)

Mode: β-
Δ: -84990 *110* keV
SpA: 2.00×10^{10} Ci/g
Prod: fission; daughter ^{114}Pd; ^{114}Cd(n,p)

Photons (^{114}Ag)

⟨γ⟩=99 *11* keV

γ_{mode}	γ(keV)	γ(%)†
γ E2	558.43 *3*	10
γ E2	576.03 *20*	0.8 *2*
γ E2+45%M1	651.1 *4*	0.40 *9*
γ M1+2.5%E2	805.9 *4*	0.28 *5*
γ E2	1209.5 *4*	0.16 *5*
γ	1304.3 *4*	0.56 *8*
γ E2	1364.3 *4*	0.19 *5*
γ E2,M1	1659.8 *5*	0.32 *8*
γ[E2]	1994.8 *6*	0.81 *6*

† 40% uncert(syst)

Continuous Radiation (^{114}Ag)

⟨β-⟩=2175 keV;⟨IB⟩=7.8 keV

E_{bin}(keV)		⟨ ⟩(keV)	(%)
0 - 10	β-	0.00257	0.051
	IB	0.061	
10 - 20	β-	0.0079	0.053
	IB	0.060	0.42
20 - 40	β-	0.0330	0.110
	IB	0.120	0.42
40 - 100	β-	0.261	0.369
	IB	0.35	0.53
100 - 300	β-	3.57	1.72
	IB	1.06	0.59
300 - 600	β-	19.0	4.12
	IB	1.34	0.31
600 - 1300	β-	161	16.4
	IB	2.3	0.26
1300 - 2500	β-	756	39.6
	IB	2.0	0.114
2500 - 5000	β-	1235	38.1
	IB	0.61	0.020
5000 - 5030	β-	0.0062	0.000124
	IB	2.3×10^{-9}	4.6×10^{-11}

$^{114}_{48}$Cd(stable)

Δ: -90023 *3* keV
%: 28.73 *21*

$^{114}_{49}$In(1.1983 *17* min)

Mode: β-(98.1 %), ε(1.9 %)
Δ: -88570 *4* keV
SpA: 1.3698×10^{9} Ci/g
Prod: daughter ^{114}In (49.51 d); ^{113}In(n,γ)

Photons (^{114}In)

⟨γ⟩=2.03 *24* keV

γ_{mode}	γ(keV)	γ(%)†
γ_ϵE2	558.43 *3*	<0.07
γ_ϵE2	576.03 *20*	0.0039 *4*
γ_ϵ[E2]	747.8 *2*	$<18\times10^{-5}$
$\gamma_{\beta-}$E2	1299.89 *5*	0.140 *11*

† uncert(syst): 30% for ε, 7.1% for β-

Continuous Radiation (^{114}In)

⟨β-⟩=773 keV;⟨IB⟩=1.47 keV

E_{bin}(keV)		⟨ ⟩(keV)	(%)
0 - 10	β-	0.0192	0.381
	IB	0.031	
10 - 20	β-	0.059	0.39
	IB	0.030	0.21
20 - 40	β-	0.243	0.81
	IB	0.059	0.21
40 - 100	β-	1.86	2.64
	IB	0.163	0.25
100 - 300	β-	22.9	11.2
	IB	0.43	0.24
300 - 600	β-	100	21.9
	IB	0.40	0.096
600 - 1300	β-	454	49.2
	IB	0.33	0.041
1300 - 1986	β-	194	13.1
	IB	0.021	0.00149

$^{114}_{49}$In(49.51 *1* d)

Mode: IT(95.7 *3* %), ε(4.3 *3* %)
Δ: -88380 *4* keV
SpA: 2.3135×10^{4} Ci/g
Prod: ^{113}In(n,γ)

Photons (^{114}In)

⟨γ⟩=94 *3* keV

γ_{mode}	γ(keV)	γ(%)†
Cd L_ℓ	2.767	0.0045 *8*
In L_ℓ	2.905	0.076 *14*
Cd L_η	2.957	0.0021 *3*
In L_η	3.112	0.052 *8*
Cd L_α	3.133	0.122 *19*
In L_α	3.286	2.1 *3*
Cd L_β	3.392	0.086 *14*
In L_β	3.550	1.9 *3*
Cd L_γ	3.782	0.0086 *16*
In L_γ	3.949	0.19 *3*
Cd $K_{\alpha2}$	22.984	0.90 *3*
Cd $K_{\alpha1}$	23.174	1.69 *6*
In $K_{\alpha2}$	24.002	9.8 *4*
In $K_{\alpha1}$	24.210	18.3 *8*
Cd $K_{\beta1}$'	26.085	0.431 *15*
Cd $K_{\beta2}$'	26.801	0.089 *3*
In $K_{\beta1}$'	27.265	4.72 *20*
In $K_{\beta2}$'	28.022	0.99 *4*
γ_{IT} E4	190.28 *3*	15.4
γ_ϵE2	558.43 *3*	4.4
γ_ϵE2	725.24 *3*	4.3

† uncert(syst): 7.0% for ε, 2.5% for IT

Atomic Electrons (^{114}In)

⟨e⟩=143.1 *25* keV

e_{bin}(keV)	⟨e⟩(keV)	e(%)
4	2.20	58 *6*
19	0.146	0.76 *8*
20	0.73	3.6 *4*
22	0.0177	0.080 *8*
23	0.207	0.89 *9*
24	0.200	0.84 *9*
25	0.00247	0.0097 *10*
26	0.0128	0.049 *5*
27	0.027	0.101 *11*
162	63.9	39.4 *13*
186	36.4	19.6 *6*
187	23.6	12.6 *4*
189	1.33	0.703 *22*
190	14.2	7.46 *24*
532	0.099	0.0186 *14*
554	0.0113	0.00204 *15*
555	0.00204	0.00037 *3*
558	0.00313	0.00056 *4*
699	0.065	0.0092 *7*
721	0.0074	0.00102 *7*
722	0.00094	0.000130 *10*
724	0.00142	0.000196 *14*
725	0.00051	7.0 *5* $\times10^{-5}$

Continuous Radiation (^{114}In)

⟨IB⟩=0.0020 keV

E_{bin}(keV)		⟨ ⟩(keV)	(%)
10 - 20	IB	0.000169	0.00091
20 - 40	IB	0.00144	0.0060
40 - 100	IB	0.000104	0.000166
100 - 300	IB	0.00028	0.000168
300 - 359	IB	3.6×10^{-6}	1.15×10^{-6}

$^{114}_{50}$Sn(stable)

Δ: -90557 *3* keV
%: 0.65 *1*

$^{114}_{51}$Sb(3.49 *3* min)

Mode: ε
Δ: -84670 *200* keV
SpA: 4.72×10^{8} Ci/g
Prod: ^{114}Sn(p,n); ^{115}Sn(p,2n); ^{115}In(^{3}He,4n)

Photons (^{114}Sb)

⟨γ⟩=1868 *20* keV

γ_{mode}	γ(keV)	γ(%)†
Sn L_ℓ	3.045	0.024 *3*
Sn L_η	3.272	0.0123 *13*
Sn L_α	3.443	0.65 *7*
Sn L_β	3.742	0.58 *7*
Sn L_γ	4.220	0.068 *10*
Sn $K_{\alpha2}$	25.044	4.86 *16*
Sn $K_{\alpha1}$	25.271	9.1 *3*
Sn $K_{\beta1}$'	28.474	2.37 *8*
Sn $K_{\beta2}$'	29.275	0.499 *17*
γ [M1+E2]	215.08 *19*	0.39 *8*
γ	290.75 *13*	0.049 *10*
γ	320.7 *3*	0.22 *4*
γ	327.22 *7*	7.1 *5*
γ E2	375.31 *14*	0.030 *6*
γ	390.32 *8*	1.18 *24*
γ	441.76 *15*	0.049 *10*
γ	450.97 *16*	0.020 *4*
γ [M1+E2]	489.56 *18*	0.15 *3*
γ	573.84 *17*	0.089 *18*
γ	593.09 *16*	0.13 *3*
γ [E2]	619.01 *21*	0.059 *12*
γ	627.33 *20*	~0.14
γ	634.0 *3*	0.030 *6*
γ [E2]	653.30 *10*	0.16 *3*
γ [E1]	668.85 *15*	1.3 *3*
γ [M1+E2]	704.64 *18*	0.049 *10*
γ	717.54 *9*	4.6 *3*
γ	771.7 *3*	0.039 *8*
γ	787.1 *3*	0.046 *10*
γ	856.9 *3*	0.049 *7*
γ E2	887.62 *5*	17.4 *7*
γ [E2]	921.97 *19*	0.118 *24*
γ	932.61 *12*	0.24 *5*
γ [M1+E2]	939.10 *13*	0.99 *20*
γ	964.16 *17*	0.13 *3*
γ [E1]	974.89 *6*	2.8 *3*
γ [E2]	990.44 *15*	0.069 *14*
γ	1010.6 *3*	0.059 *12*
γ	1019.89 *12*	0.47 *10*
γ	1071.3 *3*	0.55 *11*
γ	1121.9 *5*	0.068 *12*
γ	1131.63 *19*	0.32 *6*
γ [M1+E2]	1154.18 *14*	1.6 *3*
γ	1170.02 *22*	0.28 *6*

Photons (^{114}Sb) (continued)

γ_{mode}	γ(keV)	γ(%)†
γ [E1]	1204.11 17	0.118 24
γ [M1+E2]	1239.89 16	0.14 3
γ	1250.6 3	0.16 3
γ	1264.27 23	0.16 3
γ E2	1299.89 5	98.7
γ (E2)	1314.41 12	0.60 12
γ [M1+E2]	1327.77 20	0.069 14
γ	1337.9 3	0.079 16
γ	1364.5 8	0.027 5
γ	1377.20 18	0.18 4
γ	1403.4 3	0.083 4
γ	1415.60 21	0.030 6
γ E2	1465.74 15	0.72 14
γ	1476.78 17	0.039 8
γ [E1]	1507.06 16	0.20 4
γ	1515.17 18	0.21 4
γ [E2]	1525.69 19	0.020 4
γ	1539.0 6	0.035 7
γ [E2]	1559.98 15	0.99 20
γ	1575.82 22	0.030 6
γ [E2]	1594.34 16	0.59 12
γ	1605.15 9	0.15 3
γ	1616.0 3	<0.10
γ	1623.9 3	0.048 5
γ [M1+E2]	1643.73 14	1.3 3
γ	1677.7 3	0.026 3
γ	1716.72 18	0.109 22
γ	1725.93 20	~0.020
γ	1743.3 3	0.068 5
γ	1755.11 19	0.109 22
γ	1778.6 4	0.050 7
γ	1804.00 17	0.27 5
γ	1819.4 5	0.040 6
γ [E2]	1828.65 18	0.030 6
γ	1842.39 18	0.36 7
γ	1868.8 3	0.116 7
γ	1886.6 3	0.041 5
γ	1907.93 10	1.16 6
γ	1926.23 10	1.70 9
γ	1940.3 7	0.020 5
γ	1950.8 3	0.062 6
γ	1991.0 6	~0.039
γ	2027.3 3	0.048 5
γ	2041.1 3	0.055 7
γ	2057.6 3	0.92 18
γ	2095.7 3	0.021 3
γ [M1+E2]	2178.99 16	0.21 4
γ	2192.9 3	0.110 10
γ [E2]	2238.99 14	1.18 24
γ	2265.6 10	0.020 8
γ	2285.1 5	0.021 6
γ	2329.7 5	0.011 3
γ	2350.1 3	0.028 3
γ	2397.33 20	0.038 3
γ	2421.0 5	0.009 3
γ [E2]	2454.06 15	0.38 8
γ [M1+E2]	2481.94 16	0.18 4
γ	2718.4 3	0.058 5
γ	2729.99 18	0.16 3
γ	2829.2 6	0.0148 20
γ	2916.4 3	0.158 20
γ [E2]	2943.61 14	0.039 8
γ	3059.9 8	0.013 3
γ	3082.9 8	0.013 3
γ	3107.4 8	0.0069 20
γ	3142.6 9	0.018 3
γ	3153.5 9	0.021 3
γ	3185.5 4	0.029 3
γ	3212.8 5	0.0118 20
γ	3226.3 6	0.035 3
γ	3439.0 4	0.0158 20
γ [E2]	3478.85 17	0.020 4
γ	3494.9 5	0.058 5
γ	3562.5 4	0.108 8
γ	3650.5 4	0.074 6
γ [E2]	3781.80 17	0.0020 4
γ	3795.2 15	0.0049 20
γ	3868.7 7	0.0109 20

Photons (^{114}Sb) (continued)

γ_{mode}	γ(keV)	γ(%)†
γ	4141.3 19	0.020 5
γ	4204.6 15	0.012 3
γ	4305.0 16	0.0059 20
γ	4475 2	~0.0039
γ	4547 3	0.0030 10
γ	4947 5	~0.0020
γ	4987 5	0.0030 10

† 1.0% uncert(syst)

Atomic Electrons (^{114}Sb)

⟨e⟩=3.2 2 keV

e_{bin}(keV)	⟨e⟩(keV)	e(%)
4	0.65	16.0 16
20	0.041	0.204 21
21	0.34	1.63 17
24	0.092	0.38 4
25	0.098	0.40 4
27 - 28	0.0191	0.069 6
186 - 215	0.067	0.035 9
262 - 292	0.013	0.0045 22
298	0.30	~0.10
316 - 361	0.10	0.029 11
371 - 413	0.008	0.0022 10
422 - 460	0.0056	0.00123 23
485 - 490	0.00082	0.00017 3
545 - 593	0.0057	0.0010 3
598 - 640	0.015	0.0024 5
649 - 688	0.07	~0.009
700 - 743	0.011	0.0015 6
758 - 787	0.0007	9 4 $\times10^{-5}$
828 - 857	0.0006	7 3 $\times10^{-5}$
858	0.219	0.0255 11
883 - 933	0.052	0.0058 4
935 - 981	0.0198	0.00209 20
986 - 1020	0.0051	0.00051 22
1042 - 1071	0.0054	0.00051 23
1093 - 1141	0.022	0.0020 4
1150 - 1175	0.0033	0.00028 5
1200 - 1247	0.0040	0.00032 7
1250 - 1264	0.00020	1.6 6 $\times10^{-5}$
1271	0.811	0.0638 14
1285	0.0049	0.00038 8
1295	0.0932	0.00720 16
1296 - 1338	0.0320	0.00246 5
1348 - 1386	0.0021	0.00015 5
1399 - 1448	0.0056	0.00039 7
1461 - 1559	0.0113	0.00075 10
1565 - 1649	0.0168	0.00105 14
1673 - 1754	0.0021	0.000121 23
1774 - 1868	0.0048	0.00027 6
1878 - 1962	0.032	0.0017 4
1987 - 2067	0.0053	0.00026 8
2091 - 2189	0.0018	8.4 14 $\times10^{-5}$
2192 - 2284	0.0071	0.00032 6
2301 - 2397	0.00038	1.6 3 $\times10^{-5}$
2417 - 2481	0.0031	0.000126 17
2689 - 2729	0.00088	3.3 8 $\times10^{-5}$
2800 - 2887	0.00059	2.1 6 $\times10^{-5}$
2912 - 2943	0.00026	9.1 14 $\times10^{-6}$
3031 - 3124	0.00024	7.9 12 $\times10^{-6}$
3138 - 3226	0.00029	9.1 13 $\times10^{-6}$
3410 - 3494	0.00034	9.8 15 $\times10^{-6}$
3533 - 3621	0.00058	1.6 3 $\times10^{-5}$
3646 - 3650	3.1 $\times10^{-5}$	8.5 19 $\times10^{-7}$
3753 - 3840	5.5 $\times10^{-5}$	1.4 3 $\times10^{-6}$
3864 - 3868	4.4 $\times10^{-6}$	1.1 3 $\times10^{-7}$
4112 - 4204	9.8 $\times10^{-5}$	2.4 5 $\times10^{-6}$
4276 - 4304	1.8 $\times10^{-5}$	4.2 15 $\times10^{-7}$
4446 - 4543	2.1 $\times10^{-5}$	4.6 15 $\times10^{-7}$
4918 - 4986	1.4 $\times10^{-5}$	2.9 8 $\times10^{-7}$

Continuous Radiation (^{114}Sb)

⟨β+⟩=1171 keV;⟨IB⟩=6.0 keV

E_{bin}(keV)		⟨ ⟩(keV)	(%)
0 - 10	β+	2.33 $\times10^{-6}$	2.81 $\times10^{-5}$
	IB	0.039	
10 - 20	β+	8.5 $\times10^{-5}$	0.00051
	IB	0.038	0.27
20 - 40	β+	0.00212	0.0065
	IB	0.085	0.30
40 - 100	β+	0.092	0.119
	IB	0.22	0.33
100 - 300	β+	4.30	1.95
	IB	0.64	0.36
300 - 600	β+	32.6	7.0
	IB	0.80	0.19
600 - 1300	β+	251	26.0
	IB	1.50	0.169
1300 - 2500	β+	670	36.7
	IB	1.8	0.101
2500 - 4590	β+	214	7.6
	IB	0.84	0.028
	Σβ+		79

$^{114}_{52}$Te(15.2 7 min)

Mode: ϵ

Δ: -81760 200 keV syst.

SpA: 1.08$\times10^{8}$ Ci/g

Prod: ^{112}Sn(^{4}He,2n); ^{112}Sn(^{3}He,n); protons on Sb

Photons (^{114}Te)

γ_{mode}	γ(keV)	γ(rel)
γ	45.828 20	15 5
γ	54.568 20	27 5
γ	56.48 4	4.0 8
γ	83.768 20	67 14
γ	90.248 20	100 18
γ	147.41 5	12 2
γ	188.11 4	18 2
γ	188.22 19	4 1
γ	234.49 9	4 1
γ	244.59 4	33 7
γ	264.0 3	2.6 1
γ	346.9 3	18 2
γ	379.7 3	13 2
γ	407.7 3	4 1
γ	437.1 3	8 2
γ	461.6 3	3.4 1
γ	479.08 10	26 3
γ	544.84 24	10 2
γ	600.2 3	12 2
γ	636.0 3	6.70 13
γ	708.4 6	~3
γ	726.58 18	43 6
γ	735.4 3	5 2
γ	778.1 3	4.4 2
γ	844.8 3	7 1
γ	907.0 8	8.3 2
γ	932.97 20	11 2
γ	1099.7 4	5 2
γ	1119.6 4	5 2
γ	1138.9 3	4.2 2
γ	1161.0 3	6.50 13
γ	1199.9 3	7 2
γ	1200.5 3	7.2 2
γ	1352.4 4	7.2 2
γ	1360.7 3	6.50 13
γ	1418.4 3	32 3
γ	1432.6 4	4 1
γ	1494.7 4	5 2
γ	1652.9 3	12 2
γ	1673.9 5	3 1

Photons (^{114}Te)
(continued)

γ_{mode}	γ(keV)	γ(rel)
γ	1708.3 *4*	2.5 *1*
γ	1781.2 *5*	4 *1*
γ	1803.7 *5*	5 *2*
γ	1841.0 *3*	17 *2*
γ	1897.5 *3*	38 *5*
γ	1902.3 *10*	7.3 *2*
γ	1959.7 *4*	2.3 *1*
γ	2017.1 *4*	1.7 *1*

$^{114}_{53}$I (2.1 *2* s)

Mode: ϵ
Δ: -72860 *280* keV syst
SpA: 4.0×10^{10} Ci/g
Prod: ^{58}Ni on ^{63}Cu

Photons (^{114}I)

γ_{mode}	γ(keV)
γ	682.6 *6*
γ E2	708.8 *6*

$^{114}_{54}$Xe(10.0 *4* s)

Mode: ϵ
Δ: -66910 *290* keV syst
SpA: 9.6×10^{9} Ci/g
Prod: ^{58}Ni on ^{58}Ni

Photons (^{114}Xe)

γ_{mode}	γ(keV)	γ(rel)
γ	103.1 *2*	48 *12*
γ	161.6 *2*	64 *19*
γ	308.5 *2*	100 *20*

$^{114}_{55}$Cs(570 *20* ms)

Mode: ϵ(99.982 *6* %), ϵp(7 *2* %), $\epsilon\alpha$(0.16 *6* %), α(0.018 *6* %)
Δ: -54710 *510* keV syst
SpA: 1.00×10^{11} Ci/g
Prod: ^{58}Ni on ^{58}Ni

Alpha Particles (^{114}Cs)

α(keV)
3239 *30*

Photons (^{114}Cs)

γ_{mode}	γ(keV)	γ(rel)
γ_ϵ[E2]	449.53 *16*	100 *2*
γ_ϵ	618.3 *2*	5.0 *7*
γ_ϵ[M1+E2]	698.03 *16*	11.8 *9*
γ_ϵ	758.2 *2*	3.0 *8*
γ_ϵ[E2]	1147.57 *16*	2.7 *9*

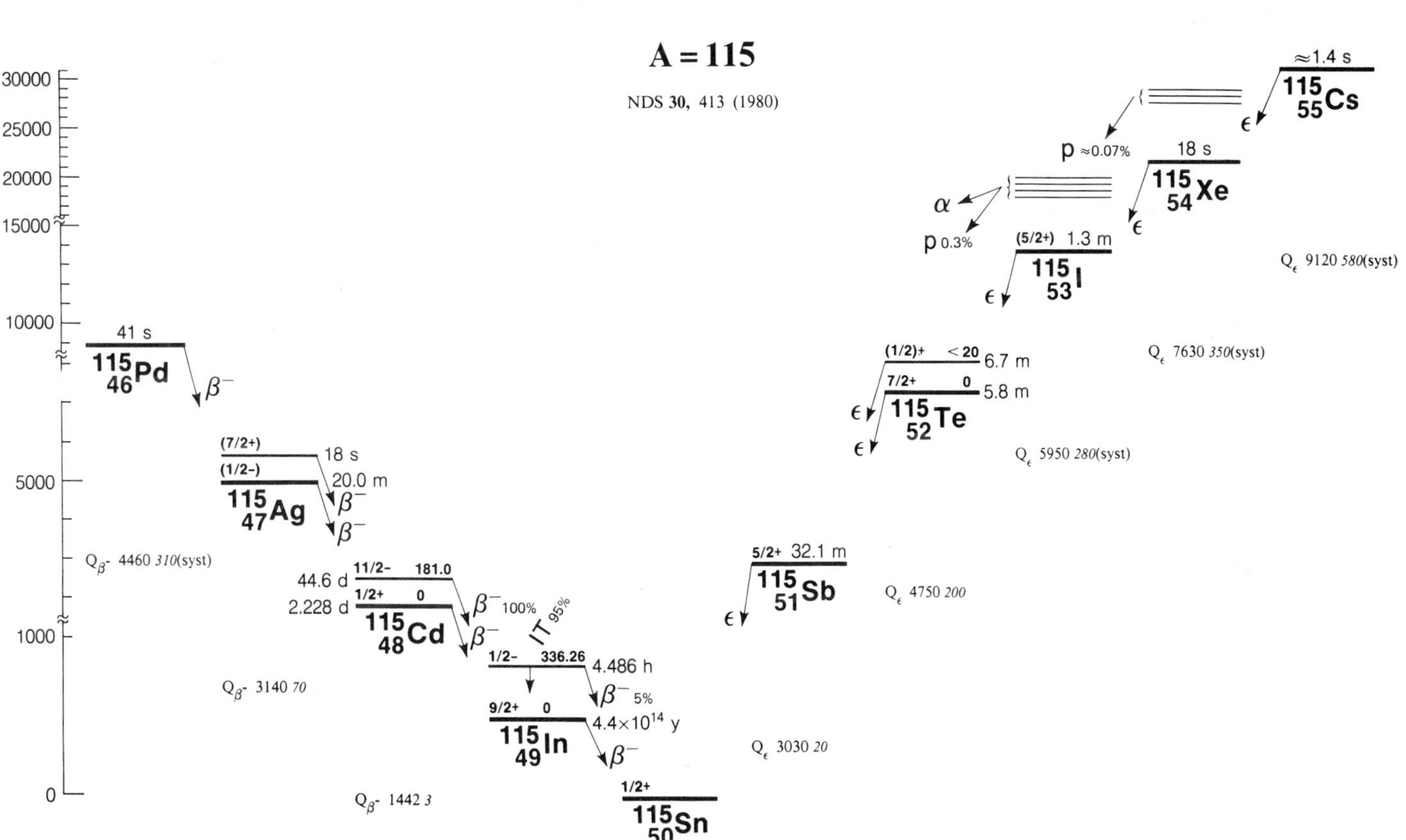

$^{115}_{46}$Pd(41 *3* s)

Mode: β-

Δ: -80490 *300* keV syst

SpA: 2.37×10^{9} Ci/g

Prod: fission

Photons (^{115}Pd)

γ_{mode}	γ(keV)
γ (E2)	47.2 *5*
γ	62.5 *10*
γ	87.7 *8*
γ	113.5 *10*
γ	255.3 *8*
γ	303 *1*
γ	326.5 *10*
γ	343.0 *8*
γ	423 *1*
γ	445 *1*
γ	594 *1*

$^{115}_{47}$Ag(20.0 *5* min)

Mode: β-

Δ: -84950 *70* keV

SpA: 8.17×10^{7} Ci/g

Prod: fission; ^{116}Cd(γ,p)

Photons (^{115}Ag)

⟨γ⟩=483 *10* keV

γ_{mode}	γ(keV)	γ(%)
Cd $K_{\alpha2}$	22.984	0.55 *5*
Cd $K_{\alpha1}$	23.174	1.03 *10*
Cd $K_{\beta1}$'	26.085	0.262 *25*
Cd $K_{\beta2}$'	26.801	0.054 *5*
γ [M1]	113.1 *3*	0.067 *13*
γ M1	131.49 *9*	2.9 *7*
γ E2(+M1)	212.73 *10*	4.4 *5*
γ E2+~17%M1	229.02 *7*	18
γ [M1+E2]	236.06 *19*	0.83 *7*
γ [M1]	243.65 *11*	0.40 *9*
γ [E1]	246.99 *21*	~0.18
γ [M1+E2]	275.7 *4*	~0.07
γ [E2]	302.64 *17*	0.70 *7*
γ [M1+E2]	325.98 *8*	2.03 *22*
γ [E2]	360.51 *10*	0.33 *4*
γ [M1+E2]	372.10 *8*	2.09 *23*
γ [M1]	388.84 *22*	0.47 *5*
γ [M1+E2]	416.10 *15*	0.23 *3*
γ [M1+E2]	420.03 *11*	0.128 *23*
γ [M1]	472.67 *9*	4.0 *5*
γ [M1+E2]	507.16 *19*	1.49 *18*
γ [M1+E2]	547.59 *14*	0.27 *4*
γ	564.9 *5*	0.117 *25*
γ [E1]	584.60 *23*	0.15 *3*
γ [E2]	602.10 *18*	0.092 *20*
γ	626.9 *5*	0.043 *11*
γ	638.4 *4*	0.09 *4*
γ [M1]	649.04 *9*	3.0 *4*
γ	653.5 *3*	~0.036
γ	670.5 *3*	0.140 *23*
γ [E2]	698.08 *8*	2.2 *3*
γ	717.0 *3*	0.07 *2*
γ [E1]	731.25 *22*	0.054 *16*
γ	751.5 *3*	0.115 *23*
γ	755.6 *5*	0.047 *14*
γ	762.3 *7*	0.045 *9*
γ	765.7 *3*	0.058 *11*

Photons (^{115}Ag) (continued)

γ_{mode}	γ(keV)	γ(%)
γ [M1+E2]	776.61 *13*	0.54 *9*
γ	798.0 *5*	0.032 *4*
γ	801.6 *5*	0.059 *11*
γ	828.9 *3*	0.146 *14*
γ	838.74 *25*	0.050 *9*
γ	844.44 *25*	0.041 *7*
γ	850.9 *4*	0.139 *16*
γ [E1]	862.74 *21*	0.128 *16*
γ	869.2 *5*	0.032 *7*
γ	879.5 *5*	0.131 *16*
γ	888.4 *5*	0.056 *9*
γ	897.1 *3*	0.09 *4*
γ	920.7 *5*	0.068 *20*
γ	931.7 *3*	0.173 *18*
γ	948.4 *5*	0.027 *5*
γ	956.59 *25*	0.041 *5*
γ [M1]	962.61 *17*	0.54 *5*
γ	973.9 *5*	0.112 *14*
γ	995.2 *3*	0.160 *18*
γ	1001.1 *3*	0.056 *7*
γ	1005.1 *5*	0.036 *5*
γ	1010.2 *5*	0.065 *9*
γ	1022.1 *5*	0.076 *9*
γ	1029.7 *3*	0.126 *16*
γ	1049.6 *6*	0.081 *13*
γ	1056.5 *6*	0.184 *20*
γ	1064.0 *6*	0.092 *14*
γ	1069.8 *6*	0.047 *11*
γ	1088.09 *24*	0.058 *11*
γ [E1]	1091.76 *20*	0.17 *5*
γ	1106.2 *7*	0.047 *11*
γ	1126.15 *25*	0.229 *22*
γ [M1]	1150.56 *21*	0.234 *22*
γ	1183.7 *5*	0.29 *9*
γ [E1]	1193.24 *20*	0.054 *7*
γ [M1+E2]	1222.51 *23*	0.117 *13*
γ	1234.2 *5*	0.056 *9*
γ	1242.1 *5*	0.038 *7*
γ	1256.34 *24*	0.126 *14*
γ	1308.7 *7*	0.047 *7*
γ	1317.10 *24*	0.086 *11*
γ [M1+E2]	1336.56 *19*	0.050 *9*
γ	1357.7 *7*	0.110 *11*
γ [E1]	1379.24 *16*	0.60 *5*
γ	1394.67 *22*	0.171 *16*
γ [E1]	1406.49 *23*	0.216 *20*
γ [M1+E2]	1436.18 *22*	0.090 *11*
γ [M1]	1464.12 *17*	0.48 *4*
γ	1485.36 *24*	0.130 *14*
γ [E1]	1506.80 *14*	1.19 *13*
γ	1528 *1*	0.020 *4*
γ	1531.1 *10*	0.045 *7*
γ	1534.7 *3*	0.063 *9*
γ [E1]	1564.9 *3*	0.040 *7*
γ [M1+E2]	1594.60 *23*	0.056 *9*
γ [M1+E2]	1606.00 *23*	0.256 *25*
γ [M1+E2]	1640.49 *17*	0.58 *9*
γ [E1]	1648.68 *21*	0.30 *4*
γ	1663.7 *3*	0.076 *23*
γ [E1]	1665.22 *20*	0.12 *3*
γ	1676.7 *5*	0.162 *16*
γ [E1]	1683.18 *14*	0.032 *7*
γ	1700.4 *7*	0.029 *7*
γ	1711.1 *3*	0.198 *20*
γ	1743.1 *5*	0.052 *9*
γ [E2]	1752.65 *17*	0.092 *11*
γ	1766.77 *22*	0.090 *13*
γ [E1]	1795.33 *14*	0.32 *3*
γ [E1]	1807.10 *23*	0.032 *5*
γ	1823.3 *3*	0.045 *7*
γ [E1]	1841.59 *19*	1.80 *16*
γ [M1+E2]	1884.14 *16*	0.34 *3*
γ	1910.67 *20*	0.277 *25*
γ [E1]	1926.82 *13*	1.35 *11*
γ	1979.26 *24*	0.058 *7*
γ	2011.3 *5*	0.036 *5*
γ	2022.83 *21*	0.104 *13*
γ	2041.4 *5*	0.043 *7*
γ	2074.0 *7*	0.016 *4*
γ	2082.0 *5*	0.126 *13*
γ	2096.2 *3*	0.018 *5*
γ [M1]	2113.15 *15*	1.17 *11*
γ	2126.9 *5*	0.049 *7*
γ	2137.4 *5*	0.0216 *18*
γ [E1]	2155.84 *12*	2.81 *23*
γ	2167.2 *7*	0.023 *4*
γ	2173.1 *3*	0.032 *4*
γ	2183.8 *3*	0.038 *4*
γ	2186.7 *3*	0.0144 *18*
γ	2208.4 *3*	0.013 *4*
γ	2257.4 *2*	0.018 *4*
γ	2265.0 *3*	0.022 *4*
γ	2296.3 *7*	0.056 *7*
γ	2302.2 *7*	0.032 *5*
γ [E1]	2314.25 *18*	0.043 *5*
γ	2339.8 *3*	0.067 *7*
γ	2374.1 *7*	0.014 *4*
γ	2383.33 *20*	0.216 *22*
γ	2414.9 *5*	0.022 *4*
γ	2430.3 *3*	0.018 *4*
γ	2435.4 *7*	0.050 *5*
γ	2442.6 *7*	0.023 *4*
γ	2451.21 *24*	0.076 *9*
γ	2486.42 *20*	0.079 *9*
γ	2494.0 *3*	0.133 *13*
γ	2526.8 *3*	0.094 *11*
γ	2568.9 *3*	0.054 *7*
γ	2635.8 *5*	0.011 *4*
γ	2659.3 *3*	0.072 *7*
γ	2680.22 *25*	0.014 *4*
γ	2713.8 *5*	0.0072 *18*
γ	2906.2 *3*	0.056 *5*

Atomic Electrons (^{115}Ag)

⟨e⟩=5.3 *4* keV

e_{bin}(keV)	⟨e⟩(keV)	e(%)
4 - 26	0.137	2.17 *25*
86	0.017	0.019 *4*
105	0.57	0.55 *14*
109 - 131	0.111	0.087 *16*
186	0.57	0.31 *9*
202	2.3	1.14 *13*
209	0.18	0.09 *3*
212 - 220	0.059	0.027 *5*
225	0.40	0.179 *20*
228 - 276	0.178	0.073 *5*
299	0.124	0.042 *7*
302 - 345	0.134	0.039 *4*
356 - 393	0.054	0.0146 *10*
412 - 446	0.122	0.027 *4*
469 - 507	0.066	0.0137 *12*
521 - 565	0.0107	0.0020 *3*
575 - 623	0.063	0.0101 *13*
626 - 671	0.047	0.0071 *7*
690 - 739	0.0098	0.00139 *16*
747 - 794	0.0105	0.00139 *21*
797 - 844	0.0050	0.00061 *13*
847 - 896	0.0038	0.00044 *10*
897 - 945	0.0089	0.00096 *11*
947 - 995	0.0053	0.00055 *9*
997 - 1046	0.0044	0.00043 *9*
1049 - 1092	0.0019	0.00018 *3*
1099 - 1147	0.0044	0.00040 *7*
1150 - 1196	0.0034	0.00029 *9*
1208 - 1256	0.0017	0.00014 *4*
1282 - 1331	0.0019	0.00015 *3*
1333 - 1380	0.0040	0.00029 *3*
1391 - 1437	0.0045	0.000317 *25*
1459 - 1538	0.0063	0.00042 *4*
1561 - 1660	0.0089	0.00055 *5*
1661 - 1752	0.0026	0.000152 *25*
1763 - 1857	0.0087	0.00048 *3*
1880 - 1979	0.0059	0.00031 *3*
1985 - 2081	0.00152	7.5 *13* ×10^{-5}
2086 - 2186	0.0156	0.00074 *4*
2204 - 2302	0.00063	2.8 *4* ×10^{-5}
2310 - 2409	0.0015	6.4 *12* ×10^{-5}

Atomic Electrons (^{115}Ag) (continued)

e_{bin}(keV)	⟨e⟩(keV)	e(%)
2411 - 2500	0.00155	6.3 9 $\times10^{-5}$
2523 - 2609	0.00028	1.10 25 $\times10^{-5}$
2632 - 2713	0.00035	1.3 3 $\times10^{-5}$
2880 - 2906	0.00019	6.6 17 $\times10^{-6}$

Continuous Radiation (^{115}Ag)

⟨β-⟩=1106 keV;⟨IB⟩=2.8 keV

E_{bin}(keV)		⟨ ⟩(keV)	(%)
0 - 10	β-	0.0194	0.387
	IB	0.039	
10 - 20	β-	0.059	0.393
	IB	0.038	0.27
20 - 40	β-	0.241	0.80
	IB	0.075	0.26
40 - 100	β-	1.79	2.55
	IB	0.21	0.33
100 - 300	β-	19.6	9.7
	IB	0.60	0.34
300 - 600	β-	69	15.4
	IB	0.66	0.155
600 - 1300	β-	304	32.2
	IB	0.85	0.099
1300 - 2500	β-	644	36.2
	IB	0.34	0.021
2500 - 3140	β-	67	2.50
	IB	0.0040	0.000153

$^{115}_{47}$Ag(18.0 7 s)

Mode: β-

Δ: >-84950 70 keV

SpA: 5.35$\times10^{9}$ Ci/g

Prod: fission

Photons (^{115}Ag)

γ_{mode}	γ(keV)
γ M1	131.49 9
γ	213.3 10
γ E2+~17%M1	229.02 7
γ [E2]	360.51 10
γ	372.5 10
γ	388.6 4
γ [M1]	472.67 9
γ	699 1
γ	749.6 10

$^{115}_{48}$Cd(2.228 4 d)

Mode: β-

Δ: -88092 3 keV

SpA: 5.0973$\times10^{5}$ Ci/g

Prod: ^{114}Cd(n,γ)

Photons (^{115}Cd)

⟨γ⟩=370 5 keV

γ_{mode}	γ(keV)	γ(%)†
In L_ℓ	2.905	0.065 11
In L_η	3.112	0.030 5
In L_α	3.286	1.8 3
In L_β	3.570	1.31 22
In L_γ	3.985	0.134 24
In $K_{\alpha2}$	24.002	11.8 5
In $K_{\alpha1}$	24.210	22.0 8
In $K_{\beta1}'$	27.265	5.68 22
In $K_{\beta2}'$	28.022	1.19 5
γ M1+2.9%E2	35.548 6	0.421 8
γ E1	231.4546 24	0.740 14
γ [E1]	251.092 24	~9 $\times10^{-5}$
γ M1+1.1%E2	260.9080 24	1.94 4
γ [E1]	267.003 6	0.092 3
γ [E1]	328.358 13	0.0033 5
γ M4	336.258 18	50.1 *
γ [E1]	344.269 21	~1 $\times10^{-5}$
γ [E1]	363.906 12	0.0061 6
γ E1	492.3620 25	8.03 14
γ E1	527.910 6	27.5 6
γ [M1+E2]	595.360 11	0.0017 2
γ	690.291 6	0.00061 6
γ [E2]	705.19 25	8 2 $\times10^{-5}$
γ [M1+E2]	856.266 11	0.0022 1
γ E2	941.432 11	7 1 $\times10^{-5}$
γ	951.198 6	0.00028 3

† 4.0% uncert(syst)

* with ^{115}In(4.486 h) in equilib

Atomic Electrons (^{115}Cd)

⟨e⟩=177 4 keV

e_{bin}(keV)	⟨e⟩(keV)	e(%)
4 - 36	3.7	52 4
204 - 251	0.177	0.0774 19
257 - 300	0.0259	0.0101 3
308	133.6	43.3 12
316 - 328	9.2 $\times10^{-6}$	2.8 3 $\times10^{-6}$
332	25.2	7.59 21
333	5.69	1.71 5
335 - 364	7.78	2.32 5
464 - 500	0.328	0.0668 14
524 - 567	0.0369	0.00703 16
591 - 595	6.2 $\times10^{-6}$	1.05 15 $\times10^{-6}$
662 - 705	1.1 $\times10^{-5}$	1.6 6 $\times10^{-6}$
828 - 856	3.5 $\times10^{-5}$	4.2 4 $\times10^{-6}$
913 - 951	3.8 $\times10^{-6}$	4.1 14 $\times10^{-7}$

Continuous Radiation (^{115}Cd)

⟨β-⟩=316 keV;⟨IB⟩=0.32 keV

E_{bin}(keV)		⟨ ⟩(keV)	(%)
0 - 10	β-	0.094	1.89
	IB	0.0149	
10 - 20	β-	0.284	1.89
	IB	0.0142	0.099
20 - 40	β-	1.14	3.81
	IB	0.026	0.092
40 - 100	β-	8.1	11.5
	IB	0.066	0.103
100 - 300	β-	70	35.9
	IB	0.128	0.076
300 - 600	β-	136	31.7
	IB	0.065	0.0161
600 - 1106	β-	100	13.5
	IB	0.0110	0.00161

$^{115}_{48}$Cd(44.6 3 d)

Mode: β-

Δ: -87911 3 keV

SpA: 2.546$\times10^{4}$ Ci/g

Prod: ^{114}Cd(n,γ)

Photons (^{115}Cd)

⟨γ⟩=32.9 4 keV

γ_{mode}	γ(keV)	γ(%)
γ E2	105.229 18	0.00442 18
γ [M1]	136.74 7	~4 $\times10^{-6}$
γ M1+0.05%E2	158.022 12	0.0170 3
γ E1	231.4546 24	0.00088 3
γ M1+1.1%E2	260.9080 24	0.00092 8
γ M1	316.211 10	0.00248 10
γ M4	336.258 18	0.00494 16 *
γ [E1]	344.269 21	~4 $\times10^{-5}$
γ M1+35%E2	353.534 14	~0.00012
γ [E2]	370.62 7	~8 $\times10^{-6}$
γ [E2]	476.826 15	0.00010 2
γ [E1]	481.01 7	~8 $\times10^{-6}$
γ E2+5.9%M1	484.411 11	0.290 6
γ E1	492.3620 25	0.00960 19
γ [E2]	507.359 12	0.00026 2
γ [M1+E2]	514.944 7	0.00010 4
γ E2	544.682 15	2.0 4 $\times10^{-5}$
γ	544.71 20	6.0 12 $\times10^{-5}$?
γ M1+0.3%E2	933.847 4	2.00 4
γ E2	941.432 11	0.00024 2
γ E2	1078.17 7	<8 $\times10^{-5}$
γ M1+21%E2	1132.579 9	0.0856 17
γ E2	1290.599 10	0.890 18
γ [M1+E2]	1418.254 11	0.00184 8
γ E2+~1.5%M1	1448.786 6	0.0170 3
γ	1478.5 3	~1 $\times10^{-5}$?
γ M1+~47%E2	1486.109 11	0.00056 2

* with ^{115}In(4.486 h) in equilib

Continuous Radiation (^{115}Cd)

⟨β-⟩=602 keV;⟨IB⟩=0.96 keV

E_{bin}(keV)		⟨ ⟩(keV)	(%)
0 - 10	β-	0.0345	0.69
	IB	0.026	
10 - 20	β-	0.104	0.69
	IB	0.025	0.17
20 - 40	β-	0.423	1.41
	IB	0.048	0.167
40 - 100	β-	3.09	4.39
	IB	0.13	0.20
100 - 300	β-	33.4	16.5
	IB	0.32	0.18
300 - 600	β-	128	28.3
	IB	0.26	0.064
600 - 1300	β-	400	45.3
	IB	0.143	0.019
1300 - 1623	β-	37.8	2.73
	IB	0.00128	9.5 $\times10^{-5}$

$^{115}_{49}$In(4.41 25 $\times10^{14}$ yr)

Mode: β-

Δ: -89534 4 keV

Prod: natural source

%: 95.7 2

Continuous Radiation (^{115}In)

⟨β-⟩=153 keV;⟨IB⟩=0.080 keV

E_{bin}(keV)		⟨ ⟩(keV)	(%)
0 - 10	β-	0.198	3.95
	IB	0.0078	
10 - 20	β-	0.59	3.92
	IB	0.0071	0.050
20 - 40	β-	2.32	7.7
	IB	0.0124	0.043
40 - 100	β-	15.3	22.0
	IB	0.026	0.041
100 - 300	β-	96	51
	IB	0.026	0.017
300 - 497	β-	38.5	10.9
	IB	0.00120	0.00036

$^{115}_{49}$In(4.486 *4* h)

Mode: IT(95.0 *7* %), β-(5.0 *7* %)

Δ: -89198 *4* keV

SpA: 6.074×10^{6} Ci/g

Prod: daughter ^{115}Cd(2.228 d); ^{115}In(n,n'); ^{115}In(p,p'); ^{115}In(α,α')

Photons (^{115}In)

⟨γ⟩=163 *7* keV

γ_{mode}	γ(keV)	γ(%)†
In L_ℓ	2.905	0.054 *10*
In L_η	3.112	0.025 *4*
In L_α	3.286	1.49 *25*
In L_β	3.570	1.09 *19*
In L_γ	3.986	0.112 *21*
In $K_{\alpha 2}$	24.002	9.8 *6*
In $K_{\alpha 1}$	24.210	18.4 *11*
In $K_{\beta 1}$'	27.265	4.7 *3*
In $K_{\beta 2}$'	28.022	0.99 *6*
γ_{IT} M4	336.258 *18*	45.8
$\gamma_{\beta\text{-}}$M1+4.3%E2	497.36 *3*	0.047

† uncert(syst): 4.8% for IT, 15% for β-

Atomic Electrons (^{115}In)

⟨e⟩=160 *7* keV

e_{bin}(keV)	⟨e⟩(keV)	e(%)
4	1.29	34 *4*
19	0.086	0.44 *5*
20	0.72	3.6 *4*
23	0.189	0.82 *9*
24	0.201	0.85 *10*
26	0.0117	0.044 *5*
27	0.027	0.102 *11*
308	122.1	39.6 *21*
332	23.0	6.9 *4*
333	5.2	1.56 *8*
335	3.99	1.19 *6*
336	3.12	0.93 *5*

Continuous Radiation (^{115}In)

⟨β-⟩=14.1 keV;⟨IB⟩=0.0123 keV

E_{bin}(keV)		⟨ ⟩(keV)	(%)
0 - 10	β-	0.00475	0.095
	IB	0.00068	
10 - 20	β-	0.0143	0.095
	IB	0.00065	0.0045
20 - 40	β-	0.058	0.193
	IB	0.00120	0.0042
40 - 100	β-	0.414	0.59
	IB	0.0029	0.0046
100 - 300	β-	3.84	1.94
	IB	0.0051	0.0031
300 - 600	β-	8.0	1.87
	IB	0.00169	0.00045
600 - 833	β-	1.76	0.265
	IB	5.5×10^{-5}	8.8×10^{-6}

$^{115}_{50}$Sn(stable)

Δ: -90032 *3* keV

%: 0.36 *1*

$^{115}_{51}$Sb(32.1 *3* min)

Mode: ε

Δ: -87002 *20* keV

SpA: 5.09×10^{7} Ci/g

Prod: ^{114}Sn(d,n); ^{116}Sn(p,2n); ^{113}In(α,2n)

Photons (^{115}Sb)

⟨γ⟩=547 *24* keV

γ_{mode}	γ(keV)	γ(%)†
Sn L_ℓ	3.045	0.071 *10*
Sn L_η	3.272	0.037 *4*
Sn L_α	3.443	1.97 *22*
Sn L_β	3.742	1.73 *21*
Sn L_γ	4.220	0.20 *3*
Sn $K_{\alpha 2}$	25.044	14.6 *4*
Sn $K_{\alpha 1}$	25.271	27.2 *8*
Sn $K_{\beta 1}$'	28.474	7.12 *22*
Sn $K_{\beta 2}$'	29.275	1.50 *5*
γ E2	115.52 *8*	0.235 *20*
γ M1+4.7%E2	136.9 *4*	~0.0007
γ M1+17%E2	293.40 *23*	~0.010
γ M1+6.6%E2	373.61 *12*	0.041 *10*
γ M1+~23%E2	430.3 *3*	~0.00033
γ M1+0.2%E2	489.13 *12*	1.3 *3*
γ M1+4.3%E2	497.36 *3*	98
γ [E2]	667.01 *22*	~0.010
γ E2,M1	747.61 *14*	0.196 *20*
γ M1+~37%E2	782.53 *20*	~0.025
γ [M1+E2]	803.9 *3*	~0.0028
γ M1+2.9%E2	919.4 *3*	~0.012
γ E2	986.49 *12*	0.353 *20*
γ [E2]	1020.93 *13*	0.043 *5*
γ	1097.29 *20*	0.028 *4*
γ [M1+E2]	1121.22 *13*	0.157 *10*
γ [M1+E2]	1136.45 *12*	0.127 *10*
γ	1206.87 *25*	0.040 *4*
γ	1212.09 *22*	0.009 *3*
γ [M1+E2]	1236.74 *13*	0.58 *3*
γ	1243.50 *22*	0.044 *4*
γ E2+17%M1	1279.89 *20*	0.284 *20*
γ	1327.61 *21*	0.024 *3*
γ E2	1360.1 *3*	0.062 *5*
γ	1378.1 *3*	0.013 *3*
γ E2	1416.8 *3*	0.040 *4*

Photons (^{115}Sb)
(continued)

γ_{mode}	γ(keV)	γ(%)†
γ	1471.3 *4*	0.016 *3*
γ	1476.4 *4*	0.0098 *20*
γ	1543.1 *4*	0.0098 *20*
γ (E2)	1562.8 *3*	0.019 *4*
γ	1580.47 *23*	0.009 *3*
γ (M1)	1633.81 *12*	0.353 *20*
γ	1658.3 *4*	0.0049 *20*
γ	1695.99 *22*	0.090 *6*
γ	1717.5 *5*	0.0049 *20*
γ	1732.62 *23*	0.049 *15*
γ [E2]	1734.10 *13*	0.055 *17*
γ	1817.3 *5*	0.0147 *20*
γ	1824.97 *21*	0.051 *4*
γ	1854.7 *6*	0.0049 *20*
γ	1857.5 *3*	~0.010
γ	1867.22 *25*	0.060 *4*
γ	1938.0 *5*	0.0088 *20*
γ	1990.0 *9*	0.0078 *20*
γ [E2]	2060.1 *3*	0.0059 *20*
γ	2193.35 *23*	0.019 *3*
γ	2229.98 *23*	0.127 *10*
γ	2364.58 *25*	~0.0039
γ	2589.8 *10*	~0.0020

† 5.1% uncert(syst)

Atomic Electrons (^{115}Sb)

⟨e⟩=7.8 *3* keV

e_{bin}(keV)	⟨e⟩(keV)	e(%)
4	1.95	48 *5*
20	0.124	0.61 *6*
21	1.03	4.9 *5*
24	0.27	1.14 *12*
25	0.29	1.19 *12*
27 - 28	0.057	0.206 *19*
86 - 133	0.210	0.226 *15*
136 - 137	6.5×10^{-6}	4.8 *22* $\times 10^{-6}$
264 - 293	0.0008	0.00031 *14*
344 - 374	0.0024	0.00069 *14*
401 - 430	1.6×10^{-5}	3.9 *17* $\times 10^{-6}$
460	0.043	0.0094 *22*
468	3.25	0.69 *4*
485 - 489	0.0069	0.0014 *3*
493	0.420	0.085 *5*
496 - 497	0.102	0.0205 *9*
638 - 667	0.00021	3.3 *15* $\times 10^{-5}$
718 - 753	0.0044	0.00061 *8*
775 - 804	0.00012	1.5 *5* $\times 10^{-5}$
890 - 919	0.00020	2.3 *10* $\times 10^{-5}$
957 - 992	0.00498	0.000517 *25*
1016 - 1021	7.2×10^{-5}	7.0 *7* $\times 10^{-6}$
1068 - 1117	0.0035	0.00032 *3*
1120 - 1136	0.000252	2.23 *23* $\times 10^{-5}$
1178 - 1214	0.0063	0.00052 *6*
1232 - 1280	0.00371	0.000297 *16*
1298 - 1331	0.00067	5.0 *6* $\times 10^{-5}$
1349 - 1388	0.00047	3.4 *4* $\times 10^{-5}$
1412 - 1447	0.00020	1.4 *4* $\times 10^{-5}$
1467 - 1562	0.00028	1.9 *3* $\times 10^{-5}$
1576 - 1667	0.0036	0.000224 *16*
1688 - 1733	0.00078	4.6 *10* $\times 10^{-5}$
1788 - 1867	0.00099	5.5 *11* $\times 10^{-5}$
1909 - 1989	8.8×10^{-5}	4.6 *13* $\times 10^{-6}$
2031 - 2059	3.7×10^{-5}	1.8 *5* $\times 10^{-6}$
2164 - 2229	0.00070	3.2 *9* $\times 10^{-5}$
2335 - 2364	1.8×10^{-5}	$\sim 8 \times 10^{-7}$
2561 - 2589	8.3×10^{-6}	$\sim 3 \times 10^{-7}$

Continuous Radiation (^{115}Sb)

⟨β+⟩=224 keV;⟨IB⟩=2.7 keV

E_{bin}(keV)		⟨ ⟩(keV)	(%)
0 - 10	β+	4.62×10^{-6}	5.6×10^{-5}
	IB	0.0097	
10 - 20	β+	0.000167	0.00101
	IB	0.0101	0.069
20 - 40	β+	0.00415	0.0128
	IB	0.048	0.18
40 - 100	β+	0.177	0.228
	IB	0.052	0.080
100 - 300	β+	7.6	3.47
	IB	0.151	0.084
300 - 600	β+	48.2	10.5
	IB	0.27	0.062
600 - 1300	β+	161	18.5
	IB	1.08	0.116
1300 - 2500	β+	6.4	0.475
	IB	1.12	0.068
2500 - 2533	IB	3.9×10^{-6}	1.53×10^{-7}
	Σβ+		33

$^{115}_{52}$Te(5.8 *2* min)

Mode: ε

Δ: -82250 *200* keV

SpA: 2.816×10^{8} Ci/g

Prod: ^{112}Sn(α,n)

Photons (^{115}Te)

⟨γ⟩=1526 *52* keV

γ_{mode}	γ(keV)	γ(%)†
Sb L_ℓ	3.189	0.053 *7*
Sb L_η	3.437	0.026 *3*
Sb L_α	3.604	1.46 *16*
Sb L_β	3.931	1.27 *16*
Sb L_γ	4.438	0.153 *22*
Sb $K_{\alpha2}$	26.111	10.2 *3*
Sb $K_{\alpha1}$	26.359	19.0 *6*
Sb $K_{\beta1}$'	29.712	4.99 *15*
Sb $K_{\beta2}$'	30.561	1.06 *3*
γ [M1]	228.23 *4*	0.77 *10*
γ [M1]	281.97 *3*	0.99 *10*
γ M1+5.5%E2	374.48 *4*	3.17 *22*
γ	386.3 *3*	0.22 *6* *
γ M1+7.6%E2	428.22 *5*	1.09 *13*
γ M1	568.39 *5*	2.8 *3*
γ M2	576.89 *8*	0.193 *15*
γ M1+E2	603.27 *4*	4.42 *13*
γ (E2)	634.01 *9*	0.93 *10*
γ M1+0.6%E2	657.01 *3*	6.82 *19*
γ M1(+E2)	723.58 *3*	30
γ	804.0 *4*	0.22 *6* *
γ	921.3 *3*	0.42 *10* *
γ	942.87 *5*	1.82 *13*
γ	996.61 *5*	1.12 *13*
γ (M1+E2)	1012.3 *4*	1.66 *13* ?
γ	1023.00 *9*	0.74 *10*
γ	1051.4 *6*	0.32 *10* *
γ	1062.22 *9*	0.61 *10*
γ [M1+E2]	1071.60 *23*	~0.24
γ	1088.61 *11*	0.80 *13*
γ M1(+E2)	1098.62 *3*	16.3 *10*
γ	1213.6 *3*	0.61 *10* ?
γ	1224.83 *5*	0.77 *10*
γ	1290.45 *8*	5.9 *4*
γ E3	1300.46 *7*	2.40 *18*
γ	1317.48 *25*	~0.13
γ E2	1326.85 *4*	22.7 *13*
γ	1369.1 *4*	0.16 *6* *
γ E2	1380.59 *3*	23.0 *13*
γ	1437.3 *3*	0.38 *6* *
γ	1446.7 *4*	0.26 *6* *

Photons (^{115}Te)

(continued)

γ_{mode}	γ(keV)	γ(%)†
γ	1589.62 *16*	0.21 *3*
γ	1599.88 *5*	2.62 *22*
γ	1616.6 *3*	0.11 *3*
γ	1665.49 *8*	0.93 *10*
γ	1685.5 *3*	0.45 *10* *
γ	1743.5 *6*	0.16 *6* *
γ	1837.3 *3*	0.51 *10* *
γ	1953.81 *22*	0.42 *6* *
γ	2019.5 *10*	~0.13 *
γ	2118.6 *8*	0.12 *3* *
γ	2130.8 *4*	0.51 *6* *
γ	2389.06 *8*	0.24 *5*
γ	2459.3 *11*	0.12 *3* *
γ	2467.5 *11*	~0.08 *
γ	2481.6 *6*	0.32 *10* *
γ	2503.2 *10*	0.11 *3* *
γ	2511.2 *11*	0.15 *3* *
γ	2688.23 *16*	0.80 *10*
γ	2717.8 *7*	0.12 *3* *
γ	2746.2 *9*	0.12 *3* *
γ	3411.6 *13*	0.048 *22* *
γ	3447.9 *8*	0.05 *3* *
γ	3529.4 *11*	~0.045 *
γ	3559.6 *9*	0.061 *13* *

† 19% uncert(syst)

* with ^{115}Te(5.8 or 6.7 min)

Atomic Electrons (^{115}Te)

⟨e⟩=5.15 *21* keV

e_{bin}(keV)	⟨e⟩(keV)	e(%)
4	1.22	29 *3*
5	0.180	3.8 *4*
21	0.187	0.88 *9*
22	0.59	2.7 *3*
25	0.189	0.75 *8*
26	0.202	0.79 *8*
29 - 30	0.040	0.137 *12*
198 - 228	0.101	0.050 *6*
251 - 282	0.095	0.037 *3*
344	0.170	0.050 *4*
356 - 398	0.086	0.0223 *20*
424 - 428	0.0080	0.00189 *19*
538 - 572	0.117	0.0215 *19*
573	0.112	0.0196 *20*
576 - 604	0.039	0.0065 *6*
627	0.169	0.0270 *9*
629 - 657	0.0302	0.00464 *12*
693	0.59	0.086 *20*
719 - 724	0.096	0.013 *3*
774 - 804	0.0033	~0.00042
891 - 939	0.027	0.0029 *11*
942 - 982	0.033	0.0034 *7*
992 - 1041	0.025	0.0024 *5*
1047 - 1067	0.009	0.0008 *4*
1068	0.19	0.0179 *25*
1071 - 1099	0.031	0.0028 *4*
1183 - 1225	0.012	0.0010 *3*
1260 - 1290	0.091	0.0072 *16*
1296	0.199	0.0154 *9*
1300 - 1339	0.0323	0.00244 *12*
1350	0.189	0.0140 *8*
1364 - 1407	0.0316	0.00229 *13*
1416 - 1447	0.0023	0.00016 *6*
1559 - 1655	0.028	0.0018 *5*
1661 - 1743	0.0022	0.00013 *4*
1807 - 1837	0.0031	0.00017 *7*
1923 - 2019	0.0031	0.00016 *5*
2088 - 2130	0.0034	0.00016 *5*
2359 - 2455	0.0033	0.00014 *3*
2458 - 2510	0.0015	5.9 *13* $\times10^{-5}$
2658 - 2745	0.0046	0.00017 *4*
3381 - 3447	0.00038	1.1 *4* $\times10^{-5}$
3499 - 3559	0.00038	1.1 *3* $\times10^{-5}$

Continuous Radiation (^{115}Te)

⟨β+⟩=615 keV;⟨IB⟩=5.1 keV

E_{bin}(keV)		⟨ ⟩(keV)	(%)
0 - 10	β+	2.27×10^{-6}	2.74×10^{-5}
	IB	0.022	
10 - 20	β+	8.4×10^{-5}	0.00051
	IB	0.022	0.154
20 - 40	β+	0.00217	0.0066
	IB	0.066	0.24
40 - 100	β+	0.097	0.124
	IB	0.124	0.19
100 - 300	β+	4.54	2.06
	IB	0.36	0.20
300 - 600	β+	33.7	7.3
	IB	0.50	0.116
600 - 1300	β+	227	23.9
	IB	1.35	0.147
1300 - 2500	β+	326	19.2
	IB	2.1	0.118
2500 - 4750	β+	23.6	0.85
	IB	0.53	0.019
	Σβ+		53

$^{115}_{52}$Te(6.7 *4* min)

Mode: ε

Δ: <-82230 *200* keV

SpA: 2.44×10^{8} Ci/g

Prod: ^{112}Sn(α,n)

Photons (^{115}Te)*

⟨γ⟩=2019 *99* keV

γ_{mode}	γ(keV)	γ(%)†
Sb L_ℓ	3.189	0.061 *8*
Sb L_η	3.437	0.030 *3*
Sb L_α	3.604	1.68 *18*
Sb L_β	3.931	1.47 *18*
Sb L_γ	4.438	0.177 *25*
Sb $K_{\alpha2}$	26.111	11.8 *4*
Sb $K_{\alpha1}$	26.359	21.9 *7*
Sb $K_{\beta1}$'	29.712	5.76 *18*
Sb $K_{\beta2}$'	30.561	1.23 *4*
γ	303.0 *4*	2.7 *9*
γ	405.50 *15*	2.4 *9*
γ	548.63 *18*	3.6 *6*
γ	555.54 *21*	1.6 *4*
γ	570.22 *23*	~2
γ	610.55 *20*	3.9 *6*
γ	689.40 *23*	3.9 *9*
γ M1(+E2)	723.58 *3*	~18
γ [E2]	770.35 *10*	34.2 *12*
γ	780.55 *15*	2.1 *6*
γ	1031.85 *20*	7.8 *9*
γ [M1+E2]	1071.60 *23*	12.9 *9*
γ M1(+E2)	1098.62 *3*	~9
γ [E2]	1116.49 *19*	2.1 *9*
γ [M1+E2]	1143.5 *3*	~2
γ	1155.72 *21*	2.7 *9*
γ	1184.8 *8*	~0.6
γ	1205.5 *3*	3.3 *9*
γ	1279.25 *20*	10.2 *12*
γ	1350.76 *18*	7.8 *9*
γ	1408.0 *3*	3.3 *6*
γ [E2]	1491.54 *19*	3.3 *6*
γ	1504.12 *15*	10.2 *9*
γ [E2]	1561.22 *18*	3.6 *9*
γ	1654.7 *4*	6.3 *9*
γ [E2]	1936.26 *18*	3.3 *6*
γ	1986.0 *3*	1.3 *5*
γ	2104.30 *18*	8.4 *9*
γ [M1+E2]	2215.10 *19*	5.7 *9*
γ [M1+E2]	2659.83 *18*	0.54 *15*

† 33% uncert(syst)

* see also ^{115}Te(5.8 min)

Atomic Electrons (^{115}Te)

⟨e⟩=5.6 *3* keV

e_{bin}(keV)	⟨e⟩(keV)	e(%)
4	1.41	34 *3*
5	0.209	4.4 *5*
21	0.216	1.02 *10*
22	0.68	3.1 *3*
25	0.218	0.87 *9*
26	0.234	0.91 *9*
29 - 30	0.046	0.158 *14*
273 - 303	0.17	~0.06
375 - 405	0.09	~0.023
518 - 566	0.17	0.032 *11*
569 - 611	0.08	0.014 *7*
659 - 689	0.07	~0.010
693	0.35	~0.05
719 - 724	0.057	0.008 *4*
740	0.55	0.074 *4*
750 - 781	0.123	0.0161 *23*
1001 - 1032	0.08	0.008 *4*
1041	0.155	0.0149 *22*
1067 - 1116	0.19	0.017 *5*
1125 - 1156	0.033	0.0029 *12*
1175 - 1205	0.031	0.0026 *12*
1249 - 1279	0.09	0.007 *3*
1320 - 1351	0.06	0.0048 *19*
1378 - 1408	0.026	0.0019 *8*
1461 - 1560	0.13	0.0090 *20*
1624 - 1654	0.042	0.0026 *10*
1906 - 1985	0.030	0.0016 *3*
2074 - 2104	0.045	0.0022 *7*
2185 - 2214	0.037	0.0017 *3*
2629 - 2659	0.0030	0.00011 *3*

Continuous Radiation (^{115}Te)

⟨β+⟩=550 keV;⟨IB⟩=4.4 keV

E_{bin}(keV)		⟨ ⟩(keV)	(%)
0 - 10	β+	1.98×10^{-6}	2.38×10^{-5}
	IB	0.020	
10 - 20	β+	7.3×10^{-5}	0.000444
	IB	0.020	0.136
20 - 40	β+	0.00188	0.0058
	IB	0.064	0.23
40 - 100	β+	0.083	0.107
	IB	0.110	0.169
100 - 300	β+	3.87	1.76
	IB	0.33	0.18
300 - 600	β+	28.0	6.0
	IB	0.48	0.110
600 - 1300	β+	183	19.3
	IB	1.30	0.142
1300 - 2500	β+	319	18.2
	IB	1.65	0.095
2500 - 4000	β+	16.2	0.62
	IB	0.45	0.0157
	Σβ+		46

$^{115}_{53}$I (1.3 *2* min)

Mode: ε

Δ: -76300 *200* keV syst

SpA: 1.25×10^{9} Ci/g

Prod: protons on Ce; daughter ^{115}Xe

Photons (^{115}I)

γ_{mode}	γ(keV)
γ	275 *1*
γ	284 *1*
γ	460 *1*
γ	709 *1*

$^{115}_{54}$Xe(18 *4* s)

Mode: ε, εα, εp(0.34 *6* %)

Δ: -68670 *280* keV syst

SpA: 5.3×10^{9} Ci/g

Prod: protons on Ce; ^{104}Pd(^{16}O,5n)

εp: 2000-6000 range

εp/εα: 1100 *300*

Photons (^{115}Xe)

γ_{mode}	γ(keV)
$\gamma_{\epsilon p}$	710 *10*

$^{115}_{55}$Cs(1.4 *8* s)

Mode: ε, εp(~0.07 %)

Δ: -59550 *500* keV syst

SpA: 6×10^{10} Ci/g

Prod: protons on La

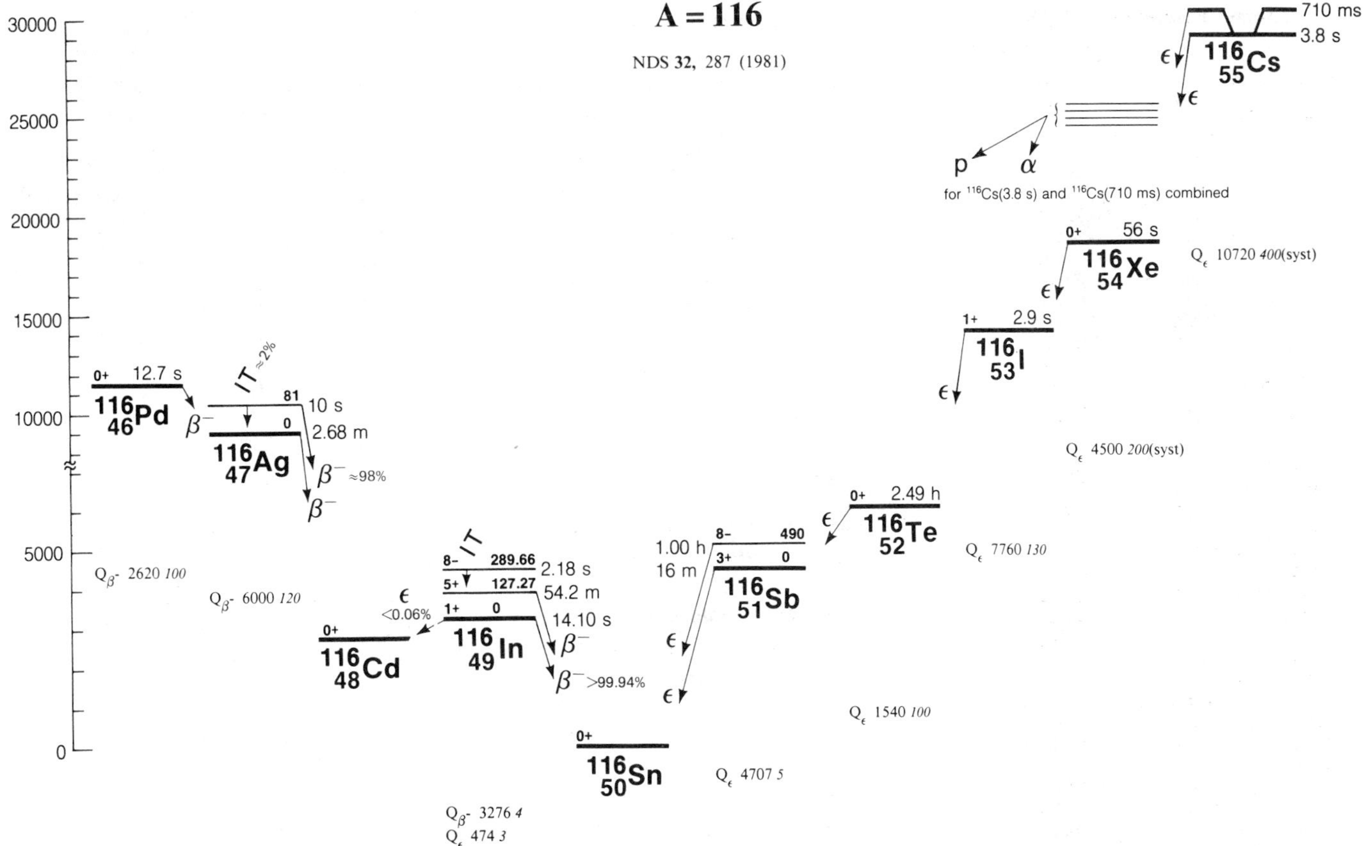

$^{116}_{46}$Pd(12.7 *4* s)

Mode: β-
Δ: -80110 *160* keV
SpA: 7.4×10^{9} Ci/g
Prod: fission

Photons (^{116}Pd)

⟨γ⟩=152 *6* keV

γ_{mode}	γ(keV)	γ(%)†
γ[M1]	101.53 *13*	8.5 *7*
γ[E1]	114.72 *10*	88
γ[M1]	177.88 *17*	11.6 *9*
γ[E1]	216.25 *15*	2.3 *4*
γ[M1]	279.41 *19*	6.2 *13*

† approximate

$^{116}_{47}$Ag(2.68 *1* min)

Mode: β-
Δ: -82720 *120* keV
SpA: 6.035×10^{8} Ci/g
Prod: fission; ^{116}Cd(n,p)

Photons (^{116}Ag)

⟨γ⟩=2149 *85* keV

γ_{mode}	γ(keV)	γ(%)†
Cd $K_{\alpha2}$	22.984	0.46 *7*
Cd $K_{\alpha1}$	23.174	0.86 *13*
Cd $K_{\beta1}$'	26.085	0.22 *3*
Cd $K_{\beta2}$'	26.801	0.046 *7*
γ[E2]	69.13 *17*	0.38 *8*
γ	422.7 *3*	0.46 *9*
γ E2	513.47 *12*	76 *4*
γ	609.8 *3*	0.70 *14*
γ	640.0 *3*	3.1 *6*
γ	643.8 *10*	1.7 *3*
γ	668.9 *3*	0.36 *7*
γ E2+30%M1	699.54 *11*	10.6 *18*
γ[M1+E2]	702.77 *18*	1.4 *3*
γ E2	705.76 *13*	2.4 *5*
γ[E1]	708.64 *22*	0.62 *12*
γ	712.0 *3*	0.52 *10*
γ[E2]	768.67 *19*	1.7 *4*
γ	781.9 *4*	0.44 *9*
γ[E2]	867.0 *3*	1.4 *3*
γ	869.2 *3*	0.91 *18*
γ	881.3 *4*	0.36 *7*
γ	993.6 *4*	1.7 *3*
γ	1080.3 *6*	0.53 *11*
γ	1128.41 *25*	2.5 *5*
γ	1133.7 *5*	0.24 *5*
γ	1152.3 *5*	0.23 *5*
γ	1173 *1*	2.5 *5*
γ[E2]	1213.01 *14*	7.3 *12*
γ	1304.2 *4*	5.5 *11*
γ[M1+E2]	1402.31 *20*	1.22 *24*
γ[E1]	1408.18 *22*	3.1 *6*
γ	1414.8 *4*	0.39 *8*
γ	1437.5 *3*	1.6 *3*
γ	1533.4 *4*	0.49 *10*
γ	1550.1 *3*	0.40 *8*

Photons (^{116}Ag) (continued)

γ_{mode}	γ(keV)	γ(%)†
γ	1569.7 *5*	0.99 *20*
γ	1575.1 *4*	0.28 *6*
γ	1604.7 *6*	1.14 *23*
γ	1631.2 *6*	0.38 *8*
γ	1641.9 *3*	2.0 *4*
γ	1649.0 *3*	0.66 *13*
γ	1681.6 *5*	1.7 *4*
γ	1705.4 *3*	0.42 *8*
γ	1801.5 *4*	0.53 *11*
γ	1836.5 *4*	0.33 *7*
γ	2003.8 *4*	1.9 *4*
γ	2091.3 *4*	1.06 *21*
γ	2134.3 *3*	1.7 *4*
γ	2151.0 *5*	0.31 *6*
γ	2246.5 *5*	2.4 *5*
γ	2289.1 *6*	1.7 *4*
γ	2315.0 *5*	0.51 *10*
γ	2348.5 *3*	1.14 *23*
γ	2478.1 *3*	11.6 *23*
γ	2501.0 *4*	0.99 *20*
γ	2661.8 *4*	4.2 *8*
γ	2703.3 *4*	0.76 *15*
γ	2703.5 *3*	7.0 *14*
γ	2759.9 *5*	0.84 *17*
γ	2828.5 *5*	1.3 *3*
γ	2833.8 *3*	2.5 *5*
γ	2919.7 *5*	0.84 *17*
γ	2958.3 *3*	0.68 *14*
γ	3217.0 *3*	0.43 *9*
γ	3347.2 *3*	0.19 *4*
γ	3398.5 *5*	0.31 *6*
γ	3916.3 *4*	0.17 *3*

† 5.3% uncert(syst)

Atomic Electrons (^{116}Ag)

⟨e⟩=4.38 *18* keV

e_{bin}(keV)	⟨e⟩(keV)	e(%)
4	0.069	1.9 *4*
19 - 26	0.060	0.29 *3*
42	0.59	1.4 *3*
65	0.21	0.33 *7*
66	0.18	0.27 *6*
68 - 69	0.098	0.143 *21*
396 - 423	0.013	~0.0032
487	1.98	0.406 *22*
509	0.226	0.0444 *24*
510 - 513	0.109	0.0214 *8*
583 - 617	0.08	0.012 *4*
636 - 669	0.016	0.0024 *7*
673	0.17	0.026 *4*
676 - 712	0.110	0.0160 *16*
742 - 782	0.033	0.0044 *7*
840 - 881	0.032	0.0038 *7*
967 - 994	0.015	0.0016 *7*
1054 - 1102	0.022	0.0020 *8*
1107 - 1152	0.023	0.0020 *7*
1169 - 1213	0.068	0.0058 *8*
1278 - 1304	0.037	0.0029 *12*
1376 - 1415	0.033	0.0023 *4*
1434 - 1533	0.0060	0.0004 *1*
1543 - 1641	0.030	0.0019 *4*
1645 - 1705	0.012	0.00072 *22*
1775 - 1836	0.0043	0.00024 *7*
1977 - 2065	0.013	0.00063 *18*
2087 - 2150	0.009	0.00045 *13*
2220 - 2314	0.019	0.00084 *19*
2322 - 2348	0.0046	0.00020 *7*
2451 - 2500	0.048	0.0020 *6*
2635 - 2733	0.046	0.0017 *4*
2756 - 2833	0.014	0.00048 *11*
2893 - 2958	0.0052	0.00018 *4*
3190 - 3216	0.0014	4.3 *12* $\times 10^{-5}$
3321 - 3398	0.0016	4.6 *9* $\times 10^{-5}$
3890 - 3916	0.00048	1.2 *3* $\times 10^{-5}$

Continuous Radiation (^{116}Ag)

⟨β-⟩=1675 keV;⟨IB⟩=5.4 keV

E_{bin}(keV)		⟨ ⟩(keV)	(%)
0 - 10	β-	0.0057	0.113
	IB	0.051	
10 - 20	β-	0.0174	0.116
	IB	0.051	0.35
20 - 40	β-	0.072	0.241
	IB	0.100	0.35
40 - 100	β-	0.57	0.80
	IB	0.29	0.44
100 - 300	β-	7.5	3.62
	IB	0.85	0.48
300 - 600	β-	37.3	8.1
	IB	1.03	0.24
600 - 1300	β-	264	27.3
	IB	1.58	0.18
1300 - 2500	β-	742	40.5
	IB	1.14	0.066
2500 - 5000	β-	619	19.1
	IB	0.32	0.0107
5000 - 5487	β-	5.4	0.106
	IB	0.000129	2.5×10^{-6}

$^{116}_{47}$Ag(10.4 *8* s)

Mode: β-(~98 %), IT(~2 %)

Δ: -82639 *120* keV

SpA: 9.0×10^{9} Ci/g

Prod: fission; ^{116}Cd(n,p)

Photons (^{116}Ag)

⟨γ⟩=1800 *150* keV

γ_{mode}	γ(keV)	γ(%)†
γ_{IT}E3	81 *1*	0.05
$\gamma_{\beta-}$	102.76 *19*	1.8 *4*
$\gamma_{\beta-}$	255.0 *3*	6.7 *13*
$\gamma_{\beta-}$	264.26 *24*	5.3 *11*
$\gamma_{\beta-}$	457.8 *5*	2.0 *4*
$\gamma_{\beta-}$E2	513.47 *12*	92
$\gamma_{\beta-}$	666.97 *24*	7.7 *15*
$\gamma_{\beta-}$E2+30%M1	699.54 *11*	8.0 *14*
$\gamma_{\beta-}$E2	705.76 *13*	61 *12*
$\gamma_{\beta-}$[E1]	708.64 *22*	0.45 *13*
$\gamma_{\beta-}$	708.96 *25*	20 *4*
$\gamma_{\beta-}$	806.78 *22*	16 *3*
$\gamma_{\beta-}$	931.24 *25*	4.3 *9*
$\gamma_{\beta-}$	974.5 *3*	1.8 *4*
$\gamma_{\beta-}$	1029.06 *25*	30 *6*
$\gamma_{\beta-}$[E2]	1213.01 *14*	5.6 *10*
$\gamma_{\beta-}$[E1]	1408.18 *22*	2.3 *5*

† approximate

Continuous Radiation (^{116}Ag)

⟨β-⟩=1811 keV;⟨IB⟩=6.1 keV

E_{bin}(keV)		⟨ ⟩(keV)	(%)
0 - 10	β-	0.00418	0.083
	IB	0.054	
10 - 20	β-	0.0128	0.085
	IB	0.053	0.37
20 - 40	β-	0.054	0.178
	IB	0.105	0.36
40 - 100	β-	0.422	0.60
	IB	0.30	0.47
100 - 300	β-	5.7	2.73
	IB	0.91	0.51
300 - 600	β-	29.0	6.3
	IB	1.12	0.26
600 - 1300	β-	221	22.8
	IB	1.8	0.20
1300 - 2500	β-	774	41.5
	IB	1.38	0.080
2500 - 5000	β-	775	24.0
	IB	0.39	0.0131
5000 - 5568	β-	6.8	0.131
	IB	0.00020	4.0×10^{-6}

$^{116}_{48}$Cd(stable)

Δ: -88721 *3* keV

%: 7.49 *9*

$^{116}_{49}$In(14.10 *3* s)

Mode: β-(>99.94 %), ϵ(<0.06 %)

Δ: -88247 *4* keV

SpA: 6.731×10^{9} Ci/g

Prod: ^{115}In(n,γ)

Photons (^{116}In)

⟨γ⟩=19.6 *4* keV

γ_{mode}	γ(keV)	γ(%)
γ [M1+E2]	113.09 *5*	$<6 \times 10^{-8}$
γ [E2]	200.1 *3*	$<5 \times 10^{-7}$
γ E2	355.42 *4*	0.00108 *17*
γ [E2]	463.33 *5*	0.25 *5*
γ [E2]	468.50 *6*	7.5 *17* $\times 10^{-5}$
γ E2+24%M1	818.74 *5*	0.0152 *23*
γ E2+28%M1	931.83 *4*	0.0110 *11*
γ	1252.7 *5*	0.031 *6*
γ E2	1293.59 *3*	1.30
γ [M1+E2]	1356.4 *3*	0.0104 *21*
γ	1497.2 *5*	0.016 *3*
γ E2	2112.32 *5*	0.021 *3*
γ [E2]	2225.41 *4*	0.0063 *6*
γ [E2]	2650.0 *3*	0.00117 *23*

Continuous Radiation (^{116}In)

⟨β-⟩=1364 keV;⟨IB⟩=3.8 keV

E_{bin}(keV)		⟨ ⟩(keV)	(%)
0 - 10	β-	0.0069	0.137
	IB	0.046	
10 - 20	β-	0.0211	0.141
	IB	0.045	0.31
20 - 40	β-	0.088	0.293
	IB	0.089	0.31
40 - 100	β-	0.69	0.97
	IB	0.26	0.39
100 - 300	β-	9.0	4.35
	IB	0.74	0.41
300 - 600	β-	44.2	9.6
	IB	0.85	0.20
600 - 1300	β-	311	32.2
	IB	1.17	0.136
1300 - 2500	β-	846	46.6
	IB	0.55	0.034
2500 - 3276	β-	153	5.7
	IB	0.0126	0.00048

$^{116}_{49}$In(54.15 *6* min)

Mode: β-

Δ: -88120 *4* keV

SpA: 2.993×10^{7} Ci/g

Prod: ^{115}In(n,γ); ^{117}Sn(γ,p)

Photons (^{116}In)

⟨γ⟩=2475 *28* keV

γ_{mode}	γ(keV)	γ(%)†
Sn $K_{\alpha2}$	25.044	0.300 *12*
Sn $K_{\alpha1}$	25.271	0.561 *22*
Sn $K_{\beta1}$'	28.474	0.147 *6*
Sn $K_{\beta2}$'	29.275	0.0309 *13*
γ E2	99.865 *14*	0.018 *7*
γ [M1+E2]	113.09 *5*	$<5 \times 10^{-7}$
γ	116.5 *10*	0.050 *20*
γ [E1]	124.74 *4*	~0.010
γ M1	138.372 *8*	3.29 *12*
γ	162.6 *5*	0.07 *2*
γ	196.5 *5*	0.050 *20*
γ [E2]	200.1 *3*	$<4 \times 10^{-6}$
γ [M1+E2]	245.22 *23*	0.037 *8*
γ [E1]	263.12 *4*	0.118 *25*
γ	272.4 *8*	0.08 *3*
γ [E2]	278.54 *3*	0.143 *17*
γ [E2]	303.83 *5*	0.118 *17*

Photons (^{116}In) (continued)

γ_{mode}	γ(keV)	γ(%)†
γ	345.2 *8*	0.030 *10*
γ E2	355.42 *4*	0.82 *4*
γ E2	416.92 *3*	29.2 *14*
γ [E1]	435.28 *15*	0.036 *14*
γ	458.5 *5*	0.07 *2*
γ [E2]	463.33 *5*	0.83 *5*
γ [E2]	468.50 *6*	0.00057 *14*
γ	500.1 *8*	0.030 *10*
γ [E1]	535.14 *14*	0.035 *13*
γ [M1+E2]	567.51 *17*	0.041 *13*
γ	639.1 *10*	0.030 *10*
γ [M1+E2]	655.61 *23*	0.11 *4*
γ	679.9 *10*	0.030 *10*
γ [E2]	688.94 *15*	0.160 *25*
γ [M1+E2]	705.88 *17*	0.169 *25*
γ [E1]	730.76 *17*	0.068 *25*
γ	736 *1*	<0.0030
γ [E1]	780.36 *23*	0.110 *20*
γ E2+24%M1	818.74 *5*	11.5 *4*
γ [E1]	830.62 *17*	0.052 *10*
γ E2+28%M1	931.83 *4*	0.084 *13*
γ E1	972.543 *21*	0.454 *16*
γ E3	1072.407 *23*	0.016 *6*
γ E2	1097.29 *4*	56.2 *11*
γ	1235.5 *10*	0.093 *17*
γ	1254.1 *10*	0.040 *19*
γ E2	1293.59 *3*	84.4 *17*
γ E2	1507.68 *14*	10.0 *3*
γ	1712.3 *10?*	
γ E2	1752.89 *23*	2.46 *8*
γ E2	2112.32 *5*	15.6 *4*
γ [E2]	2225.41 *4*	0.048 *8*

† 0.47% uncert(syst)

Atomic Electrons (^{116}In)

⟨e⟩=4.46 *8* keV

e_{bin}(keV)	⟨e⟩(keV)	e(%)
4 - 28	0.077	1.17 *11*
71 - 100	0.040	0.049 *19*
109	0.72	0.66 *3*
112 - 133	0.020	~0.016
134	0.112	0.083 *3*
137 - 171	0.037	0.026 *3*
192 - 241	0.0091	0.0042 *7*
243 - 279	0.032	0.0126 *17*
299 - 345	0.050	0.0154 *8*
351 - 355	0.0092	0.00263 *12*
388	1.22	0.314 *17*
406	0.00043	0.00011 *4*
412	0.141	0.0343 *18*
413 - 459	0.122	0.0289 *9*
462 - 506	0.0021	0.00044 *9*
531 - 568	0.0013	0.00024 *6*
610 - 656	0.0036	0.00058 *17*
660 - 707	0.0076	0.00112 *12*
726 - 751	0.00071	9.5 *16* $\times10^{-5}$
776 - 780	9.8 $\times10^{-5}$	1.26 *20* $\times10^{-5}$
790	0.171	0.0216 *9*
801 - 831	0.0274	0.00336 *10*
903 - 943	0.00338	0.000364 *20*
968 - 973	0.000323	3.34 *11* $\times10^{-5}$
1043	0.00032	3.0 *12* $\times10^{-5}$
1068	0.551	0.0516 *15*
1072 - 1097	0.0852	0.00780 *18*
1206 - 1254	0.0011	9 *3* $\times10^{-5}$
1264	0.697	0.0551 *16*
1289 - 1294	0.1058	0.00820 *19*
1478 - 1507	0.081	0.00548 *19*
1724 - 1753	0.042	0.00245 *15*
2083 - 2112	0.095	0.00454 *13*
2196 - 2225	0.00028	1.28 *18* $\times10^{-5}$

Continuous Radiation (^{116}In)

⟨β-⟩=310 keV;⟨IB⟩=0.30 keV

E_{bin}(keV)		⟨ ⟩(keV)	(%)
0 - 10	β-	0.089	1.78
	IB	0.0148	
10 - 20	β-	0.269	1.79
	IB	0.0141	0.098
20 - 40	β-	1.08	3.61
	IB	0.026	0.091
40 - 100	β-	7.7	11.0
	IB	0.065	0.102
100 - 300	β-	70	35.3
	IB	0.123	0.073
300 - 600	β-	156	36.1
	IB	0.054	0.0137
600 - 1137	β-	75	10.7
	IB	0.0051	0.00078

$^{116}_{49}$In(2.18 *4* s)

Mode: IT
Δ: -87957 *4* keV
SpA: 3.82×10^{10} Ci/g
Prod: ^{115}In(n,γ)

Photons (^{116}In)

⟨γ⟩=62.6 *6* keV

γ_{mode}	γ(keV)	γ(%)†
In L_ℓ	2.905	0.024 *4*
In L_η	3.112	0.0138 *21*
In L_α	3.286	0.65 *10*
In L_β	3.558	0.53 *9*
In L_γ	3.959	0.053 *9*
In $K_{\alpha2}$	24.002	3.64 *14*
In $K_{\alpha1}$	24.210	6.8 *3*
In $K_{\beta1}'$	27.265	1.76 *7*
In $K_{\beta2}'$	28.022	0.368 *15*
γ E3	162.390 *3*	36.6

† 1.9% uncert(syst)

Atomic Electrons (^{116}In)

⟨e⟩=94.0 *15* keV

e_{bin}(keV)	⟨e⟩(keV)	e(%)
4	1.68	44 *5*
19	0.087	0.45 *5*
20	0.73	3.6 *4*
23	0.191	0.83 *9*
24	0.204	0.86 *9*
26	0.0119	0.045 *5*
27	0.028	0.103 *11*
134	53.9	40.1 *10*
158	18.2	11.5 *3*
159	11.4	7.17 *18*
162	7.49	4.63 *12*

$^{116}_{50}$Sn(stable)

Δ: -91523 *3* keV
%: 14.53 *11*

$^{116}_{51}$Sb(15.8 *8* min)

Mode: ε
Δ: -86816 *6* keV
SpA: 1.03×10^8 Ci/g
Prod: daughter ^{116}Te; ^{115}In(α,3n); ^{116}Sn(p,n); ^{116}Sn(d,2n)

Photons (^{116}Sb)

⟨γ⟩=1717 *81* keV

γ_{mode}	γ(keV)	γ(%)†
Sn L_ℓ	3.045	0.053 *7*
Sn L_η	3.272	0.027 *3*
Sn L_α	3.443	1.47 *16*
Sn L_β	3.742	1.29 *16*
Sn L_γ	4.220	0.151 *22*
Sn $K_{\alpha2}$	25.044	10.9 *3*
Sn $K_{\alpha1}$	25.271	20.3 *6*
Sn $K_{\beta1}'$	28.474	5.31 *16*
Sn $K_{\beta2}'$	29.275	1.12 *4*
γ [M1+E2]	113.09 *5*	<14×10^{-5}
γ M1	138.372 *8*	0.0083 *17*
γ [E2]	200.1 *3*	<0.0011
γ [M1+E2]	245.22 *23*	0.00042 *12*
γ [E1]	263.12 *4*	0.00030 *9*
γ [E2]	303.83 *5*	0.00030 *7*
γ E2	355.42 *4*	0.0102 *10*
γ E2	416.92 *3*	0.074 *15*
γ [E1]	435.28 *15*	0.0015 *7*
γ [E2]	463.33 *5*	0.31 *6*
γ [E2]	468.50 *6*	0.17 *3*
γ [E1]	535.14 *14*	0.0014 *6*
γ [M1+E2]	655.61 *23*	0.0013 *5*
γ [E2]	688.94 *15*	0.0066 *17*
γ [E2]	731.7 *3*	0.093 *19*
γ [E1]	780.36 *23*	0.0013 *3*
γ E2+24%M1	818.74 *5*	0.144 *13*
γ E2+28%M1	931.83 *4*	24.8 *12*
γ E2	1293.59 *3*	85
γ E2	1507.68 *14*	0.41 *8*
γ [M1+E2]	1550.2 *5*	0.41 *8*
γ	1666.3 *4*	0.110 *22*
γ E2	1752.89 *23*	0.028 *6*
γ [M1+E2]	1794.6 *4*	0.044 *9*
γ E2	2112.32 *5*	0.194 *16*
γ [E2]	2225.41 *4*	14.2 *9*
γ [E2]	2843.8 *5*	1.0 *4*
γ	2959.8 *4*	0.20 *4*
γ [E2]	3088.1 *4*	0.102 *20*
γ [E2]	4270 *6*	~0.0022

† 7.1% uncert(syst)

Atomic Electrons (^{116}Sb)

⟨e⟩=4.07 *18* keV

e_{bin}(keV)	⟨e⟩(keV)	e(%)
4	1.45	36 *4*
20	0.093	0.46 *5*
21	0.77	3.7 *4*
24	0.205	0.85 *9*
25	0.220	0.89 *9*
27 - 28	0.043	0.154 *14*
84 - 113	0.0019	0.0017 *8*
134 - 171	0.00046	0.00032 *12*
196 - 245	9.1 $\times10^{-5}$	4.3 *16* $\times10^{-5}$
259 - 304	3.0 $\times10^{-5}$	1.1 *4* $\times10^{-5}$
326 - 355	0.00070	0.000211 *17*
388 - 435	0.0141	0.0033 *5*
439 - 469	0.0086	0.0019 *3*
506 - 535	1.5 $\times10^{-5}$	2.9 *13* $\times10^{-6}$
626 - 660	0.00015	2.3 *9* $\times10^{-5}$
684 - 732	0.0018	0.00025 *5*
751 - 790	0.00214	0.000271 *24*
814 - 819	0.000338	4.2 *3* $\times10^{-5}$
903	0.312	0.0346 *21*

Atomic Electrons (^{116}Sb) (continued)

e_{bin}(keV)	⟨e⟩(keV)	e(%)
927 - 932	0.0487	0.00525 *25*
1264	0.70	0.055 *4*
1289 - 1294	0.106	0.0082 *5*
1478 - 1549	0.0068	0.00045 *6*
1637 - 1724	0.0009	5.1 *17* $\times10^{-5}$
1748 - 1794	0.00035	2.0 *4* $\times10^{-5}$
2083 - 2112	0.00118	5.7 *4* $\times10^{-5}$
2196 - 2225	0.083	0.0038 *3*
2815 - 2843	0.0049	0.00018 *7*
2931 - 2959	0.00075	2.5 *8* $\times10^{-5}$
3059 - 3087	0.00047	1.5 *3* $\times10^{-5}$
4241 - 4269	8.3 $\times10^{-6}$	1.9 *9* $\times10^{-7}$

Continuous Radiation (^{116}Sb)

⟨β+⟩=491 keV;⟨IB⟩=3.9 keV

E_{bin}(keV)		⟨ ⟩(keV)	(%)
0 - 10	β+	3.42 $\times10^{-6}$	4.12 $\times10^{-5}$
	IB	0.019	
10 - 20	β+	0.000124	0.00075
	IB	0.019	0.131
20 - 40	β+	0.00309	0.0095
	IB	0.058	0.21
40 - 100	β+	0.133	0.172
	IB	0.103	0.158
100 - 300	β+	6.0	2.72
	IB	0.30	0.166
300 - 600	β+	41.8	9.0
	IB	0.42	0.097
600 - 1300	β+	234	25.0
	IB	1.19	0.129
1300 - 2500	β+	209	12.9
	IB	1.62	0.093
2500 - 3413	IB	0.22	0.0082
	Σβ+		50

$^{116}_{51}$Sb(1.005 *10* h)

Mode: ϵ

Δ: -86326 *56* keV

SpA: 2.69$\times10^{7}$ Ci/g

Prod: ^{115}In(α,3n); ^{113}In(α,n); ^{116}Sn(p,n); ^{116}Sn(d,2n)

Photons (^{116}Sb)

⟨γ⟩=2981 *78* keV

γ_{mode}	γ(keV)	γ(%)†
Sn L_ℓ	3.045	0.144 *20*
Sn L_η	3.272	0.077 *9*
Sn L_α	3.443	4.0 *5*
Sn L_β	3.741	3.5 *5*
Sn L_γ	4.212	0.40 *6*
Sn $K_{\alpha2}$	25.044	27.8 *14*
Sn $K_{\alpha1}$	25.271	52 *3*
Sn $K_{\beta1}$'	28.474	13.6 *7*
Sn $K_{\beta2}$'	29.275	2.86 *15*
γ E2	99.865 *14*	32 *3*
γ M1+0.2%E2	135.519 *20*	29 *3*
γ M1+0.08%E2	407.347 *17*	42 *3*
γ M1,E2	436.67 *4*	4.1 *3*
γ E2	542.865 *17*	52 *3*
γ E2	844.02 *4*	12 *1*
γ E1	972.543 *21*	72 *5*
γ E3	1072.407 *23*	28.1 *20*
γ E2	1293.59 *3*	100

Photons (^{116}Sb) (continued)

γ_{mode}	γ(keV)	γ(%)†
γ	1315.4 *4*	0.56 *10*
γ	1501.2 *4*	0.65 *10*

† 4.0% uncert(syst)

Atomic Electrons (^{116}Sb)

⟨e⟩=61 *3* keV

e_{bin}(keV)	⟨e⟩(keV)	e(%)
4	4.0	98 *11*
20	0.24	1.17 *13*
21	1.96	9.3 *10*
24 - 28	1.19	4.8 *4*
71	26.1	37 *4*
95	3.2	3.4 *3*
96	7.9	8.2 *8*
99	2.36	2.38 *22*
100	0.48	0.48 *5*
106	6.5	6.1 *7*
131 - 136	1.28	0.97 *8*
378 - 407	2.31	0.60 *4*
432 - 437	0.028	0.0064 *7*
514 - 543	1.59	0.307 *16*
815 - 844	0.187	0.0229 *17*
943 - 973	0.395	0.042 *3*
1043 - 1072	0.67	0.064 *4*
1264 - 1311	0.96	0.075 *3*
1472 - 1500	0.0044	0.00030 *12*

Continuous Radiation (^{116}Sb)

⟨β+⟩=124 keV;⟨IB⟩=2.2 keV

E_{bin}(keV)		⟨ ⟩(keV)	(%)
0 - 10	β+	4.92 $\times10^{-6}$	5.9 $\times10^{-5}$
	IB	0.0057	
10 - 20	β+	0.000177	0.00107
	IB	0.0063	0.043
20 - 40	β+	0.00438	0.0135
	IB	0.045	0.17
40 - 100	β+	0.184	0.238
	IB	0.031	0.047
100 - 300	β+	7.5	3.42
	IB	0.098	0.053
300 - 600	β+	41.3	9.1
	IB	0.24	0.053
600 - 1300	β+	75	9.4
	IB	1.06	0.114
1300 - 2288	IB	0.72	0.046
	Σβ+		22

$^{116}_{52}$Te(2.49 *4* h)

Mode: ϵ

Δ: -85280 *100* keV

SpA: 1.085$\times10^{7}$ Ci/g

Prod: protons on Sb; deuterons on Sb; ^{114}Sn(α,2n)

Photons (^{116}Te)

⟨γ⟩=76.3 *25* keV

γ_{mode}	γ(keV)	γ(%)†
Sb L_ℓ	3.189	0.19 *3*
Sb L_η	3.437	0.096 *10*
Sb L_α	3.604	5.2 *6*
Sb L_β	3.930	4.5 *6*
Sb L_γ	4.428	0.52 *8*
Sb $K_{\alpha2}$	26.111	33.1 *14*
Sb $K_{\alpha1}$	26.359	62 *3*
Sb $K_{\beta1}$'	29.712	16.2 *7*
Sb $K_{\beta2}$'	30.561	3.46 *15*
γ E2	93.74 *17*	32.1
γ M1	103.02 *17*	0.99 *20*
γ	108.5 *3*	0.016 *3*
γ	157.25 *23*	0.14 *3*
γ [M1+E2]	180.87 *23*	0.071 *14*
γ	363.0 *3*	0.019 *4*
γ [M1+E2]	366.9 *3*	0.045 *9*
γ [M1+E2]	447.85 *25*	0.017 *4*
γ [M1+E2]	457.14 *24*	0.034 *7*
γ	466.0 *3*	~0.0022
γ	471.47 *23*	0.013 *3*
γ	480.76 *24*	0.125 *25*
γ [M1+E2]	550.87 *24*	0.097 *19*
γ	574.50 *23*	0.013 *3*
γ	583.8 *3*	0.029 *6*
γ [M1+E2]	628.72 *24*	1.01 *20*
γ [M1+E2]	638.01 *24*	0.24 *5*
γ [M1+E2]	824.0 *3*	0.035 *7*
γ [E2]	917.7 *3*	0.045 *9*
γ [M1+E2]	1055.3 *3*	0.22 *4*
γ [M1+E2]	1064.6 *3*	0.077 *15*

† 6.2% uncert(syst)

Atomic Electrons (^{116}Te)

⟨e⟩=57.4 *21* keV

e_{bin}(keV)	⟨e⟩(keV)	e(%)
4	4.4	105 *11*
5 - 21	1.16	14.6 *13*
22	1.92	8.8 *9*
25 - 30	1.40	5.4 *4*
63	29.2	46 *3*
73 - 78	0.36	0.50 *10*
89	8.8	9.8 *6*
90	6.1	6.8 *4*
93	3.21	3.45 *22*
94 - 127	0.80	0.84 *5*
150 - 181	0.026	0.017 *5*
332 - 367	0.0038	0.00112 *20*
417 - 467	0.0058	0.0013 *4*
471 - 481	0.0006	0.00012 *6*
520 - 570	0.0041	0.00078 *14*
574 - 608	0.030	0.0050 *10*
624 - 638	0.0049	0.00079 *13*
793 - 824	0.00068	8.5 *18* $\times10^{-5}$
887 - 918	0.00067	7.5 *13* $\times10^{-5}$
1025 - 1065	0.0041	0.00040 *7*

Continuous Radiation (^{116}Te)

⟨β+⟩=1.18 keV;⟨IB⟩=0.67 keV

E_{bin}(keV)		⟨ ⟩(keV)	(%)
0 - 10	β+	1.64 $\times10^{-6}$	1.98 $\times10^{-5}$
	IB	0.00029	
10 - 20	β+	5.9 $\times10^{-5}$	0.000358
	IB	0.00095	0.0056
20 - 40	β+	0.00142	0.00436
	IB	0.049	0.19
40 - 100	β+	0.051	0.067
	IB	0.0043	0.0069
100 - 300	β+	0.90	0.459

Continuous Radiation (^{116}Te) (continued)

E_{bin}(keV)		⟨ ⟩(keV)	(%)
	IB	0.037	0.018
300 - 600	β+	0.231	0.069
	IB	0.17	0.038
600 - 1300	IB	0.40	0.048
1300 - 1446	IB	0.0033	0.00025
	Σβ+		0.60

$^{116}_{53}$I (2.91 *15* s)

Mode: ε
Δ: -77520 *160* keV
SpA: 2.97×10^{10} Ci/g
Prod: daughter ^{116}Xe; ^{103}Rh(^{16}O,3n)

Photons (^{116}I)

⟨γ⟩=63.4 *15* keV

γ_{mode}	γ(keV)	γ(%)
Te L_ℓ	3.335	0.0030 *4*
Te L_η	3.606	0.00147 *15*
Te L_α	3.768	0.083 *9*
Te L_β	4.123	0.073 *9*
Te L_γ	4.660	0.0089 *13*
Te $K_{\alpha2}$	27.202	0.548 *17*
Te $K_{\alpha1}$	27.472	1.02 *3*
Te $K_{\beta1}$'	30.980	0.272 *8*
Te $K_{\beta2}$'	31.877	0.0592 *19*
γ	540.2 *4*	1.20 *11*
γ (E2)	678.9 *3*	8.3

Atomic Electrons (^{116}I)

⟨e⟩=0.370 *19* keV

e_{bin}(keV)	⟨e⟩(keV)	e(%)
4	0.045	1.04 *11*
5	0.033	0.70 *7*
22	0.010	0.045 *5*
23	0.031	0.135 *14*
26	0.0100	0.038 *4*
27	0.0109	0.041 *4*
30	0.00157	0.0052 *5*
31	0.00058	0.00188 *19*
508	0.027	~0.005
535	0.0033	~0.0006
536	0.0005	~0.00010
539	0.0008	~0.00014
540	0.00019	~3×10^{-5}
647	0.167	0.0259 *12*
674	0.0218	0.00323 *15*
675	0.00139	0.000205 *10*
678	0.00467	0.00069 *3*
679	0.00111	0.000163 *8*

Continuous Radiation (^{116}I)

⟨β+⟩=3021 keV;⟨IB⟩=15.9 keV

E_{bin}(keV)		⟨ ⟩(keV)	(%)
0 - 10	β+	2.60×10^{-7}	3.13×10^{-6}
	IB	0.072	
10 - 20	β+	1.0×10^{-5}	6.0×10^{-5}
	IB	0.072	0.50
20 - 40	β+	0.000264	0.00081
	IB	0.144	0.50
40 - 100	β+	0.0125	0.0160
	IB	0.42	0.64
100 - 300	β+	0.67	0.301
	IB	1.31	0.73
300 - 600	β+	6.1	1.31
	IB	1.8	0.41
600 - 1300	β+	73	7.3
	IB	3.3	0.37
1300 - 2500	β+	477	24.5
	IB	4.0	0.22
2500 - 5000	β+	2001	55
	IB	4.0	0.120
5000 - 7500	β+	463	8.5
	IB	0.74	0.0130
7500 - 7760	IB	0.00067	8.9×10^{-6}
	Σβ+		97

$^{116}_{54}$Xe(56 *2* s)

Mode: ε
Δ: -73020 *260* keV syst
SpA: 1.73×10^{9} Ci/g
Prod: protons on Ce; ^{104}Pd(^{16}O,4n)

Photons (^{116}Xe)

γ_{mode}	γ(keV)	γ(rel)
γ	104.5 *2*	100
γ	191.6 *2*	38 *5*
γ	226.4 *2*	29 *3*
γ	247.7 *3*	40 *5*
γ	300.0 *4*	12 *1*
γ	310.7 *4*	42 *3*
γ	412.7 *2*	36 *6*
γ	923.0 *1*	25 *8*

$^{116}_{55}$Cs(3.81 *16* s)

Mode: ε, εp, εα
Δ: -62300 *310* keV
SpA: 2.334×10^{10} Ci/g
Prod: ^{90}Zr(^{32}S,p5n); ^{92}Mo(^{32}S,3p5n); protons on La

εp/β+: 0.0044 *14*
εα/β+: 0.00008 *2*
εp: 2000-6500 range

Photons (^{116}Cs)

⟨γ⟩=1826 *52* keV

γ_{mode}	γ(keV)	γ(%)†
γ	269.6 *4*	1.3 *3*
γ	322.2 *4*	3.8 *5*
γ	345.9 *4*	1.2 *3*
γ	360.2 *4*	0.9 *3*
γE2	393.47 *18*	95 *9*
γ[M1+E2]	458.39 *22*	2.8 *4*
γ	465.4 *3*	1.1 *3*
γ[E2]	517.4 *4*	2.4 *5*
γE2	524.24 *17*	76 *5*
γ	528.5 *4*	1.9 *5*
γ[E2]	541.15 *23*	5.8 *6*
γ[M1+E2]	552.8 *4*	0.85 *19*
γ[M1+E2]	556.49 *23*	1.42 *19*
γ[E2]	560.1 *3*	6.9 *6*
γ[M1+E2]	584.2 *3*	1.5 *3*
γ[E2]	611.4 *3*	5.7 *5*
γE2	615.10 *24*	30.4 *19*
γ[M1+E2]	622.34 *20*	10.4 *10*
γ[M1+E2]	639.24 *22*	7.1 *7*
γ	656.3 *6*	0.66 *19*
γ[E2]	659.4 *5*	0.9 *3*
γ[E2]	677.4 *6*	3.6 *5*
γ	684.1 *6*	0.95 *19*
γ	706.6 *6*	0.95 *19*
γ	823.0 *8*	~0.38
γ	874.7 *5*	1.5 *3*
γ	903.7 *8*	3.3 *7*
γ	906.2 *5*	2.4 *5*
γ	911.2 *4*	3.6 *5*
γ[M1+E2]	921.3 *4*	0.9 *3*
γ[M1+E2]	928.2 *4*	~0.38
γ[M1+E2]	965.6 *5*	1.5 *4*
γ	969.4 *6*	~0.6
γ	1008.5 *6*	~0.6
γ[E2]	1015.81 *24*	3.3 *4*
γ	1033.9 *8*	1.9 *5*
γ	1035.9 *8*	0.9 *3*
γ	1044.9 *6*	1.0 *3*
γ	1061.5 *4*	7.1 *7*
γ	1072.7 *6*	1.7 *4*
γ[M1+E2]	1080.73 *23*	7.1 *7*
γ	1164.3 *8*	1.5 *4*
γ	1168.0 *8*	2.8 *5*
γ	1247.0 *8*	1.9 *4*
γ[E2]	1321.6 *4*	1.9 *4*
γ	1441.0 *8*	0.9 *3*
γ[E2]	1445.6 *4*	1.5 *4*

† 9.5% uncert(syst)

Continuous Radiation (^{116}Cs)

⟨β+⟩=3767 keV;⟨IB⟩=20 keV

E_{bin}(keV)		⟨ ⟩(keV)	(%)
0 - 10	β+	1.32×10^{-7}	1.59×10^{-6}
	IB	0.081	
10 - 20	β+	5.3×10^{-6}	3.21×10^{-5}
	IB	0.081	0.56
20 - 40	β+	0.000148	0.000453
	IB	0.161	0.56
40 - 100	β+	0.0074	0.0095
	IB	0.47	0.73
100 - 300	β+	0.416	0.186
	IB	1.50	0.83
300 - 600	β+	3.91	0.83
	IB	2.0	0.48
600 - 1300	β+	48.3	4.83
	IB	4.0	0.45
1300 - 2500	β+	340	17.4
	IB	5.0	0.28
2500 - 5000	β+	1910	51
	IB	5.3	0.156
5000 - 7500	β+	1402	23.8
	IB	1.19	0.021
7500 - 9802	β+	62	0.80
	IB	0.045	0.00057
	Σβ+		99

$^{116}_{55}$Cs(710 *80* ms)

Mode: ϵ, ϵp, ϵα
Δ: -62300 *310* keV
SpA: 8.75×10^{10} Ci/g
Prod: ^{90}Zr(^{32}S,p5n); ^{92}Mo(^{32}S,3p5n); protons on La

ϵp/β+: 0.0028 *7*
ϵα/β+: 0.00049 *25*

Photons (^{116}Cs)

γ_{mode}	γ(keV)
γE2	393.47 *18*

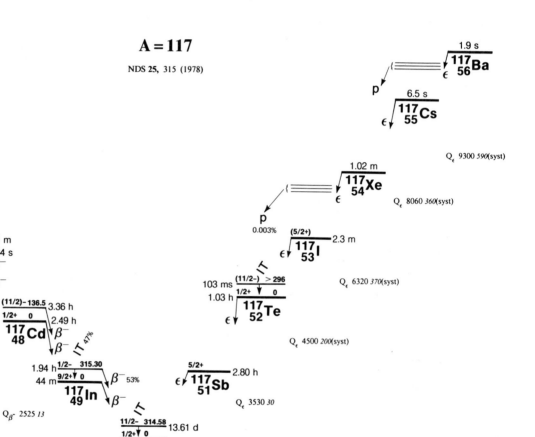

A = 117

NDS **25**, 315 (1978)

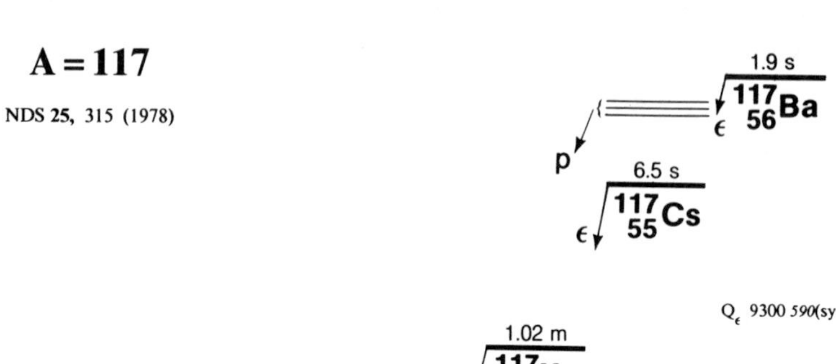
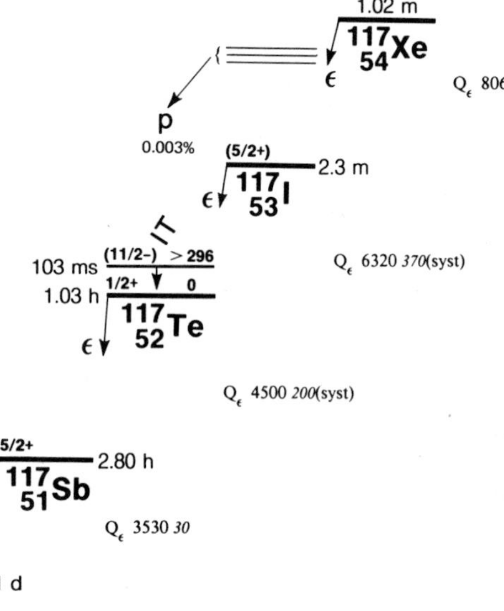

$^{117}_{46}$Pd(5.0 *6* s)

Mode: β-
SpA: 1.80×10^{10} Ci/g
Prod: fission

$^{117}_{47}$Ag(5.34 *5* s)

Mode: β-
Δ: -82250 *50* keV
SpA: 1.693×10^{10} Ci/g
Prod: fission

Photons (^{117}Ag)
⟨γ⟩=748 *20* keV

γ_{mode}	γ(keV)	γ(%)†
Cd L_ℓ	2.767	0.015 *3*
Cd L_η	2.957	0.0072 *15*
Cd L_α	3.133	0.41 *9*
Cd L_β	3.391	0.29 *6*
Cd L_γ	3.775	0.029 *6*
Cd $K_{\alpha2}$	22.984	3.1 *4*
Cd $K_{\alpha1}$	23.174	5.8 *8*
Cd $K_{\beta1}'$	26.085	1.47 *20*
Cd $K_{\beta2}'$	26.801	0.30 *4*
γ	104.77 *11*	0.19 *4*
γM1	135.36 *7*	46
γM1	141.99 *12*	4.5 *9*
γE2	156.98 *9*	5.9 *14*
γM1	184.33 *10*	6.0 *5*
γ	202.31 *9*	0.59 *14*

Photons (^{117}Ag)
(continued)

γ_{mode}	γ(keV)	γ(%)†
γE1	204.44 *11*	5.4 *9*
γ	215.2 *3*	0.64 *18*
γE1	219.44 *12*	3.1 *3*
γ	249.37 *21*	0.97 *23*
γM1,E2	297.90 *8*	20.0 *11*
γM1,E2	307.08 *8*	2.2 *4*
γM1,E2	321.98 *8*	6.7 *6*
γM1,E2	337.67 *8*	8.8 *13*
γ	341.10 *13*	1.9 *5*
γ	353.17 *24*	0.42 *7*
γ[E2]	362.56 *11*	1.6 *3*
γ	365.19 *14*	1.5 *3*
γ	377.46 *11*	1.06 *18*
γM1,E2	386.64 *8*	38.0 *20*
γ	413.54 *19*	2.6 *5*
γ	420.66 *15*	2.1 *5*
γ	442.44 *9*	1.13 *21*
γ	482.23 *10*	1.9 *3*

Photons (^{117}Ag)
(continued)

γ_{mode}	γ(keV)	γ(%)†
γ	486.67 21	0.9 3
γ	492.57 21	0.92 23
γ	500.5 4	1.0 4
γ	522.00 9	8.9 6
γ	526.43 11	1.2 5
γ	529.77 15	0.68 19
γ	543.1 3	0.28 9
γ	546.37 21	0.32 14
γ	555.48 24	0.51 24
γ	557.65 13	2.0 4
γ	569.63 16	0.6 3
γ	581.74 14	0.92 14
γ	585.8 3	0.28 9 ?
γ	591.8 3	0.28 9
γ	608.7 6	0.51 14
γ	637.22 15	1.43 14
γ	665.12 15	0.61 17
γ	684.54 9	7.5 5
γ	691.6 9	~0.23
γ	743.0 4	0.32 14
γ	754.64 18	1.1 3
γ	772.0 5	~0.18
γ	786.18 14	1.66 18
γ	801.07 21	0.51 14
γ	819.89 9	1.66 18
γ	834.20 18	~0.46
γ	895.67 21	0.32 14
γ	899.4 3	~0.14
γ	913.31 22	0.72 18
γ	1037.9 3	0.92 14
γ	1130.5 3	1.61 18
γ	1141.28 18	0.55 9
γ	1220.39 22	0.97 12
γ	1258.5 4	1.5 4
γ	1330.2 5	0.9 3
γ	1455.7 4	0.32 14
γ	1508.5 4	0.28 9

† 20% uncert(syst)

Atomic Electrons (^{117}Ag)
⟨e⟩=21.8 19 keV

e_{bin}(keV)	⟨e⟩(keV)	e(%)
4 - 26	0.76	12.0 17
78 - 105	0.11	~0.14
109	8.8	8.1 16
115	0.79	0.69 14
130	1.9	1.4 3
131	1.23	0.94 19
132 - 157	1.09	0.76 9
158	0.72	0.45 4
176 - 223	0.67	0.36 5
245 - 249	0.014	0.006 3
271	1.35	0.50 8
280 - 307	0.80	0.27 3
311	0.47	0.15 3
314 - 359	0.43	0.129 20
360	1.60	0.45 4
361 - 410	0.42	0.108 16
413 - 460	0.09	0.021 6
466 - 503	0.24	0.048 17
516 - 566	0.123	0.023 5
569 - 611	0.031	0.0052 18
633 - 681	0.12	0.018 7
684 - 728	0.019	0.0027 11
739 - 786	0.030	0.0039 12
793 - 834	0.025	0.0031 11
869 - 913	0.012	0.0013 5
1011 - 1038	0.008	0.0008 3
1104 - 1141	0.017	0.0015 5
1194 - 1232	0.016	0.0013 4
1254 - 1303	0.007	0.00052 22
1326 - 1330	0.0008	6 3 $\times 10^{-5}$
1429 - 1456	0.0020	~0.00014
1482 - 1508	0.0016	0.00011 5

Continuous Radiation (^{117}Ag)
⟨β-⟩=1810 keV;⟨IB⟩=5.3 keV

E_{bin}(keV)		⟨ ⟩(keV)	(%)
0 - 10	β-	0.0065	0.129
	IB	0.058	
10 - 20	β-	0.0199	0.132
	IB	0.058	0.40
20 - 40	β-	0.083	0.276
	IB	0.114	0.40
40 - 100	β-	0.65	0.92
	IB	0.33	0.50
100 - 300	β-	8.7	4.21
	IB	0.96	0.54
300 - 600	β-	44.3	9.6
	IB	1.13	0.27
600 - 1300	β-	330	34.1
	IB	1.65	0.19
1300 - 2500	β-	1039	56
	IB	0.95	0.057
2500 - 4035	β-	386	13.6
	IB	0.071	0.0025

$^{117}_{47}$Ag(1.21 3 min)

Mode: β-

Δ: -82250 50 keV

SpA: 1.32×10^{9} Ci/g

Prod: fission

Photons (^{117}Ag)
⟨γ⟩=1303 29 keV

γ_{mode}	γ(keV)	γ(%)†
Cd L_ℓ	2.767	0.0084 19
Cd L_η	2.957	0.0041 8
Cd L_α	3.133	0.23 5
Cd L_β	3.390	0.16 4
Cd L_γ	3.773	0.016 4
Cd $K_{\alpha2}$	22.984	1.69 23
Cd $K_{\alpha1}$	23.174	3.2 4
Cd $K_{\beta1}$'	26.085	0.81 11
Cd $K_{\beta2}$'	26.801	0.168 23
γ	104.77 11	0.160 22
γ M1	135.36 7	23
γ M1	141.99 12	1.9 3
γ E2	156.98 9	7.9 9
γ M1	184.33 10	0.25 3
γ	202.31 9	0.69 14
γ	229.6 6	~0.09
γ M1,E2	297.90 8	0.50 12
γ M1,E2	307.08 8	1.84 21
γ M1,E2	312.17 8	5.9 6
γ M1,E2	321.98 8	0.17 5
γ	327.17 9	1.13 16
γ	333.0 6	0.18 7
γ M1,E2	337.67 8	10.3 6
γ	353.17 24	0.53 7
γ	377.46 11	0.026 8
γ M1,E2	386.64 8	1.55 11
γ M1,E2	426.15 10	6.9 4
γ	442.44 9	0.93 9
γ	467.49 13	2.46 21
γ	476.9 10	~0.07
γ	482.23 10	0.047 14
γ	500.5 5	0.39 9
γ	522.00 9	0.36 4
γ	526.43 11	0.030 14
γ	529.77 15	1.27 13
γ	555.48 24	0.6 3
γ	665.12 15	1.13 11
γ	684.54 9	0.19 5
γ	701.0 5	0.32 11
γ	737.4 5	0.28 7
γ	746.60 25	0.74 9

Photons (^{117}Ag)
(continued)

γ_{mode}	γ(keV)	γ(%)†
γ	767.0 5	0.35 9
γ	779.66 13	1.79 18
γ	808.4 5	0.115 23
γ	819.89 9	0.041 7
γ	836.2 5	0.44 9
γ	839.8 3	0.35 7
γ	913.31 22	0.47 10
γ	941.5 5	0.55 16
γ	949.3 3	0.21 7
γ	1056.3 5	0.37 9
γ	1118.89 17	0.53 11
γ	1220.39 22	0.63 9
γ	1228.36 19	0.67 11
γ	1258.5 4	0.90 18
γ	1309.4 3	~0.8
γ	1341.6 3	~0.5
γ	1349.3 5	~0.6
γ	1407.48 21	~0.39
γ	1608.95 20	4.3 4
γ	1648.6 3	0.41 12
γ	1657.66 15	2.23 23
γ	1695.84 18	1.52 23
γ	1748.7 3	1.13 18
γ	1749.58 13	1.13 18
γ	1776.9 3	0.48 12
γ	1780.2 5	0.37 9
γ	1854.34 12	2.5 3
γ	1877.76 17	1.29 21
γ	1899.9 5	0.64 18
γ	1908.3 3	0.51 16
γ	1963.81 19	1.06 16
γ	1995.33 15	4.0 4
γ	2013.12 17	3.7 3
γ	2035.28 24	1.27 18
γ	2042.1 5	0.46 11
γ	2056.65 12	3.3 3
γ	2089.1 3	0.44 14
γ	2118.92 25	0.35 11
γ	2192.01 11	1.86 23
γ	2200.9 3	~0.11
γ	2216.7 4	0.92 18
γ	2247.2 3	2.8 3
γ	2341.5 4	0.51 16
γ	2413.0 3	0.94 14
γ	2417.0 3	0.94 14
γ	2478.6 4	0.51 21
γ	2505.56 25	0.25 9
γ	2514.0 3	1.45 18
γ	2663.0 4	0.48 12
γ	2681.2 4	0.74 14
γ	2738.8 4	0.23 9
γ	2850.7 3	0.69 12
γ	2861.2 6	0.21 7
γ	2888.2 3	2.55 23
γ	2896.9 3	0.35 9
γ	3029.8 5	0.60 14
γ	3599.6 6	0.37 11

† 20% uncert(syst)

Atomic Electrons (^{117}Ag)
⟨e⟩=11.7 10 keV

e_{bin}(keV)	⟨e⟩(keV)	e(%)
4 - 26	0.43	6.7 9
78 - 105	0.10	~0.11
109	4.4	4.0 9
115	0.33	0.29 5
130	2.5	1.92 22
131	0.61	0.47 10
132 - 142	0.27	0.20 3
153	0.57	0.37 4
156 - 203	0.27	0.16 3
226 - 271	0.035	0.013 4
280	0.117	0.042 8
285	0.37	0.128 21
294 - 309	0.14	0.047 10
311	0.56	0.180 23

Atomic Electrons (^{117}Ag) (continued)

e_{bin}(keV)	⟨e⟩(keV)	e(%)
312 - 360	0.200	0.059 7
373 - 423	0.314	0.078 6
425 - 474	0.08	0.018 6
476 - 523	0.031	0.0062 24
526 - 555	0.016	0.0029 14
638 - 685	0.023	0.0036 12
697 - 747	0.017	0.0024 7
753 - 793	0.024	0.0031 13
804 - 840	0.009	0.0011 4
887 - 923	0.011	0.0012 4
938 - 949	0.0010	0.00010 4
1030 - 1056	0.0031	0.00030 14
1092 - 1119	0.0042	0.00039 17
1194 - 1232	0.015	0.0012 3
1255 - 1283	0.005	~0.0004
1305 - 1349	0.007	0.00057 23
1381 - 1407	0.0025	~0.00018
1582 - 1669	0.045	0.0028 6
1692 - 1780	0.017	0.00097 19
1828 - 1908	0.024	0.0013 3
1937 - 2035	0.060	0.0030 5
2038 - 2118	0.0055	0.00026 6
2165 - 2247	0.024	0.00108 21
2315 - 2414	0.0093	0.00039 8
2416 - 2513	0.0085	0.00034 8
2636 - 2735	0.0052	0.00020 4
2738 - 2835	0.0027	1.0 3 $\times 10^{-6}$
2847 - 2896	0.0103	0.00036 8
3003 - 3029	0.0020	6.6 21 $\times 10^{-5}$
3573 - 3599	0.0011	3.1 11 $\times 10^{-5}$

Continuous Radiation (^{117}Ag)

⟨β-⟩=1299 keV;⟨IB⟩=3.6 keV

E_{bin}(keV)		⟨ ⟩(keV)	(%)
0 - 10	β-	0.0124	0.246
	IB	0.043	
10 - 20	β-	0.0378	0.252
	IB	0.043	0.30
20 - 40	β-	0.157	0.52
	IB	0.084	0.29
40 - 100	β-	1.20	1.70
	IB	0.24	0.37
100 - 300	β-	14.9	7.2
	IB	0.69	0.38
300 - 600	β-	65	14.3
	IB	0.78	0.18
600 - 1300	β-	325	34.7
	IB	1.07	0.124
1300 - 2500	β-	591	32.7
	IB	0.63	0.037
2500 - 4170	β-	302	10.3
	IB	0.073	0.0025

$^{117}_{48}$Cd(2.49 4 h)

Mode: β-
Δ: -86417 13 keV
SpA: 1.076×10^{7} Ci/g
Prod: ^{116}Cd(n,γ); ^{116}Cd(d,p)

Photons (^{117}Cd)

⟨γ⟩=1087 13 keV

γ_{mode}	γ(keV)	γ(%)†
In L_ℓ	2.905	0.0096 18
In L_η	3.112	0.0049 8
In L_α	3.286	0.26 4
In L_β	3.565	0.20 3
In L_γ	3.967	0.020 4
In $K_{\alpha2}$	24.002	1.64 10
In $K_{\alpha1}$	24.210	3.08 19
In $K_{\beta1}$'	27.265	0.79 5
In $K_{\beta2}$'	28.022	0.166 11
γ E1	71.111 11	0.39 6
γ E2+2.1%M1	89.725 9	3.26 20
γ	105.439 20	~0.022
γ [E2]	131.233 16	~0.011
γ [M1+E2]	132.656 24	~0.022
γ (E1)	160.836 13	0.25 11
γ [M1+E2]	170.987 19	0.025 11
γ [M1+E2]	172.23 4	~0.008
γ	179.38 3	0.10 3
γ [M1+E2]	220.958 14	1.17 8
γ	221.04 15	<0.11
γ M1+E2	273.350 11	28
γ [M1+E2]	279.81 3	~0.11
γ	279.83 4	<0.11
γ	284.82 3	0.084 22
γ [E1]	292.068 15	0.64 8
γ	309.68 3	0.070 14 ?
γ	314.353 22	~0.08
γ M4	315.301 12	*
γ E1	344.461 9	17.9 4
γ	385.5 4	0.036 7 ?
γ	387.98 3	0.31 6
γ	397.30 5	0.20 6
γ	416.92 13	<0.033
γ	419.792 20	0.18 4
γ E1	434.185 12	9.8 4
γ	439.39 3	0.11 4
γ	453.57 6	~0.036
γ [E1]	463.055 17	0.75 6
γ	497.74 5	~0.11
γ	500.46 5	<0.028
γ	526.44 19	<0.06
γ	526.79 8	0.14 6
γ	597.6 3	<0.028 ?
γ	626.94 9	0.11 3
γ	644.33 15	<0.033
γ	660.84 3	0.11 3
γ	687.72 10	<0.022
γ	699.62 4	0.24 4
γ	712.74 3	0.57 15
γ	716.67 9	0.20 4
γ	728.67 3	0.24 4
γ	736.11 6	~0.06
γ [M1+E2]	748.062 21	0.56 20
γ	757.6 2	~0.028
γ	787.78 9	0.056 11 ?
γ	831.823 24	2.26 8
γ [E2]	840.217 20	0.81 6
γ	850.72 8	0.12 4
γ	861.32 4	~0.28
γ	862.61 3	0.61 6
γ [E2]	880.717 13	3.96 19
γ	945.655 20	1.53 8
γ	949.62 3	0.22 4
γ	952.34 4	0.14 3
γ	963.055 24	0.61 6
γ	964.48 3	<0.11
γ	965.83 19	~0.08
γ	969.32 3	0.45 6
γ [M1+E2]	970.51 3	<0.11
γ	975.5 5	0.073 15 ?
γ	994.18 4	<0.033
γ [E2]	1011.203 21	~0.08
γ [M1+E2]	1012.44 4	~0.08
γ	1035.43 8	0.24 4
γ	1036.11 5	<0.033
γ [E2]	1051.703 21	3.79 20
γ	1052.780 24	0.73 17
γ	1061.26 15	<0.11
γ	1116.641 21	1.03 6
γ [M1+E2]	1120.03 3	0.24 4
γ	1125.16 8	0.45 6

Photons (^{117}Cd) (continued)

γ_{mode}	γ(keV)	γ(%)†
γ [M1]	1142.434 16	1.67 11
γ	1143.69 4	0.14 6
γ [M1+E2]	1183.43 4	0.13 3
γ	1229.12 4	0.61 6
γ [M1+E2]	1232.158 17	0.28 6
γ	1247.873 18	1.20 6
γ	1249.30 3	<0.06
γ	1260.005 21	1.14 6
γ [M1+E2]	1272.731 23	0.73 6
γ	1276.07 5	0.025 11
γ [M1+E2]	1291.01 3	0.67 6
γ	1294.08 3	0.45 9 ?
γ [E1]	1303.268 17	18.4 4
γ [M1+E2]	1314.66 4	0.59 6
γ [E2]	1316.08 4	<0.06
γ	1317.45 6	<0.033
γ	1337.596 18	1.62 11
γ [M1+E2]	1362.455 25	0.24 4
γ [M1+E2]	1404.38 4	0.12 3
γ	1408.706 18	1.28 6
γ [M1+E2]	1422.24 3	0.33 6
γ	1430.991 23	~0.6
γ [E1]	1433.57 3	~0.11
γ	1450.07 4	0.61 6
γ	1468.62 13	0.039 11
γ [E1]	1475.49 4	0.42 6
γ [M1+E2]	1511.97 3	~0.07
γ	1521.18 4	0.09 3
γ	1562.222 19	1.42 6
γ	1563.65 3	~0.08
γ [E1]	1576.615 18	11.19 22
γ	1578.28 5	0.14 6
γ [E1]	1583.08 3	0.053 25
γ	1596.00 5	<0.06
γ	1597.43 6	<0.11
γ	1651.945 20	0.28 11
γ	1682.053 20	0.70 6
γ	1685.73 5	0.039 17
γ [E1]	1706.91 3	1.00 6
γ	1723.055 20	2.01 8
γ	1739.12 5	0.13 3
γ [E1]	1748.84 4	0.08 3
γ	1756.84 5	~0.045
γ [E1]	1856.42 3	0.25 6
γ	1867.3 1	0.11 3
γ	2012.46 5	0.109 22
γ	2030.18 5	0.064 20

† 10% uncert(syst)
* with ^{117}In(1.94 h)

Atomic Electrons (^{117}Cd)

⟨e⟩= 9.7 5 keV

e_{bin}(keV)	⟨e⟩(keV)	e(%)
4 - 43	0.51	7.4 7
62	3.22	5.2 3
67 - 77	0.026	0.037 11
85	0.403	0.47 3
86	1.07	1.25 8
89	0.311	0.349 22
90 - 133	0.095	0.097 9
143 - 179	0.028	0.019 7
193 - 221	0.19	0.094 21
245	2.3	0.95 17
252 - 264	0.030	0.012 4
269	0.33	0.12 3
270 - 314	0.14	0.051 16
317	0.281	0.0889 25
340 - 389	0.063	0.0181 18
392 - 439	0.146	0.0356 14
449 - 498	0.0049	0.0010 4
499 - 527	0.0031	~0.0006
570 - 616	0.0022	0.00037 18
623 - 672	0.0054	0.0008 3
683 - 732	0.025	0.0036 8
735 - 784	0.0022	0.00029 8
787 - 836	0.049	0.0060 16

Atomic Electrons (^{117}Cd)
(continued)

e_{bin}(keV)	⟨e⟩(keV)	e(%)
839 - 881	0.057	0.0067 4
918 - 967	0.033	0.0036 9
968 - 1012	0.0044	0.00044 14
1024 - 1061	0.050	0.0049 4
1089 - 1125	0.034	0.0031 4
1138 - 1183	0.0042	0.00037 3
1201 - 1249	0.031	0.0025 4
1256 - 1303	0.092	0.0072 3
1310 - 1359	0.014	0.0011 4
1362 - 1409	0.016	0.0012 3
1418 - 1465	0.0066	0.00046 11
1468 - 1563	0.047	0.00301 23
1568 - 1658	0.0108	0.00067 10
1678 - 1756	0.016	0.00096 23
1828 - 1867	0.0014	7.4 15 $\times 10^{-5}$
1985 - 2030	0.00084	4.2 12 $\times 10^{-5}$

Continuous Radiation (^{117}Cd)
⟨β-⟩=434 keV;⟨IB⟩=0.70 keV

E_{bin}(keV)		⟨ ⟩(keV)	(%)
0 - 10	β-	0.133	2.66
	IB	0.018	
10 - 20	β-	0.389	2.60
	IB	0.018	0.123
20 - 40	β-	1.49	4.99
	IB	0.034	0.117
40 - 100	β-	9.3	13.4
	IB	0.088	0.137
100 - 300	β-	59	30.9
	IB	0.21	0.119
300 - 600	β-	86	20.5
	IB	0.18	0.042
600 - 1300	β-	187	20.3
	IB	0.147	0.018
1300 - 2210	β-	91	6.0
	IB	0.0129	0.00090

$^{117}_{48}$Cd(3.36 5 h)

Mode: β-

Δ: -86281 13 keV

SpA: 7.97×10^{6} Ci/g

Prod: ^{116}Cd(n,γ); ^{116}Cd(d,p)

Photons (^{117}Cd)
⟨γ⟩=2040 50 keV

γ_{mode}	γ(keV)	γ(%)†
In $K_{\alpha2}$	24.002	~0.27
In $K_{\alpha1}$	24.210	~0.5
In $K_{\beta1}'$	27.265	~0.13
In $K_{\beta2}'$	28.022	~0.028
γ	97.707 23	1.05 13
γ	99.10 4	~0.10
γ	100.80 15	~0.08
γ [E2]	131.233 16	~0.0020
γ [M1+E2]	132.656 24	~0.0040
γ	168.622 23	0.29 5
γ [M1+E2]	220.958 14	0.21 7
γ [E1]	292.068 15	0.12 4
γ	299.54 3	0.45 8
γ	310.25 6	0.50 10
γ	313.82 3	<0.047
γ	325.43 6	0.13 5
γ	366.937 19	3.33 24
γ [E2]	381.21 3	<0.047 ?

Photons (^{117}Cd)
(continued)

γ_{mode}	γ(keV)	γ(%)†
γ	407.96 6	~0.09
γ	439.39 3	0.19 7
γ	442.72 15	0.026 5 ?
γ	460.96 4	1.62 13
γ	484.802 22	1.02 13
γ	518.83 18	~0.06
γ	544.75 13	~0.16
γ	564.405 14	14.7 8
γ	597.34 20	0.13 3 ?
γ	617.46 3	0.34 8
γ	627.26 15	0.24 5 ?
γ	631.802 24	2.80 18
γ	663.50 4	0.68 8
γ [M1+E2]	684.85 3	~0.07
γ	712.74 3	0.99 13
γ	730.90 4	0.105 21 ?
γ	743.9 10	<0.026 ?
γ [M1+E2]	748.062 21	4.5 10
γ	762.720 22	1.73 13
γ	788.30 4	0.50 10
γ	827.58 10	0.26 8
γ	860.426 25	7.9 3
γ [E2]	880.717 13	0.72 24
γ	886.00 4	0.39 8
γ	891.39 9	0.079 16 ?
γ	929.34 9	0.79 13
γ	931.341 18	3.64 24
γ	957.23 6	0.39 10
γ	995.0 5	0.052 10 ?
γ	1029.047 22	11.7 4
γ	1065.978 19	23.1 5
γ	1120.0 3	<0.26 ?
γ	1170.67 6	0.65 13
γ	1196.25 7	0.39 10
γ	1205.44 15	0.13 4
γ	1208.40 10	<0.10
γ	1209.02 4	0.18 8
γ	1209.31 9	~0.13
γ	1234.598 21	11.0 3
γ	1256.77 6	0.18 8
γ	1339.29 6	2.07 24
γ	1365.516 22	1.65 10
γ	1371.2 5	0.031 6 ?
γ [M1+E2]	1432.911 17	13.4 3
γ	1442.03 6	0.018 4 ?
γ	1652.30 9	0.47 10
γ	1669.36 9	0.63 8
γ	1957.36 9	0.16 4
γ	1997.309 16	26.2
γ	2096.40 3	7.44 16
γ	2322.73 6	7.86 18
γ	2400.35 9	0.76 5
γ	2414.2 2	<0.16
γ	2417.41 9	1.02 5
γ	2440.03 15	<0.26 ?
γ	2462.5 3	0.212 24
γ	2476.19 17	0.186 18
γ	2540.83 12	0.149 18

† 8.4% uncert(syst)

Atomic Electrons (^{117}Cd)
⟨e⟩=2.7 5 keV

e_{bin}(keV)	⟨e⟩(keV)	e(%)
4 - 27	0.07	1.2 6
70	0.5	~0.7
71 - 73	0.09	~0.12
93	0.06	~0.07
94	0.13	~0.14
95 - 141	0.14	0.13 6
164 - 193	0.039	0.021 7
217 - 264	0.008	0.0035 12
272 - 314	0.060	0.021 7
321 - 325	0.0010	~0.00032
339	0.11	~0.032
353 - 381	0.024	0.006 3
404 - 443	0.043	0.010 5

Atomic Electrons (^{117}Cd)
(continued)

e_{bin}(keV)	⟨e⟩(keV)	e(%)
457 - 491	0.032	0.007 3
515 - 519	0.0030	~0.0006
536	0.25	~0.046
541 - 590	0.049	0.009 3
593 - 636	0.062	0.010 4
657 - 703	0.017	0.0024 10
709 - 716	0.0022	0.00031 13
720	0.073	0.010 3
727 - 763	0.040	0.0054 15
784 - 828	0.0041	0.00051 21
832	0.08	0.009 5
853 - 901	0.035	0.0040 9
903 - 931	0.043	0.0047 18
953 - 995	0.0010	0.00011 4
1001	0.10	0.010 5
1025 - 1029	0.015	0.0014 6
1038	0.18	0.017 8
1062 - 1092	0.029	0.0027 11
1116 - 1143	0.0048	0.00042 20
1166 - 1206	0.0077	0.00066 19
1207	0.07	0.006 3
1208 - 1257	0.013	0.0010 3
1311 - 1343	0.025	0.0019 6
1361 - 1371	0.0015	0.00011 4
1405	0.103	0.0073 7
1414 - 1442	0.0153	0.00107 10
1624 - 1669	0.0063	0.00039 10
1929 - 1957	0.0008	4.0 16 $\times 10^{-5}$
1969	0.11	0.0057 21
1993 - 2092	0.050	0.0024 6
2093 - 2096	0.0008	3.7 13 $\times 10^{-5}$
2295 - 2389	0.041	0.0018 4
2396 - 2476	0.0032	0.00013 3
2513 - 2540	0.00060	2.4 7 $\times 10^{-5}$

Continuous Radiation (^{117}Cd)
⟨β-⟩=235 keV;⟨IB⟩=0.22 keV

E_{bin}(keV)		⟨ ⟩(keV)	(%)
0 - 10	β-	0.179	3.59
	IB	0.0113	
10 - 20	β-	0.53	3.54
	IB	0.0105	0.073
20 - 40	β-	2.06	6.9
	IB	0.019	0.067
40 - 100	β-	13.3	19.1
	IB	0.045	0.070
100 - 300	β-	84	44.5
	IB	0.075	0.045
300 - 600	β-	87	21.8
	IB	0.035	0.0087
600 - 1300	β-	36.4	4.20
	IB	0.021	0.0026
1300 - 2500	β-	11.6	0.74
	IB	0.0025	0.000160
2500 - 2661	β-	0.0289	0.00114
	IB	2.3×10^{-7}	8.9×10^{-9}

$^{117}_{49}$In(43.8 7 min)

Mode: β-

Δ: -88943 5 keV

SpA: 3.67×10^{7} Ci/g

Prod: daughter ^{117}Cd

Photons (^{117}In)

⟨γ⟩=692 *14* keV

γ_{mode}	γ(keV)	γ(%)†
Sn L_ℓ	3.045	0.0150 *25*
Sn L_η	3.272	0.0079 *11*
Sn L_α	3.443	0.42 *6*
Sn L_β	3.742	0.37 *6*
Sn L_γ	4.217	0.043 *8*
Sn $K_{\alpha2}$	25.044	3.1 *3*
Sn $K_{\alpha1}$	25.271	5.8 *6*
Sn $K_{\beta1}$'	28.474	1.51 *16*
Sn $K_{\beta2}$'	29.275	0.32 *3*
γ M4	156.02 *3*	*
γ M1+0.02%E2	158.560 *12*	87 *9*
γ (M2)	396.95 *9*	0.14 *4*
γ (E2)	552.97 *9*	99.7

† 0.20% uncert(syst)
* with ^{117}Sn(13.61 d)

Atomic Electrons (^{117}In)

⟨e⟩=22.0 *16* keV

e_{bin}(keV)	⟨e⟩(keV)	e(%)
4	0.41	10.2 *15*
20	0.026	0.130 *18*
21	0.22	1.04 *15*
24	0.058	0.24 *3*
25	0.063	0.25 *4*
27	0.00107	0.0039 *6*
28	0.0111	0.040 *6*
129	15.4	11.9 *13*
154	2.29	1.49 *16*
155	0.035	0.0226 *24*
158	0.56	0.36 *4*
159	0.0103	0.0065 *7*
368	0.023	0.0063 *18*
392	0.0031	0.00079 *23*
393	0.00033	8.3 *24* ×10^{-5}
396	0.00069	0.00017 *5*
397	0.00015	3.9 *11* ×10^{-5}
524	2.51	0.480 *19*
549	0.352	0.064 *3*
552	0.070	0.0127 *5*
553	0.0155	0.00280 *11*

Continuous Radiation (^{117}In)

⟨β-⟩=245 keV;⟨IB⟩=0.19 keV

E_{bin}(keV)		⟨ ⟩(keV)	(%)
0 - 10	β-	0.110	2.20
	IB	0.0121	
10 - 20	β-	0.331	2.21
	IB	0.0114	0.079
20 - 40	β-	1.33	4.45
	IB	0.021	0.073
40 - 100	β-	9.5	13.5
	IB	0.049	0.077
100 - 300	β-	83	42.5
	IB	0.078	0.048
300 - 600	β-	138	33.4
	IB	0.019	0.0052
600 - 1139	β-	11.5	1.80
	IB	0.00020	3.1×10^{-5}

$^{117}_{49}$In(1.942 *12* h)

Mode: β-(52.9 *15* %), IT(47.1 *15* %)
Δ: -88628 *5* keV
SpA: 1.379×10^{7} Ci/g
Prod: daughter ^{117}Cd

Photons (^{117}In)

⟨γ⟩=91 *6* keV

γ_{mode}	γ(keV)	γ(%)†
In L_ℓ	2.905	0.030 *6*
In L_η	3.112	0.014 *3*
In L_α	3.286	0.83 *16*
In L_β	3.570	0.61 *12*
In L_γ	3.986	0.062 *13*
In $K_{\alpha2}$	24.002	5.4 *6*
In $K_{\alpha1}$	24.210	10.2 *11*
Sn $K_{\alpha2}$	25.044	0.54 *3*
Sn $K_{\alpha1}$	25.271	1.02 *5*
In $K_{\beta1}$'	27.265	2.6 *3*
In $K_{\beta2}$'	28.022	0.55 *6*
Sn $K_{\beta1}$'	28.474	0.266 *14*
Sn $K_{\beta2}$'	29.275	0.056 *3*
γ_{β}-M1+0.02%E2	158.560 *12*	15.9
γ_{IT} M4	315.301 *12*	19.1
γ_{β}-M1+1.7%E2	845.95 *13*	0.0017 *6*
γ_{β}-M1+2.3%E2	861.36 *5*	0.019 *3*
γ_{β}-E2+0.3%M1	1004.51 *13*	0.0063 *13*
γ_{β}-E2	1019.92 *5*	0.0066 *11*

† uncert(syst): 4.3% for IT, 10% for β-

Atomic Electrons (^{117}In)

⟨e⟩=87 *7* keV

e_{bin}(keV)	⟨e⟩(keV)	e(%)
4 - 28	1.54	24 *3*
129 - 159	3.35	2.52 *10*
287	63.0	21.9 *22*
311	12.2	3.9 *4*
312	3.0	0.97 *10*
314 - 315	3.8	1.22 *9*
817 - 861	0.00039	4.7 *6* ×10^{-5}
975 - 1020	0.000160	1.63 *18* ×10^{-5}

Continuous Radiation (^{117}In)

⟨β-⟩=183 keV;⟨IB⟩=0.31 keV

E_{bin}(keV)		⟨ ⟩(keV)	(%)
0 - 10	β-	0.0072	0.144
	IB	0.0077	
10 - 20	β-	0.0221	0.147
	IB	0.0075	0.052
20 - 40	β-	0.091	0.303
	IB	0.0144	0.050
40 - 100	β-	0.70	0.98
	IB	0.040	0.061
100 - 300	β-	8.3	4.07
	IB	0.099	0.057
300 - 600	β-	34.0	7.5
	IB	0.085	0.020
600 - 1300	β-	120	13.4
	IB	0.053	0.0069
1300 - 1769	β-	20.0	1.41
	IB	0.00113	8.3×10^{-5}

$^{117}_{50}$Sn(stable)

Δ: -90396 *3* keV
%: 7.68 *7*

$^{117}_{50}$Sn(13.61 *4* d)

Mode: IT
Δ: -90082 *3* keV
SpA: 8.200×10^{4} Ci/g
Prod: ^{116}Sn(n,γ); ^{114}Cd(α,n)

Photons (^{117}Sn)

⟨γ⟩=157.9 *7* keV

γ_{mode}	γ(keV)	γ(%)†
Sn L_ℓ	3.045	0.119 *16*
Sn L_η	3.272	0.057 *6*
Sn L_α	3.443	3.3 *4*
Sn L_β	3.747	2.8 *4*
Sn L_γ	4.230	0.33 *5*
Sn $K_{\alpha2}$	25.044	19.5 *6*
Sn $K_{\alpha1}$	25.271	36.5 *12*
Sn $K_{\beta1}$'	28.474	9.5 *3*
Sn $K_{\beta2}$'	29.275	2.01 *7*
γ M4	156.02 *3*	2.11 *4*
γ M1+0.02%E2	158.560 *12*	86.4

† 0.46% uncert(syst)

Atomic Electrons (^{117}Sn)

⟨e⟩=161 *3* keV

e_{bin}(keV)	⟨e⟩(keV)	e(%)
4	3.2	79 *8*
20	0.166	0.82 *8*
21	1.38	6.6 *7*
24	0.37	1.53 *16*
25	0.40	1.60 *16*
27	0.0068	0.0246 *25*
28	0.070	0.25 *3*
127	84.1	66.3 *19*
129	15.3	11.85 *24*
152	41.6	27.4 *8*
154	2.28	1.48 *3*
155	9.3	5.98 *17*
156	2.07	1.33 *4*
158	0.562	0.356 *7*
159	0.01029	0.00649 *13*

$^{117}_{51}$Sb(2.80 *1* h)

Mode: ϵ
Δ: -88641 *9* keV
SpA: 9.57×10^{6} Ci/g
Prod: ^{115}In(α,2n); ^{117}Sn(p,n)

Photons (^{117}Sb)

$\langle\gamma\rangle$=167.5 *9* keV

γ_{mode}	γ(keV)	γ(%)†
Sn L_ℓ	3.045	0.117 *16*
Sn L_η	3.272	0.061 *6*
Sn L_α	3.443	3.2 *4*
Sn L_β	3.742	2.9 *4*
Sn L_γ	4.220	0.34 *5*
Sn $K_{\alpha2}$	25.044	24.0 *7*
Sn $K_{\alpha1}$	25.271	44.9 *14*
Sn $K_{\beta1}$'	28.474	11.7 *4*
Sn $K_{\beta2}$'	29.275	2.47 *8*
γ M1+0.02%E2	158.560 *12*	85.9
γ (M2)	396.95 *9*	0.00011 *4*
γ (E2)	552.97 *9*	0.082 *13*
γ M1+1.7%E2	845.95 *13*	0.054 *16*
γ M1+2.3%E2	861.36 *5*	0.31 *3*
γ E2+0.3%M1	1004.51 *13*	0.205 *25*
γ E2	1019.92 *5*	0.105 *15*
γ M1+25%E2	1021.0 *5*	0.112 *17*
γ M1+3.6%E2	1287.7 *3*	0.028 *6*
γ	1339.5 *10*	0.009 *3*
γ [M1+E2]	1419.68 *24*	0.017 *6*
γ E2	1446.2 *3*	0.056 *15*
γ [M1+E2]	1578.23 *24*	0.019 *6*

† 0.47% uncert(syst)

Atomic Electrons (^{117}Sb)

$\langle e\rangle$=24.3 *5* keV

e_{bin}(keV)	⟨e⟩(keV)	e(%)
4	3.2	79 *8*
20	0.205	1.01 *10*
21	1.70	8.1 *8*
24 - 28	1.03	4.2 *3*
129	15.2	11.78 *24*
154	2.27	1.47 *3*
155	0.0345	0.0223 *5*
158	0.559	0.354 *7*
368 - 397	2.2 ×10^{-5}	6 *3* ×10^{-6}
524 - 553	0.0024	0.00046 *7*
817 - 861	0.0067	0.00081 *7*
975 - 1021	0.0055	0.00055 *4*
1258 - 1288	0.00032	2.6 *5* ×10^{-5}
1310 - 1339	7.2 ×10^{-5}	~5×10^{-6}
1390 - 1420	0.00058	4.1 *9* ×10^{-5}
1442 - 1446	6.2 ×10^{-5}	4.3 *10* ×10^{-6}
1549 - 1578	0.00016	1.0 *3* ×10^{-5}

Continuous Radiation (^{117}Sb)

⟨β+⟩=4.45 keV;⟨IB⟩=0.90 keV

E_{bin}(keV)		⟨ ⟩(keV)	(%)
0 - 10	β+	2.41×10^{-6}	2.91×10^{-5}
	IB	0.00044	
10 - 20	β+	8.6×10^{-5}	0.00052
	IB	0.00143	0.0084
20 - 40	β+	0.00205	0.0063
	IB	0.044	0.17
40 - 100	β+	0.078	0.101
	IB	0.0047	0.0073
100 - 300	β+	2.01	0.97
	IB	0.040	0.019
300 - 600	β+	2.36	0.62
	IB	0.19	0.042
600 - 1300	IB	0.58	0.067
1300 - 1596	IB	0.034	0.0025
	Σβ+		1.7

$^{117}_{52}$Te(1.03 *3* h)

Mode: ε

Δ: -85110 *30* keV

SpA: 2.59×10^7 Ci/g

Prod: ^{114}Sn(α,n); protons on Sb

Photons (^{117}Te)

$\langle\gamma\rangle$=1275 *42* keV

γ_{mode}	γ(keV)	γ(%)†
Sb L_ℓ	3.189	0.084 *11*
Sb L_η	3.437	0.042 *4*
Sb L_α	3.604	2.32 *25*
Sb L_β	3.931	2.03 *25*
Sb L_γ	4.438	0.24 *4*
Sb $K_{\alpha2}$	26.111	16.2 *5*
Sb $K_{\alpha1}$	26.359	30.2 *9*
Sb $K_{\beta1}$'	29.712	7.94 *24*
Sb $K_{\beta2}$'	30.561	1.69 *5*
γ	568.4 *8*	0.65 *13*
γ	634.7 *9*	0.45 *13*
γ E2	719.7 *5*	64.7
γ	830.6 *11*	0.52 *13*
γ	886.6 *6*	1.49 *19*
γ [M1+E2]	923.8 *6*	6.2 *7*
γ	930.3 *10*	~0.19
γ	996.6 *5*	3.9 *4*
γ	1090.7 *6*	6.9 *7*
γ	1354.4 *9*	0.52 *13*
γ [M1+E2]	1360.9 *8*	0.45 *13*
γ	1454.1 *11*	0.84 *13*
γ [M1]	1565.0 *7*	0.97 *13*
γ	1580.3 *7*	~0.19
γ	1595.2 *15*	~0.19 ?
γ	1716.3 *5*	15.9 *16*
γ	2212.9 *15*	0.32 *13* ?
γ [E2]	2284.7 *7*	0.39 *13*
γ	2300.0 *6*	11.2 *12*
γ	2379.2 *15*	~0.13 ?
γ	2884.9 *15*	~0.06 ?

† 2.2% uncert(syst)

Atomic Electrons (^{117}Te)

$\langle e\rangle$=5.95 *24* keV

e_{bin}(keV)	⟨e⟩(keV)	e(%)
4	1.94	46 *5*
5	0.29	6.1 *6*
21	0.30	1.40 *14*
22	0.94	4.3 *4*
25	0.30	1.19 *12*
26	0.32	1.25 *13*
29 - 30	0.063	0.218 *19*
538 - 568	0.015	~0.0027
604 - 635	0.009	~0.0015
689	1.14	0.166 *5*
715	0.146	0.0204 *6*
716 - 720	0.0461	0.00642 *13*
800 - 831	0.007	~0.0009
856 - 900	0.111	0.0125 *20*
919 - 966	0.053	0.0056 *20*
992 - 997	0.006	0.0006 *3*
1060 - 1091	0.07	0.007 *3*
1324 - 1361	0.0090	0.00068 *17*
1424 - 1454	0.006	0.00044 *18*
1535 - 1594	0.0123	0.00080 *11*
1686 - 1716	0.10	0.0060 *22*
2182 - 2281	0.052	0.0023 *8*
2284 - 2378	0.0077	0.00033 *9*
2854 - 2884	0.00027	~9×10^{-6}

Continuous Radiation (^{117}Te)

⟨β+⟩=198 keV;⟨IB⟩=2.4 keV

E_{bin}(keV)		⟨ ⟩(keV)	(%)
0 - 10	β+	2.77×10^{-6}	3.33×10^{-5}
	IB	0.0082	
10 - 20	β+	0.000102	0.00062
	IB	0.0086	0.059
20 - 40	β+	0.00260	0.0080
	IB	0.052	0.19
40 - 100	β+	0.113	0.146
	IB	0.046	0.071
100 - 300	β+	4.82	2.20
	IB	0.141	0.077
300 - 600	β+	29.9	6.5
	IB	0.27	0.061
600 - 1300	β+	132	14.5
	IB	0.91	0.099
1300 - 2500	β+	30.5	2.14
	IB	0.95	0.056
2500 - 2810	IB	0.0140	0.00054
	Σβ+		25

$^{117}_{52}$Te(103 *3* ms)

Mode: IT

Δ: >-84814 *30* keV

SpA: 1.39×10^{11} Ci/g

Prod: ^{114}Sn(α,n); ^{115}Sn(α,2n)

Photons (^{117}Te)

$\langle\gamma\rangle$=263 *3* keV

γ_{mode}	γ(keV)	γ(%)
γ (M1)	21.6 *2*	10.3
γ E2	274.4 *2*	95.1

$^{117}_{53}$I (2.3 *1* min)

Mode: ε

Δ: -80610 *200* keV syst

SpA: 7.0×10^8 Ci/g

Prod: ^{109}Ag(^{12}C,4n); protons on Ce

Photons (^{117}I)

γ_{mode}	γ(keV)	γ(rel)
γ	111.4 *10*	0.5 *1*
γ E2	274.4 *2*	27 *5*
γ	294.7 *10*	1.3 *3*
γ	303.2 *10*	2.0 *4*
γ	325.9 *10*	100
γ	407 *1*	0.90 *18* ?
γ	497.2 *10*	1.5 *3* ?
γ	683.1 *10*	2.5 *5* ?
γ	837.3 *10*	2.0 *4* ?

$^{117}_{54}$Xe(1.02 *3* min)

Mode: ϵ, ϵp(0.0029 *6* %)
Δ: -74290 *320* keV syst
SpA: 1.57×10^{9} Ci/g
Prod: protons on Ce; ^{16}O on ^{104}Pd

ϵp: 1700-3700 range

Photons (^{117}Xe)

γ$_{mode}$	γ(keV)	γ(rel)
γ$_ϵ$	58.4 *8*	
γ$_ϵ$	73.7 *7*	14.9 *4*
γ$_ϵ$	94.4 *6*	12.3 *4*
γ$_ϵ$	104.5 *6*	
γ$_ϵ$	112.2 *12*	5.1 *4*
γ$_ϵ$	117.0 *6*	41.0 *8*
γ$_ϵ$	155.5 *12*	11.0 *6*
γ$_ϵ$	160.8 *12*	33.8 *8*
γ$_ϵ$	198.8 *7*	
γ$_ϵ$	203.3 *8*	4.5 *7*
γ$_ϵ$	221.4 *6*	100
γ$_ϵ$	257.1 *12*	16.6 *9*
γ$_ϵ$	295.1 *7*	74.4 *15*
γ$_ϵ$	307.6 *10*	22.7 *8*
γ$_ϵ$	315.8 *6*	25.6 *9*
γ$_ϵ$[E2]	353.5 *8*	19.6 *9*
γ$_ϵ$	439.7 *7*	19.9 *8*
γ$_ϵ$	519.2 *8*	55.4 *13*
γ$_ϵ$	544.1 *7*	
γ$_ϵ$	609.8 *12*	4.5 *7*
γ$_ϵ$	639.1 *12*	50.4 *13*
γ$_ϵ$	661.1 *7*	55.6 *13*
γ$_{ϵp}$(E2)	678.9 *3*	
γ$_ϵ$	1523.3 *10*	23.7 *9*

$^{117}_{55}$Cs(6.5 *4* s)

Mode: ϵ
Δ: -66230 *180* keV
SpA: 1.41×10^{10} Ci/g
Prod: protons on La

$^{117}_{56}$Ba(1.9 *2* s)

Mode: ϵ, ϵp
Δ: -56930 *560* keV syst
SpA: 4.3×10^{10} Ci/g
Prod: ^{32}S on ^{92}Mo

ϵp: 2000-6000 range

A = 118

NDS **17**, 1 (1976)

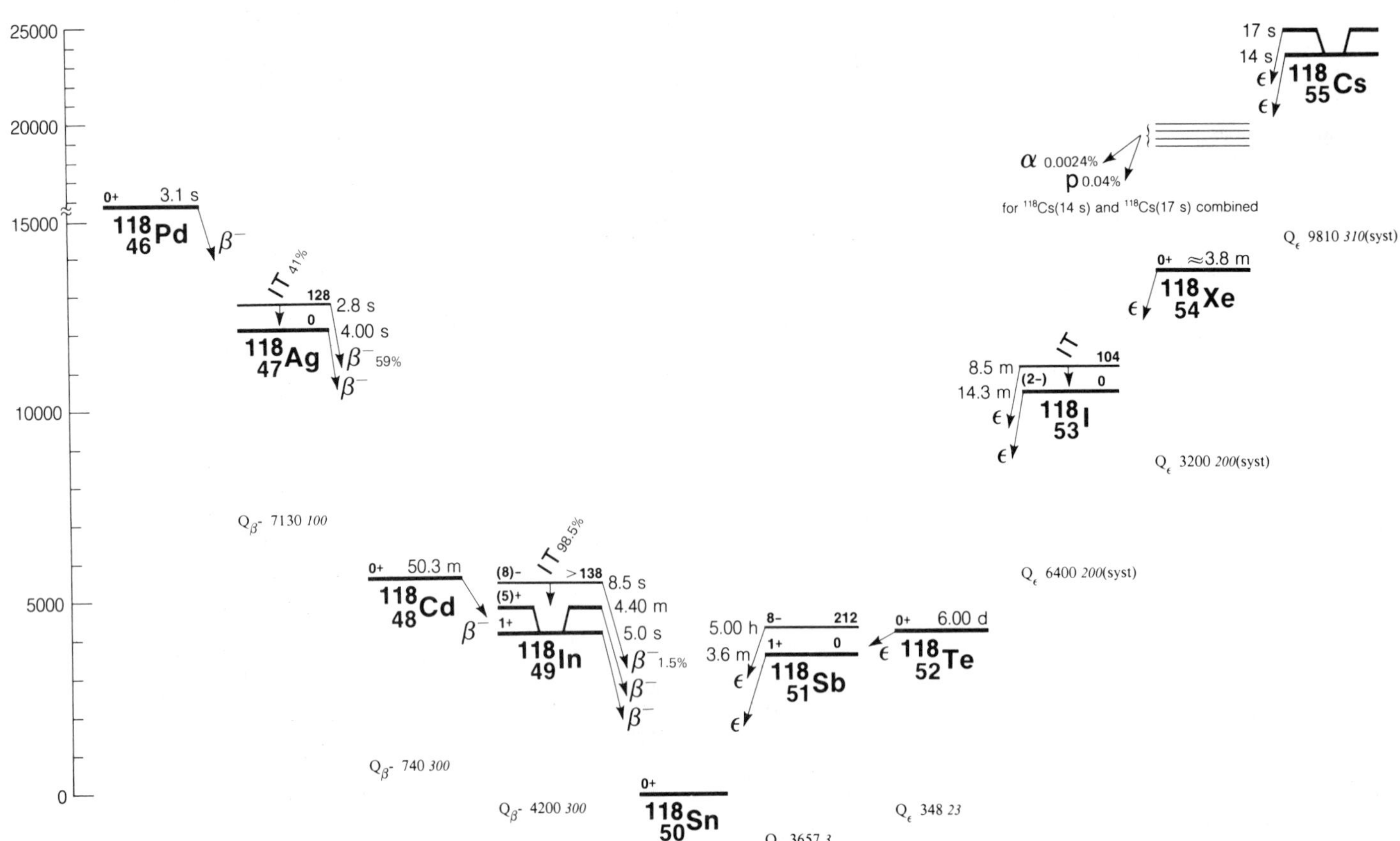

$^{118}_{46}$Pd(3.1 *3* s)

Mode: β-
SpA: 2.8×10^{10} Ci/g
Prod: fission

$^{118}_{47}$Ag(4.00 *4* s)

Mode: β-
Δ: -79580 *100* keV
SpA: 2.195×10^{10} Ci/g
Prod: fission

Photons (^{118}Ag)

⟨γ⟩=2044 *120* keV

γ_{mode}	γ(keV)	γ(%)
γ [E2]	487.77 *13*	90
γ	677.08 *13*	38 *4*
γ	770.90 *13*	0.6 *3*
γ	781.55 *19*	6.5 *5*
γ	797.81 *13*	7.8 *5*
γ	808.28 *13*	0.22 *9*
γ	1058.61 *20*	2.3 *9*
γ	1269.32 *22*	4.8 *5*
γ	1939.0 *3*	4.5 *5*
γ	2101.5 *3*	8.9 *9*
γ	2779.2 *3*	7.2 *9*
γ	2789.4 *3*	10.3 *14*
γ	3225.9 *3*	11.9 *18*

† approximate

Continuous Radiation (^{118}Ag)

⟨β-⟩=2825 keV;⟨IB⟩=11.8 keV

E_{bin}(keV)		⟨ ⟩(keV)	(%)
0 - 10	β-	0.00138	0.0274
	IB	0.070	
10 - 20	β-	0.00424	0.0282
	IB	0.070	0.48
20 - 40	β-	0.0177	0.059
	IB	0.138	0.48
40 - 100	β-	0.141	0.199
	IB	0.41	0.62
100 - 300	β-	1.96	0.94
	IB	1.26	0.70
300 - 600	β-	10.8	2.33
	IB	1.66	0.39
600 - 1300	β-	98	10.0
	IB	3.0	0.34
1300 - 2500	β-	554	28.6
	IB	3.2	0.18
2500 - 5000	β-	1867	52
	IB	1.9	0.060
5000 - 7130	β-	293	5.4
	IB	0.047	0.00088

$^{118}_{47}$Ag(2.8 *3* s)

Mode: β-(59 %), IT(41 %)
Δ: -79452 *100* keV
SpA: 3.0×10^{10} Ci/g
Prod: fission

Photons (^{118}Ag)

⟨γ⟩=1235 *80* keV

γ_{mode}	γ(keV)	γ(%)†
Ag L_ℓ	2.634	0.031 *8*
Ag L_η	2.806	0.018 *5*
Ag L_α	2.984	0.84 *22*
Ag L_β	3.209	0.64 *17*
Ag L_γ	3.553	0.061 *16*
Ag $K_{\alpha 2}$	21.990	4.8 *10*
Ag $K_{\alpha 1}$	22.163	9.1 *19*
Cd $K_{\alpha 2}$	22.984	0.131 *24*
Cd $K_{\alpha 1}$	23.174	0.25 *5*
Ag $K_{\beta 1}$'	24.934	2.3 *5*
Ag $K_{\beta 2}$'	25.603	0.45 *9*
Cd $K_{\beta 1}$'	26.085	0.063 *11*
Cd $K_{\beta 2}$'	26.801	0.0130 *24*
γ_{IT} (E3)	127.74 *16*	7.2
$\gamma_{\beta-}$[E2]	487.77 *13*	59
$\gamma_{\beta-}$	677.08 *13*	58 *6*
$\gamma_{\beta-}$	770.90 *13*	19.8 *4*
$\gamma_{\beta-}$	808.28 *13*	5.7 *6*
$\gamma_{\beta-}$	1058.61 *20*	32 *4*

† uncert(syst): 25% for IT, 17% for β-

Atomic Electrons (^{118}Ag)

⟨e⟩=42 *5* keV

e_{bin}(keV)	⟨e⟩(keV)	e(%)
3 - 26	1.50	28 *4*
102	20.3	20 *4*
124	13.9	11.2 *23*
127	2.9	2.3 *5*
128	0.51	0.40 *8*
461	1.7	0.36 *7*
484 - 488	0.29	0.060 *10*
650	0.7	~0.11
673 - 677	0.11	0.017 *7*
744 - 782	0.30	0.040 *15*
804 - 808	0.009	0.0011 *4*
1032 - 1059	0.27	0.027 *11*

Continuous Radiation (^{118}Ag)

⟨β-⟩=1340 keV;⟨IB⟩=5.0 keV

E_{bin}(keV)		⟨ ⟩(keV)	(%)
0 - 10	β-	0.00130	0.0259
	IB	0.037	
10 - 20	β-	0.00400	0.0267
	IB	0.037	0.25
20 - 40	β-	0.0168	0.056
	IB	0.072	0.25
40 - 100	β-	0.133	0.188
	IB	0.21	0.32
100 - 300	β-	1.83	0.88
	IB	0.65	0.36
300 - 600	β-	9.8	2.13
	IB	0.82	0.19
600 - 1300	β-	86	8.7
	IB	1.41	0.160
1300 - 2500	β-	427	22.3
	IB	1.28	0.073
2500 - 5000	β-	814	24.7
	IB	0.44	0.0148
5000 - 5322	β-	1.70	0.0335
	IB	2.1×10^{-5}	4.2×10^{-7}

$^{118}_{48}$Cd(50.3 *2* min)

Mode: β-
Δ: -86710 *20* keV
SpA: 3.168×10^{7} Ci/g
Prod: fission

Continuous Radiation (^{118}Cd)

⟨β-⟩=244 keV;⟨IB⟩=0.19 keV

E_{bin}(keV)		⟨ ⟩(keV)	(%)
0 - 10	β-	0.110	2.20
	IB	0.0121	
10 - 20	β-	0.332	2.21
	IB	0.0113	0.079
20 - 40	β-	1.34	4.46
	IB	0.021	0.072
40 - 100	β-	9.5	13.5
	IB	0.049	0.077
100 - 300	β-	84	42.6
	IB	0.078	0.048
300 - 600	β-	138	33.3
	IB	0.019	0.0051
600 - 740	β-	10.8	1.70
	IB	0.000151	2.4×10^{-5}

$^{118}_{49}$In(5.0 *3* s)

Mode: β-
Δ: -87450 *300* keV
SpA: 1.79×10^{10} Ci/g
Prod: daughter ^{118}Cd; ^{118}Sn(n,p)

Photons (^{118}In)

⟨γ⟩=78 *25* keV

γ_{mode}	γ(keV)	γ(%)†
γ [E2]	528.3 *4*	0.70 *10*
γ [M1+E2]	813.66 *20*	0.19 *4*
γ [E2]	827.1 *5*	0.035 *15*
γ	1096.9 *3*	0.10 *4*
γ [M1+E2]	1173.2 *4*	0.43 *9*
γ E2	1229.69 *3*	5
γ E2	1908.2 *10*	0.13 *3*
γ [E2]	2043.34 *20*	0.10 *3*

† 40% uncert(syst)

Continuous Radiation (^{118}In)

⟨β-⟩=1707 keV;⟨IB⟩=5.4 keV

E_{bin}(keV)		⟨ ⟩(keV)	(%)
0 - 10	β-	0.00467	0.093
	IB	0.053	
10 - 20	β-	0.0143	0.095
	IB	0.052	0.36
20 - 40	β-	0.060	0.198
	IB	0.103	0.36
40 - 100	β-	0.469	0.66
	IB	0.30	0.46
100 - 300	β-	6.2	3.01
	IB	0.88	0.49
300 - 600	β-	31.5	6.9
	IB	1.07	0.25
600 - 1300	β-	238	24.5
	IB	1.65	0.19
1300 - 2500	β-	829	44.4
	IB	1.15	0.067
2500 - 4200	β-	602	20.2
	IB	0.169	0.0059

$^{118}_{49}$In(4.40 *5* min)

Mode: β-

Δ: -87450 *300* keV

SpA: 3.62×10^{8} Ci/g

Prod: ^{118}Sn(n,p); ^{119}Sn(γ,p)

Photons (^{118}In)

⟨γ⟩=2718 *65* keV

γ_{mode}	γ(keV)	γ(%)†
Sn $K_{\alpha2}$	25.044	0.166 *22*
Sn $K_{\alpha1}$	25.271	0.31 *4*
Sn $K_{\beta1}$'	28.474	0.081 *11*
Sn $K_{\beta2}$'	29.275	0.0171 *22*
γ [M1+E2]	208.74 *19*	2.3 *8*
γ [M1+E2]	230.0 *6*	0.87 *19*
γ [E2]	445.83 *17*	5.9 *3*
γ [M1+E2]	474.65 *22*	3.0 *3*
γ [E2]	560.9 *4*	1.4 *5*
γ	637.2 *3*	3.5 *5*
γ [M1+E2]	683.38 *16*	55 *5*
γ [M1+E2]	813.66 *20*	3.4 *4*
γ E2	1050.75 *3*	82 *5*
γ	1096.9 *3*	3.1 *7*
γ [M1+E2]	1173.2 *4*	1.3 *2*
γ E2	1229.69 *3*	96.0
γ [E2]	1259.48 *19*	3.9 *4*
γ [E2]	1504.2 *6*	0.9 *4*
γ [E2]	1734.12 *16*	~0.5 ?
γ [E2]	2043.34 *20*	3.4 *3*
γ	2326.6 *3*	0.58 *20*

† 0.30% uncert(syst)

Atomic Electrons (^{118}In)

⟨e⟩=4.45 *23* keV

e_{bin}(keV)	⟨e⟩(keV)	e(%)
4 - 28	0.043	0.65 *10*
180	0.34	0.19 *8*
201 - 230	0.21	0.10 *3*
417	0.217	0.052 *3*
441 - 442	0.0321	0.0073 *4*
445	0.109	0.024 *3*
446 - 475	0.0189	0.0040 *5*
532 - 561	0.040	0.0076 *23*
608 - 637	0.06	0.010 *5*
654	1.10	0.167 *24*

Atomic Electrons (^{118}In) (continued)

e_{bin}(keV)	⟨e⟩(keV)	e(%)
679	0.14	0.021 *4*
682 - 683	0.034	0.0049 *7*
784 - 814	0.062	0.0079 *12*
1022	0.84	0.083 *5*
1046 - 1093	0.160	0.0152 *14*
1096 - 1144	0.0139	0.00122 *22*
1169 - 1173	0.0020	0.00017 *3*
1200	0.834	0.0695 *14*
1225 - 1259	0.165	0.0135 *4*
1475 - 1503	0.007	0.00050 *20*
2014 - 2043	0.0212	0.00105 *9*
2297 - 2326	0.0027	0.00012 *5*

Continuous Radiation (^{118}In)

⟨β-⟩=545 keV;⟨IB⟩=0.83 keV

E_{bin}(keV)		⟨ ⟩(keV)	(%)
0 - 10	β-	0.0382	0.76
	IB	0.024	
10 - 20	β-	0.116	0.77
	IB	0.023	0.159
20 - 40	β-	0.477	1.59
	IB	0.044	0.153
40 - 100	β-	3.59	5.09
	IB	0.118	0.18
100 - 300	β-	40.5	19.9
	IB	0.28	0.161
300 - 600	β-	144	32.1
	IB	0.22	0.053
600 - 1300	β-	310	36.6
	IB	0.117	0.0151
1300 - 1920	β-	47.2	3.24
	IB	0.0041	0.00029

$^{118}_{49}$In(8.5 *3* s)

Mode: IT(98.5 %), β-(1.5 %)

Δ: >-87312 *300* keV

SpA: 1.08×10^{10} Ci/g

Prod: ^{118}Sn(n,p); ^{121}Sb(n,α)

Photons (^{118}In)

⟨γ⟩=39.9 *15* keV

γ_{mode}	γ(keV)	γ(%)
In L_{ℓ}	2.905	0.078 *14*
In L_{η}	3.112	0.046 *8*
In L_{α}	3.286	2.1 *4*
In L_{β}	3.557	1.8 *3*
In L_{γ}	3.955	0.17 *3*
In $K_{\alpha2}$	24.002	11.0 *7*
In $K_{\alpha1}$	24.210	20.6 *14*
In $K_{\beta1}$'	27.265	5.3 *4*
In $K_{\beta2}$'	28.022	1.11 *8*
γ_{β}-E1	40.83 *8*	0.0058 *7*
γ_{IT} (E3)	138.2 *5*	21.6
γ_{β}-E2	253.733 *10*	0.021 *4*
γ_{β}-E2	1050.75 *3*	0.023 *5*
γ_{β}-[E3]	1091.57 *8*	0.00105 *12*
γ_{β}-E2	1229.69 *3*	0.023 *5*

Atomic Electrons (^{118}In)*

⟨e⟩=96 *4* keV

e_{bin}(keV)	⟨e⟩(keV)	e(%)
4	2.04	53 *6*
19	0.096	0.49 *6*
20	0.81	4.0 *5*
23	0.211	0.91 *11*
24	0.22	0.95 *11*
26	0.0131	0.050 *6*
27	0.030	0.114 *13*
110	48.8	44 *3*
134	35.2	26.2 *16*
137	4.3	3.11 *19*
138	4.7	3.41 *21*

* with IT

Continuous Radiation (^{118}In)

⟨β-⟩=10.1 keV;⟨IB⟩=0.017 keV

E_{bin}(keV)		⟨ ⟩(keV)	(%)
0 - 10	β-	0.000364	0.0072
	IB	0.00042	
10 - 20	β-	0.00111	0.0074
	IB	0.00041	0.0029
20 - 40	β-	0.00459	0.0153
	IB	0.00079	0.0028
40 - 100	β-	0.0351	0.0497
	IB	0.0022	0.0034
100 - 300	β-	0.424	0.207
	IB	0.0055	0.0031
300 - 600	β-	1.76	0.387
	IB	0.0048	0.00116
600 - 1300	β-	6.6	0.73
	IB	0.0032	0.00041
1300 - 1763	β-	1.32	0.092
	IB	8.0×10^{-5}	5.8×10^{-6}

$^{118}_{50}$Sn(stable)

Δ: -91652 *3* keV

%: 24.22 *11*

$^{118}_{51}$Sb(3.6 *1* min)

Mode: ϵ

Δ: -87995 *4* keV

SpA: 4.42×10^{8} Ci/g

Prod: daughter ^{118}Te; ^{115}In(α,n)

Photons (^{118}Sb)

⟨γ⟩=51 *3* keV

γ_{mode}	γ(keV)	γ(%)†
Sn L_{ℓ}	3.045	0.027 *4*
Sn L_{η}	3.272	0.0139 *14*
Sn L_{α}	3.443	0.74 *8*
Sn L_{β}	3.742	0.65 *8*
Sn L_{γ}	4.220	0.077 *11*
Sn $K_{\alpha2}$	25.044	5.52 *17*
Sn $K_{\alpha1}$	25.271	10.3 *3*
Sn $K_{\beta1}$'	28.474	2.70 *8*
Sn $K_{\beta2}$'	29.275	0.568 *18*
γ [E2]	528.3 *4*	0.385 *8*
γ [M1+E2]	813.66 *20*	0.0098 *13*
γ [E2]	827.1 *5*	0.356 *25*

Photons (^{118}Sb) (continued)

γ_{mode}	γ(keV)	γ(%)†
γ	1096.9 *3*	0.046 *7*
γ [M1+E2]	1173.2 *4*	0.047 *5*
γ E2	1229.69 *3*	2.47
γ [E2]	1267.2 *5*	0.573 *25*
γ [M1+E2]	1447.9 *5*	0.022 *7*
γ [M1+E2]	1699.9 *7*	0.057 *5*
γ [E2]	1907.8 *5*	0.040 *5*
γ [E2]	2043.34 *20*	0.0097 *11*
γ	2326.6 *3*	0.009 *3*
γ [E2]	2677.6 *5*	0.012 *5*

† 10% uncert(syst)

Atomic Electrons (^{118}Sb)

$\langle e \rangle$=1.46 *9* keV

e_{bin}(keV)	$\langle e \rangle$(keV)	e(%)
4	0.74	18.2 *19*
20	0.047	0.232 *24*
21	0.39	1.85 *19*
24	0.104	0.43 *4*
25	0.112	0.45 *5*
27 - 28	0.0217	0.078 *7*
499 - 528	0.0123	0.00245 *10*
784 - 827	0.0059	0.00073 *5*
1068 - 1097	0.00043	4.1 *18* $\times10^{-5}$
1144 - 1173	0.00055	4.8 *7* $\times10^{-5}$
1200 - 1238	0.0296	0.00245 *18*
1263 - 1267	0.00073	5.8 *3* $\times10^{-5}$
1419 - 1448	0.00021	1.5 *4* $\times10^{-5}$
1671 - 1699	0.00045	2.7 *3* $\times10^{-5}$
1879 - 1907	0.00026	1.39 *16* $\times10^{-5}$
2014 - 2043	6.1 $\times10^{-5}$	3.0 *3* $\times10^{-6}$
2297 - 2326	4.0 $\times10^{-5}$	1.7 *7* $\times10^{-6}$
2648 - 2677	6.3 $\times10^{-5}$	2.4 *8* $\times10^{-6}$

Continuous Radiation (^{118}Sb)

$\langle\beta+\rangle$=882 keV;$\langle$IB$\rangle$=4.9 keV

E_{bin}(keV)		$\langle\ \rangle$(keV)	(%)
0 - 10	β+	2.70 $\times10^{-6}$	3.25 $\times10^{-5}$
	IB	0.032	
10 - 20	β+	9.8 $\times10^{-5}$	0.00059
	IB	0.032	0.22
20 - 40	β+	0.00246	0.0075
	IB	0.073	0.26
40 - 100	β+	0.108	0.139
	IB	0.18	0.27
100 - 300	β+	5.1	2.31
	IB	0.50	0.28
300 - 600	β+	39.9	8.6
	IB	0.61	0.143
600 - 1300	β+	314	32.5
	IB	1.21	0.135
1300 - 2500	β+	521	30.7
	IB	1.7	0.097
2500 - 3657	β+	1.59	0.063
	IB	0.50	0.018
	$\Sigma\beta$+		74

$^{118}_{51}$Sb(5.00 *1* h)

Mode: ϵ
Δ: -87783 *4* keV
SpA: 5.311$\times10^{6}$ Ci/g
Prod: ^{115}In(α,n); protons on U; ^{118}Sn(p,n); ^{118}Sn(d,2n)

Photons (^{118}Sb)

$\langle\gamma\rangle$=2576 *55* keV

γ_{mode}	γ(keV)	γ(%)†
Sn L_ℓ	3.045	0.156 *22*
Sn L_η	3.272	0.081 *9*
Sn L_α	3.443	4.3 *5*
Sn L_β	3.742	3.8 *5*
Sn L_γ	4.217	0.44 *6*
Sn $K_{\alpha2}$	25.044	31.7 *15*
Sn $K_{\alpha1}$	25.271	59 *3*
Sn $K_{\beta1}$'	28.474	15.5 *8*
Sn $K_{\beta2}$'	29.275	3.26 *16*
γ E1	40.83 *8*	19.3 *20*
γ E2	253.733 *10*	99 *6*
γ [M1+E2]	813.66 *20*	~0.02
γ E2	1050.75 *3*	97 *5*
γ [E3]	1091.57 *8*	3.5 *3*
γ E2	1229.69 *3*	99.9
γ [E2]	2043.34 *20*	0.02 *1*

† 0.10% uncert(syst)

Atomic Electrons (^{118}Sb)

$\langle e \rangle$=31 *1* keV

e_{bin}(keV)	$\langle e \rangle$(keV)	e(%)
4	4.3	106 *12*
12	4.2	36 *4*
20	0.27	1.33 *14*
21	2.24	10.6 *12*
24 - 28	1.36	5.5 *4*
36	1.17	3.2 *3*
37 - 41	1.22	3.20 *23*
225	11.5	5.1 *3*
249	1.33	0.53 *3*
250 - 254	1.29	0.515 *23*
784 - 814	0.00037	4.7 *21* $\times10^{-5}$
1022	1.00	0.098 *5*
1046 - 1092	0.237	0.0225 *9*
1200 - 1230	1.002	0.0832 *15*
2014 - 2043	0.00013	6 *3* $\times10^{-6}$

Continuous Radiation (^{118}Sb)

$\langle\beta+\rangle$=0.260 keV;$\langle$IB$\rangle$=0.50 keV

E_{bin}(keV)		$\langle\ \rangle$(keV)	(%)
0 - 10	β+	1.87 $\times10^{-6}$	2.26 $\times10^{-5}$
	IB	0.00023	
10 - 20	β+	6.4 $\times10^{-5}$	0.000389
	IB	0.00126	0.0072
20 - 40	β+	0.00143	0.00440
	IB	0.044	0.17
40 - 100	β+	0.0418	0.056
	IB	0.0038	0.0060
100 - 300	β+	0.217	0.140
	IB	0.036	0.017
300 - 600	IB	0.157	0.035
600 - 1294	IB	0.25	0.032
	$\Sigma\beta$+		0.20

$^{118}_{52}$Te(6.00 *2* d)

Mode: ϵ
Δ: -87647 *23* keV
SpA: 1.844$\times10^{5}$ Ci/g
Prod: protons on Sb; ^{121}Sb(d,5n)

Photons (^{118}Te)

$\langle\gamma\rangle$=19.8 *4* keV

γ_{mode}	γ(keV)	γ(%)
Sb L_ℓ	3.189	0.110 *15*
Sb L_η	3.437	0.055 *6*
Sb L_α	3.604	3.1 *3*
Sb L_β	3.930	2.7 *3*
Sb L_γ	4.444	0.33 *5*
Sb $K_{\alpha2}$	26.111	21.1 *6*
Sb $K_{\alpha1}$	26.359	39.3 *12*
Sb $K_{\beta1}$'	29.712	10.3 *3*
Sb $K_{\beta2}$'	30.561	2.20 *7*

Atomic Electrons (^{118}Te)

$\langle e \rangle$=5.5 *3* keV

e_{bin}(keV)	$\langle e \rangle$(keV)	e(%)
4	2.5	60 *6*
5	0.42	9.0 *9*
21	0.39	1.82 *19*
22	1.22	5.6 *6*
25	0.39	1.55 *16*
26	0.42	1.63 *17*
29	0.067	0.232 *24*
30	0.0153	0.052 *5*

Continuous Radiation (^{118}Te)

$\langle$IB$\rangle$=0.059 keV

E_{bin}(keV)		$\langle\ \rangle$(keV)	(%)
10 - 20	IB	0.00093	0.0054
20 - 40	IB	0.049	0.19
40 - 100	IB	0.0031	0.0052
100 - 300	IB	0.0059	0.0036
300 - 348	IB	2.8 $\times10^{-5}$	9.2 $\times10^{-6}$

$^{118}_{53}$I (14.3 *1* min)

Mode: ϵ
Δ: -81250 *200* keV syst
SpA: 1.114$\times10^{8}$ Ci/g
Prod: protons on I; ^{12}C on Ag; protons on La; protons on Te

Photons (^{118}I)

⟨γ⟩=923 *170* keV

γ_{mode}	γ(keV)	γ(%)†
γ	496.4 *6*	~0.2
γ	544.7 *3*	11.3 *11*
γ	551.6 *5*	1.9 *2*
γ	559.3 *10*	~0.2
γ (E2)	599.9 *3*	
γ (E2)	604.96 *22*	95
γ	685.4 *5*	~0.4
γ	711.8 *6*	~0.4
γ	740.5 *4*	1.5 *2*
γ	1149.7 *3*	4.6 *5*
γ	1256.5 *5*	3.5 *4*
γ	1285.2 *5*	0.60 *12*
γ	1338.2 *5*	11.4 *11*

† approximate

$^{118}_{53}$I (8.5 *5* min)

Mode: ε, IT
Δ: -81146 *200* keV syst
SpA: 1.87×10^{8} Ci/g
Prod: ^{12}C on Ag; protons on La; protons on Te

Photons (^{118}I)

γ_{mode}	γ(keV)	γ(rel)
γ_{IT}	104 *1*	
γ_{ϵ}(E2)	599.9 *3*	100 *5*
γ_{ϵ}(E2)	604.96 *22*	~100
γ_{ϵ}(E2)	613.8 *3*	~56

$^{118}_{54}$Xe(3.8 *9* min)

Mode: ε
Δ: -78050 *280* keV syst
SpA: 4.19×10^{8} Ci/g
Prod: protons on La; protons on Ce

Photons (^{118}Xe)

γ_{mode}	γ(keV)	γ(rel)
γ	53.5 *10*	100 *20*
γ	60 *1*	88 *20*
γ	119.9 *10*	76 *8*
γ	150.5 *10*	44 *6*
γ	274 *1*	29 *12*

$^{118}_{55}$Cs(14 *2* s)

Mode: ε, εp(0.038 *7* %), εα(0.0024 *4* %) (includes ^{118}Cs(17 s))
Δ: -68240 *130* keV
SpA: 6.66×10^{9} Ci/g
Prod: protons on Ta; protons on Th; protons on La

Photons (^{118}Cs(14 + 17 s))

γ_{mode}	γ(keV)	γ(rel)
γ	52.9 *3*	23 *4*
γ	56.9 *3*	2.0 *7*
γ	117.6 *4*	8 *3*
γ	148.3 *3*	14.0 *21*
γ	300.3 *2*	3.3 *7*
γ	316.3 *3*	4.7 *13*
γ	337.2 *2*	
γ	387.3 *3*	2.5 *4*
γ	397.5 *2*	20.4 *12*
γ	407.5 *3*	7 *3*
γ	408.0 *3*	7 *3*
γ	417.9 *2*	13.7 *17*
γ	423.4 *3*	3.1 *7*
γ	437.9 *2*	21.1 *11*
γ	472.7 *2*	100 *27*
γ	487.3 *3*	8 *3*
γ	492.5 *3*	16 *7*
γ	493.0 *2*	61 *7*
γ	502 *1*	
γ	512.9 *5*	28 *8*
γ	530.5 *3*	13 *3*
γ	545.1 *3*	5.1 *7*
γ	555.9 *2*	63 *5*
γ	577.3 *3*	7.3 *15*
γ	586.2 *3*	~47
γ	586.5 *2*	73 *27*
γ	590.0 *3*	~24
γ	590.6 *2*	<35
γ	599.8 *3*	11 *3*
γ	618.8 *4*	4.0 *13*
γ	630.2 *3*	21 *4*
γ	631.1 *3*	21 *4*
γ	635.2 *3*	7 *3*
γ	638.2 *2*	9 *3*
γ	651.3 *3*	4.7 *9*
γ	660.5 *5*	3.3 *9*
γ	676.5 *2*	35 *4*
γ	700.6 *3*	~1
γ	707.7 *5*	1.7 *7*
γ	723.2 *3*	2.0 *9*
γ	733.7 *5*	1.7 *5*
γ	751.4 *5*	2.7 *8*
γ	802.0 *3*	5.3 *20*
γ	810.2 *3*	~5
γ	838.6 *3*	4.0 *13*
γ	845.1 *5*	8 *3*
γ	848.3 *4*	16 *5*
γ	891.5 *3*	14.0 *17*
γ	901.0 *4*	6.1 *8*
γ	914.9 *3*	5.6 *8*
γ	928.1 *3*	56 *7*
γ	968.2 *4*	6.3 *8*
γ	983.0 *4*	4.0 *13*
γ	1021.8 *2*	56 *8*
γ	1028.8 *2*	73 *11*
γ	1085.7 *3*	5.6 *11*
γ	1089.6 *3*	~5
γ	1112.0 *2*	20 *5*
γ	1143.7 *2*	21 *5*
γ	1161.7 *3*	4.7 *8*
γ	1184.5 *2*	56 *9*
γ	1228.3 *2*	85 *9*
γ	1281.6 *3*	6.1 *12*
γ	1307.4 *3*	3.7 *7*
γ	1342.8 *3*	2.4 *5*
γ	1353.4 *2*	35 *5*
γ	1356.6 *3*	12 *3*
γ	1364.5 *3*	12.7 *3*
γ	1393.4 *3*	13.3 *20*
γ	1435.1 *3*	4.7 *7*
γ	1469.0 *3*	3.3 *7*
γ	1491.1 *4*	3.5 *11*
γ	1500.5 *3*	9.3 *20*
γ	1520.8 *4*	6.0 *13*
γ	1532.2 *3*	5.2 *9*
γ	1580.4 *3*	4.7 *9*
γ	1598.2 *4*	7.3 *12*
γ	1640.1 *3*	31 *5*
γ	1644.6 *4*	4.8 *8*
γ	1657.2 *5*	3.3 *7*
γ	1671.5 *4*	4.8 *8*
γ	1686.5 *3*	20 *3*
γ	1765.8 *4*	4.0 *8*

Photons (^{118}Cs(14 + 17 s))
(continued)

γ_{mode}	γ(keV)	γ(rel)
γ	1780.8 *4*	4.0 *13*
γ	1806.5 *3*	19 *4*
γ	1838.3 *3*	19 *4*
γ	1846.7 *3*	4.7 *7*
γ	1853.0 *4*	3.3 *13*
γ	1873.0 *4*	7 *3*
γ	1912.3 *3*	8.7 *13*
γ	1930.7 *3*	21 *4*
γ	2051.5 *3*	6.4 *12*
γ	2067.7 *5*	4.8 *9*
γ	2121.8 *5*	4.3 *8*
γ	2170.8 *5*	3.3 *13*
γ	2190.2 *5*	7.2 *13*
γ	2286.3 *6*	4.0 *13*
γ	2325.5 *5*	4.0 *13*
γ	2359.5 *5*	5.3 *13*
γ	2434.5 *10*	4.7 *13*
γ	2459.5 *10*	3.3 *13*
γ	2531 *1*	4.4 *13*
γ	2665 *1*	7 *3*
γ	2678 *1*	8 *3*
γ	2743 *1*	6 *2*
γ	2825 *1*	5.3 *13*
γ	2974 *1*	9 *3*

$^{118}_{55}$Cs(17 *3* s)

Mode: ε, εp(0.038 *7* %), εα(0.0024 *4* %) (includes ^{118}Cs(14 s))
Δ: -68240 *130* keV
SpA: 5.51×10^{9} Ci/g
Prod: protons on Ta; protons on Th; protons on La

see ^{118}Cs(14 s) for γ rays

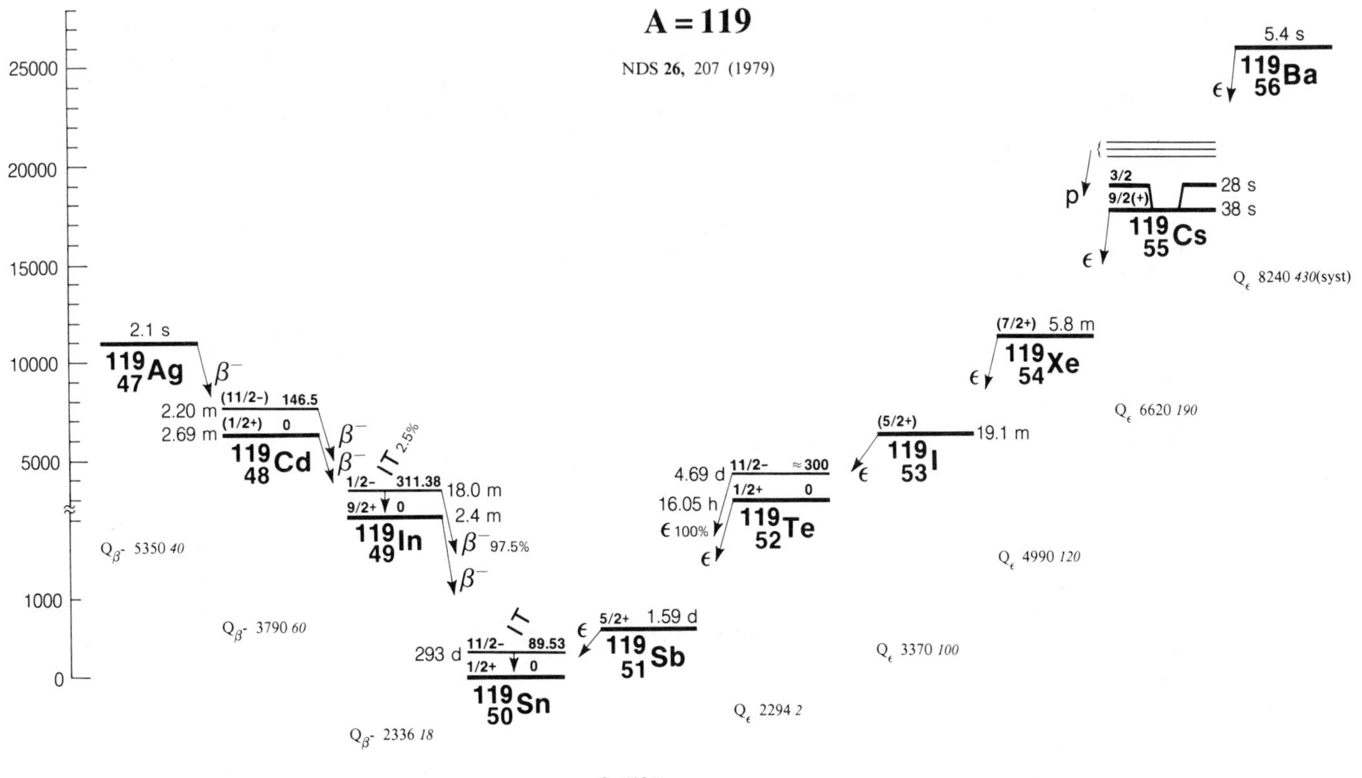

Q_ε 593 8

$^{119}_{47}$Ag(2.1 *1* s)

Mode: β-
Δ: -78590 *70* keV
SpA: 3.85×10^{10} Ci/g
Prod: fission

Photons (^{119}Ag)

⟨γ⟩=1331 *29* keV

γ_{mode}	γ(keV)	γ(%)†
Cd L_ℓ	2.767	0.069 *16*
Cd L_η	2.957	0.032 *7*
Cd L_α	3.133	1.9 *4*
Cd L_β	3.392	1.3 *3*
Cd L_γ	3.783	0.13 *3*
γ [M1]	14.33 *9*	~0.13
Cd $K_{\alpha2}$	22.984	13.5 *21*
Cd $K_{\alpha1}$	23.174	25 *4*
Cd $K_{\beta1}$'	26.085	6.5 *10*
Cd $K_{\beta2}$'	26.801	1.35 *21*
γ M1	26.88 *6*	2.2 *4*
γ M1	67.27 *8*	5.7 *10*
γ E2	81.61 *8*	2.08 *22*
γ	131.65 *9*	
γ	150.59 *14*	0.24 *6*
γ	173.3 *5*	0.27 *6*
γ	177.66 *7*	
γ	191.67 *13*	~0.05 ?
γ E1	198.93 *7*	6.5 *8*
γ E1	213.26 *8*	7.2 *9*
γ	224.97 *9*	0.11 *4*
γ	235.19 *7*	0.22 *6*
γ [M1+E2]	247.45 *5*	0.70 *11*
γ	262.26 *14*	0.11 *3*
γ	270.93 *10*	0.33 *10*
γ	280.48 *9*	0.48 *9*
γ	324.65 *9*	0.15 *3*
γ M1,E2	366.17 *6*	10.0 *6*
γ	370.49 *15*	2.81 *20*
γ	372.09 *6*	1.26 *12*
γ [M1+E2]	378.88 *6*	0.22 *4*
γ	393.04 *7*	0.70 *10*
γ M1+E2	398.96 *8*	8.9 *14*
γ [E2]	400.14 *7*	1.28 *19*
γ	406.93 *6*	2.27 *19*
γ	412.85 *5*	0.98 *12*
γ	431.07 *16*	0.29 *4*
γ	439.10 *10*	0.87 *9*
γ	472.41 *9*	0.51 *6*
γ	482.64 *6*	1.89 *14*
γ	497.82 *8*	3.16 *22*
γ	515.67 *11*	0.086 *21*
γ	525.07 *12*	0.27 *5*
γ	528.64 *9*	0.43 *9*
γ	530.98 *10*	0.76 *10*
γ	543.83 *6*	3.10 *21*
γ	561.68 *12*	0.24 *4*
γ	570.70 *7*	0.62 *10*
γ [E1]	577.81 *7*	0.35 *6*
γ [M1+E2]	595.58 *6*	0.68 *6*
γ	605.04 *22*	0.28 *4* ?
γ [M1+E2]	626.33 *7*	10.7
γ	628.43 *14*	2.2 *3*
γ	638.02 *12*	0.24 *5*
γ	654.38 *7*	3.4 *5*
γ	655.31 *15*	1.53 *24*
γ	660.29 *6*	5.8 *5*
γ	693.33 *10*	0.44 *8*
γ	720.32 *10*	0.44 *9* ?
γ	727.0 *3*	0.27 *5*
γ	731.58 *11*	0.13 *4*
γ	732.61 *19*	0.24 *5*
γ	737.50 *9*	1.07 *13*
γ	746.16 *14*	0.47 *18*
γ	753.61 *11*	0.58 *14*
γ [M1+E2]	779.01 *5*	4.5 *3*
γ [E2]	805.89 *7*	0.12 *4*
γ [E1]	825.25 *7*	2.16 *17*
γ	830.76 *7*	0.52 *6*
γ	846.05 *13*	0.43 *5* ?
γ	851.29 *8*	1.94 *16*
γ	871.94 *13*	0.33 *9*
γ	876.77 *9*	0.20 *6*
γ	885.26 *10*	0.46 *9*
γ	897.15 *11*	0.31 *13*
γ	926.8 *5*	0.32 *9*
γ	950.3 *7*	0.33 *13*
γ [M1+E2]	974.45 *7*	0.49 *10*
γ	1002.50 *7*	0.57 *9*
γ	1008.42 *6*	2.31 *21*
γ	1013.81 *19*	0.25 *6*
γ [M1+E2]	1026.46 *5*	6.2 *5*
γ	1044.14 *20*	~0.09 ?
γ [E2]	1053.34 *7*	0.45 *6*
γ	1064.7 *5*	0.41 *10*
γ	1111.49 *19*	0.24 *5*
γ	1139.54 *19*	0.10 *3*
γ	1145.46 *19*	0.20 *4*
γ [E1]	1173.38 *8*	1.07 *15*
γ	1243.06 *21*	0.24 *4* ?
γ	1251.43 *9*	0.82 *15*
γ	1257.40 *20*	0.21 *4* ?
γ	1274.75 *19*	0.32 *10*
γ	1333.1 *3*	0.49 *9*
γ [M1+E2]	1374.59 *6*	1.32 *17*
γ [E2]	1401.46 *7*	0.89 *8*
γ	1426.4 *5*	0.14 *5*
γ	1511.62 *19*	0.21 *4*
γ	1526.31 *13*	1.40 *16*
γ	1532.23 *13*	0.32 *10*
γ	1688.90 *15*	0.87 *12*
γ	1694.82 *16*	0.59 *11*
γ	1799.84 *24*	0.43 *11* ?
γ	1824.6 *6*	0.48 *18*
γ	1851.8 *5*	0.62 *13*

Photons (^{119}Ag) (continued)

γ_{mode}	γ(keV)	γ(%)†
γ	1898.39 12	2.46 21
γ	1925.26 13	0.93 14
γ	1970.2 6	0.43 15
γ	1996.81 18	0.67 8
γ	2028.1 3	1.07 10
γ	2043.2 3	0.11 4
γ	2049.2 3	0.11 4
γ	2060.98 16	4.4 5
γ	2087.85 16	0.62 13
γ	2151.51 19	0.44 10
γ	2195.73 19	0.20 5
γ	2206.76 24	0.13 4 ?
γ	2212.68 24	0.12 5 ?
γ	2301.9 7	0.37 17
γ	2334.6 9	0.21 5
γ	2344.6 3	0.16 5 ?
γ	2376.0 7	0.34 10
γ	2386.19 19	0.43 10
γ	2391.97 25	0.20 8 ?
γ	2415.3 3	~0.32
γ	2435.0 4	0.33 13
γ	2442.2 3	0.64 16
γ	2470.33 25	0.93 22 ?
γ	2529.3 7	0.52 20
γ	2554.1 22	~0.11
γ	2563.7 9	~0.13
γ	2571.1 7	0.20 6
γ	2633.9 4	0.21 5
γ	2649.32 19	0.20 6
γ	2655.53 25	0.43 9 ?
γ	2676.19 19	~0.12
γ	2706.8 12	~0.26
γ	2722.0 7	0.43 16
γ	2732.3 10	~0.17
γ	2757.1 5	0.61 10
γ	2768.1 6	0.29 8
γ	2786.32 18	2.16 21
γ	2813.19 19	0.18 9
γ	2928.5 10	0.10 4
γ	2938.3 5	0.50 6
γ	2951.7 5	1.47 14

† 21% uncert(syst)

Atomic Electrons (^{119}Ag)

⟨e⟩=16.5 8 keV

e_{bin}(keV)	⟨e⟩(keV)	e(%)
4	1.6	45 8
5 - 14	0.70	10.7 10
19	0.93	4.8 9
20 - 22	0.50	2.4 3
23	1.5	6.4 12
25 - 26	0.35	1.34 23
41	2.9	7.2 12
55	2.5	4.5 5
63	0.53	0.84 14
64 - 67	0.189	0.29 4
78	1.21	1.56 17
81 - 124	0.35	0.41 5
147 - 195	0.57	0.32 4
198 - 247	0.18	0.084 13
254 - 298	0.042	0.016 7
321 - 325	0.0011	~0.00035
339	0.46	0.136 15
344 - 393	0.75	0.206 25
395 - 439	0.115	0.029 4
446 - 494	0.12	0.026 8
497 - 544	0.10	0.020 6
551 - 600	0.24	0.040 8
601 - 650	0.22	0.035 8
651 - 700	0.031	0.0047 10
705 - 754	0.098	0.0133 15
775 - 824	0.033	0.0042 5
825 - 873	0.035	0.0042 12
876 - 924	0.0069	0.00076 25
926 - 974	0.0067	0.00071 13
976 - 1023	0.100	0.0101 13

Atomic Electrons (^{119}Ag) (continued)

e_{bin}(keV)	⟨e⟩(keV)	e(%)
1026 - 1065	0.0100	0.00097 16
1085 - 1119	0.0039	0.00036 10
1136 - 1173	0.0048	0.00042 5
1216 - 1257	0.011	0.00089 23
1271 - 1306	0.0031	0.00024 10
1329 - 1375	0.0177	0.00130 12
1397 - 1426	0.0017	0.00012 3
1485 - 1532	0.011	0.00075 21
1662 - 1694	0.0077	0.00046 12
1773 - 1872	0.018	0.00097 23
1894 - 1993	0.0108	0.00056 11
1996 - 2087	0.028	0.0014 3
2125 - 2212	0.0038	0.00018 4
2275 - 2372	0.0065	0.00028 5
2375 - 2470	0.0089	0.00037 7
2503 - 2570	0.0036	0.00014 4
2607 - 2706	0.0063	0.00024 4
2718 - 2813	0.0117	0.00042 8
2902 - 2951	0.0070	0.00024 5

Continuous Radiation (^{119}Ag)

⟨β-⟩=1854 keV;⟨IB⟩=6.2 keV

E_{bin}(keV)		⟨ ⟩(keV)	(%)
0 - 10	β-	0.00394	0.078
	IB	0.055	
10 - 20	β-	0.0121	0.080
	IB	0.055	0.38
20 - 40	β-	0.0505	0.168
	IB	0.108	0.37
40 - 100	β-	0.398	0.56
	IB	0.31	0.48
100 - 300	β-	5.3	2.58
	IB	0.94	0.52
300 - 600	β-	27.6	6.0
	IB	1.15	0.27
600 - 1300	β-	215	22.1
	IB	1.8	0.21
1300 - 2500	β-	806	43.0
	IB	1.41	0.082
2500 - 5000	β-	798	25.5
	IB	0.33	0.0112
5000 - 5350	β-	1.74	0.0343
	IB	2.5×10^{-5}	5.0×10^{-7}

$^{119}_{48}$Cd(2.69 2 min)

Mode: β-
Δ: -83940 60 keV
SpA: 5.86×10^{8} Ci/g
Prod: fission

Photons (^{119}Cd)

⟨γ⟩=1782 95 keV

γ_{mode}	γ(keV)	γ(%)
In L_ℓ	2.905	0.0070 13
In L_η	3.112	0.0035 6
In L_α	3.286	0.19 3
In L_β	3.567	0.14 3
In L_γ	3.972	0.014 3
In $K_{\alpha2}$	24.002	1.3 1
In $K_{\alpha1}$	24.210	2.44 19
In $K_{\beta1}'$	27.265	0.63 5
In $K_{\beta2}'$	28.022	0.132 11
γ [E1]	50.04 6	0.41 4

Photons (^{119}Cd) (continued)

γ_{mode}	γ(keV)	γ(%)
γ	130.7 4	0.86 8
γ E2	133.99 7	7.5 5
γ	143.17 18	0.13 5
γ	153.35 11	0.082 16
γ	153.4 3	0.082 16
γ [E1]	184.03 7	0.23 4
γ	245.8 4	0.21 4 *
γ	287.34 11	0.49 12
γ M1,E2	292.80 6	41
γ M4	311.38 3	#
γ	331.6 4	0.21 8 *
γ	337.38 11	0.62 8
γ E1	342.85 7	18.9 12
γ	348.2 5	0.29 8 *
γ	355.0 3	~0.41 *
γ	416.98 7	0.57 8
γ	437.75 15	<0.07
γ	440.18 20	0.45 12 *
γ [E1]	446.06 8	2.05 21
γ	446.91 10	0.25 8
γ	473.6 4	0.45 8 *
γ [E1]	476.83 8 *	~0.8
γ	494.2 4	0.21 8 *
γ	597.8 3	0.57 8 *
γ	708.9 5	0.74 25 *
γ	732.98 20	1.72 25
γ	742.86 20	0.49 8
γ	784.6 3	1.15 16 *
γ	828.28 15	0.86 8
γ	836.81 19	~0.8
γ	864.96 15	0.53 16
γ	870.73 8	0.57 8
γ	902.6 3	1.0 4 *
γ	928.9 10	0.94 12 *
γ	933.8 5	0.082 16
γ	941.57 11	3.5 3
γ	970.80 19	0.90 25
γ	981.5 4	0.66 21 *
γ	1013.90 18	0.74 16
γ	1018.31 13	1.35 12
γ [E2]	1050.25 7	7.4 4
γ	1115.62 15	1.48 21
γ	1122.57 19	0.38 8
γ	1132.76 9	1.02 12
γ	1152.29 13	0.94 12
γ	1222.8 5	0.37 12 *
γ	1266.75 8	~0.11
γ	1287.71 7	3.0 5
γ	1316.79 7	9.8 12
γ	1317.64 11	~2
γ	1332.1 4	1.19 16 *
γ	1360.96 14	<0.34
γ	1409.91 18	0.45 12
γ	1426.31 13	0.38 8
γ	1473.87 21	
γ	1539.62 15	<0.19
γ	1549.74 9	1.02 25
γ	1609.59 7	12.2 6
γ	1623.07 17	<0.39
γ	1683.72 8	0.82 8
γ	1713.65 10	2.58 21
γ	1733.77 7	9 2
γ	1752.75 18	0.69 8
γ	1763.69 8	10.3 20
γ	1773.0 3	0.74 25 *
γ	1869.89 20	0.45 16
γ	1910.41 15	<0.41
γ	1919.93 19	0.33 12
γ	1956.5 4	0.74 16 *
γ	1960.45 15	<0.24 ?
γ	2026.57 7	1.56 21
γ	2056.49 10	2.46 20
γ	2063.7 3	0.45 8
γ	2356.5 3	7 3
γ	2394.1 10	~0.7 *
γ	2466.6 4	0.57 8 *

Photons (^{119}Cd)
(continued)

γ_{mode}	γ(keV)	γ(%)
γ	2525.5 6	0.41 12 *
γ	2548.05 10	2.0 8
γ	2593.9 6	0.49 16 *
γ	2688.2 4	0.11 5 *
γ	2701.4 3	0.33 8 *
γ	2748.0 7	0.16 4 *
γ	2781.0 7	~0.08 *

* with ^{119}Cd(2.69 or 2.20 min)
# with ^{119}In(18.0 min)

Atomic Electrons (^{119}Cd)
$\langle e\rangle$=10.3 8 keV

e_{bin}(keV)	$\langle e\rangle$(keV)	e(%)
4 - 50	0.46	5.8 6
103	0.24	~0.23
106	3.40	3.2 2
115 - 127	0.13	0.10 5
130	0.96	0.73 6
131 - 180	0.265	0.196 11
183 - 218	0.015	~0.007
242 - 259	0.029	~0.011
265	3.0	1.1 3
283 - 287	0.005	~0.0018
289	0.46	0.16 6
292 - 309	0.14	0.048 12
315	0.299	0.095 7
320 - 355	0.083	0.025 4
389 - 438	0.059	0.014 3
439 - 477	0.030	0.0066 17
490 - 494	0.0007	~0.00015
570 - 598	0.011	0.0018 9
681 - 729	0.040	0.0057 19
732 - 781	0.016	0.0021 9
784 - 833	0.020	0.0025 10
836 - 875	0.023	0.0026 9
898 - 943	0.057	0.0062 19
954 - 990	0.025	0.0026 7
1010 - 1050	0.085	0.0083 5
1088 - 1133	0.031	0.0029 7
1148 - 1195	0.0035	0.00030 12
1219 - 1267	0.021	0.0016 7
1283 - 1332	0.09	0.0073 23
1333 - 1382	0.0038	0.00028 11
1398 - 1426	0.0029	0.00021 8
1512 - 1609	0.08	0.0051 16
1619 - 1713	0.063	0.0037 12
1725 - 1772	0.070	0.0040 12
1842 - 1933	0.0088	0.00046 11
1952 - 2036	0.020	0.00100 23
2052 - 2063	0.0017	8.4 22 $\times10^{-5}$
2329 - 2393	0.034	0.0015 6
2439 - 2525	0.011	0.00045 15
2544 - 2593	0.0030	0.00012 4
2660 - 2753	0.0026	9.6 20 $\times10^{-5}$
2777 - 2780	3.8 $\times10^{-5}$	~1 $\times10^{-6}$

$^{119}_{48}$Cd(2.20 2 min)

Mode: β-
Δ: -83794 60 keV
SpA: 7.16$\times10^8$ Ci/g
Prod: fission

Photons (^{119}Cd)*
$\langle\gamma\rangle$=2394 70 keV

γ_{mode}	γ(keV)	γ(%)†
γ	98.00 14	0.77 5
γ	105.54 7	3.3 3
γ	178.66 6	0.93 13
γ	284.85 15	0.77 10
γ	304.44 11	0.17 8 ?
γ	318.0 3	0.43 8
γ	355.0 3	0.32 13
γ	360.23 13	1.02 12
γ	363.35 16	0.65 8
γ	411.41 8	1.90 15
γ	422.40 9	9.5 8
γ	437.75 15	~0.30
γ	584.93 8	4.8 3
γ	632.99 15	1.5 5
γ	667.78 18	~0.50
γ	688.3 5	0.75 10
γ	709.0 5	1.2 4
γ	720.58 11	18.0 12
γ	817.67 7	1.35 10
γ	878.37 12	0.38 8
γ	902.5 5	1.13 25
γ	923.22 7	6.9 4
γ	983.91 11	0.98 15
γ	996.33 7	2.0 5
γ	999.69 20	0.75 15
γ	1025.01 7	25
γ	1036.07 16	0.95 17
γ	1101.88 7	9.8 8
γ	1142.98 11	0.93 13
γ	1185.44 11	2.92 20
γ	1203.67 7	13.5 10
γ	1216.52 22	~0.13
γ	1222.9 5	0.50 17
γ	1235.7 3	0.25 10
γ	1279.52 17	0.25 10
γ	1334.49 22	2.5 10
γ	1344.13 14	6.8 4
γ	1353.05 12	0.82 17
γ	1360.96 14	1.50 18
γ	1364.10 11	5.3 5
γ	1388.35 16	0.88 13
γ	1397.48 18	~1
γ	1436.42 9	5.2 3
γ	1468.4 3	1.05 25
γ	1474.2 4	2.03 22
γ	1529.1 3	0.52 25
γ	1539.62 15	0.85 17
γ	1623.07 17	
γ	1624.5 4	0.4 1
γ	1633.6 4	0.35 10
γ	1638.92 22	0.55 12
γ	1668.54 12	3.08 25
γ	1701.91 16	2.83 23
γ	1772.9 5	1.30 25
γ	1806.0 4	0.90 13
γ	1845.7 4	0.55 12
γ	1910.41 15	
γ	1950.1 3	0.95 20 ?
γ	1960.45 15	1.08 17
γ	2021.34 8	22.3 10
γ	2104.25 15	5.9 4
γ	2389.10 11	0.73 8
γ	2394.2 10	~1
γ	2422.48 17	4.5 4
γ	2520.3 3	1.13 15
γ	2527.4 12	1.0 3
γ	2672.1 3	0.70 8
γ	2701.2 4	0.35 8

† approximate
* see also ^{119}Cd(2.69 min)

$^{119}_{49}$In(2.4 1 min)

Mode: β-
Δ: -87730 18 keV
SpA: 6.6$\times10^8$ Ci/g
Prod: ^{120}Sn(γ,p); fission; daughter ^{119}In(18.0 min)

Photons (^{119}In)
$\langle\gamma\rangle$=768.8 18 keV

γ_{mode}	γ(keV)	γ(%)†
Sn L_ℓ	3.045	~0.040
Sn L_η	3.272	~0.029
Sn L_α	3.443	~1
Sn L_β	3.731	~3
Sn L_γ	4.327	~0.5
γ M1	23.869 6	16.1 4
Sn $K_{\alpha2}$	25.044	0.0532 16
Sn $K_{\alpha1}$	25.271	0.099 3
Sn $K_{\beta1}'$	28.474	0.0260 8
Sn $K_{\beta2}'$	29.275	0.00547 18
γ M4	65.662 9	*
γ (M2)	697.48 3	0.49 4
γ E2	763.14 3	99.08
γ	1214.9 1	0.44 11

† 0.15% uncert(syst)
* with ^{119}Sn(293 d)

Atomic Electrons (^{119}In)
$\langle e\rangle$=20 6 keV

e_{bin}(keV)	$\langle e\rangle$(keV)	e(%)
4	1.6	~37
19	11.9	~61
20	1.2	~6
21	0.0038	0.0179 18
23	3.1	~13
24	0.7	3.0 15
25	0.00108	0.0044 4
27	1.84 $\times10^{-5}$	6.7 7 $\times10^{-5}$
28	0.000191	0.00068 7
668	0.028	0.0043 4
693	0.0037	0.00054 5
694	5.5 $\times10^{-5}$	8.0 7 $\times10^{-6}$
697	0.00091	0.000131 12
734	1.53	0.208 4
759	0.201	0.0265 5
762	0.0397	0.00521 10
763	0.00805	0.001055 21
1186	0.0032	~0.00027
1210	0.00037	~3$\times10^{-5}$
1211	2.1 $\times10^{-5}$	~2$\times10^{-6}$
1214	7.7 $\times10^{-5}$	~6$\times10^{-6}$
1215	1.6 $\times10^{-5}$	~1$\times10^{-6}$

Continuous Radiation (^{119}In)
$\langle\beta-\rangle$=612 keV;$\langle$IB$\rangle$=0.99 keV

E_{bin}(keV)		$\langle\ \rangle$(keV)	(%)
0 - 10	β-	0.0297	0.59
	IB	0.026	
10 - 20	β-	0.090	0.60
	IB	0.025	0.18
20 - 40	β-	0.373	1.24
	IB	0.049	0.169
40 - 100	β-	2.83	4.02
	IB	0.132	0.20
100 - 300	β-	33.3	16.3
	IB	0.32	0.19

Continuous Radiation (^{119}In)
(continued)

E_{bin}(keV)		⟨ ⟩(keV)	(%)
300 - 600	β-	131	28.9
	IB	0.27	0.065
600 - 1300	β-	392	44.7
	IB	0.155	0.020
1300 - 2246	β-	53	3.57
	IB	0.0065	0.00045

$^{119}_{49}$In(18.0 *3* min)

Mode: β-(97.5 *5* %), IT(2.5 *5* %)
Δ: -87419 *18* keV
SpA: 8.78×10^7 Ci/g
Prod: ^{120}Sn(γ,p); fission

Photons (^{119}In)
⟨γ⟩=7.0 *7* keV

γ_{mode}	γ(keV)	γ(%)†
$\gamma_{\beta-}$M1	23.869 *6*	0.029 *10*
γ_{IT} M4	311.38 *3*	0.99
$\gamma_{\beta-}$(M1)	897.03 *16*	0.018 *4*
$\gamma_{\beta-}$E2+2.3%M1	920.89 *16*	0.018 *4*
$\gamma_{\beta-}$M1+E2	1065.65 *18*	0.13
$\gamma_{\beta-}$(E2)	1089.52 *18*	0.026 *6*
$\gamma_{\beta-}$M1	1163.86 *21*	0.054 *6*
$\gamma_{\beta-}$[M1+E2]	1187.73 *21*	0.010 *4*
$\gamma_{\beta-}$[M1+E2]	1225.84 *18*	0.020 *5*
$\gamma_{\beta-}$[M1]	1249.71 *18*	0.073 *6*
$\gamma_{\beta-}$M1+E2	1330 *3*	0.0038 *13*
$\gamma_{\beta-}$E2	1354 *3*	0.0011 *4*

† uncert(syst): 20% for IT, approximate for β-

Continuous Radiation (^{119}In)
⟨β-⟩=1041 keV;⟨IB⟩=2.4 keV

E_{bin}(keV)		⟨ ⟩(keV)	(%)
0 - 10	β-	0.0105	0.209
	IB	0.038	
10 - 20	β-	0.0322	0.214
	IB	0.037	0.26
20 - 40	β-	0.134	0.445
	IB	0.073	0.25
40 - 100	β-	1.04	1.47
	IB	0.21	0.32
100 - 300	β-	13.4	6.5
	IB	0.58	0.33
300 - 600	β-	63	13.8
	IB	0.62	0.146
600 - 1300	β-	389	40.8
	IB	0.72	0.085
1300 - 2500	β-	573	34.0
	IB	0.18	0.0113
2500 - 2624	β-	1.31	0.052
	IB	6.9×10^{-6}	2.7×10^{-7}

$^{119}_{50}$Sn(stable)

Δ: -90066 *3* keV
%: 8.58 *4*

$^{119}_{50}$Sn(293.0 *13* d)

Mode: IT
Δ: -89976 *3* keV
SpA: 3745 Ci/g
Prod: ^{118}Sn(n,γ)

Photons (^{119}Sn)
⟨γ⟩=11.44 *19* keV

γ_{mode}	γ(keV)	γ(%)†
Sn L_ℓ	3.045	0.155 *23*
Sn L_η	3.272	0.060 *6*
Sn L_α	3.443	4.3 *6*
Sn L_β	3.750	4.8 *8*
Sn L_γ	4.307	0.74 *13*
γ M1	23.869 *6*	16.1
Sn $K_{\alpha2}$	25.044	8.1 *3*
Sn $K_{\alpha1}$	25.271	15.1 *6*
Sn $K_{\beta1}$'	28.474	3.95 *14*
Sn $K_{\beta2}$'	29.275	0.83 *3*
γ M4	65.662 *9*	0.0200 *4*

† 2.5% uncert(syst)

Atomic Electrons (^{119}Sn)
⟨e⟩=78.3 *10* keV

e_{bin}(keV)	⟨e⟩(keV)	e(%)
4	4.4	106 *11*
19	11.9	61.4 *20*
20	1.32	6.68 *24*
21	0.57	2.7 *3*
23	3.07	13.4 *4*
24	0.86	3.63 *16*
25	0.164	0.66 *7*
27	0.0028	0.0102 *11*
28	0.029	0.104 *11*
36	11.8	32.4 *9*
61	9.7	15.9 *5*
62	24.0	39.0 *11*
65	8.59	13.2 *4*
66	1.76	2.68 *8*

$^{119}_{51}$Sb(1.587 *8* d)

Mode: ε
Δ: -89472 *8* keV
SpA: 6.91×10^5 Ci/g
Prod: daughter ^{119}Te; ^{119}Sn(p,n); ^{118}Sn(d,n)

Photons (^{119}Sb)
⟨γ⟩=23.2 *4* keV

γ_{mode}	γ(keV)	γ(%)†
Sn L_ℓ	3.045	0.143 *21*
Sn L_η	3.272	0.083 *9*
Sn L_α	3.443	4.0 *5*
Sn L_β	3.736	5.5 *8*
Sn L_γ	4.289	0.82 *14*
γ M1	23.869 *6*	16.1
Sn $K_{\alpha2}$	25.044	21.3 *6*
Sn $K_{\alpha1}$	25.271	39.7 *12*
Sn $K_{\beta1}$'	28.474	10.4 *3*
Sn $K_{\beta2}$'	29.275	2.19 *7*

† 2.5% uncert(syst)

Atomic Electrons (^{119}Sb)
⟨e⟩=24.0 *6* keV

e_{bin}(keV)	⟨e⟩(keV)	e(%)
4	4.5	108 *11*
19	11.9	61.4 *20*
20	1.43	7.2 *3*
21	1.50	7.1 *7*
23	3.07	13.4 *4*
24	1.11	4.7 *3*
25	0.43	1.74 *18*
27	0.0073	0.027 *3*
28	0.076	0.27 *3*

Continuous Radiation (^{119}Sb)
⟨IB⟩=0.081 keV

E_{bin}(keV)		⟨ ⟩(keV)	(%)
10 - 20	IB	0.00126	0.0072
20 - 40	IB	0.044	0.17
40 - 100	IB	0.0033	0.0053
100 - 300	IB	0.018	0.0094
300 - 569	IB	0.0135	0.0037

$^{119}_{52}$Te(16.05 *5* h)

Mode: ε
Δ: -87178 *8* keV
SpA: 1.641×10^6 Ci/g
Prod: ^{121}Sb(p,3n); ^{116}Sn(α,n)

Photons (^{119}Te)
⟨γ⟩=750 *7* keV

γ_{mode}	γ(keV)	γ(%)†
Sb L_ℓ	3.189	0.110 *15*
Sb L_η	3.437	0.055 *6*
Sb L_α	3.604	3.0 *3*
Sb L_β	3.931	2.7 *3*
Sb L_γ	4.438	0.32 *5*
Sb $K_{\alpha2}$	26.111	21.2 *6*
Sb $K_{\alpha1}$	26.359	39.6 *12*
Sb $K_{\beta1}$'	29.712	10.4 *3*
Sb $K_{\beta2}$'	30.561	2.22 *7*
γ [E1]	148.97 *12*	0.030 *6*
γ M1+1.6%E2	270.502 *25*	0.118 *25*
γ E2	429.42 *5*	0.085 *25*
γ [E1]	627.41 *11*	0.020 *4*

Photons (^{119}Te) (continued)

γ_{mode}	γ(keV)	γ(%)†
γ E2	644.07 3	84.5
γ [E1]	683.27 10	0.110 25
γ [M1+E2]	694.67 10	0.09 3
γ M1,E2	699.93 5	10.1 5
γ [E1]	713.35 8	0.059 17
γ [E1]	769.20 7	0.110 25
γ [E1]	787.78 7	0.27 4
γ	794.9 4	0.059 17
γ [E1]	843.64 7	0.30 4
γ [M1+E2]	1049.78 7	
γ [M1+E2]	1105.64 6	0.56 7
γ [M1+E2]	1121.30 7	0.20 4
γ [M1]	1177.16 7	0.72 9
γ	1217.21 7	0.00042 9
γ	1327.33 10	0.0127 25
γ [M1+E2]	1338.74 9	0.23 4
γ [E1]	1413.27 7	1.10 8
γ [E2]	1479.20 6	~0.034
γ	1487.71 7	0.00084 17
γ	1700.8 4	0.025 8
γ M1,E2	1749.70 6	3.89 25
γ [E2]	1821.22 8	~0.034
γ [M1+E2]	1875.35 20	0.042 17

† 0.59% uncert(syst)

Atomic Electrons (^{119}Te)

⟨e⟩=7.8 3 keV

e_{bin}(keV)	⟨e⟩(keV)	e(%)
4	2.5	61 6
5	0.38	8.1 8
21	0.39	1.84 19
22	1.23	5.6 6
25	0.39	1.57 16
26	0.42	1.64 17
29 - 30	0.083	0.286 24
118 - 149	0.0023	0.0019 3
240 - 270	0.0121	0.0050 9
399 - 429	0.0041	0.0010 3
597	0.00016	2.7 5 $\times10^{-5}$
614	1.76	0.286 6
623 - 664	0.304	0.0475 7
669	0.21	0.031 4
679 - 713	0.035	0.0050 6
739 - 788	0.0035	0.00046 7
790 - 840	0.0020	0.00025 3
843 - 844	5.1 $\times10^{-5}$	6.0 9 $\times10^{-6}$
1075 - 1121	0.0101	0.00094 12
1147 - 1187	0.0099	0.00086 9
1213 - 1217	5.1 $\times10^{-7}$	4.2 18 $\times10^{-8}$
1297 - 1339	0.0026	0.00020 4
1383 - 1413	0.0048	0.00034 3
1449 - 1488	0.00031	2.1 9 $\times10^{-5}$
1670 - 1749	0.032	0.00186 18
1791 - 1875	0.00057	3.1 9 $\times10^{-5}$

Continuous Radiation (^{119}Te)

⟨β+⟩=5.8 keV;⟨IB⟩=0.91 keV

E_{bin}(keV)		⟨ ⟩(keV)	(%)
0 - 10	β+	2.18 $\times10^{-6}$	2.63 $\times10^{-5}$
	IB	0.00052	
10 - 20	β+	8.0 $\times10^{-5}$	0.000481
	IB	0.00117	0.0071
20 - 40	β+	0.00196	0.0060
	IB	0.049	0.18
40 - 100	β+	0.077	0.101
	IB	0.0053	0.0084
100 - 300	β+	2.21	1.05
	IB	0.039	0.019

Continuous Radiation (^{119}Te) (continued)

E_{bin}(keV)		⟨ ⟩(keV)	(%)
300 - 600	β+	3.56	0.90
	IB	0.18	0.040
600 - 1300	β+	0.0060	0.00099
	IB	0.59	0.067
1300 - 1650	IB	0.049	0.0036
	$\Sigma\beta$+		2.1

$^{119}_{52}$Te(4.69 4 d)

Mode: ϵ

Δ: ~-86878 8 keV

SpA: 2.340$\times10^5$ Ci/g

Prod: ^{121}Sb(p,3n); ^{121}Sb(d,4n); ^{116}Sn(α,n)

Photons (^{119}Te)

⟨γ⟩=1512 14 keV

γ_{mode}	γ(keV)	γ(%)†
Sb L_ℓ	3.189	0.117 16
Sb L_η	3.437	0.058 6
Sb L_α	3.604	3.2 4
Sb L_β	3.931	2.8 4
Sb L_γ	4.439	0.34 5
Sb $K_{\alpha2}$	26.111	22.5 7
Sb $K_{\alpha1}$	26.359	42.0 13
Sb $K_{\beta1}$'	29.712	11.1 3
Sb $K_{\beta2}$'	30.561	2.35 8
γ E1	116.59 4	0.45 3
γ E1	153.57 3	67 3
γ M1(+19%E2)	164.31 4	1.31 5
γ	184.09 20	~0.027
γ	190.52 20	0.033 13
γ	201.15 20	~0.013
γ M1(+E2)	241.83 4	0.060 13
γ M1+1.6%E2	270.502 25	28.2 6
γ	369.7 3	0.033 13
γ E1	395.40 4	0.33 3
γ M1,E2	700.37 5	0.47 5
γ	760.3 5	0.047 20
γ M1,E2	777.89 4	~0.07
γ [E1]	818.70 7	0.11 3
γ M1(+44%E2)	859.69 5	0.16 2
γ E1	871.56 7	0.39 3
γ M1(+14%E2)	912.56 5	6.30 13
γ [E1]	917.03 10	0.09 4
γ M1+38%E2	942.20 3	5.13 10
γ [E1]	952.85 10	0.09 4
γ E2(+M1)	970.87 5	0.23 3
γ	972.22 20	~0.10 ?
γ E1	976.28 5	2.7 6
γ M1(+14%E2)	979.18 4	3.03 7
γ	1013.21 19	~0.9 ?
γ [E1]	1013.26 5	~2
γ M1(+48%E2)	1048.39 4	3.21 6
γ	1066.3 3	0.10 3
γ E1	1081.34 10	1.61 3
γ M2(+2.3%E3)	1095.77 4	2.25 5
γ [E1]	1110.43 10	~0.011 ?
γ E2(+M1)	1136.76 6	7.72 15
γ E2(+M1)	1212.70 3	66.7
γ M1,E2	1249.68 5	0.175 13
γ [E1]	1255.09 6	~0.015
γ E1	1311.71 10	0.123 14
γ E3	1366.26 4	1.074 21
γ	1391.9 10	0.033 7
γ	1407.41 15	~0.13
γ	1700.8 10	0.020 7
γ [E1]	1859.22 10	0.14 4
γ	1955.45 5	<0.027 ?
γ E1	2013.2 4	0.32 2

Photons (^{119}Te) (continued)

γ_{mode}	γ(keV)	γ(%)†
γ E1(+0.3%M2)	2089.60 9	4.72 9
γ	2126.3 4	~0.027
γ	2225.6 4	<0.027 ?
γ	2242.0 10	<0.013
γ	2360.38 20	<0.07

† 0.75% uncert(syst)

Atomic Electrons (^{119}Te)

⟨e⟩=15.3 4 keV

e_{bin}(keV)	⟨e⟩(keV)	e(%)
4	2.7	64 7
5 - 21	0.83	10.7 9
22	1.31	6.0 6
25	0.42	1.66 17
26	0.45	1.74 18
29 - 30	0.088	0.30 3
86 - 117	0.050	0.056 3
123	4.02	3.27 15
134	0.268	0.200 18
149	0.61	0.411 18
153 - 201	0.226	0.145 9
211 - 238	0.0082	0.0038 9
240	2.45	1.02 3
241 - 270	0.429	0.161 4
339 - 370	0.0061	0.0017 3
391 - 395	0.00077	0.000196 13
670 - 700	0.0112	0.00166 24
730 - 778	0.0021	0.00028 11
788 - 829	0.0033	0.00040 4
841 - 887	0.103	0.0117 5
908 - 953	0.164	0.0177 5
966 - 1013	0.029	0.0029 10
1018 - 1066	0.123	0.0118 3
1077 - 1110	0.099	0.0090 10
1132 - 1137	0.0134	0.00118 13
1182	0.70	0.059 7
1208 - 1255	0.109	0.0090 10
1281 - 1312	0.00056	4.4 4 $\times10^{-5}$
1336 - 1377	0.0207	0.00154 6
1387 - 1407	0.00017	1.2 6 $\times10^{-5}$
1670 - 1700	0.00013	8 4 $\times10^{-6}$
1829 - 1925	0.00058	3.1 8 $\times10^{-5}$
1951 - 2012	0.00112	5.7 3 $\times10^{-5}$
2059 - 2126	0.0164	0.00079 3
2195 - 2241	0.00010	~5 $\times10^{-6}$
2330 - 2360	0.00017	~7 $\times10^{-6}$

Continuous Radiation (^{119}Te)

⟨β+⟩=2.47 keV;⟨IB⟩=0.43 keV

E_{bin}(keV)		⟨ ⟩(keV)	(%)
0 - 10	β+	8.7 $\times10^{-8}$	1.05 $\times10^{-6}$
	IB	0.00035	
10 - 20	β+	3.18 $\times10^{-6}$	1.93 $\times10^{-5}$
	IB	0.00102	0.0060
20 - 40	β+	7.9 $\times10^{-5}$	0.000244
	IB	0.049	0.19
40 - 100	β+	0.00329	0.00425
	IB	0.0044	0.0071
100 - 300	β+	0.123	0.057
	IB	0.031	0.0150
300 - 600	β+	0.66	0.145
	IB	0.120	0.027
600 - 1300	β+	1.68	0.202
	IB	0.19	0.023
1300 - 2323	β+	8.3 $\times10^{-8}$	6.4 $\times10^{-9}$
	IB	0.036	0.0022
	$\Sigma\beta$+		0.41

$^{119}_{53}$I (19.1 *4* min)

Mode: ϵ
Δ: -83810 *100* keV
SpA: 8.27×10^{7} Ci/g
Prod: ^{14}N on Pd; protons on I; protons on La; ^{109}Ag(^{12}C,2n)

Photons (^{119}I)

⟨γ⟩=312 *5* keV

γ_{mode}	γ(keV)	γ(%)†
Te L_ℓ	3.335	0.064 *10*
Te L_η	3.606	0.032 *4*
Te L_α	3.768	1.78 *24*
Te L_β	4.123	1.57 *23*
Te L_γ	4.659	0.19 *3*
Te $K_{\alpha2}$	27.202	11.6 *6*
Te $K_{\alpha1}$	27.472	21.6 *11*
Te $K_{\beta1}$'	30.980	5.7 *3*
Te $K_{\beta2}$'	31.877	1.25 *6*
γ	63.19 *19*	0.59 *9*
γ	207.5 *4*	0.29 *5*
γ	240.4 *5*	0.50 *5*
γ(M1)	257.59 *16*	90.0
γ	320.78 *19*	1.98 *18*
γ	349.3 *4*	0.21 *5*
γ	363.0 *5*	0.27 *4*
γ	378.19 *19*	0.37 *4*
γ	404.2 *5*	0.31 *4*
γ	556.8 *4*	1.80 *18*
γ	561.0 *5*	0.36 *13*
γ	568.4 *5*	0.22 *4*
γ	604.4 *17*	0.24 *5*
γ	631.6 *5*	0.90 *18*
γ	635.78 *19*	2.6 *5*
γ	643.2 *5*	
γ	706.4 *5*	1.26 *18*
γ	746.38 *21*	0.25 *5*
γ	973.0 *5*	0.36 *9*
γ	1003.97 *25*	0.50 *11*

† 1.1% uncert(syst)

Atomic Electrons (^{119}I)

⟨e⟩=15.4 *8* keV

e_{bin}(keV)	⟨e⟩(keV)	e(%)
4	0.97	22 *3*
5	0.71	15.1 *19*
22	0.212	0.96 *10*
23	0.65	2.8 *3*
26 - 58	1.0	3.5 *12*
59	0.5	~0.9
62 - 63	0.16	~0.26
176 - 209	0.08	0.043 *18*
226	9.1	4.03 *17*
235 - 240	0.010	0.0043 *19*
253	1.30	0.513 *21*
257 - 289	0.42	0.160 *20*
316 - 363	0.057	0.017 *5*
372 - 404	0.015	0.0040 *16*
525 - 573	0.065	0.012 *4*
599 - 636	0.08	0.012 *5*
675 - 715	0.027	0.0040 *17*
741 - 746	0.0006	8 *4* ×10^{-5}
941 - 973	0.010	0.0010 *4*
999 - 1004	0.0008	8 *4* ×10^{-5}

Continuous Radiation (^{119}I)

⟨β+⟩=495 keV;⟨IB⟩=4.1 keV

E_{bin}(keV)		⟨ ⟩(keV)	(%)
0 - 10	β+	2.89×10^{-6}	3.48×10^{-5}
	IB	0.019	
10 - 20	β+	0.000110	0.00067
	IB	0.019	0.133
20 - 40	β+	0.00289	0.0089
	IB	0.062	0.22
40 - 100	β+	0.132	0.170
	IB	0.106	0.163
100 - 300	β+	6.3	2.85
	IB	0.30	0.168
300 - 600	β+	46.1	9.9
	IB	0.41	0.094
600 - 1300	β+	278	29.7
	IB	1.17	0.127
1300 - 2500	β+	164	10.8
	IB	1.8	0.104
2500 - 3112	IB	0.147	0.0056
	Σβ+		53

$^{119}_{54}$Xe(5.8 *3* min)

Mode: ϵ
Δ: -78820 *160* keV
SpA: 2.72×10^{8} Ci/g
Prod: protons on La; protons on Ce

Photons (^{119}Xe)

γ_{mode}	γ(keV)	γ(rel)
γ	88.1 *6*	43 *3*
γ	91.0 *6*	16 *1*
γ	95.9 *7*	38 *3*
γ	100.2 *7*	95 *6*
γ	141.7 *6*	11 *2*
γ	146.9 *8*	8 *1*
γ(M1+3.8%E2)	208.4 *7*	60 *5*
γ	221.1 *8*	~3
γ	232.7 *5*	100
γ	236.5 *7*	10 *2*
γ	278.0 *8*	~3
γ	295.1 *8*	8 *1*
γ	308.5 *7*	11 *2*
γ	320.8 *6*	12 *2*
γ	438.0 *8*	13 *3*
γ	461.4 *7*	91 *7*
γ	536.8 *8*	9 *2*
γ	693.3 *8*	12 *3*
γ	737.3 *8*	

$^{119}_{55}$Cs(37.7 *10* s)

Mode: ϵ
Δ: -72200 *100* keV
SpA: 2.49×10^{9} Ci/g
Prod: protons on Ta

Photons (^{119}Cs)

γ_{mode}	γ(keV)	γ(rel)
γ	169 *1*	~67
γ	176 *1*	~81
γ	224 *1*	100
γ	246 *1*	34 *9*
γ	257 *1*	57 *12*
γ	314 *1*	38 *11*
γ	390 *1*	<33
γ	667 *1*	17 *8*

$^{119}_{55}$Cs(28 *1* s)

decay not observed
Δ: -72200 *100* keV
SpA: 3.34×10^{9} Ci/g
Prod: protons on La

$^{119}_{56}$Ba(5.4 *3* s)

Mode: ϵ, ϵp
Δ: -63960 *410* keV syst
SpA: 1.66×10^{10} Ci/g
Prod: ^{106}Cd(^{16}O,3n); ^{90}Zr(^{32}S,3n)

ϵp/β+: 0.25 *2*
ϵp: ~2000-6000 range

A = 120

NDS **17**, 39 (1976)

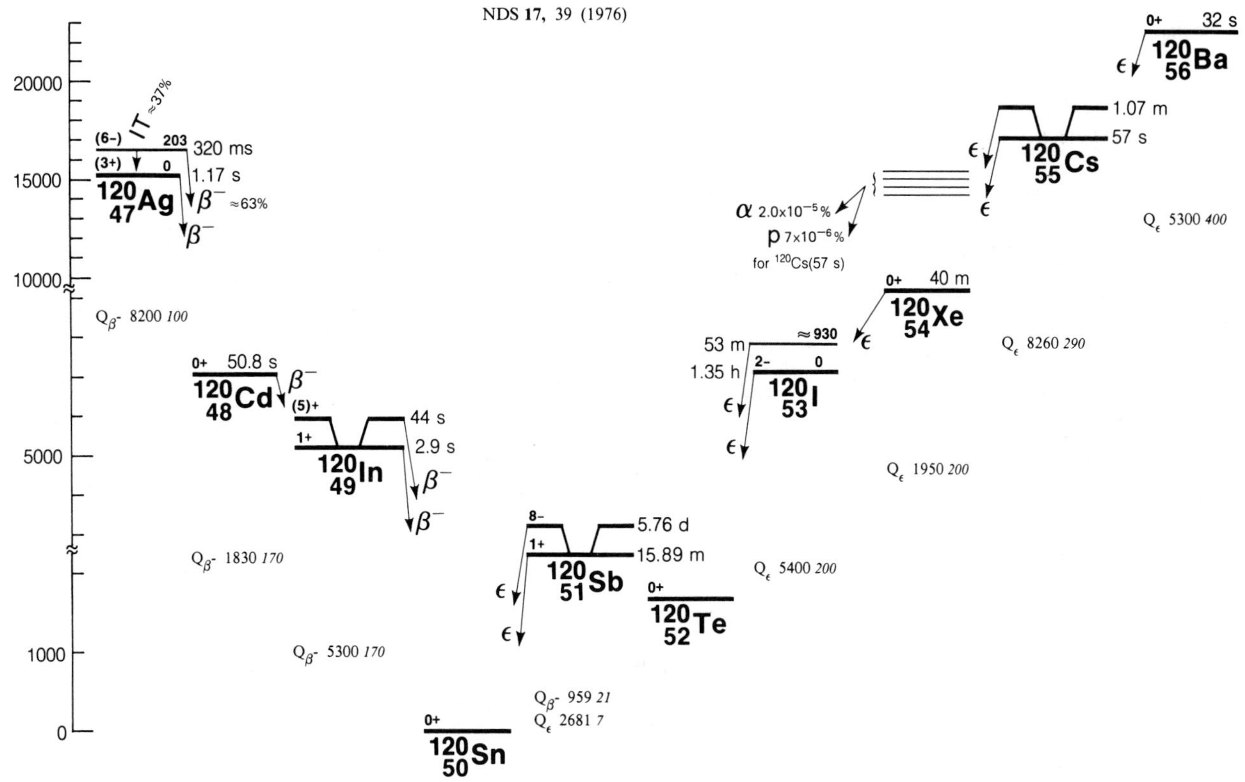

$^{120}_{47}$Ag(1.17 *5* s)

Mode: β-
Δ: -75770 *100* keV
SpA: 6.1×10^{10} Ci/g
Prod: fission

Photons (^{120}Ag)

γ_{mode}	γ(keV)
γ [E2]	505.93 *16*
γ	697.8 *2*
γ	817.13 *16*
γ	1323.07 *16*

$^{120}_{47}$Ag(320 *40* ms)

Mode: β-(~63 %), IT(~37 %)
Δ: -75567 *100* keV
SpA: 1.20×10^{11} Ci/g
Prod: fission

Photons (^{120}Ag)

γ_{mode}	γ(keV)	γ(%)
γ_{IT} E3	203.0 *5*	~23
$\gamma_{\beta-}$[E2]	505.93 *16*	
$\gamma_{\beta-}$	697.8 *2*	
$\gamma_{\beta-}$	817.13 *16*	
$\gamma_{\beta-}$	830.0 *2*	
$\gamma_{\beta-}$	925.8 *2*	
$\gamma_{\beta-}$	1323.07 *16*	

$^{120}_{48}$Cd(50.80 *21* s)

Mode: β-
Δ: -83973 *19* keV
SpA: 1.838×10^{9} Ci/g
Prod: protons on Sn; fission

see ^{120}In(2.9 s) for γ rays

Continuous Radiation (^{120}Cd)

⟨β-⟩=708 keV;⟨IB⟩=1.26 keV

E_{bin}(keV)		⟨ ⟩(keV)	(%)
0 - 10	β-	0.0223	0.444
	IB	0.029	
10 - 20	β-	0.068	0.454
	IB	0.028	0.20
20 - 40	β-	0.282	0.94
	IB	0.055	0.19
40 - 100	β-	2.16	3.07

Continuous Radiation (^{120}Cd)
(continued)

E_{bin}(keV)		⟨ ⟩(keV)	(%)
	IB	0.151	0.23
100 - 300	β-	26.4	12.9
	IB	0.39	0.22
300 - 600	β-	112	24.6
	IB	0.35	0.083
600 - 1300	β-	450	49.5
	IB	0.25	0.032
1300 - 1830	β-	117	8.1
	IB	0.0088	0.00063

$^{120}_{49}$In(2.9 *1* s)

Mode: β-
Δ: -85800 *170* keV
SpA: 2.884×10^{10} Ci/g
Prod: ^{120}Sn(n,p); ^{123}Sb(n,α); fission; daughter ^{120}Cd

Photons (^{120}In)

⟨γ⟩=347 *19* keV

γ_{mode}	γ(keV)	γ(%)†
γ	251.5 *10*	0.19 *6* ? *
γ [E2]	703.8 *3*	1.42 *19*
γ E2+34%M1	925.9 *5*	0.52 *21*
γ [E2]	988.6 *7*	0.13 *6*
γ E2	1171.44 *20*	19.0 *15*
γ E2+16%M1	1184.1 *5*	0.91 *13*

Photons (^{120}In) (continued)

γ_{mode}	γ(keV)	γ(%)†
γ	1207.5 *9*	0.59 *11* ? *
γ	1249.6 *6*	0.23 *6*
γ	2039.8 *10*	1.86 *17* ?
γ [E2]	2097.4 *6*	0.42 *10*
γ	2149.2 *16*	0.09 *4* ?
γ [E2]	2355.5 *6*	0.30 *11*
γ	2390.2 *10*	1.14 *19* ?
γ	2421.0 *6*	0.16 *7*

† 7.9% uncert(syst)
* with ^{120}Cd or ^{120}In(2.9 s)

Continuous Radiation (^{120}In)

$\langle\beta-\rangle$=2210 keV;$\langle$IB$\rangle$=8.1 keV

E_{bin}(keV)		$\langle\ \rangle$(keV)	(%)
0 - 10	β-	0.00256	0.0509
	IB	0.061	
10 - 20	β-	0.0079	0.052
	IB	0.061	0.42
20 - 40	β-	0.0329	0.109
	IB	0.120	0.42
40 - 100	β-	0.260	0.367
	IB	0.35	0.54
100 - 300	β-	3.53	1.70
	IB	1.07	0.59
300 - 600	β-	18.7	4.05
	IB	1.36	0.32
600 - 1300	β-	158	16.1
	IB	2.3	0.26
1300 - 2500	β-	733	38.4
	IB	2.1	0.118
2500 - 5000	β-	1292	39.1
	IB	0.71	0.024
5000 - 5300	β-	4.51	0.089
	IB	5.2×10^{-5}	1.03×10^{-6}

$^{120}_{49}$In(44.4 *10* s)

Mode: β-
Δ: -85800 *170* keV
SpA: 2.10×10^{9} Ci/g
Prod: ^{120}Sn(n,p); fission

Photons (^{120}In)

$\langle\gamma\rangle$=2977 *48* keV

γ_{mode}	γ(keV)	γ(%)†
Sn L_ℓ	3.045	0.0034 *6*
Sn L_η	3.272	0.0018 *3*
Sn L_α	3.443	0.094 *15*
Sn L_β	3.742	0.081 *14*
Sn L_γ	4.210	0.0092 *17*
Sn $K_{\alpha2}$	25.044	0.68 *8*
Sn $K_{\alpha1}$	25.271	1.26 *15*
Sn $K_{\beta1}'$	28.474	0.33 *4*
Sn $K_{\beta2}'$	29.275	0.070 *8*
γ E1	89.83 *20*	6.3 *15*
γ	177 *3*	0.29 *10*
γ E2	197.3 *3*	7.9 *6*
γ [E2]	267.7 *4*	1.5 *3*
γ	354.6 *16*	1.4 *3*
γ	401.1 *19*	0.8 *3*
γ [M1+E2]	414.8 *4*	2.7 *4*
γ [M1+E2]	448.8 *4*	0.58 *19*
γ	465.4 *22*	0.78 *19*
γ [E2]	546.0 *5*	1.6 *3*

Photons (^{120}In) (continued)

γ_{mode}	γ(keV)	γ(%)†
γ [M1+E2]	592.2 *3*	1.5 *3*
γ	609.9 *19*	1.5 *3*
γ	637.0 *5*	1.6 *3*
γ	697 *3*	1.7 *4*
γ [E2]	702.5 *6*	2.4 *5*
γ [M1+E2]	713.2 *3*	6.9 *4*
γ [M1+E2]	863.59 *24*	30.1 *19*
γ E2+34%M1	925.9 *5*	1.5 *5*
γ [E1]	965.2 *4*	7.9 *5*
γ [M1+E2]	984.6 *3*	2.3 *5*
γ E2	1023.06 *19*	60.2 *19*
γ [E3]	1112.9 *3*	0.103 *25*
γ [E1]	1162.4 *4*	1.9 *4*
γ E2	1171.44 *20*	97.1
γ E2+16%M1	1184.1 *5*	2.6 *5*
γ	1245.0 *11*	~0.5
γ	1249.6 *6*	1.3 *4*
γ [E2]	1294.5 *3*	10.8 *5*
γ [E2]	1471.9 *4*	4.4 *5*
γ [M1+E2]	1581.9 *14*	~0.29 ?
γ [E2]	1886.6 *3*	3.9 *15*
γ [E2]	2007.7 *3*	6.3 *6*
γ [E2]	2097.4 *6*	1.2 *3*
γ	2178.2 *12*	2.4 *3*
γ	2268.0 *11*	1.4 *3*
γ [E2]	2355.5 *6*	0.9 *3*
γ	2421.0 *6*	0.9 *3*
γ [E2]	2604.9 *14*	1.9 *5*

† 0.51% uncert(syst)

Atomic Electrons (^{120}In)

$\langle e\rangle$=6.72 *25* keV

e_{bin}(keV)	$\langle e\rangle$(keV)	e(%)
4 - 28	0.176	2.7 *4*
61	0.81	1.3 *3*
85 - 90	0.18	0.21 *3*
148	0.04	~0.029
168	1.58	0.94 *7*
173 - 177	0.012	~0.007
193	0.34	0.174 *13*
196 - 197	0.083	0.0423 *23*
238	0.15	0.063 *13*
263 - 268	0.033	0.0127 *16*
325 - 372	0.08	0.024 *10*
386 - 420	0.162	0.041 *5*
436 - 465	0.025	0.006 *3*
517 - 563	0.083	0.0153 *20*
581 - 610	0.058	0.010 *3*
633 - 673	0.070	0.0105 *25*
684 - 713	0.161	0.0234 *24*
834	0.44	0.053 *7*
859 - 897	0.087	0.0101 *12*
921 - 965	0.075	0.0080 *8*
980 - 985	0.0045	0.00045 *9*
994	0.639	0.0643 *24*
1019 - 1023	0.100	0.0098 *3*
1084 - 1113	0.0023	0.00021 *9*
1133	0.0080	0.00071 *14*
1142	0.888	0.0777 *16*
1155 - 1184	0.165	0.0142 *4*
1216 - 1265	0.103	0.0082 *6*
1290 - 1294	0.0135	0.00105 *5*
1443 - 1472	0.037	0.0025 *3*
1553 - 1581	0.0025	0.00016 *7*
1857 - 1886	0.026	0.0014 *5*
1978 - 2068	0.046	0.00232 *19*
2093 - 2177	0.013	0.00059 *18*
2239 - 2326	0.011	0.00047 *12*
2351 - 2420	0.0045	0.00019 *7*
2576 - 2604	0.0101	0.00039 *9*

Continuous Radiation (^{120}In)

$\langle\beta-\rangle$=944 keV;$\langle$IB$\rangle$=2.1 keV

E_{bin}(keV)		$\langle\ \rangle$(keV)	(%)
0 - 10	β-	0.0148	0.294
	IB	0.035	
10 - 20	β-	0.0453	0.301
	IB	0.035	0.24
20 - 40	β-	0.188	0.62
	IB	0.068	0.24
40 - 100	β-	1.45	2.05
	IB	0.19	0.29
100 - 300	β-	18.1	8.8
	IB	0.52	0.29
300 - 600	β-	81	17.8
	IB	0.53	0.127
600 - 1300	β-	420	44.9
	IB	0.57	0.068
1300 - 2500	β-	405	24.5
	IB	0.132	0.0084
2500 - 3105	β-	17.6	0.66
	IB	0.00099	3.8×10^{-5}

$^{120}_{50}$Sn(stable)

Δ: -91102 *3* keV
%: 32.59 *10*

$^{120}_{51}$Sb(15.89 *4* min)

Mode: ϵ
Δ: -88421 *8* keV
SpA: 9.857×10^{7} Ci/g
Prod: ^{120}Sn(p,n); ^{120}Sn(d,2n); ^{119}Sn(d,n); ^{121}Sb(γ,n); ^{121}Sb(n,2n)

Photons (^{120}Sb)

$\langle\gamma\rangle$=39.1 *5* keV

γ_{mode}	γ(keV)	γ(%)†
Sn L_ℓ	3.045	0.059 *8*
Sn L_η	3.272	0.030 *3*
Sn L_α	3.443	1.63 *18*
Sn L_β	3.742	1.43 *18*
Sn L_γ	4.220	0.168 *24*
Sn $K_{\alpha2}$	25.044	12.1 *4*
Sn $K_{\alpha1}$	25.271	22.5 *7*
Sn $K_{\beta1}'$	28.474	5.89 *18*
Sn $K_{\beta2}'$	29.275	1.24 *4*
γ [E2]	703.8 *3*	0.269 *14*
γ [E2]	988.6 *7*	0.105 *12*
γ E2	1171.44 *20*	2.16 *3*

† 4.6% uncert(syst)

Atomic Electrons (^{120}Sb)

$\langle e\rangle$=3.11 *19* keV

e_{bin}(keV)	$\langle e\rangle$(keV)	e(%)
4	1.61	40 *4*
20	0.103	0.51 *5*
21	0.85	4.1 *4*
24	0.227	0.94 *10*
25	0.244	0.99 *10*
27	0.0042	0.0152 *15*

Atomic Electrons (^{120}Sb) (continued)

e_{bin}(keV)	⟨e⟩(keV)	e(%)
28	0.043	0.155 *16*
675	0.0046	0.00069 *5*
699	0.00054	7.7 *5* $\times10^{-5}$
700	8.0 $\times10^{-5}$	1.14 *7* $\times10^{-5}$
703	0.000123	1.75 *11* $\times10^{-5}$
704	2.48 $\times10^{-5}$	3.53 *23* $\times10^{-6}$
959	0.00116	0.000121 *15*
984	0.000142	1.45 *18* $\times10^{-5}$
985	5.3 $\times10^{-6}$	5.3 *7* $\times10^{-7}$
988	3.4 $\times10^{-5}$	3.5 *4* $\times10^{-6}$
989	5.1 $\times10^{-7}$	5.2 *6* $\times10^{-8}$
1142	0.0197	0.00173 *4*
1167	0.00238	0.000204 *5*
1168	7.21 $\times10^{-5}$	6.18 *15* $\times10^{-6}$
1171	0.000579	4.95 *12* $\times10^{-5}$

Continuous Radiation (^{120}Sb)

⟨β+⟩=326 keV;⟨IB⟩=3.0 keV

E_{bin}(keV)		⟨ ⟩(keV)	(%)
0 - 10	β+	4.87 $\times10^{-6}$	5.9 $\times10^{-5}$
	IB	0.0136	
10 - 20	β+	0.000176	0.00107
	IB	0.0139	0.096
20 - 40	β+	0.00439	0.0135
	IB	0.051	0.19
40 - 100	β+	0.188	0.242
	IB	0.073	0.113
100 - 300	β+	8.3	3.76
	IB	0.21	0.115
300 - 600	β+	55	11.9
	IB	0.31	0.071
600 - 1300	β+	233	25.9
	IB	1.04	0.112
1300 - 2500	β+	30.1	2.16
	IB	1.26	0.075
2500 - 2681	IB	0.0045	0.00018
	Σβ+		44

$^{120}_{51}$Sb(5.76 *2* d)

Mode: ε

Δ: -88421 *8* keV

SpA: 1.889$\times10^{5}$ Ci/g

Prod: ^{119}Sn(d,n); ^{120}Sn(d,2n); ^{120}Sn(p,n); ^{121}Sb(n,2n); ^{121}Sb(γ,n)

Photons (^{120}Sb)

⟨γ⟩=2469 *22* keV

γ_{mode}	γ(keV)	γ(%)
Sn L_ℓ	3.045	0.137 *18*
Sn L_η	3.272	0.072 *8*
Sn L_α	3.443	3.8 *4*
Sn L_β	3.742	3.4 *4*
Sn L_γ	4.225	0.41 *6*
Sn $K_{\alpha2}$	25.044	27.6 *9*
Sn $K_{\alpha1}$	25.271	51.5 *16*
Sn $K_{\beta1}$'	28.474	13.5 *4*
Sn $K_{\beta2}$'	29.275	2.83 *10*
γ E1	89.83 *20*	80 *3*
γ E2	197.3 *3*	88 *3*
γ E2	1023.06 *19*	99 *2*

Photons (^{120}Sb) (continued)

γ_{mode}	γ(keV)	γ(%)
γ [E3]	1112.9 *3*	1.3 *1*
γ E2	1171.44 *20*	99.9

Atomic Electrons (^{120}Sb)

⟨e⟩=44.4 *10* keV

e_{bin}(keV)	⟨e⟩(keV)	e(%)
4	3.8	93 *10*
20	0.235	1.16 *12*
21	1.95	9.3 *10*
24	0.52	2.15 *22*
25	0.56	2.26 *23*
27	0.0095	0.035 *4*
28	0.099	0.35 *4*
61	10.3	17.0 *7*
85	1.43	1.67 *7*
86	0.444	0.518 *22*
89	0.381	0.429 *18*
90	0.084	0.093 *4*
168	17.6	10.5 *4*
193	3.75	1.94 *8*
196	0.401	0.204 *8*
197	0.529	0.269 *11*
994	1.05	0.106 *3*
1019	0.133	0.0130 *4*
1022	0.0261	0.00255 *7*
1023	0.00530	0.000518 *15*
1084	0.0252	0.00232 *20*
1108	0.0030	0.000267 *23*
1109	0.00042	3.8 *3* $\times10^{-5}$
1112	0.00067	6.0 *5* $\times10^{-5}$
1113	0.000135	1.22 *11* $\times10^{-5}$
1142	0.914	0.0800 *16*
1167	0.1100	0.00943 *19*
1168	0.00334	0.000286 *6*
1171	0.0268	0.00229 *5*

Continuous Radiation (^{120}Sb)

⟨IB⟩=0.048 keV

E_{bin}(keV)		⟨ ⟩(keV)	(%)
10 - 20	IB	0.00130	0.0074
20 - 40	IB	0.044	0.17
40 - 100	IB	0.0020	0.0035
100 - 199	IB	0.00054	0.00044

$^{120}_{52}$Te(stable)

Δ: -89380 *19* keV

%: 0.096 *2*

$^{120}_{53}$I (1.35 *1* h)

Mode: ε

Δ: -83980 *200* keV

SpA: 1.934$\times10^{7}$ Ci/g

Prod: protons on I; daughter ^{120}Xe; protons on Te; ^{11}B on Cd; protons on La; ^{14}N on Pd

Photons (^{120}I)*

⟨γ⟩=1735 *55* keV

γ_{mode}	γ(keV)	γ(%)†
Te L_ℓ	3.335	0.023 *3*
Te L_η	3.606	0.0113 *12*
Te L_α	3.768	0.63 *7*
Te L_β	4.123	0.56 *7*
Te L_γ	4.660	0.068 *10*
Te $K_{\alpha2}$	27.202	4.19 *13*
Te $K_{\alpha1}$	27.472	7.82 *25*
Te $K_{\beta1}$'	30.980	2.08 *7*
Te $K_{\beta2}$'	31.877	0.453 *15*
γ [E2]	542.3 *3*	1.10 *15*
γ E2	560.12 *25*	73
γ E2	601.3 *3*	5.8 *12*
γ	613.6 *8*	0.66 *23*
γ	640.77 *22*	9.1 *4*
γ	652.6 *9*	0.22 *8*
γ	658.6 *9*	
γ	661.7 *3*	1.12 *11*
γ	701.1 *3*	0.30 *6*
γ	712.6 *10*	0.15 *5*
γ	728.8 *6*	0.30 *6*
γ	729.6 *10*	
γ	734.9 *6*	0.35 *7*
γ	742.7 *6*	0.35 *7*
γ	751.6 *10*	0.29 *10*
γ	763.6 *10*	0.29 *10*
γ	852.9 *7*	0.17 *6*
γ	907.8 *4*	0.28 *6* ?
γ	968.7 *11*	0.19 *4*
γ	974.7 *6*	1.61 *15*
γ	979.2 *7*	0.39 *7*
γ	1038.6 *14*	0.51 *18*
γ	1085.5 *10*	~0.15
γ	1100.6 *8*	0.41 *7*
γ	1168.4 *8*	0.58 *7*
γ	1200.9 *3*	1.93 *18*
γ	1221.6 *17*	
γ	1255.0 *8*	0.80 *7*
γ	1283.0 *10*	0.36 *7*
γ	1299.0 *10*	0.42 *7*
γ	1302.4 *3*	0.82 *10*
γ	1382.6 *19*	0.51 *18*
γ	1410.5 *7*	1.31 *15*
γ	1418.6 *19*	
γ	1422.5 *7*	0.80 *7*
γ	1427.6 *19*	
γ	1451.3 *10*	0.51 *7*
γ	1491.6 *10*	0.44 *7*
γ	1522.4 *3*	11.2 *7*
γ	1534.5 *7*	2.04 *22*
γ	1542.6 *14*	1.17 *22*
γ	1548.6 *4*	0.91 *10*
γ	1551.8 *14*	0.36 *7*
γ	1604.6 *22*	0.7 *3*
γ	1663.2 *14*	0.36 *7*
γ	1673.6 *23*	0.51 *18*
γ	1763.6 *24*	0.51 *18*
γ	1768.6 *24*	0.22 *8*
γ	1789.6 *24*	1.3 *5*
γ	1832.6 *25*	
γ	1874.3 *14*	0.58 *15*
γ	1894.6 *14*	0.58 *7*
γ	1911 *3*	0.58 *20*
γ	1935 *3*	0.29 *10*
γ	1983.0 *14*	0.44 *7*
γ	2034 *3*	0.36 *13*
γ	2045 *3*	0.22 *8*
γ	2082.5 *4*	0.9 *3* ?
γ	2108.7 *5*	0.55 *11*
γ	2129.0 *14*	0.80 *7*
γ	2142 *3*	0.36 *13*
γ	2158 *3*	0.36 *13*
γ	2172 *3*	0.7 *3*
γ	2181 *3*	0.51 *18*
γ	2187.6 *14*	1.39 *15*
γ	2218 *3*	0.22 *8*
γ	2374.9 *20*	0.36 *7*
γ	2376.9 *8*	0.48 *10* ?
γ	2404 *3*	1.0 *4*
γ	2454.4 *7*	2.04 *22*
γ	2491.4 *14*	1.02 *7*
γ	2510 *3*	0.29 *10*

Photons (^{120}I)* (continued)

γ_{mode}	γ(keV)	γ(%)†
γ	2526 3	0.29 10
γ	2563.8 8	1.96 18
γ	2569 4	
γ	2638 4	0.22 8
γ	2654 4	0.22 8
γ	2696.8 20	0.36 7
γ	2740 4	0.44 15
γ	2747 4	0.29 10
γ	2778 4	0.36 13
γ	2800 4	0.29 10
γ	2829 4	0.22 8
γ	2937.0 9	0.70 10 ?
γ	2939 4	0.29 10
γ	2987 4	0.44 15
γ	3029 4	0.36 13
γ	3047 4	1.3 5
γ	3082 4	0.29 10
γ	3098 4	0.36 13
γ	3160 4	0.22 8
γ	3182 4	0.66 23
γ	3334 5	0.22 8
γ	3395 5	0.29 10
γ	3442 5	0.22 8
γ	3545 5	0.29 10
γ	3580 5	0.29 10
γ	3608 5	0.66 23
γ	3694 5	0.29 10
γ	3742 5	0.29 10
γ	4120 6	0.29 10
γ	4134 6	0.29 10
γ	4148 6	0.29 10
γ	4188 6	0.29 10
γ	4283 6	0.22 8
γ	4288 6	0.22 8
γ	4413 6	0.15 5

† approximate
* see also ^{120}I(53 min)

Atomic Electrons (^{120}I)

$\langle e \rangle$=4.29 23 keV

e_{bin}(keV)	$\langle e \rangle$(keV)	e(%)
4	0.35	8.0 8
5	0.25	5.4 6
22	0.077	0.35 4
23	0.236	1.03 11
26	0.077	0.29 3
27	0.083	0.31 3
30 - 31	0.0164	0.054 4
511	0.031	0.0061 9
528	1.98	0.37 4
537 - 542	0.0057	0.00106 10
555	0.233	0.042 4
556 - 560	0.124	0.0222 15
570	0.14	0.025 5
582 - 601	0.037	0.0063 15
609	0.17	~0.027
613 - 662	0.056	0.009 3
669 - 713	0.023	0.0034 9
720 - 764	0.012	0.0017 5
821 - 853	0.0025	~0.00030
876 - 908	0.0038	0.00043 21
937 - 979	0.027	0.0029 10
1007 - 1054	0.007	0.0007 3
1069 - 1101	0.0047	0.00043 19
1137 - 1169	0.022	0.0019 7
1196 - 1223	0.009	0.0007 3
1250 - 1299	0.015	0.0012 3
1301 - 1351	0.0040	~0.00029
1378 - 1422	0.022	0.0016 4
1446 - 1544	0.12	0.0083 23
1547 - 1642	0.011	0.00070 20
1658 - 1737	0.0050	0.00029 10
1758 - 1843	0.012	0.00070 24
1863 - 1951	0.0118	0.00062 13
1978 - 2078	0.012	0.00058 14
2081 - 2180	0.023	0.00106 18

Atomic Electrons (^{120}I) (continued)

e_{bin}(keV)	$\langle e \rangle$(keV)	e(%)
2183 - 2217	0.0022	0.00010 3
2343 - 2423	0.018	0.00077 17
2449 - 2532	0.017	0.00070 13
2559 - 2653	0.0033	0.00013 3
2665 - 2746	0.0065	0.00024 5
2768 - 2828	0.0025	9.0 24 $\times 10^{-5}$
2905 - 2997	0.0076	0.00026 5
3015 - 3097	0.0085	0.00028 8
3128 - 3181	0.0036	0.00011 4
3302 - 3394	0.0020	6.0 17 $\times 10^{-5}$
3410 - 3441	0.0009	2.5 10 $\times 10^{-5}$
3513 - 3607	0.0047	0.00013 3
3662 - 3741	0.0022	5.9 16 $\times 10^{-5}$
4088 - 4187	0.0041	0.000099 18
4251 - 4287	0.0015	3.5 9 $\times 10^{-5}$
4381 - 4412	0.00050	1.1 4 $\times 10^{-5}$

$^{120}_{53}$I (53 *4* min)

Mode: ϵ
Δ: ~-83050 kev
SpA: 2.96$\times 10^7$ Ci/g
Prod: alphas on Sb; protons on I; protons on Te; ^{11}B on Cd; protons on La; ^{14}N on Pd

Photons (^{120}I)

$\langle \gamma \rangle$=4471 180 keV

γ_{mode}	γ(keV)	γ(%)†
Te L_ℓ	3.335	0.025 3
Te L_η	3.606	0.0122 13
Te L_α	3.768	0.68 8
Te L_β	4.123	0.60 8
Te L_γ	4.660	0.073 10
Te $K_{\alpha2}$	27.202	4.52 15
Te $K_{\alpha1}$	27.472	8.4 3
Te $K_{\beta1}$'	30.980	2.24 8
Te $K_{\beta2}$'	31.877	0.488 18
γ	425.3 4	2.8 5 ?
γ	433.0 7	2.1 4 &
γ	477.9 7	1.2 3 &
γ	485.1 5	1.2 3 &
γ E2	560.12 25	99.4
γ E2	601.3 3	87 9
γ E2	614.7 4	67 13
γ	651.9 6	0.7 1 ?
γ	654.5 5	2.1 4
γ	694.4 7	0.56 20 &
γ	703.9 5	1.9 4
γ	728.5 7	10 2
γ	763.2 4	3.5 5
γ	874.7 5	1.0 3 &
γ	881.8 5	2.3 4 &
γ	921.0 4	4.3 5 &
γ	976.0 10	35 7
γ	1031.5 6	~2
γ	1039.9 5	6.5 7 ?
γ	1054.0 11	10 2
γ	1059.2 5	5.1 5
γ	1158.0 6	2.7 5 &
γ	1197.3 6	2.3 4
γ	1261.3 7	1.7 3
γ	1328.0 13	6.0 12
γ	1334.6 7	4.4 9
γ	1346.2 4	18.9 13
γ	1363.5 10	4.2 9 &
γ	1402.1 7	3.6 7
γ	1405.0 8	9.3 9
γ	1441.1 14	1.3 3 &

Photons (^{120}I) (continued)

γ_{mode}	γ(keV)	γ(%)†
γ	1453.0 15	12.0 24
γ	1761.4 10	3.8 7 &
γ	1775.8 10	5.1 7 &
γ	1851.4 20	1.6 3 &
γ	1868.3 10	3.8 8 &
γ	1922.8 15	2.0 4 &
γ	1988.2 10	2.3 5
γ	2094.3 20	2.2 4 &
γ	2305.4 20	2.0 4 &
γ	2403.2 10	6.7 7
γ	2462.8 20	4.1 8 &
γ	2560.6 10	3.7 8
γ	2602.5 20	3.2 7 &
γ	2811.2 20	4.1 6 &
γ	2864.3 20	2.3 7 &
γ	2932.9 15	4.3 6 &
γ	3105.1 20	2.1 4 &

† approximate
& with ^{120}I(1.35 h or 53 min)

Atomic Electrons (^{120}I)

$\langle e \rangle$=11.0 7 keV

e_{bin}(keV)	$\langle e \rangle$(keV)	e(%)
4	0.37	8.6 9
5	0.27	5.8 6
22 - 31	0.53	2.20 13
393 - 433	0.18	0.045 16
446 - 485	0.07	0.016 6
528	2.7	0.51 10
555	0.32	0.058 12
556 - 560	0.170	0.030 4
570	2.10	0.37 4
583	1.6	0.27 5
596 - 623	0.70	0.116 9
647 - 694	0.049	0.007 3
697 - 731	0.23	0.033 13
758 - 763	0.008	0.0011 5
843 - 889	0.10	0.011 4
916 - 921	0.008	0.0009 4
944	0.38	~0.040
971 - 1008	0.14	0.014 5
1022 - 1059	0.18	0.018 6
1126 - 1165	0.047	0.0041 14
1192 - 1229	0.017	0.0014 6
1256 - 1303	0.08	0.0063 22
1314 - 1363	0.21	0.016 5
1370 - 1409	0.12	0.0085 25
1421 - 1453	0.10	0.007 3
1730 - 1820	0.068	0.0039 10
1836 - 1922	0.038	0.0020 6
1956 - 1987	0.014	0.0007 3
2062 - 2093	0.013	0.00061 22
2274 - 2371	0.040	0.0017 5
2398 - 2462	0.025	0.0010 3
2529 - 2602	0.033	0.0013 3
2779 - 2863	0.029	0.00102 24
2901 - 2932	0.019	0.00064 19
3073 - 3104	0.009	0.00029 9

$^{120}_{54}$Xe(40 *1* min)

Mode: ϵ
Δ: -82030 280 keV
SpA: 3.917$\times 10^7$ Ci/g
Prod: protons on I; protons on Ce

Photons (^{120}Xe)

⟨γ⟩=404 6 keV

γmode	γ(keV)	γ(%)†
I L_ℓ	3.485	0.22 3
I L_η	3.780	0.111 13
I L_α	3.937	6.3 8
I L_β	4.317	6.0 9
I L_γ	4.908	0.80 13
γ E1	25.05 8	30 3
I $K_{\alpha2}$	28.317	30.0 13
I $K_{\alpha1}$	28.612	55.7 23
I $K_{\beta1}'$	32.278	15.0 6
I $K_{\beta2}'$	33.225	3.40 15
γ [M1+E2]	40.47 10	0.15 5
γ M1	40.86 10	0.67 20
γ	47.52 9	0.054 9
γ [M1+E2]	49.36 10	0.081 18
γ M1	51.50 10	0.43 5
γ [M1+E2]	53.73 12	0.072 18
γ [M1+E2]	56.31 12	0.090 18
γ [M1+E2]	58.54 10	0.099 18
γ	64.67 10	0.072 9
γ	66.04 11	0.063 9
γ M1(+E2)	69.57 10	0.95 14
γ E1	72.57 9	9.0
γ M1+14%E2	77.15 9	4.0 3
γ (M1)+E2	81.13 9	0.52 5
γ M1+E2	85.94 11	0.59 5
γ	88.39 9	0.18 3
γ (M1)+E2	89.72 10	1.79 18
γ	96.90 12	0.19 3
γ M1+E2	98.80 10	1.6 3
γ	99.20 10	0.54 11
γ [M1+E2]	100.86 11	0.117 18
γ	104.9 3	0.027 9
γ	106.51 11	0.045 9
γ	111.12 12	0.099 18
γ	113.43 9	0.036 9
γ	124.58 10	0.144 18
γ (M1)+E2	128.65 8	1.58 13
γ [M1+E2]	133.46 11	0.25 3
γ	139.67 10	0.66 5
γ	141.84 17	0.072 9
γ [M1+E2]	146.72 9	0.28 3
γ [E1]	153.70 9	0.045 9
γ	156.88 16	0.081 9
γ [E1]	158.51 11	0.027 9
γ	164.85 10	0.30 3
γ [E1]	171.77 9	0.36 7
γ M1,E2	172.12 11	0.99 20
γ [M1+E2]	174.33 13	0.40 13
γ M1,E2	175.80 11	4.5 9
γ	175.94 13	0.95 19
γ M1,E2	178.01 9	6.8 5
γ	182.22 20	0.054 9
γ	183.90 13	0.054 9
γ	188.56 11	0.063 9
γ	195.30 14	0.126 9
γ	197.35 15	0.027 9
γ [E1]	200.84 11	0.31 4
γ [M1+E2]	203.40 12	0.22 3
γ	205.71 10	0.34 4
γ	210.61 16	0.072 18
γ [M1+E2]	221.48 11	0.50 5
γ	224.37 13	0.054 9
γ	232.25 13	0.081 9
γ	236.2 3	0.054 9
γ	242.44 13	0.036 9
γ [M1+E2]	246.1 1	0.23 3
γ	253.23 10	0.099 9
γ	261.88 12	0.063 9
γ	270.94 23	0.144 18
γ	277.39 14	0.37 4
γ	279.40 16	0.38 4
γ	282.71 13	0.135 18
γ	285.45 12	0.108 9
γ [M1+E2]	295.46 10	1.13 14
γ	300.64 15	0.045 9
γ	302.22 14	0.108 9
γ	309.40 12	0.59 9
γ	311.21 22	0.081 9
γ [M1+E2]	315.74 15	0.171 18
γ [M1+E2]	317.12 14	0.207 18
γ	322.44 22	0.126 18

Photons (^{120}Xe) (continued)

γmode	γ(keV)	γ(%)†
γ	323.57 13	0.15 3
γ	331.12 16	0.162 18
γ	335.73 10	1.04 14
γ	342.81 19	0.25 4
γ [M1+E2]	346.96 11	0.54 9
γ [M1+E2]	350.13 11	0.43 5
γ	359.44 10	0.95 14
γ	365.6 3	0.108 9
γ [E1]	375.18 11	0.036 9
γ [M1+E2]	375.66 14	0.23 7
γ	376.59 10	0.13 4
γ [M1+E2]	384.90 14	0.95 11
γ	390.5 3	0.072 9
γ	396.14 13	0.063 18
γ	399.91 15	0.090 18
γ [M1+E2]	401.40 12	0.22 3
γ	403.71 17	0.135 18
γ	407.99 16	0.099 9
γ	412.3 3	0.045 9
γ [M1+E2]	424.11 9	1.21 13
γ [M1+E2]	426.80 13	0.35 5
γ	429.26 14	0.22 3
γ [M1+E2]	436.04 14	0.081 9
γ	439.64 14	0.189 18
γ	446.27 18	0.108 18
γ [E1]	449.16 9	1.66 18
γ	451.24 16	0.27 5
γ	457.66 18	0.108 9
γ	461.69 15	0.081 9
γ	463.91 16	0.135 18
γ	465.41 20	0.32 4
γ	467.06 14	0.53 5
γ	472.3 3	0.045 9
γ	475.76 15	0.58 6
γ [M1+E2]	478.29 14	0.33 5
γ	481.3 3	0.153 18
γ	489.63 14	0.22 3
γ [M1+E2]	493.92 15	0.090 9
γ [M1+E2]	495.30 14	0.054 18
γ [M1+E2]	504.31 14	0.40 13
γ [M1+E2]	504.48 14	0.27 8
γ [M1+E2]	506.70 16	0.18 5
γ	516.12 14	0.30 5
γ	518.75 17	0.072 9
γ [E1]	529.36 14	1.38 14
γ [M1+E2]	535.77 14	0.189 18
γ	540.75 22	0.25 5
γ	551.32 15	0.135 18
γ [M1+E2]	555.45 13	1.48 18
γ	562.55 15	0.24 3
γ	568.78 18	0.38 5
γ	572.28 12	0.33 5
γ [M1+E2]	574.11 12	0.22 5
γ	576.8 4	0.09 3
γ [E1]	580.50 13	0.77 8
γ [M1+E2]	590.22 12	1.58 18
γ	594.11 14	0.54 6
γ	596.28 16	0.25 4
γ	604.68 17	0.32 4
γ	619.34 14	0.090 9
γ	627.13 14	0.180 18
γ	630.95 13	1.04 13
γ [M1+E2]	638.33 12	0.30 4
γ [M1+E2]	647.51 12	0.39 5
γ	652.21 16	0.108 18
γ	656.6 3	0.24 3
γ [M1+E2]	663.79 18	0.12 3
γ	664.75 15	0.44 5
γ [M1+E2]	678.79 11	1.64 15
γ [M1+E2]	682.49 13	0.56 6
γ [M1+E2]	685.38 14	0.24 3
γ	688.83 14	0.135 18
γ [M1+E2]	693.50 15	0.063 18
γ [M1+E2]	694.56 14	0.13 3
γ [M1+E2]	696.86 11	0.171 18
γ	704.51 17	0.099 18
γ	706.9 3	0.036 9
γ [M1+E2]	725.85 14	0.50 5
γ	737.13 12	0.072 9
γ [M1+E2]	743.92 14	0.14 3
γ	745.00 12	0.16 3
γ [M1+E2]	748.36 12	1.06 11
γ [M1+E2]	753.16 11	1.45 12
γ [M1+E2]	762.34 11	4.5 4

Photons (^{120}Xe) (continued)

γmode	γ(keV)	γ(%)†
γ	777.99 12	0.063 9
γ	779.64 18	0.081 9
γ [M1+E2]	793.62 10	1.30 15
γ	803.3 4	0.081 9
γ	807.78 17	0.081 9
γ [M1+E2]	811.69 10	0.73 8
γ [M1+E2]	820.22 12	0.22 4
γ [M1+E2]	822.43 14	0.24 4
γ [M1+E2]	825.52 11	0.35 5
γ [E1]	850.56 11	0.198 18
γ	851.96 10	0.44 5
γ [M1+E2]	854.86 18	0.099 18
γ [M1+E2]	863.19 11	0.61 6
γ [M1+E2]	867.19 17	0.23 4
γ [M1+E2]	869.58 11	0.108 18
γ [M1+E2]	872.57 13	0.153 18
γ	875.67 11	0.85 9
γ [M1+E2]	880.45 17	0.054 18
γ	883.9 4	0.108 18
γ [M1+E2]	885.26 16	0.32 5
γ	892.82 10	0.045 9
γ [E1]	897.62 13	0.036 9
γ	900.0 4	0.045 9
γ [M1+E2]	904.22 18	0.036 9
γ	909.84 12	0.081 9
γ [M1+E2]	921.07 12	0.29 4
γ	925.53 17	0.045 9
γ [M1+E2]	930.43 15	0.22 3
γ	933.55 12	0.045 9
γ [M1+E2]	940.34 9	0.34 5
γ	944.49 18	0.063 9
γ [E1]	965.39 10	1.20 12
γ [M1+E2]	970.89 14	0.31 3
γ	983.86 17	0.045 9
γ [M1+E2]	988.96 14	0.59 5
γ [M1+E2]	998.23 11	0.171 18
γ	1013.49 17	0.027 9
γ [E1]	1023.27 11	0.33 5
γ	1029.23 15	0.27 4
γ [M1+E2]	1032.87 18	0.47 5
γ [E1]	1057.92 18	~0.018
γ	1061.01 16	0.090 9
γ	1086.06 17	0.027 9
γ [M1+E2]	1117.61 14	0.054 9
γ [E1]	1142.66 14	0.081 18

† 7.8% uncert(syst)

Atomic Electrons (^{120}Xe)

⟨e⟩=44.8 22 keV

e_{bin}(keV)	⟨e⟩(keV)	e(%)
5	5.9	125 15
7 - 18	1.04	10.1 18
20	7.4	37 4
21	0.08	0.39 14
23	1.04	4.5 6
24	3.0	12.5 13
25 - 35	1.93	7.0 5
36	2.0	~6
39	1.5	3.9 8
40 - 43	0.5	~1
44	2.73	6.2 4
45 - 56	2.0	4.0 9
57	1.3	2.3 8
58 - 65	1.1	~2
66	1.2	1.9 8
67 - 114	5.6	6.8 9
119 - 142	0.82	0.62 12
143	1.2	0.8 3
145	1.5	1.06 20
146 - 195	1.13	0.66 14
196 - 245	0.147	0.067 10
246 - 295	0.27	0.101 12
296 - 345	0.24	0.077 13
346 - 395	0.190	0.051 4
396 - 445	0.142	0.033 4
446 - 495	0.075	0.016 3

Atomic Electrons (^{120}Xe) (continued)

e_{bin}(keV)	⟨e⟩(keV)	e(%)
496 - 544	0.121	0.0234 *24*
546 - 595	0.110	0.0196 *23*
596 - 643	0.065	0.0106 *22*
646 - 695	0.099	0.0151 *13*
696 - 745	0.163	0.0225 *25*
746 - 794	0.085	0.0111 *9*
798 - 847	0.047	0.0057 *9*
849 - 898	0.0259	0.0030 *3*
899 - 944	0.0211	0.00228 *18*
951 - 1000	0.027	0.0028 *3*
1008 - 1058	0.0033	0.00032 *7*
1060 - 1086	0.00078	7.2 *15* $\times10^{-5}$
1109 - 1143	0.00057	5.1 *8* $\times10^{-5}$

Continuous Radiation (^{120}Xe)

⟨β+⟩=10.4 keV;⟨IB⟩=0.96 keV

E_{bin}(keV)		⟨ ⟩(keV)	(%)
0 - 10	β+	1.09×10^{-6}	1.31×10^{-5}
	IB	0.00076	
10 - 20	β+	4.22×10^{-5}	0.000255
	IB	0.00104	0.0066
20 - 40	β+	0.00110	0.00338
	IB	0.059	0.21
40 - 100	β+	0.0479	0.062
	IB	0.0072	0.0117
100 - 300	β+	1.73	0.81
	IB	0.041	0.020
300 - 600	β+	6.3	1.44
	IB	0.17	0.038
600 - 1300	β+	2.33	0.342
	IB	0.55	0.062
1300 - 1925	IB	0.124	0.0086
	Σβ+		2.7

$^{120}_{55}$Cs(57 *6* s)

Mode: ε, εα(2.0 *4* $\times10^{-5}$ %), εp(7 *3* $\times10^{-6}$ %)
Δ: -73770 *80* keV
SpA: 1.64×10^{9} Ci/g
Prod: protons on Ta; protons on Th; protons on U; protons on La; ^{107}Ag(^{18}O,5n); ^{107}Ag(^{16}O,3n)

see ^{120}Cs(1.07 min) for γ rays

$^{120}_{55}$Cs(1.07 *5* min)

Mode: ε
Δ: -73770 *80* keV
SpA: 1.46×10^{9} Ci/g
Prod: protons on La

Photons(^{120}Cs(57 s + 1.07 min))

γ_{mode}	γ(keV)	γ(rel)
γ	273.4 *3*	0.36 *1*
γ	322.4 *1*	100
γ	348.6 *3*	0.48 *6*
γ	365.6 *2*	1.39 *12*
γ	395.5 *3*	2.7 *3*
γ	398.3 *4*	0.48 *9*
γ	416.0 *5*	0.21 *6*
γ	437.3 *3*	0.76 *8*
γ	451.0 *3*	1.39 *15*
γ	451.9 *3*	0.76 *9*
γ	453.7 *3*	0.88 *9*
γ	473.5 *1*	30 *3*
γ	475.6 *3*	1.4 *3*
γ	478.5 *2*	1.51 *21*
γ	525.2 *1*	2.85 *24*
γ	539.5 *5*	0.45 *15*
γ	545.2 *3*	1.5 *3*
γ	553.4 *1*	19.1 *15*
γ	560.8 *3*	0.91 *15*
γ	584.4 *3*	1.8 *4*
γ	585.8 *3*	4.5 *9*
γ	588.3 *3*	0.7 *2*
γ	601.2 *2*	10.9 *9*
γ	605.1 *2*	3.6 *3*
γ	631.7 *3*	0.3 *1*
γ	643.7 *3*	1.06 *11*
γ	649.8 *3*	0.27 *6*
γ	655.4 *5*	0.12 *3*
γ	668.2 *3*	0.94 *9*
γ	701.8 *2*	2.4 *2*
γ	713.3 *5*	0.15 *3*
γ	722.9 *8*	0.12 *3*
γ	736.0 *3*	0.39 *3*
γ	747.2 *3*	0.48 *3*
γ	801.0 *7*	0.24 *6*
γ	817.0 *7*	0.27 *6*
γ	831.0 *7*	0.24 *6*
γ	834.0 *5*	0.36 *6*
γ	849.2 *3*	0.61 *6*
γ	869.5 *3*	0.61 *6*
γ	875.8 *3*	6.1 *6*
γ	913.5 *7*	0.18 *6*
γ	949.1 *3*	7.3 *12*
γ	951.7 *3*	~2
γ	956.9 *5*	0.15 *6*
γ	970.5 *5*	0.24 *6*
γ	1021.0 *3*	1.06 *12*
γ	1064.8 *3*	0.70 *9*
γ	1085.5 *4*	0.24 *6*
γ	1098.3 *3*	0.85 *9*
γ	1130.3 *4*	0.48 *6*
γ	1142.4 *4*	0.45 *6*
γ	1147.3 *4*	0.30 *6*
γ	1174.8 *7*	0.15 *6*
γ	1195.9 *3*	0.39 *6*
γ	1212.5 *7*	~0.12
γ	1224.2 *7*	0.21 *6*
γ	1239.7 *7*	0.15 *3*
γ	1254.0 *6*	0.15 *6*
γ	1274.1 *4*	7.3 *9*
γ	1275.8 *4*	
γ	1290.4 *6*	0.33 *6*
γ	1300.7 *6*	0.27 *6*
γ	1310.5 *6*	0.30 *6*
γ	1329.3 *8*	0.39 *6*
γ	1365.7 *6*	0.58 *9*
γ	1368.8 *6*	0.61 *9*
γ	1389.4 *3*	2.0 *2*
γ	1402.8 *3*	1.0 *1*
γ	1408.0 *9*	0.21 *3*
γ	1422.5 *3*	0.76 *9*
γ	1433.8 *5*	0.24 *6*
γ	1445.1 *3*	2.40 *24*
γ	1452.7 *3*	1.18 *15*
γ	1471.0 *6*	0.36 *9*
γ	1477.5 *5*	0.45 *6*
γ	1490.0 *6*	0.21 *6*
γ	1522.0 *6*	0.30 *6*
γ	1526.0 *8*	0.30 *6*
γ	1533.0 *6*	0.27 *6*
γ	1538.5 *6*	0.42 *9*
γ	1563.0 *8*	0.21 *6*
γ	1593.2 *4*	0.42 *6*

Photons(^{120}Cs(57 s + 1.07 min)) (continued)

γ_{mode}	γ(keV)	γ(rel)
γ	1618.9 *4*	0.24 *6*
γ	1632.7 *8*	0.27 *9*
γ	1637.1 *8*	0.24 *6*
γ	1660.2 *3*	1.06 *15*
γ	1672.7 *8*	2.1 *3*
γ	1728.5 *3*	1.7 *2*
γ	1744.5 *4*	0.39 *6*
γ	1749.0 *5*	0.24 *6*
γ	1778.1 *6*	0.21 *6*
γ	1801.5 *8*	0.18 *6*
γ	1807.0 *8*	0.42 *9*
γ	1827.8 *6*	0.30 *6*
γ	1842.5 *4*	0.45 *6*
γ	1864.5 *4*	0.48 *6*
γ	1919.5 *4*	0.96 *12*
γ	1931.7 *4*	1.15 *12*
γ	1960.5 *8*	0.30 *6*
γ	1964.2 *8*	0.21 *6*
γ	1982.0 *5*	1.51 *24*
γ	1996.0 *5*	0.27 *6*
γ	2013.0 *5*	0.60 *15*
γ	2056.0 *8*	0.91 *15*
γ	2059.0 *8*	0.55 *15*
γ	2079.2 *5*	0.45 *9*
γ	2095.4 *6*	0.30 *9*
γ	2135.6 *4*	0.73 *9*
γ	2187.3 *4*	0.37 *6*
γ	2214.7 *6*	0.30 *6*
γ	2315.4 *4*	0.88 *12*
γ	2365.2 *6*	0.27 *6*
γ	2429.2 *6*	0.30 *6*
γ	2467.3 *4*	0.70 *9*
γ	2495.4 *4*	0.61 *15*
γ	2516.9 *6*	0.18 *6*
γ	2629.2 *6*	0.24 *6*
γ	2681.8 *8*	0.36 *6*
γ	2688.3 *8*	0.30 *6*
γ	2736.7 *6*	0.18 *3*
γ	2787.3 *4*	0.6 *1*
γ	2845.7 *6*	0.30 *6*
γ	2927.2 *4*	0.6 *1*
γ	3007.0 *6*	0.9 *1*
γ	3031.9 *8*	0.36 *9*
γ	3035.7 *8*	0.9 *2*
γ	3109.9 *4*	0.4 *1*
γ	3149.8 *8*	0.67 *13*
γ	3277.2 *4*	0.5 *1*
γ	3358.1 *4*	0.40 *6*
γ	3916.7 *8*	0.36 *6*
γ	4021.0 *8*	0.15 *6*
γ	4026.0 *8*	0.09 *3*
γ	4068.5 *8*	0.15 *6*
γ	4314.5 *8*	0.15 *6*

$^{120}_{56}$Ba(32 *5* s)

Mode: ε
Δ: -68470 *410* keV syst
SpA: 2.9×10^{9} Ci/g
Prod: ^{106}Cd(^{16}O,2n); ^{32}S on ^{96}Ru

Photons (^{120}Ba)

γ_{mode}	γ(keV)
γ	51
γ	182

A = 121

NDS **26**, 385 (1979)

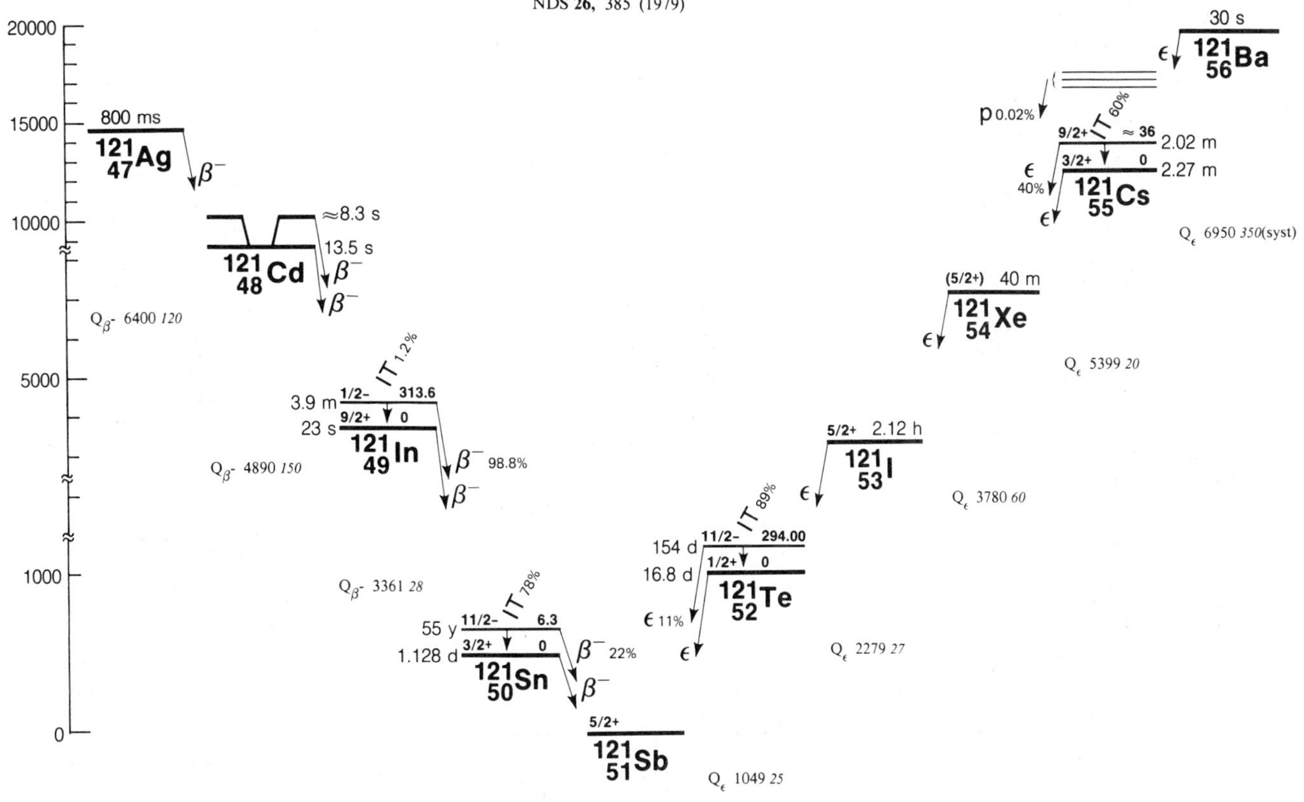

Q_{β^-} 388.9 *24*

$^{121}_{47}$Ag(800 *100* ms)

Mode: β-

Δ: -74550 *190* keV

SpA: 7.80×10^{10} Ci/g

Prod: fission

Photons (^{121}Ag)

γ$_{mode}$	γ(keV)	γ(rel)
γ M1	115.01 *10*	21.1 *21*
γ	146.4 *4*	2.0 *6*
γ	178.48 *10*	10.4 *10*
γ	203.0 *3*	2.0 *4* ?
γ	293.63 *10*	7.8 *8*
γ M1,E2	314.81 *10*	100 *7*
γ M1,E2	353.68 *10*	57 *4*
γ	362.16 *10*	11.4 *11*
γ	369.57 *10*	17.1 *12*
γ	372.04 *10*	9.3 *9*
γ	415.7 *3*	3.4 *7*
γ	430.57 *10*	12.0 *12*
γ	436.06 *10*	6.3 *6*
γ	439.72 *15*	4.4 *4*
γ	500.75 *10*	24.1 *17*
γ	547.5 *4*	1.5 *4*
γ	550.45 *20*	4.6 *5*
γ	583.1 *4*	2.3 *5*
γ	602.54 *20*	5.5 *6*
γ	605.2 *3*	3.3 *7*
γ	620.6 *3*	2.3 *5*

Photons (^{121}Ag) (continued)

γ$_{mode}$	γ(keV)	γ(rel)
γ	630.4 *3*	2.3 *5*
γ	636.72 *20*	4.1 *8*
γ	641.0 *3*	2.1 *4*
γ	645.2 *5*	1.0 *3*
γ	650.3 *3*	2.1 *4*
γ	667.50 *15*	8.3 *8*
γ	717.71 *20*	4.7 *5*
γ	725.60 *15*	6.2 *6*
γ	732.8 *3*	2.3 *5*
γ	740.27 *20*	2.8 *6*
γ	772.51 *20*	8.3 *8*
γ	785.20 *15*	4.9 *5*
γ	802.0 *2*	5.5 *6*
γ	817.21 *15*	10.9 *11*
γ	831.1 *4*	2.3 *5*
γ	856.2 *3*	4.1 *8*
γ	864.0 *2*	4.4 *9*
γ	1063.49 *20*	2.8 *6*
γ	1156.97 *20*	7.8 *8*
γ	1170.5 *3*	6.2 *6*
γ	1195.90 *15*	15.4 *15*
γ	1209.3 *4*	2.3 *5*
γ	1260.6 *3*	3.3 *7*
γ	1371.06 *20*	4.2 *8*
γ	1375.8 *5*	1.3 *4*
γ	1510.51 *15*	16.6 *17*
γ	1812.0 *3*	4.6 *9*
γ	1818.7 *5*	2.1 *4*
γ	1862.1 *5*	2.6 *5*
γ	1937.2 *5*	3.1 *6*
γ	2205.2 *5*	6.8 *14*
γ	2519.4 *5*	3.7 *7*

$^{121}_{48}$Cd(13.5 *3* s)

Mode: β-

Δ: -80950 *150* keV

SpA: 6.73×10^{9} Ci/g

Prod: fission

Photons (^{121}Cd)

γ$_{mode}$	γ(keV)	γ(rel)
γ	175.3 *3*	0.70 *7*
γ	192.86 *20*	0.96 *10*
γ E2	210.22 *10*	4.7 *3*
γ	275.03 *10*	4.4 *3*
γ	299.28 *10*	2.1 *1*
γ	317.9 *3*	1.54 *20*
γ E2,M1	324.42 *10*	100 *7*
γ	327.93 *10*	3.7 *4*
γ E2,M1	349.42 *10*	24.3 *17*
γ	375.87 *20*	1.8 *2*
γ	402.78 *10*	6.0 *4*
γ	441.6 *10*	2.2 *4*
γ	559.48 *20*	2.6 *3*
γ	567.5 *6*	0.7 *1*
γ	572.6 *2*	3.0 *3*
γ	650.8 *10*	9.5 *19*
γ	673.7 *10*	8.5 *17*
γ	677.55 *10*	6.0 *6*
γ	764.46 *10*	7.5 *8*
γ	781.84 *10*	3.3 *7*
γ	878.05 *20*	2.2 *4*
γ	910.1 *10*	3.4 *7*
γ	943.3 *6*	0.80 *16*
γ	978.7 *1*	8.0 *7*

Photons (^{121}Cd)
(continued)

γ_{mode}	γ(keV)	γ(rel)
γ	987.95 *10*	19.8 *14*
γ	1001.0 *3*	3.2 *3*
γ	1040.55 *10*	29 *6*
γ	1052.2 *4*	0.5 *2*
γ	1057.7 *4*	2.1 *4*
γ	1096.3 *2*	8.2 *8*
γ	1149.8 *3*	3.6 *4*
γ	1266.9 *3*	1.4 *3*
γ	1277.87 *20*	5.6 *6*
γ	1296.32 *20*	5.4 *5*
γ	1303.2 *5*	2.6 *5*
γ	1315.42 *10*	13.7 *14*
γ	1323.65 *20*	3.6 *7*
γ	1327.84 *20*	4.1 *8*
γ	1342.5 *2*	3.0 *7*
γ	1349.7 *10*	1.4 *3*
γ	1400.4 *10*	1.5 *3*
γ	1451.1 *3*	2.1 *4*
γ	1483.5 *2*	11.8 *12*
γ	1584.38 *20*	7.7 *8*
γ	1627.4 *4*	3.2 *6*
γ	1647.71 *10*	7.8 *8*
γ	1661.98 *20*	3.3 *6*
γ	1698.98 *20*	9.5 *10*
γ	1823.4 *3*	2.7 *5*
γ	1835.38 *20*	5.1 *10*
γ	1853.51 *20*	6.3 *6*
γ	1885.5 *3*	4.3 *9*
γ	1909.3 *6*	1.4 *3*
γ	1974.4 *10*	1.4 *3*
γ	2023.3 *4*	1.4 *3*
γ	2030.3 *4*	1.7 *3*
γ	2035.6 *7*	1.10 *22*
γ	2115.87 *5*	3.0 *6*
γ	2159.8 *5*	1.2 *2*
γ	2176.9 *5*	0.8 *2*

$^{121}_{48}$Cd(8.3 *8* s)

Mode: β-
Δ: -80950 *150* keV
SpA: 1.08×10^{10} Ci/g
Prod: fission

Photons (^{121}Cd)

γ_{mode}	γ(keV)	γ(rel)
γ	100.59 *10*	24.0 *17*
γ E2,M1	112.54 *10*	5.8 *4*
γ E2	160.3 *3*	4.0 *4*
γ	289.35 *10*	5.1 *10*
γ E2,M1	420.33 *10*	24.6 *17*
γ	447.35 *10*	19.0 *19*
γ	466.4 *3*	10.5 *10*
γ	899.2 *3*	9.8 *20*
γ	952.75 *10*	23.3 *24*
γ	988.05 *10*	88 *6*
γ	1020.89 *10*	95 *7*
γ	1069.3 *3*	10.5 *21*
γ	1121.81 *20*	13 *3*
γ	1130.1 *3*	11.7 *24*
γ	1139.62 *10*	36 *4*
γ	1181.56 *10*	66 *7*
γ	1271.27 *20*	17 *3*
γ	1336.81 *20*	12.3 *25*
γ	1382.05 *20*	17.8 *18*
γ	1433.3 *3*	9.8 *20*
γ	1456.7 *3*	6.8 *13*
γ	1467.65 *20*	17 *3*
γ	1487.39 *20*	15 *3*
γ	1504.3 *2*	21.5 *22*
γ	1535.7 *5*	4.3 *9*
γ	2059.99 *10*	100 *10*

Photons (^{121}Cd)
(continued)

γ_{mode}	γ(keV)	γ(rel)
γ	2159.8 *5*	5.5 *11*
γ	2292.8 *3*	14 *3*
γ	2332.8 *3*	17 *3*
γ	2365.9 *3*	42 *8*
γ	2456.2 *3*	27 *5*
γ	2511.8 *8*	12.3 *25*
γ	2563.6 *8*	15 *3*
γ	2870.2 *8*	1.8 *4*

$^{121}_{49}$In(23.1 *6* s)

Mode: β-
Δ: -85840 *28* keV
SpA: 3.98×10^{9} Ci/g
Prod: ^{122}Sn(γ,p); fission

Photons (^{121}In)
⟨γ⟩=927 *56* keV

γ_{mode}	γ(keV)	γ(%)†
Sn L_ℓ	3.045	0.0022 *7*
Sn L_η	3.272	0.00056 *7*
Sn L_α	3.443	0.061 *20*
Sn L_β	3.789	0.031 *6*
Sn L_γ	4.218	0.0030 *5*
Sn $K_{\alpha2}$	25.044	0.217 *16*
Sn $K_{\alpha1}$	25.271	0.41 *3*
Sn $K_{\beta1}$'	28.474	0.106 *8*
Sn $K_{\beta2}$'	29.275	0.0223 *17*
γ [M1]	56.334 *20*	0.19 *2*
γ M1	60.343 *20*	0.064 *13*
γ E1	261.95 *3*	7.9 *5*
γ (M1,E2)	657.31 *5*	7.1 *5*
γ	808.89 *6*	0.22 *4*
γ E2+M1	869.23 *6*	1.1 *1*
γ [M2]	919.26 *5*	4.2 *3*
γ E2	925.56 *6*	87 *6*
γ	1092.8 *4*	0.34 *3*

† approximate

Atomic Electrons (^{121}In)
⟨e⟩=2.15 *8* keV

e_{bin}(keV)	⟨e⟩(keV)	e(%)
2	0.014	~0.6
4	0.050	1.3 *4*
5 - 25	0.045	0.46 *17*
27	0.133	0.49 *6*
28 - 60	0.096	0.24 *3*
233	0.198	0.085 *6*
257 - 262	0.0332	0.0129 *6*
628	0.149	0.024 *3*
653 - 657	0.024	0.0037 *5*
780 - 809	0.0030	0.00038 *18*
840 - 869	0.0184	0.0022 *3*
890	0.156	0.0175 *14*
896	1.04	0.116 *8*
915 - 919	0.0250	0.00273 *18*
921	0.128	0.0139 *10*
922 - 926	0.0367	0.00397 *25*
1064 - 1093	0.0032	0.00030 *13*

Continuous Radiation (^{121}In)
⟨β-⟩=983 keV;⟨IB⟩=2.2 keV

E_{bin}(keV)		⟨ ⟩(keV)	(%)
0 - 10	β-	0.0125	0.249
	IB	0.037	
10 - 20	β-	0.0384	0.255
	IB	0.036	0.25
20 - 40	β-	0.159	0.53
	IB	0.071	0.25
40 - 100	β-	1.24	1.75
	IB	0.20	0.31
100 - 300	β-	15.8	7.6
	IB	0.55	0.31
300 - 600	β-	73	16.0
	IB	0.57	0.135
600 - 1300	β-	424	44.7
	IB	0.61	0.074
1300 - 2435	β-	469	28.8
	IB	0.109	0.0073

$^{121}_{49}$In(3.88 *10* min)

Mode: β-(98.8 *2* %), IT(1.2 *2* %)
Δ: -85526 *28* keV
SpA: 4.00×10^{8} Ci/g
Prod: ^{122}Sn(γ,p); fission

Photons (^{121}In)
⟨γ⟩=64 *5* keV

γ_{mode}	γ(keV)	γ(%)†
Sn L_ℓ	3.045	0.052 *13*
Sn L_η	3.272	0.027 *6*
Sn L_α	3.443	1.5 *3*
Sn L_β	3.742	1.3 *3*
Sn L_γ	4.217	0.15 *4*
In $K_{\alpha2}$	24.002	0.140 *24*
In $K_{\alpha1}$	24.210	0.26 *5*
Sn $K_{\alpha2}$	25.044	10.8 *22*
Sn $K_{\alpha1}$	25.271	20 *4*
In $K_{\beta1}$'	27.265	0.068 *12*
In $K_{\beta2}$'	28.022	0.0142 *25*
Sn $K_{\beta1}$'	28.474	5.3 *11*
Sn $K_{\beta2}$'	29.275	1.11 *22*
$\gamma_{\beta-}$M1	60.343 *20*	20
γ_{IT} M4	313.6 *1*	0.48
$\gamma_{\beta-}$E2+M1	909.1 *5*	0.33 *15*
$\gamma_{\beta-}$	1041.8 *4*	1.12 *20*
$\gamma_{\beta-}$(E2,M1)	1102.1 *4*	0.92 *25*
$\gamma_{\beta-}$[M1+E2]	1120.64 *20*	0.510
$\gamma_{\beta-}$[M1+E2]	1403.2 *10*	~0.051
$\gamma_{\beta-}$	2804.1 *5*	0.11 *4*
$\gamma_{\beta-}$	2864.5 *5*	0.102 *25*
$\gamma_{\beta-}$	3059.9 *14*	0.087 *25* ?
$\gamma_{\beta-}$	3228.2 *20*	0.015 *5* ?

† uncert(syst): 17% for IT, approximate for β-

Atomic Electrons (^{121}In)
⟨e⟩=22 *3* keV

e_{bin}(keV)	⟨e⟩(keV)	e(%)
4	1.5	36 *8*
19 - 20	0.103	0.51 *11*
21	0.76	3.6 *8*
23 - 28	0.47	1.9 *3*
31	13.4	43 *9*
56	3.1	5.6 *11*

Atomic Electrons (^{121}In) (continued)

e_{bin}(keV)	⟨e⟩(keV)	e(%)
59 - 60	0.80	1.34 *21*
286	1.6	0.56 *10*
309 - 314	0.49	0.158 *17*
880 - 909	0.0052	0.00059 *25*
1013 - 1042	0.011	0.0011 *5*
1073 - 1121	0.018	0.0016 *3*
1374 - 1403	0.00049	3.6 *16* $\times10^{-5}$
2775 - 2864	0.00085	3.0 *8* $\times10^{-5}$
3031 - 3059	0.00032	1.1 *4* $\times10^{-5}$
3199 - 3228	5.5 $\times10^{-5}$	1.7 *7* $\times10^{-6}$

Continuous Radiation (^{121}In)

⟨β-⟩=1503 keV;⟨IB⟩=4.4 keV

E_{bin}(keV)		⟨ ⟩(keV)	(%)
0 - 10	β-	0.0057	0.113
	IB	0.048	
10 - 20	β-	0.0175	0.116
	IB	0.048	0.33
20 - 40	β-	0.073	0.242
	IB	0.095	0.33
40 - 100	β-	0.57	0.80
	IB	0.27	0.42
100 - 300	β-	7.3	3.55
	IB	0.80	0.45
300 - 600	β-	36.2	7.9
	IB	0.94	0.22
600 - 1300	β-	268	27.6
	IB	1.38	0.159
1300 - 2500	β-	869	47.0
	IB	0.79	0.048
2500 - 3675	β-	322	11.5
	IB	0.049	0.0017

$^{121}_{50}$Sn(1.1275 *17* d)

Mode: β-
Δ: -89202 *3* keV
SpA: 9.571$\times10^5$ Ci/g
Prod: ^{120}Sn(n,γ); ^{123}Sb(d,α); fission

Continuous Radiation (^{121}Sn)

⟨β-⟩=115 keV;⟨IB⟩=0.047 keV

E_{bin}(keV)		⟨ ⟩(keV)	(%)
0 - 10	β-	0.275	5.5
	IB	0.0060	
10 - 20	β-	0.81	5.4
	IB	0.0053	0.037
20 - 40	β-	3.12	10.4
	IB	0.0087	0.031
40 - 100	β-	19.4	28.1
	IB	0.0161	0.026
100 - 300	β-	85	48.5
	IB	0.0105	0.0076
300 - 389	β-	6.9	2.14
	IB	5.6$\times10^{-5}$	1.8$\times10^{-5}$

$^{121}_{50}$Sn(55 *5* yr)

Mode: IT(77.6 *20* %), β-(22.4 *20* %)
Δ: -89195 *3* keV
SpA: 54 Ci/g
Prod: ^{120}Sn(n,γ); fission

Photons (^{121}Sn)

⟨γ⟩=5.02 *25* keV

γ_{mode}	γ(keV)	γ(%)†
Sn L_ℓ	3.045	0.094 *23*
Sb L_ℓ	3.189	0.023 *4*
Sn L_η	3.272	0.00014 *3*
Sb L_η	3.437	0.0117 *16*
Sn L_α	3.443	2.6 *6*
Sb L_α	3.604	0.64 *9*
Sb L_β	3.930	0.56 *9*
Sn L_β	4.035	0.41 *10*
Sn L_γ	4.336	0.0031 *9*
Sb L_γ	4.434	0.067 *11*
γ_{IT} (M4)	6.31 *8*	8.6 *18*$\times10^{-10}$
Sb $K_{\alpha2}$	26.111	4.5 *4*
Sb $K_{\alpha1}$	26.359	8.4 *8*
Sb $K_{\beta1}$'	29.712	2.20 *21*
Sb $K_{\beta2}$'	30.561	0.47 *5*
$\gamma_{\beta-}$M1	37.134 *8*	1.85

† uncert(syst): 2.6% for IT, 8.9% for β-

Atomic Electrons (^{121}Sn)

⟨e⟩=7.9 *7* keV

e_{bin}(keV)	⟨e⟩(keV)	e(%)
2	1.16	49 *10*
4	2.3	59 *12*
5	0.106	2.2 *3*
6	1.6	28 *6*
7	1.18	17.8 *16*
21	0.083	0.39 *5*
22	0.26	1.19 *16*
25	0.083	0.33 *5*
26	0.089	0.35 *5*
29	0.0143	0.049 *7*
30	0.0032	0.0110 *15*
32	0.69	2.12 *19*
33	0.069	0.210 *19*
36	0.168	0.46 *4*
37	0.040	0.108 *10*

Continuous Radiation (^{121}Sn)

⟨β-⟩=27.2 keV;⟨IB⟩=0.0111 keV

E_{bin}(keV)		⟨ ⟩(keV)	(%)
0 - 10	β-	0.055	1.11
	IB	0.00142	
10 - 20	β-	0.161	1.07
	IB	0.00126	0.0087
20 - 40	β-	0.63	2.09
	IB	0.0021	0.0074
40 - 100	β-	4.07	5.8
	IB	0.0039	0.0063
100 - 300	β-	21.3	11.9
	IB	0.0025	0.0018
300 - 358	β-	1.05	0.332
	IB	4.4$\times10^{-6}$	1.40$\times10^{-6}$

$^{121}_{51}$Sb(stable)

Δ: -89591 *3* keV
%: 57.3 *9*

$^{121}_{52}$Te(16.8 *4* d)

Mode: ε
Δ: -88542 *25* keV
SpA: 6.43$\times10^4$ Ci/g
Prod: daughter ^{121}I; ^{121}Sb(d,2n); ^{121}Sb(p,n); ^{120}Te(n,γ); daughter ^{121}Te(154 d)

Photons (^{121}Te)

⟨γ⟩=577 *10* keV

γ_{mode}	γ(keV)	γ(%)†
Sb L_ℓ	3.189	0.114 *16*
Sb L_η	3.437	0.057 *6*
Sb L_α	3.604	3.2 *4*
Sb L_β	3.931	2.8 *4*
Sb L_γ	4.441	0.34 *5*
Sb $K_{\alpha2}$	26.111	21.9 *11*
Sb $K_{\alpha1}$	26.359	40.8 *20*
Sb $K_{\beta1}$'	29.712	10.7 *5*
Sb $K_{\beta2}$'	30.561	2.29 *12*
γ M1	37.134 *8*	0.117 *3*
γ M1+4.0%E2	65.542 *10*	0.259 *9*
γ (E2)	470.457 *9*	1.41 *3*
γ M1+8.2%E2	507.591 *8*	17.7 *4*
γ E2	573.133 *9*	80.3 *17*

† 2.2% uncert(syst)

Atomic Electrons (^{121}Te)

⟨e⟩=9.1 *3* keV

e_{bin}(keV)	⟨e⟩(keV)	e(%)
4	2.6	62 *7*
5	0.41	8.7 *10*
7	0.0745	1.12 *4*
21	0.40	1.89 *21*
22	1.27	5.8 *6*
25	0.41	1.61 *18*
26	0.44	1.69 *19*
29	0.070	0.24 *3*
30	0.0159	0.054 *6*
32	0.0433	0.134 *5*
33	0.00434	0.0132 *5*
35	0.174	0.50 *3*
36	0.0106	0.0292 *10*
37	0.00252	0.00681 *23*
61	0.052	0.086 *22*
65	0.013	0.021 *5*
66	0.00026	0.00039 *8*
440	0.0489	0.0111 *5*
466	0.0073	0.00156 *7*
470	0.00180	0.000384 *18*
477	0.615	0.129 *4*
503	0.081	0.0160 *6*
507	0.0194	0.00383 *14*
508	0.000404	8.0 *3* $\times10^{-5}$
543	2.00	0.369 *11*
568	0.235	0.0413 *12*
569	0.0483	0.00849 *25*
572	0.0567	0.0099 *3*
573	0.0131	0.00228 *7*

Continuous Radiation (^{121}Te)

⟨IB⟩=0.073 keV

E_{bin}(keV)		⟨ ⟩(keV)	(%)
10 - 20	IB	0.00092	0.0053
20 - 40	IB	0.049	0.19
40 - 100	IB	0.0035	0.0058
100 - 300	IB	0.0143	0.0076
300 - 541	IB	0.0050	0.00143

$^{121}_{52}$Te(154 *7* d)

Mode: IT(88.6 *11* %), ϵ(11.4 *11* %)
Δ: -88248 *25* keV
SpA: 7007 Ci/g
Prod: ^{121}Sb(d,2n); ^{121}Sb(p,n); ^{120}Te(n,γ)

Photons (^{121}Te)

⟨γ⟩=216.9 *22* keV

γ_{mode}	γ(keV)	γ(%)†
Sb L_ℓ	3.189	0.024 *4*
Te L_ℓ	3.335	0.115 *16*
Sb L_η	3.437	0.0123 *15*
Sb L_α	3.604	0.68 *9*
Te L_η	3.606	0.036 *4*
Te L_α	3.768	3.2 *4*
Sb L_β	3.930	0.60 *9*
Te L_β	4.151	2.2 *3*
Sb L_γ	4.438	0.072 *11*
Te L_γ	4.692	0.27 *4*
Sb $K_{\alpha2}$	26.111	4.7 *4*
Sb $K_{\alpha1}$	26.359	8.8 *7*
Te $K_{\alpha2}$	27.202	9.8 *4*
Te $K_{\alpha1}$	27.472	18.2 *7*
Sb $K_{\beta1}$'	29.712	2.31 *18*
Sb $K_{\beta2}$'	30.561	0.49 *4*
Te $K_{\beta1}$'	30.980	4.85 *18*
Te $K_{\beta2}$'	31.877	1.06 *4*
γ_ϵM1	37.134 *8*	0.94 *10*
γ_{IT} M4	81.788 *15*	0.0478 *14*
γ_ϵ[M1+E2]	103.859 *20*	0.0009 *3*
γ_{IT} M1+5.1%E2	212.22 *3*	81.4
γ_ϵ[M1+E2]	909.860 *13*	0.0703 *15*
γ_ϵ[E2]	946.994 *13*	0.00822 *24*
γ_ϵ[M1+E2]	998.300 *11*	0.0796 *18*
γ_ϵ[M1+E2]	1024.01 *25*	~8×10⁻⁵
γ_ϵ[E2]	1035.434 *13*	0.00056 *24*
γ_ϵ[E2]	1102.158 *18*	2.54 *6*
γ_ϵ	1107.53 *4*	0.00040 *16*
γ_ϵ	1144.66 *4*	0.00108 *5*

† uncert(syst): 10% for ϵ, 1.2% for IT

Atomic Electrons (^{121}Te)

⟨e⟩=76.9 *13* keV

e_{bin}(keV)	⟨e⟩(keV)	e(%)
4	2.4	55 *6*
5 - 37	3.79	38.8 *25*
50	16.2	32.5 *11*
73	0.00049	~0.0007
77	31.4	40.7 *14*
81	8.0	9.9 *4*
82 - 104	1.83	2.24 *8*
180	11.2	6.19 *14*
207 - 212	2.13	1.023 *21*
879 - 917	0.00129	0.000146 *16*
942 - 968	0.00106	0.000109 *14*

Atomic Electrons (^{121}Te) (continued)

e_{bin}(keV)	⟨e⟩(keV)	e(%)
994 - 1035	0.000171	1.72 *21* $\times10^{-5}$
1072 - 1114	0.0303	0.00282 *11*
1140 - 1145	1.4 $\times10^{-6}$	1.2 *5* $\times10^{-7}$

Continuous Radiation (^{121}Te)

⟨β+⟩=0.00309 keV;⟨IB⟩=0.031 keV

E_{bin}(keV)		⟨ ⟩(keV)	(%)
0 - 10	β+	9.9$\times10^{-9}$	1.20$\times10^{-7}$
	IB	2.8$\times10^{-5}$	
10 - 20	β+	3.59$\times10^{-7}$	2.17$\times10^{-6}$
	IB	0.000105	0.00061
20 - 40	β+	8.6$\times10^{-6}$	2.65$\times10^{-5}$
	IB	0.0056	0.021
40 - 100	β+	0.000311	0.000407
	IB	0.00041	0.00067
100 - 300	β+	0.00277	0.00166
	IB	0.0024	0.00120
300 - 600	IB	0.0083	0.0019
600 - 1300	IB	0.0141	0.0017
1300 - 1306	IB	3.7$\times10^{-11}$	2.7$\times10^{-12}$
	Σβ+		0.0021

$^{121}_{53}$I (2.12 *1* h)

Mode: ϵ
Δ: -86263 *19* keV
SpA: 1.222$\times10^{7}$ Ci/g
Prod: ^{121}Sb(α,4n); ^{122}Te(d,3n); protons on La

Photons (^{121}I)

⟨γ⟩=285 *19* keV

γ_{mode}	γ(keV)	γ(%)†
Te L_ℓ	3.335	0.112 *19*
Te L_η	3.606	0.055 *8*
Te L_α	3.768	3.1 *5*
Te L_β	4.123	2.7 *4*
Te L_γ	4.661	0.33 *6*
Te $K_{\alpha2}$	27.202	20.5 *21*
Te $K_{\alpha1}$	27.472	38 *4*
Te $K_{\beta1}$'	30.980	10.2 *11*
Te $K_{\beta2}$'	31.877	2.21 *23*
γ[M1]	56.77 *17*	0.029 *6*
γ M1+8.3%E2	144.43 *20*	0.093 *19* ?
γ M1+5.1%E2	212.22 *3*	84
γ E2	230.9 *4*	0.29 *6*
γ[M1]	244.5 *4*	0.109 *22*
γ[M1]	262.84 *21*	0.093 *19*
γ[M1]	278.9 *4*	0.15 *3*
γ[M1]	293.23 *20*	0.059 *12* ?
γ[M1]	319.6 *2*	1.04 *21*
γ[M1]	367.6 *4*	0.051 *10*
γ[M1]	381.90 *18*	0.49 *10*
γ E2	471.5 *5*	0.86 *17*
γ E2	475.05 *21*	1.04 *21*
γ	531.82 *19*	6.1 *12*
γ (E2)	594.12 *18*	0.37 *7*
γ	594.5 *4*	0.37 *7*
γ	598.5 *4*	1.5 *3*
γ	639.1 *4*	0.051 *10*
γ	673.8 *4*	0.067 *13*
γ	678.4 *7*	0.076 *15*
γ	687.6 *5*	0.017 *3*

Photons (^{121}I) (continued)

γ_{mode}	γ(keV)	γ(%)†
γ	695.1 *6*	0.17 *3*
γ	699.6 *7*	0.19 *4*
γ	712.5 *7*	0.025 *5*
γ	751.9 *6*	0.034 *7*
γ	769.4 *6*	0.034 *7*
γ	782.2 *6*	0.084 *17*
γ	793.7 *10*	0.0084 *17*
γ	801.5 *8*	0.017 *3*
γ	806.7 *4*	0.24 *5*
γ	865.8 *9*	0.059 *12*
γ	875.0 *9*	0.051 *10*
γ	888.5 *6*	0.059 *12*
γ	911.8 *7*	0.025 *5*
γ	936.6 *4*	0.19 *4*
γ	958.7 *4*	0.025 *5*
γ	994.4 *6*	0.059 *12*
γ	1014.7 *6*	0.29 *6*
γ	1043.5 *6*	0.034 *7*
γ	1087.6 *9*	0.034 *7*
γ	1094.5 *8*	0.067 *13*
γ	1128.6 *9*	0.059 *12*
γ	1136.9 *5*	0.017 *3*
γ	1151.3 *8*	0.084 *17*
γ	1170.9 *4*	0.034 *7*
γ	1199.2 *4*	0.059 *12*
γ	1227.6 *12*	0.0084 *17*
γ	1256.0 *5*	0.017 *3*
γ	1274.3 *14*	0.017 *3*
γ	1306.6 *6*	0.084 *17*
γ	1340.8 *9*	0.017 *3*
γ	1363.6 *6*	0.034 *7*
γ	1414.1 *8*	0.017 *3*
γ	1439.8 *12*	0.0042 *8*
γ	1469.5 *9*	0.017 *3*
γ	1486.5 *14*	0.025 *5*
γ	1518.8 *5*	0.017 *3*
γ	1550.4 *10*	0.025 *5*
γ	1681.7 *9*	0.025 *5*
γ	1731.0 *5*	0.0034 *7*
γ	1841.8 *15*	0.017 *3*

† 10% uncert(syst)

Atomic Electrons (^{121}I)

⟨e⟩=19.7 *12* keV

e_{bin}(keV)	⟨e⟩(keV)	e(%)
4	1.69	39 *6*
5	1.24	26 *4*
22	0.37	1.70 *24*
23	1.15	5.0 *7*
25 - 26	0.40	1.52 *21*
27	0.41	1.52 *22*
30 - 57	0.088	0.28 *3*
113 - 144	0.029	0.025 *4*
180	11.5	6.4 *7*
199	0.044	0.022 *4*
207	1.57	0.76 *8*
208 - 247	0.68	0.321 *24*
258 - 293	0.087	0.030 *6*
315 - 363	0.044	0.0132 *18*
367 - 382	0.0048	0.00127 *21*
440 - 475	0.081	0.0182 *22*
500 - 532	0.17	~0.033
562 - 607	0.056	0.010 *3*
634 - 683	0.012	0.0018 *5*
687 - 720	0.0020	0.00028 *8*
738 - 782	0.0055	0.00072 *25*
789 - 834	0.0013	0.00016 *6*
843 - 888	0.0019	0.00023 *7*
905 - 954	0.0029	0.00032 *14*
958 - 994	0.0066	0.00068 *23*
1009 - 1056	0.0016	0.00016 *4*
1063 - 1105	0.0022	0.00021 *7*
1117 - 1166	0.0031	0.00027 *6*
1167 - 1199	0.0007	6.2 *23* $\times10^{-5}$
1222 - 1270	0.00033	2.7 *8* $\times10^{-5}$
1273 - 1309	0.0009	7 *3* $\times10^{-5}$

Atomic Electrons (^{121}I) (continued)

e_{bin}(keV)	⟨e⟩(keV)	e(%)
1332 - 1364	0.00031	2.3 10 ×10^{-5}
1382 - 1414	0.00017	1.2 5 ×10^{-5}
1435 - 1482	0.00033	2.3 7 ×10^{-5}
1486 - 1550	0.00033	2.2 6 ×10^{-5}
1650 - 1730	0.00020	1.2 4 ×10^{-5}
1810 - 1841	0.00011	5.9 23 ×10^{-6}

Continuous Radiation (^{121}I)

⟨β+⟩=61 keV;⟨IB⟩=1.65 keV

E_{bin}(keV)		⟨ ⟩(keV)	(%)
0 - 10	β+	3.87 ×10^{-6}	4.66 ×10^{-5}
	IB	0.0030	
10 - 20	β+	0.000146	0.00088
	IB	0.0033	0.023
20 - 40	β+	0.00377	0.0115
	IB	0.053	0.19
40 - 100	β+	0.163	0.210
	IB	0.018	0.028
100 - 300	β+	6.3	2.92
	IB	0.065	0.034
300 - 600	β+	29.3	6.5
	IB	0.21	0.047
600 - 1300	β+	25.3	3.49
	IB	0.91	0.099
1300 - 2067	IB	0.38	0.025
	Σβ+		13.2

$^{121}_{54}$Xe(40.1 *20* min)

Mode: ϵ

Δ: -82490 *60* keV

SpA: 3.87×10^{7} Ci/g

Prod: ^{127}I(p,7n); protons on Ce

Photons (^{121}Xe)

⟨γ⟩=1366 *24* keV

$γ_{mode}$	γ(keV)	γ(%)†
I $L_ℓ$	3.485	0.112 24
I $L_η$	3.780	0.055 11
I $L_α$	3.937	3.1 6
I $L_β$	4.320	2.8 6
I $L_γ$	4.884	0.35 8
I $K_{α2}$	28.317	18 3
I $K_{α1}$	28.612	34 5
I $K_{β1}$'	32.278	9.2 14
I $K_{β2}$'	33.225	2.1 3
γ M1(+E2)	57.88 10	0.74 5
γ	72.7 3	0.21 4
γ [M1]	76.89 11	0.106 18
γ M1	80.16 11	3.3 3
γ E2	83.97 15	0.088 18
γ E2	95.73 10	6.2 5
γ M1	132.90 9	15.2 12
γ M1	134.77 11	0.92 12
γ M1(+E2)	157.06 10	0.44 5
γ M1(+E2)	175.90 11	5.9 9
γ M1(+E2)	177.77 11	1.50 18
γ [M1]	192.58 11	0.070 18
γ [M1]	204.45 15	0.14 4
γ [M1]	214.94 11	0.81 11
γ M1(+E2)	252.79 9	18
γ M1+E2	300.77 13	3.6 3
γ M1,E2	310.67 9	7.6 7
γ [M1]	314.57 17	0.26 5

Photons (^{121}Xe) (continued)

$γ_{mode}$	γ(keV)	γ(%)†
γ (M1)	396.44 14	1.27 11
γ (E2)	433.66 12	1.48 12
γ (M1+E2)	445.37 10	10.8 10
γ [M1]	485.50 20	0.37 5
γ [M1]	516.71 16	0.25 7 ?
γ (M1)	516.92 15	0.16 7 ?
γ (E2)	529.34 14	1.9 3
γ [M1]	538.97 22	0.070 18
γ [M1]	548.22 18	0.158 18
γ [M1]	560.29 20	0.176 18
γ [M1]	585.81 24	0.158 18
γ [M1]	601.06 20	0.123 18
γ [M1]	626.83 16	0.070 18
γ [M1]	631.18 22	0.176 18
γ (E2)	649.82 14	2.15 21
γ [M1]	660.69 22	0.53 5
γ	685.54 15	0.088 18
γ	695.97 16	0.26 4
γ	724.79 15	0.158 18
γ	743.3 3	0.123 18
γ	753.85 16	0.83 9
γ	762.44 16	0.39 5
γ	773.21 19	0.35 5
γ	782.67 15	0.79 11
γ	805.43 16	0.11 3
γ	809.26 22	0.72 11
γ	818.88 18	0.123 18
γ	824.77 16	0.21 4
γ	831.28 14	0.58 7
γ	841.56 21	<0.25 ?
γ	842.60 15	1.02 12
γ	857.4 3	0.19 4
γ	860.8 5	0.21 4
γ	866.34 16	0.67 9
γ	875.75 20	0.123 18
γ	894.04 19	0.176 18
γ	899.4 5	0.070 18
γ	902.56 16	0.16 4
γ	906.10 19	0.21 4
γ	910.91 17	0.26 5
γ	913.52 21	0.14 4
γ	918.46 21	0.14 3
γ	930.86 20	1.9 3
γ	938.33 15	0.33 7
γ	939.72 16	0.55 11
γ	944.66 16	0.42 7
γ	947.45 21	0.16 4
γ	958.07 20	1.34 14
γ	962.07 15	0.21 4
γ	965.98 14	0.26 5
γ	978.8 5	0.16 4
γ	981.82 17	0.26 5
γ	992.1 3	0.26 4
γ	996.65 18	0.55 7
γ	1006.64 17	0.23 5
γ	1010.53 21	0.33 7
γ	1023.86 13	1.55 18
γ	1035.46 15	1.48 18
γ	1046.42 20	0.76 9
γ	1077.56 16	0.23 5
γ	1083.1 3	0.70 11
γ	1094.35 21	0.30 7
γ	1096.4 5	0.18 5
γ	1113.3 3	0.19 5
γ	1114.93 19	0.070 18
γ	1121.72 22	0.23 5
γ	1128.2 4	0.23 4
γ	1143.09 20	0.25 4
γ	1146.14 23	0.49 9
γ	1170.90 21	0.23 5
γ	1180.91 14	0.46 7
γ	1186.42 20	0.93 14
γ	1202.19 24	0.088 18
γ	1210.2 3	0.14 4
γ	1217.18 23	0.16 4
γ	1229.82 23	0.37 5
γ	1254.62 22	0.26 9
γ	1258.57 21	0.23 7
γ	1261.74 24	0.18 5
γ	1273.4 4	0.21 4
γ	1276.64 13	0.23 4
γ	1286.70 24	0.16 4
γ	1306.9 4	0.18 4
γ	1314.54 22	0.14 4

Photons (^{121}Xe) (continued)

$γ_{mode}$	γ(keV)	γ(%)†
γ	1339.41 17	0.63 9
γ	1348.22 20	0.30 5
γ	1358.5 4	0.25 4
γ	1364.7 3	0.30 7
γ	1369.49 25	0.16 4
γ	1382.7 3	0.21 4
γ	1402.65 24	0.25 4
γ	1427.37 25	0.25 4
γ	1441.53 23	0.18 4
γ	1459.52 23	0.088 18
γ	1462.9 3	0.088 18
γ	1470.7 3	0.21 5
γ	1474.16 23	0.106 18
γ	1482.92 20	0.25 4
γ	1508.2 3	0.21 5
γ	1540.80 20	0.77 11
γ	1553.61 20	0.58 7
γ	1560.47 24	0.28 5
γ	1566.1 5	0.23 5
γ	1574.00 23	0.35 7
γ	1597.15 22	0.19 4
γ	1616.4 3	0.23 4
γ	1631.63 16	0.79 12
γ	1654.04 20	0.11 3
γ	1655.03 22	0.26 9 ?
γ	1668.16 23	0.088 18
γ	1691.3 2	0.14 4
γ	1695.6 5	0.11 4
γ	1703.38 21	0.21 4
γ	1710.3 4	0.30 5
γ	1728.69 25	0.12 4
γ	1731.92 22	0.26 5
γ	1740.5 4	0.32 5
γ	1766.33 16	0.25 5
γ	1778.45 22	0.35 7
γ	1786.98 19	0.23 5
γ	1801.03 24	0.79 12
γ	1807.8 3	0.16 4
γ	1815.4 3	0.26 5
γ	1820.7 3	0.14 4
γ	1824.21 16	0.070 18
γ	1833.3 3	0.25 5
γ	1835.63 23	0.44 11
γ	1855.11 23	0.14 4
γ	1858.2 3	0.21 5 ?
γ	1864.7 3	0.11 4
γ	1868.97 20	0.18 5
γ	1877.79 23	0.14 4
γ	1890.1 3	0.14 4
γ	1891.8 3	0.40 9
γ	1901.10 17	0.11 3
γ	1909.97 18	0.19 4
γ	1917.87 19	0.19 4
γ	1927.27 25	0.26 5
γ	1980.47 24	0.49 9
γ	1990.54 23	0.12 3
γ	2012.57 23	0.35 7
γ	2023.01 25	0.21 5
γ	2047.91 22	0.44 9
γ	2077.00 15	0.30 7
γ	2087.74 17	0.63 12
γ	2092.73 23	0.23 7
γ	2116.96 21	0.18 5
γ	2122.06 25	0.25 5
γ	2159.06 21	0.12 3
γ	2212.8 3	0.35 9
γ	2228.56 21	0.14 4
γ	2256.35 23	0.25 5
γ	2265.8 4	0.14 4
γ	2278.39 18	0.35 9
γ	2302.8 5	0.12 4
γ	2321.1 3	0.28 5
γ	2336.27 18	0.35 9
γ	2347.26 21	0.14 4
γ	2365.9 6	0.12 4
γ	2369.9 3	0.16 4
γ	2379.34 22	0.088 18
γ	2413.16 19	0.32 5
γ	2418.27 23	0.18 4
γ	2428.68 23	0.99 14
γ	2434.3 6	0.19 5
γ	2463.0 3	0.18 4
γ	2486.56 23	0.35 7
γ	2493.32 19	0.12 4

Photons (^{121}Xe)
(continued)

γ_{mode}	γ(keV)	γ(%)†
γ	2495.8 6	0.11 4
γ	2510.3 3	0.46 11
γ	2514.11 23	0.35 9
γ	2528.13 20	0.19 5
γ	2538.74 23	0.28 7
γ	2544.91 21	1.50 18
γ	2548.8 3	0.28 5
γ	2569.6 4	0.30 5
γ	2577.9 3	0.12 4
γ	2589.05 18	0.42 11
γ	2593.9 3	0.51 11
γ	2605.02 20	0.09 3
γ	2607.7 6	0.09 3
γ	2621.80 21	1.09 18
γ	2634.2 3	0.26 12 ?
γ	2634.42 23	0.26 12 ?
γ	2637.54 22	0.83 16
γ	2643.62 23	3.1 4
γ	2651.8 3	0.25 5
γ	2664.80 21	0.25 5
γ	2667.4 3	0.23 5
γ	2671.7 3	0.19 5
γ	2677.2 3	0.30 7
γ	2689.5 4	0.28 5
γ	2690.01 22	<0.11 ?
γ	2697.8 3	0.26 5
γ	2700.6 3	0.14 4
γ	2714.3 3	0.18 5
γ	2718.4 6	0.19 5
γ	2721.52 20	0.11 4
γ	2728.7 3	0.09 4
γ	2751.9 3	0.16 4
γ	2757.41 21	0.070 18
γ	2760.1 3	0.16 4
γ	2763.3 6	0.09 4
γ	2797.69 20	1.23 18
γ	2808.8 3	0.09 3
γ	2815.29 21	0.053 18
γ	2818.3 3	0.053 18
γ	2830.7 3	0.18 4
γ	2847.6 3	~0.04 ?
γ	2847.6 3	~0.04 ?
γ	2849.4 3	0.053 18
γ	2926.7 5	0.09 3
γ	2936.9 3	0.09 3
γ	2941.71 18	0.16 5
γ	2950.9 6	0.07 3
γ	2953.33 24	0.12 4
γ	2964.5 3	0.26 5
γ	2977.2 3	0.11 4
γ	2979.9 3	0.053 18
γ	2984.2 3	0.09 4
γ	2994.2 3	0.053 18
γ	2996.7 6	0.053 18
γ	3025.4 3	0.07 3
γ	3033.98 20	0.16 5
γ	3052.2 5	0.11 4
γ	3071.14 20	0.09 3
γ	3097.1 3	~0.04 ?
γ	3097.4 3	~0.04 ?
γ	3126.5 5	0.09 3
γ	3145.90 24	0.11 3
γ	3160.1 3	0.30 7
γ	3230.0 3	~0.04

† upper limit

Atomic Electrons (^{121}Xe)
⟨e⟩=33.2 15 keV

e_{bin}(keV)	⟨e⟩(keV)	e(%)
5	2.9	62 12
23 - 24	1.33	5.7 7
25	0.81	3.3 10
27 - 44	0.98	3.3 4
47	1.94	4.1 3
51	0.091	0.18 4
53	1.2	~2
57 - 58	0.33	~0.6
63	5.3	8.5 7

Atomic Electrons (^{121}Xe)
(continued)

e_{bin}(keV)	⟨e⟩(keV)	e(%)
68 - 84	0.80	1.08 21
91	3.1	3.4 3
95 - 96	0.85	0.90 7
100	4.5	4.5 4
102 - 135	1.39	1.14 7
143	1.4	0.96 22
145 - 193	1.01	0.62 11
199 - 215	0.0259	0.0124 12
220	2.1	0.96 21
248 - 296	1.49	0.56 7
300 - 315	0.145	0.047 7
363 - 412	0.65	0.160 17
428 - 452	0.119	0.027 3
480 - 529	0.108	0.0216 21
534 - 581	0.0126	0.00226 16
585 - 631	0.073	0.0119 9
645 - 692	0.021	0.0032 5
695 - 743	0.027	0.0038 12
749 - 798	0.043	0.0055 14
800 - 843	0.038	0.0047 12
852 - 901	0.041	0.0046 14
902 - 949	0.052	0.0056 11
953 - 1002	0.054	0.0055 13
1005 - 1050	0.025	0.0024 6
1061 - 1110	0.017	0.0016 3
1112 - 1153	0.022	0.0019 5
1166 - 1213	0.0097	0.00082 18
1216 - 1262	0.012	0.00101 21
1268 - 1315	0.011	0.00088 23
1325 - 1369	0.0113	0.00084 16
1378 - 1426	0.0047	0.00034 8
1427 - 1475	0.0071	0.00049 10
1478 - 1573	0.020	0.00129 24
1583 - 1677	0.016	0.00101 19
1686 - 1782	0.019	0.00109 17
1786 - 1885	0.0186	0.00102 13
1886 - 1980	0.0083	0.00043 9
1985 - 2084	0.0122	0.00060 10
2087 - 2180	0.0044	0.00021 5
2195 - 2288	0.0072	0.00032 5
2298 - 2396	0.0120	0.00051 9
2401 - 2495	0.0106	0.00043 6
2505 - 2604	0.030	0.00117 15
2607 - 2700	0.029	0.00111 19
2709 - 2808	0.0102	0.00037 7
2810 - 2909	0.0021	7.3 15 $\times10^{-5}$
2918 - 3001	0.0048	0.000163 21
3019 - 3113	0.0021	6.7 13 $\times10^{-5}$
3121 - 3197	0.0016	4.9 14 $\times10^{-5}$
3225 - 3229	2.0 $\times10^{-5}$	~6 $\times10^{-7}$

Continuous Radiation (^{121}Xe)
⟨β+⟩=461 keV;⟨IB⟩=2.7 keV

E_{bin}(keV)		⟨ ⟩(keV)	(%)
0 - 10	β+	0.00053	0.0067
	IB	0.017	
10 - 20	β+	0.00193	0.0145
	IB	0.017	0.119
20 - 40	β+	0.00396	0.0123
	IB	0.067	0.23
40 - 100	β+	0.129	0.169
	IB	0.096	0.148
100 - 300	β+	5.00	2.29
	IB	0.28	0.155
300 - 600	β+	32.3	7.0
	IB	0.36	0.085
600 - 1300	β+	199	21.0
	IB	0.75	0.084
1300 - 2500	β+	225	13.7
	IB	0.94	0.053
2500 - 3684	β+	0.206	0.0081
	IB	0.18	0.0065
	Σβ+		44

$^{121}_{55}$Cs(2.27 5 min)

Mode: ε
Δ: -77090 60 keV
SpA: 6.83×10^8 Ci/g
Prod: ^{12}C on ^{115}In; protons on Ta; protons on Th; protons on U; ^{124}Xe(p,4n)

Photons (^{121}Cs)

γ_{mode}	γ(keV)	γ(rel)
γ	38.38 2	7.5 10
γ	159.8 3	7.8 20
γ	234.5 1	39.2 20
γ	235.2 1	11.8 20
γ	280.4 5	31.4 20
γ	281.0 5	11.8 20
γ	287 1	~4
γ	414.6 2	37 10
γ	427.3 1	76 8
γ	459.8 1	100 6
γ	554.0 2	15.7 20
γ	684.5 3	3.7 10
γ	701.0 5	2.9 10
γ	706.6 3	5.9 20
γ	733 1	~4
γ	1418 1	~2

$^{121}_{55}$Cs(2.02 5 min)

Mode: IT(60 4 %), ε(40 4 %)
Δ: ~-77054 kev
SpA: 7.67×10^8 Ci/g
Prod: ^{12}C on ^{115}In; protons on Ta; protons on Th; protons on U; ^{124}Xe(p,4n)

Photons (^{121}Cs)

γ_{mode}	γ(keV)	γ(rel)
γ	85.85 5	15.6 8
γ	153.700 5	100 3
γ	179.4 1	14.1 4
γ	196.1 1	0.45 22
γ	210.2 5	14 3
γ	239.6 1	68.5 14
γ	270.5 5	15.5 4
γ	296.2 1	14.5 7
γ	321.5 1	13.7 14
γ	451 5	12.3 14
γ	563 1	18 3
γ	620.0 5	4.1 14
γ	836 1	~0.5
γ	915.1 2	2.5 11
γ	1070 1	~1.0

$^{121}_{56}$Ba(29.7 15 s)

Mode: ε, εp(0.02 1 %)
Δ: -70140 350 keV syst
SpA: 3.10×10^9 Ci/g
Prod: ^{92}Mo(^{32}S,2pn); ^{93}Nb(^{32}S,p3n)

εp: 2000-3600 range

A = 122

NDS **B7**, 419 (1972)

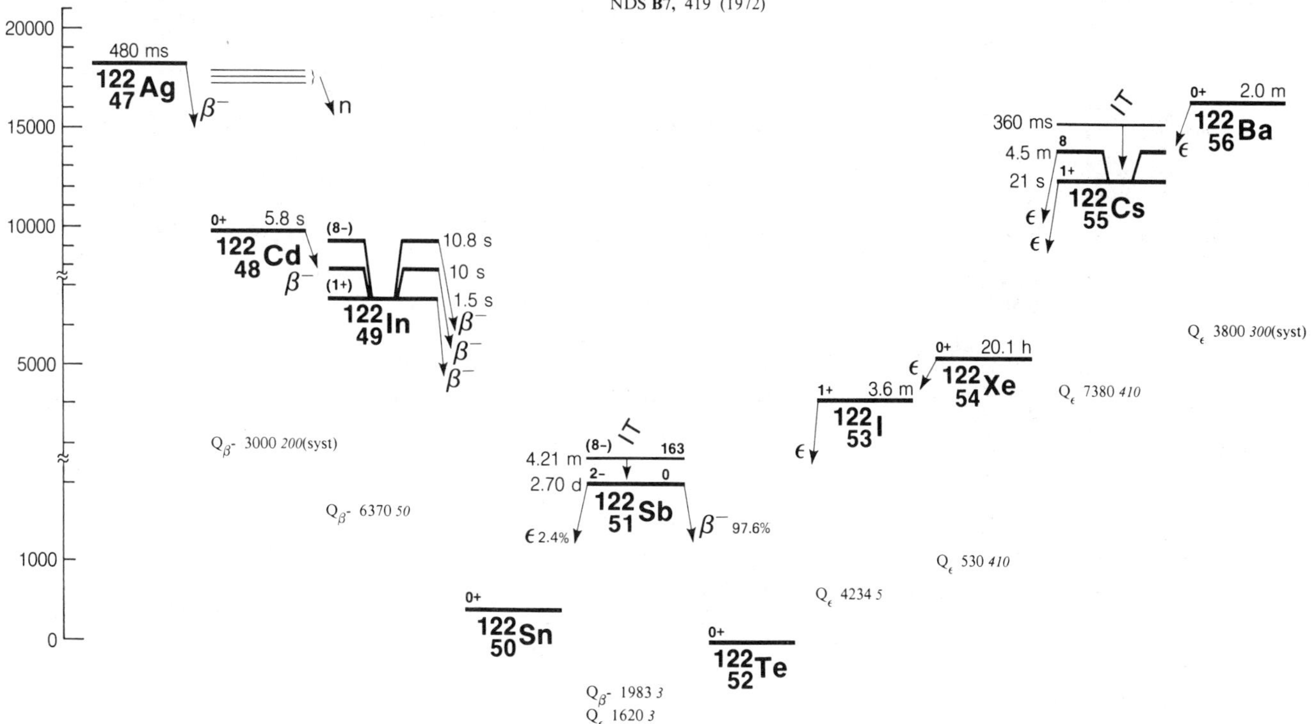

$^{122}_{47}$Ag(480 *80* ms)

Mode: β-, β-n
SpA: 1.02×10^{11} Ci/g
Prod: fission

Photons (^{122}Ag)

⟨γ⟩=1123 *130* keV

γ_{mode}	γ(keV)	γ(%)†
γ[E2]	569.45 *8*	96
γ	650.20 *12*	21 *3*
γ[E2]	759.70 *8*	32 *3*
γ	798.4 *3*	12 *4*
γ	1367.8 *5*	4.1 *18*
γ	1423.1 *9*	~3

† 4.2% uncert(syst)

$^{122}_{48}$Cd(5.78 *9* s)

Mode: β-
Δ: -80580 *210* keV syst
SpA: 1.508×10^{10} Ci/g
Prod: fission; protons on Sn

$^{122}_{49}$In(1.5 *3* s)

Mode: β-
Δ: -83580 *50* keV
SpA: 4.94×10^{10} Ci/g
Prod: ^{122}Sn(n,p); daughter ^{122}Cd

Photons (^{122}In)

⟨γ⟩=629 *35* keV

γ_{mode}	γ(keV)	γ(%)†
γ	400.27 *25*	0.9 *1*
γ[E2]	948.03 *20*	0.49 *9*
γ[M1+E2]	1013.42 *9*	2.4 *5*
γ[E2]	1140.59 *5*	29
γ[M1+E2]	1275.21 *14*	0.26 *3*
γ[E1]	1352.46 *15*	0.43 *6*
γ[E2]	1390.01 *10*	1.80 *15*
γ[M1+E2]	1635.29 *14*	0.52 *6*
γ[M1+E2]	1830.9 *4*	0.20 *3*
γ[M1+E2]	2065.93 *15*	1.97 *17*
γ[E2]	2154.00 *10*	0.13 *3*
γ[M1+E2]	2165.5 *3*	0.85 *10*
γ[M1+E2]	2408.54 *15*	0.43 *6*
γ[E2]	2415.79 *13*	0.55 *6*
γ[M1+E2]	2734.9 *3*	0.26 *3*
γ[E2]	2759.44 *15*	3.1 *3*
γ[M1+E2]	2775.86 *15*	0.23 *3*
γ[M1+E2]	2966.4 *3*	0.131 *17*
γ[M1+E2]	2976.0 *4*	0.84 *9*
γ[M1+E2]	3039.7 *4*	~0.12
γ[M1+E2]	3583.1 *4*	~0.12
γ[M1+E2]	3820.0 *3*	0.26 *3*
γ[M1+E2]	4004.3 *7*	0.12 *3*
γ[M1+E2]	4106.9 *3*	0.15 *3*

† 10% uncert(syst)

$^{122}_{49}$In(10.3 *6* s)

Mode: β-
Δ: -83580 *50* keV
SpA: 8.7×10^{9} Ci/g
Prod: fission; ^{124}Sn(d,α)

Photons (^{122}In)*

⟨γ⟩=3130 *140* keV

γ_{mode}	γ(keV)	γ(%)†
Sn $K_{\alpha2}$	25.044	0.073 *4*
Sn $K_{\alpha1}$	25.271	0.137 *7*
Sn $K_{\beta1}$'	28.474	0.0358 *19*
Sn $K_{\beta2}$'	29.275	0.0075 *4*
γ[M1+E2]	643.77 *20*	2.9 *4*
γ[M1+E2]	819.69 *12*	8.8 *10*
γ[M1+E2]	831.65 *12*	8.0 *10*
γ[M1+E2]	902.77 *12*	5.6 *6*
γ[M1+E2]	974.77 *5*	14.2 *15*
γ[E2]	1001.45 *5*	54 *5*
γ[M1+E2]	1013.42 *9*	13 *3*
γ[M1+E2]	1091.96 *12*	9.6 *10*
γ[E2]	1140.59 *5*	99
γ[M1+E2]	1163.96 *8*	26 *3*
γ[E2]	1190.65 *7*	28 *3*
γ[M1+E2]	1301.34 *10*	7.1 *6*
γ[M1+E2]	1699.01 *20*	2.48 *20*
γ[M1+E2]	2093.41 *11*	4.3 *4*
γ[E2]	2154.00 *10*	0.69 *17*

† 10% uncert(syst)
* see also ^{122}In(10.8 s)

Atomic Electrons (^{122}In)

⟨e⟩=3.47 *14* keV

e_{bin}(keV)	⟨e⟩(keV)	e(%)
4 - 28	0.0187	0.28 *3*
615 - 644	0.072	0.0117 *18*
790	0.138	0.017 *3*
802	0.123	0.015 *3*
815 - 832	0.041	0.0050 *5*
874 - 903	0.089	0.0101 *14*
946	0.18	0.019 *3*
970 - 971	0.022	0.0023 *4*
972	0.58	0.060 *6*
974 - 975	0.0053	0.00054 *8*
984	0.16	0.016 *4*
997 - 1013	0.116	0.0116 *9*
1063	0.106	0.0100 *16*
1087 - 1092	0.0161	0.00148 *19*
1111	0.93	0.084 *9*
1135	0.27	0.024 *4*
1136	0.113	0.0099 *11*
1137 - 1160	0.064	0.0055 *5*
1161	0.25	0.0216 *25*
1163 - 1191	0.046	0.0039 *3*
1272 - 1301	0.074	0.0058 *7*
1670 - 1698	0.0195	0.00117 *12*
2064 - 2153	0.032	0.00153 *15*

$^{122}_{49}$In(10.8 *4* s)

Mode: β-

Δ: -83580 *50* keV

SpA: 8.3×10^9 Ci/g

Prod: fission

Photons (^{122}In)

⟨γ⟩=3607 *170* keV

γ_{mode}	γ(keV)	γ(%)†
Sn L_ℓ	3.045	0.036 *6*
Sn L_η	3.272	0.0194 *25*
Sn L_α	3.443	1.01 *13*
Sn L_β	3.741	0.87 *12*
Sn L_γ	4.208	0.099 *16*
Sn $K_{\alpha2}$	25.044	7.1 *6*
Sn $K_{\alpha1}$	25.271	13.3 *10*
Sn $K_{\beta1}$'	28.474	3.5 *3*
Sn $K_{\beta2}$'	29.275	0.73 *6*
γ[E1]	103.68 *5*	82 *8*
γ	113.70 *14*	0.5 *2*
γ[E2]	163.30 *4*	69 *7*
γ	239.18 *20*	0.8 *2* •
γ[M1+E2]	244.00 *4*	5.8 *5*
γ[E3]	266.98 *6*	~0.12
γ[E2]	281.05 *5*	5.5 *5*
γ	309.58 *20*	1.0 *1* •
γ	332.73 *20*	0.34 *7* •
γ	350.78 *20*	1.1 *1* •
γ[M1+E2]	407.29 *4*	7.8 *7*
γ	440.58 *20*	0.6 *1* •
γ	592.27 *15*	2.6 *2*
γ[E2]	692.95 *15*	3.0 *3*
γ	813.6 *3*	0.9 *1* •
γ	840.56 *6*	1.6 *2*
γ[M1+E2]	877.61 *4*	11.5 *10*
γ[E2]	1001.45 *5*	97 *10*
γ[M1+E2]	1007.90 *14*	4.6 *4*
γ[M1+E2]	1013.37 *11*	3.4 *2*
γ	1057.53 *15*	1.8 *2* •
γ[M1+E2]	1060.2 *3*	1.2 *2*
γ[E3]	1105.14 *7*	1.2 *1*
γ[M1+E2]	1121.61 *4*	68 *7*
γ[E2]	1140.59 *5*	100
γ	1157.0 *4*	0.40 *4* •
γ[M1+E2]	1294.42 *10*	7.5 *7*

Photons (^{122}In) (continued)

γ_{mode}	γ(keV)	γ(%)†
γ	1516.2 *4*	0.45 *7* •
γ	1593.85 *20*	0.9 *1* •
γ[M1+E2]	1941.86 *15*	1.6 *2*
γ	2742.0 *4*	0.55 *6*

† 10% uncert(syst)

• with ^{122}In(10.3 or 10.8 s)

Atomic Electrons (^{122}In)

⟨e⟩=45.7 *25* keV

e_{bin}(keV)	⟨e⟩(keV)	e(%)
4 - 28	1.87	29 *3*
74	8.7	11.6 *12*
85	0.18	~0.22
99	1.15	1.16 *12*
100 - 114	0.76	0.74 *6*
134	20.9	15.6 *17*
159	5.1	3.2 *4*
162 - 210	1.34	0.82 *7*
215 - 263	1.35	0.58 *6*
266 - 310	0.19	0.066 *12*
322 - 351	0.052	0.016 *7*
378 - 411	0.42	0.109 *10*
436 - 441	0.0028	0.0006 *3*
563 - 592	0.052	0.009 *4*
664 - 693	0.062	0.0093 *9*
784 - 814	0.030	0.0037 *13*
836 - 878	0.193	0.023 *3*
972 - 1013	1.33	0.136 *12*
1028 - 1076	0.057	0.0054 *8*
1092 - 1141	1.93	0.174 *14*
1153 - 1157	0.00047	4.1 *17* ×10^{-5}
1265 - 1294	0.079	0.0062 *8*
1487 - 1565	0.0081	0.00052 *16*
1589 - 1593	0.0007	4.6 *16* ×10^{-5}
1913 - 1941	0.0111	0.00058 *8*
2723 - 2751	0.0022	8.1 *23* ×10^{-5}

$^{122}_{50}$Sn(stable)

Δ: -89945 *3* keV

%: 4.63 *3*

$^{122}_{51}$Sb(2.70 *1* d)

Mode: β-(97.62 *13* %), ϵ(2.38 *13* %)

Δ: -88326 *3* keV

SpA: 3.964×10^5 Ci/g

Prod: ^{121}Sb(n,γ)

Photons (^{122}Sb)

⟨γ⟩=433 *4* keV

γ_{mode}	γ(keV)	γ(%)†
Sn L_ℓ	3.045	0.0025 *3*
Sn L_η	3.272	0.00129 *13*
Te L_ℓ	3.335	0.00051 *7*
Sn L_α	3.443	0.069 *8*
Te L_η	3.606	0.00025 *3*
Sn L_β	3.742	0.061 *8*
Te L_α	3.768	0.0141 *15*

Photons (^{122}Sb) (continued)

γ_{mode}	γ(keV)	γ(%)†
Te L_β	4.123	0.0123 *15*
Sn L_γ	4.222	0.0072 *10*
Te L_γ	4.654	0.00148 *21*
Sn $K_{\alpha2}$	25.044	0.507 *15*
Sn $K_{\alpha1}$	25.271	0.95 *3*
Te $K_{\alpha2}$	27.202	0.092 *3*
Te $K_{\alpha1}$	27.472	0.172 *5*
Sn $K_{\beta1}$'	28.474	0.248 *8*
Sn $K_{\beta2}$'	29.275	0.0521 *17*
Te $K_{\beta1}$'	30.980	0.0457 *14*
Te $K_{\beta2}$'	31.877	0.0100 *3*
$\gamma_{\beta-}$E2	564.37 *10*	70.0
$\gamma_{\beta-}$[E2]	616.7 *6*	0.011 *4*
$\gamma_{\beta-}$E2+7.7%M1	692.90 *11*	3.82 *13*
$\gamma_{\beta-}$[E2]	793.5 *3*	0.016 *4*
γ_{ϵ}[E2]	1140.59 *5*	0.75
$\gamma_{\beta-}$	1188.8 *4*	0.0042 *7* ?
$\gamma_{\beta-}$[E2]	1257.26 *11*	0.80 *4*
$\gamma_{\beta-}$	1753.1 *4*	0.0091 *7*

† uncert(syst): 13.3% for ϵ, 5.7% for β-

Atomic Electrons (^{122}Sb)

⟨e⟩=2.46 *4* keV

e_{bin}(keV)	⟨e⟩(keV)	e(%)
4 - 31	0.154	2.30 *19*
533	1.87	0.352 *8*
559	0.221	0.0395 *9*
560 - 585	0.1172	0.0208 *3*
612 - 617	4.6 ×10^{-5}	7.5 *23* ×10^{-6}
661	0.076	0.0115 *4*
688 - 693	0.0130	0.00189 *6*
762 - 793	0.00030	4.0 *9* ×10^{-5}
1157 - 1189	4.2 ×10^{-5}	3.6 *16* ×10^{-6}
1225 - 1257	0.0089	0.00072 *4*
1721 - 1752	6.1 ×10^{-5}	3.5 *13* ×10^{-6}

Continuous Radiation (^{122}Sb)

⟨β-⟩=563 keV;⟨IB⟩=0.91 keV

E_{bin}(keV)		⟨ ⟩(keV)	(%)
0 - 10	β-	0.0370	0.74
	IB	0.024	
10 - 20	β-	0.112	0.75
	IB	0.023	0.162
20 - 40	β-	0.461	1.53
	IB	0.046	0.160
40 - 100	β-	3.43	4.87
	IB	0.121	0.19
100 - 300	β-	37.5	18.5
	IB	0.29	0.169
300 - 600	β-	130	29.0
	IB	0.24	0.058
600 - 1300	β-	329	38.0
	IB	0.153	0.019
1300 - 1983	β-	63	4.18
	IB	0.0077	0.00054

$^{122}_{51}$Sb(4.21 *2* min)

Mode: IT

Δ: -88163 *3* keV

SpA: 3.656×10^8 Ci/g

Prod: ^{121}Sb(n,γ); ^{123}Sb(n,2n)

Photons (^{122}Sb)

$\langle\gamma\rangle$=70.4 *16* keV

γ_{mode}	γ(keV)	γ(%)
Sb L_{ℓ}	3.189	0.25 *7*
Sb L_{η}	3.437	0.14 *4*
Sb L_{α}	3.604	6.8 *19*
Sb L_{β}	3.927	5.7 *16*
Sb L_{γ}	4.388	0.61 *16*
γ [E3]	26 *1*	~0.0031
Sb $K_{\alpha2}$	26.111	22.0 *9*
Sb $K_{\alpha1}$	26.359	41.0 *16*
Sb $K_{\beta1}$'	29.712	10.8 *4*
Sb $K_{\beta2}$'	30.561	2.30 *9*
γ E1	61.45 *3*	57.4 *23*
γ E2	76.08 *3*	18.5 *7*

Atomic Electrons (^{122}Sb)

$\langle e\rangle$=92 *9* keV

e_{bin}(keV)	$\langle e\rangle$(keV)	e(%)
4	6.3	149 *42*
5	0.32	6.7 *7*
21	0.44	2.1 *3*
22	18.2	84 *40*
25	5.0	20 *9*
26	1.4	5.5 *21*
29	0.070	0.243 *25*
30	0.0159	0.054 *6*
31	11.3	36.4 *16*
46	23.2	50.9 *23*
57	2.82	4.95 *22*
61	0.73	1.20 *5*
71	3.18	4.46 *20*
72	14.3	19.9 *9*
75	3.81	5.06 *23*
76	0.84	1.11 *5*

$^{122}_{52}$Te(stable)

Δ: -90309 *3* keV
%: 2.60 *1*

$^{122}_{53}$I (3.62 *6* min)

Mode: ϵ
Δ: -86075 *6* keV
SpA: 4.25$\times 10^{8}$ Ci/g
Prod: ^{121}Sb(α,3n); ^{122}Te(p,n); protons on La; ^{122}Te(d,2n); daughter ^{122}Xe

Photons (^{122}I)

$\langle\gamma\rangle$=160 *20* keV

γ_{mode}	γ(keV)	γ(%)†
Te L_{ℓ}	3.335	0.028 *4*
Te L_{η}	3.606	0.0136 *14*
Te L_{α}	3.768	0.77 *8*
Te L_{β}	4.123	0.68 *8*
Te L_{γ}	4.661	0.083 *12*
Te $K_{\alpha2}$	27.202	5.07 *15*
Te $K_{\alpha1}$	27.472	9.5 *3*
Te $K_{\beta1}$'	30.980	2.51 *8*
Te $K_{\beta2}$'	31.877	0.547 *18*

Photons (^{122}I)
(continued)

γ_{mode}	γ(keV)	γ(%)†
γ E2	564.37 *10*	18 *4*
γ [E2]	616.7 *6*	0.030 *6*
γ	683.87 *21*	0.85 *17*
γ E2+7.7%M1	692.90 *11*	1.28 *18*
γ [E2]	793.5 *3*	1.24 *25*
γ	954.0 *4*	~0.012
γ	1039.0 *5*	0.0071 *14* ?
γ	1075.4 *10*	
γ	1130.9 *7*	0.014 *3* ?
γ	1188.8 *4*	0.028 *5* ?
γ [E2]	1257.26 *11*	0.27 *4*
γ	1336.3 *4*	0.030 *6*
γ	1358.0 *18*	0.0124 *25*
γ	1376.76 *21*	0.039 *8*
γ	1500.10 *24*	0.15 *3*
γ	1534.9 *4*	0.057 *11*
γ	1723.7 *4*	0.042 *9* ?
γ	1734 *4*	0.023 *5*
γ	1747.6 *3*	0.30 *6*
γ	1753.1 *4*	0.062 *9*
γ	1793.0 *5*	0.014 *3* ?
γ [M1+E2]	1844.2 *3*	0.096 *19*
γ	1941.12 *22*	0.0089 *18* ?
γ	1945.5 *4*	0.030 *6* ?
γ	2029.2 *4*	0.041 *8*
γ	2155.8 *4*	0.027 *5*
γ	2192.98 *25*	0.24 *5*
γ	2227.1 *5*	0.0124 *25*
γ [E2]	2233.9 *12*	0.0071 *14*
γ	2288.1 *4*	0.0106 *21* ?
γ	2311.9 *3*	0.0035 *7* ?
γ [M1+E2]	2334.5 *12*	0.0035 *7* ?
γ	2347.4 *13*	0.0124 *25*
γ	2481.3 *5*	0.016 *3*
γ	2488.9 *6*	0.0071 *14*
γ	2586.5 *5*	0.037 *7*
γ	2593.6 *4*	0.0053 *11* ?
γ	2720.2 *4*	0.027 *5*
γ	2739.0 *4*	0.0089 *18*
γ	2794 *7*	0.0018 *4*
γ	2920.0 *5*	0.014 *3*
γ	2983 *4*	0.0089 *18*
γ	3045.6 *5*	0.0053 *11*
γ	3208 *4*	0.0053 *11*
γ	3291 *4*	0.0035 *7*

† 10% uncert(syst)

Atomic Electrons (^{122}I)

$\langle e\rangle$=1.96 *11* keV

e_{bin}(keV)	$\langle e\rangle$(keV)	e(%)
4	0.42	9.6 *10*
5	0.31	6.5 *7*
22	0.093	0.42 *4*
23	0.29	1.25 *13*
26	0.093	0.35 *4*
27	0.101	0.38 *4*
30 - 31	0.0198	0.066 *5*
533	0.47	0.089 *18*
559	0.056	0.0100 *20*
560 - 585	0.030	0.0054 *7*
612 - 661	0.040	0.0061 *14*
679 - 693	0.0067	0.00098 *18*
762 - 793	0.023	0.0031 *5*
922 - 954	0.00016	~2$\times 10^{-5}$
1007 - 1039	8.2 $\times 10^{-5}$	8 *4* $\times 10^{-6}$
1099 - 1131	0.00015	1.4 *6* $\times 10^{-5}$
1157 - 1189	0.00028	2.4 *11* $\times 10^{-5}$
1225 - 1257	0.0030	0.00024 *3*
1305 - 1354	0.00066	5.0 *14* $\times 10^{-5}$
1357 - 1377	4.7 $\times 10^{-5}$	3.4 *14* $\times 10^{-6}$
1468 - 1534	0.0016	0.00011 *4*
1692 - 1789	0.0030	0.00017 *5*
1792 - 1843	0.00080	4.4 *9* $\times 10^{-5}$
1909 - 1997	0.00045	2.3 *6* $\times 10^{-5}$
2024 - 2028	3.1 $\times 10^{-5}$	1.5 *5* $\times 10^{-6}$

Atomic Electrons (^{122}I)
(continued)

e_{bin}(keV)	$\langle e\rangle$(keV)	e(%)
2124 - 2223	0.0016	7.3 *22* $\times 10^{-5}$
2226 - 2316	0.00016	6.9 *14* $\times 10^{-6}$
2330 - 2347	1.1 $\times 10^{-5}$	4.8 *12* $\times 10^{-7}$
2449 - 2488	0.00011	4.6 *12* $\times 10^{-6}$
2555 - 2593	0.00020	8.0 *24* $\times 10^{-6}$
2688 - 2762	0.00017	6.3 *16* $\times 10^{-6}$
2789 - 2793	1.0 $\times 10^{-6}$	3.7 *12* $\times 10^{-8}$
2888 - 2982	0.000100	3.4 *8* $\times 10^{-6}$
3014 - 3045	2.2 $\times 10^{-5}$	7.5 *23* $\times 10^{-7}$
3177 - 3260	3.4 $\times 10^{-5}$	1.06 *24* $\times 10^{-6}$
3286 - 3291	1.8 $\times 10^{-6}$	5.5 *16* $\times 10^{-8}$

Continuous Radiation (^{122}I)

$\langle\beta+\rangle$=1087 keV;$\langle$IB$\rangle$=6.4 keV

E_{bin}(keV)		$\langle\ \rangle$(keV)	(%)
0 - 10	β+	1.53$\times 10^{-6}$	1.84$\times 10^{-5}$
	IB	0.037	
10 - 20	β+	5.9$\times 10^{-5}$	0.000354
	IB	0.036	0.25
20 - 40	β+	0.00154	0.00473
	IB	0.084	0.29
40 - 100	β+	0.072	0.092
	IB	0.21	0.32
100 - 300	β+	3.60	1.62
	IB	0.60	0.34
300 - 600	β+	29.4	6.3
	IB	0.75	0.18
600 - 1300	β+	257	26.4
	IB	1.48	0.165
1300 - 2500	β+	696	38.6
	IB	2.1	0.119
2500 - 4234	β+	100	3.72
	IB	1.06	0.036
	$\Sigma\beta$+		77

$^{122}_{54}$Xe(20.1 *1* h)

Mode: ϵ
Δ: -85540 *410* keV
SpA: 1.278$\times 10^{6}$ Ci/g
Prod: ^{127}I(p,6n); ^{120}Te(α,2n); protons on Ce

Photons (^{122}Xe)

$\langle\gamma\rangle$=68 *6* keV

γ_{mode}	γ(keV)	γ(%)†
I L_{ℓ}	3.485	0.130 *18*
I L_{η}	3.780	0.063 *7*
I L_{α}	3.937	3.6 *4*
I L_{β}	4.320	3.2 *4*
I L_{γ}	4.892	0.41 *6*
I $K_{\alpha2}$	28.317	22.6 *7*
I $K_{\alpha1}$	28.612	42.1 *13*
I $K_{\beta1}$'	32.278	11.3 *4*
I $K_{\beta2}$'	33.225	2.57 *9*
γ	52.99 *20*	0.023 *8*
γ M1(+E2)	58.01 *15*	0.070 *16*
γ M1	61.74 *14*	0.37 *4*
γ	66.71 *15*	0.039 *16*
γ M1	72.52 *14*	0.195 *16*
γ M1(+E2)	90.63 *13*	0.61 *5*
γ	104.0 *3*	0.047 *16*
γ	110.1 *3*	0.047 *16*

Photons (^{122}Xe)
(continued)

γ_{mode}	γ(keV)	γ(%)†
γ	116.2 *3*	0.101 *16*
γ M1	148.64 *13*	3.1 *4*
γ	163.16 *12*	0.148 *16*
γ M1,E2	174.6 *4*	0.16 *3*
γ M1,E2	175.6 *4*	0.33 *4*
γ	186.94 *13*	0.62 *6*
γ	201.45 *14*	0.133 *16*
γ	253.64 *14*	0.117 *16*
γ	288.35 *14*	0.47 *5*
γ M1,E2	350.09 *12*	7.8 *16*
γ	355.06 *16*	0.179 *23*
γ	416.80 *13*	1.76 *16*

† 14% uncert(syst)

Atomic Electrons (^{122}Xe)
⟨e⟩= 9.7 *5* keV

e_{bin}(keV)	⟨e⟩(keV)	e(%)
5	3.4	71 *8*
20	0.019	~0.09
23	0.71	3.1 *3*
24	0.92	3.9 *4*
25	0.08	0.31 *11*
27	0.36	1.32 *14*
28	0.48	1.72 *18*
29	0.29	1.03 *11*
31 - 53	0.42	1.02 *23*
57	0.53	0.9 *3*
58 - 106	0.47	0.60 *19*
109 - 112	0.016	~0.015
115	0.77	0.67 *9*
116 - 163	0.40	0.27 *5*
168 - 201	0.079	0.045 *13*
220 - 255	0.040	0.016 *7*
283 - 288	0.007	0.0023 *10*
317	0.53	0.17 *4*
322 - 355	0.112	0.033 *7*
384 - 417	0.07	~0.019

Continuous Radiation (^{122}Xe)
⟨IB⟩=0.085 keV

E_{bin}(keV)		⟨ ⟩(keV)	(%)
10 - 20	IB	0.00059	0.0035
20 - 40	IB	0.059	0.21
40 - 100	IB	0.0042	0.0072
100 - 300	IB	0.0140	0.0074
300 - 530	IB	0.0067	0.0019

$^{122}_{55}$Cs(21.0 *7* s)

Mode: ε
Δ: -78160 *70* keV
SpA: 4.33×10^{9} Ci/g
Prod: protons on Ta; protons on Th; protons on U; ^{113}In(^{12}C,3n); ^{109}Ag(^{18}O,5n); protons on La

Photons (^{122}Cs)
⟨γ⟩=726 *13* keV

γ_{mode}	γ(keV)	γ(%)†
γ [E2]	331.43 *17*	95.0
γ [M1+E2]	371.35 *17*	0.64 *9*
γ [M1+E2]	385.66 *18*	0.20 *5*
γ [E2]	497.49 *15*	2.0 *3*
γ [M1+E2]	511.80 *16*	9.0 *12*
γ	647.5 *6*	0.57 *10*
γ [E2]	665.76 *22*	0.57 *10*
γ	759.3 *10*	~0.07
γ [E2]	817.24 *20*	6.2 *4*
γ	821.4 *10*	0.12 *6*
γ	827.2 *10*	0.19 *6*
γ	839.8 *8*	0.38 *10*
γ [E2]	843.22 *18*	3.8 *4*
γ	850.51 *24*	0.13 *6*
γ	872.59 *23*	0.34 *4*
γ [M1+E2]	882.2 *4*	1.6 *3*
γ [M1+E2]	883.15 *16*	1.56 *23*
γ	944.7 *8*	0.28 *10*
γ	1035.2 *3*	1.14 *19*
γ	1038.2 *4*	0.50 *10*
γ	1147.7 *11*	~0.19
γ	1163.25 *22*	0.75 *12*
γ	1176.3 *11*	0.14 *6*
γ	1193.3 *13*	~0.05
γ	1221.86 *25*	0.34 *10*
γ	1236.17 *25*	0.32 *10*
γ	1384.39 *23*	1.39 *19*
γ	1420.8 *13*	0.12 *6*
γ	1427.5 *10*	0.32 *12*
γ	1440.6 *13*	~0.04
γ	1456.5 *11*	0.21 *4*
γ	1459.3 *8*	0.18 *4*
γ	1485.9 *10*	0.48 *12*
γ	1494.67 *24*	1.23 *19*
γ	1513.8 *4*	0.26 *4*
γ	1550.2 *13*	0.20 *4*
γ	1733.65 *24*	1.04 *19*
γ	1798.0 *6*	0.75 *12*
γ	1812.3 *6*	0.44 *9*
γ	1835.4 *11*	0.38 *12*
γ	1889.1 *13*	0.44 *9*
γ	1932.3 *5*	0.88 *12*
γ	2011.2 *4*	0.75 *12*
γ	2161.8 *13*	0.38 *10*
γ	2198.5 *3*	1.04 *19*

† 1.0% uncert(syst)

$^{122}_{55}$Cs(4.5 *2* min)

Mode: ε
Δ: -78160 *70* keV
SpA: 3.42×10^{8} Ci/g
Prod: protons on Ta; protons on Th; protons on U; ^{113}In(^{12}C,3n); ^{109}Ag(^{16}O,3n); protons on La

Photons (^{122}Cs)
⟨γ⟩=2599 *70* keV

γ_{mode}	γ(keV)	γ(%)†
γ	278.3 *4*	1.22 *19*
γ	280.6 *4*	1.87 *19*
γ [M1+E2]	307.75 *23*	1.50 *19*
γ [E2]	331.43 *17*	93.5
γ [M1+E2]	371.35 *17*	3.0 *5*
γ [M1+E2]	385.66 *18*	0.96 *20*
γ	459.2 *4*	0.7 *3*
γ	467.2 *4*	0.7 *3*
γ [E2]	497.49 *15*	79 *5*
γ [M1+E2]	511.80 *16*	8.7 *11*
γ [E2]	560.11 *17*	14.0 *19*
γ [M1+E2]	574.42 *16*	5.6 *7*
γ [E2]	589.83 *19*	2.3 *5*

Photons (^{122}Cs)
(continued)

γ_{mode}	γ(keV)	γ(%)†
γ [E2]	638.69 *18*	63 *5*
γ [E2]	654.10 *24*	7.5 *14*
γ [E2]	685.8 *6*	4.7 *7*
γ	738.5 *4*	0.7 *3*
γ [E2]	750.85 *24*	11.2 *19*
γ	781.6 *4*	0.7 *3*
γ	790.7 *3*	1.7 *3*
γ	797.7 *4*	0.7 *3*
γ [E2]	813.2 *5*	1.40 *19*
γ	815.7 *4*	2.4 *4*
γ	828.5 *4*	0.7 *3*
γ [E2]	843.22 *18*	3.7 *5*
γ [M1+E2]	883.15 *16*	7.3 *11*
γ	922.1 *4*	0.7 *3*
γ	946.44 *24*	4.2 *7*
γ	971.4 *3*	1.40 *19*
γ	993.5 *6*	2.0 *4*
γ	1024.8 *3*	0.9 *4*
γ	1059.0 *4*	0.9 *3*
γ	1089.0 *3*	3.7 *9*
γ	1098.08 *22*	11.7 *19*
γ	1123.7 *6*	0.9 *4*
γ	1126.8 *8*	0.8 *4*
γ	1165.0 *5*	1.3 *3*
γ	1202.2 *6*	1.0 *5*
γ	1228.52 *22*	~0.7
γ	1298.3 *5*	4.7 *7*
γ	1334.2 *6*	~0.7
γ	1380.5 *3*	3.6 *9*
γ	1405.8 *3*	8.7 *19*
γ	1411.7 *8*	~0.7
γ	1460.2 *4*	4.3 *7*
γ	1495.7 *4*	3.3 *12*
γ	1561.7 *5*	1.0 *4*
γ	1572.8 *4*	1.0 *5*
γ	1589.0 *6*	~0.7
γ	1634.2 *4*	0.9 *4*
γ	1684.8 *4*	1.8 *8*
γ	1697.7 *6*	~0.9
γ	1766.1 *8*	1.0 *4*
γ	2052.3 *5*	2.5 *3*
γ	2102.3 *5*	2.3 *5*
γ	2162.4 *8*	0.56 *19*
γ	2173.2 *8*	0.47 *19*
γ	2208.6 *8*	0.47 *19*
γ	2224.8 *6*	0.8 *4*

† 1.1% uncert(syst)

$^{122}_{55}$Cs(360 *20* ms)

Mode: IT
Δ: >-78160 keV
SpA: 1.14×10^{11} Ci/g
Prod: ^{3}He on La; ^{3}He on Ce

Photons (^{122}Cs)

γ_{mode}	γ(keV)	γ(rel)
γ E2(+9.0%M1)	45.85 *15*	50 *13*
γ M1(+9.1%E2)	81.2 *1*	100 *21*

$^{122}_{56}$Ba(1.96 *15* min)

Mode: ε
Δ: -74360 *310* keV syst
SpA: 7.8×10^{8} Ci/g
Prod: ^{32}S on ^{96}Ru

A = 123

NDS **29**, 453 (1980)

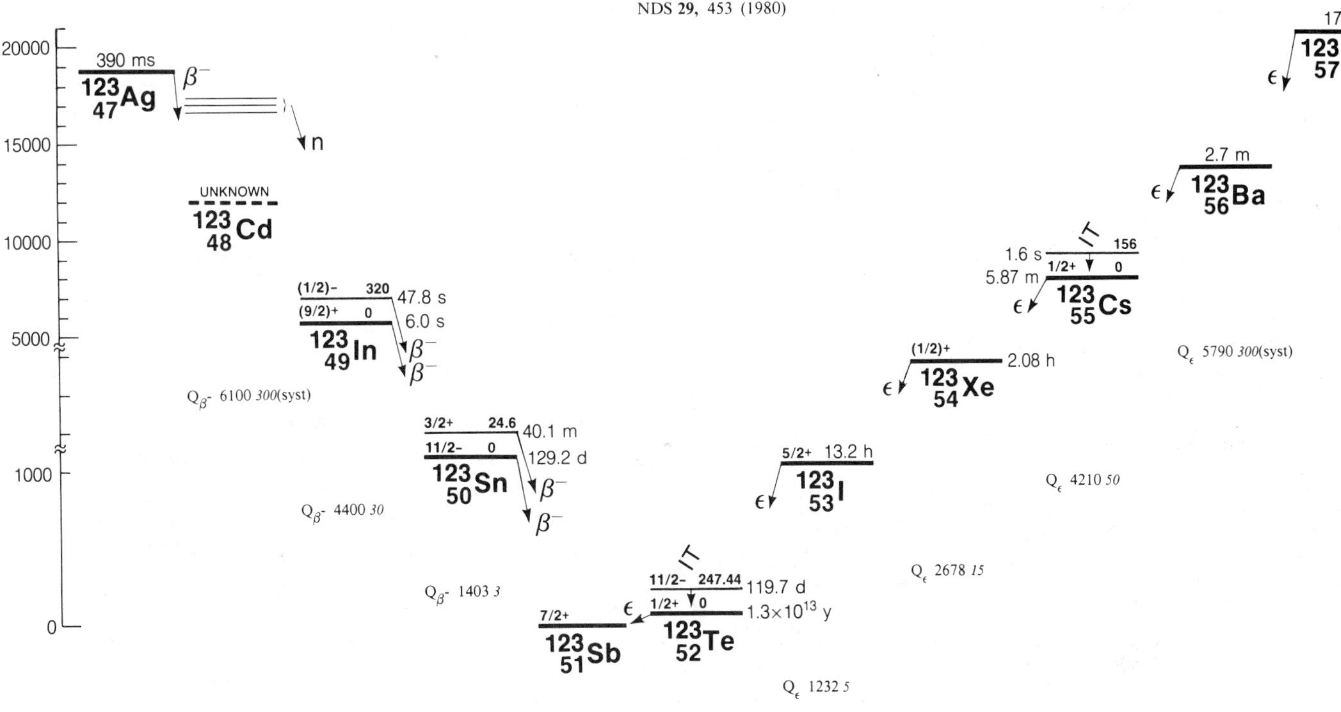

$^{123}_{47}$Ag(390 *30* ms)

Mode: β-, β-n
SpA: 1.10×10^{11} Ci/g
Prod: fission

$^{123}_{49}$In(5.98 *6* s)

Mode: β-
Δ: -83420 *30* keV
SpA: 1.448×10^{10} Ci/g
Prod: fission

Photons (^{123}In)

⟨γ⟩=1102 *50* keV

$γ_{mode}$	γ(keV)	γ(%)
γ [M1]	174.17 *6*	0.19 *3*
γ	175.01 *8*	0.13 *3*
γ [E1]	223.6 *4*	0.12 *4*
γ [E2]	284.87 *15*	0.17 *4*
γ [E1]	425.52 *25*	0.17 *8*
γ [E1]	536.22 *23*	0.90 *8*
γ E2	618.81 *23*	2.6 *2*
γ (E2)	845.56 *13*	1.3 *2*
γ E2+M1	931.4 *5*	0.3 *1*
γ	956.9 *3*	0.4 *1*
γ (E2)	1019.73 *13*	32 *2*
γ E2,M1	1130.42 *15*	63 *4*
γ	1131.0 *3*	<0.2

Photons (^{123}In)
(continued)

$γ_{mode}$	γ(keV)	γ(%)
γ [M2]	1155.0 *3*	~0.04
γ	1382.38 *18*	1.12 *7*
γ	2001.18 *25*	0.27 *6*

Continuous Radiation (^{123}In)

⟨β-⟩=1371 keV;⟨IB⟩=3.8 keV

E_{bin}(keV)		⟨ ⟩(keV)	(%)
0 - 10	β-	0.0066	0.132
	IB	0.046	
10 - 20	β-	0.0204	0.136
	IB	0.045	0.32
20 - 40	β-	0.085	0.282
	IB	0.089	0.31
40 - 100	β-	0.67	0.94
	IB	0.26	0.39
100 - 300	β-	8.8	4.23
	IB	0.74	0.42
300 - 600	β-	43.7	9.5
	IB	0.85	0.20
600 - 1300	β-	312	32.2
	IB	1.18	0.137
1300 - 2500	β-	849	46.7
	IB	0.56	0.034
2500 - 3781	β-	157	5.8
	IB	0.0136	0.00052

$^{123}_{49}$In(47.8 *5* s)

Mode: β-
Δ: -83100 *36* keV
SpA: 1.905×10^{9} Ci/g
Prod: ^{124}Sn(γ,p); fission

Photons (^{123}In)

⟨γ⟩=66 *25* keV

$γ_{mode}$	γ(keV)	γ(%)†
Sn $K_{α2}$	25.044	~2
Sn $K_{α1}$	25.271	~5
Sn $K_{β1}$'	28.474	~1
Sn $K_{β2}$'	29.275	~0.26
γM1	125.76 *4*	~38
γE2+M1	896.5 *5*	0.075 *24*
γM1+E2	1170 *1*	0.10
γ	2469 *3*	~0.017 ?
γ	2598 *3*	0.039 *15* ?
γ	2695 *3*	0.050 *18*
γ	3064 *2*	0.056 *17*
γ	3103 *3*	0.036 *12* ?
γ	3127 *2*	0.089 *25*
γ	3155 *3*	0.033 *12* ?
γ[M1+E2]	3234 *3*	0.12 *3*

† 31% uncert(syst)

Atomic Electrons (^{123}In)

⟨e⟩=12 *5* keV

e_{bin}(keV)	⟨e⟩(keV)	e(%)
4	0.33	~8
20 - 28	0.3	1.4 *5*
97	9.6	~10
121	1.4	~1
122 - 126	0.50	0.40 *14*
867 - 896	0.0012	0.00014 *4*
1141 - 1170	0.0011	1.0 *3* $\times10^{-4}$
2440 - 2468	7.6 $\times10^{-5}$	~3 $\times10^{-6}$
2569 - 2666	0.00034	1.3 *4* $\times10^{-5}$
2691 - 2694	2.6 $\times10^{-5}$	1.0 *4* $\times10^{-6}$
3035 - 3126	0.00078	2.5 *5* $\times10^{-5}$
3151 - 3233	0.00054	1.7 *4* $\times10^{-5}$

Continuous Radiation (^{123}In)

⟨β-⟩=2009 keV;⟨IB⟩=6.9 keV

E_{bin}(keV)		⟨ ⟩(keV)	(%)
0 - 10	β-	0.00304	0.060
	IB	0.058	
10 - 20	β-	0.0093	0.062
	IB	0.058	0.40
20 - 40	β-	0.0390	0.130
	IB	0.114	0.40
40 - 100	β-	0.307	0.434
	IB	0.33	0.51
100 - 300	β-	4.15	2.0
	IB	1.00	0.56
300 - 600	β-	21.8	4.72
	IB	1.25	0.29
600 - 1300	β-	180	18.3
	IB	2.1	0.23
1300 - 2500	β-	804	42.2
	IB	1.68	0.097
2500 - 4695	β-	999	32.0
	IB	0.39	0.0133

$^{123}_{50}$Sn(129.2 *4* d)

Mode: β-

Δ: -87820 *3* keV

SpA: 8217 Ci/g

Prod: ^{122}Sn(n,γ)

Photons (^{123}Sn)

⟨γ⟩=6.9 *11* keV

γ_{mode}	γ(keV)	γ(%)†
γ M1+0.6%E2	160.32 *18*	0.00191 *9*
γ	1020.95 *17*	0.00193 *10*
γ M1+11%E2	1030.22 *21*	0.0310 *12*
γ	1088.63 *21*	0.6
γ	1100.5 *3*	~8 $\times10^{-5}$
γ	1177.08 *17*	0.000228 *24*
γ	1181.3 *3*	0.00029 *3*
γ	1260.9 *4*	~5 $\times10^{-5}$
γ	1337.4 *3*	0.00076 *4*

† 17% uncert(syst)

Continuous Radiation (^{123}Sn)

⟨β-⟩=523 keV;⟨IB⟩=0.76 keV

E_{bin}(keV)		⟨ ⟩(keV)	(%)
0 - 10	β-	0.0477	0.95
	IB	0.023	
10 - 20	β-	0.144	0.96
	IB	0.022	0.154
20 - 40	β-	0.58	1.93
	IB	0.042	0.147
40 - 100	β-	4.15	5.9
	IB	0.114	0.18
100 - 300	β-	41.0	20.5
	IB	0.27	0.154
300 - 600	β-	133	29.8
	IB	0.20	0.050
600 - 1300	β-	340	39.7
	IB	0.087	0.0117
1300 - 1403	β-	3.28	0.247
	IB	2.0 $\times10^{-5}$	1.54 $\times10^{-6}$

$^{123}_{50}$Sn(40.08 *7* min)

Mode: β-

Δ: -87796 *3* keV

SpA: 3.814$\times10^{7}$ Ci/g

Prod: ^{122}Sn(n,γ); ^{124}Sn(n,2n)

Photons (^{123}Sn)

⟨γ⟩=141 *3* keV

γ_{mode}	γ(keV)	γ(%)†
Sb L_ℓ	3.189	0.0163 *22*
Sb L_η	3.437	0.0082 *9*
Sb L_α	3.604	0.45 *5*
Sb L_β	3.930	0.39 *5*
Sb L_γ	4.434	0.047 *7*
Sb $K_{\alpha2}$	26.111	3.16 *12*
Sb $K_{\alpha1}$	26.359	5.89 *22*
Sb $K_{\beta1}'$	29.712	1.55 *6*
Sb $K_{\beta2}'$	30.561	0.330 *13*
γM1+0.6%E2	160.32 *18*	85.6
γ[M1+E2]	171.0 *4*	~0.007
γ[M1+E2]	381.5 *3*	0.042 *3*
γ[E2]	541.8 *3*	0.020 *3*
γ[E2]	552.5 *3*	0.0103 *17*

† 2.3% uncert(syst)

Atomic Electrons (^{123}Sn)

⟨e⟩=20.2 *5* keV

e_{bin}(keV)	⟨e⟩(keV)	e(%)
4	0.38	9.1 *10*
5	0.053	1.13 *12*
21	0.058	0.27 *3*
22	0.184	0.84 *9*
25	0.059	0.233 *24*
26	0.063	0.24 *3*
29	0.0101	0.035 *4*
30	0.00229	0.0077 *8*
130	16.3	12.5 *4*
141	0.0015	~0.0011
156	2.52	1.62 *5*
159	0.468	0.294 *9*
160	0.164	0.102 *4*
166	0.00019	~0.00012
167	0.00014	<16 $\times10^{-5}$
170	6.9 $\times10^{-5}$	~4 $\times10^{-5}$
171	1.5 $\times10^{-5}$	~9 $\times10^{-6}$

Atomic Electrons (^{123}Sn) (continued)

e_{bin}(keV)	⟨e⟩(keV)	e(%)
351	0.00218	0.00062 *6*
377	0.00032	8.5 *18* $\times10^{-5}$
381	7.9 $\times10^{-5}$	2.1 *5* $\times10^{-5}$
511	0.00054	0.000105 *19*
522	0.00027	5.2 *9* $\times10^{-5}$
537	7.1 $\times10^{-5}$	1.32 *24* $\times10^{-5}$
538	5.9 $\times10^{-6}$	1.1 *2* $\times10^{-6}$
541	1.5 $\times10^{-5}$	2.9 *5* $\times10^{-6}$
542	3.5 $\times10^{-6}$	6.6 *12* $\times10^{-7}$
548	3.9 $\times10^{-5}$	7.1 *12* $\times10^{-6}$
552	9.6 $\times10^{-6}$	1.7 *3* $\times10^{-6}$

Continuous Radiation (^{123}Sn)

⟨β-⟩=459 keV;⟨IB⟩=0.59 keV

E_{bin}(keV)		⟨ ⟩(keV)	(%)
0 - 10	β-	0.0456	0.91
	IB	0.021	
10 - 20	β-	0.138	0.92
	IB	0.020	0.139
20 - 40	β-	0.57	1.89
	IB	0.038	0.132
40 - 100	β-	4.25	6.0
	IB	0.100	0.155
100 - 300	β-	47.1	23.2
	IB	0.22	0.13
300 - 600	β-	160	35.9
	IB	0.148	0.036
600 - 1267	β-	246	31.1
	IB	0.040	0.0057

$^{123}_{51}$Sb(stable)

Δ: -89222.9 *23* keV

%: 42.7 *9*

$^{123}_{52}$Te(1.3$\times10^{13}$ yr)

Mode: ϵ

Δ: -89171.7 *20* keV

Prod: natural source

%: 0.908 *3*

Photons (^{123}Te)

⟨γ⟩=0.217 *15* keV

γ_{mode}	γ(keV)	γ(%)
Sb L_ℓ	3.189	0.140 *18*
Sb L_η	3.437	0.00115 *12*
Sb L_α	3.604	3.9 *4*
Sb L_β	4.190	0.74 *9*
Sb L_γ	4.555	0.024 *5*
Sb $K_{\alpha2}$	26.111	0.0437 *13*
Sb $K_{\alpha1}$	26.359	0.0815 *24*
Sb $K_{\beta1}'$	29.712	0.0214 *7*
Sb $K_{\beta2}'$	30.561	0.00457 *15*

Atomic Electrons (^{123}Te)

⟨e⟩=2.6 *3* keV

e_{bin}(keV)	⟨e⟩(keV)	e(%)
4	2.6	62 *6*
5	0.066	1.41 *14*
21	0.00080	0.0038 *4*
22	0.0025	0.0116 *12*
25	0.00081	0.0032 *3*
26	0.00087	0.0034 *3*
29	0.000139	0.00048 *5*
30	3.2 $\times 10^{-5}$	0.000107 *11*

Continuous Radiation (^{123}Te)

⟨IB⟩=0.039 keV

E_{bin}(keV)		⟨ ⟩(keV)	(%)
10 - 20	IB	0.00118	0.0070
20 - 40	IB	0.037	0.141
40 - 51	IB	4.3×10^{-5}	0.000099

$^{123}_{52}$Te(119.7 *1* d)

Mode: IT
Δ: -88924.3 *20* keV
SpA: 8869 Ci/g
Prod: ^{122}Te(n,γ); ^{123}Sb(d,2n)

Photons (^{123}Te)

⟨γ⟩=148.1 *7* keV

γ_{mode}	γ(keV)	γ(%)†
Te L_ℓ	3.335	0.147 *21*
Te L_η	3.606	0.050 *6*
Te L_α	3.768	4.1 *5*
Te L_β	4.145	2.9 *4*
Te L_γ	4.687	0.36 *6*
Te $K_{\alpha2}$	27.202	14.5 *7*
Te $K_{\alpha1}$	27.472	27.0 *12*
Te $K_{\beta1}'$	30.980	7.2 *3*
Te $K_{\beta2}'$	31.877	1.57 *7*
γ M4	88.45 *3*	0.090 *4*
γ M1+1.2%E2	158.989 *21*	84.0
γ E5	247.44 *4*	0.00034 *3*

† 0.42% uncert(syst)

Atomic Electrons (^{123}Te)

⟨e⟩=102.0 *24* keV

e_{bin}(keV)	⟨e⟩(keV)	e(%)
4	2.28	53 *6*
5	1.25	26 *3*
22	0.27	1.20 *13*
23	0.82	3.6 *4*
26	0.27	1.02 *11*
27	0.29	1.07 *11*
30	0.041	0.138 *15*
31	0.0153	0.050 *5*
57	24.6	43.4 *22*
84	39.0	46.5 *24*
87	3.18	3.64 *19*
88	8.8	10.1 *5*
127	17.6	13.9 *3*
154	2.73	1.77 *4*
155	0.0668	0.0432 *14*

Atomic Electrons (^{123}Te) (continued)

e_{bin}(keV)	⟨e⟩(keV)	e(%)
158	0.576	0.364 *7*
159	0.138	0.0870 *18*
216	0.00230	0.00107 *10*
242	0.0033	0.00134 *13*
246	0.00110	0.00045 *4*

$^{123}_{53}$I (13.2 *1* h)

Mode: ε
Δ: -87939 *5* keV
SpA: 1.930×10^{6} Ci/g
Prod: ^{121}Sb(α,2n)

Photons (^{123}I)

⟨γ⟩=172.8 *9* keV

γ_{mode}	γ(keV)	γ(%)†
Te L_ℓ	3.335	0.137 *19*
Te L_η	3.606	0.068 *7*
Te L_α	3.768	3.8 *4*
Te L_β	4.123	3.4 *4*
Te L_γ	4.661	0.41 *6*
Te $K_{\alpha2}$	27.202	25.1 *9*
Te $K_{\alpha1}$	27.472	46.8 *16*
Te $K_{\beta1}'$	30.980	12.4 *4*
Te $K_{\beta2}'$	31.877	2.71 *10*
γM1+1.2%E2	158.989 *21*	83.3
γ[M1]	173.36 *9*	0.00083 *25*
γ[M1]	182.61 *4*	0.0129 *6*
γ	190.69 *10*	0.00050 *17*
γ[M1]	192.18 *4*	0.0198 *9*
γ[M1]	197.24 *7*	~0.00033
γ[M1]	198.23 *7*	0.0033 *8*
γ[M1]	206.81 *6*	0.0033 *8*
γ[M1]	207.79 *8*	0.0011 *3*
γ[M1]	247.95 *4*	0.0711 *25*
γ[E2]	257.51 *5*	0.0015 *4*
γ	258.99 *20*	0.0009 *4*
γ[M1]	278.26 *4*	0.0022 *4*
γM1+12%E2	281.00 *3*	0.0791 *25*
γ[E2]	285.31 *11*	0.0042 *4* ?
γ	295.14 *16*	0.0016 *3*
γ[M1]	329.25 *14*	0.0026 *6*
γ[E2]	330.72 *7*	0.0116 *6*
γ[M1]	343.59 *4*	0.0042 *4*
γM1+1.0%E2	346.34 *3*	0.126 *4*
γ	405.04 *8*	0.0029 *6*
γ	437.00 *18*	<0.0015
γE2+19%M1	439.99 *3*	0.428 *14*
γ	454.76 *6*	0.0039 *5*
γ	505.33 *3*	0.316 *10*
γ	528.95 *3*	1.39 *4*
γ[E2]	538.52 *4*	0.382 *12*
γ[E1]	556.05 *11*	0.0031 *4*
γ	562.78 *7*	0.0011 *4*
γ	578.40 *9*	0.0015 *4*
γ[M1]	599.61 *15*	0.0026 *9*
γ	610.26 *14*	0.0011 *3*
γE2+40%M1	624.59 *4*	0.0833 *25*
γ	628.12 *7*	0.00158 *25*
γ	687.94 *3*	0.0267 *15*
γ	735.76 *6*	0.0616 *22*
γ	760.84 *20*	0.00062 *21*
γ	783.58 *4*	0.0594 *22*
γ[E1]	837.06 *11*	0.00050 *8*
γ	877.61 *12*	~0.0011
γ	894.75 *6*	0.00092 *25*
γ	898.19 *20*	~0.0006
γ	909.12 *7*	0.00133 *25*
γ	1036.60 *12*	0.00100 *25*
γ	1068.11 *7*	0.00142 *8*

† 0.48% uncert(syst)

Atomic Electrons (^{123}I)

⟨e⟩=27.6 *5* keV

e_{bin}(keV)	⟨e⟩(keV)	e(%)
4	2.07	48 *5*
5	1.52	32 *3*
22	0.46	2.07 *21*
23	1.41	6.2 *6*
26 - 31	1.06	3.9 *3*
127	17.4	13.7 *3*
142 - 151	0.00233	0.00155 *9*
154	2.71	1.76 *4*
155	0.0663	0.0428 *14*
158	0.571	0.361 *7*
159 - 208	0.143	0.0895 *18*
216 - 263	0.0171	0.00735 *22*
273 - 315	0.0110	0.00357 *11*
324 - 373	0.00177	0.000517 *20*
400 - 440	0.0215	0.00521 *17*
450 - 497	0.040	0.008 *4*
500 - 547	0.020	0.0039 *5*
551 - 600	0.00225	0.000381 *18*
605 - 628	0.000364	5.9 *3* $\times 10^{-5}$
656 - 704	0.0014	0.00021 *8*
729 - 761	0.0010	0.00013 *6*
779 - 805	0.00014	1.8 *8* $\times 10^{-5}$
832 - 878	4.9 $\times 10^{-5}$	5.7 *19* $\times 10^{-6}$
890 - 909	5.4 $\times 10^{-6}$	6.0 *17* $\times 10^{-7}$
1005 - 1037	2.5 $\times 10^{-5}$	2.5 *9* $\times 10^{-6}$
1063 - 1068	2.2 $\times 10^{-6}$	2.0 *9* $\times 10^{-7}$

Continuous Radiation (^{123}I)

⟨IB⟩=0.30 keV

E_{bin}(keV)		⟨ ⟩(keV)	(%)
10 - 20	IB	0.00071	0.0042
20 - 40	IB	0.055	0.20
40 - 100	IB	0.0044	0.0073
100 - 300	IB	0.033	0.0158
300 - 600	IB	0.116	0.026
600 - 1073	IB	0.089	0.0123

$^{123}_{54}$Xe(2.08 *2* h)

Mode: ε
Δ: -85261 *16* keV
SpA: 1.225×10^{7} Ci/g
Prod: ^{127}I(p,5n)

Photons (^{123}Xe)

⟨γ⟩=411 *10* keV

γ_{mode}	γ(keV)	γ(%)†
I L_ℓ	3.485	0.127 *21*
I L_η	3.780	0.062 *9*
I L_α	3.937	3.5 *5*
I L_β	4.320	3.1 *5*
I L_γ	4.887	0.39 *7*
I $K_{\alpha2}$	28.317	21.7 *22*
I $K_{\alpha1}$	28.612	40 *4*
I $K_{\beta1}'$	32.278	10.8 *11*
I $K_{\beta2}'$	33.225	2.46 *25*
γ(E2)	39.95 *16*	0.0029 *10*
γM1	138.05 *15*	0.245 *25*
γE2	148.91 *9*	49
γ(M1+E2)	177.99 *10*	14.9 *7*
γ[E2]	192.14 *16*	0.083 *15*
γM1,E2	330.19 *9*	8.6 *5*
γ	474.06 *14*	0.103 *15*
γ	671.66 *14*	0.049 *10*

Photons (^{123}Xe)
(continued)

γ_{mode}	γ(keV)	γ(%)†
γ[M1+E2]	680.91 16	0.201 15
γ	691.76 18	0.113 15
γ[M1+E2]	718.44 12	0.171 15
γ	728.26 20	0.122 15
γ	752.33 14	0.059 10
γ	782.84 12	0.45 4
γ[M1+E2]	802.52 18	0.020 5
γ[M1]	816.19 15	0.073 10
γ	820.12 16	0.059 10
γ[M1+E2]	823.31 16	0.015 5 ?
γ[M1+E2]	833.11 16	0.039 5
γ	842.99 16	0.039 5
γ[M1+E2]	853.72 17	0.029 5
γ	859.67 18	0.044 5 ?
γ[M1+E2]	862.19 15	0.034 5
γ[M1+E2]	870.64 13	0.28 3
γ[M1]	899.72 13	2.45 24
γ	908.93 17	0.088 10
γ[M1+E2]	912.12 16	0.083 10
γ	935.04 13	0.31 3
γ	943.5 3	0.049 10
γ	949.39 14	0.029 5
γ	964.12 13	0.54 5
γ[M1+E2]	979.91 15	0.28 3
γ[M1+E2]	1004.59 16	0.073 10 ?
γ[M1+E2]	1011.10 16	0.44 5
γ	1013.79 15	0.118 15
γ	1040.95 18	0.073 10 ?
γ[E2]	1048.63 13	0.137 15
γ	1060.57 14	0.78 10
γ[M1+E2]	1064.32 16	0.66 7
γ[M1]	1093.40 15	2.79 25
γ	1113.03 13	1.57 15
γ[M1+E2]	1132.10 15	0.059 5
γ[M1+E2]	1153.49 16	0.098 10 ?
γ[M1+E2]	1161.18 15	0.103 10
γ	1189.86 18	0.054 5 ?
γ[M1+E2]	1201.32 18	0.093 10
γ[M1]	1236.85 22	0.098 15
γ	1241.85 14	0.45 10
γ[E2]	1242.31 15	~0.11
γ	1264.7 4	0.029 5
γ[M1+E2]	1274.38 21	0.044 5
γ	1296.3 3	0.029 5
γ[M1+E2]	1310.09 14	0.132 10
γ[M1+E2]	1326.85 19	0.0098 24 ?
γ	1334.3 3	0.025 5
γ	1390.76 14	0.118 10
γ[M1+E2]	1508.12 19	0.049 5 ?
γ[M1+E2]	1534.63 13	0.30 3
γ	1603.87 15	0.171 15
γ[M1+E2]	1625.82 14	0.59 5
γ[M1+E2]	1657.03 20	0.132 15 ?
γ[M1+E2]	1686.82 13	0.61 7
γ[M1]	1715.91 13	0.191 25
γ	1732.23 12	0.142 20
γ	1756.07 15	0.093 10
γ[M1+E2]	1778.02 14	0.098 10
γ	1785.15 15	0.029 5
γ[M1+E2]	1807.10 14	1.25 12
γ	1822.04 18	0.122 15
γ[E2]	1864.81 12	0.064 10
γ	1871.04 21	0.059 10
γ	1884.42 12	0.64 7
γ	1913.51 12	0.073 10
γ[M1+E2]	1919.76 18	0.025 5
γ	1934.06 15	0.221 24
γ[M1+E2]	1956.01 14	0.098 15
γ	1974.24 18	0.137 20
γ	1992.36 25	0.044 5
γ	2003.32 19	0.186 25
γ	2037.51 22	0.24 3
γ	2052.32 22	0.044 5
γ[M1+E2]	2059.09 25	0.073 10
γ	2062.41 12	0.049 10
γ[M1+E2]	2071.95 18	0.167 20
γ[M1+E2]	2101.03 18	0.157 20
γ[M1+E2]	2107.48 21	0.020 5
γ[M1+E2]	2125.14 24	0.0098 24 ?
γ[M1]	2136.56 21	0.025 5
γ	2144.55 25	0.034 5
γ	2151.7 5	0.015 5
γ	2173.6 3	0.064 10
γ	2178.6 5	0.039 5

Photons (^{123}Xe)
(continued)

γ_{mode}	γ(keV)	γ(%)†
γ	2189.70 22	0.020 5
γ	2201.23 21	0.044 10
γ[M1+E2]	2211.29 25	0.039 10
γ	2218.78 22	0.034 5
γ[M1+E2]	2249.94 18	0.0123 24
γ[M1+E2]	2277.33 24	0.0073 20 ?
γ	2280.8 5	0.0073 20
γ[M1]	2306.42 24	0.0098 24 ?
γ	2327.5 5	0.034 5
γ	2367.69 22	0.0098 20
γ[M1+E2]	2389.28 25	0.0049 15
γ	2411.4 4	0.0049 15 ?
γ	2419.6 5	0.0049 15
γ	2431.1 4	0.0049 15 ?
γ[E2]	2455.32 24	0.0147 25 ?
γ	2560.3 4	0.0147 25 ?
γ	2580.0 4	0.0049 15 ?

† 10% uncert(syst)

Atomic Electrons (^{123}Xe)
$\langle e \rangle$=37.4 22 keV

e_{bin}(keV)	$\langle e \rangle$(keV)	e(%)
5	3.3	70 10
7 - 40	2.51	10.0 7
105	0.068	0.065 7
116	18.6	16.1 16
133 - 138	0.0142	0.0106 9
144	5.9	4.1 4
145	3.4	2.3 4
148	1.27	0.86 9
149 - 192	1.3	0.75 19
297 - 330	0.77	0.255 19
441 - 474	0.0035	~0.0008
638 - 687	0.0129	0.00195 25
691 - 728	0.0040	0.00057 18
747 - 790	0.011	0.0015 5
797 - 843	0.0087	0.00106 13
849 - 895	0.057	0.0066 6
899 - 947	0.018	0.0020 4
948 - 981	0.0106	0.00109 15
999 - 1049	0.022	0.0021 5
1055 - 1099	0.068	0.0064 8
1108 - 1157	0.0060	0.00053 10
1160 - 1209	0.008	0.0007 3
1231 - 1277	0.0036	0.00029 4
1291 - 1334	0.00062	4.8 9 $\times 10^{-5}$
1358 - 1391	0.0011	8 3 $\times 10^{-5}$
1475 - 1571	0.0049	0.00033 4
1593 - 1686	0.0150	0.00092 7
1699 - 1789	0.0136	0.00077 8
1802 - 1901	0.0086	0.00047 9
1908 - 2004	0.0048	0.00024 4
2019 - 2119	0.0046	0.000222 15
2120 - 2218	0.00162	7.5 10 $\times 10^{-5}$
2244 - 2335	0.00042	1.8 3 $\times 10^{-5}$
2356 - 2454	0.000216	9.0 10 $\times 10^{-6}$
2527 - 2579	0.00010	4.0 10 $\times 10^{-6}$

Continuous Radiation (^{123}Xe)
$\langle\beta+\rangle$=151 keV;$\langle$IB$\rangle$=2.4 keV

E_{bin}(keV)		$\langle\ \rangle$(keV)	(%)
0 - 10	β+	2.49 $\times 10^{-6}$	3.00 $\times 10^{-5}$
	IB	0.0066	
10 - 20	β+	9.7 $\times 10^{-5}$	0.00059
	IB	0.0067	0.046
20 - 40	β+	0.00259	0.0079
	IB	0.059	0.21
40 - 100	β+	0.118	0.152
	IB	0.037	0.058

Continuous Radiation (^{123}Xe)
(continued)

E_{bin}(keV)		$\langle\ \rangle$(keV)	(%)
100 - 300	β+	5.3	2.41
	IB	0.114	0.062
300 - 600	β+	33.2	7.2
	IB	0.24	0.054
600 - 1300	β+	108	12.5
	IB	1.00	0.107
1300 - 2500	β+	3.79	0.280
	IB	0.97	0.059
2500 - 2529	IB	1.64 $\times 10^{-6}$	6.5 $\times 10^{-8}$
	$\Sigma\beta$+		23

$^{123}_{55}$Cs(5.87 5 min)

Mode: ϵ

Δ: -81050 40 keV

SpA: 2.602$\times 10^{8}$ Ci/g

Prod: ^{115}In(^{12}C,4n); ^{127}I(α,8n); ^{124}Xe(p,2n)

Photons (^{123}Cs)
$\langle\gamma\rangle$=185 12 keV

γ_{mode}	γ(keV)	γ(%)†
Xe L_{ℓ}	3.634	0.068 11
Xe L_{η}	3.955	0.031 5
Xe L_{α}	4.104	1.9 3
Xe L_{β}	4.522	1.7 3
Xe L_{γ}	5.124	0.21 4
Xe $K_{\alpha2}$	29.461	11.0 12
Xe $K_{\alpha1}$	29.782	20.5 21
Xe $K_{\beta1}'$	33.606	5.5 6
Xe $K_{\beta2}'$	34.606	1.31 14
γ[M2]	71.25 13	0.19 4
γ[M1]	83.24 9	3.0 6
γ(M1)	97.36 9	14.5
γ[M1]	180.60 9	0.57 11
γ[M1]	209.54 10	0.20 4
γ	261.80 11	1.8 4
γ	303.97 11	0.61 12
γ	306.91 11	3.3 7
γ	345.04 13	0.32 6
γ	404.78 19	0.29 6
γ	430.27 13	0.25 5
γ	434.40 16	0.88 18
γ	442.40 13	0.61 12
γ	488.02 19	0.25 5
γ	499.17 20	1.17 23
γ	513.51 13	
γ	541.46 16	0.25 5
γ	596.54 20	8.3 17
γ	608.31 24	0.36 7
γ	610.88 13	2.6 5
γ	643.94 16	1.8 4
γ	667.77 16	0.81 16
γ	711.3 4	0.51 10
γ	725.0 3	0.20 4
γ	741.31 16	2.4 5
γ	751.01 16	0.61 12
γ	819.1 4	0.15 3
γ	915.2 3	0.20 4
γ	945.4 4	0.25 5
γ	1126.0 4	0.28 6

† 14% uncert(syst)

Atomic Electrons (^{123}Cs)

⟨e⟩=17.7 *12* keV

e_{bin}(keV)	⟨e⟩(keV)	e(%)
5	1.70	34 *5*
24	0.34	1.41 *20*
25	0.43	1.74 *25*
28 - 33	0.45	1.55 *14*
37	1.6	4.4 *9*
49	1.8	3.7 *8*
63	7.1	11.3 *16*
66	0.69	1.05 *21*
70 - 83	0.69	0.90 *11*
92	1.35	1.47 *21*
93 - 97	0.39	0.40 *5*
146 - 181	0.17	0.110 *17*
204 - 227	0.14	~0.06
256 - 306	0.31	0.11 *5*
307 - 345	0.021	0.007 *3*
370 - 408	0.074	0.019 *6*
425 - 465	0.053	0.012 *5*
483 - 507	0.014	0.0028 *11*
536 - 576	0.26	0.045 *21*
591 - 639	0.10	0.017 *5*
643 - 690	0.017	0.0025 *9*
706 - 751	0.06	0.009 *4*
784 - 819	0.0026	~0.00033
881 - 915	0.0063	0.0007 *3*
940 - 945	0.00052	6 *3* $\times 10^{-5}$
1091 - 1126	0.0034	0.00031 *15*

Continuous Radiation (^{123}Cs)

⟨β+⟩=1001 keV;⟨IB⟩=6.7 keV

E_{bin}(keV)		⟨ ⟩(keV)	(%)
0 - 10	β+	1.29×10^{-6}	1.54×10^{-5}
	IB	0.034	
10 - 20	β+	5.2×10^{-5}	0.000311
	IB	0.034	0.23
20 - 40	β+	0.00143	0.00436
	IB	0.084	0.29
40 - 100	β+	0.070	0.089
	IB	0.19	0.29
100 - 300	β+	3.64	1.64
	IB	0.56	0.31
300 - 600	β+	30.0	6.4
	IB	0.71	0.166
600 - 1300	β+	259	26.6
	IB	1.50	0.166
1300 - 2500	β+	636	35.6
	IB	2.4	0.134
2500 - 4210	β+	72	2.70
	IB	1.12	0.038
	Σβ+		73

$^{123}_{55}$Cs(1.60 *15* s)

Mode: IT
Δ: -80894 *40* keV
SpA: 4.7×10^{10} Ci/g
Prod: ^{115}In(^{12}C,4n); ^{109}Ag(^{18}O,4n); ^{124}Xe(p,2n)

Photons (^{123}Cs)

γ_{mode}	γ(keV)	γ(rel)
γ	30.7 *5* ?	
γ[E3]	61.7 *20*	1.2
γ	63.9 *4*	15
γ(E2)	94.6 *4*	100 *3*

$^{123}_{56}$Ba(2.7 *4* min)

Mode: ε
Δ: -75260 *300* keV syst
SpA: 5.6×10^{8} Ci/g
Prod: ^{16}O on In; ^{16}O on Sn; ^{14}N on In; ^{12}C on Sn

Photons (^{123}Ba)

γ_{mode}	γ(keV)	γ(rel)
γ	30.7 *5*	56 *11* ?
γ	58.3 *10*	2.2 *4*
γ	63.9 *4*	14 *4*
γ	92.8 *5*	51 *5*
γ(E2)	94.6 *4*	100
γ	116.1 *6*	54 *8*
γ	120.0 *6*	27 *4*
γ	123.4 *5*	69 *6*
γ	137.0 *6*	23 *7*

$^{123}_{57}$La(17 *3* s)

Mode: ε
SpA: 5.3×10^{9} Ci/g
Prod: ^{96}Ru(^{32}S,2n3p); ^{98}Ru(^{32}S,4n3p)

Photons (^{123}La)

γ_{mode}	γ(keV)
γ	93 *1*

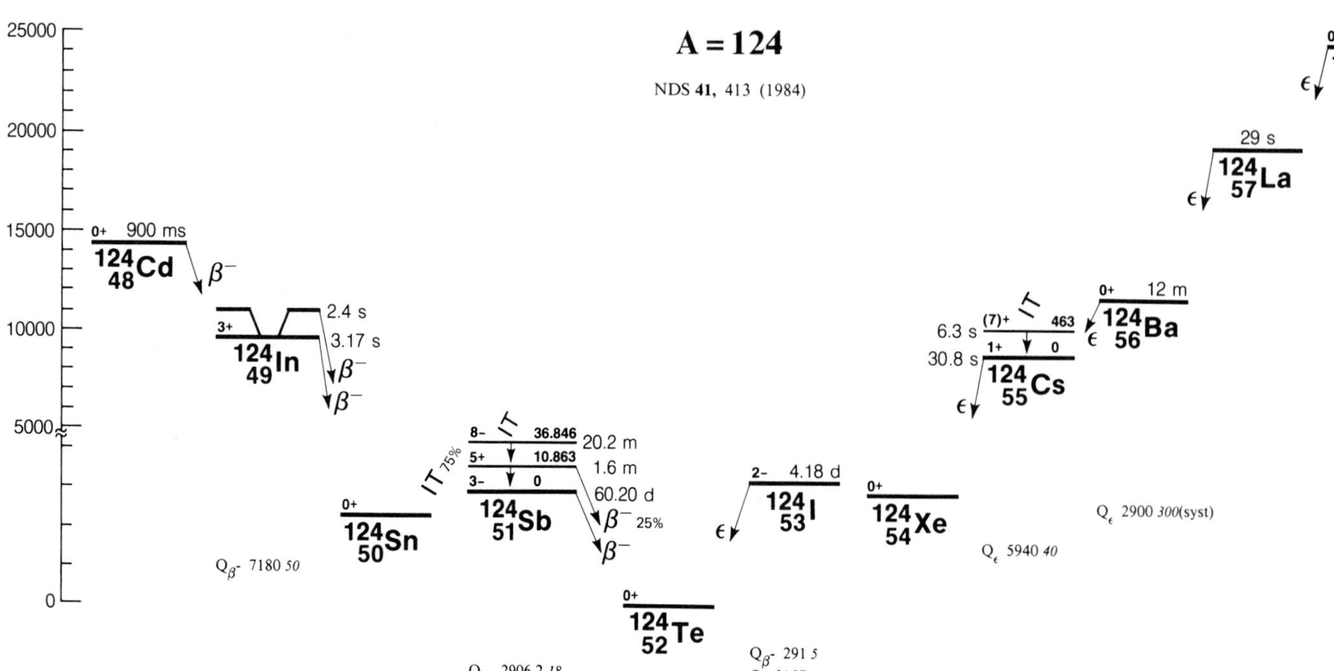

$^{124}_{48}$Cd(900 *200* ms)

Mode: β-
SpA: 7.0×10^{10} Ci/g
Prod: fission

Photons (^{124}Cd)

⟨γ⟩=136 *10* keV

γ$_{mode}$	γ(keV)	γ(%)†
In L$_ℓ$	2.905	0.067 *13*
In L$_η$	3.112	0.032 *6*
In L$_α$	3.286	1.8 *3*
In L$_β$	3.569	1.3 *3*
In L$_γ$	3.979	0.14 *3*
In K$_{α2}$	24.002	13.1 *14*
In K$_{α1}$	24.210	25 *3*
In K$_{β1}$'	27.265	6.3 *7*
In K$_{β2}$'	28.022	1.33 *14*
γ M1	36.58 *4*	4.6 *6*
γ E1	62.88 *10*	23 *3*
γ E1	143.41 *4*	12.9 *16*
γ E1	179.99 *4*	50 *5*

† approximate

Atomic Electrons (^{124}Cd)

⟨e⟩=17.7 *9* keV

e$_{bin}$(keV)	⟨e⟩(keV)	e(%)
4	1.61	42 *6*
9	3.3	38 *5*
19	0.114	0.59 *8*
20	0.96	4.8 *7*
23	0.25	1.09 *15*
24	0.27	1.13 *16*
26	0.0157	0.059 *8*

Atomic Electrons (^{124}Cd) (continued)

e$_{bin}$(keV)	⟨e⟩(keV)	e(%)
27	0.036	0.135 *19*
32	1.44	4.5 *6*
33	0.143	0.44 *6*
35	4.4	12.6 *17*
36	0.41	1.13 *15*
37	0.0125	0.034 *5*
59	0.98	1.67 *22*
62	0.20	0.32 *4*
63	0.042	0.067 *9*
115	0.81	0.70 *9*
139	0.109	0.078 *10*
140	0.0121	0.0086 *11*
143	0.029	0.020 *3*
152	2.18	1.44 *15*
176	0.31	0.176 *18*
179	0.061	0.034 *4*
180	0.0132	0.0073 *8*

$^{124}_{49}$In(3.17 *5* s)

Mode: β-
Δ: -81060 *50* keV
SpA: 2.58×10^{10} Ci/g
Prod: ^{124}Sn(n,p); fission

Photons (^{124}In)

⟨γ⟩=2649 *96* keV

γ$_{mode}$	γ(keV)	γ(%)
γ	339.9 *3* &	
γ	409.59 *20* &	
γ	449.32 *20* &	
γ	496.5 *6* &	
γ	549.0 *3* &	
γ	574.1 *3* &	
γ	614.81 *9*	0.64 *5*
γ	706.91 *8*	2.00 *20*
γ	820.3 *4* &	
γ	849.72 *20* &	
γ [E2]	969.94 *4*	3.0 *5*
γ	977.14 *15* &	
γ [M1+E2]	997.75 *4*	21.1 *15*
γ	1042.03 *12*	1.20 *10*
γ	1060.37 *15*	0.80 *7*
γ	1089.86 *5*	3.2 *3*
γ E2	1131.63 *4*	68 *6*
γ	1138.60 *19*	0.47 *15*
γ	1164.1 *7*	0.21 *4*
γ [M1+E2]	1203.77 *19*	0.19 *4*
γ [E2]	1214.18 *11*	0.65 *5*
γ	1234.8 *5*	0.12 *3*
γ [M1+E2]	1294.73 *10*	0.73 *7*
γ	1314.73 *5*	4.5 *4*
γ	1329.99 *18*	0.25 *3*
γ	1352.8 *3*	0.25 *3*
γ	1449.91 *17*	0.25 *3*
γ	1452.6 *4* &	
γ	1470.74 *7*	6.0 *5*
γ [M1+E2]	1490.74 *12*	0.190 *20*
γ	1519.48 *19*	1.00 *10*
γ	1526.1 *3*	0.210 *20*
γ [E2]	1571.28 *9*	2.40 *20*
γ	1611.59 *19*	0.36 *3*
γ	1639.40 *19*	0.18 *5*
γ	1672.3 *6* &	
γ	1695.61 *8*	0.38 *4*
γ	1704.66 *8*	0.90 *8*
γ [M1+E2]	1734.71 *14*	0.39 *4*
γ	1743.44 *13*	2.10 *20*
γ	1746.69 *20*	1.30 *20*
γ	1762.7 *9*	0.08 *3*

Photons (^{124}In)
(continued)

γ_{mode}	γ(keV)	γ(%)
γ [M1+E2]	1787.72 *8*	0.44 *4*
γ [E2]	1815.53 *8*	0.19 *4*
γ	1856.0 *4* &	
γ	1907.2 *6* &	
γ [M1+E2]	2082.53 *11*	3.4 *3*
γ [E2]	2129.37 *6*	0.59 *5*
γ [M1+E2]	2201.51 *19*	0.58 *5*
γ	2419.83 *17*	0.55 *5*
γ [E2]	2426.35 *10*	1.30 *10*
γ [M1+E2]	2593.29 *24*	0.19 *4*
γ	2609.33 *19*	0.60 *6*
γ	2628.49 *20*	0.52 *6*
γ	2699.6 *4* &	
γ [M1+E2]	2732.44 *13*	0.44 *4*
γ	2781.3 *5* &	
γ [M1+E2]	3024.3 *3*	0.19 *4*
γ	3109.5 *5*	0.15 *3*
γ [E2]	3214.14 *11*	21.5 *20*
γ [E2]	3264.09 *20*	1.30 *10*
γ [E2]	3333.12 *19*	0.22 *4*
γ [E2]	3396.5 *8*	0.11 *3*
γ [E2]	3710.3 *3*	0.34 *4*
γ [E2]	3724.90 *24*	0.18 *4*
γ [E2]	3761.5 *3*	1.00 *10*
γ	3834.2 *7*	0.070 *20*
γ [E2]	3864.05 *14*	0.57 *5*
γ	3887.9 *8*	0.090 *20*
γ	3910.6 *9*	0.080 *20*
γ [E2]	3917.06 *8*	1.90 *20*
γ [E2]	4043.7 *5*	0.16 *4*
γ [E2]	4155.9 *3*	0.56 *6*
γ [E2]	4227.8 *3*	0.18 *4*
γ [E2]	4264.0 *3*	0.46 *5*
γ [E2]	4331.3 *4*	0.20 *3*
γ [E2]	4470.2 *4*	0.19 *3*
γ [E2]	4528.7 *4*	0.41 *5*
γ [E2]	4604.5 *7*	0.35 *4*

& with ^{124}In(3.17 or 2.4 s)

Continuous Radiation (^{124}In)
⟨β-⟩=1968 keV;⟨IB⟩=6.9 keV

E_{bin}(keV)		⟨ ⟩(keV)	(%)
0 - 10	β-	0.00370	0.073
	IB	0.057	
10 - 20	β-	0.0114	0.076
	IB	0.056	0.39
20 - 40	β-	0.0474	0.158
	IB	0.111	0.39
40 - 100	β-	0.373	0.53
	IB	0.32	0.50
100 - 300	β-	4.99	2.41
	IB	0.97	0.54
300 - 600	β-	25.8	5.6
	IB	1.21	0.28
600 - 1300	β-	203	20.8
	IB	2.0	0.23
1300 - 2500	β-	781	41.5
	IB	1.61	0.093
2500 - 5000	β-	924	28.4
	IB	0.53	0.017
5000 - 6048	β-	29.3	0.56
	IB	0.0024	4.2×10^{-5}

${}^{124}_{49}$In(2.4 *4* s)

Mode: β-
Δ: -81060 *50* keV
SpA: 3.3×10^{10} Ci/g
Prod: fission

Photons (^{124}In)#
⟨γ⟩=3536 *120* keV

γ_{mode}	γ(keV)	γ(%)
Sn L_{ℓ}	3.045	0.043 *6*
Sn L_{η}	3.272	0.023 *3*
Sn L_{α}	3.443	1.20 *15*
Sn L_{β}	3.741	1.03 *14*
Sn L_{γ}	4.203	0.116 *17*
Sn $K_{\alpha2}$	25.044	8.0 *5*
Sn $K_{\alpha1}$	25.271	14.9 *10*
Sn $K_{\beta1}'$	28.474	3.9 *3*
Sn $K_{\beta2}'$	29.275	0.82 *6*
γ E1	102.92 *4*	45 *3*
γ E2	120.38 *4*	38 *3*
γ M1(+E2)	243.14 *4*	10.6 *10*
γ	253.46 *5*	4.4 *4*
γ M1(+E2)	363.52 *4*	17.0 *13*
γ	403.02 *20*	1.40 *20*
γ	432.7 *3*	1.10 *20*
γ	569.12 *15*	1.80 *20*
γ	783.9 *3*	1.40 *20*
γ	915.36 *20*	2.7 *3*
γ	955.87 *7*	12.4 *10*
γ [E2]	969.94 *4*	52 *4*
γ	1037.4 *3*	2.00 *20*
γ [E3]	1072.86 *5*	47 *4*
γ	1106.45 *9*	1.00 *20*
γ	1116.76 *7*	15.5 *15*
γ E2	1131.63 *4*	100 *8*
γ	1186.68 *11*	0.80 *20*
γ	1199.01 *7*	8.8 *8*
γ	1359.90 *7*	39 *3*
γ	1440.13 *10*	9.2 *8*
γ	1484.70 *20*	2.4 *3*
γ	1762.7 *9*	0.50 *20* &

see also ^{124}In(3.17 s)

Atomic Electrons (^{124}In)
⟨e⟩=46.0 *19* keV

e_{bin}(keV)	⟨e⟩(keV)	e(%)
4 - 28	2.16	34 *4*
74	4.8	6.5 *5*
91	21.9	24.0 *19*
98 - 103	1.02	1.03 *5*
116	7.3	6.3 *5*
119	0.52	0.43 *4*
120	1.32	1.10 *9*
214 - 253	1.8	0.81 *14*
311 - 360	1.07	0.32 *3*
363 - 410	0.17	0.044 *11*
420 - 467	0.09	0.020 *7*
492 - 540	0.060	0.011 *4*
544 - 574	0.028	0.0051 *19*
755 - 791	0.030	0.0039 *14*
816 - 850	0.027	0.0033 *14*
886 - 927	0.15	0.016 *7*
941 - 977	0.71	0.076 *6*
1008 - 1044	0.98	0.094 *9*
1068 - 1117	1.26	0.115 *9*
1127 - 1170	0.22	0.019 *3*
1182 - 1199	0.011	0.0009 *3*
1331 - 1360	0.29	0.022 *9*
1411 - 1455	0.09	0.0064 *19*
1480 - 1485	0.0021	0.00015 *5*
1643 - 1733	0.0062	0.00037 *13*
1758 - 1855	0.008	0.00045 *15*
1878 - 1906	0.0060	0.00032 *12*
2670 - 2752	0.016	0.00060 *14*
2777 - 2781	0.00050	1.8 *6* $\times10^{-5}$

Continuous Radiation (^{124}In)
⟨β-⟩=1670 keV;⟨IB⟩=5.2 keV

E_{bin}(keV)		⟨ ⟩(keV)	(%)
0 - 10	β-	0.00462	0.092
	IB	0.052	
10 - 20	β-	0.0142	0.094
	IB	0.051	0.36
20 - 40	β-	0.059	0.197
	IB	0.101	0.35
40 - 100	β-	0.466	0.66
	IB	0.29	0.45
100 - 300	β-	6.2	3.00
	IB	0.87	0.49
300 - 600	β-	31.9	6.9
	IB	1.05	0.25
600 - 1300	β-	246	25.3
	IB	1.60	0.18
1300 - 2500	β-	875	46.9
	IB	1.05	0.062
2500 - 5000	β-	510	17.2
	IB	0.148	0.0051
5000 - 5166	β-	0.115	0.00228
	IB	5.5×10^{-7}	1.09×10^{-8}

${}^{124}_{50}$Sn(stable)

Δ: -88237.1 *16* keV
%: 5.79 *5*

${}^{124}_{51}$Sb(60.20 *3* d)

Mode: β-
Δ: -87619.0 *23* keV
SpA: 1.7492×10^{4} Ci/g
Prod: ^{123}Sb(n,γ)

Photons (^{124}Sb)
⟨γ⟩=1852 *26* keV

γ_{mode}	γ(keV)	γ(%)†
Te $K_{\alpha2}$	27.202	0.130 *4*
Te $K_{\alpha1}$	27.472	0.243 *7*
Te $K_{\beta1}'$	30.980	0.0645 *19*
Te $K_{\beta2}'$	31.877	0.0141 *4*
γ	185.50 *6*	~0.039
γ [E1]	254.421 *4*	<0.028
γ [E1]	335.809 *10*	0.079 *20*
γ	370.48 *8*	0.020 *7*
γ	385.9 *1*	~0.08
γ [M1+E2]	399.966 *5*	0.147 *23*
γ	444.01 *3*	0.193 *22*
γ	468.89 *6*	0.029 *12*
γ	476.85 *7*	0.034 *12*
γ [M1+E2]	482.03 *4*	0.064 *23*
γ E2	498.4 *1*	<0.08 ?
γ	525.40 *3*	0.159 *23*
γ	553.8 *1*	<0.020
γ	592.30 *6*	0.00048 *18*
γ E2	602.731 *3*	97.8
γ [E2]	632.395 *10*	0.098 *23*
γ E2	645.8563 *18*	7.4 *4*
γ	662.35 *4*	0.017 *3*
γ M1+0.2%E2	709.325 *10*	1.35 *6*
γ E2+29%M1	713.783 *4*	2.28 *10*
γ E2+8.4%M1	722.7864 *24*	10.9 *4*

Photons (^{124}Sb)
(continued)

γ_{mode}	γ(keV)	γ(%)†
γ	735.70 6	0.127 23
γ	735.774 11	0.137 23
γ	743.74 4	0.0051 16
γ [M1+E2]	766.107 23	<0.06
γ	775 2	$\sim$0.09
γ M1,E2	790.713 4	0.74 4
γ	817.09 6	0.078 23
γ	846.72 6	$\sim3\times10^{-5}$
γ	857.02 4	0.020 7
γ	899.28 6	0.016 5
γ	928.11 6	$\sim3\times10^{-5}$
γ	937.9 8	0.0064 13
γ E1	968.202 3	1.87 6
γ	976.21 6	0.075 12
γ	997.69 8	0.010 4
γ	1014.5 8	0.008 3
γ E1	1045.1319 23	1.86 8
γ [E2]	1053.95 17	<0.018
γ	1086.50 8	0.033 12
γ	1097 1	$\sim$0.004
γ	1128.57 13	0.00042 14
γ	1163.43 8	0.0186 23
γ [M1+E2]	1195.81 4	0.006 3
γ	1205.50 13	0.00018 6
γ	1234.72 3	$\sim$0.029
γ	1253.4 4	$\sim$0.008
γ	1263.1 2	0.029 6
γ	1269 1	0.012 5
γ	1301.2 3	0.034 7
γ E2	1325.514 3	1.59 7
γ M1,E2	1355.177 10	1.05 9
γ	1356.08 6	$\sim$0.0028
γ E1	1368.165 5	2.66 11
γ E1	1376.13 4	0.52 8
γ [E2]	1385.29 7	0.052 19
γ	1428.7 5	<0.010
γ	1433.01 6	$\sim$0.00023
γ E2+14%M1	1436.565 3	1.25 11
γ	1445.094 5	0.33 7
γ	1453.06 4	0.018 8
γ E2+8.9%M1	1488.888 23	0.69 7
γ	1505.6 2	<0.006
γ M1,E2	1526.41 6	0.40 7
γ	1557 1	0.022 5
γ	1560.50 6	0.0019 7
γ	1579.80 4	0.41 7
γ	1622.06 6	0.033 9
γ	1637.43 6	0.0023 7
γ E1	1690.9824 23	47.1 14
γ	1720.47 8	0.083 13
γ	1851.35 13	0.0020 7
γ [M1+E2]	1918.58 4	0.049 15
γ	1950.4 2	<0.008
γ	2015 1	0.0088 18
γ [E2]	2039.288 4	0.068 23
γ M1,E2	2078.86 6	0.013 5
γ	2090.942 5	5.49 19
γ M1,E2	2091.612 23	$\sim$0.06
γ	2098.91 4	0.044 8
γ [E2]	2108.07 7	$\sim$0.039
γ	2151.5 5	<0.0018
γ	2172.26 6	0.0010 5
γ	2182.52 4	$\sim$0.039
γ	2203 1	$\sim$0.0008
γ	2224.79 6	0.000245 23
γ	2253.9 3	0.00147 20
γ	2274 1	0.00049 20 ?
γ	2283.27 6	0.0078 23
γ	2293.705 3	0.0311 23
γ	2323.20 8	0.0023 5
γ	2454.07 13	0.0006 3
γ	2483.29 3	$\sim$0.13
γ	2681.58 6	0.0014 3
γ	2693.663 5	$\sim$0.15
γ	2871.7 10	<0.0008 ?

† 0.51% uncert(syst)

Atomic Electrons (^{124}Sb)
$\langle e\rangle$=3.74 5 keV

e_{bin}(keV)	$\langle e\rangle$(keV)	e(%)
4 - 31	0.0339	0.48 3
154 - 185	0.007	$\sim$0.005
223 - 254	0.0005	$\sim$0.00021
304 - 339	0.0025	0.00078 19
354 - 400	0.012	0.0034 7
412 - 450	0.011	0.0025 9
464 - 498	0.006	0.0012 5
520 - 560	0.0009	0.00018 7
571	2.36	0.413 9
587 - 592	1.7×10^{-6}	$\sim3\times10^{-7}$
598	0.335	0.0560 12
601	0.0022	0.00036 9
602	0.0678	0.01126 23
603	0.0160	0.00266 6
614	0.160	0.0261 14
627 - 662	0.0287	0.00448 18
678 - 682	0.079	0.0116 4
691	0.206	0.0298 13
704 - 744	0.062	0.0087 5
759 - 791	0.0174	0.0023 3
812 - 857	0.00045	5.5 22 $\times10^{-5}$
867 - 906	0.00029	3.3 14 $\times10^{-5}$
923 - 972	0.0125	0.00133 6
975 - 1022	0.0096	0.00095 5
1040 - 1086	0.00184	0.000176 19
1092 - 1132	0.00017	1.5 7 $\times10^{-5}$
1158 - 1205	0.00034	$\sim3\times10^{-5}$
1222 - 1269	0.00076	6.1 16 $\times10^{-5}$
1294 - 1336	0.0379	0.00288 13
1344 - 1385	0.0062	0.00046 3
1397 - 1445	0.0209	0.00148 14
1448 - 1529	0.0108	0.00073 6
1548 - 1637	0.0033	0.00021 8
1659	0.167	0.0101 4
1686 - 1720	0.0246	0.00146 5
1820 - 1919	0.00042	2.2 6 $\times10^{-5}$
1945 - 2038	0.00054	2.7 7 $\times10^{-5}$
2047 - 2140	0.032	0.0016 5
2147 - 2242	0.00023	$\sim1\times10^{-5}$
2249 - 2322	0.00039	1.73 13 $\times10^{-5}$
2422 - 2482	0.0006	$\sim3\times10^{-5}$
2650 - 2693	0.0018	$\sim7\times10^{-5}$
2840 - 2871	1.8×10^{-6}	$\sim6\times10^{-8}$

Continuous Radiation (^{124}Sb)
$\langle\beta-\rangle$=386 keV;$\langle$IB$\rangle$=0.61 keV

E_{bin}(keV)		$\langle\ \rangle$(keV)	(%)
0 - 10	β-	0.152	3.05
	IB	0.0165	
10 - 20	β-	0.442	2.96
	IB	0.0158	0.110
20 - 40	β-	1.69	5.6
	IB	0.030	0.104
40 - 100	β-	10.2	14.8
	IB	0.077	0.120
100 - 300	β-	62	32.7
	IB	0.18	0.101
300 - 600	β-	85	20.5
	IB	0.146	0.035
600 - 1300	β-	136	14.8
	IB	0.133	0.0161
1300 - 2303	β-	91	5.8
	IB	0.017	0.00116

$^{124}_{51}$Sb(1.55 *8* min)

Mode: IT(75 5 %), β-(25 5 %)
Δ: -87608.1 23 keV
SpA: 9.7×10^{8} Ci/g
Prod: ^{123}Sb(n,γ)

Photons (^{124}Sb)
$\langle\gamma\rangle$=440 90 keV

γ_{mode}	γ(keV)	γ(%)†
Sb L_{ℓ}	3.189	0.068 11
Sb L_{η}	3.437	0.0154 18
Sb L_{α}	3.604	1.9 3
Sb L_{β}	3.951	1.9 3
Sb L_{γ}	4.553	0.31 6
γ_{IT} [M2]	10.8630 11	0.00305
$\gamma_{\beta-}$E2	498.4 1	25
$\gamma_{\beta-}$E2	602.731 3	25
$\gamma_{\beta-}$E2	645.8563 18	25
$\gamma_{\beta-}$	1101 1	$\sim$0.50

† uncert(syst): 6.7% for IT, 20% for β-

Atomic Electrons (^{124}Sb)*
$\langle e\rangle$=7.14 22 keV

e_{bin}(keV)	$\langle e\rangle$(keV)	e(%)
4	0.9	21.8 25
5	0.85	18.1 21
6	2.29	37.1 21
7	1.47	21.8 12
10	1.31	13.1 7
11	0.319	2.97 17

* with IT

Continuous Radiation (^{124}Sb)
$\langle\beta-\rangle$=105 keV;$\langle$IB$\rangle$=0.129 keV

E_{bin}(keV)		$\langle\ \rangle$(keV)	(%)
0 - 10	β-	0.0134	0.267
	IB	0.0048	
10 - 20	β-	0.0406	0.270
	IB	0.0046	0.032
20 - 40	β-	0.166	0.55
	IB	0.0088	0.031
40 - 100	β-	1.23	1.74
	IB	0.023	0.036
100 - 300	β-	13.0	6.5
	IB	0.050	0.029
300 - 600	β-	40.6	9.2
	IB	0.031	0.0076
600 - 1300	β-	49.6	6.4
	IB	0.0076	0.00108
1300 - 1669	β-	0.62	0.0444
	IB	2.6×10^{-5}	1.9×10^{-6}

$^{124}_{51}$Sb(20.2 *2* min)

Mode: IT
Δ: -87582.2 23 keV
SpA: 7.50×10^{7} Ci/g
Prod: ^{123}Sb(n,p)

Photons (^{124}Sb)

⟨γ⟩=0.216 16 keV

γ_{mode}	γ(keV)	γ(%)
Sb L_ℓ	3.189	0.106 15
Sb L_η	3.437	0.062 7
Sb L_α	3.604	2.9 4
Sb L_β	3.923	2.4 3
Sb L_γ	4.350	0.24 3
γ E3	25.981 3	0.00310

Atomic Electrons (^{124}Sb)

⟨e⟩=25.4 6 keV

e_{bin}(keV)	⟨e⟩(keV)	e(%)
4	2.9	68 7
5	0.0038	0.081 8
21	0.0335	0.158 5
22	16.9	77.8 22
25	4.59	18.2 5
26	0.98	3.78 11

$^{124}_{52}$Te(stable)

Δ: -90525.2 17 keV

%: 4.816 8

$^{124}_{53}$I (4.18 2 d)

Mode: ϵ

Δ: -87368 4 keV

SpA: 2.519×10^5 Ci/g

Prod: ^{121}Sb(α,n); ^{123}Sb(α,3n); protons on Te

Photons (^{124}I)

⟨γ⟩=852 37 keV

γ_{mode}	γ(keV)	γ(%)†
Te L_ℓ	3.335	0.091 12
Te L_η	3.606	0.045 5
Te L_α	3.768	2.5 3
Te L_β	4.123	2.3 3
Te L_γ	4.662	0.28 4
Te $K_{\alpha2}$	27.202	16.7 5
Te $K_{\alpha1}$	27.472	31.2 9
Te $K_{\beta1}$'	30.980	8.29 25
Te $K_{\beta2}$'	31.877	1.81 6
γ	185.50 6	~0.06
γ [E1]	254.421 4	<0.006
γ	307.35 8	~0.018
γ [E1]	335.809 10	0.017 5
γ	370.48 8	0.0020 7
γ	381.7 5	0.0169 12
γ [M1+E2]	399.966 5	0.0152 25
γ	444.01 3	0.036 9
γ	468.89 6	0.0030 12
γ	476.85 7	0.11 4
γ	478.7 5	0.0260 24
γ [M1+E2]	482.03 4	~0.2
γ	525.40 3	0.030 8
γ	541.28 3	0.188 6
γ [E2]	553.4 6	~0.06 ?
γ	592.30 6	0.041 9
γ E2	602.731 3	61

Photons (^{124}I)
(continued)

γ_{mode}	γ(keV)	γ(%)†
γ	631.5 3	~0.006
γ [E2]	632.395 10	0.0038 13
γ	635 1	
γ E2	645.8563 18	0.94 3
γ	662.35 4	0.055 4
γ [E2]	695.1 7	~0.12 ?
γ	707.32 8	0.07 3
γ M1+0.2%E2	709.325 10	0.052 13
γ E2+29%M1	713.783 4	0.106 10
γ E2+8.4%M1	722.7864 24	9.96 19
γ	735.774 11	0.0142 25
γ	743.74 4	0.016 4
γ [M1+E2]	766.107 23	<0.016
γ	775.1 2	0.013 4
γ M1,E2	790.713 4	0.034 4
γ	795.70 3	0.044 3 ?
γ	846.72 6	~0.0024 ?
γ	877.09 3	0.026 6 ?
γ	899.28 6	0.022 7
γ	928.11 6	0.0021 9 ?
γ	961.74 8	0.0224 24
γ E1	968.202 3	0.412 8
γ	976.21 6	0.107 12
γ [M1]	984.17 22	0.014 3
γ	997.69 8	0.021 9 ?
γ E1	1045.1319 23	0.412 19
γ [E2]	1053.95 17	0.121 6
γ	1086.50 8	0.018 6
γ	1128.57 13	0.0436 9
γ	1160 1	
γ	1163.43 8	0.010 4
γ [M1+E2]	1195.81 4	0.019 7
γ	1205.50 13	0.0188 6
γ	1234.72 3	~0.005
γ [M1+E2]	1315.33 16	0.034 4
γ E2	1325.514 3	1.45 4
γ M1,E2	1355.177 10	0.041 3
γ	1356.08 6	0.073 24 ?
γ E1	1368.165 5	0.277 12
γ E1	1376.13 4	1.66 3
γ	1408 1	
γ [E2]	1417.8 7	
γ	1433.01 6	~0.006 ?
γ E2+14%M1	1436.565 3	0.058 7
γ	1445.094 5	0.034 7
γ	1453.06 4	0.06 3
γ	1479 1	
γ E2+8.9%M1	1488.888 23	0.182 6
γ E1	1509.48 3	2.99 6
γ	1560.50 6	0.163 24
γ	1622.06 6	0.048 12
γ	1637.43 6	0.194 12
γ	1662.0 8	~0.06
γ	1675.51 8	0.109 24
γ	1685 1	
γ E1	1690.9824 23	10.41 20
γ	1705.5 3	0.0139 24
γ	1720.47 8	0.170 12
γ	1726 1	
γ	1739.5 9	0.024 6
γ	1752.44 8	0.050 6
γ	1847 1	
γ	1851.35 13	0.206 24
γ [M1+E2]	1918.58 4	0.157 18
γ	2021 2	<0.006
γ [M1+E2]	2038.11 16	0.339 18
γ [E2]	2039.288 4	0.0032 14
γ M1,E2	2078.86 6	0.344 12
γ	2090.942 5	0.570 11
γ M1,E2	2091.612 23	~0.015
γ	2094.5 10	
γ	2098.91 4	0.139 6
γ	2102 1	
γ	2140 1	
γ [E1]	2144.323 10	0.109 6
γ	2214.7 5	0.010 5
γ	2224.79 6	0.00035 4
γ	2232.25 3	0.569 12
γ	2275.8 5	~0.006
γ	2283.27 6	0.66 3
γ	2287 1	
γ	2293.705 3	0.0069 6
γ	2323.20 8	0.0048 16

Photons (^{124}I)
(continued)

γ_{mode}	γ(keV)	γ(%)†
γ	2385.07 24	0.0194 24
γ	2454.07 13	0.067 18
γ	2483.29 3	~0.024 ?
γ [M1]	2640.83 16	~0.005 ?
γ	2681.58 6	0.037 11
γ	2693.663 5	~0.016 ?
γ	2734 1	
γ [E1]	2747.044 10	0.460 18
γ	2834.97 3	~0.0048 ?
γ	2987.79 24	0.008 4
γ	3001.01 8	~0.005 ?

† 8% uncert(syst)

Atomic Electrons (^{124}I)

⟨e⟩=6.5 3 keV

e_{bin}(keV)	⟨e⟩(keV)	e(%)
4	1.38	32 3
5	1.01	21.6 22
22	0.31	1.38 14
23	0.94	4.1 4
26	0.31	1.17 12
27	0.33	1.24 13
30 - 31	0.065	0.216 17
154 - 185	0.010	~0.006
223 - 254	7.2×10^{-5}	3.2 14 $\times10^{-5}$
276 - 307	0.0015	~0.0005
331 - 377	0.0016	0.00046 15
381 - 412	0.0012	~0.00030
437 - 482	0.018	0.0039 10
494 - 541	0.008	0.0016 5
548 - 560	0.0011	0.00020 9
571	1.46	0.26 3
587 - 592	0.00014	2.4 11 $\times10^{-5}$
598	0.207	0.035 4
600 - 646	0.077	0.0127 8
657 - 706	0.197	0.0285 8
707 - 744	0.0329	0.00457 10
759 - 796	0.0015	0.00020 5
815 - 860	0.0062	0.00074 21
867 - 899	0.00038	4.4 19 $\times10^{-5}$
923 - 972	0.0046	0.00048 7
975 - 1022	0.00356	0.000350 14
1040 - 1086	0.00073	7 1 $\times10^{-5}$
1097 - 1132	0.0100	0.00089 20
1151 - 1201	0.0019	0.00016 4
1203 - 1235	5.6×10^{-5}	$\sim5\times10^{-6}$
1284 - 1325	0.0165	0.00127 4
1336 - 1376	0.0092	0.000683 15
1401 - 1449	0.0019	0.000137 23
1452 - 1529	0.0158	0.00107 4
1556 - 1652	0.0033	0.00020 4
1657 - 1752	0.0442	0.00266 7
1820 - 1918	0.0026	0.00014 3
1989 - 2087	0.0091	0.00045 6
2090 - 2183	0.00061	2.9 3 $\times10^{-5}$
2193 - 2291	0.0067	0.00030 7
2293 - 2384	0.00010	4.4 13 $\times10^{-6}$
2422 - 2482	0.00045	1.9 6 $\times10^{-5}$
2609 - 2693	0.00039	1.5 6 $\times10^{-5}$
2715 - 2803	0.00147	5.4 3 $\times10^{-5}$
2830 - 2834	2.8×10^{-6}	10 5 $\times10^{-8}$
2956 - 3000	5.7×10^{-5}	1.9 7 $\times10^{-6}$

Continuous Radiation (^{124}I)

$\langle\beta+\rangle$=188 keV;$\langle$IB$\rangle$=2.6 keV

E_{bin}(keV)		$\langle\ \rangle$(keV)	(%)
0 - 10	β+	2.12×10^{-6}	2.55×10^{-5}
	IB	0.0077	
10 - 20	β+	8.1×10^{-5}	0.000487
	IB	0.0079	0.054
20 - 40	β+	0.00210	0.0064
	IB	0.056	0.20
40 - 100	β+	0.093	0.120
	IB	0.043	0.067
100 - 300	β+	4.08	1.86
	IB	0.132	0.073
300 - 600	β+	25.8	5.6
	IB	0.24	0.055
600 - 1300	β+	110	12.1
	IB	0.85	0.091
1300 - 2500	β+	48.4	3.14
	IB	1.13	0.065
2500 - 3157	IB	0.089	0.0033
	Σβ+		23

$^{124}_{54}$Xe(stable)

Δ: -87659.6 *21* keV

%: 0.10 *1*

$^{124}_{55}$Cs(30.8 *5* s)

Mode: ε

Δ: -81720 *40* keV

SpA: 2.92×10^{9} Ci/g

Prod: protons on Ta; protons on Th; protons on U; ^{115}In(^{12}C,3n)

Photons (^{124}Cs)

$\langle\gamma\rangle$=315 *20* keV

γ_{mode}	γ(keV)	γ(%)†
Xe L_ℓ	3.634	0.0133 *18*
Xe L_η	3.955	0.0061 *7*
Xe L_α	4.104	0.37 *4*
Xe L_β	4.522	0.33 *4*
Xe L_γ	5.123	0.042 *6*
Xe $K_{\alpha2}$	29.461	2.19 *8*
Xe $K_{\alpha1}$	29.782	4.08 *15*
Xe $K_{\beta1}$'	33.606	1.10 *4*
Xe $K_{\beta2}$'	34.606	0.261 *10*
γ E2	354.05 *15*	40
γ [E2]	359.66 *21*	~0.24
γ [M1+E2]	368.9 *3*	~0.08
γ	380.8 *5*	~0.040
γ	388.2 *10*	~0.7
γ M1+9.4%E2	401.15 *21*	0.087 *16*
γ [E2]	422.21 *18*	0.37 *6*
γ M1+15%E2	492.58 *13*	3.6 *6*
γ E2	524.84 *22*	0.40 *12*
γ [E2]	749.61 *22*	0.20 *3*
γ [M1+E2]	781.87 *17*	0.20 *4*
γ [E2]	846.63 *15*	1.19 *12*
γ	865.1 *4*	0.40 *12*
γ	866.5 *8*	0.226 *12*
γ	879.9 *5*	~0.07
γ M1+35%E2	893.73 *21*	0.159 *24*
γ [E2]	914.78 *16*	4.0 *4*
γ	925.2 *5*	0.24 *4*
γ	957.7 *4*	~0.012
γ	1003.3 *6*	0.032 *8*
γ	1020.1 *10*	~0.040
γ	1073.7 *9*	0.028 *8*

Photons (^{124}Cs)
(continued)

γ_{mode}	γ(keV)	γ(%)†
γ	1099.8 *4*	0.095 *16*
γ	1132.1 *4*	0.143 *24*
γ	1268.8 *6*	0.091 *24*
γ [M1+E2]	1274.44 *17*	0.44 *6*
γ	1291.6 *10*	~0.032
γ [E2]	1336.1 *4*	0.48 *8*
γ	1358.9 *4*	0.16 *4*
γ	1424.6 *6*	0.024 *8*
γ	1528.1 *10*	0.032 *8*
γ	1544.6 *10*	0.036 *12*
γ [E2]	1628.49 *18*	0.95 *12*
γ	1673.5 *5*	0.032 *8*
γ	1689.7 *8*	0.56 *8*
γ	1759.4 *10*	0.032 *8*
γ	1851.4 *4*	0.28 *6*
γ	1979.6 *10*	0.099 *24*
γ	2019.5 *8*	0.64 *10*
γ	2126.1 *10*	0.79 *20*
γ	2382.1 *10*	~0.16

† 13% uncert(syst)

Atomic Electrons (^{124}Cs)

$\langle$e$\rangle$=4.3 *4* keV

e_{bin}(keV)	$\langle$e$\rangle$(keV)	e(%)
5	0.34	6.8 *7*
24 - 33	0.242	0.93 *5*
319	2.6	0.82 *11*
325 - 346	0.022	0.0067 *25*
349	0.47	0.134 *18*
353	0.098	0.028 *4*
354 - 401	0.09	0.024 *8*
417 - 422	0.0036	0.00087 *12*
458	0.16	0.036 *6*
487 - 525	0.044	0.0089 *11*
715 - 750	0.0088	0.00120 *17*
776 - 812	0.0204	0.0025 *3*
831 - 879	0.018	0.0021 *5*
880 - 925	0.074	0.0084 *8*
952 - 999	0.0009	9 *4* $\times10^{-5}$
1002 - 1039	0.00040	3.8 *18* $\times10^{-5}$
1065 - 1100	0.0027	0.00025 *9*
1127 - 1132	0.00024	2.1 *10* $\times10^{-5}$
1234 - 1274	0.0074	0.00059 *9*
1286 - 1335	0.0069	0.00053 *8*
1336 - 1359	0.00024	1.8 *7* $\times10^{-5}$
1390 - 1425	0.00023	1.6 *8* $\times10^{-5}$
1494 - 1544	0.00059	3.9 *13* $\times10^{-5}$
1594 - 1689	0.0139	0.00086 *12*
1725 - 1817	0.0020	0.00011 *5*
1846 - 1945	0.0009	4.5 *15* $\times10^{-5}$
1974 - 2019	0.0043	0.00022 *8*
2092 - 2125	0.0051	0.00024 *10*
2348 - 2381	0.0009	~4×10^{-5}

Continuous Radiation (^{124}Cs)

$\langle\beta+\rangle$=1950 keV;$\langle$IB$\rangle$=10.5 keV

E_{bin}(keV)		$\langle\ \rangle$(keV)	(%)
0 - 10	β+	5.5×10^{-7}	6.6×10^{-6}
	IB	0.055	
10 - 20	β+	2.20×10^{-5}	0.000133
	IB	0.055	0.38
20 - 40	β+	0.00061	0.00187
	IB	0.115	0.40
40 - 100	β+	0.0301	0.0386
	IB	0.32	0.49
100 - 300	β+	1.64	0.73
	IB	0.97	0.54
300 - 600	β+	14.5	3.09
	IB	1.24	0.29

Continuous Radiation (^{124}Cs)
(continued)

E_{bin}(keV)		$\langle\ \rangle$(keV)	(%)
600 - 1300	β+	153	15.5
	IB	2.3	0.26
1300 - 2500	β+	743	39.0
	IB	2.8	0.158
2500 - 5000	β+	1037	32.6
	IB	2.6	0.077
5000 - 5940	IB	0.071	0.00136
	Σβ+		91

$^{124}_{55}$Cs(6.3 *2* s)

Mode: IT

Δ: -81257 *40* keV

SpA: 1.37×10^{10} Ci/g

Prod: ^{3}He on La; ^{3}He on Ce

Photons (^{124}Cs)

$\langle\gamma\rangle$=305 *6* keV

γ_{mode}	γ(keV)	γ(%)†
Cs L_ℓ	3.795	0.30 *5*
Cs L_η	4.142	0.132 *18*
Cs L_α	4.285	8.2 *12*
Cs L_β	4.732	7.3 *11*
Cs L_γ	5.369	0.97 *16*
Cs $K_{\alpha2}$	30.625	43 *4*
Cs $K_{\alpha1}$	30.973	80 *7*
Cs $K_{\beta1}$'	34.967	22.0 *20*
Cs $K_{\beta2}$'	36.006	5.4 *5*
γ M1(+E2)	53.9 *3*	5.6 *5*
γ E1	58.2 *3*	16.5 *8*
γ M2	64.9 *3*	2.0 *4*
γ E1	89.5 *4*	41.5 *13*
γ M1(+E2)	96.6 *3*	40.9 *13*
γ (E3)	161.5 *6*	~0.7
γ M1(+E2)	169.5 *1*	4.6 *7*
γ M1+43%E2	189.0 *3*	21.4 *13*
γ E2	211.6 *4*	44.2 *13*
γ (M1+E2)	270.3 *1*	8.3 *7*

† approximate

Atomic Electrons (^{124}Cs)

$\langle$e$\rangle$=114 *5* keV

e_{bin}(keV)	$\langle$e$\rangle$(keV)	e(%)
5	6.3	123 *17*
6	0.90	15.7 *23*
18	4.7	26 *3*
22	3.02	13.6 *7*
25 - 26	2.9	11.6 *11*
29	21.3	74 *15*
30 - 53	4.47	10.6 *6*
54	5.79	10.8 *4*
57 - 58	0.290	0.507 *22*
59	8.1	14 *3*
60	2.8	4.8 *10*
61	21.6	35.6 *14*
64 - 89	4.8	6.9 *8*
91	4.26	4.69 *18*
92 - 134	3.3	2.9 *4*
153	4.7	3.08 *21*
156 - 170	1.4	0.9 *3*
176	8.4	4.80 *17*
183 - 212	3.99	2.00 *8*
234 - 270	1.20	0.50 *5*

$^{124}_{56}$Ba(11.9 *10* min)

Mode: ε
Δ: -78820 *300* keV syst
SpA: 1.27×10^{8} Ci/g
Prod: ^{115}In(^{14}N,3n); ^{116}Sn(^{12}C,4n); protons on La

Photons (^{124}Ba)

⟨γ⟩=348 *33* keV

γ_{mode}	γ(keV)	γ(%)†
γ	156.8 *14*	3.2 *6*
γ M1,E2	169.7 *6*	20 *4*
γ M1+43%E2	189.0 *3*	10.2 *20*
γ	211.6 *11*	4.0 *8*
γ	252.9 *6*	4.8 *10*
γ	272.2 *7*	7.8 *16*
γ	752 *3*	3.6 *7*
γ	933 *3*	3.2 *6*
γ	1047.2 *15*	2.8 *6* ?
γ	1098 *4*	1.20 *24* ?
γ	1216.8 *15*	12.2 *24* ?

† approximate

$^{124}_{57}$La(29 *3* s)

Mode: ε
SpA: 3.1×10^{9} Ci/g
Prod: ^{32}S on ^{96}Ru; ^{32}S on ^{98}Ru

$^{124}_{58}$Ce(6 *2* s)

Mode: ε
SpA: 1.4×10^{10} Ci/g
Prod: ^{32}S on ^{96}Ru; ^{32}S on ^{98}Ru

A = 125

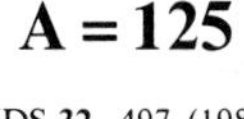

NDS **32**, 497 (1981)

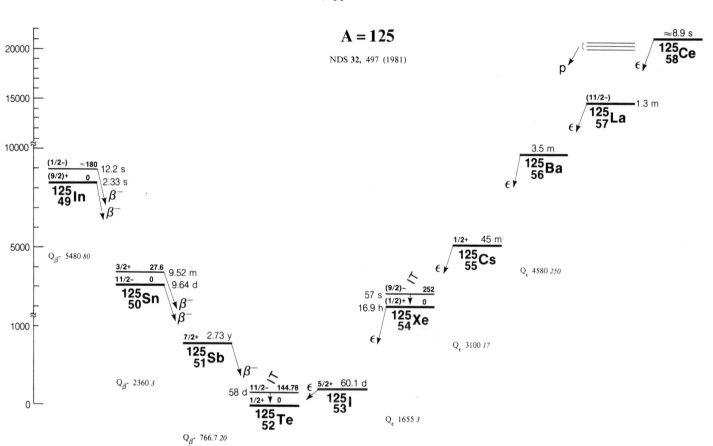

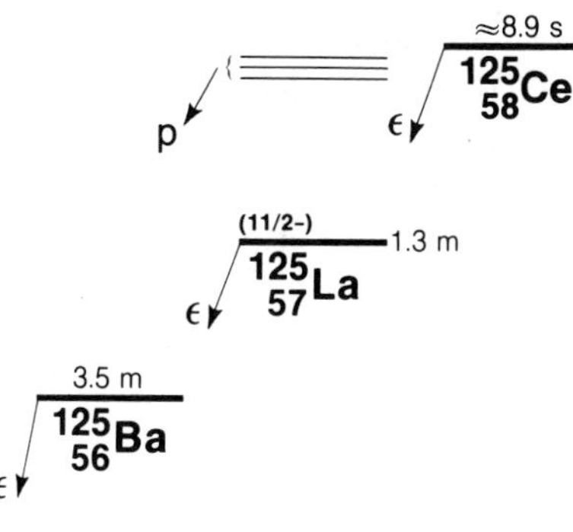

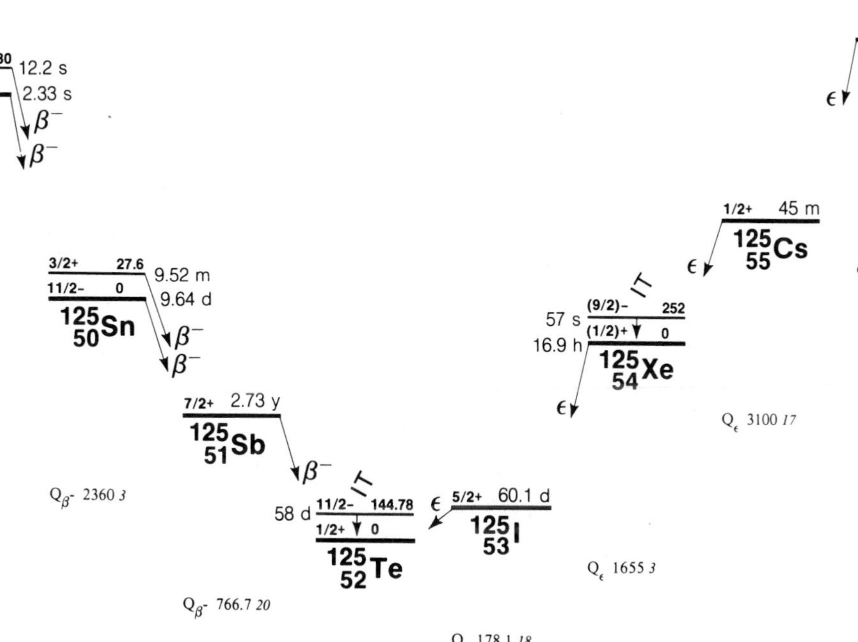

$^{125}_{49}$In(2.33 *4* s)

Mode: β-
Δ: -80420 *80* keV
SpA: 3.35×10^{10} Ci/g
Prod: fission

Photons (^{125}In)

⟨γ⟩=1293 *54* keV

γ_{mode}	γ(keV)	γ(%)
γ	318.8 *4*	0.12 *3*
γ [E1]	426.13 *9*	2.5 *2*
γ	507.87 *12*	0.5 *1*
γ M1,E2	618.00 *9*	8.0 *4*
γ	744.74 *9*	5.6 *5*
γ M1,E2	827.28 *9*	2.5 *2*
γ M1,E2	936.60 *9*	3.0 *2*
γ [E2]	1031.87 *10*	10.3 *6*
γ M1,E2	1335.14 *9*	76 *4*
γ	1362.73 *10*	0.25 *5*
γ	1558.3 *4*	1.0 *1*

$^{125}_{49}$In(12.2 *1* s)

Mode: β-
Δ: ~-80240 keV
SpA: 7.19×10^{9} Ci/g
Prod: fission

Photons (^{125}In)

γ_{mode}	γ(keV)
γ M1+46%E2	187.63 *3*

$^{125}_{50}$Sn(9.64 *3* d)

Mode: β-
Δ: -85898.2 *21* keV
SpA: 1.084×10^5 Ci/g
Prod: ^{124}Sn(n,γ); ^{124}Sn(d,p)

Photons (^{125}Sn)

⟨γ⟩=312 *4* keV

γ_{mode}	γ(keV)	γ(%)†
γ	234.63 *10*	0.0325 *18*
γ	257.80 *4*	0.0181 *9*
γ [E2]	260.20 *5*	0.0181 *9*
γ	270.49 *4*	0.099 *4*
γ [M1+E2]	282.35 *4*	0.0172 *18*
γ	286.07 *9*	0.0054 *9*
γ	311.19 *6*	0.0081 *9*
γ E2+18%M1	332.04 *4*	1.31 *5*
γ	350.88 *3*	0.246 *9*
γ	363.57 *5*	0.0027 *5*
γ [M1+E2]	386.60 *6*	0.0045 *9*
γ	434.06 *5*	0.0226 *18*
γ E1	469.77 *4*	1.38 *5*
γ	487.13 *20*	0.0118 *18*
γ	524.25 *4*	0.0090 *9*
γ [E1]	562.85 *5*	0.0145 *18*
γ	652.38 *6*	0.0380 *18*
γ	684.03 *8*	0.0099 *18*
γ [E1]	800.23 *3*	0.99 *4*
γ [E1]	822.38 *3*	3.99 *14*
γ	890.91 *5*	0.0081 *18*
γ E1	893.31 *3*	0.271 *18*
γ	903.59 *6*	0.012 *3*
γ	912.58 *6*	0.0063 *18*
γ E1	915.46 *4*	3.85 *14*
γ	921.36 *5*	0.0768 *18*
γ	934.73 *5*	0.194 *7*
γ [M1+E2]	1017.34 *4*	0.298 *11*
γ M1+42%E2	1067.03 *3*	9.04
γ [E2]	1087.60 *5*	1.11 *5*
γ [E2]	1089.17 *4*	4.28 *15*
γ	1111.41 *6*	0.0127 *18*
γ	1151.11 *3*	0.107 *4*
γ	1163.79 *4*	0.0289 *18*
γ	1173.26 *3*	0.169 *6*
γ	1186.13 *8*	0.0081 *9*
γ	1198.66 *8*	0.0145 *9*
γ	1208.28 *8*	0.0072 *18*
γ	1220.81 *8*	0.250 *9*
γ	1259.23 *5*	0.0289 *18*
γ	1291.23 *20*	0.0045 *14*
γ [M1+E2]	1349.38 *4*	0.055 *3*
γ [M1+E2]	1419.63 *4*	0.454 *16*
γ	1557.36 *4*	0.0038 *9*
γ	1591.27 *5*	0.0235 *18*
γ [M1+E2]	1806.22 *4*	0.138 *5*
γ	1889.40 *4*	0.069 *5*
γ	1982.47 *4*	0.0030 *9*
γ	2001.75 *5*	1.79 *6*
γ	2038.23 *20*	0.0027 *5*
γ	2200.58 *6*	0.0362 *18*
γ	2275.29 *7*	0.170 *6*

† 29% uncert(syst)

Continuous Radiation (^{125}Sn)

⟨β-⟩=811 keV;⟨IB⟩=1.8 keV

E_{bin}(keV)		⟨ ⟩(keV)	(%)
0 - 10	β-	0.063	1.28
	IB	0.031	
10 - 20	β-	0.182	1.21
	IB	0.030	0.21
20 - 40	β-	0.67	2.26
	IB	0.059	0.20
40 - 100	β-	4.01	5.8
	IB	0.164	0.25
100 - 300	β-	29.1	15.0
	IB	0.44	0.25
300 - 600	β-	73	16.4
	IB	0.45	0.106
600 - 1300	β-	328	35.0
	IB	0.49	0.058
1300 - 2360	β-	376	23.0
	IB	0.094	0.0061

$^{125}_{50}$Sn(9.52 *5* min)

Mode: β-
Δ: -85870.6 *21* keV
SpA: 1.579×10^8 Ci/g
Prod: ^{124}Sn(n,γ)

Photons (^{125}Sn)

⟨γ⟩=346 *7* keV

γ_{mode}	γ(keV)	γ(%)†
Sb $K_{\alpha2}$	26.111	0.547 *19*
Sb $K_{\alpha1}$	26.359	1.02 *4*
Sb $K_{\beta1}$'	29.712	0.269 *9*
Sb $K_{\beta2}$'	30.561	0.0572 *21*
γ [E2]	260.20 *5*	0.0059 *12*
γ [M1+E2]	279.0 *5*	~0.07
γ [M1+E2]	282.35 *4*	0.0056 *12*
γ [M1+E2]	310.7 *3*	~0.07
γ E2+18%M1	332.04 *4*	97.0 *19*
γ [E2]	386.4 *3*	0.087 *10*
γ [E2]	589.8 *4*	0.20 *4*
γ [E2]	642.8 *3*	0.16 *4*
γ	662.1 *10*	
γ	779.2 *6*	0.013 *5*
γ	840.9 *4*	0.068 *19*
γ [M1+E2]	1017.34 *4*	0.098 *19*
γ	1058.3 *5*	<0.039 ?
γ [M1+E2]	1093.0 *4*	0.039 *19*
γ	1151.7 *3*	~0.029 ?
γ	1294.1 *10*	
γ [M1+E2]	1304.7 *5*	<0.019 ?
γ [M1+E2]	1349.38 *4*	0.018 *4*
γ	1369.0 *4*	0.097 *19*
γ [M1+E2]	1403.7 *3*	0.70 *3*
γ	1483.7 *3*	0.18 *3*
γ	1581.7 *5*	<0.012 ?
γ [M1+E2]	1615.4 *4*	0.116 *19*
γ	1634.1 *10*	<0.039 ?
γ [E2]	1735.7 *3*	0.029 *10*
γ	1913.7 *5*	~0.019
γ [E2]	1947.4 *4*	~0.010
γ	2113.1 *10*	<0.0039 ?

†<0.1% uncert(syst)

Atomic Electrons (^{125}Sn)

⟨e⟩=8.03 *18* keV

e_{bin}(keV)	⟨e⟩(keV)	e(%)
4 - 30	0.142	2.10 *17*
230 - 279	0.009	~0.0035
280 - 282	0.005	~0.0018
302	6.53	2.16 *6*
306 - 311	0.0010	~0.00032
327	0.773	0.236 *6*
328	0.283	0.086 *6*
331 - 356	0.267	0.081 *3*
382 - 386	0.00087	0.000227 *22*
559 - 590	0.0057	0.00101 *17*
612 - 643	0.0038	0.00062 *13*
749 - 779	0.00019	~3×10^{-5}
810 - 841	0.0009	~0.00012
987 - 1028	0.0016	0.00016 *4*
1054 - 1093	0.00056	5.2 *22* ×10^{-5}
1121 - 1152	0.00028	~2×10^{-5}
1274 - 1319	0.00028	2.1 *8* ×10^{-5}
1339 - 1373	0.0070	0.00051 *6*
1399 - 1404	0.00095	6.8 *7* ×10^{-5}
1453 - 1551	0.0014	1.0 *4* ×10^{-4}
1577 - 1633	0.00117	7.4 *13* ×10^{-5}
1705 - 1735	0.00022	1.3 *4* ×10^{-5}
1883 - 1947	0.00018	9 *4* ×10^{-6}
2083 - 2112	1.1 ×10^{-5}	~5×10^{-7}

Continuous Radiation (^{125}Sn)

⟨β-⟩=798 keV;⟨IB⟩=1.55 keV

E_{bin}(keV)		⟨ ⟩(keV)	(%)
0 - 10	β-	0.0198	0.394
	IB	0.032	
10 - 20	β-	0.060	0.402
	IB	0.031	0.22
20 - 40	β-	0.249	0.83
	IB	0.060	0.21
40 - 100	β-	1.90	2.69
	IB	0.168	0.26
100 - 300	β-	22.7	11.1
	IB	0.44	0.25
300 - 600	β-	96	21.1
	IB	0.42	0.101
600 - 1300	β-	448	48.3
	IB	0.37	0.045
1300 - 2056	β-	229	15.2
	IB	0.029	0.0020

$^{125}_{51}$Sb(2.73 *3* yr)

Mode: β-
Δ: -88258 *3* keV
SpA: 1048 Ci/g
Prod: daughter ^{125}Sn

Photons (^{125}Sb)

⟨γ⟩=443 *4* keV

γ_{mode}	γ(keV)	γ(%)†
Te L_{ℓ}	3.335	0.131 *19*
Te L_{η}	3.606	0.062 *7*
Te L_{α}	3.768	3.6 *5*
Te L_{β}	4.126	3.1 *4*
Te L_{γ}	4.663	0.38 *6*
γ [M1]	19.881 *22*	~0.02
Te $K_{\alpha2}$	27.202	21.7 *10*
Te $K_{\alpha1}$	27.472	40.4 *18*
Te $K_{\beta1}$'	30.980	10.7 *5*
Te $K_{\beta2}$'	31.877	2.34 *11*
γ M1+0.07%E2	35.4919 *5*	6.01 *18* *
γ M4	109.287 *13*	0.071 *12* *
γ [E1]	110.798 *25*	0.0009 *1*
γ (E1)	116.931 *10*	0.255 *5*
γ [E1]	146.188 *19*	0.00062 *4*
γ M1	172.626 *21*	0.182 *4*
γ M1+26%E2	176.316 *10*	6.79 *14*
γ [M1]	178.759 *16*	0.027 *4*

Photons (^{125}Sb)
(continued)

γ_{mode}	γ(keV)	γ(%)†
γ [E2]	198.640 *23*	0.013 *3*
γ E2+29%M1	204.100 *14*	0.323 *7*
γ M1+1.1%E2	208.015 *15*	0.236 *5*
γ (M1,E2)	227.897 *21*	0.132 *4*
γ [E1]	314.90 *3*	0.0042 *4*
γ E1	321.031 *15*	0.410 *8*
γ (E2)	380.416 *13*	1.52 *3*
γ E2+31%M1	407.994 *20*	0.183 *6*
γ M1+22%E2	427.875 *10*	29.4 *6*
γ E2+16%M1	443.486 *20*	0.303 *7*
γ E2	463.367 *10*	10.45 *21*
γ [M2]	497.346 *14*	0.0036 *4*
γ E2	600.500 *21*	17.8 *4*
γ E2	606.633 *14*	5.02 *10*
γ M1+10%E2	635.890 *12*	11.32 *23*
γ (M1,E2)	671.381 *12*	1.80 *4*

† 1.0% uncert(syst)
* with ^{125}Te(58 d) in equilib

Atomic Electrons (^{125}Sb)
$\langle e\rangle$=39.7 *22* keV

e_{bin}(keV)	$\langle e\rangle$(keV)	e(%)
4	4.6	117 *8*
5	1.39	30 *3*
15 - 22	0.43	2.02 *21*
23	1.22	5.3 *6*
26 - 30	0.89	3.32 *24*
31	2.98	9.7 *4*
34 - 35	0.833	2.40 *7*
77	9.6	12.4 *22*
79 - 85	0.0239	0.0280 *12*
104	3.8	3.6 *6*
105	5.9	5.6 *10*
106	1.57 $\times10^{-5}$	1.49 *18* $\times10^{-5}$
108	2.3	2.1 *4*
109 - 142	0.62	0.56 *9*
145	1.37	0.95 *3*
146 - 194	0.429	0.249 *7*
196 - 228	0.044	0.0218 *13*
283 - 321	0.00968	0.00331 *8*
349 - 380	0.107	0.0302 *11*
396	1.39	0.351 *9*
403 - 443	0.646	0.151 *3*
458 - 497	0.0750	0.0163 *3*
569 - 607	0.960	0.165 *3*
631 - 671	0.099	0.0156 *9*

Continuous Radiation (^{125}Sb)
$\langle\beta-\rangle$=86 keV;$\langle$IB$\rangle$=0.038 keV

E_{bin}(keV)		$\langle\ \rangle$(keV)	(%)
0 - 10	β-	0.60	12.2
	IB	0.0044	
10 - 20	β-	1.61	10.8
	IB	0.0037	0.026
20 - 40	β-	5.2	17.8
	IB	0.0060	0.021
40 - 100	β-	19.1	29.4
	IB	0.0112	0.018
100 - 300	β-	42.8	25.5
	IB	0.0109	0.0071
300 - 600	β-	17.1	4.35
	IB	0.00133	0.00037
600 - 622	β-	0.0181	0.00298
	IB	1.43 $\times10^{-8}$	2.4 $\times10^{-9}$

$^{125}_{52}$Te(stable)

Δ: -89025.0 *24* keV
%: 7.14 *1*

$^{125}_{52}$Te(58 *1* d)

Mode: IT
Δ: -88880.2 *24* keV
SpA: 1.80$\times10^4$ Ci/g
Prod: daughter ^{125}Sb; ^{124}Te(n,γ)

Photons (^{125}Te)
$\langle\gamma\rangle$=36.0 *7* keV

γ_{mode}	γ(keV)	γ(%)
Te L_ℓ	3.335	0.24 *3*
Te L_η	3.606	0.102 *11*
Te L_α	3.768	6.5 *7*
Te L_β	4.131	5.4 *7*
Te L_γ	4.670	0.65 *10*
Te $K_{\alpha2}$	27.202	33.7 *11*
Te $K_{\alpha1}$	27.472	62.9 *21*
Te $K_{\beta1}'$	30.980	16.7 *6*
Te $K_{\beta2}'$	31.877	3.64 *13*
γ M1+0.07%E2	35.4919 *5*	6.67 *18*
γ M4	109.287 *13*	0.282 *8*

Atomic Electrons (^{125}Te)
$\langle e\rangle$=110.6 *19* keV

e_{bin}(keV)	$\langle e\rangle$(keV)	e(%)
4	6.6	164 *11*
5	2.37	50 *5*
22	0.62	2.8 *3*
23	1.90	8.3 *9*
26	0.62	2.36 *24*
27	0.67	2.5 *3*
30	0.096	0.32 *3*
31	3.37	11.0 *4*
34	0.666	1.93 *7*
35	0.274	0.78 *3*
77	40.7	52.5 *18*
104	16.0	15.3 *5*
105	24.7	23.5 *8*
108	9.7	8.9 *3*
109	2.45	2.24 *8*

$^{125}_{53}$I (60.14 *11* d)

Mode: ϵ
Δ: -88847 *3* keV
SpA: 1.737$\times10^4$ Ci/g
Prod: ^{123}Sb(α,2n); daughter ^{125}Xe; deuterons on Te; protons on Te

Photons (^{125}I)
$\langle\gamma\rangle$=42.3 *8* keV

γ_{mode}	γ(keV)	γ(%)†
Te L_ℓ	3.335	0.22 *3*
Te L_η	3.606	0.112 *12*
Te L_α	3.768	6.2 *7*
Te L_β	4.122	5.7 *7*
Te L_γ	4.668	0.71 *10*
Te $K_{\alpha2}$	27.202	40.5 *13*
Te $K_{\alpha1}$	27.472	75.6 *25*
Te $K_{\beta1}'$	30.980	20.1 *7*
Te $K_{\beta2}'$	31.877	4.38 *15*
γ M1+0.07%E2	35.4919 *5*	6.66

† 2.0% uncert(syst)

Atomic Electrons (^{125}I)
$\langle e\rangle$=17.9 *6* keV

e_{bin}(keV)	$\langle e\rangle$(keV)	e(%)
4	6.3	158 *10*
5	2.6	55 *6*
22	0.74	3.4 *4*
23	2.28	10.0 *10*
26	0.74	2.8 *3*
27	0.80	3.0 *3*
30	0.116	0.39 *4*
31	3.37	11.0 *3*
34	0.665	1.93 *6*
35	0.274	0.781 *23*

Continuous Radiation (^{125}I)
$\langle$IB$\rangle$=0.056 keV

E_{bin}(keV)		$\langle\ \rangle$(keV)	(%)
10 - 20	IB	0.00078	0.0046
20 - 40	IB	0.053	0.19
40 - 100	IB	0.0018	0.0034
100 - 143	IB	5.5 $\times10^{-5}$	5.0 $\times10^{-5}$

$^{125}_{54}$Xe(16.9 *2* h)

Mode: ϵ
Δ: -87191.5 *21* keV
SpA: 1.483$\times10^6$ Ci/g
Prod: ^{127}I(p,3n); ^{124}Xe(n,γ)

Photons (^{125}Xe)
$\langle\gamma\rangle$=261 *3* keV

γ_{mode}	γ(keV)	γ(%)†
I L_ℓ	3.485	0.169 *23*
I L_η	3.780	0.081 *9*
I L_α	3.937	4.7 *5*
I L_β	4.320	4.2 *5*
I L_γ	4.891	0.53 *8*
I $K_{\alpha2}$	28.317	29.5 *10*
I $K_{\alpha1}$	28.612	54.8 *18*
I $K_{\beta1}'$	32.278	14.7 *5*
I $K_{\beta2}'$	33.225	3.34 *12*
γ M1+0.05%E2	54.966 *14*	5.9 *3*
γ E2	74.865 *18*	0.121 *24*
γ M1+1.5%E2	113.573 *21*	0.47 *9*
γ M1+11%E2	188.438 *20*	54.9

Photons (^{125}Xe)
(continued)

γ_{mode}	γ(keV)	γ(%)†
γ [M1]	210.43 3	0.08 3
γ E2	243.404 21	28.8 9
γ [M1]	372.08 6	0.247 11
γ M1	453.84 3	4.23 16
γ [M1+E2]	553.89 23	0.033 6 ?
γ	635.64 23	~0.12
γ [M1+E2]	636.01 20	~0.11
γ	717.1 7	<0.016
γ	727.01 21	0.051 6
γ	809.4 7	<0.011
γ [M1+E2]	819.29 23	0.014 6
γ [M1]	846.45 20	1.03 3
γ	865.1 6	
γ [M1+E2]	901.41 20	0.538 22
γ	920.1 6	
γ	937.44 21	0.115 11
γ	988.7 10	~0.0033 ?
γ	992.41 21	0.104 6
γ [M1+E2]	1007.72 23	0.143 11
γ	1020.4 4	0.022 6
γ	1069.7 5	<0.037
γ	1075.4 4	0.055 6
γ [E2]	1089.85 20	0.049 6
γ	1108.5 6	~0.0033 ?
γ [M1+E2]	1138.7 3	0.285 16
γ	1180.84 21	0.63 3
γ [M1+E2]	1193.6 3	0.077 6
γ	1253.3 5	0.0016 6
γ [E2]	1268.5 3	~0.0011
γ	1318.3 4	0.0016 6
γ	1328.2 5	~0.0011
γ [M1+E2]	1382.1 3	0.0038 6
γ	1441.8 5	0.0060 11
γ	1561.7 4	0.0013 4
γ	1625 1	
γ	1772 1	

† 1.8% uncert(syst)

Atomic Electrons (^{125}Xe)
⟨e⟩=32.3 6 keV

e_{bin}(keV)	⟨e⟩(keV)	e(%)
5	4.4	93 10
22	4.76	21.9 11
23	0.93	4.0 4
24	1.20	5.0 5
27 - 42	1.37	4.7 3
50	1.45	2.91 15
54 - 80	0.75	1.21 7
108 - 114	0.041	0.037 6
155	10.0	6.42 17
177	0.012	0.0069 23
183	1.39	0.756 20
184 - 209	0.721	0.387 8
210	3.95	1.88 7
238 - 243	1.043	0.436 10
339 - 372	0.0186	0.0054 3
421 - 454	0.239	0.0562 21
521 - 554	0.00127	0.00024 5
602 - 636	0.0061	0.0010 4
684 - 727	0.0012	0.00017 7
776 - 819	0.0220	0.00271 14
841 - 868	0.0126	0.00146 17
896 - 937	0.0031	0.00034 9
956 - 1003	0.0040	0.00041 8
1007 - 1042	0.0009	8 3 $\times 10^{-5}$
1057 - 1105	0.0045	0.00041 5
1107 - 1148	0.006	0.00056 25
1160 - 1194	0.0020	0.00017 4
1220 - 1268	2.9 $\times 10^{-5}$	2.4 8 $\times 10^{-6}$
1285 - 1328	2.6 $\times 10^{-5}$	2.0 8 $\times 10^{-6}$
1349 - 1382	4.6 $\times 10^{-5}$	3.4 6 $\times 10^{-6}$
1409 - 1442	5.3 $\times 10^{-5}$	3.7 16 $\times 10^{-6}$
1529 - 1561	1.0 $\times 10^{-5}$	7 3 $\times 10^{-7}$

Continuous Radiation (^{125}Xe)
⟨β+⟩=1.34 keV;⟨IB⟩=0.63 keV

E_{bin}(keV)		⟨ ⟩(keV)	(%)
0 - 10	β+	1.72 $\times 10^{-6}$	2.07 $\times 10^{-5}$
	IB	0.00034	
10 - 20	β+	6.5 $\times 10^{-5}$	0.000393
	IB	0.00064	0.0038
20 - 40	β+	0.00163	0.00500
	IB	0.060	0.21
40 - 100	β+	0.061	0.080
	IB	0.0052	0.0087
100 - 300	β+	1.04	0.53
	IB	0.037	0.018
300 - 600	β+	0.234	0.070
	IB	0.165	0.037
600 - 1300	IB	0.36	0.044
1300 - 1467	IB	0.0023	0.00017
	$\Sigma\beta$+		0.69

$^{125}_{54}$Xe(57 1 s)

Mode: IT
Δ: -86939.4 22 keV
SpA: 1.57$\times 10^{9}$ Ci/g
Prod: daughter ^{125}Cs; ^{127}I(p,3n); ^{122}Te(α,n)

Photons (^{125}Xe)
⟨γ⟩=116.1 19 keV

γ_{mode}	γ(keV)	γ(%)
Xe L_ℓ	3.634	0.152 21
Xe L_η	3.955	0.084 9
Xe L_α	4.104	4.2 5
Xe L_β	4.511	4.0 5
Xe L_γ	5.089	0.49 6
Xe $K_{\alpha2}$	29.461	18.1 6
Xe $K_{\alpha1}$	29.782	33.7 10
Xe $K_{\beta1}$'	33.606	9.1 3
Xe $K_{\beta2}$'	34.606	2.15 7
γ M1	111.7 5	61.8 12
γ E3	140.4 5	19.7 5

Atomic Electrons (^{125}Xe)
⟨e⟩=137.3 20 keV

e_{bin}(keV)	⟨e⟩(keV)	e(%)
5	4.0	81 8
24	0.56	2.31 24
25	0.71	2.8 3
28	0.196	0.69 7
29	0.42	1.45 15
30	0.056	0.188 19
32	0.0213	0.066 7
33	0.048	0.147 15
77	25.1	32.5 9
106	44.6	42.2 13
107	0.366	0.343 10
111	1.17	1.05 3
112	0.0492	0.0440 12
135	26.6	19.7 6
136	20.2	14.9 5
139	10.6	7.61 24
140	2.61	1.86 6

$^{125}_{55}$Cs(45 1 min)

Mode: ϵ
Δ: -84092 17 keV
SpA: 3.34$\times 10^{7}$ Ci/g
Prod: ^{127}I(α,6n): daughter ^{125}Ba

Photons (^{125}Cs)
⟨γ⟩=349 28 keV

γ_{mode}	γ(keV)	γ(%)†
Xe L_ℓ	3.634	0.090 16
Xe L_η	3.955	0.041 6
Xe L_α	4.104	2.5 4
Xe L_β	4.522	2.2 4
Xe L_γ	5.124	0.29 5
Xe $K_{\alpha2}$	29.461	14.8 17
Xe $K_{\alpha1}$	29.782	28 3
Xe $K_{\beta1}$'	33.606	7.4 8
Xe $K_{\beta2}$'	34.606	1.76 20
γ M1	111.7 5	8.6 17
γ	334.9 10	2.0 4
γ	413.1 9	5.3 11
γ	428.3 8	1.6 3
γ	524.8 9	24 5
γ	539.9 8	3.1 6
γ	600.1 9	3.1 6
γ	653.9 10	0.43 9
γ	711.8 9	3.5 7
γ	779.9 10	0.22 4
γ	807.9 10	0.077 15
γ	864.9 10	0.30 6
γ	921.9 10	0.83 17
γ	994.9 10	0.53 11
γ	1059.9 10	0.18 4
γ	1158.2 9	0.43 9
γ	1199.6 9	0.19 4
γ	1213.1 9	0.34 7
γ	1227.9 10	0.15 3
γ	1311.3 9	0.22 4
γ	1324.8 9	0.093 19
γ	1467.6 9	0.26 5
γ	1579.3 9	0.28 6
γ	1586.5 8	
γ	1698.1 8	0.28 6
γ	1782.9 10	0.22 4
γ	1824.9 10	0.124 25
γ	1854.9 10	0.19 4
γ	2043.1 9	0.062 12
γ	2115.9 10	0.80 16
γ	2154.8 9	0.19 4
γ	2200.9 10	0.124 25
γ	2268.9 10	0.124 25
γ	2370.9 10	
γ	2412.6 9	0.046 9
γ	2432.1 9	0.062 12
γ	2524.3 9	0.124 25
γ	2543.8 9	0.025 5
γ	2622.9 10	0.046 9
γ	2725.9 10	0.046 9

† approximate

Atomic Electrons (^{125}Cs)
⟨e⟩= 9.7 9 keV

e_{bin}(keV)	⟨e⟩(keV)	e(%)
5	2.3	46 7
24	0.46	1.9 3
25	0.58	2.3 4
28	0.160	0.57 8
29	0.34	1.19 18
30 - 33	0.103	0.33 3
77	3.5	4.5 9
106	0.58	0.55 11
107 - 112	0.22	0.20 3
300 - 335	0.13	~0.041
379	0.20	~0.05

Atomic Electrons (^{125}Cs) (continued)

e_{bin}(keV)	⟨e⟩(keV)	e(%)
394 - 428	0.11	0.027 *11*
490	0.7	~0.13
505 - 540	0.22	0.042 *15*
566 - 600	0.08	~0.014
619 - 654	0.010	~0.0016
677 - 712	0.07	~0.011
745 - 780	0.0053	0.0007 *3*
802 - 830	0.0045	~0.0005
859 - 887	0.012	~0.0013
916 - 960	0.008	0.0009 *4*
989 - 1025	0.0031	0.00030 *12*
1054 - 1060	0.00033	3.1 *14* $\times 10^{-5}$
1124 - 1165	0.007	0.00061 *23*
1179 - 1228	0.0058	0.00049 *16*
1277 - 1325	0.0032	0.00025 *9*
1433 - 1468	0.0024	0.00017 *7*
1545 - 1578	0.0024	0.00015 *7*
1664 - 1748	0.0036	0.00021 *7*
1777 - 1854	0.0025	0.00014 *4*
2009 - 2081	0.0049	0.00024 *9*
2110 - 2200	0.0026	0.00012 *3*
2234 - 2268	0.0007	3.3 *12* $\times 10^{-5}$
2378 - 2431	0.00062	2.6 *7* $\times 10^{-5}$
2490 - 2588	0.0010	4.1 *10* $\times 10^{-5}$
2617 - 2691	0.00024	9 *3* $\times 10^{-6}$
2720 - 2725	3.1 $\times 10^{-5}$	1.2 *4* $\times 10^{-6}$

Continuous Radiation (^{125}Cs)

⟨β+⟩=327 keV;⟨IB⟩=3.8 keV

E_{bin}(keV)		⟨ ⟩(keV)	(%)
0 - 10	β+	2.14 $\times 10^{-6}$	2.57 $\times 10^{-5}$
	IB	0.0131	
10 - 20	β+	8.6 $\times 10^{-5}$	0.00052
	IB	0.0130	0.090
20 - 40	β+	0.00235	0.0072
	IB	0.065	0.22
40 - 100	β+	0.112	0.144
	IB	0.073	0.113
100 - 300	β+	5.4	2.45
	IB	0.21	0.117
300 - 600	β+	38.2	8.3
	IB	0.34	0.078
600 - 1300	β+	196	21.2
	IB	1.22	0.131
1300 - 2500	β+	87	5.8
	IB	1.7	0.101
2500 - 3100	IB	0.096	0.0036
	Σβ+		38

$^{125}_{56}$Ba(3.5 *4* min)

Mode: ϵ

Δ: -79510 *250* keV

SpA: 4.3$\times 10^{8}$ Ci/g

Prod: ^{115}In(^{14}N,4n); ^{12}C on Sn; ^{16}O on In

Photons (^{125}Ba)

γ_{mode}	γ(keV)	γ(rel)
γ	45.1 *4*	~3
γ[M1+E2]	55.3 *4*	48 *4*
γ	63.2 *4*	~8
γ	77.8 *4*	100
γ(E2)	85.7 *4*	82 *8*
γ	100.3 *4*	~6
γ	108.3 *4*	8 *2*
γ[M1+E2]	141.0 *4*	86 *8*

$^{125}_{57}$La(1.27 *10* min)

Mode: ϵ

SpA: 1.18$\times 10^{9}$ Ci/g

Prod: ^{16}O on In; ^{32}S on ^{96}Ru; ^{32}S on ^{98}Ru

Photons (^{125}La)

γ_{mode}	γ(keV)	γ(rel)
γ	43.6 *5*	10.3 *21*
γ	67.6 *5*	100 *12*

$^{125}_{58}$Ce(8.9 *7* s)

Mode: ϵ, ϵp

SpA: 9.8$\times 10^{9}$ Ci/g

Prod: ^{90}Zr(^{40}Ca,α,n); ^{32}S on ^{96}Ru; ^{32}S on ^{98}Ru

ϵp: 2000-4700 range

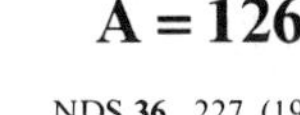

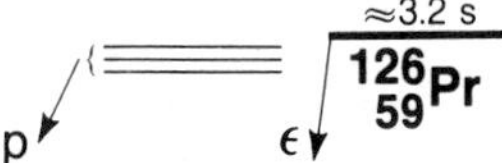

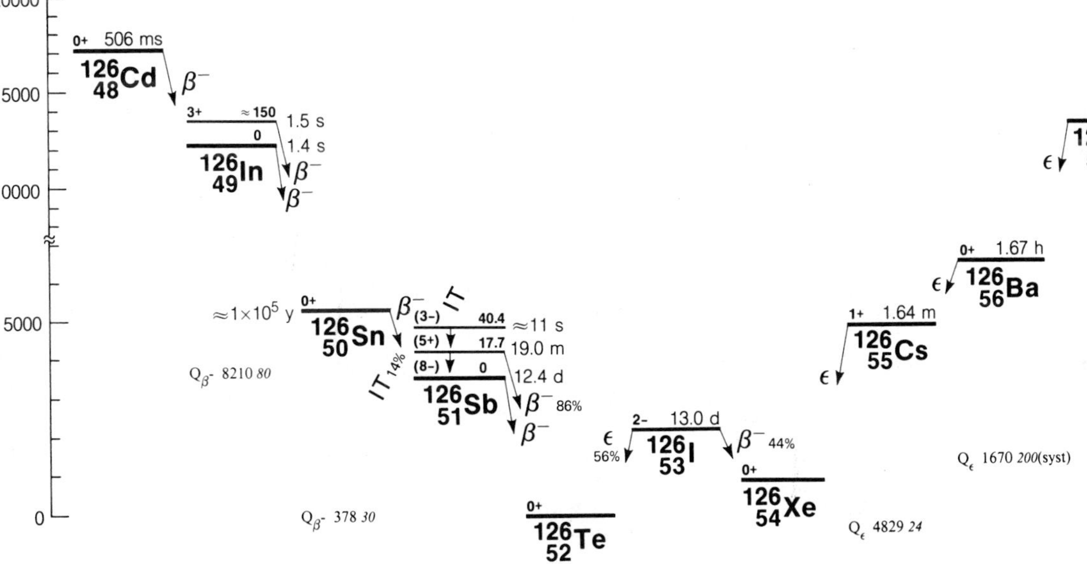

$^{126}_{48}$Cd(506 *15* ms)

Mode: β-
SpA: 9.6×10^{10} Ci/g
Prod: fission

Photons (^{126}Cd)

γ_{mode}	γ(keV)	γ(rel)
γ	62.56 17	1.6 3
γ	102.7 3	1.2 4
γ[M1]	260.00 10	100 4
γ	277.3 7	0.6 2
γ	325.3 3	~0.6
γ	365.45 17	2.3 6
γ	428.01 7	84 3
γ	555.25 13	4.8 6
γ	585.3 3	0.9 3
γ	652.93 22	1.2 4
γ	688.01 10	5.9 4

$^{126}_{49}$In(1.45 *22* s)

Mode: β-
Δ: -77810 *80* keV
SpA: 4.9×10^{10} Ci/g
Prod: fission

Photons (^{126}In)

⟨γ⟩=4306 *110* keV

γ_{mode}	γ(keV)	γ(%)
Sn L_ℓ	3.045	0.081 16
Sn L_η	3.272	0.045 8
Sn L_α	3.443	2.2 4
Sn L_β	3.739	1.9 3
Sn L_γ	4.187	0.21 4
Sn $K_{\alpha2}$	25.044	11.4 14
Sn $K_{\alpha1}$	25.271	21 3
Sn $K_{\beta1}$'	28.474	5.5 7
Sn $K_{\beta2}$'	29.275	1.17 15
γE2	57.45 3	5.5 9
γE1	111.79 5	88 8
γ	171.05 8	0.20 4
γ	175.3 3	0.16 4 •
γ	212.32 13	0.61 7 •
γ	251.75 25	0.22 5 •
γM1(+E2)	258.52 3	9.2 7
γ	266.08 15	0.31 6 •
γ	269.25 5	6.3 5
γM1(+E2)	315.97 4	11.6 10
γ	323.9 4	0.25 10 •
γ	362.73 5	2.5 2
γ	387.59 8	1.2 1
γ	402.8 2	0.5 1 •
γ	417.90 13	0.66 10 •
γ	433.31 25	0.51 10 •
γ	443.98 3	2.0 2
γ	478.0 3	0.37 10 •
γ	501.44 4	6.3 5
γ	515.79 20	1.0 2 •
γ	525.46 15	0.64 10 •
γ	571.71 4	3.0 2
γ	595.84 19	0.61 10 •
γ	708.0 3	0.7 2 •
γ	717.7 4	0.64 20 •
γ	776.85 19	1.1 2 •
γ	788.26 4	8.2 6
γ	848.30 4	0.83 10
γ	905.75 4	11.4 10
γ[E2]	908.62 5	99 7
γ	945.18 20	1.0 2 •
γ	957.9 4	0.55 20 •

Photons (^{126}In) (continued)

γ_{mode}	γ(keV)	γ(%)
γ	962.82 10	2.0 3
γ	977.37 7	2.7 3
γ	1020.41 6	0.58 15
γ	1064.84 4	4.7 3
γE2	1141.15 4	100 7
γ	1192.57 4	4.4 3
γ	1224.20 25	0.89 20 •
γ	1235.88 7	2.5 2
γ	1280.06 19	1.2 2 •
γ	1314.41 12	1.7 2
γ	1367.30 5	2.8 2
γ	1378.03 4	23.2 20
γ	1406.80 10	3.5 3
γ	1495.33 19	1.9 6
γ	1507.2 3	0.74 20 •
γ	1564.41 10	2.3 2
γ	1590.21 15	1.6 2
γ	1611.75 10	5.6 4
γ	1636.54 3	29.6 20
γ	1731.3 5	1.7 2
γ	1758.39 12	4.9 4
γ	2035.2 3	2.2 2 •
γ	2123.3 3	2.6 2 •
γ	2560.15 19	3.8 3
γ	2828.6 3	4.9 3

• with ^{126}In(1.4 or 1.5 s)

Atomic Electrons (^{126}In)

⟨e⟩=47.8 *25* keV

e_{bin}(keV)	⟨e⟩(keV)	e(%)
4	2.2	55 9
4 - 28	1.35	6.2 7
28	9.6	34 6
53	6.4	11.9 20
54	6.5	12.1 20
57	3.4	6.0 10
83	8.3	10.1 9
107 - 146	1.78	1.63 10
167 - 212	0.11	0.060 24
223 - 269	1.6	0.68 12
287 - 334	1.04	0.35 4
358 - 404	0.12	0.032 9
413 - 449	0.08	0.018 7
472 - 522	0.21	0.044 16
525 - 572	0.08	0.014 6
591 - 632	0.032	0.0052 23
679 - 718	0.025	0.0037 12
748 - 788	0.13	0.017 7
819 - 848	0.011	0.0013 6
877 - 879	0.17	0.019 7
879	1.22	0.138 11
901 - 948	0.47	0.051 4
953 - 991	0.052	0.0054 6
1016 - 1065	0.060	0.0058 19
1067 - 1115	1.48	0.133 7
1131 - 1174	0.264	0.0232 16
1188 - 1236	0.072	0.0059 8
1246 - 1285	0.026	0.0020 5
1298 - 1338	0.028	0.0021 6
1349 - 1378	0.20	0.015 5
1402 - 1407	0.0033	0.00024 9
1461 - 1560	0.066	0.0044 6
1561 - 1658	0.27	0.017 4
1683 - 1758	0.045	0.0026 6
2006 - 2105	0.056	0.0027 3
2106 - 2203	0.0179	0.00083 7
2341 - 2370	0.0106	0.00045 5
2531 - 2607	0.021	0.00083 18
2632 - 2718	0.0075	0.00028 3
2741 - 2828	0.027	0.00095 18
2845 - 2873	0.0011	4.0 11 ×10^{-5}
3217 - 3316	0.096	0.00292 12
3340 - 3434	0.0131	0.000391 14
3475 - 3504	0.00207	6.0 5 ×10^{-5}

Atomic Electrons (^{126}In) (continued)

e_{bin}(keV)	⟨e⟩(keV)	e(%)
3789 - 3886	0.0200	0.00052 5
3935 - 3963	0.0094	0.000239 19
4212 - 4303	0.0166	0.00039 3
4326 - 4330	6.1 ×10^{-5}	1.4 3 ×10^{-6}
4627 - 4699	0.00111	2.4 3 ×10^{-5}
4768 - 4796	0.00046	10 3 ×10^{-6}

Continuous Radiation (^{126}In)

⟨β-⟩=2471 keV;⟨IB⟩=9.7 keV

E_{bin}(keV)		⟨ ⟩(keV)	(%)
0 - 10	β-	0.00212	0.0421
	IB	0.065	
10 - 20	β-	0.0065	0.0434
	IB	0.064	0.45
20 - 40	β-	0.0272	0.091
	IB	0.128	0.44
40 - 100	β-	0.216	0.304
	IB	0.37	0.57
100 - 300	β-	2.94	1.42
	IB	1.15	0.64
300 - 600	β-	15.7	3.40
	IB	1.48	0.35
600 - 1300	β-	135	13.8
	IB	2.6	0.29
1300 - 2500	β-	664	34.7
	IB	2.5	0.144
2500 - 5000	β-	1468	43.0
	IB	1.28	0.040
5000 - 7219	β-	185	3.34
	IB	0.036	0.00067

$^{126}_{49}$In(1.5 *2* s)

Mode: β-
Δ: ~-77660 keV
SpA: 4.8×10^{10} Ci/g
Prod: fission

Photons (^{126}In)

⟨γ⟩=2812 *58* keV

γ_{mode}	γ(keV)	γ(%)
γ	175.34 25	0.034 10 •
γ	212.36 10	0.13 2 •
γ	251.79 20	0.045 10 •
γ	266.12 15	0.065 15 •
γ	323.9 4	0.053 20 •
γ	402.84 20	0.10 2 •
γ	417.94 10	0.14 2 •
γ	433.35 20	0.11 2 •
γ	478.02 25	0.076 20 •
γ	503.97 10	0.21 4
γ	515.83 20	0.20 4 •
γ	525.50 15	0.13 2 •
γ	595.88 15	0.13 2 •
γ	631.81 5	1.6 1
γ	708.07 25	0.15 4 •
γ	717.7 4	0.13 4 •
γ	776.89 15	0.22 4 •
γ[E2]	908.62 5	4.3 5
γ	945.22 20	0.20 4 •
γ	957.9 4	0.11 4 •
γ[M1+E2]	969.66 4	14.9 10
γ	988.97 20	0.27 5
γ	1053.10 5	0.50 4
γ	1068.01 8	0.44 4

Photons (^{126}In) (continued)

γ_{mode}	γ(keV)	γ(%)
γ	1077.79 12	0.32 4
γ	1115.40 20	0.25 4
γ	1135.69 7	2.0 2
γE2	1141.15 4	56 4
γ[M1+E2]	1174.40 9	0.31 4
γ	1224.24 20	0.18 4 *
γ[M1+E2]	1229.32 5	1.7 1
γ	1250.56 25	0.31 6
γ[M1+E2]	1252.01 10	1.7 1
γ	1280.10 15	0.24 3 *
γ[M1+E2]	1327.56 9	0.58 5
γ	1330.81 15	0.32 4
γ	1489.91 10	0.43 4
γ[M1+E2]	1495.45 10	1.1 2
γ	1507.2 3	0.15 3 *
γ[M1+E2]	1571.00 9	2.6 2
γ[M1+E2]	1593.69 7	1.1 1
γ	1601.46 5	1.4 1
γ	1643.54 20	0.37 4
γ	1687.32 7	2.2 2
γ	1745.22 13	0.74 6
γ	2035.21 25	0.45 4 *
γ[M1+E2]	2105.43 10	2.0 2
γ[E2]	2110.81 5	3.1 2
γ	2123.37 25	0.53 4 *
γ[M1+E2]	2203.69 8	2.2 2
γ[E2]	2370.47 6	1.9 2
γ[E2]	2636.59 10	1.1 1
γ	2719.1 3	0.63 6
γ[M1+E2]	2745.38 11	0.95 10
γ	2776.1 5	0.38 4
γ[M1+E2]	2822.99 7	1.0 1
γ	2874.0 9	0.29 4 *
γ[E2]	3246.56 10	3.2 3
γ[E2]	3344.82 9	21.6 4
γ[E2]	3434.9 6	0.29 4
γ[E2]	3504.5 3	0.49 4
γ[E2]	3817.9 4	0.35 3
γ[E2]	3886.51 11	4.7 5
γ[E2]	3964.12 7	2.4 2
γ[E2]	4240.96 15	1.5 2
γ[E2]	4257.0 3	0.33 6
γ[E2]	4303.23 15	2.5 3
γ[E2]	4330.8 6	0.13 3
γ[E2]	4656.4 5	0.21 3
γ[E2]	4699.4 6	0.10 3
γ[E2]	4797.0 6	0.13 4

* with ^{126}In(1.5 or 1.4 s)

Continuous Radiation (^{126}In)

$\langle\beta-\rangle$=1963 keV;$\langle$IB$\rangle$=6.7 keV

E_{bin}(keV)		$\langle\ \rangle$(keV)	(%)
0 - 10	β-	0.00324	0.064
	IB	0.057	
10 - 20	β-	0.0099	0.066
	IB	0.057	0.39
20 - 40	β-	0.0415	0.138
	IB	0.112	0.39
40 - 100	β-	0.327	0.463
	IB	0.33	0.50
100 - 300	β-	4.43	2.13
	IB	0.98	0.55
300 - 600	β-	23.2	5.03
	IB	1.22	0.29
600 - 1300	β-	190	19.4
	IB	2.0	0.23
1300 - 2500	β-	813	42.9
	IB	1.60	0.093
2500 - 5000	β-	929	29.7
	IB	0.38	0.0128
5000 - 5872	β-	3.45	0.066
	IB	0.00018	3.3×10^{-6}

$^{126}_{50}$Sn($\sim1\times10^5$ yr)

Mode: β-
Δ: -86021 11 keV
SpA: $\sim$0.03 Ci/g
Prod: fission

Photons (^{126}Sn)

$\langle\gamma\rangle$=57 3 keV

γ_{mode}	γ(keV)	γ(%)†
Sb L_ℓ	3.189	0.15 3
Sb L_η	3.437	0.069 11
Sb L_α	3.604	4.2 7
Sb L_β	3.929	5.1 9
Sb L_γ	4.518	0.80 16
γ E1	21.647 8	1.24 13
γ M2	22.656 12	0.100 11 *
γ M1	23.282 8	6.4 6
Sb $K_{\alpha2}$	26.111	8.5 5
Sb $K_{\alpha1}$	26.359	15.8 10
Sb $K_{\beta1}'$	29.712	4.2 3
Sb $K_{\beta2}'$	30.561	0.88 6
γ M1	42.637 8	0.50 5
γ E1	64.284 7	9.6 10
γ E2	86.940 10	8.9 9
γ E1	87.566 8	37

† 11% uncert(syst)
* with ^{126}Sb($\sim$11 s) in equilib

Atomic Electrons (^{126}Sn)

$\langle e\rangle$=54.9 20 keV

e_{bin}(keV)	$\langle e\rangle$(keV)	e(%)
4	2.5	61 9
5	1.79	38 5
12	0.39	3.2 3
13	0.6	$\sim$4
14	0.9	$\sim$7
17	0.7	4.2 15
18	7.8	43 5
19	9.2	50 5
21	0.25	1.19 13
22	4.7	21.5 23
23	1.04	4.6 5
25	0.157	0.62 7
26	0.169	0.65 8
29	0.027	0.093 11
30	0.0061	0.0207 24
34	1.81	5.4 6
38	0.156	0.41 4
39	0.0029	0.0076 8
42	0.041	0.099 10
43	0.00152	0.0036 4
56	9.2	16.2 16
57	5.0	8.8 9
60	0.43	0.73 7
63	0.073	0.116 12
64	0.038	0.060 6
82	1.19	1.45 15
83	5.1	6.1 6
86	1.15	1.34 14
87	0.48	0.56 6
88	0.0047	0.0054 6

Continuous Radiation (^{126}Sn)

$\langle\beta-\rangle$=70 keV;$\langle$IB$\rangle$=0.018 keV

E_{bin}(keV)		$\langle\ \rangle$(keV)	(%)
0 - 10	β-	0.471	9.5
	IB	0.0036	
10 - 20	β-	1.35	9.0
	IB	0.0029	0.021
20 - 40	β-	4.94	16.6
	IB	0.0043	0.0154
40 - 100	β-	25.6	37.9
	IB	0.0058	0.0098
100 - 250	β-	37.8	27.0
	IB	0.00132	0.00109

$^{126}_{51}$Sb(12.4 1 d)

Mode: β-
Δ: -86400 30 keV
SpA: 8.36×10^4 Ci/g
Prod: descendant ^{126}Sn; ^{128}Te(d,α); ^{126}Te(n,p); ^{124}Sn(α,pn)

Photons (^{126}Sb)

$\langle\gamma\rangle$=2749 55 keV

γ_{mode}	γ(keV)	γ(%)
Te $K_{\alpha2}$	27.202	0.59 3
Te $K_{\alpha1}$	27.472	1.10 5
Te $K_{\beta1}'$	30.980	0.293 14
Te $K_{\beta2}'$	31.877	0.064 3
γ	149.3 4	$\sim$0.40
γ E2	208.5 16	0.50 20
γ	223.6 3	1.40 10
γ E2	278.6 3	2.4 6
γ	296.1 4	0.50 20
γ M1+6.6%E2	296.9 3	4.5 4
γ E2	414.6 3	83.3 21
γ	414.7 3	1.0 3
γ	554.8 3	1.69 20
γ	573.1 3	6.7 3
γ [E2]	593.0 3	7.5 4
γ	620.1 3	0.88 8
γ	639.5 3	0.90 10
γ	656.3 3	2.19 10
γ E2	666.370 11	99.7
γ	674.7 3	3.7 10
γ E2	694.9 4	99.7
γ E2(+M1)	696.9 4	29 7
γ (E1)	720.4 3	53.8 24
γ (E1)	856.5 3	17.6 9
γ	953.9 3	1.20 10
γ	959.5 4	0.50 10
γ E2	989.1 3	6.8 3
γ E2	1034.7 3	1.00 5
γ [M1+E2]	1063.5 5	0.51 18
γ	1212.8 3	2.39 20
γ [E2]	1478.1 5	0.29 3

Atomic Electrons (^{126}Sb)

$\langle e\rangle$=12.7 3 keV

e_{bin}(keV)	$\langle e\rangle$(keV)	e(%)
4 - 31	0.155	2.21 15
117 - 149	0.12	$\sim$0.10
177 - 224	0.29	0.15 5
247 - 264	0.27	0.11 3
265	0.37	0.139 13
274 - 297	0.128	0.045 5
383	3.84	1.00 4

Atomic Electrons (^{126}Sb) (continued)

e_{bin}(keV)	$\langle e\rangle$(keV)	e(%)
410	0.607	0.148 *6*
414 - 415	0.153	0.0370 *13*
523 - 572	0.39	0.071 *15*
573 - 620	0.069	0.0117 *24*
625	0.038	~0.006
635	2.07	0.326 *8*
638 - 662	0.36	0.054 *6*
663	1.94	0.293 *7*
665	0.69	0.10 *3*
666 - 675	0.034	0.0050 *8*
689	0.383	0.056 *3*
690 - 720	0.50	0.072 *4*
825 - 856	0.122	0.0147 *8*
922 - 959	0.105	0.0111 *9*
984 - 1032	0.0334	0.00333 *25*
1034 - 1063	0.0014	0.00013 *3*
1181 - 1213	0.023	0.0020 *8*
1446 - 1478	0.0027	0.000186 *18*

Continuous Radiation (^{126}Sb)

$\langle\beta-\rangle$=340 keV;$\langle$IB$\rangle$=0.45 keV

E_{bin}(keV)		$\langle\ \rangle$(keV)	(%)
0 - 10	β-	0.128	2.56
	IB	0.0152	
10 - 20	β-	0.379	2.53
	IB	0.0145	0.101
20 - 40	β-	1.49	4.98
	IB	0.027	0.095
40 - 100	β-	9.8	14.1
	IB	0.069	0.108
100 - 300	β-	68	35.8
	IB	0.146	0.085
300 - 600	β-	95	22.5
	IB	0.104	0.025
600 - 1300	β-	130	15.1
	IB	0.067	0.0083
1300 - 1895	β-	34.8	2.36
	IB	0.0033	0.00023

$^{126}_{51}$Sb(19.0 *3* min)

Mode: β-(86 *4* %), IT(14 *4* %)
Δ: -86382 *30* keV
SpA: 7.85×10^7 Ci/g
Prod: ^{126}Te(n,p); daughter ^{126}Sn

Photons (^{126}Sb)

$\langle\gamma\rangle$=1548 *45* keV

γ_{mode}	γ(keV)	γ(%)†
Sb L_ℓ	3.189	0.0151 *22*
Sb L_η	3.437	0.0085 *9*
Sb L_α	3.604	0.42 *5*
Sb L_β	3.926	0.33 *4*
Sb L_γ	4.351	0.032 *4*
γ_{IT} E3	17.7 *3*	4.34 *14* ×10^{-5}
Te $K_{\alpha2}$	27.202	0.393 *17*
Te $K_{\alpha1}$	27.472	0.73 *3*
Te $K_{\beta1}$'	30.980	0.195 *9*
Te $K_{\beta2}$'	31.877	0.0424 *19*
$\gamma_{\beta-}$E2	414.6 *3*	86 *4*
$\gamma_{\beta-}$	620.1 *3*	1.57 *14*
$\gamma_{\beta-}$E2	666.370 *11*	86
$\gamma_{\beta-}$E2	694.9 *4*	82 *4*
$\gamma_{\beta-}$	928.4 *10*	1.3 *3* ?

Photons (^{126}Sb) (continued)

γ_{mode}	γ(keV)	γ(%)†
$\gamma_{\beta-}$E2	1034.7 *3*	1.78 *13*
$\gamma_{\beta-}$[M1+E2]	1063.5 *5*	0.54 *9*
$\gamma_{\beta-}$[E2]	1478.1 *5*	0.30 *4*

† uncert(syst): 29% for IT, 4.7% for β-

Atomic Electrons (^{126}Sb)

$\langle e\rangle$=11.3 *3* keV

e_{bin}(keV)	$\langle e\rangle$(keV)	e(%)
4	0.43	10.2 *11*
5	0.025	0.53 *6*
13	0.582	4.37 *17*
14	0.89	6.56 *25*
17	0.453	2.68 *10*
18 - 31	0.144	0.746 *24*
383	3.91	1.02 *6*
410	0.62	0.151 *8*
414 - 415	0.156	0.0377 *20*
588 - 620	0.034	0.006 *3*
635	1.77	0.280 *15*
661	0.209	0.0317 *17*
662	0.0370	0.0056 *3*
663	1.60	0.242 *13*
665 - 695	0.336	0.0490 *18*
897 - 928	0.017	0.0019 *9*
1003 - 1035	0.0311	0.00308 *22*
1059 - 1063	0.00110	0.000104 *19*
1446 - 1478	0.0028	0.000195 *21*

Continuous Radiation (^{126}Sb)

$\langle\beta-\rangle$=621 keV;$\langle$IB$\rangle$=1.12 keV

E_{bin}(keV)		$\langle\ \rangle$(keV)	(%)
0 - 10	β-	0.0201	0.399
	IB	0.025	
10 - 20	β-	0.061	0.408
	IB	0.025	0.17
20 - 40	β-	0.253	0.84
	IB	0.048	0.166
40 - 100	β-	1.92	2.73
	IB	0.132	0.20
100 - 300	β-	23.0	11.2
	IB	0.34	0.19
300 - 600	β-	94	20.8
	IB	0.31	0.074
600 - 1300	β-	376	41.2
	IB	0.23	0.029
1300 - 1912	β-	125	8.6
	IB	0.0116	0.00084

$^{126}_{51}$Sb(~11 s)

Mode: IT
Δ: -86360 *30* keV
SpA: ~8×10^9 Ci/g
Prod: daughter ^{126}Sn; fission

Photons (^{126}Sb)

$\langle\gamma\rangle$=0.249 *23* keV

γ_{mode}	γ(keV)	γ(%)
Sb L_ℓ	3.189	0.083 *13*
Sb L_η	3.437	0.026 *3*
Sb L_α	3.604	2.3 *3*
Sb L_β	3.939	2.8 *5*
Sb L_γ	4.549	0.47 *9*
γM2	22.656 *12*	0.134

Atomic Electrons (^{126}Sb)

$\langle e\rangle$=21.4 *6* keV

e_{bin}(keV)	$\langle e\rangle$(keV)	e(%)
4	1.05	25 *3*
5	1.26	27 *3*
18	10.1	56.3 *25*
19	4.34	23.4 *10*
22	3.77	17.3 *8*
23	0.89	3.95 *18*

$^{126}_{52}$Te(stable)

Δ: -90067.3 *24* keV
%: 18.95 *1*

$^{126}_{53}$I (13.02 *7* d)

Mode: ϵ(56.3 *20* %), β-(43.7 *20* %)
Δ: -87912 *5* keV
SpA: 7.96×10^4 Ci/g
Prod: ^{123}Sb(α,n); ^{125}Te(d,n); ^{126}Te(d,2n); ^{127}I(γ,n); protons on La; protons on Ce; ^{127}I(n,2n)

Photons (^{126}I)

$\langle\gamma\rangle$=421 *5* keV

γ_{mode}	γ(keV)	γ(%)†
Te L_ℓ	3.335	0.065 *9*
Te L_η	3.606	0.032 *3*
Te L_α	3.768	1.81 *20*
Te L_β	4.123	1.61 *20*
Te L_γ	4.662	0.20 *3*
Te $K_{\alpha2}$	27.202	11.9 *4*
Te $K_{\alpha1}$	27.472	22.3 *7*
Xe $K_{\alpha2}$	29.461	0.143 *5*
Xe $K_{\alpha1}$	29.782	0.265 *9*
Te $K_{\beta1}$'	30.980	5.92 *18*
Te $K_{\beta2}$'	31.877	1.29 *4*
Xe $K_{\beta1}$'	33.606	0.0718 *25*
Xe $K_{\beta2}$'	34.606	0.0170 *6*
$\gamma_{\beta-}$E2	388.635 *9*	34.1 *7*
$\gamma_{\beta-}$E2+0.5%M1	491.244 *9*	2.85 *6*
γ_ϵE2	666.370 *11*	33.1 *7*
γ_ϵE2	694.9 *4*	~0.00023 ?
γ_ϵ(E2+M1)	753.859 *12*	4.16 *9*
$\gamma_{\beta-}$E2	879.878 *10*	0.755 *16*
γ_ϵ[E2]	1206.8 *3*	0.00046 *13*
γ_ϵ	1378.78 *4*	0.00238 *17*
γ_ϵ[E2]	1420.224 *15*	0.295 *6*
γ_ϵ	2045.14 *4*	0.0046 *3*

† uncert(syst): 7.2% for ϵ, 7.3% for β-

Atomic Electrons (^{126}I)

$\langle e \rangle$=6.48 *16* keV

e_{bin}(keV)	$\langle e \rangle$(keV)	e(%)
4	0.99	22.7 *23*
5	0.75	15.8 *16*
22	0.219	0.99 *10*
23	0.67	2.9 *3*
24 - 25	0.0100	0.041 *3*
26	0.219	0.84 *9*
27	0.237	0.88 *9*
28 - 33	0.053	0.175 *13*
354	1.89	0.534 *15*
383	0.227	0.0592 *17*
384 - 389	0.179	0.0464 *8*
457 - 491	0.125	0.0270 *7*
635	0.686	0.108 *3*
661 - 695	0.1191	0.0180 *4*
722 - 754	0.096	0.0133 *16*
845 - 880	0.0139	0.00164 *4*
1175 - 1207	5.3 $\times 10^{-6}$	4.5 *11* $\times 10^{-7}$
1347 - 1388	0.00252	0.000181 *8*
1415 - 1420	0.000386	2.72 *10* $\times 10^{-5}$
2013 - 2044	2.7 $\times 10^{-5}$	1.3 *5* $\times 10^{-6}$

Continuous Radiation (^{126}I)

$\langle\beta-\rangle$=134 keV;$\langle\beta+\rangle$=5.5 keV;$\langle$IB$\rangle$=0.67 keV

E_{bin}(keV)		$\langle\ \rangle$(keV)	(%)
0 - 10	β-	0.0436	0.87
	β+	6.4 $\times 10^{-7}$	7.7 $\times 10^{-6}$
	IB	0.0067	
10 - 20	β-	0.131	0.87
	β+	2.37 $\times 10^{-5}$	0.000143
	IB	0.0067	0.046
20 - 40	β-	0.52	1.74
	β+	0.00059	0.00182
	IB	0.042	0.150
40 - 100	β-	3.63	5.2
	β+	0.0234	0.0304
	IB	0.032	0.050
100 - 300	β-	30.4	15.5
	β+	0.64	0.308
	IB	0.076	0.043
300 - 600	β-	64	14.8
	β+	1.94	0.436
	IB	0.118	0.027
600 - 1300	β-	35.1	4.71
	β+	2.85	0.371
	IB	0.29	0.033
1300 - 2156	IB	0.101	0.0065
	$\Sigma\beta$+		1.15

$^{126}_{54}$Xe(stable)

Δ: -89162 *7* keV
%: 0.09 *1*

$^{126}_{55}$Cs(1.64 *2* min)

Mode: ϵ
Δ: -84334 *24* keV
SpA: 9.07$\times 10^{8}$ Ci/g
Prod: daughter ^{126}Ba; protons on Ta; protons on Th; protons on U

Photons (^{126}Cs)

$\langle\gamma\rangle$=307 *14* keV

γ_{mode}	γ(keV)	γ(%)†
Xe L_ℓ	3.634	0.026 *4*
Xe L_η	3.955	0.0117 *14*
Xe L_α	4.104	0.72 *9*
Xe L_β	4.522	0.63 *9*
Xe L_γ	5.124	0.082 *13*
Xe $K_{\alpha2}$	29.461	4.2 *3*
Xe $K_{\alpha1}$	29.782	7.9 *5*
Xe $K_{\beta1}$'	33.606	2.13 *14*
Xe $K_{\beta2}$'	34.606	0.50 *4*
γ	213.5 *3*	0.18 *8* ?
γ [E2]	364.64 *19*	0.45 *12*
γ E2	388.635 *9*	41
γ [E2]	434.01 *13*	1.14 *16*
γ E2+0.5%M1	491.244 *9*	5.2 *5*
γ E2	553.4 *3*	0.28 *8*
γ	713.1 *5*	~0.08
γ [E2]	736.5 *3*	0.24 *8*
γ [M1+E2]	798.65 *18*	0.53 *8*
γ E2	879.878 *10*	1.39 *13*
γ E2	925.25 *13*	4.8 *5*
γ	1033.4 *3*	0.28 *8*
γ [M1+E2]	1289.89 *18*	0.37 *8*
γ	1467.4 *3*	~0.08
γ	1608.0 *5*	0.12 *4*
γ	1623.1 *5*	0.24 *4*
γ	1674.5 *5*	0.20 *8*
γ [E2]	1678.52 *18*	0.65 *12*
γ	1958.6 *3*	0.16 *4*
γ	2067.4 *5*	0.28 *8*
γ	2155 *1*	0.024 *8*
γ	2178.2 *7*	0.090 *16*
γ	2408 *1*	0.110 *20*
γ	2456.0 *5*	0.065 *12*
γ	2502.9 *5*	0.037 *12*
γ	2566.9 *7*	0.033 *12*

† 7.9% uncert(syst)

Atomic Electrons (^{126}Cs)

$\langle e \rangle$=4.37 *21* keV

e_{bin}(keV)	$\langle e \rangle$(keV)	e(%)
5	0.65	13.1 *15*
24	0.130	0.54 *6*
25	0.165	0.67 *8*
28 - 33	0.173	0.60 *5*
179 - 214	0.025	~0.014
330	0.028	0.0085 *23*
354	2.26	0.64 *5*
359 - 365	0.0062	0.0017 *3*
383	0.271	0.071 *6*
384 - 433	0.276	0.071 *4*
434	0.00043	0.000100 *15*
457	0.191	0.042 *4*
486 - 519	0.046	0.0094 *7*
548 - 553	0.0016	0.00030 *6*
679 - 713	0.0066	0.0009 *3*
731 - 764	0.0120	0.0016 *3*
793 - 799	0.0019	0.00023 *4*
845 - 891	0.097	0.0110 *9*
920 - 925	0.0119	0.00130 *11*
999 - 1033	0.0038	~0.00038
1255 - 1290	0.0051	0.00040 *9*
1433 - 1467	0.0007	~5 $\times 10^{-5}$
1573 - 1670	0.0099	0.00061 *10*
1673 - 1678	0.00084	5.0 *9* $\times 10^{-5}$
1924 - 1958	0.0011	5.8 *24* $\times 10^{-5}$
2033 - 2120	0.0020	10 *4* $\times 10^{-5}$
2144 - 2177	0.00058	2.7 *10* $\times 10^{-5}$
2373 - 2468	0.0012	4.9 *11* $\times 10^{-5}$
2497 - 2566	0.00020	8 *3* $\times 10^{-6}$

Continuous Radiation (^{126}Cs)

$\langle\beta+\rangle$=1323 keV;$\langle$IB$\rangle$=7.6 keV

E_{bin}(keV)		$\langle\ \rangle$(keV)	(%)
0 - 10	β+	1 $\times 10^{-6}$	1.20 $\times 10^{-5}$
	IB	0.042	
10 - 20	β+	4.01 $\times 10^{-5}$	0.000242
	IB	0.042	0.29
20 - 40	β+	0.00111	0.00339
	IB	0.094	0.33
40 - 100	β+	0.054	0.070
	IB	0.24	0.37
100 - 300	β+	2.87	1.29
	IB	0.71	0.40
300 - 600	β+	24.1	5.2
	IB	0.89	0.21
600 - 1300	β+	224	22.9
	IB	1.7	0.19
1300 - 2500	β+	762	41.2
	IB	2.4	0.131
2500 - 4829	β+	311	11.0
	IB	1.53	0.050
	$\Sigma\beta$+		81

$^{126}_{56}$Ba(1.67 *3* h)

Mode: ϵ
Δ: -82660 *200* keV syst
SpA: 1.49$\times 10^{7}$ Ci/g
Prod: ^{115}In(^{14}N,3n); ^{118}Sn(^{12}C,4n); ^{16}O on Sn; B on Sb; daughter ^{126}La

Photons (^{126}Ba)

$\langle\gamma\rangle$=571 *14* keV

γ_{mode}	γ(keV)	γ(%)†
Cs L_ℓ	3.795	0.147 *25*
Cs L_η	4.142	0.066 *10*
Cs L_α	4.285	4.0 *6*
Cs L_β	4.732	3.6 *6*
Cs L_γ	5.369	0.48 *8*
Cs $K_{\alpha2}$	30.625	22.6 *24*
Cs $K_{\alpha1}$	30.973	42 *4*
Cs $K_{\beta1}$'	34.967	11.4 *12*
Cs $K_{\beta2}$'	36.006	2.8 *3*
γ	94.66 *14*	0.24 *8*
γ	106.87 *19*	0.49 *12*
γ	126.52 *20*	0.45 *12*
γ	129.96 *19*	0.77 *16*
γ	192.48 *16*	
γ	201.05 *17*	0.61 *16*
γ	203.8 *3*	0.33 *8*
γ	208.19 *14*	0.53 *12*
γ	213.5 *3*	0.18 *8*
γ [M1]	217.89 *9*	4.1 *4*
γ	231.76 *14*	0.90 *12* ?
γ M1,E2	233.64 *8*	19.6 *10*
γ	239.25 *17*	0.45 *12*
γ E1	241.02 *9*	6.0 *5*
γ M1,E2	257.62 *8*	7.6 *4*
γ	269.3 *3*	0.18 *8*
γ [E2]	281.20 *11*	3.1 *3*
γ	285.08 *12*	0.41 *12*
γ	290.68 *21*	0.53 *12*
γ	303.05 *22*	0.12 *4*
γ	309.06 *12*	0.33 *8*
γ	320.9 *3*	~0.08
γ	324.8 *5*	0.24 *8*
γ	328.30 *13*	2.08 *20*
γ	348.47 *19*	0.73 *16*
γ	353.56 *16*	0.53 *12*
γ	392.42 *16*	0.77 *16*
γ	400.67 *13*	1.14 *16*
γ [M1+E2]	415.68 *15*	0.61 *12*

Photons (^{126}Ba) (continued)

γ_{mode}	γ(keV)	γ(%)†
γ [E1]	440.84 14	0.41 8
γ	452.84 22	0.18 8
γ	457.14 15	0.69 12
γ	475.98 18	0.33 8
γ	489.38 13	2.9 3
γ	508.57 20	
γ	535.6 3	0.37 8
γ	538.90 17	1.95 20
γ	542.70 11	0.90 16
γ	548.7 3	0.65 12
γ	551.1 3	0.73 12
γ	558.64 20	0.12 4
γ	560.76 19	0.28 12
γ	583.43 19	0.41 12
γ	608.31 18	0.41 8
γ [M1+E2]	611.19 16	0.49 8
γ	640.41 18	0.57 8
γ	642.8 3	0.37 8
γ	668.08 16	0.28 12
γ [M1+E2]	681.86 12	4.4 5
γ	685.6 3	0.45 8
γ	691.68 13	0.81 8
γ	698.97 17	0.28 8
γ	702.76 22	~0.08
γ	709.62 17	1.5 3
γ	745.00 15	0.57 8
γ	750.35 15	0.12 4
γ	779.0 5	~0.16
γ	781.54 22	0.45 8
γ	835.91 19	~0.16
γ [M1+E2]	839.92 14	0.49 12
γ	841.95 17	1.06 20
γ [E1]	856.52 14	0.73 12
γ [M1+E2]	863.90 13	1.5 3
γ	877.0 3	0.28 8
γ	882.48 18	0.28 8
γ	899.6 3	0.20 8
γ	903.5 3	0.24 8
γ	906.08 16	0.37 8
γ	910.0 5	0.12 4
γ	913.36 19	0.45 8
γ	929.58 17	0.61 12
γ	953.18 14	0.49 12
γ	964.75 17	0.20 8
γ [M1+E2]	976.76 11	1.79 16
γ [M1+E2]	977.14 14	0.65 20
γ [M1+E2]	984.04 16	1.14 24
γ [E1]	993.35 13	2.4 4
γ [M1+E2]	1000.74 12	0.24 8
γ [M1+E2]	1008.02 17	0.49 8
γ	1011.85 16	0.73 16
γ [M1+E2]	1035.43 13	1.6 3
γ [E1]	1052.03 13	1.22 20
γ [M1+E2]	1059.41 14	0.41 8
γ [M1+E2]	1097.54 12	0.69 16
γ [M1+E2]	1210.78 14	1.8 4
γ [M1+E2]	1234.38 10	2.0 4
γ [M1+E2]	1241.66 16	1.06 20
γ [M1+E2]	1293.05 12	3.7 4

† 7.9% uncert(syst)

Atomic Electrons (^{126}Ba)

⟨e⟩=15.5 7 keV

e_{bin}(keV)	⟨e⟩(keV)	e(%)
5	3.1	61 9
6	0.44	7.8 11
25	0.68	2.7 4
26	0.85	3.3 5
29	0.22	0.76 11
30	0.52	1.75 25
31 - 71	0.48	1.0 3
89 - 130	0.8	0.8 3
165 - 178	0.20	0.12 5
182	0.67	0.37 4
195 - 196	0.11	~0.05
198	3.00	1.52 12
199 - 218	0.41	0.200 21
222	0.98	0.44 3
226 - 227	0.020	~0.009
228	0.50	0.22 5
229 - 241	0.31	0.13 4
245	0.33	0.133 15
249 - 298	0.56	0.21 3
302 - 349	0.11	0.036 11
352 - 401	0.14	0.037 11
405 - 453	0.15	0.035 13
456 - 503	0.09	0.017 7
507 - 556	0.109	0.021 5
557 - 607	0.053	0.0089 20
608 - 656	0.16	0.024 4
662 - 710	0.08	0.012 3
714 - 750	0.015	0.0020 7
773 - 821	0.037	0.0045 13
828 - 877	0.071	0.0083 13
881 - 930	0.023	0.0025 8
941 - 988	0.113	0.0118 14
992 - 1035	0.049	0.0048 7
1046 - 1093	0.014	0.0013 3
1096 - 1097	0.00034	3.1 8 $\times 10^{-5}$
1175 - 1211	0.069	0.0058 8
1229 - 1257	0.053	0.0043 7
1287 - 1293	0.0075	0.00058 9

Continuous Radiation (^{126}Ba)

⟨β+⟩=2.96 keV;⟨IB⟩=0.53 keV

E_{bin}(keV)		⟨ ⟩(keV)	(%)
0 - 10	β+	8.6×10^{-7}	1.03×10^{-5}
	IB	0.00047	
10 - 20	β+	3.44×10^{-5}	0.000207
	IB	0.00056	0.0035
20 - 40	β+	0.00092	0.00283
	IB	0.071	0.23
40 - 100	β+	0.0395	0.051
	IB	0.0068	0.0117
100 - 300	β+	1.11	0.53
	IB	0.030	0.0147
300 - 600	β+	1.79	0.446
	IB	0.114	0.025
600 - 1300	β+	0.0132	0.00216
	IB	0.29	0.034
1300 - 1670	IB	0.020	0.00141
	$\Sigma\beta$+		1.04

$^{126}_{57}$La(1.0 3 min)

Mode: ϵ

SpA: 1.5×10^{9} Ci/g

Prod: ^{115}In(^{16}O,5n); ^{121}Sb(^{12}C,7n)

Photons (^{126}La)

γ_{mode}	γ(keV)
γ E2	256 1
γ [M1+E2]	340 1
γ E2	460 1
γ E2	625 1

$^{126}_{58}$Ce(50 6 s)

Mode: ϵ

SpA: 1.78×10^{9} Ci/g

Prod: ^{32}S on ^{96}Ru; ^{32}S on ^{98}Ru

$^{126}_{59}$Pr(3.2 6 s)

Mode: ϵ, ϵp

SpA: 2.5×10^{10} Ci/g

Prod: ^{90}Zr(^{40}Ca,p3n)

ϵp: 2100-5000 range

A = 127

NDS **35**, 181 (1982)

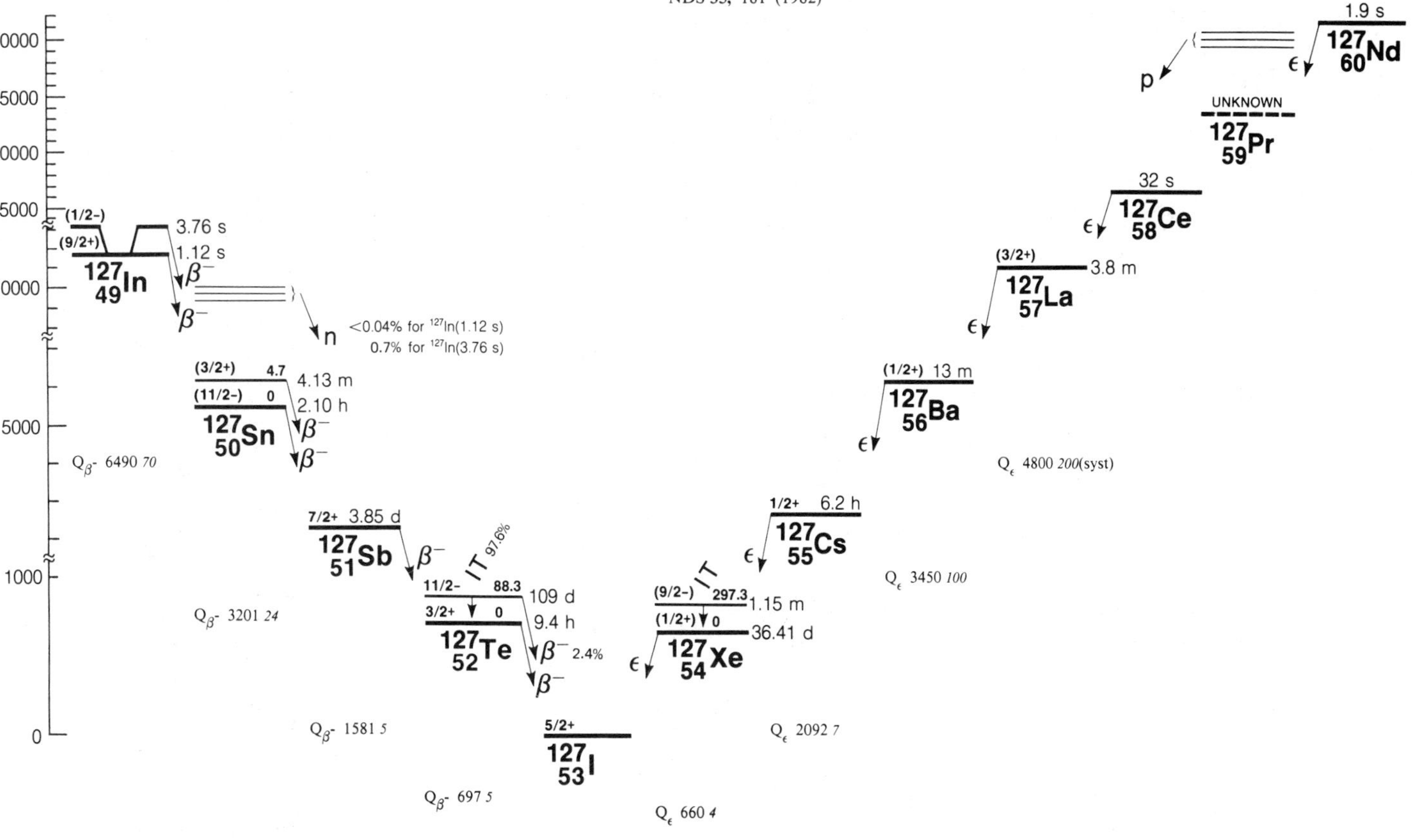

$^{127}_{49}$In(1.12 *2* s)

Mode: β-, β-n(<0.04 %)
Δ: -77010 *70* keV
SpA: 5.91×10^{10} Ci/g
Prod: fission

Photons (^{127}In)

⟨γ⟩=1766 *220* keV

γ_{mode}	γ(keV)	γ(%)†
γ	104.0 *3*	~0.14
γ	144.0 *3*	~0.14
γ	184.7 *3*	<0.14
γ	218.9 *3*	<0.14
γ	243.50 *16*	0.47 *7*
γM1	252.29 *21*	
γ	270.9 *3*	~0.14
γ	317.62 *19*	~0.14
γ	321.5 *3*	<0.14
γ	353.48 *19*	0.54 *7*
γ	359.6 *3*	~0.14
γ	376.0 *3*	~0.14
γ	406.7 *3*	0.20 *7*
γ	421.41 *21*	~0.14
γ	424.2 *3*	0.34 *7*
γ	468.01 *20*	1.35 *14*
γ	501.6 *3*	<0.14
γ	523.3 *3*	<0.14
γ[M1+E2]	549.09 *18*	0.34 *7*
γ	566.0 *3*	0.47 *7*

Photons (^{127}In) (continued)

γ_{mode}	γ(keV)	γ(%)†
γ	591.9 *3*	~0.14
γ	638.67 *20*	3.5 *3*
γ	646.1 *2*	8.6 *9*
γ	696.10 *22*	
γ	715.4 *3*	2.03 *20*
γ	731.9 *3*	0.41 *7*
γ	746.00 *18*	0.81 *7*
γ	792.59 *17*	2.17 *20*
γ	805.10 *17*	7.8 *8*
γ	808.0 *10*	0.41 *7*
γ	822.1 *3*	0.27 *7*
γ	855.97 *19*	1.15 *14*
γ	892.55 *23*	1.22 *14*
γ	909.69 *19*	0.54 *7*
γ	945.56 *21*	0.47 *7*
γ	948.39 *20*	
γ	956.29 *20*	6.4 *7*
γ	963.72 *21*	5.0 *5*
γ	970.50 *19*	0.20 *7*
γ	977.0 *10*	0.20 *7*
γ	980.0 *10*	~0.14
γ	989.1 *3*	0.20 *7*
γ[E2]	1048.60 *18*	7.4 *7*
γ	1070.72 *22*	0.95 *14*
γ	1094.7 *3*	5.1 *5*
γ	1099.47 *18*	0.61 *7*
γ	1214.00 *18*	1.15 *14*
γ	1242.7 *3*	0.68 *7*
γ	1293.7 *3*	0.20 *7*
γ	1326.6 *3*	0.88 *7*
γ	1389.3 *4*	0.61 *7*
γ	1410.8 *3*	0.20 *7*
γ	1436.6 *3*	0.41 *7*
γ	1555.79 *21*	1.76 *20*
γ[E2]	1597.69 *19*	68

Photons (^{127}In) (continued)

γ_{mode}	γ(keV)	γ(%)†
γ	1619.2 *5*	0.20 *7*
γ	1632.8 *4*	~0.14
γ	1697.64 *23*	0.34 *7*
γ	1771.0 *10*	~0.14
γ	1778.7 *5*	1.29 *14*
γ	1819.6 *5*	0.41 *7*
γ	1905.0 *10*	0.27 *7*
γ	1982.0 *10*	0.54 *7*
γ	2019.09 *20*	0.47 *7*
γ	2464.0 *10*	0.54 *7*
γ	2511.0 *10*	0.68 *7*

† 1.5% uncert(syst)

Continuous Radiation (^{127}In)

⟨β-⟩=2151 keV;⟨IB⟩=7.7 keV

E_{bin}(keV)		⟨ ⟩(keV)	(%)
0 - 10	β-	0.00258	0.051
	IB	0.060	
10 - 20	β-	0.0079	0.053
	IB	0.060	0.42
20 - 40	β-	0.0331	0.110
	IB	0.119	0.41
40 - 100	β-	0.262	0.370
	IB	0.35	0.53
100 - 300	β-	3.56	1.72
	IB	1.05	0.59
300 - 600	β-	18.9	4.10

Continuous Radiation (^{127}In)
(continued)

E_{bin}(keV)		⟨ ⟩(keV)	(%)
	IB	1.33	0.31
600 - 1300	β-	161	16.4
	IB	2.2	0.25
1300 - 2500	β-	763	39.9
	IB	1.9	0.112
2500 - 5000	β-	1201	37.3
	IB	0.58	0.019
5000 - 5844	β-	3.16	0.061
	IB	0.000123	2.4×10^{-6}

$^{127}_{49}$In(3.76 *3* s)

Mode: β-, β-n(0.68 *6* %)
Δ: -77010 *70* keV
SpA: 2.158×10^{10} Ci/g
Prod: fission

Photons (^{127}In)
⟨γ⟩=501 *26* keV

γ_{mode}	γ(keV)	γ(%)†
Sn L_{ℓ}	3.045	0.0037 *6*
Sn L_{η}	3.272	0.0019 *3*
Sn L_{α}	3.443	0.104 *15*
Sn L_{β}	3.742	0.090 *15*
Sn L_{γ}	4.217	0.0105 *19*
Sn $K_{\alpha2}$	25.044	0.77 *8*
Sn $K_{\alpha1}$	25.271	1.44 *15*
Sn $K_{\beta1}'$	28.474	0.38 *4*
Sn $K_{\beta2}'$	29.275	0.079 *8*
γ	136.7 *3*	
γM1	252.29 *21*	77 *8*
γ	696.10 *22*	0.76 *25*
γ	832.78 *23*	4.0 *5*
γ	948.39 *20*	5.5 *5*
γ	1085.07 *23*	3.28 *25*
γ	3074.0 *10*	5.8 *5*

† 9.9% uncert(syst)

Atomic Electrons (^{127}In)
⟨e⟩=8.3 *7* keV

e_{bin}(keV)	⟨e⟩(keV)	e(%)
4 - 28	0.196	2.9 *4*
223	6.8	3.1 *3*
248	0.95	0.38 *4*
251	0.177	0.070 *7*
667 - 696	0.012	~0.0018
804 - 833	0.052	0.006 *3*
919 - 948	0.06	0.007 *3*
1056 - 1085	0.031	0.0030 *12*
3045 - 3073	0.022	0.00071 *18*

Continuous Radiation (^{127}In)
⟨β-⟩=2665 keV;⟨IB⟩=10.8 keV

E_{bin}(keV)		⟨ ⟩(keV)	(%)
0 - 10	β-	0.00180	0.0358
	IB	0.068	
10 - 20	β-	0.0055	0.0368
	IB	0.067	0.47
20 - 40	β-	0.0232	0.077
	IB	0.133	0.46
40 - 100	β-	0.183	0.259
	IB	0.39	0.60
100 - 300	β-	2.50	1.20
	IB	1.21	0.67
300 - 600	β-	13.4	2.89
	IB	1.58	0.37
600 - 1300	β-	116	11.8
	IB	2.8	0.32
1300 - 2500	β-	586	30.5
	IB	2.9	0.164
2500 - 5000	β-	1752	49.5
	IB	1.63	0.051
5000 - 6485	β-	196	3.67
	IB	0.021	0.00039

$^{127}_{50}$Sn(2.10 *4* h)

Mode: β-
Δ: -83504 *25* keV
SpA: 1.175×10^{7} Ci/g
Prod: fission; ^{130}Te(n,α)

Photons (^{127}Sn)
⟨γ⟩=1863 *68* keV

γ_{mode}	γ(keV)	γ(%)†
Sb L_{ℓ}	3.189	0.011 *3*
Sb L_{η}	3.437	0.0059 *14*
Sb L_{α}	3.604	0.30 *7*
Sb L_{β}	3.928	0.26 *6*
Sb L_{γ}	4.410	0.029 *7*
Sb $K_{\alpha2}$	26.111	1.52 *15*
Sb $K_{\alpha1}$	26.359	2.8 *3*
Sb $K_{\beta1}'$	29.712	0.74 *7*
Sb $K_{\beta2}'$	30.561	0.158 *16*
γ	34.9 *3*	0.038 *8*
γ	46.9 *3*	0.080 *15*
γ	51.5 *3*	0.019 *4*
γ	52.8 *3*	0.038 *8*
γ	56.9 *3*	0.057 *11*
γ[M1]	65.90 *20*	0.14 *3*
γ[E2]	70.36 *23*	0.38 *8*
γ	83.4 *3*	0.19 *4*
γ	88.1 *3*	0.042 *8*
γ[M1]	97.28 *18*	0.45 *8*
γ[M1]	104.12 *22*	0.19 *4*
γ[M1]	110.26 *19*	0.38 *4*
γ[M1]	119.7 *3*	2.16 *23*
γ[M1]	124.09 *19*	~0.08
γ[M1]	142.2 *3*	0.42 *4*
γ[M1]	143.58 *24*	0.49 *4*
γ[E1]	155.93 *16*	0.23 *4*
γ[M1]	157.13 *20*	0.27 *4*
γ[M1]	169.24 *19*	2.01 *19*
γ[M1]	170.47 *24*	<0.15
γ[M1]	177.89 *22*	0.11 *4*
γ[M1]	181.13 *25*	0.15 *4*
γ[M1]	184.02 *19*	0.45 *8*
γ[M1]	184.66 *22*	1.10 *23*
γ[M1]	190.08 *22*	0.57 *8*
γ[M1]	195.0 *3*	~0.08 ?
γ[M1]	202.86 *18*	0.76 *8*
γ	204.16 *18*	0.23 *4*
γ[M1]	205.31 *20*	0.23 *4*
γ[M1]	208.03 *24*	0.15 *4*

Photons (^{127}Sn)
(continued)

γ_{mode}	γ(keV)	γ(%)†
γ	211.6 *3*	0.11 *4*
γ	213.06 *20*	0.11 *4*
γ	215.2 *3*	~0.08
γ[M1]	220.22 *19*	0.30 *4*
γ[M1]	228.3 *3*	0.19 *4*
γ[M1]	232.21 *24*	0.83 *8*
γ[M1]	234.35 *19*	0.53 *4*
γ[M1]	235.45 *24*	0.27 *4*
γ	248.26 *20*	~0.08
γ	255.01 *23*	0.11 *4*
γ[M1]	262.34 *24*	2.31 *23*
γ[M1]	265.90 *23*	2.12 *23*
γ	271.28 *24*	0.11 *4*
γ[M1]	279.15 *20*	0.57 *8*
γ	281.95 *22*	0.53 *4*
γ[M1]	284.07 *17*	2.7 *3*
γ[M1]	292.79 *22*	1.25 *11*
γ	301.44 *18*	0.11 *4* ?
γ	306.19 *21*	~0.08 ?
γ	331.62 *19*	0.45 *4*
γ	348.41 *23*	0.49 *4*
γ	353.21 *19*	0.11 *4*
γ	357.1 *3*	0.19 *4*
γ	360.37 *20*	0.19 *4*
γ	362.49 *18*	0.42 *4*
γ	365.23 *19*	0.19 *4*
γ	379.13 *21*	0.19 *4*
γ	390.60 *19*	1.25 *11*
γ	396.74 *22*	0.34 *4*
γ	405.39 *19*	0.45 *8*
γ	407.29 *19*	1.52 *15*
γ	420.7 *4*	0.15 *4*
γ	425.53 *17*	0.23 *4*
γ	438.4 *3*	6.1 *6*
γ	444.6 *3*	0.45 *8*
γ	446.45 *20*	0.23 *8*
γ	452.42 *19*	0.38 *4*
γ	468.74 *21*	0.45 *4*
γ	487.7 *3*	0.45 *4*
γ[M1+E2]	491.24 *21*	5.3 *5*
γ	493.37 *18*	3.1 *3*
γ	500.86 *19*	1.52 *15*
γ	509.14 *20*	1.4 *3*
γ	509.9 *3*	0.76 *15*
γ	514.15 *22*	0.27 *8*
γ	518.42 *17*	0.19 *4*
γ	528.4 *5*	~0.11
γ	530.6 *3*	~0.11
γ	539.5 *3*	0.23 *8*
γ	545.66 *20*	2.27 *23*
γ	563.22 *18*	~0.15
γ	566.26 *24*	~0.11
γ	570.3 *3*	0.57 *8*
γ	583.41 *21*	3.2 *3*
γ	592.19 *17*	2.01 *19*
γ	609.55 *18*	0.30 *4*
γ	616.19 *24*	0.23 *4*
γ	621.76 *21*	0.45 *4*
γ	631.74 *21*	0.53 *11*
γ	634.7 *4*	0.27 *8*
γ	649.30 *17*	0.80 *8*
γ	668.68 *20*	0.19 *4* ?
γ	702.9 *3*	0.15 *4*
γ	708.7 *3*	0.19 *4*
γ	758.72 *24*	0.15 *4*
γ	773.84 *24*	0.42 *4*
γ	805.83 *18*	8.2 *8*
γ	823.19 *15*	10.6 *23*
γ	824.65 *17*	6.1 *11*
γ	847.7 *3*	0.19 *4* ?
γ	859.58 *18*	8.0 *8*
γ	865.1 *3*	0.34 *4*
γ	889.09 *20*	0.34 *4*
γ	896.9 *4*	0.19 *4* ?
γ	912.21 *25*	0.11 *4*
γ	916.43 *21*	1.17 *11*
γ	929.77 *24*	0.34 *4*
γ	976.0 *4*	0.76 *15*
γ	979.13 *17*	7.2 *15*
γ	980.2 *3*	0.76 *15*
γ	997.95 *17*	1.93 *19*
γ	1002.51 *21*	1.74 *19*
γ	1036.25 *20*	1.97 *19*
γ	1044.87 *18*	0.27 *4*

Photons (^{127}Sn)
(continued)

γ_{mode}	γ(keV)	γ(%)†
γ	1055.07 *20*	0.23 *8*
γ	1064.5 *4*	0.38 *8*
γ	1093.1 *3*	3.8 *8*
γ	1095.52 *17*	19 *4*
γ	1114.35 *17*	38
γ	1134.6 *3*	0.11 *4*
γ	1142.0 *4*	0.19 *4*
γ	1159.3 *5*	0.91 *19*
γ	1160.37 *22*	2.4 *5*
γ	1179.19 *22*	0.49 *4*
γ	1220.6 *3*	0.53 *4*
γ	1237.49 *21*	0.11 *4*
γ	1291.93 *21*	0.76 *8*
γ	1310.75 *21*	∼0.08 ?
γ	1360.44 *17*	0.15 *4*
γ	1368.4 *4*	0.53 *4*
γ	1434.20 *17*	0.30 *4*
γ	1458.20 *25*	0.27 *8*
γ	1471.4 *3*	0.76 *15*
γ	1472.49 *18*	1.3 *3*
γ	1542.97 *21*	∼0.08 ?
γ	1584.33 *21*	1.78 *19*
γ	1600.2 *3*	0.15 *4*
γ	1611.1 *3*	0.15 *4*
γ	1647.8 *2*	1.02 *11*
γ	1666.63 *19*	0.49 *4*
γ	1709.75 *24*	0.27 *4*
γ	1720.1 *4*	0.19 *4*
γ	1750.8 *4*	0.19 *8*
γ	1752.96 *25*	0.27 *8*
γ	1812.9 *4*	0.11 *4*
γ	1937.53 *17*	∼0.08
γ	2003.43 *22*	5.3 *5*
γ	2093.47 *18*	∼0.08
γ	2102.4 *3*	0.49 *4*
γ	2150.59 *22*	<0.07
γ	2160.1 *4*	0.30 *4*
γ	2304.1 *4*	0.11 *4*
γ	2317.5 *3*	1.10 *11*
γ	2389.7 *4*	0.11 *4* ?
γ	2447.4 *3*	0.34 *4*
γ	2470.0 *5*	0.11 *4*
γ	2513.9 *5*	0.11 *4*
γ	2584.9 *5*	1.55 *15* ?
γ	2695.7 *3*	1.63 *15*
γ	2805.26 *24*	0.38 *4*
γ	2846.3 *3*	0.95 *11*
γ	2880.9 *4*	0.27 *4*

† 12% uncert(syst)

Atomic Electrons (^{127}Sn)
⟨e⟩=10.1 *5* keV

e_{bin}(keV)	⟨e⟩(keV)	e(%)
4	0.29	6.8 *18*
5 - 30	0.39	2.2 *5*
31	0.3	<2
34 - 35	0.18	0.51 *23*
40	0.52	1.3 *3*
42	0.016	∼0.04
43	0.20	<0.9
46 - 52	0.20	0.41 *18*
53	0.19	∼0.4
56 - 65	0.07	0.12 *5*
66	0.47	0.71 *15*
67 - 88	0.60	0.82 *10*
89	0.63	0.71 *8*
93 - 138	0.523	0.467 *22*
139	0.37	0.27 *3*
140 - 153	0.074	0.050 *11*
154	0.24	0.15 *3*
155 - 204	0.68	0.378 *19*
205 - 231	0.089	0.041 *4*
232	0.210	0.091 *10*
233 - 234	0.00202	0.00086 *6*
235	0.189	0.080 *9*
241 - 251	0.088	0.036 *9*

Atomic Electrons (^{127}Sn)
(continued)

e_{bin}(keV)	⟨e⟩(keV)	e(%)
254	0.214	0.084 *9*
255 - 302	0.267	0.099 *6*
305 - 353	0.076	0.023 *5*
356 - 405	0.15	0.04 *1*
406 - 452	0.25	0.060 *24*
457 - 506	0.43	0.091 *13*
508 - 553	0.14	0.026 *8*
559 - 605	0.088	0.015 *4*
609 - 649	0.023	0.0036 *12*
664 - 709	0.0064	0.0009 *3*
728 - 774	0.009	0.0012 *4*
775	0.41	0.053 *6*
793	0.13	∼0.016
794	0.20	0.025 *5*
801 - 848	0.22	0.027 *6*
855 - 899	0.040	0.0046 *12*
908 - 950	0.09	0.009 *4*
967 - 1014	0.076	0.0077 *17*
1025 - 1065	0.21	0.020 *9*
1084	0.33	0.030 *15*
1088 - 1138	0.11	0.0101 *24*
1141 - 1190	0.013	0.0011 *3*
1207 - 1237	0.0017	0.00014 *5*
1261 - 1311	0.007	0.00056 *21*
1330 - 1368	0.0055	0.00041 *13*
1404 - 1442	0.017	0.0012 *3*
1454 - 1542	0.0028	0.00019 *5*
1554 - 1647	0.024	0.0015 *4*
1662 - 1752	0.0062	0.00036 *7*
1782 - 1812	0.0007	3.9 *18* $\times 10^{-5}$
1907 - 2003	0.030	0.0015 *5*
2063 - 2159	0.0048	0.00023 *5*
2274 - 2359	0.0064	0.00028 *8*
2385 - 2483	0.0027	0.000109 *24*
2509 - 2584	0.0071	0.00028 *8*
2665 - 2695	0.0071	0.00027 *8*
2775 - 2850	0.0065	0.00023 *4*
2876 - 2880	0.00014	4.8 *13* $\times 10^{-6}$

Continuous Radiation (^{127}Sn)
⟨β-⟩=509 keV;⟨IB⟩=1.02 keV

E_{bin}(keV)		⟨ ⟩(keV)	(%)
0 - 10	β-	0.100	1.99
	IB	0.020	
10 - 20	β-	0.298	1.99
	IB	0.019	0.135
20 - 40	β-	1.19	3.96
	IB	0.037	0.129
40 - 100	β-	8.1	11.7
	IB	0.098	0.152
100 - 300	β-	63	32.7
	IB	0.24	0.136
300 - 600	β-	100	23.8
	IB	0.22	0.052
600 - 1300	β-	117	13.4
	IB	0.26	0.031
1300 - 2500	β-	183	10.0
	IB	0.123	0.0075
2500 - 3201	β-	36.0	1.33
	IB	0.0030	0.000110

$^{127}_{50}$Sn(4.13 *3* min)

Mode: β-
Δ: -83499 *25* keV
SpA: 3.58×10^{8} Ci/g
Prod: fission; ^{130}Te(n,α)

Photons (^{127}Sn)

γ_{mode}	γ(keV)	γ(rel)
γ[M1+E2]	491.24 *21*	100
γ	860.3 *10*	
γ	979 *1*	
γ	1095.6 *10*	
γ	1348 *1*	5.3 *14*
γ	1564 *2*	4.4 *18*
γ	1584.5 *10*	

$^{127}_{51}$Sb(3.85 *5* d)

Mode: β-
Δ: -86705 *6* keV
SpA: 2.67×10^{5} Ci/g
Prod: fission; daughter ^{127}Sn; ^{128}Te(γ,p)

Photons (^{127}Sb)
⟨γ⟩=664 *49* keV

γ_{mode}	γ(keV)	γ(%)†
Te L_{ℓ}	3.335	0.0062 *9*
Te L_{η}	3.606	0.0031 *4*
Te L_{α}	3.768	0.171 *22*
Te L_{β}	4.123	0.151 *21*
Te L_{γ}	4.656	0.018 *3*
Te $K_{\alpha2}$	27.202	1.12 *8*
Te $K_{\alpha1}$	27.472	2.10 *15*
Te $K_{\beta1}$'	30.980	0.56 *4*
Te $K_{\beta2}$'	31.877	0.121 *9*
γ[M1]	61.43 *10*	1.38 *11*
γ[M1+E2]	152.6 *6*	
γ[M1]	154.7 *4*	∼0.14
γ E2+30%M1	252.75 *25*	8.2 *3*
γ[M1]	280.7 *4*	0.64 *14*
γ(E2+24%M1)	291.0 *3*	1.94 *11*
γ[E1]	293.1 *5*	∼0.28
γ[M1]	310.4 *3*	0.25 *11*
γ M1+E2	391.9 *4*	0.92 *7*
γ[E2]	412.19 *25*	3.7 *4*
γ[M1]	441.8 *3*	0.7 *3*
γ M1+48%E2	445.7 *3*	4.17 *11*
γ[M1+E2]	451.5 *5*	0.18 *7*
γ	455.9 *5*	∼0.11 ?
γ M1+8.0%E2	473.61 *24*	24.7 *7*
γ[M1]	503.2 *3*	0.74 *25*
γ[E2]	543.7 *3*	2.8 *4*
γ	584.1 *5*	∼0.32
γ(M1+E2)	604.1 *4*	4.27 *11*
γ	624.3 *4*	0.064 *21* ?
γ[M1+E2]	638.3 *4*	0.42 *14*
γ	652.6 *6*	0.35 *7*
γ[M1+E2]	668.0 *4*	0.71 *7*
γ	682.3 *6*	0.53 *25*
γ	685.7 *4*	35.3
γ M1+E2	698.5 *3*	3.49 *7* ?
γ[E2]	722.6 *3*	1.80 *11*
γ	764.0 *8*	∼0.07 ?
γ M1+4.2%E2	784.0 *3*	14.5 *3*
γ[M1+E2]	817.2 *5*	0.39 *18*
γ	820.9 *6*	∼0.21 ?
γ[E2]	925.1 *5*	0.49 *7*
γ(M1+E2)	1141.6 *4*	0.35 *7*
γ	1155.9 *6*	∼0.039 ?
γ M1+E2	1290.8 *5*	0.35 *11*
γ	1378.2 *9*	∼0.07

† 2.3% uncert(syst)

Atomic Electrons (^{127}Sb)

⟨e⟩=6.4 *4* keV

e_{bin}(keV)	⟨e⟩(keV)	e(%)
4 - 27	0.288	4.1 *3*
30	0.99	3.3 *3*
31	0.00119	0.0039 *5*
56	0.226	0.40 *4*
57 - 61	0.087	0.146 *8*
123 - 155	0.037	0.029 *13*
221	0.96	0.434 *19*
248 - 253	0.277	0.111 *7*
259	0.175	0.068 *4*
261 - 310	0.078	0.028 *3*
360 - 408	0.254	0.067 *5*
410 - 412	0.038	0.009 *4*
414	0.179	0.0432 *15*
420 - 441	0.041	0.0094 *11*
442	1.03	0.232 *8*
445 - 474	0.208	0.0444 *23*
498 - 544	0.100	0.0193 *25*
552 - 600	0.140	0.024 *3*
603 - 652	0.049	0.0078 *14*
653	8.6 $\times10^{-6}$	~1 $\times10^{-6}$
654	0.6	~0.09
663 - 698	0.22	0.033 *7*
718 - 732	0.0067	0.00092 *11*
752	0.307	0.0408 *12*
759 - 789	0.058	0.0075 *5*
812 - 821	0.0016	0.00019 *7*
893 - 925	0.0077	0.00085 *11*
1110 - 1156	0.0053	0.00047 *10*
1259 - 1291	0.0042	0.00034 *10*
1346 - 1378	0.0006	~4 $\times10^{-5}$

Continuous Radiation (^{127}Sb)

⟨β-⟩=308 keV;⟨IB⟩=0.30 keV

E_{bin}(keV)		⟨ ⟩(keV)	(%)
0 - 10	β-	0.093	1.86
	IB	0.0147	
10 - 20	β-	0.281	1.87
	IB	0.0140	0.097
20 - 40	β-	1.13	3.76
	IB	0.026	0.091
40 - 100	β-	7.9	11.4
	IB	0.065	0.101
100 - 300	β-	70	35.7
	IB	0.122	0.073
300 - 600	β-	151	35.0
	IB	0.055	0.0140
600 - 1300	β-	77	10.6
	IB	0.0081	0.00117
1300 - 1493	β-	0.330	0.0244
	IB	5.3 $\times10^{-6}$	4.0 $\times10^{-7}$

$^{127}_{52}$Te(9.35 *7* h)

Mode: β-
Δ: -88286 *4* keV
SpA: 2.639 $\times10^{6}$ Ci/g
Prod: ^{126}Te(n,γ); daughter ^{127}Te(109 d); fission

Photons (^{127}Te)

⟨γ⟩=4.8 *4* keV

γ_{mode}	γ(keV)	γ(%)†
γ M1+0.7%E2	57.609 *11*	0.030 *3*
γ E2	145.250 *9*	0.00335 *16*
γ M1+0.7%E2	172.131 *8*	~0.00034
γ M1+21%E2	202.859 *8*	0.0579 *20*
γ M1+4.0%E2	215.07 *6*	0.0387 *17*
γ M1+3.6%E2	360.32 *6*	0.135 *3*
γ E2	374.989 *9*	~0.00023
γ M1+0.7%E2	417.93 *6*	0.99
γ M1+E2	618.4 *3*	0.000129 *20*

† 10% uncert(syst)

Continuous Radiation (^{127}Te)

⟨β-⟩=224 keV;⟨IB⟩=0.164 keV

E_{bin}(keV)		⟨ ⟩(keV)	(%)
0 - 10	β-	0.126	2.52
	IB	0.0112	
10 - 20	β-	0.379	2.53
	IB	0.0104	0.073
20 - 40	β-	1.52	5.06
	IB	0.019	0.066
40 - 100	β-	10.6	15.1
	IB	0.044	0.069
100 - 300	β-	87	44.5
	IB	0.066	0.041
300 - 600	β-	121	29.6
	IB	0.0132	0.0037
600 - 697	β-	4.15	0.66
	IB	3.3 $\times10^{-5}$	5.3 $\times10^{-6}$

$^{127}_{52}$Te(109 *2* d)

Mode: IT(97.6 *2* %), β-(2.4 *2* %)
Δ: -88198 *4* keV
SpA: 9433 Ci/g
Prod: ^{126}Te(n,γ); fission

Photons (^{127}Te)

⟨γ⟩=11.1 *3* keV

γ_{mode}	γ(keV)	γ(%)†
Te L_ℓ	3.335	0.120 *17*
Te L_η	3.606	0.038 *4*
Te L_α	3.768	3.3 *4*
Te L_β	4.150	2.3 *3*
Te L_γ	4.693	0.28 *5*
Te $K_{\alpha2}$	27.202	10.3 *5*
Te $K_{\alpha1}$	27.472	19.2 *8*
I $K_{\alpha2}$	28.317	0.41 *4*
I $K_{\alpha1}$	28.612	0.76 *8*
Te $K_{\beta1}$'	30.980	5.11 *22*
Te $K_{\beta2}$'	31.877	1.11 *5*
I $K_{\beta1}$'	32.278	0.204 *20*
I $K_{\beta2}$'	33.225	0.046 *5*
$\gamma_{\beta-}$M1+0.7%E2	57.609 *11*	0.50 *5*
γ_{IT} M4	88.25 *8*	0.084
$\gamma_{\beta-}$M1+5.1%E2	593.30 *9*	0.00225 *19*
$\gamma_{\beta-}$M1+E2	628.6 *3*	8.4 *19* $\times10^{-5}$
$\gamma_{\beta-}$E2	650.91 *9*	0.00028 *9*
$\gamma_{\beta-}$E2	658.88 *10*	0.0122 *9*

† uncert(syst): 0.20% for IT, 13% for β-

Atomic Electrons (^{127}Te)

⟨e⟩=76.0 *14* keV

e_{bin}(keV)	⟨e⟩(keV)	e(%)
4 - 53	4.7	72 *5*
56	23.1	40.9 *15*
57 - 58	0.033	0.058 *6*
83	12.5	15.0 *6*
84	24.3	29.0 *11*
87	9.0	10.3 *4*
88	2.34	2.65 *10*
560 - 595	8.9 $\times10^{-5}$	1.57 *11* $\times10^{-5}$
618 - 659	0.000325	5.2 *4* $\times10^{-5}$

Continuous Radiation (^{127}Te)

⟨β-⟩=6.1 keV;⟨IB⟩=0.0049 keV

E_{bin}(keV)		⟨ ⟩(keV)	(%)
0 - 10	β-	0.00303	0.061
	IB	0.00030	
10 - 20	β-	0.0088	0.059
	IB	0.00028	0.0020
20 - 40	β-	0.0339	0.113
	IB	0.00052	0.0018
40 - 100	β-	0.222	0.318
	IB	0.00123	0.0019
100 - 300	β-	1.85	0.94
	IB	0.0020	0.00124
300 - 600	β-	3.61	0.86
	IB	0.00053	0.000145
600 - 728	β-	0.330	0.052
	IB	4.2 $\times10^{-6}$	6.8 $\times10^{-7}$

$^{127}_{53}$I (stable)

Δ: -88984 *4* keV
%: 100

$^{127}_{54}$Xe(36.41 *2* d)

Mode: ϵ
Δ: -88323 *6* keV
SpA: 2.8238 $\times10^{4}$ Ci/g
Prod: ^{127}I(p,n); ^{126}Xe(n,γ)

Photons (^{127}Xe)

⟨γ⟩=271 *8* keV

γ_{mode}	γ(keV)	γ(%)†
I L_ℓ	3.485	0.146 *20*
I L_η	3.780	0.070 *7*
I L_α	3.937	4.1 *5*
I L_β	4.320	3.7 *5*
I L_γ	4.895	0.47 *7*
I $K_{\alpha2}$	28.317	25.3 *8*
I $K_{\alpha1}$	28.612	46.9 *15*
I $K_{\beta1}$'	32.278	12.6 *4*
I $K_{\beta2}$'	33.225	2.86 *10*
γ M1+0.7%E2	57.609 *11*	1.13 *6*
γ E2	145.250 *9*	3.94 *11*
γ M1+0.7%E2	172.131 *8*	23.5 *7*
γ M1+21%E2	202.859 *8*	68
γ E2	374.989 *9*	15.9 *5*
γ M1+E2	618.4 *3*	0.0131 *8*

† 5.0% uncert(syst)

Atomic Electrons (^{127}Xe)

⟨e⟩=30.7 *8* keV

e_{bin}(keV)	⟨e⟩(keV)	e(%)
5	3.8	80 *8*
23	0.80	3.4 *4*
24	1.92	7.9 *7*
27 - 58	1.39	4.40 *25*
112	1.58	1.41 *5*
139	4.73	3.40 *12*
140 - 168	1.39	0.904 *18*
170	11.3	6.7 *4*
171 - 172	0.192	0.112 *3*
198	1.96	0.99 *7*
202 - 203	0.51	0.251 *15*
342	0.91	0.266 *10*
370 - 375	0.192	0.0519 *16*
585 - 618	0.00043	7.3 *9* $\times 10^{-5}$

Continuous Radiation (^{127}Xe)

⟨IB⟩=0.074 keV

E_{bin}(keV)		⟨ ⟩(keV)	(%)
10 - 20	IB	0.00060	0.0035
20 - 40	IB	0.059	0.21
40 - 100	IB	0.0040	0.0069
100 - 300	IB	0.0079	0.0045
300 - 457	IB	0.00136	0.00041

$^{127}_{54}$Xe(1.153 *15* min)

Mode: IT

Δ: -88026 *6* keV

SpA: 1.277×10^{9} Ci/g

Prod: ^{127}I(p,n); daughter ^{127}Cs; protons on Ce

Photons (^{127}Xe)

⟨γ⟩=168 *4* keV

γ_{mode}	γ(keV)	γ(%)†
Xe L_ℓ	3.634	0.119 *17*
Xe L_η	3.955	0.064 *7*
Xe L_α	4.104	3.3 *4*
Xe L_β	4.512	3.1 *4*
Xe L_γ	5.095	0.39 *5*
Xe $K_{\alpha2}$	29.461	15.7 *6*
Xe $K_{\alpha1}$	29.782	29.1 *12*
Xe $K_{\beta1}$'	33.606	7.9 *3*
Xe $K_{\beta2}$'	34.606	1.86 *8*
γ M1+0.8%E2	124.76 *11*	69.0
γ E3	172.5 *3*	38.0 *21*

† 4.3% uncert(syst)

Atomic Electrons (^{127}Xe)

⟨e⟩=129 *3* keV

e_{bin}(keV)	⟨e⟩(keV)	e(%)
5	3.1	63 *7*
24	0.48	2.00 *21*
25	0.61	2.5 *3*
28	0.169	0.60 *6*
29	0.36	1.25 *13*
30	0.048	0.162 *17*
32	0.0184	0.057 *6*

Atomic Electrons (^{127}Xe)
(continued)

e_{bin}(keV)	⟨e⟩(keV)	e(%)
33	0.042	0.127 *13*
90	24.1	26.7 *6*
119	3.87	3.24 *7*
120	0.42	0.35 *3*
124	0.91	0.733 *22*
125	0.230	0.184 *5*
138	47.5	34.5 *20*
167	22.4	13.4 *8*
168	14.3	8.5 *5*
171	5.0	2.90 *17*
172	5.3	3.07 *18*

$^{127}_{55}$Cs(6.25 *10* h)

Mode: ε

Δ: -86231 *8* keV

SpA: 3.95×10^{6} Ci/g

Prod: ^{127}I(α,4n); protons on Ce

Photons (^{127}Cs)

⟨γ⟩=363 *24* keV

γ_{mode}	γ(keV)	γ(%)†
Xe L_ℓ	3.634	0.139 *19*
Xe L_η	3.955	0.063 *7*
Xe L_α	4.104	3.9 *4*
Xe L_β	4.522	3.4 *4*
Xe L_γ	5.125	0.44 *6*
Xe $K_{\alpha2}$	29.461	22.9 *8*
Xe $K_{\alpha1}$	29.782	42.6 *14*
Xe $K_{\beta1}$'	33.606	11.5 *4*
Xe $K_{\beta2}$'	34.606	2.72 *10*
γ [M1]	90.44 *16*	~0.18
γ M1+0.8%E2	124.76 *11*	15.6 *12*
γ [M1]	175.24 *16*	<0.012
γ M1,E2	196.66 *15*	0.210 *18*
γ	217.9 *6*	~0.018
γ [M1]	265.69 *17*	0.088 *12*
γ M1,E2	287.10 *13*	3.4 *3*
γ M1,E2	321.42 *13*	1.24 *6*
γ [M1]	344.11 *21*	0.058 *6*
γ M1,E2	411.86 *13*	58
γ M1,E2	462.34 *13*	4.2 *3*
γ [M1]	519.35 *21*	~0.012
γ [M1]	555.7 *3*	0.149 *12*
γ M1,E2	587.11 *13*	3.50 *23*
γ	595.6 *5*	0.015 *6*
γ [M1]	806.45 *20*	0.339 *23*
γ	823.0 *3*	0.098 *8*
γ	875.7 *3*	0.030 *3*
γ	894.2 *4*	<0.013 ?
γ	894.64 *22*	<0.013 ?
γ	931.21 *19*	0.315 *18*
γ	947.50 *24*	0.0029 *12*
γ	965.1 *5*	0.0088 *18*
γ	985.09 *22*	0.055 *6*
γ	990.9 *3*	0.0245 *23*
γ	995.71 *24*	0.0257 *23*
γ	1025.1 *5*	0.0111 *18*
γ	1072.4 *3*	0.0315 *23*
γ	1081.4 *3*	0.0146 *18*
γ	1111.3 *6*	0.0111 *18*
γ	1159.9 *4*	0.032 *4*
γ	1170.95 *24*	0.041 *5*
γ	1181.74 *22*	0.079 *9*
γ	1197.2 *3*	0.169 *18*
γ	1213.18 *25*	0.033 *4*
γ	1237.0 *4*	0.0099 *12*
γ	1261.40 *22*	0.071 *6*
γ	1290.6 *5*	0.0117 *23*
γ	1306.5 *2*	0.152 *12*

Photons (^{127}Cs)
(continued)

γ_{mode}	γ(keV)	γ(%)†
γ	1341.3 *5*	0.0076 *12*
γ	1362.8 *3*	0.0134 *18*
γ	1368.7 *5*	0.054 *5*
γ	1394.7 *3*	0.0111 *18*
γ	1402.8 *3*	0.040 *4*
γ	1409.84 *23*	<0.09 ?
γ	1410.1 *3*	<0.09 ?
γ	1418.8 *4*	0.046 *4*
γ	1431.6 *6*	0.0029 *6*
γ	1453.3 *3*	0.016 *3*
γ	1485.1 *3*	0.0175 *23*
γ	1534.60 *23*	0.066 *6*
γ	1558.5 *4*	0.0041 *6*
γ	1582.81 *22*	0.032 *3*
γ	1649.9 *3*	0.0070 *6*
γ	1681.8 *3*	0.0164 *18*
γ	1716.7 *5*	0.0169 *18*
γ	1774.7 *3*	0.0117 *12*
γ	1806.5 *3*	0.0016 *4*
γ	1830.7 *4*	0.0020 *5*
γ	1973.4 *5*	0.0105 *12*

† 16% uncert(syst)

Atomic Electrons (^{127}Cs)

⟨e⟩=17.6 *8* keV

e_{bin}(keV)	⟨e⟩(keV)	e(%)
5	3.5	71 *7*
24	0.70	2.9 *3*
25	0.89	3.6 *4*
28	0.248	0.88 *9*
29	0.53	1.83 *19*
30 - 56	0.25	0.68 *9*
85 - 89	0.023	0.027 *11*
90	5.4	6.0 *5*
119	0.87	0.73 *6*
120 - 162	0.394	0.312 *17*
170 - 218	0.014	0.0072 *19*
231 - 266	0.35	0.140 *14*
282 - 321	0.202	0.070 *7*
339 - 344	0.00080	0.000236 *21*
377	3.2	0.85 *13*
406 - 428	0.80	0.194 *24*
457 - 485	0.036	0.0078 *13*
514 - 561	0.120	0.022 *3*
582 - 596	0.020	0.0034 *6*
772 - 818	0.0112	0.00143 *14*
822 - 871	0.0007	9 *4* $\times 10^{-5}$
875 - 913	0.0042	~0.00047
926 - 965	0.0021	0.00022 *6*
980 - 1025	0.00036	3.7 *10* $\times 10^{-5}$
1038 - 1081	0.00071	6.8 *20* $\times 10^{-5}$
1106 - 1155	0.0016	0.00014 *4*
1159 - 1208	0.0026	0.00022 *7*
1212 - 1261	0.0009	7 *3* $\times 10^{-5}$
1272 - 1307	0.0017	0.00013 *5*
1328 - 1375	0.0019	0.00014 *6*
1384 - 1432	0.00077	5.5 *15* $\times 10^{-5}$
1448 - 1534	0.00078	5.3 *16* $\times 10^{-5}$
1548 - 1647	0.00044	2.8 *8* $\times 10^{-5}$
1649 - 1740	0.00023	1.3 *4* $\times 10^{-5}$
1769 - 1830	3.9 $\times 10^{-5}$	2.2 *5* $\times 10^{-6}$
1939 - 1972	7.2 $\times 10^{-5}$	3.7 *13* $\times 10^{-6}$

Continuous Radiation (^{127}Cs)

⟨β+⟩=14.2 keV;⟨IB⟩=1.19 keV

E_{bin}(keV)		⟨ ⟩(keV)	(%)
0 - 10	β+	1.45×10^{-6}	1.74×10^{-5}
	IB	0.00095	
10 - 20	β+	5.7×10^{-5}	0.000346
	IB	0.00111	0.0073
20 - 40	β+	0.00153	0.00469
	IB	0.064	0.22
40 - 100	β+	0.067	0.087
	IB	0.0087	0.0142
100 - 300	β+	2.38	1.11
	IB	0.045	0.022
300 - 600	β+	7.6	1.77
	IB	0.20	0.044
600 - 1300	β+	4.11	0.57
	IB	0.72	0.081
1300 - 2092	IB	0.150	0.0103
	Σβ+		3.5

$^{127}_{56}$Ba(12.7 *4* min)

Mode: ϵ

Δ: -82780 *100* keV

SpA: 1.17×10^8 Ci/g

Prod: ^{16}O on In; ^{14}N on In; ^{12}C on Sn; ^{16}O on Sn; ^{133}Cs(d,8n); daughter ^{127}La

Photons (^{127}Ba)

⟨γ⟩=173 *5* keV

γ_{mode}	γ(keV)	γ(%)†
Cs L_ℓ	3.795	0.107 *16*
Cs L_η	4.142	0.050 *6*
Cs L_α	4.285	3.0 *4*
Cs L_β	4.731	2.6 *4*
Cs L_γ	5.355	0.34 *5*
Cs $K_{\alpha2}$	30.625	14.7 *7*
Cs $K_{\alpha1}$	30.973	27.3 *13*
Cs $K_{\beta1}$'	34.967	7.5 *3*
Cs $K_{\beta2}$'	36.006	1.82 *9*
γ E2	66.23 *15*	2.12 *21*
γ M1	72.67 *18*	0.75 *6*
γ M1+E2	114.70 *18*	9.3 *4*
γ (M1)	138.90 *18*	0.3 *1*
γ M1+E2	180.93 *17*	12.4
γ [M1]	428.68 *24*	0.27 *5*
γ	440.7 *6*	~0.050
γ	450.8 *6*	0.087 *25*
γ	523.5 *6*	0.43 *10*
γ	532.1 *4*	0.050 *12*
γ	567.58 *22*	0.35 *4*
γ	574.1 *3*	0.087 *25*
γ	578.0 *3*	0.64 *6*
γ	619.0 *10*	<0.025
γ	621.6 *6*	~0.025
γ	625.5 *7*	0.050 *12*
γ	646.8 *4*	0.050 *12*
γ	691.5 *4*	0.050 *12*
γ	713.0 *4*	<0.025
γ	872.5 *3*	0.099 *12*
γ	1012.1 *3*	0.112 *12*
γ	1020.0 *3*	0.161 *12*
γ	1062.0 *3*	0.050 *12*
γ	1084.8 *3*	0.43 *4*
γ	1108.3 *3*	0.112 *25*
γ	1134.7 *3*	<0.025
γ	1150.4 *3*	0.19 *3*
γ	1200.92 *25*	1.61 *19*
γ	1223.0 *3*	~0.025
γ	1289.3 *3*	0.124 *12*
γ	1385.32 *23*	0.099 *25*
γ	1437.1 *3*	<0.025
γ	1448.6 *3*	0.050 *12*

Photons (^{127}Ba) (continued)

γ_{mode}	γ(keV)	γ(%)†
γ	1500.02 *21*	0.37 *4*
γ	1511.7 *4*	0.124 *12*
γ	1522.1 *3*	0.074 *12*
γ	1566.25 *21*	0.38 *4*
γ	1576.2 *6*	0.062 *12*
γ	1618.0 *3*	0.25 *4*
γ	1697.0 *8*	~0.050
γ	1753.52 *23*	0.25 *4*
γ	1800.6 *3*	0.074 *25*
γ	1842.6 *3*	0.112 *25*
γ	1915.3 *3*	0.11 *4*
γ	1920.6 *8*	0.037 *12*
γ	1950.8 *3*	0.136 *25*
γ	1962.8 *6*	0.062 *12*
γ	1981.52 *23*	0.223 *25*
γ	1991.9 *6*	0.124 *12*
γ	2028.2 *7*	0.062 *12*
γ	2057.4 *4*	0.124 *25*
γ	2074.7 *5*	0.112 *25*
γ	2089.7 *3*	0.161 *25*
γ	2099.4 *4*	0.136 *25*
γ	2140.16 *25*	0.037 *12*
γ	2172.1 *4*	0.136 *25*
γ	2182.19 *22*	0.223 *25*
γ	2189.4 *5*	0.037 *12*
γ	2222.4 *7*	0.062 *12*
γ	2238.3 *4*	0.050 *12*
γ	2321.09 *22*	0.149 *25*
γ	2467.8 *7*	0.149 *25*

† 9.7% uncert(syst)

Atomic Electrons (^{127}Ba)

⟨e⟩=24.5 *15* keV

e_{bin}(keV)	⟨e⟩(keV)	e(%)
5	2.4	46 *6*
6 - 29	1.40	9.0 *6*
30	2.8	9.4 *10*
31 - 37	0.65	1.81 *13*
61	4.5	7.4 *8*
65	1.05	1.60 *16*
66 - 73	0.43	0.65 *4*
79	4.8	6.1 *12*
103	0.10	0.09 *3*
109	1.3	1.1 *5*
110	0.6	~0.5
113 - 139	0.5	0.46 *23*
145	3.0	2.1 *3*
175 - 181	0.9	0.50 *15*
393 - 441	0.025	0.0062 *10*
445 - 488	0.013	~0.0027
496 - 542	0.032	0.0060 *21*
562 - 611	0.0082	0.0014 *3*
613 - 656	0.0016	0.00024 *10*
677 - 713	0.0005	$\sim7\times10^{-5}$
836 - 872	0.0018	0.00021 *10*
976 - 1020	0.0040	0.00041 *14*
1026 - 1072	0.007	0.0007 *3*
1079 - 1115	0.0034	0.00030 *14*
1129 - 1165	0.017	~0.0015
1187 - 1223	0.0030	0.00025 *9*
1253 - 1289	0.0014	0.00011 *5*
1349 - 1385	0.0010	8 *4* $\times10^{-5}$
1401 - 1449	0.00062	4.4 *18* $\times10^{-5}$
1464 - 1561	0.0094	0.00062 *14*
1565 - 1661	0.0027	0.00017 *6*
1691 - 1765	0.0026	0.00015 *5*
1795 - 1885	0.0019	0.00010 *3*
1910 - 1992	0.0045	0.00023 *4*
2021 - 2104	0.0040	0.00019 *4*
2134 - 2233	0.0034	0.00016 *3*
2237 - 2320	0.0009	4.1 *14* $\times10^{-5}$
2432 - 2467	0.0009	3.7 *12* $\times10^{-5}$

Continuous Radiation (^{127}Ba)

⟨β+⟩=575 keV;⟨IB⟩=4.9 keV

E_{bin}(keV)		⟨ ⟩(keV)	(%)
0 - 10	β+	1.60×10^{-6}	1.91×10^{-5}
	IB	0.022	
10 - 20	β+	6.6×10^{-5}	0.000395
	IB	0.021	0.147
20 - 40	β+	0.00185	0.0056
	IB	0.074	0.25
40 - 100	β+	0.091	0.117
	IB	0.120	0.19
100 - 300	β+	4.70	2.12
	IB	0.34	0.19
300 - 600	β+	36.4	7.8
	IB	0.47	0.108
600 - 1300	β+	257	26.9
	IB	1.27	0.138
1300 - 2500	β+	277	17.2
	IB	2.2	0.122
2500 - 3450	IB	0.39	0.0144
	Σβ+		54

$^{127}_{57}$La(3.8 *5* min)

Mode: ϵ

Δ: -77980 *220* keV syst

SpA: 3.9×10^8 Ci/g

Prod: ^{12}C on Sb; ^{16}O on In; ^{32}S on ^{96}Ru; ^{32}S on ^{98}Ru

Photons (^{127}La)

γ_{mode}	γ(keV)	γ(rel)
γ	25.0 *5*	24 *5*
γ	56.2 *5*	100 *10*

$^{127}_{58}$Ce(32 *4* s)

Mode: ϵ

SpA: 2.7×10^9 Ci/g

Prod: ^{32}S on ^{96}Ru; ^{32}S on ^{98}Ru

Photons (^{127}Ce)

γ_{mode}	γ(keV)
γ	58.4 *5*

$^{127}_{60}$Nd(1.9 *4* s)

Mode: ϵ, ϵp

SpA: 3.9×10^{10} Ci/g

Prod: ^{92}Mo(^{40}Ca,αn)

ϵp: 2000-6400 range

A = 128

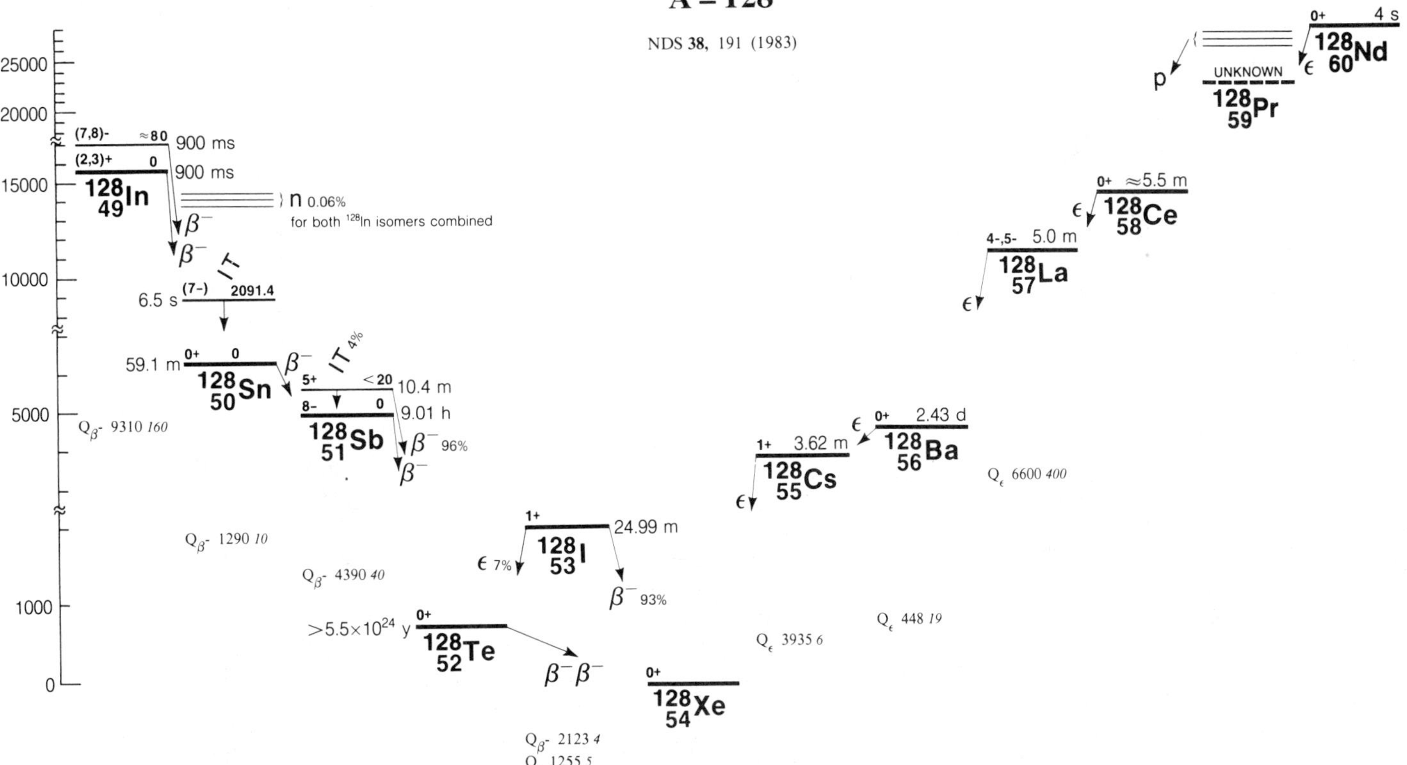

$^{128}_{49}$In(900 *100* ms)

Mode: β-, β-n(0.059 *8* %, includes other ^{128}In isomer)
Δ: -74000 *170* keV
SpA: 6.8×10^{10} Ci/g
Prod: fission

Photons (^{128}In)

⟨γ⟩=3072 *90* keV

γ_{mode}	γ(keV)	γ(%)†
γ	310.57 *20*	0.18 *4*
γ	384.12 *25*	0.15 *4*
γ	449.76 *6*	0.55 *7*
γ	468.1 *3*	0.11 *4*
γ	474.59 *15*	0.26 *6*
γ	538.27 *4*	1.20 *8*
γ	583.4 *3*	0.24 *6*
γ	704.15 *15*	0.42 *6*
γ	760.3 *3*	0.22 *7*
γ	886.97 *15*	0.45 *10*
γ	935.31 *4*	8.0 *5*
γ	1045.31 *12*	0.36 *10*
γ	1082.28 *20*	0.42 *7*
γ	1089.60 *6*	7.4 *5*
γ	1105.31 *9*	1.5 *1*
γ	1123.22 *15*	0.48 *6*
γ	1129.70 *17*	0.26 *6*
γ[E2]	1168.83 *4*	50 *5*
γ	1236.55 *25*	0.31 *7*
γ	1241.14 *9*	0.9 *1*
γ	1281.51 *13*	0.59 *7*
γ	1409.88 *8*	1.1 *1*
γ	1464.40 *10*	2.5 *2*
γ	1473.57 *5*	1.7 *1*
γ	1514.88 *25*	0.42 *1*
γ	1587.82 *10*	2.4 *2*

Photons (^{128}In) (continued)

γ_{mode}	γ(keV)	γ(%)†
γ	1593.6 *3*	0.34 *7*
γ	1678.5 *3*	0.36 *7*
γ	1696.53 *8*	1.3 *1*
γ	1739.36 *8*	2.0 *1*
γ	1783.65 *9*	1.5 *1*
γ	1816.72 *9*	2.4 *2*
γ	1893.3 *3*	0.4 *1*
γ	1923.33 *8*	1.0 *1*
γ	1945.85 *12*	0.38 *10*
γ	1961.56 *8*	1.2 *1*
γ	1967.9 *4*	0.31 *10*
γ	2104.13 *5*	6.5 *4*
γ	2205.3 *5*	0.37 *10*
γ	2258.42 *6*	3.1 *2*
γ	2351.04 *11*	1.4 *1*
γ	2578.70 *9*	1.0 *1*
γ	2786.12 *9*	0.7 *1*
γ	2906.30 *9*	0.72 *10*
γ	3051.14 *9*	1.7 *2*
γ	3058.4 *3*	0.62 *20*
γ	3128.91 *14*	1.1 *1*
γ	3225.8 *3*	1.0 *1*
γ	3519.84 *11*	16.6 *15*
γ	3886.31 *14*	3.6 *3*
γ	3954.92 *9*	3.9 *3*
γ	4038.11 *14*	1.7 *2*
γ	4227.2 *3*	0.29 *7*
γ	4297.71 *14*	11.8 *8*
γ	4509.8 *10*	0.20 *7*

† 10% uncert(syst)

Continuous Radiation (^{128}In)

⟨β-⟩=2848 keV;⟨IB⟩=12.0 keV

E_{bin}(keV)		⟨ ⟩(keV)	(%)
0 - 10	β-	0.00152	0.0302
	IB	0.070	
10 - 20	β-	0.00467	0.0311
	IB	0.070	0.48
20 - 40	β-	0.0196	0.065
	IB	0.138	0.48
40 - 100	β-	0.155	0.219
	IB	0.41	0.62
100 - 300	β-	2.14	1.03
	IB	1.26	0.70
300 - 600	β-	11.6	2.51
	IB	1.66	0.39
600 - 1300	β-	104	10.5
	IB	3.0	0.34
1300 - 2500	β-	563	29.1
	IB	3.2	0.18
2500 - 5000	β-	1731	49.0
	IB	2.1	0.064
5000 - 7500	β-	432	7.6
	IB	0.136	0.0025
7500 - 8221	β-	5.4	0.071
	IB	0.000161	2.1×10^{-6}

$^{128}_{49}$In(900 *100* ms)

Mode: β-, β-n(0.059 *8* %, includes other ^{128}In isomer)
Δ: ~-73920 keV
SpA: 6.8×10^{10} Ci/g
Prod: fission

Photons (^{128}In)

⟨γ⟩=1571 45 keV

γ_{mode}	γ(keV)	γ(%)†
Sn $K_{\alpha2}$	25.044	0.8 1
Sn $K_{\alpha1}$	25.271	1.50 20
Sn $K_{\beta1}$'	28.474	0.39 5
Sn $K_{\beta2}$'	29.275	0.082 11
γ[M1]	79.23 16	1.8 4
γE3	91.13 9	*
γE1	120.44 6	11.1 10
γ[M1]	207.23 14	0.46 10
γ[M1]	257.07 12	4.4 3
γ	310.38 20	0.43 10
γ	321.12 8	10.5 7
γ	383.93 25	0.36 10
γ[M1]	426.09 9	1.6 2
γ	457.59 8	2.1 2
γ	467.9 3	0.26 10
γ	474.40 15	0.63 15
γ	546.49 25	0.60 15
γ	609.50 13	0.87 15
γ	703.96 15	1.0 1
γ	760.1 3	0.53 15
γ	762.96 13	1.1 2
γ	811.7 3	0.87 20
γ[E2]	831.44 6	*
γ	886.78 15	1.1 2
γ	904.19 10	3.0 3
γ	1054.77 10	5.8 5
γ	1061.29 18	1.5 2
γ	1067.09 13	1.3 2
γ	1082.09 20	1.0 2
γ	1123.03 15	1.2 2
γ[E2]	1168.83 4	*
γ	1236.36 25	0.8 2
γ	1262.00 15	0.9 2
γ	1264.51 25	1.4 2
γ	1356.26 18	1.4 2
γ	1514.69 25	1.0 2
γ	1573.40 18	0.9 2
γ	1593.4 3	0.8 2
γ	1678.3 3	0.9 2
γ	1779.87 12	3.4 3
γ	1866.94 12	32.3 20
γ	1893.1 3	1.0 2
γ	1967.7 4	0.8 2
γ	1973.74 12	19.5 10
γ	2122.01 12	3.8 3
γ	2205.1 5	0.9 2

† 10% uncert(syst)
* with ^{128}Sn(6.5 s)

Atomic Electrons (^{128}In)

⟨e⟩=4.3 4 keV

e_{bin}(keV)	⟨e⟩(keV)	e(%)
4	0.107	2.6 4
20 - 28	0.098	0.44 5
50	0.86	1.7 4
75	0.17	0.22 5
78 - 79	0.042	0.054 10
91	0.94	1.03 10
116	0.134	0.115 11
117 - 120	0.055	0.046 3
178 - 207	0.064	0.035 7
228	0.38	0.166 13
253 - 281	0.085	0.033 5
292	0.5	~0.16
306 - 355	0.11	0.033 16
379 - 428	0.13	0.032 7
439 - 474	0.034	0.0075 24
517 - 546	0.013	~0.0026
580 - 610	0.017	0.0029 14
675 - 704	0.016	0.0024 11
731 - 763	0.024	0.0032 12
782 - 812	0.012	0.0015 7
858 - 904	0.049	0.0056 19
1026 - 1067	0.09	0.0090 25
1078 - 1123	0.012	0.0011 4
1207 - 1236	0.023	0.0019 5
1258 - 1264	0.0025	0.00019 6
1327 - 1356	0.011	0.0008 3
1485 - 1573	0.017	0.0011 3
1589 - 1678	0.0061	0.00037 14
1751 - 1779	0.020	0.0011 4
1838	0.16	0.009 3
1862 - 1945	0.12	0.0063 18
1963 - 1973	0.013	0.00068 21
2093 - 2176	0.023	0.0011 3
2201 - 2204	0.00054	2.5 9 $\times10^{-5}$

Continuous Radiation (^{128}In)

⟨β-⟩=2616 keV;⟨IB⟩=10.5 keV

E_{bin}(keV)		⟨ ⟩(keV)	(%)
0 - 10	β-	0.00175	0.0349
	IB	0.067	
10 - 20	β-	0.0054	0.0359
	IB	0.067	0.46
20 - 40	β-	0.0226	0.075
	IB	0.132	0.46
40 - 100	β-	0.179	0.253
	IB	0.39	0.59
100 - 300	β-	2.46	1.18
	IB	1.20	0.66
300 - 600	β-	13.3	2.87
	IB	1.56	0.36
600 - 1300	β-	118	11.9
	IB	2.8	0.31
1300 - 2500	β-	620	32.2
	IB	2.8	0.158
2500 - 5000	β-	1652	47.7
	IB	1.50	0.048
5000 - 7299	β-	211	3.79
	IB	0.046	0.00085

$^{128}_{50}$Sn(59.1 5 min)

Mode: β-
Δ: -83310 40 keV
SpA: 2.485×10^7 Ci/g
Prod: fission

Photons (^{128}Sn)

⟨γ⟩=602 22 keV

γ_{mode}	γ(keV)	γ(%)†
Sb L_ℓ	3.189	0.20 3
Sb L_η	3.437	0.101 11
Sb L_α	3.604	5.5 6
Sb L_β	3.930	4.9 6
Sb L_γ	4.435	0.58 8
Sb $K_{\alpha2}$	26.111	38.6 15
Sb $K_{\alpha1}$	26.359	72 3
Sb $K_{\beta1}$'	29.712	18.9 8
Sb $K_{\beta2}$'	30.561	4.03 17
γ M1	32.5 3	4.01 24
γ M1	46.2 4	13.0 6
γ M1	75.38 25	27.7 18
γ	81.6 3	0.18 6
γ	116.6 3	0.18 6
γ[M1]	153.3 3	6.5 6
γ	404.83 25	5.9 6
γ	437.4 3	4.1 6
γ[M1]	482.71 24	59
γ[E2]	558.09 23	16.5 18
γ	680.98 17	15.9 18

† 10% uncert(syst)

Atomic Electrons (^{128}Sn)

⟨e⟩=53.2 14 keV

e_{bin}(keV)	⟨e⟩(keV)	e(%)
2	1.04	51 3
4	4.7	111 12
5	0.67	14.2 15
16	10.3	66 3
21	0.71	3.3 4
22	2.24	10.2 11
25 - 26	1.48	5.8 4
28	2.08	7.5 5
29 - 33	0.73	2.35 12
42	3.56	8.6 4
45	16.0	35.5 23
46 - 51	0.30	0.62 19
71	3.10	4.4 3
74 - 117	0.98	1.28 10
123	1.31	1.07 11
149 - 153	0.255	0.171 14
374 - 407	0.33	0.09 3
433 - 437	0.022	~0.005
452	2.22	0.49 3
478 - 483	0.361	0.075 4
528 - 558	0.51	0.095 9
650 - 681	0.29	0.044 21

Continuous Radiation (^{128}Sn)

⟨β-⟩=199 keV;⟨IB⟩=0.133 keV

E_{bin}(keV)		⟨ ⟩(keV)	(%)
0 - 10	β-	0.148	2.96
	IB	0.0100	
10 - 20	β-	0.443	2.96
	IB	0.0093	0.065
20 - 40	β-	1.77	5.9
	IB	0.0167	0.058
40 - 100	β-	12.1	17.3
	IB	0.038	0.059
100 - 300	β-	91	47.3
	IB	0.051	0.032
300 - 600	β-	93	23.5
	IB	0.0080	0.0022
600 - 655	β-	0.77	0.126
	IB	2.5×10^{-6}	4.1×10^{-7}

$^{128}_{50}$Sn(6.5 5 s)

Mode: IT
Δ: -81219 40 keV
SpA: 1.286×10^{10} Ci/g
Prod: fission

Photons (^{128}Sn)

$\langle\gamma\rangle$=2011 *14* keV

γ_{mode}	γ(keV)	γ(%)
Sn L_ℓ	3.045	0.096 *14*
Sn L_η	3.272	0.063 *7*
Sn L_α	3.443	2.7 *3*
Sn L_β	3.731	2.4 *3*
Sn L_γ	4.163	0.25 *3*
Sn $K_{\alpha2}$	25.044	8.7 *4*
Sn $K_{\alpha1}$	25.271	16.2 *7*
Sn $K_{\beta1}$'	28.474	4.24 *17*
Sn $K_{\beta2}$'	29.275	0.89 *4*
γ E3	91.13 *9*	3.6 *1*
γ [E2]	831.44 *6*	100
γ [E2]	1168.83 *4*	100

Atomic Electrons (^{128}Sn)

$\langle e\rangle$=82.2 *17* keV

e_{bin}(keV)	$\langle e\rangle$(keV)	e(%)
4	2.7	68 *7*
20	0.074	0.36 *4*
21	0.61	2.9 *3*
24	0.164	0.68 *7*
25	0.175	0.71 *8*
27	0.0030	0.0109 *12*
28	0.031	0.112 *12*
62	21.3	34.5 *12*
87	43.0	49.4 *17*
90	9.5	10.5 *4*
91	2.00	2.20 *8*
802	1.37	0.171 *7*
827	0.170	0.0206 *9*
828	0.0078	0.00094 *4*
831	0.0423	0.00509 *21*
1140	0.92	0.080 *3*
1164	0.106	0.0091 *4*
1165	0.0081	0.00069 *3*
1168	0.0224	0.00191 *8*
1169	0.00454	0.000388 *16*

$^{128}_{51}$Sb(9.01 *3* h)

Mode: β-

Δ: -84600 *40* keV

SpA: 2.717×10^6 Ci/g

Prod: fission; ^{128}Te(n,p); ^{130}Te(d,α)

Photons (^{128}Sb)

$\langle\gamma\rangle$=3108 *72* keV

γ_{mode}	γ(keV)	γ(%)†
Te L_ℓ	3.335	0.0061 *12*
Te L_η	3.606	0.0031 *5*
Te L_α	3.768	0.17 *3*
Te L_β	4.123	0.15 *3*
Te L_γ	4.653	0.018 *4*
Te $K_{\alpha2}$	27.202	1.09 *14*
Te $K_{\alpha1}$	27.472	2.0 *3*
Te $K_{\beta1}$'	30.980	0.54 *7*
Te $K_{\beta2}$'	31.877	0.118 *16*
γ [M1]	102.90 *18*	0.4 *1*
γ	118.5 *3*	0.6 *1*
γ [M1]	152.78 *22*	0.5 *1*
γ [E2]	204.35 *12*	1.0 *2*
γ [M1]	214.90 *16*	1.0 *2*
γ	227.36 *18*	1.5 *3*

Photons (^{128}Sb)
(continued)

γ_{mode}	γ(keV)	γ(%)†
γ	235.06 *10*	0.3 *1*
γ	249.81 *14*	0.6 *1* ?
γ	278.4 *3*	0.6 *1*
γ E2	314.15 *4*	61 *3*
γ	317.80 *15*	3 *1*
γ [E1]	322.16 *9*	3 *1*
γ	356.64 *13*	1.5 *3*
γ	366.11 *24*	1.5 *3*
γ	404.55 *25*	1.0 *2*
γ	445.85 *21*	1.5 *3*
γ	454.6 *3*	1.5 *3*
γ	459.37 *24*	1.5 *3*
γ E1+M2	526.51 *9*	45 *2*
γ	582.91 *19*	1.0 *2*
γ	594.29 *8*	1.01 *13*
γ	603.02 *25*	1.7 *3*
γ M1,E2	628.77 *9*	31 *2*
γ E1+M2	636.31 *8*	36 *2*
γ M1,E2	654.30 *15*	17 *1*
γ	667.04 *23*	2.5 *3*
γ	683.91 *24*	3 *1*
γ	692.90 *21*	~2
γ	727.84 *24*	4 *1*
γ E2	743.41 *4*	100
γ E2	754.06 *4*	100
γ	773.87 *23*	1.5 *3*
γ	802.8 *3*	1.2 *2*
γ E1	813.61 *15*	13 *2*
γ	835.04 *22*	<2
γ	844.09 *14*	<4
γ	845.68 *20*	2.5 *3*
γ	860.8 *3*	0.4 *1*
γ	878.1 *4*	3.5 *4*
γ	908.43 *8*	0.72 *18*
γ	972.10 *18*	<2
γ	1047.6 *4*	3.5 *4*
γ	1078.72 *17*	~2
γ	1112.76 *22*	~2
γ	1129.7 *4*	0.8 *2*
γ	1158.24 *14*	1.5 *3*
γ	1181.63 *20*	4.5 *5*
γ	1250.2 *3*	<2
γ	1259.46 *21*	<2
γ	1340.12 *15*	<2
γ	1378.1 *4*	1.8 *4*
γ	1593.3 *5*	0.5 *1*
γ	1686.33 *20*	0.5 *1*
γ	1708.13 *20*	0.3 *1*
γ	1785.96 *21*	0.4 *1*

† 5.0% uncert(syst)

Atomic Electrons (^{128}Sb)

$\langle e\rangle$=20 *3* keV

e_{bin}(keV)	$\langle e\rangle$(keV)	e(%)
4 - 31	0.29	4.1 *4*
71 - 118	0.49	0.57 *20*
121 - 153	0.132	0.106 *19*
173 - 218	0.62	0.33 *6*
222 - 250	0.09	0.038 *12*
273 - 278	0.008	0.0031 *13*
282	4.79	1.70 *9*
286 - 290	0.21	0.07 *4*
309	0.56	0.183 *10*
310 - 357	0.67	0.212 *19*
361 - 405	0.052	0.014 *6*
414 - 459	0.15	0.036 *11*
495	2.8	~0.6
522 - 562	0.5	~0.10
571 - 594	0.040	0.007 *3*
597	0.79	0.133 *19*
598 - 603	0.006	0.0010 *5*
604	1.6	~0.26
622	0.41	0.066 *10*
624 - 667	0.61	0.10 *3*
679 - 696	0.07	0.011 *5*
712	1.77	0.249 *13*

Atomic Electrons (^{128}Sb)
(continued)

e_{bin}(keV)	$\langle e\rangle$(keV)	e(%)
722	1.74	0.241 *13*
723 - 771	0.643	0.086 *3*
773 - 814	0.17	0.021 *4*
829 - 878	0.075	0.009 *3*
903 - 940	0.013	~0.0014
967 - 1016	0.037	0.0036 *18*
1043 - 1081	0.046	0.0044 *18*
1098 - 1130	0.025	0.0022 *7*
1150 - 1182	0.047	0.0041 *17*
1218 - 1259	0.020	~0.0016
1308 - 1346	0.023	0.0017 *8*
1373 - 1378	0.0021	0.00015 *6*
1561 - 1655	0.0067	0.00042 *13*
1676 - 1754	0.0048	0.00028 *9*
1781 - 1785	0.00034	1.9 *8* ×10^{-5}

Continuous Radiation (^{128}Sb)

$\langle\beta-\rangle$=486 keV;$\langle$IB$\rangle$=0.72 keV

E_{bin}(keV)		$\langle\ \rangle$(keV)	(%)
0 - 10	β-	0.056	1.11
	IB	0.021	
10 - 20	β-	0.169	1.12
	IB	0.020	0.142
20 - 40	β-	0.69	2.28
	IB	0.039	0.136
40 - 100	β-	5.01	7.1
	IB	0.103	0.160
100 - 300	β-	50.6	25.3
	IB	0.24	0.138
300 - 600	β-	141	32.0
	IB	0.18	0.044
600 - 1300	β-	235	27.5
	IB	0.111	0.0140
1300 - 2052	β-	54	3.63
	IB	0.0060	0.00043

$^{128}_{51}$Sb(10.4 *2* min)

Mode: β-(96.4 *10* %), IT(3.6 *10* %)

Δ: >-84580 keV

SpA: 1.41×10^8 Ci/g

Prod: daughter ^{128}Sn; ^{128}Te(n,p); ^{130}Te(d,α)

Photons (^{128}Sb)

$\langle\gamma\rangle$=1909 *43* keV

γ_{mode}	γ(keV)	γ(%)†
Te L_ℓ	3.335	0.0044 *6*
Te L_η	3.606	0.00222 *25*
Te L_α	3.768	0.122 *14*
Te L_β	4.122	0.107 *14*
Te L_γ	4.651	0.0128 *19*
Te $K_{\alpha2}$	27.202	0.78 *4*
Te $K_{\alpha1}$	27.472	1.46 *8*
Te $K_{\beta1}$'	30.980	0.388 *21*
Te $K_{\beta2}$'	31.877	0.085 *5*
$\gamma_{\beta-}$[M1]	193.48 *10*	~1.0
$\gamma_{\beta-}$	249.81 *14*	0.69 *18*
$\gamma_{\beta-}$E2	314.15 *4*	89 *5*
$\gamma_{\beta-}$	594.29 *8*	3.3 *4*
$\gamma_{\beta-}$E2	743.41 *4*	96
$\gamma_{\beta-}$E2	754.06 *4*	96.4 *19*
$\gamma_{\beta-}$	787.76 *7*	7.1 *10*
$\gamma_{\beta-}$	844.09 *14*	2.2 *4*

Photons (^{128}Sb) (continued)

γ_{mode}	γ(keV)	γ(%)†
$\gamma_{\beta-}$	908.43 *8*	2.3 *3*
$\gamma_{\beta-}$	1040.9 *3*	0.96 *19*
$\gamma_{\beta-}$	1098.6 *8*	~0.29
$\gamma_{\beta-}$	1101.91 *8*	~0.39
$\gamma_{\beta-}$	1141.9 *3*	0.77 *19*
$\gamma_{\beta-}$	1158.24 *14*	1.74 *19*
$\gamma_{\beta-}$	1355.1 *3*	0.58 *19*
$\gamma_{\beta-}$	1585.4 *10*	~0.29
$\gamma_{\beta-}$	1608.7 *10*	0.48 *19*

† 5.1% uncert(syst)

Atomic Electrons (^{128}Sb)

$\langle e \rangle$=13.6 *4* keV

e_{bin}(keV)	$\langle e \rangle$(keV)	e(%)
4 - 31	0.55	8.9 *12*
162 - 193	0.18	0.11 *5*
218 - 250	0.07	~0.030
282	7.0	2.47 *14*
309	0.82	0.265 *15*
310	0.409	0.132 *8*
313 - 314	0.312	0.100 *5*
562 - 594	0.08	0.013 *7*
712	1.71	0.240 *13*
722	1.68	0.232 *7*
738 - 788	0.69	0.092 *8*
812 - 844	0.033	0.0040 *19*
877 - 908	0.032	0.0036 *17*
1009 - 1041	0.011	0.0011 *5*
1067 - 1110	0.014	0.0013 *5*
1126 - 1158	0.019	0.0017 *7*
1323 - 1355	0.0050	0.00038 *19*
1554 - 1608	0.0056	0.00036 *15*

Continuous Radiation (^{128}Sb)

$\langle\beta-\rangle$=944 keV;$\langle$IB$\rangle$=2.1 keV

E_{bin}(keV)		$\langle\ \rangle$(keV)	(%)
0 - 10	$\beta-$	0.0133	0.264
	IB	0.035	
10 - 20	$\beta-$	0.0406	0.270
	IB	0.035	0.24
20 - 40	$\beta-$	0.169	0.56
	IB	0.068	0.24
40 - 100	$\beta-$	1.30	1.84
	IB	0.19	0.29
100 - 300	$\beta-$	16.3	7.9
	IB	0.52	0.30
300 - 600	$\beta-$	74	16.1
	IB	0.54	0.129
600 - 1300	$\beta-$	397	42.2
	IB	0.59	0.071
1300 - 2500	$\beta-$	455	27.4
	IB	0.128	0.0083
2500 - 2579	$\beta-$	0.288	0.0114
	IB	7.7×10^{-7}	3.1×10^{-8}

$^{128}_{52}$Te($>5.5\times10^{24}$ yr)

Mode: β-β-
Δ: -88993 *3* keV
Prod: natural source
%: 31.69 *2*

$^{128}_{53}$I (24.99 *2* min)

Mode: β-(93.1 *8* %), ϵ(6.9 *8* %)
Δ: -87738 *4* keV
SpA: 5.877×10^{7} Ci/g
Prod: ^{127}I(n,γ)

Photons (^{128}I)

$\langle\gamma\rangle$=90 *8* keV

γ_{mode}	γ(keV)	γ(%)†
Te L_ℓ	3.335	0.0081 *14*
Te L_η	3.606	0.0040 *6*
Te L_α	3.768	0.23 *3*
Te L_β	4.123	0.20 *3*
Te L_γ	4.662	0.025 *4*
Te $K_{\alpha2}$	27.202	1.49 *16*
Te $K_{\alpha1}$	27.472	2.8 *3*
Te $K_{\beta1}$'	30.980	0.74 *8*
Te $K_{\beta2}$'	31.877	0.161 *17*
$\gamma_{\beta-}$E2	442.917 *9*	16.9
$\gamma_{\beta-}$E2+10%M1	526.575 *10*	1.57 *5*
$\gamma_{\beta-}$E2	613.523 *12*	0.00309 *14*
γ_{ϵ}E2	743.41 *4*	0.165 *7*
$\gamma_{\beta-}$[E2]	907.86 *4*	~0.00013
$\gamma_{\beta-}$E2	969.490 *12*	0.407 *11*
$\gamma_{\beta-}$E2	1140.095 *13*	0.0103 *4*
$\gamma_{\beta-}$[E2]	1434.43 *4*	0.00064 *5*

† uncert(syst): 19% for ϵ, 10% for β-

Atomic Electrons (^{128}I)

$\langle e \rangle$=1.36 *8* keV

e_{bin}(keV)	$\langle e \rangle$(keV)	e(%)
4	0.123	2.8 *4*
5	0.098	2.1 *3*
22	0.027	0.123 *18*
23	0.084	0.37 *5*
24 - 33	0.068	0.255 *22*
408	0.74	0.180 *18*
437	0.088	0.0202 *21*
438	0.031	0.0071 *7*
442 - 443	0.031	0.0070 *6*
492	0.0523	0.0106 *5*
521 - 527	0.0099	0.00190 *8*
579 - 614	9.4×10^{-5}	1.61 *7* $\times10^{-5}$
712 - 743	0.00342	0.000478 *20*
873 - 908	2.4×10^{-6}	2.7 *12* $\times10^{-7}$
935 - 969	0.00666	0.000709 *21*
1106 - 1140	0.000140	1.27 *5* $\times10^{-5}$
1400 - 1434	7.0×10^{-6}	5.0 *4* $\times10^{-7}$

Continuous Radiation (^{128}I)

$\langle\beta-\rangle$=737 keV;$\langle\beta+\rangle$=0.00313 keV;$\langle$IB$\rangle$=1.46 keV

E_{bin}(keV)		$\langle\ \rangle$(keV)	(%)
0 - 10	$\beta-$	0.0185	0.368
	$\beta+$	3.29×10^{-8}	3.97×10^{-7}
	IB	0.029	
10 - 20	$\beta-$	0.056	0.376
	$\beta+$	1.17×10^{-6}	7.1×10^{-6}
	IB	0.029	0.20
20 - 40	$\beta-$	0.234	0.78
	$\beta+$	2.67×10^{-5}	8.2×10^{-5}
	IB	0.059	0.21
40 - 100	$\beta-$	1.79	2.53
	$\beta+$	0.00075	0.00100
	IB	0.155	0.24
100 - 300	$\beta-$	21.6	10.6
	$\beta+$	0.00235	0.00166
	IB	0.41	0.23

Continuous Radiation (^{128}I) (continued)

E_{bin}(keV)		$\langle\ \rangle$(keV)	(%)
300 - 600	$\beta-$	92	20.2
	IB	0.40	0.095
600 - 1300	$\beta-$	409	44.3
	IB	0.35	0.043
1300 - 2123	$\beta-$	212	13.9
	IB	0.029	0.0020
	$\Sigma\beta+$		0.0028

$^{128}_{54}$Xe(stable)

Δ: -89860.8 *16* keV
%: 1.91 *3*

$^{128}_{55}$Cs(3.62 *2* min)

Mode: ϵ
Δ: -85926 *6* keV
SpA: 4.051×10^{8} Ci/g
Prod: daughter ^{128}Ba; ^{128}Xe(p,n)

Photons (^{128}Cs)

$\langle\gamma\rangle$=188.1 *24* keV

γ_{mode}	γ(keV)	γ(%)†
Xe L_ℓ	3.634	0.042 *6*
Xe L_η	3.955	0.0192 *20*
Xe L_α	4.104	1.17 *13*
Xe L_β	4.522	1.03 *13*
Xe L_γ	5.124	0.134 *19*
Xe $K_{\alpha2}$	29.461	6.94 *21*
Xe $K_{\alpha1}$	29.782	12.9 *4*
Xe $K_{\beta1}$'	33.606	3.49 *11*
Xe $K_{\beta2}$'	34.606	0.83 *3*
γ E2	442.917 *9*	26.8 *5*
γ M1+E2	460.083 *25*	0.032 *3*
γ E2+10%M1	526.575 *10*	2.42 *5*
γ E2	590.249 *17*	0.072 *8*
γ E2	613.523 *12*	0.351 *10*
γ	897 *1*	~0.005 ?
γ [E2]	907.86 *4*	~0.0027
γ [E2]	966.507 *23*	0.035 *3*
γ E2	969.490 *12*	0.627 *16*
γ E2+39%M1	986.656 *24*	0.029 *3*
γ	1001 *1*	~0.005 ?
γ (M1)	1030.180 *18*	0.217 *11*
γ [M1+E2]	1081.15 *4*	0.0161 *11*
γ	1118 *1*	0.0032 *8* ?
γ E2	1140.095 *13*	1.169 *22*
γ	1157.58 *3*	0.0072 *11*
γ	1162.02 *4*	0.0096 *13*
γ [M1+E2]	1203.44 *4*	0.0072 *11*
γ [E2]	1239.71 *3*	0.0121 *16*
γ	1283.43 *6*	0.0096 *13*
γ (M1)	1303.380 *24*	0.117 *4*
γ	1376.3 *3*	0.0080 *13*
γ	1392.32 *4*	0.0137 *16*
γ [M1+E2]	1407.90 *10*	0.0035 *8*
γ [E2]	1434.43 *4*	0.0129 *8*
γ	1461.22 *3*	0.0289 *19*
γ	1474.45 *16*	0.0048 *11*
γ [E2]	1477.56 *4*	0.0088 *13*
γ	1488.20 *19*	0.0013 *5*
γ	1513.02 *3*	0.0523 *19*
γ	1514.5 *10*	~0.0013 ?
γ [M1+E2]	1541.23 *3*	0.0177 *13*
γ	1551.87 *18*	0.0016 *5*
γ [M1+E2]	1556.751 *20*	0.0134 *8*
γ [E2]	1599.84 *4*	0.0046 *8*

Photons (^{128}Cs)
(continued)

γ_{mode}	γ(keV)	γ(%)†
γ (E2)	1629.10 3	0.135 3
γ [M1+E2]	1663.52 3	0.0244 13
γ	1678.52 20	0.0043 11
γ (M1)	1684.15 3	0.115 3
γ	1749.02 6	0.0032 11
γ	1756.73 15	~0.0008
γ	1795.6 4	0.0013 5
γ [E2]	1804.30 10	0.0056 8
γ	1810.00 6	0.0070 5
γ [M1+E2]	1829.949 25	0.0064 8
γ	1837.53 17	0.0035 8
γ	1858.9 4	0.0021 8
γ [M1+E2]	1867.98 10	0.0064 8
γ	1918.89 4	0.0378 19
γ	1978.17 4	0.0472 19
γ	1987.79 3	0.0265 13
γ [E2]	1999.662 21	0.0008 3
γ	2001.02 16	~0.0005
γ	2039.59 3	0.0319 24
γ	2063 1	~0.0008 ?
γ [M1+E2]	2067.80 3	0.0035 5
γ	2078.44 18	0.0121 11
γ	2090.68 16	0.0027 5
γ	2107.75 18	0.0021 5
γ	2121.86 15	~0.0011
γ	2130.12 6	0.0021 5
γ	2141.02 7	0.0070 5
γ	2148.66 4	0.047 3
γ E2	2155.67 3	0.161 11
γ [M1+E2]	2190.09 3	0.055 4
γ	2232.42 18	0.0027 5
γ	2255.22 20	0.0013 5
γ	2275.59 6	0.0147 11
γ	2283.30 15	0.0032 5
γ	2314.9 4	0.0008 3
γ	2326.8 4	~0.0005
γ	2348.4 3	0.0011 3
γ	2361.80 4	0.047 4
γ	2364.09 17	0.011 3
γ	2380.4 3	0.0008 3
γ [M1+E2]	2394.54 10	0.0214 11
γ	2416.59 4	0.0335 21
γ	2430.70 3	0.0059 5
γ	2467.0 3	0.0011 3
γ	2482.50 3	0.0029 3
γ	2494.89 11	0.00024 8
γ [E2]	2510.71 3	0.0032 3
γ	2550.66 18	0.0008 3
γ	2564.77 15	0.0021 3
γ	2591.57 4	0.0056 5
γ	2617.25 16	0.0013 3
γ [E2]	2632.99 3	0.0062 5
γ	2656.69 6	0.0016 3
γ	2662.0 3	~0.0005
γ	2667.59 7	0.0032 3
γ	2683.5 4	~0.00019
γ	2718.50 6	0.0008 3
γ	2723.9 3	0.0008 3
γ	2747 1	0.0021 3 ?
γ	2796.8 4	0.00043 11
γ	2814.8 4	0.00035 13
γ	2823.3 3	0.00029 11
γ [E2]	2837.45 10	0.0052 4
γ	2859.50 4	0.0056 4
γ	2876.7 5	0.00021 8
γ	2893 1	~8×10^{-5} ?
γ	2937.80 11	0.0029 3
γ	2963.71 19	0.00019 8
γ	3060.16 16	~0.00011
γ	3099.59 6	0.00016 5
γ	3104.9 3	0.00134 11
γ	3110.49 7	~3×10^{-5}
γ	3125 1	~2×10^{-5} ?
γ	3167.22 10	0.00024 3
γ	3204 1	<4×10^{-5} ?
γ	3406.62 19	0.00021 5
γ	3493 1	8 3×10^{-5} ?

† 4.8% uncert(syst)

Atomic Electrons (^{128}Cs)
⟨e⟩=3.39 12 keV

e_{bin}(keV)	⟨e⟩(keV)	e(%)
5	1.06	21.4 22
24	0.213	0.89 9
25	0.27	1.09 11
28	0.075	0.27 3
29	0.160	0.56 6
30 - 33	0.048	0.154 10
408	1.17	0.287 8
426	0.00148	0.00035 5
437	0.141	0.0322 9
438 - 460	0.0987	0.0224 4
492	0.081	0.0164 7
521 - 556	0.0173	0.00329 12
579 - 614	0.0111	0.00189 6
862 - 908	0.00013	1.5 7 ×10^{-5}
932 - 981	0.0114	0.00122 3
982 - 1030	0.00447	0.000447 25
1047 - 1083	0.00031	2.9 4 ×10^{-5}
1106 - 1153	0.0162	0.00146 4
1156 - 1205	0.00026	2.16 23 ×10^{-5}
1234 - 1283	0.00167	0.000132 7
1298 - 1342	0.00031	2.4 3 ×10^{-5}
1358 - 1407	0.00031	2.2 4 ×10^{-5}
1408 - 1456	0.00041	2.8 8 ×10^{-5}
1460 - 1556	0.00087	5.8 13 ×10^{-5}
1565 - 1663	0.00277	0.000171 5
1673 - 1770	0.000257	1.51 10 ×10^{-5}
1775 - 1867	0.00022	1.23 15 ×10^{-5}
1884 - 1983	0.00077	4.0 8 ×10^{-5}
1987 - 2086	0.00036	1.8 4 ×10^{-5}
2087 - 2185	0.00203	9.5 6 ×10^{-5}
2189 - 2282	0.00015	6.6 14 ×10^{-6}
2292 - 2390	0.00068	2.9 5 ×10^{-5}
2393 - 2490	0.000105	4.3 6 ×10^{-6}
2494 - 2591	5.6 ×10^{-5}	2.2 4 ×10^{-6}
2598 - 2689	7.8 ×10^{-5}	3.0 3 ×10^{-6}
2712 - 2810	4.5 ×10^{-5}	1.63 16 ×10^{-6}
2814 - 2903	4.6 ×10^{-5}	1.6 3 ×10^{-6}
2929 - 3026	3.2 ×10^{-6}	1.10 24 ×10^{-7}
3055 - 3133	8.4 ×10^{-6}	2.7 6 ×10^{-7}
3162 - 3203	2.3 ×10^{-7}	7 3 ×10^{-9}
3372 - 3458	1.2 ×10^{-6}	3.7 10 ×10^{-8}
3488 - 3492	4.5 ×10^{-8}	1.3 5 ×10^{-9}

Continuous Radiation (^{128}Cs)
⟨β+⟩=869 keV;⟨IB⟩=5.9 keV

E_{bin}(keV)		⟨ ⟩(keV)	(%)
0 - 10	β+	1.47×10^{-6}	1.77×10^{-5}
	IB	0.031	
10 - 20	β+	5.9×10^{-5}	0.000356
	IB	0.030	0.21
20 - 40	β+	0.00163	0.00499
	IB	0.080	0.27
40 - 100	β+	0.079	0.102
	IB	0.17	0.26
100 - 300	β+	4.08	1.84
	IB	0.50	0.28
300 - 600	β+	32.9	7.1
	IB	0.63	0.146
600 - 1300	β+	268	27.7
	IB	1.38	0.152
1300 - 2500	β+	541	31.0
	IB	2.2	0.123
2500 - 3935	β+	22.5	0.86
	IB	0.82	0.029
	$\Sigma\beta$+		69

$^{128}_{56}$Ba(2.43 5 d)

Mode: ϵ
Δ: -85478 18 keV
SpA: 4.20×10^{5} Ci/g
Prod: ^{133}Cs(p,6n); ^{133}Cs(d,7n); protons on Pr

Photons (^{128}Ba)
⟨γ⟩=67.3 20 keV

γ_{mode}	γ(keV)	γ(%)†
Cs L_ℓ	3.795	0.149 20
Cs L_η	4.142	0.067 7
Cs L_α	4.285	4.1 5
Cs L_β	4.731	3.8 5
Cs L_γ	5.374	0.50 7
Cs $K_{\alpha2}$	30.625	22.8 7
Cs $K_{\alpha1}$	30.973	42.2 13
Cs $K_{\beta1}$'	34.967	11.5 4
Cs $K_{\beta2}$'	36.006	2.82 9
γ [M1]	101.551 22	0.019 4
γ [M1]	129.10 19	0.016 4
γ [M1]	143.63 5	0.010 4
γ [M1]	159.52 4	0.013 4
γ [M1]	186.94 21	~0.038
γ [M1]	215.29 4	0.070 6
γ [M1]	229.33 6	0.106 9
γ M1(+E2)	273.266 10	14.5
γ [M1]	316.98 10	0.032 6
γ [M1]	358.93 4	0.096 9
γ [M1]	374.816 20	0.309 15

† 4.8% uncert(syst)

Atomic Electrons (^{128}Ba)
⟨e⟩=8.3 4 keV

e_{bin}(keV)	⟨e⟩(keV)	e(%)
5	3.1	61 6
6	0.49	8.6 9
25	0.69	2.8 3
26	0.85	3.3 3
29	0.224	0.77 8
30	0.53	1.76 18
31 - 66	0.175	0.52 3
93 - 142	0.0164	0.0156 24
143 - 187	0.022	0.013 3
193 - 229	0.0216	0.0109 8
237	1.68	0.71 6
268	0.30	0.11 3
272 - 317	0.083	0.030 8
323 - 370	0.0352	0.0104 5
374 - 375	0.00085	0.000227 11

Continuous Radiation (^{128}Ba)
⟨IB⟩=0.093 keV

E_{bin}(keV)		⟨ ⟩(keV)	(%)
10 - 20	IB	0.00046	0.0028
20 - 40	IB	0.074	0.24
40 - 100	IB	0.0056	0.0100
100 - 300	IB	0.0107	0.0059
300 - 448	IB	0.0018	0.00054

$^{128}_{57}$La(5.0 *3* min)

Mode: ε

Δ: -78880 *400* keV

SpA: 2.93×10^8 Ci/g

Prod: ^{121}Sb(^{12}C,5n); ^{123}Sb(^{12}C,7n); ^{115}In(^{16}O,3n); ^{130}Ba(p,3n); ^{118}Sn(^{14}N,4n)

Photons (^{128}La)

⟨γ⟩=2122 *26* keV

γ_{mode}	γ(keV)	γ(%)†
Ba L_{ℓ}	3.954	0.037 *5*
Ba L_{η}	4.331	0.0163 *17*
Ba L_{α}	4.465	1.02 *11*
Ba L_{β}	4.945	0.92 *11*
Ba L_{γ}	5.616	0.121 *17*
Ba $K_{\alpha2}$	31.817	5.37 *16*
Ba $K_{\alpha1}$	32.194	9.9 *3*
Ba $K_{\beta1}$'	36.357	2.74 *8*
Ba $K_{\beta2}$'	37.450	0.693 *23*
γ E2	284.09 *8*	87.2
γ	315.4 *5*	0.55 *10*
γ [E2]	386.05 *19*	1.65 *13*
γ	412.0 *3*	~0.5
γ [M1]	427.21 *15*	0.78 *13*
γ [M1]	439.93 *15*	2.08 *17*
γ	451.6 *3*	0.55 *12*
γ [M1]	475.12 *16*	2.0 *4*
γ E2	479.30 *8*	54 *3*
γ	483.0 *3*	0.89 *13*
γ E2	487.85 *13*	10.0 *5*
γ	494.12 *22*	1.1 *3*
γ	531.8 *3*	0.89 *9*
γ [M1]	561.08 *13*	1.08 *9*
γ	567.02 *17*	3.8 *3*
γ	570.6 *6*	0.30 *9*
γ	587.5 *3*	0.66 *17*
γ	591.72 *19*	1.00 *13*
γ M1+E2	600.45 *11*	10.5 *6*
γ	606.92 *22*	1.95 *17*
γ	609.00 *12*	8.2 *5*
γ	625.92 *14*	3.82 *22*
γ M1,E2	632.52 *16*	5.6 *4*
γ E2	643.55 *16*	14.7 *9*
γ	658.0 *6*	0.41 *17*
γ	673.0 *4*	0.60 *15*
γ	675.7 *4*	0.60 *12*
γ	681.9 *4*	0.71 *17*
γ	715.3 *3*	0.6 *1*
γ	774.6 *3*	1.37 *17*
γ	781.6 *4*	0.34 *10*
γ	793.5 *3*	1.0 *3*
γ	827.7 *3*	0.89 *10*
γ	838.89 *24*	0.83 *9*
γ	884.54 *12*	8.1 *5*
γ	915.05 *16*	3.5 *3*
γ	938.95 *22*	2.6 *3*
γ	988.5 *3*	1.13 *11*
γ (E1)	1005.7 *3*	1.45 *13*
γ	1036.20 *14*	2.05 *17*
γ	1040.38 *13*	9.9 *6*
γ	1045.7 *5*	0.93 *12*
γ	1049.1 *4*	0.72 *14*
γ	1053.12 *13*	10.2 *6*
γ	1070.40 *16*	4.5 *3*
γ	1079.03 *25*	1.72 *17*
γ	1088.29 *11*	8.9 *6*
γ	1101.04 *16*	4.6 *4*
γ	1124.9 *5*	0.55 *10*
γ	1144.2 *4*	1.55 *15*
γ	1164.7 *4*	0.89 *9*
γ	1168.00 *21*	1.61 *13*
γ	1176.01 *20*	1.2 *4*
γ (M1)	1276.07 *18*	4.9 *5*
γ	1302.8 *3*	1.30 *17*
γ	1318.9 *3*	1.12 *17*
γ	1348.7 *4*	0.59 *18*
γ	1412.3 *3*	3.5 *3*
γ	1440.08 *25*	1.81 *19*
γ	1483.4 *5*	0.37 *10*
γ	1505.96 *22*	3.6 *3*
γ	1515.49 *14*	0.58 *11*
γ	1549.70 *17*	1.4 *3*
γ	1605.7 *3*	1.48 *17*
γ	1653.6 *3*	0.50 *13*
γ	1662.11 *14*	0.98 *17*
γ	1688.0 *3*	0.43 *10*
γ	1710.7 *10*	0.37 *9*
γ	1722.8 *9*	0.31 *11*
γ	1726.6 *7*	0.62 *17*
γ	1755.7 *3*	1.23 *14*
γ	1908.5 *6*	0.57 *10*
γ	1919.37 *25*	1.23 *13*
γ	1957.6 *4*	0.52 *9*
γ	2025.5 *8*	0.34 *10*
γ	2177.6 *7*	0.44 *9*
γ	2191.0 *8*	0.32 *12*
γ	2212.0 *6*	0.85 *12*

† 0.57% uncert(syst)

Atomic Electrons (^{128}La)

⟨e⟩=20.5 *4* keV

e_{bin}(keV)	⟨e⟩(keV)	e(%)
5	0.51	9.8 *10*
6 - 36	0.94	8.5 *7*
247	9.39	3.81 *9*
278	1.65	0.594 *21*
279 - 315	0.938	0.333 *5*
349 - 390	0.197	0.054 *6*
402 - 440	0.31	0.073 *7*
442	2.21	0.50 *3*
446 - 494	1.08	0.232 *10*
524 - 572	0.87	0.16 *3*
582 - 627	0.84	0.139 *13*
631 - 681	0.144	0.022 *3*
682 - 715	0.0023	0.00033 *14*
737 - 782	0.057	0.008 *3*
787 - 834	0.037	0.0047 *15*
838 - 884	0.21	0.024 *9*
902 - 939	0.055	0.0060 *25*
951 - 1000	0.056	0.0057 *17*
1003 - 1052	0.53	0.052 *12*
1053 - 1101	0.11	0.010 *3*
1107 - 1144	0.066	0.0058 *15*
1159 - 1176	0.0070	0.00061 *16*
1239 - 1281	0.114	0.0091 *10*
1297 - 1343	0.011	0.0008 *3*
1347 - 1375	0.033	0.0024 *12*
1403 - 1446	0.028	0.0020 *6*
1469 - 1549	0.057	0.0038 *11*
1568 - 1661	0.031	0.0019 *5*
1673 - 1755	0.023	0.0013 *3*
1871 - 1957	0.018	0.00098 *23*
1988 - 2024	0.0026	0.00013 *6*
2140 - 2211	0.011	0.00052 *12*

Continuous Radiation (^{128}La)

⟨β+⟩=1202 keV;⟨IB⟩=7.0 keV

E_{bin}(keV)		⟨ ⟩(keV)	(%)
0 - 10	β+	8.4×10^{-7}	1.00×10^{-5}
	IB	0.039	
10 - 20	β+	3.53×10^{-5}	0.000212
	IB	0.038	0.27
20 - 40	β+	0.00102	0.00312
	IB	0.090	0.31
40 - 100	β+	0.053	0.067
	IB	0.22	0.34
100 - 300	β+	2.89	1.29
	IB	0.65	0.36
300 - 600	β+	24.6	5.2
	IB	0.81	0.19
600 - 1300	β+	226	23.1
	IB	1.57	0.18
1300 - 2500	β+	688	37.6
	IB	2.2	0.124
2500 - 5000	β+	261	8.9
	IB	1.36	0.044
5000 - 5837	IB	0.0043	8.3×10^{-5}
	Σβ+		76

$^{128}_{58}$Ce(5.5 *20* min)

Mode: ε

SpA: 2.7×10^8 Ci/g

Prod: protons on Nd

$^{128}_{60}$Nd(4 *2* s)

Mode: ε, εp

SpA: 2.0×10^{10} Ci/g

Prod: ^{92}Mo(^{40}Ca,α)

εp: 2100-4500 range

A = 129

NDS 39, 551 (1983)

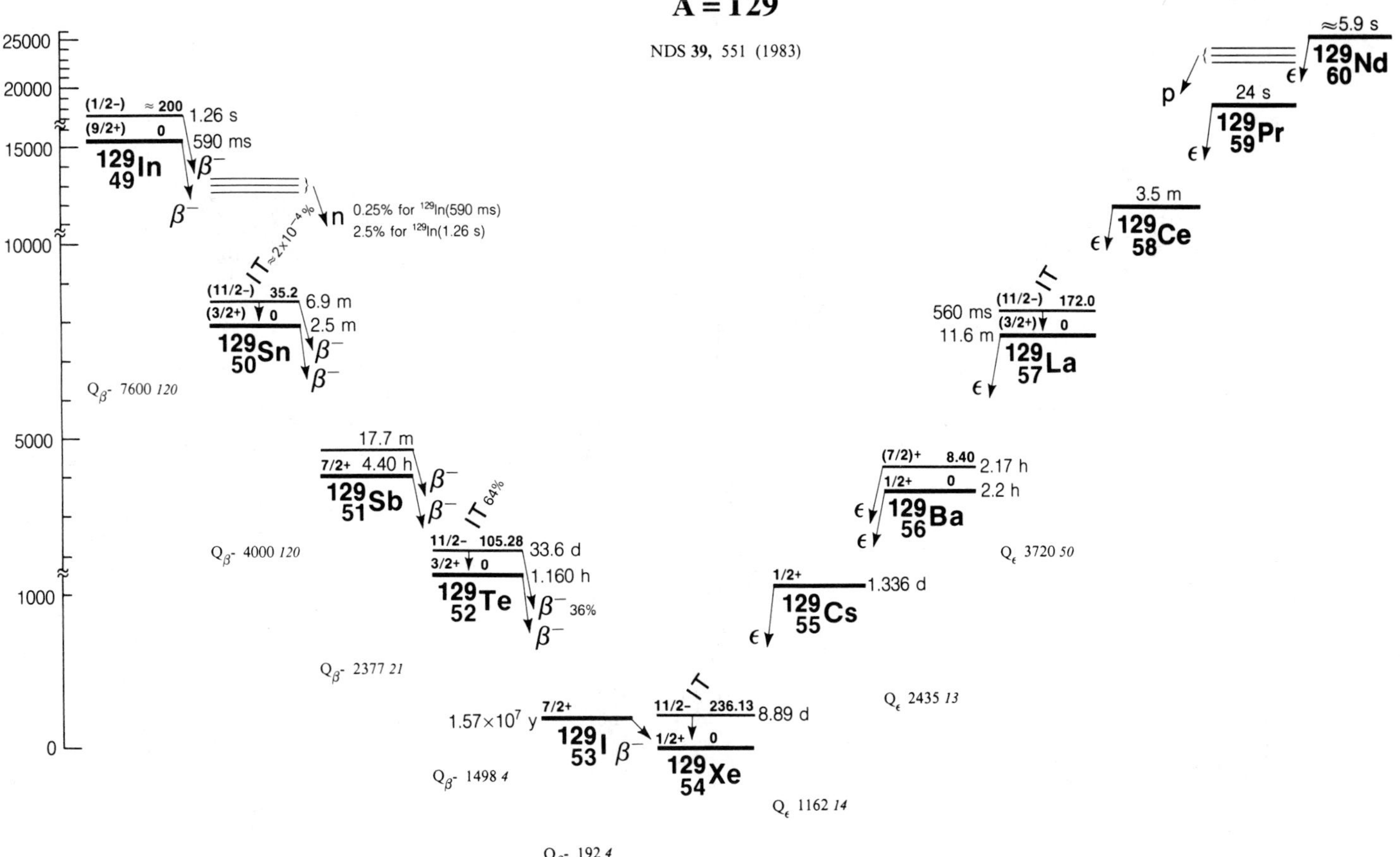

$^{129}_{49}$In(590 *20* ms)

Mode: β-, β-n(0.25 *5* %)
Δ: -73030 *170* keV
SpA: 8.7×10^{10} Ci/g
Prod: fission

Photons (^{129}In)

⟨γ⟩=2169 *110* keV

γ_{mode}	γ(keV)	γ(%)†
γ	212.9 *3*	0.18 *4* •
γ	252.9 *3*	0.27 *5* •
γ[M1]	279.44 *19*	0.36 *4*
γ[M1]	285.30 *21*	1.02 *9*
γ	382.6 *3*	1.25 *13* •
γ	417.2 *3*	~0.09 *
γ	474.0 *3*	0.27 *5* •
γ	480.1 *3*	0.31 *5* •
γ	501.3 *3*	0.36 *4* •
γ	570.3 *3*	1.69 *18* •
γ	663.1 *3*	0.31 *5* •
γ	728.88 *21*	4.6 *4*
γ	769.34 *18*	9.1 *9*
γ	821.58 *20*	0.49 *5*
γ	937.1 *3*	0.22 *4* •
γ	1008.32 *21*	6.0 *6*
γ	1045.3 *3*	<0.09 *
γ	1054.64 *22*	3.6 *4*
γ	1063.6 *3*	0.13 *4* •
γ	1074.58 *20*	2.36 *22*
γ	1095.74 *20*	3.5 *4*
γ	1101.02 *20*	1.47 *13*
γ	1136.1 *3*	1.87 *18* •
γ	1172.9 *3*	0.13 *4* •
γ	1308.8 *3*	0.22 *4* •
γ	1323.8 *3*	1.51 *13* •
γ	1348.74 *20*	2.09 *22*
γ	1354.02 *20*	0.80 *9*
γ	1427.4 *3*	0.44 *4* •
γ	1455.2 *3*	0.67 *9* •
γ	1498.6 *3*	0.36 *4* •
γ	1577.6 *3*	0.27 *5* •
γ	1716.2 *3*	0.22 *4* •
γ	1781.38 *25*	1.87 *18*
γ	1865.07 *20*	32 *3*
γ	1906.4 *3*	0.40 *4* •
γ	1977.1 *3*	0.44 *4* •
γ	2022.1 *3*	0.40 *4* •
γ	2066.7 *3*	0.93 *9*
γ	2118.07 *20*	45 *4*
γ	2189.6 *3*	1.65 *18* •
γ	2213.0 *5*	0.62 *5*
γ	2302.1 *10*	0.53 *9* •
γ	2367.1 *10*	0.67 *9* •
γ	2546.1 *10*	1.56 *18*

† 3.4% uncert(syst)
• with ^{129}In(590 ms or 1.26 s)

Continuous Radiation (^{129}In)

⟨β-⟩=2482 keV;⟨IB⟩=9.7 keV

E_{bin}(keV)		⟨ ⟩(keV)	(%)
0 - 10	β-	0.00189	0.0376
	IB	0.065	
10 - 20	β-	0.0058	0.0387
	IB	0.065	0.45
20 - 40	β-	0.0243	0.081
	IB	0.129	0.45
40 - 100	β-	0.193	0.272
	IB	0.38	0.58
100 - 300	β-	2.64	1.27
	IB	1.16	0.64
300 - 600	β-	14.3	3.09
	IB	1.50	0.35
600 - 1300	β-	126	12.8
	IB	2.6	0.30
1300 - 2500	β-	654	34.0
	IB	2.6	0.146
2500 - 5000	β-	1621	47.3
	IB	1.16	0.038
5000 - 6836	β-	65	1.22
	IB	0.0062	0.000117

$^{129}_{49}$In(1.26 *2* s)

Mode: β-, β-n(2.5 *5* %)
Δ: ~-72830 keV
SpA: 5.34×10^{10} Ci/g
Prod: fission

Photons (^{129}In) *

γ_{mode}	γ(keV)	γ(rel)
$\gamma_{\beta-}$[M1]	315.30 *21*	100 *10*
$\gamma_{\beta-}$	906.70 *24*	5.7 *7*
$\gamma_{\beta-}$	973.20 *24*	2.12 *24*
$\gamma_{\beta-}$	1222.00 *24*	9.0 *9*
$\gamma_{\beta-}$	1288.50 *24*	3.5 *5*

* see also ^{129}In(590 ms)

$^{129}_{50}$Sn(2.5 *1* min)

Mode: β-
Δ: -80630 *120* keV
SpA: 5.82×10^{8} Ci/g
Prod: fission

Photons (^{129}Sn)

γ_{mode}	γ(keV)
γ	645.6 *3*

$^{129}_{50}$Sn(6.9 *1* min)

Mode: β-, IT(~0.0002 %)
Δ: -80595 *120* keV
SpA: 2.11×10^{8} Ci/g
Prod: fission

Photons (^{129}Sn)

γ_{mode}	γ(keV)	γ(rel)
γ	381.9 *4*	78 *16*
γ	569.4 *4*	89 *18*
γ	1134.0 *4*	100
γ	1161.1 *3*	
γ	1321.5 *4*	75 *15*

$^{129}_{51}$Sb(4.40 *1* h)

Mode: β-
Δ: -84631 *22* keV
SpA: 5.521×10^{6} Ci/g
Prod: fission; ^{130}Te(γ,p)

Photons (^{129}Sb)

$\langle\gamma\rangle$=1356 *24* keV

γ_{mode}	γ(keV)	γ(%)†
Te $K_{\alpha2}$	27.202	0.27 *3*
Te $K_{\alpha1}$	27.472	0.50 *6*
Te $K_{\beta1}$'	30.980	0.132 *17*
Te $K_{\beta2}$'	31.877	0.029 *4*
γ	95.9 *7*	0.17 *4*
γ M4	105.28 *5*	*
γ	116.0 *7*	0.17 *4*
γ	124.9 *10*	
γ	136.6 *10*	

Photons (^{129}Sb) (continued)

γ_{mode}	γ(keV)	γ(%)†
γ	146.4 *7*	0.22 *4*
γ	164.8 *10*	0.034 *9*
γ[M1]	180.3 *3*	2.54 *13*
γ	197.2 *10*	0.065 *21*
γ	217.0 *10*	~0.013
γ	226.1 *10*	0.022 *9*
γ	231.9 *10*	0.301 *9*
γ	244.3 *6*	0.52 *9* ?
γ[M1]	268.2 *3*	0.26 *4*
γ[M1]	295.0 *4*	1.03 *9*
γ	313.3 *10*	0.84 *9*
γ[M1+E2]	332.5 *4*	
γ[M1]	359.3 *4*	2.8 *3*
γ[E2]	364.0 *4*	~0.43 ?
γ	404.7 *6*	1.38 *13*
γ	434.7 *10*	~0.09
γ[M1]	453.4 *4*	0.77 *22*
γ	499.0 *4*	0.22 *9*
γ	524.0 *4*	1.59 *13*
γ[M1]	544.25 *23*	17.9 *9*
γ[M1+E2]	633.7 *3*	2.75 *22*
γ	654.3 *4*	3.01 *22*
γ	669.6 *10*	0.82 *17* ?
γ	683.4 *3*	0.69 *22*
γ	683.5 *3*	5.1 *3*
γ[M1+E2]	736.6 *7*	0.39 *9* ?
γ[E2]	760.6 *4*	3.8 *3*
γ	772.9 *4*	2.75 *22*
γ[E2]	785.9 *4*	1.89 *17*
γ	812.4 *3*	43.0
γ	875.8 *10*	
γ	876.0 *5*	2.58 *21*
γ	914.3 *4*	20.0 *11*
γ	939.3 *4*	0.73 *17*
γ	950.4 *10*	~0.022
γ[M1+E2]	966.1 *3*	7.7 *5*
γ	984.1 *10*	<0.28
γ	995.2 *11*	~0.13
γ	1029.7 *5*	12.6 *8*
γ	1066.6 *10*	~0.05
γ	1083.6 *7*	0.52 *17*
γ	1103.7 *6*	0.22 *9*
γ	1121.2 *10*	~0.043
γ	1125.2 *10*	0.108 *22*
γ	1139.0 *10*	0.17 *4*
γ	1146.7 *5*	0.022 *4* ?
γ	1154.8 *10*	
γ	1161.6 *10*	0.108 *22*
γ	1167.8 *7*	0.26 *9*
γ[M1+E2]	1207.5 *4*	0.90 *13*
γ	1223.1 *10*	0.17 *4*
γ	1237.2 *15*	0.26 *9*
γ	1257.4 *6*	~0.34
γ	1262.2 *5*	0.73 *17*
γ	1272.2 *11*	~0.26 ?
γ[M1+E2]	1280.9 *7*	0.56 *13*
γ	1300.4 *5*	0.26 *9* ?
γ	1317.1 *4*	0.34 *9*
γ	1326.1 *6*	0.52 *9*
γ	1418.2 *7*	0.52 *9*
γ	1436.1 *6*	0.30 *13*
γ	1479.2 *5*	0.47 *22* ?
γ	1499.1 *10*	~0.09
γ	1525.5 *6*	0.43 *13*
γ	1540.4 *11*	0.13 *4* ?
γ	1568.6 *5*	0.69 *9*
γ	1580.9 *10*	0.077 *9*
γ	1598.4 *7*	0.52 *13*
γ[E1]	1621.5 *4*	~0.26
γ	1654.4 *10*	0.99 *21*
γ	1690.1 *6*	0.065 *21*
γ	1723.9 *20*	~0.26
γ	1736.3 *10*	6.0 *7* ?
γ	1752.1 *10*	<0.11
γ	1779.3 *10*	<0.10
γ	1841.6 *10*	0.22 *9*
γ	1870.4 *6*	0.30 *9*
γ	1916.3 *12*	~0.04 ?
γ	1919.0 *10*	0.026 *9*
γ	1919.0 *11*	
γ	1935.9 *10*	0.034 *9*
γ	1974.8 *10*	0.073 *13*
γ	2010.9 *10*	~0.0043
γ	2030.3 *10*	~0.009
γ	2041.8 *10*	~0.0043
γ	2069.8 *6*	0.56 *13*
γ	2071.4 *10*	0.43 *7*
γ	2091.3 *10*	0.017 *4*
γ	2112.8 *5*	0.34 *13*
γ	2133.0 *15*	~0.043
γ	2198.7 *15*	0.056 *13* ?
γ	2262.3 *10*	
γ	2265.6 *15*	~0.017 ?

† 2.3% uncert(syst)
* with ^{129}Te(33.6 d)

Atomic Electrons (^{129}Sb)

$\langle e\rangle$=4.1 *4* keV

e_{bin}(keV)	$\langle e\rangle$(keV)	e(%)
4 - 31	0.071	1.02 *12*
64 - 112	0.22	0.28 *12*
115 - 146	0.08	~0.06
148	0.44	0.295 *19*
160 - 200	0.126	0.070 *10*
212 - 244	0.08	0.038 *13*
263	0.089	0.034 *3*
264 - 313	0.07	0.023 *9*
327	0.176	0.054 *5*
332 - 373	0.11	0.030 *8*
400 - 449	0.051	0.012 *3*
452 - 499	0.045	0.009 *4*
512	0.62	0.121 *8*
519 - 544	0.108	0.0200 *11*
602 - 650	0.15	0.025 *5*
652	0.10	~0.015
653 - 683	0.020	0.0030 *12*
705 - 754	0.144	0.020 *3*
756 - 773	0.017	0.0023 *4*
781	0.6	~0.07
782 - 812	0.09	0.012 *5*
844 - 876	0.037	0.0043 *20*
883	0.23	~0.026
908 - 919	0.046	0.0050 *18*
934	0.113	0.0121 *19*
935 - 984	0.021	0.0022 *3*
990 - 995	0.00022	~2×10^{-5}
998	0.13	~0.013
1025 - 1072	0.028	0.0027 *9*
1079 - 1125	0.0044	0.0004 *1*
1130 - 1176	0.0141	0.00121 *21*
1191 - 1240	0.017	0.0014 *3*
1249 - 1296	0.017	0.00137 *24*
1299 - 1326	0.0011	8.2 *23* $\times10^{-5}$
1386 - 1435	0.0067	0.00048 *16*
1436 - 1479	0.0044	0.00030 *15*
1494 - 1590	0.014	0.00093 *20*
1593 - 1692	0.010	0.00059 *19*
1704 - 1778	0.041	0.0024 *9*
1810 - 1904	0.0038	0.00021 *6*
1911 - 2010	0.00061	3.1 *9* $\times10^{-5}$
2025 - 2112	0.0080	0.00039 *9*
2128 - 2198	0.00034	1.6 *5* $\times10^{-5}$
2234 - 2265	9.2 $\times10^{-5}$	~4×10^{-6}

Continuous Radiation (^{129}Sb)

$\langle\beta-\rangle$=390 keV;$\langle$IB$\rangle$=0.57 keV

E_{bin}(keV)		$\langle\ \rangle$(keV)	(%)
0 - 10	β-	0.108	2.17
	IB	0.017	
10 - 20	β-	0.323	2.16
	IB	0.0163	0.114
20 - 40	β-	1.28	4.28
	IB	0.031	0.107

Continuous Radiation (^{129}Sb)
(continued)

E_{bin}(keV)		⟨ ⟩(keV)	(%)
40 - 100	β-	8.7	12.4
	IB	0.080	0.124
100 - 300	β-	65	33.7
	IB	0.18	0.103
300 - 600	β-	100	23.8
	IB	0.136	0.033
600 - 1300	β-	158	17.8
	IB	0.100	0.0124
1300 - 2272	β-	57	3.65
	IB	0.0100	0.00067

$^{129}_{51}$Sb(17.7 *1* min)

Mode: β-
Δ: >-84631 keV
SpA: 8.23×10^{7} Ci/g
Prod: fission
decay data not yet evaluated

$^{129}_{52}$Te(1.160 *5* h)

Mode: β-
Δ: -87008 *4* keV
SpA: 2.094×10^{7} Ci/g
Prod: daughter ^{129}Te(33.6 d); fission; ^{128}Te(n,γ); ^{130}Te(n,2n)

Photons (^{129}Te)
⟨γ⟩=62.4 *12* keV

γ_{mode}	γ(keV)	γ(%)†
I L_ℓ	3.485	0.054 *15*
I L_η	3.780	0.036 *7*
I L_α	3.937	1.5 *4*
I L_β	4.297	3.3 *7*
I L_γ	4.999	0.61 *14*
γ M1+0.2%E2	27.791 *21*	16.3 *16*
γ M1+3.2%E2	208.94 *3*	0.180 *5*
γ [M1]	210.48 *4*	~0.0013
γ [E2]	242.22 *4*	1.24 *25* $\times10^{-6}$
γ M1+25%E2	250.576 *23*	0.383 *11*
γ [M1]	270.30 *4*	0.0046 *3*
γ E2	278.367 *24*	0.567 *17*
γ M1+0.8%E2	281.23 *3*	0.165 *5*
γ [M1+E2]	281.41 *4*	~7$\times10^{-5}$
γ [M1]	281.42 *3*	2.04 *25* $\times10^{-5}$
γ	281.70 *3*	0.0015 *3*
γ [M1+E2]	320.60 *4*	2.4 *4* $\times10^{-5}$
γ [M1]	342.59 *4*	0.0493 *15*
γ M1+41%E2	342.89 *4*	0.0085 *8*
γ [E2]	382.07 *3*	0.00062 *23*
γ	415.86 *14*	~0.0006
γ M1+1.6%E2	459.52 *3*	7.70 *23*
γ	462.02 *5*	<0.00023
γ M1+22%E2	487.31 *3*	1.42 *5*
γ [E2]	490.35 *3*	<0.00033
γ	491.90 *4*	0.00115 *23*
γ [E2]	531.81 *3*	0.088 *3*
γ	551.54 *3*	0.0035 *4*
γ [E2]	552.01 *3*	0.00139 *23*
γ	560.02 *4*	0.0061 *4*
γ [M1+E2]	562.82 *4*	<1.9$\times10^{-5}$
γ M1+0.3%E2	624.30 *3*	0.097 *3*
γ	701.01 *4*	0.0013 *3*
γ [E2]	701.74 *3*	4.6 *7* $\times10^{-5}$

Photons (^{129}Te)
(continued)

γ_{mode}	γ(keV)	γ(%)†
γ [E2]	722.45 *9*	<0.00023
γ M1+11%E2	729.53 *3*	0.00129 *20*
γ	732.32 *5*	0.00131 *23*
γ M1+7.4%E2	740.93 *3*	0.0376 *13*
γ [M1+E2]	768.72 *3*	0.0041 *3*
γ	768.97 *4*	0.00072 *7*
γ [E2]	771.77 *3*	1.17 *16* $\times10^{-5}$
γ	773.30 *4*	~0.00023
γ	802.11 *3*	0.192 *6*
γ	804.61 *5*	0.0216 *23*
γ	816.98 *20*	<6$\times10^{-5}$
γ	829.90 *3*	0.00639 *23*
γ [M1+E2]	833.24 *3*	0.0454 *14*
γ	918.27 *16*	0.00062 *15*
γ [M1+E2]	931.40 *9*	0.00021 *9*
γ	982.25 *3*	0.0160 *5*
γ	1013.56 *4*	0.0013 *3*
γ	1019.54 *4*	0.0022 *5*
γ M1+0.08%E2	1022.34 *3*	0.00069 *6*
γ [M1+E2]	1050.13 *3*	0.00070 *6*
γ M1+27%E2	1083.81 *3*	0.493 *15*
γ M1+0.6%E2	1111.60 *3*	0.191 *8*
γ	1168.78 *20*	<5$\times10^{-5}$
γ [E2]	1181.97 *9*	0.00012 *5*
γ	1232.82 *3*	0.0075 *3*
γ	1260.61 *3*	0.0112 *5*
γ	1264.13 *4*	0.0082 *3*
γ	1291.92 *4*	0.00028 *4*

† 6.5% uncert(syst)

Atomic Electrons (^{129}Te)
⟨e⟩=22.2 *23* keV

e_{bin}(keV)	⟨e⟩(keV)	e(%)
5	1.9	37 *8*
23	15.1	67 *10*
24	0.00176	0.0074 *8*
27	3.6	13.6 *20*
28	0.91	3.3 *5*
29 - 32	0.000193	0.00062 *4*
176 - 217	0.0778	0.0389 *9*
237 - 282	0.1017	0.0406 *10*
287 - 321	0.00415	0.00134 *6*
337 - 383	0.00080	0.000234 *11*
411 - 416	4.2 $\times10^{-6}$	~1$\times10^{-6}$
426	0.365	0.086 *3*
429 - 462	0.121	0.0266 *9*
482 - 531	0.0137	0.00281 *9*
532 - 563	6.8 $\times10^{-5}$	1.2 *3* $\times10^{-5}$
591 - 624	0.00352	0.000592 *19*
668 - 717	0.00100	0.000142 *6*
718 - 767	0.00027	3.7 *4* $\times10^{-5}$
768 - 817	0.0046	0.00059 *20*
825 - 833	0.000154	1.86 *25* $\times10^{-5}$
885 - 931	1.3 $\times10^{-5}$	1.4 *5* $\times10^{-6}$
949 - 989	0.00026	2.8 *10* $\times10^{-5}$
1008 - 1051	0.0071	0.00068 *3*
1078 - 1112	0.00441	0.000408 *13*
1136 - 1182	1.7 $\times10^{-6}$	1.5 *5* $\times10^{-7}$
1200 - 1233	0.00024	2.0 *5* $\times10^{-5}$
1255 - 1292	2.9 $\times10^{-5}$	2.3 *7* $\times10^{-6}$

Continuous Radiation (^{129}Te)
⟨β-⟩=520 keV;⟨IB⟩=0.74 keV

E_{bin}(keV)		⟨ ⟩(keV)	(%)
0 - 10	β-	0.0413	0.82
	IB	0.023	
10 - 20	β-	0.125	0.83
	IB	0.022	0.154
20 - 40	β-	0.51	1.70
	IB	0.042	0.147
40 - 100	β-	3.79	5.4
	IB	0.113	0.18
100 - 300	β-	41.1	20.3
	IB	0.26	0.153
300 - 600	β-	144	32.2
	IB	0.20	0.048
600 - 1300	β-	323	38.2
	IB	0.078	0.0110
1300 - 1470	β-	6.8	0.506
	IB	8.8$\times10^{-5}$	6.6$\times10^{-6}$

$^{129}_{52}$Te(33.6 *1* d)

Mode: IT(64 *7* %), β-(36 *7* %)
Δ: -86903 *4* keV
SpA: 3.013×10^{4} Ci/g
Prod: ^{128}Te(n,γ); fission

Photons (^{129}Te)
⟨γ⟩=37.0 *5* keV

γ_{mode}	γ(keV)	γ(%)†
Te L_ℓ	3.335	0.079 *11*
Te L_η	3.606	0.029 *3*
Te L_α	3.768	2.2 *3*
Te L_β	4.141	1.65 *23*
Te L_γ	4.686	0.20 *3*
Te $K_{\alpha2}$	27.202	8.2 *4*
Te $K_{\alpha1}$	27.472	15.3 *8*
γ_βM1+0.2%E2	27.791 *21*	0.0272 *5*
Te $K_{\beta1}$'	30.980	4.08 *22*
Te $K_{\beta2}$'	31.877	0.89 *5*
γ_β[M1]	76.04 *3*	0.00032 *7*
γ_{IT} M4	105.28 *5*	0.147 *6*
γ_β[M1]	115.22 *4*	0.00027 *8*
γ_βM1+3.2%E2	208.94 *3*	2.9 *3* $\times10^{-5}$
γ_β[E2]	242.22 *4*	0.00067 *9*
γ_βM1+25%E2	250.576 *23*	0.00039 *6*
γ_βE2	278.367 *24*	0.00058 *8*
γ_β[M1+E2]	281.41 *4*	~5$\times10^{-5}$
γ_β[M1]	281.42 *3*	0.00051 *5*
γ_β[M1+E2]	320.60 *4*	0.00061 *9*
γ_β[M1+E2]	357.45 *4*	<1.4$\times10^{-4}$
γ_βM1+1.6%E2	459.52 *3*	0.00122 *14*
γ_βM1+22%E2	487.31 *3*	0.000225 *25*
γ_β[E2]	490.35 *3*	<0.00023
γ_β[M1+E2]	552.37 *4*	0.00028 *9*
γ_β	556.57 *3*	0.118 *4*
γ_β[M1+E2]	562.82 *4*	<0.00047
γ_β[E1]	671.80 *3*	0.0248 *9*
γ_β[E2]	695.84 *5*	2.99 *6*
γ_β[E2]	701.74 *3*	0.0248 *9*
γ_β	705.49 *5*	0.0051 *5*
γ_β	716.30 *10*	<0.00023
γ_βM1+11%E2	729.53 *3*	0.70 *3*
γ_βM1+7.4%E2	740.93 *3*	0.0270 *9*
γ_β[M1+E2]	768.72 *3*	0.00292 *22*
γ_β[E2]	771.77 *3*	0.00029 *3*
γ_β[M1+E2]	794.59 *4*	0.00056 *14*
γ_βM1+17%E2	816.97 *3*	0.091 *3*
γ_β[M1+E2]	844.76 *3*	0.0342 *19*
γ_β	925.24 *10*	<6$\times10^{-5}$
γ_β[E2]	1003.54 *4*	0.00070 *14*
γ_βM1+0.08%E2	1022.34 *3*	0.0174 *9*

Photons (^{129}Te) (continued)

γ_{mode}	γ(keV)	γ(%)†
$\gamma_{\beta-}$[M1+E2]	1050.13 *3*	0.0177 *12*
$\gamma_{\beta-}$	1175.82 *10*	~9×10^{-5}
$\gamma_{\beta-}$	1203.61 *10*	0.00023 *5*
$\gamma_{\beta-}$[M1+E2]	1254.11 *4*	0.00042 *5*
$\gamma_{\beta-}$[M1+E2]	1281.90 *4*	0.00022 *4*
$\gamma_{\beta-}$	1373.53 *3*	0.00027 *3*
$\gamma_{\beta-}$	1401.32 *3*	0.00346 *9*

† uncert(syst): 11% for IT, 36% for β-

Atomic Electrons (^{129}Te)

⟨e⟩=60.3 *15* keV

e_{bin}(keV)	⟨e⟩(keV)	e(%)
4 - 43	2.93	47 *4*
71	4.3 ×10^{-5}	6.0 *14* ×10^{-5}
73	23.9	32.5 *16*
75 - 82	0.00011	0.00013 *4*
100	9.9	9.8 *5*
101	15.9	15.7 *8*
104	6.1	5.9 *3*
105	1.56	1.49 *7*
110 - 115	2.2 ×10^{-5}	2.0 *5* ×10^{-5}
176 - 217	0.000144	6.8 *7* ×10^{-5}
237 - 281	0.000173	7.0 *5* ×10^{-5}
287 - 324	6.2 ×10^{-5}	2.1 *3* ×10^{-5}
352 - 357	9.1 ×10^{-7}	~3×10^{-7}
426 - 460	8.1 ×10^{-5}	1.87 *18* ×10^{-5}
482 - 530	0.0028	~0.0005
547 - 563	0.00050	9 *4* ×10^{-5}
639 - 683	0.062	0.0093 *4*
691 - 740	0.0317	0.00454 *13*
741 - 790	0.00193	0.000246 *9*
794 - 840	0.00102	0.000125 *14*
844 - 892	2.1 ×10^{-5}	2.5 *4* ×10^{-6}
920 - 925	6.5 ×10^{-8}	~7×10^{-9}
970 - 1018	0.00058	5.8 *5* ×10^{-5}
1021 - 1050	4.8 ×10^{-5}	4.7 *5* ×10^{-6}
1143 - 1176	3.2 ×10^{-6}	2.7 *11* ×10^{-7}
1198 - 1221	5.2 ×10^{-6}	4.3 *7* ×10^{-7}
1249 - 1282	3.6 ×10^{-6}	2.8 *5* ×10^{-7}
1340 - 1373	2.9 ×10^{-5}	2.2 *9* ×10^{-6}
1396 - 1401	4.2 ×10^{-6}	3.0 *11* ×10^{-7}

Continuous Radiation (^{129}Te)

⟨β-⟩=206 keV;⟨IB⟩=0.33 keV

E_{bin}(keV)		⟨ ⟩(keV)	(%)
0 - 10	β-	0.0161	0.322
	IB	0.0088	
10 - 20	β-	0.0485	0.323
	IB	0.0085	0.059
20 - 40	β-	0.196	0.65
	IB	0.0164	0.057
40 - 100	β-	1.40	1.99
	IB	0.044	0.069
100 - 300	β-	13.9	6.9
	IB	0.108	0.062
300 - 600	β-	44.8	10.0
	IB	0.089	0.021
600 - 1300	β-	131	14.6
	IB	0.053	0.0066
1300 - 1604	β-	15.1	1.09
	IB	0.00051	3.8×10^{-5}

$^{129}_{53}$I (1.57 *4* ×10^7 yr)

Mode: β-
Δ: -88506 *4* keV
SpA: 0.000177 Ci/g
Prod: fission

Photons (^{129}I)

⟨γ⟩=24.8 *5* keV

γ_{mode}	γ(keV)	γ(%)
Xe L_ℓ	3.634	0.124 *17*
Xe L_η	3.955	0.057 *6*
Xe L_α	4.104	3.4 *4*
Xe L_β	4.522	3.0 *4*
Xe L_γ	5.121	0.39 *6*
Xe $K_{\alpha2}$	29.461	20.4 *8*
Xe $K_{\alpha1}$	29.782	37.8 *15*
Xe $K_{\beta1}$'	33.606	10.2 *4*
Xe $K_{\beta2}$'	34.606	2.42 *10*
γ M1+0.08%E2	39.5710 *22*	7.5 *2*

Atomic Electrons (^{129}I)

⟨e⟩=14.1 *5* keV

e_{bin}(keV)	⟨e⟩(keV)	e(%)
5	7.1	143 *9*
24	0.63	2.6 *3*
25	0.79	3.2 *3*
28	0.220	0.78 *8*
29	0.47	1.63 *17*
30	0.063	0.211 *22*
32	0.0239	0.074 *8*
33	0.055	0.165 *17*
34	3.61	10.6 *4*
35	0.106	0.30 *4*
38	0.755	1.96 *7*
39	0.290	0.74 *4*
40	0.0284	0.072 *3*

Continuous Radiation (^{129}I)

⟨β-⟩=41.5 keV;⟨IB⟩=0.0065 keV

E_{bin}(keV)		⟨ ⟩(keV)	(%)
0 - 10	β-	0.80	16.3
	IB	0.0020	
10 - 20	β-	2.18	14.6
	IB	0.00144	0.0102
20 - 40	β-	7.2	24.5
	IB	0.0017	0.0062
40 - 100	β-	25.0	39.1
	IB	0.00124	0.0023
100 - 154	β-	6.3	5.5
	IB	3.2×10^{-5}	3.0×10^{-5}

$^{129}_{54}$Xe(stable)

Δ: -88697.4 *19* keV
%: 26.4 *6*

$^{129}_{54}$Xe(8.89 *2* d)

Mode: IT
Δ: -88461.3 *19* keV
SpA: 1.139×10^5 Ci/g
Prod: ^{128}Xe(n,γ)

Photons (^{129}Xe)

⟨γ⟩=51.3 *8* keV

γ_{mode}	γ(keV)	γ(%)
Xe L_ℓ	3.634	0.25 *3*
Xe L_η	3.955	0.110 *12*
Xe L_α	4.104	6.9 *8*
Xe L_β	4.524	6.0 *8*
Xe L_γ	5.129	0.78 *11*
Xe $K_{\alpha2}$	29.461	36.6 *12*
Xe $K_{\alpha1}$	29.782	68.0 *23*
Xe $K_{\beta1}$'	33.606	18.4 *6*
Xe $K_{\beta2}$'	34.606	4.35 *15*
γ M1+0.08%E2	39.5710 *22*	7.5 *2*
γ M4	196.56 *3*	4.59 *14*

Atomic Electrons (^{129}Xe)

⟨e⟩=183 *4* keV

e_{bin}(keV)	⟨e⟩(keV)	e(%)
5	10.2	205 *16*
24	1.12	4.7 *5*
25	1.43	5.8 *6*
28	0.40	1.40 *14*
29	0.84	2.9 *3*
30	0.112	0.38 *4*
32	0.043	0.133 *14*
33	0.098	0.30 *3*
34	3.61	10.6 *4*
35	0.106	0.30 *4*
38	0.755	1.96 *7*
39	0.290	0.74 *4*
40	0.0284	0.072 *3*
162	102.6	63.3 *23*
191	30.7	16.0 *6*
192	16.8	8.8 *3*
195	5.58	2.86 *10*
196	7.8	4.00 *15*
197	0.314	0.160 *6*

$^{129}_{55}$Cs(1.3358 *25* d)

Mode: ϵ
Δ: -87536 *14* keV
SpA: 7.577×10^5 Ci/g
Prod: ^{127}I(α,2n); ^{130}Ba(p,2p)

Photons (^{129}Cs)

⟨γ⟩=281 *3* keV

γ_{mode}	γ(keV)	γ(%)†
Xe L_ℓ	3.634	0.184 *25*
Xe L_η	3.955	0.084 *9*
Xe L_α	4.104	5.1 *6*
Xe L_β	4.522	4.5 *6*
Xe L_γ	5.125	0.59 *8*
Xe $K_{\alpha2}$	29.461	30.2 *10*
Xe $K_{\alpha1}$	29.782	56.1 *18*
Xe $K_{\beta1}$'	33.606	15.2 *5*

Photons (^{129}Cs)
(continued)

γ_{mode}	γ(keV)	γ(%)†
Xe K$_{\beta2}$'	34.606	3.59 *13*
γ M1+0.08%E2	39.5710 *22*	2.99 *9*
γ [E2]	89.783 *5*	0.0025 *6*
γ M1	93.3097 *18*	0.656 *18*
γ M1+17%E2	177.023 *4*	0.271 *5*
γ M1+E2	266.806 *5*	0.274 *6*
γ M1(+E2)	270.333 *4*	0.214 *4*
γ M1+8.3%E2	278.5917 *22*	1.33 *3*
γ E2	282.118 *5*	0.243 *5*
γ	302.72 *14*	<0.00031
γ E2+18%M1	318.1626 *17*	2.46 *5*
γ E2	321.689 *5*	0.071 *6*
γ	357.494 *8*	0.0059 *9*
γ M1+48%E2	371.9012 *18*	30.8 *6*
γ	373.33 *3*	<0.025
γ M1	411.4720 *17*	22.5 *5*
γ [M1+E2]	492.820 *8*	0.0114 *9*
γ [M1+E2]	533.08 *3*	0.0095 *6*
γ	534.517 *8*	0.0213 *9*
γ M1(+E2)	548.924 *4*	3.42 *7*
γ [E2]	572.65 *3*	<0.005
γ [M1+E2]	582.603 *9*	~0.0009
γ	584.84 *14*	~0.0005
γ [M1+E2]	586.129 *8*	0.0129 *12*
γ M1(+E2)	588.495 *4*	0.607 *12*
γ	624.300 *7*	0.0283 *6*
γ	627.826 *8*	0.0017 *4*
γ [M1+E2]	864.720 *8*	0.0323 *9*
γ [M1+E2]	904.290 *8*	0.0083 *6*
γ	906.416 *8*	0.221 *4*
γ	945.987 *8*	0.0699 *14*

† 5.2% uncert(syst)

Atomic Electrons (^{129}Cs)
⟨e⟩=16.6 *6* keV

e$_{bin}$(keV)	⟨e⟩(keV)	e(%)
5	6.2	125 *11*
24	0.93	3.9 *4*
25	1.18	4.8 *5*
28	0.33	1.16 *12*
29	0.70	2.42 *25*
30 - 33	0.209	0.67 *4*
34	1.44	4.21 *16*
35 - 55	0.472	1.23 *4*
59	0.338	0.575 *20*
84 - 93	0.0872	0.098 *3*
142 - 177	0.0729	0.0495 *16*
232 - 281	0.267	0.1082 *19*
282 - 323	0.256	0.0888 *19*
337	2.00	0.594 *15*
339 - 373	0.390	0.1063 *23*
377	1.35	0.358 *10*
406 - 411	0.233	0.0573 *13*
458 - 500	0.0015	0.00030 *7*
514 - 554	0.162	0.031 *4*
567 - 593	0.0041	0.00071 *11*
619 - 628	0.00011	1.8 *8* ×10^{-5}
830 - 872	0.0039	0.00045 *18*
899 - 946	0.0016	0.00017 *6*

Continuous Radiation (^{129}Cs)
⟨IB⟩=0.23 keV

E$_{bin}$(keV)		⟨ ⟩(keV)	(%)
10 - 20	IB	0.00049	0.0029
20 - 40	IB	0.065	0.22
40 - 100	IB	0.0054	0.0092
100 - 300	IB	0.028	0.0139
300 - 600	IB	0.077	0.018
600 - 1162	IB	0.053	0.0072

$^{129}_{56}$Ba(2.23 *11* h)

Mode: ϵ
Δ: -85100 *14* keV
SpA: 1.09×10^7 Ci/g
Prod: ^{133}Cs(p,5n); ^{130}Ba(n,2n); ^{130}Ba(γ,n)

Photons (^{129}Ba(2.2 + 2.17 h))

γ_{mode}	γ(keV)	γ(rel)
γE2	53.26 *6*	0.23 *2*
γ[M1]	73.26 *6*	0.21 *2*
γ[M1]	85.22 *6*	0.09 *1*
γM1+4.0%E2	129.07 *5*	14.3 *8*
γM1	135.65 *5*	2.06 *14*
γ(E1)	149.10 *7*	0.95 *10*
γ[M1]	151.82 *7*	0.25 *2*
γ	164.57 *7*	0.82 *8*
γ[M1]	176.92 *7*	7.7 *4*
γM1+5.9%E2	182.34 *4*	100
γ(M1+6.9%E2)	202.34 *5*	35.6 *18*
γM1(+25%E2)	214.30 *5*	19.8 *12*
γ[M1]	220.87 *5*	13.1 *8*
γ(M1+E2)	237.58 *5*	2.84 *20*
γ[M1+E2]	263.81 *5*	1.69 *16*
γ	286.55 *8*	0.66 *7*
γ	328.73 *6*	0.49 *5*
γ	334.06 *6*	0.95 *10*
γ	343.49 *8*	0.25 *3*
γ	346.03 *8*	0.20 *2*
γ	366.03 *6*	1.94 *25*
γ	382.53 *7*	0.47 *6*
γ[E2]	386.68 *7*	0.43 *4*
γ	392.28 *8*	21.3 *12*
γ	394.50 *6*	0.92 *10*
γ	414.50 *6*	0.94 *10*
γ	416.10 *11*	0.60 *6*
γ	419.8 *1*	25.0 *13*
γ	425.77 *8*	1.19 *12*
γ	436.99 *8*	1.35 *15*
γ(E2)	459.52 *6*	13.8 *8*
γ	467.76 *6*	3.4 *3*
γ(E2)	481.39 *6*	8.2 *5*
γ	491.83 *11*	0.22 *3*
γ(M1+E2)	501.38 *6*	4.7 *3*
γ	525.32 *9*	1.22 *12*
γ	534.35 *6*	2.55 *21*
γ	542.63 *6*	3.09 *25*
γ	546.31 *6*	10.6 *6*
γ	548.36 *6*	3.49 *25*
γ	551.75 *12*	0.48 *5*
γ	556.85 *10*	2.14 *23*
γ	566.31 *6*	2.14 *12*
γ(E3)	569.02 *8*	1.32 *16*
γ	596.83 *5*	14.2 *8*
γ	601.18 *11*	0.54 *6*
γ	619.58 *6*	0.62 *6*
γ	656.03 *8*	3.28 *25*
γ	658.18 *9*	0.60 *6*
γ	670.14 *9*	0.80 *8*
γ	678.83 *7*	12.6 *7*
γ	689.52 *10*	1.86 *12*
γ	690.14 *8*	2.08 *18*
γ	700.74 *8*	2.42 *19*

Photons (^{129}Ba(2.2 + 2.17 h))
(continued)

γ_{mode}	γ(keV)	γ(rel)
γ	712.32 *9*	2.4 *3*
γ	729.63 *11*	
γ	743.40 *9*	0.19 *3*
γ	748.65 *5*	5.8 *3*
γ	760.21 *7*	1.01 *15*
γ	768.90 *8*	2.48 *20*
γ	780.21 *6*	6.1 *4*
γ	803.01 *7*	8.0 *5*
γ	820.60 *9*	2.30 *17*
γ	829.19 *8*	0.48 *5*
γ	833.47 *7*	1.21 *10*
γ	872.48 *8*	4.7 *3*
γ	892.73 *6*	18.9 *12*
γ	926.22 *9*	0.99 *12*
γ	933.47 *10*	1.94 *13*
γ	935.25 *11*	2.03 *18*
γ	947.21 *11*	1.10 *14*
γ	957.66 *7*	4.55 *9*
γ	962.55 *6*	2.18 *18*
γ	991.15 *9*	1.59 *16*
γ	999.52 *7*	6.8 *5*
γ	1027.18 *19*	0.25 *3*
γ	1034.80 *8*	8.5 *5*
γ	1044.55 *7*	17 *1*
γ	1046.76 *8*	5.4 *5*
γ	1072.36 *9*	0.54 *10*
γ	1078.04 *9*	1.13 *14*
γ	1122.22 *8*	6.0 *4*
γ	1126.51 *9*	2.3 *3*
γ	1164.09 *8*	1.83 *18*
γ	1209.11 *8*	6.8 *4*
γ	1221.46 *6*	6.2 *5*
γ	1236.93 *10*	0.86 *9*
γ	1250.54 *15*	~1
γ	1292.40 *16*	1.77 *18*
γ	1370.67 *18*	0.56 *8*
γ	1386.03 *8*	0.35 *5*
γ	1443.51 *18*	0.60 *8*
γE1	1459.04 *6*	48.5 *25*
γ	1472.53 *9*	0.88 *9*
γ	1492.53 *8*	0.55 *7*
γ	1603.61 *8*	0.22 *3*
γ	1623.60 *7*	10.4 *6*
γ	1641.37 *6*	0.41 *8*
γ	1751.92 *15*	0.72 *5*
γ	1805.94 *8*	0.42 *4*
γ	1830.18 *17*	0.46 *5*
γ	1947.19 *24*	0.19 *2*
γ	1953.76 *24*	0.45 *5*

$^{129}_{56}$Ba(2.17 *4* h)

Mode: ϵ
Δ: -85092 *14* keV
SpA: 1.119×10^7 Ci/g
Prod: ^{121}Sb(^{11}B,3n); ^{133}Cs(p,5n)

see ^{129}Ba(2.2 h) for γ rays

$^{129}_{57}$La(11.6 *2* min)

Mode: ϵ
Δ: -81380 *50* keV
SpA: 1.256×10^8 Ci/g
Prod: ^{12}C on Sb; ^{16}O on In; ^{130}Ba(p,2n)

Photons (^{129}La)

$\langle\gamma\rangle$=363 5 keV

γ_{mode}	γ(keV)	γ(%)†
Ba L_{ℓ}	3.954	0.082 12
Ba L_{η}	4.331	0.036 4
Ba L_{α}	4.465	2.3 3
Ba L_{β}	4.946	2.0 3
Ba L_{γ}	5.618	0.27 4
Ba $K_{\alpha2}$	31.817	12.1 6
Ba $K_{\alpha1}$	32.194	22.2 11
Ba $K_{\beta1}$'	36.357	6.1 3
Ba $K_{\beta2}$'	37.450	1.55 8
γ E1	64.62 5	0.222 25
γ [M1]	85.22 7	<0.025 ?
γ [E2]	102.13 5	<0.025
γ M1	110.53 4	16.9 8
γ [E1]	138.63 5	0.40 5
γ M1,E2	143.22 5	1.26 5
γ E2+45%M1	168.04 5	1.28 5
γ	173.58 10	0.49 10
γ [M1]	178.43 6	0.124 25
γ [M1]	203.24 6	0.20 7
γ [M1]	205.49 9	0.27 5
γ [E1]	207.84 5	0.47 10
γ	244.78 20	0.148 25
γ M1,E2	253.75 4	8.0 4
γ	254.89 7	0.32 7
γ	270.82 8	0.074 25
γ [M1+E2]	278.57 5	24.7
γ	307.17 9	0.35 5 ?
γ E1	318.37 5	2.03 12
γ	339.22 7	0.22 5
γ	341.57 8	0.30 5
γ	346.39 8	<0.07
γ M1(+E2)	346.47 5	5.11 25
γ [M1+E2]	348.71 8	1.58 15
γ	349.37 10	~0.10
γ	381.37 8	0.099 25
γ	393.51 6	~0.10
γ [M1+E2]	406.19 7	0.49 5
γ	413.98 9	0.12 5
γ	431.69 7	0.40 5
γ	433.32 7	0.25 5
γ E2	448.59 5	5.2 3
γ M1(+E2)	457.00 5	8.0 6
γ [M1+E2]	458.13 6	2.10 15
γ	507.26 7	0.96 10
γ	531.02 8	0.12 5
γ	533.82 7	0.42 5
γ	549.41 8	0.54 7
γ	570.24 6	0.17 7
γ	588.29 8	0.099 25
γ	601.35 6	1.01 7
γ	609.38 7	0.22 5
γ	610.04 7	0.124 25
γ	617.79 7	1.04 7
γ	622.0 3	0.25 7
γ	628.10 9	0.074 25
γ	632.78 20	~0.049
γ	651.54 8	0.30 5
γ	652.91 9	~0.049
γ [M1+E2]	674.82 9	~0.049
γ	703.48 6	0.64 5
γ	711.88 5	0.124 25
γ	738.86 8	0.272 25
γ	744.25 10	~0.049
γ	760.76 9	<0.07
γ	771.72 10	0.074 25
γ	776.55 8	<0.049
γ	778.08 6	0.64 7
γ	808.86 9	0.099 25
γ	814.3 3	~0.049
γ	816.36 7	0.15 5
γ	831.78 20	0.15 5
γ	841.17 8	0.20 5
γ	841.25 12	0.32 5
γ	866.07 12	<0.049
γ	880.21 6	0.40 5
γ	888.61 6	0.49 5
γ	901.33 24	0.074 25
γ [M1]	928.57 8	0.272 25
γ	965.95 24	0.124 25
γ	984.39 8	0.148 25
γ	1004.3 3	0.074 25

Photons (^{129}La)
(continued)

γ_{mode}	γ(keV)	γ(%)†
γ	1017.57 9	0.222 25
γ	1061.89 12	0.099 25
γ	1067.93 7	0.247 25
γ	1071.14 9	<0.07
γ	1086.52 8	0.247 25
γ	1094.92 8	0.074 25
γ	1119.82 12	0.074 25
γ	1135.75 8	0.124 25
γ	1150.91 10	~0.049
γ	1160.68 8	0.148 25
γ	1185.49 7	0.124 25
γ	1236.48 10	0.099 25
γ	1291.78 7	0.40 5
γ	1321.24 10	0.074 25
γ	1328.72 8	0.173 25
γ	1356.39 7	0.124 25
γ	1356.79 10	0.124 25
γ	1381.60 10	~0.025
γ	1409.27 8	0.124 25
γ	1439.25 7	0.074 25
γ	1459.87 10	0.124 25
γ	1486.39 17	<0.049
γ	1499.67 10	0.099 25
γ	1524.48 10	0.099 25
γ	1533.47 11	~0.025
γ	1547.89 8	~0.025
γ	1551.01 17	0.124 25
γ	1587.69 9	0.074 25
γ	1610.15 7	0.074 25
γ	1672.09 11	~0.025
γ	1711.90 11	~0.049 ?
γ	1736.71 11	0.074 25
γ	1755.73 8	~0.025 ?
γ	1778.23 10	0.099 25
γ	1785.5 5	~0.049
γ	1792.99 17	0.124 25
γ	1866.26 8	~0.049
γ	1910.08 20	0.074 25
γ	1966.89 17	<0.025
γ	1990.46 11	~0.049
γ	2071.55 17	<0.049
γ	2285.26 17	<0.049

† 2.8% uncert(syst)

Atomic Electrons (^{129}La)

$\langle e\rangle$=21.9 6 keV

e_{bin}(keV)	⟨e⟩(keV)	e(%)
5	1.14	21.7 24
6	0.82	14.3 16
25 - 65	1.35	4.76 25
73	8.0	11.0 5
79 - 102	0.042	0.042 8
105	1.55	1.48 7
106 - 143	1.57	1.34 9
162 - 208	0.31	0.184 22
216	1.11	0.52 4
217 - 240	0.037	~0.017
241	2.92	1.21 10
243 - 271	0.30	0.121 23
273	0.53	0.19 6
277 - 318	0.79	0.262 23
333 - 381	0.170	0.049 6
388 - 433	0.81	0.195 20
443 - 470	0.184	0.041 6
494 - 534	0.047	0.009 3
543 - 591	0.076	0.013 4
595 - 637	0.024	0.0039 10
646 - 675	0.018	0.0027 12
697 - 744	0.025	0.0035 11
755 - 804	0.019	0.0025 8
808 - 851	0.017	0.0020 7
860 - 901	0.0102	0.00115 14
923 - 967	0.0062	0.00066 18
978 - 1024	0.0053	0.00053 19
1030 - 1071	0.0085	0.00082 24
1081 - 1130	0.0057	0.00051 13

Atomic Electrons (^{129}La)
(continued)

e_{bin}(keV)	⟨e⟩(keV)	e(%)
1131 - 1180	0.0020	0.00018 7
1184 - 1231	0.0013	0.00010 5
1235 - 1284	0.0049	0.00039 17
1286 - 1329	0.0052	0.00040 12
1344 - 1382	0.0018	0.00013 5
1402 - 1449	0.0023	0.00016 5
1454 - 1550	0.0046	0.00030 6
1573 - 1671	0.0010	6.4 24 $\times10^{-5}$
1674 - 1773	0.0034	0.00020 4
1777 - 1873	0.0011	6.2 20 $\times10^{-5}$
1904 - 1989	0.00056	2.9 12 $\times10^{-5}$
2034 - 2070	0.00019	~9 $\times10^{-6}$
2248 - 2284	0.00017	~8 $\times10^{-6}$

Continuous Radiation (^{129}La)

$\langle\beta+\rangle$=675 keV;$\langle$IB$\rangle$=4.7 keV

E_{bin}(keV)		⟨ ⟩(keV)	(%)
0 - 10	β+	1.68 $\times10^{-6}$	2.01 $\times10^{-5}$
	IB	0.025	
10 - 20	β+	7.1 $\times10^{-5}$	0.000424
	IB	0.025	0.17
20 - 40	β+	0.00203	0.0062
	IB	0.077	0.26
40 - 100	β+	0.103	0.132
	IB	0.139	0.21
100 - 300	β+	5.3	2.41
	IB	0.40	0.22
300 - 600	β+	41.2	8.9
	IB	0.51	0.118
600 - 1300	β+	287	30.1
	IB	1.21	0.133
1300 - 2500	β+	340	20.6
	IB	1.9	0.107
2500 - 3720	β+	0.82	0.0323
	IB	0.39	0.0140
	Σβ+		62

$^{129}_{57}$La(560 *50* ms)

Mode: IT

Δ: -81208 *50* keV

SpA: 9.0$\times10^{10}$ Ci/g

Prod: ^{121}Sb(^{12}C,4n)

Photons (^{129}La)

$\langle\gamma\rangle$=50.7 7 keV

γ_{mode}	γ(keV)	γ(%)†
La L_{ℓ}	4.121	0.27 4
La L_{η}	4.529	0.141 15
La L_{α}	4.649	7.5 9
La L_{β}	5.154	7.2 8
La L_{γ}	5.842	0.94 12
La $K_{\alpha2}$	33.034	24.4 8
La $K_{\alpha1}$	33.442	44.8 14
La $K_{\beta1}$'	37.777	12.5 4
La $K_{\beta2}$'	38.927	3.27 11
γ M1	67.5 3	23.8
γ E3	104.5 3	4.69

† 2.5% uncert(syst)

Atomic Electrons (^{129}La)

$\langle e \rangle$=125.3 *18* keV

e_{bin}(keV)	$\langle e \rangle$(keV)	e(%)
5	3.7	67 *7*
6	2.9	48 *5*
26	0.187	0.71 *7*
27	0.53	1.95 *20*
28	0.86	3.1 *3*
29	19.9	69.6 *20*
31	0.135	0.43 *4*
32	0.57	1.77 *18*
33	0.164	0.50 *5*
36	0.029	0.079 *8*
37	0.048	0.130 *13*
38	0.0161	0.043 *4*
61	5.32	8.69 *25*
62	0.513	0.832 *24*
66	17.1	26.1 *7*
67	0.366	0.544 *15*

Atomic Electrons (^{129}La)
(continued)

e_{bin}(keV)	$\langle e \rangle$(keV)	e(%)
67	0.00549	0.00814 *23*
98	1.93	1.96 *6*
99	54.0	54.6 *15*
103	13.4	12.9 *4*
104	3.74	3.58 *10*
105	0.00352	0.00336 *10*

$^{129}_{58}$Ce(3.5 *5* min)

Mode: ϵ
SpA: 4.2×10^{8} Ci/g
Prod: protons on Pr; ^{114}Cd(^{20}Ne,5n)

$^{129}_{59}$Pr(24 *5* s)

Mode: ϵ
SpA: 3.6×10^{9} Ci/g
Prod: ^{32}S on ^{102}Pd

$^{129}_{60}$Nd(5.9 *6* s)

Mode: ϵ, ϵp
SpA: 1.40×10^{10} Ci/g
Prod: ^{32}S on ^{102}Pd

A = 130

NDS **13**, 133 (1974)

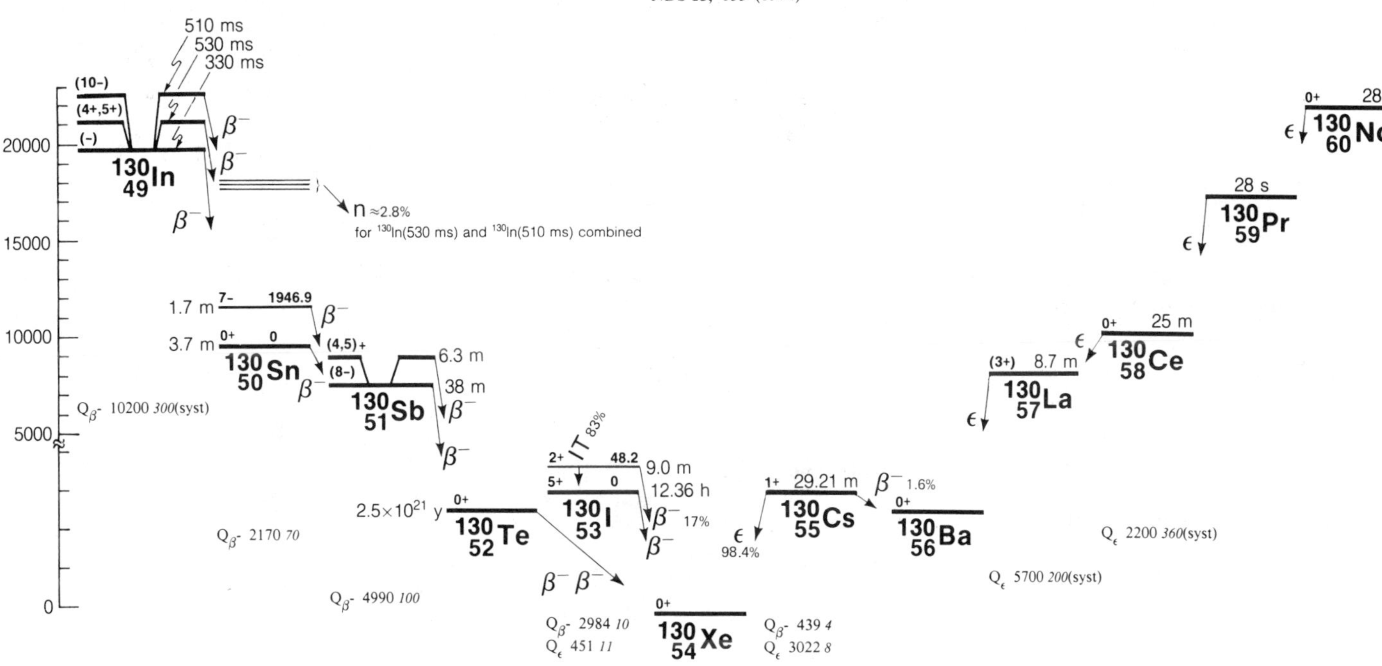

$^{130}_{49}$In(330 *30* ms)

Mode: β-
Δ: -69990 *320* keV syst
SpA: 1.099×10^{11} Ci/g
Prod: fission

decay data not yet evaluated

$^{130}_{49}$In(530 *30* ms)

Mode: β-, β-n(2.8 *15* %, includes ^{130}In(510 ms))
Δ: -69990 *320* keV syst
SpA: 9.1×10^{10} Ci/g
Prod: fission

decay data not yet evaluated

$^{130}_{49}$In(510 *20* ms)

Mode: β-, β-n(2.8 *15* %, includes ^{130}In(530 ms))
Δ: -69990 *320* keV syst
SpA: 9.3×10^{10} Ci/g
Prod: fission

decay data not yet evaluated

$^{130}_{50}$Sn(3.72 *11* min)

Mode: β-
Δ: -80190 *120* keV
SpA: 3.88×10^8 Ci/g
Prod: fission

Photons (^{130}Sn)

⟨γ⟩=964 *26* keV

γ_{mode}	γ(keV)	γ(%)†
Sb L_ℓ	3.189	0.083 *12*
Sb L_η	3.437	0.042 *5*
Sb L_α	3.604	2.3 *3*
Sb L_β	3.930	2.0 *3*
Sb L_γ	4.433	0.24 *4*
Sb $K_{\alpha2}$	26.111	16.0 *9*
Sb $K_{\alpha1}$	26.359	29.8 *17*
Sb $K_{\beta1}$'	29.712	7.8 *5*
Sb $K_{\beta2}$'	30.561	1.67 *10*
γ M1	70.19 *9*	35.9 *18*
γ	149.6 *1*	6.1 *3* *
γ M1,E2	192.64 *13*	71 *4*
γ M1,E2	229.31 *12*	23.6 *12*
γ	316.67 *15*	1.42 *14*
γ	341.45 *14*	2.13 *21*
γ	384.69 *15*	1.42 *14*
γ	434.88 *20*	14.2 *7*
γ	472.04 *14*	0.71 *7*
γ	550.66 *12*	3.2 *4*
γ	726.14 *16*	0.71 *7*
γ	743.30 *9*	18.7 *9*
γ	779.98 *9*	59 *3*
γ	852.6 *3*	0.70 *14* *
γ	872.3 *3*	1.4 *3* *

† 9.9% uncert(syst)
* with ^{130}(3.7 or 1.7 min)

Atomic Electrons (^{130}Sn)

⟨e⟩=53 *3* keV

e_{bin}(keV)	⟨e⟩(keV)	e(%)
4	1.95	46 *6*
5 - 30	2.17	13.9 *8*
40	21.3	54 *3*
65	4.19	6.4 *3*
66	0.384	0.58 *3*
69	0.96	1.39 *7*
70	0.227	0.324 *17*
162	12.8	7.9 *17*
188	2.1	1.1 *3*
189 - 193	1.1	0.55 *25*
199	3.1	1.54 *25*
225 - 229	0.67	0.30 *8*
286 - 317	0.17	0.056 *21*
337 - 385	0.07	0.021 *8*
404 - 442	0.49	0.12 *6*
467 - 472	0.0032	0.0007 *3*
520 - 551	0.08	0.015 *7*
696 - 743	0.31	0.044 *19*
749 - 780	0.9	0.12 *6*

Continuous Radiation (^{130}Sn)

⟨β-⟩=422 keV;⟨IB⟩=0.51 keV

E_{bin}(keV)		⟨ ⟩(keV)	(%)
0 - 10	β-	0.053	1.05
	IB	0.019	
10 - 20	β-	0.160	1.07
	IB	0.019	0.129
20 - 40	β-	0.65	2.18
	IB	0.035	0.122
40 - 100	β-	4.86	6.9
	IB	0.092	0.143
100 - 300	β-	52	25.9
	IB	0.20	0.116
300 - 600	β-	165	37.3
	IB	0.122	0.030
600 - 1300	β-	197	25.5
	IB	0.029	0.0042
1300 - 1473	β-	1.21	0.090
	IB	1.61×10^{-5}	1.21×10^{-6}

$^{130}_{50}$Sn(1.7 *1* min)

Mode: β-
Δ: -78243 *120* keV
SpA: 8.5×10^8 Ci/g
Prod: fission

Photons (^{130}Sn)*

γ_{mode}	γ(keV)	γ(%)†
γ	43.8 *1*	3.1 *7*
γ [M1]	60.20 *8*	4.8 *10*
γ	63.1 *1*	4.8 *10*
γ E2	84.70 *8*	14.3 *17*
γ M1	144.90 *8*	34 *3*
γ	311.3 *1*	13.9 *14*
γ	543.6 *2*	9.9 *10*
γ	899.2 *2*	16.7 *17*

† upper limit
* see also ^{130}Sn(3.7 min)

Photons (^{130}Sn(3.7 or 1.7 min))

γ_{mode}	γ(keV)	γ(rel)*
γ	91.9 *1*	2.4 *7*
γ	96.7 *1*	2.4 *7*
γ	912.6 *5*	2.7 *7*
γ	937.7 *5*	5.1 *10*
γ	962.4 *5*	3.4 *7*
γ	972.2 *5*	3.1 *7*
γ	1159.0 *5*	3.1 *7*
γ	1221.1 *8*	~1
γ	1456.3 *8*	3.1 *7*
γ	1478.5 *8*	3.1 *7*
γ	1660.7 *8*	4.1 *7*
γ	2153.2 *8*	~1

* relative to 34 for 145γ with ^{130}Sn(1.7 min)

$^{130}_{51}$Sb(38.4 *9* min)

Mode: β-
Δ: -82360 *100* keV
SpA: 3.77×10^7 Ci/g
Prod: ^{130}Te(n,p); fission

Photons (^{130}Sb)

⟨γ⟩=3265 *54* keV

γ_{mode}	γ(keV)	γ(%)
Te L_ℓ	3.335	0.018 *3*
Te L_η	3.606	0.0094 *11*
Te L_α	3.768	0.51 *6*
Te L_β	4.122	0.45 *6*
Te L_γ	4.646	0.052 *8*
Te $K_{\alpha2}$	27.202	3.12 *19*
Te $K_{\alpha1}$	27.472	5.8 *4*
Te $K_{\beta1}$'	30.980	1.55 *9*
Te $K_{\beta2}$'	31.877	0.338 *21*
γ E2	182.46 *9*	65 *4*
γ M1(+E2)	258.11 *22*	4.0 *8*
γ M1,E2	285.50 *19*	3.5 *4*
γ M1,E2	303.46 *18*	5.8 *6*
γ E1	331.08 *10*	78 *4*
γ M1,E2	455.55 *17*	4.8 *5*
γ	462.6 *3*	0.8 *2*
γ E1	468.19 *9*	18 *1*
γ	483.88 *22*	2.2 *3*
γ	506.9 *3*	2.0 *2*
γ	595.7 *3*	1.0 *2*
γ	626.91 *21*	2.8 *3*
γ	635.83 *21*	1.6 *4*
γ	654.9 *3*	2.0 *2*
γ	658.47 *20*	1.7 *4*
γ	669.4 *3*	1.1 *3*
γ	681.18 *22*	6.5 *7*
γ	686.80 *21*	3.2 *4*
γ M1(+E2)	732.17 *9*	22 *2*
γ E2	793.62 *9*	100 *5*
γ	829.84 *20*	1.8 *4*
γ E2	839.58 *10*	100
γ	856.1 *3*	1.6 *4*
γ	883.5 *3*	1.2 *3*
γ	915.4 *8*	1.8 *4*
γ	926.2 *3*	0.40 *8*
γ M1,E2	935.10 *15*	19 *1*
γ	992.3 *10*	1.9 *4*
γ	1000.5 *7*	2.3 *5*
γ	1030.7 *7*	1.5 *3*
γ	1075.3 *6*	0.40 *8*
γ	1089.5 *4*	3.7 *7*
γ	1133.5 *3*	0.40 *8*
γ	1137.6 *3*	0.30 *6*
γ	1141.6 *3*	2.0 *4*
γ	1146.4 *5*	0.6 *2*
γ	1239.0 *3*	1.8 *4*
γ	1258.6 *7*	1.0 *2*
γ	1292.4 *4*	3.7 *7*
γ	1369.5 *7*	1.10 *22*
γ	1418.97 *22*	1.20 *24*
γ	1444.2 *8*	2.5 *5*
γ	1472.7 *3*	0.60 *12*
γ	1488.6 *8*	0.6 *2*
γ	1499.8 *8*	~0.4
γ	1521.8 *4*	0.80 *16*
γ	1533.6 *4*	0.90 *18*
γ	1562.01 *21*	0.60 *12*
γ	1582.1 *6*	1.9 *4*
γ	1617.2 *8*	0.9 *2*
γ	1626.8 *8*	0.6 *2*
γ	1655.2 *7*	0.80 *16*
γ	1750.05 *23*	0.30 *6*
γ	1762.9 *7*	2.5 *5*
γ	1884.6 *8*	0.7 *2*
γ	1948.2 *8*	1.2 *2*
γ	1997.6 *5*	2.1 *2*
γ	2024.6 *4*	0.40 *8*

Atomic Electrons (^{130}Sb)

⟨e⟩=31.1 *11* keV

e_{bin}(keV)	⟨e⟩(keV)	e(%)
4 - 31	0.85	12.3 *9*
151	16.0	10.6 *7*
178	4.0	2.23 *14*
181	0.379	0.209 *14*
182	0.64	0.354 *23*
226 - 272	1.33	0.53 *5*
281 - 285	0.065	0.023 *6*
299	1.57	0.52 *3*
302 - 331	0.270	0.083 *4*
424 - 468	0.56	0.130 *11*
475 - 507	0.07	0.014 *6*
564 - 604	0.10	0.018 *6*
622 - 669	0.27	0.042 *11*
676 - 700	0.48	0.069 *11*
727 - 732	0.074	0.0102 *14*
762	1.62	0.213 *11*
789 - 798	0.294	0.0372 *23*
808	1.51	0.186 *10*
824 - 856	0.291	0.0348 *22*
879 - 926	0.32	0.035 *5*
930 - 969	0.090	0.0095 *19*
987 - 1031	0.025	0.0025 *9*
1043 - 1089	0.045	0.0043 *18*
1102 - 1146	0.034	0.0031 *9*
1207 - 1254	0.026	0.0022 *7*
1258 - 1292	0.034	0.0027 *12*
1338 - 1369	0.009	0.0007 *3*
1387 - 1419	0.028	0.0020 *7*
1439 - 1489	0.015	0.0010 *3*
1490 - 1585	0.038	0.0025 *5*
1595 - 1654	0.011	0.00067 *19*
1718 - 1762	0.019	0.0011 *4*
1853 - 1947	0.012	0.00062 *17*
1966 - 2024	0.015	0.00075 *23*

Continuous Radiation (^{130}Sb)

⟨β-⟩=672 keV;⟨IB⟩=1.32 keV

E_{bin}(keV)		⟨ ⟩(keV)	(%)
0 - 10	β-	0.0393	0.78
	IB	0.027	
10 - 20	β-	0.119	0.79
	IB	0.026	0.18
20 - 40	β-	0.482	1.61
	IB	0.050	0.17
40 - 100	β-	3.51	4.99
	IB	0.138	0.21
100 - 300	β-	35.2	17.5
	IB	0.35	0.20
300 - 600	β-	104	23.5
	IB	0.33	0.080
600 - 1300	β-	299	33.2
	IB	0.32	0.039
1300 - 2500	β-	226	13.5
	IB	0.072	0.0045
2500 - 2844	β-	4.02	0.155
	IB	9.7×10^{-5}	3.8×10^{-6}

$^{130}_{51}$Sb(6.3 *2* min)

Mode: β-
Δ: -82360 *100* keV
SpA: 2.29×10^{8} Ci/g
Prod: ^{130}Te(n,p); fission; daughter ^{130}Sn(3.7 min)

Photons (^{130}Sb)

⟨γ⟩=2653 *45* keV

γ_{mode}	γ(keV)	γ(%)
Te L_ℓ	3.335	0.0110 *16*
Te L_η	3.606	0.0057 *6*
Te L_α	3.768	0.31 *4*
Te L_β	4.121	0.27 *3*
Te L_γ	4.645	0.032 *5*
Te $K_{\alpha2}$	27.202	1.87 *10*
Te $K_{\alpha1}$	27.472	3.50 *19*
Te $K_{\beta1}$'	30.980	0.93 *5*
Te $K_{\beta2}$'	31.877	0.203 *11*
γ E2	182.46 *9*	41 *2*
γ [M1+E2]	348.61 *18*	5.1 *5*
γ	369.7 *3*	2.3 *5* ?
γ	370.13 *20*	2.0 *4*
γ	405.44 *17*	0.5 *2*
γ E1	468.19 *9*	3.1 *3*
γ	481.73 *20*	1.9 *4*
γ	502.54 *20*	1.9 *4*
γ	627.2 *3*	5.1 *5*
γ	647.9 *3*	2.4 *5* ?
γ	658.1 *3*	0.7 *2*
γ	697.62 *18*	4.4 *4*
γ [M1+E2]	748.74 *22*	4.0 *4*
γ E2	793.62 *9*	86 *4*
γ	816.52 *18*	12 *1*
γ E2	839.58 *10*	100
γ	861.40 *25*	~0.4
γ [E2]	920.6 *2*	4 *1*
γ	942.3 *4*	2.8 *3*
γ	949.92 *22*	1.0 *2*
γ	985.5 *4*	1.6 *3*
γ	1017.70 *16*	30 *2*
γ	1039.7 *4*	1.0 *2*
γ	1046.6 *4*	2.8 *3*
γ	1071.8 *4*	2.2 *2* ?
γ [M1+E2]	1103.06 *19*	3.7 *4*
γ	1132.37 *22*	1.3 *3*
γ [E2]	1142.23 *19*	5.6 *6*
γ	1177.3 *3*	2.2 *2*
γ	1200.16 *17*	3.6 *4*
γ	1232.4 *4*	1.3 *2*
γ	1299.0 *5*	0.8 *2*
γ	1323.2 *5*	~0.4 ?
γ	1491.24 *19*	1.3 *3*
γ	1597.8 *3*	2.6 *3*
γ [E2]	1896.68 *20*	1.3 *3*
γ	1925.99 *23*	~0.4 ?

Atomic Electrons (^{130}Sb)

⟨e⟩=19.3 *6* keV

e_{bin}(keV)	⟨e⟩(keV)	e(%)
4 - 31	0.51	7.4 *5*
151	10.1	6.7 *4*
178	2.50	1.40 *7*
181	0.239	0.132 *7*
182	0.406	0.224 *12*
317 - 366	0.58	0.18 *3*
369 - 405	0.027	0.007 *3*
436 - 482	0.15	0.033 *9*
498 - 503	0.009	~0.0017
595 - 644	0.20	0.033 *15*
647 - 693	0.08	0.013 *6*
697 - 744	0.093	0.0130 *20*
748 - 749	0.0026	0.00034 *6*
762	1.39	0.183 *9*
785 - 794	0.39	0.050 *11*
808	1.51	0.186 *4*
812 - 861	0.281	0.0337 *16*
889 - 938	0.109	0.0121 *25*
941 - 986	0.33	0.034 *16*
1008 - 1047	0.11	0.011 *3*
1067 - 1110	0.127	0.0116 *11*
1127 - 1176	0.064	0.0055 *15*
1177 - 1201	0.016	0.0013 *5*
1227 - 1267	0.008	0.0006 *3*
1291 - 1323	0.0045	0.00035 *17*

Atomic Electrons (^{130}Sb) (continued)

e_{bin}(keV)	⟨e⟩(keV)	e(%)
1459 - 1491	0.010	0.0007 *3*
1566 - 1597	0.019	0.0012 *5*
1865 - 1925	0.0122	0.00065 *13*

Continuous Radiation (^{130}Sb)

⟨β-⟩=994 keV;⟨IB⟩=2.3 keV

E_{bin}(keV)		⟨ ⟩(keV)	(%)
0 - 10	β-	0.0140	0.279
	IB	0.037	
10 - 20	β-	0.0428	0.285
	IB	0.036	0.25
20 - 40	β-	0.178	0.59
	IB	0.071	0.25
40 - 100	β-	1.37	1.94
	IB	0.20	0.31
100 - 300	β-	17.1	8.3
	IB	0.55	0.31
300 - 600	β-	77	16.8
	IB	0.57	0.135
600 - 1300	β-	404	43.0
	IB	0.64	0.076
1300 - 2500	β-	468	27.8
	IB	0.18	0.0110
2500 - 3357	β-	26.5	0.99
	IB	0.0017	6.7×10^{-5}

$^{130}_{52}$Te(2.5 *3* $\times10^{21}$ yr)

Mode: β-β-
Δ: -87348 *4* keV
Prod: natural source
%: 33.80 *2*

$^{130}_{53}$I (12.36 *1* h)

Mode: β-
Δ: -86897 *10* keV
SpA: 1.9503×10^{6} Ci/g
Prod: ^{129}I(n,γ); ^{133}Cs(n,α); ^{130}Te(p,n); ^{130}Te(d,2n)

Photons (^{130}I)

⟨γ⟩=2139 *21* keV

γ_{mode}	γ(keV)	γ(%)†
Xe L_ℓ	3.634	0.0027 *4*
Xe L_η	3.955	0.00126 *14*
Xe L_α	4.104	0.076 *9*
Xe L_β	4.522	0.067 *9*
Xe L_γ	5.118	0.0085 *12*
Xe $K_{\alpha2}$	29.461	0.446 *19*
Xe $K_{\alpha1}$	29.782	0.83 *4*
Xe $K_{\beta1}$'	33.606	0.224 *10*
Xe $K_{\beta2}$'	34.606	0.0530 *24*
γ	158.83 *18*	0.020 *7*
γ	190.455 *17*	<0.0049
γ	227.559 *21*	0.012 *5*
γ	246.435 *19*	0.047 *5*
γ	280.13 *4*	0.024 *7*

Photons (^{130}I) (continued)

γ_{mode}	γ(keV)	γ(%)†
γ	293.51 *20*	<0.0049
γ[E1]	302.51 *5*	0.013 *5*
γ	363.478 *16*	0.089 *20*
γ[M1+E2]	418.013 *14*	34.2 *7*
γ[M1+E2]	427.935 *14*	0.082 *9*
γ	457.732 *16*	0.237 *15*
γ[M1+E2]	510.390 *14*	0.852 *20*
γ[E2]	536.101 *17*	99
γ	539.107 *16*	1.4 *3*
γ[M1+E2]	553.932 *8*	0.662 *16*
γ[M1+E2]	586.108 *11*	1.69 *3*
γ[M1+E2]	603.564 *7*	0.615 *21*
γ	622.90 *9*	~0.017
γ[E2]	668.563 *8*	96.1 *19*
γ[E2]	686.019 *9*	1.069 *21*
γ[E2]	729.561 *16*	~0.011
γ[E2]	739.482 *14*	82.3 *16*
γ	748.76 *14*	0.012 *5*
γ	771.0 *5*	~0.004
γ	800.366 *19*	0.101 *5*
γ	808.315 *19*	0.236 *6*
γ	814.17 *8*	0.025 *5*
γ	821.208 *19*	0.043 *5*
γ[E1]	854.99 *5*	0.035 *5*
γ	867.84 *10*	0.043 *6*
γ	877.37 *4*	0.191 *8*
γ	896.71 *8*	0.021 *5*
γ	944.233 *24*	0.062 *14*
γ	967.040 *15*	0.877 *18*
γ	996.836 *16*	0.028 *5*
γ	1060.30 *14*	0.017 *5*
γ	1094.32 *20*	0.028 *8*
γ[M1+E2]	1096.496 *15*	0.553 *11*
γ[E2]	1122.207 *18*	0.254 *8*
γ[M1+E2]	1157.493 *8*	11.31 *23*
γ	1222.58 *3*	0.179 *5*
γ[E2]	1272.124 *10*	0.749 *15*
γ	1305.04 *3*	0.00485 *20*
γ	1403.926 *18*	0.345 *12*
γ	1417.73 *8*	0.0119 *20*
γ	1424.768 *18*	0.0208 *20*
γ	1488.23 *14*	0.0119 *20*
γ	1500.27 *8*	0.0396 *20*
γ	1545.93 *4*	0.023 *4*
γ	1547.793 *24*	0.018 *4*
γ	1607.32 *10*	0.045 *3*
γ	1689.77 *10*	0.0054 *10*

† 2.0% uncert(syst)

Atomic Electrons (^{130}I)

⟨e⟩=11.0 *3* keV

e_{bin}(keV)	⟨e⟩(keV)	e(%)
5 - 33	0.118	1.58 *15*
124 - 159	0.006	~0.0044
185 - 228	0.006	0.0027 *13*
241 - 289	0.0035	0.0014 *5*
292 - 329	0.004	~0.0012
358 - 363	0.0009	~0.00024
383	1.83	0.48 *5*
393 - 428	0.36	0.087 *16*
452 - 476	0.035	0.0074 *10*
502	3.14	0.63 *3*
505 - 519	0.066	0.013 *5*
531	0.477	0.090 *4*
534 - 581	0.215	0.0395 *20*
585 - 623	0.0057	0.00096 *12*
634	2.18	0.344 *15*
651	0.0234	0.00359 *16*
663	0.292	0.0440 *20*
664 - 695	0.105	0.0157 *4*
705	1.62	0.23 *1*
714 - 749	0.284	0.0387 *13*
766 - 814	0.0072	0.0009 *3*
816 - 863	0.0040	0.00048 *18*
867 - 910	0.0013	0.00015 *6*
932 - 967	0.013	0.0014 *6*

Atomic Electrons (^{130}I) (continued)

e_{bin}(keV)	⟨e⟩(keV)	e(%)
991 - 1026	0.00024	2.4 *11* $\times10^{-5}$
1055 - 1096	0.0126	0.00117 *12*
1117 - 1157	0.177	0.0157 *21*
1188 - 1223	0.0020	0.00017 *7*
1238 - 1272	0.0092	0.00074 *3*
1300 - 1305	6.9 $\times10^{-6}$	5.3 *22* $\times10^{-7}$
1369 - 1418	0.0036	0.00026 *10*
1419 - 1466	0.00043	2.9 *10* $\times10^{-5}$
1483 - 1573	0.00073	4.8 *12* $\times10^{-5}$
1602 - 1689	9.2 $\times10^{-5}$	5.6 *16* $\times10^{-6}$

Continuous Radiation (^{130}I)

⟨β-⟩=285 keV;⟨IB⟩=0.27 keV

E_{bin}(keV)		⟨ ⟩(keV)	(%)
0 - 10	β-	0.107	2.13
	IB	0.0137	
10 - 20	β-	0.320	2.14
	IB	0.0129	0.090
20 - 40	β-	1.28	4.28
	IB	0.024	0.084
40 - 100	β-	8.9	12.8
	IB	0.059	0.092
100 - 300	β-	75	38.4
	IB	0.108	0.064
300 - 600	β-	131	31.0
	IB	0.048	0.0121
600 - 1300	β-	68	9.2
	IB	0.0084	0.00120
1300 - 1779	β-	1.29	0.090
	IB	8.3 $\times10^{-5}$	6.0 $\times10^{-6}$

$^{130}_{53}$I (9.0 *1* min)

Mode: IT(83 *3* %), β-(17 *3* %)

Δ: -86849 *10* keV

SpA: 1.606 $\times10^{8}$ Ci/g

Prod: ^{129}I(n,γ); ^{133}Cs(n,α); ^{130}Te(p,n)

Photons (^{130}I)

⟨γ⟩=121 *18* keV

γ_{mode}	γ(keV)	γ(%)†
I L_{ℓ}	3.485	0.108 *16*
I L_{η}	3.780	0.029 *4*
I L_{α}	3.937	3.0 *4*
I L_{β}	4.354	2.1 *3*
I L_{γ}	4.942	0.27 *5*
I $K_{\alpha2}$	28.317	7.5 *5*
I $K_{\alpha1}$	28.612	13.9 *10*
Xe $K_{\alpha2}$	29.461	0.029 *6*
Xe $K_{\alpha1}$	29.782	0.055 *11*
I $K_{\beta1}'$	32.278	3.7 *3*
I $K_{\beta2}'$	33.225	0.85 *6*
Xe $K_{\beta1}'$	33.606	0.015 *3*
Xe $K_{\beta2}'$	34.606	0.0035 *7*
γ_{IT} (M3)	48.2 *3*	0.042 *8*
$\gamma_{\beta-}$	352.2 *5*	0.00119 *24*
$\gamma_{\beta-}$[M1+E2]	427.935 *14*	0.00038 *6*
$\gamma_{\beta-}$[M1+E2]	510.390 *14*	0.0040 *5*
$\gamma_{\beta-}$[E2]	536.101 *17*	17
$\gamma_{\beta-}$[M1+E2]	586.108 *11*	1.18 *16*
$\gamma_{\beta-}$[M1+E2]	603.564 *7*	0.00022 *3*
$\gamma_{\beta-}$[E2]	668.563 *8*	0.0117 *23*

Photons (^{130}I) (continued)

γ_{mode}	γ(keV)	γ(%)†
$\gamma_{\beta-}$[E2]	686.019 *9*	0.00038 *5*
$\gamma_{\beta-}$	836.9 *6*	0.00085 *17*
$\gamma_{\beta-}$[E2]	946.0 *5*	0.0008 *3* *
$\gamma_{\beta-}$[M1+E2]	1028.3 *4*	0.039 *6*
$\gamma_{\beta-}$[M1+E2]	1096.496 *15*	0.0026 *3*
$\gamma_{\beta-}$[E2]	1122.207 *18*	0.178 *24*
$\gamma_{\beta-}$	1174.2 *6*	0.0014 *3*
$\gamma_{\beta-}$	1263.6 *5*	0.0022 *4*
$\gamma_{\beta-}$[E2]	1272.124 *10*	0.00027 *3*
$\gamma_{\beta-}$	1380.5 *4*	0.041 *7*
$\gamma_{\beta-}$	1440.5 *6*	0.0116 *23*
$\gamma_{\beta-}$[M1+E2]	1614.4 *4*	0.51 *8*
$\gamma_{\beta-}$	1760.3 *6*	0.032 *6*
$\gamma_{\beta-}$	1849.7 *5*	0.0019 *3*
$\gamma_{\beta-}$	1958.3 *7*	0.020 *4*
$\gamma_{\beta-}$	1966.6 *4*	0.057 *9*
$\gamma_{\beta-}$	1989.25 *20*	0.0026 *7*
$\gamma_{\beta-}$	2008.3 *7*	0.0049 *10*
$\gamma_{\beta-}$	2029.4 *4*	0.00046 *20*
$\gamma_{\beta-}$	2092.7 *7*	0.0054 *11*
$\gamma_{\beta-}$	2101.6 *10*	0.0107 *21*
$\gamma_{\beta-}$	2109.0 *6*	0.0110 *22*
$\gamma_{\beta-}$[E2]	2150.5 *4*	0.024 *4*
$\gamma_{\beta-}$	2296.4 *6*	0.0041 *8*
$\gamma_{\beta-}$	2307.9 *10*	0.0024 *5* ?
$\gamma_{\beta-}$	2502.7 *4*	0.0132 *21*
$\gamma_{\beta-}$	2544.4 *7*	0.00051 *10*
$\gamma_{\beta-}$	2762.7 *10*	0.00095 *19*

† uncert(syst): 3.6% for IT, 18% for β-

* doublet

Atomic Electrons (^{130}I)

⟨e⟩=32.2 *10* keV

e_{bin}(keV)	⟨e⟩(keV)	e(%)
5	2.5	53 *6*
15	4.4	29.3 *20*
23 - 33	0.85	3.41 *21*
43	7.6	17.6 *12*
44	10.4	23.9 *16*
47	4.6	9.7 *7*
48	1.18	2.45 *16*
318 - 352	6.8 $\times10^{-5}$	~2 $\times10^{-5}$
393 - 428	2.4 $\times10^{-5}$	5.9 *9* $\times10^{-6}$
476 - 510	0.54	0.108 *22*
531 - 569	0.141	0.026 *4*
581 - 604	0.0067	0.00115 *24*
634 - 681	0.00032	5.1 *9* $\times10^{-5}$
685 - 686	3.0 $\times10^{-7}$	4.3 *5* $\times10^{-8}$
802 - 837	1.5 $\times10^{-5}$	~2 $\times10^{-6}$
911 - 946	7.2 $\times10^{-6}$	~8 $\times10^{-7}$
994 - 1028	0.00070	7.0 *14* $\times10^{-5}$
1062 - 1096	0.0022	0.00020 *3*
1117 - 1140	0.00036	3.2 *4* $\times10^{-5}$
1169 - 1174	2.2 $\times10^{-6}$	1.9 *9* $\times10^{-7}$
1229 - 1272	2.7 $\times10^{-5}$	2.2 *9* $\times10^{-6}$
1346 - 1380	0.00040	3.0 *13* $\times10^{-5}$
1406 - 1440	0.00011	8 *3* $\times10^{-6}$
1580 - 1613	0.0055	0.00035 *6*
1726 - 1815	0.00026	1.5 *6* $\times10^{-5}$
1844 - 1932	0.00047	2.4 *8* $\times10^{-5}$
1953 - 2029	0.000123	6.3 *13* $\times10^{-6}$
2058 - 2150	0.00036	1.70 *23* $\times10^{-5}$
2262 - 2307	3.9 $\times10^{-5}$	1.7 *5* $\times10^{-6}$
2468 - 2543	7.6 $\times10^{-5}$	3.1 *10* $\times10^{-6}$
2728 - 2762	4.9 $\times10^{-6}$	1.8 *6* $\times10^{-7}$

Continuous Radiation (^{130}I)

⟨β-⟩=159 keV;⟨IB⟩=0.35 keV

E_{bin}(keV)		⟨ ⟩(keV)	(%)
0 - 10	β-	0.00297	0.059
	IB	0.0060	
10 - 20	β-	0.0090	0.060
	IB	0.0059	0.041
20 - 40	β-	0.0371	0.124
	IB	0.0115	0.040
40 - 100	β-	0.278	0.395
	IB	0.032	0.050
100 - 300	β-	3.18	1.56
	IB	0.088	0.050
300 - 600	β-	12.9	2.84
	IB	0.091	0.022
600 - 1300	β-	67	7.1
	IB	0.099	0.0118
1300 - 2496	β-	76	4.61
	IB	0.019	0.00125

$^{130}_{54}$Xe(stable)

Δ: -89881.1 *15* keV
%: 4.1 *1*

$^{130}_{55}$Cs(29.21 *4* min)

Mode: ε(98.4 *2* %), β-(1.6 *2* %)
Δ: -86859 *8* keV
SpA: 4.951×10^7 Ci/g
Prod: ^{127}I(α,n)

Photons (^{130}Cs)

⟨γ⟩=59 *3* keV

γ_{mode}	γ(keV)	γ(%)†
Xe L_ℓ	3.634	0.071 *10*
Xe L_η	3.955	0.032 *3*
Xe L_α	4.104	1.98 *22*
Xe L_β	4.522	1.75 *21*
Xe L_γ	5.124	0.23 *3*
Xe $K_{\alpha2}$	29.461	11.8 *4*
Xe $K_{\alpha1}$	29.782	21.9 *7*
Xe $K_{\beta1}$'	33.606	5.91 *18*
Xe $K_{\beta2}$'	34.606	1.40 *5*
γ	352.2 *5*	0.00020 *5*
γ [E2]	536.101 *17*	4.0
γ [M1+E2]	586.108 *11*	0.50 *7*
γ	671.2 *7*	0.013 *3*
γ	894.6 *7*	0.41 *8*
γ [M1+E2]	1028.3 *4*	0.020 *3*
γ [E2]	1122.207 *18*	0.075 *10*
γ	1257.3 *7*	0.081 *16*
γ	1263.6 *5*	0.039 *7*
γ	1380.5 *4*	0.0069 *11*
γ	1480.7 *7*	0.025 *5*
γ [M1+E2]	1614.4 *4*	0.26 *4*
γ	1686.9 *10*	0.20 *4*
γ	1706.5 *10*	0.15 *3*
γ	1849.7 *5*	0.033 *6*
γ	1958.3 *7*	0.017 *4*
γ	1966.6 *4*	0.0095 *15*
γ	1996.8 *10*	0.17 *4*
γ	2092.7 *7*	0.014 *3*
γ [E2]	2150.5 *4*	0.0120 *19*
γ	2502.7 *4*	0.0022 *4*

† 12% uncert(syst)

Atomic Electrons (^{130}Cs)

⟨e⟩=3.28 *20* keV

e_{bin}(keV)	⟨e⟩(keV)	e(%)
5	1.79	36 *4*
24	0.36	1.50 *15*
25	0.46	1.85 *19*
28	0.127	0.45 *5*
29	0.27	0.94 *10*
30	0.036	0.122 *12*
32 - 33	0.045	0.138 *11*
318 - 352	1.1 ×10^{-5}	~4×10^{-6}
502	0.128	0.026 *3*
531 - 552	0.041	0.0075 *8*
581 - 586	0.0028	0.00049 *10*
637 - 671	0.00030	~5×10^{-5}
860 - 895	0.007	0.0008 *4*
911 - 946	3.7 ×10^{-6}	4.0 *17* ×10^{-7}
994 - 1028	0.00036	3.6 *7* ×10^{-5}
1088 - 1122	0.00104	9.6 *12* ×10^{-5}
1223 - 1264	0.0013	0.00011 *4*
1346 - 1380	6.7 ×10^{-5}	5.0 *21* ×10^{-6}
1446 - 1481	0.00022	1.5 *7* ×10^{-5}
1580 - 1672	0.0052	0.00032 *6*
1681 - 1706	0.00036	2.1 *6* ×10^{-5}
1815 - 1849	0.00024	1.3 *5* ×10^{-5}
1924 - 1996	0.0014	6.9 *23* ×10^{-5}
2058 - 2150	0.00018	8.7 *18* ×10^{-6}
2468 - 2502	1.2 ×10^{-5}	4.9 *16* ×10^{-7}

Continuous Radiation (^{130}Cs)

⟨β-⟩=2.10 keV;⟨β+⟩=394 keV;⟨IB⟩=4.1 keV

E_{bin}(keV)		⟨ ⟩(keV)	(%)
0 - 10	β-	0.00380	0.076
	β+	2.09×10^{-6}	2.51×10^{-5}
	IB	0.0158	
10 - 20	β-	0.0112	0.075
	β+	8.4×10^{-5}	0.000505
	IB	0.0156	0.108
20 - 40	β-	0.0438	0.146
	β+	0.00230	0.0070
	IB	0.066	0.22
40 - 100	β-	0.280	0.404
	β+	0.110	0.142
	IB	0.087	0.134
100 - 300	β-	1.47	0.81
	β+	5.5	2.46
	IB	0.25	0.138
300 - 600	β-	0.297	0.088
	β+	40.2	8.7
	IB	0.37	0.084
600 - 1300	β+	235	25.2
	IB	1.21	0.130
1300 - 2500	β+	113	7.6
	IB	2.0	0.112
2500 - 3022	IB	0.113	0.0043
	Σβ+		44

$^{130}_{56}$Ba(stable)

Δ: -87299 *8* keV
%: 0.106 *2*

$^{130}_{57}$La(8.7 *1* min)

Mode: ε
Δ: -81600 *200* keV syst
SpA: 1.661×10^8 Ci/g
Prod: ^{130}Ba(p,n); C on Sb

Photons (^{130}La)

⟨γ⟩=1394 *15* keV

γ_{mode}	γ(keV)	γ(%)†
Ba L_ℓ	3.954	0.041 *6*
Ba L_η	4.331	0.0181 *19*
Ba L_α	4.465	1.13 *12*
Ba L_β	4.946	1.02 *13*
Ba L_γ	5.618	0.135 *19*
Ba $K_{\alpha2}$	31.817	6.03 *19*
Ba $K_{\alpha1}$	32.194	11.1 *3*
Ba $K_{\beta1}$'	36.357	3.07 *10*
Ba $K_{\beta2}$'	37.450	0.78 *3*
γ	196.46 *6*	0.20 *3* ?
γ	234.41 *5*	0.11 *3*
γ	267.75 *5*	0.59 *5*
γ	325.36 *6*	
γ	327.47 *10*	1.97 *12*
γ E2	357.31 *6*	81.0
γ	367.09 *5*	0.28 *3*
γ	397.51 *10*	0.44 *8*
γ	452.93 *5*	3.8 *7*
γ	459.16 *5*	0.87 *7*
γ	463.94 *5*	0.62 *5*
γ	483.60 *5*	1.04 *5*
γ	502.16 *5*	0.178 *10*
γ	521.82 *6*	1.50 *6*
γ (E2)	544.43 *5*	17.6 *4*
γ [M1+E2]	550.65 *4*	27.3 *6*
γ	566.56 *10*	0.35 *3*
γ	569.44 *5*	3.33 *7*
γ	575.67 *5*	2.6 *5* *
γ	576.17 *7*	
γ	591.02 *10*	0.50 *4*
γ	601.50 *5*	0.31 *3*
γ	604.02 *10*	0.16 *3*
γ	649.39 *5*	1.66 *11*
γ	655.62 *5*	0.17 *3*
γ	692.68 *7*	0.32 *4*
γ	703.02 *10*	0.34 *5*
γ	718.01 *5*	2.89 *6*
γ	789.30 *5*	0.35 *5*
γ	800.02 *10*	1.01 *10*
γ	818.02 *10*	0.19 *3*
γ	821.02 *10*	0.78 *7*
γ	840.32 *5*	0.23 *4*
γ	869.25 *5*	1.58 *14*
γ [E2]	907.96 *6*	17.4 *4*
γ	936.53 *5*	0.87 *7*
γ	942.76 *5*	1.03 *8*
γ	956.83 *5*	0.48 *4*
γ	974.75 *5*	3.09 *15*
γ	985.76 *5*	0.28 *8* ?
γ	1003.59 *5*	8.11 *16*
γ	1017.02 *10*	0.32 *4*
γ	1120.09 *5*	2.05 *6*
γ	1151.83 *8*	0.373 *23*
γ	1170.94 *5*	3.89 *17*
γ	1177.17 *5*	2.17 *11*
γ	1200.05 *5*	3.14 *19*
γ	1349.02 *10*	0.99 *12*
γ	1409.76 *5*	0.45 *12*
γ	1415.98 *5*	0.09 *4*
γ	1438.69 *5*	2.39 *16*
γ	1444.91 *5*	0.95 *9*
γ	1487.18 *5*	0.79 *15*
γ	1525.40 *5*	6.9 *3*
γ	1557.35 *7*	
γ	1648.02 *10*	0.43 *10*
γ	1654.02 *10*	0.30 *7*
γ	1691.02 *10*	0.42 *7*
γ	1721.59 *5*	2.07 *8*
γ	1948.02 *10*	0.66 *16*
γ	1960.40 *5*	
γ	1989.34 *5*	

† 1.2% uncert(syst)
* combined intensity for doublet

Atomic Electrons (^{130}La)

$\langle e \rangle$=12.6 *3* keV

e_{bin}(keV)	$\langle e \rangle$(keV)	e(%)
5	0.57	10.9 *11*
6	0.41	7.2 *7*
25 - 36	0.64	2.28 *11*
159 - 197	0.048	0.028 *12*
228 - 268	0.06	0.027 *13*
290	0.12	~0.04
320	5.61	1.75 *4*
321 - 330	0.042	0.013 *5*
351	0.685	0.195 *5*
352	0.366	0.1041 *24*
356 - 398	0.309	0.087 *4*
415 - 465	0.29	0.067 *24*
478 - 502	0.06	~0.011
507	0.59	0.117 *5*
513	1.09	0.21 *4*
516 - 565	0.54	0.100 *17*
566 - 612	0.07	0.011 *4*
618 - 666	0.026	0.0040 *11*
681 - 718	0.07	0.011 *5*
752 - 800	0.046	0.0059 *18*
803 - 840	0.033	0.0040 *18*
863 - 908	0.375	0.0428 *20*
919 - 969	0.18	0.019 *7*
970 - 1017	0.026	0.0026 *9*
1083 - 1120	0.033	0.0031 *13*
1134 - 1177	0.12	0.010 *3*
1194 - 1200	0.0058	0.00048 *21*
1312 - 1349	0.011	0.0009 *4*
1372 - 1416	0.037	0.0026 *8*
1433 - 1482	0.013	0.0009 *3*
1486 - 1524	0.07	0.0047 *19*
1611 - 1690	0.026	0.0016 *5*
1716 - 1721	0.0024	0.00014 *6*
1911 - 1947	0.0052	0.00027 *11*

Continuous Radiation (^{130}La)

$\langle\beta+\rangle$=1193 keV;$\langle$IB$\rangle$=7.7 keV

E_{bin}(keV)		$\langle\ \rangle$(keV)	(%)
0 - 10	β+	8.7×10^{-7}	1.04×10^{-5}
	IB	0.038	
10 - 20	β+	3.67×10^{-5}	0.000221
	IB	0.038	0.26
20 - 40	β+	0.00106	0.00324
	IB	0.093	0.32
40 - 100	β+	0.055	0.070
	IB	0.22	0.33
100 - 300	β+	2.99	1.34
	IB	0.65	0.36
300 - 600	β+	25.2	5.4
	IB	0.82	0.19
600 - 1300	β+	226	23.1
	IB	1.68	0.19
1300 - 2500	β+	661	36.2
	IB	2.6	0.143
2500 - 5000	β+	278	9.5
	IB	1.59	0.052
5000 - 5343	IB	0.00117	2.3×10^{-5}
	$\Sigma\beta$+		76

$^{130}_{58}$Ce(25 *2* min)

Mode: ϵ

Δ: -79400 *300* keV syst

SpA: 5.8×10^{7} Ci/g

Prod: ^{139}La(p,10n); ^{130}Ba(^{3}He,3n)

Photons (^{130}Ce)

γ_{mode}	γ(keV)
γ	107.9
γ	130.8 *5*
γ	134.4
γ	162.2
γ	181.3
γ	209.1
γ	219.6
γ	307.3

$^{130}_{59}$Pr(28 *6* s)

Mode: ϵ

SpA: 3.1×10^{9} Ci/g

Prod: ^{32}S on ^{106}Cd

$^{130}_{60}$Nd(28 *3* s)

Mode: ϵ

SpA: 3.1×10^{9} Ci/g

Prod: ^{32}S on ^{106}Cd

A = 131

NDS **17**, 573 (1976)

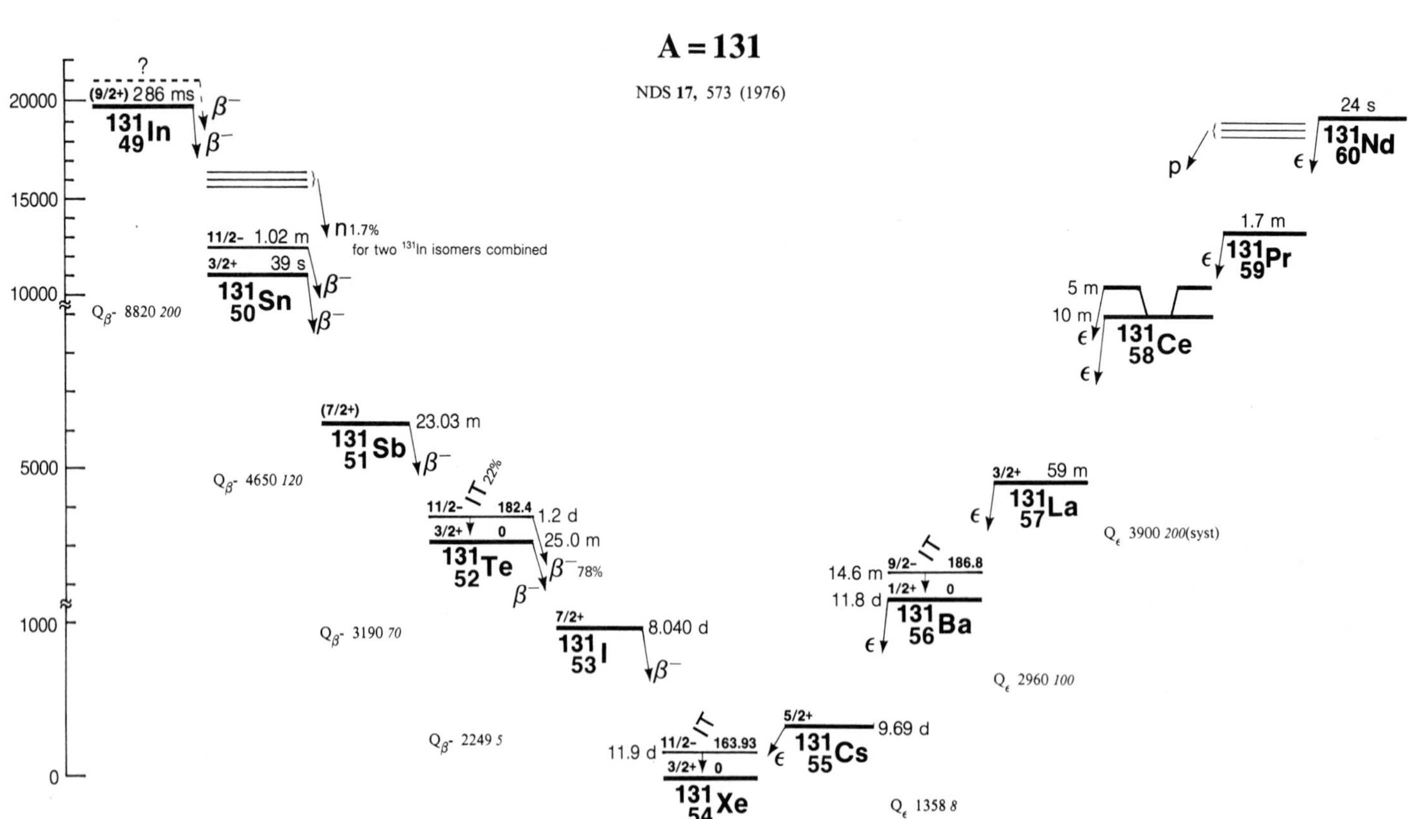

$^{131}_{49}$In(286 *9* ms)

Mode: β-, β-n(1.72 *23* %, includes other ^{131}In isomer)
Δ: -68550 *240* keV
SpA: 1.13×10^{11} Ci/g
Prod: fission

Photons (^{131}In(specific isomer unknown))

γ$_{mode}$	γ(keV)	γ(rel)
γ	332.8 *2*	~10
γ	2433 *1*	100

$^{131}_{49}$In($t_{1/2}$ unknown)

Mode: β-, β-n(1.72 *23* %, includes ^{131}In(286 ms))
Δ: -68550 *240* keV
Prod: fission

see ^{131}In(286 ms) for γ rays

$^{131}_{50}$Sn(39 *2* s)

Mode: β-
Δ: -77370 *140* keV
SpA: 2.19×10^{9} Ci/g
Prod: fission

decay data not yet evaluated

$^{131}_{50}$Sn(1.02 *5* min)

Mode: β-
Δ: >-77370 keV
SpA: 1.40×10^{9} Ci/g
Prod: fission

decay data not yet evaluated

$^{131}_{51}$Sb(23.03 *4* min)

Mode: β-
Δ: -82020 *70* keV
SpA: 6.231×10^{7} Ci/g
Prod: fission

Photons (^{131}Sb)

⟨γ⟩=1809 *63* keV

γ$_{mode}$	γ(keV)	γ(%)†
γ	134.56 *11*	2.4 *9*
γ	159.4 *3*	0.44 *13*
γ M4	182.436 *20*	*
γ	274.17 *11*	<2
γ [M1+E2]	295.67 *19*	<3
γ	301.10 *11*	2.2 *4*
γ	324.22 *17*	1.1 *4*
γ	326.6 *3*	~1
γ	433.71 *22*	<4 ?
γ	457.2 *4*	<1
γ	620.03 *22*	1.5 *3*
γ	625.6 *3*	<2 ?
γ	642.24 *9*	22 *4*
γ	657.9 *4*	<7
γ	669.04 *14*	1.8 *3*
γ	726.27 *12*	3.8 *4*
γ	824.9 *2*	2.5 *3*
γ	854.2 *3*	3.1 *4* ?
γ	865.1 *3*	0.44 *9* ?
γ	911.67 *21*	0.66 *13*
γ	933.04 *10*	24.6 *18*
γ	943.34 *9*	44
γ	958.62 *11*	0.57 *18*
γ	991.5 *5*	1.3 *4*
γ	1050.97 *18*	~0.6
γ	1123.48 *22*	8.3 *7*
γ	1191.4 *4*	<1.0 ?
γ	1207.34 *10*	3.8 *3*
γ	1234.14 *13*	2.2 *4*
γ	1248.89 *20*	0.48 *22* ?
γ	1267.56 *16*	2.77 *22*
γ	1283.8 *8*	0.48 *9* ?
γ	1331.3 *3*	0.79 *9*
γ	1360.3 *3*	~0.9
γ	1392.1 *3*	0.7 *3*
γ	1398.82 *23*	1.28 *13*
γ	1455.07 *10*	~0.44
γ	1470.37 *20*	1.45 *13*
γ	1517.2 *3*	1.14 *13*
γ	1537.5 *3*	~0.4
γ	1544.56 *23*	0.8 *4* ?
γ	1553.07 *22*	~0.5
γ	1558.6 *3*	0.40 *18*
γ	1573.47 *20*	0.97 *22*
γ	1608.69 *22*	1.3 *3*
γ	1722.5 *4*	2.29 *13*
γ	1756.17 *13*	1.06 *13*
γ	1822.1 *4*	1.14 *22*
γ	1854.4 *3*	4.0 *5*
γ	1916.2 *8*	0.9 *4*
γ	1956.6 *3*	0.7 *4*
γ	1965.7 *4*	1.2 *6*
γ	1984.9 *4*	0.40 *18*
γ	2017.0 *7*	~0.6
γ	2031.0 *10*	0.22 *9*
γ	2116.9 *4*	~0.18
γ	2149.8 *6*	~0.5
γ	2167.5 *10*	0.31 *13*
γ	2179.8 *3*	2.1 *3*
γ	2256.3 *3*	0.66 *9*
γ	2335.4 *3*	1.76 *9*
γ	2354.5 *3*	0.31 *13*
γ	2398.40 *12*	1.06 *9*
γ	2496.40 *22*	0.62 *9*
γ	2552.02 *23*	0.35 *9*
γ	2662.07 *20*	1.01 *9*

† 9.1% uncert(syst)
* with ^{131}Te(1.2 d)

Continuous Radiation (^{131}Sb)

⟨β-⟩=582 keV;⟨IB⟩=1.05 keV

E$_{bin}$(keV)		⟨ ⟩(keV)	(%)
0 - 10	β-	0.053	1.06
	IB	0.024	
10 - 20	β-	0.160	1.07
	IB	0.023	0.161
20 - 40	β-	0.65	2.16
	IB	0.045	0.155
40 - 100	β-	4.68	6.7
	IB	0.120	0.19
100 - 300	β-	45.0	22.6
	IB	0.30	0.170
300 - 600	β-	121	27.4
	IB	0.26	0.062
600 - 1300	β-	254	29.2
	IB	0.23	0.027
1300 - 2500	β-	151	9.0
	IB	0.053	0.0034
2500 - 3008	β-	6.1	0.231
	IB	0.00029	1.09×10^{-5}

$^{131}_{52}$Te(25.0 *1* min)

Mode: β-
Δ: -85206 *4* keV
SpA: 5.740×10^{7} Ci/g
Prod: ^{130}Te(n,γ); daughter ^{131}Te (1.2 d)

Photons (^{131}Te)

⟨γ⟩=421 *3* keV

γ$_{mode}$	γ(keV)	γ(%)†
I L_ℓ	3.485	0.023 *3*
I L_η	3.780	0.0114 *12*
I L_α	3.937	0.65 *7*
I L_β	4.320	0.58 *7*
I L_γ	4.886	0.072 *10*
I $K_{\alpha2}$	28.317	4.01 *12*
I $K_{\alpha1}$	28.612	7.46 *23*
I $K_{\beta1}'$	32.278	2.01 *6*
I $K_{\beta2}'$	33.225	0.455 *15*
γ	78.560 *21*	3.3 *8* ×10^{-5}
γ [M1+E2]	109.379 *4*	0.064 *7*
γ	141.188 *19*	0.028 *5*
γ M1+10%E2	149.717 *3*	68.9
γ [M1+E2]	151.272 *20*	0.17 *6*
γ	221.535 *6*	0.033 *5*
γ	267.5 *3*	~0.004
γ	274.686 *5*	<0.007 ?
γ	278.199 *25*	0.099 *5*
γ [M1+E2]	280.192 *9*	0.017 *5*
γ	294.731 *19*	<0.0048 ?
γ	296.72 *3*	0.0085 *19*
γ	297.092 *13*	0.050 *5*
γ	300.00 *5*	0.039 *5*
γ [M1+E2]	342.951 *3*	0.703 *14*
γ	345.793 *12*	0.014 *4*
γ	351.67 *3*	0.023 *4*
γ [M1+E2]	353.666 *12*	0.019 *4*
γ [M1+E2]	384.065 *3*	0.896 *18*
γ	402.367 *10*	~0.007
γ	403.733 *18*	0.0067 *20*
γ	421.380 *19*	0.042 *8*
γ	438.280 *21*	~0.007
γ M1+E2	452.3299 *19*	18.2 *4*
γ [M1+E2]	469.76 *4*	0.015 *6*
γ [M1+E2]	492.667 *4*	4.84 *10*
γ	494.854 *20*	0.076 *7*
γ	496.220 *4*	0.034 *7*
γ [M1+E2]	544.921 *8*	0.427 *14*
γ	546.914 *25*	0.0066 *17*

Photons (^{131}Te) (continued)

γ_{mode}	γ(keV)	γ(%)†
γ	550.427 5	0.028 7
γ	567.327 11	0.103 6
γ	574.922 18	0.031 5
γ [M1+E2]	602.046 3	4.19 8
γ	605.599 5	0.117 7
γ [M1+E2]	654.300 9	1.53 3
γ	696.193 19	0.179 14
γ	702.36 5	~0.008
γ [E2]	702.519 17	0.00084 10
γ [E2]	727.015 4	0.469 9
γ	744.44 4	~0.008
γ	805.571 19	0.014 6
γ	825.111 5	0.028 7
γ	842.012 11	0.200 7
γ M1,E2	852.235 17	0.044 5
γ	853.82 4	0.096 5
γ	856.062 18	0.131 7
γ	881.15 8	0.025 4
γ	898.585 9	0.138 7
γ	934.490 4	0.875 18
γ	948.548 4	2.26 5
γ	951.390 11	0.331 7
γ	997.249 8	3.34 7
γ	999.242 25	0.029 4
γ	1005.777 18	0.015 3
γ	1007.964 9	0.799 16
γ	1035.05 9	~0.0028
γ	1066.87 10	~0.006
γ	1098.264 5	0.172 7
γ	1146.964 8	4.96 10
γ	1148.520 19	0.110 7
γ	1148.957 25	0.062 7
γ	1155.83 8	~0.0041
γ	1184.788 20	0.0055 21
γ	1198.58 5	0.0055 14
γ	1265.21 8	0.0048 14
γ	1277.438 5	0.118 5
γ	1294.338 11	0.482 10
γ	1298.235 19	0.0048 21
γ	1307.96 5	0.0069 7
γ	1350.912 9	0.061 3
γ	1427.153 6	0.105 3
γ	1500.627 9	0.115 3
γ	1527.736 20	0.057 3
γ	1548.16 9	~0.0009
γ	1579.98 10	0.0083 7
γ	1650.90 5	0.0124 7
γ	1800.62 5	0.0034 7
γ	1891.10 9	~0.0028
γ	1922.93 10	0.0034 7
γ	1973.1 4	0.0021 7
γ	2040.82 9	0.0069 7
γ	2072.64 10	0.0062 14

† 1.3% uncert(syst)

Atomic Electrons (^{131}Te)

⟨e⟩=24.9 4 keV

e_{bin}(keV)	⟨e⟩(keV)	e(%)
5 - 45	1.06	14.6 13
73 - 109	0.057	0.066 16
117	17.9	15.3 3
118 - 141	0.055	0.047 19
145	3.28	2.27 5
146 - 188	0.875	0.588 12
216 - 264	0.013	0.0053 18
266 - 313	0.055	0.0181 12
319 - 354	0.065	0.0186 14
369 - 417	0.0123	0.0032 4
419	0.81	0.194 19
420 - 470	0.34	0.075 7
487 - 534	0.052	0.0104 10
540 - 575	0.128	0.023 3
597 - 621	0.061	0.0099 11
649 - 698	0.0195	0.0029 3
701 - 744	0.00171	0.000237 15
772 - 821	0.0055	0.00068 20

Atomic Electrons (^{131}Te) (continued)

e_{bin}(keV)	⟨e⟩(keV)	e(%)
823 - 865	0.0049	0.00059 16
876 - 918	0.042	0.0046 17
929 - 975	0.054	0.0056 21
992 - 1035	0.008	0.0008 3
1062 - 1098	0.0020	0.00019 8
1114 - 1156	0.057	0.0051 21
1165 - 1199	6.7×10^{-5}	5.7 24 $\times10^{-6}$
1232 - 1277	0.0054	0.00043 16
1289 - 1318	0.0011	9 3 $\times10^{-5}$
1346 - 1394	0.0009	6 3 $\times10^{-5}$
1422 - 1467	0.0010	7 3 $\times10^{-5}$
1495 - 1579	0.00067	4.4 15 $\times10^{-5}$
1618 - 1650	9.4×10^{-5}	5.8 22 $\times10^{-6}$
1767 - 1858	4.0×10^{-5}	2.2 8 $\times10^{-6}$
1886 - 1972	3.8×10^{-5}	2.0 6 $\times10^{-6}$
2008 - 2072	8.1×10^{-5}	4.0 10 $\times10^{-6}$

Continuous Radiation (^{131}Te)

⟨β-⟩=696 keV;⟨IB⟩=1.27 keV

E_{bin}(keV)		⟨ ⟩(keV)	(%)
0 - 10	β-	0.0294	0.59
	IB	0.028	
10 - 20	β-	0.089	0.60
	IB	0.028	0.19
20 - 40	β-	0.367	1.22
	IB	0.053	0.19
40 - 100	β-	2.74	3.90
	IB	0.147	0.23
100 - 300	β-	30.8	15.2
	IB	0.38	0.21
300 - 600	β-	114	25.2
	IB	0.34	0.081
600 - 1300	β-	386	42.7
	IB	0.28	0.034
1300 - 2099	β-	162	10.7
	IB	0.021	0.00146

$^{131}_{52}$Te(1.25 *8* d)

Mode: β-(77.8 16 %), IT(22.2 16 %)

Δ: -85024 4 keV

SpA: 8.0×10^{5} Ci/g

Prod: ^{130}Te(n,γ); fission

Photons (^{131}Te(1.2 d + 25.0 min))

⟨γ⟩=1423 10 keV

γ_{mode}	γ(keV)	γ(%)†*
Te L_ℓ	3.335	0.026 4
I L_ℓ	3.485	0.021 3
Te L_η	3.606	0.0116 13
Te L_α	3.768	0.71 8
I L_η	3.780	0.0101 13
I L_α	3.937	0.57 7
Te L_β	4.128	0.61 8
I L_β	4.320	0.51 7
Te L_γ	4.672	0.075 11
I L_γ	4.883	0.063 10
Te $K_{\alpha2}$	27.202	3.71 15
Te $K_{\alpha1}$	27.472	6.9 3
I $K_{\alpha2}$	28.317	3.41 12
I $K_{\alpha1}$	28.612	6.34 23
Te $K_{\beta1}$'	30.980	1.84 8
Te $K_{\beta2}$'	31.877	0.401 17

Photons (^{131}Te(1.2 d + 25.0 min)) (continued)

γ_{mode}	γ(keV)	γ(%)†*
I $K_{\beta1}$'	32.278	1.71 6
I $K_{\beta2}$'	33.225	0.387 15
$\gamma_{\beta-}$	36.85 3	0.0120 15 ?
$\gamma_{\beta-}$	51.02 4	0.0062 15 ?
$\gamma_{\beta-}$	52.60 6	0.0054 15
$\gamma_{\beta-}$	53.933 25	~0.0012
$\gamma_{\beta-}$	55.81 10	0.0027 12
$\gamma_{\beta-}$	60.85 7	0.0058 15
$\gamma_{\beta-}$	62.383 20	0.0363 23 ?
$\gamma_{\beta-}$	63.21 10	0.0043 15
$\gamma_{\beta-}$	64.99 5	0.0081 19 ?
$\gamma_{\beta-}$	66.96 5	0.023 4
$\gamma_{\beta-}$	73.30 4	0.027 3 ?
$\gamma_{\beta-}$	78.560 21	0.015 3
$\gamma_{\beta-}$	79.11 3	0.128 4
$\gamma_{\beta-}$M1	81.130 14	4.06 8
$\gamma_{\beta-}$	86.482 24	0.147 4
$\gamma_{\beta-}$	95.01 12	~0.0039
$\gamma_{\beta-}$	96.41 20	0.0058 23
$\gamma_{\beta-}$	98.19 4	0.014 3
$\gamma_{\beta-}$	100.03 9	0.073 4
$\gamma_{\beta-}$	101.91 4	0.170 15
$\gamma_{\beta-}$M1	102.064 9	7.93 16
$\gamma_{\beta-}$	103.16 4	0.046 8 ?
$\gamma_{\beta-}$	104.99 5	0.027 4 ?
$\gamma_{\beta-}$[M1+E2]	109.379 4	0.0052 11 ?
$\gamma_{\beta-}$	111.906 21	0.031 8
$\gamma_{\beta-}$	113.33 4	0.012 4
$\gamma_{\beta-}$	123.7 5	~0.0039
$\gamma_{\beta-}$	125.2 3	0.009 3
$\gamma_{\beta-}$	126.21 9	~0.006
$\gamma_{\beta-}$	127.488 21	0.023 8 ?
$\gamma_{\beta-}$	130.51 10	0.070 8
$\gamma_{\beta-}$	132.21 10	~0.005
$\gamma_{\beta-}$	134.863 18	0.708 23
$\gamma_{\beta-}$	137.64 9	~0.08
$\gamma_{\beta-}$	149.21 6	0.077 19
$\gamma_{\beta-}$M1+10%E2	149.717 3	5.1 7
$\gamma_{\beta-}$	151.082 15	0.08 3
$\gamma_{\beta-}$	156.41 9	~0.039 ?
$\gamma_{\beta-}$	159.70 3	0.128 15
$\gamma_{\beta-}$	169.70 5	0.031 8
$\gamma_{\beta-}$	172.01 20	0.012 4
$\gamma_{\beta-}$	177.22 14	0.066 12 ?
$\gamma_{\beta-}$	182.27 3	0.73 19
γ_{IT} M4	182.436 20	0.855 24
$\gamma_{\beta-}$	183.195 15	0.155 19
$\gamma_{\beta-}$	188.15 3	0.213 12
$\gamma_{\beta-}$	189.65 4	0.50 4
$\gamma_{\beta-}$	190.47 3	0.116 15
$\gamma_{\beta-}$E1	200.637 16	7.54 15
$\gamma_{\beta-}$	203.97 4	0.019 8 ?
$\gamma_{\beta-}$	207.476 22	0.039 12
$\gamma_{\beta-}$	210.3 3	0.015 4
$\gamma_{\beta-}$	212.79 17	0.012 4 ?
$\gamma_{\beta-}$	213.970 21	0.425 19
$\gamma_{\beta-}$	227.55 5	~0.015 ?
$\gamma_{\beta-}$	230.67 4	0.193 12
$\gamma_{\beta-}$	232.36 6	0.093 12
$\gamma_{\beta-}$	234.69 6	~0.015 ?
$\gamma_{\beta-}$E1	240.934 10	7.58 15
$\gamma_{\beta-}$	253.146 14	0.650 13
$\gamma_{\beta-}$	255.465 25	0.309 12
$\gamma_{\beta-}$	261.44 4	0.015 4
$\gamma_{\beta-}$	267.42 15	~0.015 ?
$\gamma_{\beta-}$	269.10 22	<0.11 ?
$\gamma_{\beta-}$	278.570 17	1.78 4
$\gamma_{\beta-}$	280.70 4	~0.035
$\gamma_{\beta-}$	283.25 5	0.39 4
$\gamma_{\beta-}$	290.35 8	0.077 12
$\gamma_{\beta-}$	296.72 3	0.050 8
$\gamma_{\beta-}$	302.701 18	0.039 12
$\gamma_{\beta-}$	304.03 6	0.039 8
$\gamma_{\beta-}$	309.40 3	0.38 4
$\gamma_{\beta-}$	323.7 4	~0.015
$\gamma_{\beta-}$	330.832 23	0.031 12
$\gamma_{\beta-}$M1,E2	334.277 9	9.55 19
$\gamma_{\beta-}$	335.500 23	0.135 23
$\gamma_{\beta-}$[M1+E2]	342.951 3	0.0203 5 ?
$\gamma_{\beta-}$	342.998 13	0.39 12
$\gamma_{\beta-}$	346.06 8	0.10 3
$\gamma_{\beta-}$	351.33 4	0.209 19
$\gamma_{\beta-}$	354.00 5	0.08 4

Photons (^{131}Te(1.2 d + 25.0 min))
(continued)

γ_{mode}	γ(keV)	γ(%)†*
$\gamma_{\beta-}$	355.05 4	0.228 12
$\gamma_{\beta-}$	357.56 16	0.019 8 ?
$\gamma_{\beta-}$	362.13 21	~0.08 ?
$\gamma_{\beta-}$	365.052 22	1.20 15
$\gamma_{\beta-}$	375.797 20	0.012 4
$\gamma_{\beta-}$	377.86 14	<0.039 ?
$\gamma_{\beta-}$	379.57 9	0.019 8
$\gamma_{\beta-}$	383.831 20	0.20 3
$\gamma_{\beta-}$	403.733 18	0.032 10
$\gamma_{\beta-}$	408.18 24	~0.06 ?
$\gamma_{\beta-}$	417.44 4	0.278 19
$\gamma_{\beta-}$	432.46 4	0.66 3
$\gamma_{\beta-}$M1+E2	452.3299 19	1.5 3
$\gamma_{\beta-}$	462.941 24	1.82 4
$\gamma_{\beta-}$	468.22 3	0.31 3
$\gamma_{\beta-}$[M1+E2]	492.667 4	0.140 3 ?
$\gamma_{\beta-}$	506.81 20	0.089 15
$\gamma_{\beta-}$	524.75 3	0.135 15
$\gamma_{\beta-}$	530.65 9	0.104 19
$\gamma_{\beta-}$	541.50 3	0.112 23
$\gamma_{\beta-}$	546.914 25	0.039 8
$\gamma_{\beta-}$	558.07 3	0.023 8 ?
$\gamma_{\beta-}$	572.56 4	~0.043
$\gamma_{\beta-}$	580.20 9	0.077 23
$\gamma_{\beta-}$	586.296 19	1.97 8
$\gamma_{\beta-}$	597.21 13	0.050 19
$\gamma_{\beta-}$[M1+E2]	602.046 3	0.34 6
$\gamma_{\beta-}$	609.40 3	0.139 15
$\gamma_{\beta-}$	637.32 5	<0.031 ?
$\gamma_{\beta-}$	657.21 20	~0.031
$\gamma_{\beta-}$	665.108 23	4.33 9
$\gamma_{\beta-}$	681.83 21	0.031 8
$\gamma_{\beta-}$	685.88 4	0.155 12
$\gamma_{\beta-}$	695.58 4	0.40 3
$\gamma_{\beta-}$[E2]	702.519 17	0.391 19
$\gamma_{\beta-}$	713.130 22	1.43 15
$\gamma_{\beta-}$	738.77 4	0.066 12
$\gamma_{\beta-}$	744.217 22	1.59 4
$\gamma_{\beta-}$	748.27 4	~0.015 ?
$\gamma_{\beta-}$M1,E2	773.676 19	38.1 8
$\gamma_{\beta-}$	774.14 4	0.54 8
$\gamma_{\beta-}$	782.479 19	7.77 16
$\gamma_{\beta-}$	793.771 17	13.8 3
$\gamma_{\beta-}$	801.80 13	0.019 8
$\gamma_{\beta-}$	822.776 21	6.11 12
$\gamma_{\beta-}$	844.80 5	0.15 4
$\gamma_{\beta-}$	849.05 16	0.039 12 ?
$\gamma_{\beta-}$	852.10 4	~0.39
$\gamma_{\beta-}$M1,E2	852.235 17	20.6 4
$\gamma_{\beta-}$	856.062 18	0.62 4
$\gamma_{\beta-}$	864.865 24	0.19 4
$\gamma_{\beta-}$	872.330 18	0.101 12
$\gamma_{\beta-}$	881.95 4	0.035 12
$\gamma_{\beta-}$	909.994 18	3.29 8
$\gamma_{\beta-}$	920.571 20	1.20 8
$\gamma_{\beta-}$	923.35 5	0.116 23
$\gamma_{\beta-}$	930.29 6	~0.019 ?
$\gamma_{\beta-}$	941.35 4	0.78 3
$\gamma_{\beta-}$	987.84 9	0.155 12
$\gamma_{\beta-}$	995.28 4	0.089 15
$\gamma_{\beta-}$	999.242 25	0.169 18
$\gamma_{\beta-}$	1003.73 4	~0.027 ?
$\gamma_{\beta-}$	1005.777 18	0.072 14
$\gamma_{\beta-}$	1023.412 18	0.062 8
$\gamma_{\beta-}$	1028.01 9	~0.008 ?
$\gamma_{\beta-}$	1035.49 3	0.104 8
$\gamma_{\beta-}$	1059.710 18	1.55 4
$\gamma_{\beta-}$	1072.340 23	0.023 4
$\gamma_{\beta-}$	1108.72 4	0.023 8
$\gamma_{\beta-}$	1114.05 3	0.012 4
$\gamma_{\beta-}$	1125.475 17	11.41 23
$\gamma_{\beta-}$	1128.046 18	0.97 8
$\gamma_{\beta-}$	1134.16 21	~0.008 ?
$\gamma_{\beta-}$	1148.82 4	1.5 3
$\gamma_{\beta-}$	1148.957 25	0.36 9
$\gamma_{\beta-}$	1150.899 23	0.66 8
$\gamma_{\beta-}$	1162.65 4	0.027 8
$\gamma_{\beta-}$	1165.458 22	0.139 12
$\gamma_{\beta-}$	1182.01 4	~0.012
$\gamma_{\beta-}$	1206.605 17	9.74 19
$\gamma_{\beta-}$	1211.20 4	0.062 12
$\gamma_{\beta-}$	1227.19 22	<0.008 ?
$\gamma_{\beta-}$	1237.380 23	0.66 3
$\gamma_{\beta-}$	1254.13 13	0.027 4

Photons (^{131}Te(1.2 d + 25.0 min))
(continued)

γ_{mode}	γ(keV)	γ(%)†*
$\gamma_{\beta-}$	1315.173 22	0.70 8
$\gamma_{\beta-}$	1316.19 4	0.10 4
$\gamma_{\beta-}$	1318.52 4	0.039 8
$\gamma_{\beta-}$	1334.02 5	0.054 8
$\gamma_{\beta-}$	1340.54 3	0.101 12
$\gamma_{\beta-}$	1376.90 22	0.043 8
$\gamma_{\beta-}$	1389.49 4	0.015 4
$\gamma_{\beta-}$	1394.75 4	0.108 8
$\gamma_{\beta-}$	1403.85 13	~0.012
$\gamma_{\beta-}$	1496.286 17	0.058 8
$\gamma_{\beta-}$	1547.81 4	0.070 8
$\gamma_{\beta-}$	1646.001 17	1.24 4
$\gamma_{\beta-}$	1697.03 5	0.015 4
$\gamma_{\beta-}$	1830.559 17	~0.008 ?
$\gamma_{\beta-}$	1880.24 9	0.062 8
$\gamma_{\beta-}$	1887.72 3	1.35 4
$\gamma_{\beta-}$	1924.567 22	~0.0039
$\gamma_{\beta-}$	1936.06 5	0.073 8
$\gamma_{\beta-}$	1980.273 17	0.031 8
$\gamma_{\beta-}$	2001.05 3	2.01 4
$\gamma_{\beta-}$	2168.42 4	0.348 19
$\gamma_{\beta-}$	2270.62 8	0.383 19
$\gamma_{\beta-}$	2332.75 24	0.0027 4

† uncert(syst): 7.2% for IT, 2.9% for β-
* with ^{131}Te(25.0 min) in equilib

Atomic Electrons (^{131}Te(1.2 d + 25.0 min))
$\langle e \rangle$=52.3 9 keV

e_{bin}(keV)	$\langle e \rangle$(keV)	e(%)*
4 - 47	2.33	30.4 23
48	2.33	4.86 16
49 - 68	0.26	0.45 15
69	3.46	5.03 23
70 - 116	1.98	2.25 18
117	1.31	1.13 15
118 - 150	0.62	0.44 8
151	22.1	14.7 5
152 - 176	0.46	0.28 4
177	5.29	2.98 11
178	4.81	2.70 10
179 - 181	1.21	0.67 4
182	1.67	0.92 3
183 - 232	0.46	0.226 19
234 - 283	0.27	0.11 3
285 - 334	0.96	0.311 19
335 - 384	0.046	0.013 3
399 - 448	0.17	0.040 9
451 - 497	0.029	0.0063 13
502 - 547	0.0090	0.0017 4
553 - 602	0.067	0.012 5
604 - 653	0.08	~0.013
656 - 706	0.057	0.0084 23
708 - 749	0.94	0.127 19
761 - 802	0.48	0.062 16
812 - 860	0.47	0.058 8
864 - 910	0.074	0.008 3
915 - 962	0.0070	0.00075 17
966 - 1006	0.0058	0.00059 14
1018 - 1067	0.019	0.0019 8
1068 - 1116	0.14	0.013 5
1118 - 1165	0.031	0.0028 7
1173 - 1223	0.11	0.009 4
1226 - 1254	0.0010	8 3 $\times10^{-5}$
1282 - 1329	0.009	0.00072 23
1333 - 1377	0.0016	0.00012 3
1384 - 1404	0.00016	1.2 4 $\times10^{-5}$
1463 - 1547	0.0011	7 2 $\times10^{-5}$
1613 - 1696	0.010	0.00059 22
1797 - 1891	0.010	0.00051 17
1903 - 2000	0.013	0.00068 22
2135 - 2168	0.0021	10 3 $\times10^{-5}$
2237 - 2332	0.0022	10 3 $\times10^{-5}$

* with ^{131}Te(25.0 min) in equilib

$^{131}_{53}$I (8.040 1 d)

Mode: β-
Δ: -87455 4 keV
SpA: 1.23974$\times10^5$ Ci/g
Prod: fission

Photons (^{131}I)
$\langle \gamma \rangle$=382 6 keV

γ_{mode}	γ(keV)	γ(%)†
Xe L_ℓ	3.634	0.0089 12
Xe L_η	3.955	0.0041 4
Xe L_α	4.104	0.25 3
Xe L_β	4.521	0.22 3
Xe L_γ	5.116	0.028 4
Xe $K_{\alpha2}$	29.461	1.42 5
Xe $K_{\alpha1}$	29.782	2.64 10
Xe $K_{\beta1}'$	33.606	0.71 3
Xe $K_{\beta2}'$	34.606	0.169 7
γ M1(+<3.8%E2)	80.185 7	2.62 5
γ [M1+E2]	85.918 12	~9$\times10^{-5}$
γ M4	163.932 7	*
γ E2(+6.4%M1)	177.211 8	0.265 5
γ [E2]	232.167 13	~0.0014
γ [M1+E2]	272.492 11	0.0564 11
γ E2	284.299 8	6.06 12
γ [E1]	295.833 14	~0.0007
γ [E1]	302.442 14	0.0045 9
γ [M1+E2]	318.084 11	0.080 3
γ [M1+E2]	324.624 11	0.022 4
γ E2+7.0%M1	325.782 9	0.251 5
γ [M1+E2]	358.41 1	0.00917 20
γ E2+4.7%M1	364.483 8	81.2 16
γ [M1+E2]	404.809 9	0.0564 20
γ E2	502.993 9	0.361 7
γ E2	636.975 9	7.27 15
γ [E2]	642.708 8	0.220 4
γ M1+4.1%E2	722.892 8	1.80 4

† 0.99% uncert(syst)
* with ^{131}Xe(11.9 d)

Atomic Electrons (^{131}I)
$\langle e \rangle$=10.04 16 keV

e_{bin}(keV)	$\langle e \rangle$(keV)	e(%)
5	0.224	4.5 5
24 - 33	0.157	0.60 3
46	1.66	3.65 15
51	7.0 $\times10^{-5}$	~0.00014
75	0.439	0.587 25
79 - 86	0.120	0.151 5
143 - 177	0.096	0.0639 23
198 - 238	0.0065	0.00276 17
250	0.618	0.248 7
261 - 302	0.184	0.0656 11
313 - 326	0.0069	0.00215 11
330	5.12	1.55 4
353 - 358	0.000125	3.5 5 $\times10^{-5}$
359	0.770	0.214 6
360 - 405	0.352	0.0971 16
468 - 503	0.0152	0.00321 8
602 - 643	0.215	0.0355 8
688 - 723	0.0576	0.00832 20

Continuous Radiation (^{131}I)

⟨β-⟩=182 keV;⟨IB⟩=0.113 keV

E_{bin}(keV)		⟨ ⟩(keV)	(%)
0 - 10	β-	0.173	3.46
	IB	0.0092	
10 - 20	β-	0.51	3.43
	IB	0.0085	0.059
20 - 40	β-	2.03	6.8
	IB	0.0150	0.053
40 - 100	β-	13.4	19.3
	IB	0.033	0.052
100 - 300	β-	90	47.4
	IB	0.042	0.027
300 - 600	β-	75	19.6
	IB	0.0053	0.00150
600 - 807	β-	0.166	0.0253
	IB	4.4×10^{-6}	7.0×10^{-7}

$^{131}_{54}$Xe(stable)

Δ: -88426 4 keV

%: 21.2 4

$^{131}_{54}$Xe(11.9 *1* d)

Mode: IT

Δ: -88262 4 keV

SpA: 8.38×10^4 Ci/g

Prod: ^{130}Xe(n,γ)

Photons (^{131}Xe)

⟨γ⟩=20.0 5 keV

γ_{mode}	γ(keV)	γ(%)†
Xe L_ℓ	3.634	0.130 18
Xe L_η	3.955	0.052 6
Xe L_α	4.104	3.6 4
Xe L_β	4.530	3.0 4
Xe L_γ	5.140	0.40 6
Xe $K_{\alpha2}$	29.461	15.5 7
Xe $K_{\alpha1}$	29.782	28.9 13
Xe $K_{\beta1}$'	33.606	7.8 4
Xe $K_{\beta2}$'	34.606	1.85 9
γ M4	163.932 7	1.96

† 3.0% uncert(syst)

Atomic Electrons (^{131}Xe)

⟨e⟩=143 4 keV

e_{bin}(keV)	⟨e⟩(keV)	e(%)
5	3.2	64 7
24	0.48	1.98 21
25	0.61	2.4 3
28	0.168	0.60 6
29	0.36	1.24 13
30	0.048	0.161 17
32	0.0183	0.056 6
33	0.042	0.126 14
129	78.5	60.7 25
158	22.2	14.0 6
159	24.0	15.1 6
163	10.9	6.7 3
164	2.69	1.64 7

$^{131}_{55}$Cs(9.69 *1* d)

Mode: ϵ

Δ: -88079 6 keV

SpA: 1.0286×10^5 Ci/g

Prod: daughter ^{131}Ba

Photons (^{131}Cs)

⟨γ⟩=22.9 4 keV

γ_{mode}	γ(keV)	γ(%)
Xe L_ℓ	3.634	0.131 18
Xe L_η	3.955	0.060 6
Xe L_α	4.104	3.6 4
Xe L_β	4.521	3.3 4
Xe L_γ	5.131	0.43 6
Xe $K_{\alpha2}$	29.461	21.3 6
Xe $K_{\alpha1}$	29.782	39.7 12
Xe $K_{\beta1}$'	33.606	10.7 3
Xe $K_{\beta2}$'	34.606	2.54 8

Atomic Electrons (^{131}Cs)

⟨e⟩=5.7 4 keV

e_{bin}(keV)	⟨e⟩(keV)	e(%)
5	3.3	67 7
24	0.66	2.7 3
25	0.83	3.4 3
28	0.231	0.82 8
29	0.49	1.71 17
30	0.066	0.221 23
32	0.025	0.078 8
33	0.057	0.174 18

Continuous Radiation (^{131}Cs)

⟨IB⟩=0.076 keV

E_{bin}(keV)		⟨ ⟩(keV)	(%)
10 - 20	IB	0.00050	0.0030
20 - 40	IB	0.065	0.22
40 - 100	IB	0.0045	0.0079
100 - 300	IB	0.0058	0.0036
300 - 347	IB	2.2×10^{-5}	7.2×10^{-6}

$^{131}_{56}$Ba(11.8 *2* d)

Mode: ϵ

Δ: -86721 8 keV

SpA: 8.45×10^4 Ci/g

Prod: ^{130}Ba(n,γ); ^{133}Cs(p,3n)

Photons (^{131}Ba)

⟨γ⟩=458 23 keV

γ_{mode}	γ(keV)	γ(%)†
Cs L_ℓ	3.795	0.189 25
Cs L_η	4.142	0.086 9
Cs L_α	4.285	5.2 6
Cs L_β	4.731	4.7 6
Cs L_γ	5.364	0.61 9
Cs $K_{\alpha2}$	30.625	28.0 9

Photons (^{131}Ba) (continued)

γ_{mode}	γ(keV)	γ(%)†
Cs $K_{\alpha1}$	30.973	51.8 16
Cs $K_{\beta1}$'	34.967	14.2 4
Cs $K_{\beta2}$'	36.006	3.47 12
γ E2(+<6%M1)	54.864 16	0.096 6
γ M1+0.4%E2	78.718 12	0.754 24
γ [M1+E2]	82.466 15	~0.020
γ M1+3.7%E2	92.271 9	0.64 4
γ E2	123.777 10	29.1 6
γ	128.13 9	0.016 4 ?
γ M1+21%E2	133.582 12	2.19 8
γ (E2)	137.330 15	0.032 4
γ [M1+E2]	157.146 15	0.199 20
γ M1	216.048 11	20 4
γ M1	239.612 16	2.41 6
γ M1,E2	246.842 19	0.60 4
γ M1	249.417 14	2.81 8
γ M1,E2	294.476 17	0.159 20
γ	323.233 24	<0.020 ?
γ M1,E2	351.158 24	0.119 20
γ	368.92 3	0.030 10
γ M1,E2	373.194 14	13.3 14
γ M1,<20%E2	403.988 16	1.29 8
γ M1,E2	427.549 21	0.099 6
γ M1,E2	451.39 3	0.042 4
γ	461.190 25	~0.06
γ	462.62 3	~0.040 ?
γ (M1)	480.378 21	0.34 4
γ E2	486.454 18	1.89 20
γ M1	496.258 16	44 4
γ	546.26 9	0.0060 20
γ	562.844 24	0.0060 20
γ M1,E2	572.649 21	0.159 10
γ M1,E2	584.967 25	1.23 8
γ (E2)	620.035 16	1.57 8
γ [M1+E2]	674.39 2	0.129 8
γ	696.425 21	0.147 10
γ	703.40 9	0.0068 10
γ	797.44 6	0.020 5
γ M1+39%E2	831.535 18	0.219 20
γ [M1+E2]	914.001 20	0.044 4
γ	919.45 9	~0.008 ?
γ M1	923.805 18	0.70 6
γ	954.59 6	0.034 4
γ M1,E2	968.865 20	0.034 4
γ	1046.86 6	~0.20
γ M1,E2	1047.582 18	~1
γ	1126.07 17	0.0030 8
γ	1170.63 6	0.0016 6 ?
γ	1208.54 18	0.0016 6 ?
γ	1342.12 17	0.0010 4 ?

† 2.0% uncert(syst)

Atomic Electrons (^{131}Ba)

⟨e⟩=44.6 10 keV

e_{bin}(keV)	⟨e⟩(keV)	e(%)
5	4.1	80 8
6 - 25	1.48	13.3 10
26	1.05	4.1 4
29 - 79	2.69	6.8 3
81 - 87	0.089	0.103 12
88	15.8	18.0 5
91 - 101	0.85	0.87 3
118	4.52	3.82 11
119	2.70	2.27 7
121 - 122	0.062	0.051 9
123	1.61	1.31 4
124 - 157	0.670	0.530 12
180	3.3	1.9 4
204 - 249	1.61	0.76 5
258 - 294	0.020	0.0077 10
315 - 364	0.92	0.27 4
367 - 415	0.29	0.078 9
422 - 458	0.096	0.0215 19
460	2.19	0.48 4
461 - 510	0.39	0.080 6
527 - 573	0.049	0.0090 14

Atomic Electrons (^{131}Ba) (continued)

e_{bin}(keV)	⟨e⟩(keV)	e(%)
579 - 620	0.057	0.0097 6
638 - 674	0.0074	0.0011 3
691 - 703	0.0005	8 4 $\times 10^{-5}$
761 - 797	0.0052	0.00066 9
826 - 832	0.00081	9.8 12 $\times 10^{-5}$
878 - 924	0.0192	0.00216 16
933 - 969	0.00077	8.2 14 $\times 10^{-5}$
1011 - 1048	0.025	0.0025 10
1090 - 1135	5.6 $\times 10^{-5}$	5.1 20 $\times 10^{-6}$
1165 - 1208	2.2 $\times 10^{-5}$	1.9 9 $\times 10^{-6}$
1306 - 1342	1.1 $\times 10^{-5}$	~8 $\times 10^{-7}$

Continuous Radiation (^{131}Ba)

⟨β+⟩=0.00071 keV;⟨IB⟩=0.23 keV

E_{bin}(keV)		⟨ ⟩(keV)	(%)
0 - 10	β+	1.50 $\times 10^{-8}$	1.80 $\times 10^{-7}$
	IB	0.00033	
10 - 20	β+	5.5 $\times 10^{-7}$	3.33 $\times 10^{-6}$
	IB	0.00042	0.0025
20 - 40	β+	1.21 $\times 10^{-5}$	3.76 $\times 10^{-5}$
	IB	0.071	0.23
40 - 100	β+	0.000251	0.000347
	IB	0.0062	0.0108
100 - 300	β+	0.000443	0.000330
	IB	0.028	0.0139
300 - 600	IB	0.076	0.018
600 - 1234	IB	0.047	0.0064
	Σβ+		0.00072

$^{131}_{56}$Ba(14.6 *2* min)

Mode: IT

Δ: -86534 8 keV

SpA: 9.83 $\times 10^{7}$ Ci/g

Prod: ^{133}Cs(p,3n)

Photons (^{131}Ba)

⟨γ⟩=76.6 23 keV

γ_{mode}	γ(keV)	γ(%)†
Ba L_ℓ	3.954	0.20 3
Ba L_η	4.331	0.113 12
Ba L_α	4.465	5.6 7
Ba L_β	4.932	5.4 6
Ba L_γ	5.567	0.66 8
Ba $K_{\alpha2}$	31.817	13.6 5
Ba $K_{\alpha1}$	32.194	25.0 10
Ba $K_{\beta1}$'	36.357	6.9 3
Ba $K_{\beta2}$'	37.450	1.75 7
γ E3	78.73 12	1.19 5
γ M1+2.1%E2	108.12 14	55

† 3.6% uncert(syst)

Atomic Electrons (^{131}Ba)

⟨e⟩=109.4 24 keV

e_{bin}(keV)	⟨e⟩(keV)	e(%)
5	2.9	55 6
6	2.21	39 4
25	0.105	0.41 4
26	0.30	1.15 12
27	0.49	1.85 19
30	0.110	0.36 4
31	0.31	1.01 11
32	0.067	0.211 22
35	0.034	0.096 10
36	0.0181	0.050 5
41	5.89	14.3 6
71	27.0	38.2 16
73	48.0	66 3
77	0.174	0.225 9
78	11.7	15.1 6
79	2.95	3.75 15
102	5.32	5.20 22
103	0.273	0.265 11
107	1.22	1.14 5
108	0.330	0.306 13

$^{131}_{57}$La(59 *2* min)

Mode: ϵ

Δ: -83760 100 keV

SpA: 2.43 $\times 10^{7}$ Ci/g

Prod: ^{130}Ba(d,n); ^{123}Sb(^{12}C,4n); ^{133}Cs(α,6n)

Photons (^{131}La)

⟨γ⟩=421 25 keV

γ_{mode}	γ(keV)	γ(%)†
Ba L_ℓ	3.954	0.149 20
Ba L_η	4.331	0.066 7
Ba L_α	4.465	4.1 5
Ba L_β	4.946	3.7 5
Ba L_γ	5.619	0.49 7
Ba $K_{\alpha2}$	31.817	22.0 8
Ba $K_{\alpha1}$	32.194	40.6 14
Ba $K_{\beta1}$'	36.357	11.2 4
Ba $K_{\beta2}$'	37.450	2.84 10
γ M1+~3.8%E2	79.95 16	0.78 16
γ M1+2.1%E2	108.12 14	23.1 11
γ [M1+E2]	157.23 23	0.055 11
γ	159.6 3	
γ (M1)	160.72 21	2.0 4
γ M1+E2	177.11 18	0.18 4
γ M1+E2	208.64 24	2.9 6
γ [M1+E2]	209.1 3	~0.5
γ (M1)	240.67 21	1.4 3
γ M1,E2	245.09 25	0.36 7
γ M1,E2	257.06 20	3.5 7
γ [M1+E2]	276.62 25	<0.13 ?
γ M1,E2	285.23 20	13 3
γ E2	316.76 24	1.13 24
γ	352.8 3	0.91 18 *
γ M1	365.18 22	16 3
γ M1,E2	402.3 3	1.0 2
γ M1,E2	417.78 23	18 4
γ M1,E2	433.85 23	0.80 16
γ	438.7 10	0.27 6 ?
γ M1,E2	453.73 24	6.4 13
γ M1,E2	525.90 24	9.8 20
γ [M1+E2]	561.85 25	1.3 3
γ M1,E2	593.4 3	1.6 3
γ [M1+E2]	610.96 23	0.98 20
γ	628.1 10	0.16 3
γ	644.7 10	0.036 7 ?
γ	657.4 6	0.38 8
γ	660.1 10	0.38 8

Photons (^{131}La) (continued)

γ_{mode}	γ(keV)	γ(%)†
γ [M1+E2]	719.08 24	0.20 4
γ	750.7 10	0.036 7 ?
γ	770.5 3	0.13 6
γ	794.7 10	0.036 7 ?
γ	841.0 10	0.25 5
γ	866.0 6	1.16 24
γ	878.7 3	~0.15
γ	926.7 10	0.055 11 ?
γ	974.1 6	0.73 15
γ	1135.7 10	0.091 18
γ	1178.0 8	0.29 6
γ	1291.7 20	0.127 25
γ	1330.7 20	0.036 7 ?
γ	1351.7 20	0.036 7 ?
γ	1366.7 20	0.073 15
γ	1386.6 8	0.13 6
γ	1441.7 20	0.055 11
γ	1494.7 8	~0.07
γ	1500.7 20	0.091 18
γ	1697.7 20	0.109 22
γ	1718.7 20	0.036 7
γ	1794.7 20	0.036 7 ?
γ	1823.7 20	0.091 18
γ	1876.7 20	0.073 15
γ	1954.7 20	0.073 15
γ	2164.7 20	0.018 4 ?
γ	2216.7 20	0.036 7 ?

† 7.1% uncert(syst)

* doublet

Atomic Electrons (^{131}La)

⟨e⟩=29.9 9 keV

e_{bin}(keV)	⟨e⟩(keV)	e(%)
5	2.08	40 4
6	1.50	26 3
25 - 26	0.65	2.53 20
27	0.80	3.0 3
30 - 43	1.43	4.1 3
71	11.4	16.1 8
74 - 80	0.20	0.27 4
102	2.24	2.19 11
103 - 152	1.40	1.23 9
155 - 204	1.17	0.66 9
207 - 245	0.62	0.28 5
248	1.5	0.62 13
251 - 285	0.56	0.20 4
311 - 317	0.07	0.024 10
328	1.4	0.42 8
347 - 365	0.32	0.090 12
380	1.1	0.29 7
396 - 439	0.64	0.15 3
448 - 488	0.48	0.100 24
520 - 562	0.20	0.037 6
574 - 623	0.073	0.0124 22
627 - 660	0.0035	0.00053 17
682 - 719	0.0072	0.00104 23
733 - 770	0.0036	0.00049 24
789 - 836	0.024	0.0029 14
840 - 879	0.0062	0.0007 3
889 - 937	0.011	~0.0012
968 - 974	0.0017	0.00018 9
1098 - 1141	0.0046	0.00041 17
1172 - 1178	0.00055	4.7 21 $\times 10^{-5}$
1254 - 1293	0.0019	0.00015 6
1314 - 1361	0.0025	0.00019 7
1365 - 1404	0.0007	5.2 20 $\times 10^{-5}$
1436 - 1463	0.0015	0.00010 4
1489 - 1500	0.00023	1.5 5 $\times 10^{-5}$
1660 - 1757	0.0016	9 3 $\times 10^{-5}$
1786 - 1876	0.0014	7.7 22 $\times 10^{-5}$
1917 - 1954	0.00057	3.0 12 $\times 10^{-5}$
2127 - 2216	0.00038	1.8 5 $\times 10^{-5}$

Continuous Radiation (^{131}La)

⟨β+⟩=179 keV;⟨IB⟩=3.1 keV

E_{bin}(keV)		⟨ ⟩(keV)	(%)
0 - 10	β+	1.73×10^{-6}	2.07×10^{-5}
	IB	0.0077	
10 - 20	β+	7.3×10^{-5}	0.000436
	IB	0.0076	0.052
20 - 40	β+	0.00207	0.0063
	IB	0.073	0.23
40 - 100	β+	0.103	0.132
	IB	0.045	0.070
100 - 300	β+	4.98	2.26
	IB	0.132	0.072
300 - 600	β+	32.8	7.1
	IB	0.28	0.063
600 - 1300	β+	121	13.7
	IB	1.22	0.131
1300 - 2500	β+	19.6	1.35
	IB	1.35	0.081
2500 - 2960	IB	0.0152	0.00058
	Σβ+		25

$^{131}_{58}$Ce(10 *1* min)

Mode: ϵ

Δ: -79860 *220* keV syst

SpA: 1.43×10^{8} Ci/g

Prod: ^{130}Ba(α,3n); ^{130}Ba(^{3}He,2n); ^{139}La(p,9n)

Photons (^{131}Ce)

⟨γ⟩=788 *20* keV

γ_{mode}	γ(keV)	γ(%)†
γ	19.5 *10*	0.20 *4*
γ	24.9 *10*	0.60 *12*
γ [M1+E2]	26.0 *6*	8.6 *17*
γ M2	108 *1*	
γ M1,E2	119.0 *6*	7.4 *7*
γ	145.0 *6*	1.20 *12*
γ	145.8 *10*	0.20 *2*
γ M1(+E2)	169.3 *6*	20 *2*
γ [E2]	195.3 *6*	1.0 *1*
γ	244.7 *7*	4.2 *4*
γ	263.4 *10*	0.60 *6*
γ	270.9 *8*	1.20 *12*
γ	302.6 *7*	0.40 *4*
γ	326.1 *10*	0.60 *6*
γ	389.9 *8*	1.0 *1*
γ	392.0 *8*	2.20 *22*
γ	401.3 *10*	0.40 *4*
γ	403.8 *10*	1.20 *12*
γ	414.0 *7*	10.6 *11*

Photons (^{131}Ce) (continued)

γ_{mode}	γ(keV)	γ(%)†
γ	432.8 *10*	1.20 *12*
γ	442.3 *8*	2.20 *22*
γ	470.2 *10*	1.20 *12*
γ	475.5 *10*	3.2 *3*
γ	477.8 *10*	2.0 *2*
γ	547.3 *6*	2.4 *4*
γ	582.1 *10*	1.40 *21*
γ	597.6 *7*	1.40 *21*
γ	601.6 *10*	2.0 *3*
γ	613.2 *10*	0.80 *12*
γ	638.1 *10*	0.80 *12*
γ	643.1 *10*	0.80 *12*
γ	651.2 *10*	0.40 *6*
γ	656 *1*	1.00 *15*
γ	678.9 *10*	0.40 *6*
γ	686.8 *10*	0.20 *3*
γ	692.3 *10*	0.20 *3*
γ	694.3 *10*	0.20 *3*
γ	701.5 *10*	0.40 *6*
γ	714 *1*	0.60 *9*
γ	718.3 *10*	0.40 *6*
γ	726.9 *10*	0.80 *12*
γ	730.8 *10*	0.20 *3*
γ	742.6 *7*	0.20 *3*
γ	748.8 *10*	1.20 *18* ?
γ	799.3 *10*	0.60 *9*
γ	812.8 *10*	1.40 *21*
γ	817.8 *10*	0.40 *6*
γ	834.8 *10*	2.2 *3*
γ	864.8 *10*	2.2 *3*
γ	878.2 *10*	0.60 *9*
γ	884 *1*	2.4 *4*
γ	902.5 *10*	0.60 *9*
γ	909.9 *10*	1.40 *21*
γ	928.2 *10*	0.60 *9*
γ	963.1 *10*	1.20 *18*
γ	973.1 *10*	0.60 *9*
γ	997.8 *10*	2.0 *3*
γ	1058.1 *10*	1.20 *18*
γ	1068.6 *10*	0.40 *6*
γ	1072.8 *10*	0.60 *9*
γ	1129 *1*	1.40 *21*
γ	1164.8 *10*	2.0 *3*
γ	1236.6 *10*	1.40 *21*
γ	1299.9 *10*	1.40 *21*
γ	1356.6 *10*	0.80 *12*
γ	1380.8 *10*	0.80 *12*
γ	1412.2 *10*	0.60 *9*
γ	1417.3 *10*	0.60 *9*
γ	1427.9 *10*	0.40 *6*
γ	1448.8 *10*	1.00 *15*
γ	1468.9 *10*	6.0 *9* ?
γ	1480.6 *10*	0.80 *12*
γ	1487.8 *10*	1.60 *24*
γ	1529.3 *10*	0.80 *12*
γ	1694.5 *10*	1.60 *24*
γ	1714.6 *10*	0.60 *9*
γ	1775.3 *10*	0.80 *12*
γ	1805.7 *10*	0.40 *6*
γ	1872.3 *10*	0.40 *6*
γ	1893.6 *10*	0.40 *6*

† 15% uncert(syst)

$^{131}_{58}$Ce(5 *1* min)

Mode: ϵ

Δ: -79860 *220* keV syst

SpA: 2.9×10^{8} Ci/g

Prod: ^{130}Ba(α,3n); ^{130}Ba(^{3}He,2n); ^{139}La(p,9n)

Photons (^{131}Ce)

γ_{mode}	γ(keV)	γ(rel)
γ	230.4 *10*	36 *4*
γ	395.5 *10*	100 *11*
γ	421.3 *10*	54 *7*

$^{131}_{59}$Pr(1.7 *3* min)

Mode: ϵ

SpA: 8.6×10^{8} Ci/g

Prod: ^{32}S on Mo-Sn targets

$^{131}_{60}$Nd(24 *3* s)

Mode: ϵ, ϵp

SpA: 3.5×10^{9} Ci/g

Prod: ^{32}S on Mo-Sn targets

A = 132

NDS 17, 225 (1976)

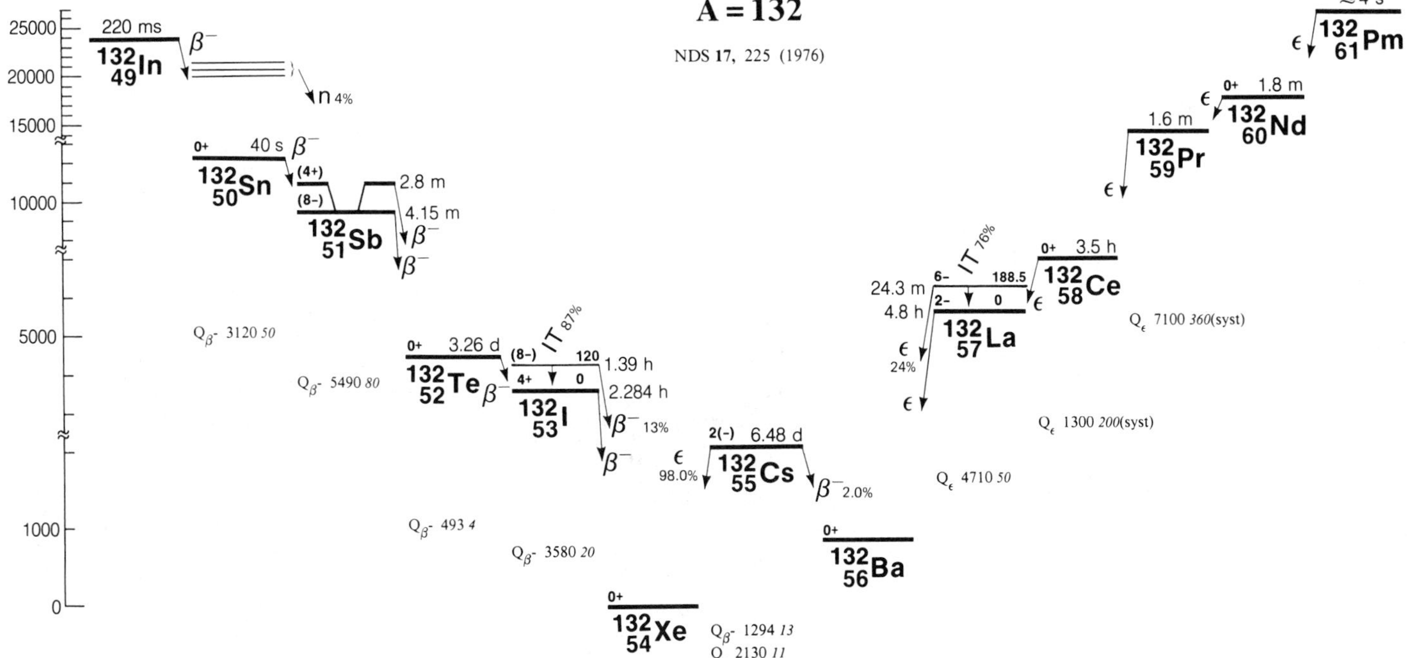

$^{132}_{49}$In(220 *30* ms)

Mode: β-, β-n(4.2 *9* %)
SpA: 1.18×10^{11} Ci/g
Prod: fission

Photons (^{132}In)

γ_{mode}	γ(keV)
$\gamma_{\beta-}$	4041 *2*

$^{132}_{50}$Sn(40 *1* s)

Mode: β-
Δ: -76610 *80* keV
SpA: 2.12×10^{9} Ci/g
Prod: fission

Photons (^{132}Sn)

⟨γ⟩=1278 *32* keV

γ_{mode}	γ(keV)	γ(%)†
Sb L_ℓ	3.189	0.058 *9*
Sb L_η	3.437	0.029 *4*
Sb L_α	3.604	1.61 *20*
Sb L_β	3.930	1.41 *19*
Sb L_γ	4.434	0.17 *3*
Sb $K_{\alpha2}$	26.111	11.2 *7*
Sb $K_{\alpha1}$	26.359	20.9 *13*
Sb $K_{\beta1}$'	29.712	5.5 *4*

Photons (^{132}Sn) (continued)

γ_{mode}	γ(keV)	γ(%)†
Sb $K_{\beta2}$'	30.561	1.17 *8*
γ M1	85.5 *9*	49 *3*
γ M1,E2	246.7 *8*	42.0 *21*
γ M1,E2	340.3 *8*	42.9 *21*
γ	528.8 *15*	2.0 *2* ?
γ	548.9 *15*	1.90 *19* ?
γ	651.9 *9*	1.90 *19*
γ	898.6 *8*	42.0
γ	992.2 *8*	38.1 *21*
γ	1077.7 *11*	2.0 *2*
γ [E2]	1238.9 *9*	13.4 *8*

† 7.0% uncert(syst)

Atomic Electrons (^{132}Sn)

⟨e⟩=41.6 *17* keV

e_{bin}(keV)	⟨e⟩(keV)	e(%)
4	1.36	32 *4*
5 - 30	1.52	9.8 *6*
55	22.7	41 *3*
81	4.3	5.3 *3*
85 - 86	1.11	1.31 *8*
216	5.0	2.3 *3*
242 - 247	1.06	0.44 *8*
310	2.78	0.90 *7*
336 - 340	0.52	0.16 *3*
498 - 545	0.10	0.019 *7*
548 - 549	0.0014	~0.00026
621 - 652	0.038	0.006 *3*
868 - 899	0.54	0.06 *3*
962 - 992	0.44	0.045 *20*
1047 - 1078	0.021	0.0020 *9*
1208 - 1239	0.142	0.0118 *7*

Continuous Radiation (^{132}Sn)

⟨β-⟩=690 keV;⟨IB⟩=1.20 keV

E_{bin}(keV)		⟨ ⟩(keV)	(%)
0 - 10	β-	0.0236	0.470
	IB	0.028	
10 - 20	β-	0.072	0.480
	IB	0.028	0.19
20 - 40	β-	0.298	0.99
	IB	0.054	0.19
40 - 100	β-	2.28	3.23
	IB	0.148	0.23
100 - 300	β-	27.5	13.4
	IB	0.38	0.21
300 - 600	β-	115	25.3
	IB	0.33	0.080
600 - 1300	β-	445	49.1
	IB	0.23	0.029
1300 - 1796	β-	101	7.0
	IB	0.0068	0.00049

$^{132}_{51}$Sb(4.15 *5* min)

Mode: β-
Δ: -79730 *80* keV
SpA: 3.43×10^{8} Ci/g
Prod: fission

Photons (^{132}Sb)

$\langle\gamma\rangle$=2583 *130* keV

γ_{mode}	γ(keV)	γ(%)†
Te L_ℓ	3.335	0.069 *10*
Te L_η	3.606	0.036 *4*
Te L_α	3.768	1.92 *23*
Te L_β	4.121	1.67 *22*
Te L_γ	4.636	0.19 *3*
Te $K_{\alpha2}$	27.202	10.5 *6*
Te $K_{\alpha1}$	27.472	19.6 *12*
Te $K_{\beta1}'$	30.980	5.2 *3*
Te $K_{\beta2}'$	31.877	1.14 *7*
γ E2	103.37 *9*	35 *2*
γ E1	150.76 *9*	66 *3*
γ	276.14 *19*	4 *1*
γ	293.13 *20*	4 *1*
γ	368.63 *16*	7 *1*
γ E1	382.26 *9*	~7
γ	496.76 *15*	13 *1*
γ E2,M1	696.90 *9*	100 *10*
γ	882.08 *25*	6 *1*
γ (E2)	973.87 *10*	100 *10*
γ	1041.64 *25*	18 *1*
γ	1167.0 *3*	10 *1*
γ	1378.8 *3*	4 *1*
γ	1763.8 *8*	4 *1*
γ	1854.7 *8*	~2
γ	2664.1 *10*	4 *1*

† 0.30% uncert(syst)

Atomic Electrons (^{132}Sb)

$\langle e\rangle$=56.6 *18* keV

e_{bin}(keV)	$\langle e\rangle$(keV)	e(%)
4 - 31	3.05	46 *3*
72	26.6	37.2 *23*
98	3.36	3.42 *21*
99	9.2	9.3 *6*
102	0.68	0.66 *4*
103	2.68	2.61 *16*
119	4.22	3.54 *18*
146 - 151	0.82	0.562 *23*
244 - 293	0.59	0.23 *9*
337 - 382	0.45	0.13 *5*
465 - 497	0.39	0.08 *4*
665	2.2	0.33 *6*
692 - 697	0.36	0.053 *8*
850 - 882	0.08	0.010 *5*
942	1.25	0.133 *5*
969 - 1010	0.38	0.039 *9*
1037 - 1042	0.028	0.0027 *12*
1135 - 1167	0.10	0.009 *4*
1347 - 1379	0.034	0.0025 *11*
1732 - 1823	0.038	0.0021 *8*
1850 - 1854	0.0016	~9 $\times10^{-5}$
2632 - 2663	0.019	0.00071 *25*

Continuous Radiation (^{132}Sb)

$\langle\beta-\rangle$=1280 keV;$\langle IB\rangle$=3.4 keV

E_{bin}(keV)		$\langle\ \rangle$(keV)	(%)
0 - 10	β-	0.0086	0.172
	IB	0.044	
10 - 20	β-	0.0265	0.176
	IB	0.043	0.30
20 - 40	β-	0.110	0.366
	IB	0.085	0.29
40 - 100	β-	0.86	1.21
	IB	0.24	0.37
100 - 300	β-	11.0	5.3
	IB	0.69	0.39
300 - 600	β-	53	11.5
	IB	0.78	0.18
600 - 1300	β-	338	35.3
	IB	1.05	0.122
1300 - 2500	β-	717	40.2
	IB	0.49	0.029
2500 - 3819	β-	161	5.8
	IB	0.022	0.00080

$^{132}_{51}$Sb(2.8 *1* min)

Mode: β-

Δ: -79730 *80* keV

SpA: 5.08$\times10^8$ Ci/g

Prod: fission

Photons (^{132}Sb)

$\langle\gamma\rangle$=2602 *53* keV

γ_{mode}	γ(keV)	γ(%)†
Te L_ℓ	3.335	0.026 *4*
Te L_η	3.606	0.0138 *17*
Te L_α	3.768	0.72 *9*
Te L_β	4.120	0.63 *9*
Te L_γ	4.635	0.073 *11*
Te $K_{\alpha2}$	27.202	3.9 *3*
Te $K_{\alpha1}$	27.472	7.3 *5*
Te $K_{\beta1}'$	30.980	1.95 *14*
Te $K_{\beta2}'$	31.877	0.42 *3*
γ E2	103.37 *9*	13.8 *10*
γ	138.47 *10*	0.69 *20*
γ	311.97 *20*	0.69 *20*
γ	353.82 *16*	3.0 *10*
γ E1	382.26 *9*	7.9 *10*
γ	436.75 *16*	3.0 *10*
γ	447.16 *16*	~2
γ	609.88 *18*	~2
γ	635.60 *15*	9.9 *10*
γ E2,M1	696.90 *9*	86 *4*
γ	814.20 *19*	4.9 *10*
γ	816.57 *19*	10.9 *10*
γ	930.07 *17*	0.99 *20*
γ (E2)	973.87 *10*	98.9
γ	989.41 *14*	14.8 *10*
γ	1093.09 *19*	4.9 *10*
γ	1133.64 *17*	5.9 *10*
γ	1152.1 *3*	3.0 *10*
γ	1183.0 *4*	1.3 *3*
γ	1196.46 *19*	3.0 *10*
γ	1213.3 *4*	~2
γ	1274.6 *4*	1.19 *20*
γ	1306.78 *19*	0.99 *20*
γ	1436.57 *17*	~2
γ	1454.3 *3*	0.59 *20*
γ	1513.47 *20*	~2
γ	1539.94 *17*	0.99 *20*
γ	1575.04 *19*	1.3 *3*
γ	1634.0 *8*	0.99 *20*
γ	1644.5 *8*	~2
γ	1787.7 *3*	3.5 *4*
γ	1891.0 *3*	0.99 *20*
γ	1893.35 *21*	0.89 *20*
γ	2280.63 *21*	0.99 *20*
γ	2587.9 *3*	1.5 *3*
γ	2633.8 *8*	0.49 *20*
γ	2913.2 *8*	0.49 *20*

† 0.20% uncert(syst)

Atomic Electrons (^{132}Sb)

$\langle e\rangle$=23.4 *9* keV

e_{bin}(keV)	$\langle e\rangle$(keV)	e(%)
4	0.40	9.3 *11*
5 - 31	0.74	8.0 *8*
72	10.5	14.7 *11*
98	1.33	1.35 *10*
99	3.7	3.7 *3*
102	0.268	0.262 *19*
103	1.06	1.03 *8*
107 - 138	0.24	~0.21
280 - 322	0.16	0.05 *3*
349 - 382	0.169	0.048 *6*
405 - 447	0.17	0.042 *18*
578 - 610	0.22	0.037 *17*
631 - 636	0.031	0.0049 *24*
665	1.9	0.29 *4*
692 - 697	0.31	0.045 *6*
782 - 817	0.24	0.031 *11*
898 - 930	0.013	0.0014 *7*
942	1.24	0.132 *5*
958 - 989	0.38	0.040 *9*
1061 - 1102	0.11	0.010 *3*
1120 - 1165	0.076	0.0066 *20*
1178 - 1213	0.025	0.0021 *10*
1243 - 1275	0.019	0.0015 *5*
1302 - 1307	0.0012	9 *4* $\times10^{-5}$
1405 - 1454	0.021	0.0015 *7*
1482 - 1574	0.032	0.0021 *7*
1602 - 1644	0.021	0.0013 *5*
1756 - 1787	0.023	0.0013 *5*
1859 - 1893	0.012	0.00063 *17*
2249 - 2280	0.0052	0.00023 *8*
2556 - 2633	0.009	0.00037 *10*
2881 - 2912	0.0022	8 *3* $\times10^{-5}$

Continuous Radiation (^{132}Sb)

$\langle\beta-\rangle$=1220 keV;$\langle IB\rangle$=3.2 keV

E_{bin}(keV)		$\langle\ \rangle$(keV)	(%)
0 - 10	β-	0.0097	0.193
	IB	0.042	
10 - 20	β-	0.0297	0.198
	IB	0.042	0.29
20 - 40	β-	0.124	0.411
	IB	0.082	0.28
40 - 100	β-	0.96	1.36
	IB	0.23	0.36
100 - 300	β-	12.2	5.9
	IB	0.66	0.37
300 - 600	β-	58	12.6
	IB	0.74	0.17
600 - 1300	β-	352	37.0
	IB	0.96	0.112
1300 - 2500	β-	669	37.8
	IB	0.42	0.025
2500 - 3819	β-	128	4.56
	IB	0.020	0.00074

$^{132}_{52}$Te(3.26 *3* d)

Mode: β-

Δ: -85217 *21* keV

SpA: 3.04$\times10^5$ Ci/g

Prod: fission

Photons (^{132}Te)

⟨γ⟩=234 *4* keV

γ_{mode}	γ(keV)	γ(%)†
I L_ℓ	3.485	0.118 *18*
I L_η	3.780	0.057 *7*
I L_α	3.937	3.3 *4*
I L_β	4.320	2.9 *4*
I L_γ	4.886	0.36 *6*
I $K_{\alpha2}$	28.317	20.4 *15*
I $K_{\alpha1}$	28.612	38 *3*
I $K_{\beta1}$'	32.278	10.2 *8*
I $K_{\beta2}$'	33.225	2.32 *17*
γ M1(+0.03%E2)	49.82 *13*	14.4 *10*
γ M1(+E2)	111.86 *10*	1.85 *18*
γ M1(+E2)	116.4 *1*	1.94 *18*
γ E2	228.26 *16*	88.2 *18*

† 0.45% uncert(syst)

Atomic Electrons (^{132}Te)

⟨e⟩=42.8 *12* keV

e_{bin}(keV)	⟨e⟩(keV)	e(%)
5	3.0	65 *8*
17	11.8	71 *6*
23	0.64	2.8 *3*
24	0.83	3.5 *4*
27	0.32	1.19 *15*
28	0.43	1.55 *19*
29	0.0117	0.041 *5*
31	0.053	0.168 *20*
32	0.027	0.084 *10*
45	4.2	9.4 *7*
49	0.93	1.90 *15*
50	0.232	0.47 *4*
79	1.0	1.2 *4*
83	0.9	1.1 *3*
107	0.34	~0.32
111	0.20	0.18 *7*
112	0.20	~0.18
115	0.046	0.04 *2*
116	0.04	~0.03
195	13.8	7.09 *20*
223	2.36	1.06 *3*
224	0.658	0.294 *8*
227	0.636	0.280 *8*
228	0.152	0.0666 *19*

Continuous Radiation (^{132}Te)

⟨β-⟩=59 keV;⟨IB⟩=0.0131 keV

E_{bin}(keV)		⟨ ⟩(keV)	(%)
0 - 10	β-	0.56	11.3
	IB	0.0030	
10 - 20	β-	1.58	10.6
	IB	0.0024	0.0167
20 - 40	β-	5.7	19.0
	IB	0.0033	0.0118
40 - 100	β-	26.6	39.9
	IB	0.0038	0.0065
100 - 215	β-	24.9	19.2
	IB	0.00053	0.00045

$^{132}_{53}$I (2.284 *2* h)

Mode: β-

Δ: -85710 *20* keV

SpA: 1.0394×10^7 Ci/g

Prod: daughter ^{132}Te

Photons (^{132}I)

⟨γ⟩=2290 *21* keV

γ_{mode}	γ(keV)	γ(%)†
Xe L_ℓ	3.634	0.0019 *3*
Xe L_η	3.955	0.00089 *10*
Xe L_α	4.104	0.054 *6*
Xe L_β	4.522	0.047 *6*
Xe L_γ	5.117	0.0060 *9*
Xe $K_{\alpha2}$	29.461	0.314 *14*
Xe $K_{\alpha1}$	29.782	0.583 *25*
Xe $K_{\beta1}$'	33.606	0.158 *7*
Xe $K_{\beta2}$'	34.606	0.0373 *17*
γ M1,E2	136.35 *16*	0.079 *10*
γ	147.19 *7*	0.237 *20*
γ M1,E2	183.5 *3*	0.16 *3*
γ E2,M1	254.95 *13*	0.19 *3*
γ M1,E2	262.75 *9*	1.44 *9*
γ	278.8 *3*	~0.039
γ E2,M1	284.76 *8*	0.79 *7*
γ	302.00 *19*	~0.0049
γ	306.37 *9*	0.11 *4* ?
γ	310.0 *3*	0.09 *4*
γ	316.1 *3*	0.16 *4*
γ	343.2 *3*	0.099 *20*
γ	351.7 *3*	0.079 *20*
γ [M1+E2]	363.50 *8*	0.45 *6*
γ	387.84 *12*	0.17 *3*
γ	416.8 *4*	0.46 *9*
γ	431.95 *9*	0.45 *9*
γ M1+E2	445.29 *17*	0.67 *8*
γ	473.75 *14*	0.27 *5*
γ	478.52 *12*	0.10 *4*
γ	487.88 *25*	0.18 *5*
γ [M1+E2]	505.91 *9*	5.02 *19*
γ M1,E2	522.68 *7*	16.1 *6*
γ	535.26 *19*	0.52 *8*
γ	547.02 *11*	1.25 *9*
γ	591.13 *10*	~0.06
γ (E1)	600.0 *4*	0.09 *3* ?
γ	620.94 *14*	~0.39
γ	621.18 *15*	~2
γ E2(+M1)	630.27 *6*	13.8 *6*
γ M1(+E2)	650.59 *11*	2.66 *20*
γ	659 *1*	0.39 *8*
γ E2	667.73 *5*	98.7
γ M1,E2	669.86 *7*	4.9 *8*
γ E2,M1	671.6 *3*	5.2 *4*
γ (E2)	726.8 *4*	2.2 *6* ?
γ E2+M1	726.9 *3*	3.2 *6*
γ	728.70 *11*	1.1 *3*
γ	764.5 *10*	0.39 *8*
γ	771.8 *5*	<0.039 ?
γ E2	772.68 *5*	76.2 *18*
γ	780.12 *13*	1.23 *6*
γ [M1+E2]	784.89 *12*	0.42 *5*
γ	791.48 *13*	0.09 *3*
γ M1(+E2)	809.77 *11*	2.9 *3*
γ E2(+M1)	812.28 *8*	5.6 *5*
γ	863.31 *18*	0.59 *5*
γ	877.23 *16*	1.08 *5*
γ	889.46 *20*	~0.04
γ	910.51 *11*	0.92 *5*
γ	927.83 *13*	0.44 *8*
γ	948.04 *17*	~0.08
γ M1(+E2)	954.62 *7*	18.1 *6*
γ	984.67 *14*	0.56 *6*
γ	1001.9 *5*	~0.039 ?
γ	1010.9 *6*	~0.08
γ	1016.0 *6*	~0.05
γ	1035.07 *11*	0.57 *5*
γ	1050.5 *5*	0.044 *15*
γ	1065.5 *7*	0.034 *11*
γ	1087.01 *13*	0.07 *3*
γ	1097.03 *10*	0.035 *12*
γ	1113.05 *18*	0.059 *20*
γ	1127.08 *15*	0.051 *24*
γ M1(+E2)	1136.18 *8*	3.02 *17*
γ	1137.4 *6*	<0.30 ?
γ (M1)	1143.61 *12*	1.38 *10*
γ [M1+E2]	1148.38 *11*	0.21 *5*
γ M1,E2	1173.26 *10*	1.09 *10*
γ	1254.41 *18*	~0.05
γ	1263.9 *4*	~0.023
γ	1272.76 *23*	0.15 *3*

Photons (^{132}I) (continued)

γ_{mode}	γ(keV)	γ(%)†
γ [M1+E2]	1290.79 *12*	1.14 *6*
γ	1295.36 *7*	1.97 *10*
γ [E2]	1298.00 *8*	0.86 *8*
γ	1314.15 *12*	0.059 *20*
γ [M1+E2]	1317.82 *18*	0.107 *14*
γ M1(+E2)	1372.08 *9*	2.47 *10*
γ	1392.5 *20*	~0.24
γ M1(+E2)	1398.56 *9*	7.1 *3*
γ	1410.6 *3*	0.059 *20*
γ M1+E2	1442.54 *7*	1.42 *6*
γ	1456.56 *12*	0.048 *10*
γ	1476.55 *17*	0.138 *20*
γ	1503.6 *6*	0.009 *3*
γ	1519.72 *20*	0.051 *6*
γ	1542.31 *16*	~0.010
γ	1592.91 *12*	0.044 *6*
γ	1617.90 *17*	0.015 *3*
γ	1618.96 *17*	0.008 *3*
γ	1637.7 *3*	0.017 *4*
γ	1661.3 *5*	0.017 *4*
γ	1671.7 *5*	~0.018 ?
γ	1716.0 *3*	0.052 *5*
γ	1720.8 *6*	0.055 *5*
γ	1727.30 *8*	0.062 *9*
γ	1752.50 *15*	0.030 *10*
γ	1757.35 *14*	0.38 *3*
γ	1760.31 *18*	~0.020
γ	1778.66 *24*	0.059 *15*
γ	1786.6 *4*	~0.008
γ	1814.2 *5*	0.010 *4*
γ	1830.0 *7*	0.029 *9*
γ	1879.3 *6*	0.016 *3*
γ	1913.8 *5*	~0.06
γ [M1+E2]	1921.06 *10*	1.18 *9*
γ [E2]	1985.55 *18*	0.0121 *22*
γ [M1+E2]	2002.35 *9*	1.09 *10*
γ	2086.83 *12*	0.25 *4*
γ	2172.58 *15*	0.19 *3*
γ	2187.45 *20*	0.007 *3*
γ	2223.18 *11*	0.118 *20*
γ	2249.22 *17*	0.030 *10*
γ	2291.5 *5*	~0.0039
γ	2390.58 *17*	0.168 *20*
γ	2408.93 *23*	0.010 *3*
γ	2444.4 *5*	~0.0039
γ	2455.2 *6*	~0.0030
γ	2488.7 *3*	<0.0020 ?
γ	2525.17 *15*	0.037 *7*
γ	2546.8 *3*	~0.0020
γ	2569.7 *6*	0.0030 *10*
γ	2592.4 *3*	<0.00049 ?
γ	2652.0 *6*	<0.00049 ?
γ	2717.4 *7*	0.0030 *10*
γ	2927 *1*	0.00039 *8*

† 0.20% uncert(syst)

Atomic Electrons (^{132}I)

⟨e⟩=8.18 *17* keV

e_{bin}(keV)	⟨e⟩(keV)	e(%)
5 - 33	0.083	1.12 *11*
102 - 149	0.15	0.13 *4*
178 - 220	0.033	0.0161 *24*
228 - 275	0.312	0.130 *9*
278 - 317	0.040	0.014 *3*
329 - 363	0.045	0.0134 *19*
382 - 431	0.074	0.018 *4*
432 - 469	0.024	0.0054 *15*
471	0.20	0.043 *6*
473 - 487	0.0019	0.00039 *15*
488	0.61	0.125 *19*
500 - 547	0.20	0.039 *5*
557 - 595	0.05	~0.008
596	0.40	0.067 *11*
599 - 630	0.159	0.026 *4*
633	2.24	0.354 *7*
635 - 659	0.28	0.045 *6*

Atomic Electrons (^{132}I) (continued)

e_{bin}(keV)	⟨e⟩(keV)	e(%)
662	0.269	0.0406 8
663 - 694	0.32	0.047 5
721 - 737	0.031	0.0042 8
738	1.42	0.192 6
746 - 791	0.46	0.059 4
804 - 843	0.053	0.0064 12
855 - 893	0.023	0.0026 9
905 - 913	0.0030	0.00033 13
920	0.30	0.033 6
922 - 967	0.058	0.0061 9
976 - 1016	0.010	0.0010 4
1030 - 1078	0.0034	0.00032 8
1082 - 1131	0.073	0.0066 7
1132 - 1173	0.022	0.00193 22
1220 - 1268	0.043	0.0034 7
1272 - 1318	0.0085	0.00066 10
1338 - 1376	0.112	0.0082 8
1387 - 1422	0.0280	0.00200 17
1437 - 1485	0.0041	0.00029 4
1498 - 1592	0.00069	4.4 11 $\times10^{-5}$
1603 - 1693	0.0016	9.5 19 $\times10^{-5}$
1711 - 1809	0.0042	0.00024 6
1813 - 1909	0.0099	0.00053 6
1913 - 2001	0.0110	0.00056 6
2052 - 2138	0.0026	0.00013 3
2153 - 2248	0.0011	5.0 13 $\times10^{-5}$
2257 - 2356	0.0009	3.7 13 $\times10^{-5}$
2374 - 2454	0.00023	9.5 21 $\times10^{-6}$
2483 - 2569	0.00023	9 3 $\times10^{-6}$
2587 - 2683	1.5 $\times10^{-5}$	5.6 24 $\times10^{-7}$
2712 - 2716	2.0 $\times10^{-6}$	7 3 $\times10^{-8}$
2892 - 2926	1.9 $\times10^{-6}$	6.7 21 $\times10^{-8}$

Continuous Radiation (^{132}I)

⟨β-⟩=486 keV;⟨IB⟩=0.72 keV

E_{bin}(keV)		⟨ ⟩(keV)	(%)
0 - 10	β-	0.059	1.17
	IB	0.021	
10 - 20	β-	0.177	1.18
	IB	0.020	0.142
20 - 40	β-	0.72	2.38
	IB	0.039	0.135
40 - 100	β-	5.2	7.3
	IB	0.103	0.160
100 - 300	β-	50.0	25.1
	IB	0.24	0.138
300 - 600	β-	137	31.2
	IB	0.18	0.044
600 - 1300	β-	238	27.9
	IB	0.111	0.0141
1300 - 2140	β-	55	3.61
	IB	0.0072	0.00051

$^{132}_{53}$I (1.39 3 h)

Mode: IT(86.8 20 %), β-(13.2 20 %)
Δ: -85590 20 keV
SpA: 1.70$\times10^{7}$ Ci/g
Prod: fission; ^{130}Te(α,pn)

Photons (^{132}I)

⟨γ⟩=319 28 keV

γ_{mode}	γ(keV)	γ(%)†
I L_ℓ	3.485	~0.20
Xe L_ℓ	3.634	0.0030 5
I L_η	3.780	~0.12
I L_α	3.937	~6
Xe L_η	3.955	0.00146 18
Xe L_α	4.104	0.084 11
I L_β	4.311	~5
Xe L_β	4.520	0.074 10
I L_γ	4.820	~0.6
Xe L_γ	5.106	0.0092 14
I $K_{\alpha2}$	28.317	6.6 8
I $K_{\alpha1}$	28.612	12.4 15
Xe $K_{\alpha2}$	29.461	0.45 3
Xe $K_{\alpha1}$	29.782	0.84 6
I $K_{\beta1}$'	32.278	3.3 4
I $K_{\beta2}$'	33.225	0.75 10
Xe $K_{\beta1}$'	33.606	0.228 17
Xe $K_{\beta2}$'	34.606	0.054 4
γ_{IT} E3	98 1	3.7 5
$\gamma_{\beta-}$(E2)	175.1 5	8.3 5
$\gamma_{\beta-}$M1+E2	310.1 8	0.61 12 ?
$\gamma_{\beta-}$(E1)	600.0 4	13.2 7
$\gamma_{\beta-}$	610.1 10	1.4 3
$\gamma_{\beta-}$	614.1 7	2.4 7
$\gamma_{\beta-}$E2	667.73 5	13 3
$\gamma_{\beta-}$E2	772.68 5	13 3

† uncert(syst): 2.3% for IT, 16% for β-

Atomic Electrons (^{132}I)

⟨e⟩=97 14 keV

e_{bin}(keV)	⟨e⟩(keV)	e(%)
5	5.5	~117
17	11.9	~69
21	3.1	~15
22 - 33	1.5	~6
65	16.9	26 3
93	41.8	45 6
97	9.7	10.0 12
98 - 141	4.5	3.9 3
170 - 175	0.83	0.48 3
276 - 310	0.065	0.023 4
565 - 614	0.24	0.043 7
633 - 668	0.35	0.055 10
738 - 773	0.29	0.039 7

Continuous Radiation (^{132}I)

⟨β-⟩=64 keV;⟨IB⟩=0.090 keV

E_{bin}(keV)		⟨ ⟩(keV)	(%)
0 - 10	β-	0.0065	0.130
	IB	0.0028	
10 - 20	β-	0.0197	0.131
	IB	0.0027	0.019
20 - 40	β-	0.080	0.268
	IB	0.0052	0.018
40 - 100	β-	0.59	0.84
	IB	0.0139	0.022
100 - 300	β-	6.2	3.09
	IB	0.032	0.018
300 - 600	β-	19.2	4.34
	IB	0.023	0.0056
600 - 1300	β-	36.6	4.32
	IB	0.0101	0.00137
1300 - 1660	β-	1.44	0.105
	IB	3.9$\times10^{-5}$	2.9$\times10^{-6}$

$^{132}_{54}$Xe(stable)

Δ: -89290 4 keV
%: 26.9 5

$^{132}_{55}$Cs(6.475 10 d)

Mode: ϵ(98.0 1 %), β-(2.0 1 %)
Δ: -87160 12 keV
SpA: 1.5277$\times10^{5}$ Ci/g
Prod: ^{133}Cs(p,pn); ^{132}Xe(p,n); ^{133}Cs(n,2n)

Photons (^{132}Cs)

⟨γ⟩=712.1 14 keV

γ_{mode}	γ(keV)	γ(%)†
Xe L_ℓ	3.634	0.127 17
Xe L_η	3.955	0.058 6
Xe L_α	4.104	3.5 4
Xe L_β	4.522	3.1 4
Xe L_γ	5.126	0.41 6
Xe $K_{\alpha2}$	29.461	21.0 6
Xe $K_{\alpha1}$	29.782	39.0 12
Xe $K_{\beta1}$'	33.606	10.5 3
Xe $K_{\beta2}$'	34.606	2.49 8
γ_ϵ[M1+E2]	363.50 8	0.074 9
$\gamma_{\beta-}$E2	464.59 4	1.87 10
γ_ϵ[M1+E2]	505.91 9	0.82 6
$\gamma_{\beta-}$E2(+M1)	567.169 20	0.244 21
γ_ϵE2(+M1)	630.27 6	0.99 7
$\gamma_{\beta-}$E2,M1	663.106 20	0.063 19
γ_ϵE2	667.73 5	97.47
γ_ϵE2	772.68 5	0.074 10
$\gamma_{\beta-}$E2(+M1)	1031.75 3	0.121 9
γ_ϵM1(+E2)	1136.18 8	0.49 3
γ_ϵ[E2]	1298.00 8	0.062 7
γ_ϵ[M1+E2]	1317.82 18	0.60 5
γ_ϵ[E2]	1985.55 18	0.068 9

† uncert(syst): 0.11% for ϵ, 5.0% for β-

Atomic Electrons (^{132}Cs)

⟨e⟩=8.3 4 keV

e_{bin}(keV)	⟨e⟩(keV)	e(%)
5	3.2	65 7
6	0.00038	0.0066 7
24	0.64	2.7 3
25	0.82	3.3 3
26 - 28	0.227	0.81 8
29	0.48	1.68 17
30 - 36	0.146	0.46 3
329 - 363	0.0059	0.00178 23
427 - 471	0.131	0.0297 15
500 - 530	0.0136	0.00263 21
561 - 596	0.030	0.0051 8
625 - 630	0.0069	0.00111 18
633	2.21	0.350 7
657 - 658	0.00027	4.1 15 $\times10^{-5}$
662	0.266	0.0401 8
663 - 668	0.1334	0.0201 3
738 - 773	0.00162	0.000218 25
994 - 1032	0.00206	0.000206 13
1102 - 1136	0.0079	0.00071 10
1263 - 1313	0.0086	0.00067 9
1317 - 1318	0.00022	1.66 24 $\times10^{-5}$
1951 - 1985	0.00055	2.8 4 $\times10^{-5}$

Continuous Radiation (^{132}Cs)

⟨β-⟩=4.63 keV;⟨β+⟩=3.09 keV;⟨IB⟩=0.68 keV

E_{bin}(keV)		⟨ ⟩(keV)	(%)
0 - 10	β-	0.00360	0.072
	β+	2.86×10^{-6}	3.44×10^{-5}
	IB	0.00067	
10 - 20	β-	0.0105	0.070
	β+	0.000111	0.00067
	IB	0.00083	0.0053
20 - 40	β-	0.0399	0.134
	β+	0.00287	0.0088
	IB	0.064	0.22
40 - 100	β-	0.239	0.347
	β+	0.113	0.147
	IB	0.0070	0.0116
100 - 300	β-	1.36	0.72
	β+	2.24	1.13
	IB	0.038	0.019
300 - 600	β-	2.45	0.58
	β+	0.74	0.219
	IB	0.167	0.037
600 - 1300	β-	0.52	0.078
	IB	0.40	0.047
1300 - 1462	IB	0.0042	0.00032
	Σβ+		1.50

$^{132}_{56}$Ba(stable)

Δ: -88453 *8* keV

%: 0.101 *2*

$^{132}_{57}$La(4.8 *2* h)

Mode: ϵ

Δ: -83740 *50* keV

SpA: 4.95×10^{6} Ci/g

Prod: ^{132}Ba(d,2n); protons on Ba; protons on La; protons on Ce

Photons (^{132}La)

⟨γ⟩=1734 *33* keV

γ_{mode}	γ(keV)	γ(%)†
Ba L_ℓ	3.954	0.097 *13*
Ba L_η	4.331	0.043 *5*
Ba L_α	4.465	2.7 *3*
Ba L_β	4.946	2.4 *3*
Ba L_γ	5.620	0.32 *5*
Ba $K_{\alpha2}$	31.817	14.3 *4*
Ba $K_{\alpha1}$	32.194	26.4 *8*
Ba $K_{\beta1}$'	36.357	7.30 *23*
Ba $K_{\beta2}$'	37.450	1.85 *6*
γ M1,E2	192.91 *3*	~1 ?
γ M1,E2	305.90 *3*	0.51 *7*
γ	360.56 *4*	0.20 *6*
γ	382.81 *3*	0.43 *6*
γ	430 *2*	0.12 *5*
γ E2	464.59 *4*	77 *5*
γ [E2]	472.09 *8*	0.36 *3* ?
γ E2,M1	479.47 *3*	2.23 *17*
γ	498.80 *4*	0.54 *5*
γ E1	515.83 *4*	5.1 *5*
γ M1(+E2)	540.411 *23*	7.8 *6*
γ	553.46 *3*	0.208 *15*
γ E2(+M1)	567.169 *20*	15.9 *11*
γ	601.77 *3*	0.347 *23*
γ	645.10 *3*	0.316 *23*
γ	654.10 *3*	0.347 *23*
γ E2,M1	663.106 *20*	9.2 *6*
γ	688.705 *25*	0.270 *23*

Photons (^{132}La)
(continued)

γ_{mode}	γ(keV)	γ(%)†
γ E2,M1	697.71 *3*	0.95 *6*
γ	787.24 *5*	0.023 *8*
γ	801 *2*	~0.15
γ	838.00 *3*	<0.11 ?
γ	847 *2*	~0.15
γ	856.53 *6*	0.100 *15*
γ	859.36 *3*	0.270 *23*
γ E1	881.61 *3*	0.95 *6*
γ E1	899.36 *3*	4.7 *3*
γ	912.37 *10*	0.054 *8*
γ	918.68 *5*	0.200 *23*
γ	929.74 *4*	0.200 *15*
γ	940.97 *3*	0.27 *3*
γ	966.52 *3*	0.400 *23*
γ	976 *2*	~0.31
γ [E1]	995.30 *3*	~0.18 ?
γ	1014.62 *5*	0.046 *15*
γ E2(+M1)	1031.75 *3*	7.9 *5*
γ	1036.91 *3*	0.32 *3*
γ M1,E2	1046.63 *3*	3.47 *23*
γ	1096.11 *9*	0.031 *8*
γ	1150.82 *6*	0.089 *20* ?
γ	1170.05 *5*	0.062 *15*
γ	1173.11 *6*	0.123 *15*
γ	1188.44 *5*	0.293 *23* ?
γ	1198.56 *4*	0.116 *15*
γ	1208.49 *7*	0.231 *23*
γ E2,M1	1221.265 *23*	2.97 *18*
γ	1242.16 *3*	0.208 *15*
γ	1246.863 *24*	0.354 *23*
γ	1264.872 *25*	0.285 *23*
γ	1285 *2*	~0.19
γ	1308 *2*	~0.15
γ	1342.80 *3*	0.36 *3*
γ	1355.33 *6*	0.092 *15*
γ	1366 *1*	
γ	1370 *2*	0.27 *13*
γ	1390 *2*	~0.15
γ	1396.94 *6*	0.185 *15*
γ	1407 *2*	~0.15
γ	1416.80 *4*	0.054 *8*
γ	1439.77 *3*	0.28 *3*
γ [E1]	1467.96 *7*	0.05 *2*
γ	1516.62 *7*	0.039 *15*
γ	1533.63 *5*	1.49 *9*
γ	1535.71 *3*	<0.23 ?
γ	1556 *2*	~0.15
γ E2,M1	1581.79 *4*	0.89 *6*
γ E1	1604.07 *3*	3.70 *23*
γ	1617.19 *10*	0.054 *8*
γ	1699.60 *9*	0.054 *8*
γ	1704 *2*	~0.15
γ	1738.14 *6*	0.069 *15*
γ	1755 *2*	~0.15
γ	1766 *2*	~0.39
γ	1800.32 *3*	0.270 *23*
γ	1824.14 *4*	0.55 *4*
γ	1846 *2*	~0.39
γ	1870 *2*	~0.15
γ	1877.14 *6*	0.246 *23* ?
γ E1	1909.963 *23*	9.1 *6*
γ	1949.45 *11*	0.154 *15* ?
γ	1983.78 *7*	0.039 *8*
γ	1998.26 *4*	0.47 *3*
γ	2041 *2*	~0.6
γ	2082.41 *9*	0.116 *15*
γ E1	2102.87 *3*	5.9 *4*
γ	2187.72 *5*	0.154 *23*
γ	2257.04 *10*	~0.015
γ	2264.72 *8*	0.031 *8*
γ	2296.29 *6*	0.131 *15*
γ	2304 *2*	~0.31
γ [E2]	2367.31 *6*	0.223 *15*
γ E1	2391.30 *4*	0.99 *7*
γ	2410 *2*	~0.42
γ	2453 *2*	~0.46
γ [M1+E2]	2463.24 *6*	0.89 *5* ?
γ	2557 *2*	~0.19
γ	2631.80 *9*	0.246 *15*
γ	2659 *2*	~0.15
γ	2684 *2*	~0.15
γ E1	2690.8 *9*	0.55 *11* ?
γ	2709 *2*	~0.19
γ	2744.18 *8*	0.154 *15*

Photons (^{132}La)
(continued)

γ_{mode}	γ(keV)	γ(%)†
γ E1	2754.88 *5*	1.62 *10*
γ	2776 *2*	~0.12
γ	2812 *2*	~0.15
γ	2872 *2*	~0.23
γ E1	2959.38 *6*	0.95 *7*
γ	2970 *2*	~0.12
γ [M1+E2]	3030.40 *6*	0.162 *15* ?
γ E1	3098.38 *6*	0.49 *3*
γ	3125 *2*	~0.12
γ	3143 *2*	~0.08
γ	3155.3 *9*	0.108 *22* ?
γ [E1]	3170.70 *11*	0.277 *23*
γ	3185 *2*	~0.08
γ E1	3198.96 *9*	0.72 *5*
γ	3269 *2*	~0.15
γ	3284 *2*	~0.12
γ	3303.65 *9*	0.077 *8*
γ	3354 *2*	~0.27
γ	3363 *2*	~0.19
γ	3382 *2*	~0.15
γ	3390 *2*	~0.06
γ	3437 *2*	~0.08
γ	3477 *2*	~0.08
γ [E2]	3494.98 *7*	0.085 *17* ?
γ	3506 *2*	~0.12
γ	3562.96 *7*	0.039 *8*
γ	3605 *2*	~0.12
γ E1	3635.27 *11*	0.34 *3*
γ	3663.53 *9*	0.013 *3* ?
γ	3775.91 *8*	0.046 *8*

† 5.2% uncert(syst)

Atomic Electrons (^{132}La)

⟨e⟩=10.5 *3* keV

e_{bin}(keV)	⟨e⟩(keV)	e(%)
5	1.35	26 *3*
6	0.98	17.0 *17*
25	0.111	0.44 *5*
26	0.31	1.21 *12*
27	0.52	1.95 *20*
30	0.116	0.38 *4*
31	0.33	1.07 *11*
32 - 36	0.126	0.38 *3*
155 - 193	0.33	0.21 *9*
268 - 306	0.063	0.023 *3*
323 - 361	0.036	0.010 *5*
377 - 425	0.011	0.0027 *12*
427	3.34	0.78 *5*
429 - 442	0.107	0.0243 *17*
459	0.56	0.122 *8*
461 - 499	0.254	0.054 *3*
503	0.32	0.063 *13*
510 - 516	0.017	0.0033 *7*
530	0.50	0.095 *7*
534 - 567	0.167	0.0302 *23*
596 - 645	0.249	0.040 *3*
648 - 698	0.078	0.0118 *7*
750 - 796	0.0037	~0.0005
800 - 847	0.017	0.0021 *4*
851 - 899	0.047	0.0054 *4*
904 - 941	0.016	0.0017 *6*
958 - 999	0.122	0.0123 *8*
1009 - 1059	0.090	0.0089 *11*
1090 - 1136	0.0032	0.00028 *11*
1145 - 1193	0.051	0.0043 *7*
1197 - 1247	0.017	0.00144 *24*
1248 - 1285	0.0044	0.00035 *14*
1302 - 1350	0.008	0.00061 *20*
1353 - 1401	0.0064	0.00047 *13*
1402 - 1440	0.0034	0.00024 *9*
1462 - 1555	0.027	~0.0018
1567 - 1662	0.0215	0.00136 *7*
1667 - 1765	0.0087	0.00051 *14*
1787 - 1876	0.048	0.00261 *21*
1904 - 1997	0.0106	0.00055 *7*
2004 - 2102	0.032	0.00154 *15*

Atomic Electrons (^{132}La) (continued)

e_{bin}(keV)	⟨e⟩(keV)	e(%)
2150 - 2227	0.0014	6.3 *19* $\times10^{-5}$
2251 - 2330	0.0045	0.00020 *5*
2354 - 2452	0.016	0.00065 *25*
2457 - 2556	0.0021	8 *4* $\times10^{-5}$
2594 - 2690	0.0065	0.00025 *4*
2703 - 2775	0.0087	0.00032 *3*
2806 - 2871	0.0014	5.0 *24* $\times10^{-5}$
2922 - 2993	0.0052	0.000176 *14*
3024 - 3124	0.0034	0.000111 *14*
3133 - 3232	0.0049	0.000153 *15*
3247 - 3345	0.0037	0.00011 *3*
3348 - 3440	0.0014	4.2 *10* $\times10^{-5}$
3458 - 3557	0.0013	3.7 *9* $\times10^{-5}$
3558 - 3634	0.0018	4.9 *8* $\times10^{-5}$
3658 - 3738	0.00019	5.1 *15* $\times10^{-6}$
3770 - 3775	2.8 $\times10^{-5}$	7.4 *20* $\times10^{-7}$

Continuous Radiation (^{132}La)

⟨β+⟩=543 keV;⟨IB⟩=4.9 keV

E_{bin}(keV)		⟨ ⟩(keV)	(%)
0 - 10	β+	1.19 $\times10^{-6}$	1.42 $\times10^{-5}$
	IB	0.019	
10 - 20	β+	4.97 $\times10^{-5}$	0.000299
	IB	0.019	0.128
20 - 40	β+	0.00142	0.00434
	IB	0.085	0.28
40 - 100	β+	0.071	0.091
	IB	0.108	0.166
100 - 300	β+	3.47	1.57
	IB	0.32	0.18
300 - 600	β+	24.2	5.2
	IB	0.48	0.111
600 - 1300	β+	146	15.4
	IB	1.3	0.143
1300 - 2500	β+	302	16.9
	IB	1.7	0.099
2500 - 4710	β+	67	2.40
	IB	0.77	0.026
	Σβ+		42

$^{132}_{57}$La(24.3 *5* min)

Mode: IT(76 %), ϵ(24 %)
Δ: -83552 *50* keV
SpA: 5.86$\times10^{7}$ Ci/g
Prod: ^{139}La(p,p7n); ^{133}Cs(α,5n)

Photons (^{132}La)

⟨γ⟩=491 *31* keV

γ_{mode}	γ(keV)	γ(%)
γ_{IT} M3	53 *1*	0.043 *9*
γ_{IT} M1(+E2)	135.2 *10*	44 *9*
γ_ϵM1,E2	237.6 *10*	3.9 *8*
γ_ϵM1	285 *1*	7.0 *13*
γ_ϵ	390 *1*	4.8 *9*
γ_ϵE2	464.59 *4*	22
γ_ϵE2,M1	479.47 *3*	3.1 *9*
γ_ϵE2(+M1)	567.169 *20*	4.4 *11*
γ_ϵ	601.77 *3*	1.3 *4*
γ_ϵE2,M1	663.106 *20*	5.0 *15*
γ_ϵE2,M1	697.71 *3*	4.0 *9*
γ_ϵE1	899.36 *3*	6.6 *9*
γ_ϵ	991 *1*	2.6 *5*
γ_ϵE2(+M1)	1031.75 *3*	2.6 *5*
γ_ϵM1,E2	1046.63 *3*	5.7 *9*

$^{132}_{58}$Ce(3.51 *11* h)

Mode: ϵ
Δ: -82440 *210* keV syst
SpA: 6.76$\times10^{6}$ Ci/g
Prod: protons on Ce; ^{139}La(p,8n)

Photons (^{132}Ce)

⟨γ⟩=273 *9* keV

γ_{mode}	γ(keV)	γ(%)
La L_ℓ	4.121	0.173 *25*
La L_η	4.529	0.073 *8*
La L_α	4.649	4.7 *6*
La L_β	5.166	4.3 *6*
La L_γ	5.880	0.58 *9*
La $K_{\alpha2}$	33.034	23.5 *8*
La $K_{\alpha1}$	33.442	43.3 *14*
La $K_{\beta1}$'	37.777	12.1 *4*
La $K_{\beta2}$'	38.927	3.16 *11*
γ	61.4 *10*	0.24 *8*
γ	88.2 *10*	0.14 *6*
γ M1,E2	142.2 *10*	0.64 *11*
γ E1	155.4 *10*	11.5 *7*
γ	176.8 *10*	~0.24
γ E1	182.1 *10*	82 *4*
γ M1	190.1 *10*	2.8 *3*
γ	199.2 *10*	0.15 *7*
γ	205.3 *10*	0.15 *7*
γ E1	216.7 *10*	5.3 *4*
γ M1,E2	251.4 *10*	2.6 *3*
γ	268.1 *10*	0.57 *16*
γ	279.9 *10*	2.1 *3*
γ	302.7 *10*	2.4 *3*
γ M1,E2	329.6 *10*	2.8 *5*
γ M1,E2	368.1 *10*	1.3 *3*
γ	424.4 *10*	1.2 *3*
γ	431.5 *10*	1.05 *25*
γ	451.5 *10*	1.6 *4*
γ	576.2 *10*	1.2 *4*

Atomic Electrons (^{132}Ce)

⟨e⟩=16.9 *6* keV

e_{bin}(keV)	⟨e⟩(keV)	e(%)
5	2.30	42 *5*
6	1.67	28 *3*
22 - 26	0.32	1.3 *5*
27	0.51	1.89 *19*
28	0.83	3.0 *3*
31	0.131	0.42 *4*
32	0.55	1.71 *18*
33 - 82	0.9	1.9 *7*
83 - 103	0.30	0.30 *6*
116	0.80	0.69 *4*
136 - 142	0.14	0.10 *4*
143	4.52	3.16 *17*
149 - 150	0.138	0.092 *5*
151	0.66	0.44 *5*
154 - 175	0.092	0.057 *15*
176	0.66	0.376 *21*
177 - 211	0.696	0.383 *14*
212	0.39	0.18 *3*
215 - 264	0.51	0.21 *7*
267 - 303	0.36	0.124 *24*
323 - 368	0.18	0.055 *10*
385 - 432	0.19	0.047 *16*
445 - 452	0.014	0.0030 *15*
537 - 576	0.042	~0.008

Continuous Radiation (^{132}Ce)

⟨IB⟩=0.33 keV

E_{bin}(keV)		⟨ ⟩(keV)	(%)
10 - 20	IB	0.00033	0.0020
20 - 40	IB	0.083	0.25
40 - 100	IB	0.0091	0.0166
100 - 300	IB	0.033	0.0160
300 - 600	IB	0.110	0.025
600 - 1118	IB	0.091	0.0124

$^{132}_{59}$Pr(1.6 *3* min)

Mode: ϵ
Δ: -75340 *300* keV syst
SpA: 8.9$\times10^{8}$ Ci/g
Prod: protons on Ta

Photons (^{132}Pr)

γ_{mode}	γ(keV)	γ(rel)
γ	325.0 *3*	100 *20*
γ	496.1 *10*	23 *5*
γ	533.1 *10*	19 *4*

$^{132}_{60}$Nd(1.75 *17* min)

Mode: ϵ
SpA: 8.1$\times10^{8}$ Ci/g
Prod: ^{32}S on ^{106}Cd

$^{132}_{61}$Pm(4 *2* s)

Mode: ϵ
SpA: 1.96$\times10^{10}$ Ci/g
Prod: ^{32}S on ^{106}Cd

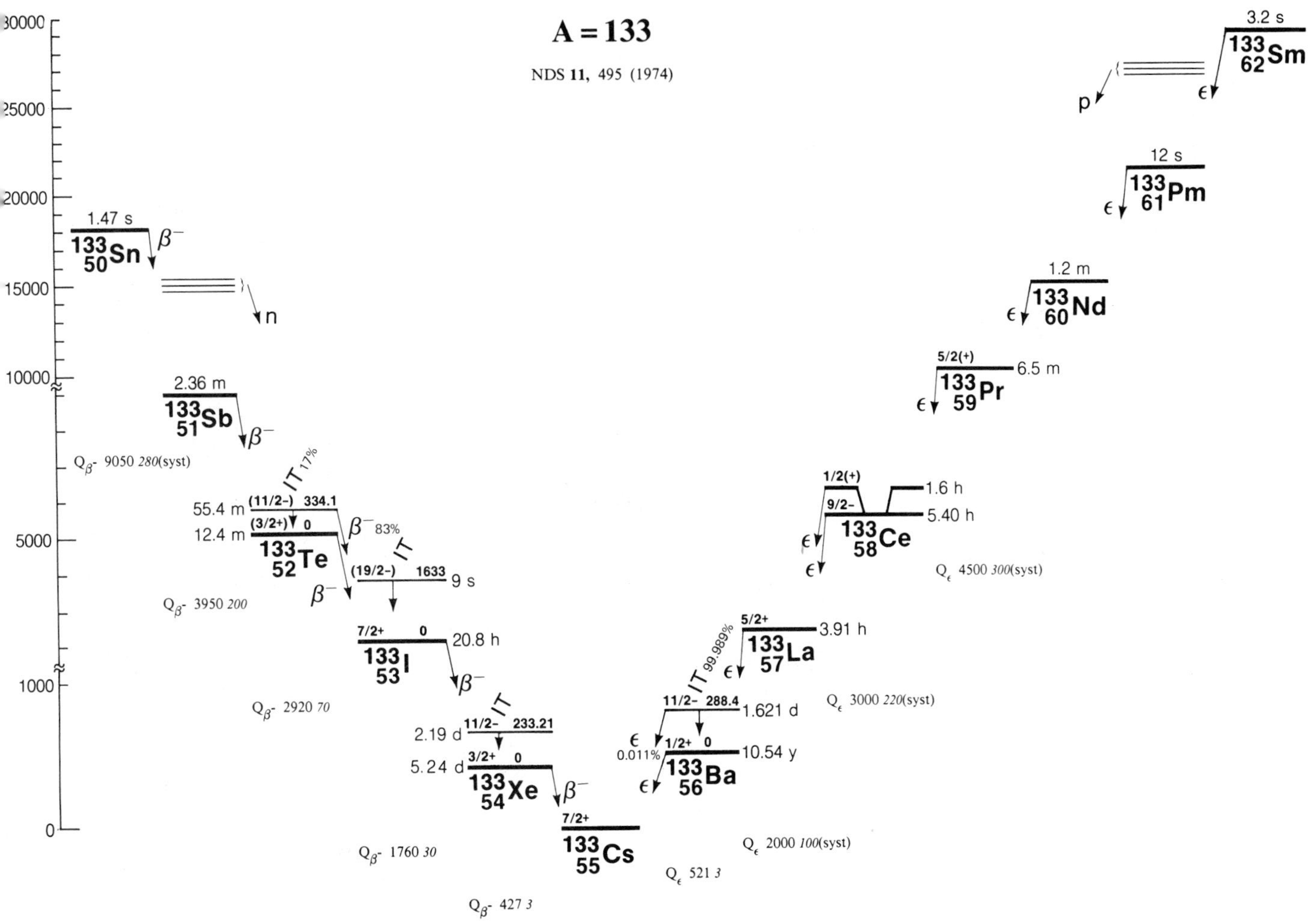

$^{133}_{50}$Sn(1.47 *4* s)

Mode: β-, β-n
Δ: -69990 *210* keV syst
SpA: 4.60×10^{10} Ci/g
Prod: fission

Photons (^{133}Sn)

γ_{mode}	γ(keV)
γ	963

$^{133}_{51}$Sb(2.36 *5* min)

Mode: β-
Δ: -79040 *210* keV
SpA: 5.98×10^{8} Ci/g
Prod: fission

Photons (^{133}Sb)

γ_{mode}	γ(keV)	γ(rel)
γ	160.48 *5*	1.04 *7*
γ	258.04 *7*	1.7 *5*
γ	260.98 *12*	1.0 *2*
γ	266.04 *8*	0.9 *1*
γ	276.56 *4*	1.7 *4*
γ	279.48 *15*	0.6 *2*
γ	290.67 *14*	~0.5 ?
γ	308.242 *11*	10.1 *2*
γ	356.83 *13*	0.7 *1*
γ	363.87 *17*	0.5 *1*
γ	403.19 *15*	~0.7
γ	404.35 *6*	5.3 *2*
γ	412.93 *13*	0.9 *2*
γ	422.17 *11*	2.0 *3*
γ	423.40 *3*	8.2 *3*
γ	440.96 *8*	1.5 *3*
γ	523.36 *15*	0.6 *2*
γ	529.40 *19*	~0.5 ?
γ	538.76 *4*	4.4 *5*
γ	558.38 *9*	1.0 *3*
γ	560.87 *15*	0.7 *2*
γ	572.30 *21*	0.4 *1*
γ	591.08 *11*	0.9 *3*
γ	632.422 *25*	9.0 *6*
γ	679.59 *22*	1.4 *4*
γ	687.47 *4*	3.8 *5*
γ	691.08 *3*	6.5 *4*
γ	808.93 *7*	2.3 *3*
γ	822.4 *10*	0.90 *18* ?
γ	836.886 *8*	25.8 *7*

Photons (^{133}Sb)
(continued)

γ_{mode}	γ(keV)	γ(rel)
γ	889.73 *15*	0.6 *2*
γ	927.67 *8*	2.7 *5*
γ	930 *1*	3.9 *11*
γ	936.34 *8*	2.0 *5*
γ	939.55 *25*	~0.7 ?
γ	964.06 *20*	0.8 *2*
γ	987.18 *4*	5.0 *4*
γ	1014.41 *9*	1.9 *3*
γ	1026.83 *3*	12.8 *5*
γ	1056 *1*	1.0 *3* ?
γ	1065.49 *3*	6.4 *5*
γ	1079.68 *10*	1.2 *3*
γ	1083.75 *14*	0.6 *2*
γ	1088.19 *11*	1.3 *2*
γ	1096.203 *15*	100 *2*
γ	1110.82 *5*	4.1 *2*
γ	1113.08 *7*	4.5 *2*
γ	1115.181 *23*	11.2 *3*
γ	1180.28 *18*	1.8 *3*
γ	1183.47 *3*	7.5 *5*
γ	1202.77 *5*	2.5 *3*
γ	1218.66 *4*	4.2 *3*
γ	1235.98 *5*	4.9 *5*
γ	1249.77 *12*	2.0 *4*
γ	1265.316 *19*	12.7 *8*
γ	1271.545 *20*	9.8 *6*
γ	1293.5 *3*	0.5 *2*
γ	1300.02 *11*	1.8 *3*
γ	1305.228 *20*	13.2 *4*
γ	1309.68 *9*	1.6 *3* ?

Photons (^{133}Sb)
(continued)

γ_{mode}	γ(keV)	γ(rel)
γ	1344.07 25	0.6 2
γ	1354.28 21	0.6 2 ?
γ	1393.61 11	1.3 2
γ	1410.06 16	1.0 2 ?
γ	1419.2 4	0.6 2
γ	1421.3 2	0.7 2
γ	1425.17 13	1.1 3
γ	1443.29 25	~0.4
γ	1484.35 5	3.0 2
γ	1489.68 3	6.0 4
γ	1496.56 5	2.5 3
γ	1529.13 10	1.6 3
γ	1552.130 22	9.7 5
γ	1558.67 11	1.1 3
γ	1579.15 16	3.8 7
γ	1580.9 4	2.0 7
γ	1641.51 3	9.1 4
γ	1654.23 5	3.7 3
γ	1658.64 4	5.2 3
γ	1697.96 5	3.0 3
γ	1705.50 9	2.2 2
γ	1728.625 14	16.9 3
γ	1775.78 6	2.7 4
γ	1794.9 4	0.4 1
γ	1877.19 5	3.6 3
γ	1886.9 3	0.5 2
γ	1896.7 3	1.2 4
γ	1904.59 19	1.2 4
γ	1931.0 5	~0.3
γ	1933.0 8	<0.10 ?
γ	1934.2 5	~0.3
γ	1944.10 6	5.3 3
γ	1946.45 16	1.7 3
γ	1976.44 8	2.2 3
γ	1992.01 21	0.7 2
γ	2018.45 16	0.7 2
γ	2416 1	20 4
γ	2447.2 10	3.0 6
γ	2755.4 10	30 6

$^{133}_{52}$Te(12.4 *3* min)

Mode: β-
Δ: -82990 *80* keV
SpA: 1.14×10^{8} Ci/g
Prod: fission; daughter ^{133}Te(55.4 min)

Photons (^{133}Te)
⟨γ⟩=952 *18* keV

γ_{mode}	γ(keV)	γ(%)†
I L_ℓ	3.485	0.0039 8
I L_η	3.780	0.0019 4
I L_α	3.937	0.110 20
I L_β	4.320	0.097 19
I L_γ	4.885	0.0121 25
I $K_{\alpha2}$	28.317	0.68 9
I $K_{\alpha1}$	28.612	1.26 17
I $K_{\beta1}'$	32.278	0.34 5
I $K_{\beta2}'$	33.225	0.077 10
γ [M1+E2]	312.17 22	72.6
γ	384.5 3	0.29 7
γ	392.9 6	0.58 22
γ [M1+E2]	407.86 23	30.9 7
γ	475.0 3	1.2 3
γ	546.3 3	0.58 22
γ	587.0 4	0.51 15
γ	613.4 3	~0.29
γ	720.0 3	6.8 5
γ	787.1 3	5.7 4
γ	844.4 5	3.3 3
γ	930.8 4	4.6 6

Photons (^{133}Te)
(continued)

γ_{mode}	γ(keV)	γ(%)†
γ	1000.8 6	6.4 7
γ	1021.2 3	2.8 3
γ	1062.0 5	1.31 22
γ	1252.2 5	1.16 15
γ	1307.7 8	0.9 3
γ	1313.0 6	0.8 3
γ	1333.4 3	10.2 6
γ	1405.8 4	0.58 15
γ	1474.0 6	0.36 15
γ	1518.6 8	0.51 7
γ	1588.2 9	~0.29
γ	1717.9 3	3.5 3
γ	1824.8 8	0.58 22
γ	1881.8 6	1.45 22
γ	2137.0 8	0.29 7
γ	2228.2 10	~0.29
γ	2540.4 10	0.073 22

† 1.1% uncert(syst)

Atomic Electrons (^{133}Te)
⟨e⟩=10.1 *10* keV

e_{bin}(keV)	⟨e⟩(keV)	e(%)
5 - 32	0.18	2.5 4
279	6.0	2.1 3
307	0.87	0.28 7
308 - 351	0.35	0.11 3
360	0.022	~0.006
375	1.62	0.43 4
379 - 393	0.007	0.0017 8
403	0.24	0.060 13
407 - 442	0.097	0.023 6
470 - 513	0.021	0.0041 20
541 - 587	0.021	0.0038 15
608 - 613	0.0010	~0.00017
687 - 720	0.13	0.019 9
754 - 787	0.10	0.013 6
811 - 844	0.053	0.007 3
898 - 931	0.06	0.007 3
968 - 1017	0.12	0.012 4
1020 - 1062	0.017	0.0016 7
1219 - 1252	0.012	0.0010 4
1275 - 1313	0.10	0.008 3
1328 - 1373	0.018	0.0013 4
1401 - 1441	0.0034	0.00024 11
1469 - 1555	0.0066	0.00044 14
1583 - 1587	0.00030	$\sim2\times10^{-5}$
1685 - 1717	0.025	0.0015 6
1792 - 1881	0.014	0.00075 22
2104 - 2195	0.0032	0.00015 5
2223 - 2227	0.00022	$\sim10\times10^{-6}$
2507 - 2539	0.00038	1.5 6 $\times10^{-5}$

Continuous Radiation (^{133}Te)
⟨β-⟩=787 keV;⟨IB⟩=1.58 keV

E_{bin}(keV)		⟨ ⟩(keV)	(%)
0 - 10	β-	0.0254	0.507
	IB	0.031	
10 - 20	β-	0.077	0.52
	IB	0.030	0.21
20 - 40	β-	0.317	1.06
	IB	0.059	0.20
40 - 100	β-	2.38	3.38
	IB	0.163	0.25
100 - 300	β-	26.8	13.2
	IB	0.43	0.24
300 - 600	β-	102	22.6
	IB	0.41	0.098
600 - 1300	β-	383	42.0
	IB	0.39	0.047

Continuous Radiation (^{133}Te)
(continued)

E_{bin}(keV)		⟨ ⟩(keV)	(%)
1300 - 2500	β-	272	16.8
	IB	0.065	0.0043
2500 - 2608	β-	0.247	0.0098
	IB	1.06×10^{-6}	4.2×10^{-8}

$^{133}_{52}$Te(55.4 *4* min)

Mode: β-(83 %), IT(17 %)
Δ: -82656 *80* keV
SpA: 2.552×10^{7} Ci/g
Prod: fission

Photons (^{133}Te)
⟨γ⟩=1696 *76* keV

γ_{mode}	γ(keV)	γ(%)†
$\gamma_{\beta-}$(M2)	74.1 2	0.50 8 *
$\gamma_{\beta-}$	81.5 2	0.50 8
$\gamma_{\beta-}$	88.0 2	1.9 3
$\gamma_{\beta-}$	94.9 2	3.8 6
$\gamma_{\beta-}$	164.34 9	1.01 15
$\gamma_{\beta-}$	168.87 9	7.6 11
$\gamma_{\beta-}$	177.1 2	0.94 14
$\gamma_{\beta-}$	178.2 2	0.63 9
$\gamma_{\beta-}$	184.45 10	0.25 4
$\gamma_{\beta-}$	193.22 10	0.76 11
$\gamma_{\beta-}$	198.2 2	0.38 6
$\gamma_{\beta-}$	213.36 8	2.6 4
$\gamma_{\beta-}$	220.94 13	0.31 5
$\gamma_{\beta-}$	224.03 13	0.25 4
$\gamma_{\beta-}$	244.28 10	0.44 7
$\gamma_{\beta-}$	251.49 10	0.38 6
$\gamma_{\beta-}$	257.64 9	0.63 9
$\gamma_{\beta-}$	261.55 7	8.8 13
$\gamma_{\beta-}$	285.7 5	0.63 13
γ_{IT} M4	334.14 7	7.99 16
$\gamma_{\beta-}$	344.5 2	0.94 14
$\gamma_{\beta-}$	347.22 9	0.82 12
$\gamma_{\beta-}$	355.57 14	0.38 6
$\gamma_{\beta-}$	362.81 15	0.69 10
$\gamma_{\beta-}$	376.83 14	0.38 6
$\gamma_{\beta-}$	396.96 9	1.07 16
$\gamma_{\beta-}$	429.02 11	2.3 3
$\gamma_{\beta-}$	435.4 7	0.76 25
$\gamma_{\beta-}$	444.90 9	2.8 4
$\gamma_{\beta-}$	462.11 16	2.1 3
$\gamma_{\beta-}$	471.85 9	1.45 22
$\gamma_{\beta-}$	478.59 10	1.13 17
$\gamma_{\beta-}$	519.6 2	0.31 5
$\gamma_{\beta-}$	534.85 11	1.26 19
$\gamma_{\beta-}$	574.04 10	2.3 3
$\gamma_{\beta-}$	622.03 16	1.01 15
$\gamma_{\beta-}$[E2]	647.40 8	21 3 *
$\gamma_{\beta-}$	702.75 12	2.7 4
$\gamma_{\beta-}$	731.69 15	1.07 16
$\gamma_{\beta-}$	733.89 10	2.1 3
$\gamma_{\beta-}$	779.75 10	2.5 4
$\gamma_{\beta-}$	795.7 4	0.94 14
$\gamma_{\beta-}$	800.51 12	1.39 21
$\gamma_{\beta-}$	863.91 13	18 3
$\gamma_{\beta-}$	882.83 12	3.0 5
$\gamma_{\beta-}$	897.7 4	0.31 5
$\gamma_{\beta-}$[E2]	912.58 10	63 *
$\gamma_{\beta-}$	914.72 13	12.0 18
$\gamma_{\beta-}$	934.4 3	0.94 14
$\gamma_{\beta-}$	978.19 9	5.9 9
$\gamma_{\beta-}$	980.4 2	1.7 3
$\gamma_{\beta-}$	982.9 2	0.82 12
$\gamma_{\beta-}$	1007.5 10	0.76 25
$\gamma_{\beta-}$	1029.8 2	1.13 17
$\gamma_{\beta-}$	1061.83 11	2.0 3
$\gamma_{\beta-}$	1348.9 2	1.8 3

Photons (^{133}Te)
(continued)

γ_{mode}	γ(keV)	γ(%)†
$\gamma_{\beta-}$	1459.1 2	1.57 24
$\gamma_{\beta-}$	1516.1 3	0.69 10
$\gamma_{\beta-}$	1531.6 4	0.63 9
$\gamma_{\beta-}$	1587.4 2	1.39 21
$\gamma_{\beta-}$	1683.3 2	4.2 6
$\gamma_{\beta-}$	1704.4 3	0.69 10
$\gamma_{\beta-}$	1885.7 3	0.82 12
$\gamma_{\beta-}$	2004.9 3	3.4 5
$\gamma_{\beta-}$	2027.7 4	0.88 13
$\gamma_{\beta-}$	2049.2 4	1.07 16

† uncert(syst): 13% for IT, 11% for β-
* with ^{133}I(9 s) in equilib

Continuous Radiation (^{133}Te)
⟨β-⟩=600 keV;⟨IB⟩=1.11 keV

E_{bin}(keV)		⟨ ⟩(keV)	(%)
0 - 10	β-	0.0203	0.404
	IB	0.024	
10 - 20	β-	0.062	0.412
	IB	0.024	0.165
20 - 40	β-	0.256	0.85
	IB	0.046	0.160
40 - 100	β-	1.95	2.76
	IB	0.127	0.20
100 - 300	β-	23.0	11.3
	IB	0.33	0.19
300 - 600	β-	93	20.6
	IB	0.30	0.071
600 - 1300	β-	339	37.6
	IB	0.24	0.030
1300 - 2342	β-	143	9.1
	IB	0.026	0.0017

$^{133}_{53}$I (20.8 1 h)

Mode: β-
Δ: -85910 30 keV
SpA: 1.133×10^{6} Ci/g
Prod: fission

Photons (^{133}I)
⟨γ⟩=607 110 keV

γ_{mode}	γ(keV)	γ(%)†
Xe L_ℓ	3.634	0.00118 22
Xe L_η	3.955	0.00054 9
Xe L_α	4.104	0.033 6
Xe L_β	4.522	0.029 5
Xe L_γ	5.119	0.0037 7
Xe $K_{\alpha2}$	29.461	0.192 25
Xe $K_{\alpha1}$	29.782	0.36 5
Xe $K_{\beta1}$'	33.606	0.097 12
Xe $K_{\beta2}$'	34.606	0.023 3
γ [M1+E2]	150.392 13	0.029 6
γ	176.96 7	0.078 17 ?
γ	203.79 3	0.0043 9
γ M4	233.207 16	*
γ [M1+E2]	245.935 20	0.035 9
γ [M1+E2]	262.689 10	0.356 10
γ [E2]	267.163 12	0.117 6
γ [M1+E2]	345.464 11	0.104 17
γ [M1+E2]	361.092 13	0.11 4
γ [M1+E2]	372.035 22	~0.009
γ [M1+E2]	381.58 3	0.045 4
γ [E2]	386.78 3	0.059 4

Photons (^{133}I)
(continued)

γ_{mode}	γ(keV)	γ(%)†
γ [M1+E2]	417.555 15	0.153 10
γ [M1+E2]	422.898 10	0.309 9
γ [M1+E2]	438.92 3	0.040 4
γ [M1+E2]	510.516 11	1.81 4
γ	510.802 16	<0.009 ?
γ	522.427 20	<0.09 ?
γ [M1+E2]	529.852 8	86
γ	537.63 3	0.035 7
γ	554.491 21	<0.0009 ?
γ [E2]	556.164 14	0.020 3
γ	567.1 4	~0.003
γ [E2]	617.969 12	0.539 12
γ [M1+E2]	648.74 3	0.056 13
γ [M1+E2]	670.108 21	0.042 5
γ	678.48 4	0.022 7
γ [M1+E2]	680.244 11	0.645 16
γ [M1+E2]	706.555 10	1.49 4
γ [M1+E2]	768.361 10	0.457 12
γ [E2]	789.589 21	0.050 4
γ [M1+E2]	820.500 18	0.154 5
γ [M1+E2]	856.265 12	1.23 4
γ [E2]	875.314 9	4.47 10
γ	909.67 3	0.212 8
γ [M1+E2]	911.43 3	0.046 6
γ	1018.1 5	0.006 3
γ	1035.523 13	0.0086 17
γ [M1+E2]	1052.277 19	0.552 13
γ	1060.06 3	0.137 6
γ [E2]	1087.661 20	0.0121 17
γ [E2]	1236.404 9	1.49 4
γ [E2+5.1%M1]	1298.210 8	2.33 5
γ	1327.22 3	<0.00043 ?
γ [M1+E2]	1350.348 18	0.148 4
γ [E2]	1386.113 14	0.009 3
γ	1589.90 3	0.0029 4

† 2.3% uncert(syst)
* with ^{133}Xe(2.19 d)

Atomic Electrons (^{133}I)
⟨e⟩=4.4 5 keV

e_{bin}(keV)	⟨e⟩(keV)	e(%)
5 - 33	0.051	0.68 10
116 - 150	0.025	0.019 7
169 - 211	0.009	0.0049 17
228 - 267	0.069	0.0296 15
311 - 360	0.025	0.0076 9
361 - 404	0.0279	0.0072 6
412 - 439	0.0050	0.00119 12
476 - 488	0.073	0.0153 22
495	3.2	0.65 10
503 - 522	0.0146	0.0029 4
524	0.40	0.076 13
525	0.06	~0.011
529	0.094	0.018 4
530 - 567	0.024	0.0045 10
583 - 618	0.0178	0.00302 13
636 - 680	0.059	0.0089 10
701 - 734	0.0164	0.0023 3
755 - 790	0.0059	0.00076 8
815 - 856	0.099	0.0119 6
870 - 911	0.0163	0.00186 19
984 - 1031	0.0100	0.00098 15
1034 - 1083	0.00175	0.000167 21
1087 - 1088	4.9 ×10^{-6}	4.5 7 ×10^{-7}
1202 - 1236	0.0188	0.00156 6
1264 - 1298	0.0283	0.00223 9
1316 - 1352	0.00203	0.000154 18
1381 - 1386	1.3 ×10^{-5}	9.6 24 ×10^{-7}
1555 - 1589	2.5 ×10^{-5}	1.6 7 ×10^{-6}

Continuous Radiation (^{133}I)
⟨β-⟩=406 keV;⟨IB⟩=0.50 keV

E_{bin}(keV)		⟨ ⟩(keV)	(%)
0 - 10	β-	0.068	1.37
	IB	0.019	
10 - 20	β-	0.205	1.37
	IB	0.018	0.124
20 - 40	β-	0.82	2.75
	IB	0.034	0.118
40 - 100	β-	5.8	8.3
	IB	0.088	0.136
100 - 300	β-	54	27.2
	IB	0.19	0.111
300 - 600	β-	150	34.0
	IB	0.119	0.029
600 - 1300	β-	195	25.0
	IB	0.030	0.0042
1300 - 1527	β-	0.257	0.0189
	IB	5.5 ×10^{-6}	4.1 ×10^{-7}

$^{133}_{53}$I (9 2 s)

Mode: IT
Δ: -84277 30 keV
SpA: 9.1×10^{9} Ci/g
Prod: fission

Photons (^{133}I)
⟨γ⟩=1578 130 keV

γ_{mode}	γ(keV)	γ(%)
I L_ℓ	3.485	0.118 17
I L_η	3.780	0.056 7
I L_α	3.937	3.3 4
I L_β	4.321	3.0 4
I L_γ	4.900	0.39 6
I $K_{\alpha2}$	28.317	19.1 11
I $K_{\alpha1}$	28.612	35.6 21
I $K_{\beta1}$'	32.278	9.6 6
I $K_{\beta2}$'	33.225	2.17 13
γ (M2)	74.1 2	4.0 1
γ [E2]	647.40 8	99.6 20
γ [E2]	912.58 10	99.8 20

Atomic Electrons (^{133}I)
⟨e⟩=55.1 18 keV

e_{bin}(keV)	⟨e⟩(keV)	e(%)
5	3.1	65 7
23	0.60	2.6 3
24	0.78	3.3 4
27	0.30	1.12 13
28	0.40	1.45 17
29	0.0110	0.038 4
31	0.049	0.158 18
32	0.025	0.079 9
41	30.5	74 4
69	9.9	14.4 8
70	1.83	2.64 15
73	2.65	3.63 20
74	0.65	0.88 5
614	2.26	0.368 15
642	0.269	0.0419 17
643	0.0520	0.0081 3
646	0.0607	0.0094 4
647	0.0206	0.00319 13
879	1.43	0.162 7
907	0.169	0.0187 8
908	0.0205	0.00226 9
912	0.0467	0.00513 21
913	0.00114	0.000125 5

$^{133}_{54}$Xe(5.245 *6* d)

Mode: β-

Δ: -87665 *7* keV

SpA: 1.8718×10^5 Ci/g

Prod: fission; ^{132}Xe(n,γ)

Photons (^{133}Xe)

⟨γ⟩=45.9 *9* keV

γ_{mode}	γ(keV)	γ(%)†
Cs L_ℓ	3.795	0.092 *13*
Cs L_η	4.142	0.042 *5*
Cs L_α	4.285	2.5 *3*
Cs L_β	4.732	2.3 *3*
Cs L_γ	5.362	0.30 *4*
Cs $K_{\alpha2}$	30.625	14.1 *6*
Cs $K_{\alpha1}$	30.973	26.0 *10*
Cs $K_{\beta1}$'	34.967	7.1 *3*
Cs $K_{\beta2}$'	36.006	1.74 *7*
γ M1+1.7%E2	79.612 *8*	0.239 *20*
γ M1+2.7%E2	80.989 *5*	37.0
γ [M1+E2]	160.601 *8*	0.045 *3*
γ [M1+E2]	223.24 *1*	0.000110 *10*
γ M1(+0.2%E2)	302.851 *9*	0.0044 *4*
γ E2	383.841 *9*	0.00214 *18*

† 2.7% uncert(syst)

Atomic Electrons (^{133}Xe)

⟨e⟩=36.3 *9* keV

e_{bin}(keV)	⟨e⟩(keV)	e(%)
5	2.00	39 *4*
6 - 44	1.93	10.4 *6*
45	24.3	54.1 *18*
74	0.037	0.050 *5*
75	4.95	6.57 *22*
76	1.32	1.74 *5*
78 - 79	0.0107	0.0136 *10*
80	1.38	1.72 *6*
81 - 125	0.373	0.456 *15*
155 - 187	0.0044	0.0028 *8*
218 - 267	0.00043	0.000163 *18*
297 - 303	9.2 ×10^{-5}	3.1 *6* ×10^{-5}
348 - 384	0.000154	4.4 *3* ×10^{-5}

Continuous Radiation (^{133}Xe)

⟨β-⟩=100 keV;⟨IB⟩=0.036 keV

E_{bin}(keV)		⟨ ⟩(keV)	(%)
0 - 10	β-	0.322	6.5
	IB	0.0052	
10 - 20	β-	0.94	6.3
	IB	0.0045	0.032
20 - 40	β-	3.59	12.0
	IB	0.0073	0.026
40 - 100	β-	21.4	31.1
	IB	0.0124	0.020
100 - 300	β-	73	43.8
	IB	0.0061	0.0048
300 - 346	β-	1.30	0.417
	IB	3.7×10^{-6}	1.19×10^{-6}

$^{133}_{54}$Xe(2.19 *3* d)

Mode: IT

Δ: -87432 *7* keV

SpA: 4.48×10^5 Ci/g

Prod: fission; ^{132}Xe(n,γ); ^{130}Te(α,n)

Photons (^{133}Xe)

⟨γ⟩=41.4 *8* keV

γ_{mode}	γ(keV)	γ(%)†
Xe L_ℓ	3.634	0.118 *16*
Xe L_η	3.955	0.051 *5*
Xe L_α	4.104	3.3 *4*
Xe L_β	4.525	2.9 *4*
Xe L_γ	5.134	0.38 *6*
Xe $K_{\alpha2}$	29.461	16.2 *6*
Xe $K_{\alpha1}$	29.782	30.1 *11*
Xe $K_{\beta1}$'	33.606	8.1 *3*
Xe $K_{\beta2}$'	34.606	1.92 *7*
γ M4	233.207 *16*	10.3

† 3.0% uncert(syst)

Atomic Electrons (^{133}Xe)

⟨e⟩=192 *4* keV

e_{bin}(keV)	⟨e⟩(keV)	e(%)
5	2.9	59 *6*
24	0.50	2.06 *21*
25	0.63	2.5 *3*
28	0.175	0.62 *6*
29	0.37	1.29 *13*
30	0.050	0.168 *17*
32	0.019	0.059 *6*
33	0.043	0.131 *14*
199	125.4	63.1 *17*
228	47.9	21.0 *6*
232	10.8	4.67 *13*
233	2.78	1.19 *3*

$^{133}_{55}$Cs(stable)

Δ: -88093 *6* keV

%: 100

$^{133}_{56}$Ba(10.54 *3* yr)

Mode: ϵ

Δ: -87572 *7* keV

SpA: 255.0 Ci/g

Prod: ^{132}Ba(n,γ); ^{133}Cs(p,n)

Photons (^{133}Ba)

⟨γ⟩=404 *5* keV

γ_{mode}	γ(keV)	γ(%)†
Cs L_ℓ	3.795	0.24 *3*
Cs L_η	4.142	0.111 *12*
Cs L_α	4.285	6.7 *8*
Cs L_β	4.728	6.5 *9*
Cs L_γ	5.389	0.91 *14*
Cs $K_{\alpha2}$	30.625	35.6 *14*
Cs $K_{\alpha1}$	30.973	65.7 *25*
Cs $K_{\beta1}$'	34.967	18.0 *7*
Cs $K_{\beta2}$'	36.006	4.39 *18*
γ M1+1.0%E2	53.148 *11*	2.17 *4*
γ M1+1.7%E2	79.612 *8*	3.18 *25*
γ M1+2.7%E2	80.989 *5*	34.2 *19*
γ [M1+E2]	160.601 *8*	0.60 *3*
γ [M1+E2]	223.24 *1*	0.460 *12*
γ E2	276.388 *9*	7.09 *14*
γ M1(+0.2%E2)	302.851 *9*	18.4 *4*
γ E2	355.999 *10*	62.2
γ E2	383.841 *9*	8.92 *18*

† 1.6% uncert(syst)

Atomic Electrons (^{133}Ba)

⟨e⟩=54.7 *15* keV

e_{bin}(keV)	⟨e⟩(keV)	e(%)
5	5.0	98 *10*
6	1.03	18.0 *19*
17	1.83	10.7 *3*
25	1.07	4.3 *5*
26	1.33	5.2 *5*
29 - 35	1.43	4.7 *3*
44	2.12	4.9 *4*
45	22.5	50 *3*
47 - 74	1.49	2.72 *8*
75	4.6	6.1 *4*
76 - 125	3.15	3.94 *13*
155 - 187	0.134	0.078 *11*
218 - 240	0.805	0.336 *10*
267	1.87	0.70 *3*
271 - 303	0.551	0.191 *5*
320	4.20	1.31 *4*
348 - 384	1.609	0.457 *6*

Continuous Radiation (^{133}Ba)

⟨IB⟩=0.067 keV

E_{bin}(keV)		⟨ ⟩(keV)	(%)
10 - 20	IB	0.00054	0.0033
20 - 40	IB	0.064	0.21
40 - 100	IB	0.00168	0.0035
100 - 137	IB	6.8×10^{-6}	6.2×10^{-6}

$^{133}_{56}$Ba(1.621 *4* d)

Mode: IT(99.989 %), ϵ(0.011 %)
Δ: -87284 *7* keV
SpA: 6.057×10^{5} Ci/g
Prod: ^{133}Cs(p,n); ^{133}Cs(d,2n); ^{132}Ba(n,γ)

Photons (^{133}Ba)

⟨γ⟩=66.8 *20* keV

γ_{mode}	γ(keV)	γ(%)
Ba L_ℓ	3.954	0.19 *3*
Ba L_η	4.331	0.098 *10*
Ba L_α	4.465	5.3 *7*
Ba L_β	4.923	7.9 *12*
Ba L_γ	5.706	1.39 *24*
γ_{IT} M1	12.29 *4*	1.38 *6*
Ba $K_{\alpha2}$	31.817	15.3 *6*
Ba $K_{\alpha1}$	32.194	28.1 *11*
Ba $K_{\beta1}$'	36.357	7.8 *3*
Ba $K_{\beta2}$'	37.450	1.97 *8*
γ_{IT} M4	276.09 *15*	17.5 *7*
γ_ϵ[E2]	632.5 *10*	0.0110 *22*

Atomic Electrons (^{133}Ba)

⟨e⟩=219 *5* keV

e_{bin}(keV)	⟨e⟩(keV)	e(%)
5	1.74	33 *4*
6	7.9	130 *9*
7	0.526	7.8 *3*
11	1.78	16.1 *6*
12	0.529	4.38 *15*
25	0.119	0.47 *5*
26	0.34	1.29 *14*
27	0.55	2.08 *22*
30	0.124	0.41 *4*
31	0.35	1.14 *12*
32	0.076	0.237 *25*
35	0.038	0.108 *11*
36	0.0203	0.056 *6*
239	140.8	59.0 *19*
270	36.0	13.3 *4*
271	13.4	4.96 *16*
275	11.4	4.14 *14*
276	3.07	1.11 *4*

$^{133}_{57}$La(3.912 *8* h)

Mode: ϵ
Δ: -85570 *100* keV syst
SpA: 6.023×10^{6} Ci/g
Prod: ^{133}Cs(α,4n); ^{134}Ba(p,2n); daughter ^{133}Ce(5.40 h)

Photons (^{133}La)

⟨γ⟩=81.6 *5* keV

γ_{mode}	γ(keV)	γ(%)†
γ M1	12.29 *4*	*
Ba $K_{\alpha2}$	31.817	21.1 *6*
Ba $K_{\alpha1}$	32.194	38.9 *12*
Ba $K_{\beta1}$'	36.357	10.8 *3*
Ba $K_{\beta2}$'	37.450	2.73 *9*
γ	113.6 *3*	0.0040 *6*

Photons (^{133}La) (continued)

γ_{mode}	γ(keV)	γ(%)†
γ	136.7 *2*	0.0015 *6*
γ	158.4 *3*	0.0013 *6*
γ	210.72 *12*	0.0047 *10*
γ	227.85 *10*	0.0057 *10*
γ	256.57 *6*	0.0173 *11*
γ M4	276.09 *15*	*
γ [M1+E2]	278.83 *3*	1.90 *4*
γ	281.8 *2*	0.0095 *4*
γ [M1+E2]	286.34 *5*	0.0232 *21*
γ [M1+E2]	290.06 *5*	1.074 *21*
γ [E2]	291.17 *5*	0.331 *7*
γ	293.17 *11*	0.019 *4*
γ [M1+E2]	302.38 *4*	1.243 *25*
γ	309.5 *1*	0.0078 *21*
γ	324.63 *12*	0.0059 *15*
γ	328.17 *7*	0.0201 *15*
γ	339.41 *8*	0.040 *6*
γ	345.1 *4*	~0.0011 ?
γ	347.1 *3*	0.0025 *11*
γ	353.26 *9*	0.0144 *13*
γ	374.13 *15*	0.0046 *13*
γ	385.25 *9*	0.0515 *15*
γ	428.7 *2*	0.0032 *11*
γ	432.3 *3*	~0.0017 ?
γ	435.77 *10*	0.0198 *21*
γ	441.9 *4*	~0.0013 ?
γ	445.5 *3*	~0.0032
γ	465.53 *11*	0.0080 *11*
γ	469.3 *3*	0.0165 *13*
γ	481.75 *8*	0.0333 *19*
γ	494.5 *3*	~0.0023
γ	519.1 *4*	0.017 *6*
γ [M1+E2]	527.51 *8*	0.0659 *17*
γ	534.83 *12*	0.0418 *13*
γ	541.3 *3*	~0.0017
γ	556.28 *8*	0.0978 *20*
γ	560.25 *15*	0.0171 *15*
γ	565.29 *7*	0.511 *10*
γ [E2]	567.22 *7*	0.163 *4*
γ	571.9 *3*	0.0188 *19*
γ	581.3 *2*	0.0091 *11*
γ	584.77 *8*	0.156 *4*
γ	592.18 *15*	0.0237 *15*
γ [M1+E2]	595.99 *14*	0.359 *8*
γ	604.9 *3*	0.0025 *11*
γ [M1+E2]	618.29 *12*	0.802 *16*
γ [M1+E2]	621.61 *9*	0.490 *11*
γ	630.60 *8*	0.125 *4*
γ [M1+E2]	632.8 *4*	0.878 *18*
γ	652.99 *15*	0.0074 *13*
γ	664.21 *13*	0.0716 *19*
γ	671.80 *12*	0.0310 *15*
γ	676.50 *14*	0.0268 *13*
γ	682.0 *5*	~0.0017
γ	684.3 *5*	~0.0015
γ	689.5 *3*	~0.0019
γ	721.9 *3*	0.0034 *8*
γ	733.5 *10*	0.0021 *10*
γ	751.8 *3*	0.0431 *15*
γ	775.3 *3*	0.0032 *8*
γ	791.4 *3*	0.0036 *10*
γ	802.3 *4*	~0.0023
γ	810.3 *2*	0.042 *3*
γ	821.1 *3*	0.0131 *11*
γ	846.16 *13*	0.475 *13*
γ	848.4 *3*	0.0065 *13* ?
γ	850.5 *2*	0.0270 *15*
γ [M1+E2]	858.50 *13*	0.384 *11*
γ	874.8 *2*	0.0553 *17*
γ	887.1 *2*	0.0270 *13*
γ	891.9 *3*	0.0023 *8*
γ	909.5 *3*	0.0078 *17*
γ	911.6 *2*	0.133 *4*
γ	920.7 *2*	0.0169 *10*
γ	923.9 *2*	0.0171 *10*
γ	933.0 *2*	0.0110 *13*
γ	981.0 *3*	0.0040 *8*
γ	992.8 *3*	0.0051 *11*
γ	1021.6 *2*	0.0057 *10*
γ	1038.1 *2*	0.0074 *10*
γ	1043.0 *2*	0.0080 *10*
γ	1061.48 *10*	0.0794 *25*
γ	1080.9 *4*	0.0013 *6*
γ	1093 *1*	0.0716 *21*
γ	1099.93 *13*	0.190 *6*
γ	1111.9 *4*	0.0017 *8*
γ	1175.7 *2*	0.0047 *8*
γ	1182.2 *3*	0.0019 *8*
γ	1192.3 *3*	~0.0011
γ	1199.5 *3*	0.0234 *11*
γ	1211.8 *2*	0.0427 *13*
γ	1219.5 *3*	0.0013 *6*
γ	1229.9 *3*	0.0032 *8*
γ	1241.0 *3*	0.0021 *8*
γ	1261.0 *3*	0.0237 *13*
γ	1284.0 *3*	0.0640 *23*
γ	1317.2 *3*	0.0249 *15*
γ	1329.4 *3*	0.0066 *8*
γ	1340.2 *3*	0.0034 *6*
γ	1387.3 *3*	0.0044 *8*
γ	1398.5 *3*	0.0038 *8*
γ	1404.7 *4*	0.0021 *6*
γ	1415.9 *3*	0.0036 *8*
γ	1467.8 *4*	~0.0013
γ	1478.5 *4*	~0.0013
γ	1516.1 *5*	~0.0006
γ	1528.5 *4*	0.0038 *8*
γ	1540.0 *4*	0.0017 *6*
γ	1551.1 *4*	0.0106 *8*
γ	1563.4 *3*	0.0087 *6*
γ	1581.7 *4*	~0.0008
γ	1595.6 *5*	~0.0006 ?
γ	1608.35 *13*	0.0437 *17*
γ	1620.9 *4*	0.0009 *4*
γ	1659.6 *5*	0.0009 *4*
γ	1677.3 *3*	0.0036 *6*
γ	1694.6 *4*	0.0047 *6*
γ	1706.7 *4*	0.0009 *4*
γ	1720.2 *2*	~0.0006
γ	1757.2 *4*	0.0015 *4*
γ	1769.4 *3*	0.0068 *8*
γ	1782.9 *5*	~0.0009
γ	1805.6 *5*	~0.00038
γ	1818.1 *4*	0.00095 *19*
γ	1830.2 *3*	0.0011 *4*
γ	1851.7 *6*	~0.00038 ?
γ	1886.7 *4*	0.00057 *19* ?

† 11% uncert(syst)
* with ^{133}Ba(1.621 d)

Atomic Electrons (^{133}La)

⟨e⟩=6.5 *3* keV

e_{bin}(keV)	⟨e⟩(keV)	e(%)
5	1.99	38 *4*
6	1.44	25 *3*
25	0.164	0.64 *7*
26	0.46	1.78 *18*
27	0.77	2.9 *3*
30	0.171	0.57 *6*
31	0.49	1.57 *16*
32 - 76	0.187	0.56 *4*
99 - 137	0.0018	0.0017 *8*
152 - 190	0.0013	0.0007 *3*
205 - 228	0.0019	~0.0009
241	0.224	0.093 *8*
244 - 293	0.376	0.143 *7*
296 - 346	0.033	0.0110 *18*
347 - 395	0.0034	0.0010 *5*
398 - 446	0.0033	0.00077 *23*
457 - 504	0.0052	0.00106 *20*
513 - 562	0.046	0.0085 *19*
564 - 613	0.084	0.0142 *14*
616 - 664	0.0131	0.0021 *3*
666 - 714	0.0012	0.00017 *7*
716 - 765	0.00033	4.5 *13* ×10^{-5}
769 - 816	0.010	0.0012 *6*
820 - 869	0.0128	0.00154 *22*
870 - 918	0.0034	0.00039 *14*
919 - 955	0.00017	1.9 *6* ×10^{-5}

Atomic Electrons (^{133}La)
(continued)

e_{bin}(keV)	⟨e⟩(keV)	e(%)
975 - 1024	0.0013	0.00013 5
1032 - 1081	0.0035	0.00033 12
1087 - 1112	0.00054	4.9 17 $\times 10^{-5}$
1138 - 1187	0.0009	7.3 24 $\times 10^{-5}$
1191 - 1240	0.00044	3.6 11 $\times 10^{-5}$
1241 - 1284	0.0011	9 3 $\times 10^{-5}$
1292 - 1340	0.00016	1.2 3 $\times 10^{-5}$
1350 - 1399	0.00015	1.1 3 $\times 10^{-5}$
1403 - 1441	3.0 $\times 10^{-5}$	2.1 9 $\times 10^{-6}$
1462 - 1558	0.00026	1.7 4 $\times 10^{-5}$
1562 - 1659	0.00050	3.2 11 $\times 10^{-5}$
1669 - 1768	0.000106	6.1 15 $\times 10^{-6}$
1777 - 1851	2.6 $\times 10^{-5}$	1.5 4 $\times 10^{-6}$
1881 - 1886	6.2 $\times 10^{-7}$	3.3 15 $\times 10^{-8}$

$^{133}_{58}$Ce(5.40 *5* h)

Mode: ϵ
Δ: -82570 *200* keV syst
SpA: 4.36$\times 10^{6}$ Ci/g
Prod: ^{139}La(p,7n); alphas on Ba

Photons (^{133}Ce)

γ_{mode}	γ(keV)	γ(rel)
γ(E1)	58.2 6	47 10
γM1	87.7 7	18 4 *
γM1+E2	130.8 5	93 19
γ	177.3 5	3.3 7
γ	178.6 7	2.3 5
γ	210.8 10	1.5 3
γ	228.2 10	1.8 4 ?
γ	235.8 10	0.88 18 ?
γ	248.5 6	5.5 11
γ	261.0 6	6.6 13
γ	282.4 10	1.21 24
γ	286.5 10	1.8 4
γ	293.9 10	2.0 4
γ	306.7 5	4.6 9
γ	338.5 7	3.5 7
γ	346.1 6	12.5 25
γ	364.0 7	5.1 10
γ	397.2 10	1.8 4
γ	404.3 6	6.4 13
γ	409.9 10	2.5 5
γ	421.9 10	2.1 4
γ	430.6 10	2.6 5
γ	432.6 7	9.2 18
γ	436.1 10	1.10 22
γ	443.5 10	7.0 14
γ	453 1	3.0 6
γ	459.4 10	3.3 7
γ	476.9 5	100 20
γ	494.8 7	1.6 3
γ	509.6 6	36 7
γ	523.4 5	7.7 15 *
γ	533.9 10	1.8 4
γ	541.1 10	7.9 16
γ	557.3 3	
γ	567.7 6	1.6 3
γ	584.5 10	1.5 3
γ	595.5 10	2.4 5
γ	611.8 7	4.3 9 ?
γ	621.2 7	4.4 9
γ	643.5 10	1.6 3
γ	645.2 7	3.2 6
γ	654.2 3	1.5 3
γ	677.5 10	1.3 3
γ	684.4 10	0.88 18
γ	689.2 7	8.9 18
γ	783.6 6	5.3 11
γ	784.1 7	15 3

Photons (^{133}Ce)
(continued)

γ_{mode}	γ(keV)	γ(rel)
γ	819.6 10	3.0 6
γ	829.5 10	2.6 5
γ	858.2 10	3.0 6
γ	943.1 10	2.9 6
γ	950.3 7	3.2 6
γ	989.9 8	6.3 13
γ	1022.1 10	2.7 6
γ	1090.9 10	2.6 5
γ	1098.1 10	3.9 8
γ	1182.9 10	2.1 4
γ	1189.9 10	0.88 18
γ	1199.9 10	2.2 4
γ	1376.9 10	3.2 6
γ	1431.3 8	2.3 5
γ	1494.4 10	6.6 13
γ	1499.4 8	8.8 18
γ	1526.0 7	5.7 11
γ	1534.1 10	1.10 22
γ	1583.9 7	5.7 11
γ	1710.9 10	2.2 4
γ	1720.1 8	3.5 7
γ	1851.4 10	1.8 4
γ	2016.5 6	2.9 6
γ	2042.4 10	1.5 3
γ	2055.4 10	0.77 15
γ	2074.9 10	0.66 13
γ	2108.4 10	1.6 3
γ	2119.4 10	1.9 4
γ	2147.3 6	0.99 20

* with ^{133}Ce(5.40 or 1.6 h)

$^{133}_{58}$Ce(1.62 *7* h)

Mode: ϵ
Δ: -82570 *200* keV syst
SpA: 1.46$\times 10^{7}$ Ci/g
Prod: ^{139}La(p,7n); alphas on Ba

Photons (^{133}Ce)*

γ_{mode}	γ(keV)	γ(rel)
γ	76.6 4	35 5
γ	96.897 10	100 15
γ	173.5 4	0.9 2
γ	376.3 10	2.0 4
γ	557.3 3	25 5
γ	580.2 10	10 2 ?
γ	607.3 10	

* see also ^{133}Ce(5.40 h)

$^{133}_{59}$Pr(6.5 *3* min)

Mode: ϵ
Δ: -78070 *220* keV syst
SpA: 2.17$\times 10^{8}$ Ci/g
Prod: protons on Gd

Photons (^{133}Pr)

γ_{mode}	γ(keV)	γ(rel)
γ	73.6 10	74 15
γ[M1+E2]	134.2 6	100 20
γ	148.8 10	8.0 16
γ	183 1	3.5 7
γ	222.8 10	5 1
γ	241.5 10	40 8
γ	276.3 10	6.4 13
γ	314.9 7	85 17
γ	315.2 10	85 17
γ	330.2 6	43 9
γ	362.1 7	12.0 24
γ	370.2 7	2.2 4
γ	435.7 10	2.2 4
γ	460.1 10	17 3
γ	464.3 6	50 10
γ	486.9 10	1.4 3
γ	496.3 7	5.5 11
γ	522.1 10	1.10 22
γ	530.3 10	1.4 3
γ	537.2 10	5.3 11
γ	617 1	3.5 7
γ	620.6 10	
γ	645.0 7	10 2
γ	656.8 10	1.4 3
γ	700.4 7	1.9 4
γ	704.3 10	2.0 4
γ	779.2 7	1.5 3
γ	814 1	
γ	826.3 8	2.8 6
γ	834.5 7	2.0 4
γ	839.4 10	1.0 2
γ	843.3 10	1.4 3
γ	853.2 10	3.2 6
γ	887.5 10	
γ	897.6 10	1.3 3
γ	972 1	1.3 3
γ	975.9 10	2.5 5
γ	1046.4 10	1.4 3
γ	1147.3 10	2.5 5
γ	1320.5 10	1.4 3
γ	1322.5 8	0.90 18
γ	1343.1 10	1.8 4
γ	1406.1 10	1.4 3
γ	1451.5 10	1.9 4
γ	1460.6 10	3.2 6
γ	1494.4 10	7.8 16
γ	1625.2 10	1.8 4
γ	1638.7 10	1.4 3
γ	1802.9 10	3.2 6
γ	1812.1 10	2.8 6
γ	1830.8 10	3.5 7
γ	1863.7 10	5 1
γ	1875.1 10	4.2 8
γ	1949.3 10	2.2 4
γ	2020.7 10	0.70 14

$^{133}_{60}$Nd(1.17 *17* min)

Mode: ϵ
SpA: 1.20$\times 10^{9}$ Ci/g
Prod: ^{32}S on ^{106}Cd

Photons (^{133}Nd)

γ_{mode}	γ(keV)
γ	61
γ	106
γ	166
γ	227
γ	251
γ	369

$^{133}_{61}$Pm(12 *3* s)

Mode: ϵ
SpA: 6.9×10^{9} Ci/g
Prod: ^{32}S on ^{106}Cd

$^{133}_{62}$Sm(3.2 *4* s)

Mode: ϵ, ϵp
SpA: 2.4×10^{10} Ci/g
Prod: ^{32}S on ^{106}Cd

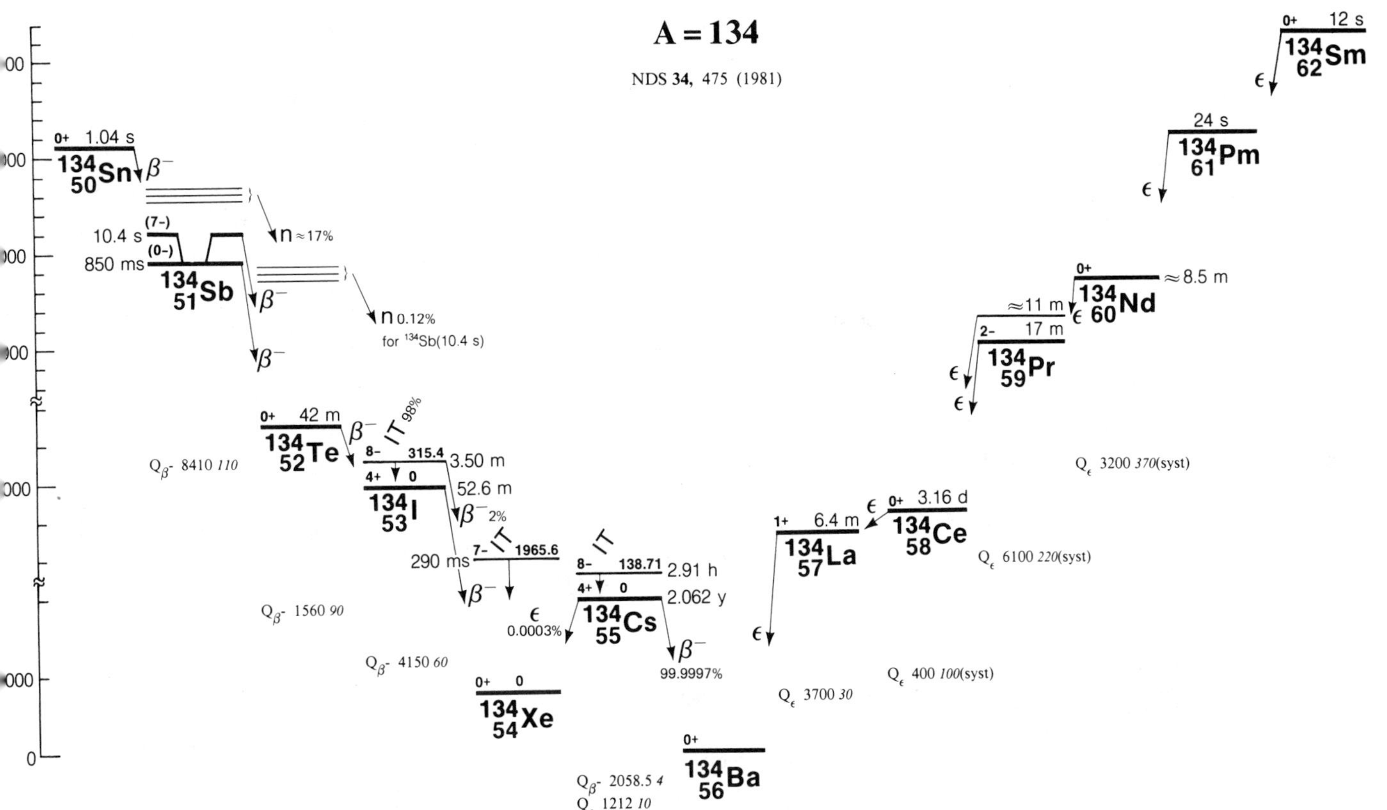

$^{134}_{50}$Sn(1.04 *2* s)

Mode: β-, β-n(∼17 %)
SpA: 5.91×10^{10} Ci/g
Prod: fission

β-n: 320, 435, 500, 760, 860, 1020

$^{134}_{51}$Sb(850 *100* ms)

Mode: β-
Δ: -74000 *150* keV
SpA: 6.8×10^{10} Ci/g
Prod: fission

Continuous Radiation (^{134}Sb)

⟨β-⟩=3791 keV;⟨IB⟩=18 keV

E_{bin}(keV)		⟨ ⟩(keV)	(%)
0 - 10	β-	0.00071	0.0141
	IB	0.081	
10 - 20	β-	0.00218	0.0145
	IB	0.081	0.56
20 - 40	β-	0.0091	0.0303
	IB	0.161	0.56
40 - 100	β-	0.073	0.102
	IB	0.48	0.73
100 - 300	β-	1.01	0.485
	IB	1.50	0.83
300 - 600	β-	5.6	1.21
	IB	2.0	0.48
600 - 1300	β-	54	5.4
	IB	4.0	0.44
1300 - 2500	β-	341	17.5
	IB	4.8	0.27
2500 - 5000	β-	1857	49.6
	IB	4.5	0.135
5000 - 7500	β-	1475	24.9
	IB	0.66	0.0117
7500 - 8410	β-	58	0.75
	IB	0.0027	3.2×10^{-5}

$^{134}_{51}$Sb(10.43 *14* s)

Mode: β-, β-n(0.120 *8* %)
Δ: -74000 *150* keV
SpA: 7.81×10^{9} Ci/g
Prod: fission

Photons (^{134}Sb)

⟨γ⟩=2037 *69* keV

γ_{mode}	γ(keV)	γ(%)
Te L_ℓ	3.335	0.065 *10*
Te L_η	3.606	0.034 *4*
Te L_α	3.768	1.80 *22*
Te L_β	4.121	1.58 *21*
Te L_γ	4.638	0.18 *3*
Te $K_{\alpha2}$	27.202	10.2 *6*
Te $K_{\alpha1}$	27.472	19.0 *12*
Te $K_{\beta1}$'	30.980	5.0 *3*
Te $K_{\beta2}$'	31.877	1.10 *7*
γ E2	115.2 *1*	49 *3*
γ E2	297.11 *10*	97 *5*
γ M1,E2	706.5 *6*	57 *3*
γ [E2]	1279.67 *19*	100 *5*

Atomic Electrons (^{134}Sb)

⟨e⟩=62.2 21 keV

e_{bin}(keV)	⟨e⟩(keV)	e(%)
4	1.01	23 3
5	0.71	15.1 18
22	0.186	0.84 10
23	0.57	2.5 3
26	0.186	0.71 8
27	0.202	0.75 9
30	0.029	0.097 11
31	0.0107	0.035 4
83	30.6	36.7 24
110	3.79	3.43 22
111	8.6	7.8 5
114	2.66	2.33 15
115	0.61	0.53 3
265	8.5	3.22 18
292	1.30	0.444 25
293	0.249	0.085 5
296	0.319	0.108 6
297	0.075	0.0251 14
675	1.24	0.18 3
702	0.16	0.023 4
705	0.030	0.0042 7
706	0.010	0.0014 4
1248	0.94	0.075 5
1275	0.117	0.0092 6
1279	0.0281	0.00220 14
1280	0.000354	2.77 18 $\times10^{-5}$

Continuous Radiation (^{134}Sb)

⟨β-⟩=2804 keV;⟨IB⟩=11.6 keV

E_{bin}(keV)		⟨ ⟩(keV)	(%)
0 - 10	β-	0.00145	0.0289
	IB	0.070	
10 - 20	β-	0.00447	0.0298
	IB	0.069	0.48
20 - 40	β-	0.0187	0.062
	IB	0.138	0.48
40 - 100	β-	0.148	0.209
	IB	0.40	0.62
100 - 300	β-	2.03	0.98
	IB	1.26	0.70
300 - 600	β-	11.0	2.39
	IB	1.65	0.38
600 - 1300	β-	100	10.1
	IB	3.0	0.34
1300 - 2500	β-	556	28.7
	IB	3.1	0.18
2500 - 5000	β-	1872	53
	IB	1.9	0.059
5000 - 6719	β-	263	4.86
	IB	0.036	0.00068

$^{134}_{52}$Te(41.8 *8* min)

Mode: β-

Δ: -82410 *110* keV

SpA: 3.36$\times10^{7}$ Ci/g

Prod: fission

Photons (^{134}Te)

⟨γ⟩=858 15 keV

γ_{mode}	γ(keV)	γ(%)†
I L_ℓ	3.485	0.054 8
I L_η	3.780	0.026 3
I L_α	3.937	1.49 18
I L_β	4.320	1.33 17
I L_γ	4.884	0.165 24
I $K_{\alpha2}$	28.317	9.1 4
I $K_{\alpha1}$	28.612	16.8 8
γ	29.575 19	<0.030
I $K_{\beta1}'$	32.278	4.53 21
I $K_{\beta2}'$	33.225	1.03 5
γ [M1+E2]	76.731 14	0.270 23
γ M1(+<3.8%E2)	79.451 11	20.9 6
γ	101.420 14	0.41 6
γ	130.995 17	0.17 6
γ	136.92 18	~0.09 ?
γ M1,E2	180.870 13	18.0 9
γ [M1+E2]	183.042 22	~0.6 ?
γ M1,E2	201.216 12	8.7 3
γ M1,E2	210.445 15	22.3 9
γ [M1+E2]	259.773 24	0.46 9
γ M1,E2	277.947 8	20.9 9
γ M1,E2	435.000 18	18.6 12
γ [M1+E2]	460.989 21	9.9 6
γ	464.575 16	4.6 3
γ [M1+E2]	565.994 11	18.3 9
γ	636.216 20	1.65 20
γ [E2]	645.445 14	0.87 9
γ	665.791 19	1.16 17
γ	712.946 19	4.6 3
γ	742.521 17	15.1 9
γ M1,E2	767.210 14	29.0 12
γ [E2]	843.940 13	1.2 3
γ	895.99 3	0.43 12
γ	925.562 25	1.45 17
γ [E2]	1026.981 23	0.43 12

† 6.9% uncert(syst)

Atomic Electrons (^{134}Te)

⟨e⟩=38.1 12 keV

e_{bin}(keV)	⟨e⟩(keV)	e(%)
5	1.39	30 3
23 - 44	1.27	4.7 3
46	12.7	27.4 13
68 - 72	0.34	~0.5
74	2.41	3.24 15
75 - 104	1.95	2.49 11
126 - 137	0.031	~0.024
148	4.0	2.7 5
150	0.13	~0.08
168	1.58	0.94 13
176	0.9	0.49 20
177	3.7	2.1 3
178 - 227	1.6	0.81 13
245	2.11	0.86 7
255 - 278	0.46	0.17 4
402	0.88	0.22 3
428 - 465	0.83	0.192 24
533 - 566	0.68	0.127 16
603 - 645	0.082	0.013 4
661 - 709	0.33	0.048 20
712 - 743	0.65	0.088 13
762 - 811	0.118	0.0153 20
839 - 863	0.009	0.0010 4
891 - 926	0.022	0.0024 11
994 - 1027	0.0063	0.00063 15

Continuous Radiation (^{134}Te)

⟨β-⟩=205 keV;⟨IB⟩=0.141 keV

E_{bin}(keV)		⟨ ⟩(keV)	(%)
0 - 10	β-	0.144	2.88
	IB	0.0103	
10 - 20	β-	0.431	2.88
	IB	0.0096	0.067
20 - 40	β-	1.72	5.7
	IB	0.017	0.060
40 - 100	β-	11.8	16.9
	IB	0.039	0.062
100 - 300	β-	89	46.4
	IB	0.055	0.034
300 - 600	β-	99	24.8
	IB	0.0099	0.0027
600 - 713	β-	2.75	0.439
	IB	2.7$\times10^{-5}$	4.3$\times10^{-6}$

$^{134}_{53}$I (52.6 *4* min)

Mode: β-

Δ: -83970 *60* keV

SpA: 2.667$\times10^{7}$ Ci/g

Prod: fission

Photons (^{134}I)

⟨γ⟩=2611 25 keV

γ_{mode}	γ(keV)	γ(%)†
Xe L_ℓ	3.634	0.0041 16
Xe L_η	3.955	0.0020 8
Xe L_α	4.104	0.12 4
Xe L_β	4.521	0.10 4
Xe L_γ	5.110	0.013 5
Xe $K_{\alpha2}$	29.461	0.64 20
Xe $K_{\alpha1}$	29.782	1.2 4
Xe $K_{\beta1}'$	33.606	0.32 10
Xe $K_{\beta2}'$	34.606	0.076 24
γ	135.406 16	4.1 3
γ	139.067 22	0.77 5
γ	152.00 15	0.106 11
γ	162.50 7	0.30 4 ?
γ M1,E2	188.448 19	0.79 6
γ [E2]	217.015 23	0.25 3
γ M1(+E2)	235.497 19	2.19 19
γ [M1+E2]	278.935 24	0.153 19
γ	319.827 25	0.47 4
γ	351.1 1	0.50 6
γ M1+42%E2	405.462 16	7.3 4
γ	410.96 3	0.56 6
γ M1(+1.0%E2)	433.372 20	4.20 19
γ [M1+E2]	458.893 24	1.34 10
γ	465.52 10	0.36 4
γ [E2]	488.910 22	1.43 10
γ [M1+E2]	514.431 20	2.21 13
γ E2+21%M1	540.868 16	7.6 4
γ	565.54 4	0.94 6
γ	571.13 5	0.34 8
γ M1(+11%E2)	595.382 16	11.2 6
γ M1+37%E2	621.819 16	10.6 6
γ	627.976 22	2.20 13
γ M1+9.3%E2	677.357 20	7.8 5
γ	706.58 7	0.83 6
γ [M1+E2]	730.788 19	1.81 11
γ [E2]	739.21 3	0.71 7
γ E2+15%M1	766.733 22	4.1 3
γ	816.422 23	0.59 5
γ E2	847.061 22	95.4 19
γ M1+29%E2	857.315 19	6.96 19
γ	864.0 3	0.19 3
γ E2	884.127 18	64.9 13
γ [M1+E2]	922.74 4	0.14 3
γ M1+14%E2	947.801 22	4.01 19
γ	966.92 4	0.36 5

Photons (^{134}I) (continued)

γ_{mode}	γ(keV)	γ(%)†
γ [E2]	974.708 25	4.8 3
γ M1+0.9%E2	1040.13 4	1.91 19
γ	1052.2 3	0.067 19 ?
γ	1058.8 3	0.10 3 ?
γ M1+2.5%E2	1072.573 21	15.0 8
γ	1087.02 20	0.086 19
γ	1100.09 12	0.69 6
γ	1103.41 4	0.73 6
γ M1+19%E2	1136.248 18	9.2 5
γ	1159.15 6	0.33 3
γ	1164.19 6	0.13 3
γ	1183.2 5	<0.12 ?
γ	1190.06 8	0.35 3
γ	1225.5 3	0.067 19
γ	1238.81 4	0.21 6
γ	1243.8 3	0.076 19
γ	1269.51 5	0.56 4
γ	1322.4 3	0.10 4
γ	1336.16 17	0.14 3
γ	1352.63 5	0.40 4
γ	1394.87 21	0.076 19
γ	1407.42 20	0.095 19
γ	1414.3 5	0.22 6
γ	1428.2 3	0.17 4
γ	1431.37 25	0.17 4
γ	1455.26 5	2.29 19
γ	1470.02 5	0.75 5
γ	1505.5 4	0.11 4 ?
γ	1541.53 7	0.51 4
γ [E2]	1613.79 3	4.29 19
γ	1629.32 6	0.20 4
γ	1644.27 4	0.40 5
γ	1655.21 10	0.23 3
γ [E2]	1741.435 23	2.67 19
γ M1+1.1%E2	1806.86 4	5.5 3
γ	1868.52 20	0.067 19 ?
γ	1893.2 3	0.057 10
γ	1925.87 6	0.181 19
γ	1947.3 3	0.095 19
γ [E2]	2020.366 24	0.172 19
γ	2159.9 3	0.21 3
γ	2236.75 5	0.53 14
γ	2262.5 3	0.095 19
γ	2312.42 20	0.24 3
γ	2408.72 17	0.075 11
γ	2452.9 3	0.064 10
γ	2467.43 21	0.143 19
γ	2513.43 6	0.067 10
γ	2629.93 21	0.067 8
γ	2645.30 9	~0.019
γ	2699.5 5	0.032 8
γ	2840 4	~0.019 ?

† 0.31% uncert(syst)

Atomic Electrons (^{134}I)

⟨e⟩= 9.6 8 keV

e_{bin}(keV)	⟨e⟩(keV)	e(%)
5 - 33	0.17	2.4 8
101	1.1	~1
105 - 128	0.22	0.20 9
130	0.26	~0.20
131 - 162	0.46	0.32 12
182 - 188	0.091	0.049 7
201	0.31	0.156 18
212 - 244	0.104	0.045 7
273 - 320	0.058	0.019 7
346 - 351	0.005	~0.0015
371	0.42	0.114 8
376	0.021	~0.006
399	0.234	0.059 4
400 - 433	0.197	0.047 3
453 - 489	0.163	0.035 3
506	0.254	0.050 3
509 - 541	0.094	0.018 3
560	0.0031	~0.0006
561	0.39	0.070 5
564 - 571	0.0024	0.00042 16
587	0.326	0.055 3
590 - 628	0.18	0.029 5
643 - 677	0.284	0.044 3
696 - 739	0.15	0.0209 14
761 - 782	0.023	0.0030 7
811	0.0012	~0.00014
812	1.58	0.194 6
815 - 847	0.414	0.0496 10
850	1.02	0.120 3
852 - 888	0.198	0.0226 5
913 - 962	0.160	0.0172 7
966 - 1006	0.044	0.0045 3
1018 - 1035	0.0061	0.00059 9
1038	0.250	0.0241 13
1039 - 1087	0.057	0.0054 6
1095 - 1136	0.165	0.0150 7
1149 - 1191	0.0062	0.00053 17
1204 - 1244	0.008	0.00069 23
1264 - 1302	0.0030	0.00023 7
1317 - 1360	0.0050	0.00038 13
1373 - 1421	0.024	0.0017 6
1423 - 1507	0.015	0.00103 24
1536 - 1628	0.048	0.00305 17
1639 - 1737	0.0244	0.00143 10
1740 - 1834	0.059	0.00332 16
1859 - 1946	0.0024	0.00013 3
1986 - 2019	0.00138	6.9 7 $\times 10^{-5}$
2125 - 2202	0.0042	0.00019 6
2228 - 2311	0.0024	0.000107 24
2374 - 2466	0.0016	6.6 13 $\times 10^{-5}$
2479 - 2512	0.00037	1.5 5 $\times 10^{-5}$
2595 - 2695	0.00062	2.4 5 $\times 10^{-5}$
2698 - 2699	3.7 $\times 10^{-6}$	1.4 5 $\times 10^{-7}$
2805 - 2839	9.6 $\times 10^{-5}$	~3 $\times 10^{-6}$

Continuous Radiation (^{134}I)

⟨β-⟩=608 keV;⟨IB⟩=1.01 keV

E_{bin}(keV)		⟨ ⟩(keV)	(%)
0 - 10	β-	0.0350	0.70
	IB	0.025	
10 - 20	β-	0.106	0.71
	IB	0.025	0.17
20 - 40	β-	0.437	1.45
	IB	0.048	0.166
40 - 100	β-	3.27	4.64
	IB	0.130	0.20
100 - 300	β-	36.5	18.0
	IB	0.32	0.18
300 - 600	β-	130	29.0
	IB	0.27	0.065
600 - 1300	β-	344	39.5
	IB	0.18	0.023
1300 - 2419	β-	93	6.0
	IB	0.0158	0.00108

$^{134}_{53}$I (3.50 2 min)

Mode: IT(97.7 10 %), β-(2.3 10 %)
Δ: -83655 60 keV
SpA: 4.003$\times 10^{8}$ Ci/g
Prod: fission

Photons (^{134}I)

⟨γ⟩=286 9 keV

γ_{mode}	γ(keV)	γ(%)†
I L_ℓ	3.485	0.131 18
I L_η	3.780	0.066 7
I L_α	3.937	3.7 4
I L_β	4.319	3.3 4
I L_γ	4.884	0.41 6
I $K_{\alpha2}$	28.317	22.5 8
I $K_{\alpha1}$	28.612	41.8 14
Xe $K_{\alpha2}$	29.461	0.120 4
Xe $K_{\alpha1}$	29.782	0.224 7
I $K_{\beta1}$'	32.278	11.2 4
I $K_{\beta2}$'	33.225	2.55 9
Xe $K_{\beta1}$'	33.606	0.0605 19
Xe $K_{\beta2}$'	34.606	0.0143 5
γ_{IT} M1	43.95 18	10.3 8
$\gamma_{\beta\text{-}}$E3	234.4 5	1.5 6
γ_{IT} E3	271.5 3	79
γ_{IT} (M4)	315.4 3	<0.47 ?
$\gamma_{\beta\text{-}}$E2	847.061 22	2.2 7
$\gamma_{\beta\text{-}}$E2	884.127 18	2.2 7

† uncert(syst): 3.8% for IT, 43% for β-

Atomic Electrons (^{134}I)

⟨e⟩=67.0 14 keV

e_{bin}(keV)	⟨e⟩(keV)	e(%)
5	3.4	73 8
11	8.06	74.7 21
23 - 33	2.56	10.3 6
39	3.85	9.9 3
43 - 44	1.078	2.50 6
200 - 234	1.283	0.610 10
238	31.0	13.0 3
266	3.64	1.37 3
267	7.60	2.85 7
270	0.733	0.271 6
271	2.27	0.839 19
282 - 315	1.4	~0.5
812 - 850	0.0698	0.00840 14
879 - 884	0.00528	0.000601 12

Continuous Radiation (^{134}I)

⟨β-⟩=23.1 keV;⟨IB⟩=0.052 keV

E_{bin}(keV)		⟨ ⟩(keV)	(%)
0 - 10	β-	0.000284	0.0057
	IB	0.00086	
10 - 20	β-	0.00087	0.0058
	IB	0.00085	0.0059
20 - 40	β-	0.00361	0.0120
	IB	0.00165	0.0057
40 - 100	β-	0.0280	0.0396
	IB	0.0047	0.0072
100 - 300	β-	0.353	0.171
	IB	0.0129	0.0073
300 - 600	β-	1.63	0.357
	IB	0.0135	0.0032
600 - 1300	β-	9.5	1.01
	IB	0.0150	0.0018
1300 - 2500	β-	11.6	0.70
	IB	0.0031	0.00020
2500 - 2501	β-	1.90$\times 10^{-8}$	7.6$\times 10^{-10}$
	IB	7.3$\times 10^{-17}$	2.9$\times 10^{-18}$

$^{134}_{54}$Xe(stable)

Δ: -88125 *7* keV
%: 10.4 *2*

$^{134}_{54}$Xe(290 *17* ms)

Mode: IT
Δ: -86159 *7* keV
SpA: 1.10×10^{11} Ci/g
Prod: ^{134}Xe(n,n')

Photons (^{134}Xe)

⟨γ⟩=1896 *56* keV

γ_{mode}	γ(keV)	γ(%)
Xe L_ℓ	3.634	0.039 *5*
Xe L_η	3.955	0.0217 *23*
Xe L_α	4.104	1.09 *12*
Xe L_β	4.511	1.04 *12*
Xe L_γ	5.094	0.129 *17*
Xe $K_{\alpha2}$	29.461	5.23 *17*
Xe $K_{\alpha1}$	29.782	9.7 *3*
Xe $K_{\beta1}'$	33.606	2.63 *8*
Xe $K_{\beta2}'$	34.606	0.622 *21*
γ E3	234.4 *5*	68 *8*
γ E2	847.061 *22*	100.0
γ E2	884.127 *18*	100

Atomic Electrons (^{134}Xe)

⟨e⟩=69.5 *10* keV

e_{bin}(keV)	⟨e⟩(keV)	e(%)
5	1.03	20.9 *21*
24	0.161	0.67 *7*
25	0.204	0.82 *9*
28	0.057	0.201 *21*
29	0.121	0.42 *4*
30	0.0161	0.054 *6*
32	0.0062	0.0190 *19*
33	0.0140	0.043 *4*
200	40.1	20.1 *5*
229	12.9	5.63 *13*
230	5.97	2.60 *6*
233	4.16	1.78 *4*
234	1.011	0.431 *10*
812	1.65	0.203 *5*
842	0.225	0.0267 *6*
846	0.0460	0.00543 *12*
847	0.01007	0.00119 *3*
850	1.56	0.184 *4*
879	0.212	0.0241 *5*
883	0.0432	0.00490 *11*
884	0.00948	0.001072 *24*

$^{134}_{55}$Cs(2.062 *5* yr)

Mode: β-(99.9997 *1* %), ϵ(0.0003 *1* %)
Δ: -86913 *6* keV
SpA: 1294 Ci/g
Prod: ^{133}Cs(n,γ)

Photons (^{134}Cs)

⟨γ⟩=1555 *5* keV

γ_{mode}	γ(keV)	γ(%)
Ba L_ℓ	3.954	0.00166 *22*
Ba L_η	4.331	0.00074 *8*
Ba L_α	4.465	0.046 *5*
Ba L_β	4.946	0.041 *5*
Ba L_γ	5.613	0.0054 *7*
Ba $K_{\alpha2}$	31.817	0.243 *6*
Ba $K_{\alpha1}$	32.194	0.447 *11*
Ba $K_{\beta1}'$	36.357	0.124 *3*
Ba $K_{\beta2}'$	37.450	0.0313 *9*
$\gamma_{\beta-}$[M1+E2]	242.726 *21*	0.0210 *8*
$\gamma_{\beta-}$[M1+E2]	326.595 *22*	0.0144 *6*
$\gamma_{\beta-}$E2+1.3%M1	475.357 *20*	1.46 *4*
$\gamma_{\beta-}$E2+1.9%M1	563.237 *8*	8.38 *15*
$\gamma_{\beta-}$M1+7.8%E2	569.321 *13*	15.4 *3*
$\gamma_{\beta-}$E2	604.710 *8*	97.6 *3*
$\gamma_{\beta-}$E2	795.867 *14*	85.4 *4*
$\gamma_{\beta-}$E2	801.951 *15*	8.73 *17*
γ_ϵE2	847.061 *22*	0.00030 *10*
$\gamma_{\beta-}$E2+23%M1	1038.592 *19*	1.00 *2*
$\gamma_{\beta-}$E2	1167.944 *11*	1.80 *4*
$\gamma_{\beta-}$E2	1365.184 *14*	3.04 *6*

Atomic Electrons (^{134}Cs)

⟨e⟩=6.86 *7* keV

e_{bin}(keV)	⟨e⟩(keV)	e(%)
5 - 36	0.065	0.82 *5*
195 - 243	0.0039	0.00187 *14*
289 - 327	0.00158	0.00054 *6*
438 - 475	0.0742	0.0168 *5*
526	0.270	0.0513 *14*
532	0.674	0.127 *4*
557 - 564	0.145	0.0259 *5*
567	2.80	0.494 *10*
568 - 569	0.0237	0.00418 *10*
599	0.427	0.0713 *14*
603 - 605	0.1109	0.0184 *3*
758	1.68	0.222 *5*
765	0.170	0.0223 *6*
790	0.226	0.0286 *6*
791 - 802	0.1046	0.01316 *17*
1001 - 1039	0.0184	0.00183 *4*
1131 - 1168	0.0269	0.00237 *6*
1328 - 1365	0.0388	0.00291 *7*

Continuous Radiation (^{134}Cs)

⟨β-⟩=157 keV;⟨IB⟩=0.104 keV

E_{bin}(keV)		⟨ ⟩(keV)	(%)
0 - 10	β-	0.477	9.8
	IB	0.0079	
10 - 20	β-	1.22	8.2
	IB	0.0072	0.050
20 - 40	β-	3.62	12.4
	IB	0.0128	0.045
40 - 100	β-	10.9	16.9
	IB	0.029	0.046
100 - 300	β-	65	33.8
	IB	0.041	0.025
300 - 600	β-	75	18.8
	IB	0.0068	0.0018
600 - 1300	β-	1.00	0.158
	IB	2.0×10^{-5}	3.0×10^{-6}
1300 - 1454	β-	0.000460	3.44×10^{-5}
	IB	5.1×10^{-9}	3.9×10^{-10}

$^{134}_{55}$Cs(2.91 *1* h)

Mode: IT
Δ: -86774 *6* keV
SpA: 8.04×10^{6} Ci/g
Prod: ^{133}Cs(n,γ)

Photons (^{134}Cs)

⟨γ⟩=26.8 *5* keV

γ_{mode}	γ(keV)	γ(%)†
Cs L_ℓ	3.795	0.17 *3*
Cs L_η	4.142	0.105 *12*
Cs L_α	4.285	4.6 *7*
Cs L_β	4.705	6.9 *11*
Cs L_γ	5.425	1.12 *20*
γ M1	11.219 *16*	0.94 *9*
Cs $K_{\alpha2}$	30.625	9.0 *3*
Cs $K_{\alpha1}$	30.973	16.7 *6*
Cs $K_{\beta1}'$	34.967	4.56 *16*
Cs $K_{\beta2}'$	36.006	1.11 *4*
γ E3	127.491 *2*	12.7
γ M4	138.710 *15*	0.00394 *25*

† 2.4% uncert(syst)

Atomic Electrons (^{134}Cs)

⟨e⟩=109.1 *19* keV

e_{bin}(keV)	⟨e⟩(keV)	e(%)
5	3.2	61 *6*
6	5.2	93 *10*
10	1.30	13.0 *13*
11	0.37	3.4 *3*
25	0.27	1.09 *11*
26	0.34	1.31 *14*
29	0.089	0.30 *3*
30	0.208	0.70 *7*
31	0.031	0.101 *10*
34	0.025	0.073 *8*
35	0.0098	0.028 *3*
92	31.7	34.7 *10*
103	0.294	0.286 *19*
122	51.0	41.7 *12*
126	11.8	9.3 *3*
127	3.03	2.38 *7*
133	0.127	0.096 *7*
134	0.112	0.084 *6*
137	0.0241	0.0176 *12*
138	0.040	0.0292 *20*
139	0.0095	0.0068 *5*

$^{134}_{56}$Ba(stable)

Δ: -88972 *6* keV
%: 2.417 *27*

$^{134}_{57}$La(6.45 *16* min)

Mode: ϵ
Δ: -85270 *30* keV
SpA: 2.17×10^{8} Ci/g
Prod: daughter ^{134}Ce; ^{133}Cs(α,3n); ^{134}Ba(p,n)

Photons (^{134}La)

$\langle\gamma\rangle$=73.8 9 keV

γ_{mode}	γ(keV)	γ(%)†
Ba L_ℓ	3.954	0.055 8
Ba L_η	4.331	0.0244 25
Ba L_α	4.465	1.54 17
Ba L_β	4.946	1.39 17
Ba L_γ	5.620	0.18 3
Ba $K_{\alpha2}$	31.817	8.20 25
Ba $K_{\alpha1}$	32.194	15.1 5
Ba $K_{\beta1}$'	36.357	4.18 13
Ba $K_{\beta2}$'	37.450	1.06 4
γ [E2]	232.632 15	<18 ×10^{-9}
γ [M1+E2]	242.726 21	7.9 7 ×10^{-5}
γ E2+1.3%M1	475.357 20	0.0055 4
γ E2+1.9%M1	563.237 8	0.359 9
γ [E2]	592.593 23	0.0165 12
γ E2	604.710 8	5.04 14
γ [M1+E2]	659.68 3	0.00257 25
γ	718.716 23	0.0120 8
γ E2	795.867 14	0.0076 5
γ	861.278 21	0.0041 4
γ [M1+E2]	920.312 22	0.0179 9
γ [E2]	991.72 3	0.0057 4
γ E2+23%M1	1038.592 19	0.00376 9
γ [M1+E2]	1104.64 3	~0.0008 ?
γ E2	1155.827 23	0.0194 10
γ E2	1167.944 11	0.0771 24
γ E2	1168.90 4	0.022 4
γ	1185.20 4	0.00174 25
γ E2	1211.144 21	0.118 3
γ	1243.85 12	0.0022 5 ?
γ	1255.65 8	0.0014 6 ?
γ	1260.1 6	0.0014 6 ?
γ	1308.33 13	~0.0007
γ E2	1320.70 3	0.0820 21
γ [E2]	1347.36 3	0.00489 25
γ	1368.95 5	0.0031 3
γ M1	1396.740 24	0.0354 9
γ	1402.91 3	0.0102 5
γ M1,E2	1424.511 20	0.188 5
γ	1431.91 4	0.0022 5 ?
γ M1	1483.545 22	0.149 4
γ M1	1488.28 8	0.0043 9
γ	1528.60 5	0.00265 25
γ E2(+M1)	1554.95 3	0.414 10
γ M1,E2	1561.26 5	0.0049 14 ?
γ [M1+E2]	1579.990 25	0.0047 10
γ	1591.0 3	0.0009 3 ?
γ	1649.69 16	0.00120 21
γ	1660.55 4	0.0030 3
γ	1674.11 13	0.00058 25 ?
γ	1683.31 6	0.0032 3
γ	1719.06 4	0.0062 3
γ E2	1732.13 4	0.234 6
γ	1749.66 7	0.00141 21
γ	1765.54 19	0.0022 5
γ E2	1774.375 22	0.0478 12
γ	1800.7 5	0.00035 13 ?
γ	1836.44 12	0.00166 25
γ	1859.42 6	0.00356 25
γ [E2]	1883.93 3	0.0033 3
γ	1893.23 5	0.00232 25
γ	1918.75 13	0.0007 3
γ	1932.18 5	0.0044 3
γ M1,E2	1959.971 24	0.0975 24
γ M1,E2	1966.14 3	0.0043 6
γ M1,E2	1995.14 4	0.0893 22
γ M1,E2	2029.215 21	0.0393 10
γ	2051.51 8	0.0059 6
γ	2064.4 5	0.00062 21 ?
γ [E2]	2088.248 23	0.058 3
γ	2091.83 5	0.0099 10
γ	2104.13 5	0.00311 21
γ	2124.49 5	0.0310 8
γ M1,E2	2143.220 25	0.0160 8
γ	2156.04 8	0.0042 4
γ [E0+E2]	2159.65 3	<0.00025 ?
γ M1,E2	2223.78 4	0.0181 5
γ	2240.89 19	0.00050 17
γ	2246.54 6	0.00054 17
γ	2282.29 4	0.0088 4
γ	2312.89 7	0.00224 17

Photons (^{134}La) (continued)

γ_{mode}	γ(keV)	γ(%)†
γ	2334.29 14	0.0019 4
γ	2345.57 21	0.00064 10
γ	2442.7 3	0.00043 9
γ	2464.14 13	0.00072 11
γ	2481.98 12	0.00060 9
γ M1,E2	2564.671 25	0.0056 3
γ	2570.84 3	0.0189 5
γ	2599.84 4	0.00282 21
γ	2656.21 8	0.00048 7
γ	2667.35 5	0.00120 8
γ	2696.53 5	0.00302 21
γ	2722.51 12	0.00045 7
γ	2729.19 5	<0.0029 ?
γ [E2]	2747.92 3	<1.7 ×10^{-4} ?
γ	2758.9 3	0.00046 7
γ	2764.17 8	0.00138 14
γ	2788.73 21	0.00031 5
γ	2804.12 19	0.00029 4
γ	2824.09 25	0.00034 7
γ	2828.48 4	0.00034 7
γ	2851.24 6	0.00111 8
γ	2866.45 24	0.00024 4
γ	2894.93 14	0.00041 4
γ	2938.99 14	0.00282 21
γ	3027.35 6	0.00061 6
γ	3061.15 5	0.00061 6
γ	3086.68 12	0.00034 5
γ	3160.05 15	0.00056 5
γ	3245.85 19	0.00023 5
γ	3327.21 12	0.00058 6
γ	3449.47 18	0.00059 6

† 3.4% uncert(syst)

Atomic Electrons (^{134}La)

$\langle e\rangle$=2.41 11 keV

e_{bin}(keV)	$\langle e\rangle$(keV)	e(%)
5	0.77	14.7 15
6	0.56	9.7 10
25	0.064	0.250 25
26	0.180	0.69 7
27	0.30	1.11 11
30	0.067	0.220 22
31	0.189	0.61 6
32 - 36	0.072	0.216 15
195 - 243	1.5 ×10^{-5}	7 3 ×10^{-6}
438 - 475	0.000279	6.3 4 ×10^{-5}
526 - 563	0.0143	0.00269 7
567	0.145	0.0255 9
587 - 622	0.0280	0.00466 13
654 - 681	0.00026	~4 ×10^{-5}
713 - 758	0.000193	2.6 3 ×10^{-5}
790 - 824	9.4 ×10^{-5}	1.2 5 ×10^{-5}
855 - 883	0.00037	4.2 8 ×10^{-5}
914 - 954	0.000147	1.57 13 ×10^{-5}
986 - 1033	8.19 ×10^{-5}	8.18 20 ×10^{-6}
1037 - 1067	1.5 ×10^{-5}	1.4 7 ×10^{-6}
1099 - 1148	0.00153	0.000136 8
1150 - 1185	0.00171	0.000146 4
1205 - 1254	0.00033	2.7 3 ×10^{-5}
1255 - 1303	0.00094	7.31 24 ×10^{-5}
1307 - 1347	0.000245	1.86 12 ×10^{-5}
1359 - 1403	0.0030	0.000219 24
1419 - 1451	0.00238	0.000165 6
1478 - 1575	0.0058	0.00038 4
1579 - 1678	0.00010	6.4 16 ×10^{-6}
1682 - 1773	0.00297	0.000174 4
1795 - 1892	0.000099	5.4 8 ×10^{-6}
1895 - 1994	0.00229	0.000117 7
2014 - 2106	0.00103	5.0 4 ×10^{-5}
2118 - 2218	0.000255	1.18 10 ×10^{-5}
2219 - 2312	9.4 ×10^{-5}	4.2 9 ×10^{-6}
2328 - 2427	8.6 ×10^{-6}	3.6 8 ×10^{-7}
2437 - 2533	0.000172	6.8 4 ×10^{-6}
2559 - 2655	5.2 ×10^{-5}	2.01 23 ×10^{-6}

Atomic Electrons (^{134}La) (continued)

e_{bin}(keV)	$\langle e\rangle$(keV)	e(%)
2659 - 2758	4.3 ×10^{-5}	1.6 4 ×10^{-6}
2759 - 2857	1.54 ×10^{-5}	5.5 8 ×10^{-7}
2860 - 2938	1.6 ×10^{-5}	5.5 15 ×10^{-7}
2990 - 3086	8.3 ×10^{-6}	2.8 5 ×10^{-7}
3123 - 3208	3.9 ×10^{-6}	1.2 3 ×10^{-7}
3240 - 3326	3.0 ×10^{-6}	9.2 23 ×10^{-8}
3412 - 3448	2.9 ×10^{-6}	8.5 22 ×10^{-8}

Continuous Radiation (^{134}La)

$\langle\beta+\rangle$=756 keV;$\langle$IB$\rangle$=5.9 keV

E_{bin}(keV)		$\langle\ \rangle$(keV)	(%)
0 - 10	β+	1.21 ×10^{-6}	1.45 ×10^{-5}
	IB	0.027	
10 - 20	β+	5.08 ×10^{-5}	0.000306
	IB	0.027	0.19
20 - 40	β+	0.00147	0.00448
	IB	0.081	0.27
40 - 100	β+	0.075	0.096
	IB	0.152	0.23
100 - 300	β+	4.03	1.81
	IB	0.44	0.25
300 - 600	β+	32.8	7.0
	IB	0.57	0.132
600 - 1300	β+	263	27.3
	IB	1.37	0.150
1300 - 2500	β+	454	26.6
	IB	2.4	0.135
2500 - 3700	β+	2.85	0.112
	IB	0.76	0.027
	$\Sigma\beta$+		63

$^{134}_{58}$Ce(3.16 4 d)

Mode: ϵ

Δ: -84870 110 keV syst

SpA: 3.08×10^{5} Ci/g

Prod: ^{139}La(p,6n); protons on Ta; protons on Gd; ^{134}Ba(α,4n); protons on Pr

Photons (^{134}Ce)

$\langle\gamma\rangle$=27.8 5 keV

γ_{mode}	γ(keV)	γ(%)
La L_ℓ	4.121	0.160 22
La L_η	4.529	0.068 7
La L_α	4.649	4.4 5
La L_β	5.165	4.1 5
La L_γ	5.887	0.56 8
γ	22.706 19	
γ	31.897 16	
La $K_{\alpha2}$	33.034	21.7 7
La $K_{\alpha1}$	33.442	39.9 12
La $K_{\beta1}$'	37.777	11.1 3
La $K_{\beta2}$'	38.927	2.91 10
γ	39.093 16	
γ	54.603 14	0.0169 24
γ	59.05 20	0.00021 10 ?
γ	61.799 20	0.0036 8
γ	66.27 20	~0.00012 ?
γ	68.622 16	0.00076 25
γ	70.86 12	0.00121 20
γ	90.171 12	0.0055 5
γ	93.696 16	0.00093 25

Photons (^{134}Ce)
(continued)

γ_{mode}	γ(keV)	γ(%)
γ	103.002 *9*	0.0253 *25*
γ	104.42 *4*	0.0021 *3*
γ	107.247 *16*	0.0061 *4*
γ	116.17 *4*	0.0027 *3*
γ	130.422 *14*	0.209 *15*
γ	131.949 *14*	0.0171 *17*
γ	150.21 *3*	0.0038 *4*
γ	158.793 *10*	0.039 *3*
γ	162.318 *9*	0.230 *16*
γ	168.470 *14*	0.0122 *9*
γ	187.021 *11*	0.0218 *15*
γ	193.172 *10*	0.040 *3*
γ	197.886 *13*	0.0136 *10*
γ	200.572 *19*	0.00161 *24*
γ	205.28 *3*	0.0042 *5*
γ	220.593 *17*	0.0046 *6*
γ	239.665 *15*	0.0131 *10*
γ	252.489 *14*	0.0037 *5*
γ	261.795 *13*	<0.0034 ?
γ	262.371 *19*	0.0034 *5*
γ	265.55 *7*	0.0039 *5*
γ	294.268 *12*	0.054 *4*
γ	300.888 *12*	0.088 *7*
γ	323.594 *16*	0.0156 *16*
γ	355.491 *11*	0.0088 *9*

Atomic Electrons (^{134}Ce)
⟨e⟩=6.3 *3* keV

e_{bin}(keV)	⟨e⟩(keV)	e(%)
5	2.10	38 *4*
6	1.58	26 *3*
16 - 23	0.011	~0.07
26	0.167	0.63 *7*
27	0.47	1.74 *18*
28	0.76	2.8 *3*
30 - 31	0.121	0.39 *4*
32	0.51	1.58 *16*
33 - 77	0.31	0.83 *10*
84 - 132	0.18	0.16 *6*
144 - 193	0.042	0.026 *8*
194 - 240	0.0035	0.0017 *5*
246 - 295	0.014	0.0054 *19*
300 - 324	0.0011	0.00036 *15*
349 - 355	0.00012	3.3 *15* $\times 10^{-5}$

Continuous Radiation (^{134}Ce)
⟨IB⟩=0.101 keV

E_{bin}(keV)		⟨ ⟩(keV)	(%)
10 - 20	IB	0.00034	0.0021
20 - 40	IB	0.083	0.25
40 - 100	IB	0.0080	0.0150
100 - 300	IB	0.0089	0.0052
300 - 400	IB	0.00044	0.000138

$^{134}_{59}$Pr(17 *2* min)

Mode: ϵ
Δ: -78770 *200* keV syst
SpA: 8.25×10^{7} Ci/g
Prod: ^{127}I(^{12}C,5n); protons on Pr; protons on Gd

Photons (^{134}Pr(17 + ~11 min))

γ_{mode}	γ(keV)	γ(rel)
γ	169.09 *18*	0.50 *8*
γ	184.7 *3*	1.9 *3*
γ	189.5 *4*	1.4 *2*
γ	206.8 *5*	~0.3
γ	215.32 *15*	9.7 *9*
γ	231.6 *3*	1.00 *25*
γ	293.6 *5*	1.2 *3*
γ	299.1 *5*	0.5 *2* ?
γ	309.02 *24*	3.8 *5*
γ	331.7 *3*	7.4 *11*
γ	333.9 *4*	7.3 *10*
γ	384.42 *17*	4.9 *5*
γ	392.0 *6*	0.4 *1*
γ E2	409.35 *9*	100 *11*
γ	417.55 *18*	2.5 *4*
γ	429.24 *19*	2.40 *24*
γ	446.7 *6*	~0.4
γ	480.3 *4*	1.3 *3*
γ	517.8 *3*	3.1 *10*
γ [M1+E2]	556.21 *13*	14.0 *14*
γ	594.20 *16*	1.7 *3*
γ E2	639.71 *15*	25 *3*
γ	644.57 *18*	1.8 *3*
γ	667.24 *18*	2.5 *5*
γ	677.70 *15*	5.1 *7*
γ	685.7 *5*	1.0 *4*
γ	718.3 *4*	1.1 *5* ?
γ	763.29 *14*	4.4 *5*
γ	786.8 *4*	1.0 *3*
γ	794.5 *5*	~0.4
γ	809.5 *4*	0.40 *19* ?
γ E2	814.4 *4*	1.1 *3*
γ	846.79 *18*	2.8 *3*
γ [E2]	965.56 *13*	14.1 *14*
γ	973.76 *15*	9.6 *13*
γ	978.61 *14*	5.7 *9*
γ	1000.0 *5*	1.0 *2*
γ	1062.11 *17*	0.50 *18* ?
γ	1125.58 *19*	4.2 *6*
γ	1162.4 *3*	~0.50 ?
γ	1196.9 *5*	1.2 *3*
γ	1213.3 *6*	0.5 *1*
γ	1233.91 *17*	1.10 *18*
γ	1312.5 *4*	1.7 *7* ?
γ	1365.7 *4*	1.1 *3*
γ	1494.6 *3*	4.0 *6*
γ	1579.9 *3*	3.5 *5*
γ	1904.5 *3*	7.4 *10*
γ	1964.1 *10*	~1 ?
γ	2136.1 *3*	8 *1*
γ	2233.0 *25*	1.5 *3*
γ	2331.0 *25*	2.1 *4*

$^{134}_{59}$Pr(~11 min)

Mode: ϵ
Δ: >-78770 *200* keV syst
SpA: ~1.3×10^{8} Ci/g
Prod: protons on Gd

see ^{134}Pr(17 min) for γ rays

$^{134}_{60}$Nd(8.5 *15* min)

Mode: ϵ
Δ: -75570 *320* keV syst
SpA: 1.6×10^{8} Ci/g
Prod: protons on Gd

Photons (^{134}Nd)
⟨γ⟩=328 *24* keV

γ_{mode}	γ(keV)	γ(%)
Pr L_ℓ	4.453	0.19 *3*
Pr L_η	4.929	0.075 *9*
Pr L_α	5.031	5.0 *6*
Pr L_β	5.619	4.5 *6*
Pr L_γ	6.410	0.63 *9*
Pr $K_{\alpha2}$	35.550	22.0 *8*
Pr $K_{\alpha1}$	36.026	40.0 *15*
Pr $K_{\beta1}$'	40.720	11.3 *4*
Pr $K_{\beta2}$'	41.981	3.16 *13*
γ M1,E2	90.2 *6*	2.2 *4*
γ M1,E2	93.2 *7*	0.29 *6*
γ E1	101.1 *8*	2.3 *5*
γ M1,E2	104.2 *7*	1.7 *4*
γ M1,E2	115.6 *8*	1.7 *4*
γ M1,E2	119.2 *7*	0.58 *12*
γ [M1+E2]	125.9 *6*	0.52 *10*
γ	131.0 *7*	0.52 *10*
γ	144.3 *10*	1.6 *3*
γ [M1+E2]	147.5 *7*	1.3 *3*
γ E1	163.1 *5*	58 *12*
γ [M1+E2]	183.4 *7*	0.81 *16*
γ	189.3 *7*	0.99 *20*
γ [E1]	216.7 *7*	12.4 *25*
γ E1	289.0 *6*	13 *3*
γ	295.7 *8*	1.7 *3*
γ [M1+E2]	309.3 *7*	1.7 *3*
γ [M1+E2]	320.3 *6*	1.04 *21*
γ [E1]	335.9 *7*	0.41 *8*
γ	352.4 *7*	0.87 *17*
γ [E1]	379.1 *6*	1.04 *21*
γ	458.8 *8*	0.64 *13*
γ	467.9 *10*	2.8 *6*
γ [E1]	483.4 *6*	2.3 *5*
γ	583.0 *10*	1.16 *23*
γ	673.0 *10*	1.16 *23*
γ	992.0 *10*	1.8 *4*
γ	1000.0 *10*	4.1 *8*

Atomic Electrons (^{134}Nd)
⟨e⟩=23.9 *16* keV

e_{bin}(keV)	⟨e⟩(keV)	e(%)
6	3.5	58 *7*
7 - 29	1.06	8.6 *7*
30	0.71	2.39 *25*
34 - 41	0.83	2.37 *15*
48	1.7	3.5 *7*
51 - 59	0.50	0.90 *13*
62	1.12	1.8 *4*
74	0.98	1.3 *3*
77 - 83	0.62	0.78 *13*
84	1.2	~1
86 - 120	3.4	3.4 *7*
121	3.9	3.3 *7*
124 - 147	0.98	0.70 *17*
156	0.55	0.35 *7*
157 - 163	0.34	0.22 *3*
175	0.56	0.32 *7*
177 - 217	0.26	0.13 *3*
247 - 296	0.98	0.38 *6*
302 - 351	0.17	0.054 *14*
352 - 379	0.0047	0.00127 *21*
417 - 462	0.22	0.051 *21*
466 - 483	0.012	0.0026 *6*
541 - 583	0.05	~0.009
631 - 673	0.038	~0.006
950 - 999	0.12	0.012 *5*
1000	0.0006	~6 $\times 10^{-5}$

Continuous Radiation (^{134}Nd)

$\langle\beta+\rangle$=151 keV;$\langle$IB$\rangle$=4.7 keV

E_{bin}(keV)		$\langle\ \rangle$(keV)	(%)
0 - 10	$\beta+$	5.5×10^{-7}	6.5×10^{-6}
	IB	0.0063	
10 - 20	$\beta+$	2.46×10^{-5}	0.000148
	IB	0.0061	0.042
20 - 40	$\beta+$	0.00076	0.00230
	IB	0.087	0.25
40 - 100	$\beta+$	0.0410	0.052
	IB	0.047	0.081
100 - 300	$\beta+$	2.21	1.00
	IB	0.123	0.067
300 - 600	$\beta+$	16.5	3.57
	IB	0.30	0.067
600 - 1300	$\beta+$	93	9.68
	IB	1.52	0.161
1300 - 2500	$\beta+$	40	2.69
	IB	2.5	0.144
2500 - 3084	IB	0.116	0.0044
	$\Sigma\beta+$		17

$^{134}_{61}$Pm(24 *2* s)

Mode: ϵ
SpA: 3.5×10^{9} Ci/g
Prod: ^{32}S on ^{112}Sn; ^{32}S on ^{106}Cd

Photons (^{134}Pm)

γ_{mode}	γ(keV)	γ(rel)
γ E2	294 *1*	100 *20*
γ	460 *1*	15 *3*
γ E2	495 *1*	60 *12*
γ E2	632 *1*	10 *2*

$^{134}_{62}$Sm(12 *3* s)

Mode: ϵ
SpA: 6.8×10^{9} Ci/g
Prod: ^{32}S on ^{112}Sn; ^{32}S on ^{106}Cd

A = 135

NDS **14**, 191 (1975)

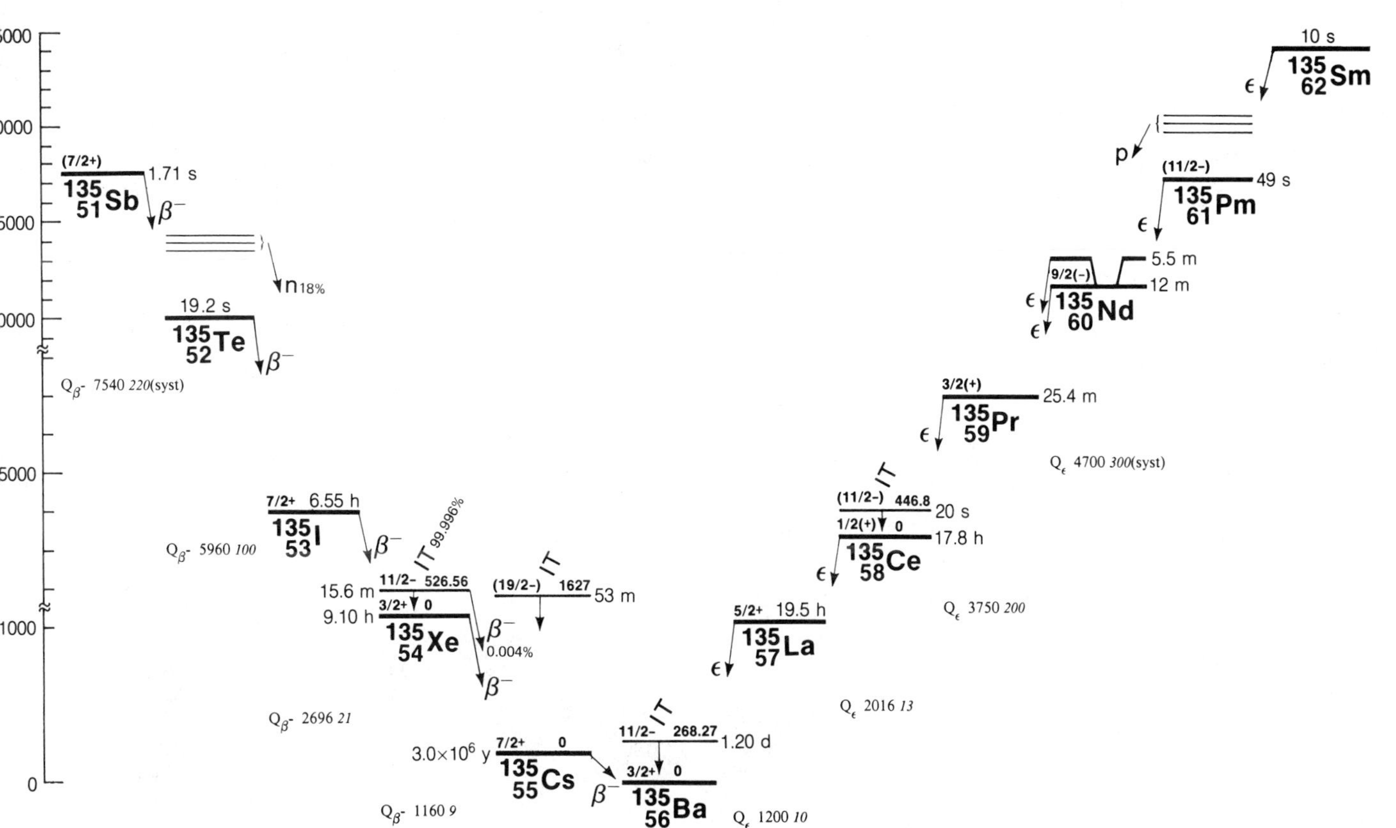

$^{135}_{51}$Sb(1.706 *14* s)

Mode: β-, β-n(17.5 *20* %)
Δ: -70310 *200* keV syst
SpA: 4.03×10^{10} Ci/g
Prod: fission

Delayed Neutrons (^{135}Sb)

n(keV)	n(%)†
162 *3*	0.4
458 *2*	0.6
499 *3*	0.5
549 *3*	0.5
623 *3*	0.4
738 *3*	0.4
848 *2*	0.8
977 *2*	1.0
1042 *2*	3
1201 *2*	2.1
1251 *4*	0.4
1322 *4*	0.9
1384 *4*	0.7
1458 *2*	4
1549 *3*	1.0
1618 *5*	0.4

† 11% uncert(syst)

Photons (^{135}Sb)

γ_{mode}	γ(keV)
$\gamma_{\beta\text{-n}}$	115
$\gamma_{\beta\text{-n}}$	297
$\gamma_{\beta\text{-n}}$	706
$\gamma_{\beta\text{-n}}$	1279

$^{135}_{52}$Te(19.2 *2* s)

Mode: β-
Δ: -77850 *100* keV
SpA: 4.27×10^{9} Ci/g
Prod: fission

Photons (^{135}Te)

γ_{mode}	γ(keV)	γ(rel)
γ [M1+E2]	266.8 *8*	15 *3*
γ [M1+E2]	603.5 *8*	100 *20*
γ [M1+E2]	870.3 *8*	23 *5*

$^{135}_{53}$I (6.55 *3* h)

Mode: β-
Δ: -83813 *22* keV
SpA: 3.544×10^{6} Ci/g
Prod: fission

Photons (^{135}I)

⟨γ⟩=1647 *17* keV

γ_{mode}	γ(keV)	γ(%)†
Xe L_ℓ	3.634	0.0051 *22*
Xe L_η	3.955	0.0023 *10*
Xe L_α	4.104	0.14 *6*
Xe L_β	4.521	0.13 *6*
Xe L_γ	5.124	0.016 *7*
Xe $K_{\alpha2}$	29.461	0.8 *3*
Xe $K_{\alpha1}$	29.782	1.5 *6*
Xe $K_{\beta1}'$	33.606	0.40 *17*
Xe $K_{\beta2}'$	34.606	0.09 *4*
γ [M1+E2]	112.771 *21*	~0.013
γ [M1+E2]	113.14 *3*	~0.007
γ [M1+E2]	162.53 *3*	0.010 *3*
γ [M1+E2]	165.66 *3*	0.031 *3*
γ	184.49 *7*	0.023 *3*
γ [M1+E2]	197.155 *19*	0.033 *3*
γ [M1+E2]	220.512 *13*	1.75 *6*
γ	229.73 *3*	0.232 *9*
γ	247.5 *10*	
γ [M1+E2]	254.67 *4*	0.023 *9*
γ	264.30 *3*	0.184 *7*
γ [M1+E2]	288.462 *15*	3.09 *11*
γ	290.258 *17*	0.303 *20*
γ	304.896 *23*	0.032 *3*
γ	305.75 *3*	0.095 *4*
γ	308.69 *7*	~0.009
γ	326.053 *16*	<0.0046
γ	333.66 *3*	0.037 *3*
γ	342.87 *4*	<0.0017
γ	361.99 *3*	0.19 *3*
γ	403.029 *22*	0.232 *9*
γ	414.875 *22*	0.301 *17*
γ	417.667 *16*	3.52 *11*
γ	429.954 *25*	0.303 *23*
γ	433.793 *19*	0.553 *23*
γ	451.655 *25*	0.315 *17*
γ M4	526.563 *11*	13.3 *4* *
γ	530.81 *3*	0.031 *14*
γ (M1,E2)	546.564 *12*	7.13 *23*
γ	575.86 *4*	0.129 *23*
γ	588.33 *3*	0.052 *14*
γ	617.44 *5*	0.037 *17*
γ	649.88 *3*	0.46 *3*
γ	656.10 *8*	0.074 *14*
γ	679.17 *3*	0.054 *14*
γ	684.72 *9*	0.023 *9*
γ	690.18 *3*	0.129 *14*
γ	707.924 *20*	0.66 *6*
γ	785.48 *3*	0.152 *20*
γ	795.780 *25*	<0.046 ?
γ	797.92 *3*	0.17 *3*
γ	807.13 *4*	0.046 *17*
γ	836.821 *14*	6.67 *23*
γ	960.31 *5*	~0.03
γ	961.437 *23*	0.15 *3*
γ [E2]	971.960 *20*	0.89 *6*
γ	972.63 *3*	1.20 *6*
γ	995.07 *3*	0.15 *3*
γ E1	1038.754 *19*	7.9 *3*
γ	1096.83 *3*	0.089 *14*
γ	1101.529 *24*	1.60 *5*
γ	1123.969 *24*	3.61 *11*
γ (E2)	1131.523 *12*	22.5 *7*
γ	1151.524 *16*	<0.0029 ?
γ	1159.90 *3*	0.103 *23*
γ	1169.114 *21*	0.87 *3*
γ	1180.47 *9*	0.063 *9*
γ	1225.73 *3*	0.043 *17*
γ	1240.481 *20*	0.90 *4*
γ	1254.84 *3*	<0.011 ?
γ (E2,M1)	1260.420 *14*	28.6 *9*
γ	1277.84 *12*	~0.06
γ	1308.71 *20*	0.034 *9*
γ	1315.81 *8*	0.066 *17*
γ	1334.60 *8*	0.031 *9*
γ	1343.58 *5*	0.077 *11*
γ	1367.90 *4*	0.61 *3*
γ	1416.3 *4*	0.031 *9* ?
γ	1448.36 *3*	0.31 *3*
γ (E2,M1)	1457.573 *16*	8.6 *3*
γ	1502.766 *25*	1.07 *4*

Photons (^{135}I)
(continued)

γ_{mode}	γ(keV)	γ(%)†
γ	1522.00 *7*	0.037 *17*
γ	1566.396 *21*	1.29 *6*
γ (E2)	1678.082 *13*	9.5 *4*
γ	1706.487 *23*	4.09 *17*
γ	1728.926 *25*	0.00094 *19*
γ	1791.224 *24*	7.70 *23*
γ	1830.69 *3*	0.58 *3*
γ	1845.438 *25*	0.006 *3*
γ	1927.296 *23*	0.295 *14*
γ	1948.53 *5*	0.063 *6*
γ	1968.337 *17*	0.016 *4*
γ	2045.89 *3*	0.87 *4*
γ	2112.39 *6*	0.069 *6*
γ	2151.51 *10*	0.022 *3*
γ	2179.7 *5*	0.0040 *17*
γ	2189.41 *20*	0.013 *3*
γ	2255.482 *23*	0.61 *3*
γ	2357 *1*	<0.0007
γ	2408.66 *4*	0.95 *4*
γ	2466.11 *8*	0.072 *3*
γ	2477.1 *4*	0.0014 *3* ?

† 1.1% uncert(syst)
* with ^{135}Xe(15.6 min) in equilib

Atomic Electrons (^{135}I)

⟨e⟩=19 *7* keV

e_{bin}(keV)	⟨e⟩(keV)	e(%)
5 - 33	0.22	2.9 *11*
78 - 113	0.015	0.017 *4*
128 - 166	0.024	0.0170 *22*
179 - 225	0.38	0.200 *17*
229 - 274	0.35	0.14 *1*
283 - 329	0.084	0.029 *4*
333 - 380	0.022	0.0059 *23*
383 - 430	0.21	0.053 *21*
433 - 452	0.0028	0.00064 *22*
492	12.7	~3
496 - 512	0.25	0.050 *8*
521	2.3	~0.44
522	0.32	~0.06
525	0.41	~0.08
526 - 575	0.33	0.063 *25*
576 - 622	0.012	0.0020 *9*
644 - 690	0.019	0.0028 *11*
702 - 751	0.0045	0.00062 *22*
761 - 807	0.10	~0.013
831 - 837	0.017	0.0020 *8*
926 - 973	0.036	0.0039 *9*
990 - 1039	0.0515	0.00511 *17*
1062 - 1101	0.326	0.0298 *23*
1117 - 1164	0.061	0.0054 *5*
1168 - 1206	0.009	0.0008 *4*
1220 - 1260	0.41	0.033 *4*
1272 - 1316	0.0021	0.00016 *4*
1329 - 1368	0.0061	0.00046 *19*
1382 - 1423	0.093	0.0065 *8*
1443 - 1487	0.023	0.0016 *3*
1497 - 1565	0.012	0.0008 *3*
1644 - 1728	0.122	0.0074 *8*
1757 - 1844	0.062	0.0035 *12*
1893 - 1967	0.0026	0.00014 *4*
2011 - 2108	0.0062	0.00031 *10*
2111 - 2188	0.00026	$1.2\ 3\times10^{-5}$
2221 - 2255	0.0037	0.00017 *5*
2322 - 2408	0.0055	0.00023 *7*
2432 - 2476	0.00041	$1.7\ 5\times10^{-5}$

Continuous Radiation (^{135}I)

$\langle\beta\text{-}\rangle$=358 keV;$\langle$IB$\rangle$=0.42 keV

E_{bin}(keV)		$\langle\ \rangle$(keV)	(%)
0 - 10	β-	0.096	1.91
	IB	0.0164	
10 - 20	β-	0.286	1.90
	IB	0.0157	0.109
20 - 40	β-	1.14	3.79
	IB	0.030	0.103
40 - 100	β-	7.8	11.1
	IB	0.076	0.118
100 - 300	β-	62	31.7
	IB	0.158	0.093
300 - 600	β-	133	30.5
	IB	0.097	0.024
600 - 1300	β-	150	18.8
	IB	0.031	0.0041
1300 - 2169	β-	3.98	0.268
	IB	0.00051	3.4×10^{-5}

$^{135}_{54}$Xe(9.104 *20* h)

Mode: β-

Δ: -86509 *11* keV

SpA: 2.550×10^{6} Ci/g

Prod: ^{134}Xe(n,γ); fission; ^{138}Ba(n,α)

Photons (^{135}Xe)

$\langle\gamma\rangle$=249 *8* keV

γ_{mode}	γ(keV)	γ(%)†
Cs L_ℓ	3.795	0.0100 *18*
Cs L_η	4.142	0.0046 *8*
Cs L_α	4.285	0.28 *4*
Cs L_β	4.731	0.25 *4*
Cs L_γ	5.358	0.032 *6*
Cs $K_{\alpha2}$	30.625	1.5 *1*
Cs $K_{\alpha1}$	30.973	2.77 *19*
Cs $K_{\beta1}$'	34.967	0.76 *5*
Cs $K_{\beta2}$'	36.006	0.185 *13*
γ [M1+E2]	158.196 *13*	0.289 *10*
γ [M1+E2]	200.197 *19*	0.012 *5*
γ [M1+E2]	249.792 *12*	90
γ [M1+E2]	358.393 *17*	0.221 *8*
γ [E2]	373.128 *25*	0.015 *3*
γ [E2]	407.988 *13*	0.359 *13*
γ [M1+E2]	454.233 *20*	0.0036 *7*
γ [M1+E2]	573.324 *23*	0.0048 *7*
γ [M1+E2]	608.185 *15*	2.90 *9*
γ [M1+E2]	654.429 *13*	0.045 *3*
γ [E2]	731.520 *19*	0.055 *3*
γ [M1+E2]	812.625 *15*	0.070 *3*
γ [E2]	1062.416 *14*	0.0041 *8*

† 3.3% uncert(syst)

Atomic Electrons (^{135}Xe)

$\langle e\rangle$=15.7 *10* keV

e_{bin}(keV)	$\langle e\rangle$(keV)	e(%)
5 - 35	0.41	5.3 *7*
122 - 164	0.119	0.092 *12*
194 - 200	0.00064	0.00033 *12*
214	12.2	5.7 *4*
244	2.0	0.81 *18*
245	0.3	~0.12
249	0.48	0.19 *7*
322 - 368	0.0205	0.0063 *6*

Atomic Electrons (^{135}Xe) (continued)

e_{bin}(keV)	$\langle e\rangle$(keV)	e(%)
372 - 418	0.0232	0.0061 *3*
449 - 454	3.4×10^{-5}	7.6 *20* $\times10^{-6}$
537 - 573	0.095	0.017 *3*
602 - 649	0.0182	0.0030 *4*
653 - 696	0.00120	0.000173 *11*
726 - 732	0.000208	2.86 *14* $\times10^{-5}$
777 - 813	0.0018	0.00023 *4*
1026 - 1062	6.3×10^{-5}	6.1 *11* $\times10^{-6}$

Continuous Radiation (^{135}Xe)

$\langle\beta\text{-}\rangle$=303 keV;$\langle$IB$\rangle$=0.29 keV

E_{bin}(keV)		$\langle\ \rangle$(keV)	(%)
0 - 10	β-	0.087	1.73
	IB	0.0146	
10 - 20	β-	0.261	1.74
	IB	0.0138	0.096
20 - 40	β-	1.05	3.51
	IB	0.026	0.090
40 - 100	β-	7.5	10.7
	IB	0.064	0.100
100 - 300	β-	71	35.7
	IB	0.117	0.070
300 - 600	β-	163	37.8
	IB	0.047	0.0122
600 - 910	β-	60	8.8
	IB	0.0029	0.00047

$^{135}_{54}$Xe(15.65 *10* min)

Mode: IT(99.996 %), β-(0.004 %)

Δ: -85982 *11* keV

SpA: 8.90×10^{7} Ci/g

Prod: daughter ^{135}I; ^{134}Xe(n,γ); fission; ^{138}Ba(n,α)

Photons (^{135}Xe)

$\langle\gamma\rangle$=432 *5* keV

γ_{mode}	γ(keV)	γ(%)†
Xe L_ℓ	3.634	0.026 *4*
Xe L_η	3.955	0.0118 *12*
Xe L_α	4.104	0.71 *8*
Xe L_β	4.521	0.64 *8*
Xe L_γ	5.126	0.083 *12*
Xe $K_{\alpha2}$	29.461	4.02 *13*
Xe $K_{\alpha1}$	29.782	7.47 *24*
Xe $K_{\beta1}$'	33.606	2.02 *7*
Xe $K_{\beta2}$'	34.606	0.478 *17*
γ_{IT} M4	526.563 *11*	81.2
$\gamma_{\beta\text{-}}$[E2]	786.91 *10*	0.0040 *8*
$\gamma_{\beta\text{-}}$	1133 *1*	0.00020 *4* ?
$\gamma_{\beta\text{-}}$	1192 *1*	3.0 *6* $\times10^{-5}$?
$\gamma_{\beta\text{-}}$	1358 *1*	0.00020 *4* ?

† 1.2% uncert(syst)

Atomic Electrons (^{135}Xe)

$\langle e\rangle$=98.4 *18* keV

e_{bin}(keV)	$\langle e\rangle$(keV)	e(%)
5	0.65	13.1 *13*
24	0.123	0.51 *5*
25	0.157	0.63 *7*
28	0.043	0.154 *16*
29	0.093	0.32 *3*
30	0.0124	0.042 *4*
32	0.0047	0.0146 *15*
33	0.0108	0.033 *3*
492	77.2	15.7 *4*
521	13.9	2.66 *6*
522	1.93	0.370 *9*
525	2.53	0.481 *11*
526	1.61	0.307 *7*
527	0.1046	0.0199 *5*

Continuous Radiation (^{135}Xe)

$\langle\beta\text{-}\rangle$=0.0127 keV;$\langle$IB$\rangle$=$1.17\times10^{-5}$ keV

E_{bin}(keV)		$\langle\ \rangle$(keV)	(%)
0 - 10	β-	4.43×10^{-6}	8.8×10^{-5}
	IB	6.1×10^{-7}	
10 - 20	β-	1.33×10^{-5}	8.9×10^{-5}
	IB	5.8×10^{-7}	4.1×10^{-6}
20 - 40	β-	5.3×10^{-5}	0.000178
	IB	1.08×10^{-6}	3.8×10^{-6}
40 - 100	β-	0.000372	0.00053
	IB	2.7×10^{-6}	4.1×10^{-6}
100 - 300	β-	0.00320	0.00163
	IB	4.8×10^{-6}	2.9×10^{-6}
300 - 600	β-	0.0068	0.00157
	IB	1.9×10^{-6}	4.7×10^{-7}
600 - 900	β-	0.00233	0.000343
	IB	1.07×10^{-7}	1.7×10^{-8}

$^{135}_{55}$Cs(3.0 *3* $\times10^{6}$ yr)

Mode: β-

Δ: -87668 *8* keV

SpA: 0.00090 Ci/g

Prod: daughter ^{135}Xe; fission

Continuous Radiation (^{135}Cs)

$\langle\beta\text{-}\rangle$=56 keV;$\langle$IB$\rangle$=0.0118 keV

E_{bin}(keV)		$\langle\ \rangle$(keV)	(%)
0 - 10	β-	0.59	12.0
	IB	0.0029	
10 - 20	β-	1.67	11.1
	IB	0.0022	0.0156
20 - 40	β-	5.9	19.9
	IB	0.0030	0.0108
40 - 100	β-	26.8	40.3
	IB	0.0033	0.0057
100 - 205	β-	21.3	16.7
	IB	0.00038	0.00033

$^{135}_{55}$Cs(53 *2* min)

Mode: IT
Δ: -86041 *8* keV
SpA: 2.628×10^{7} Ci/g
Prod: ^{134}Xe(d,n); ^{132}Xe(α,p); ^{135}Ba(n,p); protons on Ba

Photons (^{135}Cs)

⟨γ⟩=1590 *25* keV

γ_{mode}	γ(keV)	γ(%)
Cs L_ℓ	3.795	0.0064 *9*
Cs L_η	4.142	0.0029 *3*
Cs L_α	4.285	0.175 *20*
Cs L_β	4.731	0.159 *20*
Cs L_γ	5.367	0.021 *3*
Cs $K_{\alpha2}$	30.625	0.97 *4*
Cs $K_{\alpha1}$	30.973	1.79 *7*
Cs $K_{\beta1}$'	34.967	0.489 *20*
Cs $K_{\beta2}$'	36.006	0.119 *5*
γ [E2]	786.91 *10*	99.70
γ M4	840 *1*	96 *3*

Atomic Electrons (^{135}Cs)

⟨e⟩=36.6 *10* keV

e_{bin}(keV)	⟨e⟩(keV)	e(%)
5	0.137	2.7 *3*
6	0.0189	0.33 *4*
25	0.029	0.117 *12*
26	0.036	0.140 *15*
29	0.0095	0.032 *3*
30	0.0223	0.075 *8*
31	0.0033	0.0108 *11*
34	0.0026	0.0078 *8*
35	0.00105	0.0030 *3*
751	1.90	0.253 *13*
781	0.229	0.0294 *15*
782	0.0374	0.00478 *24*
786	0.055	0.0070 *4*
787	0.0143	0.00182 *9*
804	28.1	3.49 *13*
834	4.00	0.479 *17*
835	0.77	0.092 *3*
839	1.01	0.120 *4*
840	0.263	0.0314 *11*

$^{135}_{56}$Ba(stable)

Δ: -87873 *6* keV
%: 6.592 *18*

$^{135}_{56}$Ba(1.196 *8* d)

Mode: IT
Δ: -87605 *6* keV
SpA: 8.09×10^{5} Ci/g
Prod: ^{134}Ba(n,γ)

Photons (^{135}Ba)

⟨γ⟩=60 *3* keV

γ_{mode}	γ(keV)	γ(%)
Ba L_ℓ	3.954	0.124 *17*
Ba L_η	4.331	0.053 *6*
Ba L_α	4.465	3.4 *4*
Ba L_β	4.947	3.1 *4*
Ba L_γ	5.628	0.42 *6*
Ba $K_{\alpha2}$	31.817	15.5 *7*
Ba $K_{\alpha1}$	32.194	28.6 *13*
Ba $K_{\beta1}$'	36.357	7.9 *4*
Ba $K_{\beta2}$'	37.450	2.00 *10*
γ M4	268.272 *11*	15.6

Atomic Electrons (^{135}Ba)

⟨e⟩=207 *6* keV

e_{bin}(keV)	⟨e⟩(keV)	e(%)
5	1.71	33 *4*
6	1.24	21.6 *23*
25	0.120	0.47 *5*
26	0.34	1.31 *14*
27	0.56	2.11 *23*
30	0.126	0.42 *5*
31	0.36	1.15 *12*
32	0.077	0.24 *3*
35	0.039	0.110 *12*
36	0.0206	0.057 *6*
231	138.3	59.9 *24*
262	29.6	11.3 *5*
263	20.3	7.7 *3*
267	11.5	4.31 *17*
268	3.11	1.16 *5*

$^{135}_{57}$La(19.48 *16* h)

Mode: ε
Δ: -86673 *12* keV
SpA: 1.1916×10^{6} Ci/g
Prod: ^{133}Cs(α,2n); ^{134}Ba(d,n); ^{138}Ba(p,4n); ^{135}Ba(p,n)

Photons (^{135}La)

⟨γ⟩=35.7 *5* keV

γ_{mode}	γ(keV)	γ(%)†
Ba L_ℓ	3.954	0.148 *20*
Ba L_η	4.331	0.066 *7*
Ba L_α	4.465	4.1 *5*
Ba L_β	4.946	3.7 *5*
Ba L_γ	5.621	0.50 *7*
Ba $K_{\alpha2}$	31.817	21.9 *7*
Ba $K_{\alpha1}$	32.194	40.4 *12*
Ba $K_{\beta1}$'	36.357	11.2 *3*
Ba $K_{\beta2}$'	37.450	2.83 *9*
γ	107.308 *11*	0.00100 *5*
γ M1+E2	220.972 *9*	0.0554 *11*
γ [E2]	259.576 *11*	0.00459 *20*
γ	267.193 *11*	<0.0009
γ M4	268.272 *11* ?	*
γ	366.884 *9*	0.0326 *7*
γ	374.500 *9*	0.0191 *4*
γ	394.015 *20*	0.0046 *4*
γ [M1+E2]	480.548 *9*	1.54
γ	587.855 *9*	0.1129 *23*
γ	634.076 *9*	0.0227 *5*
γ	759.040 *24*	0.00075 *22*
γ	787.92 *10*	0.00012 *5* ?

Photons (^{135}La)
(continued)

γ_{mode}	γ(keV)	γ(%)†
γ	855.046 *10*	0.0183 *6*
γ	874.561 *19*	0.165 *3*
γ	980.011 *22*	0.00513 *10*
γ	1008.89 *10*	~2×10^{-5} ?

† approximate
* with ^{135}Ba(1.20 d)

Atomic Electrons (^{135}La)

⟨e⟩=6.0 *3* keV

e_{bin}(keV)	⟨e⟩(keV)	e(%)
5	2.06	39 *4*
6	1.50	26 *3*
25	0.170	0.67 *7*
26	0.48	1.85 *19*
27	0.80	3.0 *3*
30	0.178	0.59 *6*
31	0.51	1.63 *17*
32 - 70	0.193	0.58 *4*
101 - 107	0.00029	~0.00028
184 - 230	0.0130	0.0068 *3*
254 - 267	0.000181	7.1 *4* ×10^{-5}
329 - 375	0.0034	0.0010 *4*
388 - 394	4.5 ×10^{-5}	~1×10^{-5}
443 - 481	0.089	0.020 *3*
550 - 597	0.0041	0.0007 *3*
628 - 634	0.00010	1.6 *8* ×10^{-5}
722 - 759	1.9 ×10^{-5}	2.6 *13* ×10^{-6}
782 - 818	0.00031	~4×10^{-5}
837 - 874	0.0032	0.00038 *18*
943 - 980	8.5 ×10^{-5}	9 *4* ×10^{-6}
1003 - 1009	3.9 ×10^{-8}	~4×10^{-9}

Continuous Radiation (^{135}La)

⟨β+⟩=0.0105 keV;⟨IB⟩=0.42 keV

E_{bin}(keV)		⟨ ⟩(keV)	(%)
0 - 10	β+	2.19×10^{-7}	2.63×10^{-6}
	IB	0.00035	
10 - 20	β+	8.4×10^{-6}	5.07×10^{-5}
	IB	0.00037	0.0022
20 - 40	β+	0.000197	0.00061
	IB	0.077	0.24
40 - 100	β+	0.00478	0.0065
	IB	0.0075	0.0134
100 - 300	β+	0.0056	0.00452
	IB	0.035	0.0170
300 - 600	IB	0.135	0.030
600 - 1200	IB	0.160	0.021
	Σβ+		0.0117

$^{135}_{58}$Ce(17.8 *3* h)

Mode: ε
Δ: -84657 *18* keV
SpA: 1.307×10^{6} Ci/g
Prod: ^{139}La(p,5n); ^{139}La(d,6n); protons on Er; ^{136}Ce(γ,n); ^{134}Ba(α,3n)

Photons (^{135}Ce)

⟨γ⟩=815 8 keV

γ_{mode}	γ(keV)	γ(%)†
La L_ℓ	4.121	0.178 25
La L_η	4.529	0.076 8
La L_α	4.649	4.9 6
La L_β	5.163	4.6 6
La L_γ	5.894	0.65 10
La $K_{\alpha2}$	33.034	23.5 7
La $K_{\alpha1}$	33.442	43.2 13
γ M1(+0.03%E2)	34.509 7	1.85 8
La $K_{\beta1}$'	37.777	12.0 4
La $K_{\beta2}$'	38.927	3.15 11
γ M1	59.046 7	0.023 4
γ	65.0 3	0.0059 21
γ M1+E2	86.981 10	0.286 17
γ M1(+E2)	88.719 7	0.374 21
γ	115.71 20	0.021 4
γ M1	118.041 9	0.315 21
γ M1+3.8%E2	119.531 8	1.10 4
γ	123.8 3	~0.005
γ M1	132.905 22	0.097 21
γ	156.003 13	~0.008
γ	162.815 12	0.021 8
γ	177.98 6	0.050 8
γ	179.043 9	0.029 8
γ	187.047 11	0.0134 21
γ	200.776 11	0.044 6
γ	202.91 20	~0.017 ?
γ M1+E2	206.512 6	7.85 25
γ	210.145 24	~0.013
γ	223.816 11	~0.021
γ M1(+E2)	265.558 6	42.0
γ	267.762 9	0.66 4
γ	281.10 11	~0.0042 ?
γ	299.104 8	1.30 13
γ E2	300.067 7	22.7 7
γ	304.500 8	0.056 4
γ	312.9 4	~0.038
γ	318.818 14	0.035 4
γ	326.18 9	~0.013
γ	343.050 10	0.105 8
γ	379.819 12	1.50 5
γ	387.824 7	0.588 21
γ	398.054 9	0.496 21
γ	400.01 1	0.336 13
γ	400.970 25	~0.008
γ	433.962 23	~0.008
γ	459.056 11	~0.12
γ	459.08 9	~0.013
γ	465.268 13	0.109 25
γ	483.542 7	1.87 7
γ	485.035 11	0.210 21
γ	495.30 3	0.021 8
γ	505.865 11	0.052 4
γ	518.051 6	13.4 8
γ	528.316 10	0.122 8
γ	546.036 12	0.655 25
γ	560.74 3	~0.008
γ	562.825 9	0.147 13
γ	566.866 8	0.609 25
γ	572.261 7	10.5 3
γ	577.097 7	5.08 17
γ [E2]	583.9 5	0.042 17 ?
γ	604.566 8	2.90 13
γ	606.770 6	19.3 6
γ	611.115 14	0.021 8 ?
γ	621.870 8	0.428 17
γ	651.31 3	0.0088 17
γ	655.888 13	~0.0042
γ	664.077 10	0.088 21
γ	665.567 10	0.21 4
γ	665.816 8	3.02 21
γ	666.044 10	~0.042
γ	684.318 11	0.378 21
γ	693.38 6	~0.008
γ	696.61 20	0.034 13 ?
γ	712.35 9	0.029 6 ?
γ	718.827 11	0.395 17
γ	726.94 3	~0.013
γ	727.15 9	~0.008
γ	727.89 6	~0.06
γ	728.46 3	~0.021
γ	738.460 23	0.039 4

Photons (^{135}Ce)
(continued)

γ_{mode}	γ(keV)	γ(%)†
γ	750.81 20	0.0076 21
γ	770.88 3	0.038 13
γ	772.969 23	~0.13
γ	773.929 15	~0.22
γ	777.872 12	0.328 25
γ	782.601 19	~0.021
γ	783.608 6	10.5 3
γ	815.66 3	0.038 8
γ	828.381 8	5.12 21
γ	832.014 23	0.038 13
γ	834.930 14	0.113 4
γ	845.086 10	0.147 8
γ	871.364 7	3.19 6
γ	875.12 3	~0.06
γ	894.657 16	0.040 4
γ	905.873 6	1.60 8
γ	933.70 3	0.034 8
γ	938.603 17	0.016 4
γ	964.918 8	0.328 25
γ	983.376 16	~0.008
γ	984.383 11	0.076 13
γ	993.45 6	~0.013
γ	994.70 3	~0.008
γ	1038.525 23	0.042 8
γ	1051.899 10	0.067 8
γ	1067.14 14	0.0067 17
γ	1101.417 18	0.025 6
γ	1139.428 13	0.088 8
γ	1149.583 8	0.66 3
γ	1171.429 7	0.185 8
γ	1173.937 13	0.185 13
γ	1179.62 3	0.034 8
γ	1184.092 10	1.10 5
γ	1214.13 3	0.097 21
γ	1232.982 13	0.105 13
γ	1243.137 11	~0.0042
γ	1258.4 6	~0.038
γ	1273.17 3	~0.017
γ	1299.20 3	0.071 8
γ	1333.71 3	~0.038
γ	1360.15 3	~0.038
γ	1376.7 4	0.0067 21
γ	1392.75 3	0.0076 21
γ	1439.492 13	0.055 8
γ	1449.647 10	0.0071 17
γ	1466.915 15	0.214 13
γ	1479.68 3	0.0059 21
γ	1497.09 7	~0.0025
γ	1501.423 16	0.059 8
γ	1531.60 7	0.0210 21
γ	1541.9 7	0.006 3
γ	1550.68 14	0.0097 21
γ	1560.468 16	0.071 8
γ	1585.19 14	0.0231 21
γ	1599.26 3	0.0080 13
γ	1766.978 16	0.122 13
γ	1797.15 7	0.0147 17
γ	1850.74 14	0.00046 13

† 2.4% uncert(syst)

Atomic Electrons (^{135}Ce)

⟨e⟩=25.6 9 keV

e_{bin}(keV)	⟨e⟩(keV)	e(%)
5	2.29	42 4
6	1.82	30 3
20 - 27	0.71	2.68 21
28	2.19	7.8 6
29 - 77	2.09	5.90 24
79	0.147	0.186 13
81	0.67	0.83 17
82 - 132	0.53	0.56 18
133 - 164	0.027	0.018 6
168	1.64	0.98 4
171 - 218	0.48	0.24 6
222 - 224	0.00017	~8 $\times 10^{-5}$
227	5.7	2.51 24

Atomic Electrons (^{135}Ce)
(continued)

e_{bin}(keV)	⟨e⟩(keV)	e(%)
229 - 242	0.06	~0.025
259	0.76	0.29 5
260	0.4	~0.16
261	2.26	0.87 3
262 - 307	0.96	0.34 3
312 - 361	0.15	0.044 16
362 - 401	0.034	0.009 3
420 - 467	0.09	0.021 10
477 - 524	0.6	0.12 6
527 - 576	1.2	0.22 7
577 - 625	0.13	0.022 9
627 - 673	0.11	0.017 8
678 - 727	0.016	0.0023 8
728 - 776	0.23	~0.030
777 - 826	0.15	0.019 7
827 - 875	0.10	0.012 4
888 - 937	0.011	0.0012 4
938 - 987	0.0027	0.00028 8
988 - 1037	0.0018	0.00017 6
1038 - 1067	0.00051	4.8 20 $\times 10^{-5}$
1095 - 1144	0.016	0.0014 4
1145 - 1194	0.019	0.0017 6
1204 - 1253	0.0011	9 3 $\times 10^{-5}$
1257 - 1299	0.0014	0.00011 4
1321 - 1370	0.0007	5.1 23 $\times 10^{-5}$
1371 - 1411	0.0006	4.4 20 $\times 10^{-5}$
1428 - 1474	0.0031	0.00022 7
1478 - 1560	0.0015	0.000099 23
1579 - 1598	4.3 $\times 10^{-5}$	2.7 9 $\times 10^{-6}$
1728 - 1812	0.0013	7 3 $\times 10^{-5}$
1844 - 1850	5.5 $\times 10^{-7}$	3.0 12 $\times 10^{-8}$

Continuous Radiation (^{135}Ce)

⟨β+⟩=3.18 keV;⟨IB⟩=0.48 keV

E_{bin}(keV)		⟨ ⟩(keV)	(%)
0 - 10	β+	7.1 $\times 10^{-6}$	8.5 $\times 10^{-5}$
	IB	0.00053	
10 - 20	β+	0.000270	0.00164
	IB	0.00047	0.0030
20 - 40	β+	0.0062	0.0191
	IB	0.083	0.25
40 - 100	β+	0.129	0.178
	IB	0.0096	0.017
100 - 300	β+	0.98	0.498
	IB	0.035	0.017
300 - 600	β+	1.99	0.479
	IB	0.13	0.029
600 - 1300	β+	0.082	0.0132
	IB	0.21	0.026
1300 - 1750	IB	0.0121	0.00087
	Σβ+		1.19

$^{135}_{58}$Ce(20 2 s)

Mode: IT

Δ: -84210 18 keV

SpA: 4.1 $\times 10^{9}$ Ci/g

Prod: ^{128}Te(^{12}C,5n)

Photons (^{135}Ce)

⟨γ⟩=54.7 *18* keV

γ_{mode}	γ(keV)	γ(%)†
Ce L_ℓ	4.289	0.128 *19*
Ce L_η	4.730	0.073 *9*
Ce L_α	4.838	3.5 *4*
Ce L_β	5.367	3.6 *5*
Ce L_γ	6.088	0.47 *6*
Ce $K_{\alpha2}$	34.279	8.9 *5*
Ce $K_{\alpha1}$	34.720	16.3 *10*
Ce $K_{\beta1}'$	39.232	4.6 *3*
Ce $K_{\beta2}'$	40.437	1.24 *8*
γ M1	82.67 *4*	1.1 *3*
γ E3	150.7 *3*	21.4
γ M1+E2	213.43 *4*	3.66 *16*
γ E2	296.10 *4*	0.79 *11*

† 5.0% uncert(syst)

Atomic Electrons (^{135}Ce)

⟨e⟩=109 *4* keV

e_{bin}(keV)	⟨e⟩(keV)	e(%)
6 - 42	4.8	57 *6*
76 - 83	0.26	0.34 *7*
110	34.8	31.6 *18*
144	4.23	2.93 *16*
145	48.4	33.5 *19*
149	12.5	8.3 *5*
150	2.95	1.96 *11*
151 - 173	1.23	0.75 *4*
207 - 256	0.30	0.14 *3*
290 - 296	0.024	0.0082 *9*

$^{135}_{59}$Pr(25.4 *5* min)

Mode: ϵ

Δ: -80910 *200* keV

SpA: 5.48×10^{7} Ci/g

Prod: ^{136}Ce(p,2n); protons on Gd; ^{127}I(^{12}C,4n)

Photons (^{135}Pr)

γ_{mode}	γ(keV)	γ(rel)
γ M1	82.67 *4*	50 *5*
γ M1+E2	213.43 *4*	48 *5*
γ E2	296.10 *4*	10.4 *17* ?
γ M1(+E2)	296.11 *4*	100 *12*
γ	324.81 *12*	1.4 *3* *
γ	324.82 *12*	1.4 *3*
γ	484.12 *17*	~2
γ	484.13 *17*	~4
γ M1	538.25 *12*	30 *3*
γ	593.1 *4*	1.19 *24*
γ M1	614.3 *4*	6.5 *10*
γ E2	620.92 *12*	5.7 *10*
γ	697.56 *17*	~5
γ M1,E3	720.6 *8*	1.7 *7*
γ	724.21 *19*	~1
γ	746.5 *4*	~0.5
γ	806.88 *19*	4.3 *7*
γ E2,M1	934.15 *14*	2.9 *5*
γ	1016.82 *14*	2.4 *3*
γ	1107.8 *4*	0.33 *14*
γ	1131.0 *9*	0.62 *14*
γ	1143.7 *8*	~0.14
γ	1213.7 *6*	0.48 *14*
γ	1284.7 *4*	0.48 *14*
γ	1367.4 *4*	0.76 *14*
γ	1432.6 *4*	1.05 *19*

Photons (^{135}Pr) (continued)

γ_{mode}	γ(keV)	γ(rel)
γ	1460.8 *9*	0.62 *14*
γ	1538.7 *3*	1.8 *4* *
γ	1538.7 *3*	1.8 *4*
γ	1646.0 *4*	0.67 *14*
γ	1678.9 *9*	1.57 *24*
γ	1707.4 *12*	0.29 *10*
γ	1752.1 *3*	1.7 *3*
γ	1755 *3*	0.33 *14*
γ	1784.0 *15*	0.33 *10*
γ	1845 *3*	~0.29
γ	1860 *3*	0.38 *14*
γ	1867.9 *8*	0.62 *24*
γ	1937.9 *6*	0.29 *10*
γ	1950.6 *8*	0.67 *19*
γ	1973.0 *15*	0.81 *14*
γ	2020.6 *6*	0.33 *10*
γ	2084 *3*	~0.48
γ	2107 *3*	~0.7
γ	2322 *4*	1.0 *5*
γ	2356 *4*	~0.29

* doublet

$^{135}_{60}$Nd(12.4 *6* min)

Mode: ϵ

Δ: -76210 *360* keV syst

SpA: 1.12×10^{8} Ci/g

Prod: ^{136}Ce(^{3}He,4n); ^{138}Ce(α,7n)

Photons (^{135}Nd)

⟨γ⟩=596 *23* keV

γ_{mode}	γ(keV)	γ(%)†
Pr L_ℓ	4.453	0.21 *4*
Pr L_η	4.929	0.095 *15*
Pr L_α	5.031	5.7 *9*
Pr L_β	5.597	7.4 *15*
Pr L_γ	6.487	1.3 *3*
Pr $K_{\alpha2}$	35.550	18.8 *12*
Pr $K_{\alpha1}$	36.026	34.2 *22*
Pr $K_{\beta1}'$	40.720	9.7 *6*
γ M1	41.38 *5*	23 *5*
Pr $K_{\beta2}'$	41.981	2.70 *18*
γ M2	112.52 *5*	4.6 *5*
γ E2+M1	164.60 *5*	4.1 *3*
γ E2	184.98 *7*	2.8 *3*
γ M1	203.98 *5*	51
γ (E2)	205.98 *7*	3.06 *25*
γ E2+M1	220.8 *3*	0.71 *25*
γ E2+M1	233.29 *24*	1.27 *15*
γ E2	245.36 *6*	3.5 *3*
γ [M1+E2]	247.84 *17*	~0.2
γ M1	256.03 *13*	2.7 *3*
γ	259.7 *3*	0.76 *25*
γ M1	271.75 *14*	2.5 *3*
γ E3	316.50 *6*	1.12 *20*
γ	322.5 *10*	0.61 *20*
γ	351.3 *4*	1.02 *20*
γ E2	372.72 *19*	2.29 *20*
γ E2	385.80 *18*	1.48 *20*
γ E2+M1	414.8 *3*	1.12 *15*
γ E2+M1	441.01 *13*	14.7 *10*
γ	443.0 *4*	~1
γ E2+M1	451.82 *16*	4.33 *25*
γ E2	475.74 *14*	8.4 *5*
γ M1	482.4 *4*	1.7 *4*
γ M1	490.1 *3*	0.87 *15*
γ E2	493.20 *17*	1.48 *15*
γ E1	501.49 *9*	9.9 *6*
γ E2	531.8 *3*	2.0 *5*
γ	572.1 *4*	0.56 *20*
γ M1	593.52 *24*	3.9 *5*

Photons (^{135}Nd) (continued)

γ_{mode}	γ(keV)	γ(%)†
γ	616.4 *3*	1.9 *3* ?
γ	670.9 *3*	0.51 *20*
γ	708.3 *5*	0.82 *15*
γ	738.8 *4*	0.82 *15* ?
γ	746.2 *3*	0.76 *25*
γ	778.1 *4*	~0.6 ?
γ	966.5 *7*	2.7 *8* ?
γ	1172.0 *7*	1.1 *4*
γ	1480.5 *6*	1.2 *4*
γ	1585.9 *7*	0.76 *25* ?
γ	1752.3 *7*	~2

† 2.0% uncert(syst)

Atomic Electrons (^{135}Nd)

⟨e⟩=85 *5* keV

e_{bin}(keV)	⟨e⟩(keV)	e(%)
6	3.2	52 *7*
7 - 34	3.2	30 *5*
35	16.6	48 *11*
36 - 39	0.096	0.258 *22*
40	4.0	10.0 *22*
41	1.2	2.8 *6*
71	21.3	30 *3*
106	6.2	5.9 *7*
107 - 143	5.1	4.35 *23*
158 - 159	0.39	0.25 *10*
162	12.7	7.8 *16*
163 - 191	1.48	0.86 *7*
197	2.0	0.99 *20*
198 - 247	2.85	1.35 *7*
248 - 281	0.58	0.22 *3*
309 - 351	0.54	0.168 *19*
366 - 415	1.48	0.37 *6*
434 - 483	1.18	0.265 *17*
484 - 532	0.158	0.032 *5*
552 - 594	0.31	0.055 *9*
610 - 629	0.026	0.0042 *20*
664 - 708	0.067	0.010 *3*
732 - 778	0.024	0.0032 *16*
925 - 966	0.06	~0.006
1130 - 1172	0.019	~0.0016
1439 - 1480	0.015	~0.0010
1544 - 1585	0.009	~0.0006
1710 - 1751	0.022	~0.0013

Continuous Radiation (^{135}Nd)

⟨β+⟩=910 keV;⟨IB⟩=7.1 keV

E_{bin}(keV)		⟨ ⟩(keV)	(%)
0 - 10	β+	7.3×10^{-7}	8.7×10^{-6}
	IB	0.031	
10 - 20	β+	3.28×10^{-5}	0.000197
	IB	0.030	0.21
20 - 40	β+	0.00102	0.00309
	IB	0.091	0.30
40 - 100	β+	0.056	0.071
	IB	0.18	0.28
100 - 300	β+	3.20	1.43
	IB	0.51	0.29
300 - 600	β+	26.7	5.7
	IB	0.67	0.156
600 - 1300	β+	225	23.2
	IB	1.56	0.17
1300 - 2500	β+	563	31.3
	IB	2.7	0.150
2500 - 4494	β+	91	3.36
	IB	1.35	0.046
	Σβ+		65

$^{135}_{60}$Nd(5.5 *5* min)

Mode: ϵ
Δ: -76210 *360* keV syst
SpA: 2.53×10^{8} Ci/g
Prod: protons on Gd

$^{135}_{61}$Pm(49 *7* s)

Mode: ϵ
SpA: 1.69×10^{9} Ci/g
Prod: ^{141}Pr(α,10n); ^{32}S on ^{106}Cd

Photons (^{135}Pm)

γ_{mode}	γ(keV)	γ(rel)
γ	129 *1*	
γ	198.7 *10*	100
γ	271 *1*	
γ	362.2 *10*	<20
γ	465 *1*	

$^{135}_{62}$Sm(10 *2* s)

Mode: ϵ, ϵp
SpA: 8.1×10^{9} Ci/g
Prod: ^{32}S on ^{106}Cd

A = 136

NDS **26**, 473 (1979)

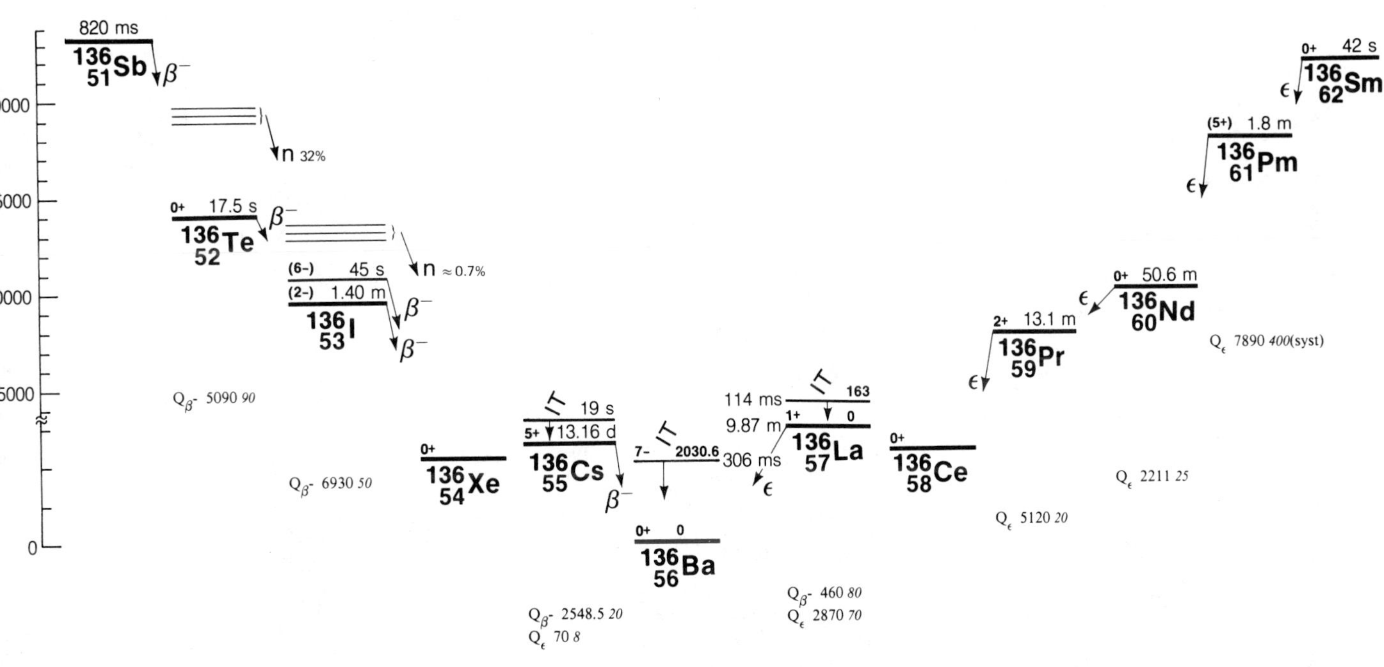

$^{136}_{51}$Sb(820 *20* ms)

Mode: β-, β-n(32 *14* %)
SpA: 6.83×10^{10} Ci/g
Prod: fission

$^{136}_{52}$Te(17.5 *4* s)

Mode: β-, β-n(0.7 *4* %)
Δ: -74410 *100* keV
SpA: 4.65×10^{9} Ci/g
Prod: fission

β-n: 250, 315, 430, 525, 595, 690, 765

Photons (^{136}Te)

⟨γ⟩=2172 *170* keV

γ$_{mode}$	γ(keV)	γ(%)†
I L$_ℓ$	3.485	0.021 *5*
I L$_η$	3.780	0.0101 *21*
I L$_α$	3.937	0.58 *12*
I L$_β$	4.320	0.52 *11*
I L$_γ$	4.887	0.064 *15*
I K$_{α2}$	28.317	3.6 *6*
I K$_{α1}$	28.612	6.7 *12*
I K$_{β1}$'	32.278	1.8 *3*
I K$_{β2}$'	33.225	0.41 *7*
γ M1	87.43 *15*	13 *3*
γ [M1+E2]	135.18 *25*	3.0 *6*
γ [M1+E2]	297.93 *23*	0.60 *12*
γ [M1+E2]	332.87 *18*	20 *4*
γ [M1+E2]	356.3 *3*	2.2 *4*
γ [M1+E2]	491.44 *17*	2.6 *5*
γ [M1+E2]	543.36 *19*	2.4 *5*
γ [M1+E2]	578.87 *16*	20 *4*
γ [M1+E2]	630.79 *16*	11.2 *22*
γ	645.0 *5*	0.60 *12*
γ	685.1 *5*	0.80 *16*
γ [M1+E2]	738.38 *19*	5.8 *12*
γ	1341.4 *5*	2.0 *4*
γ	1567.1 *5*	1.4 *3*
γ [E1]	2078.12 *23*	24 *5*
γ [E1]	2497.0 *3*	5.2 *10*
γ [E1]	2569.56 *23*	16 *3*
γ [E1]	2604.6 *3*	1.20 *24*
γ [E1]	2656.5 *3*	1.4 *3*
γ [E1]	2804.2 *4*	2.4 *5*
γ [E1]	3049.7 *4*	2.2 *4*
γ [E1]	3235.4 *3*	16 *3*

† approximate

Atomic Electrons (^{136}Te)

⟨e⟩=14.6 *15* keV

e$_{bin}$(keV)	⟨e⟩(keV)	e(%)
5	0.54	11.4 *24*
23 - 32	0.41	1.65 *17*
54	6.7	12.3 *25*
82	1.22	1.5 *3*
83 - 87	0.46	0.54 *7*
102	1.1	1.1 *3*
130 - 135	0.40	0.30 *11*
265 - 298	0.065	0.024 *4*
300	1.5	0.49 *10*
323 - 356	0.46	0.14 *3*
458 - 491	0.120	0.026 *5*
510 - 543	0.095	0.018 *4*
546	0.60	0.11 *3*
574 - 579	0.104	0.018 *4*
598	0.30	0.051 *13*
612 - 652	0.079	0.013 *3*
680 - 705	0.13	0.018 *5*
733 - 738	0.021	0.0029 *6*
1308 - 1341	0.019	0.0014 *6*
1534 - 1566	0.011	0.0007 *3*
2045 - 2077	0.091	0.0044 *8*
2464 - 2536	0.067	0.0027 *4*
2564 - 2656	0.0159	0.00061 *6*
2771 - 2803	0.0079	0.00029 *5*
3016 - 3049	0.0070	0.00023 *4*
3202 - 3234	0.050	0.0016 *3*

Continuous Radiation (^{136}Te)

⟨β-⟩=1206 keV;⟨IB⟩=3.3 keV

E$_{bin}$(keV)		⟨ ⟩(keV)	(%)
0 - 10	β-	0.0127	0.252
	IB	0.041	
10 - 20	β-	0.0387	0.258
	IB	0.040	0.28
20 - 40	β-	0.161	0.53
	IB	0.079	0.27
40 - 100	β-	1.24	1.75
	IB	0.22	0.35
100 - 300	β-	15.4	7.5
	IB	0.64	0.36
300 - 600	β-	69	15.1
	IB	0.71	0.167
600 - 1300	β-	359	38.3
	IB	0.93	0.109
1300 - 2500	β-	470	27.1
	IB	0.54	0.032
2500 - 4757	β-	291	9.3
	IB	0.115	0.0040

$^{136}_{53}$I (1.400 *17* min)

Mode: β-
Δ: -79510 *50* keV
SpA: 9.83×10^{8} Ci/g
Prod: fission

Photons (^{136}I)

⟨γ⟩=2461 *180* keV

γ$_{mode}$	γ(keV)	γ(%)†
γ [M1+E2]	219.44 *11*	0.83 *7*
γ	240.52 *20*	0.23 *5*
γ	270.36 *10*	0.21 *5*
γ	309.12 *11*	0.34 *3*
γ [M1+E2]	344.65 *8*	2.43 *20*
γ	362.50 *21*	0.128 *20*
γ E2	381.32 *5*	
γ [M1+E2]	396.19 *13*	0.43 *5*
γ [M1+E2]	431.35 *6*	0.21 *6*
γ	434.19 *7*	0.80 *6*
γ [M1+E2]	597.91 *14*	0.36 *4*
γ	682.7 *3*	0.19 *3* *
γ [E2]	812.67 *6*	0.87 *23*
γ	865.54 *8*	0.65 *5*
γ [M1+E2]	976.50 *8*	2.7 *2*
γ [M1+E2]	994.09 *13*	1.63 *9*
γ	1057.45 *21*	0.29 *5*
γ [M1+E2]	1101.71 *11*	0.49 *7*
γ [M1+E2]	1178.82 *18*	0.22 *3*
γ	1222.6 *4*	0.16 *3* *
γ	1246.86 *7*	2.29 *12*
γ E2	1313.041 *10*	68 *14*
γ [M1+E2]	1321.15 *7*	25.1 *18*
γ	1400.35 *13*	0.11 *3*
γ	1536.38 *9*	1.31 *7*
γ	1555.98 *10*	0.47 *3*
γ [E1]	1583.64 *13*	0.26 *3*
γ	1624.63 *22*	0.24 *3*
γ [E1]	1635.18 *10*	0.38 *4*
γ	1639.8 *5*	0.19 *4* *
γ	1663.1 *8*	0.07 *3* *
γ [M1+E2]	1666.04 *21*	0.18 *3*
γ	1686.1 *3*	0.31 *3* *
γ	1689.0 *3*	0.26 *3* *
γ	1709.47 *10*	0.70 *5*
γ	1738.1 *3*	0.16 *3* *
γ [E1]	1819.91 *16*	0.22 *3*
γ [E1]	1910.8 *3*	0.095 *20*
γ [E1]	1962.24 *13*	2.31 *13*
γ [E1]	1968.3 *3*	0.17 *3*
γ [E1]	1979.82 *11*	0.135 *20*
γ [E1]	2039.35 *18*	0.16 *3*
γ	2167.57 *22*	~0.08
γ	2227.9 *5*	0.11 *3* *
γ [E2]	2289.53 *8*	10.5 *5*
γ [E1]	2313.24 *24*	0.067 *14*
γ	2382.81 *21*	0.22 *3*
γ [E2]	2414.74 *11*	6.9 *3*
γ	2427.8 *3*	0.18 *3* *
γ	2480.4 *4*	0.13 *3* *
γ	2548.3 *3*	0.13 *3*
γ	2581.9 *7*	0.054 *14* *
γ	2602.24 *23*	~0.12
γ [E2]	2634.18 *7*	6.8 *3*
γ [E1]	2657.89 *24*	0.095 *14*
γ	2828.6 *3*	0.101 *13*
γ	2849.41 *9*	0.034 *14*
γ	2869.01 *10*	4.0 *3*
γ	2889.7 *7*	0.041 *13* *
γ [E1]	2956.31 *9*	0.73 *4*
γ [E2]	2979.07 *21*	0.31 *3*
γ	3132.1 *5*	0.049 *10* *
γ [E1]	3141.04 *16*	0.70 *4*
γ	3195.32 *24*	0.169 *20*
γ	3200.3 *3*	0.047 *20*
γ	3211.9 *2*	0.52 *3*
γ	3272.2 *6*	0.088 *20*
γ [M1+E2]	3348.65 *23*	0.196 *20*
γ	3482.5 *3*	0.092 *9* *
γ	3489.4 *5*	0.047 *8* *
γ	3626.3 *3*	0.169 *14* *
γ [E1]	3634.37 *23*	0.122 *13*
γ	3674.4 *3*	0.169 *14* *
γ	3774.49 *22*	0.027 *12*
γ	3794.3 *6*	0.049 *12*
γ	3925.0 *4*	0.082 *13* *
γ	3967.8 *5*	0.095 *12* *
γ	4005.0 *8*	0.054 *14* *
γ	4064.01 *20*	0.169 *20*
γ [E1]	4209.16 *22*	0.047 *11*
γ	4269.32 *9*	0.358 *20*
γ	4454.05 *16*	0.041 *10*
γ	4474.01 *22*	0.135 *14*
γ	4519.1 *6*	0.015 *7*
γ [E1]	4544.9 *3*	0.058 *11*
γ	4560.0 *4*	0.095 *12* *
γ	4614.4 *6*	0.050 *10* *
γ	4711.1 *4*	0.054 *14*
γ	4723.0 *14*	~0.013 *
γ	4729.6 *8*	0.027 *7* *
γ	4739.5 *4*	0.108 *14*
γ	4772.0 *8*	0.037 *10* *
γ	4864.4 *6*	0.024 *6* *
γ	4873.1 *3*	0.019 *8* *
γ	4889.3 *4*	0.149 *20* *
γ [E1]	4929.53 *20*	0.115 *12*
γ	5016.93 *21*	0.088 *9*

Photons (^{136}I)
(continued)

γ_{mode}	γ(keV)	γ(%)†
γ	5091.2 *6*	0.036 *8* *
γ	5128.9 *11*	0.028 *10* *
γ	5187.0 *4*	0.070 *12* *
γ	5217.7 *4*	0.027 *10*
γ	5255.4 *9*	0.039 *10*
γ	5320.98 *24*	0.074 *14*
γ	5608.1 *3*	0.15 *3*
γ	5800.0 *3*	0.13 *3*
γ	5870.7 *12*	~0.027
γ	5968.4 *10*	~0.014
γ	6012.9 *10*	~0.015
γ	6052.5 *4*	0.054 *14*
γ	6103.8 *3*	0.13 *3*
γ	6114.4 *7*	0.067 *20*
γ	6126.3 *5*	0.09 *3*
γ	6169.7 *8*	0.014 *4*
γ	6199.9 *13*	0.011 *5*
γ	6253.3 *8*	0.034 *10*
γ	6408.9 *8*	0.019 *7*

† 1.5% uncert(syst)
* with ^{136}I(1.40 min or 45 s)

Continuous Radiation (^{136}I)
$\langle\beta-\rangle$=1958 keV;$\langle$IB$\rangle$=6.9 keV

E_{bin}(keV)		$\langle\ \rangle$(keV)	(%)
0 - 10	β-	0.0076	0.151
	IB	0.056	
10 - 20	β-	0.0228	0.152
	IB	0.056	0.39
20 - 40	β-	0.092	0.307
	IB	0.110	0.38
40 - 100	β-	0.66	0.94
	IB	0.32	0.49
100 - 300	β-	6.7	3.32
	IB	0.96	0.54
300 - 600	β-	27.9	6.1
	IB	1.20	0.28
600 - 1300	β-	191	19.7
	IB	2.0	0.23
1300 - 2500	β-	730	38.7
	IB	1.65	0.095
2500 - 5000	β-	977	30.2
	IB	0.53	0.017
5000 - 6930	β-	24.6	0.459
	IB	0.0032	6.0×10^{-5}

$^{136}_{53}$I (45 *1* s)

Mode: β-
Δ: >-79510 keV
SpA: 1.83×10^{9} Ci/g
Prod: fission

Photons (^{136}I)*
$\langle\gamma\rangle$=2136 *130* keV

γ_{mode}	γ(keV)	γ(%)†
Xe L_ℓ	3.634	0.021 *3*
Xe L_η	3.955	0.0102 *12*
Xe L_α	4.104	0.59 *7*
Xe L_β	4.520	0.52 *7*
Xe L_γ	5.108	0.065 *9*
Xe $K_{\alpha2}$	29.461	3.26 *18*
Xe $K_{\alpha1}$	29.782	6.1 *3*
Xe $K_{\beta1}$'	33.606	1.64 *9*

Photons (^{136}I)*
(continued)

γ_{mode}	γ(keV)	γ(%)†
Xe $K_{\beta2}$'	34.606	0.388 *22*
γ	164.13 *8*	0.43 *9*
γ	182.64 *7*	0.74 *9*
γ E2	197.30 *4*	78 *5*
γ	318.69 *7*	0.51 *4*
γ	339.36 *12*	0.26 *5*
γ	346.76 *7*	3.01 *18*
γ [M1+E2]	370.11 *5*	17.5 *10*
γ E2	381.32 *5*	99.8 *6*
γ [M1+E2]	431.35 *6*	0.62 *16*
γ	482.81 *7*	1.75 *9*
γ	552.74 *6*	0.84 *6*
γ	716.87 *7*	0.98 *7*
γ	750.04 *6*	5.8 *4*
γ	770.71 *11*	1.28 *8*
γ [E2]	812.67 *6*	2.6 *8*
γ	914.16 *7*	3.5 *2*
γ E2	1313.041 *10*	100
γ	1385.70 *18*	0.18 *3*
γ	1592.70 *15*	0.25 *3*
γ	1796.01 *15*	0.69 *5*
γ	1938.44 *18*	0.21 *4*
γ	2135.73 *18*	0.69 *5*
γ	2166.11 *15*	~0.07
γ	2178.4 *2*	0.88 *6*
γ	2363.40 *15*	0.41 *4*
γ	3599.9 *4*	0.062 *13*
γ	3736.0 *4*	0.089 *14*
γ	4396.3 *8*	0.030 *9*

† 10% uncert(syst)
* see also ^{136}I(1.40 min)

Atomic Electrons (^{136}I)
$\langle$e$\rangle$=33.8 *11* keV

e_{bin}(keV)	$\langle$e$\rangle$(keV)	e(%)
5 - 33	0.90	12.3 *12*
130 - 159	0.22	0.15 *7*
163	17.0	10.4 *6*
164 - 183	0.037	0.021 *10*
192	3.22	1.67 *10*
193	1.11	0.58 *4*
196 - 197	1.15	0.59 *3*
284 - 319	0.19	0.06 *3*
334 - 335	0.0023	~0.0007
336	1.14	0.34 *4*
338 - 346	0.031	0.009 *4*
347	5.7	1.65 *10*
365 - 397	1.50	0.399 *19*
426 - 448	0.06	~0.013
477 - 518	0.031	0.006 *3*
547 - 553	0.0039	0.0007 *3*
682 - 717	0.12	0.017 *8*
736 - 778	0.087	0.011 *3*
807 - 813	0.0079	0.00098 *20*
880 - 914	0.054	0.006 *3*
1278 - 1313	1.18	0.092 *8*
1351 - 1386	0.0018	0.00013 *6*
1558 - 1592	0.0021	0.00013 *5*
1761 - 1795	0.0052	0.00029 *11*
1904 - 1937	0.0015	8 *3* $\times10^{-5}$
2101 - 2177	0.0104	0.00049 *11*
2329 - 2362	0.0024	0.00010 *3*
3565 - 3599	0.00026	7.4 *21* $\times10^{-6}$
3701 - 3735	0.00037	10 *3* $\times10^{-6}$
4362 - 4395	0.00011	2.6 *9* $\times10^{-6}$

Continuous Radiation (^{136}I)
$\langle\beta-\rangle$=2153 keV;$\langle$IB$\rangle$=7.8 keV

E_{bin}(keV)		$\langle\ \rangle$(keV)	(%)
0 - 10	β-	0.00284	0.056
	IB	0.060	
10 - 20	β-	0.0087	0.058
	IB	0.060	0.41
20 - 40	β-	0.0363	0.121
	IB	0.118	0.41
40 - 100	β-	0.285	0.403
	IB	0.34	0.53
100 - 300	β-	3.81	1.84
	IB	1.05	0.58
300 - 600	β-	19.7	4.28
	IB	1.33	0.31
600 - 1300	β-	162	16.5
	IB	2.2	0.25
1300 - 2500	β-	744	39.0
	IB	2.0	0.113
2500 - 5000	β-	1222	37.5
	IB	0.62	0.021
5000 - 5188	β-	0.98	0.0194
	IB	5.7×10^{-6}	1.13×10^{-7}

$^{136}_{54}$Xe(stable)

Δ: -86431 *7* keV
%: 8.9 *1*

$^{136}_{55}$Cs(13.16 *3* d)

Mode: β-
Δ: -86361 *6* keV
SpA: 7.296×10^{4} Ci/g
Prod: ^{139}La(n,α); ^{138}Ba(d,α); fission; protons on Ce

Photons (^{136}Cs)
$\langle\gamma\rangle$=2171 *20* keV

γ_{mode}	γ(keV)	γ(%)†
Ba L_ℓ	3.954	0.041 *6*
Ba L_η	4.331	0.0204 *22*
Ba L_α	4.465	1.13 *13*
Ba L_β	4.939	1.05 *13*
Ba L_γ	5.592	0.134 *18*
Ba $K_{\alpha2}$	31.817	5.03 *20*
Ba $K_{\alpha1}$	32.194	9.3 *4*
Ba $K_{\beta1}$'	36.357	2.57 *10*
Ba $K_{\beta2}$'	37.450	0.65 *3*
γ E1	66.94 *3*	12.5 *10*
γ E1	86.36 *4*	6.3 *3*
γ [E2]	109.71 *4*	0.410 *20*
γ E2	153.29 *3*	7.47 *16*
γ E3	163.98 *4*	4.62 *10* *
γ	166.62 *6*	0.63 *3*
γ E1	176.64 *4*	13.6 *3*
γ [M1+E2]	187.33 *4*	0.60 *6*
γ	233.56 *6*	0.080 *10*
γ E1	273.68 *3*	12.69 *25*
γ [M1+E2]	302.70 *14*	0.030 *10*
γ [E2]	315.41 *9*	<0.006 ?
γ	319.92 *6*	0.60 *6*
γ E2	340.62 *3*	48.6 *10*
γ [M1+E2]	490.03 *14*	0.080 *20*
γ	507.25 *6*	0.98 *5*
γ [M1+E2]	732.68 *8*	0.016 *3*
γ E2	818.56 *3*	99.760
γ E2	1048.09 *5*	79.7 *16*
γ E2	1235.42 *4*	19.8 *8*

Photons (^{136}Cs)
(continued)

γ_{mode}	γ(keV)	γ(%)†
γ	1321.78 5	<0.020 ?
γ [E2]	1538.12 14	0.100 20
γ [E2]	1551.24 8	0.016 3

† <0.1% uncert(syst)
* with ^{136}Ba(306 ms) in equilib

Atomic Electrons (^{136}Cs)
⟨e⟩=35.4 4 keV

e_{bin}(keV)	⟨e⟩(keV)	e(%)
5 - 27	1.34	19.8 14
29	2.15	7.3 6
30 - 36	0.201	0.64 4
49	0.90	1.84 10
61 - 110	1.57	2.22 8
116	2.73	2.36 7
127	6.43	5.08 15
129 - 153	2.31	1.60 7
158	3.94	2.49 7
159	2.66	1.68 5
161	0.04	~0.023
163	1.53	0.94 3
164 - 196	0.605	0.361 11
228 - 274	0.451	0.187 5
278 - 302	0.038	~0.014
303	3.68	1.21 3
309 - 341	0.906	0.270 6
453 - 502	0.041	0.009 4
506 - 507	0.0013	~0.00025
695 - 733	0.00052	7.4 18 $\times10^{-5}$
781	1.90	0.243 5
813 - 818	0.336	0.0413 7
1011	1.14	0.113 3
1042 - 1048	0.193	0.0185 4
1198 - 1235	0.278	0.0231 9
1284 - 1322	0.00026	~2 $\times10^{-5}$
1501 - 1550	0.00132	8.7 14 $\times10^{-5}$

Continuous Radiation (^{136}Cs)
⟨β-⟩=100.0 keV;⟨IB⟩=0.036 keV

E_{bin}(keV)		⟨ ⟩(keV)	(%)
0 - 10	β-	0.331	6.6
	IB	0.0052	
10 - 20	β-	0.96	6.4
	IB	0.0045	0.031
20 - 40	β-	3.65	12.2
	IB	0.0072	0.025
40 - 100	β-	21.4	31.1
	IB	0.0124	0.020
100 - 300	β-	70	42.3
	IB	0.0067	0.0051
300 - 600	β-	3.17	0.86
	IB	0.00022	6.1 $\times10^{-5}$
600 - 682	β-	0.0479	0.0077
	IB	2.8 $\times10^{-7}$	4.6 $\times10^{-8}$

$^{136}_{55}$Cs(19 2 s)

Mode: IT
Δ: >-86361 keV
SpA: 4.3$\times10^{9}$ Ci/g
Prod: protons on La

$^{136}_{56}$Ba(stable)

Δ: -88909 6 keV
%: 7.854 39

$^{136}_{56}$Ba(306 3 ms)

Mode: IT
Δ: -86878 6 keV
SpA: 1.073$\times10^{11}$ Ci/g
Prod: daughter ^{136}Cs

Photons (^{136}Ba)
⟨γ⟩=1923 11 keV

γ_{mode}	γ(keV)	γ(%)
Ba L_ℓ	3.954	0.096 13
Ba L_η	4.331	0.057 6
Ba L_α	4.465	2.7 3
Ba L_β	4.928	2.7 3
Ba L_γ	5.572	0.34 4
Ba $K_{\alpha2}$	31.817	8.8 3
Ba $K_{\alpha1}$	32.194	16.3 5
Ba $K_{\beta1}$'	36.357	4.51 15
Ba $K_{\beta2}$'	37.450	1.14 4
γ E3	163.98 4	30.7 6
γ [E2]	315.41 9	<0.006 ?
γ E2	818.56 3	99.76
γ E2	1048.09 5	99.8 10

Atomic Electrons (^{136}Ba)
⟨e⟩=106.7 13 keV

e_{bin}(keV)	⟨e⟩(keV)	e(%)
5	1.36	26 3
6	1.12	19.8 20
25	0.069	0.27 3
26	0.194	0.75 8
27	0.32	1.20 12
30	0.072	0.237 24
31	0.204	0.66 7
32	0.044	0.137 14
35	0.0220	0.063 6
36	0.0118	0.033 3
127	42.7	33.8 8
158	26.2	16.5 4
159	17.7	11.1 3
163	10.19	6.25 15
164	2.58	1.58 4
781	1.90	0.243 5
813	0.268	0.0330 7
817	0.0558	0.00682 14
818	0.01266	0.00155 3
1011	1.43	0.141 3
1042	0.185	0.0177 4
1043	0.00752	0.000721 14
1047	0.0399	0.00381 8
1048	0.00904	0.000863 17

$^{136}_{57}$La(9.87 3 min)

Mode: ϵ
Δ: -86040 70 keV
SpA: 1.400$\times10^{8}$ Ci/g
Prod: ^{133}Cs(α,n); ^{135}Ba(d,n); ^{136}Ba(d,2n); ^{136}Ba(p,n)

Photons (^{136}La)
⟨γ⟩=46 4 keV

γ_{mode}	γ(keV)	γ(%)†
Ba L_ℓ	3.954	0.095 13
Ba L_η	4.331	0.042 4
Ba L_α	4.465	2.6 3
Ba L_β	4.946	2.4 3
Ba L_γ	5.620	0.32 4
Ba $K_{\alpha2}$	31.817	14.1 4
Ba $K_{\alpha1}$	32.194	26.0 8
Ba $K_{\beta1}$'	36.357	7.19 22
Ba $K_{\beta2}$'	37.450	1.82 6
γ	541.55 10	0.0046 18
γ [M1+E2]	732.68 8	0.0112 14
γ [E2]	760.54 4	0.289 6
γ	767 1	~0.0037 ?
γ E2	818.56 3	2.3
γ	894 1	0.0046 9 ?
γ	906.59 13	<0.0012
γ	934.45 15	~0.0021
γ	981.3 10	<12 $\times10^{-5}$
γ	1262.15 10	0.0299 21
γ	1310.47 6	0.099 4
γ	1323.04 4	0.264 8
γ	1466 1	0.0028 12
γ	1496.96 9	0.0428 18
γ	1514.55 20	0.0023 9
γ [E2]	1551.24 8	0.0109 13
γ	1667.12 13	0.0110 14
γ	1713.2 10	<0.00023
γ	1791.4 3	0.0067 18
γ	1822.0 5	0.0034 14
γ	1955 1	0.0016 5
γ	2080.70 10	0.0193 11
γ	2129.02 6	0.0469 18
γ	2286 2	~0.00046 ?
γ	2332.5 10	<0.0046
γ	2485.67 13	0.0032 5

† 43% uncert(syst)

Atomic Electrons (^{136}La)
⟨e⟩=3.84 18 keV

e_{bin}(keV)	⟨e⟩(keV)	e(%)
5	1.33	25 3
6	0.96	16.7 17
25	0.109	0.43 4
26	0.31	1.19 12
27	0.51	1.92 20
30	0.114	0.38 4
31	0.32	1.05 11
32 - 36	0.124	0.371 25
504 - 542	0.00016	~3 $\times10^{-5}$
695 - 733	0.0065	0.00089 4
755 - 781	0.045	0.0057 11
813 - 857	0.0078	0.00096 16
869 - 907	5.4 $\times10^{-5}$	~6 $\times10^{-6}$
928 - 976	6.2 $\times10^{-6}$	~7 $\times10^{-7}$
980 - 981	3.0 $\times10^{-8}$	~3 $\times10^{-9}$
1225 - 1273	0.0014	0.00011 4
1286 - 1323	0.0033	0.00025 10
1429 - 1477	0.00043	2.9 12 $\times10^{-5}$
1491 - 1550	0.00019	1.24 18 $\times10^{-5}$
1630 - 1712	0.00010	6.3 25 $\times10^{-6}$
1754 - 1821	8.6 $\times10^{-5}$	4.9 18 $\times10^{-6}$
1918 - 1954	1.3 $\times10^{-5}$	7 3 $\times10^{-7}$
2043 - 2128	0.00048	2.3 6 $\times10^{-5}$
2249 - 2331	1.9 $\times10^{-5}$	~8 $\times10^{-7}$
2448 - 2485	2.0 $\times10^{-5}$	8 3 $\times10^{-7}$

Continuous Radiation (^{136}La)

⟨β+⟩=299 keV;⟨IB⟩=3.9 keV

E_{bin}(keV)		⟨ ⟩(keV)	(%)
0 - 10	β+	1.70×10^{-6}	2.03×10^{-5}
	IB	0.0122	
10 - 20	β+	7.1×10^{-5}	0.000429
	IB	0.0120	0.083
20 - 40	β+	0.00205	0.0063
	IB	0.072	0.24
40 - 100	β+	0.103	0.132
	IB	0.069	0.107
100 - 300	β+	5.2	2.35
	IB	0.20	0.109
300 - 600	β+	37.6	8.1
	IB	0.33	0.075
600 - 1300	β+	196	21.3
	IB	1.26	0.134
1300 - 2500	β+	59	4.10
	IB	1.9	0.109
2500 - 2870	IB	0.047	0.0018
	Σβ+		36

$^{136}_{57}$La(114 *6* ms)

Mode: IT
Δ: -85877 *70* keV
SpA: 1.19×10^{11} Ci/g
Prod: ^{136}Ba(p,n); ^{137}Ba(p,2n); ^{133}Cs(α,n)

Photons (^{136}La)

γ_{mode}	γ(keV)	γ(rel)
γ	66.8 *10*	~10
γ	96.2 *5*	100

$^{136}_{58}$Ce(stable)

Δ: -86500 *50* keV
%: 0.19 *1*

$^{136}_{59}$Pr(13.1 *1* min)

Mode: ε
Δ: -81380 *50* keV
SpA: 1.055×10^{8} Ci/g
Prod: daughter ^{136}Nd; ^{136}Ce(p,n); ^{136}Ce(d,2n)

Photons (^{136}Pr)

⟨γ⟩=1515 *35* keV

γ_{mode}	γ(keV)	γ(%)†
Ce L_ℓ	4.289	0.077 *10*
Ce L_η	4.730	0.032 *3*
Ce L_α	4.838	2.10 *23*
Ce L_β	5.390	1.89 *23*
Ce L_γ	6.142	0.26 *4*
Ce $K_{\alpha2}$	34.279	9.9 *3*
Ce $K_{\alpha1}$	34.720	18.2 *6*
Ce $K_{\beta1}$'	39.232	5.13 *16*

Photons (^{136}Pr)
(continued)

γ_{mode}	γ(keV)	γ(%)†
Ce $K_{\beta2}$'	40.437	1.38 *5*
γ	221.9 *10*	0.137 *15* *
γ	276.5 *16*	0.21 *3* *
γ E2	460.94 *16*	7.7 *4*
γ [E2]	523.9 *4*	0.342 *23* *
γ E2	539.86 *13*	52 *3*
γ E2	552.04 *15*	76
γ	590.4 *10*	0.129 *15* *
γ	672.79 *18*	0.236 *23* *
γ E2	761.45 *21*	1.5 *3*
γ [E2]	841.41 *21*	0.076 *15* *
γ	855.98 *18*	0.144 *15* *
γ	974.89 *25*	0.34 *5* *
γ	990.9 *4*	0.167 *23* *
γ E2	1000.80 *18*	5.02 *23*
γ	1012.22 *25*	0.220 *23* *
γ	1032.0 *4*	0.106 *23* *
γ [M1+E2]	1042.3 *3*	0.160 *15* *
γ [M1+E2]	1063.00 *18*	0.205 *23*
γ E2	1091.90 *17*	18.5 *9*
γ	1203.6 *3*	0.205 *23*
γ [E2]	1281.6 *4*	0.129 *15*
γ	1359.1 *3*	0.99 *8*
γ	1368.3 *4*	0.175 *23*
γ	1425.17 *24*	0.94 *11*
γ	1489.0 *22*	0.114 *15* *
γ [M1+E2]	1503.2 *3*	0.25 *3*
γ	1514.75 *23*	1.92 *21*
γ	1547.1 *22*	0.099 *15*
γ	1589.8 *4*	<0.15
γ [M1+E2]	1602.86 *17*	3.9 *3*
γ [E2]	1628.4 *5*	0.106 *15*
γ	1632.8 *19*	0.152 *15* *
γ	1639 *3*	0.129 *15* *
γ	1646.8 *25*	0.091 *8* *
γ	1677.6 *5*	0.144 *15*
γ	1735.78 *23*	0.43 *5*
γ	1748.7 *13*	0.160 *15*
γ	1773.8 *3*	0.243 *23*
γ	1789.8 *4*	0.091 *8*
γ	1812.1 *4*	<0.11 ?
γ [E2]	1887.1 *5*	0.129 *15*
γ	1899.0 *3*	0.94 *21*
γ	1918.98 *23*	0.106 *15*
γ	1965.03 *24*	0.160 *15*
γ	1971 *3*	0.076 *8* ?
γ	2021.1 *25*	0.106 *8*
γ [M1+E2]	2043.1 *3*	0.74 *5*
γ	2058.8 *19*	0.182 *15*
γ	2066.79 *23*	2.99 *16*
γ	2082.7 *4*	0.182 *23*
γ [M1+E2]	2108.7 *5*	0.129 *15*
γ	2129.7 *4*	0.213 *23*
γ	2141.2 *4*	0.182 *23*
γ [E2]	2154.90 *19*	0.34 *4*
γ [M1+E2]	2171.0 *5*	0.213 *23*
γ	2188.9 *5*	0.213 *15*
γ	2204.9 *5*	0.076 *8*
γ	2216.2 *22*	0.182 *23*
γ	2240.7 *3*	0.68 *5*
γ	2270.1 *3*	0.51 *4*
γ	2275.64 *22*	0.24 *5*
γ	2291.6 *22*	0.160 *15*
γ	2313.7 *3*	0.63 *5*
γ	2351.9 *4*	0.30 *3*
γ	2368.8 *16*	0.144 *15*
γ	2379.7 *4*	0.29 *3*
γ [M1+E2]	2389.8 *5*	0.106 *8*
γ	2439.0 *5*	0.129 *8*
γ	2450.99 *25*	0.71 *8*
γ	2458.83 *23*	0.129 *15*
γ [M1+E2]	2470.3 *3*	0.144 *15*
γ	2559 *3*	0.152 *15*
γ [E2]	2595.1 *3*	0.160 *15*
γ	2613.5 *6*	0.106 *8*
γ	2622.5 *4*	0.129 *8*
γ [M1+E2]	2648.6 *5*	0.053 *8*
γ	2681.1 *4*	0.129 *15* ?
γ	2728.8 *4*	0.129 *15*
γ	2792.7 *4*	0.106 *8*
γ	2810.0 *4*	0.182 *23*
γ	2842.1 *19*	0.099 *8*
γ	2931.8 *4*	0.084 *8*
γ [E2]	2941.8 *5*	0.053 *8*

Photons (^{136}Pr)
(continued)

γ_{mode}	γ(keV)	γ(%)†
γ	2982 *3*	0.076 *8*
γ	3027.3 *7*	<0.05 ?
γ	3037 *3*	<0.05 ?
γ	3153.3 *6*	0.053 *8*
γ	3174.5 *4*	0.061 *8*
γ [E2]	3200.6 *5*	0.076 *8*
γ [E2]	3262.8 *5*	0.45 *5*
γ	3280.8 *5*	0.061 *8*
γ	3362.0 *4*	0.061 *8*
γ [M1+E2]	3471.0 *3*	0.114 *8*
γ	3579.3 *7*	0.076 *8*
γ	3709 *3*	0.076 *8*

† 6.6% uncert(syst)
* with ^{136}Pr or ^{136}Nd

Atomic Electrons (^{136}Pr)

⟨e⟩= 9.6 *3* keV

e_{bin}(keV)	⟨e⟩(keV)	e(%)
6	1.50	26 *3*
7 - 28	0.49	4.1 *3*
29	0.34	1.17 *12*
32 - 39	0.384	1.14 *7*
181 - 222	0.023	~0.012
236 - 277	0.024	~0.010
420	0.363	0.086 *5*
454 - 483	0.095	0.0206 *6*
499	1.94	0.387 *24*
512	2.71	0.53 *4*
517 - 540	0.409	0.077 *3*
545	0.337	0.062 *4*
546 - 590	0.232	0.0423 *19*
632 - 673	0.007	~0.0011
721 - 761	0.040	0.0055 *8*
801 - 850	0.0050	0.00061 *20*
855 - 856	0.00010	$\sim1\times10^{-5}$
934 - 975	0.096	0.0100 *6*
984 - 1032	0.0247	0.00247 *16*
1036 - 1042	0.00053	5.1 *10* $\times10^{-5}$
1051	0.281	0.0267 *14*
1056 - 1092	0.0493	0.00454 *17*
1163 - 1203	0.0031	0.00026 *12*
1241 - 1282	0.00195	0.000157 *17*
1319 - 1368	0.015	0.0011 *4*
1385 - 1425	0.012	0.0008 *4*
1449 - 1546	0.028	0.0019 *7*
1549 - 1646	0.063	0.0040 *5*
1671 - 1768	0.0092	0.00054 *12*
1772 - 1859	0.009	0.00051 *20*
1879 - 1970	0.0044	0.00023 *4*
1981 - 2077	0.038	0.0019 *5*
2081 - 2176	0.0122	0.00057 *6*
2182 - 2274	0.017	0.00075 *13*
2285 - 2384	0.0072	0.00031 *5*
2388 - 2469	0.0083	0.00034 *7*
2518 - 2617	0.0044	0.00017 *2*
2621 - 2688	0.0025	9.6 *23* $\times10^{-5}$
2722 - 2809	0.0025	9.1 *17* $\times10^{-5}$
2836 - 2931	0.00095	3.3 *6* $\times10^{-5}$
2935 - 3031	0.00084	2.8 *8* $\times10^{-5}$
3036 - 3134	0.00058	1.9 *4* $\times10^{-5}$
3148 - 3240	0.0036	0.000113 *10*
3256 - 3355	0.00080	2.4 *3* $\times10^{-5}$
3356 - 3431	0.00069	2.02 *18* $\times10^{-5}$
3465 - 3539	0.00041	1.2 *3* $\times10^{-5}$
3573 - 3669	0.00048	1.3 *3* $\times10^{-5}$
3702 - 3708	5.3×10^{-5}	1.4 *4* $\times10^{-6}$

Continuous Radiation (^{136}Pr)

$\langle\beta+\rangle$=740 keV;$\langle$IB$\rangle$=6.3 keV

E_{bin}(keV)		$\langle\ \rangle$(keV)	(%)
0 - 10	β+	8.7×10^{-7}	1.05×10^{-5}
	IB	0.026	
10 - 20	β+	3.86×10^{-5}	0.000232
	IB	0.025	0.18
20 - 40	β+	0.00116	0.00354
	IB	0.088	0.29
40 - 100	β+	0.062	0.079
	IB	0.147	0.23
100 - 300	β+	3.41	1.53
	IB	0.43	0.24
300 - 600	β+	27.4	5.9
	IB	0.58	0.134
600 - 1300	β+	213	22.1
	IB	1.47	0.160
1300 - 2500	β+	454	25.6
	IB	2.5	0.139
2500 - 4568	β+	42.2	1.57
	IB	1.01	0.035
	Σβ+		57

$^{136}_{60}$Nd(50.6 *3* min)

Mode: ε

Δ: -79170 *60* keV

SpA: 2.729×10^{7} Ci/g

Prod: protons on Gd; ^{136}Ce(^{3}He,3n)

Photons (^{136}Nd) *

$\langle\gamma\rangle$=240 *9* keV

γ_{mode}	γ(keV)	γ(%)†
Pr L_ℓ	4.453	0.30 *5*
Pr L_η	4.929	0.132 *16*
Pr L_α	5.031	8.2 *11*
Pr L_β	5.602	9.6 *15*
Pr L_γ	6.471	1.6 *3*
Pr $K_{\alpha2}$	35.550	30.4 *22*
Pr $K_{\alpha1}$	36.026	55 *4*
γ M1	40.1 *5*	21.0 *17*
Pr $K_{\beta1}'$	40.720	15.7 *11*
Pr $K_{\beta2}'$	41.981	4.4 *3*
γ M1	100.6 *7*	1.33 *11*
γ M1+2.8%E2	109.2 *6*	35 *7*
γ	131.1 *10*	0.68 *7*
γ [M1+E2]	144.6 *5*	2.14 *10*
γ M1(+E2)	149.3 *6*	9.1 *5*
γ [M1+E2]	184.6 *5*	0.53 *4*
γ	204.1 *10*	0.20 *4*
γ	218.8 *7*	0.0030 *6*
γ [M1+E2]	291.9 *6*	~0.0035
γ M1,E2	294.5 *8*	1.02 *5*
γ [M1+E2]	335.9 *7*	0.32 *5*
γ M1,E2	390.3 *6*	1.01 *5*
γ	425.0 *10*	0.32 *8*
γ [M1+E2]	436.5 *6*	0.32 *5*
γ [M1+E2]	476.5 *6*	1.66 *17*
γ	488.0 *6*	0.25 *5*
γ	523.3 *7*	0.29 *5*
γ	525.9 *10*	0.29 *5*
γ M1(+E2)	534.9 *6*	
γ M1,E2	574.9 *6*	12.8 *7*
γ	605.8 *8*	0.55 *7*
γ	632.6 *6*	0.23 *5*
γ	644.4 *7*	0.57 *10*
γ	653.1 *8*	0.27 *5* ?
γ	672.6 *6*	0.46 *6*
γ [M1+E2]	755.8 *6*	0.46 *6*
γ [M1+E2]	900.3 *6*	0.36 *6*
γ [M1+E2]	940.4 *6*	0.52 *6*
γ [M1+E2]	972.4 *10*	1.16 *14*

† 8.6% uncert(syst)

* see also ^{136}Pr

Atomic Electrons (^{136}Nd)

$\langle$e$\rangle$=67 *5* keV

e_{bin}(keV)	$\langle$e$\rangle$(keV)	e(%)
6	5.0	81 *10*
7	1.96	29 *4*
28 - 30	1.84	6.3 *5*
33	14.3	43 *4*
34 - 36	2.51	7.4 *6*
39	3.9	10.2 *9*
40 - 59	2.04	4.4 *3*
67	20.7	31 *6*
89 - 101	0.44	0.48 *15*
102	4.0	3.9 *8*
103	1.63	1.59 *22*
107	3.5	3.3 *3*
108 - 149	3.5	2.8 *5*
162 - 204	0.09	0.052 *15*
250 - 295	0.193	0.073 *9*
329 - 348	0.089	0.026 *5*
383 - 431	0.063	0.016 *4*
435 - 484	0.15	0.035 *7*
486 - 533	0.59	0.11 *3*
564 - 611	0.16	0.028 *5*
626 - 673	0.021	0.0033 *13*
714 - 756	0.017	0.0023 *5*
858 - 900	0.023	0.0026 *5*
930 - 972	0.033	0.0036 *7*

Continuous Radiation (^{136}Nd)

$\langle\beta+\rangle$=24.4 keV;$\langle$IB$\rangle$=1.68 keV

E_{bin}(keV)		$\langle\ \rangle$(keV)	(%)
0 - 10	β+	7.4×10^{-7}	8.8×10^{-6}
	IB	0.00153	
10 - 20	β+	3.31×10^{-5}	0.000199
	IB	0.00134	0.0091
20 - 40	β+	0.00100	0.00305
	IB	0.088	0.25
40 - 100	β+	0.052	0.066
	IB	0.023	0.045
100 - 300	β+	2.35	1.08
	IB	0.052	0.026
300 - 600	β+	11.7	2.61
	IB	0.21	0.047
600 - 1300	β+	10.3	1.42
	IB	0.93	0.102
1300 - 2070	IB	0.37	0.025
	Σβ+		5.2

$^{136}_{61}$Pm(1.78 *10* min)

Mode: ε

Δ: -71280 *400* keV syst

SpA: 7.7×10^{8} Ci/g

Prod: ^{121}Sb(^{20}Ne,5n)

Photons (^{136}Pm)

$\langle\gamma\rangle$=1803 *21* keV

γ_{mode}	γ(keV)	γ(%)†
Nd L_ℓ	4.633	0.031 *3*
Nd L_η	5.146	0.0122 *10*
Nd L_α	5.228	0.83 *6*
Nd L_β	5.855	0.75 *7*
Nd L_γ	6.688	0.106 *11*
Nd $K_{\alpha2}$	36.847	3.54 *13*
Nd $K_{\alpha1}$	37.361	6.43 *23*
Nd $K_{\beta1}'$	42.240	1.83 *7*
Nd $K_{\beta2}'$	43.562	0.535 *21*
γ [M1+E2]	303.1 *5*	13.7 *3*
γ [M1+E2]	370.0 *3*	10.17 *20*
γ [E2]	373.9 *3*	90
γ [M1+E2]	488.3 *3*	9.18 *18*
γ [E2]	603.0 *3*	49.9 *10*
γ [E2]	678.3 *8*	6.93 *14*
γ [M1+E2]	693.3 *10*	2.61 *9*
γ [E2]	696.3 *10*	10.08 *20*
γ [E2]	770.7 *3*	18.0 *4*
γ [E2]	815.2 *3*	31.0 *6*
γ [M1+E2]	858.3 *3*	31.5 *6*
γ [E2]	862.1 *4*	7.74 *18*
γ	1060.0 *5*	13.8 *3*
γ [M1+E2]	1070.6 *4*	2.61 *9*

† 5.4% uncert(syst)

Atomic Electrons (^{136}Pm)

$\langle$e$\rangle$=18.9 *7* keV

e_{bin}(keV)	$\langle$e$\rangle$(keV)	e(%)
6 - 42	0.99	11.0 *8*
260	1.7	0.67 *13*
296 - 303	0.42	0.14 *4*
326	0.95	0.29 *6*
330	6.5	1.96 *13*
363 - 364	0.16	0.045 *13*
367	1.13	0.308 *21*
368 - 374	0.63	0.171 *8*
445	0.56	0.13 *3*
481 - 488	0.113	0.023 *5*
559	1.70	0.303 *14*
596 - 635	0.558	0.0915 *24*
650 - 696	0.50	0.075 *4*
727 - 771	0.53	0.072 *3*
772	0.71	0.093 *4*
808 - 814	0.131	0.0161 *5*
815	0.90	0.11 *3*
819 - 862	0.36	0.043 *4*
1016 - 1064	0.34	0.033 *13*
1069 - 1070	0.0020	0.00019 *4*

Continuous Radiation (^{136}Pm)

$\langle\beta+\rangle$=2063 keV;$\langle$IB$\rangle$=13.4 keV

E_{bin}(keV)		$\langle\ \rangle$(keV)	(%)
0 - 10	β+	2.37×10^{-7}	2.82×10^{-6}
	IB	0.057	
10 - 20	β+	1.10×10^{-5}	6.6×10^{-5}
	IB	0.056	0.39
20 - 40	β+	0.000348	0.00106
	IB	0.120	0.41
40 - 100	β+	0.0200	0.0254
	IB	0.33	0.51
100 - 300	β+	1.23	0.55
	IB	1.00	0.56
300 - 600	β+	11.5	2.44
	IB	1.31	0.31
600 - 1300	β+	128	12.9
	IB	2.5	0.28
1300 - 2500	β+	678	35.3
	IB	3.5	0.19
2500 - 5000	β+	1228	37.4
	IB	4.2	0.125
5000 - 6914	β+	17.0	0.324
	IB	0.26	0.0048
	$\Sigma\beta$+		89

$^{136}_{62}$Sm(42 *4* s)

Mode: ϵ
SpA: 1.96×10^{9} Ci/g
Prod: protons on W; protons on Ta

Photons (^{136}Sm)

γ_{mode}	γ(keV)	γ(rel)
γ	114.5 *10*	100 *20*
γ	123.6 *10*	8.0 *16*
γ	154.8 *10*	9.0 *18*
γ	185.2 *10*	9.0 *18*
γ	234.5 *10*	5 *1*
γ	256 *1*	13 *3*
γ	287 *1*	13 *3*
γ	291.7 *10*	13 *3* ?
γ	313.3 *10*	11.0 *22*
γ	404.3 *10*	15 *3*
γ	411.5 *10*	9.0 *18*
γ	434.3 *10*	16 *3*
γ	666 *1*	8.0 *16*
γ	747.4 *10*	12.0 *24*
γ	762.9 *10*	9.0 *18*
γ	874.5 *10*	16 *3*
γ	975 *1*	8.0 *16*
γ	1010 *1*	1.0 *2*

A = 137

NDS **38**, 87 (1983)

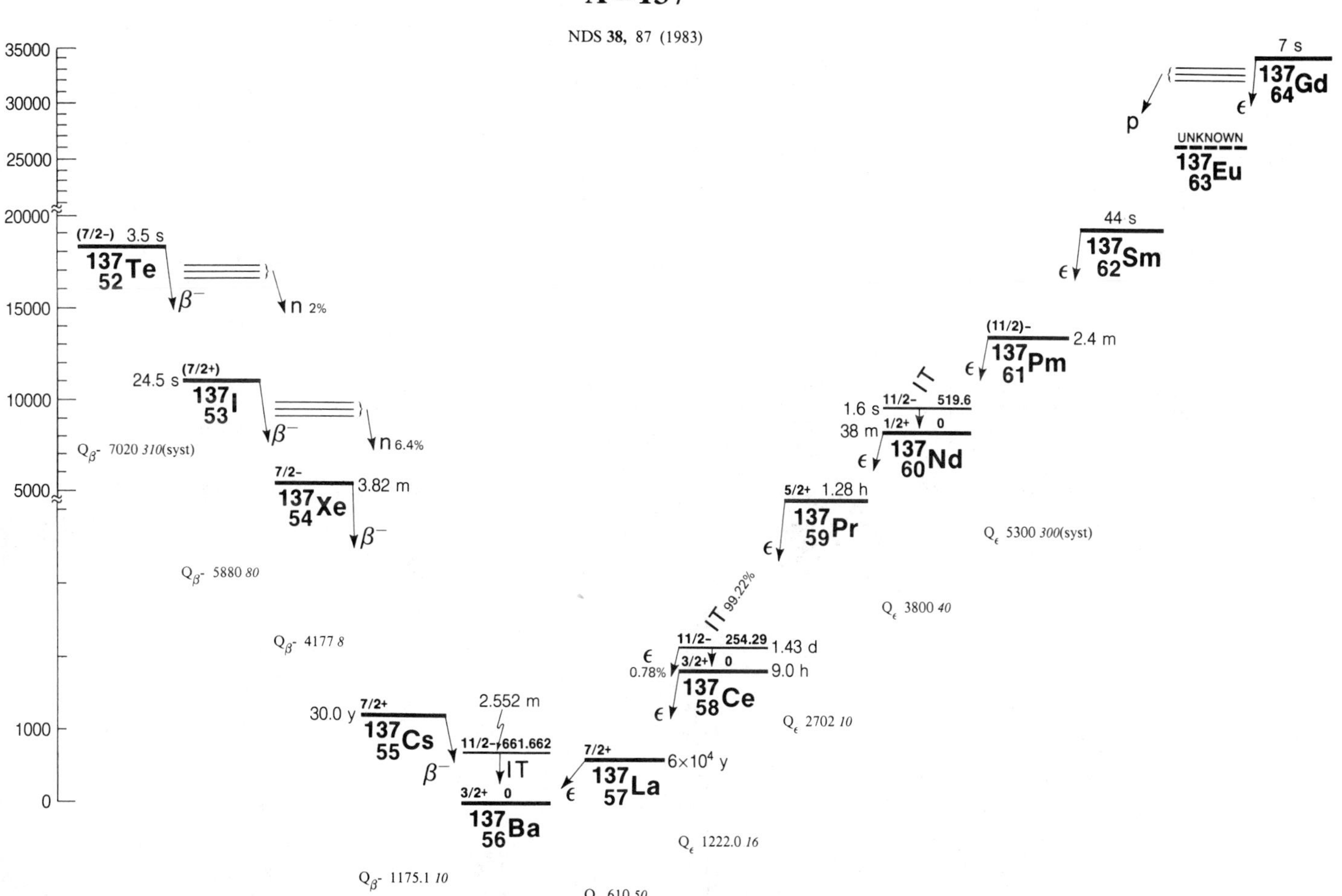

$^{137}_{52}$Te(3.5 *5* s)

Mode: β-, β-n(2.0 *7* %)
Δ: -69480 *300* keV syst
SpA: 2.1×10^{10} Ci/g
Prod: fission

Photons (^{137}Te)

γ_{mode}	γ(keV)
$\gamma_{\beta-}$[M1+E2]	243.6 *3*

$^{137}_{53}$I (24.5 *2* s)

Mode: β-, β-n(6.4 *4* %)
Δ: -76500 *80* keV
SpA: 3.31×10^{9} Ci/g
Prod: fission

β-n(avg): 540 *40*

Delayed Neutrons (^{137}I)

n(keV)	n(%)†
27 *10*	0.31 *6*
76 *10*	0.13 *6*
149 *12*	~0.030
166 *12*	~0.04
268 *10*	0.43 *5*
318 *10*	~0.08
380 *10*	1.09 *8*
413 *10*	0.47 *6*
478 *10*	0.56 *9*
503 *10*	0.70 *11*
523 *10*	0.37 *9*
566 *10*	0.59 *10*
590 *10*	0.40 *8*
607 *10*	0.24 *6*
749 *10*	0.22 *5*
855 *10*	0.31 *6*
958 *11*	0.14 *5*
993 *12*	0.11 *5*
1059 *11*	~0.08
1219 *11*	0.11 *3*

† 6% uncert(syst)

Photons (^{137}I)

⟨γ⟩=1138 *41* keV

γ_{mode}	γ(keV)	γ(%)
$\gamma_{\beta-}$	252.85 *5*	0.14 *3*
$\gamma_{\beta-}$	283.84 *7*	0.073 *15*
$\gamma_{\beta-}$[E2]	316.41 *8*	0.087 *17*
$\gamma_{\beta-}$	377.8 *3*	0.032 *6*
$\gamma_{\beta-}$[M1+E2]	385.13 *6*	0.46 *9*
$\gamma_{\beta-}$	394.47 *5*	0.46 *9*
$\gamma_{\beta-}$	408.29 *20*	0.052 *10*
$\gamma_{\beta-}$	412.93 *9*	0.098 *20*
$\gamma_{\beta-}$	431.80 *15*	0.052 *10*
$\gamma_{\beta-}$	435.27 *6*	0.34 *7*
$\gamma_{\beta-}$	442.4 *3*	0.027 *5*
$\gamma_{\beta-}$	463.50 *6*	0.25 *5*
$\gamma_{\beta-}$	477.93 *6*	0.39 *8*

Photons (^{137}I) (continued)

γ_{mode}	γ(keV)	γ(%)
$\gamma_{\beta-}$	527.9 *3*	0.075 *15*
$\gamma_{\beta-}$	532.48 *9*	0.18 *4*
$\gamma_{\beta-}$[M1+E2]	538.83 *20*	0.080 *16*
$\gamma_{\beta-}$	546.96 *6*	0.24 *5*
$\gamma_{\beta-}$	565.71 *7*	0.110 *22*
$\gamma_{\beta-}$	570.49 *5*	0.24 *5*
$\gamma_{\beta-}$	576.00 *6*	0.53 *11*
$\gamma_{\beta-}$	578.04 *7*	0.40 *8*
$\gamma_{\beta-}$[E2]	601.05 *4*	4.8 *10*
$\gamma_{\beta-}$	614.05 *16*	0.036 *7*
$\gamma_{\beta-}$	620.7 *3*	0.032 *6*
$\gamma_{\beta-}$[M1+E2]	633.42 *5*	0.27 *5*
$\gamma_{\beta-}$	647.9 *4*	0.057 *11*
$\gamma_{\beta-}$	655.11 *7*	0.14 *3*
$\gamma_{\beta-}$	659.19 *10*	0.40 *8*
$\gamma_{\beta-}$	673.34 *20*	0.043 *9*
$\gamma_{\beta-}$	678.05 *20*	0.05 *1*
$\gamma_{\beta-}$	681.97 *7*	0.20 *4*
$\gamma_{\beta-}$	694.47 *7*	0.16 *3*
$\gamma_{\beta-}$[M1+E2]	701.54 *5*	0.43 *9*
$\gamma_{\beta-}$	709.67 *6*	0.28 *6*
$\gamma_{\beta-}$	725.46 *11*	0.110 *22*
$\gamma_{\beta-}$	727.23 *6*	0.070 *14*
$\gamma_{\beta-}$	746.2 *3*	0.066 *13*
$\gamma_{\beta-}$	773.20 *5*	0.94 *19*
$\gamma_{\beta-}$	777.9 *4*	0.055 *11*
$\gamma_{\beta-}$	786.1 *5*	0.048 *10*
$\gamma_{\beta-}$	796.07 *15*	0.10 *2*
$\gamma_{\beta-}$	811.85 *5*	0.23 *5*
$\gamma_{\beta-}$	830.68 *20*	0.057 *11*
$\gamma_{\beta-}$	833.43 *10*	0.043 *9*
$\gamma_{\beta-}$	852.47 *12*	0.052 *10*
$\gamma_{\beta-}$	863.25 *20*	0.089 *18*
$\gamma_{\beta-}$	867.80 *15*	0.120 *24*
$\gamma_{\beta-}$	869.84 *15*	0.14 *3*
$\gamma_{\beta-}$	877.15 *11*	0.073 *15*
$\gamma_{\beta-}$	882.09 *6*	0.31 *6*
$\gamma_{\beta-}$	888.42 *10*	0.10 *2*
$\gamma_{\beta-}$[M1+E2]	893.40 *7*	0.34 *7*
$\gamma_{\beta-}$	909.31 *15*	0.110 *22*
$\gamma_{\beta-}$	927.01 *17*	0.17 *3*
$\gamma_{\beta-}$	937.09 *12*	0.18 *4*
$\gamma_{\beta-}$	941.43 *5*	0.67 *13*
$\gamma_{\beta-}$	950.86 *17*	0.082 *16*
$\gamma_{\beta-}$	973.04 *11*	0.120 *24*
$\gamma_{\beta-}$	978.16 *20*	0.087 *17*
$\gamma_{\beta-}$	1019.30 *15*	0.073 *15*
$\gamma_{\beta-}$	1067.10 *9*	0.30 *6*
$\gamma_{\beta-}$	1099.95 *18*	0.046 *9*
$\gamma_{\beta-}$	1103.53 *17*	0.057 *11*
$\gamma_{\beta-}$	1123.9 *5*	0.032 *6*
$\gamma_{\beta-}$	1127.63 *8*	0.17 *3*
$\gamma_{\beta-}$	1136.0 *3*	0.043 *9*
$\gamma_{\beta-}$	1150.32 *9*	0.20 *4*
$\gamma_{\beta-}$	1158.35 *13*	0.094 *19*
$\gamma_{\beta-}$	1165.03 *6*	0.096 *19*
$\gamma_{\beta-}$	1205.1 *6*	0.17 *3* ?
$\gamma_{\beta-}$	1217.96 *4*	13 *3*
$\gamma_{\beta-}$[M1+E2]	1220.01 *6*	3.5 *7*
$\gamma_{\beta-}$	1223.99 *20*	0.24 *5*
$\gamma_{\beta-}$	1236.60 *17*	0.066 *13*
$\gamma_{\beta-}$	1242.9 *7*	0.018 *4*
$\gamma_{\beta-}$	1248.49 *5*	0.51 *10*
$\gamma_{\beta-}$	1256.82 *12*	0.089 *18*
$\gamma_{\beta-}$	1268.5 *4*	0.039 *8*
$\gamma_{\beta-}$	1275.3 *3*	0.048 *10*
$\gamma_{\beta-}$[M1+E2]	1302.58 *4*	4.4 *9*
$\gamma_{\beta-}$	1330.4 *3*	0.062 *12*
$\gamma_{\beta-}$[M1+E2]	1334.96 *5*	0.42 *8*
$\gamma_{\beta-}$	1346.7 *5*	0.034 *7*
$\gamma_{\beta-}$	1351.0 *1*	0.14 *3*
$\gamma_{\beta-}$	1357.27 *9*	0.18 *4*
$\gamma_{\beta-}$	1366.62 *12*	0.13 *3*
$\gamma_{\beta-}$	1396.06 *8*	0.15 *3*
$\gamma_{\beta-}$	1400.8 *3*	0.064 *13*
$\gamma_{\beta-}$	1409.73 *15*	0.15 *3*
$\gamma_{\beta-}$	1439.0 *7*	0.048 *10*
$\gamma_{\beta-}$	1446.1 *3*	0.05 *1*
$\gamma_{\beta-}$	1456.37 *10*	0.21 *4*
$\gamma_{\beta-}$	1488.65 *17*	0.16 *3*
$\gamma_{\beta-}$	1497.17 *20*	0.110 *22*
$\gamma_{\beta-}$	1506.8 *4*	0.078 *16*

Photons (^{137}I) (continued)

γ_{mode}	γ(keV)	γ(%)
$\gamma_{\beta-}$	1512.16 *6*	1.24 *25*
$\gamma_{\beta-}$[M1+E2]	1534.34 *5*	3.2 *7*
$\gamma_{\beta-}$	1543.47 *14*	0.10 *2*
$\gamma_{\beta-}$	1554.00 *12*	0.21 *4*
$\gamma_{\beta-}$	1580.2 *4*	0.05 *1*
$\gamma_{\beta-}$	1600.0 *3*	0.046 *9*
$\gamma_{\beta-}$	1617.0 *4*	0.032 *6*
$\gamma_{\beta-}$	1628.33 *15*	0.10 *2*
$\gamma_{\beta-}$	1680.52 *15*	0.120 *24*
$\gamma_{\beta-}$	1696.76 *20*	0.10 *2*
$\gamma_{\beta-}$	1704.7 *4*	0.055 *11*
$\gamma_{\beta-}$	1715.51 *8*	0.42 *8*
$\gamma_{\beta-}$	1720.28 *10*	0.20 *4*
$\gamma_{\beta-}$	1755.22 *11*	0.23 *5*
$\gamma_{\beta-}$	1766.08 *6*	1.24 *25*
$\gamma_{\beta-}$	1773.82 *20*	0.066 *13*
$\gamma_{\beta-}$	1788.05 *11*	0.120 *24*
$\gamma_{\beta-}$	1804.90 *9*	0.32 *6*
$\gamma_{\beta-}$	1808.74 *6*	0.79 *16*
$\gamma_{\beta-}$	1820.54 *10*	0.23 *5*
$\gamma_{\beta-}$	1832.40 *18*	0.10 *2*
$\gamma_{\beta-}$	1835.0 *4*	0.057 *11*
$\gamma_{\beta-}$	1840.49 *20*	0.089 *18*
$\gamma_{\beta-}$[E2]	1841.41 *20*	
$\gamma_{\beta-}$	1849.53 *5*	0.20 *4*
$\gamma_{\beta-}$	1853.3 *5*	0.055 *11*
$\gamma_{\beta-}$	1856.4 *7*	0.05 *1*
$\gamma_{\beta-}$	1859.40 *12*	0.10 *2*
$\gamma_{\beta-}$	1873.07 *6*	1.5 *3*
$\gamma_{\beta-}$	1879.19 *11*	0.092 *18*
$\gamma_{\beta-}$	1898.3 *3*	0.078 *16*
$\gamma_{\beta-}$	1922.33 *12*	0.094 *19*
$\gamma_{\beta-}$	1926.4 *3*	0.15 *3*
$\gamma_{\beta-}$	1930.3 *7*	0.069 *14*
$\gamma_{\beta-}$[M1+E2]	1936.00 *5*	0.37 *7*
$\gamma_{\beta-}$	1954.29 *20*	0.075 *15*
$\gamma_{\beta-}$	1974.10 *9*	0.15 *3*
$\gamma_{\beta-}$	1991.15 *6*	0.23 *5*
$\gamma_{\beta-}$	1997.04 *7*	0.93 *19*
$\gamma_{\beta-}$	1999.19 *13*	0.10 *2*
$\gamma_{\beta-}$	2005.08 *12*	0.066 *13*
$\gamma_{\beta-}$	2013.0 *3*	0.120 *24*
$\gamma_{\beta-}$	2029.81 *5*	1.8 *4*
$\gamma_{\beta-}$	2036.0 *3*	0.120 *24*
$\gamma_{\beta-}$	2058.72 *10*	0.25 *5*
$\gamma_{\beta-}$	2069.6 *2*	0.071 *14*
$\gamma_{\beta-}$	2079.5 *3*	0.048 *10*
$\gamma_{\beta-}$	2092.74 *11*	0.21 *4*
$\gamma_{\beta-}$	2100.04 *6*	0.44 *9*
$\gamma_{\beta-}$	2110.8 *7*	0.055 *11*
$\gamma_{\beta-}$	2114.0 *4*	0.110 *22*
$\gamma_{\beta-}$	2123.16 *11*	0.28 *6*
$\gamma_{\beta-}$	2133.5 *5*	0.037 *7*
$\gamma_{\beta-}$	2139.0 *6*	0.031 *6*
$\gamma_{\beta-}$	2147.00 *17*	0.10 *2*
$\gamma_{\beta-}$	2155.67 *11*	0.14 *3*
$\gamma_{\beta-}$	2190.99 *10*	0.16 *3*
$\gamma_{\beta-}$	2220.81 *10*	0.23 *5*
$\gamma_{\beta-}$	2229.95 *15*	0.10 *2*
$\gamma_{\beta-}$	2237.61 *11*	0.05 *1*
$\gamma_{\beta-}$	2243.46 *10*	0.26 *5*
$\gamma_{\beta-}$	2288.4 *5*	0.030 *6*
$\gamma_{\beta-}$	2300.2 *4*	0.046 *9*
$\gamma_{\beta-}$	2312.7 *3*	0.048 *10*
$\gamma_{\beta-}$	2320.5 *4*	0.064 *13*
$\gamma_{\beta-}$	2330.0 *6*	0.057 *11*
$\gamma_{\beta-}$	2332.9 *4*	0.078 *16*
$\gamma_{\beta-}$	2345.59 *9*	0.14 *3*
$\gamma_{\beta-}$	2351.60 *15*	0.16 *3*
$\gamma_{\beta-}$	2356.26 *11*	0.19 *4*
$\gamma_{\beta-}$	2359.9 *6*	0.034 *7*
$\gamma_{\beta-}$	2368.27 *10*	0.085 *17*
$\gamma_{\beta-}$	2380.28 *20*	0.092 *18*
$\gamma_{\beta-}$	2405.7 *6*	0.027 *5*
$\gamma_{\beta-}$	2422.68 *10*	0.21 *4*
$\gamma_{\beta-}$	2452.55 *8*	0.26 *5*
$\gamma_{\beta-}$	2463.5 *4*	0.034 *7*
$\gamma_{\beta-}$	2474.77 *12*	0.10 *2*
$\gamma_{\beta-}$	2487.2 *6*	0.046 *9*
$\gamma_{\beta-}$	2491.33 *8*	0.17 *3*
$\gamma_{\beta-}$	2498.7 *5*	0.027 *5*
$\gamma_{\beta-}$	2508.6 *4*	0.052 *10*

Photons (^{137}I) (continued)

γ_{mode}	γ(keV)	γ(%)
$\gamma_{\beta-}$	2532.4 3	0.073 15
$\gamma_{\beta-}$	2562.8 6	0.064 13
$\gamma_{\beta-}$	2566.9 6	0.092 18
$\gamma_{\beta-}$	2589.8 11	0.020 4
$\gamma_{\beta-}$	2594.8 3	0.078 16
$\gamma_{\beta-}$	2629.61 7	0.78 16
$\gamma_{\beta-}$	2657.0 3	0.048 10
$\gamma_{\beta-}$	2671.57 20	0.34 7
$\gamma_{\beta-}$	2676.36 12	0.059 12
$\gamma_{\beta-}$	2693.64 11	0.120 24
$\gamma_{\beta-}$	2705.3 10	0.057 11
$\gamma_{\beta-}$	2726.12 20	0.096 19
$\gamma_{\beta-}$	2741.8 3	0.048 10
$\gamma_{\beta-}$	2766.88 9	0.31 6
$\gamma_{\beta-}$	2807.70 8	0.19 4
$\gamma_{\beta-}$	2829.8 3	0.077 15
$\gamma_{\beta-}$	2851.2 5	0.039 8
$\gamma_{\beta-}$	2899.6 4	0.062 12
$\gamma_{\beta-}$	2909.8 4	0.069 14
$\gamma_{\beta-}$	2916.4 8	0.029 6
$\gamma_{\beta-}$	2922.6 3	0.092 18
$\gamma_{\beta-}$	2943.03 21	0.05 1
$\gamma_{\beta-}$	2943.1 3	0.05 1
$\gamma_{\beta-}$	2952.8 5	0.043 9
$\gamma_{\beta-}$	2960.39 21	0.092 18
$\gamma_{\beta-}$	2983.5 3	0.073 15
$\gamma_{\beta-}$	3022.84 10	0.043 9
$\gamma_{\beta-}$	3100.4 5	0.043 9
$\gamma_{\beta-}$	3117.6 3	0.14 3
$\gamma_{\beta-}$	3194.34 11	0.75 15
$\gamma_{\beta-}$	3212.9 8	0.032 6
$\gamma_{\beta-}$	3261.15 19	0.059 12
$\gamma_{\beta-}$	3276.66 10	0.066 13
$\gamma_{\beta-}$	3353.0 4	0.14 3
$\gamma_{\beta-}$	3391.6 5	0.036 7
$\gamma_{\beta-}$	3417.1 5	0.036 7
$\gamma_{\beta-}$	3458.54 14	0.31 6
$\gamma_{\beta-}$	3500.6 4	0.015 3
$\gamma_{\beta-}$	3570.11 15	0.57 11
$\gamma_{\beta-}$	3670.6 4	0.059 12
$\gamma_{\beta-}$	3682.7 10	0.023 5
$\gamma_{\beta-}$	3729.54 18	0.42 8
$\gamma_{\beta-}$	3795.37 11	0.85 17
$\gamma_{\beta-}$	3800.7 3	0.14 3
$\gamma_{\beta-}$	3862.18 19	0.25 5
$\gamma_{\beta-}$	3866.19 17	0.23 5
$\gamma_{\beta-}$	3911.23 17	0.45 9
$\gamma_{\beta-}$	3986.86 8	0.25 5
$\gamma_{\beta-}$	3996.20 11	0.49 10
$\gamma_{\beta-}$	4016.2 8	0.021 4
$\gamma_{\beta-}$	4028.71 11	0.13 3
$\gamma_{\beta-}$	4038.7 4	0.05 1
$\gamma_{\beta-}$	4084.1 4	0.087 17
$\gamma_{\beta-}$	4105.0 6	0.031 6
$\gamma_{\beta-}$	4130.3 4	0.073 15
$\gamma_{\beta-}$	4140.9 4	0.064 13
$\gamma_{\beta-}$	4160.97 21	0.28 6
$\gamma_{\beta-}$	4172.9 4	0.05 1
$\gamma_{\beta-}$	4189.0 7	0.018 4
$\gamma_{\beta-}$	4211.4 6	0.032 6
$\gamma_{\beta-}$	4276.1 3	0.27 5
$\gamma_{\beta-}$	4318.6 7	0.017 3
$\gamma_{\beta-}$	4332.7 4	0.120 24
$\gamma_{\beta-}$	4348.0 20	~0.02
$\gamma_{\beta-}$	4380.2 5	0.036 7
$\gamma_{\beta-}$	4402.6 4	0.24 5
$\gamma_{\beta-}$	4478.0 7	0.025 5
$\gamma_{\beta-}$	4502.6 12	0.014 3
$\gamma_{\beta-}$	4542.1 8	0.021 4
$\gamma_{\beta-}$	4561.2 7	0.030 6
$\gamma_{\beta-}$	4757.3 10	0.010 2
$\gamma_{\beta-}$	4784.7 10	0.013 3
$\gamma_{\beta-}$	4880.4 10	0.018 4

Continuous Radiation (^{137}I)

⟨β-⟩=1967 keV;⟨IB⟩=7.1 keV

E_{bin}(keV)		⟨ ⟩(keV)	(%)
0 - 10	β-	0.0073	0.145
	IB	0.055	
10 - 20	β-	0.0223	0.148
	IB	0.055	0.38
20 - 40	β-	0.092	0.306
	IB	0.109	0.38
40 - 100	β-	0.70	0.99
	IB	0.32	0.48
100 - 300	β-	8.4	4.09
	IB	0.95	0.53
300 - 600	β-	35.5	7.8
	IB	1.18	0.28
600 - 1300	β-	197	20.6
	IB	2.0	0.22
1300 - 2500	β-	625	33.2
	IB	1.7	0.100
2500 - 5000	β-	1058	31.5
	IB	0.71	0.023
5000 - 5880	β-	43.5	0.83
	IB	0.0027	4.8×10^{-5}

$^{137}_{54}$Xe(3.818 13 min)

Mode: β-

Δ: -82385 7 keV

SpA: 3.589×10^{8} Ci/g

Prod: ^{136}Xe(n,γ); fission

Photons (^{137}Xe)

⟨γ⟩=190.8 24 keV

γ_{mode}	γ(keV)	γ(%)†
Cs $K_{\alpha2}$	30.625	0.098 13
Cs $K_{\alpha1}$	30.973	0.181 24
Cs $K_{\beta1}'$	34.967	0.049 7
Cs $K_{\beta2}'$	36.006	0.0121 16
γ	298.02 5	0.119 9
γ	393.37 4	0.140 9
γ M1+E2	455.459 3	31.2
γ	482.18 10	0.015 3
γ	594.63 5	0.084 9
γ	633.24 22	~0.0025
γ	683.13 5	0.0203 22
γ	715.22 7	0.0066 16
γ	750.60 8	0.0209 22
γ	802.34 7	0.0041 12
γ	848.82 4	0.62 3
γ	933.75 5	0.084 6
γ	982.17 4	0.209 12
γ	1009.87 20	0.0041 9
γ	1066.56 6	0.054 9
γ	1067.40 7	0.049 9
γ	1102.33 7	0.0165 16
γ	1108.59 6	0.051 5
γ [E2]	1114.32 4	0.092 6
γ	1119.30 6	0.107 6
γ [E2]	1184.67 5	0.084 9
γ	1195.71 6	0.048 3
γ	1219.13 7	0.0028 6
γ	1231.78 20	0.0016 6
γ	1236.26 24	0.0034 6
γ	1250.54 9	0.0066 9
γ [E2]	1273.16 6	0.228 22
γ	1280.18 5	0.0094 9
γ	1327.95 4	0.0293 25
γ	1460.76 5	0.0172 22
γ	1518.96 11	0.0019 6
γ	1564.04 6	0.0100 12
γ [M1+E2]	1569.78 4	0.085 6
γ	1574.76 6	0.072 9
γ	1576.79 5	0.103 9

Photons (^{137}Xe) (continued)

γ_{mode}	γ(keV)	γ(%)†
γ	1593.98 13	0.0031 6
γ [M1+E2]	1612.49 6	0.125 9
γ	1643.90 9	~0.0009
γ	1651.16 6	0.0044 6
γ	1665.29 5	0.053 3
γ	1677.2 6	0.0010 4
γ	1713.2 8	~0.0007
γ	1720.9 6	0.0011 5
γ	1726.22 13	0.0024 4
γ	1761.26 22	0.0059 16
γ	1783.40 4	0.415 19
γ	1843.0 4	0.0014 4
γ	1867.79 4	0.0162 16
γ	1907.64 19	0.0053 6
γ	1916.21 5	0.097 12
γ	1932.85 20	0.0013 4
γ	1946.99 19	0.0027 3
γ	1974.72 20	0.0012 4
γ	2000.19 12	0.0178 19
γ	2003.41 17	0.0066 12
γ	2043.6 3	0.0018 4
γ [E2]	2067.94 6	0.0103 9
γ	2084.49 7	0.0137 9
γ	2096.26 17	0.0072 9
γ	2099.35 9	0.0134 12
γ	2119.4 4	0.0017 4
γ	2188.39 7	0.0081 9
γ	2212.05 17	0.0041 6
γ	2216.71 22	0.0024 9
γ	2255.22 24	0.0025 3
γ	2287.2 4	0.00140 25
γ	2304.4 6	0.00056 25
γ	2310.84 20	0.00097 25
γ	2367.78 11	0.0066 9
γ	2393.55 12	0.081 6
γ	2463.5 5	0.0007 3
γ	2489.62 17	0.0027 3
γ	2528.55 13	0.0015 3
γ	2581.75 7	0.0234 22
γ	2639.18 20	0.0009 3
γ	2735.18 19	0.0014 3
γ	2849.94 4	0.184 9
γ	2921.91 12	0.0178 16
γ	3037.19 7	0.0044 6
γ	3135.6 7	0.00037 16
γ	3159.38 20	0.0119 12
γ	3194.0 4	0.00090 19
γ	3250.0 4	0.00106 19
γ	3377.36 12	0.00200 25
γ	3451.7 4	0.00047 22
γ	3458.3 4	0.00128 22
γ	3476.3 4	0.00087 19
γ	3583.98 19	0.0020 3
γ	3694.0 3	0.0062 6
γ	3736.6 5	0.00031 9
γ	3907.1 4	0.00134 22
γ	3938.3 4	~0.00025
γ	3955.5 6	0.00034 12
γ	3976.4 8	0.00031 12

† 1.6% uncert(syst)

Atomic Electrons (^{137}Xe)

⟨e⟩=1.93 22 keV

e_{bin}(keV)	⟨e⟩(keV)	e(%)
5 - 35	0.026	0.34 5
262 - 298	0.010	~0.0036
357 - 393	0.007	~0.0020
419	1.55	0.37 5
446	0.0005	~0.00011
450	0.23	0.052 12
454	0.048	0.011 3
455 - 482	0.013	0.0028 7
559 - 597	0.0025	~0.00044
628 - 647	0.00043	$\sim7\times10^{-5}$
677 - 715	0.00060	9 3 $\times10^{-5}$
745 - 766	0.00013	1.8 7 $\times10^{-5}$

Atomic Electrons (^{137}Xe)
(continued)

e_{bin}(keV)	⟨e⟩(keV)	e(%)
797 - 844	0.011	0.0014 *7*
848 - 849	0.00034	4.0 *19* $\times10^{-5}$
898 - 946	0.0041	0.00044 *17*
974 - 1010	0.00052	5.3 *20* $\times10^{-5}$
1031 - 1078	0.0034	0.00032 *7*
1083 - 1119	0.0017	0.00016 *6*
1149 - 1196	0.0018	0.000156 *25*
1200 - 1249	0.0027	0.000222 *22*
1250 - 1292	0.00070	5.5 *11* $\times10^{-5}$
1322 - 1328	4.4 $\times10^{-5}$	3.3 *13* $\times10^{-6}$
1425 - 1461	0.00017	1.2 *5* $\times10^{-5}$
1483 - 1577	0.0040	0.00026 *3*
1588 - 1685	0.00072	4.4 *11* $\times10^{-5}$
1690 - 1782	0.0034	0.00019 *7*
1807 - 1906	0.0008	4.4 *15* $\times10^{-5}$
1907 - 2002	0.00030	1.5 *3* $\times10^{-5}$
2008 - 2098	0.00035	1.69 *25* $\times10^{-5}$
2114 - 2212	9.7 $\times10^{-5}$	4.5 *10* $\times10^{-6}$
2215 - 2310	3.5 $\times10^{-5}$	1.6 *3* $\times10^{-6}$
2332 - 2428	0.00054	2.3 *7* $\times10^{-5}$
2454 - 2546	0.00014	5.6 *16* $\times10^{-6}$
2576 - 2638	2.3 $\times10^{-5}$	8.8 *23* $\times10^{-7}$
2699 - 2734	7.6 $\times10^{-6}$	2.8 *9* $\times10^{-7}$
2814 - 2886	0.0011	3.8 *10* $\times10^{-5}$
2916 - 3001	3.1 $\times10^{-5}$	1.06 *24* $\times10^{-6}$
3031 - 3131	5.5 $\times10^{-5}$	1.8 *5* $\times10^{-6}$
3134 - 3214	1.7 $\times10^{-5}$	5.2 *10* $\times10^{-7}$
3244 - 3341	8.8 $\times10^{-6}$	2.6 *7* $\times10^{-7}$
3372 - 3471	1.32 $\times10^{-5}$	3.9 *7* $\times10^{-7}$
3475 - 3548	7.8 $\times10^{-6}$	2.2 *7* $\times10^{-7}$
3578 - 3658	2.5 $\times10^{-5}$	6.8 *18* $\times10^{-7}$
3688 - 3736	5.0 $\times10^{-6}$	1.3 *3* $\times10^{-7}$
3871 - 3955	9.3 $\times10^{-6}$	2.4 *5* $\times10^{-7}$
3971 - 3975	1.7 $\times10^{-7}$	4.3 *18* $\times10^{-9}$

Continuous Radiation (^{137}Xe)
⟨β-⟩=1696 keV;⟨IB⟩=5.3 keV

E_{bin}(keV)		⟨ ⟩(keV)	(%)
0 - 10	β-	0.0052	0.103
	IB	0.052	
10 - 20	β-	0.0158	0.105
	IB	0.052	0.36
20 - 40	β-	0.065	0.217
	IB	0.102	0.36
40 - 100	β-	0.502	0.71
	IB	0.30	0.45
100 - 300	β-	6.4	3.09
	IB	0.88	0.49
300 - 600	β-	31.4	6.8
	IB	1.06	0.25
600 - 1300	β-	233	24.0
	IB	1.64	0.19
1300 - 2500	β-	856	45.7
	IB	1.12	0.066
2500 - 4177	β-	568	19.3
	IB	0.144	0.0052

$^{137}_{55}$Cs(30.0 *2* yr)

Mode: β-
Δ: -86561 *6* keV
SpA: 87.0 Ci/g
Prod: fission

Photons (^{137}Cs)
⟨γ⟩=566 *11* keV

γ_{mode}	γ(keV)	γ(%)†*
Ba L_ℓ	3.954	0.0144 *20*
Ba L_η	4.331	0.0064 *7*
Ba L_α	4.465	0.40 *5*
Ba L_β	4.944	0.37 *5*
Ba L_γ	5.620	0.049 *7*
Ba $K_{\alpha2}$	31.817	2.05 *7*
Ba $K_{\alpha1}$	32.194	3.77 *14*
Ba $K_{\beta1}$'	36.357	1.04 *4*
Ba $K_{\beta2}$'	37.450	0.264 *10*
γ M4	661.660 *3*	85.21

†0.08% uncert(syst)
* with ^{137}Ba(2.552 min) in equilib

Atomic Electrons (^{137}Cs)
⟨e⟩=61.9 *10* keV

e_{bin}(keV)	⟨e⟩(keV)	e(%)
5	0.199	3.8 *4*
6	0.147	2.6 *3*
25	0.0159	0.062 *6*
26	0.045	0.173 *18*
27	0.074	0.28 *3*
30	0.0166	0.055 *6*
31	0.047	0.152 *16*
32	0.0101	0.032 *3*
35	0.0051	0.0145 *15*
36	0.0027	0.0076 *8*
624	49.4	7.91 *16*
656	9.46	1.44 *3*
660	1.59	0.241 *5*
661	0.918	0.139 *3*
662	0.0279	0.00422 *9*

Continuous Radiation (^{137}Cs)
⟨β-⟩=188 keV;⟨IB⟩=0.124 keV

E_{bin}(keV)		⟨ ⟩(keV)	(%)
0 - 10	β-	0.170	3.42
	IB	0.0095	
10 - 20	β-	0.499	3.33
	IB	0.0087	0.061
20 - 40	β-	1.95	6.5
	IB	0.0155	0.054
40 - 100	β-	12.8	18.4
	IB	0.035	0.054
100 - 300	β-	94	48.8
	IB	0.045	0.029
300 - 600	β-	68	18.2
	IB	0.0089	0.0023
600 - 1175	β-	10.7	1.40
	IB	0.00137	0.00020

$^{137}_{56}$Ba(stable)

Δ: -87736 *6* keV
%: 11.23 *4*

$^{137}_{56}$Ba(2.552 *1* min)

Mode: IT
Δ: -87074 *6* keV
SpA: 5.3658$\times10^{8}$ Ci/g
Prod: daughter ^{137}Cs

Photons (^{137}Ba)
⟨γ⟩=599 *12* keV

γ_{mode}	γ(keV)	γ(%)†
Ba L_ℓ	3.954	0.0151 *20*
Ba L_η	4.331	0.0068 *7*
Ba L_α	4.465	0.42 *5*
Ba L_β	4.944	0.39 *5*
Ba L_γ	5.620	0.051 *7*
Ba $K_{\alpha2}$	31.817	2.16 *7*
Ba $K_{\alpha1}$	32.194	3.97 *12*
Ba $K_{\beta1}$'	36.357	1.10 *3*
Ba $K_{\beta2}$'	37.450	0.278 *9*
γ M4	661.660 *3*	90.1

†<0.1% uncert(syst)

Atomic Electrons (^{137}Ba)
⟨e⟩=65.2 *11* keV

e_{bin}(keV)	⟨e⟩(keV)	e(%)
5	0.210	4.0 *4*
6	0.155	2.7 *3*
25	0.0167	0.066 *7*
26	0.047	0.182 *19*
27	0.078	0.29 *3*
30	0.0175	0.058 *6*
31	0.050	0.161 *16*
32	0.0107	0.033 *3*
35	0.0054	0.0153 *16*
36	0.0029	0.0080 *8*
624	52.0	8.33 *17*
656	9.96	1.52 *3*
660	1.68	0.254 *5*
661	0.967	0.146 *3*
662	0.0294	0.00445 *9*

$^{137}_{57}$La(6 *2* $\times10^{4}$ yr)

Mode: ε
Δ: -87130 *50* keV
SpA: 0.043 Ci/g
Prod: daughter ^{137}Ce

Photons (^{137}La)
⟨γ⟩=25.3 *5* keV

γ_{mode}	γ(keV)	γ(%)
Ba L_ℓ	3.954	0.148 *20*
Ba L_η	4.331	0.065 *7*
Ba L_α	4.465	4.1 *5*
Ba L_β	4.945	3.7 *5*
Ba L_γ	5.623	0.50 *7*
Ba $K_{\alpha2}$	31.817	21.8 *7*
Ba $K_{\alpha1}$	32.194	40.1 *12*
Ba $K_{\beta1}$'	36.357	11.1 *3*
Ba $K_{\beta2}$'	37.450	2.81 *9*

Atomic Electrons (^{137}La)

⟨e⟩=5.9 *3* keV

e_{bin}(keV)	⟨e⟩(keV)	e(%)
5	2.05	39 *4*
6	1.51	26 *3*
25	0.169	0.66 *7*
26	0.48	1.83 *19*
27	0.79	3.0 *3*
30	0.177	0.58 *6*
31	0.50	1.62 *17*
32	0.108	0.34 *3*
35	0.054	0.154 *16*
36	0.029	0.080 *8*

Continuous Radiation (^{137}La)

⟨IB⟩=0.123 keV

E_{bin}(keV)		⟨ ⟩(keV)	(%)
10 - 20	IB	0.00037	0.0022
20 - 40	IB	0.077	0.24
40 - 100	IB	0.0071	0.0127
100 - 300	IB	0.020	0.0105
300 - 600	IB	0.018	0.0048
600 - 610	IB	2.1×10^{-8}	3.4×10^{-9}

$^{137}_{58}$Ce(9.0 *3* h)

Mode: ε

Δ: -85910 *50* keV

SpA: 2.54×10^{6} Ci/g

Prod: daughter ^{137}Pr; ^{139}La(p,3n); ^{136}Ce(n,γ); ^{134}Ba(α,n)

Photons (^{137}Ce)

⟨γ⟩=40.8 *7* keV

γ_{mode}	γ(keV)	γ(%)†
La L_ℓ	4.121	0.24 *4*
La L_η	4.529	0.115 *12*
La L_α	4.649	6.4 *8*
La L_β	5.143	9.1 *14*
La L_γ	5.960	1.6 *3*
La $K_{\alpha2}$	33.034	22.0 *7*
La $K_{\alpha1}$	33.442	40.4 *12*
La $K_{\beta1}$'	37.777	11.3 *3*
La $K_{\beta2}$'	38.927	2.95 *10*
γ [M1+E2]	148.91 *6*	0.0011 *5*
γ [M1+E2]	217.10 *5*	0.0049 *7*
γ E2	433.31 *6*	0.0652 *13*
γ E2	436.66 *6*	0.334 *11*
γ M1+E2	447.23 *6*	2.24
γ E2+M1	479.21 *6*	0.0150 *7*
γ M1+E2	482.56 *5*	0.0576 *20*
γ [E2]	493.13 *6*	0.0132 *7*
γ [M1]	529.45 *10*	~0.00045 ?
γ E2	631.46 *5*	0.0168 *9*
γ [M1+E2]	678.35 *9*	0.0011 *5*
γ M1+E2	698.77 *6*	0.0392 *20*
γ [E2]	709.35 *6*	0.00134 *22*
γ [E2]	724.25 *11*	~0.0009
γ M1+E2	771.06 *8*	0.0076 *5*
γ [M1+E2]	781.63 *8*	0.0038 *5*
γ (M1+E2)	915.87 *5*	0.0647 *22*
γ M1+E2	926.45 *5*	0.0426 *16*
γ [E2]	1160.91 *10*	0.00188 *18*

† 4.5% uncert(syst)

Atomic Electrons (^{137}Ce)

⟨e⟩=13.8 *5* keV

e_{bin}(keV)	⟨e⟩(keV)	e(%)
4	2.97	68.8 *24*
5	2.57	48 *4*
6	3.8	63 *7*
9	1.47	15.9 *6*
10 - 26	0.631	5.11 *15*
27	0.48	1.76 *18*
28	0.78	2.8 *3*
31	0.122	0.39 *4*
32	0.51	1.60 *16*
33 - 38	0.232	0.68 *5*
110 - 149	0.00056	0.00048 *16*
178 - 217	0.00120	0.00065 *9*
394 - 444	0.178	0.043 *6*
446 - 493	0.0068	0.0015 *3*
523 - 529	4.2×10^{-6}	8 *3* $\times10^{-7}$
593 - 639	0.00060	0.000100 *5*
660 - 709	0.00148	0.00022 *4*
718 - 766	0.00034	4.6 *7* $\times10^{-5}$
770 - 782	2.5×10^{-5}	3.2 *5* $\times10^{-6}$
877 - 926	0.0027	0.00030 *4*
1122 - 1161	3.0×10^{-5}	2.64 *24* $\times10^{-6}$

Continuous Radiation (^{137}Ce)

⟨β+⟩=0.0130 keV;⟨IB⟩=0.43 keV

E_{bin}(keV)		⟨ ⟩(keV)	(%)
0 - 10	β+	1.98×10^{-7}	2.37×10^{-6}
	IB	0.00038	
10 - 20	β+	7.8×10^{-6}	4.72×10^{-5}
	IB	0.00033	0.0020
20 - 40	β+	0.000191	0.00059
	IB	0.083	0.25
40 - 100	β+	0.00507	0.0068
	IB	0.0091	0.0167
100 - 300	β+	0.0078	0.0061
	IB	0.035	0.017
300 - 600	IB	0.136	0.030
600 - 1211	IB	0.165	0.022
	Σβ+		0.0136

$^{137}_{58}$Ce(1.433 *13* d)

Mode: IT(99.22 *3* %), ε(0.78 *3* %)

Δ: -85656 *50* keV

SpA: 6.65×10^{5} Ci/g

Prod: ^{139}La(p,3n); ^{136}Ce(n,γ); ^{134}Ba(α,n)

Photons (^{137}Ce)

⟨γ⟩=55.0 *14* keV

γ_{mode}	γ(keV)	γ(%)†
Ce L_ℓ	4.289	0.149 *21*
Ce L_η	4.730	0.058 *6*
Ce L_α	4.838	4.0 *5*
Ce L_β	5.392	3.6 *5*
Ce L_γ	6.153	0.51 *7*
Ce $K_{\alpha2}$	34.279	15.7 *6*
Ce $K_{\alpha1}$	34.720	28.7 *10*
Ce $K_{\beta1}$'	39.232	8.1 *3*
Ce $K_{\beta2}$'	40.437	2.18 *8*
γ_ϵ[E1]	87.01 *12*	0.0088 *13*
γ_ϵE1	169.06 *4*	0.44 *3*
γ_{IT} M4	254.29 *5*	11.0 *4*
γ_ϵE2	762.10 *10*	0.192 *9*

Photons (^{137}Ce)

(continued)

γ_{mode}	γ(keV)	γ(%)†
γ_ϵE2	824.61 *7*	0.44 *9*
γ_ϵE2	835.18 *7*	0.103 *4*
γ_ϵ[E2]	906.65 *10*	0.0028 *5*
γ_ϵM1+E2	917.23 *10*	0.0128 *22*
γ_ϵE3	993.66 *8*	0.0020 *3*
γ_ϵM2(+E3)	1004.24 *8*	0.022 *3*

† uncert(syst): 6.0% for ε, <0.1% for IT

Atomic Electrons (^{137}Ce)*

⟨e⟩=202 *4* keV

e_{bin}(keV)	⟨e⟩(keV)	e(%)
6	2.8	48 *5*
7	0.44	6.8 *7*
27	0.119	0.43 *5*
28	0.33	1.19 *12*
29	0.53	1.84 *19*
32	0.052	0.160 *17*
33	0.29	0.88 *9*
34	0.176	0.52 *5*
35	0.026	0.075 *8*
38	0.042	0.110 *11*
39	0.0170	0.044 *5*
214	127.8	59.8 *16*
248	36.9	14.9 *4*
249	16.4	6.58 *18*
253	12.7	5.00 *14*
254	3.49	1.37 *4*

* with IT

$^{137}_{59}$Pr(1.28 *3* h)

Mode: ε

Δ: -83200 *50* keV

SpA: 1.79×10^{7} Ci/g

Prod: protons on Ce; ^{136}Ce(d,n)

Photons (^{137}Pr)

⟨γ⟩=113 *3* keV

γ_{mode}	γ(keV)	γ(%)†
Ce L_ℓ	4.289	0.129 *17*
Ce L_η	4.730	0.053 *6*
Ce L_α	4.838	3.5 *4*
Ce L_β	5.390	3.2 *4*
Ce L_γ	6.142	0.44 *6*
Ce $K_{\alpha2}$	34.279	16.6 *5*
Ce $K_{\alpha1}$	34.720	30.5 *9*
Ce $K_{\beta1}$'	39.232	8.6 *3*
Ce $K_{\beta2}$'	40.437	2.31 *8*
γ M1	160.44 *9*	0.97 *11*
γ	249.05 *11*	0.022 *7*
γ [E2]	273.81 *10*	0.044 *9*
γ	310.8 *5*	0.024 *9*
γ M1	329.32 *10*	0.18 *3*
γ	337.94 *20*	0.033 *11*
γ M1,E2	354.09 *10*	0.58 *6*
γ M1,E2	402.88 *10*	0.064 *7*
γ	416.55 *12*	0.081 *9*
γ M1,E2	434.25 *8*	1.28 *11*
γ M1,E2	514.53 *9*	1.08 *11*
γ	573.6 *5*	0.040 *4*
γ	584.5 *4*	~0.009
γ [M1+E2]	591.04 *15*	0.053 *4*
γ	603.14 *11*	0.310 *24*
γ	610.05 *19*	0.026 *4*

Photons (^{137}Pr)
(continued)

γ_{mode}	γ(keV)	γ(%)†
γ	647.3 4	0.018 4
γ	654.8 3	0.031 4
γ [M1+E2]	665.60 11	0.117 9
γ [M1+E2]	671.31 14	0.033 4
γ [E2]	676.69 11	~0.013
γ	695.71 16	0.013 4
γ	698.21 15	~0.013
γ	706.81 12	0.044 7
γ	713.72 16	0.055 4
γ	734.5 3	0.0088 22 ?
γ [M1+E2]	745.87 10	0.189 9
γ [M1+E2]	754.36 14	0.020 4
γ	763.57 9	0.189 9
γ	825.92 21	0.051 4
γ M1	837.13 9	1.8
γ	856.76 20	0.018 4
γ	867.25 12	0.246 11
γ	921.65 10	0.169 11
γ	934.02 13	0.062 7
γ [M1+E2]	945.13 15	~0.009
γ [M1+E2]	954.26 14	0.0220 22
γ	962.77 19	<0.0044
γ	973.4 3	0.013 4
γ	1001.93 10	0.114 9
γ [E2]	1019.69 11	0.044 7
γ [M1+E2]	1028.82 17	0.0110 22 ?
γ	1057.1 3	0.029 4
γ	1067.23 15	0.013 4
γ M1,E2	1089.02 14	0.418 22
γ M1	1097.35 11	0.202 13
γ [M1+E2]	1105.56 13	0.090 9
γ [M1+E2]	1111.37 17	0.086 9
γ	1124.69 18	0.059 11
γ	1128.7 3	0.042 9
γ	1170.90 13	0.059 7
γ [M1+E2]	1180.12 10	0.081 11
γ [M1+E2]	1200.03 25	0.013 4
γ	1247.09 16	0.026 4
γ	1259.63 21	~0.009
γ	1260.17 22	0.011 4
γ [M1+E2]	1271.80 18	0.018 4
γ	1277.20 13	0.059 7
γ	1282.5 4	0.013 4
γ	1286.74 12	0.046 4
γ	1289.2 3	0.013 4
γ	1300.8 4	~0.0044
γ	1350.76 15	0.015 4
γ	1367.02 13	0.013 4
γ [M1+E2]	1373.74 18	0.013 4
γ [M1+E2]	1419.95 13	0.075 4
γ	1436.17 12	<0.020 ?
γ [M1+E2]	1437.73 14	0.246 18
γ [M1+E2]	1454.01 17	0.051 7
γ [M1+E2]	1468.46 23	<0.009
γ	1477.29 18	0.0132 22
γ	1499.90 20	0.081 9
γ	1506.1 5	0.022 7
γ	1512.23 17	0.040 9
γ [M1+E2]	1518.01 14	0.062 9
γ	1533.44 21	0.0066 22
γ	1542.1 6	~0.009
γ	1560.7 6	~0.007
γ	1571.6 3	0.0088 22
γ	1613.7 4	0.013 4
γ [M1+E2]	1619.86 13	0.073 11
γ	1639.0 3	0.037 4
γ	1680.08 14	0.051 7
γ	1693.88 20	0.029 4
γ [M1+E2]	1700.13 13	0.128 13
γ	1719.3 3	~0.009
γ [M1+E2]	1727.82 17	0.033 7
γ	1742.7 7	0.011 4
γ	1765.70 18	0.013 4
γ [M1+E2]	1774.04 13	~0.007 ?
γ [M1+E2]	1791.82 16	0.024 7
γ	1801.26 12	0.163 13
γ	1819.7 4	0.022 9
γ	1833.5 3	~0.013
γ [M1+E2]	1841.55 16	~0.013
γ	1853.2 5	0.015 4
γ	1864.5 9	~0.007
γ [M1+E2]	1871.33 23	0.020 4
γ [M1+E2]	1888.26 17	0.077 15

Photons (^{137}Pr)
(continued)

γ_{mode}	γ(keV)	γ(%)†
γ	1894.2 4	0.022 7
γ	1913.8 3	0.026 9
γ [M1+E2]	1934.47 12	0.020 4
γ [M1+E2]	1952.25 15	0.029 9
γ	1966.4 4	0.013 4
γ [M1+E2]	1973.94 15	0.0066 22 ?
γ	1993.1 3	~0.0044
γ	2114.32 14	0.062 11
γ [M1+E2]	2115.36 18	0.029 9
γ	2128.3 5	0.018 4
γ [M1+E2]	2134.38 14	0.051 11
γ [M1+E2]	2145.14 24	~0.009
γ	2153.5 3	0.024 7
γ	2187.6 3	0.0066 22
γ [M1+E2]	2275.79 17	0.015 4
γ [M1+E2]	2305.58 23	0.011 4
γ	2320.5 4	0.0066 22
γ	2348.1 3	0.020 7
γ	2480.9 4	0.0066 22
γ	2518.4 6	0.0066 22

† 17% uncert(syst)

Atomic Electrons (^{137}Pr)
⟨e⟩=5.4 3 keV

e_{bin}(keV)	⟨e⟩(keV)	e(%)
6	2.5	43 4
7	0.34	5.2 5
27	0.126	0.46 5
28	0.35	1.26 13
29	0.56	1.96 20
32	0.055	0.170 17
33	0.31	0.94 10
34	0.188	0.55 6
35 - 39	0.090	0.243 15
120	0.32	0.26 3
154 - 160	0.071	0.046 4
209 - 249	0.0083	0.0037 9
267 - 314	0.078	0.026 4
323 - 362	0.020	0.0058 10
376 - 417	0.089	0.023 5
428 - 474	0.071	0.015 3
508 - 551	0.0141	0.0027 5
563 - 610	0.013	0.0023 10
614 - 660	0.0078	0.00124 19
664 - 713	0.0092	0.00133 24
714 - 763	0.0066	0.0009 4
785 - 831	0.069	0.0086 14
836 - 881	0.0058	0.00067 20
894 - 939	0.0027	0.00030 8
944 - 988	0.0029	0.00030 11
995 - 1029	0.0011	0.00011 3
1049 - 1097	0.0190	0.00179 17
1099 - 1140	0.0029	0.00026 5
1160 - 1207	0.00095	8.0 18 $\times10^{-5}$
1219 - 1266	0.0023	0.00018 4
1270 - 1310	0.00045	3.5 9 $\times10^{-5}$
1327 - 1374	0.00042	3.1 9 $\times10^{-5}$
1380 - 1428	0.0054	0.00039 5
1430 - 1478	0.0031	0.00021 4
1493 - 1579	0.00170	0.000110 15
1599 - 1694	0.0035	0.000210 25
1699 - 1796	0.0026	0.00015 4
1800 - 1894	0.0022	0.000120 15
1907 - 1992	0.00062	3.2 6 $\times10^{-5}$
2074 - 2152	0.0018	8.7 13 $\times10^{-5}$
2181 - 2280	0.00030	1.3 3 $\times10^{-5}$
2299 - 2347	0.00017	7 3 $\times10^{-6}$
2440 - 2517	9.4 $\times10^{-5}$	3.8 12 $\times10^{-6}$

Continuous Radiation (^{137}Pr)
⟨β+⟩=190 keV;⟨IB⟩=3.4 keV

E_{bin}(keV)		⟨ ⟩(keV)	(%)
0 - 10	β+	1.25 $\times10^{-6}$	1.49 $\times10^{-5}$
	IB	0.0081	
10 - 20	β+	5.5 $\times10^{-5}$	0.000330
	IB	0.0079	0.055
20 - 40	β+	0.00165	0.00501
	IB	0.081	0.25
40 - 100	β+	0.086	0.110
	IB	0.050	0.081
100 - 300	β+	4.44	2.0
	IB	0.139	0.076
300 - 600	β+	31.0	6.7
	IB	0.28	0.063
600 - 1300	β+	135	15.0
	IB	1.23	0.131
1300 - 2500	β+	19.5	1.39
	IB	1.58	0.094
2500 - 2702	IB	0.0067	0.00026
	$\Sigma\beta$+		25

$^{137}_{60}$Nd(38.5 15 min)

Mode: ϵ

Δ: -79400 60 keV

SpA: 3.56$\times10^{7}$ Ci/g

Prod: protons on Gd; ^{141}Pr(α,7np)

Photons (^{137}Nd)
⟨γ⟩=838 23 keV

γ_{mode}	γ(keV)	γ(%)†
Pr L_ℓ	4.453	0.28 8
Pr L_η	4.929	0.12 4
Pr L_α	5.031	7.5 20
Pr L_β	5.616	6.8 19
Pr L_γ	6.396	0.9 3
Pr $K_{\alpha2}$	35.550	28.4 18
Pr $K_{\alpha1}$	36.026	52 3
Pr $K_{\beta1}$'	40.720	14.6 9
Pr $K_{\beta2}$'	41.981	4.1 3
γ M1+E2	75.44 6	17.3 19
γ	110.8 3	<0.26
γ [M1]	144.35 10	<0.07
γ	149.0 4	0.14 4
γ [M1+E2]	167.64 10	<0.20
γ	170.2 5	~0.09
γ [M1+E2]	180.91 9	0.13 3
γ M1,E2	198.51 6	0.65 7
γ	227.9 3	0.21 5
γ M1+E2	230.53 9	1.30 13
γ M1,E2	238.16 6	3.6 3
γ	245.4 5	0.08 3
γ	252.0 3	0.12 3
γ	258.0 4	0.16 5
γ M1,E2	267.05 6	0.98 13
γ M1(+E2)	276.38 7	0.65 10
γ	288.51 24	<0.20
γ M1,E2	306.70 6	10.1 5
γ M1,E2	313.60 7	0.78 13
γ	317.6 3	0.12 4
γ	323.2 4	0.12 4
γ	325.5 3	0.10 4
γ	327.1 3	0.10 4
γ	342.88 14	<0.26
γ E2,M1	348.54 8	<0.26
γ [E2]	350.12 10	0.39 13
γ	361.0 4	0.19 4
γ M1,E2	382.14 6	1.04 13
γ	384.6 3	0.12 4
γ	386.2 4	0.12 4
γ	395.1 4	0.10 4
γ [M1+E2]	447.95 9	<0.13

Photons (^{137}Nd) (continued)

γ_{mode}	γ(keV)	γ(%)†
γ M1(+E2)	474.89 *7*	1.17 *20*
γ	494.55 *25*	0.10 *3*
γ M1,E2	505.21 *6*	9.1 *13*
γ	525.36 *13*	0.39 *7*
γ (E2)	531.02 *10*	0.78 *13*
γ	540.7 *5*	<0.26
γ	547.05 *8*	0.39 *8*
γ	576.5 *3*	0.32 *3*
γ M1,E2	580.65 *6*	13
γ	598.17 *20*	0.30 *7*
γ M1,E2	615.59 *8*	1.05 *10*
γ M1,E2	619.24 *8*	0.59 *7*
γ	623.79 *14*	0.31 *7*
γ	627.56 *15*	<0.18
γ	631.8 *3*	0.052 *19*
γ	635.1 *3*	0.060 *20*
γ	637.6 *3*	0.10 *3*
γ	644.5 *3*	0.09 *3*
γ	646.9 *3*	0.12 *4*
γ M1(+E2)	649.58 *14*	<0.20
γ	654.0 *4*	<0.05
γ	661.2 *3*	0.10 *3*
γ	667.08 *25*	<0.26
γ	678.2 *3*	0.10 *4*
γ E2(+M1)	686.12 *7*	1.8 *3*
γ [E2]	687.78 *10*	<0.13
γ	696.8 *3*	
γ	708.4 *3*	<0.05
γ	711.5 *3*	0.09 *3*
γ [M1+E2]	725.02 *14*	<0.26
γ	727.1 *3*	0.08 *3*
γ	748.1 *3*	0.12 *4*
γ M1,E2	761.56 *8*	9.1 *13*
γ	776.4 *3*	0.14 *5*
γ M1(+E2)	781.59 *7*	9.5 *10*
γ	785.0 *3*	<0.20 ?
γ	797.4 *3*	0.07 *3*
γ	799.8 *3*	0.08 *3*
γ	811.4 *4*	0.09 *3*
γ	814.5 *5*	0.17 *5*
γ	847.7 *4*	0.08 *3*
γ	849.5 *4*	~0.05
γ [E2]	857.03 *7*	<0.23
γ	860.2 *3*	0.21 *7*
γ	863.61 *13*	<0.34
γ	879.8 *3*	0.10 *3*
γ	883.78 *14*	<0.32
γ M1(+E2)	925.94 *9*	7.4 *8*
γ M1(+E2)	929.19 *8*	3.0 *4*
γ	940.1 *3*	0.09 *3*
γ	942.9 *3*	0.11 *3*
γ	959.22 *14*	<0.07
γ	967.3 *4*	0.08 *3*
γ	971.2 *4*	0.09 *3*
γ [E2]	1001.38 *9*	<0.26
γ	1029.6 *3*	0.16 *5*
γ	1044.52 *11*	0.98 *10*
γ	1048.3 *3*	0.09 *4*
γ	1102.44 *15*	0.30 *7*
γ	1116.7 *3*	0.09 *3*
γ [M1]	1119.92 *10*	0.59 *13*
γ	1166.8 *5*	0.13 *4*
γ	1176.2 *4*	0.13 *4*
γ	1179.23 *15*	<0.13
γ	1182.1 *5*	0.10 *3*
γ	1191.1 *5*	0.08 *3*
γ	1204.3 *3*	0.09 *3*
γ	1213.1 *4*	0.13 *4*
γ	1218.25 *17*	<0.20 ?
γ	1234.54 *18*	0.52 *8*
γ	1243.02 *12*	1.43 *13*
γ	1247.73 *24*	<0.13
γ	1253.5 *3*	0.07 *3*
γ	1256.6 *3*	0.18 *7*
γ	1263.0 *4*	0.08 *3*
γ	1268.1 *4*	0.09 *3*
γ	1293.69 *16*	0.39 *7* ?
γ	1296.7 *3*	<0.026 ?
γ	1309.98 *18*	0.46 *7*
γ	1320.2 *4*	0.09 *3*
γ	1340.5 *4*	0.08 *3*
γ	1351.3 *4*	0.10 *4*
γ	1360.13 *13*	0.65 *13*
γ [M1+E2]	1365.11 *11*	<0.26

Photons (^{137}Nd) (continued)

γ_{mode}	γ(keV)	γ(%)†
γ	1372.3 *5*	0.09 *3*
γ	1388.40 *13*	<0.32
γ [E2]	1401.64 *13*	0.32 *7*
γ	1409.7 *4*	0.12 *4*
γ	1427.98 *12*	<0.07
γ	1434.1 *4*	0.17 *5*
γ	1444.0 *4*	0.11 *3*
γ	1464.00 *12*	0.75 *13*
γ	1472.3 *4*	0.09 *3*
γ	1484.58 *14*	0.62 *10*
γ	1493.5 *4*	0.12 *5*
γ	1542.5 *4*	0.09 *3*
γ [M1+E2]	1546.01 *10*	0.39 *8*
γ	1551.7 *5*	0.07 *3*
γ	1574.5 *4*	0.10 *4*
γ	1586.90 *14*	0.32 *7*
γ	1594.81 *10*	0.46 *7*
γ	1607.8 *3*	0.17 *5*
γ	1626.48 *12*	0.91 *7*
γ	1646.5 *7*	0.17 *7*
γ	1662.51 *13*	<0.13
γ	1700.6 *4*	0.10 *3*
γ	1731.05 *13*	<0.26
γ	1744.52 *10*	<0.07
γ	1765.0 *4*	0.30 *5*
γ	1769.7 *4*	0.32 *10*
γ	1779.5 *4*	0.12 *4*
γ	1791.3 *5*	<0.07
γ [E2]	1813.06 *10*	0.85 *10*
γ	1845.60 *24*	0.08 *3*
γ	1865.34 *13*	0.36 *7*
γ	1893.9 *3*	0.32 *10*
γ [M1+E2]	1901.51 *9*	0.78 *7*
γ	1933.18 *12*	0.39 *7* ?
γ	1940.77 *14*	<0.026
γ	1964.6 *5*	0.10 *3*
γ	1969.04 *13*	0.32 *6*
γ [E2]	1976.94 *10*	<0.09
γ	1999.7 *5*	0.08 *3*
γ	2008.62 *12*	0.32 *7*
γ [M1+E2]	2051.22 *9*	0.98 *7*
γ	2091.2 *5*	0.07 *3*
γ	2096.5 *3*	<0.09 ?
γ [E2]	2126.65 *10*	<0.17
γ	2205.2 *5*	~0.05
γ	2230.3 *4*	<0.07 ?
γ	2276.8 *3*	<0.07
γ	2288.9 *4*	<0.20 ?
γ	2334.7 *3*	<0.039 ?
γ	2364.3 *4*	<0.013 ?
γ	2514.9 *3*	<0.09
γ	2543.9 *4*	<0.08 ?
γ	2570.5 *5*	<0.039
γ	2590.4 *3*	<0.013
γ	2639.5 *10*	<0.039
γ	2753.8 *10*	<0.013
γ	2801 *1*	<0.013
γ	2831 *1*	<0.013

† 15% uncert(syst)

Atomic Electrons (^{137}Nd)

⟨e⟩=54 *16* keV

e_{bin}(keV)	⟨e⟩(keV)	e(%)
6	5.4	89 *25*
7 - 30	2.26	13.8 *11*
33	14.8	44 *5*
34 - 41	1.07	3.06 *21*
69	17.9	~26
74	4.3	~6
75 - 111	1.3	~2
126 - 175	0.30	0.20 *3*
179 - 228	1.22	0.61 *5*
229 - 261	0.33	0.141 *22*
265	1.20	0.45 *8*
266 - 314	0.52	0.18 *3*
316 - 361	0.143	0.042 *8*

Atomic Electrons (^{137}Nd) (continued)

e_{bin}(keV)	⟨e⟩(keV)	e(%)
375 - 406	0.027	0.0072 *15*
433 - 475	0.59	0.13 *3*
483 - 531	0.17	0.035 *6*
534 - 582	0.80	0.15 *3*
586 - 635	0.045	0.0074 *14*
636 - 685	0.094	0.014 *3*
686 - 734	0.29	0.040 *11*
740 - 785	0.40	0.053 *10*
791 - 838	0.017	0.0020 *7*
841 - 887	0.26	0.029 *5*
898 - 943	0.052	0.0057 *8*
952 - 1001	0.006	0.0006 *3*
1003 - 1048	0.021	0.0020 *9*
1060 - 1110	0.020	0.0019 *4*
1111 - 1160	0.0097	0.00086 *20*
1161 - 1207	0.033	0.0027 *9*
1211 - 1257	0.016	0.0013 *3*
1261 - 1310	0.011	0.0009 *3*
1313 - 1360	0.019	0.0014 *4*
1364 - 1410	0.0064	0.00046 *12*
1421 - 1466	0.019	0.0014 *4*
1471 - 1569	0.020	0.00131 *23*
1573 - 1661	0.016	0.0010 *3*
1689 - 1785	0.019	0.00106 *16*
1790 - 1888	0.0174	0.00094 *14*
1891 - 1976	0.013	0.00068 *18*
1993 - 2091	0.0136	0.00067 *8*
2095 - 2188	0.0008	3.5 *16* $\times10^{-5}$
2198 - 2293	0.0013	$<6\times10^{-5}$
2322 - 2363	7.6 $\times10^{-5}$	$<3\times10^{-6}$
2473 - 2569	0.0008	$<3\times10^{-5}$
2584 - 2638	0.00015	$<6\times10^{-6}$
2712 - 2800	0.00013	$<5\times10^{-6}$
2824 - 2830	6.2 $\times10^{-6}$	$<2\times10^{-7}$

Continuous Radiation (^{137}Nd)

⟨β+⟩=292 keV;⟨IB⟩=5.1 keV

E_{bin}(keV)		⟨ ⟩(keV)	(%)
0 - 10	β+	6.3 $\times10^{-7}$	7.6 $\times10^{-6}$
	IB	0.0112	
10 - 20	β+	2.85 $\times10^{-5}$	0.000171
	IB	0.0109	0.076
20 - 40	β+	0.00088	0.00267
	IB	0.087	0.26
40 - 100	β+	0.0478	0.061
	IB	0.073	0.119
100 - 300	β+	2.62	1.18
	IB	0.20	0.108
300 - 600	β+	20.3	4.36
	IB	0.36	0.082
600 - 1300	β+	133	14.1
	IB	1.43	0.153
1300 - 2500	β+	135	8.2
	IB	2.5	0.141
2500 - 3725	β+	0.70	0.0273
	IB	0.39	0.0142
	Σβ+		28

$^{137}_{60}$Nd(1.60 *15* s)

Mode: IT

Δ: -78880 *60* keV

SpA: 4.2×10^{10} Ci/g

Prod: ^{119}Sn(^{22}Ne,4n); ^{122}Sn(^{20}Ne,5n); ^{122}Sn(^{22}Ne,7n); ^{126}Te(^{18}O,7n)

Photons (^{137}Nd)

⟨γ⟩=370 *13* keV

γ_{mode}	γ(keV)	γ(%)†
Nd L_ℓ	4.633	0.19 *3*
Nd L_η	5.146	0.087 *15*
Nd L_α	5.228	5.1 *7*
Nd L_β	5.844	4.9 *8*
Nd L_γ	6.662	0.67 *12*
Nd $K_{\alpha2}$	36.847	17.5 *7*
Nd $K_{\alpha1}$	37.361	31.9 *12*
Nd $K_{\beta1}$'	42.240	9.1 *3*
Nd $K_{\beta2}$'	43.562	2.65 *11*
γ M1+E2	108.52 *14*	34 *4*
γ M1+E2	177.48 *14*	57 *5*
γ E3	233.58 *23*	63.7
γ E2	286.00 *14*	20.8 *24*

† 1.1% uncert(syst)

Atomic Electrons (^{137}Nd)

⟨e⟩=144 *8* keV

e_{bin}(keV)	⟨e⟩(keV)	e(%)
6 - 42	5.8	69 *8*
65	20.3	31.3 *16*
101	3.5	3.5 *6*
102	7.7	~8
107 - 109	3.4	~3
134	18.5	13.8 *12*
170 - 177	6.6	3.9 *14*
190	38.2	20.1 *5*
226	4.89	2.16 *5*
227	23.8	10.46 *25*
232	6.76	2.91 *7*
233 - 280	4.8	1.99 *14*
284 - 286	0.170	0.060 *5*

$^{137}_{61}$Pm(2.4 *1* min)

Mode: ϵ

Δ: -74100 *310* keV syst

SpA: 5.70×10^{8} Ci/g

Prod: ^{141}Pr(α,8n); daughter ^{137}Sm

Photons (^{137}Pm)

⟨γ⟩=1020 *45* keV

γ_{mode}	γ(keV)	γ(%)†
Nd L_ℓ	4.633	0.27 *5*
Nd L_η	5.146	0.115 *22*
Nd L_α	5.228	7.2 *12*
Nd L_β	5.850	6.7 *13*
Nd L_γ	6.673	0.93 *18*
Nd $K_{\alpha2}$	36.847	26.8 *25*
Nd $K_{\alpha1}$	37.361	49 *5*
Nd $K_{\beta1}$'	42.240	13.9 *13*
Nd $K_{\beta2}$'	43.562	4.1 *4*
γ	87.0 *2*	5.7 *11*
γ M1+E2	108.52 *14*	35
γ M1,E2	160.35 *22*	1.6 *3*
γ M1+E2	177.48 *14*	42 *7* *
γ	192.8 *3*	0.56 *11*
γ	199.0 *4*	<0.7
γ	213.0 *5*	<0.35
γ	220.2 *5*	<0.35
γ E3	233.58 *23*	30 *6* *
γ M1,E2	268.87 *21*	8.5 *17*
γ E2	286.00 *14*	15.4 *25* *
γ	293.1 *5*	<0.35
γ	325.1 *5*	2.9 *6*
γ M1,E2	328.7 *3*	3.3 *7*
γ	340.3 *4*	0.46 *9*
γ M1,E2	352.3 *3*	1.30 *4*

Photons (^{137}Pm) (continued)

γ_{mode}	γ(keV)	γ(%)†
γ M1,E2	370.6 *3*	2.91 *6*
γ	379.5 *4*	0.46 *9*
γ M1,E2	389.04 *20*	2.9 *6*
γ	397.72 *23*	1.19 *24*
γ M1,E2	410.1 *4*	7.00 *14*
γ M1,E2	414.0 *4*	2.2 *4*
γ	419.1 *5*	<0.35
γ	433.98 *25*	0.66 *13* ?
γ [M1+E2]	457.2 *3*	4.2 *9*
γ M1,E2	459.2 *5*	1.9 *4*
γ	463.7 *5*	<0.35
γ M1,E2	470.48 *23*	3.64 *7*
γ [E2]	506.2 *3*	7.0 *14*
γ	511.9 *4*	
γ	525.03 *24*	1.4 *3*
γ	529.0 *3*	2.5 *5*
γ [M1+E2]	533.71 *24*	5.18 *7*
γ [M1+E2]	548.66 *22*	4.0 *8*
γ	565.76 *21*	2.62 *4*
γ M1+E2	580.8 *4*	13 *3*
γ	586.9 *4*	0.63 *13*
γ	591.5 *4*	<0.35
γ	611.6 *5*	<0.35
γ	645.7 *3*	0.66 *13*
γ	658.68 *23*	0.52 *10*
γ	669.2 *5*	<0.35
γ	672.7 *4*	0.39 *8*
γ M2	690.73 *21*	2.91 *7*
γ	695.4 *5*	<0.35
γ	703.5 *5*	<0.35
γ	712.8 *4*	0.70 *14*
γ	714.3 *4*	0.39 *8*
γ	722.9 *4*	0.52 *10*
γ	734.7 *4*	<0.35 ?
γ [E2]	743.24 *23*	0.66 *13* ?
γ	749.1 *5*	<0.35
γ	759.8 *3*	1.08 *22*
γ	788.8 *5*	0.84 *17*
γ [E2]	799.1 *4*	0.49 *10* ?
γ	818 *1*	<0.35
γ	822.7 *3*	0.77 *15*
γ	829 *1*	<0.35
γ	836.8 *6*	0.98 *20*
γ	854 *1*	<0.35
γ	871 *1*	<0.35
γ	880 *1*	<0.35
γ	895 *1*	<0.35
γ	921.2 *6*	0.66 *13*
γ [M1+E2]	922.75 *23*	2.4 *5*
γ	939 *1*	<0.35
γ	946 *1*	<0.35
γ	968 *1*	<0.35
γ	977 *1*	<0.35
γ	991 *1*	<0.35
γ	995.5 *3*	0.42 *8* ?
γ	1013.7 *5*	0.70 *14*
γ	1028 *1*	<0.35
γ	1038 *1*	<0.35
γ [E1]	1047.72 *24*	0.70 *14*
γ [E1]	1064.8 *3*	0.98 *20*
γ	1090 *1*	0.42 *8*
γ	1092.0 *4*	4.4 *9*
γ	1123.6 *5*	0.39 *8*
γ	1136 *1*	0.42 *8*
γ	1141 *1*	<0.35
γ	1145 *1*	<0.35
γ	1153 *1*	<0.35
γ	1171 *1*	<0.35
γ	1189.6 *4*	1.15 *23*
γ	1201 *1*	<0.35
γ	1278.9 *8*	0.56 *11*
γ [E1]	1284.8 *3*	6.8 *14*
γ	1293.5 *3*	~1
γ	1381 *1*	<0.35
γ	1409 *1*	<0.35
γ	1461 *1*	<0.35
γ	1468 *1*	<0.35
γ	1518 *1*	0.42 *8*
γ	1551 *1*	<0.35
γ	1605 *1*	<0.35

† 29% uncert(syst)

* with ^{137}Nd(1.6 s) in equilib

Atomic Electrons (^{137}Pm)

⟨e⟩=118 *11* keV

e_{bin}(keV)	⟨e⟩(keV)	e(%)
6	3.3	53 *10*
7 - 43	7.7	50 *9*
65	21.9	34 *7*
80 - 87	4.8	~6
101	3.8	3.7 *10*
102	8.3	~8
107 - 117	4.2	~4
134	12.8	9.6 *17*
149 - 187	5.0	3.0 *10*
190	17.8	9.3 *19*
191 - 226	3.6	1.60 *25*
227	11.0	4.9 *10*
232	3.1	1.4 *3*
233 - 282	3.7	1.48 *16*
284 - 334	1.09	0.36 *5*
336 - 383	1.23	0.34 *5*
387 - 434	0.88	0.21 *3*
450 - 499	0.9	0.19 *6*
500 - 549	1.12	0.21 *5*
559 - 605	0.18	0.031 *7*
610 - 659	0.490	0.076 *4*
660 - 708	0.159	0.023 *3*
711 - 760	0.074	0.010 *3*
774 - 823	0.063	0.0080 *25*
827 - 874	0.018	0.0022 *9*
878 - 924	0.100	0.0113 *25*
932 - 977	0.030	0.0031 *11*
984 - 1032	0.024	0.0024 *6*
1036 - 1086	0.10	0.009 *4*
1088 - 1138	0.024	0.0022 *7*
1139 - 1188	0.024	0.0021 *9*
1189 - 1235	0.008	~0.0007
1241 - 1287	0.068	0.0054 *11*
1292 - 1337	0.0028	~0.00021
1365 - 1409	0.0031	~0.00022
1417 - 1467	0.005	~0.00036
1468 - 1561	0.010	0.00066 *25*
1598 - 1604	0.0003	~2×10^{-5}

Continuous Radiation (^{137}Pm)

⟨β+⟩=777 keV;⟨IB⟩=7.3 keV

E_{bin}(keV)		⟨ ⟩(keV)	(%)
0 - 10	β+	6.2×10^{-7}	7.4×10^{-6}
	IB	0.026	
10 - 20	β+	2.85×10^{-5}	0.000171
	IB	0.026	0.18
20 - 40	β+	0.00090	0.00273
	IB	0.084	0.27
40 - 100	β+	0.0506	0.064
	IB	0.165	0.27
100 - 300	β+	2.90	1.30
	IB	0.44	0.25
300 - 600	β+	23.8	5.1
	IB	0.61	0.142
600 - 1300	β+	187	19.4
	IB	1.62	0.18
1300 - 2500	β+	439	24.4
	IB	2.9	0.160
2500 - 4780	β+	124	4.40
	IB	1.41	0.047
	Σβ+		55

$^{137}_{62}$Sm(44 *8* s)

Mode: ϵ

SpA: 1.9×10^{9} Ci/g

Prod: protons on Gd

$^{137}_{64}$Gd(7 *3* s)

Mode: ϵ, ϵp
SpA: 1.1×10^{10} Ci/g
Prod: ^{50}Cr(^{92}Mo,αn);

ϵp: 2200-6600 range

A = 138

NDS **36**, 289 (1982)

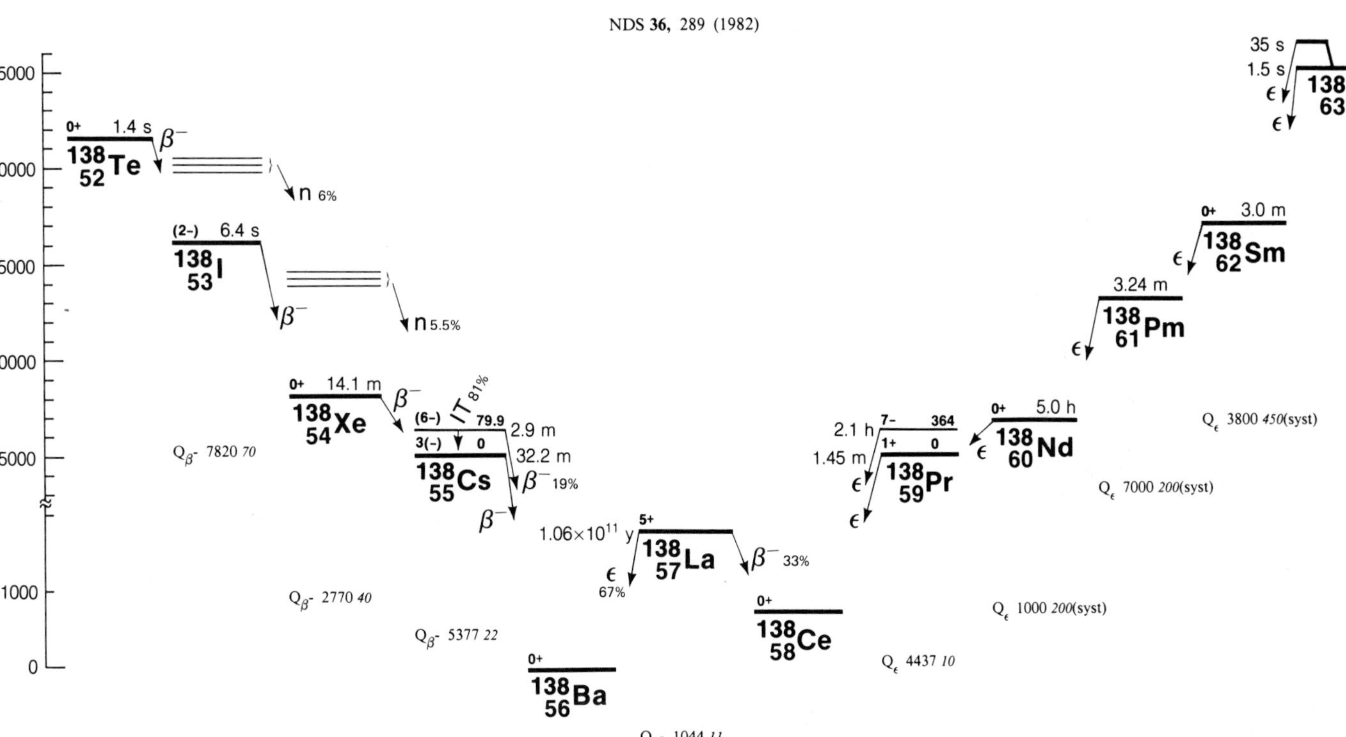

$^{138}_{52}$Te(1.4 *4* s)

Mode: β-, β-n(6 *4* %)
SpA: 4.6×10^{10} Ci/g
Prod: fission

$^{138}_{53}$I (6.41 *6* s)

Mode: β-, β-n(5.5 *4* %)
Δ: -72310 *80* keV
SpA: 1.209×10^{10} Ci/g
Prod: fission

Photons (^{138}I)

⟨γ⟩=1883 *25* keV

γ_{mode}	γ(keV)	γ(%)†
Xe $K_{\alpha 2}$	29.461	0.124 *7*
Xe $K_{\alpha 1}$	29.782	0.231 *13*
Xe $K_{\beta 1}$'	33.606	0.062 *3*
Xe $K_{\beta 2}$'	34.606	0.0148 *8*
$\gamma_{\beta-}$	430.7 *3*	0.39 *16*
$\gamma_{\beta-}$[E2]	483.673 *24*	4.91 *23*
$\gamma_{\beta-}$[E2]	588.801 *17*	78 *2*

Photons (^{138}I) (continued)

γ_{mode}	γ(keV)	γ(%)†
$\gamma_{\beta-}$	650.64 *14*	0.33 *8* ?
$\gamma_{\beta-}$	738.5 *5*	0.35 *8*
$\gamma_{\beta-}$	779.0 *5*	0.27 *8*
$\gamma_{\beta-}$	830.63 *6*	2.25 *14*
$\gamma_{\beta-}$	837.86 *19*	0.18 *6* ?
$\gamma_{\beta-}$	849.78 *16*	0.19 *4* ?
$\gamma_{\beta-}$	869.8 *3*	4.5 *7*
$\gamma_{\beta-}$	875.19 *8*	12.9 *11*
$\gamma_{\beta-}$[E1]	942.91 *8*	0.92 *5*
$\gamma_{\beta-}$	987.27 *16*	0.22 *6* ?
$\gamma_{\beta-}$	1108.33 *12*	0.55 *4*
$\gamma_{\beta-}$	1258.84 *14*	1.01 *6*
$\gamma_{\beta-}$	1277.44 *8*	3.31 *16*
$\gamma_{\beta-}$	1314.30 *6*	1.40 *5*
$\gamma_{\beta-}$	1332.24 *13*	0.226 *16*
$\gamma_{\beta-}$	1355.41 *13*	0.55 *5* ?
$\gamma_{\beta-}$	1371.62 *19*	0.37 *8*
$\gamma_{\beta-}$[E1]	1426.58 *8*	1.6 *5*
$\gamma_{\beta-}$	1444.87 *17*	0.359 *23*
$\gamma_{\beta-}$	1463.98 *8*	0.86 *9*
$\gamma_{\beta-}$	1500.42 *10*	1.27 *7*
$\gamma_{\beta-}$	1525.83 *12*	1.18 *9*
$\gamma_{\beta-}$	1528.34 *15*	0.92 *6*
$\gamma_{\beta-}$	1567.76 *17*	0.32 *8* ?
$\gamma_{\beta-}$	1623.65 *13*	0.73 *5*
$\gamma_{\beta-}$	1673.29 *7*	1.71 *6*
$\gamma_{\beta-}$	1678.59 *19*	0.37 *4*
$\gamma_{\beta-}$	1697.30 *19*	0.38 *5*
$\gamma_{\beta-}$	1707.07 *20*	0.34 *6*
$\gamma_{\beta-}$	1745.0 *3*	1.49 *16*
$\gamma_{\beta-}$	1809.3 *1*	2.85 *14*
$\gamma_{\beta-}$	1835.49 *17*	0.34 *3*
$\gamma_{\beta-}$	1866.24 *9*	0.47 *4*

Photons (^{138}I) (continued)

γ_{mode}	γ(keV)	γ(%)†
$\gamma_{\beta-}$	1889.01 *19*	0.53 *5*
$\gamma_{\beta-}$	1919.85 *12*	0.195 *16*
$\gamma_{\beta-}$	1946.25 *13*	0.569 *23*
$\gamma_{\beta-}$	1954.85 *9*	1.03 *3*
$\gamma_{\beta-}$	2032.81 *11*	1.20 *8*
$\gamma_{\beta-}$	2058.82 *14*	1.12 *6*
$\gamma_{\beta-}$	2085.38 *11*	1.08 *5*
$\gamma_{\beta-}$	2114.62 *12*	0.21 *4*
$\gamma_{\beta-}$	2151.47 *20*	0.148 *23*
$\gamma_{\beta-}$	2157.72 *18*	0.234 *16*
$\gamma_{\beta-}$	2262.09 *7*	5.38 *16*
$\gamma_{\beta-}$	2301.56 *15*	1.43 *6*
$\gamma_{\beta-}$	2307.6 *4*	0.20 *3*
$\gamma_{\beta-}$	2363.67 *15*	0.390 *23*
$\gamma_{\beta-}$	2375.60 *11*	0.218 *23* ?
$\gamma_{\beta-}$	2398.09 *10*	0.93 *4*
$\gamma_{\beta-}$	2543.64 *9*	1.11 *3*
$\gamma_{\beta-}$	2572.31 *11*	1.67 *5*
$\gamma_{\beta-}$	2644.57 *20*	0.35 *5*
$\gamma_{\beta-}$	2670.4 *4*	0.117 *16*
$\gamma_{\beta-}$	2674.17 *11*	0.164 *23*
$\gamma_{\beta-}$	2685.34 *15*	0.257 *16*
$\gamma_{\beta-}$	2700.07 *20*	0.140 *16*
$\gamma_{\beta-}$	2793.9 *3*	0.273 *23*
$\gamma_{\beta-}$	2806.5 *3*	0.125 *23*
$\gamma_{\beta-}$	2835.59 *18*	1.73 *6*
$\gamma_{\beta-}$	2842.6 *5*	0.086 *23*
$\gamma_{\beta-}$	2927.67 *11*	0.164 *8*
$\gamma_{\beta-}$	2964.39 *11*	0.218 *16*
$\gamma_{\beta-}$	3025.8 *5*	0.15 *4*
$\gamma_{\beta-}$	3141.47 *20*	0.296 *16*
$\gamma_{\beta-}$	3310.24 *10*	1.52 *6*
$\gamma_{\beta-}$	3366.9 *3*	0.11 *3*

Photons (^{138}I) (continued)

γ_{mode}	γ(keV)	γ(%)†
$\gamma_{\beta-}$	3372.9 6	0.19 3
$\gamma_{\beta-}$	3452.3 3	0.172 16
$\gamma_{\beta-}$	3458.57 20	0.250 16
$\gamma_{\beta-}$	3496.62 15	1.02 6
$\gamma_{\beta-}$	3516.46 11	0.226 23
$\gamma_{\beta-}$	3578.4 3	0.265 23
$\gamma_{\beta-}$	3584.67 20	0.421 23
$\gamma_{\beta-}$	3593.12 11	0.343 16
$\gamma_{\beta-}$	3888.07 20	0.30 3
$\gamma_{\beta-}$	3899.03 11	0.062 16
$\gamma_{\beta-}$	3905.9 3	0.125 16
$\gamma_{\beta-}$	4014.1 3	0.125 16
$\gamma_{\beta-}$	4089.97 20	0.312 16
$\gamma_{\beta-}$	4099.67 20	0.195 16
$\gamma_{\beta-}$	4128.1 3	0.140 16
$\gamma_{\beta-}$	4181.90 11	0.87 6
$\gamma_{\beta-}$	4306.8 3	0.226 16
$\gamma_{\beta-}$	4318.87 20	0.57 4
$\gamma_{\beta-}$	4496.1 3	0.226 16
$\gamma_{\beta-}$	4515.2 5	0.117 16
$\gamma_{\beta-}$	4697.0 3	0.335 23
$\gamma_{\beta-}$	4720.0 4	0.070 8
$\gamma_{\beta-}$	4752.67 25	0.117 8
$\gamma_{\beta-}$	4973.9 3	0.140 8
$\gamma_{\beta-}$	5004.8 4	0.164 16
$\gamma_{\beta-}$	5261.1 4	0.086 16
$\gamma_{\beta-}$	5329.7 6	0.109 16
$\gamma_{\beta-}$	5341.45 25	0.172 23

† 2.6% uncert(syst)

Atomic Electrons (^{138}I)

⟨e⟩=3.53 16 keV

e_{bin}(keV)	⟨e⟩(keV)	e(%)
5 - 33	0.033	0.44 4
396 - 431	0.017	~0.0041
449	0.185	0.041 3
478 - 484	0.0365	0.0076 3
554	2.13	0.385 20
583	0.256	0.0439 23
584	0.060	0.0102 5
588	0.065	0.0111 6
589 - 616	0.023	0.0038 7
645 - 651	0.0012	0.00018 8
704 - 744	0.012	0.0016 6
774 - 815	0.040	0.0050 24
825 - 838	0.07	~0.008
841	0.18	~0.022
844 - 875	0.041	0.0047 17
908 - 953	0.0091	0.00099 17
982 - 987	0.00043	~4 $\times10^{-5}$
1074 - 1108	0.007	0.0006 3
1224 - 1273	0.045	0.0036 13
1276 - 1321	0.022	0.0017 5
1327 - 1372	0.0047	0.00035 12
1392 - 1440	0.018	0.0013 3
1444 - 1491	0.020	0.0014 4
1494 - 1589	0.019	0.0012 3
1618 - 1710	0.033	0.0020 4
1740 - 1835	0.028	0.0016 5
1854 - 1954	0.017	0.00088 17
1998 - 2084	0.024	0.0012 3
2109 - 2157	0.0026	0.00012 3
2228 - 2307	0.042	0.0019 5
2329 - 2397	0.0089	0.00038 8
2509 - 2571	0.015	0.00060 13
2610 - 2699	0.0054	0.00021 3
2759 - 2842	0.0112	0.00040 9
2893 - 2991	0.0025	8.5 15 $\times10^{-5}$
3020 - 3107	0.0013	4.2 11 $\times10^{-5}$
3136 - 3141	0.00018	5.7 15 $\times10^{-6}$
3276 - 3372	0.0082	0.00025 5
3418 - 3516	0.0072	0.00021 3
3544 - 3592	0.0044	0.000123 17
3854 - 3905	0.0020	5.1 8 $\times10^{-5}$
3980 - 4065	0.0022	5.5 8 $\times10^{-5}$

Atomic Electrons (^{138}I) (continued)

e_{bin}(keV)	⟨e⟩(keV)	e(%)
4085 - 4181	0.0042	0.000101 18
4272 - 4318	0.0030	7.1 11 $\times10^{-5}$
4462 - 4514	0.00129	2.9 5 $\times10^{-5}$
4662 - 4752	0.0019	4.1 6 $\times10^{-5}$
4939 - 4973	0.00050	1.02 19 $\times10^{-5}$

Continuous Radiation (^{138}I)

⟨β-⟩=2588 keV;⟨IB⟩=10.5 keV

E_{bin}(keV)		⟨ ⟩(keV)	(%)
0 - 10	β-	0.00296	0.059
	IB	0.066	
10 - 20	β-	0.0091	0.060
	IB	0.065	0.45
20 - 40	β-	0.0378	0.126
	IB	0.130	0.45
40 - 100	β-	0.294	0.417
	IB	0.38	0.58
100 - 300	β-	3.82	1.85
	IB	1.17	0.65
300 - 600	β-	18.8	4.08
	IB	1.52	0.35
600 - 1300	β-	136	14.0
	IB	2.7	0.31
1300 - 2500	β-	568	29.7
	IB	2.8	0.157
2500 - 5000	β-	1554	44.2
	IB	1.67	0.052
5000 - 7231	β-	308	5.5
	IB	0.068	0.00126

$^{138}_{54}$Xe(14.08 8 min)

Mode: β-

Δ: -80130 50 keV

SpA: 9.67$\times10^{7}$ Ci/g

Prod: fission

Photons (^{138}Xe)

⟨γ⟩=1126 9 keV

γ_{mode}	γ(keV)	γ(%)†
γ M1	4.90 3	0.19 3
γ M1	10.89 3	0.70 14
Cs $K_{\alpha2}$	30.625	1.10 8
Cs $K_{\alpha1}$	30.973	2.03 14
Cs $K_{\beta1}$'	34.967	0.55 4
Cs $K_{\beta2}$'	36.006	0.136 10
γ	68.38 15	0.047 16
γ	137.31 15	~0.07
γ M1,E2	153.904 3	5.95 13
γ	197.10 5	
γ M1,E2	242.656 10	3.50 7
γ M1,E2	258.446 18	31.5 6
γ	282.61 3	0.428 13
γ	324.47 8	0.023 8
γ	329.4 5	0.015 8
γ	335.37 8	0.107 10
γ	371.52 4	0.501 25
γ M1,E2	396.560 10	6.30 13
γ M1	401.46 3	2.17 10
γ	403.75 15	
γ M1,E2	434.61 4	20.3 4
γ	500.22 3	0.362 13
γ	530.16 4	0.252 13
γ	533.38 18	0.015 6
γ	537.88 10	0.117 16

Photons (^{138}Xe) (continued)

γ_{mode}	γ(keV)	γ(%)†
γ	541.05 4	~0.022
γ	556.02 9	0.117 13
γ	568.62 4	0.306 16
γ	579.73 14	0.076 13
γ	585.85 21	0.019 7
γ	588.88 8	0.123 9
γ	619.93 7	~0.022
γ	647.07 13	~0.016
γ	654.13 3	0.145 13
γ	675.44 10	0.072 13
γ	680.34 11	0.054 13
γ	691.23 11	0.032 13
γ	693.72 6	0.088 13
γ	697.32 4	0.022 9
γ	703.60 12	0.057 10
γ	733.9 4	0.032 10
γ	745.11 6	
γ	754.83 12	~0.025
γ	774.30 8	0.066 10
γ	778.11 13	0.044 10
γ	792.89 12	0.022 9
γ	799.74 21	~0.016
γ	816.21 16	0.072 13
γ	848.78 12	0.044 13
γ	851.22 4	0.069 13
γ	865.86 6	0.296 19
γ	869.39 5	0.62 4
γ	869.87 14	
γ	896.78 3	0.132 13
γ	902.54 7	0.044 13
γ	912.57 3	0.328 19
γ	917.18 4	0.920 18
γ	936.37 6	0.135 13
γ	941.27 6	0.230 16
γ	946.79 12	0.063 13
γ	953.09 23	0.028 13
γ	996.96 21	0.063 16
γ	1076.61 10	0.088 16
γ	1093.88 4	0.410 25
γ	1098.77 4	0.214 16
γ	1102.25 8	0.107 13
γ	1114.27 5	1.47 7
γ	1141.67 6	0.51 3
γ	1145.19 7	0.132 19
γ	1153.57 7	~0.032
γ	1160.98 7	0.098 13
γ	1189.45 12	0.082 13
γ	1194.34 12	0.088 13
γ	1205.24 12	0.035 13
γ	1218.7 5	0.038 16
γ	1228.08 8	0.063 22
γ	1311.07 8	0.085 16
γ	1356.43 17	0.050 16
γ	1361.33 17	0.035 16
γ	1381.76 14	0.069 16
γ	1385.58 9	0.076 16
γ	1473.17 10	0.069 13
γ	1548.74 13	0.076 19
γ	1571.93 15	0.265 25
γ	1578.53 14	0.050 19
γ	1614.49 4	0.236 25
γ	1646.51 12	0.066 13
γ	1768.39 4	16.7 4
γ	1783.21 14	0.038 16
γ	1799.86 17	0.035 13
γ	1812.82 7	0.180 19
γ	1850.88 6	1.42 5
γ	1887.34 8	0.069 13
γ	1925.39 7	0.56 3
γ	2004.78 6	5.35 13
γ [E1]	2015.94 5	12.25 25
γ	2040.69 14	0.032 10
γ	2079.29 7	1.44 4
γ [E1]	2252.33 7	2.29 7
γ	2266.8 5	0.038 13
γ [E1]	2321.95 7	0.62 3
γ [E1]	2326.84 7	0.057 10
γ [E1]	2475.30 14	0.312 16
γ [E1]	2492.69 14	0.054 6
γ [E1]	2497.59 14	0.173 13

† 3.5% uncert(syst)

Atomic Electrons (^{138}Xe)

⟨e⟩=11.0 5 keV

e_{bin}(keV)	⟨e⟩(keV)	e(%)
5 - 35	0.33	4.1 5
63 - 101	0.08	~0.11
118	1.9	1.61 23
132 - 137	0.009	~0.007
148	0.260	0.175 11
149 - 154	0.39	~0.26
207	0.50	0.241 15
222	4.02	1.81 11
237 - 247	0.15	0.062 12
253	0.74	0.29 9
257 - 299	0.21	0.082 20
319 - 336	0.025	~0.008
361	0.39	0.108 13
365 - 397	0.249	0.066 4
399	1.09	0.27 4
400 - 435	0.21	0.050 8
464 - 505	0.025	0.0052 18
520 - 569	0.021	0.0039 11
574 - 620	0.0051	0.0008 3
639 - 689	0.0080	0.0012 3
690 - 738	0.0027	0.00037 12
742 - 788	0.0031	0.00040 12
792 - 833	0.016	0.0019 7
843 - 892	0.023	0.0027 9
896 - 942	0.010	0.0012 3
946 - 992	0.0010	0.00011 5
996 - 1041	0.0011	~0.00010
1058 - 1106	0.033	0.0030 9
1109 - 1156	0.0078	0.00070 19
1158 - 1205	0.0027	0.00023 6
1213 - 1228	0.00017	1.4 5 $\times10^{-5}$
1275 - 1320	0.0014	0.00011 4
1325 - 1361	0.0017	0.00013 4
1376 - 1386	0.00021	1.5 5 $\times10^{-5}$
1437 - 1536	0.0034	0.00022 7
1543 - 1641	0.0035	0.00022 6
1645 - 1732	0.12	0.007 3
1747 - 1846	0.031	0.0017 4
1850 - 1924	0.0050	0.00026 8
1969 - 2043	0.100	0.0051 7
2074 - 2078	0.0013	6.4 22 $\times10^{-5}$
2216 - 2291	0.0119	0.000533 20
2316 - 2326	0.000347	1.50 7 $\times10^{-5}$
2439 - 2497	0.00208	8.5 4 $\times10^{-5}$

Continuous Radiation (^{138}Xe)

⟨β-⟩=628 keV;⟨IB⟩=1.25 keV

E_{bin}(keV)		⟨ ⟩(keV)	(%)
0 - 10	β-	0.073	1.46
	IB	0.025	
10 - 20	β-	0.219	1.46
	IB	0.024	0.167
20 - 40	β-	0.87	2.91
	IB	0.046	0.161
40 - 100	β-	6.0	8.6
	IB	0.126	0.19
100 - 300	β-	48.5	24.9
	IB	0.32	0.18
300 - 600	β-	90	20.9
	IB	0.31	0.074
600 - 1300	β-	228	24.3
	IB	0.33	0.040
1300 - 2500	β-	253	15.4
	IB	0.068	0.0043
2500 - 2759	β-	1.37	0.054
	IB	2.1 $\times10^{-5}$	8.4 $\times10^{-7}$

$^{138}_{55}$Cs(32.2 *1* min)

Mode: β-

Δ: -82900 23 keV

SpA: 4.231$\times10^{7}$ Ci/g

Prod: fission; ^{138}Ba(n,p)

Photons (^{138}Cs)

⟨γ⟩=2361 25 keV

γ_{mode}	γ(keV)	γ(%)†
Ba $K_{\alpha2}$	31.817	0.429 25
Ba $K_{\alpha1}$	32.194	0.79 5
Ba $K_{\beta1}$'	36.357	0.219 13
Ba $K_{\beta2}$'	37.450	0.055 3
γ [M1]	107.87 7	~0.06
γ [M1]	112.58 8	0.130 23
γ M1,E2	138.04 4	1.49 8
γ [E2]	191.94 5	0.50 4
γ [M1+E2]	193.89 5	0.328 23
γ E2,M1	212.30 7	0.172 13
γ M1,E2	227.72 4	1.51 4
γ [M1]	324.88 6	0.298 18
γ [M1+E2]	333.78 6	0.089 15
γ [M1+E2]	363.96 6	0.244 23
γ	365.22 7	0.191 23
γ	368.8 3	0.022 8
γ M1(+20%E2)	408.95 4	4.66 9
γ [M1+E2]	421.61 5	0.427 23
γ E2	462.782 5	30.7 6
γ [M1]	516.81 6	0.43 5
γ M1+0.6%E2	546.987 5	10.76 23
γ [E2]	575.71 9	0.021 8
γ [M1+E2]	596.14 22	0.026 10
γ [M1+E2]	683.57 7	0.108 14
γ	702.77 11	0.084 13
γ	717.89 11	0.040 12
γ	754.6 3	0.034 12
γ [M1+E2]	766.07 11	0.146 14
γ [M1+E2]	773.25 7	0.233 18
γ [M1+E2]	782.05 4	0.33 3
γ	797.7 5	0.053 23
γ	802.6 3	~0.038
γ	813.01 15	0.060 18
γ	842.33 12	0.082 11
γ	855.93 11	0.023 9
γ E2	871.73 4	5.12 13
γ [M1+E2]	880.77 6	0.11 3
γ	934.98 10	0.181 16
γ	945.61 11	0.031 13
γ	952.96 10	0.053 14
γ M1(+E2)	1009.767 7	29.8 6
γ [M1+E2]	1041.60 18	0.063 17
γ [E1]	1054.28 13	0.159 19
γ	1147.27 7	1.24 7
γ	1199.14 24	0.17 3
γ [M1+E2]	1203.66 5	0.40 4
γ	1264.88 11	0.137 17
γ [E2]	1343.55 6	1.14 5
γ	1359.01 24	0.048 19
γ	1386.34 11	0.076 11
γ	1415.74 10	0.37 3
γ E2	1435.798 10	76.3 15
γ [E1]	1445.06 13	0.97 19
γ [M1+E2]	1495.61 20	0.18 4
γ [M1+E2]	1555.30 6	0.366 23
γ	1614.08 17	0.137 23
γ [M1+E2]	1717.28 14	0.107 23
γ	1727.66 11	0.111 13
γ [E1]	1748.34 16	0.07 3
γ	1778.18 14	0.137 23
γ	1806.71 10	0.092 11
γ	1821.78 24	0.045 10
γ [M1+E2]	1903.20 19	0.046 14
γ	1941.0 3	0.079 15
γ [E1]	2023.87 16	0.118 15
γ	2062.37 14	0.111 11
γ	2105.9 3	0.055 10
γ	2113.6 3	0.021 9
γ [E1]	2211.11 16	0.21 6
γ [E2]	2217.84 4	15.2 3
γ [E1]	2486.64 16	0.023 8
γ [M1+E2]	2499.33 14	0.17 5

Photons (^{138}Cs) (continued)

γ_{mode}	γ(keV)	γ(%)†
γ	2510.5 8	0.015 7
γ	2583.06 7	0.239 15
γ	2609.35 14	0.034 5
γ [E2]	2639.44 5	7.63 23
γ	2731.13 13	0.120 8
γ	2806.56 17	0.100 8
γ [M1]	2931.39 20	0.020 4
γ	3049.86 17	0.031 5
γ	3072.12 14	0.019 4
γ	3180.4 7	0.0084 23
γ [E2]	3338.97 19	0.151 9
γ	3352.6 3	0.035 4
γ [E2]	3366.97 25	0.227 13
γ	3437.5 6	0.011 3
γ	3442.6 6	0.011 3
γ [E2]	3643.4 3	0.023 3
γ	3652.5 8	0.0053 15
γ [E2]	3935.10 14	0.018 3
γ	4080.1 5	0.0175 23

† 0.39% uncert(syst)

Atomic Electrons (^{138}Cs)

⟨e⟩=6.37 16 keV

e_{bin}(keV)	⟨e⟩(keV)	e(%)
5 - 36	0.119	1.52 15
70 - 75	0.087	0.119 24
101	0.59	0.59 9
102 - 113	0.021	0.020 4
132	0.14	0.10 4
133 - 175	0.33	0.22 4
186 - 189	0.051	0.027 3
190	0.253	0.133 8
191 - 228	0.088	0.041 9
287 - 334	0.074	0.0241 24
358 - 369	0.0062	0.0017 4
372	0.319	0.086 6
384 - 422	0.089	0.0224 16
425	1.34	0.316 9
457	0.200	0.0438 12
458 - 479	0.106	0.0227 6
510	0.508	0.100 3
511 - 559	0.0925	0.0171 4
570 - 596	0.00029	5.0 12 $\times10^{-5}$
646 - 683	0.0064	0.00097 21
697 - 745	0.0187	0.0025 3
749 - 798	0.0061	0.00080 18
801 - 851	0.095	0.0114 4
855 - 898	0.0189	0.00217 17
908 - 953	0.0019	0.00021 7
972	0.54	0.055 10
1004 - 1053	0.091	0.0091 14
1110 - 1147	0.017	0.0015 7
1162 - 1204	0.0089	0.00076 13
1227 - 1265	0.0017	0.00014 6
1306 - 1354	0.0161	0.00123 7
1358 - 1386	0.0036	0.00026 12
1398	0.799	0.0572 16
1408 - 1445	0.136	0.00948 21
1458 - 1554	0.0071	0.00048 5
1577 - 1613	0.0013	8 4 $\times10^{-5}$
1680 - 1777	0.0044	0.00026 4
1784 - 1866	0.00090	4.9 12 $\times10^{-5}$
1897 - 1986	0.0011	5.9 13 $\times10^{-5}$
2018 - 2113	0.0014	7.1 17 $\times10^{-5}$
2174 - 2217	0.128	0.00585 23
2449 - 2546	0.0028	0.000112 22
2572 - 2638	0.0562	0.00216 9
2694 - 2769	0.0012	4.4 10 $\times10^{-5}$
2801 - 2894	0.00020	7.0 12 $\times10^{-6}$
2925 - 3012	0.00016	5.5 16 $\times10^{-6}$
3035 - 3071	0.00012	4.1 11 $\times10^{-6}$
3302 - 3400	0.00256	7.7 4 $\times10^{-5}$
3405 - 3442	6.3 $\times10^{-5}$	1.9 6 $\times10^{-6}$
3606 - 3651	0.000159	4.4 5 $\times10^{-6}$
4043 - 4079	0.000172	4.2 5 $\times10^{-6}$

Continuous Radiation (^{138}Cs)

⟨β-⟩=1237 keV;⟨IB⟩=3.2 keV

E_{bin}(keV)		⟨ ⟩(keV)	(%)
0 - 10	β-	0.0094	0.186
	IB	0.043	
10 - 20	β-	0.0286	0.191
	IB	0.042	0.29
20 - 40	β-	0.119	0.395
	IB	0.083	0.29
40 - 100	β-	0.92	1.30
	IB	0.24	0.36
100 - 300	β-	11.5	5.6
	IB	0.67	0.38
300 - 600	β-	53	11.7
	IB	0.76	0.18
600 - 1300	β-	339	35.4
	IB	0.99	0.116
1300 - 2500	β-	737	41.4
	IB	0.41	0.025
2500 - 3941	β-	94	3.44
	IB	0.0109	0.00040

$^{138}_{55}$Cs(2.9 *1* min)

Mode: IT(81 *3* %), β-(19 *3* %)

Δ: -82820 *23* keV

SpA: 4.69×10^{8} Ci/g

Prod: ^{138}Ba(n,p); fission

Photons (^{138}Cs)

⟨γ⟩=419 *44* keV

γ_{mode}	γ(keV)	γ(%)†
Cs L_{ℓ}	3.795	0.114 *16*
Ba L_{ℓ}	3.954	0.007 *1*
Cs L_{η}	4.142	0.041 *5*
Cs L_{α}	4.285	3.2 *4*
Ba L_{η}	4.331	0.0032 *4*
Ba L_{α}	4.465	0.194 *23*
Cs L_{β}	4.746	2.6 *4*
Ba L_{β}	4.944	0.175 *23*
Cs L_{γ}	5.396	0.35 *6*
Ba L_{γ}	5.605	0.023 *3*
Cs $K_{\alpha2}$	30.625	11.5 *5*
Cs $K_{\alpha1}$	30.973	21.2 *10*
Ba $K_{\alpha2}$	31.817	0.97 *5*
Ba $K_{\alpha1}$	32.194	1.78 *10*
Cs $K_{\beta1}$'	34.967	5.8 *3*
Cs $K_{\beta2}$'	36.006	1.42 *7*
Ba $K_{\beta1}$'	36.357	0.49 *3*
Ba $K_{\beta2}$'	37.450	0.125 *7*
γ_{IT} M3	79.9 *3*	0.370
$\gamma_{\beta-}$[M1]	107.87 *7*	~0.19
$\gamma_{\beta-}$[M1]	112.58 *8*	1.52 *19*
$\gamma_{\beta-}$[E2]	191.94 *5*	15.4 *6*
$\gamma_{\beta-}$E2,M1	212.30 *7*	0.58 *6*
$\gamma_{\beta-}$[M1]	324.88 *6*	1.01 *10*
$\gamma_{\beta-}$M1(+20%E2)	408.95 *4*	0.109 *20*
$\gamma_{\beta-}$E2	462.782 *5*	18.7 *10*
$\gamma_{\beta-}$[M1]	516.81 *6*	1.45 *20*
$\gamma_{\beta-}$E2	871.73 *4*	0.120 *22*
$\gamma_{\beta-}$E2	1435.798 *10*	19

† uncert(syst): 3.7% for IT, 16% for β-

Atomic Electrons (^{138}Cs)

⟨e⟩=58.5 *11* keV

e_{bin}(keV)	⟨e⟩(keV)	e(%)
5	2.33	46 *5*
6 - 36	1.83	13.6 *10*
44	19.5	44.5 *18*
70	0.09	~0.13
74	9.9	13.3 *6*
75	11.7	15.6 *7*
79	5.15	6.5 *3*
80	1.33	1.67 *7*
102 - 113	0.195	0.182 *21*
154	3.63	2.35 *13*
175 - 212	1.48	0.79 *3*
287 - 325	0.119	0.041 *4*
372 - 409	0.0089	0.0024 *4*
425 - 463	0.99	0.230 *11*
479 - 517	0.087	0.0179 *22*
834 - 872	0.0025	0.00029 *5*
1398 - 1436	0.23	0.0165 *23*

Continuous Radiation (^{138}Cs)

⟨β-⟩=261 keV;⟨IB⟩=0.72 keV

E_{bin}(keV)		⟨ ⟩(keV)	(%)
0 - 10	β-	0.00132	0.0262
	IB	0.0087	
10 - 20	β-	0.00403	0.0269
	IB	0.0086	0.060
20 - 40	β-	0.0168	0.056
	IB	0.017	0.059
40 - 100	β-	0.131	0.185
	IB	0.049	0.075
100 - 300	β-	1.70	0.82
	IB	0.141	0.079
300 - 600	β-	8.3	1.81
	IB	0.163	0.038
600 - 1300	β-	59	6.1
	IB	0.23	0.026
1300 - 2500	β-	161	8.8
	IB	0.109	0.0066
2500 - 3366	β-	31.9	1.17
	IB	0.0029	0.000112

$^{138}_{56}$Ba(stable)

Δ: -88276 *6* keV

%: 71.70 *7*

$^{138}_{57}$La(1.06 *3* ×10^{11} yr)

Mode: ϵ(66.7 *14* %), β-(33.3 *14* %)

Δ: -86531 *5* keV

Prod: natural source

%: 0.09 *1*

Photons (^{138}La)

⟨γ⟩=1237 *23* keV

γ_{mode}	γ(keV)	γ(%)†
Ba L_{ℓ}	3.954	0.098 *13*
Ba L_{η}	4.331	0.043 *5*
Ba L_{α}	4.465	2.7 *3*
Ba L_{β}	4.944	2.5 *3*
Ba L_{γ}	5.628	0.34 *5*
Ba $K_{\alpha2}$	31.817	14.3 *4*

Photons (^{138}La) (continued)

γ_{mode}	γ(keV)	γ(%)†
Ba $K_{\alpha1}$	32.194	26.3 *8*
Ba $K_{\beta1}$'	36.357	7.27 *22*
Ba $K_{\beta2}$'	37.450	1.84 *6*
$\gamma_{\beta-}$E2	788.742 *8*	33.3
γ_{ϵ}E2	1435.798 *10*	66.7

† uncert(syst): 2.1% for ϵ, 4.2% for β-

Atomic Electrons (^{138}La)

⟨e⟩=4.68 *19* keV

e_{bin}(keV)	⟨e⟩(keV)	e(%)
5	1.34	26 *3*
6	1.02	17.7 *18*
25	0.111	0.43 *4*
26	0.31	1.20 *12*
27	0.52	1.94 *20*
30	0.116	0.38 *4*
31	0.33	1.06 *11*
32	0.071	0.221 *23*
35	0.035	0.101 *10*
36	0.0189	0.053 *5*
1398	0.699	0.0500 *14*
1430	0.0879	0.00615 *18*
1431	0.00249	0.000174 *5*
1435	0.0186	0.00130 *4*
1436	0.00423	0.000295 *9*

Continuous Radiation (^{138}La)

⟨β-⟩=23.8 keV;⟨IB⟩=0.064 keV

E_{bin}(keV)		⟨ ⟩(keV)	(%)
0 - 10	β-	0.154	3.11
	IB	0.00150	
10 - 20	β-	0.442	2.96
	IB	0.00126	0.0086
20 - 40	β-	1.62	5.5
	IB	0.052	0.165
40 - 100	β-	8.5	12.5
	IB	0.0061	0.0109
100 - 300	β-	13.1	9.3
	IB	0.0031	0.0022
300 - 309	IB	2.0×10^{-8}	6.6×10^{-9}

$^{138}_{58}$Ce(stable)

Δ: -87575 *11* keV

%: 0.25 *1*

$^{138}_{59}$Pr(1.45 *5* min)

Mode: ϵ

Δ: -83138 *15* keV

SpA: 9.4×10^{8} Ci/g

Prod: daughter ^{138}Nd; ^{138}Ce(p,n)

Photons (^{138}Pr)

⟨γ⟩=49 *3* keV

γ_{mode}	γ(keV)	γ(%)†
Ce L_ℓ	4.289	0.043 *6*
Ce L_η	4.730	0.0176 *18*
Ce L_α	4.838	1.17 *13*
Ce L_β	5.390	1.05 *13*
Ce L_γ	6.142	0.145 *20*
Ce $K_{\alpha2}$	34.279	5.57 *17*
Ce $K_{\alpha1}$	34.720	10.2 *3*
Ce $K_{\beta1}$'	39.232	2.87 *9*
Ce $K_{\beta2}$'	40.437	0.77 *3*
γ E2	688.20 *11*	0.82 *4*
γ [M1+E2]	721.70 *20*	0.077 *10*
γ E2	788.742 *8*	2.4
γ	1082.3 *3*	0.043 *7*
γ [E1]	1348.3 *4*	~0.011
γ	1426.3 *3*	~0.012
γ	1447.82 *19*	0.130 *17*
γ [E2]	1510.44 *20*	0.086 *12*
γ (E2)	1551.14 *14*	0.42 *5*
γ	1682.27 *18*	0.036 *7*
γ	1804.0 *3*	0.034 *7*
γ [M1+E2]	1853.7 *4*	0.024 *5*
γ	2114.5 *3*	0.038 *7*
γ	2236.56 *19*	0.079 *12*
γ	2471.00 *18*	0.053 *10*
γ [E2]	2642.4 *4*	0.008 *4*
γ	3177.4 *10*	0.009 *4*

† 17% uncert(syst)

Atomic Electrons (^{138}Pr)

⟨e⟩=1.90 *10* keV

e_{bin}(keV)	⟨e⟩(keV)	e(%)
6	0.84	14.3 *15*
7	0.112	1.72 *18*
27	0.042	0.154 *16*
28	0.118	0.42 *4*
29	0.188	0.65 *7*
32	0.0184	0.057 *6*
33	0.104	0.31 *3*
34	0.063	0.186 *19*
35 - 39	0.0301	0.081 *5*
648 - 688	0.0280	0.00426 *20*
715 - 748	0.053	0.0071 *12*
782 - 789	0.0097	0.00124 *15*
1042 - 1082	0.0007	7 *3* $\times10^{-5}$
1308 - 1348	7.0 $\times10^{-5}$	5 *3* $\times10^{-6}$
1386 - 1426	0.0015	0.00011 *5*
1436	0.25	0.0172 *21*
1441 - 1511	0.045	0.0030 *6*
1545 - 1642	0.00105	6.7 *11* $\times10^{-5}$
1676 - 1764	0.00033	1.9 *8* $\times10^{-5}$
1797 - 1852	0.00034	1.8 *3* $\times10^{-5}$
2074 - 2113	0.00032	1.5 *6* $\times10^{-5}$
2196 - 2235	0.00062	2.8 *10* $\times10^{-5}$
2431 - 2470	0.00038	1.6 *6* $\times10^{-5}$
2602 - 2641	6.9 $\times10^{-5}$	2.6 *10* $\times10^{-6}$
3137 - 3176	5.2 $\times10^{-5}$	1.6 *8* $\times10^{-6}$

Continuous Radiation (^{138}Pr)

⟨β+⟩=1159 keV;⟨IB⟩=8.1 keV

E_{bin}(keV)		⟨ ⟩(keV)	(%)
0 - 10	β+	6.4 $\times10^{-7}$	7.7 $\times10^{-6}$
	IB	0.038	
10 - 20	β+	2.84 $\times10^{-5}$	0.000171
	IB	0.037	0.26
20 - 40	β+	0.00086	0.00262
	IB	0.096	0.32
40 - 100	β+	0.0466	0.059
	IB	0.22	0.33

Continuous Radiation (^{138}Pr)
(continued)

E_{bin}(keV)		⟨ ⟩(keV)	(%)
100 - 300	β+	2.67	1.20
	IB	0.64	0.35
300 - 600	β+	23.3	4.96
	IB	0.81	0.19
600 - 1300	β+	220	22.5
	IB	1.68	0.19
1300 - 2500	β+	724	39.3
	IB	2.8	0.153
2500 - 4437	β+	189	6.9
	IB	1.8	0.061
	Σβ+		75

$^{138}_{59}$Pr(2.1 *1* h)

Mode: ε

Δ: -82774 *15* keV

SpA: 1.08$\times10^{7}$ Ci/g

Prod: ^{140}Ce(p,3n); ^{138}Ce(p,n); ^{139}La(^{3}He,4n); ^{141}Pr(p,p3n); protons on Ta; protons on Er

Photons (^{138}Pr)

⟨γ⟩=2238 *54* keV

γ_{mode}	γ(keV)	γ(%)
Ce L_ℓ	4.289	0.161 *22*
Ce L_η	4.730	0.068 *7*
Ce L_α	4.838	4.4 *5*
Ce L_β	5.387	4.0 *5*
Ce L_γ	6.137	0.55 *8*
Ce $K_{\alpha2}$	34.279	20.2 *7*
Ce $K_{\alpha1}$	34.720	36.9 *13*
Ce $K_{\beta1}$'	39.232	10.4 *4*
Ce $K_{\beta2}$'	40.437	2.8 *1*
γ	75.5 *10*	0.11 *5*
γ [E2]	80.4 *4*	0.12 *4*
γ	158.0 *14*	0.11 *5*
γ	170.0 *10*	~0.14
γ	177.5 *14*	0.064 *14*
γ	184.0 *14*	0.053 *16*
γ	196.0 *14*	0.083 *17*
γ	206.0 *14*	0.16 *7*
γ	231.0 *14*	~0.12
γ E3	302.68 *12*	80 *5*
γ	351.0 *14*	0.105 *21*
γ	359.41 *14*	~0.24
γ E1	390.89 *13*	6.1 *3*
γ	457.9 *6*	~0.5
γ M1+E2	547.50 *13*	5.2 *3*
γ (E2+M1)	635.71 *13*	1.78 *10*
γ	681.6 *4*	0.21 *4*
γ	769.8 *4*	0.66 *8*
γ E2	788.742 *8*	100.00
γ	940 *2*	0.50 *15*
γ E2	1037.77 *14*	101 *5*
γ	1083.1 *14*	0.20 *5*
γ	1202.4 *8*	~0.06
γ	1238.6 *5*	0.120 *14*
γ	1257.0 *14*	<0.030
γ	1279.7 *14*	0.090 *14*
γ [E1]	1348.3 *4*	0.33 *4*
γ	1392.7 *7*	0.120 *14*
γ	1416.0 *14*	<0.02
γ	1453.1 *4*	0.260 *19*
γ	1509.3 *10*	0.056 *12*
γ	1527.6 *6*	0.170 *17*
γ	1541.3 *5*	0.160 *14*
γ	1583.1 *5*	0.121 *13*
γ	1631.5 *10*	~0.04
γ	1671.3 *5*	0.102 *10*

Photons (^{138}Pr)
(continued)

γ_{mode}	γ(keV)	γ(%)
γ	1709.3 *8*	0.088 *11*
γ	1726.5 *10*	0.023 *8*
γ	1797.5 *8*	0.10 *1*
γ	1808.1 *7*	0.251 *18*
γ	1851.0 *12*	0.028 *7*
γ	1864.4 *7*	0.214 *14*
γ	1884.1 *7*	0.080 *8*
γ	1914.0 *14*	<0.010
γ	1927.0 *14*	<0.010
γ	1947.0 *14*	<0.010
γ	1956.0 *14*	<0.010
γ	1957.0 *8*	0.094 *9*
γ	1969.0 *14*	<0.010
γ	1979.0 *14*	<0.010
γ	2010.2 *14*	0.034 *6*
γ	2028.4 *7*	0.190 *14*
γ	2030.7 *9*	~0.06
γ	2066.1 *14*	~0.02
γ	2111.7 *14*	0.03 *1*
γ	2118.9 *9*	0.038 *7*
γ	2119.3 *12*	0.040 *15*
γ	2198.5 *14*	0.012 *4*
γ	2222.5 *14*	0.06 *2*
γ	2236.4 *14*	0.03 *1*

† <0.1% uncert(syst)

Atomic Electrons (^{138}Pr)

⟨e⟩=51.8 *18* keV

e_{bin}(keV)	⟨e⟩(keV)	e(%)
6	3.2	54 *6*
7 - 40	2.61	13.3 *7*
69 - 118	0.36	0.48 *13*
130 - 178	0.11	0.074 *22*
183 - 231	0.032	0.016 *7*
262	26.0	9.9 *7*
296	3.27	1.10 *7*
297	8.0	2.68 *18*
301	2.55	0.85 *6*
302 - 351	0.83	0.270 *14*
353 - 391	0.025	0.0066 *8*
417 - 458	0.027	~0.006
507 - 547	0.28	0.055 *10*
595 - 641	0.083	0.014 *3*
675 - 682	0.0010	~0.00014
729	0.014	~0.0020
748	2.19	0.293 *6*
763 - 789	0.408	0.0521 *7*
900 - 940	0.010	~0.0011
997	1.61	0.162 *9*
1031 - 1077	0.285	0.0276 *11*
1082 - 1083	0.00010	~9 $\times10^{-6}$
1162 - 1202	0.0024	0.00020 *8*
1217 - 1257	0.0016	0.00013 *5*
1273 - 1308	0.00203	0.000156 *19*
1342 - 1387	0.0019	0.00014 *5*
1391 - 1416	0.0028	0.00020 *10*
1447 - 1543	0.0061	0.00041 *13*
1577 - 1670	0.0024	0.00015 *4*
1686 - 1768	0.0033	0.00019 *6*
1791 - 1887	0.0035	0.00019 *5*
1907 - 2004	0.0032	0.00016 *4*
2009 - 2106	0.0013	6.1 *13* $\times10^{-5}$
2110 - 2197	0.00080	3.7 *11* $\times10^{-5}$
2216 - 2235	9.6 $\times10^{-5}$	4.3 *14* $\times10^{-6}$

Continuous Radiation (^{138}Pr)

$\langle\beta+\rangle$=176 keV;$\langle$IB$\rangle$=3.4 keV

E_{bin}(keV)		$\langle\ \rangle$(keV)	(%)
0 - 10	β+	1.27×10^{-6}	1.52×10^{-5}
	IB	0.0076	
10 - 20	β+	5.6×10^{-5}	0.000336
	IB	0.0074	0.051
20 - 40	β+	0.00167	0.00510
	IB	0.083	0.25
40 - 100	β+	0.088	0.112
	IB	0.048	0.077
100 - 300	β+	4.46	2.02
	IB	0.132	0.072
300 - 600	β+	30.6	6.6
	IB	0.28	0.063
600 - 1300	β+	126	14.0
	IB	1.27	0.135
1300 - 2500	β+	15.2	1.09
	IB	1.54	0.092
2500 - 2672	IB	0.0038	0.000149
	Σβ+		24

$^{138}_{60}$Nd(5.04 *9* h)

Mode: ϵ

Δ: -82140 *200* keV syst

SpA: 4.51×10^{6} Ci/g

Prod: protons on Ta; protons on Er; ^{136}Ce(α,2n); ^{141}Pr(p,4n); protons on Gd

Photons (^{138}Nd)

$\langle\gamma\rangle$=43.6 *6* keV

γ_{mode}	γ(keV)	γ(%)†
Pr L_ℓ	4.453	0.185 *25*
Pr L_η	4.929	0.074 *8*
Pr L_α	5.031	5.0 *6*
Pr L_β	5.619	4.5 *6*
Pr L_γ	6.415	0.64 *9*
Pr $K_{\alpha2}$	35.550	22.4 *7*
Pr $K_{\alpha1}$	36.026	40.8 *12*
Pr $K_{\beta1}$'	40.720	11.6 *4*
Pr $K_{\beta2}$'	41.981	3.22 *11*
γ (M1)	62.71 *20*	0.065 *7*
γ M1,E2	116.4 *2*	0.086 *10*
γ M1(+E2)	126.27 *4*	0.110 *14*
γ M1,E2	132.82 *4*	0.175 *22*
γ	168.0 *5*	
γ M1,E2	194.30 *4*	0.264 *14*
γ M1,E2	199.61 *4*	0.56 *3*
γ M1+E2	214.22 *5*	0.072 *24*
γ M1,E2	215.46 *4*	0.29 *3*
γ	234.1 *10*	
γ	284.4 *5*	0.043 *12*
γ	294.4 *10*	
γ M1(+E2)	325.88 *4*	2.93 *7*
γ	327.12 *5*	0.026 *5*
γ M1,E2	341.73 *4*	0.41 *4*
γ (M1)	541.34 *5*	0.041 *12*

† 17% uncert(syst)

Atomic Electrons (^{138}Nd)

$\langle$e$\rangle$=7.3 *4* keV

e_{bin}(keV)	$\langle$e$\rangle$(keV)	e(%)
6	3.5	57 *6*
7	0.48	7.0 *7*
21	0.058	0.28 *3*
28	0.167	0.59 *6*
29	0.46	1.60 *16*
30	0.72	2.43 *25*
34	0.33	0.96 *10*
35	0.37	1.05 *11*
36 - 84	0.285	0.59 *3*
91 - 133	0.18	0.170 *25*
152 - 200	0.342	0.207 *14*
207 - 242	0.026	0.012 *4*
278 - 283	0.0010	~0.0004
284	0.31	0.111 *20*
285 - 327	0.116	0.037 *6*
335 - 342	0.0092	0.0027 *7*
499 - 541	0.0029	0.00058 *15*

Continuous Radiation (^{138}Nd)

$\langle$IB$\rangle$=0.29 keV

E_{bin}(keV)		$\langle\ \rangle$(keV)	(%)
10 - 20	IB	0.00027	0.00163
20 - 40	IB	0.091	0.26
40 - 100	IB	0.018	0.037
100 - 300	IB	0.032	0.0157
300 - 600	IB	0.096	0.022
600 - 1000	IB	0.050	0.0072

$^{138}_{61}$Pm(3.24 *5* min)

Mode: ϵ

Δ: -75140 *280* keV syst

SpA: 4.20×10^{8} Ci/g

Prod: ^{141}Pr(α,7n); protons on Gd

Photons (^{138}Pm)

γ_{mode}	γ(keV)	γ(rel)
γ M1+E2	437.56 *13*	10.4 *6*
γ E2	493.16 *13*	21.6 *13*
γ E2	520.99 *16*	100
γ	593.21 *23*	0.9 *1*
γ	699.3 *3*	0.5 *1*
γ E2	729.10 *15*	37.8 *23*
γ E1	740.44 *22*	6.4 *5*
γ	786.04 *17*	0.9 *2*
γ M1,E2	810.26 *22*	3.1 *3*
γ	818.6 *4*	1.1 *3*
γ M1+E2	829.15 *21*	7.1 *5*
γ E2	884.5 *4*	0.8 *2*
γ [M1+E2]	930.72 *13*	5.1 *3*
γ	944.36 *23*	0.8 *2*
γ	970.76 *24*	0.8 *3*
γ M1,E2	972.2 *3*	4.5 *3*
γ	1011.87 *22*	3.8 *5*
γ E2	1014.15 *17*	7.4 *7*
γ	1033.4 *3*	0.3 *1*
γ	1091.6 *3*	~0.8
γ	1097.8 *3*	1.2 *4*
γ	1117.99 *24*	0.7 *2*
γ	1134.71 *21*	2.5 *3*
γ	1140.9 *3*	0.8 *2*
γ	1161.11 *22*	0.7 *2*
γ	1214.6 *4*	0.5 *1*
γ	1258.9 *4*	0.3 *1*
γ	1259.1 *3*	0.5 *2*
γ	1279.19 *17*	11.0 *8*
γ	1318.1 *4*	0.6 *2*
γ	1322.1 *4*	0.6 *2*

Photons (^{138}Pm) (continued)

γ_{mode}	γ(keV)	γ(rel)
γ	1360.1 *4*	0.6 *2*
γ	1373.4 *4*	1.3 *3*
γ	1375.9 *4*	1.1 *3*
γ	1460.5 *4*	~0.6
γ	1470.9 *3*	0.8 *2*
γ	1483.19 *20*	2.5 *3*
γ	1508.9 *3*	~0.4
γ	1509.58 *20*	~0.8
γ	1576.7 *4*	0.9 *2*
γ	1675.4 *3*	3.2 *4*
γ	1711.20 *20*	1.5 *3*
γ	1736.6 *4*	0.7 *2*
γ	1744.8 *3*	1.1 *2*
γ	1789.9 *5*	0.6 *2*
γ	1800.18 *21*	0.3 *1* ?
γ	1803.0 *3*	1.7 *3*
γ	1851.2 *4*	0.5 *1*
γ	1951.2 *4*	1.1 *2*
γ	1984.2 *3*	0.6 *2*
γ	2029.6 *5*	0.6 *2*
γ	2036.1 *5*	~0.4
γ	2138.2 *3*	0.3 *1*
γ	2242.1 *10*	1.4 *5*
γ	2273.2 *3*	0.6 *2*
γ	2303.1 *5*	0.8 *3*
γ	2332.7 *4*	0.3 *1*
γ	2369.6 *4*	0.7 *3*
γ	2403.5 *3*	1.3 *4*
γ	2605.2 *3*	3.0 *5*
γ	2731.4 *3*	1.1 *3*
γ	2754.4 *5*	~0.2
γ	2770.1 *10*	~0.8
γ	2841.1 *3*	0.5 *2*
γ	2962.8 *4*	0.9 *3*
γ	2966.1 *10*	0.5 *2*
γ	3016.1 *10*	0.6 *2*
γ	3139.1 *10*	~0.4
γ	3460.5 *3*	2.9 *4*
γ	3480.0 *4*	1.0 *2*

$^{138}_{62}$Sm(3.0 *3* min)

Mode: ϵ

Δ: -71340 *450* keV syst

SpA: 4.5×10^{8} Ci/g

Prod: protons on Gd

Photons (^{138}Sm)

γ_{mode}	γ(keV)
γ	53.6 *10*
γ	74.7 *10*

$^{138}_{63}$Eu(1.5 *4* s)

Mode: ϵ

SpA: 4.4×10^{10} Ci/g

Prod: ^{32}S on ^{112}Sn

$^{138}_{63}$Eu(35 *6* s)

Mode: ϵ

SpA: 2.3×10^{9} Ci/g

Prod: ^{32}S on ^{112}Sn

A = 139

NDS 32, 1 (1981)

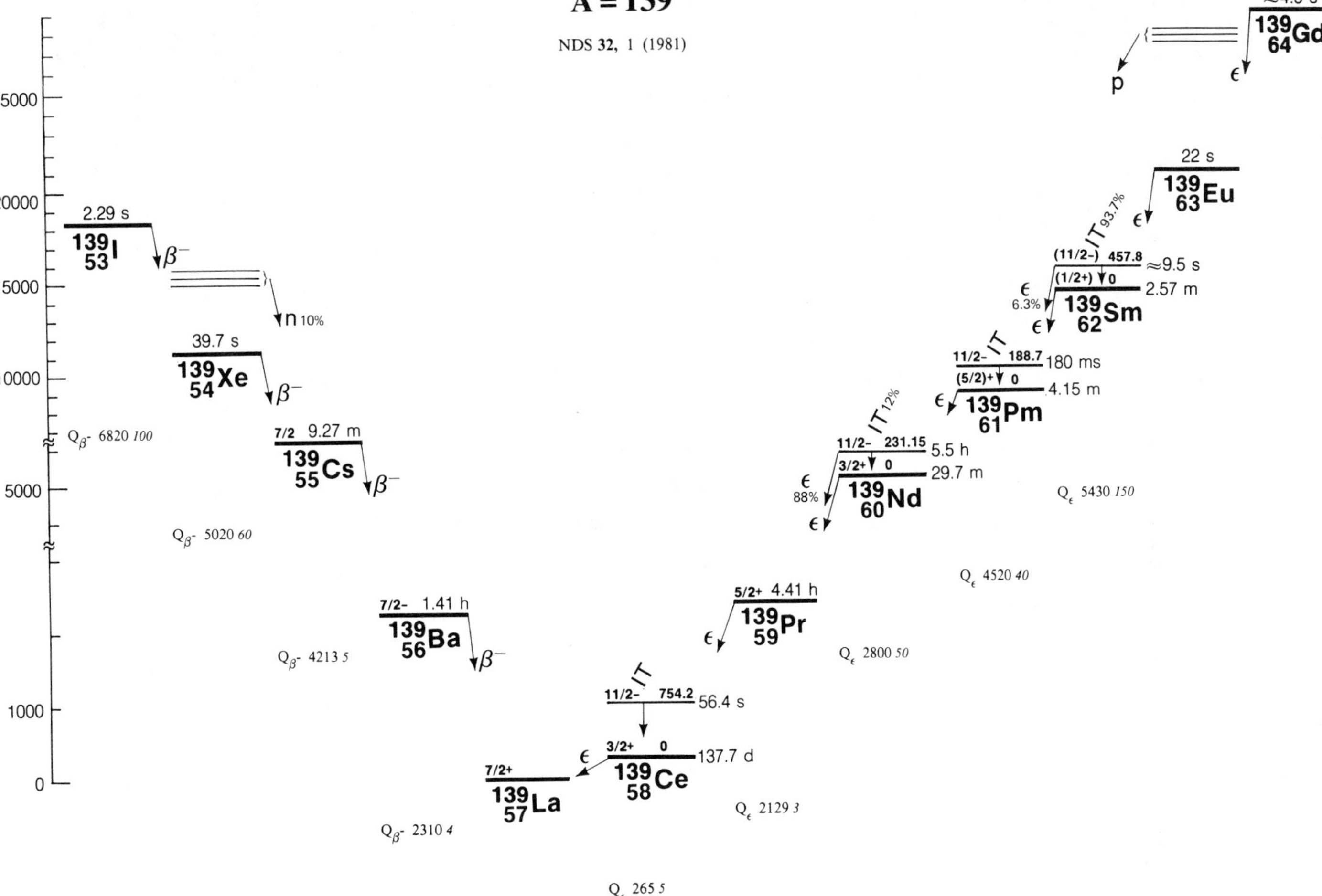

$^{139}_{53}$I (2.29 *2* s)

Mode: β-, β-n(9.5 *6*%)
Δ: -68880 *120* keV
SpA: 3.06×10^{10} Ci/g
Prod: fission
β-n: 130, 190, 290, 485, 565

Photons (^{139}I)

γ_{mode}	γ(keV)
γ	71 *1* ?
γ	192 *1* ?
γ	198 *1* ?
γ	273 *1* ?
γ	382 *1* ?
γ	386 *1* ?
γ	468 *1* ?
γ	653 *1* ?
γ	683 *1* ?
γ	1313 *1* ?

$^{139}_{54}$Xe(39.68 *14* s)

Mode: β-
Δ: -75700 *60* keV
SpA: 2.028×10^{9} Ci/g
Prod: fission

Photons (^{139}Xe)

⟨γ⟩=888 *11* keV

γ_{mode}	γ(keV)	γ(%)†
Cs L_ℓ	3.795	0.020 *4*
Cs L_η	4.142	0.0094 *16*
Cs L_α	4.285	0.56 *9*
Cs L_β	4.730	0.49 *9*
Cs L_γ	5.352	0.063 *12*
Cs $K_{\alpha2}$	30.625	2.83 *21*
Cs $K_{\alpha1}$	30.973	5.2 *4*
Cs $K_{\beta1}$'	34.967	1.43 *11*
Cs $K_{\beta2}$'	36.006	0.35 *3*
γ	55.79 *15*	0.10 *3*
γ	71.17 *4*	0.24 *8*
γ M1+E2	103.82 *4*	0.291 *21*
γ	119.07 *11*	0.062 *21*
γ	121.62 *4*	0.36 *3*
γ M1+E2	174.98 *3*	18.5 *11*
γ E2,M1	218.650 *25*	52 *3*
γ	225.44 *4*	2.7 *3*
γ M1,E2	289.82 *4*	8.5 *5*
γ E2,M1	296.61 *4*	20.2 *11*
γ	326.47 *21*	0.062 *21*

Photons (^{139}Xe) (continued)

γ_{mode}	γ(keV)	γ(%)†
γ E2,M1	338.88 *4*	0.56 *3*
γ	356.75 *5*	0.46 *3*
γ	393.63 *3*	6.2 *3*
γ	427.39 *6*	0.057 *21*
γ	440.96 *11*	0.083 *21*
γ	442.69 *5*	0.146 *21*
γ	446.82 *12*	0.062 *21*
γ	454.58 *10*	0.182 *21*
γ	466.79 *19*	0.068 *16*
γ	491.49 *6*	1.34 *7*
γ	498.08 *6*	0.047 *21*
γ	505.12 *6*	0.30 *3*
γ	513.86 *4*	0.78 *8*
γ	515.25 *4*	0.48 *8*
γ	549.02 *5*	0.54 *3*
γ	565.55 *9*	0.057 *16*
γ	569.62 *11*	0.083 *16*
γ	590.21 *8*	0.057 *21*
γ	595.52 *12*	0.182 *21*
γ	601.89 *6*	0.48 *4*
γ	613.12 *6*	5.1 *3*
γ	623.97 *18*	~0.09
γ	626.74 *6*	0.73 *6*
γ	646.57 *5*	0.55 *4*
γ	652.83 *6*	0.22 *3*
γ	673.06 *6*	0.130 *21*
γ	675.96 *7*	0.151 *21*
γ	699.66 *10*	0.083 *21*
γ	710.24 *6*	0.17 *3*
γ	716.93 *6*	0.16 *3*
γ	720.34 *11*	0.06 *3*
γ	724.00 *5*	1.67 *9*
γ	730.56 *6*	0.21 *5*

Photons (^{139}Xe)
(continued)

γ_{mode}	γ(keV)	γ(%)†
γ	732.51 4	1.63 10
γ	745.25 7	0.49 3
γ	761.18 7	0.19 3
γ	773.17 12	0.09 3
γ	775.69 9	0.09 3
γ	782.82 18	0.057 21
γ	786.42 19	~0.20
γ	788.10 6	3.13 24
γ	801.72 6	0.52 4
γ	818.29 14	0.26 3
γ	820.65 10	0.08 3
γ	832.47 17	0.099 21
γ	847.46 8	0.234 21
γ	879.93 13	0.140 21
γ	888.83 15	~0.06
γ	891.71 6	0.20 3
γ	896.28 19	0.20 3
γ	924.97 11	~0.10
γ	926.96 24	~0.09
γ	937.7 3	0.068 16
γ	942.65 5	0.109 21
γ	946.07 10	0.088 21
γ	957.32 24	0.068 21
γ	961.52 11	0.073 21
γ	967.44 19	0.062 21
γ	970.34 16	0.073 21
γ	980.76 13	0.151 21
γ	986.04 8	0.31 3
γ	996.26 10	0.30 3
γ	1001.55 13	0.073 21
γ	1006.32 6	0.218 16
γ	1017.88 20	0.12 3
γ	1020.94 23	~0.042
γ	1036.41 18	0.094 21
γ	1046.44 12	0.25 3
γ	1067.69 10	0.13 3
γ	1099.26 15	0.06 3
γ	1105.37 13	0.109 21
γ	1114.57 8	0.31 3
γ	1137.63 5	0.37 3
γ	1149.05 23	0.114 21
γ	1171.51 10	0.073 21
γ	1176.53 13	0.08 3
γ	1178.78 10	0.47 4
γ	1190.91 15	0.047 21
γ	1199.53 16	0.125 21
γ	1206.42 8	0.58 5
γ	1214.91 10	0.068 21
γ	1219.48 16	0.146 21
γ	1229.04 12	0.057 21
γ	1243.06 7	0.55 4
γ	1259.25 5	0.48 4
γ	1273.11 21	0.06 3
γ	1289.55 8	0.41 6
γ	1291.16 24	0.16 5
γ	1298.03 10	0.37 6
γ	1300.40 10	~0.08
γ	1310.24 8	~0.08
γ	1316.22 15	0.10 3
γ	1324.57 16	0.17 3
γ	1345.01 6	1.06 3
γ	1352.35 18	0.09 3
γ	1363.07 5	0.27 3
γ	1367.20 11	0.172 21
γ	1386.20 7	0.54 4
γ	1404.21 10	0.114 21
γ	1417.02 14	0.146 21
γ	1428.61 17	0.17 3
γ	1434.23 5	0.16 3
γ	1437.84 15	0.07 3
γ	1448.82 7	0.099 21
γ	1453.20 7	0.45 4
γ	1459.48 13	0.15 3
γ	1490.14 10	0.23 6
γ	1503.15 16	0.13 6
γ	1519.99 6	0.76 5
γ	1540.82 18	0.062 21
γ	1543.81 19	0.057 16
γ	1579.35 9	0.19 8
γ	1584.45 11	0.146 21
γ	1608.71 16	0.094 21
γ	1612.82 15	0.14 3
γ	1615.27 17	0.15 3
γ	1641.84 19	0.14 3

Photons (^{139}Xe)
(continued)

γ_{mode}	γ(keV)	γ(%)†
γ	1652.88 5	0.10 3
γ	1666.32 23	~0.042
γ	1670.45 7	1.04 6
γ	1681.25 15	0.099 21
γ	1700.01 18	0.099 21
γ	1711.55 15	0.213 21
γ	1722.39 13	0.047 21
γ	1765.61 25	0.042 16
γ	1774.02 9	0.31 3
γ	1777.19 15	0.187 21
γ	1785.95 18	0.057 21
γ	1790.76 11	0.35 5
γ	1793.16 19	0.11 4
γ	1804.20 20	0.114 21
γ	1813.89 13	0.11 3
γ	1816.5 3	0.11 3
γ	1831.47 15	0.073 21
γ	1851.83 17	0.10 3
γ	1854.86 11	0.12 3
γ	1857.96 11	0.10 3
γ	1862.44 17	~0.15
γ	1863.94 10	0.23 10
γ	1895.88 7	0.57 4
γ	1911.49 18	0.114 16
γ	1935.21 6	0.052 16
γ	1939.52 9	0.078 16
γ	1967.05 7	0.120 21
γ	1979.58 10	0.49 4
γ	1994.15 17	0.10 3
γ	2006.91 19	0.094 21
γ	2014.96 10	0.146 16
γ	2021.63 16	0.068 16
γ	2025.02 12	0.068 16
γ	2039.02 6	0.068 21
γ	2063.84 9	0.39 3
γ	2086.12 9	0.62 4
γ	2099.70 11	0.120 16
γ	2103.70 13	0.052 16
γ	2110.19 6	0.270 21
γ	2116.87 8	0.302 21
γ	2192.46 7	0.31 3
γ	2205.11 15	~0.042
γ	2227.35 22	0.34 8
γ	2238.89 16	0.135 21
γ	2249.8 4	0.062 16
γ	2255.41 16	0.083 16
γ	2291.85 8	0.37 3
γ	2304.77 9	0.27 3
γ	2328.83 6	0.59 4
γ	2367.43 8	0.125 16
γ	2403.87 10	0.244 21
γ	2423.75 15	0.042 10
γ	2430.16 19	0.036 10
γ	2437.85 16	0.088 16
γ	2452.41 11	0.042 16
γ	2464.32 16	0.104 16
γ	2507.68 11	0.073 21
γ	2510.50 9	0.25 3
γ	2535.49 16	0.057 16
γ	2574.03 11	0.32 3
γ	2578.85 10	0.057 16
γ	2613.00 17	~0.031
γ	2633.75 16	0.099 16
γ	2640.30 19	~0.031
γ	2673.5 5	0.052 16
γ	2693.50 14	0.073 21
γ	2736.80 22	0.109 21
γ	2754.13 16	0.062 16
γ	2761.86 24	0.062 16
γ	2769.35 12	0.276 21
γ	2790.95 14	0.250 21
γ	2815.12 14	0.208 21
γ	2832.77 23	0.057 10
γ	2853.83 19	0.083 16
γ	2872.70 21	0.114 16
γ	2886.7 4	0.078 16
γ	2904.61 24	0.073 16
γ	2911.78 22	0.062 16
γ	2918.23 23	0.114 21
γ	2936.79 17	0.052 16
γ	2941.9 3	0.073 16
γ	2989.45 16	0.068 16
γ	3028.7 4	0.062 16
γ	3111.07 16	~0.036

Photons (^{139}Xe)
(continued)

γ_{mode}	γ(keV)	γ(%)†
γ	3130.42 22	0.07 3
γ	3147.11 15	0.057 10
γ	3156.02 16	0.042 10
γ	3168.77 22	0.057 10
γ	3214.88 16	0.036 10
γ	3375.25 12	0.140 16
γ	3424.9 5	0.068 21
γ	3504.69 16	0.062 10

† 12% uncert(syst)

Atomic Electrons (^{139}Xe)

⟨e⟩=24.5 13 keV

e_{bin}(keV)	⟨e⟩(keV)	e(%)
5 - 51	1.22	11.8 15
55 - 104	0.74	1.0 3
113 - 122	0.08	0.06 3
139	4.7	3.4 4
169	0.63	0.374 25
170 - 175	0.8	~0.5
183	8.9	4.9 4
189	0.28	~0.15
213	1.5	0.71 18
214 - 225	0.8	0.39 15
254	0.89	0.35 3
261	2.03	0.78 7
284 - 334	0.72	0.25 3
338 - 358	0.28	~0.08
388 - 437	0.08	0.020 7
438 - 486	0.10	0.021 7
490 - 534	0.029	0.0057 19
543 - 591	0.16	0.028 12
594 - 642	0.052	0.0084 21
645 - 695	0.048	0.007 3
697 - 745	0.061	0.009 3
747 - 796	0.09	0.011 4
797 - 844	0.0079	0.0010 3
846 - 895	0.011	0.0013 4
896 - 945	0.0105	0.00113 23
946 - 995	0.016	0.0017 4
996 - 1045	0.0074	0.00073 20
1046 - 1094	0.0058	0.00054 19
1098 - 1144	0.014	0.0012 3
1148 - 1195	0.012	0.0010 3
1198 - 1243	0.013	0.0011 3
1254 - 1300	0.016	0.0012 3
1305 - 1352	0.022	0.0016 4
1357 - 1404	0.0073	0.00052 10
1411 - 1459	0.0095	0.00067 16
1467 - 1548	0.012	0.00083 22
1573 - 1669	0.017	0.00102 22
1676 - 1773	0.0111	0.00064 11
1776 - 1863	0.0123	0.00067 12
1876 - 1975	0.0080	0.00041 8
1978 - 2074	0.0119	0.00058 10
2080 - 2169	0.0051	0.00024 5
2187 - 2286	0.0082	0.00037 6
2287 - 2368	0.0061	0.00026 6
2388 - 2475	0.0038	0.000154 25
2500 - 2598	0.0034	0.000133 25
2604 - 2701	0.0015	5.6 11 $\times 10^{-5}$
2718 - 2814	0.0050	0.00018 3
2818 - 2917	0.0034	0.000118 14
2931 - 3028	0.00075	2.5 5 $\times 10^{-5}$
3075 - 3168	0.00131	4.2 7 $\times 10^{-5}$
3179 - 3214	0.00018	5.6 20 $\times 10^{-6}$
3339 - 3424	0.00097	2.9 6 $\times 10^{-5}$
3469 - 3504	0.00028	8.2 23 $\times 10^{-6}$

Continuous Radiation (^{139}Xe)

⟨β-⟩=1752 keV;⟨IB⟩=5.8 keV

E_{bin}(keV)		⟨ ⟩(keV)	(%)
0 - 10	β-	0.0064	0.127
	IB	0.053	
10 - 20	β-	0.0195	0.130
	IB	0.052	0.36
20 - 40	β-	0.081	0.269
	IB	0.103	0.36
40 - 100	β-	0.62	0.88
	IB	0.30	0.46
100 - 300	β-	7.8	3.77
	IB	0.89	0.49
300 - 600	β-	35.7	7.8
	IB	1.08	0.25
600 - 1300	β-	230	23.9
	IB	1.7	0.19
1300 - 2500	β-	734	39.3
	IB	1.30	0.075
2500 - 5000	β-	744	23.8
	IB	0.30	0.0102
5000 - 5020	β-	0.000424	8.5×10^{-6}
	IB	8.6×10^{-11}	1.7×10^{-12}

$^{139}_{55}$Cs(9.27 *5* min)

Mode: β-

Δ: -80715 *7* keV

SpA: 1.458×10^{8} Ci/g

Prod: fission; daughter ^{139}Xe

Photons (^{139}Cs)

⟨γ⟩=329 *6* keV

γ_{mode}	γ(keV)	γ(%)†
γ	188.91 *6*	0.0085 *15*
γ	196.66 *5*	0.0092 *15*
γ [E1]	230.86 *9*	0.033 *3*
γ	233.52 *6*	0.0100 *23*
γ	249.85 *11*	0.0108 *23*
γ	260.10 *5*	0.0077 *23*
γ	267.61 *22*	0.010 *3*
γ [M1+E2]	312.59 *7*	0.0085 *15*
γ	339.18 *9*	0.0062 *23*
γ	357.24 *5*	0.0146 *23*
γ [E1]	376.07 *7*	0.042 *3*
γ	397.15 *5*	0.017 *5*
γ	401.25 *8*	0.0108 *23*
γ	404.75 *25*	0.0085 *23*
γ	416.46 *6*	0.013 *3*
γ	419.33 *11*	0.010 *3*
γ	430.18 *6*	0.039 *3*
γ	434.50 *11*	0.015 *3*
γ	448.86 *8*	0.032 *4*
γ [M1+E2]	454.79 *5*	0.140 *9*
γ [M1+E2]	466.93 *5*	0.0216 *23*
γ	505.77 *11*	0.010 *3*
γ	515.98 *5*	0.055 *5*
γ	528.29 *7*	0.039 *13*
γ	532.12 *3*	0.229 *12*
γ	538.30 *8*	0.014 *3*
γ [M1+E2]	542.95 *7*	0.025 *3*
γ	558.14 *9*	0.009 *3*
γ [M1+E2]	567.84 *5*	0.105 *9*
γ	567.87 *4*	0.039 *9*
γ [M1+E2]	594.15 *4*	0.075 *5*
γ [E2]	598.32 *9*	0.0123 *23*
γ	601.63 *4*	0.069 *4*
γ [M1+E2]	604.33 *4*	0.045 *3*
γ	613.82 *11*	0.016 *5*
γ	617.33 *5*	0.025 *5*
γ	619.8 *3*	0.017 *5*
γ [E2]	627.383 *22*	1.65 *9*
γ	651.25 *5*	0.049 *4*

Photons (^{139}Cs)
(continued)

γ_{mode}	γ(keV)	γ(%)†
γ	656.68 *6*	0.033 *4*
γ [M1+E2]	666.27 *6*	0.031 *3*
γ	669.18 *4*	0.045 *4*
γ	672.38 *6*	0.021 *3*
γ	690.16 *6*	0.0231 *23*
γ	715.06 *5*	0.076 *5*
γ	728.44 *5*	0.043 *4*
γ	735.78 *5*	0.069 *5*
γ	737.66 *10*	0.046 *5*
γ	770.94 *18*	0.018 *3*
γ	773.6 *3*	0.010 *3*
γ	788.04 *5*	0.009 *3*
γ [M1+E2]	793.41 *4*	0.081 *5*
γ [M1+E2]	798.40 *8*	0.028 *4*
γ	806.62 *6*	0.0131 *23*
γ	827.67 *5*	0.117 *8*
γ	831.54 *8*	0.013 *5*
γ	849.44 *11*	0.011 *3*
γ	858.62 *19*	0.012 *3*
γ	883.52 *19*	0.014 *4*
γ	890.72 *5*	0.079 *5*
γ	925.10 *5*	0.071 *5*
γ	929.27 *4*	0.246 *13*
γ [M1+E2]	932.50 *9*	0.014 *4*
γ [M1+E2]	946.63 *7*	0.105 *8*
γ	955.29 *9*	0.031 *5*
γ [M1+E2]	966.64 *24*	0.018 *5*
γ	972.83 *21*	0.014 *5*
γ	1041.11 *7*	0.034 *3*
γ	1059.82 *8*	0.016 *5*
γ	1063.85 *9*	~0.011
γ [E2]	1066.95 *12*	0.025 *5*
γ	1076.99 *16*	0.028 *5*
γ	1092.69 *9*	0.0447 *23*
γ	1109.12 *15*	0.043 *6*
γ [M1+E2]	1111.21 *6*	0.021 *5*
γ [M1+E2]	1121.05 *5*	0.050 *5*
γ [M1+E2]	1159.48 *10*	0.028 *4*
γ	1178.50 *7*	0.072 *6*
γ	1185.20 *5*	0.032 *5*
γ	1190.56 *4*	0.197 *12*
γ	1216.22 *11*	0.022 *4*
γ	1241.11 *11*	0.016 *3*
γ	1249.47 *18*	0.016 *3*
γ [M1+E2]	1283.39 *3*	7.7 *4*
γ [M1+E2]	1306.22 *5*	0.113 *10*
γ [E2]	1308.28 *5*	0.400 *23*
γ	1316.57 *9*	0.014 *4*
γ	1321.92 *5*	0.250 *14*
γ	1344.71 *21*	0.013 *4*
γ	1353.43 *9*	0.023 *4*
γ	1386.99 *24*	0.020 *4*
γ	1393.66 *10*	0.016 *4*
γ	1410.74 *4*	0.161 *9*
γ [M1+E2]	1420.79 *3*	0.85 *5*
γ	1462.63 *6*	0.039 *6*
γ	1472.83 *11*	0.012 *5*
γ	1500.73 *14*	0.015 *4*
γ [M1+E2]	1529.74 *8*	0.028 *6*
γ	1531.77 *16*	0.023 *6*
γ	1539.23 *14*	0.031 *3*
γ	1546.73 *5*	0.032 *4*
γ	1563.78 *14*	0.011 *3*
γ	1564.50 *10*	0.032 *4*
γ	1573.86 *9*	0.027 *3*
γ [E2]	1591.80 *7*	0.056 *5*
γ	1601.00 *12*	0.021 *8*
γ [M1+E2]	1620.87 *6*	0.447 *23*
γ [M1+E2]	1677.63 *8*	0.095 *7*
γ	1680.88 *4*	0.65 *3*
γ	1689.17 *9*	0.021 *4*
γ	1698.81 *5*	0.189 *11*
γ	1711.25 *9*	0.084 *6*
γ	1714.07 *8*	0.017 *5*
γ	1722.68 *3*	0.080 *5*
γ	1738.05 *14*	0.032 *3*
γ [E2]	1748.43 *5*	0.016 *3*
γ	1768.72 *17*	0.0146 *23*
γ	1793.94 *12*	0.023 *3*
γ	1814.60 *17*	0.013 *4*
γ	1818.84 *16*	0.016 *4*
γ [E2]	1851.22 *5*	0.011 *3*
γ [M1+E2]	1877.54 *4*	0.364 *19*
γ [M1+E2]	1887.72 *4*	0.235 *13*

Photons (^{139}Cs)
(continued)

γ_{mode}	γ(keV)	γ(%)†
γ [M1+E2]	1904.61 *5*	0.132 *8*
γ [M1+E2]	1933.60 *5*	0.261 *14*
γ	1949.30 *5*	0.035 *4*
γ	1998.44 *5*	0.031 *3*
γ	2003.51 *17*	0.015 *3*
γ	2021.04 *10*	0.14 *5*
γ	2022.09 *7*	~0.07
γ	2038.11 *4*	0.045 *4*
γ	2079.22 *18*	0.045 *5*
γ	2090.00 *5*	0.146 *9*
γ	2100.21 *11*	0.050 *6*
γ	2111.05 *5*	0.70 *4*
γ [E2]	2157.12 *8*	0.044 *5*
γ	2166.90 *19*	0.013 *4*
γ	2174.10 *5*	0.214 *12*
γ [M1+E2]	2219.18 *7*	0.022 *3*
γ [E2]	2230.01 *8*	0.015 *3*
γ [E2]	2250.02 *24*	0.012 *3*
γ	2269.6 *3*	0.015 *3*
γ [M1+E2]	2305.01 *8*	0.032 *4*
γ	2330.19 *13*	0.011 *5*
γ	2339.5 *5*	0.030 *7*
γ	2350.05 *3*	0.59 *3*
γ	2352.54 *21*	0.024 *9*
γ	2376.07 *9*	0.074 *6*
γ	2380.81 *7*	0.200 *12*
γ	2418.93 *16*	0.013 *3*
γ	2422.30 *18*	0.031 *4*
γ	2524.49 *11*	0.030 *5*
γ	2529.89 *16*	0.085 *23*
γ [M1+E2]	2531.99 *5*	0.45 *3*
γ	2605.97 *4*	0.260 *15*
γ	2649.47 *7*	0.179 *10*
γ	2674.12 *18*	0.037 *5*
γ [M1+E2]	2774.18 *13*	0.032 *3*
γ [M1+E2]	2837.10 *7*	0.029 *3*
γ	2847.87 *10*	0.106 *6*
γ	2979.13 *24*	0.0139 *15*
γ	2997.44 *8*	0.092 *6*
γ	3047.44 *13*	0.032 *3*
γ	3096.92 *14*	0.0092 *23*
γ	3171.71 *23*	0.0193 *23*
γ	3270.49 *19*	0.0108 *15*
γ	3323.66 *11*	0.053 *5*
γ	3364.37 *11*	0.085 *6*
γ	3418.90 *13*	0.042 *4*
γ [M1+E2]	3464.47 *7*	0.116 *7*
γ	3645.84 *13*	0.0293 *23*
γ	3665.76 *8*	0.146 *9*
γ	3724.28 *14*	0.0277 *23*
γ	3769.30 *11*	0.048 *3*
γ	3820.13 *24*	0.0115 *15*
γ	3839.85 *15*	0.0193 *15*
γ	3854.01 *16*	0.0208 *15*
γ	3887.9 *3*	0.0085 *15*
γ	3912.35 *21*	0.0131 *15*

† 39% uncert(syst)

Continuous Radiation (^{139}Cs)

⟨β-⟩=1650 keV;⟨IB⟩=5.2 keV

E_{bin}(keV)		⟨ ⟩(keV)	(%)
0 - 10	β-	0.0078	0.155
	IB	0.051	
10 - 20	β-	0.0236	0.157
	IB	0.051	0.35
20 - 40	β-	0.096	0.321
	IB	0.100	0.35
40 - 100	β-	0.72	1.02
	IB	0.29	0.44
100 - 300	β-	8.2	4.02
	IB	0.85	0.48
300 - 600	β-	36.2	7.9
	IB	1.03	0.24
600 - 1300	β-	241	25.0
	IB	1.58	0.18

Continuous Radiation (^{139}Cs) (continued)

E_{bin}(keV)		⟨ ⟩(keV)	(%)
1300 - 2500	β-	785	42.0
	IB	1.10	0.064
2500 - 4213	β-	579	19.4
	IB	0.164	0.0057

$^{139}_{56}$Ba(1.410 *7* h)

Mode: β-

Δ: -84928 *6* keV

SpA: 1.599×10^{7} Ci/g

Prod: ^{138}Ba(n,γ); fission

Photons (^{139}Ba)

⟨γ⟩=43.5 *17* keV

γ_{mode}	γ(keV)	γ(%)†
La L_ℓ	4.121	0.0095 *14*
La L_η	4.529	0.0040 *5*
La L_α	4.649	0.26 *3*
La L_β	5.166	0.23 *3*
La L_γ	5.877	0.032 *5*
La $K_{\alpha2}$	33.034	1.31 *7*
La $K_{\alpha1}$	33.442	2.40 *13*
La $K_{\beta1}$'	37.777	0.67 *4*
La $K_{\beta2}$'	38.927	0.175 *10*
γ M1	165.853 *7*	22
γ E2	1053.2 *3*	0.00036 *15*
γ E2+M1	1090.8 *2*	0.0094 *9*
γ	1215.6 *3*	0.0036 *3*
γ E2+4.1%M1	1219.0 *3*	0.0045 *6*
γ [M1+E2]	1254.68 *14*	0.030 *3*
γ E2+M1	1256.66 *20*	0.0031 *4*
γ	1310.56 *17*	0.0163 *15*
γ M1+38%E2	1370.48 *21*	0.0033 *3*
γ	1381.4 *3*	~9×10^{-5}
γ [M1+E2]	1392.4 *3*	~9×10^{-5}
γ E2+M1	1420.53 *14*	0.30 *3*
γ	1476.41 *17*	0.00184 *4*
γ [M1+E2]	1517.3 *3*	<6×10^{-5} ?
γ M1+44%E2	1536.33 *21*	0.0030 *3*
γ [M1+E2]	1558.2 *3*	0.00027 *12*
γ [M1+E2]	1578.2 *4*	0.00060 *15*
γ	1595.4 *3*	0.0028 *5*
γ [M1+E2]	1601.6 *7*	0.00015 *3*
γ [M1+E2]	1683.2 *3*	0.0036 *3*
γ [M1+E2]	1691.2 *10*	0.00033 *3*
γ	1754.7 *3*	~6×10^{-5}
γ	1761.2 *3*	9 *3*×10^{-5}
γ E2	1767.4 *7*	~0.00030
γ [M1+E2]	1797.4 *10*	~6×10^{-5}
γ	1894.4 *4*	~6×10^{-5}
γ	1920.5 *3*	~0.00012
γ	2060.2 *4*	~0.00015

† 4.6% uncert(syst)

Atomic Electrons (^{139}Ba)

⟨e⟩=8.1 *3* keV

e_{bin}(keV)	⟨e⟩(keV)	e(%)
5 - 38	0.35	4.3 *3*
127	6.4	5.03 *25*
160	1.08	0.68 *3*
164	0.213	0.130 *7*
165 - 166	0.084	0.0509 *21*
1014 - 1053	0.00017	1.6 *3* ×10^{-5}
1085 - 1091	2.7 ×10^{-5}	2.5 *5* ×10^{-6}

Atomic Electrons (^{139}Ba) (continued)

e_{bin}(keV)	⟨e⟩(keV)	e(%)
1177 - 1219	0.00062	5.1 *8* ×10^{-5}
1248 - 1272	0.00026	2.1 *7* ×10^{-5}
1304 - 1353	7.8 ×10^{-5}	5.9 *10* ×10^{-6}
1364 - 1392	0.0039	0.00028 *5*
1414 - 1437	0.00066	4.6 *7* ×10^{-5}
1470 - 1563	8.3 ×10^{-5}	5.5 *8* ×10^{-6}
1572 - 1652	4.8 ×10^{-5}	2.9 *4* ×10^{-6}
1677 - 1766	1.19 ×10^{-5}	7 *1* ×10^{-7}
1791 - 1889	1.5 ×10^{-6}	8 *4* ×10^{-8}
1893 - 1919	1.5 ×10^{-7}	8 *4* ×10^{-9}
2021 - 2059	1.2 ×10^{-6}	~6×10^{-8}

Continuous Radiation (^{139}Ba)

⟨β-⟩=891 keV;⟨IB⟩=1.9 keV

E_{bin}(keV)		⟨ ⟩(keV)	(%)
0 - 10	β-	0.0160	0.318
	IB	0.034	
10 - 20	β-	0.0488	0.325
	IB	0.034	0.23
20 - 40	β-	0.202	0.67
	IB	0.066	0.23
40 - 100	β-	1.55	2.20
	IB	0.18	0.28
100 - 300	β-	19.1	9.3
	IB	0.49	0.28
300 - 600	β-	84	18.5
	IB	0.50	0.118
600 - 1300	β-	436	46.5
	IB	0.49	0.059
1300 - 2310	β-	349	22.2
	IB	0.064	0.0043

$^{139}_{57}$La(stable)

Δ: -87238 *4* keV

%: 99.91 *1*

$^{139}_{58}$Ce(137.66 *13* d)

Mode: ε

Δ: -86973 *7* keV

SpA: 6824 Ci/g

Prod: ^{138}Ce(n,γ); ^{139}La(d,2n); ^{139}La(p,n)

Photons (^{139}Ce)

⟨γ⟩=160.2 *14* keV

γ_{mode}	γ(keV)	γ(%)†
La L_ℓ	4.121	0.178 *25*
La L_η	4.529	0.077 *8*
La L_α	4.649	4.9 *6*
La L_β	5.160	5.0 *7*
La L_γ	5.912	0.73 *11*
La $K_{\alpha2}$	33.034	22.8 *6*
La $K_{\alpha1}$	33.442	41.9 *12*
La $K_{\beta1}$'	37.777	11.7 *3*
La $K_{\beta2}$'	38.927	3.06 *10*
γ M1	165.853 *7*	79.9

† 1.0% uncert(syst)

Atomic Electrons (^{139}Ce)

⟨e⟩=33.1 *3* keV

e_{bin}(keV)	⟨e⟩(keV)	e(%)
5	2.21	40 *4*
6	1.97	33 *3*
26	0.175	0.66 *7*
27	0.49	1.83 *19*
28	0.80	2.9 *3*
31	0.127	0.40 *4*
32	0.53	1.66 *17*
33	0.153	0.47 *5*
36	0.027	0.074 *8*
37	0.045	0.122 *12*
38	0.0150	0.040 *4*
127	21.82	17.19 *7*
160	3.685	2.309 *10*
164	1.012	0.615 *3*

Continuous Radiation (^{139}Ce)

⟨IB⟩=0.080 keV

E_{bin}(keV)		⟨ ⟩(keV)	(%)
10 - 20	IB	0.00043	0.0026
20 - 40	IB	0.075	0.23
40 - 99	IB	0.0036	0.0077

$^{139}_{58}$Ce(56.4 *5* s)

Mode: IT

Δ: -86219 *7* keV

SpA: 1.429×10^{9} Ci/g

Prod: ^{138}Ce(n,γ); ^{139}La(p,n); ^{140}Ce(n,2n); ^{138}Ba(^{3}He,2n)

Photons (^{139}Ce)

⟨γ⟩=699.7 *23* keV

γ_{mode}	γ(keV)	γ(%)†
Ce L_ℓ	4.289	0.0129 *17*
Ce L_η	4.730	0.0054 *6*
Ce L_α	4.838	0.35 *4*
Ce L_β	5.388	0.32 *4*
Ce L_γ	6.142	0.044 *6*
Ce $K_{\alpha2}$	34.279	1.61 *5*
Ce $K_{\alpha1}$	34.720	2.94 *9*
Ce $K_{\beta1}$'	39.232	0.828 *25*
Ce $K_{\beta2}$'	40.437	0.223 *8*
γ M4	754.21 *9*	92.5

† 0.32% uncert(syst)

Atomic Electrons (^{139}Ce)

⟨e⟩=54.7 *9* keV

e_{bin}(keV)	⟨e⟩(keV)	e(%)
6	0.25	4.3 *4*
7	0.035	0.53 *5*
27	0.0122	0.044 *5*
28	0.034	0.122 *12*
29	0.054	0.189 *19*
32	0.0053	0.0164 *17*
33	0.030	0.091 *9*
34	0.0181	0.054 *6*
35	0.0026	0.0077 *8*
38	0.0043	0.0113 *12*
39	0.00174	0.0045 *5*
714	43.7	6.13 *12*

Atomic Electrons (^{139}Ce) (continued)

e_{bin}(keV)	⟨e⟩(keV)	e(%)
748	8.25	1.103 *22*
753	1.80	0.240 *5*
754	0.420	0.0558 *11*

$^{139}_{59}$Pr(4.41 *4* h)

Mode: ϵ
Δ: -84844 *7* keV
SpA: 5.11×10^6 Ci/g
Prod: daughter ^{139}Nd; ^{140}Ce(p,2n); ^{141}Pr(n,3n); ^{141}Pr(γ,2n); ^{140}Pr(γ,n)

Photons (^{139}Pr)

⟨γ⟩=41.3 *6* keV

γ_{mode}	γ(keV)	γ(%)†
Ce L_ℓ	4.289	0.159 *21*
Ce L_η	4.730	0.065 *7*
Ce L_α	4.838	4.3 *5*
Ce L_β	5.389	3.9 *5*
Ce L_γ	6.142	0.54 *7*
Ce $K_{\alpha2}$	34.279	20.5 *6*
Ce $K_{\alpha1}$	34.720	37.5 *11*
Ce $K_{\beta1}$'	39.232	10.5 *3*
Ce $K_{\beta2}$'	40.437	2.84 *10*
γ M1(+E2)	255.077 *18*	0.189 *5*
γ	354.23 *8*	0.0095 *19*
γ	587.41 *3*	0.0057 *19*
γ	664.65 *8*	0.0030 *8*
γ [M1+E2]	696.01 *5*	~0.0034
γ M4	754.21 *9*	0.0114 *19* *
γ	823.97 *20*	0.0106 *19*
γ [E2]	1065.14 *3*	~0.0027
γ [E2]	1088.67 *12*	0.0084 *19*
γ M1	1320.219 *22*	0.0559 *11*
γ	1341.48 *3*	~0.0038
γ E2	1347.304 *12*	0.38
γ M1(+E2)	1375.561 *22*	0.124 *6*
γ E2	1563.347 *22*	0.0334 *19*
γ M1,E2	1596.554 *23*	0.0274 *23*
γ E2	1630.636 *20*	0.276 *8*
γ	1652.547 *21*	0.0312 *19*
γ	1710.32 *24*	0.0014 *3*
γ	1729.79 *7*	0.0072 *11*
γ [M1+E2]	1818.42 *3*	0.0247 *15*
γ	1907.62 *3*	0.0137 *15*
γ	1965.39 *24*	0.00046 *19*
γ	1984.86 *8*	0.00061 *19*
γ [E2]	2016.22 *4*	0.0095 *11*

† 5.3% uncert(syst)
* with ^{139}Ce(56.4 s) in equilib

Atomic Electrons (^{139}Pr)

⟨e⟩=5.6 *3* keV

e_{bin}(keV)	⟨e⟩(keV)	e(%)
6	3.1	53 *5*
7	0.42	6.4 *7*
27	0.155	0.57 *6*
28	0.43	1.55 *16*
29	0.69	2.41 *25*
32	0.068	0.209 *21*
33	0.38	1.15 *12*
34	0.231	0.68 *7*
35 - 39	0.111	0.299 *19*
215 - 255	0.036	0.0164 *17*
314 - 354	0.0007	~0.00023
547 - 587	0.00021	$\sim4\times10^{-5}$

Atomic Electrons (^{139}Pr) (continued)

e_{bin}(keV)	⟨e⟩(keV)	e(%)
624 - 665	0.00021	3.2 *13* $\times10^{-5}$
689 - 714	0.0054	0.00076 *13*
748 - 784	0.00150	0.00020 *3*
817 - 824	3.8×10^{-5}	4.6 *21* $\times10^{-6}$
1025 - 1065	0.00018	1.7 *4* $\times10^{-5}$
1082 - 1089	2.2×10^{-5}	2.0 *3* $\times10^{-6}$
1280 - 1320	0.0059	0.000451 *21*
1335 - 1375	0.0029	0.000213 *23*
1523 - 1612	0.00396	0.000250 *11*
1624 - 1723	0.00058	3.58 *23* $\times10^{-5}$
1724 - 1817	0.00031	1.73 *21* $\times10^{-5}$
1867 - 1964	0.00013	7 *3* $\times10^{-6}$
1976 - 2015	9.7×10^{-5}	4.9 *5* $\times10^{-6}$

Continuous Radiation (^{139}Pr)

⟨β+⟩=39.6 keV;⟨IB⟩=2.0 keV

E_{bin}(keV)		⟨ ⟩(keV)	(%)
0 - 10	β+	1.06×10^{-6}	1.27×10^{-5}
	IB	0.0022	
10 - 20	β+	4.65×10^{-5}	0.000279
	IB	0.0020	0.0138
20 - 40	β+	0.00138	0.00420
	IB	0.085	0.25
40 - 100	β+	0.070	0.090
	IB	0.020	0.035
100 - 300	β+	3.20	1.46
	IB	0.059	0.030
300 - 600	β+	16.9	3.74
	IB	0.22	0.049
600 - 1300	β+	19.4	2.60
	IB	1.06	0.114
1300 - 2129	IB	0.54	0.036
	Σβ+		7.9

$^{139}_{60}$Nd(29.7 *5* min)

Mode: ϵ
Δ: -82040 *50* keV
SpA: 4.55×10^7 Ci/g
Prod: daughter ^{139}Nd(5.5 h); ^{141}Pr(d,4n); ^{141}Pr(p,3n); daughter ^{139}Pm; protons on Gd

Photons (^{139}Nd)

⟨γ⟩=159 *9* keV

γ_{mode}	γ(keV)	γ(%)†
Pr L_ℓ	4.453	0.138 *19*
Pr L_η	4.929	0.055 *6*
Pr L_α	5.031	3.7 *4*
Pr L_β	5.620	3.4 *4*
Pr L_γ	6.414	0.47 *7*
Pr $K_{\alpha2}$	35.550	16.8 *5*
Pr $K_{\alpha1}$	36.026	30.6 *10*
Pr $K_{\beta1}$'	40.720	8.7 *3*
Pr $K_{\beta2}$'	41.981	2.42 *8*
γ M1+E2	113.95 *5*	0.58 *12*
γ [M1+E2]	183.75 *15*	0.64 *6*
γ	221.0 *3*	0.064 *16*
γ	368.1 *3*	0.11 *3*
γ M1	405.12 *9*	5.8
γ	411.62 *18*	0.13 *3*
γ M1	474.92 *17*	1.10 *10*
γ [E1]	485.30 *23*	0.38 *6*
γ	511.88 *23*	<3
γ [M1+E2]	588.87 *16*	0.67 *10*
γ	621.8 *3*	1.02 *11*
γ [E1]	669.05 *21*	1.28 *16*
γ	696.3 *3*	0.32 *6* ?

Photons (^{139}Nd) (continued)

γ_{mode}	γ(keV)	γ(%)†
γ	916.99 *22*	1.28 *13*
γ	923.50 *22*	1.12 *11*
γ [E1]	1074.17 *22*	2.13 *22*
γ	1096.3 *3*	0.32 *6*
γ	1221.3 *8*	0.24 *8*
γ	1233.4 *8*	0.11 *5* ?
γ	1246.8 *8*	0.16 *6*
γ	1311.9 *4*	
γ	1328.61 *22*	0.24 *6*
γ	1405.6 *5*	0.51 *8*
γ	1449.1 *10*	0.13 *5*
γ	1463.5 *6*	0.27 *6*
γ	1501.4 *3*	0.21 *6*
γ	1532.1 *10*	0.11 *5*

† 19% uncert(syst)

Atomic Electrons (^{139}Nd)

⟨e⟩=6.4 *3* keV

e_{bin}(keV)	⟨e⟩(keV)	e(%)
6	2.6	43 *4*
7	0.35	5.1 *5*
28	0.125	0.44 *5*
29	0.35	1.20 *12*
30	0.54	1.83 *19*
34	0.245	0.72 *7*
35	0.27	0.79 *8*
36 - 41	0.115	0.304 *18*
72	0.33	0.46 *9*
107 - 114	0.21	~0.19
142	0.177	0.125 *16*
177 - 221	0.071	0.039 *10*
326 - 362	0.008	~0.0025
363	0.53	0.15 *3*
367 - 412	0.108	0.027 *4*
433 - 479	0.16	0.035 *12*
484 - 512	0.012	~0.0023
547 - 589	0.067	0.012 *4*
615 - 663	0.030	0.0048 *10*
668 - 696	0.0021	0.00030 *11*
875 - 923	0.053	0.0060 *22*
1032 - 1074	0.022	0.0021 *3*
1089 - 1096	0.0008	8 *4* $\times10^{-5}$
1179 - 1227	0.008	0.00064 *23*
1232 - 1247	0.00041	3.3 *16* $\times10^{-5}$
1287 - 1328	0.0034	~0.00027
1364 - 1407	0.008	0.00060 *24*
1422 - 1463	0.0060	0.00041 *15*
1490 - 1531	0.0017	0.00011 *5*

Continuous Radiation (^{139}Nd)

⟨β+⟩=201 keV;⟨IB⟩=3.6 keV

E_{bin}(keV)		⟨ ⟩(keV)	(%)
0 - 10	β+	1.04×10^{-6}	1.24×10^{-5}
	IB	0.0085	
10 - 20	β+	4.66×10^{-5}	0.000280
	IB	0.0082	0.057
20 - 40	β+	0.00143	0.00435
	IB	0.083	0.25
40 - 100	β+	0.077	0.098
	IB	0.057	0.095
100 - 300	β+	4.03	1.82
	IB	0.146	0.080
300 - 600	β+	28.6	6.2
	IB	0.29	0.066
600 - 1300	β+	137	15.0
	IB	1.25	0.133
1300 - 2500	β+	31.1	2.18
	IB	1.7	0.101
2500 - 2800	IB	0.023	0.00090
	Σβ+		25

$^{139}_{60}$Nd(5.5 *2* h)

Mode: ε(88 *2* %), IT(12 *2* %)
Δ: -81809 *50* keV
SpA: 4.10×10^6 Ci/g
Prod: ^{141}Pr(p,3n); ^{142}Nd(γ,3n); protons on Gd

Photons (^{139}Nd)

⟨γ⟩=1336 *27* keV

γ_{mode}	γ(keV)	γ(%)†
Pr L_ℓ	4.453	0.23 *4*
Nd L_ℓ	4.633	0.022 *5*
Pr L_η	4.929	0.095 *17*
Pr L_α	5.031	6.2 *10*
Nd L_η	5.146	0.0077 *17*
Nd L_α	5.228	0.58 *12*
Pr L_β	5.618	5.6 *10*
Nd L_β	5.861	0.51 *11*
Pr L_γ	6.407	0.78 *15*
Nd L_γ	6.707	0.074 *17*
Pr $K_{\alpha2}$	35.550	26.6 *14*
Pr $K_{\alpha1}$	36.026	48.5 *25*
Nd $K_{\alpha2}$	36.847	1.9 *4*
Nd $K_{\alpha1}$	37.361	3.5 *7*
Pr $K_{\beta1}$'	40.720	13.7 *7*
Pr $K_{\beta2}$'	41.981	3.83 *21*
Nd $K_{\beta1}$'	42.240	0.99 *20*
Nd $K_{\beta2}$'	43.562	0.29 *6*
γ_ϵM1+E2	92.95 *6*	0.87 *18*
γ_ϵ[M1+E2]	101.17 *7*	0.15 *5*
γ_ϵM1+E2	113.95 *5*	34 *4*
γ_ϵE2+M1	147.90 *7*	0.53 *7*
γ_ϵ	151.45 *20*	<0.06
γ_ϵ	172.15 *20*	<0.06
γ_ϵM1+E2	209.72 *5*	1.53 *15*
γ_ϵE2+M1	214.62 *6*	0.42 *6*
γ_{IT} M4	231.15 *5*	0.76
γ_ϵ[M1+E2]	254.69 *8*	1.02 *12*
γ_ϵ[M1+E2]	302.67 *7*	0.48 *8*
γ_ϵ	326.1 *4*	0.31 *5*
γ_ϵM1+E2	340.53 *9*	0.52 *8*
γ_ϵ(E2)	362.52 *6*	1.98 *15*
γ_ϵ(M1)	403.84 *6*	2.04 *18*
γ_ϵM1+E2	424.33 *7*	0.65 *8*
γ_ϵ	475.6 *4*	0.53 *8*
γ_ϵM1	547.74 *8*	2.01 *15*
γ_ϵ(M1)	572.24 *7*	0.66 *10*
γ_ϵ[E1]	600.47 *22*	0.48 *8*
γ_ϵM1+E2	673.41 *9*	0.53 *6*
γ_ϵM1+20%E2	701.25 *8*	3.5 *4*
γ_ϵM2	708.17 *8*	22.4 *10*
γ_ϵ	732.5 *5*	0.49 *8*
γ_ϵE2	738.17 *13*	30
γ_ϵE1	796.51 *18*	3.6 *4*
γ_ϵ(E2)	802.43 *8*	5.9 *6*
γ_ϵE1	810.18 *22*	5.4 *5*
γ_ϵE3	822.12 *9*	1.47 *18*
γ_ϵE2	828.03 *19*	8.8 *5*
γ_ϵ	852.12 *14*	<0.15 ?
γ_ϵ	894.9 *4*	0.30 *5*
γ_ϵ	900.1 *4*	0.33 *6*
γ_ϵE2+M1	910.13 *22*	6.5 *3*
γ_ϵE1	982.14 *12*	22.4 *7*
γ_ϵE1	1006.23 *18*	2.73 *21*
γ_ϵE2+M1	1012.14 *8*	2.34 *18*
γ_ϵE1	1024.80 *22*	1.29 *9*
γ_ϵE1	1075.09 *13*	2.97 *21*
γ_ϵE2+M1	1105.09 *8*	2.31 *18*
γ_ϵ	1165.9 *5*	0.29 *5*
γ_ϵE1	1220.84 *19*	1.41 *15*
γ_ϵM1+E2	1226.76 *9*	1.14 *12*
γ_ϵ	1234.7 *5*	0.36 *6*
γ_ϵ	1245.8 *5*	0.25 *5*
γ_ϵE1	1322.66 *14*	1.59 *15*
γ_ϵ	1344.66 *13*	0.39 *12*
γ_ϵ	1364.9 *5*	0.29 *5*
γ_ϵ	1374.66 *9*	0.36 *12*
γ_ϵ	1464.0 *4*	0.27 *9*
γ_ϵ	1469.9 *4*	0.50 *12*

Photons (^{139}Nd) (continued)

γ_{mode}	γ(keV)	γ(%)†
γ_ϵ	1510.59 *10*	0.12 *5*
γ_ϵ	1680.8 *5*	0.27 *6*
γ_ϵE1	2060.83 *13*	4.1 *4*
γ_ϵ	2085.1 *7*	0.045 *18*
γ_ϵ	2201.3 *8*	0.108 *21*

† uncert(syst): 9.1% for ε, 17% for IT

Atomic Electrons (^{139}Nd)

⟨e⟩=78 *8* keV

e_{bin}(keV)	⟨e⟩(keV)	e(%)
6	5.3	86 *14*
7 - 51	4.7	23.1 *17*
59	0.11	0.19 *6*
72	22.2	31 *4*
86 - 106	1.0	1.1 *5*
107	3.6	3.4 *6*
108	7.0	~6
109 - 151	3.3	~3
165 - 173	0.49	0.29 *3*
188	13.5	7.2 *15*
203 - 215	0.33	0.159 *24*
224	4.5	2.0 *4*
225	2.4	1.08 *22*
230 - 261	2.3	0.99 *16*
284 - 326	0.27	0.088 *9*
334 - 382	0.321	0.089 *7*
397 - 434	0.077	0.019 *4*
469 - 506	0.142	0.0281 *22*
530 - 572	0.081	0.0150 *14*
594 - 631	0.023	0.0037 *9*
659	0.157	0.024 *3*
666	2.93	0.439 *21*
667 - 708	1.46	0.209 *12*
726 - 768	0.410	0.055 *3*
780 - 828	0.415	0.052 *3*
845 - 895	0.20	0.023 *5*
899 - 940	0.224	0.0239 *10*
964 - 1012	0.137	0.0140 *13*
1018 - 1063	0.077	0.0073 *10*
1068 - 1105	0.0126	0.00116 *14*
1124 - 1166	0.006	0.00049 *24*
1179 - 1228	0.048	0.0040 *5*
1229 - 1246	0.0008	6.7 *25* ×10^{-5}
1281 - 1323	0.021	0.00159 *24*
1333 - 1375	0.007	0.00052 *22*
1422 - 1509	0.017	0.0012 *3*
1639 - 1680	0.0034	0.00021 *9*
2019 - 2084	0.0246	0.00122 *11*
2159 - 2200	0.0010	4.8 *19* ×10^{-5}

Continuous Radiation (^{139}Nd)

⟨β+⟩=30.0 keV;⟨IB⟩=0.88 keV

E_{bin}(keV)		⟨ ⟩(keV)	(%)
0 - 10	β+	2.57×10^{-7}	3.07×10^{-6}
	IB	0.00160	
10 - 20	β+	1.15×10^{-5}	6.9×10^{-5}
	IB	0.00143	0.0097
20 - 40	β+	0.000349	0.00106
	IB	0.079	0.23
40 - 100	β+	0.0181	0.0232
	IB	0.022	0.042
100 - 300	β+	0.85	0.388
	IB	0.047	0.024
300 - 600	β+	5.3	1.15
	IB	0.137	0.031
600 - 1300	β+	18.6	2.09
	IB	0.34	0.038
1300 - 2500	β+	5.3	0.363
	IB	0.24	0.0147
2500 - 2917	IB	0.0054	0.00021
	Σβ+		4.0

$^{139}_{61}$Pm(4.15 *5* min)

Mode: ε
Δ: -77520 *60* keV
SpA: 3.26×10^8 Ci/g
Prod: ^{142}Nd(p,4n); ^{142}Nd(d,5n)

Photons (^{139}Pm)

⟨γ⟩=210 *7* keV

γ_{mode}	γ(keV)	γ(%)†
Nd L_ℓ	4.633	0.067 *7*
Nd L_η	5.146	0.0263 *20*
Nd L_α	5.228	1.82 *12*
Nd L_β	5.856	1.63 *14*
Nd L_γ	6.691	0.232 *24*
Nd $K_{\alpha2}$	36.847	7.77 *24*
Nd $K_{\alpha1}$	37.361	14.1 *5*
Nd $K_{\beta1}$'	42.240	4.03 *13*
Nd $K_{\beta2}$'	43.562	1.18 *4*
γ M1	95.27 *8*	1.54 *10*
γ	272.66 *19*	0.101 *25*
γ	307.45 *12*	0.30 *4*
γ M1	367.76 *12*	3.04 *19*
γ M1,E2	402.72 *11*	12.5
γ	412.95 *19*	~0.13
γ M1,E2	463.03 *12*	3.51 *23*
γ	530.08 *16*	0.300 *25*
γ	578.65 *21*	0.049 *12*
γ	580.11 *18*	0.36 *4*
γ	603.0 *3*	0.125 *13*
γ	631.7 *4*	0.100 *11*
γ	661.12 *18*	0.35 *4*
γ	663.4 *4*	0.101 *20*
γ	675.38 *17*	0.162 *12*
γ	702.2 *3*	0.071 *19*
γ	720.9 *10*	0.047 *13*
γ	756.39 *17*	1.71 *14*
γ	815.67 *17*	0.975 *25*
γ	819.0 *4*	~0.050 ?
γ	856.3 *4*	0.083 *11*
γ	879.3 *4*	0.047 *8*
γ	981.37 *20*	0.91 *6*
γ	1079.8 *4*	0.125 *25*
γ	1104.9 *3*	0.035 *8*
γ	1126.04 *21*	0.250 *25*
γ	1174.62 *19*	0.150 *12*
γ	1193.2 *3*	0.175 *25*
γ	1205.67 *21*	0.046 *10*
γ	1264.94 *20*	0.076 *11*
γ	1291.75 *21*	0.237 *25*
γ	1299.6 *3*	0.067 *8*
γ	1351.02 *19*	0.179 *16*
γ	1358.58 *22*	0.045 *15*
γ	1426.41 *25*	0.080 *19*
γ	1473.2 *3*	0.033 *11*
γ	1484.05 *10*	0.175 *14*
γ	1558.30 *19*	0.111 *17*
γ	1602.8 *3*	0.36 *4*
γ	1618.61 *18*	~0.17
γ	1644.38 *16*	0.262 *25*
γ	1664.56 *22*	0.033 *5*
γ	1704.69 *15*	0.61 *6*
γ	1798.14 *21*	0.075 *10*
γ	1825.83 *25*	0.039 *6*
γ	1849.4 *5*	0.036 *6*
γ	1858.45 *20*	0.213 *25*
γ	1876.91 *21*	0.138 *25*
γ	1886.1 *3*	0.036 *6*
γ	1956.48 *24*	0.124 *25*
γ	2021.33 *17*	0.150 *12*
γ	2107.40 *14*	0.31 *4*
γ	2261.16 *20*	0.188 *25*
γ	2350.9 *4*	0.031 *6*
γ	2359.20 *23*	0.138 *13*

† 12% uncert(syst)

Atomic Electrons (^{139}Pm)

⟨e⟩=5.8 *3* keV

e_{bin}(keV)	⟨e⟩(keV)	e(%)
6	0.81	13.1 *13*
7	0.59	8.6 *9*
29 - 30	0.216	0.73 *6*
31	0.244	0.79 *8*
35 - 42	0.290	0.80 *5*
52	1.13	2.18 *15*
88	0.246	0.279 *19*
89 - 95	0.101	0.109 *5*
229 - 273	0.04	0.016 *8*
300 - 307	0.007	0.0023 *11*
324	0.346	0.107 *7*
359	1.0	0.28 *7*
361 - 369	0.074	0.0203 *19*
396	0.16	0.040 *13*
397 - 413	0.062	0.015 *5*
419	0.23	0.055 *14*
456 - 487	0.060	0.013 *3*
523 - 572	0.022	0.0041 *17*
573 - 620	0.020	0.0034 *12*
625 - 674	0.011	0.0017 *6*
675 - 721	0.05	~0.006
749 - 775	0.032	0.0042 *19*
809 - 856	0.0075	0.0009 *3*
872 - 879	0.00018	2.1 *10* $\times10^{-5}$
938 - 981	0.020	~0.0021
1036 - 1082	0.007	0.00065 *24*
1098 - 1131	0.0030	0.00027 *11*
1150 - 1199	0.0042	0.00036 *13*
1204 - 1248	0.0043	0.00035 *15*
1256 - 1300	0.0018	0.00014 *4*
1307 - 1352	0.0033	0.00025 *10*
1357 - 1383	0.0010	~7×10^{-5}
1419 - 1467	0.0026	0.00018 *7*
1472 - 1559	0.0056	0.00037 *13*
1575 - 1663	0.012	0.00076 *23*
1698 - 1797	0.0022	0.00012 *3*
1806 - 1885	0.0045	0.00025 *6*
1913 - 1978	0.0025	0.00013 *4*
2014 - 2106	0.0031	0.00015 *5*
2218 - 2316	0.0029	0.00013 *3*
2344 - 2358	0.00020	8.3 *25* $\times10^{-6}$

Continuous Radiation (^{139}Pm)

⟨β+⟩=1054 keV;⟨IB⟩=8.6 keV

E_{bin}(keV)		⟨ ⟩(keV)	(%)
0 - 10	β+	5.2×10^{-7}	6.2×10^{-6}
	IB	0.034	
10 - 20	β+	2.38×10^{-5}	0.000143
	IB	0.034	0.24
20 - 40	β+	0.00075	0.00229
	IB	0.091	0.30
40 - 100	β+	0.0428	0.054
	IB	0.21	0.33
100 - 300	β+	2.53	1.13
	IB	0.58	0.32
300 - 600	β+	22.0	4.69
	IB	0.76	0.18
600 - 1300	β+	203	20.7
	IB	1.7	0.19
1300 - 2500	β+	650	35.4
	IB	3.1	0.169
2500 - 4520	β+	177	6.4
	IB	2.1	0.069
	Σβ+		68

$^{139}_{61}$Pm(180 *20* ms)

Mode: IT

Δ: -77331 *60* keV

SpA: 1.15×10^{11} Ci/g

Prod: daughter ^{139}Sm(9.5 s); ^{141}Pr(α,6n)

Photons (^{139}Pm)

⟨γ⟩=85.1 *11* keV

γ_{mode}	γ(keV)	γ(%)
Pm L_ℓ	4.809	0.110 *12*
Pm L_η	5.363	0.061 *5*
Pm L_α	5.430	2.96 *21*
Pm L_β	6.069	3.2 *3*
Pm L_γ	6.927	0.45 *4*
Pm $K_{\alpha2}$	38.171	7.05 *22*
Pm $K_{\alpha1}$	38.725	12.8 *4*
Pm $K_{\beta1}$'	43.793	3.69 *12*
Pm $K_{\beta2}$'	45.183	1.09 *4*
γ E3	188.7 *3*	39.7

Atomic Electrons (^{139}Pm)

⟨e⟩=103.0 *14* keV

e_{bin}(keV)	⟨e⟩(keV)	e(%)
6	1.31	20.2 *21*
7	1.19	16.9 *17*
30	0.052	0.170 *17*
31	0.142	0.46 *5*
32	0.214	0.67 *7*
36	0.055	0.153 *16*
37	0.127	0.34 *4*
38	0.047	0.123 *13*
39	0.0044	0.0114 *12*
42	0.0167	0.040 *4*
43	0.0066	0.0152 *16*
44	0.0025	0.0058 *6*
144	38.0	26.5 *6*
181	4.93	2.72 *6*
182	42.3	23.3 *5*
187	11.4	6.10 *14*
188	2.72	1.44 *3*
189	0.403	0.214 *5*

$^{139}_{62}$Sm(2.57 *1* min)

Mode: ε

Δ: -72090 *160* keV

SpA: 5.252×10^{8} Ci/g

Prod: ^{142}Nd(α,7n); ^{144}Sm(α,α5n); protons on Gd

Photons (^{139}Sm)

γ_{mode}	γ(keV)	γ(rel)
γ E2+M1	273.4 *5*	100 *24*
γ	306.3 *5*	92 *23*
γ	597 *1*	23 *7*

$^{139}_{62}$Sm(9.5 *10* s)

Mode: IT(93.7 *5* %), ε(6.3 *5* %)

Δ: -71632 *160* keV

SpA: 8.2×10^{9} Ci/g

Prod: ^{142}Nd(α,7n); ^{144}Sm(α,α5n)

Photons (^{139}Sm)

⟨γ⟩=283 *15* keV

γ_{mode}	γ(keV)	γ(%)†
Pm L_ℓ	4.809	0.0102 *12*
Sm L_ℓ	4.993	0.23 *4*
Pm L_η	5.363	0.0051 *5*
Pm L_α	5.430	0.274 *25*
Sm L_η	5.589	0.109 *16*
Sm L_α	5.633	6.0 *9*
Pm L_β	6.077	0.28 *3*
Sm L_β	6.324	6.1 *9*
Pm L_γ	6.942	0.040 *5*
Sm L_γ	7.231	0.87 *14*
Pm $K_{\alpha2}$	38.171	0.80 *5*
Pm $K_{\alpha1}$	38.725	1.45 *8*
Sm $K_{\alpha2}$	39.522	16.8 *14*
Sm $K_{\alpha1}$	40.118	30.4 *25*
Pm $K_{\beta1}$'	43.793	0.418 *24*
Pm $K_{\beta2}$'	45.183	0.124 *8*
Sm $K_{\beta1}$'	45.379	8.8 *7*
Sm $K_{\beta2}$'	46.819	2.61 *22*
γ_{IT} E2+M1	112.0 *3*	24 *3*
γ_{IT} E2+M1	155.5 *3*	34 *4*
γ_ϵE3	188.7 *3*	2.51 *
γ_{IT} E3	190.3 *3*	38 *4*
γ_{IT} (E2)	267.5 *3*	37 *4*

† uncert(syst): 7.9% for ε, 2.6% for IT

* with ^{139}Pm(180 ms) in equilib

Atomic Electrons (^{139}Sm)

⟨e⟩=164 *9* keV

e_{bin}(keV)	⟨e⟩(keV)	e(%)
6 - 45	6.6	77 *11*
65	15.0	23 *4*
104	2.7	2.6 *8*
105	6.0	~6
109	14.0	12.9 *22*
110 - 112	2.7	2.4 *11*
143	35.3	25 *3*
144 - 182	11.7	7.5 *16*
183	29.6	16.2 *17*
184	17.8	9.7 *10*
187 - 188	0.89	0.48 *3*
189	11.7	6.2 *7*
190	3.1	1.62 *17*
221	5.1	2.3 *3*
260 - 268	1.88	0.72 *5*

$^{139}_{63}$Eu(22 *3* s)

Mode: ε

SpA: 3.6×10^{9} Ci/g

Prod: protons on Gd

Photons (^{139}Eu)

γ_{mode}	γ(keV)
γ	111

$^{139}_{64}$Gd(4.9 *10* s)

Mode: ε, εp

SpA: 1.5×10^{10} Ci/g

Prod: ^{50}Cr(^{92}Mo,2pn)

εp:2100-5500

A = 140

NDS **28**, 267 (1979)

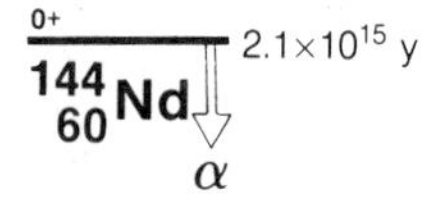

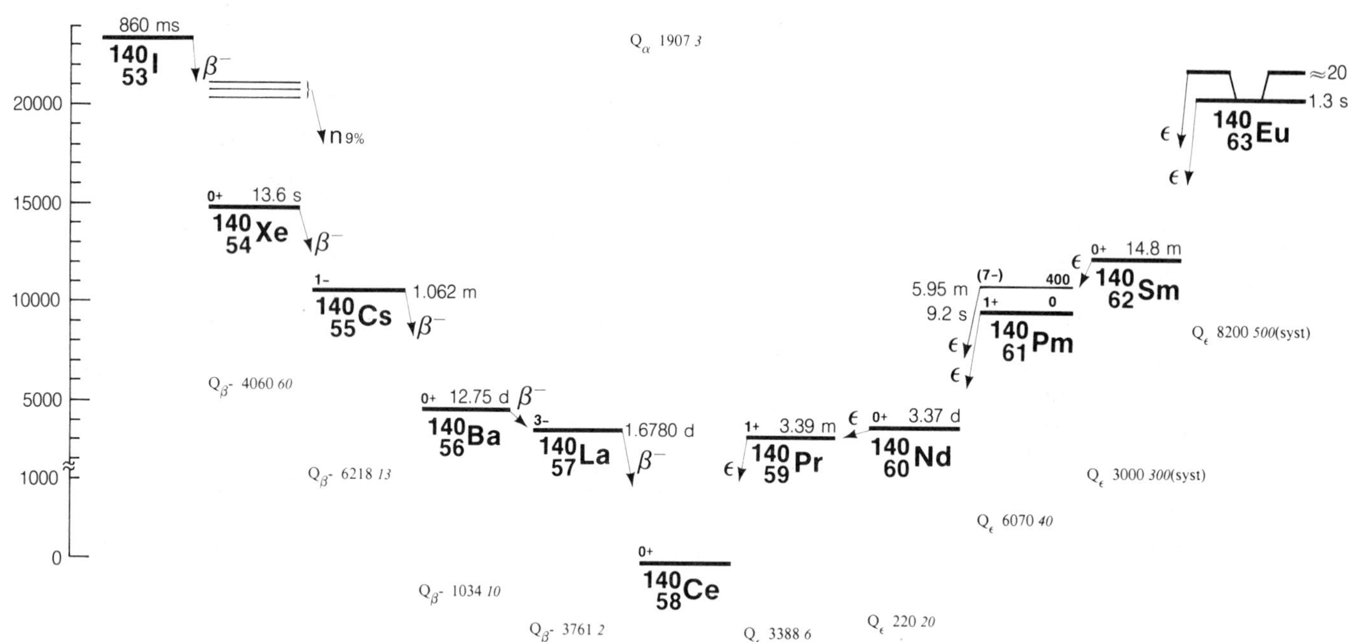

$^{140}_{53}$I (860 *40* ms)

Mode: β-, β-n(9.2 *6* %)
SpA: 6.4×10^{10} Ci/g
Prod: fission

Photons (^{140}I)

γ_{mode}	γ(keV)	γ(%)
$\gamma_{\beta-}$	372.35 *19*	~60
$\gamma_{\beta-}$[E2]	377.1 *5*	
$\gamma_{\beta-}$[E2]	457.7 *5*	

$^{140}_{54}$Xe(13.6 *1* s)

Mode: β-
Δ: -73020 *60* keV
SpA: 5.78×10^{9} Ci/g
Prod: fission

Photons (^{140}Xe)

⟨γ⟩=1152 *42* keV

γ_{mode}	γ(keV)	γ(%)†
Cs L_ℓ	3.795	0.083 *16*
Cs L_η	4.142	0.044 *7*
Cs L_α	4.285	2.3 *4*
Cs L_β	4.712	3.3 *7*
Cs L_γ	5.444	0.55 *13*
γ (M1)	13.930 *21*	0.92 *21*
Cs $K_{\alpha2}$	30.625	8.1 *4*
Cs $K_{\alpha1}$	30.973	15.0 *6*
Cs $K_{\beta1}$'	34.967	4.09 *18*
Cs $K_{\beta2}$'	36.006	1.00 *5*
γ [M1]	38.328 *23*	0.27 *4*
γ	45.89 *9*	~0.042
γ [M1+E2]	47.761 *24*	0.063 *21*
γ (M1)	50.823 *24*	1.83 *13*
γ M1	80.119 *23*	4.62 *25*
γ	84.24 *9*	0.23 *6*
γ [M1+E2]	89.168 *22*	0.46 *4*
γ [M1+E2]	93.61 *3*	0.126 *21*
γ [M1+E2]	99.55 *4*	0.063 *21*
γ M1+~50%E2	103.098 *22*	1.11 *6*
γ M1+~50%E2	104.517 *22*	1.43 *8*
γ [M1+E2]	108.96 *3*	0.084 *21*
γ E2+~41%M1	112.514 *23*	4.12 *25*
γ M1	118.447 *19*	4.68 *25*
γ [M1+E2]	119.49 *8*	0.13 *4*
γ	121.54 *15*	~0.042
γ [E2]	128.91 *8*	0.040 *8*
γ	132.97 *7*	0.065 *10*
γ [M1+E2]	138.38 *8*	0.034 *8*
γ [M1+E2]	147.31 *4*	0.029 *8*

Photons (^{140}Xe) (continued)

γ_{mode}	γ(keV)	γ(%)†
γ	158.73 *15*	0.113 *15*
γ M1	167.25 *8*	1.30 *13*
γ [M1+E2]	176.47 *6*	0.122 *15*
γ [M1+E2]	182.40 *7*	0.252 *25*
γ M1,E2	196.04 *10*	0.23 *5*
γ M1,E2	198.13 *3*	0.59 *6*
γ	202.84 *8*	0.101 *13*
γ M1,E2	212.06 *3*	2.41 *25*
γ [M1+E2]	214.80 *7*	0.33 *5*
γ [M1+E2]	218.08 *8*	0.116 *25*
γ [M1+E2]	219.9 *1*	0.063 *21*
γ [M1+E2]	226.57 *6*	0.032 *13*
γ	226.58 *7*	0.032 *13*
γ	232.51 *7*	0.073 *17*
γ	241.93 *7*	0.151 *21*
γ [M1+E2]	252.95 *8*	0.052 *13*
γ M1,E2	276.98 *7*	0.59 *6*
γ M1,E2	280.99 *6*	1.45 *15*
γ [E2]	282.81 *11*	~0.08
γ [M1+E2]	290.51 *6*	0.46 *4*
γ [M1+E2]	294.92 *6*	0.050 *21*
γ [M1+E2]	320.19 *6*	0.078 *21*
γ [M1+E2]	326.12 *6*	0.071 *21*
γ	331.10 *7*	0.36 *4*
γ [M1+E2]	335.54 *6*	0.23 *4*
γ	343.75 *9*	0.059 *21*
γ [M1+E2]	358.46 *9*	0.118 *21*
γ [M1+E2]	373.88 *6*	0.57 *6*
γ E2	390.00 *8*	1.53 *15*
γ [M1+E2]	396.37 *8*	0.80 *8*
γ [M1+E2]	409.95 *5*	0.204 *21*
γ [M1+E2]	429.42 *5*	0.69 *6*
γ [M1+E2]	435.35 *6*	0.22 *4*

Photons (^{140}Xe) (continued)

γ_{mode}	γ(keV)	γ(%)†
γ M1	438.63 5	2.7 3
γ [M1+E2]	441.32 7	0.55 15
γ	445.08 14	0.74 8
γ	455.19 14	~0.06
γ	461.82 8	1.55 13
γ [M1+E2]	483.11 6	0.18 3
γ [M1+E2]	503.56 5	0.18 4
γ	505.30 14	0.14 4
γ [M1+E2]	509.49 5	0.88 13
γ [M1+E2]	514.82 8	1.09 10
γ [M1+E2]	518.91 5	1.07 11
γ	524.56 13	0.48 4
γ [M1+E2]	547.87 5	1.20 8
γ M1	557.26 5	5.3 4
γ	562.08 8	0.30 5
γ	568.21 15	0.10 4
γ	570.9 7	0.12 5
γ [M1+E2]	573.26 7	0.16 5
γ	588.16 13	~0.042
γ [M1+E2]	608.08 4	2.35 25
γ E2+~41%M1	622.01 4	8.4 6
γ	627.36 13	1.03 10
γ [M1+E2]	639.45 7	1.39 13
γ M1+E2	653.38 6	~0.42
γ (E1)	653.44 7	4.6 4
γ	655.69 7	0.74 19
γ	670.84 10	0.32 10
γ	686.2 3	0.18 4
γ	690.96 13	~0.08
γ	697.12 13	0.19 4
γ	721.4 4	0.08 3
γ	733.75 11	0.30 5
γ	735.46 13	0.22 4
γ	774.14 7	3.89 25
γ [E1]	774.20 7	~0.32
γ	786.29 13	0.43 5
γ	800.22 13	0.57 21
γ E1	805.57 5	21
γ	820.9 5	0.046 21
γ [E1]	842.15 8	0.61 6
γ	847.31 8	0.21 6
γ [E1]	850.56 9	0.44 6
γ	857.0 4	0.111 25
γ	862.66 8	0.17 5
γ	864.05 14	0.23 6
γ [E1]	879.71 6	2.88 19
γ	889.09 13	0.41 5
γ	901.00 8	0.21 6
γ	903.02 13	0.42 6
γ [E1]	912.75 8	0.92 8
γ [E1]	925.00 6	1.51 13
γ	935.9 3	0.088 21
γ	945.5 4	0.086 25
γ	951.83 8	0.92 8
γ	963.6 5	0.15 6
γ	966.5 7	~0.10
γ	982.50 14	0.48 8
γ [E1]	988.94 6	3.19 25
γ	999.8 5	0.11 5
γ [E1]	1018.61 6	0.30 5
γ [E1]	1024.55 6	0.27 5
γ [E1]	1077.13 8	~0.17
γ	1079.8 5	0.29 10
γ	1086.7 5	0.27 11
γ	1089.8 10	~0.13
γ [E1]	1123.13 6	~0.17
γ [E1]	1132.66 7	0.69 17
γ [E1]	1137.06 6	2.25 21
γ	1141.5 4	0.32 8
γ	1154.5 3	0.32 4
γ	1168.6 10	~0.13
γ [E1]	1170.75 8	~0.10
γ [E1]	1176.68 8	1.15 15
γ	1180.2 5	0.15 6
γ	1189.5 10	0.25 6
γ	1192.8 10	0.17 6
γ [E1]	1209.07 8	1.43 17
γ [E1]	1215.51 4	0.29 6
γ [E1]	1289.19 8	0.21 6
γ [E1]	1309.12 4	6.7 6
γ [E1]	1315.06 4	8.6 8
γ [E1]	1347.45 4	0.15 4
γ [E1]	1413.64 4	12.8 13
γ [E1]	1427.57 3	1.20 13

Photons (^{140}Xe) (continued)

γ_{mode}	γ(keV)	γ(%)†
γ	1585.7 3	0.19 4
γ	1661.2 3	0.25 6
γ [E1]	1885.80 18	0.36 8
γ [E1]	2073.97 21	0.23 6
γ [E1]	2112.38 18	0.38 11
γ [E1]	2211.92 17	0.10 4
γ [E1]	2286.03 21	0.17 6

† 14% uncert(syst)

Atomic Electrons (^{140}Xe)

⟨e⟩=27.9 8 keV

e_{bin}(keV)	⟨e⟩(keV)	e(%)
2	0.082	3.5 6
5	1.36	26 3
6	0.95	17 4
8	2.4	29 7
9 - 12	0.35	3.9 8
13	0.85	6.7 15
14	0.24	1.7 4
15	1.51	10.2 8
25 - 43	1.49	4.8 6
44	3.03	6.9 4
45	0.65	1.45 21
46 - 64	0.87	1.7 3
67	0.68	1.01 6
69	0.85	1.25 8
73 - 75	0.73	0.99 5
77	2.28	2.98 19
79 - 80	0.30	0.38 14
82	1.88	2.28 13
83 - 132	3.47	3.31 23
133 - 182	0.90	0.53 4
184 - 233	0.24	0.117 18
236 - 286	0.362	0.144 10
289 - 339	0.110	0.035 5
343 - 391	0.185	0.051 3
393 - 441	0.38	0.092 10
444 - 489	0.174	0.037 4
498 - 547	0.324	0.062 4
548 - 591	0.40	0.069 5
602 - 651	0.203	0.033 3
652 - 699	0.022	0.0032 8
716 - 764	0.09	0.012 6
768 - 816	0.21	0.027 4
820 - 867	0.046	0.0054 7
874 - 920	0.039	0.0043 8
924 - 967	0.034	0.0036 5
977 - 1024	0.0083	0.00085 8
1041 - 1090	0.012	0.0011 3
1097 - 1144	0.036	0.0032 3
1149 - 1193	0.0162	0.00139 19
1203 - 1215	0.00140	0.000116 10
1253 - 1289	0.077	0.0060 4
1303 - 1347	0.0125	0.00095 6
1378 - 1426	0.075	0.0055 5
1550 - 1625	0.0036	0.00023 8
1655 - 1660	0.00029	1.7 7 $\times 10^{-5}$
1850 - 1885	0.0016	8.6 18 $\times 10^{-5}$
2038 - 2111	0.0026	0.000124 22
2176 - 2250	0.0010	4.6 12 $\times 10^{-5}$
2280 - 2285	8.7 $\times 10^{-5}$	3.8 12 $\times 10^{-6}$

Continuous Radiation (^{140}Xe)

⟨β-⟩=1194 keV;⟨IB⟩=3.1 keV

E_{bin}(keV)		⟨ ⟩(keV)	(%)
0 - 10	β-	0.0098	0.196
	IB	0.042	
10 - 20	β-	0.0301	0.201
	IB	0.041	0.29
20 - 40	β-	0.125	0.416
	IB	0.081	0.28
40 - 100	β-	0.97	1.37
	IB	0.23	0.35
100 - 300	β-	12.3	6.0
	IB	0.65	0.37
300 - 600	β-	58	12.7
	IB	0.72	0.17
600 - 1300	β-	362	37.9
	IB	0.92	0.108
1300 - 2500	β-	653	37.5
	IB	0.37	0.022
2500 - 4060	β-	107	3.72
	IB	0.023	0.00080

$^{140}_{55}$Cs(1.062 5 min)

Mode: β-
Δ: -77076 16 keV
SpA: 1.258×10^9 Ci/g
Prod: fission

Photons (^{140}Cs)

⟨γ⟩=2252 32 keV

γ_{mode}	γ(keV)	γ(%)†
Ba L_ℓ	3.954	0.00085 12
Ba L_η	4.331	0.00038 4
Ba L_α	4.465	0.023 3
Ba L_β	4.946	0.021 3
Ba L_γ	5.612	0.0027 4
Ba $K_{\alpha2}$	31.817	0.124 6
Ba $K_{\alpha1}$	32.194	0.228 11
Ba $K_{\beta1}'$	36.357	0.063 3
Ba $K_{\beta2}'$	37.450	0.0160 8
γ	401.04 14	~0.04
γ	411.06 16	~0.10
γ [E1]	413.32 12	0.22 4
γ E2	528.32 4	4.08 8
γ E2	602.39 4	70 4
γ	627.75 11	0.190 21
γ	643.34 15	0.084 14
γ E1,M1	672.24 5	1.55 8
γ	693.33 10	0.16 5
γ [M1+E2]	695.62 10	0.43 5
γ [E1]	726.53 10	0.11 5
γ	728.99 15	0.11 5
γ	736.28 13	0.65 6
γ	740.9 3	~0.06
γ	759.5 10	0.056 11
γ [M1+E2]	760.27 20	0.08 4
γ	794.00 16	~0.07
γ	799.03 11	0.17 3
γ	809.39 21	0.11 3
γ M1+E2	820.81 13	0.303 21
γ [M1+E2]	828.56 21	~0.035
γ	861.92 17	0.049 14
γ	874.46 18	0.063 21
γ	880.50 17	0.10 4
γ	893.0 10	0.14 3
γ	902.0 10	0.56 11
γ E2+45%M1	908.44 6	11.3 6
γ [M1+E2]	918.87 11	0.26 3
γ [E1]	939.17 12	0.063 21
γ	944.34 25	0.056 21
γ	950.24 21	0.08 3
γ	969.5 5	0.070 21

Photons (^{140}Cs) (continued)

γ_{mode}	γ(keV)	γ(%)†
γ	980.67 18	~0.13
γ	984.86 15	0.15 7
γ	999.71 17	0.127 14
γ	1007.87 11	1.11 11
γ	1011.09 14	0.61 11
γ[M1+E2]	1040.37 11	0.25 3
γ	1058.11 21	0.08 3
γ	1064.00 19	0.17 4
γ	1068.05 16	0.13 4
γ	1073.28 13	0.27 4
γ	1099.08 24	0.17 4
γ	1100.5 10	0.16 3
γ	1102.36 18	0.12 4
γ	1105.48 18	~0.08
γ	1114.7 10	~0.04
γ[M1+E2]	1130.03 8	3.10 15
γ	1147.77 17	0.127 21
γ	1172.73 21	0.070 21
γ	1181.6 3	~0.06
γ	1190.18 14	0.43 4
γ E1,M1	1200.56 5	6.1 3
γ E2	1221.65 9	2.97 15
γ	1263.7 3	0.077 21
γ	1271.08 15	0.13 3
γ	1276.98 15	0.17 3
γ	1280.9 3	0.11 4
γ	1288.9 3	0.17 5
γ	1292.58 21	0.24 5
γ	1299.4 2	0.31 4
γ	1322.3 3	0.07 3
γ	1337.85 23	0.11 5
γ	1363.48 14	0.29 3
γ	1376.10 17	0.134 21
γ M1+2.8%E2	1391.41 10	2.18 11
γ	1397.3 10	0.099 20
γ	1411.5 5	0.34 6
γ[M1+E2]	1419.01 13	0.46 10
γ E1+37%M2	1422.15 9	1.08 11
γ	1443.00 22	0.14 4
γ	1454.9 3	0.19 3
γ	1459.21 17	0.19 3
γ	1474.1 10	0.077 15
γ	1479.2 5	0.084 21
γ	1493.7 5	0.11 3
γ[E1]	1514.4 2	0.12 3
γ	1517.79 15	0.19 3
γ	1527.5 5	0.08 4
γ	1536.19 10	0.65 5
γ	1542.33 23	0.084 21
γ	1601.60 13	0.62 5
γ	1607.85 22	0.25 3
γ	1614.30 12	0.82 6
γ	1627.4 10	0.099 20
γ(E1)	1634.97 8	3.18 16
γ	1648.37 18	0.08 4
γ	1651.20 15	0.21 4
γ	1663.49 13	0.23 4
γ	1702.2 3	0.19 3
γ	1707.46 10	1.57 11
γ	1718.49 14	0.32 4
γ	1726.7 10	0.18 4
γ[M1+E2]	1735.98 13	0.61 8
γ	1737.1 10	0.42 8
γ	1756.1 10	~0.08
γ	1766.3 10	0.127 25
γ	1770.34 15	0.14 4
γ	1784.31 20	0.11 4
γ	1794.69 24	0.13 4
γ	1799.48 20	0.24 5
γ M1+20%E2	1827.30 11	0.49 7
γ	1834.77 14	0.21 5
γ[M1+E2]	1853.42 13	4.02 25
γ	1857.86 14	0.87 21
γ	1899.48 21	0.084 21
γ	1911.74 19	~0.07
γ	1918.3 10	~0.25
γ	1929.1 4	0.12 3
γ	1940.48 18	0.15 6
γ	1943.3 6	~0.11
γ[E1]	1950.29 19	0.46 14
γ[E2]	1993.80 10	0.51 11
γ	2010.4 5	~0.30
γ[E1]	2021.83 16	0.15 4
γ	2038.24 24	0.20 4

Photons (^{140}Cs) (continued)

γ_{mode}	γ(keV)	γ(%)†
γ	2048.71 12	0.41 6
γ	2061.59 17	0.21 5
γ	2067.67 23	0.44 7
γ[E1]	2086.39 16	0.17 4
γ	2090.1 3	~0.07
γ M1+E2	2101.70 12	3.94 25
γ[E1]	2120.05 19	0.11 3
γ	2146.94 25	0.49 5
γ[M1+E2]	2170.39 11	0.92 9
γ	2179.51 14	0.10 4
γ	2185.41 14	0.36 4
γ	2193.1 10	0.077 15
γ[M1+E2]	2235.10 21	~0.28
γ[E1]	2237.35 9	3.9 3
γ[M1+E2]	2250.78 19	0.11 4
γ	2268.59 15	1.72 12
γ[M1+E2]	2277.25 13	0.79 6
γ	2309.84 11	0.38 7
γ	2312.77 24	0.20 7
γ E1	2330.58 8	4.6 3
γ	2340.82 12	~0.28
γ	2362.0 5	0.25 10
γ[E1]	2372.20 22	~0.08
γ	2386.1 3	~0.08
γ[E1]	2401.84 21	0.15 4
γ[M1]	2429.68 11	1.68 12
γ	2453.4 8	~0.08
γ	2456.5 5	0.23 7
γ	2460.6 4	~0.14
γ[E1]	2462.51 11	0.56 11
γ	2477.6 3	0.11 3
γ	2496.88 15	0.36 7
γ	2514.28 23	0.33 11
γ	2521.90 14	4.1 3
γ	2554.5 10	0.042 8
γ	2664.0 10	0.056 11
γ	2674.6 10	0.092 18
γ	2704.08 13	0.84 6
γ	2765.5 10	0.120 24
γ	2787.79 15	0.106 21
γ	2848.91 17	0.76 6
γ	2874.29 14	0.56 6
γ	2969.93 21	0.52 6
γ[M1+E2]	2998.94 19	0.30 3
γ	3023.34 25	0.20 3
γ[E1]	3053.96 12	1.50 13
γ	3067.69 21	0.190 21
γ	3099.25 15	0.13 3
γ	3116.7 3	0.22 3
γ	3167.1 10	0.049 10
γ[E1]	3189.94 19	0.28 3
γ	3243.50 16	0.34 5
γ	3249.25 11	0.25 5
γ	3267.6 10	~0.08
γ	3285.9 2	~0.11
γ	3304.0 10	0.056 21
γ	3316.7 10	0.028 6
γ[E1]	3341.68 17	0.39 4
γ[E1]	3370.93 11	0.71 7
γ[E1]	3394.76 21	~0.11
γ	3408.4 10	0.028 6
γ[E1]	3435.64 20	0.44 11
γ	3451.28 18	0.65 7
γ[E1]	3477.78 13	0.25 4
γ	3526.2 3	0.084 21
γ	3545.0 6	0.084 21
γ	3564.52 21	0.113 21
γ	3602.39 20	0.25 4
γ	3629.8 5	0.11 3
γ	3636.4 5	0.099 21
γ	3658.7 10	~0.042
γ	3671.16 19	0.113 21
γ	3698.5 8	<0.042
γ	3756.4 4	0.15 3
γ[E1]	3786.1 3	0.10 4
γ[E1]	3793.23 20	0.30 4
γ	3825.5 15	~0.028
γ	3829.6 10	0.035 14
γ	3845.87 17	0.070 21
γ	3851.62 12	~0.028
γ	3918.7 10	0.035 7
γ[E1]	3944.06 17	0.45 9
γ	4003.8 10	0.014 3
γ	4053.02 19	~0.035

Photons (^{140}Cs) (continued)

γ_{mode}	γ(keV)	γ(%)†
γ	4075.8 6	0.063 21
γ	4108.2 8	0.070 21
γ	4171.1 10	0.021 4
γ	4210.2 8	0.07 3
γ	4237.7 8	0.049 21
γ	4381.5 8	0.042 14
γ	4405.1 10	0.021 4
γ[E1]	4416.58 20	0.084 21
γ	4472.9 8	0.070 21
γ	4499.7 10	0.042 14
γ	4525.4 8	0.09 3
γ	4531.5 6	0.070 21
γ[E1]	4571.49 23	~0.028
γ[E1]	4786.14 19	~0.028
γ	4813.3 10	~0.028
γ[E1]	4982.5 4	0.049 21
γ	5228.3 15	<0.028

† 1.4% uncert(syst)

Atomic Electrons (^{140}Cs)

⟨e⟩=3.81 12 keV

e_{bin}(keV)	⟨e⟩(keV)	e(%)
5 - 36	0.0334	0.42 3
364 - 413	0.012	0.0031 12
491	0.144	0.0293 8
522 - 528	0.0289	0.00552 10
565	2.03	0.360 19
590	0.005	~0.0008
596	0.248	0.0416 22
597 - 643	0.181	0.030 3
656 - 703	0.042	0.0062 13
721 - 762	0.011	0.0015 4
772 - 821	0.0121	0.00154 25
823 - 869	0.015	0.0017 6
871	0.227	0.0261 14
873 - 919	0.049	0.0054 3
932 - 981	0.031	0.0033 10
984 - 1031	0.0135	0.00133 24
1034 - 1077	0.013	0.0012 3
1093 - 1142	0.060	0.0055 8
1143 - 1153	0.006	0.00049 22
1163	0.102	0.0088 6
1167 - 1216	0.0494	0.00416 19
1220 - 1270	0.014	0.00113 22
1271 - 1317	0.0035	0.00027 7
1321 - 1370	0.0366	0.00271 18
1371 - 1419	0.036	0.0026 6
1421 - 1469	0.0055	0.00038 9
1473 - 1570	0.018	0.00120 25
1577 - 1670	0.045	0.0028 4
1681 - 1779	0.024	0.00138 16
1783 - 1881	0.061	0.0033 3
1892 - 1989	0.0127	0.00066 11
1993 - 2089	0.044	0.00214 19
2096 - 2192	0.0208	0.00098 9
2198 - 2293	0.059	0.00264 21
2303 - 2401	0.0209	0.00088 8
2416 - 2513	0.034	0.0014 3
2516 - 2553	0.0036	0.00014 4
2627 - 2703	0.0058	0.00022 6
2728 - 2811	0.0050	0.00018 5
2837 - 2932	0.0062	0.00021 4
2961 - 3053	0.0100	0.000331 23
3062 - 3161	0.0031	0.000101 14
3162 - 3248	0.0040	0.000124 23
3262 - 3357	0.0045	0.000135 11
3365 - 3450	0.0060	0.00018 3
3472 - 3565	0.0025	7.0 11 $\times 10^{-5}$
3592 - 3670	0.0020	5.4 8 $\times 10^{-5}$
3692 - 3788	0.0022	5.8 7 $\times 10^{-5}$
3792 - 3881	0.00078	2.0 4 $\times 10^{-5}$
3907 - 4003	0.0016	4.1 7 $\times 10^{-5}$
4016 - 4107	0.00074	1.8 4 $\times 10^{-5}$
4134 - 4232	0.00060	1.4 4 $\times 10^{-5}$
4236 - 4237	4.7 $\times 10^{-6}$	1.1 5 $\times 10^{-7}$

Atomic Electrons (^{140}Cs)
(continued)

e_{bin}(keV)	⟨e⟩(keV)	e(%)
4344 - 4435	0.00080	1.8 *3* $\times10^{-5}$
4462 - 4534	0.00097	2.2 *4* $\times10^{-5}$
4566 - 4570	1.2 $\times10^{-5}$	2.5 *11* $\times10^{-7}$
4749 - 4812	0.00020	4.2 *14* $\times10^{-6}$
4945 - 4981	0.00015	3.1 *12* $\times10^{-6}$

Continuous Radiation (^{140}Cs)
⟨β-⟩=1743 keV;⟨IB⟩=5.9 keV

E_{bin}(keV)		⟨ ⟩(keV)	(%)
0 - 10	β-	0.0087	0.172
	IB	0.051	
10 - 20	β-	0.0264	0.176
	IB	0.051	0.35
20 - 40	β-	0.109	0.363
	IB	0.100	0.35
40 - 100	β-	0.83	1.18
	IB	0.29	0.45
100 - 300	β-	10.0	4.88
	IB	0.86	0.48
300 - 600	β-	42.8	9.4
	IB	1.05	0.25
600 - 1300	β-	243	25.5
	IB	1.66	0.19
1300 - 2500	β-	633	34.4
	IB	1.35	0.078
2500 - 5000	β-	778	23.3
	IB	0.52	0.0169
5000 - 6218	β-	36.2	0.69
	IB	0.0031	5.6 $\times10^{-5}$

$^{140}_{56}$Ba(12.746 *10* d)

Mode: β-
Δ: -83294 *10* keV
SpA: 7.317$\times10^{4}$ Ci/g
Prod: fission

Photons (^{140}Ba)
⟨γ⟩=182.6 *16* keV

γ_{mode}	γ(keV)	γ(%)†
La L_ℓ	4.121	0.120 *23*
La L_η	4.529	0.073 *9*
La L_α	4.649	3.3 *6*
La L_β	5.126	7.7 *14*
La L_γ	5.998	1.5 *3*
γ M1	13.87 *3*	1.20 *12*
γ M1+~0.1%E2	29.99 *3*	13.6 *12*
La $K_{\alpha2}$	33.034	0.60 *4*
La $K_{\alpha1}$	33.442	1.11 *7*
La $K_{\beta1}'$	37.777	0.309 *18*
La $K_{\beta2}'$	38.927	0.081 *5*
γ [E2]	43.85 *4*	0.015 *3*
γ [M1]	113.59 *3*	0.018 *2* ?
γ M1	118.85 *3*	0.051 *5*
γ [M1]	132.72 *3*	0.202 *17*
γ M1(+E2)	162.70 *3*	6.21 *12*
γ M1(+E2)	304.867 *19*	4.30 *9*
γ M1	423.720 *25*	3.12 *15*
γ [M1]	437.585 *24*	1.93 *4*
γ [E2]	467.57 *3*	0.146 *12*
γ M1	537.31 *4*	24.39

† 0.90% uncert(syst)

Atomic Electrons (^{140}Ba)
⟨e⟩=34.5 *15* keV

e_{bin}(keV)	⟨e⟩(keV)	e(%)
5	0.25	4.5 *6*
6	3.4	55 *7*
8	4.0	52 *5*
13	1.36	10.9 *11*
14	0.41	3.0 *3*
24	14.2	60 *6*
25 - 28	0.64	2.60 *23*
29	3.7	13.0 *12*
30	1.07	3.6 *3*
31 - 80	0.27	0.67 *9*
94 - 119	0.088	0.093 *8*
124	1.94	1.57 *12*
126 - 163	0.70	0.45 *15*
266 - 305	0.56	0.208 *22*
385 - 432	0.431	0.109 *3*
436 - 468	0.00655	0.00148 *5*
498	1.28	0.257 *6*
531 - 537	0.224	0.0421 *8*

Continuous Radiation (^{140}Ba)
⟨β-⟩=276 keV;⟨IB⟩=0.26 keV

E_{bin}(keV)		⟨ ⟩(keV)	(%)
0 - 10	β-	0.122	2.44
	IB	0.0133	
10 - 20	β-	0.364	2.43
	IB	0.0125	0.087
20 - 40	β-	1.45	4.83
	IB	0.023	0.081
40 - 100	β-	9.8	14.1
	IB	0.057	0.089
100 - 300	β-	73	38.1
	IB	0.104	0.062
300 - 600	β-	122	28.4
	IB	0.047	0.0119
600 - 1004	β-	70	9.8
	IB	0.0051	0.00077

$^{140}_{57}$La(1.6780 *3* d)

Mode: β-
Δ: -84328 *4* keV
SpA: 5.55825$\times10^{5}$ Ci/g
Prod: ^{139}La(n,γ); fission; daughter ^{140}Ba

Photons (^{140}La)
⟨γ⟩=2315 *17* keV

γ_{mode}	γ(keV)	γ(%)†
Ce L_ℓ	4.289	0.0053 *8*
Ce L_η	4.730	0.0023 *3*
Ce L_α	4.838	0.144 *18*
Ce L_β	5.388	0.128 *17*
Ce L_γ	6.124	0.0171 *25*
Ce $K_{\alpha2}$	34.279	0.56 *3*
Ce $K_{\alpha1}$	34.720	1.03 *6*
Ce $K_{\beta1}'$	39.232	0.291 *17*
Ce $K_{\beta2}'$	40.437	0.078 *5*
γ M1	64.122 *8*	~0.010
γ M1	68.906 *6*	0.061 *15*
γ M1	109.407 *6*	0.200 *14*
γ	131.113 *7*	0.44 *9*
γ M1,E2	173.529 *9*	0.129 *20*
γ	241.955 *8*	0.47 *3*
γ E1+13%M2	266.550 *8*	0.452 *25*
γ E2	306.85 *13*	~0.021
γ M1+0.2%E2	328.758 *9*	20.7 *4*
γ E2,M1	397.663 *9*	0.10 *3*
γ M1+23%E2	432.52 *3*	2.99 *6*
γ [E2]	438.165 *11*	0.020 *10*
γ [E2]	444.81 *13*	~0.024
γ E2	487.026 *18*	45.9 *9*
γ [E2]	618.34 *13*	0.06 *3*
γ M1+10%E2	751.661 *21*	4.3 *3*
γ M1+0.2%E2	815.782 *20*	23.6 *5*
γ E1(+0.1%M2)	867.83 *13*	5.59 *11*
γ E2	919.55 *3*	2.68 *5*
γ M1+2.6%E2	925.19 *2*	7.03 *14*
γ	936.6 *3*	0.057 *17*
γ M1+1.1%E2	950.86 *21*	0.541 *19*
γ	1087.1 *7*	~0.0029
γ	1415.9 *7*	~0.006
γ E2	1596.54 *14*	95.4
γ	2083.56 *14*	0.0116 *7*
γ E2	2348.19 *14*	0.849 *17*
γ	2464.36 *19*	0.017 *6*
γ	2521.72 *14*	3.44 *8*
γ	2533.1 *3*	0.0039 *14*
γ M1,E2	2547.39 *24*	0.104 *3*
γ E2	2899.8 *5*	0.0658 *19*
γ M1(+E2)	3119.2 *7*	0.0258 *10*
γ M1,E2	3319.7 *6*	0.0045 *14*

†<0.1% uncert(syst)

Atomic Electrons (^{140}La)
⟨e⟩=9.20 *15* keV

e_{bin}(keV)	⟨e⟩(keV)	e(%)
6 - 39	0.33	2.89 *22*
58 - 104	0.28	0.36 *10*
108 - 133	0.12	0.10 *5*
167 - 202	0.06	~0.033
226 - 267	0.086	0.037 *5*
288	2.37	0.823 *23*
300 - 307	0.00057	0.00019 *6*
322	0.330	0.103 *3*
323 - 357	0.128	0.0389 *11*
391 - 439	0.258	0.0650 *19*
443 - 445	5.9 $\times10^{-5}$	1.3 *4* $\times10^{-5}$
447	1.99	0.446 *13*
480	0.247	0.0515 *15*
481 - 487	0.188	0.0388 *7*
578 - 618	0.0022	0.00037 *14*
711 - 752	0.174	0.0243 *16*
775	0.760	0.098 *3*
809 - 827	0.174	0.0214 *4*
861 - 910	0.262	0.0296 *6*
913 - 951	0.0433	0.00471 *9*
1047 - 1087	4.8 $\times10^{-5}$	~5 $\times10^{-6}$
1375 - 1416	7.1 $\times10^{-5}$	~5 $\times10^{-6}$
1556	1.013	0.0651 *15*
1590 - 1595	0.161	0.01011 *19*
1863 - 1898	0.170	0.0091 *11*
2043 - 2082	9.8 $\times10^{-5}$	4.8 *17* $\times10^{-6}$
2308 - 2347	0.00760	0.000329 *8*
2424 - 2521	0.025	0.0010 *3*
2527 - 2546	0.000129	5.1 *4* $\times10^{-6}$
2859 - 2899	0.000504	1.76 *5* $\times10^{-5}$
3079 - 3118	0.000193	6.3 *4* $\times10^{-6}$
3279 - 3318	3.2 $\times10^{-5}$	1.0 *3* $\times10^{-6}$

Continuous Radiation (^{140}La)

⟨β-⟩=525 keV;⟨IB⟩=0.77 keV

E_{bin}(keV)		⟨ ⟩(keV)	(%)
0 - 10	β-	0.0398	0.79
	IB	0.023	
10 - 20	β-	0.121	0.81
	IB	0.022	0.154
20 - 40	β-	0.496	1.65
	IB	0.043	0.148
40 - 100	β-	3.71	5.3
	IB	0.114	0.18
100 - 300	β-	41.2	20.3
	IB	0.27	0.154
300 - 600	β-	145	32.4
	IB	0.20	0.049
600 - 1300	β-	309	36.9
	IB	0.095	0.0125
1300 - 2500	β-	25.3	1.72
	IB	0.0026	0.00018
2500 - 3761	β-	0.00320	0.000113
	IB	5.7×10^{-7}	2.1×10^{-8}

$^{140}_{58}$Ce(stable)

Δ: -88089 *4* keV
%: 88.48 *10*

$^{140}_{59}$Pr(3.39 *1* min)

Mode: ε
Δ: -84701 *7* keV
SpA: 3.955×10^8 Ci/g
Prod: daughter ^{140}Nd; ^{140}Ce(p,n); ^{141}Pr(γ,n)

Photons (^{140}Pr)

⟨γ⟩=23.5 *7* keV

γ_{mode}	γ(keV)	γ(%)†
Ce L_ℓ	4.289	0.085 *11*
Ce L_η	4.730	0.035 *4*
Ce L_α	4.838	2.30 *25*
Ce L_β	5.390	2.07 *25*
Ce L_γ	6.142	0.28 *4*
Ce $K_{\alpha2}$	34.279	10.9 *3*
Ce $K_{\alpha1}$	34.720	20.0 *6*
Ce $K_{\beta1}$'	39.232	5.62 *17*
Ce $K_{\beta2}$'	40.437	1.52 *5*
γ M1	109.407 *6*	0.00078 *6* ?
γ M1,E2	173.529 *9*	0.00050 *8* ?
γ E2	306.85 *13*	0.185 *20*
γ [E2]	438.165 *11*	$\sim8\times10^{-5}$
γ [E2]	444.81 *13*	~0.00017
γ [E2]	618.34 *13*	0.00023 *11*
γ M1+10%E2	751.661 *21*	0.0309 *19*
γ E1(+0.1%M2)	867.83 *13*	0.0020 *8*
γ M1+2.6%E2	925.19 *2*	0.0275 *18*
γ	936.6 *3*	0.0035 *9*
γ M1+1.1%E2	950.86 *21*	0.00146 *17*
γ [E2]	1420.3 *5*	0.009 *2*
γ E2	1596.54 *14*	0.50
γ E2	2348.19 *14*	0.0061 *6*
γ	2464.36 *19*	$\sim6\times10^{-6}$
γ	2521.72 *14*	0.0135 *9*
γ	2533.1 *3*	0.00024 *8*
γ M1,E2	2547.39 *24*	0.00028 *3*
γ E2	2899.8 *5*	<0.0011
γ M1(+E2)	3119.2 *7*	0.00100 *25*
γ M1,E2	3319.7 *6*	~0.00014 ?

† 8.0% uncert(syst)

Atomic Electrons (^{140}Pr)

⟨e⟩=4.07 *20* keV

e_{bin}(keV)	⟨e⟩(keV)	e(%)
6	1.64	28 *3*
7	0.221	3.4 *3*
27	0.083	0.30 *3*
28	0.231	0.83 *8*
29	0.37	1.28 *13*
32	0.036	0.111 *11*
33	0.204	0.62 *6*
34 - 69	0.201	0.55 *4*
103 - 133	0.0052	0.0050 *19*
167 - 174	5.0×10^{-5}	3.0 *10* $\times10^{-5}$
266 - 307	0.0231	0.0084 *8*
398 - 445	1.6×10^{-5}	3.8 *12* $\times10^{-6}$
578 - 618	8.5×10^{-6}	1.5 *6* $\times10^{-6}$
711 - 752	0.00126	0.000176 *10*
827 - 868	1.9×10^{-5}	2.3 *7* $\times10^{-6}$
885 - 931	0.00098	0.000110 *7*
935 - 951	8.6×10^{-6}	9.1 *14* $\times10^{-7}$
1380 - 1420	0.000124	8.9 *17* $\times10^{-6}$
1556 - 1595	0.0062	0.00039 *3*
1863	0.90	0.049 *5*
1897	0.14	0.0075 *15*
2308 - 2347	5.5×10^{-5}	2.38 *20* $\times10^{-6}$
2424 - 2521	9.9×10^{-5}	4.0 *12* $\times10^{-6}$
2527 - 2546	5.7×10^{-7}	2.2 *4* $\times10^{-8}$
2859 - 2899	4.2×10^{-6}	$\sim1\times10^{-7}$
3079 - 3118	7.5×10^{-6}	2.4 *6* $\times10^{-7}$
3279 - 3318	1.0×10^{-6}	3.0 *13* $\times10^{-8}$

Continuous Radiation (^{140}Pr)

⟨β+⟩=544 keV;⟨IB⟩=5.5 keV

E_{bin}(keV)		⟨ ⟩(keV)	(%)
0 - 10	β+	1.07×10^{-6}	1.28×10^{-5}
	IB	0.020	
10 - 20	β+	4.74×10^{-5}	0.000285
	IB	0.020	0.139
20 - 40	β+	0.00143	0.00435
	IB	0.082	0.26
40 - 100	β+	0.076	0.098
	IB	0.117	0.18
100 - 300	β+	4.19	1.88
	IB	0.33	0.18
300 - 600	β+	33.5	7.2
	IB	0.46	0.106
600 - 1300	β+	243	25.5
	IB	1.38	0.149
1300 - 2500	β+	263	16.3
	IB	2.6	0.145
2500 - 3388	IB	0.48	0.017
	Σβ+		51

$^{140}_{60}$Nd(3.37 *2* d)

Mode: ε
Δ: -84481 *19* keV
SpA: 2.768×10^5 Ci/g
Prod: ^{141}Pr(p,2n); ^{141}Pr(d,3n); protons on Gd

Photons (^{140}Nd)

⟨γ⟩=28.6 *5* keV

γ_{mode}	γ(keV)	γ(%)
Pr L_ℓ	4.453	0.181 *25*
Pr L_η	4.929	0.073 *8*
Pr L_α	5.031	4.9 *5*
Pr L_β	5.618	4.5 *6*
Pr L_γ	6.419	0.64 *9*
Pr $K_{\alpha2}$	35.550	21.9 *7*
Pr $K_{\alpha1}$	36.026	39.9 *12*
Pr $K_{\beta1}$'	40.720	11.3 *3*
Pr $K_{\beta2}$'	41.981	3.15 *11*

Atomic Electrons (^{140}Nd)

⟨e⟩=6.0 *4* keV

e_{bin}(keV)	⟨e⟩(keV)	e(%)
6	3.4	55 *6*
7	0.50	7.3 *8*
28	0.163	0.57 *6*
29	0.45	1.56 *16*
30	0.71	2.38 *24*
34	0.32	0.94 *10*
35	0.36	1.03 *10*
36	0.068	0.191 *19*
39	0.043	0.110 *11*
40	0.030	0.075 *8*
41	0.0081	0.0199 *20*

Continuous Radiation (^{140}Nd)

⟨IB⟩=0.106 keV

E_{bin}(keV)		⟨ ⟩(keV)	(%)
10 - 20	IB	0.00030	0.0018
20 - 40	IB	0.089	0.25
40 - 100	IB	0.0151	0.032
100 - 220	IB	0.00138	0.00099

$^{140}_{61}$Pm(9.2 *2* s)

Mode: ε
Δ: -78410 *40* keV
SpA: 8.44×10^9 Ci/g
Prod: daughter ^{140}Sm; ^{141}Pr(^{3}He,4n); ^{142}Nd(p,3n)

Photons (^{140}Pm)

⟨γ⟩=142 *12* keV

γ_{mode}	γ(keV)	γ(%)†
Nd L_ℓ	4.633	0.0224 *24*
Nd L_η	5.146	0.0087 *7*
Nd L_α	5.228	0.60 *4*
Nd L_β	5.856	0.54 *5*
Nd L_γ	6.691	0.077 *9*
Nd $K_{\alpha2}$	36.847	2.57 *10*
Nd $K_{\alpha1}$	37.361	4.67 *18*
Nd $K_{\beta1}$'	42.240	1.33 *5*
Nd $K_{\beta2}$'	43.562	0.388 *17*
γ	159.6 *3*	0.95 *25*
γ	476.9 *3*	2.5 *5*
γ E2	639.37 *19*	0.62 *6*
γ E2(+M1)	716.2 *3*	0.93 *15*

Photons (^{140}Pm)
(continued)

γ_{mode}	γ(keV)	γ(%)†
γ E2	773.57 18	5
γ	1013.6 3	0.7 3
γ	1057.5 3	0.160 25
γ	1138.5 3	1.5 6
γ	1204.6 3	1.9 4
γ	1366.3 5	0.07 2
γ (E2)	1489.8 3	0.91 5
γ	1558.6 5	0.075 20
γ	1585.0 6	0.140 25
γ	1773.7 4	0.140 25
γ	1941.7 7	0.120 25
γ	2332.2 5	0.14 3
γ	2467.0 12	0.065 15

† 20% uncert(syst)

Atomic Electrons (^{140}Pm)
⟨e⟩=1.56 20 keV

e_{bin}(keV)	⟨e⟩(keV)	e(%)
6	0.27	4.3 5
7	0.194	2.8 3
29	0.0189	0.065 7
30	0.052	0.175 18
31	0.081	0.26 3
35	0.027	0.076 8
36	0.048	0.133 14
37 - 42	0.0213	0.055 4
116	0.22	~0.19
152	0.04	~0.023
153	0.05	~0.04
158 - 160	0.026	~0.017
433	0.12	~0.028
470 - 477	0.025	~0.005
596 - 639	0.0233	0.0039 3
673 - 716	0.040	0.0058 15
730	0.123	0.017 3
766 - 773	0.024	0.0031 4
970 - 1014	0.017	~0.0018
1050 - 1095	0.024	~0.0022
1131 - 1161	0.031	~0.0027
1197 - 1204	0.0046	0.00039 18
1323 - 1366	0.0010	~8 $\times10^{-5}$
1369	0.058	0.0043 10
1406 - 1446	0.021	0.00146 20
1483 - 1579	0.0046	0.00030 6
1583 - 1584	4.2 $\times10^{-5}$	~3 $\times10^{-6}$
1730 - 1772	0.0016	9 4 $\times10^{-5}$
1898 - 1940	0.0012	6 3 $\times10^{-5}$
2289 - 2331	0.0012	5.2 20 $\times10^{-5}$
2423 - 2466	0.00053	2.2 8 $\times10^{-5}$

Continuous Radiation (^{140}Pm)
⟨β+⟩=2037 keV;⟨IB⟩=13 keV

E_{bin}(keV)		⟨ ⟩(keV)	(%)
0 - 10	β+	2.42 $\times10^{-7}$	2.89 $\times10^{-6}$
	IB	0.056	
10 - 20	β+	1.12 $\times10^{-5}$	6.7 $\times10^{-5}$
	IB	0.056	0.39
20 - 40	β+	0.000356	0.00108
	IB	0.119	0.41
40 - 100	β+	0.0205	0.0260
	IB	0.33	0.51
100 - 300	β+	1.26	0.56
	IB	0.99	0.55
300 - 600	β+	11.7	2.49
	IB	1.30	0.30
600 - 1300	β+	130	13.1
	IB	2.5	0.28
1300 - 2500	β+	683	35.6
	IB	3.4	0.19

Continuous Radiation (^{140}Pm)
(continued)

E_{bin}(keV)		⟨ ⟩(keV)	(%)
2500 - 5000	β+	1212	37.2
	IB	4.0	0.120
5000 - 6070	β+	0.0265	0.00053
	IB	0.20	0.0039
	Σβ+		89

$^{140}_{61}$Pm(5.95 *5* min)

Mode: ε
Δ: -78010 *130* keV
SpA: 2.255 $\times10^{8}$ Ci/g
Prod: ^{142}Nd(p,3n); protons on ^{144}Sm; ^{142}Nd(d,4n); ^{141}Pr(^{3}He,4n)

Photons (^{140}Pm)
⟨γ⟩=2350 46 keV

γ_{mode}	γ(keV)	γ(%)†
Nd L_ℓ	4.633	0.074 8
Nd L_η	5.146	0.0298 23
Nd L_α	5.228	2.00 13
Nd L_β	5.854	1.82 16
Nd L_γ	6.688	0.26 3
Nd $K_{\alpha2}$	36.847	8.45 25
Nd $K_{\alpha1}$	37.361	15.4 5
Nd $K_{\beta1}$'	42.240	4.38 13
Nd $K_{\beta2}$'	43.562	1.28 4
γ E1	90.44 9	0.19 3
γ E1	144.75 15	0.35 4
γ	257.9 4	~1
γ E3	419.90 16	92 2
γ E1,E2,M1	474.22 16	1.0 2
γ [E2]	564.65 16	~0.2
γ	566.1 5	~0.29
γ	635.1 4	0.40 8
γ	651.8 4	~0.5
γ	667.5 3	~0.2
γ [E1]	695.15 23	0.20 4
γ	721.9 3	~0.2
γ E2	773.57 18	100 5
γ [M1+E2]	839.90 19	0.5 1
γ	880.4 4	~0.5
γ	930.7 5	0.15 6
γ [M1+E2]	1017.8 4	0.45 10
γ E2	1028.32 20	100 2
γ	1110.3 5	0.20 4
γ [M1+E2]	1197.52 20	3.8 2
γ	1261.9 4	1.0 3
γ	1306.9 3	0.11 3
γ	1397.4 3	0.13 3
γ	1451.7 3	0.24 5
γ	1486.4 5	0.12 3
γ	1733.8 4	0.3 1
γ	1837.8 4	~0.2
γ	1907.1 5	0.40 5
γ	1957.4 5	0.22 4
γ	2145.7 5	0.75 10
γ	2240.3 5	0.20 4
γ	2247.6 5	0.14 3
γ	2407.7 10	~0.02

† 5.0% uncert(syst)

Atomic Electrons (^{140}Pm)
⟨e⟩=29.4 5 keV

e_{bin}(keV)	⟨e⟩(keV)	e(%)
6 - 47	2.40	26.7 18
83 - 101	0.037	0.038 3
138 - 145	0.0069	0.0050 4
214 - 258	0.21	~0.09
376	15.2	4.03 12
413	3.94	0.95 3
414	0.93	0.225 7
418 - 467	1.44	0.344 8
468 - 474	0.0022	0.00046 20
521 - 566	0.022	0.0042 19
592 - 635	0.037	0.006 3
645 - 694	0.012	0.0018 7
695 - 722	0.0010	~0.00014
730	2.46	0.337 18
766 - 796	0.489	0.0636 24
833 - 880	0.015	~0.0018
887 - 931	0.0035	~0.00039
974	0.010	0.0011 3
985	1.78	0.181 5
1011 - 1028	0.323	0.0316 6
1067 - 1110	0.0038	~0.00035
1154 - 1197	0.085	0.0073 13
1218 - 1263	0.018	0.0014 7
1300 - 1307	0.00025	1.9 9 $\times10^{-5}$
1354 - 1397	0.0019	0.00014 7
1408 - 1452	0.0047	0.00033 12
1479 - 1486	0.00023	1.6 7 $\times10^{-5}$
1690 - 1733	0.0034	~0.0002
1794 - 1864	0.0057	0.00031 11
1900 - 1956	0.0028	0.00015 5
2102 - 2197	0.008	0.00040 13
2204 - 2246	0.0015	6.7 23 $\times10^{-5}$
2364 - 2406	0.00017	~7 $\times10^{-6}$

Continuous Radiation (^{140}Pm)
⟨β+⟩=982 keV;⟨IB⟩=7.8 keV

E_{bin}(keV)		⟨ ⟩(keV)	(%)
0 - 10	β+	5.9 $\times10^{-7}$	7.0 $\times10^{-6}$
	IB	0.033	
10 - 20	β+	2.73 $\times10^{-5}$	0.000163
	IB	0.033	0.23
20 - 40	β+	0.00086	0.00262
	IB	0.088	0.29
40 - 100	β+	0.0489	0.062
	IB	0.20	0.31
100 - 300	β+	2.88	1.29
	IB	0.55	0.31
300 - 600	β+	24.8	5.3
	IB	0.71	0.166
600 - 1300	β+	222	22.8
	IB	1.63	0.18
1300 - 2500	β+	629	34.7
	IB	2.9	0.160
2500 - 4248	β+	102	3.79
	IB	1.66	0.056
	Σβ+		68

$^{140}_{62}$Sm(14.82 *10* min)

Mode: ε
Δ: -75410 *300* keV syst
SpA: 9.06 $\times10^{7}$ Ci/g
Prod: protons on Er; ^{142}Nd(α,6n); protons on Gd

Photons (^{140}Sm)

γ_{mode}	γ(keV)	γ(rel)
γ	75.7 *4*	4.0 *13*
γ E2+M1	84.8 *7*	13.3 *13*
γ M1	109.7 *3*	6.0 *7*
γ M1	114.3 *3*	11.3 *7*
γ E1	120.5 *3*	20.7 *13*
γ M1(+E2)	136.1 *3*	4.7 *7*
γ E1	139.76 *24*	61 *3*
γ E2	145.8 *3*	4.0 *7*
γ E2	150.3 *4*	1.3 *3*
γ M1	158.2 *3*	6.7 *7*
γ	163.3 *3*	1.3 *3*
γ E1	199.9 *3*	6.7 *7*
γ	205.1 *3*	1.0 *3*
γ M1+E2	220.8 *3*	14.7 *13*
γ M1	225.4 *3*	100 *3*
γ E2+M1	237.2 *4*	1.00 *13*
γ E2+M1	255.5 *3*	2.13 *20*
γ	260.2 *3*	2.13 *20*
γ E2+M1	279.1 *3*	4.5 *3*
γ E2(+M1)	306.8 *5*	1.40 *13*
γ M1(+E2)	312.1 *3*	3.1 *3*
γ E2(+M1)	339.7 *3*	16.7 *13*
γ E2(+M1)	344.82 *23*	8.7 *7*
γ E2(+M1)	370.0 *5*	1.93 *20*
γ E2+M1	409.6 *4*	1.60 *13*
γ	415.4 *3*	1.2 *4*
γ	425.9 *3*	1.53 *13*
γ	444.8 *5*	1.07 *13*
γ E2+M1	468.0 *5*	0.80 *7*
γ M1(+E2)	480.9 *3*	0.87 *7* ?
γ M1+E2	503.0 *3*	4.7 *7*
γ E2(+M1)	534.0 *5*	2.27 *20*

Photons (^{140}Sm)
(continued)

γ_{mode}	γ(keV)	γ(rel)
γ E2+M1	565.66 *25*	1.93 *20*
γ	604.2 *5*	1.87 *20*
γ	608.5 *5*	1.27 *13*
γ E2	668.1 *5*	2.4 *2*
γ M1(+E2)	700.8 *5*	1.73 *20*
γ M1(+E2)	725.6 *5*	1.73 *20*
γ E2(+M1)	761.3 *5*	4.1 *3*
γ	805.2 *5*	0.47 *7*
γ	814.1 *4*	2.0 *2*
γ M1+E2	825.3 *3*	5.1 *5*
γ	844.7 *3*	4.0 *7*
γ	855.9 *4*	0.7 *3* ?
γ	862.2 *5*	2.0 *2*
γ	873.9 *5*	3.3 *3*
γ	904.4 *5*	0.40 *7*
γ	1000.2 *4*	0.67 *7*
γ	1022.7 *7*	1.87 *20*
γ	1057.9 *7*	2.07 *20*
γ	1097.6 *3*	2.9 *3*
γ	1116.2 *4*	1.73 *20*
γ	1138.1 *7*	11.3 *13*
γ	1167.0 *3*	4.1 *4*
γ	1189.0 *7*	0.87 *7*
γ	1249.8 *5*	2.5 *3*
γ	1254.9 *5*	1.47 *13*
γ	1274.4 *4*	9.7 *10*
γ	1283.4 *7*	3.0 *3*
γ	1325.1 *3*	2.07 *20*
γ [M1+E2]	1393.8 *4*	6.3 *6*
γ	1409.7 *3*	0.33 *7*
γ [M1+E2]	1444.6 *3*	0.60 *7*
γ [E1]	1479.4 *4*	~0.33
γ [E1]	1530.2 *3*	8.0 *7*
γ	1577.5 *7*	2.6 *3*
γ [M1+E2]	1669.96 *25*	2.7 *3*

$^{140}_{63}$Eu(1.3 *2* s)

Mode: ϵ
Δ: -67210 *400* keV syst
SpA: 4.8×10^{10} Ci/g
Prod: protons on Gd

Photons (^{140}Eu)

γ_{mode}	γ(keV)
γ (E2)	531.0 *5*

$^{140}_{63}$Eu(~20 s)

Mode: ϵ
Δ: -67210 *400* keV syst
SpA: 4×10^{9} Ci/g
Prod: alphas on ^{144}Sm

Photons (^{140}Eu)

γ_{mode}	γ(keV)
γ (E2)	531.0 *5*
γ (E2)	714 *1*

A = 141

NDS **23**, 529 (1978)

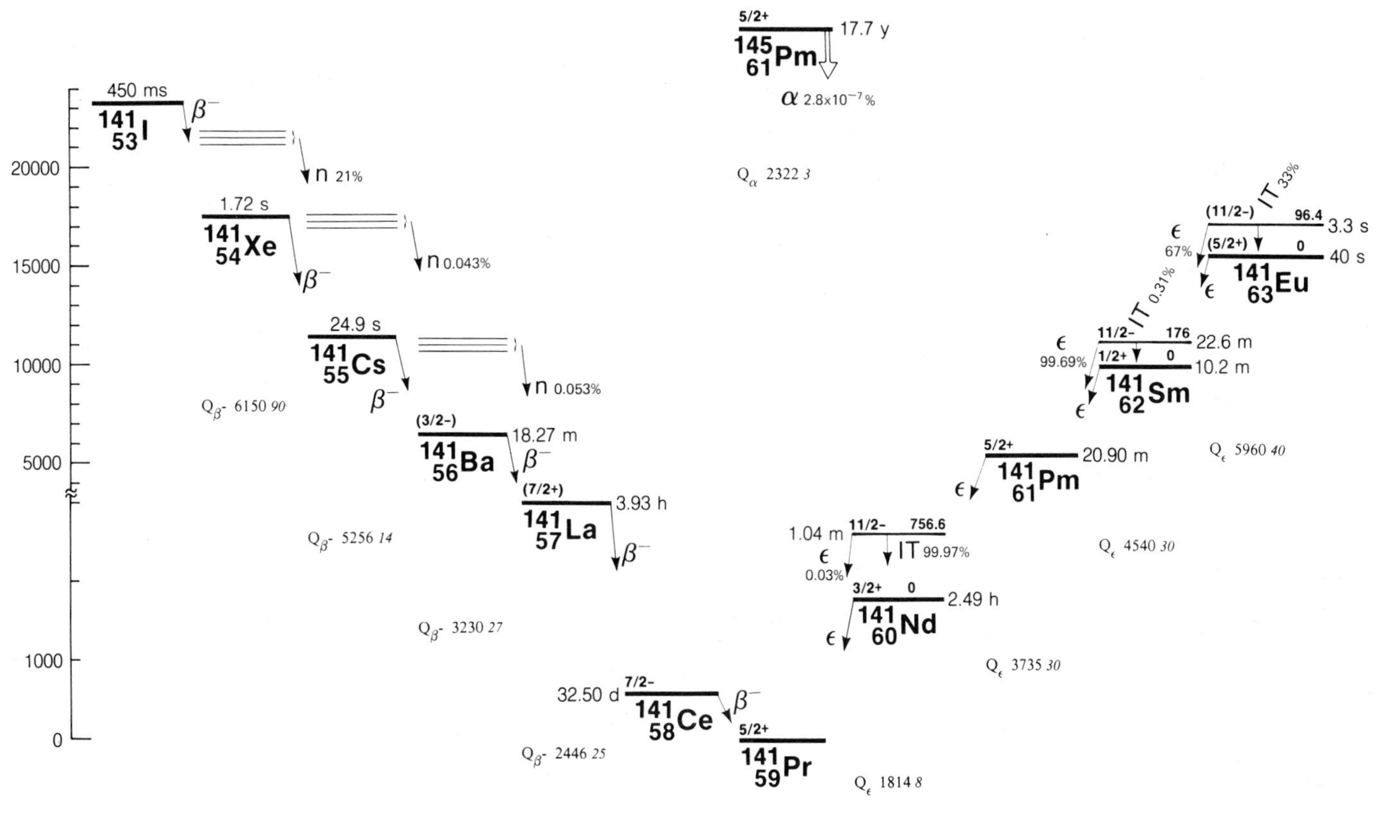

$^{141}_{53}$I (450 *30* ms)

Mode: β-, β-n(21 *3* %)
SpA: 9.1×10^{10} Ci/g
Prod: fission
β-n: 160, 225?, 300, 340?, 395, 450, 550, 610?, 685?

$^{141}_{54}$Xe(1.72 *3* s)

Mode: β-, β-n(0.043 *3* %)
Δ: -68360 *90* keV
SpA: 3.83×10^{10} Ci/g
Prod: fission

Photons (^{141}Xe)

γ_{mode}	γ(keV)	γ(rel)
γ	69.01 *3*	19 *3*
γ	81.805 *20*	9.9 *13*
γ	89.34 *7*	2.9 *3*
γ	93.7 *7*	0.08 *2*
γ	96.29 *19*	0.38 *6*
γ	100.781 *24*	11.0 *12*
γ	105.93 *3*	37 *4*
γ(E1)	118.719 *19*	52 *6*
γ	137.695 *20*	3.3 *3*
γ	149.41 *10*	0.43 *6*
γ	151.05 *8*	0.10 *2*
γ	168.27 *15*	0.13 *1*
γ	177.11 *12*	0.14 *1*
γM1,E2	187.73 *3*	11.6 *5*
γ	234.77 *11*	0.23 *2*
γ	246.18 *25*	0.06 *2*
γ	248.57 *17*	0.08 *2*
γ	254.25 *18*	0.54 *4*
γ	280.09 *6*	~0.2
γ	283.14 *5*	1.81 *15*
γ	286.07 *6*	0.99 *5*
γ	313.59 *16*	0.10 *3*
γ	314.78 *12*	0.12 *2*
γ	317.56 *9*	0.09 *2*
γ	318.00 *14*	0.09 *2*
γ	333.18 *19*	0.35 *3*
γ	335.43 *10*	0.24 *4*
γM1,E2	361.90 *6*	3.9 *7*
γ	369.43 *5*	7.2 *7*
γM1,E2	388.97 *6*	6.8 *5*
γ	421.74 *21*	1.56 *24*
γ	423.88 *14*	5.5 *4*
γ	432.07 *13*	0.21 *3*
γ	434.33 *21*	0.80 *15*
γ	435.70 *10*	0.30 *5*
γ	437.97 *17*	0.39 *6*
γ	451.24 *5*	0.65 *8*
γ	452.73 *8*	2.25 *8*
γM1,E2	459.48 *10*	20.1 *15*
γM1,E2	467.82 *6*	12.9 *8*
γ	492.86 *10*	3.03 *14*
γ	507.3 *6*	0.86 *16*
γ	509.81 *17*	3.47 *14*
γ	532.66 *14*	0.26 *5*
γ	538.53 *8*	2.9 *5*
γ	539.03 *22*	2.9 *5*
γM1,E2	540.02 *6*	23.5 *13*
γ	551.69 *20*	2.09 *22*
γM1,E2	557.17 *5*	21 *4*
γ	570.40 *24*	0.41 *7*
γ	576.77 *9*	1.62 *10*
γ	578.2 *4*	0.57 *6*
γ	594.65 *10*	1.68 *12*
γ	599.65 *15*	0.58 *10*
γ	604.6 *5*	1.05 *20*
γ	613.18 *14*	3.7 *6*
γ	629.36 *7*	2.17 *10*
γ	644.46 *8*	3.00 *21*
γ	649.2 *4*	1.07 *19*
γ	677.35 *17*	1.1 *3*
γ	677.68 *6*	1.1 *3*

Photons (^{141}Xe) (continued)

γ_{mode}	γ(keV)	γ(rel)
γ	677.88 10	1.1 3
γ	722.1 7	0.33 9
γ	728.96 11	0.74 6
γ	731.92 5	3.29 11
γ	736.78 23	0.34 7
γ	739.42 8	0.39 9
γ	745.41 16	0.95 13
γ	755.30 3	5.24 21
γ	769.98 12	0.15 5
γ	770.92 18	0.15 5
γ	773.18 6	10.0 4
γ	778.06 11	1.2 3
γ	784.44 25	~0.45
γ	792.16 7	1.49 13
γ	801.26 20	0.62 16
γ	805.4 3	1.24 12
γ	807.66 16	0.97 13
γ	818.97 24	0.27 8
γ	823.4 3	0.94 9
γ	825.7 3	0.44 9
γ	843.02 14	1.86 14
γ	854.79 12	1.11 20
γ	854.80 12	1.11 20
γ	864.4 3	~0.12
γ	869.58 7	1.30 13
γ	870.5 4	<0.4
γ	874.46 22	1.5 3
γ	880.73 23	0.36 17
γ	894.77 12	3.6 5
γ	898.2 7	0.83 21
γ	909.45 4	100 3
γ	913.68 18	0.81 8
γ	914.25 5	0.81 8
γ	933.23 5	0.62 8
γ	942.8 8	1.13 13
γ	943.90 20	0.96 16
γ	944.14 13	0.58 14
γ	977.35 14	0.55 18
γ	977.63 18	0.55 18
γ	979.89 6	4.47 19
γ	986.02 24	1.67 10
γ	990.07 11	0.85 9
γ	992.13 13	0.93 14
γ	999.49 10	0.73 11
γ	1009.05 11	1.9 4
γ	1015.03 5	1.49 8
γ	1025.3 4	0.99 9
γ	1028.14 3	7.6 9
γ	1051.95 5	4.67 18
γ	1061.92 20	0.85 14
γ	1090.85 11	1.00 8
γ	1092.73 14	0.85 11
γ	1097.18 5	2.0 4
γ	1099.7 9	~2
γ	1104.2 3	0.72 11
γ	1112.16 20	0.48 12
γ	1120.96 5	3.27 11
γ	1131.3 6	0.38 11
γ	1134.6 3	1.36 11
γ	1140.4 3	0.50 12
γ	1177.71 14	0.66 12
γ	1196.78 11	0.85 24
γ	1205.25 12	0.22 8
γ	1208.46 19	0.62 7
γ	1214.55 16	1.27 17
γ	1216.92 14	2.00 11
γ	1219.19 16	0.71 8
γ	1233.0 9	0.80 14
γ	1246.18 22	1.30 13
γ	1253.33 21	0.94 23
γ	1276.57 20	0.28 9
γ	1312.21 21	0.58 8
γ	1317.49 13	0.54 12
γ	1323.9 3	0.57 15
γ	1330.84 20	0.46 12
γ	1351.64 14	1.05 18
γ	1360.36 17	0.45 10
γ	1360.39 13	0.45 10
γ	1368.92 9	3.50 11
γ	1372.11 17	0.81 10
γ	1387.46 15	0.59 7
γ	1393.3 8	0.43 8
γ	1400.91 24	0.18 8
γ	1406.84 13	0.26 11

Photons (^{141}Xe) (continued)

γ_{mode}	γ(keV)	γ(rel)
γ	1412.99 21	0.41 12
γ	1421.2 17	0.23 9
γ	1427.13 18	0.35 10
γ	1436.21 10	2.6 9
γ	1439.5 9	0.79 18
γ	1489.21 13	1.08 12
γ	1497.9 6	0.49 9
γ	1502.4 4	1.71 13
γ	1510.55 25	0.40 8
γ	1526.17 22	0.55 16
γ	1540.99 15	0.51 15
γ	1546.91 21	0.54 8
γ	1551.0 7	0.66 12
γ	1556.65 9	13.1 10
γ	1579.06 17	0.60 11
γ	1600.21 17	0.58 7
γ	1621.49 13	0.31 11
γ	1654.88 15	0.66 12
γ	1688.1 3	1.18 10
γ	1739.27 20	0.57 21
γ	1748.9 7	0.48 14
γ	1755.59 24	2.41 15
γ	1770.0 4	1.74 13
γ	1795.95 15	0.66 10
γ	1799.79 13	1.3 1
γ	1829.71 16	0.67 9
γ	1860.83 6	0.95 9
γ	1882.60 12	0.56 14
γ	1897.75 6	0.62 15
γ	1917.08 14	0.76 12
γ	1921.24 18	0.38 11
γ	1933.9 6	0.83 13
γ	1960.00 22	0.35 7
γ	2016.3 10	0.88 13
γ	2020.29 12	0.92 11
γ	2059.3 3	0.59 12
γ	2109.64 19	0.55 11
γ	2125.9 3	0.37 11
γ	2141.55 21	0.36 11
γ	2141.58 17	0.36 11
γ	2168.7 11	0.28 11
γ	2173.59 23	0.56 12
γ	2210.4 5	0.76 8
γ	2218.67 15	0.81 10
γ	2231.95 14	0.44 10
γ	2236.5 3	0.53 10
γ	2268.86 13	0.45 12
γ	2282.7 8	0.71 8
γ	2286.27 20	0.38 8
γ	2336.6 11	0.4 1
γ	2372.62 18	0.25 7
γ	2394.2 6	1.02 13
γ	2410.36 23	0.57 7
γ	2429.14 14	0.56 10
γ	2448.8 14	0.31 8
γ	2476.04 21	0.30 7
γ	2488.14 10	0.27 7
γ	2488.17 21	0.27 7
γ	2546.62 14	1.75 13
γ	2577.26 16	0.5 1
γ	2601.04 16	0.69 11
γ	2628.42 14	0.47 11
γ	2635.74 21	0.49 11
γ	2665.33 14	0.37 7
γ	2683.39 17	0.49 12
γ	2709.23 13	0.54 9
γ	2709.62 15	0.54 9
γ	2734.34 14	0.64 16
γ	2790.81 18	~0.15
γ	2791.03 13	~0.15
γ	2826.9 3	0.37 9
γ	2827.94 13	0.37 9
γ	2838.17 13	0.51 7
γ	2875.08 13	0.39 7
γ	2896.85 21	0.21 6
γ	2896.96 13	0.21 6
γ	2910.36 23	0.18 7
γ	2944.09 13	~0.28
γ	2944.50 17	~0.28
γ	2983.13 18	0.23 8
γ	3103.55 21	0.32 8
γ	3221.68 15	0.22 5

$^{141}_{55}$Cs(24.94 6 s)

Mode: β-, β-n(0.053 4 %)
Δ: -74515 22 keV
SpA: 3.164×10^9 Ci/g
Prod: fission
β-n(avg): 240 50

Photons (^{141}Cs)

$\langle\gamma\rangle$=999 21 keV

γ_{mode}	γ(keV)	γ(%)†
Ba L_ℓ	3.954	0.124 20
Ba L_η	4.331	0.056 7
Ba L_α	4.465	3.4 5
Ba L_β	4.945	3.1 5
Ba L_γ	5.615	0.41 7
Ba $K_{\alpha2}$	31.817	18.2 13
Ba $K_{\alpha1}$	32.194	33.5 23
Ba $K_{\beta1}$'	36.357	9.3 7
Ba $K_{\beta2}$'	37.450	2.35 17
γ[M1]	48.64 3	9.9 5
γ[M1+E2]	55.09 4	0.206 17
γ	340.6 3	~0.04
γ	439.9 3	~0.04
γ	448.62 15	0.09 4
γ	502.10 21	0.18 3
γ	549.87 18	0.08 4
γ	555.21 8	4.71 25
γ	561.64 8	5.8 3
γ	569.81 22	0.10 4
γ	585.52 13	0.36 5
γ	588.74 8	5.1 3
γ	591.95 13	0.17 5
γ	605.27 9	1.24 8
γ	612.81 12	0.42 5
γ	638.80 11	0.25 4
γ	642.39 19	0.12 4
γ	646.33 13	0.84 8
γ	647.13 24	0.84 8
γ	649.16 16	0.50 10
γ	654.32 10	0.91 7
γ	660.82 9	0.99 7
γ	691.94 7	3.73 21
γ	698.40 7	0.64 7
γ	709.46 9	0.19 4
γ	728.27 20	0.10 4
γ	771.86 20	0.33 5
γ	778.57 20	0.33 5
γ	806.2 4	~0.12
γ	827.0 5	0.16 5
γ	895.2 3	0.17 5
γ	902.2 3	0.31 6
γ	938.54 25	0.26 8
γ	953.8 3	0.11 5
γ	972.81 25	0.28 6
γ	985.8 3	0.19 5
γ	1007.67 13	0.55 7
γ	1018.43 19	0.21 6
γ	1024.63 22	0.21 7
γ	1043.0 3	0.11 4
γ	1056.30 13	0.54 6
γ	1061.89 12	1.30 8 ?
γ	1068.20 18	0.58 6 ?
γ	1072.47 14	0.49 6
γ	1097.92 16	0.41 6
γ	1117.3 4	0.19 5
γ	1140.56 14	1.36 11
γ	1147.16 11	3.83 22
γ	1153.72 16	1.17 10
γ	1165.87 14	0.37 10
γ	1171.68 21	1.08 12
γ	1177.80 13	1.32 13
γ	1182.13 16	0.42 10
γ	1194.01 10	5.4 4
γ	1195.80 11	<0.44 ?
γ	1200.3 3	0.21 9
γ	1209.22 16	0.16 7
γ	1214.51 13	0.88 8
γ	1226.44 13	0.81 9
γ	1230.0 3	0.35 8
γ	1232.7 4	0.30 7

Photons (^{141}Cs)
(continued)

γ_{mode}	γ(keV)	γ(%)†
γ	1264.1 *3*	0.24 *6*
γ	1277.5 *5*	0.16 *6*
γ	1290.1 *8*	0.11 *5*
γ	1315.8 *3*	0.19 *5*
γ	1344.0 *3*	0.11 *5*
γ	1360.52 *23*	0.36 *7*
γ	1400.1 *3*	~0.10
γ	1432.47 *22*	0.58 *8*
γ	1448.74 *24*	0.51 *9*
γ	1453.6 *3*	~0.14
γ	1497.28 *24*	0.29 *8*
γ	1503.71 *24*	~0.10
γ	1517.7 *4*	0.41 *8*
γ	1524.5 *7*	~0.14
γ	1539.6 *3*	~0.05
γ	1540.9 *3*	~0.05
γ	1572.4 *5*	0.31 *10*
γ	1575.5 *4*	~0.13
γ	1606.2 *10*	~0.15
γ	1625.2 *4*	0.23 *8*
γ	1630.5 *4*	0.30 *9*
γ	1653.6 *3*	0.13 *6*
γ	1661.73 *18*	0.58 *7*
γ	1715.0 *3*	0.57 *10*
γ	1751.1 *3*	0.22 *8*
γ	1757.20 *23*	0.23 *8*
γ	1765.0 *5*	~0.15
γ	1773.2 *3*	0.23 *9*
γ	1783.4 *4*	~0.16
γ	1789.34 *25*	0.47 *9*
γ	1809.3 *3*	0.19 *8*
γ	1819.50 *14*	0.54 *9*
γ	1826.8 *3*	0.24 *8*
γ	1843.7 *3*	~0.12
γ	1852.2 *3*	0.17 *7*
γ	1868.82 *15*	~0.13
γ	1885.8 *3*	~0.17
γ	1894.40 *23*	0.50 *10*
γ	1906.33 *24*	0.36 *9*
γ	1918.4 *3*	0.37 *9*
γ	1933.6 *3*	0.45 *9*
γ	1940.8 *3*	0.64 *10*
γ	1957.5 *4*	~0.15
γ	1964.3 *7*	0.21 *9*
γ	1989.46 *23*	0.25 *6*
γ	1994.3 *3*	0.38 *8*
γ	1998.4 *9*	0.21 *8*
γ	2046.4 *3*	0.27 *9*
γ	2060.33 *24*	0.49 *9*
γ	2066.1 *4*	0.50 *9*
γ	2088.63 *22*	0.41 *8*
γ	2095.23 *21*	0.28 *9*
γ	2143.0 *6*	0.34 *9*
γ	2221.7 *3*	~0.14
γ	2386.1 *3*	0.29 *8*
γ	2394.9 *3*	~0.16
γ	2399.4 *3*	0.23 *9*
γ	2411.37 *23*	~0.13
γ	2489.5 *8*	0.19 *6*
γ	2504.3 *6*	~0.13
γ	2615.4 *3*	0.12 *3*
γ	2615.5 *4*	0.12 *3*
γ	2638.5 *5*	~0.11
γ	2672.08 *18*	~0.11
γ	2709.4 *3*	~0.16
γ	2728.4 *5*	~0.14
γ	2820.93 *21*	0.28 *7*
γ	2846.9 *8*	0.21 *7*
γ	2950.1 *3*	0.23 *7*
γ	2962.3 *3*	0.26 *7*
γ	2976.6 *6*	0.19 *5*
γ	3031.8 *3*	0.12 *6*
γ	3039.50 *25*	0.30 *6*
γ	3057.8 *6*	0.25 *6*
γ	3072.18 *21*	0.68 *8*
γ	3078.6 *3*	0.37 *7*
γ	3099.3 *3*	~0.10
γ	3116.1 *3*	0.36 *7*
γ	3120.82 *21*	~0.11
γ	3133.8 *3*	0.44 *6*
γ	3170.5 *7*	0.12 *6*
γ	3183.5 *3*	~0.09
γ	3192.3 *4*	~0.45
γ	3194.3 *7*	~0.25

Photons (^{141}Cs)
(continued)

γ_{mode}	γ(keV)	γ(%)†
γ	3205.25 *13*	~0.10
γ	3218.1 *3*	~0.07
γ	3225.8 *3*	0.21 *5*
γ	3239.1 *4*	~0.07
γ	3252.9 *3*	0.18 *5*
γ	3260.2 *3*	0.23 *5*
γ	3273.3 *3*	0.17 *4*
γ	3304.43 *17*	0.13 *4*
γ	3314.9 *9*	0.10 *4*
γ	3331.7 *3*	0.21 *4*
γ	3349.2 *3*	0.19 *4*
γ	3378.6 *4*	0.09 *4*
γ	3385.5 *5*	~0.06
γ	3395.1 *7*	~0.07
γ	3416.1 *5*	~0.05
γ	3475.4 *5*	~0.05
γ	3495.8 *4*	~0.05
γ	3530.3 *3*	0.044 *20*
γ	3935.9 *7*	0.030 *12*

† 10% uncert(syst)

Atomic Electrons (^{141}Cs)

⟨e⟩=20.0 *8* keV

e_{bin}(keV)	⟨e⟩(keV)	e(%)
5	1.74	33 *4*
6	1.25	22 *3*
11	7.7	69 *5*
18 - 26	0.73	3.18 *24*
27	0.66	2.5 *3*
30	0.147	0.49 *6*
31	0.42	1.35 *16*
32 - 36	0.159	0.48 *4*
43	4.0	9.4 *6*
47	0.83	1.76 *12*
48 - 55	1.0	2.0 *7*
303 - 341	0.0030	~0.0010
402 - 449	0.006	0.0015 *7*
465 - 512	0.009	0.0019 *9*
518 - 565	0.51	0.10 *3*
568 - 617	0.16	0.027 *6*
623 - 672	0.14	0.022 *8*
686 - 734	0.026	0.0037 *12*
741 - 790	0.013	0.0017 *6*
800 - 827	0.0009	0.00010 *4*
858 - 902	0.013	0.0014 *5*
916 - 954	0.009	0.0010 *3*
967 - 1013	0.017	0.0018 *5*
1017 - 1066	0.049	0.0047 *11*
1067 - 1116	0.08	0.0071 *23*
1117 - 1166	0.12	0.010 *3*
1170 - 1214	0.044	0.0037 *8*
1220 - 1264	0.0083	0.00068 *18*
1272 - 1316	0.0037	0.00029 *10*
1323 - 1363	0.0052	0.00039 *15*
1394 - 1443	0.013	0.0009 *3*
1447 - 1496	0.009	0.00061 *17*
1497 - 1593	0.012	0.00075 *17*
1600 - 1678	0.012	0.00071 *19*
1709 - 1808	0.021	0.00122 *20*
1814 - 1913	0.022	0.00119 *19*
1917 - 2009	0.0122	0.00063 *11*
2023 - 2106	0.015	0.00072 *13*
2137 - 2221	0.0013	6 *3* $\times 10^{-5}$
2349 - 2410	0.0054	0.00023 *6*
2452 - 2503	0.0020	8 *3* $\times 10^{-5}$
2578 - 2672	0.0036	0.00014 *4*
2691 - 2783	0.0023	8 *3* $\times 10^{-5}$
2809 - 2846	0.0014	5.0 *18* $\times 10^{-5}$
2913 - 3002	0.0057	0.00019 *3*
3020 - 3120	0.0123	0.00040 *5*
3128 - 3225	0.0091	0.00029 *6*
3233 - 3331	0.0042	0.000128 *19*
3341 - 3438	0.0017	5.1 *12* $\times 10^{-5}$
3458 - 3529	0.00047	1.3 *5* $\times 10^{-5}$
3898 - 3935	0.00013	3.4 *14* $\times 10^{-6}$

Continuous Radiation (^{141}Cs)

⟨β-⟩=1864 keV;⟨IB⟩=6.5 keV

E_{bin}(keV)		⟨ ⟩(keV)	(%)
0 - 10	β-	0.0119	0.238
	IB	0.054	
10 - 20	β-	0.0361	0.241
	IB	0.053	0.37
20 - 40	β-	0.147	0.490
	IB	0.105	0.36
40 - 100	β-	1.08	1.54
	IB	0.30	0.47
100 - 300	β-	11.3	5.6
	IB	0.91	0.51
300 - 600	β-	38.1	8.5
	IB	1.13	0.26
600 - 1300	β-	180	18.9
	IB	1.9	0.21
1300 - 2500	β-	648	34.2
	IB	1.60	0.092
2500 - 5000	β-	985	30.3
	IB	0.50	0.0168
5000 - 5207	β-	1.11	0.0221
	IB	7.4×10^{-6}	1.48×10^{-7}

$^{141}_{56}$Ba(18.27 *1* min)

Mode: β-

Δ: -79771 *24* keV

SpA: 7.297×10^{7} Ci/g

Prod: fission

Photons (^{141}Ba)

⟨γ⟩=845 *11* keV

γ_{mode}	γ(keV)	γ(%)†
La L_ℓ	4.121	0.019 *3*
La L_η	4.529	0.0082 *13*
La L_α	4.649	0.52 *8*
La L_β	5.166	0.47 *8*
La L_γ	5.874	0.063 *11*
La $K_{\alpha2}$	33.034	2.56 *25*
La $K_{\alpha1}$	33.442	4.7 *5*
La $K_{\beta1}$'	37.777	1.31 *13*
La $K_{\beta2}$'	38.927	0.34 *3*
γ	112.92 *6*	0.93 *6*
γ	163.04 *4*	0.44 *4*
γ	180.63 *4*	0.49 *4*
γ(M1)	190.31 *4*	46 *3*
γ	234.79 *11*	0.055 *18*
γ	242.49 *11*	0.078 *18*
γ	276.91 *3*	23.3 *12*
γ	281.49 *5*	~0.08
γ	304.18 *4*	25.2 *13*
γ	343.68 *4*	14.2 *8*
γ	349.20 *7*	0.29 *5*
γ	364.42 *8*	0.58 *5*
γ	381.32 *21*	0.115 *23*
γ	389.83 *6*	1.32 *7*
γ	418.58 *9*	~0.05
γ	457.54 *4*	4.8 *3*
γ	462.12 *4*	4.8 *3*
γ	467.22 *4*	5.5 *3*
γ	486.29 *10*	0.06 *3*
γ	522.07 *12*	0.43 *6*
γ	523.98 *11*	0.40 *6*
γ	527.46 *8*	0.38 *4*
γ	541.26 *18*	0.08 *3*

Photons (^{141}Ba)
(continued)

γ_{mode}	γ(keV)	γ(%)†
γ	551.68 *22*	~0.06
γ	561.59 *11*	0.10 *3*
γ	572.1 *3*	0.25 *4*
γ	588.74 *19*	~0.05
γ	599.21 *8*	0.23 *3*
γ	608.98 *17*	0.24 *3*
γ	625.17 *4*	3.27 *18*
γ	635.93 *12*	0.28 *4*
γ	641.33 *7*	0.36 *5*
γ	647.85 *4*	5.6 *3*
γ	660.8 *3*	0.28 *5*
γ	669.90 *12*	0.18 *3*
γ	675.29 *16*	<0.21
γ	675.7 *5*	0.22 *11*
γ	685.50 *21*	0.13 *5*
γ	687.6 *5*	0.10 *5*
γ	698.68 *9*	0.28 *11*
γ	700.69 *14*	~0.21
γ	704.61 *11*	0.30 *3*
γ	739.03 *5*	4.28 *23*
γ	753.92 *15*	0.051 *14*
γ	754.0 *3*	0.051 *14*
γ	762.26 *9*	0.14 *3*
γ	778.15 *25*	0.11 *4*
γ	801.4 *5*	0.08 *3*
γ	806.09 *12*	0.10 *3*
γ	826.25 *12*	0.33 *5*
γ	831.64 *7*	1.52 *9*
γ	833.2 *3*	~0.16
γ	845.9 *3*	~0.06
γ	867.65 *11*	0.15 *4*
γ	876.12 *8*	3.40 *18*
γ	880.6 *3*	0.20 *4*
γ	884.94 *18*	~0.06
γ	909.15 *15*	0.12 *4*
γ	929.34 *5*	0.69 *5*
γ	943.18 *10*	0.73 *6*
γ	958.78 *25*	~0.06
γ	981.52 *11*	0.78 *7*
γ	996.41 *12*	0.12 *4*
γ	1012.46 *15*	0.10 *4*
γ	1034.32 *11*	0.29 *5*
γ	1040.88 *12*	0.10 *4*
γ	1046.27 *13*	0.35 *5*
γ	1094.11 *13*	0.22 *4*
γ	1160.65 *15*	0.24 *9*
γ	1160.80 *8*	0.92 *9*
γ	1197.36 *11*	4.6 *3*
γ	1224.66 *10*	0.41 *5*
γ	1235.69 *25*	0.14 *4*
γ	1263.96 *14*	0.82 *7*
γ	1273.57 *14*	0.52 *6*
γ	1277.90 *12*	0.66 *7*
γ	1309.2 *3*	0.23 *11*
γ	1311.22 *11*	0.60 *14*
γ	1323.84 *9*	0.94 *6*
γ	1345.61 *13*	0.22 *4*
γ	1357.5 *5*	0.16 *5*
γ	1376.88 *14*	0.70 *6*
γ	1405.30 *10*	0.27 *5*
γ	1436.61 *14*	0.82 *7*
γ	1458.53 *12*	0.68 *6*
γ	1501.53 *11*	0.31 *4*
γ	1539.92 *14*	0.09 *3*
γ	1550.48 *14*	0.31 *4*
γ	1568.34 *10*	0.25 *4*
γ	1599.88 *22*	0.060 *23*
γ	1609.1 *3*	~0.037
γ	1621.57 *12*	0.055 *23*
γ	1642.66 *22*	0.074 *23*
γ	1653.79 *14*	0.75 *6*
γ	1682.20 *10*	1.34 *9*
γ	1712.80 *21*	0.170 *23*
γ	1727.8 *3*	0.06 *3*
γ	1735.43 *12*	0.18 *4*
γ	1740.79 *14*	0.31 *4*
γ	1795.5 *3*	0.51 *5*
γ	1821.06 *20*	0.078 *23*
γ	1860.6 *3*	0.078 *23*
γ	1875.84 *21*	~0.037
γ	1912.58 *17*	0.129 *23*
γ	1918.7 *4*	0.046 *18*
γ	1989.70 *21*	0.18 *3*
γ	2026.44 *17*	0.38 *5*

Photons (^{141}Ba)
(continued)

γ_{mode}	γ(keV)	γ(%)†
γ	2058.8 *9*	0.051 *18*
γ	2081.7 *4*	~0.028
γ	2121.1 *9*	~0.037
γ	2136.5 *3*	0.110 *18*
γ	2164.73 *19*	0.156 *23*
γ	2195.6 *4*	0.092 *18*
γ	2216.75 *17*	0.060 *18*
γ	2278.59 *20*	0.097 *23*
γ	2468.90 *19*	0.18 *4*

† 6.5% uncert(syst)

Atomic Electrons (^{141}Ba)
⟨e⟩=22.4 *20* keV

e_{bin}(keV)	⟨e⟩(keV)	e(%)
5 - 38	0.70	8.6 *8*
74 - 113	0.6	~0.7
124 - 142	0.17	0.13 *6*
151	10.9	7.2 *5*
157 - 181	0.07	0.043 *16*
184	1.76	0.95 *7*
185 - 234	0.53	0.278 *15*
235 - 237	0.0020	~0.0008
238	1.9	~0.8
241 - 243	0.007	~0.003
265	1.8	~0.7
271	0.4	~0.15
275 - 304	0.58	0.20 *8*
305	0.9	~0.28
310 - 359	0.33	0.10 *3*
363 - 412	0.020	0.0052 *22*
413 - 462	0.72	0.17 *5*
466 - 513	0.055	0.011 *3*
516 - 562	0.028	0.0052 *14*
566 - 609	0.26	0.043 *17*
619 - 666	0.090	0.014 *3*
669 - 715	0.10	~0.014
723 - 773	0.026	0.0035 *12*
777 - 826	0.044	0.0055 *21*
827 - 876	0.08	0.010 *4*
879 - 928	0.026	0.0029 *10*
929 - 976	0.019	0.0020 *7*
980 - 1029	0.012	0.0012 *4*
1033 - 1055	0.0041	0.00039 *16*
1088 - 1122	0.015	~0.0013
1154 - 1197	0.07	0.0063 *25*
1218 - 1268	0.026	0.0021 *6*
1270 - 1319	0.027	0.0021 *6*
1322 - 1371	0.012	0.0009 *3*
1376 - 1420	0.015	0.0011 *4*
1430 - 1463	0.0053	0.00036 *11*
1495 - 1594	0.0087	0.00057 *13*
1599 - 1697	0.025	0.0015 *4*
1702 - 1794	0.0087	0.00050 *12*
1815 - 1913	0.0026	0.00014 *3*
1917 - 1989	0.0042	0.00021 *7*
2020 - 2116	0.0020	0.000099 *25*
2120 - 2216	0.0024	0.00011 *3*
2240 - 2277	0.0007	3.1 *12* $\times 10^{-5}$
2430 - 2468	0.0012	5.1 *19* $\times 10^{-5}$

Continuous Radiation (^{141}Ba)
⟨β-⟩=945 keV;⟨IB⟩=2.1 keV

E_{bin}(keV)		⟨ ⟩(keV)	(%)
0 - 10	β-	0.0187	0.372
	IB	0.035	
10 - 20	β-	0.057	0.379
	IB	0.035	0.24
20 - 40	β-	0.235	0.78
	IB	0.067	0.23
40 - 100	β-	1.78	2.52
	IB	0.19	0.29
100 - 300	β-	20.8	10.2
	IB	0.52	0.29

Continuous Radiation (^{141}Ba)
(continued)

E_{bin}(keV)		⟨ ⟩(keV)	(%)
300 - 600	β-	84	18.5
	IB	0.53	0.127
600 - 1300	β-	376	40.4
	IB	0.60	0.071
1300 - 2500	β-	446	26.3
	IB	0.163	0.0102
2500 - 3230	β-	16.5	0.62
	IB	0.00099	3.8 $\times 10^{-5}$

$^{141}_{57}$La(3.93 *5* h)

Mode: β-
Δ: -83000 *25* keV
SpA: 5.66×10^{6} Ci/g
Prod: fission

Photons (^{141}La)

γ_{mode}	γ(keV)	γ(rel)
γ	324.60 *20*	0.08 *3*
γ	434.98 *18*	~0.04
γ[M1+E2]	474.9 *9*	~0.04 ?
γ	547.5 *3*	~0.06
γ	561.8 *7*	~0.06
γ	580.9 *3*	0.07 *3*
γ	589.3 *3*	~0.06
γ[E2]	662.02 *6*	1.58 *9*
γ	676.99 *17*	0.08 *3*
γ	694.7 *3*	~0.07
γ	710.38 *14*	0.19 *4*
γ[M1+E2]	834.95 *10*	0.12 *4*
γ	852.88 *15*	0.19 *4*
γ	964.5 *3*	~0.09
γ[M1+E2]	1354.47 *9*	100 *5*
γ	1368.68 *18*	0.30 *3*
γ[M1+E2]	1496.97 *9*	1.11 *7*
γ	1511.93 *15*	0.56 *5*
γ	1604.86 *14*	0.52 *5*
γ	1693.27 *10*	4.50 *25*
γ	1738.97 *10*	0.95 *6*
γ	1943.8 *3*	0.20 *3*
γ	2030.17 *20*	0.31 *3*
γ	2049.2 *3*	0.14 *2*
γ	2171.1 *3*	1.24 *12*
γ	2173.95 *15*	1.00 *12*
γ	2207.34 *13*	0.48 *4*
γ	2266.88 *14*	2.52 *15*
γ	2328.9 *11*	~0.03

$^{141}_{58}$Ce(32.50 *1* d)

Mode: β-
Δ: -85446 *4* keV
SpA: 2.8494×10^{4} Ci/g
Prod: ^{140}Ce(n,γ); ^{141}Pr(n,p)

Photons (^{141}Ce)

$\langle\gamma\rangle$=77.0 *14* keV

γ_{mode}	γ(keV)	γ(%)†
Pr L_ℓ	4.453	0.041 *6*
Pr L_η	4.929	0.0165 *18*
Pr L_α	5.031	1.11 *12*
Pr L_β	5.619	1.00 *12*
Pr L_γ	6.410	0.139 *19*
Pr $K_{\alpha2}$	35.550	5.01 *18*
Pr $K_{\alpha1}$	36.026	9.1 *3*
Pr $K_{\beta1}$'	40.720	2.58 *9*
Pr $K_{\beta2}$'	41.981	0.72 *3*
γ M1+0.5%E2	145.440 *3*	48.4

† 0.8% uncert(syst)

Atomic Electrons (^{141}Ce)

$\langle e\rangle$=25.7 *6* keV

e_{bin}(keV)	$\langle e\rangle$(keV)	e(%)
6	0.78	12.8 *13*
7	0.098	1.43 *15*
28	0.037	0.131 *14*
29	0.104	0.36 *4*
30	0.162	0.54 *6*
34	0.073	0.215 *22*
35	0.082	0.235 *24*
36	0.0157	0.044 *5*
39	0.0099	0.025 *3*
40	0.0068	0.0171 *18*
41	0.00184	0.0046 *5*
103	19.6	19.0 *5*
139	3.65	2.63 *7*
144	0.801	0.556 *16*
145	0.226	0.156 *4*

Continuous Radiation (^{141}Ce)

$\langle\beta-\rangle$=145 keV;$\langle$IB$\rangle$=0.074 keV

E_{bin}(keV)		$\langle\ \rangle$(keV)	(%)
0 - 10	β-	0.218	4.37
	IB	0.0075	
10 - 20	β-	0.65	4.31
	IB	0.0067	0.047
20 - 40	β-	2.53	8.4
	IB	0.0116	0.041
40 - 100	β-	16.3	23.5
	IB	0.024	0.038
100 - 300	β-	92	50.1
	IB	0.023	0.0152
300 - 581	β-	33.4	9.3
	IB	0.00142	0.00042

$^{141}_{59}$Pr(stable)

Δ: -86027 *4* keV
%: 100

$^{141}_{60}$Nd(2.49 *3* h)

Mode: ϵ
Δ: -84213 *9* keV
SpA: 8.93$\times10^{6}$ Ci/g
Prod: ^{141}Pr(p,n); ^{141}Pr(d,2n); protons on Gd; ^{142}Nd(n,2n)

Photons (^{141}Nd)

$\langle\gamma\rangle$=50.0 *9* keV

γ_{mode}	γ(keV)	γ(%)†
Pr L_ℓ	4.453	0.179 *24*
Pr L_η	4.929	0.071 *7*
Pr L_α	5.031	4.8 *5*
Pr L_β	5.619	4.4 *5*
Pr L_γ	6.414	0.61 *9*
Pr $K_{\alpha2}$	35.550	21.8 *7*
Pr $K_{\alpha1}$	36.026	39.7 *12*
Pr $K_{\beta1}$'	40.720	11.2 *3*
Pr $K_{\beta2}$'	41.981	3.13 *11*
γ M1+0.5%E2	145.440 *3*	0.241 *22*
γ [E2]	981.57 *15*	0.0217 *23*
γ M1+E2	1127.01 *15*	0.80
γ E2(+M1)	1147.25 *14*	0.306 *12*
γ [E2]	1289.5 *3*	0.0098 *15*
γ (E2)	1292.69 *14*	0.46 *4*
γ E2	1298.60 *21*	0.127 *14*
γ	1306 *1*	<0.00032 ?
γ	1310.7 *5*	~0.00040
γ [M1+E2]	1434.9 *3*	~0.006
γ	1434.93 *15*	0.0167 *24*
γ	1456.1 *5*	0.00080 *24*
γ	1580.37 *15*	0.0060 *9*
γ	1608.35 *19*	0.0183 *24*
γ [E2]	1657.0 *4*	0.00096 *24*

† 3.8% uncert(syst)

Atomic Electrons (^{141}Nd)

$\langle e\rangle$=6.1 *4* keV

e_{bin}(keV)	$\langle e\rangle$(keV)	e(%)
6	3.4	55 *6*
7	0.46	6.7 *7*
28	0.162	0.57 *6*
29	0.45	1.55 *16*
30	0.70	2.36 *24*
34	0.32	0.94 *10*
35	0.36	1.02 *10*
36 - 41	0.149	0.394 *24*
103 - 145	0.082	0.075 *7*
940 - 982	0.00046	4.8 *5* $\times10^{-5}$
1085 - 1127	0.024	0.0022 *3*
1140 - 1147	0.00097	8.5 *14* $\times10^{-5}$
1247 - 1293	0.0094	0.00074 *5*
1297 - 1311	6.1 $\times10^{-5}$	4.7 *4* $\times10^{-6}$
1393 - 1435	0.00032	2.3 *10* $\times10^{-5}$
1449 - 1538	6.3 $\times10^{-5}$	4.1 *20* $\times10^{-6}$
1566 - 1656	0.00023	1.5 *6* $\times10^{-5}$

Continuous Radiation (^{141}Nd)

$\langle\beta+\rangle$=9.1 keV;$\langle$IB$\rangle$=1.28 keV

E_{bin}(keV)		$\langle\ \rangle$(keV)	(%)
0 - 10	β+	7.0$\times10^{-7}$	8.3$\times10^{-6}$
	IB	0.00086	
10 - 20	β+	3.10$\times10^{-5}$	0.000186
	IB	0.00068	0.0045
20 - 40	β+	0.00093	0.00284
	IB	0.090	0.26
40 - 100	β+	0.0467	0.060
	IB	0.020	0.040
100 - 300	β+	1.85	0.86
	IB	0.044	0.022
300 - 600	β+	6.1	1.41
	IB	0.20	0.044
600 - 1300	β+	1.11	0.170
	IB	0.78	0.086
1300 - 1814	IB	0.151	0.0106
	$\Sigma\beta$+		2.5

$^{141}_{60}$Nd(1.040 *15* min)

Mode: IT(99.968 *8* %), ϵ(0.032 *8* %)
Δ: -83456 *9* keV
SpA: 1.275$\times10^{9}$ Ci/g
Prod: ^{141}Pr(p,n); ^{142}Nd(n,2n); ^{142}Nd(γ,n)

Photons (^{141}Nd)

$\langle\gamma\rangle$=695 *7* keV

γ_{mode}	γ(keV)	γ(%)†
Nd L_ℓ	4.633	0.0164 *17*
Nd L_η	5.146	0.0065 *5*
Nd L_α	5.228	0.44 *3*
Nd L_β	5.854	0.40 *4*
Nd L_γ	6.693	0.058 *6*
Nd $K_{\alpha2}$	36.847	1.83 *6*
Nd $K_{\alpha1}$	37.361	3.32 *10*
Nd $K_{\beta1}$'	42.240	0.95 *3*
Nd $K_{\beta2}$'	43.562	0.276 *10*
γ_ϵM1+0.5%E2	145.440 *3*	0.019
γ_{IT} M4	756.63 *8*	91.5
γ_ϵM2	971 *1*	0.028
γ_ϵE3	1116.4 *10*	~0.0038

† uncert(syst): 25% for ϵ, <0.1% for IT

Atomic Electrons (^{141}Nd)

$\langle e\rangle$=62.0 *11* keV

e_{bin}(keV)	$\langle e\rangle$(keV)	e(%)
6 - 42	0.52	5.9 *4*
103 - 145	0.010	0.009 *4*
713	49.1	6.89 *15*
750	9.67	1.29 *3*
755	2.15	0.285 *6*
756 - 757	0.507	0.0670 *15*
929 - 971	0.0023	0.00024 *4*
1074 - 1116	0.00014	1.3 *6* $\times10^{-5}$

$^{141}_{61}$Pm(20.90 *5* min)

Mode: ϵ
Δ: -80480 *30* keV
SpA: 6.379$\times10^{7}$ Ci/g
Prod: ^{141}Pr(α,4n); ^{142}Nd(p,2n); ^{142}Nd(d,3n)

Photons (^{141}Pm)

$\langle\gamma\rangle$=218 *12* keV

γ_{mode}	γ(keV)	γ(%)†
Nd L_ℓ	4.633	0.095 *10*
Nd L_η	5.146	0.037 *3*
Nd L_α	5.228	2.56 *16*
Nd L_β	5.856	2.30 *20*
Nd L_γ	6.692	0.33 *3*
Nd $K_{\alpha2}$	36.847	11.0 *3*
Nd $K_{\alpha1}$	37.361	19.9 *6*
Nd $K_{\beta1}$'	42.240	5.68 *16*
Nd $K_{\beta2}$'	43.562	1.66 *6*
γ	180.16 *7*	0.028 *9*
γ M1	193.69 *3*	1.61 *8*
γ	289.08 *6*	0.17 *3*
γ [M1+E2]	402.89 *7*	0.021 *4*
γ	431.62 *10*	0.014 *3* ?
γ	537.82 *9*	0.078 *18*
γ	544.88 *6*	0.064 *18*
γ	597.14 *5*	0.055 *14*

Photons (^{141}Pm) (continued)

γ_{mode}	γ(keV)	γ(%)†
γ [M1+E2]	622.03 4	0.87 5
γ	646.89 10	0.064 5
γ	706.79 9	0.046 14
γ [M1+E2]	739.00 8	0.027 3
γ [M1+E2]	744.24 5	0.0414 9
γ M4	756.63 8	0.092 22 *
γ E2	886.23 4	2.41 12
γ	901.05 6	0.051 23
γ [M1+E2]	958.13 8	0.064 5
γ	966.19 10	0.087 9
γ	1023.26 5	0.129 23
γ [E2]	1029.60 4	0.359 23
γ [M1+E2]	1043.03 9	0.037 4
γ [E2]	1051.73 9	0.097 9
γ [M1+E2]	1080.34 8	0.046 9
γ [M1+E2]	1088.36 8	0.015 3
γ	1117.95 9	0.012 3
γ E2	1223.28 3	4.6
γ	1235.46 8	0.0069 14
γ	1282.06 8	0.020 4
γ [E2]	1345.49 4	1.27 6
γ	1363.09 10	0.0037 9
γ [E2]	1370.94 5	0.10 3
γ E2+M1	1403.15 5	0.74 5
γ	1474.55 18	0.0060 14
γ [M1+E2]	1564.62 5	0.82 5
γ	1581.33 17	0.0101 18
γ	1596.84 5	0.74 5
γ	1626.74 5	0.27 3
γ	1703.55 8	0.055 9
γ	1808.35 9	0.0014 3 ?
γ	1820.42 5	0.074 18
γ	1872.69 7	0.023 9
γ	1880.04 7	0.32 3
γ	1897.23 8	0.051 14
γ [E2]	1967.51 5	0.166 18
γ	2052.85 5	0.120 18
γ	2066.37 7	0.060 9
γ	2073.72 7	0.62 5
γ	2109.50 5	0.083 9
γ	2145.29 21	0.014 5
γ	2160.65 15	0.0083 14
γ	2246.53 5	0.069 18
γ	2265.19 21	0.032 5
γ [E2]	2303.62 7	0.110 9
γ	2311.65 8	0.023 5
γ	2335.99 21	0.0115 23
γ	2354.33 15	0.046 9
γ [E2]	2388.51 10	0.060 9
γ	2418.59 20	0.0074 9
γ	2429.59 21	0.028 5
γ	2463.43 10	0.0124 25
γ	2505.33 8	0.028 9
γ	2514.79 21	0.0041 9
γ	2601.69 20	0.012 3
γ	2610.3 4	0.014 5
γ	2618.99 21	0.018 5
γ	2732.49 21	0.0032 6
γ	2750.72 16	0.0028 5
γ	2804.0 4	0.023 5
γ	2865.3 11	0.0023 5
γ	2944.40 16	0.0055 9
γ	2985.5 10	0.041 9
γ [E2]	3055.86 8	0.0023 5

† approximate
* with ^{141}Nd(1.04 min) in equilib

Atomic Electrons (^{141}Pm)

$\langle e \rangle$=4.00 16 keV

e_{bin}(keV)	$\langle e \rangle$(keV)	e(%)
6	1.15	18.5 19
7	0.83	12.1 12
29	0.081	0.28 3
30	0.224	0.75 8
31	0.34	1.11 11
35	0.114	0.32 3

Atomic Electrons (^{141}Pm) (continued)

e_{bin}(keV)	$\langle e \rangle$(keV)	e(%)
36	0.204	0.57 6
37 - 42	0.091	0.235 16
137	0.005	~0.004
150	0.466	0.310 16
173 - 194	0.105	0.0559 24
246 - 289	0.020	~0.008
359 - 403	0.0028	0.00076 20
424 - 432	0.00017	4.0 19 $\times 10^{-5}$
494 - 544	0.007	0.0014 6
545 - 591	0.040	0.0069 17
596 - 641	0.0096	0.0016 3
645 - 663	0.0014	~0.00021
695 - 744	0.052	0.0074 16
750 - 757	0.0124	0.0017 3
843 - 886	0.061	0.0072 3
894 - 923	0.0034	0.00037 11
951 - 999	0.0101	0.00102 15
1008 - 1052	0.0050	0.00049 4
1073 - 1118	0.00046	4.2 12 $\times 10^{-5}$
1180 - 1229	0.081	0.0068 12
1234 - 1282	0.00033	2.6 13 $\times 10^{-5}$
1302 - 1345	0.0219	0.00167 9
1356 - 1403	0.0139	0.00102 16
1431 - 1521	0.0117	0.00077 13
1538 - 1625	0.014	0.0009 3
1660 - 1702	0.0006	3.8 17 $\times 10^{-5}$
1765 - 1854	0.0043	0.00024 8
1866 - 1961	0.0024	0.000127 14
1966 - 2065	0.0068	0.00034 11
2066 - 2159	0.0018	8.6 21 $\times 10^{-5}$
2203 - 2302	0.0023	0.000101 14
2305 - 2387	0.00127	5.4 8 $\times 10^{-5}$
2411 - 2509	0.00039	1.6 4 $\times 10^{-5}$
2513 - 2612	0.00034	1.3 3 $\times 10^{-5}$
2613 - 2707	4.2 $\times 10^{-5}$	1.6 4 $\times 10^{-6}$
2725 - 2822	0.00019	6.7 21 $\times 10^{-6}$
2858 - 2943	0.00028	10 3 $\times 10^{-6}$
2978 - 3055	5.8 $\times 10^{-5}$	1.9 5 $\times 10^{-6}$

Continuous Radiation (^{141}Pm)

$\langle \beta+ \rangle$=630 keV;$\langle$IB$\rangle$=6.2 keV

E_{bin}(keV)		$\langle\ \rangle$(keV)	(%)
0 - 10	β+	7.0×10^{-7}	8.3×10^{-6}
	IB	0.023	
10 - 20	β+	3.22×10^{-5}	0.000193
	IB	0.022	0.154
20 - 40	β+	0.00101	0.00309
	IB	0.080	0.25
40 - 100	β+	0.057	0.073
	IB	0.145	0.24
100 - 300	β+	3.25	1.45
	IB	0.37	0.21
300 - 600	β+	26.5	5.7
	IB	0.52	0.120
600 - 1300	β+	212	21.9
	IB	1.43	0.155
1300 - 2500	β+	384	22.3
	IB	2.7	0.150
2500 - 3735	β+	3.84	0.150
	IB	0.90	0.032
	Σβ+		52

$^{141}_{62}$Sm(10.2 2 min)

Mode: ε
Δ: -75942 13 keV
SpA: 1.31×10^{8} Ci/g
Prod: protons on Dy; daughter ^{141}Eu

Photons (^{141}Sm)

$\langle\gamma\rangle$=865 37 keV

γ_{mode}	γ(keV)	γ(%)†
Pm L_ℓ	4.809	0.101 10
Pm L_η	5.363	0.038 3
Pm L_α	5.430	2.72 18
Pm L_β	6.097	2.44 21
Pm L_γ	6.979	0.35 4
Pm $K_{\alpha2}$	38.171	11.1 4
Pm $K_{\alpha1}$	38.725	20.1 6
Pm $K_{\beta1}$'	43.793	5.78 19
Pm $K_{\beta2}$'	45.183	1.71 6
γ (E2)	324.44 15	2.51 25
γ M1+34%E2	403.93 8	42.5
γ E2	438.35 8	37.7 13
γ	728.37 16	1.27 25
γ	767.39 17	1.15 9
γ	854.32 14	1.32 9
γ	888.74 15	0.64 9
γ	1046.48 20	1.49 13
γ	1057.42 13	3.27 21
γ	1091.84 12	2.59 17
γ	1292.66 13	6.8 4
γ	1336.6 4	0.51 17
γ	1352.8 4	0.38 17
γ	1446.70 22	0.30 9
γ	1464.26 21	1.9 4
γ	1481.12 23	0.51 9
γ	1495.76 12	1.79 13
γ	1498.68 21	0.60 9
γ	1515.4 3	0.68 9
γ	1566.38 21	0.21 9
γ	1588.3 3	~0.25
γ	1599.70 21	0.60 25
γ	1600.79 20	4.0 4
γ	1634.12 20	0.34 9
γ	1885.05 22	0.72 9
γ	1902.60 21	0.89 9
γ	1992.2 3	0.68 9
γ	2004.72 20	0.89 9
γ	2038.04 20	2.80 21

† 1.4% uncert(syst)

Atomic Electrons (^{141}Sm)

$\langle e \rangle$=11.5 8 keV

e_{bin}(keV)	$\langle e \rangle$(keV)	e(%)
6	1.19	18.4 19
7	0.86	12.1 12
30 - 44	1.05	3.12 15
279 - 324	0.31	0.108 9
359	4.0	1.10 22
393	2.14	0.544 21
397	0.65	0.16 3
402 - 404	0.18	0.045 7
431	0.362	0.084 3
432 - 438	0.174	0.0399 10
683 - 728	0.07	0.011 5
760 - 809	0.037	~0.0046
844 - 889	0.023	0.0027 11
1001 - 1050	0.15	0.014 5
1051 - 1092	0.010	0.0010 3
1247 - 1293	0.12	0.010 4
1308 - 1353	0.007	~0.0006
1402 - 1451	0.056	0.0039 12
1453 - 1543	0.032	0.0022 5
1555 - 1633	0.067	0.0043 15
1840 - 1901	0.018	0.0010 3
1947 - 2037	0.045	0.0023 6

Continuous Radiation (^{141}Sm)

⟨β+⟩=690 keV;⟨IB⟩=6.7 keV

E_{bin}(keV)		⟨ ⟩(keV)	(%)
0 - 10	β+	6.2×10^{-7}	7.4×10^{-6}
	IB	0.024	
10 - 20	β+	2.93×10^{-5}	0.000176
	IB	0.024	0.164
20 - 40	β+	0.00094	0.00287
	IB	0.075	0.24
40 - 100	β+	0.054	0.069
	IB	0.164	0.28
100 - 300	β+	3.16	1.42
	IB	0.40	0.22
300 - 600	β+	25.9	5.6
	IB	0.56	0.129
600 - 1300	β+	198	20.6
	IB	1.53	0.167
1300 - 2500	β+	420	23.6
	IB	2.8	0.155
2500 - 4136	β+	42.1	1.58
	IB	1.15	0.039
	Σβ+		53

$^{141}_{62}$Sm(22.6 *2* min)

Mode: ε(99.69 *3* %), IT(0.31 *3* %)
Δ: -75766 *13* keV
SpA: 5.90×10^7 Ci/g
Prod: ^{142}Nd(α,5n); protons on Er; ^{144}Sm(γ,3n); ^{144}Sm(p,p3n)

Photons (^{141}Sm)

⟨γ⟩=1634 *27* keV

γ_{mode}	γ(keV)	γ(%)†
Pm L_ℓ	4.809	0.182 *19*
Pm L_η	5.363	0.069 *6*
Pm L_α	5.430	4.9 *3*
Pm L_β	6.097	4.4 *4*
Pm L_γ	6.978	0.63 *7*
Pm $K_{\alpha2}$	38.171	19.8 *8*
Pm $K_{\alpha1}$	38.725	36.0 *15*
Pm $K_{\beta1}$'	43.793	10.4 *4*
Pm $K_{\beta2}$'	45.183	3.07 *14*
γ_ϵ	108.42 *15*	0.20 *4*
γ_ϵ	149.20 *23*	0.29 *8* ?
γ_{IT} M4	174.9 *7*	0.00438
γ_ϵM1	196.86 *11*	75 *7*
γ_ϵ	247.76 *11*	0.78 *16*
γ_ϵM2	431.86 *9*	40.9 *20*
γ_ϵM1	538.25 *10*	8.5 *6*
γ_ϵ	577.65 *18*	0.90 *25* ?
γ_ϵ	583.37 *16*	0.29 *8* ?
γ_ϵ	607.87 *19*	1.02 *12*
γ_ϵ[E3]	628.73 *9*	2.70 *8*
γ_ϵ	648.77 *13*	0.37 *8*
γ_ϵ	676.95 *12*	1.39 *20*
γ_ϵE2,M1	684.48 *10*	8.0 *6*
γ_ϵ	704.35 *10*	0.45 *8*
γ_ϵ	725.79 *24*	1.47 *25* ?
γ_ϵ	750.30 *14*	1.59 *25*
γ_ϵ	764.3 *3*	0.16 *4*
γ_ϵ	768.2 *5*	0.16 *4*
γ_ϵ	777.22 *15*	20.6 *8*
γ_ϵ	786.01 *8*	6.9 *4*
γ_ϵ	805.88 *9*	3.6 *7*
γ_ϵ	820.7 *3*	0.16 *4*
γ_ϵ	837.09 *17*	3.64 *12* ?
γ_ϵ	874.99 *9*	1.27 *4*
γ_ϵ	882.0 *3*	0.16 *4*
γ_ϵ	896.53 *9*	1.47 *16*
γ_ϵ	911.43 *14*	9.32 *25*
γ_ϵ	924.71 *9*	2.3 *3*

Photons (^{141}Sm) (continued)

γ_{mode}	γ(keV)	γ(%)†
γ_ϵ	952.11 *11*	0.90 *4*
γ_ϵ	955.22 *16*	0.69 *4*
γ_ϵ	974.08 *15*	0.20 *4*
γ_ϵ(E1)	983.41 *14*	7.4 *3*
γ_ϵ	997.0 *3*	0.37 *8* ?
γ_ϵ	1009.20 *17*	2.94 *25*
γ_ϵ	1029.3 *3*	0.53 *12* ?
γ_ϵ	1108.29 *13*	1.27 *12*
γ_ϵ	1117.61 *14*	3.27 *25*
γ_ϵ	1145.01 *13*	8.8 *3*
γ_ϵ	1287.72 *17*	0.29 *12* ?
γ_ϵ	1380.9 *6*	0.20 *8*
γ_ϵ	1434.78 *12*	0.37 *12*
γ_ϵ	1462.96 *11*	1.8 *3*
γ_ϵ	1490.36 *8*	9.4 *6*
γ_ϵ	1786.41 *15*	11.1 *5*
γ_ϵ	1897.71 *25*	0.38 *6*
γ_ϵ	1979.62 *20*	0.40 *5*
γ_ϵ	2073.72 *15*	1.4 *5*

† uncert(syst): 9.8% for ε, 9.7% for IT

Atomic Electrons (^{141}Sm)

⟨e⟩=54.9 *25* keV

e_{bin}(keV)	⟨e⟩(keV)	e(%)
1 - 2	0.004	~0.3
6	2.13	33 *4*
7	1.56	21.9 *23*
8 - 45	1.88	5.6 *3*
63 - 108	0.25	0.29 *12*
128 - 149	0.25	0.191 *20*
152	23.0	15.2 *15*
167 - 175	0.238	0.141 *4*
189	3.7	1.95 *19*
190 - 203	1.54	0.79 *6*
240 - 248	0.033	0.013 *7*
387	13.4	3.47 *19*
424	2.07	0.49 *3*
425 - 432	0.93	0.217 *7*
493 - 538	0.77	0.153 *10*
563 - 608	0.28	0.049 *5*
621 - 670	0.45	0.071 *15*
675 - 724	0.17	0.025 *6*
725 - 771	0.9	0.12 *5*
776 - 821	0.17	0.021 *7*
830 - 876	0.29	0.034 *15*
880 - 929	0.14	0.015 *4*
938 - 984	0.15	0.016 *4*
990 - 1029	0.013	0.0013 *5*
1063 - 1111	0.24	0.022 *8*
1116 - 1145	0.028	0.0024 *10*
1243 - 1288	0.005	~0.00039
1336 - 1381	0.0032	~0.00024
1390 - 1435	0.029	0.0020 *9*
1445 - 1490	0.14	0.010 *4*
1741 - 1785	0.13	0.007 *3*
1853 - 1934	0.0079	0.00042 *13*
1972 - 2067	0.015	0.0007 *3*
2072	0.00035	$\sim2\times10^{-5}$

Continuous Radiation (^{141}Sm)

⟨β+⟩=376 keV;⟨IB⟩=4.7 keV

E_{bin}(keV)		⟨ ⟩(keV)	(%)
0 - 10	β+	7.5×10^{-7}	8.9×10^{-6}
	IB	0.0140	
10 - 20	β+	3.53×10^{-5}	0.000212
	IB	0.0136	0.094
20 - 40	β+	0.00113	0.00344
	IB	0.066	0.21
40 - 100	β+	0.064	0.082
	IB	0.119	0.21

Continuous Radiation (^{141}Sm) (continued)

E_{bin}(keV)		⟨ ⟩(keV)	(%)
100 - 300	β+	3.57	1.61
	IB	0.24	0.131
300 - 600	β+	26.6	5.7
	IB	0.39	0.089
600 - 1300	β+	149	16.0
	IB	1.39	0.149
1300 - 2500	β+	175	10.1
	IB	2.1	0.118
2500 - 4519	β+	22.2	0.81
	IB	0.42	0.0146
	Σβ+		34

$^{141}_{63}$Eu(40.0 *7* s)

Mode: ε
Δ: -69980 *40* keV
SpA: 1.98×10^9 Ci/g
Prod: protons on Gd; alphas on ^{144}Sm; ^{144}Sm(p,4n)

Photons (^{141}Eu)

⟨γ⟩=318 *9* keV

γ_{mode}	γ(keV)	γ(%)†
Sm L_ℓ	4.993	0.035 *4*
Sm L_η	5.589	0.0131 *11*
Sm L_α	5.633	0.94 *7*
Sm L_β	6.343	0.85 *8*
Sm L_γ	7.271	0.125 *14*
Sm $K_{\alpha2}$	39.522	3.67 *16*
Sm $K_{\alpha1}$	40.118	6.7 *3*
Sm $K_{\beta1}$'	45.379	1.93 *8*
Sm $K_{\beta2}$'	46.819	0.57 *3*
γ	202.33 *16*	0.26 *7*
γ	213.42 *15*	0.18 *8*
γ	234.68 *22*	0.086 *16*
γ	354.5 *3*	0.22 *6*
γ	369.41 *18*	2.47 *18*
γ	382.85 *13*	4.5 *3*
γ	384.44 *13*	8.5 *5*
γ	393.95 *15*	13.7
γ	395.54 *15*	2.49 *20*
γ	395.75 *25*	0.25 *6*
γ	433.83 *22*	0.86 *8*
γ	593.10 *14*	4.5 *3*
γ	594.69 *14*	0.63 *7*
γ	596.28 *14*	0.77 *7*
γ	597.86 *14*	1.86 *16*
γ	605.87 *17*	1.44 *16*
γ	606.0 *3*	0.48 *7*
γ	687.87 *17*	0.29 *4*
γ	698.97 *15*	0.29 *3*
γ	723.9 *3*	0.25 *8* ?
γ	764.9 *4*	0.33 *14*
γ	776.0 *4*	~0.41
γ	799.6 *6*	0.41 *10*
γ	817.4 *3*	0.12 *3*
γ	882.87 *20*	1.11 *10*
γ	893.67 *25*	0.12 *3*
γ	935.7 *3*	0.19 *4*
γ	976.2 *3*	0.26 *4*
γ	990.0 *5*	0.22 *4*
γ	996.1 *3*	0.56 *6*
γ	999.82 *19*	0.38 *4* ?
γ	1052.0 *3*	0.30 *4*
γ	1053.4 *4*	~0.14
γ	1081.82 *16*	0.29 *4*
γ	1083.41 *18*	0.12 *3*
γ	1234.32 *23*	0.27 *6*
γ	1245.42 *23*	0.44 *7*
γ	1300.38 *21*	0.30 *6*

Photons (^{141}Eu) (continued)

γ_{mode}	γ(keV)	γ(%)†
γ	1382.0 3	0.33 7
γ	1392.5 3	0.21 6
γ	1510.63 21	0.53 10
γ	1560.7 5	0.14 6
γ	1676.0 6	0.23 11
γ	1691.6 7	~0.21
γ	1744.9 4	0.41 12
γ	1766.4 3	0.62 12
γ	1826.3 3	0.36 16
γ	1839.0 5	0.33 10
γ	2221.9 3	0.26 10

† 21% uncert(syst)

Atomic Electrons (^{141}Eu)

$\langle e \rangle$=4.8 9 keV

e_{bin}(keV)	$\langle e \rangle$(keV)	e(%)
7	0.62	8.9 10
8 - 32	0.179	1.34 11
33	0.107	0.32 4
37 - 45	0.132	0.34 2
155 - 202	0.11	0.07 3
206 - 235	0.016	0.008 4
308	0.018	~0.006
323	0.19	~0.06
336	0.34	~0.10
338	0.6	~0.19
347	1.0	~0.28
348	0.0004	~0.00011

Atomic Electrons (^{141}Eu) (continued)

e_{bin}(keV)	$\langle e \rangle$(keV)	e(%)
349	0.19	~0.06
353 - 384	0.28	0.074 25
386	0.14	~0.04
387 - 434	0.21	0.053 22
546	0.18	~0.033
548 - 597	0.27	0.048 14
598 - 641	0.025	0.0041 16
652 - 699	0.021	0.0031 13
716 - 765	0.036	0.0049 19
768 - 817	0.008	0.0011 4
836 - 883	0.035	0.0041 20
886 - 935	0.011	0.0012 5
936 - 983	0.026	0.0027 10
988 - 1037	0.020	0.0020 6
1044 - 1083	0.0029	0.00027 7
1187 - 1234	0.012	0.0010 4
1238 - 1254	0.006	0.00047 22
1293 - 1335	0.005	0.00041 20
1346 - 1392	0.0042	0.00031 13
1464 - 1559	0.010	0.0007 3
1629 - 1720	0.017	0.0010 3
1737 - 1832	0.010	0.00057 18
1837 - 1838	0.00010	~5×10^{-6}
2175 - 2220	0.0026	0.00012 6

Continuous Radiation (^{141}Eu)

$\langle \beta+ \rangle$=1819 keV;$\langle$IB$\rangle$=12.6 keV

E_{bin}(keV)		$\langle\ \rangle$(keV)	(%)
0 - 10	$\beta+$	2.33×10^{-7}	2.78×10^{-6}
	IB	0.052	
10 - 20	$\beta+$	1.13×10^{-5}	6.8×10^{-5}
	IB	0.051	0.36
20 - 40	$\beta+$	0.000374	0.00114
	IB	0.110	0.38
40 - 100	$\beta+$	0.0225	0.0286
	IB	0.31	0.48
100 - 300	$\beta+$	1.43	0.64
	IB	0.91	0.50
300 - 600	$\beta+$	13.3	2.82
	IB	1.18	0.28
600 - 1300	$\beta+$	143	14.4
	IB	2.3	0.26
1300 - 2500	$\beta+$	694	36.4
	IB	3.5	0.19
2500 - 5000	$\beta+$	967	30.4
	IB	4.1	0.121
5000 - 5958	IB	0.122	0.0023
	$\Sigma\beta+$		85

$^{141}_{63}$Eu(3.3 3 s)

Mode: ϵ(67 %), IT(33 %)
Δ: -69884 40 keV
SpA: 2.19×10^{10} Ci/g
Prod: ^{144}Sm(p,4n)

Photons (^{141}Eu)

$\langle \gamma \rangle$=58 7 keV

γ_{mode}	γ(keV)	γ(%)†
γ_{IT}E3	96.4 2	0.67
γ_ϵ	116.7 5	0.074 22
γ_ϵM4	174.9 7	*
γ_ϵ	234.68 22	0.042 17
γ_ϵ	369.41 18	0.29 5
γ_ϵ	393.95 15	0.60 23
γ_ϵ	395.54 15	0.11 4
γ_ϵ	433.83 22	0.42 15
γ_ϵ	518.2 3	~0.43
γ_ϵ	605.87 17	
γ_ϵ	606.0 4	0.14 5 ‡
γ_ϵ	635.0 5	0.044 15
γ_ϵ	804.4 5	0.43 6
γ_ϵ	883 25	0.53 7
γ_ϵ	887.6 3	0.28 6
γ_ϵ	1225.8 4	~0.015
γ_ϵ	1595.2 4	0.39 7

† 9.1% uncert(syst)
* with ^{141}Sm(22.6 min)
‡ combined intensity for doublet

A = 142

NDS **25**, 53 (1978)

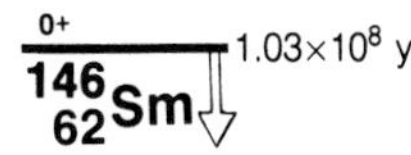

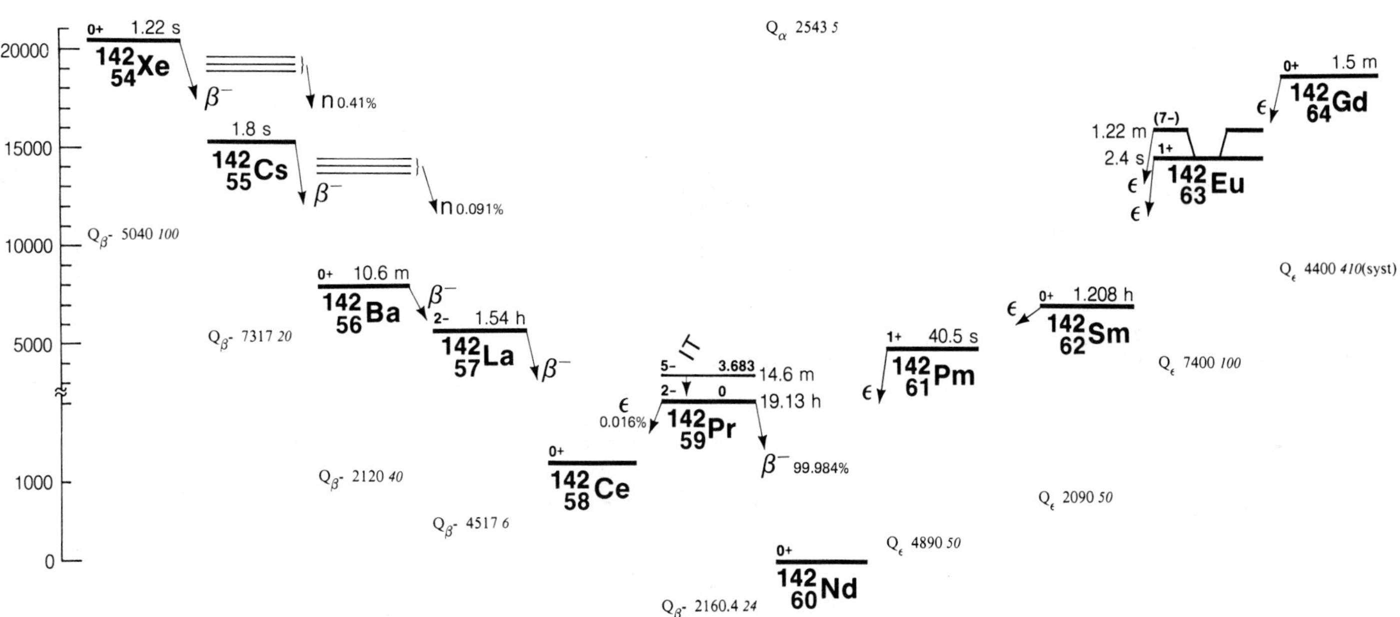

$^{142}_{54}$Xe(1.22 *2* s)

Mode: β-, β-n(0.41 *3* %)
Δ: -65550 *100* keV
SpA: 4.97×10^{10} Ci/g
Prod: fission

Photons (^{142}Xe)

γ_{mode}	γ(keV)	γ(rel)
γ	12.19 *5*	~0.15
γ	19.23 *18*	3.2 *9*
γ	20.86 *9*	13.5 *12*
γ	24.01 *5*	4.0 *12*
γ	33.02 *9*	~0.032
γ	38.87 *5*	23.3 *17*
γ	38.92 *4*	2.2 *3*
γ	46.24 *5*	5.0 *8*
γ	57.10 *16*	~0.014
γ	58.26 *15*	~0.014
γ	70.46 *15*	2.4 *10*
γ	72.96 *4*	27.4 *17*
γ	94.42 *9*	~2
γ	100.09 *12*	0.37 *12*
γ	100.28 *7*	0.37 *12*
γ	105.59 *14*	2.0 *3*
γ	113.4 *3*	1.8 *3*
γ	117.6 *4*	2.3 *5*
γ	119.32 *15*	2.5 *4*
γ	124.58 *8*	8.6 *18*
γ	157.60 *5*	17.4 *11*
γ	161.8 *3*	3.3 *5*
γ	164.96 *5*	21.8 *13*
γ	167.41 *16*	1.7 *6*
γ	170.82 *8*	1.5 *4*
γ	191.69 *5*	35.7 *20*
γ	197.54 *8*	6.8 *8*

Photons (^{142}Xe) (continued)

γ_{mode}	γ(keV)	γ(rel)
γ	203.84 *5*	65 *3*
γ	203.89 *4*	27.6 *14*
γ	211.73 *7*	3.9 *9*
γ	219.00 *6*	5.4 *10*
γ	242.76 *5*	2.7 *13*
γ	250.65 *6*	32.0 *22*
γ	265.24 *7*	1.6 *4*
γ	286.73 *7*	17.1 *9*
γ	291.97 *6*	17.5 *10*
γ	304.16 *7*	1.8 *5*
γ	309.12 *8*	27.2 *15*
γ	313.69 *12*	3.1 *5*
γ	330.3 *3*	2.3 *3*
γ	334.79 *12*	12.3 *8*
γ	337.2 *6*	2.1 *5*
γ	349.34 *16*	5.8 *12*
γ	352.93 *6*	12.7 *14*
γ	373.5 *6*	0.9 *4*
γ	379.96 *8*	10.3 *6*
γ	394.21 *9*	17.5 *12*
γ	404.65 *21*	2.0 *5*
γ	406.44 *6*	11.0 *8*
γ	414.34 *5*	46.8 *25*
γ	418.88 *9*	3.6 *6*
γ	421.9 *13*	~1
γ	428.41 *11*	5.5 *7*
γ	432.76 *13*	11.6 *12*
γ	438.25 *12*	6.3 *7*
γ	447.36 *8*	7.0 *7*
γ	453.21 *4*	19.9 *12*
γ	468.21 *7*	20.3 *12*
γ	468.25 *7*	3.4 *2*
γ	497.6 *4*	2.3 *5*
γ	524.63 *13*	7.6 *6*
γ	538.34 *6*	77 *4*
γ	547.45 *14*	6.7 *11*
γ	557.88 *22*	7.5 *9*

Photons (^{142}Xe) (continued)

γ_{mode}	γ(keV)	γ(rel)
γ	562.3 *5*	2.5 *7*
γ	571.94 *4*	100 *5*
γ	578.0 *4*	3.4 *7*
γ	582.34 *13*	4.4 *7*
γ	586.64 *15*	3.0 *8*
γ	605.62 *9*	22.0 *14*
γ	618.17 *4*	72 *4*
γ	627.5 *22*	<2
γ	644.90 *4*	63 *4*
γ	657.09 *4*	79 *4*
γ	662.12 *17*	3.0 *8*
γ	664.64 *13*	16.8 *12*
γ	669.2 *5*	2.8 *6*
γ	672.03 *12*	7.4 *7*
γ	693.65 *9*	5.2 *4*
γ	709.2 *4*	2.0 *4*
γ	718.26 *12*	0.71 *24*
γ	718.86 *9*	0.71 *24*
γ	724.4 *8*	1.3 *6*
γ	727.2 *5*	1.9 *6*
γ	735.6 *5*	6.4 *14*
γ	737.49 *14*	12.6 *17*
γ	741.1 *4*	3.3 *5*
γ	744.99 *12*	3.1 *4*
γ	761.8 *5*	1.8 *4*
γ	765.88 *14*	3.4 *4*
γ	776.41 *15*	2.8 *4*
γ	792.3 *4*	2.3 *5*
γ	801.34 *13*	6.2 *7*
γ	807.46 *17*	5.7 *9*
γ	816.04 *16*	3.6 *6*
γ	823.58 *22*	3.4 *4*
γ	829.8 *5*	1.3 *4*
γ	862.99 *16*	2.7 *4*
γ	891.28 *8*	11.2 *9*
γ	917.62 *23*	3.0 *6*
γ	930.4 *4*	2.4 *5*

Photons (^{142}Xe)
(continued)

γ_{mode}	γ(keV)	γ(rel)
γ	943.82 8	2.9 10
γ	957.39 9	6.7 10
γ	983.08 16	2.5 6
γ	991.55 7	8.8 7
γ	996.42 16	3.6 5
γ	1020.3 4	2.6 5
γ	1040.67 24	1.6 4
γ	1068.24 16	2.7 5
γ	1089.85 13	1.3 4
γ	1108.46 5	0.71 3
γ	1156.85 11	6.6 5
γ	1164.8 3	3.1 4
γ	1183.24 7	2.7 5
γ	1187.75 15	5.8 6
γ	1195.43 7	4.5 7
γ	1218.61 12	6.2 9
γ	1227.19 5	18.9 11
γ	1233.05 19	4.8 5
γ	1258.02 13	7.7 6
γ	1300.15 5	30.6 16
γ	1304.22 14	2.4 4
γ	1312.34 5	21.4 12
γ	1338.25 19	2.9 6 ?
γ	1363.22 15	1.8 4
γ	1376.58 17	2.2 3 ?
γ	1384.50 25	1.7 3
γ	1395.02 24	1.7 3 ?
γ	1410.59 8	5.9 5
γ	1431.85 21	1.2 4
γ	1456.82 14	1.2 4
γ	1487.0 12	~0.7
γ	1511.8 9	0.7 3
γ	1520.5 6	1.2 3
γ	1595.2 6	1.4 4
γ	1602.28 9	1.9 4
γ	1607.1 3	3.5 4
γ	1616.4 7	1.0 3
γ	1625.05 12	0.8 3
γ	1632.94 13	1.4 4
γ	1710.66 14	2.1 4
γ	1718.55 14	0.9 4
γ	1773.4 8	1.3 5
γ	1781.23 16	1.7 6
γ	1790.54 13	1.1 4
γ	1805.24 18	1.2 4
γ	1836.78 12	1.9 5
γ	1844.78 17	2.0 4 ?
γ	1862.6 3	1.5 5 ?
γ	1875.70 12	1.7 5
γ	1901.99 14	7.6 7
γ	1972.29 24	~0.6
γ	2077.62 25	1.7 5 ?

$^{142}_{55}$Cs(1.80 *8* s)

Mode: β-, β-n(0.091 *3* %)

Δ: -70590 *30* keV

SpA: 3.66×10^{10} Ci/g

Prod: fission

β-n(avg): 240 *60*

Photons (^{142}Cs)

γ_{mode}	γ(keV)	γ(rel)
γ	100.6 5	
γ	209.3 8	0.51 25
γ	325.1 3	0.26 8
γE2	359.63 5	100 5
γ	400.98 22	0.20 7
γ	459.1 8	0.54 25
γ[E2]	475.20 21	1.14 17
γ	510.74 16	7.8 9
γ	608.3 3	1.20 22 ?
γ	858.14 16	0.92 12

Photons (^{142}Cs)
(continued)

γ_{mode}	γ(keV)	γ(rel)
γ	932.70 11	2.47 18
γ	966.88 5	28.3 15
γ	986.98 21	1.35 17
γ	1015.26 12	0.62 13
γ	1064.49 10	2.45 18
γ	1098.9 3	~0.5
γ	1101.2 4	1.0 3
γ	1118.8 3	0.42 15
γ	1136.98 23	0.66 12
γ	1175.96 6	10.3 6
γ	1192.5 3	1.52 21
γ	1243.27 12	1.8 2
γ	1279.99 7	6.2 4
γ	1326.51 4	34.3 18
γ	1333.48 17	1.80 19
γ	1422.25 18	3.1 5
γ	1424.04 17	2.7 5
γ	1559.84 15	1.62 16
γ	1610.95 18	1.26 13
γ	1768.57 16	1.71 16
γ	1817.96 16	1.22 13
γ	1899.11 15	2.9 3
γ	1915.9 4	0.59 12
γ	1935.26 18	1.26 14
γ	1956.80 17	0.93 16
γ	1960.7 8	0.41 16
γ	1982.13 12	4.4 3
γ	2051.0 7	0.30 12
γ	2056.1 5	0.46 12
γ	2246.77 18	1.40 16
γ	2254.1 3	0.96 14
γ	2341.76 12	0.65 10
γ	2351.3 3	0.47 10 ?
γ	2394.1 3	~0.18
γ	2397.85 22	2.63 23
γ	2412.2 9	0.26 12
γ	2508.9 7	0.39 13
γ	2522.92 20	0.90 14
γ	2575.9 6	0.38 11
γ	2613.2 3	0.72 11 ?
γ	2655.9 4	0.39 12
γ	2677.4 4	~0.23
γ	2725.77 20	1.62 13
γ	2757.3 3	1.25 14
γ	2784.83 16	0.47 11
γ	2796.6 6	0.32 10
γ	2839.6 8	0.25 10
γ	2882.55 20	1.36 12
γ	2923.66 17	1.71 16
γ	2938.6 3	0.45 10
γ	2988.6 7	0.34 12
γ	3079.6 4	~0.13
γ	3144.45 16	0.56 12
γ	3167.5 3	0.40 12 ?
γ	3261.75 19	0.62 10
γ	3283.29 17	2.06 20
γ	3369.4 4	0.21 9
γ	3426.5 4	0.87 12
γ	3573.26 18	2.32 17
γ	3661.2 4	0.3 1 ?
γ	3786.3 4	0.69 10
γ	3797.4 3	0.76 10
γ	3835.0 4	0.93 11
γ	3870.5 5	0.71 12
γ	3897.9 7	0.34 9
γ	3931.7 4	0.94 11
γ	4009.3 7	0.26 7
γ[E2]	4028.4 4	0.50 7
γ	4037.2 3	0.37 7
γ	4085.7 6	0.32 7
γ	4145.7 11	0.17 8
γ	4178.5 5	0.33 8 ?
γ	4198.4 4	0.18 7
γ	4205.8 4	0.27 7 ?
γ	4217.4 5	0.15 7 ?
γ	4238.3 3	0.78 9 ?
γ	4250.6 8	0.26 7
γ	4277.3 3	0.51 8 ?
γ	4362.9 4	~0.10 ?
γ	4369.3 4	1.06 10
γ	4418.2 4	0.86 9
γ	4494.30 22	0.31 6
γ	4537.2 10	0.13 5
γ	4549.6 8	0.18 5

Photons (^{142}Cs)
(continued)

γ_{mode}	γ(keV)	γ(rel)
γ	4564.8 7	0.24 5
γ	4578.2 7	0.23 6
γ	4609.8 7	0.19 5
γ	4647.8 5	0.36 5
γ	4670.2 13	0.07 3
γ	4681.8 9	0.10 3
γ	4694.7 7	0.14 4
γ	4730.4 11	0.059 24
γ	4738.4 11	0.059 24
γ	4812.2 8	0.095 24
γ[E2]	4863.2 4	0.062 25
γ	4891.1 8	0.14 3
γ	4897.2 5	0.10 3 ?
γ	4937.1 12	0.047 19
γ	4955.5 14	0.039 18
γ	4993.6 13	0.043 18
γ	5006.2 9	0.078 19
γ	5028.2 12	0.046 19
γ	5374.16 14	0.040 13

$^{142}_{56}$Ba(10.6 *2* min)

Mode: β-

Δ: -77910 *40* keV

SpA: 1.249×10^{8} Ci/g

Prod: fission

Photons (^{142}Ba)
⟨γ⟩=1038 *39* keV

γ_{mode}	γ(keV)	γ(%)
γ	69.83 10	~0.41
γ	77.28 9	~1
γ (M1)	77.57 4	~11
γ	122.98 6	1.07 14
γ	154.29 9	0.60 8
γ	161.74 10	~0.12
γ	176.90 6	1.71 21
γ	216.45 9	0.23 6
γ	222.67 10	0.31 6
γ	231.57 3	11.8 17
γ	242.75 17	0.19 8
γ	255.23 3	20.6 19
γ	269.37 7	0.78 10
γ	283.92 17	0.21 8
γ	286.28 4	1.07 16
γ	309.14 3	2.6 4
γ	334.86 7	1.44 21
γ	337.17 16	0.29 6
γ	347.51 7	~0.16
γ	363.85 4	4.6 6
γ	379.05 5	0.54 8
γ	417.9 3	0.39 8
γ	425.08 6	5.7 8
γ	432.12 6	1.13 16
γ	434.83 17	0.35 6
γ	449.39 12	0.25 6
γ	457.44 17	0.45 8
γ	473.47 20	0.35 6
γ	487.84 23	~0.12
γ	513.10 23	0.27 10
γ	532.09 18	~0.10
γ	537.07 18	~0.12
γ	557.81 16	0.35 6
γ	590.68 23	0.29 6
γ	599.89 7	1.85 21
γ	604.23 7	0.37 8
γ	769.38 4	0.70 8
γ	786.5 3	0.29 8
γ	792.30 18	0.25 8
γ	823.29 5	0.47 8
γ	840.27 4	3.5 5
γ	876.7 9	~0.08
γ	894.98 4	12.7 13

Photons (^{142}Ba)
(continued)

γ_{mode}	γ(keV)	γ(%)
γ	948.89 *4*	10.3 *14*
γ	1000.95 *4*	9.1 *13*
γ	1032.48 *7*	0.56 *8*
γ	1078.52 *4*	10.8 *15*
γ	1093.71 *5*	2.6 *4*
γ	1122.70 *8*	0.35 *6*
γ	1126.55 *4*	1.77 *25*
γ	1148.43 *5*	0.45 *6*
γ	1202.34 *5*	6.2 *8*
γ	1204.12 *4*	15.8 *18*
γ	1283.5 *5*	0.19 *8*
γ	1379.99 *5*	3.9 *5*

Continuous Radiation (^{142}Ba)
$\langle\beta-\rangle$=378 keV;$\langle$IB$\rangle$=0.47 keV

E_{bin}(keV)		$\langle\ \rangle$(keV)	(%)
0 - 10	β-	0.075	1.50
	IB	0.017	
10 - 20	β-	0.227	1.52
	IB	0.0165	0.115
20 - 40	β-	0.92	3.07
	IB	0.031	0.108
40 - 100	β-	6.6	9.5
	IB	0.080	0.124
100 - 300	β-	63	31.7
	IB	0.167	0.098
300 - 600	β-	149	34.4
	IB	0.104	0.026
600 - 1300	β-	136	16.8
	IB	0.049	0.0063
1300 - 2042	β-	22.0	1.50
	IB	0.0022	0.000154

$^{142}_{57}$La(1.542 *8* h)

Mode: β-
Δ: -80025 *7* keV
SpA: 1.431×10^{7} Ci/g
Prod: fission; ^{142}Ce(n,p)

Photons (^{142}La)
$\langle\gamma\rangle$=2485 *70* keV

γ_{mode}	γ(keV)	γ(%)†
γ	105.96 *19*	~0.14
γ	119.5 *6*	<0.047
γ	142.24 *22*	<0.047
γ	169.6 *7*	<0.047
γ	173.42 *17*	~0.09
γ	298.0 *3*	<0.09
γ	332.16 *12*	<0.09
γ	353.7 *6*	<0.047
γ	355.48 *8*	<0.047
γ	367.47 *14*	~0.09
γ	393.51 *6*	~0.09
γ	408.5 *4*	<0.09
γ	420.87 *10*	0.24 *10*
γ	428.0 *5*	<0.09
γ [E1]	433.34 *5*	0.38 *14*
γ	514.79 *11*	~0.14
γ	532.07 *20*	~0.14
γ	538.44 *7*	<0.09
γ	545.9 *7*	<0.047
γ	571.7 *5*	<0.09
γ [E2]	578.14 *4*	1.23 *19*
γ	597.7 *5*	<0.09
γ	601.9 *5*	<0.09
γ	619.57 *10*	0.14 *5*

Photons (^{142}La)
(continued)

γ_{mode}	γ(keV)	γ(%)†
γ E2	641.25 *3*	47
γ	792.62 *11*	<0.09
γ	861.60 *5*	1.8 *3*
γ	877.74 *11*	~0.19
γ [M1+E2]	894.93 *3*	8.5 *10*
γ	916.81 *12*	<0.09
γ	946.43 *13*	~0.09
γ	962.27 *11*	0.38 *14*
γ	991.3 *3*	~0.09
γ	1006.53 *6*	0.24 *10*
γ (E1)	1011.47 *4*	3.9 *5*
γ	1020.77 *18*	<0.09
γ	1039.3 *3*	~0.09
γ	1043.72 *5*	2.7 *3*
γ	1056.15 *14*	<0.09
γ	1061.67 *8*	~0.14
γ	1070.55 *12*	~0.14
γ	1074.3 *3*	~0.09
γ	1088.96 *12*	0.24 *10*
γ	1100.2 *5*	<0.09
γ	1105.61 *17*	<0.09
γ	1112.7 *3*	~0.09
γ	1116.8 *3*	~0.09
γ	1130.65 *8*	0.47 *14*
γ	1144.57 *20*	~0.14
γ	1160.27 *4*	1.75 *24*
γ	1174.51 *18*	~0.14
γ	1190.97 *20*	0.38 *14*
γ	1205.51 *12*	<0.09
γ	1231.6 *5*	0.28 *10*
γ	1233.22 *8*	1.8 *3*
γ	1242.4 *3*	~0.19
γ	1264.8 *3*	~0.09
γ	1270.2 *4*	~0.09
γ	1279.27 *18*	<0.09
γ	1282.67 *15*	<0.09
γ	1288.24 *16*	~0.09
γ	1323.27 *20*	0.33 *10*
γ	1332.4 *4*	~0.09
γ	1341.3 *6*	<0.09
γ	1354.92 *20*	~0.09
γ [M1+E2]	1363.01 *5*	2.1 *3*
γ	1373.7 *7*	~0.19
γ (E2)	1389.37 *10*	0.43 *14*
γ	1395.37 *20*	~0.19
γ	1402.27 *20*	~0.14
γ	1445.64 *16*	~0.14
γ	1455.18 *9*	~0.09
γ	1493.91 *13*	~0.14
γ	1500.33 *14*	~0.09
γ	1516.37 *20*	0.43 *14*
γ	1535.56 *18*	0.24 *10*
γ	1540.4 *1*	0.47 *10*
γ	1545.97 *7*	3.0 *4*
γ	1618.27 *20*	0.28 *10*
γ	1651.5 *3*	~0.19
γ	1688.2 *3*	0.24 *10*
γ	1723.04 *11*	1.52 *24*
γ	1752.5 *7*	~0.09
γ	1756.52 *5*	3.0 *4*
γ	1767.71 *10*	~0.19
γ	1771.1 *5*	~0.19
γ	1788.49 *19*	<0.09
γ	1794.05 *20*	~0.09
γ	1806.71 *8*	~0.14
γ	1817.17 *14*	~0.09
γ	1885.5 *7*	0.52 *14*
γ	1901.45 *6*	7.9 *5*
γ	1923.26 *8*	0.24 *10*
γ	1933.72 *14*	~0.14
γ	1948.3 *4*	0.47 *14*
γ	1954.1 *10*	<0.09
γ	1960.52 *12*	~0.14
γ [E2]	2004.26 *5*	0.95 *19*
γ	2025.57 *8*	1.23 *19*
γ	2038.77 *13*	1.00 *19*
γ	2050.47 *20*	0.47 *14*
γ	2055.19 *4*	2.7 *3*
γ	2077.07 *11*	0.66 *14*
γ	2086.17 *20*	0.38 *14*
γ	2100.43 *12*	0.95 *19*
γ	2126.3 *3*	0.33 *14*
γ	2139.39 *11*	0.52 *14*
γ	2180.37 *20*	0.52 *14*

Photons (^{142}La)
(continued)

γ_{mode}	γ(keV)	γ(%)†
γ	2187.22 *7*	5.3 *7*
γ	2290.6 *6*	0.33 *14*
γ	2358.60 *14*	0.76 *14*
γ	2364.28 *12*	0.43 *14*
γ	2397.77 *5*	14.8 *16*
γ	2419.6 *4*	~0.19
γ	2459.5 *4*	0.38 *14*
γ	2513.3 *6*	~0.14
γ	2532.4 *7*	~0.09
γ	2539.5 *5*	0.71 *14*
γ	2542.69 *6*	10.1 *13*
γ	2663.6 *3*	0.71 *14*
γ	2666.81 *8*	1.71 *24*
γ	2672.7 *4*	~0.19
γ	2779.18 *10*	<0.09
γ	2782.4 *4*	0.28 *10*
γ	2800.9 *4*	0.57 *14*
γ	2818.17 *8*	0.76 *19*
γ	2828.63 *14*	0.24 *10*
γ	2970.87 *18*	0.71 *14*
γ	2971.98 *11*	3.0 *3*
γ	2991.59 *16*	~0.09
γ	2999.84 *14*	0.47 *14*
γ	3007.2 *5*	~0.19
γ	3012.97 *20*	0.66 *14*
γ	3022.4 *7*	~0.09
γ	3034.30 *11*	0.52 *14*
γ	3046.97 *20*	0.38 *14*
γ	3062.5 *7*	~0.14
γ	3075.94 *17*	~0.14
γ	3155.1 *3*	~0.19
γ	3181.1 *3*	0.28 *10*
γ	3236.77 *20*	0.28 *10*
γ	3242.5 *3*	~0.19
γ	3273.3 *3*	~0.14
γ	3314.77 *20*	1.23 *19*
γ	3334.44 *18*	<0.09
γ	3401.76 *13*	0.28 *10*
γ	3420.42 *10*	<0.09
γ	3459.41 *8*	0.33 *14*
γ	3469.87 *14*	<0.09
γ	3612.11 *18*	0.81 *19*
γ	3632.83 *15*	1.04 *19*
γ	3719.18 *17*	0.28 *10*
γ	3746.4 *8*	<0.09
γ	3850.5 *3*	0.24 *10*
γ	3975.67 *18*	<0.09
γ	4045.3 *3*	<0.09
γ	4192.4 *3*	

† 11% uncert(syst)

Continuous Radiation (^{142}La)
$\langle\beta-\rangle$=842 keV;$\langle$IB$\rangle$=1.9 keV

E_{bin}(keV)		$\langle\ \rangle$(keV)	(%)
0 - 10	β-	0.0309	0.62
	IB	0.032	
10 - 20	β-	0.093	0.62
	IB	0.031	0.21
20 - 40	β-	0.381	1.27
	IB	0.060	0.21
40 - 100	β-	2.79	3.97
	IB	0.167	0.26
100 - 300	β-	29.1	14.4
	IB	0.45	0.25
300 - 600	β-	97	21.7
	IB	0.45	0.106
600 - 1300	β-	347	37.8
	IB	0.49	0.058
1300 - 2500	β-	274	16.5
	IB	0.19	0.0116
2500 - 4517	β-	91	2.98
	IB	0.031	0.00109

$^{142}_{58}$Ce(>5×10^{16} yr)

Δ: -84542 *4* keV
%: 11.08 *10*

$^{142}_{59}$Pr(19.13 *4* h)

Mode: β-(99.9836 *8* %), ϵ(0.0164 *8* %)
Δ: -83799 *4* keV
SpA: 1.1536×10^6 Ci/g
Prod: ^{141}Pr(n,γ); ^{139}La(α,n); ^{142}Ce(d,2n); ^{141}Pr(d,p)

Photons (^{142}Pr)

⟨γ⟩=58.4 *6* keV

γmode	γ(keV)	γ(%)†
γβ-[E1]	508.7 *5*	0.0229 *18*
γϵE2	641.25 *3*	0.00205
γβ-[E2]	1575.5 *3*	3.70

† uncert(syst): 4.9% for ϵ, 14% for β-

Continuous Radiation (^{142}Pr)

⟨β-⟩=809 keV;⟨IB⟩=1.65 keV

Ebin(keV)		⟨ ⟩(keV)	(%)
0 - 10	β-	0.0295	0.59
	IB	0.031	
10 - 20	β-	0.089	0.59
	IB	0.031	0.21
20 - 40	β-	0.360	1.20
	IB	0.060	0.21
40 - 100	β-	2.61	3.71
	IB	0.167	0.26
100 - 300	β-	26.7	13.3
	IB	0.44	0.25
300 - 600	β-	89	19.9
	IB	0.44	0.104
600 - 1300	β-	383	41.0
	IB	0.43	0.052
1300 - 2160	β-	307	19.7
	IB	0.055	0.0036

$^{142}_{59}$Pr(14.6 *5* min)

Mode: IT
Δ: -83795 *4* keV
SpA: 9.1×10^7 Ci/g
Prod: ^{141}Pr(n,γ)

Photons (^{142}Pr)

γmode	γ(keV)	γ(%)
γ[M3]	3.683 *4*	7.9×10^{-9}

$^{142}_{60}$Nd(stable)

Δ: -85959 *3* keV
%: 27.13 *10*

$^{142}_{61}$Pm(40.5 *5* s)

Mode: ϵ
Δ: -81070 *50* keV
SpA: 1.945×10^9 Ci/g
Prod: daughter ^{142}Sm; ^{142}Nd(p,n); protons on Gd

Photons (^{142}Pm)

⟨γ⟩=72 *10* keV

γmode	γ(keV)	γ(%)
Nd L_ℓ	4.633	0.043 *4*
Nd L_η	5.146	0.0168 *13*
Nd L_α	5.228	1.16 *7*
Nd L_β	5.856	1.04 *9*
Nd L_γ	6.691	0.148 *15*
Nd $K_{\alpha2}$	36.847	4.97 *14*
Nd $K_{\alpha1}$	37.361	9.0 *3*
Nd $K_{\beta1}$'	42.240	2.58 *7*
Nd $K_{\beta2}$'	43.562	0.752 *25*
γ[E2]	641.2 *5*	0.65 *3*
γ	808.7 *6*	0.020 *4*
γ	1007.3 *5*	0.021 *3*
γ	1552.1 *6*	0.031 *7*
γ[E2]	1575.5 *3*	3.3
γ	1781.9 *14*	~0.020 ?
γ	2384.2 *5*	0.112 *10*
γ	2582.8 *5*	0.046 *3*
γ[E2]	2845.6 *8*	0.079 *7*
γ	3045.4 *10*	0.0214 *16*
γ	3127.6 *7*	0.0155 *20*
γ	3357.4 *14*	0.0076 *15*

Atomic Electrons (^{142}Pm)

⟨e⟩=1.45 *7* keV

ebin(keV)	⟨e⟩(keV)	e(%)
6	0.52	8.4 *9*
7	0.37	5.5 *6*
29	0.037	0.125 *13*
30	0.102	0.34 *4*
31	0.156	0.51 *5*
35	0.052	0.147 *15*
36	0.093	0.26 *3*
37 - 42	0.041	0.107 *7*
598 - 641	0.0244	0.00404 *22*
765 - 809	0.0006	~7×10^{-5}
964 - 1007	0.00044	~5×10^{-5}
1509 - 1574	0.046	0.0030 *5*
1738 - 1781	0.00022	~1×10^{-5}
2341 - 2383	0.0009	4.0 *14* ×10^{-5}
2539 - 2581	0.00036	1.4 *5* ×10^{-5}
2802 - 2844	0.00069	2.44 *20* ×10^{-5}
3002 - 3084	0.00023	7.7 *17* ×10^{-6}
3120 - 3126	1.4 ×10^{-5}	4.5 *13* ×10^{-7}
3314 - 3356	4.7 ×10^{-5}	1.4 *5* ×10^{-6}

Continuous Radiation (^{142}Pm)

⟨β+⟩=1364 keV;⟨IB⟩=9.8 keV

Ebin(keV)		⟨ ⟩(keV)	(%)
0 - 10	β+	4.20×10^{-7}	5.01×10^{-6}
	IB	0.042	
10 - 20	β+	1.94×10^{-5}	0.000117
	IB	0.042	0.29
20 - 40	β+	0.00062	0.00187
	IB	0.099	0.33
40 - 100	β+	0.0351	0.0447
	IB	0.25	0.39

Continuous Radiation (^{142}Pm) (continued)

Ebin(keV)		⟨ ⟩(keV)	(%)
100 - 300	β+	2.11	0.94
	IB	0.72	0.40
300 - 600	β+	18.8	4.02
	IB	0.93	0.22
600 - 1300	β+	187	19.0
	IB	1.9	0.21
1300 - 2500	β+	740	39.5
	IB	3.1	0.17
2500 - 4890	β+	417	14.4
	IB	2.7	0.085
	Σβ+		78

$^{142}_{62}$Sm(1.2082 *8* h)

Mode: ϵ
Δ: -78986 *16* keV
SpA: 1.8265×10^7 Ci/g
Prod: ^{142}Nd(α,4n)

Photons (^{142}Sm)

⟨γ⟩=36.3 *9* keV

γmode	γ(keV)	γ(%)
Pm L_ℓ	4.809	0.194 *19*
Pm L_η	5.363	0.074 *6*
Pm L_α	5.430	5.2 *3*
Pm L_β	6.097	4.7 *4*
Pm L_γ	6.979	0.68 *7*
Pm $K_{\alpha2}$	38.171	21.2 *6*
Pm $K_{\alpha1}$	38.725	38.4 *11*
Pm $K_{\beta1}$'	43.793	11.1 *3*
Pm $K_{\beta2}$'	45.183	3.27 *11*
γE1	679 *1*	0.099 *20*
γ	849 *1*	0.079 *16*
γ	954 *1*	
γ	1243 *3*	0.26
γ	1345 *2*	0.132 *20*
γ	1830 *1*	

Atomic Electrons (^{142}Sm)

⟨e⟩=5.9 *3* keV

ebin(keV)	⟨e⟩(keV)	e(%)
6	2.27	35 *4*
7	1.65	23.2 *24*
30	0.155	0.51 *5*
31	0.43	1.38 *14*
32	0.64	2.01 *21*
36	0.166	0.46 *5*
37	0.38	1.03 *10*
38 - 44	0.231	0.59 *4*
634 - 679	0.00135	0.00021 *4*
804 - 849	0.0022	~0.00028
1198 - 1243	0.0047	~0.00039
1300 - 1345	0.0021	0.00016 *8*

Continuous Radiation (^{142}Sm)

$\langle\beta+\rangle$=27.7 keV;$\langle$IB$\rangle$=1.9 keV

E_{bin}(keV)		$\langle\ \rangle$(keV)	(%)
0 - 10	$\beta+$	6.3×10^{-7}	7.6×10^{-6}
	IB	0.0017	
10 - 20	$\beta+$	2.97×10^{-5}	0.000178
	IB	0.00144	0.0098
20 - 40	$\beta+$	0.00094	0.00286
	IB	0.060	0.17
40 - 100	$\beta+$	0.051	0.065
	IB	0.070	0.154
100 - 300	$\beta+$	2.44	1.11
	IB	0.055	0.028
300 - 600	$\beta+$	12.6	2.81
	IB	0.22	0.049
600 - 1300	$\beta+$	12.6	1.71
	IB	1.04	0.113
1300 - 2090	IB	0.48	0.032
	$\Sigma\beta+$		5.7

$^{142}_{63}$Eu(2.4 *2* s)

Mode: ϵ

Δ: -71590 *100* keV

SpA: 2.88×10^{10} Ci/g

Prod: ^{144}Sm(p,3n)

Photons (^{142}Eu)

$\langle\gamma\rangle$=194 *18* keV

γ_{mode}	γ(keV)	γ(%)
Sm L_ℓ	4.993	0.020 *8*
Sm L_η	5.589	0.008 *3*
Sm L_α	5.633	0.54 *20*
Sm L_β	6.338	0.50 *19*
Sm L_γ	7.253	0.07 *3*
Sm $K_{\alpha2}$	39.522	1.7 *3*
Sm $K_{\alpha1}$	40.118	3.1 *6*
Sm $K_{\beta1}$'	45.379	0.90 *18*
Sm $K_{\beta2}$'	46.819	0.27 *5*
γ	68.6 *7*	0.62 *9*
γ E2	768.38 *19*	11.2
γ [M1+E2]	890.0 *3*	1.49 *13*
γ [M1+E2]	1287.8 *3*	1.53 *13*
γ	1405.6 *4*	0.80 *9*
γ [E2]	1658.4 *3*	1.5 *3*
γ	1754.5 *4*	1.46 *11*
γ [E2]	2056.1 *3*	0.55 *7*

Continuous Radiation (^{142}Eu)

$\langle\beta+\rangle$=2717 keV;$\langle$IB$\rangle$=16.3 keV

E_{bin}(keV)		$\langle\ \rangle$(keV)	(%)
0 - 10	$\beta+$	1.21×10^{-7}	1.44×10^{-6}
	IB	0.067	
10 - 20	$\beta+$	5.9×10^{-6}	3.51×10^{-5}
	IB	0.067	0.46
20 - 40	$\beta+$	0.000195	0.00059
	IB	0.136	0.47
40 - 100	$\beta+$	0.0118	0.0150
	IB	0.39	0.61
100 - 300	$\beta+$	0.76	0.339
	IB	1.22	0.68
300 - 600	$\beta+$	7.3	1.56
	IB	1.62	0.38
600 - 1300	$\beta+$	87	8.7
	IB	3.1	0.35
1300 - 2500	$\beta+$	530	27.3
	IB	4.0	0.22
2500 - 5000	$\beta+$	1840	52

Continuous Radiation (^{142}Eu) (continued)

E_{bin}(keV)		$\langle\ \rangle$(keV)	(%)
	IB	4.8	0.141
5000 - 7400	$\beta+$	252	4.71
	IB	0.93	0.0166
	$\Sigma\beta+$		94

$^{142}_{63}$Eu(1.22 *2* min)

Mode: ϵ

Δ: -71590 *100* keV

SpA: 1.080×10^{9} Ci/g

Prod: ^{144}Sm(d,4n); ^{144}Sm(p,3n); alphas on ^{144}Sm

Photons (^{142}Eu)

$\langle\gamma\rangle$=2586 *160* keV

γ_{mode}	γ(keV)	γ(%)
Sm L_ℓ	4.993	0.24 *3*
Sm L_η	5.589	0.104 *10*
Sm L_α	5.633	6.4 *6*
Sm L_β	6.336	5.7 *6*
Sm L_γ	7.201	0.75 *8*
Sm $K_{\alpha2}$	39.522	4.20 *14*
Sm $K_{\alpha1}$	40.118	7.62 *25*
Sm $K_{\beta1}$'	45.379	2.21 *7*
Sm $K_{\beta2}$'	46.819	0.655 *24*
γ E2,M1	201.20 *22*	1.1 *2*
γ E2,M1	273.9 *4*	1.2 *2*
γ	475.1 *4*	0.75 *10*
γ M1	540.34 *16*	5.0 *4*
γ E1	556.94 *15*	87 *3*
γ M1,E2	564.10 *17*	8.3 *4*
γ	581.08 *19*	0.44 *10* ?
γ	629.10 *17*	4.1 *2*
γ	741.54 *18*	1.7 *2*
γ E2	768.38 *19*	100
γ	832.98 *20*	0.42 *9*
γ	848.4 *3*	0.40 *8* ?
γ	887.01 *17*	0.69 *7*
γ	906.75 *20*	0.50 *12*
γ	954.77 *17*	0.58 *8*
γ	982.89 *25*	0.24 *5*
γ	1016.50 *17*	11.0 *6*
γ E2	1023.66 *17*	92 *3*
γ	1151.27 *23*	0.35 *7*
γ	1199.29 *23*	0.39 *10*
γ	1212.4 *3*	0.47 *10*
γ	1342.28 *20*	2.98 *14*
γ	1427.34 *20*	0.78 *15*
γ	1652.52 *21*	0.29 *6*
γ	1700.54 *21*	0.83 *7*
γ	1724.69 *23*	0.12 *4*
γ	1728.9 *3*	0.20 *5*
γ	1839.0 *3*	0.44 *5*
γ	1889.63 *25*	0.15 *3*
γ	1937.65 *23*	0.51 *6*
γ	2258.78 *20*	0.64 *6*

Atomic Electrons (^{142}Eu)

$\langle$e$\rangle$=32.2 *13* keV

e_{bin}(keV)	$\langle$e$\rangle$(keV)	e(%)
7	4.8	69 *8*
8 - 16	0.124	1.42 *13*
17	12.6	73 *5*
22	0.0091	0.041 *3*
23	3.8	17.0 *12*
24	1.04	4.4 *3*

Atomic Electrons (^{142}Eu) (continued)

e_{bin}(keV)	$\langle$e$\rangle$(keV)	e(%)
31 - 45	0.388	1.12 *5*
154 - 201	0.40	0.25 *5*
227 - 274	0.26	0.11 *3*
428 - 475	0.05	~0.012
494	0.38	0.078 *6*
510	1.29	0.253 *10*
517 - 564	0.89	0.17 *3*
573 - 622	0.18	~0.031
627 - 629	0.007	0.0011 *5*
695	0.05	~0.007
722	2.7	0.38 *8*
734 - 768	0.56	0.073 *11*
786 - 833	0.023	0.0029 *12*
840 - 887	0.033	0.0039 *15*
899 - 948	0.022	0.0024 *10*
953 - 976	0.23	~0.023
977	1.82	0.186 *7*
981 - 1024	0.378	0.0372 *22*
1104 - 1152	0.014	0.0012 *5*
1166 - 1212	0.010	0.0009 *4*
1295 - 1342	0.052	0.0040 *18*
1381 - 1427	0.013	0.0009 *4*
1606 - 1699	0.019	0.0011 *3*
1717 - 1792	0.0052	0.00029 *12*
1831 - 1930	0.0082	0.00044 *13*
1931 - 1936	0.00016	8 *4* $\times10^{-6}$
2212 - 2257	0.0063	0.00029 *11*

Continuous Radiation (^{142}Eu)

$\langle\beta+\rangle$=1454 keV;$\langle$IB$\rangle$=8.8 keV

E_{bin}(keV)		$\langle\ \rangle$(keV)	(%)
0 - 10	$\beta+$	3.91×10^{-7}	4.66×10^{-6}
	IB	0.045	
10 - 20	$\beta+$	1.89×10^{-5}	0.000113
	IB	0.045	0.31
20 - 40	$\beta+$	0.00063	0.00190
	IB	0.098	0.33
40 - 100	$\beta+$	0.0374	0.0475
	IB	0.27	0.42
100 - 300	$\beta+$	2.32	1.04
	IB	0.77	0.43
300 - 600	$\beta+$	20.8	4.43
	IB	0.97	0.23
600 - 1300	$\beta+$	202	20.6
	IB	1.9	0.21
1300 - 2500	$\beta+$	762	40.8
	IB	2.6	0.145
2500 - 5000	$\beta+$	467	16.0
	IB	2.1	0.066
5000 - 5052	IB	3.0×10^{-7}	5.9×10^{-9}
	$\Sigma\beta+$		83

$^{142}_{64}$Gd(1.5 *3* min)

Mode: ϵ

Δ: -67190 *400* keV syst

SpA: 8.8×10^{8} Ci/g

Prod: ^{144}Sm(α,6n)

Photons (^{142}Gd)

γ_{mode}	γ(keV)
γ	179 *1*

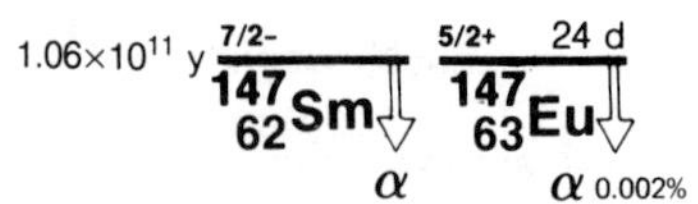

Q_α 2310.6 *11* Q_α 2989 *3*

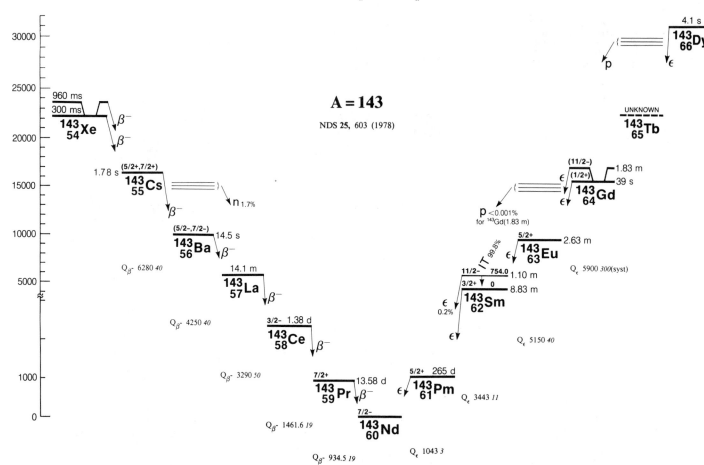

30000
25000
20000
15000
10000
5000
1000
0

A = 143

NDS **25**, 603 (1978)

960 ms
300 ms
$^{143}_{54}$Xe β⁻ β⁻

1.78 s (5/2+,7/2+) $^{143}_{55}$Cs β⁻ n 1.7%

Q_{β^-} 6280 *40*

(5/2-,7/2-) 14.5 s $^{143}_{56}$Ba β⁻

14.1 m $^{143}_{57}$La β⁻

Q_{β^-} 4250 *40*

3/2- 1.38 d $^{143}_{58}$Ce β⁻

Q_{β^-} 3290 *50*

7/2+ 13.58 d $^{143}_{59}$Pr β⁻

Q_{β^-} 1461.6 *19*

7/2- $^{143}_{60}$Nd

Q_{β^-} 934.5 *19*

5/2+ 265 d $^{143}_{61}$Pm ε

Q_ϵ 3443 *11*

Q_ϵ 1043 *3*

11/2- 754.0 1.10 m IT 99.8%
3/2+ 0 8.83 m
$^{143}_{62}$Sm ε 0.2% ε

Q_ϵ 5150 *40*

5/2+ 2.63 m $^{143}_{63}$Eu ε

Q_ϵ 5900 *300*(syst)

(11/2-) 1.83 m
(1/2+) 39 s
$^{143}_{64}$Gd ε ε
p <0.001% for ^{143}Gd(1.83 m)

UNKNOWN $^{143}_{65}$Tb

4.1 s $^{143}_{66}$Dy ε p

$^{143}_{54}$Xe(300 *30* ms)

Mode: β-
SpA: 1.03×10^{11} Ci/g
Prod: fission

$^{143}_{54}$Xe(960 *20* ms)

Mode: β-
SpA: 5.85×10^{10} Ci/g
Prod: fission

$^{143}_{55}$Cs(1.78 *1* s)

Mode: β-, β-n(1.68 *17* %)
Δ: -67790 *40* keV
SpA: 3.671×10^{10} Ci/g
Prod: fission

β-n(avg): 350 *10*
β-n: 125, 180, 225, 310, 350

Photons (^{143}Cs)

⟨γ⟩=690 *10* keV

γ_{mode}	γ(keV)	γ(%)†
Ba L_ℓ	3.954	0.092 *18*
Ba L_η	4.331	0.048 *8*
Ba L_α	4.465	2.6 *5*
Ba L_β	4.939	2.3 *4*
Ba L_γ	5.549	0.26 *5*
Ba $K_{\alpha2}$	31.817	2.02 *19*
Ba $K_{\alpha1}$	32.194	3.7 *4*
$\gamma_{\beta-}$E2	33.61 *17*	0.43 *6*
Ba $K_{\beta1}$'	36.357	1.03 *10*
Ba $K_{\beta2}$'	37.450	0.261 *25*
$\gamma_{\beta-}$	74.03 *19*	0.18 *3*
$\gamma_{\beta-}$	77.73 *17*	0.55 *8*
$\gamma_{\beta-}$[M1]	117.1 *3*	1.85 *9*
$\gamma_{\beta-}$	146.0 *3*	0.55 *8*
$\gamma_{\beta-}$	160.06 *25*	0.68 *10*
$\gamma_{\beta-}$M1,E2	195.06 *14*	18.5 *9*
$\gamma_{\beta-}$	198.75 *20*	0.074 *11*
$\gamma_{\beta-}$	203.4 *3*	0.074 *11*
$\gamma_{\beta-}$M1,E2	228.67 *18*	3.70 *18*
$\gamma_{\beta-}$M1,E2	232.37 *16*	13.9 *7*
$\gamma_{\beta-}$	234.1 *3*	1.48 *22*
$\gamma_{\beta-}$	237.8 *3*	0.61 *9*
$\gamma_{\beta-}$M1,E2	263.09 *16*	5.5 *3*
$\gamma_{\beta-}$M1,E2	272.78 *15*	5.5 *3*
$\gamma_{\beta-}$	298.9 *3*	1.48 *22*
$\gamma_{\beta-}$	302.4 *3*	0.74 *11*
$\gamma_{\beta-}$M1,E2	306.39 *15*	9.6 *5*
$\gamma_{\beta-}$	388.93 *19*	0.92 *14*

Photons (^{143}Cs) (continued)

γ_{mode}	γ(keV)	γ(%)†
$\gamma_{\beta-}$	407.3 *3*	0.37 *6*
$\gamma_{\beta-}$	417.7 *3*	0.18 *3*
$\gamma_{\beta-}$	466.65 *15*	7.4 *4*
$\gamma_{\beta-}$	524.0 *5*	0.18 *3*
$\gamma_{\beta-}$	527.31 *24*	4.25 *21*
$\gamma_{\beta-}$	534.8 *3*	1.85 *9*
$\gamma_{\beta-}$	553.4 *3*	0.37 *6*
$\gamma_{\beta-}$	570.61 *25*	2.03 *10*
$\gamma_{\beta-}$	589.8 *5*	0.37 *6*
$\gamma_{\beta-}$	595.7 *5*	0.37 *6*
$\gamma_{\beta-}$	605.0 *3*	2.40 *12*
$\gamma_{\beta-}$	612.1 *4*	1.11 *17*
$\gamma_{\beta-}$	618.0 *4*	0.28 *4*
$\gamma_{\beta-}$	626.38 *22*	4.25 *21*
$\gamma_{\beta-}$	653.3 *5*	0.28 *4*
$\gamma_{\beta-}$	659.99 *17*	6.8 *3*
$\gamma_{\beta-}$	661.71 *15*	6.8 *3*
$\gamma_{\beta-}$	670.4 *3*	0.37 *6*
$\gamma_{\beta-}$	681.9 *5*	0.37 *6*
$\gamma_{\beta-}$	711.8 *5*	0.092 *14*
$\gamma_{\beta-}$	729.2 *3*	1.85 *9*
$\gamma_{\beta-}$	743.3 *5*	0.37 *6*
$\gamma_{\beta-}$	753.0 *5*	0.18 *3*
$\gamma_{\beta-}$	756.9 *3*	0.55 *8*
$\gamma_{\beta-}$	778.0 *4*	0.74 *11*
$\gamma_{\beta-}$	787.5 *5*	0.18 *3*
$\gamma_{\beta-}$	792.2 *3*	0.92 *14*
$\gamma_{\beta-}$	793.9 *3*	0.37 *6*
$\gamma_{\beta-}$	822.3 *5*	0.74 *11*
$\gamma_{\beta-}$	833.70 *23*	1.85 *9*

Photons (^{143}Cs)
(continued)

γ_{mode}	γ(keV)	γ(%)†
$\gamma_{\beta-}$	837.3 *3*	0.87 *13*
$\gamma_{\beta-}$	855.6 *5*	0.55 *8*
$\gamma_{\beta-}$	868.0 *3*	1.05 *16*
$\gamma_{\beta-}$	871.7 *3*	0.67 *10*
$\gamma_{\beta-}$	911.3 *5*	0.28 *4*
$\gamma_{\beta-}$	969.6 *5*	0.55 *8*
$\gamma_{\beta-}$	985.3 *5*	0.37 *6*
$\gamma_{\beta-}$	1021.3 *5*	0.46 *7*
$\gamma_{\beta-}$	1024.8 *5*	0.18 *3*
$\gamma_{\beta-}$	1083.2 *5*	0.37 *6*
$\gamma_{\beta-}$	1153.2 *5*	0.18 *3*
$\gamma_{\beta-}$	1208.1 *5*	0.48 *7*
$\gamma_{\beta-}$	1219.8 *3*	0.37 *6*
$\gamma_{\beta-}$	1312.0 *5*	0.37 *6*
$\gamma_{\beta-}$	1804.4 *3*	0.92 *14*
$\gamma_{\beta-}$	1908.9 *4*	0.92 *14*
$\gamma_{\beta-}$	1978.2 *3*	2.59 *13*
$\gamma_{\beta-}$	2634.6 *5*	0.74 *11*
$\gamma_{\beta-}$	2648.3 *5*	0.74 *11*
$\gamma_{\beta-}$	2662.5 *5*	0.37 *6*
$\gamma_{\beta-}$	2674.4 *5*	0.55 *8*
$\gamma_{\beta-}$	2683.5 *5*	0.74 *11*

† approximate

Atomic Electrons (^{143}Cs)
$\langle e \rangle$=34.6 *20* keV

e_{bin}(keV)	$\langle e \rangle$(keV)	e(%)
5	1.36	26 *5*
6 - 27	1.04	17 *3*
28	11.9	42 *6*
30 - 31	0.063	0.205 *22*
32	1.27	3.9 *6*
33	2.4	7.4 *11*
34 - 80	2.0	3.4 *8*
109 - 155	0.57	0.48 *12*
158	4.1	2.58 *22*
159 - 194	1.8	0.93 *15*
195	2.29	1.18 *10*
197 - 223	0.33	0.16 *6*
226	1.02	0.45 *4*
227 - 268	1.49	0.62 *7*
269	0.97	0.36 *4*
271 - 306	0.28	0.096 *14*
351 - 389	0.08	0.021 *9*
401 - 429	0.28	~0.06
461 - 497	0.25	0.051 *19*
516 - 565	0.13	0.025 *7*
568 - 617	0.23	0.040 *13*
618 - 665	0.41	0.065 *21*
669 - 716	0.052	0.008 *3*
719 - 757	0.062	0.0084 *21*
772 - 821	0.078	0.010 *3*
822 - 871	0.042	0.0051 *16*
872 - 911	0.0050	0.0006 *3*
932 - 980	0.015	0.0016 *6*
984 - 1025	0.010	0.0010 *4*
1046 - 1083	0.005	0.00052 *25*
1116 - 1153	0.0025	0.00023 *11*
1171 - 1220	0.011	0.0009 *3*
1275 - 1312	0.0044	0.00034 *15*
1767 - 1803	0.008	0.00044 *18*
1871 - 1941	0.025	0.0013 *4*
1972 - 1977	0.0027	0.00014 *5*
2597 - 2682	0.019	0.00071 *11*

Continuous Radiation (^{143}Cs)
$\langle \beta- \rangle$=2721 keV;$\langle$IB$\rangle$=11.1 keV

E_{bin}(keV)		$\langle\ \rangle$(keV)	(%)
0 - 10	β-	0.00165	0.0328
	IB	0.069	
10 - 20	β-	0.00507	0.0337
	IB	0.068	0.47
20 - 40	β-	0.0212	0.070
	IB	0.135	0.47
40 - 100	β-	0.167	0.236
	IB	0.40	0.61
100 - 300	β-	2.26	1.09
	IB	1.23	0.68
300 - 600	β-	12.1	2.61
	IB	1.61	0.38
600 - 1300	β-	107	10.8
	IB	2.9	0.33
1300 - 2500	β-	577	29.9
	IB	3.0	0.169
2500 - 5000	β-	1823	52
	IB	1.69	0.053
5000 - 6280	β-	200	3.76
	IB	0.021	0.00038

$^{143}_{56}$Ba(14.5 *5* s)

Mode: β-

Δ: -74070 *50* keV

SpA: 5.31×10^{9} Ci/g

Prod: fission

Photons (^{143}Ba)
$\langle \gamma \rangle$=226 *5* keV

γ_{mode}	γ(keV)	γ(%)†
La L_{ℓ}	4.121	0.030 *7*
La L_{η}	4.529	0.015 *3*
La L_{α}	4.649	0.82 *19*
La L_{β}	5.160	0.72 *16*
La L_{γ}	5.809	0.087 *19*
γ E2+37%M1	29.84 *4*	0.110 *22*
La $K_{\alpha2}$	33.034	0.35 *4*
La $K_{\alpha1}$	33.442	0.64 *8*
La $K_{\beta1}$'	37.777	0.179 *22*
La $K_{\beta2}$'	38.927	0.047 *6*
γ	174.57 *6*	0.19 *3*
γ (M1)	177.24 *7*	0.29 *5*
γ (M1)	178.56 *7*	0.90 *5*
γ M1,E2	181.67 *4*	0.214 *20*
γ	208.40 *8*	0.22 *3*
γ M1,E2	211.51 *4*	6.5
γ	218.76 *9*	0.35 *3*
γ M1,E2	254.33 *6*	0.68 *5*
γ	257.44 *9*	0.019 *7*
γ M1,E2	261.43 *5*	0.40 *4*
γ	281.95 *6*	0.065 *13*
γ M1,E2	291.27 *4*	2.20 *13*
γ	297.04 *8*	0.24 *3*
γ	310.89 *9*	0.104 *20*
γ	351.81 *6*	0.091 *20*
γ	365.06 *10*	0.045 *13*
γ	367.22 *5*	0.52 *3*
γ	397.63 *4*	0.38 *3*
γ	408.14 *9*	0.136 *20*
γ	424.32 *10*	0.124 *19*
γ M1,E2	431.57 *5*	0.78 *5*
γ	436.00 *6*	0.56 *4*
γ	459.19 *6*	0.071 *20*
γ	465.84 *6*	0.28 *3*
γ	482.90 *5*	0.234 *19*
γ	487.90 *9*	0.045 *13*
γ	544.46 *5*	0.27 *3*
γ	577.04 *9*	0.20 *3*
γ	601.65 *20*	0.15 *3*

Photons (^{143}Ba)
(continued)

γ_{mode}	γ(keV)	γ(%)†
γ	613.24 *6*	0.18 *3*
γ	619.38 *9*	0.21 *3*
γ	633.76 *5*	0.33 *3*
γ	643.08 *5*	0.26 *3*
γ	667.09 *15*	0.189 *20*
γ	669.57 *9*	0.175 *19*
γ	713.52 *5*	0.12 *3*
γ	719.03 *4*	1.07 *7*
γ	764.85 *6*	0.38 *4*
γ	798.79 *4*	3.63 *23*
γ	853.90 *19*	0.35 *7*
γ	859.09 *20*	0.28 *3*
γ	884.24 *11*	0.143 *19*
γ	895.19 *4*	0.94 *7*
γ	925.03 *4*	1.13 *8*
γ	942.09 *6*	0.21 *3*
γ	973.1 *4*	0.085 *20*
γ	980.46 *4*	2.46 *16*
γ	1000.2 *6*	0.091 *20*
γ	1010.30 *4*	2.11 *16*
γ	1037.64 *19*	0.143 *19*
γ	1063.36 *21*	0.058 *13*
γ	1116.66 *5*	0.20 *3*
γ	1196.42 *4*	1.55 *11*
γ	1261.1 *5*	0.065 *20*
γ	1367.14 *11*	0.136 *20*
γ	1378.08 *5*	0.097 *20*
γ	1401.9 *6*	0.071 *20*
γ	1407.92 *5*	0.150 *19*
γ	1443.23 *23*	0.13 *2*
γ	1649.09 *11*	0.22 *3*
γ	2000.89 *11*	0.091 *20*
γ	2016.1 *5*	0.117 *20*
γ	2055.87 *24*	0.208 *20*
γ	2203.4 *9*	0.065 *20*
γ	2297.1 *3*	0.12 *3*
γ	2347.14 *24*	0.19 *3*
γ	2386.9 *9*	0.071 *20*
γ	2392.1 *7*	0.078 *20*

† 25% uncert(syst)

Atomic Electrons (^{143}Ba)
$\langle e \rangle$=8.6 *8* keV

e_{bin}(keV)	$\langle e \rangle$(keV)	e(%)
5	0.42	7.7 *17*
6	0.28	4.8 *10*
24	3.3	14 *3*
26 - 28	0.037	0.135 *18*
29	0.87	3.0 *6*
30 - 38	0.25	0.83 *16*
136 - 172	0.49	0.34 *3*
173	1.3	0.76 *16*
174 - 223	0.62	0.30 *6*
243 - 251	0.025	0.0100 *25*
252	0.25	0.101 *12*
253 - 297	0.109	0.040 *7*
305 - 352	0.040	0.012 *6*
359 - 408	0.115	0.030 *6*
418 - 466	0.043	0.0099 *23*
477 - 506	0.011	~0.0021
538 - 580	0.024	0.0042 *13*
595 - 643	0.030	0.0049 *13*
661 - 708	0.028	~0.0042
712 - 760	0.08	0.011 *6*
763 - 799	0.013	0.0016 *8*
815 - 859	0.032	0.0037 *12*
878 - 925	0.028	0.0031 *12*
934 - 980	0.08	0.008 *3*
994 - 1038	0.009	0.0009 *3*
1057 - 1078	0.0028	~0.00026
1110 - 1157	0.019	~0.0017
1190 - 1222	0.0038	0.00032 *11*
1255 - 1261	0.00012	10 *5* ×10^{-6}
1328 - 1377	0.0051	0.00037 *9*
1378 - 1408	0.0016	0.00012 *5*
1437 - 1443	0.00021	1.4 *6* ×10^{-5}

Atomic Electrons (^{143}Ba) (continued)

e_{bin}(keV)	⟨e⟩(keV)	e(%)
1610 - 1648	0.0022	0.00014 *6*
1962 - 2055	0.0034	0.00017 *4*
2164 - 2258	0.0012	5.5 *17* $\times 10^{-5}$
2291 - 2387	0.0025	0.000106 *24*
2391	1.2 $\times 10^{-5}$	5.2 *25* $\times 10^{-7}$

Continuous Radiation (^{143}Ba)

⟨β-⟩=1708 keV;⟨IB⟩=5.4 keV

E_{bin}(keV)		⟨ ⟩(keV)	(%)
0 - 10	β-	0.00486	0.097
	IB	0.053	
10 - 20	β-	0.0149	0.099
	IB	0.052	0.36
20 - 40	β-	0.062	0.206
	IB	0.103	0.36
40 - 100	β-	0.484	0.68
	IB	0.30	0.46
100 - 300	β-	6.3	3.06
	IB	0.88	0.49
300 - 600	β-	31.5	6.9
	IB	1.07	0.25
600 - 1300	β-	235	24.2
	IB	1.65	0.19
1300 - 2500	β-	836	44.7
	IB	1.15	0.068
2500 - 4250	β-	599	20.1
	IB	0.169	0.0059

$^{143}_{57}$La(14.14 *16* min)

Mode: β-

Δ: -78320 *50* keV

SpA: 9.30×10^{7} Ci/g

Prod: fission

Photons (^{143}La)

⟨γ⟩=93 *3* keV

γ_{mode}	γ(keV)	γ(%)†
γ	300.02 *23*	0.017 *4*
γ	432.64 *21*	0.041 *8*
γ	454.28 *22*	
γ	475.4 *10*	<0.050 ?
γ	476.72 *21*	0.076 *15*
γ	527.36 *20*	0.064 *13*
γ	560.13 *22*	0.088 *18*
γ	581.77 *22*	0.082 *17*
γ	620.48 *18*	1.0
γ	621.40 *16*	0.48 *5*
γ	643.84 *16*	0.72 *7*
γ	774.74 *19*	0.15 *3*
γ	798.10 *19*	0.48 *5*
γ	808.0 *3*	
γ	860.15 *22*	0.051 *10*
γ	919.7 *3*	0.096 *20*
γ	942.6 *5*	0.030 *6*
γ	1053.11 *18*	0.26 *3*
γ	1064.12 *21*	0.028 *6*
γ	1076.48 *17*	0.15 *3*
γ	1086.56 *21*	0.026 *5*
γ	1122.66 *19*	0.14 *3*
γ	1139.48 *24*	0.079 *16*
γ	1146.03 *20*	0.31 *3*
γ	1148.76 *19*	0.40 *4*
γ	1164.86 *21*	0.14 *3*
γ	1167.58 *25*	0.042 *8* ?

Photons (^{143}La) (continued)

γ_{mode}	γ(keV)	γ(%)†
γ	1201.0 *3*	0.097 *20*
γ	1211.76 *24*	0.041 *8*
γ	1240.0 *3*	0.053 *10*
γ	1298.8 *4*	<0.02
γ	1346.7 *3*	
γ	1401.7 *3*	0.023 *5*
γ	1423.32 *25*	0.023 *5*
γ	1475.4 *3*	0.079 *16*
γ	1556.7 *3*	0.42 *4*
γ	1592.4 *3*	0.045 *9*
γ	1611.77 *24*	0.035 *7*
γ	1664.1 *3*	
γ	1707.96 *19*	0.16 *3*
γ	1740.9 *3*	0.049 *10*
γ	1838.0 *3*	0.084 *17*
γ	1876.0 *3*	0.096 *20*
γ	1878.39 *22*	0.087 *18*
γ	1938.03 *19*	0.16 *3*
γ	1961.40 *20*	0.40 *4*
γ	1980.22 *21*	0.14 *3*
γ	2003.5 *3*	0.102 *20*
γ	2056.9 *4*	0.027 *6* ?
γ	2066.0 *3*	0.026 *5*
γ	2384.8 *3*	0.072 *14*
γ	2499.79 *22*	0.29 *3*
γ	2624.9 *3*	0.13 *3*
γ	2709.9 *3*	0.035 *7*
γ	2825.5 *5*	0.051 *10* ?

† 40% uncert(syst)

Continuous Radiation (^{143}La)

⟨β-⟩=1310 keV;⟨IB⟩=3.6 keV

E_{bin}(keV)		⟨ ⟩(keV)	(%)
0 - 10	β-	0.0096	0.191
	IB	0.044	
10 - 20	β-	0.0292	0.195
	IB	0.044	0.30
20 - 40	β-	0.121	0.402
	IB	0.086	0.30
40 - 100	β-	0.92	1.31
	IB	0.25	0.38
100 - 300	β-	11.1	5.4
	IB	0.71	0.40
300 - 600	β-	48.9	10.7
	IB	0.81	0.19
600 - 1300	β-	313	32.6
	IB	1.10	0.128
1300 - 2500	β-	799	44.1
	IB	0.51	0.031
2500 - 3269	β-	137	5.05
	IB	0.0109	0.00042

$^{143}_{58}$Ce(1.375 *8* d)

Mode: β-

Δ: -81615 *4* keV

SpA: 6.64×10^{5} Ci/g

Prod: ^{142}Ce(n,γ)

Photons (^{143}Ce)

⟨γ⟩=274 *14* keV

γ_{mode}	γ(keV)	γ(%)†
Pr L_ℓ	4.453	0.150 *25*
Pr L_η	4.929	0.061 *9*
Pr L_α	5.031	4.1 *6*
Pr L_β	5.619	3.6 *6*
Pr L_γ	6.410	0.51 *9*
Pr $K_{\alpha2}$	35.550	18.2 *19*
Pr $K_{\alpha1}$	36.026	33 *4*
Pr $K_{\beta1}'$	40.720	9.4 *10*
Pr $K_{\beta2}'$	41.981	2.6 *3*
γ M1+0.2%E2	57.3817 *10*	11.8 *13*
γ E2	122.50 *16*	0.0105 *25*
γ (E2)	139.73 *3*	0.10 *3*
γ	168.7 *7*	<0.6
γ M1,E2	215.83 *8*	0.206 *21*
γ M1(+E2)	231.58 *3*	2.02 *21*
γ M1+37%E2	293.273 *18*	42
γ	337.5 *3*	0.29 *3*
γ E2	350.655 *18*	3.4 *3*
γ M1(+E2)	371.30 *4*	0.021 *4*
γ	376.6 *7*	0.063 *8*
γ E1	389.49 *18*	0.029 *8*
γ	400.3 *7*	0.097 *8*
γ	412.8 *7*	0.030 *3*
γ M1(+E2)	433.00 *3*	0.13 *3*
γ	438.52 *15*	0.118 *13*
γ E2,M1	447.40 *8*	0.067 *13*
γ M1	490.38 *3*	1.97 *21*
γ	497.88 *16*	0.034 *8*
γ	549.2 *3*	0.071 *8*
γ (E2)	556.88 *21*	0.025 *8*
γ M1+E2	587.13 *8*	0.24 *4*
γ	657.6 *3*	0.0126 *13*
γ M1+E2	664.58 *4*	5.3 *6*
γ	709.63 *15*	0.0084 *21*
γ (M1)	721.96 *4*	5.1 *5*
γ	791.16 *16*	0.017 *4*
γ	802.3 *7*	0.037 *4*
γ	806.48 *23*	0.025 *8*
γ M1	809.82 *14*	0.025 *8*
γ M1	880.40 *8*	0.92 *8*
γ	891.57 *21*	0.013 *4* ?
γ E2	937.79 *8*	0.034 *8*
γ M1	1002.90 *15*	0.067 *13*
γ	1031.29 *21*	0.017 *4*
γ	1047.11 *25*	~0.008
γ M1,E2	1060.28 *15*	0.034 *8*
γ E2,M1	1103.09 *14*	0.37 *5*
γ	1314.4 *7*	0.034 *3*
γ	1324.56 *21*	0.013 *4*
γ	1340.38 *25*	0.0038 *13*

† 9.5% uncert(syst)

Atomic Electrons (^{143}Ce)

⟨e⟩=30.0 *14* keV

e_{bin}(keV)	⟨e⟩(keV)	e(%)
6	2.9	47 *7*
7	0.36	5.2 *8*
15	10.2	66 *7*
28 - 41	1.79	5.7 *4*
51	4.8	9.4 *10*
56	1.11	1.99 *22*
57 - 98	0.36	0.61 *6*
116 - 163	0.13	~0.09
167 - 216	0.44	0.24 *3*
225 - 232	0.11	0.047 *13*
251	5.6	2.21 *22*
286	0.77	0.27 *3*
287 - 336	0.80	0.268 *15*
337 - 384	0.079	0.0229 *18*
388 - 437	0.024	0.0060 *13*
438 - 484	0.160	0.035 *3*
489 - 515	0.0092	0.0018 *4*
542 - 587	0.013	0.0024 *6*
616 - 665	0.23	0.037 *8*
668 - 716	0.238	0.035 *3*

Atomic Electrons (^{143}Ce) (continued)

e_{bin}(keV)	⟨e⟩(keV)	e(%)
720 - 768	0.0105	0.00143 *13*
784 - 810	0.00047	5.9 *13* $\times10^{-5}$
838 - 886	0.034	0.0041 *3*
890 - 938	0.00075	8.4 *18* $\times10^{-5}$
961 - 1005	0.0025	0.00026 *4*
1018 - 1061	0.0080	0.00076 *17*
1096 - 1103	0.00122	0.000112 *22*
1272 - 1319	0.00071	5.5 *19* $\times10^{-5}$
1323 - 1340	1.3 $\times10^{-5}$	10 *4* $\times10^{-7}$

Continuous Radiation (^{143}Ce)

⟨β-⟩=409 keV;⟨IB⟩=0.50 keV

E_{bin}(keV)		⟨ ⟩(keV)	(%)
0 - 10	β-	0.071	1.42
	IB	0.019	
10 - 20	β-	0.208	1.39
	IB	0.018	0.124
20 - 40	β-	0.81	2.70
	IB	0.034	0.118
40 - 100	β-	5.6	8.0
	IB	0.088	0.137
100 - 300	β-	55	27.4
	IB	0.19	0.111
300 - 600	β-	153	34.7
	IB	0.120	0.030
600 - 1300	β-	194	24.4
	IB	0.035	0.0048
1300 - 1404	β-	0.75	0.056
	IB	4.5 $\times10^{-6}$	3.5 $\times10^{-7}$

$^{143}_{59}$Pr(13.58 *3* d)

Mode: β-
Δ: -83077 *4* keV
SpA: 6.724$\times10^{4}$ Ci/g
Prod: daughter ^{143}Ce; fission

Photons (^{143}Pr)

γ_{mode}	γ(keV)	γ(%)
γ E2	741.98 *4*	1.2 *4* $\times10^{-6}$

Continuous Radiation (^{143}Pr)

⟨β-⟩=315 keV;⟨IB⟩=0.31 keV

E_{bin}(keV)		⟨ ⟩(keV)	(%)
0 - 10	β-	0.080	1.60
	IB	0.0151	
10 - 20	β-	0.242	1.62
	IB	0.0144	0.100
20 - 40	β-	0.98	3.28
	IB	0.027	0.093
40 - 100	β-	7.1	10.1
	IB	0.067	0.104
100 - 300	β-	69	34.5
	IB	0.125	0.075
300 - 600	β-	167	38.5
	IB	0.054	0.0137
600 - 935	β-	71	10.3
	IB	0.0039	0.00062

$^{143}_{60}$Nd(stable)

Δ: -84012 *3* keV
%: 12.18 *5*

$^{143}_{61}$Pm(265 *7* d)

Mode: ϵ
Δ: -82969 *4* keV
SpA: 3446 Ci/g
Prod: daughter ^{143}Sm; ^{141}Pr(α,2n); ^{143}Nd(p,n)

Photons (^{143}Pm)

⟨γ⟩=316 *18* keV

γ_{mode}	γ(keV)	γ(%)†
Nd L_ℓ	4.633	0.193 *19*
Nd L_η	5.146	0.075 *6*
Nd L_α	5.228	5.2 *3*
Nd L_β	5.855	4.7 *4*
Nd L_γ	6.697	0.68 *7*
Nd $K_{\alpha2}$	36.847	22.0 *6*
Nd $K_{\alpha1}$	37.361	40.0 *12*
Nd $K_{\beta1}$'	42.240	11.4 *3*
Nd $K_{\beta2}$'	43.562	3.33 *11*
γ E2	741.98 *4*	38.5

† 6.2% uncert(syst)

Atomic Electrons (^{143}Pm)

⟨e⟩=7.3 *3* keV

e_{bin}(keV)	⟨e⟩(keV)	e(%)
6	2.30	37 *4*
7	1.71	25 *3*
29	0.162	0.55 *6*
30	0.45	1.50 *15*
31	0.69	2.24 *23*
35	0.229	0.65 *7*
36	0.41	1.14 *12*
37	0.102	0.27 *3*
40	0.0077	0.0190 *19*
41	0.050	0.122 *12*
42	0.0237	0.057 *6*
698	1.00	0.143 *9*
735	0.144	0.0195 *13*
736	0.0102	0.00139 *9*
740	0.0264	0.00357 *23*
741	0.0066	0.00090 *6*
742	0.0077	0.00104 *7*

Continuous Radiation (^{143}Pm)

⟨IB⟩=0.25 keV

E_{bin}(keV)		⟨ ⟩(keV)	(%)
10 - 20	IB	0.00025	0.00154
20 - 40	IB	0.074	0.21
40 - 100	IB	0.042	0.096
100 - 300	IB	0.022	0.0111
300 - 600	IB	0.066	0.0149
600 - 1043	IB	0.042	0.0060

$^{143}_{62}$Sm(8.83 *2* min)

Mode: ϵ
Δ: -79526 *11* keV
SpA: 1.488$\times10^{8}$ Ci/g
Prod: ^{142}Nd(α,3n); ^{144}Sm(n,2n); ^{144}Sm(γ,n); ^{142}Nd(^{3}He,2n)

Photons (^{143}Sm)

⟨γ⟩=62.7 *23* keV

γ_{mode}	γ(keV)	γ(%)†
Pm L_ℓ	4.809	0.112 *11*
Pm L_η	5.363	0.042 *3*
Pm L_α	5.430	3.01 *19*
Pm L_β	6.097	2.70 *23*
Pm L_γ	6.979	0.39 *4*
Pm $K_{\alpha2}$	38.171	12.2 *4*
Pm $K_{\alpha1}$	38.725	22.2 *6*
Pm $K_{\beta1}$'	43.793	6.40 *19*
Pm $K_{\beta2}$'	45.183	1.89 *6*
γ [M1+E2]	271.90 *18*	0.28 *3*
γ	458.49 *23*	0.045 *5*
γ	797.2 *5*	0.011 *4*
γ	1056.48 *18*	1.75
γ [E2]	1173.1 *6*	0.38 *4*
γ	1243.07 *16*	0.200 *21*
γ	1342.0 *5*	0.026 *4*
γ	1403.22 *20*	0.30 *3*
γ	1514.97 *16*	0.61 *7*
γ	1544.37 *18*	0.045 *5*
γ	1752.8 *5*	0.0030 *10*
γ	1808.5 *3*	0.0030 *10*
γ	1816.27 *23*	0.021 *4*
γ	1853.7 *5*	0.0049 *12*
γ	2000 *1*	<0.0018 ?
γ	2009.0 *7*	0.0114 *18* ?
γ	2080.4 *3*	0.0105 *18*
γ	2171.5 *9*	0.0149 *3*
γ	2190.6 *14*	0.0032 *7*
γ	2344 *2*	0.0035 *9*
γ	2443.4 *9*	0.0021 *7*
γ	2462.4 *14*	~0.0014
γ	2634.5 *7*	0.0044 *9*
γ	2781.4 *10*	~0.0007
γ	2906.4 *7*	0.0012 *4*
γ	3053.3 *10*	0.00052 *17*

† 10% uncert(syst)

Atomic Electrons (^{143}Sm)

⟨e⟩=3.58 *18* keV

e_{bin}(keV)	⟨e⟩(keV)	e(%)
6	1.31	20.4 *21*
7	0.95	13.4 *14*
30	0.089	0.29 *3*
31	0.247	0.80 *8*
32	0.37	1.16 *12*
36	0.096	0.26 *3*
37	0.220	0.59 *6*
38 - 71	0.140	0.347 *24*
109 - 128	0.016	0.013 *6*
166 - 185	0.007	0.0040 *17*
223 - 272	0.057	0.024 *5*
413 - 458	0.0030	~0.0007
509 - 554	0.0009	~0.00017
739 - 785	0.0017	0.00023 *11*
790 - 797	5.0 $\times10^{-5}$	~6$\times10^{-6}$
856 - 901	0.0013	~0.00015
979 - 1024	0.033	~0.0033
1049 - 1098	0.007	0.0006 *3*
1102 - 1149	0.0071	0.00063 *7*
1153 - 1198	0.0041	0.00035 *14*
1236 - 1243	0.00052	4.2 *19* $\times10^{-5}$
1297 - 1342	0.00042	3.3 *15* $\times10^{-5}$
1358 - 1403	0.0046	0.00034 *16*

Atomic Electrons (^{143}Sm) (continued)

e_{bin}(keV)	⟨e⟩(keV)	e(%)
1470 - 1543	0.009	0.0006 *3*
1708 - 1807	0.00028	1.6 *6* $\times10^{-5}$
1809 - 1852	8.9 $\times10^{-5}$	4.9 *22* $\times10^{-6}$
1955 - 2035	0.00022	1.1 *3* $\times10^{-5}$
2073 - 2170	0.00018	9 *3* $\times10^{-6}$
2183 - 2189	4.2 $\times10^{-6}$	1.9 *8* $\times10^{-7}$
2299 - 2398	4.7 $\times10^{-5}$	2.0 *7* $\times10^{-6}$
2417 - 2461	1.5 $\times10^{-5}$	6 *3* $\times10^{-7}$
2589 - 2633	3.5 $\times10^{-5}$	1.4 *5* $\times10^{-6}$
2736 - 2780	5.4 $\times10^{-6}$	~2 $\times10^{-7}$
2861 - 2905	9.1 $\times10^{-6}$	3.2 *12* $\times10^{-7}$
3008 - 3052	3.7 $\times10^{-6}$	1.2 *5* $\times10^{-7}$

Continuous Radiation (^{143}Sm)

⟨β+⟩=497 keV;⟨IB⟩=5.9 keV

E_{bin}(keV)		⟨ ⟩(keV)	(%)
0 - 10	β+	7.0 $\times10^{-7}$	8.3 $\times10^{-6}$
	IB	0.019	
10 - 20	β+	3.31 $\times10^{-5}$	0.000198
	IB	0.018	0.126
20 - 40	β+	0.00107	0.00323
	IB	0.069	0.22
40 - 100	β+	0.061	0.078
	IB	0.137	0.24
100 - 300	β+	3.56	1.59
	IB	0.31	0.17
300 - 600	β+	29.0	6.2
	IB	0.45	0.103
600 - 1300	β+	213	22.3
	IB	1.45	0.156
1300 - 2500	β+	251	15.4
	IB	2.9	0.159
2500 - 3443	IB	0.58	0.021
	$\Sigma\beta$+		46

$^{143}_{62}$Sm(1.10 *3* min)

Mode: IT(99.76 *6* %), ϵ(0.24 *6* %)
Δ: -78772 *11* keV
SpA: 1.19$\times10^{9}$ Ci/g
Prod: ^{142}Nd(^{3}He,2n); ^{144}Sm(n,2n); ^{144}Sm(γ,n); ^{144}Sm(p,pn)

Photons (^{143}Sm)

⟨γ⟩=684 *27* keV

γ_{mode}	γ(keV)	γ(%)†
Pm L_ℓ	4.809	0.00034 *4*
Sm L_ℓ	4.993	0.0210 *23*
Pm L_η	5.363	0.000130 *14*
Pm L_α	5.430	0.0092 *9*
Sm L_η	5.589	0.0080 *7*
Sm L_α	5.633	0.56 *4*
Pm L_β	6.097	0.0082 *9*
Sm L_β	6.340	0.52 *5*
Pm L_γ	6.978	0.00118 *15*
Sm L_γ	7.273	0.077 *9*
Pm $K_{\alpha2}$	38.171	0.037 *3*
Pm $K_{\alpha1}$	38.725	0.067 *5*
Sm $K_{\alpha2}$	39.522	2.1 *1*
Sm $K_{\alpha1}$	40.118	3.81 *19*
Pm $K_{\beta1}$'	43.793	0.0194 *14*
Pm $K_{\beta2}$'	45.183	0.0057 *4*

Photons (^{143}Sm) (continued)

γ_{mode}	γ(keV)	γ(%)†
Sm $K_{\beta1}$'	45.379	1.10 *5*
Sm $K_{\beta2}$'	46.819	0.328 *17*
γ_ϵ[M1+E2]	271.90 *18*	0.190 *8*
γ_ϵM2	690.1 *11*	0.202
γ_{IT} M4	754.01 *16*	90
γ_ϵE3	961.9 *11*	0.032 *7*

† uncert(syst): 25% for ϵ, <0.1% for IT

Atomic Electrons (^{143}Sm)

⟨e⟩=71.2 *25* keV

e_{bin}(keV)	⟨e⟩(keV)	e(%)
6 - 45	0.62	6.6 *6*
227 - 272	0.039	0.017 *7*
645 - 690	0.0337	0.00517 *20*
707	55.7	7.9 *4*
746	9.1	1.22 *5*
747	2.41	0.322 *14*
752 - 754	3.32	0.441 *15*
917 - 962	0.0017	0.00018 *3*

Continuous Radiation (^{143}Sm)

⟨β+⟩=0.96 keV;⟨IB⟩=0.0129 keV

E_{bin}(keV)		⟨ ⟩(keV)	(%)
0 - 10	β+	1.79 $\times10^{-9}$	2.14 $\times10^{-8}$
	IB	3.7 $\times10^{-5}$	
10 - 20	β+	8.4 $\times10^{-8}$	5.06 $\times10^{-7}$
	IB	3.6 $\times10^{-5}$	0.00025
20 - 40	β+	2.72 $\times10^{-6}$	8.3 $\times10^{-6}$
	IB	0.000158	0.00049
40 - 100	β+	0.000156	0.000199
	IB	0.00030	0.00052
100 - 300	β+	0.0090	0.00403
	IB	0.00061	0.00034
300 - 600	β+	0.072	0.0154
	IB	0.00095	0.00022
600 - 1300	β+	0.487	0.051
	IB	0.0034	0.00037
1300 - 2500	β+	0.391	0.0250
	IB	0.0066	0.00037
2500 - 3235	IB	0.00081	3.0 $\times10^{-5}$
	$\Sigma\beta$+		0.096

$^{143}_{63}$Eu(2.63 *5* min)

Mode: ϵ
Δ: -74380 *40* keV
SpA: 4.99$\times10^{8}$ Ci/g
Prod: ^{144}Sm(d,3n); ^{144}Sm(p,2n)

Photons (^{143}Eu)

⟨γ⟩=365 *9* keV

γ_{mode}	γ(keV)	γ(%)†
Sm L_ℓ	4.993	0.068 *8*
Sm L_η	5.589	0.025 *3*
Sm L_α	5.633	1.82 *16*
Sm L_β	6.342	1.65 *18*
Sm L_γ	7.269	0.24 *3*
Sm $K_{\alpha2}$	39.522	6.9 *3*

Photons (^{143}Eu) (continued)

γ_{mode}	γ(keV)	γ(%)†
Sm $K_{\alpha1}$	40.118	12.6 *5*
Sm $K_{\beta1}$'	45.379	3.64 *14*
Sm $K_{\beta2}$'	46.819	1.08 *5*
γ[M1+E2]	107.71 *7*	2.09 *23*
γ[E1]	203.18 *17*	0.143 *22*
γ	429.58 *13*	0.114 *15*
γ	458.70 *15*	0.043 *15*
γ	551.4 *3*	0.060 *15*
γ[E2]	556.56 *20*	0.067 *15*
γ	607.72 *13*	0.262 *15*
γ	691.17 *22*	0.058 *15*
γ	733.23 *20*	0.063 *15*
γ M4	754.01 *16*	*
γ[M1+E2]	798.2 *3*	0.075 *23*
γ	805.32 *13*	1.01 *5*
γ[E2]	999.68 *10*	0.548 *23*
γ[M1+E2]	1107.39 *9*	7.5
γ[M1+E2]	1163.40 *22*	0.028 *8*
γ	1272.2 *3*	0.060 *11*
γ[E2]	1369.27 *16*	0.90 *3*
γ	1429.26 *12*	0.353 *23*
γ	1458.38 *13*	1.16 *5*
γ	1536.96 *12*	3.29 *9*
γ	1566.09 *13*	0.600 *23*
γ	1578.52 *19*	0.035 *8*
γ	1607.40 *12*	1.01 *5*
γ	1668.84 *20*	0.097 *15*
γ	1680.9 *3*	0.024 *6*
γ	1715.10 *12*	0.173 *22*
γ	1778.4 *3*	0.026 *11*
γ	1805.00 *12*	1.67 *9*
γ	1912.70 *12*	2.13 *11*
γ	1955.54 *25*	0.060 *19*
γ	1962.64 *15*	0.232 *15*
γ	2001.6 *5*	0.029 *11*
γ	2070.34 *15*	0.51 *3*
γ	2102.50 *17*	0.95 *5*
γ	2131.7 *3*	0.036 *5*
γ[E2]	2167.43 *25*	0.052 *5*
γ	2228.12 *22*	0.40 *5*
γ[E2]	2270.78 *22*	0.120 *15*
γ	2280.0 *3*	0.026 *5*
γ	2303.04 *24*	0.034 *5*
γ	2312.4 *3*	0.022 *3*
γ	2323.6 *3*	0.035 *5*
γ	2410.74 *24*	0.017 *3*
γ	2450.44 *22*	0.035 *5*
γ	2479.84 *22*	0.085 *8*
γ	2558.14 *22*	0.034 *5*
γ	2570.8 *3*	0.016 *4*
γ	2578.19 *20*	0.011 *5*
γ	2587.54 *22*	0.048 *5*
γ	2646.8 *3*	0.023 *4*
γ	2685.89 *19*	0.041 *5*
γ	2708.9 *4*	0.016 *3*
γ	2842.1 *6*	0.011 *3*
γ	3031.2 *6*	0.014 *3*
γ	3154.0 *6*	0.011 *3*
γ	3342.9 *7*	0.0097 *23*
γ	4093 *3*	0.0053 *15*

† 9.3% uncert(syst)
* with ^{143}Sm(1.10 min)

Atomic Electrons (^{143}Eu)

⟨e⟩=5.0 *6* keV

e_{bin}(keV)	⟨e⟩(keV)	e(%)
7	1.20	17.3 *22*
8	0.150	1.94 *21*
31 - 32	0.187	0.59 *5*
33	0.203	0.61 *6*
37 - 45	0.250	0.64 *4*
61	1.37	2.3 *4*
100	0.6	~0.6
101	0.3	~0.3
106	0.21	~0.20

Atomic Electrons (^{143}Eu)
(continued)

e_{bin}(keV)	⟨e⟩(keV)	e(%)
107 - 156	0.06	~0.06
195 - 203	0.00172	0.00088 *9*
383 - 430	0.011	0.0029 *14*
451 - 459	0.0005	~0.00012
505 - 551	0.0065	0.0013 *4*
555 - 601	0.012	~0.0021
606 - 644	0.0024	~0.00037
683 - 732	0.0027	~0.00038
733 - 758	0.030	~0.0040
790 - 805	0.006	0.0007 *3*
953 - 1000	0.0132	0.00137 *7*
1061	0.18	0.017 *4*
1100 - 1117	0.032	0.0029 *6*
1156 - 1163	0.00011	9 *3* $\times10^{-6}$
1225 - 1272	0.0011	~9 $\times10^{-5}$
1322 - 1369	0.0159	0.00120 *5*
1382 - 1429	0.021	0.0015 *6*
1451 - 1536	0.059	0.0040 *15*
1558 - 1634	0.017	0.0011 *4*
1661 - 1758	0.021	0.0012 *5*
1771 - 1866	0.024	0.0013 *5*
1905 - 2000	0.0072	0.00037 *10*
2024 - 2121	0.016	0.00080 *21*
2124 - 2221	0.0041	0.00019 *7*
2224 - 2322	0.0026	0.000116 *12*
2364 - 2449	0.0011	4.7 *12* $\times10^{-5}$
2472 - 2571	0.00100	4.0 *8* $\times10^{-5}$
2576 - 2662	0.00066	2.5 *6* $\times10^{-5}$
2678 - 2708	6.7 $\times10^{-5}$	2.5 *6* $\times10^{-6}$
2795 - 2841	8.4 $\times10^{-5}$	3.0 *12* $\times10^{-6}$
2984 - 3030	0.00011	3.6 *12* $\times10^{-6}$
3107 - 3153	8.2 $\times10^{-5}$	2.6 *10* $\times10^{-6}$
3296 - 3342	6.8 $\times10^{-5}$	2.1 *7* $\times10^{-6}$
4046 - 4092	3.1 $\times10^{-5}$	8 *3* $\times10^{-7}$

Continuous Radiation (^{143}Eu)
⟨β+⟩=1287 keV;⟨IB⟩=10.4 keV

E_{bin}(keV)		⟨ ⟩(keV)	(%)
0 - 10	β+	3.45 $\times10^{-7}$	4.11 $\times10^{-6}$
	IB	0.040	
10 - 20	β+	1.67 $\times10^{-5}$	0.000100
	IB	0.039	0.27
20 - 40	β+	0.00055	0.00167
	IB	0.094	0.31
40 - 100	β+	0.0329	0.0418
	IB	0.25	0.39
100 - 300	β+	2.03	0.91
	IB	0.68	0.38
300 - 600	β+	18.1	3.86
	IB	0.90	0.21
600 - 1300	β+	173	17.6
	IB	2.0	0.22
1300 - 2500	β+	641	34.3
	IB	3.4	0.19
2500 - 5000	β+	453	15.3
	IB	3.0	0.095
5000 - 5150	IB	0.00049	9.8 $\times10^{-6}$
	Σβ+		72

$^{143}_{64}$Gd(39 *2* s)

Mode: ε
Δ: -68480 *300* keV syst
SpA: 2.01$\times10^{9}$ Ci/g
Prod: ^{144}Sm(α,5n)

Photons (^{143}Gd)
⟨γ⟩=365 *10* keV

γ_{mode}	γ(keV)	γ(%)†
Eu L_ℓ	5.177	0.075 *9*
Eu L_η	5.817	0.027 *3*
Eu L_α	5.843	1.98 *16*
Eu L_β	6.594	1.80 *19*
Eu L_γ	7.569	0.27 *3*
Eu $K_{\alpha2}$	40.902	7.3 *4*
Eu $K_{\alpha1}$	41.542	13.1 *6*
Eu $K_{\beta1}$'	46.999	3.82 *18*
Eu $K_{\beta2}$'	48.496	1.16 *6*
γ[M1+E2]	204.88 *5*	19.4 *14*
γ(M1)	258.92 *3*	74.8
γ[E2]	463.79 *5*	9.9 *8*
γ	554.10 *10*	~0.7
γ	813.02 *10*	5.4 *5*
γ	1284.3 *4*	1.0 *4*
γ	1464.9 *4*	0.9 *3*

† 1.3% uncert(syst)

$^{143}_{64}$Gd(1.83 *3* min)

Mode: ε, εp(<0.001 %)
Δ: -68480 *300* keV syst
SpA: 7.16$\times10^{8}$ Ci/g
Prod: ^{144}Sm(α,5n)

εp: 2500-5000 range

Photons (^{143}Gd)
⟨γ⟩=1417 *20* keV

γ_{mode}	γ(keV)	γ(%)†
Eu L_ℓ	5.177	0.25 *3*
Eu L_η	5.817	0.091 *8*
Eu L_α	5.843	6.6 *5*
Eu L_β	6.593	6.1 *6*
Eu L_γ	7.578	0.92 *11*
Eu $K_{\alpha2}$	40.902	23.6 *13*
Eu $K_{\alpha1}$	41.542	42.5 *24*
Eu $K_{\beta1}$'	46.999	12.4 *7*
Eu $K_{\beta2}$'	48.496	3.74 *22*
γM2	117.54 *5*	6.5 *5*
γ	130.94 *9*	0.37 *6*
γM1	210.92 *6*	1.10 *8*
γM1	271.87 *4*	84.3
γM1+E2	303.73 *16*	1.01 *8*
γE3	389.40 *6*	3.46 *25*
γ	428.0 *3*	0.25 *8*
γ	497.27 *12*	0.59 *8*
γ	545.12 *10*	0.59 *8*
γM1+E2	587.92 *4*	15.7 *11*
γ	590.7 *3*	~0.34
γ	594.04 *9*	0.58 *5*
γ	625.10 *9*	1.18 *8*
γM1+E2	668.03 *5*	9.7 *7*
γ	698.73 *16*	0.38 *5*
γ	776.68 *12*	0.84 *8*
γE2	785.44 *8*	5.5 *4*
γE2	798.84 *6*	10.7 *8*
γE2	824.36 *14*	5.0 *3*
γ	830.32 *8*	0.54 *5*
γ	835.44 *17*	0.56 *5*
γ	846.0 *4*	0.25 *8*
γM1+E2	890.57 *8*	1.77 *17*
γE2	906.88 *8*	2.11 *25*
γE2	916.50 *7*	4.3 *3*
γ	926.0 *4*	0.55 *8*
γ(M1+E2)	984.87 *7*	2.02 *17*
γ	993.0 *5*	0.46 *5*
γ	1008.22 *6*	1.35 *8*
γ	1041.24 *7*	3.03 *25*
γ	1059.23 *16*	0.84 *8*

Photons (^{143}Gd)
(continued)

γ_{mode}	γ(keV)	γ(%)†
γ	1087.16 *14*	1.60 *8*
γ	1139.16 *9*	0.81 *8*
γ	1158.12 *11*	0.56 *8*
γ	1162.63 *15*	0.76 *8*
γ	1196.83 *15*	0.89 *8*
γ	1213.02 *9*	0.56 *8*
γ	1219.14 *6*	4.1 *3*
γ	1225.7 *3*	0.25 *8*
γ	1231.7 *5*	0.67 *8*
γ	1276.81 *17*	0.25 *8*
γ	1293.2 *3*	0.84 *8*
γ	1298.08 *14*	0.35 *6*
γ	1329.2 *8*	0.25 *8*
γ	1354.3 *3*	0.51 *8*
γ	1373.55 *14*	1.10 *8*
γ	1386.62 *11*	1.26 *8*
γ	1404.49 *11*	2.87 *25*
γ	1489.7 *3*	0.66 *8*
γ	1503.33 *15*	1.18 *8*
γ	1629.16 *7*	1.94 *17*
γ	1633.3 *8*	~0.08
γ	1675.6 *1*	0.48 *8*
γ	1702.58 *10*	1.10 *8*
γ	1746.69 *9*	0.76 *8*
γ	1793.14 *9*	2.61 *17*
γ	1807.06 *6*	7.7 *6*
γ	1820.12 *9*	3.03 *25*
γ	1886.00 *14*	0.76 *8*
γ	2338.7 *8*	0.25 *8*

† 0.59% uncert(syst)

Atomic Electrons (^{143}Gd)
⟨e⟩=87 *3* keV

e_{bin}(keV)	⟨e⟩(keV)	e(%)
7	2.7	39 *5*
8 - 47	4.2	32 *3*
69	34.4	50 *4*
82	0.13	~0.16
109	10.1	9.2 *7*
110 - 111	2.88	2.61 *15*
116	3.14	2.71 *22*
117 - 162	1.58	1.24 *9*
173 - 211	0.22	0.11 *3*
223	18.8	8.40 *24*
251 - 259	0.18	0.070 *17*
264	3.08	1.17 *3*
265 - 304	0.957	0.353 *9*
341 - 389	1.11	0.313 *17*
420 - 449	0.037	~0.008
489 - 538	0.041	~0.008
539 - 588	1.2	0.21 *5*
589 - 625	0.47	0.076 *23*
650 - 699	0.106	0.016 *4*
728 - 777	0.62	0.082 *4*
778 - 827	0.138	0.0173 *19*
828 - 877	0.236	0.027 *3*
883 - 926	0.045	0.0050 *4*
936 - 985	0.111	0.012 *3*
986 - 1034	0.10	0.010 *4*
1039 - 1087	0.044	0.0042 *20*
1091 - 1139	0.059	0.0053 *15*
1142 - 1190	0.12	0.010 *4*
1195 - 1232	0.022	0.0018 *6*
1245 - 1293	0.028	0.0022 *7*
1296 - 1338	0.045	0.0034 *11*
1346 - 1386	0.050	0.0037 *17*
1396 - 1441	0.017	0.0012 *4*
1455 - 1502	0.021	0.0014 *6*
1581 - 1674	0.050	0.0031 *9*
1695 - 1792	0.17	0.010 *3*
1799 - 1885	0.030	0.0016 *4*
2290 - 2337	0.0026	0.00011 *5*

Continuous Radiation (^{143}Gd)

$\langle\beta+\rangle$=1103 keV;$\langle$IB$\rangle$=9.7 keV

E_{bin}(keV)		$\langle\ \rangle$(keV)	(%)
0 - 10	β+	3.56×10^{-7}	4.23×10^{-6}
	IB	0.035	
10 - 20	β+	1.76×10^{-5}	0.000105
	IB	0.035	0.24
20 - 40	β+	0.00059	0.00180
	IB	0.087	0.29
40 - 100	β+	0.0362	0.0460
	IB	0.23	0.37
100 - 300	β+	2.27	1.01
	IB	0.60	0.33
300 - 600	β+	20.2	4.31
	IB	0.79	0.18
600 - 1300	β+	188	19.2
	IB	1.8	0.20
1300 - 2500	β+	589	32.1
	IB	3.5	0.19
2500 - 5000	β+	305	10.2
	IB	2.6	0.083
5000 - 5611	IB	0.0123	0.00024
	$\Sigma\beta$+		67

$^{143}_{66}$Dy(4.1 *3* s)

Mode: ϵ, ϵp
SpA: 1.77×10^{10} Ci/g
Prod: ^{92}Mo(^{58}Ni,α2pn); ^{92}Mo(^{56}Fe,αn)

ϵp: 2000-6400 range

A = 144

NDS **27**, 97 (1979)

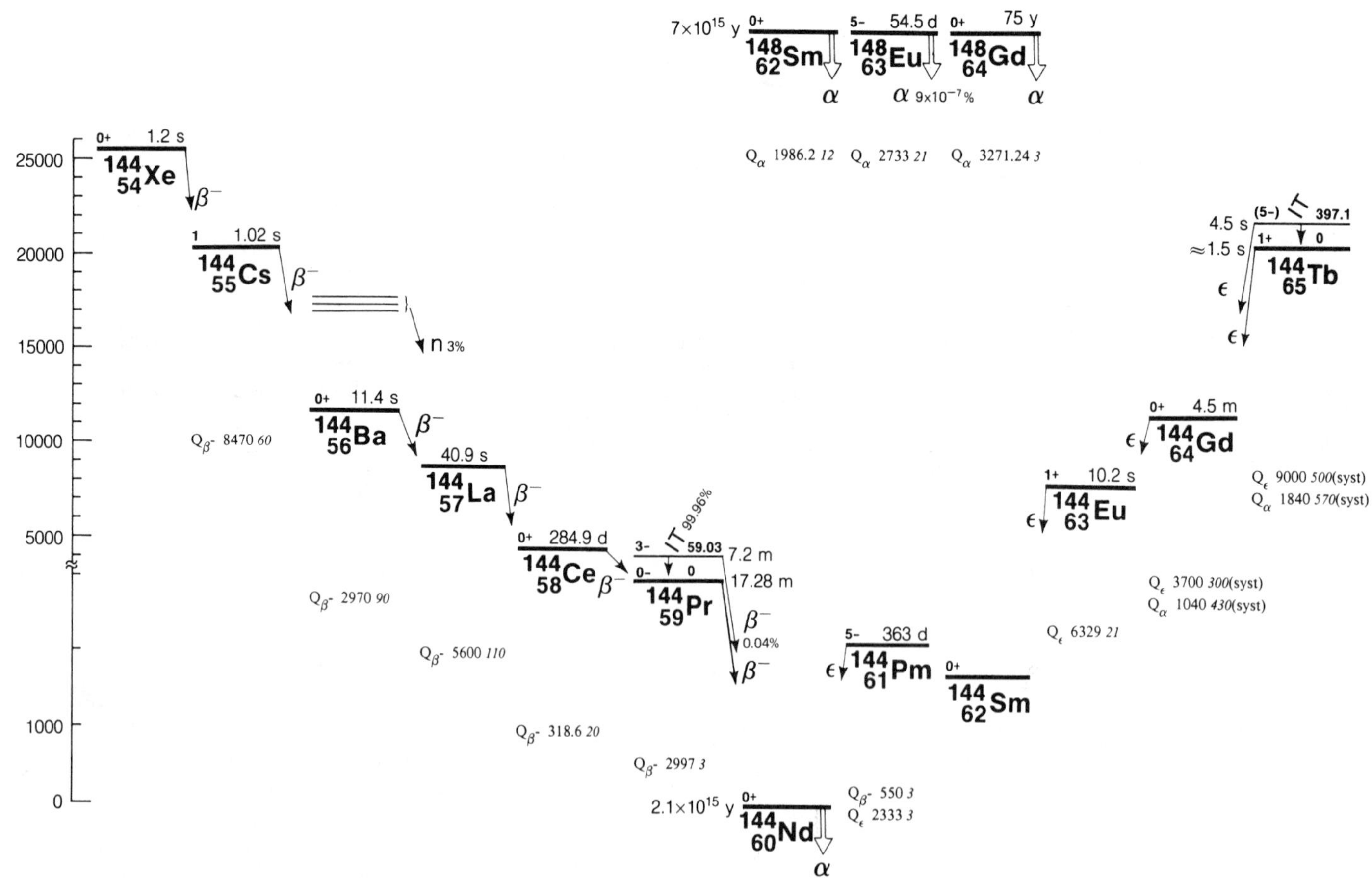

$^{144}_{54}$Xe(1.15 *20* s)

Mode: β-

SpA: 5.1×10^{10} Ci/g

Prod: fission

$^{144}_{55}$Cs(1.02 *3* s)

Mode: β-, β-n(3.0 *6* %)

Δ: -63410 *60* keV

SpA: 5.57×10^{10} Ci/g

Prod: fission

β-n(avg): 290 *20*

Photons (^{144}Cs)

⟨γ⟩=1078 *33* keV

γ_{mode}	γ(keV)	γ(%)†
Ba L_ℓ	3.954	0.013 *3*
Ba L_η	4.331	0.0061 *13*
Ba L_α	4.465	0.36 *8*
Ba L_β	4.943	0.33 *7*
Ba L_γ	5.601	0.042 *10*
Ba $K_{\alpha2}$	31.817	1.7 *3*
Ba $K_{\alpha1}$	32.194	3.2 *6*
Ba $K_{\beta1}$'	36.357	0.89 *17*
Ba $K_{\beta2}$'	37.450	0.23 *4*
$\gamma_{\beta-}$E2	199.20 *9*	47
$\gamma_{\beta-}$	273.5 *3*	0.33 *9*
$\gamma_{\beta-}$	308.24 *15*	2.21 *23*
$\gamma_{\beta-}$E2,M1	330.84 *15*	4.6 *4*
$\gamma_{\beta-}$M1,E2	348.07 *17*	1.50 *9*
$\gamma_{\beta-}$	359.6 *3*	0.24 *5*
$\gamma_{\beta-}$	444.0 *3*	0.24 *5*
$\gamma_{\beta-}$	476.92 *22*	0.38 *5*
$\gamma_{\beta-}$	559.66 *13*	9.2 *5*
$\gamma_{\beta-}$	639.07 *14*	9.6 *5*
$\gamma_{\beta-}$	758.85 *14*	9.6 *5*
$\gamma_{\beta-}$	820.3 *3*	1.88 *19*
$\gamma_{\beta-}$	896.52 *25*	0.61 *9*
$\gamma_{\beta-}$	1009.4 *3*	1.41 *14*
$\gamma_{\beta-}$	1021.4 *4*	0.33 *9*
$\gamma_{\beta-}$	1025.37 *22*	1.03 *19*
$\gamma_{\beta-}$	1078.5 *4*	1.08 *14*
$\gamma_{\beta-}$	1088.8 *3*	1.41 *19*
$\gamma_{\beta-}$	1104.79 *22*	0.61 *9*
$\gamma_{\beta-}$	1115.99 *23*	1.69 *14*
$\gamma_{\beta-}$	1195.5 *5*	~0.28
$\gamma_{\beta-}$	1275.7 *5*	~0.28
$\gamma_{\beta-}$	1318.9 *5*	1.6 *3*
$\gamma_{\beta-}$	1373.44 *22*	1.32 *23*
$\gamma_{\beta-}$	1452.86 *22*	0.99 *19*
$\gamma_{\beta-}$	1648.5 *3*	0.94 *19*
$\gamma_{\beta-}$	1664.44 *21*	1.13 *19*
$\gamma_{\beta-}$	1804.3 *6*	0.75 *14*
$\gamma_{\beta-}$	1836.7 *6*	1.93 *23*
$\gamma_{\beta-}$	2012.51 *21*	2.6 *3*
$\gamma_{\beta-}$	2053.9 *8*	0.99 *19*
$\gamma_{\beta-}$	2117.9 *8*	1.27 *19*
$\gamma_{\beta-}$	2139.4 *8*	2.6 *3*
$\gamma_{\beta-}$	2176.2 *6*	2.2 *3*
$\gamma_{\beta-}$	2371.3 *6*	1.50 *24*
$\gamma_{\beta-}$	2391.2 *10*	1.41 *24*
$\gamma_{\beta-}$	2409.4 *10*	3.2 *4*
$\gamma_{\beta-}$	2472.1 *12*	1.32 *23*
$\gamma_{\beta-}$	2710.9 *15*	2.2 *4*
$\gamma_{\beta-}$	2752.7 *20*	0.94 *24*
$\gamma_{\beta-}$	3014.5 *6*	0.61 *19*
$\gamma_{\beta-}$	3057 *3*	1.1 *3*

† 20% uncert(syst)

Atomic Electrons (^{144}Cs)

⟨e⟩=16.6 *21* keV

e_{bin}(keV)	⟨e⟩(keV)	e(%)
5 - 36	0.50	6.5 *9*
162	10.3	6.4 *13*
193	1.26	0.65 *13*
194	1.6	0.83 *17*
198	0.63	0.32 *6*
199 - 236	0.19	0.093 *18*
268 - 274	0.15	~0.06
293	0.40	0.137 *19*
302 - 348	0.28	0.089 *10*
354 - 360	0.0028	~0.0008
407 - 444	0.024	0.0057 *24*
471 - 477	0.0026	~0.0006
522 - 560	0.31	~0.06
602 - 639	0.27	~0.045
721 - 759	0.22	0.030 *15*
783 - 820	0.039	0.0049 *24*
859 - 896	0.011	0.0013 *7*
972 - 1021	0.043	0.0044 *14*
1024 - 1073	0.041	0.0039 *12*
1077 - 1116	0.029	0.0026 *10*
1158 - 1195	0.0037	~0.00032
1238 - 1281	0.019	0.0015 *7*
1313 - 1336	0.015	0.0011 *5*
1367 - 1415	0.011	0.0008 *3*
1447 - 1453	0.0014	0.00010 *4*
1611 - 1663	0.019	0.0012 *4*
1767 - 1836	0.022	0.0013 *4*
1975 - 2053	0.028	0.0014 *4*
2080 - 2175	0.044	0.0021 *4*
2334 - 2408	0.040	0.0017 *4*
2435 - 2471	0.008	0.00034 *12*
2673 - 2752	0.018	0.00069 *18*
2977 - 3056	0.0091	0.00030 *8*

$^{144}_{56}$Ba(11.4 *5* s)

Mode: β-

Δ: -71870 *80* keV

SpA: 6.7×10^{9} Ci/g

Prod: fission

Photons (^{144}Ba)

γ_{mode}	γ(keV)	γ(rel)
γ M1	16.26 *8*	2.7 *5*
γ	22.46 *20*	1.5 *5*
γ M1(+E2)	42.34 *6*	8.0 *15*
γ	67.76 *10*	2.3 *3*
γ M1(+E2)	68.99 *6*	15.5 *15*
γ	71.26 *20*	1.2 *4*
γ M1,E2	81.88 *8*	27.0 *25*
γ	87.76 *20*	0.5 *2*
γ	92.66 *20*	0.8 *2*
γ	101.16 *10*	3.3 *5*
γ M1(+E2)	103.93 *7*	100 *6*
γ	108.16 *20*	0.5 *1*
γ	109.16 *20*	1.6 *2*
γ M1(+E2)	111.33 *7*	24 *2*
γ	115.09 *9*	11.2 *8*
γ	137.26 *20*	1.2 *3*
γ	138.77 *8*	7.0 *8*
γ	153.06 *20*	3.5 *5*
γ M1,E2	156.66 *8*	65 *4*
γ	163.27 *12*	2.8 *4*
γ	167.55 *12*	3.5 *5*
γ M1,E2	172.91 *6*	63 *4*
γ	180.84 *15*	4.5 *5*
γ	202.16 *20*	4.0 *5*
γ	207.75 *9*	9.5 *10*
γ	215.25 *8*	2.0 *4* ?
γ	219.7 *3*	3.0 *5*

Photons (^{144}Ba) (continued)

γ_{mode}	γ(keV)	γ(rel)
γ M1,E2	228.26 *20*	8.0 *8*
γ	234.50 *16*	2.0 *4*
γ	239.1 *3*	1.5 *3*
γ	259.70 *12*	14.5 *15*
γ	261.0 *3*	2.0 *5*
γ	279.8 *3*	2.3 *5*
γ	289.14 *12*	9.4 *15*
γ	291.80 *10*	13.2 *20*
γ	294.11 *15*	6.3 *10*
γ	302.59 *12*	4.5 *8*
γ	309.37 *15*	1.8 *4*
γ	321.6 *3*	3.0 *5*
γ	329.66 *16*	1.0 *3*
γ	335.2 *3*	3.0 *5*
γ	354.9 *3*	3.2 *5*
γ	373.54 *13*	8.1 *10*
γ	382.80 *13*	4.4 *6*
γ	388.23 *7*	67 *4*
γ	417.67 *9*	13.6 *12*
γ	430.56 *7*	85 *6*
γ	444.6 *3*	1.0 *3*
γ	467.9 *3*	3.0 *5*
γ	474.1 *3*	3.0 *5*
γ	482.52 *15*	3.5 *6*
γ	499.55 *8*	3.5 *6*
γ	515.89 *15*	34 *3*
γ	527.5 *3*	2.0 *4*
γ	541.09 *10*	20 *4*
γ	559.5 *3*	2.0 *5*
γ	570.53 *11*	15.0 *12*
γ	583.42 *10*	13.0 *15*
γ	608.04 *17*	1.5 *3*
γ	619.9 *3*	1.5 *3*
γ	652.41 *10*	2.5 *5*
γ	673.95 *16*	2.5 *5*
γ	703.20 *17*	3.0 *6*
γ	785.11 *16*	3.0 *6*
γ	791.89 *18*	2.5 *5*
γ	805.04 *15*	2.5 *5*
γ	817.2 *3*	2.5 *5*
γ	889.20 *16*	2.5 *5*

$^{144}_{57}$La(40.9 *4* s)

Mode: β-

Δ: -74850 *110* keV

SpA: 1.899×10^{9} Ci/g

Prod: fission

Photons (^{144}La)

⟨γ⟩=2170 *55* keV

γ_{mode}	γ(keV)	γ(%)†
Ce L_ℓ	4.289	0.0049 *8*
Ce L_η	4.730	0.0021 *3*
Ce L_α	4.838	0.132 *18*
Ce L_β	5.388	0.119 *18*
Ce L_γ	6.131	0.016 *3*
Ce $K_{\alpha2}$	34.279	0.61 *5*
Ce $K_{\alpha1}$	34.720	1.11 *10*
Ce $K_{\beta1}$'	39.232	0.31 *3*
Ce $K_{\beta2}$'	40.437	0.084 *8*
γ	226.2 *4*	2.3 *4*
γ	303.60 *18*	1.3 *4* ?
γ	368.0 *3*	1.8 *4*
γ E2	397.25 *16*	88
γ	431.52 *21*	3.9 *5*
γ	453.4 *3*	4.5 *7*
γ E2	541.18 *15*	37 *3*
γ	584.9 *2*	10.1 *9*

Photons (^{144}La) (continued)

γ_{mode}	γ(keV)	γ(%)†
γ	597.2 *4*	0.61 *18*
γ	662.4 *4*	0.88 *18*
γ	705.4 *3*	4.0 *4*
γ	735.11 *21*	7.71 *18*
γ	753.1 *3*	1.49 *18*
γ	813.2 *5*	0.70 *9*
γ	833.3 *5*	0.70 *9*
γ	844.78 *16*	23.0 *18*
γ	890.6 *3*	1.49 *18*
γ	952.0 *3*	5.1 *4*
γ	968.8 *3*	3.9 *4*
γ	979.1 *4*	1.49 *18*
γ	1052.7 *4*	2.19 *9*
γ	1062.9 *5*	<0.5
γ	1071.0 *4*	0.79 *9*
γ	1084.3 *5*	1.05 *18*
γ	1092.1 *4*	1.23 *18*
γ	1102.6 *3*	1.14 *18*
γ	1115.5 *6*	0.26 *9*
γ	1247.3 *5*	0.53 *9*
γ	1276.29 *21*	1.8 *3*
γ	1294.2 *3*	7.9 *9*
γ	1307.4 *6*	0.88 *18*
γ	1347.0 *5*	1.7 *4*
γ	1380.3 *6*	0.88 *18*
γ	1390.4 *6*	0.79 *18*
γ	1421.9 *4*	1.1 *3*
γ	1431.8 *3*	6.0 *7*
γ	1468.3 *4*	0.35 *9*
γ	1489.4 *4*	1.4 *3*
γ	1524.6 *4*	4.2 *5*
γ	1623.8 *6*	0.79 *18*
γ	1631.4 *6*	1.1 *3*
γ	1661.2 *6*	0.88 *18*
γ	1673.54 *24*	1.6 *3*
γ	1683.1 *6*	0.53 *9*
γ	1714.8 *5*	2.0 *4*
γ	1756.0 *5*	2.1 *4*
γ	1765.7 *7*	0.61 *18*
γ	1806.1 *7*	0.88 *18*
γ	1819.1 *4*	0.70 *18*
γ	1853.3 *8*	0.88 *18*
γ	1863.8 *8*	0.61 *18*
γ	1942.7 *5*	1.7 *4*
γ	1955.2 *6*	1.8 *4*
γ	1966.7 *8*	1.1 *3*
γ	1996.4 *5*	3.8 *6*
γ	2008.0 *8*	1.5 *4*
γ	2051.4 *8*	1.3 *3*
γ	2324.3 *6*	0.96 *18*
γ	2340.0 *5*	0.44 *9*
γ	2538.9 *10*	0.44 *9*
γ	2662.0 *6*	2.5 *4*
γ	2865.4 *6*	1.6 *4*
γ	2879 *2*	~0.5
γ	2895.4 *20*	0.6 *3*

† 10% uncert(syst)

Atomic Electrons (^{144}La)

$\langle e \rangle$=11.8 *7* keV

e_{bin}(keV)	$\langle e \rangle$(keV)	e(%)
6 - 39	0.169	2.02 *22*
186 - 226	0.36	~0.19
263 - 304	0.13	~0.05
328	0.10	~0.032
357	5.3	1.49 *15*
361 - 368	0.024	0.007 *3*
391	1.05	0.27 *5*
392 - 432	0.65	0.16 *3*
447 - 453	0.041	~0.009
501	1.35	0.27 *2*
535 - 541	0.284	0.053 *3*
544	0.31	~0.06
557 - 597	0.08	0.014 *5*
622 - 665	0.13	0.019 *9*
695 - 735	0.26	0.037 *16*

Atomic Electrons (^{144}La) (continued)

e_{bin}(keV)	$\langle e \rangle$(keV)	e(%)
747 - 793	0.034	0.0044 *16*
804	0.4	~0.06
807 - 850	0.11	0.013 *5*
884 - 928	0.15	0.017 *7*
939 - 979	0.054	0.0057 *17*
1012 - 1061	0.087	0.0085 *23*
1062 - 1110	0.031	0.0028 *9*
1114 - 1115	0.00013	~1×10^{-5}
1207 - 1254	0.12	0.010 *4*
1267 - 1307	0.050	0.0039 *11*
1340 - 1389	0.036	0.0027 *8*
1390 - 1432	0.08	0.0057 *23*
1449 - 1523	0.065	0.0044 *15*
1583 - 1682	0.069	0.0042 *8*
1708 - 1805	0.044	0.0025 *6*
1813 - 1902	0.028	0.0015 *4*
1915 - 2011	0.083	0.0043 *9*
2045 - 2050	0.0015	7 *3* $\times 10^{-5}$
2284 - 2339	0.011	0.00046 *13*
2498 - 2538	0.0031	0.00012 *4*
2622 - 2661	0.017	0.00065 *22*
2825 - 2894	0.017	0.00060 *15*

Continuous Radiation (^{144}La)

$\langle \beta - \rangle$=1596 keV;$\langle IB \rangle$=4.9 keV

E_{bin}(keV)		$\langle \ \rangle$(keV)	(%)
0 - 10	β-	0.0058	0.115
	IB	0.050	
10 - 20	β-	0.0178	0.118
	IB	0.050	0.35
20 - 40	β-	0.074	0.246
	IB	0.098	0.34
40 - 100	β-	0.58	0.81
	IB	0.28	0.43
100 - 300	β-	7.4	3.60
	IB	0.83	0.47
300 - 600	β-	36.5	8.0
	IB	0.99	0.23
600 - 1300	β-	262	27.0
	IB	1.49	0.17
1300 - 2500	β-	817	44.2
	IB	0.97	0.057
2500 - 4662	β-	473	15.9
	IB	0.133	0.0046

$^{144}_{58}$Ce(284.9 *2* d)

Mode: β-

Δ: -80442 *5* keV

SpA: 3182.8 Ci/g

Prod: fission

Photons (^{144}Ce)

$\langle \gamma \rangle$=19.2 *6* keV

γ_{mode}	γ(keV)	γ(%)†
Pr L_ℓ	4.453	0.024 *4*
Pr L_η	4.929	0.0095 *11*
Pr L_α	5.031	0.64 *9*
Pr L_β	5.612	0.65 *10*
Pr L_γ	6.453	0.101 *16*
γ M1	33.603 *13*	0.288 *22*
Pr $K_{\alpha2}$	35.550	2.40 *12*
Pr $K_{\alpha1}$	36.026	4.37 *22*
Pr $K_{\beta1}$'	40.720	1.24 *6*

Photons (^{144}Ce) (continued)

γ_{mode}	γ(keV)	γ(%)†
γ M1(+0.2%E2)	40.90 *5*	0.40 *4*
Pr $K_{\beta2}$'	41.981	0.345 *18*
γ M1	53.435 *10*	0.095 *6*
γ M3	59.03 *5*	~0.0011
γ M1	80.103 *8*	1.13 *11*
γ E2	99.936 *13*	0.039 *3*
γ M1	133.539 *8*	11.1 *4*

† 1.4% uncert(syst)

Atomic Electrons (^{144}Ce)

$\langle e \rangle$= 9.8 *3* keV

e_{bin}(keV)	$\langle e \rangle$(keV)	e(%)
6	0.41	6.7 *9*
7 - 17	0.26	2.7 *3*
27	0.289	1.08 *9*
28 - 33	0.242	0.80 *4*
34	0.33	0.97 *12*
35 - 36	0.059	0.17 *3*
38	0.92	2.42 *24*
39 - 80	1.01	1.8 *3*
92	5.05	5.51 *25*
93 - 100	0.0344	0.0362 *18*
127	0.95	0.75 *3*
128 - 134	0.286	0.216 *7*

Continuous Radiation (^{144}Ce)

$\langle \beta - \rangle$=82 keV;$\langle IB \rangle$=0.025 keV

E_{bin}(keV)		$\langle \ \rangle$(keV)	(%)
0 - 10	β-	0.424	8.5
	IB	0.0043	
10 - 20	β-	1.21	8.1
	IB	0.0036	0.025
20 - 40	β-	4.46	15.0
	IB	0.0055	0.020
40 - 100	β-	23.5	34.7
	IB	0.0086	0.0142
100 - 300	β-	53	33.6
	IB	0.0035	0.0027
300 - 319	β-	0.082	0.0271
	IB	5.8×10^{-8}	1.9×10^{-8}

$^{144}_{59}$Pr(17.28 *5* min)

Mode: β-

Δ: -80760 *5* keV

SpA: 7.554×10^7 Ci/g

Prod: daughter ^{144}Ce

Photons (^{144}Pr)

$\langle \gamma \rangle$=28.9 *6* keV

γ_{mode}	γ(keV)	γ(%)†
γ [E1]	624.66 *12*	0.00114 *20*
γ [E2]	675.02 *6*	0.00278 *20*
γ E2	696.543 *13*	1.34
γ E1	814.19 *5*	0.00329 *17*
γ M1+E2	864.56 *12*	0.00259 *16*
γ [M1+E2]	1388.05 *15*	0.00597 *21*
γ E1	1489.21 *4*	0.272 *9*
γ	1510.73 *5*	$<3 \times 10^{-6}$

Photons (^{144}Pr) (continued)

γ_{mode}	γ(keV)	γ(%)†
γ [E2]	1561.10 12	~0.00023
γ [E1]	2185.75 4	0.70 3
γ	2654.7 7	0.00019 3

† 1.0% uncert(syst)

Continuous Radiation (^{144}Pr)

⟨β-⟩=1209 keV;⟨IB⟩=3.1 keV

E_{bin}(keV)		⟨ ⟩(keV)	(%)
0 - 10	β-	0.0100	0.198
	IB	0.042	
10 - 20	β-	0.0304	0.202
	IB	0.042	0.29
20 - 40	β-	0.126	0.419
	IB	0.082	0.28
40 - 100	β-	0.97	1.37
	IB	0.23	0.36
100 - 300	β-	12.0	5.8
	IB	0.66	0.37
300 - 600	β-	55	11.9
	IB	0.74	0.17
600 - 1300	β-	347	36.2
	IB	0.95	0.112
1300 - 2500	β-	743	41.9
	IB	0.35	0.022
2500 - 2997	β-	51	1.95
	IB	0.0022	8.5×10^{-5}

$^{144}_{59}$Pr(7.2 *3* min)

Mode: IT(99.96 %), β-(0.04 %)
Δ: -80701 5 keV
SpA: 1.81×10^{8} Ci/g
Prod: daughter ^{144}Ce; ^{144}Nd(n,p)

Photons (^{144}Pr)

⟨γ⟩=12.1 13 keV

γ_{mode}	γ(keV)	γ(%)
Pr L_ℓ	4.453	0.19 5
Pr L_η	4.929	0.042 10
Pr L_α	5.031	5.2 12
Pr L_β	5.662	3.6 9
Pr L_γ	6.472	0.51 13
Pr $K_{\alpha2}$	35.550	8.5 17
Pr $K_{\alpha1}$	36.026	16 3
Pr $K_{\beta1}$'	40.720	4.4 9
Pr $K_{\beta2}$'	41.981	1.23 25
γ_{IT} M3	59.03 5	0.079
$\gamma_{\beta-}$E2	696.543 13	0.040
$\gamma_{\beta-}$E1	814.19 5	0.040

Atomic Electrons (^{144}Pr)

⟨e⟩=46 4 keV

e_{bin}(keV)	⟨e⟩(keV)	e(%)
6	3.0	49 11
7	0.63	9.2 21
17	5.5	32 7
28	0.063	0.22 5
29	0.18	0.61 14
30	0.28	0.93 21
34	0.12	0.37 8

Atomic Electrons (^{144}Pr) (continued)

e_{bin}(keV)	⟨e⟩(keV)	e(%)
35	0.14	0.40 9
36	0.027	0.075 17
39	0.017	0.043 10
40	0.012	0.029 7
41	0.0031	0.0078 17
52	10.0	19 4
53	16.9	32 6
58	7.4	13 3
59	2.1	3.5 7

$^{144}_{60}$Nd(2.1 *4* $\times10^{15}$ yr)

Mode: α
Δ: -83757 3 keV
Prod: natural source
%: 23.80 10

Alpha Particles (^{144}Nd)

α(keV)
1830 15

$^{144}_{61}$Pm(363 *14* d)

Mode: ε
Δ: -81424 4 keV
SpA: 2498 Ci/g
Prod: ^{141}Pr(α,n); ^{144}Nd(p,n); ^{146}Nd(p,3n); ^{145}Nd(p,2n)

Photons (^{144}Pm)

⟨γ⟩=1563 71 keV

γ_{mode}	γ(keV)	γ(%)
Nd L_ℓ	4.633	0.197 20
Nd L_η	5.146	0.077 6
Nd L_α	5.228	5.3 3
Nd L_β	5.855	4.8 4
Nd L_γ	6.695	0.69 7
Nd $K_{\alpha2}$	36.847	22.5 7
Nd $K_{\alpha1}$	37.361	40.9 12
Nd $K_{\beta1}$'	42.240	11.7 3
Nd $K_{\beta2}$'	43.562	3.40 11
γ E1	301.80 6	0.18 4
γ E2	476.81 3	42.2 8
γ [E2]	582.49 8	0.19 2
γ E2	618.06 3	99.1 20
γ [E1]	694.04 14	0.55 10
γ E2	696.543 13	100
γ E1+5.2%M2	778.61 6	1.52 5
γ E1	814.19 5	0.55 3
γ (E0+M1+E2)	890.17 14	<0.08
γ	1396.67 6	0.0006 2
γ [E2]	1508.22 14	<0.0007 ?
γ	1510.73 5	<0.00050 ?

Atomic Electrons (^{144}Pm)

⟨e⟩=16.1 4 keV

e_{bin}(keV)	⟨e⟩(keV)	e(%)
6	2.35	38 4
7	1.73	25 3
29 - 30	0.63	2.10 17
31	0.70	2.29 23
35 - 42	0.84	2.31 14
258 - 302	0.0064	0.0024 5
433	2.03	0.468 13
470 - 477	0.473	0.1004 20
539	0.0068	0.00126 14
574	3.26	0.567 16
575 - 582	0.00145	0.000251 18
611	0.488	0.0799 23
612 - 650	0.198	0.0321 5
653	2.8	0.43 4
687 - 735	0.58	0.084 6
771 - 814	0.0099	0.00128 10
847 - 890	0.0013	~0.00015
1353 - 1397	8.7×10^{-6}	~6×10^{-7}
1465 - 1509	8.5×10^{-6}	<6×10^{-7}

Continuous Radiation (^{144}Pm)

⟨IB⟩=0.24 keV

E_{bin}(keV)		⟨ ⟩(keV)	(%)
10 - 20	IB	0.00025	0.00152
20 - 40	IB	0.074	0.21
40 - 100	IB	0.043	0.096
100 - 300	IB	0.026	0.0130
300 - 600	IB	0.060	0.0138
600 - 1019	IB	0.033	0.0046

$^{144}_{62}$Sm(stable)

Δ: -81974 4 keV
%: 3.1 1

$^{144}_{63}$Eu(10.2 *1* s)

Mode: ε
Δ: -75645 21 keV
SpA: 7.43×10^{9} Ci/g
Prod: ^{144}Sm(p,n)

Photons (^{144}Eu)

⟨γ⟩=200 10 keV

γ_{mode}	γ(keV)	γ(%)†
Sm L_ℓ	4.993	0.028 3
Sm L_η	5.589	0.0104 8
Sm L_α	5.633	0.75 5
Sm L_β	6.343	0.68 6
Sm L_γ	7.271	0.099 10
Sm $K_{\alpha2}$	39.522	2.92 8
Sm $K_{\alpha1}$	40.118	5.29 15
Sm $K_{\beta1}$'	45.379	1.53 4
Sm $K_{\beta2}$'	46.819	0.454 15
γ [M1+E2]	763.11 21	0.045 6
γ [E2]	817.7 2	1.56 8
γ	1000.6 4	0.038 10
γ [E2]	1660.15 17	9.6
γ [E2]	2423.25 17	0.96 4

† 6.3% uncert(syst)

Atomic Electrons (^{144}Eu)

$\langle e \rangle$=1.02 *5* keV

e_{bin}(keV)	$\langle e \rangle$(keV)	e(%)
7	0.49	7.1 *7*
8	0.064	0.82 *8*
31	0.0210	0.067 *7*
32	0.058	0.180 *18*
33	0.085	0.26 *3*
37	0.0094	0.025 *3*
38	0.039	0.104 *11*
39	0.037	0.096 *10*
40 - 45	0.0189	0.045 *3*
716 - 763	0.0020	0.00028 *7*
771	0.0392	0.0051 *3*
810 - 818	0.0078	0.00096 *5*
954 - 1000	0.0009	$\sim 10 \times 10^{-5}$
1613	0.122	0.0076 *5*
1652 - 1659	0.0203	0.00123 *7*
2376 - 2422	0.0105	0.000439 *22*

Continuous Radiation (^{144}Eu)

$\langle \beta+ \rangle$=2063 keV;$\langle$IB$\rangle$=14.2 keV

E_{bin}(keV)		$\langle\ \rangle$(keV)	(%)
0 - 10	β+	1.93×10^{-7}	2.29×10^{-6}
	IB	0.056	
10 - 20	β+	9.3×10^{-6}	5.6×10^{-5}
	IB	0.056	0.39
20 - 40	β+	0.000309	0.00094
	IB	0.118	0.40
40 - 100	β+	0.0186	0.0236
	IB	0.33	0.52
100 - 300	β+	1.19	0.53
	IB	0.99	0.55
300 - 600	β+	11.1	2.36
	IB	1.31	0.30
600 - 1300	β+	123	12.4
	IB	2.6	0.29
1300 - 2500	β+	638	33.3
	IB	3.7	0.20
2500 - 5000	β+	1285	38.5
	IB	4.7	0.137
5000 - 6329	β+	5.4	0.106
	IB	0.37	0.0069
	$\Sigma\beta$+		87

$^{144}_{64}$Gd(4.5 *1* min)

Mode: ϵ
Δ: -71940 *300* keV syst
SpA: 2.90×10^{8} Ci/g
Prod: ^{144}Sm(α,4n); protons on Er

Photons (^{144}Gd)

γ_{mode}	γ(keV)	γ(rel)
γ M1+E2	270.5 *5*	6.7 *17*
γ M1+E2	273.8 *5*	9.2 *25*
γ M1	333.2 *5*	100
γ E1	347.0 *5*	33 *3*
γ	622.0 *5*	18 *3*
γ	629.8 *5*	32 *3*
γ	641.9 *5*	17 *3*
γ	867.7 *5*	18 *3*

$^{144}_{65}$Tb($\sim$1.5 s)

Mode: ϵ
Δ: -62940 *400* keV syst
SpA: 4×10^{10} Ci/g
Prod: ^{144}Sm(^{10}B,α6n); ^{90}Zr(^{58}Ni,n3p)

Photons (^{144}Tb)

γ_{mode}	γ(keV)
γ [E2]	743.0 *3*

$^{144}_{65}$Tb(4.5 *5* s)

Mode: IT, ϵ
Δ: -62543 *400* keV syst
SpA: 1.61×10^{10} Ci/g
Prod: ^{144}Sm(^{10}B,α6n); ^{90}Zr(^{58}Ni,n3p)

Photons (^{144}Tb)

γ_{mode}	γ(keV)	γ(rel)
γ_{IT} E3	113.0 *3*	3.0 *8*
γ_ϵ[E2]	168.9 *3*	9.5 *16*
γ_{IT}	284.1 *3*	100 *10*
γ_ϵ[M1+E2]	315.0 *3*	7.9 *16*
γ_ϵ[E2]	558.1 *3*	24 *3*
γ_ϵ[E1]	600.3 *3*	9.5 *16*
γ_ϵ[E2]	743.0 *3*	59 *6*
γ_ϵ[E2]	959.3 *3*	25 *3*
γ_ϵ[E1]	1001.4 *3*	33 *5*

A = 145

NDS **29**, 533 (1980)

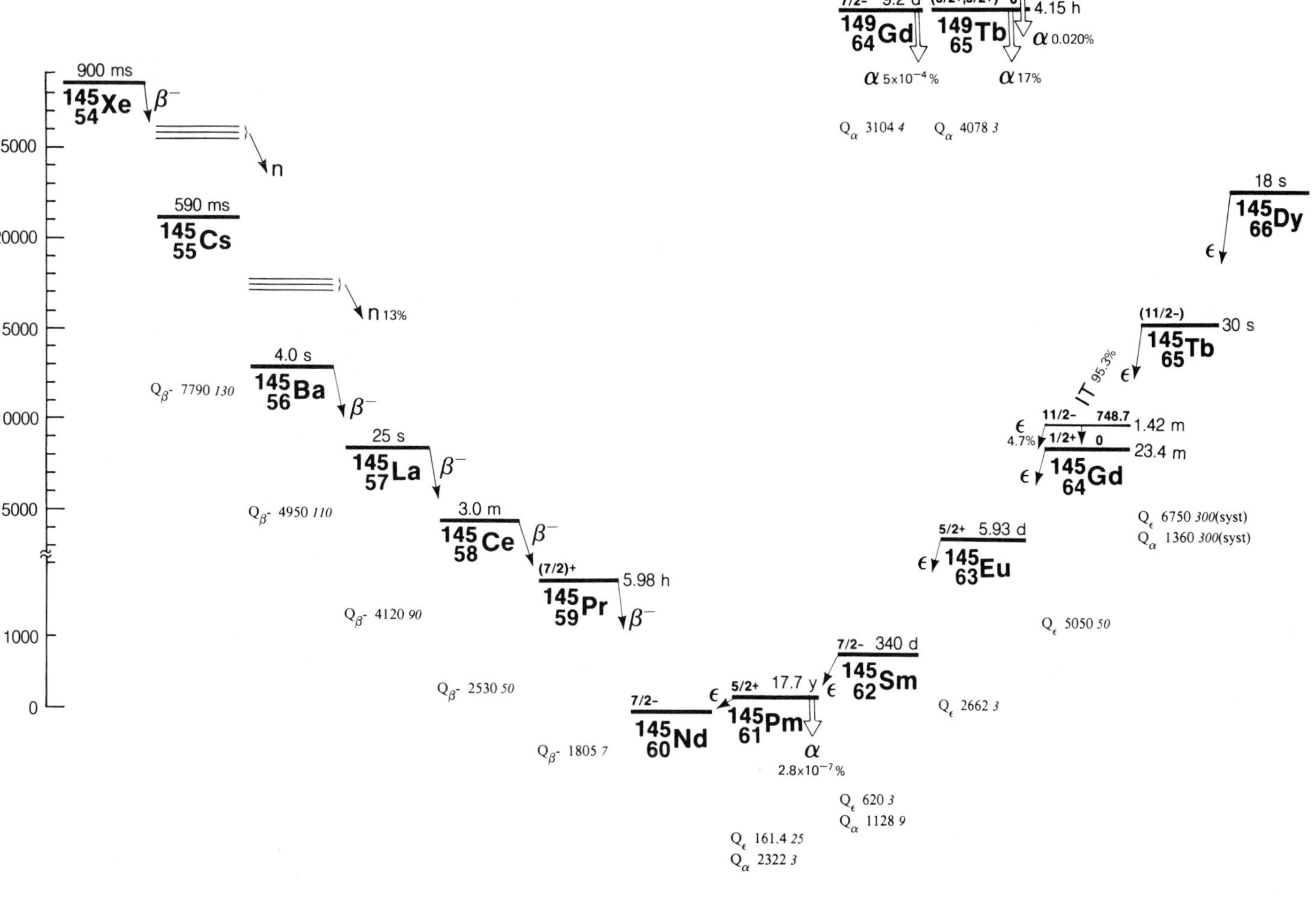

$^{145}_{54}$Xe(900 *300* ms)

Mode: β-, β-n
SpA: 6.0×10^{10} Ci/g
Prod: fission

$^{145}_{55}$Cs(590 *10* ms)

Mode: β-, β-n(13 %)
Δ: -60240 *70* keV
SpA: 7.76×10^{10} Ci/g
Prod: fission

β-n(avg): 460 *30*

Photons (^{145}Cs)

γ_{mode}	γ(keV)
γ	112.6 *10*
γ	175.1 *10*
γ	199 *1*

$^{145}_{56}$Ba(4.0 *3* s)

Mode: β-
Δ: -68040 *130* keV
SpA: 1.79×10^{10} Ci/g
Prod: fission

Photons (^{145}Ba)

⟨γ⟩=303 *9* keV

γ_{mode}	γ(keV)	γ(%)†
La L$_\ell$	4.121	0.099 *14*
La L$_\eta$	4.529	0.042 *5*
La L$_\alpha$	4.649	2.7 *3*
La L$_\beta$	5.166	2.5 *3*
La L$_\gamma$	5.877	0.33 *5*
La K$_{\alpha2}$	33.034	13.6 *8*
La K$_{\alpha1}$	33.442	25.0 *15*
La K$_{\beta1}$'	37.777	7.0 *4*
La K$_{\beta2}$'	38.927	1.83 *11*
γ (M1)	65.77 *15*	6.3 *6*
γ (M1)	92.31 *13*	7.9 *8*
γ (M1)	97.14 *16*	20.2
γ [M1]	123.68 *15*	1.21 *20*
γ [M1]	161.98 *15*	3.6 *4*
γ [M1]	189.45 *14*	1.8 *4*
γ	247.48 *17*	0.61 *20*
γ	254.29 *16*	1.8 *4*

Photons (^{145}Ba)
(continued)

γ_{mode}	γ(keV)	γ(%)†
γ	286.56 *16*	1.6 *4*
γ	303.75 *18*	3.6 *4*
γ	325.63 *17*	2.02 *20*
γ	334.74 *20*	1.01 *20*
γ	344.0 *4*	1.21 *20*
γ	351.43 *18*	1.01 *20*
γ	378.87 *17*	6.5 *6*
γ	408.00 *21*	0.81 *20*
γ	417.94 *17*	5.9 *6*
γ	477.77 *25*	2.02 *20*
γ	533.14 *17*	2.63 *20*
γ	543.5 *3*	
γ	544.54 *16*	5.1 *6* *
γ	571.64 *21*	1.6 *4*
γ	578.49 *18*	2.0 *4*
γ	598.91 *16*	3.6 *4*
γ	668.22 *17*	~0.4
γ	684.20 *19*	1.41 *20*
γ	701.39 *18*	0.61 *20*
γ	730.8 *2*	1.6 *4*
γ	733.98 *17*	0.61 *20*
γ	843.8 *3*	1.21 *20*
γ	1111.63 *21*	1.41 *20*

† 7.4% uncert(syst)
* combined intensity for doublet

Atomic Electrons (^{145}Ba)

⟨e⟩=38.1 *14* keV

e_{bin}(keV)	⟨e⟩(keV)	e(%)
5	1.32	24 *3*
6 - 26	1.05	16.3 *18*
27	5.6	20.9 *22*
28 - 38	1.02	3.38 *23*
53	5.0	9.4 *10*
58	12.1	20.7 *18*
59	1.47	2.5 *3*
60 - 87	2.24	2.86 *19*
91	2.77	3.0 *3*
92 - 122	0.95	0.97 *6*
123	1.09	0.89 *11*
124 - 162	0.67	0.44 *7*
183 - 215	0.32	0.16 *6*
241 - 287	0.63	0.24 *9*
296 - 345	0.68	0.21 *7*
346 - 379	0.39	0.10 *5*
402 - 439	0.14	0.033 *14*
472 - 506	0.26	0.053 *23*
527 - 573	0.28	0.050 *15*
577 - 599	0.022	0.0036 *17*
629 - 679	0.064	0.010 *4*
683 - 731	0.060	0.009 *4*
733 - 734	0.0005	~7×10^{-5}
805 - 844	0.026	~0.0032
1073 - 1112	0.022	0.0020 *10*

Continuous Radiation (^{145}Ba)

⟨β-⟩=1981 keV;⟨IB⟩=6.8 keV

E_{bin}(keV)		⟨ ⟩(keV)	(%)
0 - 10	β-	0.00337	0.067
	IB	0.057	
10 - 20	β-	0.0103	0.069
	IB	0.057	0.40
20 - 40	β-	0.0431	0.143
	IB	0.113	0.39
40 - 100	β-	0.338	0.478
	IB	0.33	0.50
100 - 300	β-	4.49	2.17
	IB	0.99	0.55
300 - 600	β-	23.1	5.01
	IB	1.23	0.29
600 - 1300	β-	186	19.1
	IB	2.0	0.23
1300 - 2500	β-	801	42.2
	IB	1.64	0.095
2500 - 4950	β-	966	30.8
	IB	0.39	0.0135

$^{145}_{57}$La(24.8 *20* s)

Mode: β-
Δ: -72990 *100* keV
SpA: 3.09×10^{9} Ci/g
Prod: fission

Photons (^{145}La)

⟨γ⟩=639 *9* keV

γ_{mode}	γ(keV)	γ(%)†
Ce L_ℓ	4.289	0.071 *10*
Ce L_η	4.730	0.029 *3*
Ce L_α	4.838	1.92 *24*
Ce L_β	5.389	1.72 *23*
Ce L_γ	6.137	0.24 *4*
Ce $K_{\alpha2}$	34.279	9.0 *5*
Ce $K_{\alpha1}$	34.720	16.6 *10*
Ce $K_{\beta1}$'	39.232	4.7 *3*
Ce $K_{\beta2}$'	40.437	1.26 *8*
γ E2	64.23 *18*	0.0302 *10*
γ [M1]	70.04 *16*	10.4 *5*
γ [M1]	118.08 *17*	3.50 *18*
γ	127.1 *3*	0.63 *6*
γ [M1]	163.77 *20*	2.61 *13*
γ [M1]	169.58 *21*	3.09 *15*
γ	234.4 *8*	0.55 *6*
γ	237.49 *20*	1.03 *10*
γ	287.94 *23*	0.15 *3*
γ	291.34 *21*	1.03 *10*
γ	311.9 *3*	0.15 *3*
γ	327.0 *3*	0.74 *7*
γ	355.57 *19*	3.68 *18*
γ	359.9 *3*	0.92 *9*
γ	376.9 *3*	1.25 *13*
γ	387.26 *22*	0.63 *6*
γ	403.67 *23*	0.92 *9*
γ	429.9 *3*	1.58 *16*
γ	435.30 *22*	1.62 *16*
γ	447.0 *3*	3.09 *15*
γ	451.71 *23*	0.52 *5*
γ	463.8 *8*	0.29 *6*
γ	484.14 *24*	0.70 *7*
γ	505.34 *22*	1.66 *17*
γ	514.4 *3*	0.52 *5*
γ	590.73 *24*	0.52 *5*
γ	605.90 *24*	0.92 *9*
γ	632.43 *25*	1.40 *14*
γ	644.57 *24*	1.62 *16*
γ	658.5 *3*	0.37 *4*
γ	663.7 *3*	0.48 *5*
γ	668.8 *3*	0.29 *6*
γ	671.2 *3*	1.77 *18*
γ	687.9 *3*	0.74 *7*
γ	721.63 *24*	0.74 *7*
γ	730.16 *25*	0.74 *7*
γ	743.2 *8*	1.40 *14*
γ	763.5 *4*	0.59 *6*
γ	774.0 *8*	0.59 *6*
γ	786.9 *3*	1.66 *17*
γ	799.2 *3*	0.63 *6*
γ	841.32 *22*	0.52 *5*
γ	846.2 *4*	0.22 *4*
γ	883.59 *24*	0.81 *8*
γ	889.36 *20*	0.96 *10*
γ	895.17 *21*	0.48 *5*
γ	931.63 *24*	2.72 *14*
γ	959.40 *19*	0.22 *4*
γ	1020.7 *3*	1.33 *13*
γ	1031.0 *3*	1.69 *17*
γ	1036.7 *3*	0.77 *8*
γ	1050.5 *4*	1.40 *14*
γ	1221.9 *4*	0.37 *4*
γ	1237.4 *4*	0.85 *9*
γ	1595.9 *4*	1.14 *11*
γ	1819.2 *5*	2.98 *15*
γ	1922.2 *5*	0.63 *6*
γ	1946.2 *4*	0.85 *9*
γ	2087.2 *3*	0.81 *8* ?
γ	2156.0 *5*	0.92 *9*
γ	2205.2 *4*	0.81 *8*
γ	2289.3 *3*	0.41 *4*
γ	2295.1 *3*	0.41 *4*
γ	2306.6 *3*	0.52 *5*
γ	2351.1 *12*	0.55 *6*
γ	2359.4 *3*	1.33 *13*
γ	2376.7 *3*	0.59 *6*
γ	2473.2 *5*	0.48 *5*
γ	2479.0 *5*	0.74 *7*
γ	2526.5 *21*	0.29 *6*
γ	2543.2 *5*	0.70 *7*

† 19% uncert(syst)

Atomic Electrons (^{145}La)

⟨e⟩=21.8 *7* keV

e_{bin}(keV)	⟨e⟩(keV)	e(%)
6	1.38	24 *3*
7 - 29	0.77	4.8 *3*
30	8.9	30.0 *19*
32 - 59	0.44	1.19 *9*
63	2.42	3.8 *3*
64	0.238	0.37 *3*
69	0.60	0.87 *6*
70	0.169	0.242 *15*
78	1.75	2.25 *15*
87 - 121	0.73	0.69 *17*
123	0.83	0.67 *4*
126 - 127	0.028	~0.023
129	0.93	0.72 *5*
157 - 197	0.57	0.33 *5*
228 - 271	0.17	0.068 *25*
281 - 327	0.38	0.12 *5*
336 - 386	0.24	0.070 *19*
387 - 435	0.37	0.09 *3*
440 - 484	0.15	0.033 *10*
499 - 514	0.016	0.0033 *14*
550 - 599	0.09	0.015 *5*
600 - 647	0.16	0.025 *7*
652 - 690	0.052	0.0077 *24*
703 - 746	0.10	0.014 *4*
757 - 806	0.041	0.0052 *14*
835 - 884	0.049	0.0057 *17*
888 - 932	0.06	0.007 *3*
953 - 996	0.059	0.0060 *20*
1010 - 1050	0.034	0.0034 *12*
1181 - 1231	0.017	0.0014 *5*
1232 - 1237	0.00040	3.2 *13* $\times10^{-5}$
1555 - 1595	0.012	0.0008 *3*
1779 - 1818	0.029	0.0016 *6*
1882 - 1945	0.013	0.00070 *19*
2047 - 2116	0.013	0.00064 *18*
2149 - 2249	0.010	0.00046 *12*
2255 - 2354	0.025	0.00109 *19*
2358 - 2439	0.0083	0.00034 *9*
2467 - 2542	0.0081	0.00033 *7*

Continuous Radiation (^{145}La)

⟨β-⟩=1456 keV;⟨IB⟩=4.3 keV

E_{bin}(keV)		⟨ ⟩(keV)	(%)
0 - 10	β-	0.0082	0.163
	IB	0.047	
10 - 20	β-	0.0251	0.167
	IB	0.047	0.32
20 - 40	β-	0.104	0.347
	IB	0.092	0.32
40 - 100	β-	0.80	1.14
	IB	0.26	0.40
100 - 300	β-	10.1	4.90
	IB	0.77	0.43
300 - 600	β-	46.6	10.2
	IB	0.90	0.21
600 - 1300	β-	287	30.0
	IB	1.3	0.150
1300 - 2500	β-	742	40.4
	IB	0.79	0.047
2500 - 4120	β-	370	12.7
	IB	0.084	0.0030

$^{145}_{58}$Ce(2.98 *15* min)

Mode: β-
Δ: -77100 *50* keV
SpA: 4.34×10^{8} Ci/g
Prod: fission; ^{148}Nd(n,α)

Photons (^{145}Ce)

$\langle\gamma\rangle$=773 26 keV

γ_{mode}	γ(keV)	γ(%)†
Pr L_ℓ	4.453	0.15 3
Pr L_η	4.929	0.060 9
Pr L_α	5.031	4.0 6
Pr L_β	5.619	3.6 6
Pr L_γ	6.410	0.50 9
Pr $K_{\alpha2}$	35.550	18.0 21
Pr $K_{\alpha1}$	36.026	33 4
Pr $K_{\beta1}$'	40.720	9.3 11
Pr $K_{\beta2}$'	41.981	2.6 3
γ M1	62.68 11	15.3 18
γ	125.80 15	0.59 12
γ	188.48 15	1.18 24
γ	207.70 12	1.18 24
γ	232.04 12	2.4 5
γ	284.12 11	8.9 9
γ	346.80 12	0.59 12
γ	350.86 14	4.1 8
γ	423.68 13	4.1 8
γ	435.69 14	1.8 4
γ	439.74 12	6.5 7
γ	491.82 11	1.18 24
γ	498.43 15	0.59 12
γ	554.50 12	0.59 12
γ	655.72 13	1.18 24
γ	670.57 19	1.18 24
γ	723.86 12	59 3
γ	758.69 20	0.59 12
γ	782.55 15	2.4 5
γ	859.36 14	1.8 4
γ	911.17 19	0.59 12
γ	1110.31 21	1.18 24
γ	1118.86 21	0.59 12
γ	1147.54 14	9.4 9
γ	1210.22 14	0.59 12

† 12% uncert(syst)

Atomic Electrons (^{145}Ce)

$\langle e\rangle$=31.2 21 keV

e_{bin}(keV)	$\langle e\rangle$(keV)	e(%)
6	2.8	46 7
7	0.35	5.2 8
21	13.8	67 8
28 - 41	1.77	5.7 4
56	5.1	9.2 11
57	0.080	0.141 17
61	1.20	1.96 23
62 - 84	0.52	0.77 16
119 - 166	0.50	0.35 13
182 - 231	0.53	0.27 11
232	0.005	~0.0020
242	0.8	~0.34
277 - 309	0.55	0.19 7
340 - 382	0.29	~0.08
394 - 440	0.54	0.13 6
450 - 498	0.09	0.020 8
513 - 555	0.025	~0.005
614 - 656	0.07	0.012 5
664 - 671	0.006	~0.0009
682	1.5	~0.22
717 - 759	0.35	0.048 21
776 - 817	0.046	0.006 3
853 - 869	0.018	0.0020 9
904 - 911	0.0020	0.00022 10
1068 - 1118	0.17	0.015 7
1119 - 1147	0.023	0.0020 9
1168 - 1210	0.009	0.0008 4

Continuous Radiation (^{145}Ce)

$\langle\beta\text{-}\rangle$=660 keV;$\langle$IB$\rangle$=1.15 keV

E_{bin}(keV)		$\langle\ \rangle$(keV)	(%)
0 - 10	β-	0.0293	0.58
	IB	0.027	
10 - 20	β-	0.089	0.59
	IB	0.027	0.18
20 - 40	β-	0.366	1.22
	IB	0.051	0.18
40 - 100	β-	2.76	3.92
	IB	0.141	0.22
100 - 300	β-	31.7	15.6
	IB	0.35	0.20
300 - 600	β-	121	26.9
	IB	0.31	0.074
600 - 1300	β-	390	43.8
	IB	0.22	0.028
1300 - 2467	β-	114	7.4
	IB	0.018	0.00121

$^{145}_{59}$Pr(5.98 2 h)

Mode: β-

Δ: -79636 8 keV

SpA: 3.614×10^6 Ci/g

Prod: fission; ^{146}Nd(γ,p); ^{145}Nd(n,p)

Photons (^{145}Pr)

$\langle\gamma\rangle$=14.76 11 keV

γ_{mode}	γ(keV)	γ(%)†
Nd L_ℓ	4.633	0.0016 3
Nd L_η	5.146	0.00065 10
Nd L_α	5.228	0.044 6
Nd L_β	5.855	0.039 6
Nd L_γ	6.684	0.0055 9
Nd $K_{\alpha2}$	36.847	0.178 19
Nd $K_{\alpha1}$	37.361	0.32 3
Nd $K_{\beta1}$'	42.240	0.092 10
Nd $K_{\beta2}$'	43.562	0.027 3
γ E2	67.223 16	~0.007
γ M1	72.482 7	0.202 21
γ [M1]	90.609 11	0.0060 4
γ	130.734 12	~0.00026
γ	242.893 16	0.00133 13
γ	262.905 20	0.00228 22
γ	263.044 12	0.00340 21
γ	303.169 8	0.00538 22
γ	318.653 9	0.01127 23
γ	339.7 25	<0.00022 ?
γ	352.460 8	0.0301 6
γ	353.513 18	0.0030 4
γ	364.74 6	0.00022 9
γ	401.989 11	0.00056 13
γ	424.88 20	0.00056 17
γ	448.5 5	<1.7×10^{-4}
γ	466.93 5	0.00211 21
γ	475.63 3	0.00348 22
γ	492.597 8	0.0228 5
γ	504.67 5	0.00047 17
γ	515.89 3	0.0060 4
γ	606.36 3	0.0014 3
γ	623.473 9	0.0175 4
γ	657.646 8	0.0477 10
γ [E2]	675.773 7	0.378 8
γ [M1+E2]	707.927 10	0.0082 3
γ [M1+E2]	713.187 15	0.0069 3
γ	744 3	<0.00022 ?
γ [M1+E2]	748.255 7	0.430 9
γ	778.80 3	0.00047 22
γ [E2]	780.409 11	0.0034 3
γ	848.207 10	0.0555 11
γ [M1+E2]	864.47 5	0.00099 21

Photons (^{145}Pr)
(continued)

γ_{mode}	γ(keV)	γ(%)†
γ	869.40 3	~0.0005
γ [M1+E2]	869.73 5	0.0007 3
γ	920.689 9	0.1204 24
γ [M1+E2]	936.95 5	0.0022 4
γ	978.940 9	0.191 4
γ	1011.158 19	0.0007 3
γ	1012.746 11	0.0045 3
γ	1018.006 17	0.0078 3
γ	1051.422 9	0.144 3
γ	1088.506 16	0.00464 22
γ	1089.83 5	0.00138 17
γ	1093.765 16	0.00443 13
γ	1150.241 8	0.165 3
γ	1160.988 17	0.0123 4
γ	1162.31 5	0.0072 4
γ	1177.213 21	0.00310 17
γ	1182.47 3	0.00064 13
γ	1213.01 5	0.00060 17
γ	1218.27 5	0.00056 17
γ	1249.695 21	0.0019 3
γ	1259.0 9	~0.00022 ?
γ	1266.13 6	0.00052 13
γ	1271.39 6	0.00120 17
γ	1285.49 5	0.00043 13
γ	1331.397 9	0.0054 3
γ	1336.657 15	0.00138 17
γ	1338.61 6	<0.0013?
γ	1403.879 10	0.0039 4
γ	1527.05 3	0.00129 17
γ	1532.00 9	0.00034 13

† 16% uncert(syst)

Atomic Electrons (^{145}Pr)

$\langle e\rangle$=0.406 25 keV

e_{bin}(keV)	$\langle e\rangle$(keV)	e(%)
6	0.020	0.32 5
7	0.0141	0.21 3
24	0.006	~0.024
29	0.184	0.63 7
30 - 60	0.0218	0.061 4
61	0.021	~0.034
65	0.053	0.081 9
66 - 67	0.012	0.018 5
71	0.0134	0.0188 20
72 - 91	0.0053	0.0071 6
124 - 131	5.9 ×10^{-5}	~5×10^{-5}
199 - 243	0.0008	~0.00039
256 - 303	0.0018	0.0007 3
309 - 358	0.0033	0.0010 5
359 - 405	4.5 ×10^{-5}	~1×10^{-5}
418 - 467	0.0014	~0.00030
469 - 516	0.00056	0.00012 4
563 - 606	0.0006	~0.00011
614 - 623	0.0016	~0.00026
632	0.0110	0.00174 8
651 - 676	0.00305	0.00046 3
700 - 702	4.6 ×10^{-5}	6.6 17 ×10^{-6}
705	0.015	0.0021 5
706 - 748	0.0028	0.00038 7
772 - 821	0.0013	~0.00016
826 - 870	0.00026	3.1 13 ×10^{-5}
877 - 921	0.0029	0.00033 16
930 - 979	0.0044	0.00047 22
1004 - 1051	0.0031	0.00031 14
1081 - 1119	0.0029	0.00026 13
1134 - 1182	0.00057	5.0 17 ×10^{-5}
1206 - 1253	6.8 ×10^{-5}	5.5 16 ×10^{-6}
1257 - 1295	9.5 ×10^{-5}	7 3 ×10^{-6}
1324 - 1360	6.3 ×10^{-5}	4.6 19 ×10^{-6}
1397 - 1404	8.0 ×10^{-6}	5.7 25 ×10^{-7}
1483 - 1531	2.1 ×10^{-5}	1.4 5 ×10^{-6}

Continuous Radiation (^{145}Pr)

⟨β-⟩=677 keV;⟨IB⟩=1.17 keV

E_{bin}(keV)		⟨ ⟩(keV)	(%)
0 - 10	β-	0.0264	0.53
	IB	0.028	
10 - 20	β-	0.080	0.54
	IB	0.027	0.19
20 - 40	β-	0.331	1.10
	IB	0.053	0.18
40 - 100	β-	2.51	3.55
	IB	0.145	0.22
100 - 300	β-	29.0	14.2
	IB	0.37	0.21
300 - 600	β-	116	25.5
	IB	0.32	0.078
600 - 1300	β-	431	47.6
	IB	0.22	0.028
1300 - 1805	β-	99	6.9
	IB	0.0068	0.00050

$^{145}_{60}$Nd(>6×10^{16} yr)

Δ: -81441 *3* keV

%: 8.30 *5*

$^{145}_{61}$Pm(17.7 *4* yr)

Mode: ϵ, α(2.8×10^{-7} %)

Δ: -81280 *4* keV

SpA: 139 Ci/g

Prod: daughter ^{145}Sm

Alpha Particles (^{145}Pm)

α(keV)
2240 *40*

Photons (^{145}Pm)

⟨γ⟩=33.1 *6* keV

γ_{mode}	γ(keV)	γ(%)†
Nd L_ℓ	4.633	0.215 *22*
Nd L_η	5.146	0.086 *7*
Nd L_α	5.228	5.8 *4*
Nd L_β	5.849	5.6 *5*
Nd L_γ	6.713	0.85 *9*
Nd $K_{\alpha2}$	36.847	22.5 *7*
Nd $K_{\alpha1}$	37.361	41.0 *12*
Nd $K_{\beta1}$'	42.240	11.7 *4*
Nd $K_{\beta2}$'	43.562	3.41 *12*
γE2	67.223 *16*	0.771 *21*
γM1	72.482 *7*	2.58

† 16% uncert(syst)

Atomic Electrons (^{145}Pm)

⟨e⟩=13.6 *4* keV

e_{bin}(keV)	⟨e⟩(keV)	e(%)
6	2.46	40 *4*
7	2.07	30 *3*
24	0.598	2.53 *8*
29	2.49	8.6 *4*
30	0.46	1.54 *16*
31	0.71	2.29 *23*
35	0.234	0.67 *7*

Atomic Electrons (^{145}Pm) (continued)

e_{bin}(keV)	⟨e⟩(keV)	e(%)
36	0.42	1.17 *12*
37	0.104	0.28 *3*
40	0.0079	0.0195 *20*
41	0.051	0.125 *13*
42	0.0242	0.058 *6*
60	0.133	0.221 *7*
61	2.17	3.57 *12*
65	0.67	1.03 *5*
66	0.639	0.97 *3*
67	0.150	0.224 *8*
71	0.171	0.240 *11*
72	0.0481	0.067 *3*

Continuous Radiation (^{145}Pm)

⟨IB⟩=0.108 keV

E_{bin}(keV)		⟨ ⟩(keV)	(%)
10 - 20	IB	0.00030	0.0019
20 - 40	IB	0.071	0.20
40 - 100	IB	0.036	0.084
100 - 161	IB	0.00022	0.00020

$^{145}_{62}$Sm(340 *3* d)

Mode: ϵ

Δ: -80660 *4* keV

SpA: 2649 Ci/g

Prod: ^{144}Sm(n,γ)

Photons (^{145}Sm)

⟨γ⟩=65.0 *11* keV

γ_{mode}	γ(keV)	γ(%)†
Pm L_ℓ	4.809	0.37 *4*
Pm L_η	5.363	0.143 *11*
Pm L_α	5.430	10.0 *7*
Pm L_β	6.097	9.1 *8*
Pm L_γ	6.980	1.31 *14*
Pm $K_{\alpha2}$	38.171	40.6 *13*
Pm $K_{\alpha1}$	38.725	73.8 *23*
Pm $K_{\beta1}$'	43.793	21.2 *7*
Pm $K_{\beta2}$'	45.183	6.28 *22*
γ M1(+E2)	60.91 *5*	12.36
γ [E2]	431.06 *15*	4.4 *3* ×10^{-5}
γ [M1+E2]	491.97 *14*	0.00272 *12*

† 1.9% uncert(syst)

Atomic Electrons (^{145}Sm)

⟨e⟩=29.3 *7* keV

e_{bin}(keV)	⟨e⟩(keV)	e(%)
6	4.4	68 *7*
7	3.2	45 *5*
16	11.0	69.9 *20*
30	0.30	0.98 *10*
31	0.82	2.6 *3*
32	1.23	3.9 *4*
36	0.32	0.88 *9*
37	0.73	1.97 *20*
38	0.27	0.71 *7*
39	0.025	0.066 *7*

Atomic Electrons (^{145}Sm) (continued)

e_{bin}(keV)	⟨e⟩(keV)	e(%)
42	0.096	0.228 *23*
43	0.038	0.088 *9*
44	0.0146	0.034 *4*
53	4.82	9.0 *3*
54	0.481	0.892 *25*
59	1.24	2.09 *6*
60	0.0212	0.0356 *10*
61	0.358	0.590 *17*
386	2.5 ×10^{-6}	~7×10^{-7}
424	4.3 ×10^{-7}	~1×10^{-7}
425	6.9 ×10^{-8}	~2×10^{-8}
429	6.9 ×10^{-8}	~2×10^{-8}
430	4.2 ×10^{-8}	~1×10^{-8}
431	3.0 ×10^{-8}	~7×10^{-9}
447	0.00017	3.9 *11* ×10^{-5}
485	2.6 ×10^{-5}	5.5 *18* ×10^{-6}
486	1.5 ×10^{-6}	~3×10^{-7}
490	5.7 ×10^{-6}	1.2 *4* ×10^{-6}
491	3.7 ×10^{-7}	~8×10^{-8}
492	1.7 ×10^{-6}	3.4 *12* ×10^{-7}

Continuous Radiation (^{145}Sm)

⟨IB⟩=0.157 keV

E_{bin}(keV)		⟨ ⟩(keV)	(%)
10 - 20	IB	0.00023	0.00143
20 - 40	IB	0.060	0.17
40 - 100	IB	0.066	0.150
100 - 300	IB	0.019	0.0100
300 - 600	IB	0.0106	0.0029
600 - 620	IB	4.5×10^{-8}	7.2×10^{-9}

$^{145}_{63}$Eu(5.93 *4* d)

Mode: ϵ

Δ: -77998 *5* keV

SpA: 1.519×10^5 Ci/g

Prod: daughter ^{145}Gd; ^{144}Sm(d,n); ^{144}Sm(p,γ); daughter ^{149}Tb

Photons (^{145}Eu)

γ_{mode}	γ(keV)	γ(rel)
γ M1+E2	110.86 *9*	2.4 *2*
γ M1+E2	191.35 *10*	0.86 *8*
γ	519.4 *4*	0.16 *4*
γ [M1+E2]	526.10 *17*	0.30 *6*
γ E1	542.58 *12*	6.4 *4*
γ E1	653.45 *9*	23.2 *20*
γ [M1+E2]	713.80 *17*	0.40 *8*
γ M1	764.87 *15*	2.5 *2*
γ	838.5 *4*	0.21 *7*
γ E2	893.67 *12*	100
γ	910.63 *21*	0.11 *2*
γ	949.9 *5*	0.08 *2* ?
γ (M1)	1078.51 *19*	0.67 *9*
γ (M1+E2)	1239.90 *18*	0.17 *4*
γ M1+E2	1423.3 *3*	0.72 *8*
γ M1(+E2)	1532.23 *17*	0.55 *8*
γ	1547.3 *4*	0.23 *9*
γ	1625.0 *10*	~0.027
γ M1(+E2)	1658.54 *15*	25.0 *21*
γ E1	1804.29 *19*	1.71 *17*
γ (E1)	1857.75 *20*	0.67 *11*
γ M1(+E2)	1876.75 *20*	2.21 *21*
γ [E2]	1972.17 *17*	0.08 *2*

Photons (^{145}Eu)
(continued)

γ_{mode}	γ(keV)	γ(rel)
γ M1	1996.95 20	10.7 11
γ	2110.5 3	0.07 2
γ [E2]	2133.57 16	0.36 6
γ	2155.35 20	0.19 4
γ	2193.0 3	0.06 2
γ	2276.9 5	0.17 4
γ	2291.8 18	<0.006 ?
γ	2329.3 3	0.29 4
γ [M1+E2]	2340.8 5	0.038 13
γ	2346.8 4	0.29 4
γ	2387.6 3	0.044 4
γ [M1+E2]	2425.90 19	0.21 2
γ	2482.3 3	0.051 8
γ	2508.1 4	0.046 6
γ	2513.0 4	0.036 4

$^{145}_{64}$Gd(23.4 5 min)

Mode: ϵ
Δ: -72950 50 keV
SpA: 5.54×10^7 Ci/g
Prod: ^{144}Sm(α,3n); ^{144}Sm(^{3}He,2n)

Photons (^{145}Gd)
$\langle\gamma\rangle$=1794 110 keV

γ_{mode}	γ(keV)	γ(%)
γ [M1+E2]	329.28 15	2.8 5
γ	782.1 5	0.28 6 ?
γ [E2]	808.43 16	9.0
γ	914.7 4	0.25 5
γ [M1+E2]	949.34 22	0.77 14
γ	953.2 3	1.42 17
γ [M1+E2]	1041.78 18	10.1 14
γ	1071.5 4	0.88 18 ?
γ	1072.0 3	2.8 4
γ	1567.24 20	0.94 11
γ [M1+E2]	1599.74 20	1.84 22
γ	1719.23 19	1.20 13
γ [M1+E2]	1757.77 22	34 4
γ	1784.0 4	0.43 8
γ	1806.7 10	0.24 6 ?
γ	1814.8 20	0.9 3 ?
γ	1845.2 4	0.57 10
γ	1880.4 3	33 4
γ	1891.4 3	0.44 9
γ	2203.14 20	0.20 4
γ	2451.3 4	0.32 7
γ [E2]	2494.6 5	1.30 18
γ	2581.7 3	0.27 6
γ	2642.0 5	1.94 22
γ	2662.5 4	0.60 16
γ	2665.9 4	0.65 5 ?
γ	2672.4 3	0.16 5
γ	2765.0 15	0.17 4
γ	2837.2 3	0.41 10
γ	2867.9 7	0.12 3
γ	2906.7 3	0.11 3
γ	2956.23 19	<0.27 ?
γ	3236.0 3	0.14 5
γ	3259.7 4	0.20 5
γ	3285.51 23	0.15 4
γ	3294.1 3	<0.25 ?
γ	3369.5 4	<0.14 ?
γ	3544.4 5	<0.29 ?
γ	3602.8 4	<0.18 ?
γ	3623.4 3	<0.38 ?
γ	3644.4 5	<0.16 ?
γ	3685.7 16	<0.25 ?

$^{145}_{64}$Gd(1.42 5 min)

Mode: IT(95.3 5 %), ϵ(4.7 5 %)
Δ: -72201 50 keV
SpA: 9.1×10^8 Ci/g
Prod: ^{144}Sm(^{3}He,2n); protons on Er; ^{144}Sm(α,3n); ^{148}Sm(^{3}He,6n)

Photons (^{145}Gd)
$\langle\gamma\rangle$=636 24 keV

γ_{mode}	γ(keV)	γ(%)†
Eu L_ℓ	5.177	0.0048 5
Gd L_ℓ	5.362	0.20 5
Eu L_η	5.817	0.00177 16
Eu L_α	5.843	0.129 10
Gd L_η	6.049	0.091 21
Gd L_α	6.054	5.2 14
Eu L_β	6.594	0.117 12
Gd L_β	6.815	8.0 17
Eu L_γ	7.571	0.0174 20
Gd L_γ	7.967	1.6 3
γ_{IT}[M1+1.1%E2]	27.30 10	4.8
Eu $K_{\alpha2}$	40.902	0.474 25
Eu $K_{\alpha1}$	41.542	0.85 5
Gd $K_{\alpha2}$	42.309	2.61 16
Gd $K_{\alpha1}$	42.996	4.7 3
Eu $K_{\beta1}$'	46.999	0.248 13
Eu $K_{\beta2}$'	48.496	0.075 4
Gd $K_{\beta1}$'	48.652	1.38 8
Gd $K_{\beta2}$'	50.214	0.42 3
γ_ϵ[M1+E2]	329.28 15	4.45
γ_ϵ(M2)	386.4 3	4.02
γ_{IT}[M4]	721.4 3	83

† uncert(syst): 11% for ϵ, 0.52% for IT

Atomic Electrons (^{145}Gd)
$\langle e\rangle$=113 5 keV

e_{bin}(keV)	$\langle e\rangle$(keV)	e(%)
7 - 8	4.4	56 10
19	12.5	66 10
20	3.4	17 7
25 - 48	6.4	24 3
281 - 329	0.73	0.25 6
338 - 386	2.35	0.68 3
671	65.1	9.7 6
713	13.0	1.82 10
714 - 721	5.68	0.79 3

Continuous Radiation (^{145}Gd)
$\langle\beta+\rangle$=65 keV;$\langle$IB$\rangle$=0.57 keV

E_{bin}(keV)		$\langle\ \rangle$(keV)	(%)
0 - 10	β+	1.20×10^{-8}	1.43×10^{-7}
	IB	0.0020	
10 - 20	β+	6.0×10^{-7}	3.56×10^{-6}
	IB	0.0020	0.0136
20 - 40	β+	2.01×10^{-5}	6.1×10^{-5}
	IB	0.0045	0.0153
40 - 100	β+	0.00124	0.00157
	IB	0.0124	0.020
100 - 300	β+	0.079	0.0354
	IB	0.034	0.019
300 - 600	β+	0.73	0.156
	IB	0.045	0.0105
600 - 1300	β+	7.6	0.77
	IB	0.097	0.0107
1300 - 2500	β+	33.0	1.75
	IB	0.18	0.0097

Continuous Radiation (^{145}Gd)
(continued)

E_{bin}(keV)		$\langle\ \rangle$(keV)	(%)
2500 - 5000	β+	23.9	0.81
	IB	0.19	0.0059
5000 - 5083	IB	2.3×10^{-6}	4.5×10^{-8}
	$\Sigma\beta$+		3.5

$^{145}_{65}$Tb(30 3 s)

Mode: ϵ
Δ: -66200 300 keV syst
SpA: 2.6×10^9 Ci/g
Prod: ^{144}Sm(^{10}B,α5n); ^{90}Zr(^{58}Ni,3p)

Photons (^{145}Tb)
$\langle\gamma\rangle$=1562 86 keV

γ_{mode}	γ(keV)	γ(%)
γ [M1+1.1%E2]	27.30 10	*
γ	200.68 19	7.2 7
γ	246.7 3	4.1 4
γ [E1]	257.76 24	39 4
γ [E1]	268.55 25	3.0 4
γ	371.52 21	1.4 3
γ [E2]	524.0 2	10 1
γ	537.07 20	23 2
γ	572.20 21	14 2
γ	698.08 23	5.5 6
γ [M4]	721.4 3	*
γ	908.58 19	7.3 8
γ [M1+E2]	935.19 22	5.3 6
γ [M1+E2]	987.66 20	37
γ [E2]	1014.96 20	5.0 6
γ	1109.26 19	14 2
γ [E2]	1388.05 25	6.3 7
γ	1432.58 21	9.6 9
γ	1446.8 3	15 2

* with ^{145}Gd(1.42 min)

$^{145}_{66}$Dy(18 3 s)

Mode: ϵ
SpA: 4.2×10^9 Ci/g
Prod: protons on W; protons on Ta; ^{90}Zr(^{58}Ni,n2p)

Photons (^{145}Dy)

γ_{mode}	γ(keV)	γ(%)
γ	639.7 3	12 4

A = 146

NDS **41**, 195 (1984)

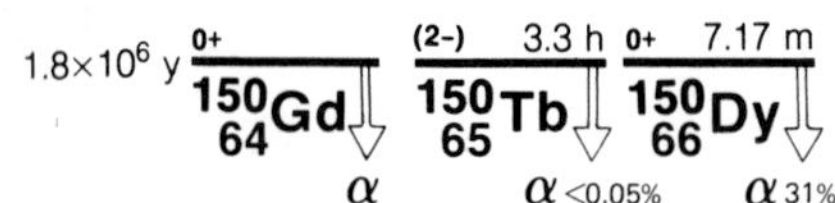

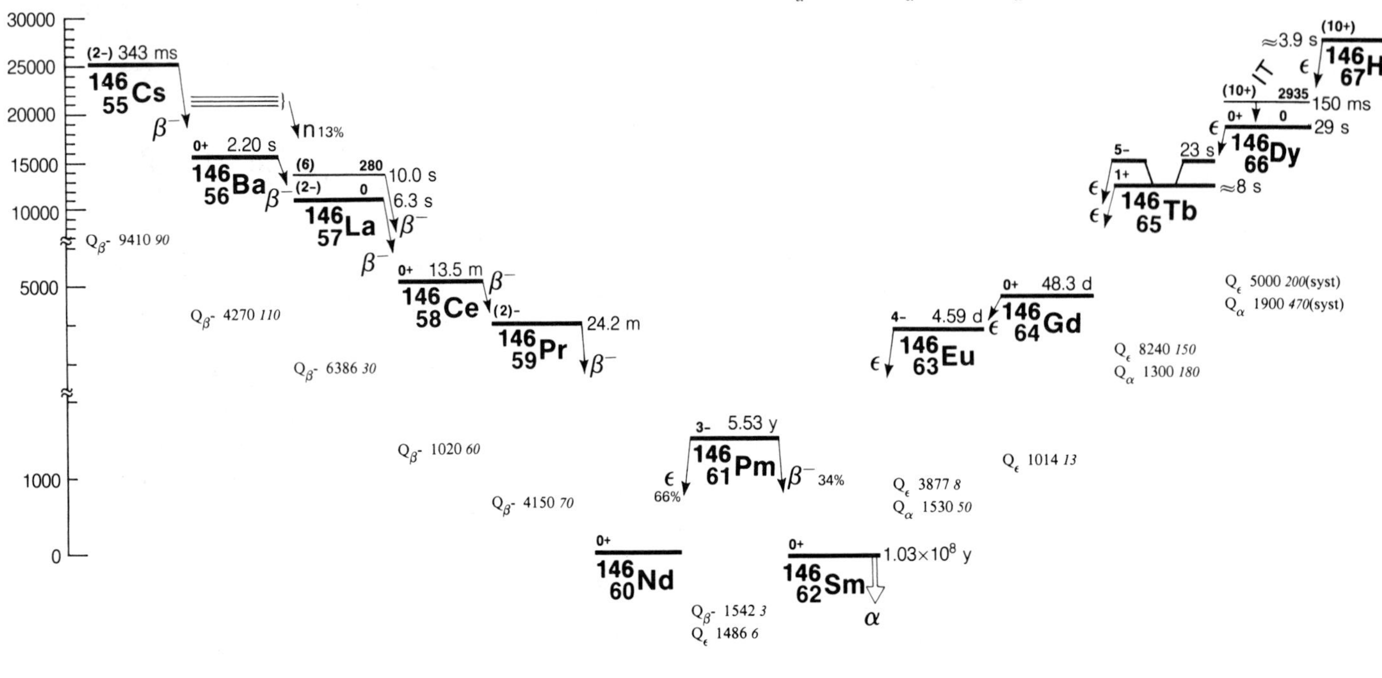

Qα 2543 5

$^{146}_{55}$Cs(343 7 ms)

Mode: β-, β-n(13.2 6 %)

Δ: -55690 100 keV

SpA: 9.67×10¹⁰ Ci/g

Prod: fission

β-n(avg): 530 70

Photons (^{146}Cs)

⟨γ⟩=817 19 keV

γmode	γ(keV)	γ(%)†
Ba L_ℓ	3.954	0.021 3
Ba L_η	4.331	0.0099 11
Ba L_α	4.465	0.58 7
Ba L_β	4.943	0.53 6
Ba L_γ	5.599	0.067 9
Ba $K_{\alpha2}$	31.817	2.78 10
Ba $K_{\alpha1}$	32.194	5.13 18
Ba $K_{\beta1}$'	36.357	1.42 5
Ba $K_{\beta2}$'	37.450	0.359 14
$\gamma_{\beta\text{-}}$E2	181.19 7	57.0 11
$\gamma_{\beta\text{-}}$[E1]	307.52 8	2.74 6
$\gamma_{\beta\text{-}}$[E2]	332.59 7	6.44 13
$\gamma_{\beta\text{-}}$[E1]	557.90 9	9.18 18
$\gamma_{\beta\text{-}}$[E1]	640.11 9	2.91 8
$\gamma_{\beta\text{-}}$[E1]	739.09 11	3.02 7
$\gamma_{\beta\text{-}}$[E1]	745.37 17	0.57 6
$\gamma_{\beta\text{-}}$	772.1 4	0.80 16
$\gamma_{\beta\text{-}}$	795.7 4	0.68 14
$\gamma_{\beta\text{-}}$	809.0 4	0.57 11

Photons (^{146}Cs) (continued)

γmode	γ(keV)	γ(%)†
$\gamma_{\beta\text{-}}$	817.0 4	0.97 19
$\gamma_{\beta\text{-}}$	821.3 4	0.68 14
$\gamma_{\beta\text{-}}$[E1]	827.58 17	0.91 7
$\gamma_{\beta\text{-}}$	847.5 4	0.40 8
$\gamma_{\beta\text{-}}$	861.8 4	0.29 6
$\gamma_{\beta\text{-}}$	868.1 4	0.74 15
$\gamma_{\beta\text{-}}$[E2]	871.6 4	1.54 8
$\gamma_{\beta\text{-}}$	893.6 6	1.37 7
$\gamma_{\beta\text{-}}$	917.3 5	1.14 6
$\gamma_{\beta\text{-}}$	934.23 17	1.77 8
$\gamma_{\beta\text{-}}$	944.3 4	0.83 17
$\gamma_{\beta\text{-}}$	976.6 3	0.63 7
$\gamma_{\beta\text{-}}$[E2]	1052.89 15	0.46 9
$\gamma_{\beta\text{-}}$	1075.25 16	0.63 7
$\gamma_{\beta\text{-}}$	1115.42 16	1.37 10
$\gamma_{\beta\text{-}}$	1158.9 4	1.3 3
$\gamma_{\beta\text{-}}$	1218.0 4	0.51 10
$\gamma_{\beta\text{-}}$	1256.44 16	1.08 10
$\gamma_{\beta\text{-}}$	1300.5 4	0.48 10
$\gamma_{\beta\text{-}}$[M1+E2]	1385.48 15	1.37 9
$\gamma_{\beta\text{-}}$	1451.5 4	0.60 12
$\gamma_{\beta\text{-}}$	1456.5 4	0.80 9
$\gamma_{\beta\text{-}}$	1488.1 4	1.14 23
$\gamma_{\beta\text{-}}$	1502.1 4	1.31 9
$\gamma_{\beta\text{-}}$	1534.5 3	0.57 9
$\gamma_{\beta\text{-}}$	1546.4 9	0.17 3
$\gamma_{\beta\text{-}}$[E2]	1566.66 16	1.6 3
$\gamma_{\beta\text{-}}$	1598.0 7	2.28 19
$\gamma_{\beta\text{-}}$	1656.7 9	1.8 4
$\gamma_{\beta\text{-}}$	1715.7 3	1.6 3
$\gamma_{\beta\text{-}}$	1787.5 5	0.68 14
$\gamma_{\beta\text{-}}$	1813.2 6	0.80 12
$\gamma_{\beta\text{-}}$	1968.7 5	2.3 5
$\gamma_{\beta\text{-}}$	1981.2 9	0.80 16
$\gamma_{\beta\text{-}}$	1983.6 9	0.51 10

Photons (^{146}Cs) (continued)

γmode	γ(keV)	γ(%)†
$\gamma_{\beta\text{-}}$	1995.8 9	0.29 6
$\gamma_{\beta\text{-}}$	2027.9 9	0.46 9
$\gamma_{\beta\text{-}}$	2163.0 9	0.51 10
$\gamma_{\beta\text{-}}$	2188 9	0.34 7
$\gamma_{\beta\text{-}}$	2344 9	1.4 3
$\gamma_{\beta\text{-}}$	2444 9	0.74 15
$\gamma_{\beta\text{-}}$	2567 9	0.46 9

† 5.3% uncert(syst)

Atomic Electrons (^{146}Cs)

⟨e⟩=22.9 4 keV

e_{bin}(keV)	⟨e⟩(keV)	e(%)
5 - 36	0.81	10.5 7
144	15.1	10.5 3
175	1.86	1.06 3
176	2.75	1.57 4
180	1.02	0.565 16
270	0.068	0.0253 12
295	0.511	0.173 8
302 - 333	0.139	0.0426 14
520 - 558	0.123	0.0234 9
603 - 640	0.0333	0.00548 23
702 - 745	0.050	0.0070 13
758 - 808	0.065	0.0084 21
809 - 857	0.080	0.0096 18
860 - 907	0.067	0.0075 21
911 - 944	0.018	0.0020 6

Atomic Electrons (^{146}Cs)
(continued)

e_{bin}(keV)	⟨e⟩(keV)	e(%)
971 - 1015	0.0080	0.00079 15
1038 - 1078	0.027	0.0025 9
1109 - 1159	0.021	0.0018 8
1181 - 1219	0.018	0.0015 6
1250 - 1299	0.008	0.00060 22
1300 - 1348	0.017	0.00127 21
1379 - 1419	0.016	0.0011 3
1445 - 1488	0.025	0.0017 5
1496 - 1593	0.048	0.0031 7
1597 - 1678	0.030	0.0018 6
1710 - 1808	0.014	0.00081 21
1931 - 2027	0.034	0.0017 4
2126 - 2187	0.0061	0.00028 8
2307 - 2343	0.009	0.00039 14
2407 - 2443	0.0048	0.00020 7
2530 - 2566	0.0028	0.00011 4

$^{146}_{56}$Ba(2.20 *3* s)

Mode: β-
Δ: -65100 *110* keV
SpA: 3.01×10^{10} Ci/g
Prod: fission

Photons (^{146}Ba)

γ_{mode}	γ(keV)	γ(rel)
γ	56.30 6	2.9 3
γ	75.93 6	0.63 7
γ	77.52 6	4.5 5
γ	97.89 6	1.83 14
γ	107.43 8	1.55 7
γ	107.99 5	0.56 7
γ	114.98 8	4.79 14
γ	121.26 5	70.4 14
γ	139.87 7	4.43 14
γ	140.88 5	100.0 22
γ	144.73 6	13 3
γ	145.41 5	1.20 21
γ	148.67 12	0.42 7
γ	159.12 7	1.83 14
γ	164.69 7	2.0 3
γ	171.69 6	1.9 3
γ	175.41 5	23.8 6
γ	182.15 9	1.3 4
γ	186.01 7	7.46 21
γ	197.18 5	61.5 12
γ	208.25 7	1.48 7
γ	218.01 7	1.48 14
γ	231.71 6	53.2 11
γ	241.91 6	14.0 4
γ	246.56 11	3.38 21
γ	247.86 8	4.50 21
γ	251.34 6	97 3
γ	254.60 7	1.06 21
γ	269.58 6	21.3 7
γ	270.08 8	6.6 4
γ	272.90 9	1.3 3
γ	274.75 6	4.5 6
γ	279.67 5	1.55 21
γ	279.84 8	19.5 8
γ	283.39 10	0.77 21
γ	284.61 7	13.5 4
γ	290.88 11	2.9 4
γ	291.55 9	5.6 4
γ	295.07 6	27.0 9
γ	296.6 3	~0.6
γ	298.21 6	21.1 5
γ	301.60 9	1.06 21
γ	302.93 8	2.32 21
γ	310.8 17	<0.14
γ	314.59 17	2.18 21
γ	316.43 7	13.2 8

Photons (^{146}Ba)
(continued)

γ_{mode}	γ(keV)	γ(rel)
γ	317.84 7	2.0 4
γ	322.48 11	1.8 3
γ	325.88 6	1.7 5
γ[M1+E2]	336.58 7	1.3 3
γ	342.25 9	4.0 3
γ	347.58 12	1.8 4
γ	349.89 9	3.5 4
γ	352.27 7	3.9 4
γ	355.44 9	5.0 3
γ	355.56 8	0.99 14
γ	359.29 8	1.8 4
γ	360.30 17	1.0 4
γ[E1]	372.59 4	4.9 7
γ	377.56 5	5.8 3
γ	380.30 8	1.7 4
γ	385.04 10	1.8 3
γ	388.75 8	1.8 4
γ	389.21 9	1.9 4
γ	392.24 8	12.0 25
γ	392.74 6	18 3
γ	413.71 6	11.3 4
γ	417.63 7	6.2 5
γ	429.34 5	30.0 8
γ	433.86 6	~0.9
γ	436.73 11	1.6 4
γ	439.09 5	6.8 3
γ	441.38 7	9.7 4
γ	443.74 11	1.5 3
γ	450.16 6	4.9 3
γ	462.84 8	1.13 21
γ	466.76 5	4.3 15
γ	466.93 9	2.2 4
γ	478.34 8	1.27 21
γ	488.23 7	10.8 4
γ	489.99 6	4.0 3
γ	500.17 7	1.8 3
γ	503.24 21	1.2 4
γ	506.46 7	0.84 14
γ[M1+E2]	507.88 6	5.2 9
γ	509.6 17	<0.21
γ	511.99 8	2.25 21
γ	525.30 10	0.7 3
γ	530.79 8	1.76 14
γ	534.64 9	1.7 3
γ	546.74 11	0.56 21
γ	555.92 17	5.3 3
γ	564.44 7	2.7 4
γ	565.1 3	0.42 14
γ	568.29 8	1.4 4
γ	574.75 5	3.4 3
γ	576.50 8	1.0 3
γ	585.40 7	1.55 21
γ	588.96 8	1.13 21
γ	597.97 6	1.3 4
γ	607.33 8	1.9 4
γ	613.0 3	0.9 3
γ	621.87 8	1.48 21
γ	635.39 6	2.8 3
γ	669.08 10	1.3 6
γ	671.99 7	0.77 21
γ	681.9 1	0.9 3
γ	683.29 6	1.3 3
γ	692.14 5	7.0 3
γ	702.19 14	1.97 21
γ	715.31 8	1.3 3
γ	722.48 9	3.7 4
γ	724.10 9	5.8 4
γ	728.6 5	0.77 21
γ	734.02 16	0.8 3
γ	735.99 10	1.7 3
γ	742.98 9	1.48 21
γ	744.29 13	0.84 21
γ	751.33 14	1.5 4
γ	752.73 8	4.4 4
γ	759.21 6	4.36 14
γ	761.08 13	0.42 7
γ	764.44 10	0.70 14
γ	769.66 7	1.1 3
γ	772.56 9	0.7 3
γ	785.13 10	3.1 9
γ	789.33 9	2.5 4
γ	794.88 10	2.3 4
γ	802.46 8	3.1 4

Photons (^{146}Ba)
(continued)

γ_{mode}	γ(keV)	γ(rel)
γ	809.48 8	2.0 4
γ	812.9 3	0.7 3
γ	817.82 13	1.27 14
γ	822.08 8	3.17 14
γ	830.12 8	0.28 7
γ	834.39 8	1.3 3
γ	841.22 9	1.34 21
γ	842.01 16	1.83 21
γ	847.51 16	2.67 14
γ	851.59 8	0.35 7
γ	851.62 9	1.41 14
γ	867.13 13	1.2 4
γ	868.91 7	3.0 6
γ	869.68 16	1.2 6
γ	876.47 9	1.13 7
γ[E1]	880.47 6	4.7 3
γ	887.35 7	2.46 14
γ	894.68 6	6.3 11
γ	896.62 8	1.55 14
γ	915.59 12	1.13 14
γ	943.48 7	1.48 7
γ	949.24 14	0.35 7
γ	955.43 9	0.70 14
γ	974.15 8	0.70 7
γ	976.89 7	1.55 14
γ	981.20 9	0.21 7
γ	993.23 13	1.76 14
γ	1002.67 7	0.42 14
γ	1013.46 14	0.9 4
γ	1021.52 17	<0.28
γ	1023.22 10	0.63 14
γ	1040.09 7	1.69 7
γ	1043.37 8	1.20 7
γ	1049.54 14	~0.42
γ	1050.91 8	1.62 21
γ	1052.39 8	1.97 7
γ	1060.81 9	0.77 7
γ	1064.73 5	4.0 8
γ	1068.41 18	4.79 10
γ	1071.06 7	6.0 3
γ	1076.68 7	2.25 14
γ	1078.62 17	2.6 3
γ	1088.99 7	1.55 7
γ	1095.60 6	6.33 21
γ	1096.83 7	1.5 6
γ	1101.16 23	0.56 7
γ	1105.35 6	3.59 7
γ	1109.13 7	2.04 7
γ	1128.33 8	0.91 7
γ	1141.94 8	0.35 14
γ	1148.58 7	0.70 7
γ	1155.8 3	0.91 14
γ	1162.09 6	1.55 7
γ	1174.35 7	1.27 21
γ	1182.07 8	0.42 14
γ	1184.28 8	1.13 7
γ	1185.67 13	1.90 7
γ	1195.42 13	1.4 3
γ	1202.84 18	0.77 14
γ	1207.13 14	0.77 14
γ	1211.95 8	1.3 4
γ	1214.0 3	0.77 14
γ	1226.7 3	0.35 7
γ	1232.02 13	0.49 7
γ	1246.47 7	0.70 7
γ	1254.3 3	0.99 7
γ	1258.30 10	0.42 7
γ	1283.54 14	0.63 7
γ	1311.67 25	0.56 14
γ	1320.8 3	1.06 7
γ	1328.54 8	0.35 7
γ	1337.50 7	0.42 14
γ	1338.50 18	0.99 14
γ	1340.85 8	1.48 21
γ	1350.04 14	1.41 21
γ	1384.84 19	0.99 14
γ	1404.99 18	0.49 14
γ	1427.57 14	0.63 7
γ	1443.22 20	1.5 4
γ	1453.86 9	0.49 7
γ	1456.71 23	0.63 14
γ	1482.52 18	0.70 14
γ	1483.88 13	1.55 21

Photons (^{146}Ba)
(continued)

γ_{mode}	γ(keV)	γ(rel)
γ	1489.57 *21*	0.49 *7*
γ	1492.9 *17*	
γ	1496.2 *3*	0.84 *14*
γ	1503.50 *13*	1.90 *14*
γ	1525.45 *14*	0.35 *7*
γ	1581.9 *3*	0.63 *7*
γ	1643.0 *3*	1.8 *4*
γ	1650.3 *10*	0.28 *14*
γ	1656.33 *19*	0.42 *7*
γ	1667.5 *5*	0.49 *14*
γ	1708.9 *5*	0.99 *14*
γ	1767.4 *17*	<0.6
γ	1773.14 *23*	0.28 *7*
γ	1870.81 *22*	0.56 *14*
γ	1899.9 *7*	0.56 *14*
γ	1903.7 *5*	0.56 *14*
γ	1919.8 *3*	0.28 *7*
γ	1939.4 *3*	0.42 *7*
γ	1964.6 *17*	<0.35
γ	1992.9 *17*	<0.35
γ	2030.7 *17*	<0.14
γ	2044.62 *22*	0.56 *14*

$^{146}_{57}$La(6.27 *10* s)

Mode: β-
Δ: -69370 *90* keV
SpA: 1.167×10^{10} Ci/g
Prod: fission

Photons (^{146}La)

⟨γ⟩=1478 *22* keV

γ_{mode}	γ(keV)	γ(%)†
Ce L_ℓ	4.289	0.0109 *18*
Ce L_η	4.730	0.0047 *7*
Ce L_α	4.838	0.29 *4*
Ce L_β	5.387	0.27 *4*
Ce L_γ	6.125	0.036 *6*
Ce $K_{\alpha2}$	34.279	1.30 *13*
Ce $K_{\alpha1}$	34.720	2.39 *24*
Ce $K_{\beta1}$'	39.232	0.67 *7*
Ce $K_{\beta2}$'	40.437	0.181 *18*
γ[M1+E2]	107.64 *6*	<0.023
γ[E1]	118.53 *6*	0.061 *15*
γ[M1+E2]	194.74 *12*	0.040 *19*
γ E2	258.46 *4*	76
γ[E1]	292.35 *6*	0.84 *5*
γ	294.87 *9*	<0.08
γ[M1+E2]	302.38 *11*	<0.08
γ	316.55 *12*	<0.08
γ	346.21 *8*	0.334 *23*
γ	366.62 *8*	0.289 *23*
γ[E2]	382.97 *10*	<0.08
γ E2	409.95 *7*	5.17 *23*
γ[E1]	421.15 *5*	0.395 *23*
γ[E1]	457.35 *5*	0.89 *5*
γ	467.11 *15*	<0.08
γ[E2]	502.93 *9*	0.42 *3*
γ[E1]	514.70 *10*	0.51 *3*
γ	533.56 *11*	0.099 *23*
γ	549.85 *16*	0.129 *15*
γ	595.83 *7*	0.357 *23*
γ	646.04 *17*	<0.08
γ[E1]	666.10 *4*	7.37 *23*
γ	693.18 *11*	~0.38
γ[E1]	702.30 *5*	7.70 *23*
γ[E2]	713.50 *6*	0.71 *3*
γ	744.60 *14*	0.12 *3*
γ	756.78 *8*	0.205 *23*
γ E2	784.63 *5*	3.57 *23*
γ	787.14 *15*	<0.08
γ	793.03 *9*	1.14 *15*

Photons (^{146}La)
(continued)

γ_{mode}	γ(keV)	γ(%)†
γ	797.35 *12*	<0.08
γ	808.75 *9*	0.182 *23*
γ	829.23 *7*	1.15 *5*
γ[E1]	832.03 *5*	0.51 *3*
γ	836.07 *9*	0.17 *3*
γ	852.23 *13*	0.48 *5*
γ	870.32 *14*	0.48 *7*
γ	881.66 *10*	~0.08
γ[M1+E2]	908.24 *13*	0.342 *23*
γ	915.04 *12*	0.099 *23*
γ[E1]	924.57 *4*	8.9 *3*
γ	927.58 *13*	0.20 *3*
γ	948.42 *13*	0.205 *23*
γ	959.18 *9*	0.289 *23*
γ	992.83 *16*	0.15 *4*
γ[M1+E2]	1015.81 *4*	4.03 *15*
γ	1028.47 *13*	0.30 *3*
γ	1036.70 *11*	0.65 *4*
γ	1043.46 *14*	0.17 *3*
γ	1064.67 *14*	~0.08
γ	1083.41 *14*	0.448 *23*
γ	1114.89 *22*	0.17 *4*
γ[M1+E2]	1123.45 *4*	0.52 *5*
γ	1128.8 *3*	0.34 *3*
γ	1134.04 *5*	0.22 *3*
γ	1140.00 *11*	0.49 *4*
γ	1167.1 *3*	0.26 *3*
γ	1172.70 *8*	1.02 *5*
γ	1187.57 *14*	<0.08
γ	1195.16 *10*	0.129 *23*
γ	1201.94 *14*	0.20 *3*
γ	1239.6 *4*	<0.08
γ	1261.92 *14*	0.091 *15*
γ[E2]	1274.27 *5*	1.52 *8*
γ[M1+E2]	1318.19 *11*	1.52 *8*
γ	1324.91 *11*	0.09 *3*
γ	1336.5 *3*	0.24 *3*
γ	1348.04 *13*	<0.08
γ	1350.20 *11*	0.18 *3*
γ	1354.58 *9*	0.106 *8*
γ	1362.64 *8*	0.198 *23*
γ	1368.8 *4*	0.114 *23*
γ	1376.60 *13*	<0.08
γ[E2]	1381.91 *5*	2.05 *8*
γ	1386.4 *1*	0.25 *3*
γ[E2]	1398.77 *10*	0.52 *3*
γ	1407.24 *11*	0.114 *23*
γ	1443.44 *10*	0.33 *3*
γ	1486.21 *10*	0.24 *4*
γ	1489.70 *11*	0.21 *3*
γ	1495.33 *8*	<0.08
γ[M1+E2]	1498.13 *5*	1.75 *8*
γ	1509.11 *23*	0.13 *3*
γ	1538.41 *5*	0.78 *3*
γ	1543.91 *6*	0.52 *4*
γ	1550.90 *11*	0.46 *3*
γ	1573.43 *14*	0.92 *5*
γ	1581.36 *20*	0.14 *3*
γ	1585.25 *9*	0.061 *15*
γ	1587.50 *11*	0.167 *15*
γ	1594.60 *21*	0.046 *15*
γ	1619.16 *14*	0.35 *3*
γ	1625.81 *12*	0.084 *15*
γ	1642.91 *10*	0.175 *15*
γ	1671.64 *16*	0.175 *15*
γ	1678.7 *3*	0.190 *23*
γ	1685.46 *20*	0.099 *15*
γ	1701.13 *10*	0.175 *15*
γ	1733.9 *3*	0.046 *8*
γ	1742.27 *14*	0.084 *15*
γ	1752.22 *17*	<0.08
γ	1752.52 *11*	0.122 *23*
γ[E2]	1756.59 *5*	0.96 *4*
γ	1772.59 *7*	0.53 *4*
γ	1793.00 *8*	0.74 *3*
γ	1803.76 *10*	0.251 *23*
γ	1813.15 *12*	<0.08
γ	1828.4 *3*	0.046 *15*
γ	1831.89 *15*	0.205 *23*
γ	1868.04 *13*	0.40 *5*
γ	1892.07 *22*	0.137 *15*
γ	1897.46 *10*	0.243 *15*
γ	1908.11 *21*	0.053 *15*
γ	1916.41 *11*	0.144 *15*

Photons (^{146}La)
(continued)

γ_{mode}	γ(keV)	γ(%)†
γ	1920.93 *21*	0.220 *15*
γ	1937.2 *1*	0.167 *15*
γ	1944.31 *21*	0.243 *23*
γ	1959.93 *9*	0.114 *8*
γ	1963.9 *3*	0.030 *8*
γ	1968.5 *3*	<0.08
γ	1974.98 *9*	0.175 *15*
γ	1981.1 *3*	0.076 *15*
γ	1992.48 *9*	0.274 *23*
γ	2028.68 *10*	0.59 *4*
γ	2052.50 *11*	0.97 *6*
γ	2054.06 *10*	0.31 *3*
γ	2060.08 *11*	0.441 *23*
γ	2072.96 *10*	0.114 *8*
γ	2109.54 *10*	0.061 *15*
γ	2127.36 *20*	0.099 *15*
γ	2140.56 *15*	0.243 *15*
γ	2155.92 *10*	0.388 *23*
γ	2179.39 *21*	0.304 *23*
γ	2183.9 *3*	0.160 *15*
γ	2188.50 *9*	1.41 *7*
γ	2233.44 *10*	<0.08
γ	2253.20 *10*	0.82 *5*
γ	2291.06 *10*	0.084 *8*
γ	2292.96 *20*	0.198 *15*
γ	2310.96 *10*	0.18 *3*
γ	2318.54 *11*	0.62 *3*
γ	2322.27 *8*	0.433 *23*
γ	2332.96 *15*	0.43 *3*
γ	2358.47 *7*	2.43 *6*
γ	2368.00 *10*	0.243 *15*
γ	2368.10 *13*	0.243 *15*
γ	2380.96 *15*	0.304 *23*
γ	2395.36 *10*	0.167 *15*
γ	2397.66 *8*	0.43 *3*
γ	2404.30 *12*	0.410 *23*
γ	2410.96 *10*	0.205 *23*
γ	2417.27 *9*	0.426 *23*
γ	2454.81 *11*	0.59 *3*
γ	2474.93 *10*	0.44 *3*
γ	2520.46 *20*	0.47 *3*
γ	2533.15 *10*	0.47 *3*
γ	2558.96 *20*	0.160 *23*
γ	2582.51 *11*	0.95 *6*
γ	2603.30 *10*	1.19 *7*
γ	2610.40 *21*	0.296 *15*
γ	2694.77 *9*	1.25 *4*
γ	2861.75 *10*	0.091 *15*
γ	2905.96 *20*	0.266 *15*
γ	2910.26 *20*	0.137 *15*
γ	2996.2 *4*	0.258 *15*
γ	2996.86 *25*	0.258 *15*
γ	2998.16 *20*	0.160 *15*
γ	3024.56 *8*	0.74 *5*
γ	3029.0 *3*	<0.08
γ	3070.39 *13*	1.04 *8*
γ	3083.36 *9*	0.49 *3*
γ	3276.6 *3*	0.190 *15*
γ	3450.1 *3*	<0.08
γ	3486.3 *3*	<0.08
γ	3698.08 *25*	0.114 *15*
γ	3765.5 *3*	<0.08
γ	4152.4 *3*	0.175 *15*
γ	4431.6 *3*	0.053 *8*
γ	4690.1 *3*	0.030 *8*

† 11% uncert(syst)

Atomic Electrons (^{146}La)

⟨e⟩=15.6 *11* keV

e_{bin}(keV)	⟨e⟩(keV)	e(%)
6 - 39	0.37	4.5 *5*
67 - 113	0.019	0.024 *11*
117 - 154	0.010	0.006 *3*
188 - 195	0.0029	0.0016 *6*
218	10.3	4.7 *5*
252	2.00	0.79 *8*
253	0.56	0.223 *23*

Atomic Electrons (^{146}La) (continued)

e_{bin}(keV)	⟨e⟩(keV)	e(%)
254	0.003	<0.0027
257	0.56	0.218 *22*
258 - 306	0.188	0.071 *8*
310 - 346	0.026	0.008 *4*
360 - 367	0.0040	~0.0011
370	0.298	0.081 *4*
376 - 421	0.0935	0.0231 *6*
427 - 474	0.0288	0.0062 *5*
493 - 534	0.066	0.0133 *23*
543 - 590	0.013	~0.0024
594 - 640	0.078	0.0125 *6*
645 - 693	0.117	0.0176 *14*
696 - 744	0.103	0.0140 *9*
747 - 796	0.080	0.010 *3*
797 - 847	0.027	0.0033 *10*
851 - 887	0.082	0.0093 *5*
902 - 948	0.022	0.0024 *4*
952 - 996	0.100	0.0102 *18*
1003 - 1043	0.027	0.0026 *5*
1058 - 1100	0.027	0.0025 *5*
1108 - 1155	0.023	0.0021 *7*
1161 - 1202	0.0067	0.00057 *17*
1221 - 1269	0.0238	0.00193 *12*
1273 - 1322	0.037	0.0029 *4*
1323 - 1371	0.0384	0.00285 *17*
1375 - 1407	0.0093	0.00067 *13*
1437 - 1486	0.030	0.0021 *3*
1488 - 1586	0.043	0.0028 *5*
1588 - 1684	0.0089	0.00054 *10*
1693 - 1792	0.031	0.00177 *22*
1797 - 1896	0.0111	0.00060 *11*
1897 - 1991	0.0148	0.00076 *12*
2012 - 2108	0.019	0.00095 *18*
2115 - 2213	0.024	0.00112 *22*
2227 - 2321	0.031	0.0013 *3*
2326 - 2416	0.024	0.00103 *12*
2434 - 2532	0.0113	0.00046 *8*
2542 - 2609	0.017	0.00066 *13*
2654 - 2694	0.0083	0.00031 *10*
2821 - 2909	0.0031	0.000108 *20*
2956 - 3043	0.0167	0.00056 *8*
3064 - 3082	0.0012	4.0 *9* $\times10^{-5}$
3236 - 3275	0.0011	3.3 *9* $\times10^{-5}$
3410 - 3485	0.0004	~1 $\times10^{-5}$
3658 - 3725	0.00076	2.1 *6* $\times10^{-5}$
3759 - 3764	2.7 $\times10^{-5}$	~7 $\times10^{-7}$
4112 - 4151	0.00084	2.0 *5* $\times10^{-5}$
4391 - 4430	0.00025	5.6 *14* $\times10^{-6}$
4650 - 4689	0.00014	3.0 *9* $\times10^{-6}$

Continuous Radiation (^{146}La)

⟨β-⟩=2147 keV;⟨IB⟩=7.9 keV

E_{bin}(keV)		⟨ ⟩(keV)	(%)
0 - 10	β-	0.00375	0.075
	IB	0.059	
10 - 20	β-	0.0115	0.076
	IB	0.059	0.41
20 - 40	β-	0.0479	0.159
	IB	0.116	0.40
40 - 100	β-	0.374	0.53
	IB	0.34	0.52
100 - 300	β-	4.86	2.35
	IB	1.03	0.57
300 - 600	β-	24.3	5.3
	IB	1.30	0.30
600 - 1300	β-	183	18.8
	IB	2.2	0.25
1300 - 2500	β-	700	37.2
	IB	2.0	0.112
2500 - 5000	β-	1155	34.1
	IB	0.86	0.028
5000 - 6386	β-	80	1.50
	IB	0.0080	0.000149

$^{146}_{57}$La(10.0 *1* s)

Mode: β-
Δ: -69090 *150* keV
SpA: 7.47×10^{9} Ci/g
Prod: fission

Photons (^{146}La)

⟨γ⟩=1311 *97* keV

γ_{mode}	γ(keV)	γ(%)
γ	183.3 *3*	8.2 *16*
γ E2	258.46 *4*	93 *19*
γ [E1]	292.35 *6*	1.14 *13*
γ	366.9 *4*	4.7 *9*
γ	379.9 *5*	7.3 *15*
γ E2	409.95 *7*	81 *16*
γ	446.0 *5*	7.3 *15*
γ [E2]	502.93 *9*	26 *5*
γ [E1]	514.70 *10*	31 *6*
γ	550.2 *4*	2.1 *4*
γ	666.4 *3*	2.6 *5*
γ [E1]	702.30 *5*	10.5 *11*
γ	773.2 *4*	2.1 *4*
γ	785.0 *4*	5.1 *10*
γ	958.8 *3*	13 *3*
γ	1142.0 *4*	6.4 *13*

$^{146}_{58}$Ce(13.52 *13* min)

Mode: β-
Δ: -75760 *80* keV
SpA: 9.52×10^{7} Ci/g
Prod: fission

Photons (^{146}Ce)

⟨γ⟩=289 *9* keV

γ_{mode}	γ(keV)	γ(%)†
Pr L_ℓ	4.453	0.18 *3*
Pr L_η	4.929	0.085 *13*
Pr L_α	5.031	4.7 *8*
Pr L_β	5.603	4.9 *8*
Pr L_γ	6.408	0.71 *12*
γ M1	12.204 *19*	0.158 *10*
γ E2	22.83 *3*	~0.010
γ M1+39%E2	35.03 *3*	1.02 *5*
Pr $K_{\alpha2}$	35.550	4.07 *17*
Pr $K_{\alpha1}$	36.026	7.4 *3*
Pr $K_{\beta1}'$	40.720	2.10 *9*
Pr $K_{\beta2}'$	41.981	0.59 *3*
γ M1+38%E2	52.18 *4*	0.479 *25*
γ E2,M1	87.21 *4*	0.71 *5*
γ M1,E2	98.50 *3*	2.9 *3*
γ E1	100.87 *5*	2.29 *15*
γ M1	106.23 *4*	0.56 *5*
γ M1,E2	133.53 *3*	7.4 *4*
γ M1+~50%E2	141.26 *4*	2.96 *15*
γ	173.82 *15*	0.209 *10*
γ E1	210.48 *4*	4.49 *20*
γ E1	218.212 *25*	18.9 *10*
γ	245.48 *10*	0.245 *15*
γ M1,E2	250.87 *5*	2.29 *10*
γ E1	264.53 *3*	8.2 *4*
γ E1	316.708 *24*	51.0 *25*
γ [E1]	351.74 *3*	2.09 *10*
γ	369.41 *5*	0.158 *10*
γ	375.63 *13*	0.092 *10*
γ	415.72 *5*	1.17 *5*
γ	462.68 *20*	0.031 *5*
γ	467.90 *5*	0.76 *5*
γ	490.73 *5*	0.056 *10*
γ [E1]	502.94 *4*	0.92 *5*
γ	526.81 *15*	0.061 *5*

† 7.8% uncert(syst)

Atomic Electrons (^{146}Ce)

⟨e⟩=43.1 *25* keV

e_{bin}(keV)	⟨e⟩(keV)	e(%)
5	0.59	11.1 *8*
6	3.6	58 *9*
7 - 28	3.9	28 *4*
29	12.2	42 *7*
30	0.131	0.44 *5*
34	3.4	10.1 *16*
35	0.97	2.8 *4*
36 - 45	0.75	1.67 *15*
46	1.2	2.7 *8*
51 - 52	0.45	0.88 *20*
57	2.00	3.5 *4*
59 - 87	1.2	1.7 *4*
92	4.2	4.6 *8*
93 - 98	0.9	~1.0
99	1.32	1.33 *7*
100 - 141	2.3	1.8 *5*
167 - 212	1.67	0.90 *4*
217 - 265	0.48	0.209 *13*
275	1.37	0.50 *3*
310 - 352	0.333	0.107 *5*
363 - 410	0.08	~0.020
414 - 463	0.061	0.014 *5*
466 - 503	0.0068	0.0014 *4*
520 - 527	0.00047	~9 $\times10^{-5}$

$^{146}_{59}$Pr(24.15 *18* min)

Mode: β-
Δ: -76780 *70* keV
SpA: 5.33×10^{7} Ci/g
Prod: fission; ^{146}Nd(n,p)

Photons (^{146}Pr)

⟨γ⟩=1018 *17* keV

γ_{mode}	γ(keV)	γ(%)†
Nd L_ℓ	4.633	0.00164 *18*
Nd L_η	5.146	0.00066 *6*
Nd L_α	5.228	0.044 *3*
Nd L_β	5.854	0.040 *4*
Nd L_γ	6.681	0.0056 *6*
Nd $K_{\alpha2}$	36.847	0.184 *10*
Nd $K_{\alpha1}$	37.361	0.335 *17*
Nd $K_{\beta1}'$	42.240	0.095 *5*
Nd $K_{\beta2}'$	43.562	0.0278 *15*
γ [E1]	146.33 *18*	0.049 *6*
γ [M1+E2]	191.08 *16*	0.015 *7*
γ	446.57 *25*	0.077 *10*
γ E2	453.89 *4*	48.0 *24*
γ [E2]	461.7 *3*	0.046 *9*
γ	481.55 *19*	~0.038
γ [M1+E2]	507.96 *7*	0.461 *24*
γ	537.57 *20*	0.28 *4*
γ	562.17 *16*	0.370 *19*
γ [E1]	587.91 *15*	0.072 *14*
γ E2	589.48 *17*	0.346 *19*
γ [E1]	597.90 *16*	0.079 *8*
γ [E1]	601.83 *5*	2.98 *14*
γ [E1]	716.08 *9*	0.064 *8*

Photons (^{146}Pr) (continued)

γ_{mode}	γ(keV)	γ(%)†
γ [E1]	727.32 *10*	0.57 *3*
γ E1	735.81 *5*	7.4 *4*
γ	766.98 *17*	0.036 *8*
γ	772.3 *8*	0.031 *7*
γ	774.50 *19*	0.259 *14*
γ [E1]	788.98 *5*	6.3 *3*
γ	817.1 *3*	0.017 *7*
γ	839.62 *17*	0.029 *8*
γ [M1+E2]	849.4 *4*	0.079 *10*
γ [E1]	922.96 *5*	2.33 *12*
γ [E1]	928.38 *17*	0.168 *11*
γ	954.12 *17*	0.016 *3*
γ	1012.5 *3*	0.13 *4*
γ [M1+E2]	1016.83 *6*	1.23 *6*
γ	1081.39 *16*	0.79 *4*
γ	1148.7 *3*	0.226 *14*
γ	1164.9 *5*	0.126 *11*
γ	1183.38 *25*	0.093 *8*
γ	1191.7 *4*	0.047 *6*
γ [E1]	1235.27 *11*	0.365 *19*
γ [E2]	1243.37 *20*	0.64 *3*
γ	1248.4 *3*	0.048 *9*
γ [E2]	1303.3 *4*	0.071 *9*
γ [M1+E2]	1323.72 *15*	0.56 *3*
γ [M1+E2]	1329.14 *10*	0.58 *3*
γ [M1+E2]	1333.71 *16*	0.69 *3*
γ	1338.7 *5*	0.130 *11*
γ [E1]	1376.84 *5*	4.37 *24*
γ	1411.2 *5*	0.081 *8*
γ	1436.1 *4*	0.159 *11*
γ [M1+E2]	1451.88 *8*	2.28 *12*
γ	1463.41 *24*	0.054 *7*
γ [E2]	1470.72 *6*	1.19 *6*
γ	1500.26 *22*	0.086 *10*
γ	1504.9 *3*	0.038 *10*
γ	1508.74 *24*	0.078 *7*
γ [M1+E2]	1516.29 *10*	0.26 *3*
γ	1518.2 *10*	~0.20
γ [M1+E2]	1524.79 *4*	15.6 *8*
γ	1529.9 *10*	0.125 *19*
γ	1555.7 *4*	0.080 *7*
γ	1594.13 *22*	0.137 *10*
γ	1614.12 *16*	0.056 *7*
γ	1650.0 *8*	0.060 *7*
γ	1689.93 *17*	0.62 *3*
γ	1741.3 *9*	0.11 *3*
γ	1744.5 *5*	0.107 *11*
γ	1766.0 *4*	0.072 *10*
γ	1781.28 *22*	0.076 *4*
γ [E2]	1787.59 *16*	0.085 *4*
γ	1812.8 *5*	0.105 *8*
γ	1831.2 *3*	0.239 *14*
γ	1882.6 *10*	0.063 *7*
γ	1897.9 *20*	~0.04
γ	1902.75 *19*	0.13 *4*
γ [E2]	1905.77 *9*	0.044 *7*
γ	1915.7 *3*	0.085 *7*
γ	1921.00 *5*	0.054 *7*
γ	1940.2 *8*	0.075 *8*
γ	1958.7 *4*	0.184 *12*
γ	1961.4 *10*	0.041 *6*
γ [E2]	1978.67 *6*	0.216 *14*
γ	1984.2 *3*	0.247 *14*
γ	1992.33 *23*	0.060 *7*
γ	2006.33 *18*	0.244 *14*
γ	2056.1 *15*	~0.029
γ	2098.21 *17*	0.097 *13*
γ	2121.0 *10*	0.089 *8*
γ	2127.0 *10*	0.090 *8*
γ	2143.81 *17*	0.174 *11*
γ	2149.2 *8*	0.024 *5*
γ	2157.5 *4*	0.055 *7*
γ	2179.48 *23*	0.212 *12*
γ	2208.0 *10*	0.058 *19*
γ	2218.0 *4*	0.082 *11*
γ	2227.68 *21*	0.456 *24*
γ	2246.0 *10*	0.040 *6*
γ [E1]	2252.09 *9*	1.03 *5*
γ	2266.7 *5*	0.062 *14*
γ	2322.1 *15*	0.030 *5*
γ	2356.63 *19*	0.83 *4*
γ	2372.1 *10*	0.043 *19*
γ	2460.21 *18*	0.499 *24*
γ	2477.7 *4*	0.042 *5*
γ	2517.08 *21*	0.331 *19*
γ	2681.56 *21*	0.432 *24*
γ	2775.7 *5*	0.216 *14*
γ	2779.1 *20*	~0.034
γ	2830.4 *3*	0.226 *14*
γ	2881.7 *4*	0.076 *5*
γ	2893.4 *8*	0.0110 *10*
γ	2915.28 *23*	0.044 *4*
γ	2937.82 *8*	0.080 *5*
γ	3080.4 *4*	0.087 *5*
γ	3140.9 *4*	0.080 *6*
γ	3164.9 *3*	0.026 *3*
γ	3255.4 *13*	0.023 *3*
γ	3289.3 *20*	<0.08
γ	3292.5 *3*	0.151 *8*
γ	3386.3 *12*	0.0149 *19*
γ	3709.3 *13*	0.064 *7*

† 31% uncert(syst)

Atomic Electrons (^{146}Pr)

⟨e⟩=4.05 *14* keV

e_{bin}(keV)	⟨e⟩(keV)	e(%)
6 - 42	0.052	0.59 *4*
103 - 148	0.0090	0.0073 *14*
184 - 191	0.0014	0.0008 *3*
403	0.004	~0.0010
410	2.50	0.61 *3*
418 - 446	0.0050	0.0012 *4*
447	0.407	0.091 *5*
448	0.058	0.0131 *7*
452	0.088	0.0195 *10*
453 - 502	0.085	0.0182 *21*
506 - 555	0.034	0.0063 *18*
556 - 602	0.0477	0.0084 *4*
673 - 721	0.083	0.0121 *6*
723 - 772	0.083	0.0111 *8*
773 - 817	0.0138	0.00175 *12*
832 - 879	0.0198	0.00226 *14*
885 - 928	0.0052	0.00057 *3*
947 - 973	0.031	0.0032 *7*
1005 - 1038	0.018	0.0018 *7*
1074 - 1121	0.0077	0.00070 *22*
1140 - 1186	0.0033	0.00029 *8*
1190 - 1237	0.0143	0.00119 *6*
1241 - 1290	0.032	0.0025 *3*
1295 - 1339	0.0340	0.00256 *16*
1368 - 1411	0.042	0.0030 *5*
1420 - 1469	0.0268	0.00187 *11*
1470 - 1475	0.0062	0.00042 *13*
1481	0.23	0.015 *3*
1486 - 1571	0.043	0.0028 *4*
1587 - 1684	0.008	0.00049 *18*
1688 - 1788	0.0086	0.00049 *8*
1806 - 1904	0.0053	0.00029 *5*
1909 - 2005	0.0105	0.00054 *8*
2013 - 2106	0.0042	0.00020 *4*
2114 - 2212	0.0126	0.00058 *8*
2216 - 2316	0.0081	0.00035 *10*
2320 - 2417	0.0048	0.00020 *6*
2434 - 2516	0.0035	0.00014 *4*
2638 - 2736	0.0048	0.00018 *4*
2769 - 2850	0.0024	8.5 *19* $\times 10^{-5}$
2872 - 2937	0.00095	3.3 *7* $\times 10^{-5}$
3037 - 3135	0.00124	4.0 *8* $\times 10^{-5}$
3139 - 3212	0.00016	5.1 *13* $\times 10^{-6}$
3246 - 3343	0.0013	4.0 *10* $\times 10^{-5}$
3379 - 3385	1.3 $\times 10^{-5}$	3.7 *10* $\times 10^{-7}$
3666 - 3708	0.00037	1.0 *3* $\times 10^{-5}$

Continuous Radiation (^{146}Pr)

⟨β-⟩=1298 keV;⟨IB⟩=3.7 keV

E_{bin}(keV)		⟨ ⟩(keV)	(%)
0 - 10	β-	0.0149	0.296
	IB	0.043	
10 - 20	β-	0.0452	0.301
	IB	0.042	0.29
20 - 40	β-	0.186	0.62
	IB	0.083	0.29
40 - 100	β-	1.39	1.98
	IB	0.24	0.36
100 - 300	β-	15.8	7.8
	IB	0.68	0.38
300 - 600	β-	62	13.8
	IB	0.77	0.18
600 - 1300	β-	305	32.4
	IB	1.09	0.126
1300 - 2500	β-	558	30.9
	IB	0.68	0.040
2500 - 4150	β-	356	11.9
	IB	0.100	0.0036

$^{146}_{60}$Nd(stable)

Δ: -80935 *3* keV
%: 17.19 *8*

$^{146}_{61}$Pm(5.53 *5* yr)

Mode: ε(66.1 *13* %), β-(33.9 *13* %)
Δ: -79450 *7* keV
SpA: 443 Ci/g
Prod: ^{146}Nd(p,n); ^{148}Nd(p,3n); ^{146}Nd(d,2n)

Photons (^{146}Pm)

⟨γ⟩=754 *21* keV

γ_{mode}	γ(keV)	γ(%)†
Nd L_ℓ	4.633	0.130 *13*
Nd L_η	5.146	0.051 *4*
Nd L_α	5.228	3.49 *22*
Nd L_β	5.855	3.2 *3*
Nd L_γ	6.697	0.46 *5*
Nd $K_{\alpha2}$	36.847	14.8 *4*
Nd $K_{\alpha1}$	37.361	26.9 *8*
Sm $K_{\alpha2}$	39.522	0.0375 *20*
Sm $K_{\alpha1}$	40.118	0.068 *4*
Nd $K_{\beta1}'$	42.240	7.66 *22*
Nd $K_{\beta2}'$	43.562	2.23 *7*
Sm $K_{\beta1}'$	45.379	0.0197 *10*
Sm $K_{\beta2}'$	46.819	0.0058 *3*
γ_ϵ[E1]	146.33 *18*	0.149 *21*
γ_ϵE2	453.89 *4*	65.1 *24*
γ_ϵE2	589.48 *17*	0.42 *11*
$\gamma_{\beta-}$E1	633.10 *7*	2.15 *20*
γ_ϵE1	735.81 *5*	22.9 *19*
$\gamma_{\beta-}$E2	747.20 *9*	33.9 *16*

† uncert(syst): 2.0% for ε, 3.8% for β-

Atomic Electrons (^{146}Pm)

$\langle e \rangle$= 9.8 *3* keV

e_{bin}(keV)	$\langle e \rangle$(keV)	e(%)
6	1.55	24.9 *25*
7	1.15	16.9 *17*
8 - 29	0.110	0.38 *4*
30	0.30	1.01 *10*
31	0.46	1.50 *15*
32 - 45	0.55	1.53 *9*
103 - 146	0.0149	0.0137 *17*
410	3.38	0.82 *4*
447	0.552	0.124 *5*
448 - 454	0.255	0.0566 *14*
546 - 589	0.046	0.0080 *9*
625 - 633	0.0050	0.00080 *5*
692	0.234	0.034 *3*
700	0.95	0.136 *7*
729 - 747	0.234	0.0316 *9*

Continuous Radiation (^{146}Pm)

$\langle \beta - \rangle$=83 keV;$\langle$IB$\rangle$=0.23 keV

E_{bin}(keV)		$\langle \ \rangle$(keV)	(%)
0 - 10	β-	0.0497	1.00
	IB	0.0044	
10 - 20	β-	0.145	0.97
	IB	0.0040	0.028
20 - 40	β-	0.55	1.85
	IB	0.056	0.165
40 - 100	β-	3.40	4.92
	IB	0.045	0.090
100 - 300	β-	25.5	13.0
	IB	0.043	0.025
300 - 600	β-	46.7	11.1
	IB	0.054	0.0125
600 - 1032	β-	7.1	1.08
	IB	0.028	0.0040

$^{146}_{62}$Sm(1.03 *5* $\times 10^8$ yr)

Mode: α

Δ: -80992 *6* keV

SpA: 2.32×10^{-5} Ci/g

Prod: ^{147}Sm(n,2n); alphas on Nd

Alpha Particles (^{146}Sm)

α(keV)
2460 *20*

$^{146}_{63}$Eu(4.59 *3* d)

Mode: ϵ

Δ: -77114 *10* keV

SpA: 1.948×10^{5} Ci/g

Prod: daughter ^{146}Gd; ^{147}Sm(p,2n)

Photons (^{146}Eu)

$\langle \gamma \rangle$=2219 *160* keV

γ_{mode}	γ(keV)	γ(%)†
γ	234.9 *5*	0.021 *4*
γ E1	271.65 *11*	0.95 *6*
γ	387.3 *5*	0.025 *5*
γ E2,M1	394.23 *20*	0.46 *5*
γ E2(+M1)	397.41 *16*	0.96 *10*
γ	403.7 *4*	0.42 *8*
γ E2	410.74 *18*	0.98 *10*
γ E2	430.48 *9*	4.5 *4*
γ	471.5 *5*	0.049 *10*
γ	482.2 *5*	0.033 *7*
γ	605.8 *5*	0.024 *5*
γ M1(+E2)	621.76 *14*	0.29 *10*
γ E1	633.10 *7*	44 *7*
γ E2	634.00 *7*	38 *6*
γ	652.5 *5*	0.088 *18*
γ E1	664.54 *9*	6.8 *8*
γ	673.3 *5*	0.029 *6*
γ E1	702.14 *9*	6.5 *12*
γ E2+M1	704.88 *21*	2.7 *5*
γ [E2]	720.19 *18*	~0.039
γ [M1+E2]	742.47 *21*	0.88 *18*
γ E2	747.20 *9*	98.0
γ	760.9 *5*	0.098 *20*
γ [E1]	775.34 *18*	0.09 *3*
γ [E2]	790.90 *13*	0.47 *5*
γ M1,E2	804.35 *20*	0.08 *3*
γ	804.8 *3*	0.08 *3*
γ	812.2 *4*	~0.05
γ	850.15 *16*	0.20 *4*
γ [E1]	865.40 *13*	0.13 *3*
γ	881.4 *5*	0.034 *7*
γ M1+E2	888.68 *12*	1.60 *10*
γ [M1+E2]	899.82 *12*	3.8 *3*
γ	914.04 *16*	0.39 *4*
γ	914.12 *13*	0.39 *4*
γ	927.9 *5*	0.022 *4*
γ	937.3 *5*	0.078 *16*
γ	941.2 *5*	0.16 *3*
γ	947.9 *5*	0.0108 *22*
γ	968.6 *5*	0.046 *9*
γ	971.4 *5*	0.069 *14*
γ	978.2 *5*	0.073 *15*
γ	1009.2 *5*	0.0108 *22*
γ	1021.9 *5*	0.030 *6*
γ	1027.1 *5*	0.068 *14*
γ	1037.0 *5*	0.074 *15*
γ	1047.5 *5*	0.058 *12*
γ	1053.2 *5*	0.127 *25*
γ E2(+M1)	1057.82 *12*	5.6 *6*
γ [M1+E2]	1086.48 *14*	0.29 *3*
γ	1094.3 *5*	0.029 *6*
γ	1107.3 *5*	0.049 *10*
γ	1116.88 *17*	0.41 *4*
γ E2+M1	1133.22 *13*	0.87 *7*
γ M1(+E2)	1150.68 *16*	2.18 *15*
γ [M1+E2]	1150.91 *20*	2.18 *18*
γ	1161.8 *5*	0.0108 *22*
γ	1166.7 *5*	0.023 *5*
γ M1,E2	1176.62 *12*	2.07 *20*
γ	1214.22 *12*	0.22 *3*
γ	1231.0 *5*	0.018 *4*
γ	1230.96 *3*	0.018 *4*
γ	1255.6 *5*	0.036 *7*
γ	1273.5 *8*	0.098 *20*
γ E2(+M1)	1297.23 *13*	5.6 *4*
γ	1333.04 *20*	0.19 *3*
γ	1345.50 *13*	0.127 *20*
γ M1+E2	1356.14 *20*	0.65 *3*
γ E2,M1	1378.64 *14*	1.03 *10*
γ	1401.7 *5*	0.046 *9*
γ (E1)	1407.91 *19*	2.54 *21*
γ E2	1408.44 *18*	3.26 *7*
γ (M1)	1445.54 *20*	0.35 *6*
γ	1488.1 *5*	0.022 *4*
γ	1488.34 *15*	~0.06
γ	1490.2 *5*	0.022 *4*
γ	1497.5 *5*	0.030 *6*
γ	1497.89 *11*	0.030 *6*
γ (M1)	1501.04 *23*	0.12 *4*
γ E2,M1	1516.96 *13*	0.32 *6*
γ [M1+E2]	1522.68 *12*	1.25 *10*
γ E2	1533.82 *12*	1.25 *10*
γ	1580.0 *5*	0.0118 *24*
γ E2,M1	1592.65 *14*	0.53 *6*
γ M1(+E2)	1633.51 *12*	0.42 *5*
γ E2	1648.12 *11*	0.79 *10*
γ	1664.2 *13*	~0.039
γ	1681.9 *5*	0.019 *4*
γ M1,E2	1686.42 *14*	0.73 *7*
γ	1691.74 *20*	0.40 *5*
γ E2,M1	1691.82 *12*	0.40 *5*
γ	1711.94 *20*	0.137 *20*
γ	1724.24 *20*	0.08 *3*
γ M1(+E2)	1756.17 *12*	0.96 *10*
γ E1	1766.31 *12*	0.71 *7*
γ	1796.7 *5*	0.036 *7*
γ [M1+E2]	1797.16 *20*	0.036 *7*
γ M1,E2	1802.75 *18*	0.16 *3*
γ	1818.94 *20*	0.12 *3*
γ	1858.48 *22*	~0.05
γ	1861.9 *5*	0.016 *3*
γ E1,E2	1878.76 *12*	0.18 *4*
γ	1897.9 *5*	0.024 *5*
γ	1898.7 *5*	0.042 *5*
γ	1902.4 *5*	0.037 *7*
γ E2	1931.23 *13*	1.37 *12*
γ	1948.9 *10*	0.029 *10*
γ	1956.9 *10*	0.069 *10*
γ	1962.9 *5*	0.056 *11*
γ	1966.0 *3*	0.25 *4*
γ	1980.3 *3*	0.18 *3*
γ E2,M1	1980.3 *3*	0.22 *3*
γ	2004.2 *5*	0.026 *5*
γ	2011.34 *20*	0.127 *20*
γ M1,E2	2011.38 *17*	0.147 *20*
γ	2017.3 *5*	0.024 *5*
γ	2038.8 *3*	0.049 *10*
γ E2,M1	2051.83 *19*	0.82 *10*
γ E2	2080.78 *14*	2.19 *20*
γ	2096.0 *5*	0.014 *3*
γ	2103.3 *3*	0.09 *3*
γ	2131.88 *10*	0.026 *5*
γ	2132.0 *5*	0.026 *5*
γ E1,E2	2136.56 *18*	0.17 *3*
γ E2	2155.63 *18*	0.54 *6*
γ	2164.5 *3*	0.07 *3*
γ	2203.8 *3*	0.11 *3*
γ	2222.2 *3*	0.11 *3*
γ [E2]	2226.64 *14*	~0.039
γ (M1)	2244.5 *3*	0.19 *4*
γ	2244.81 *23*	0.147 *20*
γ E2,M1	2267.50 *12*	0.51 *5*
γ	2310.4 *3*	0.020 *4*
γ	2310.7 *5*	0.020 *4*
γ M1,E2	2320.41 *15*	0.108 *20*
γ	2346.04 *20*	0.34 *4*
γ	2359.0 *3*	0.039 *10*
γ	2389.26 *12*	0.24 *5*
γ (E2,M1)	2401.24 *20*	0.26 *5*
γ	2401.76 *21*	0.27 *5*
γ E2,M1	2436.74 *17*	1.01 *10*
γ E2,M1	2491.57 *21*	0.26 *4*
γ	2498.07 *19*	0.049 *10*
γ [E2]	2544.35 *21*	0.049 *10*
γ	2629.6 *3*	0.059 *20*
γ E1	2644.47 *17*	0.147 *20*
γ	2644.5 *3*	0.039 *10*
γ E2,M1	2671.9 *3*	0.039 *10*
γ	2680.0 *4*	0.015 *5*
γ	2711.7 *21*	0.0127 *25*
γ	2724.4 *4*	0.029 *6*
γ	2761.5 *5*	~0.012
γ	2770.9 *10*	~0.014
γ	2798.9 *3*	0.029 *6*
γ M1,E2	2799.02 *19*	0.039 *10*
γ	2905.3 *4*	0.053 *16*
γ	2945.9 *5*	~0.009

† 1.0% uncert(syst)

$^{146}_{64}$Gd(48.27 *10* d)

Mode: ε
Δ: -76100 *11* keV
SpA: 1.853×10^4 Ci/g
Prod: ^{144}Sm(α,2n); protons on Er; protons on Ta

Photons (^{146}Gd)

⟨γ⟩=255.5 *21* keV

γ_{mode}	γ(keV)	γ(%)†
Eu L_ℓ	5.177	0.54 *5*
Eu L_η	5.817	0.198 *15*
Eu L_α	5.843	14.5 *9*
Eu L_β	6.594	13.2 *11*
Eu L_γ	7.573	1.97 *20*
Eu $K_{\alpha2}$	40.902	53.3 *15*
Eu $K_{\alpha1}$	41.542	96 *3*
Eu $K_{\beta1}$'	46.999	27.9 *8*
Eu $K_{\beta2}$'	48.496	8.5 *3*
γ	76.696 *10*	0.023 *9*
γ M1	114.865 *20*	44.0 *9*
γ M1	115.666 *20*	44.0 *9*
γ M1+<0.5%E2	154.726 *20*	46.6
γ (M1,E2)	267.96 *7*	~0.037
γ [M1+E2]	383.62 *7*	0.047 *19*
γ M1(+E2)	421.79 *7*	0.081 *9*
γ (M1,E2)	576.16 *20*	0.065 *9*

† 1.1% uncert(syst)

Atomic Electrons (^{146}Gd)

⟨e⟩=125.8 *16* keV

e_{bin}(keV)	⟨e⟩(keV)	e(%)
7	6.0	86 *9*
8	4.4	57 *6*
28 - 47	4.84	13.5 *6*
66	31.5	47.5 *13*
67	31.3	46.6 *13*
69 - 77	0.040	~0.06
106	23.0	21.7 *5*
107	7.17	6.71 *19*
108	7.18	6.67 *19*
109 - 116	4.35	3.82 *6*
147	4.49	3.06 *7*
148 - 155	1.394	0.912 *15*
219 - 268	0.009	0.0038 *18*
335 - 384	0.013	0.0037 *10*
414 - 422	0.0016	0.00040 *12*
528 - 576	0.0046	0.00085 *24*

Continuous Radiation (^{146}Gd)

⟨IB⟩=0.19 keV

E_{bin}(keV)		⟨ ⟩(keV)	(%)
10 - 20	IB	0.00020	0.00126
20 - 40	IB	0.056	0.153
40 - 100	IB	0.092	0.20
100 - 300	IB	0.022	0.0114
300 - 600	IB	0.023	0.0058
600 - 784	IB	0.00106	0.000166

$^{146}_{65}$Tb(~8 s)

Mode: ε
Δ: -67860 *150* keV
SpA: ~9×10^9 Ci/g
Prod: daughter ^{146}Dy; protons on Ta

Photons (^{146}Tb)

γ_{mode}	γ(keV)	γ(%)
γ	1059.7 *4*	
γ E2	1971.7 *3*	12 *4*

$^{146}_{65}$Tb(23 *2* s)

Mode: ε
Δ: -67860 *150* keV
SpA: 3.3×10^9 Ci/g
Prod: ^{141}Pr(^{12}C,7n)

Photons (^{146}Tb)

⟨γ⟩=2861 *89* keV

γ_{mode}	γ(keV)	γ(%)
γ [M1+E2]	116.6 *5*	
γ E2	324.3 *5*	0.8 *3*
γ [M1+E2]	440.8 *5*	13.1 *9*
γ	655.6 *6*	2.6 *3*
γ	987.4 *4*	1.0 *3*
γ	1032.3 *6*	2.9 *3*
γ E2	1078.7 *5*	51 *3*
γ [M1+E2]	1296.1 *4*	
γ [M1+E2]	1416.9 *6*	17.1 *11*
γ E3	1579.1 *6*	99.7
γ [M1+E2]	1831.7 *6*	
γ [M1+E2]	1843.5 *4*	
γ E2	1971.7 *3*	0.29 *6*
γ [M1+E2]	2060.9 *5*	
γ [M1+E2]	3139.7 *4*	11.7 *8*

$^{146}_{66}$Dy(29 *3* s)

Mode: ε
Δ: -62860 *250* keV syst
SpA: 2.6×10^9 Ci/g
Prod: ^{90}Zr(^{58}Ni,2p)

Photons (^{146}Dy)

γ_{mode}	γ(keV)	γ(%)
γ	280.1 *3*	53 *15*
γ	337.8 *3*	5 *1*

$^{146}_{66}$Dy(150 *20* ms)

Mode: IT
Δ: -59925 *250* keV syst
SpA: 1.10×10^{11} Ci/g
Prod: ^{90}Zr(^{58}Ni,2p)

Photons (^{146}Dy)

⟨γ⟩=2921 *210* keV

γ_{mode}	γ(keV)	γ(%)
Dy L_ℓ	5.743	0.127 *25*
Dy L_α	6.491	3.3 *6*
Dy L_η	6.534	0.067 *12*
Dy L_β	7.352	3.8 *7*
Dy L_γ	8.454	0.57 *11*
Dy $K_{\alpha2}$	45.208	4.4 *5*
Dy $K_{\alpha1}$	45.998	7.9 *10*
Dy $K_{\beta1}$'	52.063	2.4 *3*
Dy $K_{\beta2}$'	53.735	0.71 *9*
γ E3	126.80 *24*	3.0 *6*
γ E1	237.0 *3*	100 *20*
γ M1,E2	289.50 *24*	40 *8*
γ (E3)	416.30 *24*	54 *10*
γ [E2]	498.8 *3*	14 *3*
γ [E1]	673.5 *3*	87 *15*
γ [E2]	682.7 *3*	99.30
γ [E2]	925.1 *3*	89 *18*
γ [E1]	1099.8 *3*	16 *3*

Atomic Electrons (^{146}Dy)

⟨e⟩=91 *6* keV

e_{bin}(keV)	⟨e⟩(keV)	e(%)
8 - 52	2.9	32 *4*
73	4.7	6.4 *13*
118	17.4	15 *3*
119	13.5	11.3 *22*
125	8.1	6.5 *13*
126 - 127	2.3	1.78 *25*
183	4.9	2.7 *5*
228 - 235	1.07	0.47 *7*
236	7.6	3.2 *13*
237 - 282	1.7	0.62 *21*
287 - 290	0.49	0.17 *6*
363	10.4	2.9 *5*
407 - 445	6.8	1.64 *14*
490 - 499	0.20	0.042 *6*
620	1.25	0.20 *4*
629	3.64	0.579 *23*
664 - 683	1.07	0.159 *6*
871	2.4	0.27 *6*
916 - 925	0.48	0.053 *7*
1046 - 1092	0.17	0.016 *3*
1098 - 1100	0.0058	0.00053 *9*

$^{146}_{67}$Ho(3.9 *8* s)

Mode: ε
SpA: 1.8×10^{10} Ci/g
Prod: ^{90}Zr(^{58}Ni,pn)

Photons (^{146}Ho)

$\langle\gamma\rangle$=2251 *170* keV

γ_{mode}	γ(keV)	γ(%)*
γ E3	126.80 *24*	1.1 *4*
γ E1	237.0 *3*	54 *7*
γ M1,E2	289.50 *24*	30 *6*
γ (E3)	416.30 *24*	20 *6*
γ [E2]	498.8 *3*	9.2 *25*
γ [E1]	673.5 *3*	58 *9*
γ [E2]	682.7 *3*	103.3
γ [E2]	925.1 *3*	71 *14*
γ [E1]	1099.8 *3*	~13

* with ^{146}Dy(150 ms) in equilib

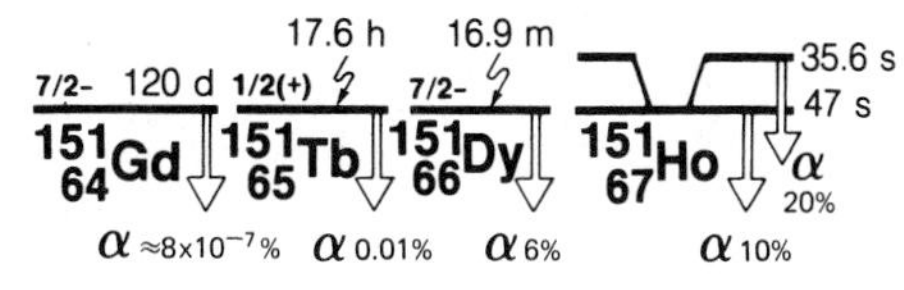

A = 147

NDS **25**, 113 (1978)

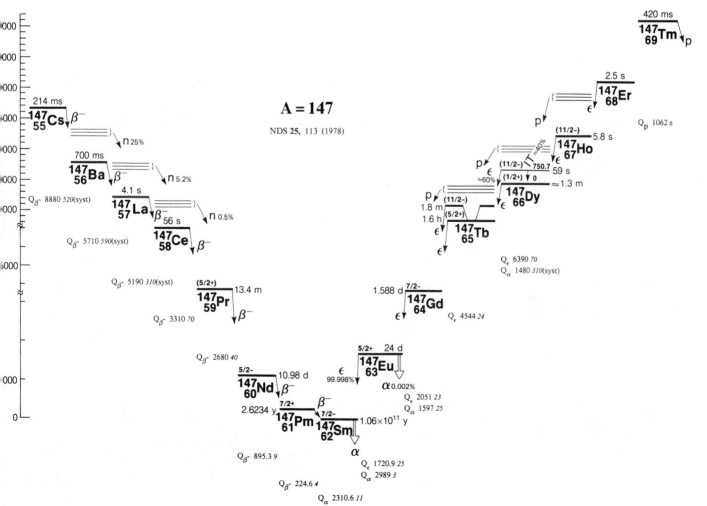

147 55 Cs 214 ms β⁻ n 25%
Qβ- 8880 520(syst)
147 56 Ba 700 ms β⁻ n 5.2%
Qβ- 5710 590(syst)
147 57 La 4.1 s β⁻ n 0.5%
Qβ- 5190 310(syst)
147 58 Ce 56 s β⁻
Qβ- 3310 70
147 59 Pr (5/2+) 13.4 m β⁻
Qβ- 2680 40
147 60 Nd 5/2- 10.98 d β⁻
Qβ- 895.3 9
147 61 Pm 7/2+ 2.6234 y β⁻
Qβ- 224.6 4
147 62 Sm 7/2- 1.06×10¹¹ y α
Qα 2310.6 11
147 63 Eu 5/2+ 24 d ε 99.998% α 0.002%
Qε 1720.9 25
Qα 2989 3
147 64 Gd 7/2- 1.588 d ε
Qε 2051 23
Qα 1597 25
147 65 Tb (11/2-) 1.8 m (5/2+) 1.6 h ε p
Qε 4544 24
147 66 Dy (11/2-) 750.7 59 s (1/2+) 0 ≈1.3 m ε ≈60% IT ≈40% ε p
Qε 6390 70
Qα 1480 310(syst)
147 67 Ho (11/2-) 5.8 s ε p
147 68 Er 2.5 s ε p
147 69 Tm 420 ms p
Qp 1062 6

$^{147}_{55}$Cs(214 *30* ms)

Mode: β-, β-n(25 *3* %)
Δ: -52380 *130* keV
SpA: 1.06×10^{11} Ci/g
Prod: fission

$^{147}_{56}$Ba(700 *40* ms)

Mode: β-, β-n(5.2 *5* %)
Δ: -61260 *510* keV syst
SpA: 7.0×10^{10} Ci/g
Prod: fission

$^{147}_{57}$La(4.10 *25* s)

Mode: β-, β-n(0.50 *17* %)
Δ: -66970 *300* keV syst
SpA: 1.72×10^{10} Ci/g
Prod: fission

Photons (^{147}La)

γ_{mode}	γ(keV)	γ(rel)
γ	117.62 *25*	100 *10*
γ	151.7 *5*	1.6 *3*
γ	156.0 *5*	7.2 *14*
γ	161.6 *3*	2.0 *4*
γ	174.7 *5*	1.20 *24*
γ	186.1 *3*	49 *5*
γ	206.9 *5*	1.5 *3*
γ	214.5 *4*	35 *4*
γ	215.3 *3*	27 *3*
γ	224.5 *4*	3.0 *6*
γ	235.2 *4*	22.0 *22*
γ	245.9 *5*	2.8 *6*
γ	255.9 *5*	3.1 *6*
γ	258.5 *8*	0.70 *14*
γ	273.4 *5*	18 *3*
γ	279.2 *5*	0.80 *16*
γ	283.0 *3*	20 *2*
γ	292.2 *4*	2.0 *4*
γ	308.3 *3*	2.0 *4*
γ	319.9 *4*	3.0 *6*
γ	332.9 *3*	5 *1*
γ	343.8 *5*	0.80 *16*
γ	352.8 *4*	13 *2*
γ	358.8 *5*	7.9 *16*
γ	376.9 *3*	4.0 *8*
γ	381.7 *8*	1.0 *2*
γ	388.2 *5*	4.9 *10*
γ	393.2 *5*	3.1 *6*
γ	399.0 *4*	13 *2*
γ	401.7 *5*	7.0 *14*
γ	410.1 *5*	1.8 *4*
γ	416.2 *5*	1.8 *4*
γ	432.7 *5*	7.4 *15*
γ	437.5 *4*	39 *4*
γ	474.1 *5*	0.20 *4*
γ	490.1 *5*	4.0 *8*
γ	494.5 *3*	11.0 *17*
γ	507.5 *3*	10.0 *15*
γ	516.6 *4*	20 *2*
γ	525.8 *5*	1.20 *24*
γ	557.2 *5*	3.8 *8*
γ	570.5 *5*	6.6 *13*
γ	598.0 *5*	6.5 *13*
γ	644.5 *8*	1.0 *2*
γ	648.7 *8*	1.0 *2*
γ	706.3 *5*	4.5 *9*
γ	709.4 *5*	3.9 *8*

$^{147}_{58}$Ce(56.4 *12* s)

Mode: β-
Δ: -72160 *80* keV
SpA: 1.35×10^{9} Ci/g
Prod: fission

Photons (^{147}Ce)

γ_{mode}	γ(keV)	γ(rel)
γ	92.9 *4*	51 *10*
γ	138.1 *3*	10 *2*
γ	178.4 *5*	5.3 *11*
γ	198.9 *5*	25 *5*
γ	218.6 *3*	32 *3*
γ	244.4 *3*	10 *2*
γ	254.4 *5*	7.9 *16*
γ	269.1 *4*	100 *20*
γ	289.9 *5*	19 *4*
γ	319.2 *3*	10 *2*
γ	362.0 *4*	6.2 *12*
γ	374.4 *4*	43 *9*
γ	439.8 *5*	
γ	449.3 *3*	8 *2*
γ	452.3 *5*	28 *6* ?
γ	464.5 *3*	12 *2*
γ	467.3 *4*	30 *6*
γ	580.2 *3*	51 *5*
γ	605.4 *3*	12 *2*
γ	668.9 *3*	7 *2*
γ	674.3 *3*	10 *2*
γ	701.1 *3*	24 *2*
γ	721.0 *3*	10 *2*
γ	749.0 *3*	9 *2*
γ	799.3 *3*	14 *2*
γ	802.8 *3*	13 *2*
γ	808.4 *3*	7 *2*
γ	811.0 *3*	6 *1*
γ	832.2 *4*	16 *3*
γ	1194.2 *4*	<3 ?

$^{147}_{59}$Pr(13.4 *3* min)

Mode: β-
Δ: -75470 *40* keV
SpA: 9.54×10^{7} Ci/g
Prod: ^{148}Nd(γ,p); fission

Photons (^{147}Pr)

⟨γ⟩=863 *27* keV

γ_{mode}	γ(keV)	γ(%)†
Nd L_{ℓ}	4.633	0.21 *4*
Nd L_{η}	5.146	0.086 *13*
Nd L_{α}	5.228	5.6 *8*
Nd L_{β}	5.853	5.1 *8*
Nd L_{γ}	6.676	0.71 *12*
Nd $K_{\alpha2}$	36.847	21 *3*
Nd $K_{\alpha1}$	37.361	39 *5*
Nd $K_{\beta1}$'	42.240	11.1 *15*
Nd $K_{\beta2}$'	43.562	3.2 *4*
γ M1+1.4%E2	49.91 *6*	4.1 *11*
γ M1+7.4%E2	77.92 *6*	10.2 *3*
γ E2	86.55 *7*	5.0 *10*
γ [M1]	100.19 *7*	0.31 *5*
γ M1(+E2)	127.83 *5*	8.78 *19*
γ [M1]	140.52 *7*	0.67 *5*
γ [M1]	186.75 *5*	1.44 *24*
γ [E2]	190.43 *7*	1.15 *7*
γ [E2]	214.38 *7*	1.27 *24*
γ [M1]	249.01 *8*	1.51 *14*
γ [E2]	264.67 *7*	0.264 *24*
γ [E1]	305.71 *8*	
γ [M1]	314.58 *6*	24
γ	328.86 *7*	5.3 *7*

Photons (^{147}Pr)
(continued)

γ_{mode}	γ(keV)	γ(%)†
γ [M1]	335.56 *6*	6.5 *5*
γ [M1+E2]	388.75 *6*	1.82 *24*
γ [E2]	413.48 *7*	1.01 *14*
γ	454.58 *10*	1.15 *19*
γ [M1+E2]	466.67 *6*	1.6 *3*
γ	477.67 *7*	5.5 *5*
γ	492.7 *10*	0.7 *3*
γ	499.79 *5*	1.15 *19* ?
γ [M1+E2]	503.51 *11*	~0.7
γ [M1+E2]	516.58 *6*	0.96 *24*
γ [E1]	554.72 *7*	7.9 *10*
γ	577.87 *7*	16.3 *14*
γ [E2]	604.71 *10*	0.67 *12*
γ	608.94 *12*	1.51 *19*
γ	620.7 *5*	
γ	627.51 *10*	0.31 *5*
γ [M1+E2]	631.33 *10*	0.77 *10*
γ [E1]	641.27 *6*	19.0 *12*
γ	699.63 *10*	0.24 *7*
γ [E1]	705.83 *9*	0.50 *7*
γ	719.19 *7*	0.60 *12* ?
γ [E1]	769.10 *6*	0.43 *12*
γ [E1]	793.96 *6*	1.32 *14*
γ	795.4 *5*	
γ	853.9 *5*	
γ	858.74 *12*	
γ [E1]	881.39 *8*	0.31 *7*
γ	886.95 *8*	0.55 *10*
γ	904.27 *15*	1.01 *5*
γ	920.1 *5*	
γ	934.2 *3*	0.19 *7*
γ	942.09 *9*	1.01 *10*
γ [E1]	949.67 *10*	0.144 *24*
γ	980.8 *5*	
γ [E1]	995.96 *6*	1.63 *19*
γ [E1]	1083.39 *8*	0.96 *10*
γ [E1]	1096.16 *8*	0.22 *7*
γ	1100.59 *17*	0.46 *12*
γ	1112.6 *5*	0.120 *24*
γ [E1]	1136.41 *9*	1.63 *24*
γ	1156.79 *15*	
γ [E1]	1182.71 *6*	1.22 *14*
γ	1198.5 *5*	
γ	1205.3 *5*	
γ	1214.33 *10*	0.53 *10* ?
γ	1230.02 *9*	0.240 *24*
γ	1261.03 *21*	5.3 *12*
γ [E1]	1264.24 *10*	0.72 *24*
γ	1300.43 *8*	3.0 *4*
γ [E1]	1310.54 *6*	0.60 *7*
γ	1324.4 *10*	1.0 *4*
γ	1350.34 *9*	0.098 *10*
γ	1358.79 *14*	0.192 *24*
γ	1391.73 *10*	0.110 *17*
γ [E1]	1397.97 *9*	0.115 *7*
γ	1416.77 *10*	0.31 *5*
γ [E1]	1465.57 *11*	0.14 *5*
γ	1471.1 *8*	
γ	1518.27 *11*	0.168 *24*
γ [E1]	1543.49 *11*	0.298 *14*
γ	1545.54 *14*	0.288 *24*
γ	1559.7 *8*	
γ [E1]	1593.40 *9*	0.288 *24*
γ	1606.40 *9*	0.120 *12*
γ	1617.16 *17*	
γ	1623.45 *14*	0.312 *24*
γ	1673.36 *14*	0.264 *24*
γ	1793.22 *11*	0.264 *24*
γ	1808.40 *9*	0.053 *7*
γ	1846.41 *9*	0.077 *7*
γ	1941.83 *10*	0.084 *10*
γ	1960.63 *10*	0.060 *7*
γ	1995.15 *8*	0.175 *12*
γ	2076.5 *6*	0.050 *5*
γ	2095.42 *11*	0.060 *7*
γ	2103.1 *5*	0.041 *5*
γ	2164.73 *10*	
γ	2309.79 *11*	0.077 *7*
γ	2314.9 *4*	0.144 *24*
γ	2335.33 *10*	0.144 *14*

† 13% uncert(syst)

Atomic Electrons (^{147}Pr)

⟨e⟩=55.4 *25* keV

e_{bin}(keV)	⟨e⟩(keV)	e(%)
6	4.9	79 *17*
7	1.8	27 *5*
29 - 31	1.27	4.2 *4*
34	8.8	25.5 *9*
35 - 42	0.80	2.21 *21*
43	6.3	15 *3*
44 - 57	1.39	2.9 *4*
71	3.00	4.2 *3*
72 - 79	2.42	3.2 *3*
80	5.5	6.8 *13*
84	4.4	5.3 *4*
85	1.5	1.7 *3*
86 - 134	3.2	2.9 *7*
139 - 187	1.20	0.77 *6*
189 - 221	0.48	0.231 *16*
242 - 265	0.075	0.0305 *21*
271	3.4	1.3 *3*
285 - 334	2.2	0.74 *11*
335 - 383	0.25	0.071 *12*
387 - 434	0.44	0.10 *4*
447 - 496	0.29	0.062 *14*
497 - 534	0.7	~0.14
548 - 588	0.25	0.045 *12*
598 - 641	0.285	0.047 *3*
656 - 706	0.031	0.0046 *17*
712 - 750	0.0198	0.0027 *3*
762 - 794	0.00284	0.00036 *3*
838 - 887	0.037	0.0044 *17*
891 - 936	0.032	0.0035 *14*
941 - 990	0.0153	0.00160 *17*
994 - 1040	0.0074	0.00071 *7*
1053 - 1100	0.025	0.0023 *5*
1106 - 1139	0.0105	0.00093 *9*
1171 - 1217	0.09	0.007 *4*
1221 - 1267	0.062	0.0050 *18*
1281 - 1324	0.026	0.0020 *7*
1343 - 1392	0.0068	0.00050 *16*
1396 - 1422	0.0015	0.00011 *3*
1458 - 1550	0.0096	0.00064 *13*
1563 - 1630	0.0082	0.00052 *14*
1666 - 1765	0.0034	0.00020 *7*
1786 - 1845	0.0013	7.2 *20* $\times10^{-5}$
1898 - 1994	0.0032	0.00016 *4*
2033 - 2102	0.0014	7.0 *15* $\times10^{-5}$
2266 - 2334	0.0031	0.00014 *3*

Continuous Radiation (^{147}Pr)

⟨β-⟩=726 keV;⟨IB⟩=1.36 keV

E_{bin}(keV)		⟨ ⟩(keV)	(%)
0 - 10	β-	0.0274	0.55
	IB	0.029	
10 - 20	β-	0.083	0.55
	IB	0.029	0.20
20 - 40	β-	0.341	1.13
	IB	0.055	0.19
40 - 100	β-	2.54	3.61
	IB	0.153	0.24
100 - 300	β-	28.3	13.9
	IB	0.39	0.22
300 - 600	β-	108	24.0
	IB	0.36	0.087
600 - 1300	β-	402	44.3
	IB	0.30	0.037
1300 - 2490	β-	184	12.0
	IB	0.029	0.0020

$^{147}_{60}$Nd(10.98 *1* d)

Mode: β-

Δ: -78156 *3* keV

SpA: 8.090×10^{4} Ci/g

Prod: ^{146}Nd(n,γ); fission

Photons (^{147}Nd)

⟨γ⟩=141 *5* keV

γ_{mode}	γ(keV)	γ(%)†
Pm L_ℓ	4.809	0.122 *17*
Pm L_η	5.363	0.047 *6*
Pm L_α	5.430	3.3 *4*
Pm L_β	6.097	2.9 *4*
Pm L_γ	6.975	0.42 *6*
Pm $K_{\alpha2}$	38.171	13.3 *14*
Pm $K_{\alpha1}$	38.725	24.1 *25*
Pm $K_{\beta1}$'	43.793	6.9 *7*
Pm $K_{\beta2}$'	45.183	2.05 *21*
γ M1+0.9%E2	91.104 *13*	28
γ M1+0.3%E2	120.496 *19*	0.40 *4*
γ M1+4.3%E2	196.644 *20*	0.204 *17*
γ M1+1.9%E2	275.385 *13*	0.80 *5*
γ M1+12%E2	319.414 *13*	1.95 *11*
γ M1+8.3%E2	398.155 *16*	0.87 *6*
γ E2	410.518 *16*	0.139 *8*
γ M1+28%E2	439.910 *16*	1.20 *8*
γ M1(+E2)	489.259 *18*	0.153 *8*
γ M1+14%E2	531.013 *16*	13.1 *7*
γ	589.36 *4*	0.046 *5*
γ E2(+2.7%M1)	594.798 *15*	0.265 *17*
γ E2(+M1)	680.46 *4*	0.020 *4*
γ M1+45%E2	685.902 *17*	0.81 *5*

† 6.4% uncert(syst)

Atomic Electrons (^{147}Nd)

⟨e⟩=36.4 *24* keV

e_{bin}(keV)	⟨e⟩(keV)	e(%)
6	1.44	22 *3*
7 - 44	2.3	18.3 *21*
46	22.5	49 *5*
75	0.236	0.31 *3*
84	5.8	6.9 *7*
85	0.21	0.24 *4*
89	1.20	1.34 *14*
90 - 120	0.65	0.70 *5*
147 - 197	0.081	0.052 *4*
230 - 275	0.463	0.179 *8*
312 - 353	0.151	0.0449 *20*
365 - 411	0.130	0.0332 *19*
432 - 482	0.032	0.0072 *7*
483 - 531	1.07	0.217 *11*
544 - 593	0.0138	0.00249 *25*
594 - 641	0.034	0.0054 *5*
673 - 686	0.0065	0.00095 *7*

Continuous Radiation (^{147}Nd)

⟨β-⟩=233 keV;⟨IB⟩=0.19 keV

E_{bin}(keV)		⟨ ⟩(keV)	(%)
0 - 10	β-	0.146	2.92
	IB	0.0115	
10 - 20	β-	0.432	2.88
	IB	0.0108	0.075
20 - 40	β-	1.70	5.7
	IB	0.020	0.068
40 - 100	β-	11.1	16.0
	IB	0.046	0.073
100 - 300	β-	78	40.5

Continuous Radiation (^{147}Nd)
(continued)

E_{bin}(keV)		⟨ ⟩(keV)	(%)
	IB	0.075	0.046
300 - 600	β-	122	28.9
	IB	0.022	0.0059
600 - 804	β-	20.1	3.08
	IB	0.00051	8.1×10^{-5}

$^{147}_{61}$Pm(2.6234 *2* yr)

Mode: β-

Δ: -79052 *3* keV

SpA: 927.00 Ci/g

Prod: daughter ^{147}Nd; fission

Photons (^{147}Pm)

⟨γ⟩=0.00439 *18* keV

γ_{mode}	γ(keV)	γ(%)†
γ M1+30%E2	76.14 *6*	1.03 *24* $\times10^{-8}$
γ M1+10%E2	121.26 *4*	0.00285
γ E2	197.39 *4*	3.4 *6* $\times10^{-7}$

† 3.9% uncert(syst)

Continuous Radiation (^{147}Pm)

⟨β-⟩=62 keV;⟨IB⟩=0.0142 keV

E_{bin}(keV)		⟨ ⟩(keV)	(%)
0 - 10	β-	0.54	10.9
	IB	0.0032	
10 - 20	β-	1.52	10.2
	IB	0.0025	0.018
20 - 40	β-	5.5	18.4
	IB	0.0036	0.0127
40 - 100	β-	26.4	39.4
	IB	0.0043	0.0073
100 - 225	β-	28.0	21.1
	IB	0.00069	0.00058

$^{147}_{62}$Sm(1.06 *2* $\times 10^{11}$ yr)

Mode: α
Δ: -79276 *3* keV
Prod: natural source
%: 15.0 *2*

Alpha Particles (^{147}Sm)

α(keV)
2233 *5*

$^{147}_{63}$Eu(24 *1* d)

Mode: ε(99.9978 *6* %), α(0.0022 *6* %)
Δ: -77555 *4* keV
SpA: 3.70$\times 10^4$ Ci/g
Prod: ^{147}Sm(p,n); ^{148}Sm(p,2n); deuterons on Sm

Alpha Particles (^{147}Eu)

α(keV)
2908 *5*

Photons (^{147}Eu)

⟨γ⟩=493 *12* keV

γ_{mode}	γ(keV)	γ(%)†
Sm L_ℓ	4.993	0.29 *3*
Sm L_η	5.589	0.107 *9*
Sm L_α	5.633	7.6 *5*
Sm L_β	6.342	7.0 *6*
Sm L_γ	7.270	1.02 *11*
Sm $K_{\alpha2}$	39.522	29.1 *11*
Sm $K_{\alpha1}$	40.118	52.6 *20*
Sm $K_{\beta1}$'	45.379	15.2 *6*
Sm $K_{\beta2}$'	46.819	4.53 *19*
γM1+30%E2	76.14 *6*	0.77 *13*
γM1+10%E2	121.26 *4*	22.7 *21*
γ	160.84 *20*	0.077 *13*
γE2	197.39 *4*	26
γ	278.24 *20*	~0.05
γ	327.9 *3*	~0.04
γ	471.6 *3*	0.12 *3*
γ	504.8 *7*	0.18 *8*
γ	549.6 *7*	0.17 *4*
γM1	601.48 *7*	6.8 *6*
γM1+18%E2	677.61 *7*	10.7 *8*
γ	738.24 *20*	0.08 *3*
γ	749.89 *25*	0.24 *4*
γE2	798.87 *7*	5.5 *5*
γ(M1)	809.3 *5*	0.046 *13*
γ	829.0 *7*	
γ	846 *1*	0.101 *10*
γE1	857.07 *8*	3.1 *3*
γ[E1]	867.9 *7*	
γ(M1)	879.80 *9*	0.19 *3*
γM1,E2	885.7 *7*	0.044 *9*
γE1	933.21 *8*	3.6 *3*
γ(E1)	942.71 *21*	0.21 *4*
γM1+2.9%E2	955.94 *9*	3.9 *3*
γ	964.0 *8*	
γ	1029.0 *3*	0.08 *3*
γ	1053.1 *8*	0.023 *8*
γ	1058.85 *20*	0.057 *10*
γ(E1)	1063.97 *21*	0.13 *3*
γM1	1077.19 *8*	6.4 *5*
γ	1106.0 *8*	0.028 *3*
γ	1120.63 *18*	0.19 *3*
γ	1158.2 *9*	~0.008
γ(E2)	1180.11 *20*	0.18 *3*
γE2+M1	1196.77 *18*	0.26 *3*
γM1+E2	1255.96 *16*	1.01 *8*
γ(M1)	1274.50 *20*	0.049 *5*
γ	1318.02 *17*	0.126 *15*
γM1(+E2)	1332.10 *17*	0.34 *5*
γ	1350.63 *20*	0.11 *4*
γ	1427.50 *25*	0.124 *13*
γ	1448.74 *20*	0.25 *3*
γ	1467.1 *12*	
γ	1482 *1*	0.0039 *10*
γ	1542.0 *12*	
γ	1604 *1*	0.0093 *18*
γ	1626 *1*	0.0039 *10*
γ	1637 *2*	0.0010 *5*
γ	1655.7 *9*	0.0034 *10*

† 10% uncert(syst)

Atomic Electrons (^{147}Eu)

⟨e⟩=40.1 *16* keV

e_{bin}(keV)	⟨e⟩(keV)	e(%)
7	5.0	72 *8*
8 - 45	4.00	18.4 *10*
68 - 69	0.68	0.99 *15*
74	14.1	19.0 *18*
75 - 76	0.17	0.23 *4*
114	3.4	3.0 *3*
115	0.45	0.39 *7*
120	0.87	0.73 *8*
121	0.24	0.200 *22*
151	6.1	4.0 *4*
153 - 161	0.011	~0.007
190	1.67	0.88 *9*
191 - 231	1.45	0.75 *5*
271 - 281	0.005	~0.0019
320 - 328	0.0010	~0.00030
425 - 472	0.017	~0.0038
497 - 543	0.011	~0.0021
548 - 595	0.51	0.091 *8*
600 - 631	0.57	0.091 *7*
670 - 703	0.112	0.0167 *12*
731 - 763	0.145	0.0193 *18*
791 - 839	0.070	0.0087 *5*
844 - 886	0.0397	0.0045 *3*
896 - 943	0.142	0.0157 *10*
948 - 982	0.0254	0.00267 *17*
1006 - 1053	0.197	0.0191 *16*
1056 - 1106	0.038	0.0036 *3*
1111 - 1158	0.0094	0.00083 *13*
1172 - 1209	0.022	0.0018 *4*
1228 - 1274	0.0069	0.00055 *10*
1285 - 1332	0.0095	0.00074 *15*
1343 - 1381	0.0020	0.00014 *7*
1402 - 1449	0.0043	0.00031 *13*
1474 - 1557	0.00012	8 *4* $\times 10^{-6}$
1579 - 1654	0.00013	8.3 *23* $\times 10^{-6}$

Continuous Radiation (^{147}Eu)

⟨β+⟩=1.01 keV;⟨IB⟩=0.62 keV

E_{bin}(keV)		⟨ ⟩(keV)	(%)
0 - 10	β+	1.66$\times 10^{-7}$	1.97$\times 10^{-6}$
	IB	0.00060	
10 - 20	β+	7.8$\times 10^{-6}$	4.68$\times 10^{-5}$
	IB	0.00026	0.00164
20 - 40	β+	0.000245	0.00075
	IB	0.059	0.165
40 - 100	β+	0.0125	0.0160
	IB	0.079	0.17
100 - 300	β+	0.403	0.192
	IB	0.034	0.017
300 - 600	β+	0.58	0.149
	IB	0.125	0.028
600 - 1300	β+	0.0128	0.00204
	IB	0.31	0.035
1300 - 1721	IB	0.018	0.00127
	Σβ+		0.36

$^{147}_{64}$Gd(1.588 *4* d)

Mode: ε
Δ: -75505 *23* keV
SpA: 5.595$\times 10^5$ Ci/g
Prod: ^{144}Sm(α,n); ^{147}Sm(α,4n)

Photons (^{147}Gd)

⟨γ⟩=1324 *17* keV

γ_{mode}	γ(keV)	γ(%)†
Eu L_ℓ	5.177	0.27 *3*
Eu L_η	5.817	0.100 *8*
Eu L_α	5.843	7.3 *5*
Eu L_β	6.594	6.7 *6*
Eu L_γ	7.574	1.00 *10*
Eu $K_{\alpha2}$	40.902	26.8 *8*
Eu $K_{\alpha1}$	41.542	48.4 *15*
Eu $K_{\beta1}$'	46.999	14.1 *4*
Eu $K_{\beta2}$'	48.496	4.26 *15*
γ	64.4 *10*	
γ	102.79 *20*	
γ M1,E2	106.519 *20*	0.09 *3*
γ M1+~14%E2	111.15 *5*	0.29 *8*
γ M1,E2	164.66 *7*	<0.07
γ M1+25%E2	166.35 *6*	0.31 *8*
γ (M1)	176.68 *9*	0.033 *8*
γ	209.19 *10*	<0.06 ‡
γ	210.39 *10*	
γ M1,E2	213.49 *20*	
γ M1,E2	214.89 *10*	0.25 *8*
γ E1	216.89 *10*	1.24 *17*
γ E1	217.13 *6*	0.50 *17*
γ M1+1.7%E2	229.309 *19*	61 *3*
γ M1(+E2)	240.58 *4*	1.48 *8*
γ M1(+E2)	249.13 *6*	0.38 *2*
γ	252.19 *10*	<0.08
γ M1(+E2)	261.19 *6*	1.90 *9*
γ	286.6 *3*	0.21 *4* ‡
γ	287.39 *10*	

Photons (^{147}Gd) (continued)

γ_{mode}	γ(keV)	γ(%)†
γ [E1]	291.35 7	0.19 4
γ	293.19 10	~0.10
γ E2	297.06 19	0.34 6
γ M1	309.97 6	4.05 17
γ M1(+E2)	318.52 5	2.10 8
γ	327.4 10	0.082 15
γ	329.83 16	~0.029
γ	341.34 10	0.165 15
γ M1	346.42 9	2.05 8
γ M1+0.6%E2	369.90 5	16.5 6
γ	376.0 5	0.18 1
γ M2	395.93 5	33.0 15
γ	404.06 16	<0.06
γ	408.22 20	<0.06
γ	417.01 19	<0.06
γ	417.83 18	<0.06
γ [E1]	431.43 7	0.15 7
γ	432.65 17	~0.08
γ	457.81 11	<0.10
γ	460.54 16	~0.13
γ M1	484.87 6	2.90 14
γ	490.6 10	0.033 10
γ	495.80 12	0.05 2
γ	506.40 19	~0.05
γ	516.73 15	0.03 1
γ	529.2 10	~0.04
γ	538.01 15	0.07 2
γ M1	547.09 8	0.29 5
γ [M1+E2]	548.71 5	~0.07
γ M1	559.10 5	6.20 22
γ	560.29 10	<0.30
γ	570.5 6	0.13 5
γ (M1)	573.45 16	0.15 5
γ	580.48 12	0.05 2
γ	583.84 18	~0.03
γ	595.6 4	0.095 20
γ E2	610.49 5	1.53 13
γ M1+36%E2	619.03 6	3.47 15
γ E3	625.24 5	4.5 3
γ M1	632.31 5	1.64 7
γ	646.8 9	0.074 8
γ M1	693.39 12	0.25 2
γ (M1)	701.12 16	0.35 4
γ	703.9 7	
γ [M1]	704.37 14	0.66 4 ‡
γ M1	714.86 10	0.31 2
γ	726.6 7	<0.030
γ	733.2 7	<0.08
γ (M1)	734.99 17	0.16 5
γ	738.04 17	0.05 2
γ (M1)	751.30 21	0.15 5
γ M1(+E2)	755.10 5	1.99 8
γ E1	765.84 5	10.9 6
γ [E1]	776.23 6	~1
γ E2	776.40 8	4.16 21
γ M1(+~20%E2)	778.02 4	4.76 21
γ E2(+M1)	782.78 11	1.15 4
γ M1+E2	788.78 11	0.78 6
γ E2(+M1)	805.64 24	0.24 2
γ E2(+M1)	810.07 11	0.50 5
γ M1,E2	820.52 20	0.19 2
γ	827.77 9	0.50 8
γ	834.64 16	0.124 13
γ	840.74 18	~0.08
γ E2+39%M1	861.62 5	1.68 6
γ	867.8 9	<0.030
γ	879.1 5	0.22 3
γ	881.21 11	0.06 2
γ M1+E2	893.51 5	7.8 4
γ	896.5 9	<0.2
γ M1,E2	910.45 11	0.54 3
γ	917.53 17	0.05 2
γ M1+28%E2	929.00 5	19.4 8
γ	936.8 10	0.016 4
γ E1	954.8 10	0.18 5
γ	966.0 10	<0.030
γ	968.4 3	0.11 2
γ	975.0 10	<0.02
γ	976.9 3	0.03 1
γ E1	983.6 1	0.165 20
γ (E2)	988.6 4	0.124 15
γ	995.72 14	0.83 5
γ E1	1006.42 4	1.31 8
γ	1017.33 15	0.11 2

Photons (^{147}Gd) (continued)

γ_{mode}	γ(keV)	γ(%)†
γ	1040.25 15	0.38 4
γ	1043.97 12	0.13 2
γ	1048.49 18	0.066 20
γ	1061.10 12	0.15 3
γ E1	1069.37 6	6.9 5
γ	1081.0 6	0.07 2
γ	1096.50 19	<0.06
γ E2	1122.81 5	0.87 5
γ	1125.28 17	~0.11
γ E1	1130.93 16	6.2 4
γ E2	1149.04 14	0.37 3
γ	1150.49 12	0.054 3
γ	1155.00 18	<0.06
γ E1	1160.28 8	0.64 4
γ M1,E2	1169.89 15	0.095 8
γ	1184.83 24	~0.025
γ	1196.9 4	~0.025
γ	1208.88 17	0.033 4
γ	1212.91 10	0.115 8
γ	1216.0 10	~0.030
γ	1219.18 21	0.045 5
γ E1	1235.73 4	1.11 7
γ	1245.34 21	0.062 5
γ	1270.2 4	0.045 5
γ	1305.7 4	0.045 5
γ E1(+M2)	1324.94 5	0.83 5
γ	1335.87 11	0.045 5
γ E2	1370.33 20	0.075 8
γ	1378.35 14	0.045 5
γ	1389.58 8	0.054 5
γ M1(+E2)	1399.20 15	0.16 1
γ	1406.7 10	<0.0050
γ	1409.5 8	<0.0050
γ	1466.95 16	0.029 5
γ	1474.65 21	0.074 8
γ	1530.7 5	0.054 5
γ	1544.43 14	0.012 4
γ	1565.17 11	0.37 3 ‡
γ	1566.04 15	
γ M1,E2	1586.90 11	0.54 3
γ	1601.5 15	~0.010
γ	1628.8 3	0.016 3
γ	1641.0 20	<0.0050
γ M1,E2	1676.28 11	0.249 13
γ	1680.80 18	0.062 4
γ	1721.27 9	0.010 3
γ	1731.92 24	<0.006 ?
γ	1735.0 25	<0.006 ?
γ	1757.59 16	0.027 3
γ	1766.26 20	0.041 4
γ	1775.0 20	<0.002
γ	1783.3 5	~0.004
γ (E1)	1795.35 15	0.77 4
γ	1807.41 16	0.041 4
γ (E2)	1816.20 12	0.140 8
γ	1824.0 5	0.015 3
γ	1844.3 3	0.031 3
γ	1858.1 3	~0.006
γ	1901.0 20	<0.002
γ	1905.59 11	0.012 4
γ	1910.10 18	0.037 4
γ	1936.56 18	0.0037 13
γ	1950.58 9	0.066 8
γ	1961.22 24	0.032 3
γ	1982.6 5	0.010 2
γ	1986.89 16	0.064 6

† 10% uncert(syst)
‡ combined intensity for doublet

Atomic Electrons (^{147}Gd)

⟨e⟩=58.0 13 keV

e_{bin}(keV)	⟨e⟩(keV)	e(%)
7	3.0	43 5
8	2.22	29 3
32 - 63	2.70	7.2 3
98 - 128	0.30	0.27 4
157 - 177	0.21	0.124 18
181	17.2	9.5 5
192 - 217	0.81	0.40 6
221	2.74	1.24 7
222 - 260	1.43	0.62 3
261	0.75	0.287 14
270 - 319	0.90	0.31 3
320	0.0005	~0.00016
321	2.34	0.73 3
322 - 346	0.085	0.025 3
347	14.8	4.26 21
356 - 384	0.492	0.135 5
388	2.65	0.68 3
389 - 436	1.18	0.292 9
442 - 491	0.072	0.0151 12
494 - 542	0.56	0.110 6
545 - 594	0.86	0.151 6
595 - 640	0.198	0.0321 10
645 - 694	0.108	0.0163 8
696 - 744	0.64	0.089 5
747 - 792	0.171	0.0223 17
798 - 845	0.33	0.039 8
848 - 896	0.75	0.086 6
902 - 948	0.164	0.0177 17
953 - 1000	0.034	0.0035 6
1005 - 1054	0.067	0.0065 5
1059 - 1106	0.087	0.0081 4
1112 - 1160	0.0231	0.00205 10
1162 - 1211	0.0140	0.00118 13
1212 - 1257	0.0032	0.00026 5
1262 - 1306	0.028	~0.0022
1312 - 1361	0.011	0.00083 24
1362 - 1409	0.00102	7.4 9 $\times 10^{-5}$
1418 - 1468	0.0017	0.00012 4
1473 - 1564	0.016	0.00106 21
1579 - 1675	0.0074	0.00045 5
1679 - 1776	0.0080	0.00046 3
1781 - 1862	0.0021	0.000117 15
1888 - 1985	0.0022	0.00011 3

Continuous Radiation (^{147}Gd)

⟨β+⟩=0.80 keV;⟨IB⟩=0.42 keV

E_{bin}(keV)		⟨ ⟩(keV)	(%)
0 - 10	β+	1.33×10^{-8}	1.58×10^{-7}
	IB	0.00064	
10 - 20	β+	6.5×10^{-7}	3.90×10^{-6}
	IB	0.00023	0.00149
20 - 40	β+	2.15×10^{-5}	6.5×10^{-5}
	IB	0.056	0.153
40 - 100	β+	0.00124	0.00158
	IB	0.093	0.20
100 - 300	β+	0.062	0.0281
	IB	0.032	0.0160
300 - 600	β+	0.335	0.074
	IB	0.092	0.021
600 - 1300	β+	0.405	0.054
	IB	0.133	0.0162
1300 - 2328	β+	2.34×10^{-7}	1.80×10^{-8}
	IB	0.0159	0.00105
	Σβ+		0.157

$^{147}_{65}$Tb(1.65 *10* h)

Mode: ϵ
Δ: -70960 *29* keV
SpA: 1.29×10^{7} Ci/g
Prod: ^{141}Pr(^{12}C,6n); protons on U; protons on Ta

Photons (^{147}Tb)

⟨γ⟩=1161 *73* keV

γ_{mode}	γ(keV)	γ(%)†
Gd L_ℓ	5.362	0.162 *18*
Gd L_η	6.049	0.058 *5*
Gd L_α	6.054	4.3 *3*
Gd L_β	6.852	3.9 *4*
Gd L_γ	7.878	0.59 *7*
Gd $K_{\alpha2}$	42.309	14.9 *5*
Gd $K_{\alpha1}$	42.996	26.7 *9*
Gd $K_{\beta1}$'	48.652	7.9 *3*
Gd $K_{\beta2}$'	50.214	2.38 *9*
γ E2+47%M1	119.7 *3*	4.4 *4*
γ E1	139.9 *3*	19.9 *20*
γ (M1)	347.4 *6*	1.7 *2*
γ (M1+E2)	407.2 *3*	1.4 *2*
γ [E1]	434.9 *4*	0.7 *2*
γ	547.0 *3*	2.0 *2*
γ [E1]	554.6 *3*	3.7 *4*
γ (M1+E2)	694.5 *3*	31
γ (E2)	1152.2 *4*	73 *6*

† 10% uncert(syst)

Atomic Electrons (^{147}Tb)

⟨e⟩=15.6 *9* keV

e_{bin}(keV)	⟨e⟩(keV)	e(%)
7	1.74	24 *3*
8	1.27	15.8 *18*
33 - 34	0.262	0.78 *6*
35	0.41	1.17 *12*
36 - 48	0.66	1.64 *8*
69	2.7	3.8 *6*
90	1.78	1.99 *20*
111	0.49	0.44 *10*
112	1.1	1.0 *5*
118 - 140	0.99	0.79 *14*
297 - 346	0.34	0.113 *12*
347 - 385	0.16	0.045 *14*
399 - 435	0.036	0.009 *3*
497 - 546	0.19	0.038 *15*
547 - 555	0.0045	0.00082 *17*
644	1.5	0.23 *8*
686 - 694	0.30	0.043 *11*
1102	1.41	0.128 *12*
1144 - 1152	0.265	0.0231 *16*

Continuous Radiation (^{147}Tb)

⟨β+⟩=518 keV;⟨IB⟩=5.9 keV

E_{bin}(keV)		⟨ ⟩(keV)	(%)
0 - 10	β+	5.1×10^{-7}	6.1×10^{-6}
	IB	0.018	
10 - 20	β+	2.59×10^{-5}	0.000155
	IB	0.018	0.124
20 - 40	β+	0.00089	0.00269
	IB	0.062	0.19
40 - 100	β+	0.055	0.069
	IB	0.167	0.30
100 - 300	β+	3.28	1.47
	IB	0.31	0.17
300 - 600	β+	26.1	5.6

Continuous Radiation (^{147}Tb) (continued)

E_{bin}(keV)		⟨ ⟩(keV)	(%)
	IB	0.47	0.108
600 - 1300	β+	170	18.0
	IB	1.49	0.160
1300 - 2500	β+	259	14.7
	IB	2.5	0.143
2500 - 4544	β+	60	2.15
	IB	0.85	0.028
	Σβ+		42

$^{147}_{65}$Tb(1.83 *6* min)

Mode: ϵ
Δ: -70960 *29* keV
SpA: 6.97×10^{8} Ci/g
Prod: ^{141}Pr(^{12}C,6n); ^{151}Eu(^{3}He,7n)

Photons (^{147}Tb)

⟨γ⟩=1449 *86* keV

γ_{mode}	γ(keV)	γ(%)†
γ E3	997.6 *4*	0.9 *2*
γ [M1]	1397.7 *2*	85 *6*
γ	1797.8 *3*	14 *1*

† 20% uncert(syst)

$^{147}_{66}$Dy(~1.3 min)

Mode: ϵ
Δ: -64570 *70* keV
SpA: ~9.8×10^{8} Ci/g
Prod: ^{142}Nd(^{12}C,7n)

Photons (^{147}Dy)

γ_{mode}	γ(keV)
γ (M1)	100.7 *5*
γ (M1)	253.4 *5*
γ (M1)	365.3 *5*

$^{147}_{66}$Dy(59 *3* s)

Mode: ϵ(~60 %), IT(~40 %), ϵp
Δ: -63819 *70* keV
SpA: 1.29×10^{9} Ci/g
Prod: ^{14}N on ^{141}Pr

ϵp: ~3800

Photons (^{147}Dy)

⟨γ⟩=242 *9* keV

γ_{mode}	γ(keV)	γ(%) †
Dy L_ℓ	5.743	0.116 *12*
Dy L_α	6.491	3.03 *22*
Dy L_η	6.534	0.040 *3*
Dy L_β	7.383	2.79 *25*
Dy L_γ	8.511	0.43 *5*
Dy $K_{\alpha2}$	45.208	9.3 *4*
Dy $K_{\alpha1}$	45.998	16.6 *7*
Dy $K_{\beta1}$'	52.063	5.03 *22*
Dy $K_{\beta2}$'	53.735	1.47 *7*
γ_{IT}M1+3.8%E2	72.02 *9*	5.48
γ_{IT}M4	678.66 *15*	32.8

† approximate

$^{147}_{67}$Ho(5.8 *4* s)

Mode: ϵ, ϵp
SpA: 1.25×10^{10} Ci/g
Prod: ^{94}Mo(^{58}Ni,2n3p); ^{92}Mo(^{58}Ni,3p)

Photons (^{147}Ho)

⟨γ⟩=839 *43* keV

γ_{mode}	γ(keV)	γ(%)
γ_ϵM1+3.8%E2	72.02 *9*	*
γ_ϵ[E1]	189.18 *23*	33 *3*
γ_ϵ	292.87 *23*	4.3 *8*
γ_ϵ[E2]	394.57 *22*	5.2 *9*
γ_ϵ	431.66 *24*	2.9 *7*
γ_ϵ	445.2 *3*	5.4 *8*
γ_ϵ	486.81 *21*	20 *2*
γ_ϵ[E1]	589.10 *25*	5.4 *8*
γ_ϵM4	678.66 *15*	*
γ_ϵ[M1+E2]	779.68 *21*	5.8 *10*
γ_ϵ[M1+E2]	884.04 *19*	33
γ_ϵ	918.47 *24*	2.0 *5*
γ_ϵ[E2]	956.06 *20*	7.4 *11*
γ_ϵ[E2]	1263.80 *25*	12 *2*

* with ^{147}Dy(59 s)

$^{147}_{68}$Er(2.5 *2* s)

Mode: ϵ, ϵp
SpA: 2.68×10^{10} Ci/g
Prod: ^{92}Mo(^{58}Ni,n2p)

ϵp: ~4300

$^{147}_{69}$Tm(420 *100* ms)

Mode: p
SpA: 8.9×10^{10} Ci/g
Prod: ^{92}Mo(^{58}Ni,2np)

p: 1055 *6*

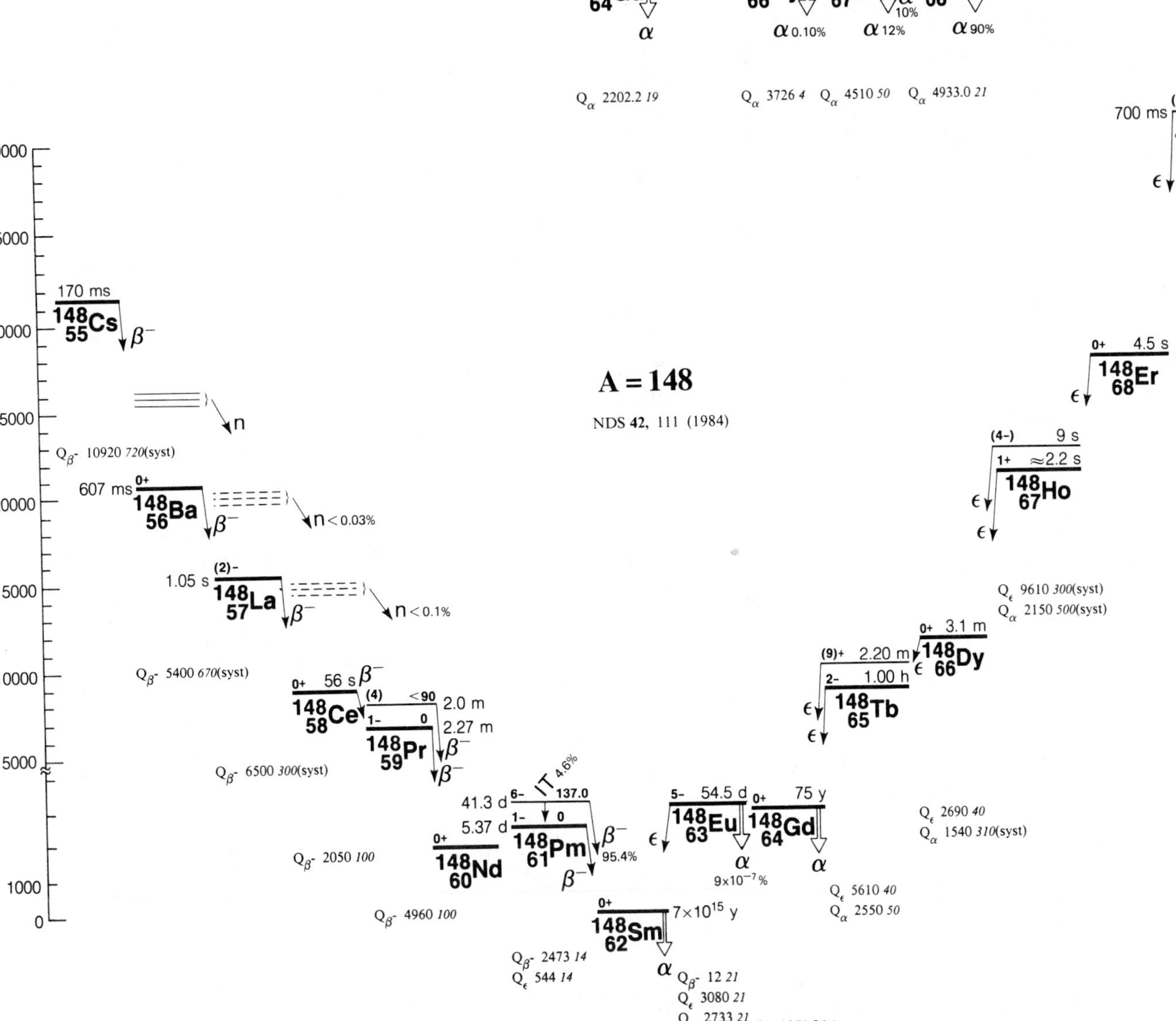

$^{148}_{55}$Cs(170 *7* ms)

Mode: β-, β-n
Δ: -47590 *370* keV
SpA: 1.08×10^{11} Ci/g
Prod: fission

Photons (^{148}Cs)

γ_{mode}	γ(keV)
$\gamma_{\beta\text{-}}$[E2]	142.5 *10*

$^{148}_{56}$Ba(607 *25* ms)

Mode: β-, β-n?(<0.03 %)
Δ: -58510 *620* keV syst
SpA: 7.5×10^{10} Ci/g
Prod: fission

Photons (^{148}Ba)

⟨γ⟩=191 *6* keV

γ_{mode}	γ(keV)	γ(%)†
γ	44.59 *6*	0.20 *3*
γ	47.08 *17*	~0.06
γM1(+E2)	47.28 *4*	0.93 *3*
γE1	48.51 *4*	1.66 *3*

Photons (^{148}Ba) (continued)

γ_{mode}	γ(keV)	γ(%)†
γ	49.41 *12*	0.15 *3*
γM1+E2	53.91 *4*	2.86 *6*
γE1	56.12 *4*	29.2 *9*
γ	58.54 *11*	0.32 *3*
γM1(+E2)	61.52 *4*	2.28 *5*
γ	72.30 *4*	1.93 *18*
γ	84.16 *6*	1.78 *15*
γM1(+E2)	86.54 *4*	2.01 *9*
γ	92.46 *6*	0.76 *3*
γ	96.92 *8*	0.58 *6*
γ	97.92 *12*	0.76 *18*
γ	98.49 *7*	2.9 *6*
γ	107.34 *6*	0.44 *6*
γ[E1]	110.03 *4*	0.67 *6*
γ	112.72 *5*	0.44 *3*
γ	120.70 *8*	0.88 *18*
γ	127.48 *6*	0.47 *6*

Photons (^{148}Ba)
(continued)

γ_{mode}	γ(keV)	γ(%)†
γE2,M1	133.81 *4*	3.88 *8*
γ	145.57 *16*	1.75 *6*
γ	153.48 *7*	0.35 *6*
γ	156.46 *6*	1.46 *9*
γ	168.84 *5*	1.34 *9*
γ	174.61 *7*	0.85 *9*
γ	177.26 *7*	0.70 *6*
γ	184.96 *19*	~0.35
γ	205.21 *13*	0.58 *6*
γ	213.13 *7*	0.73 *9*
γ	214.88 *8*	2.34 *9*
γ	217.97 *6*	0.35 *9*
γ	230.20 *9*	~0.20
γ	235.68 *6*	1.05 *6*
γ	246.74 *9*	1.20 *12*
γ	256.8 *5*	0.8 *3*
γ	259.46 *6*	1.61 *9*
γ	270.39 *15*	1.46 *15*
γ	307.90 *8*	0.96 *6*
γ	313.37 *7*	0.93 *15*
γ	317.47 *10*	0.64 *12*
γ	323.72 *17*	0.76 *6*
γ	345.23 *11*	0.23 *6*
γ	369.50 *19*	0.15 *6*
γ	390.40 *7*	~0.5
γ	404.52 *19*	0.64 *6*
γ	410.56 *9*	1.52 *6*
γ	415.96 *9*	3.59 *9*
γ	427.0 *14*	0.70 *23*
γ	444.31 *6*	2.54 *9*
γ	476.82 *19*	0.61 *23*
γ	535.28 *18*	~0.20
γ	569.6 *4*	0.44 *20*
γ	583.62 *15*	0.18 *6*
γ	600.2 *4*	0.64 *23*
γ	607.41 *15*	1.72 *15*
γ	661.31 *15*	0.23 *9*
γ	720.9 *4*	0.61 *23*
γ	757.22 *17*	0.41 *18*
γ	774.8 *4*	~0.18
γ	873.07 *25*	0.53 *6*
γ	1011.32 *20*	0.55 *23*
γ	1118.89 *10*	1.14 *6*
γ	1173.18 *19*	~0.35
γ	1196.59 *11*	0.47 *15*

† 2.7% uncert(syst)

$^{148}_{57}$La(1.05 *1* s)

Mode: β-, β-n?(<0.1 %)
Δ: -63910 *330* keV syst
SpA: 5.31×10^{10} Ci/g
Prod: fission

Photons (^{148}La)
⟨γ⟩=1252 *32* keV

γ_{mode}	γ(keV)	γ(%)†
Ce L_ℓ	4.289	0.055 *15*
Ce L_η	4.730	0.025 *7*
Ce L_α	4.838	1.5 *4*
Ce L_β	5.384	1.4 *3*
Ce L_γ	6.110	0.18 *5*
Ce $K_{\alpha2}$	34.279	5.4 *5*
Ce $K_{\alpha1}$	34.720	9.8 *9*
Ce $K_{\beta1}$'	39.232	2.8 *3*
Ce $K_{\beta2}$'	40.437	0.75 *7*
γE2	158.376 *5*	55.6 *11*
γ	252.12 *6*	1.67 *11*
γ	257.00 *9*	0.33 *6*
γ[E2]	295.00 *4*	6.67 *13*
γ	298.84 *8*	0.72 *6*
γ	369.36 *8*	0.67 *6*
γ	378.86 *4*	3.9 *4*
γ[E1]	387.85 *6*	1.39 *6*
γ	425.58 *7*	1.00 *6*
γ	433.18 *5*	1.11 *6*
γ[E2]	482.06 *5*	0.95 *6*
γ[E2]	536.38 *6*	0.50 *6*
γ[E1]	601.80 *4*	7.62 *15*
γE2	611.82 *15*	2.89 *6*
γ	654.41 *7*	0.78 *22*
γ[M1+E2]	663.12 *5*	1.50 *6*
γ[E1]	682.85 *5*	6.4 *5*
γ	713.29 *9*	0.50 *6*
γ[E1]	760.18 *4*	8.6 *4*
γ[M1+E2]	770.44 *10*	0.50 *6*
γ[M1+E2]	777.06 *4*	7.17 *14*
γ	794.34 *9*	0.72 *6*
γ	819.64 *16*	1.3 *3*
γ[M1+E2]	831.38 *4*	5.2 *3*
γ	886.95 *11*	0.44 *6*
γ	921.22 *19*	0.56 *11*
γ[M1+E2]	958.12 *5*	3.95 *17*
γ	967.99 *12*	0.39 *11*
γ[E2]	989.75 *4*	9.3 *3*
γ	1104.97 *15*	0.33 *6*
γ	1130.81 *7*	1.06 *11*
γ	1256.95 *8*	0.61 *6*
γ	1298.37 *25*	0.83 *6* ?
γ	1303.05 *15*	<0.22
γ	1316.56 *14*	0.44 *6*
γ	1338.53 *7*	1.78 *11*
γ	1425.49 *11*	0.89 *6* ?
γ	1431.46 *7*	1.33 *6*
γ	1464.27 *11*	0.89 *6* ?
γ	1496.91 *7*	0.61 *6*
γ	1569.79 *11*	0.39 *11*
γ	1589.83 *7*	0.83 *6*
γ	1732.61 *7*	0.67 *6*
γ	1769.18 *21*	0.89 *11* ?
γ	1890.98 *7*	1.22 *6*
γ	1985.89 *15*	2.45 *6*
γ	1995.12 *13*	3.28 *11*
γ	2031.1 *3*	1.17 *11*
γ	2033.86 *24*	0.67 *11* ?
γ	2093.61 *14*	7.0 *14*
γ	2153.50 *13*	0.72 *11*
γ	2219.8 *4*	1.50 *11*
γ	2391.79 *21*	3.9 *3*
γ	2550.16 *21*	~0.33

† 0.54% uncert(syst)

Atomic Electrons (^{148}La)
⟨e⟩=36.9 *16* keV

e_{bin}(keV)	⟨e⟩(keV)	e(%)
6	1.1	20 *5*
7 - 54	3.6	12 *3*
118	19.2	16.3 *5*
152	5.16	3.39 *10*
153	2.60	1.70 *5*
157	1.77	1.13 *3*
212 - 258	1.02	0.42 *6*
288 - 329	0.255	0.086 *8*
338 - 388	0.36	0.10 *4*
393 - 442	0.12	0.028 *9*
476 - 496	0.0278	0.0057 *5*
530 - 571	0.180	0.0319 *8*
595 - 642	0.171	0.027 *3*
648 - 683	0.036	0.0055 *12*
707 - 754	0.32	0.044 *6*
759 - 794	0.20	0.026 *5*
813 - 847	0.036	0.0044 *8*
880 - 928	0.105	0.0115 *21*
949 - 990	0.201	0.0210 *9*
1065 - 1105	0.020	0.0018 *8*
1124 - 1131	0.0024	0.00022 *9*
1217 - 1263	0.020	0.0016 *5*
1276 - 1316	0.028	0.0022 *8*
1332 - 1338	0.0034	0.00025 *11*
1385 - 1431	0.037	0.0026 *7*
1456 - 1549	0.020	0.0013 *4*
1563 - 1589	0.0018	0.00012 *4*
1692 - 1768	0.015	0.0009 *3*
1851 - 1945	0.030	0.0016 *5*
1955 - 2053	0.098	0.0049 *13*
2087 - 2179	0.024	0.0011 *3*
2213 - 2219	0.0016	7.2 *24* ×10^{-5}
2351 - 2391	0.029	0.0012 *4*
2510 - 2549	0.0023	~9×10^{-5}

$^{148}_{58}$Ce(56 *1* s)

Mode: β-
Δ: -70410 *140* keV
SpA: 1.353×10^{9} Ci/g
Prod: fission

Photons (^{148}Ce)
⟨γ⟩=317 *5* keV

γ_{mode}	γ(keV)	γ(%)†
γ[M1]	74.812 *13*	~0.13
γ[M1+E2]	90.775 *20*	2.05 *12*
γ[M1]	97.815 *13*	1.29 *8*
γE2	98.064 *3*	1.55 *8*
γ[M1+E2]	98.869 *20*	12.4 *7*
γE1	100.927 *4*	3.59 *18*
γ[E1]	103.45 *5*	<0.50
γ[M1+E2]	105.104 *20*	5.24 *25*
γ	116.82 *3*	3.76 *20*
γ[M1+E2]	121.067 *3*	13.2 *6*
γ[E1]	168.08 *8*	1.10 *12*
γ[M1]	168.56 *3*	0.33 *5*
γ[M1]	184.451 *23*	1.65 *8*
γ[E1]	187.96 *10*	1.50 *8*
γ[M1]	188.16 *5*	1.15 *12*
γ[M1]	191.491 *19*	1.65 *10*
γ[E1]	193.67 *12*	0.33 *7*
γ[E1]	194.604 *20*	4.01 *8*
γ[M1]	195.878 *13*	6.5 *3*
γ[M1]	231.45 *8*	0.33 *7*
γ[E1]	233.57 *4*	1.0 *1*
γ[M1]	247.41 *8*	0.94 *10*
γE1	269.415 *19*	17.0 *9*
γ[E1]	271.45 *11*	0.67 *10*
γ[M1]	273.66 *3*	5.4 *3*
γ[E1]	285.378 *23*	0.80 *8*
γ[M1]	287.03 *5*	1.84 *17*
γ[M1]	289.555 *19*	5.7 *3*
γ[E1]	291.613 *15*	16.7 *8*
γ[E1]	324.72 *4*	7.6 *3*
γ[M1]	332.64 *9*	1.50 *15*
γ[E1]	346.26 *11*	0.58 *7*
γ[M1]	352.51 *8*	1.77 *20*
γ	368.46 *12*	1.62 *22*
γ	374.69 *11*	0.68 *15*
γ[E1]	390.482 *18*	1.32 *23*
γ	399.53 *4*	0.77 *20*
γ	421.73 *4*	3.64 *18*
γ	478.14 *10*	1.34 *20*
γ	521.10 *20*	0.68 *15*

† 3.0% uncert(syst)

Continuous Radiation (^{148}Ce)

⟨β-⟩=633 keV;⟨IB⟩=1.05 keV

E_{bin}(keV)		⟨ ⟩(keV)	(%)
0 - 10	β-	0.0288	0.57
	IB	0.027	
10 - 20	β-	0.088	0.58
	IB	0.026	0.18
20 - 40	β-	0.362	1.20
	IB	0.050	0.17
40 - 100	β-	2.74	3.88
	IB	0.136	0.21
100 - 300	β-	31.7	15.6
	IB	0.34	0.19
300 - 600	β-	124	27.5
	IB	0.29	0.069
600 - 1300	β-	408	45.8
	IB	0.18	0.023
1300 - 2050	β-	67	4.65
	IB	0.0048	0.00034

$^{148}_{59}$Pr(2.27 *4* min)

Mode: β-

Δ: -72460 *100* keV

SpA: 5.583×10^{8} Ci/g

Prod: fission; ^{150}Nd(γ,pn); ^{148}Nd(n,p)

Photons (^{148}Pr)

⟨γ⟩=881 *16* keV

$γ_{mode}$	γ(keV)	γ(%) †
Nd $L_ℓ$	4.633	0.0068 *8*
Nd $L_η$	5.146	0.0028 *3*
Nd $L_α$	5.228	0.182 *16*
Nd $L_β$	5.852	0.165 *17*
Nd $L_γ$	6.675	0.023 *3*
Nd $K_{α2}$	36.847	0.73 *5*
Nd $K_{α1}$	37.361	1.32 *9*
Nd $K_{β1}$'	42.240	0.378 *24*
Nd $K_{β2}$'	43.562	0.110 *7*
γ	147.9 *3*	<0.6
γ	171.4 *3*	<0.6
γ[E1]	246.9 *3*	0.184 *23*
γE2	301.629 *16*	61 *3*
γE2	421.9 *4*	1.34 *18*
γE2	450.6 *3*	0.183 *18*
γ[M1+E2]	489.1 *4*	0.37 *6*
γ	499.0 *3*	0.98 *24* *
γE2	614.93 *23*	2.26 *18*
γ[E1]	636.48 *16*	1.46 *12*
γ[E1]	660.17 *15*	1.71 *12*
γE1	697.50 *11*	4.09 *18*
γ[E1]	721.19 *13*	4.3 *3*
γ	825.18 *19*	1.28 *12*
γM1+E2	869.13 *13*	3.72 *24*
γ	903.03 *19*	1.40 *12*
γ[M1+E2]	946.98 *15*	1.40 *12*
γ	999.1 *5*	0.61 *24* *
γ[E1]	1022.82 *13*	4.8 *4*
γ	1079.9 *3*	0.79 *24* *
γ	1132.1 *3*	1.34 *12* *
γ	1157.23 *23*	1.16 *12*
γ[E2]	1170.76 *13*	0.61 *6*
γ[E1]	1210.2 *4*	1.77 *18*
γ[E2]	1248.61 *15*	3.23 *24*
γ	1327.4 *10*	~0.12 *
γ	1343.4 *3*	0.43 *6*
γ[M1+E2]	1357.67 *12*	5.5 *4*
γ	1381.6 *3*	2.26 *18*
γ	1426.6 *4*	0.49 *12* *
γ	1520.1 *3*	0.73 *6*
γ	1580.5 *7*	~0.24 *
γ[E2]	1659.29 *12*	~0.24

Photons (^{148}Pr)
(continued)

$γ_{mode}$	γ(keV)	γ(%) †
γ[E1]	1682.13 *24*	~0.24
γ	1727.1 *6*	~0.24 *
γ	1772.16 *17*	0.79 *6*
γ[M1+E2]	1907.93 *20*	1.10 *12*
γ[M1+E2]	1931.62 *20*	0.73 *6*
γ	2079.2 *3*	1.16 *12* *
γ	2106.61 *25*	0.079 *18*
γ	2130.3 *2*	1.83 *12*
γ	2143.4 *5*	0.49 *6*
γ	2224.2 *5*	0.37 *6*
γ	2241.3 *3*	0.18 *6*
γ	2542.9 *3*	0.24 *6*
γ[E1]	2629.11 *19*	1.28 *12*
γ	2735.9 *5*	0.55 *6*
γ	2983.1 *6*	0.37 *6*

† approximate
* with ^{148}Pm(2.27 or 2.0 min)

Atomic Electrons (^{148}Pr)

⟨e⟩=10.0 *4* keV

e_{bin}(keV)	⟨e⟩(keV)	e(%)
6 - 42	0.213	2.43 *19*
104 - 148	0.21	~0.17
164 - 203	0.042	~0.025
240 - 247	0.00146	0.00061 *6*
258	6.4	2.49 *13*
295	1.54	0.52 *3*
300	0.343	0.114 *6*
301 - 302	0.093	0.0309 *14*
378 - 422	0.108	0.028 *3*
444 - 493	0.08	0.017 *7*
497 - 499	0.0020	~0.00039
571 - 617	0.127	0.0217 *12*
629 - 678	0.095	0.0143 *6*
690 - 721	0.0152	0.00216 *8*
782 - 825	0.035	~0.0044
826	0.104	0.013 *3*
859 - 903	0.088	0.0100 *23*
940 - 979	0.054	0.0055 *9*
992 - 1036	0.021	0.0021 *8*
1073 - 1114	0.041	0.0037 *14*
1125 - 1171	0.030	0.00259 *24*
1203 - 1248	0.058	0.0048 *3*
1284 - 1327	0.098	0.0075 *14*
1336 - 1383	0.055	0.0041 *12*
1420 - 1427	0.0010	7 *3* ×10^{-5}
1477 - 1574	0.013	0.0008 *3*
1579 - 1676	0.010	0.00064 *24*
1681 - 1771	0.012	0.00068 *23*
1864 - 1930	0.0243	0.00129 *14*
2036 - 2129	0.033	0.0016 *4*
2136 - 2235	0.0055	0.00025 *7*
2499 - 2586	0.0073	0.00028 *4*
2622 - 2692	0.0043	0.00016 *5*
2729 - 2735	0.00056	2.0 *7* ×10^{-5}
2940 - 2982	0.0025	9 *3* ×10^{-5}

Continuous Radiation (^{148}Pr)

⟨β-⟩=1785 keV;⟨IB⟩=5.9 keV

E_{bin}(keV)		⟨ ⟩(keV)	(%)
0 - 10	β-	0.0054	0.107
	IB	0.053	
10 - 20	β-	0.0164	0.109
	IB	0.053	0.37
20 - 40	β-	0.068	0.227
	IB	0.105	0.36
40 - 100	β-	0.53	0.75
	IB	0.30	0.46
100 - 300	β-	6.8	3.29

Continuous Radiation (^{148}Pr)
(continued)

E_{bin}(keV)		⟨ ⟩(keV)	(%)
	IB	0.90	0.50
300 - 600	β-	32.7	7.1
	IB	1.10	0.26
600 - 1300	β-	228	23.6
	IB	1.7	0.20
1300 - 2500	β-	760	40.6
	IB	1.33	0.077
2500 - 4960	β-	757	24.1
	IB	0.31	0.0106

$^{148}_{59}$Pr(2.0 *1* min)

Mode: β-

Δ: <-72370 keV

SpA: 6.3×10^{8} Ci/g

Prod: fission

Photons (^{148}Pr) *

⟨γ⟩=938 *36* keV

$γ_{mode}$	γ(keV)	γ(%)†
Nd $L_ℓ$	4.633	0.0114 *14*
Nd $L_η$	5.146	0.0047 *5*
Nd $L_α$	5.228	0.31 *3*
Nd $L_β$	5.853	0.28 *3*
Nd $L_γ$	6.676	0.039 *5*
Nd $K_{α2}$	36.847	1.23 *9*
Nd $K_{α1}$	37.361	2.24 *16*
Nd $K_{β1}$'	42.240	0.64 *5*
Nd $K_{β2}$'	43.562	0.186 *14*
$γ_{β-}$[E1]	246.9 *3*	1.80 *21*
$γ_{β-}$E2	301.629 *16*	95 *8*
$γ_{β-}$E2	450.6 *3*	50 *4*
$γ_{β-}$	522.4 *5*	0.90 *15*
$γ_{β-}$E1	697.50 *11*	40 *3*
$γ_{β-}$[M1+E2]	934.9 *6*	1.6 *1*
$γ_{β-}$	1106.1 *4*	4.1 *4*
$γ_{β-}$	1556.8 *3*	4.9 *4*

† 1.0% uncert(syst)
* see also ^{148}Pr(2.27 min)

Atomic Electrons (^{148}Pr)

⟨e⟩=17.5 *9* keV

e_{bin}(keV)	⟨e⟩(keV)	e(%)
6 - 42	0.36	4.1 *3*
203 - 247	0.085	0.041 *4*
258	10.0	3.9 *3*
295	2.39	0.81 *7*
300	0.53	0.178 *14*
301 - 302	0.145	0.048 *3*
407	2.63	0.65 *5*
444 - 479	0.67	0.149 *10*
515 - 522	0.008	0.0015 *7*
654 - 697	0.51	0.077 *5*
891 - 935	0.048	0.0053 *11*
1063 - 1106	0.08	0.007 *4*
1513 - 1555	0.06	0.0041 *18*

Continuous Radiation (^{148}Pr)

$\langle\beta-\rangle$=1715 keV;$\langle$IB$\rangle$=5.4 keV

E_{bin}(keV)		$\langle\ \rangle$(keV)	(%)
0 - 10	β-	0.00472	0.094
	IB	0.053	
10 - 20	β-	0.0144	0.096
	IB	0.052	0.36
20 - 40	β-	0.060	0.20
	IB	0.103	0.36
40 - 100	β-	0.470	0.66
	IB	0.30	0.46
100 - 300	β-	6.1	2.96
	IB	0.89	0.49
300 - 600	β-	30.6	6.7
	IB	1.08	0.25
600 - 1300	β-	232	23.9
	IB	1.66	0.19
1300 - 2500	β-	850	45.4
	IB	1.15	0.068
2500 - 4298	β-	595	20.1
	IB	0.162	0.0057

$^{148}_{60}$Nd($>2.7\times10^{18}$ yr)

Δ: -77418 *4* keV

%: 5.76 *3*

$^{148}_{61}$Pm(5.370 *9* d)

Mode: β-

Δ: -76874 *15* keV

SpA: 1.643×10^5 Ci/g

Prod: ^{148}Nd(p,n); ^{148}Nd(d,2n); ^{147}Pm(n,γ)

Photons (^{148}Pm)

$\langle\gamma\rangle$=574 *7* keV

γ_{mode}	γ(keV)	γ(%)†
Sm $K_{\alpha2}$	39.522	0.0579 *19*
Sm $K_{\alpha1}$	40.118	0.105 *3*
Sm $K_{\beta1}$'	45.379	0.0304 *10*
Sm $K_{\beta2}$'	46.819	0.0090 *3*
γ [E2]	303.581 *18*	0.038 *4*
γ	362.81 *14*	0.0022 *4* ?
γ [E1]	393.798 *20*	0.0155 *22*
γ E2	550.274 *15*	22.0 *4*
γ [M1+E2]	592.828 *19*	0.353 *7*
γ E1	611.270 *15*	1.021 *20*
γ	819.269 *20*	0.0133 *22*
γ E2	874.17 *3*	0.235 *9*
γ [M1+E2]	896.408 *19*	0.981 *20*
γ M1+E2	903.917 *19*	0.053 *6*
γ E1	914.850 *16*	11.46 *23*
γ [M1+E2]	1113.879 *19*	0.0211 *18*
γ	1152.46 *12*	0.0029 *13*
γ [E2]	1371.31 *14*	0.0138 *13*
γ E2	1454.187 *20*	0.0506 *21*
γ E1	1465.120 *18*	22.2
γ [E1]	1507.674 *17*	0.0056 *9*
γ [E2]	1664.15 *2*	0.0119 *10*
γ	1734.113 *19*	0.0386 *8*
γ	1763.73 *12*	0.0062 *7*
γ	2284.38 *2*	0.0444 *22*
γ	2313.99 *12*	<0.00022 ?

† 2.3% uncert(syst)

Atomic Electrons (^{148}Pm)

$\langle$e$\rangle$=1.53 *3* keV

e_{bin}(keV)	$\langle$e$\rangle$(keV)	e(%)
7 - 45	0.0166	0.177 *15*
257 - 304	0.0055	0.00206 *20*
316 - 363	0.00057	0.00017 *4*
386 - 394	6.7×10^{-5}	1.72 *19* $\times10^{-5}$
503	0.92	0.183 *5*
543	0.148	0.0272 *8*
544 - 546	0.035	0.0065 *10*
549	0.0364	0.00663 *19*
550 - 593	0.0275	0.00489 *17*
604 - 611	0.00247	0.000409 *9*
772 - 819	0.00042	$\sim5\times10^{-5}$
827 - 867	0.038	0.0045 *9*
868	0.105	0.0121 *3*
872 - 915	0.0241	0.00267 *14*
1067 - 1114	0.00063	5.9 *13* $\times10^{-5}$
1145 - 1152	8.9×10^{-6}	$\sim8\times10^{-7}$
1324 - 1371	0.000242	1.82 *16* $\times10^{-5}$
1407	0.00072	5.11 *24* $\times10^{-5}$
1418	0.143	0.0101 *3*
1446 - 1465	0.0237	0.00162 *4*
1500 - 1506	5.5×10^{-6}	3.7 *5* $\times10^{-7}$
1617 - 1687	0.00061	3.6 *12* $\times10^{-5}$
1717 - 1762	0.00015	8.7 *25* $\times10^{-6}$
2238 - 2313	0.00044	1.9 *7* $\times10^{-5}$

Continuous Radiation (^{148}Pm)

$\langle\beta-\rangle$=722 keV;$\langle$IB$\rangle$=1.44 keV

E_{bin}(keV)		$\langle\ \rangle$(keV)	(%)
0 - 10	β-	0.0437	0.87
	IB	0.028	
10 - 20	β-	0.130	0.87
	IB	0.028	0.19
20 - 40	β-	0.52	1.73
	IB	0.054	0.19
40 - 100	β-	3.60	5.1
	IB	0.148	0.23
100 - 300	β-	34.1	17.0
	IB	0.38	0.22
300 - 600	β-	107	24.1
	IB	0.37	0.087
600 - 1300	β-	304	33.5
	IB	0.37	0.044
1300 - 2473	β-	272	16.7
	IB	0.067	0.0043

$^{148}_{61}$Pm(41.29 *11* d)

Mode: β-(95.4 *5* %), IT(4.6 *5* %)

Δ: -76737 *15* keV

SpA: 2.137×10^4 Ci/g

Prod: ^{148}Nd(p,n); ^{148}Nd(d,2n); ^{147}Pm(n,γ)

Photons (^{148}Pm)

$\langle\gamma\rangle$=1986 *14* keV

γ_{mode}	γ(keV)	γ(%)†
Pm L_ℓ	4.809	0.0092 *11*
Sm L_ℓ	4.993	0.0175 *19*
Pm L_η	5.363	0.0039 *4*
Pm L_α	5.430	0.248 *22*
Sm L_η	5.589	0.0066 *6*
Sm L_α	5.633	0.47 *3*
Pm L_β	6.091	0.230 *24*

Photons (^{148}Pm) (continued)

γ_{mode}	γ(keV)	γ(%)†
Sm L_β	6.342	0.42 *4*
Pm L_γ	6.959	0.032 *4*
Sm L_γ	7.264	0.061 *7*
Pm $K_{\alpha2}$	38.171	0.82 *5*
Pm $K_{\alpha1}$	38.725	1.48 *9*
Sm $K_{\alpha2}$	39.522	1.78 *7*
Sm $K_{\alpha1}$	40.118	3.23 *13*
Pm $K_{\beta1}$'	43.793	0.427 *25*
Pm $K_{\beta2}$'	45.183	0.126 *8*
Sm $K_{\beta1}$'	45.379	0.94 *4*
Sm $K_{\beta2}$'	46.819	0.278 *12*
γ_{IT} E4	61.30 *5*	0.00033
γ_{IT} M1	75.7 *1*	1.02
$\gamma_{\beta-}$M1+3.8%E2	98.479 *16*	2.47 *6*
$\gamma_{\beta-}$[M1+E2]	189.645 *16*	1.091 *25*
$\gamma_{\beta-}$E1	222.53 *11*	0.071 *16*
$\gamma_{\beta-}$E2	279.30 *3*	0.0014 *5*
$\gamma_{\beta-}$[M1+E2]	288.124 *16*	12.48 *24*
$\gamma_{\beta-}$[E2]	299.08 *3*	0.089 *18*
$\gamma_{\beta-}$[E1]	300.68 *3*	0.00074 *14*
$\gamma_{\beta-}$E1+3.6%M2	311.634 *15*	3.90 *7*
$\gamma_{\beta-}$[E2]	362.093 *17*	0.177 *18*
$\gamma_{\beta-}$E1+0.03%M2	414.071 *13*	18.6 *4*
$\gamma_{\beta-}$E2	432.763 *17*	5.32 *10*
$\gamma_{\beta-}$[E2]	460.572 *17*	0.419 *18*
$\gamma_{\beta-}$E1+0.05%M2	501.279 *16*	6.74 *13*
$\gamma_{\beta-}$E2	534.16 *11*	0.28 *3*
$\gamma_{\beta-}$E2	550.274 *15*	94.4 *19*
$\gamma_{\beta-}$E2+27%M1	553.257 *14*	0.39 *5*
$\gamma_{\beta-}$E1+M2	571.948 *16*	0.214 *8*
$\gamma_{\beta-}$E1+0.06%M2	599.757 *16*	12.53 *25*
$\gamma_{\beta-}$E1	611.270 *15*	5.46 *11*
$\gamma_{\beta-}$E2	629.962 *16*	88.6
$\gamma_{\beta-}$M1+6.6%E2	714.75 *3*	0.045 *4*
$\gamma_{\beta-}$E2	725.704 *14*	32.7 *6*
$\gamma_{\beta-}$E2	915.348 *15*	17.1 *3*
$\gamma_{\beta-}$E2	1013.827 *15*	20.2 *4*
$\gamma_{\beta-}$E2	1183.216 *18*	0.0356 *20*
$\gamma_{\beta-}$E2	1344.71 *3*	0.058 *4*

† uncert(syst): 11% for IT, 0.75% for β-

Atomic Electrons (^{148}Pm)

$\langle$e$\rangle$=20.5 *5* keV

e_{bin}(keV)	$\langle$e$\rangle$(keV)	e(%)
6 - 30	0.55	7.8 *7*
31	0.97	3.16 *18*
32 - 45	0.206	0.59 *3*
52	1.95	3.77 *12*
54 - 76	0.96	1.58 *6*
91	0.512	0.564 *18*
92 - 98	0.211	0.220 *5*
143 - 190	0.45	0.30 *4*
215 - 232	0.0009	0.00042 *13*
241	1.9	0.80 *19*
252 - 301	0.61	0.22 *6*
304 - 315	0.034	0.011 *5*
354 - 386	0.717	0.191 *4*
406 - 454	0.299	0.0693 *10*
459 - 501	0.0346	0.0071 *3*
503	3.95	0.784 *22*
506 - 534	0.046	0.009 *4*
543	0.634	0.117 *3*
544 - 572	0.528	0.0957 *16*
583	3.08	0.528 *12*
592 - 630	0.712	0.1145 *15*
668	0.00229	0.00034 *3*
679	0.95	0.140 *4*
707 - 726	0.196	0.0272 *6*
869 - 915	0.454	0.0518 *12*
967 - 1014	0.477	0.0490 *12*
1136 - 1183	0.0007	6 *3* $\times10^{-5}$
1298 - 1345	0.00105	8.0 *5* $\times10^{-5}$

Continuous Radiation (^{148}Pm)

⟨β-⟩=149 keV;⟨IB⟩=0.085 keV

E_{bin}(keV)		⟨ ⟩(keV)	(%)
0 - 10	β-	0.203	4.05
	IB	0.0076	
10 - 20	β-	0.60	3.99
	IB	0.0069	0.048
20 - 40	β-	2.34	7.8
	IB	0.0120	0.042
40 - 100	β-	15.1	21.7
	IB	0.025	0.040
100 - 300	β-	85	46.1
	IB	0.029	0.018
300 - 600	β-	43.9	11.4
	IB	0.0041	0.00113
600 - 1016	β-	1.98	0.299
	IB	7.5×10^{-5}	1.14×10^{-5}

$^{148}_{62}$Sm(7 *3* $\times10^{15}$ yr)

Mode: α

Δ: -79346 *3* keV

Prod: natural source

%: 11.3 *1*

Alpha Particles (^{148}Sm)

α(keV)
1960 *20*

$^{148}_{63}$Eu(54.5 *5* d)

Mode: ε, α(9 *3* $\times10^{-7}$ %)

Δ: -76266 *21* keV

SpA: 1.619×10^{4} Ci/g

Prod: ^{148}Sm(p,n); ^{147}Sm(d,n); ^{148}Sm(d,2n); ^{149}Sm(p,2n)

Alpha Particles (^{148}Eu)

α(keV)
2630 *30*

Photons (^{148}Eu)

⟨γ⟩=2164 *30* keV

$γ_{mode}$	γ(keV)	γ(%)†
Sm $L_ℓ$	4.993	0.29 *9*
Sm $L_η$	5.589	0.11 *3*
Sm $L_α$	5.633	7.6 *20*
Sm $L_β$	6.340	7.0 *19*
Sm $L_γ$	7.265	1.0 *3*
γ	22.46 *10*	<0.049
Sm $K_{α2}$	39.522	24.2 *7*
Sm $K_{α1}$	40.118	43.8 *13*
Sm $K_{β1}$'	45.379	12.7 *4*
Sm $K_{β2}$'	46.819	3.76 *13*
γM1	66.74 *3*	~0.020
γ[E1]	92.05 *5*	~0.020
γM1+3.8%E2	98.479 *16*	0.062 *4*

Photons (^{148}Eu) (continued)

$γ_{mode}$	γ(keV)	γ(%)†
γ	114.85 *20*	<0.01
γE1	116.07 *4*	0.102 *10*
γ	125.2 *4*	<0.01
γ	152.0 *4*	<0.01
γ	157.56 *6*	~0.010
γ	158.14 *9*	~0.010
γ[M1+E2]	160.73 *8*	0.020 *4*
γ	163.7 *3*	<0.01
γ(M1)	166.10 *8*	0.059 *5*
γ	172.6 *5*	<0.01
γ	180.3 *5*	<0.01
γE1	182.80 *4*	0.126 *8*
γ[M1+E2]	189.645 *16*	0.152 *6*
γ	212.9 *5*	<0.01
γ(M1)	216.01 *5*	0.015 *5*
γM1	216.10 *4*	0.023 *5*
γE1	222.53 *11*	0.020 *4*
γ	225.2 *4*	<0.01
γ	230.4 *5*	<0.01
γ	241.628 *24*	1.19 *6*
γE2	243.68 *9*	0.28 *3*
γM1,E2	252.58 *4*	0.098 *5*
γ	253.5 *5*	<0.01
γE2	279.30 *3*	0.059 *20*
γ[M1+E2]	288.124 *16*	0.314 *16*
γM1	296.15 *6*	0.054 *6*
γ[E2]	299.08 *3*	0.0022 *5*
γ[E1]	300.68 *3*	0.028 *5*
γ(E2)	308.36 *3*	0.075 *7*
γE2,M1	310.02 *5*	0.10 *3*
γ(M1)	310.42 *10*	0.15 *3*
γE1+3.6%M2	311.634 *15*	1.43 *5*
γ(M1)	319.32 *4*	0.148 *10*
γ[E2]	322.24 *4*	~0.020
γ	361.5 *5*	~0.039 ?
γ[E2]	362.093 *17*	0.025 *3*
γM1	377.59 *3*	0.199 *14*
γ	379.2 *5*	<0.01
γ	380.8 *7*	<0.01
γ	382.2 *5*	<0.01
γ	398.6 *5*	<0.01
γE1+0.03%M2	414.071 *13*	9.8 *4*
γE2	414.075 *24*	10.1 *8*
γ[M1+E2]	423.21 *10*	~0.020
γ[M1+E2]	432.12 *6*	~0.10
γE2	432.763 *17*	2.82 *11*
γM1	437.19 *3*	0.212 *23*
γ(E2)	446.93 *3*	0.030 *6*
γ[M1+E2]	449.69 *9*	0.049 *20*
γ[E2]	460.572 *17*	0.0105 *7*
γM1+14%E2	468.46 *3*	0.433 *24*
γ	471.3 *6*	<0.01
γ(M1)	480.81 *3*	0.099 *9*
γ	486.43 *13*	0.030 *10*
γ[E1]	486.73 *5*	0.030 *10*
γ	489.41 *3*	~0.020
γ[M1+E2]	493.52 *8*	0.059 *20*
γM1	494.68 *4*	0.049 *20*
γM1	495.40 *5*	0.28 *3*
γE1+0.05%M2	501.279 *16*	0.94 *3*
γ[E1]	504.58 *4*	0.059 *20*
γ[E1]	516.78 *3*	0.446 *24*
γ	532 *1*	~0.020 ?
γE2	534.16 *11*	0.079 *20*
γ[E1]	539.33 *7*	~0.010
γE2	550.274 *15*	98.6
γE2+27%M1	553.257 *14*	17 *3*
γ(E1)	553.261 *24*	5.7 *10*
γE1+M2	571.948 *16*	9.2 *6*
γ[M1+E2]	573.65 *4*	~0.020
γ(E2)	576.01 *4*	0.046 *6*
γ[E2]	584.18 *3*	~0.020
γ[E1]	587.45 *5*	0.059 *20*
γ	587.89 *3*	0.030 *10*
γ[M1+E2]	595.13 *4*	0.17 *5*
γE1+0.06%M2	599.757 *16*	0.315 *17*
γ[M1+E2]	602.57 *4*	0.20 *4*
γ[E1]	603.05 *4*	~0.020
γE1	611.270 *15*	20.1 *11*
γE1	620.00 *3*	0.88 *6*
γ[M1+E2]	620.43 *10*	0.20 *4*
γE2	629.962 *16*	71 *4*
γ[M1+E2]	636.678 *24*	0.034 *5*
γ	643.9 *3*	0.015 *5*

Photons (^{148}Eu) (continued)

$γ_{mode}$	γ(keV)	γ(%)†
γ[E1]	646.82 *5*	~0.010
γ	651.97 *4*	~0.020
γM1+42%E2	654.26 *3*	1.58 *9*
γ[E2]	656.89 *3*	0.128 *20*
γ[E2]	662.94 *5*	0.099 *10*
γ[M1+E2]	666.94 *11*	0.108 *20*
γ[E2]	667.14 *5*	0.069 *20*
γM1(+38%E2)	683.17 *3*	1.26 *7*
γ[M1+E2]	690.74 *3*	0.122 *7*
γ	702.27 *12*	~0.020
γ[E1]	704.54 *4*	0.020 *5*
γ[M1+E2]	706.24 *3*	~0.010
γM1+6.6%E2	714.75 *3*	1.68 *8*
γ[M1+E2]	719.66 *3*	0.207 *12*
γE2	725.704 *14*	12.0 *5*
γ[E1]	732.79 *5*	0.059 *6*
γ	735.15 *8*	0.15 *3*
γ[M1+E2]	735.157 *24*	~0.039
γ[M1+E2]	746.11 *9*	~0.020
γM1,E2	756.62 *3*	0.302 *16*
γ[E1]	770.32 *3*	0.413 *22*
γ[E2]	773.98 *4*	~0.010
γ[E1]	780.09 *4*	0.108 *10*
γ	788.49 *4*	0.033 *4*
γ[M1+E2]	790.83 *5*	0.039 *8*
γ[E1]	792.70 *4*	0.102 *7*
γ[E1]	799.24 *4*	0.36 *5*
γ	810.20 *7*	0.073 *7*
γ[E2]	817.64 *4*	~0.010
γ[M1+E2]	817.69 *10*	~0.010
γ[E2]	825.96 *3*	0.019 *8*
γ[M1+E2]	828.60 *5*	0.027 *8*
γ[E2]	833.19 *10*	~0.010
γ	838.3 *7*	<0.01
γ[E1]	851.26 *3*	0.015 *7*
γ[E2]	860.16 *5*	0.033 *5*
γE2+26%M1	869.95 *3*	4.9 *3*
γ	876.5 *7*	<0.01
γM1+4.8%E2	895.883 *24*	0.59 *3*
γM1+E2	903.917 *19*	0.35 *3*
γ[M1+E2]	906.84 *4*	0.182 *11*
γE2	915.348 *15*	2.39 *8*
γM1	924.801 *23*	0.278 *16*
γ[E1]	930.01 *5*	0.79 *20*
γ[E2]	930.85 *3*	1.38 *20*
γ[M1+E2]	935.76 *4*	0.038 *10*
γ	938.23 *8*	0.085 *8*
γ[E1]	949.54 *3*	0.222 *13*
γ	961.4 *5*	~0.010
γE2	967.331 *23*	3.02 *15*
γ[E1]	976.52 *7*	0.087 *8*
γ[M1+E2]	979.86 *5*	0.193 *13*
γ(M1)	989.67 *5*	0.424 *22*
γ	1001.8 *10*	<0.01
γE2	1013.827 *15*	0.51 *3*
γM1,E2	1034.07 *3*	7.1 *4*
γ[E2]	1035.92 *4*	0.108 *20*
γ	1038.5 *10*	<0.01
γ[E1]	1046.78 *8*	~0.039
γ(M1)	1047.94 *4*	0.138 *20*
γ[E2]	1059.42 *4*	~0.010
γ[E1]	1066.63 *4*	0.286 *20*
γ[M1+E2]	1068.330 *25*	0.091 *8*
γ[E2]	1070.13 *5*	0.165 *10*
γ[E1]	1082.14 *3*	0.173 *11*
γM1,E2	1089.17 *3*	0.197 *11*
γM1	1097.248 *23*	0.114 *8*
γM1	1104.33 *4*	0.341 *20*
γ	1107.77 *10*	0.087 *9*
γ[M1+E2]	1113.879 *19*	0.119 *12*
γ[E1]	1127.63 *4*	0.068 *6*
γ[E1]	1133.09 *6*	0.041 *5*
γ	1138.4 *5*	0.020 *5* ?
γM1(+E2)	1146.86 *5*	1.80 *9*
γ	1151.47 *4*	~0.020
γ	1156.4 *5*	<0.01
γ[E1]	1165.55 *5*	0.083 *5*
γE2	1183.216 *18*	1.54 *8*
γ[E2]	1194.16 *3*	0.112 *8*
γ	1201.3 *5*	~0.010
γE1	1207.515 *24*	0.58 *3*
γ	1218.99 *6*	0.036 *5*
γ(M1,E2)	1221.32 *3*	0.129 *9*
γ	1228.77 *10*	~0.010

Photons (^{148}Eu) (continued)

γ_{mode}	γ(keV)	γ(%)†
γ[E1]	1236.433 22	0.368 21
γ	1243 1	<0.01
γ(M1,E2)	1266.82 4	0.127 9
γ[E2]	1269.39 4	~0.010
γM1,E2	1309.88 3	0.426 24
γ	1327.3 10	<0.01
γE1	1328.57 3	1.16 6
γ	1338.2 10	<0.01
γM1+3.8%E2	1344.08 5	1.08 20
γE2	1344.71 3	2.17 15
γM1+32%E2	1353.61 9	0.56 3
γE1(+1.0%M2)	1362.77 5	0.53 3
γ	1370.99 8	0.018 6
γ	1377.4 5	~0.010
γ[M1+E2]	1390.59 7	~0.010
γ[E1]	1409.28 7	0.132 10
γ	1419.6 5	0.0049 20
γ	1422.43 5	~0.010
γ	1431.3 10	<0.01
γE2	1454.187 20	0.327 22
γM1+24%E2	1460.85 8	0.85 7
γ	1492.94 6	0.091 6
γ	1503.23 3	0.160 10
γ	1511.63 6	0.036 4
γ	1521.92 3	0.140 9
γ[M1+E2]	1533.12 5	0.021 5
γ	1535.90 7	0.074 7
γ	1540.53 10	0.059 10
γM1,E2	1543.34 4	0.67 3
γ[M1+E2]	1547.16 6	0.039 10
γ	1551.7 5	<0.0049
γE2	1560.81 3	0.76 4
γ[E1]	1565.85 6	0.039 10
γ	1572.5 5	~0.010
γ	1579.1 5	<0.0049
γ	1602.4 10	<0.0049
γ	1617.7 10	<0.0049
γM1+E2	1621.582 24	4.12 21
γ(E1)	1635.39 3	0.141 9
γ	1640.3 10	<0.0049
γM1+E2	1650.500 22	3.29 17
γ[M1+E2]	1654.08 3	0.108 12
γ	1658.3 4	<0.0049
γ[E2]	1664.15 2	0.067 6
γ	1668.7 10	<0.0049
γE2	1677.90 4	0.375 20
γ[E1]	1680.88 4	0.028 3
γ[E2]	1699.57 4	0.0108 20
γ[M1+E2]	1748.53 10	0.035 3
γ[E2]	1776.82 5	0.062 4
γ	1788.8 5	0.0049 20
γ	1840.04 10	0.0168 20
γ	1843.5 5	0.0049 10
γ[E2]	1939.83 3	0.058 4
γ[E2]	1974.04 5	0.048 4
γ	2122.90 6	~0.010
γ[E2]	2163.07 5	~0.010
γ[E2]	2173.29 4	0.218 12

† <0.1% uncert(syst)

Atomic Electrons (^{148}Eu)

$\langle e \rangle$=27 3 keV

e_{bin}(keV)	$\langle e \rangle$(keV)	e(%)
7	5.1	74 24
8	0.61	7.9 8
15	1.2	<16
16	1.7	<22
20	0.019	~0.09
21	0.9	<9
22 - 32	0.9	3.2 11
33	0.71	2.14 22
37 - 85	0.95	2.38 12
90 - 136	0.073	0.066 6
143 - 190	0.093	0.060 7
195 - 244	0.37	0.18 6
245 - 294	0.14	0.053 12

Atomic Electrons (^{148}Eu) (continued)

e_{bin}(keV)	$\langle e \rangle$(keV)	e(%)
295 - 335	0.056	0.0176 23
352 - 362	0.0018	0.00049 22
367	0.85	0.233 17
370 - 416	0.424	0.107 3
421 - 470	0.166	0.0380 12
471 - 501	0.0161	0.0033 3
503	4.12	0.819 18
504 - 505	8.5 $\times10^{-6}$	1.7 5 $\times10^{-6}$
506	0.92	0.18 3
509 - 524	0.00145	0.000284 24
525	1.0	~0.20
526 - 575	1.68	0.30 3
576 - 582	0.00048	8 3 $\times10^{-5}$
583	2.45	0.420 22
584 - 630	0.682	0.110 3
635 - 684	0.535	0.080 3
686 - 735	0.115	0.0161 8
738 - 787	0.0151	0.00199 23
788 - 837	0.139	0.0169 13
838 - 884	0.165	0.0190 7
888 - 937	0.101	0.0110 4
938 - 987	0.22	0.023 5
988 - 1037	0.051	0.0050 6
1038 - 1088	0.0265	0.00251 17
1089 - 1138	0.072	0.0064 9
1139 - 1187	0.0222	0.00191 14
1190 - 1236	0.0074	0.00061 5
1241 - 1282	0.0169	0.00133 15
1291 - 1338	0.083	0.0064 4
1342 - 1390	0.0061	0.000451 22
1402 - 1447	0.0228	0.00161 14
1452 - 1550	0.035	0.00231 18
1553 - 1652	0.143	0.0089 9
1653 - 1747	0.00249	0.000147 7
1769 - 1842	0.00040	2.2 5 $\times10^{-5}$
1893 - 1973	0.00138	7.2 4 $\times10^{-5}$
2076 - 2172	0.00281	0.000132 8

Continuous Radiation (^{148}Eu)

$\langle\beta+\rangle$=0.93 keV;$\langle$IB$\rangle$=0.35 keV

E_{bin}(keV)		$\langle\ \rangle$(keV)	(%)
0 - 10	β+	3.80 $\times10^{-8}$	4.52 $\times10^{-7}$
	IB	0.00061	
10 - 20	β+	1.81 $\times10^{-6}$	1.08 $\times10^{-5}$
	IB	0.00025	0.00161
20 - 40	β+	5.8 $\times10^{-5}$	0.000177
	IB	0.058	0.162
40 - 100	β+	0.00318	0.00406
	IB	0.077	0.17
100 - 300	β+	0.141	0.065
	IB	0.027	0.0138
300 - 600	β+	0.57	0.129
	IB	0.081	0.018
600 - 1300	β+	0.214	0.0316
	IB	0.100	0.0125
1300 - 1900	IB	0.0071	0.00049
	$\Sigma\beta$+		0.23

$^{148}_{64}$Gd(75 *3* yr)

Mode: α

Δ: -76278 4 keV

SpA: 32.4 Ci/g

Prod: ^{147}Sm(α,3n); ^{151}Eu(p,4n)

Alpha Particles (^{148}Gd)

α(keV)
3182.787 24

$^{148}_{65}$Tb(1.000 *17* h)

Mode: ϵ

Δ: -70670 40 keV

SpA: 2.12$\times10^{7}$ Ci/g

Prod: ^{141}Pr(^{12}C,5n); protons on Dy; ^{139}La(^{16}O,7n); protons on Ta

Photons (^{148}Tb)

$\langle\gamma\rangle$=1851 19 keV

γ_{mode}	γ(keV)	γ(%)†
γ E1	142.889 12	0.306 13
γ	382.62 11	~0.15
γ E1	489.059 10	19.8 4
γ	589.9 22	0.60 3
γ E2	631.947 12	10.71 21
γ M1	639.44 5	2.65 10
γ E2	784.433 15	84.4 17
γ E2	808.52 6	0.442 19
γ	841.6 5	0.267 19
γ [E1]	915.18 3	0.30 3
γ E2,M1	960.11 4	1.08 9
γ E2,M1	1002.56 7	0.337 20
γ	1007.64 7	0.65 6
γ E2,M1	1050.16 4	0.81 3
γ E2+5.2%M1	1079.013 21	11.46 23
γ E1	1089.42 3	2.26 5
γ	1105.66 11	0.62 3
γ	1167 3	0.11 3
γ	1215.0 19	0.17 3
γ	1230 1	0.33 12
γ E2,M1	1230.20 5	0.82 3
γ	1248.2 25	0.20 5
γ	1341.09 4	0.17 8
γ E2,M1	1404.24 3	2.14 5
γ [E1]	1426.51 7	0.279 18
γ	1449.16 4	0.91 3
γ	1469.93 10	0.13 5
γ	1526.53 5	0.64 3
γ	1599.45 5	1.08 3
γ M1,E2	1639.67 8	0.42 6
γ	1641.99 7	0.45 6
γ	1679 3	0.28 4
γ	1719.25 5	1.46 9
γ [E1]	1721.36 3	0.35 9
γ	1737.9 19	0.17 3
γ	1802.63 24	0.51 7
γ	1816.16 7	0.44 3
γ	1830.15 4	1.83 10
γ	1848.37 8	0.59 3
γ [E2]	1863.440 23	5.65 11
γ [M1+E2]	1915.56 7	0.40 3
γ	1988.7 12	0.33 4
γ	2088.51 5	0.44 7
γ	2101.87 10	0.63 5
γ	2131.04 7	1.22 4
γ	2155.28 24	0.31 5
γ	2168.0 11	0.31 4
γ [E2]	2188.66 3	1.72 6
γ	2247.3 4	0.33 3
γ	2288.1 5	0.40 3
γ	2301.43 21	0.253 25
γ	2310.96 5	1.16 3
γ	2331.9 5	0.51 3
γ	2346.46 16	0.44 6
γ	2362.9 8	0.57 3
γ	2485.9 5	0.53 4
γ	2510.58 15	0.31 3
γ	2593.3 19	0.26 10
γ	2614.57 4	0.69 5

Photons (^{148}Tb)
(continued)

γ_{mode}	γ(keV)	γ(%)†
γ	2634.6 *6*	0.14 *3*
γ[E2]	2699.98 *8*	0.64 *3*
γ	2777.5 *6*	~0.07
γ	2794.71 *24*	0.16 *3*
γ	2858.5 *16*	0.28 *7*
γ	2871.8 *22*	0.38 *7*
γ	3089.63 *7*	0.16 *5*
γ	3125.4 *9*	0.26 *3*
γ	3130.88 *16*	0.69 *5*
γ	3266.5 *6*	~0.35
γ	3269.2 *9*	0.56 *4*
γ	3296 *3*	0.10 *3*
γ	3552.9 *16*	0.30 *3*
γ	3574.90 *21*	0.23 *3*
γ	3644.9 *16*	0.37 *4*
γ	3685.8 *16*	0.51 *4*
γ	3983.7 *16*	0.30 *3*
γ	4068.17 *24*	0.14 *3*

† approximate

$^{148}_{65}$Tb(2.20 *5* min)

Mode: ϵ
Δ: >-70670 keV
SpA: 5.76×10^{8} Ci/g
Prod: ^{141}Pr(^{12}C,5n); ^{139}La(^{16}O,7n); protons on Dy; ^{151}Eu(^{3}He,6n)

Photons (^{148}Tb)
$\langle\gamma\rangle$=2627 *160* keV

γ_{mode}	γ(keV)	γ(%)
Gd L_{ℓ}	5.362	0.190 *19*
Gd L_{η}	6.049	0.068 *5*
Gd L_{α}	6.054	5.0 *3*
Gd L_{β}	6.851	4.6 *4*
Gd L_{γ}	7.878	0.69 *7*
Gd $K_{\alpha2}$	42.309	17.5 *5*
Gd $K_{\alpha1}$	42.996	31.5 *10*
Gd $K_{\beta1}$'	48.652	9.3 *3*
Gd $K_{\beta2}$'	50.214	2.80 *10*
γ E2	129.5 *3*	2.5 *4*
γ E1	142.889 *12*	2.66 *16*
γ E2	394.51 *20*	86 *5*
γ E2	481.80 *23*	3.0 *4*
γ E1	489.059 *10*	5.2 *5*
γ E2	631.947 *12*	93 *4*
γ[E1]	752.92 *22*	1.7 *2*
γ E2	784.433 *15*	100
γ E2	808.52 *6*	2.9 *6*
γ E3	882.4 *3*	92 *4*

$^{148}_{66}$Dy(3.1 *1* min)

Mode: ϵ
Δ: -67980 *60* keV
SpA: 4.09×10^{8} Ci/g
Prod: protons on Ta; ^{142}Nd(^{12}C,6n)

Photons (^{148}Dy)

γ_{mode}	γ(keV)	γ(%)
γ E1	620.24 *1*	99.7

$^{148}_{67}$Ho(~2.2 s)

Mode: ϵ
Δ: -58370 *300* keV syst
SpA: ~3.0×10^{10} Ci/g
Prod: daughter ^{148}Er

Photons (^{148}Ho)

γ_{mode}	γ(keV)	γ(%)
γ E2	1667.7 *3*	7.4 *14*

$^{148}_{67}$Ho(9 *1* s)

Mode: ϵ
Δ: >-58370 keV syst
SpA: 8.2×10^{9} Ci/g
Prod: ^{10}B on ^{144}Sm; ^{92}Zn(^{58}Ni,np)

Photons (^{148}Ho)

γ_{mode}	γ(keV)	γ(rel)
γ	504.3 *3*	17 *3*
γ E2	661.5 *2*	69 *7*
γ E3	1688.3 *2*	100

$^{148}_{68}$Er(4.5 *4* s)

Mode: ϵ
SpA: 1.57×10^{10} Ci/g
Prod: ^{92}Mo(^{58}Ni,2p)

Photons (^{148}Er)

γ_{mode}	γ(keV)	γ(%)
γ	244.2 *3*	7.0 *14*
γ	315.2 *3*	6.0 *12*

$^{148}_{69}$Tm(700 *200* ms)

Mode: ϵ
SpA: 6.9×10^{10} Ci/g
Prod: ^{92}Mo(^{58}Ni,np)

Photons (^{148}Tm)

γ_{mode}	γ(keV)	γ(%)†
γ E2	131.1 *3*	
γ[E1]	247.1 *3*	~15 ?
γ[E2]	257.5 *3*	52 *20*
γ[E2]	283.0 *3*	46 *20* ?
γ[E2]	646.6 *3*	100
γ[E1]	730.3 *3*	35 *15*
γ[E2]	877.4 *3*	72 *25*
γ[E2]	1002.9 *3*	55 *20*

† approximate

A = 149

NDS **19**, 337 (1976)

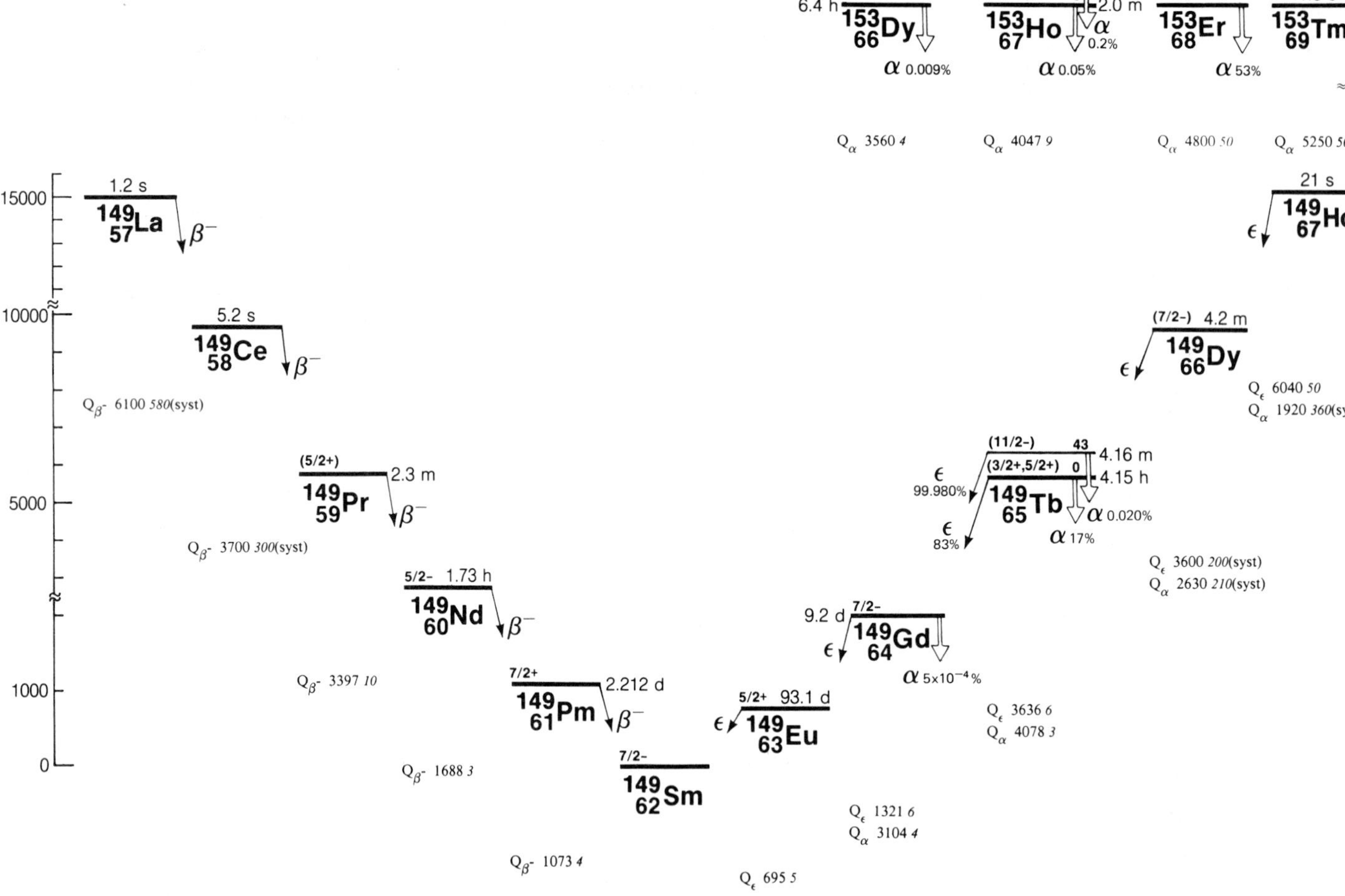

$^{149}_{57}$La(1.2 *4* s)

Mode: β-

Δ: -61190 *500* keV syst

SpA: 4.8×10^{10} Ci/g

Prod: fission

$^{149}_{58}$Ce(5.2 *4* s)

Mode: β-

Δ: -67290 *300* keV syst

SpA: 1.36×10^{10} Ci/g

Prod: fission

Photons (^{149}Ce)

γ_{mode}	γ(keV)	γ(rel)
γ	57.7 *3*	100 *20*
γ	86.4 *3*	20 *4*
γ	104.5 *3*	2.2 *4*
γ	129.2 *3*	1.8 *4*
γ	144.7 *3*	2.7 *5*
γ	172.5 *3*	2.8 *6*
γ	211.4 *3*	2.0 *4*
γ	225.7 *3*	1.3 *3*
γ	232.8 *3*	1.8 *4*
γ	258.5 *3*	1.5 *3*
γ	284.9 *3*	1.5 *3*
γ	294.0 *3*	6.3 *13*
γ	311.0 *3*	1.3 *3*
γ	322.4 *3*	7.2 *14*
γ	380.0 *3*	34 *7*
γ	390.0 *3*	2.3 *5*
γ	417.3 *3*	1.8 *4*
γ	460.0 *3*	2.0 *4*
γ	702.8 *3*	2.5 *5*
γ	864.5 *3*	7.8 *16*
γ	892.7 *3*	8.0 *16*

$^{149}_{59}$Pr(2.3 *2* min)

Mode: β-

Δ: -70988 *11* keV

SpA: 5.5×10^{8} Ci/g

Prod: ^{150}Nd(γ,p); fission

Photons (^{149}Pr)

⟨γ⟩=180 *7* keV

γ_{mode}	γ(keV)	γ(%)†
γ	108.5 *1*	5.7 *11*
γ	112.1 *1*	0.60 *12*
γ	120.3 *1*	0.84 *17*
γ	138.5 *1*	6.6 *13*
γ	156.0 *1*	1.08 *22*
γ	162.3 *1*	1.9 *4*
γ	165.1 *1*	6.0 *12*
γ	177.7 *1*	0.48 *10*
γ	182.6 *1*	0.60 *12*
γ	207.7 *1*	1.7 *3*
γ	238.7 *1*	0.60 *12*
γ	258.3 *1*	3.4 *7*
γ	312.9 *1*	0.54 *11*
γ	316.4 *1*	1.6 *3*
γ	321.3 *1*	1.5 *3*
γ	333.0 *1*	3.7 *7*

Photons (^{149}Pr)
(continued)

γ_{mode}	γ(keV)	γ(%)†
γ	341.3 2	0.54 11
γ	351.2 2	0.48 10
γ	366.0 1	1.9 4
γ	388.7 2	0.48 10
γ	390.6 2	0.42 8
γ	397.5 1	0.90 18
γ	406.3 1	1.4 3
γ	409.5 1	0.66 13
γ	433.0 1	1.4 3
γ	459.9 2	0.48 10
γ	474.6 1	1.7 3
γ	494.6 2	0.42 8
γ	517.4 1	2.9 6
γ	530.6 2	0.54 11
γ	571.1 2	0.48 10
γ	604.1 2	0.72 14
γ	623.0 1	1.08 22
γ	654.7 2	0.48 10
γ	662.5 1	1.08 22
γ	724.3 2	0.48 10
γ	755.8 2	0.54 11
γ	782.0 2	0.78 16
γ	797.4 2	0.54 11
γ	874.1 2	0.48 10

† approximate

$^{149}_{60}$Nd(1.73 1 h)

Mode: β-

Δ: -74385 4 keV

SpA: 1.216×10^7 Ci/g

Prod: ^{148}Nd(n,γ)

Photons (^{149}Nd)

⟨γ⟩=384 9 keV

γ_{mode}	γ(keV)	γ(%)†
Pm L_ℓ	4.809	0.105 12
Pm L_η	5.363	0.042 4
Pm L_α	5.430	2.83 24
Pm L_β	6.095	2.5 3
Pm L_γ	6.961	0.36 4
γ [E2]	29.952 9	0.0180 6
Pm $K_{\alpha2}$	38.171	9.5 5
Pm $K_{\alpha1}$	38.725	17.3 10
Pm $K_{\beta1}$'	43.793	5.0 3
Pm $K_{\beta2}$'	45.183	1.47 9
γ [E1]	58.857 7	1.52 3
γ [E1]	65.398 11	0.048 9
γ M1+33%E2	74.314 7	1.26 19
γ [E1]	74.620 13	1.0 3
γ M1+E2	75.682 10	0.33 11
γ	91.111 25	0.074 14
γ M1	96.985 7	1.53 11
γ	107.77 3	0.063 14
γ M1+3.0%E2	114.307 6	18.8 18
γ M1+E2	116.974 7	0.117 22
γ	122.399 16	0.232 22
γ E1	126.605 8	0.115 11
γ M1+E2	137.056 15	0.046 4
γ [E1]	139.185 10	0.48 4
γ E1	155.842 6	6.1 6
γ E1	177.817 10	0.164 11
γ	185.461 8	0.093 19
γ E2	188.621 6	1.99 22
γ E2	192.003 8	0.60 6
γ M1(+<21%E2)	198.911 9	1.46 14
γ M1,E2	208.133 7	2.9 3
γ M1+14%E2	211.293 6	27.3 14
γ M1,E2	213.945 12	0.41 4
γ	226.829 24	0.16 3
γ	229.571 13	0.50 4

Photons (^{149}Nd)
(continued)

γ_{mode}	γ(keV)	γ(%)†
γ M2	240.198 7	4.0 3
γ E2(+M1)	245.710 10	1.04 10
γ	250.80 6	0.038 8
γ	254.18 3	0.087 16
γ E1	258.054 8	0.38 3
γ M1	267.685 6	6.1 6
γ E1	270.149 6	10.7 10
γ M1,E2	273.225 11	0.23 4
γ [E1]	275.425 10	0.60 11
γ	276.943 22	0.32 7
γ M1,E2	282.446 7	0.62 6
γ M1,E2	288.181 10	0.68 7
γ E1	294.790 9	0.58 6
γ	301.116 18	0.38 3
γ	310.930 13	0.52 4
γ E1	326.542 7	4.7 4
γ	342.79 11 ?	
γ [E1]	347.844 10	0.19 9
γ E1	349.213 7	1.48 14
γ	352.73 7 ?	
γ	357.03 9 ?	
γ	360.018 10	0.164 11
γ	366.627 10	0.66 7
γ	380.91 8 ?	
γ [M1+E2]	384.659 7	0.34 3
γ	396.754 7 ?	
γ E1	423.527 6	9.4 9
γ	425.237 13 ?	
γ E1	443.515 7	1.50 14
γ	480.25 9 ?	
γ	483.54 9 ?	
γ	493.88 9 ?	
γ	498.01 10 ?	
γ	510.27 10 ?	
γ	515.622 11 ?	
γ	533.15 8	0.087 16
γ	538.13 8	0.11 3
γ E1	540.500 6	7.7 7
γ	556.41 6	1.2 4
γ	579.263 13	~0.08
γ	583.02 3	0.085 8
γ	594.40 7 ?	
γ	598.06 7 ?	
γ	630.221 24	0.221 22
γ	635.46 3	0.112 11
γ E1	654.807 6	7.3 7
γ	661.88 13 ?	
γ	673.55 10 ?	
γ	675.76 6 ?	
γ	681.9 1 ?	
γ	686.92 3	0.104 11
γ	696.236 12	0.172 16
γ	712.57 3	0.085 8
γ	718.448 15	0.030 3
γ	727.88 8	0.022 6
γ	740.54 5	0.019 3
γ	749.60 7 ?	
γ	754.27 3	0.035 6
γ	758.59 13	0.016 3
γ	761.43 7	
γ	768.153 25	0.068 6
γ	781.39 8	0.0071 8
γ	786.72 6	0.0126 11
γ	793.40 5	0.0254 25
γ	795.89 11	0.0068 25
γ	808.833 15	0.169 16
γ	813.25 13 ?	
γ	829.33 21 ?	
γ	832.09 8	0.0221 25
γ	837.316 13	0.035 6
γ	842.83 3	0.055 6
γ	849.912 17	0.025 3
γ	859.40 6 ?	
γ	861.51 5 ?	
γ	864.98 6 ?	
γ	871.35 3	0.038 6
γ	874.052 15 ?	
γ	886.57 10 ?	
γ	907.67 10 ?	
γ	915.310 17 ?	
γ	923.86 3	0.115 11
γ	929.41 6 ?	
γ	935.88 8 ?	
γ	938.76 8 ?	

Photons (^{149}Nd)
(continued)

γ_{mode}	γ(keV)	γ(%)†
γ	945.79 5	0.024 3
γ	952.046 19	0.0109 22
γ	963.920 13	0.021 3
γ	967.42 6 ?	
γ	971.69 13 ?	
γ	979.00 5	0.112 11
γ	986.65 14 ?	
γ	992.90 8	0.0150 16
γ	1022.776 12	0.120 11
γ	1027.16 5 ?	
γ	1031.75 9 ?	
γ	1041.914 16	0.027 3
γ	1051.88 16 ?	
γ	1075.82 6	0.025 3
γ	1078.74 3	0.068 8
γ	1100.770 16	0.060 6
γ	1123.441 17	0.016 3
γ	1125.30 5	0.030 4
γ	1128.53 14 ?	
γ	1135.92 11 ?	
γ	1141.75 5	0.0038 6
γ	1150.08 5	0.0033 6
γ	1171.76 5	0.0049 6
γ	1175.70 8	0.0052 6
γ	1190.32 9	0.0033 3
γ	1197.755 17	0.0076 8
γ	1202.29 13 ?	
γ	1225.59 6	~0.0027
γ	1234.067 12	0.292 22
γ	1259.58 10	0.0038 8
γ	1263.99 8	0.0071 8
γ	1280.33 5	~0.0011
γ	1284.45 6	~0.0011
γ	1290.05 8	0.0044 14
γ	1298.36 5	0.0017 3
γ	1312.061 16	0.0079 11
γ	1357.22 5	0.0025 3
γ	1381.43 6	0.00243 22
γ	1448.04 23	0.00046 14
γ	1454.20 5	0.0013 3
γ	1473.9 3 ?	
γ	1495.74 6	0.0008 3
γ	1568.51 5	0.00055 14

† 4.0% uncert(syst)

Atomic Electrons (^{149}Nd)

⟨e⟩=47.5 17 keV

e_{bin}(keV)	⟨e⟩(keV)	e(%)
6	1.26	19.5 23
7 - 28	2.38	19.5 15
29	1.42	4.9 8
30 - 51	1.36	4.1 4
52	1.21	2.34 17
57 - 68	1.38	2.1 3
69	12.2	17.7 21
72 - 106	1.19	1.47 20
107	2.7	2.48 25
108 - 156	2.77	2.21 11
163	0.72	0.44 8
166	7.3	4.38 25
169 - 192	0.55	0.302 22
195	4.3	2.18 19
197 - 202	0.40	0.20 5
204	1.29	0.63 6
205 - 254	3.72	1.65 7
256 - 304	0.68	0.251 21
309 - 358	0.13	0.040 11
359 - 398	0.234	0.062 5
416 - 444	0.0399	0.0095 6
488 - 537	0.19	0.038 7
538 - 585	0.027	0.0048 14
590 - 635	0.094	0.0155 14
642 - 690	0.029	0.0045 7
695 - 742	0.0054	0.00076 20
747 - 796	0.007	0.0010 4
798 - 843	0.0041	0.00051 14

Atomic Electrons (^{149}Nd)
(continued)

e_{bin}(keV)	⟨e⟩(keV)	e(%)
848 - 879	0.0026	~0.00030
901 - 948	0.0042	0.00046 *15*
950 - 997	0.0033	0.00034 *14*
1015 - 1056	0.0032	0.00031 *10*
1068 - 1118	0.0015	0.00014 *3*
1119 - 1168	0.00041	3.6 *9* $\times 10^{-5}$
1169 - 1218	0.0046	~0.00038
1219 - 1267	0.0011	9 *3* $\times 10^{-5}$
1273 - 1312	7.3 $\times 10^{-5}$	5.6 *16* $\times 10^{-6}$
1336 - 1381	4.4 $\times 10^{-5}$	3.3 *13* $\times 10^{-6}$
1403 - 1451	3.5 $\times 10^{-5}$	2.5 *8* $\times 10^{-6}$
1453 - 1523	8.6 $\times 10^{-6}$	5.7 *24* $\times 10^{-7}$
1561 - 1567	1.0 $\times 10^{-6}$	7 *3* $\times 10^{-8}$

Continuous Radiation (^{149}Nd)

⟨β-⟩=457 keV;⟨IB⟩=0.60 keV

E_{bin}(keV)		⟨ ⟩(keV)	(%)
0 - 10	β-	0.051	1.02
	IB	0.020	
10 - 20	β-	0.155	1.03
	IB	0.020	0.138
20 - 40	β-	0.63	2.11
	IB	0.038	0.131
40 - 100	β-	4.65	6.6
	IB	0.099	0.154
100 - 300	β-	48.6	24.1
	IB	0.22	0.129
300 - 600	β-	154	34.6
	IB	0.151	0.037
600 - 1300	β-	246	30.3
	IB	0.052	0.0071
1300 - 1574	β-	3.22	0.240
	IB	4.5 $\times 10^{-5}$	3.4 $\times 10^{-6}$

$^{149}_{61}$Pm(2.2117 *21* d)

Mode: β-
Δ: -76074 *5* keV
SpA: 3.962$\times 10^{5}$ Ci/g
Prod: daughter ^{149}Nd; fission

Photons (^{149}Pm)

⟨γ⟩=10.7 *6* keV

γ_{mode}	γ(keV)	γ(%)†
Sm $K_{\alpha2}$	39.522	0.062 *4*
Sm $K_{\alpha1}$	40.118	0.113 *8*
Sm $K_{\beta1}$'	45.379	0.0328 *23*
Sm $K_{\beta2}$'	46.819	0.0097 *7*
γ M1+E2	73.04 *19*	6 *3* $\times 10^{-6}$
γ M1,E2	208.34 *21*	~0.0009
γ	242.4 *4*	~0.0008
γ M1	254.58 *12*	0.0043 *12*
γ [E2]	263.47 *5*	0.008 *4*
γ M1	277.14 *12*	0.024 *6*
γ [M1+E2]	281.39 *20*	0.0044 *12*
γ M1	286.03 *5*	2.85
γ	305.05 *21*	0.0020 *9*
γ M1	327.62 *14*	0.0031 *8*
γ E2	350.18 *14*	0.00027 *8*
γ	350.66 *21*	~0.0018
γ	359.55 *22*	0.0020 *9*
γ	531.8 *4*	~0.0011
γ E2	535.97 *17*	0.0104 *19*
γ	544.6 *3*	0.0023 *9*

Photons (^{149}Pm)
(continued)

γ_{mode}	γ(keV)	γ(%)†
γ	547.5 *4*	~0.0017
γ E2	558.53 *17*	0.014 *3*
γ	568.52 *20*	0.017 *3*
γ	591.08 *20*	0.062 *7*
γ	614.13 *20*	0.0119 *18*
γ	636.69 *20*	0.0057 *18*
γ	808.1 *3*	0.0138 *18*
γ	830.6 *3*	0.028 *4*
γ	833.5 *4*	0.028 *4*
γ	859.4 *3*	0.094 *14*
γ	882.0 *3*	0.021 *3*

† 6.0% uncert(syst)

Atomic Electrons (^{149}Pm)

⟨e⟩=0.70 *4* keV

e_{bin}(keV)	⟨e⟩(keV)	e(%)
7 - 45	0.0207	0.210 *20*
65 - 73	1.3 $\times 10^{-5}$	~2 $\times 10^{-5}$
162 - 208	0.0014	0.00071 *17*
217 - 236	0.0068	0.0030 *6*
239	0.55	0.228 *15*
241 - 277	0.0020	0.00077 *12*
278	0.082	0.0294 *20*
279 - 281	0.0070	0.00251 *16*
284	0.0192	0.0067 *5*
285 - 328	0.0062	0.00215 *14*
342 - 360	8.7 $\times 10^{-5}$	2.5 *11* $\times 10^{-5}$
485 - 534	0.0021	0.00041 *10*
535 - 584	0.0037	0.0007 *3*
589 - 637	0.00046	8 *3* $\times 10^{-5}$
761 - 808	0.0019	0.00024 *9*
813 - 859	0.0036	0.00043 *19*
874 - 882	9.5 $\times 10^{-5}$	1.1 *5* $\times 10^{-5}$

Continuous Radiation (^{149}Pm)

⟨β-⟩=366 keV;⟨IB⟩=0.40 keV

E_{bin}(keV)		⟨ ⟩(keV)	(%)
0 - 10	β-	0.067	1.34
	IB	0.017	
10 - 20	β-	0.203	1.35
	IB	0.0164	0.114
20 - 40	β-	0.83	2.75
	IB	0.031	0.108
40 - 100	β-	6.0	8.6
	IB	0.079	0.123
100 - 300	β-	60	30.0
	IB	0.160	0.094
300 - 600	β-	168	38.2
	IB	0.085	0.021
600 - 1073	β-	131	17.9
	IB	0.0127	0.0019

$^{149}_{62}$Sm(>1$\times 10^{16}$ yr)

Δ: -77147 *3* keV
%: 13.8 *1*

$^{149}_{63}$Eu(93.1 *4* d)

Mode: ϵ
Δ: -76452 *6* keV
SpA: 9413 Ci/g
Prod: ^{149}Sm(p,n); ^{150}Sm(p,2n)

Photons (^{149}Eu)

⟨γ⟩=63.6 *11* keV

γ_{mode}	γ(keV)	γ(%)
Sm L_ℓ	4.993	0.218 *22*
Sm L_η	5.589	0.081 *6*
Sm L_α	5.633	5.8 *4*
Sm L_β	6.342	5.3 *5*
Sm L_γ	7.275	0.79 *8*
Sm $K_{\alpha2}$	39.522	22.3 *6*
Sm $K_{\alpha1}$	40.118	40.3 *12*
Sm $K_{\beta1}$'	45.379	11.7 *3*
Sm $K_{\beta2}$'	46.819	3.47 *12*
γ M1+E2	73.04 *19*	0.008 *3*
γ M1	178.4 *2*	0.017 *8*
γ M1,E2	208.34 *21*	~0.004
γ [M1+E2]	251.45 *18*	~0.014
γ M1	254.58 *12*	0.59 *6*
γ M1	277.14 *12*	3.3 *2*
γ [M1+E2]	281.39 *20*	0.020 *4*
γ M1	327.62 *14*	3.90 *20*
γ E2	350.18 *14*	0.34 *3*
γ E2	506.03 *14*	0.55 *5*
γ E2	528.59 *14*	0.53 *5*
γ E2	535.97 *17*	0.0461 *10*
γ E2	558.53 *17*	0.0620 *12*

Atomic Electrons (^{149}Eu)

⟨e⟩=8.4 *4* keV

e_{bin}(keV)	⟨e⟩(keV)	e(%)
7	3.8	55 *6*
8	0.53	6.8 *7*
15 - 31	0.34	1.53 *15*
32	0.44	1.37 *14*
33	0.65	1.97 *20*
37	0.072	0.193 *20*
38	0.30	0.79 *8*
39	0.29	0.74 *8*
40 - 73	0.160	0.37 *3*
132 - 178	0.009	0.0064 *24*
201 - 208	0.137	0.066 *7*
230	0.66	0.288 *18*
235 - 280	0.172	0.065 *3*
281	0.61	0.218 *12*
303 - 350	0.161	0.0506 *19*
459 - 506	0.058	0.0122 *7*
512 - 559	0.0091	0.00174 *8*

Continuous Radiation (^{149}Eu)

⟨IB⟩=0.19 keV

E_{bin}(keV)		⟨ ⟩(keV)	(%)
10 - 20	IB	0.00021	0.00133
20 - 40	IB	0.059	0.164
40 - 100	IB	0.078	0.17
100 - 300	IB	0.023	0.0119
300 - 600	IB	0.029	0.0071
600 - 695	IB	0.00028	4.5 $\times 10^{-5}$

$^{149}_{64}$Gd(9.25 *10* d)

Mode: ϵ, α(0.00046 *15* %)
Δ: -75131 *5* keV
SpA: 9.47×10^4 Ci/g
Prod: ^{151}Eu(p,3n); ^{147}Sm(α,2n)

Alpha Particles (^{149}Gd)

α(keV)
3018 *5*

Photons (^{149}Gd)

⟨γ⟩=418 *10* keV

γ_{mode}	γ(keV)	γ(%)†
Eu L_ℓ	5.177	0.35 *4*
Eu L_η	5.817	0.123 *10*
Eu L_α	5.843	9.4 *8*
Eu L_β	6.591	9.1 *10*
Eu L_γ	7.607	1.46 *18*
γ[M2]	37.88 *23*	0.103 *16* ?
Eu $K_{\alpha2}$	40.902	29.4 *10*
Eu $K_{\alpha1}$	41.542	53.0 *17*
Eu $K_{\beta1}'$	46.999	15.4 *5*
Eu $K_{\beta2}'$	48.496	4.67 *17*
γ	126.0 *4*	0.126 *24* ?
γ	132.3 *3*	0.063 *24* ?
γ	138.0 *7*	0.079 *24* ?
γM1+3.5%E2	149.87 *15*	41.9 *18*
γ	184.7 *6*	~0.041 ?
γE1	214.6 *4*	0.146 *18*
γ	230.9 *6*	0.12 *6* ?
γE2	252.5 *4*	0.19 *5*
γE1	260.81 *22*	1.04 *7*
γ	264.5 *6*	~0.023 ?
γM1	272.27 *17*	2.63 *11*
γM1(+E2)	298.68 *17*	22.7 *18*
γM2	346.63 *18*	18.0
γE2	404.5 *3*	0.7 *3*
γ	431.1 *6*	0.059 *9*
γM1	460.07 *23*	0.43 *4*
γE2	478.84 *23*	0.171 *18*
γE2	483.3 *6*	
γE3	496.50 *19*	1.30 *7*
γM1	516.77 *20*	2.0 *3*
γM1	534.38 *22*	2.38 *11*
γM1	601.2 *6*	0.054 *11*
γE1	645.31 *19*	1.06 *9*
γM1	663.1 *4*	0.70 *11*
γE2(+M1)	666.64 *22*	0.70 *11*
γE1	748.9 *4*	6.2 *7*
γM1+E2	789.04 *18*	5.3 *5*
γM1	812.9 *4*	~0.10
γ	863.6 *6*	0.058 *18* ?
γM1	876.1 *4*	0.162 *20*
γM1	933.5 *5*	0.40 *9*
γM1	938.91 *19*	1.62 *25*
γ(E1)	947.9 *6*	0.67 *11*

† 33% uncert(syst)

Atomic Electrons (^{149}Gd)

⟨e⟩=66.4 *19* keV

e_{bin}(keV)	⟨e⟩(keV)	e(%)
7	3.7	53 *6*
8	3.1	40 *5*
30	5.1	17 *3*
31 - 35	3.6	11.1 *10*
36	2.0	5.6 *9*
37 - 84	1.71	4.3 *3*
90	0.026	~0.029
101	21.5	21.2 *10*
118 - 138	0.09	0.070 *25*
142	4.3	3.06 *22*
143 - 185	1.53	1.03 *11*
204 - 245	0.68	0.304 *13*
250	3.5	1.4 *4*
251 - 297	1.0	0.35 *10*
298	10.5	3.53 *13*
299 - 347	2.63	0.773 *21*
356 - 405	0.062	0.017 *5*
412 - 460	0.232	0.0526 *24*
468 - 517	0.48	0.100 *6*
526 - 553	0.0427	0.0081 *3*
593 - 638	0.094	0.0153 *21*
644 - 678	0.0194	0.0029 *4*
700 - 749	0.29	0.039 *9*
764 - 813	0.044	0.0056 *13*
815 - 864	0.0087	0.00105 *17*
868 - 899	0.085	0.0096 *12*
925 - 948	0.0151	0.00162 *16*

Continuous Radiation (^{149}Gd)

⟨IB⟩=0.34 keV

E_{bin}(keV)		⟨ ⟩(keV)	(%)
10 - 20	IB	0.00020	0.00126
20 - 40	IB	0.056	0.153
40 - 100	IB	0.092	0.20
100 - 300	IB	0.027	0.0138
300 - 600	IB	0.075	0.0168
600 - 1300	IB	0.090	0.0118
1300 - 1321	IB	1.02×10^{-7}	7.8×10^{-9}

$^{149}_{65}$Tb(4.15 *5* h)

Mode: ϵ(83.3 *14* %), α(16.7 *14* %)
Δ: -71495 *6* keV
SpA: 5.07×10^6 Ci/g
Prod: ^{141}Pr(^{12}C,4n); ^{151}Eu(α,6n); protons on Ta

Alpha Particles (^{149}Tb)

⟨α⟩=662.5 *7* keV

α(keV)	α(%)
3645 *4*	0.0050 *17*
3966 *4*	16.7

Photons (^{149}Tb)

⟨γ⟩=1279 *10* keV

γ_{mode}	γ(keV)	γ(%)†
Gd L_ℓ	5.362	0.236 *24*
Gd L_η	6.049	0.085 *7*
Gd L_α	6.054	6.2 *4*
Gd L_β	6.851	5.7 *5*
Gd L_γ	7.878	0.86 *9*
Gd $K_{\alpha2}$	42.309	21.7 *7*
Gd $K_{\alpha1}$	42.996	39.1 *12*
Gd $K_{\beta1}'$	48.652	11.5 *4*
Gd $K_{\beta2}'$	50.214	3.48 *12*
γ_ϵM1	98.35 *16*	0.198 *20*
γ_ϵM1+39%E2	165.00 *8*	27.8 *8*
γ_ϵM1+E2	187.21 *3*	4.45 *13*
γ_ϵE2	352.21 *8*	33.0 *10*
γ_ϵM1	388.33 *15*	20.4 *10*
γ_ϵM1	464.47 *5*	6.44 *16*
γ_ϵM1	651.68 *5*	16.7 *4*
γ_ϵ(E1)	674.24 *13*	0.56 *5*
γ_ϵ	687.4 *4*	0.40 *8*
γ_ϵ	723.47 *24*	0.165 *14*
γ_ϵ	740.1 *2*	0.406 *20*
γ_ϵE1	772.58 *16*	1.61 *5*
γ_ϵ(E1)	792.8 *3*	0.10 *4*
γ_ϵM1+40%E2	816.68 *8*	12.3 *5*
γ_ϵ	835.3 *3*	0.37 *4*
γ_ϵM1+45%E2	852.80 *15*	16.2 *4*
γ_ϵE1+1.0%M2	861.45 *13*	8.4 *3*
γ_ϵ	913.4 *4*	0.069 *20*
γ_ϵE2	955.56 *23*	0.41 *4*
γ_ϵ	966.04 *17*	0.48 *4* ?
γ_ϵ	978.35 *21*	0.45 *4*
γ_ϵ	996.7 *6*	0.133 *10*
γ_ϵ	1001.98 *20*	0.251 *13*
γ_ϵ(M1)	1033.6 *3*	0.343 *16*
γ_ϵM1+E2	1040.01 *15*	1.43 *3*
γ_ϵ	1054.4 *4*	0.192 *21*
γ_ϵ	1117.7 *3*	0.195 *19*
γ_ϵ(M1)	1131.97 *19*	0.804 *24*
γ_ϵ	1136.2 *3*	1.25 *3*
γ_ϵ	1145.0 *3*	0.34 *4*
γ_ϵ(M2)	1166.98 *19*	0.501 *23*
γ_ϵ(M1)	1175.80 *14*	3.70 *7*
γ_ϵE2	1191.72 *20*	0.340 *24*
γ_ϵ[E2]	1205.01 *16*	0.34 *3*
γ_ϵ	1261.6 *3*	0.176 *12*
γ_ϵ	1281.1 *3*	0.12 *4*
γ_ϵ	1302.92 *18*	0.78 *16*
γ_ϵE2,M1	1341.73 *18*	2.33 *5*
γ_ϵ	1366.68 *20*	0.099 *20*
γ_ϵ	1379.6 *3*	0.406 *20*
γ_ϵ	1403.0 *3*	0.297 *20*
γ_ϵ	1449.5 *3*	0.93 *3*
γ_ϵ	1461.7 *6*	0.105 *25*
γ_ϵ	1477.2 *4*	0.119 *20*
γ_ϵ	1483.88 *21*	0.262 *20*
γ_ϵ	1491.4 *3*	0.297 *20*
γ_ϵ	1498.3 *4*	0.109 *20*
γ_ϵ	1513.9 *4*	0.168 *20*
γ_ϵ	1540.15 *25*	0.32 *3*
γ_ϵ	1545.5 *6*	0.129 *20*
γ_ϵ	1559.4 *4*	0.129 *20*
γ_ϵ	1585.6 *3*	0.198 *20*
γ_ϵ	1595.0 *4*	0.052 *10*
γ_ϵM1,E2	1640.27 *13*	3.23 *10*
γ_ϵ	1679.2 *3*	0.218 *20*
γ_ϵ	1734.19 *24*	0.10 *4*
γ_ϵ	1798.6 *3*	0.228 *20*
γ_ϵ	1806.20 *18*	0.515 *20*
γ_ϵM1	1827.48 *13*	1.20 *5*
γ_ϵ	1874.4 *3*	0.44 *3*
γ_ϵ	1909.5 *4*	0.248 *20*
γ_ϵ	1940.5 *3*	0.69 *3*
γ_ϵ	1948.34 *21*	0.69 *3*
γ_ϵ	1973.4 *3*	0.14 *3*
γ_ϵ	1992.48 *15*	0.083 *20*
γ_ϵE2	2008.40 *19*	0.78 *3*
γ_ϵ	2035.14 *22*	0.158 *20*
γ_ϵ	2107.7 *7*	0.168 *20*
γ_ϵ	2135.55 *21*	0.28 *3* ?
γ_ϵ	2183.35 *20*	0.70 *3*
γ_ϵ	2238.6 *3*	0.119 *10*
γ_ϵ	2248.5 *5*	0.089 *10*
γ_ϵ	2260.8 *2*	0.149 *20*
γ_ϵ	2284.6 *3*	0.525 *20*
γ_ϵ	2451.9 *5*	0.27 *4*
γ_ϵ	2456.9 *3*	0.104 *12*
γ_ϵ	2472.9 *7*	0.122 *19*
γ_ϵ	2561.7 *4*	0.287 *20*
γ_ϵ	2647.6 *3*	0.188 *14*
γ_ϵ	2669.7 *7*	0.109 *10*
γ_ϵ	2755.3 *6*	0.16 *4*
γ_ϵ	2774.5 *6*	0.144 *22*
γ_ϵ	2796.4 *3*	0.109 *10*
γ_ϵ	2823.2 *2*	0.099 *20*

Photons (^{149}Tb) (continued)

γ_{mode}	γ(keV)	γ(%)†
γ_ϵ	2857.1 *2*	0.146 *11*
γ_ϵ	2937.3 *2*	0.065 *15*
γ_ϵM1	2963.95 *18*	0.89 *4*
γ_ϵ	3013.3 *5*	0.11 *4*
γ_ϵ	3186.0 *4*	0.087 *14*
γ_ϵ	3202.1 *2*	0.228 *20*
γ_ϵ	3273.6 *3*	0.158 *20*

† 6.3% uncert(syst)

Atomic Electrons (^{149}Tb)

$\langle e \rangle$=36.0 *9* keV

e_{bin}(keV)	$\langle e \rangle$(keV)	e(%)
7	2.19	30 *3*
8	1.60	20.0 *21*
33 - 48	1.86	4.89 *20*
90 - 98	0.062	0.067 *5*
115	11.7	10.2 *5*
137	1.5	1.09 *25*
157	2.9	1.85 *21*
158	0.73	0.47 *11*
163	0.85	0.52 *8*
164 - 187	0.82	0.47 *12*
302	2.91	0.96 *3*
338	2.92	0.86 *5*
344 - 388	1.52	0.424 *9*
414 - 463	0.844	0.201 *6*
464	0.00683	0.00147 *5*
601	1.15	0.190 *6*
624 - 673	0.246	0.0381 *19*
674 - 723	0.037	0.0053 *14*
732 - 773	0.51	0.067 *8*
784 - 834	0.82	0.102 *7*
835 - 863	0.136	0.0160 *10*
905 - 954	0.043	0.0047 *12*
955 - 1004	0.064	0.0065 *11*
1025 - 1067	0.015	0.0014 *3*
1082 - 1131	0.218	0.0196 *15*
1132 - 1176	0.0426	0.00367 *12*
1183 - 1231	0.0076	0.00063 *19*
1253 - 1302	0.067	0.0052 *11*
1303 - 1342	0.017	0.0013 *3*
1353 - 1402	0.021	0.0015 *6*

Atomic Electrons (^{149}Tb) (continued)

e_{bin}(keV)	$\langle e \rangle$(keV)	e(%)
1403 - 1450	0.016	0.0011 *3*
1453 - 1552	0.018	0.00121 *24*
1558 - 1639	0.068	0.0043 *7*
1671 - 1756	0.011	0.00063 *20*
1777 - 1873	0.035	0.00194 *16*
1890 - 1985	0.032	0.0017 *3*
1991 - 2085	0.0064	0.00031 *7*
2099 - 2198	0.011	0.00051 *15*
2211 - 2283	0.0078	0.00035 *10*
2402 - 2471	0.0051	0.00021 *5*
2511 - 2597	0.0044	0.00017 *5*
2619 - 2705	0.0025	9.5 *25* $\times 10^{-5}$
2724 - 2822	0.0045	0.00016 *3*
2849 - 2936	0.0097	0.000332 *17*
2956 - 3012	0.0024	8.1 *14* $\times 10^{-5}$
3136 - 3223	0.0036	0.000114 *22*
3265 - 3272	0.00018	5.5 *16* $\times 10^{-6}$

$^{149}_{65}$Tb(4.16 *4* min)

Mode: ϵ(99.980 *4* %), α(0.020 *4* %)
Δ: -71452 *14* keV
SpA: 3.03×10^8 Ci/g
Prod: ^{139}La(^{16}O,6n); ^{141}Pr(^{12}C,4n)

Alpha Particles (^{149}Tb)

α(keV)
3999 *7*

Photons (^{149}Tb)

$\langle \gamma \rangle$=744 keV

γ_{mode}	γ(keV)	γ(%)†
γ	165.0 *1*	6.8 *5*
γ	630.7 *3*	~3
γ[M1+E2]	796.0 *1*	90

† approximate

$^{149}_{66}$Dy(4.2 *2* min)

Mode: ϵ
Δ: -67890 *200* keV syst
SpA: 3.00×10^8 Ci/g
Prod: ^{141}Pr(^{14}N,6n); ^{142}Nd(^{12}C,5n)

Photons (^{149}Dy)

γ_{mode}	γ(keV)	γ(rel)
γ	100.8 *1*	100
γ	106.3 *1*	51 *3*
γ	253.4 *1*	50 *5*
γ	653.6 *1*	60 *6*
γ	736.5 *1*	19 *3*
γ	741.7 *1*	17 *3*
γ	775.3 *1*	35 *4*
γ	789.4 *1*	65 *7*
γ	1274.2 *3*	18 *4*
γ	1776.5 *3*	79 *12*
γ	1806.2 *3*	64 *10*

$^{149}_{67}$Ho(21 *2* s)

Mode: ϵ
Δ: -61850 *210* keV syst
SpA: 3.5×10^9 Ci/g
Prod: ^{10}B on ^{144}Sm

Photons (^{149}Ho)

γ_{mode}	γ(keV)	γ(rel)
γ	1073.3 *1*	13 *1*
γ	1091.1 *1*	100
γ	1583.6 *2*	9 *2*

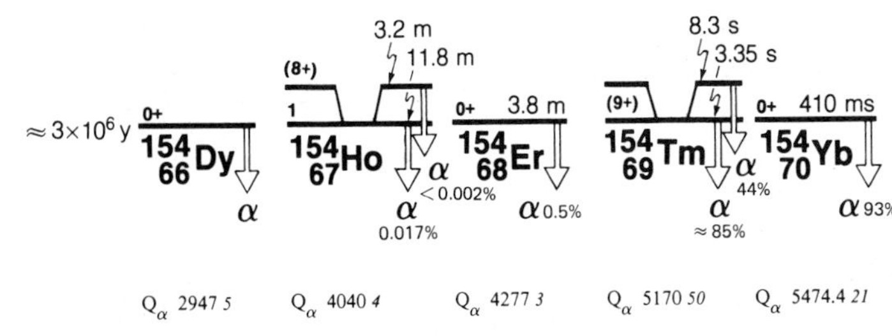

Q_α 2947 *5* Q_α 4040 *4* Q_α 4277 *3* Q_α 5170 *50* Q_α 5474.4 *21*

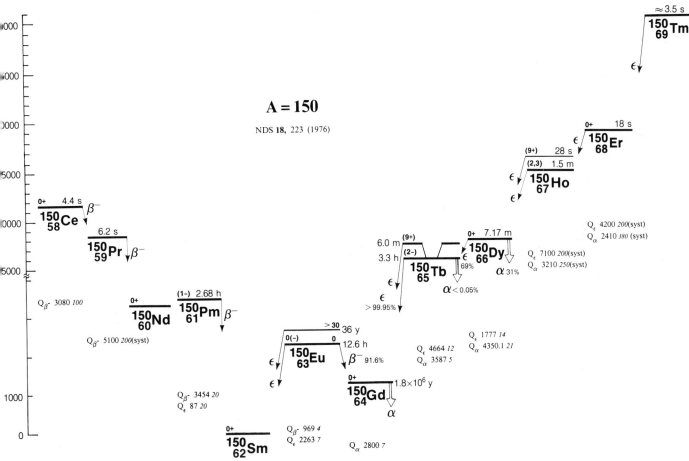

A = 150

NDS **18**, 223 (1976)

$^{150}_{58}$Ce: 0+ 4.4 s, β^-; Q_{β^-} 3080 *100*

$^{150}_{59}$Pr: 6.2 s, β^-; Q_{β^-} 5100 *200*(syst)

$^{150}_{60}$Nd: 0+

$^{150}_{61}$Pm: (1-) 2.68 h, β^-; Q_{β^-} 3454 *20*, Q_ϵ 87 *20*

$^{150}_{62}$Sm: 0+

$^{150}_{63}$Eu: >30 36 y; 0(-) 0 12.6 h, β^- 91.6%, ϵ; Q_{β^-} 969 *4*, Q_ϵ 2263 *7*

$^{150}_{64}$Gd: 0+ 1.8×10^6 y, α; Q_α 2800 *7*

$^{150}_{65}$Tb: (9+) 6.0 m; (2-) 3.3 h; ϵ >99.95%, α<0.05%; Q_ϵ 4664 *12*, Q_α 3587 *5*

$^{150}_{66}$Dy: 0+ 7.17 m; ϵ 69%, α 31%; Q_ϵ 1777 *14*, Q_α 4350.1 *21*

$^{150}_{67}$Ho: (9+) 28 s; (2,3) 1.5 m; ϵ; Q_ϵ 7100 *200*(syst), Q_α 3210 *250*(syst)

$^{150}_{68}$Er: 0+ 18 s; ϵ; Q_ϵ 4200 *200*(syst), Q_α 2410 *380* (syst)

$^{150}_{69}$Tm: ≈3.5 s; ϵ

$^{150}_{58}$Ce(4.4 *4* s)

Mode: β-
Δ: -65510 *220* keV syst
SpA: 1.58×10^{10} Ci/g
Prod: fission

Photons (^{150}Ce or ^{150}Pr)

γ_{mode}	γ(keV)	γ(rel)
γ	100.1 *3*	3.0 *6*
γ	103.7 *3*	5 *1*
γ	109.9 *3*	100 *20*
γ	141.1 *3*	3.0 *6*
γ	145.5 *3*	2.0 *4*
γ	160.7 *3*	3.0 *6*
γ	196.4 *3*	3.0 *6*
γ	199.9 *3*	3.0 *6*

Photons (^{150}Ce or ^{150}Pr) (continued)

γ_{mode}	γ(keV)	γ(rel)
γ	255.2 *3*	2.0 *4*
γ	258.7 *3*	1.0 *2*
γ	278.3 *3*	4.0 *8*
γ	289.9 *3*	7.0 *14*
γ	329.1 *3*	5 *1*
γ	348.7 *3*	3.0 *6*
γ	372.8 *3*	4.0 *8*
γ	397.7 *3*	7.0 *14*
γ	430.7 *3*	6.0 *12*
γ	469.0 *3*	3.0 *6*
γ	546.6 *3*	9.0 *18*
γ	721.9 *3*	20 *4*
γ	803.8 *3*	13 *3*
γ	851.9 *3*	21 *4*
γ	931.2 *3*	7.0 *14*
γ	946.3 *3*	6.0 *12*
γ	1060.4 *5*	7.0 *14*
γ	1073.7 *5*	7.0 *14*
γ	1140.5 *5*	22 *4*

$^{150}_{59}$Pr(6.19 *16* s)

Mode: β-
Δ: -68590 *200* keV syst
SpA: 1.15×10^{10} Ci/g
Prod: ^{150}Nd(n,p); fission

Photons (^{150}Pr)*

γ_{mode}	γ(keV)	γ(rel)
γ[E2]	130.12 *6*	100 *20*
γ	153.6 *3*	3.0 *6*
γ	211.5 *3*	3.0 *6*
γ[E2]	250.7 *3*	4.0 *8*

* see also ^{150}Ce

$^{150}_{60}$Nd(>5×10^{18} yr)

Δ: -73694 *4* keV
%: 5.64 *3*

$^{150}_{61}$Pm(2.68 *2* h)

Mode: β-
Δ: -73607 *20* keV
SpA: 7.80×10^{6} Ci/g
Prod: ^{150}Nd(p,n); protons on Ta; protons on Gd

Photons (^{150}Pm)

⟨γ⟩=1491 *21* keV

γ_{mode}	γ(keV)	γ(%)†
Sm L$_\ell$	4.993	0.0069 *8*
Sm L$_\eta$	5.589	0.00272 *25*
Sm L$_\alpha$	5.633	0.184 *15*
Sm L$_\beta$	6.339	0.169 *17*
Sm L$_\gamma$	7.256	0.024 *3*
Sm K$_{\alpha2}$	39.522	0.68 *4*
Sm K$_{\alpha1}$	40.118	1.22 *7*
Sm K$_{\beta1}$'	45.379	0.355 *21*
Sm K$_{\beta2}$'	46.819	0.105 *7*
γ [E2]	209.39 *6*	0.079 *11*
γ	218.1 *8*	~0.007
γ	225.3 *4*	~0.007 ?
γ	237.1 *4*	0.041 *7*
γ [E1]	241.10 *13*	~0.014
γ E1	251.54 *5*	0.191 *15*
γ	259.3 *10*	~0.007
γ [E2]	272.78 *5*	0.063 *16*
γ	276.4 *4*	0.10 *3*
γ E1	298.00 *4*	0.148 *10*
γ E2	305.70 *6*	0.125 *16*
γ [M1+E2]	310.69 *6*	0.024 *6*
γ E2	333.93 *5*	69
γ E1	345.92 *5*	0.44 *3*
γ	358.6 *5*	0.035 *14*
γ M1+E2+E0	371.14 *5*	0.083 *8*
γ E2	406.49 *5*	5.7 *4*
γ E2?	420.47 *5*	0.110 *21* ?
γ E1,E2	425.31 *6*	0.49 *5*
γ E2	439.41 *4*	0.78 *6*
γ E1,E2	453.39 *6*	0.15 *2*
γ E2	458.38 *6*	0.035 *4*
γ [E1]	464.54 *13*	0.048 *7*
γ [M1+E2]	492.63 *13*	0.35 *4*
γ	499.5 *4*	~0.014
γ	532.3 *8*	0.028 *14*
γ	542.8 *4*	0.041 *14*
γ [M1+E2]	547.50 *11*	0.041 *14*
γ [E1]	565.66 *11*	1.33 *10*
γ	573.0 *6*	~0.014 ?
γ [M1+E2]	587.01 *13*	1.36 *10*
γ	600.8 *10*	~0.014
γ [E1]	612.24 *13*	0.95 *7*
γ	620.51 *13*	0.130 *20*
γ	627.9 *10*	~0.014
γ	633.1 *3*	0.069 *14*
γ [E1]	652.90 *12*	0.34 *4*
γ [E1]	667.1 *1*	0.16 *3*
γ M1(+E0+E2)	712.19 *4*	4.46 *22*
γ E2	731.16 *5*	0.274 *25*
γ E1	737.42 *4*	2.22 *14*
γ	762.6 *7*	0.110 *21*
γ	812.1 *8*	0.097 *21*
γ (E1)	831.80 *5*	12.1 *5*
γ [E1]	842.58 *7*	0.41 *5*
γ	848.75 *24*	~0.014 ?
γ E2(+E0)	859.88 *4*	3.50 *14*
γ [E1]	876.35 *11*	7.4 *4*
γ	889.0 *4*	0.14 *3*
γ [M1+E2]	904.44 *12*	0.92 *6*
γ E1	910.84 *6*	0.054 *9*
γ [E1]	917.41 *9*	0.52 *6*

Photons (^{150}Pm) (continued)

γ_{mode}	γ(keV)	γ(%)†
γ [E2]	921.58 *7*	0.84 *6*
γ	943.76 *8*	~0.004
γ [E1]	972.80 *11*	0.097 *14*
γ [M1+E2]	998.82 *12*	0.062 *21* ?
γ [E1]	1004.32 *8*	0.81 *6*
γ [E1]	1024.04 *12*	0.75 *6*
γ [E2]	1046.12 *6*	0.34 *3*
γ [E1]	1066.02 *7*	0.46 *5*
γ M1+E2+E0	1083.33 *4*	0.180 *15*
γ [M1+E2]	1094.11 *8*	0.076 *14*
γ	1097.8 *4*	0.035 *14*
γ	1120.8 *8*	0.090 *14*
γ	1128.6 *8*	0.069 *14*
γ	1154.49 *13*	0.69 *5*
γ E1	1165.72 *6*	16.1 *6*
γ E2	1170.57 *5*	1.08 *10*
γ	1179.7 *3*	0.103 *21*
γ E2	1193.81 *6*	4.79 *19*
γ	1201.7 *3*	0.076 *14*
γ [E1]	1213.72 *7*	1.05 *8*
γ [E1]	1223.11 *10*	2.79 *16*
γ	1269.3 *10*	0.069 *21*
γ	1296.1 *8*	0.069 *21*
γ [E1]	1324.43 *13*	17.7 *7*
γ	1341.38 *21*	0.055 *14*
γ E1	1350.25 *6*	0.098 *15* ?
γ	1351.0 *4*	0.090 *14*
γ	1358.5 *8*	0.062 *14*
γ	1364.1 *7*	0.021 *7* ?
γ E1?	1379.29 *10*	3.19 *22*
γ [E2]	1417.26 *6*	~0.00027
γ	1435.76 *21*	0.27 *5*
γ	1452.31 *13*	0.13 *3*
γ	1485.7 *3*	0.048 *14*
γ	1499.1 *4*	0.076 *14*
γ	1507.4 *3*	~0.07 ?
γ [E1]	1519.41 *7*	0.27 *5*
γ [E1]	1629.59 *9*	0.81 *5*
γ	1647.12 *13*	0.26 *4*
γ	1672.0 *4*	0.090 *14*
γ [E1]	1713.22 *10*	0.36 *4*
γ	1727.2 *3*	0.19 *3*
γ [E1]	1736.23 *11*	7.0 *4*
γ	1766.7 *3*	0.19 *3*
γ	1788.9 *3*	~0.014
γ	1810.06 *23*	0.041 *14*
γ	1821.6 *3*	~0.035
γ	1833.8 *4*	~0.028
γ	1846.2 *3*	~0.0034 ?
γ	1865.2 *10*	0.069 *14*
γ	1874.1 *10*	0.069 *14*
γ	1893.2 *10*	0.048 *14*
γ	1906.2 *3*	0.076 *14*
γ	1914.4 *6*	0.076 *14* ?
γ [E1]	1925.90 *7*	0.35 *7*
γ	1940.6 *3*	0.069 *14*
γ [E1]	1963.52 *10*	1.51 *9*
γ	2003.8 *3*	~0.028
γ	2018.5 *4*	0.041 *14*
γ	2033.5 *3*	0.98 *7*
γ	2173.17 *21*	0.055 *21*
γ	2195.4 *3*	0.090 *21*
γ	2216.55 *23*	0.24 *4*
γ [E1]	2259.82 *8*	0.069 *21*
γ	2371.0 *8*	0.076 *21*
γ	2453.1 *10*	~0.014
γ	2478.90 *13*	0.38 *4*
γ	2507.09 *21*	0.055 *21*
γ	2529.3 *3*	0.34 *4*
γ	2546.1 *7*	0.035 *14*
γ	2550.47 *23*	0.12 *3*
γ	2623.5 *6*	0.041 *14*
γ	2651.7 *10*	0.021 *7*
γ	2679.5 *6*	0.048 *21* ?
γ	2688.5 *7*	~0.007 ?
γ	2699.8 *5*	0.055 *21*
γ	2704.2 *4*	~0.041
γ	2716.0 *3*	~0.007
γ	2803.3 *4*	0.048 *21*
γ	2855.8 *8*	~0.007
γ	2878.3 *4*	~0.028
γ	2883.3 *8*	~0.028
γ	2893.0 *3*	0.21 *3*
γ	2932.2 *10*	~0.0021

Photons (^{150}Pm) (continued)

γ_{mode}	γ(keV)	γ(%)†
γ	2941.0 *10*	~0.0021
γ	3022.4 *7*	0.028 *7*
γ	3038.1 *4*	~0.014
γ	3049.9 *3*	~0.014
γ	3080.1 *6*	~0.014
γ	3089.3 *4*	~0.007
γ	3137.2 *4*	~0.0021
γ	3187.8 *20*	~0.0034

† 5.8% uncert(syst)

Atomic Electrons (^{150}Pm)

⟨e⟩=10.7 *4* keV

e_{bin}(keV)	⟨e⟩(keV)	e(%)
7 - 45	0.200	2.15 *21*
163 - 211	0.042	0.023 *3*
212 - 259	0.043	0.018 *4*
264 - 276	0.010	0.0036 *10*
287	6.4	2.21 *14*
290 - 324	0.029	0.0096 *11*
326	0.82	0.252 *16*
327	0.70	0.214 *14*
332	0.275	0.083 *5*
333 - 359	0.167	0.0500 *22*
360	0.374	0.104 *7*
363 - 413	0.184	0.047 *3*
414 - 463	0.044	0.0101 *19*
464 - 501	0.012	0.0024 *7*
519 - 566	0.111	0.021 *4*
571 - 621	0.032	0.0055 *9*
625 - 670	0.18	0.028 *8*
675 - 724	0.072	0.0103 *12*
729 - 765	0.0079	0.00106 *23*
785 - 832	0.300	0.0372 *12*
835 - 882	0.085	0.0098 *10*
887 - 936	0.0107	0.00118 *14*
937 - 977	0.0149	0.00155 *11*
991 - 1039	0.0181	0.00179 *13*
1044 - 1094	0.0087	0.00081 *15*
1096 - 1133	0.155	0.0139 *8*
1147 - 1195	0.152	0.0131 *4*
1200 - 1249	0.0069	0.00056 *8*
1262 - 1304	0.125	0.0098 *5*
1312 - 1358	0.0435	0.00328 *16*
1362 - 1411	0.0089	0.00064 *16*
1416 - 1461	0.0034	0.00024 *7*
1473 - 1518	0.0024	0.000161 *23*
1583 - 1680	0.0146	0.00090 *13*
1689 - 1787	0.054	0.00317 *21*
1799 - 1898	0.0066	0.00036 *5*
1899 - 1997	0.020	0.00102 *21*
2002 - 2032	0.0016	8 *3* ×10^{-5}
2126 - 2215	0.0043	0.00020 *5*
2252 - 2324	0.0007	2.9 *13* ×10^{-5}
2363 - 2460	0.0036	0.00015 *5*
2471 - 2549	0.0050	0.00020 *5*
2577 - 2673	0.0021	8.1 *17* ×10^{-5}
2676 - 2756	0.00049	1.8 *7* ×10^{-5}
2796 - 2894	0.0023	7.9 *21* ×10^{-5}
2924 - 3021	0.00039	1.3 *4* ×10^{-5}
3030 - 3129	0.00020	6.5 *21* ×10^{-6}
3130 - 3186	2.5 ×10^{-5}	~8×10^{-7}

Continuous Radiation (^{150}Pm)

⟨β-⟩=781 keV;⟨IB⟩=1.61 keV

E_{bin}(keV)		⟨ ⟩(keV)	(%)
0 - 10	β-	0.0303	0.60
	IB	0.030	
10 - 20	β-	0.092	0.61
	IB	0.030	0.21
20 - 40	β-	0.373	1.24
	IB	0.058	0.20
40 - 100	β-	2.73	3.89
	IB	0.160	0.25
100 - 300	β-	28.7	14.2
	IB	0.42	0.24
300 - 600	β-	103	22.8
	IB	0.41	0.097
600 - 1300	β-	361	39.9
	IB	0.40	0.048
1300 - 2500	β-	269	16.1
	IB	0.100	0.0062
2500 - 3454	β-	17.2	0.63
	IB	0.00156	5.8×10^{-5}

$^{150}_{62}$Sm(stable)

Δ: -77061 3 keV

%: 7.4 1

$^{150}_{63}$Eu(12.62 10 h)

Mode: β-(91.6 %), ε(8.4 %)

Δ: -74798 8 keV

SpA: 1.655×10^6 Ci/g

Prod: ^{150}Sm(p,n); ^{150}Sm(d,2n)

Photons (^{150}Eu)

⟨γ⟩=39.0 14 keV

γ_{mode}	γ(keV)	γ(%)
Sm L_ℓ	4.993	0.0176 18
Sm L_η	5.589	0.0065 5
Sm L_α	5.633	0.47 3
Sm L_β	6.342	0.43 4
Sm L_γ	7.273	0.063 7
Sm $K_{\alpha 2}$	39.522	1.82 5
Sm $K_{\alpha 1}$	40.118	3.30 10
Sm $K_{\beta 1}'$	45.379	0.96 3
Sm $K_{\beta 2}'$	46.819	0.284 10
γ_ϵ[E2]	209.39 6	0.0188 24
γ_ϵ[E2]	272.78 5	0.0016 5
γ_ϵE2	305.70 6	0.0033 4
γ_ϵE2	333.93 5	3.7 3
γ_ϵE2	406.49 5	2.60 22
γ_ϵE2?	420.47 5	0.00031 6
γ_ϵE1,E2	425.31 6	0.0073 8
γ_ϵE1,E2	453.39 6	0.00043 7
γ_ϵ	620.51 13	0.028 4
γ_ϵM1(+E0+E2)	712.19 4	0.117 8
γ_ϵ(E1)	831.80 5	0.181 11
γ_ϵE2(+E0)	859.88 4	0.0100 8
γ_ϵ[E1]	917.41 9	0.036 4
γ_ϵ[E2]	921.58 7	0.199 14
γ_ϵ[E2]	1046.12 6	0.0090 9
γ_ϵE1	1165.72 6	0.242 14
γ_ϵE2	1193.81 6	0.0137 10
γ_ϵ[E1]	1223.11 10	0.190 13
γ_ϵ	1452.31 13	0.028 9
γ_ϵ[E1]	1629.59 9	0.055 5
γ_ϵ[E1]	1963.52 10	0.103 8

Atomic Electrons (^{150}Eu)

⟨e⟩=1.20 5 keV

e_{bin}(keV)	⟨e⟩(keV)	e(%)
7	0.31	4.4 5
8	0.041	0.53 5
31 - 32	0.049	0.154 12
33	0.053	0.161 16
37 - 45	0.066	0.169 9
163 - 209	0.0059	0.0034 3
226 - 273	0.00065	0.00026 5
287	0.34	0.117 10
298 - 306	0.000116	3.9 4 $\times 10^{-5}$
326	0.044	0.0134 11
327	0.037	0.0114 9
332 - 334	0.0233	0.0070 4
360	0.172	0.048 4
374 - 420	0.047	0.0117 7
424 - 453	2.0×10^{-5}	4.7 20 $\times 10^{-6}$
574 - 621	0.0013	~0.00022
665 - 712	0.0057	0.00085 21
785 - 832	0.00235	0.000297 17
852 - 875	0.0048	0.00054 4
910 - 922	0.00090	9.8 6 $\times 10^{-5}$
999 - 1046	0.000205	2.04 18 $\times 10^{-5}$
1119 - 1166	0.00236	0.000210 10
1176 - 1223	0.00166	0.000140 9
1405 - 1452	0.00044	$\sim 3 \times 10^{-5}$
1583 - 1628	0.00038	2.41 19 $\times 10^{-5}$
1917 - 1962	0.00064	3.33 25 $\times 10^{-5}$

Continuous Radiation (^{150}Eu)

⟨β-⟩=299 keV;⟨β+⟩=2.16 keV;⟨IB⟩=0.43 keV

E_{bin}(keV)		⟨ ⟩(keV)	(%)
0 - 10	β-	0.071	1.41
	β+	2.90×10^{-8}	3.46×10^{-7}
	IB	0.0144	
10 - 20	β-	0.213	1.42
	β+	1.39×10^{-6}	8.3×10^{-6}
	IB	0.0137	0.095
20 - 40	β-	0.87	2.89
	β+	4.51×10^{-5}	0.000137
	IB	0.030	0.102
40 - 100	β-	6.3	8.9
	β+	0.00255	0.00325
	IB	0.071	0.114
100 - 300	β-	61	30.6
	β+	0.129	0.059
	IB	0.127	0.075
300 - 600	β-	153	35.3
	β+	0.75	0.165
	IB	0.071	0.017
600 - 1300	β-	78	11.1
	β+	1.28	0.163
	IB	0.068	0.0077
1300 - 2263	IB	0.034	0.0022
	Σβ+		0.39

$^{150}_{63}$Eu(35.8 10 yr)

Mode: ε

Δ: >-74768 keV

SpA: 66.6 Ci/g

Prod: ^{150}Sm(p,n); ^{151}Eu(n,2n)

Photons (^{150}Eu)

⟨γ⟩=1501 14 keV

γ_{mode}	γ(keV)	γ(%)†
Sm L_ℓ	4.993	0.226 23
Sm L_η	5.589	0.084 6
Sm L_α	5.633	6.1 4
Sm L_β	6.341	5.6 5
Sm L_γ	7.277	0.83 9
Sm $K_{\alpha 2}$	39.522	23.2 7
Sm $K_{\alpha 1}$	40.118	42.1 12
Sm $K_{\beta 1}'$	45.379 –	12.2 4
Sm $K_{\beta 2}'$	46.819	3.62 12
γ E1	78.77 6	0.081 5
γ [M1+E2]	124.79 10	~0.0019 ?
γ [E1]	135.18 7	0.011 3
γ E2(+E0)	205.22 6	0.029 4
γ E1	251.54 5	0.168 8
γ [E2]	272.78 5	0.015 4
γ	272.84 11	0.023 8
γ E1(+M2)	284.99 5	0.164 8
γ [E1]	286.24 6	0.07 3
γ [E2]	286.25 5	0.097 14
γ [E1]	286.27 9	~0.06
γ E1	298.00 4	0.630 13
γ E2	305.70 6	0.030 4
γ [M1+E2]	310.69 6	0.029 6
γ [M1+E2]	314.86 6	0.017 6
γ E2	333.93 5	94.0 19
γ	335.7 4	0.036 14 ?
γ E1?	342.60 10	0.166 14 ?
γ E1	345.92 5	0.390 14
γ [M1+E2]	370.22 10	0.035 11
γ M1+E2+E0	371.14 5	0.073 7
γ E0+M1+E2	372.68 15	0.234 9
γ [M1+E2]	377.68 7	0.096 8
γ [E1]	377.80 9	0.0150 19
γ E1	381.99 5	0.110 8 ?
γ E2?	402.10 5	0.761 19
γ E2	403.02 10	0.235 9
γ E2	406.49 5	0.137 14
γ E2?	420.47 5	0.018 4
γ E1,E2	425.31 6	0.0030 4
γ E2	439.41 4	78.7 16
γ [E2]	448.78 5	0.250 8
γ E1,E2	453.39 6	0.025 3 ?
γ E2	458.38 6	0.043 4
γ E1	461.77 5	0.808 19
γ E1?	464.23 15	0.35 6
γ	464.78 8	<0.16
γ M1+E2+E0	474.49 5	0.142 7
γ	476.88 13	0.018 5 ?
γ E1?	485.93 9	0.162 7
γ E2	505.49 5	4.71 9
γ [E1]	510.00 10	0.118 8
γ E2?	515.77 6	0.968 19
γ M1(+E2)	520.08 6	0.455 9
γ [E2]	540.53 7	0.085 7
γ E2?	542.99 15	0.132 7
γ [M1+E2]	547.50 11	<0.0006
γ [E2]	553.16 6	0.032 7
γ [E1]	565.66 11	$\sim 4 \times 10^{-5}$
γ E1	571.25 5	0.404 10
γ E0+M1+E2	575.43 9	0.030 8
γ E1	584.25 4	51.5 10
γ [E2]	590.71 10	0.031 5
γ [E2]	596.47 5	0.052 9
γ [M1+E2]	596.62 9	~0.021
γ E2	607.32 5	0.164 5
γ [M1+E2]	612.58 6	0.088 8
γ E2	625.55 5	0.304 7
γ	637.82 24	0.014 7
γ [E1]	652.90 12	$\sim 9 \times 10^{-6}$
γ E1?	662.68 6	0.015 4
γ [E1]	666.98 5	0.233 7
γ M1+E2+E0	675.80 8	0.498 10
γ [E2]	676.84 6	<0.019 ?
γ [E2]	699.82 6	~0.006
γ M1(+E0+E2)	712.19 4	1.063 21
γ E2	731.16 5	0.331 8
γ E1	737.42 4	9.42 18
γ E1	741.44 6	0.840 17
γ E1	748.02 5	5.05 10
γ E1	749.76 6	0.659 13

Photons (^{150}Eu)
(continued)

γ_{mode}	γ(keV)	γ(%)†
γ E2	751.03 *7*	2.10 *5*
γ E1	756.46 *10*	0.111 *7*
γ [E1]	759.49 *6*	0.075 *7*
γ E2?	761.91 *10*	0.027 *5*
γ E2?	773.24 *5*	0.595 *12*
γ [E2]	776.61 *6*	0.007 *3* ?
γ [M1+E2]	787.11 *9*	~0.004 ?
γ M1(+E2)	816.38 *10*	0.048 *5*
γ [M1+E2]	828.52 *6*	0.572 *11*
γ E2?	830.76 *5*	0.517 *10*
γ (E1)	831.80 *5*	0.074 *5*
γ [E1]	836.58 *9*	0.298 *9*
γ [E2]	838.25 *7*	0.055 *9*
γ E2(+E0)	859.88 *4*	0.572 *11*
γ M1+E2?	869.25 *5*	1.81 *4*
γ [E1]	876.35 *11*	~0.00020
γ E1	899.08 *5*	0.921 *19*
γ [M1+E2]	904.44 *12*	~2×10^{-5}
γ E1	910.84 *6*	0.091 *7*
γ [E2]	915.35 *10*	0.016 *6*
γ E2?	923.20 *8*	0.301 *10*
γ	943.76 *8*	~0.008 ?
γ [E1]	953.23 *5*	0.044 *7*
γ [E2]	958.56 *10*	0.016 *7*
γ [E2]	978.46 *6*	0.020 *5*
γ [M1+E2]	998.82 *12*	~2×10^{-6}
γ [E1]	1024.04 *12*	~2×10^{-5}
γ	1045.85 *6*	0.90 *5*
γ [E2]	1046.12 *6*	0.081 *7*
γ E1	1049.03 *7*	5.24 *10*
γ E2?	1070.96 *5*	0.139 *8*
γ [E1]	1081.03 *9*	~0.028
γ M1+E2+E0	1083.33 *4*	0.159 *8*
γ [E2]	1115.07 *9*	0.015 *6*
γ [E1]	1122.83 *8*	0.324 *9*
γ E1	1165.72 *6*	0.098 *6*
γ E2	1170.57 *5*	1.31 *3*
γ E2	1193.81 *6*	0.781 *15*
γ M1(+E2+E0)	1197.08 *5*	1.11 *4*
γ [M1+E2]	1209.45 *8*	0.0033 *16*
γ E2	1246.93 *6*	1.87 *5*
γ M1(+E0)	1251.23 *5*	0.160 *9*
γ E1	1261.95 *9*	0.451 *9*
γ E2	1308.66 *4*	0.873 *18*
γ [M1+E2]	1321.87 *9*	0.165 *9*
γ E2	1334.01 *6*	0.407 *10*
γ M1(+E2+E0)	1343.74 *5*	2.54 *5*
γ [E1]	1346.16 *10*	0.028 *7*
γ E1	1350.25 *6*	0.165 *7*
γ	1379.12 *5*	0.039 *5*
γ [E2]	1417.26 *6*	~0.00023
γ [M1+E2]	1420.83 *9*	0.007 *3*
γ E2	1485.43 *5*	1.80 *7*
γ M1+E0+E2?	1499.30 *8*	0.0132 *9*
γ E2	1636.49 *5*	0.705 *14*
γ E2	1690.64 *5*	0.100 *3*
γ	1710.0 *10*	0.00056 *19*
γ [E1]	1713.22 *10*	<0.0049
γ [E1]	1736.23 *11*	~0.00019 ?
γ E2	1783.15 *5*	0.101 *3*
γ [E2]	1818.44 *9*	0.0038 *5*
γ M1,E2	1833.22 *10*	0.0025 *5*

† 3.2% uncert(syst)

Atomic Electrons (^{150}Eu)
⟨e⟩=26.3 *5* keV

e_{bin}(keV)	⟨e⟩(keV)	e(%)
7	3.9	57 *6*
8 - 45	2.72	13.5 *8*
71 - 118	0.0076	0.0099 *9*
123 - 158	0.0068	0.0044 *6*
197 - 245	0.10	~0.04
250 - 286	0.055	0.021 *4*
287	8.66	3.02 *9*
289 - 324	0.036	0.0118 *12*

Atomic Electrons (^{150}Eu)
(continued)

e_{bin}(keV)	⟨e⟩(keV)	e(%)
326	1.15	0.351 *12*
327	0.95	0.292 *8*
328 - 377	0.702	0.209 *4*
378 - 382	0.00025	6.5 *25* ×10^{-5}
393	4.60	1.17 *3*
394 - 430	0.076	0.0186 *14*
432	0.797	0.185 *5*
433 - 479	0.701	0.157 *3*
484 - 533	0.0887	0.0176 *4*
534 - 583	0.884	0.163 *4*
584 - 631	0.037	0.0060 *10*
636 - 684	0.058	0.0087 *18*
691 - 737	0.285	0.0406 *7*
740 - 790	0.065	0.0084 *7*
791 - 838	0.080	0.0097 *19*
852 - 899	0.035	0.0040 *5*
903 - 952	0.0034	0.00037 *6*
953 - 1002	0.060	0.0060 *16*
1016 - 1064	0.0174	0.00168 *23*
1068 - 1116	0.0040	0.00037 *3*
1119 - 1166	0.064	0.0056 *5*
1169 - 1215	0.0452	0.00377 *11*
1239 - 1287	0.0298	0.00235 *6*
1297 - 1346	0.062	0.0048 *8*
1349 - 1379	0.00024	1.8 *6* ×10^{-5}
1410 - 1452	0.0253	0.00176 *7*
1478 - 1499	0.00440	0.000297 *10*
1590 - 1689	0.0120	0.000752 *16*
1702 - 1786	0.00148	8.51 *25* ×10^{-5}
1811 - 1832	1.30 ×10^{-5}	7.1 *7* ×10^{-7}

Continuous Radiation (^{150}Eu)
⟨IB⟩=0.27 keV

E_{bin}(keV)		⟨ ⟩(keV)	(%)
10 - 20	IB	0.00022	0.00135
20 - 40	IB	0.059	0.164
40 - 100	IB	0.077	0.17
100 - 300	IB	0.024	0.0123
300 - 600	IB	0.061	0.0140
600 - 1300	IB	0.052	0.0066
1300 - 1520	IB	0.00077	5.7×10^{-5}

$^{150}_{64}$Gd(1.79 *8* ×10^{6} yr)

Mode: α
Δ: -75766 *8* keV
SpA: 0.00133 Ci/g
Prod: daughter ^{150}Eu(12.6 h); ^{151}Eu(d,3n); alphas on Sm

Alpha Particles (^{150}Gd)

α(keV)
2726 *10*

$^{150}_{65}$Tb(3.27 *10* h)

Mode: ϵ(>99.95 %), α(<0.05 %)
Δ: -71102 *11* keV
SpA: 6.39×10^{6} Ci/g
Prod: protons on Gd

Alpha Particles (^{150}Tb)

α(keV)
3492 *5*

Photons (^{150}Tb)
⟨γ⟩=1683 *37* keV

γ_{mode}	γ(keV)	γ(%)†
Gd L_ℓ	5.362	0.148 *15*
Gd L_η	6.049	0.053 *4*
Gd L_α	6.054	3.9 *3*
Gd L_β	6.852	3.6 *3*
Gd L_γ	7.880	0.54 *6*
Gd $K_{\alpha2}$	42.309	13.8 *5*
Gd $K_{\alpha1}$	42.996	24.8 *9*
Gd $K_{\beta1}$'	48.652	7.3 *3*
Gd $K_{\beta2}$'	50.214	2.21 *9*
γ[E1]	154.04 *11*	0.059 *9*
γE1	412.22 *11*	0.90 *8*
γM1+E2+E0?	436.9 *3*	0.94 *7*
γE1+M2	496.06 *8*	15.1 *7*
γ	525.2 *4*	0.68 *11*
γ	557.8 *4*	~0.36
γE2	566.26 *9*	2.0 *9*
γE2	569.18 *20*	2.52 *25*
γ[E1]	573.4 *3*	~0.36
γ	574.2 *7*	~0.36
γ	602.7 *3*	0.43 *14*
γE2	637.86 *8*	72
γE2	650.10 *13*	4.1 *4*
γ	701 *1*	0.32 *7*
γ	747.8 *7*	0.79 *14*
γ	772.7 *6*	~0.36
γ	779.3 *5*	0.54 *14*
γE2	792.05 *19*	4.7 *5*
γ	812.7 *5*	0.65 *9*
γE1	820.78 *20*	1.44 *14*
γM1+E2+E0?	879.98 *23*	3.2 *4*
γ[E1]	884.2 *3*	~0.22
γ	949.4 *4*	1.08 *11*
γ	951.7 *6*	0.94 *14*
γ[M1+E2]	954.3 *3*	1.22 *14*
γ	961.2 *6*	~0.36
γ	1045.38 *25*	1.22 *14*
γ	1073.8 *5*	0.86 *22*
γ	1167.6 *5*	0.50 *14*
γ	1175.0 *5*	0.61 *14*
γ	1275.12 *25*	~0.43
γ	1291.16 *22*	1.66 *14*
γ[M1+E2]	1316.84 *20*	~0.50
γ	1342.8 *7*	0.56 *14*
γ	1349.1 *6*	1.26 *11*
γ	1387.0 *5*	~0.36
γ	1414.2 *3*	~0.29
γ	1429.91 *20*	2.45 *22*
γ	1444.8 *6*	0.97 *11*
γE1	1453.4 *3*	3.7 *4*
γM1,E2	1517.83 *23*	2.9 *3*
γ	1524.6 *6*	0.47 *5*
γ	1541.44 *25*	~0.43
γM1	1592.1 *3*	1.80 *22*
γ	1688.6 *5*	0.72 *22*
γ	1726.3 *3*	0.216 *22*
γ	1771.18 *25*	0.58 *14*
γ	1776.0 *13*	0.50 *14*
γ	1787.22 *21*	1.80 *22*
γ	1857.7 *10*	~0.29
γ	1900.4 *4*	0.86 *9*
γ	1912.4 *5*	0.40 *6*

Photons (^{150}Tb) (continued)

γ_{mode}	γ(keV)	γ(%)†
γ	2014.9 *4*	1.15 *14*
γ	2037.1 *6*	0.79 *14*
γ[E1]	2091.2 *3*	1.51 *14*
γ	2113.8 *13*	0.43 *7*
γ	2147.1 *5*	0.94 *11*
γ	2179.29 *25*	~0.43
γ	2193.4 *8*	0.25 *4*
γ	2206.2 *5*	0.94 *14*
γ	2313 *1*	0.38 *4*
γ	2363.8 *5*	0.94 *14*
γ	2396.5 *4*	0.61 *6*
γ	2409.03 *25*	0.37 *4*
γ	2425.07 *21*	1.01 *10*
γ	2488.2 *10*	~0.50
γ	2590.5 *5*	0.259 *22*
γ	2735.2 *8*	0.58 *11*
γ	2750.2 *10*	~0.22
γ	2844.0 *5*	0.345 *22*
γ	2871.5 *5*	0.360 *22*
γ	2982.9 *6*	0.37 *4*
γ	3034.3 *4*	0.36 *7*
γ	3083.5 *4*	0.35 *4*
γ	3094.1 *7*	0.17 *4*
γ	3133 *3*	~0.14
γ	3264 *3*	~0.36
γ	3403.7 *20*	~0.36
γ	3423.7 *20*	0.22 *7*
γ	3603.7 *20*	0.07 *3*
γ	3659.7 *20*	0.08 *3*
γ	3666.7 *20*	0.058 *22*
γ	3734.7 *20*	0.12 *3*
γ	3778.7 *20*	0.115 *22*
γ	4113 *3*	0.18 *3*
γ	4213 *3*	0.072 *22*
γ	4294 *3*	0.12 *3*
γ	4323 *3*	0.08 *3*
γ	4413 *3*	0.043 *14*
γ	4454 *3*	0.065 *14*

† 5.6% uncert(syst)

Atomic Electrons (^{150}Tb)

⟨e⟩=13.3 *24* keV

e_{bin}(keV)	⟨e⟩(keV)	e(%)
7	1.59	22.0 *23*
8	1.16	14.4 *15*
33 - 34	0.244	0.72 *5*
35	0.38	1.09 *11*
36 - 48	0.61	1.52 *8*
104 - 153	0.0059	0.0053 *23*
362 - 411	0.11	0.029 *7*
412 - 437	0.019	0.0045 *11*
446	2.5	~0.6
475	0.05	~0.010
488	0.4	~0.09
489 - 518	0.26	0.052 *20*
519	0.8	~0.16
523 - 572	0.29	0.05 *3*
573 - 574	0.00023	~4 $\times10^{-5}$
588	2.65	0.45 *3*
594 - 643	0.78	0.124 *5*
648 - 698	0.084	0.012 *3*
699 - 748	0.19	0.026 *4*
763 - 811	0.110	0.014 *3*
812 - 834	0.12	0.014 *4*
872 - 911	0.12	0.014 *3*
941 - 961	0.019	0.0020 *4*
995 - 1044	0.065	0.0064 *23*
1045 - 1075	0.0063	0.00059 *22*
1117 - 1167	0.030	0.0027 *10*
1168 - 1175	0.0007	6 *3* $\times10^{-5}$
1225 - 1274	0.049	0.0039 *15*
1275 - 1317	0.059	0.0046 *12*
1334 - 1380	0.059	0.0043 *16*
1385 - 1434	0.066	0.0047 *9*
1435 - 1475	0.075	0.0051 *9*
1491 - 1591	0.063	0.0041 *5*

Atomic Electrons (^{150}Tb) (continued)

e_{bin}(keV)	⟨e⟩(keV)	e(%)
1638 - 1737	0.049	0.0029 *8*
1763 - 1862	0.024	0.0013 *3*
1892 - 1987	0.023	0.0012 *4*
2007 - 2106	0.028	0.00135 *24*
2107 - 2205	0.033	0.0015 *5*
2263 - 2359	0.039	0.0017 *4*
2362 - 2438	0.017	0.0007 *2*
2480 - 2540	0.0029	0.00011 *4*
2582 - 2589	0.00036	1.4 *5* $\times10^{-5}$
2685 - 2749	0.0073	0.00027 *8*
2794 - 2870	0.0094	0.00033 *8*
2933 - 3027	0.0061	0.00021 *5*
3032 - 3131	0.0055	0.00018 *4*
3214 - 3262	0.0028	~9 $\times10^{-5}$
3354 - 3422	0.0044	0.00013 *5*
3554 - 3653	0.0014	4.0 *10* $\times10^{-5}$
3658 - 3733	0.0016	4.3 *10* $\times10^{-5}$
3770 - 3777	0.00011	3.0 *9* $\times10^{-6}$
4063 - 4111	0.0012	2.9 *8* $\times10^{-5}$
4163 - 4244	0.0011	2.6 *7* $\times10^{-5}$
4273 - 4363	0.00084	2.0 *5* $\times10^{-5}$
4404 - 4452	0.00044	1.0 *3* $\times10^{-5}$

Continuous Radiation (^{150}Tb)

⟨β+⟩=524 keV;⟨IB⟩=6.2 keV

E_{bin}(keV)		⟨ ⟩(keV)	(%)
0 - 10	β+	4.25 $\times10^{-7}$	5.05 $\times10^{-6}$
	IB	0.018	
10 - 20	β+	2.14 $\times10^{-5}$	0.000128
	IB	0.018	0.124
20 - 40	β+	0.00073	0.00222
	IB	0.062	0.20
40 - 100	β+	0.0450	0.057
	IB	0.169	0.30
100 - 300	β+	2.72	1.22
	IB	0.31	0.17
300 - 600	β+	21.8	4.69
	IB	0.48	0.110
600 - 1300	β+	153	16.1
	IB	1.49	0.160
1300 - 2500	β+	296	16.7
	IB	2.6	0.145
2500 - 4664	β+	50.4	1.82
	IB	1.02	0.035
	Σβ+		41

$^{150}_{65}$Tb(6.0 *2* min)

Mode: ϵ

Δ: -71102 *11* keV

SpA: 2.09 $\times10^{8}$ Ci/g

Prod: ^{151}Eu(α,5n); ^{151}Eu(^{3}He,4n); ^{141}Pr(^{12}C,3n); ^{139}La(^{16}O,5n); protons on Dy

Photons (^{150}Tb)

⟨γ⟩=2383 *140* keV

γ_{mode}	γ(keV)	γ(%)†
Gd L_ℓ	5.362	0.25 *3*
Gd L_η	6.049	0.089 *7*
Gd L_α	6.054	6.6 *4*
Gd L_β	6.852	6.0 *5*
Gd L_γ	7.880	0.92 *10*

Photons (^{150}Tb) (continued)

γ_{mode}	γ(keV)	γ(%)†
Gd $K_{\alpha2}$	42.309	23.3 *8*
Gd $K_{\alpha1}$	42.996	42.0 *15*
Gd $K_{\beta1}$'	48.652	12.4 *4*
Gd $K_{\beta2}$'	50.214	3.74 *14*
γ[E1]	95.30 *11*	0.5 *1*
γ[E1]	154.04 *11*	1.0 *1*
γ M1	161.82 *17*	7.3 *5*
γ E2?	179.14 *20*	~1
γ[E1]	181.03 *18*	~1
γ E1	274.44 *21*	1.9 *7*
γ E1	342.85 *9*	25 *2*
γ E1	412.22 *11*	10.1 *10*
γ E1	415.05 *12*	4.0 *5*
γ E2	438.15 *9*	42 *3*
γ M1(+E2)	455.48 *17*	12 *3*
γ E1+M2	496.06 *8*	23.5 *10*
γ E2	510.35 *17*	26 *4* ?
γ E2	566.26 *9*	21.8 *20*
γ E2	637.86 *8*	100
γ E2	648.13 *20*	18 *2*
γ E2	650.10 *13*	70 *3*
γ	789.7 *4*	2.3 *9*
γ E2	827.27 *9*	41 *3*

† 1.0% uncert(syst)

Atomic Electrons (^{150}Tb)

⟨e⟩=36 *4* keV

e_{bin}(keV)	⟨e⟩(keV)	e(%)
7	2.7	37 *4*
8	1.97	24.5 *25*
33 - 48	2.16	5.78 *25*
87 - 104	0.104	0.105 *9*
112	3.6	3.27 *23*
129 - 178	1.70	1.14 *16*
179 - 224	0.09	0.042 *13*
266 - 293	0.73	0.25 *2*
334 - 365	0.46	0.131 *7*
388	2.61	0.67 *5*
404	0.032	0.0080 *8*
405	1.0	0.26 *10*
407 - 438	0.75	0.174 *9*
446	3.9	~0.9
447 - 455	0.23	0.051 *16*
460	1.29	0.28 *4*
488 - 510	1.2	~0.24
516	0.94	0.182 *17*
558 - 566	0.224	0.040 *3*
588	3.7	0.63 *13*
598	0.65	0.109 *12*
600	2.52	0.42 *2*
629 - 650	1.54	0.243 *17*
739	0.07	~0.010
777	1.11	0.143 *11*
781 - 827	0.242	0.0296 *19*

$^{150}_{66}$Dy(7.17 *5* min)

Mode: ϵ(69 *3* %), α(31 *3* %)

Δ: -69325 *11* keV

SpA: 1.747 $\times10^{8}$ Ci/g

Prod: ^{141}Pr(^{14}N,5n); ^{140}Ce(^{16}O,6n); ^{159}Tb(p,10n); protons on Er; protons on Ta

Alpha Particles (^{150}Dy)

α(keV)
4233 *4*

Photons (^{150}Dy)

⟨γ⟩=297 *11* keV

γ_{mode}	γ(keV)	γ(%)†
Tb L_ℓ	5.546	0.178 *18*
Tb L_α	6.269	4.7 *3*
Tb L_η	6.284	0.062 *5*
Tb L_β	7.114	4.3 *4*
Tb L_γ	8.196	0.66 *7*
Tb $K_{\alpha2}$	43.744	15.7 *5*
Tb $K_{\alpha1}$	44.482	28.1 *8*
Tb $K_{\beta1}$'	50.338	8.41 *25*
Tb $K_{\beta2}$'	51.958	2.51 *8*
γ,E1	396.75 *9*	68 *3*

† 4.3% uncert(syst)

Atomic Electrons (^{150}Dy)

⟨e⟩=6.6 *3* keV

e_{bin}(keV)	⟨e⟩(keV)	e(%)
8	2.8	36 *4*
9	0.38	4.4 *5*
35	0.28	0.79 *8*
36	0.42	1.17 *12*
37	0.138	0.37 *4*
41	0.051	0.123 *13*
42	0.172	0.41 *4*
43	0.224	0.52 *5*
44	0.046	0.103 *11*
48	0.0187	0.039 *4*
49	0.0208	0.043 *4*
50	0.0167	0.034 *3*
345	1.69	0.49 *3*
388	0.243	0.063 *4*
389	0.0199	0.0051 *3*
395	0.058	0.0147 *8*
396	0.0139	0.0035 *2*
397	0.00240	0.00060 *3*

Continuous Radiation (^{150}Dy)

⟨IB⟩=0.46 keV

E_{bin}(keV)		⟨ ⟩(keV)	(%)
10 - 20	IB	0.000122	0.00078
20 - 40	IB	0.024	0.064
40 - 100	IB	0.096	0.20
100 - 300	IB	0.027	0.0137
300 - 600	IB	0.108	0.024
600 - 1300	IB	0.20	0.025
1300 - 1380	IB	9.0×10^{-5}	6.9×10^{-6}

$^{150}_{67}$Ho(1.47 *25* min)

Mode: ε

Δ: -62220 *200* keV syst

SpA: 8.5×10^{8} Ci/g

Prod: ^{141}Pr(^{16}O,7n); ^{159}Tb(^{3}He,12n)

Photons (^{150}Ho)

γ_{mode}	γ(keV)	γ(rel)
γ[E1]	591.3 *3*	31 *3*
γ[E2]	653.44 *22*	30 *5*
γ[E2]	803.46 *22*	100

$^{150}_{67}$Ho(28 *3* s)

Mode: ε

Δ: >-62220 keV syst

SpA: 2.7×10^{9} Ci/g

Prod: ^{141}Pr(^{16}O,7n); ^{144}Sm(^{10}B,4n)

Photons (^{150}Ho)

⟨γ⟩=2356 *170* keV

γ_{mode}	γ(keV)	γ(%)
γ[E2]	393.95 *25*	93 *5*
γ[E1]	411.2 *2*	7 *2*
γ[E2]	551.15 *25*	88 *5*
γ[E2]	624.3 *2*	3 *1*
γ[E2]	653.44 *22*	100 *5*
γ[E2]	803.46 *22*	100

$^{150}_{68}$Er(18.5 *7* s)

Mode: ε

Δ: -58020 *280* keV syst

SpA: 3.99×10^{9} Ci/g

Prod: ^{94}Mo(^{58}Ni,2p); ^{92}Mo(^{60}Ni,2p); ^{144}Sm(^{12}C,6n)

Photons (^{150}Er)

γ_{mode}	γ(keV)
γ[E1]	475.8 *3*

$^{150}_{69}$Tm(3.5 *6* s)

Mode: ε

SpA: 1.9×10^{10} Ci/g

Prod: ^{92}Mo(^{60}Ni,pn)

Photons (^{150}Tm)

γ_{mode}	γ(keV)	γ(rel)
γ	207.5 *3*	96 *26*
γ	474.8 *3*	93 *29*
γ	1578.9 *3*	100 *20*
γ	1786.4 *3*	12 *3*

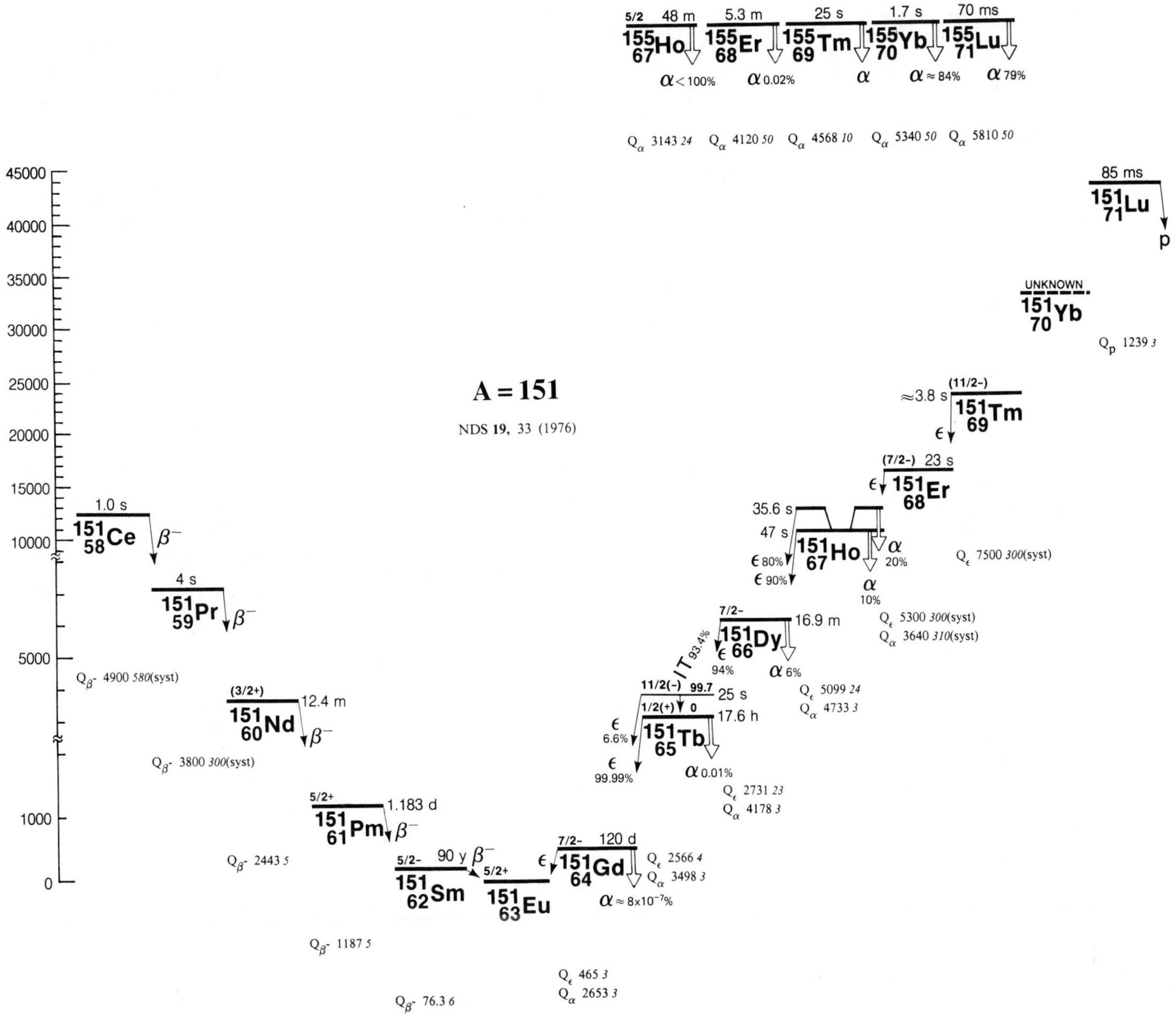

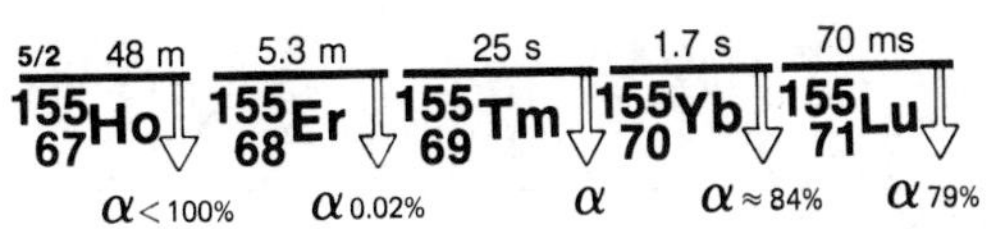

Q_α 3143 *24* Q_α 4120 *50* Q_α 4568 *10* Q_α 5340 *50* Q_α 5810 *50*

A = 151

NDS **19**, 33 (1976)

45000
40000
35000
30000
25000
20000
15000
10000
5000
1000
0

1.0 s $^{151}_{58}$Ce β^-
Q_{β^-} 4900 *580*(syst)

4 s $^{151}_{59}$Pr β^-
Q_{β^-} 3800 *300*(syst)

(3/2+) 12.4 m $^{151}_{60}$Nd β^-
Q_{β^-} 2443 *5*

5/2+ 1.183 d $^{151}_{61}$Pm β^-
Q_{β^-} 1187 *5*

5/2- 90 y $^{151}_{62}$Sm β^-
Q_{β^-} 76.3 *6*

5/2+ $^{151}_{63}$Eu

7/2- 120 d $^{151}_{64}$Gd ε α ≈ 8x10^{-7}%
Q_ϵ 465 *3*
Q_α 2653 *3*

11/2(-) 99.7 25 s
1/2(+) 0 17.6 h $^{151}_{65}$Tb
ε 6.6% ε 99.99% α 0.01%
Q_ϵ 2566 *4*
Q_α 3498 *3*

7/2- 16.9 m $^{151}_{66}$Dy IT 93.4% ε 94% α 6%
Q_ϵ 2731 *23*
Q_α 4178 *3*

35.6 s 47 s $^{151}_{67}$Ho ε 80% ε 90% α 20% α 10%
Q_ϵ 5099 *24*
Q_α 4733 *3*

(7/2-) 23 s $^{151}_{68}$Er ε
Q_ϵ 5300 *300*(syst)
Q_α 3640 *310*(syst)

(11/2-) ≈3.8 s $^{151}_{69}$Tm ε
Q_ϵ 7500 *300*(syst)

UNKNOWN $^{151}_{70}$Yb

85 ms $^{151}_{71}$Lu p
Q_p 1239 *3*

$^{151}_{58}$Ce(1.02 *6* s)

Mode: β-
Δ: -62260 *500* keV syst
SpA: 5.3×10^{10} Ci/g
Prod: fission

Photons (^{151}Ce)

γ_{mode}	γ(keV)
γ	84.79 *9*
γ	96.8 *2*
γ	118.57 *9*

$^{151}_{59}$Pr(4.0 *7* s)

Mode: β-
Δ: -67160 *300* keV syst
SpA: 1.7×10^{10} Ci/g
Prod: fission

Photons (^{151}Pr)

γ_{mode}	γ(keV)
γ	164.0 *1*

$^{151}_{60}$Nd(12.44 *7* min)

Mode: β-
Δ: -70957 *4* keV
SpA: 1.001×10^{8} Ci/g
Prod: ^{150}Nd(n,γ)

Photons (^{151}Nd)

⟨γ⟩=916 *37* keV

γ_{mode}	γ(keV)	γ(%)†
Pm L_ℓ	4.809	0.075 *9*
Pm L_η	5.363	0.030 *3*
Pm L_α	5.430	2.00 *17*
Pm L_β	6.094	1.82 *19*
Pm L_γ	6.971	0.26 *3*
γ [E1]	31.45 *10*	2.0 *10*
Pm $K_{\alpha2}$	38.171	7.1 *3*
Pm $K_{\alpha1}$	38.725	12.9 *6*
Pm $K_{\beta1}'$	43.793	3.72 *16*
Pm $K_{\beta2}'$	45.183	1.10 *5*
γ M1+1.2%E2	58.38 *9*	0.5 *1*

Photons (^{151}Nd) (continued)

γ_{mode}	γ(keV)	γ(%)†
γ M1+6.8%E2	68.85 13	1.25 12
γ M1+34%E2	85.26 7	2.72 14
γ E1	89.84 7	1.77 18
γ M1+E2	103.01 10	0.46 6
γ [M1+E2]	112.11 23	~0.46
γ E1	116.71 9	46.8 18
γ [E2]	127.06 24	~0.06
γ	130.8 5	~0.06
γ E1	138.87 11	7.8 4
γ E1	149.34 10	0.32 3
γ M1+E2	171.02 9	4.1 3
γ E1	175.09 7	7.7 5
γ	183.31 14	0.52 5
γ	192.7 8	~0.05 ?
γ [E2]	197.37 23	0.34 5
γ (M1+E2)	199.89 21	0.35 6
γ [E1]	207.72 12	~0.14 ?
γ	221.5 4	<0.046
γ [M1+E2]	230.07 25	~0.06 ?
γ	239.05 17	~0.6
γ [M1+E2]	239.17 10	~0.37
γ	248.5 4	0.28 5 ?
γ M1(+E2)	255.58 11	16.8 8
γ (E1)	263.53 13	0.80 8
γ [M1+E2]	268.74 21	0.14 6 ?
γ	275.61 16	0.179 20 ?
γ	280.32 18	<0.06
γ	291.1 4	~0.032 ?
γ E1	300.63 9	1.97 15
γ	312.71 15	0.29 3
γ	320.18 21	0.90 9
γ M1(+E2)	324.43 10	0.58 6
γ	332.65 14	~0.6
γ	333.16 15	~0.18
γ	347.17 18	0.43 5
γ	357.0 3	0.41 5
γ	363.7 4	0.11 3
γ [E2]	365.18 11	0.20 3 ?
γ [E1]	373.95 17	0.14 3
γ	384.9 4	<0.06 ?
γ	391.03 15	~0.044
γ (M1)	402.28 12	1.96 15
γ [E1]	407.61 20	0.46 5
γ	411.9 4	~0.08 ?
γ	414.30 14	0.23 6
γ M1(+E2)	423.56 8	7.3 5
γ [M1+E2]	427.44 13	0.55 9
γ [M1+E2]	439.07 20	0.38 5 ?
γ	447.00 23	0.31 5 ?
γ	455.0 3	0.08 3 ?
γ [M1+E2]	460.66 12	1.10 11
γ	487.6 3	~0.24
γ	491.21 20	0.23 3 ?
γ	507.74 14	0.061 15
γ	516.47 13	0.14 3 ?
γ [M1+E2]	524.32 20	0.60 5
γ	531.8 7	0.17 3
γ	542.0 4	0.58 5
γ	550.01 23	0.67 5
γ	562.8 5	0.24 3
γ [E1]	577.37 13	0.44 5
γ (M1)	585.32 12	1.58 14
γ [M1+E2]	589.79 18	0.31 11
γ	597.40 15	0.80 8
γ	611.6 4	0.11 5 ?
γ	618.9 3	0.32 3
γ	630.08 20	0.17 5 ?
γ	643.53 25	~0.06 ?
γ [M1+E2]	658.64 17	0.84 8
γ	665.81 11	~0.06
γ	670.9 3	0.29 5 ?
γ	677.89 13	2.6 3
γ	724.19 10	0.28 3
γ	736.27 13	7.2 5
γ [E1]	739.12 17	1.53 23
γ	755.64 12	1.44 14
γ	758.9 3	0.17 5 ?
γ	762.1 4	<0.12
γ	765.1 10	0.23 6
γ	767.72 14	0.37 6
γ	773.78 21	0.34 5 ?
γ	785.3 3	0.098 23 ?
γ	789.58 20	0.125 23
γ (E1)	797.51 15	5.5 3
γ	809.1 3	0.24 5
γ	812.8 7	0.18 3
γ	820.2 3	~0.12
γ [M1+E2]	828.96 17	0.26 6
γ	840.90 11	1.06 11
γ	852.98 14	0.47 8 ?
γ	855.0 3	~0.09 ?
γ	859.8 3	~0.09 ?
γ	862.3 13	~0.09 ?
γ [M1+E2]	869.88 19	0.26 11 ?
γ	871.8 13	~0.21
γ	876.3 3	0.49 8
γ	878.6 3	0.14 6 ?
γ	879.7 6	0.40 8
γ	881.0 17	0.11 5 ?
γ	899.6 15	0.14 6
γ	901.6 13	0.17 8
γ	905.0 4	0.28 8
γ	910.0 3	~0.21 ?
γ (M1)	914.22 17	1.21 12
γ	925.4 7	0.15 3
γ	933.4 13	~0.11
γ	935.4 13	~0.11
γ	944.2 4	0.40 5
γ	958.20 15	0.57 5
γ	968 2	0.17 5
γ [M1+E2]	972.89 17	~0.09
γ	990.6 18	~0.09
γ	995.3 3	0.12 5 ?
γ	1002.6 4	~0.12 ?
γ	1008 2	~0.09 ?
γ	1016.59 14	2.92 24
γ	1020.6 3	~0.06 ?
γ	1028.7 15	0.08 3
γ	1032.72 25	0.12 3
γ	1035.9 10	0.24 9
γ [M1+E2]	1041.74 17	0.49 6
γ	1048.04 15	0.81 5
γ	1066.3 12	0.18 8
γ	1070.0 18	0.14 5
γ	1072.7 3	0.18 5
γ	1080.6 5	0.37 6
γ	1091.7 15	~0.06
γ	1099.8 3	~0.06
γ	1107.1 3	0.47 5
γ [E1]	1122.22 16	4.6 3
γ	1126.8 3	0.21 5
γ	1133.29 15	0.17 5
γ	1145.4 3	0.23 3
γ	1156.6 12	0.23 5
γ	1169.7 3	0.29 5
γ	1173.0 18	0.11 5
γ [E1]	1180.61 15	15
γ	1188.8 7	0.24 5
γ	1201.1 3	0.24 3
γ	1223.4 3	~0.06
γ	1232.7 8	0.11 3
γ	1238.3 14	0.08 3
γ	1254.7 4	~0.06
γ	1268.2 13	~0.08
γ	1270.1 4	0.17 5
γ	1277.5 3	0.023 6
γ	1286.4 3	0.31 3
γ	1293.9 7	0.20 5 ?
γ [M1+E2]	1297.31 16	0.24 5
γ	1314.8 3	0.37 5
γ	1316.6 13	0.12 5
γ	1324.2 13	~0.05
γ	1328.5 4	0.38 5
γ	1333.35 25	0.20 5
γ	1341.0 13	0.17 5
γ	1360.0 4	~0.14 ?
γ	1362.7 3	0.37 6
γ	1365.9 3	~0.06 ?
γ	1379.2 8	0.20 5
γ	1383.1 12	0.11 5
γ	1394.5 3	~0.09
γ	1396.5 13	~0.06
γ	1403 2	~0.017 ?
γ	1407 2	0.046 15 ?
γ	1460.8 7	0.11 3 ?
γ	1464.8 3	~0.06 ?
γ	1474.9 3	0.11 3
γ	1484.8 4	0.31 3
γ	1549.19 23	0.37 3
γ	1565.5 4	0.080 23
γ	1571.5 8	0.09 3
γ	1577.7 13	0.167 24
γ	1596.9 4	0.113 24
γ	1618.04 25	0.38 3
γ	1627.5 14	0.046 15 ?
γ	1636.6 3	0.11 3
γ	1703 3	~0.031
γ	1717.1 3	0.138 23
γ	1753.3 13	0.054 15
γ	1756.91 24	0.057 15
γ	1775.5 3	0.28 5
γ	1788.36 24	0.09 3 ?
γ	1794 2	0.054 23
γ	1806.9 3	0.06 3
γ	1811 2	0.07 3
γ	1818.4 6	0.075 15
γ	1863.6 6	0.075 15
γ	1873.62 24	~0.024 ?
γ	1892.2 3	0.21 3
γ	1926.4 18	0.043 6
γ	2019 3	0.061 9

† 13% uncert(syst)

Atomic Electrons (^{151}Nd)

$\langle e \rangle$=29.2 9 keV

e_{bin}(keV)	$\langle e \rangle$(keV)	e(%)
6	0.88	13.6 16
7	0.65	9.1 11
13	0.42	3.2 7
24	1.44	6.0 11
25 - 39	0.93	3.0 4
40	2.21	5.5 3
42 - 69	2.06	3.6 3
72	4.99	7.0 3
78	1.08	1.39 7
79 - 90	1.28	1.57 7
94	0.68	0.72 4
96 - 106	0.45	0.45 15
109	0.80	0.73 3
110 - 125	0.70	0.62 4
126	1.40	1.11 15
127 - 176	1.66	1.13 13
177 - 208	0.30	0.16 6
210	3.0	1.4 3
214 - 264	1.04	0.42 9
267 - 314	0.35	0.12 3
317 - 365	0.314	0.090 7
367 - 416	0.91	0.23 4
417 - 463	0.109	0.025 5
471 - 518	0.12	0.025 6
523 - 571	0.172	0.032 4
574 - 623	0.089	0.0149 24
624 - 671	0.11	0.018 9
676 - 724	0.30	0.043 19
729 - 778	0.14	0.019 4
779 - 828	0.078	0.0097 23
829 - 878	0.100	0.0117 16
879 - 928	0.045	0.0050 11
929 - 975	0.07	0.007 3
983 - 1033	0.057	0.0056 11
1034 - 1082	0.062	0.0058 7
1084 - 1133	0.025	0.0023 5
1135 - 1182	0.139	0.0122 20
1187 - 1237	0.0091	0.00075 20
1238 - 1287	0.027	0.0021 4
1288 - 1335	0.019	0.0015 3
1338 - 1387	0.0065	0.00048 12
1388 - 1430	0.0038	0.00027 9
1440 - 1533	0.013	0.00089 22
1542 - 1635	0.0099	0.00063 15
1658 - 1756	0.0077	0.00045 10
1762 - 1857	0.0062	0.00034 7
1862 - 1925	0.00086	4.6 13 $\times 10^{-5}$
1974 - 2018	0.0006	3.2 13 $\times 10^{-5}$

Continuous Radiation (^{151}Nd)

⟨β-⟩=591 keV;⟨IB⟩=0.98 keV

E_{bin}(keV)		⟨ ⟩(keV)	(%)
0 - 10	β-	0.0407	0.81
	IB	0.025	
10 - 20	β-	0.123	0.82
	IB	0.024	0.168
20 - 40	β-	0.503	1.68
	IB	0.046	0.162
40 - 100	β-	3.71	5.3
	IB	0.126	0.19
100 - 300	β-	39.0	19.3
	IB	0.31	0.18
300 - 600	β-	128	28.7
	IB	0.26	0.062
600 - 1300	β-	326	37.3
	IB	0.18	0.022
1300 - 2326	β-	93	6.1
	IB	0.0129	0.00089

$^{151}_{61}$Pm(1.1833 *17* d)

Mode: β-

Δ: -73400 *6* keV

SpA: 7.308×10^5 Ci/g

Prod: daughter ^{151}Nd

Photons (^{151}Pm)

⟨γ⟩=321 *16* keV

γ_{mode}	γ(keV)	γ(%)†
Sm L_ℓ	4.993	0.088 9
Sm L_η	5.589	0.034 3
Sm L_α	5.633	2.37 17
Sm L_β	6.340	2.17 20
Sm L_γ	7.267	0.32 3
γ E1	25.693 15	0.93 7
γ M1+26%E2	35.124 14	0.016 3
Sm $K_{\alpha2}$	39.522	8.4 3
Sm $K_{\alpha1}$	40.118	15.1 6
Sm $K_{\beta1}'$	45.379	4.38 17
Sm $K_{\beta2}'$	46.819	1.30 6
γ [E1]	59.97 3	0.0246 22
γ E2	61.003 12	0.0090 11
γ (E1)	62.912 12	0.213 16
γ M1+0.8%E2	64.870 11	1.93 16
γ M1+3.8%E2	65.820 12	1.14 9
γ M1+2.8%E2	69.687 11	0.49 5
γ E2	76.211 14	0.206 16
γ (M1)	88.77 3	0.0123 16
γ [E1]	91.64 4	~0.007
γ	92.96 3	0.034 3
γ E1	98.036 12	0.36 3
γ (M1+E2)	98.690 18	0.058 9
γ M1	99.995 10	2.53 18
γ [M1+E2]	100.76 3	0.012 3
γ E1	101.904 12	1.29 9
γ M1,E2	102.557 19	~0.031
γ M1+1.0%E2	104.812 10	3.52 25
γ [M1+E2]	109.55 4	0.085 7
γ	113.00 5	0.0096 16
γ (M1)	121.73 3	0.099 9
γ [M1+E2]	125.46 5	0.0125 18
γ	126.8 5	0.0029 13 ?
γ [E1]	130.40 4	0.067 7
γ	134.18 4	0.0202 20
γ	134.83 4	~0.004
γ [E1]	138.37 3	0.039 4
γ [M1+E2]	139.02 3	0.027 7
γ M1(+E2)	139.278 19	0.47 4
γ	141.7 5	0.010 3 ?
γ	143.08 4	0.010 3
γ M1(+E2)	143.145 19	0.213 18
γ	146.2 4	0.017 3 ?

Photons (^{151}Pm) (continued)

γ_{mode}	γ(keV)	γ(%)†
γ [E1]	147.51 4	0.159 13
γ [E1]	148.54 4	0.054 5
γ	150.1 4	0.0090 22 ?
γ [E1]	155.532 24	0.025 5
γ (M1)	156.185 19	0.148 13
γ E1	162.907 10	0.84 7
γ M1	163.560 15	1.50 11
γ E1	167.724 9	7.8 7
γ M1	168.377 15	0.83 9
γ E1	176.495 19	0.85 7
γ M1	177.149 13	3.58 25
γ (E1)	186.555 24	0.166 13
γ	192.96 7	0.0074 22
γ	195.47 11	0.027 7
γ [E1]	201.935 25	0.85 7
γ (E2)	204.148 18	0.134 11
γ [E1]	205.74 6	0.0090 22
γ [M1+E2]	206.59 4	0.036 7
γ	207.23 7	0.0074 18
γ M1	208.965 18	1.64 13
γ [E2]	215.22 3	0.0090 22
γ E1	227.143 23	0.31 3
γ M1	227.797 21	0.049 18
γ [M1+E2]	228.94 11	0.027 7
γ M1+0.8%E2	232.396 20	1.03 9
γ	232.87 4	0.087 20
γ [E1]	236.07 5	0.094 16
γ [E1]	236.67 3	0.159 18
γ (M1)	236.74 4	0.19 5
γ [E1]	237.06 3	0.52 9
γ E1	240.061 15	3.61 25
γ [M1+E2]	247.18 4	0.018 5
γ [E1]	247.83 3	0.027 5
γ	250.5 8	~0.009
γ [E1]	254.222 20	0.161 13
γ E1	258.089 19	0.54 5
γ	261.46 3	0.011 5
γ	270.45 4	0.065 7
γ E1	275.185 15	6.6 6
γ [E1]	277.26 3	0.047 7
γ [M1+E2]	277.911 24	0.008 3
γ M1	280.09 3	0.237 20
γ [E2]	284.91 3	<0.0022
γ E1	290.708 21	0.80 7
γ	292.46 12	~0.011
γ [E2]	294.76 11	0.013 5
γ	295.24 4	0.015 5
γ	297.74 4	0.043 5
γ	298.6 5	~0.006 ?
γ	302.05 3	0.013 5
γ	302.56 4	0.022 5
γ	302.70 3	0.027 7
γ [E1]	306.75 3	0.244 20
γ M1	308.93 6	0.081 9
γ [M1+E2]	310.42 4	0.017 5
γ [M1+E2]	310.74 3	0.035 6
γ	314.86 5	0.062 7
γ [M1+E2]	315.24 4	0.0078 22
γ	321.85 4	0.096 11
γ E1	323.909 19	1.22 10
γ	325.33 4	0.015 3
γ [E1]	325.833 22	0.105 13
γ	329.0 8	~0.013
γ (E1)	329.700 22	0.233 20
γ E1	340.055 13	22
γ [E1]	340.823 25	0.072 18
γ E1	344.872 13	2.11 18
γ [M1+E2]	345.86 3	0.038 9
γ	349.64 9	0.0092 22
γ M1	349.73 3	0.132 13
γ	352.06 4	0.015 4
γ [M1+E2]	353.26 4	0.112 11
γ	356.9 5	0.008 3
γ	358.30 5	0.015 3
γ	360.83 7	0.011 3
γ	368.63 12	0.0155 25
γ	374.07 12	0.025 5
γ [M1+E2]	376.85 4	0.016 5
γ	378.73 12	0.010 5
γ [E1]	379.815 25	0.94 7
γ	381.2 3	0.020 7
γ [E1]	390.703 20	0.049 5
γ [E1]	395.52 2	0.043 5
γ	398.72 4	0.029 5

Photons (^{151}Pm) (continued)

γ_{mode}	γ(keV)	γ(%)†
γ	400.74 3	0.007 3
γ	404.61 3	0.065 7
γ [E1]	406.93 4	0.188 18
γ [M1+E2]	410.74 3	0.063 7
γ [M1+E2]	415.55 3	0.022 5
γ	416.8 4	0.016 5
γ	420.54 4	0.054 7
γ	424.41 4	0.047 5
γ	425.64 5	0.010 3
γ	427.13 6	0.063 7
γ	429.09 12	0.016 5
γ E1	440.818 24	1.50 13
γ	443.60 11	0.021 9
γ E1	445.635 24	4.0 3
γ	448.42 11	0.023 9
γ [E1]	451.29 4	0.28 3
γ	451.99 9	0.013 5
γ	454.42 7	0.013 5
γ	455.97 4	0.037 5
γ	458.62 6	~0.0045
γ	461.96 4	0.031 5
γ	463.8 4	~0.009 ?
γ	467.11 5	0.011 5
γ	470.43 3	0.018 6
γ	471.30 5	0.018 6
γ	471.77 4	0.012 5
γ	473.8 8	~0.007 ?
γ [M1+E2]	477.61 3	0.091 10
γ	487.09 20	0.017 5 ?
γ	490.23 4	0.129 13
γ	495.01 7	0.010 5
γ	495.67 7	0.012 5
γ	503.7 7	~0.0045 ?
γ	505.0 4	0.014 4
γ [E1]	507.25 5	0.048 6
γ	510.1 7	0.009 4
γ [E1]	516.16 4	0.190 18
γ [E1]	520.98 4	0.031 5
γ	532.64 3	0.030 4
γ [E1]	537.58 4	0.049 6
γ	550.68 9	0.015 3
γ	554.55 9	0.016 3
γ	562.27 5	0.019 3
γ	564.89 4	0.35 3
γ	572.50 3	0.050 11
γ	573.15 3	0.028 6
γ	574.95 4	0.118 11
γ	575.1 5	0.0029 11
γ	581.1 6	0.0045 13 ?
γ	583.09 20	0.026 3
γ	584.77 4	0.0085 16
γ	593.70 7	0.0101 20
γ	597.57 7	0.079 9
γ	598.0 10	~0.009
γ	599.1 7	0.008 3 ?
γ	602.25 4	0.011 3
γ	604.0 5	0.0081 22
γ	605.48 4	0.0096 25
γ	609.24 10	0.047 5
γ	620.37 9	0.069 7
γ	636.07 3	1.38 12
γ [E1]	654.11 3	0.235 20
γ	655.6 4	0.011 5 ?
γ	661.79 5	0.012 5
γ	663.39 7	0.088 11
γ	668.49 5	0.0034 16
γ	668.54 4	0.35 4
γ	669.04 4	0.28 4
γ	671.19 3	0.87 7
γ	678.36 3	0.042 5
γ	699.0 7	0.019 6 ?
γ	704.17 4	0.32 3
γ	709.13 3	0.128 14
γ	711.96 4	0.087 11
γ	713.9 4	0.009 3
γ (E1)	717.67 3	4.0 3
γ	718.95 3	0.011 3
γ	727.0 25	0.0067 13
γ	736.06 3	0.46 4
γ	740.88 3	0.020 3
γ (E1)	752.80 3	1.23 11
γ	755.38 4	0.0067 22
γ	758.44 4	0.0090 20
γ	769.04 4	0.086 9

Photons (^{151}Pm) (continued)

γ_{mode}	γ(keV)	γ(%)†
γ	772.70 3	0.83 7
γ	785.02 5	0.203 19
γ	792.8 7	~0.0022
γ	795.97 4	0.054 5
γ	807.82 3	0.48 4
γ	811.69 3	0.065 6
γ	817.638 20	0.17 3
γ[E1]	817.67 3	0.09 3
γ[E1]	822.48 3	0.025 3
γ	848.59 5	0.269 22
γ	856.47 4	0.0065 11
γ	860.34 4	0.0083 11
γ	867.1 7	0.0029 11 ?
γ	877.51 3	0.092 8
γ	883.71 5	0.048 4
γ	887.325 22	0.0027 9
γ	894.66 4	0.0027 9
γ	898.53 4	0.0231 25
γ	903.49 5	0.0031 9
γ	911.24 15	0.0255 25
γ	919.3 7	~0.0018 ?
γ	922.1 7	~0.0013 ?
γ	926.16 4	0.0040 7
γ	933.94 5	0.0038 7
γ	940 5	0.0038 9
γ	948.58 4	0.32 3
γ	953.40 4	0.080 9
γ	959.53 4	0.055 7
γ	964.35 4	0.0047 9
γ	968.86 18	0.0146 16
γ	1011.7 4	0.0036 9

† 12% uncert(syst)

Atomic Electrons (^{151}Pm)

⟨e⟩=24.1 5 keV

e_{bin}(keV)	⟨e⟩(keV)	e(%)
7	1.56	22.7 24
8 - 16	0.228	2.7 3
18	1.95	10.8 9
19	1.16	6.1 5
23 - 52	1.83	6.17 25
53	1.98	3.7 3
54 - 56	0.25	0.46 10
57	0.73	1.28 11
58	3.17	5.5 4
59 - 91	1.77	2.67 9
92	0.70	0.76 9
93 - 96	0.21	0.22 3
97	0.63	0.65 5
98 - 116	0.59	0.57 3
117	0.64	0.55 4
118 - 120	0.0070	0.0059 17
121	0.55	0.45 4
122 - 129	0.35	0.29 3
130	1.43	1.10 8
131 - 180	1.54	0.97 5
181 - 230	1.03	0.513 24
231 - 280	0.296	0.117 5
282 - 292	0.0154	0.0054 6
293	0.61	0.21 4
294 - 343	0.284	0.089 6
344 - 393	0.046	0.0127 17
394 - 443	0.163	0.0401 24
444 - 493	0.0186	0.0040 4
494 - 543	0.029	0.0056 21
544 - 593	0.07	0.011 6
594 - 636	0.072	0.012 4
646 - 694	0.091	0.0136 20
696 - 745	0.062	0.0086 22
746 - 795	0.029	0.0038 13
796 - 842	0.016	0.0019 6
847 - 896	0.0027	0.00031 7
897 - 946	0.012	0.0013 5
947 - 969	0.00073	7.6 22 ×10^{-5}
1004 - 1012	1.3 ×10^{-5}	~1×10^{-6}

Continuous Radiation (^{151}Pm)

⟨β-⟩=281 keV;⟨IB⟩=0.27 keV

E_{bin}(keV)		⟨ ⟩(keV)	(%)
0 - 10	β-	0.118	2.37
	IB	0.0135	
10 - 20	β-	0.352	2.35
	IB	0.0128	0.089
20 - 40	β-	1.39	4.65
	IB	0.024	0.082
40 - 100	β-	9.4	13.5
	IB	0.058	0.091
100 - 300	β-	71	36.7
	IB	0.106	0.064
300 - 600	β-	136	31.8
	IB	0.046	0.0117
600 - 1187	β-	63	8.7
	IB	0.0055	0.00082

$^{151}_{62}$Sm(90 *6* yr)

Mode: β-

Δ: -74587 3 keV

SpA: 26.3 Ci/g

Prod: fission; ^{150}Sm(n,γ)

Photons (^{151}Sm)

⟨γ⟩=0.0131 8 keV

γ_{mode}	γ(keV)	γ(%)
Eu L_ℓ	5.177	0.00103 16
Eu L_η	5.817	0.00053 6
Eu L_α	5.843	0.027 4
Eu L_β	6.544	0.060 9
Eu L_γ	7.699	0.0130 21
γ M1(+0.09%E2)	21.448 6	0.030 2

Atomic Electrons (^{151}Sm)

⟨e⟩=0.152 8 keV

e_{bin}(keV)	⟨e⟩(keV)	e(%)
7	0.0021	0.030 4
8	0.023	0.29 4
13	0.075	0.56 5
14	0.0148	0.105 9
20	0.0285	0.145 13
21	0.0085	0.040 4

Continuous Radiation (^{151}Sm)

⟨β-⟩=125 keV;⟨IB⟩=0.054 keV

E_{bin}(keV)		⟨ ⟩(keV)	(%)
0 - 10	β-	0.255	5.10
	IB	0.0064	
10 - 20	β-	0.75	5.00
	IB	0.0057	0.040
20 - 40	β-	2.91	9.7
	IB	0.0096	0.034
40 - 100	β-	18.3	26.5
	IB	0.018	0.030
100 - 300	β-	89	49.7
	IB	0.0138	0.0097
300 - 421	β-	13.4	4.06
	IB	0.00019	6.0×10^{-5}

$^{151}_{63}$Eu(stable)

Δ: -74663 3 keV

%: 47.8 5

$^{151}_{64}$Gd(120 *20* d)

Mode: ϵ, α(~8×10^{-7} %)

Δ: -74198 4 keV

SpA: 7206 Ci/g

Prod: ^{151}Eu(p,n); ^{151}Eu(d,2n)

Alpha Particles (^{151}Gd)

α(keV)
2600 30

Photons (^{151}Gd)

⟨γ⟩=64.2 18 keV

γ_{mode}	γ(keV)	γ(%)†
Eu L_ℓ	5.177	0.33 4
Eu L_η	5.817	0.132 11
Eu L_α	5.843	8.7 7
Eu L_β	6.571	10.9 12
Eu L_γ	7.644	2.0 3
γM1(+0.09%E2)	21.448 6	2.35 10
Eu $K_{\alpha2}$	40.902	23.7 8
Eu $K_{\alpha1}$	41.542	42.7 14
Eu $K_{\beta1}'$	46.999	12.4 4
Eu $K_{\beta2}'$	48.496	3.76 14
γE2	106.41 5	0.066 3
γ[M1]	110.44 11	0.0061 5
γM1	153.51 5	5.1
γM2	174.59 5	2.45 10
γ(E3)	196.04 5	~0.012
γM1+21%E2	196.37 9	~0.012
γ[E1]	221.68 5	~0.0031
γ(E2)	238.87 8	0.118 6
γE1	243.13 5	4.59 15
γM1	260.32 8	0.0428 15
γM1+8.8%E2	285.90 6	0.0688 20
γM1+14%E2	307.35 6	0.841 25
γE1	328.09 5	0.067 3
γ	332.00 7	0.0087 10
γ	353.44 7	0.104 3
γ[M1+E2]	394.74 21	0.00061 15
γ[M1+E2]	416.18 21	0.00046 15

† 16% uncert(syst)

Atomic Electrons (^{151}Gd)

⟨e⟩=30.9 *9* keV

e_{bin}(keV)	⟨e⟩(keV)	e(%)
7	2.8	41 *4*
8	3.9	50 *5*
13	6.0	45 *3*
14	1.19	8.4 *5*
20	2.29	11.6 *7*
21	0.68	3.23 *19*
32 - 62	2.19	6.06 *25*
98 - 103	0.0495	0.0500 *22*
105	2.6	2.4 *5*
106 - 110	0.00349	0.00328 *15*
126	5.9	4.65 *22*
145 - 154	0.66	0.45 *7*
167	1.45	0.87 *4*
168 - 215	0.871	0.490 *12*
220 - 260	0.215	0.086 *3*
278 - 327	0.048	0.0161 *22*
328 - 368	0.0025	0.0007 *3*
387 - 416	2.4×10^{-5}	6.0 *14* $\times10^{-6}$

Continuous Radiation (^{151}Gd)

⟨IB⟩=0.158 keV

E_{bin}(keV)		⟨ ⟩(keV)	(%)
10 - 20	IB	0.00021	0.00135
20 - 40	IB	0.055	0.152
40 - 100	IB	0.089	0.18
100 - 300	IB	0.0111	0.0064
300 - 465	IB	0.00147	0.00044

$^{151}_{65}$Tb(17.6 *1* h)

Mode: ϵ(99.9905 *15* %), α(0.0095 *15* %)
Δ: -71632 *5* keV
SpA: 1.179×10^{6} Ci/g
Prod: ^{151}Eu(α,4n); protons on Gd

Alpha Particles (^{151}Tb)

⟨α⟩=0.324 keV

α(keV)	α(%)
3184 *4*	1×10^{-5}
3408 *4*	0.0095

Photons (^{151}Tb)

⟨γ⟩=880 *41* keV

γ_{mode}	γ(keV)	γ(%)†
Gd L_ℓ	5.362	0.38 *5*
Gd L_η	6.049	0.139 *15*
Gd L_α	6.054	10.0 *10*
Gd L_β	6.850	9.2 *10*
Gd L_γ	7.874	1.39 *18*
Gd $K_{\alpha2}$	42.309	33.8 *22*
Gd $K_{\alpha1}$	42.996	61 *4*
Gd $K_{\beta1}$'	48.652	17.9 *12*
Gd $K_{\beta2}$'	50.214	5.4 *4*
γM1+40%E2	108.16 *7*	25 *5*
γE2	149.04 *12*	0.47 *9*
γ[M1]	160.81 *13*	0.42 *8*
γM1	180.32 *8*	11.4 *23*
γM1	192.09 *8*	3.8 *8*
γ	212.1 *5*	0.044 *9* ?
γ	236.24 *11*	0.047 *9* ?
γ	241.34 *20*	0.075 *15*
γM1	251.81 *9*	26
γM1	263.57 *11*	0.16 *3*

Photons (^{151}Tb)
(continued)

γ_{mode}	γ(keV)	γ(%)†
γM1	287.08 *7*	25 *5*
γM1	318.36 *12*	0.23 *5*
γE1	380.31 *12*	4.3 *9*
γM1	385.28 *14*	1.14 *23*
γE2	395.24 *8*	9.6 *19*
γM1	416.56 *10*	1.4 *3*
γM1	426.52 *12*	4.1 *8*
γM1	443.90 *10*	10.4 *21*
γM1+E2	467.40 *9*	1.04 *21*
γ(E2)	479.17 *8*	16 *3*
γ(M1)	500.19 *14*	0.39 *8*
γ[M1]	511.96 *14*	0.34 *7*
γ	536.7 *3*	0.47 *9*
γE2	587.33 *9*	17 *3*
γE1	604.78 *9*	3.2 *6*
γE1	616.55 *10*	10.4 *21*
γ	656.64 *22*	0.52 *10*
γ	661.00 *16*	0.49 *10*
γM1(+E2)	692.29 *12*	1.5 *3*
γM1(+E2)	703.64 *10*	3.8 *8*
γE2	730.97 *9*	9.1 *18*
γE1	762.64 *17*	0.34 *7*
γ(M1)	805.34 *13*	0.62 *12*
γ	811.80 *11*	~0.05 ?
γ[E2]	839.14 *10*	~0.05 ?
γ	865.47 *17*	0.099 *20*
γ	881.3 *3*	0.096 *19*
γ(E1)	884.37 *20*	0.14 *3*
γ(E2)	905.60 *13*	1.01 *20*
γM1+E2	913.51 *14*	0.13 *3*
γ	938.8 *3*	0.125 *25*
γ	947.2 *7*	0.070 *14*
γ	958.73 *19*	~0.039 ?
γ	968.1 *8*	0.039 *8*
γ	974.3 *10*	0.060 *12*
γ(M1)	979.36 *13*	0.26 *5*
γ(M1)	983.38 *19*	0.107 *21*
γ	990.08 *21*	0.026 *5* ?
γM1+E2	1010.19 *17*	0.13 *3*
γ	1025.2 *7*	0.14 *3*
γ	1030.3 *3*	0.047 *9*
γ	1050.6 *3*	0.16 *3*
γ(M1)	1061.49 *22*	0.18 *4*
γ	1098.6 *3*	0.042 *8*
γ	1110.11 *14*	0.70 *14*
γ	1132.05 *14*	0.055 *11*
γ	1157.40 *19*	0.17 *3*
γ	1163.20 *22*	0.086 *17*
γ(M1)	1171.12 *14*	0.57 *11*
γ	1182.17 *20*	0.19 *4*
γ(M1)	1191.65 *14*	0.16 *3*
γ	1195.4 *8*	0.16 *3*
γ	1214.0 *6*	0.052 *10* ?
γ	1217.6 *3*	0.078 *16* ?
γ	1222.31 *17*	0.47 *9*
γ	1231.77 *16*	0.026 *5*
γ	1259.10 *17*	0.026 *5*
γ	1281.09 *16*	0.127 *25*
γ(M1)	1312.37 *12*	0.57 *11*
γ	1348.68 *25*	0.16 *3*
γ	1352.45 *17*	0.18 *4*
γ	1362.2 *3*	0.094 *19*
γ(M1)	1383.74 *13*	0.32 *6*
γ	1397.18 *15*	0.052 *10*
γ	1448.8 *4*	0.109 *22*
γ(M1)	1483.57 *15*	0.41 *8*
γ	1495.34 *16*	0.19 *4*
γ	1510.38 *15*	0.026 *5* ?
γ	1554.8 *10*	0.026 *5*
γ	1579.2 *4*	0.026 *5*
γ	1599.44 *13*	0.19 *4*
γ	1632.4 *4*	0.057 *11*
γ	1638.52 *15*	0.062 *12*
γ(M1)	1670.81 *13*	0.60 *12*
γ	1689.70 *18*	0.065 *13*
γ	1720.3 *6*	0.042 *8* ?
γ	1746.68 *16*	0.013 *3* ?
γ	1778.97 *14*	0.062 *12*
γ	1815.5 *4*	0.101 *20*
γ	1870.1 *5*	0.091 *18*
γ	1962.74 *16*	0.013 *3* ?

† 19% uncert(syst)

Atomic Electrons (^{151}Tb)

⟨e⟩=73 *5* keV

e_{bin}(keV)	⟨e⟩(keV)	e(%)
7	4.1	57 *7*
8	3.0	37 *5*
33 - 48	3.03	8.2 *4*
58	17.7	31 *6*
99	0.17	0.17 *4*
100	7.2	7.2 *16*
101	3.4	3.4 *8*
106	1.7	1.6 *4*
107 - 111	1.8	1.67 *24*
130	4.9	3.8 *8*
141 - 190	3.3	2.08 *25*
191 - 192	0.031	0.016 *5*
202	7.0	3.5 *7*
204 - 236	0.051	0.024 *4*
237	5.6	2.4 *5*
239 - 287	2.8	1.08 *12*
310 - 345	0.99	0.29 *4*
366 - 415	2.2	0.57 *7*
416 - 466	1.40	0.33 *4*
467 - 512	0.27	0.057 *8*
528 - 566	0.89	0.16 *3*
579 - 617	0.24	0.040 *5*
642 - 691	0.56	0.084 *14*
692 - 731	0.101	0.0142 *19*
754 - 804	0.041	0.0054 *9*
805 - 839	0.008	0.0009 *3*
855 - 904	0.040	0.0047 *7*
905 - 951	0.020	0.0022 *3*
957 - 1003	0.015	0.0016 *4*
1008 - 1054	0.0099	0.00097 *15*
1060 - 1109	0.022	0.0021 *9*
1110 - 1157	0.032	0.0028 *4*
1161 - 1210	0.018	0.0015 *5*
1212 - 1259	0.0041	0.00033 *13*
1262 - 1311	0.024	0.0019 *3*
1312 - 1361	0.0118	0.00088 *15*
1362 - 1399	0.0033	0.00023 *7*
1433 - 1482	0.0145	0.00100 *18*
1483 - 1582	0.0049	0.00032 *10*
1588 - 1682	0.0168	0.00103 *16*
1688 - 1777	0.0024	0.00014 *4*
1807 - 1869	0.0014	8 *3* $\times10^{-5}$
1912 - 1961	0.00017	9 *4* $\times10^{-6}$

Continuous Radiation (^{151}Tb)

⟨β+⟩=5.6 keV;⟨IB⟩=1.13 keV

E_{bin}(keV)		⟨ ⟩(keV)	(%)
0 - 10	β+	2.09×10^{-7}	2.48×10^{-6}
	IB	0.00088	
10 - 20	β+	1.03×10^{-5}	6.2×10^{-5}
	IB	0.00043	0.0029
20 - 40	β+	0.000342	0.00104
	IB	0.046	0.124
40 - 100	β+	0.0190	0.0242
	IB	0.115	0.25
100 - 300	β+	0.77	0.358
	IB	0.042	0.021
300 - 600	β+	2.60	0.60
	IB	0.18	0.039
600 - 1300	β+	2.18	0.270
	IB	0.59	0.067
1300 - 2458	β+	0.0288	0.00215
	IB	0.158	0.0104
	Σβ+		1.26

$^{151}_{65}$Tb(25 *3* s)

Mode: IT(93.4 %), ϵ(6.6 %)
Δ: -71532 *5* keV
SpA: 2.9×10^{9} Ci/g
Prod: ^{151}Eu(α,4n); ^{151}Eu(^{3}He,3n)

Photons (^{151}Tb)

⟨γ⟩=80.3 *25* keV

γ_{mode}	γ(keV)	γ(%)
Gd L_{ℓ}	5.362	0.0167 *17*
Tb L_{ℓ}	5.546	0.42 *5*
Gd L_{η}	6.049	0.0059 *5*
Gd L_{α}	6.054	0.44 *3*
Tb L_{α}	6.269	11.1 *11*
Tb L_{η}	6.284	0.197 *16*
Gd L_{β}	6.851	0.41 *4*
Tb L_{β}	7.073	17.7 *20*
Gd L_{γ}	7.884	0.062 *7*
Tb L_{γ}	8.283	3.5 *5*
γ_{IT} M1+0.05%E2	23.04 *2*	3.18 *13*
Gd $K_{\alpha 2}$	42.309	1.54 *6*
Gd $K_{\alpha 1}$	42.996	2.78 *10*
Tb $K_{\alpha 2}$	43.744	0.075 *5*
Tb $K_{\alpha 1}$	44.482	0.134 *8*
Gd $K_{\beta 1}$'	48.652	0.82 *3*
γ_{IT} M1	49.58 *2*	24.4 *10*
Gd $K_{\beta 2}$'	50.214	0.247 *10*
Tb $K_{\beta 1}$'	50.338	0.0400 *24*
Tb $K_{\beta 2}$'	51.958	0.0119 *8*
γ_{IT} (E2)	72.62 *3*	0.121 *5*
γ_{ϵ}[M1+E2]	326.1 *4*	0.35 *11*
γ_{ϵ}[M1+E2]	379.70 *6*	6.3 *3*
γ_{ϵ}[M1+E2]	504.5 *2*	0.51 *16*
γ_{ϵ}[M1+E2]	522.4 *1*	1.51 *13*
γ_{ϵ}[M1+E2]	830.5 *1*	3.28 *22*

$^{151}_{66}$Dy(16.9 *5* min)

Mode: ϵ(94.5 *8* %), α(5.5 *8* %)
Δ: -68902 *23* keV
SpA: 7.37×10^{7} Ci/g
Prod: ^{141}Pr(^{14}N,4n); ^{140}Ce(^{16}O,5n); ^{159}Tb(p,9n); ^{142}Nd(^{12}C,3n)

Alpha Particles (^{151}Dy)

α(keV)
4067 *3*

Photons (^{151}Dy)

⟨γ⟩=230 *15* keV

γ_{mode}	γ(keV)	γ(%)†
γ_{ϵ}	22.95 *5*	2.4 *2*
γ_{ϵ}	176.1 *2*	8.2 *9*
γ_{ϵ}	386.3 *3*	10.2 *9*
γ_{ϵ}	432.9 *4*	2.0 *5*
γ_{ϵ}	464.0 *4*	1.5 *5*
γ_{ϵ}	477.1 *3*	4.7 *10*
γ_{ϵ}	547.1 *3*	9.0 *9*
γ_{ϵ}	984.9 *5*	~0.9
γ_{ϵ}	1010.8 *5*	1.4 *5*
γ_{ϵ}	1096.5 *5*	1.0 *5*
γ_{ϵ}	1115.1 *5*	1.8 *5*
γ_{ϵ}	1130.9 *5*	1.4 *5*
γ_{ϵ}	1143.1 *5*	1.5 *5*

† 0.85% uncert(syst)

$^{151}_{67}$Ho(47 *2* s)

Mode: ϵ(90 *3* %), α(10 *3* %)
Δ: -63803 *29* keV
SpA: 1.58×10^{9} Ci/g
Prod: ^{16}O on ^{142}Nd; ^{144}Sm(^{11}B,4n); ^{58}Ni on ^{107}Ag; ^{58}Ni on ^{103}Rh

Alpha Particles (^{151}Ho)

α(keV)
4607 *3*

Photons (^{151}Ho(35.6 or 47 s))

γ_{mode}	γ(keV)
γ	210.2 *2*
γ	352.2 *4*
γ	488.9 *4*
γ	527.4 *1*
γ	551.0 *1*
γ	653.8 *1*
γ	694.8 *2*
γ	776.2 *1*
γ	804.4 *1*
γ	1047.1 *1*

$^{151}_{67}$Ho(35.6 *4* s)

Mode: ϵ(80 *5* %), α(20 *5* %)
Δ: -63803 *29* keV
SpA: 2.078×10^{9} Ci/g
Prod: ^{141}Pr(^{16}O,6n); ^{144}Sm(^{11}B,4n); ^{58}Ni on ^{107}Ag; ^{58}Ni on ^{103}Rh

Alpha Particles (^{151}Ho)

α(keV)
4517 *3*

see ^{151}Ho(47 s) for γ rays

$^{151}_{68}$Er(23 *2* s)

Mode: ϵ
Δ: -58500 *300* keV syst
SpA: 3.2×10^{9} Ci/g
Prod: ^{156}Dy(^{3}He,8n)

$^{151}_{69}$Tm(3.8 *8* s)

Mode: ϵ
Δ: -51000 *430* keV syst
SpA: 1.8×10^{10} Ci/g
Prod: ^{96}Ru(^{58}Ni,3p)

Photons (^{151}Tm)

γ_{mode}	γ(keV)
γ	801.6 *5*

$^{151}_{71}$Lu(85 *10* ms)

Mode: p
SpA: 1.08×10^{11} Ci/g
Prod: ^{96}Ru(^{58}Ni,p2n)
p: 1231 *3*

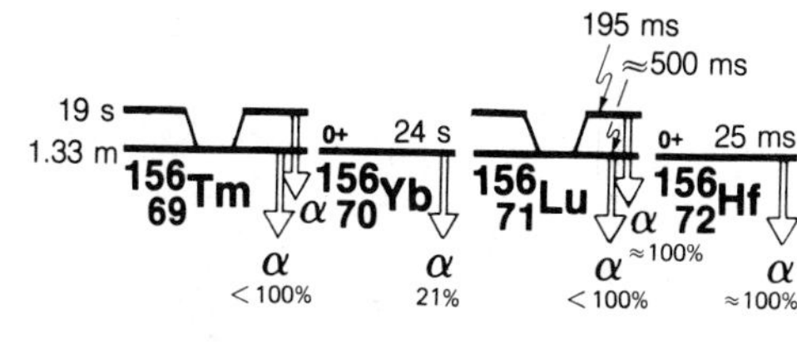

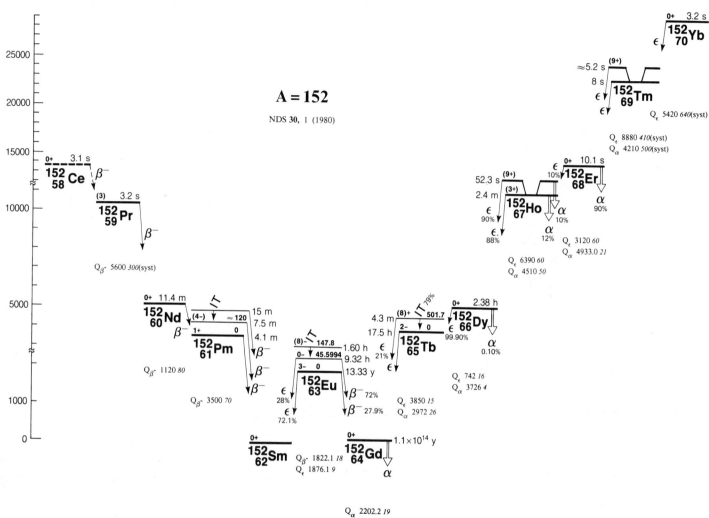

A = 152

NDS **30**, 1 (1980)

$^{152}_{58}$Ce(3.1 *3* s)

Mode: β-
SpA: 2.15×10^{10} Ci/g
Prod: fission

Photons (^{152}Ce)

γ_{mode}	γ(keV)
γ	285.0 *3*

$^{152}_{59}$Pr(3.24 *19* s)

Mode: β-
Δ: -64560 *300* keV syst
SpA: 2.06×10^{10} Ci/g
Prod: fission

Photons (^{152}Pr)

γ_{mode}	γ(keV)	γ(rel)
γ [E2]	72.6 *2*	63 *14*
γ [E2]	164.03 *11*	100 *8*

$^{152}_{60}$Nd(11.4 *2* min)

Mode: β-
Δ: -70160 *30* keV
SpA: 1.085×10^{8} Ci/g
Prod: ^{150}Nd(t,p); fission

Photons (^{152}Nd)

γ_{mode}	γ(keV)	γ(rel)
γ E1	16.03 23	25 6
γ (M1+<8.3%E2)	28.46 16	~4
γ [E1]	74.48 20	~4 ?
γ (M1)	175.5 3	<7 ?
γ [E1]	249.96 16	68 3
γ [E1]	278.43 16	100
γ [M1+E2]	294.46 19	12 1

$^{152}_{61}$Pm(4.1 *1* min)

Mode: β-

Δ: -71270 *80* keV

SpA: 3.01×10^{8} Ci/g

Prod: daughter ^{152}Nd; ^{152}Sm(n,p); ^{150}Nd(t,n)

Photons (^{152}Pm)

$\langle\gamma\rangle$=151 *3* keV

γ_{mode}	γ(keV)	γ(%)†
Sm L_ℓ	4.993	0.039 5
Sm L_η	5.589	0.0169 18
Sm L_α	5.633	1.03 10
Sm L_β	6.333	0.97 11
Sm L_γ	7.230	0.135 16
Sm $K_{\alpha2}$	39.522	2.86 23
Sm $K_{\alpha1}$	40.118	5.2 4
Sm $K_{\beta1}$'	45.379	1.50 12
Sm $K_{\beta2}$'	46.819	0.45 4
γ E2	121.7758 4	15.7 12
γ [E2]	125.759 17	0.00093 18
γ [M1+E2]	147.965 13	0.00010 3
γ [E1]	152.911 7	0.00022 11
γ [E1]	239.301 25	0.009 4
γ E2	244.6923 10	0.51 7
γ [E1]	251.618 8	0.065 6
γ [E2]	269.834 11	0.0085 11
γ [E2]	272.40 4	0.097 16
γ [M1+E2]	275.437 12	0.000138 20
γ E1(+4.1%M2)	295.929 6	0.0041 6
γ	314.6 3	0.091 17
γ [E1]	329.423 9	0.128 11
γ [E1]	357.113 11	0.0008 3
γ [E1]	416.045 6	0.017 6
γ [M1+E2]	423.402 7	$6.5\ 19\times10^{-6}$
γ E1(+0.1%M2)	443.894 13	0.026 3
γ [E2]	443.983 4	0.025 3
γ (E0+M1+E2)	482.334 9	0.028 3
γ E2+3.2%M1	488.615 9	0.0038 5
γ [M1+E2]	535.229 25	0.017 5
γ [M1+E2]	547.405 11	0.0024 5
γ E2	562.916 16	0.23 3
γ E1	564.010 14	0.07 3
γ M1+E2	566.420 6	0.00120 16
γ	662.5 3	0.100 25
γ E1	674.698 7	0.136 11
γ [M1+E2]	686.535 9	0.0030 11
γ E0+M1+E2	688.675 4	0.066 8
γ	695.92 9	1.35 8
γ [E1]	700.315 11	0.0027 8
γ [E1]	719.330 8	0.00054 10
γ (M1+E2)	719.419 12	0.00110 16
γ	735.13 17	0.30 3
γ [E1]	805.720 24	0.027 6
γ E2	810.450 4	0.025 3
γ	812.94 17	0.28 3
γ [E1]	826.074 18	~0.00018
γ [E1]	839.446 9	0.0026 9
γ E1	841.585 6	2.19 8
γ E2+2.7%M1	867.384 6	0.0101 17
γ	870.097 17	0.050 12
γ	903.88 21	0.091 25

Photons (^{152}Pm) (continued)

γ_{mode}	γ(keV)	γ(%)†
γ E1	919.390 7	0.35 3
γ E2	926.316 8	0.276 23
γ	929.1 3	0.083 25
γ [M1+E2]	958.630 24	0.048 10
γ E1	961.07 4	1.92 9
γ E1	963.360 6	1.81 7
γ E2+1.0%M1	964.110 12	0.060 9
γ	974.59 10	<0.017
γ	981.68 21	0.124 25
γ	995.855 10	0.038 10
γ	1049.7 3	~0.050
γ	1079.43 13	0.32 3
γ E2	1085.885 12	0.041 6
γ E2+0.4%M1	1112.075 6	0.033 7
γ	1127.8 3	0.066 25
γ [M1+E2]	1171.006 8	0.037 4
γ	1252.90 15	~0.033
γ [E1]	1273.76 16	<0.017
γ [E2]	1292.781 8	0.106 10
γ	1315.9 3	0.116 25
γ	1321.28 20	0.61 4
γ	1330.71 15	<0.017
γ E1+M2	1388.987 10	0.191 17
γ	1397.8 3	0.025 8
γ E1+0.2%M2	1408.002 7	0.193 25
γ [E1]	1510.761 10	0.00163 22
γ E1(+0.5%M2)	1528.116 8	0.042 15
γ	1537.50 9	<0.017
γ	1558.768 17	0.0046 12
γ	1627.4 4	0.025 8
γ M1(+E2)	1647.300 24	0.013 3
γ	1680.542 17	~0.04
γ E2	1769.074 24	0.018 4
γ	1823.26 21	<0.017
γ	1848.3 3	0.025 8
γ [E2]	1870.65 16	~0.017
γ	1905.1 3	0.025 8
γ	1921.00 13	~0.017
γ	1970.1 3	0.025 8
γ	2042.78 13	0.058 8
γ [M1+E2]	2115.34 16	0.158 17
γ	2162.86 20	<0.017
γ	2172.28 15	0.050 8
γ	2184.81 17	0.100 17
γ	2222.44 20	0.058 17
γ [E2]	2237.11 16	0.025 8
γ	2294.06 15	~0.017
γ	2306.59 17	0.050 8
γ	2329.2 5	~0.017
γ	2384.34 24	0.025 8
γ	2561.4 6	0.025 8
γ	2799.84 20	0.042 17

† 12% uncert(syst)

Atomic Electrons (^{152}Pm)

$\langle e\rangle$=18.3 *7* keV

e_{bin}(keV)	$\langle e\rangle$(keV)	e(%)
7	0.75	10.8 14
8 - 45	0.311	1.37 9
75	8.0	10.6 8
79 - 106	0.00051	0.0006 3
114	4.0	3.5 3
115	2.82	2.45 19
118 - 119	0.00036	0.00030 11
120	1.64	1.36 10
121	0.071	0.058 5
122	0.38	0.310 24
124 - 153	0.00014	0.00011 4
174 - 223	0.087	0.044 5
226 - 274	0.061	0.025 4
275 - 323	0.0067	0.0023 5
328 - 377	0.00052	0.00015 4
397 - 444	0.009	0.0022 8
475 - 520	0.0130	0.0026 3
527 - 566	0.00258	0.00047 4
616 - 663	0.06	0.010 4

Atomic Electrons (^{152}Pm) (continued)

e_{bin}(keV)	$\langle e\rangle$(keV)	e(%)
667 - 713	0.021	0.0030 14
718 - 766	0.010	0.0013 6
779 - 826	0.0247	0.00310 17
832 - 879	0.039	~0.004
882 - 929	0.045	0.0049 5
935 - 982	0.0101	0.00106 19
988 - 1033	0.007	~0.0007
1039 - 1086	0.0040	0.00037 9
1104 - 1128	0.0011	0.00010 4
1163 - 1206	0.0007	$\sim6\times10^{-5}$
1227 - 1274	0.013	0.0010 4
1284 - 1331	0.0023	0.00018 6
1342 - 1391	0.008	~0.0006
1396 - 1408	0.000226	$1.61\ 16\times10^{-5}$
1464 - 1557	0.00051	$3.4\ 10\times10^{-5}$
1581 - 1679	0.0011	$7\ 3\times10^{-5}$
1722 - 1822	0.00062	$3.5\ 10\times10^{-5}$
1824 - 1923	0.00099	$5.3\ 13\times10^{-5}$
1962 - 2041	0.0007	$3.4\ 13\times10^{-5}$
2069 - 2166	0.0036	0.00017 3
2171 - 2260	0.0016	$7.2\ 16\times10^{-5}$
2282 - 2378	0.00048	$2.1\ 6\times10^{-5}$
2515 - 2560	0.00022	$9\ 4\times10^{-6}$
2753 - 2798	0.00034	$1.2\ 6\times10^{-5}$

Continuous Radiation (^{152}Pm)

$\langle\beta-\rangle$=1377 keV;$\langle$IB$\rangle$=3.9 keV

E_{bin}(keV)		$\langle\ \rangle$(keV)	(%)
0 - 10	β-	0.0082	0.163
	IB	0.046	
10 - 20	β-	0.0250	0.167
	IB	0.045	0.31
20 - 40	β-	0.104	0.346
	IB	0.089	0.31
40 - 100	β-	0.80	1.14
	IB	0.26	0.39
100 - 300	β-	10.0	4.87
	IB	0.74	0.41
300 - 600	β-	46.6	10.2
	IB	0.85	0.20
600 - 1300	β-	300	31.3
	IB	1.20	0.139
1300 - 2500	β-	807	44.2
	IB	0.61	0.037
2500 - 3500	β-	212	7.7
	IB	0.025	0.00091

$^{152}_{61}$Pm(7.52 *8* min)

Mode: β-

Δ: ~-71150 keV

SpA: 1.644×10^{8} Ci/g

Prod: ^{152}Sm(n,p); ^{150}Nd(t,n)

Photons (^{152}Pm)

$\langle\gamma\rangle$=1508 *28* keV

γ_{mode}	γ(keV)	γ(%)†
Sm L_ℓ	4.993	0.132 16
Sm L_η	5.589	0.057 6
Sm L_α	5.633	3.5 3
Sm L_β	6.333	3.3 3
Sm L_γ	7.234	0.46 5
Sm $K_{\alpha2}$	39.522	10.2 7
Sm $K_{\alpha1}$	40.118	18.5 12

Photons (^{152}Pm) (continued)

γ_{mode}	γ(keV)	γ(%)†
Sm $K_{\beta1}'$	45.379	5.4 *4*
Sm $K_{\beta2}'$	46.819	1.59 *11*
γ E2	121.7758 *4*	45 *3*
γ [E2]	125.759 *17*	0.032 *5*
γ [M1+E2]	147.965 *13*	0.012 *3*
γ [E1]	207.679 *12*	0.0031 *9*
γ E2	244.6923 *10*	78 *4*
γ [M1+E2]	275.437 *12*	0.0041 *3*
γ [E2]	285.838 *16*	0.049 *5*
γ [E2]	316.03 *4*	~0.05
γ [E1]	330.559 *12*	0.033 *5*
γ E2	340.46 *4*	31.3 *20*
γ	385.298 *23*	0.19 *3*
γ [M1+E2]	423.402 *7*	0.00080 *19*
γ [E1]	432.31 *6*	1.64 *11*
γ [E2]	443.983 *4*	0.87 *5*
γ	493.40 *16*	0.37 *9*
γ [E1]	493.517 *14*	0.0223 *22*
γ	496.37 *4*	0.082 *23*
γ	500.0 *3*	~0.14
γ	516.42 *23*	0.86 *9*
γ	523.171 *21*	0.56 *6*
γ [M1+E2]	538.238 *10*	0.0030 *5*
γ [E1]	556.454 *11*	0.0135 *13*
γ [E2]	561.275 *11*	0.0047 *10*
γ	584.97 *17*	0.37 *11*
γ [M1+E2]	603.31 *17*	0.23 *9*
γ [E2]	616.043 *9*	0.0064 *8*
γ	644.34 *4*	0.114 *25*
γ	645.20 *16*	0.95 *9*
γ E0+M1+E2	656.483 *8*	3.44 *18*
γ [E2]	664.80 *4*	0.084 *10*
γ	671.136 *18*	0.90 *10*
γ E1	674.698 *7*	0.50 *4*
γ E0+M1+E2	688.675 *4*	2.30 *11*
γ (M1+E2)	719.419 *12*	0.0324 *25*
γ	725.09 *20*	0.77 *11*
γ	746.39 *20*	0.23 *9*
γ [E2]	762.87 *6*	0.20 *9*
γ [E1]	768.953 *8*	0.064 *8*
γ [E1]	781.08 *6*	4.2 *3*
γ E2	810.450 *4*	0.86 *4*
γ	810.59 *14*	4.2 *3*
γ	854.2 *3*	0.57 *11*
γ E2+2.7%M1	867.384 *6*	1.25 *7*
γ [E2]	901.174 *8*	2.13 *16*
γ E1	919.390 *7*	1.30 *10*
γ	932.1 *3*	0.29 *9*
γ [E2]	943.77 *17*	~0.20
γ E2+1.0%M1	964.110 *12*	1.75 *12*
γ E2+10%M1	1005.256 *10*	2.89 *15*
γ	1021.50 *14*	1.18 *11*
γ	1056.7 *3*	0.57 *14*
γ E2	1085.885 *12*	1.21 *8*
γ [E1]	1097.11 *7*	28.7 *14*
γ E2+0.4%M1	1112.075 *6*	4.02 *21*
γ	1193.37 *17*	0.95 *9*
γ E1(+M2)	1212.934 *7*	1.00 *8*
γ	1233.99 *14*	2.87 *20*
γ M1,E2	1249.947 *10*	0.82 *5*
γ	1297.50 *23*	<0.11
γ E1	1363.75 *3*	0.43 *8*
γ	1390.551 *21*	0.197 *25*
γ [E1]	1437.56 *6*	22.7 *11*
γ E1	1457.624 *7*	0.35 *3*
γ	1515.6 *3*	~0.11
γ [M1+E2]	1523.89 *15*	~0.06
γ	1572.1 *3*	0.11 *3*
γ	1591.77 *16*	0.11 *3*
γ	1608.44 *3*	0.089 *18*
γ	1613.53 *23*	0.11 *3*
γ	1635.241 *21*	0.0056 *20*
γ	1677.97 *14*	0.40 *11*
γ [E2]	1809.73 *15*	~0.14
γ	1860.20 *17*	0.23 *6*
γ	1881.86 *20*	0.09 *3*
γ	1953.98 *23*	0.26 *6*
γ	2004.89 *20*	0.23 *6*
γ	2064.8 *3*	~0.06
γ [E2]	2188.69 *15*	0.20 *6*
γ	2200.65 *17*	0.69 *11*
γ	2280.43 *16*	0.09 *3*
γ	2283.09 *20*	0.20 *6*
γ	2491.4 *3*	0.09 *3*
γ	2512.0 *3*	0.23 *6*
γ [M1+E2]	2529.14 *15*	0.29 *6*

† 7.3% uncert(syst)

Atomic Electrons (^{152}Pm)

$\langle e \rangle$=76.7 *23* keV

e_{bin}(keV)	$\langle e \rangle$(keV)	e(%)
7	2.5	37 *4*
8 - 45	1.11	4.9 *3*
75	22.8	30.5 *24*
79 - 101	0.021	0.025 *10*
114	11.4	10.0 *8*
115	8.1	7.0 *6*
118 - 119	0.013	0.011 *4*
120	4.7	3.9 *3*
121 - 161	1.29	1.06 *7*
198	12.5	6.3 *4*
200 - 229	0.0007	~0.00031
237	2.91	1.23 *7*
238 - 286	2.16	0.90 *3*
294	2.79	0.95 *6*
308 - 340	0.86	0.258 *12*
377 - 426	0.091	0.0232 *13*
431 - 476	0.11	0.025 *8*
486 - 532	0.022	0.0043 *13*
537 - 585	0.031	0.0056 *23*
596 - 645	1.57	0.253 *20*
649 - 689	0.29	0.044 *6*
700 - 746	0.064	0.0088 *10*
755 - 804	0.17	0.022 *10*
807 - 854	0.101	0.0121 *13*
860 - 901	0.038	0.0043 *6*
912 - 958	0.108	0.0115 *5*
962 - 1010	0.048	0.0049 *16*
1014 - 1057	0.256	0.0244 *14*
1065 - 1112	0.129	0.0120 *4*
1147 - 1193	0.10	0.008 *3*
1203 - 1251	0.035	0.0029 *5*
1290 - 1317	0.0031	0.00023 *4*
1344 - 1391	0.151	0.0109 *7*
1411 - 1457	0.0270	0.00189 *9*
1469 - 1567	0.0083	0.00054 *12*
1570 - 1634	0.005	0.00033 *16*
1670 - 1763	0.0025	0.00014 *6*
1802 - 1880	0.0041	0.00022 *8*
1907 - 2003	0.0055	0.00028 *9*
2018 - 2063	0.0006	~3×10^{-5}
2142 - 2236	0.012	0.00054 *14*
2273 - 2282	0.00039	1.7 *6* $\times10^{-5}$
2445 - 2528	0.0061	0.00025 *4*

Continuous Radiation (^{152}Pm)

$\langle\beta-\rangle$=808 keV;$\langle$IB$\rangle$=1.68 keV

E_{bin}(keV)		$\langle\ \rangle$(keV)	(%)
0 - 10	β-	0.0247	0.492
	IB	0.031	
10 - 20	β-	0.075	0.500
	IB	0.031	0.21
20 - 40	β-	0.309	1.03
	IB	0.059	0.21
40 - 100	β-	2.32	3.30
	IB	0.165	0.25
100 - 300	β-	26.5	13.0
	IB	0.44	0.25
300 - 600	β-	102	22.6
	IB	0.42	0.101
600 - 1300	β-	387	42.5
	IB	0.41	0.050
1300 - 2500	β-	263	15.6
	IB	0.117	0.0073
2500 - 3253	β-	26.7	0.99
	IB	0.0020	7.8×10^{-5}

$^{152}_{61}$Pm(15 *1* min)

Mode: β-, IT

Δ: >-71150 keV

SpA: 8.2×10^{7} Ci/g

Prod: ^{152}Sm(n,p); ^{150}Nd(t,n)

Photons (^{152}Pm)

γ_{mode}	γ(keV)
γ	137.4 *10*
γ	200.3 *10*
γ	229.9 *10*
γ	360.4 *10*
γ	1214 *1*
γ	1233.8 *7*
γ	1437.5 *3*

$^{152}_{62}$Sm(stable)

Δ: -74773 *3* keV

%: 26.7 *2*

$^{152}_{63}$Eu(13.33 *4* yr)

Mode: ϵ(72.08 *19* %), β-(27.92 *19* %)

Δ: -72897 *3* keV

SpA: 176.4 Ci/g

Prod: ^{151}Eu(n,γ)

Photons (^{152}Eu)

$\langle\gamma\rangle$=1162 *8* keV

γ_{mode}	γ(keV)	γ(%)
Sm L_{ℓ}	4.993	0.226 *23*
Gd L_{ℓ}	5.362	0.0029 *3*
Sm L_{η}	5.589	0.088 *7*
Sm L_{α}	5.633	6.0 *4*
Gd L_{η}	6.049	0.00111 *9*
Gd L_{α}	6.054	0.077 *5*
Sm L_{β}	6.339	5.6 *5*
Gd L_{β}	6.847	0.071 *6*
Sm L_{γ}	7.264	0.82 *8*
Gd L_{γ}	7.863	0.0106 *11*
Sm $K_{\alpha2}$	39.522	21.2 *6*
Sm $K_{\alpha1}$	40.118	38.4 *12*
Gd $K_{\alpha2}$	42.309	0.254 *8*
Gd $K_{\alpha1}$	42.996	0.457 *15*
Sm $K_{\beta1}'$	45.379	11.1 *3*
Sm $K_{\beta2}'$	46.819	3.30 *11*
Gd $K_{\beta1}'$	48.652	0.135 *4*
Gd $K_{\beta2}'$	50.214	0.0407 *15*
$\gamma_{\beta-}$E2	117.25 *4*	0.00042 *5*
γ_{ϵ}E2	121.7758 *4*	28.4 *6*

Photons (^{152}Eu)
(continued)

γ_{mode}	γ(keV)	γ(%)
γ_{ϵ}[E2]	125.759 17	0.0119 19 ?
γ_{ϵ}[M1+E2]	147.965 13	0.040 8
γ_{ϵ}[E1]	152.911 7	$1.6\,8\times10^{-5}$
$\gamma_{\beta-}$[E2]	173.102 18	~0.00042
$\gamma_{\beta-}$[E2]	175.185 4	0.00222 23
$\gamma_{\beta-}$[E1]	192.613 5	0.00688 21
$\gamma_{\beta-}$E1	195.17 3	0.0019 4
γ_{ϵ}	202.87 16	~0.0038
γ_{ϵ}[E1]	207.679 12	0.0044 13
$\gamma_{\beta-}$E0+M1+E2	209.18 3	0.00012 4
$\gamma_{\beta-}$[E1]	209.454 5	0.0044 13
γ_{ϵ}	212.562 15	0.0196 6
γ_{ϵ}	237.27 6	0.0094 8
γ_{ϵ}[E1]	239.301 25	0.0042 10
γ_{ϵ}E2	244.6923 10	7.51 15
γ_{ϵ}[E1]	251.618 8	0.0626 21
γ_{ϵ}[E2]	269.834 11	0.0081 8
$\gamma_{\beta-}$[E2]	270.54 5	0.0020 7
$\gamma_{\beta-}$E2	271.132 11	0.0730 21
γ_{ϵ}[M1+E2]	275.437 12	0.0336 10
γ_{ϵ}[E2]	285.838 16	0.0111 10
γ_{ϵ}E1(+4.1%M2)	295.929 6	0.440 10
$\gamma_{\beta-}$(E2)	315.175 11	0.0504 12
γ_{ϵ}[E2]	316.03 4	~0.0021
γ_{ϵ}	320.02 15	0.0017 6
$\gamma_{\beta-}$[M1]	324.799 4	0.0751 21
γ_{ϵ}[E1]	329.423 9	0.1230 25
γ_{ϵ}[E1]	330.559 12	0.0075 10
γ_{ϵ}E2	340.46 4	0.027 6
$\gamma_{\beta-}$E2	344.286 2	26.6 5
$\gamma_{\beta-}$E2	351.704 18	0.0124 16
γ_{ϵ}[E1]	357.113 11	0.0048 6
$\gamma_{\beta-}$E1+0.2%M2	367.798 3	0.858 17
γ_{ϵ}	379.35 18	0.00083 21
γ_{ϵ}	385.298 23	0.0050 6
$\gamma_{\beta-}$E0+M1+E2	387.79 3	0.0019 3
γ_{ϵ}	387.89 8	0.00292 21
γ_{ϵ}	391.31 14	0.00125 21
γ_{ϵ}	406.73 15	0.00083 21
$\gamma_{\beta-}$E2	411.122 4	2.23 5
γ_{ϵ}[E1]	416.045 6	0.1106 22
γ_{ϵ}[M1+E2]	423.402 7	0.0027 6
γ_{ϵ}	440.85 10	0.0108 19
$\gamma_{\beta-}$	441.003 20	~0.025
γ_{ϵ}E1(+0.1%M2)	443.894 13	2.80 6
γ_{ϵ}[E2]	443.983 4	0.322 13
γ_{ϵ}(E0+M1+E2)	482.334 9	0.0271 21
$\gamma_{\beta-}$[E1]	482.432 17	0.0016 7
γ_{ϵ}E2+3.2%M1	488.615 9	0.407 8
γ_{ϵ}[E1]	493.517 14	0.0313 21
$\gamma_{\beta-}$[E2]	493.777 11	0.0091 8
γ_{ϵ}	496.37 4	0.0046 8
$\gamma_{\beta-}$E0+M1+E2	496.443 17	0.0048 4
$\gamma_{\beta-}$E2(+<22%M1)	503.401 4	0.152 3
$\gamma_{\beta-}$[M1+E2]	520.242 4	0.0538 15
γ_{ϵ}	523.171 21	0.0147 8
$\gamma_{\beta-}$E0+M1+E2	526.889 17	0.0130 7
$\gamma_{\beta-}$[E1]	534.252 5	0.0429 10
γ_{ϵ}[M1+E2]	535.229 25	0.0079 23
γ_{ϵ}[M1+E2]	538.238 10	0.0042 6
γ_{ϵ}[E1]	556.454 11	0.0190 10
$\gamma_{\beta-}$M1+E2	557.79 4	0.0038 4
γ_{ϵ}[E2]	561.275 11	0.00104 21
γ_{ϵ}E2	562.916 16	0.025 4 ?
$\gamma_{\beta-}$[E2]	562.97 3	0.00033 8
γ_{ϵ}E1	564.010 14	0.467 12
γ_{ϵ}M1+E2	566.420 6	0.129 3
γ_{ϵ}	571.82 8	0.0048 8
$\gamma_{\beta-}$E0+M1+E2	586.307 4	0.461 9
γ_{ϵ}[E2]	616.043 9	0.0090 8
γ_{ϵ}	644.34 4	0.0064 9
γ_{ϵ}E0+M1+E2	656.483 8	0.144 3
γ_{ϵ}[E2]	664.80 4	0.0188 21
γ_{ϵ}	671.136 18	0.0235 20
γ_{ϵ}E1	674.698 7	0.167 6
$\gamma_{\beta-}$E2(+<35%M1)	675.045 17	0.0196 12
$\gamma_{\beta-}$E2,>0.1%M1	678.586 3	0.469 9
γ_{ϵ}	683.31 11	0.0031 8
γ_{ϵ}[M1+E2]	686.535 9	0.0192 17
γ_{ϵ}E0+M1+E2	688.675 4	0.849 16
$\gamma_{\beta-}$[E2]	702.96 4	0.0036 6
$\gamma_{\beta-}$[E2]	703.56 4	0.00167 4
$\gamma_{\beta-}$[E1]	712.854 4	0.0959 21

Photons (^{152}Eu)
(continued)

γ_{mode}	γ(keV)	γ(%)
γ_{ϵ}[E1]	719.330 8	0.058 8
γ_{ϵ}(M1+E2)	719.419 12	0.267 10
γ_{ϵ}	728.02 4	0.0113 10
γ_{ϵ}	735.39 10	0.0058 10
γ_{ϵ}	756.11 9	0.0054 8
$\gamma_{\beta-}$E2+5.5%M1	764.909 5	0.180 8
γ_{ϵ}[E1]	768.953 8	0.090 8
$\gamma_{\beta-}$E1	778.920 4	12.98 25
$\gamma_{\beta-}$	794.79 2	0.0248 12
γ_{ϵ}[E1]	805.720 24	0.0127 10
γ_{ϵ}E2	810.450 4	0.318 6
γ_{ϵ}[E1]	839.446 9	0.0165 10
γ_{ϵ}E1	841.585 6	0.162 3
γ_{ϵ}E2+2.7%M1	867.384 6	4.21 8
γ_{ϵ}[E2]	901.174 8	0.089 6
γ_{ϵ}	906.01 6	0.0150 13
γ_{ϵ}E1	919.390 7	0.435 10
γ_{ϵ}E2	926.316 8	0.265 8
$\gamma_{\beta-}$E2	930.592 5	0.075 3
$\gamma_{\beta-}$	937.00 6	0.0033 4
γ_{ϵ}[M1+E2]	958.630 24	0.0231 21
γ_{ϵ}E1	963.360 6	0.1340 24
γ_{ϵ}E2+1.0%M1	964.110 12	14.5 3
γ_{ϵ}	968.0 4	~0.0027
$\gamma_{\beta-}$(E0)+M1+E2	974.09 3	0.0144 10
$\gamma_{\beta-}$E2	990.219 17	0.0305 12
γ_{ϵ}	1001.1 3	0.0040 19
γ_{ϵ}E2+10%M1	1005.256 10	0.647 14
γ_{ϵ}	1084.0 10	0.244 8
γ_{ϵ}E2	1085.885 12	9.94 20
$\gamma_{\beta-}$E2,0.2%M1	1089.706 4	1.71 3
$\gamma_{\beta-}$E2	1109.193 5	0.183 4
γ_{ϵ}E2+0.4%M1	1112.075 6	13.6 3
γ_{ϵ}	1139.0 10	0.0013 4
γ_{ϵ}[M1+E2]	1171.006 8	0.0357 13
$\gamma_{\beta-}$	1205.91 2	0.0149 10
γ_{ϵ}E1(+M2)	1212.934 7	1.40 3
γ_{ϵ}M1,E2	1249.947 10	0.183 9
$\gamma_{\beta-}$E0+M1+E2	1261.349 17	0.0332 12
γ_{ϵ}[E2]	1292.781 8	0.102 6
$\gamma_{\beta-}$E1+0.06%M2	1299.158 5	1.63 3
$\gamma_{\beta-}$M1	1318.38 3	~0.0018
$\gamma_{\beta-}$E2	1348.12 6	0.0168 12
γ_{ϵ}E1	1363.75 3	0.0243 10
γ_{ϵ}	1390.551 21	0.0052 8
γ_{ϵ}E1+0.2%M2	1408.002 7	20.8 4
γ_{ϵ}E1	1457.624 7	0.496 10
γ_{ϵ}E1(+0.5%M2)	1528.116 8	0.265 6
$\gamma_{\beta-}$E2	1605.632 17	0.0074 6
γ_{ϵ}	1608.44 3	0.0050 4
γ_{ϵ}	1635.241 21	0.00015 4 ?
γ_{ϵ}M1(+E2)	1647.300 24	0.0060 8
γ_{ϵ}E2	1769.074 24	0.0087 6

Atomic Electrons (^{152}Eu)
⟨e⟩=44.5 7 keV

e_{bin}(keV)	⟨e⟩(keV)	e(%)
7	4.1	59 6
8 - 48	2.48	12.2 7
65 - 67	0.0023	0.0034 16
75	14.4	19.2 5
79 - 110	0.025	0.026 5
114	7.21	6.31 18
115	5.10	4.44 13
116 - 119	0.0047	0.0040 5
120	2.97	2.47 7
121 - 168	0.826	0.677 16
171 - 196	0.0020	0.0011 5
198	1.20	0.607 17
200 - 249	0.536	0.225 6
250 - 289	0.0356	0.0130 9
294	2.43	0.828 24
295 - 344	0.829	0.246 4
349 - 397	0.237	0.0639 16
399 - 447	0.0900	0.0214 4
450 - 496	0.0289	0.0061 3
502 - 551	0.066	0.0124 17

Atomic Electrons (^{152}Eu)
(continued)

e_{bin}(keV)	⟨e⟩(keV)	e(%)
554 - 598	0.057	0.0101 10
607 - 656	0.259	0.041 3
657 - 706	0.050	0.0073 9
709 - 758	0.161	0.0221 6
759 - 806	0.0386	0.00500 8
809 - 854	0.104	0.0126 3
859 - 906	0.054	0.0062 25
912 - 961	0.376	0.0407 10
962 - 1005	0.0154	0.00159 3
1037 - 1086	0.516	0.0489 10
1088 - 1137	0.0481	0.00435 8
1138 - 1171	0.05	~0.004
1198 - 1246	0.015	0.0013 4
1248 - 1297	0.0150	0.00120 3
1298 - 1347	0.00069	$5.3\,4\times10^{-5}$
1348 - 1390	0.139	0.0102 3
1400 - 1411	0.0262	0.00187 4
1450 - 1527	0.00249	0.000168 6
1555 - 1646	0.00030	$1.92\,24\times10^{-5}$
1722 - 1768	0.000123	$7.1\,5\times10^{-6}$

Continuous Radiation (^{152}Eu)
⟨β-⟩=83 keV;⟨β+⟩=0.087 keV;⟨IB⟩=0.25

E_{bin}(keV)		⟨ ⟩(keV)	(%)
0 - 10	β-	0.0423	0.85
	β+	8.1×10^{-9}	9.6×10^{-8}
	IB	0.0043	
10 - 20	β-	0.123	0.82
	β+	3.83×10^{-7}	2.29×10^{-6}
	IB	0.0038	0.027
20 - 40	β-	0.471	1.58
	β+	1.21×10^{-5}	3.69×10^{-5}
	IB	0.049	0.143
40 - 100	β-	2.86	4.15
	β+	0.00064	0.00081
	IB	0.073	0.150
100 - 300	β-	18.5	9.6
	β+	0.0235	0.0110
	IB	0.047	0.027
300 - 600	β-	30.6	7.2
	β+	0.058	0.0139
	IB	0.041	0.0100
600 - 1300	β-	30.1	3.59
	β+	0.00474	0.00075
	IB	0.025	0.0031
1300 - 1754	β-	0.66	0.0493
	IB	0.0018	0.000127
	Σβ+		0.027

$^{152}_{63}$Eu(9.32 1 h)

Mode: β-(72 4 %), ε(28 4 %)
Δ: -72851 3 keV
SpA: 2.2121×10^{6} Ci/g
Prod: ^{151}Eu(n,γ)

Photons (^{152}Eu)
⟨γ⟩=306 4 keV

γ_{mode}	γ(keV)	γ(%)†
Sm L_{ℓ}	4.993	0.078 8
Sm L_{η}	5.589	0.0301 24
Sm L_{α}	5.633	2.09 14
Sm L_{β}	6.340	1.92 17
Sm L_{γ}	7.265	0.28 3

Photons (^{152}Eu) (continued)

γ_{mode}	γ(keV)	γ(%)†
Sm $K_{\alpha2}$	39.522	7.55 *25*
Sm $K_{\alpha1}$	40.118	13.7 *5*
Sm $K_{\beta1}$'	45.379	3.96 *13*
Sm $K_{\beta2}$'	46.819	1.18 *4*
$\gamma_{\beta-}$E2	117.25 *4*	0.0172 *20*
γ_{ϵ}E2	121.7758 *4*	7.2 *4*
γ_{ϵ}[E2]	125.759 *17*	0.00094 *15*
γ_{ϵ}[E1]	152.911 *7*	0.0015 *7*
γ_{ϵ}	160.0 *5*	~0.0007 ?
$\gamma_{\beta-}$[E2]	175.185 *4*	~6 ×10⁻⁵
$\gamma_{\beta-}$[E2]	191.473 *10*	~0.0007
γ_{ϵ}	218.1 *3*	~4 ×10⁻⁵
γ_{ϵ}	220.84 *15*	<0.00034
γ_{ϵ}E2	244.6923 *10*	0.0255 *9*
γ_{ϵ}	256.98 *22*	0.0010 *4*
$\gamma_{\beta-}$[E1]	266.84 *4*	~0.0012
$\gamma_{\beta-}$E2	271.132 *11*	0.076 *3*
γ_{ϵ}[E2]	272.40 *4*	0.0104 *16*
$\gamma_{\beta-}$(E2)	315.175 *11*	~0.0014
γ_{ϵ}E2	340.46 *4*	~0.005
$\gamma_{\beta-}$E2	344.286 *2*	2.44 *5*
γ_{ϵ}	387.8 *3*	~0.0007
$\gamma_{\beta-}$	412.79 *13*	~0.0007
γ_{ϵ}[E2]	443.983 *4*	0.0255 *12*
γ_{ϵ}[M1+E2]	547.405 *11*	0.0095 *19*
γ_{ϵ}E2	562.916 *16*	0.226 *7*
$\gamma_{\beta-}$E0+M1+E2	586.307 *4*	~0.013
γ_{ϵ}	605.23 *15*	~0.004
$\gamma_{\beta-}$[E2]	632.78 *3*	~0.0012
$\gamma_{\beta-}$[E1]	646.79 *3*	~0.0007
γ_{ϵ}E0+M1+E2	688.675 *4*	0.0672 *24*
$\gamma_{\beta-}$[E1]	699.261 *14*	0.0717 *22*
γ_{ϵ}[E1]	700.315 *11*	0.011 *3*
$\gamma_{\beta-}$[E2]	703.56 *4*	0.0678 *23*
γ_{ϵ}	703.7 *3*	~0.0007 ?
γ_{ϵ}	796.1 *3*	0.0035 *15*
γ_{ϵ}E2	810.450 *4*	0.0252 *10*
$\gamma_{\beta-}$[E1]	825.39 *3*	~0.0007
γ_{ϵ}[E1]	826.074 *18*	~0.0007
γ_{ϵ}E1	841.585 *6*	14.6 *4*
$\gamma_{\beta-}$	845.22 *12*	0.0092 *13*
γ_{ϵ}	870.097 *17*	0.091 *3*
γ_{ϵ}	915.7 *4*	0.0102 *15*
$\gamma_{\beta-}$E2	930.592 *5*	~0.0021
γ_{ϵ}E1	961.07 *4*	0.204 *16*
γ_{ϵ}E1	963.360 *6*	12.0 *3*
$\gamma_{\beta-}$E1	970.391 *9*	0.605 *20*
γ_{ϵ}	995.855 *10*	0.0700 *23*
γ_{ϵ}	1039.2 *5*	0.0083 *18*
$\gamma_{\beta-}$	1116.35 *12*	~0.0010
γ_{ϵ}	1137.5 *3*	~0.014
γ_{ϵ}	1168.14 *15*	0.0061 *23*
γ_{ϵ}	1207.3 *6*	0.0029 *10*
γ_{ϵ}	1289.92 *15*	0.00092 *14*
$\gamma_{\beta-}$E1	1314.675 *9*	0.955 *19*
γ_{ϵ}E1+M2	1388.987 *10*	0.77 *4*
γ_{ϵ}	1406.5 *5*	~0.0007
$\gamma_{\beta-}$[E1]	1411.70 *3*	0.0452 *9*
γ_{ϵ}	1420.0 *10*	~0.0006
$\gamma_{\beta-}$	1460.63 *12*	0.0016 *4*
γ_{ϵ}[E1]	1510.761 *10*	0.0066 *7*
γ_{ϵ}	1558.768 *17*	0.0084 *7*
γ_{ϵ}	1680.542 *17*	~0.08
$\gamma_{\beta-}$[E1]	1755.98 *3*	0.00263 *15*

† uncert(syst): 14% for ϵ, 12% for β-

Atomic Electrons (^{152}Eu)

$\langle e \rangle$=10.9 *3* keV

e_{bin}(keV)	$\langle e \rangle$(keV)	e(%)
7	1.39	20.1 *21*
8 - 48	0.87	4.21 *24*
67	0.0087	0.0129 *15*
75	3.65	4.9 *3*
79 - 113	0.0106	0.0098 *8*
114	1.83	1.60 *9*
115	1.30	1.13 *6*
116 - 119	0.0032	0.00272 *25*
120	0.75	0.63 *3*
121 - 168	0.206	0.169 *8*
171 - 220	0.00443	0.00224 *11*
221 - 270	0.0180	0.00776 *24*
271 - 272	0.000320	0.000118 *9*
294	0.224	0.0760 *22*
307 - 344	0.0732	0.0217 *4*
363 - 411	0.057	0.015 *2*
412 - 444	0.0106	0.0025 *5*
501 - 547	0.0112	0.00215 *18*
555 - 604	0.045	0.0079 *8*
605 - 653	0.049	0.0077 *6*
657 - 704	0.0073	0.00107 *14*
749 - 796	0.145	0.0182 *6*
803 - 845	0.0273	0.00328 *18*
862 - 909	0.00074	8 *3* ×10⁻⁵
914 - 963	0.134	0.0145 *4*
969 - 998	0.009	~0.0009
1031 - 1066	0.0014	~0.00014
1091 - 1137	0.00039	~4 ×10⁻⁵
1160 - 1207	7.6 ×10⁻⁵	~7 ×10⁻⁶
1243 - 1290	0.00725	0.000574 *16*
1306 - 1342	0.023	~0.0017
1360 - 1406	0.0042	~0.00031
1410 - 1459	4.2 ×10⁻⁵	3.0 *11* ×10⁻⁶
1460 - 1557	0.00017	1.1 *4* ×10⁻⁵
1634 - 1706	0.0011	~6 ×10⁻⁵
1748 - 1754	2.64 ×10⁻⁶	1.51 *8* ×10⁻⁷

Continuous Radiation (^{152}Eu)

$\langle\beta-\rangle$=496 keV;$\langle\beta+\rangle$=0.0311 keV;$\langle$IB$\rangle$=0.96 keV

E_{bin}(keV)		$\langle\ \rangle$(keV)	(%)
0 - 10	β-	0.0211	0.420
	β+	1.34 ×10⁻⁹	1.60 ×10⁻⁸
	IB	0.020	
10 - 20	β-	0.064	0.424
	β+	6.4 ×10⁻⁸	3.81 ×10⁻⁷
	IB	0.020	0.139
20 - 40	β-	0.259	0.86
	β+	2.03 ×10⁻⁶	6.2 ×10⁻⁶
	IB	0.055	0.18
40 - 100	β-	1.91	2.71
	β+	0.000109	0.000139
	IB	0.128	0.21
100 - 300	β-	21.0	10.4
	β+	0.00468	0.00215
	IB	0.28	0.158
300 - 600	β-	79	17.5
	β+	0.0191	0.00434
	IB	0.27	0.063
600 - 1300	β-	307	33.7
	β+	0.0072	0.00106
	IB	0.19	0.024
1300 - 1922	β-	86	6.0
	IB	0.0081	0.00058
	$\Sigma\beta$+		0.0077

$^{152}_{63}$Eu(1.600 *17* h)

Mode: IT
Δ: -72749 *3* keV
SpA: 1.288×10⁷ Ci/g
Prod: ^{154}Sm(p,3n); ^{152}Sm(p,n); ^{151}Eu(n,γ)

Photons (^{152}Eu)

$\langle\gamma\rangle$=76.1 *7* keV

γ_{mode}	γ(keV)	γ(%)†
Eu L_{ℓ}	5.177	0.38 *5*
Eu L_{η}	5.817	0.177 *17*
Eu L_{α}	5.843	10.1 *10*
Eu L_{β}	6.568	12.0 *15*
Eu L_{γ}	7.596	2.0 *3*
γ [E1]	12.617 *14*	0.29 *4*
γ M1+0.2%E2	18.23 *4*	1.26 *21*
γ E3	39.75 *10*	0.0130 *5*
Eu $K_{\alpha2}$	40.902	7.37 *21*
Eu $K_{\alpha1}$	41.542	13.3 *4*
Eu $K_{\beta1}$'	46.999	3.86 *11*
Eu $K_{\beta2}$'	48.496	1.17 *4*
γ M1+1.0%E2	77.249 *15*	0.69 *5*
γ E1+<0.2%M2	89.866 *6*	69.9

† 0.29% uncert(syst)

Atomic Electrons (^{152}Eu)

$\langle e \rangle$=66.4 *14* keV

e_{bin}(keV)	$\langle e \rangle$(keV)	e(%)
5 - 6	0.173	3.3 *3*
7	3.7	53 *6*
8	4.2	55 *7*
10	4.0	39 *7*
11 - 29	4.6	31 *4*
32	11.0	34.3 *15*
33	13.4	41.0 *19*
34 - 35	0.208	0.61 *5*
38	7.1	18.7 *8*
39	2.21	5.6 *3*
40	0.427	1.07 *6*
41	10.39	25.1 *5*
42 - 77	0.347	0.51 *3*
82	2.78	3.40 *7*
83 - 90	1.65	1.913 *23*

$^{152}_{64}$Gd(1.08 *8* ×10¹⁴ yr)

Mode: α
Δ: -74719 *3* keV
Prod: natural source
%: 0.20 *1*

Alpha Particles (^{152}Gd)

α(keV)
2140 *30*

$^{152}_{65}$Tb(17.5 *1* h)

Mode: ϵ
Δ: -70869 *15* keV
SpA: 1.178×10⁶ Ci/g
Prod: ^{151}Eu(α,3n); protons on Gd; protons on Ta

Photons (^{152}Tb)

$\langle\gamma\rangle$=1126 40 keV

γ_{mode}	γ(keV)	γ(%)†
Gd L_ℓ	5.362	0.208 21
Gd L_η	6.049	0.074 6
Gd L_α	6.054	5.5 4
Gd L_β	6.852	5.0 4
Gd L_γ	7.881	0.77 8
Gd $K_{\alpha2}$	42.309	19.3 6
Gd $K_{\alpha1}$	42.996	34.7 11
Gd $K_{\beta1}$'	48.652	10.2 3
Gd $K_{\beta2}$'	50.214	3.09 11
γ M1+E2	113.4 3	0.0022 4
γ E2	117.25 4	0.32 8
γ M1+E2	143.79 21	0.0030 6
γ M1+E2	156.3 4	0.0040 8
γ [E2]	173.102 18	~0.007
γ [E2]	175.185 4	0.039 4
γ M1+E2	180.0 2	0.015 3
γ [E2]	191.473 10	~0.0010
γ [E1]	192.613 5	0.00263 17
γ E1	195.17 3	0.36 6
γ E0+M1+E2	209.18 3	0.023 6
γ M1+E2	248.77 9	0.097 23
γ [E1]	266.84 4	~0.0016
γ [E2]	270.54 5	0.38 11
γ E2	271.132 11	7.5 5
γ (E2)	315.175 11	0.89 5
γ [M1]	324.799 4	0.0310 19
γ E2	344.286 2	57
γ E2	351.704 18	0.216 21
γ E1+0.2%M2	367.798 3	0.328 18
γ E0+M1+E2	387.79 3	0.36 3
γ E2	411.122 4	3.57 23
γ	441.003 20	~0.16
γ	456.92 14	0.029 11
γ	466.0 3	0.029 11
γ	472.0 7	~0.011
γ [E1]	482.432 17	0.040 17
γ	490.6 5	0.057 12
γ [E2]	493.777 11	0.122 10
γ E0+M1+E2	496.443 17	0.117 9
γ E2(+<22%M1)	503.401 4	0.063 4
γ [M1+E2]	520.242 4	0.061 5
γ E0+M1+E2	526.889 17	0.226 14
γ [E1]	534.252 5	0.049 4
γ E0+M1+E2	543.67 9	0.171 17
γ [E1]	547.38 8	0.063 11
γ M1+E2	557.79 4	0.093 10
γ [E2]	562.97 3	0.063 11
γ E0+M1+E2	586.307 4	8.2 3
γ M1+39%E2	622.80 8	0.82 6
γ	633.82 14	0.017 6
γ [M1+E2]	641.17 9	0.034 11
γ	648.39 14	0.074 11
γ [M1+E2]	656.24 12	0.034 11
γ [E2]	658.88 7	~0.023
γ E2(+<35%M1)	675.045 17	0.48 3
γ E2(+>0.1%M1)	678.586 3	0.194 11
γ	693.26 14	0.029 11
γ [E1]	699.261 14	0.095 7
γ [E2]	702.96 4	0.69 10
γ [E2]	703.56 4	1.3 3
γ	709.1 6	0.040 18
γ [E1]	712.854 4	0.109 8
γ [E1]	716.44 16	0.040 17
γ [M1+E2]	730.46 16	~0.023
γ [E1]	738.85 8	0.194 17
γ E2+5.5%M1	764.909 5	2.41 12
γ E1	778.920 4	5.0 3
γ [M1+E2]	788.21 18	0.046 11
γ	794.79 2	0.157 14
γ [M1+E2]	812.82 8	0.205 17
γ [E1]	817.97 7	0.091 11
γ [M1+E2]	831.98 7	0.097 17
γ	841.00 14	0.046 11
γ E0+M1+E2	854.8 5	0.040 12
γ [M1+E2]	857.17 24	0.074 11
γ [E2]	893.33 8	0.59 5
γ	902.45 13	0.131 11
γ [M1+E2]	909.06 16	0.108 11
γ	914.7 5	0.029 12
γ M1(+>20%E2)	928.43 9	0.40 6
γ E2	930.592 5	1.33 8

Photons (^{152}Tb) (continued)

γ_{mode}	γ(keV)	γ(%)†
γ	937.00 6	0.164 16
γ M1	952.60 21	0.051 11
γ	957.6 10	~0.06
γ E1	970.391 9	0.80 6
γ (E0)+M1+E2	974.09 3	2.75 16
γ [E2]	985.01 12	0.034 11
γ E2	990.219 17	0.74 4
γ M1+E2+E0	1010.58 7	0.37 3
γ	1016.19 14	0.068 11
γ	1036.98 17	0.108 11
γ	1048.85 23	~0.034
γ	1052.28 24	0.108 17
γ	1061.6 10	~0.06
γ	1070.6 3	0.046 17
γ [E2]	1084.24 16	0.137 23
γ [E1]	1085.94 17	0.22 3
γ E2(+0.2%M1)	1089.706 4	0.71 4
γ [E2]	1106.64 8	0.37 4
γ E2	1109.193 5	2.44 10
γ [E1]	1123.60 8	0.074 17
γ	1130.8 4	0.108 17
γ (M1)	1137.61 8	0.74 5
γ	1149.4 7	0.040 18
γ	1159.98 17	0.22 3
γ	1171.9 7	0.040 12
γ E2	1185.76 7	0.21 3
γ [E1]	1190.49 12	0.39 3
γ (M1)	1202.53 25	0.040 11
γ	1205.91 2	0.094 10
γ (E1)	1209.06 17	0.26 3
γ	1236.4 3	0.063 17
γ	1247.10 17	0.103 11
γ E0+M1+E2	1261.349 17	0.81 4
γ [M1+E2]	1275.41 11	0.091 11
γ	1284.61 22	0.057 11
γ	1290.0 4	0.029 11
γ E1+0.06%M2	1299.158 5	1.84 14
γ E1	1314.675 9	1.27 9
γ [M1+E2]	1316.22 8	0.17 3
γ M1	1318.38 3	0.34 6
γ E2	1325.75 7	0.80 6
γ	1336.6 2	0.120 18
γ E2	1348.12 6	0.83 5
γ	1360.0 3	~0.06
γ [E1]	1365.8 3	0.091 17
γ [E1]	1369.09 12	0.120 23
γ	1372.4 4	~0.040
γ [E1]	1400.53 17	0.143 17
γ	1406.19 17	0.09 3
γ	1409.3 6	~0.046
γ (E2)	1411.00 17	0.63 6
γ	1420.20 17	0.046 11
γ	1427.31 14	0.091 11
γ [M1+E2]	1430.83 17	0.080 11
γ	1436.54 22	0.051 23
γ [E1]	1441.7 3	0.085 11
γ	1446.2 4	0.074 15
γ	1446.63 11	0.16 3
γ	1481.2 5	0.046 18
γ	1490.09 22	0.057 17
γ (E0+M1+E2)	1495.36 16	0.091 17
γ E0+M1+E2	1507.0 5	0.040 18
γ M1	1517.76 8	0.62 5
γ (E2)	1532.21 24	0.08 3
γ [E1]	1544.27 12	0.06 3
γ	1562.25 12	0.063 17
γ	1565.95 11	0.097 17
γ	1571.10 17	0.137 17
γ	1574.9 10	~0.040
γ E1	1586.19 11	0.85 6
γ [E1]	1596.51 14	0.16 3
γ [M1+E2]	1596.88 7	0.18 7
γ	1598.80 17	0.21 4
γ	1599.42 22	0.17 7
γ E2	1605.632 17	0.179 16
γ [E1]	1606.18 17	0.18 4
γ	1631.2 3	0.13 3
γ [E2]	1631.39 8	0.097 11
γ [M1+E2]	1640.02 17	~0.046
γ	1645.86 18	0.080 23
γ E0+M1+E2	1663.2 8	0.051 24
γ M1	1667.35 13	0.66 5
γ [E2]	1681.54 17	0.046 11
γ	1727.49 22	0.046 11

Photons (^{152}Tb) (continued)

γ_{mode}	γ(keV)	γ(%)†
γ	1735.8 10	~0.06
γ M1	1739.4 3	0.080 17
γ E2	1757.42 11	0.63 5
γ	1761.5 10	~0.029
γ (E0+M1+E2)	1771.43 11	0.262 23
γ (E0+M1+E2)	1778.80 11	0.074 17
γ (M1)	1789.12 14	0.50 4
γ (E2)	1798.79 17	0.171 23
γ [E2]	1802.8 3	0.091 17
γ	1809.4 3	0.085 17
γ (E0+M1+E2)	1818.62 17	0.046 11
γ [M1+E2]	1825.3 3	0.125 17
γ M1	1841.03 17	0.09 3
γ M1(+E2)	1857.3 4	0.154 23
γ (E2)	1862.05 8	0.43 3
γ	1871.2 10	0.131 24
γ [E1]	1887.0 3	~0.017
γ [M1+E2]	1890.7 3	~0.040
γ M1	1902.52 8	1.67 11
γ [E2]	1908.32 17	~0.029
γ	1915.1 4	~0.034
γ M1	1920.9 2	0.38 5
γ M1,E2	1932.98 22	~0.034
γ E2	1941.16 7	0.63 5
γ E1	1955.39 12	0.31 3
γ	1962.9 10	~0.06
γ	1970.3 4	0.034 11
γ	1975.4 3	0.063 11
γ	1983.50 22	0.063 11
γ [E2]	1993.80 17	0.080 17
γ M1+E2	2019.8 7	~0.023
γ E1+E2	2033.64 17	0.120 17
γ [E1]	2043.3 3	0.080 11
γ	2050.9 6	~0.013
γ	2065.3 10	~0.06
γ M1	2068.1 4	0.085 11
γ M1,E2	2075.0 3	0.068 23
γ [E1]	2078.7 3	0.063 23
γ	2085.8 10	~0.018
γ [M1+E2]	2092.6 5	0.125 25
γ [E2]	2093.98 11	0.057 11
γ E2	2103.5 4	0.080 18
γ M1,E2	2113.96 17	0.085 17
γ (M1)	2118.51 24	0.046 11
γ [E2]	2127.7 4	0.034 11
γ M1	2151.0 3	0.200 23
γ	2158.69 22	0.074 11
γ M1,E2	2168.79 24	0.085 17
γ (M1)	2179.45 17	0.057 11
γ M1	2185.11 17	0.188 23
γ (E0+M1+E2)	2195.7 4	0.068 18
γ	2211.7 10	~0.029
γ	2218 1	~0.011
γ [E1]	2251.8 3	0.085 11
γ	2254.99 22	0.068 11
γ (E0+M1+E2)	2260.7 4	0.051 11
γ (M1)	2265.20 11	0.074 11
γ	2276.6 10	0.10 3
γ	2291 1	~0.06
γ E0+M1+E2	2307 1	~0.013
γ	2324.0 8	~0.029
γ	2342.8 4	0.125 17
γ	2348.81 17	0.046 17
γ	2357 1	~0.029
γ (E0+M1+E2)	2365.10 11	0.382 23
γ (M1)	2375.42 14	0.78 5
γ (E0+M1+E2)	2385.09 17	0.160 17
γ	2398.2 4	0.091 23
γ (E2)	2404.92 17	1.31 8
γ (E0+M1+E2)	2427.4 3	0.057 11
γ [M1+E2]	2480.4 5	~0.017
γ	2495.3 3	0.080 11
γ M1,E2	2518.4 3	0.114 23
γ M1,E2	2523.73 17	0.080 17
γ (M1)	2536.33 11	0.28 3
γ	2551.9 5	~0.023
γ	2556.9 5	~0.023
γ E1	2569.80 22	0.194 17
γ	2575.9 4	0.051 17
γ	2580.5 8	0.046 18
γ	2584.6 3	0.063 17
γ (M1)	2588.3 3	0.36 3
γ	2599.27 22	~0.023
γ (M1)	2602.6 3	0.051 17

Photons (^{152}Tb) (continued)

γ_{mode}	γ(keV)	γ(%)†
γ M1(+E2)	2619.94 17	0.268 23
γ [E2]	2638.0 3	~0.023
γ (E2)	2655.6 5	0.051 17
γ E1?	2662.9 3	0.200 17
γ	2667.9 4	0.137 18
γ M1,E2	2680.0 4	0.057 17
γ	2687.1 4	0.040 11
γ	2694.3 6	0.057 17
γ M1,E2	2697.87 22	0.234 23
γ	2703.1 10	0.068 18
γ [E2]	2709.38 11	0.217 17
γ (E2)	2719.70 14	0.34 3
γ [E2]	2729.37 17	<0.017
γ (E1)	2733.9 4	0.091 18
γ (M1)	2740.8 3	~0.023
γ	2744.0 4	0.085 18
γ M1,E2	2755.09 24	0.114 11
γ	2761.9 4	~0.011
γ	2768.3 7	~0.029
γ [E2]	2778.0 3	~0.029
γ	2788.2 6	~0.029
γ	2790.7 10	0.046 6
γ E1	2795.7 3	0.097 11
γ	2799.5 10	~0.011
γ E1	2808.9 5	0.051 12
γ	2815.8 6	~0.011
γ	2820.8 7	~0.023
γ	2832.0 15	~0.017
γ E1,E2	2837 1	0.029 12
γ	2844.9 7	~0.017
γ E1	2861.1 3	0.074 11
γ	2870.0 7	0.057 12
γ	2882.1 5	0.114 12
γ	2889.5 10	0.034 12
γ	2893 1	0.074 12
γ E1	2906.6 4	0.051 11
γ	2914.08 22	~0.023
γ	2921.5 10	~0.023
γ	2927.7 8	0.046 12
γ M1,E2	2940.7 4	0.080 11
γ	2961.2 8	0.023 6
γ	2965.4 6	~0.011
γ	2971.0 15	~0.007
γ	2979.9 8	0.034 11
γ	2984.5 4	~0.017
γ	2993.9 7	~0.023
γ	2998.0 15	~0.023
γ	3016.0 15	~0.011
γ	3024.3 4	~0.017
γ	3042.15 22	0.074 11
γ	3056.6 10	0.029 6
γ [M1+E2]	3066.7 5	~0.011
γ [E2]	3085.1 3	~0.011
γ	3088.9 8	~0.0046
γ	3095 1	~0.017
γ	3106.2 4	0.029 6
γ	3134.6 11	~0.008
γ [E1]	3140.0 3	0.023 6
γ	3154.0 15	0.017 6
γ	3160.0 6	0.051 11
γ	3165.0 7	0.068 11
γ	3173.5 10	0.023 6
γ	3189.1 7	0.017 6
γ [E1]	3205.4 3	0.040 11
γ	3223.5 10	~0.023
γ	3228.9 11	0.017 6
γ	3245.0 15	~0.0046
γ [E1]	3250.9 4	~0.0046
γ	3268.0 15	0.017 6
γ	3275.5 20	~0.011
γ	3285.0 4	0.0063 23
γ	3309.7 6	0.013 3
γ	3324.1 8	0.034 6
γ	3328.8 4	0.011 4
γ	3338.2 7	~0.0046
γ	3367.0 15	~0.006
γ	3380 2	~0.006
γ [E2]	3411.0 5	0.0068 23
γ	3478.9 11	0.013 3
γ	3573.1 11	~0.0023

† 18% uncert(syst)

Atomic Electrons (^{152}Tb)

⟨e⟩=20.0 11 keV

e_{bin}(keV)	⟨e⟩(keV)	e(%)
7	2.24	31 3
8	1.64	20.4 21
33 - 34	0.341	1.01 8
35	0.53	1.52 16
36 - 67	1.03	2.38 12
94 - 142	0.27	0.24 3
143 - 192	0.053	0.034 4
193 - 220	0.078	0.037 8
221	1.03	0.47 3
240 - 285	0.550	0.208 9
294	5.2	1.8 4
301 - 335	0.0637	0.0207 8
336	1.07	0.32 6
337 - 381	1.42	0.41 3
382	1.43	0.37 5
386 - 434	0.35	0.084 14
439 - 488	0.155	0.033 3
489 - 535	0.061	0.0121 12
536	0.84	0.16 3
539 - 586	0.21	0.037 6
591 - 640	0.046	0.0075 5
641 - 689	0.080	0.0122 19
691 - 739	0.149	0.0208 8
745 - 793	0.046	0.0060 6
794 - 843	0.0215	0.00259 21
846 - 895	0.064	0.0073 6
901 - 950	0.164	0.018 3
951 - 999	0.202	0.0204 25
1002 - 1051	0.054	0.0052 7
1052 - 1101	0.096	0.0090 3
1102 - 1149	0.020	0.00176 25
1152 - 1201	0.0110	0.00094 16
1202 - 1249	0.037	0.0030 4
1253 - 1299	0.058	0.00456 22
1306 - 1353	0.017	0.0013 3
1356 - 1405	0.024	0.00174 21
1406 - 1447	0.0052	0.00037 6
1457 - 1556	0.043	0.0029 3
1558 - 1656	0.0242	0.00151 11
1659 - 1756	0.0337	0.00196 9
1759 - 1858	0.0486	0.00265 13
1860 - 1956	0.0282	0.00149 7
1961 - 2060	0.0096	0.00048 5
2061 - 2157	0.0131	0.00062 4
2160 - 2259	0.0061	0.00028 4
2263 - 2358	0.0342	0.00146 7
2363 - 2445	0.0062	0.000259 17
2468 - 2563	0.0145	0.00058 3
2568 - 2666	0.0141	0.00054 3
2669 - 2767	0.0100	0.000372 22
2770 - 2868	0.0040	0.000143 19
2871 - 2969	0.0030	0.000103 12
2971 - 3065	0.0017	5.7 10 $\times 10^{-5}$
3077 - 3173	0.0020	6.5 9 $\times 10^{-5}$
3179 - 3278	0.00086	2.7 4 $\times 10^{-5}$
3279 - 3373	0.00032	9.5 18 $\times 10^{-6}$
3378 - 3477	0.00011	3.2 10 $\times 10^{-6}$
3523 - 3572	1.7 $\times 10^{-5}$	~5 $\times 10^{-7}$

Continuous Radiation (^{152}Tb)

⟨β+⟩=206 keV;⟨IB⟩=4.2 keV

E_{bin}(keV)		⟨ ⟩(keV)	(%)
0 - 10	β+	3.14 $\times 10^{-7}$	3.73 $\times 10^{-6}$
	IB	0.0080	
10 - 20	β+	1.57 $\times 10^{-5}$	9.4 $\times 10^{-5}$
	IB	0.0075	0.052
20 - 40	β+	0.00054	0.00162
	IB	0.052	0.151
40 - 100	β+	0.0323	0.0410
	IB	0.135	0.27
100 - 300	β+	1.82	0.82
	IB	0.149	0.082
300 - 600	β+	13.1	2.84
	IB	0.29	0.066
600 - 1300	β+	78	8.2

Continuous Radiation (^{152}Tb) (continued)

E_{bin}(keV)		⟨ ⟩(keV)	(%)
	IB	1.08	0.116
1300 - 2500	β+	110	6.4
	IB	1.9	0.103
2500 - 3850	β+	3.02	0.117
	IB	0.58	0.020
	Σβ+		18

$^{152}_{65}$Tb(4.3 2 min)

Mode: IT(78.9 8 %), ϵ(21.1 8 %)

Δ: -70367 15 keV

SpA: 2.87×10^8 Ci/g

Prod: ^{151}Eu(α,3n); ^{152}Gd(p,n); ^{153}Eu(α,5n); protons on Dy; ^{139}La(^{16}O,3n)

Photons (^{152}Tb)

⟨γ⟩=747 35 keV

γ_{mode}	γ(keV)	γ(%)†
Gd L_ℓ	5.362	0.057 6
Tb L_ℓ	5.546	0.36 4
Gd L_η	6.049	0.0205 18
Gd L_α	6.054	1.51 11
Tb L_α	6.269	9.4 8
Tb L_η	6.284	0.160 14
Gd L_β	6.851	1.38 13
Tb L_β	7.094	9.8 9
Gd L_γ	7.878	0.210 24
Tb L_γ	8.166	1.50 16
Gd $K_{\alpha2}$	42.309	5.25 17
Gd $K_{\alpha1}$	42.996	9.4 3
Tb $K_{\alpha2}$	43.744	22.8 13
Tb $K_{\alpha1}$	44.482	40.8 24
γ_{IT} (M1)	47.90 9	1.18 9
Gd $K_{\beta1}$'	48.652	2.78 9
Gd $K_{\beta2}$'	50.214	0.84 3
Tb $K_{\beta1}$'	50.338	12.2 7
Tb $K_{\beta2}$'	51.958	3.64 22
γ_{IT} (M1)	58.85 15	6.8 4
γ_{IT} [E2]	64.95 17	0.185 12
γ_ϵM1+E2	92.43 14	0.310 18
γ_{IT} [E2]	106.75 16	0.31 6
γ_{IT} E3	159.59 10	16.5 7
γ_ϵ[E2]	173.102 18	~0.06
γ_ϵ[E2]	175.185 4	0.0067 8
γ_ϵ[E1]	192.613 5	3.79 20 $\times 10^{-5}$
γ_ϵ[E1]	197.66 18	0.286 18
γ_ϵM1+E2	220.84 18	0.435 24
γ_{IT} M1(+35%E2)	235.39 9	4.3 2
γ_ϵM1,E2	255.4 3	0.370 18
γ_ϵE2	271.132 11	~0.08
γ_{IT} M1(+28%E2)	277.19 10	8.5 3
γ_{IT} E2	283.29 5	60
γ_ϵ	303.93 19	~0.06
γ_ϵ	311.43 10	0.08 3 *
γ_ϵ	311.81 18	
γ_ϵ(E2)	315.175 11	0.153 11
γ_ϵ[M1]	324.799 4	0.0297 20
γ_ϵE2	344.286 2	20 1
γ_ϵE2	351.704 18	1.72 15
γ_ϵE1+0.2%M2	367.798 3	0.00473 23
γ_ϵ[E2]	385.92 9	3.26 18
γ_ϵ	396.36 17	0.13 3
γ_ϵE2	411.122 4	18.1 9
γ_ϵE2	427.64 9	0.93 5
γ_ϵ	440.22 15	~0.35
γ_ϵ(M1+E2+E0)	440.85 11	0.40 15
γ_ϵ	441.003 20	~0.18
γ_ϵ	447.72 17	0.095 24

Photons (^{152}Tb)
(continued)

γ_{mode}	γ(keV)	γ(%)†
γ_ϵ[E2]	470.7 *3*	0.46 *18*
γ_ϵE2	471.96 *8*	11.8 *6*
γ_ϵ[E2]	493.777 *11*	0.00078 *7*
γ_ϵE2(+<22%M1)	503.401 *4*	0.060 *4*
γ_ϵE2	519.45 *9*	4.7 *3*
γ_ϵE0+M1+E2	526.889 *17*	1.80 *14*
γ_ϵE2	532.65 *8*	4.27 *23*
γ_ϵ[M1+E2]	579.33 *9*	0.221 *18*
γ_ϵE0+M1+E2	586.307 *4*	1.40 *9*
γ_ϵ[M1+E2]	634.27 *10*	0.94 *7*
γ_ϵM1(+E2)	647.47 *12*	4.3 *3*
γ_ϵE1	652.9 *3*	0.340 *24*
γ_ϵE2(+>0.1%M1)	678.586 *3*	0.186 *11*
γ_ϵ[E1]	715.14 *17*	0.58 *4*
γ_ϵM1(+E2)	726.07 *12*	3.17 *16*
γ_ϵE2+5.5%M1	764.909 *5*	0.0155 *7*
γ_ϵ	770.55 *16*	0.155 *24*
γ_ϵE1	778.920 *4*	0.072 *3*
γ_ϵ	794.79 *2*	0.174 *21*
γ_ϵE2	930.592 *5*	0.229 *17*
γ_ϵ	946.07 *17*	0.48 *4*
γ_ϵ	1041.0 *4*	0.137 *25*
γ_ϵ	1052.28 *24*	~0.07
γ_ϵ	1074.49 *14*	0.66 *5*
γ_ϵE2(+0.2%M1)	1089.706 *4*	0.68 *4*
γ_ϵM1(+E2)	1106.22 *9*	2.80 *17*
γ_ϵE2	1109.193 *5*	0.0157 *5*
γ_ϵ[M1+E2]	1166.91 *9*	3.70 *21*
γ_ϵ	1205.91 *2*	0.104 *13*
γ_ϵ	1242.51 *17*	0.101 *18*

† uncert(syst): 3.8% for ϵ, 1.1% for IT
* combined intensity for doublet

Atomic Electrons (^{152}Tb)
$\langle e\rangle$=142 *3* keV

e_{bin}(keV)	$\langle e\rangle$(keV)	e(%)
4 - 6	0.3	~7
7	4.8	69 *5*
8	6.8	87 *10*
9 - 49	4.81	17.8 *9*
50	4.1	8.1 *6*
51 - 99	3.81	6.3 *3*
105 - 107	0.084	0.080 *13*
108	19.9	18.5 *8*
109 - 147	0.036	0.027 *7*
151	33.6	22.2 *10*
152	22.7	14.9 *7*
158	14.5	9.2 *4*
159 - 205	5.21	3.16 *10*
212 - 228	2.17	0.96 *4*
231	7.7	3.3 *7*
233 - 282	3.6	1.31 *15*
283 - 325	2.15	0.73 *4*
336 - 385	2.28	0.648 *22*
386 - 434	1.16	0.279 *13*
438 - 487	1.29	0.273 *19*
492 - 536	0.39	0.074 *7*
565 - 608	0.45	0.076 *13*
626 - 671	0.071	0.0112 *18*
676 - 725	0.18	0.026 *7*
726 - 772	0.009	0.0012 *5*
777 - 795	0.0011	0.00014 *7*
880 - 929	0.019	0.0021 *8*
930 - 946	0.0023	0.00025 *12*
991 - 1040	0.034	0.0033 *9*
1041 - 1090	0.081	0.0076 *19*
1098 - 1117	0.107	0.0096 *21*
1156 - 1205	0.021	0.0018 *3*
1206 - 1243	0.00035	2.9 *13* $\times10^{-5}$

Continuous Radiation (^{152}Tb)
$\langle\beta+\rangle$=2.93 keV;$\langle$IB$\rangle$=0.35 keV

E_{bin}(keV)		$\langle\ \rangle$(keV)	(%)
0 - 10	$\beta+$	7.4×10^{-8}	8.8×10^{-7}
	IB	0.00026	
10 - 20	$\beta+$	3.69×10^{-6}	2.21×10^{-5}
	IB	0.000169	0.00115
20 - 40	$\beta+$	0.000124	0.000376
	IB	0.0096	0.026
40 - 100	$\beta+$	0.0072	0.0091
	IB	0.024	0.052
100 - 300	$\beta+$	0.342	0.156
	IB	0.0104	0.0052
300 - 600	$\beta+$	1.56	0.351
	IB	0.045	0.0098
600 - 1300	$\beta+$	1.02	0.141
	IB	0.19	0.021
1300 - 2472	$\beta+$	0.00382	0.000285
	IB	0.065	0.0044
	$\Sigma\beta+$		0.66

$^{152}_{66}$Dy(2.38 *2* h)

Mode: ϵ(99.900 *7* %), α(0.100 *7* %)
Δ: -70127 *6* keV
SpA: 8.66×10^6 Ci/g
Prod: ^{141}Pr(^{14}N,3n); ^{152}Gd(α,4n); ^{142}Nd(^{12}C,2n); protons on Ta; ^{154}Gd(^{3}He,5n)

Alpha Particles (^{152}Dy)

α(keV)
3629 *4*

Photons (^{152}Dy)
$\langle\gamma\rangle$=287 *3* keV

γ_{mode}	γ(keV)	γ(%)†
Tb L_ℓ	5.546	0.26 *3*
Tb L_α	6.269	6.8 *4*
Tb L_η	6.284	0.090 *7*
Tb L_β	7.113	6.3 *6*
Tb L_γ	8.202	0.98 *10*
Tb $K_{\alpha2}$	43.744	22.5 *7*
Tb $K_{\alpha1}$	44.482	40.4 *12*
Tb $K_{\beta1}$'	50.338	12.1 *4*
Tb $K_{\beta2}$'	51.958	3.60 *12*
γE1	256.93 *13*	97.5 *10*

† <0.1% uncert(syst)

Atomic Electrons (^{152}Dy)
$\langle e\rangle$=11.7 *4* keV

e_{bin}(keV)	$\langle e\rangle$(keV)	e(%)
8	4.1	52 *5*
9	0.61	7.0 *7*
35	0.40	1.14 *12*
36	0.60	1.67 *17*
37	0.198	0.53 *6*
41	0.073	0.177 *18*
42	0.246	0.59 *6*
43	0.32	0.75 *8*
44	0.066	0.149 *15*

Atomic Electrons (^{152}Dy)
(continued)

e_{bin}(keV)	$\langle e\rangle$(keV)	e(%)
48	0.027	0.056 *6*
49	0.030	0.061 *6*
50	0.0240	0.048 *5*
205	4.17	2.03 *5*
248	0.578	0.233 *5*
249	0.138	0.0554 *12*
255	0.159	0.0625 *14*
256	0.000540	0.000211 *5*
257	0.0445	0.0173 *4*

Continuous Radiation (^{152}Dy)
$\langle$IB$\rangle$=0.19 keV

E_{bin}(keV)		$\langle\ \rangle$(keV)	(%)
10 - 20	IB	0.00019	0.00120
20 - 40	IB	0.034	0.092
40 - 100	IB	0.136	0.29
100 - 300	IB	0.0154	0.0088
300 - 485	IB	0.0033	0.00096

$^{152}_{67}$Ho(2.35 *11* min)

Mode: ϵ(88 *3* %), α(12 *3* %)
Δ: -63740 *60* keV
SpA: 5.25×10^8 Ci/g
Prod: ^{141}Pr(^{16}O,5n); ^{144}Sm(^{11}B,3n)

Alpha Particles (^{152}Ho)

α(keV)
4387 *3*

Photons (^{152}Ho)
$\langle\gamma\rangle$=760 *37* keV

γ_{mode}	γ(keV)	γ(%)†
γ_ϵE2	613.9 *1*	88
γ_ϵE2	647.5 *1*	14 *4*
γ_ϵ	755.5 *6*	6.2 *12*
γ_ϵ	930.0 *7*	8.8 *18*

† 3.4% uncert(syst)

$^{152}_{67}$Ho(52.3 *5* s)

Mode: ϵ(90 *3* %), α(10 *3* %)
Δ: -63740 *60* keV
SpA: 1.410×10^9 Ci/g
Prod: ^{141}Pr(^{16}O,5n); ^{144}Sm(^{11}B,3n)

Alpha Particles (^{152}Ho)

α(keV)
4453 *3*

Photons (^{152}Ho)

$\langle\gamma\rangle$=1990 *56* keV

γ_{mode}	γ(keV)	γ(%)†
γ_ϵE2	492.90 *14*	53 *5*
γ_ϵE2	613.9 *1*	90
γ_ϵE2	647.5 *1*	90 *5*
γ_ϵE2	683.50 *14*	77 *5*
γ_ϵM1(+E2)	758.5 *3*	10 *3*

† 3.3% uncert(syst)

$^{152}_{68}$Er(10.1 *2* s)

Mode: α(90 *3* %), ϵ(10 *3* %)
Δ: -60620 *60* keV
SpA: 7.10×10^9 Ci/g
Prod: ^{141}Pr(^{19}F,8n); ^{142}Nd(^{16}O,6n); ^{140}Ce(^{20}Ne,8n); ^{58}Ni on ^{107}Ag

Alpha Particles (^{152}Er)

α(keV)
4802 *5*

$^{152}_{69}$Tm(8 *1* s)

Mode: ϵ
Δ: -51740 *410* keV syst
SpA: 8.9×10^9 Ci/g
Prod: daughter ^{152}Yb

decay data not yet evaluated

$^{152}_{69}$Tm(5.2 *6* s)

Mode: ϵ
Δ: -51740 *410* keV syst
SpA: 1.34×10^{10} Ci/g
Prod: protons on Er

decay data not yet evaluated

$^{152}_{70}$Yb(3.2 *3* s)

Mode: ϵ
Δ: -46320 *500* keV syst
SpA: 2.09×10^{10} Ci/g
Prod: ^{96}Ru(^{58}Ni,2p); ^{107}Ag on V-Ni targets

decay data not yet evaluated

A = 153

NDS **37**, 487 (1982)

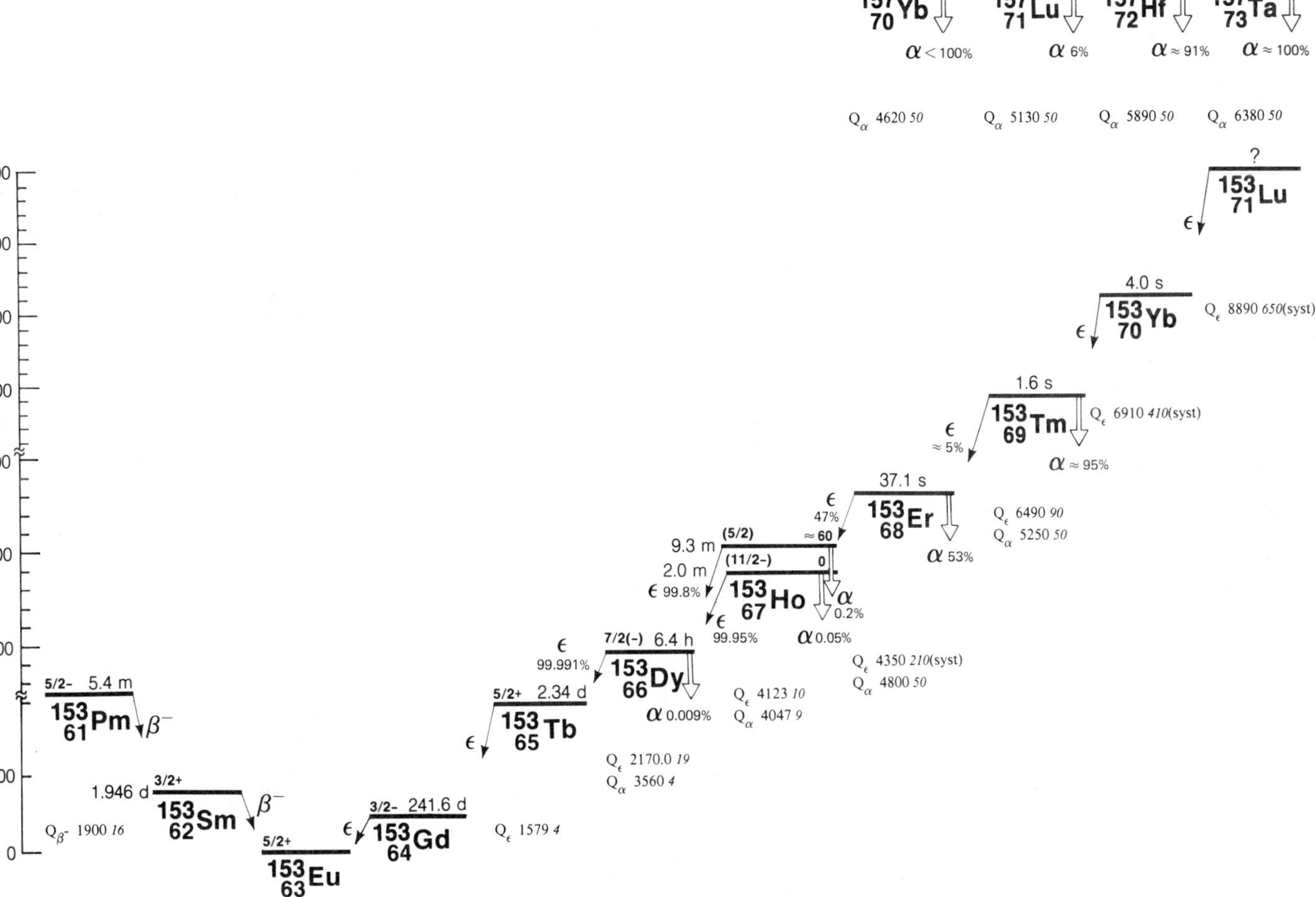

$^{153}_{61}$Pm(5.4 *2* min)

Mode: β-

Δ: -70669 *16* keV

SpA: 2.27×10^8 Ci/g

Prod: ^{154}Sm(γ,p)

Photons (^{153}Pm)

γ_{mode}	γ(keV)	γ(rel)
γ E1	28.26 *14*	32 *6*
γ E1	35.80 *14*	100 *20*
γ E1	83.4 *3*	4.4 *12*
γ E1	90.9 *3*	14 *4*
γ E1	119.57 *21*	24.0 *24*
γ E1	127.11 *21*	56 *11*
γ E1	129.39 *25*	7.2 *20*
γ M1+E2	147.1 *3*	2.0 *4*
γ E1	175.33 *25*	8.0 *12*
γ E1	182.87 *25*	10.8 *12*

$^{153}_{62}$Sm(1.946 *4* d)

Mode: β-

Δ: -72569 *3* keV

SpA: 4.386×10^5 Ci/g

Prod: ^{152}Sm(n,γ); ^{150}Nd(α,n)

Photons (^{153}Sm)

⟨γ⟩=62.6 *8* keV

γ_{mode}	γ(keV)	γ(%)
Eu L_ℓ	5.177	0.186 *19*
Eu L_η	5.817	0.069 *6*
Eu L_α	5.843	5.0 *3*
Eu L_β	6.594	4.5 *4*
Eu L_γ	7.566	0.66 *7*
γ E1	14.0639 *3*	0.00044 *15*
γ E2	19.8130 *3*	0.00012 *5*
Eu $K_{\alpha2}$	40.902	18.0 *6*
Eu $K_{\alpha1}$	41.542	32.4 *10*
Eu $K_{\beta1}$'	46.999	9.4 *3*
Eu $K_{\beta2}$'	48.496	2.85 *10*
γ (M1+E2)	54.1985 *19*	0.0016 *3*
γ	55.9 *3*	0.035 *7*
γ (E1)	68.2624 *19*	0.0012 *4*
γ M1+1.9%E2	69.67344 *21*	5.32 *22*
γ E1	75.42255 *25*	0.187 *19*
γ M1+39%E2	83.36762 *24*	0.21 *3*
γ M1+3.8%E2	89.48643 *25*	0.167 *18*
γ M1+E2	96.880 *5*	0.0070 *14*
γ E1	97.4315 *3*	0.73 *2*
γ M1+1.7%E2	103.1806 *3*	28.3 *6*
γ [E1]	118.104 *5*	0.00030 *6*
γ (E1)	151.6300 *19*	0.0115 *23*
γ [E2]	166.554 *5*	0.00060 *12*
γ [E1]	172.303 *5*	0.00040 *8*
γ M1+49%E2	172.85400 *25*	0.066 *6*
γ	412.14 *8*	0.0023 *2*
γ [M1+E2]	424.43 *8*	0.00230 *23*
γ [M1+E2]	436.86 *6*	0.00180 *18*
γ	461.76 *12*	0.0014 *3*
γ	463.63 *7*	0.0156 *16*
γ	484.82 *14*	0.0004 *1* ?
γ	509.02 *8*	0.0022 *2*
γ [M1+E2]	521.31 *8*	0.0075 *8*
γ	531.44 *12*	0.064 *6*
γ	533.31 *7*	0.032 *3*

Photons (^{153}Sm) (continued)

γ_{mode}	γ(keV)	γ(%)
γ	539.06 *7*	0.020 *2*
γ [E1]	542.53 *8*	0.0027 *3*
γ	545.74 *11*	0.0010 *1*
γ [E1]	554.96 *6*	0.0051 *5*
γ	574.31 *14*	0.00010 *3* ?
γ	578.69 *8*	0.0033 *3*
γ	584.44 *8*	0.00086 *9*
γ	587.48 *14*	0.00034 *4*
γ [M1+E2]	590.98 *8*	0.00091 *9*
γ [E1]	596.73 *8*	0.0115 *12*
γ	598.4 *3*	0.0015 *2*
γ	603.63 *13*	0.0031 *3* ?
γ [E1]	609.16 *6*	0.0127 *13*
γ	615.41 *11*	0.00080 *8*
γ	617.69 *13*	0.0011 *1*
γ	630.7 *3*	0.00010 *2*
γ	634.62 *12*	0.00030 *8*
γ	636.49 *7*	0.00190 *19*
γ	657.55 *25*	0.00040 *4*
γ	662.90 *14*	8 *3* ×10^{-5}
γ	676.96 *14*	1.0 *3* ×10^{-5}
γ	681.87 *8*	<1.4 ×10^{-5}
γ [M1+E2]	694.16 *8*	2.0 *5* ×10^{-5}
γ	701.06 *13*	2.5 *6* ×10^{-5}
γ [M1+E2]	706.59 *6*	1.5 *4* ×10^{-5}
γ	713.6 *3*	0.00019 *4*
γ	718.59 *11*	2.0 *5* ×10^{-5}
γ	760.33 *14*	2.5 *6* ×10^{-5}
γ	763.8 *6*	~3 ×10^{-5} ?

Atomic Electrons (^{153}Sm)

⟨e⟩=44.9 *8* keV

e_{bin}(keV)	⟨e⟩(keV)	e(%)
6	0.0007	~0.013
7	2.09	30 *3*
8	1.50	19.4 *20*
12 - 20	0.055	0.39 *10*
21	5.09	24.1 *11*
27 - 54	2.26	6.1 *3*
55	22.6	41.3 *12*
56 - 61	0.010	~0.017
62	2.24	3.64 *17*
63 - 90	1.30	1.86 *6*
95	5.13	5.39 *16*
96 - 145	2.63	2.63 *4*
150 - 173	0.0099	0.0060 *6*
364 - 412	0.00057	0.00015 *4*
413 - 462	0.0015	0.00035 *16*
463 - 509	0.007	0.0013 *5*
513 - 561	0.0023	0.00042 *9*
566 - 615	0.00038	6.4 *12* ×10^{-5}
616 - 665	3.4 ×10^{-5}	5.3 *15* ×10^{-6}
669 - 718	4.4 ×10^{-6}	6.3 *19* ×10^{-7}
719 - 764	3.4 ×10^{-7}	4.5 *19* ×10^{-8}

Continuous Radiation (^{153}Sm)

⟨β-⟩=225 keV;⟨IB⟩=0.166 keV

E_{bin}(keV)		⟨ ⟩(keV)	(%)
0 - 10	β-	0.129	2.57
	IB	0.0112	
10 - 20	β-	0.385	2.57
	IB	0.0105	0.073
20 - 40	β-	1.54	5.1
	IB	0.019	0.066
40 - 100	β-	10.6	15.2
	IB	0.044	0.069
100 - 300	β-	86	44.3

Continuous Radiation (^{153}Sm) (continued)

E_{bin}(keV)		⟨ ⟩(keV)	(%)
	IB	0.067	0.041
300 - 600	β-	118	29.0
	IB	0.0145	0.0040
600 - 809	β-	7.7	1.20
	IB	0.000161	2.6 ×10^{-5}

$^{153}_{63}$Eu(stable)

Δ: -73379 *4* keV

%: 52.2 *5*

$^{153}_{64}$Gd(241.6 *2* d)

Mode: ϵ

Δ: -72895 *3* keV

SpA: 3532 Ci/g

Prod: ^{152}Gd(n,γ); ^{153}Eu(d,2n)

Photons (^{153}Gd)

⟨γ⟩=101.5 *17* keV

γ_{mode}	γ(keV)	γ(%)†
Eu L_ℓ	5.177	0.36 *4*
Eu L_η	5.817	0.131 *10*
Eu L_α	5.843	9.5 *6*
Eu L_β	6.593	8.7 *8*
Eu L_γ	7.576	1.32 *14*
γ E1	14.0639 *3*	0.017 *6*
Eu $K_{\alpha2}$	40.902	34.4 *10*
Eu $K_{\alpha1}$	41.542	62.1 *18*
Eu $K_{\beta1}$'	46.999	18.1 *5*
Eu $K_{\beta2}$'	48.496	5.47 *18*
γ (M1+E2)	54.1985 *19*	~0.015
γ (E1)	68.2624 *19*	~0.011
γ M1+1.9%E2	69.67344 *21*	2.30 *8*
γ E1	75.42255 *25*	0.081 *9*
γ M1+39%E2	83.36762 *24*	0.193 *19*
γ M1+3.8%E2	89.48643 *25*	0.072 *9*
γ E1	97.4315 *3*	27.6
γ M1+1.7%E2	103.1806 *3*	19.6 *4*
γ (E1)	151.6300 *19*	~0.10
γ M1+49%E2	172.85400 *25*	0.0285 *24*

† 5.4% uncert(syst)

Atomic Electrons (^{153}Gd)

⟨e⟩=39.9 *7* keV

e_{bin}(keV)	⟨e⟩(keV)	e(%)
6	0.010	~0.17
7	3.9	56 *6*
8	2.9	38 *4*
12 - 20	0.076	0.54 *9*
21	2.20	10.4 *4*
27 - 47	3.43	9.5 *4*
49	3.46	7.1 *4*
52 - 54	0.023	~0.04
55	15.6	28.6 *8*
60 - 61	0.0007	0.0012 *4*
62	0.97	1.57 *6*
63 - 90	1.62	2.01 *6*
95	3.55	3.73 *11*

Atomic Electrons (^{153}Gd) (continued)

e_{bin}(keV)	⟨e⟩(keV)	e(%)
96 - 145	2.10	2.11 4
150 - 173	0.0047	0.0028 3

Continuous Radiation (^{153}Gd)

⟨IB⟩=0.155 keV

E_{bin}(keV)		⟨ ⟩(keV)	(%)
10 - 20	IB	0.00021	0.00132
20 - 40	IB	0.056	0.152
40 - 100	IB	0.090	0.20
100 - 300	IB	0.0086	0.0053
300 - 484	IB	0.00046	0.000138

$^{153}_{65}$Tb(2.34 1 d)

Mode: ϵ

Δ: -71316 5 keV

SpA: 3.647×10^5 Ci/g

Prod: protons on Gd; ^{153}Eu(α,4n); ^{151}Eu(α,2n)

Photons (^{153}Tb)

⟨γ⟩=308 4 keV

γ_{mode}	γ(keV)	γ(%)†
Gd L_ℓ	5.362	0.38 4
Gd L_η	6.049	0.145 12
Gd L_α	6.054	10.1 7
Gd L_β	6.844	10.0 10
Gd L_γ	7.892	1.58 18
γ M1+6.3%E2	41.564 11	3.7 4
Gd $K_{\alpha2}$	42.309	29.1 9
Gd $K_{\alpha1}$	42.996	52.3 15
Gd $K_{\beta1}$'	48.652	15.4 5
Gd $K_{\beta2}$'	50.214	4.66 16
γ M1+3.1%E2	51.81 3	0.363 24
γ [E1]	54.331 19	0.090 15
γ [E1]	66.12 3	~0.030
γ M1+5.0%E2	68.209 13	0.354 18
γ E1	82.865 12	5.0 3
γ M1+0.1%E2	87.619 14	1.27 7
γ E2+17%M1	88.37 8	0.384 24
γ E1	90.14 3	0.246 15
γ M1	91.562 21	0.171 9
γ E2	93.37 3	0.102 6
γ E1	102.275 11	5.8 3
γ E2	106.84 8	0.036 9
γ M1	109.773 10	6.2 3
γ M1	126.11 5	0.120 9
γ M1	129.183 14	0.53 3
γ M1+25%E2	132.53 4	0.168 12
γ (M1)	139.86 3	0.129 9
γ E1	141.950 15	1.05 6
γ [M1+E2]	147.60 9	0.045 15
γ	151.80 10	0.075 15
γ	152.49 20	0.045 15
γ E2	166.02 9	0.060 15 ?
γ E1	170.484 12	6.6 4
γ (E1)	174.427 20	1.52 8
γ E1	177.72 9	0.120 15
γ E1	183.514 17	0.98 5
γ M1	186.29 4	~0.024
γ E1	186.86 4	0.147 15
γ E1	193.837 22	0.324 18
γ M1	195.21 3	0.77 4
γ E1	197.33 8	0.060 15

Photons (^{153}Tb) (continued)

γ_{mode}	γ(keV)	γ(%)†
γ E1	206.27 4	0.222 21
γ M1	208.07 3	0.58 4
γ E1	210.24 3	1.50 9
γ E1	212.048 10	30.0 15
γ	216.01 15	0.048 15
γ	223.66 15	0.066 12
γ	224.6 3	~0.014
γ	229.51 25	0.027 12
γ	232.71 25	0.027 12
γ	233.98 6	0.084 21
γ	238.41 25	0.033 9
γ [M1+E2]	239.56 4	~0.021
γ	241.11 4	0.019 9
γ M1	249.63 3	2.28 12
γ M1	258.97 4	0.078 18
γ (E1)	262.046 22	0.54 5
γ M1	267.26 7	0.060 15
γ	268.2 4	~0.012
γ E2	273.91 4	0.051 15
γ M1	275.37 4	0.23 5
γ	277.6 6	~0.009
γ	278.66 15	0.066 18
γ	280.37 8	~0.011
γ	291.74 7	0.054 18
γ	292.8 3	0.021 9
γ	295.0 3	~0.015
γ	298.65 11	0.027 12
γ E2+46%M1	299.59 8	0.129 15
γ E1	303.610 22	0.85 6
γ	310.63 8	~0.018
γ	310.9 3	~0.015
γ	312.02 9	~0.012
γ M1	315.48 4	0.45 5
γ E1	316.04 4	0.40 4
γ	318.61 20	0.063 12
γ	319.95 5	0.279 24
γ E1	325.31 3	0.042 15
γ M1	327.20 5	0.19 4
γ	328.6 4	~0.021
γ E2	332.38 5	0.141 24
γ	338.9 5	~0.027
γ M1	340.50 8	0.228 24
γ M1	346.16 7	0.057 9
γ	348.78 5	0.045 9
γ	352.1 3	~0.018
γ	353.2 5	<0.018
γ M1	355.05 5	0.204 21
γ	356.9 4	~0.015
γ M1	361.23 9	0.15 4
γ	362.88 6	0.045 15
γ	364.56 20	~0.024
γ E2+33%M1	368.74 4	0.045 15
γ	371.15 8	0.093 12
γ E2+44%M1	379.3 3	0.027 9
γ	382.5 4	~0.018
γ	390.70 10	0.033 12 ?
γ	392.1 5	~0.015
γ	393.20 8	0.12 3 ?
γ	395.70 12	~0.030
γ	398.4 4	~0.030
γ	400.58 5	0.16 3
γ	404.7 5	0.024 9
γ M1	406.86 4	0.075 15
γ	410.37 14	0.033 6
γ	412.8 5	~0.015
γ	417.28 20	~0.033
γ [M1+E2]	419.56 4	~0.012
γ M1+31%E2	420.63 15	0.051 18
γ	423.4 3	~0.012
γ	433.5 4	~0.009
γ M1	436.323 22	0.372 18
γ	442.15 4	0.156 15
γ M1	448.42 4	~0.09
γ M1	455.38 3	<0.030
γ E2+48%M1	462.55 9	~0.018 ?
γ	467.25 3	0.300 18
γ	471.2 4	~0.018
γ	473.5 6	~0.009
γ	477.0 6	~0.009
γ	479.72 23	0.016 6
γ E1	481.99 5	0.057 6
γ E1	484.08 10	0.045 5
γ	488.82 8	0.072 15
γ E1	494.52 8	0.054 6

Photons (^{153}Tb) (continued)

γ_{mode}	γ(keV)	γ(%)†
γ	496.52 7	0.204 15
γ	499.80 6	0.023 6
γ	501.5 4	~0.012
γ M1	503.57 20	0.048 12
γ	504.82 6	0.048 12
γ M1	507.18 4	0.108 15
γ M1	508.83 3	0.168 24
γ	512.5 4	~0.021
γ	513.74 20	0.045 9
γ	515.26 20	0.029 6
γ	523.8 4	~0.018
γ	525.68 6	0.120 12
γ M1	530.46 5	0.099 12
γ	533.09 5	0.084 9
γ E2+31%M1	541.37 6	0.087 18
γ	542.5 5	~0.021
γ M1	548.40 11	0.120 12
γ	550.3 5	~0.018
γ E1	552.88 4	0.108 6
γ [E1]	553.98 4	~0.012
γ	555.6 5	~0.015
γ E1	557.32 8	0.057 5
γ	564.61 10	0.027 9
γ	566.2 3	0.033 9
γ	570.2 5	~0.015
γ M1	571.56 7	0.060 12
γ	574.07 9	~0.018
γ	576.8 4	~0.015
γ	579.89 4	0.123 9
γ	581.8 6	~0.015
γ	586.2 3	0.021 9
γ	591.2 4	~0.024
γ	594.6 3	0.030 12
γ	598.61 7	0.060 12 ?
γ	599.47 8	~0.012
γ E2	605.41 20	~0.024
γ	607.5 4	~0.024
γ	610.2 5	~0.012
γ	613.1 5	~0.012
γ	615.63 9	~0.018
γ M1+47%E2	618.01 8	~0.018
γ	621.85 5	0.024 6
γ	629.73 4	0.348 15
γ E2	636.00 6	0.081 18
γ	638.31 10	0.096 21
γ	646.8 3	0.024 9
γ E1	653.21 4	0.143 7
γ E1,E2	665.31 3	0.300 18
γ	667.9 3	0.048 12
γ	671.8 5	~0.021
γ E1	674.05 10	0.093 18
γ	678.61 20	0.057 9
γ	682.14 19	0.045 9
γ E1	690.06 5	0.243 12
γ [E1]	695.58 4	~0.021
γ	698.78 20	~0.021
γ	704.0 6	0.010 5
γ [E1]	705.80 5	0.037 6
γ	711.42 6	0.103 7
γ	713.6 5	~0.015
γ E2	719.23 5	0.045 5 ?
γ	721.37 7	0.087 12
γ E1	727.80 7	0.096 9
γ	731.62 5	~0.015
γ	733.0 6	0.033 12
γ E1	736.32 4	0.126 12
γ [E1]	739.76 18	0.33 3
γ	742.21 5	~0.042
γ	743.7 8	<0.012
γ	745.5 5	~0.018
γ [M1+E2]	747.81 3	~0.024
γ	750.2 5	~0.015
γ	754.0 4	0.066 12
γ E1	755.73 4	0.117 15
γ M1	761.696 25	0.078 9
γ	765.1 3	~0.018
γ	771.4 4	0.024 9
γ	774.4 4	0.042 9
γ	777.13 9	~0.024
γ	779.63 6	0.096 9
γ	782.72 5	<0.024
γ	785.65 3	0.267 12
γ	788.8 6	0.024 9
γ [M1+E2]	793.74 18	0.029 8

Photons (^{153}Tb) (continued)

γ_{mode}	γ(keV)	γ(%)†
γ	796.8 6	0.021 8
γ	798.97 10	0.067 8
γ	807.7 4	0.022 9
γ	812.2 4	0.019 9
γ E2+35%M1	816.02 3	0.261 21
γ	821.19 6	0.033 12
γ	824.5 5	~0.018
γ	826.25 4	0.060 15 ?
γ	827.65 5	0.120 21
γ E1	835.437 20	0.99 4
γ E1	841.55 17	0.035 7 ?
γ E1	845.66 4	0.345 18
γ	849.05 9	~0.011
γ E1	852.00 4	0.258 18
γ M1	857.59 3	0.177 15
γ	859.86 18	0.057 9
γ E1	865.50 4	0.204 12
γ	869.0 4	0.029 9
γ E1	870.93 9	0.027 9
γ	875.4 3	~0.009
γ	878.8 6	~0.011
γ	880.6 3	0.045 12
γ (E1)	882.19 6	0.039 12
γ	883.6 4	0.027 12
γ	885.90 10	0.036 12 ?
γ	890.8 6	~0.015
γ E1	895.86 5	0.035 5
γ	899.3 6	~0.009
γ E1,E2	903.646 22	0.60 4
γ E1	906.09 4	0.44 3
γ E1,E2	912.4 3	0.020 6
γ	914.7 6	0.014 6
γ	916.5 5	~0.021
γ M1	918.12 3	0.105 15
γ E1	925.50 4	0.117 6
γ	929.8 5	~0.012
γ	934.1 6	~0.008
γ	936.61 10	~0.021
γ E1	937.42 5	0.123 9
γ E1	945.209 20	0.83 3
γ	948.39 17	~0.021
γ	951.6 6	~0.015
γ E1	955.43 4	0.024 9
γ E1,E2	956.37 9	0.039 12
γ	958.0 6	0.014 6
γ	965.24 12	0.029 6
γ	967.2 6	0.014 6
γ E1	972.45 3	0.333 21
γ	979.34 12	0.034 6
γ	982.2 3	0.015 6
γ	989.0 5	~0.024
γ E1	991.857 24	1.01 5
γ M1	997.09 8	0.027 6
γ	1012.04 15	0.025 5
γ	1015.10 6	0.072 5
γ	1016.8 6	~0.017
γ	1019.5 6	0.013 6
γ E1	1022.05 6	0.076 6
γ	1024.98 7	~0.015
γ	1030.9 6	~0.011
γ	1033.3 8	~0.006
γ [E1]	1035.28 4	~0.015
γ E1	1036.76 17	0.039 12
γ E1,E2	1051.42 7	0.049 6
γ	1054.95 13	0.037 7
γ E1	1060.066 25	0.132 12
γ	1061.8 3	0.038 8
γ	1066.14 9	0.046 9
γ	1068.7 3	0.036 12
γ E1	1070.6 3	~0.024
γ	1076.6 4	0.016 6
γ E1	1078.42 9	0.059 9
γ E2	1085.52 11	0.019 4
γ E1	1090.25 6	0.023 5
γ	1098.9 5	~0.012
γ E1	1101.629 23	0.327 15
γ	1106.01 12	0.051 5
γ	1107.6 3	~0.011
γ	1111.41 18	0.014 4
γ E1	1118.58 10	0.049 4
γ	1131.82 6	~0.0048
γ	1136.0 6	~0.008
γ E1	1139.04 8	0.067 4
γ [M1+E2]	1144.53 9	0.0144 18

Photons (^{153}Tb) (continued)

γ_{mode}	γ(keV)	γ(%)†
γ	1153.26 25	0.0069 12
γ	1157.2 4	0.0060 15
γ	1173.25 22	0.010 3
γ E1	1179.27 8	0.0288 24
γ E1	1198.86 9	0.070 4
γ	1203.3 4	0.0045 12
γ	1210.15 10	0.0066 24
γ	1212.8 6	~0.0042
γ E1	1218.27 9	0.0348 24
γ	1231.07 7	0.039 6
γ	1233.7 8	~0.007
γ E1	1272.64 7	0.0459 24
γ	1294.54 20	0.0144 15
γ	1299.9 4	0.0054 15
γ	1313.19 14	0.0027 12
γ	1322.91 20	0.0102 21
γ	1332.91 20	0.0036 15
γ	1343.01 20	0.0042 6
γ	1350.6 3	0.0033 6
γ E1	1360.00 12	0.0246 21
γ	1380.63 10	~0.0014
γ	1401.56 12	0.0036 9

† 3.3% uncert(syst)

Atomic Electrons (^{153}Tb)

⟨e⟩=42.2 10 keV

e_{bin}(keV)	⟨e⟩(keV)	e(%)
2 - 4	0.052	3.14 21
7	3.9	55 6
8	3.3	41 5
11 - 19	0.331	1.85 10
33	4.3	13.1 12
34	6.2	18.2 18
35 - 36	1.07	3.03 25
37	1.22	3.26 19
38	0.254	0.67 4
40	2.8	7.1 7
41	1.21	2.9 3
42 - 59	1.70	3.60 12
60	5.2	8.7 5
61 - 100	2.79	3.42 7
101	1.13	1.12 6
102 - 151	1.78	1.47 5
152 - 160	0.310	0.196 11
162	1.68	1.04 6
163 - 212	1.34	0.680 22
214 - 262	0.327	0.138 9
265 - 314	0.41	0.146 12
315 - 365	0.120	0.036 5
367 - 416	0.102	0.026 3
417 - 466	0.169	0.038 5
467 - 516	0.078	0.0158 15
517 - 566	0.038	0.007 1
567 - 616	0.043	0.0073 20
617 - 666	0.025	0.0040 7
667 - 715	0.029	0.0042 5
717 - 766	0.035	0.0047 9
767 - 816	0.043	0.0054 5
817 - 866	0.029	0.0034 6
867 - 916	0.0237	0.00266 18
917 - 966	0.0211	0.00225 16
967 - 1016	0.0101	0.00102 12
1017 - 1066	0.0084	0.00080 10
1067 - 1111	0.0031	0.000282 25
1117 - 1166	0.00159	0.000140 17
1168 - 1217	0.0014	0.00012 4
1218 - 1265	0.00095	7.7 14 $\times10^{-5}$
1271 - 1316	0.00065	5 1 $\times10^{-5}$
1321 - 1360	0.00015	1.1 3 $\times10^{-5}$
1372 - 1402	1.4 $\times10^{-5}$	1.0 4 $\times10^{-6}$

Continuous Radiation (^{153}Tb)

⟨β+⟩=0.195 keV;⟨IB⟩=0.62 keV

E_{bin}(keV)		⟨ ⟩(keV)	(%)
0 - 10	β+	8.2×10^{-8}	9.8×10^{-7}
	IB	0.00064	
10 - 20	β+	3.99×10^{-6}	2.39×10^{-5}
	IB	0.00020	0.00124
20 - 40	β+	0.000127	0.000387
	IB	0.046	0.124
40 - 100	β+	0.0062	0.0080
	IB	0.115	0.25
100 - 300	β+	0.126	0.064
	IB	0.037	0.018
300 - 600	β+	0.062	0.0172
	IB	0.140	0.031
600 - 1300	IB	0.28	0.034
1300 - 1579	IB	0.0025	0.00019
	Σβ+		0.090

$^{153}_{66}$Dy(6.4 1 h)

Mode: ϵ(99.9906 14 %), α(0.0094 14 %)
Δ: -69146 5 keV
SpA: 3.20×10^{6} Ci/g
Prod: ^{152}Gd(α,3n); ^{154}Gd(^{3}He,4n)

Alpha Particles (^{153}Dy)

⟨α⟩=0.326 keV

α(keV)	α(%)
3304 4	$\sim2\times10^{-6}$
3465 4	0.0094

Photons (^{153}Dy)

⟨γ⟩=673 8 keV

γ_{mode}	γ(keV)	γ(%)†
Tb L_ℓ	5.546	0.50 5
Tb L_α	6.269	13.1 9
Tb L_η	6.284	0.174 14
Tb L_β	7.113	12.1 11
Tb L_γ	8.198	1.86 20
Tb $K_{\alpha2}$	43.744	42.7 15
Tb $K_{\alpha1}$	44.482	77 3
Tb $K_{\beta1}'$	50.338	22.9 8
Tb $K_{\beta2}'$	51.958	6.8 3
γM1	70.79 4	0.058 13
γM1	71.13 6	~0.026
γ	78.39 20	0.051 19
γM1+1.2%E2	80.79 3	9.0 5
γM2	82.50 5	0.64 6
γ	88.22 8	0.147 19
γ	88.89 20	0.27 3
γM1	93.06 3	0.76 4
γ	94.29 20	0.192 19
γ	94.89 20	0.28 3
γ	95.99 20	0.058 19
γ	96.69 20	0.058 19
γM1+2.5%E2	99.69 4	8.1 6
γ	124.39 20	0.090 19
γ	124.97 8	0.173 19
γ	127.19 10	0.243 13
γM1	128.16 5	0.218 13
γE1	132.98 4	0.205 13
γ[E1]	143.27 6	0.128 19
γ	144.19 20	0.19 3
γM1+17%E2	147.45 4	2.69 19
γM1	149.03 5	0.61 6
γ	157.79 6	0.090 13

Photons (^{153}Dy) (continued)

γ_{mode}	γ(keV)	γ(%)†
γ[M1+E2]	159.72 5	0.045 13
γ	162.29 9	~0.019 ?
γM1	173.42 4	0.269 19
γM1	182.32 7	0.205 19
γ(M1)	185.37 7	0.096 13
γ[M1+E2]	185.70 9	0.058 13
γ	188.5 5	~0.038
γM1	190.50 5	0.70 4
γ	191.54 10	0.06 3
γE1	193.79 8	0.30 3
γE2+31%M1	204.27 7	0.22 4
γ[M1+E2]	210.14 7	0.11 3
γE1	213.77 3	8.4 4
γM1	218.58 6	1.12 6
γ	235.37 8	0.154 19
γ[M1+E2]	240.51 4	0.275 19
γM1	242.09 5	0.102 13
γM1	244.20 4	3.07 19
γE1	247.39 6	0.38 3
γM1	254.21 3	6.4
γ	258.0 3	0.16 3
γ	260.76 10	~0.032 ?
γ	262.38 8	0.70 16
γ	264.38 9	~0.038 ?
γ	269.86 7	0.045 19 ?
γ	270.61 11	0.14 3
γ[E1]	272.46 6	0.134 19
γE2	274.51 5	5.3 3
γ[E1]	280.90 6	0.160 19
γ	283.28 5	0.102 19
γ(M1)	289.01 7	0.17 5
γM1	290.77 6	0.43 3
γ	295.49 10	0.13 3
γ	296.02 6	0.15 3
γ	296.60 6	0.90 9
γ	297.50 9	0.141 19
γ	299.88 8	0.051 19
γ(M1)	302.71 8	0.15 3
γ[M1+E2]	305.41 8	0.096 13
γ	305.90 14	0.032 13 ?
γ[M1+E2]	308.75 5	<0.09
γ[M1+E2]	317.65 7	0.122 13
γM1	323.72 4	0.93 10
γM1	325.00 4	0.51 6
γ	326.69 13	0.058 19 ?
γ[E2]	331.35 7	0.096 13
γ	332.88 8	0.058 13
γM1	334.48 5	0.160 19
γ	335.09 6	0.064 13
γ	337.33 7	0.070 13
γ	340.84 11	0.019 6 ?
γ	350.86 9	0.038 6 ?
γ	361.84 9	~0.013 ?
γ	362.92 7	~0.026 ?
γM1	363.91 5	0.26 3
γ[E1]	365.97 8	0.032 13
γ[E1]	367.43 8	0.051 13
γ	369.16 7	0.051 13
γ	370.29 20	0.058 13
γE2	371.56 6	0.63 5
γ	374.19 10	0.14 3
γ(M1)	376.09 7	0.128 19
γ(M1)	376.20 8	0.13 4
γ	378.51 11	0.038 13 ?
γ(M1)	383.69 5	0.160 19
γ	384.69 20	0.096 19
γM1	389.54 4	1.06 5
γ	395.77 8	0.07 3
γ	397.48 13	0.070 13
γ[E1]	400.61 5	0.090 13
γ	403.99 10	~0.019 ?
γ(E2)	405.87 6	0.32 3
γ	408.44 11	0.09 3
γ	411.18 13	0.064 19
γM1+3.7%E2	415.73 5	0.83 10
γ	418.87 8	0.051 13
γ(M1)	420.09 7	0.42 5
γ	424.6 4	0.026 6
γ	426.24 9	0.083 19
γ(M1)	429.57 6	0.21 4
γ(M1)	434.16 5	1.02 13
γ	437.92 13	0.115 19
γ(M1)	441.50 7	0.28 3
γE2	444.71 5	1.0 3

Photons (^{153}Dy) (continued)

γ_{mode}	γ(keV)	γ(%)†
γE2	448.47 7	0.83 10
γM1	451.59 7	0.36 3
γ	456.70 4	0.15 3
γM1	462.63 5	0.61 9
γ	465.17 12	0.096 19
γ	467.62 12	0.064 13
γ	468.69 20	0.17 3
γE1	471.40 5	0.96 5
γ	473.6 4	0.045 13
γ	477.89 20	0.154 19
γ	480.59 20	0.28 3
γ	481.91 8	0.166 19
γ	485.74 6	0.26 3
γ[M1+E2]	491.07 7	~0.05
γ	491.28 9	~0.04
γ(M1)	500.21 6	0.22 3
γ	503.84 9	0.07 3
γ	508.65 8	0.26 5
γE2+38%M1	511.84 5	1.9 3
γ(E1)	513.10 6	0.48 6
γ	514.87 10	0.17 5
γ	515.61 11	0.17 5
γ	518.59 20	0.115 19
γ	522.14 7	0.102 19
γ(E1)	527.04 8	0.41 3
γ	532.35 9	0.11 3
γM1	535.81 7	0.147 19
γE1	537.49 4	0.96 16
γ	543.22 7	0.19 4
γM1	544.61 7	0.38 5
γ[E1]	553.39 6	0.19 5
γ	557.58 8	0.058 13 ?
γM1	562.32 5	0.45 6
γ	571.01 12	0.083 19
γM1+E2	571.86 6	0.26 4
γ	574.76 11	0.051 19 ?
γ	576.25 7	0.17 4
γE2+22%M1	579.42 6	0.35 3
γ	581.29 10	0.083 19
γ	582.48 13	0.29 5
γM1	593.83 6	0.90 10
γ	597.39 20	0.077 19
γ	601.60 11	~0.019 ?
γ	604.6 5	0.051 19
γ	609.48 8	0.13 3
γ	610.88 10	0.051 13 ?
γ(M1)	614.61 12	0.38 4
γ	618.59 20	0.102 19
γ	623.1 4	0.045 13
γ	626.73 8	0.045 13
γ	634.96 7	0.045 13 ?
γ	637.3 4	0.032 13
γ	638.30 7	0.10 3
γ	639.56 7	0.24 3
γ	641.63 7	0.070 13
γ	643.02 10	0.070 19
γ(M1)	644.30 7	0.24 4
γ	646.88 13	0.045 13
γ(E2)	652.42 9	0.17 3
γ	653.4 3	0.077 19
γ	654.60 8	0.064 13
γ	657.7 3	0.058 13
γM1(+E2+E0)	659.93 5	0.90 18
γ	673.57 10	0.211 19
γ	681.40 11	0.128 19
γ	686.03 7	0.090 13
γ	686.16 10	0.058 19
γ	694.81 11	0.077 19
γ	697.20 9	0.102 19
γ	703.96 6	0.20 3
γ(M1)	705.75 6	0.45 13
γ	708.77 11	~0.026
γ	709.22 7	~0.026
γ	711.07 13	0.10 3
γ	713.99 20	0.18 3
γ	719.45 7	0.064 19
γ	722.42 7	0.19 4
γ	725.59 20	0.090 19
γ	726.87 6	0.19 6
γ[M1+E2]	740.72 5	0.23 4
γ	744.79 20	0.090 13
γ	746.19 7	0.18 3
γ	752.65 11	0.14 4
γ	754.6 4	0.14 4

Photons (^{153}Dy) (continued)

γ_{mode}	γ(keV)	γ(%)†
γ	758.29 7	0.115 13
γ	761.82 8	0.14 3
γ	766.2 3	0.102 19
γ	779.80 9	0.20 3
γ	781.75 7	0.122 19
γ	783.76 15	0.090 19
γ	785.59 13	~0.019 ?
γ	788.56 12	0.077 19
γ	790.02 7	0.13 3
γ	793.31 8	0.058 19
γ	793.32 10	0.058 19
γ	795.77 11	0.032 13 ?
γ	802.29 10	0.064 19
γ	804.67 10	0.13 3
γ	820.31 8	0.115 19
γ	827.68 6	0.22 3
γ	829.08 7	0.19 3
γ	831.45 9	0.141 19
γ	835.86 9	0.045 19
γ	835.91 8	0.045 19
γ	841.63 7	0.09 3
γ	842.78 8	0.14 3
γ	846.86 11	0.070 19
γ	848.49 10	0.109 19
γ	850.94 10	0.10 3
γ	857.71 8	0.141 19
γ	864.70 9	0.102 19
γ	869.52 7	0.090 19
γ	871.49 20	0.24 3
γ	873.0 3	0.102 19
γ	879.17 7	0.12 3
γ	886.35 8	0.141 19
γ	892.08 6	0.128 19
γ	896.78 11	0.15 3
γ	900.31 8	0.166 19
γ	920.00 8	0.134 19
γ	921.9 3	0.051 19
γ	928.76 13	0.045 13 ?
γ	937.60 8	0.058 19
γ	939.39 20	0.115 19
γ	941.49 20	0.14 3
γ	944.19 20	0.115 19
γ	945.8 3	0.070 19
γ	950.85 11	0.17 3
γ	952.53 8	0.115 19
γ	954.65 10	0.045 19 ?
γ	959.96 6	0.51 13
γ	963.48 7	0.102 19
γ	966.09 20	0.090 13
γ	972.25 6	0.18 3
γ	974.3 3	0.064 19
γ	978.2 3	0.077 19
γ	979.97 11	0.22 3
γ	989.08 7	0.038 13
γ	999.97 10	0.051 19
γ	1002.49 7	0.42 6
γ	1006.64 13	0.09 3
γ	1012.69 6	0.35 3
γ	1014.19 20	0.13 3
γ	1015.04 9	~0.026 ?
γ	1024.26 9	0.83 13
γ	1026.71 10	0.16 4
γ	1031.2 3	0.10 3
γ	1033.7 4	0.070 19
γ	1035.38 10	0.19 3
γ	1040.03 7	0.13 3
γ	1040.17 6	0.48 10
γ	1050.07 9	1.09 13
γ	1057.08 13	0.19 3
γ	1058.47 11	0.16 3
γ	1062.69 20	0.17 3
γ	1068.33 10	0.15 3
γ	1074.74 10	~0.038 ?
γ	1077.6 4	0.058 19
γ	1081.91 8	0.051 19 ?
γ	1087.27 11	0.19 3
γ	1097.02 8	0.058 19
γ	1099.33 8	0.16 3
γ	1102.19 6	0.10 3
γ	1102.52 10	0.10 3
γ	1104.57 6	0.83 13
γ	1110.96 6	0.13 3
γ	1117.62 10	0.09 3
γ	1119.41 12	0.13 3

Photons (^{153}Dy) (continued)

γ_{mode}	γ(keV)	γ(%)†
γ	1122.47 *8*	0.19 *3*
γ	1128.99 *20*	0.13 *3*
γ	1131.33 *9*	0.045 *19*
γ	1131.47 *10*	~0.04
γ	1131.84 *11*	0.045 *19*
γ	1132.74 *9*	0.15 *3*
γ	1140.05 *12*	0.14 *3*
γ	1142.0 *5*	~0.038
γ	1147.5 *5*	~0.032
γ	1151.40 *6*	0.17 *3*
γ	1153.5 *4*	0.070 *19*
γ	1159.0 *5*	0.045 *19*
γ	1161.00 *10*	0.11 *3*
γ	1161.14 *13*	0.11 *3*
γ	1166.59 *6*	0.13 *3*
γ	1175.36 *6*	0.06 *3*
γ	1175.85 *10*	0.12 *3*
γ	1183.8 *4*	0.032 *13*
γ	1185.62 *7*	0.038 *13* ?
γ	1186.88 *6*	0.038 *13* ?
γ	1191.4 *4*	0.051 *19*
γ	1194.03 *11*	0.17 *3*
γ	1199.20 *11*	0.096 *19*
γ	1201.88 *7*	0.35 *5*
γ	1205.53 *11*	0.13 *3*
γ	1210.98 *8*	~0.019 ?
γ	1215.80 *6*	0.13 *3*
γ	1217.50 *12*	~0.019 ?
γ	1225.17 *8*	0.35 *5*
γ	1230.4 *4*	0.064 *19*
γ	1251.88 *10*	0.109 *19*
γ	1253.92 *8*	0.35 *5*
γ	1266.28 *7*	0.083 *19*
γ	1268.1 *3*	0.11 *3*
γ	1270.1 *3*	0.10 *3*
γ	1271.32 *13*	0.10 *3*
γ	1274.2 *4*	0.070 *19*
γ	1279.41 *10*	0.058 *19*
γ	1281.04 *9*	0.32 *6*
γ	1284.37 *6*	0.16 *3*
γ	1285.17 *13*	0.22 *4*
γ	1287.44 *8*	0.058 *19*
γ	1293.36 *10*	0.070 *19*
γ	1295.67 *10*	0.14 *3*
γ	1297.37 *15*	0.122 *19*
γ	1298.63 *15*	~0.038 ?
γ	1301.81 *16*	0.102 *19*
γ	1306.67 *11*	0.096 *19*
γ	1308.4 *4*	0.051 *19*
γ	1309.52 *12*	0.064 *19*
γ	1315.04 *6*	0.64 *10*
γ	1325.57 *10*	0.032 *13* ?
γ	1340.63 *8*	0.058 *19*
γ	1344.6 *4*	0.045 *19*
γ	1347.98 *10*	0.17 *3*
γ	1375.00 *6*	0.58 *10*
γ	1380.23 *8*	0.51 *13*
γ	1383.23 *9*	0.32 *6*
γ	1389.95 *12*	0.22 *5*
γ	1398.67 *14*	0.058 *19*
γ	1402.12 *7*	0.51 *5*
γ	1410.66 *11*	0.083 *19*
γ	1422.6 *4*	0.058 *19*
γ	1426.78 *15*	0.045 *19*
γ	1431.1 *3*	0.12 *3*
γ	1433.08 *7*	0.14 *3*
γ	1446.39 *10*	0.13 *3*
γ	1453.37 *9*	0.17 *3*
γ	1454.50 *12*	0.19 *3*
γ	1461.01 *14*	0.07 *3*
γ	1466.41 *8*	0.08 *3*
γ	1467.78 *7*	0.07 *3*
γ	1486.78 *10*	0.051 *19*
γ	1495.22 *11*	~0.026 ?
γ	1497.63 *8*	0.045 *19*
γ	1500.0 *5*	~0.026
γ	1508.04 *9*	0.29 *6*
γ	1510.94 *9*	0.064 *19*
γ	1516.52 *12*	0.058 *19*
γ	1518.7 *5*	0.032 *13*
γ	1522.95 *7*	0.09 *3*
γ	1526.5 *4*	0.064 *19*
γ	1528.43 *8*	0.36 *6*
γ	1533.35 *9*	0.051 *19*

Photons (^{153}Dy) (continued)

γ_{mode}	γ(keV)	γ(%)†
γ	1537.20 *7*	0.51 *10*
γ	1548.48 *9*	~0.032
γ	1549.23 *14*	~0.026
γ	1553.8 *5*	~0.038
γ	1556.4 *4*	0.08 *3*
γ	1559.65 *8*	0.10 *3*
γ	1561.95 *8*	0.06 *3*
γ	1565.73 *12*	~0.038 ?
γ	1570.72 *8*	0.42 *5*
γ	1572.82 *9*	0.08 *3*
γ	1577.64 *7*	0.64 *13*
γ	1583.60 *13*	0.35 *6*
γ	1595.36 *9*	0.16 *4*
γ	1608.86 *7*	0.54 *12*
γ	1614.92 *10*	0.08 *3*
γ	1617.83 *10*	0.032 *13* ?
γ	1628.12 *8*	~0.026 ?
γ	1632.9 *3*	0.18 *5*
γ	1637.7 *5*	0.032 *13*
γ	1644.57 *9*	~0.026 ?
γ	1649.51 *6*	0.32 *10*
γ	1658.28 *6*	0.22 *6*
γ	1659.33 *8*	0.16 *5*
γ	1671.98 *7*	~0.038
γ	1675.8 *5*	0.058 *19*
γ	1693.90 *8*	~0.026 ?
γ	1698.72 *6*	0.10 *3*
γ	1710.89 *9*	~0.026 ?
γ	1737.9 *5*	0.051 *19*
γ	1741.83 *7*	0.096 *19*
γ	1749.20 *6*	0.19 *6*
γ	1765.2 *6*	~0.026
γ	1771.53 *13*	<0.013 ?
γ	1777.55 *9*	0.09 *3*
γ	1792.47 *10*	0.032 *13* ?
γ	1798.27 *12*	0.10 *3*
γ	1810.46 *15*	0.032 *13* ?
γ	1824.93 *8*	~0.019 ?
γ	1833.1 *5*	0.051 *19*
γ	1859.12 *10*	0.058 *13*
γ	1913.8 *8*	~0.013
γ	1924.4 *8*	~0.013
γ	1935.3 *5*	0.051 *13*
γ	1949.8 *6*	~0.026
γ	1978.4 *7*	~0.013
γ	2012.04 *12*	~0.013
γ	2024.23 *15*	0.045 *13*

† 4.7% uncert(syst)

Atomic Electrons (^{153}Dy)

$\langle e\rangle$=66.2 *14* keV

e_{bin}(keV)	$\langle e\rangle$(keV)	e(%)
7	5.1	68 *7*
8	2.7	33 *4*
9 - 26	1.21	13.3 *13*
29	9.1	31.6 *17*
31	6.3	20.7 *21*
35 - 45	4.9	12.6 *9*
48	7.4	15.5 *11*
49 - 71	0.23	0.39 *8*
72	3.07	4.3 *3*
73	0.60	0.83 *11*
74	3.7	5.0 *5*
75 - 89	4.8	6.1 *7*
91	2.16	2.38 *18*
92 - 94	0.32	0.34 *11*
95	1.54	1.61 *14*
96 - 145	2.58	2.24 *10*
146 - 194	2.28	1.30 *4*
195 - 201	0.043	0.0220 *20*
202	1.83	0.91 *5*
203 - 251	2.06	0.90 *6*
252 - 301	0.96	0.357 *18*
302 - 351	0.468	0.143 *7*
352 - 401	0.68	0.180 *9*
402 - 452	0.35	0.082 *7*

Atomic Electrons (^{153}Dy) (continued)

e_{bin}(keV)	$\langle e\rangle$(keV)	e(%)
454 - 503	0.35	0.074 *6*
504 - 553	0.237	0.045 *4*
554 - 603	0.147	0.0255 *23*
604 - 653	0.115	0.019 *4*
654 - 703	0.109	0.0161 *23*
704 - 753	0.061	0.0084 *15*
754 - 803	0.048	0.0062 *12*
804 - 851	0.050	0.0060 *12*
856 - 903	0.036	0.0041 *8*
908 - 957	0.056	0.0061 *14*
958 - 1007	0.103	0.0105 *25*
1008 - 1057	0.060	0.0057 *13*
1058 - 1107	0.039	0.0036 *6*
1109 - 1158	0.036	0.0032 *6*
1159 - 1208	0.025	0.0021 *5*
1209 - 1258	0.040	0.0033 *5*
1259 - 1308	0.025	0.0020 *6*
1309 - 1350	0.039	0.0029 *8*
1359 - 1403	0.022	0.0016 *3*
1409 - 1458	0.013	0.00092 *22*
1459 - 1558	0.063	0.0041 *7*
1560 - 1659	0.027	0.0017 *3*
1663 - 1758	0.0096	0.00056 *13*
1763 - 1862	0.0026	0.00015 *4*
1872 - 1971	0.0016	8.3 *23* $\times10^{-5}$
1972 - 2023	0.0006	3.1 *14* $\times10^{-5}$

Continuous Radiation (^{153}Dy)

$\langle\beta+\rangle$=10.4 keV;$\langle$IB$\rangle$=1.21 keV

E_{bin}(keV)		$\langle\ \rangle$(keV)	(%)
0 - 10	β+	1.84×10^{-7}	2.19×10^{-6}
	IB	0.00116	
10 - 20	β+	9.4×10^{-6}	5.6×10^{-5}
	IB	0.00064	0.0043
20 - 40	β+	0.000322	0.00098
	IB	0.034	0.093
40 - 100	β+	0.0192	0.0244
	IB	0.137	0.29
100 - 300	β+	0.95	0.435
	IB	0.040	0.020
300 - 600	β+	4.74	1.06
	IB	0.155	0.034
600 - 1300	β+	4.72	0.63
	IB	0.61	0.067
1300 - 2170	IB	0.23	0.0153
	$\Sigma\beta$+		2.2

$^{153}_{67}$Ho(2.0 *1* min)

Mode: ϵ(99.949 *25* %), α(0.051 *25* %)

Δ: -65023 *10* keV

SpA: 6.1×10^{8} Ci/g

Prod: ^{144}Nd(^{14}N,5n); ^{142}Nd(^{14}N,3n); ^{147}Sm(^{10}B,4n)

Alpha Particles (^{153}Ho)

α(keV)
3910 *10*

Photons (^{153}Ho)

γ_{mode}	γ(keV)	γ(rel)
γ	109.0 *7*	~2
γ	162.0 *7*	~3
γ	295.80 *15*	100
γ	334.60 *15*	45 *10*
γ	366.10 *15*	4.0 *10*
γ	438.10 *15*	16.0 *20*
γ	638.30 *15*	29 *5*
γ	1087.2 *3*	5.0 *20*
γ	1277.0 *15*	10 *3*

$^{153}_{67}$Ho(9.3 *5* min)

Mode: ϵ(99.82 *8* %), α(0.18 *8* %)
Δ: ~-64963 keV
SpA: 1.32×10^8 Ci/g
Prod: ^{156}Dy(p,4n); ^{141}Pr(^{16}O,4n); ^{142}Nd(^{14}N,3n); ^{147}Sm(^{10}B,4n)

Alpha Particles (^{153}Ho)

α(keV)
4010 *5*

Photons (^{153}Ho)

γ_{mode}	γ(keV)	γ(rel)
γE2	108.91 *13*	100 *4*
γM1+E2	161.77 *13*	84 *5*
γ	198.9 *3*	7.3 *18*
γM1	230.16 *15*	53 *4*
γ	259.02 *15*	
γE2	270.68 *14*	72 *4*
γ	294.95 *19*	
γ	366.02 *15*	92
γ	391.93 *18*	12 *3*
γ	420.12 *15*	
γ	456.71 *15*	
γ	553.7 *3*	27 *4*
γ	565.62 *18*	20 *6*

$^{153}_{68}$Er(37.1 *2* s)

Mode: α(53 *3* %), ϵ(47 *3* %)
Δ: -60670 *210* keV syst
SpA: 1.969×10^9 Ci/g
Prod: ^{142}Nd(^{16}O,5n); ^{141}Pr(^{19}F,7n); ^{147}Sm(^{12}C,6n); ^{58}Ni on ^{107}Ag; ^{58}Ni on Pd; ^{140}Ce(^{20}Ne,7n)

Alpha Particles (^{153}Er)

α(keV)
4674 *10*

$^{153}_{69}$Tm(1.59 *8* s)

Mode: α(~95 %), ϵ(~5 %)
Δ: -54180 *210* keV syst
SpA: 3.76×10^{10} Ci/g
Prod: ^{141}Pr(^{20}Ne,8n); ^{142}Nd(^{19}F,8n); ^{58}Ni on ^{107}Ag

Alpha Particles (^{153}Tm)

α(keV)
5109 *5*

$^{153}_{70}$Yb(4.0 *5* s)

Mode: ϵ
Δ: -47270 *450* keV syst
SpA: 1.69×10^{10} Ci/g
Prod: daughter ^{157}Hf

$^{153}_{71}$Lu($t_{1/2}$ unknown)

Mode: ϵ
Δ: -38380 *540* keV syst
Prod: ^{58}Ni on Mo-Sn targets; ^{107}Ag on V-Ni targets

A = 154

NDS **26**, 281 (1979)

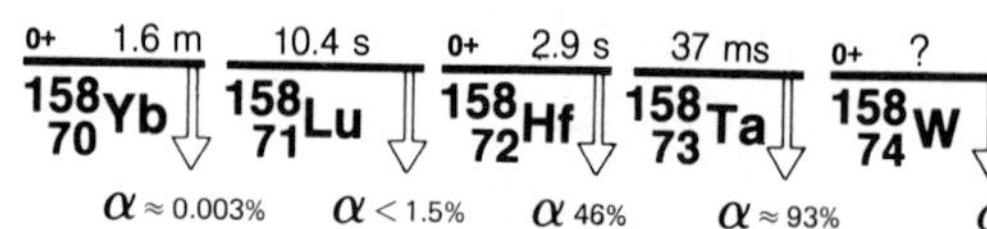

Q_α 4175 *10* Q_α 4970 *220*(syst) Q_α 5402 *4* Q_α 6210 *50* Q_α 6620 *30*

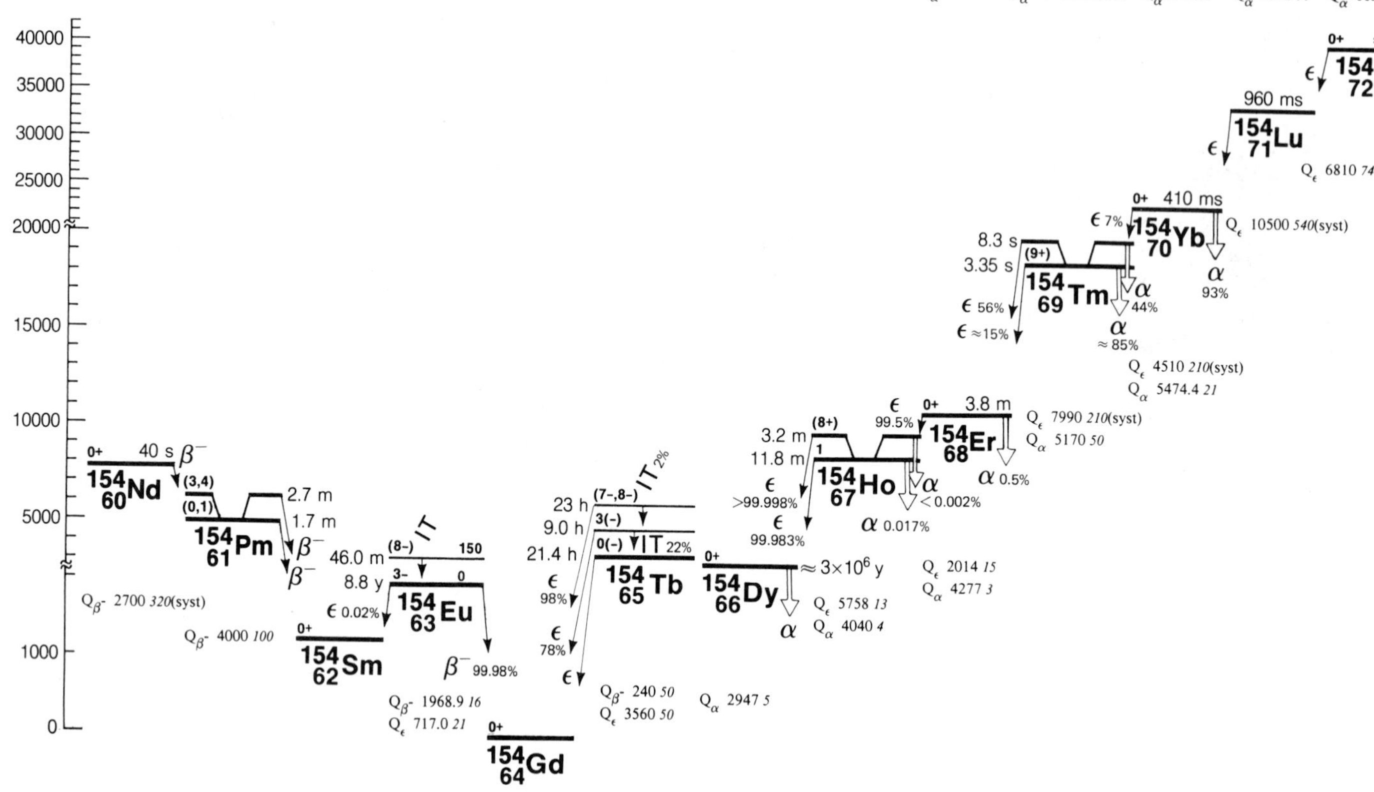

$^{154}_{60}$Nd(40 *10* s)

Mode: β-

Δ: -65770 *300* keV syst

SpA: 1.8×10^9 Ci/g

Prod: fission

Photons (^{154}Nd)

γ_{mode}	γ(keV)
γ	400 *20*
γ	700 *20*

$^{154}_{61}$Pm(1.7 *2* min)

Mode: β-

Δ: -68470 *100* keV

SpA: 7.2×10^8 Ci/g

Prod: ^{154}Sm(n,p)

Photons (^{154}Pm)

⟨γ⟩=1906 *58* keV

γ_{mode}	γ(keV)	γ(%)†
Sm L_ℓ	4.993	0.131 *24*
Sm L_η	5.589	0.06 *1*
Sm L_α	5.633	3.5 *6*
Sm L_β	6.330	3.3 *6*
Sm L_γ	7.216	0.45 *8*
Sm $K_{\alpha2}$	39.522	6.9 *10*
Sm $K_{\alpha1}$	40.118	12.6 *19*
Sm $K_{\beta1}'$	45.379	3.6 *6*
Sm $K_{\beta2}'$	46.819	1.08 *16*
γ E2	82.11 *8*	12.6 *19*
γ E2	184.90 *8*	4.8 *10*
γ	359.09 *20*	
γ	414.99 *20*	1.14 *17*
γ	664.81 *15*	1.40 *24*
γ	700.41 *16*	0.48 *12*
γ	754.5 *6*	4.7 *5*
γ [E1]	839.76 *12*	12.9 *7*
γ	891.75 *14*	6.9 *5*
γ [E2]	911.44 *15*	4.7 *4*
γ [E1]	921.87 *13*	8.6 *5*
γ	962.72 *14*	3.7 *3*
γ	970.24 *15*	5.2 *4*
γ [E2]	1017.84 *15*	10.3 *6*
γ	1033.2 *7*	1.0 *3*
γ [M1+E2]	1096.33 *13*	5.8 *4*
γ	1119.1 *6*	<0.8
γ	1148.32 *14*	9.4 *6*
γ	1162.9 *5*	0.80 *23*

Photons (^{154}Pm) (continued)

γ_{mode}	γ(keV)	γ(%)†
γ [E2]	1173.74 *14*	0.015 *7*
γ [E2]	1178.44 *14*	3.7 *3*
γ	1297.8 *8*	~0.43
γ [M1+E2]	1358.64 *13*	0.25 *6*
γ [E1]	1394.24 *15*	12.6 *19*
γ [E2]	1440.74 *13*	0.22 *5*
γ	1809.03 *24*	1.63 *25*
γ	1891.13 *25*	1.4 *3*
γ	1894.7 *6*	0.50 *23*
γ	1988.08 *12*	1.30 *19*
γ	2059.04 *11*	19.4 *21*
γ	2070.19 *13*	1.86 *19*
γ	2141.15 *12*	11.0 *6*
γ	2213.8 *10*	0.56 *16*
γ	2347.9 *3*	1.78 *17*
γ	2370.3 *5*	0.64 *12*
γ	2510.72 *19*	1.36 *16*
γ	2536.3 *3*	0.64 *16*
γ	2592.83 *20*	0.43 *10*
γ	2618.4 *3*	0.39 *16*
γ	2779.1 *10*	0.31 *10*

† 15% uncert(syst)

Atomic Electrons (^{154}Pm)

⟨e⟩=45 *4* keV

e_{bin}(keV)	⟨e⟩(keV)	e(%)
7	2.6	38 *7*
8 - 33	0.50	2.6 *3*
35	8.8	25 *4*
37 - 45	0.250	0.64 *6*
74	1.63	2.2 *3*
75	19.8	26 *4*
80	0.37	0.46 *7*
81	5.0	6.2 *10*
82	1.42	1.7 *3*
138 - 185	2.0	1.33 *19*
368 - 415	0.09	~0.025
618 - 665	0.07	0.012 *6*
693 - 708	0.14	~0.020
747 - 793	0.153	0.0196 *21*
832 - 875	0.37	0.043 *12*
884 - 923	0.26	0.028 *9*
955 - 986	0.261	0.0269 *24*
1010 - 1050	0.18	0.017 *3*
1072 - 1119	0.21	0.019 *9*
1127 - 1174	0.104	0.0091 *13*
1177 - 1178	0.00240	0.000204 *17*
1251 - 1298	0.008	~0.0006
1312 - 1359	0.089	0.0066 *10*
1387 - 1434	0.0174	0.00125 *14*
1439 - 1441	0.000113	7.9 *17* $\times 10^{-6}$
1762 - 1848	0.039	0.0022 *7*
1883 - 1981	0.017	0.0009 *3*
1986 - 2069	0.23	0.011 *4*
2094 - 2167	0.12	0.0057 *21*
2206 - 2301	0.015	0.0007 *3*
2323 - 2369	0.0084	0.00036 *11*
2464 - 2546	0.021	0.00084 *20*
2572 - 2617	0.0039	0.00015 *6*
2732 - 2778	0.0025	9 *4* $\times 10^{-5}$

Continuous Radiation (^{154}Pm)

⟨β-⟩=846 keV;⟨IB⟩=1.8 keV

E_{bin}(keV)		⟨ ⟩(keV)	(%)
0 - 10	β-	0.0200	0.397
	IB	0.033	
10 - 20	β-	0.061	0.405
	IB	0.032	0.22
20 - 40	β-	0.251	0.84
	IB	0.062	0.22
40 - 100	β-	1.91	2.71
	IB	0.17	0.27
100 - 300	β-	22.7	11.1
	IB	0.46	0.26
300 - 600	β-	95	20.9
	IB	0.46	0.108
600 - 1300	β-	416	45.1
	IB	0.45	0.054
1300 - 2500	β-	300	18.2
	IB	0.097	0.0061
2500 - 3078	β-	11.0	0.418
	IB	0.00054	2.1 $\times 10^{-5}$

$^{154}_{61}$Pm(2.7 *1* min)

Mode: β-

Δ: -68470 *100* keV

SpA: 4.51$\times 10^{8}$ Ci/g

Prod: ^{154}Sm(n,p)

Photons (^{154}Pm)

⟨γ⟩=1999 *59* keV

γ_{mode}	γ(keV)	γ(%)†
Sm L_ℓ	4.993	0.22 *4*
Sm L_η	5.589	0.101 *19*
Sm L_α	5.633	6.0 *11*
Sm L_β	6.331	5.7 *11*
Sm L_γ	7.219	0.78 *15*
Sm $K_{\alpha2}$	39.522	12.7 *20*
Sm $K_{\alpha1}$	40.118	23 *4*
Sm $K_{\beta1}$'	45.379	6.7 *10*
Sm $K_{\beta2}$'	46.819	2.0 *3*
γ E2	82.11 *8*	19 *4*
γ E2	184.90 *8*	38.9 *25*
γ	231.3 *3*	13.2 *10*
γ E2	278.0 *7*	~1.0
γ	280.28 *16*	13.3 *15*
γ [E1]	359.11 *19*	3.9 *3*
γ	364.6 *3*	1.45 *22*
γ [E2]	375.52 *24*	1.6 *3*
γ	438.92 *20*	3.5 *3*
γ [E1]	526.0 *3*	1.3 *3*
γ	547.00 *16*	13.3 *8*
γ [E1]	636.5 *7*	0.62 *25* ?
γ	659.2 *4*	0.74 *25*
γ	664.81 *15*	0.51 *9*
γ [E1]	694.54 *20*	1.06 *22*
γ	700.41 *16*	0.18 *5*
γ	743.2 *3*	3.7 *6*
γ [E1]	745.92 *17*	4.1 *6*
γ	834.30 *25*	4.6 *4*
γ [E1]	839.76 *12*	2.73 *25*
γ [E2]	911.44 *15*	0.45 *5*
γ [E1]	914.4 *3*	2.0 *4*
γ [E1]	921.87 *13*	1.82 *17*
γ [E1]	930.82 *16*	6.4 *5*
γ	954.4 *3*	1.8 *3*
γ	962.72 *14*	1.36 *13*
γ [M1+E2]	1096.33 *13*	0.56 *6*
γ	1116.6 *6*	~0.7
γ [E2]	1173.74 *14*	0.7 *3*
γ [E2]	1178.44 *14*	0.35 *5*
γ	1204.9 *6*	2.1 *3*
γ [M1+E2]	1273.0 *3*	2.2 *5*
γ [E2]	1289.93 *24*	0.49 *20*
γ	1318.1 *8*	1.3 *3*
γ [M1+E2]	1358.64 *13*	11.4 *7*
γ [M1+E2]	1394.22 *20*	3.9 *12*
γ [E1]	1394.24 *15*	~0.5
γ	1433.8 *5*	1.6 *3*
γ [M1+E2]	1440.46 *15*	15.5 *25*
γ [E2]	1440.74 *13*	9.8 *18*
γ [M1+E2]	1457.9 *3*	4.2 *5*
γ [M1+E2]	1549.26 *24*	3.4 *8*
γ	1551.5 *6*	2.1 *6*
γ	1612.8 *3*	1.1 *3*
γ [E2]	1625.35 *15*	4.8 *4*
γ	1656.27 *15*	4.8 *4*
γ	1720.74 *18*	0.49 *22*
γ [E2]	1734.15 *23*	2.6 *3*
γ	1797.7 *3*	2.6 *4*
γ	1841.17 *15*	3.8 *4*
γ	2059.04 *11*	7.1 *7*
γ	2141.15 *12*	4.0 *3*

† 12% uncert(syst)

Atomic Electrons (^{154}Pm)

⟨e⟩=87 *7* keV

e_{bin}(keV)	⟨e⟩(keV)	e(%)
7	4.4	64 *13*
8 - 33	0.92	4.8 *6*
35	13.3	38 *7*
37 - 45	0.46	1.18 *12*
74	2.5	3.3 *7*
75	30.1	40 *8*
80	0.57	0.70 *14*
81	7.6	9.4 *18*
82	2.2	2.6 *5*

Atomic Electrons (^{154}Pm)
(continued)

e_{bin}(keV)	⟨e⟩(keV)	e(%)
138	10.3	7.4 *5*
177	1.36	0.77 *5*
178	3.06	1.72 *11*
183 - 231	4.2	2.2 *8*
233 - 280	2.0	~0.8
312 - 359	0.38	0.12 *3*
363 - 392	0.26	~0.07
431 - 479	0.070	0.016 *6*
500 - 547	0.7	~0.14
590 - 636	0.053	0.009 *4*
648 - 696	0.14	~0.02
699 - 746	0.075	0.0105 *18*
787 - 834	0.17	0.022 *10*
838 - 884	0.104	0.0118 *8*
904 - 953	0.09	0.010 *3*
954 - 963	0.006	0.0006 *3*
1050 - 1096	0.028	0.0026 *11*
1109 - 1158	0.056	0.0048 *18*
1166 - 1205	0.009	0.00080 *24*
1226 - 1273	0.080	0.0064 *15*
1282 - 1318	0.22	0.017 *3*
1347 - 1394	0.56	0.040 *7*
1411 - 1505	0.24	0.0167 *21*
1542 - 1624	0.16	0.0099 *19*
1649 - 1733	0.053	0.0031 *4*
1751 - 1840	0.078	0.0044 *13*
2012 - 2094	0.11	0.0055 *17*
2133 - 2140	0.0059	0.00028 *10*

Continuous Radiation (^{154}Pm)

⟨β-⟩=828 keV;⟨IB⟩=1.67 keV

E_{bin}(keV)		⟨ ⟩(keV)	(%)
0 - 10	β-	0.0193	0.385
	IB	0.032	
10 - 20	β-	0.059	0.392
	IB	0.032	0.22
20 - 40	β-	0.244	0.81
	IB	0.062	0.21
40 - 100	β-	1.86	2.63
	IB	0.17	0.27
100 - 300	β-	22.2	10.9
	IB	0.46	0.26
300 - 600	β-	94	20.7
	IB	0.44	0.106
600 - 1300	β-	430	46.5
	IB	0.42	0.050
1300 - 2500	β-	280	17.8
	IB	0.054	0.0036
2500 - 2629	β-	0.073	0.00288
	IB	3.6 $\times 10^{-7}$	1.45 $\times 10^{-8}$

$^{154}_{62}$Sm(stable)

Δ: -72466 *3* keV

%: 22.7 *2*

$^{154}_{63}$Eu(8.8 *1* yr)

Mode: β-(99.98 *1* %), ϵ(0.02 *1* %)

Δ: -71749 *4* keV

SpA: 264 Ci/g

Prod: ^{153}Eu(n,γ)

Photons (^{154}Eu)

$\langle\gamma\rangle$=1253 11 keV

γ_{mode}	γ(keV)	γ(%)†
Gd L_ℓ	5.362	0.116 12
Gd L_η	6.049	0.050 4
Gd L_α	6.054	3.06 21
Gd L_β	6.840	2.94 24
Gd L_γ	7.836	0.43 4
Gd $K_{\alpha2}$	42.309	7.36 25
Gd $K_{\alpha1}$	42.996	13.2 5
Gd $K_{\beta1}$'	48.652	3.90 13
Gd $K_{\beta2}$'	50.214	1.18 5
$\gamma_{\beta-}$[E1]	58.68 3	0.0039 4
$\gamma_{\beta-}$[M1]	80.21 3	~0.0028
γ_ϵE2	82.11 8	~0.0033
$\gamma_{\beta-}$E2	123.100 4	40.5 8
$\gamma_{\beta-}$(M1+E2)	124.46 5	0.0019 7
$\gamma_{\beta-}$	125.4 10	0.0071 21
$\gamma_{\beta-}$	129.5 10	0.0138 21 ?
$\gamma_{\beta-}$E2+29%M1	131.546 23	0.0206 21
$\gamma_{\beta-}$[E2]	134.86 3	0.0099 11
$\gamma_{\beta-}$[M1]	146.04 3	0.026 5
$\gamma_{\beta-}$[M1]	156.25 5	0.0099 20
$\gamma_{\beta-}$	162.1 10	~0.0010
$\gamma_{\beta-}$	165.9 10	0.0023 5
$\gamma_{\beta-}$[E2]	180.745 22	0.0046 11
$\gamma_{\beta-}$	182.2 10	
γ_ϵE2	184.90 8	0.0041 11
$\gamma_{\beta-}$[E1]	188.274 12	0.227 7
$\gamma_{\beta-}$	195.5 5	~0.0020
$\gamma_{\beta-}$	209.4 4	0.0024 8
$\gamma_{\beta-}$	219.4 10	0.0023 9
$\gamma_{\beta-}$	229.0 5	0.0020 8
$\gamma_{\beta-}$E2	232.076 22	0.0213 17
$\gamma_{\beta-}$	237 1	~0.006
$\gamma_{\beta-}$E2	247.968 7	6.60 14
$\gamma_{\beta-}$	260.9 10	0.0020 8 ?
$\gamma_{\beta-}$[E2]	267.52 3	0.014 3
$\gamma_{\beta-}$[E1]	269.77 3	0.0071 11
$\gamma_{\beta-}$	274 1	0.0039 8
$\gamma_{\beta-}$[E1]	279.81 4	0.0030 6
$\gamma_{\beta-}$[E1]	290.02 4	0.0034 7
$\gamma_{\beta-}$[E1]	295.8 4	0.0024 5
$\gamma_{\beta-}$	296 1	~0.0014
$\gamma_{\beta-}$[E1]	301.25 4	0.0099 20
$\gamma_{\beta-}$[M1+E2]	305.12 6	0.018 4
$\gamma_{\beta-}$	312.3 10	0.015 3
$\gamma_{\beta-}$	315.60 3	0.0046 9
$\gamma_{\beta-}$	320 1	~0.0011
$\gamma_{\beta-}$[M1]	322.04 3	0.067 13
$\gamma_{\beta-}$	329.84 5	0.0094 9
$\gamma_{\beta-}$[M1+E2]	337.92 12	0.0020 5
$\gamma_{\beta-}$E2	346.75 4	0.030 6
$\gamma_{\beta-}$	368.2 10	0.0030 6
$\gamma_{\beta-}$[E2]	370.78 4	0.0053 14
$\gamma_{\beta-}$[E2]	375.51 10	~0.0018
$\gamma_{\beta-}$M1+E2	382.03 3	0.0069 8
$\gamma_{\beta-}$[M1+E2]	397.15 4	0.030 6
$\gamma_{\beta-}$[E1]	401.32 3	0.209 7
$\gamma_{\beta-}$[M1+E2]	403.538 23	0.027 5
$\gamma_{\beta-}$	414.3 10	0.0050 10
$\gamma_{\beta-}$	419.61 5	~0.0039
$\gamma_{\beta-}$(M1)	422.11 4	~0.003
$\gamma_{\beta-}$E2	444.518 17	0.507 13
$\gamma_{\beta-}$[E2]	448.27 3	~0.0035
$\gamma_{\beta-}$	463.9 10	0.0043 9
$\gamma_{\beta-}$[E2]	468.08 4	0.057 11
$\gamma_{\beta-}$E2	478.29 4	0.217 7
$\gamma_{\beta-}$	480.6 10	0.0050 10
$\gamma_{\beta-}$[E2]	483.75 3	0.0050 10
$\gamma_{\beta-}$	484.6 10	0.0039 8
$\gamma_{\beta-}$[E1]	489.03 10	0.007 3
$\gamma_{\beta-}$E2	506.49 5	0.0051 5
$\gamma_{\beta-}$	509.9 10	0.037 7
$\gamma_{\beta-}$[E1]	512.0 4	0.033 7
$\gamma_{\beta-}$E2	518.01 3	0.040 4
$\gamma_{\beta-}$[E1]	532.88 9	0.0110 21
$\gamma_{\beta-}$[E2]	546.03 5	0.0109 10
$\gamma_{\beta-}$E2	557.63 3	0.256 7
$\gamma_{\beta-}$	563.4 10	
$\gamma_{\beta-}$[E1]	569.24 10	0.0099 20
$\gamma_{\beta-}$E1	582.06 3	0.841 25
$\gamma_{\beta-}$E1	591.811 23	4.84 11
$\gamma_{\beta-}$[E1]	596.92 19	0.0057 11
$\gamma_{\beta-}$M1+25%E2	598.23 3	0.0086 10
$\gamma_{\beta-}$	600 1	~0.006
$\gamma_{\beta-}$E0+E2+M1	602.85 3	0.037 5
$\gamma_{\beta-}$[M1+E2]	613.34 3	0.092 18
$\gamma_{\beta-}$[E1]	620.57 10	0.0092 18
$\gamma_{\beta-}$E2	625.263 19	0.314 9
$\gamma_{\beta-}$E2	642.47 5	0.005 3
$\gamma_{\beta-}$E2	649.56 3	0.072 8
$\gamma_{\beta-}$[M1+E2]	651.0 3	0.0099 20 ?
$\gamma_{\beta-}$[M1+E2]	664.67 3	0.029 6
$\gamma_{\beta-}$[E1]	668.85 9	0.030 12
$\gamma_{\beta-}$E2+14%M1+E0	676.594 22	0.141 4
$\gamma_{\beta-}$E2+1.8%M1+E0	692.485 17	1.69 3
$\gamma_{\beta-}$M1	715.828 22	0.176 7
$\gamma_{\beta-}$[M1+E2]	722.68 5	~0.0027
$\gamma_{\beta-}$E1	723.356 22	19.7 4
$\gamma_{\beta-}$[E2]	737.71 4	0.0075 11
$\gamma_{\beta-}$E2+2.8%M1	756.808 22	4.34 10
$\gamma_{\beta-}$	774.4 10	~0.010
$\gamma_{\beta-}$[E2]	790.18 10	0.011 3
$\gamma_{\beta-}$[E1]	800.40 9	0.032 5
$\gamma_{\beta-}$[E1]	801.32 10	
$\gamma_{\beta-}$E2	815.585 17	0.464 13
$\gamma_{\beta-}$(E2)	830.30 3	0.0045 6
$\gamma_{\beta-}$[M1+E2]	845.42 3	0.550 18
$\gamma_{\beta-}$[E2]	850.69 3	0.229 7
$\gamma_{\beta-}$E2(+1.2%M1+E0)	873.230 18	11.45 22
$\gamma_{\beta-}$(E1)	880.54 4	0.063 8
$\gamma_{\beta-}$E2(+6.0%M1+E0)	892.79 3	0.462 13
$\gamma_{\beta-}$	898.6 6	0.0020 5
$\gamma_{\beta-}$E1	904.101 22	0.823 25
$\gamma_{\beta-}$	906.1 10	0.0121 24
$\gamma_{\beta-}$	919.2 10	0.0124 25
$\gamma_{\beta-}$E2	924.561 22	0.063 6
$\gamma_{\beta-}$[E2]	928.07 5	0.0025 7
$\gamma_{\beta-}$[E1]	981.14 9	0.0082 21
$\gamma_{\beta-}$	984.5 10	~0.006
$\gamma_{\beta-}$E2	996.329 18	10.29 21
$\gamma_{\beta-}$E2+1.6%M1	1004.775 21	17.9 3
$\gamma_{\beta-}$	1012.8 2	0.0029 12
$\gamma_{\beta-}$[M1+E2]	1022.25 10	0.007 3
$\gamma_{\beta-}$	1033.4 10	0.0121 24
$\gamma_{\beta-}$[E2]	1047.37 3	0.134 13
$\gamma_{\beta-}$	1049.4 1	0.017 4
$\gamma_{\beta-}$	1110 1	~0.003
$\gamma_{\beta-}$(E1)	1118.29 4	0.103 3
$\gamma_{\beta-}$	1124.2 10	0.0071 11
$\gamma_{\beta-}$[E1]	1128.51 4	0.267 7
$\gamma_{\beta-}$	1136.1 10	0.0075 11
$\gamma_{\beta-}$E2	1140.75 3	0.216 7
$\gamma_{\beta-}$	1153.1 5	0.014 4
$\gamma_{\beta-}$(E2)	1160.344 22	0.052 8
$\gamma_{\beta-}$	1170.0 5	~0.0043 ?
$\gamma_{\beta-}$(E1)	1188.6 4	0.082 16
$\gamma_{\beta-}$	1232.1 5	~0.009
$\gamma_{\beta-}$[E1]	1241.39 4	0.131 3
$\gamma_{\beta-}$(E1)	1245.84 10	0.896 21
$\gamma_{\beta-}$E1	1274.54 3	35.5 7
$\gamma_{\beta-}$[M1+E2]	1274.82 3	~0.002
$\gamma_{\beta-}$[M1+E2]	1289.93 3	0.0114 23
$\gamma_{\beta-}$(E1)	1291.46 6	0.0135 21
$\gamma_{\beta-}$E0+E2+M1	1295.34 3	0.0039 7
$\gamma_{\beta-}$	1387.0 5	0.0199 21 ?
$\gamma_{\beta-}$[M1+E2]	1399.27 5	~0.0029
$\gamma_{\beta-}$[M1+E2]	1408.310 21	0.023 3
$\gamma_{\beta-}$[E1]	1414.56 6	0.0038 6
$\gamma_{\beta-}$[E2]	1418.43 3	0.0021 4
$\gamma_{\beta-}$[M1+E2]	1419.22 17	0.0067 18
$\gamma_{\beta-}$[E1]	1425.66 9	~0.0013
$\gamma_{\beta-}$[E1]	1489.70 19	0.0030 5
$\gamma_{\beta-}$(E1)	1493.80 10	0.654 21
$\gamma_{\beta-}$	1510.0 5	0.0050 11 ?
$\gamma_{\beta-}$[E2]	1522.78 3	0.0023 8 ?
$\gamma_{\beta-}$[E2]	1531.409 21	0.0060 4
$\gamma_{\beta-}$[M1+E2]	1537.90 3	0.050 4
$\gamma_{\beta-}$	1593.0 2	1.03 11
$\gamma_{\beta-}$E1	1596.582 20	1.83 9
$\gamma_{\beta-}$[E2]	1667.18 17	0.0020 3
$\gamma_{\beta-}$[E1]	1673.62 9	~0.012
$\gamma_{\beta-}$[M1+E2]	1714.73 10	~0.0006
$\gamma_{\beta-}$	1771.8 6	~0.00028 ?
$\gamma_{\beta-}$[E2]	1837.83 10	0.00082 21
$\gamma_{\beta-}$	1894.9 6	0.00060 21 ?

† uncert(syst): 57% for ϵ, 5.6% for β-

Atomic Electrons (^{154}Eu)

$\langle e\rangle$=52.4 7 keV

e_{bin}(keV)	$\langle e\rangle$(keV)	e(%)
7 - 51	2.92	31.7 22
57 - 72	0.0011	0.0015 7
73	19.2	26.4 8
74 - 112	0.057	0.066 11
115	11.3	9.9 3
116	8.17	7.05 20
117 - 120	0.0033	~0.0028
121	2.73	2.25 7
122	2.09	1.72 5
123	1.30	1.06 3
124 - 173	0.040	0.0297 24
174 - 222	1.07	0.541 16
224 - 273	0.491	0.203 4
274 - 323	0.0091	0.0030 4
325 - 374	0.0142	0.0041 5
375 - 424	0.0376	0.0095 3
425 - 473	0.0296	0.0067 4
475 - 525	0.0158	0.0032 3
526 - 575	0.118	0.0216 6
580 - 626	0.066	0.0108 6
634 - 677	0.75	0.114 7
684 - 730	0.267	0.0380 22
736 - 783	0.0417	0.00553 10
788 - 837	0.372	0.0453 19
838 - 886	0.094	0.0109 3
890 - 934	0.00267	0.000296 16
946 - 995	0.672	0.0704 13
996 - 1046	0.0834	0.00835 16
1047 - 1091	0.00777	0.000717 22
1102 - 1146	0.0035	0.00031 3
1151 - 1196	0.0086	0.00072 3
1208 - 1246	0.277	0.0226 6
1266 - 1295	0.0471	0.00371 8
1319 - 1368	0.00092	6.8 16 $\times10^{-5}$
1369 - 1418	0.00035	2.5 4 $\times10^{-5}$
1419 - 1467	0.00464	0.000321 16
1473 - 1552	0.028	0.0018 5
1585 - 1672	0.0045	0.00028 7
1706 - 1788	1.6 $\times10^{-5}$	9.2 25 $\times10^{-7}$
1829 - 1893	9.8 $\times10^{-6}$	5.3 22 $\times10^{-7}$

Continuous Radiation (^{154}Eu)

$\langle\beta$-$\rangle$=227 keV;$\langle$IB$\rangle$=0.24 keV

E_{bin}(keV)		$\langle\ \rangle$(keV)	(%)
0 - 10	β-	0.229	4.60
	IB	0.0106	
10 - 20	β-	0.66	4.44
	IB	0.0099	0.069
20 - 40	β-	2.52	8.4
	IB	0.018	0.063
40 - 100	β-	14.7	21.5
	IB	0.043	0.068
100 - 300	β-	66	36.3
	IB	0.080	0.048
300 - 600	β-	70	17.0
	IB	0.049	0.0119
600 - 1300	β-	60	6.9
	IB	0.029	0.0036
1300 - 1846	β-	13.4	0.93
	IB	0.00104	7.5 $\times10^{-5}$

$^{154}_{63}$Eu(46.0 *3* min)

Mode: IT
Δ: -71599 *8* keV
SpA: 2.654×10^{7} Ci/g
Prod: ^{154}Sm(p,n); ^{154}Sm(d,2n); ^{153}Eu(n,γ)

Photons (^{154}Eu)

⟨γ⟩=84 *8* keV

γ_{mode}	γ(keV)	γ(%)†
γ	22.00 *6*	~0.044
γ M1+0.6%E2	27.55 *5*	0.65 *7*
γ E1	28.815 *20*	0.66 *7*
γ	31.816 *10*	4.2 *3*
γ E1	35.839 *10*	0.8 *3*
γ	35.854 *10*	9.7 *6*
Eu $K_{\alpha2}$	40.902	8.4 *14*
Eu $K_{\alpha1}$	41.542	15.1 *24*
Eu $K_{\beta1}$'	46.999	4.4 *7*
Eu $K_{\beta2}$'	48.496	1.33 *22*
γ [E1]	68.207 *10*	37
γ [E1]	100.916 *10*	28.3 *9*
γ [M1]	136.754 *14*	0.11 *4*

† 16% uncert(syst)

$^{154}_{64}$Gd(stable)

Δ: -73718 *3* keV
%: 2.18 *3*

$^{154}_{65}$Tb(21.4 *5* h)

Mode: ε
Δ: -70160 *50* keV
SpA: 9.51×10^{5} Ci/g
Prod: ^{153}Eu(α,3n); ^{155}Gd(p,2n); ^{151}Eu(α,n); protons on Gd; protons on Ta; ^{154}Gd(d,2n)

Photons (^{154}Tb)

⟨γ⟩=2364 *38* keV

γ_{mode}	γ(keV)	γ(%)†
Gd L_{ℓ}	5.362	0.32 *4*
Gd L_{η}	6.049	0.119 *11*
Gd L_{α}	6.054	8.4 *7*
Gd L_{β}	6.849	7.8 *8*
Gd L_{γ}	7.871	1.17 *14*
Gd $K_{\alpha2}$	42.309	27.2 *13*
Gd $K_{\alpha1}$	42.996	48.9 *24*
Gd $K_{\beta1}$'	48.652	14.4 *7*
Gd $K_{\beta2}$'	50.214	4.36 *22*
γ[M1]	80.21 *3*	~0.00015
γE2	123.100 *4*	28 *4*
γE2+29%M1	131.546 *23*	0.00108 *23*
γ[E2]	134.86 *3*	0.0200 *24*
γ[M1]	146.04 *3*	0.0082 *17*
γ[M1]	156.25 *5*	0.0031 *7*
γ[E2]	180.745 *22*	0.0025 *6*
γE2	232.076 *22*	0.028 *5*
γE2	247.968 *7*	~2
γ[E1]	269.77 *3*	0.0022 *4*
γ[E1]	279.81 *4*	0.012 *3*
γ[E1]	290.02 *4*	0.014 *3*
γ[E1]	295.8 *4*	0.009 *3*
γ	315.60 *3*	0.0025 *5*
γ	329.84 *5*	0.0125 *24*
γ[E1]	401.32 *3*	0.066 *5*
γ[M1+E2]	403.538 *23*	0.110 *23*
γ(M1)	422.11 *4*	~0.022
γ	429.6 *5*	0.57 *19*
γE2	444.518 *17*	1.02 *5*
γ	470.40 *21*	0.43 *5*
γ[E2]	483.75 *3*	0.020 *4*
γ	489.0 *6*	0.243 *23*
γ[E1]	512.0 *4*	0.12 *5*
γ	536.79 *24*	1.41 *19*
γE2	557.63 *3*	5.8 *4*
γE1	582.06 *3*	0.267 *20*
γ[M1+E2]	588.31 *5*	0.48 *5*
γE0+E2+M1	602.85 *3*	0.28 *3*
γE2	625.263 *19*	0.169 *5*
γ	653.8 *5*	0.10 *5*
γE2+14%M1+E0	676.594 *22*	0.19 *3*
γ	687 *1*	0.07 *3*
γE2+1.8%M1+E0	692.485 *17*	3.40 *16*
γM1	701.29 *6*	0.52 *6*
γE1	705.16 *7*	5.1 *3*
γM1	715.828 *22*	0.72 *4*
γE1	722.08 *6*	8.2 *5*
γ[E2]	737.71 *4*	0.055 *9*
γE2+2.8%M1	756.808 *22*	0.23 *4*
γ	789.5 *9*	0.28 *5*
γE2	815.585 *17*	0.93 *5*
γ[E2]	850.69 *3*	0.93 *5*
γ	864.6 *8*	0.19 *8*
γE2(+1.2%M1+E0)	873.230 *18*	6.2 *3*
γ[E1]	878.33 *6*	3.00 *18*
γ(E1)	880.54 *4*	0.048 *13*
γE2	924.561 *22*	0.084 *15*
γ	945.03 *17*	0.44 *4*
γ	956.9 *7*	0.24 *5*
γE2	996.329 *18*	5.5 *3*
γE2+1.6%M1	1004.775 *21*	0.94 *21*
γ	1016.06 *10*	0.44 *8*
γ	1032.98 *8*	0.47 *8*
γ[E2]	1047.37 *3*	0.99 *8*
γ	1053.43 *15*	~0.14
γ[E2]	1059.39 *7*	0.89 *9*
γ(E1)	1118.29 *4*	2.47 *19*
γM1,E2	1123.39 *5*	6.1 *4*
γ[E1]	1128.51 *4*	0.20 *6*
γ(E2)	1160.344 *22*	0.21 *3* ?
γ(E2)	1172.12 *14*	
γ	1175.0 *8*	0.25 *11*
γ(E1)	1188.6 *4*	0.30 *9*
γ[M1+E2]	1190.94 *7*	1.0 *3*
γ	1218.6 *9*	0.17 *5*
γ[E1]	1241.39 *4*	3.12 *23*
γE1	1274.54 *3*	11.2 *8*
γ(E1)	1291.46 *6*	7.4 *4*
γE0+E2+M1	1295.34 *3*	0.029 *6*
γ	1308.9 *3*	0.84 *6*
γ	1325.06 *9*	1.25 *9*
γ	1370.2 *10*	~0.15
γ	1374.43 *17*	0.55 *5*
γ	1391.35 *16*	0.53 *5*
γ	1405 *1*	0.30 *14*
γ[M1+E2]	1408.310 *21*	0.095 *14*
γ[E1]	1414.56 *6*	2.06 *15*
γ[E2]	1418.43 *3*	0.016 *3*
γ	1439.0 *3*	~0.14
γ	1458.53 *17*	1.66 *11*
γ	1481.2 *5*	0.24 *8*
γ[M1]	1506.54 *7*	0.30 *3*
γ	1527.53 *24*	0.21 *5*
γ[E2]	1531.409 *21*	~0.025
γ	1593.79 *20*	0.65 *7*
γ	1606.3 *3*	0.14 *5*
γ	1737.9 *3*	0.69 *7*
γ	1774.53 *20*	0.30 *5*
γ	1839.82 *24*	0.11 *5*
γ	1907.12 *9*	1.41 *11*
γ	1909.39 *20*	1.41 *11*
γ	1973.40 *16*	0.22 *11*
γ	1974.67 *24*	~0.15
γM1,E2	1996.62 *5*	8.0 *5*
γ	2041.97 *9*	2.09 *11*
γM1,E2	2064.16 *7*	7.6
γ	2108.26 *17*	0.37 *8*
γM1,E2	2119.72 *5*	4.5 *3*
γ	2174.35 *18*	~0.015
γM1,E2	2187.26 *7*	10.6 *7*
γ	2219.56 *16*	0.87 *6*
γ	2278.5 *3*	0.33 *4*
γ	2307.51 *8*	1.55 *10*
γ	2342.66 *16*	1.59 *11*
γ	2344.89 *14*	1.59 *11*
γ	2363.27 *12*	0.43 *5*
γ	2376.9 *5*	0.33 *7*
γ	2380.1 *7*	0.33 *7*
γ	2402.5 *3*	0.26 *4*
γ	2430.61 *8*	2.31 *14*
γ	2442.2 *8*	0.08 *3*
γ	2449.5 *6*	0.08 *3*
γ	2467.01 *20*	0.53 *15*
γ	2467.99 *14*	0.84 *19*
γ	2486.37 *12*	1.41 *9*
γ	2500.0 *5*	0.30 *6*
γ	2504.2 *6*	0.24 *8*
γ	2525.1 *7*	0.19 *3*
γ	2532.29 *24*	0.144 *23*
γ	2575.1 *5*	0.084 *23*
γ	2590.11 *20*	0.10 *3*
γ	2599.59 *9*	0.27 *3*
γ	2611.1 *3*	0.11 *5*
γ	2646.3 *7*	0.41 *4*
γ	2655.39 *24*	0.26 *3*
γ	2665.88 *16*	0.44 *4*
γ	2711.7 *5*	0.122 *15*
γ	2716 *1*	~0.030
γ	2722.69 *9*	~0.13
γ	2727.8 *9*	0.046 *15*
γ	2734.2 *3*	~0.030
γ	2750 *1*	0.068 *23*
γ	2770.5 *8*	0.068 *23*
γ	2788.98 *16*	0.74 *5*
γ	2811.3 *10*	0.057 *18*
γ	2820.7 *15*	0.043 *13*
γ	2826.5 *4*	0.62 *4*
γ	2851.2 *9*	0.091 *15*
γ	2866.83 *18*	0.307 *24*
γ	2873.1 *10*	0.068 *15*
γ	2886.6 *4*	0.100 *23*
γ	2900.05 *22*	~1
γ	2908.5 *15*	0.061 *15*
γ	2940.5 *10*	0.061 *23*
γ	2949.6 *4*	0.50 *3*
γ	2967.6 *15*	0.038 *15*
γ	2989.92 *18*	0.50 *3*
γ	2998.9 *15*	~0.030
γ	3009.7 *4*	0.37 *3*
γ	3023.15 *22*	0.84 *8*
γ	3032.4 *8*	0.152 *15*
γ	3039.4 *15*	0.038 *15*
γ	3061.9 *6*	0.175 *15*
γ	3085 *2*	~0.015
γ	3090.5 *10*	0.114 *15*
γ	3103 *2*	~0.015
γ	3122.2 *15*	0.046 *15*
γ	3137.5 *15*	0.061 *15*
γ	3141 *1*	0.114 *23*
γ	3163 *2*	~0.038
γ	3170.8 *10*	0.070 *11*
γ	3185.0 *6*	0.068 *11*
γ	3205 *2*	~0.008
γ	3222.9 *15*	0.031 *12*
γ	3227.6 *10*	0.084 *15*
γ	3263.8 *10*	0.076 *15*
γ	3280 *2*	~0.023
γ	3292 *1*	0.069 *10*
γ	3294.4 *10*	0.076 *15*
γ	3328.3 *15*	~0.023
γ	3345.8 *15*	0.059 *11*
γ	3350.7 *15*	0.042 *11*
γ	3381.4 *15*	~0.015
γ	3414.6 *10*	0.068 *11*
γ	3435 *2*	~0.007
γ	3467.9 *20*	~0.010

† 11% uncert(syst)

Atomic Electrons (^{154}Tb)

⟨e⟩=44 3 keV

e_{bin}(keV)	⟨e⟩(keV)	e(%)
7	3.4	47 6
8	2.5	32 4
30 - 72	2.44	6.6 3
73	13.2	18 3
78 - 106	0.015	0.017 6
115	7.8	6.7 10
116	5.6	4.8 7
121	1.9	1.54 23
122	1.43	1.17 18
123 - 156	0.90	0.73 11
172 - 220	0.30	0.15 7
224 - 273	0.14	0.056 17
278 - 323	0.0051	0.0017 7
328 - 372	0.017	0.0047 20
379 - 428	0.14	0.036 10
429 - 477	0.042	0.009 3
481 - 530	0.35	0.069 12
532 - 581	0.171	0.031 3
582 - 626	0.073	0.0118 18
637 - 686	1.39	0.214 16
687 - 736	0.0587	0.0083 3
737 - 782	0.038	0.0049 9
788 - 830	0.250	0.0305 14
842 - 881	0.0497	0.00574 22
895 - 944	0.020	0.0022 9
945 - 989	0.185	0.0194 13
994 - 1040	0.059	0.0059 4
1045 - 1078	0.19	0.017 4
1110 - 1159	0.070	0.0062 11
1160 - 1191	0.034	0.0028 3
1210 - 1259	0.163	0.0132 9
1266 - 1309	0.048	0.0038 10
1317 - 1366	0.047	0.0035 6
1367 - 1417	0.033	0.0024 10
1418 - 1459	0.015	0.00106 22
1473 - 1556	0.017	0.0011 3
1585 - 1605	0.0018	0.00011 4
1688 - 1773	0.015	0.0009 3
1790 - 1859	0.034	0.0018 6
1899 - 1995	0.162	0.0083 11
2014 - 2112	0.194	0.0095 10
2118 - 2212	0.169	0.0079 9
2218 - 2313	0.054	0.0024 5
2327 - 2423	0.052	0.0022 4
2429 - 2525	0.027	0.00108 22
2530 - 2616	0.0139	0.00054 9
2638 - 2733	0.0059	0.00022 4
2739 - 2836	0.018	0.00063 11
2843 - 2942	0.019	0.00066 22
2948 - 3040	0.015	0.00050 9
3053 - 3139	0.0035	0.000113 17
3155 - 3244	0.0028	8.8 14 $\times 10^{-5}$
3255 - 3349	0.00133	4.0 7 $\times 10^{-5}$
3364 - 3461	0.00066	2.0 5 $\times 10^{-5}$
3466	1.9 $\times 10^{-6}$	~5 $\times 10^{-8}$

Continuous Radiation (^{154}Tb)

⟨β+⟩=9.6 keV;⟨IB⟩=0.83 keV

E_{bin}(keV)		⟨ ⟩(keV)	(%)
0 - 10	β+	4.05 $\times 10^{-8}$	4.81 $\times 10^{-7}$
	IB	0.00101	
10 - 20	β+	2.03 $\times 10^{-6}$	1.21 $\times 10^{-5}$
	IB	0.00055	0.0037
20 - 40	β+	6.9 $\times 10^{-5}$	0.000208
	IB	0.046	0.125
40 - 100	β+	0.00407	0.0052
	IB	0.116	0.25
100 - 300	β+	0.218	0.098
	IB	0.042	0.021
300 - 600	β+	1.39	0.304
	IB	0.141	0.031
600 - 1300	β+	4.98	0.56
	IB	0.33	0.039
1300 - 2500	β+	2.96	0.182
	IB	0.140	0.0084
2500 - 3560	β+	0.000219	8.7 $\times 10^{-6}$
	IB	0.0082	0.00030
	Σβ+		1.15

$^{154}_{65}$Tb(9.0 5 h)

Mode: ε(78.2 7 %), IT(21.8 7 %)
Δ: >-70160 keV
SpA: 2.26$\times 10^{6}$ Ci/g
Prod: protons on Gd; ^{153}Eu(α,3n); ^{155}Gd(p,2n); protons on Ta

Photons (^{154}Tb)

γ_{mode}	γ(keV)	γ(rel)
γ_ϵ[E1]	58.68 3	0.00017 6
γ_ϵ[M1]	80.21 3	~0.006
γ_ϵE2	123.100 4	100 14
γ_ϵ(M1+E2)	124.46 5	0.8 3
γ_ϵE2+29%M1	131.546 23	0.042 5
γ_ϵ[E2]	134.86 3	0.067 8
γ_ϵ[M1]	146.04 3	0.0019 4
γ_ϵ[M1]	156.25 5	0.00072 15
γ_ϵ[E2]	180.745 22	0.012 3
γ_ϵ[E1]	188.274 12	0.010 4
γ_ϵE2	232.076 22	1.57 11
γ_ϵE2	247.968 7	73 7
γ_ϵ[E1]	269.77 3	0.00052 9
γ_ϵ[E1]	279.81 4	0.0066 17
γ_ϵ[E1]	290.02 4	0.0074 19
γ_ϵ[E1]	295.8 4	0.059 22
γ_ϵ[E1]	301.25 4	0.00043 17
γ_ϵ[M1+E2]	305.12 6	0.0008 3
γ_ϵ	315.60 3	0.0123 25
γ_ϵ[M1]	322.04 3	0.0029 12
γ_ϵ	329.84 5	0.69 6
γ_ϵ[M1+E2]	337.92 12	0.85 19
γ_ϵE2	346.75 4	5.2 7
γ_ϵ[E2]	370.78 4	0.018 6
γ_ϵ[E2]	375.51 10	~0.0022
γ_ϵM1+E2	382.03 3	3.48 25
γ_ϵ[M1+E2]	397.15 4	0.046 10
γ_ϵ[E1]	401.32 3	0.0152 12
γ_ϵ[M1+E2]	403.538 23	0.059 15
γ_ϵE1	415.76 7	7.0 5
γ_ϵ	419.61 5	~0.006
γ_ϵ(M1)	422.11 4	~0.010
γ_ϵE2	444.518 17	3.41 16
γ_ϵ[E2]	448.27 3	~0.08
γ_ϵM1	461.4 3	0.84 7
γ_ϵ[E2]	468.08 4	0.0025 10
γ_ϵE2	478.29 4	0.009 3
γ_ϵ[E2]	483.75 3	0.011 3
γ_ϵ	484.9 4	0.52 7
γ_ϵ[E1]	489.03 10	0.009 4
γ_ϵ	492.2 5	0.45 7
γ_ϵE2	506.49 5	2.2 3
γ_ϵ[E1]	512.0 4	0.8 3
γ_ϵE2	518.01 3	20.5 9
γ_ϵ[E1]	532.88 9	0.14 5
γ_ϵE1	540.22 7	64.5
γ_ϵ[E2]	546.03 5	0.24 3
γ_ϵE2	557.63 3	1.03 19
γ_ϵ[E1]	569.24 10	0.012 3
γ_ϵE1	582.06 3	0.061 5
γ_ϵE1	591.811 23	0.21 7
γ_ϵM1+25%E2	598.23 3	4.4 4
γ_ϵE0+E2+M1	602.85 3	0.128 21
γ_ϵ[M1+E2]	613.34 3	0.14 3
γ_ϵ[E1]	620.57 10	0.011 3
γ_ϵE2	625.263 19	0.84 4
γ_ϵE2	642.47 5	2.3 3
γ_ϵE2	649.56 3	36.4 21
γ_ϵ	660.3 5	0.39 13
γ_ϵ[M1+E2]	664.67 3	0.044 10
γ_ϵ[E1]	668.85 9	0.37 12
γ_ϵE2+14%M1+E0	676.594 22	10.3 5
γ_ϵE2+1.8%M1+E0	692.485 17	11.4 5
γ_ϵ[M1+E2]	714.60 12	0.084 24
γ_ϵM1	715.828 22	0.38 6
γ_ϵ[M1+E2]	722.68 5	1.2 3
γ_ϵE1	723.356 22	0.852 19
γ_ϵ[E2]	737.71 4	0.026 4
γ_ϵ[E1]	753.68 14	~0.8 ?
γ_ϵE2+2.8%M1	756.808 22	8.8 4
γ_ϵ	796.4 5	0.84 7
γ_ϵ[E1]	800.40 9	0.40 9
γ_ϵE2	815.585 17	3.13 16
γ_ϵ	826.4 5	0.45 7
γ_ϵ(E2)	830.30 3	2.27 19
γ_ϵ[M1+E2]	845.42 3	0.84 7
γ_ϵ[E2]	850.69 3	0.50 8
γ_ϵ	857.3 4	0.32 7
γ_ϵE2(+1.2%M1+E0)	873.230 18	30.5 14
γ_ϵ(E1)	880.54 4	1.20 15
γ_ϵE2(+6.0%M1+E0)	892.79 3	10.2 5
γ_ϵE1	904.101 22	0.036 3
γ_ϵ[E1]	922.25 8	1.3 4
γ_ϵE2	924.561 22	4.6 4
γ_ϵ[E2]	928.07 5	1.3 3
γ_ϵ	953.05 13	1.87 13
γ_ϵ	965.1 3	1.10 13
γ_ϵ[E1]	981.14 9	~0.10
γ_ϵ	981.4 4	1.03 19
γ_ϵ	983.90 19	1.48 19
γ_ϵE2	996.329 18	27.4 13
γ_ϵE2+1.6%M1	1004.775 21	36.1 16
γ_ϵ	1013.05 12	0.32 7
γ_ϵ	1021.0 3	0.97 13
γ_ϵ	1033.26 13	1.55 13
γ_ϵ	1041.91 10	0.71 7
γ_ϵ[E2]	1047.37 3	0.46 4
γ_ϵ	1054.15 11	0.71 19
γ_ϵ[E1]	1058.23 8	0.90 13
γ_ϵ(M1+E2)	1061.35 12	0.45 6
γ_ϵ	1072.20 7	1.16 13
γ_ϵ	1084.60 13	1.23 13
γ_ϵ	1102.0 3	0.90 13
γ_ϵ	1104.97 18	~0.19
γ_ϵ(E1)	1118.29 4	0.72 8
γ_ϵ[E1]	1128.51 4	5.1 3
γ_ϵE2	1140.75 3	4.76 25
γ_ϵ	1149.02 12	3.2 5
γ_ϵ	1152.46 17	7.2 10
γ_ϵ(E2)	1160.344 22	0.113 23
γ_ϵ	1177.88 10	0.97 13
γ_ϵ(E1)	1188.6 4	2.0 7
γ_ϵ	1208.17 7	1.68 13
γ_ϵ	1214.1 6	0.35 5
γ_ϵ	1229.24 12	1.9 3
γ_ϵ	1233.6 3	~0.39 ?
γ_ϵ	1237.5 8	0.5 3
γ_ϵ[E1]	1241.39 4	0.91 12
γ_ϵ(E1)	1245.84 10	1.09 20
γ_ϵ	1258.10 10	5.4 4
γ_ϵ	1265.34 13	~0.26
γ_ϵE1	1274.54 3	2.6 5
γ_ϵ[M1+E2]	1274.82 3	1.0 4
γ_ϵ	1280.57 12	~0.6
γ_ϵ	1288.44 17	4.6 4
γ_ϵ[M1+E2]	1289.93 3	0.017 4
γ_ϵE0+E2+M1	1295.34 3	0.014 3
γ_ϵ	1309.43 10	0.39 7
γ_ϵ	1330.8 6	0.19 7
γ_ϵ	1339.72 7	0.97 13
γ_ϵ	1346.0 6	0.13 7
γ_ϵ	1360.3 6	0.13 7
γ_ϵ	1377.6 9	~0.2 ?
γ_ϵ	1387.6 3	0.97 7
γ_ϵ[M1+E2]	1399.27 5	~1
γ_ϵ[M1+E2]	1408.310 21	0.051 11
γ_ϵ[E2]	1418.43 3	0.0073 17
γ_ϵ	1419.99 17	2.4 5
γ_ϵ[E1]	1425.66 9	~0.016
γ_ϵ	1450.4 2	0.45 5
γ_ϵ	1490.17 10	3.4 3
γ_ϵ(E1)	1493.80 10	0.79 3

Photons (^{154}Tb)
(continued)

γ_{mode}	γ(keV)	γ(rel)
γ_ϵ	1515.9 4	1.03 13
γ_ϵ	1520.46 7	1.87 13
γ_ϵ[E2]	1522.78 3	1.2 4
γ_ϵ[E2]	1531.409 21	0.013 4
γ_ϵ[M1+E2]	1537.90 3	0.076 8
γ_ϵ	1553.24 17	0.77 13
γ_ϵE1	1596.582 20	0.079 7
γ_ϵ	1619.2 3	1.29 13
γ_ϵ	1652.0 5	0.39 13
γ_ϵ[E1]	1673.62 9	~0.15
γ_ϵ	1715 1	0.13 7
γ_ϵ	1721 2	~0.5
γ_ϵ[E1]	1815.03 8	0.45 13
γ_ϵ	1858.8 3	0.71 7
γ_ϵ	1887 1	0.13 7
γ_ϵ	1894.91 20	0.55 8
γ_ϵ	1901.5 5	0.23 10
γ_ϵ	1905.83 12	0.52 7
γ_ϵ	1931.0 5	0.39 7
γ_ϵ	1934.68 10	2.39 19
γ_ϵ	1949 1	0.13 7
γ_ϵ	1964.98 7	6.4 5
γ_ϵ	1997.75 17	1.2 3
γ_ϵ	2014.9 7	0.19 7
γ_ϵ	2025.0 3	0.65 7
γ_ϵ	2035.5 4	0.13 7
γ_ϵ	2054.2 4	0.29 4
γ_ϵ	2084.9 3	0.41 4
γ_ϵ	2101.6 7	0.13 7
γ_ϵ	2106.8 3	0.52 7
γ_ϵ	2126.3 6	0.13 7
γ_ϵ	2142.88 20	0.71 7
γ_ϵ	2153.79 12	3.29 19
γ_ϵ	2182.65 10	0.58 13
γ_ϵ	2212.94 7	2.71 19
γ_ϵ	2245.72 17	1.48 13
γ_ϵ	2251.8 7	~0.26
γ_ϵ	2283.5 3	0.71 7
γ_ϵ	2295.9 3	1.03 7
γ_ϵ	2324.5 4	0.39 5
γ_ϵ	2358.3 3	0.65 7
γ_ϵ	2372.4 4	0.43 5
γ_ϵ	2389.5 2	0.65 7
γ_ϵ	2411.1 4	0.32 13
γ_ϵ	2422.2 5	0.10 3
γ_ϵ	2473 1	0.19 7 ?
γ_ϵ	2496.3 8	0.32 13
γ_ϵ	2520.8 10	~0.06
γ_ϵ	2532.3 7	0.07 3
γ_ϵ	2540 1	0.06 3 ?
γ_ϵ	2546.9 8	~0.09
γ_ϵ	2554.1 5	0.181 19
γ_ϵ	2559.6 4	0.47 3
γ_ϵ	2630.5 8	0.16 4
γ_ϵ	2634.3 8	0.16 4
γ_ϵ	2643 1	~0.26
γ_ϵ	2652 1	~0.26
γ_ϵ	2683.4 5	0.206 19
γ_ϵ	2839.2 15	0.065 19
γ_ϵ	2921.4 15	~0.013
γ_ϵ	2934.2 8	0.071 19
γ_ϵ	2942.2 10	0.09 3
γ_ϵ	3240.4 15	0.058 19
γ_ϵ	3260.0 15	~0.06

$^{154}_{65}$Tb(22.6 *6* h)

Mode: ϵ(98.2 *6* %), IT(1.8 *6* %)
Δ: >-70160 keV
SpA: 9.00×10^5 Ci/g
Prod: ^{153}Eu(α,3n); protons on Ta

Photons (^{154}Tb)
$\langle\gamma\rangle$=2285 *65* keV

γ_{mode}	γ(keV)	γ(%)†
Gd L_ℓ	5.362	0.43 5
Gd L_η	6.049	0.164 17
Gd L_α	6.054	11.3 10
Gd L_β	6.847	10.5 12
Gd L_γ	7.865	1.57 19
Gd $K_{\alpha2}$	42.309	35.5 21
Gd $K_{\alpha1}$	42.996	64 4
Gd $K_{\beta1}'$	48.652	18.8 11
Gd $K_{\beta2}'$	50.214	5.7 4
γ_ϵ[M1]	80.21 3	~0.0012
γ_ϵE2	123.100 4	47 8
γ_ϵ(M1+E2)	124.46 5	1.6 6
γ_ϵE2+29%M1	131.546 23	0.0087 11
γ_ϵM1	141.38 3	7.8 8
γ_ϵ(M1+E2)	172.04 4	4.9 4
γ_ϵ[E2]	180.745 22	0.0015 4
γ_ϵE1	225.99 3	28.9 22
γ_ϵE2	232.076 22	0.27 3
γ_ϵE2	247.968 7	85 9
γ_ϵ(E2)	265.83 4	4.2 4
γ_ϵ	267.5 3	4.2 4
γ_ϵ	304.8 10	1.12 6
γ_ϵE2	305.01 10	1.51 4
γ_ϵ	315.60 3	0.0015 3
γ_ϵE2	318.45 9	~0.02
γ_ϵ	329.84 5	0.119 17
γ_ϵ[M1+E2]	337.92 12	1.6 4
γ_ϵE2	346.75 4	75 6
γ_ϵM1+E2	382.03 3	0.69 6
γ_ϵE2	426.85 9	18.7
γ_ϵ[E2]	448.27 3	~0.039
γ_ϵE2	479.30 12	4.1 4
γ_ϵE2	506.49 5	4.3 4
γ_ϵE2	518.01 3	4.05 23
γ_ϵ[M1+E2]	545.61 10	0.62 22
γ_ϵ[E2]	546.03 5	0.120 14
γ_ϵ	565.5 2	2.80 8
γ_ϵM1+25%E2	598.23 3	0.86 8
γ_ϵE2	625.263 19	0.100 7
γ_ϵE2	642.47 5	4.4 3
γ_ϵE2+23%M1+E0	648.29 8	~0.37
γ_ϵE2	649.56 3	7.2 6
γ_ϵE2+14%M1+E0	676.594 22	1.79 21
γ_ϵ[M1+E2]	714.60 12	0.84 21
γ_ϵ[M1+E2]	722.68 5	2.3 6
γ_ϵE2+2.8%M1	756.808 22	1.83 12
γ_ϵ(E2)	830.30 3	0.45 5
γ_ϵE2(+1.2%M1+E0)	873.230 18	3.6 3
γ_ϵE2	888.89 9	1.51 22
γ_ϵE2(+6.0%M1+E0)	892.79 3	5.1 3
γ_ϵE2	924.561 22	0.80 11
γ_ϵ[E2]	928.07 5	0.25 6
γ_ϵ(E1)	993.03 10	17.5 15
γ_ϵE2	996.329 18	3.27 24
γ_ϵE2+1.6%M1	1004.775 21	7.5 6
γ_ϵ(M1+E2)	1061.35 12	4.5 6
γ_ϵ	1093.6 4	~0.37
γ_ϵE2	1140.75 3	2.39 15
γ_ϵ[M1+E2]	1193.90 5	3.2 5
γ_ϵ[E2]	1235.64 10	0.65 9
γ_ϵ[M1+E2]	1274.82 3	0.2 1
γ_ϵ	1315.0 4	~0.37
γ_ϵ[M1+E2]	1399.27 5	~2
γ_ϵE1	1419.88 6	50 3
γ_ϵ[E2]	1522.78 3	0.23 8
γ_ϵ[E2]	1540.65 5	0.45 9
γ_ϵ	1741.8 4	0.60 6

† 11% uncert(syst)

Atomic Electrons (^{154}Tb)
$\langle e\rangle$=112 *6* keV

e_{bin}(keV)	$\langle e\rangle$(keV)	e(%)
7	4.6	64 8
8	3.5	43 5
30 - 72	3.19	8.6 4
73	22.2	30 6
74 - 81	0.9	1.2 5
91	4.7	5.1 5
115	13.1	11.4 21
116	9.6	8.3 15
117	0.3	~0.3
121	3.2	2.6 5
122	4.3	3.5 7
123 - 172	3.7	2.8 4
173 - 182	1.47	0.84 7
198	13.6	6.9 8
216 - 232	1.6	0.72 21
240	3.4	1.43 16
241 - 288	3.58	1.44 10
296	0.020	~0.007
297	6.8	2.29 18
298 - 347	2.36	0.70 3
374 - 420	1.49	0.39 3
425 - 472	0.76	0.169 8
477 - 518	0.30	0.060 19
537 - 575	0.102	0.018 3
590 - 635	1.07	0.175 15
640 - 677	0.32	0.048 8
706 - 755	0.095	0.0134 11
756 - 780	0.0127	0.00162 17
822 - 871	0.301	0.0360 19
872 - 921	0.065	0.0073 4
923 - 955	0.407	0.0429 22
985 - 1011	0.20	0.020 4
1043 - 1092	0.079	0.0074 9
1093 - 1141	0.0088	0.00078 4
1144 - 1193	0.104	0.009 2
1194 - 1236	0.008	0.00061 21
1265 - 1314	0.008	~0.0007
1315 - 1349	0.05	~0.004
1370 - 1419	0.424	0.0308 18
1473 - 1539	0.0119	0.00080 12
1692 - 1740	0.009	0.00051 23

Continuous Radiation (^{154}Tb)
$\langle$IB$\rangle$=0.67 keV

E_{bin}(keV)		$\langle\ \rangle$(keV)	(%)
10 - 20	IB	0.00018	0.00115
20 - 40	IB	0.045	0.121
40 - 100	IB	0.114	0.24
100 - 300	IB	0.039	0.019
300 - 600	IB	0.155	0.034
600 - 1300	IB	0.31	0.038
1300 - 1423	IB	0.00086	6.5×10^{-5}

$^{154}_{66}$Dy(2.9 *15* $\times10^6$ yr)

Mode: α
Δ: -70395 *9* keV
SpA: 0.0008 Ci/g
Prod: ^{154}Gd(α,4n); protons on Ta

Alpha Particles (^{154}Dy)

α(keV)
2872 *5*

$^{154}_{67}$Ho(11.8 *5* min)

Mode: ϵ(99.983 *4* %), α(0.017 *4* %)
Δ: -64637 *12* keV
SpA: 1.03×10^8 Ci/g
Prod: ^{156}Dy(p,3n); daughter ^{154}Er
^{147}Sm(^{10}B,3n)

Alpha Particles (^{154}Ho)

α(keV)
3933 *5*

Photons (^{154}Ho)

γ_{mode}	γ(keV)	γ(rel)
γ[E2]	158.36 *12*	6.2 *14*
γ[E2]	334.99 *8*	100
γ[E2]	412.80 *8*	25 *6*
γ[M1+E2]	571.16 *10*	16 *3*
γ[E2]	906.15 *13*	2.4 *8*

$^{154}_{67}$Ho(3.2 *1* min)

Mode: ϵ(>99.998 %), α(<0.002 %)
Δ: -64637 *12* keV
SpA: 3.81×10^8 Ci/g
Prod: ^{148}Sm(^{11}B,5n); ^{148}Sm(^{10}B,4n);
^{147}Sm(^{10}B,3n); ^{144}Nd(^{14}N,4n)

Alpha Particles (^{154}Ho)

α(keV)
3721 *5*

Photons (^{154}Ho)

⟨γ⟩=1737 *42* keV

γ_{mode}	γ(keV)	γ(%)
γ[E2]	158.36 *12*	3.7 *3*
γ	289.62 *20*	4.0 *3*
γ	296.22 *20*	12.1 *5*
γ[M1+E2]	310.41 *23*	2.8 *3*
γ[E2]	334.99 *8*	94.2
γ[E2]	347.04 *23*	11.8 *9*
γ[E2]	407.3 *3*	23.1 *9*
γ[E2]	412.80 *8*	79 *4*
γ[M1+E2]	435.17 *23*	2.4 *3*
γ	444.6 *4*	4.8 *5*
γ	472.3 *6*	2.4 *4*
γ[E2]	477.52 *14*	52.7 *19*
γ[M1+E2]	505.40 *21*	15.3 *7*
γ[E2]	524.80 *18*	15.1 *7*
γ[M1+E2]	571.16 *10*	9.4 *16*
γ[M1+E2]	726.07 *13*	12.2 *19*
γ[M1+E2]	815.70 *22*	12 *3*
γ[E2]	906.15 *13*	1.4 *4*
γ[M1+E2]	1250.87 *13*	15.1 *19*

$^{154}_{68}$Er(3.75 *12* min)

Mode: ϵ(99.53 *13* %), α(0.47 *13* %)
Δ: -62623 *12* keV
SpA: 3.25×10^8 Ci/g
Prod: ^{147}Sm(^{12}C,5n); ^{148}Sm(^{12}C,6n)

Alpha Particles (^{154}Er)

α(keV)
4166 *3*

$^{154}_{69}$Tm(3.35 *5* s)

Mode: α(~85 %), ϵ(~15 %)
Δ: -54630 *210* keV syst
SpA: 1.98×10^{10} Ci/g
Prod: ^{141}Pr(^{20}Ne,7n); ^{142}Nd(^{19}F,7n)

Alpha Particles (^{154}Tm)

α(keV)
5030 *3*

Photons (^{154}Tm)

γ_{mode}	γ(keV)
γ_ϵ[E2]	542 *1*
γ_ϵ[E2]	560 *1*
γ_ϵ[E2]	602 *1*
γ_ϵ[E2]	625 *1*

$^{154}_{69}$Tm(8.3 *3* s)

Mode: ϵ(56 *15* %), α(44 *15* %)
Δ: -54630 *210* keV syst
SpA: 8.5×10^9 Ci/g
Prod: ^{141}Pr(^{20}Ne,7n); ^{142}Nd(^{19}F,7n)

Alpha Particles (^{154}Tm)

α(keV)
4955 *3*

$^{154}_{70}$Yb(410 *30* ms)

Mode: α(93 *2* %), ϵ(7 *2* %)
Δ: -50120 *280* keV syst
SpA: 8.6×10^{10} Ci/g
Prod: ^{144}Sm(^{16}O,6n); ^{142}Nd(^{20}Ne,8n)

Alpha Particles (^{154}Yb)

α(keV)
5318 *5*

$^{154}_{71}$Lu(960 *100* ms)

Mode: ϵ
Δ: -39630 *540* keV syst
SpA: 5.4×10^{10} Ci/g
Prod: ^{58}Ni on Mo-Sn targets; daughter ^{158}Ta
^{107}Ag on V-Ni targets

$^{154}_{72}$Hf(2 *1* s)

Mode: ϵ
Δ: -32820 *580* keV syst
SpA: 3.1×10^{10} Ci/g
Prod: ^{58}Ni on Mo-Sn targets; daughter ^{158}W
^{107}Ag on V-Ni targets

A = 155

NDS **15**, 409 (1975)

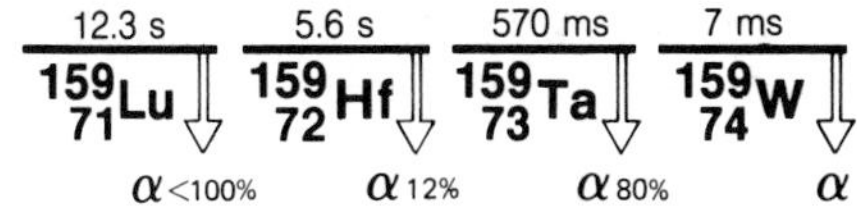

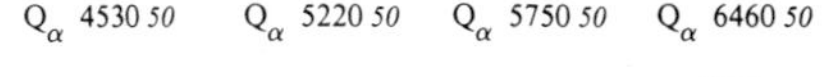

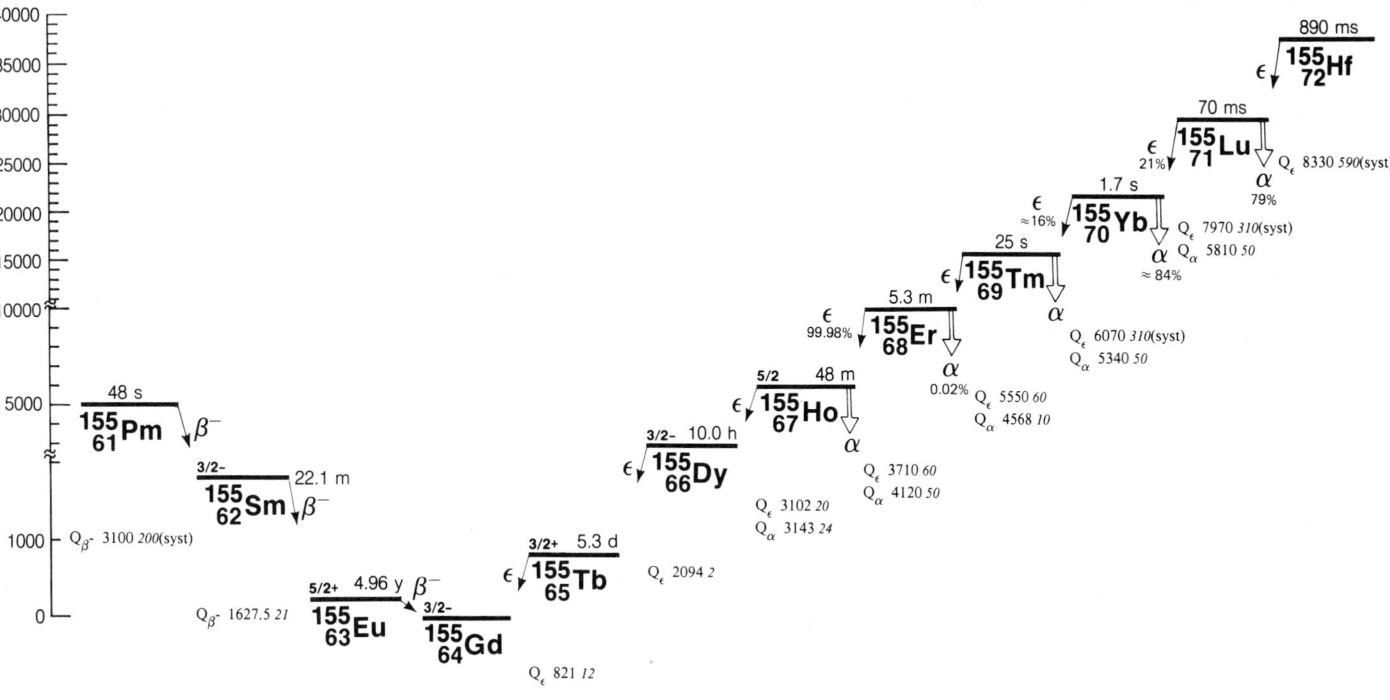

$^{155}_{61}$Pm(48 *4* s)

Mode: β-
Δ: -67100 *200* keV syst
SpA: 1.51×10^{9} Ci/g
Prod: fission

Photons (^{155}Pm)

γ_{mode}	γ(keV)	γ(rel)
γ	53.1 *5*	12 *2*
γ	409.8 *2*	28 *2*
γ	725.4 *2*	68 *3*
γ	762.0 *3*	19 *4*
γ	778.6 *2*	100

$^{155}_{62}$Sm(22.1 *2* min)

Mode: β-
Δ: -70202 *3* keV
SpA: 5.49×10^{7} Ci/g
Prod: ^{154}Sm(n,γ)

Photons (^{155}Sm)

⟨γ⟩=103 *4* keV

γ_{mode}	γ(keV)	γ(%)
Eu L_ℓ	5.177	0.057 *9*
Eu L_η	5.817	0.021 *3*
Eu L_α	5.843	1.52 *19*
Eu L_β	6.594	1.36 *20*
Eu L_γ	7.556	0.20 *3*
γ E1	25.72 *4*	0.45 *8*
Eu $K_{\alpha2}$	40.902	5.0 *3*
Eu $K_{\alpha1}$	41.542	9.0 *5*
Eu $K_{\beta1}'$	46.999	2.60 *15*
Eu $K_{\beta2}'$	48.496	0.79 *5*
γ M1+E2	61.60 *5*	0.225 *23*
γ [M1+E2]	64.69 *11*	0.0068 *15*
γ M1+E2	78.63 *4*	0.262 *15*
γ [M1+E2]	84.05 *19*	0.0023 *6*
γ E1	90.41 *12*	0.0094 *22*
γ E1	104.347 *5*	75 *4*
γ E1	138.35 *11*	0.075 *19*
γ E1	141.439 *11*	2.03 *8*
γ [E2]	167.16 *4*	0.038 *8*
γ E1	169.04 *11*	0.038 *11*
γ	178.8 *3*	0.0019 *4*
γ	183.7 *3*	0.0019 *4*
γ	195.72 *25*	0.0086 *19*
γ E1	203.04 *5*	0.038 *8*
γ	219.8 *5*	0.0021 *5*
γ [M1+E2]	228.76 *6*	0.052 *8*
γ	229.9 *3*	0.0028 *8*
γ M1(+E2)	245.786 *12*	3.8
γ [E1]	287.09 *19*	0.0011 *4*
γ [M1+E2]	307.39 *5*	0.011 *3*
γ	426.21 *17*	0.0128 *22*
γ	460.92 *9*	0.066 *9*

Photons (^{155}Sm)
(continued)

γ_{mode}	γ(keV)	γ(%)
γ	510.26 *12*	0.0120 *15*
γ	522.52 *8*	0.150 *15*
γ	571.86 *12*	0.019 *4*
γ [M1+E2]	603.80 *18*	0.011 *4*
γ	631.23 *19*	0.017 *5*
γ	648.61 *13*	0.0075 *19*
γ	663.96 *9*	0.060 *15*
γ [M1+E2]	665.40 *18*	0.0056 *15*
γ [M1+E2]	677.45 *23*	0.0068 *19*
γ	713.30 *12*	0.0060 *15*
γ	768.31 *9*	0.0056 *11*
γ	817.65 *12*	0.00094 *19*
γ [E2]	832.56 *19*	0.00094 *19*
γ	861.13 *22*	0.0049 *11*
γ	877.02 *19*	0.0028 *6*
γ [M1+E2]	911.19 *18*	0.00094 *19*
γ [E2]	923.24 *23*	0.00094 *19*
γ	932.9 *3*	0.0094 *19*
γ	997.6 *3*	0.012 *3*
γ	1002.56 *22*	0.017 *4*
γ	1018.1 *4*	0.00094 *19*
γ	1055.60 *16*	~0.0009
γ	1096.8 *5*	0.00094 *19*
γ	1132.35 *19*	0.0019 *4*
γ	1159.5 *4*	0.013 *3*
γ	1197.04 *16*	0.0056 *11*
γ	1222.75 *16*	0.023 *5*
γ	1263.9 *4*	0.0022 *8*
γ	1301.38 *16*	0.079 *8*

Atomic Electrons (^{155}Sm)

⟨e⟩=16.6 *8* keV

e_{bin}(keV)	⟨e⟩(keV)	e(%)
7	0.65	9.3 *14*
8	0.45	5.8 *9*
13 - 55	1.6	5.2 *9*
56	8.9	16.0 *9*
57 - 93	0.8	1.2 *4*
96	1.59	1.65 *9*
97	0.71	0.74 *4*
103	0.53	0.52 *3*
104 - 147	0.212	0.190 *8*
155 - 196	0.026	0.0149 *20*
197	0.78	0.40 *12*
201 - 246	0.24	0.10 *3*
259 - 307	0.0020	0.00073 *22*
378 - 426	0.005	~0.0013
453 - 502	0.009	~0.0020
503 - 523	0.0026	0.00049 *19*
555 - 604	0.0019	0.00033 *11*
615 - 665	0.0039	0.0006 *3*
669 - 713	0.00013	1.9 *8* $\times10^{-5}$
720 - 769	0.0003	~5 $\times10^{-5}$
784 - 833	0.00024	2.9 *12* $\times10^{-5}$
853 - 884	0.00032	3.6 *17* $\times10^{-5}$
903 - 949	0.00033	~3 $\times10^{-5}$
954 - 1002	0.00051	5 *3* $\times10^{-5}$
1007 - 1055	4.5 $\times10^{-5}$	4.4 *20* $\times10^{-6}$
1084 - 1132	0.00029	~3 $\times10^{-5}$
1149 - 1197	0.00056	4.8 *21* $\times10^{-5}$
1215 - 1264	0.0014	~0.00011
1293 - 1301	0.00023	1.7 *8* $\times10^{-5}$

Continuous Radiation (^{155}Sm)

⟨β-⟩=551 keV;⟨IB⟩=0.82 keV

E_{bin}(keV)		⟨ ⟩(keV)	(%)
0 - 10	β-	0.0366	0.73
	IB	0.024	
10 - 20	β-	0.111	0.74
	IB	0.023	0.161
20 - 40	β-	0.456	1.52
	IB	0.045	0.155
40 - 100	β-	3.41	4.85
	IB	0.120	0.19
100 - 300	β-	38.0	18.7
	IB	0.29	0.165
300 - 600	β-	139	30.9
	IB	0.22	0.054
600 - 1300	β-	356	41.5
	IB	0.101	0.0137
1300 - 1628	β-	13.9	1.02
	IB	0.00027	2.1 $\times10^{-5}$

$^{155}_{63}$Eu(4.96 *1* yr)

Mode: β-

Δ: -71829 *4* keV

SpA: 465.0 Ci/g

Prod: daughter ^{155}Sm

Photons (^{155}Eu)

⟨γ⟩=63 *3* keV

γ_{mode}	γ(keV)	γ(%)†
Gd L_ℓ	5.362	0.125 *18*
Gd L_η	6.049	0.049 *6*
Gd L_α	6.054	3.3 *4*
Gd L_β	6.846	3.1 *4*
Gd L_γ	7.867	0.47 *6*
γ M1+6.6%E2	18.771 *3*	0.051 *7*
γ E1	26.527 *3*	0.317 *11*
γ M1+17%E2	31.452 *3*	~0.007
Gd $K_{\alpha2}$	42.309	6.9 *4*
Gd $K_{\alpha1}$	42.996	12.4 *7*
γ E1	45.2977 *17*	1.28 *4*
Gd $K_{\beta1}$'	48.652	3.64 *20*
Gd $K_{\beta2}$'	50.214	1.10 *6*
γ E1	57.9793 *20*	0.067 *3*
γ M1+3.7%E2	60.0153 *17*	1.14 *9*
γ M1+4.9%E2	86.066 *5*	0.15 *3*
γ E1	86.5423 *20*	34
γ E1	105.313 *2*	20.6 *4*
γ E2	146.081 *5*	0.0530 *22*

† 6.5% uncert(syst)

Atomic Electrons (^{155}Eu)

⟨e⟩=17.9 *6* keV

e_{bin}(keV)	⟨e⟩(keV)	e(%)
2 - 3	0.032	1.4 *3*
7	1.35	19 *3*
8	1.03	12.9 *18*
9	0.029	0.34 *11*
10	1.01	10.2 *10*
11	0.57	5.3 *10*
12	0.83	7.2 *14*
13 - 14	0.130	0.96 *5*
17	0.55	3.2 *6*
18 - 35	0.78	3.2 *3*
36	4.6	12.7 *12*
37 - 51	0.470	1.16 *4*
52	0.75	1.45 *12*
53	0.147	0.28 *3*
55	2.42	4.40 *12*
56 - 60	0.291	0.50 *4*
78	1.01	1.29 *12*
79 - 105	1.84	2.08 *8*
138 - 146	0.0190	0.0136 *4*

Continuous Radiation (^{155}Eu)

⟨β-⟩=47.1 keV;⟨IB⟩=0.0088 keV

E_{bin}(keV)		⟨ ⟩(keV)	(%)
0 - 10	β-	0.74	15.0
	IB	0.0024	
10 - 20	β-	2.02	13.6
	IB	0.0017	0.0123
20 - 40	β-	6.8	22.9
	IB	0.0022	0.0080
40 - 100	β-	24.9	38.6
	IB	0.0022	0.0038
100 - 253	β-	12.7	9.9
	IB	0.00029	0.00024

$^{155}_{64}$Gd(stable)

Δ: -72082 *3* keV

%: 14.80 *5*

$^{155}_{65}$Tb(5.32 *6* d)

Mode: ϵ

Δ: -71260 *13* keV

SpA: 1.583×10^5 Ci/g

Prod: protons on Gd; ^{153}Eu(^{3}He,n); ^{153}Eu(α,2n); daughter ^{155}Dy; protons on Ta

Photons (^{155}Tb)

⟨γ⟩=184 *7* keV

γ_{mode}	γ(keV)	γ(%)†
Gd L_ℓ	5.362	0.61 *16*
Gd L_η	6.049	0.24 *7*
Gd L_α	6.054	16 *4*
Gd L_β	6.845	15 *4*
Gd L_γ	7.862	2.3 *6*
γ M1+6.6%E2	18.771 *3*	0.057 *11*
γ (E2)	21.036 *17*	0.00023 *5*
γ E1	26.527 *3*	0.271 *12*
γ M1+17%E2	31.452 *3*	~0.019
Gd $K_{\alpha2}$	42.309	38.7 *25*
Gd $K_{\alpha1}$	42.996	70 *4*
γ E1	45.2977 *17*	1.42 *20*
γ	47.309 *15*	~1
Gd $K_{\beta1}$'	48.652	20.5 *13*
Gd $K_{\beta2}$'	50.214	6.2 *4*
γ	51.73 *15*	~1
γ E1	57.9793 *20*	0.18 *3*
γ M1+3.7%E2	60.0153 *17*	0.94 *11*
γ E1	80.81 *12*	0.028 *6*
γ M1+4.9%E2	86.066 *5*	0.27 *7*
γ E1	86.5423 *20*	28.6 *21*
γ	88.509 *20*	~0.6
γ M1+E2	99.03 *5*	0.06 *3*
γ (E2)	101.02 *5*	0.13 *4*
γ (E1)	103.74 *25*	0.081 *7*
γ E1	105.313 *2*	23
γ	118.01 *5*	<0.07
γ E1	120.553 *20*	0.044 *9*
γ E2	146.081 *5*	0.095 *16*
γ M1+2.5%E2	148.65 *3*	2.4 *3*
γ [E2]	150.63 *3*	0.22 *4*
γ (E2)	158.53 *4*	0.35 *7*
γ M1(+E2)	160.52 *3*	0.67 *14*
γ M1+E2	161.322 *19*	2.7 *5*
γ M1(+E2)	162.80 *8*	0.23 *5*
γ M1+E2	163.313 *25*	4.3 *9*
γ E1	175.18 *3*	0.21 *4*
γ M1+E2	180.092 *19*	8.3 *18*
γ E1	181.53 *12*	0.55 *9*
γ M1(+E2)	182.08 *3*	0.13 *4*
γ [M1+E2]	191.96 *4*	0.046 *18*
γ E1	200.30 *12*	0.193 *18*
γ M1	207.99 *13*	0.237 *23*
γ E1	208.61 *3*	0.078 *16*
γ E1	216.1 *3*	0.113 *23*
γ [M1+E2]	220.16 *13*	0.22 *4*
γ M1+E2	220.67 *13*	0.39 *9*
γ M1+E2	226.83 *12*	0.14 *3*
γ [E1]	235.19 *3*	0.057 *12* ?
γ M1+E2	239.44 *13*	0.21 *4*
γ M1(+E2)?	247.9 *4*	0.021 *9* ?
γ M1	253.59 *13*	0.021 *9*
γ M1(+E2)	262.35 *5*	5.5 *11*
γ E1	268.63 *3*	0.57 *11*
γ (E2)	281.12 *5*	0.246 *25*
γ M1	286.85 *12*	0.239 *23*
γ M1+E2	321.84 *3*	0.147 *23*
γ [E1]	325.98 *13*	0.041 *9*
γ M1(+E2)	340.61 *3*	0.99 *16*

Photons (^{155}Tb)
(continued)

γ_{mode}	γ(keV)	γ(%)†
γ [M1+E2]	361.1 3	0.032 7
γ E1	367.14 3	1.22 25
γ E1	367.66 5	0.76 16
γ M1+E2	370.79 13	0.13 4
γ E1	380.3 4	0.0083 21
γ [M1+E2]	383.47 13	0.023 5
γ [M1+E2]	391.3 3	0.018 5
γ M1	402.24 13	0.064 11
γ [E1]	427.15 3	0.029 7
γ [E2]	446.45 23	0.0069 14
γ M1(+E2)	451.3 3	0.032 7
γ (E1)	454.7 4	0.016 5
γ [E1]	487.22 23	0.025 9
γ [E1]	488.79 13	0.016 3
γ M1(+E2)	501.9 3	0.026 6
γ [E1]	505.99 23	0.040 8
γ M1(+E2)	532.51 23	0.044 9
γ M1(+E2)?	542.7 3	0.0071 21 ?
γ M1(+E2)	555.5 4	0.026 5
γ M1(+E2)	560.0 4	0.120 23
γ E0+M1+E2	588.0 3	0.0081 16
γ E0+M1+E2	592.53 23	0.016 5
γ M1(+E2)	615.5 4	0.018 5
γ M1(+E2)	648.0 3	0.012 5

† 17% uncert(syst)

Atomic Electrons (^{155}Tb)
⟨e⟩=73 10 keV

e_{bin}(keV)	⟨e⟩(keV)	e(%)
1	0.09	~6
7	6.6	92 20
8	5.0	63 14
10 - 35	5.5	34.7 23
36	4.4	12.0 9
37 - 38	0.5	~1
39	6.2	~16
40	7.2	~18
41 - 43	1.4	3.3 10
44	7.5	~17
45	0.19	~0.4
46	3.4	~7
47 - 49	1.1	~2
50	2.1	~4
51 - 54	1.4	2.8 9
55	2.7	4.9 5
56 - 105	5.4	6.3 5
108 - 112	1.5	1.38 24
113	1.8	1.6 4
116 - 125	0.019	0.015 3
130	2.9	2.3 6
131 - 181	3.7	2.3 3
182 - 231	1.3	0.62 15
232 - 281	0.43	0.17 3
285 - 333	0.24	0.079 13
339 - 384	0.035	0.0098 9
389 - 439	0.0062	0.00154 22
443 - 492	0.0070	0.00150 24
494 - 543	0.027	0.0051 7
547 - 593	0.0059	0.00103 14
598 - 647	0.00093	0.00015 4
648	5.9 $\times 10^{-6}$	9 4 $\times 10^{-7}$

Continuous Radiation (^{155}Tb)
⟨IB⟩=0.22 keV

E_{bin}(keV)		⟨ ⟩(keV)	(%)
10 - 20	IB	0.00019	0.00121
20 - 40	IB	0.046	0.124
40 - 100	IB	0.114	0.25
100 - 300	IB	0.024	0.0127
300 - 600	IB	0.035	0.0085
600 - 821	IB	0.0031	0.00048

$^{155}_{66}$Dy(10.0 *3* h)

Mode: ε
Δ: -69166 12 keV
SpA: 2.02×10^{6} Ci/g
Prod: ^{159}Tb(p,5n); ^{153}Gd(α,2n); ^{154}Gd(α,3n); ^{152}Gd(α,n); ^{159}Tb(d,6n); protons on U; protons on Ta

Photons (^{155}Dy)
⟨γ⟩=594 34 keV

γ_{mode}	γ(keV)	γ(%)†
Tb L_ℓ	5.546	0.30 3
Tb L_α	6.269	7.9 6
Tb L_η	6.284	0.105 9
Tb L_β	7.113	7.3 7
Tb L_γ	8.196	1.12 12
γ M1+1.4%E2	23.08 4	0.018 4
Tb $K_{\alpha2}$	43.744	26.3 11
Tb $K_{\alpha1}$	44.482	47.1 20
Tb $K_{\beta1}$'	50.338	14.1 6
Tb $K_{\beta2}$'	51.958	4.20 19
γ M1+2.1%E2	65.47 3	1.4 3
γ M1(+E2)	67.06 9	0.022 4
γ [E1]	84.83 6	0.051 10
γ M1+1.9%E2	90.32 3	1.3 3
γ [E1]	94.22 4	0.0115 23
γ M1+2.9%E2	115.27 4	0.061 12
γ M1(+1.8%E2)	118.34 6	0.101 20
γ [M1+E2]	131.97 13	0.0115 23
γ [M1+E2]	134.61 20	0.0081 16
γ [M1+E2]	153.24 11	0.101 20
γ E2	155.80 4	0.18 4
γ E1	161.47 4	1.21 24
γ [M1+E2]	172.35 10	0.0088 18
γ [M1+E2]	179.06 6	0.0074 15
γ E1	184.54 5	3.8 8
γ [M1+E2]	195.75 13	0.032 6
γ M1+14%E2	205.60 5	0.36 7
γ [E2]	208.66 6	0.051 10
γ E1	226.94 4	68 14
γ	245.36 13	0.0121 24
γ	248.61 20	0.095 19
γ [M1+E2]	269.38 6	0.17 3
γ M1+7.9%E2	271.07 5	1.12 22
γ [M1+E2]	294.78 15	0.067 14
γ	301.32 15	0.014 3 ?
γ M1	311.1 2	0.04 1
γ M1	317.86 15	0.18 4
γ	322.66 11	0.20 4
γ M1+E2	326.41 12	0.13 3
γ E2	334.85 7	0.108 22
γ	352.63 11	0.054 11
γ	357.01 20	0.034 7
γ [E2]	377.71 10	0.014 3 ?
γ	379.31 20	0.041 8
γ	382.79 12	0.17 3
γ	393.8 1	0.034 7 ?
γ	403.90 12	0.16 3
γ	411.52 13	0.13 3
γ	420.71 20	0.027 5

Photons (^{155}Dy)
(continued)

γ_{mode}	γ(keV)	γ(%)†
γ [E1]	424.91 9	0.061 12
γ M1	433.13 9	0.60 12
γ	452.21 20	0.16 3
γ	458.91 20	0.122 24
γ	463.01 20	0.067 14
γ	467.61 20	0.041 8
γ M1	484.12 10	1.08 22
γ [M1+E2]	496.05 9	0.17 3
γ M1	498.61 9	1.4 3
γ M1	508.43 11	0.97 19
γ	517.6 2	0.074 15
γ M1	549.60 10	0.88 18
γ	559.3 3	0.061 12
γ M1,E2	570.57 13	0.18 4
γ M1	586.37 9	0.18 4
γ M1,E2	588.16 11	0.074 15
γ	596.62 13	
γ	609.71 20	0.081 16
γ [M1+E2]	610.67 17	0.041 8
γ	618.81 20	0.095 19
γ M1,E2	641.20 8	1.15 23
γ M1	653.95 12	0.13 3
γ [E1]	656.87 12	0.18 4
γ	658.8 3	0.061 12
γ M1	664.28 8	2.2 4
γ M1	678.49 11	0.18 4
γ	683.7 3	0.088 18
γ	695.1 10	0.20 4
γ	720.9 3	0.22 5
γ	724.1 3	0.20 4
γ	726.1 3	0.13 3
γ	743.96 11	0.39 8
γ	748.70 14	0.054 11
γ	750.31 18	0.054 11
γ	756.8 5	0.041 8
γ	760.31 20	0.13 3
γ E1	773.79 9	0.108 22
γ	781.9 5	0.054 11
γ	784.0 5	0.088 18
γ	796.11 20	0.088 18
γ	807.4 3	0.074 15
γ	809.75 12	0.067 14
γ [M1+E2]	812.18 11	0.34 7
γ	825.74 8	0.108 22
γ	829.3 10	~0.06
γ M1	835.25 11	0.23 5
γ [E2]	838.40 12	0.16 3
γ E1	841.45 12	0.25 5
γ	848.81 20	0.18 4
γ	855.26 12	0.061 12
γ [E1]	884.41 9	0.041 8
γ	891.22 8	0.51 10
γ M1,E2	905.46 9	2.3 5
γ	912.21 20	0.19 4
γ	921.05 11	0.095 19
γ M1	928.53 8	0.61 12
γ	938.01 20	0.074 15
γ	940.53 13	0.30 6
γ	962.21 20	0.074 15
γ	972.41 20	0.095 19
γ	981.81 20	0.25 5
γ [E1]	996.72 11	~0.34
γ E1	999.68 8	2.4 5
γ	1002.91 12	~0.34
γ E1	1013.16 9	0.32 7
γ E1	1062.19 11	0.36 7
γ E2	1068.3 2	0.52 10
γ E1	1090.00 8	2.7 6
γ [E1]	1098.43 11	0.081 16
γ	1100.20 13	0.095 19
γ E1	1115.41 11	0.37 7
γ E1	1120.20 15	0.074 15
γ	1143.4 5	0.047 10
γ E1	1155.47 9	2.0 4
γ E1	1166.40 10	1.6 3
γ	1173.3 5	0.064 13
γ	1184.4 5	0.027 5
γ	1199.0 5	0.054 11
γ	1203.7 5	0.054 11
γ	1222.5 5	0.074 15
γ	1232.2 5	0.014 3
γ M1	1242.64 10	0.088 18
γ E1	1251.67 12	0.88 18
γ	1295.11 12	0.18 4

Photons (^{155}Dy) (continued)

γ_{mode}	γ(keV)	γ(%)†
γ	1304.11 9	0.14 3
γ (E1)	1316.47 15	0.15 3
γ	1332 1	0.09 4
γ E1	1336.86 9	0.43 9
γ	1348.4 5	0.047 10
γ E1	1367.89 9	0.73 15
γ	1385.47 16	0.095 19 ?
γ	1388.94 9	0.108 22
γ [E1]	1393.93 6	0.24 5
γ	1403.1 5	0.034 7
γ	1412.02 8	0.067 14
γ [M1+E2]	1414.98 6	0.25 5
γ [E1]	1427.18 9	0.35 7
γ	1429.60 15	0.041 8
γ M1	1438.06 5	0.30 6
γ	1451.6 3	0.15 3
γ E1	1479.21 10	0.54 11
γ E1	1492.65 10	0.54 11
γ (E1)	1509.21 5	0.25 5
γ	1522.54 8	0.027 5
γ	1562.5 5	0.074 15
γ	1565.1 5	0.027 5
γ	1566.67 7	0.020 4
γ	1573.48 8	0.088 18
γ	1591.07 16	0.054 11
γ [E1]	1594.48 9	0.067 14
γ (E1)	1599.53 5	0.24 5
γ	1607.4 10	0.047 13
γ	1610.8 10	~0.027
γ	1637.81 7	0.84 17
γ	1656.54 16	0.041 8
γ E1	1665.00 6	0.91 18
γ E1	1684.80 9	0.101 20
γ	1709.8 10	0.020 4
γ	1728.13 7	0.041 8
γ [E1]	1750.27 9	0.027 5
γ [M1+E2]	1764.47 12	0.034 7
γ	1793.61 7	0.044 9
γ	1803.0 7	0.014 3
γ	1813.0 5	0.020 4
γ [E1]	1835.62 12	0.014 3
γ	1845.4 4	~0.007
γ	1866.06 15	0.0108 22
γ	1889.2 3	0.020 4
γ	1954.7 3	0.014 3

† 15% uncert(syst)

Atomic Electrons (^{155}Dy)

⟨e⟩=19.6 10 keV

e_{bin}(keV)	⟨e⟩(keV)	e(%)
8	4.8	62 7
9	0.66	7.5 8
13	1.18	8.8 18
14 - 35	0.64	2.39 18
36	0.70	1.95 21
37	0.231	0.62 7
38	1.3	3.3 7
41 - 50	0.92	2.14 12
57	0.78	1.4 3
58 - 107	1.26	1.71 15
108 - 157	0.64	0.48 5
159 - 172	0.0080	0.0049 7
175	3.4	1.9 4
176 - 217	0.160	0.083 12
218	0.48	0.22 4
219 - 268	0.71	0.31 4
269 - 318	0.099	0.035 8
319 - 360	0.057	0.017 5
369 - 417	0.123	0.032 5
419 - 468	0.43	0.096 10
475 - 519	0.179	0.036 4
534 - 581	0.058	0.0106 13
584 - 634	0.27	0.044 7
639 - 688	0.069	0.0105 17
692 - 741	0.030	0.0043 13
742 - 789	0.050	0.0065 10

Atomic Electrons (^{155}Dy) (continued)

e_{bin}(keV)	⟨e⟩(keV)	e(%)
794 - 841	0.032	0.0039 13
847 - 891	0.13	0.015 4
897 - 945	0.039	0.0042 8
948 - 997	0.043	0.0045 9
998 - 1046	0.042	0.0041 12
1048 - 1097	0.0147	0.00137 21
1098 - 1147	0.038	0.0034 4
1148 - 1197	0.0093	0.00080 12
1198 - 1244	0.012	0.00102 22
1250 - 1297	0.0108	0.00085 17
1302 - 1351	0.0128	0.00097 14
1359 - 1407	0.023	0.00166 20
1410 - 1457	0.0123	0.00086 9
1470 - 1566	0.0098	0.00064 9
1572 - 1663	0.024	0.0015 4
1676 - 1763	0.0027	0.00015 3
1784 - 1882	0.00080	4.4 10 $\times 10^{-5}$
1887 - 1953	0.00019	1.0 4 $\times 10^{-5}$

Continuous Radiation (^{155}Dy)

⟨β+⟩=5.6 keV;⟨IB⟩=1.11 keV

E_{bin}(keV)		⟨ ⟩(keV)	(%)
0 - 10	β+	1.96×10^{-7}	2.33×10^{-6}
	IB	0.00094	
10 - 20	β+	9.9×10^{-6}	5.9×10^{-5}
	IB	0.00043	0.0029
20 - 40	β+	0.000339	0.00103
	IB	0.034	0.093
40 - 100	β+	0.0198	0.0252
	IB	0.138	0.29
100 - 300	β+	0.90	0.415
	IB	0.040	0.020
300 - 600	β+	3.51	0.80
	IB	0.162	0.036
600 - 1300	β+	1.21	0.178
	IB	0.60	0.066
1300 - 2094	IB	0.138	0.0096
	Σβ+		1.42

$^{155}_{67}$Ho(48 1 min)

Mode: ε, α

Δ: -66064 24 keV

SpA: 2.53×10^{7} Ci/g

Prod: protons on Dy; protons on Ho; protons on Ta

Alpha Particles (^{155}Ho)

α(keV)
3940 20

Photons (^{155}Ho)

⟨γ⟩=186 5 keV

γ_{mode}	γ(keV)	γ(%)†
Dy L_ℓ	5.743	0.27 3
Dy L_α	6.491	7.1 5
Dy L_η	6.534	0.097 8
Dy L_β	7.379	6.9 7
Dy L_γ	8.526	1.10 13

Photons (^{155}Ho) (continued)

γ_{mode}	γ(keV)	γ(%)†
γ_ϵM1+3.8%E2	39.38 4	0.75 15
Dy $K_{\alpha2}$	45.208	20.4 7
Dy $K_{\alpha1}$	45.998	36.5 13
γ_ϵM1+2.4%E2	47.43 4	0.52 10
Dy $K_{\beta1}$'	52.063	11.1 4
Dy $K_{\beta2}$'	53.735	3.25 13
γ_ϵM1+E2	66.08 5	0.075 23
γ_ϵE2	86.81 4	0.60 12
γ_ϵM1	96.93 4	0.68 15
γ_ϵ[M1+E2]	101.15 5	0.075 23
γ_ϵE1	103.96 5	1.5 3
γ_ϵM1+E2	108.79 5	0.32 6
γ_ϵ	115.32 6	0.045 15
γ_ϵE2	115.59 5	0.45 11
γ_ϵ	121.32 10	0.105 23
γ_ϵ	121.82 10	0.023 8
γ_ϵ	124.32 10	0.045 15
γ_ϵE1	124.64 5	0.18 4
γ_ϵM1	136.32 4	3.8 8
γ_ϵM1	137.62 5	0.38 11
γ_ϵE2	146.68 5	0.53 11
γ_ϵ	149.23 7	0.11 3 ?
γ_ϵM1	163.01 5	0.75 15
γ_ϵM1	185.05 5	1.6 4
γ_ϵM1+E2	189.10 5	0.19 5
γ_ϵE1	200.47 6	0.45 11
γ_ϵM1	202.40 4	0.82 15
γ_ϵ	206.62 10	0.53 11
γ_ϵ[E1]	208.54 5	0.91 4
γ_ϵ	212.76 4	0,34 8
γ_ϵE2	216.31 6	0.22 5
γ_ϵ	218.81 7	0.82 15 *
γ_ϵ	219.20 8	
γ_ϵE1	240.28 5	7.5 15
γ_ϵM1	243.43 7	0.11 4 ?
γ_ϵE1	247.93 5	0.98 15
γ_ϵ	251.92 10	0.075 23
γ_ϵM1	259.28 7	0.120 23
γ_ϵE1	262.26 5	0.54 15
γ_ϵ	266.62 10	0.21 5
γ_ϵE1	272.21 5	0.35 11
γ_ϵ	281.31 6	0.090 23
γ_ϵM1	286.04 5	0.35 11
γ_ϵ	288.62 10	0.39 11
γ_ϵE2	304.43 5	0.26 6
γ_ϵ[E1]	309.69 4	0.50 11 ?
γ_ϵ(E1)	312.10 7	0.22 5 ?
γ_ϵE1	321.30 6	0.34 8
γ_ϵM1	325.42 5	1.6 4
γ_ϵ	336.20 8	0.29 6 ?
γ_ϵ	343.85 13	0.41 19 ?
γ_ϵ(M1)	344.32 6	0.41 19
γ_ϵE1	349.07 4	0.41 11
γ_ϵ	353.44 6	0.112 23 ?
γ_ϵ	366.83 6	<0.08
γ_ϵ	369.14 5	0.46 9 *
γ_ϵ	369.27 6	
γ_ϵ	373.22 10	0.27 5
γ_ϵ	377.09 6	0.083 23
γ_ϵ	383.23 13	0.75 15 *
γ_ϵ(E2)	383.62 7	
γ_ϵ	390.92 10	0.097 23
γ_ϵ	396.90 6	0.090 23
γ_ϵE1	408.52 5	0.82 15
γ_ϵM1	416.70 6	0.25 5
γ_ϵM1(+E2)	420.76 6	0.41 9
γ_ϵ	439.62 10	0.36 8
γ_ϵ	448.92 10	0.44 9
γ_ϵ	456.08 6	0.38 8
γ_ϵ	460.42 10	0.37 8
γ_ϵ	476.26 8	0.15 3 ?
γ_ϵ	478.08 6	0.075 15
γ_ϵ	479.07 7	0.25 5 ?
γ_ϵ	493.55 6	0.060 15
γ_ϵ	495.32 10	0.13 3
γ_ϵ	516.12 10	0.25 5
γ_ϵ	523.63 8	0.22 5 ?
γ_ϵ	529.22 10	0.052 15
γ_ϵ	533.22 10	0.052 15
γ_ϵ	536.72 10	0.052 15
γ_ϵ	542.62 10	0.13 3
γ_ϵ	554.32 10	0.14 3
γ_ϵM1	557.08 6	0.16 3

Photons (^{155}Ho)
(continued)

γ_{mode}	γ(keV)	γ(%)†
γ_ϵ	558.82 10	0.090 23
γ_ϵ(M1+E2)	566.20 5	0.17 4
γ_ϵ(E1)	569.23 6	0.18 5
γ_ϵ	576.65 6	0.060 15
γ_ϵ	598.82 10	0.060 15
γ_ϵ(M1+E2)	615.70 6	0.09 3
γ_ϵM1+E2	616.53 6	0.09 3
γ_ϵ	623.92 20	0.052 15
γ_ϵM1+E2	648.47 9	0.105 23 ?
γ_ϵM1	654.15 6	0.13 3
γ_ϵ	688.02 20	0.075 15
γ_ϵ	699.68 6	0.112 23
γ_ϵ	737.02 20	0.060 15
γ_ϵ	752.84 6	~0.045
γ_ϵ	765.76 6	0.14 3
γ_ϵ	768.62 20	0.17 3
γ_ϵ	791.82 20	0.083 23
γ_ϵ	803.12 20	0.17 4
γ_ϵ	825.52 20	0.13 3
γ_ϵ	827.22 20	0.13 3
γ_ϵ	834.47 15	0.22 5 ?
γ_ϵ	868.63 9	0.090 23 ?
γ_ϵ	872.42 20	0.052 15
γ_ϵ	875.22 20	0.052 15
γ_ϵE1	892.28 9	0.25 5
γ_ϵE1	897.01 9	0.44 9
γ_ϵ	954.52 20	0.075 15
γ_ϵE1	993.95 9	0.35 8
γ_ϵE1	1015.30 9	0.35 8
γ_ϵ	1027.92 20	0.090 23
γ_ϵE1	1033.33 8	0.56 11
γ_ϵ	1056.02 20	0.052 15
γ_ϵE1	1081.38 9	0.38 8
γ_ϵE1	1178.32 8	0.25 5
γ_ϵ	1270.62 20	0.052 15
γ_ϵ	1327.22 20	0.060 15
γ_ϵ	1333.02 20	0.060 15

† 27% uncert(syst)
* combined intensity for doublet

Atomic Electrons (^{155}Ho)
$\langle e \rangle$=25.0 9 keV

e_{bin}(keV)	$\langle e \rangle$(keV)	e(%)
8	2.7	34 4
9	2.14	25 3
12	0.043	~0.35
30	0.93	3.1 6
31	0.51	1.7 3
32	0.56	1.8 4
33 - 36	0.65	1.88 21
37	0.80	2.1 3
38	0.96	2.5 5
39 - 40	0.42	1.08 17
43	0.87	2.0 4
44 - 71	1.82	3.5 4
78	0.59	0.75 15
79	0.54	0.69 14
83	2.7	3.2 7
84 - 125	2.45	2.53 24
127	0.55	0.43 9
128 - 130	0.111	0.087 16
131	0.79	0.60 14
134 - 184	1.9	1.24 19
185 - 232	1.12	0.56 6
234 - 282	0.62	0.23 4
284 - 330	0.54	0.17 4
334 - 384	0.24	0.065 10
386 - 432	0.19	0.046 12
438 - 487	0.086	0.019 5
488 - 537	0.067	0.0132 24
541 - 590	0.028	0.005 1
591 - 640	0.024	0.0040 7
641 - 688	0.010	0.0016 6
691 - 738	0.018	0.0025 9
744 - 792	0.027	0.0035 11
794 - 843	0.019	0.0023 4

Atomic Electrons (^{155}Ho)
(continued)

e_{bin}(keV)	$\langle e \rangle$(keV)	e(%)
860 - 901	0.0048	0.00054 17
940 - 986	0.0159	0.00165 23
992 - 1033	0.0071	0.00070 12
1047 - 1081	0.00088	8.3 16 $\times 10^{-5}$
1125 - 1171	0.0026	0.00023 4
1176 - 1217	0.0012	~10 $\times 10^{-5}$
1262 - 1279	0.0025	0.00020 8
1318 - 1333	0.00042	3.2 11 $\times 10^{-5}$

Continuous Radiation (^{155}Ho)
$\langle \beta + \rangle$=201 keV;$\langle$IB$\rangle$=4.7 keV

E_{bin}(keV)		$\langle\ \rangle$(keV)	(%)
0 - 10	β+	3.62×10^{-7}	4.30×10^{-6}
	IB	0.0084	
10 - 20	β+	1.90×10^{-5}	0.000113
	IB	0.0079	0.055
20 - 40	β+	0.00068	0.00205
	IB	0.033	0.100
40 - 100	β+	0.0435	0.055
	IB	0.17	0.33
100 - 300	β+	2.67	1.20
	IB	0.151	0.083
300 - 600	β+	20.8	4.49
	IB	0.31	0.070
600 - 1300	β+	120	12.9
	IB	1.45	0.153
1300 - 2500	β+	58	3.85
	IB	2.4	0.140
2500 - 3102	IB	0.138	0.0052
	$\Sigma\beta$+		22

$^{155}_{68}$Er(5.3 3 min)

Mode: ϵ(99.978 7 %), α(0.022 7 %)
Δ: -62360 60 keV
SpA: 2.29×10^8 Ci/g
Prod: ^{156}Dy(α,5n); protons on Ta

Alpha Particles (^{155}Er)

α(keV)
4012 5

decay data not yet evaluated

$^{155}_{69}$Tm(25 4 s)

Mode: α, ϵ
Δ: -56810 30 keV
SpA: 2.9×10^9 Ci/g
Prod: ^{144}Sm(^{14}N,3n)

Alpha Particles (^{155}Tm)

α(keV)
4450 10

Photons (^{155}Tm)

γ_{mode}	γ(keV)	γ(rel)
γ_ϵ	31.5 1	5.3 7
γ_ϵ	63.8 1	3.3 5
γ_ϵ	88.1 2	17 5
γ_ϵ	94.5 2	0.13 3
γ_ϵ	98.0 2	0.47 19
γ_ϵ	152.0 1	6.67 20
γ_ϵ	171.6 1	1.87 25
γ_ϵ	196.7 2	0.27 8 ?
γ_ϵ	226.8 2	100 23
γ_ϵ	241.6 2	0.9 3
γ_ϵ	247.60 15	6.3 9
γ_ϵ	273.9 2	1.13 23
γ_ϵ	305.0 2	1.5 4
γ_ϵ	311.60 25	0.60 17
γ_ϵ	315.3 3	2.4 6
γ_ϵ	317.2 3	0.9 3
γ_ϵ	323.50 25	8.3 21
γ_ϵ	327.9 4	0.28 13
γ_ϵ	328.6 4	~0.2
γ_ϵ	331.4 4	0.40 12
γ_ϵ	379.0 3	3.7 7
γ_ϵ	380.1 3	1.3 3
γ_ϵ	385.7 5	0.47 20 ?
γ_ϵ	395.7 4	0.60 17
γ_ϵ	396.8 4	0.44 10
γ_ϵ	433.40 25	2.0 4
γ_ϵ	466.8 4	0.33 10
γ_ϵ	497.0 4	1.4 3
γ_ϵ	498.7 4	0.7 3
γ_ϵ	501.1 5	1.3 4
γ_ϵ	518.7 4	3.3 7
γ_ϵ	521.0 6	0.5 2
γ_ϵ	527.5 4	1.07 20
γ_ϵ	532.0 5	20 5
γ_ϵ	533.3 5	5.2 13
γ_ϵ	549.3 4	1.1 3
γ_ϵ	558.0 4	0.5 1
γ_ϵ	575.7 3	2.0 3
γ_ϵ	583.8 4	0.5 2
γ_ϵ	585.5 4	0.9 3
γ_ϵ	606.7 2	11.3 23
γ_ϵ	619.7 3	1.6 3

$^{155}_{70}$Yb(1.7 2 s)

Mode: α(84 10 %), ϵ(16 10 %)
Δ: -50740 310 keV syst
SpA: 3.5×10^{10} Ci/g
Prod: ^{144}Sm(^{16}O,5n); ^{142}Nd(^{20}Ne,7n)

Alpha Particles (^{155}Yb)

α(keV)
5191 5

$^{155}_{71}$Lu(70 *6* ms)

Mode: α(79 *4* %), ε(21 *4* %)
Δ: -42770 *430* keV syst
SpA: 1.05×10^{11} Ci/g
Prod: ^{144}Sm(^{19}F,8n)

Alpha Particles (^{155}Lu)

α(keV)
5630 *30*

$^{155}_{72}$Hf(890 *120* ms)

Mode: ε
Δ: -34440 *590* keV syst
SpA: 5.7×10^{10} Ci/g
Prod: ^{58}Ni on Mo-Sn targets; daughter ^{159}W; ^{107}Ag on V-Ni targets

A = 156

NDS **18**, 553 (1976)

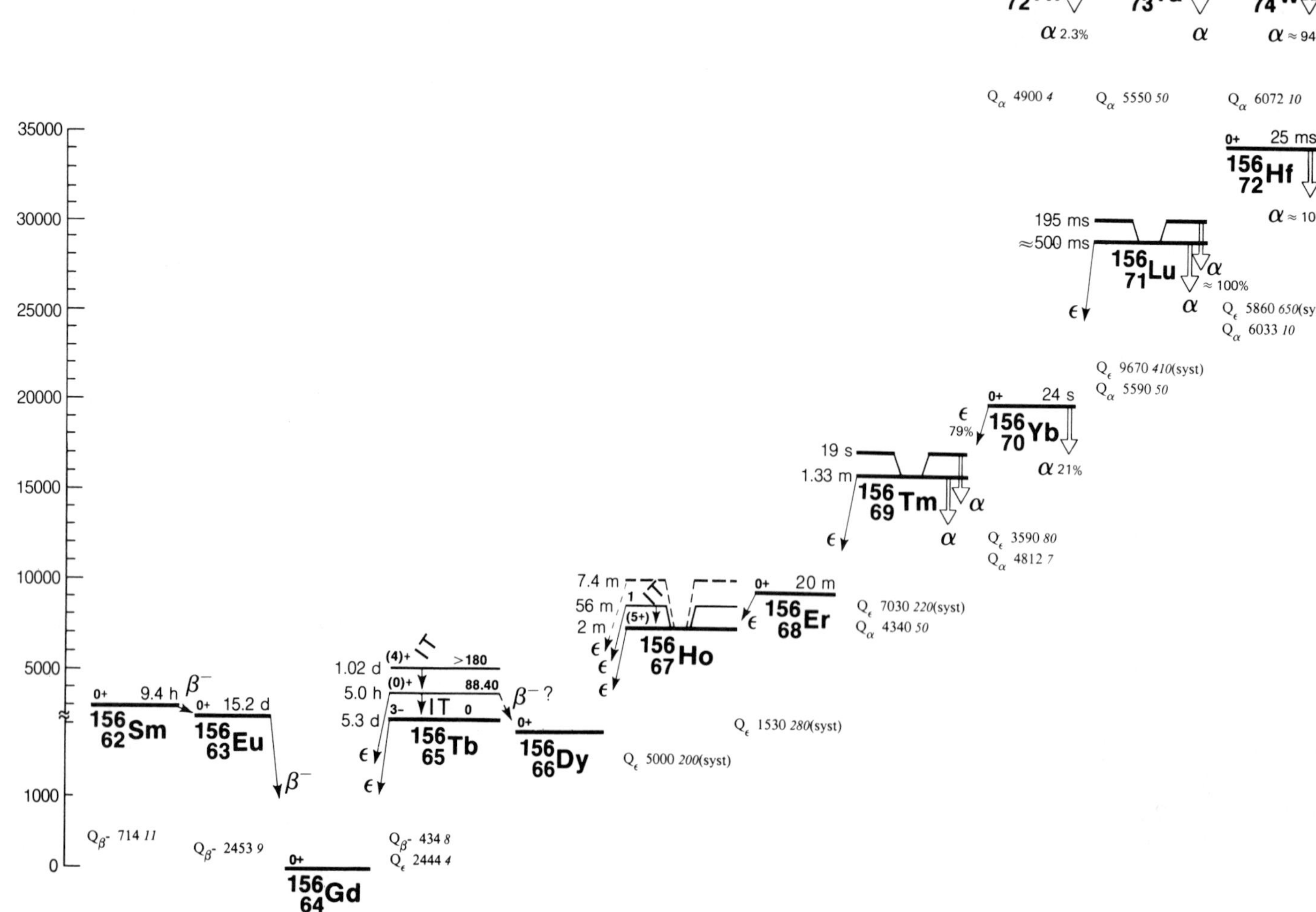

$^{156}_{62}$Sm(9.4 2 h)

Mode: β-

Δ: -69380 14 keV

SpA: 2.14×10^{6} Ci/g

Prod: fission

Photons (^{156}Sm)

γ_{mode}	γ(keV)	γ(rel)
γ (M1)	22.57 10	~32
γ (M1)	38.07 10	<14
γ E1	65.00 19	6.2 16
γ E1	87.57 17	100 14
γ E1	103.07 21	4.8 10
γ E1	165.73 13	61.7 12
γ E1	203.80 10	96.8 19
γ	219.0 10	~2
γ	243.5 4	5.6 3
γ (M1)	268.80 21	10.5 5
γ [M1]	291.37 19	12.6 6

$^{156}_{63}$Eu(15.19 6 d)

Mode: β-

Δ: -70094 10 keV

SpA: 5.510×10^{4} Ci/g

Prod: multiple n-capture from ^{154}Sm; daughter ^{156}Sm; fission

Photons (^{156}Eu)

⟨γ⟩=1324 17 keV

γ_{mode}	γ(keV)	γ(%)†
Gd L_ℓ	5.362	0.082 12
Gd L_η	6.049	0.037 5
Gd L_α	6.054	2.2 3
Gd L_β	6.837	2.1 3
Gd L_γ	7.823	0.30 4
Gd $K_{\alpha2}$	42.309	3.8 4
Gd $K_{\alpha1}$	42.996	6.9 7
Gd $K_{\beta1}'$	48.652	2.04 21
Gd $K_{\beta2}'$	50.214	0.62 7
γ E2	88.9854 24	8.9 9
γ	138.54 14	0.0083 9
γ	160.15 5	0.0108 11
γ [E1]	190.24 3	0.0173 16
γ E2	199.232 9	0.78 4
γ	215.72 20	0.013 3
γ	244.72 20	0.009 3
γ	281.42 20	0.0082 20
γ E1(+7.2%M2)	290.51 15	0.0092 20
γ	317.57 5	0.066 4
γ	335.71 11	0.0107 14
γ	348.29 9	0.0143 20
γ	354.22 9	0.0153 20
γ	434.46 6	0.216 11
γ	472.70 6	0.144 8
γ	490.31 5	0.178 12
γ	494.9 15	0.016 7
γ	498.88 5	0.061 7
γ	554.74 4	0.023 5
γ	585.92 4	0.066 6
γ E1	599.51 3	2.26 13
γ	632.69 5	0.037 6
γ E1	646.28 3	7.0 3
γ	706.95 5	0.043 14
γ	709.91 4	0.90 5

Photons (^{156}Eu) (continued)

γ_{mode}	γ(keV)	γ(%)†
γ E1	723.50 3	5.91 12
γ	768.56 7	0.088 9
γ	784.17 4	0.041 7
γ [M1]	797.77 4	0.108 13
γ [M1+E2]	811.82 3	10.2
γ [E1]	820.33 4	0.157 10
γ [M1+E2]	836.52 3	0.088 11
γ	839.07 5	0.033 8
γ E1	841.220 21	0.210 17
γ	858.44 5	0.124 12
γ E2	865.925 21	0.155 15
γ E1	867.10 4	1.38 12
γ	872.48 4	0.033 7
γ	903.49 4	0.032 8
γ	907 3	0.39 9
γ [M1]	916.45 8	~0.041
γ [M1+E2]	928.72 4	0.024 8
γ [E1]	944.32 4	1.37 8
γ	947.46 5	0.31 9
γ E2	960.52 8	1.59 10
γ	961.0 6	0.15 3
γ [E2]	969.84 3	0.379 18
γ	1011.88 3	0.336 20
γ E2	1018.58 5	0.080 9
γ	1027.48 3	0.117 10
γ	1037.33 6	0.035 6
γ [M1+E2]	1040.451 21	0.54 3
γ [M1+E2]	1065.156 20	5.16 18
γ	1076.08 5	0.37 6
γ [E2]	1079.20 3	4.8 4
γ	1101.75 4	0.037 11
γ	1115.80 3	0.056 8
γ	1129.436 21	0.139 9
γ	1140.50 3	0.291 19
γ E1	1153.472 17	7.0 7
γ [E2]	1154.141 20	5.12 24
γ	1156.01 9	0.138 20
γ	1164.03 11	0.059 9
γ	1169.07 3	0.291 20
γ	1187.3 5	0.015 7
γ E1	1230.69 3	8.8 3
γ E1(+1.7%M2)	1242.457 17	6.6 3
γ	1258.06 3	0.098 9
γ E1(+M2)	1277.460 25	3.15 12
γ [E1]	1366.444 25	1.72 7
γ	1626.22 4	0.035 7
γ	1682.07 6	0.30 4
γ	1857.44 11	0.250 17
γ [M1+E2]	1876.966 23	1.69 7
γ	1937.63 4	2.1 1
γ	1946.36 13	0.186 13
γ M1	1965.950 23	4.12 17
γ	2026.62 4	3.47 14
γ	2032.45 11	0.127 11
γ [M1+E2]	2097.78 4	4.19 15
γ	2110.54 13	0.084 8
γ	2116.52 5	0.123 8
γ	2121.44 11	0.0049 16
γ	2170.99 13	0.054 5
γ	2180.95 3	2.39 10
γ	2186.76 4	3.9 4
γ	2205.51 5	0.98 4
γ	2211.85 12	0.095 6
γ	2255.5 5	0.0063 12
γ	2259.97 13	0.0174 17
γ	2269.93 3	1.10 4
γ	2293.45 11	0.0245 20
γ	2301.02 20	0.0098 16
γ	2344.3 7	~0.0026
γ	2361.2 3	0.0181 17

† 4.9% uncert(syst)

Atomic Electrons (^{156}Eu)

⟨e⟩=30.0 18 keV

e_{bin}(keV)	⟨e⟩(keV)	e(%)
7	0.91	12.6 18
8 - 36	0.90	9.3 13
39	5.3	13.7 14
40 - 48	0.136	0.327 25
81	6.9	8.5 9
82	6.3	7.7 8
87	3.3	3.8 4
88 - 137	0.94	1.05 11
138 - 183	0.186	0.125 7
188 - 237	0.108	0.056 2
240 - 289	0.010	~0.0038
290 - 336	0.0052	0.0017 6
340 - 384	0.016	~0.004
422 - 471	0.041	0.009 3
472 - 504	0.0048	0.0010 3
536 - 585	0.039	0.0072 6
586 - 633	0.103	0.0172 7
638 - 673	0.125	0.019 3
699 - 748	0.033	0.0045 7
760 - 808	0.47	0.062 16
810 - 859	0.052	0.0063 10
860 - 909	0.027	0.0030 6
910 - 959	0.061	0.0066 4
960 - 990	0.031	0.0031 7
999	2.4	~0.24
1004 - 1052	0.67	0.065 21
1057 - 1106	0.217	0.0198 10
1107 - 1157	0.038	0.0034 3
1161 - 1208	0.131	0.0111 4
1222 - 1270	0.15	~0.012
1276 - 1316	0.017	0.00133 24
1358 - 1366	0.00216	0.000159 7
1576 - 1675	0.0049	0.00030 13
1680 - 1681	0.00011	~7×10^{-6}
1807 - 1896	0.058	0.0031 6
1916 - 1982	0.120	0.0062 9
2018 - 2115	0.074	0.0036 4
2120 - 2210	0.085	0.0040 10
2220 - 2311	0.013	0.00058 21
2336 - 2360	3.1 ×10^{-5}	1.3 5 ×10^{-6}

Continuous Radiation (^{156}Eu)

⟨β-⟩=395 keV;⟨IB⟩=0.69 keV

E_{bin}(keV)		⟨ ⟩(keV)	(%)
0 - 10	β-	0.195	3.91
	IB	0.0163	
10 - 20	β-	0.56	3.77
	IB	0.0156	0.108
20 - 40	β-	2.13	7.1
	IB	0.029	0.102
40 - 100	β-	12.4	18.0
	IB	0.077	0.120
100 - 300	β-	55	30.6
	IB	0.18	0.106
300 - 600	β-	55	13.0
	IB	0.17	0.041
600 - 1300	β-	144	15.9
	IB	0.169	0.020
1300 - 2453	β-	125	7.7
	IB	0.031	0.0020

$^{156}_{64}$Gd(stable)

Δ: -72547 3 keV

%: 20.47 4

$^{156}_{65}$Tb(5.34 9 d)

Mode: ϵ
Δ: -70103 5 keV
SpA: 1.57×10^{5} Ci/g
Prod: ^{153}Eu(α,n); ^{156}Gd(p,n)

Photons (^{156}Tb)

⟨γ⟩=1823 25 keV

γmode	γ(keV)	γ(%)†
Gd L$_ℓ$	5.362	0.44 5
Gd L$_η$	6.049	0.174 13
Gd L$_α$	6.054	11.7 8
Gd L$_β$	6.844	11.2 10
Gd L$_γ$	7.866	1.68 17
Gd K$_{α2}$	42.309	32.5 10
Gd K$_{α1}$	42.996	58.4 17
Gd K$_{β1}$'	48.652	17.2 5
Gd K$_{β2}$'	50.214	5.20 18
γ E2	88.9854 24	18.3 4
γ M1	111.950 21	3.88 21
γ M1(+35%E2)	115.637 25	0.050 12
γ M1+19%E2	155.17 3	1.50 9
γ [E1]	190.24 3	0.00151 17
γ E2	199.232 9	37.9 15
γ [M1+E2]	212.76 3	0.038 9
γ	248.85 4	0.0206 21
γ E2(+<3.8%M1)	262.558 23	5.41 21
γ [M1+E2]	267.12 3	0.06 3
γ E2	296.54 3	4.00 12
γ E2	356.451 25	12.6 7
γ [E2]	374.51 3	0.050 6
γ E2+39%M1	381.41 5	0.54 7
γ	407.1 10	0.053 12
γ E1	422.40 3	7.0 4
γ	445.47 4	0.035 12
γ [E1]	496.45 5	0.056 15
γ E1	534.35 3	61.2 18
γ [E2]	567.86 4	0.026 6
γ	576.23 9	2.00 21
γ [E1]	578.89 3	0.42 3
γ	592.74 4	0.018 6
γ E1	596.95 4	0.041 9
γ [E1]	603.84 5	0.085 12
γ [E2]	609.41 5	0.018 6
γ [M1+E2]	614.64 4	0.165 15
γ	626.83 9	0.29 4
γ	632.70 5	~0.0038
γ [E1]	636.47 4	~0.0038
γ E1	641.05 8	0.059 12
γ	651.17 7	~0.012
γ [M1+E2]	658.12 5	0.19 3
γ	668.23 5	0.053 12 ?
γ	673.64 10	0.044 12
γ	676.19 5	0.091 12 ?
γ E1+4.6%M2	676.25 4	0.091 12
γ E1	686.27 4	0.40 4
γ (E1)	689.52 4	0.16 3
γ [E2]	691.85 3	0.21 3
γ [E1]	697.73 5	0.118 18
γ (E1)	704.34 4	0.135 18
γ	706.54 7	<0.029
γ	717.04 10	0.097 18
γ	747.91 4	0.25 3
γ [E2]	770.67 4	<0.032
γ E1	780.17 3	2.23 9
γ E1	796.91 4	0.035 12
γ (E1)	804.87 3	0.229 21
γ E1	816.22 10	0.024 9
γ	827.15 5	~0.035
γ E1	841.220 21	0.232 19
γ E1	845.6 1	0.044 9
γ	855.25 9	0.29 3
γ E1	860.87 4	0.129 24
γ E2	865.925 21	0.30 3
γ	866.05 7	0.31 3
γ	877.30 6	0.044 15
γ	898.95 6	0.024 6
γ	921.93 10	0.07 3
γ [E2]	925.84 3	3.61 7
γ (E1)	949.19 4	1.39 6
γ [M1+E2]	959.82 3	1.75 7
γ [E2]	969.84 3	0.073 12
γ	973.90 4	0.091 21
γ E1	984.47 7	0.100 24
γ E1	987.97 5	0.27 4
γ [M1+E2]	1009.62 4	0.059 15
γ [M1+E2]	1037.79 4	1.00 12
γ [M1+E2]	1040.451 21	0.59 3
γ [M1+E2]	1065.156 20	10.1 3
γ [M1+E2]	1067.21 3	2.9 4
γ	1129.436 21	0.154 12
γ E1	1153.472 17	0.19 3
γ [E2]	1154.141 20	10.1 3
γ [M1+E2]	1159.05 3	6.82 24
γ	1169.07 3	0.056 9
γ	1174.07 8	0.106 15
γ	1180.31 13	0.047 12
γ E1(+2.6%M2)	1187.20 5	0.54 3
γ [E2]	1208.85 4	0.029 9
γ [M1+E2]	1222.374 23	29.4
γ E1	1230.69 3	0.76 5
γ E1(+1.7%M2)	1242.457 17	0.182 21
γ	1250.64 5	0.026 9
γ	1258.06 3	0.019 3
γ [E2]	1266.44 3	1.06 15
γ E1(+M2)	1277.460 25	0.022 3
γ [M1+E2]	1334.32 3	2.35 9
γ [E1]	1366.444 25	0.0120 18
γ	1373.30 8	0.024 6
γ [E2]	1421.604 23	11.8 7
γ	1449.87 5	0.032 12
γ [E1]	1563.65 5	0.041 12
γ E1	1646.09 3	3.44 12
γ [E1]	1762.88 5	0.08 3
γ E1	1815.11 3	0.376 21
γ E1	1845.32 3	3.8 4
γ	1886.91 5	0.065 21
γ	1893.1 7	0.038 6
γ	1944.28 7	0.051 6
γ E1	2014.34 3	1.06 4
γ	2092.4 7	0.041 9
γ	2103.32 3	0.0059 18 ?

† 5.1% uncert(syst)

Atomic Electrons (^{156}Tb)

⟨e⟩=84.7 11 keV

e$_{bin}$(keV)	⟨e⟩(keV)	e(%)
7	4.8	66 7
8	3.7	46 5
33 - 36	1.76	5.1 3
39	11.0	28.3 9
40 - 48	1.15	2.76 15
62	3.03	4.9 3
65	0.033	0.051 12
81	14.2	17.5 6
82	12.9	15.8 5
87	6.82	7.81 24
88 - 116	3.63	3.79 9
140 - 148	0.193	0.132 8
149	8.8	5.9 3
151 - 190	0.162	0.106 12
191	2.69	1.41 6
192 - 241	3.29	1.65 4
242 - 289	0.935	0.367 8
295 - 331	1.24	0.403 21
348 - 395	0.536	0.150 4
399 - 446	0.0327	0.0079 4
484 - 533	1.32	0.268 16
534 - 582	0.056	0.0100 25
584 - 634	0.025	0.0041 9
635 - 683	0.028	0.0044 5
684 - 730	0.037	0.0052 9
740 - 789	0.0107	0.00140 17
790 - 838	0.043	0.0052 16
839 - 878	0.099	0.0113 6
891 - 938	0.094	0.0104 17
941 - 990	0.062	0.0063 10
1001 - 1040	0.37	0.037 8
1057 - 1104	0.266	0.0244 15
1109 - 1158	0.25	0.022 4
1159 - 1208	0.71	0.061 15
1209 - 1258	0.150	0.0124 21
1259 - 1284	0.052	0.0040 9
1316 - 1365	0.0096	0.00073 13
1366 - 1414	0.215	0.0156 9
1420 - 1513	0.0078	0.00055 3
1555 - 1645	0.0261	0.00163 6
1713 - 1808	0.026	0.00148 14
1813 - 1894	0.0057	0.00031 5
1936 - 2013	0.0072	0.000366 13
2042 - 2102	0.00057	2.8 11 ×10^{-5}

Continuous Radiation (^{156}Tb)

⟨IB⟩=0.18 keV

E$_{bin}$(keV)		⟨ ⟩(keV)	(%)
10 - 20	IB	0.00020	0.00125
20 - 40	IB	0.045	0.123
40 - 100	IB	0.112	0.24
100 - 300	IB	0.0128	0.0074
300 - 600	IB	0.0094	0.0022
600 - 933	IB	0.0030	0.00044

$^{156}_{65}$Tb(5.0 1 h)

Mode: IT, ϵ, β-(?)
Δ: -70015 5 keV
SpA: 4.02×10^{6} Ci/g
Prod: ^{156}Gd(p,n); ^{155}Gd(d,n); ^{156}Gd(d,2n)

Photons (^{156}Tb)

⟨γ⟩=4.38 10 keV

γmode	γ(keV)	γ(%)
Tb L$_ℓ$	5.546	0.215 25
Tb L$_α$	6.269	5.7 5
Tb L$_η$	6.284	0.122 10
Tb L$_β$	7.079	6.5 6
Tb L$_γ$	8.110	0.93 8
Tb K$_{α2}$	43.744	1.56 8
Tb K$_{α1}$	44.482	2.80 14
Tb K$_{β1}$'	50.338	0.84 4
Tb K$_{β2}$'	51.958	0.250 13
γ$_{IT}$ E3	88.40 5	1.12

Atomic Electrons (^{156}Tb)

⟨e⟩=83.2 20 keV

e$_{bin}$(keV)	⟨e⟩(keV)	e(%)
8	4.4	57 6
9	0.035	0.40 4
35	0.028	0.079 9
36	2.15	5.9 3
37	0.0137	0.037 4
41	0.0051	0.0123 13
42	0.0171	0.041 5

Atomic Electrons (^{156}Tb)
(continued)

e_{bin}(keV)	⟨e⟩(keV)	e(%)
43	0.0223	0.052 *6*
44	0.0046	0.0103 *11*
48	0.00186	0.0039 *4*
49	0.00207	0.0042 *5*
50	0.00167	0.0033 *4*
80	29.5	36.8 *16*
81	27.4	33.8 *15*
86	0.172	0.199 *9*
87	15.3	17.7 *8*
88	4.14	4.69 *21*

$^{156}_{65}$Tb(1.02 *4* d)

Mode: IT
Δ: >-69923 keV
SpA: 8.2×10^5 Ci/g
Prod: ^{157}Gd(p,2n)

Photons (^{156}Tb)

γ_{mode}	γ(keV)	γ(%)
γ E1	49.63 *1*	74

$^{156}_{66}$Dy(>1.0×10^{18} yr)

Δ: -70536 *7* keV
%: 0.06 *1*

$^{156}_{67}$Ho(2 min)

Mode: ε
Δ: -65540 *200* keV syst
SpA: 6×10^8 Ci/g
Prod: ^{149}Sm(^{11}B,4n); daughter ^{156}Er

see ^{156}Ho(56 min) for γ rays

$^{156}_{67}$Ho(55.6 *6* min)

Mode: ε, IT
Δ: -65540 *200* keV syst
SpA: 2.168×10^7 Ci/g
Prod: ^{156}Dy(p,n)

Photons (^{156}Ho(2 + 56 min))

γ_{mode}	γ(keV)	γ(rel)*
γ_ϵE2	137.8 *5*	67 *5*
γ_ϵE2	266.4 *5*	100 *9*
γ_ϵE2	366.3 *5*	20.9 *20*
γ_ϵ	404.45 *23*	~2
γ_ϵ	424.8 *10*	0.8 *2*
γ_ϵ	538.7 *4*	1.1 *2*
γ_ϵ	564.7 *3*	1.1 *2*
γ_ϵ	617.5 *5*	2.6 *5*
γ_ϵ	659.5 *5*	1.0 *3*
γ_ϵ	666.8 *5*	2.0 *2*
γ_ϵ(E2)	683.8 *5*	11.5 *10*
γ_ϵ	690.83 *22*	9.5 *10*
γ_ϵ	752.9 *5*	3.6 *5*
γ_ϵ	755.28 *22*	2.4 *7*
γ_ϵ	764.11 *22*	7 *1*
γ_ϵ	884.1 *3*	16.1 *13*
γ_ϵ[E2]	890.7 *8*	8.5 *15*
γ_ϵ	930.98 *22*	7.4 *10*
γ_ϵ	950.2 *7*	2.0 *2*
γ_ϵ	965.0 *5*	1.0 *3*
γ_ϵ	1001.3 *10*	
γ_ϵ	1030.7 *7*	7.7 *10*
γ_ϵ	1121.4 *14*	7.5 *10*
γ_ϵ	1155.3 *10*	
γ_ϵ	1180.6 *5*	1.0 *3*
γ_ϵ	1205.8 *10*	
γ_ϵ	1218 *1*	
γ_ϵ	1222.61 *21*	5.9 *10*
γ_ϵ	1292.5 *10*	
γ_ϵ	1301.4 *10*	
γ_ϵ	1337.8 *10*	
γ_ϵ	1390.8 *3*	3.2 *12*
γ_ϵ	1416 *1*	
γ_ϵ	1422.6 *10*	
γ_ϵ	1432.9 *10*	
γ_ϵ	1453 *1*	
γ_ϵ	1471 *1*	
γ_ϵ	1528.1 *10*	
γ_ϵ	1535.4 *10*	
γ_ϵ	1543.6 *10*	
γ_ϵ	1633.9 *10*	
γ_ϵ	1648 *1*	
γ_ϵ	1761.7 *10*	
γ_ϵ	1820.1 *10*	
γ_ϵ	1854.6 *10*	
γ_ϵ	2020.7 *10*	
γ_ϵ	2025.9 *10*	
γ_ϵ	2042.9 *10*	
γ_ϵ	2218.5 *10*	
γ_ϵ	2252.1 *10*	
γ_ϵ	2331.3 *10*	
γ_ϵ	2392.3 *10*	
γ_ϵ	2458.3 *10*	

* with ^{156}Ho(2 min) in equilib

$^{156}_{67}$Ho(7.4 min)

Mode: ε
Δ: -65540 *200* keV syst
SpA: 1.6×10^8 Ci/g
Prod: ^{160}Dy(p,5n)

Photons (^{156}Ho)

γ_{mode}	γ(keV)
γ E2	137.8 *5*
γ E2	266.4 *5*
γ E2	366.3 *5*
γ	445.7 *20*
γ	1576.9 *20*
γ	2022.5 *20*

$^{156}_{68}$Er(19.5 *10* min)

Mode: ε
Δ: -64000 *200* keV syst
SpA: 6.2×10^7 Ci/g
Prod: protons on Ta

Photons (^{156}Er)

γ_{mode}	γ(keV)	γ(rel)
γ M1	29.8 *1*	17.0 *17*
γ E1	35.2 *1*	100

$^{156}_{69}$Tm(1.33 *5* min)

Mode: α, ε
Δ: -56970 *80* keV
SpA: 9.0×10^8 Ci/g
Prod: ^{147}Sm(^{14}N,5n); protons on Er

Alpha Particles (^{156}Tm)

α(keV)
4230 *10*

Photons (^{156}Tm)

γ_{mode}	γ(keV)	γ(rel)
γ	290.70 *15*	0.43 *8*
γ	326.0 *1*	0.28 *4*
γ	344.6 *1*	100
γ	350.0 *5*	0.20 *8*
γ	406.9 *3*	0.11 *4*
γ	420.8 *1*	1.6 *2*
γ	423.45 *20*	0.50 *8*
γ	429.90 *15*	0.15 *4*
γ	451.5 *4*	<0.15
γ	452.9 *1*	18.9 *19*
γ	475.65 *10*	1.1 *1*
γ	484.85 *15*	0.53 *7*
γ	507.4 *4*	<0.2
γ	543.50 *15*	0.32 *6*
γ	554.00 *15*	1.0 *1*
γ	586.00 *15*	15.8 *19*
γ	608.85 *15*	1.6 *2*
γ	640.5 *2*	0.57 *9*
γ	700.0 *2*	1.4 *3*
γ	749.0 *3*	0.57 *15*
γ	763.90 *25*	0.19 *6*
γ	773.00 *25*	0.23 *6*
γ	814.30 *25*	0.68 *15*
γ	826.0 *4*	0.19 *8*
γ	866.00 *15*	0.53 *8*
γ	876.2 *2*	2.6 *4*
γ	898.5 *2*	1.4 *2*
γ	930.5 *2*	6.0 *8*
γ	959.0 *2*	9.6 *9*
γ	974.9 *3*	0.23 *8*
γ	1006.80 *25*	3.6 *6*
γ	1017.00 *25*	1.2 *3*
γ	1084.4 *3*	0.25 *8*
γ	1160.5 *4*	0.19 *6*
γ	1173.30 *25*	<0.6
γ	1202.2 *3*	0.9 *2*
γ	1226.1 *4*	1.4 *2*
γ	1286.1 *2*	3.2 *8*
γ	1366.1 *3*	1.9 *5*
γ	1405.2 *3*	0.5 *1*
γ	1415.2 *3*	0.38 *9*
γ	1517.50 *25*	3.2 *6*

Photons (^{156}Tm)
(continued)

γ_{mode}	γ(keV)	γ(rel)
γ	1529.3 *4*	0.47 *13*
γ	1545.5 *6*	~0.15
γ	1565.3 *4*	1.9 *6*
γ	1573.4 *5*	0.38 *9*
γ	1664.0 *5*	1.4 *3*
γ	1670.1 *4*	1.5 *3*
γ	1677.2 *5*	0.28 *9*
γ	1711.7 *10*	0.25 *9*
γ	1722.5 *8*	0.36 *9*
γ	1738.7 *10*	0.28 *9*
γ	1760.5 *8*	0.43 *13*
γ	1767.3 *8*	0.19 *8*
γ	1779.4 *9*	0.19 *8*
γ	1825.3 *6*	0.9 *2*

$^{156}_{69}$Tm(19 *3* s)

Mode: α
Δ: -56970 *80* keV
SpA: 3.7×10^{9} Ci/g
Prod: ^{147}Sm(^{14}N,5n); ^{162}Er(^{3}He,p8n)

Alpha Particles (^{156}Tm)

α(keV)
4460 *10*

$^{156}_{70}$Yb(24 *1* s)

Mode: ϵ(79 *6* %), α(21 *6* %)
Δ: -53380 *60* keV
SpA: 2.97×10^{9} Ci/g
Prod: ^{162}Er(^{3}He,9n); ^{58}Ni on Pd

Alpha Particles (^{156}Yb)

α(keV)
4688 *7*

$^{156}_{71}$Lu(~500 ms)

Mode: α, ϵ
Δ: -43720 *410* keV syst
SpA: $\sim8\times10^{10}$ Ci/g
Prod: ^{144}Sm(^{19}F,7n); ^{58}Ni on ^{107}Ag

Alpha Particles (^{156}Lu)

α(keV)
5450 *10*

$^{156}_{71}$Lu(195 *25* ms)

Mode: α
Δ: -43720 *410* keV syst
SpA: 1.01×10^{11} Ci/g
Prod: ^{58}Ni on ^{107}Ag; ^{144}Sm(^{19}F,7n)

Alpha Particles (^{156}Lu)

α(keV)
5568 *5*

$^{156}_{72}$Hf(25 *4* ms)

Mode: α
Δ: -37860 *500* keV syst
SpA: 1.04×10^{11} Ci/g
Prod: ^{58}Ni on ^{107}Ag

Alpha Particles (^{156}Hf)

α(keV)
5878 *10*

A = 157

NDS **39**, 103 (1983)

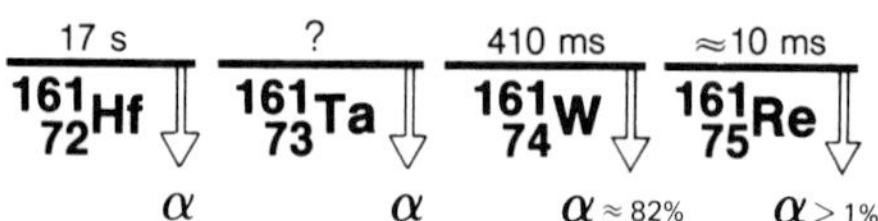

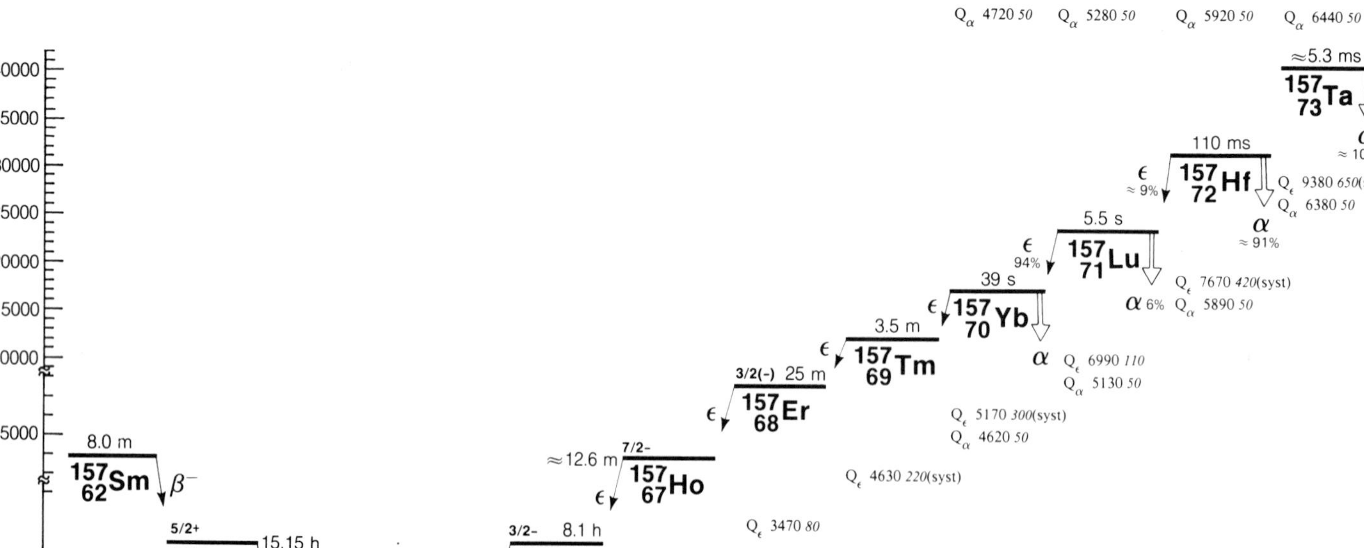

$^{157}_{62}$Sm(8.0 *5* min)

Mode: β-
Δ: -66870 *200* keV
SpA: 1.50×10^{8} Ci/g
Prod: ^{160}Gd(n,α)

Photons (^{157}Sm)

γ$_{mode}$	γ(keV)	γ(rel)
γ [M1+E2]	58.9 *3*	~2
γ (M1+E2)	76.83 *20*	2.5 *5*
γ [E1]	120.97 *17*	9.3 *10*
γ [E1]	186.24 *25*	1.2 *3*
γ [E1]	190.0 *3*	1.4 *4*
γ [E1]	196.35 *15*	32 *3*
γ [E1]	197.79 *15*	100
γ	216.9 *4*	0.5 *2*
γ [E1]	255.2 *3*	0.5 *2*
γ [E1]	263.06 *18*	2.7 *4*
γ	275.8 *4*	1.3 *3*
γ [E2]	317.32 *21*	2.3 *4*
γ [M1+E2]	394.15 *15*	24.0 *24*
γ	843.96 *20*	9.3 *9* ?
γ	988.9 *3*	2.0 *3*

$^{157}_{63}$Eu(15.15 *4* h)

Mode: β-
Δ: -69473 *7* keV
SpA: 1.318×10^{6} Ci/g
Prod: ^{160}Gd(p,α); neutrons on Gd; deuterons on ^{160}Gd; fission; ^{158}Gd(γ,p); ^{154}Sm(α,p)

Photons (^{157}Eu)

γ$_{mode}$	γ(keV)	γ(rel)
γ E1	9.373 *14*	7.2 *14*
γ M1+4.3%E2	51.802 *14*	3.3 *7*
γ M1+3.2%E2	54.529 *9*	16 *3*
γ E1	63.902 *15*	100 *20*
γ (M1+E2)	64.512 *18*	0.58 *12*
γ M1+3.2%E2	76.909 *18*	0.87 *17*
γ [M1+E2]	96.05 *4*	0.048 *10*
γ [E2]	116.314 *17*	0.17 *4*
γ E2	131.438 *18*	0.25 *5*
γ	158.389 *19*	0.106 *21*
γ [M1+E2]	161.813 *19*	0.37 *7*
γ	208.601 *19*	0.64 *13*
γ [M1+E2]	212.026 *19*	0.26 *5*
γ [M1+E2]	226.607 *24*	0.16 *3*
γ [E1]	237.83 *3*	0.067 *13*
γ	246.487 *24*	0.034 *7*
γ [M1+E2]	276.820 *22*	0.18 *4*
γ [E1]	288.040 *25*	0.42 *8*
γ	291.80 *4*	0.096 *19*
γ [M1+E2]	302.970 *21*	0.29 *6*
γ (E1)	318.704 *19*	12.6 *25*
γ [E1]	334.424 *22*	3.6 *7*
γ	340.10 *4*	0.096 *19*
γ [E2]	344.607 *19*	0.15 *3*
γ [E2]	358.907 *17*	1.3 *3*
γ (E1)	370.505 *19*	48 *10*
γ [M1+E2]	379.878 *18*	1.15 *23*
γ [M1+E2]	383.20 *3*	0.31 *6*
γ [E1]	393.385 *20*	0.53 *11*
γ [E1]	398.935 *21*	5.8 *12*
γ [M1+E2]	409.119 *17*	11.7 *23*

Photons (^{157}Eu) (continued)

γ$_{mode}$	γ(keV)	γ(rel)
γ (M1)	410.708 *16*	76 *15*
γ [E1]	420.081 *16*	4.0 *8*
γ	427.343 *23*	0.70 *14*
γ [M1+E2]	434.408 *19*	1.5 *3*
γ [E1]	450.737 *21*	5.3 *11*
γ [M1+E2]	460.921 *16*	5.3 *11*
γ [E1]	470.294 *17*	0.87 *17*
γ [E1]	474.611 *16*	11.0 *22*
γ	491.855 *22*	0.39 *8*
γ [E2]	506.42 *2*	0.36 *7*
γ [E1]	524.823 *16*	1.3 *3*
γ	543.656 *23*	0.14 *3*
γ	543.82 *3*	0.14 *3*
γ	553.029 *22*	0.15 *3*
γ [E1]	555.198 *23*	0.15 *3*
γ	567.507 *19*	0.63 *13*
γ [M1+E2]	570.932 *18*	6.8 *14*
γ	585.50 *4*	0.077 *15*
γ	591.094 *22*	0.69 *14*
γ	607.20 *24*	0.096 *19*
γ	613.29 *23*	0.072 *14* ?
γ	619.309 *18*	16 *3*
γ [M1+E2]	622.734 *18*	4.2 *9*
γ	625.71 *4*	0.063 *13*
γ	628.682 *18*	0.43 *9*
γ [E1]	632.107 *19*	0.20 *4*
γ [E2]	635.726 *22*	0.20 *4*
γ	655.606 *22*	0.81 *16*
γ	668.37 *24*	0.053 *11*
γ [M1+E2]	674.47 *23*	0.072 *14*
γ	683.211 *18*	1.0 *2*
γ	683.27 *4*	0.36 *7*
γ	685.19 *4*	0.21 *4*
γ [M1+E2]	687.528 *21*	5.1 *10*
γ	699.00 *4*	0.27 *6*
γ	700.92 *4*	1.3 *3*
γ	707.407 *21*	0.20 *4*
γ	716.780 *24*	0.120 *24*
γ [M1+E2]	729.00 *23*	0.096 *19*
γ	739.29 *20*	0.077 *15*
γ	750.80 *3*	0.56 *11*
γ [E1]	751.430 *23*	0.56 *11* ?
γ	752.72 *4*	1.11 *22*
γ	762.09 *3*	1.7 *4*
γ [M1+E2]	762.650 *25*	1.7 *4*
γ	803.8 *2*	0.077 *15*
γ	814.70 *4*	0.096 *19*
γ	816.62 *3*	0.31 *6*
γ	864.97 *20*	0.087 *17*
γ	934.05 *20*	0.17 *3*
γ	944.41 *4*	0.14 *3*
γ	985.85 *20*	0.63 *13*
γ	996.21 *4*	0.13 *3*
γ	1051.18 *20*	0.111 *22*
γ	1060.11 *3*	0.120 *24*
γ	1115.69 *20*	0.082 *16*
γ	1167.49 *20*	0.20 *4*

$^{157}_{64}$Gd(stable)

Δ: -70835 *3* keV
%: 15.65 *3*

$^{157}_{65}$Tb(150 *30* yr)

Mode: ϵ
Δ: -70773 *4* keV
SpA: 15 Ci/g
Prod: daughter ^{157}Dy; ^{157}Gd(p,n)

Photons (^{157}Tb)

⟨γ⟩=9.55 *16* keV

γ$_{mode}$	γ(keV)	γ(%)
Gd L$_{\ell}$	5.362	0.144 *17*
Gd L$_{\eta}$	6.049	0.060 *5*
Gd L$_{\alpha}$	6.054	3.8 *3*
Gd L$_{\beta}$	6.812	6.7 *8*
Gd L$_{\gamma}$	8.004	1.43 *20*
Gd K$_{\alpha2}$	42.309	5.68 *17*
Gd K$_{\alpha1}$	42.996	10.2 *3*
Gd K$_{\beta1}$'	48.652	3.01 *9*
Gd K$_{\beta2}$'	50.214	0.91 *3*
γ M1+3.2%E2	54.529 *9*	0.0210 *15*

Atomic Electrons (^{157}Tb)

⟨e⟩=3.7 *3* keV

e$_{bin}$(keV)	⟨e⟩(keV)	e(%)
4	0.0089	0.208 *15*
7	0.66	9.0 *9*
8	2.48	30 *3*
33	0.041	0.122 *12*
34	0.060	0.176 *18*
35	0.157	0.45 *5*
36	0.052	0.144 *15*
40	0.054	0.136 *14*
41	0.087	0.211 *22*
42	0.0244	0.058 *6*
43	0.0151	0.035 *4*
46	0.0147	0.0318 *25*
47	0.0189	0.040 *5*
48	0.0067	0.0139 *14*
53	0.0050	0.0095 *9*
54	0.00135	0.00249 *23*
55	7.5 ×10^{-5}	0.000138 *16*

Continuous Radiation (^{157}Tb)

⟨IB⟩=0.063 keV

E$_{bin}$(keV)		⟨ ⟩(keV)	(%)
10 - 20	IB	0.00042	0.0027
20 - 40	IB	0.023	0.066
40 - 63	IB	0.037	0.080

$^{157}_{66}$Dy(8.1 *1* h)

Mode: ϵ
Δ: -69434 *7* keV
SpA: 2.46×10^{6} Ci/g
Prod: ^{159}Tb(p,3n); ^{154}Gd(α,n); ^{159}Tb(d,4n); ^{156}Gd(α,3n)

Photons (^{157}Dy)

⟨γ⟩=341 *6* keV

γ$_{mode}$	γ(keV)	γ(%)†
Tb L$_{\ell}$	5.546	0.28 *3*
Tb L$_{\alpha}$	6.269	7.3 *5*
Tb L$_{\eta}$	6.284	0.096 *8*
Tb L$_{\beta}$	7.114	6.6 *6*
Tb L$_{\gamma}$	8.197	1.02 *11*
Tb K$_{\alpha2}$	43.744	24.2 *12*

Photons (^{157}Dy) (continued)

γ_{mode}	γ(keV)	γ(%)†
Tb $K_{\alpha1}$	44.482	43.3 21
Tb $K_{\beta1}$'	50.338	13.0 6
Tb $K_{\beta2}$'	51.958	3.87 20
γ M1+1.2%E2	60.84 5	~0.4
γ M1+<2.5%E2	83.04 4	0.60 18
γ E2	143.88 6	0.035 6
γ E1	182.29 12	2.2 5
γ [E1]	265.33 12	0.24 9
γ [E1]	297.0 7	~0.08
γ E1	326.17 12	90.0
γ E1	405.3 7	~0.015
γ E1	498.8 8	~0.010
γ M1	553.6 5	0.012 4
γ M1	576.5 6	0.029 5
γ M1	597.5 6	0.076 15
γ [M1+E2]	636.7 5	~0.019
γ [M1+E2]	637.3 6	~0.013
γ [M1+E2]	697.5 5	~0.0027
γ (E2)	745.1 7	0.049 10
γ E2	770.1 7	0.023 5
γ E2	776.7 6	0.052 11
γ M1(+E2)	897.0 15	~0.0027 ?
γ	910.4 10	0.0027 5
γ E2	931.98 20	0.009 3
γ (E0+E2)	983.7 12	<0.0009
γ (E0+E2)	992.81 19	0.0027 9
γ (M1)	1044.6 12	0.0045 18
γ (E1)	1102.9 6	~0.0018
γ	1215.3 7	~0.0018
γ	1276.1 7	0.0045 18

† 2.2% uncert(syst)

Atomic Electrons (^{157}Dy)

⟨e⟩=12.4 6 keV

e_{bin}(keV)	⟨e⟩(keV)	e(%)
8	4.4	56 6
9	0.91	10 3
31	0.60	1.9 6
35	0.43	1.22 13
36	0.65	1.80 20
37 - 83	1.76	3.7 4
92 - 136	0.17	0.133 23
142 - 182	0.039	0.023 3
213 - 258	0.015	0.0065 19
263 - 265	0.00048	0.00018 5
274	2.83	1.03 3
288 - 297	0.00061	0.00021 7
317	0.382	0.120 3
318 - 353	0.207	0.0642 10
397 - 405	7.3 ×10^{-5}	1.8 7 ×10^{-5}
447 - 491	0.00022	4.8 21 ×10^{-5}
497 - 546	0.0101	0.0019 3
552 - 598	0.0036	0.00061 24
628 - 646	0.00051	8 3 ×10^{-5}
689 - 738	0.0041	0.00058 7
743 - 777	0.00057	7.4 8 ×10^{-5}
845 - 890	0.00042	4.8 13 ×10^{-5}
895 - 941	0.0018	0.00019 6
975 - 993	0.00048	4.8 13 ×10^{-5}
1036 - 1051	4.8 ×10^{-5}	4.6 13 ×10^{-6}
1094 - 1103	2.8 ×10^{-6}	2.6 10 ×10^{-7}
1163 - 1208	4.1 ×10^{-5}	~4×10^{-6}
1213 - 1224	8.7 ×10^{-5}	~7×10^{-6}
1267 - 1276	1.5 ×10^{-5}	~1×10^{-6}

Continuous Radiation (^{157}Dy)

⟨IB⟩=0.37 keV

E_{bin}(keV)		⟨ ⟩(keV)	(%)
10 - 20	IB	0.00018	0.00114
20 - 40	IB	0.034	0.092
40 - 100	IB	0.139	0.29
100 - 300	IB	0.034	0.017
300 - 600	IB	0.099	0.023
600 - 1300	IB	0.061	0.0086
1300 - 1338	IB	1.9×10^{-7}	1.44×10^{-8}

$^{157}_{67}$Ho(12.6 6 min)

Mode: ε

Δ: -66890 50 keV

SpA: 9.5×10^{7} Ci/g

Prod: protons on Dy; protons on Ho

Photons (^{157}Ho)

⟨γ⟩=468 15 keV

γ_{mode}	γ(keV)	γ(%)†
Dy L_{ℓ}	5.743	0.50 7
Dy L_{α}	6.491	13.1 15
Dy L_{η}	6.534	0.177 24
Dy L_{β}	7.382	12.1 17
Dy L_{γ}	8.508	1.9 3
Dy $K_{\alpha2}$	45.208	39 3
Dy $K_{\alpha1}$	45.998	69 5
Dy $K_{\beta1}$'	52.063	20.9 14
Dy $K_{\beta2}$'	53.735	6.1 4
γ M1+<8.3%E2	61.18 8	5.1 8
γ	67.38 12	
γ M1+E2	78.86 8	0.052 8
γ M1+E2	86.59 8	4.9 7
γ M1+E2	98.87 15	0.116 17 ?
γ [E1]	106.57 11	0.80 12
γ E2+18%M1	109.91 10	0.58 9
γ E1	121.17 12	0.26 4
γ	126.93 10	0.22 3 ?
γ [E1]	130.00 10	0.7 1
γ	132.05 13	0.100 15 ?
γ E2	147.77 8	1.8 3
γ E1	150.04 10	0.72 11
γ E1	153.12 10	2.7 4
γ M1	162.40 11	1.18 18
γ M1(+E2)	163.86 20	0.34 5
γ E1	173.48 11	0.60 9
γ E1	188.11 10	3.9 6
γ M1	193.45 8	6.5 10
γ E2	196.5 1	0.60 9
γ E1	208.86 10	1.20 18
γ (E1)	224.26 20	0.134 20
γ (E1)	227.36 20	0.24 4
γ E1	234.66 11	0.88 13
γ	253.23 12	0.148 22 ?
γ E1	258.13 12	1.13 17 ?
γ [M1+E2]	269.39 12	0.17 3
γ M1+E2	272.32 9	4.0 6
γ	273.85 11	0.26 4 ?
γ M1	280.04 8	20 3
γ E1	297.09 11	0.72 11
γ (E1)	320.21 11	2.1 3
γ M1(+E2)	341.22 8	15.5 23
γ	347.03 10	0.060 9
γ M1+E2	353.66 20	0.19 3
γ M1	358.91 8	0.48 7
γ M1	360.55 10	0.48 7
γ	365.16 20	0.036 5
γ	367.26 20	0.23 4
γ [E2]	377.77 15	0.0060 9 ?
γ [M1+E2]	379.30 10	0.126 19
γ M1+E2	388.41 10	0.42 6
γ	394.26 20	0.054 8

Photons (^{157}Ho) (continued)

γ_{mode}	γ(keV)	γ(%)†
γ	400.16 20	0.156 23
γ	405.90 12	0.074 11 ?
γ	416.76 20	0.030 5
γ E2	420.09 9	0.27 4
γ	428.46 20	0.122 18
γ	430.57 11	0.108 16
γ M1+E2	449.46 20	0.18 3
γ M1	463.26 11	0.26 4 ?
γ M1	465.89 10	0.164 25
γ	468.06 20	0.158 24
γ M1+E2	476.64 9	0.55 8
γ M1(+E2)	508.32 10	2.8 4
γ	523.00 11	0.096 14 ?
γ	527.5 6	0.040 6
γ	540.49 10	0.118 18
γ	550.06 20	0.146 22
γ M1	555.50 8	2.6 4
γ	567.86 20	0.138 21
γ M1	570.25 10	0.51 8
γ	582.26 20	0.074 11
γ M1	597.56 20	0.19 3
γ	610.26 20	0.088 13
γ	627.07 10	0.19 3 ?
γ M1	649.11 10	0.30 5
γ (E1)	662.07 11	0.23 4
γ E1	685.5 1	0.84 13
γ M1+E2	688.25 10	0.34 5 ?
γ M1	702.93 11	0.166 25 ?
γ E1	708.62 10	1.3 2
γ (M1)	748.96 9	0.17 3
γ E1	779.11 11	0.43 7
γ M1	791.16 10	0.21 3
γ	828.16 20	0.60 9
γ M1+E2	835.55 8	1.02 15
γ (M1)	842.56 10	0.166 25
γ M1	870.03 10	0.91 14
γ M1+E2	896.73 8	4.0 6
γ M1+E2	929.15 10	0.59 9
γ	936.46 20	0.152 23
γ	946.26 20	0.18 3
γ	954.96 20	0.076 11
γ	963.06 20	0.128 19
γ	969.16 20	0.074 11
γ	1038.89 10	0.092 14 ?
γ	1044.06 20	0.028 4
γ	1053.96 20	0.084 13
γ	1063.46 20	0.106 16
γ	1072.96 20	0.046 7
γ	1082.36 20	0.078 12
γ	1090.46 20	0.052 8
γ M1+E2	1150.07 10	0.79 12
γ	1158.96 20	0.044 7
γ	1169.76 20	0.19 3
γ	1172.36 20	0.068 10
γ	1192.00 11	0.120 18 ?
γ	1202.16 20	0.064 10
γ M1+E2	1211.25 10	2.1 3
γ	1232.34 10	0.058 9 ?
γ	1239.76 20	0.056 8
γ	1274.76 20	0.084 13
γ	1298.36 20	0.078 12
γ	1303.06 20	0.088 13
γ	1318.93 10	0.17 3 ?
γ	1332.56 20	0.048 7
γ	1349.66 20	0.082 12
γ	1358.96 20	0.068 10
γ	1380.11 10	0.22 3 ?
γ	1395.06 20	0.036 5
γ	1406.46 20	0.114 17
γ	1460.16 20	0.21 3
γ	1490.96 20	0.038 6
γ	1510.36 20	0.048 7
γ	1521.96 20	0.034 5
γ	1563.76 20	0.058 9
γ	1763.26 20	0.044 7
γ	1788.66 20	0.062 9

† 10% uncert(syst)

Atomic Electrons (^{157}Ho)

⟨e⟩=56 4 keV

e_{bin}(keV)	⟨e⟩(keV)	e(%)
7	3.1	42 6
8	5.0	64 10
9	3.7	43 7
25	0.039	0.16 6
33	3.8	12 4
36 - 51	3.38	8.5 5
52	2.9	5.6 9
53	3.2	6.0 9
56 - 77	2.60	4.3 4
78	3.3	~4
79	2.3	<6
85	1.4	~2
86 - 134	3.3	3.2 4
139	0.34	0.25 4
140	3.2	2.3 3
141 - 189	1.50	0.91 6
191 - 225	1.2	0.56 14
226	5.4	2.4 4
227 - 274	1.32	0.49 6
278 - 280	0.27	0.096 12
287	2.3	0.8 3
288 - 335	0.79	0.25 5
338 - 387	0.27	0.078 15
388 - 431	0.152	0.037 6
440 - 487	0.28	0.061 20
496 - 544	0.40	0.080 9
546 - 596	0.105	0.0187 19
597 - 641	0.041	0.0064 11
647 - 695	0.050	0.0076 6
700 - 749	0.0241	0.0033 3
770 - 819	0.129	0.016 3
820 - 869	0.18	0.021 6
870 - 915	0.071	0.0080 14
920 - 969	0.0077	0.00083 16
985 - 1031	0.011	0.0011 3
1035 - 1083	0.0032	0.00030 9
1088 - 1119	0.030	0.0027 7
1138 - 1186	0.071	0.0061 15
1190 - 1239	0.013	0.00111 23
1240 - 1279	0.0079	0.00063 20
1289 - 1333	0.008	0.00064 21
1341 - 1387	0.0041	0.0003 1
1393 - 1437	0.0047	0.00033 15
1451 - 1520	0.0033	0.00022 6
1555 - 1562	0.00016	1.0 5 $\times10^{-5}$
1709 - 1787	0.0017	0.00010 3

Continuous Radiation (^{157}Ho)

⟨β+⟩=28.7 keV;⟨IB⟩=2.1 keV

E_{bin}(keV)		⟨ ⟩(keV)	(%)
0 - 10	β+	2.74 $\times10^{-7}$	3.25 $\times10^{-6}$
	IB	0.0019	
10 - 20	β+	1.43 $\times10^{-5}$	8.5 $\times10^{-5}$
	IB	0.00140	0.0096
20 - 40	β+	0.000502	0.00152
	IB	0.024	0.066
40 - 100	β+	0.0310	0.0394
	IB	0.163	0.34
100 - 300	β+	1.66	0.75
	IB	0.058	0.030
300 - 600	β+	9.6	2.11
	IB	0.22	0.048
600 - 1300	β+	17.3	2.16
	IB	1.01	0.109
1300 - 2479	β+	0.102	0.0076
	IB	0.60	0.039
	Σβ+		5.1

$^{157}_{68}$Er(25 3 min)

Mode: ε

Δ: -63420 90 keV

SpA: 4.8×10^{7} Ci/g

Prod: protons on ^{181}Ta; ^{165}Ho(p,9n); ^{149}Sm(^{12}C,4n)

Photons (^{157}Er)

γ_{mode}	γ(keV)	γ(rel)
γ(E1)	53.051 20	~100
γ(M1+E2)	66.908 20	20.9 18
γ	83.37 8	12.2 16
γ	117.4 3	3.3 7
γ	121.41 7	71 14
γ	142.0 3	0.88 22
γ	144.3 3	1.7 3
γ	150.42 9	21 4
γ	157.77 10	1.11 18
γ	160.6 3	4.5 5
γ	179.90 16	8.3 4
γ	182.4 10	
γ	198.89 13	0.40 8
γ	201.37 12	1.5 4
γ	238.3 5	1.6 3
γ	284.74 13	0.86 11
γ	302.30 16	2.4 4
γ	303.60 10	8.7 8
γ	305.13 10	3.2 5
γ	307.61 13	2.6 6
γ	317.60 20	3.6 4
γ	347.70 10	11.3 9
γ	349.5 3	1.4 4
γ	354.47 15	1.7 3
γ	356.94 13	3.5 6
γ	372.14 8	1.8 2
γ	374.62 7	0.6 1
γ(M1)	391.31 8	100 6
γ	398.82 13	1.0 3
γ	411.70 20	0.31 11
γ	422.87 14	2.1 4
γ	431.00 10	5.2 10
γ	436.71 11	0.70 23
γ	440.31 14	1.2 4
γ	442.7 3	1.9 6
γ	443.6 3	1.5 5
γ	456.20 10	1.01 25
γ	460.50 20	2.9 6
γ	474.40 20	1.3 4
γ	481.0 4	~0.50
γ	482.17 7	1.0 2
γ	482.19 13	1.0 2
γ	492.96 11	1.11 23
γ	502.5 3	1.9 4
γ	503.48 17	6.0 4
γ	517.14 9	1.8 5
γ[E1]	517.34 17	1.8 5
γ	518.5 3	2.4 7
γ	527.80 9	4.8 6
γ	538.1 3	<1
γ	549.08 7	19.2 21
γ	560.9 3	0.6 2
γ	564.4 1	<0.35
γ	569 1	
γ	574.0 1	3.2 4
γ	584.05 9	3.7 4
γ	611.0 1	4.8 5
γ	614.7 1	0.9 3
γ	651.8 1	3.3 4
γ	672.0 2	1.68 25
γ	673.5 2	2.2 3
γ	694.6 1	1.3 5
γ	719.5 5	1.0 2
γ	735.6 2	<1
γ	785.8 2	2.0 4
γ	792.6 2	1.99 23
γ	796.0 2	1.1 2
γ	816.7 2	
γ	824.8 2	
γ	889.7 10	
γ	910.6 3	1.8 5

Photons (^{157}Er) (continued)

γ_{mode}	γ(keV)	γ(rel)
γ	921.1 1	0.53 7
γ	942.0 2	1.8 4
γ	960.7 2	1.5 5
γ	972.2 2	
γ	980.5 2	0.80 17
γ	1026.31 15	1.1 3
γ	1028.79 15	~0.8
γ	1114.7 1	1.33 20
γ	1129.01 12	1.2 4
γ	1137.8 2	0.40 13
γ	1141.2 2	
γ	1142.87 12	1.9 5
γ	1148.3 3	<2
γ	1154.0 3	1.2 4
γ	1184.5 3	0.42 17
γ	1195.92 12	2.1 4
γ	1205.5 3	1.3 4
γ	1226.39 23	<0.6
γ	1228.87 22	1.7 4
γ	1238.1 1	1.7 5
γ	1242.7 1	3.3 6
γ	1246.9 2	
γ	1278.8 2	1.4 5
γ	1310.14 17	1.1 4
γ	1312.62 18	0.7 3
γ	1373.7 2	1.8 5
γ	1378.4 3	1.4 4
γ	1391.7 2	1.8 5
γ	1396.4 3	1.3 4
γ	1398.2 2	1.8 5
γ	1403.33 22	1.0 2
γ	1422.5 2	6.3 9
γ	1433.1 2	3.1 6
γ	2170.0 5	3.5 4
γ	2217.5 3	2.2 3

$^{157}_{69}$Tm(3.5 3 min)

Mode: ε

Δ: -58790 210 keV syst

SpA: 3.4×10^{8} Ci/g

Prod: protons on Er; protons on Ta

Photons (^{157}Tm)

γ_{mode}	γ(keV)	γ(rel)
γ	94.2 3	1.5 5
γ M1+E2	100.13 4	25 2
γ M1+E2	110.44 5	88 10
γ	116.41 5	9 1
γ M1+E2	131.21 5	40 8
γ	139.49 5	2.3 3
γ	141.5 3	0.9 2
γ	154.83 11	3.2 5
γ	155.46 18	
γ E3	155.5 3	1.2 4
γ	157.01 9	3.2 5
γ	159.29 16	0.80 16 ?
γ M1+E2	169.92 5	18.5 20
γ E1	175.25 12	30 5
γ M1+E2	195.59 12	29 5
γ	201.42 5	1.1 3
γ	222.61 18	1.3 4
γ M1+E2	231.34 5	10 2
γ	234.30 13	5.4 10
γ M1+E2	241.65 4	68 7
γ E1	247.62 4	30 4
γ M1+E2	250.31 5	12.5 20
γ M1+E2	257.16 10	3.5 10 ?
γ E1	270.70 4	7.5 15
γ	290.50 15	4.7 10
γ	304.43 8	7.2 20
γ M1(+E2)	308.26 9	20 4

Photons (^{157}Tm)
(continued)

γ_{mode}	γ(keV)	γ(rel)
γ	317.83 *6*	6 *1*
γ	321.80 *12*	6.5 *10*
γ	331.62 *15*	9 *1*
γ	347.75 *5*	25 *5*
γ	348.5 *3*	90 *15*
γ	357.29 *10*	72 *10*
γ	358.05 *6*	47 *9*
γ M1+E2	360.72 *11*	40 *7*
γ M1+E2	367.60 *10*	45 *6*
γ E1	370.83 *5*	54 *8*
γ E1	381.14 *5*	20 *2*
γ E1	385.55 *9*	95 *10*
γ	387.4 *3*	20 *3*
γ	406.5 *3*	4.5 *10*
γ	412.16 *18*	1.2 *3*
γ	421.9 *6*	0.8 *3*
γ	433.1 *4*	2.4 *4*
γ	439.08 *9*	5.3 *10*
γ	443.92 *9*	3.5 *10*
γ	447.75 *8*	4 *1*
γ M1+E2	449.04 *6*	14 *3*
γ M1+E2	455.10 *12*	100 *3*
γ	474.5 *4*	14 *3*
γ	479.68 *14*	2.0 *5*
γ E2	484.8 *4*	27 *4*
γ	488.6 *4*	1.2 *4*
γ E1	496.6 *5*	10 *3*
γ M1+E2	525 *3*	43 *6*
γ M1+E2	535.5 *3*	37 *6*
γ E1	549.17 *6*	55 *15*
γ M1+E2	555.7 *6*	34 *10*
γ	558.16 *11*	5.7 *10*
γ	570.2 *3*	2.6 *8*
γ	573.21 *18*	10 *1*
γ E1	575.13 *8*	26 *7*
γ	581.1 *5*	10 *3*
γ	587.7 *3*	1.0 *4*
γ	593.91 *9*	1.2 *4*
γ	596.01 *18*	4.8 *10*
γ	617.7 *18*	~1
γ	623.0 *3*	2.4 *10*
γ	630.3 *4*	1.8 *8*
γ	639.11 *18*	7.0 *15*
γ	642.72 *11*	4.7 *10*
γ	655.1 *6*	~2
γ	682.3 *6*	3.5 *7*
γ	685.57 *9*	24 *6*
γ	689.37 *11*	11 *3*
γ	702.9 *7*	2.2 *6*
γ	714.1 *7*	3.0 *8*
γ	718.2 *6*	1.5 *5*
γ	733.0 *9*	1.6 *6*
γ	735.7 *6*	1.5 *5*
γ	742.85 *11*	13 *3*
γ	748.8 *6*	13 *3*
γ	754.9 *6*	7.6 *2*
γ	763.83 *10*	9.5 *3*
γ	771.9 *6*	3.7 *10*

$^{157}_{70}$Yb(38.6 *10* s)

Mode: α, ϵ
Δ: -53620 *210* keV syst
SpA: 1.84×10^{9} Ci/g
Prod: ^{162}Er(^{3}He,8n); ^{58}Ni on ^{103}Rh; protons on Ta

Alpha Particles (^{157}Yb)

α(keV)
4504 *10*

$^{157}_{71}$Lu(5.5 *3* s)

Mode: ϵ(94 *2* %), α(6 *2* %)
Δ: -46630 *220* keV syst
SpA: 1.23×10^{10} Ci/g
Prod: ^{58}Ni on ^{107}Ag; protons on Ta; protons on W

Alpha Particles (^{157}Lu)

α(keV)
4996 *5*

$^{157}_{72}$Hf(110 *6* ms)

Mode: α(91 *7* %), ϵ(9 *7* %)
Δ: -38960 *450* keV syst
SpA: 1.03×10^{11} Ci/g
Prod: ^{144}Sm(^{20}Ne,7n); ^{58}Ni on ^{107}Ag; ^{32}S on ^{144}Sm

Alpha Particles (^{157}Hf)

α(keV)
5735 *5*

$^{157}_{73}$Ta(5.3 *18* ms)

Mode: α
Δ: -29580 *550* keV syst
SpA: 1.0×10^{11} Ci/g
Prod: ^{58}Ni on ^{107}Ag

Alpha Particles (^{157}Ta)

α(keV)
6219 *10*

A = 158

NDS **31**, 381 (1980)

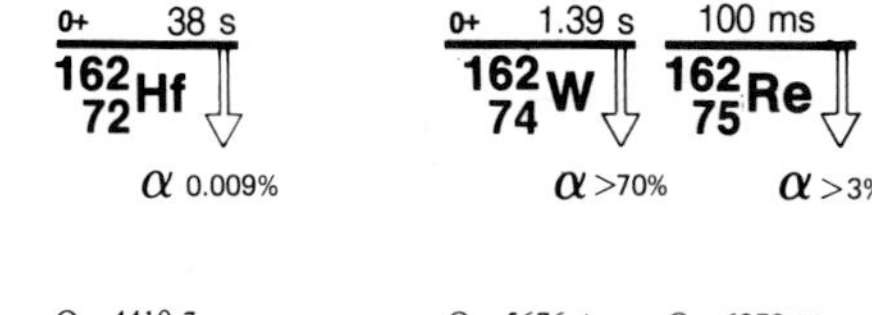

Q_α 4419 *7* Q_α 5676 *4* Q_α 6270 *50*

0+ ? 158 74 W α

37 ms 158 73 Ta ε ≈ 7% α ≈ 93% Q_ϵ 7210 *740*(syst) Q_α 6620 *30*

0+ 2.9 s 158 72 Hf ε 54% α 46% Q_ϵ 11300 *540*(syst) Q_α 6210 *50*

10.4 s 158 71 Lu ε > 98.5% α < 1.5% Q_ϵ 4940 *290*(syst) Q_α 5402 *4*

0+ 1.6 m 158 70 Yb ε ≈ 99.997% α ≈ 0.003% Q_ϵ 8780 *70* Q_α 4970 *220*(syst)

(2-) 4.0 m 158 69 Tm ε Q_ϵ 2680 *200*(syst) Q_α 4175 *10*

0+ 2.2 h 158 68 Er ε Q_ϵ 6500 *220*(syst)

27 m 2- 67.15 IT ≈ 78%; 21 m (9+); 11.3 m 5+; 158 67 Ho ε ≈ 22%; ε; ε Q_ϵ 1000 *100*(syst)

0+ 158 66 Dy Q_ϵ 4220 *30*

10.5 s 0- 110 IT; 150 y 3- 0; 158 65 Tb ε 82% β⁻ 18% Q_{β^-} 939 *3* Q_ϵ 1221.9 *16*

0+ 158 64 Gd

(1-) 46.0 m 158 63 Eu β⁻ Q_{β^-} 3450 *80*

0+ 5.5 m 158 62 Sm β⁻ Q_{β^-} 2050 *220*(syst)

5000, 0000, 5000, 0000, 5000, 0000, 5000, 0000, 5000, 0000, 5000, 1000, 0

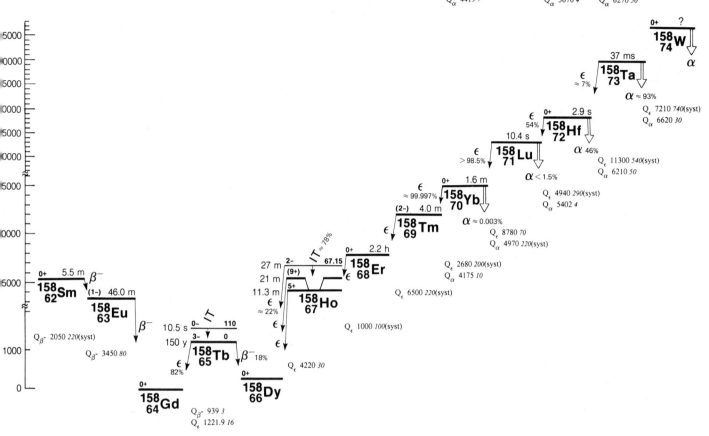

$^{158}_{62}$Sm(5.51 *9* min)

Mode: β-

Δ: -65200 *200* keV syst

SpA: 2.16×10⁸ Ci/g

Prod: fission

Photons (^{158}Sm)

γ_{mode}	γ(keV)	γ(rel)
γ	100.2 *3*	30.6 *17*
γ	108.7 *3*	7.7 *9*
γ	132.3 *3*	4.8 *7*
γ	149.0 *3*	32.5 *21*
γ	177.7 *3*	26.2 *14*
γ	189.4 *3*	100 *6*
γ	190.7 *3*	27 *3*
γ	224.1 *3*	56 *3*
γ	226.6 *3*	34 *3*
γ	229.7 *3*	44 *3*
γ	283.0 *3*	4.6 *4*
γ	285.4 *3*	11.1 *13*
γ	299.7 *3*	13.8 *15*
γ	321.3 *3*	55 *3*
γ	324.5 *3*	70 *4*

Photons (^{158}Sm) (continued)

γ_{mode}	γ(keV)	γ(rel)
γ	326.8 *3*	13.6 *9*
γ	338.6 *3*	24.5 *21*
γ	361.7 *3*	43 *3*
γ	363.6 *3*	82 *5*
γ	376.5 *3*	3.5 *3*
γ	551.2 *3*	19.9 *13*
γ	791.4 *3*	10.8 *8*
γ	1162.9 *3*	8.0 *4*
γ	1209.9 *3*	5.8 *8*
γ	1343.3 *3*	5.5 *4*
γ	1448.5 *3*	2.38 *21*

† 1.9% uncert(syst)

$^{158}_{63}$Eu(46.0 *3* min)

Mode: β-

Δ: -67250 *80* keV

SpA: 2.587×10⁷ Ci/g

Prod: ^{160}Gd(d,α); fission; ^{158}Gd(n,p)

Photons (^{158}Eu)

⟨γ⟩=1081 *49* keV

γ_{mode}	γ(keV)	γ(%)
Gd L_ℓ	5.362	0.154 *22*
Gd L_η	6.049	0.070 *9*
Gd L_α	6.054	4.1 *5*
Gd L_β	6.836	4.0 *5*
Gd L_γ	7.818	0.56 *7*
Gd $K_{\alpha2}$	42.309	6.1 *6*
Gd $K_{\alpha1}$	42.996	11.0 *11*
Gd $K_{\beta1}'$	48.652	3.2 *3*
Gd $K_{\beta2}'$	50.214	0.98 *10*
γ E2	79.55 *6*	11.1 *11*
γ E2	181.97 *7*	1.95 *20*
γ [E1]	209.99 *6*	~0.020
γ	218.5 *4*	0.038 *12*
γ [M1+E2]	245.40 *13*	0.08 *1*
γ E1	528.08 *8*	1.27 *8*
γ E1	606.43 *6*	3.30 *17*
γ M1+E2	698.71 *8*	0.88 *5*
γ M1	743.08 *6*	3.00 *15*
γ M1+E2	751.93 *7*	0.24 *4*
γ	764.02 *10*	0.52 *4*
γ E2	769.89 *7*	0.54 *4*
γ M1	777.05 *7*	0.65 *8*
γ E1	780.19 *8*	0.74 *5*
γ [M1+E2]	816.42 *7*	0.305 *20*
γ E1	824.18 *9*	1.08 *8*

Photons (^{158}Eu)
(continued)

γ_{mode}	γ(keV)	γ(%)
γ E1	827.96 *11*	0.32 *3*
γ [E1]	852.82 *8*	0.330 *22*
γ [E1]	870.72 *10*	1.05 *10*
γ [E1]	870.77 *8*	0.202 *22*
γ	879.37 *15*	0.140 *22*
γ [E1]	897.67 *5*	10.4 *5*
γ E1	906.54 *7*	1.52 *10*
γ [E1]	917.31 *8*	0.23 *3*
γ [E1]	922.45 *14*	0.25 *3*
γ (E1)	922.55 *7*	1.35 *15*
γ E2	925.69 *8*	0.100 *17*
γ [E1]	940.51 *7*	0.28 *8*
γ E1	944.21 *6*	25
γ (E1)	953.08 *7*	1.65 *10*
γ E1	962.17 *6*	1.58 *9*
γ E1	977.22 *6*	13.5 *7*
γ E1	987.05 *6*	1.13 *8*
γ E2	998.51 *10*	0.32 *4*
γ M1+E2	1004.04 *10*	0.40 *8*
γ [E1]	1005.67 *16*	1.02 *12*
γ E1	1034.56 *20*	0.120 *25*
γ [E1]	1061.76 *12*	0.270 *22*
γ E2	1107.66 *5*	4.33 *19*
γ E2	1116.54 *9*	1.05 *8*
γ	1130.3 *4*	~0.015
γ [M1+E2]	1138.31 *10*	0.178 *22*
γ E1	1141.55 *13*	0.153 *22*
γ	1166.6 *5*	~0.013
γ M1	1180.48 *10*	0.28 *5*
γ [E1]	1184.22 *12*	2.5 *3*
γ E2	1186.01 *8*	2.5 *5*
γ E2	1187.21 *7*	3.59 *21*
γ	1215.5 *3*	0.070 *15*
γ [E2]	1233.77 *12*	0.13 *3*
γ	1245.2 *4*	0.025 *10*
γ	1250.5 *4*	0.028 *10*
γ E2	1260.03 *11*	0.32 *5*
γ E1	1263.76 *12*	1.83 *15*
γ	1263.8 *3*	0.17 *8*
γ [E1]	1283.81 *10*	0.052 *8*
γ [E1]	1292.22 *13*	0.22 *3*
γ [E1]	1301.77 *9*	0.153 *15*
γ [M1+E2]	1312.12 *10*	0.222 *17*
γ E1	1323.52 *12*	0.190 *15*
γ [E1]	1348.30 *10*	1.38 *8*
γ [E1]	1353.79 *12*	0.097 *13*
γ [E1]	1371.74 *13*	0.110 *25*
γ E2	1372.06 *20*	0.063 *25*
γ	1433.84 *23*	0.052 *20*
γ [M1+E2]	1438.0 *3*	0.03 *1*
γ [E1]	1475.57 *11*	0.04 *1*
γ	1492.6 *7*	~0.02
γ E2	1517.6 *3*	0.025 *8*
γ	1532.12 *9*	0.05 *1*
γ	1552.1 *7*	0.028 *8*
γ	1597.30 *23*	0.020 *5*
γ	1643.83 *23*	0.03 *1*
γ	1693.5 *3*	0.052 *13*
γ [E2]	1702.74 *8*	0.113 *15*
γ [E1]	1714.09 *6*	0.153 *17*
γ	1738.31 *21*	0.105 *20*
γ [M1+E2]	1768.39 *10*	0.033 *8*
γ	1784.85 *21*	0.06 *1*
γ	1793.63 *8*	0.013 *5*
γ [M1+E2]	1814.98 *8*	0.030 *8*
γ	1836.0 *6*	0.030 *8*
γ [M1+E2]	1850.74 *7*	0.125 *25*
γ	1857.1 *5*	0.080 *17*
γ [M1+E2]	1884.71 *6*	1.02 *5*
γ [M1]	1930.29 *8*	0.038 *8*
γ [M1+E2]	1944.49 *9*	1.35 *8*
γ	1956.3 *3*	0.075 *10*
γ [E2]	1964.26 *8*	0.110 *13*
γ [M1]	2024.04 *10*	0.77 *5*
γ	2139.1 *4*	0.22 *6*
γ	2163.5 *4*	0.035 *5*
γ [M1+E2]	2189.88 *13*	0.023 *5*
γ	2194.3 *7*	0.028 *8*
γ	2203.9 *4*	0.07 *1*
γ	2215.4 *3*	0.093 *13*
γ [M1+E2]	2245.96 *9*	0.38 *3*
γ	2260.8 *3*	0.205 *20*
γ	2268.3 *5*	0.052 *8*

Photons (^{158}Eu)
(continued)

γ_{mode}	γ(keV)	γ(%)
γ	2273.8 *5*	0.042 *8*
γ [M1+E2]	2315.94 *13*	0.017 *8*
γ [E2]	2325.51 *10*	0.015 *5*
γ	2340.6 *10*	0.025 *8*
γ	2367.81 *24*	0.65 *4*
γ	2395.49 *14*	0.042 *8*
γ	2396.0 *3*	0.042 *8*
γ [M1+E2]	2419.77 *11*	0.013 *5*
γ	2421.1 *11*	0.013 *5*
γ	2447.36 *24*	0.63 *5*
γ	2451.0 *3*	0.188 *5*
γ	2464.1 *20*	~0.010
γ	2475.6 *3*	0.033 *5*
γ [M1]	2499.32 *12*	0.0525 *25*
γ	2514.1 *5*	0.090 *8*
γ	2520.6 *12*	~0.006
γ	2541.49 *23*	0.009 *3*
γ	2564.1 *20*	0.0075 *25*
γ	2601.1 *12*	0.024 *5*
γ	2640.1 *20*	0.009 *4*
γ	2670.7 *3*	0.010 *4*
γ	2703.1 *20*	0.028 *5*
γ	2743.9 *15*	0.052 *5*
γ	2762.05 *22*	0.016 *3*
γ	2806 *3*	0.0042 *20*
γ	2824.1 *20*	0.028 *5*
γ	2844.1 *20*	0.017 *4*
γ	2873.1 *20*	~0.004
γ	2884.1 *20*	~0.0045
γ	2967 *3*	~0.004

Atomic Electrons (^{158}Eu)
⟨e⟩=45.0 *25* keV

e_{bin}(keV)	⟨e⟩(keV)	e(%)
7	1.73	24 *3*
8	1.32	16.5 *23*
29	6.5	22.1 *22*
33 - 71	1.96	3.47 *22*
72	23.1	32 *3*
78	6.3	8.1 *8*
79	1.48	1.86 *19*
132 - 181	0.87	0.59 *4*
182 - 218	0.038	0.020 *3*
237 - 245	0.0054	0.0023 *8*
478 - 527	0.0257	0.0053 *3*
528 - 556	0.049	0.0088 *5*
598 - 606	0.0090	0.00150 *7*
648 - 697	0.221	0.0323 *24*
698 - 745	0.121	0.0167 *18*
750 - 780	0.042	0.0055 *6*
803 - 851	0.132	0.0157 *9*
852 - 899	0.31	0.034 *6*
903 - 952	0.227	0.0245 *10*
953 - 1002	0.054	0.0056 *5*
1003 - 1035	0.00305	0.000302 *23*
1053 - 1100	0.127	0.0120 *4*
1106 - 1142	0.152	0.0134 *9*
1158 - 1207	0.0311	0.00264 *16*
1208 - 1257	0.030	0.0025 *3*
1258 - 1305	0.0194	0.00151 *12*
1310 - 1354	0.0041	0.00031 *5*
1363 - 1388	0.0017	0.00013 *4*
1425 - 1474	0.0012	8.7 *20* $\times10^{-5}$
1475 - 1551	0.0017	0.00011 *3*
1589 - 1688	0.0054	0.00032 *12*
1692 - 1786	0.0031	0.00018 *3*
1792 - 1883	0.022	0.00118 *14*
1894 - 1974	0.037	0.00193 *17*
2016 - 2113	0.0046	0.00022 *6*
2131 - 2224	0.0105	0.00048 *6*
2238 - 2333	0.0079	0.00034 *11*
2339 - 2425	0.0098	0.00041 *10*
2439 - 2534	0.0031	0.000127 *20*
2540 - 2639	0.00042	1.6 *4* $\times10^{-5}$
2653 - 2742	0.00088	3.3 *7* $\times10^{-5}$
2754 - 2843	0.00052	1.9 *4* $\times10^{-5}$

Atomic Electrons (^{158}Eu)
(continued)

e_{bin}(keV)	⟨e⟩(keV)	e(%)
2865 - 2960	4.4 $\times10^{-5}$	1.5 *6* $\times10^{-6}$
2965 - 2966	8.7 $\times10^{-7}$	~3 $\times10^{-8}$

Continuous Radiation (^{158}Eu)
⟨β-⟩=923 keV;⟨IB⟩=2.1 keV

E_{bin}(keV)		⟨ ⟩(keV)	(%)
0 - 10	β-	0.0216	0.429
	IB	0.034	
10 - 20	β-	0.065	0.436
	IB	0.034	0.23
20 - 40	β-	0.270	0.90
	IB	0.066	0.23
40 - 100	β-	2.03	2.88
	IB	0.18	0.28
100 - 300	β-	23.1	11.4
	IB	0.50	0.28
300 - 600	β-	90	19.8
	IB	0.51	0.122
600 - 1300	β-	358	39.0
	IB	0.58	0.069
1300 - 2500	β-	405	23.6
	IB	0.19	0.0118
2500 - 3370	β-	44.9	1.64
	IB	0.0043	0.000164

$^{158}_{64}$Gd(stable)

Δ: -70702 *3* keV
%: 24.84 *12*

$^{158}_{65}$Tb(150 *30* yr)

Mode: ϵ(82 *2* %), β-(18 *2* %)
Δ: -69480 *4* keV
SpA: 15 Ci/g
Prod: multiple n-capture from ^{156}Dy

Photons (^{158}Tb)
⟨γ⟩=787 *27* keV

γ_{mode}	γ(keV)	$\gamma(\%)^{\dagger}$
Gd L_{ℓ}	5.362	0.35 *4*
Dy L_{ℓ}	5.743	0.034 *5*
Gd L_{η}	6.049	0.140 *13*
Gd L_{α}	6.054	9.2 *7*
Dy L_{α}	6.491	0.90 *9*
Dy L_{η}	6.534	0.0151 *16*
Gd L_{β}	6.840	9.1 *9*
Dy L_{β}	7.368	0.89 *10*
Gd L_{γ}	7.873	1.40 *16*
Dy L_{γ}	8.459	0.130 *15*
Gd $K_{\alpha2}$	42.309	22.6 *9*
Gd $K_{\alpha1}$	42.996	40.6 *17*
Dy $K_{\alpha2}$	45.208	1.47 *12*
Dy $K_{\alpha1}$	45.998	2.62 *22*
Gd $K_{\beta1}'$	48.652	12.0 *5*
Gd $K_{\beta2}'$	50.214	3.62 *16*
Dy $K_{\beta1}'$	52.063	0.80 *7*
Dy $K_{\beta2}'$	53.735	0.234 *20*
γ_{ϵ}E2	79.55 *6*	11.4 *9*
γ_{β}-E2	98.980 *10*	4.6 *4*

Photons (^{158}Tb) (continued)

γ_{mode}	γ(keV)	γ(%)†
γ_ϵE2	181.97 7	9.2 7
γ_ϵ[E1]	209.99 6	~0.010
$\gamma_{\beta-}$E2	218.280 10	0.93 7
γ_ϵE1	780.19 8	9.3 5
γ_ϵ[E1]	897.7 5	0.103 22
γ_ϵ[E1]	897.67 5	0.133 19
γ_ϵE2	925.69 8	0.048 8
γ_ϵE1	944.21 6	43
γ_ϵE1	962.17 6	19.8 12
γ_ϵE1	977.22 6	0.172 23
γ_ϵE2	1107.66 5	2.06 12
γ_ϵE2	1187.21 7	1.71 11

† uncert(syst): 5.1% for ε, 13% for β-

Atomic Electrons (^{158}Tb)

⟨e⟩=64.5 22 keV

e_{bin}(keV)	⟨e⟩(keV)	e(%)
7	3.7	51 6
8	3.4	42 5
9	0.28	3.3 4
29	6.6	22.7 18
33 - 44	2.06	5.58 25
45	2.4	5.3 4
46 - 71	1.54	2.23 16
72	23.7	33 3
78	6.4	8.3 7
79	1.51	1.91 15
80	0.220	0.276 22
90	2.96	3.3 3
91	2.54	2.78 23
97 - 99	1.79	1.84 12
132	2.48	1.88 16
160 - 209	1.84	1.05 5
210 - 236	0.084	0.0394 23
565 - 611	0.0005	~8×10^{-5}
617 - 666	0.0005	~8×10^{-5}
711 - 730	0.108	0.0148 9
772 - 780	0.0192	0.00248 12
847 - 896	0.42	0.048 3
897 - 944	0.266	0.0289 13
954 - 977	0.0335	0.00351 17
1057 - 1106	0.049	0.00460 25
1107 - 1137	0.0325	0.00286 19
1179 - 1187	0.0060	0.00051 3

Continuous Radiation (^{158}Tb)

⟨β-⟩=48.7 keV;⟨IB⟩=0.17 keV

E_{bin}(keV)		⟨ ⟩(keV)	(%)
0 - 10	β-	0.0181	0.362
	IB	0.0031	
10 - 20	β-	0.054	0.363
	IB	0.0024	0.0167
20 - 40	β-	0.220	0.73
	IB	0.039	0.109
40 - 100	β-	1.56	2.23
	IB	0.092	0.19
100 - 300	β-	13.9	7.1
	IB	0.019	0.0118
300 - 600	β-	27.0	6.4
	IB	0.0106	0.0026
600 - 1142	β-	5.9	0.89
	IB	0.0037	0.00052

$^{158}_{65}$Tb(10.5 2 s)

Mode: IT
Δ: -69370 4 keV
SpA: 6.58×10^9 Ci/g
Prod: ^{159}Tb(n,2n); ^{159}Tb(γ,n)

Photons (^{158}Tb)

⟨γ⟩=23.7 6 keV

γ_{mode}	γ(keV)	γ(%)†
Tb L_ℓ	5.546	0.25 3
Tb L_α	6.269	6.6 5
Tb L_η	6.284	0.069 6
Tb L_β	7.130	5.5 5
Tb L_γ	8.231	0.88 10
Tb $K_{\alpha2}$	43.744	13.8 6
Tb $K_{\alpha1}$	44.482	24.7 11
Tb $K_{\beta1}$'	50.338	7.4 3
Tb $K_{\beta2}$'	51.958	2.20 11
γM3	109.9 15	0.88

† 3.4% uncert(syst)

Atomic Electrons (^{158}Tb)

⟨e⟩=84.2 16 keV

e_{bin}(keV)	⟨e⟩(keV)	e(%)
8	3.6	47 5
9	0.71	8.2 9
35	0.24	0.70 8
36	0.37	1.02 11
37	0.121	0.33 4
41	0.045	0.108 12
42	0.150	0.36 4
43	0.196	0.46 5
44	0.040	0.091 10
48	0.0164	0.034 4
49	0.0182	0.037 4
50	0.0147	0.029 3
58	29.5	50.9 20
101	17.8	17.6 7
102	18.7	18.3 7
108	9.8	9.1 4
109	0.093	0.086 3
110	2.77	2.53 10

$^{158}_{66}$Dy(stable)

Δ: -70419 5 keV
%: 0.10 1

$^{158}_{67}$Ho(11.3 4 min)

Mode: ε
Δ: -66200 30 keV
SpA: 1.05×10^8 Ci/g
Prod: ^{159}Tb(α,5n); ^{148}Nd(^{14}N,4n)

Photons (^{158}Ho(11.3 or 27 min))

γ_{mode}	γ(keV)	γ(rel)
γ[M1+E2]	98.31 3	0.19 9
γE2	98.980 10	70 5
γ[M1+E2]	119.15 3	0.46 4
γ[M1+E2]	151.08 3	0.30 6
γ	186.88 10	0.30 8
γ[E2]	217.46 4	
γE2	218.280 10	100 2
γ	219.78 20	0.8 4
γ[E2]	270.22 3	0.26 5
γ	301.84 15	~0.07
γ[E1]	301.89 7	~0.07
γE2	320.60 3	11.1 19
γ[E1]	327.10 7	<3
γ	406.2 1	0.19 4
γ[E1]	425.42 7	1.8 3
γ[M1+E2]	441.59 6	0.19 6
γE2	462.15 9	0.77 8
γ[E1]	473.97 8	0.49 7
γM1	487.21 6	0.16 6
γ[E1]	508.13 14	0.61 20
γ	514.4 1	1.1 3
γE2	517.38 10	1.0 3
γ[E2]	526.04 4	~0.18
γ[E1]	533.88 10	0.13 5
γ[E1]	538.87 9	0.35 5
γ[E1]	543.68 7	0.16 5
γ(M1)	556.58 10	0.35 6
γ	560.19 14	0.17 7
γ[E1]	570.06 9	0.15 7
γ[M1+E2]	580.40 4	0.6 1
γ[M1+E2]	615.19 6	0.35 11
γ(E2)	624.68 17	~0.7
γ[E2]	629.18 3	0.9 3
γE1	630.26 13	1.4 4
γE2	642.32 5	0.45 12
γE2	660.82 6	0.53 9
γE2	677.11 3	1.34 24
γM1+E2	707.24 9	0.54 8
γE2	727.487 17	6.93 25
γE2	731.47 4	5.81 15
γM1+E2	740.65 5	0.88 12
γ[M1+E2]	766.59 9	0.63 12
γ[E2]	768.53 3	1.08 9
γ[M1+E2]	775.44 7	0.58 11
γM1	777.10 4	0.97 12
γE2	792.96 4	0.28 11
γ[M1+E2]	807.63 9	1.05 21
γ	829.68 20	0.30 7
γE2	838.94 7	1.29 9
γ	846.63 3	14.3 20
γ	847.46 3	34 4
γE2	850.62 4	21.3 6
γ[E1]	854.30 8	0.50 9
γE2	858.32 9	1.85 8
γE2	875.28 10	0.88 7
γ[E1]	890.42 6	0.68 14
γ	891.66 10	2.64 14
γ	893.93 14	0.43 9
γE2	896.25 4	2.7 1
γE2	908.88 20	0.44 10
γ	918.07 10	0.22 6
γE2	933.94 13	0.30 7
γ[M1+E2]	944.04 5	2.2 3
γ(E2)	945.766 19	37 4
γ(E2)	946.44 3	17.7 25
γE2	948.93 4	34.5 10
γ[M1+E2]	962.92 5	0.57 7
γE2	977.46 9	2.25 11
γ[M1+E2]	986.81 3	1.11 10
γ[E2]	990.02 7	1.09 7
γE2	994.56 5	8.63 22
γE2	997.709 22	5.91 17
γ[M1+E2]	1010.87 5	1.32 13
γ	1012.4 2	0.43 4
γ[M1+E2]	1038.18 20	0.57 12
γM1	1043.69 16	0.46 7
γ[M1+E2]	1047.41 10	1.57 9
γ[M1+E2]	1063.19 4	2.50 17
γE2	1064.91 3	5.38 20
γ(E1)	1080.05 6	1.04 9
γE2	1085.79 3	1.06 9
γE2	1094.96 5	0.51 10
γ[E2]	1109.18 6	0.70 11

Photons (^{158}Ho(11.3 or 27 min))
(continued)

γ_{mode}	γ(keV)	γ(rel)
γ[E2]	1124.76 *9*	0.32 *9*
γ[M1+E2]	1138.10 *13*	0.24 *7*
γ	1159.58 *20*	0.60 *13*
γ[M1+E2]	1161.50 *4*	3.11 *14*
γ[E1]	1166.88 *12*	0.40 *7*
γE2	1169.42 *8*	0.77 *8*
γE2	1181.20 *5*	1.80 *16*
γ[E1]	1196.49 *8*	0.75 *12*
γE1	1201.46 *8*	1.33 *13*
γE1	1211.02 *6*	2.27 *10*
γE2	1230.26 *9*	1.10 *13*
γE2	1234.14 *9*	0.65 *10*
γ[M1+E2]	1238.61 *12*	~0.16
γ	1246.1 *2*	0.33 *8*
γM1	1261.48 *20*	0.42 *12*
γ	1272.84 *8*	2.61 *12*
γ	1279.78 *20*	0.45 *11*
γE1	1298.33 *6*	3.36 *17*
γE1	1301.34 *15*	0.56 *20*
γ[E1]	1317.38 *18*	0.27 *13*
γ	1324.4 *5*	~0.14
γ	1332.3 *3*	0.27 *12*
γE2	1337.83 *6*	0.77 *8*
γE1	1342.88 *5*	0.9 *1*
γE1	1365.08 *6*	1.20 *11*
γ[M1+E2]	1374.83 *11*	0.22 *8*
γ(M1)	1386.58 *20*	~0.10
γE2	1392.1 *2*	0.34 *10*
γ	1402.28 *20*	0.54 *12*
γ[E1]	1414.77 *8*	1.80 *18*
γ[E2]	1417.76 *5*	0.86 *10*
γ[E2]	1432.99 *11*	0.48 *8*
γ[E2]	1436.14 *6*	0.58 *8*
γE1	1441.86 *5*	1.06 *10*
γ[M1+E2]	1452.93 *11*	~0.13
γE2	1463.49 *4*	3.36 *9*
γ	1490.35 *7*	~0.19
γ[M1+E2]	1493.98 *11*	~0.16
γ[E1]	1501.78 *9*	0.55 *9*
γ[E1]	1509.11 *10*	1.06 *12*
γE2	1521.18 *20*	2.3 *3*
γ	1523.10 *14*	0.55 *22*
γM1	1564.38 *20*	0.31 *9*
γ[E1]	1573.04 *14*	0.53 *12*
γE2	1578.10 *4*	8.3 *5*
γ	1583.88 *10*	0.97 *13*
γ[M1+E2]	1592.29 *11*	0.17 *7*
γM1	1603.17 *10*	0.57 *9*
γ[E1]	1608.09 *10*	0.28 *9*
γ[E2]	1611.49 *17*	0.50 *9*
γE2	1623.73 *4*	6.2 *3*
γ	1645.5 *4*	~0.16
γE2	1678.08 *10*	1.14 *9*
γ	1681.48 *20*	0.57 *11*
γ[E2]	1687.04 *16*	~0.13
γ	1690.58 *20*	0.35 *8*
γ	1698.30 *9*	0.82 *16*
γ[E2]	1709.37 *14*	0.29 *8*
γ	1711.57 *16*	0.35 *8*
γM1	1738.35 *5*	1.44 *19*
γ[M1+E2]	1751.25 *13*	~0.29
γM1	1753.39 *9*	0.61 *16*
γ[E2]	1772.07 *5*	~0.18
γ	1784.18 *10*	1.59 *11*
γM1	1790.67 *4*	23.4 *9*
γ[E2]	1796.38 *4*	0.83 *11*
γ[M1+E2]	1842.01 *4*	0.45 *10*
γM1+E2	1857.28 *20*	0.88 *11*
γ	1867.5 *4*	~0.14
γM1	1876.75 *9*	1.68 *9*
γ[E2]	1880.93 *11*	0.69 *8*
γM1	1886.28 *20*	0.39 *8*
γ[M1+E2]	1894.04 *10*	0.38 *8*
γ	1899.6 *3*	0.17 *8*
γ[E2]	1913.28 *20*	0.20 *8*
γ	1918.68 *20*	0.48 *8*
γ	1938.0 *3*	0.25 *8*
γ	1949.6 *3*	0.32 *10*
γE2	1956.63 *5*	1.06 *12*
γ	1969.68 *20*	0.38 *9*
γ[E2]	1975.73 *9*	0.23 *7*
γ[E2]	1998.87 *16*	0.15 *7*
γM1	2008.94 *4*	1.76 *14*

Photons (^{158}Ho(11.3 or 27 min))
(continued)

γ_{mode}	γ(keV)	γ(rel)
γ[E2]	2019.64 *13*	0.24 *8*
γM1+E2	2029.20 *13*	0.99 *10*
γ[E2]	2034.41 *10*	0.44 *9*
γM1+E2	2065.31 *6*	3.23 *19*
γ[E2]	2071.84 *12*	0.54 *9*
γM1	2076.2 *3*	0.83 *10*
γ	2092.6 *2*	1.39 *18*
γM1+E2	2095.38 *20*	0.72 *15*
γ[E2]	2106.06 *13*	0.28 *7*
γM1	2119.52 *6*	2.34 *14*
γE2	2146.4 *4*	0.22 *8*
γ	2170.68 *20*	0.63 *11*
γM1	2201.52 *11*	4.77 *23*
γM1	2221.46 *11*	4.5 *3*
γ	2236.1 *5*	~0.15
γ[E2]	2283.59 *6*	~0.08
γ(M1)	2289.8 *3*	0.29 *8*
γ	2303.28 *20*	0.35 *8*
γ	2310.6 *1*	1.01 *9*
γ(E2)	2337.80 *6*	0.24 *8*
γ[M1+E2]	2355.01 *10*	0.19 *8*
γ	2366.72 *17*	~0.10
γ	2371.78 *20*	0.38 *7*
γ(E2)	2396.08 *20*	0.64 *13*
γ	2409.8 *5*	0.19 *8*
γ[E2]	2418.51 *16*	0.21 *7*
γ[M1+E2]	2439.74 *11*	~0.08
γ	2444.18 *20*	0.33 *10*
γ	2464.5 *3*	0.25 *7*
γE2	2486.47 *18*	0.28 *6*
γ	2492.8 *3*	0.31 *7*
γ[E1]	2507.0 *1*	1.14 *14*
γ	2514.08 *20*	0.76 *8*
γM1	2525.58 *20*	0.41 *6*
γ(E2)	2545.75 *9*	3.32 *16*
γ	2559.6 *4*	0.20 *8*
γ[E2]	2573.28 *10*	0.20 *6*
γ	2584.4 *4*	~0.11
γ	2593.7 *4*	0.16 *6*
γE1	2605.98 *10*	7.0 *3*
γ	2645.9 *5*	0.13 *5*
γ	2658.78 *20*	0.55 *7*
γ[E2]	2672.27 *13*	0.32 *6*
γ	2744.6 *5*	0.12 *5*
γ	2757.68 *20*	0.29 *8*
γ	2827.8 *4*	0.22 *4*
γ	2842.9 *4*	0.12 *4*
γ	2850.6 *4*	0.15 *7*
γ	2873.8 *3*	0.21 *4*
γ	2884.5 *3*	0.13 *4*
γ	2892.75 *17*	~0.06
γ	2908.8 *3*	0.23 *6*
γ	2922.6 *4*	0.14 *4*
γ	2929.7 *4*	0.14 *4*
γ	2934.78 *20*	0.36 *4*
γ	2939.6 *5*	0.09 *4*
γ	3019.2 *5*	0.10 *4*
γ	3040.1 *6*	~0.04
γ	3103.2 *6*	~0.04
γ	3127.0 *6*	~0.04
γ	3166.4 *7*	~0.05
γ[E2]	3582.38 *16*	0.05 *2*
γ	3615.9 *10*	~0.02
γ	3651.2 *10*	~0.02
γ	3873.1 *10*	~0.02

$^{158}_{67}$Ho(21.3 *23* min)

Mode: ε
Δ: -66200 *30* keV
SpA: 5.6×10^{7} Ci/g
Prod: ^{159}Tb(α,5n); protons on Ta

Photons (^{158}Ho)

γ_{mode}	γ(keV)
γE2	98.980 *10*
γ	153.1 *10*
γ	166.5 *10*
γ	187.1 *10*
γE2	218.280 *10*
γ	266.3 *10*
γE2	320.60 *3*
γE2	406.29 *10*
γ	475.1 *10*
γ	731.6 *10*
γ	839.0 *10*
γE2	977.46 *9*
γ	1007.1 *10*
γ	1053.3 *10*
γ[M1+E2]	1484.7 *10*

$^{158}_{67}$Ho(27 *2* min)

Mode: IT(78 *12* %), ε(22 *12* %)
Δ: -66133 *30* keV
SpA: 4.4×10^{7} Ci/g
Prod: ^{159}Tb(α,5n); daughter ^{158}Er

Photons (^{158}Ho) *

γ_{mode}	γ(keV)	γ(%)†
γ_{IT} E3	67.15 *5*	0.162 *7*

† approximate
* see also ^{158}Ho(11.3 min)

$^{158}_{68}$Er(2.25 *7* h)

Mode: ε
Δ: -65200 *110* keV syst
SpA: 8.8×10^{6} Ci/g
Prod: protons on Ta; ^{165}Ho(p,8n); ^{150}Sm(^{12}C,4n); ^{151}Eu(^{11}B,4n)

Photons (^{158}Er)

⟨γ⟩=81 *3* keV

γ_{mode}	γ(keV)	γ(%)†
γM1+0.5%E2	24.45 *12*	0.17 *3*
γE1	30.66 *11*	0.22 *3*
γM1+0.4%E2	43.42 *9*	~0.5
γ[E1]	45.35 *10*	~1
γE1	50.5 *4*	1.13 *22*
γE1	69.80 *12*	0.21 *5*
γM1+0.8%E2	71.85 *4*	10.6 *11*
γM1	92.92 *9*	0.38 *16*
γE2	115.27 *8*	0.23 *8*
γM1	195.46 *8*	1.58 *16*
γM1+E2	204.30 *20*	0.22 *5*
γE2	207.6 *4*	0.13 *4*
γM1	238.88 *8*	0.22 *3*
γE1	248.47 *5*	2.57 *16*
γM1	271.47 *9*	0.22 *5*
γM1+E2	277.23 *12*	~0.08
γM1	286.36 *10*	0.16 *5*
γE1	295.92 *9*	0.95 *13*
γM1	310.73 *8*	1.65 *17*
γM1	314.89 *6*	0.37 *7*
γE1	326.6 *3*	0.42 *13*
γ	328.8 *3*	~0.21

Photons (^{158}Er)
(continued)

γ_{mode}	γ(keV)	γ(%)†
γ [E1]	336.3 4	~0.08
γ E1	341.39 8	1.12 13
γ M1	358.21 10	1.65 17
γ M1	386.74 5	6.6 5
γ M1	472.68 9	0.95 11
γ M1	516.11 12	1.43 16
γ E2	587.96 12	0.15 7

† 3.0% uncert(syst)

$^{158}_{69}$Tm(4.02 *10* min)

Mode: ε
Δ: -58700 *200* keV syst
SpA: 2.96×10^8 Ci/g
Prod: ^{162}Er(p,5n); ^{148}Sm(^{14}N,4n)

Photons (^{158}Tm)
⟨γ⟩=1087 *17* keV

γ_{mode}	γ(keV)	γ(%)†
Er L_ℓ	6.151	0.145 15
Er L_α	6.944	3.7 3
Er L_η	7.058	0.051 4
Er L_β	7.935	3.5 3
Er L_γ	9.165	0.55 6
Er $K_{\alpha2}$	48.221	9.8 4
Er $K_{\alpha1}$	49.128	17.4 7
Er $K_{\beta1}'$	55.616	5.40 20
Er $K_{\beta2}'$	57.406	1.53 6
γ	104.5 3	0.083 21
γ	171.71 12	0.083 21
γ	174.9 3	0.062 21
γ	177.8 4	0.069 21
γ [E2]	182.67 6	0.11 5
γ E2	192.08 4	69
γ [M1+E2]	223.25 4	0.138 14
γ	239.91 20	0.28 6
γ	247.99 10	0.26 4
γ [M1+E2]	256.41 7	0.041 7
γ [E2]	268.16 6	0.221 21
γ	278.87 9	0.048 7 ?
γ [E2]	286.93 9	0.028 7
γ	305.67 6	0.076 14
γ E2	335.02 4	18.7 17
γ	352.30 20	0.05 2
γ	355.95 12	0.076 21
γ	359.01 20	0.07 3
γ [E2]	363.60 5	0.30 3
γ [M1+E2]	374.09 5	0.32 3
γ	390.40 8	0.097 14
γ	394.67 11	0.076 14
γ	405.91 20	0.048 14
γ	414.9 3	0.062 14
γ	416.88 20	0.083 7
γ M1+E2	428.42 5	0.41 4
γ	430.73 11	0.055 21
γ E2	443.05 7	0.41 4
γ	446.02 7	0.159 14
γ E2	461.82 5	0.95 9
γ (E1)	482.78 7	0.097 14
γ	484.66 17	0.083 14
γ M1	500.35 6	0.48 5
γ	504.61 20	0.50 12
γ M1+E2	516.15 5	0.77 19
γ (E1)	571.08 9	0.24 3
γ E0+M1+E2	597.33 5	0.103 21
γ M1(+E2)	599.71 10	0.207 21
γ [E2]	611.08 6	0.29 4
γ E2	614.17 5	1.89 17
γ M1+E2	627.92 4	7.5 7
γ	635.4 3	~0.035
γ M1+E2	656.50 5	1.90 17

Photons (^{158}Tm)
(continued)

γ_{mode}	γ(keV)	γ(%)†
γ	667.31 15	0.131 21
γ	669.27 6	0.26 3
γ	676.71 10	0.12 4
γ [M1+E2]	684.82 7	0.37 3
γ	699.15 8	0.069 21
γ (E1)	702.47 11	0.35 7
γ	703.8 3	0.26 7
γ (E1)	706.02 6	0.83 8
γ E0(+M1+E2)	729.98 6	<0.07
γ [E2]	763.82 9	0.090 14
γ	777.56 17	0.069 21
γ	780.45 13	0.083 21
γ	788.92 21	0.103 21
γ	794.33 10	0.31 3
γ E0+E2+M1	796.84 4	1.26 10
γ (E1)	814.69 6	1.28 10
γ E2	820.00 5	3.6 3
γ	830.93 20	0.090 21
γ	834.41 11	0.12 3
γ M1+E2	851.17 5	5.2 5
γ [M1+E2]	853.74 7	0.35 6
γ	889.5 4	0.14 5
γ [E2]	890.23 6	0.30 8
γ	899.52 24	0.069 14
γ	910.82 10	0.44 5
γ	922.40 8	0.21 3
γ	949.37 13	0.38 14
γ	962.17 6	0.179 21
γ	968.2 4	0.048 14
γ	971.41 11	0.09 3
γ	978.36 15	0.097 14
γ E2	988.92 5	4.1 3
γ	998.92 7	0.41 6
γ	1008.5 4	0.138 21
γ	1011.4 3	0.15 4
γ M1+E2	1018.32 8	0.77 7
γ [E2]	1042.97 8	0.90 9
γ	1048.66 25	0.124 21
γ	1052.36 25	0.10 4
γ E2	1065.00 6	1.73 14
γ	1109.7 3	0.110 21
γ	1113.4 4	0.083 14
γ	1132.81 10	0.173 21
γ E1	1149.71 5	8.4 7
γ	1172.81 10	0.79 7
γ	1206.8 3	0.10 3
γ	1215.30 8	0.72 8
γ	1217.1 5	0.13 6
γ (E1)	1225.92 18	1.52 12
γ	1234.54 24	0.12 3
γ	1239.8 2	0.220 21
γ	1253.38 16	0.13 6
γ	1262.3 4	0.10 3
γ	1275.02 19	0.12 3
γ	1281.78 19	0.110 21
γ	1294.9 5	0.14 3
γ	1303.21 15	0.41 6
γ	1307.34 13	0.41 6
γ	1311.76 15	0.19 3
γ	1319.36 25	0.19 4
γ (E1)	1333.94 6	3.7 3
γ	1360.3 4	0.193 21
γ [M1+E2]	1377.99 8	0.29 4
γ	1407.7 3	0.19 3
γ	1418.01 18	1.54 14
γ	1428.4 5	0.09 4
γ	1438.02 19	0.21 6
γ	1448.58 11	0.33 6
γ	1453.6 3	~0.14
γ	1458.9 5	0.17 6
γ	1472.9 4	0.10 3
γ	1482.5 4	0.21 4
γ	1489.66 25	0.21 4
γ	1494.75 13	0.69 7
γ	1501.93 10	0.90 8
γ	1505.65 11	0.62 6
γ	1526.02 7	0.25 4
γ	1533.4 7	0.11 3
γ	1550.31 7	1.78 16
γ	1567.3 3	0.14 3
γ [E2]	1570.07 9	0.39 6
γ	1577.28 13	0.97 14
γ	1614.33 10	~0.06
γ	1616.80 19	0.22 6

Photons (^{158}Tm)
(continued)

γ_{mode}	γ(keV)	γ(%)†
γ	1630.11 19	0.10 3
γ	1640.66 11	0.23 4
γ	1686.83 14	0.18 3
γ	1693.81 20	0.25 4
γ	1701.39 11	0.08 3
γ	1751.31 20	0.145 21
γ	1761.31 20	0.145 21
γ	1771.2 3	0.110 21
γ	1777.86 13	0.32 4
γ	1785.22 18	0.16 3
γ	1811.71 15	0.21 3
γ	1832.3 4	0.069 21
γ	1840.11 20	0.18 4
γ	1867.55 15	0.33 4
γ	1879.0 4	0.138 21
γ	1889.16 25	0.131 21
γ	1904.08 20	0.23 4
γ	1909.8 3	0.16 3
γ	1924.9 5	0.09 4
γ	1931.0 5	0.08 4
γ	1951.38 17	0.18 4
γ	1977.30 18	0.12 3
γ	2036.4 1	0.23 4
γ	2090.7 3	0.103 21
γ	2112.87 14	0.23 4
γ	2118.7 3	0.14 4
γ	2143.47 17	0.28 4
γ	2162.8 3	0.16 3
γ	2176.11 19	0.31 4
γ	2186.7 3	0.22 4
γ	2197.3 3	0.158 14
γ	2208.7 5	0.08 3
γ	2221.2 5	0.12 5
γ	2241.1 4	0.15 3
γ	2274.0 3	0.18 4
γ	2368.19 19	0.21 4
γ	2389.4 3	0.07 3
γ	2422.06 20	0.29 4
γ	2453.7 15	0.17 7
γ	2456.9 15	~0.13
γ	2470.4 15	0.17 5
γ	2480.6 12	0.16 6
γ	2487.4 15	~0.13
γ	2548.4 15	0.20 7
γ	2643 3	~0.07
γ	2656 4	~0.06
γ	2672.6 12	0.12 4
γ	2686 4	~0.05
γ	2816.4 20	0.23 6
γ	2825.31 19	~0.06
γ	2838 4	~0.041
γ	2888 3	0.08 3
γ	3000 4	0.062 21
γ	3017.39 19	0.062 21
γ	3036 4	0.076 21
γ	3053 4	0.10 3

† 2.9% uncert(syst)

Atomic Electrons (^{158}Tm)
⟨e⟩=40.7 *11* keV

e_{bin}(keV)	⟨e⟩(keV)	e(%)
8	1.35	16.2 17
9	0.80	8.7 9
10 - 55	1.08	4.4 3
95 - 125	0.18	0.16 5
135	16.9	12.6 7
162 - 181	0.109	0.065 15
182	2.52	1.38 10
183	4.52	2.47 13
184	3.52	1.91 10
190	2.59	1.36 7
191 - 240	0.88	0.45 3
246 - 277	0.040	0.016 5
278	1.94	0.70 6
279 - 327	0.70	0.215 12
333 - 382	0.31	0.089 9

Atomic Electrons (^{158}Tm) (continued)

e_{bin}(keV)	⟨e⟩(keV)	e(%)
385 - 434	0.140	0.035 *4*
435 - 483	0.20	0.044 *10*
484 - 523	0.12	~0.023
540 - 589	0.67	0.12 *4*
590 - 635	0.32	0.052 *10*
640 - 688	0.18	~0.028
689 - 737	0.034	0.0047 *14*
739 - 789	0.27	0.035 *7*
792 - 841	0.36	0.044 *11*
842 - 891	0.047	0.0055 *14*
892 - 940	0.142	0.0153 *16*
941 - 990	0.102	0.0105 *14*
991 - 1040	0.062	0.0062 *5*
1041 - 1075	0.021	0.0019 *4*
1092 - 1141	0.118	0.0107 *13*
1148 - 1196	0.056	0.0048 *11*
1197 - 1246	0.029	0.0024 *6*
1250 - 1299	0.055	0.0044 *6*
1301 - 1350	0.024	0.0018 *3*
1351 - 1399	0.049	0.0036 *14*
1401 - 1450	0.064	0.0045 *11*
1451 - 1549	0.083	0.0055 *15*
1557 - 1644	0.026	0.0016 *3*
1677 - 1776	0.022	0.00127 *24*
1777 - 1874	0.022	0.00121 *23*
1877 - 1975	0.0068	0.00036 *10*
1979 - 2061	0.010	0.00049 *13*
2081 - 2178	0.020	0.00094 *16*
2184 - 2272	0.0056	0.00025 *7*
2311 - 2399	0.0103	0.00044 *10*
2412 - 2491	0.0092	0.00038 *9*
2539 - 2635	0.0035	0.00013 *4*
2641 - 2684	0.00043	1.6 *5* $\times10^{-5}$
2759 - 2836	0.0045	0.00016 *4*
2878 - 2960	0.0013	4.3 *14* $\times10^{-5}$
2978 - 3051	0.0020	6.7 *17* $\times10^{-5}$

Continuous Radiation (^{158}Tm)

⟨β+⟩=1545 keV;⟨IB⟩=14.4 keV

E_{bin}(keV)		⟨ ⟩(keV)	(%)
0 - 10	β+	1.37×10^{-7}	1.63×10^{-6}
	IB	0.044	
10 - 20	β+	7.5×10^{-6}	4.49×10^{-5}
	IB	0.044	0.30
20 - 40	β+	0.000282	0.00085
	IB	0.089	0.31
40 - 100	β+	0.0195	0.0246
	IB	0.31	0.50
100 - 300	β+	1.37	0.61
	IB	0.78	0.43
300 - 600	β+	12.9	2.75
	IB	1.04	0.24
600 - 1300	β+	135	13.7
	IB	2.3	0.25
1300 - 2500	β+	609	32.1
	IB	4.2	0.23
2500 - 5000	β+	786	24.6
	IB	5.4	0.162
5000 - 6308	β+	1.38	0.0273
	IB	0.22	0.0042
	Σβ+		74

$^{158}_{70}$Yb(1.55 *10* min)

Mode: ϵ(~99.997 %), α(~0.003 %)
Δ: -56023 *16* keV
SpA: 7.6×10^{8} Ci/g
Prod: ^{122}Te(^{40}Ar,4n); protons on Ta; protons on W

Alpha Particles (^{158}Yb)

α(keV)
4069 *10*

Photons (^{158}Yb)

γ_{mode}	γ(keV)	γ(rel)
γ	74.1 *1*	100
γ	147.7 *1*	1.7 *4*
γ	160.3 *1*	2.1 *4*
γ	252.6 *2*	3.3 *6*

$^{158}_{71}$Lu(10.4 *1* s)

Mode: ϵ(>98.5 %), α(<1.5 %)
Δ: -47240 *70* keV
SpA: 6.64×10^{9} Ci/g
Prod: protons on Ta; protons on W

Alpha Particles (^{158}Lu)

α(keV)
4665 *10*

Photons (^{158}Lu)

γ_{mode}	γ(keV)	γ(rel)
γ_ϵ	358.2 *1*	100
γ_ϵ	477.0 *3*	21 *5*

$^{158}_{72}$Hf(2.9 *2* s)

Mode: ϵ(54 *3* %), α(46 *3* %)
Δ: -42300 *280* keV syst
SpA: 2.19×10^{10} Ci/g
Prod: ^{144}Sm(^{20}Ne,6n); ^{58}Ni on ^{107}Ag

Alpha Particles (^{158}Hf)

α(keV)
5268 *5*

$^{158}_{73}$Ta(36.8 *16* ms)

Mode: α(93 *6* %), ϵ(7 *6* %)
Δ: -31000 *540* keV syst
SpA: 1.03×10^{11} Ci/g
Prod: ^{58}Ni on ^{107}Ag; daughter ^{162}Re

Alpha Particles (^{158}Ta)

α(keV)
6051 *6*

$^{158}_{74}$W($t_{1/2}$ unknown)

Mode: α
Δ: -23780 *580* keV syst
Prod: ^{58}Ni on Mo-Sn targets; ^{107}Ag on V-Ni targets

Alpha Particles (^{158}W)

α(keV)
6450 *30*

A = 159

NDS 27, 155 (1979)

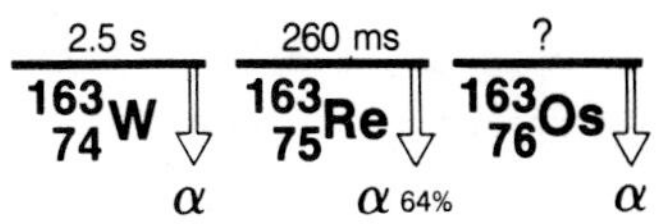

Q_α 5520 *50* Q_α 6070 *50* Q_α 6670 *50*

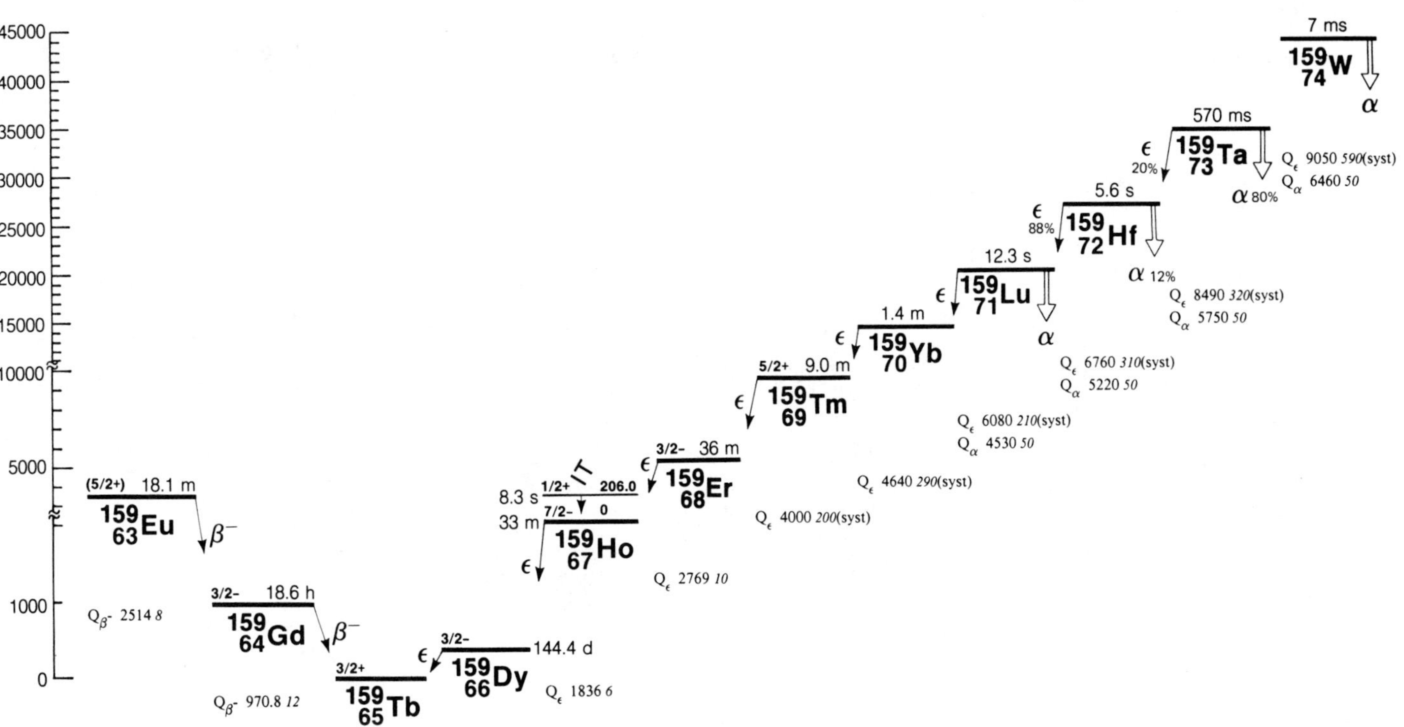

Q_ϵ 365.6 *10*

$^{159}_{63}$Eu(18.1 *1* min)

Mode: β-

Δ: -66059 *9* keV

SpA: 6.53×10^7 Ci/g

Prod: ^{160}Gd(γ,p); fission; ^{160}Gd(n,pn)

Photons (^{159}Eu)

γ_{mode}	γ(keV)	γ(rel)
γ E1	17.10 *8*	8.3 *22*
γ E1	67.81 *7*	100 *22*
γ [M1+E2]	71.16 *11*	56 *14*
γ [E1]	78.44 *7*	47 *9*
γ [M1+E2]	81.13 *10*	6.4 *17*
γ [M1+E2]	90.2 *6*	3.2 *5*
γ [M1+E2]	95.54 *7*	36 *4*
γ	102.34 *20*	34 *3*
γ [M1+E2]	105.51 *12*	3.7 *3*
γ [E1]	108.49 *15*	1.47 *22*
γ [E2]	121.87 *12*	20 *3*
γ [M1+E2]	146.25 *7*	17
γ [E1]	159.57 *9*	7.1 *5*
γ [M1+E2]	176.67 *8*	6.8 *3*
γ [E2]	227.38 *10*	8.5 *25*
γ	498.0 *6*	1.7 *5*
γ	521.2 *7*	8 *3*
γ [E1]	551.27 *18*	2.03 *17*
γ	575.3 *4*	13.6 *17*
γ	588.20 *14*	2.0 *3*
γ	595.8 *4*	17 *3*
γ [E1]	601.98 *17*	4.6 *3*
γ	613.60 *15*	6.6 *5*
γ [M1+E2]	645.18 *14*	1.86 *17*

Photons (^{159}Eu) (continued)

γ_{mode}	γ(keV)	γ(rel)
γ	659.36 *10*	6.9 *5*
γ	664.68 *8*	15.9 *9*
γ [M1+E2]	676.42 *9*	9.8 *5*
γ	681.78 *8*	12.0 *7*
γ [E1]	693.52 *11*	2.54 *17*
γ [M1+E2]	720.98 *22*	0.85 *17*
γ [M1+E2]	726.31 *14*	3.4 *3*
γ	732.49 *9*	1.27 *25* ?
γ [E1]	744.23 *10*	4.7 *3*
γ [E1]	753.66 *15*	4.7 *3*
γ [E1]	762.9 *3*	1.69 *17*
γ [E1]	804.75 *13*	13.4 *9*
γ [E1]	829.47 *22*	2.9 *3*
γ [M1+E2]	872.56 *14*	1.10 *17*
γ [E1]	880.55 *21*	1.69 *17*
γ	915.5 *6*	8.5 *17*
γ [M1+E2]	935.08 *19*	1.5 *3*
γ [M1+E2]	1016.21 *19*	2.5 *9*
γ	1038.0 *7*	10.2 *17*
γ [E1]	1043.56 *21*	2.7 *3*
γ	1060.48 *23*	1.53 *17*
γ	1077.57 *23*	1.36 *17*
γ [E1]	1094.65 *18*	6.3 *5*
γ	1108.8 *10*	14 *5*
γ	1128.28 *22*	2.7 *3*
γ	1159.2 *5*	3.2 *5*
γ	1181.4 *10*	5.9 *17*
γ	1220.5 *4*	10.2 *17*
γ	1301.3 *3*	1.69 *17*
γ	1350.6 *5*	0.63 *10*
γ	1433.5 *5*	12.7 *25*
γ	1451.84 *18*	1.02 *17*
γ	1468.94 *18*	1.53 *17*
γ	1519.65 *17*	3.4 *3*

$^{159}_{64}$Gd(18.56 *8* h)

Mode: β-

Δ: -68573 *4* keV

SpA: 1.062×10^6 Ci/g

Prod: ^{158}Gd(n,γ)

Photons (^{159}Gd)

⟨γ⟩=52.7 *21* keV

γ_{mode}	γ(keV)	γ(%)†
Tb L_ℓ	5.546	0.066 *9*
Tb L_α	6.269	1.74 *20*
Tb L_η	6.284	0.023 *3*
Tb L_β	7.113	1.58 *20*
Tb L_γ	8.189	0.24 *3*
γ [E1]	15.37 *8*	
Tb $K_{\alpha2}$	43.744	5.7 *6*
Tb $K_{\alpha1}$	44.482	10.2 *10*
Tb $K_{\beta1}'$	50.338	3.1 *3*
Tb $K_{\beta2}'$	51.958	0.91 *9*
γ M1+1.4%E2	57.945 *10*	2.27 *22*
γ M1+1.6%E2	79.53 *4*	0.042 *4*
γ E2	137.48 *4*	0.0065 *11*
γ M1(+E2)	210.65 *8*	0.018 *3*
γ E1	226.02 *4*	0.211 *11*
γ [E1]	237.40 *14*	0.0078 *12*
γ [E1]	274.12 *15*	0.0058 *14*
γ M1(+E2)	290.19 *7*	0.030 *3*
γ E1	305.56 *3*	0.059 *4*
γ [M1+E2]	348.13 *7*	0.216 *15*
γ E1+0.3%M2	363.50 *3*	11.0 *6*
γ (M1)	536.79 *16*	0.0011 *3*

Photons (^{159}Gd) (continued)

γ_{mode}	γ(keV)	γ(%)†
γ(M1)	559.56 11	0.0216 22
γ M1+0.1%E2	580.78 13	0.062 4
γ(M1)	616.33 16	0.0022 5
γ(M1)	617.51 11	0.0140 22
γ(M1)	674.27 16	0.00037 11
γ[E1]	854.91 13	0.0023 3

† 25% uncert(syst)

Atomic Electrons (^{159}Gd)

$\langle e \rangle$=5.77 24 keV

e_{bin}(keV)	$\langle e \rangle$(keV)	e(%)
6	1.24	20.8 20
8	1.06	13.7 19
9 - 35	0.27	1.92 21
36	0.153	0.42 6
37 - 48	0.235	0.57 4
49	1.38	2.8 3
50	0.38	0.76 8
56	0.44	0.78 8
57 - 85	0.152	0.254 22
129 - 174	0.0189	0.0117 11
185 - 230	0.0050	0.0024 4
235 - 283	0.0087	0.0035 8
288 - 306	0.03	0.010 3
312	0.323	0.104 7
339 - 364	0.074	0.0209 11
485 - 529	0.0074	0.00142 9
535 - 581	0.00271	0.00048 3
608 - 622	0.00027	4.5 5 $\times 10^{-5}$
666 - 674	4.9 $\times 10^{-6}$	7.4 17 $\times 10^{-7}$
803 - 847	2.9 $\times 10^{-5}$	3.5 5 $\times 10^{-6}$
853 - 855	9.8 $\times 10^{-7}$	1.15 14 $\times 10^{-7}$

Continuous Radiation (^{159}Gd)

$\langle \beta- \rangle$=305 keV;$\langle$IB$\rangle$=0.29 keV

E_{bin}(keV)		$\langle\ \rangle$(keV)	(%)
0 - 10	β-	0.089	1.77
	IB	0.0146	
10 - 20	β-	0.267	1.78
	IB	0.0139	0.097
20 - 40	β-	1.08	3.60
	IB	0.026	0.090
40 - 100	β-	7.7	11.0
	IB	0.064	0.100
100 - 300	β-	71	35.7
	IB	0.119	0.071
300 - 600	β-	156	36.3
	IB	0.051	0.0130
600 - 971	β-	69	9.9
	IB	0.0041	0.00063

$^{159}_{65}$Tb(stable)

Δ: -69544 4 keV
%: 100

$^{159}_{66}$Dy(144.4 2 d)

Mode: ϵ
Δ: -69178 4 keV
SpA: 5687 Ci/g
Prod: ^{158}Dy(n,γ); ^{159}Tb(d,2n); ^{159}Tb(p,n); protons on Ta

Photons (^{159}Dy)

$\langle \gamma \rangle$=45.6 10 keV

γ_{mode}	γ(keV)	γ(%)
Tb L_ℓ	5.546	0.31 3
Tb L_α	6.269	8.3 6
Tb L_η	6.284	0.110 9
Tb L_β	7.112	7.7 7
Tb L_γ	8.203	1.21 13
γ[E1]	15.37 8	
Tb $K_{\alpha2}$	43.744	27.2 10
Tb $K_{\alpha1}$	44.482	48.7 18
Tb $K_{\beta1}'$	50.338	14.6 5
Tb $K_{\beta2}'$	51.958	4.34 18
γ M1+1.4%E2	57.945 10	2.22 16
γ M1+1.6%E2	79.53 4	0.0040 9
γ E2	137.48 4	0.00062 16
γ M1(+E2)	210.65 8	7.7 15 $\times 10^{-5}$
γ E1	226.02 4	~4 $\times 10^{-6}$
γ M1(+E2)	290.19 7	0.000133 21
γ E1	305.56 3	~1 $\times 10^{-6}$
γ[M1+E2]	348.13 7	0.00095 11
γ E1+0.3%M2	363.50 3	0.000222 14

Atomic Electrons (^{159}Dy)

$\langle e \rangle$=11.6 6 keV

e_{bin}(keV)	$\langle e \rangle$(keV)	e(%)
6	1.21	20.4 15
8	4.9	64 7
9	0.76	8.7 9
28	0.0041	0.015 3
35	0.48	1.37 14
36	0.73	2.02 21
37 - 48	1.12	2.72 14
49	1.37	2.79 21
50	0.39	0.79 7
56	0.43	0.76 6
57 - 85	0.126	0.218 16
129 - 174	0.00032	0.00024 4
202 - 238	3.2 $\times 10^{-5}$	1.4 4 $\times 10^{-5}$
254 - 298	0.00014	4.6 15 $\times 10^{-5}$
304 - 348	3.9 $\times 10^{-5}$	1.16 25 $\times 10^{-5}$
355 - 364	1.35 $\times 10^{-6}$	3.80 22 $\times 10^{-7}$

Continuous Radiation (^{159}Dy)

$\langle$IB$\rangle$=0.18 keV

E_{bin}(keV)		$\langle\ \rangle$(keV)	(%)
10 - 20	IB	0.00019	0.00125
20 - 40	IB	0.034	0.092
40 - 100	IB	0.134	0.29
100 - 300	IB	0.0071	0.0046
300 - 366	IB	5.7 $\times 10^{-5}$	1.57 $\times 10^{-5}$

$^{159}_{67}$Ho(33 1 min)

Mode: ϵ
Δ: -67342 7 keV
SpA: 3.58 $\times 10^{7}$ Ci/g
Prod: ^{159}Tb(α,4n); ^{160}Dy(p,2n); protons on Ta

Photons (^{159}Ho)

$\langle \gamma \rangle$=329 8 keV

γ_{mode}	γ(keV)	γ(%)†
Dy L_ℓ	5.743	0.57 6
Dy L_α	6.491	14.9 12
Dy L_η	6.534	0.201 18
Dy L_β	7.380	14.1 14
Dy L_γ	8.520	2.2 3
γ M1+2.5%E2	30.58 11	0.21 3
γ M1+3.2%E2	31.39 9	0.59 5
γ E1	41.11 9	0.47 5
Dy $K_{\alpha2}$	45.208	42.0 21
Dy $K_{\alpha1}$	45.998	75 4
Dy $K_{\beta1}'$	52.063	22.8 11
Dy $K_{\beta2}'$	53.735	6.7 4
γ M1+3.9%E2	56.67 10	4.9 5
γ E2	61.97 11	0.235 24
γ[E1]	72.50 10	0.24 3
γ M1+4.7%E2	79.81 9	1.19 13
γ M1	85.75 9	0.163 16
γ	88.28 20	0.08 2
γ M1+E2	99.43 12	0.28 3
γ E1	100.59 9	3.7 4
γ[E1]	103.07 11	0.142 15
γ E1	120.92 9	33 3
γ E1	131.98 9	21.7 21
γ E2	136.48 10	0.45 4
γ E1	152.31 10	1.13 11
γ E1	155.77 11	2.08 21
γ M1(+E2)	159.41 13	0.39 4
γ	166.0 10	
γ M1(+E2)	173.09 9	2.22 23
γ E1	177.59 9	6.0 6
γ E2	179.24 14	0.17 3
γ E1	186.34 9	3.2 3
γ	195.58 20	0.08 2
γ[E1]	217.73 9	3.2 3
γ M1(+E2)	252.90 9	13.4 13
γ M1(+E2)	258.84 9	1.26 12
γ[E1]	265.50 21	0.154 15
γ	268.2 5	0.068 17
γ E1	296.08 19	0.80 8
γ M1	309.57 9	13.8 13
γ M1(+E2)	338.65 9	0.56 5
γ	352.8 5	0.064 16
γ E2	395.32 10	0.25 3
γ[M1+E2]	620.95 11	0.24 3
γ	667.88 20	0.040 11
γ[M1+E2]	694.91 13	
γ M1(+E2)	706.70 11	0.88 9
γ	714.1 10	
γ M1(+E2)	766.12 11	0.091 23
γ M1(+E2)	780.66 12	0.044 11
γ E1	807.29 11	0.91 9
γ	816.7 5	0.029 8
γ E1	838.68 10	2.8 3
γ	850.68 20	0.23 3
γ	866.0 5	0.024 7
γ	879.2 10	
γ(E1)	881.25 12	0.31 3
γ	891.8 5	0.064 16
γ	898.10 11	0.028 7
γ E1	912.64 12	0.225 23
γ	939.21 12	0.050 12
γ	992.0 5	0.050 12
γ M1(+E2)	1016.27 11	0.36 4
γ M1(+E2)	1019.02 11	0.174 13
γ E1	1024.29 14	0.150 15
γ	1035.18 20	0.043 11
γ	1062.6 5	0.047 12
γ M1(+E2)	1075.69 11	0.101 25
γ	1109.1 5	

Photons (^{159}Ho)
(continued)

γ_{mode}	γ(keV)	γ(%)†
γ	1145.20 16	
γ	1153.3 5	0.047 12
γ	1162.5 5	
γ	1193.08 20	0.046 12
γ	1198.1 5	
γ M1(+E2)	1201.87 14	0.138 13
γ	1333.1 5	
γ	1460.98 20	0.047 12

† 9.0% uncert(syst)

Atomic Electrons (^{159}Ho)
$\langle e \rangle$=52.2 17 keV

e_{bin}(keV)	$\langle e \rangle$(keV)	e(%)
3	1.48	51 5
8	5.6	72 8
9	4.4	51 6
19	0.027	0.144 17
22	1.70	7.7 9
23 - 24	1.8	7.9 13
26	1.20	4.6 5
29 - 47	5.8	15.7 7
48	4.3	9.1 11
49 - 54	2.60	5.0 5
55	1.36	2.5 3
56 - 65	0.91	1.55 11
67	3.4	5.1 5
70 - 77	0.68	0.95 10
78	2.25	2.9 3
79 - 129	4.4	4.0 3
130 - 179	1.38	0.90 11
184 - 196	0.021	0.012 3
199	3.1	1.6 5
205 - 253	1.4	0.61 15
256	3.2	1.25 12
257 - 302	0.66	0.224 21
308 - 353	0.199	0.064 5
386 - 395	0.0062	0.00160 9
567 - 614	0.020	0.0034 10
619 - 668	0.048	0.007 3
698 - 727	0.016	0.0023 5
754 - 800	0.056	0.0071 8
805 - 851	0.0153	0.00184 23
857 - 905	0.0056	0.00064 12
911 - 939	0.0017	~0.00019
962 - 1011	0.025	0.0025 5
1014 - 1063	0.0046	0.00045 12
1067 - 1099	0.0017	0.00015 7
1139 - 1185	0.0051	0.00045 11
1191 - 1202	0.00075	6.3 14 $\times 10^{-5}$
1407 - 1453	0.0009	~7$\times 10^{-5}$
1459 - 1461	3.2 $\times 10^{-5}$	~2$\times 10^{-6}$

Continuous Radiation (^{159}Ho)
$\langle$IB$\rangle$=0.85 keV

E_{bin}(keV)		$\langle\ \rangle$(keV)	(%)
10 - 20	IB	0.00017	0.00111
20 - 40	IB	0.023	0.061
40 - 100	IB	0.164	0.34
100 - 300	IB	0.041	0.021
300 - 600	IB	0.166	0.037
600 - 1300	IB	0.44	0.051
1300 - 1779	IB	0.019	0.00136

$^{159}_{67}$Ho(8.30 8 s)

Mode: IT
Δ: -67136 7 keV
SpA: 8.20$\times 10^{9}$ Ci/g
Prod: daughter ^{159}Er; ^{160}Dy(p,2n)

Photons (^{159}Ho)
$\langle\gamma\rangle$=100 4 keV

γ_{mode}	γ(keV)	γ(%)†
Ho L_{ℓ}	5.943	0.152 18
Ho L_{α}	6.716	3.9 3
Ho L_{η}	6.789	0.076 7
Ho L_{β}	7.630	4.5 4
Ho L_{γ}	8.795	0.69 7
γ M3	39.95 14	0.000239 15
Ho $K_{\alpha2}$	46.700	5.4 3
Ho $K_{\alpha1}$	47.547	9.6 6
Ho $K_{\beta1}$'	53.822	2.95 17
Ho $K_{\beta2}$'	55.556	0.85 5
γ E1	166.04 13	5.2 4
γ E3	205.99 9	39.7

† 1.5% uncert(syst)

Atomic Electrons (^{159}Ho)
$\langle e \rangle$=106 3 keV

e_{bin}(keV)	$\langle e \rangle$(keV)	e(%)
8	1.48	18.3 20
9	1.41	15.7 18
31	0.316	1.03 7
32	0.91	2.84 19
37	0.089	0.24 3
38	0.46	1.21 9
39	0.142	0.37 4
40	0.121	0.305 20
44	0.041	0.094 11
45	0.045	0.100 11
46	0.062	0.135 15
47	0.0138	0.029 3
48	0.00113	0.0024 3
51	0.00193	0.0038 4
52	0.0108	0.0208 23
53	0.0044	0.0083 9
54	0.00177	0.0033 4
110	0.40	0.36 3
150	29.3	19.5 10
157	0.073	0.047 3
158	0.0117	0.0074 5
164	0.0194	0.0118 9
165	0.000142	8.6 6 $\times 10^{-5}$
166	0.0054	0.00328 23
197	34.4	17.5 9
198	18.8	9.5 5
204	13.6	6.6 4
205	0.128	0.062 3
206	3.67	1.79 10

$^{159}_{68}$Er(36 1 min)

Mode: ε
Δ: -64573 12 keV
SpA: 3.28$\times 10^{7}$ Ci/g
Prod: ^{165}Ho(p,7n); protons on Ta; ^{153}Eu(^{11}B,5n)

Photons (^{159}Er)
$\langle\gamma\rangle$=817 52 keV

γ_{mode}	γ(keV)	γ(%)†
Ho L_{ℓ}	5.943	0.31 3
Ho L_{α}	6.716	8.0 6
Ho L_{η}	6.789	0.111 10
Ho L_{β}	7.654	7.6 8
Ho L_{γ}	8.840	1.20 14
Ho $K_{\alpha2}$	46.700	22.9 9
Ho $K_{\alpha1}$	47.547	40.8 15
Ho $K_{\beta1}$'	53.822	12.5 5
Ho $K_{\beta2}$'	55.556	3.59 14
γ M1+E2	97.56 23	0.75 3
γ E2	106.96 23	0.39 8
γ	132.17 24	~0.33
γ E1	166.04 13	4.9 10 *
γ E3	205.99 9	9.5 19 *
γ E1	211.74 18	1.14 23
γ [E1]	218.40 17	~0.9
γ [E1]	253.00 15	~3
γ [E1]	307.90 19	0.33 7
γ [E1]	314.56 19	1.14 23
γ	366.5 3	0.22 4
γ	370.1 3	0.14 3
γ E1	391.43 23	0.43 9
γ M1+E2	418.87 22	0.18 4
γ E1	436.58 21	0.24 5
γ M1	482.8 3	0.15 3
γ M1	505.83 20	1.8 4
γ E2	551.66 23	2.3 5
γ E2	562.82 23	0.32 6
γ E2	580.80 18	4.0 8
γ M1	599.1 3	0.27 5
γ	610.3 3	0.20 4
γ	613.5 3	0.16 3
γ M1	624.28 23	33 7
γ M1	649.23 20	23 5
γ	668.2 3	0.24 5
γ M1	774.2 3	0.18 4
γ M1+E2	817.75 24	0.25 5
γ M1+E2	843.98 23	0.69 14
γ	887.47 21	0.49 10
γ	918.9 3	0.111 22
γ E1,E2	942.3 3	1.05 21
γ M1+E2	947.72 21	0.15 3
γ M1	964.9 3	0.30 6
γ M1	972.55 24	0.15 3
γ M1	987.9 3	0.19 4
γ	1046.6 3	0.19 4
γ	1052.1 3	0.111 22
γ	1078.7 3	0.095 19
γ	1086.8 3	0.092 18
γ	1103.1 3	0.18 4
γ	1110.2 3	0.15 3
γ	1120.0 3	0.14 3
γ	1125.7 3	0.108 22
γ	1137.7 3	0.20 4
γ	1152.5 3	0.19 4
γ	1185.9 3	0.27 5
γ M1	1189.3 3	0.21 4
γ M1	1198.92 21	0.82 16
γ	1209.6 3	0.43 9
γ	1219.9 3	0.20 4
γ	1224.6 3	0.14 3
γ [E1]	1226.21 24	0.29 6
γ (E1)	1231.2 3	0.36 7
γ (E1)	1232.50 24	0.52 10
γ (E1)	1240.10 24	0.59 12
γ [E2]	1255.80 21	0.18 4
γ	1285.4 3	0.19 4
γ	1291.5 3	0.21 4
γ	1295.5 3	0.114 23
γ [M1+E2]	1316.34 24	0.15 3
γ [E1]	1329.15 22	0.28 6
γ	1335.0 3	0.26 5
γ E2	1355.33 19	0.95 19
γ	1365.4 3	0.15 3
γ (E1)	1392.65 22	0.56 11
γ	1402.5 3	0.28 6
γ	1418.2 3	0.15 3
γ [E1]	1427.18 20	0.75 15
γ	1443.4 3	0.16 3
γ	1445.9 3	0.20 4

Photons (^{159}Er)
(continued)

γ_{mode}	γ(keV)	γ(%)†
γ	1459.3 3	0.32 6
γ [E1]	1466.76 23	0.26 5
γ [M1+E2]	1476.93 21	0.19 4
γ [M1+E2]	1495.19 22	0.13 3
γ M1	1523.53 22	0.14 3
γ (E1)	1553.0 3	0.92 18
γ M1+E2	1558.69 22	0.46 9
γ [E1]	1567.06 18	0.26 5
γ	1594.9 3	0.14 3
γ E2	1598.6 3	0.26 5
γ	1618.2 3	0.124 25
γ	1631.9 3	0.33 7
γ	1652.8 3	0.121 24
γ	1658.8 3	0.17 3
γ [M1+E2]	1678.58 24	0.072 14
γ [M1+E2]	1680.18 19	0.14 3
γ [E2]	1685.24 21	0.124 25
γ	1708.4 3	0.13 3
γ	1748.77 23	0.118 24
γ	1775.3 3	0.24 5
γ	1786.1 3	0.17 3
γ	1792.3 3	0.118 24
γ (E1)	1838.87 23	1.21 24
γ [E1]	1891.23 21	1.01 20
γ	1906.3 3	0.13 3
γ (E1)	2001.77 23	0.95 19
γ E2	2006.5 3	0.49 10
γ	2038.2 3	0.21 4
γ	2091.87 23	0.28 6

† 4.9% uncert(syst)

* with ^{159}Ho(8.3 s) in equilib

Atomic Electrons (^{159}Er)
⟨e⟩=42 3 keV

e_{bin}(keV)	⟨e⟩(keV)	e(%)
5 - 7	0.7	~13
8	3.0	37 4
9	2.3	26 3
31 - 77	2.87	6.7 6
88 - 132	1.9	1.9 6
150	7.0	4.7 9
156 - 166	0.22	0.138 20
197	8.4	4.3 9
198	4.5	2.3 5
202 - 203	0.011	0.0054 10
204	3.2	1.6 3
205	0.030	0.015 3
206	0.88	0.43 9
209 - 253	0.062	0.026 6
259 - 308	0.049	0.019 3
311 - 358	0.059	0.018 7
361 - 410	0.038	0.010 3
411 - 450	0.24	0.054 10
473 - 507	0.18	0.036 6
525 - 563	0.26	0.049 7
569	3.0	0.52 11
571 - 591	0.051	0.0089 10
594	2.0	0.33 7
597 - 641	0.91	0.146 18
647 - 668	0.089	0.0137 22
719 - 766	0.026	0.0035 7
772 - 818	0.034	0.0043 15
832 - 879	0.029	0.0034 14
885 - 934	0.068	0.0075 13
938 - 987	0.0091	0.00095 10
988 - 1037	0.014	0.0014 5
1038 - 1087	0.021	0.0020 6
1094 - 1138	0.023	0.0020 5
1143 - 1191	0.072	0.0062 9
1197 - 1240	0.022	0.0018 4
1246 - 1295	0.014	0.0011 3
1300 - 1347	0.036	0.0028 4
1353 - 1402	0.019	0.0014 3
1404 - 1451	0.019	0.0013 3
1457 - 1557	0.035	0.00233 25
1558 - 1657	0.024	0.0015 3
1669 - 1767	0.012	0.00068 16

Atomic Electrons (^{159}Er)
(continued)

e_{bin}(keV)	⟨e⟩(keV)	e(%)
1773 - 1851	0.0194	0.00107 14
1882 - 1951	0.0146	0.00075 10
1983 - 2036	0.0088	0.00044 11
2082 - 2090	0.00060	2.9 12 $\times10^{-5}$

Continuous Radiation (^{159}Er)
⟨β+⟩=44.2 keV;⟨IB⟩=2.1 keV

E_{bin}(keV)		⟨ ⟩(keV)	(%)
0 - 10	β+	3.36×10^{-7}	3.99×10^{-6}
	IB	0.0026	
10 - 20	β+	1.79×10^{-5}	0.000107
	IB	0.0020	0.0141
20 - 40	β+	0.00064	0.00195
	IB	0.0153	0.045
40 - 100	β+	0.0410	0.052
	IB	0.18	0.37
100 - 300	β+	2.28	1.03
	IB	0.064	0.034
300 - 600	β+	13.7	3.01
	IB	0.21	0.046
600 - 1300	β+	27.3	3.33
	IB	0.96	0.103
1300 - 2500	β+	0.85	0.062
	IB	0.64	0.041
2500 - 2603	IB	1.11×10^{-5}	4.4×10^{-7}
	Σβ+		7.5

$^{159}_{69}$Tm(9.0 4 min)

Mode: ε

Δ: -60570 200 keV syst

SpA: 1.31×10^{8} Ci/g

Prod: protons on Ta; ^{162}Er(p,4n)

Photons (^{159}Tm)
⟨γ⟩=384 7 keV

γ_{mode}	γ(keV)	γ(%)†
Er L_ℓ	6.151	0.61 7
Er L_α	6.944	15.6 14
Er L_η	7.058	0.208 20
Er L_β	7.937	14.8 15
Er L_γ	9.177	2.4 3
γ E1	38.47 5	6.5 15
Er $K_{\alpha2}$	48.221	41.2 25
Er $K_{\alpha1}$	49.128	73 4
Er $K_{\beta1}'$	55.616	22.6 14
Er $K_{\beta2}'$	57.406	6.4 4
γ M1+<8.3%E2	59.26 3	4.9 8
γ	75.01 12	0.13 5
γ (M1)	76.12 4	0.31 6
γ M1	76.90 6	~0.21
γ M1+<12%E2	84.94 4	6.4 6
γ M1	86.90 6	0.36 10
γ E2	88.82 6	0.87 21
γ	91.7 5	~0.031
γ	94.1 5	~0.05
γ	105.94 9	~0.10
γ	112.56 10	0.10 3
γ M1	114.06 4	1.03 15
γ M1	119.82 6	2.8 6
γ [E1]	124.55 6	0.13 3
γ E1	127.29 6	1.1 4
γ E1	128.07 5	4.5 9
γ E2	136.81 5	0.62 10

Photons (^{159}Tm)
(continued)

γ_{mode}	γ(keV)	γ(%)†
γ M1	142.15 5	0.103 21
γ E2	144.20 4	2.21 21
γ M1	161.06 4	3.4 4
γ M1	163.02 5	1.49 23
γ M1	170.75 8	0.26 5
γ	179.73 20	0.103 21
γ	183.1 6	~0.07
γ	191.11 7	0.23 5
γ M1(+E2)	196.66 5	2.4 4
γ E2	199.00 4	1.28 18
γ	206.92 8	0.31 10
γ E1	212.23 5	1.39 21
γ	214.03 15	0.21 8
γ M1	220.32 4	5.1 3
γ	229.05 7	0.15 5
γ (E1)	243.23 6	1.03 15
γ	246.78 13	0.18 8
γ [M1+E2]	247.96 6	1.08 21
γ E1	252.82 5	2.1 3
γ	263.32 10	0.051 10
γ	267.73 8	0.21 5
γ E1	271.49 5	5.6 8
γ E2	284.81 7	0.41 8
γ E1	289.13 5	5.0 5
γ	297.05 7	0.21 5
γ	307.56 8	0.44 10
γ	313.63 8	0.62 13
γ	334.99 7	0.31 8
γ	344.63 8	~0.10
γ E1	348.39 5	3.8 5
γ	358.90 10	0.46 10
γ	361.76 15	0.154 3
γ	367.68 10	0.41 8
γ	373.53 20	0.62 15
γ	375.03 15	1.9 4
γ (E2)	395.80 7	0.31 5
γ	401.3 3	0.072 15
γ (E1)	408.64 10	2.3 3
γ	415.93 15	0.51 10
γ (M1)	422.60 12	0.51 10
γ	429.23 20	0.067 15
γ	434.49 10	0.26 10
γ	439.4 4	0.08 3
γ	445.68 9	0.46 5
γ	450.43 10	1.03 10
γ	453.93 10	0.56 8
γ	461.83 10	1.13 15
γ	468.23 15	0.31 5
γ	472.96 9	0.36 5
γ	482.53 20	0.41 10
γ	485.03 20	0.36 10
γ	496.85 9	0.10 3
γ	501.03 10	0.82 10
γ	518.43 25	0.87 21
γ	525.7 3	0.26 3
γ	532.51 15	0.36 8
γ	534.49 10	0.46 10
γ	541.78 15	1.18 21
γ	549.43 20	0.31 8
γ	558.43 15	0.67 15
γ	559.4 5	~0.15
γ	567.13 15	0.36 8
γ	572.97 9	0.13 4 ?
γ	583.53 14	0.21 8
γ	601.33 20	0.13 3
γ	605.43 15	0.31 6
γ	617.16 10	0.17 6
γ (M1)	619.26 13	0.31 10
γ	634.33 15	0.33 8
γ	643.03 20	0.077 15
γ	690.63 20	0.31 8
γ	693.73 20	0.15 3
γ	703.9 3	0.10 3
γ	713.33 15	0.13 3
γ	729.73 20	0.26 5
γ	733.43 20	0.21 5
γ	737.33 20	0.15 4
γ	740.13 20	0.15 4
γ	755.8 3	0.21 4
γ	758.03 20	0.10 3
γ	762.26 19	0.46 8
γ	770.68 14	0.26 5
γ	778.98 12	0.21 4
γ	783.83 20	0.21 4

Photons (^{159}Tm)
(continued)

γ_{mode}	γ(keV)	γ(%)†
γ	787.2 *4*	~0.08
γ	792.21 *12*	0.21 *6*
γ	822.53 *25*	0.13 *4*
γ	829.43 *25*	0.31 *8*
γ	843.3 *3*	0.21 *5*
γ	857.7 *3*	0.36 *10*
γ	888.77 *18*	0.15 *4*
γ	902.4 *3*	0.36 *10*
γ	906.27 *12*	0.21 *6*
γ	921.93 *20*	0.36 *8*
γ	933.01 *17*	0.15 *4*
γ	956.33 *25*	0.56 *13*
γ	990.99 *14*	0.26 *5*
γ	1060.09 *15*	0.15 *5*
γ	1132.00 *17*	0.10 *3*
γ	1135.68 *15*	0.62 *13*
γ	1146.2 *3*	0.15 *5*
γ	1168.4 *5*	0.08 *4*
γ	1174.15 *15*	0.15 *5*
γ	1190.9 *5*	~0.15
γ	1208.3 *4*	0.31 *8*
γ	1211.1 *4*	0.31 *8*
γ	1248.0 *3*	0.46 *10*
γ	1261.6 *3*	0.46 *10*
γ	1270.2 *3*	0.77 *10*

† 15% uncert(syst)

Atomic Electrons (^{159}Tm)
⟨e⟩=64.7 *18* keV

e_{bin}(keV)	⟨e⟩(keV)	e(%)
2	0.76	43 *7*
8	5.6	68 *8*
9	3.2	35 *4*
10 - 19	1.73	14.8 *17*
27	6.4	23.5 *23*
29 - 49	5.7	15.7 *12*
50	5.4	10.9 *17*
51	1.9	3.7 *6*
53 - 55	0.20	0.36 *9*
57	2.9	5.1 *8*
58 - 59	0.56	0.95 *15*
62	2.5	4.1 *8*
65 - 74	1.2	1.8 *4*
75	2.4	3.2 *4*
76 - 79	2.48	3.24 *24*
80	1.7	2.1 *5*
82 - 86	1.63	1.95 *17*
87	1.26	1.45 *22*
88 - 103	0.26	0.28 *6*
104	2.6	2.5 *3*
105 - 154	5.9	4.7 *3*
155 - 162	0.37	0.23 *3*
163	2.25	1.38 *8*
169 - 214	2.00	1.01 *12*
218 - 267	0.80	0.34 *4*
268 - 316	0.52	0.18 *3*
318 - 367	0.60	0.18 *6*
368 - 416	0.45	0.11 *3*
419 - 468	0.31	0.070 *19*
471 - 518	0.30	0.061 *17*
523 - 572	0.14	0.025 *5*
573 - 619	0.040	0.007 *3*
625 - 672	0.051	0.008 *3*
676 - 725	0.088	0.013 *3*
726 - 775	0.054	0.0072 *19*
777 - 823	0.028	0.0035 *15*
827 - 876	0.048	0.0056 *17*
879 - 925	0.027	0.0030 *14*
931 - 956	0.012	0.0013 *6*
981 - 1003	0.006	~0.0006
1050 - 1089	0.024	0.0022 *10*
1111 - 1160	0.030	0.0026 *8*
1164 - 1213	0.044	0.0037 *12*
1238 - 1270	0.0075	0.00060 *17*

Continuous Radiation (^{159}Tm)
⟨β+⟩=296 keV;⟨IB⟩=9.1 keV

E_{bin}(keV)		⟨ ⟩(keV)	(%)
0 - 10	β+	1.39×10^{-7}	1.65×10^{-6}
	IB	0.0110	
10 - 20	β+	7.6×10^{-6}	4.54×10^{-5}
	IB	0.0105	0.072
20 - 40	β+	0.000284	0.00086
	IB	0.027	0.089
40 - 100	β+	0.0195	0.0246
	IB	0.22	0.41
100 - 300	β+	1.31	0.58
	IB	0.20	0.112
300 - 600	β+	11.6	2.47
	IB	0.40	0.090
600 - 1300	β+	96	10.0
	IB	1.9	0.20
1300 - 2500	β+	182	10.5
	IB	4.7	0.26
2500 - 3941	β+	4.81	0.185
	IB	1.64	0.058
	Σβ+		24

$^{159}_{70}$Yb(1.4 *2* min)

Mode: ε
Δ: -55930 *210* keV syst
SpA: 8.4×10^{8} Ci/g
Prod: protons on W; protons on Ta

Photons (^{159}Yb)

γ_{mode}	γ(keV)	γ(rel)
γ	77.74 *5*	7 *1*
γ	113.14 *5*	12 *2*
γ	166.14 *5*	100
γ	176.08 *5*	14 *2*
γ	177.19 *5*	20 *4*
γ	191.92 *10*	4 *1*
γ	193.71 *5*	5 *1*
γ	197.62 *5*	5 *1*
γ	239.1 *1*	10 *2*
γ	329.7 *1*	18 *3*
γ	390.3 *1*	18 *3*
γ	497.2 *3*	9 *2*

$^{159}_{71}$Lu(12.3 *1* s)

Mode: ε, α
Δ: -49850 *60* keV
SpA: 5.61×10^{9} Ci/g
Prod: protons on W

Alpha Particles (^{159}Lu)

α(keV)
4420 *10*

Photons (^{159}Lu)

γ_{mode}	γ(keV)	γ(rel)
γ	150.51 *5*	100
γ	187.5 *1*	25 *5*
γ	369.3 *2*	19 *4*

$^{159}_{72}$Hf(5.6 *5* s)

Mode: ε(88 *1* %), α(12 *1* %)
Δ: -43090 *310* keV syst
SpA: 1.19×10^{10} Ci/g
Prod: ^{144}Sm(^{20}Ne,5n); daughter ^{163}W; ^{58}Ni on ^{107}Ag

Alpha Particles (^{159}Hf)

α(keV)
5095 *5*

$^{159}_{73}$Ta(570 *180* ms)

Mode: α(80 *5* %), ε(20 *5* %)
Δ: -34600 *430* keV syst
SpA: 7.2×10^{10} Ci/g
Prod: ^{58}Ni on ^{107}Ag

Alpha Particles (^{159}Ta)

α(keV)
5601 *6*

$^{159}_{74}$W (7 *3* ms)

Mode: α
Δ: -25550 *590* keV syst
SpA: 1.0×10^{11} Ci/g
Prod: daughter ^{163}Os

Alpha Particles (^{159}W)

α(keV)
6299 *6*

A = 160

NDS 12, 477 (1974)

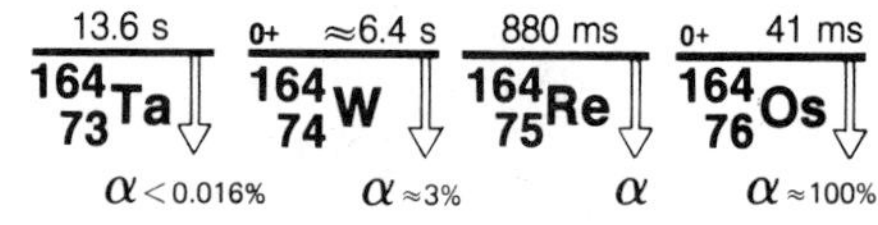

Q_α 4400 200(syst) Q_α 5278 3 Q_α 5920 50 Q_α 6478 21

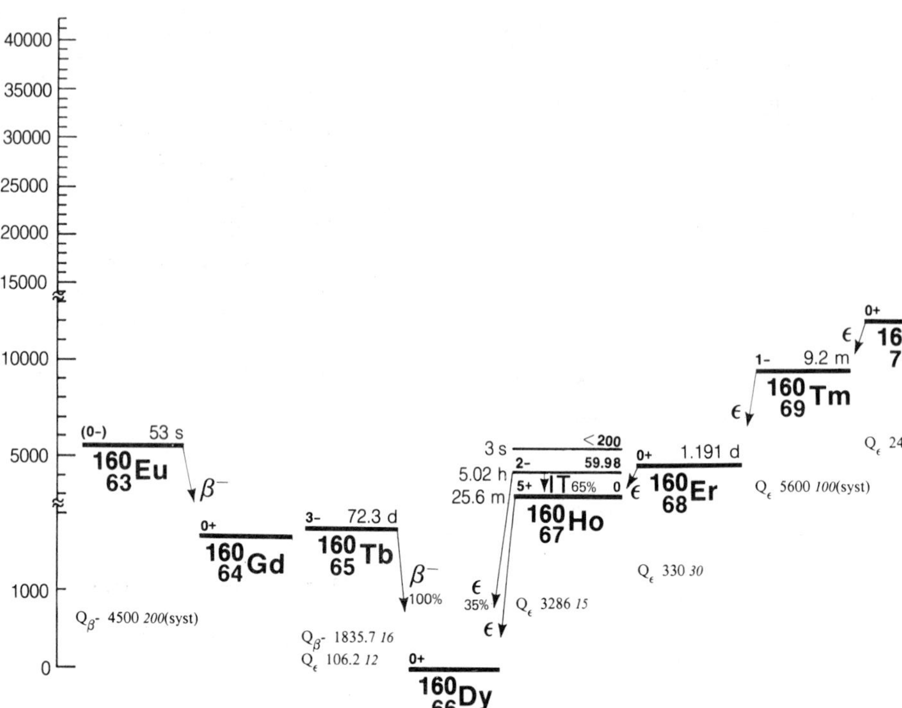

$^{160}_{63}$Eu(53 10 s)

Mode: β-

Δ: -63450 200 keV syst

SpA: 1.32×10^{9} Ci/g

Prod: ^{160}Gd(n,p)

Photons (^{160}Eu)

γ_{mode}	γ(keV)	γ(rel)
γ (E2)	74.5 5	24 5
γ (E2)	172.6 9	100
γ	411.1 8	56 8
γ	513.6 10	60 9
γ	736.4 22	11.0 16
γ	821.2 12	49 7
γ	924.7 10	19 3
γ	993.8 13	36 5
γ	1147 5	6 1
γ	1184 4	6 1
γ	1233.4 13	12.0 18
γ	1300.4 18	14.0 21
γ	1385.1 14	14.0 21
γ	1459.7 14	9.0 14 ?

$^{160}_{64}$Gd(stable)

Δ: -67954 3 keV

%: 21.86 4

$^{160}_{65}$Tb(72.3 2 d)

Mode: β-

Δ: -67848 4 keV

SpA: 1.129×10^{4} Ci/g

Prod: ^{159}Tb(n,γ); ^{159}Tb(d,p)

Photons (^{160}Tb)

⟨γ⟩=1127 15 keV

γ_{mode}	γ(keV)	γ(%)†
Dy L_ℓ	5.743	0.164 18
Dy L_α	6.491	4.3 3
Dy L_η	6.534	0.073 6
Dy L_β	7.367	4.3 4
Dy L_γ	8.454	0.62 6
Dy $K_{\alpha2}$	45.208	6.0 3
Dy $K_{\alpha1}$	45.998	10.7 5
Dy $K_{\beta1}'$	52.063	3.25 14
Dy $K_{\beta2}'$	53.735	0.95 5
γ E2	86.788 2	13.2 5
γ E2+49%M1	93.919 5	0.056 5
γ [E2]	176.487 22	0.0049 10
γ (E2)	189.642 18	<19×10^{-5}
γ E2	197.034 6	5.15 12
γ E1	215.645 6	3.95 10
γ [E1]	230.623 11	0.073 6
γ [E1]	237.605 24	0.008 3
γ [E2]	239.56 3	<0.0036
γ [E1]	243.14 3	~0.010
γ [E1]	246.488 16	0.023 4
γ E1+0.03%M2	298.576 4	26.8 8
γ E1	309.564 7	0.856 19
γ E1	337.334 15	0.324 10
γ [E1]	349.849 23	0.015 3

Photons (^{160}Tb) (continued)

γ_{mode}	γ(keV)	γ(%)†
γ [E1]	379.34 3	0.015 3
γ E1+0.06%M2	392.495 6	1.33 3
γ [E1]	432.780 22	0.017 3
γ E1	486.051 22	0.084 5
γ (E2)	682.334 9	0.57 3
γ (E2)	707.49 12	<0.023
γ E2	765.265 10	2.00 5
γ (E2)	871.975 19	0.208 23
γ E2+0.4%M1	879.367 7	29.8
γ E2+0.7%M1	962.297 9	9.8 3
γ E2	966.155 7	25.0 5
γ E1	1002.869 23	1.02 3
γ (E2)	1004.83 3	<0.10 ?
γ (E2)	1069.008 18	0.092 11
γ	1102.597 16	0.53 2
γ E1	1115.112 21	1.53 3
γ E1	1177.941 7	15.2 3
γ E1	1199.901 23	2.32 5
γ [E1]	1251.313 23	0.098 10
γ E1	1271.860 8	7.49 19
γ E1	1285.61 7	0.0137 20
γ	1299.629 16	~0.005
γ E1	1312.144 21	2.92 7

† 5.0% uncert(syst)

Atomic Electrons (^{160}Tb)

⟨e⟩=49.5 9 keV

e_{bin}(keV)	⟨e⟩(keV)	e(%)
8	1.70	21.8 24
9	1.34	15.6 17
33	6.7	20.4 9
36 - 52	0.56	1.41 7
78	12.9	16.5 7
79	12.0	15.1 6
85	6.5	7.6 3
86	1.53	1.76 9
87 - 136	0.234	0.268 15
143 - 193	2.07	1.32 3
195 - 244	0.247	0.1247 25
245 - 291	1.19	0.473 14
296 - 342	0.0915	0.0292 6
348 - 392	0.00760	0.00198 4
424 - 433	0.00179	0.000413 25

Atomic Electrons (^{160}Tb) (continued)

e_{bin}(keV)	⟨e⟩(keV)	e(%)
477 - 486	0.000334	7.0 3 ×10^{-5}
629 - 675	0.0250	0.00393 21
680 - 711	0.0655	0.0092 3
756 - 765	0.0141	0.00186 4
818 - 864	0.84	0.101 5
870 - 912	1.052	0.1163 22
949 - 997	0.191	0.0200 3
1001 - 1049	0.015	~0.0015
1060 - 1107	0.0191	0.00178 12
1113 - 1146	0.158	0.0140 4
1169 - 1218	0.0919	0.00763 18
1232 - 1278	0.0357	0.00283 6
1284 - 1312	0.00423	0.000324 8

Continuous Radiation (^{160}Tb)

⟨β-⟩=204 keV;⟨IB⟩=0.148 keV

E_{bin}(keV)		⟨ ⟩(keV)	(%)
0 - 10	β-	0.153	3.06
	IB	0.0102	
10 - 20	β-	0.455	3.03
	IB	0.0095	0.066
20 - 40	β-	1.80	6.0
	IB	0.017	0.059
40 - 100	β-	12.2	17.5
	IB	0.039	0.061
100 - 300	β-	86	45.3
	IB	0.057	0.035
300 - 600	β-	88	22.0
	IB	0.0148	0.0039
600 - 1300	β-	14.8	2.15
	IB	0.00099	0.000144
1300 - 1552	β-	0.089	0.0065
	IB	2.1 ×10^{-6}	1.56 ×10^{-7}

$^{160}_{66}$Dy(stable)

Δ: -69683 4 keV
%: 2.34 5

$^{160}_{67}$Ho(25.6 3 min)

Mode: ε
Δ: -66397 15 keV
SpA: 4.59×10^{7} Ci/g
Prod: daughter ^{160}Ho(5.02 h); ^{159}Tb(α,3n); protons on Dy

Photons(^{160}Ho(25.6 min + 5.02 h))

⟨γ⟩=1783 18 keV

γ_{mode}	γ(keV)	γ(%)$^{\&}$
γE2	86.788 2	14
γE2+49%M1	93.919 5	0.0191 21
γM1+E2	107.84 5	0.24 3
γ(M1+E2)	126.80 14	0.048 11
γ	163.20 10	0.032 7
γ[E2]	176.487 22	0.0021 8
γ(E2)	189.642 18	<0.006
γE2	197.034 6	14.0 3
γE1	215.645 6	0.154 10
γ[E2]	234.64 14	0.042 14

Photons(^{160}Ho(25.6 min + 5.02 h)) (continued)

γ_{mode}	γ(keV)	γ(%)$^{\&}$
γ[E1]	237.605 24	0.0015 6
γ[E2]	239.56 3	0.070 14
γ[E1]	243.14 3	~0.0021
γ[E1]	246.488 16	0.010 4
γ[E2]	256.12 11	<0.022 ?
γ[E2]	282.42 11	0.025 6
γE2	297.33 12	1.12 7
γ	297.90 10	0.31 7
γE1+0.03%M2	298.576 4	1.05 6
γE1	309.564 7	0.292 20
γ	337.30 4	0.10 4
γ[E1]	349.849 23	0.0032 7
γ(M1+E2)	363.96 11	0.14 4
γ[E1]	379.34 3	0.0063 24
γE1+0.06%M2	392.495 6	0.45 3
γE2	405.68 4	0.31 4
γ[E1]	432.780 22	0.0036 7
γ	446.70 10	0.14 3
γ	469.9 3	0.14 6
γE1	486.051 22	0.036 12
γ[M1+E2]	490.76 17	~0.07
γ	491.9 6	0.07 3
γ	494.6 3	0.17 7
γ(E2)	513.53 5	1.40 14
γE2	538.54 3	3.9 3
γ	546.00 20	~0.10
γ	555.4 4	~0.14 ?
γ	558.5 4	~0.14 ?
γ	564.4 4	0.042 14 ?
γ	575.40 20	0.15 7
γ	583.30 20	0.14 6
γ	596.0 10	~0.07
γ	606.60 20	~0.13 ?
γ	609.3 10	0.15 7
γ	612.2 5	~0.06
γ[M1+E2]	621.1 3	0.07 3
γ	630.3 7	~0.06 ?
γ	636.9 3	~0.028 ?
γ[M1+E2]	640.33 14	0.29 7
γ[M1+E2]	645.25 3	16.2 7
γ(E2)	646.38 5	3.1 3
γ	654.4 7	0.098 14 ?
γ	665.7 5	0.14 3 ?
γ(E2)	682.334 9	0.384 20
γ(E2)	707.49 12	0.45 7
γ	712.8 9	~0.06 ?
γE2	728.18 3	30.8 8
γE2	753.09 5	2.8 3
γ	758.2 3	0.48 7 ?
γE2	765.265 10	3.69 15
γ	773.00 20	0.14 3 ?
γ	791.8 3	0.14 3 ?
γ	815.40 10	0.10 3 ?
γ	826.20 10	0.87 10
γ	838.66 14	0.056 14
γ	843.6 5	0.21 3
γ	853.8 6	0.14 3
γ[M1+E2]	857.06 12	0.50 4 ?
γ(E2)	871.975 19	6.1 5
γE2+0.4%M1	879.367 7	20.2 6
γ	900.70 20	0.042 14 ?
γ	903.58 14	0.17 4
γ	907.3 4	0.056 14 ?
γ	912.3 4	0.056 14
γ	914.2 3	0.070 14 ?
γ	921.70 10	0.056 14 ?
γ	936.00 20	0.10 3
γ	941.00 20	0.38 4
γ	946.3 3	0.056 14 ?
γE2+0.7%M1	962.297 9	18.1 6
γE2	966.155 7	16.9 5
γ	975.90 10	0.056 14 ?
γ[E2]	987.1 3	0.056 14
γ	989.1 7	0.10 3
γ	994.40 20	0.056 14
γE1	1002.869 23	0.200 7
γ(E2)	1004.83 3	1.90 14
γ	1018.5 3	0.070 14 ?
γ	1030.90 20	0.070 14
γ	1039.70 20	0.070 14
γ(M1+E2)	1047.80 10	0.80 10
γ	1058.40 10	0.027 7 ?
γ	1062.5 3	0.070 14 ?

Photons(^{160}Ho(25.6 min + 5.02 h)) (continued)

γ_{mode}	γ(keV)	γ(%)$^{\&}$
γ(E2)	1069.008 18	2.7 4
γ	1077.30 10	0.056 14 ?
γ	1082.90 10	0.084 14
γ	1090.20 20	0.17 3
γ	1094.90 20	0.29 4
γ	1102.59 5	0.14 3
γ	1112.50 20	0.39 4
γE1	1115.112 21	0.33 3
γ	1122.10 10	0.14 3
γ[E1]	1125.0 3	0.14 3
γ(E2)	1130.60 10	1.09 11
γ	1140.90 20	0.056 14
γ	1145.5 3	0.084 14
γ[E2]	1154.39 11	0.15 3
γ	1159.9 3	0.070 14
γ	1170.80 22	
γ	1172.70 10	0.22 3 ‡
γ	1174.80 10	0.24 3
γE1	1177.941 7	0.59 4
γ	1182.80 20	0.22 3
γ	1193.20 20	0.18 3
γ	1199.89 4	0.87 17
γE1	1199.901 23	0.456 14
γ	1201.0 5	0.084 14
γ	1208.3 5	0.056 14 ?
γ	1216.90 16	0.119 17
γ	1220.20 20	0.042 10 ?
γ	1229.1 5	<0.042
γ(E1)	1234.67 16	0.27 3
γ	1240.90 20	0.14 4 ?
γ	1244.20 10	0.15 4
γ	1250.0 5	0.042 14
γ[E1]	1251.313 23	0.042 14
γ	1259.70 10	0.056 14
γ	1262.7 10	0.31 3
γE1	1271.860 8	2.56 17
γ	1273.0 10	0.15 7 ?
γM1	1278.9 3	0.070 14
γ	1281.0 3	0.13 3
γE1	1285.61 7	1.62 14
γ	1287.6 8	0.20 3
γ	1298.5 4	0.043 10
γ(E1)	1302.7 3	0.17 3
γ	1305.00 20	0.13 3
γE1	1312.144 21	0.63 5
γ	1317.60 20	0.17 3
γ(E2)	1320.9 4	0.084 14
γ	1331.90 20	0.14 3
γ	1338.40 20	0.196 14
γ(M1)	1345.31 14	0.25 3
γ[E2]	1349.48 15	0.27 3
γ	1357.4 5	0.070 14
γ	1359.40 20	0.17 3
γ[M1+E2]	1366.19 13	0.15 3
γ	1369.90 10	
γ	1370.8 7	0.84 10 ‡
γ	1375.70 20	0.28 4
γ(M1)	1389.26 18	0.17 3
γ(M1+E2)	1396.50 20	0.13 3
γ	1400.7 7	
γ	1402.40 10	0.50 7 ‡
γ	1408.2 5	
γ[M1+E2]	1410.14 18	0.50 7 ‡
γ	1414.85 16	0.084 14
γ	1419.00 20	0.14 3
γ	1421.1 8	0.056 14
γ	1428.50 10	0.098 14
γ	1432.00 10	0.83 10
γ	1433.7 8	
γ	1435.73 15	0.14 3 ‡
γM1	1438.70 20	0.20 4
γ	1441.8 9	<0.10
γ	1444.00 10	0.070 11
γ	1449.3 3	0.036 7 ?
γ	1452.5 3	0.031 6 ?
γ	1465.20 10	0.11 3
γ	1468.6 8	0.10 3
γ	1473.30 10	0.57 8
γ	1481.6 3	0.084 14
γ	1489.50 10	0.25 4
γ	1502.70 10	0.070 14 ?
γ	1509.9 3	0.027 7 ?
γ[M1+E2]	1518.35 5	0.168 14

Photons(^{160}Ho(25.6 min + 5.02 h))
(continued)

γ_{mode}	γ(keV)	γ(%)$^{\&}$
γ	1526.0 5	0.020 8
γ	1547.3 3	0.017 7 ?
γ	1556.60 10	0.112 14
γ	1561.5 6	0.022 8 ?
γ	1566.20 10	0.032 8
γ	1576.30 10	0.039 7
γ	1586.30 10	0.109 13
γ	1588.7 4	0.034 7 ?
γ[E2]	1607.54 3	0.108 11 ?
γ	1619.4 3	0.154 14 ?
γ	1621.5 3	0.182 14
γ	1635.90 10	0.032 7
γ	1655.1 4	0.087 11 ?
γE1	1664.76 13	0.32 3
γ(E2)	1670.2 3	0.31 3
γ	1695.10 10	0.112 14
γ	1698.20 10	0.035 7 ?
γ	1718.03 14	0.70 8
γ	1734.30 15	0.098 14
γ	1755.2 6	0.025 8 ?
γ	1764.5 3	0.066 13
γ	1782.95 14	0.31 3
γ	1804.81 14	0.32 3
γ	1816.40 10	0.18 3
γ	1856.2 3	0.056 14 ?
γ[E2]	1866.4 3	0.104 13
γ	1869.73 14	0.140 14
γ	1873.1 6	0.042 14 ?
γ	1919.10 10	0.070 14
γ	1922.80 20	0.43 10
γ	1925.3 8	0.039 8
γ	1929.3 5	0.042 14
γ	1963.2 6	0.027 7 ?
γ	1969.6 9	~0.020 ?
γE1	1985.1 3	0.140 14
γ	1998.15 11	0.140 14
γ	2002.50 10	0.140 14
γ	2044.10 20	0.070 14
γ	2051.30 10	0.084 14
γ	2052.5 7	0.025 10 ?
γ	2057.7 3	0.031 4 ?
γ	2068.20 20	0.49 6
γ	2077.90 10	0.010 3 ?
γ	2084.94 11	0.25 3
γ	2091.0 5	~0.028
γ	2095.50 20	0.036 7
γ	2101.5 3	~0.014
γ	2114.50 10	0.024 6
γ	2124.1 9	<0.028
γ	2169.1 3	0.070 14
γ	2176.2 8	~0.018
γ	2180.60 10	0.11 4
γE1	2182.1 3	0.62 7
γ	2190.70 10	0.027 7
γ	2204.40 20	0.027 4
γ	2236.45 16	0.053 10
γ	2276.60 10	0.031 3
γ	2283.1 9	~0.028 ?
γ	2288.6 3	0.043 3
γ	2296.7 9	~0.028 ?
γ	2300.2 7	0.042 14 ?
γ	2304.9 7	~0.028 ?
γ	2308.9 5	~0.028
γ	2321.2 9	0.049 13
γ	2325.1 5	0.0238 14
γ	2352.6 4	0.020 6
γ	2355.0 5	0.035 7
γ	2362.70 10	0.098 14
γE1	2382.8 3	0.059 13
γ	2391.30 20	0.042 7
γ	2416.79 16	~0.015
γ	2425.3 5	0.032 10 ?
γ	2427.9 5	<0.014
γ	2433.48 16	0.084 14
γ	2436.2 3	0.070 14
γ	2438.7 5	0.070 14 ?
γ	2450.50 20	0.098 14
γ	2461.0 9	~0.0028 ?
γ	2473.2 4	0.027 10
γ	2479.50 10	0.070 7
γ	2487.60 10	0.034 7 ?
γ	2494.4 3	0.084 14
γ	2504.10 10	0.041 6

Photons(^{160}Ho(25.6 min + 5.02 h))
(continued)

γ_{mode}	γ(keV)	γ(%)$^{\&}$
γ	2514.7 5	0.032 6
γ	2520.27 16	~0.010
γ	2537.30 10	0.022 8 ?
γE1	2544.12 13	1.26 13
γ	2549.30 10	0.017 7 ?
γ(E1)	2560.40 10	0.24 4
γ	2570.50 10	0.053 6
γE1	2574.60 10	0.28 4
γ	2581.10 20	0.028 7
γE1	2588.07 18	0.20 3
γ	2596.9 5	0.032 13 ?
γ	2601.3 5	0.031 13 ?
γ	2609.40 10	0.084 14
γE1	2613.66 15	1.04 11
γ	2617.7 4	0.14 3 ?
γ[E1]	2630.91 13	0.56 7
γ	2633.6 4	0.59 7
γ	2643.40 10	0.13 3
γE1	2648.00 20	0.57 7
γ	2653.0 7	0.031 10 ?
γ	2657.6 5	0.035 10 ?
γ	2658.60 20	0.028 13 ?
γ	2669.0 3	0.112 14 ?
γE1	2674.86 18	1.54 14
γ	2681.30 10	0.076 11
γ	2686.50 10	0.024 7
γ	2694.4 9	~0.014 ?
γ	2698.20 10	0.057 13
γ	2701.40 10	0.036 8
γ	2717.70 10	0.050 10
γ	2729.6 5	~0.0028 ?
γ	2735.10 20	0.17 3
γ	2742.60 10	0.027 7 ?
γ	2747.60 10	0.014 4
γ	2753.4 8	~0.011 ?
γ	2761.4 4	0.017 7
γ	2765.20 10	0.168 14
γ	2774.5 7	0.0084 14 ?
γ	2817.0 8	0.014 3
γ	2843.0 9	~0.0028 ?
γ	2851.4 5	0.042 8
γ	2860.6 9	0.017 4 ?
γ	2865.9 9	0.008 3
γ	2875.0 9	0.0056 14 ?
γ	2889.0 9	~0.006 ?
γ	2895.5 9	0.013 3
γ	2974.3 8	0.0056 14

‡ combined intensity for doublet
& per 100 decays of ^{160}Ho(25.6 min + 5.02 h)

$^{160}_{67}$Ho(5.02 5 h)

Mode: IT(65 3 %), ϵ(35 3 %)
Δ: -66337 15 keV
SpA: 3.90×10^{6} Ci/g
Prod: ^{159}Tb(α,3n); protons on Dy; daughter ^{160}Er

Photons (^{160}Ho)$^{\&}$

γ_{mode}	γ(keV)	γ(%)
γ_{IT} E3	59.98 3	0.069 5

& see also ^{160}Ho(25.6 min + 5.02 h)

$^{160}_{67}$Ho(3 s)

decay mode uncertain
Δ: -66197 15 keV
SpA: 2.1×10^{10} Ci/g
Prod: ^{159}Tb(α,3n)

$^{160}_{68}$Er(1.191 4 d)

Mode: ϵ
Δ: -66063 28 keV
SpA: 6.852×10^{5} Ci/g
Prod: protons on Er; protons on U

Photons (^{160}Er)

$\langle\gamma\rangle$=36.8 6 keV

γ_{mode}	γ(keV)	γ(%)
Ho L_{ℓ}	5.943	0.28 3
Ho L_{α}	6.716	7.1 5
Ho L_{η}	6.789	0.093 7
Ho L_{β}	7.655	6.9 6
Ho L_{γ}	8.868	1.14 12
Ho $K_{\alpha2}$	46.700	21.1 6
Ho $K_{\alpha1}$	47.547	37.4 11
Ho $K_{\beta1}'$	53.822	11.5 3
Ho $K_{\beta2}'$	55.556	3.30 11

Atomic Electrons (^{160}Er)

$\langle e\rangle$=6.5 4 keV

e_{bin}(keV)	$\langle e\rangle$(keV)	e(%)
8	2.6	32 3
9	2.12	23.3 24
37	0.35	0.94 10
38	0.193	0.51 5
39	0.52	1.34 14
44	0.162	0.37 4
45	0.175	0.39 4
46	0.241	0.53 5
47	0.054	0.114 12
48	0.0044	0.0093 10
51	0.0075	0.0146 15
52	0.042	0.081 8
53	0.0172	0.032 3
54	0.0069	0.0129 13

Continuous Radiation (^{160}Er)

$\langle IB\rangle$=0.19 keV

E_{bin}(keV)		$\langle\ \rangle$(keV)	(%)
10 - 20	IB	0.00020	0.00131
20 - 40	IB	0.013	0.036
40 - 100	IB	0.18	0.37
100 - 260	IB	0.0031	0.0027

$^{160}_{69}$Tm(9.2 *4* min)

Mode: ϵ
Δ: -60460 *100* keV syst
SpA: 1.28×10^8 Ci/g
Prod: ^{164}Er(p,5n); ^{162}Er(p,3n); protons on Ta

Photons (^{160}Tm)

γ_{mode}	γ(keV)	γ(%)
γ E2	125.7 *1*	35 *5*
γ E2	264.0 *1*	9.4 *12*
γ	389.0 *8*	0.7 *3*
γ	520.2 *8*	0.5 *2*
γ	527.0 *8*	0.8 *3*
γ	548.4 *8*	~0.6
γ	597.1 *6*	1.7 *3*
γ	617.5 *6*	1.6 *3*
γ	636.4 *8*	0.8 *3*
γ	640.1 *7*	1.6 *3*
γ	665.0 *8*	0.8 *3*
γ	681.7 *7*	0.7 *3*
γ	728.5 *5*	12.8 *12*
γ	767.8 *5*	2.9 *3*
γ	797.7 *6*	2.7 *4*
γ E2	854.4 *5*	8.1 *7*
γ E2	861.4 *5*	7.0 *6*
γ	882.0 *6*	2.2 *3*
γ	985.5 *7*	0.9 *3*
γ	1000.2 *8*	0.7 *3*
γ	1007.7 *6*	2.5 *3*
γ	1102.7 *12*	~0.6
γ	1151.8 *8*	1.1 *3*
γ	1249.1 *6*	2.8 *4*
γ	1264.1 *8*	1.3 *3*
γ	1269.7 *7*	3.0 *3*
γ	1340.5 *10*	0.5 *2*
γ	1368.5 *5*	8.6 *7*
γ	1394.7 *6*	3.6 *4*
γ	1409.4 *10*	0.7 *3*
γ	1460.6 *6*	4.3 *5*
γ	1526.4 *6*	3.8 *4*
γ	1536.6 *8*	1.1 *2*
γ	1585.9 *7*	1.0 *2*
γ	1768.5 *8*	0.8 *2*
γ	1801.1 *10*	0.6 *2*
γ	1862.6 *9*	0.8 *2*
γ	1894.4 *11*	0.9 *3*
γ	2068.5 *8*	1.1 *3*
γ	2123.4 *8*	1.1 *3*
γ	2202.2 *7*	1.9 *3*
γ	2214.1 *8*	1.1 *3*
γ	2403.9 *9*	0.7 *3*
γ	2924.3 *10*	0.8 *3*

$^{160}_{70}$Yb(4.8 *2* min)

Mode: ϵ
Δ: -58060 *200* keV syst
SpA: 2.45×10^8 Ci/g
Prod: ^{40}Ar on ^{124}Te; protons on Ta

Photons (^{160}Yb)

γ_{mode}	γ(keV)	γ(rel)
γ	34.18 *10*	3.1 *5*
γ	42.02 *10*	7.3 *6*
γ	62.05 *10*	0.46 *15*
γ	94.29 *7*	0.92 *8*
γ	98.24 *5*	2.8 *2*
γ	99.46 *5*	2.1 *1*
γ	116.44 *5*	1.96 *16*
γ	132.23 *5*	14.0 *7*
γ	140.35 *5*	22.2 *10*
γ	155.76 *7*	1.7 *2*
γ	173.74 *6*	100 *4*
γ	174.4 *1*	13.2 *15*
γ	215.78 *6*	48 *2*
γ	278.0 *3*	1.0 *2* ?
γ	320.00 *15*	3.4 *3*
γ	327.60 *15*	5.6 *4*
γ	354.6 *3*	1.1 *2*
γ	356.9 *5*	0.74 *20* ?
γ	366.2 *3*	1.05 *25* ?
γ	373.0 *1*	10.0 *5*
γ	386.3 *2*	3.0 *3*
γ	389.45 *15*	5.2 *3*
γ	395.16 *25*	1.61 *23*
γ	429.0 *4*	1.2 *3* ?
γ	465.2 *4*	1.4 *3* ?
γ	563.1 *3*	1.8 *4*
γ	582.12 *20*	3.0 *4*
γ	588.7 *3*	1.5 *4* ?

$^{160}_{71}$Lu(34.5 *15* s)

Mode: ϵ
Δ: -50260 *320* keV syst
SpA: 2.02×10^9 Ci/g
Prod: protons on W

decay data not yet evaluated

$^{160}_{72}$Hf(~12 s)

Mode: ϵ(97.7 *6* %), α(2.3 *6* %)
Δ: -46060 *60* keV
SpA: ~ 6×10^9 Ci/g
Prod: ^{144}Sm(^{20}Ne,4n); ^{58}Ni on ^{107}Ag

Alpha Particles (^{160}Hf)

α(keV)
4777 *5*

$^{160}_{73}$Ta($t_{1/2}$ unknown)

Mode: α
Δ: -35740 *410* keV syst
Prod: ^{58}Ni on ^{107}Ag

Alpha Particles (^{160}Ta)

α(keV)
5413 *5*

$^{160}_{74}$W (81 *15* ms)

Mode: α(~ 94 %), ϵ(~ 6 %)
Δ: -29360 *500* keV syst
SpA: 1.02×10^{11} Ci/g
Prod: ^{32}S on ^{144}Sm; daughter ^{164}Os

Alpha Particles (^{160}W)

α(keV)
5920 *10*

A = 161

NDS **13**, 493 (1974)

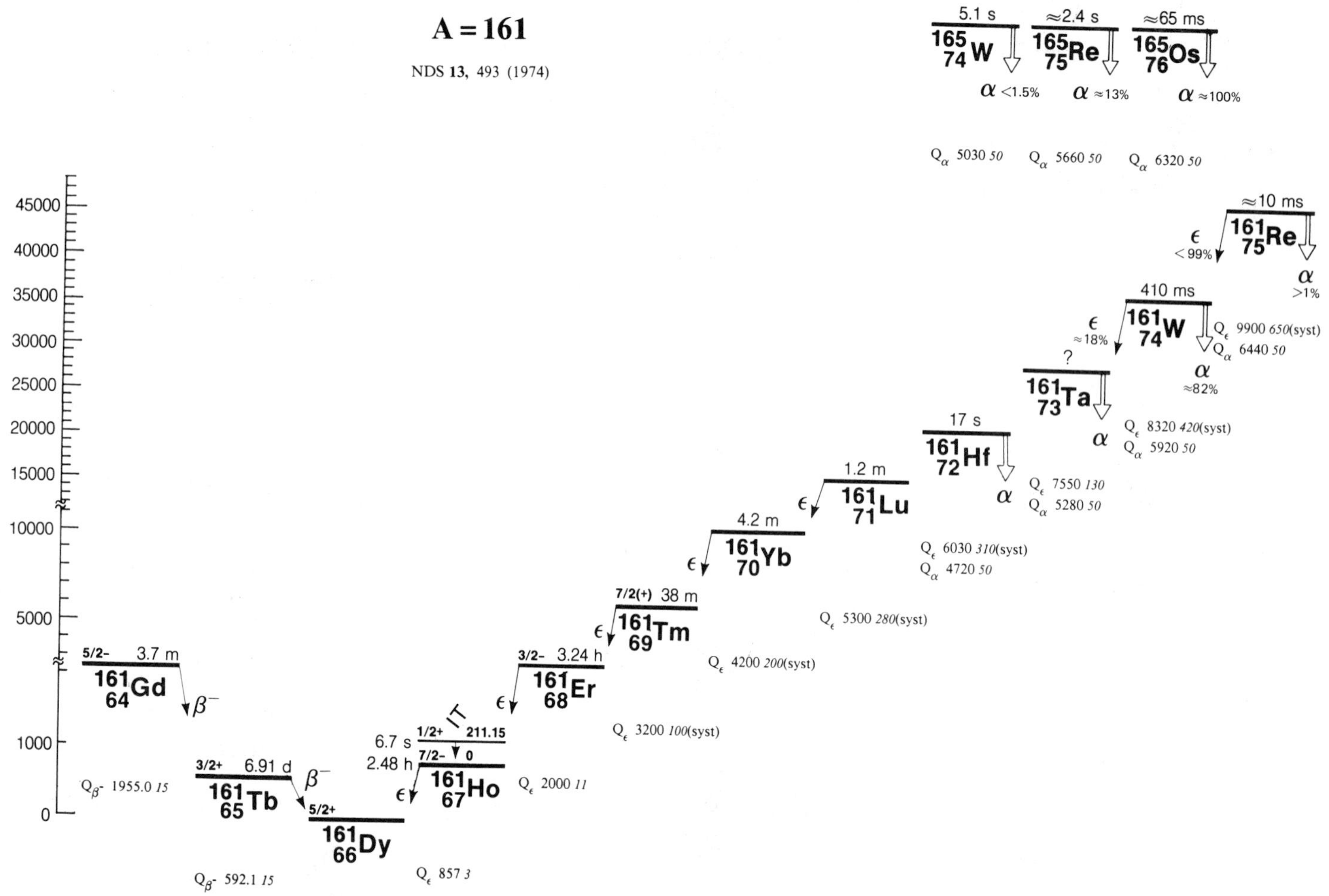

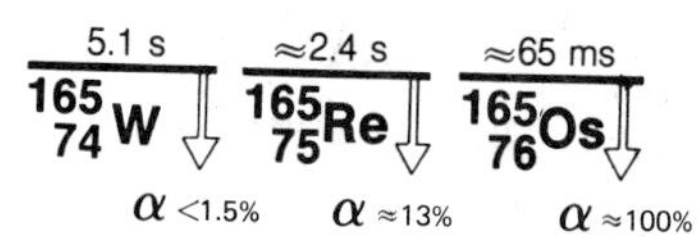

$^{161}_{64}$Gd(3.7 *1* min)

Mode: β-

Δ: -65518 *4* keV

SpA: 3.15×10^8 Ci/g

Prod: ^{160}Gd(n,γ)

Photons (^{161}Gd)

⟨γ⟩=391 *6* keV

γ_{mode}	γ(keV)	γ(%)†
Tb L_ℓ	5.546	0.160 *22*
Tb L_α	6.269	4.2 *5*
Tb L_η	6.284	0.058 *7*
Tb L_β	7.113	3.8 *5*
Tb L_γ	8.184	0.58 *8*
Tb $K_{\alpha 2}$	43.744	13.3 *7*
Tb $K_{\alpha 1}$	44.482	23.9 *12*
Tb $K_{\beta 1}$'	50.338	7.1 *4*
Tb $K_{\beta 2}$'	51.958	2.13 *11*
γ M1	56.288 *9*	3.80 *16*
γ [M1+E2]	62.900 *12*	0.060 *15*
γ	71.57 *3*	0.060 *15*
γ M1(+E2)	77.394 *8*	1.07 *10*
γ	79.41 *4*	0.06 *1*
γ [E1]	85.765 *19*	0.15 *3*
γ	89.43 *15*	0.05 *1*
γ	97.04 *7*	0.125 *20*
γ E1	102.316 *9*	14.0 *8*
γ	105.643 *14*	0.74 *10*
γ	121.7 *3*	0.010 *3*
γ [E2]	133.682 *10*	0.016 *2*

Photons (^{161}Gd) (continued)

γ_{mode}	γ(keV)	γ(%)†
γ E1	165.216 *11*	2.6 *2*
γ	168.543 *15*	0.082 *12*
γ M1+E2	181.234 *9*	0.75 *3*
γ	191.407 *18*	0.64 *3*
γ M1	258.627 *10*	0.99 *4*
γ	270.858 *14*	0.88 *4*
γ [E1]	283.549 *11*	6.00 *25*
γ M1+E2	314.915 *11*	22.9 *9*
γ [M1+E2]	338.078 *16*	1.7 *1*
γ [E1]	360.943 *11*	60.6 *15*
γ [E2]	394.366 *18*	0.22 *2*
γ [M2]	417.231 *12*	0.31 *4*
γ [E1]	423.843 *12*	0.18 *3*
γ	452.091 *15*	0.06 *1* ?
γ [E1]	480.131 *12*	2.70 *15*
γ	529.485 *14*	1.27 *7*
γ	772.18 *10*	0.0036 *3*
γ	818.9 *3*	0.0007 *2* ?
γ	832.0 *3*	0.0010 *2*
γ	835.0 *3*	0.0009 *2*
γ	857.93 *11*	0.00210 *15*
γ	866.5 *4*	0.00040 *15* ?
γ	869.30 *16*	0.00097 *15*
γ	911.53 *12*	0.0019 *2*
γ	924.55 *12*	0.0019 *2*
γ	928.42 *14*	0.0013 *2*
γ	932.85 *10*	0.0029 *2*
γ	937.53 *9*	0.0092 *5*
γ	947.75 *24*	0.00053 *6*
γ	951.10 *22*	0.00055 *6*
γ	955.35 *8*	0.0079 *4*
γ	972.3 *8*	0.00024 *8* ?
γ	979.37 *14*	0.0011 *1*
γ	1012.9 *1*	0.0030 *2*

Photons (^{161}Gd) (continued)

γ_{mode}	γ(keV)	γ(%)†
γ	1015.10 *14*	0.00120 *15*
γ	1026.25 *10*	0.0028 *2*
γ	1034.72 *8*	0.0274 *7*
γ	1038.20 *15*	0.00115 *10* ?
γ	1044.30 *22*	0.00055 *10*
γ	1048.75 *12*	0.0019 *2*
γ	1053.7 *3*	0.00045 *10* ?
γ	1057.06 *10*	0.0036 *2*
γ	1063.4 *2*	0.00073 *12*
γ	1066.22 *12*	0.00145 *15*
γ	1071.5 *4*	0.00035 *6* ?
γ	1093.52 *9*	0.0064 *4*
γ	1105.84 *13*	0.00133 *12*
γ	1112.20 *15*	0.0018 *5*
γ	1113.49 *10*	0.0053 *8*
γ	1117.15 *14*	0.0012 *2*
γ	1120.92 *10*	0.0042 *3*
γ	1126.30 *25*	0.00064 *15*
γ	1135.2 *4*	0.00038 *8*
γ	1143.15 *12*	0.00200 *15*
γ	1145.50 *18*	0.0009 *2*
γ	1149.94 *9*	0.0120 *6*
γ	1153.43 *12*	0.00155 *20*
γ	1160.09 *12*	~0.0016
γ	1171.4 *5*	0.00030 *12*
γ	1175.82 *11*	0.0020 *2*
γ	1181.4 *5*	0.00018 *6* ?
γ	1186.06 *18*	0.00073 *10*
γ	1192.42 *15*	0.00103 *15*
γ	1197.07 *13*	0.00140 *15*
γ	1204.8 *5*	0.00022 *7*
γ	1209.72 *11*	0.00187 *15*
γ	1219.6 *4*	0.00036 *10*
γ	1224.93 *20*	0.00056 *10*

Photons (^{161}Gd) (continued)

γ_{mode}	γ(keV)	γ(%)†
γ	1252.42 12	0.00181 12
γ	1258.88 13	0.00126 8
γ	1261.11 11	0.00195 10
γ	1271.8 5	0.00019 5
γ	1286.4 4	0.00039 7
γ	1297.9 2	0.00073 12
γ	1300.9 4	0.00036 9
γ	1308.27 10	0.00296 20
γ	1323.0 5	0.00032 7
γ	1341.10 12	0.00166 15
γ	1344.2 5	0.00033 7 ?
γ	1349.60 9	0.0051 3
γ	1354.9 3	0.00043 7
γ	1357.8 3	0.00045 7
γ	1364.19 13	0.00119 13
γ	1373.2 5	0.00027 6 ?
γ	1376.7 3	0.00041 6
γ	1379.74 13	0.00165 12
γ	1400.13 12	0.00175 12
γ	1421.37 15	0.00089 7
γ	1424.3 2	0.00066 5
γ	1430.7 6	0.00012 4 ?
γ	1433.82 16	0.00085 6
γ	1439.5 4	0.00035 5
γ	1459.5 5	0.00027 5 ?
γ	1472.45 20	0.00059 6
γ	1477.55 15	0.00136 10
γ	1480.9 3	0.00040 5
γ	1489.42 15	0.0014 1
γ	1495.82 16	0.00082 7
γ	1501.8 2	0.00058 5
γ	1533.87 15	0.00138 10
γ	1538.7 5	0.00024 6
γ	1544.80 14	0.00116 8
γ	1547.5 5	0.00024 6 ?
γ	1558.33 15	0.00099 8
γ	1567.0 6	0.00012 3 ?
γ	1590.5 5	0.00024 5
γ	1600.55 15	0.00420 25
γ	1622.95 15	0.00212 15
γ	1677.3 4	0.00028 5
γ	1691.7 4	0.00029 5
γ	1715.5 3	0.00035 5 ?
γ	1738.8 3	0.00033 5
γ	1754.1 5	0.000110 25 ?
γ	1768.6 5	0.00022 4 ?

† 2.5% uncert(syst)

Atomic Electrons (^{161}Gd)

$\langle e \rangle$=25.3 18 keV

e_{bin}(keV)	$\langle e \rangle$(keV)	e(%)
4	1.65	38.4 18
8	2.6	34 5
9 - 20	0.36	3.9 4
25	0.8	3.2 11
27 - 45	1.27	3.37 23
48	2.66	5.6 3
49	0.055	0.112 8
50	1.68	3.35 20
54	1.0	1.8 6
55 - 64	0.61	1.1 4
69	0.9	~1
70	0.7	~1
71 - 120	2.2	2.4 6
121 - 169	0.50	0.37 10
173 - 219	0.60	0.30 6
232 - 262	0.308	0.130 8
263	3.6	1.4 5
269 - 286	0.30	0.11 3
306	0.53	0.17 7
307	0.22	~0.07
309	1.68	0.54 3
313 - 361	0.64	0.19 3
365 - 410	0.186	0.050 6
415 - 452	0.063	0.0148 9
471 - 521	0.10	~0.020

Atomic Electrons (^{161}Gd) (continued)

e_{bin}(keV)	$\langle e \rangle$(keV)	e(%)
522 - 529	0.0044	~0.0008
720 - 767	0.00017	~2 $\times10^{-5}$
770 - 819	0.00018	2.2 7 $\times10^{-5}$
823 - 873	0.00014	1.6 6 $\times10^{-5}$
876 - 925	0.00066	7.3 25 $\times10^{-5}$
926 - 974	0.00031	3.2 8 $\times10^{-5}$
977 - 1026	0.0010	0.00010 4
1027 - 1074	0.00054	5.1 13 $\times10^{-5}$
1083 - 1129	0.00054	4.9 15 $\times10^{-5}$
1133 - 1181	0.00021	1.8 4 $\times10^{-5}$
1184 - 1225	0.00012	1.0 3 $\times10^{-5}$
1234 - 1279	0.00011	9 3 $\times10^{-6}$
1284 - 1334	0.00022	1.7 5 $\times10^{-5}$
1335 - 1382	0.000104	7.6 18 $\times10^{-6}$
1388 - 1434	6.1 $\times10^{-5}$	4.3 10 $\times10^{-6}$
1437 - 1487	8.2 $\times10^{-5}$	5.6 13 $\times10^{-6}$
1488 - 1583	0.00015	9.5 25 $\times10^{-6}$
1589 - 1687	3.4 $\times10^{-5}$	2.1 4 $\times10^{-6}$
1690 - 1767	6.7 $\times10^{-6}$	3.9 11 $\times10^{-7}$

Continuous Radiation (^{161}Gd)

$\langle\beta-\rangle$=559 keV;$\langle$IB$\rangle$=0.84 keV

E_{bin}(keV)		$\langle\ \rangle$(keV)	(%)
0 - 10	β-	0.0356	0.71
	IB	0.024	
10 - 20	β-	0.108	0.72
	IB	0.023	0.163
20 - 40	β-	0.444	1.48
	IB	0.045	0.157
40 - 100	β-	3.33	4.73
	IB	0.122	0.19
100 - 300	β-	37.4	18.4
	IB	0.29	0.167
300 - 600	β-	138	30.6
	IB	0.23	0.055
600 - 1300	β-	363	42.1
	IB	0.107	0.0144
1300 - 1640	β-	17.0	1.24
	IB	0.00039	2.9 $\times10^{-5}$

$^{161}_{65}$Tb(6.91 2 d)

Mode: β-

Δ: -67473 4 keV

SpA: 1.174$\times10^{5}$ Ci/g

Prod: daughter ^{161}Gd

Photons (^{161}Tb)

$\langle\gamma\rangle$=33.8 24 keV

γ_{mode}	γ(keV)	γ(%)†
Dy L_{ℓ}	5.743	0.27 5
Dy L_{α}	6.491	6.9 10
Dy L_{η}	6.534	0.106 15
Dy L_{β}	7.353	9.8 16
Dy L_{γ}	8.603	1.9 4
γ E1	25.667 3	21.0 21
γ M1	28.729 9	0.058 12
γ M1+4.2%E2	43.841 9	0.060 12
Dy $K_{\alpha2}$	45.208	6.1 10
Dy $K_{\alpha1}$	45.998	10.8 17
γ M1+0.4%E2	48.930 4	16 3
Dy $K_{\beta1}$'	52.063	3.3 5
Dy $K_{\beta2}$'	53.735	0.97 15

Photons (^{161}Tb) (continued)

γ_{mode}	γ(keV)	γ(%)†
γ [M1]	57.212 5	1.6 3
γ E1	59.239 13	0.020 5
γ E1	74.597 4	9.8 20
γ E2+48%M1	77.413 10	0.068 10
γ [E1]	87.968 11	0.21 3
γ E1	103.080 10	0.114 9
γ M1	106.142 6	0.088 8
γ [E2]	138.34 8	0.0108 20
γ [M1+E2]	238.44 8	0.00186 19
γ [M1+E2]	286.33 6	0.0123 13
γ [M1+E2]	292.30 7	0.064 7
γ [E2]	315.06 6	0.00039 10
γ [M1+E2]	319.56 7	0.0031 4
γ [E2]	341.23 7	0.0034 4
γ [M1+E2]	343.54 6	0.0135 14
γ [M1+E2]	348.29 7	0.00039 10
γ [M1+E2]	376.77 7	0.00049 10
γ [M1+E2]	392.47 6	0.0022 3
γ [E1]	418.62 6	0.0084 8
γ [M1+E2]	425.70 7	0.00029 10
γ [E1]	475.83 6	0.0186 19
γ [E2]	506.58 6	0.00078 10
γ [M1+E2]	550.42 6	0.039 4

† 10% uncert(syst)

Atomic Electrons (^{161}Tb)

$\langle e \rangle$=42 3 keV

e_{bin}(keV)	$\langle e \rangle$(keV)	e(%)
3 - 5	0.56	16 3
8	1.8	23 4
9	3.3	37 6
17	4.1	24.5 25
18	2.7	15.0 15
20 - 21	1.27	6.1 11
24	2.16	9.0 9
25 - 38	1.05	3.64 23
40	15.8	40 8
41 - 46	0.95	2.3 4
47	4.3	9.1 18
48	1.08	2.2 5
49	1.4	2.8 6
50 - 98	1.46	2.30 19
101 - 138	0.0133	0.0117 6
185 - 233	0.0030	0.0013 4
236 - 285	0.016	0.0064 18
286 - 335	0.0039	0.0013 4
336 - 385	0.00076	0.00022 4
390 - 426	0.00045	0.000107 10
453 - 499	0.0030	0.00062 22
505 - 550	0.00063	0.00012 3

Continuous Radiation (^{161}Tb)

$\langle\beta-\rangle$=155 keV;$\langle$IB$\rangle$=0.083 keV

E_{bin}(keV)		$\langle\ \rangle$(keV)	(%)
0 - 10	β-	0.201	4.03
	IB	0.0079	
10 - 20	β-	0.59	3.96
	IB	0.0072	0.050
20 - 40	β-	2.33	7.8
	IB	0.0125	0.044
40 - 100	β-	15.2	21.9
	IB	0.026	0.042
100 - 300	β-	94	50.3
	IB	0.027	0.018
300 - 592	β-	42.3	11.8
	IB	0.0017	0.00050

$^{161}_{66}$Dy(stable)

Δ: -68065 *3* keV
%: 18.9 *1*

$^{161}_{67}$Ho(2.48 *5* h)

Mode: ε
Δ: -67208 *4* keV
SpA: 7.85×10^6 Ci/g
Prod: ^{159}Tb(α,2n); protons on Dy

Photons (^{161}Ho)

⟨γ⟩=58.7 *18* keV

γ_{mode}	γ(keV)	γ(%)
Dy L_ℓ	5.743	0.47 *7*
Dy L_α	6.491	12.3 *13*
Dy L_η	6.534	0.169 *20*
Dy L_β	7.375	12.4 *16*
Dy L_γ	8.538	2.0 *3*
γ E1	25.667 *3*	29 *6*
γ M1+4.2%E2	43.841 *9*	0.115 *23*
Dy $K_{\alpha2}$	45.208	25.0 *9*
Dy $K_{\alpha1}$	45.998	44.6 *16*
γ M1+0.4%E2	48.930 *4*	0.078 *16*
Dy $K_{\beta1}$'	52.063	13.5 *5*
Dy $K_{\beta2}$'	53.735	3.97 *16*
γ E1	59.239 *13*	1.17 *23*
γ E1	74.597 *4*	0.048 *10*
γ E2+48%M1	77.413 *10*	2.2 *4*
γ M1+48%E2	98.1 *6*	0.47 *8*
γ E1	103.080 *10*	3.6 *4*
γ (E1)	157.3 *6*	0.47 *8*
γ E2	175.5 *6*	0.47 *8*
γ	210.4 *10*	
γ	234.5 *10*	
γ	239.7 *10*	
γ	339.7 *10*	
γ	416 *1*	
γ	760 *1*	

Atomic Electrons (^{161}Ho)

⟨e⟩=32.0 *18* keV

e_{bin}(keV)	⟨e⟩(keV)	e(%)
5	0.064	1.17 *24*
8	4.4	57 *8*
9	3.9	45 *6*
17	5.5	33 *6*
18	3.6	20 *4*
21	0.0055	0.027 *5*
24	4.4	18 *3*
25 - 68	4.83	12.4 *6*
69	1.5	2.2 *4*
70	1.6	2.3 *4*
73 - 75	0.134	0.18 *3*
76	0.82	1.09 *20*
77 - 122	1.06	1.16 *8*
148 - 175	0.118	0.071 *5*

Continuous Radiation (^{161}Ho)

⟨IB⟩=0.27 keV

E_{bin}(keV)		⟨ ⟩(keV)	(%)
10 - 20	IB	0.00017	0.00112
20 - 40	IB	0.023	0.061
40 - 100	IB	0.161	0.34
100 - 300	IB	0.029	0.0152
300 - 600	IB	0.054	0.0129
600 - 831	IB	0.0068	0.00106

$^{161}_{67}$Ho(6.73 *10* s)

Mode: IT
Δ: -66997 *4* keV
SpA: 9.89×10^9 Ci/g
Prod: daughter ^{161}Er; ^{162}Dy(p,2n); ^{161}Dy(p,n)

Photons (^{161}Ho)

⟨γ⟩=104 *5* keV

γ_{mode}	γ(keV)	γ(%)†
Ho L_ℓ	5.943	0.133 *16*
Ho L_α	6.716	3.4 *3*
Ho L_η	6.789	0.074 *7*
Ho L_β	7.622	4.2 *4*
Ho L_γ	8.788	0.66 *7*
Ho $K_{\alpha2}$	46.700	5.5 *3*
Ho $K_{\alpha1}$	47.547	9.7 *6*
Ho $K_{\beta1}$'	53.822	2.99 *18*
Ho $K_{\beta2}$'	55.556	0.86 *5*
γ E3	211.15 *3*	44.7 *22*

† 2.2% uncert(syst)

Atomic Electrons (^{161}Ho)

⟨e⟩=106 *3* keV

e_{bin}(keV)	⟨e⟩(keV)	e(%)
8	1.28	15.9 *18*
9	1.36	15.2 *17*
37	0.091	0.25 *3*
38	0.051	0.133 *15*
39	0.136	0.35 *4*
44	0.042	0.096 *11*
45	0.046	0.102 *12*
46	0.063	0.138 *16*
47	0.0141	0.030 *3*
48	0.00116	0.0024 *3*
51	0.00197	0.0038 *4*
52	0.0110	0.0212 *24*
53	0.0045	0.0085 *10*
54	0.00180	0.0034 *4*
156	31.5	20.3 *11*
202	35.2	17.4 *10*
203	18.9	9.3 *5*
209	13.7	6.6 *4*
210	0.125	0.060 *3*
211	3.72	1.76 *10*

$^{161}_{68}$Er(3.24 *4* h)

Mode: ε
Δ: -65209 *11* keV
SpA: 6.01×10^6 Ci/g
Prod: protons on Er; ^{165}Ho(p,5n); protons on Ta; ^{162}Er(γ,n); ^{162}Dy(α,5n); ^{161}Dy(α,4n)

Photons (^{161}Er)

⟨γ⟩=996 *110* keV

γ_{mode}	γ(keV)	γ(%)†
γ M1+E2	11.304 *23*	~0.12
Ho $K_{\alpha2}$	46.700	25.7 *9*
Ho $K_{\alpha1}$	47.547	45.7 *15*
Ho $K_{\beta1}$'	53.822	14.0 *5*
Ho $K_{\beta2}$'	55.556	4.02 *15*
γ	74.59 *10*	~0.10
γ M1+E2	76.211 *22*	1.13 *24*
γ M1	87.515 *22*	0.36 *5*
γ M1+E2	89.97 *6*	0.040 *8*
γ M1+E2	94.124 *23*	0.90 *7*
γ M1+E2	99.62 *3*	0.43 *11*
γ M1	102.01 *3*	0.024 *5*
γ [E2]	105.43 *3*	0.067 *13*
γ E1	105.59 *4*	0.031 *6*
γ	107.27 *9*	0.049 *10*
γ M1+E2	109.95 *6*	0.045 *10*
γ (M1)	122.27 *10*	0.018 *4*
γ [E1]	125.25 *3*	0.0122 *24*
γ E2	130.84 *3*	0.59 *3*
γ E2	148.16 *3*	~0.22
γ M1	150.82 *5*	~0.04
γ [E1]	152.66 *5*	0.126 *16*
γ [E1]	153.04 *4*	0.051 *10*
γ [E2]	162.12 *5*	0.033 *10*
γ [E2]	164.58 *6*	0.018 *4*
γ	180.83 *10*	0.0122 *24*
γ E1	201.461 *23*	1.09 *5*
γ E1	209.349 *24*	0.94 *20*
γ E3	211.15 *3*	12.2 *5* §
γ E1	212.765 *24*	0.83 *17*
γ [E1]	219.37 *5*	0.077 *12*
γ E1	236.43 *3*	0.486 *22*
γ	247.2 *7*	0.0122 *24*
γ E1	252.66 *3*	0.464 *24*
γ [E1]	270.9 *3*	0.0122 *24*
γ E1	276.06 *4*	0.107 *12*
γ E1	293.97 *3*	0.429 *22*
γ	301.43 *10*	~0.018
γ E1	303.473 *23*	0.322 *22*
γ	309.1 *3*	0.072 *21*
γ E1	314.78 *3*	2.44 *10*
γ	346.69 *20*	0.073 *15*
γ E1	350.41 *10*	0.082 *11*
γ [E1]	363.29 *11*	0.060 *10*
γ [E1]	370.6 *3*	0.083 *11*
γ E1	376.75 *17*	0.143 *12*
γ	421.69 *10*	0.381 *24*
γ E1	446.82 *4*	0.43 *3*
γ	454.47 *10*	0.050 *12*
γ	467.99 *20*	0.273 *17*
γ M1	489.0 *4*	0.087 *16*
γ	499.8 *4*	0.041 *16*
γ	503.30 *21*	0.037 *16*
γ M1	507.65 *17*	0.36 *3*
γ E1	527.86 *9*	0.406 *23*
γ E2	549.44 *10*	0.355 *24*
γ	554.09 *9*	0.094 *22*
γ [E1]	573.87 *9*	~0.037
γ (E2)	592.63 *4*	3.98 *17*
γ	625.7 *4*	0.09 *3*
γ (E2)	649.07 *10*	0.88 *6*
γ	662.6 *3*	0.139 *23*
γ	690.79 *20*	0.35 *3*
γ	718.8 *4*	0.104 *17*
γ E2	726.90 *9*	0.95 *6*
γ	737.0 *3*	0.254 *23*
γ	745.4 *5*	0.052 *20*
γ	747.88 *15*	0.043 *18*

Photons (^{161}Er) (continued)

γ_{mode}	γ(keV)	γ(%)†
γ	767.1 *6*	0.043 *18*
γ	783.77 *18*	0.059 *18*
γ M1+E2	799.19 *20*	0.16 *3*
γ (E2)	804.31 *12*	0.34 *5*
γ	808.95 *14*	0.38 *5*
γ	811.72 *15*	0.49 *5*
γ M1	826.52 *9*	67 *13*
γ	831.3 *5*	0.10 *5*
γ (M1)	839.28 *12*	0.107 *23*
γ [M1+E2]	842.01 *17*	0.067 *23*
γ	859.2 *7*	~0.033
γ M1+E2	864.99 *8*	1.44 *9*
γ M1+E2	868.53 *11*	0.37 *3*
γ	871.02 *12*	0.084 *18*
γ M1	875.52 *15*	0.38 *3*
γ	878.43 *15*	0.245 *24*
γ	880.05 *20*	~0.032
γ	884.40 *18*	0.055 *17*
γ M1	895.72 *8*	0.89 *5*
γ [M1+E2]	898.44 *15*	0.089 *21*
γ	904.4 *9*	~0.029
γ	913.89 *19*	<0.044
γ	923.19 *23*	~0.029
γ M1	931.70 *7*	2.07 *9*
γ	935.59 *11*	0.09 *3*
γ (M1)	938.07 *13*	0.08 *3*
γ	941.10 *23*	0.145 *17*
γ	948.5 *6*	0.272 *18*
γ	951.85 *20*	~0.028
γ	954.53 *22*	~0.031
γ	962.43 *9*	0.155 *23*
γ [M1+E2]	965.15 *15*	~0.033
γ	970.54 *11*	0.112 *20*
γ	973.03 *12*	0.299 *24*
γ	980.20 *14*	0.38 *3*
γ	998.75 *8*	0.176 *18*
γ	1008.8 *5*	0.041 *16*
γ	1010.41 *14*	0.102 *15*
γ	1017.86 *24*	0.034 *15*
γ	1021.72 *14*	0.085 *13*
γ	1028.61 *23*	0.037 *15*
γ	1037.60 *11*	0.056 *15*
γ	1047.72 *13*	0.083 *12*
γ	1061.35 *13*	0.066 *15*
γ	1064.44 *9*	0.118 *15*
γ	1077.88 *11*	0.102 *13*
γ	1088.24 *12*	0.044 *12*
γ	1098.28 *12*	0.264 *22*
γ E1	1102.66 *20*	0.215 *17*
γ	1106.6 *6*	~0.028
γ	1110.4 *6*	0.032 *12*
γ E1	1115.08 *9*	0.051 *4*
γ E1	1117.81 *15*	0.222 *21*
γ E1	1144.94 *11*	0.72 *5*
γ	1147.78 *17*	0.138 *24*
γ E1	1158.96 *7*	0.55 *4*
γ	1162.85 *11*	~0.023
γ	1172.00 *11*	0.38 *4*
γ	1174.49 *12*	1.32 *9*
γ	1183.31 *11*	0.33 *3*
γ	1185.79 *12*	0.42 *3*
γ	1189.69 *9*	0.078 *16*
γ E1	1193.21 *15*	0.61 *4*
γ	1199.4 *5*	0.028 *12*
γ	1202.4 *5*	~0.023
γ E1	1209.85 *11*	0.41 *3*
γ	1228.49 *14*	0.124 *15*
γ	1237.38 *23*	~0.017
γ	1239.06 *11*	1.183 *24*
γ	1246.47 *7*	0.296 *24*
γ	1250.36 *11*	0.43 *3*
γ [E1]	1268.63 *15*	0.245 *18*
γ	1277.20 *9*	0.139 *22*
γ (E1)	1279.93 *15*	0.60 *4*
γ [E1]	1283.41 *12*	0.024 *10*
γ	1287.1 *5*	0.035 *10*
γ	1293.60 *17*	0.022 *9*
γ	1298.89 *13*	~0.022
γ E1	1303.39 *12*	0.46 *3*
γ	1313.9 *9*	0.026 *10*
γ	1318.13 *13*	0.124 *16*
γ	1333.3 *9*	~0.010
γ E1	1338.11 *12*	0.27 *3*
γ	1341.69 *12*	0.072 *24*

Photons (^{161}Er) (continued)

γ_{mode}	γ(keV)	γ(%)†
γ	1342.8 *3*	0.055 *18*
γ	1352.47 *16*	~0.021
γ (E1)	1358.01 *11*	0.66 *5*
γ	1361.55 *16*	0.165 *23*
γ	1372.45 *13*	0.026 *11*
γ	1374.82 *12*	0.079 *16*
γ	1376.65 *14*	0.081 *15*
γ	1383.19 *23*	0.144 *15*
γ	1387.13 *17*	0.054 *12*
γ	1392.73 *12*	0.087 *10*
γ	1404.4 *5*	0.018 *6*
γ	1417.90 *12*	0.68 *5*
γ	1421.1 *4*	0.063 *16*
γ	1424.96 *12*	0.066 *12*
γ	1429.21 *12*	0.354 *24*
γ [E1]	1434.22 *11*	0.220 *20*
γ	1447.05 *12*	0.033 *7*
γ	1452.86 *14*	0.051 *7*
γ	1457.62 *8*	~0.016
γ	1461.51 *12*	0.126 *15*
γ	1464.16 *14*	0.270 *18*
γ	1468.94 *12*	0.106 *10*
γ	1477.72 *16*	0.033 *11*
γ	1480.25 *12*	0.067 *13*
γ	1488.35 *9*	0.190 *15*
γ	1492.39 *16*	0.184 *15*
γ	1495.4 *3*	~0.012
γ	1517.96 *17*	0.054 *9*
γ	1524.58 *12*	~0.032
γ	1525.7 *6*	~0.016
γ	1527.6 *9*	~0.016
γ	1531.35 *23*	~0.015
γ	1534.56 *12*	0.016 *6*
γ	1549.17 *23*	~0.009
γ	1553.93 *16*	0.170 *18*
γ	1565.23 *16*	0.016 *5*
γ	1596.4 *4*	0.015 *5*
γ	1613.8 *3*	0.018 *5*
γ	1625.38 *23*	0.013 *5*
γ	1640.35 *13*	0.022 *5*
γ [M1+E2]	1656.67 *12*	0.59 *4*
γ	1691.39 *12*	<0.017
γ	1714.83 *16*	0.032 *9*
γ M1	1740.41 *17*	0.43 *4*
γ	1817.90 *22*	~0.006
γ	1830.01 *23*	0.057 *6*
γ	1868.7 *3*	0.0171 *24*

† 16% uncert(syst)
§ with ^{161}Ho(6.7 s) in equilib

Atomic Electrons (^{161}Er)

$\langle e \rangle$=45.6 *21* keV

e_{bin}(keV)	$\langle e \rangle$(keV)	e(%)
19 - 38	2.0	7.3 *20*
39	1.4	3.5 *9*
44 - 66	1.45	3.1 *4*
67	1.3	~2
68 - 117	4.1	5.1 *17*
120 - 155	0.62	0.48 *3*
156	8.6	5.5 *3*
157 - 201	0.134	0.076 *7*
202	9.6	4.75 *23*
203	5.16	2.54 *12*
204 - 208	0.0044	0.00215 *25*
209	3.75	1.79 *9*
210 - 259	1.20	0.556 *24*
262 - 309	0.039	0.0130 *25*
313 - 362	0.0143	0.0044 *4*
363 - 412	0.08	0.020 *9*
413 - 460	0.071	0.0160 *14*
466 - 508	0.045	0.0092 *11*
518 - 566	0.188	0.0351 *20*
570 - 618	0.094	0.0159 *12*
624 - 671	0.065	0.0099 *19*
681 - 729	0.033	0.0047 *15*
735 - 767	0.053	0.0071 *22*

Atomic Electrons (^{161}Er) (continued)

e_{bin}(keV)	$\langle e \rangle$(keV)	e(%)
771	4.1	0.54 *11*
774 - 823	0.75	0.093 *15*
824 - 874	0.26	0.031 *3*
875 - 924	0.176	0.0199 *12*
925 - 973	0.039	0.0041 *9*
978 - 1027	0.015	0.0015 *4*
1028 - 1078	0.016	0.0016 *4*
1079 - 1128	0.067	0.0060 *18*
1130 - 1178	0.038	0.0033 *6*
1180 - 1230	0.061	0.0051 *15*
1231 - 1280	0.0152	0.00122 *19*
1281 - 1330	0.023	0.00173 *25*
1331 - 1379	0.030	0.0022 *6*
1381 - 1429	0.018	0.0013 *3*
1432 - 1480	0.012	0.00086 *20*
1483 - 1570	0.0059	0.00039 *11*
1585 - 1683	0.015	0.00096 *17*
1685 - 1774	0.0125	0.00074 *6*
1809 - 1867	0.00044	2.4 *8* $\times 10^{-5}$

Continuous Radiation (^{161}Er)

$\langle \beta+ \rangle$=0.491 keV;$\langle$IB$\rangle$=0.52 keV

E_{bin}(keV)		$\langle\ \rangle$(keV)	(%)
0 - 10	$\beta+$	2.65×10^{-8}	3.14×10^{-7}
	IB	0.00082	
10 - 20	$\beta+$	1.39×10^{-6}	8.3×10^{-6}
	IB	0.00019	0.00129
20 - 40	$\beta+$	4.91×10^{-5}	0.000149
	IB	0.0128	0.036
40 - 100	$\beta+$	0.00291	0.00370
	IB	0.19	0.38
100 - 300	$\beta+$	0.120	0.056
	IB	0.034	0.017
300 - 600	$\beta+$	0.330	0.078
	IB	0.106	0.024
600 - 1300	$\beta+$	0.0375	0.0058
	IB	0.168	0.021
1300 - 1789	IB	0.0111	0.00079
	$\Sigma\beta+$		0.144

$^{161}_{69}$Tm(38 *4* min)

Mode: ϵ

Δ: -62010 *100* keV syst

SpA: 3.1×10^{7} Ci/g

Prod: ^{162}Er(p,2n); protons on Ta

Photons (^{161}Tm)

$\langle \gamma \rangle$=1076 *32* keV

γ_{mode}	γ(keV)	γ(%)
Er L_{ℓ}	6.151	0.83 *9*
Er L_{α}	6.944	21.3 *17*
Er L_{η}	7.058	0.28 *3*
Er L_{β}	7.934	20.7 *22*
Er L_{γ}	9.188	3.4 *4*
γ [M1+0.4%E2]	16.661 *25*	0.0030 *11*
γ [E2]	23.473 *17*	0.0029 *4*
γ [M1+0.9%E2]	27.912 *20*	0.43 *5*
γ [M1+40%E2]	28.164 *18*	0.00190 *23*
γ [M1+0.5%E2]	29.226 *18*	0.057 *6*
γ [E1]	40.829 *21*	0.24 *3*
γ E1	45.520 *19*	25.0 *5*
γ [E1]	46.89 *3*	0.29 *4*
Er $K_{\alpha2}$	48.221	53.9 *24*

Photons (^{161}Tm)
(continued)

γ_{mode}	γ(keV)	γ(%)
Er $K_{\alpha1}$	49.128	95 4
Er $K_{\beta1}'$	55.616	29.6 13
Er $K_{\beta2}'$	57.406	8.4 4
γ M1+0.4%E2	59.491 24	5.4 5
γ (E2)	68.08 6	0.09 4
γ [E1]	68.993 21	0.35 6
γ [E1]	73.432 20	0.15 6
γ [E2]	78.01 3	0.90 18
γ [E2]	79.324 25	0.15 5
γ M1(+5.5%E2)	84.378 19	9.4 9
γ E1	87.20 6	~0.15
γ M1+14%E2	94.361 21	1.2 5
γ E1	99.74 3	2.37 25
γ M1+5.0%E2	105.865 17	3.4 3
γ [E2+40%M1]	107.235 24	0.51 7
γ M1	112.542 19	5.1 4
γ M1+3.2%E2	122.526 23	1.55 12
γ M1+25%E2	123.75 4	0.35 10
γ (E1)	125.58 6	1.58 13
γ E1	128.96 4	2.95 25
γ [M1+5.0%E2]	138.67 5	0.60 7
γ [M1+18%E2]	140.41 4	0.42 6
γ E2	143.87 3	3.8 8
γ M1+5.0%E2	146.63 4	4.8 4
γ [E1]	153.371 24	3.00 25
γ [M1+11%E2]	156.55 5	0.71 7
γ [E1]	157.809 25	1.80 16
γ M1	172.03 3	5.1 4
γ [E1]	172.85 4	0.55 15
γ M1	181.98 9	0.12 3
γ E2	190.242 23	3.4 3
γ [E1]	197.38 5	~0.12
γ [M1+E2]	206.903 24	0.60 18
γ [E1]	206.97 4	2.4 3
γ [E1]	212.86 3	3.2 3
γ M1	215.68 6	1.57 15
γ [E2]	218.11 3	1.05 10
γ E1	220.08 10	0.30 4
γ E2	241.9 3	0.10 3
γ [E2]	244.53 5	1.10 11
γ [M1+E2]	246.28 3	<0.10
γ E2	248.5 4	0.10 3
γ [M1+E2]	250.18 9	0.74 8
γ (E2)	252.49 4	1.55 14
γ M1	260.88 10	0.37 4
γ E2	263.88 10	0.50 6
γ M1	265.44 10	1.04 11
γ [E2]	266.39 3	0.66 7
γ M1	270.18 10	0.17 4
γ M1	272.05 10	0.75 10
γ [M1]	278.92 7	0.81 10
γ M1	280.98 10	0.22 5
γ M1+39%E2	283.36 7	0.83 10
γ [E1]	309.92 5	0.30 4
γ E1	325.78 20	0.27 3
γ E1	330.58 10	0.64 7
γ [E1]	369.41 5	1.40 12
γ	371.18 20	0.45 6
γ M1	372.71 11	1.05 11
γ M1	377.14 11	0.49 6
γ M1	400.62 11	0.71 8
γ [E1]	403.55 9	0.17 3
γ	407.6 4	0.33 4
γ	419.6 5	0.26 7
γ	425.6 5	0.23 5
γ	433.2 4	0.36 6
γ	436.73 7	0.15 5
γ	447.1 4	0.15 3
γ E1	454.3 4	0.42 8
γ	458.39 20	0.33 4
γ [E1]	463.04 9	0.41 5
γ	476.0 5	~0.08
γ	483.3 4	0.43 5
γ	489.5 5	0.61 6
γ	496.22 7	0.10 4
γ	503.8 4	0.38 6
γ	507.2 8	0.15 5
γ	523.6 4	0.81 7
γ	540.0 5	0.24 5
γ	549.6 4	0.35 5
γ	552.75 20	0.21 5
γ	560.2 4	0.57 6
γ	574.6 4	0.47 5

Photons (^{161}Tm)
(continued)

γ_{mode}	γ(keV)	γ(%)
γ	576.72 20	0.11 4
γ	580.91 20	0.15 5
γ	593.38 20	0.28 8
γ	608.9 4	0.27 8
γ	618.3 4	0.25 8
γ	622.3 4	0.43 14
γ	644.7 4	0.28 9
γ	654.2 4	0.38 9
γ	665.29 20	0.18 6
γ	671.08 20	0.20 7
γ	680.3 5	0.22 7
γ	696.6 5	0.28 10
γ	699.25 20	0.25 11
γ	702.0 6	0.15 7
γ	712.3 4	0.18 4
γ	716.8 4	0.22 4
γ	724.78 20	0.57 7
γ	752.1 4	0.35 5
γ	762.4 4	0.38 5
γ	776.0 4	0.35 5
γ	781.2 5	0.22 5
γ	783.62 20	0.42 6
γ	799.0 5	<0.15
γ	812.3 6	<0.2
γ	840.1 5	0.45 5
γ	843.11 21	0.20 5
γ	858.0 5	0.35 8
γ	889.6 5	0.33 8
γ	891.17 22	0.16 7
γ	901.7 5	<0.41
γ	912.2 5	0.16 4
γ	916.5 5	0.18 4
γ	935.8 5	0.22 5
γ	949.2 6	0.18 6
γ	964.1 4	0.28 4
γ	970.4 5	0.26 4
γ	984.96 20	0.29 7
γ	997.5 7	<0.25
γ	1003.2 4	0.69 8
γ	1057.6 5	0.20 5
γ	1089.1 4	0.35 4
γ	1098.8 4	0.27 3
γ	1112.0 8	0.10 4
γ	1117.5 7	0.15 6
γ	1156.3 5	0.40 4
γ	1185.1 6	0.12 3
γ	1214.78 20	~0.10
γ	1223.0 4	0.34 4
γ	1235.38 24	0.40 4
γ	1250.1 5	<0.17
γ	1268.31 20	0.15 7
γ	1271.5 5	0.43 8
γ	1276.2 6	0.25 8
γ	1305.3 6	<0.41
γ	1309.14 20	0.42 6
γ	1317.0 8	~0.12
γ	1322.1 5	0.25 6
γ	1337.31 20	~0.18
γ	1342.2 3	0.24 7
γ	1351.5 6	<0.11
γ	1355.1 5	0.27 6
γ	1384.2 6	0.20 4
γ	1421.68 20	<0.18
γ	1429.1 5	<0.08
γ	1437.3 6	0.18 5
γ	1461.1 5	0.55 6
γ	1481.17 20	~0.12
γ	1514.6 4	0.72 7
γ	1519.1 5	0.36 7
γ	1537.0 8	0.15 6
γ	1540.0 8	0.15 6
γ	1552.0 8	0.15 7
γ	1555.3 8	0.15 7
γ	1565.8 8	0.15 5
γ	1570.01 18	0.43 5
γ	1578.2 5	~0.10
γ	1581.3 5	0.19 6
γ	1591.2 8	~0.08
γ	1597.7 5	0.17 6
γ	1600.4 5	~0.12
γ	1611.4 4	0.46 6
γ	1628.5 5	0.21 6
γ	1633.1 5	0.34 6

Photons (^{161}Tm)
(continued)

γ_{mode}	γ(keV)	γ(%)
γ	1639.5 8	0.10 4
γ	1648.1 3	19.5 18
γ	1663.6 5	0.56 11
γ	1693.76 18	0.42 5
γ	1706.1 4	0.35 4
γ	1718.0 8	<0.12
γ	1721.6 4	0.15 4
γ	1735.3 6	0.18 4
γ	1742.85 18	0.17 5
γ	1747.29 18	0.21 6
γ	1753.0 8	<0.08
γ	1757.9 5	0.24 5
γ	1766.1 5	0.21 6
γ	1770.77 18	0.18 6
γ	1788.12 18	1.72 15
γ	1796.51 21	<0.050
γ	1799.61 19	0.15 3
γ	1816.3 4	0.20 4
γ	1827.2 4	0.77 8
γ	1830.7 5	~0.18
γ	1834.0 4	0.47 6
γ	1845.60 21	0.64 7
γ	1849.71 18	
γ	1850.04 21	1.65 15
γ	1854.15 18	0.82 8
γ	1861.6 4	0.28 4
γ	1867.4 4	0.35 4
γ	1873.51 21	~0.10
γ	1875.9 5	0.33 4
γ	1887.8 5	<0.12
γ	1890.87 21	0.4 1
γ	1894.98 18	0.71 8
γ	1923.14 18	<0.10
γ	1934.9 7	0.08 3
γ	1941.2 5	0.15 5
γ	1952.5 6	0.09 2
γ	1958.2 5	0.11 4
γ	1984.9 5	0.13 4
γ	2007.52 18	0.11 4
γ	2010.7 5	0.19 5
γ	2043.9 6	<0.08
γ	2062.2 5	0.15 4
γ	2067.01 19	<0.06
γ	2095.2 4	0.20 6
γ	2115.0 7	0.09 4
γ	2129.5 7	0.09 4
γ	2139.0 9	<0.06
γ	2154.4 4	0.23 5
γ	2174.4 8	<0.07
γ	2190.8 7	<0.07
γ	2223.5 8	<0.07
γ	2374.0 10	<0.08

Atomic Electrons (^{161}Tm)

⟨e⟩=98.1 23 keV

e_{bin}(keV)	⟨e⟩(keV)	e(%)
2 - 7	1.11	55 5
8	7.5	90 10
9	4.3	46 5
10	1.83	18.9 24
11 - 26	4.0	22 4
27	10.1	37 4
28 - 32	0.15	0.53 20
36	2.48	6.91 20
37	2.3	6.3 13
38 - 47	5.49	13.1 5
48	3.5	7.3 7
49	0.128	0.26 3
50	4.9	9.9 9
51 - 54	0.285	0.55 3
55	5.0	9.1 7
57 - 73	7.7	11.9 8
75	4.7	6.2 10
76 - 87	6.3	7.7 9
89	3.5	3.9 3
90 - 114	5.53	5.46 19

Atomic Electrons (^{161}Tm) (continued)

e_{bin}(keV)	⟨e⟩(keV)	e(%)
115	3.2	2.78 24
116 - 164	7.0	4.91 19
170 - 219	3.04	1.56 6
220 - 269	1.17	0.494 23
270 - 317	0.44	0.146 15
320 - 369	0.40	0.116 12
370 - 420	0.18	0.047 10
423 - 468	0.22	0.050 15
474 - 522	0.17	0.035 9
523 - 573	0.11	0.021 6
574 - 623	0.084	0.014 4
635 - 680	0.09	0.014 4
687 - 726	0.091	0.013 3
741 - 791	0.049	0.0064 19
797 - 844	0.045	0.0055 19
848 - 893	0.033	0.0038 11
899 - 949	0.058	0.0062 20
954 - 1003	0.016	0.0016 5
1032 - 1081	0.027	0.0026 8
1087 - 1128	0.017	0.0015 6
1147 - 1193	0.025	0.0021 7
1205 - 1252	0.038	0.0031 9
1259 - 1309	0.031	0.0024 6
1312 - 1355	0.008	0.00063 21
1364 - 1413	0.018	0.0013 5
1419 - 1462	0.025	0.0018 6
1471 - 1570	0.049	0.0032 6
1571 - 1664	0.43	0.027 12
1678 - 1776	0.080	0.0046 10
1778 - 1877	0.102	0.0056 10
1878 - 1977	0.015	0.00080 15
1983 - 2081	0.0096	0.00047 11
2085 - 2182	0.0059	0.00028 8
2189 - 2222	9.4×10^{-5}	$\sim4\times10^{-6}$
2316 - 2372	0.0006	$\sim2\times10^{-5}$

$^{161}_{70}$Yb(4.2 *2* min)

Mode: ϵ
Δ: -57810 *220* keV syst
SpA: 2.78×10^8 Ci/g
Prod: protons on Ta; ^{164}Er(^{3}He,6n)

Photons (^{161}Yb)
⟨γ⟩=812 *12* keV

γ_{mode}	γ(keV)	γ(%)†
Tm L_ℓ	6.341	0.44 6
Tm L_α	7.176	11.2 12
Tm L_η	7.310	0.149 18
Tm L_β	8.218	11.7 16
Tm L_γ	9.560	2.0 3
Tm $K_{\alpha2}$	49.772	27.8 16
Tm $K_{\alpha1}$	50.742	49 3
Tm $K_{\beta1}'$	57.444	15.4 9
Tm $K_{\beta2}'$	59.296	4.26 25
γ	70.9 1	0.58 12
γ E1	78.20 3	41 3
γ	140.25 8	3.4 3
γ	144.43 6	5.5 6
γ	159.67 19	0.70 16
γ	161.85 15	0.83 16
γ	188.28 5	4.2 3
γ	192.26 14	0.69 6
γ	197.7 5	~0.047
γ	222.37 20	0.28 5
γ	261.2 3	0.28 8
γ	266.0 5	0.9 3
γ	298.46 15	1.61 9
γ	310.3 3	0.53 9
γ M1+E2	314.70 15	3.17 12
γ	318.63 18	0.76 6
γ M1+E2	330.10 24	3.28 20
γ	344.7 3	1.72 16
γ	359.92 17	1.33 9
γ	381.03 14	1.97 9
γ	410.44 17	1.47 11
γ	443.02 24	0.48 8
γ	458.22 16	3.40 12
γ	471.0 1	1.97 11
γ	519.12 20	0.75 8
γ	532.91 23	0.75 11
γ	536.6 3	0.51 11
γ	550.0 5	0.31 9
γ	552.1 5	0.31 9
γ	555.50 15	1.73 9
γ	560.48 20	2.53 19
γ	566.92 22	0.84 11
γ	569.73 14	6.7 3
γ	599.88 10	30.7 14
γ	631.45 10	16.4 8
γ	641.22 21	0.66 8
γ	644.9 4	0.53 11
γ	659.10 14	3.84 16
γ	690.75 20	1.28 11
γ	714.9 4	0.62 11
γ	722.15 23	0.89 12
γ	730.9 5	0.31 11
γ	745.6 4	0.30 8
γ	771.2 3	0.55 9
γ	781.2 3	0.67 12
γ	789.5 3	0.67 8
γ	793.02 23	1.45 9
γ	800.5 3	1.17 16
γ	805.24 24	0.84 9
γ	813.2 3	0.47 6
γ	816.5 3	0.36 6
γ	823.5 5	0.28 9
γ	842.7 3	0.70 9
γ	959.6 4	0.50 11
γ	1007.2 4	1.09 12
γ	1018.5 4	0.97 9
γ	1022.0 4	0.61 8
γ	1038.2 5	0.55 9
γ	1042.7 4	1.20 9
γ	1117.3 5	0.56 14
γ	1145.6 5	1.14 11
γ	1167.1 6	0.70 11
γ	1182.5 5	1.00 11
γ	1364.9 5	1.14 11
γ	1517.8 5	1.12 14
γ	1805.8 15	0.72 16

† 10% uncert(syst)

Atomic Electrons (^{161}Yb)
⟨e⟩=43 *4* keV

e_{bin}(keV)	⟨e⟩(keV)	e(%)
9	5.1	59 9
10	3.5	35 5
12 - 18	1.0	6.8 21
19	4.0	21.4 17
39 - 62	3.8	7.7 18
68	1.47	2.15 17
69	0.9	~1
70 - 78	1.47	2.02 18
81	1.5	~2
85	2.4	~3
100 - 102	0.6	~0.6
129	1.4	~1
130 - 180	4.5	3.2 8
182 - 222	0.57	0.29 11
239 - 288	2.1	0.80 19
289 - 336	0.93	0.30 8
342 - 384	0.34	0.09 4
399 - 448	0.6	0.15 7
449 - 496	0.40	0.084 25
501 - 537	0.7	0.15 7
540	2.0	~0.4
541 - 570	0.20	0.035 12
572	1.0	~0.17
582 - 631	1.0	0.17 6
632 - 681	0.16	0.024 8
682 - 731	0.12	0.017 6
734 - 781	0.22	0.030 9
783 - 833	0.08	0.010 4
834 - 843	0.0015	0.00018 8
900 - 949	0.06	0.006 3
950 - 999	0.11	0.012 4
1005 - 1043	0.022	0.0022 6
1058 - 1107	0.050	0.0047 21
1108 - 1157	0.056	0.0050 19
1158 - 1182	0.0060	0.00051 22
1306 - 1355	0.029	~0.0022
1356 - 1365	0.0011	$8\ 4\times10^{-5}$
1458 - 1516	0.026	~0.0018
1746 - 1804	0.014	0.0008 4

$^{161}_{71}$Lu(1.2 *1* min)

Mode: ϵ
Δ: -52510 *220* keV syst
SpA: 9.7×10^8 Ci/g
Prod: protons on W

Photons (^{161}Lu)

γ_{mode}	γ(keV)	γ(rel)
γ	43.7 3	~70
γ	67.13 20	48 5
γ	86.79 15	17 4 ?
γ	100.32 10	95 9
γ	105.2 1	28 5
γ	110.78 10	100 9
γ	156.24 10	49 5
γ	170.08 20	14 4
γ	177.13 20	14 4
γ	204.57 20	30 6
γ	211.1 2	~20
γ	221.76 20	20 4
γ	256.24 25	49 8

$^{161}_{72}$Hf(17 *2* s)

Mode: α
Δ: -46480 *220* keV syst
SpA: 4.0×10^9 Ci/g
Prod: ^{144}Sm(^{20}Ne,3n); ^{147}Sm(^{20}Ne,6n)

Alpha Particles (^{161}Hf)

α(keV)
4600 10

$^{161}_{73}$Ta($t_{1/2}$ unknown)

Mode: α

Δ: -38920 *220* keV syst

Prod: ^{58}Ni on ^{107}Ag

Alpha Particles (^{161}Ta)

α(keV)
5148 *5*

$^{161}_{74}$W (410 *40* ms)

Mode: α(82 *26* %), ε(18 *26* %)

Δ: -30610 *460* keV syst

SpA: 8.2×10^{10} Ci/g

Prod: ^{58}Ni on ^{107}Ag; ^{144}Sm(^{24}Mg,7n); daughter ^{165}Os

Alpha Particles (^{161}W)

α(keV)
5777 *5*

$^{161}_{75}$Re(~10 ms)

Mode: ε(<99 %), α(>1 %)

Δ: -20710 *550* keV syst

SpA: ~1.0×10^{11} Ci/g

Prod: ^{58}Ni on ^{107}Ag

Alpha Particles (^{161}Re)

α(keV)
6279 *10*

A = 162

NDS **17**, 97 (1976)

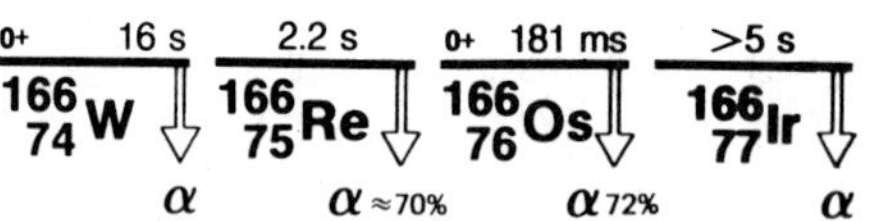

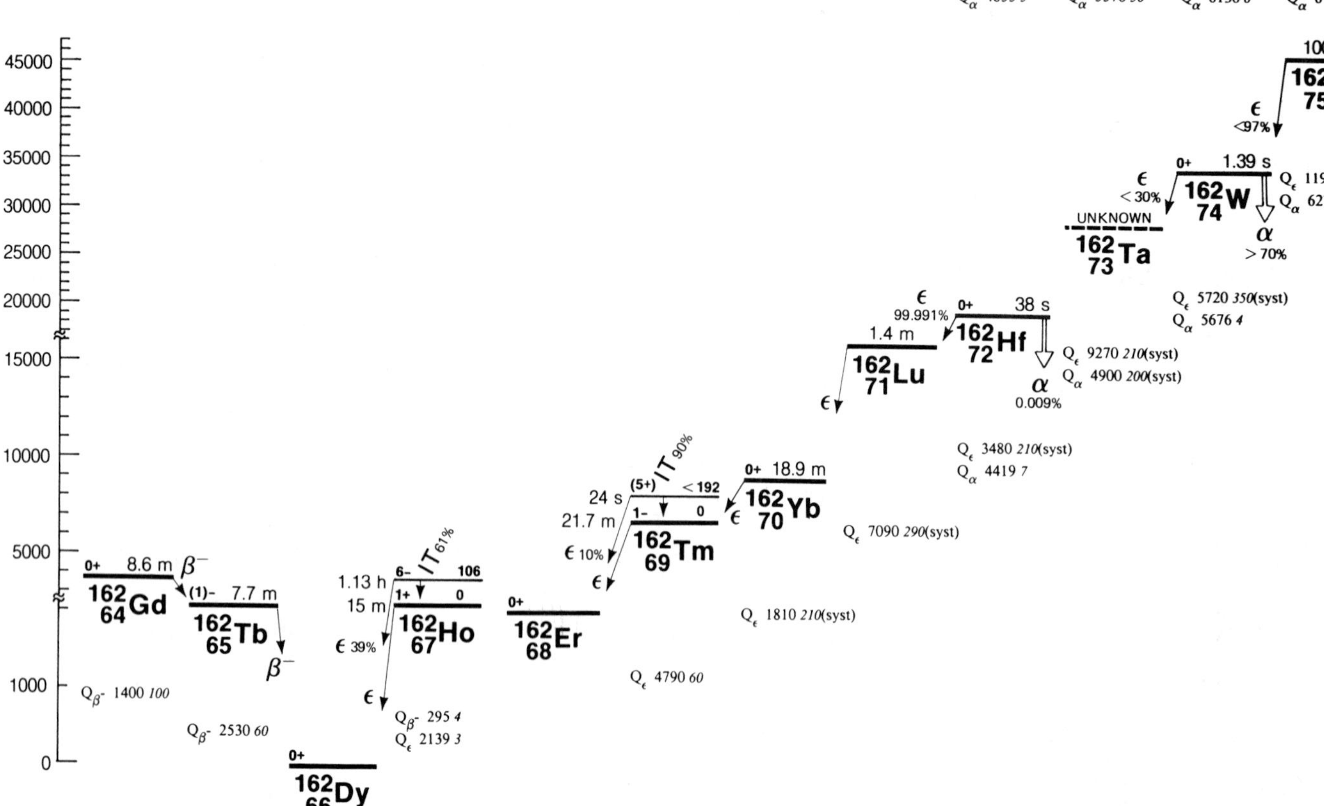

$^{162}_{64}$Gd(8.6 *3* min)

Mode: β-
Δ: -64260 *120* keV
SpA: 1.36×10^{8} Ci/g
Prod: multiple n-capture from ^{160}Gd

Photons (^{162}Gd)

⟨γ⟩=422 *24* keV

γ_{mode}	γ(keV)	γ(%)†
Tb L_ℓ	5.546	0.049 *12*
Tb L_α	6.269	1.3 *3*
Tb L_η	6.284	0.024 *5*
Tb L_β	7.061	3.0 *7*
Tb L_γ	8.330	0.67 *17*
γ (M1)	38.80 *18*	6.4 *14*
Tb $K_{\alpha2}$	43.744	0.166 *10*
Tb $K_{\alpha1}$	44.482	0.297 *18*
Tb $K_{\beta1}$'	50.338	0.089 *6*
Tb $K_{\beta2}$'	51.958	0.0265 *17*
γ [E1]	402.80 *23*	46
γ [E1]	441.60 *23*	53 *5*

† 4.3% uncert(syst)

Atomic Electrons (^{162}Gd)

⟨e⟩=15.8 *18* keV

e_{bin}(keV)	⟨e⟩(keV)	e(%)
8	0.19	2.3 *5*
9	0.98	11 *3*
30	8.0	26 *6*
31	0.84	2.7 *6*
35	0.0030	0.0086 *10*
36	0.0045	0.0126 *15*
37	2.4	6.4 *14*
38	0.58	1.5 *3*
39	0.112	0.29 *6*
41	0.00055	0.00133 *15*
42	0.00185	0.0044 *5*
43	0.0024	0.0056 *7*
44	0.00049	0.00112 *13*
48	0.000201	0.00042 *5*
49	0.00022	0.00046 *5*
50	0.000181	0.00036 *4*
351	1.14	0.325 *19*
390	1.18	0.30 *3*
394	0.151	0.0384 *23*
395	0.0258	0.0065 *4*
401	0.0390	0.0097 *6*
402	0.0086	0.00214 *13*
403	0.00236	0.00058 *3*
433	0.167	0.039 *4*
434	0.0127	0.0029 *3*
440	0.040	0.0090 *9*
441	0.0094	0.00214 *22*
442	0.00162	0.00037 *4*

Continuous Radiation (^{162}Gd)

⟨β-⟩=322 keV;⟨IB⟩=0.32 keV

E_{bin}(keV)		⟨ ⟩(keV)	(%)
0 - 10	β-	0.079	1.58
	IB	0.0154	
10 - 20	β-	0.238	1.59
	IB	0.0147	0.102
20 - 40	β-	0.97	3.22
	IB	0.027	0.095
40 - 100	β-	7.0	10.0
	IB	0.068	0.107
100 - 300	β-	67	33.9
	IB	0.13	0.078
300 - 600	β-	167	38.4
	IB	0.058	0.0147
600 - 958	β-	80	11.4
	IB	0.0049	0.00076

$^{162}_{65}$Tb(7.7 *2* min)

Mode: β-
Δ: -65660 *60* keV
SpA: 1.51×10^{8} Ci/g
Prod: ^{163}Dy(γ,p); ^{162}Dy(n,p); ^{165}Ho(n,α)

Photons (^{162}Tb)

⟨γ⟩=1097 *46* keV

γ_{mode}	γ(keV)	γ(%)†
Dy L_ℓ	5.743	0.12 *3*
Dy L_α	6.491	3.2 *7*
Dy L_η	6.534	0.055 *12*
Dy L_β	7.368	3.2 *7*
Dy L_γ	8.454	0.47 *10*
Dy $K_{\alpha2}$	45.208	4.4 *8*
Dy $K_{\alpha1}$	45.998	7.8 *14*
Dy $K_{\beta1}$'	52.063	2.4 *4*
Dy $K_{\beta2}$'	53.735	0.69 *13*
γ E2	80.5215 *20*	7.1 *16*
γ E2	185.015 *3*	2.5 *4*
γ [E1]	185.289 *5*	13.3 *17*
γ E1	260.080 *6*	79
γ [E2]	622.53 *5*	0.9 *4*
γ M1(+E2)	697.32 *5*	2.4 *4*
γ E2(+0.1%M1)	807.54 *5*	45 *3*
γ M1(+E2)	882.33 *5*	11.8 *16*
γ (E2)	888.06 *5*	39 *4*
γ [E1]	1067.62 *5*	0.63 *16*
γ E1	1196.4 *3*	0.10 *5*
γ E1	1276.9 *3*	0.09 *4*

† 5.7% uncert(syst)

Atomic Electrons (^{162}Tb)

⟨e⟩=40 *3* keV

e_{bin}(keV)	⟨e⟩(keV)	e(%)
8	1.3	17 *4*
9	1.01	12 *3*
27	3.4	13 *3*
36 - 71	1.22	2.1 *3*
72	8.1	11.2 *25*
73	8.6	12 *3*
78	0.20	0.26 *6*
79	4.4	5.6 *12*
80	1.2	1.5 *3*
131 - 177	2.09	1.48 *12*
183 - 185	0.166	0.091 *11*

Atomic Electrons (^{162}Tb) (continued)

e_{bin}(keV)	⟨e⟩(keV)	e(%)
206	3.41	1.65 *10*
251 - 260	0.77	0.303 *13*
569 - 615	0.046	0.008 *3*
620 - 644	0.13	0.020 *8*
688 - 697	0.026	0.0038 *11*
754	1.37	0.182 *16*
798 - 834	1.84	0.223 *23*
873 - 888	0.32	0.036 *3*
1014 - 1060	0.0069	0.00068 *15*
1066 - 1068	0.00024	2.2 *5* ×10^{-5}
1143 - 1189	0.0010	9 *4* ×10^{-5}
1194 - 1223	0.0008	6 *3* ×10^{-5}
1268 - 1277	0.00013	1.0 *4* ×10^{-5}

Continuous Radiation (^{162}Tb)

⟨β-⟩=504 keV;⟨IB⟩=0.70 keV

E_{bin}(keV)		⟨ ⟩(keV)	(%)
0 - 10	β-	0.0427	0.85
	IB	0.022	
10 - 20	β-	0.129	0.86
	IB	0.022	0.150
20 - 40	β-	0.53	1.76
	IB	0.041	0.144
40 - 100	β-	3.95	5.6
	IB	0.110	0.17
100 - 300	β-	43.2	21.3
	IB	0.25	0.147
300 - 600	β-	150	33.4
	IB	0.18	0.045
600 - 1300	β-	302	36.4
	IB	0.070	0.0096
1300 - 2500	β-	4.93	0.335
	IB	0.00063	4.1×10^{-5}
2500 - 2530	β-	7.9×10^{-5}	3.15×10^{-6}
	IB	5.0×10^{-11}	2.0×10^{-12}

$^{162}_{66}$Dy(stable)

Δ: -68190 *3* keV
%: 25.5 *2*

$^{162}_{67}$Ho(15 *1* min)

Mode: ϵ
Δ: -66051 *5* keV
SpA: 7.7×10^{7} Ci/g
Prod: daughter ^{162}Ho(1.13 h); ^{162}Dy(p,n)

Photons (^{162}Ho)

⟨γ⟩=125 *5* keV

γ_{mode}	γ(keV)	γ(%)†
Dy L_ℓ	5.743	0.39 *4*
Dy L_α	6.491	10.1 *7*
Dy L_η	6.534	0.145 *12*
Dy L_β	7.378	9.5 *8*
Dy L_γ	8.496	1.45 *15*
Dy $K_{\alpha2}$	45.208	25.7 *8*
Dy $K_{\alpha1}$	45.998	45.9 *15*

Photons (^{162}Ho) (continued)

γ_{mode}	γ(keV)	γ(%)†
Dy $K_{\beta1}$'	52.063	13.9 5
Dy $K_{\beta2}$'	53.735	4.09 15
γ E2	80.5215 20	7.7 4
γ E2	185.015 3	0.54 11
γ [E1]	185.289 5	~0.010
γ E1	260.080 6	~0.048
γ E2	282.726 8	~0.0037
γ (E2)	392.39 17	~0.49
γ	539.9 20	0.022 6
γ [E2]	622.53 5	~0.0010
γ M1(+E2)	697.32 5	~0.0020
γ E2+15%M1	795.36 5	~0.30
γ E2(+0.1%M1)	807.54 5	0.070 6
γ M1(+E2)	882.33 5	~0.008
γ (E2)	888.06 5	0.060 7
γ E2	980.37 5	~0.20
γ [E1]	1067.62 5	~0.00048
γ [E2]	1187.74 16	0.44 4
γ E1	1196.4 3	~0.018
γ E1	1276.9 3	~0.015
γ [E2]	1319.49 20	3.7
γ [M1+E2]	1372.76 16	0.77 15
γ [E2]	1453.28 16	0.022 7 ?
γ [M1+E2]	1668.9 10	~0.007 ?
γ [M1+E2]	1699.2 9	0.011 4 ?
γ	1736.9 20	0.008 3
γ [E2]	1779.7 9	0.031 9 ?
γ	1805.9 20	0.015 6

† 8.7% uncert(syst)

Atomic Electrons (^{162}Ho)

⟨e⟩=38.7 9 keV

e_{bin}(keV)	⟨e⟩(keV)	e(%)
8	3.9	50 5
9	3.0	34 4
27	3.75	14.0 7
36 - 71	3.14	6.9 3
72	8.8	12.2 6
73	9.4	12.9 7
78	0.223	0.284 15
79	4.78	6.1 3
80	1.34	1.66 9
131 - 177	0.22	0.15 6
183 - 229	0.027	0.015 6
251 - 283	0.00068	0.00026 7
339 - 385	0.047	0.014 6
390 - 392	0.0028	0.00070 21
486 - 532	0.0016	~0.00032
538 - 569	0.00011	1.9 9 $\times10^{-5}$
613 - 644	0.00012	~2$\times10^{-5}$
688 - 697	2.2 $\times10^{-5}$	3.2 14 $\times10^{-6}$
742 - 788	0.014	0.0019 7
793 - 834	0.0029	0.00035 4
873 - 888	0.00041	4.6 5 $\times10^{-5}$
927 - 973	0.0057	0.0006 3
978 - 1014	0.00023	2.3 8 $\times10^{-5}$
1059 - 1080	0.0005	~4$\times10^{-5}$
1125 - 1143	0.0094	0.00083 8
1179 - 1223	0.00191	0.000162 12
1266 - 1312	0.080	0.0063 5
1317 - 1365	0.024	0.0018 4
1371 - 1399	0.00112	8.1 16 $\times10^{-5}$
1444 - 1453	7.1 $\times10^{-5}$	4.9 12 $\times10^{-6}$
1615 - 1698	0.00052	3.1 8 $\times10^{-5}$
1726 - 1804	0.00079	4.6 11 $\times10^{-5}$

Continuous Radiation (^{162}Ho)

⟨β+⟩=20.7 keV;⟨IB⟩=1.9 keV

E_{bin}(keV)		⟨ ⟩(keV)	(%)
0 - 10	β+	2.94$\times10^{-7}$	3.49$\times10^{-6}$
	IB	0.00162	
10 - 20	β+	1.53$\times10^{-5}$	9.1$\times10^{-5}$
	IB	0.00108	0.0074
20 - 40	β+	0.00054	0.00163
	IB	0.024	0.065
40 - 100	β+	0.0331	0.0420
	IB	0.164	0.34
100 - 300	β+	1.73	0.79
	IB	0.053	0.027
300 - 600	β+	9.3	2.06
	IB	0.21	0.046
600 - 1300	β+	9.7	1.31
	IB	0.98	0.106
1300 - 2139	IB	0.45	0.030
	Σβ+		4.2

$^{162}_{67}$Ho(1.133 17 h)

Mode: IT(61 4 %), ε(39 4 %)
Δ: -65945 6 keV
SpA: 1.707$\times10^{7}$ Ci/g
Prod: ^{159}Tb(α,n); ^{162}Dy(p,n); protons on Dy

Photons (^{162}Ho)

⟨γ⟩=532 14 keV

γ_{mode}	γ(keV)	γ(%)†
γ_{IT} M1	38.3 1	7.3 9
γ_{IT} M1	57.8 1	4.4 7
γ_ϵE2	80.5215 20	9.1 25 &
γ_ϵ[E1]	161.14 18	0.09 3
γ_ϵE2	185.015 3	29.3 18
γ_ϵ[E1]	185.289 5	0.006 3
γ_ϵ(M1)	188.545 15	~0.23
γ_ϵ	205.4 3	0.047 15
γ_ϵE2	219.698 9	~0.035
γ_ϵ[E1]	235.882 5	0.07 3
γ_ϵE1	247.014 10	~0.06
γ_ϵ	250.9 6	~0.06
γ_ϵE1	260.080 6	0.035 15
γ_ϵ(E2)	275.451 12	0.80 18
γ_ϵ	278.3 6	~0.23
γ_ϵE2	282.726 8	11.5 4
γ_ϵ(E1)	302.766 14	0.38 4
γ_ϵE1	321.805 11	0.085 20
γ_ϵ[E1]	329.11 3	0.056 12
γ_ϵE1	333.920 10	0.32 3
γ_ϵ	347.5 5	0.029 9
γ_ϵ	424.56 20	0.474 20
γ_ϵ	467.76 20	0.243 20
γ_ϵ[E2]	622.53 5	0.0025 11
γ_ϵM1(+E2)	634.29 5	0.111 23
γ_ϵM1(+E2)	697.32 5	0.074 24
γ_ϵ[M1+E2]	775.92 18	0.097 23
γ_ϵE2+15%M1	795.36 5	0.27 3
γ_ϵE2(+0.1%M1)	807.54 5	0.119 23 &
γ_ϵ(E1)	841.74 6	0.46 6
γ_ϵM1(+E2)	882.33 5	0.37 10
γ_ϵ(E2)	888.06 5	0.103 19 &
γ_ϵM1(+E2)	917.02 5	0.59 6
γ_ϵ(E1)	937.06 5	10.9 4
γ_ϵE1	944.33 5	0.16 5
γ_ϵE2	980.37 5	0.26 5
γ_ϵ[E1]	1067.62 5	0.00028 14

Photons (^{162}Ho) (continued)

γ_{mode}	γ(keV)	γ(%)†
γ_ϵ(E1)	1124.46 6	1.19 6
γ_ϵE1	1129.35 5	0.38 4
γ_ϵ	1134.5 4	0.13 3 &
γ_ϵE1	1219.78 5	22.9 10
γ_ϵ	1805.9 10	0.026 12 &

† uncert(syst): 10% for ε, 6.6% for IT & with ^{162}Ho(1.13 h) and ^{162}Ho(15 min) in equilib

Continuous Radiation (^{162}Ho)

⟨IB⟩=0.099 keV

E_{bin}(keV)		⟨ ⟩(keV)	(%)
10 - 20	IB	6.6$\times10^{-5}$	0.00043
20 - 40	IB	0.0085	0.023
40 - 100	IB	0.061	0.127
100 - 300	IB	0.0106	0.0056
300 - 600	IB	0.0170	0.0041
600 - 1062	IB	0.00122	0.00019

$^{162}_{68}$Er(stable)

Δ: -66347 4 keV
%: 0.14 1

$^{162}_{69}$Tm(21.7 2 min)

Mode: ε
Δ: -61560 60 keV
SpA: 5.35$\times10^{7}$ Ci/g
Prod: daughter ^{162}Yb; ^{165}Ho(^{3}He,6n); ^{166}Er(p,5n); ^{164}Er(p,3n)

Photons (^{162}Tm)

⟨γ⟩=1479 13 keV

γ_{mode}	γ(keV)	γ(%)†
γ E2	102.14 3	17.5 7
γ	178.7 3	0.028 7
γ E2	227.63 3	7.1
γ [E2]	337.62 18	0.085 14
γ	380.3 5	0.071 14
γ	418.7 4	0.028 11
γ	425.07 13	~0.021
γ	432.6 4	0.050 21
γ	453.15 8	0.37 4
γ	465.22 10	0.248 21
γ	488.44 19	0.050 14
γ [M1+E2]	498.36 13	0.142 14
γ [E2]	519.77 4	0.59 5
γ	524.13 20	0.08 3
γ	533.8 4	0.050 14
γ (E1)	570.85 5	1.95 16
γ [E2]	571.17 5	~0.19
γ	634.19 10	0.170 7
γ	640.91 18	0.099 21
γ	645.82 16	0.135 14
γ E1	672.24 8	5.6 3
γ [M1+E2]	672.55 7	1.84 13
γ	695.44 13	0.092 14
γ	711.1 7	0.064 21
γ	716.6 4	0.09 4
γ	720.21 20	0.15 3
γ	733.20 25	~0.028

Photons (^{162}Tm)
(continued)

γ_{mode}	γ(keV)	γ(%)†
γ [E2]	736.79 25	0.057 14
γ	743.7 5	0.057 14
γ	759.76 17	0.099 21
γ	760.30 13	0.099 21
γ	765.07 14	0.23 6
γ (E2)	798.56 13	0.77 11
γ (E2)	798.80 4	8.4 3
γ	811.7 6	0.13 3
γ	821.72 10	0.32 3
γ	830.54 14	0.20 4
γ [E2]	841.51 4	0.65 4
γ	872.41 16	0.099 21
γ	890.43 12	0.078 14
γ (E2)	900.18 7	5.6 3
γ (E2)	900.94 5	6.5 3
γ	909.51 20	0.35 4
γ	929.25 13	0.18 4
γ	957.5 4	0.12 3
γ	960.5 3	0.12 6
γ	966.57 11	0.17 4
γ (E2)	985.22 6	1.15 5
γ	993.73 8	0.48 5
γ	1001.91 15	0.18 4
γ	1007.47 17	0.18 4
γ	1010.80 14	0.29 6
γ	1018.40 18	~0.028
γ [E2]	1026.19 13	<0.11
γ [E1]	1027.21 7	0.91 7
γ	1037.17 25	0.170 21
γ	1057.86 20	0.19 4
γ (E0+M1+E2)	1069.13 4	1.10 6
γ	1079.5 3	0.057 21
γ	1093.05 17	0.14 4
γ	1096.10 18	0.44 4
γ [E2]	1100.14 10	1.37 7
γ	1106.57 15	0.099 21
γ	1116.07 25	0.23 4
γ	1119.73 13	0.14 3
γ	1125.35 13	0.22 4
γ [E2]	1170.92 12	0.75 8
γ [E2]	1171.28 5	~0.16
γ	1199.40 23	~0.11
γ [E2]	1213.51 16	0.28 5
γ	1221.12 13	0.68 9
γ [M2]	1243.40 8	0.11 4
γ (E1)	1250.17 4	4.81 18
γ [E1]	1254.84 6	1.46 11
γ	1269.7 5	0.18 6
γ	1289.5 5	0.16 6
γ [E1]	1293.68 10	0.53 6
γ	1310.72 21	0.38 5
γ (E2)	1318.56 4	5.57 21
γ [M1+E2]	1327.77 10	0.87 7
γ	1341.48 10	0.085 21
γ [E1]	1352.32 4	3.42 14
γ	1380.8 5	0.09 3
γ	1394.0 4	0.26 5
γ [M1+E2]	1398.54 12	~0.38
γ	1404.36 5	2.84 11
γ	1411.01 15	0.28 5
γ	1412.86 21	0.28 5
γ	1416.5 4	0.135 21
γ [E2]	1429.91 11	0.44 5
γ	1447.5 3	0.170 14
γ	1451.4 4	0.21 4
γ [E1]	1471.03 8	0.60 14
γ	1475.69 9	0.135 21
γ	1478.9 5	0.043 14
γ [E1]	1493.56 18	0.16 4
γ [E2]	1500.69 12	0.12 4
γ	1506.50 5	1.38 7
γ [E1]	1521.30 10	0.65 5
γ	1533.5 4	0.31 6
γ [E2]	1535.34 25	0.39 6
γ [E1]	1543.84 18	0.078 21
γ	1549.09 16	0.28 5
γ	1556.9 6	0.099 21
γ [M2]	1573.18 8	0.18 4
γ	1575.9 5	0.14 4
γ	1581.0 4	0.135 21
γ	1596.01 14	0.43 4
γ [M1+E2]	1616.32 18	0.22 4
γ	1623.45 10	0.10 4
γ	1627.68 17	0.43 4

Photons (^{162}Tm)
(continued)

γ_{mode}	γ(keV)	γ(%)†
γ	1633.8 5	0.11 3
γ	1640.5 5	0.085 21
γ	1668.0 4	0.15 3
γ [E2]	1696.31 17	0.50 7
γ	1704.15 25	0.106 14
γ	1716.1 5	0.12 3
γ	1754.76 13	0.74 7
γ [M1+E2]	1762.97 25	0.15 3
γ	1773.1 5	0.036 14
γ [M1+E2]	1775.87 10	~0.028
γ [M1+E2]	1780.53 10	0.057 21
γ	1792.28 13	0.18 6
γ	1814.5 3	0.21 3
γ	1829.43 13	0.50 5
γ	1838.00 15	0.27 4
γ	1847.21 14	0.15 3
γ [E2]	1862.50 18	0.28 4
γ [E2]	1865.11 25	0.31 4
γ	1872.70 10	0.12 3
γ	1874.8 4	0.23 4
γ	1902.1 10	0.099 21
γ	1915.60 14	0.60 4
γ	1924.14 13	0.72 6
γ	1931.57 13	0.59 7
γ	1947.48 13	~0.028
γ	1959.46 17	0.55 4
γ [E1]	1961.57 9	0.192 21
γ	1970.32 19	0.20 4
γ	1974.84 10	1.22 9
γ	1983.5 7	0.14 4
γ	1994.28 20	0.057 14
γ	2000.6 3	0.23 3
γ	2005.3 6	0.11 4
γ [E2]	2012.30 16	0.43 4
γ	2015.87 12	1.12 7
γ	2022.5 5	0.092 14
γ	2030.9 5	0.09 3
γ	2037.09 20	0.11 3
γ	2047.91 17	0.15 4
γ	2062.66 19	0.12 3
γ	2073.3 5	0.071 14
γ	2083.5 5	0.20 3
γ [M1+E2]	2090.13 18	0.25 5
γ [E2]	2097.58 21	0.16 4
γ	2103.99 24	0.35 4
γ	2110.0 5	0.121 21
γ	2118.5 10	0.043 14
γ [E1]	2130.52 10	0.46 5
γ	2140.27 9	1.26 6
γ	2158.30 13	0.30 4
γ [M1+E2]	2176.06 17	0.08 3
γ	2185.7 3	0.06 3
γ [E2]	2192.27 18	0.29 4
γ	2202.7 10	0.036 14
γ	2206.13 24	0.078 21
γ	2212.9 8	0.064 21
γ	2216.74 10	0.55 4
γ	2227.7 10	0.10 3
γ [E1]	2231.91 8	0.84 6
γ	2250.4 5	0.13 3
γ	2257.5 6	0.114 21
γ	2260.44 14	0.11 3
γ	2265.59 14	0.071 21
γ	2269.2 3	0.078 21
γ	2302.04 21	0.057 14
γ	2305.5 11	0.050 21
γ	2311.8 3	0.071 14
γ	2318.88 10	0.13 3
γ [E1]	2324.43 14	0.071 14
γ	2329.3 9	0.078 21
γ	2334.99 23	0.11 3
γ	2338.7 5	0.078 14
γ [M1+E2]	2347.00 22	0.036 14
γ	2358.6 3	0.114 21
γ	2366.97 14	0.036 14
γ	2376.4 10	0.114 21
γ	2379.2 5	0.085 21
γ	2384.5 12	~0.043
γ [E2]	2389.95 24	0.12 3
γ	2395.2 7	0.25 4
γ	2403.8 6	0.071 21
γ	2420.2 10	0.043 14
γ	2439.4 5	0.036 14
γ	2450.02 16	0.17 3

Photons (^{162}Tm)
(continued)

γ_{mode}	γ(keV)	γ(%)†
γ	2461.5 10	0.057 14
γ	2465.2 5	0.14 3
γ	2474.5 5	0.043 14
γ	2480.1 3	0.192 21
γ	2496.18 15	0.071 21
γ	2501.7 3	0.14 3
γ [M1+E2]	2505.47 14	0.13 3
γ	2513.3 5	0.078 14
γ	2516.97 20	0.11 3
γ	2521.6 6	0.036 14
γ	2526.2 5	0.071 21
γ	2543.2 5	0.128 21
γ [E2]	2548.41 15	0.071 21
γ	2553.0 13	0.050 21
γ	2557.6 5	0.092 21
γ	2562.62 23	0.099 21
γ	2572.4 5	0.028 7
γ	2578.5 10	0.099 21
γ	2588.1 5	0.036 14
γ	2595.5 5	0.021 7
γ	2603.9 3	0.213 21
γ	2612.1 5	0.028 7
γ	2621.48 20	0.028 7
γ	2652.8 6	0.057 14
γ	2672.4 5	0.050 14
γ	2678.7 10	<0.014
γ	2687.6 3	0.050 14
γ	2698.9 5	0.106 21
γ	2709.0 3	0.071 21
γ	2712.8 4	0.14 3
γ	2727.0 5	0.021 7
γ	2735.3 5	0.021 7
γ	2739.0 5	~0.028
γ	2756.6 10	0.043 14
γ	2761.2 5	~0.028
γ [M1+E2]	2775.80 14	0.043 14
γ [E2]	2787.23 17	0.028 7
γ	2795.5 5	0.043 14
γ	2800.9 6	0.092 21
γ	2806.61 20	0.078 21
γ	2810.3 3	0.036 14
γ	2813.6 5	0.106 21
γ	2822.3 3	0.043 14
γ	2827.3 8	0.050 14
γ	2881.5 10	0.021 7
γ	2885.2 10	~0.028
γ	2888.4 8	0.050 14
γ	2901 1	0.021 7
γ	2909.6 5	0.114 21
γ	2919.5 10	0.043 14
γ	2942.6 10	~0.028
γ	2949.6 5	0.114 21
γ	2960.9 5	0.16 3
γ	2970.4 5	0.078 14
γ	2974.1 5	0.036 14
γ	3003.7 6	0.078 14
γ	3011.2 5	0.036 14
γ	3018.6 6	0.036 14
γ	3027.8 10	0.021 7
γ	3040.0 4	0.036 14
γ	3055.6 10	0.043 14
γ	3061.8 10	0.064 14
γ	3073.7 10	0.036 14
γ	3078.1 4	0.121 21
γ	3092.6 6	~0.028
γ	3103.6 10	0.050 14
γ	3120.3 7	~0.028
γ	3127.7 6	~0.028
γ	3136.1 10	~0.028
γ	3165.76 13	0.23 4
γ	3180.2 4	0.18 4
γ	3185.9 10	0.078 21
γ	3191.3 3	0.25 3
γ	3197.0 6	0.043 14
γ	3267.90 13	0.043 14
γ	3280.2 10	0.078 14
γ	3287.25 20	0.24 4
γ	3293.4 3	0.15 4
γ	3298.07 16	0.65 4
γ	3317.2 7	0.021 7
γ	3333.8 4	0.071 14
γ	3358.5 5	0.021 7
γ	3362.2 10	~0.028
γ	3368.17 12	0.036 14

Photons (^{162}Tm) (continued)

γ_{mode}	γ(keV)	γ(%)†
γ	3389.39 *20*	0.32 *4*
γ	3393.7 *10*	0.106 *14*
γ	3400.21 *16*	0.24 *4*
γ [M1+E2]	3416.12 *21*	0.26 *4*
γ	3420.6 *5*	0.064 *14*
γ	3426.1 *10*	~0.014
γ	3435.9 *4*	0.114 *14*
γ	3463.2 *10*	0.021 *7*
γ	3470.7 *8*	0.036 *14*
γ	3476.4 *7*	0.064 *14*
γ	3485.2 *10*	0.050 *14*
γ	3494.5 *4*	0.16 *3*
γ	3503.8 *7*	0.078 *14*
γ	3515.1 *10*	0.036 *14*
γ [E2]	3518.26 *21*	0.078 *14*
γ	3533.6 *8*	0.114 *21*
γ	3536.4 *10*	0.099 *21*
γ	3549.4 *8*	~0.028
γ	3567.9 *10*	0.036 *14*
γ [M1+E2]	3574.58 *13*	0.42 *4*
γ	3587.8 *3*	0.149 *21*
γ	3597.6 *10*	~0.014
γ	3618.9 *10*	~0.028
γ	3619.9 *10*	~0.028
γ	3670.9 *10*	~0.007
γ	3699.6 *10*	~0.014
γ	3723.7 *10*	0.021 *7*
γ	3736.7 *10*	~0.014
γ	3745.1 *3*	0.114 *14*
γ	3765.4 *8*	~0.007
γ	3784 *1*	0.011 *4*
γ	3797 *1*	0.021 *7*
γ	3801.8 *7*	0.028 *7*
γ	3861.1 *10*	0.021 *7*
γ	3877.7 *10*	~0.014

† 7.0% uncert(syst)

$^{162}_{69}$Tm(24.3 *17* s)

Mode: IT(89.8 *22* %), ϵ(10.2 *22* %)
Δ: <-61368 keV
SpA: 2.83×10^{9} Ci/g
Prod: ^{165}Ho(^{3}He,6n); ^{166}Er(p,5n); ^{164}Er(p,3n)

Photons (^{162}Tm)

$\langle\gamma\rangle$=265 *14* keV

γ_{mode}	γ(keV)	γ(%)†
Er L_ℓ	6.151	0.047 *12*
Tm L_ℓ	6.341	0.40 *14*
Er L_α	6.944	1.2 *3*
Er L_η	7.058	0.018 *5*
Tm L_α	7.176	10 *3*
Tm L_η	7.310	0.18 *8*
Er L_β	7.931	1.2 *3*
Tm L_β	8.207	11 *4*
Er L_γ	9.151	0.18 *5*
Tm L_γ	9.474	1.7 *7*
Er $K_{\alpha2}$	48.221	2.6 *4*
Er $K_{\alpha1}$	49.128	4.7 *7*
Tm $K_{\alpha2}$	49.772	13.3 *18*
Tm $K_{\alpha1}$	50.742	23 *3*
Er $K_{\beta1}'$	55.616	1.45 *23*
Er $K_{\beta2}'$	57.406	0.41 *7*
Tm $K_{\beta1}'$	57.444	7.3 *10*
Tm $K_{\beta2}'$	59.296	2.0 *3*
γ_{IT} M1+<39%E2	66.9 *1*	7.3 *7*
γ_ϵE2	102.14 *3*	~3
γ_ϵE2	227.63 *3*	~5
γ_ϵ[E2]	337.62 *18*	1.6 *3*

Photons (^{162}Tm) (continued)

γ_{mode}	γ(keV)	γ(%)†
γ_ϵ	345.5 *6*	0.66 *20*
γ_ϵ	354.7 *6*	0.59 *20*
γ_ϵ	453.1 *6*	0.7 *3*
γ_ϵ	478.0 *6*	0.7 *3*
γ_ϵ[E2]	571.17 *5*	~0.12
γ_ϵ[M1+E2]	584.22 *16*	~0.33
γ_ϵ[M1+E2]	672.55 *7*	0.9 *3*
γ_ϵ[M1+E2]	710.22 *10*	3.6 *3*
γ_ϵ	713.3 *7*	0.7 *3*
γ_ϵ(E2)	798.56 *13*	<0.7
γ_ϵ(E2)	798.80 *4*	5.3 *6*
γ_ϵ[E2]	811.61 *9*	6.6 *5*
γ_ϵ(E2)	900.18 *7*	2.7 *9*
γ_ϵ(E2)	900.94 *5*	4.1 *5*
γ_ϵ[E2]	1026.19 *13*	<0.05

† uncert(syst): 22% for ϵ, 12% for IT

Atomic Electrons (^{162}Tm)

$\langle e\rangle$=102 *29* keV

e_{bin}(keV)	$\langle e\rangle$(keV)	e(%)
8	3.7	48 *5*
9	3.9	45 *15*
10	3.3	34 *14*
38 - 56	2.5	5.6 *14*
57	11.7	20.4 *19*
58	8.8	15.1 *14*
65	5.6	8.6 *8*
66	3.6	~5
67 - 102	5.7	6.5 *12*
115	21.1	<37
116	15.8	<27
123	10.0	<16
124 - 170	3.8	~3
218 - 228	0.61	0.28 *9*
280 - 329	0.38	0.13 *3*
335 - 355	0.061	0.018 *6*
396 - 445	0.12	~0.030
451 - 478	0.016	0.0035 *17*
514 - 563	0.033	~0.006
569 - 615	0.06	0.010 *5*
653 - 702	0.30	0.046 *13*
704 - 741	0.20	0.028 *4*
754 - 803	0.297	0.0388 *24*
809 - 843	0.21	0.025 *5*
890 - 901	0.043	0.0048 *5*
969 - 1018	0.00018	1.9 *8* $\times10^{-5}$
1024 - 1026	7.3 $\times10^{-6}$	7 *3* $\times10^{-7}$

Continuous Radiation (^{162}Tm)

$\langle\beta+\rangle$=26.3 keV;$\langle$IB$\rangle$=0.62 keV

E_{bin}(keV)		$\langle\ \rangle$(keV)	(%)
0 - 10	β+	2.48×10^{-8}	2.94×10^{-7}
	IB	0.00106	
10 - 20	β+	1.35×10^{-6}	8.1×10^{-6}
	IB	0.00099	0.0069
20 - 40	β+	5.04×10^{-5}	0.000152
	IB	0.0025	0.0084
40 - 100	β+	0.00342	0.00432
	IB	0.021	0.039
100 - 300	β+	0.225	0.100
	IB	0.019	0.0103
300 - 600	β+	1.88	0.403
	IB	0.035	0.0080
600 - 1300	β+	13.1	1.37
	IB	0.160	0.0169
1300 - 2500	β+	11.2	0.71
	IB	0.33	0.019
2500 - 3270	IB	0.044	0.00164
	$\Sigma\beta$+		2.6

$^{162}_{70}$Yb(18.9 *2* min)

Mode: ϵ
Δ: -59750 *200* keV syst
SpA: 6.14×10^{7} Ci/g
Prod: protons on Ta; ^{164}Er(^{3}He,5n); ^{166}Er(^{3}He,7n); ^{169}Tm(p,8n); alphas on Er

Photons (^{162}Yb)

$\langle\gamma\rangle$=135 *12* keV

γ_{mode}	γ(keV)	γ(%)
Tm L_ℓ	6.341	0.41 *5*
Tm L_α	7.176	10.3 *10*
Tm L_η	7.310	0.139 *16*
Tm L_β	8.220	10.4 *13*
Tm L_γ	9.543	1.76 *25*
γ (M1+7.8%E2)	44.68 *4*	2.2 *6*
Tm $K_{\alpha2}$	49.772	24.2 *8*
Tm $K_{\alpha1}$	50.742	42.6 *14*
Tm $K_{\beta1}'$	57.444	13.4 *4*
Tm $K_{\beta2}'$	59.296	3.71 *13*
γ E1	118.78 *4*	24.8 *13*
γ E1	163.46 *4*	36

Atomic Electrons (^{162}Yb)

$\langle e\rangle$=26.1 *18* keV

e_{bin}(keV)	$\langle e\rangle$(keV)	e(%)
9	3.5	41 *5*
10	3.0	31 *4*
35	5.0	14 *4*
36	2.5	6.9 *18*
39 - 41	0.99	2.45 *16*
42	0.93	2.2 *5*
43	1.4	3.3 *9*
44 - 57	1.39	3.0 *3*
59	2.60	4.38 *24*
104	2.9	2.8 *6*
109	0.63	0.58 *3*
110 - 155	1.02	0.74 *6*
161 - 163	0.19	0.120 *17*

Continuous Radiation (^{162}Yb)

$\langle$IB$\rangle$=1.16 keV

E_{bin}(keV)		$\langle\ \rangle$(keV)	(%)
10 - 20	IB	0.00017	0.00120
20 - 40	IB	0.0063	0.018
40 - 100	IB	0.22	0.44
100 - 300	IB	0.045	0.023
300 - 600	IB	0.19	0.041
600 - 1300	IB	0.63	0.071
1300 - 1810	IB	0.070	0.0050

$^{162}_{71}$Lu(1.4 min)

Mode: ϵ
Δ: -52660 *210* keV syst
SpA: 8×10^{8} Ci/g
Prod: ^{148}Sm(^{19}F,5n); protons on W

decay data not yet evaluated

$^{162}_{72}$Hf(37.6 *8* s)

Mode: ϵ(99.991 *3* %), α(0.009 *3* %)
Δ: -49179 *17* keV
SpA: 1.84×10^{9} Ci/g
Prod: ^{147}Sm(^{20}Ne,5n); ^{142}Nd(^{24}Mg,4n)

Alpha Particles (^{162}Hf)

α(keV)
4308 *10*

Photons (^{162}Hf)

γ$_{mode}$	γ(keV)	γ(rel)
γ	173.70 *5*	100 *5*
γ	196.34 *5*	25 *5*
γ	410.12 *10*	16.8 *10*

† 34% uncert(syst)

$^{162}_{74}$W (1.39 *4* s)

Mode: α(>70 %), ϵ(<30 %)
Δ: -34200 *280* keV syst
SpA: 3.95×10^{10} Ci/g
Prod: ^{144}Sm(^{24}Mg,6n); ^{156}Dy(^{16}O,10n)

Alpha Particles (^{162}W)

α(keV)
5535 *4*

$^{162}_{75}$Re(100 *30* ms)

Mode: ϵ(<97 %), α(>3 %)
Δ: -22300 *550* keV syst
SpA: 1.0×10^{11} Ci/g
Prod: ^{107}Ag(^{58}Ni,3n)

Alpha Particles (^{162}Re)

α(keV)
6119 *6*

A = 163

NDS **29**, 653 (1980)

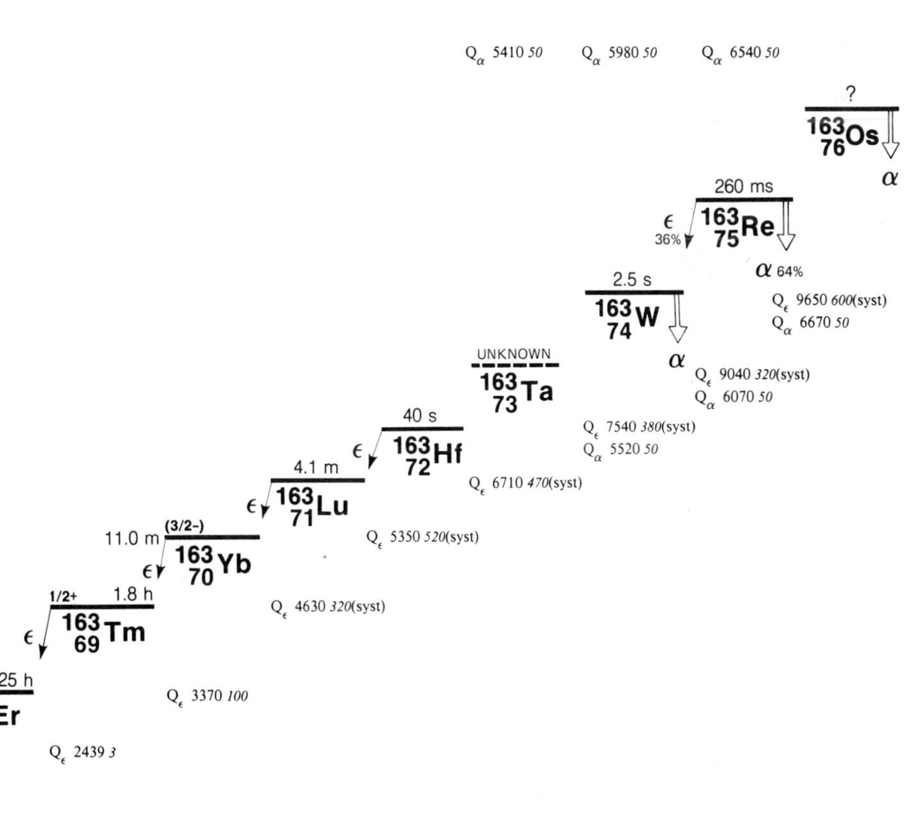

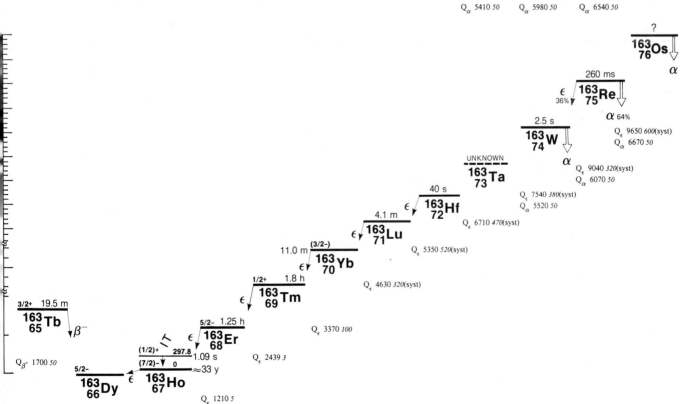

$^{163}_{65}$Tb(19.5 *3* min)

Mode: β-
Δ: -64690 *50* keV
SpA: 5.91×10^{7} Ci/g
Prod: ^{164}Dy(γ,p); ^{163}Dy(n,p)

Photons (^{163}Tb)

⟨γ⟩=785 *19* keV

γ_{mode}	γ(keV)	γ(%)†
Dy L_ℓ	5.743	0.067 *13*
Dy L_α	6.491	1.7 *3*
Dy L_η	6.534	0.029 *5*
Dy L_β	7.369	1.7 *3*
Dy L_γ	8.464	0.25 *5*
γ [M1]	37.84 *11*	~0.0045
γ [M1]	38.56 *10*	0.029 *13*
Dy $K_{\alpha2}$	45.208	2.6 *3*
Dy $K_{\alpha1}$	45.998	4.6 *5*
Dy $K_{\beta1}$'	52.063	1.41 *16*
Dy $K_{\beta2}$'	53.735	0.41 *5*
γ	68.8 *5*	~0.07 ?
γ [M1]	70.64 *11*	0.08 *4*
γ (E2)	73.40 *9*	2.0 *5*
γ [E2]	76.40 *12*	0.08 *4*
γ [M1]	77.72 *17*	~0.031
γ E2+14%M1	93.98 *12*	0.11 *5*
γ [E1]	118.22 *12*	0.09 *3*
γ	123.6 *5*	0.06 *3* ?
γ [E2]	124.83 *12*	0.13 *5*
γ [M1]	146.75 *12*	0.20 *6*
γ [M1]	154.10 *15*	0.27 *7*
γ E2	167.39 *11*	0.56 *11*
γ E1	177.44 *10*	0.65 *13*
γ E1	212.20 *11*	0.67 *13*
γ E1	250.85 *10*	6.7 *7*
γ E2	260.26 *12*	0.60 *11*
γ E1	266.42 *15*	0.20 *6*
γ E1	285.61 *11*	0.83 *16*
γ E2	316.41 *10*	8.3 *8*
γ [M1]	321.32 *13*	~0.11
γ E1	338.60 *13*	4.5 *5*
γ [E1]	344.36 *12*	0.45 *11*
γ M1+26%E2	347.25 *15*	0.34 *11*
γ E1	347.72 *11*	6.1 *6*
γ E2	351.25 *10*	26 *3*
γ [E1]	353.41 *18*	~0.11
γ E2+35%M1	354.24 *11*	4.6 *5*
γ E1	376.43 *12*	0.63 *13*
γ [E1]	383.80 *17*	~0.22
γ [M1+E2]	385.70 *17*	~0.11
γ E1	386.29 *11*	4.5 *5*
γ E2+38%M1	389.81 *10*	24.2 *25*
γ E1	391.25 *16*	0.45 *11*
γ [E1]	396.29 *15*	0.47 *9*
γ M1+32%E2	401.97 *12*	2.49 *25*
γ E1	414.99 *12*	5.4 *5*
γ E2+48%M1	421.89 *11*	11.4 *11*
γ E2+19%M1	427.65 *10*	3.5 *4*
γ E1	434.74 *13*	1.41 *22*
γ [E1]	437.29 *13*	0.34 *11*
γ [E1]	440.35 *24*	~0.22
γ E2+34%M1	441.24 *12*	0.67 *13*
γ [E1]	459.79 *13*	1.14 *18*
γ [E1]	462.40 *12*	2.2 *3*
γ M1(+30%E2)	475.37 *12*	2.9 *3*
γ M1+50%E2	479.68 *14*	0.34 *9*
γ E2	486.69 *12*	1.10 *18*
γ E1	494.48 *11*	22.4
γ E1	507.52 *11*	4.6 *5*
γ [M1+E2]	514.64 *13*	<0.45 ?
γ [M1+E2]	515.40 *13*	0.56 *22* ?
γ [E1]	527.50 *13*	0.27 *7*
γ E1	533.04 *11*	9.5 *9*
γ [E1]	545.35 *10*	1.68 *25*
γ [M1+E2]	553.09 *15*	~0.09

Photons (^{163}Tb)
(continued)

γ_{mode}	γ(keV)	γ(%)†
γ E1	559.57 *12*	2.0 *3*
γ [E2]	573.56 *14*	0.16 *5*
γ	578.2 *5*	0.18 *6* ?
γ E1	583.92 *11*	7.0 *7*
γ (M1+35%E2)	608.33 *13*	3.7 *4*
γ [M1+E2]	630.12 *19*	1.12 *18*
γ [E2]	633.44 *11*	0.29 *7*
γ [E1]	636.97 *15*	0.20 *7*
γ [E2]	649.56 *12*	0.11 *5*
γ [E1]	656.66 *15*	~0.06
γ [E1]	662.42 *13*	~0.20
γ [M1+E2]	663.77 *15*	~0.16
γ [M1+E2]	664.88 *22*	~0.22
γ [E1]	669.04 *13*	0.60 *16*
γ [M1+E2]	684.32 *11*	0.29 *13*
γ [E1]	694.50 *13*	~0.06
γ [M1+E2]	698.54 *13*	~0.09
γ [E1]	707.61 *13*	0.31 *13*
γ	722.4 *5*	~0.09 ?
γ	725.1 *5*	0.16 *7* ?
γ [E1]	733.06 *13*	0.25 *11*
γ [E2]	808.01 *14*	~0.18
γ [M1+E2]	833.46 *13*	1.01 *18*
γ	844.2 *5*	~0.11 ?
γ [E2]	896.42 *20*	~0.13

† 6.3% uncert(syst)

Atomic Electrons (^{163}Tb)

⟨e⟩=28.5 *13* keV

e_{bin}(keV)	⟨e⟩(keV)	e(%)
8	0.69	8.8 *17*
9 - 17	0.65	6.9 *13*
20	0.84	4.3 *10*
23 - 64	0.90	2.0 *3*
65	3.2	4.9 *11*
66	3.5	5.3 *12*
67 - 71	0.47	0.67 *17*
72	1.8	2.5 *6*
73 - 122	1.33	1.54 *18*
123 - 170	0.35	0.235 *18*
175 - 213	0.41	0.208 *18*
232 - 260	0.145	0.059 *4*
263	0.91	0.34 *4*
264 - 294	0.43	0.147 *12*
297	2.42	0.81 *9*
300 - 335	1.12	0.36 *3*
336	2.7	0.80 *11*
337 - 367	1.55	0.448 *24*
368	1.23	0.33 *6*
369 - 415	1.52	0.39 *3*
419 - 468	1.23	0.283 *20*
471 - 520	0.415	0.085 *5*
524 - 574	0.44	0.081 *12*
575 - 624	0.21	0.036 *5*
625 - 671	0.049	0.0076 *19*
675 - 724	0.0109	0.0016 *3*
725 - 754	0.006	0.0008 *4*
780 - 826	0.055	0.0070 *21*
831 - 844	0.0063	0.00075 *23*
887 - 896	0.00076	9 *3* ×10^{-5}

Continuous Radiation (^{163}Tb)

⟨β-⟩=292 keV;⟨IB⟩=0.28 keV

E_{bin}(keV)		⟨ ⟩(keV)	(%)
0 - 10	β-	0.093	1.86
	IB	0.0140	
10 - 20	β-	0.280	1.87
	IB	0.0133	0.093
20 - 40	β-	1.13	3.78
	IB	0.025	0.086
40 - 100	β-	8.1	11.5
	IB	0.061	0.095
100 - 300	β-	73	36.8
	IB	0.111	0.067
300 - 600	β-	149	35.0
	IB	0.046	0.0118
600 - 1278	β-	60	8.3
	IB	0.0060	0.00088

$^{163}_{66}$Dy(stable)

Δ: -66390 *3* keV
%: 24.9 *2*

$^{163}_{67}$Ho(33 *23* yr)

Mode: ϵ
Δ: -66387 *4* keV
SpA: 66 Ci/g
Prod: daughter ^{163}Er

$^{163}_{67}$Ho(1.09 *3* s)

Mode: IT
Δ: -66089 *4* keV
SpA: 4.70×10^{10} Ci/g
Prod: ^{165}Ho(γ,2n); ^{163}Dy(p,n)

Photons (^{163}Ho)

⟨γ⟩=237 *7* keV

γ_{mode}	γ(keV)	γ(%)†
Ho L_ℓ	5.943	0.057 *6*
Ho L_α	6.716	1.48 *11*
Ho L_η	6.789	0.0287 *23*
Ho L_β	7.629	1.72 *15*
Ho L_γ	8.804	0.27 *3*
Ho $K_{\alpha2}$	46.700	3.23 *14*
Ho $K_{\alpha1}$	47.547	5.75 *24*
Ho $K_{\beta1}$'	53.822	1.77 *7*
Ho $K_{\beta2}$'	55.556	0.506 *23*
γ E3	297.78 *9*	77.5

† 1.9% uncert(syst)

Atomic Electrons (^{163}Ho)

⟨e⟩=61.1 *12* keV

e_{bin}(keV)	⟨e⟩(keV)	e(%)
8	0.54	6.7 *7*
9	0.54	6.0 *6*
37	0.053	0.143 *15*
38	0.030	0.077 *8*
39	0.079	0.204 *22*
44	0.025	0.056 *6*
45	0.027	0.059 *6*
46	0.037	0.080 *9*
47	0.0082	0.0174 *18*
48	0.00067	0.00142 *15*
51	0.00115	0.00223 *24*
52	0.0064	0.0124 *13*
53	0.0026	0.0049 *5*
54	0.00105	0.00196 *21*
242	28.6	11.8 *4*
288	4.34	1.51 *6*
289	12.8	4.42 *16*
290	6.54	2.26 *8*
296	5.91	2.00 *7*
297	1.39	0.467 *17*
298	0.203	0.0681 *25*

$^{163}_{68}$Er(1.250 *7* h)

Mode: ϵ

Δ: -65177 *6* keV

SpA: 1.538×10^{7} Ci/g

Prod: ^{165}Ho(p,3n); ^{162}Er(n,γ); ^{164}Er(γ,n); daughter ^{163}Tm

Photons (^{163}Er)

⟨γ⟩=40.3 *7* keV

γ_{mode}	γ(keV)	γ(%)†
Ho L_ℓ	5.943	0.29 *3*
Ho L_α	6.716	7.4 *5*
Ho L_η	6.789	0.096 *7*
Ho L_β	7.659	6.8 *6*
Ho L_γ	8.851	1.09 *11*
Ho $K_{\alpha2}$	46.700	22.5 *7*
Ho $K_{\alpha1}$	47.547	40.1 *12*
Ho $K_{\beta1}'$	53.822	12.3 *4*
Ho $K_{\beta2}'$	55.556	3.53 *11*
γ	80.5 *5*	~0.00046 ?
γ[M1]	100.00 *9*	0.00169 *9*
γ[E2]	123.59 *15*	0.00021 *6*
γ	164.58 *20*	0.00046 *6*
γ	192.58 *20*	0.00051 *5*
γ	230.0 *5*	~0.0010 ?
γ[E1]	253.88 *20*	0.00038 *6*
γ E3	297.78 *9*	0.0115 §
γ[E1]	331.11 *14*	0.00021 *6*
γ[E1]	339.83 *11*	0.00334 *11*
γ	348.0 *5*	0.00024 *10* ?
γ	417.08 *20*	0.00087 *8*
γ E1	431.11 *13*	0.00144 *10*
γ[M1]	436.05 *8*	0.0285 *6*
γ E1	439.83 *8*	0.0276 *6*
γ[M1]	444.78 *13*	0.00091 *10*
γ	452.28 *20*	0.00069 *7*
γ[M1]	484.0 *3*	0.00028 *6*
γ	552.0 *3*	0.00032 *8*
γ	558.48 *20*	0.00059 *10*
γ[M1]	568.37 *15*	0.00086 *9*
γ[E2]	578.10 *12*	0.00133 *10*

Photons (^{163}Er) (continued)

γ_{mode}	γ(keV)	γ(%)†
γ	583 *4*	0.0063 *24*
γ[E2]	614.28 *10*	0.00334 *11*
γ	711.28 *20*	0.00123 *10*
γ[E1]	875.88 *9*	0.0069 *3*
γ[E2]	1013.53 *22*	0.00086 *10*
γ[M1+E2]	1113.53 *22*	0.0490 *14*

† 15% uncert(syst)

§ with ^{163}Ho(1.09 s) in equilib

Atomic Electrons (^{163}Er)

⟨e⟩=6.7 *4* keV

e_{bin}(keV)	⟨e⟩(keV)	e(%)
8	2.7	34 *4*
9	2.06	22.7 *23*
25	0.00027	~0.0011
37	0.37	1.01 *10*
38	0.207	0.54 *6*
39	0.56	1.43 *15*
44 - 45	0.36	0.81 *6*
46	0.26	0.56 *6*
47 - 92	0.142	0.284 *16*
98 - 137	0.00062	0.00055 *14*
155 - 198	0.00041	~0.00024
221 - 254	0.0043	0.0018 *3*
275 - 323	0.0048	0.00164 *13*
329 - 375	0.00015	4.1 *17* $\times 10^{-5}$
380 - 429	0.00594	0.00154 *5*
430 - 476	0.000381	8.8 *3* $\times 10^{-5}$
482 - 527	0.0006	~0.00011
543 - 583	0.00027	4.8 *9* $\times 10^{-5}$
605 - 614	3.58 $\times 10^{-5}$	5.90 *23* $\times 10^{-6}$
656 - 703	6.4 $\times 10^{-5}$	~1.0×10^{-5}
709 - 711	2.6 $\times 10^{-6}$	~4×10^{-7}
820 - 868	9.3 $\times 10^{-5}$	1.13 *6* $\times 10^{-5}$
874 - 876	3.28 $\times 10^{-6}$	3.76 *19* $\times 10^{-7}$
958 - 1005	2.5 $\times 10^{-5}$	2.6 *3* $\times 10^{-6}$
1011 - 1058	0.0016	0.00015 *4*
1104 - 1113	0.00030	2.7 *6* $\times 10^{-5}$

Continuous Radiation (^{163}Er)

⟨β+⟩=0.00393 keV;⟨IB⟩=0.52 keV

E_{bin}(keV)		⟨ ⟩(keV)	(%)
0 - 10	β+	2.97×10^{-8}	3.53×10^{-7}
	IB	0.00078	
10 - 20	β+	1.46×10^{-6}	8.7×10^{-6}
	IB	0.00017	0.00112
20 - 40	β+	4.40×10^{-5}	0.000134
	IB	0.0128	0.035
40 - 100	β+	0.00146	0.00195
	IB	0.19	0.39
100 - 300	β+	0.00242	0.00191
	IB	0.038	0.020
300 - 600	IB	0.13	0.029
600 - 1210	IB	0.151	0.020
	Σβ+		0.0040

$^{163}_{69}$Tm(1.81 *6* h)

Mode: ϵ

Δ: -62738 *6* keV

SpA: 1.06×10^{7} Ci/g

Prod: ^{164}Er(p,2n); ^{166}Er(d,5n); protons on Ta

Photons (^{163}Tm)

⟨γ⟩=1223 *24* keV

γ_{mode}	γ(keV)	γ(%)†
Er L_ℓ	6.151	0.57 *6*
Er L_α	6.944	14.6 *11*
Er L_η	7.058	0.188 *16*
Er L_β	7.940	13.5 *13*
Er L_γ	9.179	2.16 *24*
Er $K_{\alpha2}$	48.221	41.7 *20*
Er $K_{\alpha1}$	49.128	74 *4*
Er $K_{\beta1}'$	55.616	22.9 *11*
Er $K_{\beta2}'$	57.406	6.5 *3*
γ M1+4.9%E2	60.122 *23*	1.14 *20*
γ (E1)	63.69 *5*	0.103 *16*
γ E1	69.23 *3*	10.9
γ (E1)	72.85 *4*	0.09 *2*
γ E1	78.00 *5*	0.085 *15*
γ M1	80.449 *24*	0.47 *5*
γ E2+~14%M1	83.978 *25*	0.65 *8*
γ M1+4.1%E2	85.14 *3*	0.35 *4*
γ E1	91.56 *3*	0.33 *4*
γ (E2)	93.92 *4*	0.023 *5*
γ E1	96.37 *10*	0.040 *14*
γ (M1)	97.43 *7*	0.107 *18*
γ E1	98.31 *7*	0.091 *17*
γ M1	104.31 *3*	19.4 *20*
γ M1	111.12 *15*	~0.013 ?
γ E1	118.67 *5*	0.12 *2*
γ E1	129.21 *5*	0.10 *1*
γ E2	145.26 *3*	0.13 *2*
γ (M1)	147.72 *20*	0.016 *6*
γ E1	152.72 *10*	0.074 *13*
γ E2	153.39 *10*	0.095 *16*
γ E1	161.31 *6*	0.169 *18*
γ M1	164.43 *3*	0.97 *10*
γ M1	165.59 *4*	~0.04
γ M1(+E2)	190.07 *4*	1.28 *10*
γ M1,E2	225.4 *3*	0.035 *14*
γ M1	239.67 *4*	4.1 *3*
γ M1	241.42 *4*	9.3 *9*
γ M1	249.57 *4*	0.074 *13*
γ M1	275.21 *4*	2.4 *3*
γ	287.6 *4*	~0.04
γ	289.8 *3*	~0.07
γ E1	298.11 *11*	0.33 *4*
γ M1	299.79 *4*	4.1 *4*
γ E2	320.12 *4*	0.29 *6*
γ E1	331.28 *10*	0.21 *4*
γ M1(+E2)	335.34 *4*	0.56 *6*
γ M1	338.15 *14*	0.07 *2*
γ E2	345.72 *5*	1.05 *12*
γ M1(+E2)	355.66 *4*	0.47 *5*
γ (E1)	358.24 *11*	0.72 *6*
γ [E2]	370.98 *11*	0.06 *2*
γ M1	376.9 *3*	0.14 *4*
γ E1	389.65 *10*	0.29 *4*
γ M1	393.31 *11*	1.24 *20*
γ M1(+E2)	404.10 *4*	0.99 *20*
γ M1(+E2)	411.8 *3*	0.16 *3*
γ	422.4 *4*	0.16 *4*
γ M1,E2	433.3 *3*	0.09 *4*
γ (M1)	434.82 *20*	0.49 *6*
γ M1	439.64 *4*	0.31 *4*
γ E2	454.7 *4*	0.22 *3*
γ (M1)	456.72 *20*	0.12 *4*
γ (M1)	462.02 *20*	0.56 *6*
γ E1	469.82 *20*	0.52 *10*
γ E2	471.29 *18*	3.8 *4*
γ M1,E2	473.72 *20*	0.23 *6*
γ M1	484.02 *20*	0.23 *8*
γ	491.0 *6*	0.05 *2*
γ	493.6 *5*	0.13 *5*
γ	500.4 *8*	0.08 *3*
γ M1	504.92 *20*	0.85 *10*
γ E1	515.13 *21*	0.7 *1*
γ E2,E1	528.3 *4*	0.13 *6*
γ M1	529.9 *3*	0.27 *6*
γ M1,E2	539.7 *5*	0.05 *2*
γ M1(+E2)	550.21 *21*	1.47 *16*
γ E2	552.9 *3*	0.9 *1*
γ M1	573.2 *3*	0.25 *8*
γ M1	579.57 *13*	1.72 *18*
γ	595.7 *3*	0.20 *6*
γ E1	598.7 *3*	0.20 *6*

Photons (^{163}Tm) (continued)

γ_{mode}	γ(keV)	γ(%)†
γ E1	607.1 4	0.12 4
γ E2	612.9 3	0.83 14
γ E2	614.9 3	0.35 8
γ (E1)	633.4 4	0.15 6
γ M1	655.6 3	0.73 12
γ	662.2 6	0.28 7
γ M1	666.15 11	1.69 18
γ	674.8 5	0.16 6
γ E2	683.88 14	0.43 12
γ E2	687.6 4	0.20 6
γ E2	691.5 3	0.49 16
γ	697.0 6	0.12 4
γ	710.7 4	0.07 2
γ M1(+E2)	714.1 5	0.07 2
γ	717.4 3	0.17 3
γ	727.3 3	0.07 3
γ M1	732.0 4	0.08 3
γ [E1]	735.38 11	0.12 3
γ M1(+E2)	742.5 5	0.06 2
γ M1	747.0 5	0.06 2
γ M1(+E2)	751.52 20	0.41 6
γ M1(+E2)	755.4 3	0.25 3
γ E1,E2	758.9 3	0.25 3
γ E2	779.82 20	0.74 12
γ M1	782.3 5	0.21 8
γ	790.0 3	0.28 6
γ M1	798.6 3	0.16 4
γ M1	803.6 3	0.25 4
γ E2	813.0 5	0.13 3
γ M1(+E2)	829.0 7	0.05 2
γ M1	834.3 4	0.40 5
γ	837.6 7	0.05 2
γ E1,E2	846.5 5	0.14 4
γ E2	859.02 20	0.29 6
γ	872.5 5	0.06 2
γ E1	885.52 20	0.40 5
γ M1	894.22 20	0.36 4
γ	901.7 3	0.09 3
γ E2	906.92 20	0.27 4
γ E2	916.0 3	0.18 3
γ M1	928.7 6	0.10 2
γ E2	936.4 4	0.10 2
γ E2,E1	940.62 20	0.45 6
γ E2	945.22 20	0.70 8
γ M1	950.4 4	0.15 4
γ	957.3 8	~0.02
γ	961.0 6	0.06 2
γ E2	975.5 4	0.33 6
γ M1	986.98 13	0.20 3
γ E2,E1	990.2 6	0.10 2
γ M1	995.3 4	0.18 3
γ E1	1004.6 4	0.18 2
γ E1,E2	1011.3 6	0.08 2
γ M1(+E2)	1014.8 6	0.08 2
γ	1018.8 6	0.07 2
γ M1,E2	1022.5 7	0.05 2
γ M1	1028.9 5	0.16 3
γ E1	1033.1 5	0.16 3
γ E2	1036.2 6	0.15 3
γ E2	1039.2 6	0.10 2
γ M1(+E2)	1042.6 5	0.15 3
γ E2,E1	1046.5 5	0.14 3
γ M1	1052.46 19	0.14 2
γ M1	1066.16 12	0.20 4
γ E1	1075.02 10	0.8 1
γ E1,E2	1090.6 4	0.27 3
γ E1	1099.12 10	0.50 6
γ (E2)	1108.0 7	0.043 18
γ	1113.0 7	0.09 3
γ E1	1130.07 7	1.94 22
γ E1	1136.0 3	0.58 6
γ E1	1142.32 10	0.74 10
γ E1	1153.33 7	0.95 10
γ E1	1168.56 9	0.52 8
γ E2,E1	1175.82 20	0.45 6
γ E1	1188.88 7	0.12 3
γ M1(+E2)	1192.2 4	0.12 3
γ E1	1204.79 7	2.4 3
γ M1(+E2)	1212.4 6	0.07 3
γ E1	1223.99 7	2.02 22
γ E1	1239.8 6	0.11 2
γ E1	1247.25 7	0.82 10
γ M1(+E2)	1251.7 6	0.16 3
γ M1	1259.81 14	0.21 6

Photons (^{163}Tm) (continued)

γ_{mode}	γ(keV)	γ(%)†
γ M1	1261.02 19	0.21 6
γ E1	1264.91 7	4.9 5
γ E1,E2	1272.6 5	0.15 3
γ M1,E2	1285.6 3	0.33 4
γ M1	1299.99 7	0.45 6
γ E1,E2	1302.9 5	0.21 4
γ E2	1306.6 6	0.14 3
γ E1	1318.26 9	1.40 16
γ E1,E2	1322.6 7	0.087 2
γ M1,E2	1332.0 5	0.11 2
γ	1338.5 5	0.09 2
γ E1	1349.86 9	0.41 6
γ	1365.0 9	0.047 2
γ E1	1374.26 7	4.2 3
γ E1	1386.9 3	1.0 2
γ E1	1397.44 7	7.1 10
γ E1	1405.29 7	0.9 3
γ E1	1409.98 9	0.78 20
γ M1(+E2)	1422.8 5	0.06 2
γ E1	1434.38 7	7.6 8
γ	1446.0 5	0.07 2
γ E1	1455.81 7	3.4 4
γ E1	1465.41 7	1.79 20
γ M1	1469.45 7	2.7 3
γ M1	1480.82 20	0.55 6
γ [E1]	1488.67 8	0.05 2
γ M1	1500.49 8	0.31 4
γ [E1]	1514.29 9	0.033 10
γ E1	1526.1 6	0.83 18
γ	1532.5 7	0.083 20
γ E1,E2	1561.5 4	0.19 3
γ [E1]	1569.71 7	0.064 12
γ	1578.0 6	0.045 10
γ M1	1584.0 6	0.37 5
γ [E1]	1592.97 8	0.045 10
γ [E1]	1618.04 9	0.045 10
γ	1627.2 6	0.025 8
γ	1637.8 5	0.07 2
γ	1649.6 6	0.06 2
γ	1654.6 5	0.075 16
γ E1	1662.12 10	0.92 10
γ	1673.2 4	0.065 12
γ	1689.12 20	0.31 4
γ E1	1697.23 7	0.43 6
γ	1709.2 3	0.11 2
γ E1	1722.35 9	0.47 6
γ (M1)	1732.8 3	0.13 2
γ M1,E2	1742.1 4	0.08 1
γ (E1)	1749.22 10	0.93 12
γ	1753.4 4	0.16 2
γ	1757.1 6	0.07 2
γ	1767.6 3	0.15 2
γ	1784.12 20	0.36 4
γ	1789.0 7	0.13 2
γ	1792.8 7	0.19 3
γ M1	1803.72 10	1.15 12
γ	1813.5 5	0.033 8
γ M1	1825.2 3	0.17 3
γ	1836.1 5	0.03 1
γ	1848.5 5	0.06 2
γ	1853.3 6	0.03 1
γ	1876.4 4	0.18 3
γ	1880.2 6	0.03 1
γ	1889.5 4	0.03 1
γ	1913.5 5	0.03 1
γ E1	1936.32 20	0.31 4
γ	1948.2 4	0.06 2
γ	1957.0 6	0.04 1
γ	1970.7 6	0.013 3
γ	1983.1 5	~0.04
γ	2017.6 4	0.035 8
γ	2038.5 6	0.006 2
γ	2041.4 4	0.018 6
γ	2052.9 7	0.014 5
γ	2079.8 7	0.012 4

† 14% uncert(syst)

Atomic Electrons (^{163}Tm)

⟨e⟩=63.3 24 keV

e_{bin}(keV)	⟨e⟩(keV)	e(%)
3 - 6	0.32	12.0 21
8	5.3	63 7
9	2.9	31 4
10 - 46	6.0	30.3 17
47	20.3	43 5
48 - 94	5.9	9.4 5
95	6.0	6.3 7
96 - 101	0.110	0.115 14
102	1.45	1.42 15
103 - 152	1.70	1.49 16
153 - 181	0.33	0.20 3
182	1.62	0.89 9
184	3.6	1.93 19
188 - 237	1.88	0.84 6
238 - 286	1.72	0.70 5
287 - 336	0.81	0.264 19
337 - 385	0.40	0.109 18
387 - 436	0.52	0.127 9
437 - 485	0.266	0.058 4
489 - 538	0.43	0.084 11
539 - 587	0.147	0.0264 22
589 - 634	0.302	0.050 4
640 - 688	0.095	0.0144 16
689 - 739	0.108	0.0152 22
740 - 790	0.088	0.0116 9
794 - 838	0.049	0.0059 5
844 - 893	0.070	0.0080 7
894 - 943	0.045	0.0049 4
944 - 993	0.042	0.0043 4
994 - 1043	0.042	0.0041 4
1044 - 1091	0.042	0.0039 3
1096 - 1145	0.035	0.0031 3
1147 - 1196	0.065	0.0055 4
1202 - 1252	0.105	0.0086 6
1253 - 1301	0.0377	0.00297 19
1302 - 1350	0.120	0.0090 7
1352 - 1402	0.124	0.0090 6
1403 - 1447	0.141	0.0099 7
1454 - 1552	0.049	0.00328 20
1553 - 1652	0.027	0.00165 22
1653 - 1752	0.064	0.0037 3
1755 - 1852	0.0178	0.00100 11
1856 - 1955	0.0061	0.00032 5
1960 - 2051	0.0014	7.3 17 $\times 10^{-5}$
2070 - 2078	2.8 $\times 10^{-5}$	1.3 6 $\times 10^{-6}$

Continuous Radiation (^{163}Tm)

⟨β+⟩=11.2 keV;⟨IB⟩=1.26 keV

E_{bin}(keV)		⟨ ⟩(keV)	(%)
0 - 10	β+	8.6 $\times 10^{-8}$	1.03 $\times 10^{-6}$
	IB	0.00129	
10 - 20	β+	4.69 $\times 10^{-6}$	2.80 $\times 10^{-5}$
	IB	0.00065	0.0044
20 - 40	β+	0.000172	0.00052
	IB	0.0092	0.027
40 - 100	β+	0.0111	0.0141
	IB	0.20	0.40
100 - 300	β+	0.62	0.279
	IB	0.044	0.023
300 - 600	β+	3.63	0.80
	IB	0.140	0.031
600 - 1300	β+	6.9	0.85
	IB	0.54	0.059
1300 - 2370	β+	0.00194	0.000148
	IB	0.33	0.021
	Σβ+		1.9

$^{163}_{70}$Yb(11.05 *25* min)

Mode: ε

Δ: -59370 *100* keV

SpA: 1.043×10^{8} Ci/g

Prod: ^{169}Tm(p,7n); protons on Ta; ^{164}Er(^{3}He,4n); ^{166}Er(^{3}He,6n)

Photons (^{163}Yb)

⟨γ⟩=520 *5* keV

γ_{mode}	γ(keV)	γ(%)
Tm L_ℓ	6.341	0.45 *10*
Tm L_α	7.176	11.4 *21*
Tm L_η	7.310	0.15 *4*
γ [M1]	7.679 *20*	0.0069 *21*
Tm L_β	8.212	13 *4*
Tm L_γ	9.586	2.4 *7*
γ E2	9.74 *4*	0.000099 *4*
γ M1+~0.2%E2	13.514 *18*	~0.24
Tm $K_{\alpha2}$	49.772	20.9 *6*
Tm $K_{\alpha1}$	50.742	36.8 *11*
Tm $K_{\beta1}'$	57.444	11.5 *4*
Tm $K_{\beta2}'$	59.296	3.20 *11*
γ E1	63.62 *3*	6.9 *3*
γ M1+30%E2	79.98 *3*	0.108 *22*
γ M1+<16%E2	87.65 *3*	0.216 *22*
γ M1(+20%E2)	113.96 *4*	0.292 *22*
γ M1(+18%E2)	121.63 *4*	0.216 *22*
γ M1+<5.0%E2	123.189 *16*	2.12 *8*
γ M1+<12%E2	130.867 *16*	1.41 *4*
γ M1+47%E2	136.703 *20*	0.242 *11*
γ	141.18 *6*	0.119 *11*
γ E2	144.381 *19*	0.529 *22*
γ	151.8 *4*	0.043 *16*
γ E2	161.47 *3*	1.14 *4*
γ E1	181.83 *3*	0.259 *16*
γ E1	189.50 *3*	0.28 *5*
γ [E1]	194.1 *4*	0.049 *16*
γ E1	203.62 *3*	0.58 *3*
γ E1	217.13 *3*	0.46 *4*
γ (E2)	221.96 *6*	0.100 *16*
γ E1	234.44 *6*	0.83 *4*
γ [M1]	274.20 *5*	0.24 *3*
γ M1+28%E2	304.80 *5*	0.140 *22*
γ E2+~20%M1	312.48 *4*	0.35 *5*
γ [E1]	312.69 *3*	0.21 *5*
γ E1	326.21 *3*	1.67 *7*
γ M1(+E2)	352.83 *6*	0.78 *4*
γ [E1]	353.73 *15*	0.22 *5*
γ E1	361.41 *15*	0.259 *22*
γ M1+42%E2	366.35 *6*	0.77 *3*
γ E2+35%M1	384.57 *9*	0.241 *16*
γ E2	407.70 *7*	0.153 *11*
γ [M1+E2]	415.17 *9*	0.052 *8*
γ	416.8 *4*	0.065 *9*
γ [M1+E2]	422.85 *9*	0.032 *11*
γ M1(+E2)	435.67 *4*	0.65 *4*
γ	447.35 *24*	0.076 *4*
γ	457.4 *12*	0.032 *11* ?
γ	481.96 *24*	0.093 *11*
γ	484.70 *21*	0.097 *22*
γ	492.38 *21*	0.065 *22*
γ	520.7 *8*	0.069 *16*
γ	539.33 *14*	0.36 *3*
γ	547.01 *14*	0.13 *3*
γ	561.5 *4*	0.084 *12*
γ	567.70 *24*	0.108 *12*
γ	571.27 *21*	0.076 *12*
γ	588.85 *18*	0.069 *18* ?
γ [M1+E2]	599.05 *7*	0.28 *3*
γ	601 *4*	0.022 *7*
γ	606.01 *18*	0.59 *3*
γ	619.3 *4*	0.130 *22*
γ	622.2 *5*	0.108 *22*
γ	643.8 *4*	0.076 *11*
γ	648.95 *9*	0.238 *22*
γ	661.74 *8*	0.201 *21*
γ	670.0 *6*	0.11 *3*
γ (E1)	687.17 *7*	1.73 *9*
γ	688.96 *6*	0.22 *5*

Photons (^{163}Yb) (continued)

γ_{mode}	γ(keV)	γ(%)
γ	694.6 *4*	0.26 *3*
γ	709.66 *24*	0.12 *3*
γ	730.04 *24*	0.173 *22*
γ	737.06 *8*	0.80 *4*
γ	739.9 *12*	0.097 *22* ?
γ	743.7 *6*	0.084 *16*
γ	759.1 *4*	0.108 *22*
γ	772.66 *12*	0.274 *15*
γ	797.0 *4*	0.184 *22*
γ	802.91 *5*	0.38 *3*
γ	804.62 *20*	0.032 *5* ?
γ	806.12 *8*	0.032 *5* ?
γ	810.59 *5*	0.66 *3*
γ	817.9 *6*	0.100 *16*
γ	841.1 *14*	0.16 *8*
γ	848.5 *8*	0.29 *8*
γ	853.0 *6*	0.43 *8*
γ M1,E2	860.42 *6*	10.8 *4*
γ	867.5 *9*	0.25 *8*
γ	872.26 *13*	0.19 *8*
γ	881.4 *12*	0.060 *16* ?
γ	886.20 *7*	0.162 *22*
γ	904.3 *3*	0.30 *3*
γ	913.70 *20*	0.108 *22*
γ	913.71 *17*	0.108 *22*
γ	920.27 *24*	0.30 *3*
γ	934.9 *6*	0.16 *3*
γ	942.19 *24*	0.52 *4*
γ	948.5 *7*	0.151 *22*
γ	959.13 *18*	0.367 *22*
γ	969.99 *24*	0.324 *22*
γ	973.93 *11*	0.22 *3*
γ	983.0 *12*	0.13 *3*
γ	986.44 *19*	0.27 *3*
γ	994.12 *19*	0.21 *3*
γ	996.55 *17*	0.25 *3*
γ	1002.67 *18*	0.270 *22*
γ	1006.00 *8*	0.443 *22*
γ	1009.56 *9*	0.300 *17*
γ	1015.3 *6*	0.043 *16*
γ	1019.8 *6*	0.054 *22*
γ	1023.27 *16*	0.305 *18*
γ	1035.9 *4*	0.184 *22*
γ	1040.8 *4*	0.238 *22*
γ	1045.10 *24*	0.324 *22*
γ	1052.2 *8*	0.054 *22*
γ [E1]	1059.45 *8*	0.205 *22*
γ	1065.3 *5*	0.140 *18*
γ	1095.0 *3*	0.119 *22*
γ	1102.11 *20*	0.108 *22*
γ	1108.0 *12*	0.040 *11* ?
γ	1113.9 *6*	0.097 *22*
γ	1115.9 *12*	~0.024 ?
γ	1124.23 *24*	0.132 *16*
γ	1158.0 *8*	0.046 *14*
γ	1164.79 *24*	0.14 *3*
γ	1172.0 *12*	~0.022
γ	1178.0 *18*	0.065 *22*
γ	1218.51 *17*	0.065 *11*
γ	1226.19 *17*	0.162 *22*
γ	1235.3 *4*	0.135 *16*
γ	1240.9 *3*	0.140 *16*
γ	1252.5 *6*	0.108 *22*
γ	1284.0 *6*	0.043 *16*
γ	1291.0 *9*	0.024 *11*
γ	1303.6 *5*	0.054 *22*
γ	1315.3 *8*	0.054 *11*
γ	1331.79 *6*	0.70 *3*
γ	1335.38 *15*	0.201 *14*
γ	1345.30 *6*	0.616 *22*
γ	1364.1 *4*	0.184 *22*
γ	1370.36 *12*	0.173 *22*
γ [E1]	1384.31 *7*	0.130 *22*
γ	1414.2 *4*	0.141 *18*
γ	1453.84 *9*	0.66 *5*
γ	1465.6 *12*	0.09 *3* ?
γ	1493.34 *11*	0.378 *22*
γ	1498.0 *12*	0.054 *22* ?
γ	1506.6 *4*	0.127 *14*
γ	1511.8 *6*	0.086 *22*
γ	1514.0 *12*	~0.032 ?
γ	1520.5 *5*	0.080 *15*
γ	1544.9 *4*	0.043 *9*

Photons (^{163}Yb) (continued)

γ_{mode}	γ(keV)	γ(%)
γ	1560.5 *4*	0.108 *22*
γ	1575.16 *6*	0.381 *16*
γ	1591.4 *4*	0.081 *11*
γ	1598.5 *12*	0.032 *13* ?
γ	1602.42 *12*	0.134 *15* ?
γ	1603.49 *14*	0.134 *15* ?
γ	1611.47 *18*	0.212 *16*
γ	1619.2 *6*	0.055 *11*
γ	1625.6 *4*	0.076 *22*
γ	1651.4 *9*	0.065 *22*
γ [E1]	1658.50 *6*	0.42 *3*
γ	1676.24 *13*	0.378 *22*
γ	1683.92 *13*	0.244 *15*
γ [E1]	1689.11 *5*	0.81 *4*
γ [E1]	1696.79 *5*	0.54 *4*
γ	1708.3 *4*	0.103 *12*
γ	1734.6 *4*	0.097 *14*
γ [M1+E2]	1746.62 *6*	1.84 *11*
γ	1766.53 *8*	0.43 *3*
γ	1788.54 *24*	0.130 *16*
γ [E1]	1819.98 *5*	0.218 *16*
γ	1843.7 *5*	0.073 *9*
γ	1857.1 *12*	0.043 *9* ?
γ	1861.0 *12*	0.043 *11* ?
γ	1871.2 *5*	0.076 *11*
γ	1883.0 *12*	0.050 *9* ?
γ	1907.83 *10*	1.64 *7*
γ	2014.4 *7*	0.033 *7*
γ	2026.0 *7*	0.032 *7*
γ	2052.8 *7*	0.027 *7*
γ	2060.6 *7*	0.026 *7*

Atomic Electrons (^{163}Yb)

⟨e⟩=24.4 *16* keV

e_{bin}(keV)	⟨e⟩(keV)	e(%)
3	1.1	~32
4	0.60	15 *5*
5 - 8	0.8	14 *5*
9	3.6	41 *7*
10	3.9	40 *11*
11	1.0	~9
12 - 57	3.5	11.9 *19*
61 - 63	0.38	0.61 *6*
64	1.95	3.05 *12*
70	0.09	~0.13
71	1.23	1.71 *12*
77 - 123	2.45	2.41 *11*
127 - 175	1.15	0.80 *3*
179 - 226	0.128	0.060 *6*
232 - 274	0.176	0.068 *6*
293 - 343	0.38	0.123 *22*
344 - 388	0.17	0.048 *10*
398 - 447	0.056	0.013 *3*
448 - 492	0.047	0.010 *5*
502 - 551	0.10	0.018 *6*
552 - 601	0.049	0.0085 *24*
602 - 652	0.088	0.014 *3*
653 - 701	0.08	0.011 *4*
707 - 757	0.08	0.011 *4*
758 - 800	0.052	0.0066 *21*
801	0.54	0.067 *24*
802 - 851	0.13	0.016 *4*
852 - 900	0.096	0.0110 *23*
902 - 950	0.093	0.0100 *20*
956 - 1005	0.057	0.0058 *13*
1006 - 1055	0.024	0.0023 *5*
1056 - 1104	0.008	0.0007 *3*
1105 - 1155	0.008	0.0007 *3*
1156 - 1193	0.016	0.0014 *4*
1208 - 1256	0.0072	0.00058 *15*
1272 - 1322	0.046	0.0036 *11*
1323 - 1370	0.0100	0.00075 *18*
1374 - 1414	0.016	~0.0012
1434 - 1483	0.019	0.0013 *4*
1484 - 1583	0.027	0.0018 *4*
1588 - 1687	0.076	0.0046 *6*

Atomic Electrons (^{163}Yb) (continued)

e_{bin}(keV)	⟨e⟩(keV)	e(%)
1688 - 1787	0.022	0.0013 3
1798 - 1881	0.030	0.0016 7
1898 - 1993	0.0058	0.00030 11
2001 - 2059	0.00067	3.3 11 $\times 10^{-5}$

Continuous Radiation (^{163}Yb)

⟨β+⟩=283 keV;⟨IB⟩=5.1 keV

E_{bin}(keV)		⟨ ⟩(keV)	(%)
0 - 10	β+	3.02×10^{-7}	3.58×10^{-6}
	IB	0.0113	
10 - 20	β+	1.68×10^{-5}	0.000100
	IB	0.0107	0.074
20 - 40	β+	0.00063	0.00191
	IB	0.025	0.085
40 - 100	β+	0.0430	0.054
	IB	0.23	0.42
100 - 300	β+	2.70	1.21
	IB	0.20	0.109
300 - 600	β+	21.1	4.54
	IB	0.35	0.079
600 - 1300	β+	136	14.4
	IB	1.39	0.148
1300 - 2500	β+	123	7.7
	IB	2.5	0.142
2500 - 3370	IB	0.37	0.0135
	Σβ+		28

$^{163}_{71}$Lu(4.1 2 min)

Mode: ϵ
Δ: -54740 300 keV syst
SpA: 2.81×10^{8} Ci/g
Prod: protons on W

decay data not yet evaluated

$^{163}_{72}$Hf(40.0 6 s)

Mode: ϵ
Δ: -49390 420 keV syst
SpA: 1.72×10^{9} Ci/g
Prod: ^{147}Sm(^{20}Ne,4n); ^{142}Nd(^{24}Mg,3n)

Photons (^{163}Hf)

γ_{mode}	γ(keV)	γ(rel)
γ	45.39 8	48 2
γ	62.14 5	64 5
γ	70.98 8	100
γ	84.9 1	<2
γ	133.08 10	24 1
γ	162.25 15	16 1
γ	233.35 10	17 1
γ	496.07 10	13 1
γ	520.32 10	19 1
γ	535.25 20	4 1
γ	688.25 10	33 4

$^{163}_{74}$W (2.5 3 s)

Mode: α
Δ: -35150 310 keV syst
SpA: 2.4×10^{10} Ci/g
Prod: ^{144}Sm(^{24}Mg,5n); ^{147}Sm(^{24}Mg,8n)

Alpha Particles (^{163}W)

α(keV)
5385 5

$^{163}_{75}$Re(260 40 ms)

Mode: α(64 18 %), ϵ(36 18 %)
Δ: -26110 430 keV syst
SpA: 9.3×10^{10} Ci/g
Prod: daughter ^{167}Ir; ^{107}Ag(^{58}Ni,2n); ^{109}Ag(^{58}Ni,4n)

Alpha Particles (^{163}Re)

α(keV)
5918 6

$^{163}_{76}$Os($t_{1/2}$ unknown)

Mode: α
Δ: -16450 590 keV syst
Prod: ^{58}Ni on Mo-Sn targets; ^{107}Ag on V-Ni targets

Alpha Particles (^{163}Os)

α(keV)
6510 30

A = 164

NDS 11, 327 (1974)

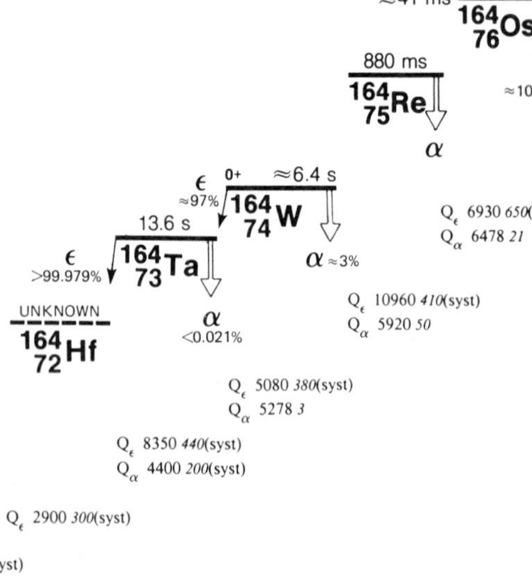

$^{164}_{65}$Tb(3.0 *1* min)

Mode: β-

Δ: -62120 *150* keV

SpA: 3.81×10^{8} Ci/g

Prod: ^{164}Dy(n,p)

Photons (^{164}Tb)

⟨γ⟩=2177 *39* keV

γ_{mode}	γ(keV)	γ(%)†
Dy L_ℓ	5.743	0.32 *8*
Dy L_α	6.491	8.4 *21*
Dy L_η	6.534	0.14 *4*
Dy L_β	7.368	8.3 *20*
Dy L_γ	8.452	1.2 *3*
γ	36.8 *4*	~0.2
γ	39.9 *3*	~0.2
Dy $K_{\alpha2}$	45.208	8.0 *7*
Dy $K_{\alpha1}$	45.998	14.3 *12*
Dy $K_{\beta1}$'	52.063	4.3 *4*
Dy $K_{\beta2}$'	53.735	1.27 *11*
γ E2	73.356 *5*	7.8 *10*
γ	84.3 *3*	0.4 *1*
γ	86.7 *3*	0.3 *1*
γ [E1]	98.118 *6*	0.3 *1*
γ	98.71 *10*	~0.2
γ	104.3 *5*	0.3 *1*
γ [E1]	123.309 *9*	1.4 *2*
γ [M1+E2]	131.0 *4*	0.3 *1* ?
γ	141.0 *5*	0.4 *1*
γ [E2]	145.875 *13*	0.10 *2* ?
γ [E1]	148.688 *9*	3.7 *3*
γ	152.6 *5*	~0.8
γ [E2]	154.178 *17*	0.054 *10* ?
γ	159.38 *18*	0.4 *1* ?
γ [E2]	168.829 *6*	24.0 *12*
γ	174.68 *23*	0.3 *1*
γ	176.7 *3*	~0.2
γ [E2]	185.831 *21*	0.3 *1*
γ [E2]	196.444 *14*	~0.2 ?
γ [E1]	200.490 *21*	~0.4
γ [E1]	206.768 *14*	1.58 *20*
γ [E1]	211.103 *13*	5.7 *6*
γ [E1]	215.071 *20*	20
γ	246.5 *5*	0.4 *1*
γ [E2]	259.06 *14*	4 *1*
γ [E1]	277.487 *16*	8.0 *6*
γ [E1]	294.563 *13*	6.7 *4*
γ [E1]	309.140 *21*	1.7 *3*
γ [E1]	344.7 *4*	5 *1*
γ	386.3 *5*	0.8 *2*
γ	410.39 *16*	5.8 *4*
γ [E2]	414.68 *14*	~0.2
γ	424.9 *4*	0.88 *20*
γ	434.9 *10*	0.44 *10*
γ [M1+E2]	465.05 *10*	0.88 *16*
γ	480.8 *5*	0.3 *1*
γ [E1]	484.9 *3*	0.4 *1*
γ [E2]	519.56 *7*	0.4 *1*
γ [M1+E2]	523.33 *14*	0.6 *1*
γ [M1+E2]	548.51 *10*	8.0 *6*
γ	559.41 *20*	0.84 *10*
γ [M1+E2]	566.3 *4*	~0.4
γ [M1+E2]	583.0 *3*	0.6 *2* ?
γ [M1+E2]	585.95 *7*	8.0 *6* ?
γ [E2]	610.93 *10*	19.0 *12*
γ [E1]	617.80 *16*	11.4 *8*
γ [M1+E2]	632.7 *4*	~0.4
γ	646.8 *3*	5.8 *6*
γ [M1+E2]	654.4 *4*	0.30 *4* ?
γ [E1]	671.82 *10*	0.9 *2*
γ [M1+E2]	673.74 *7*	8.8 *8*
γ [M1+E2]	688.39 *7*	20.0 *12*
γ	695.49 *20*	0.5 *1*
γ	701.0 *4*	0.44 *10*
γ [E1]	707.4 *4*	0.3 *1*

Photons (^{164}Tb) (continued)

γ_{mode}	γ(keV)	γ(%)†
γ	715.5 *10*	0.4 *1*
γ [E1]	723.82 *14*	~0.4
γ	744.9 *3*	0.3 *1* ?
γ [M1+E2]	754.77 *7*	22.1 *14*
γ [E2]	761.75 *7*	16.2 *10*
γ [E1]	770.0 *5*	0.5 *1*
γ [M1+E2]	779.4 *3*	1.3 *1*
γ [M1+E2]	782.39 *7*	5.4 *5*
γ	790.3 *10*	1.7 *3*
γ [E1]	797.05 *7*	0.8 *3*
γ [E2]	799.9 *5*	0.6 *2* ?
γ	808.3 *4*	2.0 *6*
γ [E1]	809.7 *4*	0.57 *20*
γ [E2]	811.1 *5*	0.57 *20* ?
γ	821.0 *10*	~0.2
γ	827.0 *5*	0.78 *20*
γ	835.2 *10*	0.58 *20*
γ [E2]	842.57 *7*	3.0 *4*
γ [E2]	845.8 *3*	5.5 *8*
γ	848.0 *10*	0.9 *3*
γ	856.3 *4*	0.4 *1*
γ	875.44 *17*	0.4 *1*
γ	882.0 *20*	~0.2
γ	889.0 *20*	0.3 *1*
γ [E1]	903.46 *7*	0.3 *1*
γ [E2]	913.4 *4*	0.3 *1* ?
γ	934.0 *5*	0.44 *10*
γ	945.2 *5*	0.44 *10*
γ	952.5 *10*	0.8 *3*
γ [E1]	965.88 *7*	1.4 *4*
γ [E1]	969.3 *3*	0.40 *8* ?
γ	976.5 *10*	~0.4
γ [E1]	982.88 *8*	0.8 *1*
γ [M1+E2]	1016.5 *4*	0.7 *2*
γ	1021.31 *17*	0.44 *10* ?
γ	1029.0 *10*	0.4 *1*
γ	1034.81 *21*	0.5 *1*
γ	1050.12 *23*	~0.2
γ [M1+E2]	1104.3 *4*	1.1 *2*
γ	1106.0 *10*	0.4 *1*
γ	1113.0 *10*	~0.2
γ	1123.2 *4*	1.4 *4*
γ	1124.9 *4*	0.4 *1* ?
γ	1133.57 *23*	0.3 *1*
γ	1148.23 *23*	1.7 *3*
γ [E2]	1152.2 *4*	0.86 *20*
γ [E1]	1155.1 *3*	1.04 *20*
γ [E1]	1166.31 *16*	1.9 *4*
γ [M1+E2]	1169.7 *3*	1.1 *4*
γ	1170.00 *17*	1.2 *4*
γ [M1+E2]	1180.97 *16*	~0.8
γ	1189.6 *4*	0.3 *1*
γ	1195.99 *23*	0.3 *1* ?
γ	1217.2 *5*	0.9 *1*
γ	1223.0 *4*	~0.2 ?
γ	1228.5 *10*	~0.2
γ	1233.1 *10*	~0.2
γ	1256.88 *23*	~0.2
γ	1270.7 *4*	0.6 *2* ?
γ [M1+E2]	1278.4 *3*	2.2 *4*
γ	1288.2 *5*	1.6 *6*
γ [M1+E2]	1289.62 *16*	4.0 *6*
γ	1301.2 *10*	~0.2
γ	1307.6 *10*	0.3 *1*
γ [M1+E2]	1321.1 *4*	1.0 *4*
γ	1329.38 *21*	0.5 *1*
γ	1334.3 *20*	~0.8
γ [M1+E2]	1366.2 *3*	1.9 *5*
γ	1372.0 *10*	~0.2
γ [M1+E2]	1377.41 *16*	5.0 *8*
γ [E2]	1394.4 *4*	~0.4
γ	1395.76 *21*	0.3 *1* ?
γ	1411.06 *23*	~0.2 ?
γ	1426.1 *4*	0.5 *1*
γ [E2]	1443.80 *16*	8.1 *8*
γ	1485.8 *4*	0.3 *1* ?
γ	1576.0 *15*	~0.2
γ	1652.2 *4*	0.4 *1*
γ	1656.26 *25*	0.9 *3* ?
γ	1665.0 *15*	0.6 *2*
γ	1740.0 *20*	0.3 *1*
γ	1877.9 *4*	0.3 *1*

Photons (^{164}Tb) (continued)

γ_{mode}	γ(keV)	γ(%)†
γ	1891.1 *4*	~0.2
γ	1898.0 *20*	~0.2
γ	1905.0 *20*	~0.2
γ	1915.32 *22*	~0.2
γ	1924.77 *18*	~0.2 ?
γ	1930.62 *24*	0.3 *1*
γ [M1+E2]	1952.1 *3*	0.44 *10*
γ [M1+E2]	1963.36 *17*	0.7 *1*
γ	1983.0 *20*	0.3 *1*
γ	1989.3 *4*	<0.2
γ	2048.0 *20*	0.3 *1*
γ	2085.3 *11*	0.3 *1* ?
γ	2099.44 *24*	0.3 *1* ?
γ [E2]	2132.18 *17*	0.8 *1*
γ	2174.2 *5*	0.4 *1* ?
γ	2219.0 *20*	~0.2
γ	2240.5 *20*	0.76 *20*
γ	2311.5 *20*	~0.2
γ	2502 *3*	0.5 *1*
γ	2512.6 *4*	1.3 *3*
γ	2627.0 *20*	0.3 *1*
γ	2763 *3*	0.4 *1* ?
γ	2786 *3*	0.3 *1*

† 5.0% uncert(syst)

Atomic Electrons (^{164}Tb)

⟨e⟩=90 *6* keV

e_{bin}(keV)	⟨e⟩(keV)	e(%)
8	3.3	43 *11*
9	2.6	30 *8*
20	3.2	16.6 *22*
28	2.2	<15
29	2.6	<18
30 - 64	8.9	25 *9*
65	12.4	19.1 *25*
66	13.6	21 *3*
70 - 71	0.39	0.56 *6*
72	6.9	9.6 *12*
73 - 114	4.6	5.7 *8*
115	7.2	6.3 *4*
116 - 159	1.40	0.99 *15*
160	3.04	1.90 *12*
161	2.80	1.74 *11*
166 - 215	2.70	1.49 *10*
224 - 270	0.99	0.413 *24*
275 - 309	0.23	0.081 *11*
333 - 381	0.8	~0.22
384 - 433	0.28	0.067 *18*
434 - 483	0.095	0.020 *5*
484 - 532	1.3	0.25 *6*
539 - 586	1.27	0.228 *15*
593 - 642	2.2	0.35 *8*
645 - 694	0.41	0.061 *10*
695 - 744	2.0	0.28 *6*
745 - 794	0.86	0.112 *13*
795 - 844	0.128	0.0155 *23*
845 - 895	0.049	0.0056 *16*
896 - 945	0.072	0.0078 *22*
950 - 996	0.072	0.0074 *18*
1007 - 1052	0.055	0.0053 *12*
1059 - 1106	0.13	0.012 *3*
1111 - 1159	0.14	0.012 *3*
1161 - 1209	0.057	0.0049 *12*
1214 - 1263	0.22	0.018 *3*
1267 - 1313	0.13	0.0105 *21*
1318 - 1366	0.15	0.011 *3*
1368 - 1418	0.176	0.0127 *12*
1424 - 1444	0.032	0.0022 *3*
1477 - 1574	0.0046	~0.00030
1598 - 1686	0.038	0.0023 *7*
1731 - 1824	0.0046	~0.00026
1837 - 1935	0.043	0.0023 *3*
1943 - 2040	0.012	0.00058 *15*
2046 - 2131	0.021	0.00101 *16*
2165 - 2258	0.015	0.00068 *22*

Atomic Electrons (^{164}Tb) (continued)

e_{bin}(keV)	⟨e⟩(keV)	e(%)
2302 - 2310	0.00036	~2×10^{-5}
2448 - 2511	0.020	0.0008 3
2573 - 2625	0.0032	0.00013 6
2710 - 2784	0.0072	0.00026 8

Continuous Radiation (^{164}Tb)

⟨β-⟩=682 keV;⟨IB⟩=1.23 keV

E_{bin}(keV)		⟨ ⟩(keV)	(%)
0 - 10	β-	0.0308	0.61
	IB	0.028	
10 - 20	β-	0.093	0.62
	IB	0.027	0.19
20 - 40	β-	0.383	1.27
	IB	0.053	0.18
40 - 100	β-	2.86	4.06
	IB	0.144	0.22
100 - 300	β-	31.6	15.6
	IB	0.37	0.21
300 - 600	β-	115	25.6
	IB	0.33	0.079
600 - 1300	β-	390	43.2
	IB	0.26	0.032
1300 - 2500	β-	142	9.2
	IB	0.024	0.00158
2500 - 2835	β-	0.277	0.0107
	IB	5.5×10^{-6}	2.2×10^{-7}

$^{164}_{66}$Dy(stable)

Δ: -65977 *3* keV
%: 28.2 *2*

$^{164}_{67}$Ho(29.0 *5* min)

Mode: ε(58 %), β-(42 %)
Δ: -64939 *4* keV
SpA: 3.95×10^{7} Ci/g
Prod: protons on Dy; ^{165}Ho(γ,n); ^{165}Ho(n,2n); ^{164}Dy(d,2n)

Photons (^{164}Ho)

⟨γ⟩=29.5 *5* keV

γ_{mode}	γ(keV)	γ(%)†
Dy L_ℓ	5.743	0.203 21
Er L_ℓ	6.151	0.030 5
Dy L_α	6.491	5.3 4
Dy L_η	6.534	0.074 6
Er L_α	6.944	0.76 11
Er L_η	7.058	0.0131 19
Dy L_β	7.379	5.0 4
Er L_β	7.921	0.79 11
Dy L_γ	8.504	0.76 8
Er L_γ	9.113	0.118 17
Dy $K_{\alpha2}$	45.208	14.2 5
Dy $K_{\alpha1}$	45.998	25.3 8
Er $K_{\alpha2}$	48.221	0.89 11
Er $K_{\alpha1}$	49.128	1.58 19
Dy $K_{\beta1}'$	52.063	7.68 24
Dy $K_{\beta2}'$	53.735	2.25 8
Er $K_{\beta1}'$	55.616	0.49 6
Er $K_{\beta2}'$	57.406	0.139 17
γ_ϵE2	73.356 5	1.97 13
$\gamma_{\beta-}$E2	91.347 9	2.5
γ_ϵ[M1+E2]	688.39 7	~0.0010
γ_ϵ[E2]	761.75 7	~0.0010

† uncert(syst): 12% for ε, 12% for β-

Atomic Electrons (^{164}Ho)

⟨e⟩=22.5 *7* keV

e_{bin}(keV)	⟨e⟩(keV)	e(%)
8	2.3	30 3
9	1.76	20.1 23
10	0.0161	0.17 3
20	0.82	4.2 3
34	1.11	3.3 4
36 - 64	1.55	3.65 15
65	3.13	4.8 3
66	3.43	5.2 4
71	0.063	0.088 6
72	1.74	2.43 17
73	0.48	0.66 5
82	2.4	2.9 4
83	2.2	2.6 3
89	0.62	0.70 9
90 - 91	0.91	1.02 9
635 - 681	6.4 ×10^{-5}	~1.0×10^{-5}
686 - 708	3.5 ×10^{-5}	4.9 23 ×10^{-6}
753 - 762	7.1 ×10^{-6}	9 4 ×10^{-7}

Continuous Radiation (^{164}Ho)

⟨β-⟩=139 keV;⟨IB⟩=0.36 keV

E_{bin}(keV)		⟨ ⟩(keV)	(%)
0 - 10	β-	0.0323	0.64
	IB	0.0070	
10 - 20	β-	0.097	0.65
	IB	0.0064	0.044
20 - 40	β-	0.395	1.32
	IB	0.025	0.077
40 - 100	β-	2.85	4.07
	IB	0.125	0.24
100 - 300	β-	27.7	13.9
	IB	0.077	0.044
300 - 600	β-	70	16.0
	IB	0.083	0.020
600 - 1038	β-	38.3	5.4
	IB	0.034	0.0048

$^{164}_{67}$Ho(37.5 *10* min)

Mode: IT
Δ: -64799 *4* keV
SpA: 3.06×10^{7} Ci/g
Prod: ^{164}Dy(d,2n); ^{165}Ho(n,2n); ^{165}Ho(γ,n)

Photons (^{164}Ho)

⟨γ⟩=47.7 *13* keV

γ_{mode}	γ(keV)	γ(%)†
Ho L_ℓ	5.943	0.65 11
Ho L_α	6.716	16.9 24
Ho L_η	6.789	0.28 4
Ho L_β	7.632	21 3
Ho L_γ	8.893	3.9 6
γ M1	37.34 5	11.7 6
γ (E3)	46 1	0.022 5
Ho $K_{\alpha2}$	46.700	21.4 12
Ho $K_{\alpha1}$	47.547	38.0 21
Ho $K_{\beta1}'$	53.822	11.7 6
Ho $K_{\beta2}'$	55.556	3.35 19
γ M1	56.64 5	6.7 3
γ [E2]	93.99 6	0.141 15

† 10% uncert(syst)

Atomic Electrons (^{164}Ho)

⟨e⟩=87 *5* keV

e_{bin}(keV)	⟨e⟩(keV)	e(%)
8	5.3	65 11
9	6.9	75 11
28	19.5	70 4
29	0.276	0.94 5
35	5.4	15.4 9
36	0.088	0.248 14
37	15.1	41 8
38	15.0	40 8
39	0.53	1.36 15
44	8.6	20 4
45	0.64	1.4 3
46	2.7	5.9 12
47	5.1	10.7 6
48	0.455	0.95 5
49	0.072	0.149 8
51	0.0076	0.0148 17
52	0.043	0.082 9
53	0.0175	0.033 4
54	0.0070	0.0130 15
55	1.40	2.57 13
85	0.115	0.135 16
86	0.102	0.119 14
92	0.056	0.061 7
93	0.00051	0.00055 6
94	0.0152	0.0162 19

$^{164}_{68}$Er(stable)

Δ: -65952 *4* keV
%: 1.61 *1*

$^{164}_{69}$Tm(2.0 *1* min)

Mode: ε
Δ: -61990 *20* keV
SpA: 5.7×10^{8} Ci/g
Prod: ^{164}Er(p,n); ^{166}Er(p,3n); daughter ^{164}Yb

Photons (^{164}Tm)

$\langle\gamma\rangle$=317 5 keV

γ_{mode}	γ(keV)	γ(%)†
Er L_ℓ	6.151	0.26 3
Er L_α	6.944	6.7 5
Er L_η	7.058	0.094 8
Er L_β	7.934	6.4 6
Er L_γ	9.160	1.00 11
Er $K_{\alpha2}$	48.221	15.9 6
Er $K_{\alpha1}$	49.128	28.3 10
Er $K_{\beta1}$'	55.616	8.8 3
Er $K_{\beta2}$'	57.406	2.48 10
γ E2	91.347 9	6.7
γ E2	208.05 3	1.19 5
γ E1	315.41 7	0.121 7
γ [E1]	524.35 11	0.040 13
γ [E2]	560.77 6	0.034 13
γ	595.09 9	0.422 13
γ	635.2 3	0.0871 3
γ	637.0 3	0.141 13
γ [M1+E2]	646.88 9	0.061 5
γ	689.59 6	0.107 13
γ	695.06 10	0.067 7 ?
γ [M1+E2]	758.80 16	0.047 7
γ E2	768.82 4	1.43 5
γ	841.92 7	0.402 20
γ [M1+E2]	854.93 9	0.274 13
γ (E2)	860.16 5	1.12 4
γ E2	905.53 7	0.335 20
γ (M1+E2+E0)	928.07 10	0.054 7
γ	935.6 6	0.047 13 ?
γ	963.4 6	0.034 7
γ [E2]	966.85 16	0.0115 19
γ [E2]	1015.06 8	0.161 7
γ	1058.0 3	0.456 20
γ [E1]	1134.51 9	0.074 7
γ E2	1154.55 10	1.67 5
γ (E2)	1165.7 4	0.75 3 ?
γ [E2]	1184.32 7	0.167 20
γ (M1+E2+E0)	1223.11 8	0.630 20
γ E1	1295.33 6	0.98 3
γ E2	1312.67 10	0.44 4
γ (E2)	1314.45 8	0.31 5
γ (E2)	1325.3 3	0.771 20
γ [E1]	1342.56 9	0.141 7
γ	1361.47 10	0.075 3 ?
γ	1373.9 7	0.054 13
γ [E2]	1378.4 8	~0.027
γ [E1]	1386.67 6	0.65 3
γ [M1+E2]	1392.37 7	0.134 7
γ	1416.6 3	0.027 7
γ [M1+E2]	1417.90 9	0.161 13
γ	1459.9 6	0.094 13 ?
γ	1483.3 10	<0.09
γ	1486.39 10	0.42 17
γ [E2]	1488.84 10	0.41 16
γ [E2]	1533.92 7	0.154 7
γ	1577.74 10	0.094 7
γ	1584.4 6	0.018 3
γ	1610.73 6	1.11 3
γ	1623.54 18	0.046 8 ?
γ E2	1674.34 7	1.02 3
γ (M1+E2+E0)	1696.89 9	0.275 13
γ	1713.9 3	0.052 3 ?
γ (M1+E2+E0)	1741.97 7	0.221 7
γ	1783.41 16	0.060 7 ?
γ [E2]	1788.23 9	0.047 7
γ (M1+E2+E0)	1819.67 10	0.409 20
γ [E2]	1833.32 7	0.101 20
γ (M1+E2+E0)	1862.6 7	0.52 3
γ	1874.76 16	0.060 7 ?
γ	1900.3 6	0.034 6 ?
γ [E2]	1911.01 10	0.050 5
γ	1934.5 4	0.181 7 ?
γ	1955.24 8	0.094 7
γ	1977.8 3	0.101 7 ?
γ	2010.9 4	0.017 3
γ	2050.58 15	0.034 7 ?
γ E2	2081.48 9	0.64 3
γ	2170.50 22	~0.04
γ M1+E2+E0	2186.71 10	0.034 7
γ [E2]	2278.06 10	0.074 7

Photons (^{164}Tm) (continued)

γ_{mode}	γ(keV)	γ(%)†
γ	2353.0 3	0.127 7
γ	2383.3 3	0.422 20
γ	2421.5 3	0.054 7
γ	2449.30 19	0.067 7
γ	2474.6 3	0.013 3
γ	2484.04 10	0.101 20
γ	2489.63 20	0.060 13
γ	2521.70 14	0.047 7
γ	2530.5 4	0.047 7
γ	2571.3 5	0.040 7
γ	2642.4 4	0.040 7
γ	2690.43 20	0.021 4
γ	2762.9 5	0.034 13
γ	2881.1 5	0.025 5
γ	2958.6 5	0.020 7
γ	3002.3 3	0.036 3
γ	3021.8 5	0.020 4
γ	3043.5 6	0.020 7 ?
γ	3081.3 4	0.015 3
γ	3107.8 6	0.017 3 ?
γ	3121.8 3	0.020 7
γ	3134.9 6	0.029 3 ?
γ	3315.9 6	0.020 7 ?

† 7.5% uncert(syst)

Atomic Electrons (^{164}Tm)

$\langle e\rangle$=26.0 9 keV

e_{bin}(keV)	$\langle e\rangle$(keV)	e(%)
8	2.5	29 3
9	1.49	16.1 18
10	0.39	4.0 4
34	2.96	8.8 7
38 - 55	1.31	3.10 13
82	6.5	7.9 7
83	5.8	7.0 6
89	1.67	1.87 15
90	1.58	1.76 15
91	0.87	0.96 8
151 - 200	0.400	0.243 9
206 - 208	0.0443	0.0215 9
258 - 307	0.0051	0.00192 10
313 - 315	0.000215	6.9 3 $\times10^{-5}$
467 - 516	0.0026	0.00053 15
522 - 561	0.026	~0.005
578 - 626	0.023	0.0040 13
627 - 647	0.012	0.0018 7
680 - 711	0.0541	0.0076 3
749 - 797	0.040	0.0052 15
803 - 852	0.056	0.0068 3
853 - 897	0.0078	0.00089 16
903 - 936	0.0027	0.00029 9
954 - 1000	0.018	0.0018 8
1005 - 1050	0.0029	0.00028 11
1056 - 1097	0.0403	0.00367 14
1108 - 1157	0.0318	0.00283 11
1164 - 1184	0.021	0.0018 5
1213 - 1257	0.0283	0.00227 13
1268 - 1317	0.0269	0.00210 11
1321 - 1370	0.0159	0.00118 11
1372 - 1418	0.0045	0.00032 8
1426 - 1475	0.017	0.0012 5
1476 - 1576	0.029	0.0019 7
1582 - 1673	0.0306	0.00188 14
1684 - 1782	0.018	0.00104 11
1786 - 1877	0.0183	0.00100 13
1891 - 1976	0.0041	0.00021 5
1993 - 2080	0.0114	0.00056 3
2113 - 2185	0.0013	6 3 $\times10^{-5}$
2221 - 2295	0.0026	0.00012 3
2326 - 2420	0.0075	0.00032 10
2427 - 2522	0.0038	0.00015 3
2528 - 2585	0.00050	1.9 7 $\times10^{-5}$
2633 - 2705	0.00065	2.4 8 $\times10^{-5}$
2753 - 2824	0.00029	1.0 4 $\times10^{-5}$
2871 - 2964	0.00076	2.6 6 $\times10^{-5}$

Atomic Electrons (^{164}Tm) (continued)

e_{bin}(keV)	$\langle e\rangle$(keV)	e(%)
2986 - 3080	0.00102	3.4 6 $\times10^{-5}$
3098 - 3133	0.000100	3.2 7 $\times10^{-6}$
3258 - 3314	0.00019	6.0 25 $\times10^{-6}$

Continuous Radiation (^{164}Tm)

$\langle\beta+\rangle$=523 keV;$\langle$IB$\rangle$=8.0 keV

E_{bin}(keV)		$\langle\ \rangle$(keV)	(%)
0 - 10	β+	2.24 $\times10^{-7}$	2.65 $\times10^{-6}$
	IB	0.019	
10 - 20	β+	1.22 $\times10^{-5}$	7.3 $\times10^{-5}$
	IB	0.018	0.125
20 - 40	β+	0.000457	0.00138
	IB	0.040	0.136
40 - 100	β+	0.0311	0.0393
	IB	0.23	0.40
100 - 300	β+	2.08	0.93
	IB	0.32	0.18
300 - 600	β+	18.1	3.86
	IB	0.49	0.112
600 - 1300	β+	151	15.6
	IB	1.63	0.17
1300 - 2500	β+	335	19.0
	IB	3.6	0.20
2500 - 3962	β+	16.9	0.65
	IB	1.59	0.055
	Σβ+		40

$^{164}_{69}$Tm(5.1 *1* min)

Mode: IT(~80 %), ϵ(~20 %)

Δ: >-61990 keV

SpA: 2.25$\times10^{8}$ Ci/g

Prod: ^{166}Er(p,3n); ^{165}Ho(α,5n)

Photons (^{164}Tm)

$\langle\gamma\rangle$=379 9 keV

γ_{mode}	γ(keV)	γ(%)†
γ_ϵE2	80.28 5	0.68 7
γ_ϵE2	91.347 9	4.8 7
γ_ϵ[M1+E2]	101.10 7	~0.19
γ_ϵE2	139.38 7	2.93 11
γ_ϵ	199.2 6	0.09 3
γ_ϵE2	208.05 3	16.7 5
γ_ϵM1	240.48 5	8.5 4
γ_ϵ	256.5 15	0.077 22
γ_ϵE2	314.85 5	11.0
γ_ϵ[E1]	385.46 12	0.32 3
γ_ϵE2	410.31 11	1.62 12
γ_ϵE1	546.98 7	5.1 3
γ_ϵM1(+E2)	583.04 7	0.81 6
γ_ϵ[E1]	625.94 13	0.30 3
γ_ϵ[M1+E2]	646.88 9	0.197 17
γ_ϵ	736.7 6	0.176 22
γ_ϵ[M1+E2]	744.56 13	0.121 11
γ_ϵ[M1+E2]	758.80 16	0.50 4
γ_ϵE1	820.81 11	1.50 8
γ_ϵ[M1+E2]	854.93 9	0.89 6
γ_ϵE2	897.89 6	4.87 22
γ_ϵ[M1+E2]	929.9 5	0.143 11
γ_ϵ[E1]	960.19 12	0.198 22
γ_ϵ[E2]	966.85 16	0.121 11
γ_ϵ[E1]	1049.74 6	1.75 9
γ_ϵ[E1]	1130.02 6	0.37 3
γ_ϵ	1140.0 8	0.187 22
γ_ϵ[E1]	1231.12 7	4.60 22

Photons (^{164}Tm)
(continued)

γ_{mode}	γ(keV)	γ(%)†
γ_ϵ[E1]	1364.59 *6*	4.68 *23*
γ_ϵ[E1]	1370.50 *7*	1.07 *11*
γ_ϵ	1498.0 *4*	0.17 *8*

† approximate

Continuous Radiation (^{164}Tm)
$\langle\beta+\rangle$=2.06 keV;$\langle$IB$\rangle$=0.34 keV

E_{bin}(keV)		$\langle\ \rangle$(keV)	(%)
0 - 10	β+	3.70×10^{-8}	4.38×10^{-7}
	IB	0.00025	
10 - 20	β+	2.0×10^{-6}	1.19×10^{-5}
	IB	0.000126	0.00087
20 - 40	β+	7.3×10^{-5}	0.000221
	IB	0.0018	0.0053
40 - 100	β+	0.00465	0.0059
	IB	0.041	0.082
100 - 300	β+	0.240	0.109
	IB	0.0100	0.0051
300 - 600	β+	1.13	0.254
	IB	0.042	0.0092
600 - 1300	β+	0.68	0.097
	IB	0.18	0.020
1300 - 1977	IB	0.061	0.0041
	$\Sigma\beta$+		0.47

$^{164}_{70}$Yb(1.26 *3* h)

Mode: ϵ
Δ: -60990 *100* keV syst
SpA: 1.51×10^{7} Ci/g
Prod: ^{169}Tm(p,6n); ^{166}Er(^{3}He,5n); protons on Ta

Photons (^{164}Yb)

γ_{mode}	γ(keV)	γ(rel)
γ M1(+E2)	97.30 *4*	41 *9*
γ	154.52 *3*	16 *3*
γ	362.92 *5*	38 *3*
γ	390.64 *5*	78 *3*
γ	401.90 *5*	31 *3*
γ	415.71 *5*	28 *3*
γ	445.44 *5*	94 *3*
γ	491.43 *17*	~9
γ	543.53 *20*	13 *3* ?
γ	588.73 *17*	28 *3*
γ	602.00 *7*	41 *3*
γ	675.33 *5*	100 *3*
γ	887.43 *20*	13 *3*

$^{164}_{71}$Lu(3.17 *3* min)

Mode: ϵ
Δ: -54690 *200* keV syst
SpA: 3.61×10^{8} Ci/g
Prod: ^{150}Sm(^{19}F,5n); ^{155}Gd(^{14}N,5n)

Photons (^{164}Lu)

γ_{mode}	γ(keV)	γ(rel)
γ[E2]	123.8 *2*	100 *2*
γ[E2]	262.1 *2*	31.9 *15*
γ[E2]	374.2 *3*	1.6 *3*
γ	412.0 *3*	1.7 *2*
γ	552.2 *2*	11.6 *8*
γ	605.0 *4*	<1.0
γ	608.2 *2*	6.1 *7*
γ[M1+E2]	618.3 *3*	2.8 *4*
γ[E2]	688.0 *3*	5.7 *7*
γ[M1+E2]	740.4 *2*	37.6 *14*
γ	747.9 *2*	15.9 *8*
γ	758.9 *3*	2.4 *5*
γ	852.1 *3*	9.2 *6*
γ[E2]	863.9 *2*	28.7 *11*
γ[M1+E2]	880.4 *2*	20.8 *10*
γ	937.9 *4*	2.8 *5*
γ[M1+E2]	949.4 *3*	7.3 *7*
γ	979.6 *4*	3.3 *9*
γ[E2]	1073.8 *5*	10.2 *9*
γ	1114.8 *4*	4.0 *9*
γ	1199.4 *4*	7.6 *7*
γ	1212.4 *3*	13.7 *9*
γ	1292.5 *4*	6.1 *9*
γ	1335.6 *6*	11.3 *8*
γ	1376.0 *4*	6.0 *6*
γ	1389.6 *4*	6.2 *6*
γ	1513.4 *5*	6.1 *11*

$^{164}_{73}$Ta(13.6 *2* s)

Mode: ϵ(>99.979 %), α(<0.021 %)
Δ: -43440 *380* keV syst
SpA: 4.93×10^{9} Ci/g
Prod: ^{127}I(^{40}Ca,3n)

Alpha Particles (^{164}Ta)

α(keV)
4625 *15*

$^{164}_{74}$W (6.4 *8* s)

Mode: ϵ(97.4 *17* %), α(2.6 *17* %)
Δ: -38360 *60* keV
SpA: 1.02×10^{10} Ci/g
Prod: ^{144}Sm(^{24}Mg,4n); ^{147}Sm(^{24}Mg,7n); ^{156}Dy(^{16}O,8n); ^{58}Ni on Pd

Alpha Particles (^{164}W)

α(keV)
5148 *5*

$^{164}_{75}$Re(880 *240* ms)

Mode: α
Δ: -27390 *410* keV syst
SpA: 5.4×10^{10} Ci/g
Prod: daughter ^{168}Ir

Alpha Particles (^{164}Re)

α(keV)
5778 *10*

$^{164}_{76}$Os(41 *20* ms)

Mode: α
Δ: -20460 *500* keV syst
SpA: 1×10^{11} Ci/g
Prod: daughter ^{168}Pt

Alpha Particles (^{164}Os)

α(keV)
6320 *20*

A = 165

NDS **11**, 189 (1974)

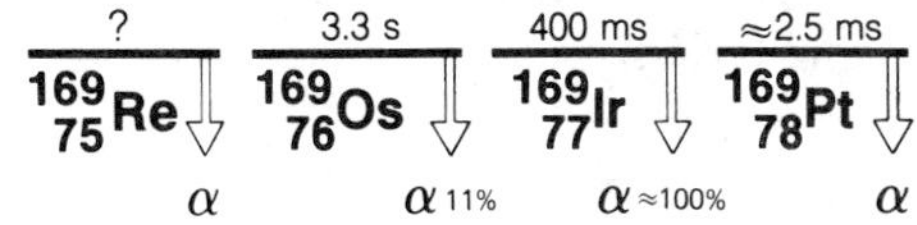

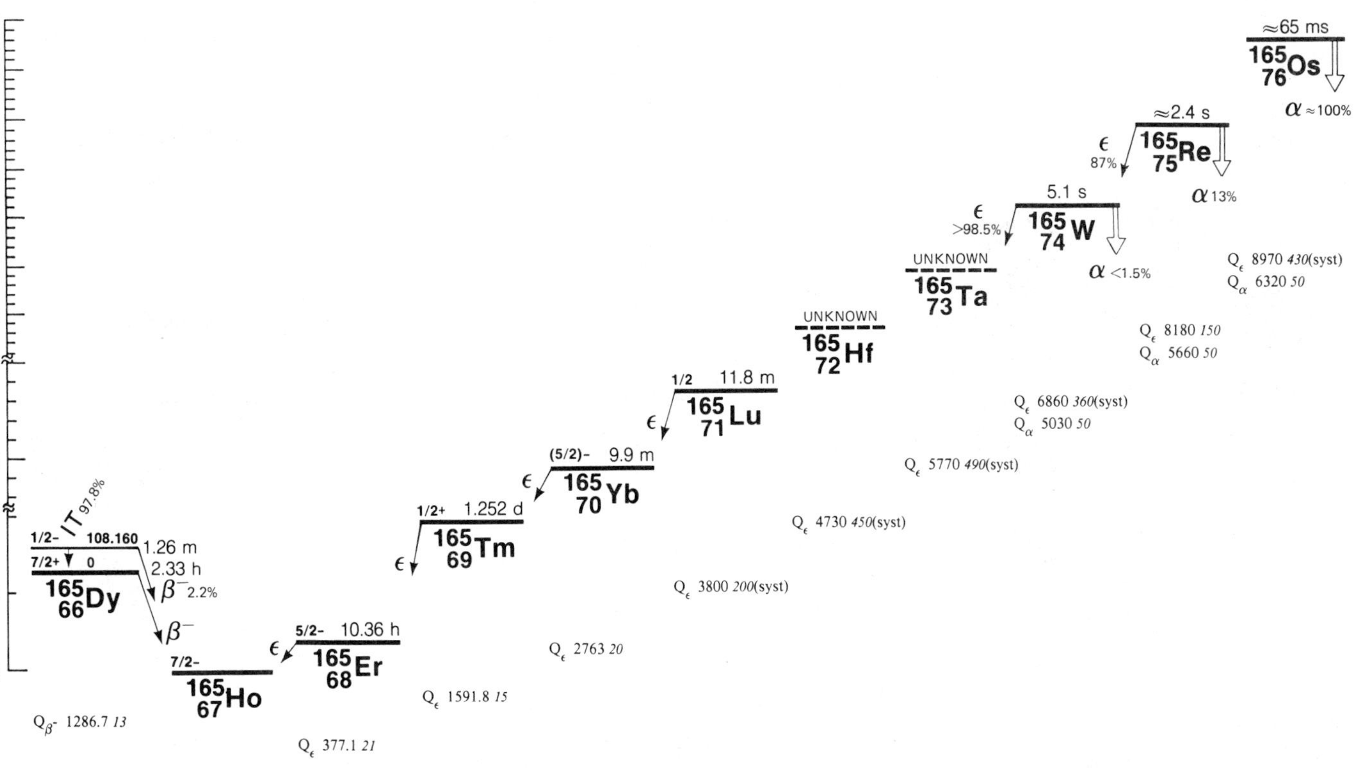

$^{165}_{66}$Dy(2.334 *6* h)

Mode: β-

Δ: -63622 *3* keV

SpA: 8.137×10^6 Ci/g

Prod: ^{164}Dy(n,γ)

Photons (^{165}Dy)

⟨γ⟩=26.0 *4* keV

γ_{mode}	γ(keV)	γ(%)†
Ho L$_\ell$	5.943	0.035 *5*
Ho L$_\alpha$	6.716	0.91 *9*
Ho L$_\eta$	6.789	0.0119 *13*
Ho L$_\beta$	7.659	0.84 *10*
Ho L$_\gamma$	8.845	0.131 *17*
γ M1+<0.2%E2	29.719 *3*	0.0041 *19*
Ho K$_{\alpha2}$	46.700	2.73 *22*
Ho K$_{\alpha1}$	47.547	4.9 *4*
Ho K$_{\beta1}$'	53.822	1.49 *12*
Ho K$_{\beta2}$'	55.556	0.43 *4*
γ M1+1.4%E2	57.869 *3*	0.0139 *21*
γ M1+<7.3%E2	67.716 *4*	~0.014
γ M1+<17%E2	71.507 *9*	0.0024 *3*
γ M1+<8.3%E2	87.588 *3*	0.0143 *14*
γ M1+<25%E2	89.756 *7*	0.0029 *6*
γ [M1]	94.704 *3*	3.58 *25*
γ [E1]	95.9346 *4*	0.00093 *7*
γ [M1+E2]	98.751 *18*	0.00092 *17*
γ [E2]	109.628 *8*	0.00059 *17*
γ [M1]	115.108 *10*	0.0071 *5*
γ M1+<35%E2	119.474 *7*	0.0071 *5*
γ [E2]	129.377 *10*	0.00050 *17*
γ [E2]	140.540 *15*	0.00210 *25*
γ [E1]	153.804 *3*	0.0057 *3*
γ [M1+E2]	170.258 *15*	0.0030 *3*
γ [E2]	174.983 *16*	0.00109 *25*
γ [E2]	209.811 *10*	~0.0010
γ [E2]	228.127 *15*	~0.00042
γ [M1+E2]	259.514 *23*	0.0146 *8*
γ [E3]	266.967 *9*	0.0011 *3*
γ M1(+E2)	279.763 *9*	0.498 *25*
γ	356.90 *25*	~0.0008
γ M2	361.670 *8*	0.84 *4*
γ [M1+E2]	405.298 *16*	0.0107 *5*
γ [M1+E2]	456.082 *10*	0.0425 *21*
γ [E2]	471.019 *20*	0.00143 *25*
γ [E1]	479.622 *8*	0.0440 *22*
γ [M1+E2]	489.830 *22*	~0.003
γ [M1+E2]	504.049 *12*	0.00109 *25*
γ [E1]	513.90 *3*	0.0032 *7* ?
γ (E2)	515.474 *8*	0.0363 *24*
γ [M1+E2]	540.614 *19*	0.0056 *6*
γ [M1+E2]	545.837 *8*	0.162 *5*
γ [E2]	565.710 *8*	0.13 *4*
γ [M1+E2]	575.556 *8*	0.0787 *24*
γ [M1+E2]	588.580 *19*	0.0033 *3*
γ [E1]	610.302 *18*	0.0053 *5*
γ [E1]	620.629 *10*	0.097 *4*
γ M1	633.425 *8*	0.568 *11*
γ [M1+E2]	660.087 *17*	0.0266 *13*
γ [E1]	694.091 *21*	0.0116 *6*
γ (E1)	715.332 *10*	0.534 *11*
γ [E1]	725.409 *16*	0.0140 *14*
γ [E1]	820.112 *15*	0.0081 *6*
γ	900.391 *5*	0.00252 *25*
γ [E1]	976.78 *6*	0.00023 *4*
γ [E1]	984.922 *17*	0.0056 *4*
γ [E1]	995.094 *5*	0.0546 *11*
γ [M1+E2]	1045.66 *5*	0.00050 *8*
γ [M1+E2]	1055.759 *21*	0.0312 *16*
γ [E1]	1079.625 *17*	0.092 *3*
γ [E1]	1091.89 *6*	0.00101 *8*
γ [M1+E2]	1140.36 *5*	0.00134 *8*
γ [E1]	1186.59 *6*	0.00046 *5*

† 12% uncert(syst)

Atomic Electrons (^{165}Dy)

⟨e⟩=7.3 *3* keV

e_{bin}(keV)	⟨e⟩(keV)	e(%)
2	0.0033	0.149 *23*
8	0.34	4.2 *6*
9	0.25	2.8 *4*
12 - 38	0.109	0.35 *3*
39	3.8	9.7 *8*
40 - 82	0.229	0.43 *5*
85	1.10	1.29 *10*
86 - 91	0.111	0.129 *9*
93	0.290	0.313 *25*
94 - 141	0.092	0.096 *6*
144 - 175	0.0014	0.00084 *18*
200 - 228	0.11	0.049 *17*
250 - 280	0.034	0.012 *3*
301	9.8 $\times 10^{-5}$	~3 $\times 10^{-5}$
306	0.60	0.195 *11*
348 - 396	0.162	0.0458 *17*
397 - 434	0.0058	0.0014 *4*
447 - 496	0.017	0.0035 *10*
502 - 546	0.016	0.0031 *6*
555 - 602	0.0554	0.0096 *3*
604 - 652	0.0124	0.00199 *11*
658 - 707	0.0090	0.00135 *5*
713 - 725	0.000349	4.89 *16* $\times 10^{-5}$
764 - 812	0.000116	1.50 *11* $\times 10^{-5}$
818 - 845	0.00035	4.2 *4* $\times 10^{-5}$
891 - 939	0.00072	7.7 *3* $\times 10^{-5}$
967 - 1000	0.0012	0.00012 *3*
1024 - 1072	0.00126	0.000122 *6*
1077 - 1092	7.9 $\times 10^{-5}$	7.3 *11* $\times 10^{-6}$
1131 - 1179	1.28 $\times 10^{-5}$	1.13 *21* $\times 10^{-6}$
1184 - 1186	1.62 $\times 10^{-7}$	1.37 *12* $\times 10^{-8}$

Continuous Radiation (^{165}Dy)

⟨β-⟩=442 keV;⟨IB⟩=0.56 keV

E_{bin}(keV)		⟨ ⟩(keV)	(%)
0 - 10	β-	0.058	1.15
	IB	0.020	
10 - 20	β-	0.174	1.16
	IB	0.019	0.134
20 - 40	β-	0.70	2.34
	IB	0.037	0.127
40 - 100	β-	5.01	7.1
	IB	0.096	0.149
100 - 300	β-	49.2	24.5
	IB	0.21	0.124
300 - 600	β-	156	35.0
	IB	0.139	0.034
600 - 1287	β-	231	29.1
	IB	0.038	0.0054

$^{165}_{66}$Dy(1.258 *6* min)

Mode: IT(97.76 *14* %), β-(2.24 *14* %)
Δ: -63514 *3* keV
SpA: 9.02×10^{8} Ci/g
Prod: ^{164}Dy(n,γ)

Photons (^{165}Dy)

⟨γ⟩=19.2 *7* keV

γ_{mode}	γ(keV)	γ(%)†
Dy L_ℓ	5.743	0.217 *25*
Ho L_ℓ	5.943	0.0016 *4*
Dy L_α	6.491	5.7 *5*
Dy L_η	6.534	0.125 *11*
Ho L_α	6.716	0.042 *9*
Ho L_η	6.789	0.00061 *15*
Dy L_β	7.346	6.8 *6*
Ho L_β	7.648	0.042 *10*
Dy L_γ	8.435	1.00 *9*
Ho L_γ	8.848	0.0068 *17*
$\gamma_{\beta-}$M1+<0.2%E2	29.719 *3*	0.0033 *15*
Dy $K_{\alpha2}$	45.208	2.63 *13*
Dy $K_{\alpha1}$	45.998	4.70 *23*
Ho $K_{\alpha2}$	46.700	0.092 *12*
Ho $K_{\alpha1}$	47.547	0.163 *21*
Dy $K_{\beta1}$'	52.063	1.43 *7*
Dy $K_{\beta2}$'	53.735	0.418 *22*
Ho $K_{\beta1}$'	53.822	0.050 *6*
Ho $K_{\beta2}$'	55.556	0.0144 *18*
$\gamma_{\beta-}$M1+1.4%E2	57.869 *3*	0.0082 *12*
$\gamma_{\beta-}$M1+<7.3%E2	67.716 *4*	~0.012
$\gamma_{\beta-}$M1+<8.3%E2	87.588 *3*	0.0113 *12*
$\gamma_{\beta-}$M1+<25%E2	89.756 *7*	0.0011 *3*
$\gamma_{\beta-}$[E1]	95.9346 *4*	0.0397 *23*
γ_{IT} E3	108.160 *3*	3.01 *12*
$\gamma_{\beta-}$[E2]	109.628 *8*	0.00023 *8*
$\gamma_{\beta-}$M1+<35%E2	119.474 *7*	0.0027 *6* ?
$\gamma_{\beta-}$[E1]	153.804 *3*	0.246 *7*
$\gamma_{\beta-}$	251.73 *10*	~0.018 ?
$\gamma_{\beta-}$[E3]	266.967 *9*	0.0007 *3*
$\gamma_{\beta-}$M2	361.670 *8*	0.54 *16*
$\gamma_{\beta-}$(E2)	515.474 *8*	1.56 *5*
$\gamma_{\beta-}$	676 *1*	~0.00046 ?

† uncert(syst): 0.14% for IT, 8.1% for β-

Atomic Electrons (^{165}Dy)

⟨e⟩=98 *3* keV

e_{bin}(keV)	⟨e⟩(keV)	e(%)
2 - 8	2.23	29 *3*
9	2.23	26 *3*
12 - 53	0.274	0.74 *4*
54	5.28	9.7 *4*
56 - 99	1.25	1.32 *6*
100	63.7	64 *3*
101 - 102	0.00018	0.00018 *7*
106	17.1	16.1 *7*
107	0.347	0.325 *15*
108	4.66	4.32 *19*
109 - 154	0.00672	0.00476 *17*
196 - 244	0.005	~0.0023
250 - 267	0.0007	0.00029 *11*
306 - 354	0.47	0.15 *4*
360 - 362	0.024	0.0066 *16*
460 - 507	0.102	0.0219 *9*
513 - 515	0.00525	0.00102 *3*
620 - 668	2.6 $\times 10^{-5}$	~4 $\times 10^{-6}$
674 - 676	1.0 $\times 10^{-6}$	~2 $\times 10^{-7}$

Continuous Radiation (^{165}Dy)

⟨β-⟩=6.9 keV;⟨IB⟩=0.0064 keV

E_{bin}(keV)		⟨ ⟩(keV)	(%)
0 - 10	β-	0.00208	0.0415
	IB	0.00033	
10 - 20	β-	0.0063	0.0417
	IB	0.00032	0.0022
20 - 40	β-	0.0253	0.084
	IB	0.00059	0.0020
40 - 100	β-	0.181	0.259

Continuous Radiation (^{165}Dy)
(continued)

E_{bin}(keV)		⟨ ⟩(keV)	(%)
	IB	0.00145	0.0023
100 - 300	β-	1.68	0.85
	IB	0.0026	0.00158
300 - 600	β-	3.73	0.87
	IB	0.00104	0.00026
600 - 1033	β-	1.28	0.188
	IB	6.4 $\times 10^{-5}$	9.8 $\times 10^{-6}$

$^{165}_{67}$Ho(stable)

Δ: -64908 *3* keV
%: 100

$^{165}_{68}$Er(10.36 *4* h)

Mode: ϵ
Δ: -64531 *4* keV
SpA: 1.833×10^{6} Ci/g
Prod: ^{165}Ho(d,2n); ^{165}Ho(p,n); ^{164}Er(n,γ); ^{164}Er(d,p); ^{166}Er(d,t)

Photons (^{165}Er)

⟨γ⟩=37.9 *6* keV

γ_{mode}	γ(keV)	γ(%)
Ho L_ℓ	5.943	0.28 *3*
Ho L_α	6.716	7.3 *5*
Ho L_η	6.789	0.094 *7*
Ho L_β	7.657	6.9 *6*
Ho L_γ	8.860	1.12 *12*
Ho $K_{\alpha2}$	46.700	21.7 *6*
Ho $K_{\alpha1}$	47.547	38.6 *11*
Ho $K_{\beta1}$'	53.822	11.9 *3*
Ho $K_{\beta2}$'	55.556	3.40 *11*

Atomic Electrons (^{165}Er)

⟨e⟩=6.6 *4* keV

e_{bin}(keV)	⟨e⟩(keV)	e(%)
8	2.6	33 *3*
9	2.09	23.0 *24*
37	0.36	0.97 *10*
38	0.200	0.52 *5*
39	0.54	1.38 *14*
44	0.167	0.38 *4*
45	0.180	0.40 *4*
46	0.249	0.54 *6*
47	0.055	0.117 *12*
48	0.0046	0.0096 *10*
51	0.0077	0.0151 *15*
52	0.043	0.083 *9*
53	0.0178	0.033 *3*
54	0.0071	0.0133 *14*

Continuous Radiation (^{165}Er)

⟨IB⟩=0.21 keV

E_{bin}(keV)		⟨ ⟩(keV)	(%)
10 - 20	IB	0.00019	0.00123
20 - 40	IB	0.0129	0.036
40 - 100	IB	0.18	0.37
100 - 300	IB	0.0091	0.0059
300 - 377	IB	0.000117	3.7×10^{-5}

$^{165}_{69}$Tm(1.2525 *12* d)

Mode: ϵ

Δ: -62939 *4* keV

SpA: 6.318×10^{5} Ci/g

Prod: protons on U; ^{166}Er(p,2n); protons on Ta

Photons (^{165}Tm)

⟨γ⟩=596 *27* keV

γ_{mode}	γ(keV)	γ(%)†
γ E1	47.154 *8*	17 *3*
Er $K_{\alpha2}$	48.221	29.3 *12*
Er $K_{\alpha1}$	49.128	51.9 *22*
γ M1+<3.1%E2	53.183 *8*	~0.09
γ M1	54.437 *8*	7.2 *14*
Er $K_{\beta1}$'	55.616	16.1 *7*
Er $K_{\beta2}$'	57.406	4.56 *21*
γ M1+45%E2	59.149 *8*	0.045 *9*
γ M1	60.403 *8*	0.71 *14*
γ (E1)	62.667 *13*	0.51 *10*
γ M1+~0.2%E2	70.636 *10*	0.14 *4*
γ E2+~18%M1	77.247 *8*	0.56 *14*
γ M1+~0.5%E2	113.587 *9*	1.7 *5*
γ M1(+E2)	141.382 *11*	0.042 *14*
γ E1	150.839 *19*	0.53 *16*
γ E1	156.03 *3*	0.065 *16*
γ [E1]	162.528 *22*	0.086 *16*
γ E2	165.603 *21*	0.14 *4*
γ M1+E2	181.59 *3*	~0.021
γ E1	195.696 *21*	0.70 *14*
γ E1	205.369 *21*	0.56 *14*
γ E1	209.988 *19*	~0.6
γ M1+~5.0%E2	218.786 *21*	2.4 *7*
γ (M1+E2)	224.10 *5*	0.042 *14*
γ E1	233.164 *19*	0.13 *3*
γ M1	234.77 *3*	0.14 *3*
γ E1	238.40 *6*	0.049 *9*
γ M1+1.4%E2	242.850 *21*	35
γ E1	248.879 *21*	~1
γ (M1)	249.92 *8*	~0.09
γ E1	264.425 *20*	0.53 *4*
γ E2	279.189 *22*	0.60 *4*
γ [E1]	292.314 *20*	1.7 *4*
γ M1+<8.3%E2	296.033 *22*	23 *4*
γ E2	297.287 *22*	0.094 *19*
γ M1	306.984 *22*	0.18 *7*
γ (M1)	312.29 *5*	0.79 *16*
γ M1	330.78 *13*	0.25 *5*
γ [M1+E2]	346.75 *8*	3.9 *4*
γ M1+E2	356.436 *22*	3.7 *11*
γ (M1+E2)	365.48 *5*	3.7 *7*
γ M1+E2	384.231 *23*	0.18 *5*
γ E1	389.19 *8*	2.7 *5*
γ M1	400.37 *4*	0.19 *5*
γ	409.9 *8*	0.028 *14*
γ [E1]	413.11 *11*	0.08 *3*
γ (E1)	414.95 *4*	0.049 *21*
γ (E2)	421.02 *3*	0.28 *4*
γ	427.6 *8*	0.049 *21*
γ E1	430.47 *4*	0.28 *4*
γ E2	442.76 *11*	0.86 *17*
γ E1	448.34 *8*	2.6 *5*
γ M1(+E2)	456.30 *3*	0.75 *15*
γ E2	460.12 *3*	3.7 *6*
γ M1+<30%E2	471.81 *3*	0.40 *8*
γ M1(+E2)	477.62 *4*	0.40 *7*
γ	480.4 *8*	0.049 *14*
γ [M1+E2]	484.39 *12*	0.12 *3*
γ	492.6 *4*	0.12 *3*
γ	495.41 *15*	0.065 *16*
γ M1+E2	514.14 *15*	0.59 *12*
γ E2	526.93 *3*	0.91 *18*
γ (E2)	531.08 *5*	<0.24
γ [E1]	534.43 *6*	0.049 *19* ?
γ [E2]	536.15 *11*	0.10 *3* ?
γ M1+E2	542.45 *3*	1.7 *4*
γ E1	557.78 *13*	0.118 *24*
γ M1+E2	558.08 *17*	~0.2
γ M1(+E2)	563.94 *11*	2.4 *5*
γ M1+E2	573.31 *15*	0.59 *12*
γ (M1+E2)	577.78 *12*	0.22 *5*
γ E2	589.60 *8*	2.3 *5*
γ (E2)	605.57 *12*	0.15 *3*
γ (M1+E2)	608.33 *5*	0.34 *7*
γ E1	610.97 *13*	0.34 *7*
γ M1	623.09 *11*	0.14 *3*
γ (M1+E2)	664.72 *12*	0.57 *11*
γ M1+E2	677.53 *11*	0.00113 *23*
γ (M1+E2)	680.76 *19*	0.13 *5*
γ E2	698.47 *8*	0.94 *19*
γ	746.7 *4*	0.23 *5*
γ	748.5 *4*	0.11 *3*
γ [E2]	791.15 *13*	0.73 *15* ?
γ M1(+E2)	806.66 *13*	8.3 *17*
γ	821.4 *6*	0.13 *3*
γ M1(+E2)	827.3 *3*	0.113 *23*
γ M1	837.53 *15*	0.49 *10*
γ [E1]	853.81 *13*	0.13 *3* ?
γ	882.4 *3*	0.035 *14*
γ (M1)	892.93 *20*	0.065 *22*
γ M1(+E2)	908.16 *15*	0.054 *22*
γ [E2]	920.38 *11*	~0.027
γ (M1)	932.65 *19*	0.091 *22*
γ (E2)	939.2 *3*	~0.035
γ	953.6 *8*	0.16 *3*
γ M1(+E2)	993.06 *19*	0.049 *14* ?
γ M1(+E2)	993.2 *4*	0.049 *14* ?
γ E1	1042.89 *15*	0.124 *25*
γ M1+E2	1046.24 *19*	0.10 *3*
γ E1	1131.09 *15*	1.4 *3*
γ E1	1184.27 *15*	2.6 *5*
γ	1231.6 *4*	0.042 *14*
γ (M1+E2)	1284.8 *4*	0.070 *16*
γ (M1+E2)	1289.09 *19*	0.19 *4*
γ	1339.6 *3*	0.030 *14* ?
γ (M1+E2)	1364.74 *20*	0.10 *3*
γ M1	1379.97 *15*	0.5 *1*
γ	1416.9 *3*	0.065 *13*
γ E1	1427.12 *15*	0.89 *18*

† 15% uncert(syst)

Atomic Electrons (^{165}Tm)

⟨e⟩=48 *3* keV

e_{bin}(keV)	⟨e⟩(keV)	e(%)
3 - 5	0.24	8.0 *15*
8	3.7	45 *5*
9	2.03	21.9 *24*
10 - 55	4.27	17.6 *10*
56	1.6	2.9 *9*
58 - 105	3.2	4.4 *5*
108 - 157	0.37	0.30 *4*
160 - 182	1.1	0.68 *17*
185	13.2	7.1 *14*
186 - 232	0.54	0.26 *3*
233	2.3	0.97 *19*
234 - 238	0.80	0.34 *6*
239	6.3	2.6 *5*
240 - 284	1.23	0.49 *5*
286	1.02	0.36 *7*
287 - 332	2.6	0.87 *16*
337 - 385	0.66	0.19 *3*
387 - 435	0.54	0.133 *15*
438 - 487	0.41	0.087 *16*
490 - 536	0.44	0.086 *19*
540 - 590	0.149	0.027 *4*
596 - 641	0.093	0.015 *3*
655 - 698	0.034	0.0050 *15*
734 - 783	0.50	0.066 *23*
789 - 838	0.10	0.013 *4*
844 - 893	0.0106	0.00122 *22*
896 - 945	0.012	0.0013 *5*
951 - 993	0.0059	0.00060 *15*
1033 - 1074	0.015	0.0014 *3*
1121 - 1131	0.028	0.0025 *5*
1174 - 1223	0.0057	0.00048 *9*
1227 - 1276	0.0078	0.00063 *16*
1279 - 1322	0.021	0.0016 *3*
1330 - 1379	0.0125	0.00091 *16*
1380 - 1427	0.00175	0.000123 *17*

Continuous Radiation (^{165}Tm)

⟨IB⟩=0.47 keV

E_{bin}(keV)		⟨ ⟩(keV)	(%)
10 - 20	IB	0.00018	0.00119
20 - 40	IB	0.0086	0.024
40 - 100	IB	0.20	0.41
100 - 300	IB	0.034	0.017
300 - 600	IB	0.100	0.023
600 - 1300	IB	0.118	0.0153
1300 - 1349	IB	1.51×10^{-6}	1.15×10^{-7}

$^{165}_{70}$Yb(9.9 *3* min)

Mode: ϵ

Δ: -60176 *20* keV

SpA: 1.15×10^{8} Ci/g

Prod: ^{169}Tm(p,5n); ^{159}Tb(^{11}B,5n)

Photons (^{165}Yb)

⟨γ⟩=235 *4* keV

γ_{mode}	γ(keV)	γ(%)
Tm $K_{\alpha2}$	49.772	30.0 *10*
Tm $K_{\alpha1}$	50.742	52.8 *18*
Tm $K_{\beta1}$'	57.444	16.6 *6*
Tm $K_{\beta2}$'	59.296	4.59 *17*
γ [E2]	68.99 *8*	6.29 *13*
γ [E1]	80.18 *8*	37.0 *7*
γ [M1+E2]	91.81 *8*	0.44 *4*
γ [M1+E2]	104.39 *9*	0.15 *4*
γ [M1+E2]	116.80 *21*	
γ [M1+E2]	118.17 *7*	1.78 *4*
γ [M1+E2]	130.06 *9*	0.41 *4*
γ [M1+E2]	132.29 *10*	~0.07
γ	134.70 *21*	0.067 *13*
γ [E2]	147.40 *7*	0.67 *4*
γ [M1+E2]	156.70 *7*	0.107 *21*
γ	158.30 *21*	0.048 *10*
γ [E1]	170.43 *10*	0.41 *4*
γ [M1+E2]	185.94 *9*	0.24 *5*
γ [M1+E2]	190.03 *16*	
γ [M1+E2]	203.35 *9*	0.20 *4*
γ [E2]	208.61 *22*	
γ [E2]	232.59 *11*	0.14 *3*

Photons (^{165}Yb)
(continued)

γ_{mode}	γ(keV)	γ(%)
γ [M1+E2]	235.11 9	0.048 10
γ [E2]	255.16 20	0.037 7
γ [M1+E2]	261.09 10	0.048 10
γ [E1]	263.81 21	
γ [M1+E2]	269.1 5	
γ [E1]	275.63 18	0.11 4
γ	282.6 10	
γ [M1+E2]	290.33 12	0.07 2
γ	292.3 10	0.015 3
γ [M1+E2]	304.11 7	0.59 4
γ	312.1 21	0.055 11
γ	320.9 21	0.13 3
γ [E2]	332.34 17	
γ [M1+E2]	361.57 16	0.126 25
γ [M1+E2]	363.3 3	0.052 10
γ	404.40 19	0.096 19
γ	416.0 4	0.100 20
γ	422.3 6	0.018 4
γ	427.1 21	0.022 4
γ	431.40 14	
γ	433.17 9	0.115 23
γ	462.41 11	0.044 9
γ [M1+E2]	479.74 16	0.19 4
γ [M1+E2]	491.56 19	
γ	544.5 6	0.026 5
γ [M1+E2]	566.7 3	0.059 12
γ	578.5 10	0.037 7
γ	589.3 3	
γ [M1+E2]	595.9 3	
γ [M1+E2]	597.87 19	0.015 3
γ	605.9 4	0.018 4
γ	636.62 17	0.122 24 ?
γ	655.9 3	0.21 4 ?
γ	675.20 21	0.048 10
γ	729 1	0.026 5
γ	736.9 10	0.033 7
γ	743.1 10	0.0111 22
γ	772.1 10	0.033 7
γ	786.09 12	0.19 4 ?
γ	826.48 20	0.13 3
γ	831.8 3	0.107 21
γ	838.9 10	0.041 8
γ	853.6 21	0.022 4
γ	854.9 3	0.018 4
γ	878.9 3	0.14 3
γ	896.43 24	
γ	920.1 21	0.015 3
γ [E1]	935.26 13	0.26 5
γ	938.1 10	0.078 16
γ	944.1 21	0.022 4
γ	948.1 21	0.018 4
γ [E2]	956.63 19	0.63 13
γ	963.1 21	0.018 4
γ	972.8 10	0.026 5
γ	976.9 21	0.015 3
γ	989.1 21	0.033 7
γ	999.27 9	0.41 4 ?
γ	1002.6 10	0.015 3
γ	1009.6 21	0.022 4
γ	1012.7 10	0.037 7
γ	1015.7 4	0.063 13
γ [E1]	1029.30 17	0.45 9
γ [M1+E2]	1073.43 9	0.53 11
γ	1090.19 10	2.59 5 ?
γ	1102.01 15	0.037 7 ?
γ	1102.1 7	<0.07 ?
γ	1117.44 11	0.13 3 ?
γ [E1]	1121.20 14	0.089 18
γ	1125.1 7	0.059 12 ?
γ	1128.3 5	0.059 12
γ	1145.7 10	0.0048 10
γ	1149.5 5	0.0048 10
γ	1154.4 6	0.25 5
γ [E1]	1161.59 14	0.070 14
γ [M1+E2]	1165.24 11	0.14 3
γ	1188.25 19	0.081 16
γ	1192.93 20	0.044 9 ?
γ	1202.43 14	0.085 17
γ	1203.5 7	0.085 17
γ	1209.4 6	0.041 8
γ	1211.27 22	0.041 8
γ [E1]	1219.33 14	0.31 6
γ	1222.16 21	
γ [E1]	1239.37 13	0.16 3
γ	1249.08 13	0.026 5
γ	1253.3 21	0.0037 7
γ [E1]	1265.98 12	0.11 4
γ	1269.4 6	0.085 17
γ [M1+E2]	1289.11 19	0.100 20
γ	1295.0 3	0.18 4
γ	1306.1 21	0.015 3
γ	1309.1 21	0.026 5
γ	1312.5 6	0.063 13
γ [M1+E2]	1329.10 9	0.18 4
γ	1341.6 10	0.015 3
γ	1353.6 10	0.0111 22
γ	1367.3 4	0.055 11
γ [E1]	1371.03 12	0.085 17
γ	1390.3 6	0.048 10
γ	1399.6 21	
γ	1402.9 10	0.037 7
γ	1405.79 13	0.026 5
γ [M1+E2]	1420.91 10	0.17 3
γ [E1]	1422.68 12	0.17 3
γ	1426.7 6	0.081 16
γ	1435.02 14	0.052 10
γ [E1]	1451.91 13	0.122 24
γ [E1]	1501.09 10	0.31 6
γ	1530.8 10	0.034 7

Atomic Electrons (^{165}Yb)
$\langle e \rangle$=73.3 19 keV

e_{bin}(keV)	$\langle e \rangle$(keV)	e(%)
9	5.8	68 7
10	5.6	58 6
21	3.77	18.1 8
32 - 57	2.88	6.9 6
59	18.0	30.4 21
60	17.3	28.6 13
67	9.3	13.9 6
68	0.0494	0.073 3
69	2.53	3.68 16
70	1.30	1.85 8
71 - 118	4.5	5.3 7
120 - 169	1.16	0.89 11
170 - 216	0.15	0.083 18
222 - 267	0.20	0.083 23
273 - 321	0.080	0.027 7
345 - 394	0.047	0.013 4
395 - 433	0.052	0.013 5
452 - 485	0.013	0.0027 8
507 - 557	0.011	0.0022 6
558 - 606	0.021	~0.0036
616 - 665	0.007	0.0012 4
666 - 713	0.0052	0.0008 3
719 - 767	0.015	~0.0021
770 - 819	0.016	0.0021 7
820 - 869	0.010	0.0012 5
870 - 920	0.051	0.0057 21
925 - 975	0.056	0.0059 20
976 - 1021	0.028	0.0028 8
1027 - 1073	0.11	~0.010
1080 - 1129	0.035	0.0032 9
1134 - 1183	0.015	0.0013 3
1184 - 1231	0.012	0.00098 22
1236 - 1285	0.015	0.0012 3
1286 - 1333	0.0052	0.00040 9
1339 - 1388	0.0109	0.00080 15
1389 - 1435	0.0030	0.00021 3
1442 - 1491	0.0039	0.00027 5
1492 - 1529	0.00021	1.4 4 $\times 10^{-5}$

Continuous Radiation (^{165}Yb)
$\langle\beta+\rangle$=70 keV;$\langle$IB$\rangle$=3.3 keV

E_{bin}(keV)		$\langle\ \rangle$(keV)	(%)
0 - 10	β+	2.09×10^{-7}	2.48×10^{-6}
	IB	0.0037	
10 - 20	β+	1.16×10^{-5}	6.9×10^{-5}
	IB	0.0030	0.021
20 - 40	β+	0.000439	0.00133
	IB	0.0113	0.036
40 - 100	β+	0.0298	0.0377
	IB	0.22	0.42
100 - 300	β+	1.85	0.83
	IB	0.083	0.044
300 - 600	β+	13.3	2.88
	IB	0.25	0.055
600 - 1300	β+	50.9	5.8
	IB	1.28	0.136
1300 - 2500	β+	3.76	0.274
	IB	1.44	0.087
2500 - 2602	IB	0.00027	1.09×10^{-5}
	$\Sigma\beta$+		9.8

$^{165}_{71}$Lu(11.8 5 min)

Mode: ϵ
Δ: -56380 200 keV syst
SpA: 9.7×10^{7} Ci/g
Prod: ^{169}Tm(^{3}He,7n)

Photons (^{165}Lu)
$\langle\gamma\rangle$=858 24 keV

γ_{mode}	γ(keV)	γ(%)[†]
γ	39.3 1	7.8 12
γ	120.58 5	25
γ	127.52 20	1.50 25
γ	132.43 10	23.0 13
γ	174.22 10	12.5 13
γ	203.58 5	10.5 13
γ	217.40 5	4.8 5
γ	253.44 10	4.0 5
γ	268.67 12	1.50 3
γ	271.03 6	5.1 4
γ	312.9 4	0.60 8
γ	319.51 20	1.13 25
γ	324.7 2	1.25 13
γ	356.52 10	5.2 4
γ	360.51 10	8.2 5
γ	372.45 10	3.3 4
γ	443.0 3	0.38 12
γ	454.7 3	0.75 15
γ	458.12 12	3.10 25
γ	519.36 15	1.80 25
γ	532.69 20	1.15 25
γ	544.04 25	0.95 25
γ	552.35 15	2.05 25
γ	605.96 20	1.20 25
γ	609.11 15	2.05 25
γ	629.75 20	1.40 25
γ	655.87 20	1.25 25
γ	659.65 25	1.50 25
γ	662.63 20	1.65 25
γ	686.60 15	2.55 25
γ	727.0 3	1.10 25
γ	753.48 15	2.2 2
γ	770.7 4	0.93 18
γ	815.3 3	1.17 25
γ	1029.93 25	1.63 25
γ	1050.1 3	1.35 25
γ	1073.75 25	1.88 25
γ	1181.9 5	0.70 25
γ	1240.6 4	1.50 25
γ	1280.89 25	1.48 25
γ	1290.1 4	1.00 25

Photons (^{165}Lu)
(continued)

γ_{mode}	γ(keV)	γ(%)†
γ	1329.1 *4*	1.55 *25*
γ	1438.7 *7*	0.57 *18*
γ	1479.0 *5*	0.65 *20*
γ	1560.0 *3*	1.92 *25*
γ	1572.3 *6*	0.63 *17*
γ	1601.56 *20*	4.0 *4*
γ	1613.51 *20*	4.0 *4*
γ	1734.4 *3*	2.3 *4*
γ	1801.9 *4*	1.9 *4*
γ	1808.1 *5*	1.1 *4*
γ	1862.3 *5*	1.1 *4*
γ	1887.1 *5*	1.1 *4*
γ	2005.1 *5*	1.1 *4*

† approximate

$^{165}_{74}$W (5.1 *5* s)

Mode: ϵ(>98.5 %), α(<1.5 %)
Δ: -39020 *220* keV syst
SpA: 1.25×10^{10} Ci/g
Prod: ^{156}Dy(^{16}O,7n)

Alpha Particles (^{165}W)

α(keV)
4909 *5*

$^{165}_{75}$Re(2.4 *6* s)

Mode: ϵ(87 *3* %), α(13 *3* %)
Δ: -30840 *230* keV syst
SpA: 2.5×10^{10} Ci/g
Prod: daughter ^{169}Ir

Alpha Particles (^{165}Re)

α(keV)
5506 *10*

$^{165}_{76}$Os(~65 ms)

Mode: α
Δ: -21870 *460* keV syst
SpA: ~1×10^{11} Ci/g
Prod: daughter ^{169}Pt; ^{106}Cd(^{63}Cu,p3n)

Alpha Particles (^{165}Os)

α(keV)
6164 *10*

A = 166

NDS **14**, 471 (1975)

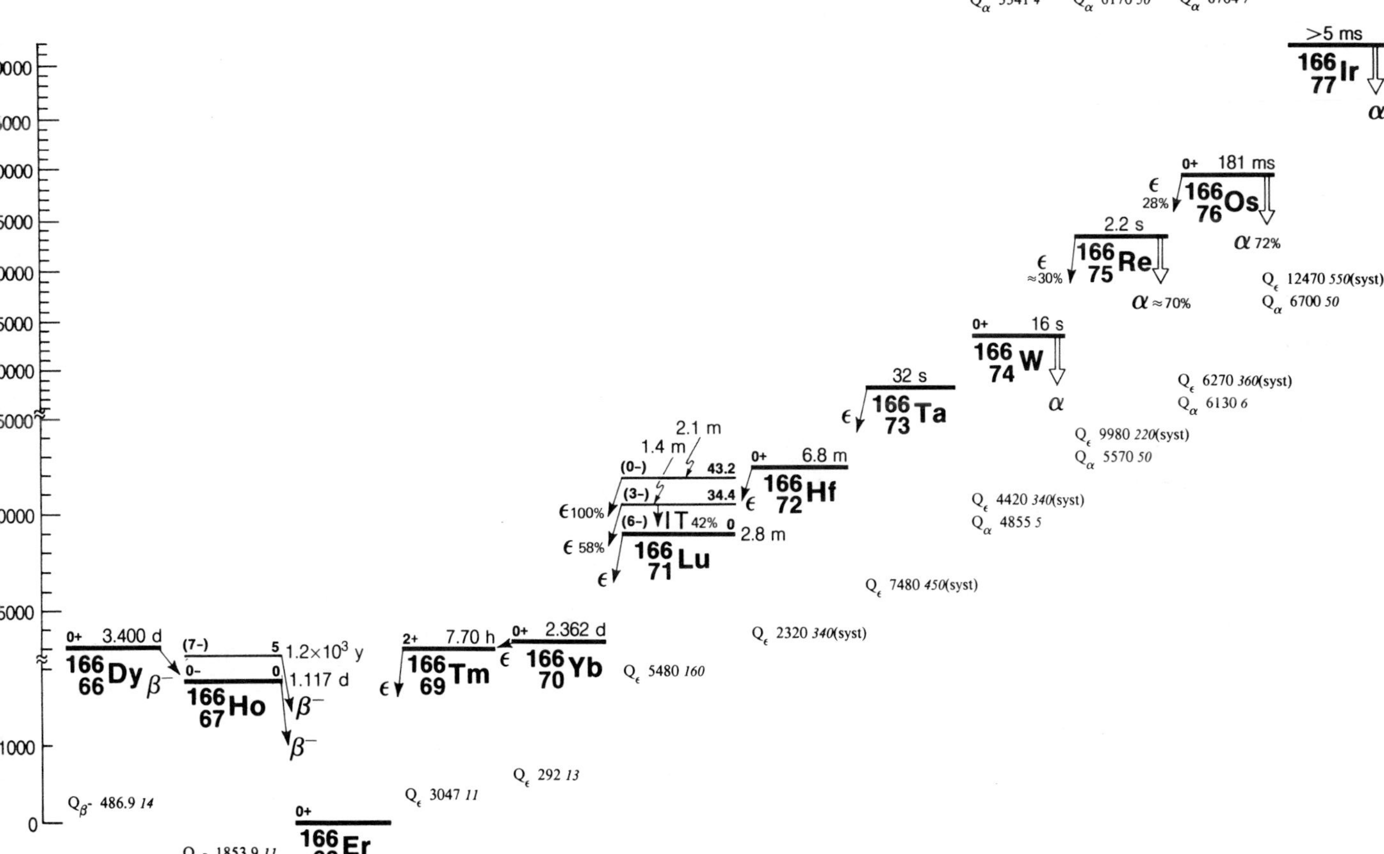

$^{166}_{66}$Dy(3.400 *4* d)

Mode: β-
Δ: -62594 *3* keV
SpA: 2.314×10^5 Ci/g
Prod: multiple n-capture from ^{164}Dy

Photons (^{166}Dy)

⟨γ⟩=40 *4* keV

γ_{mode}	γ(keV)	γ(%)
Ho L_ℓ	5.943	0.25 *6*
Ho L_α	6.716	6.6 *15*
Ho L_η	6.789	0.097 *21*
Ho L_β	7.643	7.2 *18*
Ho L_γ	8.872	1.2 *3*
γ M1	28.2308 *19*	~1.0
Ho $K_{\alpha2}$	46.700	14 *3*
Ho $K_{\alpha1}$	47.547	24 *5*
Ho $K_{\beta1}$'	53.822	7.5 *15*
γ E2	54.2431 *7*	0.67 *13*
Ho $K_{\beta2}$'	55.556	2.2 *4*
γ M1	82.4739 *19*	13 *3*
γ (M1)	290.66 *10*	~0.013
γ (E1)	343.519 *17*	0.051 *13*
γ E1	371.750 *17*	0.46 *4*
γ E1	425.993 *17*	0.54 *4*

Atomic Electrons (^{166}Dy)

⟨e⟩=40 *4* keV

e_{bin}(keV)	⟨e⟩(keV)	e(%)
8	2.3	28 *6*
9	2.2	25 *6*
19	2.5	~13
20 - 26	0.8	3.1 *15*
27	13.6	51 *10*
28 - 44	1.03	2.9 *5*
45	3.7	8.1 *16*
46	4.2	9.1 *18*
47 - 54	2.8	5.2 *6*
73	5.0	6.9 *14*
74	0.5	0.68 *14*
80	1.21	1.5 *3*
81 - 82	0.52	0.64 *10*
235 - 283	0.0041	0.0017 *8*
288 - 335	0.0152	0.0049 *4*
341 - 372	0.0159	0.0043 *3*
417 - 426	0.00266	0.00064 *4*

Continuous Radiation (^{166}Dy)

⟨β-⟩=119 keV;⟨IB⟩=0.051 keV

E_{bin}(keV)		⟨ ⟩(keV)	(%)
0 - 10	β-	0.285	5.7
	IB	0.0062	
10 - 20	β-	0.82	5.5
	IB	0.0055	0.038
20 - 40	β-	3.08	10.3
	IB	0.0091	0.032
40 - 100	β-	18.7	27.0
	IB	0.017	0.028
100 - 300	β-	85	48.3
	IB	0.0126	0.0086
300 - 487	β-	10.7	3.25
	IB	0.000146	4.7×10^{-5}

$^{166}_{67}$Ho(1.117 *2* d)

Mode: β-
Δ: -63081 *3* keV
SpA: 7.042×10^5 Ci/g
Prod: ^{165}Ho(n,γ); daughter ^{166}Dy

Photons (^{166}Ho)

⟨γ⟩=29.0 *8* keV

γ_{mode}	γ(keV)	γ(%)
Er L_ℓ	6.151	0.120 *15*
Er L_α	6.944	3.1 *3*
Er L_η	7.058	0.054 *5*
Er L_β	7.921	3.2 *3*
Er L_γ	9.108	0.47 *5*
Er $K_{\alpha2}$	48.221	2.8 *2*
Er $K_{\alpha1}$	49.128	5.0 *4*
Er $K_{\beta1}$'	55.616	1.54 *11*
Er $K_{\beta2}$'	57.406	0.44 *3*
γ E2	80.573 *7*	6.2 *4*
γ E2	184.413 *14*	0.0020 *5*
γ E2	520.901 *22*	~0.0003
γ [E2]	674.01 *4*	0.020 *2*
γ E2(+0.2%M1)	705.313 *22*	0.0147 *16*
γ E2	785.886 *22*	0.0133 *15*
γ E2	1379.32 *4*	0.93 *5*
γ E1	1581.88 *6*	0.181 *9*
γ E1	1662.45 *6*	0.116 *6*
γ [E1]	1749.96 *8*	0.025 *2*
γ [E1]	1830.53 *8*	0.008 *1*

Atomic Electrons (^{166}Ho)

⟨e⟩=28.6 *9* keV

e_{bin}(keV)	⟨e⟩(keV)	e(%)
8	1.17	13.9 *17*
9	0.90	9.7 *12*
10	0.051	0.53 *6*
23	2.37	10.3 *7*
38 - 55	0.230	0.54 *3*
71	9.1	12.7 *9*
72	8.7	12.1 *8*
78	0.175	0.223 *15*
79	4.6	5.8 *4*
80	1.13	1.41 *10*
81 - 127	0.149	0.184 *12*
175 - 184	0.00045	0.00026 *4*
463 - 513	2.2 ×10^{-5}	4.7 *20* ×10^{-6}
519 - 521	1.2 ×10^{-6}	2.3 *9* ×10^{-7}
617 - 666	0.00152	0.000240 *17*
672 - 705	0.000178	2.58 *17* ×10^{-5}
728 - 778	0.00053	7.2 *7* ×10^{-5}
784 - 786	2.31 ×10^{-5}	2.9 *3* ×10^{-6}
1322 - 1371	0.0214	0.00161 *8*
1377 - 1379	0.00079	5.8 *3* ×10^{-5}
1524 - 1605	0.00259	0.000167 *6*
1653 - 1748	0.000370	2.20 *11* ×10^{-5}
1773 - 1829	6.8 ×10^{-5}	3.8 *4* ×10^{-6}

Continuous Radiation (^{166}Ho)

⟨β-⟩=711 keV;⟨IB⟩=1.3 keV

E_{bin}(keV)		⟨ ⟩(keV)	(%)
0 - 10	β-	0.0311	0.62
	IB	0.029	
10 - 20	β-	0.094	0.62
	IB	0.028	0.20
20 - 40	β-	0.381	1.27
	IB	0.055	0.19

Continuous Radiation (^{166}Ho)
(continued)

E_{bin}(keV)		⟨ ⟩(keV)	(%)
40 - 100	β-	2.77	3.94
	IB	0.150	0.23
100 - 300	β-	28.9	14.3
	IB	0.39	0.22
300 - 600	β-	105	23.2
	IB	0.35	0.085
600 - 1300	β-	415	45.2
	IB	0.28	0.035
1300 - 1960	β-	159	10.8
	IB	0.0154	0.00109

$^{166}_{67}$Ho(1200 *180* yr)

Mode: β-
Δ: -63076 *3* keV
SpA: 1.8 Ci/g
Prod: ^{165}Ho(n,γ)

Photons (^{166}Ho)

⟨γ⟩=1748 *34* keV

γ_{mode}	γ(keV)	γ(%)†
Er L_ℓ	6.151	0.34 *4*
Er L_α	6.944	8.7 *7*
Er L_η	7.058	0.147 *12*
Er L_β	7.923	8.9 *8*
Er L_γ	9.118	1.33 *13*
Er $K_{\alpha2}$	48.221	11.2 *4*
Er $K_{\alpha1}$	49.128	19.9 *8*
Er $K_{\beta1}$'	55.616	6.16 *24*
Er $K_{\beta2}$'	57.406	1.74 *7*
γ E2	73.474 *17*	0.003 *1*
γ E2	80.573 *7*	12.8 *6*
γ (M1)	94.658 *23*	0.143 *10*
γ E2	96.829 *22*	
γ [M1+E2]	119.026 *19*	0.184 *20*
γ (E2)	121.165 *24*	0.27 *3*
γ [M1+E2]	135.242 *24*	0.102 *10*
γ [M1+E2]	140.685 *19*	0.044 *10*
γ [M1+E2]	160.055 *19*	0.100 *10*
γ [M1+E2]	161.75 *3*	0.112 *10*
γ E2	170.303 *25*	0.0163 *20*
γ E2	184.413 *14*	75 *4*
γ [E2]	190.713 *22*	0.224 *20*
γ (E2)	214.78 *4*	0.56 *7* ?
γ E2	215.855 *22*	2.65 *20*
γ [E2]	231.298 *25*	0.245 *20*
γ [E2]	259.711 *17*	1.12 *6*
γ E2	280.41 *3*	30.4 *15*
γ [E2]	300.740 *15*	3.83 *19*
γ [E2]	339.75 *5*	0.173 *20*
γ E2	365.740 *22*	2.57 *13*
γ E2	410.791 *22*	0.030 *7*
γ [E1]	410.934 *16*	11.8 *6*
γ [E1]	451.518 *18*	3.12 *15*
γ [M1+E2]	464.817 *25*	1.25 *8*
γ E2	529.817 *18*	10.4 *5*
γ [E1]	570.988 *19*	5.9 *3*
γ M1(+E2)	594.375 *19*	0.70 *4*
γ [E1]	611.572 *22*	1.42 *8*
γ [E1]	639.99 *3*	0.16 *5*
γ [M1+E2]	644.51 *6*	0.184 *20*
γ [M1+E2]	670.502 *21*	5.9 *3*
γ E2+10%M1	691.204 *22*	1.55 *8*
γ [E1]	711.673 *18*	60 *3*
γ [E1]	736.81 *3*	0.10 *3*
γ [E1]	752.257 *19*	13.4 *7*
γ M1(+E2)	778.787 *20*	3.44 *17*
γ E2+~0.1%M1	810.229 *23*	64 *3*
γ [E2]	830.556 *18*	10.8 *6*

Photons (^{166}Ho) (continued)

γ_{mode}	γ(keV)	γ(%)†
γ E2	875.616 23	0.82 5
γ [E2]	950.913 24	3.10 15
γ [E2]	1010.25 5	0.092 10
γ [E1]	1120.32 3	0.235 20
γ [E1]	1146.83 3	0.224 20
γ [E1]	1241.487 20	1.02 5
γ [E1]	1282.071 22	0.235 20
γ [E1]	1400.73 3	0.56 3
γ [E1]	1427.24 3	0.60 3

† 4.9% uncert(syst)

Atomic Electrons (^{166}Ho)

⟨e⟩=113.9 20 keV

e_{bin}(keV)	⟨e⟩(keV)	e(%)
8	3.3	39 4
9 - 16	2.6	28 3
23	4.87	21.1 11
37 - 65	1.32	2.98 13
71	18.7	26.2 14
72	18.0	24.9 13
73 - 78	0.42	0.54 5
79	9.4	12.0 6
80 - 126	3.22	3.83 16
127	19.4	15.3 8
131 - 174	0.85	0.55 4
175	8.4	4.8 3
176	4.44	2.52 14
181 - 222	4.76	2.57 8
223	4.16	1.87 10
229 - 279	2.48	0.93 3
280 - 308	0.562	0.190 6
330 - 366	0.406	0.115 6
394 - 443	0.28	0.070 14
449 - 472	0.60	0.127 8
514 - 563	0.358	0.068 4
569 - 613	0.42	0.068 24
630 - 679	1.04	0.159 9
681 - 728	0.56	0.079 10
735 - 779	2.53	0.335 15
800 - 831	0.573	0.0712 20
866 - 893	0.092	0.0103 6
941 - 953	0.0208	0.00220 9
1000 - 1010	0.00050	5.0 4 $\times10^{-5}$
1063 - 1112	0.0050	0.00046 3
1118 - 1147	0.00050	4.4 3 $\times10^{-5}$
1184 - 1233	0.0131	0.00110 6
1239 - 1282	0.00076	6.1 3 $\times10^{-5}$
1343 - 1392	0.0107	0.00079 4
1399 - 1427	0.00110	7.8 3 $\times10^{-5}$

Continuous Radiation (^{166}Ho)

⟨β-⟩=16.5 keV;⟨IB⟩=0.00111 keV

E_{bin}(keV)		⟨ ⟩(keV)	(%)
0 - 10	β-	1.91	41.0
	IB	0.00066	
10 - 20	β-	3.80	26.1
	IB	0.00028	0.0021
20 - 40	β-	7.1	25.0
	IB	0.000156	0.00060
40 - 72	β-	3.72	7.7
	IB	1.29 $\times10^{-5}$	2.9 $\times10^{-5}$

$^{166}_{68}$Er(stable)

Δ: -64935 3 keV

%: 33.6 2

$^{166}_{69}$Tm(7.70 3 h)

Mode: ε

Δ: -61888 12 keV

SpA: 2.4517 $\times10^{6}$ Ci/g

Prod: ^{165}Ho(α,3n); protons on Er; daughter ^{166}Yb

Photons (^{166}Tm)

⟨γ⟩=1894 54 keV

γ_{mode}	γ(keV)	γ(%)†
Er L_ℓ	6.151	0.62 7
Er L_α	6.944	16.0 12
Er L_η	7.058	0.240 21
Er L_β	7.930	15.7 14
Er L_γ	9.147	2.42 25
Er $K_{\alpha2}$	48.221	30.8 11
Er $K_{\alpha1}$	49.128	54.6 20
Er $K_{\beta1}'$	55.616	17.0 6
Er $K_{\beta2}'$	57.406	4.80 19
γ E2	73.474 17	0.013 4
γ E2	80.573 7	15.2 9
γ M1	84.11 3	~0.33
γ	90.7 20	
γ E2	96.829 22	
γ	112.73 3	0.84 8
γ [M1+E2]	119.026 19	0.0032 5
γ E1	131.06 3	0.88 9
γ	147.28 4	0.31 4
γ M1+E2	154.43 5	0.210 22
γ E2	170.303 25	0.077 9
γ E2	184.413 14	15.8 9
γ M1	194.77 6	0.80 18
γ E1	215.17 4	5.3 3
γ E2	215.855 22	0.047 6
γ	228.12 7	
γ	237.36 11	
γ E2	280.41 3	0.28 6
γ	293.08 10	
γ (M1+E2)	298.10 9	0.35 7
γ	319.78 10	
γ M1+E2	345.59 6	0.34 3
γ	372.8 8	
γ	385.48 10	
γ [E1]	389.33 9	~0.06
γ (M1)	403.88 10	0.82 8
γ E2	410.791 22	0.14 3
γ	413.48 10	
γ M1(+E2)	429.70 6	0.22 6
γ M1	459.56 7	2.74 22
γ	471.58 10	~0.09
γ	487.99 10	
γ (E1)	496.89 5	0.25 4
γ	501.0 15	
γ E2	520.901 22	~0.23
γ E2	529.817 18	0.183 20
γ	537.0 20	
γ	543.67 8	
γ [E1]	557.63 12	0.24 4
γ [E1]	560.76 7	0.21 5
γ M1(+E2)	594.375 19	3.08 25
γ (E1)	598.77 9	1.55 15
γ [M1+E2]	604.50 9	
γ [E1]	615.92 5	0.084 22
γ	631.0 10	0.18 4
γ [E1]	654.46 12	0.243 22
γ E1	672.25 9	7.0 9
γ (E1)	674.73 8	~2
γ E2+10%M1	691.204 22	7.4 5
γ M1+E2	701.98 13	~0.9 ?
γ E2(+0.2%M1)	705.313 22	10.4 9
γ [E1]	712.74 5	

Photons (^{166}Tm) (continued)

γ_{mode}	γ(keV)	γ(%)†
γ [E1]	727.93 12	0.40 4
γ (M1)	757.66 11	2.32 23
γ M1(+E2)	778.787 20	15.1 12
γ E2	785.886 22	9.4 9
γ E2+~0.1%M1	810.229 23	1.12 11
γ	815.5 15	~0.035
γ	846.0 10	0.88 9
γ E2	875.616 23	3.9 3
γ	902.5 15	0.14 7
γ	930.05 21	~0.6
γ	1022.25 9	
γ	1045.62 7	
γ [E2]	1057.65 7	
γ	1070.0 20	
γ	1078.74 9	<0.8
γ [E2]	1084.84 18	
γ	1097.6 11	
γ	1119.08 9	~0.22
γ	1132.0 15	
γ M1	1152.21 9	1.22 13
γ M1+E2	1161.8 4	0.33 3 ?
γ M1	1176.68 7	8.4 8
γ	1187.5 12	
γ	1192.55 9	
γ M1(+E2)	1203.87 18	~1
γ	1215.92 7	0.62 6 ?
γ E1	1235.2 4	1.46 15 ?
γ	1250.3 15	~0.11
γ M1+E2	1263.37 11	1.02 10
γ M1	1273.50 7	14.4 14
γ (M1)	1300.70 18	1.19 12
γ	1315.2 15	~0.29
γ M1	1346.98 7	0.88 9
γ	1358.0 14	
γ	1364.0 20	
γ (M1)	1374.17 18	5.2 5
γ	1431.1 8	~0.49
γ	1437.3 14	0.73 7
γ E2+E0	1447.79 11	0.60 6
γ	1462.3 15	
γ	1475.4 20	0.099 10
γ	1498.0 20	
γ (M1)	1504.6 8	0.82 8
γ	1523.1 15	0.150 15
γ [E2]	1528.36 11	0.00088 9 ?
γ	1542.1 15	
γ	1576.4 16	
γ	1591.7 16	0.221 22
γ	1603.0 20	
γ	1607.4 16	<0.051
γ	1623.1 16	0.44 4
γ	1628.6 16	
γ E1	1652.70 7	1.06 11
γ	1704.3 17	
γ	1721.4 17	
γ	1731.8 17	
γ	1736.82 7	
γ	1784.3 18	
γ	1814.4 18	
γ	1824.8 18	
γ	1829.0 20	
γ (E1)	1837.12 7	0.53 5
γ M1+E2	1867.87 7	4.2 4
γ (M1)	1895.07 18	0.82 8
γ	1908.5 19	0.53 5
γ	1967.0 20	
γ	1980.0 20	0.139 14
γ	1986.7 20	0.066 7
γ	1994.0 20	0.066 7
γ	2008.0 20	0.221 22
γ	2020.5 20	
γ M1+E2	2052.28 7	20 2
γ (M1+E2)	2079.48 18	7.1 7
γ	2093.2 21	1.81 18
γ	2128 3	0.0221 22
γ	2136 3	0.066 7
γ	2162.4 22	0.042 4
γ	2177.3 22	0.027 3
γ	2184.8 22	
γ	2192.4 22	0.159 16
γ	2205.5 22	
γ	2209.9 8	~0.033 ?
γ	2298.8 23	~0.031
γ	2313.1 23	

Photons (^{166}Tm) (continued)

γ_{mode}	γ(keV)	γ(%)†
γ	2378.9 24	
γ	2388.17 19	
γ	2396.3 24	
γ	2412.5 20	
γ	2425.5 24	~0.022
γ	2464.5 25	0.033 3
γ	2519.9 25	0.029 3
γ	2563 3	0.0110 11
γ	2601 3	0.053 5
γ	2649 3	~0.018
γ	2670.0 20	
γ	2681 3	
γ	2741 3	
γ	2783 3	

† 9.9% uncert(syst)

Atomic Electrons (^{166}Tm)

⟨e⟩=95.1 23 keV

e_{bin}(keV)	⟨e⟩(keV)	e(%)
8	5.9	70 8
9	3.9	42 5
10 - 16	0.78	7.9 8
23	5.8	25.2 15
27 - 65	3.4	8.2 10
71	22.3	31.3 19
72	21.5	29.8 18
73 - 78	0.67	0.87 11
79	11.3	14.4 9
80	2.79	3.47 21
81 - 123	1.3	1.4 3
127	4.11	3.24 19
129 - 176	3.64	2.18 11
182 - 223	1.09	0.586 20
241 - 290	0.16	0.062 17
296 - 345	0.022	0.0066 18
346 - 396	0.212	0.060 6
401 - 450	0.51	0.126 9
451 - 500	0.050	0.011 3
503 - 552	0.27	0.051 17
555 - 599	0.072	0.012 3
606 - 654	0.91	0.143 10
662 - 705	0.386	0.056 3
718 - 758	1.2	0.16 4
769 - 818	0.40	0.050 7
836 - 876	0.057	0.0066 20
893 - 930	0.0049	~0.0005
1021 - 1070	0.020	~0.0019
1077 - 1119	0.41	0.037 3
1142 - 1178	0.137	0.0118 18
1193 - 1242	0.58	0.047 5
1243 - 1292	0.191	0.0151 9
1298 - 1347	0.182	0.0138 14
1364 - 1390	0.068	0.0049 9
1418 - 1467	0.036	0.0025 3
1471 - 1566	0.017	0.0011 3
1582 - 1651	0.0118	0.00073 7
1780 - 1866	0.124	0.0068 10
1885 - 1985	0.0119	0.00062 11
1986 - 2085	0.57	0.029 4
2091 - 2190	0.0047	0.00022 5
2191 - 2290	0.0005	$\sim 2\times10^{-5}$
2291 - 2368	0.00029	$\sim 1\times10^{-5}$
2407 - 2505	0.00090	3.7 9 $\times10^{-5}$
2510 - 2599	0.0009	3.5 11 $\times10^{-5}$
2639 - 2647	3.2×10^{-5}	$\sim 1\times10^{-6}$

Continuous Radiation (^{166}Tm)

⟨β+⟩=14.1 keV;⟨IB⟩=0.76 keV

E_{bin}(keV)		⟨ ⟩(keV)	(%)
0 - 10	β+	2.84×10^{-8}	3.37×10^{-7}
	IB	0.00138	
10 - 20	β+	1.55×10^{-6}	9.2×10^{-6}
	IB	0.00072	0.0050
20 - 40	β+	5.7×10^{-5}	0.000173
	IB	0.0095	0.027
40 - 100	β+	0.00381	0.00482
	IB	0.20	0.41
100 - 300	β+	0.234	0.105
	IB	0.042	0.022
300 - 600	β+	1.74	0.376
	IB	0.097	0.022
600 - 1300	β+	8.7	0.95
	IB	0.17	0.019
1300 - 2500	β+	3.38	0.229
	IB	0.22	0.0128
2500 - 2966	IB	0.0081	0.00031
	Σβ+		1.66

$^{166}_{70}$Yb(2.362 4 d)

Mode: ϵ

Δ: -61595 8 keV

SpA: 3.329×10^5 Ci/g

Prod: ^{169}Tm(p,4n); protons on U; protons on Th

Photons (^{166}Yb)

⟨γ⟩=86.5 18 keV

γ_{mode}	γ(keV)	γ(%)
Tm L_ℓ	6.341	0.58 6
Tm L_α	7.176	14.6 10
Tm L_η	7.310	0.186 15
Tm L_β	8.224	14.2 13
Tm L_γ	9.546	2.4 3
Tm $K_{\alpha2}$	49.772	39.7 16
Tm $K_{\alpha1}$	50.742	70 3
Tm $K_{\beta1}$'	57.444	22.0 9
Tm $K_{\beta2}$'	59.296	6.1 3
γ M1	82.29 2	15.2 7

Atomic Electrons (^{166}Yb)

⟨e⟩=38.9 11 keV

e_{bin}(keV)	⟨e⟩(keV)	e(%)
9	5.0	58 6
10	4.1	42 4
23	16.3	71 3
39	0.27	0.70 7
40	0.42	1.05 11
41	0.94	2.28 24
42	0.30	0.71 8
47	0.38	0.80 8
48	0.25	0.51 5
49	0.43	0.89 9
50	0.084	0.167 18
51	0.0198	0.039 4
55	0.074	0.134 14
56	0.0214	0.038 4
57	0.041	0.072 8
72	7.1	9.8 5

Atomic Electrons (^{166}Yb) (continued)

e_{bin}(keV)	⟨e⟩(keV)	e(%)
73	0.62	0.86 4
74	0.093	0.126 6
80	1.93	2.41 12
81	0.00190	0.00235 11
82	0.56	0.68 3

Continuous Radiation (^{166}Yb)

⟨IB⟩=0.21 keV

E_{bin}(keV)		⟨ ⟩(keV)	(%)
10 - 20	IB	0.00023	0.00159
20 - 40	IB	0.0067	0.019
40 - 100	IB	0.20	0.40
100 - 210	IB	0.0019	0.00153

$^{166}_{71}$Lu(2.8 2 min)

Mode: ϵ

Δ: -56120 160 keV

SpA: 4.0×10^8 Ci/g

Prod: daughter ^{166}Hf; ^{152}Sm(^{19}F,5n); ^{169}Tm(^{3}He,6n); ^{170}Yb(p,5n)

Photons (^{166}Lu(2.8 or 1.4 or 2.1 min))

γ_{mode}	γ(keV)	γ(rel)*
γ	308.8 6	~0.24
γ	312.9 4	~0.33
γ	401.6 3	0.45 12
γ	416.1 5	~0.24
γ	549.6 6	~0.33
γ	612.1 6	0.65 16
γ	671.6 4	0.61 16
γ	697.3 6	0.37 12
γ	735 25	0.37 12
γ	769.4 8	~0.16
γ	915.9 6	~0.28
γ	942.6 6	0.37 16
γ	948.0 6	0.41 16
γ	962.1 6	~0.33
γ	1011.6 6	0.37 16
γ	1171.0 6	0.41 16
γ	1316.6 10	~0.24
γ	1389.8 6	~0.49
γ	1548.2 6	~0.20
γ	1594.5 6	~0.24
γ	1620.2 6	~0.24
γ	1654.0 6	~0.33
γ	1693.9 6	~0.24
γ	1809.3 6	~0.24
γ	1888.1 6	~0.24
γ	2149.2 6	~0.24
γ	2259.0 6	0.41 12
γ	2262.8 6	0.41 12
γ	2362.6 10	0.41 12
γ	2448.5 6	0.49 12
γ	2481.5 6	~0.20
γ	2489.6 6	~0.20
γ	2547.5 6	0.20 8
γ	2762.5 5	~0.16

* relative to 40.7 for 337γ with ^{166}Lu(2.8 min)

Photons (^{166}Lu(2.8 min))

⟨γ⟩=1864 *22* keV

γ_{mode}	γ(keV)	γ(%)†
γ E1	67.55 *4*	3.9 *4*
γ (M1+E2)	75.02 *7*	0.90 *12*
γ [M1+E2]	93.52 *5*	0.20 *4*
γ	99.50 *20*	0.45 *4*
γ E2	102.34 *3*	25.0 *12*
γ	138.96 *21*	0.41 *12*
γ	160.0 *6*	~0.24
γ	191.7 *3*	0.49 *8*
γ	195.51 *15*	0.85 *12*
γ (M1+E2)	208.65 *5*	3.7 *4*
γ (E2)	212.83 *7*	1.14 *12*
γ [M1+E2]	219.76 *16*	0.33 *4*
γ E2	228.09 *3*	76.7 *15*
γ (E2)	248.59 *5*	4.80 *24*
γ	268.13 *16*	0.81 *8*
γ [E2]	272.23 *14*	1.63 *20*
γ (M1)	274.40 *4*	9.9 *6*
γ (E1+M2)	276.20 *3*	13.6 *8*
γ [E2]	288.74 *4*	1.90 *5*
γ [M1+E2]	294.78 *15*	0.39 *8*
γ	319.34 *15*	0.75 *10*
γ [E2]	330.23 *15*	0.45 *8*
γ (E2)	337.47 *3*	40.7
γ	353.93 *20*	0.54 *12*
γ (M1)	360.04 *6*	3.6 *3*
γ (M1)	367.92 *3*	31.2 *9*
γ	377.4 *4*	0.37 *8*
γ (E1+M2)	382.95 *3*	3.05 *20*
γ [E1]	386.49 *15*	0.28 *12*
γ [E2]	396.97 *10*	1.47 *4*
γ [E2]	430.26 *3*	5.0 *3*
γ [E2]	442.94 *7*	0.53 *12*
γ	445.70 *22*	0.22 *9*
γ [E1]	454.07 *6*	1.57 *10*
γ	467.7 *5*	0.37 *11*
γ [M1+E2]	474.67 *4*	2.73 *16*
γ	487.1 *3*	0.63 *15*
γ	490.23 *21*	0.44 *12*
γ [E2]	494.70 *5*	~0.24
γ	523.9 *5*	0.50 *5*
γ	533.88 *19*	0.53 *16* ?
γ (E1+M2)	537.47 *4*	8.1 *3*
γ [E1]	577.62 *4*	4.03 *24*
γ [M1+E2]	626.59 *11*	0.41 *12*
γ (M1+E2)	629.19 *4*	7.0 *4*
γ	648.1 *6*	0.41 *12*
γ [M1+E2]	659.89 *3*	3.66 *24*
γ [M1+E2]	705.10 *6*	0.53 *11*
γ [M1+E2]	708.62 *4*	1.12 *11*
γ [M1+E2]	714.37 *14*	0.61 *6*
γ	735.2 *6*	0.37 *12*
γ	760.9 *6*	~0.24
γ [E2]	794.38 *4*	2.97 *20*
γ [E2]	811.84 *5*	1.22 *24*
γ [M1+E2]	814.41 *4*	6.7 *4*
γ [M1+E2]	829.97 *5*	
γ [M1+E2]	832.17 *5*	6.0 *4*
γ [E1]	837.32 *7*	2.73 *16*
γ [E1]	860.62 *6*	3.26 *20*
γ [E1]	902.58 *15*	0.41 *16*
γ [E2]	932.31 *5*	
γ [M1+E2]	936.71 *3*	5.71 *23*
γ	975.0 *6*	0.33 *12*
γ (E2)	997.36 *3*	17.9 *7*
γ	1021.2 *5*	0.55 *11*
γ [M1+E2]	1056.85 *10*	2.1 *5*
γ [E2]	1060.26 *5*	1.30 *8*
γ	1067.46 *6*	2.5 *3*
γ [E1]	1122.34 *6*	4.03 *20*
γ [E2]	1144.63 *14*	0.49 *12*
γ [E2]	1151.88 *4*	0.45 *12*
γ	1165.2 *6*	0.41 *16*
γ [E1]	1174.80 *7*	4.4 *4*
γ	1185.2 *6*	0.81 *24*
γ	1186.9 *6*	0.41 *16*
γ [E1]	1197.36 *4*	0.57 *8*
γ	1201.5 *4*	0.41 *8*
γ	1234.2 *3*	0.85 *16*
γ [E1]	1240.05 *15*	1.34 *16*
γ	1261.7 *6*	~0.33
γ [E1]	1290.88 *5*	9.7 *7*
γ	1301.9 *4*	0.65 *12*

Photons (^{166}Lu(2.8 min)) (continued)

γ_{mode}	γ(keV)	γ(%)†
γ	1306.0 *5*	0.49 *12*
γ	1310.8 *7*	0.53 *8*
γ	1349.4 *6*	~0.33
γ [M1+E2]	1354.32 *13*	1.7 *4*
γ	1398.0 *9*	0.73 *20*
γ [E1]	1459.81 *7*	7.8 *4*
γ	1487.59 *21*	1.06 *20*
γ	1497.72 *5*	0.73 *16*
γ	1505.1 *4*	0.73 *16*
γ [E2]	1582.41 *13*	0.26 *12*
γ [E2]	1626.54 *4*	0.94 *16*
γ	1640.3 *6*	0.37 *12*
γ	1645.4 *6*	0.28 *12*
γ	1685.75 *19*	0.49 *8*
γ	1720.3 *6*	~0.24

† 4.9% uncert(syst)

$^{166}_{71}$Lu(1.41 *10* min)

Mode: ε(58 *5* %), IT(42 *5* %)
Δ: -56086 *160* keV
SpA: 8.0×10^{8} Ci/g
Prod: ^{169}Tm(^{3}He,6n); ^{170}Yb(p,5n)

Photons (^{166}Lu)*

⟨γ⟩=1396 *49* keV

γ_{mode}	γ(keV)	γ(%)†
Yb L_{ℓ}	6.545	0.36 *6*
Lu L_{ℓ}	6.753	0.19 *3*
Yb L_{α}	7.411	9.0 *14*
Yb L_{η}	7.580	0.140 *23*
Lu L_{α}	7.650	4.6 *6*
Lu L_{η}	7.857	0.0066 *9*
Yb L_{β}	8.509	9.2 *15*
Lu L_{β}	9.008	1.7 *3*
Yb L_{γ}	9.837	1.46 *25*
Lu L_{γ}	10.401	0.22 *4*
γ_{IT} (M3)	34.4 *1*	0.00047 *6*
Yb $K_{\alpha2}$	51.354	16.4 *16*
Yb $K_{\alpha1}$	52.389	29 *3*
Yb $K_{\beta1}'$	59.316	9.2 *9*
Yb $K_{\beta2}'$	61.229	2.52 *25*
γ_{ϵ}E2	102.34 *3*	22 *5*
γ_{ϵ}	152.44 *13*	2.47 *19*
γ_{ϵ}E2	228.09 *3*	26 *8*
γ_{ϵ}E1	285.01 *5*	19.0 *10*
γ_{ϵ}[E1]	344.40 *15*	0.76 *19*
γ_{ϵ}	346.89 *10*	0.76 *19*
γ_{ϵ}	407.0 *6*	~0.8
γ_{ϵ}	412.21 *19*	2.09 *19*
γ_{ϵ}	421.26 *9*	3.61 *19*
γ_{ϵ}	464.19 *6*	1.3 *4*
γ_{ϵ}	470.4 *5*	1.0 *4*
γ_{ϵ}[E2]	494.70 *5*	~0.18
γ_{ϵ}	525.91 *7*	5.1 *6*
γ_{ϵ}	568.85 *11*	1.3 *6*
γ_{ϵ}	570.93 *6*	5.5 *6*
γ_{ϵ}	581.0 *6*	2.1 *6*
γ_{ϵ}	625.3 *6*	1.1 *4*
γ_{ϵ}	643.22 *9*	6.1 *6*
γ_{ϵ}	680.8 *4*	1.1 *4*
γ_{ϵ}	701.9 *3*	1.71 *19*
γ_{ϵ}[M1+E2]	705.10 *6*	7.6 *8*
γ_{ϵ}[M1+E2]	708.62 *4*	2.8 *3*
γ_{ϵ}	747.1 *5*	0.76 *19*
γ_{ϵ}[E2]	811.84 *5*	16.9 *10*
γ_{ϵ}[M1+E2]	829.97 *5*	17.7 *10*
γ_{ϵ}[M1+E2]	832.17 *5*	4.5 *11*
γ_{ϵ}[E1]	866.55 *8*	2.1 *4*
γ_{ϵ}[E2]	932.31 *5*	13.9 *10*

Photons (^{166}Lu)* (continued)

γ_{mode}	γ(keV)	γ(%)†
γ_{ϵ}[M1+E2]	936.71 *3*	14.2 *9*
γ_{ϵ}	984.61 *13*	3.8 *8*
γ_{ϵ}	1023.8 *6*	~1
γ_{ϵ}	1055.51 *11*	~2
γ_{ϵ}[E2]	1060.26 *5*	0.97 *24*
γ_{ϵ}	1277.46 *11*	2.1 *6*
γ_{ϵ}	1283.6 *1*	6.7 *13*
γ_{ϵ}	1348.92 *19*	1.0 *4*
γ_{ϵ}[M1+E2]	1354.32 *13*	~2
γ_{ϵ}	1505.55 *11*	2.1 *6*
γ_{ϵ}[E2]	1582.41 *13*	~0.3
γ_{ϵ}[E1]	1698.72 *7*	2.3 *6*
γ_{ϵ}	1801.3 *6*	1.7 *6*
γ_{ϵ}	1974.0 *6*	~1

† uncert(syst): 18% for ε, 12% for IT
* see also ^{166}Lu(2.8 min)

Atomic Electrons (^{166}Lu)

⟨e⟩=91 *6* keV

e_{bin}(keV)	⟨e⟩(keV)	e(%)
9	4.9	55 *10*
10	2.6	26 *5*
11 - 24	1.77	8.8 *9*
25	6.0	24 *3*
32	2.9	8.9 *12*
33 - 40	1.05	3.1 *3*
41	8.7	21 *5*
42 - 91	2.1	3.4 *10*
92	16.8	18 *4*
93	14.1	15 *4*
100	8.2	8.1 *19*
101 - 150	3.1	2.8 *6*
151 - 152	0.04	~0.03
167	5.0	3.0 *9*
218 - 228	4.4	2.0 *3*
275 - 286	0.33	0.12 *3*
334 - 360	0.8	0.22 *11*
396 - 433	0.43	0.11 *4*
454 - 495	0.5	~0.11
508 - 526	0.8	0.15 *7*
558 - 582	0.6	0.11 *5*
615 - 647	1.0	0.15 *4*
670 - 709	0.23	0.033 *7*
737 - 771	1.8	0.24 *5*
801 - 832	0.43	0.052 *8*
856 - 875	1.11	0.13 *3*
922 - 962	0.41	0.044 *12*
974 - 1023	0.11	0.011 *5*
1024 - 1060	0.016	0.0015 *6*
1216 - 1222	0.23	~0.019
1267 - 1293	0.13	~0.010
1338 - 1354	0.016	~0.0012
1495 - 1580	0.015	0.0010 *5*
1637 - 1697	0.022	0.0014 *3*
1740 - 1799	0.035	~0.0020
1913 - 1972	0.021	~0.0011

Continuous Radiation (^{166}Lu)

⟨β+⟩=198 keV;⟨IB⟩=4.1 keV

E_{bin}(keV)		⟨ ⟩(keV)	(%)
0 - 10	β+	1.14×10^{-7}	1.35×10^{-6}
	IB	0.0076	
10 - 20	β+	6.5×10^{-6}	3.87×10^{-5}
	IB	0.0073	0.050
20 - 40	β+	0.000253	0.00076
	IB	0.0161	0.055
40 - 100	β+	0.0180	0.0227
	IB	0.138	0.25
100 - 300	β+	1.24	0.55

Continuous Radiation (^{166}Lu)
(continued)

E_{bin}(keV)		⟨ ⟩(keV)	(%)
	IB	0.133	0.074
300 - 600	β+	10.8	2.31
	IB	0.23	0.052
600 - 1300	β+	82	8.6
	IB	0.96	0.101
1300 - 2500	β+	104	6.3
	IB	2.2	0.120
2500 - 3485	IB	0.48	0.017
	Σβ+		18

$^{166}_{71}$Lu(2.12 *10* min)

Mode: ε
Δ: -56077 *160* keV
SpA: 5.33×10^8 Ci/g
Prod: ^{169}Tm(^{3}He,6n); ^{170}Yb(p,5n)

Photons (^{166}Lu)*
⟨γ⟩=1744 *70* keV

γ_{mode}	γ(keV)	γ(%)
Yb L_ℓ	6.545	0.30 *6*
Yb L_α	7.411	7.4 *14*
Yb L_η	7.580	0.106 *24*
Yb L_β	8.514	7.3 *15*
Yb L_γ	9.854	1.17 *25*
Yb $K_{\alpha2}$	51.354	16.1 *17*
Yb $K_{\alpha1}$	52.389	28 *3*
Yb $K_{\beta1}$'	59.316	9.0 *10*
Yb $K_{\beta2}$'	61.229	2.5 *3*
γ E2	102.34 *3*	11 *5*
γ E2	228.09 *3*	<8
γ [E1]	518.8 *3*	1.0 *5*
γ [M1+E2]	1067.92 *16*	5.6 *9*
γ [E2]	1249.96 *24*	1.5 *6*
γ [E1]	1257.16 *7*	15.0 *15*
γ [E1]	1359.50 *7*	13.2 *16*
γ [E1]	1427.92 *9*	22.6 *23*
γ [M1+E2]	1478.06 *24*	2.7 *5*
γ [E1]	1530.27 *9*	10.9 *8*
γ [E2]	1580.40 *24*	1.0 *5*
γ	1821.0 *6*	~0.9
γ	1923.4 *6*	2.4 *3*
γ [E1]	1996.86 *12*	3.3 *9*
γ [E1]	2099.21 *12*	15.9 *19*
γ [E1]	2325.07 *17*	9.3 *8*
γ [E1]	2427.41 *17*	~0.6

* see also ^{166}Lu(2.8 min)

Atomic Electrons (^{166}Lu)
⟨e⟩=33 *6* keV

e_{bin}(keV)	⟨e⟩(keV)	e(%)
9	2.6	29 *6*
10	2.0	20 *5*
40	0.110	0.27 *4*
41	4.5	11 *5*
42 - 59	1.01	2.18 *15*
92	8.5	9 *4*
93	7.1	8 *4*
100	4.1	4.1 *20*
101	0.037	0.037 *18*
102	1.1	1.1 *5*
218 - 228	0.6	~0.26
508 - 519	0.0047	0.0009 *3*
1057 - 1068	0.045	0.0043 *12*

Atomic Electrons (^{166}Lu)
(continued)

e_{bin}(keV)	⟨e⟩(keV)	e(%)
1189 - 1196	0.189	0.0158 *18*
1239 - 1257	0.035	0.00281 *25*
1349 - 1367	0.234	0.0172 *17*
1417 - 1428	0.11	0.0080 *18*
1468 - 1528	0.149	0.0101 *8*
1570 - 1578	0.0038	0.00024 *9*
1760 - 1819	0.018	~0.0010
1862 - 1936	0.071	0.0037 *12*
1986 - 2038	0.122	0.0060 *7*
2089 - 2097	0.0198	0.00095 *10*
2264 - 2323	0.076	0.0033 *3*
2366 - 2425	0.0048	0.00020 *9*

$^{166}_{72}$Hf(6.8 *3* min)

Mode: ε
Δ: -53790 *300* keV syst
SpA: 1.67×10^8 Ci/g
Prod: protons on Ta; ^{170}Yb(^{3}He,7n)

Photons (^{166}Hf)
⟨γ⟩=188 *8* keV

γ_{mode}	γ(keV)	γ(%)
Lu $K_{\alpha2}$	52.965	27.0 *9*
Lu $K_{\alpha1}$	54.070	47.2 *16*
Lu $K_{\beta1}$'	61.219	15.2 *5*
Lu $K_{\beta2}$'	63.200	4.06 *15*
γ E1	78.13 *12*	41 *2*
γ (E1)	92.18 *22*	3.3 *4*
γ	169.3 *6*	0.45 *16*
γ	243.9 *4*	1.6 *5*
γ	283.24 *24*	1.6 *6*
γ	298.23 *23*	1.3 *4*
γ	306.12 *25*	1.7 *4*
γ	338.21 *16*	1.2 *6*
γ	341.12 *12*	4.7 *4*
γ	355.18 *25*	1.1 *5*
γ	376.36 *25*	4.0 *11*
γ [M1+E2]	407.19 *12*	4.5 *9*
γ	430.06 *10*	1.3 *3*
γ [E2]	482.41 *12*	4.1 *7*

Atomic Electrons (^{166}Hf)
⟨e⟩=21.6 *12* keV

e_{bin}(keV)	⟨e⟩(keV)	e(%)
9	3.6	39 *4*
10	1.94	18.8 *20*
11	0.73	6.7 *7*
15	3.30	22.3 *12*
29 - 61	2.47	5.77 *24*
67	1.54	2.28 *12*
68	0.52	0.77 *4*
69	0.63	0.91 *5*
76	0.67	0.88 *5*
77 - 106	0.61	0.70 *16*
158 - 181	0.5	~0.3
220 - 244	1.1	0.47 *21*
272 - 275	0.29	~0.11
278	0.8	~0.28
281 - 306	0.39	0.13 *6*
313	0.6	~0.18
327 - 341	0.29	0.09 *4*
344	0.7	~0.21
345 - 376	0.34	0.09 *4*
396 - 430	0.50	0.122 *23*
472 - 482	0.090	0.0189 *24*

$^{166}_{73}$Ta(32 *3* s)

Mode: ε
Δ: -46310 *340* keV syst
SpA: 2.10×10^9 Ci/g
Prod: protons on Hg; ^{159}Tb(^{16}O,9n); protons on Re

Photons (^{166}Ta)

γ_{mode}	γ(keV)	γ(rel)
γ	158.7 *2*	100 *4*
γ	311.7 *3*	53.6 *21*
γ	536.0 *4*	7.5 *12*
γ	552.4 *4*	5.6 *18*
γ	594.5 *3*	6.7 *10*
γ	651.4 *4*	16.1 *11*
γ	693.2 *5*	3.2 *9*
γ	742.8 *4*	13.3 *12*
γ	750.0 *5*	10.4 *18*
γ	810.1 *4*	18.6 *16*
γ	847.4 *4*	14 *3*
γ	851.7 *6*	3.4 *14*
γ	862.2 *6*	7.1 *20*
γ	864.1 *5*	9.2 *23*
γ	906.2 *6*	11.5 *15*
γ	977.0 *8*	4.7 *11*
γ	1054.4 *10*	8.3 *13*
γ	1173.8 *10*	10 *5*
γ	1288.3 *12*	5.8 *21*
γ	1447 *2*	6.3 *16*

$^{166}_{74}$W (16 *3* s)

Mode: α
Δ: -41899 *18* keV
SpA: 4.2×10^9 Ci/g
Prod: ^{156}Dy(^{16}O,6n)

Alpha Particles (^{166}W)

α(keV)
4739 *5*

$^{166}_{75}$Re(2.2 *4* s)

Mode: α(~70 %), ε(~30 %)
Δ: -31910 *220* keV syst
SpA: 2.6×10^{10} Ci/g
Prod: ^{93}Nb(^{84}Kr,α7n); ^{89}Y(^{84}Kr,7n)

Alpha Particles (^{166}Re)

α(keV)
5495 *10*

$^{166}_{76}$Os(181 *38* ms)

Mode: α(72 *13* %), ε(28 *13* %)
Δ: -25640 *280* keV syst
SpA: 9.6×10^{10} Ci/g
Prod: daughter ^{170}Pt; ^{106}Cd(^{63}Cu,p2n); ^{107}Ag(^{63}Cu,4n)

Alpha Particles (^{166}Os)

α(keV)
5981 *6*

$^{166}_{77}$Ir(>5 ms)

Mode: α
Δ: -13170 *550* keV syst
SpA: <9.8×10^{10} Ci/g
Prod: ^{58}Ni on Mo-Sn targets; ^{107}Ag on V-Ni targets

Alpha Particles (^{166}Ir)

α(keV)
6541 *20*

A = 167

NDS **17**, 143 (1976)

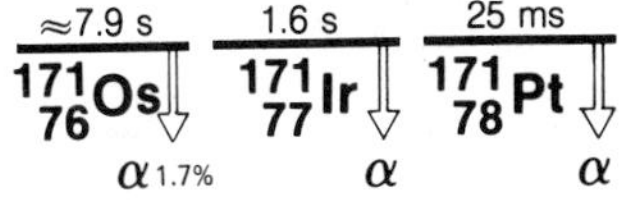

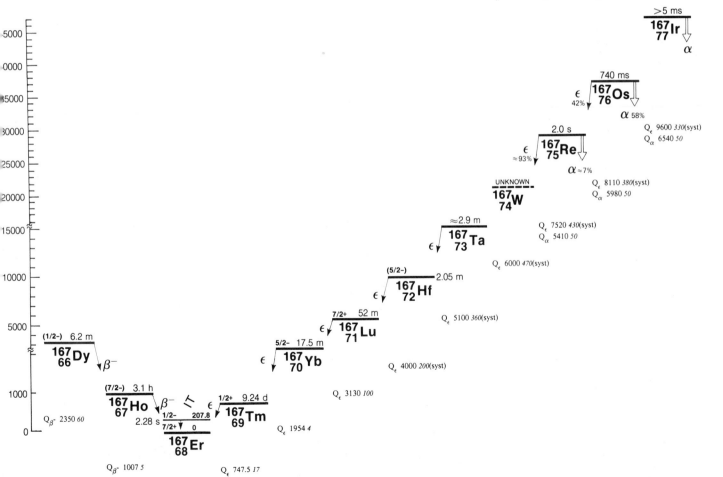

$^{167}_{66}$Dy(6.20 *8* min)

Mode: β-

Δ: -59940 *60* keV

SpA: 1.814×10^{8} Ci/g

Prod: ^{170}Er(n,α)

Photons (^{167}Dy)

⟨γ⟩=534 *14* keV

γ_{mode}	γ(keV)	γ(%)†
Ho L_ℓ	5.943	0.128 *17*
Ho L_α	6.716	3.3 *4*
Ho L_η	6.789	0.045 *6*
Ho L_β	7.656	3.1 *4*
Ho L_γ	8.843	0.49 *7*
Ho $K_{\alpha2}$	46.700	9.5 *7*
Ho $K_{\alpha1}$	47.547	16.9 *13*
Ho $K_{\beta1}$'	53.822	5.2 *4*
Ho $K_{\beta2}$'	55.556	1.49 *12*
γ M1	60.41 *5*	0.91 *14*
γ [E2]	72.72 *7*	~0.14
γ [M1+E2]	90.22 *6*	0.43 *6*
γ [M1+E2]	133.13 *6*	3.12 *24*
γ [M1+E2]	150.63 *6*	0.67 *10*
γ	159.71 *6*	~0.48
γ	249.93 *7*	9.6 *5*
γ M2	259.33 *11*	27.8 *19*
γ	310.34 *7*	25.0 *14*
γ	352.30 *17*	1.01 *10*
γ	569.66 *12*	48.0
γ	579.28 *24*	0.23 *4*
γ	599.13 *19*	0.82 *10*
γ	662.64 *17*	0.34 *4*
γ	689.53 *17*	~0.24
γ	707.03 *17*	0.96 *19*
γ	738.99 *24*	0.58 *14*
γ	745.99 *20*	0.413 *8*
γ	799.0 *4*	~0.38
γ	830.6 *3*	~0.34
γ	848.1 *3*	~0.48
γ	909.46 *19*	~0.38
γ	920.8 *3*	~0.24
γ	981.2 *3*	~0.24
γ	996.99 *20*	0.46 *5*
γ	1080.3 *3*	0.30 *4*
γ	1095.2 *3*	0.24 *10*
γ	1272.4 *3*	0.31 *4*
γ	1405.5 *3*	0.23 *3*

† 4.0% uncert(syst)

Atomic Electrons (^{167}Dy)

⟨e⟩=72 *5* keV

e_{bin}(keV)	⟨e⟩(keV)	e(%)
5 - 54	4.5	39 *3*
58 - 104	3.9	5.1 *10*
124 - 160	1.6	1.3 *5*
194	1.8	~0.9
204	38.4	18.9 *14*
241 - 249	0.7	~0.27
250	8.7	3.5 *3*
251	0.58	0.229 *16*
255	3.5	~1
257	2.03	0.79 *6*
258 - 302	1.7	0.61 *20*
308 - 352	0.28	~0.09
514	2.9	~0.6
524 - 571	0.7	0.13 *6*
577 - 607	0.027	0.0046 *22*
634 - 683	0.08	0.013 *6*
687 - 737	0.034	0.0049 *19*
738 - 775	0.028	~0.0038
790 - 839	0.025	~0.0032
840 - 865	0.021	~0.0025
900 - 941	0.024	0.0026 *11*
972 - 997	0.0040	0.00040 *16*
1025 - 1072	0.015	0.0015 *7*
1078 - 1095	0.0015	0.00014 *7*
1217 - 1264	0.008	~0.0006
1270 - 1272	0.00027	~2×10^{-5}
1350 - 1397	0.005	~0.00037
1403 - 1405	0.00017	1.2 *6* ×10^{-5}

Continuous Radiation (^{167}Dy)

⟨β-⟩=666 keV;⟨IB⟩=1.14 keV

E_{bin}(keV)		⟨ ⟩(keV)	(%)
0 - 10	β-	0.0301	0.60
	IB	0.028	
10 - 20	β-	0.091	0.61
	IB	0.027	0.19
20 - 40	β-	0.374	1.24
	IB	0.052	0.18
40 - 100	β-	2.80	3.97
	IB	0.143	0.22
100 - 300	β-	31.2	15.4
	IB	0.36	0.20
300 - 600	β-	118	26.2
	IB	0.31	0.075
600 - 1300	β-	414	46.0
	IB	0.22	0.027
1300 - 2091	β-	99	6.8
	IB	0.0081	0.00058

$^{167}_{67}$Ho(3.1 *1* h)

Mode: β-

Δ: -62292 *6* keV

SpA: 6.05×10^{6} Ci/g

Prod: ^{170}Er(p,α); ^{167}Er(n,p); ^{168}Er(γ,p); ^{164}Dy(α,p)

Photons (^{167}Ho)

⟨γ⟩=365 *40* keV

γ_{mode}	γ(keV)	γ(%)†
Er L_ℓ	6.151	0.121 *19*
Er L_α	6.944	3.1 *4*
Er L_η	7.058	0.047 *7*
Er L_β	7.927	3.2 *5*
Er L_γ	9.160	0.50 *8*
Er $K_{\alpha2}$	48.221	6.7 *7*
Er $K_{\alpha1}$	49.128	11.8 *13*
Er $K_{\beta1}$'	55.616	3.7 *4*
Er $K_{\beta2}$'	57.406	1.04 *11*
γ E2	73.77 *14*	0.46 *17*
γ M1+~9.3%E2	79.27 *20*	2.2 *5*
γ M1+~13%E2	83.47 *15*	1.6 *3*
γ [M1]	106.4 *4*	0.09 *4*
γ [M1+E2]	131.59 *18*	~0.11 ‡
γ [E2]	148.26 *18*	~0.11
γ E3	207.79 *15*	5.0 *3* *
γ [M1]	208.7 *4*	~0.17
γ M1	237.84 *15*	5.2 *3*
γ [M1]	254.64 *16*	0.21 *6*
γ [M1+E2]	303.0 *10*	<0.06 ?
γ [M1]	315.02 *22*	0.74 *17*
γ M1	321.31 *13*	24.1 *9*
γ [M1+E2]	331.8 *3*	~0.17
γ E1	346.48 *13*	57
γ M1	386.22 *14*	3.4 *2*
γ [M1]	398.49 *20*	0.92 *17*
γ [M1]	402.90 *11*	3.3 *2*
γ [E1]	429.94 *17*	0.13 *3*
γ [E2]	460.00 *11*	2.1 *2*
γ [M1]	463.40 *24*	0.46 *17*
γ [E2]	480.08 *22*	0.15 *3*
γ E2	531.5 *3*	<0.02 ?
γ [E1]	667.78 *15*	0.23 *9*
γ [E1]	744.96 *22*	0.17 *6*

† approximate

* with ^{167}Er(2.28 s) in equilib

‡ doublet

Atomic Electrons (^{167}Ho)

⟨e⟩=42.3 *13* keV

e_{bin}(keV)	⟨e⟩(keV)	e(%)
8	1.11	13.3 *21*
9 - 16	1.08	10.9 *13*
22	2.2	10.0 *23*
26	1.6	6.1 *12*
38 - 64	2.4	5.0 *9*
65	1.8	2.7 *10*
70	1.3	1.9 *4*
71 - 106	3.1	4.1 *4*
122 - 148	0.080	0.060 *18*
150	3.57	2.38 *15*
151	0.08	~0.05
180	2.04	1.13 *7*
192 - 198	0.66	0.332 *22*
199	6.4	3.24 *21*
200	0.00017	~8×10^{-5}
206	1.82	0.89 *6*
207 - 255	1.01	0.461 *17*
257 - 258	0.19	0.075 *18*
264	6.1	2.33 *10*
265 - 274	0.030	~0.011
289	1.8	0.64 *13*
293 - 341	2.58	0.80 *3*
344 - 394	0.97	0.273 *14*
395 - 430	0.25	0.063 *7*
450 - 480	0.060	0.0131 *11*
522 - 532	0.00015	~3×10^{-5}
610 - 659	0.0042	0.00067 *24*
666 - 687	0.0026	0.00037 *13*
735 - 745	0.00046	6.2 *16* ×10^{-5}

Continuous Radiation (^{167}Ho)

⟨β-⟩=188 keV;⟨IB⟩=0.144 keV

E_{bin}(keV)		⟨ ⟩(keV)	(%)
0 - 10	β-	0.214	4.30
	IB	0.0093	
10 - 20	β-	0.63	4.19
	IB	0.0086	0.060
20 - 40	β-	2.42	8.1
	IB	0.0153	0.053
40 - 100	β-	14.9	21.5
	IB	0.035	0.055
100 - 300	β-	72	39.8
	IB	0.054	0.033

Continuous Radiation (^{167}Ho)
(continued)

E_{bin}(keV)		⟨ ⟩(keV)	(%)
300 - 600	β-	73	17.3
	IB	0.020	0.0051
600 - 1007	β-	25.2	3.59
	IB	0.00164	0.00025

$^{167}_{68}$Er(stable)

Δ: -63299 *3* keV
%: 22.95 *13*

$^{167}_{68}$Er(2.28 *3* s)

Mode: IT
Δ: -63091 *3* keV
SpA: 2.55×10^{10} Ci/g
Prod: daughter ^{167}Tm; daughter ^{167}Ho; ^{166}Er(n,γ); ^{168}Er(n,2n)

Photons (^{167}Er)

⟨γ⟩=97 *3* keV

γ_{mode}	γ(keV)	γ(%)†
Er L_ℓ	6.151	0.15 *3*
Er L_α	6.944	3.8 *8*
Er L_η	7.058	0.083 *18*
Er L_β	7.902	4.8 *10*
Er L_γ	9.118	0.75 *16*
Er $K_{\alpha2}$	48.221	5.4 *11*
Er $K_{\alpha1}$	49.128	9.6 *19*
Er $K_{\beta1}'$	55.616	3.0 *6*
Er $K_{\beta2}'$	57.406	0.84 *17*
γ E3	207.79 *15*	41.7

† 2.9% uncert(syst)

Atomic Electrons (^{167}Er)

⟨e⟩=111 *13* keV

e_{bin}(keV)	⟨e⟩(keV)	e(%)
8	1.4	16 *4*
9	1.4	15 *3*
10	0.12	1.2 *3*
38	0.089	0.23 *5*
39	0.048	0.12 *3*
40	0.087	0.22 *5*
41	0.042	0.103 *23*
46	0.073	0.16 *4*
47	0.058	0.12 *3*
48	0.016	0.033 *8*
49	0.013	0.026 *6*
53	0.0064	0.012 *3*
54	0.0066	0.012 *3*
55	0.0056	0.0101 *23*
150	29.7	20 *4*
198	4.8	2.4 *5*
199	53.5	27 *5*
206	15.1	7.3 *15*
207	3.6	1.7 *3*
208	0.51	0.25 *5*

$^{167}_{69}$Tm(9.24 *2* d)

Mode: ϵ
Δ: -62552 *4* keV
SpA: 8.462×10^{4} Ci/g
Prod: ^{165}Ho(α,2n); protons on Er; protons on U

Photons (^{167}Tm)

⟨γ⟩=146 *13* keV

γ_{mode}	γ(keV)	γ(%)†
Er L_ℓ	6.151	0.49 *7*
Er L_α	6.944	12.6 *14*
Er L_η	7.058	0.200 *25*
Er L_β	7.922	13.6 *18*
Er L_γ	9.171	2.2 *3*
Er $K_{\alpha2}$	48.221	27.4 *13*
Er $K_{\alpha1}$	49.128	48.5 *24*
Er $K_{\beta1}'$	55.616	15.1 *7*
γ M1+13%E2	57.10 *4*	~3
Er $K_{\beta2}'$	57.406	4.27 *22*
γ E2	73.77 *14*	
γ E3	207.79 *15*	41 *6* *
γ [E1]	250.0 *3*	0.0022 *5*
γ [E1]	266.6 *3*	0.0022 *5*
γ [E1]	323.7 *3*	0.0021 *5*
γ E1	346.48 *13*	0.025 *3*
γ E2	531.5 *3*	1.6 *3*

† 2.4% uncert(syst)
* with ^{167}Er(2.28 s) in equilib

Atomic Electrons (^{167}Tm)

⟨e⟩=126 *10* keV

e_{bin}(keV)	⟨e⟩(keV)	e(%)
8	4.4	53 *7*
9 - 57	16.0	68 *7*
150	29.3	19 *3*
192	0.000105	5.5 *13* $\times10^{-5}$
198	4.8	2.4 *4*
199	52.8	27 *4*
206	14.9	7.2 *11*
207	3.5	1.69 *25*
208 - 257	0.51	0.24 *4*
258 - 289	0.00088	0.00031 *3*
314 - 346	0.000189	5.6 *5* $\times10^{-5}$
474 - 523	0.104	0.022 *4*
529 - 532	0.0054	0.00103 *13*

Continuous Radiation (^{167}Tm)

⟨IB⟩=0.24 keV

E_{bin}(keV)		⟨ ⟩(keV)	(%)
10 - 20	IB	0.00018	0.00122
20 - 40	IB	0.0087	0.024
40 - 100	IB	0.20	0.41
100 - 300	IB	0.019	0.0106
300 - 540	IB	0.0061	0.0017

$^{167}_{70}$Yb(17.5 *2* min)

Mode: ϵ
Δ: -60598 *5* keV
SpA: 6.43×10^{7} Ci/g
Prod: daughter ^{167}Lu; ^{169}Tm(p,3n); ^{164}Er(α,n); ^{168}Yb(d,t); ^{168}Yb(γ,n)

Photons (^{167}Yb)

⟨γ⟩=257 *5* keV

γ_{mode}	γ(keV)	γ(%)†
γ M1+9.3%E2	37.066 *13*	0.10 *4*
Tm $K_{\alpha2}$	49.772	54 *3*
Tm $K_{\alpha1}$	50.742	95 *5*
Tm $K_{\beta1}'$	57.444	29.9 *17*
Tm $K_{\beta2}'$	59.296	8.3 *5*
γ M1+0.6%E2	62.899 *14*	4.9 *8*
γ (M1+E2)	90.85 *6*	0.018 *8*
γ E2	98.252 *21*	0.082 *8*
γ M1	105.189 *16*	0.59 *6*
γ M1+1.0%E2	106.167 *14*	22.5 *10*
γ E1	113.336 *15*	55 *2*
γ E2	116.577 *19*	2.82 *6*
γ [M1+20%E2]	116.68 *7*	~0.06
γ E2	132.001 *14*	2.78 *8*
γ E1	143.466 *18*	2.10 *6*
γ E1	150.402 *15*	0.037 *10*
γ [E1]	161.34 *6*	0.035 *10*
γ E2	169.066 *15*	0.157 *14*
γ [E1]	171.75 *6*	0.037 *10*
γ E1	176.236 *16*	20.4
γ E1	177.214 *19*	2.7 *1*
γ [M1]	184.08 *18*	~0.014
γ [E1]	272.06 *14*	0.0027 *6*
γ	280.51 *20*	0.0061 *14*
γ [E1]	282.47 *14*	0.0083 *16*
γ	321.1 *5*	0.0022 *10*
γ	323.5 *5*	0.0035 *10*
γ	343.30 *8*	0.034 *4*
γ	351.8 *4*	0.0033 *12*
γ [M1+E2]	354.10 *6*	~0.0024
γ	375.91 *20*	0.0067 *16*
γ [M1+E2]	379.75 *8*	0.0043 *14*
γ	387.0 *4*	0.0022 *10*
γ	398.11 *20*	0.0047 *10*
γ [M1+E2]	405.59 *8*	0.014 *2*
γ [M1+E2]	415.44 *6*	0.004 *1*
γ	421.41 *20*	0.0041 *10*
γ [M1+E2]	441.27 *6*	0.011 *2*
γ	446.8 *3*	0.0024 *8*
γ	457.01 *10*	0.0067 *14*
γ [M1+E2]	460.27 *6*	0.026 *3*
γ [M1+E2]	470.68 *6*	0.023 *3*
γ	486.61 *20*	0.0067 *16*
γ	541.41 *20*	0.0045 *12*
γ [M1+E2]	547.44 *6*	0.012 *2*
γ	561.8 *4*	0.0029 *10*
γ	571.70 *17*	0.0065 *14*
γ	590.9 *4*	0.0047 *16*
γ	600.2 *4*	0.0041 *12*
γ	664.91 *20*	0.0090 *24*
γ	671.98 *11*	0.008 *2*
γ	680.23 *16*	0.0043 *14*
γ	686.9 *5*	0.005 *3*
γ	688.38 *16*	0.012 *3*
γ	694.5 *6*	~0.004
γ	697.1 *6*	~0.004
γ	707.66 *12*	~0.003
γ	719.5 *3*	0.0039 *12*
γ	733.2 *3*	0.007 *2* ?
γ	791.51 *20*	0.013 *2*
γ	829.4 *3*	0.0069 *18*
γ	846.16 *11*	0.013 *2*
γ	903.31 *20*	0.0067 *18*
γ	920.37 *7*	0.116 *18*
γ	923.71 *6*	0.006 *2*
γ	927.1 *8*	0.0041 *18*
γ	933.66 *12*	0.005 *2*
γ	937.01 *10*	0.007 *2*

Photons (^{167}Yb)
(continued)

γ_{mode}	γ(keV)	γ(%)†
γ	977.9 3	0.0043 14
γ	998.3 3	0.0043 14
γ	1008.6 5	0.0037 14
γ	1023.05 8	0.011 2
γ	1025.9 3	0.0045 16
γ	1037.05 6	0.61 7
γ	1048.64 11	~0.012
γ	1050.35 10	0.039 9
γ	1068.2 4	0.007 3
γ	1070.3 6	~0.0035
γ	1110.22 7	0.011 2
γ	1139.49 10	0.039 6
γ	1165.5 4	0.0031 12
γ	1213.89 9	0.006 2
γ	1217.11 20	0.0067 20
γ	1234.62 6	0.157 18
γ	1242.01 10	0.017 3
γ	1254.41 19	0.003 1
γ	1288.09 6	0.034 5
γ	1298.42 15	~0.002
γ	1304.75 8	0.033 5
γ	1320.91 10	0.012 2
γ	1332.51 20	0.0055 14
γ	1336.91 14	0.003 1
γ	1339.80 6	0.004 1
γ	1342.4 4	0.0039 14
γ	1361.48 8	0.018 3
γ	1366.5 7	0.0031 12
γ	1370.21 10	0.012 2
γ	1385.02 6	0.008 2
γ	1393.27 6	0.006 1
γ	1401.42 6	0.003 1
γ	1410.85 6	0.003 1
γ	1427.8 3	0.0031 10
γ	1433.7 3	0.0029 10
γ	1438.49 6	0.022 3
γ	1455.15 8	0.022 3
γ	1464.32 6	0.006 1
γ	1480.98 8	~0.002
γ	1487.31 14	0.009 2
γ	1498.2 3	0.0041 12
γ	1511.88 8	0.013 2
γ	1517.02 6	0.009 2
γ	1525.7 3	0.0020 6
γ	1533.1 4	0.0016 6
γ	1537.72 8	0.003 1
γ	1542.0 5	~0.0010
γ	1549.5 4	~0.0012
γ	1570.49 6	0.029 4
γ	1587.15 8	0.028 4
γ	1619.31 14	0.011 2
γ	1631.7 3	0.0014 6
γ	1643.88 8	0.015 2
γ	1675.0 7	~0.0006
γ	1680.7 6	~0.0010
γ	1693.6 5	~0.0008
γ	1793.4 6	~0.0006
γ	1807.8 5	~0.0012

† 5.0% uncert(syst)

Atomic Electrons (^{167}Yb)
⟨e⟩=78.3 18 keV

e_{bin}(keV)	⟨e⟩(keV)	e(%)
4	1.8	50 8
9	7.0	82 9
10	5.3	54 6
27 - 46	4.2	11.1 9
47	24.2	52 3
48 - 51	1.07	2.19 16
53	4.1	7.8 13
54	6.1	11.3 5
55 - 95	4.40	6.8 3
96	6.7	7.0 3
97 - 103	2.17	2.16 6
104	2.12	2.04 10
105 - 153	8.45	7.41 10

Atomic Electrons (^{167}Yb)
(continued)

e_{bin}(keV)	⟨e⟩(keV)	e(%)
159 - 184	0.531	0.317 9
213 - 262	0.0020	~0.0009
263 - 312	0.007	~0.0024
313 - 362	0.0070	0.0021 4
366 - 415	0.0091	0.0023 5
419 - 468	0.0024	0.00055 13
469 - 512	0.0023	0.00046 14
531 - 572	0.0011	0.00020 7
581 - 629	0.0023	0.00037 12
635 - 684	0.0016	0.00024 7
685 - 733	0.0008	0.00012 6
770 - 819	0.0010	0.00013 6
820 - 868	0.005	~0.0006
874 - 924	0.0016	0.00018 6
925 - 969	0.0009	9 3 $\times10^{-5}$
976 - 1024	0.021	~0.0022
1025 - 1070	0.0045	0.00043 20
1080 - 1129	0.0014	~0.00013
1130 - 1175	0.0044	~0.00038
1183 - 1229	0.0020	0.00016 6
1232 - 1280	0.0018	0.00014 5
1283 - 1332	0.0013	0.00010 3
1333 - 1379	0.0010	7.6 23 $\times10^{-5}$
1383 - 1432	0.0010	6.8 23 $\times10^{-5}$
1433 - 1483	0.00084	5.7 14 $\times10^{-5}$
1485 - 1584	0.0019	0.00012 3
1585 - 1684	0.00016	10 3 $\times10^{-6}$
1685 - 1784	3.1 $\times10^{-5}$	~2 $\times10^{-6}$
1785 - 1806	3.8 $\times10^{-6}$	~2 $\times10^{-7}$

Continuous Radiation (^{167}Yb)
⟨β+⟩=1.51 keV;⟨IB⟩=1.08 keV

E_{bin}(keV)		⟨ ⟩(keV)	(%)
0 - 10	β+	1.02 $\times10^{-7}$	1.20 $\times10^{-6}$
	IB	0.00093	
10 - 20	β+	5.6 $\times10^{-6}$	3.31 $\times10^{-5}$
	IB	0.00025	0.00170
20 - 40	β+	0.000203	0.00061
	IB	0.0064	0.018
40 - 100	β+	0.0125	0.0158
	IB	0.22	0.44
100 - 300	β+	0.507	0.236
	IB	0.045	0.023
300 - 600	β+	0.99	0.246
	IB	0.18	0.040
600 - 1300	β+	0.00452	0.00074
	IB	0.57	0.066
1300 - 1661	IB	0.046	0.0033
	Σβ+		0.50

$^{167}_{71}$Lu(51.5 *10* min)

Mode: ϵ
Δ: -57470 *100* keV
SpA: 2.19$\times10^{7}$ Ci/g
Prod: ^{168}Yb(p,2n); daughter ^{167}Hf; ^{170}Yb(p,4n)

Photons (^{167}Lu)
⟨γ⟩=980 20 keV

γ_{mode}	γ(keV)	γ(%)†
Yb L_ℓ	6.545	0.59 6
Yb L_α	7.411	14.6 11
Yb L_η	7.580	0.201 16
Yb L_β	8.512	15.1 14
Yb L_γ	9.892	2.6 3
γ [M1]	19.673 16	0.021 4
γ M1+1.9%E2	24.631 9	0.094 19
γ M1	26.23 2	0.039 8
γ E2	28.882 9	0.0035 7
γ E1	29.652 9	15 3
γ E1	33.903 11	3.2 6
γ E1	44.763 12	1.20 24
γ [E1]	49.013 11	1.11 22
Yb $K_{\alpha2}$	51.354	28.8 10
Yb $K_{\alpha1}$	52.389	50.7 18
γ [M1]	57.61 3	0.16 3
Yb $K_{\beta1}'$	59.316	16.1 6
γ [M1]	59.410 18	0.124 25
Yb $K_{\beta2}'$	61.229	4.41 17
γ M1+~6.3%E2	67.366 10	0.54 11
γ (M1+50%E2)	69.826 16	0.089 18
γ M1+2.8%E2	78.350 19	0.54 11
γ E2	78.665 11	1.5 3
γ E2	89.498 16	0.090 18
γ M1	95.263 18	0.27 5
γ E2	100.202 14	0.43 9
γ M1	102.080 18	0.45 9
γ (M1)	102.551 18	0.18 4
γ E1	120.334 15	0.54 11
γ M1(+E2)	123.19 2	0.63 13
γ	127.35 10	0.36 7
γ [M1]	133.84 5	0.45 9
γ [M1]	138.621 25	0.18 4
γ	139.56 3	<0.09
γ E1	144.965 15	2.2 4
γ (M1+E2)	151.945 25	0.22 4
γ (M1+E2)	160.473 19	0.24 5
γ [M1+50%E2]	162.46 3	0.058 12
γ	169.25 25	0.18 4
γ E2	178.868 15	2.6 5
γ [M1]	179.729 24	0.16 3
γ E2	182.13 3	1.8 4
γ [E1]	183.561 23	<0.10
γ E2	188.72 3	2.0 4
γ (M1)	194.60 4	0.090 18
γ E2	198.96 4	1.10 22
γ (M1+E2)	201.540 16	0.090 18
γ [E1]	205.236 20	0.32 6
γ [E1]	209.487 20	0.76 15
γ M1	213.213 23	3.5 7
γ E2	222.83 5	1.20 24
γ M1	229.810 22	1.05 21
γ (E2)	232.159 25	0.19 4
γ E2	235.92 5	0.83 17
γ M1	238.823 23	0.90 18
γ M1	239.139 19	8.2 16
γ [M1]	240.70 3	0.108 22
γ (E1)	242.96 5	0.81 16
γ E1	248.57 3	0.72 14
γ M1	258.544 25	1.4 3
γ M1	261.811 23	1.3 3
γ	270.02 5	0.090 18
γ E2,E1	274.573 23	0.20 4
γ [M1]	278.22 3	1.6 3
γ E1	278.823 22	3.8 8
γ	282.7 3	0.110 22
γ E2,E1	298.31 4	0.27 5
γ M1	308.476 22	0.31 6
γ M1	317.489 23	1.5 3
γ	330.26 12	0.070 14
γ (M1+E2)	332.361 24	0.120 24
γ	339.00 20	0.120 24
γ M1	340.903 25	0.38 8
γ [E1]	352.13 5	0.31 6
γ	356.18 12	0.19 4
γ [M1+E2]	362.013 21	0.30 6
γ	368.65 9	0.090 18 ?
γ M1+M2	372.26 5	0.27 5
γ (M1)	374.54 6	0.14 3
γ [E1]	377.124 24	0.63 13
γ [E1]	381.374 24	0.54 11

Photons (^{167}Lu) (continued)

γ_{mode}	γ(keV)	γ(%)†
γ[E1]	385.67 3	0.54 11
γ M1	392.65 3	0.54 11
γ[M1+E2]	396.89 5	0.72 14
γ[M1]	398.51 4	0.36 7
γ	398.91 10	0.27 5
γ M1	401.14 5	2.5 5
γ[E1]	406.776 22	0.47 9
γ(M1+E2)	411.026 24	0.57 11
γ(M1,E2)	417.74 11	0.42 8
γ	427.51 6	0.14 3
γ	436.2 5	0.060 12
γ	437.98 9	0.060 12
γ	443.0 9	0.14 3
γ E1	445.61 3	0.95 19
γ	463.91 9	0.17 3
γ	467.2 3	0.090 18
γ	470.76 7	0.29 6
γ	474.05 20	0.120 24
γ[E2]	477.18 4	0.060 12
γ	479.74 7	0.030 6 ?
γ	485.16 20	0.14 3
γ	487.53 20	0.18 4
γ	494.44 18	0.15 3
γ	512.98 3	1.10 22
γ	528.2 3	0.090 18
γ	534.44 12	0.23 5
γ	539.74 12	0.20 4
γ	549.11 7	0.090 18
γ	569.86 9	0.45 9
γ	574.10 17	0.16 3
γ	588.1 3	0.120 24
γ	591.46 9	0.78 16
γ	594.49 9	0.29 6
γ	598.74 9	0.52 10
γ	602.07 10	0.63 13
γ	609.38 7	0.27 5
γ	633.34 20	0.32 6
γ	642.02 8	0.23 5
γ	646.08 20	0.14 3
γ	651.64 25	0.110 22
γ	663.6 3	0.13 3
γ	671.0 4	0.120 24
γ	673.97 11	0.18 4
γ	677.12 9	0.13 3 ?
γ	679.86 25	0.14 3
γ	688.8 5	0.070 14
γ	696.3 4	0.120 24
γ	709.58 7	0.33 7
γ	715.88 12	0.42 8
γ	719.74 25	0.16 3
γ	730.27 15	0.27 5
γ	734.54 20	0.26 5
γ	763.2 3	0.36 7
γ	779.68 20	0.27 5
γ	784.82 10	0.63 13
γ	788.24 7	0.18 4
γ	803.9 4	0.14 3
γ	808.62 20	0.28 6
γ	815.1 3	0.15 3
γ	830.73 16	0.27 5
γ	833.53 8	0.18 4
γ	847.18 18	0.14 3
γ	855.7 4	0.18 4
γ	858.4 6	0.090 18
γ	867.78 24	0.14 3
γ	873.94 19	0.16 3
γ	883.5 9	0.18 4
γ	908.4 4	0.14 3
γ	919.97 15	0.24 5
γ	925.23 10	0.090 18
γ	963.72 7	0.33 7
γ	967.3 4	0.10 2
γ	988.35 7	0.83 17
γ	990.94 14	0.090 18
γ	1016.51 20	0.22 4
γ	1049.8 4	0.090 18
γ	1067.7 4	0.14 3
γ	1069.7 3	0.18 4
γ	1085.19 17	0.47 9
γ	1092.25 7	0.110 22
γ	1108.97 20	0.24 5
γ	1126.60 7	0.51 10
γ	1161.41 17	0.47 9
γ	1164.16 20	0.28 6
γ	1188.53 6	1.20 24
γ	1196.6 3	0.24 5
γ	1227.22 20	1.18 24
γ	1255.14 10	0.30 6
γ	1267.20 6	3.3 7
γ	1275.35 3	0.54 11
γ	1300.9 4	0.18 4
γ	1305.46 7	0.67 13
γ	1319.6 3	0.19 4
γ	1375.95 12	0.63 13
γ	1394.07 8	0.45 9
γ	1397.64 7	0.95 19
γ	1403.75 8	0.54 11
γ	1414.1 6	0.080 16
γ	1423.5 4	0.23 5
γ	1426.82 12	0.72 14
γ	1444.9 3	0.22 4
γ	1470.30 7	0.15 3
γ	1506.80 6	2.4 5
γ	1510.37 7	0.67 13
γ	1515.8 5	0.17 3
γ	1521.41 23	0.31 6
γ	1531.48 7	0.32 6
γ	1534.58 8	0.43 9
γ	1542.18 11	0.55 11
γ	1548.63 10	0.62 12
γ	1554.45 9	0.090 18 ?
γ	1558.1 3	0.23 5
γ	1562.7 5	0.15 3
γ	1578.86 15	0.35 7
γ	1607.4 3	0.26 5
γ	1610.8 3	0.25 5
γ	1629.99 6	0.15 3
γ	1633.56 7	0.95 19
γ	1644.51 11	1.23 25
γ	1654.1 5	0.14 3
γ	1656.53 8	0.33 7
γ	1665.54 8	0.41 8
γ	1675.3 4	0.30 6
γ	1677.86 12	0.27 5
γ	1680.81 25	0.36 7
γ	1696.1 3	0.27 5
γ	1702.03 13	0.14 3
γ	1713.62 15	0.65 13
γ	1719.8 7	0.090 18
γ	1730.7 3	0.18 4
γ	1735.18 10	0.42 8
γ	1740.29 10	0.31 6
γ	1747.5 4	0.16 3
γ	1752.3 7	0.070 14
γ	1759.6 4	0.20 4
γ	1772.17 8	0.21 4
γ	1801.4 4	0.120 24
γ	1819.5 3	0.21 4
γ	1824.46 12	0.060 12
γ	1833.46 18	0.26 5
γ	1849.6 6	0.14 3
γ	1868.34 20	0.49 10
γ	1872.94 10	0.21 4
γ	1879.1 6	0.14 3
γ	1889.74 9	0.18 4
γ	1893.30 20	0.27 5
γ	1895.35 8	0.54 11
γ	1899.63 10	0.27 5
γ	1910.85 9	0.19 4
γ	1917.82 6	0.53 11
γ	1926.77 10	0.26 5
γ	1933.66 18	0.45 9
γ	1936.81 11	0.51 10
γ	1941.35 8	1.3 3
γ	1945.52 10	0.05 1 ?
γ	1951.04 8	0.45 9
γ	1954.2 6	0.14 3
γ	1961.44 11	0.77 15
γ	1964.81 23	0.38 8
γ	1974.01 8	0.60 12 ?
γ	1979.28 12	0.47 9 ?
γ	1979.42 10	0.47 9 ?
γ	1983.53 12	0.24 5
γ	1989.47 15	0.77 15
γ	1995.7 7	0.080 16
γ	2000.7 3	0.24 5
γ	2013.18 12	0.60 12 ?
γ	2025.9 14	0.05 1
γ	2047.8 6	0.110 22
γ	2062.6 7	0.060 12
γ	2064.6 7	0.060 12
γ	2148.8 4	0.110 22
γ	2198.6 3	0.27 5
γ	2204.52 9	0.16 3
γ	2247.5 4	0.16 3
γ	2257.2 7	0.05 1
γ	2266.0 5	0.39 8
γ	2269.8 7	0.18 4
γ	2271.88 9	0.65 13
γ	2307.7 10	0.05 1

† 8.5% uncert(syst)

Atomic Electrons (^{167}Lu)

⟨e⟩=57.8 17 keV

e_{bin}(keV)	⟨e⟩(keV)	e(%)
6 - 8	0.32	5.1 10
9	5.0	56 7
10	4.3	43 5
11 - 18	1.89	11.8 14
19	1.7	8.8 18
20	1.4	7.0 14
21	1.6	7.9 16
22 - 40	4.4	15.5 10
41	1.01	2.5 4
42 - 68	4.7	8.8 6
69	2.6	3.8 8
70	2.6	3.7 8
73 - 121	6.6	7.3 5
122 - 151	1.39	1.05 11
152	1.9	1.2 3
154 - 177	2.51	1.47 11
178	3.8	2.1 4
179 - 228	3.79	1.88 12
229 - 278	2.13	0.87 8
279 - 328	0.59	0.19 3
329 - 377	1.18	0.34 4
379 - 428	0.39	0.098 12
429 - 478	0.18	0.041 17
479 - 528	0.09	0.019 6
529 - 578	0.21	0.040 12
579 - 628	0.12	0.020 5
631 - 680	0.10	0.016 4
686 - 735	0.09	0.012 4
743 - 788	0.068	0.0089 23
793 - 838	0.043	0.0053 14
845 - 883	0.026	0.0030 10
898 - 930	0.053	0.0058 25
953 - 991	0.021	0.0021 7
1006 - 1050	0.039	0.0038 13
1057 - 1107	0.045	0.0041 14
1108 - 1155	0.048	0.0042 21
1159 - 1206	0.14	0.011 5
1214 - 1258	0.062	0.0050 14
1265 - 1311	0.011	0.0009 3
1315 - 1362	0.068	0.0051 15
1365 - 1415	0.037	0.0027 9
1416 - 1461	0.08	0.0056 23
1468 - 1561	0.090	0.006 1
1568 - 1667	0.117	0.0073 14
1668 - 1763	0.045	0.0027 5
1770 - 1869	0.062	0.0034 6
1870 - 1969	0.14	0.0075 14
1970 - 2063	0.0106	0.00053 13
2087 - 2186	0.010	0.00047 13
2188 - 2270	0.022	0.00100 25
2297 - 2306	0.00012	5.1 21 $\times 10^{-6}$

$^{167}_{72}$Hf(2.05 *5* min)

Mode: ϵ
Δ: -53470 *220* keV syst
SpA: 5.48×10^{8} Ci/g
Prod: protons on Ta; daughter ^{167}Ta; ^{170}Yb(^{3}He,6n)

Photons (^{167}Hf)

⟨γ⟩=297 *51* keV

γ_{mode}	γ(keV)	γ(%)
Lu L_ℓ	6.753	0.220 *24*
Lu L_α	7.650	5.4 *4*
Lu L_η	7.857	0.070 *6*
Lu L_β	8.829	5.2 *5*
Lu L_γ	10.245	0.87 *10*
Lu $K_{\alpha2}$	52.965	14.2 *7*
Lu $K_{\alpha1}$	54.070	24.8 *12*
Lu $K_{\beta1}$'	61.219	8.0 *4*
Lu $K_{\beta2}$'	63.200	2.13 *11*
γ [M1+E2]	139.84 *15*	3.1 *7*
γ [E1]	175.34 *15*	4.9 *8*
γ E1	315.18 *9*	81

Atomic Electrons (^{167}Hf)

⟨e⟩=12.3 *15* keV

e_{bin}(keV)	⟨e⟩(keV)	e(%)
9	1.86	20.1 *24*
10	1.02	9.8 *12*
11	0.40	3.7 *4*
42 - 61	1.12	2.42 *10*
77	2.0	~3
112	0.37	0.33 *6*
129	0.9	~0.7
131 - 175	0.8	0.6 *3*
252	3.2	1.26 *25*
304	0.46	0.15 *3*
305 - 315	0.28	0.091 *10*

Continuous Radiation (^{167}Hf)

⟨β+⟩=523 keV;⟨IB⟩=7.6 keV

E_{bin}(keV)		⟨ ⟩(keV)	(%)
0 - 10	β+	2.03×10^{-7}	2.40×10^{-6}
	IB	0.019	
10 - 20	β+	1.18×10^{-5}	7.0×10^{-5}
	IB	0.019	0.128
20 - 40	β+	0.000467	0.00141
	IB	0.038	0.132
40 - 100	β+	0.0342	0.0430
	IB	0.25	0.43
100 - 300	β+	2.43	1.08
	IB	0.32	0.18
300 - 600	β+	21.7	4.64
	IB	0.48	0.110
600 - 1300	β+	179	18.5
	IB	1.62	0.17
1300 - 2500	β+	316	18.4
	IB	3.7	0.20
2500 - 4000	β+	4.22	0.162
	IB	1.21	0.043
	Σβ+		43

$^{167}_{73}$Ta(2.9 *15* min)

Mode: ϵ
Δ: -48370 *320* keV syst
SpA: 3.9×10^{8} Ci/g
Prod: protons on Hg; protons on Re

$^{167}_{75}$Re(2.0 *3* s)

Mode: ϵ(~93 %), α(~7 %)
Δ: -34850 *210* keV syst
SpA: 2.9×10^{10} Ci/g
Prod: ^{93}Nb(^{84}Kr,α6n); ^{89}Y(^{84}Kr,6n)

Alpha Particles (^{167}Re)

α(keV)
5330 *10*

$^{167}_{76}$Os(740 *120* ms)

Mode: α(58 *12* %), ϵ(42 *12* %)
Δ: -26740 *320* keV syst
SpA: 5.93×10^{10} Ci/g
Prod: daughter ^{171}Pt; ^{108}Cd(^{63}Cu,p3n); ^{106}Cd(^{63}Cu,pn)

Alpha Particles (^{167}Os)

α(keV)
5838 *5*

$^{167}_{77}$Ir(>5 ms)

Mode: α
Δ: -17140 *440* keV syst
SpA: <9.7×10^{10} Ci/g
Prod: ^{58}Ni on Mo-Sn targets; ^{107}Ag on V-Ni targets

Alpha Particles (^{167}Ir)

α(keV)
6386 *10*

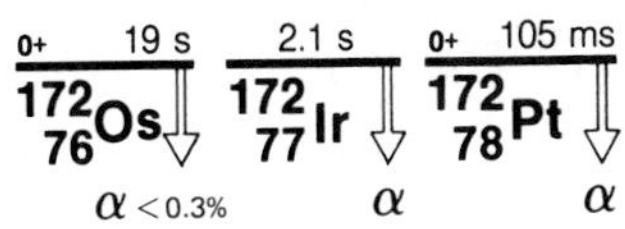

A = 168

NDS **11**, 385 (1974)

Q_α 5227 *10* Q_α 5970 *50* Q_α 6465 *4*

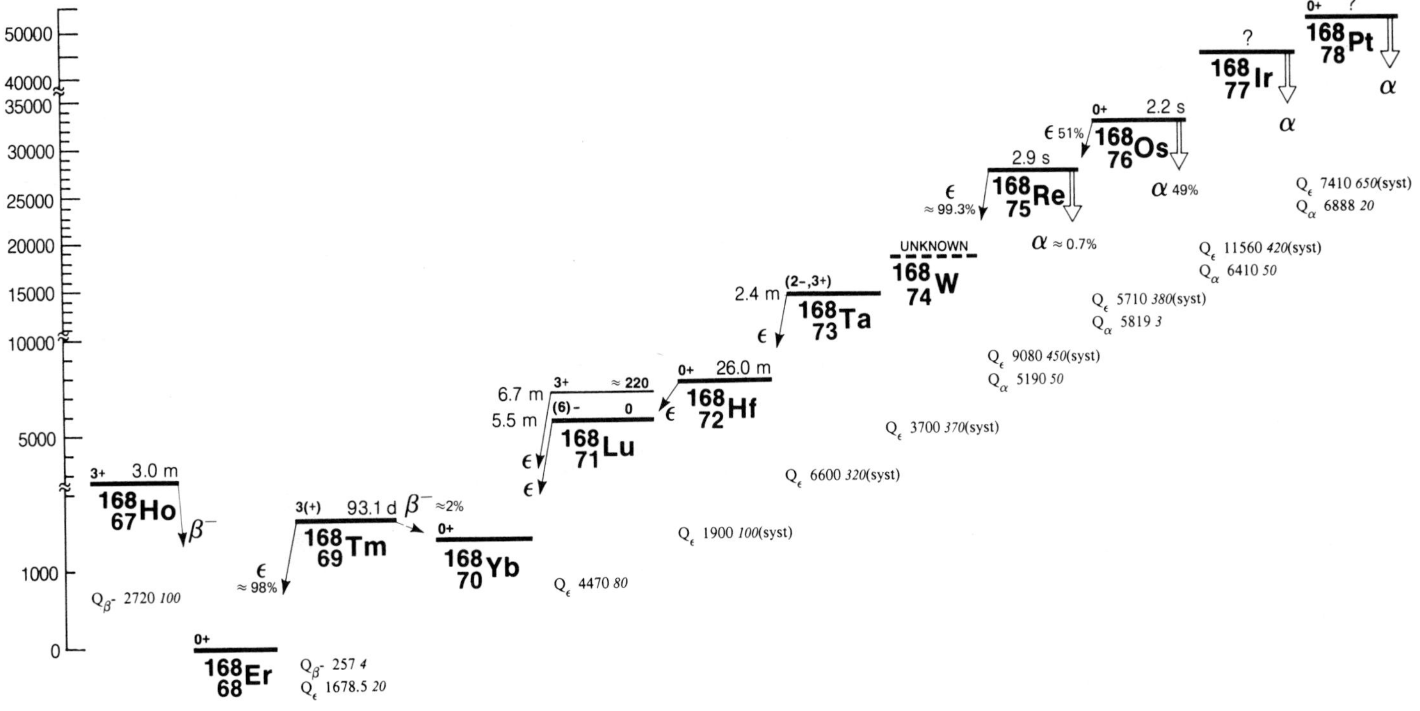

$^{168}_{67}$Ho(3.0 *1* min)

Mode: β-
Δ: -60280 *100* keV
SpA: 3.72×10^8 Ci/g
Prod: ^{168}Er(n,p); ^{170}Er(γ,np)

Photons (^{168}Ho)

⟨γ⟩=870 *12* keV

γ_{mode}	γ(keV)	γ(%)†
Er L_ℓ	6.151	0.21 *5*
Er L_α	6.944	5.4 *11*
Er L_η	7.058	0.093 *20*
Er L_β	7.921	5.5 *12*
Er L_γ	9.110	0.82 *18*
Er $K_{\alpha2}$	48.221	5.2 *10*
Er $K_{\alpha1}$	49.128	9.2 *17*
Er $K_{\beta1}$'	55.616	2.8 *5*
Er $K_{\beta2}$'	57.406	0.81 *15*
γ E2	79.8071 *10*	9.9 *20*
γ E1	99.2961 *10*	0.28 *7*
γ [M1]	123.07 *3*	0.22 *5*
γ	162.45 *8*	~0.037
γ [E2]	173.564 *17*	0.028 *11*
γ E2	184.274 *3*	6.1 *7*
γ E1	198.228 *3*	1.70 *6*
γ	209.21 *19*	~0.031
γ E2	217.45 *10*	0.021 *5*
γ (M2)	272.860 *17*	0.0034 *10*

Photons (^{168}Ho) (continued)

γ_{mode}	γ(keV)	γ(%)†
γ E2	284.15 *8*	0.042 *7* ?
γ	297.205 *4*	0.014 *3* ?
γ (E2)	348.43 *3*	0.0205 *25*
γ	365.72 *7*	0.041 *10*
γ	381.68 *19*	0.019 *9*
γ	383.63 *8*	0.050 *9*
γ M1	422.30 *8*	0.221 *20*
γ	429.79 *4*	0.131 *13*
γ M1+E2	447.41 *3*	1.44 *6*
γ	457.60 *3*	0.299 *22*
γ E1	546.70 *3*	0.159 *6*
γ [E2]	557.023 *15*	0.64 *3*
γ	559.2 *3*	~0.022
γ M1+E2	569.51 *8*	0.079 *7* ?
γ	579.93 *10*	0.050 *13*
γ	596.03 *9*	0.062 *14*
γ E2	631.654 *12*	3.13 *11*
γ E1	645.63 *3*	0.095 *5*
γ E1	673.65 *4*	0.024 *5*
γ	679.09 *15*	0.030 *10*
γ E1	720.26 *3*	0.728 *25*
γ E2	730.586 *12*	1.43 *6*
γ E2	741.297 *15*	35.9 *8*
γ E1	748.29 *4*	0.129 *8*
γ E2	815.928 *12*	18.5 *6*
γ (E2)	821.104 *15*	34.7 *8*
γ E1	829.882 *12*	0.210 *8*
γ (E2)	853.65 *3*	0.063 *6*
γ	904.96 *18*	0.015 *6*
γ E2	914.860 *12*	0.90 *4*
γ E1	928.859 *12*	0.036 *5*
γ	952.48 *4*	0.054 *4*
γ	994.94 *11*	0.029 *6*
γ (E2)	1012.10 *15*	0.032 *4* ?
γ M2,E3	1014.155 *12*	0.0022 *4*

Photons (^{168}Ho) (continued)

γ_{mode}	γ(keV)	γ(%)†
γ (E1)	1027.11 *5*	0.023 *4*
γ [M1+E2]	1034.50 *6*	0.022 *4*
γ	1076.67 *10*	0.031 *5*
γ	1104.56 *22*	0.019 *7*
γ [M1+E2]	1109.13 *7*	0.015 *4*
γ	1137.03 *4*	0.022 *8*
γ	1139.0 *5*	~0.014
γ [E1]	1173.45 *12*	0.14 *6*
γ	1176.47 *16*	0.027 *7*
γ [M1+E2]	1196.37 *15*	0.012 *3* ?
γ	1260.10 *3*	0.090 *7*
γ	1264.0 *4*	~0.011
γ	1267.95 *11*	0.053 *9*
γ	1273.60 *4*	0.19 *4*
γ E1	1277.28 *3*	0.106 *5*
γ	1297.25 *3*	0.340 *17*
γ	1341.6 *3*	0.020 *7*
γ [E1]	1351.16 *8*	0.0083 *20* ?
γ M1+E2	1359.03 *3*	0.192 *9*
γ	1371.88 *3*	1.3 *3*
γ	1433.66 *3*	0.065 *17*
γ E1	1461.56 *3*	0.083 *4* ?
γ	1488.5 *4*	~0.10
γ [M1+E2]	1529.13 *6*	0.039 *4*
γ [M1+E2]	1603.76 *6*	0.115 *7*
γ	1651.34 *19*	0.115 *7*
γ	1651.38 *21*	0.016 *5*
γ	1711.87 *21*	0.015 *4*
γ	1768.40 *4*	0.255 *10*
γ	1835.62 *19*	0.109 *4*
γ	1848.21 *4*	0.241 *14*
γ [M1+E2]	1850.43 *6*	0.173 *10*
γ [E1]	1914.74 *12*	0.045 *3*
γ [E2]	1930.23 *6*	0.122 *5*
γ	2094.8 *3*	0.015 *5*

Photons (^{168}Ho)
(continued)

γ_{mode}	γ(keV)	γ(%)†
γ	2220.49 *15*	0.031 *7*
γ [M1+E2]	2345.05 *6*	0.205 *7*
γ	2379.8 *6*	~0.005
γ	2390.2 *4*	0.010 *4*
γ	2404.76 *15*	0.0411 *24*
γ [E2]	2424.86 *6*	0.090 *4*

† 15% uncert(syst)

Atomic Electrons (^{168}Ho)
⟨e⟩=55 *6* keV

e_{bin}(keV)	⟨e⟩(keV)	e(%)
8	2.0	24 *5*
9 - 18	1.7	18 *4*
22	3.7	17 *3*
38 - 70	1.82	3.1 *4*
71	28.2	40 *8*
72 - 75	0.0064	0.0088 *12*
78	7.9	10.1 *21*
79	1.8	2.3 *5*
80 - 123	0.38	0.44 *7*
127	1.59	1.25 *14*
128 - 176	1.19	0.70 *5*
182 - 227	0.38	0.204 *17*
240 - 289	0.0072	0.0027 *8*
291 - 340	0.016	0.0050 *20*
346 - 390	0.22	0.057 *17*
400 - 449	0.082	0.019 *6*
455 - 502	0.038	0.0077 *6*
512 - 561	0.024	0.0045 *9*
567 - 616	0.140	0.0243 *10*
622 - 671	0.0470	0.0074 *4*
672 - 721	1.36	0.200 *6*
722 - 764	2.04	0.269 *8*
772 - 822	0.391	0.0482 *10*
828 - 871	0.0277	0.00324 *14*
895 - 944	0.0088	0.00096 *15*
950 - 995	0.0024	0.00024 *4*
1002 - 1052	0.0023	0.00022 *7*
1067 - 1116	0.0027	0.00024 *8*
1119 - 1168	0.0015	0.00014 *4*
1171 - 1220	0.009	0.0008 *3*
1240 - 1289	0.011	0.0009 *4*
1294 - 1343	0.033	~0.0025
1349 - 1376	0.007	0.00054 *19*
1404 - 1453	0.0029	~0.00021
1459 - 1546	0.0040	0.00027 *5*
1594 - 1654	0.0033	0.00021 *8*
1702 - 1793	0.013	0.00076 *17*
1826 - 1922	0.0040	0.000215 *18*
2037 - 2093	0.00023	~1 $\times 10^{-5}$
2163 - 2219	0.00044	2.0 *9* $\times 10^{-5}$
2288 - 2382	0.0054	0.000232 *22*
2388 - 2423	0.00029	1.19 *14* $\times 10^{-5}$

Continuous Radiation (^{168}Ho)
⟨β-⟩=670 keV;⟨IB⟩=1.18 keV

E_{bin}(keV)		⟨ ⟩(keV)	(%)
0 - 10	β-	0.0325	0.65
	IB	0.028	
10 - 20	β-	0.098	0.65
	IB	0.027	0.19
20 - 40	β-	0.399	1.33
	IB	0.052	0.18
40 - 100	β-	2.91	4.15
	IB	0.143	0.22
100 - 300	β-	30.4	15.1
	IB	0.36	0.21
300 - 600	β-	111	24.5
	IB	0.32	0.077

Continuous Radiation (^{168}Ho)
(continued)

E_{bin}(keV)		⟨ ⟩(keV)	(%)
600 - 1300	β-	408	45.0
	IB	0.23	0.030
1300 - 1899	β-	118	8.1
	IB	0.0101	0.00073

$^{168}_{68}$Er(stable)

Δ: -62999 *3* keV
%: 26.8 *2*

$^{168}_{69}$Tm(93.1 *1* d)

Mode: ε(~98 %), β-?(~2 %)
Δ: -61321 *4* keV
SpA: 8348 Ci/g
Prod: ^{170}Er(p,3n); ^{165}Ho(α,n); ^{168}Er(p,n); ^{168}Er(d,2n)

Photons (^{168}Tm)
⟨γ⟩=1146 *9* keV

γ_{mode}	γ(keV)	γ(%)†
Er L_ℓ	6.151	0.54 *6*
Er L_α	6.944	14.0 *10*
Er L_η	7.058	0.206 *16*
Er L_β	7.929	14.0 *12*
Er L_γ	9.165	2.23 *23*
Er $K_{\alpha2}$	48.221	28.1 *9*
Er $K_{\alpha1}$	49.128	49.9 *16*
Er $K_{\beta1}'$	55.616	15.5 *5*
Er $K_{\beta2}'$	57.406	4.38 *15*
γ_ϵE2	79.8071 *10*	10.8 *4*
γ_ϵ[M1]	98.9776 *20*	0.61 *12*
γ_ϵE1	99.2961 *10*	8.2 *19* ?
γ_ϵ[M1]	123.07 *3*	0.137 *24* ?
γ_ϵ[M1+E2]	137.98 *6*	0.0073 *15* ?
γ_ϵ[E2]	173.564 *17*	0.09 *3*
γ_ϵE2	184.274 *3*	16.1 *3*
γ_ϵE1	198.228 *3*	49.1 *10*
γ_ϵ	232.71 *18*	0.13 *4*
γ_ϵ	246.67 *5*	~0.28
γ_ϵ	253.75 *9*	0.11 *4*
γ_ϵ(M2)	272.860 *17*	0.10 *3*
γ_ϵE2	284.15 *8*	0.09 *3* ?
γ_ϵ	297.205 *4*	0.020 *8*
γ_ϵ(E2)	348.43 *3*	0.30 *3* ?
γ_ϵM1	422.30 *8*	0.26 *3*
γ_ϵM1+E2	447.41 *3*	21.3 *6*
γ_ϵE1	546.70 *3*	2.36 *6*
γ_ϵ[E2]	557.023 *15*	0.199 *11*
γ_ϵM1+E2	569.51 *8*	0.050 *20*
γ_ϵE2	631.654 *12*	7.65 *18*
γ_ϵE1	645.63 *3*	1.41 *7*
γ_ϵE1	673.65 *4*	0.127 *15* ?
γ_ϵE1	720.26 *3*	10.79 *23*
γ_ϵE2	730.586 *12*	4.43 *9*
γ_ϵE2	741.297 *15*	11.17 *22*
γ_ϵE1	748.29 *4*	0.328 *20*
γ_ϵ	768.67 *19*	~0.07
γ_ϵ	789.73 *23*	0.09 *4*
γ_ϵE2	815.928 *12*	45.2 *9*
γ_ϵ(E2)	821.104 *15*	10.80 *21*
γ_ϵE1	829.882 *12*	6.06 *14*
γ_ϵ(E2)	853.65 *3*	0.039 *10*
γ_ϵ	864.27 *11*	~0.27
γ_ϵ	870.80 *6*	~0.12
γ_ϵ	878.6 *3*	~0.04
γ_ϵ	885.5 *3*	~0.15

Photons (^{168}Tm)
(continued)

γ_{mode}	γ(keV)	γ(%)†
γ_ϵE2	914.860 *12*	2.81 *6*
γ_ϵE1	928.859 *12*	0.054 *15*
γ_ϵ	963.7 *3*	~0.04
γ_ϵM2,E3	1014.155 *12*	0.064 *10*
γ_ϵ(E1)	1167.33 *5*	0.069 *5* ?
γ_ϵ	1262.60 *23*	0.041
γ_ϵE1	1277.28 *3*	1.57 *5*
γ_ϵ	1312.05 *6*	0.029 *11*
γ_ϵ	1324.48 *23*	0.029 *11*
γ_ϵ[E1]	1351.16 *8*	0.0099 *19* ?
γ_ϵ[E1]	1351.60 *5*	0.069 *14* ?
γ_ϵE1	1461.56 *3*	1.22 *7* ?
γ_ϵ	1488.8 *3*	<0.004

† 6.0% uncert(syst)

Atomic Electrons (^{168}Tm)
⟨e⟩=80.1 *18* keV

e_{bin}(keV)	⟨e⟩(keV)	e(%)
8	5.0	60 *6*
9	3.2	35 *4*
10 - 18	0.94	9.5 *10*
22	4.04	18.1 *8*
38 - 70	5.3	11.3 *7*
71	30.8	43.4 *18*
72 - 75	0.0147	0.020 *3*
78	8.6	11.0 *5*
79 - 123	3.08	3.73 *14*
127	4.19	3.30 *10*
128 - 138	0.0031	0.0023 *9*
141	3.10	2.21 *6*
164 - 198	4.60	2.57 *6*
215 - 265	0.18	0.08 *3*
271 - 297	0.048	0.017 *3*
339 - 365	0.057	0.0158 *16*
390	2.4	0.62 *25*
413 - 447	0.59	0.13 *4*
489 - 538	0.066	0.0134 *6*
544 - 588	0.361	0.0628 *18*
616 - 665	0.248	0.0383 *8*
671 - 721	0.632	0.0924 *19*
722 - 769	1.94	0.256 *6*
772 - 822	0.516	0.0642 *17*
828 - 877	0.096	0.0112 *8*
878 - 927	0.0194	0.00214 *18*
928 - 964	0.0060	0.00062 *25*
1004 - 1014	0.0013	0.00012 *4*
1110 - 1159	0.00077	6.9 *5* $\times 10^{-5}$
1165 - 1205	0.0005	~4 $\times 10^{-5}$
1220 - 1269	0.0179	0.00146 *7*
1275 - 1324	0.00151	0.000117 *13*
1341 - 1352	0.000124	9.2 *13* $\times 10^{-6}$
1404 - 1453	0.0118	0.00084 *4*
1459 - 1489	0.000408	2.79 *13* $\times 10^{-5}$

Continuous Radiation (^{168}Tm)
⟨IB⟩=0.22 keV

E_{bin}(keV)		⟨ ⟩(keV)	(%)
10 - 20	IB	0.00021	0.00142
20 - 40	IB	0.0086	0.024
40 - 100	IB	0.18	0.37
100 - 300	IB	0.0129	0.0071
300 - 600	IB	0.0120	0.0030
600 - 857	IB	0.0013	0.00020

${}^{168}_{70}$Yb(stable)

Δ: -61578 *5* keV
%: 0.13 *1*

${}^{168}_{71}$Lu(5.5 *1* min)

Mode: ε
Δ: -57110 *80* keV
SpA: 2.03×10^8 Ci/g
Prod: ^{169}Tm(α,5n)

See ^{168}Lu(6.7 min) for γ rays

Continuous Radiation (^{168}Lu)

⟨β+⟩=50.0 keV;⟨IB⟩=3.0 keV

E_{bin}(keV)		⟨ ⟩(keV)	(%)
0 - 10	β+	1.76×10^{-7}	2.08×10^{-6}
	IB	0.0029	
10 - 20	β+	9.9×10^{-6}	5.9×10^{-5}
	IB	0.0022	0.0156
20 - 40	β+	0.000382	0.00115
	IB	0.0086	0.027
40 - 100	β+	0.0264	0.0333
	IB	0.24	0.44
100 - 300	β+	1.62	0.73
	IB	0.073	0.039
300 - 600	β+	11.0	2.39
	IB	0.24	0.053
600 - 1300	β+	33.3	3.86
	IB	1.24	0.132
1300 - 2500	β+	4.11	0.285
	IB	1.15	0.071
2500 - 3025	IB	0.0074	0.00029
	Σβ+		7.3

${}^{168}_{71}$Lu(6.7 *4* min)

Mode: ε
Δ: ~-56890 *80* keV
SpA: 1.669×10^8 Ci/g
Prod: ^{168}Yb(p,n); ^{169}Tm(α,5n); daughter ^{168}Hf, protons on Ta; protons on W

Photons (^{168}Lu(5.5 + 6.7 min))

γ_{mode}	γ(keV)	γ(rel)
γ [M1]	68.9 *11*	<2
γ	75 *2*	7 *3*
γ	83 *2*	~4
γ E2	87.77 *11*	85 *13*
γ [E1]	99.62 *16*	5 *1*
γ	101.3 *8*	5 *1*
γ (M1)	111.8 *4*	98 *15*
γ (M1)	111.8 *5*	
γ	114 *2*	5 *2*
γ	120 *2*	~4
γ [M1]	122.94 *13*	~4
γ	126 *2*	~4
γ	130.2 *3*	3.7 *9*
γ [E2]	131.8 *11*	~4
γ	135.7 *3* ?	
γ [E1]	145.18 *17*	4.7 *8*
γ	148.2 *5*	5 *1*
γ (M1)	156.66 *16*	31 *4*
γ (E2)	179.51 *16*	25 *3*
γ E2	198.86 *10*	180 *20*
γ (M1)	223.6 *2*	36 *4*
γ (M1)	228.63 *14*	70 *7*
γ [E2]	246.2 *6*	6 *2*
γ	268.0 *4*	8.6 *15*
γ	280 *2*	8 *2*
γ	286.6 *10* ?	
γ E2	298.61 *7*	17 *3*
γ	309 *2*	~4
γ E1+M2	324.69 *13*	30 *3*
γ	348.26 *16*	9 *3*
γ [E2]	371.84 *15*	5.3 *9*
γ [M1]	374.17 *22*	4.8 *9*
γ [E2]	379.89 *14*	~4
γ E2	384.82 *20*	~4
γ	393.2 *10*	5.4 *10*
γ [E1]	397.02 *22*	7.5 *15*
γ (M1)	401.03 *17*	26 *3*
γ	402 *2*	~2
γ M1+E2	467.77 *17*	7 *2*
γ	479.25 *18*	10.0 *15*
γ [M1+E2]	484.17 *15*	~2
γ	538 *2*	~2
γ E1	539.88 *16*	47 *5*
γ E1	583.53 *15*	29 *4*
γ	606.06 *19*	8 *2*
γ	624 *2*	9 *2*
γ	652.66 *15*	4.3 *9*
γ	697 *2*	<6
γ [M1+E2]	717.11 *13*	<4
γ M1+E2	730.51 *15*	12 *3*
γ	752.27 *18*	13 *3*
γ (E2)	780.45 *12*	24 *3*
γ [E2]	806.0 *3*	6 *2*
γ M1+E2	853.45 *14*	31 *4*
γ M1+E2	859.96 *16*	13 *2*
γ (E2)	884.73 *12*	90 *9*
γ (E2)	896.10 *11*	100
γ	902.1 *5*	8 *2*
γ	947 *1*	10 *2*
γ [E2]	966.01 *14*	5 *1*
γ (E2)	979.32 *12*	130 *21*
γ (E2)	983.87 *13*	77 *13*
γ	987.9 *4*	12 *2*
γ	988.6 *5*	10 *2*
γ (M2)	1012.61 *15*	7.1 *10*
γ (E2)	1015.72 *13*	12 *2*
γ	1025.7 *4*	5.8 *12*
γ M1+E2	1032.55 *14*	70 *8*
γ M1+E2	1071.7 *3*	24 *3*
γ (E2)	1083.59 *12*	35 *4*
γ	1088.95 *12*	9.0 *17*
γ M1+E2	1136.83 *13*	89 *9*
γ [E2]	1158.58 *16*	3.4 *5*
γ [M1+E2]	1165.01 *15*	7 *2*
γ E1	1185.02 *13*	46 *4*
γ	1193.2 *3*	5 *2*
γ	1215.4 *4*	9.0 *15*
γ (E2)	1220.05 *14*	69 *7*
γ M1+E2	1233.7 *6*	21 *2*
γ (E2)	1256.55 *25*	9 *2*
γ [M1+E2]	1264.62 *12*	19 *3*
γ	1289.0 *11*	<4
γ [E2]	1303.5 *3*	~3
γ [E1]	1311.23 *14*	7 *2*
γ	1320.3 *4*	3.5 *8*
γ (E2)	1337.62 *14*	27 *4*
γ	1360.8 *3*	~3
γ [M1+E2]	1363.87 *14*	25 *4*
γ	1380.2 *5*	2.8 *9*
γ	1387.57 *11*	6 *2*
γ	1392.1 *3*	6 *2*
γ [E1]	1413.64 *12*	16.4 *18*
γ M1+E2	1420.84 *14*	67 *7*
γ	1449.6 *5*	11 *2*
γ [E2]	1463.48 *13*	16 *2*
γ E1	1483.63 *13*	72 *9*
γ	1510.09 *16*	12 *4*
γ	1516.0 *7*	~2
γ	1525.4 *4*	4.5 *10*
γ	1532.74 *17*	2.9 *9*
γ	1594.5 *10*	<2 ?
γ	1631.5 *10*	<2 ?
γ	1669.2 *10*	~2
γ	1686.0 *5*	20 *4*
γ [E1]	1712.25 *12*	~2
γ	1849.2 *7*	~3
γ	1872.3 *8*	1.3 *4*
γ	1897.6 *10*	~0.6
γ	1902.5 *10*	
γ	1917.28 *14*	7 *3*
γ	1956.5 *3*	~0.50
γ	1967.9 *5*	~3
γ	1977.1 *10*	
γ	2031.5 *12*	~0.50
γ	2042.3 *15*	~0.50
γ	2048.0 *7*	~0.50
γ	2054.0 *15*	~0.50
γ	2071.5 *15*	~0.50
γ	2093.0 *8*	1.3 *5*
γ	2116.14 *13*	8 *2*
γ	2128.2 *10*	
γ	2141.3 *3*	18 *3*
γ	2188.5 *15*	<0.50
γ	2272.9 *11*	5 *2*
γ	2340.1 *3*	7 *3*
γ	2355 *2*	7 *3*

Continuous Radiation (^{168}Lu)

⟨β+⟩=108 keV;⟨IB⟩=3.8 keV

E_{bin}(keV)		⟨ ⟩(keV)	(%)
0 - 10	β+	1.84×10^{-7}	2.18×10^{-6}
	IB	0.0050	
10 - 20	β+	1.05×10^{-5}	6.2×10^{-5}
	IB	0.0043	0.030
20 - 40	β+	0.000402	0.00121
	IB	0.0124	0.041
40 - 100	β+	0.0280	0.0353
	IB	0.24	0.44
100 - 300	β+	1.77	0.79
	IB	0.104	0.056
300 - 600	β+	12.9	2.79
	IB	0.27	0.060
600 - 1300	β+	55	6.1
	IB	1.34	0.142
1300 - 2500	β+	38.3	2.32
	IB	1.66	0.098
2500 - 3706	β+	0.118	0.00462
	IB	0.17	0.0062
	Σβ+		12.0

${}^{168}_{72}$Hf(25.95 *20* min)

Mode: ε
Δ: -55210 *130* keV syst
SpA: 4.31×10^7 Ci/g
Prod: ^{175}Lu(p,8n); ^{170}Yb(α,6n); ^{170}Yb(^{3}He,5n); daughter ^{168}Ta

Photons (^{168}Hf)

γ_{mode}	γ(keV)	γ(rel)
γ	24.3 10	
γ	38 1	
γ	40.2 10	
γ	49 1	
γ	55 1	
γ	97.5 10	
γ	117 1	
γ	157.2 10	100
γ	159.5 10	
γ	183.8 10	147 15
γ	189.3 10	
γ	192.1 10	
γ	199.3 10	
γ	202.8 10	
γ	206.4 10	
γ	223.2 10	
γ	230.5 10	
γ	240.6 10	
γ	248.4 10	
γ	324.1 10	

$^{168}_{73}$Ta(2.4 *4* min)

Mode: ϵ

Δ: -48610 *310* keV syst

SpA: 4.6×10^{8} Ci/g

Prod: protons on Hg; protons on Re; ^{159}Tb(^{16}O,7n)

Photons (^{168}Ta)

$\langle\gamma\rangle$=722 *27* keV

γ_{mode}	γ(keV)	γ(%)†
Hf L_ℓ	6.960	0.30 3
Hf L_α	7.893	7.2 5
Hf L_η	8.139	0.119 9
Hf L_β	9.130	7.9 6
Hf L_γ	10.573	1.27 11
Hf $K_{\alpha2}$	54.611	12.1 5
Hf $K_{\alpha1}$	55.790	21.1 9
Hf $K_{\beta1}$'	63.166	6.9 3
Hf $K_{\beta2}$'	65.211	1.79 8
γ [E2]	123.9 2	37.0 16
γ [E2]	247.9 2	1.5 7
γ [E2]	261.5 2	28.4 10
γ [E2]	371.1 4	3.9 6
γ [E2]	462.9 6	1.26 22
γ [E2]	527.0 6	2.7 3
γ [M1+E2]	646.4 8	3.1 5
γ [E2]	672.7 6	1.6 6
γ [M1+E2]	750.2 6	9.7 7
γ [E2]	773.0 8	5.6 13
γ [E2]	815.6 8	2.0 4
γ [M1+E2]	834.0 8	1.9 4
γ [E2]	874.1 8	5.7 9
γ [M1+E2]	879.5 10	2.0 4
γ [M1+E2]	907.2 10	6.5 6
γ [M1+E2]	934.2 12	2.8 4
γ [E2]	986.6 10	5.4 12
γ [E2]	1058.1 14	2.4 5
γ [E2]	1159.7 20	1.1 3
γ	1179.6 14	0.9 3
γ [M1+E2]	1248.1 20	1.8 4
γ [M1+E2]	1282.4 20	2.0 5
γ [M1+E2]	1406.3 20	1.6 5
γ	1441.4 22	1.1 4
γ [E2]	1668 3	1.1 4

† 2.7% uncert(syst)

Atomic Electrons (^{168}Ta)

$\langle e\rangle$=72.5 *17* keV

e_{bin}(keV)	$\langle e\rangle$(keV)	e(%)
10	2.5	26 3
11	2.10	19.4 21
43 - 56	0.90	1.91 9
59	12.8	21.9 13
60 - 63	0.0411	0.067 4
113	18.0	16.0 9
114	13.5	11.8 7
121	0.61	0.50 3
122	7.8	6.4 4
123 - 124	2.33	1.88 8
183	0.26	0.14 6
196	4.45	2.27 12
237 - 262	3.19	1.27 4
360 - 398	0.254	0.068 6
452 - 463	0.198	0.043 4
516 - 527	0.053	0.0102 8
581 - 607	0.35	0.059 23
635 - 673	0.083	0.013 3
685 - 708	0.9	0.13 4
739 - 773	0.41	0.054 10
804 - 842	0.74	0.089 19
863 - 907	0.30	0.035 8
921 - 934	0.21	0.022 5
975 - 993	0.113	0.0115 16
1047 - 1094	0.046	0.0043 9
1114 - 1160	0.037	~0.0033
1168 - 1217	0.15	0.012 3
1237 - 1282	0.029	0.0023 5
1341 - 1376	0.08	0.0060 22
1395 - 1441	0.016	0.0011 3
1603 - 1666	0.027	0.0017 6

$^{168}_{75}$Re(2.9 *3* s)

Mode: ϵ(~99.3 %), α(~0.7 %)

Δ: -35820 *380* keV syst

SpA: 2.06×10^{10} Ci/g

Prod: ^{93}Nb(^{84}Kr,α5n); ^{89}Y(^{84}Kr,5n); ^{108}Cd(^{63}Cu,2pn); ^{107}Ag(^{63}Cu,pn); ^{109}Ag(^{63}Cu,p3n); ^{110}Pd(^{63}Cu,5n)

Alpha Particles (^{168}Re)

α(keV)
5140 10

$^{168}_{76}$Os(2.2 *1* s)

Mode: ϵ(51 *3* %), α(49 *3* %)

Δ: -30110 *60* keV

SpA: 2.62×10^{10} Ci/g

Prod: ^{93}Nb(^{84}Kr,p8n); daughter ^{172}Pt; ^{108}Cd(^{63}Cu,p2n); ^{110}Cd(^{63}Cu,p4n); ^{107}Ag(^{63}Cu,2n); ^{109}Ag(^{63}Cu,4n)

Alpha Particles (^{168}Os)

α(keV)
5680 3

$^{168}_{77}$Ir($t_{1/2}$ unknown)

Mode: α

Δ: -18560 *420* keV syst

Prod: ^{108}Cd(^{63}Cu,3n)

Alpha Particles (^{168}Ir)

α(keV)
6220 20

$^{168}_{78}$Pt($t_{1/2}$ unknown)

Mode: α

Δ: -11150 *500* keV syst

Prod: ^{58}Ni on Sn

Alpha Particles (^{168}Pt)

α(keV)
6824 20

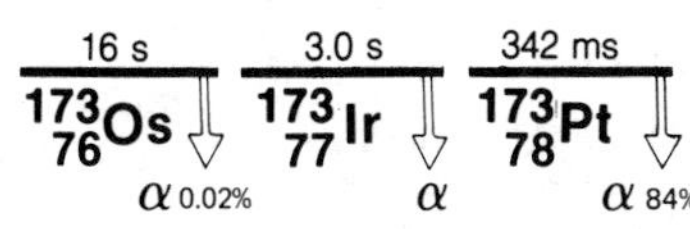

A = 169

NDS **36**, 443 (1982)

Q$_\alpha$ 5060 *50*   Q$_\alpha$ 5810 *50*   Q$_\alpha$ 6350 *50*

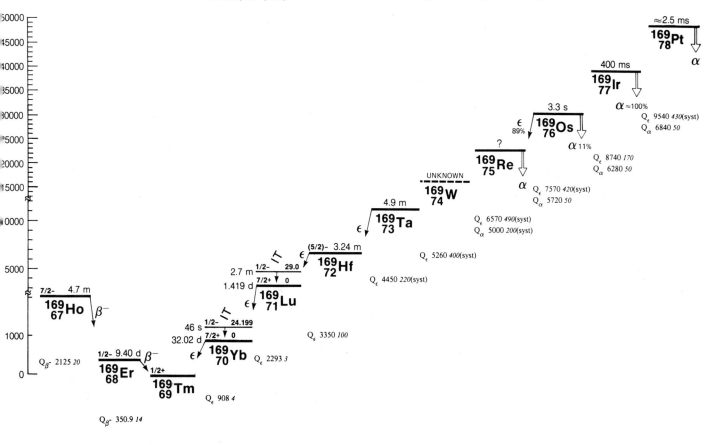

$^{169}_{67}$Ho(4.7 *1* min)

Mode: β-
Δ: -58806 *20* keV
SpA: 2.36×10^8 Ci/g
Prod: ^{170}Er(γ,p)

Photons (^{169}Ho)

⟨γ⟩=670 *25* keV

γ$_{mode}$	γ(keV)	γ(%)†
γ M1+31%E2	64.388 *20*	3.5 *7* *
γ E1	66.87 *17*	
γ [M1+E2]	73.4 *5*	
γ (E2)	74.42 *7*	1.4 *3* §
γ M1	84.67 *9*	1.4 *3*
γ [M1]	149.40 *9*	1.5 *3*
γ E1	151.54 *16*	5.8 *7*
γ [E2]	159.43 *8*	1.4 *4*
γ [E2]	167.26 *10*	0.79 *22*
γ	579.4 *5*	0.77 *8*
γ	609.4 *5*	0.56 *5*
γ [M1+E2]	628.82 *10*	2.9 *4*
γ	656.2 *5*	0.56 *5*
γ	663.2 *5*	0.19 *4*
γ [M1+E2]	676.25 *14*	4.3 *4*
γ [M1+E2]	694.81 *11*	1.4 *4*

Photons (^{169}Ho) (continued)

γ$_{mode}$	γ(keV)	γ(%)†
γ [E1]	697.36 *22*	~0.43 ?
γ [M1+E2]	698.95 *15*	1.0 *3*
γ [M1+E2]	704.84 *9*	1.4 *3*
γ [M1+E2]	716.80 *13*	3.26 *22*
γ [M1+E2]	760.93 *14*	10.7
γ [M1+E2]	764.23 *18*	0.51 *16*
γ	773.1 *5*	1.3 *3*
γ [M1+E2]	778.23 *8*	10.5 *6*
γ [M1+E2]	788.26 *7*	22.0 *22*
γ [M1+E2]	848.90 *18*	1.08 *14*
γ [E2]	852.65 *7*	11.7 *14*
γ [M1+E2]	866.20 *12*	4.6 *6*
γ [E2]	876.24 *12*	2.2 *4*
γ	1372.8 *10*	0.56 *5*
γ	1441.8 *10*	0.171 *21*
γ	1516.8 *10*	0.15 *3*
γ	1677.8 *10*	0.14 *3*
γ	1768.8 *10*	0.096 *21*
γ	1849.8 *10*	0.225 *21*

† approximate
* 64.4γ + 66.9γ
§ 73.4γ + 74.4γ

$^{169}_{68}$Er(9.40 *2* d)

Mode: β-
Δ: -60931 *3* keV
SpA: 8.219×10^4 Ci/g
Prod: ^{168}Er(n,γ)

Photons (^{169}Er)

⟨γ⟩=0.015 *3* keV

γ$_{mode}$	γ(keV)	γ(%)†
γ M1+0.1%E2	8.4099 *8*	0.16 *4*
γ M1+2.3%E2	109.7802 *3*	0.001292 *22*
γ E2	118.1901 *8*	0.0001399 *24*

† 19% uncert(syst)

Continuous Radiation (^{169}Er)

⟨β-⟩=99.6 keV;⟨IB⟩=0.035 keV

E_{bin}(keV)		⟨ ⟩(keV)	(%)
0 - 10	β-	0.327	6.6
	IB	0.0052	
10 - 20	β-	0.95	6.4
	IB	0.0045	0.031
20 - 40	β-	3.63	12.1
	IB	0.0072	0.025
40 - 100	β-	21.4	31.2
	IB	0.0123	0.020
100 - 300	β-	72	43.3
	IB	0.0062	0.0048
300 - 351	β-	1.35	0.432
	IB	4.2×10^{-6}	1.36×10^{-6}

$^{169}_{69}$Tm(stable)

Δ: -61282 3 keV

%: 100

$^{169}_{70}$Yb(32.022 8 d)

Mode: ϵ

Δ: -60374 5 keV

SpA: 2.4128×10^{4} Ci/g

Prod: ^{168}Yb(n,γ); ^{169}Tm(d,2n); daughter ^{169}Lu; protons on Ta

Photons (^{169}Yb)

⟨γ⟩=312.1 22 keV

γ_{mode}	γ(keV)	γ(%)†
Tm L_ℓ	6.341	0.81 8
Tm L_α	7.176	20.5 13
Tm L_η	7.310	0.268 20
Tm L_β	8.224	19.8 17
γ M1+0.1%E2	8.4099 8	0.332 10
Tm L_γ	9.536	3.3 3
γ M1+0.08%E2	20.7437 4	0.192 14
Tm $K_{\alpha2}$	49.772	53.5 14
Tm $K_{\alpha1}$	50.742	94.3 25
Tm $K_{\beta1}'$	57.444	29.6 8
Tm $K_{\beta2}'$	59.296	8.21 25
γ E1	63.12080 19	43.7 11
γ M1+3.3%E2	93.6151 4	2.66 6
γ M1+2.3%E2	109.7802 3	17.4 3
γ [M1]	117.25 5	0.019 7
γ E2	118.1901 8	1.88 3
γ E2	130.5239 3	11.1 4
γ [E1]	156.7359 4	0.031 7
γ M1+15%E2	177.2144 4	21.5 4
γ M1+9.3%E2	197.9581 4	34.9
γ E1+0.9%M2	240.3351 4	0.128 7
γ E1+0.1%M2	261.0788 4	1.90 4
γ [M1]	284.98 14	0.0028 5
γ [M1]	294.47 5	0.0070 7
γ [M1]	304.0 1	0.084 7 ?
γ E2	307.7382 4	10.80 22
γ [M3]	316.1481 8	~3×10^{-5} ?
γ	328.07 9	0.007 3 ?
γ	336.48 9	0.009 3 ?
γ [M2]	354.6938 6	0.020 5
γ [M2]	370.8588 4	0.0066 4
γ	424.99 5	~0.0007
γ	452.8 4	4.9 21 $\times10^{-5}$
γ	465.5 3	0.000234 21
γ	474.6 3	0.000196 21
γ [M1+E2]	494.10 12 ?	

Photons (^{169}Yb) (continued)

γ_{mode}	γ(keV)	γ(%)†
γ (M1)	514.84 12	0.00409 17
γ	562.3 3	0.000133 14
γ (M1)	579.45 14	0.00185 10
γ (M1)	600.19 14	0.00112 7
γ	614.1 5	9.1 10 $\times10^{-5}$
γ (M1)	624.62 12	0.0049 4
γ	642.5 3	7.7 7 $\times10^{-5}$
γ	709.9 3	3.8 4 $\times10^{-5}$

† 0.14% uncert(syst)

Atomic Electrons (^{169}Yb)

⟨e⟩=111.7 12 keV

e_{bin}(keV)	⟨e⟩(keV)	e(%)
4	1.47	39.3 13
9	7.1	82 8
10	5.7	58 6
11 - 49	8.30	28.8 8
50	17.9	35.4 9
51 - 63	6.06	10.89 18
71	4.26	6.0 3
83 - 97	1.61	1.88 4
100	5.44	5.45 14
101 - 117	3.92	3.61 5
118	12.3	10.5 3
120	0.71	0.593 25
121	2.98	2.46 11
122 - 131	4.72	3.78 10
139	17.9	12.9 3
147 - 181	4.07	2.41 4
188	3.77	2.01 5
189 - 238	1.49	0.760 17
239 - 286	1.34	0.541 15
292 - 336	0.623	0.208 4
345 - 393	0.0074	0.0021 3
406 - 455	0.00102	0.00023 4
456 - 505	0.000107	2.13 18 $\times10^{-5}$
506 - 555	0.000378	7.2 4 $\times10^{-5}$
560 - 605	0.00059	0.000105 8
612 - 651	0.000108	1.75 11 $\times10^{-5}$
700 - 710	4.2×10^{-7}	~6×10^{-8}

Continuous Radiation (^{169}Yb)

⟨IB⟩=0.25 keV

E_{bin}(keV)		⟨ ⟩(keV)	(%)
10 - 20	IB	0.00019	0.00131
20 - 40	IB	0.0065	0.019
40 - 100	IB	0.22	0.43
100 - 300	IB	0.019	0.0109
300 - 592	IB	0.0057	0.00160

$^{169}_{70}$Yb(46 2 s)

Mode: IT

Δ: -60350 5 keV

SpA: 1.44×10^{9} Ci/g

Prod: ^{168}Yb(n,γ)

Photons (^{169}Yb)

⟨γ⟩=1.30 18 keV

γ_{mode}	γ(keV)	γ(%)†
Yb L_ℓ	6.545	0.30 7
Yb L_α	7.411	7.5 16
Yb L_η	7.580	0.12 3
Yb L_β	8.510	7.2 15
Yb L_γ	9.784	1.07 23
γ E3	24.1994 22	0.000380

† 5.0% uncert(syst)

Atomic Electrons (^{169}Yb)

⟨e⟩=21.8 20 keV

e_{bin}(keV)	⟨e⟩(keV)	e(%)
9	2.8	31 7
10	2.0	20 5
14	4.5	32 6
15	6.0	39 8
22	4.5	20 4
23	0.54	2.4 5
24	1.5	6.1 12

$^{169}_{71}$Lu(1.4192 21 d)

Mode: ϵ

Δ: -58081 6 keV

SpA: 5.444×10^{5} Ci/g

Prod: protons on Yb; daughter ^{169}Hf; protons on U; protons on Th; protons on Ta; ^{169}Tm(α,4n)

Photons (^{169}Lu)

⟨γ⟩=1302 8 keV

γ_{mode}	γ(keV)	γ(%)†
γ M1+0.08%E2	12.3216 10	~0.018
γ [M1]	14.344 3	~0.0008
γ M1(+<0.3%E2)	20.4372 22	~0.008
γ M1	34.7813 20	~0.007
Yb $K_{\alpha2}$	51.354	33.2 10
Yb $K_{\alpha1}$	52.389	58.5 17
Yb $K_{\beta1}'$	59.316	18.6 6
Yb $K_{\beta2}'$	61.229	5.09 17
γ M1+26%E2	62.7188 8	0.66 3
γ M1+7.1%E2	70.8816 11	1.71 3
γ E2	75.0404 9	0.305 6
γ M1+5.1%E2	87.3837 11	2.47 5
γ [M1+E2]	89.788 13	~0.00016 ?
γ M1+6.4%E2	90.7690 11	0.559 12
γ M1+2.2%E2	91.9735 14	0.602 12
γ M1+E2	104.2950 15	0.479 16
γ M1+50%E2	108.005 3	0.083 4
γ M1+45%E2	110.9293 12	1.76 4
γ [M1+E2]	129.29 4	~7×10^{-5}
γ M1+4.8%E2	133.543 4	0.198 5
γ M1+35%E2	144.5758 10	0.566 14
γ E2	156.8974 11	1.41 3
γ E2	161.6506 13	0.174 9
γ E2	165.0131 20	1.98 4
γ M1+39%E2	166.43 13	0.128 6
γ [E2]	167.0138 15	~0.016
γ [M1+E2]	171.151 24	~0.0007
γ [M1+E2]	179.3571 18	~0.0047
γ E1	191.2132 17	20.7 5
γ [E2]	197.56 8	~0.009
γ E2	198.3130 14	0.77 3
γ [E2]	198.773 3	~0.042

Photons (^{169}Lu)
(continued)

γ_{mode}	γ(keV)	γ(%)†
γ (E1)	207.7153 19	0.432 14
γ [M1+E2]	222.691 4	~0.033
γ [M1+E2]	225.88 3	~0.07
γ [E1]	227.8756 21	0.22 4
γ (E2)	243.128 4	0.22 5
γ (E2)	244.472 4	0.19 4
γ (M1)	247.738 24	~0.09
γ E2+30%M1	258.309 3	0.341 14
γ [M1+E2]	272.052 13	~0.028
γ [E1]	272.97 6	~0.06
γ (E1)	278.5968 18	0.132 9
γ M1	291.233 3	0.447 16
γ (E1)	318.6445 21	0.090 4
γ [M1+E2]	326.015 4	~0.014
γ	357.1 2	~0.047
γ (E2+33%M1)	359.43 3	0.167 9
γ M1	369.238 4	0.841 21
γ M1	378.617 4	2.12 5
γ [M1]	383.582 4	0.075 12
γ (E1)	389.5259 19	0.149 7
γ (M1)	404.020 4	0.125 9
γ (M1)	406.02 14	0.043 4
γ (M1)	419.35 3	0.036 3
γ M1	423.387 21	0.028 3
γ M1	427.81 3	0.059 4
γ (E2+42%M1)	431.97 6	0.028 3
γ [M1+E2]	443.73 8	~0.0023
γ	452.42 8	0.068 9
γ (M1)	456.622 4	0.72 3
γ E2	466.17 8	<0.16
γ (E2)	466.93 21	0.047 12
γ M1	470.36 3	0.55 3
γ [M1+E2]	476.27 13	0.031 7
γ M1	478.51 8	~0.021
γ (M1)	479.99 3	0.160 19
γ M1	482.912 4	0.148 7
γ (M1)	484.70 3	0.165 7
γ [E2]	485.635 14	~0.021
γ M1	489.13 3	0.141 16
γ	492.2 3	~0.047
γ [M1]	502.78 4	0.17 5
γ M1	505.14 3	0.16 6
γ E2	519.805 13	0.069 6
γ [M1+E2]	530.361 14	~0.047
γ [M1+E2]	531.11 8	~0.0023
γ [M1+E2]	539.62 5	0.094 14
γ (M1)	542.803 15	0.073 9
γ M1	545.54 3	0.369 23
γ M1	548.595 4	0.343 19
γ	550.2 3	~0.07
γ M1+33%E2	560.378 13	0.125 4
γ (E2)	563.241 15	0.402 24
γ [E1]	569.830 4	0.146 7
γ M1	572.699 13	0.108 14
γ M1	576.403 14	0.71 6
γ M1	587.44 6	0.063 12
γ M1	590.686 13	0.71 4
γ	613.9 3	~0.035
γ [E1]	617.70 3	0.237 19
γ M1	623.09 8	0.167 14
γ [E1]	633.06 3	<0.07 ?
γ [M1+E2]	633.073 21	0.080 9
γ M1	635.41 8	0.639 21
γ (M1+~45%E2)	636.31 3	~0.50
γ (M1)	642.70 3	0.056 9
γ M1	647.285 14	0.289 21
γ [E2]	649.71 3	0.056 7
γ M1(+E2)	655.634 14	0.162 19
γ	657.38 6	0.19 8
γ	660.5 5	
γ (E2)	664.69 8	0.122 12
γ M1	667.854 21	0.073 12
γ E2	670.41 3	0.212 21
γ M1	676.071 14	0.075 9
γ	682.026 18	0.021 9
γ [E1]	687.92 3	0.270 14
γ (E2)	690.955 20	0.482 23
γ [M1+E2]	700.80 4	0.059 12
γ M1	703.42 4	0.122 14
γ M1	707.816 15	0.348 21
γ E2	719.98 3	0.209 19
γ E2+47%M1	725.15 5	0.329 16
γ (M1)	728.73 4	0.244 14
γ M1+37%E2	761.18 3	0.73 6
γ (E1)	767.49 5	0.34 3
γ	782.6 3	~0.047
γ	792.5 5	~0.024
γ E2	796.90 6	0.080 16
γ [E1]	802.47 3	0.266 21
γ [M1+E2]	812.429 21	~0.028
γ M1+32%E2	815.92 5	0.214 9
γ	817.33 8	0.212 9
γ M1	821.15 3	0.317 12
γ (M1)	824.751 21	0.047 7
γ (M1)	832.06 3	0.068 9
γ	847.9 7	~0.024
γ	856.97 8	~0.035
γ [M1+E2]	861.88 3	~0.035
γ	875.72 10	~0.035
γ M1	879.927 23	0.343 19
γ [M1+E2]	884.04 10	0.082 16
γ E1	889.740 18	5.38 14
γ [E2]	895.85 4	0.14 5
γ	903.371 21	0.040 16
γ [E2]	909.021 24	0.122 16
γ E1	916.70 3	0.87 3
γ (E2)	920.41 21	0.052 12
γ [E2]	926.688 24	~0.024
γ	934.18 9	~0.07
γ	939.85 8	~0.12
γ E1	960.621 18	23.5 5
γ	979.79 7	0.122 12
γ	984.09 14	0.176 19
γ (M1)	994.17 14	0.054 16
γ (E2)	999.789 24	0.42 3
γ E1	1007.46 3	1.81 5
γ (M1)	1013.58 4	0.080 14
γ M1	1017.62 5	0.261 19
γ	1025.72 7	0.085 12
γ M1	1031.78 3	0.132 7
γ	1037.27 8	~0.08
γ	1037.8 3	~0.047
γ (M1)	1043.12 8	0.153 12
γ	1055.94 8	0.08 3
γ M1	1060.230 23	1.92 6
γ M1	1065.04 5	0.47 4
γ	1068.383 21	0.320 19
γ (M1)	1073.87 3	1.14 6
γ E1	1078.34 3	1.08 4
γ	1088.23 4	0.101 16
γ [M1+E2]	1099.82 5	0.064 7
γ M1	1106.34 22	0.153 14
γ M1	1110.15 6	0.186 9
γ	1117.61 20	0.031 9
γ	1122.74 14	0.155 19
γ (E1)	1133.44 5	0.188 16
γ M1	1139.28 5	0.091 5
γ M1	1142.08 5	0.041 5
γ (M1)	1146.82 4	0.073 14
γ	1148.20 8	~0.05
γ M1	1151.16 5	0.21 3
γ [M1+E2]	1156.13 16	0.052 9
γ M1	1162.52 5	0.181 12
γ M1	1165.32 3	0.172 12
γ M1	1171.159 23	0.81 3
γ	1176.57 16	~0.14
γ M1	1180.48 6	0.190 21
γ M1	1184.80 3	2.23 9
γ M1	1199.14 3	0.226 19
γ M1	1205.940 23	0.51 3
γ E1	1212.44 13	0.48 4
γ	1215.12 5	0.106 9
γ (M1)	1219.58 3	0.30 4
γ	1222.98 6	0.11 3
γ M1	1244.40 5	0.075 9
γ (E2)	1251.87 9	0.066 21
γ M1	1258.542 23	0.357 9
γ	1260.82 4	0.320 19
γ	1266.98 5	~0.06
γ E2	1267.23 10	0.134 24
γ (E1)	1272.76 5	0.68 3
γ M1	1276.25 3	0.176 16
γ E1	1283.32 13	2.11 9
γ M1+41%E2	1290.59 3	1.15 6
γ M1+49%E2	1296.87 5	0.167 9
γ (M1)	1301.35 6	0.155 9
γ E2	1307.09 5	0.113 19
γ M1	1311.03 3	0.066 12
γ (M1)	1317.97 4	0.129 12
γ [M1+E2]	1321.14 16	0.077 7
γ E1	1326.91 4	0.710 18
γ E1	1338.82 4	1.63 5
γ [E1]	1343.64 5	0.148 12
γ M1	1350.514 23	0.195 5
γ M1+45%E2	1355.11 5	0.133 7
γ E2	1363.63 3	0.071 9
γ	1367.56 7	0.150 9
γ (M1)	1374.39 6	0.212 9
γ E1	1378.872 23	3.20 7
γ E2	1392.24 3	1.31 3
γ [E1]	1406.33 5	0.226 9
γ (M1)	1412.69 5	0.115 12
γ	1419.68 13	0.042 9
γ	1425.54 22	0.063 14
γ (M1)	1429.91 5	0.313 19
γ (E1)	1437.50 9	0.627 21
γ E1	1449.753 23	9.96 21
γ (E1)	1463.40 3	1.52 4
γ (E1)	1466.85 4	3.34 9
γ E2	1483.97 9	0.202 14
γ	1487.25 4	0.036 5
γ (E2)	1498.13 4	0.282 12
γ (M1)	1503.17 3	0.216 12
γ E2	1517.30 5	0.55 3
γ E1	1524.88 9	0.51 3
γ	1529.87 4	0.451 14
γ [E1]	1540.68 5	0.040 9
γ	1547.69 18	0.055 6
γ	1554.45 9	0.11 3
γ	1557.26 5	0.056 11
γ	1568.66 18	0.026 7
γ	1575.76 7	0.086 7
γ	1584.70 9	0.141 12
γ (M1)	1590.55 3	0.482 19
γ E1	1595.18 8	~0.08
γ	1595.89 23	0.023 5
γ E1	1607.50 8	0.068 5
γ E1	1618.46 4	0.714 16
γ	1626.12 14	0.023 5
γ	1630.05 4	0.066 12
γ	1636.90 9	0.221 9
γ M1	1645.21 9	0.080 5
γ M1	1658.06 4	0.797 19
γ	1671.6 1	0.055 5
γ M1	1676.19 14	0.089 4
γ M1	1682.52 3	0.29 3
γ E1	1689.34 4	0.52 3
γ (M1)	1694.42 8	0.044 3
γ	1702 1	~0.024
γ (M1+E2)	1707.78 9	0.44 6
γ	1709.94 4	0.28 7
γ	1717.44 4	0.113 9
γ	1726.34 7	0.061 7
γ	1730.38 4	~0.021
γ	1737.0 3	0.042 5
γ	1746.78 14	0.063 9
γ	1751.44 5	0.014 3
γ M1	1763.58 14	0.186 9
γ E1	1781.76 3	0.94 3
γ	1790.55 10	0.062 3
γ [E1]	1810.61 8	0.024 4
γ	1817.11 7	0.0341 24
γ	1822.42 11	0.0353 24
γ	1833.41 11	0.0313 24
γ	1838.82 5	0.0353 21
γ	1850.87 10	0.024 5
γ	1862.44 9	0.147 6
γ	1867.06 12	0.0202 16
γ	1897.6 1	0.0322 19
γ [E1]	1901.38 8	~0.042
γ	1903.04 5	0.072 4
γ (M1+E2)	1908.65 4	0.083 5
γ	1916.1 4	0.0099 24
γ	1920.81 17	0.0244 21
γ	1947.33 22	0.0118 19
γ	1954.79 14	0.043 3
γ (E1)	1959.15 5	0.277 9
γ	1969.8 2	0.0343 23
γ	1974.19 4	0.284 9
γ	1985.08 12	0.100 3
γ	2014.06 9	0.032 4
γ	2018.4 3	0.015 5

Photons (^{169}Lu)
(continued)

γ_{mode}	γ(keV)	γ(%)†
γ	2025.46 *11*	0.116 *5*
γ	2030.03 *5*	0.679 *19*
γ	2048.99 *8*	0.079 *5*
γ	2056.17 *5*	0.289 *9*
γ	2065.03 *11*	0.0139 *12*
γ	2070.85 *11*	0.0305 *12*
γ	2088.69 *14*	0.0101 *7*
γ	2095.90 *7*	0.129 *5*
γ	2101.09 *13*	0.0125 *9*
γ	2112.0 *4*	0.008 *3*
γ	2114.3 *3*	0.0165 *9*
γ	2122.47 *10*	0.197 *9*
γ	2135.4 *4*	0.0078 *12*
γ	2139.39 *17*	0.073 *9*
γ	2141.88 *20*	0.0148 *19*
γ	2148.27 *17*	0.0249 *14*
γ	2158.05 *25*	0.027 *6*
γ	2161.18 *10*	0.070 *8*
γ	2191.49 *20*	0.0160 *9*

† 3.0% uncert(syst)

Atomic Electrons (^{169}Lu)
$\langle e\rangle$=31.2 *7* keV

e_{bin}(keV)	$\langle e\rangle$(keV)	e(%)
14	0.0667	0.487 *14*
26	2.68	10.3 *3*
28	0.00012	~0.0004
29	0.610	2.07 *6*
31	0.677	2.21 *6*
40 - 41	0.58	1.42 *11*
42	0.77	1.81 *19*
43	0.61	1.4 *5*
47 - 49	0.51	1.05 *8*
50	1.7	3.4 *5*
51 - 52	0.61	1.17 *6*
53	0.71	1.35 *11*
54	0.75	1.39 *12*
57 - 59	0.114	0.198 *14*
60	1.18	1.96 *9*
61	0.81	1.3 *4*
62 - 63	0.48	0.8 *3*
65	0.662	1.02 *3*
66	0.654	0.99 *3*
68 - 75	1.25	1.76 *16*
77	1.34	1.75 *7*
78 - 103	4.1	4.5 *4*
104	0.61	0.59 *4*
105 - 129	0.54	0.49 *10*
130	1.41	1.08 *3*
131 - 179	1.97	1.32 *4*
181 - 230	0.94	0.488 *18*
233 - 282	0.127	0.051 *4*
289 - 328	0.762	0.243 *6*
343 - 391	0.242	0.0666 *19*
394 - 442	0.386	0.094 *4*
443 - 493	0.238	0.050 *3*
494 - 543	0.269	0.0519 *20*
544 - 591	0.239	0.042 *4*
594 - 643	0.133	0.0215 *20*
645 - 694	0.108	0.0164 *7*
697 - 741	0.084	0.0119 *11*
751 - 800	0.075	0.0098 *10*
801 - 848	0.126	0.0152 *6*
851 - 900	0.335	0.0374 *11*
901 - 950	0.091	0.0097 *6*
951 - 1000	0.175	0.0177 *7*
1003 - 1051	0.348	0.0343 *22*
1054 - 1101	0.104	0.0097 *7*
1104 - 1153	0.211	0.0188 *6*
1154 - 1202	0.087	0.0074 *5*
1203 - 1252	0.106	0.0086 *5*
1256 - 1305	0.0647	0.00506 *13*
1306 - 1355	0.086	0.00647 *20*
1357 - 1406	0.171	0.01229 *25*
1409 - 1458	0.0559	0.00387 *9*
1461 - 1560	0.058	0.0038 *4*

Atomic Electrons (^{169}Lu)
(continued)

e_{bin}(keV)	$\langle e\rangle$(keV)	e(%)
1565 - 1663	0.079	0.0049 *3*
1665 - 1762	0.0267	0.00157 *9*
1771 - 1865	0.0112	0.00062 *9*
1886 - 1983	0.025	0.0013 *3*
1988 - 2087	0.017	0.00084 *15*
2091 - 2190	0.0029	0.00014 *3*

Continuous Radiation (^{169}Lu)
$\langle\beta+\rangle$=1.11 keV;$\langle$IB$\rangle$=0.55 keV

E_{bin}(keV)		$\langle\ \rangle$(keV)	(%)
0 - 10	$\beta+$	9.5×10^{-8}	1.13×10^{-6}
	IB	0.00100	
10 - 20	$\beta+$	5.2×10^{-6}	3.11×10^{-5}
	IB	0.00024	0.0017
20 - 40	$\beta+$	0.000186	0.00056
	IB	0.0051	0.0148
40 - 100	$\beta+$	0.0102	0.0131
	IB	0.24	0.46
100 - 300	$\beta+$	0.234	0.116
	IB	0.035	0.018
300 - 600	$\beta+$	0.53	0.123
	IB	0.092	0.021
600 - 1300	$\beta+$	0.335	0.0463
	IB	0.159	0.019
1300 - 2102	IB	0.018	0.00121
	$\Sigma\beta+$		0.30

$^{169}_{71}$Lu(2.67 *17* min)

Mode: IT
Δ: -58052 *6* keV
SpA: 4.2×10^{8} Ci/g
Prod: ^{170}Yb(p,2n)

Photons (^{169}Lu)
$\langle\gamma\rangle$=1.43 *19* keV

γ_{mode}	γ(keV)	γ(%)†
Lu L_ℓ	6.753	0.31 *7*
Lu L_α	7.650	7.7 *16*
Lu L_η	7.857	0.13 *3*
Lu L_β	8.816	7.9 *17*
Lu L_γ	10.150	1.18 *25*
γ E3	29.0 *5*	0.00100

† 5.0% uncert(syst)

Atomic Electrons (^{169}Lu)
$\langle e\rangle$=26.4 *25* keV

e_{bin}(keV)	$\langle e\rangle$(keV)	e(%)
9	2.8	30 *7*
10	2.1	20 *5*
11	0.025	0.23 *5*
18	0.101	0.55 *11*
19	6.1	32 *7*
20	7.7	39 *8*
27	6.0	22 *5*
28	0.0105	0.037 *7*
29	1.7	6.0 *12*

$^{169}_{72}$Hf(3.24 *4* min)

Mode: ϵ
Δ: -54730 *100* keV
SpA: 3.43×10^{8} Ci/g
Prod: protons on Ta; daughter ^{169}Ta; ^{170}Yb(^{3}He,4n)

Photons (^{169}Hf)
$\langle\gamma\rangle$=504 *82* keV

γ_{mode}	γ(keV)	γ(%)†
Lu $K_{\alpha2}$	52.965	23.1 *19*
Lu $K_{\alpha1}$	54.070	40 *3*
Lu $K_{\beta1}$'	61.219	13.0 *11*
Lu $K_{\beta2}$'	63.200	3.5 *3*
γ (E1)	68.2 *6*	1.9 *4*
γ	72.9 *5*	0.58 *12* ?
γ M1+E2	123.36 *22*	7.6 *15*
γ [E1]	369.26 *22*	10.7 *21*
γ E1	492.6 *4*	83

† approximate

Atomic Electrons (^{169}Hf)
$\langle e\rangle$=24 *4* keV

e_{bin}(keV)	$\langle e\rangle$(keV)	e(%)
5	0.071	1.5 *3*
9	3.2	34 *5*
10	2.0	20 *5*
11 - 59	2.6	10.1 *8*
60	5.4	~9
61 - 62	0.23	~0.4
63	0.7	<2
64	0.8	<2
66 - 73	0.6	~0.9
112	1.3	~1
113	1.6	~1
114	1.3	<2
121	1.1	~0.9
122 - 123	0.3	~0.25
358 - 369	0.078	0.022 *3*
429	2.0	0.46 *9*
482 - 493	0.41	0.086 *13*

Continuous Radiation (^{169}Hf)
$\langle\beta+\rangle$=131 keV;$\langle$IB$\rangle$=4.8 keV

E_{bin}(keV)		$\langle\ \rangle$(keV)	(%)
0 - 10	$\beta+$	1.66×10^{-7}	1.96×10^{-6}
	IB	0.0060	
10 - 20	$\beta+$	9.6×10^{-6}	5.7×10^{-5}
	IB	0.0053	0.037
20 - 40	$\beta+$	0.000379	0.00114
	IB	0.0134	0.045
40 - 100	$\beta+$	0.0273	0.0344
	IB	0.25	0.45
100 - 300	$\beta+$	1.84	0.82
	IB	0.118	0.065
300 - 600	$\beta+$	14.6	3.16
	IB	0.29	0.064
600 - 1300	$\beta+$	81	8.8
	IB	1.51	0.159
1300 - 2500	$\beta+$	33.6	2.23
	IB	2.5	0.144
2500 - 3350	IB	0.110	0.0041
	$\Sigma\beta+$		15.0

$^{169}_{73}$Ta(4.9 *4* min)

Mode: ϵ
Δ: -50280 *200* keV syst
SpA: 2.27×10^{8} Ci/g
Prod: protons on Hg; protons on Re; ^{16}O on Tb

Photons (^{169}Ta)

γ_{mode}	γ(keV)	γ(rel)
γ E1	28.80 *4*	230 *23*
γ (E1)	38.18 *4*	57 *6*
γ	68.5 *1*	38 *8* ?
γ [M1+E2]	77.7 *1*	16 *3*
γ M1+32%E2	99.30 *14*	
γ	132.8 *1*	20 *4*
γ	153.5 *1*	80 *8*
γ	170.4 *1*	18 *4* ?
γ E2	177.0 *1*	24 *5*
γ	187.8 *2*	12 *2*
γ	192.4 *1*	100
γ	230.0 *1*	28 *6*
γ	394.5 *1*	35 *7*
γ	404.0 *2*	21 *4*
γ	440.8 *1*	38 *8*
γ	520.4 *2*	20 *4*
γ	529.0 *2*	26 *5*
γ	547.4 *3*	20 *4*
γ	595.0 *2*	59 *6*

$^{169}_{75}$Re($t_{1/2}$ unknown)

Mode: α
Δ: -38450 *350* keV syst
Prod: ^{108}Cd(^{63}Cu,2p); ^{110}Pd(^{63}Cu,4n); ^{109}Ag(^{63}Cu,p2n)

Alpha Particles (^{169}Re)

α(keV)
5050 *10*

$^{169}_{76}$Os(3.3 *2* s)

Mode: ϵ(89 *1* %), α(11 *1* %)
Δ: -30880 *230* keV syst
SpA: 1.82×10^{10} Ci/g
Prod: ^{156}Dy(^{20}Ne,7n); ^{93}Nb(^{84}Kr,p7n); daughter ^{173}Pt; ^{110}Cd(^{63}Cu,p3n); ^{108}Cd(^{63}Cu,pn); ^{109}Ag(^{63}Cu,3n)

Alpha Particles (^{169}Os)

⟨α⟩=611 keV

α(keV)	α(%)
5505 *35*	2.8
5571 *7*	8

$^{169}_{77}$Ir(400 *100* ms)

Mode: α
Δ: -22140 *240* keV syst
SpA: 7.9×10^{10} Ci/g
Prod: ^{93}Nb(^{84}Kr,8n); ^{110}Cd(^{63}Cu,4n); ^{108}Cd(^{63}Cu,2n)

Alpha Particles (^{169}Ir)

α(keV)
6090 *20*

$^{169}_{78}$Pt(~2.5 ms)

Mode: α
Δ: -12600 *460* keV syst
SpA: ~9.6×10^{10} Ci/g
Prod: ^{58}Ni on Sn

Alpha Particles (^{169}Pt)

α(keV)
6678 *15*

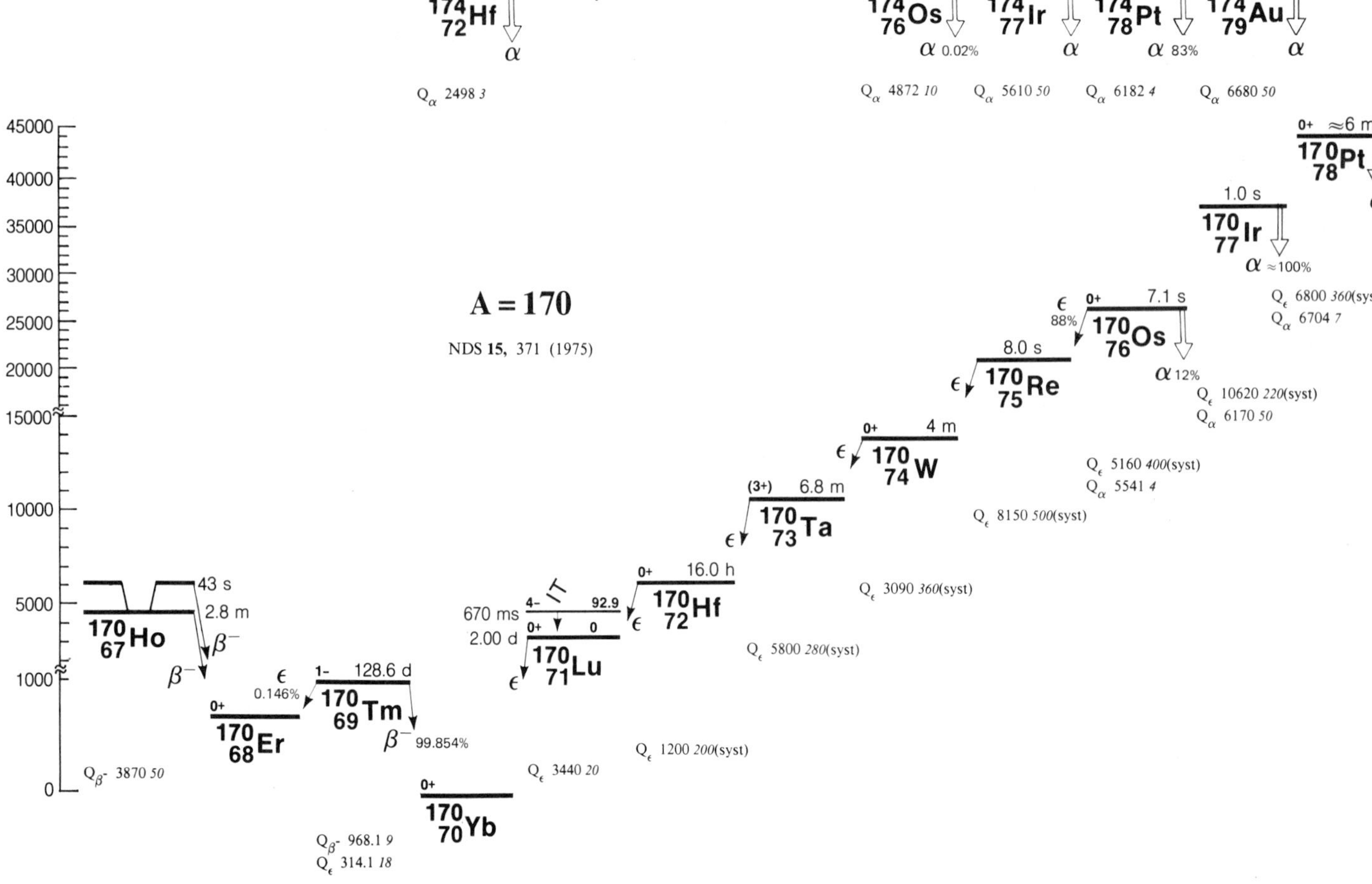

$^{170}_{67}$Ho(2.8 *2* min)

Mode: β-
Δ: -56250 *50* keV
SpA: 3.9×10^{8} Ci/g
Prod: ^{170}Er(n,p)

Photons (^{170}Ho)

⟨γ⟩=2001 *58* keV

γ_{mode}	γ(keV)	γ(%)†
Er L_ℓ	6.151	0.43 *6*
Er L_α	6.944	11.1 *12*
Er L_η	7.058	0.179 *21*
Er L_β	7.926	11.1 *13*
Er L_γ	9.128	1.69 *21*
Er $K_{\alpha2}$	48.221	16.5 *10*
Er $K_{\alpha1}$	49.128	29.3 *18*
γ (E1)	51.3 *1*	2.90 *20*
Er $K_{\beta1}$'	55.616	9.1 *6*
Er $K_{\beta2}$'	57.406	2.57 *16*
γ E2	78.65 *8*	13.2 *15*
γ M1	87.16 *9*	1.22 *15*
γ M1	94.67 *8*	2.82 *23*
γ M1	103.54 *8*	5.2 *4*
γ M1+E2	123.90 *14*	4.1 *8*
γ [E1]	141.50 *9*	1.98 *25*
γ (E1)	165.36 *8*	4.3 *4*
γ E2	181.57 *8*	27 *3*

Photons (^{170}Ho) (continued)

γ_{mode}	γ(keV)	γ(%)†
γ	218.69 *10*	1.32 *25*
γ	227.41 *9*	4.1 *5*
γ [E1]	258.17 *9*	43 *3*
γ E2	280.44 *11*	3.0 *5*
γ [M1+E2]	283.42 *10*	3.0 *5*
γ	413.2 *2*	3.63 *23*
γ	477.4 *2*	3.91 *25*
γ	662.9 *3* ?	1.40 *18* ?
γ	746.0 *2* ?	1.78 *25* ?
γ [M1+E2]	750.4 *2*	6.1 *3*
γ	786.3 *5*	5.6 *10*
γ	832.5 *10* ?	~0.8 ?
γ [M1+E2]	843.5 *2*	2.8 *8*
γ	854.7 *5*	14.0 *18*
γ [M1+E2]	867.0 *2*	2.46 *20*
γ	872.6 *3* ?	0.43 *10* ?
γ [E1]	890.2 *2*	25
γ [M1+E2]	932.1 *2*	41.7 *23*
γ	934.6 *5* ?	4.3 *10* ?
γ [M1+E2]	941.4 *2*	23.9 *5*
γ [M1+E2]	957.4 *3*	4.32 *23*
γ	976.5 *3*	3.35 *20*
γ	1024.7 *4*	1.7 *8*
γ [M1+E2]	1044.2 *2*	7.4 *4*
γ [E2]	1048.7 *8*	~0.51
γ [E1]	1111.8 *3*	2.39 *18*
γ [M1+E2]	1138.7 *2*	23.6 *10*
γ	1153.0 *3*	2.26 *20*
γ [E1]	1225.4 *3*	3.6 *5*
γ	1306.9 *3* ?	0.51 *10* ?

† 12% uncert(syst)

Atomic Electrons (^{170}Ho)

⟨e⟩=117 *4* keV

e_{bin}(keV)	⟨e⟩(keV)	e(%)
8	4.1	49 *7*
9 - 10	3.3	35 *4*
21	4.8	23 *3*
30	1.33	4.5 *6*
37	3.03	8.2 *7*
38 - 43	1.15	2.88 *14*
46	5.6	12.1 *9*
47 - 66	3.1	4.9 *16*
69	21.0	30 *4*
70	20.4	29 *3*
76	0.38	0.50 *6*
77	11.2	14.6 *17*
78 - 123	9.5	10.3 *9*
124	7.4	6.0 *6*
132 - 170	1.4	~0.9
172	3.2	1.88 *18*
173 - 219	5.7	3.05 *17*
223 - 272	1.8	0.76 *14*
274 - 283	0.25	0.09 *3*
356 - 405	0.4	~0.12
411 - 420	0.35	~0.08
468 - 477	0.08	~0.017
605 - 655	0.09	~0.014
661 - 693	0.42	0.061 *20*
729 - 778	0.39	0.053 *24*
784 - 833	1.1	0.14 *5*
834 - 882	2.1	0.24 *7*
884 - 933	1.8	0.20 *4*
934 - 977	0.14	0.015 *4*
987 - 1036	0.34	0.034 *9*
1039 - 1081	0.82	0.076 *22*

Atomic Electrons (^{170}Ho)
(continued)

e_{bin}(keV)	⟨e⟩(keV)	e(%)
1096 - 1145	0.22	0.020 5
1151 - 1168	0.036	0.0031 4
1216 - 1249	0.017	0.0014 6
1297 - 1307	0.0021	0.00016 8

Continuous Radiation (^{170}Ho)
⟨β-⟩=713 keV;⟨IB⟩=1.31 keV

E_{bin}(keV)		⟨ ⟩(keV)	(%)
0 - 10	β-	0.0262	0.52
	IB	0.029	
10 - 20	β-	0.080	0.53
	IB	0.028	0.20
20 - 40	β-	0.328	1.09
	IB	0.055	0.19
40 - 100	β-	2.48	3.51
	IB	0.150	0.23
100 - 300	β-	28.5	14.0
	IB	0.39	0.22
300 - 600	β-	112	24.8
	IB	0.35	0.084
600 - 1300	β-	411	45.5
	IB	0.28	0.035
1300 - 2500	β-	154	9.8
	IB	0.035	0.0023
2500 - 3329	β-	3.36	0.124
	IB	0.00030	1.12×10^{-5}

$^{170}_{67}$Ho(43 2 s)

Mode: β-
Δ: -56250 50 keV
SpA: 1.53×10^{9} Ci/g
Prod: ^{170}Er(n,p)

Photons (^{170}Ho)

γ_{mode}	γ(keV)	γ(rel)
γ E2	78.65 8	100 24
γ E2	181.57 8	8.5 9
γ	481.8 3	7.1 3
γ	540.73 20	12.8 5
γ [E2]	699.57 19	7.6 4
γ (E2)	812.15 20	58.8 18
γ [M1+E2]	881.00 16	11.6 5
γ [E2]	959.53 19	7.3 7
γ	1022.5 4	8.9 4
γ	1187.3 3	15.0 6
γ	1226.0 3	7.9 8
γ	1245.0 4	1.5 3
γ	1337.2 3	3.4 4
γ	1415.4 3	2.9 3
γ	1663.6 8	0.8 3
γ	1836.4 5	0.41 18
γ	1876.1 5	0.47 18
γ	1893.87 22	26.6 9
γ	1939.9 3	6.2 3
γ	1960.6 3	1.59 18
γ	1972.40 22	21.5 8
γ	1992.3 5	2.82 24
γ	2039.1 3	1.71 18
γ	2132.6 6	0.71 18
γ	2606.0 3	2.53 24
γ	2621.2 6	0.47 18
γ	2646.3 4	2.24 18
γ	2684.5 3	2.65 18
γ	2715.1 4	1.47 18

Photons (^{170}Ho)
(continued)

γ_{mode}	γ(keV)	γ(rel)
γ	2759.3 12	~0.5
γ	2789.0 15	0.71 18

$^{170}_{68}$Er(stable)

Δ: -60118 4 keV
%: 14.9 1

$^{170}_{69}$Tm(128.6 3 d)

Mode: β-(99.854 2 %), ϵ(0.146 2 %)
Δ: -59804 3 keV
SpA: 5973 Ci/g
Prod: ^{169}Tm(n,γ); ^{170}Er(p,n)

Photons (^{170}Tm)
⟨γ⟩=5.42 16 keV

γ_{mode}	γ(keV)	γ(%)†
Er L_ℓ	6.151	0.00051 8
Yb L_ℓ	6.545	0.064 7
Er L_α	6.944	0.0132 17
Er L_η	7.058	0.000178 25
Yb L_α	7.411	1.59 12
Yb L_η	7.580	0.0287 22
Er L_β	7.933	0.0129 19
Yb L_β	8.501	1.71 14
Er L_γ	9.185	0.0021 3
Yb L_γ	9.800	0.263 22
Er $K_{\alpha2}$	48.221	0.033 3
Er $K_{\alpha1}$	49.128	0.058 6
Yb $K_{\alpha2}$	51.354	1.24 7
Yb $K_{\alpha1}$	52.389	2.19 12
Er $K_{\beta1}$'	55.616	0.0181 19
Er $K_{\beta2}$'	57.406	0.0051 5
Yb $K_{\beta1}$'	59.316	0.70 4
Yb $K_{\beta2}$'	61.229	0.190 11
γ_ϵE2	78.65 8	0.0039
γ_βE2	84.3049 14	3.26

† uncert(syst): 19% for ϵ, 4.9% for β-

Atomic Electrons (^{170}Tm)
⟨e⟩=14.7 4 keV

e_{bin}(keV)	⟨e⟩(keV)	e(%)
8	0.0047	0.056 9
9	0.58	6.5 7
10	0.49	4.9 6
21	0.0014	0.0068 14
23	1.04	4.53 24
38 - 70	0.115	0.246 11
74	4.77	6.4 3
75	4.47	5.9 3
76 - 79	0.0042	0.0055 9
82	2.49	3.03 16
83	0.0252	0.0304 16
84	0.68	0.81 4

Continuous Radiation (^{170}Tm)
⟨β-⟩=315 keV;⟨IB⟩=0.31 keV

E_{bin}(keV)		⟨ ⟩(keV)	(%)
0 - 10	β-	0.082	1.64
	IB	0.0151	
10 - 20	β-	0.248	1.65
	IB	0.0143	0.100
20 - 40	β-	1.00	3.35
	IB	0.027	0.093
40 - 100	β-	7.2	10.3
	IB	0.067	0.104
100 - 300	β-	69	34.5
	IB	0.126	0.075
300 - 600	β-	164	37.8
	IB	0.055	0.0139
600 - 968	β-	74	10.6
	IB	0.0044	0.00067

$^{170}_{70}$Yb(stable)

Δ: -60772 3 keV
%: 3.05 5

$^{170}_{71}$Lu(2.00 3 d)

Mode: ϵ
Δ: -57332 20 keV
SpA: 3.84×10^{5} Ci/g
Prod: ^{169}Tm(α,3n); daughter ^{170}Hf; protons on Yb

Photons (^{170}Lu)
⟨γ⟩=2679 20 keV

γ_{mode}	γ(keV)	γ(%)†
Yb L_ℓ	6.545	0.51 8
Yb L_α	7.411	12.7 17
Yb L_η	7.580	0.184 24
Yb L_β	8.513	12.7 18
Yb L_γ	9.857	2.1 3
Yb $K_{\alpha2}$	51.354	26 4
Yb $K_{\alpha1}$	52.389	46 7
Yb $K_{\beta1}$'	59.316	14.6 21
Yb $K_{\beta2}$'	61.229	4.0 6
γ E2	84.3049 14	8.7 5
γ	118.85 6	0.032 3
γ	119.92 6	0.0067 7 ?
γ [M1+E2]	133.96 10	0.0125 13 ?
γ	142.50 15	0.0094 9
γ M1	152.62 3	0.273 9
γ	166.83 8	0.0060 7 ?
γ	170.79 12	0.0031 5 ?
γ E2	193.19 4	2.07 7
γ	199.65 15	0.0090 9
γ [E1]	201.86 8	0.0157 13 ?
γ	205.58 5	0.0078 7 ?
γ	209.9 2	0.0074 7
γ (M1)	220.80 7	0.0190 7
γ M1+E2	222.40 15	0.0403 13
γ M1+E2	223.40 15	0.0202 7
γ	225.59 7	0.0058 9 ?
γ (E0+E2)	228.16 6	0.0358 22
γ	231.50 11	0.0058 7 ?
γ M1+E2	235.82 7	0.039 4 ?
γ	238.17 9	0.0166 18 ?
γ M1	241.55 4	0.228 7
γ	249.94 8	0.0038 11 ?
γ	251.79 6	0.0470 22 ?
γ [M1+E2]	272.54 6	0.0092 9 ?
γ	275.4 2	0.0045 5

Photons (^{170}Lu)
(continued)

γ_{mode}	γ(keV)	γ(%)[†]
γ	279.47 8	0.0211 13 ?
γ M1(+E2)	282.97 9	0.199 7
γ E1	286.70 4	0.452 13
γ	292.72 5	0.0049 5 ?
γ	295.15 20	0.0045 5
γ	296.7 2	0.0076 7
γ	297.7 2	0.0038 5
γ M1	300.6 2	0.0045 5
γ M1	301.85 20	0.0058 7
γ	303.2 2	0.0040 5
γ	311.80 10	0.0072 7 ?
γ M1(+E2)	323.62 4	0.345 11
γ	329.3 2	0.0112 9
γ	339.45 20	0.0031 5
γ	340.90 15	0.0152 7
γ	366.35 15	0.0242 9
γ	368.23 9	0.0090 5 ?
γ	369.80 15	0.0260 13
γ	372.06 7	0.0305 18
γ	374.55 20	0.0045 5
γ	382.30 7	0.0582 22 ?
γ	384.85 15	0.0143 7
γ	386.56 15	0.0090 7 ?
γ M1	388.86 5	0.090 3
γ	390.40 15	0.0560 22
γ (E0+E2)	396.01 5	0.188 5
γ	401.28 15	0.009 3 ?
γ M1	404.00 15	0.0143 7
γ	406.25 15	0.0233 13
γ	407.55 20	0.0090 5
γ	410.55 15	0.0099 22
γ	416.64 9	0.0060 7 ?
γ M1(+E2)	419.70 5	0.502 13
γ	427.21 13	0.0085 13 ?
γ M1,E2	443.40 15	0.0408 13
γ M1	447.65 10	0.0703 22
γ M1	449.25 20	0.0072 7
γ (M1)	455.74 5	0.130 5
γ	457.9 1	0.0215 18 ?
γ M1	461.20 15	0.0121 18
γ	465.50 15	0.0108 9
γ (M1)	467.42 9	0.0220 13 ?
γ	472.50 15	0.0112 5
γ	478.89 8	0.056 6 ?
γ	479.58 13	0.0300 13 ?
γ	480.50 15	0.0197 9
γ	486.80 15	0.0188 9
γ	490.95 15	0.0224 7
γ E1	492.62 5	0.569 18
γ	497.50 15	0.0139 5
γ	500.58 10	0.0099 5 ?
γ	518.90 15	0.0099 5
γ	525.05 15	0.0112 13
γ (E2)	530.53 9	0.094 5
γ	534.65 15	0.0099 5
γ	535.95 15	0.0094 5
γ	539.05 15	0.0242 22
γ (M1)	540.23 6	0.206 9
γ M1(+E2)	544.42 6	0.829 22
γ	547.28 8	0.0385 18 ?
γ E1	558.90 15	0.0157 18
γ M1	560.55 15	0.0166 22
γ M1,E2	563.00 15	0.0430 13
γ	565.86 14	0.0125 7 ?
γ E1	572.27 4	1.25 3
γ	575.95 25	0.0195 9
γ (E1)	579.42 4	0.448 13
γ	584.35 15	0.0119 7
γ M1,E2	585.80 15	0.0152 9
γ M1,E2	587.15 15	0.030 5
γ M1,E2	587.15 15	12.7 18
γ	590.85 15	0.0363 13
γ	590.85 15	14.6 21
γ	595.70 15	0.0314 9
γ	598.15 15	0.0323 13
γ (E2)	612.31 11	0.0417 13 ?
γ	613.89 9	0.0090 5 ?
γ M1+E2	618.98 7	0.0739 22 ?
γ M1	621.40 15	0.043 5
γ M1	622.75 20	0.0246 18
γ	633.76 6	0.0090 5 ?
γ	636.8 2	0.022 4
γ	645.8 2	0.0134 7
γ	649.60 15	0.045 3

Photons (^{170}Lu)
(continued)

γ_{mode}	γ(keV)	γ(%)[†]
γ	652.65 20	0.0166 13
γ	655.1 2	0.0101 5
γ	656.77 15	0.0125 9 ?
γ	658.2 2	0.0099 9
γ	659.79 9	0.0108 9 ?
γ	670.54 11	0.0376 18 ?
γ	675.45 20	0.0108 9
γ	681.54 13	0.0078 5 ?
γ (M1)	688.09 6	0.197 7
γ	691.75 20	0.0166 9
γ	693.70 5	0.0237 22 ?
γ	700.15 20	0.0206 9
γ	700.90 6	0.0314 9 ?
γ (M1)	703.94 6	0.0762 22 ?
γ	706.5 5	0.074 7
γ	707.10 15	0.134 5
γ (M1)	711.65 12	0.0717 22 ?
γ	723.08 9	0.0197 9 ?
γ	728.76 9	0.043 9 ?
γ	742.14 11	0.0435 13
γ	747.04 8	0.0305 9 ?
γ [M1+E2]	751.07 13	0.0372 13 ?
γ	756.15 20	0.0202 9
γ (M1)	757.64 11	0.114 5 ?
γ	762.61 9	0.0278 9 ?
γ	785.75 20	0.028 3
γ	787.60 15	0.054 4
γ	791.96 13	0.105 5 ?
γ	801.25 20	0.0358 18
γ	802.4 2	0.0327 18
γ	805.85 25	0.018 5
γ	809.25 20	0.0278 13
γ	813.55 20	0.040 4
γ	815.7 2	0.0233 13
γ [E1]	819.75 6	0.0314 9
γ	822.30 15	0.110 5
γ E1	829.55 5	0.486 13
γ	834.45 10	0.100 4
γ (M1)	839.34 6	0.703 20
γ	850.18 6	0.0470 22
γ (M1)	851.72 6	0.081 5 ?
γ (M1)	855.31 6	0.96 3
γ	859.41 10	0.058 5 ?
γ	864.96 15	0.0358 18 ?
γ M1+E2	868.28 5	0.076 9
γ	873.76 11	0.0134 13 ?
γ	877.01 9	0.0269 13 ?
γ	879.65 25	0.0224 13
γ E1	884.22 7	0.345 20
γ	895.29 8	0.0242 13 ?
γ	901.4 2	0.067 3
γ	910.8 3	0.0412 22
γ	916.66 9	0.099 9
γ	916.85 14	0.067 7
γ (E2)	926.49 6	0.260 8
γ (M1)	938.86 5	1.58 4
γ (E2)	942.45 6	0.211 7
γ M1+E2	947.93 6	0.157 5
γ	952.55 25	0.0417 22
γ (M1)	954.23 9	0.224 7 ?
γ	962.85 25	0.0076 9
γ M1+E2	967.04 13	0.143 5 ?
γ (M1)	969.03 7	0.058 3 ?
γ (M1)	970.21 7	0.112 4 ?
γ M1+E2	980.44 6	0.130 13
γ (M1)	983.83 6	0.314 22
γ (E2)	985.12 5	5.38 18
γ (M1)	987.21 6	1.66 5
γ [E1]	987.60 5	0.134 13
γ (M1)	999.58 5	1.52 4
γ	1002.32 9	0.134 13
γ M1+E2	1003.17 5	3.45 11
γ	1009.22 11	0.0394 22 ?
γ	1012.61 14	0.0130 13 ?
γ (E2)	1028.97 6	0.81 3
γ	1034.2 3	0.027 9
γ [E1]	1046.57 9	0.087 5 ?
γ (E2)	1050.41 8	0.99 3
γ	1053.60 13	0.112 22
γ (E2)	1054.32 3	4.61 13
γ	1055.17 8	0.22 4
γ [E1]	1057.71 6	0.213 7
γ (M1)	1060.55 11	0.246 22
γ	1061.30 7	0.22 4

Photons (^{170}Lu)
(continued)

γ_{mode}	γ(keV)	γ(%)[†]
γ (E2)	1061.47 4	2.11 7
γ	1068.8 4	0.0054 5
γ (M1)	1070.98 6	0.0524 18
γ	1078.55 15	0.034 9 ?
γ	1082.1 3	0.026 3
γ	1086.9 3	0.0336 13
γ (E2)	1101.88 5	0.95 3
γ	1110.37 12	0.0121 9 ?
γ [E1]	1113.29 9	0.101 5
γ (E1)	1119.64 13	0.179 5
γ	1122.49 20	0.0157 5 ?
γ	1124.7 3	0.0381 13
γ [E1]	1132.67 6	0.067 7
γ (M1)	1133.68 6	1.03 3
γ [E1]	1135.25 9	
γ M1+E2	1136.88 6	0.157 5
γ (E2)	1138.62 3	3.49 11
γ M1+E2	1141.12 6	0.511 16
γ (E2)	1144.62 8	1.67 5
γ (E2)	1145.77 4	1.75 7
γ	1155.3 3	0.0336 22
γ	1158.4 3	0.0206 13
γ	1162.3 3	0.0403 22
γ	1173.2 4	0.031 13
γ	1180.8 3	0.0112 13
γ [E1]	1181.77 9	0.448 9 ?
γ [E1]	1187.34 7	0.0448 22
γ	1202.9 3	0.0202 13
γ	1204.8 3	0.0179 9
γ [E1]	1206.08 7	0.134 7
γ	1211.2 3	0.0358 18
γ [E1]	1213.59 6	0.052 3 ?
γ M1,E2	1217.3 2	0.202 7
γ E1	1218.41 6	1.36 4
γ (E0+E2)	1222.16 5	0.641 22
γ E1	1225.56 5	4.84 14
γ	1230.49 10	0.112 5
γ	1234.51 23	0.0224 13 ?
γ (M1)	1235.76 6	0.228 7
γ	1240.7 3	0.0166 9
γ	1241.98 5	0.0493 22
γ (E2)	1257.14 5	1.37 4
γ (M1)	1263.31 8	0.309 9
γ [E1]	1268.26 8	0.116 5
γ E1	1280.29 4	7.93 22
γ	1290.50 9	0.085 16
γ E1	1294.76 6	2.84 9
γ [E1]	1294.77 11	0.045 5
γ [E1]	1304.91 9	0.099 4
γ [E2]	1306.47 5	0.493 22
γ M1+E2	1307.46 5	1.08 4
γ	1308.03 12	0.116 13
γ [E1]	1312.83 13	0.314 18
γ [E1]	1313.05 10	0.045 5
γ (M1)	1322.90 6	0.175 13
γ	1330.7 3	0.0358 18
γ (E1)	1341.01 4	3.16 9
γ	1350.45 8	0.057 3
γ [E1]	1360.83 6	0.112 11
γ E1	1364.60 4	4.48
γ	1370.4 3	0.0233 13
γ [E1]	1373.38 8	0.166 16 ?
γ	1380.59 11	0.121 16
γ [M1+E2]	1383.62 6	0.188 7
γ	1385.5 3	0.0448 22
γ [E1]	1395.07 8	0.40 4
γ (E2)	1395.65 6	2.20 7
γ	1398.33 9	0.067 13
γ	1403.85 8	0.202 22
γ E1	1405.21 6	2.53 8
γ	1410.3 4	0.128 13
γ (E2)	1413.35 7	0.220 16
γ	1418.57 13	0.0314 18 ?
γ [E1]	1426.87 7	0.45 4
γ [M1+E2]	1427.38 6	0.33 4
γ (E1)	1428.16 4	3.38 11
γ (M1)	1435.52 7	0.246 9
γ [E1]	1438.37 9	0.0493 22
γ	1445.1 3	0.0358 18
γ [E1]	1449.74 9	0.134 18
γ (E0+E2)	1450.32 4	1.57 4
γ (E2)	1455.32 5	1.14 3
γ	1457.08 12	0.170 18
γ E1	1459.88 7	1.05 3

Photons (^{170}Lu)
(continued)

γ_{mode}	γ(keV)	γ(%)†
γ	1462.93 10	0.072 9 ?
γ	1467.53 10	0.067 7
γ[E1]	1468.02 8	0.090 9
γ	1469.30 9	0.090 5
γ(E2)	1482.15 7	0.605 22
γ[M1]	1485.74 9	0.0448 22 ?
γ	1490.4 3	0.0237 13
γ[E2]	1498.84 15	0.0340 18 ?
γ	1504.38 8	0.0090 9 ?
γ	1507.91 16	0.045 7
γ E1	1512.46 4	2.48 7
γ(M1)	1514.52 6	0.547 22
γ[E1]	1519.29 10	0.0582 22 ?
γ[E1]	1521.52 6	0.036 9
γ[E1]	1528.67 5	0.072 7
γ	1531.38 12	0.179 7
γ(E2)	1534.63 4	0.91 3
γ	1540.37 8	0.085 5
γ	1550.07 10	0.112 11
γ(E0+E2)	1550.62 8	0.448 13
γ	1560.3 3	0.0125 13
γ[M1+E2]	1565.03 9	0.202 9
γ[M1+E2]	1565.10 8	0.090 9
γ	1573.77 10	0.090 5
γ(M1)	1575.24 6	0.502 13
γ	1583.3 3	0.0582 22
γ	1585.8 4	0.0090 9
γ(E2)	1592.10 12	0.139 5
γ[E1]	1597.87 7	0.072 5
γ	1601.2 3	0.116 5
γ[E1]	1602.45 6	0.103 5
γ[E1]	1609.60 6	0.215 11
γ	1610.79 10	0.430 22
γ	1614.63 11	0.0367 18 ?
γ M1+E2	1619.70 10	0.090 5
γ M1+E2	1631.07 10	0.0986 22
γ M1+E2	1633.37 7	0.052 9 ?
γ(E2)	1634.93 8	0.094 4
γ(M1)	1636.80 11	0.0538 18 ?
γ(E1)	1641.51 8	0.309 9
γ	1645.18 13	0.0193 9 ?
γ	1648.7 3	0.0148 13
γ[E1]	1651.28 15	0.0305 13 ?
γ	1653.24 10	0.0211 13 ?
γ	1663.07 13	0.064 4
γ	1667.35 13	0.0309 18 ?
γ(M1)	1674.30 12	0.157 5
γ[E1]	1678.80 7	0.224 7
γ[M1+E2]	1683.20 13	0.054 18 ?
γ(E2)	1685.75 11	0.058 7
γ	1687.9 4	0.0224 22
γ[E1]	1700.77 10	0.134 5
γ	1703.3 3	0.085 3
γ	1705.91 14	0.047 7
γ[M2]	1714.42 7	0.018 4 ?
γ[E1]	1719.05 10	0.146 5
γ	1723.8 3	0.0269 18
γ	1731.28 13	0.0094 9 ?
γ[M1]	1736.77 11	0.039 5 ?
γ(M1)	1740.38 10	0.081 3
γ	1746.46 12	0.0300 18 ?
γ	1747.62 15	0.0112 13 ?
γ	1753.8 3	0.0448 22
γ(M1)	1758.82 14	0.081 3
γ	1761.44 11	0.042 5 ?
γ	1767.2 3	0.081 5
γ	1770.34 13	0.0112 13 ?
γ(M1)	1776.17 14	0.258 9
γ[E1]	1778.72 20	0.0242 22 ?
γ	1783.3 4	0.0242 22
γ	1784.66 11	0.039 7 ?
γ	1791.7 4	0.0349 9
γ(E2)	1793.19 10	0.090 5
γ	1796.30 5	0.0179 9
γ	1799.25 5	0.0125 9
γ[E1]	1802.20 7	0.157 5
γ E1	1809.35 7	0.771 22
γ	1818.7 5	0.0211 22
γ	1820.6 5	0.0157 18
γ[E1]	1824.76 15	0.030 3 ?
γ	1830.1 5	0.0193 18
γ	1832.4 4	0.0237 9
γ	1836.77 10	0.058 6 ?
γ	1838.52 15	0.0421 13 ?

Photons (^{170}Lu)
(continued)

γ_{mode}	γ(keV)	γ(%)†
γ	1842.79 11	0.052 3
γ	1843.3 3	0.116 13
γ	1855.0 5	0.0157 18
γ(E2)	1859.23 10	0.20 3
γ E1	1860.28 8	0.542 22
γ	1870.8 3	0.058 7
γ	1874.7 5	0.0273 13
γ	1876.2 3	0.146 9
γ[E1]	1878.55 7	0.551 18
γ	1887.1 5	0.034 5
γ	1888.7 5	0.0358 18
γ	1893.7 5	0.0426 22
γ	1896.5 3	0.055 3
γ E1	1901.30 9	0.591 22
γ	1904.5 5	0.0197 9
γ	1909.7 5	0.0202 13
γ	1917.7 5	0.0224 13
γ	1920.75 10	0.094 4 ?
γ	1936.77 12	0.213 7
γ[M1+E2]	1953.88 9	0.161 9
γ(M1)	1955.70 9	1.34 4
γ[M1+E2]	1961.03 9	0.287 9
γ	1962.4 3	0.096 3
γ(M1)	1966.73 15	0.0291 22 ?
γ	1974.0 3	0.0538 18
γ	1977.4 5	0.031 7
γ	1983.9 5	0.0255 13
γ(E2)	1985.46 14	0.076 3 ?
γ[E1]	1992.61 14	0.0179 9 ?
γ(E2)	1995.87 14	0.081 3
γ	1998.4 5	0.018 5
γ	2007.3 5	0.0125 18
γ M1+E2	2019.93 10	0.060 5
γ	2025.8 3	0.0560 22
γ E1	2027.08 10	0.164 7
γ(M1)	2030.23 9	0.287 18
γ E1	2031.67 6	0.365 11
γ M1+E2	2040.00 9	2.54 9
γ E1	2041.91 5	5.91 18
γ	2046.5 5	0.0260 13
γ	2054.4 3	0.125 5
γ[E1]	2057.03 13	0.0385 13 ?
γ(E1)	2061.81 15	0.0139 9 ?
γ	2063.2 3	0.0708 22
γ	2086.4 5	0.0202 9
γ	2094.5 5	0.0273 13
γ	2096.28 10	0.139 5
γ[E1]	2115.98 6	0.157 18
γ	2116.63 7	0.493 18
γ E1	2126.21 5	4.97 16
γ	2143.5 3	0.072 3
γ	2148.5 5	0.0336 13
γ	2152.9 5	0.0193 9
γ	2157.7 5	0.0099 5
γ	2165.7 5	0.0130 9
γ	2178.0 5	0.0188 9
γ	2183.95 20	0.0394 22
γ E1	2191.19 5	1.59 4
γ	2200.9 3	0.0538 22
γ	2205.07 10	0.0340 13
γ	2223.9 5	0.0157 18
γ	2232.7 5	0.0157 9
γ	2243.7 4	0.0323 22
γ	2246.8 5	0.0112 9
γ	2255.4 6	0.0076 9
γ	2257.4 4	0.0314 13
γ	2266.8 5	0.0161 9
γ	2268.25 20	0.188 5
γ(E1)	2275.49 5	0.87 3
γ[E1]	2279.86 4	0.190 7
γ	2284.2 5	0.014 5
γ(M1)	2289.38 10	0.0426 22
γ	2315.1 4	0.0358 18
γ E1	2315.90 6	0.206 7
γ	2325.0 4	0.0314 22
γ	2330.6 6	0.0058 9
γ	2333.9 5	0.0108 9
γ	2344.81 8	0.0448 18
γ	2352.3 5	0.0493 18
γ E1	2364.17 4	1.45 4
γ	2398.1 3	0.045 9
γ(E1)	2400.21 6	0.405 13
γ E1	2411.98 5	0.80 3
γ	2419.9 5	0.019 3

Photons (^{170}Lu)
(continued)

γ_{mode}	γ(keV)	γ(%)†
γ	2424.4 3	0.121 5
γ	2429.11 8	0.047 5
γ(M1)	2438.73 14	0.103 5 ?
γ(E2)	2452.64 8	0.134 5
γ	2459.9 5	0.0121 13
γ E1	2496.28 5	0.739 22
γ(M1)	2523.03 14	0.134 5 ?
γ	2534.0 6	0.008 3
γ(E2)	2536.94 8	0.063 5
γ	2542.8 6	0.0112 13
γ	2546.1 6	0.0067 9
γ	2558.0 5	0.0358 22
γ	2561.1 6	0.0134 13
γ	2575.3 7	~0.027
γ	2576.87 16	0.076 13 ?
γ(E1)	2582.98 4	0.139 5
γ	2599.0 5	0.031 3
γ	2637.0 6	0.0085 9
γ	2642.1 4	0.085 5
γ	2652.0 4	0.0202 22
γ	2653.0 6	0.036 4
γ M1+E2	2661.18 16	0.224 13 ?
γ E1	2663.90 5	1.22 4
γ(M1)	2667.28 4	0.081 5
γ	2677.3 7	0.0067 9
γ	2680.3 7	0.0076 9
γ E1	2691.45 8	2.22 9
γ M1+E2	2698.85 13	0.591 22
γ	2718.3 6	0.0157 18
γ	2720.9 5	0.0426 22
γ	2726.6 6	0.0112 13
γ	2729.3 7	0.0090 9
γ	2735.60 4	0.0246 22 ?
γ	2737.2 4	0.056 9
γ E1	2748.21 5	2.07 9
γ	2775.7 3	0.110 5
γ M1+E2	2783.16 13	1.00 4
γ	2793.1 7	0.0116 13
γ	2805.0 6	0.0291 13
γ	2813.7 6	0.0202 22
γ E1	2845.38 7	1.67 9
γ E2	2849.5 3	0.206 18
γ(E1)	2855.52 5	0.318 13
γ E1	2863.66 6	0.129 5
γ(M1)	2872.38 12	0.075 4
γ(M1)	2881.38 8	0.73 3
γ(E1)	2885.22 13	0.291 11
γ	2897.6 5	0.045 3
γ E1	2923.33 19	0.179 9
γ E1	2929.68 7	0.58 3
γ E1	2939.82 5	1.50 9
γ E1	2947.96 6	0.58 3
γ	2953.1 5	0.034 7
γ M1+E2	2956.69 12	0.085 3
γ	2958.22 17	0.0448 22
γ(M1)	2965.68 8	1.25 7
γ	2969.7 5	0.027 3
γ	2981.5 5	0.031 3
γ	2983.13 13	0.076 5
γ	2985.9 4	0.054 4
γ(E2)	3007.63 19	0.137 7
γ[M1+E2]	3015.34 9	0.246 11
γ	3018.5 6	0.0143 13
γ E1	3030.92 12	1.28 7
γ	3036.9 3	0.206 9
γ	3042.52 17	0.067 4
γ[E1]	3046.91 14	0.034 4
γ	3053.1 3	0.108 9
γ	3061.86 9	0.103 9 ?
γ M1+E2	3064.94 10	0.251 11
γ	3067.43 13	0.116 9
γ	3085.40 15	0.0148 9 ?
γ	3091.9 3	0.152 9
γ	3095.49 12	0.323 18
γ[M1]	3099.64 9	0.193 11
γ	3102.44 11	0.0148 13 ?
γ(E1)	3111.34 13	0.175 9
γ E1	3115.23 12	0.73 4
γ	3119.2 6	0.020 7
γ	3123.0 6	0.0188 18
γ	3128.1 5	0.040 4
γ[E1]	3131.21 14	0.0112 18
γ	3139.6 8	0.0027 9
γ	3146.16 9	0.112 9 ?

Photons (^{170}Lu)
(continued)

γ_{mode}	γ(keV)	γ(%)†
γ	3149.24 *10*	0.101 *9*
γ	3157.0 *8*	0.0040 *5*
γ	3161.1 *5*	0.045 *5*
γ	3165.68 *9*	0.099 *9* ?
γ	3169.71 *15*	0.0045 *9* ?
γ	3173.4 *7*	0.0134 *13*
γ	3179.8 *7*	0.0166 *18*
γ	3183.6 *5*	0.063 *6*
γ E1	3190.19 *24*	0.056 *5* ?
γ (E1)	3195.64 *13*	0.090 *9*
γ	3202.4 *5*	0.067 *7*
γ	3206.8 *8*	0.0134 *13*
γ	3212.2 *8*	0.0067 *9*
γ	3218.3 *6*	0.0022 *5* ?
γ	3229.75 *14*	0.0067 *9* ?
γ	3255.9 *7*	0.0134 *13*
γ	3258.2 *8*	0.0112 *13*
γ E1	3274.49 *24*	0.045 *5* ?
γ	3282.1 *8*	0.0022 *5*
γ	3291.4 *7*	0.0045 *5*
γ M1+E2	3302.6 *6*	0.0116 *13* ?
γ M1+E2	3314.05 *14*	0.0125 *13* ?
γ	3338.9 *8*	0.0018 *5*
γ	3385.0 *8*	0.0018 *5*

† 5.6% uncert(syst)

Atomic Electrons (^{170}Lu)
⟨e⟩=52.4 *14* keV

e_{bin}(keV)	⟨e⟩(keV)	e(%)
9	4.5	50 *8*
10	3.6	36 *6*
23	2.79	12.2 *7*
40 - 41	0.45	1.11 *14*
42	0.60	1.42 *25*
43 - 73	1.06	2.16 *16*
74	12.8	17.2 *10*
75	12.0	15.9 *9*
81	0.005	~0.006
82	6.7	8.1 *5*
83	0.068	0.082 *5*
84	1.82	2.17 *12*
91 - 141	0.762	0.656 *22*
142 - 191	0.726	0.410 *11*
192 - 241	0.170	0.078 *11*
242 - 291	0.11	0.043 *12*
292 - 341	0.117	0.036 *4*
343 - 391	0.12	0.033 *9*
392 - 441	0.088	0.0215 *23*
442 - 491	0.14	0.029 *7*
492 - 540	0.078	0.0151 *17*
542 - 591	0.039	0.0069 *8*
593 - 643	0.043	0.0069 *5*
644 - 692	0.035	0.0052 *11*
693 - 742	0.032	0.0045 *7*
745 - 794	0.158	0.0202 *9*
795 - 845	0.046	0.0056 *5*
846 - 895	0.155	0.0177 *8*
899 - 948	0.58	0.062 *6*
950 - 999	0.314	0.0318 *16*
1000 - 1049	0.139	0.0136 *9*
1050 - 1099	0.292	0.0271 *9*
1100 - 1148	0.0638	0.00565 *17*
1150 - 1197	0.138	0.0118 *4*
1201 - 1248	0.200	0.0163 *9*
1251 - 1299	0.077	0.00601 *21*
1302 - 1351	0.150	0.0113 *4*
1352 - 1401	0.158	0.0114 *4*
1402 - 1450	0.0470	0.00330 *16*
1451 - 1550	0.39	0.026 *8*
1551 - 1649	0.083	0.0053 *15*
1651 - 1750	0.0354	0.00207 *9*
1751 - 1850	0.0364	0.00202 *13*
1851 - 1951	0.064	0.00335 *16*
1952 - 2051	0.134	0.0068 *5*
2052 - 2151	0.069	0.00331 *20*
2155 - 2255	0.0173	0.00078 *6*

Atomic Electrons (^{170}Lu)
(continued)

e_{bin}(keV)	⟨e⟩(keV)	e(%)
2256 - 2355	0.0247	0.00107 *3*
2359 - 2458	0.0141	0.00059 *3*
2462 - 2559	0.0081	0.000326 *23*
2565 - 2662	0.0408	0.00156 *5*
2665 - 2752	0.0368	0.00136 *7*
2765 - 2863	0.0368	0.00131 *4*
2868 - 2968	0.0518	0.00178 *5*
2970 - 3067	0.0350	0.00116 *6*
3070 - 3169	0.0093	0.00030 *3*
3170 - 3266	0.00134	4.2 *3* $\times10^{-5}$
3272 - 3337	0.000105	3.2 *3* $\times10^{-6}$
3375 - 3383	2.9 $\times10^{-6}$	8 *3* $\times10^{-8}$

Continuous Radiation (^{170}Lu)
⟨β+⟩=2.42 keV;⟨IB⟩=0.52 keV

E_{bin}(keV)		⟨ ⟩(keV)	(%)
0 - 10	β+	3.07 $\times10^{-9}$	3.63 $\times10^{-8}$
	IB	0.00107	
10 - 20	β+	1.74 $\times10^{-7}$	1.04 $\times10^{-6}$
	IB	0.00029	0.0021
20 - 40	β+	6.7 $\times10^{-6}$	2.02 $\times10^{-5}$
	IB	0.0052	0.0153
40 - 100	β+	0.000465	0.00059
	IB	0.24	0.46
100 - 300	β+	0.0291	0.0130
	IB	0.031	0.0169
300 - 600	β+	0.210	0.0456
	IB	0.073	0.0166
600 - 1300	β+	1.07	0.114
	IB	0.116	0.0139
1300 - 2500	β+	1.11	0.068
	IB	0.042	0.0025
2500 - 3440	IB	0.0052	0.00019
	Σβ+		0.24

$^{170}_{71}$Lu(670 *100* ms)

Mode: IT
Δ: -57239 *20* keV
SpA: 6.2×10^{10} Ci/g
Prod: daughter ^{170}Hf; ^{170}Yb(p,n)

Photons (^{170}Lu)
⟨γ⟩=3.6 *3* keV

γ_{mode}	γ(keV)	γ(%)†
Lu L_ℓ	6.753	0.59 *10*
Lu L_α	7.650	14.5 *21*
Lu L_η	7.857	0.20 *3*
Lu L_β	8.812	17 *3*
Lu L_γ	10.302	3.3 *6*
γ E2	44.49 *6*	0.85
γ M2	48.41 *10*	0.44 *3*

† 4.7% uncert(syst)

Atomic Electrons (^{170}Lu)
⟨e⟩=87 *5* keV

e_{bin}(keV)	⟨e⟩(keV)	e(%)
9	4.3	47 *8*
10	2.6	25 *5*
11	2.3	21 *3*
34	12.2	36 *7*
35	14.3	41 *8*
38	21.4	57 *4*
39	7.7	19.8 *14*
42	7.9	19 *4*
43	0.106	0.25 *5*
44	2.2	5 *1*
46	8.9	19.3 *14*
47	0.037	0.079 *6*
48	2.66	5.5 *4*

$^{170}_{72}$Hf(16.01 *13* h)

Mode: ε
Δ: -56130 *200* keV syst
SpA: 1.151×10^{6} Ci/g
Prod: ^{175}Lu(p,6n); ^{171}Yb(α,5n); ^{3}He on Yb

Photons (^{170}Hf)
⟨γ⟩=579 *38* keV

γ_{mode}	γ(keV)	γ(%)†
Lu L_ℓ	6.753	0.82 *17*
Lu L_α	7.650	20 *4*
Lu L_η	7.857	0.29 *6*
Lu L_β	8.814	23 *4*
Lu L_γ	10.274	4.1 *8*
γ M1+0.5%E2	28.34 *7*	0.25 *5*
γ M1+0.8%E2	32.30 *7*	0.084 *17*
γ (E2)	39.04 *6*	0.0031 *7*
γ E2	44.49 *6*	0.32 *6* §
γ M1+0.4%E2	47.76 *7*	3.7 *7*
γ M2	48.41 *10*	0.023 *5* §
Lu $K_{\alpha2}$	52.965	35 *7*
γ M1	53.98 *6*	1.4 *3*
Lu $K_{\alpha1}$	54.070	62 *13*
γ M1+1.3%E2	55.12 *7*	1.4 *3*
Lu $K_{\beta1}$'	61.219	20 *4*
γ (M1)	62.77 *15*	0.14 *3*
Lu $K_{\beta2}$'	63.200	5.3 *11*
γ (M1)	70.37 *7*	0.28 *6*
γ (M1)	71.43 *7*	0.28 *6*
γ E2	71.52 *7*	0.112 *22*
γ	71.9 *3*	0.039 *8* ?
γ	74.74 *10*	0.22 *4*
γ M1	80.06 *6*	0.78 *16*
γ M1	98.47 *7*	4.2 *8*
γ M1+26%E2	99.85 *7*	2.5 *5*
γ	112.7 *3*	~0.11 ?
γ	113.78 *9*	0.22 *5* ?
γ	114.9 *3*	~0.25 ?
γ (M1+E2)	115.92 *7*	0.81 *16*
γ	116.8 *3*	~0.47 ?
γ	117.7 *3*	~0.6 ?
γ M1(+E2)	119.10 *3*	1.06 *21*
γ E1	120.17 *6*	19 *4*
γ	123.60 *11*	0.036 *7* ?
γ	127.3 *3*	0.039 *8* ?
γ	132.12 *18*	0.056 *11*
γ	139.1 *3*	0.022 *5* ?
γ	143.5 *3*	
γ E1	146.26 *9*	1.5 *3*
γ	147.6 *3*	
γ (M1+E2)	153.83 *8*	0.084 *17*
γ M1(+E2)	162.64 *12*	1.7 *4*
γ E1	164.66 *7*	33
γ	167.93 *9*	0.5 *1*

Photons (^{170}Hf) (continued)

γ_{mode}	γ(keV)	γ(%)†
γ	168.90 *8*	0.75 *15*
γ	183.8 *3*	0.039 *8* ?
γ	185.29 *7*	0.27 *5*
γ	186.37 *12*	0.084 *17*
γ (M1)	187.83 *21*	0.112 *22* ?
γ	189.2 *3*	0.070 *14* ?
γ	190.86 *11*	0.084 *17*
γ (M1)	198.32 *9*	0.17 *4*
γ M1+30%E2	208.01 *22*	3.4 *7* ?
γ	208.10 *24*	3.4 *7* ?
γ	209.04 *13*	0.64 *13*
γ	218.2 *3*	0.098 *20* ?
γ M1(+E2)	225.41 *12*	1.09 *22*
γ (M1)	242.70 *11*	0.10 *2*
γ (M1+E2)	257.71 *13*	0.112 *22*
γ	261.9 *3*	0.112 *22* ?
γ	268.9 *3*	0.14 *3* ?
γ	278.7 *3*	0.112 *22* ?
γ E1	291.45 *13*	1.3 *3*
γ M1	304.0 *3*	0.38 *8*
γ E1	308.89 *12*	2.6 *5*
γ	310.4 *3*	0.120 *24* ?
γ [E1]	315.20 *14*	0.14 *3*
γ (M1)	348.9 *3*	1.4 *3* ?
γ	377.97 *13*	0.16 *3*
γ E1	425.64 *12*	1.10 *22*
γ M1(+E2)	462.0 *3*	0.056 *11*
γ E1	470.13 *12*	0.67 *13*
γ	481.30 *24*	4.7 *9* ?
γ E1	481.38 *20*	4.7 *9* ?
γ M1	494.6 *3*	0.05 *1*
γ E1	501.57 *10*	4.7 *9*
γ (M1)	510.68 *22*	0.21 *4* ?
γ	533.4 *3*	0.05 *1* ?
γ E1	540.61 *10*	3.1 *6*
γ	554.0 *3*	0.078 *16* ?
γ E1	572.91 *11*	18 *4*
γ (M1)	587.01 *11*	0.33 *7*
γ	602.1 *3*	0.112 *22* ?
γ	605.1 *3*	
γ	608.72 *19*	0.21 *4* ?
γ (M1)	615.34 *11*	0.47 *10*
γ E1	620.67 *9*	23 *5*
γ	632.6 *3*	0.042 *8* ?
γ	639.29 *15*	0.022 *5* ?
γ	654.4 *3*	0.017 *3* ?
γ	660.9 *10*	0.028 *6* ?
γ [M1+E2]	669.41 *11*	0.21 *4* ?
γ	673.9 *10*	0.042 *8* ?
γ M1+E2	686.70 *18*	0.33 *7* ?
γ	692.7 *3*	0.031 *6* ?
γ	711.3 *3*	0.017 *3* ?
γ	723.9 *10*	0.0084 *17* ?
γ M1+E2	740.84 *10*	0.23 *5*
γ M1+E2	746.44 *20*	0.106 *21*
γ E1	757.07 *18*	0.51 *10* ?
γ	770.14 *19*	0.16 *3*
γ [M1]	785.33 *10*	0.056 *11* ?
γ	801.56 *18*	0.33 *7*
γ	808.19 *15*	0.020 *4* ?
γ	814.63 *19*	0.075 *15*
γ	878.56 *14*	0.036 *7* ?
γ	923.05 *14*	0.014 *3* ?

† 11% uncert(syst)

§ with ^{170}Lu(670 ms) in equilib

Atomic Electrons (^{170}Hf)

⟨e⟩=92 *5* keV

e_{bin}(keV)	⟨e⟩(keV)	e(%)
6 - 8	0.73	10.4 *13*
9	6.4	69 *15*
10	4.0	39 *8*
11	3.0	27 *9*
14 - 32	3.2	16.6 *19*
34	4.6	14 *3*

Atomic Electrons (^{170}Hf) (continued)

e_{bin}(keV)	⟨e⟩(keV)	e(%)
35	10.3	29 *6*
37	8.4	23 *5*
38 - 39	1.8	4.8 *7*
42	3.6	8.5 *17*
43	2.0	4.5 *9*
44	3.1	7.0 *14*
45	2.2	4.8 *10*
46 - 91	19.0	30 *4*
96 - 100	2.2	2.2 *5*
101	2.7	2.6 *3*
102 - 144	4.9	4.4 *10*
145	2.7	1.9 *9*
146 - 194	2.6	1.7 *3*
196 - 243	1.7	0.81 *18*
246 - 295	0.60	0.22 *3*
298 - 347	0.148	0.046 *7*
348 - 378	0.075	0.021 *5*
399 - 447	0.8	~0.18
451 - 500	0.25	0.053 *15*
501 - 545	0.46	0.090 *15*
552 - 601	0.59	0.105 *16*
602 - 651	0.148	0.024 *3*
652 - 701	0.042	0.0062 *12*
702 - 751	0.041	0.0057 *20*
755 - 804	0.0075	0.0010 *3*
805 - 815	0.0019	~0.00023
860 - 879	0.0009	0.00011 *5*
912 - 923	0.00012	~1×10^{-5}

Continuous Radiation (^{170}Hf)

⟨IB⟩=0.33 keV

E_{bin}(keV)		⟨ ⟩(keV)	(%)
10 - 20	IB	0.00024	0.0018
20 - 40	IB	0.0042	0.0122
40 - 100	IB	0.26	0.47
100 - 300	IB	0.023	0.013
300 - 600	IB	0.035	0.0081
600 - 1085	IB	0.0157	0.0022

$^{170}_{73}$Ta(6.76 *6* min)

Mode: ε

Δ: -50330 *200* keV syst

SpA: 1.635×10^{8} Ci/g

Prod: ^{159}Tb(^{16}O,5n); ^{165}Ho(^{12}C,7n); ^{175}Lu(^{3}He,8n)

Photons (^{170}Ta)

⟨γ⟩=309 *5* keV

γ_{mode}	γ(keV)	γ(%)†
γ	100.8 *2*	21.0 *6*
γ	221.2 *2*	15.7 *5*
γ	320.9 *2*	~0.7
γ	512.5 *5*	
γ	576.5 *6*	0.29 *4*
γ	584.3 *6*	0.34 *4*
γ	639.4 *2*	0.48 *6*
γ	665.0 *3*	1.11 *13*
γ	765.5 *2*	0.97 *15*
γ	778.8 *2*	0.90 *6*
γ	834.8 *4*	1.53 *13*
γ	860.4 *2*	7.39 *25*
γ	886.2 *5*	0.38 *8*
γ	897.6 *5*	0.15 *4*
γ	905.3 *6*	0.40 *6*
γ	961.3 *4*	0.46 *19*
γ	986.9 *3*	3.36 *13*
γ	987.0 *3*	5.88 *21*
γ	1054.3 *12*	0.29 *6*

Photons (^{170}Ta) (continued)

γ_{mode}	γ(keV)	γ(%)†
γ	1119.0 *6*	1.11 *8*
γ	1126.5 *8*	0.50 *4*
γ	1344.2 *6*	1.43 *11*
γ	1444.8 *14*	0.13 *4*

† 4.8% uncert(syst)

Continuous Radiation (^{170}Ta)

⟨β+⟩=1349 keV;⟨IB⟩=15.8 keV

E_{bin}(keV)		⟨ ⟩(keV)	(%)
0 - 10	β+	8.5×10^{-8}	1.01×10^{-6}
	IB	0.040	
10 - 20	β+	5.07×10^{-6}	3.02×10^{-5}
	IB	0.039	0.27
20 - 40	β+	0.000206	0.00062
	IB	0.078	0.27
40 - 100	β+	0.0156	0.0197
	IB	0.31	0.51
100 - 300	β+	1.19	0.53
	IB	0.70	0.39
300 - 600	β+	11.7	2.48
	IB	0.94	0.22
600 - 1300	β+	125	12.7
	IB	2.2	0.25
1300 - 2500	β+	574	30.2
	IB	4.8	0.26
2500 - 5000	β+	637	20.5
	IB	6.6	0.20
5000 - 5729	IB	0.108	0.0021
	Σβ+		66

$^{170}_{74}$W (4 *1* min)

Mode: ε

Δ: -47240 *300* keV syst

SpA: 2.8×10^{8} Ci/g

Prod: ^{155}Gd(^{20}Ne,5n); ^{156}Gd(^{20}Ne,6n); ^{155}Gd(^{22}Ne,7n)

$^{170}_{75}$Re(8.0 *5* s)

Mode: ε

Δ: -39090 *400* keV syst

SpA: 7.9×10^{9} Ci/g

Prod: ^{159}Tb(^{20}Ne,9n); protons on Tl

Photons (^{170}Re)

γ_{mode}	γ(keV)	γ(rel)
γ [E2]	156.0 *2*	89 *5*
γ [E2]	305.5 *2*	100
γ [E2]	412.5 *2*	74 *5*

$^{170}_{76}$Os(7.1 *2* s)

Mode: ϵ(88 *1* %), α(12 *1* %)
Δ: -33934 *18* keV
SpA: 8.91×10^{9} Ci/g
Prod: ^{156}Dy(^{20}Ne,6n); ^{93}Nb(^{84}Kr,p6n); daughter ^{174}Pt

Alpha Particles (^{170}Os)

α(keV)
5411 *4*

$^{170}_{77}$Ir(1.05 *15* s)

Mode: α
Δ: -23320 *220* keV syst
SpA: 4.6×10^{10} Ci/g
Prod: ^{93}Nb(^{84}Kr,7n); ^{110}Cd(^{63}Cu,3n)

Alpha Particles (^{170}Ir)

α(keV)
6030 *10*

$^{170}_{78}$Pt(~ 6 ms)

Mode: α
Δ: -16510 *280* keV syst
SpA: ~ 9.6×10^{10} Ci/g
Prod: ^{144}Sm(^{32}S,6n); ^{58}Ni on Sn

Alpha Particles (^{170}Pt)

α(keV)
6548 *7*

A = 171

NDS **43**, 127 (1984)

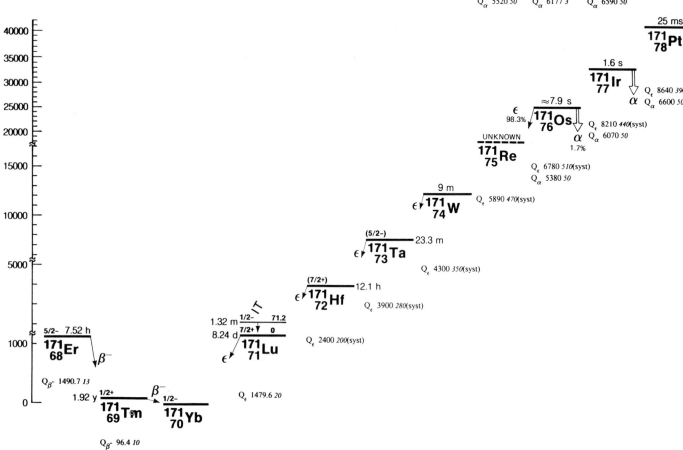

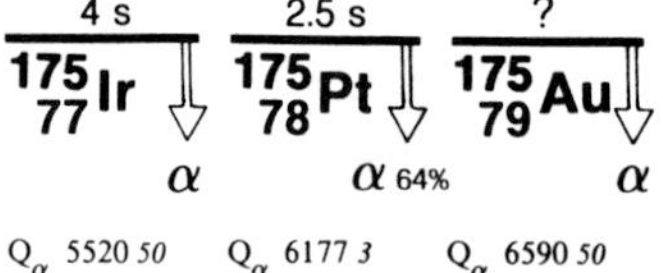

$^{171}_{68}$Er(7.52 *3* h)

Mode: β-
Δ: -57728 *4* keV
SpA: 2.4370×10^{6} Ci/g
Prod: ^{170}Er(n,γ)

Photons (^{171}Er)

⟨γ⟩=373 *6* keV

γ_{mode}	γ(keV)	γ(%)†
Tm L_ℓ	6.341	0.226 *24*
Tm L_α	7.176	5.7 *4*
Tm L_η	7.310	0.080 *6*
Tm L_β	8.217	5.9 *6*
Tm L_γ	9.540	1.00 *11*
γ M1+0.03%E2	12.391 *4*	0.0300 *18*
Tm $K_{\alpha 2}$	49.772	13.4 *6*
Tm $K_{\alpha 1}$	50.742	23.6 *10*
Tm $K_{\beta 1}'$	57.444	7.4 *3*
Tm $K_{\beta 2}'$	59.296	2.05 *9*
γ M1+5.1%E2	85.58 *6*	0.060 *4*
γ M1+2.5%E2	111.661 *3*	20.5 *8*
γ E2	116.689 *5*	2.30 *6*
γ E2	124.052 *4*	9.1 *3*
γ	166.4 *3* ?	
γ [M1]	175.66 *4*	0.089 *9*
γ [M1]	197.88 *10*	0.027 *5*
γ [E2]	210.27 *10*	~0.007
γ E1	210.65 *3*	0.642 *19*
γ M1+1.9%E2	237.17 *4*	0.302 *10*
γ [M1+E2]	261.24 *7*	<0.02
γ (M1+8.8%E2)	277.47 *4*	0.58 *2*
γ [M1]	286.40 *9*	~0.008
γ E1	295.939 *11*	28.9 *8*
γ E1	308.331 *11*	64.4 *16*
γ [M1]	363.05 *7*	0.0197 *11*
γ M1+7.3%E2	371.98 *7*	0.257 *10*
γ M2	419.991 *11*	0.083 *4*
γ E3	425.020 *12*	0.0224 *23*
γ	455.64 *20*	0.006 *2*
γ [E1]	488.11 *4*	0.005 *2*
γ [M1+E2]	495.55 *15*	~0.002
γ [M1]	506.58 *3*	0.0227 *20*
γ [M1]	518.98 *3*	0.0177 *16*
γ [M1+E2]	547.64 *8*	0.017 *4*
γ M1	559.27 *5*	0.0466 *19*
γ [E1]	573.69 *7*	0.0098 *15*
γ [E2]	586.18 *10*	~0.004
γ [M1+E2]	608.39 *5*	~0.037
γ [M1+E2]	609.15 *8*	~0.02
γ M1	620.79 *5*	0.089 *3*
γ [E2]	630.64 *3*	0.005 *1*
γ M1+1.6%E2	670.93 *5*	0.252 *5*
γ [M1+E2]	671.76 *11*	0.022 *5*
γ M1+1.9%E2	675.96 *5*	0.285 *6*
γ [M1+E2]	693.42 *14*	0.0150 *16*
γ [M1+E2]	705.81 *14*	0.012 *4*
γ M1	732.45 *5*	0.0976 *24*
γ	745.0 *5*	0.0066 *8*
γ	767.84 *20*	0.0045 *5*
γ M1+11%E2	784.05 *4*	0.240 *5*
γ M1+25%E2	796.44 *4*	0.640 *13*
γ [E1]	860.09 *8*	0.00150 *24*
γ (M1+5.9%E2)	869.63 *7*	0.055 *5*
γ	871.54 *20*	0.020 *5*
γ M1+1.3%E2	882.02 *7*	0.0385 *19*
γ M1+10%E2	908.10 *4*	0.635 *13*
γ [E2]	913.13 *4*	0.077 *5*
γ	966.1 *4*	0.0264 *8*
γ [E1]	975.65 *25*	0.0007 *3*
γ [E2]	993.68 *7*	~0.0006
γ	1051.0 *5*	~0.0004

Photons (^{171}Er)
(continued)

γ_{mode}	γ(keV)	γ(%)†
γ	1096.7 *4*	0.00106 *19*
γ	1109.1 *4*	0.00679 *21*
γ [M1+E2]	1156.03 *8*	0.00060 *15*
γ [M1+E2]	1168.42 *8*	0.00184 *15*
γ	1172.9 *5*	0.0008 *3*
γ	1182.0 *5*	~0.0003
γ	1220.7 *4*	0.0028 *2*
γ [M1+E2]	1271.59 *25*	0.00034 *15*
γ [M1+E2]	1280.08 *8*	0.0025 *2*
γ [E2]	1285.11 *8*	0.0024 *2*
γ [M1+E2]	1395.64 *25*	0.0028 *8*
γ [E2]	1400.67 *25*	0.0025 *1*

† 2.0% uncert(syst)

Atomic Electrons (^{171}Er)

⟨e⟩=58.1 *10* keV

e_{bin}(keV)	⟨e⟩(keV)	e(%)
2	0.120	5.3 *3*
3	2.05	72 *5*
4 - 5	0.95	20.5 *13*
9	1.94	22.4 *24*
10	1.86	19.0 *19*
11 - 51	1.16	3.13 *11*
52	20.7	39.6 *17*
55 - 57	0.99	1.74 *6*
65	3.62	5.60 *22*
75 - 86	0.047	0.060 *3*
102	6.2	6.1 *3*
103 - 112	4.24	3.92 *7*
114	3.58	3.13 *12*
115	3.09	2.68 *10*
116 - 117	0.191	0.164 *9*
122	1.59	1.30 *5*
123 - 167	0.509	0.402 *14*
173 - 218	0.334	0.167 *6*
227 - 237	1.18	0.501 *17*
249	2.44	0.98 *3*
251 - 300	0.738	0.252 *4*
304 - 353	0.184	0.0598 *13*
354 - 396	0.068	0.0189 *8*
410 - 456	0.0201	0.00481 *16*
460 - 509	0.0109	0.00223 *17*
510 - 559	0.0065	0.0012 *4*
561 - 611	0.0123	0.00215 *9*
612 - 661	0.0581	0.0094 *3*
662 - 708	0.0160	0.00238 *6*
722 - 768	0.061	0.0083 *4*
774 - 823	0.0187	0.00236 *11*
849 - 898	0.0472	0.00552 *13*
899 - 934	0.0032	0.00035 *7*
956 - 994	0.00020	2.1 *10* ×10^{-5}
1037 - 1087	0.00024	~2×10^{-5}
1088 - 1123	0.00015	1.4 *3* ×10^{-5}
1146 - 1182	9.5 ×10^{-5}	8 *4* ×10^{-6}
1211 - 1226	0.000155	1.27 *20* ×10^{-5}
1261 - 1285	2.7 ×10^{-5}	2.1 *3* ×10^{-6}
1336 - 1386	0.00014	1.05 *23* ×10^{-5}
1387 - 1400	1.34 ×10^{-5}	9.6 *8* ×10^{-7}

Continuous Radiation (^{171}Er)

⟨β-⟩=359 keV;⟨IB⟩=0.39 keV

E_{bin}(keV)		⟨ ⟩(keV)	(%)
0 - 10	β-	0.072	1.44
	IB	0.0168	
10 - 20	β-	0.217	1.45
	IB	0.0161	0.112
20 - 40	β-	0.88	2.93
	IB	0.030	0.105
40 - 100	β-	6.3	9.0
	IB	0.077	0.120
100 - 300	β-	61	30.4
	IB	0.156	0.092
300 - 600	β-	162	37.1
	IB	0.083	0.021
600 - 1300	β-	128	17.4
	IB	0.0131	0.0019
1300 - 1491	β-	0.216	0.0160
	IB	3.3×10^{-6}	2.5×10^{-7}

$^{171}_{69}$Tm(1.92 *1* yr)

Mode: β-
Δ: -59219 *3* keV
SpA: 1089 Ci/g
Prod: daughter ^{171}Er

Photons (^{171}Tm)

⟨γ⟩~0.62 keV

γ_{mode}	γ(keV)	γ(%)
Yb L_ℓ	6.545	~0.006
Yb L_α	7.411	~0.15
Yb L_η	7.580	~0.0022
Yb L_β	8.512	~0.15
Yb L_γ	9.842	~0.024
Yb $K_{\alpha 2}$	51.354	~0.27
Yb $K_{\alpha 1}$	52.389	~0.47
Yb $K_{\beta 1}'$	59.316	~0.15
Yb $K_{\beta 2}'$	61.229	~0.041
γ M1+32%E2	66.7408 *17*	~0.14

Atomic Electrons (^{171}Tm)

⟨e⟩~0.69 keV

e_{bin}(keV)	⟨e⟩(keV)	e(%)
5	0.05	~1.0
9	0.05	~0.6
10	0.042	~0.42
40	0.0019	~0.0046
41	0.0029	~0.007
42	0.006	~0.015
43	0.0020	~0.0045
48	0.0009	~0.0018
49	0.0028	~0.006
50	0.0027	~0.005
51	0.0009	~0.0018
52	0.0006	~0.0012
56	0.08	~0.14
57	0.15	~0.26
58	0.16	~0.27
59	0.00023	~0.00039
64	0.020	~0.031
65	0.09	~0.13
66	0.025	~0.038
67	0.0037	~0.006

Continuous Radiation (^{171}Tm)

$\langle\beta-\rangle$=24.8 keV;$\langle$IB$\rangle$=0.0024 keV

E_{bin}(keV)		$\langle\ \rangle$(keV)	(%)
0 - 10	β-	1.32	27.4
	IB	0.00111	
10 - 20	β-	3.20	21.6
	IB	0.00062	0.0045
20 - 40	β-	8.6	29.6
	IB	0.00051	0.0019
40 - 96	β-	11.6	21.4
	IB	0.000122	0.00026

$^{171}_{70}$Yb(stable)

Δ: -59315 3 keV

%: 14.3 2

$^{171}_{71}$Lu(8.24 3 d)

Mode: ϵ

Δ: -57836 4 keV

SpA: 9.27×10^4 Ci/g

Prod: ^{169}Tm(α,2n); ^{171}Yb(p,n)

Photons (^{171}Lu)

$\langle\gamma\rangle$=656 8 keV

γ_{mode}	γ(keV)	γ(%)†
Yb L_ℓ	6.545	0.98 10
Yb L_α	7.411	24.3 16
Yb L_η	7.580	0.348 23
Yb L_β	8.512	24.9 20
γ M1+0.04%E2	9.1583 10	0.159 19
Yb L_γ	9.874	4.2 4
γ E1	19.3972 11	13.8 8
γ E1	27.1420 11	0.779 16
γ	28.5 10	
γ M1+1.6%E2	46.5392 14	0.167 7
Yb $K_{\alpha2}$	51.354	36.1 17
Yb $K_{\alpha1}$	52.389	64 3
γ M1+0.3%E2	55.6974 15	1.212 24
Yb $K_{\beta1}'$	59.316	20.2 9
Yb $K_{\beta2}'$	61.229	5.5 3
γ M1+32%E2	66.7408 17	2.48 5
γ M1+7.3%E2	72.3875 23	2.00 4
γ E2	75.8990 17	6.08 12
γ M1+5.6%E2	85.6114 22	1.077 22
γ M1+7.4%E2	91.417 4	0.443 11
γ	93.61 10	0.394 24
γ	95.3 10	
γ	99.29 10	0.240 14
γ	103.91 20	0.013 3
γ M1+4.8%E2	109.296 3	0.601 14
γ [E2]	122.37 3	0.0115 10
γ (M1)	132.151 3	0.0390 24
γ [E2]	141.309 3	0.048 14
γ M1+30%E2	154.699 7	0.0467 14
γ E2	163.805 4	~0.12
γ E2	163.857 7	~0.13
γ E2	170.744 15	0.0688 24
γ E2	194.907 4	0.144 14
γ (E2)	241.446 4	0.0279 20
γ (E2)	256.70 5	0.0221 14
γ M1+10%E2	498.799 21	0.101 4
γ M1+22%E2	517.779 7	0.341 7
γ	566.45 24	~0.014
γ M1	604.526 10	0.011 3
γ M1+49%E2	627.074 7	0.837 17
γ E2+42%M1	631.062 7	0.134 3
γ E1	667.440 7	11.06 24

Photons (^{171}Lu)
(continued)

γ_{mode}	γ(keV)	γ(%)†
γ M1	676.207 13	0.019 4
γ E1	689.306 7	2.37 5
γ	697.6 3	0.091 10
γ [M1+E2]	707.304 19	0.010 3
γ E2+30%M1	712.685 7	1.135 23
γ (M1)	720.16 10	0.216 19
γ (M1)	725.00 3	0.074 4
γ	727.1 5	0.053 14
γ E1	739.827 7	48.1 10
γ	747.0 5	~0.005
γ E1	753.50 3	0.0077 19
γ E2+18%M1	759.224 7	0.0207 14
γ M1+11%E2	767.623 12	0.702 14
γ M1+~45%E2	778.006 22	0.048 14
γ E1	780.723 6	4.35 9
γ (M1)	794.051 19	0.0726 19
γ M1	816.41 3	0.0332 19
γ M2	821.900 24	0.0034 10
γ E2	825.968 6	0.156 8
γ (M1)	834.36 3	0.024 5
γ	835.0 5	0.087 19
γ (E2)	835.99 11	0.101 9
γ M1+22%E2	840.011 12	3.04 8
γ M1	850.403 24	0.058 5
γ E1	853.110 6	2.55 5
γ (E1)	862.75 5	0.0337 19
γ	867.4 4	<0.029
γ E2+25%M1	868.50 3	0.0303 19
γ M1	872.951 24	0.0082 10
γ (M1)	877.66 3	0.0178 24
γ (M1)	877.72 13	~0.008
γ (E2)	881.09 5	0.0192 14
γ	884.73 7	0.0091 10
γ M1	888.80 3	0.0154 19
γ [M1+E2]	893.76 13	0.0072 19
γ	897.31 6	0.0091 19
γ M1	902.275 23	0.147 3
γ	911.85 15	0.0207 19
γ (M1)	922.06 5	0.0173 10
γ M1+27%E2	925.66 5	0.0375 14
γ	929.41 5	0.0115 10
γ	934.34 8	0.0048 10
γ	937.52 20	0.0034 10
γ	944.40 3	0.0067 14
γ M1+24%E2	948.749 18	0.0875 19
γ (M1)	953.48 14	0.0096 24
γ [E2]	958.43 11	0.0067 14
γ	968.98 13	0.0038 14
γ [E1]	985.704 24	0.020 4
γ M1	998.05 5	0.0255 10
γ M1+21%E2	1005.101 24	0.0322 10
γ (E2)	1013.59 6	0.0125 19
γ	1020.4 6	0.024 5
γ (M1)	1093.73 5	0.0226 24
γ M1	1169.472 22	0.0048 10
γ	1202.63 9	0.0029 10
γ E1	1209.837 22	0.0664 14
γ M1	1255.082 22	0.0063 5
γ E1	1282.224 21	0.315 8
γ	1311.33 5	0.0106 5

† 2.5% uncert(syst)

Atomic Electrons (^{171}Lu)

$\langle$e$\rangle$=89.3 17 keV

e_{bin}(keV)	$\langle$e$\rangle$(keV)	e(%)
2 - 8	2.5	40 4
9	12.0	133 13
10	9.9	97 9
11 - 15	2.95	23.7 5
17	2.18	12.6 8
18 - 56	10.4	29.3 9
57	2.69	4.75 23
58	2.72	4.71 23
59 - 65	4.57	7.19 17
66	12.5	18.9 6
67	12.6	18.8 5

Atomic Electrons (^{171}Lu)
(continued)

e_{bin}(keV)	$\langle$e$\rangle$(keV)	e(%)
70 - 72	0.719	1.02 3
74	6.94	9.4 3
75 - 123	4.5	5.6 4
130 - 171	0.201	0.136 16
180 - 195	0.0402	0.0215 9
231 - 257	0.00537	0.00225 8
437 - 456	0.059	0.0131 7
488 - 518	0.0141	0.00280 22
543 - 570	0.079	0.0140 9
594 - 636	0.252	0.0413 11
646 - 695	0.887	0.131 3
696 - 745	0.295	0.0409 6
747 - 794	0.285	0.0366 13
801 - 851	0.0705	0.00848 24
852 - 901	0.0139	0.00158 7
902 - 951	0.00572	0.000612 23
952 - 998	0.0018	0.00019 6
1003 - 1032	0.00150	0.000146 16
1083 - 1108	0.00046	4.2 5 $\times10^{-5}$
1141 - 1169	0.00082	7.1 5 $\times10^{-5}$
1192 - 1221	0.00358	0.000294 9
1245 - 1282	0.00090	7.1 13 $\times10^{-5}$
1301 - 1311	5.2 $\times10^{-5}$	4.0 20 $\times10^{-6}$

Continuous Radiation (^{171}Lu)

$\langle\beta+\rangle$=0.0145 keV;$\langle$IB$\rangle$=0.39 keV

E_{bin}(keV)		$\langle\ \rangle$(keV)	(%)
0 - 10	β+	7.0×10^{-9}	8.3×10^{-8}
	IB	0.00098	
10 - 20	β+	3.84×10^{-7}	2.29×10^{-6}
	IB	0.00020	0.00143
20 - 40	β+	1.37×10^{-5}	4.15×10^{-5}
	IB	0.0051	0.0147
40 - 100	β+	0.00074	0.00095
	IB	0.24	0.46
100 - 300	β+	0.0129	0.0068
	IB	0.029	0.0157
300 - 600	β+	0.00085	0.000270
	IB	0.049	0.0114
600 - 1300	IB	0.062	0.0077
1300 - 1384	IB	2.1×10^{-5}	1.63×10^{-6}
	$\Sigma\beta$+		0.0081

$^{171}_{71}$Lu(1.32 3 min)

Mode: IT

Δ: -57765 4 keV

SpA: 8.31×10^8 Ci/g

Prod: daughter ^{171}Hf; ^{171}Yb(p,n)

Photons (^{171}Lu)

$\langle\gamma\rangle$=1.77 8 keV

γ_{mode}	γ(keV)	γ(%)
Lu L_ℓ	6.753	0.29 3
Lu L_α	7.650	7.0 6
Lu L_η	7.857	0.150 11
Lu L_β	8.798	8.6 7
Lu L_γ	10.150	1.34 11
Lu $K_{\alpha2}$	52.965	0.086 5
Lu $K_{\alpha1}$	54.070	0.151 9
Lu $K_{\beta1}'$	61.219	0.049 3
Lu $K_{\beta2}'$	63.200	0.0130 8
γ E3	71.19 15	0.20 1

Atomic Electrons (^{171}Lu)

⟨e⟩=66.3 *19* keV

e_{bin}(keV)	⟨e⟩(keV)	e(%)
8	0.0246	0.313 *17*
9	2.5	27 *3*
10	2.4	23 *3*
11	0.029	0.26 *3*
42	0.00144	0.0034 *4*
43	0.00072	0.00168 *19*
44	0.00127	0.0029 *3*
45	0.00061	0.00136 *15*
50	0.00083	0.00165 *19*
51	0.00039	0.00076 *9*
52	0.00098	0.00188 *21*
53	9.3 $\times10^{-5}$	0.000177 *20*
54	0.000195	0.00036 *4*
58	3.1 $\times10^{-5}$	5.3 *6* $\times10^{-5}$
59	0.000158	0.00027 *3*
60	0.383	0.63 *3*
61	22.4	36.8 *20*
62	21.3	34.4 *19*
69	13.0	18.8 *10*
70	0.50	0.72 *4*
71	3.74	5.3 *3*

$^{171}_{72}$Hf(12.1 *4* h)

Mode: ε

Δ: -55440 *200* keV syst

SpA: 1.51×10^{6} Ci/g

Prod: ^{175}Lu(p,5n); ^{171}Yb(^{3}He,3n); alphas on Yb

Photons (^{171}Hf)

γ_{mode}	γ(keV)	γ(rel)
γ M1+0.08%E2	31.18 *13*	
γ M1+2.7%E2	31.57 *8*	
γ E3	71.19 *15*	0.35 *7* &
γ M1	74.83 *13*	1.3 *3*
γ E2	86.37 *13*	6.0 *15*
γ M1+3.8%E2	99.13 *7*	10.9 *15*
γ M1+7.6%E2	113.13 *9*	7.5 *19*
γ (E2)	117.31 *20*	8 *3*
γ M1+20%E2	122.02 *5*	~120
γ E2	124.63 *9*	8.6 *17*
γ M1	124.64 *13*	6.0 *12*
γ [E2]	125.7 *2*	0.75 *15*
γ M1	126.27 *9*	2.3 *5*
γ M1+11%E2	133.48 *14*	3.5 *7*
γ [M1+E2]	135.16 *16*	~0.9
γ E1	137.02 *10*	51 *10*
γ E2	144.31 *13*	7.5 *15*
γ M1+16%E2	147.11 *6*	16.2 *15*
γ M1+2.8%E2	149.1 *3*	1.5 *6* ?
γ M1+0.7%E2	151.54 *17*	1.8 *4*
γ M1+21%E2	171.12 *11*	1.5 *6*
γ [E2]	173.03 *15*	~2
γ [E2]	173.58 *7*	~0.45
γ E2	177.28 *17*	2.0 *6*
γ E2+41%M1	188.36 *9*	7.6 *9*
γ E2	193.05 *8*	~8
γ (E2)	194.3 *3*	~5
γ [M1+E2]	194.68 *14*	
γ E1	200.08 *6*	7.5 *8*
γ (E2)	212.32 *15*	~1
γ M1+3.3%E2	220.14 *13*	6.8 *8*
γ (E2)	223.75 *8*	~0.9
γ E2	225.4 *3*	~3
γ E2	240.61 *16*	1.9 *8*
γ E2	247.41 *10*	2.6 *8*
γ E2	269.12 *6*	15.2 *15*
γ	272.7 *10*	<0.6
γ [M1+E2]	283.48 *17*	1.4 *6*
γ [E1]	292.22 *16*	~5
γ M1+27%E2	295.60 *5*	52 *11*
γ E2+6.2%M1	306.51 *9*	8.1 *11*
γ E2	318.22 *9*	2.7 *6*
γ M1	340.37 *13*	1.5 *6*
γ E1	347.18 *6*	56 *8*
γ (M1)	372.11 *20*	1.9 *4*
γ M1	394.72 *7*	4.2 *11*
γ	397.51 *17*	<2
γ (M1)	440.3 *3*	1.0 *3*
γ	449.01 *16*	1.1 *4*
γ E2+34%M1	460.75 *17*	~2
γ [E1]	461.6 *3*	~4 ?
γ E1	469.20 *5*	38 *4*
γ [E1]	471.83 *13*	~0.9
γ (M1)	472.12 *16*	~2
γ [M1+E2]	519.35 *7*	~2
γ [E1]	519.40 *7*	4.1 *15*
γ	526.5 *3*	1.5 *6*
γ [E1]	530.69 *17*	0.8 *3*
γ	533.0 *5*	~0.8
γ E1	540.23 *7*	13.0 *15*
γ	547.0 *3*	1.0 *4*
γ	557.6 *5*	0.75 *15*
γ [E1]	568.42 *14*	2.0 *6*
γ [M1+E2]	576.16 *19*	~1
γ M1	591.60 *9*	5.6 *8*
γ [E1]	610.72 *20*	2.8 *6*
γ M1	624.13 *15*	3.5 *11*
γ E2+34%M1	650.98 *18*	1.8 *6*
γ E1	662.25 *7*	100 *11*
γ E1	666.50 *7*	19 *6*
γ M1	674.48 *7*	4.8 *8*
γ [E1]	693.06 *7*	4.3 *8*
γ [E1]	703.72 *18*	4.0 *8*
γ	706.1 *3*	2.1 *8*
γ [E2]	722.02 *12*	
γ	724.62 *7*	5.3 *8*
γ [E1]	735.36 *20*	1.7 *6*
γ [E1]	783.37 *18*	3.8 *11*
γ [E1]	788.52 *7*	10.0 *11*
γ M1	799.10 *8*	6.8 *8*
γ [M1+E2]	802.52 *10*	~1
γ [E1]	837.20 *17*	~1
γ	842.4 *3*	2.8 *11*
γ E1	852.67 *8*	29 *4*
γ	858.5 *5*	3.0 *8*
γ	861.5 *6*	1.5 *6*
γ [E1]	869.73 *16*	6.0 *12*
γ (M1)	881.63 *12*	3.6 *11*
γ	884.0 *7*	1.8 *6*
γ M1	893.14 *8*	4.5 *9*
γ (M1)	896.0 *3*	3.4 *11*
γ	924.70 *7*	2.7 *8*
γ	958.55 *16*	0.8 *3*
γ	966.4 *3*	1.5 *6*
γ	976.5 *5*	1.2 *6*
γ	999.9 *3*	2.6 *11*
γ	1002.81 *14*	2.6 *8*
γ (E2)	1017.10 *16*	1.9 *5*
γ	1026.5 *4*	~1
γ	1036.7 *10*	1.9 *5*
γ E0+M1	1040.24 *7*	4.1 *11*
γ M1	1061.5 *3*	4.3 *11*
γ E2+22%M1	1071.81 *6*	56 *6*
γ	1076.5 *5*	3.6 *15*
γ	1080.97 *18*	5.3 *23*
γ	1084.5 *10*	1.5 *6*
γ	1150.4 *3*	2.8 *8*
γ	1154.7 *5*	3.0 *8*
γ E2	1162.26 *7*	12.4 *15*
γ	1168.3 *3*	0.9 *4*
γ	1177.04 *19*	1.1 *4*
γ (M1)	1193.82 *6*	4.9 *6*
γ [M1+E2]	1199.85 *9*	4.5 *6*
γ	1205.60 *18*	~4
γ	1219.3 *3*	3.0 *8*
γ	1226.08 *16*	1.5 *6*
γ	1229.79 *12*	4.9 *11*
γ	1236.42 *16*	2.8 *8*
γ	1241.43 *19*	2.7 *8*
γ	1248.9 *6*	1.5 *6*
γ	1256 *1*	0.8 *3*
γ	1266.42 *19*	1.1 *4*
γ	1276.0 *5*	2.6 *8*
γ	1278.4 *5*	2.3 *8*
γ	1286.90 *17*	2.6 *8*
γ	1293.0 *3*	5.6 *11*
γ	1301.68 *16*	8.1 *11*
γ	1304.72 *18*	~2
γ	1309.32 *13*	5.6 *19*
γ	1314.0 *5*	0.9 *5*
γ [E2]	1321.87 *9*	2.8 *6*
γ	1325.21 *16*	1.1 *4*
γ	1333.2 *3*	1.7 *4*
γ	1340.61 *20*	9.8 *19*
γ	1356.06 *13*	~1
γ	1361.5 *3*	3.8 *11*
γ	1372.22 *15*	5.9 *8*
γ	1383.53 *16*	2.0 *4*
γ	1388 *1*	~0.6
γ	1398.33 *23*	1.7 *8*
γ	1401.18 *18*	~2
γ	1410 *1*	~0.8 ?
γ	1436.10 *17*	2.6 *8*
γ	1460.2 *4*	2.3 *8*
γ	1470.5 *4*	1.7 *6*
γ	1478.5 *4*	1.4 *5*
γ	1484.4 *4*	1.1 *4*
γ	1489 *1*	0.75 *23*
γ	1498.78 *15*	1.5 *6*
γ	1505.54 *15*	7.6 *9*
γ	1528.7 *5*	1.1 *4*
γ [M1+E2]	1534.19 *13*	0.8 *3*
γ	1538.5 *5*	1.5 *6*
γ	1549.11 *12*	2.6 *9*
γ	1558.12 *17*	9.0 *11*
γ	1572.29 *16*	~0.8
γ	1582.9 *3*	2.0 *4*
γ	1620.80 *15*	1.5 *4*
γ	1640 *1*	0.9 *3*
γ	1657.0 *4*	2.1 *4*
γ	1706.4 *5*	1.1 *4*
γ	1712.6 *3*	3.2 *6*
γ	1719.40 *16*	~0.45
γ	1749.19 *12*	10.9 *11*
γ	1753.50 *21*	3.9 *11*
γ	1762.2 *3*	0.8 *3*
γ	1785.2 *5*	~0.8
γ	1804.4 *4*	2.1 *4*
γ	1813.7 *5*	~0.8
γ	1836.0 *3*	7.4 *11*
γ	1841.42 *16*	3.6 *11*
γ	1859.5 *3*	4.9 *8*
γ	1863.9 *10*	1.2 *4*
γ	1867.5 *15*	0.8 *4*
γ	1896.29 *12*	5.5 *8*
γ	1911.5 *6*	2.3 *6*
γ	1932.5 *25*	0.34 *15*
γ	1962 *1*	0.6 *3*
γ	1967 *1*	~0.8
γ	1971 *1*	0.6 *3*
γ	1981 *1*	0.56 *19*
γ	1986.5 *10*	0.64 *23*
γ	2011.5 *20*	0.8 *3*
γ	2018.31 *12*	1.0 *3*
γ	2022.62 *21*	12.0 *13*
γ	2134.7 *15*	0.41 *15*
γ	2141.5 *15*	0.45 *15*
γ	2177.8 *10*	0.8 *3*

& with ^{171}Lu(1.32 min) in equilib

$^{171}_{73}$Ta(23.3 *3* min)

Mode: ϵ
Δ: -51540 *200* keV syst
SpA: 4.72×10^{7} Ci/g
Prod: ^{165}Ho(^{12}C,6n); ^{159}Tb(^{16}O,4n)

Photons (^{171}Ta)

γ_{mode}	γ(keV)	γ(rel)
γ [E3]	22.2 *3*	
γ (E1)	49.55 *17*	100 *10*
γ (M1+∼3.8%E2)	61.82 *17*	9.1 *9*
γ (M1+E2)	66.60 *21*	4.5 *5*
γ [E2]	80.5 *2*	4.2 *4*
γ M1+4.8%E2	84.04 *19*	1.6 *2*
γ [M1]	92.03 *20*	10.9 *11*
γ M1+2.5%E2	116.96 *20*	5.5 *6*
γ M1+1.1%E2	140.34 *23*	0.80 *8*
γ [M1+E2]	152.23 *17*	5.8 *6*
γ E2	166.13 *20*	19.2 *19*
γ E2	175.18 *20*	16.0 *16*
γ [E2]	208.99 *20*	2.9 *3*
γ	240.6 *3*	0.60 *6*
γ	247.5 *3* ?	
γ E2	253.61 *24*	1.30 *13*
γ E2	257.30 *23*	1.10 *11*
γ E2	259.05 *24*	0.50 *5*
γ	266.9 *3*	0.50 *5*
γ	282.0 *3*	0.70 *7*
γ	352.21 *23*	3.1 *3*
γ [M1+E2]	369.84 *20*	1.30 *13*
γ	376.9 *3*	
γ	392.7 *3*	0.80 *8*
γ	406.45 *24*	4.6 *5*
γ	408.62 *21* ?	
γ	423.2 *3*	3.5 *4*
γ [M1+E2]	431.70 *21*	1.20 *12*
γ	444.19 *20*	15.6 *16*
γ	446.25 *23*	3.4 *3*
γ [M1+E2]	454.65 *20*	4.5 *5*
γ [M1+E2]	457.29 *21*	1.90 *19*
γ	461.4 *3*	2.0 *2*
γ [E2]	464.1 *3*	
γ	467.2 *3*	
γ [M1+E2]	471.19 *21*	9.0 *9*
γ	492.45 *24*	<15
γ [M1+E2]	492.51 *24*	∼2
γ	492.66 *20*	<15
γ	501.37 *20*	22.6 *23*
γ	506.01 *18*	54 *6*
γ [M1+E2]	522.06 *20*	11.0 *11*
γ	526.0 *3*	1.70 *17*
γ	530.20 *23*	3.9 *4*
γ [M1+E2]	535.97 *21*	7.7 *8*
γ [M1+E2]	537.79 *21*	14.9 *15*
γ [M1+E2]	554.3 *3*	6.9 *7*

Photons (^{171}Ta)
(continued)

γ_{mode}	γ(keV)	γ(rel)
γ	570.74 *24*	3.2 *3*
γ	573.2 *3*	0.40 *4*
γ [M1+E2]	606.87 *20*	3.9 *4*
γ [E2]	620.78 *24*	3.6 *4*
γ	630.3 *3*	2.50 *25*
γ	664.79 *25*	2.50 *25*
γ	678.2 *3*	2.9 *3*
γ	702.6 *3*	1.50 *15*
γ	717.97 *22*	2.30 *23*
γ [M1+E2]	723.18 *21*	2.50 *25*
γ	726.92 *21*	4.1 *4*
γ	731.4 *3*	1.50 *15*
γ	736.7 *3*	1.30 *13*
γ [M1+E2]	746.13 *21*	4.3 *4*
γ	767.51 *21*	9.0 *9*
γ	782.1 *3*	2.10 *21*
γ	788.74 *20*	4.5 *5*
γ	796.5 *3*	3.3 *3*
γ	802.1 *3*	3.3 *3*
γ	836.6 *3*	1.80 *18*
γ	861.52 *21*	1.70 *17* ?
γ	876.6 *3*	1.70 *17*
γ	898.90 *24*	1.60 *16*
γ	906.5 *3*	2.0 *2*
γ	919.9 *3*	1.20 *12*
γ	957.83 *20*	1.80 *18* ?
γ	986.9 *3*	8.6 *9*
γ	996.8 *3*	3.1 *3*
γ [E1]	1001.07 *25*	2.7 *3*
γ	1007.38 *18*	3.4 *3*
γ	1027.20 *24*	1.80 *18*
γ	1086.8 *10*	
γ	1434.8 *3*	
γ	1526.8 *3*	

$^{171}_{74}$W (9.0 *15* min)

Mode: ϵ
Δ: -47240 *280* keV syst
SpA: 1.22×10^{8} Ci/g
Prod: ^{155}Gd(^{20}Ne,4n); ^{156}Gd(^{20}Ne,5n); ^{155}Gd(^{22}Ne,6n)

$^{171}_{76}$Os(7.9 *6* s)

Mode: ϵ(98.3 *3* %), α(1.7 *3* %)
Δ: -34570 *380* keV syst
SpA: 8.0×10^{9} Ci/g
Prod: ^{156}Dy(^{20}Ne,5n); ^{93}Nb(^{84}Kr,p5n)

Alpha Particles (^{171}Os)

α(keV)
5240 *10*

$^{171}_{77}$Ir(1.6 *1* s)

Mode: α
Δ: -26360 *220* keV syst
SpA: 3.35×10^{10} Ci/g
Prod: ^{162}Er(^{19}F,10n); ^{93}Nb(^{84}Kr,6n); ^{110}Cd(^{63}Cu,2n)

Alpha Particles (^{171}Ir)

α(keV)
5910 *10*

$^{171}_{78}$Pt(25 *9* ms)

Mode: α
Δ: -17710 *320* keV syst
SpA: 1×10^{11} Ci/g
Prod: ^{112}Sn(^{63}Cu,p3n); ^{144}Sm(^{32}S,5n); ^{58}Ni on Sn

Alpha Particles (^{171}Pt)

α(keV)
6451 *3*

A = 172

NDS 15, 497 (1975)

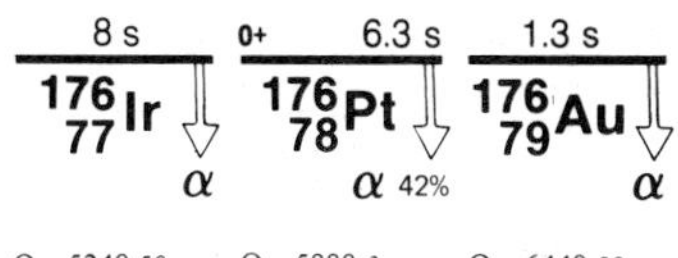

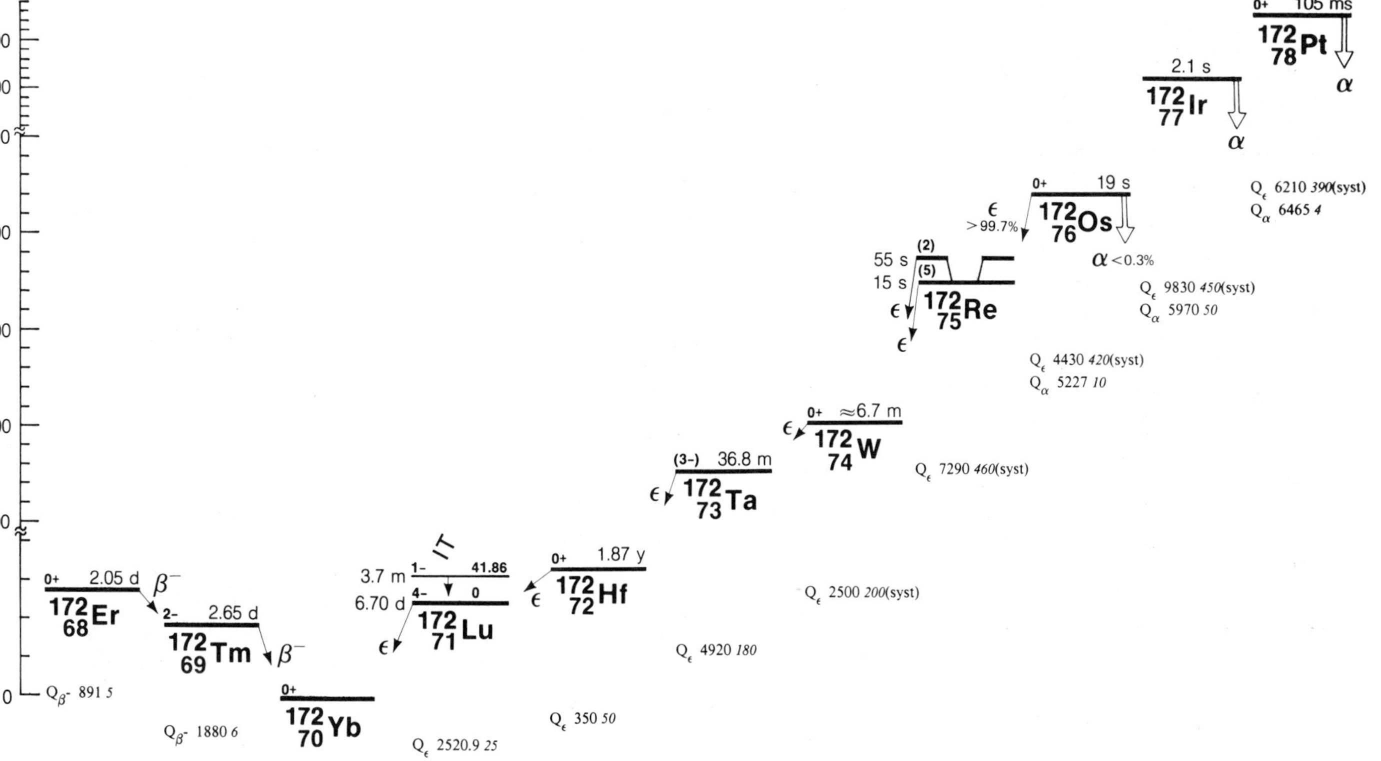

$^{172}_{68}$Er(2.054 *21* d)

Mode: β-

Δ: -56493 *5* keV

SpA: 3.70×10^{5} Ci/g

Prod: multiple n-capture from ^{170}Er

Photons (^{172}Er)

⟨γ⟩=513 *7* keV

γ_{mode}	γ(keV)	γ(%)†
Tm L_ℓ	6.341	0.19 *3*
Tm L_α	7.176	4.8 *5*
Tm L_η	7.310	0.066 *8*
Tm L_β	8.220	4.8 *6*
Tm L_γ	9.538	0.81 *11*
γ E2	29.418 *5*	0.0077 *22*
γ M1+E2	38.691 *5*	0.0073 *22*
Tm $K_{\alpha2}$	49.772	10.9 *5*
Tm $K_{\alpha1}$	50.742	19.3 *8*
Tm $K_{\beta1}'$	57.444	6.0 *3*
Tm $K_{\beta2}'$	59.296	1.68 *8*
γ M1	59.697 *3*	2.75 *9*
γ M1+E2	62.526 *4*	0.211 *4*
γ M1	68.1085 *23*	3.36 *7*
γ E1	74.920 *3*	0.120 *4*
γ	80.19 *6*	~0.0043
γ [M1]	83.30 *5*	0.0030 *9*

Photons (^{172}Er) (continued)

γ_{mode}	γ(keV)	γ(%)†
γ [M1]	89.115 *5*	0.0047 *4*
γ	113.72 *12*	~0.0013
γ	118.90 *23*	~0.0009
γ M1+E2	127.805 *3*	2.17 *5*
γ [E1]	134.6170 *25*	0.0103 *4*
γ [E2]	145.83 *5*	~0.0013
γ E1	164.035 *5*	0.69 *7*
γ [E2]	167.41 *5*	0.0073 *9*
γ [M1]	177.41 *5*	<40×10^{-5}
γ	179.36 *9*	<0.005
γ	187.4 *5*	~0.0009 ?
γ [E1]	202.726 *3*	1.05 *3*
γ [M1]	206.10 *5*	0.0030 *4*
γ	239.75 *9*	0.0043 *9*
γ [M1]	239.93 *5*	0.0116 *9*
γ	295.21 *5*	~0.0026 ?
γ [E2]	300.20 *5*	0.0211 *9*
γ	307.16 *9*	0.0103 *21*
γ [E2]	344.813 *4*	0.641 *17*
γ	370.30 *23*	<0.004
γ (M1)	383.503 *4*	2.40 *5*
γ M1	407.3385 *21*	43.0 *9*
γ	417.16 *9*	0.009 *3*
γ (M1)	446.029 *5*	3.02 *7*
γ [E2]	472.618 *5*	0.0318 *17*
γ [E2]	475.4469 *18*	1.062 *21*
γ	479.69 *9*	0.0116 *13*
γ	496.34 *12*	~0.0026
γ	526.2 *5*	~0.0030
γ [M1+E2]	535.144 *3*	0.299 *7*
γ E1	610.0635 *19*	45.2 *10*
γ	714.50 *9*	~0.0017

Photons (^{172}Er) (continued)

γ_{mode}	γ(keV)	γ(%)†
γ	734.90 *12*	<13×10^{-5}
γ	797.42 *10*	0.0108 *4*
γ	831.0 *12*	~0.00022 ?
γ	894.2 *10*	~0.0013 ?

† 2.3% uncert(syst)

Atomic Electrons (^{172}Er)

⟨e⟩=31.3 *11* keV

e_{bin}(keV)	⟨e⟩(keV)	e(%)
3	0.04	~1
9	4.0	47 *3*
10	1.41	14.4 *19*
16 - 49	2.5	9.0 *11*
50	2.49	5.02 *19*
51 - 57	1.6	2.8 *11*
58	2.46	4.24 *12*
59 - 67	1.16	1.8 *3*
68	1.6	2.3 *8*
70 - 119	1.0	~0.9
120 - 169	0.39	0.30 *13*
170 - 206	0.0240	0.0126 *6*
230 - 248	0.0064	0.0027 *6*
285 - 335	0.593	0.186 *7*
336 - 345	0.0103	0.00303 *9*
348	8.39	2.41 *7*

Atomic Electrons (^{172}Er)
(continued)

e_{bin}(keV)	⟨e⟩(keV)	e(%)
358 - 387	0.630	0.164 6
397	1.30	0.328 9
398 - 446	0.697	0.170 3
463 - 496	0.050	0.0106 25
516 - 535	0.0065	0.0012 4
551	0.806	0.146 4
600 - 610	0.158	0.0262 6
655 - 704	0.00011	$\sim 2\times10^{-5}$
705 - 738	0.0005	$\sim 7\times10^{-5}$
772 - 821	0.00011	1.4 6 $\times10^{-5}$
822 - 835	5.0 $\times10^{-5}$	$\sim 6\times10^{-6}$
884 - 894	10.0 $\times10^{-6}$	$\sim 1\times10^{-6}$

Continuous Radiation (^{172}Er)
⟨β-⟩=98 keV;⟨IB⟩=0.035 keV

E_{bin}(keV)		⟨ ⟩(keV)	(%)
0 - 10	β-	0.353	7.1
	IB	0.0051	
10 - 20	β-	1.02	6.8
	IB	0.0044	0.030
20 - 40	β-	3.85	12.9
	IB	0.0070	0.025
40 - 100	β-	22.0	32.2
	IB	0.0119	0.019
100 - 300	β-	66	40.3
	IB	0.0068	0.0048
300 - 484	β-	4.30	1.30
	IB	5.5 $\times10^{-5}$	1.7 $\times10^{-5}$

$^{172}_{69}$Tm(2.650 8 d)

Mode: β-
Δ: -57383 6 keV
SpA: 2.865$\times10^{5}$ Ci/g
Prod: daughter ^{172}Er; ^{173}Yb(γ,p)

Photons (^{172}Tm)
⟨γ⟩=470 7 keV

γ_{mode}	γ(keV)	γ(%)†
Yb L_ℓ	6.545	0.171 21
Yb L_α	7.411	4.3 4
Yb L_η	7.580	0.077 7
Yb L_β	8.501	4.6 4
Yb L_γ	9.798	0.70 7
Yb $K_{\alpha2}$	51.354	2.94 21
Yb $K_{\alpha1}$	52.389	5.2 4
Yb $K_{\beta1}$'	59.316	1.65 12
Yb $K_{\beta2}$'	61.229	0.45 3
γ E2	78.792 7	6.5 5
γ E2+25%M1	90.665 17	0.021 3
γ[M1]	131.85 3	0.0101 12
γ[M1]	142.596 18	0.100 5
γ[M1]	151.46 5	0.0014 5
γ E2	181.562 8	2.72 14
γ	186.2 3	0.0021 7
γ[M1]	196.91 3	0.0119 12
γ	267.2 3	0.0021 7
γ E2	279.74 3	0.0047 12
γ	286.3 3	0.0062 12
γ[E2]	293.55 3	0.0113 12
γ[E1]	321.77 4	0.0036 6
γ[M1+E2]	348.10 3	0.0166 12
γ[M1+E2]	358.84 4	0.0095 12
γ[M1+E2]	399.792 19	0.116 5

Photons (^{172}Tm)
(continued)

γ_{mode}	γ(keV)	γ(%)†
γ[E2]	423.05 3	0.0154 12
γ[M1+E2]	431.36 4	0.0059 6
γ[M1+E2]	436.15 3	0.244 9
γ(M1)	437.61 3	0.0072 7
γ(M1)	490.457 13	0.420 18
γ(M1)	528.279 21	0.127 6
γ	544.9 3	0.0055 12
γ[E2]	565.64 3	0.0409 24
γ[E2]	857.57 3	0.136 8
γ M1+E2	912.111 13	1.39 4
γ[E2]	964.18 4	0.335 17
γ M1+E2	1002.776 19	0.022 3
γ	1026.2 4	0.0056 24
γ(M1)	1039.13 3	0.136 7
γ[E1]	1076.20 3	0.78 3
γ M1+E2	1093.672 13	6.00 19
γ[E2]	1117.92 3	0.051 7
γ	1119.78 9	0.242 14
γ[E1]	1154.99 3	0.160 12
γ[E2]	1205.662 24	0.154 7
γ[E2]	1216.41 3	0.034 5
γ(M1)	1288.93 4	0.492 23
γ[E2]	1348.26 3	0.173 9
γ(E2)	1387.223 23	5.47 17
γ[M1+E2]	1397.97 3	0.80 4
γ(E2)	1402.566 17	0.179 10
γ(E2)	1440.387 23	0.0160 18
γ(E2)	1466.014 24	4.47 18
γ(E2)	1470.49 4	1.86 7
γ[E2]	1476.76 3	0.301 15
γ[M1+E2]	1529.818 25	5.1 3
γ(E2)	1584.126 17	0.568 24
γ[E2]	1608.61 3	4.05 19
γ[M1+E2]	1621.948 23	0.070 3

† 17% uncert(syst)

Atomic Electrons (^{172}Tm)
⟨e⟩=39.3 14 keV

e_{bin}(keV)	⟨e⟩(keV)	e(%)
9	1.55	17.3 22
10	1.31	13.1 16
17	1.72	9.8 8
29 - 68	0.95	1.59 9
69	11.2	16.3 12
70	11.5	16.5 13
71 - 76	0.194	0.255 18
77	6.2	8.1 6
78	1.52	1.94 15
79 - 125	1.06	1.03 4
129 - 177	0.604	0.357 11
179 - 225	0.188	0.104 4
232 - 280	0.0024	0.00100 12
283 - 322	0.0056	0.0019 6
338 - 376	0.054	0.015 5
389 - 438	0.081	0.0191 12
467 - 504	0.0354	0.0075 3
518 - 566	0.00459	0.00087 4
847 - 858	0.070	0.008 3
902 - 941	0.025	0.0028 5
954 - 1003	0.0102	0.00105 6
1015 - 1058	0.26	0.025 8
1066 - 1111	0.052	0.0048 11
1116 - 1155	0.0053	0.00047 3
1195 - 1228	0.0215	0.00175 10
1278 - 1326	0.127	0.0096 5
1337 - 1386	0.051	0.0038 5
1387 - 1431	0.145	0.0103 4
1438 - 1528	0.20	0.0134 24
1547 - 1620	0.097	0.0062 3

Continuous Radiation (^{172}Tm)
⟨β-⟩=493 keV;⟨IB⟩=0.81 keV

E_{bin}(keV)		⟨ ⟩(keV)	(%)
0 - 10	β-	0.122	2.44
	IB	0.021	
10 - 20	β-	0.355	2.37
	IB	0.020	0.139
20 - 40	β-	1.36	4.55
	IB	0.038	0.133
40 - 100	β-	8.3	12.0
	IB	0.103	0.159
100 - 300	β-	43.7	23.6
	IB	0.25	0.143
300 - 600	β-	86	19.5
	IB	0.22	0.052
600 - 1300	β-	274	30.2
	IB	0.158	0.020
1300 - 1880	β-	79	5.4
	IB	0.0063	0.00045

$^{172}_{70}$Yb(stable)

Δ: -59264 3 keV
%: 21.9 3

$^{172}_{71}$Lu(6.70 3 d)

Mode: ϵ
Δ: -56743 4 keV
SpA: 1.133$\times10^{5}$ Ci/g
Prod: ^{172}Yb(p,n); ^{169}Tm(α,n)

Photons (^{172}Lu)
⟨γ⟩=1935 17 keV

γ_{mode}	γ(keV)	γ(%)
Yb L_ℓ	6.545	0.75 8
Yb L_α	7.411	18.5 15
Yb L_η	7.580	0.284 22
Yb L_β	8.510	18.9 16
Yb L_γ	9.845	3.0 3
Yb $K_{\alpha2}$	51.354	32.1 22
Yb $K_{\alpha1}$	52.389	57 4
Yb $K_{\beta1}$'	59.316	18.0 12
Yb $K_{\beta2}$'	61.229	4.9 4
γ E2	78.792 7	11.0 4
γ E2+25%M1	90.665 17	5.1 3
γ M1+E2	112.757 20	1.47 10
γ	119.0 3	0.020 4
γ[M1]	131.85 3	0.00025 4
γ[M1+E2]	134.42 4	0.034 4
γ	137.98 23	0.036 8
γ[M1]	142.596 18	0.00252 24
γ	142.67 7	0.096 4
γ(M1)	145.66 17	0.072 20
γ[M1]	151.46 5	0.044 16
γ	163.2 5	0.08 3
γ	174.77 21	0.12 1
γ E2	181.562 8	20 1
γ[E1]	196.24 8	0.094 8
γ[M1]	196.91 3	0.053 6
γ E2	203.423 17	4.85 20
γ[M1]	210.31 8	0.060 12
γ	233.5 2	0.33 9
γ	241.07 14	0.066 10
γ[E2]	247.18 4	0.45 6
γ	254.99 16	0.076 12
γ(E1)	264.806 21	0.32 3
γ(M1)	269.93 4	1.83 10
γ E2	279.74 3	1.11 10

Photons (^{172}Lu)
(continued)

γmode	γ(keV)	γ(%)
γ [E2]	293.55 3	0.00165 18
γ (M1)	319.08 15	0.15 2
γ (M1)	323.92 4	1.39 6
γ E1+M2	330.66 8	0.59 7
γ	337.7 3	0.06 1
γ	348.0 4	0.03 1
γ [M1+E2]	348.10 3	0.00243 18
γ	352.3 4	0.64 9
γ (M1)	358.50 9	0.11 1
γ E1	366.70 8	0.31 4
γ (M1+E2)	372.510 25	2.56 17
γ E1	377.564 18	3.18 10
γ [E2]	382.64 17	0.05 2
γ [M1+E2]	399.792 19	0.516 24
γ [M1+E2]	410.332 19	1.96 7
γ M1+E2	416.64 16	0.076 10
γ [M1+E2]	422.59 9	0.16 2
γ [E2]	423.05 3	0.00226 18
γ (M1+E2)	427.44 4	0.14 2
γ [M1+E2]	431.36 4	0.00190 23
γ (E1)	432.561 24	1.54 9
γ [M1+E2]	436.15 3	0.0061 6
γ (M1)	437.61 3	0.22 2
γ	443.42 8	0.15 2
γ (M1)	482.21 5	0.70 5
γ (M1)	486.20 4	0.72 8
γ (M1)	490.457 13	1.87 9
γ (M1)	512.78 17	0.15 2
γ (M1+E2)	524.36 18	0.23 5
γ (M1)	528.279 21	3.89 11
γ (M1)	536.20 6	0.69 4
γ M1	540.20 4	1.30 9
γ [M1+E2]	551.15 9	0.42 3
γ	562.1 3	0.11 3
γ [E2]	565.64 3	0.00103 11
γ [M1+E2]	566.55 7	~0.17
γ (M1)	576.87 4	0.35 3
γ [M1+E2]	584.79 5	0.37 5
γ (M1+E2)	594.64 5	0.58 7
γ [E2]	607.24 3	0.65 5
γ [M1+E2]	625.88 8	0.29 4
γ [M1+E2]	630.86 4	0.30 6
γ	644.85 5	0.24 6
γ (M1)	681.91 6	0.75 6
γ M1	697.367 22	5.8 3
γ (M1+E2)	709.15 7	0.72 4
γ	723.07 12	0.48 4
γ M1	810.123 20	16.0 4
γ (M1)	816.33 5	1.10 7
γ	836.6 3	0.17 5
γ	862.4 5	0.32 15
γ M1	900.788 19	28.8 7
γ M1+E2	912.111 13	14.8 3
γ M1+E2	929.09 5	3.15 11
γ	953.05 20	0.41 7
γ	960.4 3	0.11 3
γ	963.7 4	0.17 3
γ (M1)	968.09 4	0.19 4
γ [M1+E2]	970.21 5	0.10 3
γ	979.7 3	0.14 2
γ	990.73 23	0.11 4
γ M1+E2	1002.776 19	5.26 12
γ [E2]	1019.75 5	~0.17
γ M1	1022.41 5	1.48 6
γ	1038.6 8	0.21 10
γ (M1)	1041.06 17	0.38 5
γ	1054.53 13	0.086 12
γ	1061.87 22	0.10 2
γ (M1)	1080.84 4	1.14 3
γ M1+E2	1093.672 13	63.5 12
γ M1+E2	1113.07 5	1.88 9
γ [M1+E2]	1115.532 20	0.20 4
γ	1142.0 7	0.10 4
γ	1145.42 19	0.49 7
γ	1153.01 25	0.19 3
γ E1	1166.46 8	0.14 3
γ [E2]	1205.662 24	0.0225 10
γ (M1)	1288.93 4	0.158 11
γ (E2)	1322.85 8	0.15 3
γ [E2]	1348.26 3	0.0043 4
γ (E2)	1387.223 23	0.800 16
γ [M1+E2]	1397.32 17	0.34 5
γ (E2)	1402.566 17	0.80 5

Photons (^{172}Lu)
(continued)

γmode	γ(keV)	γ(%)
γ (E2)	1440.387 23	0.49 6
γ (E2)	1466.014 24	0.65 3
γ (E2)	1470.49 4	0.60 3
γ (E2)	1488.98 4	1.10 4
γ [M1+E2]	1529.818 25	0.127 11
γ (E2)	1542.97 4	0.96 5
γ [E2]	1578.88 17	0.20 2
γ (E2)	1584.126 17	2.52 11
γ	1592.92 12	0.08 2
γ M1+E2	1602.59 8	0.29 3
γ [E2]	1608.61 3	0.102 8
γ [M1+E2]	1621.948 23	2.14 7
γ [M1+E2]	1666.68 9	0.17 7
γ M1+E2	1670.54 4	0.55 7
γ M1+E2	1724.53 4	0.43 2
γ (E2)	1812.894 21	0.18 1
γ (M1)	1914.80 7	0.58 3
γ	1920.34 10	0.10 4
γ [M1+E2]	1931.86 5	0.04 1
γ [E2]	1994.454 21	0.16 2
γ [M1+E2]	2025.18 5	0.05 2
γ (M1)	2083.61 4	0.30 6
γ M1+E2	2096.36 7	0.11 3
γ	2205.1 4	0.03 1

Atomic Electrons (^{172}Lu)
⟨e⟩=117.6 21 keV

e_{bin}(keV)	⟨e⟩(keV)	e(%)
9	6.5	73 8
10	5.4	53 6
17	2.92	16.8 7
29 - 68	7.7	19.3 12
69	19.1	27.7 12
70	19.6	28.0 12
71 - 76	0.336	0.442 22
77	10.6	13.8 6
78 - 80	3.88	4.91 16
81	4.1	5.1 3
82	3.98	4.9 3
84 - 119	4.6	4.9 5
120	5.2	4.34 23
121 - 166	1.30	0.92 4
171 - 218	7.54	4.19 12
223 - 271	1.1	0.42 11
276 - 324	0.87	0.28 7
327 - 376	0.69	0.20 4
377 - 426	0.36	0.088 9
427 - 476	1.06	0.233 8
477 - 526	0.54	0.108 6
527 - 576	0.21	0.038 5
577 - 626	0.122	0.020 3
628 - 673	0.66	0.104 6
680 - 723	0.136	0.0196 9
749 - 775	1.36	0.181 5
800 - 839	2.25	0.270 7
851 - 899	1.3	0.15 3
900 - 944	0.47	0.050 10
950 - 994	0.165	0.0170 15
1000 - 1046	2.6	0.25 8
1051 - 1094	0.62	0.057 12
1103 - 1151	0.023	0.0020 4
1153 - 1197	0.00042	3.6 6 $\times 10^{-5}$
1203 - 1228	0.0066	0.00054 4
1262 - 1289	0.0048	0.00038 6
1312 - 1348	0.046	0.0034 3
1377 - 1409	0.0453	0.00325 13
1428 - 1470	0.0332	0.00231 11
1478 - 1577	0.154	0.0100 9
1582 - 1669	0.044	0.0027 3
1714 - 1811	0.0056	0.000319 23
1853 - 1933	0.0224	0.00120 8
1964 - 2035	0.0103	0.00051 8
2073 - 2144	0.0020	9.4 15 $\times 10^{-5}$
2195 - 2203	7.5 $\times 10^{-5}$	~3 $\times 10^{-6}$

Continuous Radiation (^{172}Lu)
⟨IB⟩=0.34 keV

E_{bin}(keV)		⟨ ⟩(keV)	(%)
10 - 20	IB	0.00021	0.00150
20 - 40	IB	0.0051	0.0148
40 - 100	IB	0.24	0.45
100 - 300	IB	0.018	0.0106
300 - 600	IB	0.024	0.0055
600 - 1300	IB	0.039	0.0046
1300 - 2261	IB	0.0124	0.00080

$^{172}_{71}$Lu(3.7 5 min)

Mode: IT
Δ: -56701 4 keV
SpA: 2.9×10^{8} Ci/g
Prod: daughter ^{172}Hf

Photons (^{172}Lu)
⟨γ⟩=1.31 8 keV

γmode	γ(keV)	γ(%)
Lu L_ℓ	6.753	0.43 5
Lu L_α	7.650	10.5 8
Lu L_η	7.857	0.0198 16
Lu L_β	8.976	4.5 5
Lu L_γ	10.397	0.65 10
γ M3	41.86 4	0.00376 20

Atomic Electrons (^{172}Lu)
⟨e⟩=39.1 12 keV

e_{bin}(keV)	⟨e⟩(keV)	e(%)
9	3.9	42 5
10	0.110	1.06 12
11	0.74	6.8 8
31	5.2	16.8 10
32	0.53	1.69 10
33	17.6	54 3
39	2.01	5.1 3
40	6.5	16.4 9
41	0.61	1.47 8
42	1.90	4.6 3

$^{172}_{72}$Hf(1.87 3 yr)

Mode: ε
Δ: -56390 50 keV
SpA: 1123 Ci/g
Prod: ^{175}Lu(p,4n); alphas on Yb

Photons (^{172}Hf)

⟨γ⟩=105 *8* keV

γ_{mode}	γ(keV)	γ(%)†
Lu L_ℓ	6.753	1.9 *7*
Lu L_α	7.650	46 *16*
Lu L_η	7.857	0.55 *25*
Lu L_β	8.834	42 *17*
Lu L_γ	10.248	7 *3*
γ (M1+E2)	12.41 *20*	~0.011
γ	19.94 *10*	~0.045
γ E1	23.99 *5*	20.3 *11*
γ M1	41.13 *10*	0.268 *22*
γ M3	41.86 *4*	0.0037 *11* §
γ E2	44.17 *10*	0.32 *6*
γ M1+3.8%E2	48.17 *20*	~0.09
Lu $K_{\alpha2}$	52.965	33 *6*
Lu $K_{\alpha1}$	54.070	57 *11*
Lu $K_{\beta1}'$	61.219	18 *4*
Lu $K_{\beta2}'$	63.200	4.9 *10*
γ E1	67.35 *10*	5.3 *6*
γ (E1)	68.0 *1*	0.69 *7*
γ M1+1.9%E2	69.99 *10*	0.84 *9*
γ M1+E2	81.75 *5*	4.52 *23*
γ M1	114.06 *10*	2.6 *3*
γ M1+E2	122.92 *10*	1.14 *11*
γ M1	125.82 *5*	11.3 *6*
γ [E1]	127.91 *10*	1.46 *15*
γ M1	154.62 *10*	0.147 *23*

† 5.3% uncert(syst)
§ with ^{172}Lu(3.7 m) in equilib

Atomic Electrons (^{172}Hf)

⟨e⟩=147 *17* keV

e_{bin}(keV)	⟨e⟩(keV)	e(%)
2 - 7	0.72	14.2 *14*
9	15.9	173 *66*
10	12.8	~128
11	10.1	~94
12 - 17	7.23	51.9 *18*
18	7.9	~44
19 - 30	5.3	24 *8*
31	5.2	17 *5*
32	0.53	1.7 *5*
33	17.5	54 *16*
34	4.6	13.6 *25*
35	5.5	16 *3*
37 - 39	2.4	6.0 *17*
40	6.5	16 *5*
41	0.65	1.6 *5*
42	5.5	13 *3*
43 - 61	7.9	15.6 *12*
63	11.7	18.7 *10*
65 - 70	0.57	0.86 *5*
71	5.4	~8
73	3.8	~5
79 - 128	9.7	9.7 *22*
144 - 155	0.039	0.027 *3*

Continuous Radiation (^{172}Hf)

⟨IB⟩=0.22 keV

E_{bin}(keV)		⟨ ⟩(keV)	(%)
10 - 20	IB	0.00036	0.0028
20 - 40	IB	0.0046	0.0136
40 - 100	IB	0.22	0.40
100 - 241	IB	0.00079	0.00071

$^{172}_{73}$Ta(36.8 *3* min)

Mode: ε
Δ: -51470 *190* keV
SpA: 2.970×10^7 Ci/g
Prod: ^{175}Lu(^{3}He,6n); protons on Hf; protons on Ta

Photons (^{172}Ta)

⟨γ⟩=1429 *29* keV

γ_{mode}	γ(keV)	γ(%)†
Hf L_ℓ	6.960	0.55 *9*
Hf L_α	7.893	13.3 *19*
Hf L_η	8.139	0.21 *3*
Hf L_β	9.133	14.2 *21*
Hf L_γ	10.581	2.3 *4*
Hf $K_{\alpha2}$	54.611	23 *4*
Hf $K_{\alpha1}$	55.790	40 *6*
Hf $K_{\beta1}'$	63.166	13.1 *21*
Hf $K_{\beta2}'$	65.211	3.4 *5*
γ (E2)	95.18 *5*	16.8 *13*
γ [E1]	113.82 *13*	~0.21 ?
γ (E2)	213.96 *5*	52
γ (M1)	221.10 *7*	1.26 *5*
γ	224.9 *10*	~0.17
γ E1	237.62 *8*	1.89 *4*
γ	260.5 *10*	~0.13
γ	280.32 *14*	~0.16 ?
γ [E1]	288.94 *11*	1.60 *4*
γ E2	318.75 *17*	4.96 *10*
γ	334.92 *12*	0.66 *5*
γ	366.17 *11*	0.307 *21*
γ	379.72 *12*	0.84 *6*
γ [E2]	383.27 *12*	0.38 *4*
γ	403.00 *24*	0.33 *5*
γ [E2]	406.0 *4*	0.31 *3* ?
γ	409.1 *10*	~0.07
γ	420.42 *10*	<0.12 ?
γ	426.63 *25*	0.98 *19*
γ	429.3 *6*	~0.31
γ [M1+E2]	445.07 *9*	0.56 *8*
γ	458.72 *9*	0.53 *4*
γ [E2]	501.54 *20*	0.73 *11*
γ	503.52 *9*	1.26 *11*
γ	547.9 *3*	0.780 *21*
γ	564.30 *10*	0.59 *4*
γ	576.0 *7*	~0.08
γ	595.5 *6*	0.42 *5*
γ	598.0 *4*	0.23 *5*
γ	620.86 *24*	0.35 *5*
γ	636.18 *15*	~0.39 ?
γ [E2]	643.14 *8*	2.16 *8*
γ	653.39 *18*	~0.47 ?
γ	721.75 *18*	0.52 *6*
γ	735.2 *4*	0.21 *5*
γ	742.4 *5*	~0.5
γ	748.6 *10*	<0.13
γ (E2)	776.00 *11*	2.39 *5*
γ	790.8 *6*	~0.08
γ	804.7 *4*	0.50 *6*
γ (M1+E2)	820.29 *10*	3.02 *12*
γ	824.8 *6*	~0.30
γ	827.0 *6*	~0.12
γ	835.0 *3*	~0.47 ?
γ	838.58 *19*	0.40 *6*
γ	843.64 *18*	0.90 *7*
γ E0(+M1+E2)	857.11 *7*	4.15 *14*
γ	871.61 *8*	1.4 *5*
γ	946.5 *7*	0.27 *5*
γ [E2]	952.29 *8*	1.85 *8*
γ	980.00 *8*	3.68 *7*
γ	987.7 *3*	0.23 *5*
γ	995.42 *11*	2.06 *12*
γ	997.3 *5*	<0.44
γ [E2]	1034.25 *10*	1.92 *9*
γ	1042.6 *5*	~0.09
γ	1050.01 *13*	2.20 *7*
γ M1+E2	1075.18 *8*	3.46 *8*
γ (E2)	1085.58 *7*	7.6 *3*
γ	1109.23 *6*	14.0 *6*

Photons (^{172}Ta) (continued)

γ_{mode}	γ(keV)	γ(%)†
γ	1146.0 *3*	0.20 *5*
γ	1153.72 *20*	1.25 *9*
γ	1162.40 *6*	1.08 *12*
γ	1172.99 *7*	0.55 *6*
γ	1186.46 *5*	2.54 *9*
γ	1194.19 *5*	1.42 *9*
γ [E2]	1200.3 *3*	0.28 *4*
γ	1209.38 *12*	1.6 *6*
γ [E2]	1240.37 *10*	2.00 *9*
γ	1263.98 *14*	1.96 *12*
γ	1265.61 *23*	2.46 *15*
γ	1277.54 *5*	2.69 *6* ?
γ	1330.33 *5*	7.6 *3*
γ	1370.6 *10*	~0.06
γ	1375.14 *5*	1.97 *11*
γ	1386.96 *5*	2.55 *15*
γ [E2]	1397.36 *7*	~0.08 ?
γ	1408.15 *7*	~0.06
γ	1419.6 *3*	~0.5
γ	1479.57 *22*	2.25 *13*
γ	1481.72 *19*	0.33 *5*
γ	1523.0 *5*	0.32 *5*
γ	1544.29 *7*	6.2 *3*
γ	1637.7 *4*	0.58 *10*
γ	1695.68 *18*	0.82 *11*
γ	1714.6 *8*	~0.19
γ	1877.8 *5*	0.24 *6*
γ	2008.6 *6*	0.30 *8*
γ	2026.5 *24*	0.17 *7*
γ	2031.5 *10*	~0.17
γ	2126.6 *15*	0.22 *9*
γ	2141.4 *3*	0.27 *9*
γ	2154.6 *10*	0.19 *7*
γ	2194.3 *7*	0.31 *9*
γ	2355.4 *3*	0.56 *8*
γ	2725.1 *11*	~0.17
γ	3046.8 *7*	0.21 *5*
γ	3195.1 *9*	~0.09
γ	3512.3 *20*	0.19 *7*
γ	3815.1 *9*	~0.09

† 7.7% uncert(syst)

Atomic Electrons (^{172}Ta)

⟨e⟩=136 *6* keV

e_{bin}(keV)	⟨e⟩(keV)	e(%)
10	4.6	48 *8*
11	3.8	35 *6*
30	5.2	17.4 *15*
43 - 63	1.81	3.8 *3*
84	19.2	22.7 *20*
86	16.8	19.6 *17*
93	9.8	10.5 *9*
94 - 114	2.78	2.93 *25*
149	10.7	7.2 *15*
156 - 195	0.92	0.58 *5*
203	5.1	2.5 *5*
204 - 253	5.8	2.7 *3*
258 - 307	0.29	0.10 *4*
308 - 357	0.52	0.16 *4*
359 - 407	0.92	0.24 *4*
409 - 458	0.37	0.084 *24*
459 - 504	0.17	0.035 *14*
511 - 556	0.11	0.021 *7*
562 - 610	0.19	0.033 *7*
611 - 656	0.078	0.012 *4*
670 - 719	0.15	0.022 *4*
720 - 766	0.28	0.037 *11*
770 - 819	0.82	0.103 *20*
820 - 869	0.16	0.018 *4*
870 - 915	0.23	0.025 *12*
922 - 970	0.20	0.022 *7*
976 - 1025	0.51	0.051 *8*
1031 - 1081	0.6	~0.06
1083 - 1129	0.34	0.031 *8*
1135 - 1184	3.4	~0.3
1185 - 1215	0.8	~0.07

Atomic Electrons (^{172}Ta)
(continued)

e_{bin}(keV)	⟨e⟩(keV)	e(%)
1226	5.0	0.40 *6*
1229 - 1231	1.65	0.134 *20*
1237	23.1	1.86 *19*
1238 - 1286	4.8	0.37 *4*
1291	4.2	0.33 *7*
1293 - 1332	0.77	0.058 *10*
1343 - 1388	0.040	0.0029 *12*
1395 - 1420	0.07	~0.0048
1458 - 1542	0.19	0.013 *6*
1572 - 1649	0.036	0.0023 *8*
1684 - 1712	0.0039	0.00023 *9*
1812 - 1876	0.005	~0.00030
1943 - 2029	0.013	0.00065 *21*
2061 - 2153	0.018	0.00085 *24*
2183 - 2192	0.0009	4.1 *19* $\times 10^{-5}$
2290 - 2353	0.010	0.00042 *18*
2660 - 2723	0.0024	$\sim 9 \times 10^{-5}$
2981 - 3045	0.0027	9 *4* $\times 10^{-5}$
3130 - 3193	0.0012	$\sim 4 \times 10^{-5}$
3447 - 3510	0.0021	6 *3* $\times 10^{-5}$
3750 - 3813	0.0009	$\sim 2 \times 10^{-5}$

Continuous Radiation (^{172}Ta)

⟨β+⟩=288 keV;⟨IB⟩=6.7 keV

E_{bin}(keV)		⟨ ⟩(keV)	(%)
0 - 10	β+	1.33×10^{-7}	1.58×10^{-6}
	IB	0.0111	
10 - 20	β+	7.9×10^{-6}	4.69×10^{-5}
	IB	0.0105	0.073
20 - 40	β+	0.000318	0.00096
	IB	0.023	0.077
40 - 100	β+	0.0238	0.0299
	IB	0.24	0.42
100 - 300	β+	1.70	0.75
	IB	0.20	0.109
300 - 600	β+	14.9	3.18
	IB	0.35	0.080
600 - 1300	β+	113	11.8
	IB	1.52	0.161
1300 - 2500	β+	156	9.3
	IB	3.5	0.19
2500 - 3968	β+	2.23	0.086
	IB	0.85	0.031
	Σβ+		25

$^{172}_{74}$W (6.7 *10* min)

Mode: ε
Δ: -48970 *270* keV syst
SpA: 1.63×10^{8} Ci/g
Prod: ^{176}Hf(^{3}He,7n); ^{155}Gd(^{22}Ne,5n); ^{156}Gd(^{20}Ne,4n)

Photons (^{172}W)

γ_{mode}	γ(keV)	γ(rel)
γ E1	35.79 *18*	142 *31*
γ	39.49 *25*	35 *3*
γ	83.5 *6*	6.9 *12*
γ	90.1 *8*	16 *2*
γ	93.2 *7*	8.1 *19*
γ	109.25 *20*	23 *2*
γ	114.8 *10*	~0.50 ?
γ M1	130.18 *15*	100 *5*
γ	145.2 *5*	17 *1*
γ	153.7 *3*	37 *4*
γ	165.96 *19*	<27 ?

Photons (^{172}W)
(continued)

γ_{mode}	γ(keV)	γ(rel)
γ	169.0 *8*	9.2 *10*
γ	174.83 *25*	77 *6*
γ	191.7 *5*	3.3 *10*
γ	197.1 *10*	13 *4*
γ	221.4 *7*	~8
γ	227.8 *10*	3.9 *10* ?
γ	240.6 *5*	~12
γ	273.8 *10*	~2
γ	278.4 *10*	2.7 *11* ?
γ	393.06 *25*	6.5 *15*
γ	406.0 *4*	3.5 *11*
γ	414.4 *3*	9.9 *18*
γ	423.35 *19*	24 *2*
γ	425.19 *20*	6.5 *13*
γ (M1)	457.54 *15*	367 *50*
γ	493.33 *19*	13 *2*
γ	494.7 *5*	13 *2*
γ	576.2 *4*	3.7 *13*
γ	623.50 *17*	102 *20*
γ	630.9 *4*	~12
γ	635.9 *3*	~19
γ	674.6 *5*	9.9 *21*
γ	677.5 *6*	~10
γ	770.8 *6*	15 *2*

$^{172}_{75}$Re(15 *3* s)

Mode: ε
Δ: -41680 *370* keV syst
SpA: 4.3×10^{9} Ci/g
Prod: protons on Tl

Photons (^{172}Re)

γ_{mode}	γ(keV)	γ(rel)
γ [E2]	123.4 *1*	45 *5*
γ [E2]	253.7 *2*	100
γ [E2]	350.4 *5*	55 *13*
γ [E2]	419.4 *3*	~10

$^{172}_{75}$Re(55 *5* s)

Mode: ε
Δ: -41680 *370* keV syst
SpA: 1.19×10^{9} Ci/g
Prod: daughter ^{172}Os; protons on Tl

Photons (^{172}Re)

γ_{mode}	γ(keV)	γ(rel)
γ [E2]	123.4 *1*	135 *5*
γ [E2]	253.7 *2*	100
γ [E2]	350.4 *5*	<5 ?
γ	743.0 *2*	26 *7*

$^{172}_{76}$Os(19 *2* s)

Mode: ε(>99.7 %), α(<0.3 %)
Δ: -37260 *240* keV syst
SpA: 3.4×10^{9} Ci/g
Prod: ^{164}Er(^{16}O,8n)

Alpha Particles (^{172}Os)

α(keV)
5105 *10*

$^{172}_{77}$Ir(2.1 *1* s)

Mode: α
Δ: -27430 *380* keV syst
SpA: 2.66×10^{10} Ci/g
Prod: ^{162}Er(^{19}F,9n); ^{164}Er(^{19}F,11n); ^{93}Nb(^{84}Kr,5n)

Alpha Particles (^{172}Ir)

α(keV)
5811 *5*

$^{172}_{78}$Pt(105 *15* ms)

Mode: α
Δ: -21220 *60* keV
SpA: 9.4×10^{10} Ci/g
Prod: ^{112}Sn(^{63}Cu,p2n); ^{144}Sm(^{32}S,4n)

Alpha Particles (^{172}Pt)

α(keV)
6314 *4*

A = 173

NDS **14**, 297 (1975)

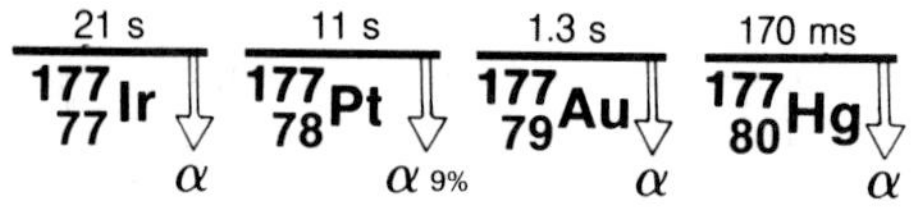

Q_α 5130 *50* Q_α 5646 *10* Q_α 6290 *50* Q_α 6730 *50*

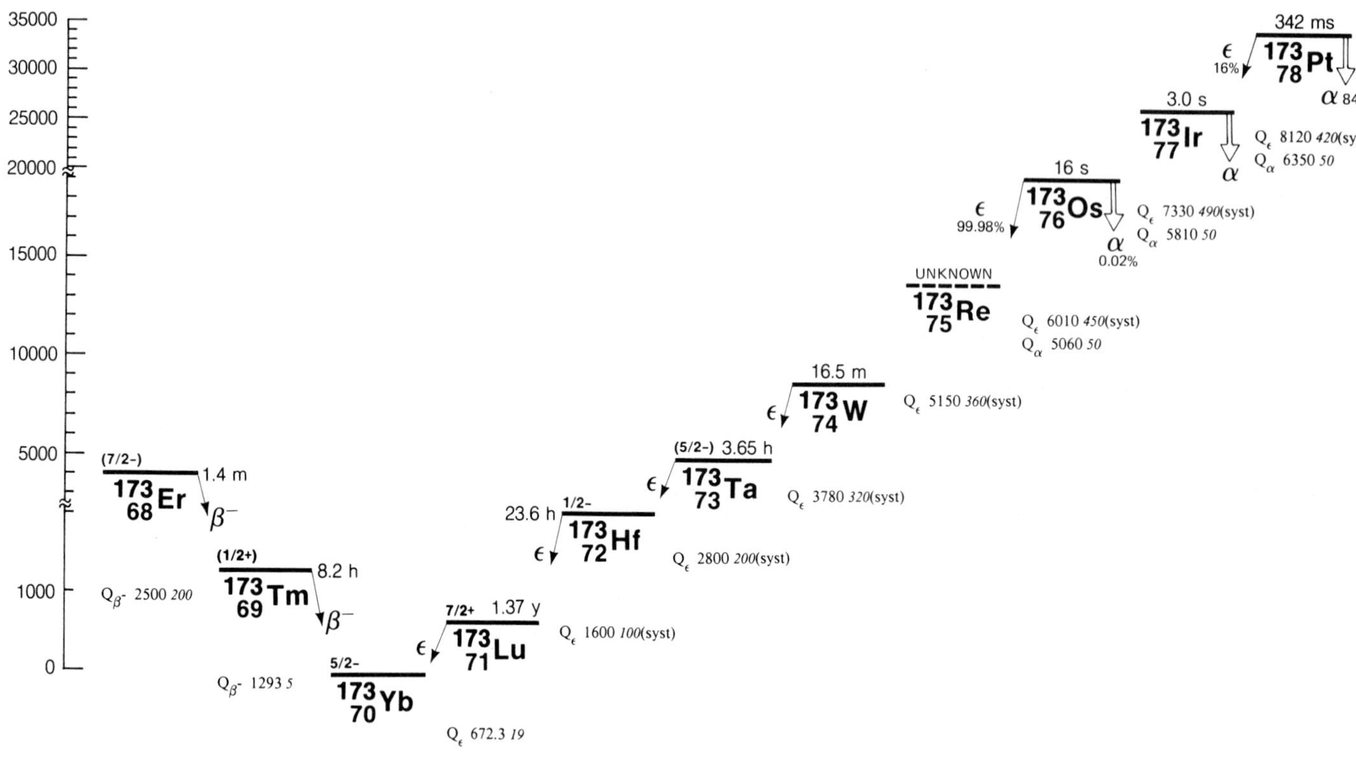

$^{173}_{68}$Er(1.4 *1* min)

Mode: β-
Δ: -53770 *200* keV
SpA: 7.7×10^8 Ci/g
Prod: ^{176}Yb(n,α)

Photons (^{173}Er)

⟨γ⟩=832 *45* keV

γ_{mode}	γ(keV)	γ(%)†
Tm L_ℓ	6.341	0.31 *4*
Tm L_α	7.176	7.7 *9*
Tm L_η	7.310	0.109 *13*
Tm L_β	8.222	7.4 *10*
Tm L_γ	9.508	1.18 *16*
Tm $K_{\alpha2}$	49.772	18.5 *18*
Tm $K_{\alpha1}$	50.742	33 *3*
Tm $K_{\beta1}'$	57.444	10.3 *10*
Tm $K_{\beta2}'$	59.296	2.8 *3*
γ M1(+E2)	94.40 *19*	4.8 *10*
γ M1(+E2)	116.34 *4*	19 *3*
γ [E2]	118.8 *2*	2.5 *6*
γ (E2)	122.60 *4*	20.6 *24*
γ [E1]	193.07 *14*	47 *5*
γ [E1]	199.33 *14*	48
γ [M1]	801.0 *4*	10 *3*
γ [M1]	895.4 *3*	54 *4*

† 6.3% uncert(syst)

Atomic Electrons (^{173}Er)

⟨e⟩=84 *4* keV

e_{bin}(keV)	⟨e⟩(keV)	e(%)
9	2.8	32 *5*
10	2.1	22 *3*
35	5.3	15 *3*
39 - 56	1.48	3.41 *19*
57	18.6	33 *5*
59	1.02	1.7 *4*
63	8.3	13.1 *16*
84 - 94	2.6	3.0 *5*
106	4.8	4.5 *7*
107 - 112	3.9	3.5 *4*
113	6.9	6.1 *8*
114	7.5	6.6 *9*
115 - 120	1.35	1.15 *12*
121	3.4	2.8 *4*
122 - 123	1.01	0.83 *9*
134	3.1	2.3 *3*
140	3.06	2.19 *16*
183 - 199	1.64	0.87 *4*
742 - 791	0.8	0.11 *4*
792 - 801	0.032	0.0041 *11*
836	3.4	0.41 *3*
885 - 895	0.67	0.075 *5*

Continuous Radiation (^{173}Er)

⟨β-⟩=585 keV;⟨IB⟩=0.96 keV

E_{bin}(keV)		⟨ ⟩(keV)	(%)
0 - 10	β-	0.0388	0.77
	IB	0.025	
10 - 20	β-	0.117	0.78
	IB	0.024	0.167
20 - 40	β-	0.481	1.60
	IB	0.046	0.160
40 - 100	β-	3.58	5.08
	IB	0.125	0.19
100 - 300	β-	38.8	19.2
	IB	0.30	0.17
300 - 600	β-	134	30.0
	IB	0.25	0.061
600 - 1300	β-	312	36.4
	IB	0.17	0.021
1300 - 2182	β-	95	6.2
	IB	0.0141	0.00098

$^{173}_{69}$Tm(8.24 *8* h)

Mode: β-
Δ: -56267 *5* keV
SpA: 2.198×10^6 Ci/g
Prod: ^{176}Yb(p,α); ^{170}Er(α,p); ^{173}Yb(n,p); ^{174}Yb(γ,p)

Photons (^{173}Tm)

$\langle\gamma\rangle$=387 *9* keV

γ_{mode}	γ(keV)	γ(%)†
Yb L_ℓ	6.545	0.044 *14*
Yb L_α	7.411	1.1 *3*
Yb L_η	7.580	0.016 *5*
Yb L_β	8.510	1.1 *4*
Yb L_γ	9.860	0.18 *6*
Yb $K_{\alpha2}$	51.354	2.0 *5*
Yb $K_{\alpha1}$	52.389	3.5 *9*
Yb $K_{\beta1}$'	59.316	1.1 *3*
Yb $K_{\beta2}$'	61.229	0.31 *8*
γ M1+23%E2	62.5 *2*	0.9 *4*
γ E2	398.8 *5*	88
γ M1	461.3 *5*	6.9 *3*

† 2.3% uncert(syst)

Atomic Electrons (^{173}Tm)

$\langle e\rangle$=15.5 *7* keV

e_{bin}(keV)	$\langle e\rangle$(keV)	e(%)
9 - 51	0.85	7.7 *17*
52	0.60	1.1 *5*
53	0.8	1.6 *7*
54	0.9	1.6 *7*
57 - 59	0.0069	0.0119 *24*
60	0.38	0.6 *3*
61 - 63	0.42	0.69 *21*
337	7.36	2.18 *7*
388	1.05	0.271 *8*
389	0.671	0.173 *5*
390 - 399	1.011	0.256 *4*
400	1.21	0.302 *13*
451 - 461	0.260	0.0575 *20*

Continuous Radiation (^{173}Tm)

$\langle\beta-\rangle$=293 keV;$\langle$IB$\rangle$=0.27 keV

E_{bin}(keV)		$\langle\ \rangle$(keV)	(%)
0 - 10	β-	0.092	1.83
	IB	0.0141	
10 - 20	β-	0.276	1.84
	IB	0.0134	0.093
20 - 40	β-	1.11	3.71
	IB	0.025	0.087
40 - 100	β-	7.9	11.3
	IB	0.061	0.096
100 - 300	β-	73	36.9
	IB	0.110	0.066
300 - 600	β-	158	36.8
	IB	0.043	0.0109
600 - 1293	β-	52	7.6
	IB	0.0030	0.00044

$^{173}_{70}$Yb(stable)

Δ: -57560 *3* keV
%: 16.12 *18*

$^{173}_{71}$Lu(1.37 *1* yr)

Mode: ϵ
Δ: -56887 *4* keV
SpA: 1508 Ci/g
Prod: ^{173}Yb(p,n); daughter ^{173}Hf

Photons (^{173}Lu)

$\langle\gamma\rangle$=103 *8* keV

γ_{mode}	γ(keV)	γ(%)†
Yb L_ℓ	6.545	0.39 *6*
Yb L_α	7.411	9.6 *12*
Yb L_η	7.580	0.123 *15*
Yb L_β	8.518	9.3 *13*
Yb L_γ	9.891	1.57 *24*
Yb $K_{\alpha2}$	51.354	25 *3*
Yb $K_{\alpha1}$	52.389	45 *5*
Yb $K_{\beta1}$'	59.316	14.2 *16*
Yb $K_{\beta2}$'	61.229	3.9 *4*
γ (M1+8.5%E2)	62.15 *4*	0.014 *4*
γ M1+5.3%E2	78.603 *18*	0.78 *3*
γ M1+4.1%E2	100.659 *12*	3.12 *10*
γ [E1]	111.07 *13*	0.043 *4*
γ [M1]	122.40 *12*	~0.018
γ E1	171.32 *3*	1.77 *8*
γ E2	179.262 *21*	0.83 *3*
γ [E2]	223.05 *12*	~0.010
γ E1	223.13 *5*	~0.08
γ E1	233.47 *5*	0.346 *7*
γ E1	271.98 *4*	13
γ E1	285.29 *4*	0.352 *7*
γ E1	334.13 *5*	0.0654 *19*
γ E1	350.58 *4*	0.186 *7*
γ M1+37%E2	456.60 *3*	0.078 *3*
γ E2+20%M1	557.26 *3*	0.299 *10*
γ E2+46%M1	635.863 *20*	0.87 *3*

† approximate

Atomic Electrons (^{173}Lu)

$\langle e\rangle$=16.3 *7* keV

e_{bin}(keV)	$\langle e\rangle$(keV)	e(%)
9	3.3	37 *6*
10	2.6	26 *4*
17	0.76	4.40 *17*
39	3.43	8.7 *3*
40 - 41	0.44	1.09 *12*
42	0.58	1.39 *21*
43 - 62	1.06	2.16 *14*
68	0.416	0.611 *24*
69 - 79	0.389	0.535 *15*
90	1.09	1.21 *5*
91 - 122	1.10	1.08 *4*
161 - 179	0.305	0.179 *4*
211	0.59	0.28 *6*
213 - 262	0.120	0.047 *7*
263 - 289	0.056	0.0205 *21*
324 - 351	0.00191	0.000566 *12*
446 - 457	0.0026	0.00057 *11*
547 - 575	0.072	0.013 *3*
625 - 636	0.0149	0.0024 *4*

Continuous Radiation (^{173}Lu)

$\langle$IB$\rangle$=0.27 keV

E_{bin}(keV)		$\langle\ \rangle$(keV)	(%)
10 - 20	IB	0.00020	0.00144
20 - 40	IB	0.0050	0.0144
40 - 100	IB	0.23	0.44
100 - 300	IB	0.021	0.0118
300 - 600	IB	0.0134	0.0035
600 - 672	IB	2.3×10^{-5}	3.7×10^{-6}

$^{173}_{72}$Hf(23.6 *1* h)

Mode: ϵ
Δ: -55290 *100* keV syst
SpA: 7.68×10^5 Ci/g
Prod: ^{175}Lu(p,3n); alphas on Yb

Photons (^{173}Hf)

$\langle\gamma\rangle$=397 *7* keV

γ_{mode}	γ(keV)	γ(%)†
Lu L_ℓ	6.753	0.50 *9*
Lu L_α	7.650	12.3 *21*
Lu L_η	7.857	0.16 *3*
Lu L_β	8.829	11.9 *21*
Lu L_γ	10.245	2.0 *4*
Lu $K_{\alpha2}$	52.965	31 *5*
Lu $K_{\alpha1}$	54.070	55 *9*
Lu $K_{\beta1}$'	61.219	18 *3*
Lu $K_{\beta2}$'	63.200	4.7 *8*
γ (M1)	117.16 *4*	0.066 *13*
γ E1	123.667 *19*	83
γ E2+28%M1	134.951 *23*	4.78 *13*
γ M1+14%E2	139.63 *3*	12.3 *10*
γ E1	162.02 *3*	6.54 *18*
γ E1	171.61 *3*	0.173 *8*
γ	229.70 *25*	0.0025 *8*
γ	239.76 *20*	0.0042 *17*
γ [E1]	288.77 *4*	0.014 *5*
γ E1	296.972 *22*	33.9 *7*
γ E1	306.562 *21*	6.33 *17*
γ E1	311.241 *23*	10.74 *22*
γ M1	356.98 *3*	0.46 *5*
γ	377.69 *25*	~0.0008
γ M1	423.04 *4*	0.066 *3*
γ	426.93 *15*	0.0100 *17*
γ [E1]	428.40 *4*	0.0124 *17*
γ (M1)	444.80 *6*	0.30 *10*
γ E2+~34%M1	451.29 *5*	0.0091 *17*
γ	458.19 *15*	~0.0033
γ	492.4 *3*	0.0012 *6*
γ M1	540.205 *22*	0.373 *12*
γ M1	546.87 *3*	0.0307 *17*
γ M1	549.796 *24*	0.444 *17*
γ E0+M1	556.46 *3*	0.0133 *17*
γ E0+M1	568.46 *3*	0.0091 *17*
γ M1	578.05 *3*	0.024 *3*
γ	593.20 *15*	0.0066 *17*
γ M1(+E2)	618.13 *3*	0.0249 *17*
γ M1	625.90 *4*	0.0307 *17*
γ	694.42 *24*	0.0100 *25*
γ E1	718.484 *25*	0.300 *16*
γ	734.51 *20*	0.0033 *8*
γ [E1]	740.07 *3*	0.025 *3*
γ M1	760.85 *4*	0.067 *3*
γ (M1)	765.53 *4*	0.0149 *8*
γ	807.1 *3*	0.0012 *6*
γ	811.57 *4*	0.0166 *17*
γ (E0+M1)	821.16 *5*	0.0042 *8*
γ	828.37 *25*	~0.0033
γ	834.12 *9*	0.0058 *8*
γ	845.4 *4*	0.0012 *6*
γ E1	853.43 *3*	0.317 *21*
γ	866.31 *7*	~0.0008

Photons (^{173}Hf) (continued)

γ_{mode}	γ(keV)	γ(%)†
γ E1	875.02 3	0.228 14
γ E1	879.70 3	0.39 3
γ	889.50 5	0.0033 8
γ M1	899.10 6	1.01 7
γ	905.79 20	0.0018 5
γ	929.16 10	0.0050 8
γ	933.9 4	0.0012 6
γ	969.07 8	0.0058 17
γ [E1]	977.02 5	0.0100 17
γ	983.18 5	0.0042 8
γ	989.9 3	~0.0008
γ	991.8 10	~0.0008
γ	1001.26 7	0.0108 17
γ	1005.94 7	0.0199 25
γ M1	1034.05 6	0.43 3
γ M1	1038.73 6	0.333 23
γ	1064.11 10	0.0050 8
γ M1	1070.70 5	0.079 8
γ	1085.7 3	~0.00033
γ	1096.00 20	0.0012 6
γ	1100.0 3	0.00050 17
γ	1118.13 5	0.0249 25
γ	1145.2 4	~0.00017
γ M1	1205.65 5	0.297 24
γ M1	1210.33 5	0.090 8
γ	1230.8 3	0.00075 17
γ	1235.48 20	~0.00017
γ	1280.6 5	~0.00017
γ	1286.58 20	0.00075 17
γ	1315.9 3	0.00058 17
γ	1332.85 20	~0.00017
γ	1363.9 5	<17 ×10^{-5}
γ	1488.9 3	0.00033 8
γ	2042.9 10	0.00025 8
γ	2127.7 10	~0.00017

† 1.2% uncert(syst)

Atomic Electrons (^{173}Hf)

⟨e⟩=46.7 14 keV

e_{bin}(keV)	⟨e⟩(keV)	e(%)
7 - 8	0.20	~3
9	4.3	46 9
10	2.3	22 4
11 - 59	3.4	13.7 16
60	8.37	13.9 3
61	0.021	0.034 6
72	2.50	3.49 15
76	10.4	13.7 12
99 - 108	0.569	0.573 19
113	2.10	1.86 4
114 - 124	1.78	1.48 3
125	1.01	0.81 4
126	0.84	0.67 3
129	3.1	2.39 23
130 - 176	2.47	1.82 8
219 - 231	0.0013	0.00056 20
234	1.42	0.606 18
237 - 287	0.914	0.358 6
288 - 314	0.389	0.131 5
346 - 395	0.105	0.029 5
412 - 458	0.017	0.0040 8
477 - 515	0.171	0.0352 13
529 - 578	0.0422	0.0078 3
582 - 631	0.0020	0.00033 8
655 - 702	0.0133	0.00196 7
708 - 756	0.0032	0.00043 8
758 - 807	0.0075	0.00097 8
809 - 857	0.085	0.0102 6
864 - 906	0.0171	0.00193 9
914 - 960	0.0016	0.00017 6
966 - 1007	0.051	0.0053 3
1022 - 1071	0.0109	0.00106 6
1075 - 1118	0.00018	1.6 7 ×10^{-5}
1134 - 1172	0.0190	0.00167 11
1195 - 1235	0.00372	0.000310 16
1253 - 1301	2.7 ×10^{-5}	2.2 9 ×10^{-6}

Atomic Electrons (^{173}Hf) (continued)

e_{bin}(keV)	⟨e⟩(keV)	e(%)
1305 - 1354	4.2 ×10^{-6}	3.2 13 ×10^{-7}
1355 - 1364	1.1 ×10^{-7}	~8 ×10^{-9}
1478 - 1489	1.5 ×10^{-6}	10 5 ×10^{-8}
1980 - 2064	7.3 ×10^{-6}	3.6 15 ×10^{-7}
2117 - 2126	4.6 ×10^{-7}	~2 ×10^{-8}

Continuous Radiation (^{173}Hf)

⟨β+⟩=0.486 keV;⟨IB⟩=0.59 keV

E_{bin}(keV)		⟨ ⟩(keV)	(%)
0 - 10	β+	1.98 ×10^{-6}	2.35 ×10^{-5}
	IB	0.00098	
10 - 20	β+	0.000103	0.00062
	IB	0.00024	0.0018
20 - 40	β+	0.00323	0.0099
	IB	0.0041	0.0121
40 - 100	β+	0.099	0.133
	IB	0.26	0.48
100 - 300	β+	0.368	0.220
	IB	0.040	0.021
300 - 600	β+	0.0161	0.00473
	IB	0.124	0.028
600 - 1300	IB	0.157	0.020
1300 - 1472	IB	6.4 ×10^{-5}	4.8 ×10^{-6}
	Σβ+		0.37

$^{173}_{73}$Ta(3.65 5 h)

Mode: ε

Δ: -52490 220 keV syst

SpA: 4.96×10^{6} Ci/g

Prod: ^{175}Lu(α,6n); daughter ^{173}W; protons on ^{181}Ta; ^{165}Ho(^{12}C,4n)

Photons (^{173}Ta)

⟨γ⟩=432 13 keV

γ_{mode}	γ(keV)	γ(%)
Hf L_ℓ	6.960	0.68 12
Hf L_α	7.893	17 3
Hf L_η	8.139	0.25 4
Hf L_β	9.129	19 3
Hf L_γ	10.627	3.3 6
γ M1	25.63 7	0.14 4
γ M1+0.4%E2	37.43 7	1.2 1
Hf $K_{\alpha2}$	54.611	29 5
Hf $K_{\alpha1}$	55.790	51 9
γ M1+6.9%E2	57.98 10	0.48 12
Hf $K_{\beta1}$'	63.166	17 3
Hf $K_{\beta2}$'	65.211	4.3 8
γ M1+44%E2	69.72 7	5.8 6
γ E2	81.51 7	1.4 4
γ (M1+E2)	90.12 7	~0.37
γ (E1)	90.28 10	~5
γ M1(+E2)	115.00 8	0.37 9
γ [M1+E2]	139.34 11	
γ M1+33%E2	160.40 6	4.8 5
γ E2	172.19 6	17 4
γ E2	180.58 7	2.1 2
γ [E2]	205.12 9	0.15 4
γ	214.28 11	0.050 13
γ [M1+E2]	246.74 8	0.12 3
γ [E2]	254.34 9	0.034 8
γ [E2]	266.92 8	0.32 9
γ	276.98 11	0.025 6

Photons (^{173}Ta) (continued)

γ_{mode}	γ(keV)	γ(%)
γ	380.28 21	0.060 15
γ	413.28 11	0.090 23
γ	434.20 9	0.050 13
γ	438.28 11	0.30 8
γ	463.11 21	0.030 8
γ (E1)	529.78 21	0.45 11
γ	530.18 21	0.17 4
γ	549.57 12	0.37 9
γ	557.08 21	0.070 17
γ	558.98 21	0.100 25
γ	566.48 21	0.050 13
γ	569.75 11	0.070 17
γ	578.11 21	0.04 1
γ (M1+E2)	587.78 21	0.21 5
γ	615.28 21	0.15 4
γ	659.81 12	0.04 1 ?
γ	662.68 21	0.19 5
γ	668.23 22	0.33 8
γ	675.08 21	0.54 13
γ	680.94 11	0.090 23
γ	685.55 12	0.12 3
γ	691.58 21	0.22 6
γ (M1+E2)	701.12 11	1.20 12
γ	702.58 21	0.29 7
γ	707.84 11	0.050 13
γ (M1)	730.14 11	0.60 15
γ	739.63 12	~0.10
γ	741.93 12	~0.10
γ	744.7 4	~0.07
γ	747.1 4	~0.10
γ	749.5 4	~0.10
γ	753.98 21	0.12 3
γ	771.38 21	0.070 17
γ	778.21 10	0.51 13
γ	783.38 21	0.070 17
γ	789.08 21	0.20 5
γ	795.18 21	0.20 5
γ	799.15 12	0.44 11
γ	801.38 21	0.20 5
γ	811.66 12	0.4 1
γ	814.48 21	0.08 2
γ	822.85 11	0.32 8
γ	836.18 21	0.24 6
γ	842.78 21	0.100 25
γ	845.94 12	0.46 11
γ [E1]	851.10 12	0.12 3
γ	857.74 12	0.28 7
γ	861.52 11	0.090 23
γ (E2)	864.74 11	0.33 8
γ	873.31 11	0.70 17
γ (E1)	876.73 11	0.35 9
γ (E1)	888.53 11	0.18 5
γ	892.6 4	0.017 4
γ	905.68 21	0.070 17
γ	914.15 11	0.28 7
γ	917.38 21	0.034 9
γ	931.58 21	0.050 13
γ	938.60 10	0.100 25
γ	941.33 12	0.18 5 ?
γ	947.18 21	0.034 9
γ	950.39 10	0.24 6
γ (E1)	958.25 12	0.51 13
γ	986.37 11	0.070 17
γ	989.58 21	0.090 23
γ (M1+E2)	995.37 15	0.26 7
γ (M1)	1006.55 10	0.56 14
γ	1009.88 21	0.100 25
γ	1013.08 21	0.070 17
γ (E1)	1029.90 12	1.60 16
γ (M1)	1045.32 11	0.30 8
γ	1057.11 11	0.050 13
γ	1067.48 21	0.100 25
γ	1070.28 21	0.100 25
γ	1085.49 15	0.30 8
γ	1088.68 21	0.100 25
γ	1104.78 21	0.070 17
γ	1126.83 11	0.100 25
γ	1166.94 10	0.090 23
γ	1176.08 21	0.14 4
γ	1178.73 10	0.26 7
γ (E1)	1208.25 11	2.7 3
γ	1216.18 21	0.12 3
γ	1252.89 11	0.16 4

Photons (^{173}Ta) (continued)

γ_{mode}	γ(keV)	γ(%)
γ	1277.0 *4*	~0.050
γ	1279.6 *4*	0.070 *17*
γ	1327.1 *4*	0.070 *17*
γ	1332.18 *17*	0.090 *23*
γ	1343.01 *11*	0.18 *5* §
γ	1349.3 *4*	0.070 *17*
γ	1368.64 *11*	0.38 *10*
γ	1375.3 *4*	0.21 *5*
γ	1380.44 *10*	0.54 *13*
γ	1390.4 *4*	0.100 *25*
γ	1393.40 *14*	0.72 *18*
γ	1405.18 *17*	0.100 *25*
γ	1409.18 *21*	0.050 *13*
γ	1413.58 *13*	0.12 *3*
γ	1425.36 *17*	0.14 *4*
γ	1432.41 *20*	0.56 *14*
γ	1434.7 *5*	0.090 *23*
γ	1445.6 *3*	0.16 *4*
γ	1486.7 *3*	0.17 *4*
γ	1492.58 *16*	0.12 *3*
γ	1499.0 *3*	0.18 *5*
γ	1504.37 *16*	0.070 *17*
γ	1547.6 *3*	0.26 *7*
γ	1573.97 *13*	0.33 *8* §
γ	1585.75 *17*	0.100 *25*
γ	1585.76 *13*	0.100 *25*
γ	1597.54 *17*	0.28 *7*
γ	1612.98 *20*	0.47 *12*
γ	1633.9 *3*	0.050 *13*
γ	1648.2 *3*	0.100 *25*
γ	1691.0 *3*	0.070 *17*
γ	1702.9 *5*	0.030 *8*
γ	1709.7 *5*	0.070 *17*
γ	1717.2 *5*	0.050 *13*
γ	1757.8 *5*	0.070 *17*
γ	1882.2 *5*	0.030 *8*
γ	1913.3 *5*	0.030 *8*
γ	1960.6 *5*	0.030 *8*
γ	2001.3 *5*	0.100 *25*
γ	2022.7 *6*	
γ	2048.5 *5*	0.050 *13*
γ	2066.7 *5*	0.030 *8*
γ	2077.3 *5*	0.050 *13*
γ	2086.9 *5*	0.030 *8*
γ	2161.5 *5*	0.12 *3*
γ	2182.0 *4*	0.070 *17*
γ	2188.3 *4*	0.100 *25*
γ	2199.8 *4*	0.100 *25*
γ	2244.9 *4*	0.100 *25*
γ	2258.1 *4*	0.030 *8*
γ	2269.4 *4*	0.12 *3*
γ	2281.0 *5*	0.030 *8*
γ	2293.6 *5*	0.030 *8*
γ	2319.9 *5*	0.030 *8*
γ	2328.3 *5*	0.030 *8*
γ	2360.2 *5*	0.050 *13*
γ	2460.0 *4*	0.050 *13*
γ	2475.0 *4*	0.08 *2*
γ	2493.0 *5*	0.030 *8*
γ	2526.5 *5*	0.050 *13*
γ	2557.5 *5*	0.017 *4*
γ	2567.8 *5*	0.030 *8*
γ	2585.7 *5*	0.030 *8*
γ	2595.3 *5*	0.017 *4*
γ	2638.1 *4*	0.070 *17*
γ	2644.1 *5*	0.12 *3*
γ	2656.5 *5*	0.050 *13*
γ	2674.9 *5*	0.030 *8*
γ	2680.1 *4*	0.08 *2*
γ	2694.1 *5*	0.017 *4*
γ	2722.3 *5*	0.017 *4*
γ	2736.7 *5*	0.017 *4*
γ	2749.1 *5*	0.017 *4*
γ	2784.5 *5*	0.0070 *18*

§ doublet

Atomic Electrons (^{173}Ta)

⟨e⟩=76 *3* keV

e_{bin}(keV)	⟨e⟩(keV)	e(%)
4	1.55	36 *4*
10	5.3	56 *10*
11	5.1	47 *8*
14 - 25	1.9	10.1 *23*
26	2.74	10.5 *9*
27 - 57	5.7	13.7 *9*
58	3.0	5.2 *6*
59	8.2	13.9 *19*
60	8.4	13.9 *19*
61 - 63	0.074	0.120 *15*
67	3.1	4.5 *6*
68	2.4	3.6 *5*
69 - 70	1.68	2.4 *3*
71	2.5	3.6 *10*
72	2.5	3.5 *10*
79	2.0	2.5 *11*
80 - 90	1.1	1.3 *5*
95	3.2	3.4 *4*
104 - 105	0.27	~0.26
107	4.6	4.3 *11*
112 - 160	2.32	1.66 *11*
161	3.1	1.9 *5*
163 - 212	4.2	2.5 *3*
213 - 257	0.044	0.018 *3*
264 - 277	0.0099	0.0037 *8*
315 - 348	0.021	~0.006
369 - 413	0.05	~0.014
423 - 465	0.039	0.009 *3*
484 - 530	0.09	0.018 *6*
538 - 588	0.031	0.0057 *19*
594 - 642	0.22	0.036 *9*
649 - 697	0.143	0.021 *3*
698 - 747	0.14	0.020 *5*
748 - 797	0.10	0.013 *4*
799 - 849	0.09	0.011 *3*
850 - 896	0.049	0.0056 *13*
903 - 950	0.074	0.0079 *14*
956 - 1005	0.062	0.0063 *7*
1006 - 1055	0.026	0.0025 *8*
1056 - 1105	0.012	0.0011 *3*
1111 - 1157	0.049	0.0043 *7*
1164 - 1212	0.015	0.0013 *3*
1214 - 1262	0.0052	0.00042 *17*
1266 - 1315	0.041	0.0032 *10*
1316 - 1365	0.037	0.0028 *10*
1366 - 1415	0.031	0.0022 *8*
1416 - 1446	0.018	0.0012 *4*
1475 - 1574	0.042	0.0028 *7*
1575 - 1652	0.0107	0.00067 *15*
1680 - 1756	0.0025	0.00015 *6*
1817 - 1911	0.0019	0.00010 *3*
1936 - 2022	0.0048	0.00025 *7*
2037 - 2134	0.0066	0.00031 *8*
2150 - 2249	0.0062	0.00028 *6*
2255 - 2351	0.0024	0.000105 *24*
2358 - 2450	0.0023	10 *3* $\times10^{-5}$
2457 - 2557	0.0024	9.7 *20* $\times10^{-5}$
2558 - 2657	0.0056	0.00021 *4*
2664 - 2747	0.00091	3.4 *7* $\times10^{-5}$
2773 - 2782	1.6 $\times10^{-5}$	5.6 *23* $\times10^{-7}$

Continuous Radiation (^{173}Ta)

⟨β+⟩=162 keV;⟨IB⟩=2.6 keV

E_{bin}(keV)		⟨ ⟩(keV)	(%)
0 - 10	β+	1.14 $\times10^{-5}$	0.000136
	IB	0.0074	
10 - 20	β+	0.00053	0.00319
	IB	0.0068	0.047
20 - 40	β+	0.0121	0.0376
	IB	0.0153	0.052
40 - 100	β+	0.120	0.173
	IB	0.24	0.42
100 - 300	β+	4.59	2.09
	IB	0.126	0.070
300 - 600	β+	29.7	6.5
	IB	0.24	0.054

Continuous Radiation (^{173}Ta) (continued)

E_{bin}(keV)		⟨ ⟩(keV)	(%)
600 - 1300	β+	114	12.8
	IB	0.95	0.102
1300 - 2500	β+	12.7	0.91
	IB	1.05	0.063
2500 - 2719	IB	0.0023	8.9 $\times10^{-5}$
	$\Sigma\beta$+		23

$^{173}_{74}$W (16.5 *5* min)

Mode: ϵ

Δ: -48710 *220* keV syst

SpA: 6.58$\times10^{7}$ Ci/g

Prod: ^{181}Ta(p,9n); ^{176}Hf(^{3}He,6n)

$^{173}_{76}$Os(16 *5* s)

Mode: ϵ(99.979 *9* %), α(0.021 *9* %)

Δ: -37540 *350* keV syst

SpA: 4.0$\times10^{9}$ Ci/g

Prod: ^{164}Er(^{16}O,7n)

Alpha Particles (^{173}Os)

α(keV)
4940 *10*

Photons (^{173}Os)

γ_{mode}	γ(keV)	γ(rel)
γ	177 *1*	100 *20*
γ	187 *1*	50 *10*
γ	276 *1*	25 *5*
γ	285 *1*	30 *6*

$^{173}_{77}$Ir(3.0 *1* s)

Mode: α

Δ: -30220 *350* keV syst

SpA: 1.94$\times10^{10}$ Ci/g

Prod: ^{162}Er(^{19}F,8n); ^{164}Er(^{19}F,10n)

Alpha Particles (^{173}Ir)

α(keV)
5665 *5*

$^{173}_{78}$Pt(342 *18* ms)

Mode: α(84 *6* %), ϵ(16 *6* %)

Δ: -22100 *240* keV syst

SpA: 8.2$\times10^{10}$ Ci/g

Prod: ^{162}Er(^{20}Ne,9n); ^{142}Nd(^{40}Ar,9n); ^{112}Sn(^{63}Cu,pn); ^{144}Sm(^{32}S,3n)

Alpha Particles (^{173}Pt)

α(keV)
6213 *8*

A = 174

NDS **41**, 511 (1984)

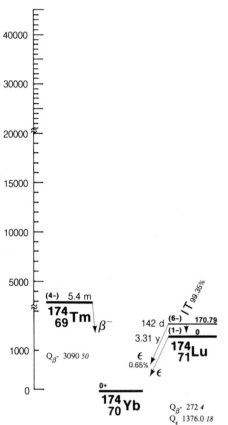

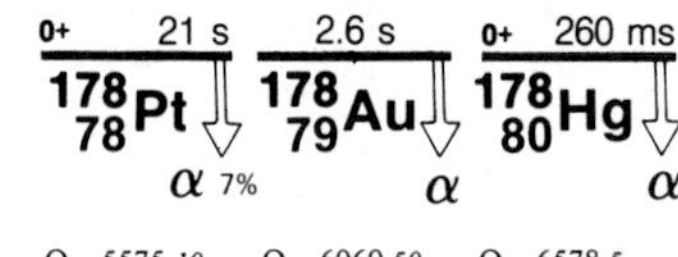

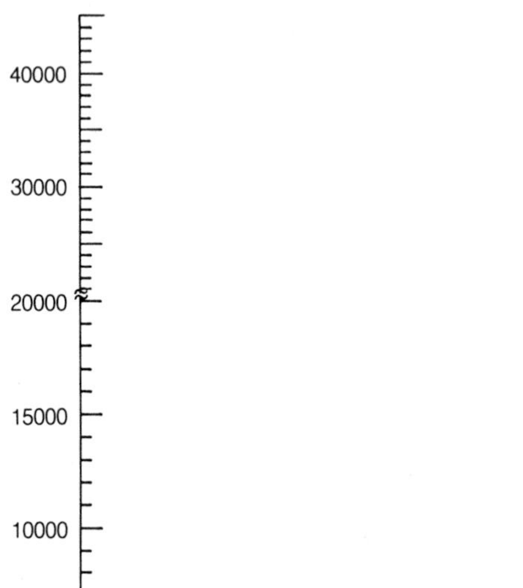

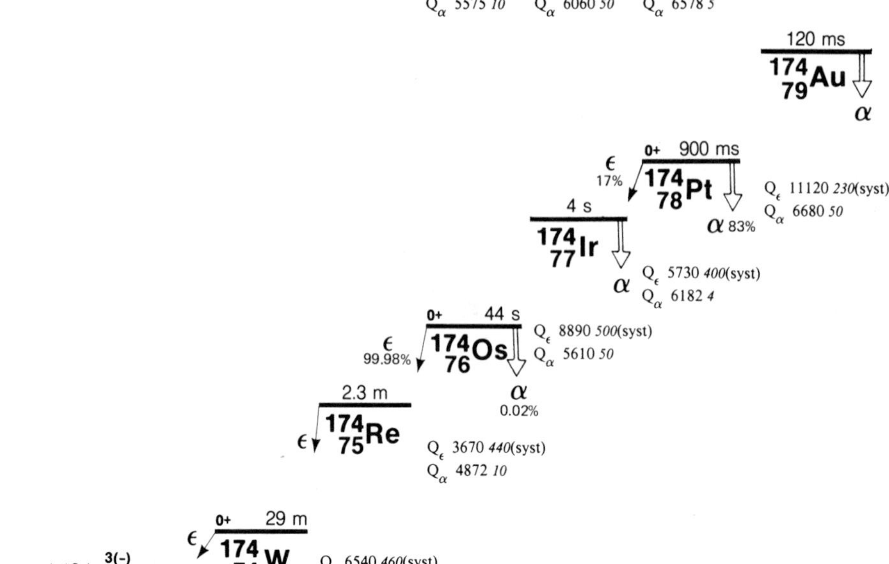

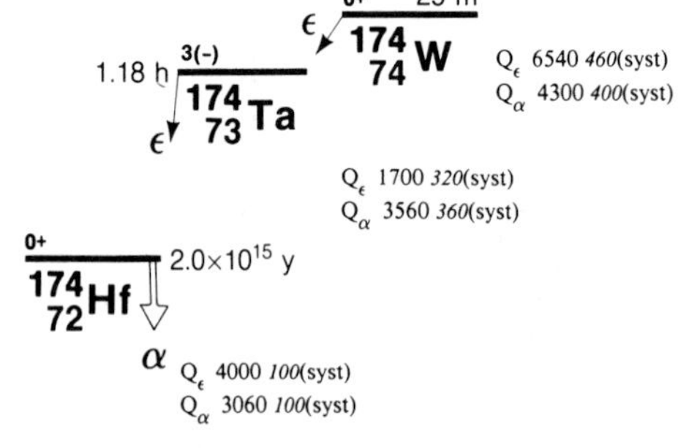

$^{174}_{69}$Tm(5.4 *1* min)

Mode: β-
Δ: -53860 *50* keV
SpA: 2.00×10^{8} Ci/g
Prod: ^{174}Yb(n,p); ^{176}Yb(γ,np)

Photons (^{174}Tm)

⟨γ⟩=1779 *38* keV

γ_{mode}	γ(keV)	γ(%)†
Yb L_ℓ	6.545	0.39 *5*
Yb L_α	7.411	9.7 *9*
Yb L_η	7.580	0.168 *15*
Yb L_β	8.503	10.2 *10*
Yb L_γ	9.810	1.59 *16*
Yb $K_{\alpha2}$	51.354	10.5 *5*
Yb $K_{\alpha1}$	52.389	18.5 *9*
Yb $K_{\beta1}$'	59.316	5.9 *3*
Yb $K_{\beta2}$'	61.229	1.61 *8*
γ E2	76.64 *4*	9.1 *9*
γ [M1]	95.17 *13*	0.10 *3*
γ [E1]	136.06 *17*	0.06 *3*
γ [E1]	138.26 *20*	~0.07
γ [E2]	149.87 *19*	~0.026
γ [M1]	153.27 *19*	0.31 *6*
γ [E2]	176.69 *4*	66.3 *21*
γ [E1]	224.65 *23*	0.30 *4*
γ [E1]	233.43 *21*	~0.09
γ [E2]	273.32 *8*	86 *3*
γ [E1]	288.13 *17*	1.54 *7*
γ	316.31 *25*	~0.09
γ [E1]	319.82 *23*	0.27 *5*
γ [E1]	348.65 *16*	~0.26
γ [E1]	349.39 *23*	0.35 *13*
γ [M1]	358.26 *15*	0.52 *5*
γ [E2]	363.60 *12*	2.7 *3*
γ E1	366.6 *1*	92 *4*
γ [E1]	387.40 *17*	0.33 *6*
γ [E1]	443.82 *10*	1.09 *10*
γ	452.37 *19*	0.30 *7*
γ [E1]	459.30 *15*	0.56 *9*
γ [E1]	482.57 *19*	0.14 *5*
γ M1	494.33 *9*	11.4 *6*
γ [E1]	502.66 *16*	0.24 *9*
γ [E1]	554.47 *15*	0.28 *7*
γ [E2]	628.45 *9*	2.70 *13*
γ [E1]	860.92 *8*	1.62 *9*
γ (E2+28%M1)	992.05 *8*	87 *3*
γ M2(+<27%E3)	1065.20 *11*	0.0078 *14*
γ [E1]	1128.68 *19*	0.18 *4*
γ [E2]	1175.18 *16*	0.14 *4*
γ [E1]	1215.08 *17*	0.18 *3*
γ E1+E3+M2	1241.89 *11*	1.72 *9*
γ [E2]	1265.37 *8*	2.2 *1*
γ [E1]	1305.37 *19*	0.65 *4*
γ [E1]	1316.7 *10*	~0.09
γ M2	1318.54 *11*	0.0105 *9*
γ [M1+E2]	1353.33 *15*	0.12 *3*
γ [E1]	1358.65 *12*	0.06 *3*
γ [M1+E2]	1448.51 *15*	0.16 *4*
γ [M1+E2]	1530.02 *15*	0.16 *4*
γ [E1]	1631.97 *12*	0.18 *4*

† 0.92% uncert(syst)

Atomic Electrons (^{174}Tm)

⟨e⟩=124 *3* keV

e_{bin}(keV)	⟨e⟩(keV)	e(%)
9	3.5	39 *5*
10 - 59	6.1	46 *4*
66	1.01	1.53 *15*
67	17.5	26 *3*
68	18.1	27 *3*
74	5.0	6.7 *7*
75	5.1	6.8 *7*
76 - 95	3.04	3.9 *3*
115	17.9	15.5 *8*
126 - 166	2.87	1.73 *9*

Atomic Electrons (^{174}Tm) (continued)

e_{bin}(keV)	⟨e⟩(keV)	e(%)
167	6.9	4.15 *21*
168	5.4	3.22 *16*
172 - 177	4.88	2.79 *10*
212	12.5	5.9 *3*
214 - 258	0.095	0.041 *6*
263	3.94	1.50 *8*
264 - 311	6.42	2.24 *6*
314 - 363	0.651	0.183 *6*
364 - 398	0.222	0.059 *6*
421 - 459	1.84	0.426 *24*
472 - 503	0.395	0.081 *4*
544 - 567	0.128	0.0226 *14*
618 - 628	0.0337	0.00543 *25*
850 - 861	0.00417	0.000489 *24*
982 - 1004	0.68	0.069 *3*
1055 - 1067	0.0022	0.00020 *4*
1114 - 1154	0.0058	0.00051 *8*
1165 - 1214	0.10	0.009 *3*
1215 - 1264	0.030	0.0024 *4*
1265 - 1314	0.0063	0.00049 *10*
1315 - 1359	0.00089	6.6 *15* $\times 10^{-5}$
1438 - 1528	0.0058	0.00039 *11*
1571 - 1630	0.0018	0.000116 *20*

Continuous Radiation (^{174}Tm)

⟨β-⟩=389 keV;⟨IB⟩=0.46 keV

E_{bin}(keV)		⟨ ⟩(keV)	(%)
0 - 10	β-	0.067	1.33
	IB	0.018	
10 - 20	β-	0.202	1.35
	IB	0.017	0.120
20 - 40	β-	0.82	2.73
	IB	0.033	0.113
40 - 100	β-	5.9	8.5
	IB	0.084	0.131
100 - 300	β-	58	29.1
	IB	0.18	0.104
300 - 600	β-	157	35.8
	IB	0.104	0.026
600 - 1206	β-	167	21.8
	IB	0.024	0.0033

$^{174}_{70}$Yb(stable)

Δ: -56953 *3* keV
%: 31.8 *4*

$^{174}_{71}$Lu(3.31 *5* yr)

Mode: ϵ
Δ: -55577 *4* keV
SpA: 621 Ci/g
Prod: ^{174}Yb(p,n); ^{174}Yb(d,2n)

Photons (^{174}Lu)

⟨γ⟩=132 *17* keV

γ_{mode}	γ(keV)	γ(%)
Yb L_ℓ	6.545	0.49 *11*
Yb L_α	7.411	12.2 *25*
Yb L_η	7.580	0.18 *4*
Yb L_β	8.511	12 *3*
Yb L_γ	9.865	2.1 *5*
Yb $K_{\alpha2}$	51.354	24 *5*
Yb $K_{\alpha1}$	52.389	42 *8*
Yb $K_{\beta1}$'	59.316	13 *3*
Yb $K_{\beta2}$'	61.229	3.6 *7*
γ E2	76.64 *4*	5.8 *12*
γ [E2]	176.69 *4*	0.021 *4*
γ M2(+<27%E3)	1065.20 *11*	0.029 *5*
γ E1+E3+M2	1241.89 *11*	6.5
γ M2	1318.54 *11*	0.040 *3*

Atomic Electrons (^{174}Lu)

⟨e⟩=43 *4* keV

e_{bin}(keV)	⟨e⟩(keV)	e(%)
9	4.2	47 *11*
10	3.6	35 *8*
15	1.4	9.2 *18*
40 - 66	2.55	5.2 *4*
67	11.2	17 *3*
68	11.6	17 *3*
74	3.2	4.3 *9*
75	3.2	4.3 *9*
76	1.5	2.0 *4*
77 - 115	0.20	0.26 *5*
166 - 177	0.0063	0.0037 *4*
1055 - 1065	0.00053	5.1 *7* $\times 10^{-5}$
1231 - 1257	0.047	0.0038 *16*
1308 - 1319	0.00076	5.8 *3* $\times 10^{-5}$

Continuous Radiation (^{174}Lu)

⟨β+⟩=0.0405 keV;⟨IB⟩=0.63 keV

E_{bin}(keV)		⟨ ⟩(keV)	(%)
0 - 10	β+	2.57×10^{-8}	3.04×10^{-7}
	IB	0.00087	
10 - 20	β+	1.39×10^{-6}	8.3×10^{-6}
	IB	0.00018	0.00129
20 - 40	β+	4.92×10^{-5}	0.000149
	IB	0.0047	0.0135
40 - 100	β+	0.00256	0.00330
	IB	0.23	0.43
100 - 300	β+	0.0364	0.0199
	IB	0.039	0.020
300 - 600	β+	0.00151	0.000481
	IB	0.136	0.030
600 - 1300	IB	0.22	0.028
1300 - 1376	IB	3.1×10^{-5}	2.4×10^{-6}
	Σβ+		0.024

$^{174}_{71}$Lu(142 *2* d)

Mode: IT(99.35 *4* %), ϵ(0.65 *4* %)
Δ: -55406 *4* keV
SpA: 5285 Ci/g
Prod: ^{174}Yb(p,n); ^{174}Yb(d,2n); ^{175}Lu(n,2n)

Photons (^{174}Lu)

⟨γ⟩=60.1 *10* keV

γ_{mode}	γ(keV)	γ(%)†
Yb L_ℓ	6.545	0.0027 *6*
Lu L_ℓ	6.753	0.84 *9*
Yb L_α	7.411	0.068 *15*
Yb L_η	7.580	0.0012 *3*
Lu L_α	7.650	20.5 *14*
Lu L_η	7.857	0.211 *17*
Yb L_β	8.503	0.072 *16*
Lu L_β	8.824	22.4 *24*
Yb L_γ	9.810	0.0113 *25*
Lu L_γ	10.345	4.4 *6*
γ_{IT} M1+0.2%E2	44.681 *23*	12.5 *3*
Yb $K_{\alpha2}$	51.354	0.071 *11*
Yb $K_{\alpha1}$	52.389	0.124 *20*
Lu $K_{\alpha2}$	52.965	19.0 *7*
Lu $K_{\alpha1}$	54.070	33.2 *13*
γ_{IT} M3	59.056 *20*	0.0270 *12*
Yb $K_{\beta1}$'	59.316	0.040 *6*
Lu $K_{\beta1}$'	61.219	10.7 *4*
Yb $K_{\beta2}$'	61.229	0.0108 *17*
Lu $K_{\beta2}$'	63.200	2.85 *12*
γ_{IT} M1+0.8%E2	67.055 *8*	6.80
γ_ϵE2	76.64 *4*	0.066 *18*
γ_{IT} E2	111.736 *24*	0.322 *18*
γ_{IT} [E4]	126.111 *21*	0.039 *15*
γ_ϵ[E2]	176.69 *4*	0.48 *12*
γ_ϵ[E2]	273.32 *8*	0.57 *6*
γ_ϵ[E2]	363.60 *12*	
γ_ϵ[E2]	628.45 *9*	0.019 *4*
γ_ϵ(E2+28%M1)	992.05 *8*	0.62
γ_ϵ[E2]	1265.37 *8*	0.016 *3*

† uncert(syst): 6.2% for ϵ, 2.7% for IT

Atomic Electrons (^{174}Lu)

⟨e⟩=117 *3* keV

e_{bin}(keV)	⟨e⟩(keV)	e(%)
4	2.56	68.4 *22*
9	6.1	66 *7*
10	2.3	22 *3*
11	4.0	37 *4*
15	0.016	0.10 *3*
34	22.1	65.3 *19*
35 - 41	0.787	2.22 *6*
42	6.54	15.5 *5*
43 - 45	2.75	6.26 *19*
48	9.9	20.6 *10*
49	1.26	2.59 *13*
50	21.9	44.0 *22*
51 - 54	0.364	0.70 *5*
56	5.42	9.6 *3*
57	11.5	20.3 *10*
58 - 102	6.48	10.3 *3*
109 - 115	0.41	0.36 *8*
116	5.1	4.4 *17*
117	3.6	3.1 *12*
124 - 168	3.6	2.8 *9*
174 - 212	0.118	0.059 *6*
263 - 273	0.046	0.0175 *12*
618 - 628	0.00024	3.9 *14* $\times 10^{-5}$
982 - 992	0.0049	0.00050 *3*
1255 - 1265	7.5 $\times 10^{-5}$	6.0 *9* $\times 10^{-6}$

$^{174}_{72}$Hf(2.0 *4* $\times 10^{15}$ yr)

Mode: α
Δ: -55849 *4* keV
Prod: natural source
%: 0.162 *2*

Alpha Particles (^{174}Hf)

α(keV)
2500 30

$^{174}_{73}$Ta(1.18 5 h)

Mode: ε
Δ: -51850 100 keV syst
SpA: 1.53×10^{7} Ci/g
Prod: ^{175}Lu(α,5n); protons on Hf; ^{175}Lu(^{3}He,4n); protons on ^{181}Ta; daughter ^{174}W

Photons (^{174}Ta)

⟨γ⟩=662 11 keV

γ_{mode}	γ(keV)	γ(%)†
Hf L_ℓ	6.960	0.62 9
Hf L_α	7.893	15.1 19
Hf L_η	8.139	0.24 3
Hf L_β	9.132	16.1 20
Hf L_γ	10.580	2.6 4
Hf $K_{\alpha2}$	54.611	25 3
Hf $K_{\alpha1}$	55.790	44 6
Hf $K_{\beta1}$'	63.166	14.4 19
Hf $K_{\beta2}$'	65.211	3.7 5
γ E2	90.898 19	15.9 12
γ E2	206.38 3	57.7
γ	211.7 4	~0.06
γ	220.3 5	0.083 11
γ [E2]	222.04 8	0.019 5
γ	259.3 5	0.061 11
γ	302.7 4	0.034 9
γ E2	310.80 4	1.05 17
γ	339.49 24	0.21 3
γ [E1]	363.02 8	0.046 12
γ	365.40 18	~0.010
γ [M1]	371.86 16	0.020 8
γ [E1]	408.41 10	0.046 14
γ E0+M1+E2	419.12 5	0.167 23
γ [M1]	441.11 6	0.202 23
γ [E2]	453.92 5	0.237 23
γ	471.16 21	0.036 9
γ [E2]	491.23 23	0.039 11
γ	560.07 15	0.121 17
γ	574.01 20	0.076 13
γ [M1]	596.04 10	0.147 17
γ E2	602.81 4	0.46 4
γ	614.2 5	0.046 17
γ	621.6 5	0.047 15
γ	628.8 5	0.062 10
γ	657.0 3	0.080 10
γ	703.1 5	0.050 14
γ [E2]	737.09 23	0.050 13
γ M1+14%E0+E2	764.72 4	1.26 10
γ	771.6 16	0.028 12
γ M1+>5.4%E0+E2	809.19 4	0.62 7
γ [E1]	834.33 11	0.19 4
γ	835.11 13	0.19 4
γ [E2]	840.56 6	0.08 3
γ	890.9 6	0.057 9
γ [E2]	900.09 4	0.47 6
γ	928.65 25	0.028 11
γ	933.83 22	0.024 10
γ [E2]	971.10 5	1.22 10
γ [E2]	979.20 11	0.24 4
γ M1+E2	996.58 12	0.38 5
γ [M1]	1005.86 8	0.07 3
γ [E2]	1017.7 3	0.121 23
γ E2	1021.93 4	0.80 7
γ	1029.72 12	~0.018
γ [M1]	1039.01 7	0.156 23
γ [M1]	1050.11 7	0.18 4
γ E2	1066.23 7	0.26 5

Photons (^{174}Ta) (continued)

γ_{mode}	γ(keV)	γ(%)†
γ E0+M1+E2	1083.30 7	0.35 4
γ M1+E2	1097.13 7	0.45 5
γ [E1]	1102.39 12	0.048 9
γ	1105.1 3	0.033 8
γ	1111.63 14	0.024 11
γ (E1)	1127.74 7	0.58 5
γ (E2)	1135.69 6	0.66 6
γ	1138.2 3	0.054 10
γ [E1]	1145.12 10	0.27 4
γ (E2)	1151.35 5	1.11 9
γ [M1]	1166.6 3	0.033 8
γ (E2)	1175.92 7	0.62 6
γ [M1]	1185.75 14	0.242 12
γ [M1]	1192.2 4	0.13 4
γ [E2]	1198.85 10	0.17 4
γ M1+E2	1205.82 4	4.8 4
γ E0+M1+E2	1210.49 17	0.138 23
γ [M1]	1212.24 8	0.62 8
γ [E1]	1217.60 10	0.42 5
γ	1221.20 18	0.030 6
γ [E2]	1226.59 6	0.6 3
γ E0+M1+E2	1228.31 4	1.4 4
γ	1233.42 13	0.208 17
γ M1+E2	1245.39 6	0.63 6
γ	1264.68 18	0.150 23
γ	1288.87 22	0.023 7
γ	1291.67 24	0.098 23
γ [M1]	1296.2 3	0.056 8
γ [E2]	1303.50 7	0.37 4
γ	1316.6 5	<0.022
γ [E2]	1319.21 5	0.104 17
γ [M1]	1328.5 3	0.058 16
γ	1331.2 4	0.104 23
γ	1350.83 17	0.17 3
γ	1356.98 20	0.40 14
γ [E2]	1357.73 6	0.83 12
γ [M1]	1360.90 7	0.92 9
γ [M1]	1405.23 10	0.054 17
γ [E2]	1412.20 5	0.21 4
γ [E2]	1421.96 15	~0.022
γ [M1]	1429.4 3	0.045 8
γ [M1]	1435.4 3	0.092 23
γ	1439.5 3	0.12 3
γ	1477.5 10	0.052 23
γ	1482.38 20	0.115 23
γ [E2]	1496.13 10	0.08 3
γ [M1+E2]	1502.42 7	0.115 23
γ [E2]	1534.9 3	0.121 23
γ	1564.25 15	0.133 23
γ [E2]	1591.5 3	0.054 16
γ	1597.1 6	<0.046
γ	1599.58 17	0.144 23
γ [E2]	1607.0 3	0.14 3
γ	1624.3 17	~0.035
γ [M1+E2]	1629.61 17	0.16 3
γ	1640.24 21	0.24 4
γ	1651.9 11	0.066 10
γ	1654.9 3	0.115 23
γ	1688.76 20	0.063 23
γ [M1+E2]	1732.76 15	0.27 4
γ	1742.1 5	0.13 3
γ	1759.9 12	0.10 3
γ	1762.3 12	0.09 3
γ	1770.63 15	0.16 3
γ	1774.1 7	0.07 3
γ [M1+E2]	1784.5 3	~0.035
γ	1816.3 10	0.069 23
γ	1829.46 11	0.34 4
γ	1853.29 17	0.063 16
γ	1887.6 5	0.08 3
γ	1898.2 5	0.10 3
γ	1928.48 16	0.069 23
γ [E2]	1939.13 15	0.22 4
γ	1944.55 16	0.21 3
γ	1953.8 11	0.08 3
γ	1957.2 11	0.042 7
γ	1968.6 7	0.015 5
γ	1974.2 3	0.100 17
γ	1993.7 7	~0.012
γ	2001.4 7	<0.030
γ	2007.9 4	0.104 17
γ	2021.10 16	0.021 6
γ [M1+E2]	2031.40 25	~0.11
γ	2040.93 13	0.19 6

Photons (^{174}Ta) (continued)

γ_{mode}	γ(keV)	γ(%)†
γ	2056.39 24	0.058 23
γ	2086.2 8	0.046 13
γ [E2]	2103.5 3	0.047 13
γ [E1]	2124.31 11	0.179 23
γ	2144.1 3	0.13 3
γ	2148.7 7	0.049 9
γ	2158.0 11	0.027 8
γ [E2]	2188.6 3	0.052 23
γ	2194.2 3	0.29 4
γ [E2]	2207.67 14	0.036 8
γ [E2]	2232.41 17	0.115 17
γ	2246.0 10	0.069 23
γ	2247.31 13	0.09 3
γ	2262.76 24	0.040 12
γ	2294.72 21	0.087 23
γ	2296.8 9	0.087 23
γ [E1]	2330.69 11	0.017 5
γ [M1+E2]	2343.6 4	0.054 17
γ	2350.5 3	0.084 17
γ	2376.5 12	0.029 11
γ	2390.8 14	0.027 13
γ	2400.6 3	0.21 4
γ [M1+E2]	2414.05 14	0.017 5
γ [M1+E2]	2438.79 17	0.21 3
γ	2443.1 9	0.023 6
γ [E2]	2479.5 3	0.15 4
γ [E2]	2485.8 3	~0.046
γ	2493.2 7	0.036 6
γ	2501.10 21	0.127 23
γ [E2]	2504.95 15	0.028 13
γ [E2]	2529.69 17	0.020 6
γ	2543.9 16	~0.023
γ [E2]	2550.0 4	0.033 7
γ	2557.4 5	0.052 9
γ	2584.2 10	~0.017
γ	2607.5 11	0.032 6
γ [E2]	2634.20 25	0.032 7
γ	2655.6 6	0.063 14
γ	2699.6 7	0.048 13
γ	2706.5 25	0.021 9
γ	2711.0 17	0.022 5
γ	2721.1 17	0.029 6
γ	2726.9 12	0.058 17
γ	2737.1 15	0.048 8
γ	2772.5 14	0.036 13
γ	2779.7 18	0.028 5
γ [M1+E2]	2790.3 3	0.046 6
γ	2796.7 12	0.14 4
γ	2808.3 5	0.077 23
γ	2817.8 15	0.070 19
γ [M1+E2]	2840.58 25	0.046 13
γ	2874.3 13	0.046 10
γ	2889.5 21	0.017 5
γ	2893.5 5	0.058 16
γ	2899.9 17	0.023 5
γ	2917.0 11	~0.06
γ	2927.6 17	0.028 6
γ [E2]	2931.48 25	0.074 16
γ	2938.8 21	0.011 4
γ	2953.0 17	0.022 8
γ	2965.1 14	0.023 8
γ [E2]	2996.7 3	0.014 6
γ	3014.7 5	0.018 6
γ	3080.5 15	0.027 8
γ	3099.8 5	0.018 6
γ	3170.0 15	0.043 12
γ	3206.3 22	0.027 8
γ	3294.2 18	~0.012
γ	3332.4 22	0.016 6
γ	3340.9 18	~0.008
γ	3644 3	0.013 6

† 2.6% uncert(syst)

Atomic Electrons (^{174}Ta)

⟨e⟩=99 *3* keV

e_{bin}(keV)	⟨e⟩(keV)	e(%)
10	5.2	54 *8*
11	4.3	39 *6*
26	4.5	17.8 *13*
43 - 63	1.97	4.1 *3*
80	21.1	26.4 *20*
81	18.9	23.2 *18*
88	0.39	0.44 *3*
89	10.6	12.0 *9*
90 - 91	3.05	3.36 *18*
141	12.5	8.8 *3*
146 - 195	2.03	1.05 *4*
196	4.29	2.19 *7*
197	3.04	1.55 *5*
200 - 250	3.21	1.56 *4*
257 - 303	0.10	0.033 *11*
307 - 356	0.117	0.034 *6*
360 - 410	0.084	0.0218 *20*
417 - 462	0.0212	0.0049 *4*
469 - 509	0.018	~0.0037
531 - 574	0.082	0.0152 *16*
585 - 629	0.024	0.0041 *8*
638 - 672	0.007	0.0010 *4*
692 - 737	0.72	0.102 *13*
744 - 775	0.49	0.066 *8*
798 - 841	0.088	0.0109 *18*
863 - 906	0.048	0.0053 *5*
914 - 962	0.068	0.0072 *9*
964 - 1012	0.044	0.0044 *4*
1015 - 1065	0.053	0.0052 *9*
1066 - 1112	0.083	0.0076 *5*
1116 - 1165	0.92	0.080 *11*
1166 - 1215	0.136	0.0114 *17*
1216 - 1265	0.101	0.0082 *17*
1266 - 1315	0.084	0.0065 *10*
1316 - 1364	0.028	0.00207 *20*
1370 - 1419	0.013	0.00094 *22*
1420 - 1468	0.0099	0.00069 *16*
1470 - 1564	0.024	0.00155 *23*
1575 - 1667	0.022	0.0014 *3*
1677 - 1776	0.025	0.0014 *3*
1777 - 1876	0.016	0.00085 *19*
1877 - 1976	0.022	0.00114 *21*
1982 - 2079	0.0082	0.00040 *7*
2083 - 2182	0.013	0.00059 *13*
2183 - 2278	0.0062	0.00028 *5*
2283 - 2381	0.0104	0.00044 *8*
2388 - 2485	0.0087	0.00036 *5*
2490 - 2582	0.0026	0.000102 *18*
2590 - 2689	0.0040	0.00015 *3*
2690 - 2788	0.0066	0.00024 *4*
2794 - 2891	0.0048	0.00017 *3*
2897 - 2995	0.00118	4.0 *7* $\times 10^{-5}$
3003 - 3098	0.00062	2.0 *6* $\times 10^{-5}$
3105 - 3204	0.0009	2.8 *8* $\times 10^{-5}$
3229 - 3323	0.00040	1.2 *4* $\times 10^{-5}$
3330 - 3339	2.0 $\times 10^{-5}$	~6 $\times 10^{-7}$
3579 - 3642	0.00014	3.9 *19* $\times 10^{-6}$

Continuous Radiation (^{174}Ta)

⟨β+⟩=295 keV;⟨IB⟩=7.6 keV

E_{bin}(keV)		⟨ ⟩(keV)	(%)
0 - 10	β+	1.13 $\times 10^{-7}$	1.34 $\times 10^{-6}$
	IB	0.0112	
10 - 20	β+	6.7 $\times 10^{-6}$	3.98 $\times 10^{-5}$
	IB	0.0106	0.074
20 - 40	β+	0.000270	0.00081
	IB	0.023	0.078
40 - 100	β+	0.0202	0.0254
	IB	0.28	0.48
100 - 300	β+	1.45	0.64
	IB	0.21	0.114
300 - 600	β+	12.7	2.72
	IB	0.38	0.086
600 - 1300	β+	101	10.4
	IB	1.64	0.17
1300 - 2500	β+	177	10.3
	IB	3.8	0.21
2500 - 3909	β+	2.91	0.112
	IB	1.25	0.044
	Σβ+		24

$^{174}_{74}$W (29 *1* min)

Mode: ϵ

Δ: -50150 *300* keV syst

SpA: 3.73×10^{7} Ci/g

Prod: ^{12}C on Er; ^{176}Hf(^{3}He,5n); ^{181}Ta(p,8n)

Photons (^{174}W)

γ_{mode}	γ(keV)	γ(rel)
γ	35.42 *8*	148 *12*
γ	49.84 *8*	27.0 *19*
γ	61.9 *4*	20.5 *17* ?
γ	73.36 *9*	5.9 *6*
γ	75.88 *12*	2.2 *4*
γ	96.44 *8*	10.8 *11* ?
γ	125.18 *5*	81 *7*
γ	136.52 *5*	78 *7*
γ	143.73 *8*	23.9 *22*
γ	162.69 *8*	14.3 *16* ?
γ	173.96 *16*	4.6 *9*
γ	181.41 *9*	<21 ?
γ	193.04 *8*	56 *5*
γ	202.04 *8*	41 *4*
γ	216.36 *19*	7.0 *13*
γ	233.37 *8*	32 *3*
γ	239.51 *13*	12.7 *13* ?
γ	289.81 *16*	9.7 *13* ?
γ	328.68 *7*	100
γ	339.76 *9*	36 *4*
γ	354.97 *11*	20.6 *22*
γ	364.52 *10*	37 *4*
γ	377.04 *10*	57 *7*
γ	378.54 *9*	84 *9*
γ	400.5 *10*	~9 ?
γ	428.83 *7*	123 *11*
γ	472.2 *9*	4.3 *8*
γ	547.5 *4*	4.3 *9*
γ	567.6 *4*	4.4 *10*
γ	835.0 *7*	5.7 *17*

$^{174}_{75}$Re(2.3 *1* min)

Mode: ϵ

Δ: -43610 *350* keV syst

SpA: 4.69×10^{8} Ci/g

Prod: daughter ^{174}Os; ^{159}Tb(^{20}Ne,5n)

Photons (^{174}Re)

γ_{mode}	γ(keV)	γ(rel)
γ E2	112.4 *3*	26.5 *15*
γ E2	243.7 *4*	49
γ E2	349.1 *3*	6.4 *7*
γ	533.8 *5*	1.7 *3*
γ	739.3 *2*	6.2 *7*
γ	759.8 *5*	2.0 *3*
γ	777.2 *2*	3.5 *5*
γ	863.4 *5*	2.4 *3*
γ	900 *1*	~4
γ	903 *1*	5 *2*
γ	981.8 *9*	1.7 *3*
γ	1002.9 *2*	7.50 *15*
γ	1088.2 *5*	3.4 *5*

$^{174}_{76}$Os(44 *4* s)

Mode: ϵ(99.98 *1* %), α(0.02 *1* %)

Δ: -39950 *300* keV syst

SpA: 1.46×10^{9} Ci/g

Prod: ^{164}Er(^{16}O,6n); protons on Hg

Alpha Particles (^{174}Os)

α(keV)
4760 *10*

Photons (^{174}Os)

γ_{mode}	γ(keV)	γ(rel)
γ	118 *1*	100 *20*
γ	138 *1*	25 *5* ?
γ	158 *1*	15 *3*
γ	302 *1*	26 *5*
γ	325 *1*	43 *9*
γ	372 *1*	20 *4*
γ	387.5 *10*	10.0 *20* ?

$^{174}_{77}$Ir(4 *1* s)

Mode: α

Δ: -31060 *400* keV syst

SpA: 1.5×10^{10} Ci/g

Prod: ^{169}Tm(^{16}O,11n); ^{164}Er(^{19}F,9n); ^{162}Er(^{19}F,7n)

Alpha Particles (^{174}Ir)

α(keV)
5478 *6*

$^{174}_{78}$Pt(900 *10* ms)

Mode: α(83 *5* %), ϵ(17 *5* %)
Δ: -25326 *19* keV
SpA: 5.02×10^{10} Ci/g
Prod: ^{162}Er(^{20}Ne,8n); ^{164}Er(^{20}Ne,10n); ^{142}Nd(^{40}Ar,8n); ^{144}Sm(^{32}S,2n)

Alpha Particles (^{174}Pt)

α(keV)
6040 *4*

$^{174}_{79}$Au(120 *20* ms)

Mode: α
Δ: -14210 *230* keV syst
SpA: 9.3×10^{10} Ci/g
Prod: ^{92}Mo on Mo; ^{92}Mo on Rb

Alpha Particles (^{174}Au)

α(keV)
6530 *20*

A = 175

NDS **18**, 331 (1976)

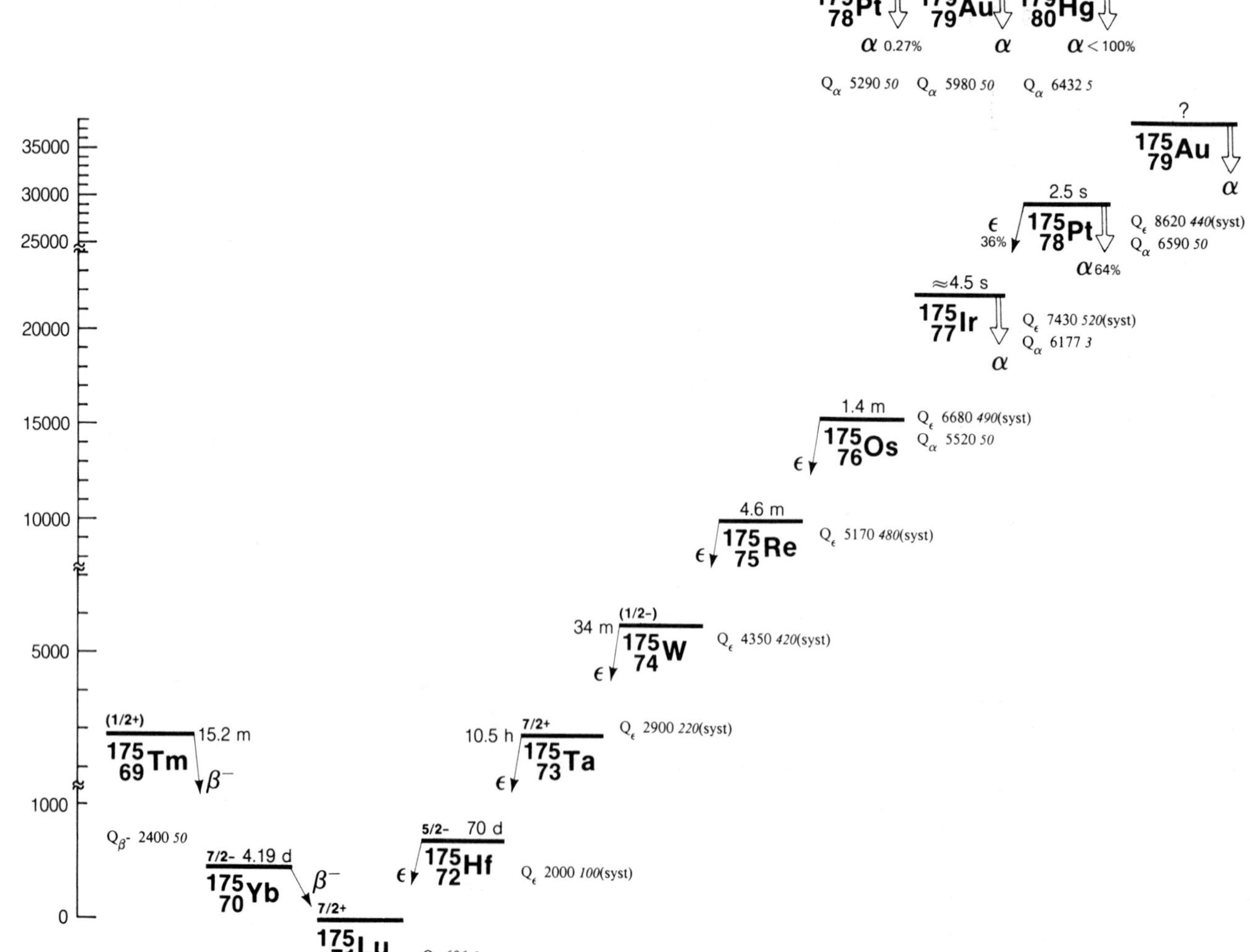

$^{175}_{69}$Tm(15.2 *5* min)

Mode: β-

Δ: -52300 *50* keV

SpA: 7.07×10^{7} Ci/g

Prod: ^{176}Yb(γ,p)

Photons (^{175}Tm)

⟨γ⟩=1074 *42* keV

γmode	γ(keV)	γ(%)†
Yb L_ℓ	6.545	~2
Yb L_α	7.411	~42
Yb L_η	7.580	~0.7
Yb L_β	8.504	~44
Yb L_γ	9.809	~7
γ [M1+E2]	41.2211 *15*	~6
γ [M1+E2]	46.7526 *18*	~0.6
Yb $K_{\alpha2}$	51.354	5.9 *7*
Yb $K_{\alpha1}$	52.389	10.3 *12*
Yb $K_{\beta1}$'	59.316	3.3 *4*
Yb $K_{\beta2}$'	61.229	0.90 *10*
γ [E2]	87.9737 *17*	~0.7
γ [M1+E2]	95.2755 *25*	~0.10
γ [M1+E2]	104.5312 *20*	1.4 *4*
γ	159.7 *3*	~0.4
γ [M1+E2]	172.170 *4*	0.24 *9*
γ [E1]	196.19 *4*	<0.14
γ	296.27 *19*	0.40 *10*
γ [M1+E2]	311.274 *7*	0.7 *3*
γ	324.85 *20*	0.28 *9*
γ [M1+E2]	363.957 *8*	12.8 *13*
γ	394.56 *20*	3.3 *10*
γ [M1]	405.178 *9*	0.26 *11*
γ	423.15 *20*	0.34 *9*
γ [E1]	428.628 *12*	<0.4
γ [M1+E2]	436.178 *11*	2.3 *3*
γ [M1+E2]	477.399 *11*	2.8 *3*
γ [E1]	487.95 *5*	0.72 *11*
γ [E1]	505.07 *5*	<0.9
γ M3	514.867 *10*	65 *7*
γ [E2]	534.731 *11*	2.0 *3*
γ [E2]	556.088 *10*	0.27 *6*
γ [E1]	577.30 *5*	1.24 *16*
γ [M1+E2]	602.841 *10*	0.13 *4*
γ [E1]	625.64 *6*	0.31 *7*
γ [M1+E2]	639.262 *11*	6.1 *7*
γ [E1]	657.32 *7*	0.11 *4*
γ	669.78 *25*	0.18 *6*
γ [E1]	685.91 *5*	<0.23 ?
γ [E2]	767.17 *4*	0.34 *7*
γ	801.34 *19*	0.20 *6*
γ [E2]	811.431 *11*	4.3 *4*
γ [E1]	858.08 *5*	5.7 *6*
γ [M1+E2]	871.70 *4*	0.54 *7*
γ [E1]	894.50 *4*	7.8 *9*
γ [E1]	941.25 *4*	14.2
γ	944.82 *20*	~0.37
γ [E1]	953.89 *7*	1.8 *3*
γ	982.17 *19*	0.9 *3*
γ [E1]	982.47 *4*	9.9 *11*
γ	993.16 *20*	~0.37
γ	1053.43 *20*	0.14 *4*
γ	1064.4 *21*	<0.09 ?
γ	1088.2 *3*	<0.14
γ	1105.6 *21*	<0.11
γ	1126.5 *4*	0.34 *9*
γ	1134.9 *3*	0.28 *9*
γ	1154.34 *19*	0.34 *16*
γ	1167.7 *4*	0.13 *4*
γ	1176.2 *3*	0.31 *4*
γ	1237.52 *19*	~0.28
γ	1252.64 *20*	0.23 *4*
γ	1262.02 *20*	0.27 *6*
γ	1269.8 *5*	<0.09 ?
γ	1278.74 *19*	0.07 *3*
γ	1289.06 *20*	<0.06

Photons (^{175}Tm) (continued)

γmode	γ(keV)	γ(%)†
γ	1308.78 *20*	0.43 *6*
γ	1335.81 *20*	0.47 *6*
γ	1350.00 *20*	0.11 *3*
γ	1377.03 *20*	3.0 *3*
γ	1454.4 *6*	0.13 *3*
γ	1511.9 *7*	0.11 *3*
γ	1525.1 *5*	1.26 *16*
γ	1537.2 *12*	0.10 *3*
γ	1558.7 *7*	0.17 *4*
γ	1566.3 *5*	0.064 *14*
γ	1578.8 *6*	~0.04
γ	1584.0 *12*	0.07 *3*
γ	1599.9 *7*	0.17 *3*
γ	1625.2 *12*	~0.06

† 2.8% uncert(syst)

Atomic Electrons (^{175}Tm)

⟨e⟩=311 *93* keV

e_{bin}(keV)	⟨e⟩(keV)	e(%)
9	15.1	~169
10	12.4	~124
27	0.23	~0.8
31	55.7	~179
32	61.4	<380
34 - 38	8.5	~23
39	35.0	~89
40	0.5	~1
41	9.8	~24
42 - 88	6.9	13 *6*
93 - 135	2.0	2.0 *8*
149 - 196	0.20	0.13 *5*
235 - 264	0.27	0.11 *5*
286 - 333	2.7	0.9 *4*
344 - 393	1.11	0.31 *7*
394 - 436	0.45	0.11 *4*
444	0.010	<0.0043
454	73.2	16.1 *18*
467 - 503	0.21	0.045 *7*
504	14.2	2.8 *3*
505 - 554	9.5	1.87 *10*
555 - 603	0.49	0.09 *4*
608 - 657	0.12	0.019 *6*
659 - 706	0.016	0.0022 *4*
740 - 767	0.165	0.0220 *24*
791 - 833	0.246	0.0302 *23*
848 - 894	0.257	0.0292 *18*
921 - 954	0.21	0.022 *4*
972 - 1007	0.043	0.0044 *9*
1027 - 1074	0.026	0.0024 *10*
1078 - 1127	0.027	0.0025 *9*
1133 - 1176	0.013	~0.0011
1191 - 1238	0.019	0.0016 *5*
1242 - 1289	0.039	0.0031 *10*
1298 - 1348	0.08	~0.006
1350 - 1393	0.017	0.0012 *5*
1444 - 1476	0.034	0.0023 *11*
1497 - 1591	0.019	0.0013 *3*
1597 - 1623	0.00031	1.9 *9* ×10^{-5}

Continuous Radiation (^{175}Tm)

⟨β-⟩=430 keV;⟨IB⟩=0.61 keV

E_{bin}(keV)		⟨ ⟩(keV)	(%)
0 - 10	β-	0.081	1.63
	IB	0.019	
10 - 20	β-	0.243	1.62
	IB	0.018	0.127
20 - 40	β-	0.97	3.24
	IB	0.035	0.121
40 - 100	β-	6.7	9.6
	IB	0.091	0.141
100 - 300	β-	56	28.8
	IB	0.20	0.118
300 - 600	β-	127	29.1
	IB	0.150	0.036
600 - 1300	β-	203	23.6
	IB	0.086	0.0111
1300 - 1885	β-	36.0	2.48
	IB	0.0029	0.00021

$^{175}_{70}$Yb(4.19 *1* d)

Mode: β-

Δ: -54704 *3* keV

SpA: 1.781×10^{5} Ci/g

Prod: ^{174}Yb(n,γ)

Photons (^{175}Yb)

⟨γ⟩=40 *3* keV

γmode	γ(keV)	γ(%)†
Lu L_ℓ	6.753	0.0183 *18*
Lu L_α	7.650	0.45 *3*
Lu L_η	7.857	0.0061 *4*
Lu L_β	8.826	0.44 *3*
Lu L_γ	10.233	0.072 *7*
Lu $K_{\alpha2}$	52.965	1.10 *4*
Lu $K_{\alpha1}$	54.070	1.93 *8*
Lu $K_{\beta1}$'	61.219	0.622 *24*
Lu $K_{\beta2}$'	63.200	0.166 *7*
γ M1+18%E2	113.806 *4*	1.91 *6*
γ M1+17%E2	137.659 *5*	0.117 *7*
γ E1(+0.07%M2)	144.863 *5*	0.332 *19*
γ E2	251.465 *6*	0.085 *7*
γ E1(+0.4%M2)	282.522 *6*	3.05 *9*
γ E1+1.1%M2	396.328 *7*	6.5

† 12% uncert(syst)

Atomic Electrons (^{175}Yb)

⟨e⟩=4.15 *8* keV

e_{bin}(keV)	⟨e⟩(keV)	e(%)
9	0.156	1.68 *18*
10 - 45	0.171	1.26 *10*
50	1.87	3.71 *13*
51 - 82	0.154	0.218 *9*
103	0.75	0.73 *3*
105	0.171	0.164 *7*
111	0.126	0.113 *4*
112 - 145	0.230	0.198 *5*
188	0.0139	0.0074 *6*
219	0.151	0.069 *3*
241 - 283	0.0474	0.0177 *5*
333	0.25	0.075 *9*
385 - 396	0.058	0.0150 *13*

Continuous Radiation (^{175}Yb)

⟨β-⟩=127 keV;⟨IB⟩=0.060 keV

E_{bin}(keV)		⟨ ⟩(keV)	(%)
0 - 10	β-	0.379	7.7
	IB	0.0065	
10 - 20	β-	1.01	6.8
	IB	0.0058	0.041
20 - 40	β-	3.28	11.1
	IB	0.0099	0.035
40 - 100	β-	15.8	23.2
	IB	0.020	0.032
100 - 300	β-	83	45.5
	IB	0.018	0.0120
300 - 469	β-	23.0	6.7
	IB	0.00056	0.000167

$^{175}_{71}$Lu(stable)

Δ: -55173 *3* keV
%: 97.41 *2*

$^{175}_{72}$Hf(70 *2* d)

Mode: ε
Δ: -54486 *4* keV
SpA: 1.07×10^4 Ci/g
Prod: ^{174}Hf(n,γ); ^{175}Lu(d,2n); ^{175}Lu(p,n)

Photons (^{175}Hf)

⟨γ⟩=364 *6* keV

γ_{mode}	γ(keV)	γ(%)†
Lu L_ℓ	6.753	0.42 *8*
Lu L_α	7.650	10.4 *19*
Lu L_η	7.857	0.135 *24*
Lu L_β	8.825	10.5 *20*
Lu L_γ	10.262	1.8 *4*
Lu $K_{\alpha2}$	52.965	27 *5*
Lu $K_{\alpha1}$	54.070	47 *8*
Lu $K_{\beta1}$'	61.219	15 *3*
Lu $K_{\beta2}$'	63.200	4.0 *7*
γ M1+1.0%E2	89.356 *10*	2.3 *3*
γ M1+18%E2	113.806 *4*	0.31 *4*
γ [M1+E2]	143.9 *6*	~0.007
γ M1	161.3 *6*	0.023 *9*
γ E2	229.59 *8*	0.77 *16*
γ M1+2.0%E2	318.95 *8*	0.17 *4*
γ M1+7.8%E2	343.40 *8*	87.0
γ E1	353.6 *6*	0.23 *4*
γ M1	432.75 *8*	1.57 *17*

† 0.57% uncert(syst)

Atomic Electrons (^{175}Hf)

⟨e⟩=43.9 *11* keV

e_{bin}(keV)	⟨e⟩(keV)	e(%)
9	3.5	38 *8*
10	1.9	18 *4*
11	0.96	8.8 *18*
26	2.7	10.3 *12*
42 - 61	2.42	5.2 *4*
78	1.13	1.45 *16*
79 - 114	0.82	0.92 *6*
133 - 166	0.15	0.094 *18*

Atomic Electrons (^{175}Hf)
(continued)

e_{bin}(keV)	⟨e⟩(keV)	e(%)
219 - 256	0.167	0.072 *8*
280	24.0	8.57 *20*
290 - 319	0.021	0.0070 *9*
333	4.33	1.30 *3*
334 - 369	1.73	0.501 *13*
422 - 433	0.072	0.0170 *15*

Continuous Radiation (^{175}Hf)

⟨IB⟩=0.26 keV

E_{bin}(keV)		⟨ ⟩(keV)	(%)
10 - 20	IB	0.00026	0.0020
20 - 40	IB	0.0043	0.0126
40 - 100	IB	0.25	0.46
100 - 300	IB	0.0078	0.0054
300 - 343	IB	1.44×10^{-5}	4.7×10^{-6}

$^{175}_{73}$Ta(10.5 *2* h)

Mode: ε
Δ: -52490 *100* keV syst
SpA: 1.71×10^6 Ci/g
Prod: ^{175}Lu(α,4n); ^{176}Hf(p,2n); daughter ^{175}W; protons on ^{181}Ta

Photons (^{175}Ta)

⟨γ⟩=1049 *24* keV

γ_{mode}	γ(keV)	γ(%)
Hf L_ℓ	6.960	0.64 *12*
Hf L_α	7.893	16 *3*
Hf L_η	8.139	0.21 *4*
Hf L_β	9.139	16 *3*
Hf L_γ	10.629	2.8 *5*
γ [E2]	35.91 *16*	
γ M1+11%E2	50.58 *16*	0.014 *3*
Hf $K_{\alpha2}$	54.611	37 *6*
Hf $K_{\alpha1}$	55.790	64 *10*
Hf $K_{\beta1}$'	63.166	21 *3*
Hf $K_{\beta2}$'	65.211	5.5 *9*
γ M1+45%E2	70.54 *21*	0.0065 *13*
γ [M1+E2]	77.22 *20*	0.0057 *11*
γ M1+5.9%E2	81.56 *14*	5.7 *10*
γ [E1]	90.40 *16*	0.38 *11*
γ M1+5.0%E2	104.58 *16*	3.0 *4*
γ E2	126.09 *20*	2.5 *8*
γ [E1]	126.14 *14*	5.5 *11*
γ [M1+E2]	126.61 *17*	0.34 *11*
γ [M1+E2]	126.73 *16*	0.27 *11*
γ (E1)	140.97 *15*	2.2 *3*
γ (M1+E2)	162.41 *19*	~0.8
γ M1+16%E2	162.52 *16*	1.4 *3*
γ [M1+E2]	162.64 *17*	0.16 *3*
γ [E1]	176.71 *17*	0.27 *11*
γ (E2)	179.27 *21*	1.22 *19*
γ (E2)	186.14 *16*	0.61 *8*
γ (E2)	192.9 *3*	0.35 *5*
γ [M1+E2]	196.63 *19*	0.15 *4*
γ (E1)	207.70 *15*	13.3 *4*
γ [E1]	217.12 *18*	0.179 *23*
γ (E2)	231.19 *16*	0.67 *4*
γ	256.3 *10*	~0.08 ?
γ [M1+E2]	260.42 *25*	0.49 *11*
γ (M1)	267.11 *15*	10.3 *11*
γ [E1]	267.70 *17*	0.42 *15*

Photons (^{175}Ta)
(continued)

γ_{mode}	γ(keV)	γ(%)
γ	275.3 *10*	~0.046 ?
γ (M1)	280.8 *4*	0.65 *4*
γ (M1)	289.25 *18*	1.41 *15*
γ [M1+E2]	294.34 *19*	0.16 *4*
γ (M1)	308.98 *23*	0.16 *5*
γ [E1]	331.72 *23*	0.053 *23*
γ (E2+47%M1)	348.67 *15*	11.4 *6*
γ	357.3 *10*	~0.030
γ (M1)	361.7 *3*	0.32 *4*
γ	366.0 *5*	0.20 *6*
γ	375.3 *10*	~0.011
γ	380.3 *10*	~0.03
γ (M1)	386.20 *21*	0.61 *6*
γ (M1)	393.84 *17*	2.01 *15*
γ [E1]	400.38 *25*	<0.046
γ	404.4 *6*	0.019 *4*
γ	433.1 *10*	0.049 *19*
γ (M1)	436.78 *21*	3.8 *4*
γ (M1)	443.4 *4*	0.15 *3*
γ	448.7 *10*	~0.015
γ	450.8 *7*	~0.06
γ [M1+E2]	462.55 *24*	0.22 *4*
γ	467.7 *7*	0.046 *19*
γ	470.9 *5*	0.160 *19*
γ (E2)	475.40 *18*	1.94 *19*
γ	482.1 *7*	0.057 *19*
γ [E1]	485.30 *25*	0.13 *3*
γ	502.3 *8*	0.12 *4*
γ [M1+E2]	525.3 *4*	0.32 *5*
γ	533.3 *4*	0.20 *3*
γ (M1)	539.78 *21*	0.91 *11*
γ [M1+E2]	546.3 *4*	0.12 *3*
γ [M1+E2]	549.5 *4*	0.07 *3*
γ [E1]	561.95 *22*	0.14 *3*
γ	568.7 *9*	0.057 *19*
γ	572.5 *4*	0.19 *3*
γ	588.6 *6*	0.11 *3*
γ	592.1 *8*	~0.04
γ (M1)	600.1 *4*	0.22 *4*
γ	609.6 *9*	0.16 *8*
γ [E1]	619.84 *24*	0.42 *8*
γ	661.7 *4*	~0.11
γ	676.5 *5*	~0.08
γ	697.8 *9*	~0.23
γ [M1+E2]	701.5 *4*	~0.19
γ	720.4 *5*	<0.046
γ (M1)	731.03 *18*	0.49 *6*
γ	739.7 *4*	0.18 *4*
γ (M1)	750.08 *22*	0.23 *4*
γ	759.3 *10*	0.11 *4* ?
γ [M1+E2]	773.74 *24*	<0.11
γ	776.6 *7*	0.13 *3*
γ	784.3 *5*	0.25 *6*
γ [M1+E2]	789.2 *4*	0.12 *4*
γ [M1+E2]	802.2 *4*	~0.08
γ (M1)	808.76 *25*	0.61 *8*
γ	812.3 *9*	0.08 *3*
γ [M1+E2]	819.1 *3*	~0.05
γ (M1)	842.8 *3*	~0.12
γ (E2)	849.58 *23*	0.46 *6*
γ [M1+E2]	852.8 *4*	~0.11
γ (M1)	857.76 *16*	3.0 *3*
γ [M1+E2]	866.5 *3*	0.51 *5*
γ (M1)	873.23 *24*	0.30 *4*
γ (M1)	876.81 *21*	0.72 *8*
γ	887.5 *7*	~0.13
γ [M1+E2]	893.67 *19*	0.14 *5*
γ (M1)	900.46 *22*	0.65 *8*
γ [M1+E2]	917.0 *3*	~0.06
γ	926.3 *4*	0.11 *4*
γ	934.1 *11*	<0.13
γ	938.2 *6*	0.14 *7*
γ [E1]	948.16 *19*	~0.19
γ	949.3 *12*	~0.18
γ	959.7 *9*	~0.23
γ [M1+E2]	962.33 *24*	1.37 *19*
γ [E1]	967.20 *23*	0.19 *8* ?
γ	986.1 *7*	0.22 *7*
γ [M1+E2]	991.5 *3*	0.46 *8*
γ [E2]	992.94 *22*	0.22 *7*
γ (M1)	998.73 *17*	2.4 *3*
γ	1010.9 *4*	0.78 *10*
γ [M1+E2]	1020.28 *19*	0.38 *15*

Photons (^{175}Ta)
(continued)

γ_{mode}	γ(keV)	γ(%)
γ[M1+E2]	1020.7 *3*	<0.11
γ	1028.7 *4*	0.40 *8*
γ[E2]	1035.65 *22*	0.76 *11*
γ	1052.2 *7*	~0.46
γ	1053.7 *15*	<0.30
γ[M1+E2]	1062.99 *24*	<0.34
γ	1068.0 *16*	0.49 *24*
γ	1071.4 *7*	0.16 *7*
γ	1083.3 *8*	0.15 *7*
γ[E2]	1085.8 *4*	~0.11
γ[M1+E2]	1091.0 *3*	0.17 *4*
γ[M1+E2]	1096.3 *3*	0.30 *8*
γ	1107.9 *5*	0.38 *11*
γ	1114.7 *6*	0.28 *5*
γ[M1+E2]	1118.18 *25*	0.76 *19*
γ[M1+E2]	1120.3 *5*	<0.30
γ[M1+E2]	1124.87 *18*	0.14 *3*
γ(M1)	1143.92 *21*	1.1 *3*
γ	1172.1 *4*	0.42 *11*
γ(M1)	1174.3 *3*	0.19 *8*
γ	1177.9 *9*	<0.15
γ	1195.8 *6*	0.22 *8*
γ	1199.4 *11*	~0.08
γ[E2]	1206.43 *18*	0.46 *11*
γ[E1]	1208.6 *3*	0.53 *11*
γ[M2]	1210.7 *5*	0.57 *11* ?
γ[M1+E2]	1225.48 *21*	2.4 *3*
γ	1231.3 *15*	~0.10
γ	1240.8 *7*	0.11 *5*
γ[E2]	1249.13 *22*	0.57 *19*
γ[M1+E2]	1249.9 *3*	2.3 *6*
γ[E1]	1259.2 *3*	0.46 *11*
γ[E1]	1261.3 *5*	0.19 *8*
γ[M1+E2]	1271.31 *25*	0.57 *15*
γ	1279.6 *7*	0.26 *9*
γ	1283.4 *3*	0.19 *7*
γ[E1]	1294.05 *24*	0.42 *6*
γ	1312.0 *11*	0.15 *3* ?
γ	1325.0 *10*	0.21 *5*
γ[M1+E2]	1348.53 *23*	0.27 *11*
γ[M1+E2]	1385.3 *3*	0.11 *4*
γ(E2)	1399.11 *23*	0.300 *23*
γ[E1]	1420.66 *25*	0.09 *4*
γ[E1]	1444.99 *25*	0.17 *7*
γ	1445.8 *3*	0.17 *7*
γ	1451.6 *3*	0.34 *8*
γ	1462.7 *3*	0.28 *7*
γ[M1+E2]	1466.8 *3*	~0.19
γ(M1)	1469.0 *5*	0.53 *15*
γ	1473.8 *8*	0.14 *7*
γ(M1)	1483.29 *25*	0.32 *5*
γ(E2)	1490.6 *3*	0.76 *11*
γ[E1]	1506.0 *3*	0.46 *11*
γ	1516.4 *7*	0.34 *8*
γ[E1]	1525.24 *24*	0.13 *6*
γ	1536.5 *6*	0.17 *5*
γ	1544.4 *5*	0.26 *8*
γ[M1+E2]	1560.51 *24*	0.29 *7*
γ	1577.7 *3*	0.14 *5*
γ	1578.9 *22*	~0.06
γ[E1]	1581.6 *3*	<0.5
γ[M1+E2]	1585.96 *24*	1.52 *23*
γ	1590.5 *13*	~0.11
γ(M1)	1611.08 *24*	0.25 *6*
γ[M1+E2]	1617.2 *3*	<0.7
γ(M1)	1618.56 *24*	1.25 *15*
γ	1620.4 *6*	0.38 *15*
γ[E1]	1632.6 *3*	<0.7
γ(M1)	1636.12 *24*	1.60 *23*
γ	1642.1 *5*	<0.11
γ	1650.3 *4*	0.29 *7*
γ(E1)	1659.29 *25*	1.03 *11*
γ	1670.1 *5*	0.14 *3*
γ	1680.5 *3*	0.22 *6*
γ	1682.5 *12*	0.08 *3*
γ[M1+E2]	1686.70 *23*	~0.11
γ	1695.6 *5*	<0.23
γ[E1]	1708.2 *3*	0.46 *11*
γ[E1]	1712.09 *24*	1.10 *15*
γ[M1+E2]	1721.8 *3*	1.10 *15*
γ	1733.4 *13*	~0.06
γ(E1)	1737.22 *23*	0.87 *11*
γ(E1)	1744.70 *24*	1.29 *15*

Photons (^{175}Ta)
(continued)

γ_{mode}	γ(keV)	γ(%)
γ	1760.2 *7*	~0.038
γ	1767.8 *4*	0.21 *5*
γ(E1)	1793.65 *23*	4.4 *5*
γ[E1]	1812.83 *25*	0.34 *8*
γ(E1)	1826.26 *24*	1.18 *19*
γ	1836.3 *5*	0.07 *3*
γ	1849.6 *6*	0.057 *23*
γ	1881.1 *9*	~0.06
γ	1888.2 *3*	0.36 *7*
γ	1892.1 *5*	0.29 *6*

Atomic Electrons (^{175}Ta)
⟨e⟩=60.0 *22* keV

e_{bin}(keV)	⟨e⟩(keV)	e(%)
5	0.052	1.00 *20*
10	5.2	54 *10*
11	4.3	39 *7*
12	0.08	~0.7
16	5.5	34 *6*
22 - 35	0.085	0.33 *6*
39	3.9	9.9 *14*
40 - 60	4.5	9.7 *6*
61	2.1	3.5 *12*
62 - 69	0.7	1.0 *5*
70	3.4	4.9 *8*
71 - 78	2.26	3.1 *3*
79	1.29	1.6 *3*
80 - 95	2.02	2.24 *18*
97	1.6	1.7 *6*
98 - 114	0.82	0.77 *7*
115	1.4	1.2 *4*
116 - 163	4.0	3.1 *4*
165 - 198	1.19	0.67 *5*
202	4.6	2.3 *3*
203 - 251	1.16	0.52 *3*
254 - 281	1.36	0.52 *4*
283	2.28	0.80 *5*
284 - 332	0.87	0.272 *18*
335 - 384	1.75	0.49 *3*
385 - 435	0.43	0.103 *8*
436 - 485	0.32	0.067 *8*
491 - 540	0.120	0.023 *4*
544 - 592	0.074	0.013 *4*
596 - 636	0.058	0.009 *3*
650 - 700	0.103	0.0154 *20*
701 - 750	0.118	0.0161 *22*
754 - 803	0.34	0.043 *4*
806 - 855	0.218	0.0264 *16*
856 - 902	0.14	0.016 *4*
906 - 955	0.28	0.030 *4*
956 - 1003	0.14	0.014 *3*
1006 - 1055	0.121	0.0116 *22*
1057 - 1106	0.082	0.0076 *17*
1107 - 1145	0.147	0.0130 *19*
1160 - 1209	0.25	0.022 *4*
1210 - 1259	0.065	0.0053 *11*
1260 - 1302	0.024	0.0019 *6*
1309 - 1355	0.0152	0.00114 *16*
1374 - 1421	0.068	0.0048 *8*
1425 - 1474	0.048	0.0033 *5*
1479 - 1577	0.228	0.0148 *14*
1579 - 1678	0.114	0.0070 *7*
1679 - 1771	0.082	0.0048 *4*
1782 - 1881	0.026	0.0014 *3*
1882 - 1890	0.00043	2.3 *9* $\times10^{-5}$

Continuous Radiation (^{175}Ta)
⟨β+⟩=2.14 keV;⟨IB⟩=0.75 keV

E_{bin}(keV)		⟨ ⟩(keV)	(%)
0 - 10	β+	8.6×10^{-8}	1.01×10^{-6}
	IB	0.00130	
10 - 20	β+	4.97×10^{-6}	2.96×10^{-5}
	IB	0.00044	0.0034
20 - 40	β+	0.000193	0.00058
	IB	0.0038	0.0114
40 - 100	β+	0.0127	0.0161
	IB	0.27	0.49
100 - 300	β+	0.55	0.256
	IB	0.032	0.017
300 - 600	β+	1.35	0.324
	IB	0.101	0.023
600 - 1300	β+	0.226	0.0337
	IB	0.30	0.035
1300 - 1919	IB	0.035	0.0025
	Σβ+		0.63

$^{175}_{74}$W (34 *1* min)

Mode: ϵ
Δ: -49590 *200* keV syst
SpA: 3.16×10^{7} Ci/g
Prod: ^{181}Ta(p,7n)

Photons (^{175}W)

γ_{mode}	γ(keV)
γ[E1]	14.978 *16*
γ M1+E2	36.408 *16*
γ E1(+M2)	51.385 *16*
γ(E1)	121.14 *8*
γ(M1+E2)	149.15 *8*
γ(M1+E2)	166.70 *10*
γ E1(+M2)?	270.29 *8*

$^{175}_{75}$Re(4.6 *4* min)

Mode: ϵ
Δ: -45240 *360* keV syst
SpA: 2.33×10^{8} Ci/g
Prod: ^{165}Ho(^{16}O,6n); ^{159}Tb(^{22}Ne,6n); ^{181}Ta(α,10n)

Photons (^{175}Re)

γ_{mode}	γ(keV)
γ	185.4 *3*

$^{175}_{76}$Os(1.4 *1* min)

Mode: ϵ
Δ: -40070 *310* keV syst
SpA: 7.6×10^{8} Ci/g
Prod: protons on Hg

Photons (^{175}Os)

γ_{mode}	γ(keV)	γ(rel)
γ	125 *1*	100
γ	170.1 *5*	6.2 *12*
γ	181 *1*	10.8 *22*
γ	226 *1*	4.5 *9*
γ	248 *1*	8.6 *17*
γ	308 *1*	3.4 *7*
γ	361 *1* ?	
γ	410 *1* ?	5 *1*

$^{175}_{77}$Ir(4.5 *10* s)

Mode: α
Δ: -33400 *380* keV syst
SpA: 1.3×10^{10} Ci/g
Prod: ^{16}O on Tm; ^{19}F on Er

Alpha Particles (^{175}Ir)

α(keV)
5393 *5*

$^{175}_{78}$Pt(2.52 *8* s)

Mode: α(64 *5* %), ϵ(36 *5* %)
Δ: -25960 *380* keV syst
SpA: 2.24×10^{10} Ci/g
Prod: ^{16}O on Yb; ^{20}Ne on Er; daughter ^{179}Hg; ^{142}Nd(^{40}Ar,7n)

Alpha Particles (^{175}Pt)

⟨α⟩=3801 *320* keV

α(keV)	α(%)
5831 *10*	4.7 *10*
5960 *3*	54 *5*
6038 *10*	4.7 *8*

Photons (^{175}Pt)

γ_{mode}	γ(keV)
γ_α	77.4 *8*
γ_α	135.4 *8*
γ_α	212.8 *8*

$^{175}_{79}$Au($t_{1/2}$ unknown)

Mode: α
Δ: -17340 *230* keV syst
Prod: ^{141}Pr(^{40}Ca,6n)

Alpha Particles (^{175}Au)

α(keV)
6440 *10*

A = 176

NDS **19**, 383 (1976)

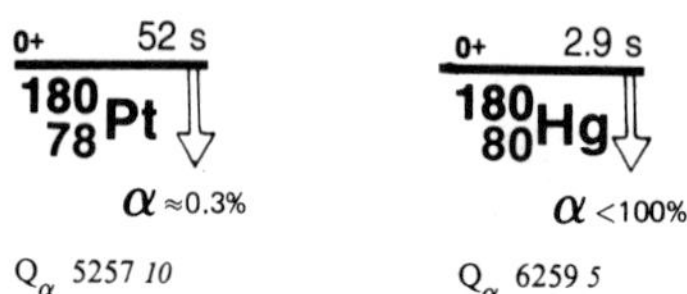

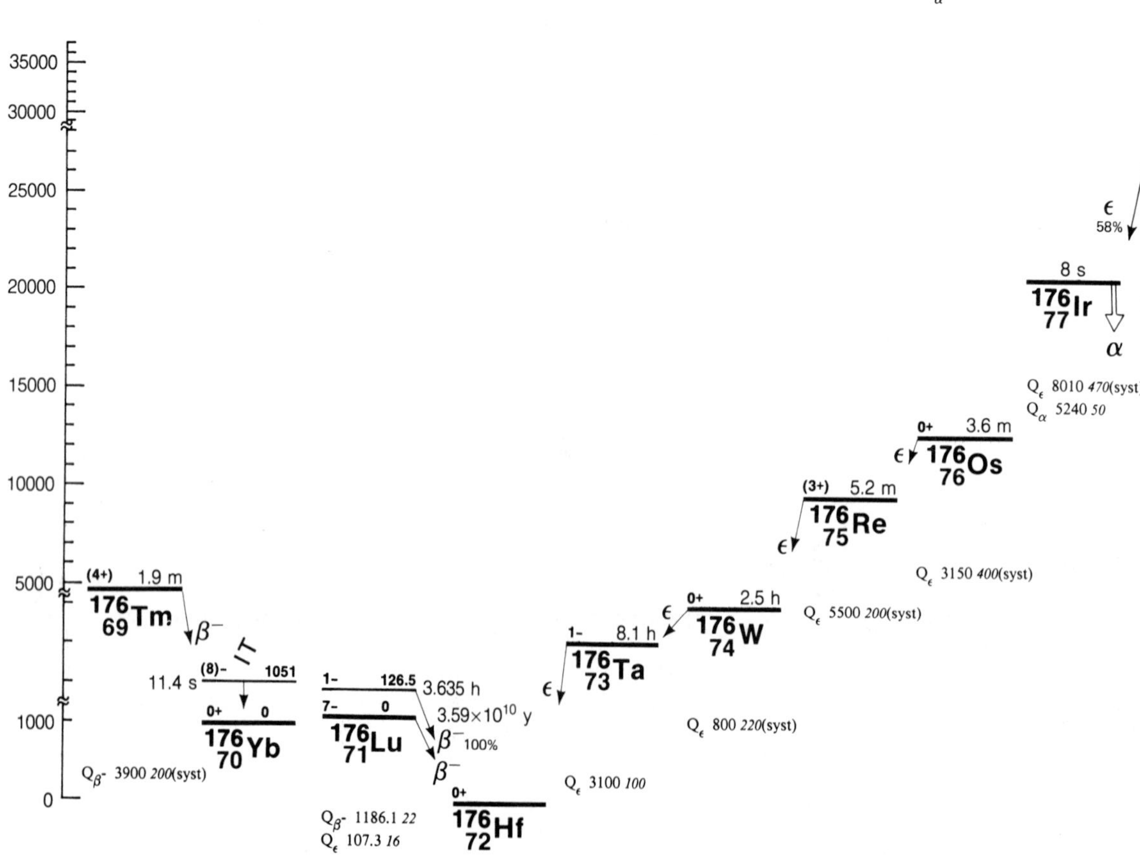

$^{176}_{69}$Tm(1.9 *1* min)

Mode: β-
Δ: -49600 *200* keV syst
SpA: 5.6×10^{8} Ci/g
Prod: ^{176}Yb(n,p)

Photons (^{176}Tm)

⟨γ⟩=1949 *86* keV

γ_{mode}	γ(keV)	γ(%)†
Yb L_{ℓ}	6.545	0.35 *5*
Yb L_{α}	7.411	8.7 *10*
Yb L_{η}	7.580	0.148 *16*
Yb L_{β}	8.504	9.1 *10*
Yb L_{γ}	9.813	1.42 *17*
Yb $K_{\alpha2}$	51.354	10.1 *8*
Yb $K_{\alpha1}$	52.389	17.7 *13*
Yb $K_{\beta1}$'	59.316	5.6 *4*
Yb $K_{\beta2}$'	61.229	1.54 *12*
γ E2	82.17 *9*	11.6 *13*
γ [M1]	95.92 *9*	0.86 *10*
γ	101.1 *3*	0.92 *10*
γ	111.2 *3*	0.10 *3*
γ [M1]	172.78 *9*	0.96 *10*
γ E2	189.80 *9*	44 *4*
γ	215.3 *3*	0.40 *7*
γ [E1]	234.40 *13*	3.2 *3*
γ [M1]	238.27 *16*	2.5 *3*
γ [M1]	239.78 *12*	7.9 *10*
γ	241.9 *3*	1.09 *13*
γ	255.14 *16*	1.09 *13*
γ	289.07 *14*	1.42 *13*
γ E2	292.87 *18*	3.5 *3*
γ [M1]	299.60 *14*	3.2 *3*
γ	305.28 *15*	0.20 *7*
γ	330.32 *11*	8.6 *10*
γ	343.51 *15*	6.9 *7*
γ	347.8 *4*	0.9 *3*
γ	366.60 *13*	<2
γ	381.87 *14*	23.1 *23*
γ	392.1 *5*	0.69 *13*
γ	410.52 *14*	<0.10
γ	410.58 *18*	4.6 *3*
γ	423.12 *17*	0.82 *17*
γ	436.81 *17*	<0.6
γ	440.9 *7*	0.30 *7*
γ	449.0 *7*	0.69 *13*
γ	451.5 *4*	1.22 *23*
γ	457.05 *16*	2.8 *3*
γ	478.05 *15*	<1
γ	482.05 *24*	2.21 *23*
γ	498.14 *15*	0.92 *10*
γ	520.2 *5*	0.82 *10*
γ	539.38 *13*	0.69 *10*
γ	554.65 *15*	0.50 *7*
γ	571.5 *3*	0.69 *10*
γ	621.65 *17*	3.4 *3*
γ	654.82 *25*	0.59 *13*
γ	712.19 *15*	0.69 *13*
γ	754.3 *9*	0.59 *13*
γ	774.8 *8*	1.12 *13*
γ	809.3 *4*	1.91 *20*
γ	852.8 *8*	1.02 *10*
γ	900.5 *3*	2.6 *3*
γ	921.41 *17*	0.50 *10*
γ	1005.97 *17*	1.02 *10*
γ	1011.3 *3*	1.52 *17*
γ	1023.2 *9*	0.59 *13*
γ	1049.88 *16*	7.0 *7*
γ	1069.14 *15*	33
γ	1088.14 *18*	5.7 *7*
γ	1111.21 *16*	5.5 *7*
γ	1121.4 *8*	<1.0
γ [E2]	1163.5 *5*	1.02 *10*
γ [E2]	1178.68 *20*	2.9 *3*
γ	1254.1 *3*	2.01 *20*

Photons (^{176}Tm)
(continued)

γ_{mode}	γ(keV)	γ(%)†
γ	1258.95 *15*	<6
γ [E2]	1260.85 *21*	2.31 *23*
γ	1273.2 *8*	0.59 *7*
γ	1282.4 *3*	1.82 *20*
γ	1349.48 *16*	1.52 *17*
γ [E2]	1353.3 *5*	0.69 *13*
γ	1358.22 *18*	0.30 *7*
γ	1493.34 *16*	0.89 *20*
γ	1521.3 *9*	1.02 *10*
γ	1589.26 *14*	2.8 *3*
γ	1612.7 *3*	1.02 *10*
γ	1748.0 *11*	0.69 *7*
γ	1756.3 *6*	0.50 *10*
γ	1845.1 *5*	0.69 *7*
γ	1881.5 *3*	0.50 *10*
γ	1891.8 *9*	<0.8
γ	1971.13 *18*	2.44 *23*
γ	2071.3 *3*	0.50 *10*
γ	2265.8 *6*	0.50 *10*
γ	2403.2 *13*	<0.50
γ	2455.6 *6*	0.92 *10*
γ	2614.1 *13*	0.92 *10*
γ	2621.4 *5*	2.9 *3*
γ	2677.7 *6*	0.92 *10*
γ	2681.8 *3*	1.32 *13*
γ	2780.3 *3*	0.59 *7*
γ	2867.5 *6*	1.82 *20*
γ	2871.6 *3*	2.11 *20*
γ	2914.2 *5*	4.3 *3*

† 21% uncert(syst)

Atomic Electrons (^{176}Tm)

⟨e⟩=107 *5* keV

e_{bin}(keV)	⟨e⟩(keV)	e(%)
9	3.1	35 *5*
10	2.6	26 *4*
21	3.5	16.7 *19*
35 - 59	2.4	6.2 *13*
72	18.5	26 *3*
73	17.5	24 *3*
80	9.7	12.1 *14*
81 - 111	5.1	5.8 *8*
128	10.9	8.5 *8*
154 - 177	1.63	0.94 *11*
178	3.5	2.0 *3*
179	1.65	0.92 *9*
180	3.6	2.02 *20*
181	3.0	1.7 *3*
187 - 236	4.7	2.32 *18*
237 - 287	4.3	1.6 *5*
288 - 320	0.7	0.24 *8*
321	3.1	~1.0
328 - 375	1.9	0.53 *17*
379 - 428	1.2	0.30 *8*
430 - 480	0.40	0.089 *24*
481 - 529	0.14	0.028 *12*
530 - 572	0.25	~0.05
593 - 622	0.09	0.015 *6*
644 - 693	0.08	0.012 *6*
702 - 748	0.16	0.022 *10*
752 - 800	0.07	0.009 *4*
807 - 853	0.12	~0.014
860 - 900	0.041	0.0047 *20*
911 - 950	0.09	0.010 *5*
962 - 1011	1.4	~0.13
1013 - 1060	0.59	0.056 *19*
1067 - 1112	0.15	0.013 *4*
1117 - 1164	0.080	0.0071 *7*
1168 - 1212	0.23	0.019 *8*
1221 - 1264	0.09	0.007 *3*
1271 - 1297	0.070	0.0054 *19*
1339 - 1358	0.012	0.0009 *3*
1432 - 1460	0.041	0.0028 *12*
1483 - 1580	0.09	0.0061 *23*
1587 - 1611	0.0055	0.00035 *13*

Atomic Electrons (^{176}Tm)
(continued)

e_{bin}(keV)	⟨e⟩(keV)	e(%)
1687 - 1784	0.037	0.0021 *6*
1820 - 1910	0.058	0.0031 *11*
1961 - 2010	0.014	0.00072 *25*
2061 - 2069	0.0013	6 *3* ×10^{-5}
2205 - 2264	0.008	0.00036 *16*
2342 - 2401	0.015	0.0006 *3*
2445 - 2454	0.0020	8 *3* ×10^{-5}
2553 - 2620	0.078	0.0030 *7*
2667 - 2719	0.011	0.00041 *11*
2770 - 2869	0.094	0.0033 *8*
2870 - 2912	0.008	0.00027 *10*

Continuous Radiation (^{176}Tm)

⟨β-⟩=743 keV;⟨IB⟩=1.54 keV

E_{bin}(keV)		⟨ ⟩(keV)	(%)
0 - 10	β-	0.0390	0.78
	IB	0.029	
10 - 20	β-	0.117	0.78
	IB	0.028	0.20
20 - 40	β-	0.478	1.59
	IB	0.055	0.19
40 - 100	β-	3.48	4.95
	IB	0.151	0.23
100 - 300	β-	35.0	17.5
	IB	0.39	0.22
300 - 600	β-	105	23.7
	IB	0.38	0.090
600 - 1300	β-	316	34.7
	IB	0.39	0.046
1300 - 2500	β-	256	15.3
	IB	0.109	0.0067
2500 - 3628	β-	27.1	0.97
	IB	0.0041	0.000144

$^{176}_{70}$Yb(stable)

Δ: -53502 *3* keV
%: 12.7 *1*

$^{176}_{70}$Yb(11.4 *5* s)

Mode: IT
Δ: -52452 *3* keV
SpA: 5.46×10^{9} Ci/g
Prod: ^{176}Y(n,n'); ^{176}Yb(d,d')

Photons (^{176}Yb)

⟨γ⟩=897 *23* keV

γ_{mode}	γ(keV)	γ(%)†
Yb L_{ℓ}	6.545	0.52 *8*
Yb L_{α}	7.411	12.9 *18*
Yb L_{η}	7.580	0.21 *3*
Yb L_{β}	8.507	13.3 *19*
Yb L_{γ}	9.818	2.1 *3*
Yb $K_{\alpha2}$	51.354	18.2 *14*
Yb $K_{\alpha1}$	52.389	32.0 *24*
Yb $K_{\beta1}$'	59.316	10.2 *8*
Yb $K_{\beta2}$'	61.229	2.79 *21*
γ E2	82.17 *9*	14 *3*

Photons (^{176}Yb)
(continued)

γ_{mode}	γ(keV)	γ(%)†
γ E1	96.0 *4*	73 *7*
γ E2	189.80 *9*	81 *6*
γ E2	292.87 *18*	92.60
γ E2	389.6 *5*	91 *5*

† 0.22% uncert(syst)

Atomic Electrons (^{176}Yb)
⟨e⟩=153 *7* keV

e_{bin}(keV)	⟨e⟩(keV)	e(%)
9	4.7	52 *9*
10 - 21	8.0	58 *7*
35	8.0	23.0 *24*
40 - 59	1.46	3.25 *16*
72	22.2	31 *6*
73	21.1	29 *6*
80	11.7	15 *3*
81 - 96	7.7	9.0 *9*
128	20.0	15.6 *11*
179	3.04	1.69 *12*
180	6.7	3.7 *3*
181	5.1	2.80 *20*
187 - 190	4.76	2.53 *12*
232	12.12	5.23 *11*
282 - 293	6.23	2.189 *21*
328	7.8	2.39 *13*
379 - 390	2.98	0.780 *21*

$^{176}_{71}$Lu(3.59 *5* $\times 10^{10}$ yr)

Mode: β-

Δ: -53395 *3* keV

Prod: natural source

%: 2.59 *2*

Photons (^{176}Lu)
⟨γ⟩=490 *10* keV

γ_{mode}	γ(keV)	γ(%)†
Hf L_ℓ	6.960	0.36 *5*
Hf L_α	7.893	8.8 *8*
Hf L_η	8.139	0.160 *15*
Hf L_β	9.125	10.0 *10*
Hf L_γ	10.550	1.58 *16*
Hf $K_{\alpha2}$	54.611	9.3 *5*
Hf $K_{\alpha1}$	55.790	16.2 *9*
Hf $K_{\beta1}$'	63.166	5.3 *3*
Hf $K_{\beta2}$'	65.211	1.38 *8*
γ E2	88.372 *9*	13.1 *13*
γ E2	201.87 *3*	84 *5*
γ E2	306.91 *5*	93.00
γ E2	401.13 *20*	0.84 *19*

† 0.20% uncert(syst)

Atomic Electrons (^{176}Lu)
⟨e⟩=115 *3* keV

e_{bin}(keV)	⟨e⟩(keV)	e(%)
10 - 56	9.9	74 *5*
60 - 77	1.29	1.68 *16*
78	17.9	23.0 *23*
79	17.4	22.0 *22*
86	10.0	11.6 *12*
87 - 88	2.9	3.3 *3*
137	18.7	13.7 *9*
191	9.7	5.1 *3*
192	4.8	2.50 *16*
199 - 202	4.74	2.37 *10*
242	11.64	4.82 *10*
296	3.66	1.237 *25*
297 - 336	2.74	0.907 *11*
390 - 401	0.029	0.0075 *11*

$^{176}_{71}$Lu(3.635 *3* h)

Mode: β-

Δ: -53268 *3* keV

SpA: 4.898$\times 10^{6}$ Ci/g

Prod: ^{175}Lu(n,γ)

Photons (^{176}Lu)
⟨γ⟩=14.4 *4* keV

γ_{mode}	γ(keV)	γ(%)†
Hf L_ℓ	6.960	0.175 *19*
Hf L_α	7.893	4.2 *3*
Hf L_η	8.139	0.080 *6*
Hf L_β	9.123	4.9 *4*
Hf L_γ	10.538	0.76 *6*
Hf $K_{\alpha2}$	54.611	2.86 *16*
Hf $K_{\alpha1}$	55.790	5.0 *3*
Hf $K_{\beta1}$'	63.166	1.63 *9*
Hf $K_{\beta2}$'	65.211	0.424 *24*
γ E2	88.372 *9*	8.9 *4*
γ E2	201.87 *3*	<0.0007
γ E2	936.43 *5*	0.000220 *13*
γ E3(+M2)	957.48 *4*	4.4 *6* $\times 10^{-5}$
γ E2	1061.60 *6*	0.000762 *25*
γ E0+E2+M1	1138.30 *4*	0.000234 *15*
γ E1+M2	1159.35 *3*	0.00139 *4*
γ (E2)	1204.81 *7*	9.3 *7* $\times 10^{-5}$
γ (E2)	1226.67 *4*	0.000132 *7*
γ M2	1247.72 *3*	2.1 *4* $\times 10^{-5}$

† 1.7% uncert(syst)

Atomic Electrons (^{176}Lu)
⟨e⟩=38.9 *10* keV

e_{bin}(keV)	⟨e⟩(keV)	e(%)
10	1.48	15.4 *17*
11	1.30	12.1 *14*
23	2.39	10.4 *5*
43 - 77	1.08	1.57 *6*
78	12.1	15.6 *8*
79	11.8	15.0 *8*
86	6.8	7.9 *4*
87	0.068	0.079 *4*
88	1.89	2.15 *11*
191 - 202	8.0 $\times 10^{-5}$	~4$\times 10^{-5}$
871 - 892	1.29 $\times 10^{-5}$	1.5 *3* $\times 10^{-6}$
925 - 957	3.0 $\times 10^{-6}$	3.3 *4* $\times 10^{-7}$

Atomic Electrons (^{176}Lu)
(continued)

e_{bin}(keV)	⟨e⟩(keV)	e(%)
1050 - 1094	0.00011	~1$\times 10^{-5}$
1127 - 1161	2.9 $\times 10^{-5}$	2.5 *12* $\times 10^{-6}$
1182 - 1227	3.7 $\times 10^{-6}$	3.1 *4* $\times 10^{-7}$
1236 - 1248	5.2 $\times 10^{-7}$	4.2 *6* $\times 10^{-8}$

$^{176}_{72}$Hf(stable)

Δ: -54581 *3* keV

%: 5.206 *4*

$^{176}_{73}$Ta(8.08 *7* h)

Mode: ε

Δ: -51480 *100* keV

SpA: 2.204$\times 10^{6}$ Ci/g

Prod: ^{175}Lu(α,3n); ^{176}Hf(p,n)

Photons (^{176}Ta)
⟨γ⟩=2131 *31* keV

γ_{mode}	γ(keV)	γ(%)†
Hf L_ℓ	6.960	0.59 *11*
Hf L_α	7.893	14.4 *25*
Hf L_η	8.139	0.22 *4*
Hf L_β	9.133	15 *3*
Hf L_γ	10.599	2.6 *5*
Hf $K_{\alpha2}$	54.611	26 *5*
Hf $K_{\alpha1}$	55.790	45 *9*
Hf $K_{\beta1}$'	63.166	15 *3*
Hf $K_{\beta2}$'	65.211	3.8 *8*
γ E2	88.372 *9*	11.4 *9*
γ M1(+E2)	91.27 *3*	0.057 *5*
γ	110.13 *20*	0.019 *3*
γ	111.33 *20*	0.016 *3*
γ	117.53 *20*	0.012 *3*
γ [E1]	118.958 *20*	0.0114 *21*
γ	125.91 *9*	<0.2 ?
γ	130.77 *7*	0.021 *5*
γ	140.9 *10*	0.050 *5*
γ M1(+E2)?	146.69 *5*	0.203 *16*
γ E2	156.95 *4*	0.34 *3*
γ M1	158.23 *5*	0.218 *17*
γ [E1]	173.01 *4*	0.0146 *21*
γ M1(+E2)?	175.53 *5*	0.41 *3*
γ	179.13 *6*	0.037 *4*
γ	185.75 *6*	0.026 *3*
γ M1+E2	190.48 *5*	0.40 *3*
γ	192.83 *8*	0.0125 *21*
γ	196.81 *9*	0.024 *6*
γ [E1]	198.13 *5*	0.036 *8*
γ E2	201.87 *3*	5.5 *4*
γ [M1+E2]	207.44 *6*	<0.08
γ M1(+E2)?	213.53 *4*	0.41 *8*
γ	216.01 *5*	0.114 *9*
γ	230.91 *8*	0.0255 *21*
γ M1(+E2)	236.17 *5*	0.078 *6*
γ M1	239.69 *4*	0.52 *4*
γ	248.32 *8*	0.027 *3*
γ [E1]	264.28 *4*	0.073 *6* ?
γ	271.61 *9*	0.0125 *21*
γ	277.77 *8*	0.0104 *21*
γ	280.80 *7*	0.0114 *21*
γ [M1+E2]	292.96 *5*	0.038 *4* ?
γ	303.56 *10*	0.0218 *21*
γ	306.82 *20*	0.026 *3*
γ	314.56 *20*	0.030 *4*
γ	315.59 *6*	0.078 *10*
γ	318.86 *11*	0.0109 *21*

Photons (^{176}Ta)
(continued)

γ_{mode}	γ(keV)	γ(%)†
γ	327.1 3	0.0135 21
γ	337.54 20	0.0120 16
γ	343.41 20	0.036 4
γ	346.94 6	0.109 9
γ M1(+E2)	350.17 5	0.078 6
γ	358.67 6	0.094 8
γ	361.68 10	0.032 5
γ [E2]	362.90 7	0.020 5
γ	366.23 25	0.0125 16
γ [E1]	380.57 5	0.125 10
γ	382.74 25	0.023 4
γ	383.80 6	0.050 5
γ	386.32 9	0.023 3
γ [E1]	388.14 7	0.029 3
γ [E1]	401.42 7	0.0187 21
γ	411.7 2	0.018 3
γ	414.26 6	0.073 15
γ	421.1 3	0.017 4
γ	423.2 3	0.017 4
γ [E1]	424.67 5	0.048 5
γ [E1]	428.91 8	0.0140 21
γ	433.54 9	0.042 5
γ	434.88 7	0.046 5
γ	440.04 8	0.021 3
γ [M1+E2]	445.72 5	0.052 10
γ	450.90 6	0.016 3
γ [E2]	452.30 5	<0.023
γ	452.34 6	<0.023
γ	454.72 7	0.017 3
γ	459.13 9	0.031 4
γ	461.30 7	0.057 10
γ M1	466.23 5	1.07 9
γ [M1+E2]	473.34 6	0.265 21
γ	474.65 7	0.083 17
γ	479.17 8	0.029 4
γ	480.81 7	0.028 4
γ [M1+E2]	483.44 5	0.026 3
γ	495.01 13	0.0135 21
γ M1	507.75 8	1.39 11
γ [E1]	512.40 5	0.38 4
γ	517.4 4	~0.031
γ	519.95 6	~0.31
γ	521.16 6	~0.26
γ (M1)	521.52 5	~2
γ	524.93 11	0.057 13
γ	529.11 17	0.014 5
γ [M1+E2]	532.65 5	0.23 4
γ	533.21 7	0.062 21
γ	540.30 8	0.057 10
γ	541.27 12	0.088 10
γ [M1+E2]	543.26 5	0.078 6
γ	545.77 11	0.21 4
γ (M1)	546.29 6	0.51 4
γ	550.4 5	0.042 10
γ	551.35 9	0.018 3
γ [E2]	553.58 5	0.021 3
γ [M1+E2]	555.06 6	0.014 3
γ	560.03 20	0.027 4
γ	561.6 3	0.013 3
γ	566.63 20	0.0120 21
γ [M1]	569.70 7	0.109 16
γ [M1+E2]	570.73 5	0.44 9
γ [E2]	571.25 6	0.255 20 ?
γ	577.16 8	0.043 5
γ [E1]	578.82 5	0.057 5
γ	583.27 6	0.0125 21
γ	584.93 20	0.019 3
γ	586.75 9	0.083 7
γ	589.93 10	0.0156 21
γ	594.93 20	0.0120 21
γ [E1]	598.67 5	0.024 4
γ	604.44 7	0.025 5
γ [M1+E2]	609.25 4	0.073 10
γ M1	611.21 5	1.22 10
γ [E1]	615.15 5	0.099 16
γ E1	616.90 4	0.97 8
γ	625.98 7	0.016 3
γ	632.15 9	0.068 5
γ	636.34 5	0.049 5
γ	638.68 8	0.192 15
γ	642.88 8	0.094 8
γ M1	644.84 4	0.96 8
γ	656.83 10	0.033 4
γ	660.69 6	0.114 9

Photons (^{176}Ta)
(continued)

γ_{mode}	γ(keV)	γ(%)†
γ	663.98 7	0.083 10
γ	664.97 5	0.057 16
γ	670.23 20	0.011 3
γ	676.89 6	0.31 7
γ M1	678.91 5	0.198 16
γ M1	685.41 5	0.114 9
γ	693.23 10	0.020 3
γ	702.03 5	0.068 5
γ M1	710.53 4	5.2 4
γ	717.48 8	0.062 5
γ	723.08 5	0.125 10
γ	730.73 5	0.031 4
γ	735.93 20	0.016 3
γ	741.00 9	0.13 1
γ [M1+E2]	760.63 6	0.016 3
γ	766.53 10	0.029 4
γ	774.0 3	0.012 3
γ	779.33 10	0.028 3
γ	782.73 10	0.032 4
γ	784.23 20	0.018 4
γ	787.13 10	0.028 3
γ	789.43 20	0.0135 21
γ [E1]	798.52 6	0.045 8
γ	799.5 3	~0.02
γ	801.73 20	0.014 3
γ	803.83 10	0.034 4
γ	808.63 10	0.035 4
γ [E1]	819.49 5	0.250 20
γ	833.50 6	
γ	833.52 7	0.073 10 ‡
γ	837.7 3	0.018 5
γ	839.47 11	0.068 10
γ	842.00 8	0.041 9
γ	842.6 5	~0.020
γ	857.69 10	0.135 11
γ [E2]	860.67 5	0.039 5
γ	863.22 10	0.114 9
γ [E1]	867.46 7	0.033 4
γ	872.33 20	0.016 3
γ	876.63 20	0.024 3
γ	878.43 20	0.023 3
γ	884.7 3	0.014 5
γ	886.33 20	0.037 5
γ [E1]	893.23 5	0.025 6
γ	900.33 10	0.036 4
γ	907.33 10	0.046 5
γ E1	923.99 5	0.70 6
γ E2	936.43 5	0.54 4
γ [M1+E2]	951.94 4	0.068 10
γ E3(+M2)	957.48 4	0.55 4
γ	960.64 6	0.073 10
γ	962.57 7	0.052 10
γ	967.06 5	0.125 16
γ	971.83 10	0.046 5
γ	975.13 20	0.042 5
γ	977.03 20	0.047 6
γ	979.97 22	0.057 5
γ	981.17 8	0.048 18
γ	986.73 20	0.031 6
γ	994.47 6	0.052 10
γ	998.33 10	0.094 16
γ	1002.65 11	0.068 10
γ [M1+E2]	1010.86 7	0.030 10
γ [M1+E2]	1017.62 4	0.114 16
γ	1021.41 8	0.034 16
γ E1	1023.17 4	2.57 21
γ	1035.03 20	0.024 5
γ	1043.32 11	0.057 10
γ [E2]	1051.11 4	0.104 16
γ [E1]	1052.94 8	0.042 6
γ E2	1061.60 6	0.52 4
γ	1064.12 8	0.083 10
γ [E2]	1066.24 5	0.62 5
γ [E2]	1089.19 4	0.192 15
γ [E1]	1091.02 7	0.073 10
γ	1097.27 10	0.062 10
γ	1107.84 9	0.244 20
γ	1112.97 7	0.049 5
γ E1	1114.43 4	0.48 10 ‡
γ	1115.03 9	
γ	1122.75 7	0.099 16
γ	1125.48 9	0.135 11
γ E0+E2+M1	1138.30 4	0.65 5
γ	1148.30 7	0.044 8

Photons (^{176}Ta)
(continued)

γ_{mode}	γ(keV)	γ(%)†
γ (M1+E2)	1155.62 5	0.62 8
γ M1	1157.50 5	3.3 3
γ E1+M2	1159.35 3	24.6
γ	1174.17 5	0.198 16
γ	1178.76 8	0.036 6
γ [M1+E2]	1184.64 7	0.104 16
γ M1	1190.23 6	4.4 4
γ [M1+E2]	1198.24 7	0.062 10
γ [M1+E2]	1201.53 7	0.35 3
γ (E2)	1204.81 7	0.317 25
γ	1211.33 13	0.078 10
γ	1213.18 7	0.140 11
γ M1+E2+E0	1223.19 5	1.92 15
γ E1	1225.03 3	5.5 4
γ (E2)	1226.67 4	0.36 4
γ	1234.29 15	0.062 10
γ [E1]	1239.92 6	0.109 22
γ M2	1247.72 3	0.44 5
γ	1250.04 18	0.120 16
γ M1+E2	1252.98 3	2.97 24
γ [M1+E2]	1258.82 6	0.18 3
γ M1+E2	1268.87 6	1.28 10
γ	1277.93 11	0.151 12
γ	1281.23 20	0.045 7
γ [M1+E2]	1287.44 4	0.088 7
γ (E2+E2+M1)	1291.06 4	1.28 10
γ [M1+E2]	1301.09 5	0.073 6
γ	1308.33 12	0.062 5
γ [M1+E2]	1325.66 6	0.078 10
γ	1332.99 8	0.036 9
γ (M1+E2)	1341.35 4	3.2 3
γ	1345.89 7	0.068 16
γ M1+E2	1357.48 4	1.92 15
γ [E1]	1366.57 5	0.208 17
γ	1371.76 6	0.146 12
γ [E2]	1379.43 4	0.052 16 ?
γ	1412.94 5	0.109 22
γ (E1)	1420.04 5	0.44 4
γ	1427.67 11	0.114 9
γ	1432.53 8	0.083 7
γ	1438.19 6	0.029 6
γ	1450.39 6	0.35 3
γ	1462.63 20	0.025 5
γ	1467.53 20	0.042 5
γ	1470.03 20	0.048 10
γ E2	1476.27 5	0.46 4
γ	1482.8 3	0.028 7
γ (E2)	1489.31 4	0.70 6
γ	1495.93 8	0.182 15
γ	1503.45 6	<0.10 ?
γ	1504.21 5	0.73 10
γ	1537.03 8	0.37 3
γ [E1]	1540.89 7	0.34 3
γ	1544.02 7	0.244 20
γ E1	1555.07 4	3.9 3
γ	1563.84 7	0.19 3
γ	1564.98 7	0.40 3
γ	1573.33 20	0.034 8
γ M1+E2	1579.74 8	0.27 3
γ M1+E2	1584.02 4	5.1 4
γ	1603.49 18	0.052 16
γ	1608.71 11	0.140 11
γ	1612.62 8	0.166 13
γ (M1)	1616.27 5	1.24 10
γ [E2]	1621.84 4	0.56 11 ‡
γ [E1]	1621.91 5	
γ	1628.6 3	0.13 3
γ M1	1630.84 5	1.71 14
γ E1	1633.71 5	2.82 23
γ	1637.85 7	0.078 16
γ E1	1643.44 4	2.31 18
γ	1659.24 11	0.104 8
γ	1665.03 20	0.047 7
γ [M1]	1672.39 4	1.14 9
γ	1673.37 8	0.43 10
γ E1	1679.20 7	1.16 9
γ [E1]	1693.65 8	0.50 4
γ M1	1696.53 5	4.5 4
γ	1697.80 8	0.31 10
γ (E2)	1704.64 5	1.35 11
γ	1705.31 6	<0.16 ?
γ	1712.0 3	0.043 10
γ	1718.1 4	0.09 3
γ [M1+E2]	1721.38 6	

Photons (^{176}Ta) (continued)

γ_{mode}	γ(keV)	γ(%)†
γ E1	1722.08 5	3.15 25
γ	1725.9 4	0.062 21
γ	1736.73 20	0.037 4
γ	1745.32 14	0.109 9
γ	1751.1 3	0.027 5
γ	1754.97 16	0.068 5
γ	1765.71 7	0.46 4
γ	1768.25 16	0.177 14
γ M1(+E2)	1774.50 4	1.50 12
γ	1793.20 15	0.192 15
γ	1820.0 3	0.083 16
γ M1	1823.70 4	4.3 4
γ (E1)	1836.24 5	0.208 17
γ	1855.72 16	0.114 9
γ	1861.37 5	0.25 6
γ M1(+E2)	1862.87 4	3.8 3
γ [E1]	1869.87 4	0.078 6 ?
γ	1875.1 3	0.024 5
γ [E2]	1912.07 4	0.012 3
γ	1937.93 20	0.023 4
γ	1948.43 18	0.11 3
γ	1949.74 5	0.12 3
γ M1+E2	1956.48 6	0.83 7
γ	1960.63 16	0.057 5
γ	1970.63 20	0.030 4
γ M1+E2	1977.94 9	0.84 7
γ	2042.7 5	0.034 11
γ M1+E2	2044.85 6	1.3 1
γ	2049.2 4	0.027 6
γ	2057.4 3	0.017 3
γ	2066.32 9	0.068 5
γ	2071.03 20	0.016 3
γ	2077.03 20	0.040 5
γ	2090.6 3	0.014 3
γ	2140.13 20	0.037 4
γ	2162.12 10	0.037 4
γ	2192.49 10	0.218 17
γ	2219.49 6	0.281 22
γ	2246.95 20	0.125 10
γ	2257.9 4	0.023 6
γ	2260.4 3	0.030 5
γ	2272.1 3	0.017 3
γ	2278.6 3	0.025 4
γ	2280.86 10	0.172 14
γ	2304.5 4	0.026 11
γ	2307.86 6	0.192 15
γ	2314.8 5	~0.026
γ	2317.05 7	0.239 19
γ	2361.53 20	0.198 16
γ	2374.2 3	0.018 4
γ	2394.57 6	0.120 10
γ	2405.42 7	0.47 4
γ	2421.7 3	0.019 3
γ	2460.3 3	0.028 4
γ	2480.5 4	0.042 5
γ	2482.95 6	0.083 7
γ	2506.2 3	0.027 5
γ	2513.85 9	0.64 5
γ	2531.6 5	0.021 6
γ	2534.2 3	0.034 6
γ	2548.4 3	0.033 5
γ	2571.63 20	0.044 5
γ	2586.1 3	0.033 5
γ	2602.22 9	0.34 4
γ	2674.20 8	0.177 14
γ	2681.6 3	0.031 8
γ	2689.7 3	0.044 10
γ	2703.4 3	0.068 16
γ	2705.6 3	0.023 9
γ	2729.33 20	0.034 5
γ	2744.5 3	0.025 4
γ	2755.3 3	0.013 4
γ	2762.57 8	0.047 6
γ	2769.1 3	0.044 5
γ	2773.83 20	0.109 16
γ	2790.01 20	0.078 6
γ	2797.18 7	0.062 5
γ	2817.40 7	0.044 6
γ	2823.93 5	0.052 10
γ E1	2831.93 7	4.2 3
γ	2845.1 3	0.0062 16
γ	2854.1 9	~0.005
γ [E1]	2855.86 5	0.011 5
γ	2863.91 20	0.104 8

Photons (^{176}Ta) (continued)

γ_{mode}	γ(keV)	γ(%)†
γ	2882.5 4	0.030 6
γ	2885.56 7	0.104 8
γ	2890.3 4	0.008 3
γ	2905.77 7	0.021 3
γ E1	2920.31 7	2.11 17
γ	2940.7 3	0.0177 21
γ	2952.43 20	0.036 4
γ	2971.6 3	0.0109 16
γ	2978.7 3	0.0177 16
γ	2995.4 3	0.0048 7

† 15% uncert(syst)
‡ combined intensity for doublet

Atomic Electrons (^{176}Ta)

⟨e⟩=72 3 keV

e_{bin}(keV)	⟨e⟩(keV)	e(%)
10	4.8	51 10
11	4.1	38 7
23	3.07	13.4 11
26 - 65	2.17	4.6 5
76 - 77	1.12	1.46 12
78	15.6	20.1 17
79	15.2	19.2 16
80 - 82	0.22	0.27 13
86	8.7	10.1 8
87	0.088	0.101 8
88	2.44	2.77 23
89 - 138	2.22	1.82 16
139 - 188	1.03	0.64 8
189 - 238	1.46	0.75 4
239 - 285	0.085	0.032 9
290 - 339	0.067	0.022 5
341 - 390	0.062	0.017 4
391 - 440	0.29	0.072 7
441 - 490	0.95	0.21 5
492 - 541	0.29	0.057 9
542 - 591	0.379	0.068 4
592 - 641	0.151	0.025 3
642 - 691	0.62	0.096 8
692 - 741	0.139	0.0198 12
743 - 793	0.027	0.0034 9
794 - 843	0.021	0.0026 6
846 - 895	0.11	0.012 3
896 - 942	0.035	0.0038 7
946 - 994	0.071	0.0074 8
996 - 1043	0.065	0.0064 7
1048 - 1092	0.304	0.0280 21
1094	1.8	~0.16
1095 - 1140	0.290	0.0258 19
1144 - 1194	0.80	0.069 21
1195 - 1243	0.95	0.077 14
1245 - 1292	0.36	0.028 5
1297 - 1346	0.041	0.0031 5
1347 - 1385	0.027	0.0020 5
1397 - 1441	0.057	0.0040 9
1448 - 1496	0.065	0.0044 5
1498 - 1598	0.39	0.025 3
1599 - 1696	0.362	0.0221 10
1700 - 1798	0.31	0.0174 15
1805 - 1903	0.069	0.0037 3
1905 - 2001	0.058	0.0030 4
2006 - 2097	0.0083	0.00040 6
2127 - 2217	0.015	0.00071 15
2236 - 2329	0.014	0.00063 12
2340 - 2420	0.011	0.00048 14
2441 - 2539	0.018	0.00072 19
2546 - 2640	0.0053	0.00020 5
2663 - 2760	0.0086	0.00032 4
2763 - 2862	0.052	0.00188 10
2871 - 2970	0.0039	0.000135 10
2976 - 2993	1.7×10^{-5}	5.6 16 $\times 10^{-7}$

Continuous Radiation (^{176}Ta)

⟨β+⟩=7.2 keV;⟨IB⟩=0.83 keV

E_{bin}(keV)		⟨ ⟩(keV)	(%)
0 - 10	β+	1.55×10^{-8}	1.84×10^{-7}
	IB	0.00142	
10 - 20	β+	9.1×10^{-7}	5.4×10^{-6}
	IB	0.00061	0.0046
20 - 40	β+	3.62×10^{-5}	0.000109
	IB	0.0041	0.0124
40 - 100	β+	0.00257	0.00324
	IB	0.27	0.48
100 - 300	β+	0.153	0.069
	IB	0.035	0.019
300 - 600	β+	0.95	0.208
	IB	0.099	0.022
600 - 1300	β+	4.15	0.447
	IB	0.26	0.029
1300 - 2500	β+	1.94	0.130
	IB	0.158	0.0093
2500 - 3100	IB	0.0071	0.00027
	Σβ+		0.86

$^{176}_{74}$W (2.5 2 h)

Mode: ϵ

Δ: -50680 200 keV syst

SpA: 7.1×10^6 Ci/g

Prod: ^{181}Ta(p,6n)

Photons (^{176}W)

γ_{mode}	γ(keV)	γ(rel)
γ E2	33.58 4	0.079 16
γ M1	50.56 4	0.87 17
γ M1	61.29 4	9.2 18
γ M1	84.14 4	4.6 9
γ M1	94.87 4	9.3 19
γ E1	100.20 5	100

$^{176}_{75}$Re(5.2 4 min)

Mode: ϵ

Δ: -45180 280 keV syst

SpA: 2.05×10^8 Ci/g

Prod: ^{165}Ho(^{16}O,5n); ^{180}W(p,5n); ^{159}Tb(^{22}Ne,5n); ^{181}Ta(α,9n)

Photons (^{176}Re)

γ_{mode}	γ(keV)	γ(rel)
γ E2	108.9 3	54 6
γ E2	240.6 3	100

$^{176}_{76}$Os(3.6 *5* min)

Mode: ϵ
Δ: -42030 *280* keV syst
SpA: 3.0×10^{8} Ci/g
Prod: ^{180}W(^{3}He,7n); protons on Au

Photons (^{176}Os)*

γ_{mode}	γ(keV)	γ(rel)
γ	81.55 *11*	36 *7*
γ	775.78 *12*	98 *20*
γ	857.33 *12*	69 *14*
γ	1209.33 *12*	71 *14*
γ	1290.88 *12*	100

* see also ^{177}Os

$^{176}_{77}$Ir(8 *1* s)

Mode: α
Δ: -34020 *370* keV syst
SpA: 7.68×10^{9} Ci/g
Prod: ^{16}O on Tm; ^{19}F on Er

Alpha Particles (^{176}Ir)

α(keV)
5118 *8*

$^{176}_{78}$Pt(6.33 *15* s)

Mode: ϵ(58 *4* %), α(42 *4* %)
Δ: -28940 *240* keV syst
SpA: 9.59×10^{9} Ci/g
Prod: ^{16}O on Yb; ^{20}Ne on Er; daughter ^{180}Hg; ^{142}Nd(^{40}Ar,6n)

Alpha Particles (^{176}Pt)

⟨α⟩=2414 keV

α(keV)	α(%)
5528 *15*	0.58 *13*
5750 *10*	41

Photons (^{176}Pt)

γ_{mode}	γ(keV)	γ(%)
γ[E2]	227 *1*	0.48 *12*

$^{176}_{79}$Au(1.3 *3* s)

Mode: α
Δ: -18570 *390* keV syst
SpA: 3.9×10^{10} Ci/g
Prod: ^{141}Pr(^{40}Ca,5n)

Alpha Particles (^{176}Au)

α(keV)	α(rel)
6260 *10*	~80
6290 *10*	~20

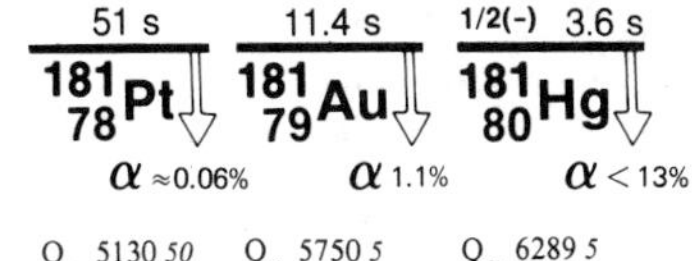

Qα 5130 *50*   Qα 5750 *5*   Qα 6289 *5*

A = 177

NDS **16**, 135 (1975)

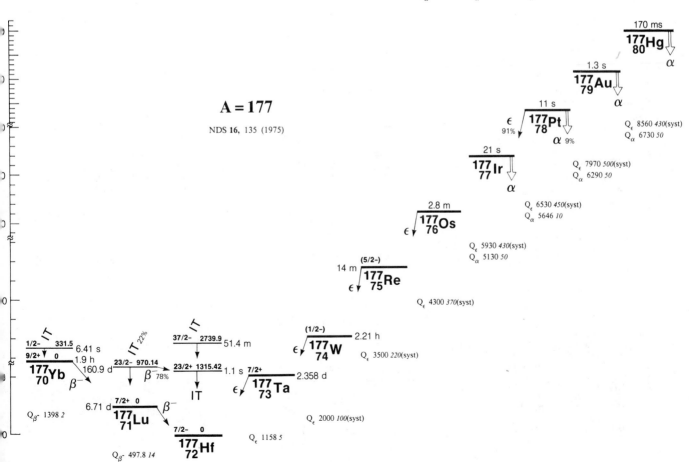

177 80 Hg 170 ms α
177 79 Au 1.3 s α — Qϵ 8560 *430*(syst), Qα 6730 *50*
177 78 Pt 11 s ϵ 91%, α 9% — Qϵ 7970 *500*(syst), Qα 6290 *50*
177 77 Ir 21 s α — Qϵ 6530 *450*(syst), Qα 5646 *10*
177 76 Os 2.8 m ϵ — Qϵ 5930 *430*(syst), Qα 5130 *50*
177 75 Re (5/2-) 14 m ϵ — Qϵ 4300 *370*(syst)
177 74 W (1/2-) 2.21 h ϵ — Qϵ 3500 *220*(syst)
177 73 Ta 7/2+ 2.358 d ϵ — Qϵ 2000 *100*(syst)
177 72 Hf 7/2- 0 — Qϵ 1158 *5*
177 71 Lu 7/2+ 0 6.71 d β⁻; 23/2- 970.14 160.9 d IT 22%, β⁻ 78% — Qβ- 497.8 *14*
177 Hf 37/2- 2739.9 51.4 m IT; 23/2+ 1315.42 1.1 s IT
177 70 Yb 9/2+ 0 1.9 h β⁻; 1/2- 331.5 6.41 s IT — Qβ- 1398 *2*

$^{177}_{70}$Yb(1.9 1 h)

Mode: β-
Δ: -50997 4 keV
SpA: 9.3×10^{6} Ci/g
Prod: ^{176}Yb(n,γ)

Photons (^{177}Yb)

⟨γ⟩=191 5 keV

γ_{mode}	γ(keV)	γ(%)†
Lu L_ℓ	6.753	0.064 8
Lu L_α	7.650	1.56 16
Lu L_η	7.857	0.0207 22
Lu L_β	8.827	1.53 18
Lu L_γ	10.243	0.25 3
Lu $K_{\alpha2}$	52.965	3.9 3
Lu $K_{\alpha1}$	54.070	6.8 6
Lu $K_{\beta1}$'	61.219	2.18 19
Lu $K_{\beta2}$'	63.200	0.58 5
γ M1+23%E2	121.6242 24	3.4 3
γ M1+5.7%E2	138.609 5	1.33 13
γ M1+24%E2	147.167 3	0.181 21
γ E1+4.3%M2	150.395 3	20.0 17
γ M1+23%E2	162.495 4	0.060 17
γ E2	268.791 4	0.171 16
γ M1	458.1 5	0.041 8 ?
γ	760.44 21	0.055 8
γ [E1]	779.14 13	0.121 16
γ (M1+E2)	899.05 21	0.64 4
γ	912.8 5	0.033 11 ?
γ	927.82 21	0.039 11 ?
γ [E1]	941.64 13	1.01 6
γ [M1+E2]	961.85 13	0.019 6
γ	967.4 5	0.033 6
γ	1028.04 21	0.63 4
γ	1063.2 5	0.017 6 ?
γ E1	1080.25 13	5.5 3
γ [M1+E2]	1109.02 13	0.176 16
γ	1119.69 21	0.54 4
γ	1149.66 21	0.64 4
γ M1+E2	1214.8 3	0.028 4
γ [E2]	1230.64 13	0.37 3
γ M1+E2	1241.31 21	3.35 22
γ M1+E2	1336.4 3	0.0132 22

† 11% uncert(syst)

Atomic Electrons (^{177}Yb)

⟨e⟩=17.5 11 keV

e_{bin}(keV)	⟨e⟩(keV)	e(%)
9 - 54	1.24	10.4 9
58	3.1	5.3 6
59 - 61	0.0118	0.0198 17
75	1.21	1.61 17
84	0.137	0.164 21
87	6.0	6.9 11
99	0.044	0.045 15
111	1.18	1.07 14
112 - 139	1.28	1.05 7
140	1.9	1.38 23
141 - 147	0.29	0.21 3
148	0.55	0.37 6
149 - 162	0.18	0.120 18
258 - 269	0.0156	0.0060 4
447 - 458	0.00171	0.00038 5
697 - 716	0.0049	0.0007 3
750 - 779	0.0010	0.00013 5
836 - 878	0.051	0.0060 16
888 - 932	0.0121	0.0013 3
939 - 967	0.024	~0.0025
1000 - 1046	0.078	0.0076 6

Atomic Electrons (^{177}Yb) (continued)

e_{bin}(keV)	⟨e⟩(keV)	e(%)
1052 - 1100	0.052	0.0048 17
1107 - 1151	0.009	0.0008 3
1167 - 1215	0.13	0.011 3
1220 - 1241	0.026	0.0021 5
1326 - 1336	8.6 ×10^{-5}	6.4 17 ×10^{-6}

Continuous Radiation (^{177}Yb)

⟨β-⟩=421 keV;⟨IB⟩=0.56 keV

E_{bin}(keV)		⟨ ⟩(keV)	(%)
0 - 10	β-	0.135	2.72
	IB	0.019	
10 - 20	β-	0.380	2.54
	IB	0.018	0.126
20 - 40	β-	1.37	4.61
	IB	0.034	0.120
40 - 100	β-	7.0	10.4
	IB	0.091	0.141
100 - 300	β-	41.5	21.0
	IB	0.21	0.119
300 - 600	β-	132	29.5
	IB	0.143	0.035
600 - 1300	β-	238	29.1
	IB	0.050	0.0067
1300 - 1398	β-	0.95	0.072
	IB	5.3×10^{-6}	4.0×10^{-7}

$^{177}_{70}$Yb(6.41 2 s)

Mode: IT
Δ: -50666 4 keV
SpA: 9.42×10^{9} Ci/g
Prod: ^{176}Yb(n,γ)

Photons (^{177}Yb)

⟨γ⟩=149.4 20 keV

γ_{mode}	γ(keV)	γ(%)
Yb L_ℓ	6.545	0.37 4
Yb L_α	7.411	9.2 5
Yb L_η	7.580	0.113 7
Yb L_β	8.522	8.6 7
Yb L_γ	9.892	1.45 14
Yb $K_{\alpha2}$	51.354	21.4 7
Yb $K_{\alpha1}$	52.389	37.7 12
Yb $K_{\beta1}$'	59.316	12.0 4
Yb $K_{\beta2}$'	61.229	3.28 12
γ E1	104.5 2	76.6 15
γ M3	227.0 2	12.3 3

Atomic Electrons (^{177}Yb)

⟨e⟩=180 4 keV

e_{bin}(keV)	⟨e⟩(keV)	e(%)
9	3.2	35 4
10	2.42	23.9 25
40	0.147	0.36 4
41	0.226	0.55 6
42	0.49	1.17 12
43	8.5	19.7 6

Atomic Electrons (^{177}Yb) (continued)

e_{bin}(keV)	⟨e⟩(keV)	e(%)
48	0.069	0.142 15
49	0.219	0.45 5
50	0.213	0.42 4
51	0.072	0.142 15
52	0.049	0.094 10
57	0.047	0.083 9
58	0.0077	0.0133 14
59	0.0184	0.031 3
94	1.92	2.05 6
95	0.503	0.532 15
96	0.584	0.612 18
102	0.574	0.562 16
103	0.159	0.155 4
104	0.203	0.195 6
104	0.00134	0.00128 4
166	97.1	58.6 20
217	36.0	16.6 6
218	11.8	5.43 18
225	12.4	5.51 18
227	3.51	1.55 5

$^{177}_{71}$Lu(6.71 1 d)

Mode: β-
Δ: -52395 3 keV
SpA: 1.0994×10^{5} Ci/g
Prod: ^{176}Lu(n,γ)

Photons (^{177}Lu)

⟨γ⟩=35.1 10 keV

γ_{mode}	γ(keV)	γ(%)†
Hf L_ℓ	6.960	0.053 6
Hf L_α	7.893	1.28 9
Hf L_η	8.139	0.0227 17
Hf L_β	9.126	1.43 11
Hf L_γ	10.556	0.228 20
Hf $K_{\alpha2}$	54.611	1.63 8
Hf $K_{\alpha1}$	55.790	2.85 14
Hf $K_{\beta1}$'	63.166	0.93 5
Hf $K_{\beta2}$'	65.211	0.242 12
γ E1+0.03%M2	71.658 15	0.154 6
γ E2+4.4%M1	112.949 5	6.4 3
γ E2+11%M1	136.705 15	0.0483 18
γ E1+0.5%M2	208.363 6	11.0
γ E2	249.654 16	0.210 8
γ E1+2.8%M2	321.312 6	0.222 10

† 4.0% uncert(syst)

Atomic Electrons (^{177}Lu)

⟨e⟩=14.5 3 keV

e_{bin}(keV)	⟨e⟩(keV)	e(%)
6 - 46	0.91	8.5 7
48	2.46	5.2 3
51 - 72	0.089	0.152 6
102	4.11	4.03 21
103	3.17	3.07 16
110	0.138	0.125 6
111	1.82	1.64 8
112	0.0348	0.0309 16
113	0.51	0.451 23
125 - 137	0.0370	0.0289 10
143	0.87	0.61 7
184 - 208	0.301	0.153 13

Atomic Electrons (^{177}Lu) (continued)

e_{bin}(keV)	⟨e⟩(keV)	e(%)
238 - 256	0.0417	0.0169 5
310 - 321	0.00451	0.00144 8

Continuous Radiation (^{177}Lu)

⟨β-⟩=133 keV;⟨IB⟩=0.066 keV

E_{bin}(keV)		⟨ ⟩(keV)	(%)
0 - 10	β-	0.277	5.6
	IB	0.0068	
10 - 20	β-	0.80	5.3
	IB	0.0061	0.043
20 - 40	β-	3.01	10.1
	IB	0.0105	0.037
40 - 100	β-	17.4	25.4
	IB	0.021	0.034
100 - 300	β-	83	45.3
	IB	0.020	0.0134
300 - 498	β-	28.9	8.2
	IB	0.00090	0.00027

$^{177}_{71}$Lu(160.9 *3* d)

Mode: β-(78 *3* %), IT(22 *3* %)
Δ: -51425 *3* keV
SpA: 4585 Ci/g
Prod: ^{176}Lu(n,γ)

Photons (^{177}Lu)

⟨γ⟩=998 *10* keV

γ_{mode}	γ(keV)	γ(%)†
Lu L_ℓ	6.753	0.142 17
Hf L_ℓ	6.960	0.66 7
Lu L_α	7.650	3.5 3
Lu L_η	7.857	0.062 6
Hf L_α	7.893	15.9 10
Hf L_η	8.139	0.234 15
Lu L_β	8.809	4.0 4
Hf L_β	9.137	16.4 12
Lu L_γ	10.196	0.64 7
Hf L_γ	10.595	2.70 24
Lu $K_{\alpha 2}$	52.965	5.5 3
Lu $K_{\alpha 1}$	54.070	9.7 5
Hf $K_{\alpha 2}$	54.611	33.5 12
$\gamma_{\beta-}$[E1]	55.149 20	1.20 24 &
Hf $K_{\alpha 1}$	55.790	58.5 22
Lu $K_{\beta 1}$'	61.219	3.12 15
Hf $K_{\beta 1}$'	63.166	19.1 7
Lu $K_{\beta 2}$'	63.200	0.83 4
Hf $K_{\beta 2}$'	65.211	4.97 19
$\gamma_{\beta-}$[E1]	69.06 7	0.011 4 &
$\gamma_{\beta-}$E1+0.03%M2	71.658 15	0.85 4 §
$\gamma_{\beta-}$[E1]	88.41 4	0.038 10 &
$\gamma_{\beta-}$M1+10%E2	105.345 5	11.5 &
$\gamma_{\beta-}$E2+4.4%M1	112.949 5	21.5 10 §
γ_{IT} E3	115.83 4	0.64 6
$\gamma_{\beta-}$[E1]	117.10 4	0.236 23 &
γ_{IT} M1+23%E2	121.6242 24	6.3 4
$\gamma_{\beta-}$M1+10%E2	128.497 4	15.2 6 &
$\gamma_{\beta-}$E2+11%M1	136.705 15	1.40 7 §
$\gamma_{\beta-}$[E1]	145.76 3	0.95 5 §
γ_{IT} M1+24%E2	147.167 3	3.83 25
$\gamma_{\beta-}$M1+12%E2	153.290 3	17.8 8 &
$\gamma_{\beta-}$[M1+E2]	159.74 3	0.60 7 &
γ_{IT} M1+30%E2	171.861 5	5.2 3
$\gamma_{\beta-}$M1+11%E2	174.403 5	12.7 5 &

Photons (^{177}Lu) (continued)

γ_{mode}	γ(keV)	γ(%)†
$\gamma_{\beta-}$[E1]	177.003 16	3.42 13 &
$\gamma_{\beta-}$[M1+E2]	181.95 5	0.096 24 &
γ_{IT} M1+17%E2	195.567 5	0.92 9
$\gamma_{\beta-}$M1+12%E2	204.099 5	14.4 7 &
$\gamma_{\beta-}$E1+0.5%M2	208.363 6	60.9 21 §
$\gamma_{\beta-}$M1+8.3%E2	214.432 6	6.6 3 &
γ_{IT} M1+21%E2	218.092 6	3.2 4
$\gamma_{\beta-}$E2	228.47 5	37.2 16 &
$\gamma_{\beta-}$E2	233.842 6	5.60 22 &
$\gamma_{\beta-}$[M1+E2]	242.38 8	0.038 8 &
$\gamma_{\beta-}$E2	249.654 16	6.1 3 §
γ_{IT} E2	268.791 4	3.61 18
$\gamma_{\beta-}$E2	281.787 5	14.1 6 &
$\gamma_{\beta-}$[E1]	283.49 7	0.35 5 &
$\gamma_{\beta-}$[E1]	291.50 4	1.02 9 &
$\gamma_{\beta-}$[E1]	292.51 4	0.81 8 &
$\gamma_{\beta-}$[E2]	296.45 3	5.4 4 &
$\gamma_{\beta-}$[E1]	299.05 3	1.56 8 &
$\gamma_{\beta-}$[E1]	305.499 16	1.71 7 &
$\gamma_{\beta-}$[E1]	313.708 7	1.25 5 &
γ_{IT} E2	319.028 5	11.0 6
$\gamma_{\beta-}$E1+2.8%M2	321.312 6	1.23 7 §
$\gamma_{\beta-}$E2	327.693 5	17.4 6 &
$\gamma_{\beta-}$[E2]	341.69 4	1.79 16 &
γ_{IT} E2	367.428 6	3.16 20
$\gamma_{\beta-}$E2	378.502 6	27.7 12 &
$\gamma_{\beta-}$[E2]	385.04 4	2.94 19 &
γ_{IT} E2	413.658 7	17.5 9
$\gamma_{\beta-}$E2	418.530 7	20.1 9 &
$\gamma_{\beta-}$[E2]	426.55 7	0.41 5 &
$\gamma_{\beta-}$[E2]	465.83 5	2.33 16 &

† uncert(syst): 28% for IT, 8.7% for β-
§ with ^{177}Hf(1.1 s) in equilib
& with ^{177}Lu(6.71 d) and ^{177}Hf(1.1 s) in equilib

Atomic Electrons (^{177}Lu)

⟨e⟩=236 *3* keV

e_{bin}(keV)	⟨e⟩(keV)	e(%)
4 - 9	1.25	13.7 16
10	6.4	66 7
11 - 23	4.5	41 4
40	12.8	32 3
42 - 46	1.95	4.4 3
48	8.3	17.4 8
50 - 56	1.89	3.58 22
58	5.7	9.7 7
59 - 62	0.185	0.304 16
63	15.1	24.0 11
66 - 87	3.58	4.4 3
88	14.8	16.8 8
94 - 96	6.6	7.0 7
102	13.8	13.5 7
103	12.3	11.9 7
104	0.0052	0.0050 5
105	8.9	8.5 9
106	0.0067	0.0063 7
107	6.5	6.1 6
108	0.00153	0.00142 15
109	12.4	11.3 8
110	0.465	0.421 21
111	8.3	7.5 5
112 - 138	18.7	15.8 5
139	8.5	6.1 3
142 - 162	17.3	11.8 5
163	9.1	5.6 3
164 - 213	10.00	5.31 11
214 - 262	14.17	6.15 10
266 - 314	5.35	1.80 4
316 - 366	4.74	1.39 3
367 - 416	2.54	0.655 10
417 - 466	0.097	0.0219 8

Continuous Radiation (^{177}Lu)

⟨β-⟩=31.7 keV;⟨IB⟩=0.0049 keV

E_{bin}(keV)		⟨ ⟩(keV)	(%)
0 - 10	β-	0.64	13.0
	IB	0.00156	
10 - 20	β-	1.73	11.6
	IB	0.00109	0.0077
20 - 40	β-	5.7	19.3
	IB	0.00129	0.0047
40 - 100	β-	19.2	30.2
	IB	0.00091	0.00167
100 - 153	β-	4.44	3.92
	IB	2.1×10^{-5}	2.0×10^{-5}

$^{177}_{72}$Hf(stable)

Δ: -52893 *3* keV
%: 18.606 *3*

$^{177}_{72}$Hf(1.08 *6* s)

Mode: IT
Δ: -51578 *3* keV
SpA: 4.36×10^{10} Ci/g
Prod: daughter ^{177}Lu(160.9 d); daughter ^{177}Hf(51.4 m)

Photons (^{177}Hf)

⟨γ⟩=1073 *18* keV

γ_{mode}	γ(keV)	γ(%)†
Hf L_ℓ	6.960	0.84 8
Hf L_α	7.893	20.4 13
Hf L_η	8.139	0.300 20
Hf L_β	9.137	21.1 16
Hf L_γ	10.595	3.5 3
Hf $K_{\alpha 2}$	54.611	43.1 15
γ[E1]	55.149 20	1.5 5
Hf $K_{\alpha 1}$	55.790	75 3
Hf $K_{\beta 1}$'	63.166	24.6 9
Hf $K_{\beta 2}$'	65.211	6.39 24
γ[E1]	69.06 7	0.014 5
γ E1+0.03%M2	71.658 15	1.11 7
γ[E1]	88.41 4	0.049 13
γ M1+10%E2	105.345 5	14.6
γ E2+4.4%M1	112.949 5	26.9 20
γ[E1]	117.10 4	0.30 4
γ M1+10%E2	128.497 4	20.0 11
γ E2+11%M1	136.705 15	1.86 12
γ[E1]	145.76 3	1.25 8
γ M1+12%E2	153.290 3	22.9 14
γ[M1+E2]	159.74 3	0.77 10
γ M1+11%E2	174.403 5	16.3 9
γ[E1]	177.003 16	4.33 25
γ[M1+E2]	181.95 5	0.15 4
γ M1+12%E2	204.099 5	19.3 12
γ E1+0.5%M2	208.363 6	79 5
γ M1+8.3%E2	214.432 6	8.3 4
γ E2	228.47 5	48 4
γ E2	233.842 6	7.4 4
γ[M1+E2]	242.38 8	0.046 10
γ E2	249.654 16	8.1 6
γ E2	281.787 5	18.1 12
γ[E1]	283.49 7	0.45 6
γ[E1]	291.50 4	1.31 12
γ[E1]	292.51 4	1.08 12
γ[E2]	296.45 3	6.9 4
γ[E1]	299.05 3	2.01 14
γ[E1]	305.499 16	2.25 13
γ[E1]	313.708 7	1.58 10

Photons (^{177}Hf)
(continued)

γ_{mode}	γ(keV)	γ(%)†
γ E1+2.8%M2	321.312 6	1.60 11
γ E2	327.693 5	22.4 9
γ [E2]	341.69 4	2.7 3
γ E2	378.502 6	37.1 23
γ [E2]	385.04 4	4.0 4
γ E2	418.530 7	25.4 14
γ [E2]	426.55 7	0.59 8
γ [E2]	465.83 5	2.78 24

† 7.9% uncert(syst)

Atomic Electrons (^{177}Hf)
⟨e⟩=240 4 keV

e_{bin}(keV)	⟨e⟩(keV)	e(%)
4 - 6	0.050	0.80 5
10	6.9	73 8
11 - 23	5.6	51 6
40	16.2	41 3
43 - 46	2.17	4.9 3
48	10.3	21.7 16
51 - 62	1.54	2.85 15
63	19.9	31.5 18
66 - 87	0.90	1.25 11
88	19.0	21.6 14
94	5.8	6.2 8
95 - 96	2.61	2.74 17
102	17.3	16.9 13
103	15.4	14.9 11
104 - 108	0.60	0.58 5
109	11.8	10.8 6
110	0.58	0.53 4
111	7.7	6.9 5
112 - 137	13.5	11.4 4
139	11.4	8.2 5
142	4.3	3.03 20
143	7.1	5.0 6
144 - 160	7.3	4.91 24
163	11.6	7.1 5
164 - 213	11.3	6.01 16
214 - 262	16.2	7.12 17
271 - 320	7.31	2.43 8
321 - 369	3.72	1.05 4
374 - 419	1.53	0.385 10
424 - 466	0.074	0.0163 10

$^{177}_{72}$Hf(51.4 5 min)

Mode: IT
Δ: -50153 3 keV
SpA: 2.067×10^{7} Ci/g
Prod: ^{176}Yb(α,3n); protons on ^{186}W

Photons (^{177}Hf)*
⟨γ⟩=1163 37 keV

γ_{mode}	γ(keV)	γ(%)
Hf L_ℓ	6.960	0.38 4
Hf L_α	7.893	9.1 7
Hf L_η	8.139	0.170 15
Hf L_β	9.120	11 1
Hf L_γ	10.577	1.83 19
Hf $K_{\alpha2}$	54.611	16.6 8
Hf $K_{\alpha1}$	55.790	28.9 14
Hf $K_{\beta1}'$	63.166	9.4 5
Hf $K_{\beta2}'$	65.211	2.45 12
γ	120.48 10	0.95 15

Photons (^{177}Hf)*
(continued)

γ_{mode}	γ(keV)	γ(%)
γ E3	213.98 10	40 4
γ	254.78 10	1.32 12
γ M1(+7.8%E2)	277.30 8	75 6
γ M1(+14%E2)	295.06 7	68 6
γ M1	311.46 7	58 4
γ M1	326.70 8	65 7
γ E2	572.36 8	7.1 6
γ E2	606.52 8	11.4 9
γ E2	638.16 8	20.0 15

* additionally, all photons with ^{177}Hf(1.1 s)

Atomic Electrons (^{177}Hf)*
⟨e⟩=251 8 keV

e_{bin}(keV)	⟨e⟩(keV)	e(%)
10 - 56	7.9	63 5
60 - 110	0.44	~0.44
111 - 120	0.38	~0.34
149	25.5	17.1 16
189	0.3	~0.18
203	46.0	22.6 22
204	23.1	11.3 11
211	1.28	0.61 6
212	47.3	22.3 20
213 - 214	5.1	2.40 21
230	24.1	10.5 10
244 - 245	0.10	~0.04
246	20.8	8.5 6
252 - 255	0.032	~0.013
261	21.7	8.3 9
266	5.3	1.98 17
267 - 316	17.9	6.08 22
317 - 327	1.23	0.38 3
507 - 541	0.99	0.188 11
561 - 607	1.28	0.222 14
627 - 638	0.275	0.0437 25

* additionally, all electrons with ^{177}Hf(1.1 s)

$^{177}_{73}$Ta(2.358 4 d)

Mode: ε
Δ: -51735 6 keV
SpA: 3.128×10^{5} Ci/g
Prod: ^{175}Lu(α,2n); daughter ^{177}W; protons on Hf

Photons (^{177}Ta)
⟨γ⟩=67.3 13 keV

γ_{mode}	γ(keV)	γ(%)†
Hf L_ℓ	6.960	0.42 4
Hf L_α	7.893	10.2 6
Hf L_η	8.139	0.139 9
Hf L_β	9.140	10.4 8
Hf L_γ	10.617	1.74 17
Hf $K_{\alpha2}$	54.611	24.1 8
Hf $K_{\alpha1}$	55.790	42.0 14
Hf $K_{\beta1}'$	63.166	13.7 5
Hf $K_{\beta2}'$	65.211	3.57 12
γ E1+0.03%M2	71.658 15	0.0134 8
γ M1+33%E2	96.36 5	0.090 14
γ M1+10%E2	105.345 5	0.0066 6
γ E2+4.4%M1	112.949 5	7.2 7
γ [E1]	129.85 6	0.00102 19

Photons (^{177}Ta)
(continued)

γ_{mode}	γ(keV)	γ(%)†
γ E2+11%M1	136.705 15	0.0074 4
γ	141.4 7	0.00054 7
γ [E1]	177.003 16	0.00195 15
γ [E1]	197.10 7	0.0029 3
γ E1+0.5%M2	208.363 6	0.96 5
γ E2	249.654 16	0.0321 18
γ [M1]	256.92 6	0.00086 8
γ [M1+E2]	268.49 7	0.00013 4 ?
γ (E1)	283.17 4	0.00055 6
γ (E2+49%M1)	297.60 8	0.00130 9
γ [E1]	311.84 6	0.00048 8
γ [E1]	313.708 7	0.00071 5
γ [E2]	319.25 5	0.0023 3
γ E1+2.8%M2	321.312 6	0.0193 13
γ [E2]	354.83 4	0.0019 3
γ [M1+E2]	364.85 7	0.00016 4 ?
γ (E2)	395.18 5	0.0052 5
γ [E1]	398.34 6	0.00056 8
γ M1+30%E2	420.74 5	0.031 3
γ M1	424.60 5	0.102 9
γ	439.0 7	0.00036 7
γ (M1)	453.26 5	0.0022 3
γ M1	491.54 4	0.031 3
γ	494.70 6	0.0042 4
γ M1+37%E2	508.13 5	0.070 6
γ M1+35%E2	526.09 5	0.0167 19
γ (M1+E2)	549.62 6	0.0060 5
γ (E1)	597.75 5	~0.009
γ M1(+E2)	604.49 4	0.021 3
γ E1	632.96 5	0.029 3
γ	681.52 5	0.00074 9
γ (E1)	734.45 5	0.039 4
γ (E1)	736.43 5	0.0158 19
γ E1	745.91 5	0.206 17
γ [E2]	760.03 6	0.00066 7
γ [E2]	805.73 7	0.0027 3
γ E1	847.40 5	0.027 3
γ [M1+E2]	872.98 6	0.00086 8
γ M1	944.79 5	0.055 5
γ	1002.83 5	0.00102 9
γ M1+25%E2	1057.74 5	0.29 3

† 22% uncert(syst)

Atomic Electrons (^{177}Ta)
⟨e⟩=22.2 9 keV

e_{bin}(keV)	⟨e⟩(keV)	e(%)
6	0.00060	0.0095 7
10	3.4	36 4
11	2.7	25 3
31 - 46	1.20	2.79 16
48	2.8	5.8 6
51 - 96	0.90	1.58 7
102	4.6	4.5 5
103	3.6	3.4 4
104 - 110	0.155	0.140 15
111	2.04	1.84 19
112	0.039	0.035 4
113	0.57	0.50 5
119 - 167	0.081	0.057 6
174 - 218	0.0290	0.0148 11
232 - 282	0.00607	0.00248 8
283 - 330	0.00134	0.000429 21
333 - 374	0.030	0.0082 7
384 - 429	0.0136	0.00325 18
436 - 485	0.014	0.0030 7
489 - 538	0.0034	0.00067 12
539 - 588	0.0027	0.00049 18
593 - 633	0.00065	0.00011 3
669 - 695	0.0044	0.00065 4
723 - 760	0.00097	0.000132 6
782 - 808	0.00048	$6.1\ 6\times10^{-5}$
836 - 879	0.0041	0.00047 4
934 - 945	0.00086	$9.2\ 7\times10^{-5}$
992 - 1003	0.016	0.0016 3
1046 - 1058	0.0032	0.00031 4

Continuous Radiation (^{177}Ta)

⟨IB⟩=0.53 keV

E_{bin}(keV)		⟨ ⟩(keV)	(%)
10 - 20	IB	0.00029	0.0023
20 - 40	IB	0.0035	0.0102
40 - 100	IB	0.28	0.50
100 - 300	IB	0.040	0.021
300 - 600	IB	0.112	0.025
600 - 1158	IB	0.098	0.0133

$^{177}_{74}$W (2.21 *4* h)

Mode: ϵ

Δ: -49730 *100* keV syst

SpA: 8.01×10^{6} Ci/g

Prod: ^{181}Ta(p,5n)

Photons (^{177}W)

⟨γ⟩=908 *11* keV

γ_{mode}	γ(keV)	γ(%)†
Ta L_ℓ	7.173	1.01 *15*
Ta L_α	8.140	24 *3*
Ta L_η	8.428	0.34 *5*
Ta L_β	9.459	25 *3*
Ta L_γ	10.983	4.1 *6*
Ta $K_{\alpha2}$	56.280	44.6 *18*
Ta $K_{\alpha1}$	57.535	78 *3*
Ta $K_{\beta1}$'	65.140	25.7 *10*
Ta $K_{\beta2}$'	67.254	6.6 *3*
γ M1+25%E2	70.02 *5*	6.3 *3*
γ	73.13 *9*	0.22 *5*
γ M1(+E2)	101.74 *5*	0.47 *11*
γ (E1)	115.05 *4*	59 *3* ‡
γ E1	115.67 *4*	
γ M1+3.1%E2	142.54 *3*	1.35 *10*
γ (M1)	149.14 *9*	0.55 *8*
γ [E2]	152.35 *4*	0.19 *6*
γ M1+~33%E2	155.94 *4*	3.9 *2*
γ [M1]	159.06 *8*	0.13 *3*
γ	172.48 *20*	0.13 *3*
γ (E1)	185.69 *7*	16.1 *8* ‡
γ (M1+~20%E2)	186.41 *3*	
γ	215.28 *20*	0.13 *3*
γ E2+~34%M1	223.21 *3*	2.2 *1*
γ E2+~15%M1	224.97 *3*	0.58 *8*
γ	237.68 *20*	0.18 *3*
γ M1+~45%E2	259.28 *9*	0.92 *6*
γ E1	271.00 *5*	3.94 *13*
γ	277.83 *20*	0.35 *5*
γ (E1)	280.81 *5*	0.87 *10*
γ [E1]	304.85 *12*	0.19 *6*
γ	305.83 *20*	0.56 *8*
γ M1	308.19 *8*	0.92 *13*
γ (E1)	311.27 *4*	1.37 *14*
γ	316.28 *20*	0.10 *3*
γ M1(+E2)	317.70 *8*	0.56 *6*
γ M1	367.51 *3*	4.2 *2*
γ M1	377.32 *4*	4.6 *2*
γ M1	382.27 *8*	0.68 *5*
γ E1	388.08 *9*	1.67 *10*
γ E2	417.13 *4*	6.1 *3*
γ M1(+E2)	418.34 *8*	0.87 *16*
γ	423.98 *10*	0.9 *2*
γ M1	426.94 *4*	13.1
γ [M1]	431.47 *8*	0.76 *8*
γ	436.58 *10*	0.53 *5*
γ M1	450.73 *8*	1.29 *3*
γ	457.2 *4*	0.27 *2*
γ M1	467.54 *11*	0.71 *3*
γ	471.78 *9*	0.020 *4*
γ [M1]	473.64 *8*	1.29 *3*
γ	497.7 *3*	0.13 *3*
γ	500.6 *4*	0.048 *16*

Photons (^{177}W) (continued)

γ_{mode}	γ(keV)	γ(%)†
γ E2	502.49 *13*	0.80 *8*
γ [M1+E2]	504.10 *8*	0.13 *3*
γ E2	528.39 *12*	2.36 *5*
γ	551.5 *6*	0.08 *3*
γ (E0+M1+E2)	562.75 *11*	0.14 *3*
γ	568.1 *3*	0.80 *3*
γ	577.7 *4*	0.24 *6*
γ E2	611.62 *12*	5.90 *12*
γ	619.4 *4*	0.30 *6*
γ	642.3 *3*	0.16 *6*
γ (E1)	647.35 *11*	2.54 *6*
γ M1	672.19 *10*	1.95 *8*
γ [E1]	678.49 *12*	0.56 *5*
γ	694.7 *3*	0.21 *3*
γ	707.0 *3*	0.18 *3*
γ	711.2 *5*	0.26 *5*
γ	714.0 *5*	0.18 *5*
γ M1	721.38 *10*	1.17 *8*
γ	755.0 *5*	0.14 *3*
γ M1	759.0 *5*	0.56 *8*
γ	771.5 *5*	0.14 *3*
γ	775.5 *5*	~0.06
γ M1	785.96 *11*	1.03 *11*
γ	789.0 *6*	~0.06
γ M1	794.45 *4*	0.58 *6*
γ [M1+E2]	822.03 *12*	0.11 *3*
γ M1	828.13 *10*	1.03 *6*
γ	836.59 *12*	1.01 *6*
γ M1	858.59 *10*	0.69 *5*
γ (M1)	877.32 *10*	1.21 *16*
γ (M1)	880.45 *9*	1.21 *16*
γ	889.5 *5*	0.18 *5*
γ [E1]	903.46 *12*	0.39 *5*
γ	939.32 *13*	0.76 *5*
γ	979.13 *12*	0.34 *2*
γ	990.27 *16*	0.80 *5*
γ	1000.08 *16*	0.53 *2*
γ	1004.7 *3*	0.37 *5*
γ E1	1014.86 *11*	4.77 *10*
γ M1	1036.39 *9*	10.2 *2*
γ [E1]	1046.00 *12*	0.42 *3*
γ	1052.5 *5*	0.18 *3*
γ [E1]	1055.81 *12*	0.64 *5*
γ M1(+E2)	1066.85 *9*	3.09 *8*
γ	1082.9 *4*	0.14 *5*
γ	1090.1 *4*	0.16 *5*
γ	1103.66 *9*	0.27 *3*
γ	1115.13 *16*	0.13 *5*
γ (M1)	1139.73 *10*	0.79 *3*
γ	1166.8 *6*	0.13 *5*
γ (M1)	1170.86 *12*	0.53 *5*
γ E1	1182.53 *9*	3.70 *8*
γ	1213.1 *4*	0.06 *2*
γ	1220.1 *4*	0.21 *2*
γ	1245.6 *4*	0.14 *2*
γ	1253.7 *4*	0.14 *2*
γ	1259.60 *9*	0.13 *2*
γ (E0+M1)	1271.07 *16*	0.05 *2*
γ	1276.3 *4*	0.13 *3*
γ	1290.06 *9*	0.06 *2*
γ (E0+M1)	1295.67 *11*	0.48 *2*
γ	1301.53 *16*	0.24 *2*
γ	1322.0 *5*	0.05 *2*
γ [M1+E2]	1326.80 *12*	0.53 *3*
γ [M1+E2]	1357.27 *12*	0.18 *2*
γ	1406.07 *12*	0.27 *3*

† 15% uncert(syst)

‡ combined intensity for doublet

Atomic Electrons (^{177}W)

⟨e⟩=92 *4* keV

e_{bin}(keV)	⟨e⟩(keV)	e(%)
2 - 6	1.46	54 *3*
10	8.0	81 *13*
11	4.9	44 *7*
12 - 14	1.43	12.2 *14*
19	2.6	~13
21	3.2	~16
28	2.0	~7
29 - 46	2.7	6.9 *12*
48	6.1	12.7 *7*
53 - 57	1.26	2.30 *12*
58	4.40	7.5 *4*
59	5.6	9.5 *7*
60	5.4	9.0 *7*
61 - 67	2.0	3.1 *8*
68	3.15	4.7 *3*
69 - 85	3.5	4.7 *4*
89	2.84	3.21 *17*
90 - 118	4.4	4.2 *5*
119	5.1	~4
131 - 177	5.0	3.3 *4*
183 - 228	1.67	0.84 *11*
235 - 281	0.92	0.37 *6*
293 - 321	3.08	1.01 *3*
350 - 358	1.02	0.29 *4*
360	3.2	0.90 *14*
364 - 413	1.68	0.436 *20*
414 - 464	1.26	0.295 *21*
465 - 510	0.23	0.047 *13*
517 - 566	0.430	0.079 *4*
567 - 612	0.424	0.071 *3*
617 - 662	0.249	0.038 *4*
667 - 714	0.131	0.0189 *21*
719 - 766	0.298	0.0405 *22*
769 - 818	0.38	0.048 *6*
819 - 869	0.079	0.0093 *13*
871 - 912	0.06	0.007 *3*
923 - 969	0.86	0.089 *4*
976 - 1025	0.32	0.032 *6*
1027 - 1073	0.133	0.0126 *12*
1078 - 1115	0.085	0.0077 *5*
1128 - 1173	0.032	0.0028 *4*
1178 - 1223	0.12	0.010 *3*
1228 - 1276	0.243	0.0197 *19*
1278 - 1327	0.049	0.0038 *5*
1339 - 1357	0.009	~0.0007
1394 - 1406	0.0015	0.00011 *5*

Continuous Radiation (^{177}W)

⟨β+⟩=0.71 keV;⟨IB⟩=0.74 keV

E_{bin}(keV)		⟨ ⟩(keV)	(%)
0 - 10	β+	2.42×10^{-8}	2.86×10^{-7}
	IB	0.00092	
10 - 20	β+	1.43×10^{-6}	8.5×10^{-6}
	IB	0.00052	0.0044
20 - 40	β+	5.7×10^{-5}	0.000171
	IB	0.0031	0.0092
40 - 100	β+	0.00383	0.00485
	IB	0.30	0.52
100 - 300	β+	0.173	0.080
	IB	0.036	0.020
300 - 600	β+	0.477	0.113
	IB	0.095	0.021
600 - 1300	β+	0.057	0.0089
	IB	0.27	0.031
1300 - 1783	IB	0.030	0.0022
	Σβ+		0.21

$^{177}_{75}$Re(14 *1* min)

Mode: ϵ
Δ: -46230 *200* keV syst
SpA: 7.6×10^{7} Ci/g
Prod: ^{181}Ta(^{3}He,7n); protons on W

Photons (^{177}Re)

γ_{mode}	γ(keV)	γ(rel)
γ M1+~0.6%E2	33.9 *2*	4.8 *10* ?
γ	49.8 *2*	
γ	76.1 *5*	30 *10*
γ M1+50%E2	79.65 *11*	85 *15*
γ (E1)	84.3 *2*	75 *15*
γ E2	94.90 *10*	46 *10*
γ M1+39%E2	101.4 *2*	35 *7* ?
γ	116.9 *3*	
γ [M1+E2]	181.60 *19*	1.6 *8*
γ E2	196.85 *17*	100 *20*
γ E2	209.8 *3*	33 *6*
γ	600.2 *6*	20 *4*
γ	708.1 *4*	30 *6* ?
γ	723.4 *4*	25 *5* ?
γ	1118.4 *8*	15 *4*
γ	1196.5 *8*	15 *4*
γ	1551.7 *15*	7.1 *24*
γ	1770.5 *8*	26 *6*
γ	1861.1 *8*	9.5 *24*
γ	1886.1 *8*	9.5 *24*
γ	1911.2 *8*	15 *6*
γ	1944.9 *8*	8 *4*
γ	1964.6 *8*	35 *10*
γ	1986.1 *8*	12 *4*

$^{177}_{76}$Os(2.8 *3* min)

Mode: ϵ
Δ: -41930 *320* keV syst
SpA: 4×10^{8} Ci/g
Prod: protons on Au; protons on Hg

Photons (^{177}Os or ^{176}Os)

γ_{mode}	γ(keV)	γ(rel)
γ	83 *1*	60 *12*
γ	157.7 *10*	30 *6*
γ	195.5 *10*	100 *20*
γ	300 *1*	33 *7*
γ	422 *1*	31 *6*
γ	456 *1*	20 *4*

$^{177}_{77}$Ir(21 *2* s)

Mode: α
Δ: -36000 *290* keV syst
SpA: 3.0×10^{9} Ci/g
Prod: ^{19}F on Er; ^{16}O on Tm

Alpha Particles (^{177}Ir)

α(keV)
5011 *10*

$^{177}_{78}$Pt(11 *2* s)

Mode: ϵ(91.0 *6* %), α(9.0 *6* %)
Δ: -29470 *350* keV syst
SpA: 5.6×10^{9} Ci/g
Prod: ^{16}O on Yb; ^{20}Ne on Er; daughter ^{181}Hg

Alpha Particles (^{177}Pt)

⟨α⟩=496 *23* keV

α(keV)	α(%)
5485 *20*	3.1 *3*
5525 *20*	5.9 *5*

$^{177}_{79}$Au(1.3 *4* s)

Mode: α
Δ: -21500 *350* keV syst
SpA: 3.8×10^{10} Ci/g
Prod: ^{168}Yb(^{19}F,10n); ^{141}Pr(^{40}Ca,4n)

Alpha Particles (^{177}Au)

α(keV)	α(rel)
6115 *10*	65
6150 *10*	35

$^{177}_{80}$Hg(170 *50* ms)

Mode: α
Δ: -12940 *240* keV syst
SpA: 9×10^{10} Ci/g
Prod: ^{142}Nd(^{40}Ca,5n)

Alpha Particles (^{177}Hg)

α(keV)
6570 *20*

A = 178

NDS 13, 549 (1974)

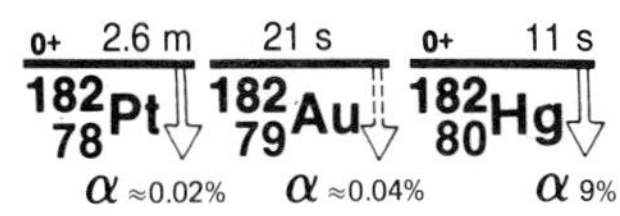

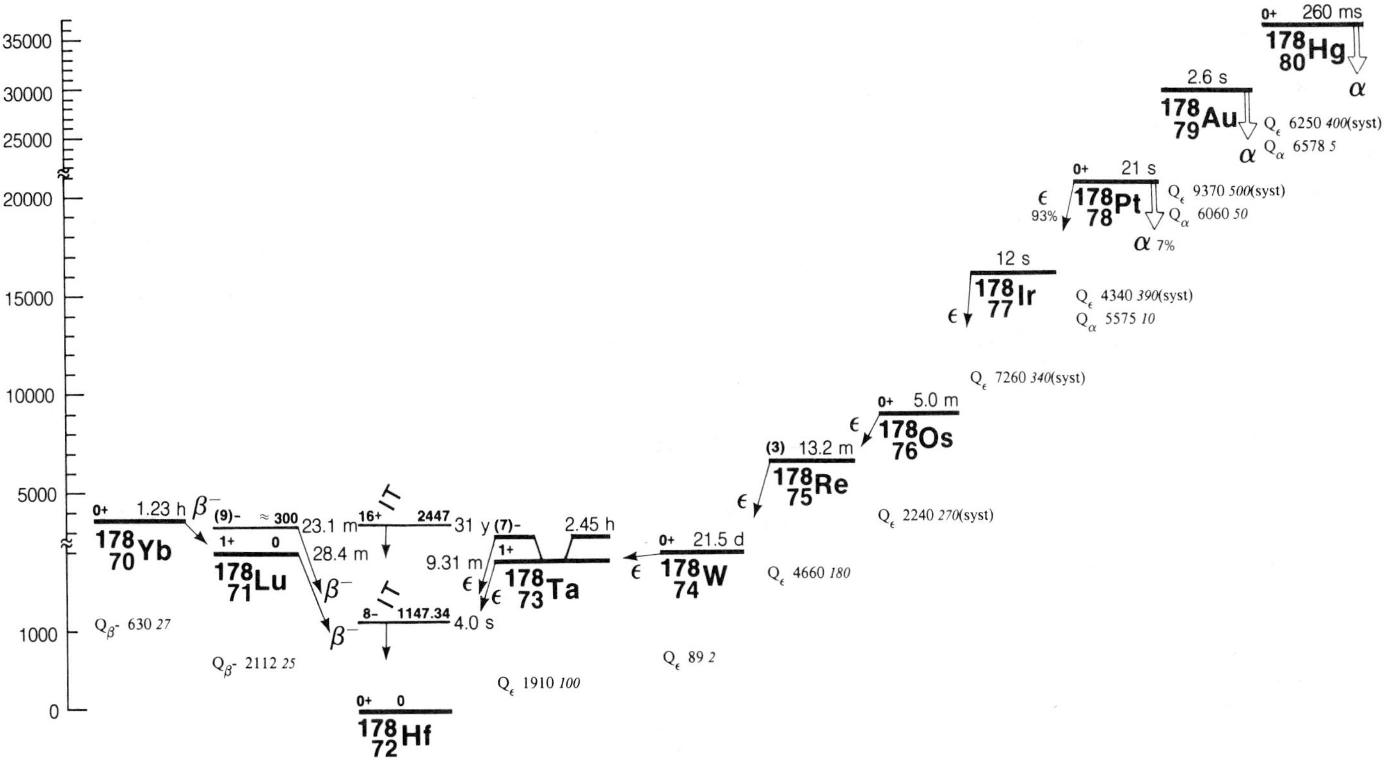

$^{178}_{70}$Yb(1.23 5 h)

Mode: β-
Δ: -49706 11 keV
SpA: 1.43×10^7 Ci/g
Prod: ^{176}Yb(t,p)

Photons (^{178}Yb)

γ_{mode}	γ(keV)	γ(rel)
γ	42.4 10	6.7 13
γ	348.4 10	64 13
γ	390.8 10	100

$^{178}_{71}$Lu(28.4 2 min)

Mode: β-
Δ: -50336 25 keV
SpA: 3.72×10^7 Ci/g
Prod: ^{179}Hf(γ,p); ^{181}Ta(n,α); ^{178}Hf(n,p); daughter ^{178}Yb

Photons (^{178}Lu)

⟨γ⟩=151 3 keV

γ_{mode}	γ(keV)	γ(%)†
Hf L_ℓ	6.960	0.11 3
Hf L_α	7.893	2.6 7
Hf L_η	8.139	0.049 13
Hf L_β	9.123	3.0 8
Hf L_γ	10.541	0.47 12
Hf $K_{\alpha2}$	54.611	2.1 5
Hf $K_{\alpha1}$	55.790	3.6 9
Hf $K_{\beta1}$'	63.166	1.2 3
Hf $K_{\beta2}$'	65.211	0.31 8
γ E2	93.159 3	6.8 17
γ (E2)	151.26 10	0.15 3 ?
γ M1+E2	203.74 9	0.31 8 ?
γ (M1+E2)	204.08 12	0.087 21 ?
γ E2	213.422 14	0.075 20
γ (M1+E2)	256.56 10	0.118 20
γ E1	1166.8 6	0.130 7
γ [E2]	1189.50 6	0.0360 14 ?
γ [E1]	1216.79 6	0.238 8
γ (E2)	1254.76 7	0.249 8
γ E1	1269.27 8	1.05 4
γ E1	1309.95 6	1.58 4
γ [E2]	1340.87 9	5.00 15
γ (E2)	1350.58 9	0.057 4
γ (M1+E2+E0)	1402.92 6	0.644 17
γ	1420.53 7	0.137 5
γ M1+E2	1468.18 7	0.095 3
γ	1473.35 9	0.106 4

Photons (^{178}Lu)
(continued)

γ_{mode}	γ(keV)	γ(%)†
γ (E2)	1496.08 6	0.357 10
γ	1513.69 7	0.144 5
γ [E2]	1561.34 7	0.0757 25
γ [E2]	1678.85 17	0.306 10
γ	1725.5 8	0.0305 15

† 50% uncert(syst)

Atomic Electrons (^{178}Lu)

⟨e⟩=26 3 keV

e_{bin}(keV)	⟨e⟩(keV)	e(%)
10	0.92	10 3
11	0.81	7.5 20
28	2.0	7.3 18
43 - 63	0.161	0.34 4
82	8.3	10.1 25
84	7.3	8.7 22
86	0.045	0.053 11
91	4.3	4.7 12
93	1.2	1.3 3
138 - 151	0.27	0.20 6
191 - 213	0.13	0.068 18
245 - 257	0.014	0.0055 19
1101 - 1124	0.00250	0.000225 9

Atomic Electrons (^{178}Lu) (continued)

e_{bin}(keV)	⟨e⟩(keV)	e(%)
1151 - 1190	0.0097	0.00082 *3*
1204 - 1253	0.0304	0.00247 *6*
1254 - 1300	0.130	0.0102 *5*
1307 - 1355	0.051	0.0039 *5*
1369 - 1418	0.12	0.009 *4*
1419 - 1468	0.032	0.0023 *7*
1471 - 1559	0.0044	0.000292 *22*
1613 - 1677	0.0081	0.00050 *3*
1714 - 1723	0.00012	7 *3* $\times10^{-6}$

Continuous Radiation (^{178}Lu)

⟨β-⟩=728 keV;⟨IB⟩=1.38 keV

E_{bin}(keV)		⟨ ⟩(keV)	(%)
0 - 10	β-	0.0337	0.67
	IB	0.029	
10 - 20	β-	0.101	0.68
	IB	0.028	0.20
20 - 40	β-	0.412	1.37
	IB	0.055	0.19
40 - 100	β-	2.99	4.25
	IB	0.153	0.24
100 - 300	β-	30.0	14.9
	IB	0.39	0.22
300 - 600	β-	98	22.0
	IB	0.37	0.089
600 - 1300	β-	394	42.7
	IB	0.32	0.040
1300 - 2112	β-	202	13.3
	IB	0.026	0.0019

$^{178}_{71}$Lu(23.1 *4* min)

Mode: β-

Δ: ~-50036 keV

SpA: 4.57×10^{7} Ci/g

Prod: ^{181}Ta(n,α); ^{179}Hf(γ,p); ^{178}Hf(n,p)

Photons (^{178}Lu)

⟨γ⟩=1056 *11* keV

γ_{mode}	γ(keV)	γ(%)†
Hf L_ℓ	6.960	0.55 *5*
Hf L_α	7.893	13.3 *8*
Hf L_η	8.139	0.217 *13*
Hf L_β	9.131	14.4 *10*
Hf L_γ	10.573	2.31 *18*
Hf $K_{\alpha2}$	54.611	20.8 *6*
Hf $K_{\alpha1}$	55.790	36.3 *10*
Hf $K_{\beta1}'$	63.166	11.9 *3*
Hf $K_{\beta2}'$	65.211	3.08 *9*
γ E1	88.857 *15*	62.1 *19* §
γ E2	93.159 *3*	17.4 *5* §
γ E2	213.422 *14*	81.3 *16* §
γ [M1+E2]	216.653 *14*	2.53 *19*
γ E2	325.546 *15*	94.4 *19* §
γ M1	331.66 *6*	11.8 *7*
γ E2	426.352 *15*	97.4 *19* §
γ [E2]	454.035 *23*	~0.023 ?

† 0.31% uncert(syst)

§ with ^{178}Hf(4.0 s) in equilib

Atomic Electrons (^{178}Lu)

⟨e⟩=149.0 *15* keV

e_{bin}(keV)	⟨e⟩(keV)	e(%)
10	4.6	48 *5*
11	3.8	35 *4*
24	8.6	36.7 *11*
28	5.21	18.8 *7*
43 - 63	1.63	3.40 *15*
78	4.31	5.56 *17*
82	21.4	26.0 *9*
84	18.9	22.6 *8*
86	1.59	1.85 *6*
91	11.1	12.2 *4*
93	3.06	3.29 *11*
148	16.8	11.4 *3*
202	2.62	1.30 *4*
203	5.41	2.67 *8*
204 - 217	8.05	3.88 *7*
260	10.9	4.18 *12*
266 - 315	7.1	2.48 *9*
316 - 332	3.19	0.993 *18*
361	7.80	2.16 *6*
389 - 426	2.97	0.711 *10*
443 - 454	0.00062	0.00014 *5*

Continuous Radiation (^{178}Lu)

⟨β-⟩=419 keV;⟨IB⟩=0.52 keV

E_{bin}(keV)		⟨ ⟩(keV)	(%)
0 - 10	β-	0.057	1.13
	IB	0.019	
10 - 20	β-	0.171	1.14
	IB	0.018	0.128
20 - 40	β-	0.70	2.33
	IB	0.035	0.121
40 - 100	β-	5.1	7.3
	IB	0.091	0.141
100 - 300	β-	53	26.2
	IB	0.20	0.115
300 - 600	β-	159	35.8
	IB	0.124	0.030
600 - 1265	β-	202	25.8
	IB	0.031	0.0045

$^{178}_{72}$Hf(stable)

Δ: -52447 *3* keV

%: 27.297 *3*

$^{178}_{72}$Hf(4.0 *2* s)

Mode: IT

Δ: -51300 *3* keV

SpA: 1.45×10^{10} Ci/g

Prod: daughter ^{178}Ta; ^{177}Hf(n,γ); ^{176}Yb(α,2n)

Photons (^{178}Hf)

⟨γ⟩=1007 *11* keV

γ_{mode}	γ(keV)	γ(%)†
Hf L_ℓ	6.960	0.54 *5*
Hf L_α	7.893	13.0 *8*
Hf L_η	8.139	0.213 *13*
Hf L_β	9.131	14.0 *10*

Photons (^{178}Hf) (continued)

γ_{mode}	γ(keV)	γ(%)†
Hf L_γ	10.572	2.26 *18*
Hf $K_{\alpha2}$	54.611	20.1 *6*
Hf $K_{\alpha1}$	55.790	35.0 *10*
Hf $K_{\beta1}'$	63.166	11.4 *3*
Hf $K_{\beta2}'$	65.211	2.97 *9*
γ E1	88.857 *15*	61.9 *19*
γ E2	93.159 *3*	17.4 *5*
γ E2	213.422 *14*	81.1 *16*
γ E2	325.546 *15*	94.1 *19*
γ E2	426.352 *15*	97.1 *19*

† 0.31% uncert(syst)

Atomic Electrons (^{178}Hf)

⟨e⟩=143.1 *14* keV

e_{bin}(keV)	⟨e⟩(keV)	e(%)
10	4.5	47 *5*
11	3.7	35 *4*
24	8.6	36.5 *11*
28	5.20	18.7 *6*
43 - 63	1.57	3.28 *14*
78	4.30	5.54 *17*
82	21.3	25.9 *9*
84	18.8	22.5 *8*
86	1.59	1.84 *6*
91	11.0	12.1 *4*
93	3.05	3.28 *11*
148	16.8	11.3 *3*
202	2.61	1.29 *4*
203	5.40	2.66 *8*
204 - 213	7.62	3.67 *7*
260	10.8	4.17 *12*
314 - 326	5.49	1.732 *24*
361	7.78	2.16 *6*
415 - 426	2.96	0.708 *10*

$^{178}_{72}$Hf(31 *1* yr)

Mode: IT

Δ: -50000 *3* keV

SpA: 64.8 Ci/g

Prod: ^{177}Hf(n,γ); ^{181}Ta(γ,p2n)

Photons (^{178}Hf)

⟨γ⟩=2200 *23* keV

γ_{mode}	γ(keV)	γ(%)†
Hf L_ℓ	6.960	0.65 *8*
Hf L_α	7.893	15.9 *14*
Hf L_η	8.139	0.253 *22*
Hf L_β	9.132	16.9 *16*
Hf L_γ	10.578	2.7 *3*
Hf $K_{\alpha2}$	54.611	26.9 *24*
Hf $K_{\alpha1}$	55.790	47 *4*
Hf $K_{\beta1}'$	63.166	15.4 *14*
Hf $K_{\beta2}'$	65.211	4.0 *4*
γ E1	88.857 *15*	62.1 *18* §
γ E2	93.159 *3*	17.3 *4* §
γ E2	213.422 *14*	81.1 *20* §
γ [M1+E2]	216.653 *14*	63.8 *14*
γ [M1+E2]	237.382 *20*	8.8 *5*
γ [M1+E2]	257.606 *17*	16.6 *5*
γ [M1+E2]	277.38 *3*	1.45 *9*
γ [M1+E2]	296.80 *3*	9.9 *4*
γ E2	325.546 *15*	94.1 *19* §
γ E2	426.352 *15*	97.1 *24* §

Photons (^{178}Hf) (continued)

γ_{mode}	γ(keV)	γ(%)†
γ [E2]	454.035 23	16.4 5
γ [E2]	494.988 22	68.9 21
γ [E2]	534.99 3	8.9 3
γ [E2]	574.18 3	83.7 25

† 0.32% uncert(syst)
§ with ^{178}Hf(4.0 s) in equilib

Atomic Electrons (^{178}Hf)

$\langle e \rangle$=212 14 keV

e_{bin}(keV)	$\langle e \rangle$(keV)	e(%)
10	5.4	57 7
11	4.5	42 5
24	8.61	36.6 11
28 - 63	7.28	23.0 6
78	4.31	5.55 16
82	21.2	25.8 8
84	18.8	22.4 7
86	1.59	1.85 5
91	11.0	12.1 4
93	3.04	3.27 10
148	16.8	11.3 4
151	25.4	17 8
172	3.1	1.8 9
192	5.2	2.7 13
202	2.61	1.29 4
203	5.40	2.66 8
204 - 248	24.3	11.4 19
255 - 258	0.46	~0.18
260	10.8	4.17 12
266 - 315	4.2	1.37 12
316 - 326	2.26	0.705 13
361	7.78	2.16 7
389 - 430	8.74	2.08 6
443 - 492	2.29	0.484 13
493 - 535	5.05	0.99 5
563 - 574	1.39	0.246 9

$^{178}_{73}$Ta(9.31 3 min)

Mode: ϵ
Δ: -50540 100 keV
SpA: 1.134×10^{8} Ci/g
Prod: daughter ^{178}W; protons on ^{181}Ta

Photons (^{178}Ta)

$\langle \gamma \rangle$=108.7 23 keV

γ_{mode}	γ(keV)	γ(%)†
Hf L_{ℓ}	6.960	0.47 6
Hf L_{α}	7.893	11.4 10
Hf L_{η}	8.139	0.162 16
Hf L_{β}	9.138	11.7 12
Hf L_{γ}	10.605	1.94 23
Hf $K_{\alpha2}$	54.611	24.3 10
Hf $K_{\alpha1}$	55.790	42.5 17
Hf $K_{\beta1}$'	63.166	13.9 6
Hf $K_{\beta2}$'	65.211	3.61 15
γ E2	93.159 3	6.7 13
γ (E2)	151.26 10	0.0046 9
γ M1+E2	203.74 9	0.010 3
γ (M1+E2)	204.08 12	0.0026 6
γ E2	213.422 14	0.099 4
γ (M1+E2)	256.56 10	0.0035 6
γ (E2)	970.03 4	0.0533 23
γ (E2)	1081.53 5	0.0179 9
γ (E2)	1106.14 7	0.531 17

Photons (^{178}Ta) (continued)

γ_{mode}	γ(keV)	γ(%)†
γ (E2)	1174.69 5	0.0157 9
γ (M1+E2+E0)	1183.45 4	0.166 7
γ [E2]	1189.50 6	0.0271 7
γ [E1]	1216.79 6	0.0063 3
γ (E2)	1254.76 7	0.0343 10
γ E1	1269.27 8	0.0292 12
γ (M1+E2)	1276.61 4	0.0334 12
γ E1	1309.95 6	0.0420 14
γ [E2]	1340.87 9	1.02 3
γ (E2)	1350.58 9	1.17 4
γ (M1+E2+E0)	1402.92 6	0.48
γ	1420.53 7	0.00427 15
γ M1+E2	1468.18 7	0.0130 4
γ	1473.35 9	0.00312 13
γ (E2)	1496.08 6	0.269 7
γ	1513.69 7	0.00450 15
γ [E2]	1561.34 7	0.0104 4
γ [E2]	1678.85 17	0.0034 3

† 21% uncert(syst)

Atomic Electrons (^{178}Ta)

$\langle e \rangle$=32.2 25 keV

e_{bin}(keV)	$\langle e \rangle$(keV)	e(%)
10	3.9	40 5
11	3.1	28 4
28	2.0	7.2 15
43 - 63	1.90	3.98 18
82	8.2	10 2
84	7.3	8.7 18
86	0.0014	0.0017 3
91	4.3	4.7 9
93	1.18	1.3 3
138 - 151	0.029	0.0196 19
191 - 213	0.0227	0.0112 6
245 - 257	0.00040	0.00016 6
959 - 970	0.000395	4.11 18 $\times 10^{-5}$
1016 - 1041	0.0160	0.00154 8
1070 - 1118	0.0107	0.00096 20
1124 - 1173	0.039	0.0034 17
1174 - 1217	0.010	0.0008 3
1243 - 1285	0.055	0.00430 15
1299 - 1348	0.027	0.0020 4
1349 - 1393	0.15	0.011 5
1400 - 1448	0.033	0.0023 9
1457 - 1552	0.00160	0.000108 4
1559 - 1613	7.7×10^{-5}	4.8 5 $\times 10^{-6}$
1668 - 1677	1.32×10^{-5}	7.9 7 $\times 10^{-7}$

Continuous Radiation (^{178}Ta)

$\langle \beta+ \rangle$=4.47 keV;$\langle$IB$\rangle$=1.47 keV

E_{bin}(keV)		$\langle\ \rangle$(keV)	(%)
0 - 10	β+	8.2×10^{-8}	9.7×10^{-7}
	IB	0.00114	
10 - 20	β+	4.80×10^{-6}	2.86×10^{-5}
	IB	0.00048	0.0036
20 - 40	β+	0.000189	0.00057
	IB	0.0038	0.0113
40 - 100	β+	0.0130	0.0164
	IB	0.28	0.50
100 - 300	β+	0.68	0.310
	IB	0.049	0.026
300 - 600	β+	2.80	0.64
	IB	0.19	0.042
600 - 1300	β+	0.98	0.145
	IB	0.77	0.085
1300 - 1910	IB	0.18	0.0125
	$\Sigma\beta$+		1.11

$^{178}_{73}$Ta(2.45 5 h)

Mode: ϵ
Δ: -50540 100 keV
SpA: 7.19×10^{6} Ci/g
Prod: ^{175}Lu(α,n); deuterons on Hf; protons on Hf; protons on ^{181}Ta

Photons (^{178}Ta)

$\langle \gamma \rangle$=1160 54 keV

γ_{mode}	γ(keV)	γ(%)†
Hf L_{ℓ}	6.960	0.92 9
Hf L_{α}	7.893	22.2 13
Hf L_{η}	8.139	0.333 20
Hf L_{β}	9.135	23.3 17
Hf L_{γ}	10.596	3.8 3
Hf $K_{\alpha2}$	54.611	43.2 14
Hf $K_{\alpha1}$	55.790	75.4 24
Hf $K_{\beta1}$'	63.166	24.7 8
Hf $K_{\beta2}$'	65.211	6.41 22
γ E1	88.857 15	61.9 19 §
γ E2	93.159 3	17.4 5 §
γ E2	213.422 14	81.1 16 §
γ E2	325.546 15	94.1 19 §
γ M1	331.66 6	~32
γ E2	426.352 15	97.1 19 §

† 0.31% uncert(syst)
§ with ^{178}Hf(4.0 s) in equilib

Atomic Electrons (^{178}Ta)

$\langle e \rangle$=163 6 keV

e_{bin}(keV)	$\langle e \rangle$(keV)	e(%)
10	7.5	79 8
11	6.2	57 6
24	8.6	36.5 11
28	5.20	18.7 6
43 - 63	3.38	7.1 3
78	4.30	5.54 17
82	21.3	25.9 9
84	18.8	22.5 8
86	1.59	1.84 6
91	11.0	12.1 4
93	3.05	3.28 11
148	16.8	11.3 3
202	2.61	1.29 4
203	5.40	2.66 8
204 - 213	7.62	3.67 7
260	10.8	4.17 12
266	10.4	~4
314 - 332	8.0	2.5 3
361	7.78	2.16 6
415 - 426	2.96	0.708 10

Continuous Radiation (^{178}Ta)

⟨IB⟩=0.34 keV

E_{bin}(keV)		⟨ ⟩(keV)	(%)
10 - 20	IB	0.00030	0.0024
20 - 40	IB	0.0035	0.0104
40 - 100	IB	0.28	0.49
100 - 300	IB	0.027	0.0153
300 - 600	IB	0.031	0.0075
600 - 762	IB	0.00141	0.00022

$^{178}_{74}$W (21.5 *1* d)

Mode: ϵ
Δ: -50450 *100* keV
SpA: 3.412×10^4 Ci/g
Prod: ^{181}Ta(p,4n)

Photons (^{178}W)

⟨γ⟩=16.6 *3* keV

γ_{mode}	γ(keV)	γ(%)
Ta L_ℓ	7.173	0.241 *25*
Ta L_α	8.140	5.7 *4*
Ta L_η	8.428	0.087 *5*
Ta L_β	9.421	10.7 *11*
Ta L_γ	11.138	2.5 *3*
Ta $K_{\alpha2}$	56.280	7.24 *21*
Ta $K_{\alpha1}$	57.535	12.6 *4*
Ta $K_{\beta1}'$	65.140	4.17 *12*
Ta $K_{\beta2}'$	67.254	1.07 *3*

Atomic Electrons (^{178}W)

⟨e⟩=4.6 *3* keV

e_{bin}(keV)	⟨e⟩(keV)	e(%)
10	0.97	9.8 *10*
11	0.71	6.4 *7*
12	2.41	20.6 *21*
44	0.049	0.112 *11*
45	0.078	0.174 *18*
46	0.154	0.33 *3*
48	0.048	0.101 *10*
53	0.042	0.078 *8*
54	0.056	0.104 *11*
55	0.070	0.127 *13*
56	0.0206	0.037 *4*
57	0.0160	0.028 *3*
62	0.0063	0.0101 *10*
63	0.0107	0.0170 *17*
64	0.0037	0.0057 *6*
65	0.0038	0.0059 *6*

Continuous Radiation (^{178}W)

⟨IB⟩=0.132 keV

E_{bin}(keV)		⟨ ⟩(keV)	(%)
10 - 20	IB	0.00155	0.0138
20 - 40	IB	0.0037	0.0114
40 - 89	IB	0.124	0.22

$^{178}_{75}$Re(13.2 *2* min)

Mode: ϵ
Δ: -45790 *210* keV
SpA: 8.00×10^7 Ci/g
Prod: ^{181}Ta(^{3}He,6n); protons on W; protons on Re; daughter ^{178}Os

Photons (^{178}Re)

⟨γ⟩=1268 *32* keV

γ_{mode}	γ(keV)	γ(%)†
W L_ℓ	7.387	0.67 *7*
W L_α	8.391	15.7 *11*
W L_η	8.724	0.249 *19*
W L_β	9.781	17.2 *15*
W L_γ	11.365	2.9 *3*
W $K_{\alpha2}$	57.981	26.8 *10*
W $K_{\alpha1}$	59.318	46.4 *17*
W $K_{\beta1}'$	67.155	15.4 *6*
W $K_{\beta2}'$	69.342	4.00 *16*
γ E2	106.06 *22*	23.1 *19*
γ [E2]	181.0 *3*	0.62 *19*
γ (E2)	237.19 *16*	45 *3*
γ [E2]	352.0 *3*	2.6 *5*
γ	500.9 *3*	1.7 *4*
γ	521.3 *5*	0.48 *14*
γ	539.5 *7*	0.48 *19*
γ	608.4 *3*	1.06 *19*
γ	635.5 *5*	0.7 *3*
γ	650.9 *5*	0.77 *19*
γ [E2]	685.7 *5*	0.87 *19*
γ	739.5 *3*	0.29 *10*
γ [E2]	767.5 *3*	0.48 *19*
γ E1+M2	778.0 *4*	3.8 *5*
γ	882.7 *3*	0.96 *19*
γ	938.9 *3*	8.9 *8*
γ	962.8 *5*	0.38 *14*
γ	976.7 *3*	3.6 *6*
γ [M1+E2]	1004.7 *3*	0.58 *19*
γ [M1+E2]	1037.7 *4*	1.0 *3*
γ [E2]	1106.4 *4*	0.58 *19*
γ [E2]	1110.7 *3*	2.7 *5*
γ	1130.6 *4*	3.3 *4*
γ [E2]	1169.3 *4*	<0.8 ?
γ [E2]	1169.7 *4*	0.77 *15* ?
γ	1229.9 *4*	0.67 *19*
γ	1255.2 *3*	1.4 *3*
γ [E2]	1274.9 *4*	1.06 *19*
γ	1289.0 *10*	~0.5
γ	1311.5 *5*	1.15 *19*
γ [M1+E2]	1343.6 *4*	~0.5
γ	1361.0 *10*	~0.5
γ	1377.2 *10*	~0.38
γ	1417.8 *5*	0.67 *19*
γ [E2]	1449.7 *4*	1.1 *3*
γ	1492.4 *3*	1.4 *3*
γ	1499.4 *5*	0.58 *19*
γ [M1+E2]	1521.3 *4*	~0.38
γ	1580.0 *10*	0.48 *19*
γ	1598.4 *4*	1.3 *3*
γ	1608.5 *4*	0.67 *19*
γ	1708.2 *4*	0.29 *10*
γ	1744.6 *5*	0.48 *19*
γ [E2]	1758.4 *4*	0.67 *19*
γ	1795.6 *7*	~0.19
γ	1833.9 *8*	0.67 *19*
γ	1836.0 *15*	~0.19
γ	1893.3 *8*	0.48 *19*
γ	1924.7 *8*	~0.38
γ	2016.3 *8*	0.29 *10*
γ	2036.4 *5*	0.29 *10*
γ	2053.0 *8*	0.38 *10*
γ	2133.1 *8*	0.58 *10*
γ	2248.2 *4*	0.19 *7*
γ	2263.7 *8*	0.19 *7*
γ	2286.2 *4*	0.29 *9* ?
γ	2306.6 *8*	0.19 *7*
γ	2312.1 *8*	0.38 *10*
γ	2324.4 *4*	0.19 *6* ?

Photons (^{178}Re) (continued)

γ_{mode}	γ(keV)	γ(%)†
γ	2455.9 *7*	0.20 *7*
γ	2468.6 *5*	~0.19
γ	2958.2 *4*	0.87 *19*
γ	3011.7 *5*	0.14 *5*
γ [E2]	3026.1 *3*	0.43 *14*
γ	3112.5 *4*	0.29 *10*
γ	3116.1 *4*	0.29 *10*
γ	3133.6 *5*	0.29 *10*
γ	3156.3 *4*	0.58 *19*
γ	3164.0 *6*	0.48 *19*
γ [E2]	3168.9 *3*	0.87 *19*
γ	3172.0 *5*	0.32 *12*
γ	3182.0 *6*	0.26 *8*
γ	3188.1 *6*	0.52 *14*
γ	3195.4 *4*	0.26 *8*
γ	3208.1 *4*	0.70 *19*
γ	3217.2 *6*	0.26 *8*
γ	3232.5 *6*	0.20 *7*
γ	3237.3 *5*	0.30 *10*
γ	3242.6 *5*	0.27 *10*
γ	3247.5 *6*	0.17 *7*
γ	3251.9 *5*	0.39 *13*
γ	3254.2 *6*	0.35 *12*
γ	3257.5 *6*	0.35 *12*
γ [M1+E2]	3263.3 *3*	0.39 *13*
γ	3277.3 *4*	0.38 *13*
γ	3291.6 *4*	0.22 *8*
γ	3363.6 *5*	0.19 *6*
γ [E2]	3369.4 *3*	0.15 *5*
γ	3376.0 *6*	0.10 *3*
γ	3383.3 *4*	0.13 *4*
γ	3393.5 *4*	0.17 *6*
γ	3399.8 *5*	0.33 *9*
γ [M1+E2]	3406.1 *3*	0.46 *12*
γ	3409.2 *5*	0.17 *6*
γ	3417.2 *6*	0.57 *7*
γ	3445.2 *4*	0.90 *10*
γ	3464.5 *4*	0.087 *19*
γ	3468.1 *5*	0.087 *19*
γ	3474.5 *5*	0.10 *3*
γ	3479.8 *5*	0.10 *3*
γ	3489.1 *5*	0.077 *19*
γ	3505.9 *5*	~0.038
γ [E2]	3512.2 *4*	0.10 *3*
γ	3525.7 *8*	0.10 *3*
γ	3528.7 *4*	0.20 *7*

† 8.1% uncert(syst)

Atomic Electrons (^{178}Re)

⟨e⟩=106 *3* keV

e_{bin}(keV)	⟨e⟩(keV)	e(%)
10	5.1	50 *6*
12	4.3	37 *4*
37	6.6	18.1 *16*
45 - 94	3.91	6.04 *22*
95	19.7	20.9 *18*
96	17.6	18.3 *16*
103	5.8	5.6 *5*
104	5.0	4.8 *4*
105	0.123	0.116 *10*
106	2.94	2.78 *24*
111	0.15	0.14 *4*
168	8.0	4.8 *4*
169 - 181	0.23	0.134 *24*
225 - 237	7.3	3.18 *12*
340 - 352	0.142	0.042 *5*
431 - 470	0.31	~0.07
489 - 538	0.08	0.016 *7*
539 - 581	0.22	0.039 *19*
596 - 641	0.091	0.015 *3*
648 - 686	0.036	0.0053 *25*
698 - 740	0.6	~0.09
755 - 778	0.15	~0.019
813	0.05	~0.006
863	13.4	1.55 *22*
869 - 907	0.7	~0.07

Atomic Electrons (^{178}Re) (continued)

e_{bin}(keV)	⟨e⟩(keV)	e(%)
920 - 968	2.8	0.30 6
974 - 1005	0.015	0.0016 6
1026 - 1061	0.25	0.023 9
1094 - 1131	0.085	0.0077 20
1157 - 1205	0.11	0.009 3
1218 - 1265	0.07	0.0060 24
1272 - 1311	0.059	0.0045 17
1332 - 1380	0.056	0.0041 12
1406 - 1452	0.08	0.0056 21
1480 - 1578	0.08	0.0051 17
1586 - 1675	0.027	0.0017 6
1689 - 1785	0.045	0.0026 7
1793 - 1891	0.024	0.0013 5
1913 - 2006	0.021	0.0011 3
2013 - 2064	0.013	0.0006 3
2121 - 2217	0.016	0.0007 3
2236 - 2323	0.020	0.00088 22
2386 - 2466	0.007	0.00030 12
2889 - 2957	0.021	0.00072 20
3000 - 3094	0.024	0.00079 19
3099 - 3198	0.079	0.0025 3
3205 - 3300	0.019	0.00057 10
3306 - 3405	0.040	0.00119 16
3406 - 3504	0.0100	0.00029 5
3509 - 3526	0.00063	1.8 6 $\times 10^{-5}$

Continuous Radiation (^{178}Re)

⟨β+⟩=155 keV;⟨IB⟩=11.7 keV

E_{bin}(keV)		⟨ ⟩(keV)	(%)
0 - 10	β+	2.94×10^{-8}	3.48×10^{-7}
	IB	0.0057	
10 - 20	β+	1.81×10^{-6}	1.08×10^{-5}
	IB	0.0057	0.041
20 - 40	β+	7.6×10^{-5}	0.000229
	IB	0.0123	0.041
40 - 100	β+	0.0060	0.0075
	IB	0.31	0.52
100 - 300	β+	0.451	0.199
	IB	0.133	0.073
300 - 600	β+	4.12	0.88
	IB	0.33	0.072
600 - 1300	β+	35.5	3.66
	IB	1.8	0.19
1300 - 2500	β+	93	5.1
	IB	5.5	0.30
2500 - 4554	β+	21.9	0.79
	IB	3.5	0.117
	Σβ+		10.7

$^{178}_{76}$Os(5.0 *4* min)

Mode: ε
Δ: -43550 *220* keV syst
SpA: 2.11×10^{8} Ci/g
Prod: ^{12}C on Yb; protons on Au; N on Tm; protons on Hg

Photons (^{178}Os or ^{179}Os)

γ_{mode}	γ(keV)	γ(rel)
γ	320 1	5 1
γ	350.8 10 ?	30 6 ?
γ	533.1 8	72 14
γ	548 1	~17
γ	551.8 10	40 8
γ	594.6 8	100
γ	600.6 10	56 11
γ	613 1	31 6
γ	632.5 10 ?	56 11 ?
γ	685.0 12	90 18
γ	745 2 ?	~40
γ	968.7 8	138 28
γ	1331.1 12	130 26

$^{178}_{77}$Ir(12 *2* s)

Mode: ε
Δ: -36290 *290* keV syst
SpA: 5.1×10^{9} Ci/g
Prod: ^{169}Tm(^{16}O,7n)

Photons (^{178}Ir)

γ_{mode}	γ(keV)	γ(rel)
γ[E2]	132.0 5	71 4
γ[E2]	266.7 3	100 5
γ	270.0 4	5.2 5
γ[E2]	363.3 4	35.2 18
γ	372.8 4	3.4 4
γ	398.7 4	12.7 13
γ[E2]	432.9 5	4.7 5
γ	532.9 5	4.7 5 ?
γ	546.9 5	6.1 6 ?
γ	625.0 5	15.1 8 ?
γ	639.5 4	13.3 13
γ	700.2 4	7.9 8 ?
γ	732.9 4	4.6 5
γ	864.9 4	8.6 9
γ	900.0 4	13.0 13
γ	970.2 4	1.5 2
γ	1017.6 5	5.5 6
γ	1201.5 4	7.1 7

$^{178}_{78}$Pt(21.0 *7* s)

Mode: ε(92.6 *8* %), α(7.4 *8* %)
Δ: -31950 *300* keV syst
SpA: 2.969×10^{9} Ci/g
Prod: ^{16}O on Yb; ^{20}Ne on Er; ^{19}F on Tm; daughter ^{182}Hg

Alpha Particles (^{178}Pt)

⟨α⟩=402 *1* keV

α(keV)	α(%)
5286 9	0.20 7
5442 8	7.2 8

$^{178}_{79}$Au(2.6 *5* s)

Mode: α
Δ: -22580 *400* keV syst
SpA: 2.1×10^{10} Ci/g
Prod: ^{168}Yb(^{19}F,9n); ^{169}Tm(^{20}Ne,11n)

Alpha Particles (^{178}Au)

α(keV)
5920 10

$^{178}_{80}$Hg(260 *30* ms)

Mode: α
Δ: -16323 *19* keV
SpA: 8.51×10^{10} Ci/g
Prod: protons on Pb

Alpha Particles (^{178}Hg)

α(keV)
6429 6

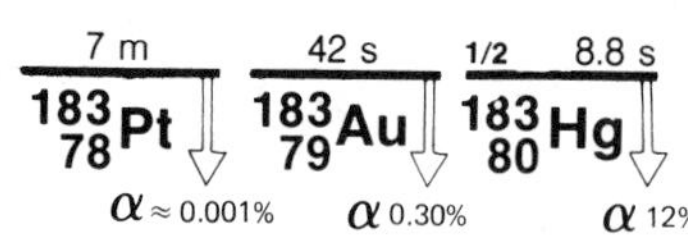

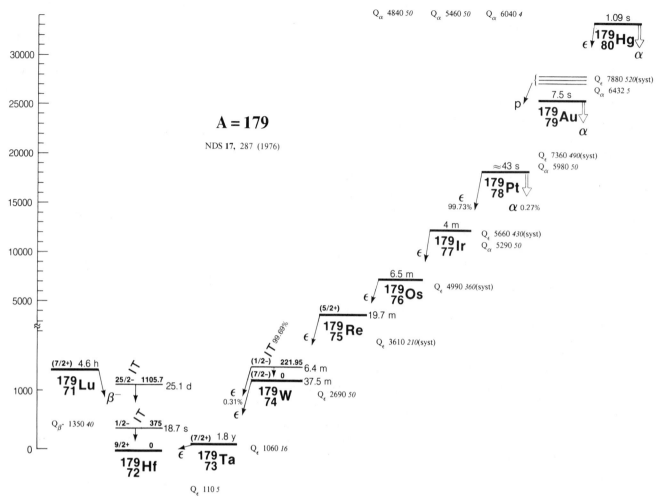

Q_α 4840 *50* Q_α 5460 *50* Q_α 6040 *4*

1.09 s
$^{179}_{80}$Hg ϵ α

Q_ϵ 7880 *520*(syst)
Q_α 6432 *5*

p
7.5 s
$^{179}_{79}$Au α

A = 179

NDS **17**, 287 (1976)

Q_ϵ 7360 *490*(syst)
Q_α 5980 *50*

≈43 s
$^{179}_{78}$Pt
ϵ 99.73% α 0.27%

4 m
$^{179}_{77}$Ir ϵ
Q_ϵ 5660 *430*(syst)
Q_α 5290 *50*

6.5 m
$^{179}_{76}$Os ϵ
Q_ϵ 4990 *360*(syst)

(5/2+) 19.7 m
$^{179}_{75}$Re ϵ
Q_ϵ 3610 *210*(syst)

IT 99.69%
(1/2-) 221.95 6.4 m
(7/2-) 0 37.5 m
$^{179}_{74}$W
ϵ 0.31%
ϵ
Q_ϵ 2690 *50*

(7/2+) 4.6 h
$^{179}_{71}$Lu β⁻
Q_{β^-} 1350 *40*

IT
25/2- 1105.7 25.1 d
IT
1/2- 375 18.7 s
9/2+ 0
$^{179}_{72}$Hf

(7/2+) 1.8 y
$^{179}_{73}$Ta ϵ
Q_ϵ 1060 *16*
Q_ϵ 110 *5*

30000
25000
20000
15000
10000
5000
1000
0

$^{179}_{71}$Lu(4.59 *6* h)

Mode: β-
Δ: -49130 *40* keV
SpA: 3.81×10^6 Ci/g
Prod: ^{180}Hf(γ,p); ^{179}Hf(n,p)

Photons (^{179}Lu)

⟨γ⟩=31.6 *25* keV

γ_{mode}	γ(keV)	γ(%)†
Hf L_ℓ	6.960	0.0069 *22*
Hf L_α	7.893	0.17 *5*
Hf L_η	8.139	0.0023 *8*
Hf L_β	9.141	0.17 *6*
Hf L_γ	10.601	0.028 *9*
Hf $K_{\alpha2}$	54.611	0.38 *9*
Hf $K_{\alpha1}$	55.790	0.67 *16*
Hf $K_{\beta1}$'	63.166	0.22 *5*
Hf $K_{\beta2}$'	65.211	0.057 *13*
γ M1+6.9%E2	122.72 *5*	0.15 *3*
γ [M1+E2]	123.36 *3*	0.47 *5*
γ E1	214.31 *3*	12.0 *11*

Photons (^{179}Lu)
(continued)

γ_{mode}	γ(keV)	γ(%)†
γ [E1]	214.94 *5*	0.48 *17*
γ [M1+E2]	279.17 *20*	0.0020 *6*
γ [M1+E2]	304.04 *15*	0.0066 *14*
γ [E1]	337.66 *4*	0.191 *20*
γ	532.47 *9*	0.0043 *9*
γ	655.82 *8*	0.029 *6*
γ	681.09 *23*	~0.0006
γ	735.77 *6*	0.018 *3*
γ	789.01 *15*	~0.00023
γ	830.54 *12*	0.0029 *9*
γ	859.13 *6*	0.106 *11*
γ	870.13 *9*	0.060 *7*
γ	891.57 *9*	0.0023 *6*
γ	953.89 *12*	0.0014 *4*
γ	983.16 *9*	0.0094 *17*
γ	998.4 *3*	~0.00029
γ	1003.32 *15*	0.0123 *23*
γ	1045.48 *12*	0.0040 *9*
γ	1076.72 *19*	0.0086 *17*
γ	1105.88 *9*	0.028 *3*
γ	1121.1 *3*	0.0009 *3*
γ	1168.20 *12*	0.0015 *4*
γ	1199.43 *18*	0.0051 *11*

† 25% uncert(syst)

Atomic Electrons (^{179}Lu)

⟨e⟩=2.1 *3* keV

e_{bin}(keV)	⟨e⟩(keV)	e(%)
10	0.057	0.60 *19*
11 - 56	0.073	0.47 *14*
57	0.16	0.28 *5*
58	0.34	~0.6
60 - 63	0.00131	0.0021 *3*
111	0.045	0.040 *8*
112	0.10	~0.09
113	0.11	~0.10
114	0.09	~0.08
120	0.0130	0.0108 *21*
121	0.07	~0.06
122 - 123	0.025	~0.021
149	0.74	0.50 *5*
150	0.029	0.020 *7*
203	0.117	0.058 *6*
204 - 239	0.098	0.047 *3*
268 - 304	0.0078	0.0028 *3*
326 - 338	0.00165	0.00050 *4*
521 - 532	9.8 ×10^{-5}	~2×10^{-5}
590 - 616	0.0021	~0.0004
645 - 681	0.0015	~0.00023
724 - 765	0.00040	5.4 *24* ×10^{-5}
778 - 826	0.008	~0.0010
828 - 870	0.0017	0.00020 *9*
880 - 918	0.0005	~5×10^{-5}

Atomic Electrons (^{179}Lu) (continued)

e_{bin}(keV)	⟨e⟩(keV)	e(%)
933 - 981	0.0008	8 *4* ×10^{-5}
983 - 1011	0.00043	~4×10^{-5}
1034 - 1077	0.0011	~0.00011
1095 - 1134	0.00043	3.8 *15* ×10^{-5}
1157 - 1199	4.3 ×10^{-5}	3.6 *15* ×10^{-6}

Continuous Radiation (^{179}Lu)

⟨β-⟩=464 keV;⟨IB⟩=0.61 keV

E_{bin}(keV)		⟨ ⟩(keV)	(%)
0 - 10	β-	0.0501	1.00
	IB	0.021	
10 - 20	β-	0.151	1.01
	IB	0.020	0.140
20 - 40	β-	0.62	2.05
	IB	0.038	0.133
40 - 100	β-	4.54	6.5
	IB	0.101	0.157
100 - 300	β-	47.5	23.6
	IB	0.23	0.132
300 - 600	β-	154	34.6
	IB	0.154	0.038
600 - 1300	β-	256	31.7
	IB	0.049	0.0069
1300 - 1350	β-	0.197	0.0150
	IB	3.9×10^{-7}	3.0×10^{-8}

$^{179}_{72}$Hf(stable)

Δ: -50476 *3* keV
%: 13.629 *5*

$^{179}_{72}$Hf(18.68 *6* s)

Mode: IT
Δ: -50101 *3* keV
SpA: 3.312×10^{9} Ci/g
Prod: ^{178}Hf(n,γ); ^{180}Hf(n,2n); γ rays on Hf

Photons (^{179}Hf)

⟨γ⟩=243 *4* keV

γ_{mode}	γ(keV)	γ(%)†
Hf L_ℓ	6.960	0.38 *8*
Hf L_α	7.893	9.2 *18*
Hf L_η	8.139	0.103 *20*
Hf L_β	9.150	8.8 *18*
Hf L_γ	10.657	1.5 *3*
Hf $K_{\alpha2}$	54.611	16 *3*
Hf $K_{\alpha1}$	55.790	29 *5*
Hf $K_{\beta1}'$	63.166	9.4 *18*
Hf $K_{\beta2}'$	65.211	2.4 *5*
γ M3	160.8 *7*	2.9 *6*
γ E1	214.31 *3*	95.20
γ [M4]	375.1 *7*	0.0048 *10*

† 0.21% uncert(syst)

Atomic Electrons (^{179}Hf)

⟨e⟩=135 *13* keV

e_{bin}(keV)	⟨e⟩(keV)	e(%)
10 - 56	6.6	55 *8*
60 - 63	0.056	0.091 *10*
95	53.4	56 *11*
149	5.88	3.95 *8*
150	31.9	21 *4*
151	17.6	11.6 *23*
158	8.5	5.4 *11*
159	5.1	3.2 *7*
160	3.4	2.1 *4*
161 - 205	1.84	0.98 *7*
212 - 214	0.378	0.178 *3*
364 - 375	0.023	0.0062 *8*

$^{179}_{72}$Hf(25.1 *3* d)

Mode: IT
Δ: -49370 *3* keV
SpA: 2.91×10^{4} Ci/g
Prod: ^{176}Yb(α,n)

Photons (^{179}Hf)

⟨γ⟩=891 *14* keV

γ_{mode}	γ(keV)	γ(%)†
Hf L_ℓ	6.960	0.50 *6*
Hf L_α	7.893	12.1 *11*
Hf L_η	8.139	0.166 *17*
Hf L_β	9.141	12.2 *13*
Hf L_γ	10.610	2.02 *25*
Hf $K_{\alpha2}$	54.611	29.3 *20*
Hf $K_{\alpha1}$	55.790	51 *4*
Hf $K_{\beta1}'$	63.166	16.7 *12*
Hf $K_{\beta2}'$	65.211	4.4 *3*
γ M1+6.9%E2	122.72 *5*	26.9 *11*
γ M1+14%E2	146.12 *8*	26.3 *11*
γ M1+E2	169.79 *9*	18.9 *9*
γ M1+E2	192.63 *11*	20.9 *18*
γ M1+E2	217.08 *13*	8.8 *7*
γ M1+E2	236.38 *14*	18.3 *5*
γ E3	257.39 *15*	3.2 *5*
γ (E2)	268.84 *9*	11.0 *7*
γ (E2)	315.91 *11*	19.7 *4*
γ (E2)	362.42 *13*	38.5 *9*
γ (E2)	409.72 *15*	20.9 *5*
γ E2	453.47 *17*	66 *3*

† 6.1% uncert(syst)

Atomic Electrons (^{179}Hf)

⟨e⟩=154 *9* keV

e_{bin}(keV)	⟨e⟩(keV)	e(%)
10	4.1	43 *6*
11 - 56	5.4	34 *4*
57	28.4	49.6 *23*
60 - 63	0.100	0.163 *10*
81	22.6	28.0 *14*
104	10.2	10 *5*
111	7.8	7.0 *3*
112 - 123	5.5	4.67 *23*
127	9.7	8 *4*
135	6.6	4.9 *3*
137 - 170	12.7	8.3 *21*
171	6.5	3.8 *19*
181 - 227	11.5	5.8 *10*
234 - 269	7.5	2.99 *19*
297	3.83	1.29 *6*

Atomic Electrons (^{179}Hf) (continued)

e_{bin}(keV)	⟨e⟩(keV)	e(%)
305 - 353	4.32	1.29 *3*
360 - 362	0.410	0.114 *4*
388	4.89	1.26 *6*
398 - 444	2.04	0.477 *9*
451 - 453	0.414	0.092 *3*

$^{179}_{73}$Ta(1.79 *8* yr)

Mode: ϵ
Δ: -50366 *6* keV
SpA: 1116 Ci/g
Prod: ^{176}Lu(α,n); protons on ^{181}Ta

Photons (^{179}Ta)

⟨γ⟩=28.3 *5* keV

γ_{mode}	γ(keV)	γ(%)
Hf L_ℓ	6.960	0.29 *3*
Hf L_α	7.893	6.9 *4*
Hf L_η	8.139	0.096 *6*
Hf L_β	9.118	9.6 *9*
Hf L_γ	10.714	1.97 *23*
Hf $K_{\alpha2}$	54.611	13.5 *4*
Hf $K_{\alpha1}$	55.790	23.6 *7*
Hf $K_{\beta1}'$	63.166	7.70 *22*
Hf $K_{\beta2}'$	65.211	2.00 *6*

Atomic Electrons (^{179}Ta)

⟨e⟩=5.6 *3* keV

e_{bin}(keV)	⟨e⟩(keV)	e(%)
10	1.78	18.6 *19*
11	2.8	24.9 *25*
43	0.227	0.53 *5*
44	0.0101	0.0231 *24*
45	0.30	0.66 *7*
46	0.093	0.20 *2*
51	0.044	0.085 *9*
52	0.095	0.183 *19*
53	0.111	0.209 *21*
54	0.103	0.191 *19*
55	0.026	0.047 *5*
56	0.0068	0.0121 *12*
60	0.0116	0.0193 *20*
61	0.0179	0.029 *3*
62	0.0091	0.0147 *15*
63	0.0073	0.0116 *12*

Continuous Radiation (^{179}Ta)

⟨IB⟩=0.19 keV

E_{bin}(keV)		⟨ ⟩(keV)	(%)
10 - 20	IB	0.00065	0.0054
20 - 40	IB	0.0042	0.0127
40 - 100	IB	0.19	0.34
100 - 110	IB	3.1×10^{-5}	2.3×10^{-5}

$^{179}_{74}$W (37.5 *5* min)

Mode: ϵ

Δ: -49307 *16* keV

SpA: 2.80×10^{7} Ci/g

Prod: ^{181}Ta(p,3n); daughter ^{179}Re

Photons (^{179}W)

γ_{mode}	γ(keV)	γ(rel)
γ E1	30.5 *5*	100
γ [M1+E2]	133.9 *2*	0.59 *6*

Continuous Radiation (^{179}W)

⟨IB⟩=0.49 keV

E_{bin}(keV)		⟨ ⟩(keV)	(%)
10 - 20	IB	0.00047	0.0041
20 - 40	IB	0.0030	0.0089
40 - 100	IB	0.30	0.52
100 - 300	IB	0.040	0.022
300 - 600	IB	0.096	0.022
600 - 1029	IB	0.052	0.0074

$^{179}_{74}$W (6.4 *1* min)

Mode: IT(99.69 *5* %), ϵ(0.31 *5* %)

Δ: -49085 *16* keV

SpA: 1.64×10^{8} Ci/g

Prod: ^{181}Ta(p,3n); daughter ^{179}Re

Photons (^{179}W)

⟨γ⟩=55.0 *22* keV

γ_{mode}	γ(keV)	γ(%)†
W L_{ℓ}	7.387	0.36 *5*
W L_{α}	8.391	8.5 *8*
W L_{η}	8.724	0.105 *11*
W L_{β}	9.790	8.7 *10*
W L_{γ}	11.427	1.57 *21*
W $K_{\alpha2}$	57.981	15.6 *13*
W $K_{\alpha1}$	59.318	26.9 *23*
W $K_{\beta1}$'	67.155	9.0 *8*
W $K_{\beta2}$'	69.342	2.32 *20*
γ_{IT} E2+13%M1	119.913 *25*	0.32 *3*
γ_{ϵ}	213.9 *5*	
γ_{IT} M3	221.95 *3*	8.6 *7*
γ_{ϵ}	222.5 *5*	
γ_{ϵ}	230.1 *5*	
γ_{ϵ}M1+E2	238.7 *3*	0.220 *11*
γ_{ϵ}[E2]	281.7 *3*	0.187 *18*
γ_{ϵ}[M1+E2]	288.9 *3*	0.029 *7*

† uncert(syst): 17% for ϵ, 8.1% for IT

Atomic Electrons (^{179}W)*

⟨e⟩=164 *8* keV

e_{bin}(keV)	⟨e⟩(keV)	e(%)
10 - 59	6.1	47 *4*
64 - 110	1.4	1.5 *3*
117 - 120	0.112	0.095 *6*
152	85.5	56 *5*
210	38.1	18.2 *15*
212	14.1	6.7 *6*
219	9.9	4.5 *4*
220	4.1	1.87 *15*
221 - 222	4.18	1.89 *11*

* with IT

Continuous Radiation (^{179}W)

⟨IB⟩=0.00109 keV

E_{bin}(keV)		⟨ ⟩(keV)	(%)
10 - 20	IB	1.49×10^{-6}	1.29×10^{-5}
20 - 40	IB	9.1×10^{-6}	2.7×10^{-5}
40 - 100	IB	0.00089	0.00155
100 - 300	IB	8.6×10^{-5}	4.9×10^{-5}
300 - 600	IB	8.9×10^{-5}	2.2×10^{-5}
600 - 1043	IB	1.8×10^{-5}	2.6×10^{-6}

$^{179}_{75}$Re(19.7 *5* min)

Mode: ϵ

Δ: -46620 *50* keV

SpA: 5.33×10^{7} Ci/g

Prod: ^{180}W(p,2n); ^{181}Ta(^{3}He,5n); ^{181}Ta(α,6n)

Photons (^{179}Re)

⟨γ⟩=1074 *16* keV

γ_{mode}	γ(keV)	γ(%)†
W L_{ℓ}	7.387	1.1 *5*
W L_{α}	8.391	26 *12*
W L_{η}	8.724	~0.39
W L_{β}	9.785	27 *13*
W L_{γ}	11.367	4.6 *20*
γ [M1+E2]	24.31 *7*	<0.11
γ M1+15%E2	53.50 *4*	<0.6
W $K_{\alpha2}$	57.981	32.8 *16*
W $K_{\alpha1}$	59.318	57 *3*
W $K_{\beta1}$'	67.155	18.9 *9*
W $K_{\beta2}$'	69.342	4.90 *25*
γ [M1+E2]	78.73 *6*	~0.17
γ E2	82.83 *3*	1.46 *11*
γ E2	96.45 *3*	0.39 *3*
γ [M1+E2]	101.19 *5*	0.30 *3*
γ M1	111.85 *4*	0.056 *11*
γ E2+13%M1	119.913 *25*	4.79 *25*
γ M1	125.48 *4*	0.087 *11*
γ (E2)	135.05 *6*	0.11 *3*
γ E2	144.91 *7*	0.059 *8*
γ	160.56 *8*	0.14 *2*
γ M1+23%E2	168.36 *3*	2.24 *11*
γ	185.22 *12*	0.21 *4*
γ	186.4 *3*	0.08 *3*
γ E1	189.09 *3*	7.5 *4*
γ [M1+E2]	190.59 *5*	<0.14
γ E2	204.21 *4*	0.53 *3*
γ [E2]	208.30 *4*	~0.14
γ (E2)	214.90 *7*	0.25 *3*
γ	217.53 *10*	0.252 *17*
γ M3	221.95 *3*	3.6 *6* §

Photons (^{179}Re) (continued)

γ_{mode}	γ(keV)	γ(%)†
γ [E1]	222.44 *5*	0.36 *14*
γ	238.3 *5*	0.11 *4*
γ E1	241.77 *11*	1.0 *3*
γ E1	242.29 *6*	1.7 *3*
γ (E2)	242.87 *4*	0.8 *3*
γ	252.93 *7*	0.28 *3*
γ [M1+E2]	255.93 *8*	~0.027
γ E2	264.82 *7*	0.59 *6*
γ E1	289.98 *3*	26.9 *14*
γ M1	296.37 *4*	8.9 *5*
γ E1	309.00 *4*	3.3 *5*
γ	309.3 *4*	1.5 *5*
γ [E2]	310.34 *4*	0.6 *3*
γ [E1]	321.99 *8*	0.13 *4*
γ [E1]	323.84 *7*	~0.06
γ	329.17 *10*	0.21 *3*
γ	335.47 *8*	0.27 *3*
γ [E1]	343.48 *4*	0.062 *22*
γ [E1]	346.30 *7*	0.08 *3*
γ [E1]	357.45 *4*	1.33 *8*
γ M1	377.92 *5*	0.41 *3*
γ M1	384.20 *10*	0.266 *25*
γ E1	401.83 *4*	7.2 *4*
γ M1	411.53 *5*	1.12 *8*
γ E1	415.45 *4*	10.6 *6*
γ M1	430.25 *4*	28
γ (E1)	455.33 *4*	1.48 *8*
γ E2	464.73 *4*	3.75 *20*
γ M1,E2	467.02 *10*	0.39 *6*
γ E1	477.36 *3*	9.2 *5*
γ M1	482.59 *8*	0.36 *3*
γ E1	498.28 *3*	5.7 *3*
γ	518.27 *25*	0.10 *3*
γ	520.8 *6*	0.031 *11*
γ M1	531.44 *5*	0.39 *4*
γ [E1]	534.05 *11*	0.087 *20*
γ M1	546.28 *6*	0.98 *6*
γ	551.77 *8*	0.50 *6*
γ	557.21 *11*	0.22 *3*
γ [E2]	565.42 *7*	0.050 *17*
γ	580.61 *10*	0.210 *22*
γ	584.28 *25*	0.067 *14*
γ [E1]	594.97 *11*	0.109 *17*
γ	599.86 *14*	0.028 *8*
γ	608.94 *10*	0.28 *3*
γ	620.8 *4*	0.048 *11*
γ	625.07 *9*	0.059 *14*
γ	627.7 *3*	0.064 *20*
γ	652.8 *3*	0.067 *22*
γ	665.2 *4*	0.045 *14*
γ	674.17 *25*	0.059 *14*
γ	684.07 *20*	0.073 *17*
γ	691.32 *20*	0.126 *25*
γ	704.83 *25*	0.076 *17*
γ	720.13 *9*	0.120 *22*
γ [E1]	735.37 *6*	0.151 *22*
γ	742.59 *20*	0.154 *22*
γ	744.87 *12*	0.38 *3*
γ	761.90 *15*	0.190 *22*
γ [E1]	773.73 *4*	0.050 *17*
γ	781.2 *3*	0.101 *17*
γ [M1+E2]	787.37 *7*	0.69 *6*
γ	798.85 *20*	0.16 *3*
γ	802.57 *12*	0.25 *3*
γ	812.13 *9*	0.55 *5*
γ	815.22 *15*	0.23 *3*
γ	827.61 *16*	0.29 *4*
γ M1	832.65 *5*	3.00 *22*
γ	836.52 *10*	0.36 *6*
γ [E2]	850.38 *11*	0.18 *4*
γ [E1]	855.28 *6*	0.22 *6*
γ	862.61 *20*	0.36 *6*
γ	874.5 *5*	0.048 *14*
γ E2	886.15 *5*	2.52 *17*
γ	903.13 *10*	0.50 *17*
γ	935.20 *18*	0.34 *3*
γ	947.06 *8*	0.67 *6*
γ [M1+E2]	953.62 *9*	0.25 *3*
γ	962.98 *13*	0.218 *22*
γ	971.37 *8*	0.52 *4*
γ	983.1 *6*	0.039 *11*
γ	989.0 *3*	0.084 *22*
γ	1030.12 *12*	0.27 *3*
γ	1035.7 *7*	0.053 *14*

Photons (^{179}Re)
(continued)

γ_{mode}	γ(keV)	γ(%)†
γ	1042.3 *3*	0.101 *22*
γ	1061.37 *15*	0.118 *22*
γ	1068.4 *6*	0.048 *14*
γ[M1+E2]	1072.33 *11*	0.062 *14*
γ	1110.29 *25*	0.17 *3*
γ	1120.1 *3*	0.12 *2*
γ[E1]	1148.84 *7*	0.042 *14*
γ[E1]	1171.30 *7*	~0.050
γ[M1+E2]	1202.92 *6*	0.090 *17*
γ	1236.7 *5*	0.056 *17*
γ	1243.8 *4*	0.078 *17*
γ[E1]	1250.03 *6*	0.084 *17*
γ	1271.5 *7*	0.062 *17*
γ[E1]	1277.46 *8*	0.073 *22*
γ[E1]	1287.98 *6*	0.22 *3*
γ[E1]	1301.60 *6*	0.11 *3*
γ	1306.4 *8*	0.10 *3*
γ	1320.09 *12*	0.20 *3*
γ[M1+E2]	1331.54 *8*	0.084 *20*
γ	1340.02 *11*	0.134 *25*
γ	1349.8 *4*	0.106 *22*
γ	1362.4 *5*	0.076 *20*
γ[M1+E2]	1371.28 *5*	1.08 *6*
γ[E1]	1384.43 *6*	0.050 *17*
γ	1398.1 *6*	0.056 *17*
γ	1402.2 *3*	0.126 *25*
γ	1431.95 *12*	~0.041
γ	1486.01 *13*	0.073 *14*
γ[M1+E2]	1499.90 *8*	0.20 *3*
γ	1529.11 *11*	0.132 *20*
γ[E1]	1560.36 *5*	3.22 *20*
γ	1599.9 *4*	0.112 *22*
γ	1608.7 *6*	0.101 *22*
γ	1625.55 *25*	0.115 *20*
γ	1649.03 *11*	0.46 *5*
γ	1663.1 *9*	0.062 *20*
γ	1667.8 *6*	0.101 *20*
γ E1	1680.28 *5*	13.0 *7*
γ[E1]	1688.99 *7*	0.44 *5*
γ	1697.25 *14*	0.042 *8*
γ	1728.87 *13*	0.090 *20*
γ	1804.8 *4*	0.17 *4*
γ[E1]	1808.90 *7*	2.23 *17*

† 3.6% uncert(syst)
§ with ^{179}W(6.4 min) in equilib

Atomic Electrons (^{179}Re)
$\langle e \rangle$=147 *12* keV

e_{bin}(keV)	$\langle e \rangle$(keV)	e(%)
9	0.07	~0.8
10	8.5	~83
12	7.0	60 *29*
13	4.6	~36
14	6.0	<84
21	0.08	<0.7
22	4.2	<38
23 - 71	11.7	25 *6*
73 - 121	16.2	17.3 *6*
123 - 150	0.50	0.36 *7*
152	35.5	23 *4*
153 - 202	1.75	1.03 *7*
203 - 208	0.092	0.045 *11*
210	15.8	7.5 *12*
211	0.00064	0.00030 *12*
212	5.9	2.8 *4*
213 - 218	0.018	0.008 *4*
219	4.1	1.9 *3*
220 - 226	4.7	2.12 *18*
227	3.97	1.75 *10*
228 - 277	1.0	0.39 *13*
278 - 327	1.82	0.63 *3*
329 - 357	0.90	0.263 *10*
361	7.4	2.04 *8*
366 - 415	0.97	0.243 *10*
418 - 467	2.13	0.501 *13*
470 - 520	0.37	0.076 *10*

Atomic Electrons (^{179}Re)
(continued)

e_{bin}(keV)	$\langle e \rangle$(keV)	e(%)
521 - 570	0.114	0.021 *4*
572 - 621	0.032	0.0054 *15*
622 - 666	0.030	0.0047 *18*
671 - 720	0.12	0.017 *5*
723 - 772	0.43	0.057 *6*
773 - 822	0.213	0.0263 *23*
824 - 873	0.072	0.009 *3*
874 - 923	0.12	0.014 *4*
924 - 973	0.038	0.0040 *11*
977 - 1026	0.014	0.0014 *4*
1027 - 1072	0.015	0.0015 *6*
1079 - 1120	0.0036	0.00033 *9*
1133 - 1181	0.010	0.00087 *25*
1191 - 1240	0.012	0.00098 *24*
1241 - 1290	0.019	0.0015 *5*
1291 - 1340	0.054	0.0042 *11*
1347 - 1396	0.0114	0.00084 *17*
1397 - 1432	0.009	0.00066 *18*
1460 - 1558	0.054	0.0036 *3*
1580 - 1679	0.180	0.0111 *7*
1685 - 1739	0.026	0.00148 *17*
1793 - 1807	0.0045	0.000248 *25*

Continuous Radiation (^{179}Re)
$\langle \beta + \rangle$=8.3 keV;$\langle$IB$\rangle$=1.56 keV

E_{bin}(keV)		$\langle\ \rangle$(keV)	(%)
0 - 10	β+	5.3×10^{-8}	6.2×10^{-7}
	IB	0.00100	
10 - 20	β+	3.20×10^{-6}	1.90×10^{-5}
	IB	0.00106	0.0088
20 - 40	β+	0.000132	0.000397
	IB	0.0032	0.0099
40 - 100	β+	0.0097	0.0122
	IB	0.31	0.53
100 - 300	β+	0.57	0.258
	IB	0.052	0.028
300 - 600	β+	3.08	0.68
	IB	0.18	0.039
600 - 1300	β+	4.61	0.57
	IB	0.70	0.076
1300 - 2468	β+	0.055	0.00408
	IB	0.32	0.021
	$\Sigma\beta$+		1.53

$^{179}_{76}$Os(6.5 *5* min)

Mode: ϵ
Δ: -43010 *200* keV syst
SpA: 1.61×10^{8} Ci/g
Prod: ^{171}Yb(^{12}C,4n); ^{169}Tm(^{14}N,4n); protons on Au

Photons (^{179}Os)*

γ_{mode}	γ(keV)	γ(rel)
γ	165.6 *9*	34 *4*
γ	218.9 *6*	100
γ	309.6 *10*	~40
γ	759.6 *10*	~68
γ	1311.6 *12*	~83

* see also ^{178}Os

$^{179}_{77}$Ir(4 *1* min)

Mode: ϵ
Δ: -38020 *310* keV syst
SpA: 2.6×10^{8} Ci/g
Prod: Ne on ^{165}Ho

$^{179}_{78}$Pt(~ 43 s)

Mode: ϵ(99.73 *4* %), α(0.27 *4* %)
Δ: -32350 *300* keV syst
SpA: ~ 1.5×10^{9} Ci/g
Prod: ^{16}O on Yb; ^{20}Ne on Er; ^{19}F on Tm; daughter ^{183}Hg

Alpha Particles (^{179}Pt)

α(keV)
5150 *10*

$^{179}_{79}$Au(7.5 *4* s)

Mode: α
Δ: -24990 *380* keV syst
SpA: 8.0×10^{9} Ci/g
Prod: ^{20}Ne on Tm

Alpha Particles (^{179}Au)

α(keV)
5848 *5*

$^{179}_{80}$Hg(1.09 *4* s)

Mode: α, ϵ, ϵp
Δ: -17110 *380* keV syst
SpA: 4.28×10^{10} Ci/g
Prod: protons on Pb
ϵp/α: 0.0028 *4*

Alpha Particles (^{179}Hg)

α(keV)
6270 *15*

A = 180

NDS 15, 559 (1975)

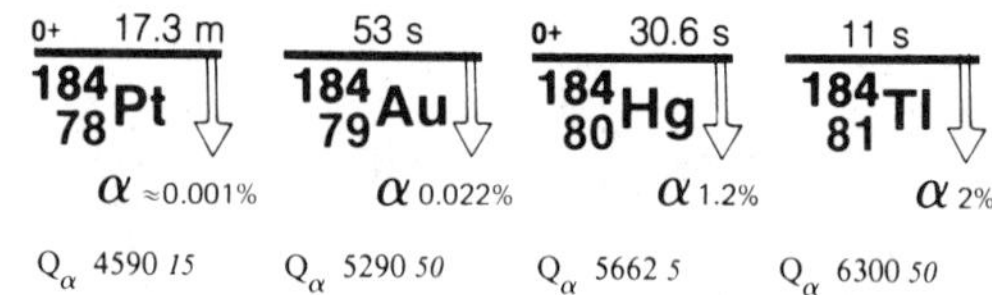

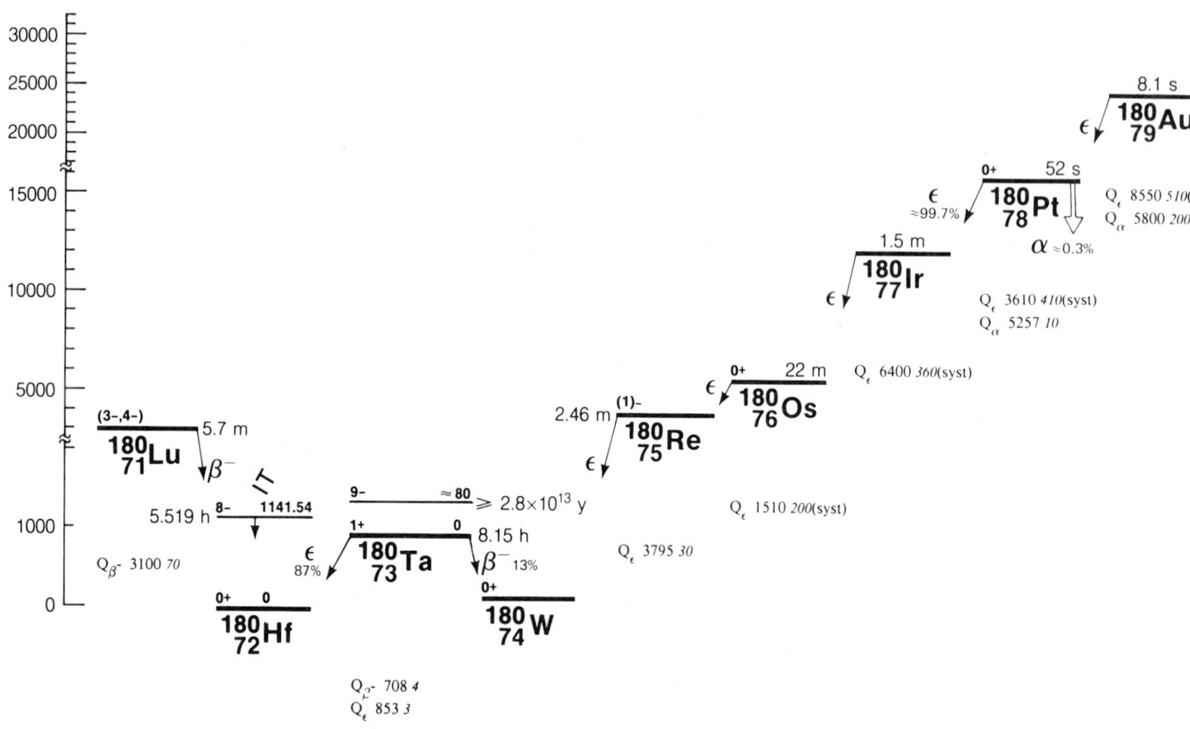

$^{180}_{71}$Lu(5.7 *1* min)

Mode: β-
Δ: -46690 *70* keV
SpA: 1.83×10^{8} Ci/g
Prod: ^{180}Hf(n,p)

Photons (^{180}Lu)

⟨γ⟩=1567 *52* keV

γ_{mode}	γ(keV)	γ(%)†
γ	68.9 *3*	1.8 *8*
γE2	93.331 *6*	13.3 *20*
γ(E1+M2)	134.6 *3*	1.65 *20*
γ	198.38 *24*	1.25 *20*
γE2	215.248 *13*	21 *3*
γM1+E2	234.6 *3*	1.05 *15*
γ	316.51 *5*	14.9 *15*
γ	333.8 *4*	1.0 *3*
γ	407.95 *5*	50
γ	424.6 *3*	1.3 *2*
γ	451.9 *3*	1.05 *20*
γ	891.56 *8*	0.70 *15*
γ	982.99 *10*	2.20 *25*
γ	1065.7 *3*	1.4 *2*
γ	1090.1 *3*	1.15 *15*
γ	1101.12 *24*	1.7 *2*
γ	1106.80 *8*	23.5 *23*
γ	1198.24 *10*	15.1 *15*
γ	1200.13 *8*	26 *3*
γ	1230.6 *3*	~0.50

Photons (^{180}Lu) (continued)

γ_{mode}	γ(keV)	γ(%)†
γ	1280.9 *3*	~0.50
γ	1299.50 *9*	14.2 *14*
γ	1316.37 *24*	1.6 *5*
γ	1434.9 *3*	2.00 *25*
γ	1445.8 *3*	0.75 *25*
γ	1514.75 *9*	7.9 *8*
γ	1874.5 *9*	0.70 *15*
γ	1888.5 *9*	1.2 *2*

† 4.0% uncert(syst)

$^{180}_{72}$Hf(stable)

Δ: -49793 *3* keV
%: 35.100 *6*

$^{180}_{72}$Hf(5.519 *4* h)

Mode: IT
Δ: -48651 *3* keV
SpA: 3.1545×10^{6} Ci/g
Prod: ^{179}Hf(n,γ)

Photons (^{180}Hf)

⟨γ⟩=998 *16* keV

γ_{mode}	γ(keV)	γ(%)†
Hf L_{ℓ}	6.960	0.46 *5*
Hf L_{α}	7.893	11.1 *9*
Hf L_{η}	8.139	0.187 *13*
Hf L_{β}	9.123	13.3 *13*
Hf L_{γ}	10.614	2.3 *3*
Hf $K_{\alpha2}$	54.611	10.1 *3*
Hf $K_{\alpha1}$	55.790	17.6 *5*
γ E1+0.3%M2	57.549 *5*	48.4 *19*
Hf $K_{\beta1}$'	63.166	5.75 *17*
Hf $K_{\beta2}$'	65.211	1.49 *5*
γ E2	93.331 *6*	17.0 *5*
γ E2	215.248 *13*	81.7 *19*
γ E2	332.269 *20*	94.4 *24*
γ E2	443.142 *17*	85 *3*
γ [M2]	500.691 *17*	12.8 *11*

† 0.21% uncert(syst)

Atomic Electrons (^{180}Hf)

⟨e⟩=147.9 *24* keV

e_{bin}(keV)	⟨e⟩(keV)	e(%)
10 - 11	7.2	71 *6*
28	5.10	18.2 *6*
43 - 45	0.40	0.90 *6*
46	7.4	16 *4*
47 - 82	11.1	20.6 *18*
83	19.2	23.2 *8*

Atomic Electrons (^{180}Hf) (continued)

e_{bin}(keV)	⟨e⟩(keV)	e(%)
84	18.3	21.8 8
91	10.6	11.6 4
92 - 93	3.06	3.30 11
150	16.7	11.1 3
204	2.59	1.27 4
205	5.29	2.59 8
206	3.68	1.79 6
213 - 215	3.77	1.77 4
267	10.6	3.96 13
321 - 332	5.23	1.614 25
378	6.51	1.72 7
432 - 434	1.82	0.422 14
435	6.9	1.60 15
441 - 490	1.94	0.409 25
491 - 501	0.47	0.095 7

$^{180}_{73}$Ta(8.152 6 h)

Mode: ϵ(87 2 %), β-(13 2 %)
Δ: -48940 3 keV
SpA: 2.1356×10^6 Ci/g
Prod: ^{180}Hf(d,2n); ^{181}Ta(n,2n); ^{181}Ta(γ,n); ^{181}Ta(d,p2n)

Photons (^{180}Ta)

⟨γ⟩=48 12 keV

γ_{mode}	γ(keV)	γ(%)†
Hf L_ℓ	6.960	~0.39
W L_ℓ	7.387	0.0096 18
Hf L_α	7.893	~9
Hf L_η	8.139	~0.13
W L_α	8.391	0.22 4
W L_η	8.724	0.0044 7
Hf L_β	9.138	~10
W L_β	9.770	0.27 5
Hf L_γ	10.611	~2
W L_γ	11.318	0.044 8
Hf $K_{\alpha2}$	54.611	~21
Hf $K_{\alpha1}$	55.790	~36
W $K_{\alpha2}$	57.981	0.17 3
W $K_{\alpha1}$	59.318	0.29 5
Hf $K_{\beta1}$'	63.166	~12
Hf $K_{\beta2}$'	65.211	~3
W $K_{\beta1}$'	67.155	0.097 16
W $K_{\beta2}$'	69.342	0.025 4
γ_ϵE2	93.331 6	~5
$\gamma_{\beta-}$E2	103.40 15	0.74 12

† uncert(syst): 2.3% for ϵ, 15% for β-

Atomic Electrons (^{180}Ta)

⟨e⟩=26 5 keV

e_{bin}(keV)	⟨e⟩(keV)	e(%)
10	3.3	~34
11	2.6	~24
12	0.068	0.59 11
28	1.4	~5
34 - 82	2.3	4.6 8
83	5.3	~6
84	5.1	~6
91	3.0	3.3 16
92	0.72	0.78 14
93	1.4	1.5 6
101 - 103	0.49	0.48 6

Continuous Radiation (^{180}Ta)

⟨β-⟩=27.8 keV;⟨IB⟩=0.35 keV

E_{bin}(keV)		⟨ ⟩(keV)	(%)
0 - 10	β-	0.0179	0.357
	IB	0.00139	
10 - 20	β-	0.053	0.355
	IB	0.00130	0.0090
20 - 40	β-	0.213	0.71
	IB	0.0023	0.0082
40 - 100	β-	1.46	2.09
	IB	0.0054	0.0085
100 - 300	β-	11.4	5.9
	IB	0.0078	0.0048
300 - 600	β-	14.1	3.50
	IB	0.00147	0.00041
600 - 853	β-	0.487	0.078
	IB	4.5×10^{-6}	7.3×10^{-7}

$^{180}_{73}$Ta(>2.8×10^{13} yr)

Δ: ~-48860 keV
%: 0.012 2

$^{180}_{74}$W (stable)

Δ: -49648 5 keV
%: 0.13 3

$^{180}_{75}$Re(2.46 5 min)

Mode: ϵ
Δ: -45850 30 keV
SpA: 4.24×10^8 Ci/g
Prod: daughter ^{180}Os; ^{181}Ta(α,5n); ^{181}Ta(^{3}He,4n); protons on W

Photons (^{180}Re)

⟨γ⟩=1105 180 keV

γ_{mode}	γ(keV)	γ(%)
W L_ℓ	7.387	0.73 15
W L_α	8.391	17 3
W L_η	8.724	0.27 5
W L_β	9.781	19 3
W L_γ	11.365	3.1 6
W $K_{\alpha2}$	57.981	29 6
W $K_{\alpha1}$	59.318	50 10
W $K_{\beta1}$'	67.155	16 3
W $K_{\beta2}$'	69.342	4.3 8
γ (M1+E2)	75.80 11	~0.7
γ E2	103.40 15	22.5 25
γ E2	234.6 6	0.49 10
γ	742.0 10	0.69 20
γ	748.8 10	1.4 4
γ E1+1.0%M2+E3	902.2 3	98
γ (M2)	1005.6 4	0.59 10
γ	1012.8 10	1.2 3
γ	1024.7 13	0.15 5
γ	1061.2 15	0.20 7
γ	1115.8 15	0.69 20
γ	1129.1 15	0.59 20
γ	1203.8 20	0.26 8
γ	1227.8 20	0.29 10
γ	1298.8 10	0.46 9

Photons (^{180}Re) (continued)

γ_{mode}	γ(keV)	γ(%)
γ	1319.3 15	0.29 7
γ	1352.7 6	0.36 8
γ	1409.0 15	1.18 10
γ	1430.6 10	0.25 8
γ	1483.6 15	0.28 6
γ	1528.8 15	0.31 6
γ	1545 3	<0.27
γ	1767.3 10	0.15 4
γ	1802.2 15	~0.20
γ	1820.3 15	~0.10
γ	1838.2 15	0.13 4
γ	1867.2 15	~0.05
γ	1877.3 10	0.24 3
γ	2025.2 20	~0.039
γ	2033.4 20	0.049 20
γ	2153.8 15	0.108 20
γ	2204.8 20	0.127 20
γ	2242 3	~0.029
γ	2260 3	0.049 20
γ	2333 3	<0.14
γ	2341 3	<0.14 ?
γ	2408 3	<0.039

Atomic Electrons (^{180}Re)

⟨e⟩=80 4 keV

e_{bin}(keV)	⟨e⟩(keV)	e(%)
6	0.22	~4
10	5.5	54 11
12	4.7	40 8
34	6.3	18.6 21
45 - 76	5.4	9 3
91	1.85	2.02 22
92	21.0	22.9 25
93	18.9	20.3 22
101	11.4	11.3 13
102	0.108	0.107 12
103	3.3	3.2 4
222 - 235	0.082	0.036 4
672 - 679	0.14	~0.021
730 - 749	0.032	0.0043 20
890 - 936	0.46	0.051 6
943 - 992	0.07	~0.007
993 - 1025	0.038	0.0038 8
1046 - 1061	0.05	~0.005
1104 - 1134	0.020	0.0018 7
1158 - 1204	0.013	~0.0011
1216 - 1250	0.027	0.0022 10
1283 - 1319	0.017	0.0013 6
1339 - 1361	0.045	~0.0034
1397 - 1431	0.017	0.0012 5
1459 - 1542	0.017	0.0011 5
1698 - 1792	0.014	0.0008 3
1798 - 1875	0.008	0.00047 19
1956 - 2031	0.0020	0.00010 4
2084 - 2172	0.0051	0.00024 8
2190 - 2271	0.0039	0.00017 8
2321 - 2405	0.0008	~4×10^{-5}

Continuous Radiation (^{180}Re)

⟨β+⟩=66 keV;⟨IB⟩=4.0 keV

E_{bin}(keV)		⟨ ⟩(keV)	(%)
0 - 10	β+	8.5×10^{-8}	1.01×10^{-6}
	IB	0.0033	
10 - 20	β+	5.2×10^{-6}	3.11×10^{-5}
	IB	0.0033	0.024
20 - 40	β+	0.000218	0.00066
	IB	0.0075	0.025
40 - 100	β+	0.0168	0.0211
	IB	0.31	0.52
100 - 300	β+	1.17	0.52

Continuous Radiation (^{180}Re) (continued)

E_{bin}(keV)		⟨ ⟩(keV)	(%)
	IB	0.088	0.047
300 - 600	β+	9.3	2.00
	IB	0.26	0.057
600 - 1300	β+	45.7	5.00
	IB	1.41	0.149
1300 - 2500	β+	100	0.70
	IB	1.9	0.115
2500 - 2789	IB	0.019	0.00073
	Σβ+		8.2

$^{180}_{76}$Os(21.7 *6* min)

Mode: ϵ
Δ: -44350 *200* keV syst
SpA: 4.81×10^{7} Ci/g
Prod: ^{169}Tm(^{15}N,4n); ^{12}C on Yb; alphas on W

Photons (^{180}Os)

γ_{mode}	γ(keV)
γE1	20.13 *8*
γ	31.63 *10*
γ	48.20 *10*
γM1,E2	49.85 *10*
γ(M1)	54.44 *8*
γE1	74.58 *8*
γ	104.01 *20*
γ	107.01 *20*
γ	113.71 *20*
γ	183.7 *3*
γ	200.03 *19*
γ	218.26 *19*
γ	249.88 *19*
γ	319.5 *4*
γ	329.0 *3*
γ	349.2 *3*
γ	402.0 *3*
γ	450.4 *4*
γ	485.7 *7*
γ	667.0 *4*
γ	716.9 *4*

$^{180}_{77}$Ir(1.5 *1* min)

Mode: ϵ
Δ: -37950 *300* keV syst
SpA: 6.9×10^{8} Ci/g
Prod: Ne on ^{165}Ho; ^{169}Tm(^{16}O,5n); ^{175}Lu(^{12}C,7n); protons on Au

Photons (^{180}Ir)

⟨γ⟩=687 *13* keV

γ_{mode}	γ(keV)	γ(%)†
γE2	132.1 *3*	39.9 *21*
γE2	276.5 *3*	42.0 *21*
γE2	386.6 *3*	2.60 *25*
γ	492.8 *3*	2.9 *3*
γ	614.1 *4*	2.14 *21*
γ	644.5 *5*	7.0 *4*
γ	699.0 *5*	9.4 *5* ?
γ	788.3 *4*	3.8 *4*
γ	846.3 *5*	2.44 *25*
γ	870.3 *5*	8.6 *5* ?
γ	890.5 *4*	9.1 *5*
γ	968.9 *5*	3.1 *3* ?
γ	1014.3 *5*	1.09 *13* ?
γ	1064.8 *4*	6.0 *6*
γ	1330.6 *5*	4.37 *25* ?

† 7.1% uncert(syst)

$^{180}_{78}$Pt(52 *3* s)

Mode: α(~0.3 %), ϵ(~99.7 %)
Δ: -34350 *280* keV syst
SpA: 1.20×10^{9} Ci/g
Prod: ^{16}O on Yb; ^{20}Ne on Er

Alpha Particles (^{180}Pt)

α(keV)
5140 *10*

$^{180}_{79}$Au(8.1 *3* s)

Mode: ϵ
Δ: -25800 *420* keV syst
SpA: 7.4×10^{9} Ci/g
Prod: protons on Pb

Photons (^{180}Au)

γ_{mode}	γ(keV)	γ(rel)
γ	152.2 *3*	100
γ	256.4 *3*	30 *6*
γ	324.0 *3*	18 *3*
γ	343.4 *3*	14 *3*
γ	450.5 *5*	~7
γ	524.2 *3*	44 *7*
γ	552.4 *4*	~7
γ	676.5 *4*	20 *4*
γ	707.7 *5*	~4
γ	808.4 *4*	30 *6*
γ	859.7 *6*	35 *7*
γ	1032.1 *7*	23 *5*

$^{180}_{80}$Hg(2.9 *3* s)

Mode: α, ϵ
Δ: -20260 *240* keV syst
SpA: 1.92×10^{10} Ci/g
Prod: protons on Pb

Alpha Particles (^{180}Hg)

α(keV)
6118 *15*

Photons (^{180}Hg)

γ_{mode}	γ(keV)	γ(rel)
γ_ϵ	125.0 *4*	9.7 *20*
γ_ϵ	300.5 *3*	100
γ_ϵ	381.2 *4*	69 *14*
γ_ϵ	405.0 *5*	~17
γ_ϵ	450.5 *5*	~16
γ_ϵ	479.9 *4*	23 *5*

A = 181

NDS **43**, 289 (1984)

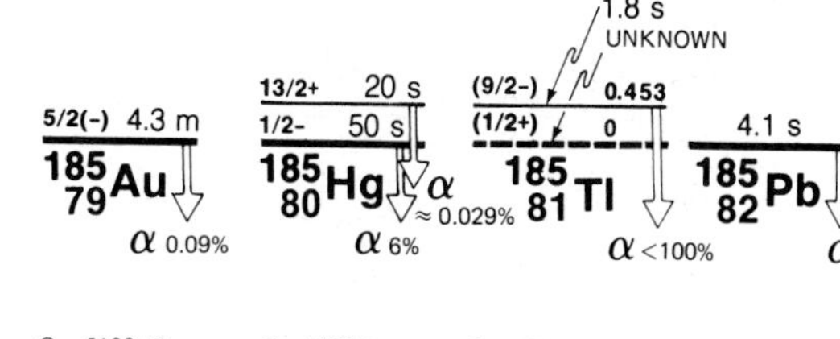

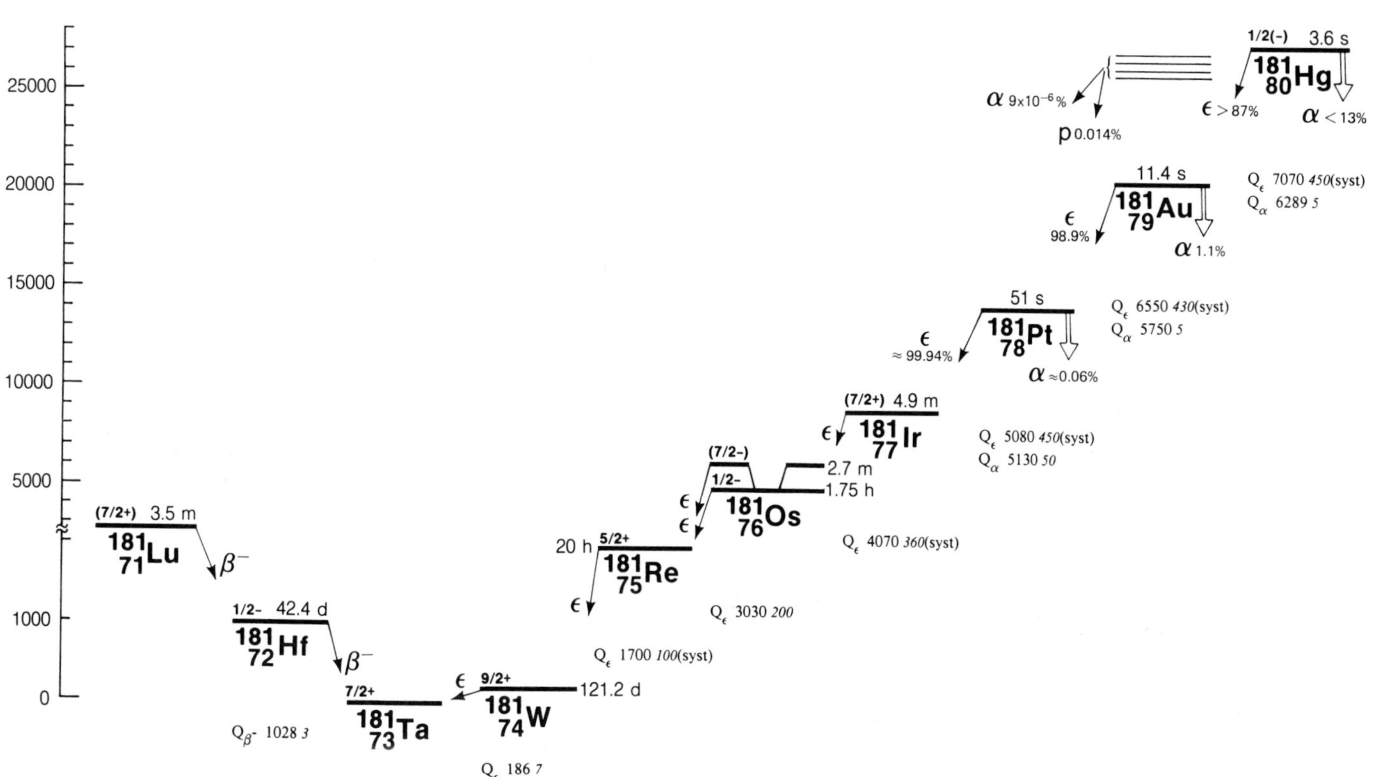

$^{181}_{71}$Lu(3.5 *3* min)

Mode: β-
SpA: 2.96×10^{8} Ci/g
Prod: ^{136}Xe on W; ^{136}Xe on Ta

Photons (^{181}Lu)

⟨γ⟩=542 *32* keV

γ_{mode}	γ(keV)	γ(%)
γ M1+~3.8%E2	45.77 *17*	6.5 *11*
γ M1+~8.3%E2	52.99 *22*	3.9 *9*
γ E2	98.76 *18*	3.5 *13*
γ [M1]	105.6 *3*	3.9 *7*
γ [M1]	125.0 *4*	3.2 *7*
γ [M1]	153.0 *3*	2.6 *11*
γ [E2]	158.5 *4*	1.5 *7*
γ [M1]	205.9 *3*	15.8 *13*
γ [E1]	240.4 *4*	4.5 *13*
γ [M1]	251.7 *3*	~2
γ [E2]	329.3 *3*	5.0 *11*
γ [M1]	334.4 *4*	3.7 *13*
γ [M1]	341.8 *4*	3.2 *9*
γ [E1]	463.7 *5*	4.5 *13*

Photons (^{181}Lu)
(continued)

γ_{mode}	γ(keV)	γ(%)
γ [E1]	574.9 *3*	15.1 *13*
γ	589.9 *10*	3.2 *11*
γ [E1]	652.5 *4*	21.6 *13*
γ [E1]	699.9 *4*	4.1 *15*
γ [E1]	805.4 *3*	8.6 *22*
γ [E1]	858.4 *3*	7.6 *22*

$^{181}_{72}$Hf(42.39 *6* d)

Mode: β-
Δ: -47417 *3* keV
SpA: 1.7018×10^{4} Ci/g
Prod: ^{180}Hf(n,γ)

Photons (^{181}Hf)

⟨γ⟩=518 *5* keV

γ_{mode}	γ(keV)	γ(%)†
γ [M1]	3.65 *11*	~0.0016 ?
γ E1	6.134 *20*	0.0101 *23*
Ta L_{ℓ}	7.173	0.215 *21*
Ta L_{α}	8.140	5.1 *3*
Ta L_{η}	8.428	0.089 *5*
Ta L_{β}	9.448	5.7 *4*
Ta L_{γ}	10.950	0.94 *7*
Ta $K_{\alpha2}$	56.280	7.8 *3*
Ta $K_{\alpha1}$	57.535	13.5 *4*
Ta $K_{\beta1}'$	65.140	4.47 *15*
Ta $K_{\beta2}'$	67.254	1.15 *4*
γ E2	132.94 *7*	35.9 *8*
γ M1+14%E2	136.17 *7*	5.80 *24*
γ (M1+<0.2%E2)	136.59 *9*	<0.4 ?
γ E2	345.83 *7*	15.1 *10*
γ M2+20%E3	475.86 *5*	0.61 *22*
γ E2+2.6%M1	482.00 *5*	80.6
γ M3+32%E4	614.94 *8*	0.258 *16*
γ (E2)	618.58 *9*	0.040 *8*

† 0.87% uncert(syst)

Atomic Electrons (^{181}Hf)

⟨e⟩=68.2 *9* keV

e_{bin}(keV)	⟨e⟩(keV)	e(%)
2 - 48	3.60	32.9 *22*
53 - 65	0.245	0.442 *21*
66	11.5	17.6 *5*
69	5.8	8.5 *7*
121	2.24	1.85 *6*
122	12.9	10.6 *3*
123	10.7	8.7 *3*
124 - 127	2.14	1.71 *10*
130	3.96	3.03 *9*
131	2.96	2.27 *7*
132 - 137	2.63	1.98 *5*
334 - 346	0.81	0.240 *8*
408	0.33	0.08 *3*
415	5.97	1.44 *3*
464 - 482	2.16	0.457 *7*
548 - 551	0.186	0.034 *4*
603 - 619	0.070	0.0116 *9*

Continuous Radiation (^{181}Hf)

⟨β-⟩=131 keV;⟨IB⟩=0.061 keV

E_{bin}(keV)		⟨ ⟩(keV)	(%)
0 - 10	β-	0.251	5.03
	IB	0.0067	
10 - 20	β-	0.74	4.91
	IB	0.0060	0.042
20 - 40	β-	2.84	9.5
	IB	0.0102	0.036
40 - 100	β-	17.8	25.7
	IB	0.020	0.032
100 - 300	β-	88	48.7
	IB	0.018	0.0116
300 - 547	β-	21.5	6.1
	IB	0.00074	0.00022

$^{181}_{73}$Ta(stable)

Δ: -48445 *3* keV
%: 99.988 *2*

$^{181}_{74}$W (121.2 *2* d)

Mode: ϵ
Δ: -48259 *7* keV
SpA: 5952 Ci/g
Prod: ^{181}Ta(d,2n); ^{181}Ta(p,n); ^{180}W(n,γ)

Photons (^{181}W)

⟨γ⟩=40 *8* keV

γ_{mode}	γ(keV)	γ(%)
γ E1	6.134 *20*	1.07 *3*
Ta L_ℓ	7.173	0.35 *13*
Ta L_α	8.140	8 *3*
Ta L_η	8.428	0.11 *4*
Ta L_β	9.451	10 *4*
Ta L_γ	11.054	1.8 *7*
Ta $K_{\alpha2}$	56.280	19 *7*
Ta $K_{\alpha1}$	57.535	33 *12*
Ta $K_{\beta1}'$	65.140	11 *4*
Ta $K_{\beta2}'$	67.254	2.8 *10*

Photons (^{181}W) (continued)

γ_{mode}	γ(keV)	γ(%)
γ M1+14%E2	136.17 *7*	0.0318 *10*
γ M1+21%E2	152.214 *20*	0.084 *3*

Atomic Electrons (^{181}W)

⟨e⟩=8.2 *11* keV

e_{bin}(keV)	⟨e⟩(keV)	e(%)
3	0.078	2.28 *8*
4	1.02	24.7 *9*
6	0.384	6.60 *23*
10	2.5	25 *9*
11	1.4	13 *5*
12	1.2	10 *4*
44	0.13	0.29 *11*
45	0.20	0.45 *16*
46	0.40	0.9 *3*
48 - 85	0.82	1.46 *20*
124 - 152	0.048	0.035 *3*

Continuous Radiation (^{181}W)

⟨IB⟩=0.27 keV

E_{bin}(keV)		⟨ ⟩(keV)	(%)
10 - 20	IB	0.00075	0.0066
20 - 40	IB	0.0034	0.0102
40 - 100	IB	0.26	0.46
100 - 186	IB	0.00157	0.00161

$^{181}_{75}$Re(19.9 *7* h)

Mode: ϵ
Δ: -46560 *100* keV syst
SpA: 8.7×10^5 Ci/g
Prod: ^{181}Ta(α,4n); ^{182}W(p,2n); ^{181}Ta(^{3}He,3n)

Photons (^{181}Re)

⟨γ⟩=813 *36* keV

γ_{mode}	γ(keV)	γ(%)†
W L_ℓ	7.387	0.79 *21*
W L_α	8.391	18 *5*
W L_η	8.724	0.26 *7*
W L_β	9.783	20 *5*
W L_γ	11.411	3.6 *9*
γ E2	19.63 *11*	~0.0025
γ M1	31.04 *11*	0.017 *7*
γ M1	38.19 *13*	0.060 *15*
γ M1+1.1%E2	43.65 *13*	0.46 *9*
W $K_{\alpha2}$	57.981	35 *8*
W $K_{\alpha1}$	59.318	60 *14*
γ M1+10%E2	64.96 *12*	2.4 *5*
W $K_{\beta1}'$	67.155	20 *5*
W $K_{\beta2}'$	69.342	5.2 *12*
γ M1+8.4%E2	71.57 *13*	0.11 *5*
γ M1	72.65 *12*	0.22 *9*
γ M1+13%E2	93.61 *15*	0.12 *5*
γ E2	102.62 *12*	0.26 *10*
γ E2	103.15 *12*	0.17 *7*
γ M1+13%E2	109.97 *11*	2.7 *5*
γ M1+3.2%E2	110.31 *12*	0.97 *22*

Photons (^{181}Re) (continued)

γ_{mode}	γ(keV)	γ(%)†
γ M1	113.40 *15*	~0.7
γ	125.79 *21*	0.10 *4*
γ M1	137.19 *20*	0.07 *3*
γ M1	144.22 *11*	0.37 *15*
γ E2	154.47 *14*	0.27 *10*
γ M1+~39%E2	163.85 *11*	0.13 *5*
γ M1+~39%E2	164.60 *14*	0.13 *5*
γ M1+~39%E2	165.80 *11*	0.13 *5*
γ [M1]	167.28 *14*	0.10 *3*
γ E2	175.27 *10*	~0.45
γ E1	177.35 *14*	1.6 *7*
γ E2	186.09 *14*	0.19 *8*
γ [E2]	192.66 *16*	0.08 *3*
γ M1	194.90 *11*	0.16 *7*
γ M1	196.85 *12*	0.52 *22*
γ	201.0 *3*	0.19 *6*
γ	211.7 *3*	0.09 *3*
γ	213.0 *3*	0.11 *4*
γ [M1]	237.93 *14*	0.10 *3*
γ [E1]	239.52 *19*	0.09 *3*
γ	245.5 *3*	0.11 *4*
γ E3	252.19 *16*	0.99 *25*
γ M1+46%E2	262.60 *18*	0.22 *8*
γ M1+39%E2	276.12 *13*	0.64 *16*
γ [M1]	277.24 *14*	0.16 *5*
γ	279.5 *3*	0.10 *4*
γ M1+~39%E2	296.06 *15*	0.21 *7*
γ	297.0 *3*	0.11 *4*
γ [E2]	317.05 *16*	0.09 *3*
γ M1	318.55 *18*	1.1 *3*
γ M1	331.87 *14*	1.3 *4*
γ [M1]	341.07 *12*	0.037 *15*
γ	350.4 *3*	0.35 *12*
γ (M1)	353.73 *21*	~0.45
γ [E2]	356.22 *16*	1.7 *8*
γ E2+38%M1	360.70 *11*	20 *4*
γ M2	365.59 *13*	57 *6*
γ	376.8 *3*	0.13 *5*
γ M1+~39%E2	382.33 *21*	0.27 *9*
γ M1	398.19 *16*	0.65 *16*
γ [E1]	409.24 *16*	0.25 *8*
γ [M1]	412.16 *16*	1.0 *3*
γ	420.0 *3*	0.10 *4*
γ (M1)	441.47 *17*	~0.45
γ (M1)	441.84 *13*	~1
γ [E1]	475.56 *14*	1.0 *3*
γ M1	487.07 *19*	~0.7
γ M1	488.78 *20*	~0.7
γ M1	515.67 *19*	0.16 *5*
γ M1	522.61 *18*	0.23 *8*
γ [M1]	522.7 *4*	0.19 *7*
γ [E1]	533.82 *17*	~0.06
γ [M1]	544.73 *20*	0.29 *10*
γ E1	558.16 *15*	2.2 *5*
γ [E1]	570.30 *19*	0.46 *12*
γ E2	587.25 *18*	0.69 *17*
γ [M1]	627.76 *16*	0.13 *5*
γ [E1]	633.07 *20*	0.16 *5*
γ E1	639.30 *14*	6.5 *13*
γ [E1]	643.79 *16*	0.58 *14*
γ E1	651.44 *19*	1.0 *3*
γ [M1]	658.81 *16*	0.23 *8*
γ E1	661.65 *17*	3.0 *6*
γ [M1]	668.5 *4*	0.31 *10*
γ	691.8 *4*	0.12 *4*
γ E2+~31%M1	693.6 *3*	0.25 *8*
γ	696.5 *3*	~0.06
γ [M1]	699.88 *17*	0.13 *5*
γ	719.9 *4*	0.13 *5*
γ [M1]	730.38 *17*	0.07 *3*
γ M1	738.07 *16*	0.30 *10*
γ M1	769.6 *4*	0.16 *5*
γ [M1]	773.34 *17*	0.07 *3*
γ	789.4 *4*	0.07 *3*
γ M1	791.6 *4*	0.12 *4*
γ M1+~50%E2	803.03 *15*	~1
γ E1	805.11 *13*	~3
γ [E1]	817.24 *18*	0.13 *5*
γ M1	822.66 *15*	0.16 *5*
γ	826.7 *4*	0.17 *6*
γ [E1]	836.15 *13*	0.46 *11*
γ E2	840.03 *18*	0.28 *10*
γ [E1]	848.28 *19*	0.13 *5*
γ M1	854.6 *4*	0.19 *6*

Photons (^{181}Re) (continued)

γ$_{mode}$	γ(keV)	γ(%)†
γ [E1]	862.74 19	0.18 6
γ [E1]	877.23 15	~0.45
γ M1	880.0 3	0.52 13
γ M1	883.31 16	0.26 9
γ [E1]	889.36 19	0.120 22
γ	891.7 4	0.10 4
γ E1	907.72 15	1.0 3
γ	947.2 3	0.22 8
γ M1	953.43 16	3.6 9
γ [E1]	964.93 19	0.21 7
γ [E2]	973.37 22	0.13 5
γ [E1]	980.37 13	0.19 6
γ M1	989.7 3	0.90 22
γ [M1+E2]	993.4 3	~0.08
γ [E1]	993.53 19	0.22 10
γ E1	1000.00 12	3.4 7
γ [E2]	1009.38 17	2.5 6
γ [E1]	1018.92 21	0.13 5
γ	1057.2 3	0.30 10
γ	1068.3 5	0.16 5
γ E1	1074.90 19	1.0 3
γ E2	1086.77 21	0.57 14
γ E1	1103.50 19	0.69 17
γ	1127.3 5	0.12 4
γ [E1]	1132.56 20	0.22 8
γ	1172.3 5	0.19 7
γ	1200.8 5	0.20 7
γ [E2]	1271.98 18	0.11 4
γ E2	1384.76 22	0.23 8
γ E2	1440.50 19	1.9 4
γ E2	1469.10 19	0.82 22
γ [M1]	1498.16 20	0.08 3
γ (M1)	1538.0 5	0.19 7

† 5.1% uncert(syst)

Atomic Electrons (^{181}Re)

⟨e⟩=135 7 keV

e$_{bin}$(keV)	⟨e⟩(keV)	e(%)
2 - 9	1.5	20 6
10	5.7	56 15
12	5.1	43 11
17 - 49	10.0	28 3
53	3.7	7.1 15
55 - 103	9.3	12.9 11
106 - 155	2.4	1.97 23
156 - 203	1.25	0.70 9
207 - 252	2.32	0.98 11
260 - 287	1.07	0.39 7
291	3.8	1.3 5
293 - 295	0.0065	0.0022 7
296	63.5	21.5 22
297 - 346	0.94	0.29 3
348 - 352	1.0	0.28 10
353	13.0	3.7 4
354 - 402	7.9	2.19 13
406 - 453	0.58	0.14 3
463 - 513	0.22	0.045 8
514 - 560	0.091	0.0170 24
564 - 599	0.32	0.055 6
616 - 662	0.135	0.021 3
666 - 710	0.078	0.0114 21
717 - 763	0.23	0.031 9
767 - 815	0.169	0.021 3
816 - 866	0.035	0.0042 8
867 - 911	0.35	0.039 9
920 - 969	0.28	0.030 3
970 - 1019	0.098	0.0099 15
1034 - 1077	0.027	0.0026 5
1084 - 1133	0.019	0.0017 7
1160 - 1202	0.0062	0.00052 14
1260 - 1272	0.00068	5.4 14 $\times 10^{-5}$
1371 - 1400	0.072	0.0052 8
1428 - 1469	0.027	0.00186 25
1486 - 1536	0.0023	0.00015 3

Continuous Radiation (^{181}Re)

⟨IB⟩=0.58 keV

E$_{bin}$(keV)		⟨ ⟩(keV)	(%)
10 - 20	IB	0.00081	0.0072
20 - 40	IB	0.0027	0.0082
40 - 100	IB	0.33	0.55
100 - 300	IB	0.035	0.019
300 - 600	IB	0.084	0.019
600 - 1300	IB	0.124	0.0157
1300 - 1434	IB	0.000143	1.08 $\times 10^{-5}$

$^{181}_{76}$Os(1.75 5 h)

Mode: ε

Δ: -43530 220 keV syst

SpA: 9.9×10^{6} Ci/g

Prod: protons on Au; ^{182}W(α,5n); ^{182}W(^{3}He,4n); ^{12}C on Yb; ^{175}Lu(^{11}B,5n)

Photons (^{181}Os)

⟨γ⟩=1355 34 keV

γ$_{mode}$	γ(keV)	γ(%)†
Re L$_{ℓ}$	7.604	0.72 18
Re L$_{α}$	8.646	17 4
Re L$_{η}$	9.027	0.23 6
Re L$_{β}$	10.125	17 4
Re L$_{γ}$	11.788	3.0 8
γ E2	33.7 3	0.0011 4 ?
Re K$_{α2}$	59.717	32 7
Re K$_{α1}$	61.141	55 13
Re K$_{β1}$'	69.214	19 4
Re K$_{β2}$'	71.470	4.7 11
γ E2	75.65 4	1.8 6
γ [M1]	99.8 3	0.3 1
γ [M1]	104.7 3	0.32 10
γ M1+4.9%E2	117.94 4	12.9 14
γ E1	144.93 6	1.4 3
γ M1+<1.7%E2	148.32 20	0.16 6
γ	153.3 8	~0.06
γ	157.9 8	0.24 10
γ E2+~45%M1	167.12 5	3.0 4
γ (M1)	210.84 19	0.16 6
γ (M1)	223.01 22	0.26 10
γ E2+43%M1	228.68 11	1.6 4
γ E2+37%M1	233.53 10	1.8 6
γ E1	238.68 10	44 4
γ M1+22%E2	242.77 6	6.1 14
γ M1+44%E2	267.50 11	1.01 20
γ [M1]	309.9 3	0.36 10
γ (M1)	324.37 18	0.48 10
γ (M1)	326.30 18	~0.28
γ [M1]	334.6 3	0.20 8
γ M1+32%E2	344.25 19	0.36 6
γ (E1)	356.63 10	1.6 3
γ (M1)	393.9 8	0.51 10
γ M1	434.62 12	1.9 4
γ	460.4 12	~0.3
γ [M1]	508.40 23	~0.20
γ	533.4 11	0.9 3
γ [M1]	567.2 3	~0.44
γ	569.7 11	~0.4
γ (M1)	591.87 20	0.7 3
γ (M1)	675.52 23	1.6 4
γ	687.1 11	0.51 16
γ M1	728.6 4	1.03 24
γ	749.5 12	<2
γ (E2)	751.17 23	3.2 8
γ M1	758.99 20	2.4 4
γ [M1]	785.6 6	0.77 24
γ (E2)	787.5 4	5.3 6
γ	791.9 15	0.61 20

Photons (^{181}Os) (continued)

γ$_{mode}$	γ(keV)	γ(%)†
γ (E2)	796.8 5	1.2 3 ?
γ E2	826.74 22	20.2
γ M1	831.59 23	7.7 10
γ [M1]	834.7 3	~0.2
γ [M1]	843.03 23	0.83 16
γ	867.9 8	2.0 3
γ M1	872.1 8	2.4 5
γ	908.4 15	~0.40
γ [M1]	920.4 5	1.3 3
γ (E2)	931.4 3	0.83 22
γ [M1]	941.7 4	1.0 3
γ (M1)	954.9 5	5.1 8
γ	959.9 15	0.46 20
γ	970.9 8	1.2 3
γ E2+44%M1	980.7 3	0.83 24
γ	990.5 8	1.5 3
γ [M1]	1000.4 6	~0.5 ?
γ [M1]	1010.15 23	0.83 24
γ [M1]	1028.4 6	~0.3
γ M1	1030.5 5	1.66 12
γ	1044.4 15	~0.3
γ	1048.4 15	~0.30
γ E2+38%M1	1060.27 22	5.7 8
γ [M1]	1064.2 3	~0.30
γ [M1]	1077.47 25	~0.4
γ E2+45%M1	1085.79 23	1.3 4
γ [M1]	1110.7 5	2.1 4
γ (E2)	1131.6 6	0.73 20
γ [M1]	1159.4 5	0.53 20
γ	1164.4 15	~0.36
γ	1176.4 15	~0.20
γ M1	1181.4 8	1.03 22
γ	1233.9 15	~0.22
γ	1243.9 15	~0.24
γ [M1]	1259.2 5	~0.26
γ (E2)	1305.1 3	1.8 3
γ M1	1325.1 4	0.40 14
γ E2+~28%M1	1346.5 5	1.09 6
γ	1368.9 15	~0.14
γ (E2)	1385.2 9	1.2 3
γ	1394.1 15	~0.06
γ	1405.1 15	~0.10
γ	1412.6 15	~0.2
γ	1419.3 15	~0.14
γ [E1]	1434.1 3	0.53 20
γ [E1]	1442.42 25	0.69 24
γ M1	1492.3 4	1.0 3
γ [M1]	1513.6 5	~0.16
γ [M1]	1538.2 3	0.26 10
γ [E1]	1550.5 4	~0.14
γ (M1)	1567.9 4	1.0 3
γ (E2)	1572.6 3	1.1 3
γ M1	1589.2 5	0.67 20
γ M1	1619.2 12	0.57 20
γ M1+39%E2	1624.3 12	0.36 16
γ	1653.6 15	~0.10
γ	1658.9 15	~0.06
γ	1666.9 15	~0.06
γ	1684.5 19	~0.06
γ M1	1705.3 3	1.4 3
γ M1+~50%E2	1739.7 3	1.25 22
γ M1	1758.4 4	0.89 20
γ M1	1781.0 3	0.40 14
γ	1794.4 19	~0.16
γ [M1]	1826.3 8	0.16 6
γ	1836 3	~0.06
γ	1846 3	~0.06
γ	1867 3	~0.16
γ	1913 3	~0.06
γ	1926.9 23	~0.10
γ E2	1937.5 5	0.20 8
γ E1	1945.8 5	0.63 20
γ	1981.6 15	1.2 4
γ [M1]	1993.4 8	0.18 8
γ [M1]	1999.3 4	~0.18
γ [M1]	2015.2 5	~0.18
γ [M1]	2069.0 8	~0.04
γ	2096 3	~0.06
γ	2115 3	~0.16
γ E1	2137.6 3	0.77 20
γ E1	2156.6 23	0.30 12
γ	2225 3	~0.16
γ E1	2257.7 22	0.26 10
γ [M1]	2266.8 4	~0.04

Photons (^{181}Os) (continued)

γ_{mode}	γ(keV)	γ(%)†
γ	2284 *3*	~0.04
γ	2302.8 *22*	0.40 *14*
γ	2355 *3*	~0.04
γ	2396 *3*	~0.04
γ[M1]	2433.9 *4*	~0.10
γ	2465 *3*	~0.04
γ[E1]	2481.8 *3*	~0.04
γ	2528.1 *23*	0.14 *6*
γ	2647 *3*	~0.04

† 6.4% uncert(syst)

Atomic Electrons (^{181}Os)

$\langle e\rangle$=77 *4* keV

e_{bin}(keV)	$\langle e\rangle$(keV)	e(%)
4	0.068	1.7 *6*
11	5.1	48 *12*
12	3.1	26 *7*
13 - 34	2.0	11.2 *21*
46	16.1	35 *4*
47 - 63	2.6	4.9 *5*
64	5.5	9 *3*
65	5.4	8 *3*
66 - 69	0.111	0.165 *22*
73	3.3	4.5 *15*
74 - 94	1.7	2.1 *4*
95	1.79	1.87 *25*
97 - 104	0.123	0.124 *23*
105	5.3	5.1 *6*
106 - 155	4.8	3.96 *24*
156 - 166	1.8	1.15 *21*
167	2.62	1.57 *13*
171	3.2	1.9 *5*
196 - 243	3.06	1.37 *10*
253 - 299	0.75	0.28 *5*
307 - 356	0.32	0.100 *14*
357 - 394	0.62	0.17 *3*
422 - 462	0.29	0.066 *21*
496 - 533	0.29	0.057 *16*
555 - 604	0.29	0.048 *9*
615 - 664	0.22	0.033 *8*
665 - 714	0.64	0.093 *15*
716 - 763	2.22	0.296 *18*
771 - 820	0.88	0.110 *13*
821 - 870	0.47	0.056 *6*
871 - 921	0.71	0.080 *12*
928 - 977	0.45	0.047 *6*
978 - 1027	0.48	0.048 *7*
1028 - 1077	0.28	0.027 *3*
1083 - 1131	0.17	0.0154 *24*
1147 - 1188	0.061	0.0052 *14*
1221 - 1259	0.083	0.0067 *10*
1275 - 1323	0.096	0.0074 *8*
1325 - 1373	0.043	0.0032 *6*
1375 - 1424	0.056	0.0040 *10*
1430 - 1479	0.023	0.0016 *4*
1480 - 1579	0.175	0.0115 *14*
1582 - 1674	0.113	0.0069 *9*
1682 - 1779	0.090	0.0053 *6*
1782 - 1874	0.021	0.00112 *24*
1900 - 1997	0.054	0.0028 *9*
2003 - 2094	0.016	0.00076 *16*
2103 - 2195	0.0087	0.00040 *11*
2212 - 2300	0.011	0.0005 *2*
2324 - 2423	0.0048	0.00020 *7*
2431 - 2526	0.0029	~0.00012
2575 - 2645	0.0007	~3×10^{-5}

Continuous Radiation (^{181}Os)

$\langle\beta+\rangle$=19.6 keV;$\langle$IB$\rangle$=2.1 keV

E_{bin}(keV)		$\langle\ \rangle$(keV)	(%)
0 - 10	β+	5.2×10^{-8}	6.1×10^{-7}
	IB	0.00130	
10 - 20	β+	3.24×10^{-6}	1.92×10^{-5}
	IB	0.0018	0.0146
20 - 40	β+	0.000137	0.000411
	IB	0.0038	0.0123
40 - 100	β+	0.0105	0.0132
	IB	0.34	0.56
100 - 300	β+	0.68	0.307
	IB	0.059	0.032
300 - 600	β+	4.57	1.00
	IB	0.19	0.041
600 - 1300	β+	13.3	1.55
	IB	0.85	0.091
1300 - 2500	β+	0.99	0.071
	IB	0.68	0.042
2500 - 2668	IB	0.00045	1.8×10^{-5}
	$\Sigma\beta$+		2.9

$^{181}_{76}$Os(2.7 *1* min)

Mode: ϵ

Δ: -43530 *220* keV syst

SpA: 3.84×10^{8} Ci/g

Prod: protons on Re; protons on Au; ^{182}W(^{3}He,4n); alphas on W

Photons (^{181}Os)

γ_{mode}	γ(keV)	γ(rel)
γ M1+4.9%E2	117.94 *4*	28 *3*
γ E1	144.93 *6*	100
γ	162.9 *10*	~0.8
γ (E2)	220.9 *10*	~0.2
γ	237.9 *10*	~1
γ (E2)	252.9 *10*	~0.2
γ	262.9 *10*	~0.50
γ	665.9 *10*	0.4 *1*
γ	1118.7 *10*	4.2 *8*
γ	1206.9 *15*	0.8 *2*
γ	1427.9 *15*	0.4 *1*
γ	1467.9 *10*	1.3 *2*

$^{181}_{77}$Ir(4.90 *15* min)

Mode: ϵ

Δ: -39460 *320* keV syst

SpA: 2.12×10^{8} Ci/g

Prod: Ne on ^{165}Ho; protons on ^{197}Au; ^{169}Tm(^{16}O,4n)

Photons (^{181}Ir)

γ_{mode}	γ(keV)	γ(rel)
γ[E1]	19.6 *2*	6.0 *2*
γ	65.3 *2*	~20
γ	93.8 *2*	29 *5*
γ	102.5 *2*	25 *8*
γ	107.6 *2*	100 *5*
γ	117.9 *2*	
γ	123.5 *2*	28 *4*
γ	166 *1*	

Photons (^{181}Ir) (continued)

γ_{mode}	γ(keV)	γ(rel)
γ	184.6 *2*	28 *3*
γ	189.9 *2*	4 *1*
γ	218.5 *5*	14 *4*
γ	227.0 *2*	58 *6*
γ	231.6 *2*	30 *4*
γ	239.2 *2*	
γ	309.0 *2*	14 *3*
γ	318.9 *2*	46 *5*
γ	350.5 *2*	7 *1*
γ	352.8 *2*	4 *1*
γ	375.2 *2*	16 *2*
γ	576.5 *2*	9 *2*
γ	700.1 *2*	9 *4*
γ	871.2 *2*	4 *1*
γ	1182.3 *3*	9 *1*
γ	1192.6 *3*	11 *2*
γ	1347.1 *3*	13 *4*
γ	1381.0 *3*	13 *2*
γ	1528.8 *3*	29 *2*
γ	1545.0 *3*	6 *1*
γ	1565.6 *3*	13 *2*
γ	1593.4 *3*	9 *1*
γ	1639.6 *3*	52 *4*
γ	1646.4 *3*	27 *3*
γ	1652.5 *3*	17 *3*
γ	1714.9 *3*	6 *1*

$^{181}_{78}$Pt(51 *5* s)

Mode: ϵ(~99.94 %), α(~0.06 %)

Δ: -34380 *320* keV syst

SpA: 1.21×10^{9} Ci/g

Prod: ^{16}O on Yb; ^{20}Ne on Er

Alpha Particles (^{181}Pt)

α(keV)
5020 *20*

$^{181}_{79}$Au(11.4 *5* s)

Mode: ϵ(98.90 *25* %), α(1.10 *25* %)

Δ: -27830 *290* keV syst

SpA: 5.30×10^{9} Ci/g

Prod: ^{169}Tm(^{20}Ne,8n); ^{168}Yb(^{19}F,6n); ^{147}Sm(^{40}Ar,p5n); protons on Pb

Alpha Particles (^{181}Au)

$\langle\alpha\rangle$=61.16 keV

α(keV)	α(%)
5482 *5*	0.5
5623 *5*	0.6

$^{181}_{80}$Hg(3.6 *3* s)

Mode: ϵ(>87 %), α(<13 %), ϵp(0.014 *4* %), ϵα(9 *3* ×10^{-6} %)
Δ: -20760 *350* keV syst
SpA: 1.57×10^{10} Ci/g
Prod: protons on Pb
ϵp: 3500-5500 (weak)
ϵα: weak

Alpha Particles (^{181}Hg)

α(keV)	α(rel)
5920 *30*	15
6003 *15*	100 *15*

A = 182

NDS **14**, 559 (1975)

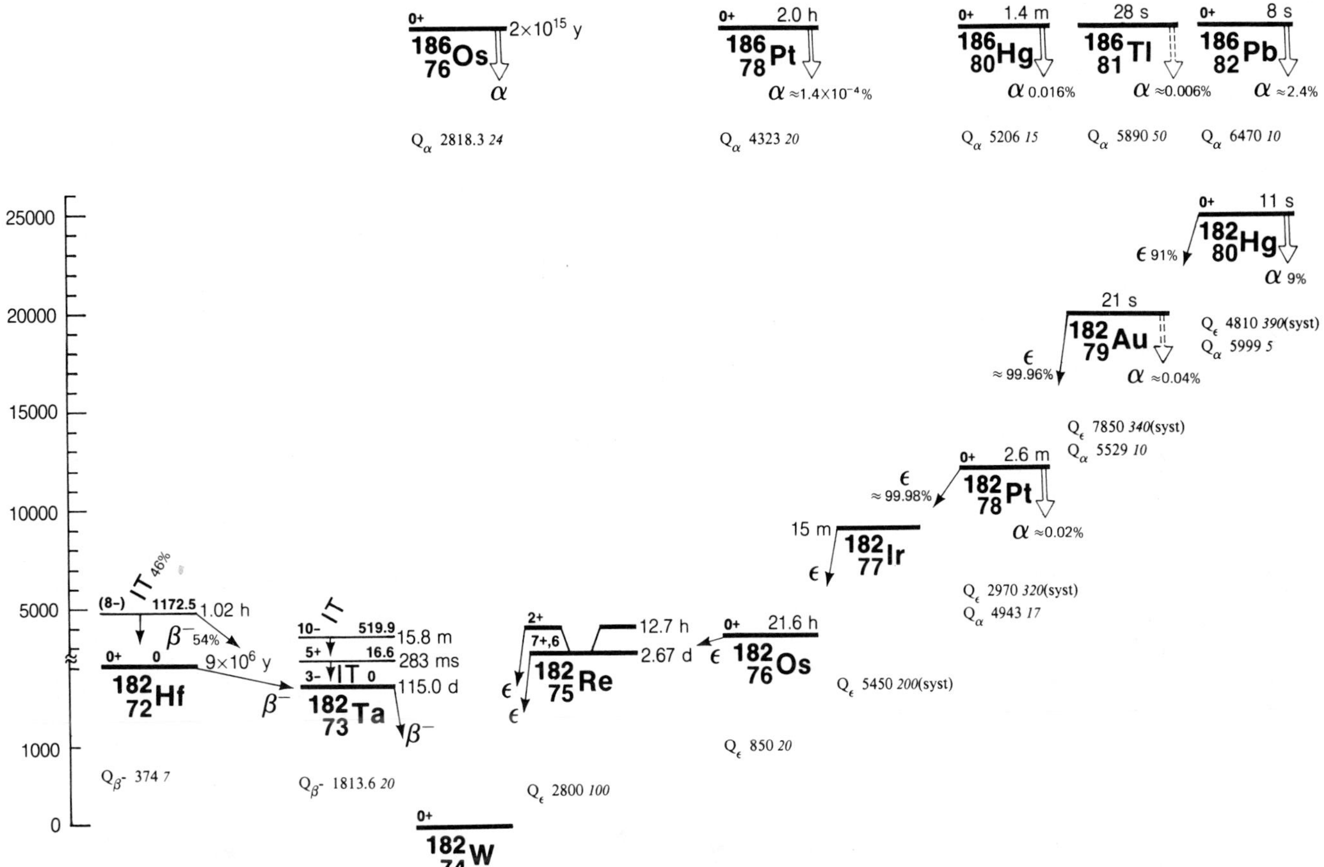

$^{182}_{72}$Hf(9.0 *20* ×10^6 yr)

Mode: β-
Δ: -46063 *7* keV
SpA: 0.00022 Ci/g
Prod: multiple n-capture from ^{180}Hf

Photons (^{182}Hf)
⟨γ⟩=239 *14* keV

γ_{mode}	γ(keV)	γ(%)†
Ta L_ℓ	7.173	0.085 *10*
Ta L_α	8.140	2.02 *19*
Ta L_η	8.428	0.031 *3*
Ta L_β	9.455	2.16 *23*
Ta L_γ	10.974	0.36 *4*
Ta $K_{\alpha2}$	56.280	4.1 *3*
Ta $K_{\alpha1}$	57.535	7.2 *6*
Ta $K_{\beta1}$'	65.140	2.38 *20*
Ta $K_{\beta2}$'	67.254	0.61 *5*
γ (E2)	97.84 *6*	0.08 *2*
γ (M1)	114.313 *16*	2.6 *4*
γ [E2]	156.073 *16*	7 *1*
γ [E2]	172.55 *7*	0.20 *4*
γ (E2)	270.386 *13*	80 *5*

† 6.0% uncert(syst)

Atomic Electrons (^{182}Hf)
⟨e⟩=32.8 *12* keV

e_{bin}(keV)	⟨e⟩(keV)	e(%)
10 - 46	1.41	12.1 *10*
47	3.1	6.7 *11*
48 - 88	0.325	0.48 *4*
89	2.0	2.3 *3*
95 - 98	0.059	0.062 *9*
103	1.07	1.04 *17*
104 - 144	0.77	0.62 *6*
145	1.45	1.00 *15*
146	1.14	0.78 *12*
153 - 173	1.08	0.70 *7*
203	12.0	5.9 *4*
259	4.6	1.76 *13*
261	1.68	0.64 *5*
268	1.57	0.59 *4*
269 - 270	0.45	0.167 *12*

Continuous Radiation (^{182}Hf)

⟨β-⟩=49.7 keV;⟨IB⟩=0.0089 keV

E_{bin}(keV)		⟨ ⟩(keV)	(%)
0 - 10	β-	0.64	13.2
	IB	0.0025	
10 - 20	β-	1.73	11.6
	IB	0.0019	0.0132
20 - 40	β-	6.3	21.1
	IB	0.0024	0.0086
40 - 100	β-	29.0	43.8
	IB	0.0021	0.0037
100 - 162	β-	12.1	10.4
	IB	8.6×10^{-5}	7.9×10^{-5}

$^{182}_{72}$Hf(1.025 *25* h)

Mode: β-(54 *2* %), IT(46 *2* %)
Δ: -44890 *7* keV
SpA: 1.68×10^7 Ci/g
Prod: ^{186}W(p,pα)

Photons (^{182}Hf)

⟨γ⟩=986 *31* keV

γ_{mode}	γ(keV)	γ(%)†
Hf L_ℓ	6.960	0.19 *3*
Ta L_ℓ	7.173	0.46 *7*
Hf L_α	7.893	4.5 *7*
Hf L_η	8.139	0.079 *13*
Ta L_α	8.140	11.1 *14*
Ta L_η	8.428	0.162 *25*
Hf L_β	9.125	5.2 *8*
Ta L_β	9.447	13.2 *20*
Hf L_γ	10.577	0.86 *14*
Ta L_γ	11.040	2.4 *4*
γ_{IT} (E1)	50.80 *16*	13.4 *15*
Hf $K_{\alpha2}$	54.611	5.2 *6*
Hf $K_{\alpha1}$	55.790	9.0 *10*
Ta $K_{\alpha2}$	56.280	18.2 *13*
Ta $K_{\alpha1}$	57.535	31.7 *23*
$\gamma_{\beta-}$(M1)	59.15 *11*	5.0 *8*
Hf $K_{\beta1}'$	63.166	2.9 *3*
Ta $K_{\beta1}'$	65.140	10.5 *8*
Hf $K_{\beta2}'$	65.211	0.77 *8*
Ta $K_{\beta2}'$	67.254	2.69 *20*
$\gamma_{\beta-}$(M1)	75.62 *11*	1.4 *3*
γ_{IT} (E2)	97.7 *2*	~9
$\gamma_{\beta-}$(E2)	97.84 *6*	~5
$\gamma_{\beta-}$(M1)	114.313 *16*	7.1 *7*
$\gamma_{\beta-}$(M1)	132.79 *20*	3.4 *8*
$\gamma_{\beta-}$(M1)	143.15 *15*	4.9 *5*
$\gamma_{\beta-}$M1	146.792 *15*	5.4 *6*
$\gamma_{\beta-}$M1	171.591 *15*	6.0*
$\gamma_{\beta-}$(E2)	173.46 *11*	3.2 *10*
$\gamma_{\beta-}$(M1)	178.74 *14*	2.5 *9*
$\gamma_{\beta-}$E3	184.954 *16*	3.3 *13**
$\gamma_{\beta-}$(M1)	195.72 *14*	1.2 *5*
$\gamma_{\beta-}$(M1)	220.79 *20*	1.1 *3*
γ_{IT} (E2)	224.3 *2*	38.3
$\gamma_{\beta-}$E2	318.383 *20*	0.88 *19**
$\gamma_{\beta-}$(M1)	339.60 *14*	6.5 *6*
γ_{IT} (E2)	344.0 *2*	46 *5*
$\gamma_{\beta-}$M4	356.545 *21*	0.036 *16**
$\gamma_{\beta-}$[E2]	374.46 *14*	
γ_{IT} (E2)	455.70 *16*	20.3 *15*
γ_{IT} M2	506.50 *16*	23.8 *19*
$\gamma_{\beta-}$[E2]	603.19 *15*	6.0 *10*
$\gamma_{\beta-}$[E1]	613.29 *16*	1.3 *3*
$\gamma_{\beta-}$[M1+E2]	627.49 *14*	1.2 *3*
$\gamma_{\beta-}$[M1+E2]	799.65 *15*	10.8 *12*
$\gamma_{\beta-}$[E2]	823.21 *14*	3.0 *6*
$\gamma_{\beta-}$[E2]	942.79 *12*	21.6 *20*
$\gamma_{\beta-}$[E1]	952.89 *16*	0.28 *12*

† uncert(syst): 4.4% for IT, 10% for β-
* with ^{182}Ta(15.8 min) in equilib

Atomic Electrons (^{182}Hf)

⟨e⟩=177 *8* keV

e_{bin}(keV)	⟨e⟩(keV)	e(%)
5 - 8	1.59	23 *3*
10	4.9	50 *8*
11 - 46	12.5	68 *7*
47	14.4	31 *4*
48 - 75	9.8	16.4 *16*
76	5.1	6.7 *8*
79	5.4	6.8 *7*
86	1.2	1.4 *4*
87	13.7	16 *5*
88	12.9	15 *5*
95	4.0	4.2 *13*
96 - 103	8.6	8.8 *14*
104	5.1	4.9 *6*
106 - 153	13.0	10.6 *12*
159	7.4	4.67 *19*
160 - 173	3.9	2.38 *25*
174	6.8	3.9 *16*
175	4.1	2.3 *9*
176 - 224	10.9	5.4 *5*
251 - 272	2.32	0.85 *8*
279	5.0	1.78 *19*
289 - 338	2.67	0.82 *7*
339 - 357	0.82	0.239 *23*
390	1.49	0.38 *3*
441	12.6	2.85 *24*
444 - 456	0.535	0.120 *5*
495 - 536	3.66	0.73 *4*
546 - 593	0.22	0.038 *11*
600 - 627	0.056	0.0091 *19*
732 - 756	0.9	0.12 *5*
788 - 823	0.20	0.025 *7*
875 - 885	0.76	0.087 *9*
931 - 953	0.178	0.0190 *12*

Continuous Radiation (^{182}Hf)

⟨β-⟩=50.7 keV;⟨IB⟩=0.033 keV

E_{bin}(keV)		⟨ ⟩(keV)	(%)
0 - 10	β-	0.058	1.15
	IB	0.0025	
10 - 20	β-	0.170	1.13
	IB	0.0023	0.0163
20 - 40	β-	0.67	2.22
	IB	0.0041	0.0144
40 - 100	β-	4.31	6.2
	IB	0.0090	0.0143
100 - 300	β-	25.6	13.8
	IB	0.0119	0.0074
300 - 600	β-	16.0	4.12
	IB	0.0030	0.00078
600 - 952	β-	3.82	0.55
	IB	0.00022	3.4×10^{-5}

$^{182}_{73}$Ta(115.0 *2* d)

Mode: β-
Δ: -46437 *3* keV
SpA: 6239 Ci/g
Prod: ^{181}Ta(n,γ)

Photons (^{182}Ta)

⟨γ⟩=1301 *10* keV

γ_{mode}	γ(keV)	γ(%)†
W L_ℓ	7.387	0.39 *4*
W L_α	8.391	9.1 *6*
W L_η	8.724	0.158 *10*
W L_β	9.770	11.2 *9*
W L_γ	11.389	2.00 *18*
γ E1	31.7359 *4*	0.63 *18*
γ E1	42.7133 *6*	0.243 *14*
W $K_{\alpha2}$	57.981	10.5 *3*
W $K_{\alpha1}$	59.318	18.2 *6*
γ M1+1.0%E2	65.7207 *2*	2.77 *13*
W $K_{\beta1}'$	67.155	6.05 *20*
γ E1	67.7485 *2*	41.2 *21*
W $K_{\beta2}'$	69.342	1.57 *5*
γ M1+8.3%E2	84.6792 *6*	2.65 *12*
γ E2	100.1054 *3*	14.0 *4*
γ [E1]	110.375 *14*	0.091 *18* ?
γ M1+8.8%E2	113.6671 *7*	1.92 *10*
γ E1	116.4152 *7*	0.43 *5*
γ	146.0 *10*	<0.010
γ E1	152.4277 *6*	7.18 *14*
γ E1	156.3803 *8*	2.73 *6*
γ M1+47%E2	179.3878 *7*	3.14 *10*
γ E2	198.3463 *9*	1.52 *5*
γ E1	222.1010 *8*	7.60 *20*
γ E2	229.315 *3*	3.63 *10*
γ E2	264.0669 *9*	3.62 *9*
γ E2	351.069 *19*	0.0129 *10*
γ E2	891.988 *12*	0.052 *7*
γ E2	928.000 *12*	0.61 *3*
γ M2+>5.9%E3	959.736 *12*	0.364 *20*
γ E2+1.4%M1	1001.702 *12*	2.00 *7*
γ (E1+17%M2)	1044.415 *12*	0.228 *15*
γ E2	1113.427 *18*	0.437 *17*
γ E2+0.1%M1	1121.302 *12*	34.7
γ E2	1157.314 *12*	0.67 *10*
γ (E1)	1158.081 *12*	0.33 *7*
γ [M1+E2]	1180.81 *8*	0.084 *7*
γ E1+19%M2+E3	1189.050 *12*	16.5 *3*
γ E2	1221.406 *12*	27.3 *5*
γ [E1]	1223.802 *12*	0.20 *3*
γ E2	1231.015 *12*	11.63 *22*
γ E2	1257.419 *12*	1.52 *3*
γ E1+15%M2	1273.728 *12*	0.664 *14*
γ M2	1289.154 *12*	1.398 *23*
γ E2	1342.741 *17*	0.260 *5*
γ	1373.833 *12*	0.231 *7*
γ	1387.395 *12*	0.076 *3*
γ (M1+~50%E2)	1410.12 *8*	0.0418 *19*
γ	1435 *1*	<0.0017
γ	1453.115 *12*	0.043 *3*

† 1.7% uncert(syst)

Atomic Electrons (^{182}Ta)

⟨e⟩=81.3 *9* keV

e_{bin}(keV)	⟨e⟩(keV)	e(%)
10	2.8	27 *3*
12	2.9	25 *3*
15	2.49	16.4 *9*
20 - 30	0.22	1.00 *18*
31	3.78	12.4 *4*
32 - 49	2.81	6.3 *3*
54	3.37	6.3 *4*
55	0.107	0.194 *20*
56	2.92	5.2 *3*
57 - 68	3.65	5.86 *17*
73	2.29	3.15 *20*
74 - 88	3.38	4.05 *11*
89	14.6	16.5 *5*
90	13.3	14.8 *5*
97	0.312	0.321 *10*
98	7.78	7.97 *25*
99	0.00067	0.00068 *14*
100	2.30	2.30 *7*
102 - 151	3.81	3.43 *13*
152 - 198	3.19	1.87 *3*

Atomic Electrons (^{182}Ta) (continued)

e_{bin}(keV)	⟨e⟩(keV)	e(%)
210 - 254	1.107	0.486 6
261 - 282	0.1075	0.0410 8
339 - 351	0.00071	0.000208 8
822 - 858	0.0245	0.00287 13
880 - 928	0.0366	0.00409 20
932 - 975	0.089	0.0095 3
990 - 1034	0.0180	0.00180 5
1042 - 1089	1.14	0.108 3
1101 - 1148	0.89	0.080 4
1152 - 1189	1.323	0.1143 22
1204 - 1247	0.456	0.0374 5
1255 - 1289	0.0526	0.00412 7
1304 - 1343	0.013	0.0010 4
1362 - 1410	0.0035	0.00025 8
1423 - 1453	0.00025	1.8 8 $\times10^{-5}$

Continuous Radiation (^{182}Ta)

⟨β-⟩=126 keV;⟨IB⟩=0.061 keV

E_{bin}(keV)		⟨ ⟩(keV)	(%)
0 - 10	β-	0.289	5.8
	IB	0.0065	
10 - 20	β-	0.84	5.6
	IB	0.0058	0.040
20 - 40	β-	3.19	10.6
	IB	0.0098	0.034
40 - 100	β-	18.8	27.3
	IB	0.019	0.031
100 - 300	β-	77	43.4
	IB	0.018	0.0121
300 - 600	β-	26.1	7.3
	IB	0.00139	0.00038
600 - 1300	β-	0.58	0.068
	IB	0.00019	2.5 $\times10^{-5}$
1300 - 1811	β-	0.0428	0.00307
	IB	1.8 $\times10^{-6}$	1.35 $\times10^{-7}$

$^{182}_{73}$Ta(283 2 ms)

Mode: IT

Δ: -46420 3 keV

SpA: 8.17$\times10^{10}$ Ci/g

Prod: ^{181}Ta(n,γ); daughter ^{182}Ta(15.8 min)

Photons (^{182}Ta)

⟨γ⟩=1.69 13 keV

γ_{mode}	γ(keV)	γ(%)
Ta L_ℓ	7.173	0.34 4
Ta L_α	8.140	8.1 7
Ta L_η	8.428	0.044 4
Ta L_β	9.470	8.5 11
Ta L_γ	11.176	1.9 3
γ (M2)	16.57 20	0.00240 10

Atomic Electrons (^{182}Ta)

⟨e⟩=12.2 5 keV

e_{bin}(keV)	⟨e⟩(keV)	e(%)
5	2.28	46 3
7	1.87	28.0 19
10	2.09	21 3
11	0.129	1.15 14
12	2.1	18.1 22
14	2.77	19.7 14
15	0.0182	0.123 9
16	0.79	4.9 3
17	0.141	0.86 6

$^{182}_{73}$Ta(15.84 10 min)

Mode: IT

Δ: -45917 3 keV

SpA: 6.52$\times10^{7}$ Ci/g

Prod: ^{181}Ta(n,γ)

Photons (^{182}Ta)

⟨γ⟩=254 6 keV

γ_{mode}	γ(keV)	γ(%)†
Ta L_ℓ	7.173	0.96 11
Ta L_α	8.140	22.8 19
Ta L_η	8.428	0.308 23
Ta L_β	9.452	26 3
Ta L_γ	11.047	4.8 6
Ta $K_{\alpha2}$	56.280	26.8 11
Ta $K_{\alpha1}$	57.535	46.8 19
Ta $K_{\beta1}$'	65.140	15.5 6
Ta $K_{\beta2}$'	67.254	3.97 17
γ M1	146.792 15	35.1 18
γ M1	171.591 15	45.9
γ E3	184.954 16	23.1 14
γ E2	318.383 20	6.5 5
γ M4	356.545 21	0.28 5

† 4.3% uncert(syst)

Atomic Electrons (^{182}Ta)

⟨e⟩=252 5 keV

e_{bin}(keV)	⟨e⟩(keV)	e(%)
5 - 54	23.2	233 12
55 - 65	0.49	0.85 5
79	35.2	44.4 25
104	38.9	37.4 20
118	17.2	14.6 9
135	8.6	6.3 4
136 - 147	3.81	2.67 10
160	9.3	5.8 3
162 - 173	7.1	4.11 17
174	47.0	27.1 17
175	28.4	16.2 10
182	13.7	7.5 5
183	8.0	4.4 3
184 - 185	6.1	3.30 15
251 - 289	3.2	1.16 14
307 - 356	2.20	0.65 5
357	0.0064	0.0018 3

$^{182}_{74}$W (stable)

Δ: -48250 3 keV

%: 26.3 2

$^{182}_{75}$Re(2.667 21 d)

Mode: ϵ

Δ: -45450 100 keV

SpA: 2.690$\times10^{5}$ Ci/g

Prod: ^{181}Ta(α,3n); ^{182}W(p,n)

Photons (^{182}Re)

⟨γ⟩=1916 39 keV

γ_{mode}	γ(keV)	γ(%)†
W L_ℓ	7.387	1.4 3
W L_α	8.391	32 6
W L_η	8.724	0.47 8
W L_β	9.779	36 6
W L_γ	11.407	6.4 12
γ [M1]	19.839 15	~0.10
γ E1	31.7359 4	0.36 10
γ M1+<0.2%E2	39.088 10	~0.48
γ E1	42.7133 6	0.32 4
W $K_{\alpha2}$	57.981	60 11
W $K_{\alpha1}$	59.318	103 19
γ [M1]	60.588 11	~0.06
γ M1+1.0%E2	65.72072 20	3.09 18
W $K_{\beta1}$'	67.155	34 6
γ E1	67.7485 2	23.4 13
W $K_{\beta2}$'	69.342	8.9 16
γ M1+8.3%E2	84.6792 6	3.5 3
γ E2	100.1054 3	16.1 12
γ M1+43%E2	107.150 9	~1
γ M1+13%E2	108.565 11	~0.7
γ [E1]	110.375 14	0.101 21
γ M1+8.8%E2	113.6671 7	5.6 3
γ E1	116.4152 7	0.57 8
γ M1+18%E2	130.813 12	~8
γ [E2]	131.316 14	~0.17
γ M1+9.0%E2	133.782 9	2.65 7
γ E1	145.381 15	0.74 11
γ M1+39%E2	147.653 12	~0.7
γ [M1]	148.839 13	1.6 3
γ M1+~8.3%E2	149.442 15	~0.8
γ M1+33%E2	151.155 17	0.67 13
γ E1	152.4277 6	9.4 10
γ	153.950 20	~0.35
γ E1	156.3803 8	7.94 22
γ [M1]	160.110 20	<0.23
γ M1+1.5%E2	169.153 10	12.6 3
γ M1+7.8%E2	172.871 9	3.75 16
γ [E1]	178.437 13	~2
γ M1+47%E2	179.3878 7	3.50 17
γ	187.5 5	
γ	188.4 3	
γ [M1]	189.613 14	<0.46
γ M1	191.401 13	8.28 18
γ E2	198.3463 9	4.41 15
γ (E2)	203.330 20	0.48 5
γ E1	205.969 15	0.53 7
γ (M1+E2)	208.241 10	~0.09
γ [E2]	209.427 13	~1
γ M1+4.6%E2	214.297 20	1.36 9
γ E2	215.715 10	0.67 7
γ E1	217.525 13	4.3 3
γ (E1)	221.617 16	~10
γ E1	222.1010 8	8.5 4
γ M1+6.9%E2	226.172 20	3.50 16
γ E2	229.315 3	29.0 9
γ E2	247.449 9	5.5 5
γ M1	256.463 16	10.8 5
γ E2	264.0669 9	4.04 9
γ E2	276.303 9	9.71 23
γ M1+<6.8%E2	281.435 10	6.23 18
γ M1+<3.8%E2	286.538 9	8.1 3
γ [E2]	299.966 12	2.2 4

Photons (^{182}Re)
(continued)

γ_{mode}	γ(keV)	γ(%)†
γ (M1)	300.468 14	~1
γ E2	313.900 16	0.64 5
γ E2	323.395 16	2.05 12
γ E2	339.054 11	6.03 16
γ E2	342.023 9	0.97 21
γ [E2]	345.40 2	0.35 7
γ E2	351.069 19	11.11 22
γ	357.080 13	0.55 5
γ E2	891.988 12	0.035 6
γ E2	928.000 12	0.425 24
γ E2	943.07 3	0.23 5
γ M2+>5.9%E3	959.736 12	0.207 12
γ E2+1.4%M1	1001.702 12	2.65 8
γ (E1+17%M2)	1044.415 12	0.299 11
γ E2	1076.26 3	11.0 3
γ E2	1088.13 4	0.207 18
γ E2	1113.427 18	4.90 10
γ E2+0.1%M1	1121.302 12	23
γ E2	1157.314 12	0.47 7
γ (E1)	1158.081 12	0.97 22
γ [M1+E2]	1180.81 8	0.594 23
γ E1+19%M2+E3	1189.050 12	9.40 17
γ E2	1221.406 12	18.2 3
γ [E1]	1223.802 12	0.23 4
γ E2	1231.015 12	15.4 3
γ E2	1257.419 12	1.06 4
γ E1+15%M2	1273.728 12	0.87 7
γ	1279.848 25	0.064 7
γ M2	1289.154 12	0.796 13
γ [E1]	1291.863 15	0.244 23
γ E2	1294.135 20	1.70 5
γ (E1)	1330.951 15	0.38 3
γ E2	1342.741 17	2.92 6
γ	1373.833 12	0.304 8
γ	1387.395 12	0.222 12
γ (M1+~50%E2)	1410.12 8	0.294 14
γ (E2)	1427.325 21	10.17 20
γ (E2)	1439.20 3	0.145 23
γ	1453.115 12	0.047 4
γ	1521.176 15	0.085 14
γ	1560.264 14	0.069 7
γ	1630.914 17	0.0131 3

† 17% uncert(syst)

Atomic Electrons (^{182}Re)
⟨e⟩=213 6 keV

e_{bin}(keV)	⟨e⟩(keV)	e(%)
8	0.6	~8
10	9.8	96 18
12	9.1	78 15
15 - 43	12.1	50 4
44	6.5	14.8 9
45 - 60	10.6	20.1 9
61	7.8	~13
62 - 88	16.6	22.5 10
89	16.8	19.0 14
90	15.3	17.0 13
91 - 97	2.1	2.2 6
98	9.0	9.2 7
99	0.0007	~0.0008
100	14.0	14.1 6
102 - 121	13.7	12.6 13
122	7.5	6.17 22
124 - 156	7.8	5.6 5
157	4.74	3.02 13
158 - 159	0.312	0.198 13
160	5.42	3.39 13
161 - 186	5.91	3.43 9
187	6.3	3.39 16
188 - 216	7.66	3.73 8
217	4.56	2.10 8
218 - 266	9.3	4.00 13
269 - 314	4.57	1.65 6
321 - 357	1.119	0.332 6
822 - 858	0.0171	0.00200 14
874 - 918	0.0291	0.00328 23

Atomic Electrons (^{182}Re)
(continued)

e_{bin}(keV)	⟨e⟩(keV)	e(%)
925 - 960	0.102	0.0109 4
975 - 1019	0.399	0.0397 12
1032 - 1078	0.97	0.092 9
1085 - 1121	0.61	0.055 3
1145 - 1189	1.103	0.0952 16
1204 - 1247	0.400	0.0329 5
1255 - 1294	0.1240	0.00973 20
1304 - 1343	0.045	0.0034 6
1358 - 1407	0.273	0.0201 9
1408 - 1453	0.0585	0.00412 16
1491 - 1561	0.0029	0.00020 8
1619 - 1629	6.2×10^{-5}	$\sim4\times10^{-6}$

Continuous Radiation (^{182}Re)
⟨IB⟩=1.06 keV

E_{bin}(keV)		⟨ ⟩(keV)	(%)
10 - 20	IB	0.00090	0.0081
20 - 40	IB	0.0032	0.0095
40 - 100	IB	0.39	0.66
100 - 300	IB	0.051	0.027
300 - 600	IB	0.128	0.029
600 - 1300	IB	0.26	0.030
1300 - 2500	IB	0.22	0.0131
2500 - 2800	IB	0.0021	8.4×10^{-5}

$^{182}_{75}$Re(12.7 2 h)

Mode: ε
Δ: -45450 100 keV
SpA: 1.356×10^{6} Ci/g
Prod: ^{181}Ta(α,3n); ^{182}W(p,n); daughter ^{182}Os

Photons (^{182}Re)
⟨γ⟩=1174 44 keV

γ_{mode}	γ(keV)	γ(%)†
W L_ℓ	7.387	0.73 15
W L_α	8.391	17 3
W L_η	8.724	0.26 5
W L_β	9.782	19 4
W L_γ	11.384	3.2 6
γ E1	31.7359 4	0.57 16
γ	41.97 10	~0.2
γ E1	42.7133 6	0.209 14
W $K_{\alpha2}$	57.981	30 7
W $K_{\alpha1}$	59.318	52 12
γ M1+1.0%E2	65.72072 20	0.224 16
W $K_{\beta1}'$	67.155	17 4
γ E1	67.7485 2	37.0 22
W $K_{\beta2}'$	69.342	4.5 10
γ M1+8.3%E2	84.6792 6	2.28 13
γ E2	100.1054 3	14.3 10
γ M1+8.8%E2	113.6671 7	0.253 23
γ E1	116.4152 7	0.37 4
γ E1	152.4277 6	6.17 25
γ E1	156.3803 8	0.360 25
γ M1+47%E2	179.3878 7	0.253 16
γ E2	198.3463 9	0.200 15
γ E1	222.1010 8	0.61 4
γ E2	229.315 3	2.1 5
γ E2	264.0669 9	0.292 14
γ M1	470.3 5	1.97 19
γ (M1)	536.0 5	0.21 3
γ (E2)	556.4 3	0.11 3

Photons (^{182}Re)
(continued)

γ_{mode}	γ(keV)	γ(%)†
γ (M1)	598.6 7	0.40 4
γ (M1)	649.7 5	0.35 6
γ (M1)	734.4 5	0.38 4
γ (M1)	786.9 6	0.25 4
γ	800.0 10	0.15 4
γ (M1)	810.2 4	0.38 6
γ [E1]	836.0 3	0.48 6
γ E2	891.988 12	0.048 11
γ (M1)	894.9 4	2.10 19
γ (M1)	900.6 6	0.35 6
γ E2	928.000 12	0.55 3
γ	943.0 10	~0.02
γ M2+>5.9%E3	959.736 12	0.327 20
γ E2+1.4%M1	1001.702 12	0.225 11
γ (E1+17%M2)	1044.415 12	0.196 13
γ E2+0.1%M1	1121.302 12	31.8
γ E2	1157.314 12	0.62 9
γ (E1)	1158.081 12	0.044 10
γ [M1+E2]	1180.81 8	0.079 4
γ E1+19%M2+E3	1189.050 12	14.8 4
γ E2	1221.406 12	25 3
γ [E1]	1223.802 12	0.016 3
γ E2	1231.015 12	1.30 6
γ E2	1257.419 12	1.39 5
γ E1+15%M2	1273.728 12	0.571 23
γ M2	1289.154 12	1.26 3
γ	1294.2 10	0.175 19
γ	1373.833 12	0.199 8
γ	1387.395 12	0.0101 8
γ (M1+~50%E2)	1410.12 8	0.0392 21
γ	1453.115 12	0.0034 3
γ	1523.0 20	~0.016
γ	1537.0 20	~0.016
γ	1543.0 20	~0.016
γ	1558.0 20	0.076 10
γ [M1+E2]	1755.9 7	<0.06
γ	1756.9 5	<0.06
γ E1	1770.8 5	0.29 4
γ	1811.2 10	<0.022
γ	1818.1 5	0.105 10
γ	1857.0 5	0.032 6
γ E1	1870.9 5	0.293 22
γ	1877.8 3	0.060 19
γ E1	1879.4 4	0.054 16
γ [E2]	1911.4 4	0.041 6
γ [E1]	1957.3 3	0.46 3
γ	2009.5 3	0.095 13
γ	2016.2 5	0.78 6
γ	2033.0 5	~0.022
γ	2047.4 5	0.118 16
γ (M2)	2057.4 3	0.83 13
γ	2073.5 5	0.041 6
γ	2084.0 4	0.064 6
γ (E1)	2099 3	~0.025
γ	2107.1 3	<0.3
γ [E1]	2108.7 4	<0.4
γ	2109.6 3	<0.3
γ	2131 3	<0.006
γ [M1+E2]	2140.7 4	0.038 6
γ	2147.0 10	0.016 6
γ	2159 3	<0.006
γ (E1)	2174.3 6	0.045 6
γ	2189 3	0.017 5
γ	2207.2 3	0.102 10
γ	2215.9 21	~0.022
γ	2230 3	0.011 3
γ	2316.0 21	0.0080 16

† 2.5% uncert(syst)

Atomic Electrons (^{182}Re)

$\langle e \rangle$=73.6 *22* keV

e_{bin}(keV)	$\langle e \rangle$(keV)	e(%)
10	5.4	53 *11*
12	4.7	40 *8*
15	2.14	14.1 *8*
20 - 30	0.20	0.90 *16*
31	3.9	12.6 *9*
32 - 55	2.30	4.8 *4*
56	2.6	4.7 *4*
57 - 88	7.49	10.4 *3*
89	14.9	16.9 *12*
90	13.6	15.1 *11*
97	0.318	0.327 *23*
98	7.9	8.1 *6*
100	2.34	2.35 *16*
102 - 151	0.583	0.494 *18*
152 - 198	0.63	0.38 *6*
210 - 254	0.42	0.189 *19*
261 - 264	0.0085	0.00326 *10*
458 - 487	0.151	0.0326 *22*
524 - 556	0.076	0.0144 *13*
580 - 599	0.065	0.0112 *16*
638 - 665	0.058	0.0087 *8*
717 - 766	0.096	0.0131 *15*
775 - 824	0.0196	0.0025 *3*
825 - 858	0.247	0.0298 *24*
880 - 928	0.080	0.0090 *4*
932 - 975	0.0243	0.00256 *9*
990 - 1034	0.00319	0.000317 *11*
1042 - 1089	1.02	0.097 *3*
1109 - 1158	1.53	0.135 *10*
1161 - 1210	0.334	0.0281 *14*
1211 - 1257	0.226	0.0185 *7*
1262 - 1304	0.046	0.0036 *3*
1318 - 1364	0.0027	0.00020 *5*
1368 - 1410	0.08	~0.006
1426 - 1473	0.017	~0.0012
1488 - 1556	0.0027	0.00018 *9*
1686 - 1769	0.0077	0.00045 *10*
1787 - 1877	0.0075	0.00041 *6*
1888 - 1978	0.025	0.0013 *4*
1988 - 2087	0.073	0.0036 *5*
2089 - 2187	0.0035	0.00016 *5*
2195 - 2247	0.00058	2.6 *8* $\times 10^{-5}$
2304 - 2314	2.5 $\times 10^{-5}$	1.1 *5* $\times 10^{-6}$

Continuous Radiation (^{182}Re)

$\langle \beta+ \rangle$=1.42 keV;$\langle$IB$\rangle$=1.60 keV

E_{bin}(keV)		$\langle\ \rangle$(keV)	(%)
0 - 10	β+	1.94 $\times 10^{-9}$	2.29 $\times 10^{-8}$
	IB	0.00072	
10 - 20	β+	1.19 $\times 10^{-7}$	7.1 $\times 10^{-7}$
	IB	0.00079	0.0069
20 - 40	β+	4.97 $\times 10^{-6}$	1.50 $\times 10^{-5}$
	IB	0.0027	0.0082
40 - 100	β+	0.000381	0.000479
	IB	0.33	0.55
100 - 300	β+	0.0266	0.0118
	IB	0.050	0.026
300 - 600	β+	0.208	0.0449
	IB	0.17	0.038
600 - 1300	β+	0.99	0.109
	IB	0.61	0.068
1300 - 2500	β+	0.191	0.0134
	IB	0.43	0.026
2500 - 2760	IB	0.0028	0.000111
	Σβ+		0.18

$^{182}_{76}$Os(21.6 *4* h)

Mode: ε

Δ: -44600 *100* keV

SpA: 7.97$\times 10^{5}$ Ci/g

Prod: ^{185}Re(p,4n); ^{182}W(α,4n); ^{182}W(^{3}He,3n)

Photons (^{182}Os)

$\langle \gamma \rangle$=435 *14* keV

γ_{mode}	γ(keV)	γ(%)†
Re L_ℓ	7.604	0.63 *10*
Re L_α	8.646	14.6 *21*
Re L_η	9.027	0.20 *3*
Re L_β	10.104	19 *3*
Re L_γ	11.884	4.0 *7*
γ M1(+0.2%E2)	27.54 *3*	0.62 *13*
γ M1	55.54 *4*	5.9 *6*
Re $K_{\alpha2}$	59.717	24 *3*
Re $K_{\alpha1}$	61.141	42 *6*
Re $K_{\beta1}$'	69.214	14.0 *19*
Re $K_{\beta2}$'	71.470	3.6 *5*
γ (M1)	110.470 *25*	0.23 *4*
γ (M1)	111.36 *5*	0.036 *10*
γ M1	115.933 *24*	0.65 *7*
γ	122.30 *10*	0.41 *13*
γ E1	130.85 *3*	3.32 *21*
γ	136.9 *5*	~0.10
γ M1(+E2)	143.47 *3*	0.083 *16*
γ	170.45 *5*	0.21 *3*
γ M1	172.43 *6*	0.35 *4*
γ [M1]	175.03 *5*	0.32 *4*
γ E1	180.22 *4*	34.7 *21*
γ	190.0 *10*	~0.10
γ M1	202.56 *5*	0.062 *10*
γ M1(+E2)	216.93 *5*	0.75 *6*
γ	223.0 *5*	~0.10 ?
γ (E1+M2)	235.75 *3*	0.41 *10*
γ (E1+M2)	241.32 *3*	0.88 *10*
γ E1(+M2)	246.78 *3*	0.59 *5*
γ (M1+E2)	261.59 *5*	0.08 *3*
γ E1	263.29 *3*	6.6 *4*
γ E1	274.32 *3*	1.71 *10*
γ M1	286.38 *5*	0.062 *10*
γ (M1)	373.2 *5*	~0.36
γ (E1+M2)	379.22 *3*	0.78 *10*
γ [E1]	438.31 *6*	0.062 *16*
γ M1(+E2)	454.53 *4*	0.29 *4*
γ	458.25 *5*	~0.021
γ	486.0 *20*	0.047 *10*
γ M1	499.04 *8*	0.34 *3*
γ M1	510.07 *3*	52
γ M1	554.58 *8*	0.31 *4*
γ (M1)	561.11 *25*	0.13 *3*
γ	632.37 *10*	~0.021
γ (M1)	727.00 *6*	0.13 *3*

† 4.8% uncert(syst)

Atomic Electrons (^{182}Os)

$\langle e \rangle$=53.2 *16* keV

e_{bin}(keV)	$\langle e \rangle$(keV)	e(%)
11	3.4	32 *6*
12	2.2	18 *3*
13	2.8	22 *4*
15	2.7	18 *4*
16 - 17	0.57	3.6 *6*
25	1.2	4.9 *10*
26 - 40	0.75	2.4 *3*
43	9.2	21.4 *21*
44	1.72	3.9 *4*
45 - 51	1.5	3.1 *6*
53	2.8	5.4 *5*
54 - 101	2.73	4.29 *25*
103 - 108	0.65	0.63 *7*

Atomic Electrons (^{182}Os) (continued)

e_{bin}(keV)	$\langle e \rangle$(keV)	e(%)
109	2.63	2.42 *15*
110 - 158	1.5	1.20 *23*
160 - 208	1.88	1.06 *4*
210 - 259	0.214	0.093 *7*
260 - 308	0.23	0.078 *21*
312 - 361	0.031	0.009 *3*
363 - 408	0.048	0.0128 *16*
414 - 436	0.088	0.0206 *23*
438	11.5	2.63 *14*
440 - 487	0.088	0.0185 *16*
489 - 497	0.029	0.0058 *11*
498	2.01	0.404 *22*
499 - 544	0.64	0.126 *5*
549 - 561	0.0108	0.0019 *4*
620 - 655	0.018	0.0028 *6*
714 - 727	0.0039	0.00055 *8*

Continuous Radiation (^{182}Os)

$\langle$IB$\rangle$=0.35 keV

E_{bin}(keV)		$\langle\ \rangle$(keV)	(%)
10 - 20	IB	0.00124	0.0112
20 - 40	IB	0.0025	0.0074
40 - 100	IB	0.33	0.54
100 - 300	IB	0.0160	0.0104
300 - 600	IB	0.0038	0.00105
600 - 611	IB	1.45 $\times 10^{-9}$	2.4 $\times 10^{-9}$

$^{182}_{77}$Ir(15 *1* min)

Mode: ε

Δ: -39150 *230* keV syst

SpA: 6.9$\times 10^{7}$ Ci/g

Prod: ^{175}Lu(^{12}C,5n); ^{169}Tm(^{16}O,3n); Ne on ^{165}Ho; protons on Au

Photons (^{182}Ir)

$\langle \gamma \rangle$=845 *13* keV

γ_{mode}	γ(keV)	γ(%)†
Os L_ℓ	7.822	0.54 *11*
Os L_α	8.904	12.1 *22*
Os L_η	9.337	0.20 *3*
Os L_β	10.462	13.3 *23*
Os L_γ	12.171	2.3 *4*
Os $K_{\alpha2}$	61.485	18 *4*
Os $K_{\alpha1}$	63.000	32 *7*
Os $K_{\beta1}$'	71.313	10.9 *25*
Os $K_{\beta2}$'	73.643	2.8 *6*
γ E2	126.92 *20*	34.4 *7*
γ	136.2 *2*	0.47 *4*
γ	142.6 *2*	0.95 *4*
γ	154.7 *2*	0.73 *4*
γ	166.0 *2*	0.65 *4*
γ	179.8 *2*	1.16 *9*
γ	197.2 *2*	0.43 *4*
γ	228.2 *3*	0.65 *4*
γ	236.3 *3*	9.0 *4*
γ [M1+E2]	251.97 *22*	0.86 *4*
γ	264.8 *4*	0.56 *9*
γ E2	273.09 *16*	43.0 *17*
γ	281.8 *3*	0.47 *4*
γ	289.0 *3*	1.08 *4*
γ	295.3 *3*	1.12 *4*
γ	306.8 *2*	0.30 *4*
γ	335.8 *2*	1.20 *4*

Photons (^{182}Ir) (continued)

γ_{mode}	γ(keV)	γ(%)†
γ	343.2 *3*	0.34 *4*
γ	351.6 *2*	0.30 *4*
γ E2	393.12 *18*	3.18 *17*
γ [E2]	397.0 *3*	1.03 *9*
γ [E2]	400.04 *22*	3.10 *17*
γ	405.0 *5*	0.34 *4*
γ	415.5 *3*	0.56 *4*
γ	429.8 *3*	1.16 *4*
γ	432.7 *4*	0.95 *4*
γ	464.7 *4*	0.26 *4*
γ E2	483.7 *5*	0.52 *4*
γ [E2]	491.06 *21*	~0.17
γ [E2]	498.0 *3*	0.47 *9*
γ	545.5 *5*	0.30 *4*
γ	549.2 *3*	0.34 *4*
γ	559.4 *7*	~0.17
γ	581.0 *5*	0.65 *4*
γ	602.3 *3*	0.43 *4*
γ	632.0 *5*	0.86 *4*
γ [M1+E2]	639.14 *21*	0.99 *4*
γ	647.0 *5*	0.77 *4*
γ	690.2 *2*	0.34 *4*
γ	747.1 *2*	0.39 *4*
γ	749.2 *8*	0.43 *9*
γ [M1+E2]	764.15 *15*	5.6 *4*
γ	779.9 *3*	0.69 *4*
γ [M1+E2]	790.15 *22*	3.14 *6*
γ	837.8 *3*	0.43 *4*
γ [E2]	891.07 *17*	5.7 *3*
γ [M1+E2]	912.22 *16*	8.7 *6*
γ	932.5 *2*	0.39 *4*
γ	938.9 *2*	1.51 *17*
γ	952.5 *7*	0.60 *9*
γ	977.1 *2*	0.60 *4*
γ [M1+E2]	999.1 *2*	1.51 *13*
γ	1032.9 *3*	0.86 *9*
γ [E2]	1063.24 *22*	2.19 *13*
γ	1111.0 *4*	0.47 *4*
γ	1118.1 *4*	2.5 *4*
γ	1121.4 *4*	0.43 *9*
γ	1130.3 *5*	0.56 *4*
γ	1158.0 *8*	1.12 *17*
γ	1160.0 *5*	1.38 *9*
γ	1218 *1*	1.42 *9*
γ	1227.0 *3*	0.56 *9*
γ	1251.6 *5*	1.89 *17*
γ	1266.1 *4*	1.55 *9*
γ	1375.1 *3*	1.03 *13*
γ	1546.0 *6*	1.16 *17*
γ	1549 *1*	0.60 *9*
γ	1652.0 *6*	2.5 *3*

† 23% uncert(syst)

Atomic Electrons (^{182}Ir)

⟨e⟩=97 *4* keV

e_{bin}(keV)	⟨e⟩(keV)	e(%)
11	3.6	33 *7*
12	2.7	22 *4*
13 - 52	1.27	5.1 *9*
53	9.4	17.7 *5*
58 - 106	2.9	3.9 *12*
114	2.28	2.01 *6*
115	18.7	16.4 *5*
116	15.2	13.1 *4*
123	0.3	~0.24
124	10.0	8.1 *3*
125 - 155	4.7	3.6 *4*
162	3.4	~2
163 - 197	1.1	0.64 *21*
199	6.5	3.25 *15*
208 - 254	2.8	1.2 *4*
260 - 307	6.1	2.31 *13*
319 - 359	1.36	0.41 *7*
380 - 429	0.61	0.153 *15*
430 - 479	0.11	0.025 *10*
480 - 528	0.16	0.031 *14*

Atomic Electrons (^{182}Ir) (continued)

e_{bin}(keV)	⟨e⟩(keV)	e(%)
533 - 581	0.32	0.056 *19*
589 - 638	0.10	0.017 *5*
639 - 688	0.08	0.011 *5*
690 - 738	0.8	0.12 *4*
744 - 790	0.23	0.030 *7*
817 - 865	1.0	0.12 *3*
878 - 927	0.39	0.043 *8*
928 - 977	0.06	0.007 *3*
986 - 1033	0.110	0.0111 *10*
1037 - 1086	0.31	0.030 *10*
1098 - 1147	0.11	0.010 *4*
1148 - 1192	0.17	0.014 *6*
1205 - 1254	0.042	0.0034 *10*
1255 - 1301	0.040	~0.0031
1362 - 1375	0.008	~0.0006
1472 - 1547	0.06	0.0043 *18*
1578 - 1650	0.08	~0.005

Continuous Radiation (^{182}Ir)

⟨β+⟩=765 keV;⟨IB⟩=13.1 keV

E_{bin}(keV)		⟨ ⟩(keV)	(%)
0 - 10	β+	6.1×10^{-8}	7.2×10^{-7}
	IB	0.024	
10 - 20	β+	3.91×10^{-6}	2.32×10^{-5}
	IB	0.024	0.168
20 - 40	β+	0.000171	0.00051
	IB	0.047	0.164
40 - 100	β+	0.0141	0.0177
	IB	0.31	0.48
100 - 300	β+	1.14	0.501
	IB	0.43	0.24
300 - 600	β+	11.1	2.35
	IB	0.62	0.143
600 - 1300	β+	110	11.2
	IB	1.9	0.20
1300 - 2500	β+	409	21.9
	IB	4.8	0.26
2500 - 5000	β+	233	8.0
	IB	4.9	0.157
5000 - 5173	IB	0.00040	7.9×10^{-6}
	Σβ+		44

$^{182}_{78}$Pt(2.6 *1* min)

Mode: ε(~99.98 %), α(~0.02 %)
Δ: -36180 *230* keV syst
SpA: 3.96×10^8 Ci/g
Prod: ^{16}O on Yb; ^{20}Ne on Er; protons on Ir

Alpha Particles (^{182}Pt)

α(keV)
4840 *20*

Photons (^{182}Pt)

γ_{mode}	γ(keV)	γ(rel)
γ_ϵ	136.0 *15*	100 *5*
γ_ϵ	146.0 *15*	15.4 *16*
γ_ϵ	186.0 *15*	7.0 *17*
γ_ϵ	210.0 *15*	12 *4*

$^{182}_{79}$Au(21 *2* s)

Mode: ε(~99.96 %), α(~0.04 %)
Δ: -28330 *290* keV syst
SpA: 2.9×10^9 Ci/g
Prod: daughter ^{182}Hg

Alpha Particles (^{182}Au)

α(keV)
5460 *20* ?

Photons (^{182}Au)

γ_{mode}	γ(keV)	γ(rel)
γ_ϵ[E2]	154.9 *10*	100
γ_ϵ[E2]	263.8 *10*	50 *10*

$^{182}_{80}$Hg(11.2 *10* s)

Mode: ε(91 *2* %), α(9 *2* %)
Δ: -23520 *300* keV syst
SpA: 5.4×10^9 Ci/g
Prod: ^{147}Sm(^{40}Ar,5n); ^{170}Yb(^{20}Ne,8n); protons on Pb

Alpha Particles (^{182}Hg)

⟨α⟩=527 *2* keV

α(keV)	α(%)
5700 *15*	0.036 *9*
5865 *15*	8.6 *18*

Photons (^{182}Hg)

γ_{mode}	γ(keV)	γ(rel)
γ_ϵ	128.9 *10*	100
γ_ϵ	216.8 *10*	75 *15*
γ_ϵ	412.6 *10*	53 *10*

A = 183

NDS **16**, 267 (1975)

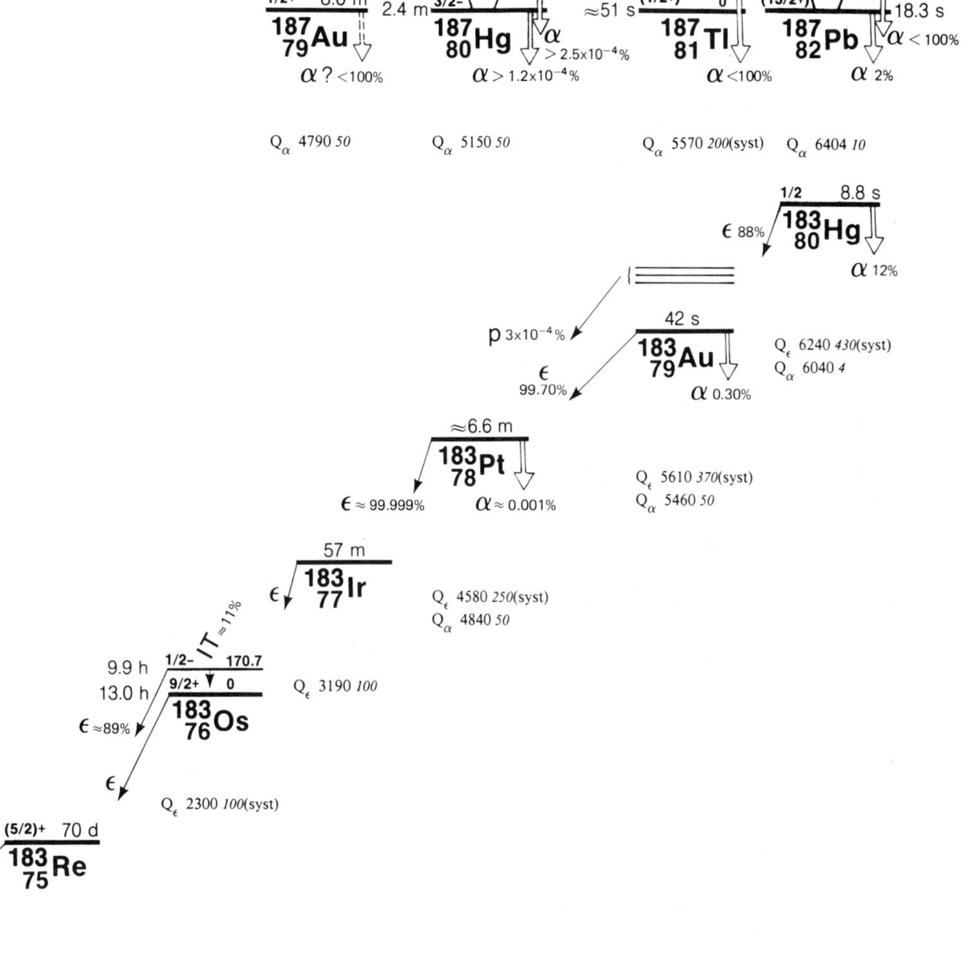

$^{183}_{72}$Hf(1.067 *17* h)

Mode: β-

Δ: -43290 *30* keV

SpA: 1.605×10^7 Ci/g

Prod: ^{186}W(n,α); ^{182}Hf(n,γ)

Photons (^{183}Hf)

⟨γ⟩=751 *53* keV

γ_{mode}	γ(keV)	γ(%)†
Ta L_ℓ	7.173	0.137 *19*
Ta L_α	8.140	3.3 *4*
Ta L_η	8.428	0.043 *5*
Ta L_β	9.468	3.1 *4*
Ta L_γ	10.981	0.51 *7*
Ta $K_{\alpha2}$	56.280	7.5 *8*
Ta $K_{\alpha1}$	57.535	13.1 *14*
Ta $K_{\beta1}'$	65.140	4.3 *5*
Ta $K_{\beta2}'$	67.254	1.12 *12*
γE1	73.173 *14*	38 *4*

Photons (^{183}Hf)
(continued)

γ_{mode}	γ(keV)	γ(%)†
γ[M1+E2]	113.74 *3*	0.14 *1*
γ[M1+E2]	143.200 *16*	0.47 *5*
γ	225.01 *10*	0.15 *2* ?
γ[E1]	284.12 *3*	0.36 *4*
γ	295.22 *8*	0.17 *2* ?
γ[E2]	315.870 *16*	1.22 *12*
γ	375.83 *12*	0.09 *2* ?
γ[E1]	397.858 *17*	2.9 *3*
γ[M1+E2]	459.070 *15*	27 *3*
γ	594.77 *14*	0.21 *10* ?
γ[M1+E2]	686.48 *7*	0.24 *2*
γ	691.74 *10*	0.30 *3*
γ	735.06 *8*	0.88 *9* ?
γ[E2]	783.753 *21*	65
γ	805.44 *12*	0.14 *3* ?
γ[E1]	856.926 *20*	0.11 *3* ?
γ	875.46 *12*	~0.06 ?
γ[E2]	1470.23 *7*	2.7 *3*
γ[E1]	1543.41 *7*	<0.04
γ	1784.31 *20*	0.07 *3* ?

† 10% uncert(syst)

Atomic Electrons (^{183}Hf)

⟨e⟩=17.8 *21* keV

e_{bin}(keV)	⟨e⟩(keV)	e(%)
6	1.45	25 *3*
10	1.10	11.2 *16*
11	0.61	5.5 *8*
12 - 57	0.83	2.8 *3*
61	1.59	2.6 *3*
62	0.62	0.99 *11*
63	0.74	1.17 *13*
64 - 65	0.0078	0.0121 *12*
70	0.40	0.56 *6*
71 - 114	1.05	1.38 *25*
132 - 158	0.31	0.23 *10*
213 - 248	0.23	0.095 *16*
272 - 316	0.115	0.038 *5*
330 - 376	0.096	0.029 *3*
386 - 388	0.0161	0.0042 *4*
392	4.0	~1
395 - 398	0.0048	0.00121 *10*
447	0.6	~0.14
448 - 459	0.42	0.093 *25*
583 - 624	0.046	0.007 *3*
668 - 692	0.07	~0.010
716	2.7	0.38 *4*
723 - 773	0.51	0.066 *6*
774 - 808	0.206	0.0264 *18*

Atomic Electrons (^{183}Hf)
(continued)

e_{bin}(keV)	⟨e⟩(keV)	e(%)
845 - 875	0.0010	0.00012 *5*
1459 - 1541	0.0134	0.00092 *8*
1717 - 1782	0.0018	~0.00010

Continuous Radiation (^{183}Hf)
⟨β-⟩=423 keV;⟨IB⟩=0.54 keV

E_{bin}(keV)		⟨ ⟩(keV)	(%)
0 - 10	β-	0.059	1.18
	IB	0.019	
10 - 20	β-	0.178	1.19
	IB	0.018	0.128
20 - 40	β-	0.72	2.41
	IB	0.035	0.122
40 - 100	β-	5.2	7.4
	IB	0.092	0.142
100 - 300	β-	51	25.6
	IB	0.20	0.117
300 - 600	β-	149	33.7
	IB	0.132	0.033
600 - 1300	β-	211	26.4
	IB	0.043	0.0059
1300 - 1550	β-	4.91	0.360
	IB	0.000113	8.5×10^{-6}

$^{183}_{73}$Ta(5.1 *1* d)

Mode: β-
Δ: -45299 *3* keV
SpA: 1.40×10^{5} Ci/g
Prod: multiple n-capture from ^{181}Ta

Photons (^{183}Ta)
⟨γ⟩=295 *6* keV

γ_{mode}	γ(keV)	γ(%)†
W L_ℓ	7.387	0.71 *7*
W L_α	8.391	16.6 *11*
W L_η	8.724	0.260 *17*
W L_β	9.763	23.5 *23*
W L_γ	11.468	4.7 *6*
γ M1+~0.2%E2	40.9764 *7*	0.47 *5*
γ M1+0.6%E2	46.4847 *6*	4.9 *6*
γ M1+1.2%E2	52.5961 *6*	5.3 *3*
W $K_{\alpha2}$	57.981	24.9 *7*
W $K_{\alpha1}$	59.318	43.1 *12*
W $K_{\beta1}$'	67.155	14.3 *4*
W $K_{\beta2}$'	69.342	3.71 *11*
γ M1+28%E2	82.9175 *10*	0.368 *12*
γ M1+2.7%E2	84.7132 *10*	1.27 *4*
γ E2	99.0808 *7*	6.5 *4*
γ (M1)	101.9363 *17*	0.311 *16*
γ M2	102.482 *3*	0.136 *13**
γ (M1)	103.1483 *17*	0.090 *17*
γ M1+5.5%E2	107.9332 *9*	10.7 *4*
γ M1+1.3%E2	109.7289 *10*	0.597 *19*
γ E2+21%M1	120.3714 *13*	0.065 *4*
γ M1+7.4%E2	142.271 *6*	0.33 *4*
γ M1+0.6%E2	144.1246 *16*	2.47 *5*
γ E2	160.5293 *10*	2.91 *8*
γ M1+4.8%E2	161.3478 *12*	8.8 *3*
γ M1+12%E2	162.3250 *10*	4.82 *13*
γ M1+28%E2	192.6464 *11*	0.364 *9*
γ E2	203.2889 *13*	0.382 *10*
γ M1+8.1%E2	205.0846 *12*	0.883 *21*
γ M1+22%E2	208.8096 *10*	0.612 *17*

Photons (^{183}Ta)
(continued)

γ_{mode}	γ(keV)	γ(%)†
γ E2	209.8695 *17*	4.55 *12*
γ E2	244.2651 *12*	8.67 *18*
γ M1+~28%E2	245.2424 *11*	0.370 *12*
γ M1+0.6%E2	246.0609 *11*	28.0 *10*
γ [M1]	286.396 *6*	~0.011
γ E2	291.7270 *12*	3.80 *11*
γ M1+4.8%E2	313.0177 *13*	3.32 *11*
γ (M1)	313.28 *3*	~4
γ M1+3.5%E2	353.9939 *12*	11.36 *25*
γ (E2)	365.6137 *13*	0.528 *22*
γ (E2)	406.5899 *12*	0.514 *18*

† 1.9% uncert(syst)
* with ^{183}W(5.15 s)

Atomic Electrons (^{183}Ta)
⟨e⟩=154.7 *22* keV

e_{bin}(keV)	⟨e⟩(keV)	e(%)
10	4.3	42 *4*
12	6.3	53 *6*
13 - 33	6.11	25.3 *8*
34	9.3	27 *3*
35 - 36	1.93	5.5 *6*
38	13.4	35.0 *15*
39	0.0116	0.030 *3*
40	9.1	22.4 *14*
41 - 87	18.2	34.3 *12*
88	7.0	8.0 *5*
89	6.4	7.2 *5*
90 - 91	1.97	2.18 *12*
92	8.4	9.2 *4*
93	4.22	4.55 *16*
96	5.7	6.0 *3*
97	3.73	3.86 *24*
98 - 144	7.92	6.94 *14*
148 - 176	7.94	5.08 *10*
177	16.0	9.1 *4*
181 - 222	2.11	1.035 *15*
232 - 282	8.8	3.7 *3*
284	3.85	1.35 *4*
285 - 313	0.85	0.28 *5*
337 - 366	1.020	0.296 *7*
394 - 407	0.0199	0.00501 *14*

Continuous Radiation (^{183}Ta)
⟨β-⟩=187 keV;⟨IB⟩=0.119 keV

E_{bin}(keV)		⟨ ⟩(keV)	(%)
0 - 10	β-	0.161	3.21
	IB	0.0095	
10 - 20	β-	0.477	3.18
	IB	0.0087	0.061
20 - 40	β-	1.89	6.3
	IB	0.0156	0.054
40 - 100	β-	12.8	18.3
	IB	0.034	0.055
100 - 300	β-	91	47.8
	IB	0.045	0.028
300 - 600	β-	80	20.7
	IB	0.0058	0.00163
600 - 776	β-	0.200	0.0312
	IB	3.4×10^{-6}	5.4×10^{-7}

$^{183}_{74}$W (stable)

Δ: -46370 *3* keV
%: 14.3 *1*

$^{183}_{74}$W (5.15 *3* s)

Mode: IT
Δ: -46061 *3* keV
SpA: 1.120×10^{10} Ci/g
Prod: daughter ^{183}Ta; ^{182}W(n,γ); γ rays on W; ^{184}W(n,2n)

Photons (^{183}W)
⟨γ⟩=127.8 *16* keV

γ_{mode}	γ(keV)	γ(%)
W L_ℓ	7.387	1.04 *10*
W L_α	8.391	24.3 *15*
W L_η	8.724	0.363 *21*
W L_β	9.764	34 *3*
W L_γ	11.473	6.8 *8*
γ M1+0.6%E2	46.4847 *6*	6.93 *17*
γ M1+1.2%E2	52.5961 *6*	8.02 *15*
W $K_{\alpha2}$	57.981	36.8 *12*
W $K_{\alpha1}$	59.318	63.7 *21*
W $K_{\beta1}$'	67.155	21.2 *7*
W $K_{\beta2}$'	69.342	5.49 *19*
γ E2	99.0808 *7*	6.57 *19*
γ M2	102.482 *3*	2.42 *10*
γ M1+5.5%E2	107.9332 *9*	18.4 *4*
γ E2	160.5293 *10*	5.08 *8*

Atomic Electrons (^{183}W)
⟨e⟩=173.2 *20* keV

e_{bin}(keV)	⟨e⟩(keV)	e(%)
10	6.3	62 *6*
12	9.1	77 *8*
30	1.72	5.82 *21*
33	22.3	68 *3*
34	13.2	38.3 *12*
35 - 36	2.76	7.8 *5*
38	22.4	58.2 *17*
40	12.4	30.6 *9*
41 - 42	3.33	8.0 *3*
44	4.63	10.6 *5*
45 - 87	10.68	21.0 *4*
88	7.11	8.1 *3*
89	6.49	7.3 *3*
90	15.3	16.9 *8*
91 - 92	6.63	7.24 *21*
96	9.7	10.1 *4*
97 - 99	5.71	5.87 *19*
100	5.59	5.60 *25*
101 - 149	6.44	5.79 *12*
150 - 161	1.514	0.982 *16*

$^{183}_{75}$Re(70.0 *11* d)

Mode: ε
Δ: -45814 *9* keV
SpA: 1.019×10^{4} Ci/g
Prod: ^{181}Ta(α,2n); daughter ^{183}Os

Photons (^{183}Re)

$\langle\gamma\rangle$=150 *6* keV

γ_{mode}	γ(keV)	γ(%)†
W L_ℓ	7.387	0.88 *14*
W L_α	8.391	20 *3*
W L_η	8.724	0.29 *3*
W L_β	9.767	28 *4*
W L_γ	11.482	5.7 *9*
γ M1+~0.2%E2	40.9764 *7*	0.0227 *24*
γ M1+0.6%E2	46.4847 *6*	8.18 *16*
γ M1+1.2%E2	52.5961 *6*	2.27 *7*
W $K_{\alpha 2}$	57.981	31 *4*
W $K_{\alpha 1}$	59.318	54 *8*
W $K_{\beta 1}$'	67.155	18.0 *25*
W $K_{\beta 2}$'	69.342	4.7 *7*
γ M1+28%E2	82.9175 *10*	0.261 *5*
γ M1+2.7%E2	84.7132 *10*	0.906 *18*
γ E2	99.0808 *7*	2.77 *5*
γ (M1)	101.9363 *17*	0.0184 *10*
γ (M1)	103.1483 *17*	0.0115 *21*
γ M1+5.5%E2	107.9332 *9*	2.23 *4*
γ M1+1.3%E2	109.7289 *10*	2.97 *6*
γ E2+21%M1	120.3714 *13*	0.0084 *5*
γ M1+0.6%E2	144.1246 *16*	0.1196 *21*
γ E2	160.5293 *10*	0.604 *11*
γ M1+4.8%E2	161.3478 *12*	0.426 *18*
γ M1+12%E2	162.3250 *10*	24.0 *4*
γ M1+28%E2	192.6464 *11*	0.259 *5*
γ E2	203.2889 *13*	0.0491 *15*
γ M1+8.1%E2	205.0846 *12*	0.113 *3*
γ M1+22%E2	208.8096 *10*	3.05 *6*
γ E2	209.8695 *17*	0.269 *6*
γ E2	244.2651 *12*	0.420 *8*
γ M1+~28%E2	245.2424 *11*	0.263 *5*
γ M1+0.6%E2	246.0609 *11*	1.35 *4*
γ E2	291.7270 *12*	2.7 *1*
γ M1+4.8%E2	313.0177 *13*	0.426 *10*
γ M1+3.5%E2	353.9939 *12*	0.550 *13*
γ (E2)	365.6137 *13*	0.0678 *25*
γ (E2)	406.5899 *12*	0.0249 *9*

† 4.9% uncert(syst)

Atomic Electrons (^{183}Re)

$\langle e\rangle$=105.0 *16* keV

e_{bin}(keV)	$\langle e\rangle$(keV)	e(%)
10	5.3	52 *9*
12	7.4	63 *9*
13 - 32	1.89	10.02 *20*
34	15.6	45.2 *13*
35 - 39	5.99	16.3 *6*
40	7.24	17.9 *6*
41 - 42	0.94	2.27 *10*
44	5.47	12.5 *5*
45 - 87	7.79	14.4 *4*
88	2.99	3.42 *10*
89	2.73	3.08 *9*
90 - 92	0.581	0.634 *22*
93	21.0	22.7 *6*
96 - 144	8.00	7.49 *12*
148 - 149	0.247	0.166 *5*
150	4.97	3.31 *9*
151 - 200	5.27	3.22 *5*
201 - 246	1.036	0.458 *9*
280 - 313	0.479	0.168 *3*
337 - 366	0.0515	0.0149 *3*
394 - 407	0.00096	0.000243 *7*

Continuous Radiation (^{183}Re)

$\langle$IB$\rangle$=0.32 keV

E_{bin}(keV)		$\langle\ \rangle$(keV)	(%)
10 - 20	IB	0.00098	0.0088
20 - 40	IB	0.0028	0.0085
40 - 100	IB	0.30	0.51
100 - 300	IB	0.0105	0.0073
300 - 555	IB	0.00053	0.000154

$^{183}_{76}$Os(13.0 *5* h)

Mode: ϵ

Δ: -43510 *100* keV syst

SpA: 1.32$\times 10^6$ Ci/g

Prod: ^{185}Re(p,3n); alphas on W; daughter ^{183}Ir

Photons (^{183}Os)

$\langle\gamma\rangle$=531 *33* keV

γ_{mode}	γ(keV)	γ(%)
Re L_ℓ	7.604	0.84 *19*
Re L_α	8.646	19 *4*
Re L_η	9.027	0.25 *5*
Re L_β	10.129	19 *4*
Re L_γ	11.802	3.4 *8*
Re $K_{\alpha 2}$	59.717	43 *9*
Re $K_{\alpha 1}$	61.141	73 *15*
Re $K_{\beta 1}$'	69.214	25 *5*
Re $K_{\beta 2}$'	71.470	6.3 *13*
γ M1+3.8%E2	114.43 *5*	20.7 *23*
γ M1+1.4%E2	145.43 *5*	1.5 *4*
γ M1+20%E2	151.02 *5*	0.31 *8*
γ M1+2.3%E2	167.93 *4*	7.7 *8*
γ M1+2.3%E2	175.50 *14*	~0.5
γ M1+2.0%E2	197.23 *10*	0.12 *3*
γ E1	236.34 *6*	2.2 *4*
γ E2	259.86 *6*	0.23 *8*
γ [E2]	320.93 *15*	~0.015
γ [E1]	338.00 *14*	~0.15
γ (M1+50%E2)	355.53 *18*	0.50 *12*
γ E1	381.78 *7*	77 *8*
γ [E1]	404.27 *7*	~0.15
γ M2	496.20 *8*	0.54 *15*
γ [M1+E2]	591.25 *13*	~0.023
γ [M1]	639.88 *17*	~0.15
γ (M1)	736.68 *12*	0.31 *8*
γ [M1+E2]	742.27 *13*	~0.031
γ (M1)	807.81 *17*	0.38 *12*
γ (M1)	851.11 *12*	3.8 *5*
γ (M1+50%E2)	887.70 *13*	<1
γ (M1+50%E2)	889.60 *23*	<1
γ (M1+50%E2)	1057.52 *23*	0.46 *8*
γ [E1]	1090.6 *6*	0.12 *4*
γ (M1)	1163.34 *22*	1.15 *15*
γ E1	1266.0 *6*	0.10 *4*
γ	1285.2 *5*	0.18 *5*
γ [E1]	1293.86 *24*	0.10 *4*
γ	1349.3 *7*	0.08 *3*
γ E1	1365.6 *7*	0.046 *15*
γ E1	1411.5 *6*	0.14 *4*
γ E1	1439.30 *24*	0.54 *8*
γ	1453.1 *5*	0.061 *23*
γ	1463.7 *7*	0.023 *8*
γ E1	1533.5 *7*	0.23 *5*

Atomic Electrons (^{183}Os)

$\langle e\rangle$=73 *4* keV

e_{bin}(keV)	$\langle e\rangle$(keV)	e(%)
11	5.9	56 *13*
12	3.3	27 *6*
13	1.3	10.3 *24*
43	26.1	61 *7*
47 - 79	5.2	8.9 *9*
96	7.4	7.7 *8*
102	10.5	10.3 *12*
104	1.1	1.0 *3*
111	2.25	2.02 *23*
112 - 157	4.25	3.16 *18*
163 - 197	0.99	0.60 *6*
224 - 266	0.076	0.032 *3*
284 - 309	0.12	0.042 *10*
310	2.7	0.87 *9*
318 - 356	0.044	0.0129 *17*
369 - 404	0.64	0.173 *12*
484 - 520	0.106	0.022 *5*
568 - 591	0.025	0.0045 *22*
627 - 671	0.048	0.0073 *16*
724 - 742	0.054	0.0073 *18*
779 - 818	0.50	0.063 *9*
839 - 888	0.104	0.0123 *14*
889 - 890	0.0004	~5$\times 10^{-5}$
986 - 1019	0.028	0.0028 *5*
1045 - 1092	0.085	0.0078 *10*
1151 - 1194	0.0175	0.00151 *16*
1214 - 1256	0.008	~0.0006
1263 - 1294	0.0050	0.00039 *18*
1337 - 1381	0.0104	0.00076 *13*
1392 - 1441	0.0025	0.00017 *4*
1443 - 1531	0.0032	0.00022 *4*

Continuous Radiation (^{183}Os)

$\langle$IB$\rangle$=1.60 keV

E_{bin}(keV)		$\langle\ \rangle$(keV)	(%)
10 - 20	IB	0.00099	0.0089
20 - 40	IB	0.0023	0.0069
40 - 100	IB	0.34	0.56
100 - 300	IB	0.052	0.028
300 - 600	IB	0.19	0.042
600 - 1300	IB	0.78	0.086
1300 - 2286	IB	0.23	0.0159

$^{183}_{76}$Os(9.9 *3* h)

Mode: ϵ(~89 %), IT(~11 %)

Δ: -43339 *100* keV syst

SpA: 1.73$\times 10^6$ Ci/g

Prod: ^{185}Re(p,3n); alphas on W; daughter ^{183}Ir

Photons (^{183}Os)

$\langle\gamma\rangle$=1033 *45* keV

γ_{mode}	γ(keV)	γ(%)†
Re L_ℓ	7.604	0.41 *8*
Os L_ℓ	7.822	0.055 *12*
Re L_α	8.646	9.5 *17*
Os L_α	8.904	1.2 *3*
Re L_η	9.027	0.122 *22*
Os L_η	9.337	0.0090 *19*
Re L_β	10.128	9.6 *18*
Os L_β	10.515	0.92 *20*
Re L_γ	11.807	1.7 *3*

Photons (^{183}Os)
(continued)

γ_{mode}	γ(keV)	γ(%)†
Os L_γ	12.268	0.16 4
Re $K_{\alpha2}$	59.717	21 4
Re $K_{\alpha1}$	61.141	36 6
Os $K_{\alpha2}$	61.485	0.91 18
Os $K_{\alpha1}$	63.000	1.6 3
Re $K_{\beta1}$'	69.214	12.2 22
Os $K_{\beta1}$'	71.313	0.54 11
Re $K_{\beta2}$'	71.470	3.1 6
Os $K_{\beta2}$'	73.643	0.14 3
γ_ϵM1+3.8%E2	114.43 5	0.51 5
γ_ϵ(E1)	126.30 10	0.20 7 ?
γ_ϵ(M1+50%E2)	147.14 6	0.53 20
γ_ϵ(M1)	163.32 10	~0.13
γ_{IT} M4	170.72 9	0.051 10
γ_ϵ[M1+50%E2]	229.6 4	0.60 13
γ_ϵ(M1)	245.76 8	0.40 13
γ_ϵ(M1)	251.88 8	0.40 13
γ_ϵ[M1]	273.4 3	0.13 5 ?
γ_ϵ[M1]	399.0 3	0.79 20
γ_ϵE1	484.61 10	1.66 20
γ_ϵ	797.3 4	0.53 13
γ_ϵ	803.4 4	0.33 13
γ_ϵ[E2]	840.44 16	0.53 13
γ_ϵ(E2)	878.6 5	1.66 20
γ_ϵ[M1+E2]	885.1 6	0.23 7
γ_ϵ(M1+E2)	954.86 15	1.13 20
γ_ϵ(M1)	1027.1 7	0.10 3
γ_ϵ(M1+50%E2)	1034.8 3	6.6
γ_ϵ(M1+E2)	1041.3 5	0.46 7
γ_ϵ(E2)	1102.00 16	52 3
γ_ϵ(M1+26%E2)	1108.13 16	23.8 20
γ_ϵ[E1]	1227.58 19	0.033 13
γ_ϵ[E1]	1341.2 8	0.066 20
γ_ϵ(M1+E2)	1353.88 16	0.16 3
γ_ϵ[E1]	1626.1 7	0.09 3
γ_ϵ(E2)	1680.3 8	0.040 13 ?
γ_ϵ(E2)	1805.5 6	0.086 20
γ_ϵ(M1)	1825.8 8	0.053 20
γ_ϵ(E2)	1905.4 4	0.19 4
γ_ϵ(E2)	1919.9 6	0.046 13

† 5.8% uncert(syst)

Atomic Electrons (^{183}Os)
⟨e⟩=30.9 18 keV

e_{bin}(keV)	⟨e⟩(keV)	e(%)
11	3.2	31 6
12	1.7	14 3
13	0.77	6.1 13
43 - 92	2.9	5.4 8
97	3.2	3.3 7
102 - 151	0.73	0.61 14
153	0.00032	~0.00021
158	4.2	2.7 5
160	4.9	3.1 6
161 - 163	0.0035	0.0021 9
168	2.7	1.6 3
169 - 218	1.55	0.88 11
219 - 263	0.190	0.08 1
270 - 273	0.0045	0.0017 4
386 - 413	0.106	0.027 3
472 - 485	0.0102	0.00215 18
726 - 769	0.08	0.011 5
785 - 830	0.101	0.0125 17
838 - 885	0.09	0.011 4
942 - 970	0.44	0.045 8
1015 - 1029	0.070	0.0069 11
1030	1.75	0.170 13
1031 - 1035	0.019	0.0018 3
1036	1.50	0.145 13
1038 - 1041	0.0013	0.00012 4
1089 - 1108	0.71	0.065 3
1215 - 1227	8.1×10^{-5}	$6.6\ 20 \times 10^{-6}$
1269 - 1282	0.007	0.00058 20
1329 - 1354	0.0015	0.00011 3
1554 - 1624	0.0020	0.00013 3

Atomic Electrons (^{183}Os)
(continued)

e_{bin}(keV)	⟨e⟩(keV)	e(%)
1668 - 1754	0.0041	0.00023 5
1793 - 1848	0.0058	0.00031 5
1893 - 1918	0.00099	$5.2\ 8 \times 10^{-5}$

Continuous Radiation (^{183}Os)
⟨IB⟩=0.52 keV

E_{bin}(keV)		⟨ ⟩(keV)	(%)
10 - 20	IB	0.00093	0.0083
20 - 40	IB	0.0021	0.0063
40 - 100	IB	0.31	0.51
100 - 300	IB	0.039	0.021
300 - 600	IB	0.095	0.022
600 - 1300	IB	0.075	0.0103
1300 - 1546	IB	6.9×10^{-5}	5.1×10^{-6}

$^{183}_{77}$Ir(57 4 min)

Mode: ϵ
Δ: -40320 140 keV syst
SpA: 1.80×10^7 Ci/g
Prod: ^{175}Lu(^{12}C,4n); ^{165}Ho(^{22}Ne,4n)

Photons (^{183}Ir)

γ_{mode}	γ(keV)	γ(rel)
γ	30.8 5	
γ	87.7 5	63 13
γ	96.2 5	6.0 12
γ	102.2 5	18 4
γ	107.7 5	4.0 8
γ	136.8 5	17 3
γ	165.7 5	14 3
γ	194.5 5	23 5
γ	228.5 5	100 20
γ	236.7 5	29 6
γ	239.7 5	26 5
γ	250.6 5	10 2
γ	254.4 5	23 5
γ	282.4 5	70 14
γ	286.1 5	2.0 4
γ	296.3 5	5 1
γ	314.4 5	11.0 22
γ	319.1 5	
γ	342.2 5	29 6
γ	347.7 5	37 7
γ	392.4 5	
γ	411.2 5	19 4
γ	457.7 5	6.0 12
γ	461.9 5	5 1
γ	498.4 5	15 3
γ	617.4 5	8.0 16
γ	655.1 5	20 4
γ	670.8 5	10 2
γ	692.2 5	33 7
γ	706.1 5	5 1
γ	724.8 5	5 1
γ	800.1 5	33 7
γ	896.6 5	18 4

$^{183}_{78}$Pt(6.6 9 min)

Mode: ϵ(~99.9987 %), α(~0.0013 %)
Δ: -35740 210 keV syst
SpA: 1.56×10^8 Ci/g
Prod: ^{174}Yb(^{16}O,7n); protons on Ir

Alpha Particles (^{183}Pt)

α(keV)
4730 20

$^{183}_{79}$Au(42.0 12 s)

Mode: ϵ(99.70 5 %), α(0.30 5 %)
Δ: -30130 310 keV syst
SpA: 1.46×10^9 Ci/g
Prod: ^{169}Tm(^{20}Ne,6n); ^{168}Yb(^{19}F,4n); ^{175}Lu(^{16}O,8n); daughter ^{183}Hg

Alpha Particles (^{183}Au)

α(keV)
5343 5

$^{183}_{80}$Hg(8.8 5 s)

Mode: ϵ(88 2 %), α(12 2 %), ϵp(0.00027 6 %)
Δ: -23890 300 keV syst
SpA: 6.7×10^9 Ci/g
Prod: protons on Pb
ϵp: 5100(weak)
ϵp/α: $2.2\ 3 \times 10^{-5}$

Alpha Particles (^{183}Hg)
⟨α⟩=707.8 24 keV

α(keV)	α(%)
5830 15	1.13 20
5905 15	10.9 21

A = 184

NDS **21**, 1 (1977)

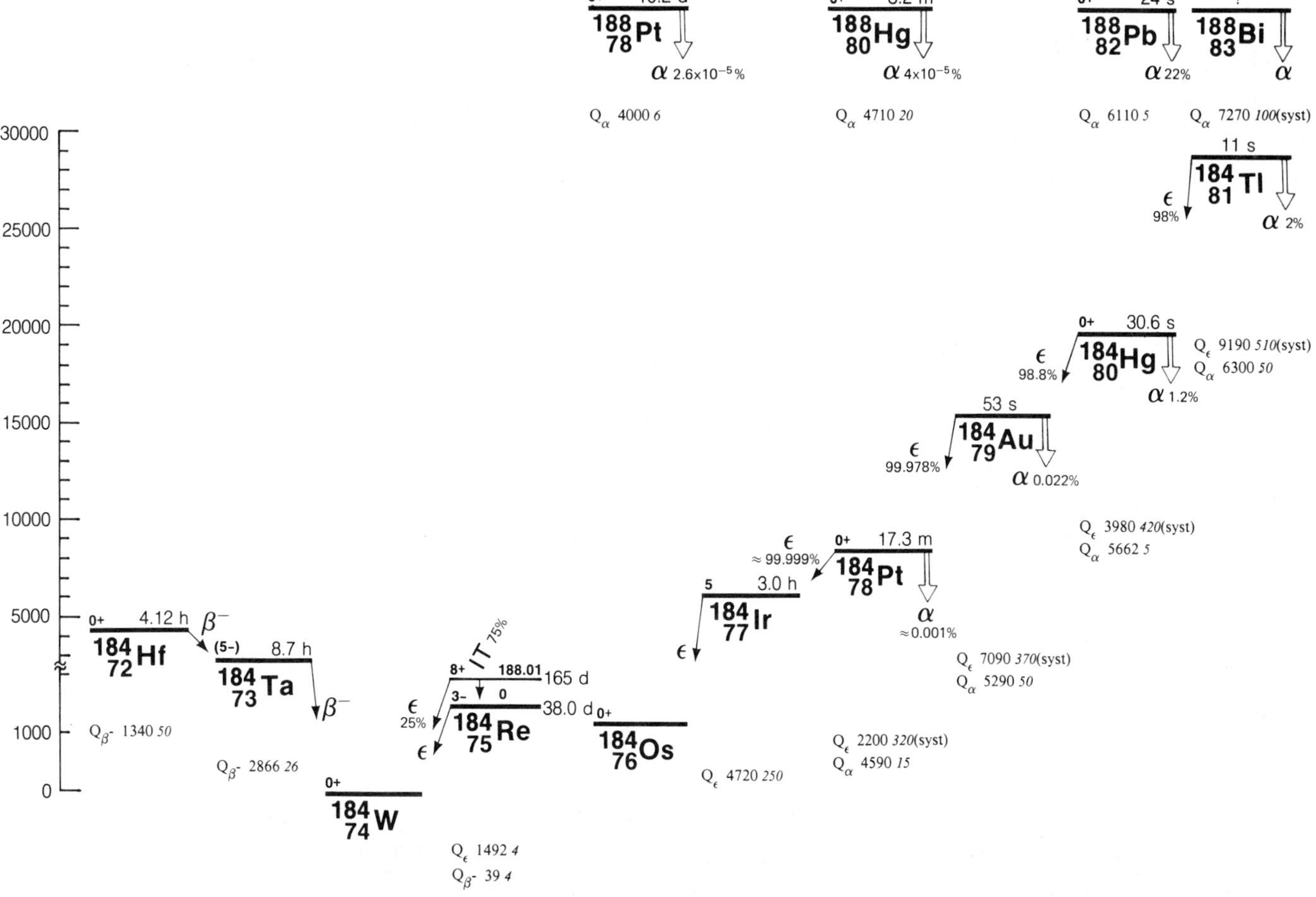

$^{184}_{72}$Hf(4.12 *5* h)

Mode: β-
Δ: -41500 *60* keV
SpA: 4.13×10^6 Ci/g
Prod: ^{186}W(p,3p)

Photons (^{184}Hf)

⟨γ⟩=251 *12* keV

γ_{mode}	γ(keV)	γ(%)
Ta L_ℓ	7.173	0.89 *17*
Ta L_α	8.140	21 *4*
Ta L_η	8.428	0.40 *6*
Ta L_β	9.423	35 *6*
Ta L_γ	11.076	7.2 *12*
γ[M1]	41.4 *2*	9.9 *12*
γ[M1]	43.9 *2*	6.1 *8*
γ[E2]	47.9 *2*	1.2 *3*
Ta $K_{\alpha2}$	56.280	6.2 *6*
Ta $K_{\alpha1}$	57.535	10.8 *10*
Ta $K_{\beta1}$'	65.140	3.6 *3*
Ta $K_{\beta2}$'	67.254	0.92 *8*
γ[E2]	139.1 *2*	48 *4*
γ[E1]	181.0 *2*	14.8 *17*
γ[E1]	344.9 *2*	38 *3*

Atomic Electrons (^{184}Hf)

⟨e⟩=168 *8* keV

e_{bin}(keV)	⟨e⟩(keV)	e(%)
10 - 11	9.8	94 *14*
12	5.2	45 *6*
30	22.9	77 *10*
32	12.0	37 *5*
33 - 36	1.46	4.4 *5*
37	15.3	42 *11*
38	17.5	46 *12*
39	6.8	17.7 *23*
40	0.0109	0.027 *4*
41	5.9	14.2 *19*
42 - 45	6.0	13 *3*
46	5.5	12 *3*
47 - 65	3.1	6.5 *13*
72	15.1	21.0 *19*
114 - 127	3.9	3.19 *24*
128	14.8	11.5 *11*
129	12.1	9.4 *9*
136	0.68	0.50 *5*
137	7.2	5.3 *5*
139 - 181	2.56	1.80 *15*
333 - 345	0.333	0.099 *6*

Continuous Radiation (^{184}Hf)

⟨β-⟩=293 keV;⟨IB⟩=0.29 keV

E_{bin}(keV)		⟨ ⟩(keV)	(%)
0 - 10	β-	0.091	1.82
	IB	0.0140	
10 - 20	β-	0.274	1.83
	IB	0.0133	0.093
20 - 40	β-	1.11	3.68
	IB	0.025	0.086
40 - 100	β-	7.8	11.2
	IB	0.061	0.096
100 - 300	β-	70	35.3
	IB	0.114	0.068
300 - 600	β-	143	33.4
	IB	0.051	0.0129
600 - 1112	β-	71	9.8
	IB	0.0068	0.00100

$^{184}_{73}$Ta(8.7 *1* h)

Mode: β-
Δ: -42844 *26* keV
SpA: 1.958×10^6 Ci/g
Prod: ^{186}W(d,α); ^{184}W(n,p)

Photons (^{184}Ta)

⟨γ⟩=1610 *13* keV

γ_{mode}	γ(keV)	γ(%)†
W L_{ℓ}	7.387	0.53 *6*
W L_{α}	8.391	12.3 *10*
W L_{η}	8.724	0.230 *16*
W L_{β}	9.773	14.5 *11*
W L_{γ}	11.333	2.40 *21*
γ M1+0.3%E2	55.281 *5*	0.41 *10*
W $K_{\alpha2}$	57.981	10.5 *9*
W $K_{\alpha1}$	59.318	18.1 *16*
γ E2	63.698 *10*	1.77 *6*
W $K_{\beta1}$'	67.155	6.0 *5*
W $K_{\beta2}$'	69.342	1.56 *14*
γ E1	87.454 *8*	0.98 *5*
γ M1+28%E2	91.269 *7*	1.07 *6*
γ E2	111.208 *6*	24.3 *10*
γ [E1]	124.056 *10*	0.63 *4*
γ E2(+<11%M1)	127.871 *11*	0.0022 *6*
γ [E1]	151.152 *11*	0.187 *16*
γ M1+22%E2	161.270 *10*	3.3 *9*
γ	162.37 *16*	1.7 *7*
γ [E1]	191.20 *18*	0.037 *15*
γ [E1]	203.71 *3*	~0.030
γ E1	215.325 *9*	11.4 *4*
γ E2	216.551 *9*	1.77 *22*
γ E1+<0.3%M2+E3	226.746 *8*	6.4 *3*
γ [E2]	230.56 *1*	0.0195 *24*
γ	244.49 *3*	3.6 *4*
γ	252.77 *4*	5.0 *15*
γ E2	252.849 *8*	44 *3*
γ	274.03 *5*	0.44 *4*
γ [E1]	294.97 *3*	0.53 *7*
γ [M1+E2]	296.47 *9*	0.71 *10*
γ	299.77 *3*	0.48 *5*
γ [E1]	315.37 *16*	0.30 *7*
γ E1+0.09%M2	318.014 *7*	23.5 *5*
γ [E1]	331.05 *11*	0.13 *3*
γ [E1]	339.49 *9*	0.19 *4*
γ	353.57 *10*	0.15 *6*
γ	359.32 *12*	0.11 *3*
γ	371.09 *12*	0.27 *4*
γ [M1+E2]	381.5 *3*	~0.17
γ	381.712 *11*	0.25 *3*
γ E2	384.256 *10*	12.8 *4*
γ	414.04 *4*	73.9
γ	461.04 *3*	10.9 *3*
γ	516.59 *25*	0.31 *4*
γ	528.29 *6*	1.02 *11*
γ E1+1.5%M2+E3	536.681 *11*	13.2 *3*
γ E2	539.225 *10*	0.127 *5*
γ	576.3 *6*	0.10 *3*
γ E2+1.4%M1	641.914 *10*	1.37 *4*
γ	655.30 *25*	0.259 *22*
γ E2+3.1%M1	769.785 *10*	0.89 *4*
γ	785.8 *6*	0.081 *22*
γ E2+0.4%M1	792.073 *9*	15.0 *3*
γ	807.69 *9*	0.50 *6*
γ	851.1 *5*	~0.05
γ E1	857.238 *9*	0.668 *21*
γ E2+0.6%M1	894.762 *10*	11.0 *3*
γ E2	903.280 *10*	15.3 *3*
γ E1+2.1%M2+E3	920.936 *10*	32.6 *7*
γ [M1+E2]	930.9 *3*	0.081 *22*
γ	942.89 *11*	0.103 *22*
γ [E1]	981.41 *9*	0.044 *15*
γ [E1]	1018.818 *10*	0.39 *3*
γ E2	1022.632 *10*	0.67 *3*
γ	1043.68 *14*	~0.007
γ	1046.40 *10*	0.044 *15*
γ [M1+E2]	1060.94 *3*	0.072 *14*
γ	1094.2 *4*	~0.022
γ	1104.15 *11*	~0.044
γ E1+0.7%M2	1110.086 *8*	2.32 *7*
γ	1115.8 *11*	0.022 *7*
γ [M1+E2]	1172.61 *16*	0.25 *7*
γ	1173.783 *11*	4.85 *22*
γ	1207.67 *10*	0.30 *3*
γ	1221.293 *10*	0.085 *6*
γ [M1+E2]	1312.45 *14*	0.10 *3*
γ M1+>50%E2	1313.79 *3*	0.31 *4*
γ	1334.97 *4*	0.052 *15*
γ [E2]	1425.46 *16*	0.170 *15*

† 1.2% uncert(syst)

Atomic Electrons (^{184}Ta)

⟨e⟩=142 *10* keV

e_{bin}(keV)	⟨e⟩(keV)	e(%)
10	4.0	40 *5*
12	3.7	32 *4*
18 - 22	1.03	4.8 *3*
42	7.3	17.4 *8*
43 - 49	1.15	2.6 *3*
52	9.4	18.0 *8*
53	9.5	17.7 *7*
55 - 59	0.42	0.75 *7*
61	5.42	8.9 *3*
62 - 99	9.0	10.8 *12*
100	17.6	17.7 *8*
101	15.5	15.3 *7*
108	0.474	0.438 *21*
109	9.1	8.4 *4*
111 - 160	6.1	4.7 *3*
161 - 181	1.3	~0.7
183	8.8	4.8 *10*
188 - 234	1.6	0.72 *13*
241 - 290	7.9	3.2 *3*
292 - 341	1.64	0.52 *3*
342 - 343	0.0019	~0.0006
345	11.4	~3
347 - 392	2.1	~0.5
402 - 451	3.5	~0.9
458 - 507	0.65	0.14 *3*
514 - 539	0.130	0.025 *4*
564 - 586	0.098	0.017 *3*
630 - 655	0.027	0.0042 *4*
700 - 738	0.73	0.101 *4*
758 - 807	0.200	0.0256 *7*
808 - 857	1.64	0.195 *9*
861 - 911	0.354	0.0396 *16*
912 - 953	0.061	0.0065 *4*
969 - 1019	0.0113	0.00114 *23*
1020 - 1061	0.039	0.0037 *3*
1082 - 1116	0.20	~0.018
1138 - 1174	0.054	0.0047 *19*
1196 - 1244	0.020	0.00161 *23*
1265 - 1314	0.0052	0.00040 *10*
1323 - 1356	0.0047	0.00035 *3*
1413 - 1425	0.00092	6.5 *4* $\times 10^{-5}$

Continuous Radiation (^{184}Ta)

⟨β-⟩=394 keV;⟨IB⟩=0.46 keV

E_{bin}(keV)		⟨ ⟩(keV)	(%)
0 - 10	β-	0.063	1.26
	IB	0.018	
10 - 20	β-	0.190	1.27
	IB	0.017	0.122
20 - 40	β-	0.77	2.58
	IB	0.033	0.115
40 - 100	β-	5.6	8.0
	IB	0.085	0.133
100 - 300	β-	56	28.0
	IB	0.18	0.106
300 - 600	β-	162	36.6
	IB	0.106	0.026
600 - 1300	β-	170	22.4
	IB	0.022	0.0031
1300 - 1580	β-	0.199	0.0146
	IB	4.9×10^{-6}	3.7×10^{-7}

$^{184}_{74}$W (stable)

Δ: -45710 *3* keV

%: 30.67 *15*

$^{184}_{75}$Re(38.0 *5* d)

Mode: ϵ

Δ: -44218 *5* keV

SpA: 1.867×10^{4} Ci/g

Prod: ^{181}Ta(α,n); ^{184}W(p,n); ^{185}Re(n,2n); ^{186}W(p,3n); deuterons on W

Photons (^{184}Re)

⟨γ⟩=893 *8* keV

γ_{mode}	γ(keV)	γ(%)†
W L_{ℓ}	7.387	0.58 *6*
W L_{α}	8.391	13.4 *9*
W L_{η}	8.724	0.201 *14*
W L_{β}	9.782	14.6 *12*
W L_{γ}	11.386	2.52 *25*
W $K_{\alpha2}$	57.981	25.5 *13*
W $K_{\alpha1}$	59.318	44.2 *22*
W $K_{\beta1}$'	67.155	14.7 *8*
W $K_{\beta2}$'	69.342	3.8 *2*
γ E2	111.208 *6*	17.1 *6*
γ [E1]	124.056 *10*	0.00179 *19*
γ E2(+<11%M1)	127.871 *11*	0.0016 *5*
γ [E1]	203.71 *3*	~0.0011
γ E1+<0.3%M2+E3	226.746 *8*	0.0182 *17*
γ [E2]	230.56 *1*	0.0147 *17*
γ E2	252.849 *8*	3.02 *25*
γ [E1]	294.97 *3*	0.019 *3*
γ E2+37%M1	380.353 *20*	0.0050 *13*
γ M1(+E2+E0)	483.042 *20*	0.018 *3*
γ E2	539.225 *10*	0.315 *12*
γ E2+1.4%M1	641.914 *10*	1.94 *4*
γ [E2]	757.383 *21*	0.062 *4*
γ E2+3.1%M1	769.785 *10*	0.671 *19*
γ E2+0.4%M1	792.073 *9*	37.5 *6*
γ E2+0.6%M1	894.762 *10*	15.5 *3*
γ E2	903.280 *10*	38.1 *6*
γ E2(+18%M1+E0)	1010.231 *21*	0.091 *6*
γ [E1]	1018.818 *10*	0.00112 *12*
γ E2	1022.632 *10*	0.505 *22*
γ [M1+E2]	1060.94 *3*	0.0026 *4*
γ E2	1121.438 *21*	0.0352 *25*
γ M1(+E2+E0)	1275.113 *19*	0.119 *6*
γ M1+>50%E2	1313.79 *3*	0.0114 *8*
γ M1+E2+E0	1319.80 *7*	0.00226 *25*
γ E2	1386.320 *19*	0.103 *5*
γ [E2]	1431.01 *7*	0.0024 *3*

† 1.7% uncert(syst)

Atomic Electrons (^{184}Re)

⟨e⟩=54.1 *10* keV

e_{bin}(keV)	⟨e⟩(keV)	e(%)
10	4.3	42 *5*
12	3.7	31 *3*
42	5.12	12.3 *5*
45 - 67	1.97	3.88 *15*
99	1.31	1.32 *5*
100	12.4	12.5 *5*
101	10.9	10.8 *4*
108	0.334	0.308 *12*
109	6.42	5.91 *23*
111	1.92	1.73 *7*
112 - 161	0.0059	0.0041 *4*
183 - 231	0.50	0.273 *23*
241 - 285	0.405	0.166 *9*
292 - 311	0.0009	0.00030 *13*
368 - 414	0.0095	0.0023 *6*
470 - 483	0.0222	0.00472 *21*
527 - 572	0.112	0.0197 *5*
630 - 642	0.0306	0.00484 *9*
688 - 700	0.0337	0.00482 *21*
723	1.62	0.224 *6*
745 - 792	0.427	0.0546 *8*

Atomic Electrons (^{184}Re) (continued)

e_{bin}(keV)	⟨e⟩(keV)	e(%)
825 - 834	2.05	0.247 *5*
883 - 903	0.498	0.0559 *7*
941 - 953	0.0211	0.00222 *10*
991 - 1023	0.00498	0.000493 *17*
1049 - 1061	0.00114	0.000108 *8*
1109 - 1121	0.000248	2.23 *11* $\times10^{-5}$
1206 - 1250	0.0056	0.00046 *14*
1263 - 1312	0.0011	9.0 *21* $\times10^{-5}$
1313 - 1361	0.00279	0.000211 *11*
1374 - 1421	0.000581	4.22 *15* $\times10^{-5}$
1428 - 1431	2.9 $\times10^{-6}$	2.06 *23* $\times10^{-7}$

Continuous Radiation (^{184}Re)

⟨IB⟩=0.37 keV

E_{bin}(keV)		⟨ ⟩(keV)	(%)
10 - 20	IB	0.00081	0.0072
20 - 40	IB	0.0027	0.0080
40 - 100	IB	0.31	0.53
100 - 300	IB	0.026	0.0153
300 - 600	IB	0.0165	0.0042
600 - 1300	IB	0.0124	0.00154
1300 - 1385	IB	4.8 $\times10^{-6}$	3.6 $\times10^{-7}$

$^{184}_{75}$Re(165 *5* d)

Mode: IT(74.7 *6* %), ϵ(25.3 *6* %)
Δ: -44030 *5* keV
SpA: 4301 Ci/g
Prod: ^{181}Ta(α,n); ^{184}W(p,n); ^{185}Re(n,2n); ^{186}W(p,3n); deuterons on W

Photons (^{184}Re)

⟨γ⟩=390 *3* keV

γ_{mode}	γ(keV)	γ(%)†
W L_ℓ	7.387	0.27 *4*
Re L_ℓ	7.604	0.68 *7*
W L_α	8.391	6.2 *8*
Re L_α	8.646	15.7 *10*
W L_η	8.724	0.102 *12*
Re L_η	9.027	0.103 *7*
W L_β	9.768	8.1 *11*
Re L_β	10.187	10.5 *9*
W L_γ	11.428	1.53 *23*
Re L_γ	11.841	1.65 *18*
γ_ϵM1+0.3%E2	55.281 *5*	2.36 *24*
W $K_{\alpha2}$	57.981	8.6 *10*
W $K_{\alpha1}$	59.318	14.8 *18*
Re $K_{\alpha2}$	59.717	14.2 *6*
Re $K_{\alpha1}$	61.141	24.6 *10*
γ_ϵE2	63.698 *10*	0.38 *6*
W $K_{\beta1}$'	67.155	4.9 *6*
Re $K_{\beta1}$'	69.214	8.3 *3*
W $K_{\beta2}$'	69.342	1.28 *16*
Re $K_{\beta2}$'	71.470	2.11 *9*
γ_{IT} M4	83.28 *4*	0.00545 *22*
γ_ϵE1	87.454 *8*	0.244 *11*
γ_ϵM1+28%E2	91.269 *7*	0.260 *12*
γ_{IT} M1+4.6%E2	104.729 *7*	13.3 *4*
γ_ϵE2	111.208 *6*	5.9 *3*
γ_ϵ[E1]	124.056 *10*	0.149 *7*
γ_ϵE2(+<11%M1)	127.871 *11*	0.00059 *17*
γ_ϵ[E1]	151.152 *11*	0.048 *4*

Photons (^{184}Re) (continued)

γ_{mode}	γ(keV)	γ(%)†
γ_ϵM1+22%E2	161.270 *10*	6.64 *13*
γ_{IT} [E5]	188.01 *4*	0.00022 *4*
γ_ϵE1	215.325 *9*	2.84 *8*
γ_ϵE2	216.551 *9*	9.63 *20*
γ_ϵE1+<0.3%M2+E3	226.746 *8*	1.52 *4*
γ_ϵ[E2]	230.56 *1*	0.0053 *6*
γ_ϵE2	252.849 *8*	10.9 *3*
γ_ϵE1+0.09%M2	318.014 *7*	5.87 *10*
γ_ϵ	381.712 *11*	0.064 *8*
γ_ϵE2	384.256 *10*	3.20 *6*
γ_ϵE1+1.5%M2+E3	536.681 *11*	3.37 *6*
γ_ϵE2	539.225 *10*	0.0316 *12*
γ_ϵE2+1.4%M1	641.914 *10*	0.352 *9*
γ_ϵE2+3.1%M1	769.785 *10*	0.241 *12*
γ_ϵE2+0.4%M1	792.073 *9*	3.75 *6*
γ_ϵE1	857.238 *9*	0.167 *4*
γ_ϵE2+0.6%M1	894.762 *10*	2.81 *7*
γ_ϵE2	903.280 *10*	3.81 *6*
γ_ϵE1+2.1%M2+E3	920.936 *10*	8.35 *15*
γ_ϵ[E1]	1018.818 *10*	0.094 *7*
γ_ϵE2	1022.632 *10*	0.182 *10*
γ_ϵE1+0.7%M2	1110.086 *8*	0.579 *18*
γ_ϵ	1173.783 *11*	1.24 *5*
γ_ϵ	1221.293 *10*	0.0211 *15*

† uncert(syst): 2.8% for ϵ, 0.80% for IT

Atomic Electrons (^{184}Re)

⟨e⟩=139.8 *20* keV

e_{bin}(keV)	⟨e⟩(keV)	e(%)
10	1.8	18 *3*
11	5.1	48 *5*
12 - 22	4.7	38 *4*
33	16.6	50.1 *17*
42 - 69	14.3	29.3 *14*
71	8.8	12.4 *6*
73	29.2	40.2 *18*
75 - 80	2.93	3.65 *16*
81	11.8	14.5 *7*
82	0.0042	0.0051 *5*
83	4.37	5.28 *24*
85 - 91	0.123	0.138 *7*
92	12.1	13.2 *5*
93 - 99	2.26	2.40 *6*
100	4.29	4.30 *25*
101	3.76	3.72 *21*
102 - 151	9.85	8.23 *16*
157 - 206	4.10	2.20 *4*
213 - 253	2.27	0.962 *14*
306 - 318	0.380	0.121 *4*
370 - 384	0.145	0.0387 *8*
467 - 470	0.110	0.023 *3*
525 - 572	0.044	0.0081 *5*
630 - 642	0.00554	0.000876 *20*
700 - 723	0.173	0.0240 *6*
758 - 792	0.0474	0.00607 *9*
825 - 857	0.415	0.0495 *23*
883 - 921	0.095	0.0106 *4*
949 - 953	0.0075	0.00079 *4*
1007 - 1041	0.0099	0.00095 *5*
1098 - 1110	0.05	~0.004
1152 - 1174	0.010	~0.0009
1209 - 1221	0.00016	~1 $\times10^{-5}$

Continuous Radiation (^{184}Re)

⟨IB⟩=0.072 keV

E_{bin}(keV)		⟨ ⟩(keV)	(%)
10 - 20	IB	0.00031	0.0028
20 - 40	IB	0.00077	0.0023
40 - 100	IB	0.070	0.119
100 - 182	IB	0.00046	0.00050

$^{184}_{76}$Os(stable)

Δ: -44257 *3* keV
%: 0.02 *1*

$^{184}_{77}$Ir(3.02 *6* h)

Mode: ϵ
Δ: -39540 *250* keV
SpA: 5.64$\times10^{6}$ Ci/g
Prod: ^{175}Lu(^{12}C,3n); ^{185}Re(^{3}He,4n)

Photons (^{184}Ir)

⟨γ⟩=1773 *24* keV

γ_{mode}	γ(keV)	γ(%)†
Os L_ℓ	7.822	0.76 *23*
Os L_α	8.904	17 *5*
Os L_η	9.337	0.27 *7*
Os L_β	10.465	18 *5*
Os L_γ	12.177	3.2 *9*
Os $K_{\alpha2}$	61.485	27 *10*
Os $K_{\alpha1}$	63.000	47 *17*
Os $K_{\beta1}$'	71.313	16 *6*
Os $K_{\beta2}$'	73.643	4.1 *15*
γ [M1]	76.77 *10*	~0.07
γ [M1]	97.43 *8*	0.22 *4*
γ [M1]	114.60 *9*	0.634 *13*
γ E2	119.77 *9*	30.3 *21*
γ	127.5 *3*	0.09 *3*
γ	131.8 *3*	0.18 *3*
γ	153.54 *20*	0.42 *3*
γ	158.23 *20*	0.51 *4*
γ [M1+E2]	163.61 *11*	0.189 *20*
γ	167.78 *20*	0.27 *2*
γ [E2]	174.20 *10*	0.162 *13*
γ [M1+E2]	185.81 *7*	0.95 *19*
γ	197.35 *15*	0.43 *3*
γ [M1+E2]	203.27 *9*	0.169 *20*
γ	208.85 *9*	0.45 *3*
γ [E2]	212.03 *7*	1.86 *13*
γ	219.66 *9*	0.68 *7*
γ	242.32 *20*	0.142 *14*
γ	245.12 *20*	0.182 *13*
γ E2	263.95 *7*	67.5
γ	272.1 *4*	0.12 *3*
γ [E2]	282.25 *9*	0.175 *13*
γ	308.0 *3*	0.101 *20*
γ	337.73 *20*	0.40 *3*
γ [E2]	347.22 *9*	0.27 *2*
γ	348.90 *20*	0.31 *3*
γ	361.08 *25*	0.216 *20*
γ [M1]	364.68 *7*	1.12 *8*
γ	368.00 *20*	0.162 *13*
γ	376.95 *14*	0.189 *20*
γ	378.62 *25*	0.128 *13*
γ	381.67 *15*	0.55 *4*
γ E2	390.37 *7*	25.7 *18*
γ	394.66 *9*	0.55 *5*
γ	400.0 *3*	0.108 *14*
γ [E1]	404.49 *10*	0.38 *3*
γ	406.47 *8*	0.89 *6*

Photons (^{184}Ir) (continued)

γmode	γ(keV)	γ(%)†
γ	410.18 25	0.209 20
γ	412.12 9	0.79 5
γ	419.60 12	0.067 20
γ	420.50 25	0.209 20
γ	427.0 3	0.101 13
γ	431.16 20	0.324 20
γ	444.9 3	0.108 14
γ	449.33 15	0.162 20
γ	464.39 20	0.32 3
γ	482.50 11	0.10 3
γ	484.21 18	0.16 5
γ	488.31 16	0.45 9
γ E1	493.09 7	5.8 4
γ E2	500.70 15	0.93 7
γ E2	502.91 9	2.94 20
γ	522.6 3	0.122 20
γ	530.21 20	0.142 14
γ E1	539.68 7	6.8 5
γ E2?	550.49 7	0.72 14
γ	557.83 12	0.28 8
γ [E2]	559.01 8	0.38 7
γ	563.38 20	0.38 3
γ	566.2 4	0.21 9
γ	567.2 5	~0.13
γ	571.16 20	0.37 3
γ E1	601.14 8	3.23 22
γ	606.22 10	0.51 4
γ	611.21 9	0.78 5
γ [E1]	613.85 8	1.32 10
γ E1	626.52 9	2.44 17
γ [M1+E2]	654.09 8	0.74 5
γ	657.85 20	0.50 4
γ	667.57 20	0.44 5
γ	682.11 10	0.63 5
γ	684.3 3	0.162 20
γ	691.44 11	0.83 6
γ [M1+E2]	697.24 6	1.67 11
γ	716.3 3	0.175 20
γ	726.1 3	0.39 3
γ [E1]	728.45 9	0.19 3
γ	738.1 3	0.155 14
γ	767.47 8	1.15 11
γ	778.29 9	1.07 7
γ	781.8 4	0.155 20
γ	786.93 25	0.53 4
γ	803.4 4	0.081 14
γ M1	815.07 8	0.73 6
γ E2	822.96 8	3.8 3
γ [E1]	825.88 8	1.28 10
γ	832.9 4	0.26 3
γ [M1+E2]	839.01 13	1.44 10
γ E2	841.26 7	7.9 5
γ	857.36 8	0.47 4
γ	868.7 3	0.169 20
γ	887.07 18	0.29 3
γ	896.32 18	0.175 20
γ	905.1 4	0.229 20
γ [E2]	942.73 11	3.6 3
γ [E1]	943.98 9	2.71 22
γ	953.29 9	0.76 8
γ E2	961.19 8	12.4 9
γ	970.75 10	0.29 3
γ	997.1 4	0.196 20
γ	1001.67 18	0.209 20
γ E2	1044.46 8	5.3 4
γ [E1]	1058.58 10	~0.07
γ	1061.92 7	2.4 3
γ	1062.10 11	0.43 9
γ	1066.21 9	1.52 11
γ	1072.6 3	0.202 20
γ	1085.8 4	0.149 20
γ	1096.1 5	0.101 20
γ	1103.42 15	0.90 13
γ E2	1105.21 9	5.3 5
γ M1	1116.84 10	0.88 7
γ	1133.7 4	0.128 20
γ	1138.4 4	0.101 20
γ	1142.22 17	0.72 5
γ	1154.28 17	0.72 5
γ [E1]	1160.15 9	0.81 5
γ	1197.6 3	0.175 20
γ	1205.8 3	0.202 20
γ	1217.2 3	0.27 3
γ	1225.3 3	0.24 3

Photons (^{184}Ir) (continued)

γmode	γ(keV)	γ(%)†
γ [E2]	1229.38 11	1.21 9
γ [E1]	1236.92 6	2.09 15
γ	1247.73 6	2.65 18
γ	1276.0 4	0.128 13
γ	1281.4 5	0.108 14
γ	1291.57 18	0.182 13
γ	1301.50 25	0.24 3
γ	1311.4 3	0.33 5
γ [E2]	1314.19 14	0.24 5
γ	1323.76 9	0.70 7
γ	1325.86 8	0.84 7
γ [E1]	1334.35 7	2.32 16
γ	1361.5 3	0.38 4
γ	1378.7 5	0.22 5
γ	1380.79 12	0.34 7
γ	1396.8 4	0.122 20
γ	1402.1 4	0.074 20
γ	1412.7 3	0.22 3
γ [E1]	1424.09 9	0.223 20
γ	1436.4 4	0.135 20
γ	1452.47 9	0.81 5
γ	1457.86 15	1.37 10
γ	1469.8 4	0.169 20
γ	1493.79 14	0.52 4
γ	1504.69 25	0.36 3
γ	1514.9 2	0.68 5
γ	1524.0 5	0.067 14
γ	1532.0 5	0.081 14
γ	1544.6 3	0.46 3
γ	1550.63 25	0.63 5
γ	1570.2 5	0.155 20
γ [M1+E2]	1578.13 14	0.49 4
γ	1607.67 25	0.51 4
γ	1625.92 20	0.86 6
γ	1635.5 3	0.32 4
γ (E2)	1672.43 9	3.7 3
γ [E2]	1697.90 16	0.40 3
γ	1718.3 5	0.095 14
γ	1746.2 6	0.17 4
γ	1774.5 7	0.088 20
γ	1818.0 5	0.24 4
γ	1849.7 5	0.34 3
γ	1861.2 6	0.095 20
γ	1895.6 6	0.236 20
γ	1899.8 6	0.149 20
γ	1914.6 6	0.182 20
γ	1930.4 5	0.088 14
γ	1940.78 17	0.128 13
γ	1945.65 18	0.243 20
γ	1962.3 6	0.135 14
γ	1992.7 6	0.108 14
γ	2005.3 7	0.149 20
γ	2015.20 9	0.169 20
γ	2028.5 7	0.142 14
γ (E2)	2062.79 8	4.5 4
γ	2080.1 6	0.22 3
γ	2134.3 6	0.59 5
γ	2166.4 6	0.28 3
γ	2243.0 6	0.95 7
γ	2257.1 6	0.209 20
γ	2336.02 18	0.088 14
γ	2373.5 6	0.149 20
γ	2399.4 7	0.22 3
γ	2469.8 6	0.149 14
γ	2478.7 6	0.162 13
γ	2665.3 5	0.236 20
γ	3178.3 12	0.155 20

† 1.0% uncert(syst)

Atomic Electrons (^{184}Ir)

⟨e⟩=112 3 keV

e_{bin}(keV)	⟨e⟩(keV)	e(%)
3	0.02	~0.7
11	5.0	46 14
12	3.6	29 8
13 - 41	1.8	8.6 21
46	8.1	17.6 13
48 - 97	3.4	5.5 8
100 - 104	0.384	0.378 15
107	22.3	20.8 15
109	16.7	15.4 11
112 - 115	0.7	~0.6
117	10.8	9.2 7
118	0.095	0.081 6
119	2.8	2.33 23
120 - 168	2.5	1.8 3
171 - 186	0.51	0.29 11
190	10.60	5.58 13
191 - 240	0.93	0.46 6
242 - 251	1.76	0.702 16
252	3.25	1.29 3
253 - 297	5.00	1.91 5
303 - 352	3.4	1.06 9
353 - 402	1.63	0.430 25
403 - 452	0.64	0.152 17
454 - 503	0.60	0.124 14
510 - 559	0.40	0.074 14
560 - 609	0.29	0.048 14
610 - 658	0.36	0.058 17
664 - 714	0.32	0.046 14
715 - 764	0.329	0.044 3
765 - 814	0.61	0.079 10
815 - 858	0.168	0.0202 21
866 - 905	0.74	0.084 6
923 - 971	0.394	0.0412 22
984 - 1034	0.50	0.050 12
1041 - 1090	0.22	0.021 4
1091 - 1140	0.085	0.0077 8
1141 - 1187	0.20	0.017 6
1193 - 1240	0.083	0.0067 14
1245 - 1292	0.11	0.0087 23
1298 - 1339	0.056	0.0043 10
1349 - 1397	0.09	0.0067 25
1399 - 1447	0.064	0.0044 14
1449 - 1548	0.097	0.0065 13
1552 - 1644	0.153	0.0096 12
1659 - 1744	0.035	0.0021 3
1762 - 1859	0.030	0.0016 4
1867 - 1960	0.029	0.0015 3
1980 - 2078	0.136	0.0068 6
2092 - 2183	0.031	0.0015 5
2230 - 2325	0.013	0.00056 14
2333 - 2405	0.0067	0.00028 8
2457 - 2476	0.0010	4.2 13 $\times 10^{-5}$
2591 - 2663	0.0045	0.00017 7
3104 - 3176	0.0024	8 3 $\times 10^{-5}$

Continuous Radiation (^{184}Ir)

⟨β+⟩=155 keV;⟨IB⟩=7.0 keV

E_{bin}(keV)		⟨ ⟩(keV)	(%)
0 - 10	β+	4.15×10^{-8}	4.89×10^{-7}
	IB	0.0056	
10 - 20	β+	2.65×10^{-6}	1.57×10^{-5}
	IB	0.0063	0.046
20 - 40	β+	0.000115	0.000347
	IB	0.0121	0.041
40 - 100	β+	0.0094	0.0118
	IB	0.33	0.53
100 - 300	β+	0.72	0.317
	IB	0.134	0.074
300 - 600	β+	6.4	1.36
	IB	0.31	0.069

Continuous Radiation (^{184}Ir) (continued)

E_{bin}(keV)		⟨ ⟩(keV)	(%)
600 - 1300	β+	48.6	5.07
	IB	1.60	0.168
1300 - 2500	β+	89	5.05
	IB	3.5	0.19
2500 - 4336	β+	10.5	0.388
	IB	1.17	0.040
	Σβ+		12.2

$^{184}_{78}$Pt(17.3 *2* min)

Mode: ϵ(~99.999 %), α(~0.001 %)
Δ: -37330 *200* keV syst
SpA: 5.90×10^{7} Ci/g
Prod: ^{174}Yb(^{16}O,6n); ^{193}Ir(p,10n)

Alpha Particles (^{184}Pt)

α(keV)
4490 *15*

Photons (^{184}Pt)

γ_{mode}	γ(keV)	γ(rel)
γ	85.1 *5*	~2
γ	92.6 *3*	16.0 *20*
γ	117.0 *4*	7.5 *7*
γ	134.9 *6*	~2
γ	139.5 *3*	5.6 *5*
γ	144.5 *4*	4.1 *10*
γ	149.5 *6*	4.5 *10*
γ	154.9 *3*	100 *5*
γ	161.6 *4*	6.5 *6*
γ	169.9 *5*	2.9 *10*
γ	176.0 *4*	2.1 *7*
γM2	182.9 *4*	8.3 *8*
γM1	191.9 *3*	94 *9*
γ	209.3 *4*	6.2 *12*
γ	211.6 *5*	5.4 *10*
γ	216.5 *3*	17 *4*
γ	225.7 *4*	6.3 *15*
γ	230.6 *4*	7.4 *10*
γ	236.8 *5*	6.1 *20*
γ	278.5 *9*	3.5 *10*
γ	315.2 *9*	2.0 *6*
γ	328.9 *6*	4.0 *15*
γ	336.5 *9*	3.0 *10*
γ	394.6 *3*	16 *3*
γ	415.9 *9*	3.5 *10*
γ	471.5 *10*	~4
γ	499.3 *10*	5.0 *20*
γ	532.1 *10*	5.0 *20*
γE1	548.4 *3*	77 *7*
γ	610.7 *9*	12.0 *20*
γ	632 *9*	~6
γ	709.6 *15*	6.0 *20*
γ	731.2 *4*	43 *5*

$^{184}_{79}$Au(53.0 *14* s)

Mode: ϵ(99.978 %), α(0.022 %)
Δ: -30240 *310* keV syst
SpA: 1.15×10^{9} Ci/g
Prod: daughter ^{184}Hg

Alpha Particles (^{184}Au)

⟨α⟩=1.127 *4* keV

α(keV)	α(%)
4990	~0.0020
5066 *20*	~0.0031
5108 *15*	~0.006
5172 *15*	0.011 *3*

Photons (^{184}Au)

γ_{mode}	γ(keV)	γ(rel)
γ	111.90 *20*	0.80 *13*
γ[M1+E2]	133.61 *17*	0.87 *13*
γE2	162.82 *8*	100
γ	221.98 *15*	4.1 *6*
γ	229.2 *3*	0.64 *15*
γ	251.41 *20*	1.10 *20*
γE2	272.83 *8*	80 *8*
γM1,M2	291.9 *4*	0.73 *10*
γ	311.80 *20*	0.61 *10*
γ	315.0 *3*	0.90 *20*
γE2	328.94 *16*	4.5 *6*
γ[E2]	352.01 *15*	1.10 *15*
γE2	362.28 *9*	35 *4*
γ	366.76 *19*	~1.0
γ(M1+E2)	378.82 *15*	2.8 *6*
γE2	390.57 *17*	1.7 *3*
γ[E2]	408.12 *13*	1.3 *4*
γ	424.01 *20*	1.00 *20* ?
γE2	432.1 *3*	3.8 *5*
γ	434.87 *18*	3.6 *15*
γ	441.10 *20*	0.64 *9*
γ	479.41 *20*	0.63 *9*
γE0,<3.8%M1,E2	485.73 *8*	11.9 *15*
γE0+E2	524.35 *18*	0.80 *15*
γ	530.3 *3*	~0.40
γM1+E2	585.79 *16*	2.3 *3*
γM1(+E2)?	591.72 *14*	6.5 *8*
γ	600.1 *3*	0.50 *7*
γ	609.5 *3*	1.72 *20*
γ	611.7 *3*	1.35 *15*
γ	627.3 *4*	0.60 *8*
γ	631.1 *4*	1.40 *18*
γ	634.9 *4*	0.90 *13*
γE2	648.55 *11*	6.1 *7*
γ	652.20 *20*	0.85 *10*
γE2	664.31 *14*	3.9 *5*
γ	672.70 *20*	0.70 *10*
γE2+E0	681.1 *21*	1.6 *2*
γ	691.00 *20*	1.10 *10*
γ	700.81 *20*	0.50 *5*
γ	752.81 *20*	1.00 *10*
γE2+<5.9%M1	776.93 *13*	13.2 *15*
γ	783.41 *20*	1.80 *20*
γE0+E2	798.69 *14*	3.4 *4*
γ	806.50 *20*	0.50 *7*
γ	811.0 *4*	0.80 *15*
γ	821.97 *17*	1.30 *15*
γ	826.50 *20*	1.60 *20*
γM1(+E2+E0)?	831.01 *22*	4.3 *5*
γE2	843.77 *12*	10.7 *12*
γM1(+E2)?	864.55 *14*	2.5 *3*
γ	867.94 *17*	1.30 *20*
γE2	870.86 *17*	7.5 *8*
γ	891.9 *3*	0.60 *8*
γ	898.8 *3*	0.70 *9*
γ	917.9 *3*	0.50 *6*
γ	923.21 *22*	1.90 *20*
γ	932.25 *19*	1.05 *10*
γ	938.8 *3*	1.50 *15*
γ	949.6 *3*	0.60 *7*
γ	962.3 *3*	0.90 *10*
γ	981.6 *3*	0.40 *5*
γ	996.4 *3*	0.65 *8*
γ	1001.2 *3*	1.90 *20*
γE1,E2	1009.4 *3*	5.1 *6*
γ	1026.59 *14*	2.6 *3*
γ	1033.8 *3*	0.70 *8*
γE2	1071.52 *15*	4.7 *6*

Photons (^{184}Au) (continued)

γ_{mode}	γ(keV)	γ(rel)
γ	1073.8 *3*	1.20 *20*
γ	1084.6 *3*	1.00 *15*
γ	1089.92 *16*	3.7 *4*
γ	1100.10 *25*	1.15 *20*
γ	1155.5 *3*	0.95 *15*
γ	1161.5 *3*	0.80 *10*
γ	1168.1 *3*	1.20 *20*
γ[E2]	1172.90 *19*	0.70 *20*
γ	1229.5 *3*	0.80 *15*
γ	1239.5 *3*	1.05 *15*
γ	1245.65 *19*	4.6 *6*
γ	1274.3 *3*	0.50 *6*
γ	1290.7 *3*	0.90 *10*
γ	1293.0 *3*	0.60 *7*
γ	1304.8 *3*	1.20 *20*
γ	1308.5 *3*	2.8 *3*
γ	1322.7 *3*	0.55 *6*
γ	1356.8 *3*	0.90 *15*
γ	1362.7 *3*	0.95 *15*
γ	1390.43 *18*	0.70 *10* ?
γ	1397.41 *22*	3.4 *4* ?
γ	1416.8 *3*	1.9 *3*
γ	1448.7 *3*	0.70 *10*
γ	1459.2 *3*	1.10 *20*
γ	1505.0 *3*	0.60 *8*
γ	1519.3 *3*	1.60 *20*
γ	1524.79 *19*	2.3 *3*
γ	1532.24 *15*	2.0 *3* ?
γ	1545.8 *3*	0.55 *7*
γ	1551.1 *3*	1.9 *3*
γ	1576.1 *3*	0.60 *8*
γ	1610.8 *3*	0.50 *7*
γ	1614.3 *3*	0.60 *7*
γ	1644.2 *4*	0.70 *8*
γ	1663.4 *4*	1.40 *20*
γ	1691.2 *4*	0.95 *15*
γ	1698.3 *4*	1.9 *3*
γ	1713.6 *4*	2.8 *3*
γ	1723.2 *4*	1.10 *20*
γ	1739.2 *4*	0.90 *20*
γ	1754.22 *14*	6.4 *8*
γ	1805.1 *4*	1.35 *20*
γ	1814.0 *3*	5.3 *7*
γ	1848.7 *4*	1.50 *20*
γ	1982.3 *5*	0.80 *10*
γ	1989.2 *5*	1.80 *25*
γ	2039.3 *5*	0.50 *6*
γ	2117.6 *5*	1.80 *25*
γ	2196.10 *22*	2.8 *4*
γ	2202.1 *3*	1.50 *20*
γ	2468.0 *5*	0.60 *7*
γ	2474.9 *3*	3.9 *5*
γ	2490.7 *3*	2.0 *3* ?

$^{184}_{80}$Hg(30.6 *3* s)

Mode: ϵ(98.75 *20* %), α(1.25 *20* %)
Δ: -26260 *280* keV syst
SpA: 1.981×10^{9} Ci/g
Prod: protons on Pb

Alpha Particles (^{184}Hg)

⟨α⟩=69.46 keV

α(keV)	α(%)
5380 *15*	0.005 *1*
5535 *15*	1.25

Photons (^{184}Hg)

γ_{mode}	γ(keV)	γ(rel)
γ_ϵ	91.9 *3*	4.7 *8*
γ_ϵ	126.73 *24*	1.4 *5*
γ_ϵ	126.9 *3*	3.4 *7*
γ_ϵ	142.5 *3*	1.9 *3*
γ_ϵ	156.6 *3*	91 *9*
γ_ϵ	159.5 *3*	1.0 *3*
γ_ϵM1	159.55 *22*	6 *1*
γ_ϵ	170.6 *3*	2.1 *3*
γ_ϵ	236.7 *3*	100
γ_ϵ	259.4 *3*	8.4 *10*
γ_ϵ	262.91 *23*	6.7 *8*
γ_ϵ	295.73 *24*	16 *2*
γ_ϵ	422.46 *21*	5.9 *7*

$^{184}_{81}$Tl(11 *1* s)

Mode: ε(97.9 *7* %), α(2.1 *7* %)
Δ: -17070 *430* keV syst
SpA: 5.4×10^9 Ci/g
Prod: ^{180}W(^{14}N,10n)

Alpha Particles (^{184}Tl)

α(keV)
5988 *5*
6162 *5*

Photons (^{184}Tl)

γ_{mode}	γ(keV)	γ(rel)
γ_ϵ[E2]	118.89 *24*	
γ_ϵ[E2]	159.34 *25*	<3
γ_ϵ[E2]	286.76 *24*	39 *4*
γ_ϵ[E2]	339.9 *3*	25 *3*
γ_ϵ[E2]	366.54 *23*	100
γ_ϵ[E2]	418.8 *3*	9 *1*
γ_ϵ[E2]	534.41 *23*	16.8 *20*
γ_ϵ[E2]	554.24 *24*	5.4 *8*
γ_ϵ	608.31 *25*	<11
γ_ϵ	616.84 *25*	<8
γ_ϵ[E2]	722.11 *24*	3.3 *5*

A = 185

NDS **33**, 557 (1981)

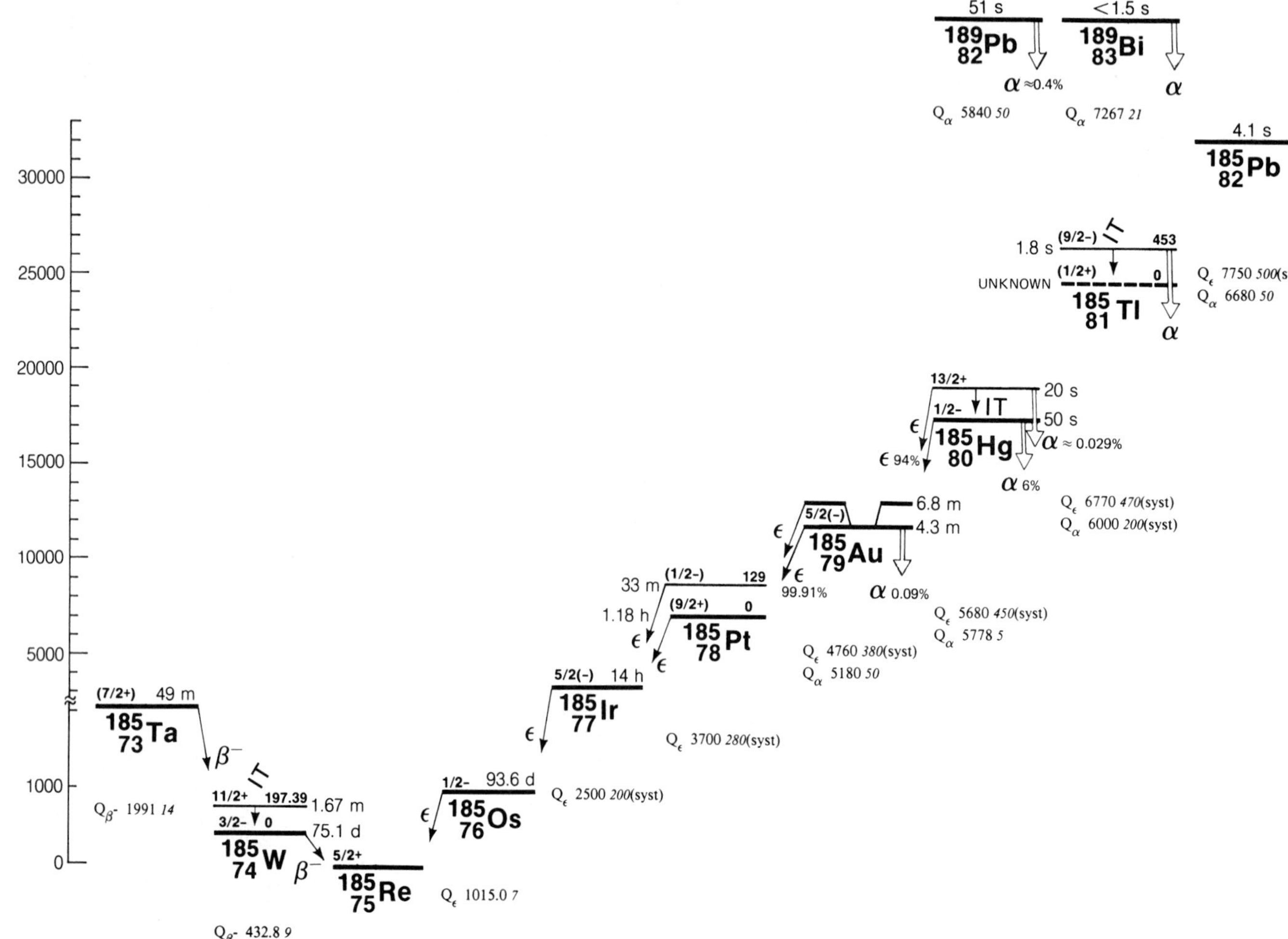

$^{185}_{73}$Ta(49 *2* min)

Mode: β-
Δ: -41403 *14* keV
SpA: 2.07×10^{7} Ci/g
Prod: ^{186}W(γ,p); ^{186}W(n,pn)

Photons (^{185}Ta)

⟨γ⟩=144 *3* keV

γ_{mode}	γ(keV)	γ(%)†
W L_ℓ	7.387	0.46 *11*
W L_α	8.391	10.8 *23*
W L_η	8.724	0.19 *4*
W L_β	9.769	13 *3*
W L_γ	11.384	2.4 *5*
γ (M1+2.3%E2)	23.546 *22*	0.13 *4*
γ E2	42.30 *3*	0.040 *15*
W $K_{\alpha2}$	57.981	8.5 *8*
W $K_{\alpha1}$	59.318	14.7 *14*
γ E2+46%M1	65.843 *18*	3.7 *7*
W $K_{\beta1}$'	67.155	4.9 *5*
W $K_{\beta2}$'	69.342	1.27 *12*
γ M1+11%E2	69.56 *7*	2.0 *3*
γ [M1+E2]	69.74 *4*	~0.031
γ [M1+E2]	93.28 *4*	0.020 *9*
γ [M1+E2]	94.59 *4*	0.018 *4* ?
γ E2+39%M1	107.839 *17*	2.71 *12*
γ [M1+E2]	122.032 *23*	0.018 *4*
γ (M1+35%E2)	147.08 *10*	1.13 *5*
γ [E2]	149.95 *8*	0.12 *3* ?
γ [E2]	164.328 *18*	0.104 *21*
γ E2	173.681 *16*	22.0 *5*
γ M1+17%E2	177.39 *7*	25.6 *5*
γ [M1+E2]	187.874 *17*	0.14 *3*
γ [E2]	243.24 *7*	3.7 *3*
γ [M1]	394.3 *4*	0.8 *3*
γ [M1]	541.4 *4*	0.8 *3*
γ	580.3 *10*	0.54 *18*
γ	588.5 *10*	0.8 *3*
γ	913 *3*	0.12 *4*
γ [E2]	965.2 *17*	0.09 *3*
γ [M1]	992.6 *17*	0.23 *8*
γ [E2]	1058.5 *17*	0.24 *8*
γ	1122 *3*	0.0020 *8*
γ	1138 *3*	0.008 *3*
γ	1147 *3*	0.008 *3*

† 10% uncert(syst)

Atomic Electrons (^{185}Ta)

⟨e⟩=98 *6* keV

e_{bin}(keV)	⟨e⟩(keV)	e(%)
5	0.00453019	0.09
10	3.3	32 *8*
11	0.61	5.3 *16*
12	4.0	34 *10*
13 - 53	6.3	22 *5*
54	11.4	21 *7*
55	0.086	0.157 *21*
56	9.6	17 *6*
57 - 60	3.9	6.7 *10*
63	3.2	5.1 *16*
64	2.8	4.5 *15*
65 - 98	7.3	9.3 *11*
104	5.64	5.41 *16*
105 - 107	1.07	1.01 *16*
108	19.6	18 *3*
110 - 154	0.79	0.58 *12*
162	4.47	2.76 *8*
163	2.57	1.57 *5*
164	0.0030	0.0018 *4*

Atomic Electrons (^{185}Ta) (continued)

e_{bin}(keV)	⟨e⟩(keV)	e(%)
165	4.3	2.6 *4*
166 - 188	6.4	3.7 *5*
231 - 243	0.559	0.239 *11*
382 - 394	0.054	0.014 *4*
472 - 519	0.27	0.055 *18*
529 - 578	0.058	0.011 *3*
580 - 588	0.005	~0.0008
896 - 923	0.023	0.0025 *7*
953 - 993	0.013	0.0013 *3*
1046 - 1077	0.0025	0.00024 *5*
1110 - 1147	0.00014	1.3 *5* ×10^{-5}

Continuous Radiation (^{185}Ta)

⟨β-⟩=639 keV;⟨IB⟩=1.07 keV

E_{bin}(keV)		⟨ ⟩(keV)	(%)
0 - 10	β-	0.0300	0.60
	IB	0.027	
10 - 20	β-	0.091	0.60
	IB	0.026	0.18
20 - 40	β-	0.373	1.24
	IB	0.050	0.17
40 - 100	β-	2.80	3.98
	IB	0.137	0.21
100 - 300	β-	31.6	15.6
	IB	0.34	0.20
300 - 600	β-	120	26.6
	IB	0.29	0.071
600 - 1300	β-	407	45.4
	IB	0.19	0.024
1300 - 1817	β-	78	5.4
	IB	0.0048	0.00035

$^{185}_{74}$W (75.1 *3* d)

Mode: β-
Δ: -43393 *3* keV
SpA: 9398 Ci/g
Prod: ^{184}W(n,γ); ^{187}Re(d,α)

Photons (^{185}W)

⟨γ⟩=0.050 *6* keV

γ_{mode}	γ(keV)	γ(%)
Re L_ℓ	7.604	0.00023 *5*
Re L_α	8.646	0.0053 *11*
Re L_η	9.027	6.9 *15* ×10^{-5}
Re L_β	10.129	0.0053 *11*
Re L_γ	11.796	0.00092 *21*
Re $K_{\alpha2}$	59.717	0.0117 *24*
Re $K_{\alpha1}$	61.141	0.020 *4*
Re $K_{\beta1}$'	69.214	0.0068 *14*
Re $K_{\beta2}$'	71.470	0.0017 *4*
γ M1+3.5%E2	125.354 *3*	0.019 *4*

Atomic Electrons (^{185}W)

⟨e⟩=0.038 *5* keV

e_{bin}(keV)	⟨e⟩(keV)	e(%)
11	0.0016	0.015 *4*
12	0.00092	0.0077 *17*
13	0.00033	0.0027 *6*
47	0.00021	0.00044 *10*
48	8.5 ×10^{-6}	1.8 *4* ×10^{-5}
49	0.00024	0.00050 *11*
51	7.5 ×10^{-5}	0.00015 *3*
54	0.023	0.042 *9*
56	7.0 ×10^{-5}	0.00012 *3*
57	8.9 ×10^{-5}	0.00016 *4*
58	5.8 ×10^{-5}	0.000100 *22*
59	8.9 ×10^{-5}	0.00015 *3*
60	2.3 ×10^{-6}	3.9 *9* ×10^{-6}
61	2.6 ×10^{-5}	4.3 *10* ×10^{-5}
66	1.5 ×10^{-5}	2.3 *5* ×10^{-5}
67	1.3 ×10^{-5}	2.0 *4* ×10^{-5}
68	6.3 ×10^{-6}	9.2 *21* ×10^{-6}
69	6.3 ×10^{-6}	9.1 *21* ×10^{-6}
113	0.0079	0.0070 *14*
115	0.00035	0.00031 *11*
122	0.0017	0.0014 *3*
123	0.00036	0.00029 *8*
125	0.00063	0.00050 *11*

Continuous Radiation (^{185}W)

⟨β-⟩=127 keV;⟨IB⟩=0.057 keV

E_{bin}(keV)		⟨ ⟩(keV)	(%)
0 - 10	β-	0.252	5.04
	IB	0.0066	
10 - 20	β-	0.74	4.93
	IB	0.0059	0.041
20 - 40	β-	2.87	9.6
	IB	0.0099	0.035
40 - 100	β-	18.0	26.1
	IB	0.019	0.031
100 - 300	β-	89	49.6
	IB	0.0152	0.0101
300 - 433	β-	15.8	4.73
	IB	0.00025	8.1 ×10^{-5}

$^{185}_{74}$W (1.67 *3* min)

Mode: IT
Δ: -43196 *3* keV
SpA: 6.07×10^{8} Ci/g
Prod: ^{184}W(n,γ); ^{186}W(γ,n); ^{186}W(n,2n)

Photons (^{185}W)

⟨γ⟩=27.5 *6* keV

γ_{mode}	γ(keV)	γ(%)†
γ (M1+2.3%E2)	23.546 *22*	0.13 *5*
γ E2	42.30 *3*	0.062 *24*
W $K_{\alpha2}$	57.981	2.37 *15*
W $K_{\alpha1}$	59.318	4.1 *3*
γ E2+46%M1	65.843 *18*	5.8 *3*
W $K_{\beta1}$'	67.155	1.36 *9*
W $K_{\beta2}$'	69.342	0.353 *23*
γ [M1+E2]	69.74 *4*	~0.046
γ [M1+E2]	93.28 *4*	0.026 *9*
γ [M1+E2]	94.59 *4*	0.104 *9*

Photons (^{185}W)
(continued)

γmode	γ(keV)	γ(%)†
γ E2+39%M1	107.839 17	0.405 19
γ [M1+E2]	122.032 23	0.100 9
γ E3	131.545 20	4.3 3
γ [E2]	164.328 18	0.58 3
γ E2	173.681 16	3.29 11
γ [M1+E2]	187.874 17	0.81 4

† 5.5% uncert(syst)

Atomic Electrons (^{185}W)
⟨e⟩=165 8 keV

e_{bin}(keV)	⟨e⟩(keV)	e(%)
12	6.7	57 12
13 - 53	6.8	26 6
54	17.8	33 9
55	0.024	0.044 5
56	14.8	27 8
57 - 62	3.9	6.4 5
63	5.1	8.0 21
64 - 112	9.2	13.2 22
118 - 119	2.1	1.77 22
120	41.3	34.4 22
121	29.2	24.0 15
122	0.004	~0.003
129	20.2	15.7 10
130	0.41	0.316 20
131	5.6	4.3 3
132 - 178	2.19	1.38 7
185 - 188	0.06	0.031 15

$^{185}_{75}$Re(stable)

Δ: -43826 3 keV
%: 37.40 2

$^{185}_{76}$Os(93.6 5 d)

Mode: ϵ
Δ: -42811 3 keV
SpA: 7541 Ci/g
Prod: ^{185}Re(d,2n); ^{184}Os(n,γ); ^{185}Re(p,n)

Photons (^{185}Os)
⟨γ⟩=713 7 keV

γmode	γ(keV)	γ(%)†
Re L_ℓ	7.604	0.44 4
Re L_α	8.646	10.0 6
Re L_η	9.027	0.129 8
Re L_β	10.122	10.8 9
Re L_γ	11.832	2.00 21
Re $K_{\alpha2}$	59.717	20.9 7
Re $K_{\alpha1}$	61.141	36.2 13
Re $K_{\beta1}'$	69.214	12.2 4
γ M1+1.0%E2	71.3081 19	~0.25
Re $K_{\beta2}'$	71.470	3.11 11
γ (E2)	122.81 6	0.024 8 ?
γ M1+3.5%E2	125.354 3	0.349 7
γ M1	162.846 5	0.561 11
γ M1	234.155 6	0.416 8
γ E2	592.065 6	1.33 3

Photons (^{185}Os)
(continued)

γmode	γ(keV)	γ(%)†
γ E2	646.110 6	81
γ E2+28%M1	717.418 6	4.12 8
γ [E2]	749.456 13	0.0032 4
γ [M1+E2]	768.92 6	0.0036 3
γ [E2]	805.702 20	~4×10^{-5}
γ M1	874.809 13	6.61 13
γ E2	880.264 8	5.00 10
γ M1	931.055 20	0.0494 16

† 1.2% uncert(syst)

Atomic Electrons (^{185}Os)
⟨e⟩=16.6 4 keV

e_{bin}(keV)	⟨e⟩(keV)	e(%)
11	2.9	27 3
12	1.63	13.6 14
13	1.00	8.0 9
47 - 58	1.77	3.46 14
59	0.45	0.8 3
60 - 71	0.24	0.37 7
91	0.571	0.626 18
110 - 152	0.393	0.311 9
160 - 163	0.323	0.199 5
222 - 234	0.0780	0.0348 8
520	0.0792	0.0152 4
574	4.42	0.769 18
580 - 592	0.0254	0.00436 9
634	0.911	0.144 3
636 - 678	0.72	0.112 8
697 - 747	0.075	0.0107 12
748 - 795	7.4 ×10^{-5}	10 3 ×10^{-6}
803	0.679	0.0846 24
804 - 809	0.203	0.0251 7
859 - 880	0.200	0.0231 4
919 - 931	0.00098	0.000107 3

Continuous Radiation (^{185}Os)
⟨IB⟩=0.34 keV

E_{bin}(keV)		⟨ ⟩(keV)	(%)
10 - 20	IB	0.00138	0.0125
20 - 40	IB	0.0026	0.0077
40 - 100	IB	0.32	0.53
100 - 300	IB	0.0111	0.0078
300 - 600	IB	0.00109	0.00026
600 - 1015	IB	0.00058	8.3×10^{-5}

$^{185}_{77}$Ir(14.0 9 h)

Mode: ϵ
Δ: -40310 200 keV syst
SpA: 1.21×10^{6} Ci/g
Prod: ^{185}Re(α,4n); ^{186}Os(p,2n)

Photons (^{185}Ir)

γmode	γ(keV)	γ(rel)
γ M1+1.7%E2	24.20 4	0.14 7
γ (M1)	30.43 7	~0.16
γ (M1)	33.85 5	~0.09
γ M1+0.7%E2	37.41 8	26 3
γ M1+3.1%E2	60.01 7	~43
γ M1+6.9%E2	90.44 8	9.8 10
γ M1+41%E2	94.51 9	3.0 3
γ E2	97.42 10	32 6
γ M1+6.2%E2	100.74 8	18.3 18
γ M1+20%E2	119.68 9	0.40 4
γ [M1]	124.94 8	~0.26
γ M1+~14%E2	126.9 2	6 3
γ M1	127.85 10	~5
γ [E1]	142.10 16	0.32 5
γ M1+35%E2	153.53 8	15.2 15
γ M1+20%E2	158.20 16	18.1 18
γ E2	160.75 9	13.2 14
γ	177.3 5	0.89 18
γ M1	184.95 8	6.5 6
γ	188.7 2	0.89 18
γ	189.4 5	1.6 3
γ	193.6 5	0.53 11
γ	203.3 5	0.53 11
γ	205.8 5	0.53 11
γ	217.5 5	0.95 19
γ E2	220.42 9	8.8 9
γ (E2)	222.36 11	12.6 13
γ E2	223.84 9	15.9 3
γ	228.9 5	0.63 13
γ	248.8 5	0.84 17
γ M1+8.9%E2	254.27 9	100 10
γ	266.5 4	
γ	276.5 5	1.4 3
γ	282.1 5	0.74 15
γ	283.1 5	0.74 15
γ E1	300.30 16	4.9 5
γ	307.1 5	
γ	308.6 5	0.63 13
γ [E2]	314.28 9	6.8 7
γ M1	321.4 2	2.1 4
γ (M1+E2)	339.2 5	1.05 21
γ	346.8 5	0.47 10
γ	352.4 6	
γ	358.4 5	0.53 11
γ	367.3 5	0.63 13
γ	370.6 5	0.63 13
γ (M1)	377.65 13	1.58 16
γ	382.2 5	0.53 11
γ	394.4 5	0.58 12
γ (M1)	398.5 2	2.4 5
γ	402.6 5	1.21 24
γ (M1)	406.9 2	3.4 4
γ (M1+27%E2)	419.0 2	2.0 4
γ	426.8 5	1.05 21
γ (M1+E2)	431.4 7	2.8 6
γ	446.1 5	0.47 10
γ	449.8 5	0.79 16
γ	453.0 5	0.37 7
γ	464.9 5	0.79 16
γ	486.4 5	0.63 13
γ	489.0 5	1.5 3
γ	501.0 5	0.95 19
γ (M1)	506.98 13	5.1 5
γ (E2+3.2%M1)	513.1 2	6.9 7
γ	517.4 5	1.6 3
γ	533.0 5	0.58 12
γ (M1+E2)	539.2 2	9.8 10
γ	544.9 5	1.05 21
γ	550.4 2	8.8 9
γ	576.1 5	1.05 21
γ	590.0 5	1.16 23
γ	601.50 14	1.05 21
γ	627.2 5	4.2 8
γ	638.5 5	2.2 4
γ	642.8 5	2.2 4
γ	646.2 5	9.3 19
γ	666.1 2	3.8 8
γ	670.9 5	3.2 6
γ	681.5 5	1.7 3
γ	691.94 14	3.2 6
γ	695.4 5	1.11 22
γ	710.6 5	0.89 18
γ	726.5 5	0.58 12

Photons (^{185}Ir) (continued)

γ_{mode}	γ(keV)	γ(rel)
γ	743.2 *5*	1.9 *4*
γ	745.7 *5*	4.3 *9*
γ	759.2 *5*	1.05 *21*
γ	761.2 *5*	1.16 *23*
γ	785.4 *5*	0.74 *15*
γ	796.4 *5*	1.05 *21*
γ	798.2 *5*	1.05 *21*
γ	807.3 *5*	5.6 *11*
γ	817.1 *5*	0.37 *7*
γ	823.9 *5*	0.26 *5*
γ	828.3 *5*	1.05 *21*
γ	850.8 *5*	1.3 *3*
γ	855.7 *5*	0.58 *12*
γ	860.6 *5*	0.26 *5*
γ	870.8 *5*	0.47 *10*
γ	913.9 *5*	5.6 *11*
γ	925.0 *5*	2.3 *5*
γ	954.9 *5*	0.79 *16*
γ	959.0 *5*	1.7 *3*
γ	966.2 *5*	1.3 *3*
γ	978.2 *5*	0.84 *17*
γ	997.2 *5*	0.89 *18*
γ	1016.8 *5*	0.26 *5*
γ	1038.7 *5*	
γ	1064.2 *5*	1.5 *3*
γ	1076.2 *5*	0.63 *13*
γ	1079.3 *5*	0.74 *15*
γ	1094.4 *5*	0.63 *13*
γ	1101.8 *5*	0.47 *10*
γ	1127.9 *5*	1.05 *21*
γ	1153.9 *5*	0.53 *11*
γ	1157.5 *5*	0.53 *11*
γ	1165.9 *5*	0.53 *11*
γ	1178.18 *25*	0.53 *11*
γ	1190.4 *5*	1.05 *21*
γ	1205.3 *5*	0.21 *4*
γ	1221.2 *5*	0.63 *13*
γ	1226.1 *5*	1.21 *24*
γ	1230.6 *5*	0.95 *19*
γ	1237.0 *5*	0.63 *13*
γ	1247.2 *5*	0.95 *19*
γ	1255.6 *5*	0.63 *13*
γ	1264.4 *5*	1.6 *3*
γ	1266.6 *5*	1.6 *3*
γ	1299.7 *5*	3.3 *7*
γ	1310.9 *5*	6.5 *13*
γ	1318.2 *5*	2.3 *5*
γ	1325.2 *5*	2.2 *4*
γ	1345.9 *5*	0.95 *19*
γ	1359.1 *5*	1.4 *3*
γ	1361.8 *5*	1.21 *24*
γ	1366.6 *5*	1.3 *3*
γ	1384.6 *5*	1.4 *3*
γ	1390.9 *5*	0.63 *13*
γ	1409.1 *5*	0.84 *17*
γ	1418.08 *23*	4.1 *8*
γ	1441.3 *5*	4.1 *8*
γ	1463.1 *5*	1.21 *24*
γ	1465.8 *5*	1.11 *22*
γ	1478.3 *5*	0.95 *19*
γ	1512.0 *5*	0.26 *5*
γ	1568.3 *5*	1.9 *4*
γ	1571.61 *23*	2.0 *4*
γ	1579.9 *5*	0.74 *15*
γ	1625.9 *5*	1.5 *3*
γ	1641.92 *23*	8.7 *17*
γ	1651.93 *25*	0.37 *7*
γ	1668.0 *3*	27 *6*
γ	1685.16 *23*	2.3 *5*
γ	1698.5 *5*	1.21 *24*
γ	1701.0 *5*	1.5 *3*
γ	1709.37 *23*	2.6 *5*
γ	1732.36 *23*	21 *4*
γ	1738.3 *3*	18 *4*
γ	1757.6 *5*	6.8 *14*
γ	1763.1 *5*	0.74 *15*
γ	1768.7 *3*	0.53 *11* ?
γ	1779.68 *23*	2.8 *6*
γ	1794.2 *5*	0.26 *5*
γ	1805.46 *25*	3.1 *6* ?
γ	1828.8 *3*	76 *15*
γ	1854.7 *5*	0.53 *11*
γ	1870.12 *23*	9.1 *18*
γ	1875.77 *25*	3.4 *7*
γ	1882.5 *5*	1.11 *22*
γ	1893.2 *5*	0.53 *11*
γ	1901.3 *5*	0.26 *5*
γ	1920.3 *5*	0.26 *5*
γ	1924.0 *5*	0.53 *11*
γ	1938.0 *5*	0.26 *5*
γ	1942.3 *5*	0.26 *5*
γ	1948.4 *5*	0.89 *18*
γ	1966.20 *25*	0.74 *15*
γ	1978.4 *5*	0.26 *5*
γ	1982.4 *5*	0.105 *21*
γ	1996.6 *5*	0.26 *5*
γ	2014.4 *5*	0.105 *21*
γ	2026.2 *5*	0.37 *7*
γ	2044.4 *5*	0.74 *15*
γ	2049.7 *5*	2.6 *5*

$^{185}_{78}$Pt(1.18 *4* h)

Mode: ϵ

Δ: -36610 *200* keV syst

SpA: 1.43×10^{7} Ci/g

Prod: descendant ^{185}Hg(50 s); ^{14}N on Ta

Photons (^{185}Pt(1.18 h + 33 min))

γ_{mode}	γ(keV)	γ(rel)
γ[M1+E2]	23.51 *11*	0.7 *2*
γ(M1+4.0%E2)	83.46 *11*	0.65 *15*
γ(M1+4.0%E2)	85.98 *9*	4.0 *5*
γE1	94.34 *7*	1.5 *3*
γM1(+E2)	103.16 *8*	2.0 *4*
γM1+35%E2	105.69 *8*	6.2 *6*
γ(E2+45%M1)	106.97 *8*	3.0 *4*
γ[E2]	109.49 *10*	5.8 *7*
γ(M1)	113.83 *10*	2.4 *5*
γM1+1.7%E2	119.93 *8*	14.7 *3*
γE2	135.37 *7*	80 *10*
γ(M1+33%E2)	137.35 *14*	~4
γ(M1)	140.87 *16*	~2
γ	152.8 *1*	6.1 *10*
γ[M1+E2]	161.59 *13*	1.4 *4*
γ	168.9 *3*	1.4 *4*
γ	187.2 *3*	~0.4
γ(E2)	191.47 *16*	2.2 *4*
γ	195.3 *4*	1.3 *2*
γE1	197.50 *8*	74 *10*
γ[M1+E2]	200.51 *15*	1.0 *2*
γ	202.67 *16*	0.7 *2*
γ(M1)	206.78 *14*	6.5 *10*
γ(M1)	212.66 *8*	12 *2*
γE1	229.70 *7*	100
γ	238.8 *3*	3.4 *5*
γ	243.07 *16*	6 *2*
γ(M1)	251.64 *16*	8 *1*
γ(M1)	253.1 *5*	3 *1*
γM1+13%E2	255.30 *9*	51 *5*
γ(M1)	264.61 *11*	8.4 *15*
γ[E2]	267.28 *14*	1.7 *4*
γ[E2]	278.22 *16*	1.5 *3*
γ(M1+36%E2)	294.31 *9*	6.7 *10*
γ(M1)	298.17 *22*	2.0 *4*
γM1+32%E2	300.16 *10*	7.8 *10*
γ(M1)	307.15 *14*	3.4 *4*
γ(M1)	314.45 *17*	2.0 *4*
γ	326.4 *6*	<2
γ	335.4 *2*	13.5 *20*
γ(M1)	341.70 *16*	3.5 *4*
γ	356.7 *4*	1.7 *4*
γ(M1)	361.68 *16*	0.7 *3*
γ	370.1 *5*	3.3 *4*
γ(M1+21%E2)	384.54 *11*	14.8 *15*
γ	414.5 *6*	
γ	416.2 *6*	
γ[E1]	418.85 *11*	6 *2*
γ(E2+34%M1)	426.32 *17*	3.3 *4*
γ(M1)	434.63 *20*	1.4 *2*
γ[E1]	442.36 *10*	1.5 *3*
γM1+33%E2	460.01 *14*	7 *1*
γM1(+<2.5%E2)	465.12 *13*	25 *5*
γ	573.3 *5*	0.5 *2*
γ	576.9 *5*	0.9 *3*
γM1	585.05 *12*	17 *3*
γ	596.87 *22*	2.4 *8*
γ	612.0 *6*	2.1 *4*
γ	620.6 *6*	2.4 *3*
γ	625.87 *21*	1.8 *4*
γ(M1)	640.88 *20*	7 *1*
γ	689.9 *6*	4.8 *10*
γ	691.22 *21*	2.8 *5* ?
γ	699.1 *6*	3.0 *9*
γ	704.19 *22*	3 *1* ?
γ	706.27 *22*	2.0 *8*
γ	720.5 *2*	20 *4*
γ	726.4 *5*	5 *1*
γ(E2+41%M1)	735.36 *19*	8 *1*
γ(E2+49%M1)	741.78 *18*	3 *1*
γ	745.9 *6*	2.1 *8*
γ	751.3 *5*	1.6 *4*
γ	773.6 *6*	2.5 *5*
γ	784.9 *6*	1.0 *3*
γ	788.3 *3*	1.1 *3*
γ	795.48 *21*	1.4 *4*
γ	801.33 *22*	2.3 *6*
γ	810.0 *8*	1.5 *4*
γ	827.6 *8*	1.3 *4*
γ	837.6 *3*	7.0 *15*
γ	842.7 *6*	~3
γ	868.0 *6*	1.5 *5*
γ(M1)	880.5 *5*	1.2 *3*
γ	891.2 *6*	2.4 *5*
γ(E2)	894.64 *19*	6.6 *10*
γ	900.9 *6*	3.2 *6*
γ	909.4 *8*	1.7 *4*
γ	955.9 *8*	3.0 *6*
γ	962.6 *8*	5 *1*
γ	1039.9 *3*	3.1 *7*
γ	1092.17 *24*	0.9 *3*
γ	1105.8 *3*	1.2 *4*
γ	1140.5 *3*	0.8 *3*
γ	1164.0 *3*	~0.6
γ	1215.7 *8*	2.8 *7*
γ	1247.5 *3*	1.4 *3*
γ	1292.9 *3*	4 *1*
γ	1370.9 *10*	2.3 *8*
γ	1396.0 *3*	4.1 *10*
γ	1418.1 *8*	2.3 *8*
γ	1490.1 *8*	1.5 *5*

$^{185}_{78}$Pt(33.0 *8* min)

Mode: ϵ

Δ: -36481 *200* keV syst

SpA: 3.08×10^{7} Ci/g

Prod: descendant ^{185}Hg(50 s)

see ^{185}Pt(1.18 h) for γ rays

$^{185}_{79}$Au(4.3 *1* min)

Mode: ϵ(99.907 *20* %), α(0.093 *20* %)
Δ: -31850 *320* keV syst
SpA: 2.36×10^8 Ci/g
Prod: protons on Pt; ^{175}Lu(^{16}O,6n); ^{169}Tm(^{22}Ne,6n)

Alpha Particles (^{185}Au)

α(keV)
5067 *5*

$^{185}_{79}$Au(6.8 *3* min)

Mode: ϵ
Δ: -31850 *320* keV syst
SpA: 1.49×10^8 Ci/g
Prod: daughter ^{185}Hg(50 s)

$^{185}_{80}$Hg(50 *2* s)

Mode: ϵ(94.5 *7* %), α(5.5 *7* %)
Δ: -26170 *320* keV syst
SpA: 1.21×10^9 Ci/g
Prod: ^{197}Au(p,13n); ^{197}Au(d,14n); ^{170}Yb(^{20}Ne,5n)

Alpha Particles (^{185}Hg)

⟨α⟩=310.7 *4* keV

α(keV)	α(%)
5569 *5*	0.22 *6*
5653 *5*	5.28 *6*

$^{185}_{80}$Hg(20 *2* s)

Mode: α, ϵ, IT(~0.029 %)
Δ: >-26170 keV syst
SpA: 3.0×10^9 Ci/g
Prod: protons on Pb

Alpha Particles (^{185}Hg)

α(keV)	α(rel)
5371 *10*	100
5408 *10*	~25
5430 ?	

$^{185}_{81}$Tl(1.8 *2* s)

Mode: IT, α
Δ: -18947 *350* keV syst
SpA: 2.8×10^{10} Ci/g
Prod: ^{180}W(^{14}N,9n); ^{182}W(^{14}N,11n)

Alpha Particles (^{185}Tl)

α(keV)
5975 *5*

Photons (^{185}Tl)

γ_{mode}	γ(keV)	γ(rel)
γ_{IT} (E3)	168.8 *5*	100 *4*
γ_{IT} M1+44%E2	284.0 *5*	13.3 *5*

$^{185}_{82}$Pb(4.1 *3* s)

Mode: α
Δ: -11650 *350* keV syst
SpA: 1.37×10^{10} Ci/g
Prod: ^{150}Sm(^{40}Ca,5n)

Alpha Particles (^{185}Pb)

α(keV)	α(rel)
6400 *10*	~72
6480 *20*	~28

A = 186

NDS **13**, 267 (1974)

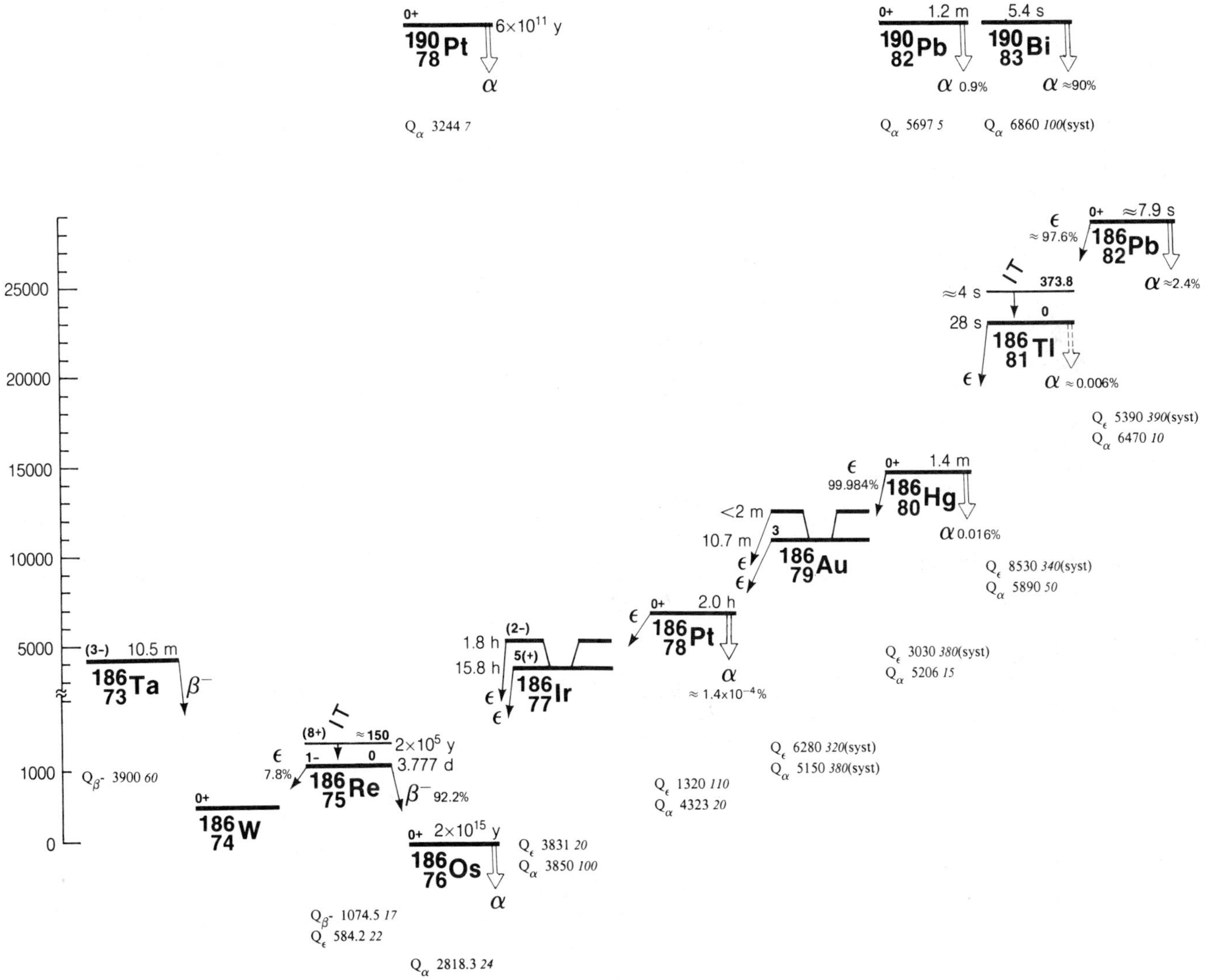

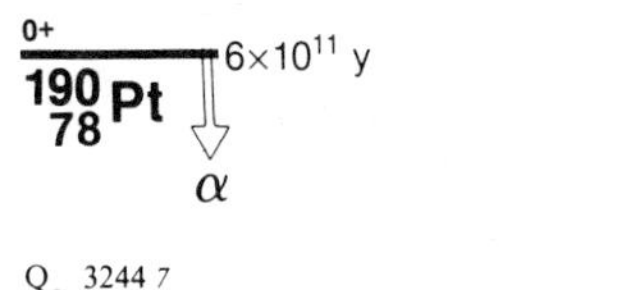

Q_α 3244 *7*

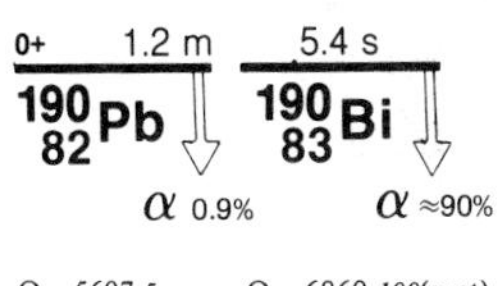

Q_α 5697 *5* Q_α 6860 *100*(syst)

0+ ≈7.9 s $^{186}_{82}$Pb: ϵ ≈ 97.6%; α ≈2.4%

$^{186}_{81}$Tl: 373.8 (≈4 s) IT; 0 (28 s); ϵ; α ≈ 0.006%

Q_ϵ 5390 *390*(syst)
Q_α 6470 *10*

0+ 1.4 m $^{186}_{80}$Hg: ϵ 99.984%; α 0.016%

Q_ϵ 8530 *340*(syst)
Q_α 5890 *50*

$^{186}_{79}$Au: <2 m; 3 10.7 m; ϵ; ϵ

Q_ϵ 3030 *380*(syst)
Q_α 5206 *15*

0+ 2.0 h $^{186}_{78}$Pt: ϵ; α ≈ 1.4×10⁻⁴%

Q_ϵ 6280 *320*(syst)
Q_α 5150 *380*(syst)

$^{186}_{77}$Ir: (2−) 1.8 h; 5(+) 15.8 h; ϵ; ϵ

Q_ϵ 1320 *110*
Q_α 4323 *20*

(3−) 10.5 m $^{186}_{73}$Ta: β⁻

Q_{β^-} 3900 *60*

0+ $^{186}_{74}$W

$^{186}_{75}$Re: (8+) ≈150 2×10⁵ y IT; 1− 0 3.777 d; ϵ 7.8%; β⁻ 92.2%

Q_{β^-} 1074.5 *17*
Q_ϵ 584.2 *22*

0+ 2×10¹⁵ y $^{186}_{76}$Os: α

Q_ϵ 3831 *20*
Q_α 3850 *100*

Q_α 2818.3 *24*

$^{186}_{73}$Ta(10.5 *5* min)

Mode: β-
Δ: -38620 *60* keV
SpA: 9.6×10⁷ Ci/g
Prod: ^{186}W(n,p)

Photons (^{186}Ta)

⟨γ⟩=1563 *39* keV

γ_{mode}	γ(keV)	γ(%)†
W L_ℓ	7.387	0.26 *4*
W L_α	8.391	6.2 *7*
W L_η	8.724	0.104 *12*
W L_β	9.778	6.9 *8*
W L_γ	11.350	1.14 *15*
W $K_{\alpha2}$	57.981	9.6 *8*
W $K_{\alpha1}$	59.318	16.5 *13*
W $K_{\beta1}$'	67.155	5.5 *5*
W $K_{\beta2}$'	69.342	1.43 *12*
γ [E1]	90.73 *13*	2.2 *9*
γ [M1]	92.65 *11*	1.7 *3*
γ E2	122.43 *7*	23 *3*

Photons (^{186}Ta) (continued)

γ_{mode}	γ(keV)	γ(%)†
γ [E1]	183.39 *9*	3.5 *6*
γ [M1]	184.17 *25*	~0.6
γ E1	198.05 *10*	59
γ E1	215.01 *9*	49.9 *10*
γ [M1+E2]	268.88 *16*	0.9 *3*
γ E2	274.30 *10*	8.0 *3*
γ	277.25 *9*	1.5 *3*
γ [E2]	291.7 *3*	4.1 *6*
γ [E2]	294.1 *3*	2.4
γ [E1]	307.66 *9*	11.4 *6*
γ	309.32 *10*	2.7 *3*
γ [M1+E2]	315.74 *18*	1.8 *3*
γ	326.5 *3*	0.9 *3*
γ [M1+E2]	339.1 *3*	~0.6
γ [E2]	340.99 *16*	~0.29
γ	383.9 *5*	~0.6
γ [E2]	402.1 *5*	0.9 *3*
γ E2	412.13 *20*	~0.6
γ	418.03 *15*	14.8 *12*
γ [E2]	440 *1*	0.94 *18*
γ	448.44 *18*	~0.6
γ	456.81 *21*	2.5 *3*
γ [M1+E2]	460.34 *23*	~0.29 ?
γ [M1+E2]	465.27 *17*	1.3 *3*
γ [E2]	488.4 *4*	~0.6
γ	510.68 *17*	44 *3*
γ	541.5 *3*	~0.29
γ [M1+E2]	546.5 *4*	0.59 *12*

Photons (^{186}Ta) (continued)

γ_{mode}	γ(keV)	γ(%)†
γ [E1]	567.50 *24*	4.0 *6*
γ	583.33 *18*	1.30 *18*
γ	596.9 *3*	~0.29
γ	601.41 *16*	~0.6
γ [E2]	609.87 *21*	4.0 *3*
γ M1+E2	615.29 *16*	33.0 *12*
γ	618.24 *17*	~0.6
γ	622.11 *23*	
γ [M1+E2]	635.1 *3*	~0.6
γ	642.1 *5*	~0.41
γ	646.49 *19*	~0.18
γ [E1]	648.65 *17*	~0.18
γ	654.9 *3*	1.8 *6*
γ	703.2 *7*	~0.6
γ [E1]	708.73 *19*	1.2 *3*
γ	725.69 *15*	1.2 *3*
γ E2	737.72 *18*	34.2 *24*
γ [M1+E2]	739.57 *17*	11.8 *12*
γ	745.6 *3*	~0.29
γ [E2]	759.7 *4*	2.1 *3*
γ [M1+E2]	799.47 *18*	2.8 *3*
γ	814.5 *3*	~0.29
γ [M1+E2]	822.8 *3*	~0.29
γ [E1]	830.30 *17*	1.8 *3*
γ	869.9 *3*	~0.29
γ [M1+E2]	884.17 *21*	2.3 *3*
γ	892.54 *17*	0.88 *12*
γ [E2]	909.4 *3*	~0.6

Photons (^{186}Ta)
(continued)

γ_{mode}	γ(keV)	γ(%)†
γ[E1]	922.95 *16*	1.4 *3*
γ	948.5 *5*	~0.29
γ	1046.1 *10*	~0.29
γ[M1+E2]	1091.7 *3*	~0.6
γ[E2]	1123.5 *3*	~0.6
γ[M1+E2]	1161.8 *4*	~0.29
γ[M1+E2]	1175.9 *3*	~0.29 ?
γ[M1+E2]	1199.91 *25*	~0.29
γ	1210.9 *3*	~0.29
γ	1213.9 *7*	~0.24
γ	1231.98 *23*	~0.18
γ	1239.0 *6*	~0.24
γ[E2]	1284.2 *4*	~0.29
γ[E2]	1298.3 *4*	~0.29
γ	1318.1 *5*	~0.29
γ[E2]	1322.3 *3*	~0.35
γ[M1+E2]	1397.8 *3*	~0.47
γ	1408.8 *5*	~0.6
γ	1428.9 *7*	~0.29
γ	1485.2 *3*	~0.29
γ	1506.28 *23*	
γ[E2]	1520.2 *3*	~0.29

† 17% uncert(syst)

Atomic Electrons (^{186}Ta)
⟨e⟩=86 *6* keV

e_{bin}(keV)	⟨e⟩(keV)	e(%)
10	2.0	20 *3*
12 - 49	4.3	25 *3*
53	7.1	13.4 *17*
55 - 93	2.2	2.9 *5*
110	1.58	1.43 *19*
111	12.0	10.8 *14*
112	10.1	9.0 *12*
114 - 115	0.75	0.66 *22*
120	6.4	5.3 *7*
121 - 122	1.88	1.54 *19*
129	4.1	3.2 *6*
132 - 144	0.43	~0.32
145	3.16	2.17 *6*
171 - 215	4.3	2.16 *21*
222 - 271	3.6	1.5 *3*
272 - 316	1.31	0.45 *6*
324 - 343	0.21	0.063 *13*
349	2.2	~0.6
371 - 419	1.6	0.42 *13*
428 - 440	0.082	0.019 *6*
441	5.1	~1
445 - 488	0.28	0.061 *17*
498 - 545	1.8	0.36 *16*
546	3.5	0.6 *3*
547 - 595	0.38	0.066 *25*
596 - 645	1.0	0.17 *5*
646 - 693	2.8	0.41 *7*
697 - 746	0.91	0.125 *17*
748 - 797	0.120	0.016 *3*
798 - 840	0.26	0.031 *10*
853 - 899	0.087	0.0100 *21*
907 - 948	0.0084	0.00091 *23*
977 - 1022	0.043	~0.0043
1034 - 1082	0.026	0.0025 *9*
1089 - 1130	0.047	0.0042 *13*
1141 - 1190	0.042	0.0036 *13*
1197 - 1239	0.024	0.0020 *5*
1249 - 1298	0.023	0.0018 *7*
1306 - 1339	0.040	0.0030 *13*
1359 - 1409	0.016	0.0012 *6*
1416 - 1451	0.017	0.0012 *5*
1473 - 1518	0.0031	0.00021 *8*

Continuous Radiation (^{186}Ta)
⟨β-⟩=881 keV;⟨IB⟩=1.9 keV

E_{bin}(keV)		⟨ ⟩(keV)	(%)
0 - 10	β-	0.0176	0.351
	IB	0.034	
10 - 20	β-	0.054	0.357
	IB	0.033	0.23
20 - 40	β-	0.221	0.74
	IB	0.064	0.22
40 - 100	β-	1.68	2.39
	IB	0.18	0.28
100 - 300	β-	19.9	9.7
	IB	0.49	0.28
300 - 600	β-	84	18.5
	IB	0.49	0.116
600 - 1300	β-	411	44.0
	IB	0.50	0.060
1300 - 2500	β-	361	22.4
	IB	0.088	0.0057
2500 - 2885	β-	2.84	0.110
	IB	7.2×10^{-5}	2.8×10^{-6}

$^{186}_{74}$W (stable)

Δ: -42517 *3* keV
%: 28.6 *2*

$^{186}_{75}$Re(3.777 *4* d)

Mode: β-(92.2 %), ϵ(7.8 %)
Δ: -41933 *3* keV
SpA: 1.8588×10^{5} Ci/g
Prod: ^{185}Re(n,γ)

Photons (^{186}Re)
⟨γ⟩=19.3 *14* keV

γ_{mode}	γ(keV)	γ(%)†
W L_ℓ	7.387	0.036 *8*
Os L_ℓ	7.822	0.046 *7*
W L_α	8.391	0.84 *16*
W L_η	8.724	0.0116 *22*
Os L_α	8.904	1.02 *13*
Os L_η	9.337	0.0202 *25*
W L_β	9.786	0.89 *18*
Os L_β	10.451	1.24 *16*
W L_γ	11.403	0.16 *3*
Os L_γ	12.144	0.22 *3*
W $K_{\alpha2}$	57.981	1.8 *4*
W $K_{\alpha1}$	59.318	3.1 *6*
Os $K_{\alpha2}$	61.485	1.01 *12*
Os $K_{\alpha1}$	63.000	1.74 *20*
W $K_{\beta1}$'	67.155	1.04 *20*
W $K_{\beta2}$'	69.342	0.27 *5*
Os $K_{\beta1}$'	71.313	0.60 *7*
Os $K_{\beta2}$'	73.643	0.152 *17*
γ_ϵE2	122.43 *7*	0.66 *6*
γ_{β}-E2	137.144 *24*	8.5 *9*
γ_{β}-E2	296.901 *23*	4.6 *21* $\times10^{-5}$
γ_{β}-E2	333.46 *3*	0.00051 *18*
γ_{β}-E2+<0.2%M1	630.357 *23*	0.0235 *10*
γ_{β}-E2	767.501 *22*	0.0263 *6*
γ_{β}-E2	773.266 *24*	1.8 *6* $\times10^{-5}$

† uncert(syst): 10% for ϵ, 20% for β-

Atomic Electrons (^{186}Re)
⟨e⟩=14.0 *6* keV

e_{bin}(keV)	⟨e⟩(keV)	e(%)
10 - 11	0.58	5.5 *7*
12	0.50	4.2 *7*
13 - 62	0.428	0.94 *5*
63	2.3	3.7 *4*
64 - 110	0.055	0.055 *4*
111	0.34	0.31 *3*
112 - 122	0.52	0.45 *3*
124	0.52	0.42 *5*
125	3.5	2.8 *3*
126	2.8	2.22 *25*
130 - 132	6.6×10^{-8}	5.1 *23* $\times10^{-8}$
134	1.09	0.81 *9*
135	0.79	0.59 *7*
136	0.034	0.025 *3*
137	0.51	0.38 *4*
140 - 143	2.2×10^{-8}	1.5 *6* $\times10^{-8}$
223 - 260	6.6×10^{-5}	2.6 *8* $\times10^{-5}$
284 - 333	4.1×10^{-5}	1.30 *19* $\times10^{-5}$
463 - 476	5.7×10^{-8}	1.2 *3* $\times10^{-8}$
617 - 630	0.000430	6.93 *18* $\times10^{-5}$
694 - 699	0.00126	0.000182 *5*
755 - 773	0.000350	4.62 *10* $\times10^{-5}$

Continuous Radiation (^{186}Re)
⟨β-⟩=323 keV;⟨IB⟩=0.37 keV

E_{bin}(keV)		⟨ ⟩(keV)	(%)
0 - 10	β-	0.067	1.34
	IB	0.0152	
10 - 20	β-	0.203	1.35
	IB	0.0145	0.101
20 - 40	β-	0.82	2.74
	IB	0.027	0.095
40 - 100	β-	6.0	8.5
	IB	0.069	0.108
100 - 300	β-	58	29.0
	IB	0.138	0.082
300 - 600	β-	152	34.7
	IB	0.070	0.018
600 - 1075	β-	106	14.6
	IB	0.0097	0.00146

$^{186}_{75}$Re(2×10^{5} yr)

Mode: IT
Δ: ~-41783 keV
SpA: 0.010 Ci/g
Prod: ^{185}Re(n,γ)

Photons (^{186}Re)
⟨γ⟩=21 *3* keV

γ_{mode}	γ(keV)	γ(%)
Re L_ℓ	7.604	0.78 *23*
Re L_α	8.646	18 *5*
Re L_η	9.027	0.32 *9*
Re L_β	10.083	32 *8*
Re L_γ	11.896	7.0 *17*
γ M1+~1.0%E2	40.35 *8*	5.6 *11*
γ M1	59.04 *8*	20 *4*
Re $K_{\alpha2}$	59.717	0.28 *6*
Re $K_{\alpha1}$	61.141	0.48 *10*
Re $K_{\beta1}$'	69.214	0.16 *3*
Re $K_{\beta2}$'	71.470	0.041 *8*
γ E2	99.39 *8*	1.19 *24*

Atomic Electrons (^{186}Re)

⟨e⟩=134 *17* keV

e_{bin}(keV)	⟨e⟩(keV)	e(%)
11	3.1	29 *12*
12	3.4	28 *10*
13	5.2	42 *7*
28	17.4	62 *13*
30	1.9	6.4 *13*
37	4.4	11.7 *23*
38	13.6	36 *17*
40	14.1	36 *16*
47	36.6	78 *20*
48	19.2	~40
49	0.33	0.68 *14*
51	0.0018	0.0035 *8*
56	8.4	15 *3*
57	0.107	0.19 *4*
58	2.0	3.4 *7*
59	0.67	1.15 *23*
60	5.5×10^{-5}	9.2 *21* $\times 10^{-5}$
61	0.00061	0.00101 *23*
66	0.00035	0.00053 *12*
67	0.00031	0.00046 *10*
68	0.00015	0.00022 *5*
69	0.00015	0.00022 *5*
87	1.5	1.7 *3*
89	1.25	1.4 *3*
96	0.027	0.028 *6*
97	0.74	0.76 *15*
98	0.0033	0.0034 *7*
99	0.22	0.22 *5*

$^{186}_{76}$Os(2 *1* $\times 10^{15}$ yr)

Mode: α

Δ: -43007 *3* keV

Prod: natural source

%: 1.58 *10*

Alpha Particles (^{186}Os)

α(keV)
2800 *100*

$^{186}_{77}$Ir(15.8 *3* h)

Mode: ε

Δ: -39176 *20* keV

SpA: 1.066×10^{6} Ci/g

Prod: ^{185}Re(α,3n); ^{187}Re(α,7n); ^{187}Re(^{3}He,4n)

Photons (^{186}Ir)

⟨γ⟩=1581 *21* keV

γ_{mode}	γ(keV)	γ(%)†
Os L_ℓ	7.822	0.73 *14*
Os L_α	8.904	16 *3*
Os L_η	9.337	0.25 *4*
Os L_β	10.467	17 *3*
Os L_γ	12.185	3.0 *5*
Os $K_{\alpha2}$	61.485	29 *5*
Os $K_{\alpha1}$	63.000	49 *9*
Os $K_{\beta1}'$	71.313	17 *3*
Os $K_{\beta2}'$	73.643	4.3 *8*

Photons (^{186}Ir) (continued)

γ_{mode}	γ(keV)	γ(%)†
γ E2	137.144 *24*	41.3
γ M1	142.91 *3*	0.27 *3* ?
γ (E2)	208.04 *17*	0.54 *7*
γ [M1+E2]	215.68 *4*	0.097 *12*
γ M1	224.12 *16*	0.19 *4*
γ M1	269.02 *12*	0.19 *8*
γ E1	276.56 *13*	1.54 *8*
γ M1	281.3 *4*	0.08 *3*
γ E1	284.26 *15*	~0.07
γ (M1)	288.79 *12*	~0.07
γ E2	296.901 *23*	62.1 *15*
γ (E2)	302.93 *4*	0.37 *7*
γ M1+E2	309.63 *12*	0.46 *15*
γ M1+E2	330.21 *17*	0.20 *4*
γ E2	333.46 *3*	0.10 *4* ?
γ E2	351.70 *12*	1.89 *12*
γ (E2)	365.139 *24*	0.73 *12*
γ M1	403.28 *16*	0.19 *5*
γ (E2)	406.66 *3*	0.19 *8*
γ E2	420.80 *3*	2.7 *4*
γ E2	434.846 *25*	33.8 *9*
γ E2	441.51 *11*	1.62 *19*
γ M1	446.7 *10*	~0.31
γ E2	476.366 *24*	0.90 *18*
γ (E2)	489.53 *16*	1.20 *12*
γ M1	551.4 *3*	~0.27
γ E2	551.99 *5*	2.90 *15*
γ E1	558.05 *16*	0.58 *19*
γ [M1+E2]	565.4 *4*	0.39 *12* ?
γ M1+E2	565.4 *4*	0.39 *12*
γ (E2)	569.78 *18*	0.42 *12*
γ E2	584.42 *11*	5.4 *4*
γ E2	622.34 *3*	3.0 *4*
γ E2+<0.2%M1	630.357 *23*	4.8 *4*
γ E2	636.38 *3*	7.2 *9*
γ E2	649.55 *16*	1.35 *15*
γ M1+E2	661.9 *7*	~0.31
γ M1	680.55 *25*	~0.18
γ M1	684.8 *4*	~0.31
γ	700.4 *10*	~0.23
γ E2	705.12 *15*	0.89 *19*
γ E2	712.69 *18*	1.57 *19*
γ M1+33%E2	729.51 *17*	0.58 *12*
γ E2+1.0%M1	760.0 *4*	0.50 *8*
γ E2	767.501 *22*	5.4 *4*
γ E2	773.266 *24*	8.9 *4*
γ E2+14%M1	804.64 *24*	1.16 *8*
γ E2	841.504 *23*	5.1 *7*
γ	846.6 *10*	6.3 *4*
γ M1	885.0 *10*	0.14 *4*
γ E2	933.28 *3*	5.2 *4*
γ (M1)	943.52 *12*	0.86 *4*
γ M1+33%E2	959.4 *4*	0.22 *4*
γ M1	1011.00 *14*	0.73 *8*
γ (M1)	1026.67 *15*	1.19 *4*
γ [M1+E2]	1046.15 *18*	0.31 *4*
γ E2	1057.1 *4*	3.1 *3*
γ E2	1070.3 *5*	0.13 *4*
γ E2	1107.56 *24*	0.75 *9*
γ (E1)	1122.0 *10*	0.48 *9*
γ E2+34%M1	1171.02 *14*	1.47 *23*
γ E2+20%M1	1187.67 *20*	1.97 *23*
γ E2	1265.4 *5*	0.77 *8*
γ [E2]	1313.93 *15*	1.97 *23*
γ E2	1323.57 *15*	1.16 *12*
γ (M1)	1332.4 *5*	0.15 *4*
γ [E2]	1343.05 *18*	0.234 *23*
γ	1363.5 *10*	0.147 *19*
γ	1378.1 *10*	0.62 *15*
γ [M1+E2]	1393.6 *5*	0.220 *23*
γ M1+50%E2	1441.02 *24*	0.77 *7*
γ E2	1466.9 *3*	0.50 *12*
γ M1	1508.3 *3*	0.93 *8*
γ M1	1597.1 *8*	0.85 *8*
γ E2	1622.51 *20*	0.89 *8*
γ E2	1647.38 *14*	4.67 *23*
γ	1690.2 *10*	0.28 *4*
γ M1	1701.4 *4*	2.08 *8*
γ (E2)	1737.92 *24*	0.75 *7*
γ M1	1751.0 *3*	0.83 *4*
γ (M1)	1788.5 *4*	0.174 *19*
γ	1829.2 *10*	0.12 *5*
γ	1842.6 *10*	0.13 *4*
γ	1869.0 *10*	0.39 *5*
γ [E2]	1893.9 *3*	0.23 *6*
γ	1997.1 *10*	0.27 *5*
γ	2138.2 *5*	0.081 *15*
γ	2144.3 *10*	0.32 *4*
γ	2165.2 *10*	0.59 *10*
γ [E2]	2172.1 *5*	0.22 *4*
γ	2185.8 *10*	0.63 *7*
γ [M1+E2]	2241.3 *5*	1.4 *4*
γ	2315.6 *10*	0.20 *3*
γ	2339.7 *10*	0.38 *5*
γ [M1+E2]	2358.3 *4*	0.28 *4*
γ [E2]	2384.2 *5*	0.37 *4*
γ [E2]	2399.8 *4*	0.40 *5*
γ	2544.9 *5*	0.26 *5*
γ	2580.3 *10*	0.077 *23*
γ (M1)	2733.7 *10*	0.058 *19*
γ	2770.7 *10*	0.035 *12*
γ	2780.4 *10*	0.32 *5*
γ	2790.2 *10*	0.18 *3*
γ	2805.8 *10*	0.058 *15*
γ [M1+E2]	2834.6 *4*	0.77 *7*
γ	2853.1 *10*	0.22 *3*
γ	2866.5 *10*	0.058 *19*
γ	2912.5 *10*	0.10 *3*
γ	2967.0 *10*	0.112 *23*
γ	2979.7 *5*	0.073 *15*
γ	2994.8 *10*	0.066 *12*
γ	3007.3 *10*	0.124 *19*
γ	3040.3 *10*	0.039 *12*
γ	3074.6 *10*	0.025 *8*
γ [E2]	3131.5 *4*	0.046 *8*

† 5.2% uncert(syst)

Atomic Electrons (^{186}Ir)

⟨e⟩=95.8 *18* keV

e_{bin}(keV)	⟨e⟩(keV)	e(%)
11	4.8	45 *9*
12	3.3	26 *5*
13 - 62	2.8	9.7 *13*
63	11.3	17.9 *10*
68 - 71	0.43	0.62 *7*
124	2.51	2.03 *11*
125	17.3	13.9 *8*
126	13.6	10.8 *6*
130 - 132	0.100	0.077 *8*
134	5.4	4.03 *22*
135	3.85	2.86 *15*
136 - 150	2.89	2.10 *10*
195 - 222	0.52	0.26 *3*
223	8.4	3.78 *12*
224 - 272	0.32	0.13 *4*
273 - 323	6.54	2.28 *4*
327 - 354	0.48	0.140 *11*
361	2.90	0.80 *3*
362 - 410	0.41	0.108 *14*
416 - 464	1.50	0.354 *8*
465 - 511	0.78	0.157 *17*
538 - 584	1.18	0.210 *12*
588 - 637	0.47	0.075 *7*
639 - 688	0.199	0.030 *4*
689 - 731	0.80	0.114 *4*
747 - 794	0.9	0.11 *4*
802 - 847	0.18	0.022 *7*
859 - 886	0.318	0.0368 *20*
920 - 959	0.246	0.0261 *10*
972 - 1016	0.167	0.0169 *14*
1024 - 1070	0.070	0.0067 *4*
1095 - 1122	0.156	0.0141 *13*
1158 - 1192	0.057	0.0048 *3*
1240 - 1269	0.116	0.0093 *7*
1290 - 1332	0.060	0.0046 *12*
1340 - 1383	0.039	0.0029 *3*
1391 - 1439	0.068	0.0048 *4*
1440 - 1467	0.0033	0.00023 *3*
1495 - 1595	0.199	0.0127 *5*
1610 - 1699	0.210	0.0128 *6*

Atomic Electrons (^{186}Ir)
(continued)

e_{bin}(keV)	⟨e⟩(keV)	e(%)
1715 - 1795	0.034	0.0020 4
1816 - 1891	0.0094	0.00051 9
1923 - 1995	0.007	~0.00038
2064 - 2163	0.042	0.0020 5
2167 - 2266	0.056	0.0025 6
2284 - 2382	0.0259	0.00112 10
2387 - 2471	0.0058	0.00024 9
2506 - 2578	0.0023	9 3 $\times 10^{-5}$
2660 - 2758	0.011	0.00039 10
2760 - 2856	0.025	0.00092 11
2863 - 2956	0.0058	0.00020 4
2964 - 3062	0.0025	8.3 13 $\times 10^{-5}$
3064 - 3129	0.000156	5.0 7 $\times 10^{-6}$

Continuous Radiation (^{186}Ir)

⟨β+⟩=14.4 keV;⟨IB⟩=3.9 keV

E_{bin}(keV)		⟨ ⟩(keV)	(%)
0 - 10	β+	1.14 $\times 10^{-8}$	1.35 $\times 10^{-7}$
	IB	0.00084	
10 - 20	β+	7.3 $\times 10^{-7}$	4.32 $\times 10^{-6}$
	IB	0.0018	0.0149
20 - 40	β+	3.16 $\times 10^{-5}$	9.5 $\times 10^{-5}$
	IB	0.0031	0.0099
40 - 100	β+	0.00255	0.00320
	IB	0.36	0.58
100 - 300	β+	0.188	0.084
	IB	0.063	0.034
300 - 600	β+	1.56	0.335
	IB	0.22	0.049
600 - 1300	β+	9.0	0.97
	IB	1.26	0.133
1300 - 2500	β+	3.63	0.246
	IB	1.9	0.107
2500 - 3694	IB	0.18	0.0065
	Σβ+		1.64

$^{186}_{77}$Ir(1.75 15 h)

Mode: ε
Δ: -39176 20 keV
SpA: 9.6×10^{6} Ci/g
Prod: ^{191}Ir(p,p5n); ^{187}Re(α,5n); ^{187}Re(^{3}He,4n); ^{186}Os(p,n)

Photons (^{186}Ir)

γ_{mode}	γ(keV)	γ(rel)
γ E2	137.144 24	100
γ M1	142.91 3	1.0 3
γ E2	296.901 23	40 7
γ +17%E2	302.93 4	0.29 7
γ E2	333.46 3	1.2 4
γ E2	441.51 11	~0.6
γ E2	476.366 24	3.4 8
γ (E2)	569.78 18	2.9 9
γ E2	584.42 11	~2
γ E2+<0.2%M1	630.357 23	56 7
γ E2	636.38 3	5.8 9
γ E2	712.69 18	10.9 22
γ E2	767.501 22	62 7
γ E2	773.266 24	34 5
γ E2	933.28 3	4.2 5
γ	986.5 6	28 10
γ (M1)	1026.67 15	~3
γ [M1+E2]	1046.15 18	2.2 5 ?
γ [E2]	1343.05 18	1.6 3

Photons (^{186}Ir)
(continued)

γ_{mode}	γ(keV)	γ(rel)
γ	1616.8 6	10 3
γ	1753.9 6	11 3

$^{186}_{78}$Pt(2.0 1 h)

Mode: ε, α(~ 0.00014 %)
Δ: -37850 110 keV
SpA: 8.4×10^{6} Ci/g
Prod: protons on Ir

Alpha Particles (^{186}Pt)

α(keV)
4230 20

Photons (^{186}Pt)

γ_{mode}	γ(keV)	γ(rel)
γ	109.1 4	0.7 2
γ	126.9 4	0.4 1
γ	149.6 4	0.6 1
γ	180.5 4	1.7 2
γ	186.5 4	0.8 2
γ	205.4 3	2.2 3
γ	210.3 3	2.6 3
γ	252.8 8	0.9 2
γ	256.0 4	2.0 3
γ	276.6 6	0.9 2
γ	280.8 4	2.4 4
γ	366.7 4	3.3 7
γ	611.5 4	8.6 8
γ	635.6 4	<5
γ	689.2 3	100

$^{186}_{79}$Au(10.7 5 min)

Mode: ε
Δ: -31580 300 keV syst
SpA: 9.4×10^{7} Ci/g
Prod: daughter ^{186}Hg; protons on Pt

Photons (^{186}Au)

γ_{mode}	γ(keV)	γ(rel)
γ E2	191.57 4	100 7
γ M1	205.12 20	3.3 2
γ (M1+E2)	225.20 20	0.70 5
γ	231.7 3	0.93 7
γ	257.9 3	1.40 8
γ [E2]	266.08 18	0.42 4 ?
γ E2	279.96 21	2.2 1
γ E2	298.85 9	41 2
γ (M1+E2)	308.12 17	0.66 5
γ (M1+E2)	327.02 22	1.3 1
γ E2	349.53 15	2.10 13
γ E2	384.46 20	3.0 5
γ [E2]	387.0 3	1.7 5
γ (M1)	415.57 10	13.7 6
γ E2	424.21 19	1.0 1
γ E2	430.31 22	2.9 2

Photons (^{186}Au)
(continued)

γ_{mode}	γ(keV)	γ(rel)
γ	440.3 3	2.0 1
γ	461.8 3	0.16 3
γ [E2]	466.25 15	2.1 1
γ E0	471.52 21	<0.4
γ M1	501.17 20	3.5 3
γ (E0+E2)	606.97 16	<9
γ [E2]	607.14 10	8.5 20
γ [E1]	609.39 22	2.4 6
γ [E2]	615.61 16	0.50 5
γ E1	676.47 18	5.3 2
γ	704.4 3	0.9 1
γ (E0+M1+E2)	732.32 16	1.8 1
γ E2	765.10 14	17.0 9
γ	791.0 3	2.0 1
γ	796.4 4	0.5 1
γ [E2]	798.54 17	8.6 10
γ	800.4 4	2.1 2
γ	810.7 3	0.61 6
γ	873.1 10	2.0 1
γ	881.58 21	3.4 2
γ	905.14 3	0.6 1
γ	907.7 4	0.5 1
γ	917.5 4	0.7 1
γ	927.3 4	0.4 1
γ [E2]	956.66 14	~0.10 ?
γ	984.5 4	2.1 2
γ [E2]	1031.18 15	2.2 1
γ	1098.0 3	0.68 7
γ	1121.9 3	0.7 1
γ	1142.71 18	0.67 7
γ	1176.1 5	1.0 1
γ	1181.5 5	1.0 1
γ	1203.0 3	1.4 1
γ	1216.2 3	6.0 3
γ	1226.1 3	1.6 1
γ	1271.1 5	0.8 1
γ	1289.2 5	3.3 2
γ	1323.7 4	0.8 1
γ	1345.5 4	0.8 1
γ	1400.0 4	0.55 6
γ	1441.56 18	0.76 7
γ	1532.7 4	1.6 2
γ	1589.5 4	0.7 1
γ	1725.9 4	2.0 2
γ	1737.6 4	3.0 3
γ	2024.6 5	3.2 3
γ	2035.6 5	4.6 3

$^{186}_{79}$Au(<2 min)

Mode: ε
Δ: -31580 300 keV syst
SpA: >5×10^{8} Ci/g
Prod: daughter ^{186}Hg

$^{186}_{80}$Hg(1.38 10 min)

Mode: ε(99.984 5 %), α(0.016 5 %)
Δ: -28550 230 keV syst
SpA: 7.3×10^{8} Ci/g
Prod: ^{197}Au(p,12n); protons on Pb

Alpha Particles (^{186}Hg)

α(keV)
5094 15

Photons (^{186}Hg)

γ_{mode}	γ(keV)	γ(rel)
γ	111.9 *4*	90 *5*
γ	227.8 *3*	3.7 *10*
γ(M1)	251.8 *4*	100 *5*
γ	349.5 *5*	~3

$^{186}_{81}$Tl(28 *2* s)

Mode: ϵ, α?(~0.006 %)
Δ: -20020 *290* keV syst
SpA: 2.14×10^9 Ci/g
Prod: ^{159}Tb(^{32}S,5n); ^{182}W(^{14}N,10n); ^{197}Au(^{3}He,14n); ^{181}Ta(^{16}O,11n)

Alpha Particles (^{186}Tl)

α(keV)
5760 ?

Photons (^{186}Tl)

γ_{mode}	γ(keV)	γ(rel)
γ(E0+M1+E2)	215.5 *3*	4.5 *5*
γ	287.9 *3*	<2 ?
γE2	356.7 *3*	32 *1*
γE2+M1	397.8 *3*	1.6 *2*
γE2	402.6 *3*	50 *1*
γE2	405.3 *2*	100
γ	412.6 *3*	1.5 *2*
γ	421.3 *3*	0.7 *1*
γE2	424.1 *2*	14.0 *7*
γ(E2)	459.2 *3*	2.7 *5*
γ	477.9 *3*	1.8 *2*
γ	497.6 *3*	<0.8
γE2+M1	573.5 *3*	1.5 *5*
γ(E2)	597.5 *2*	4.6 *3*
γ	607.5 *3*	
γ	726.5 *3*	1.7 *4*
γ	770.2 *3*	5.0 *3*
γ	788.4 *4*	2.9 *3*
γ(E2)	811.1 *4*	2.3 *5*
γ	826.4 *4*	2.4 *3*
γ	870.0 *4*	1.0 *2*
γ	1177.1 *4*	0.8 *2*
γ	1209.7 *5*	1.1 *2*
γ	1247.6 *5*	1.4 *2*
γ	1272.6 *5*	1.4 *2*

$^{186}_{81}$Tl(~4 s)

Mode: IT
Δ: -19646 *290* keV syst
SpA: $\sim 1.4 \times 10^{10}$ Ci/g
Prod: ^{182}W(^{14}N,10n); ^{197}Au(^{3}He,14n); ^{181}Ta(^{16}O,11n)

Photons (^{186}Tl)

$\langle\gamma\rangle$=304 *12* keV

γ_{mode}	γ(keV)	γ(%)
Tl L_ℓ	8.953	0.099 *10*
Tl L_α	10.259	2.02 *14*
Tl L_η	10.994	0.053 *4*
Tl L_β	12.275	2.99 *22*
Tl L_γ	14.374	0.61 *5*
Tl $K_{\alpha2}$	70.832	2.25 *11*
Tl $K_{\alpha1}$	72.873	3.81 *19*
Tl $K_{\beta1}'$	82.434	1.34 *7*
Tl $K_{\beta2}'$	85.185	0.377 *19*
γ E3	373.8 *3*	80 *3*

$^{186}_{82}$Pb(7.9 *16* s)

Mode: ϵ(~97.6 %), α(~2.4 %)
Δ: -14630 *300* keV syst
SpA: 7.4×10^9 Ci/g
Prod: ^{155}Gd(^{40}Ar,9n)

Alpha Particles (^{186}Pb)

α(keV)
6320 *10*

A = 187

NDS **36**, 559 (1982)

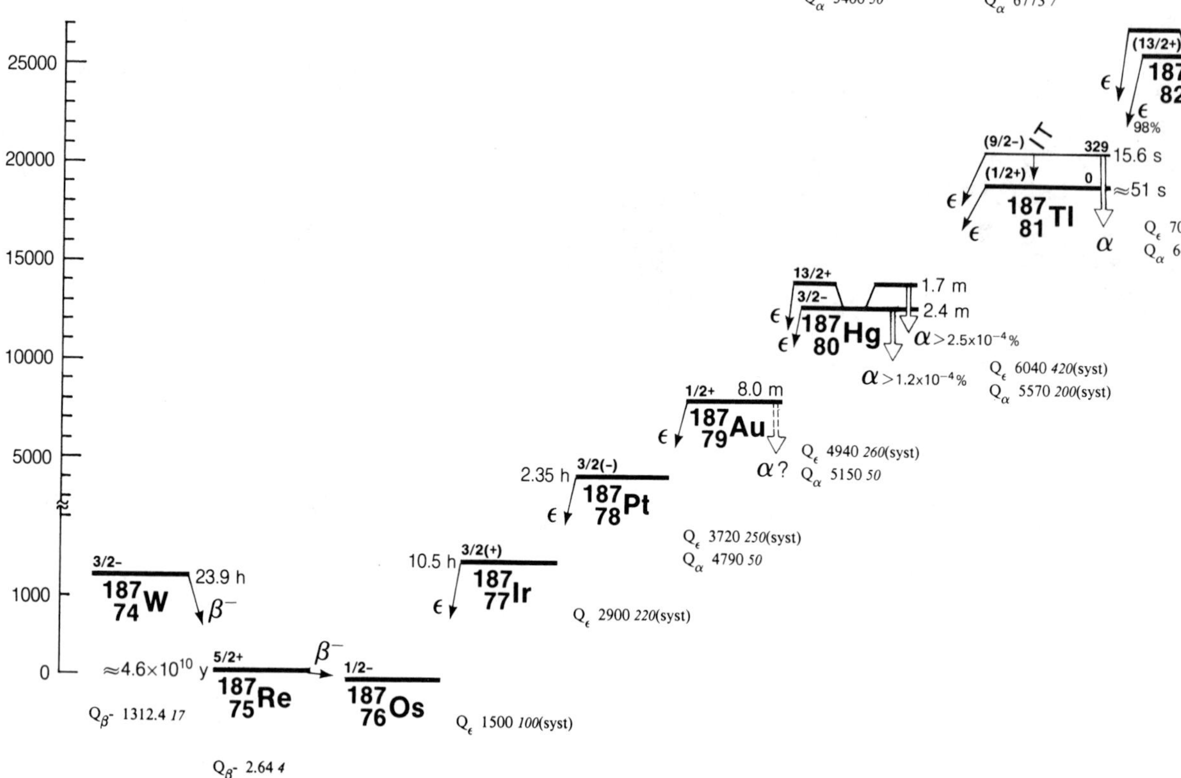

$^{187}_{74}$W (23.9 *1* h)

Mode: β-
Δ: -39912 *3* keV
SpA: 7.01×10^{5} Ci/g
Prod: ^{186}W(n,γ)

Photons (^{187}W)

⟨γ⟩=430 *5* keV

γ_{mode}	γ(keV)	γ(%)†
Re L_ℓ	7.604	~0.4
Re L_α	8.646	~10
Re L_η	9.027	~0.15
Re L_β	10.123	~10
Re L_γ	11.758	~2
γ	16.591 *20*	0.0058 *8*
γ [M1]	29.212 *20*	0.0037 *8*
γ	40.90 *5*	~0.0017
γ [M1]	43.65 *4*	~0.0017
Re $K_{\alpha2}$	59.717	5.9 *6*
Re $K_{\alpha1}$	61.141	10.2 *10*
Re $K_{\beta1}$'	69.214	3.4 *3*
Re $K_{\beta2}$'	71.470	0.88 *8*

Photons (^{187}W) (continued)

γ_{mode}	γ(keV)	γ(%)†
γ E1+1.0%M2	71.983 *4*	10.77 *22*
γ (M1)	77.374 *21*	0.0067 *17*
γ	93.20 *5*	0.0058 *8*
γ	100.121 *20*	0.0084 *17*
γ M1+13%E2	106.586 *12*	0.0246 *8*
γ M1	113.724 *8*	0.0743 *25*
γ	123.77 *11*	0.0025 *8*
γ M1+3.1%E2	134.228 *7*	8.56 *18*
γ	138.48 *6*	0.0042 *17*
γ	165.7 *4*	0.0008 *3*
γ M1+2.7%E2	168.95 *11*	0.0025 *9*
γ [M1+E2]	198.21 *3*	0.0017 *4*
γ M2	206.211 *7*	0.138 *5*
γ [E2]	208.33 *13*	0.00067 *25*
γ M1+38%E2	239.148 *18*	0.083 *4*
γ M1+33%E2	246.285 *17*	0.115 *4*
γ [M1+E2]	275.497 *25*	0.0020 *6*
γ [E2]	303.18 *11*	~0.00050
γ [M1+E2]	352.871 *18*	0.0015 *6*
γ	374.29 *14*	0.0025 *8*
γ	375.91 *13*	0.0033 *8*
γ (E2)	454.896 *17*	0.0284 *17*
γ E2	479.534 *22*	21.1 *5*
γ [E2]	484.108 *19*	0.0167 *8*
γ	492.78 *20*	0.025 *8*
γ E2	511.749 *22*	0.624 *12*
γ E1	551.517 *22*	4.92 *10*
γ	564.96 *5*	0.012 *4*
γ	573.69 *14*	0.00050 *17*
γ [E2]	576.31 *8*	0.0064 *10*

Photons (^{187}W) (continued)

γ_{mode}	γ(keV)	γ(%)†
γ	578.70 *11*	0.0009 *3*
γ (M1+25%E2)	589.123 *18*	0.1177 *25*
γ	611.75 *5*	~0.002
γ M1+23%E2	618.335 *20*	6.07 *13*
γ E2	625.473 *23*	1.052 *25*
γ [E2]	638.67 *4*	0.0031 *10*
γ [M1]	647.3 *3*	0.0008 *3*
γ [M1+E2]	682.316 *20*	<0.013
γ E1	685.744 *22*	26.4 *6*
γ [M1+E2]	692.43 *13*	~0.0013
γ [E2]	730.392 *25*	<0.017 ?
γ (M1)	745.26 *9*	0.288 *7*
γ	767.4 *8*	0.0015 *6*
γ M1+16%E2	772.89 *4*	3.98 *8*
γ [M1+E2]	816.543 *19*	0.0095 *8*
γ	825.92 *5*	0.00023 *3*
γ [M1+E2]	826.66 *13*	0.00023 *3*
γ	844.8 *4*	~0.00023
γ (M1)	864.620 *25*	0.325 *8*
γ (M1)	879.48 *9*	0.137 *3*
γ	960.15 *5*	0.00128 *8*
γ	1056.20 *5*	0.00022 *6*
γ	1086.6 *3*	$<8\times10^{-5}$?
γ	1095.86 *4*	$<7\times10^{-5}$?
γ	1190.43 *5*	0.000209 *25*
γ	1220.8 *3*	$1.7\,_{5}\times10^{-5}$
γ	1230.08 *4*	0.00128 *15*

† 3.5% uncert(syst)

Atomic Electrons (^{187}W)

⟨e⟩=37 *4* keV

e_{bin}(keV)	⟨e⟩(keV)	e(%)
2 - 4	0.10	3.3 *10*
5	1.0	<43
6	1.6	~27
7	0.018	~0.28
11	3.0	~29
12	2.0	~17
13	0.37	2.9 *13*
14	1.7	<24
15 - 58	1.2	~5
59	2.5	~4
60 - 61	1.3	2.1 *9*
63	10.1	16.1 *5*
65 - 114	1.5	2.2 *9*
121	0.0005	~0.0004
122	3.22	2.64 *8*
123 - 169	1.73	1.304 *25*
175 - 208	0.285	0.148 *10*
227 - 275	0.034	0.015 *3*
281 - 304	0.0015	0.00051 *24*
340 - 383	0.00271	0.00072 *6*
408	1.58	0.388 *12*
412 - 455	0.049	0.0112 *7*
467 - 512	0.747	0.1583 *23*
517 - 566	0.97	0.177 *14*
567 - 616	0.735	0.120 *3*
618 - 659	0.0156	0.00251 *18*
670 - 718	0.59	0.084 *7*
720 - 767	0.083	0.0109 *11*
770 - 817	0.070	0.0089 *3*
823 - 869	0.0095	0.00111 *3*
877 - 888	0.00075	8.5 *6* $\times10^{-5}$
948 - 985	2.5 $\times10^{-5}$	2.6 *12* $\times10^{-6}$
1015 - 1056	5.8 $\times10^{-6}$	~6 $\times10^{-7}$
1074 - 1119	9.1 $\times10^{-6}$	~8 $\times10^{-7}$
1149 - 1190	5.1 $\times10^{-5}$	~4 $\times10^{-6}$
1208 - 1230	1.0 $\times10^{-5}$	~8 $\times10^{-7}$

Continuous Radiation (^{187}W)

⟨β-⟩=274 keV;⟨IB⟩=0.27 keV

E_{bin}(keV)		⟨ ⟩(keV)	(%)
0 - 10	β-	0.122	2.44
	IB	0.013	
10 - 20	β-	0.365	2.43
	IB	0.0123	0.086
20 - 40	β-	1.45	4.85
	IB	0.023	0.079
40 - 100	β-	10.0	14.3
	IB	0.055	0.087
100 - 300	β-	77	40.0
	IB	0.101	0.061
300 - 600	β-	108	26.1
	IB	0.051	0.0126
600 - 1300	β-	76	9.5
	IB	0.0143	0.0019
1300 - 1313	β-	0.00113	8.7 $\times10^{-5}$
	IB	2.8 $\times10^{-10}$	2.2 $\times10^{-11}$

$^{187}_{75}$Re(4.6 *8* $\times10^{10}$ yr)

Mode: β-

Δ: -41224 *3* keV

SpA: Ci/g

Prod: natural source

%: 62.60 *2*

Continuous Radiation (^{187}Re)

E_{bin}(keV)		⟨ ⟩(keV)	(%)
0 - 3	β-	0.66	100
	IB	1.65 $\times10^{-6}$	

$^{187}_{76}$Os(stable)

Δ: -41227 *3* keV

%: 1.6 *1*

$^{187}_{77}$Ir(10.5 *3* h)

Mode: ϵ

Δ: -39730 *100* keV syst

SpA: 1.60$\times10^{6}$ Ci/g

Prod: ^{185}Re(α,2n); ^{187}Re(α,4n)

Photons (^{187}Ir)

γ_{mode}	γ(keV)	γ(rel)
γ M1	9.721 *20*	3.4 *11*
γ M1	25.53 *6*	9.9 *4*
γ M1+2.1%E2	64.545 *22*	11.8 *19*
γ E2(+<31%M1)	65.259 *22*	7.3 *13*
γ M1	74.266 *22*	23 *4*
γ E2	74.980 *22*	9.8 *10*
γ M1+4.6%E2	84.87 *4*	2.2 *2*
γ	87.58 *10*	<0.10
γ E2	90.79 *6*	0.17 *7*
γ E2+40%M1	112.39 *4*	1.1 *3*
γ E2+38%M1	113.10 *4*	5.8 *10*
γ M1+45%E2	115.58 *7*	1.2 *2*
γ [M1+E2]	146.09 *8*	0.7 *1*
γ	150.5 *10*	<0.10
γ M2	156.64 *9*	0.87 *10*
γ	159.04 *20*	0.31 *5*
γ E2+49%M1	162.72 *8*	2.8 *6*
γ (M1)	163.46 *15*	1.4 *3*
γ	170.4 *10*	<0.10
γ M1+21%E2	177.64 *4*	56 *4*
γ E2	180.83 *7*	4.2 *5*
γ (M1+E2)	181.81 *15*	0.69 *14*
γ E2	187.37 *4*	36 *3*
γ	189.2 *10*	<0.8
γ	198.36 *10*	1.46 *16*
γ (M1)	206.4 *3*	0.12 *3*
γ	209.70 *15*	0.24 *4*
γ	212.17 *10*	0.29 *5*
γ (M1)	224.24 *8*	0.6 *2*
γ	232.34 *15*	0.46 *8*
γ	240.25 *25*	1.9 *2*
γ	249.13 *25*	2.0 *2*
γ (M1)	252.86 *9*	1.9 *2*
γ M1(+E2)	258.48 *8*	5.2 *6*
γ (M1)	261.47 *7*	4.2 *4*
γ	263.4 *10*	<0.44
γ (E2)	265.93 *10*	1.8 *2*
γ (M1)	275.76 *10*	0.8 *2*
γ (E1)	277.45 *14*	1.3 *2* ?
γ M1	299.63 *14*	2.8 *7*
γ M1	314.08 *6*	19.3 *15*
γ	317.56 *15*	0.70 *13*
γ [M1]	322.98 *8*	5.7 *4*
γ	344.56 *20*	0.6 *2*
γ (M1)	348.73 *11*	0.8 *2*
γ [E1]	355.30 *12*	0.65 *7*
γ (M1)	370.55 *20*	1.6 *2*
γ	377.08 *20*	1.17 *16*
γ (E2)	384.84 *13*	6.5 *5*
γ M1	395.76 *8*	2.5 *3*
γ M1	398.95 *6*	5.1 *20*
γ E2	400.94 *6*	83 *4*

Photons (^{187}Ir)
(continued)

γ_{mode}	γ(keV)	γ(rel)
γ	412.16 *15*	5.0 *4*
γ	422.26 *15*	0.98 *10*
γ [M1]	426.47 *5*	5.1 *10*
γ M1	427.18 *5*	88 *4*
γ	433.76 *15*	1.01 *10*
γ [E1]	440.17 *12*	1.21 *18*
γ (M1)	447.85 *11*	2.6 *3*
γ [E1]	456.27 *14*	0.94 *13*
γ	462.2 *4*	1.2 *3*
γ	463.8 *4*	2.3 *5*
γ [M1+E2]	485.81 *6*	15.7 *18*
γ M1	491.73 *4*	29 *3*
γ M1	501.45 *4*	31 *3*
γ [M1+E2]	511.34 *5*	2.4 *4* ?
γ (M1)	515.70 *11*	7.8 *6*
γ (M1)	522.06 *11*	5.5 *5*
γ	564.38 *25*	1.4 *2*
γ (M1)	576.60 *5*	17.1 *15*
γ (M1)	586.60 *11*	6.2 *5*
γ (M1+36%E2)	589.19 *7*	5.9 *5*
γ M1	610.88 *8*	80 *7*
γ [M1]	636.41 *9*	3.8 *4*
γ (M1+33%E2)	651.42 *7*	8.0 *7*
γ [M1+E2]	654.45 *7*	8.5 *8*
γ [M1+E2]	664.17 *7*	2.9 *3*
γ	672.83 *20*	2.5 *3*
γ	677.06 *15*	0.67 *10*
γ [M1+E2]	701.67 *9*	1.18 *24* ?
γ [E2]	711.39 *9*	1.7 *2*
γ [M1+E2]	715.96 *7*	4.9 *4*
γ	723.56 *20*	3.0 *4*
γ (E2)	725.68 *7*	10.3 *8*
γ	732.46 *20*	0.36 *7*
γ	742.00 *20*	1.02 *14*
γ (M1)	747.68 *10*	5.1 *4*
γ (M1)	756.79 *10*	2.9 *3*
γ	760.3 *10*	<0.4
γ	787.95 *10*	1.4 *3*
γ [E2]	796.59 *9*	0.26 *5* ?
γ M1	799.78 *6*	17.8 *16*
γ	813.93 *15*	1.41 *19*
γ (E1)	841.11 *12*	6.4 *3*
γ [M1+E2]	860.78 *10*	2.0 *2*
γ [E2]	886.64 *7*	2.6 *3*
γ [M1+E2]	899.69 *10*	2.7 *3*
γ (M1)	902.88 *9*	6.4 *5*
γ M1	912.88 *6*	100
γ [M1+E2]	924.66 *9*	0.86 *12*
γ [E2]	935.05 *10*	0.76 *10*
γ M1	977.43 *6*	61 *5*
γ M1	987.15 *6*	55 *5*
γ	1012.46 *20*	0.81 *14*
γ (M1)	1015.98 *8*	1.83 *19*
γ	1027.6 *25*	1.0 *2*
γ [M1+E2]	1037.76 *9*	4.7 *6*
γ	1040.2 *3*	1.7 *3*
γ	1048.9 *3*	0.73 *25*
γ [M1+E2]	1080.52 *8*	2.7 *4*
γ [E2]	1090.24 *8*	0.42 *6*
γ (M1+~50%E2)	1102.30 *9*	2.9 *4*
γ (M1+~20%E2)	1112.02 *9*	10.6 *12*

$^{187}_{78}$Pt(2.35 *3* h)

Mode: ϵ

Δ: -36830 *200* keV syst

SpA: 7.13$\times10^{6}$ Ci/g

Prod: protons on Ir; protons on Au

Photons (^{187}Pt)

γ_{mode}	γ(keV)	γ(rel)
γ M2+4.8%E3	76.22 3	0.64 7
γ [M1]	79.56 3	1.4 4
γ M1+2.9%E2	83.17 3	23 2
γ E1	91.59 5	8.7 9
γ (M1+E2)	97.68 5	3.2 5
γ E2	106.57 3	100
γ M1+29%E2	110.18 3	65 6
γ M1+2.7%E2	122.15 3	33 3
γ (M1)	159.67 7	4.6 8
γ (M1)	162.54 15	~1
γ (M1)	166.40 15	0.5 2
γ (M1)	175.13 5	8.0 8
γ E3	186.40 4	16 7
γ E2	187.18 7	43 5
γ (M1+32%E2)	189.74 3	10.8 10
γ E1	199.21 7	8.5 10 ?
γ	200.94 20	
γ [M1]	201.71 4	8.1 14
γ [E1]	201.77 5	74 18
γ E2	205.32 4	9.5 10
γ (M1)	244.94 20	9 4
γ (M1)	245.25 17	8 3
γ M1	247.75 8	42 4
γ [E2]	281.82 7	~5
γ E1	282.38 7	27 4
γ (M1)	284.85 7	63 6
γ (M1)	300.44 10	12.5 10
γ E2	304.89 8	49 5
γ (M1)	311.88 4	21 2
γ (M1+30%E2)	329.71 10	8 1
γ [E1]	332.91 9	2.9 6
γ (M1)	361.38 7	8.2 10
γ (M1)	376.52 12	4.1 3
γ (M1)	385.29 20	4.6 3
γ E1	388.95 7	8 1
γ	400.94 10	18 3
γ [M1]	427.39 9	25 4
γ M1	480.74 15	13 1
γ E1	486.63 7	14 1
γ	492.1 4	4 1
γ M1	507.65 7	15 2
γ	511.8 3	<16
γ (M1)	530.11 17	7.4 10
γ (M1)	536.6 4	6.3 15
γ	622.64 20	15 2
γ M1	629.80 6	31 3
γ	696.1 7	10 3
γ M1	709.35 6	59 5
γ (M1)	712.96 7	11.6 16
γ	732.1 4	~6
γ	790.3 4	10 3
γ	792.8 4	13 2
γ	796.8 3	10 2
γ	816.6 4	12 2
γ M1	819.53 6	42 4
γ	861.2 4	11 2
γ	891.9 3	7.5 20
γ	895.5 3	20 3
γ	912.8 3	20 3
γ	978.0 4	18 7
γ	1145.57 20	19 4
γ	1149.18 20	12 4
γ	1157.1 7	13 4
γ	1255.74 20	27 4
γ	1269.6 5	11 2

$^{187}_{79}$Au(8.0 *4* min)

Mode: ϵ, α?

Δ: -33110 *150* keV syst

SpA: 1.26×10^{8} Ci/g

Prod: ^{174}Yb(^{19}F,6n)

Alpha Particles (^{187}Au)

α(keV)
4690 20 ?

Photons (^{187}Au)

γ_{mode}	γ(keV)	γ(rel)
γM1	31.68 9	2.5 5
γM1+~5.9%E2	48.93 5	5.4 16
γM1+1.0%E2	51.29 7	16 5
γM1+3.5%E2	65.22 6	19 4
γM1+1.0%E2	74.50 5	5.9 12
γ	84.81 13	0.61 12
γE2+~41%M1	97.67 6	1.6 3
γM1+~20%E2	115.86 6	3.5 7
γM2	117.31 13	0.95 19
γ	129.8 7	~1
γ	131.6 7	~1
γM1+E2	133.10 10	~1 ?
γM1+E2	138.14 7	~0.6 ?
γM1+E2	139.07 8	~0.9 ?
γ	140.1 7	1.00 20
γ	142.0 7	1.21 24
γM1+~50%E2	161.31 13	1.13 23
γM1+20%E2	164.78 6	4.3 9
γM1(+<26%E2)	168.24 11	0.52 22
γ	172.61 13	1.3 3
γ	178.6 7	1.17 22
γE2	181.08 6	16 3
γ(E2)	185.48 11	13 4
γE2+~41%M1	185.79 6	26 4
γE2	190.36 6	15 3
γ	191.01 13	1.7 4
γ(M1+~20%E2)	194.31 9	0.61 12 ?
γ	204.61 13	3.7 7
γ	206.01 13	1.3 3
γ(M1+14%E2)	208.36 9	1.6 3 ?
γM1(+E2)	209.00 8	5.2 10
γE2	213.52 6	7.4 15
γM1(+E2)	235.81 6	8.7 22
γ(E2)	236.74 7	11 4
γ(M1+25%E2)	247.35 9	15 3
γM1(+E2)	251.02 6	16 5
γ	251.37 22	5.6 17 ?
γ(E0+M1+E2)	260.29 6	4.8 10
γ(E0+M1+E2)	262.45 6	1.6 3 ?
γ	265.11 13	2.5 5
γM1+E2	278.75 6	3.0 6 ?
γM1+E2	288.03 6	1.6 3 ?
γ	294.81 13	5.6 11
γ	345.81 13	3.0 6
γM1+33%E2	351.67 6	14 3
γM1+E2	355.42 20	3.0 6 ?
γM1(+E2)	368.91 10	11.3 23
γ	374.88 8	9.1 18
γE2	390.31 10	22 4
γ(M1+50%E2)	400.59 6	10.4 22
γ	405.5 7	3.0 6
γ	413.5 7	5.2 10
γ	416.89 7	4.8 10 ?
γ	422.8 7	2.6 5
γM1	426.17 6	27 5
γ	429.0 7	7.8 16
γ	448.6 4	4.8 10
γ(M1)	456.36 11	6.1 12 ?
γM1(+E2)	468.1 3	6.9 14
γ(M1)	470.99 21	9.1 18 ?
γ	473.7 4	4.3 9
γ	482.07 10	4.3 9 ?
γ	484.8 4	5.2 10
γ(E0+M1+E2)	498.5 5	3 1 ?
γ	541.9 4	6.1 12
γ(M1+E2)	545.98 10	11.3 23
γM1(+E2)	559.8 4	~11 ?
γ(E2)	560.02 10	~23
γ	563.22 12	5.2 10
γ(M1+E2)	591.23 19	7.4 15
γ(M1+E2)	594.90 10	4.3 9
γ(M1+14%E2)	608.95 10	4.8 10 ?
γ	614.4 4	3.5 7
γ(M1+E2)	620.48 10	26 5

Photons (^{187}Au) (continued)

γ_{mode}	γ(keV)	γ(rel)
γ(M1+14%E2)	634.52 10	16 3
γ	705.4 4	8.7 17
γ(E1)	706.83 18	31 6
γ[E1]	720.87 19	3.5 7
γ(E1)	721.3 4	17 4
γ(M1+E2)	730.30 19	7.8 16
γ(E2)	833.8 4	17 3
γE1	915.18 17	43
γ	924.8 4	6.1 12
γ	1189.9 4	16 3
γE1	1266.85 18	35 7
γE1	1319.1 13	28 6
γE1	1332.1 3	100 20
γ	1358.9 4	10.0 20
γ	1379.7 4	4.8 10
γE1	1408.1 4	39 8
γ	1417.3 4	9.5 19
γ	1426.3 4	10.0 20
γ	1433.4 4	13 3
γ	1452.1 13	24 5
γ	1491.4 13	6.9 14
γ	1726.5 13	6.1 12
γ	1754.7 13	11.7 23
γ	1766.7 13	6.1 12
γ	1776.2 13	6.5 13
γ	1907.6 13	7.8 16
γ	1960.5 13	14 3
γ	1983.9 13	12.1 24
γ	1988.1 13	17 4
γ	2004.6 13	7.8 16
γ	2013.2 13	9.5 19
γ	2027.0 13	16 3
γ	2052.8 13	11.3 23
γ	2081.3 13	10.8 22

$^{187}_{80}$Hg(2.4 *3* min)

Mode: ϵ, α(>0.00012 %)

Δ: -28170 *210* keV syst

SpA: 4.2×10^{8} Ci/g

Prod: protons on Pb

Alpha Particles (^{187}Hg)

α(keV)
4870 20

Photons (^{187}Hg(2.4 min + 1.7 min))

γ_{mode}	γ(keV)	γ(rel)
γE3	101.12 18	6.5 13
γM1(+<20%E2)	103.42 18	32 7
γ[M1+E2]	127.40 15	0.5 1
γ[M1+E2]	133.74 23	0.80 16
γM1	142.95 12	2.7 5
γ(E2+46%M1)	153.82 20	8.8 18
γ[E1]	185.80 24	2.2 4
γM1+27%E2	203.50 19	19 4
γ(M1+20%E2)	205.52 20	10.4 21
γE2+28%M1	220.81 13	24 5
γE2	233.41 14	100
γE2	240.28 15	33 7
γ[E2]	252.72 17	8.0 16
γM1+13%E2	255.10 15	8.9 18
γ(M1+20%E2)	257.47 25	3.4 7
γ(M1)	259.08 21	1.8 4
γM1+25%E2	271.46 17	31 6
γ	284.22 20	7.0 14
γE1	298.52 20	10.6 21

Photons (^{187}Hg(2.4 min + 1.7 min)) (continued)

γ_{mode}	γ(keV)	γ(rel)
γE1	305.87 *15*	4.0 *8*
γ(E2)	319.15 *14*	6.7 *13*
γ	322.92 *20*	11.8 *24*
γE2	334.70 *15*	16 *3*
γ(M1+>42%E2)	336.60 *22*	2.4 *5*
γ(M1)	349.42 *20*	5.6 *11*
γ(M1)	363.61 *23*	2.9 *6*
γM1+38%E2	376.36 *14*	38 *8*
γ	388.1 *3*	2.5 *5*
γM1	393.47 *18*	8.3 *17*
γ(M1)	410.32 *20*	4.6 *9*
γ(M1+25%E2)	430.02 *20*	7.1 *14*
γ(E2+<22%M1)	438.72 *20*	11.3 *23*
γ	447.22 *20*	5.8 *12*
γE2	449.51 *20*	29 *6*
γ(M1)	459.62 *19*	4.2 *8*
γM1+22%E2	462.10 *13*	10.3 *21*
γE2	470.33 *18*	29 *6*
γ(E2+<31%M1)	472.70 *18*	10.6 *21*
γE2+38%M1	475.91 *15*	22 *5*
γ(M1)	476.56 *18*	11.1 *22*
γ	484.8 *3*	2.6 *5*
γM1	499.38 *16*	19 *4*
γ[E1]	501.99 *19*	3.5 *7*
γ(M1)	525.42 *20*	30 *6*
γ	537.32 *20*	2.1 *4*
γ(E1)	564.95 *18*	4.7 *9*
γ	571.42 *20*	4.5 *9*
γ(M1)	602.58 *19*	2.1 *4*
γ(M1)	616.32 *23*	1.8 *4*
γE1	625.02 *13*	14 *3*
γ(M1)	639.32 *20*	7.5 *15*
γ(E2+35%M1)	652.4 *4*	1.6 *3*
γ(M1)	657.42 *20*	6.8 *14*
γ	668.81 *25*	4.3 *9*
γ(E2)	709.6 *3*	6.4 *13*
γ[E1]	767.97 *13*	3.5 *7*
γ	835.99 *19*	2.4 *5*
γ	963.22 *20*	3.4 *7*
γ	1180.92 *20*	3.5 *7*
γ	1197.22 *20*	3.4 *7*
γ	1579.7 *4*	3.8 *8*
γ	1647.3 *5*	4.7 *9*
γ	1803.8 *6*	3.3 *7*
γ	1998.1 *8*	10.8 *22*
γ	2012.6 *8*	10.3 *21*
γ	2028.1 *8*	6.4 *13*
γ	2074.5 *9*	13 *3*
γ	2079.7 *9*	4.6 *9*
γ	2176.5 *10*	20 *4*

$^{187}_{80}$Hg(1.7 *2* min)

Mode: ϵ, α(>0.00025 %)
Δ: -28170 *210* keV syst
SpA: 5.9×10^{8} Ci/g
Prod: protons on Pb

see ^{187}Hg(2.4 min) for γ rays

Alpha Particles (^{187}Hg)

α(keV)
5035 *20*

$^{187}_{81}$Tl(~51 s)

Mode: ϵ
Δ: -22130 *370* keV syst
SpA: $\sim1.2\times10^{9}$ Ci/g
Prod: ^{180}W(^{14}N,7n)

$^{187}_{81}$Tl(15.60 *12* s)

Mode: IT, ϵ, α
Δ: -21801 *370* keV syst
SpA: 3.78×10^{9} Ci/g
Prod: ^{180}W(^{14}N,7n); ^{182}W(^{14}N,9n)

Alpha Particles (^{187}Tl)

α(keV)
5520 *10*

Photons (^{187}Tl)

γ_{mode}	γ(keV)
γ_{IT} (E2+21%M1)	299.33 *24*

$^{187}_{82}$Pb(18.3 *3* s)

Mode: ϵ(98 %), α(2 %)
Δ: -15060 *300* keV syst
SpA: 3.24×10^{9} Ci/g
Prod: ^{142}Nd(^{48}Ti,3n); ^{107}Ag(^{84}Kr,p3n); ^{155}Gd(^{40}Ar,8n); ^{150}Sm(^{40}Ca,3n)

Alpha Particles (^{187}Pb)

α(keV)
6073 *10*

Photons (^{187}Pb)

γ_{mode}	γ(keV)	γ(rel)
γ_ϵ	193.0 *3*	15.0 *10*
γ_ϵ[E1]	331.4 *3*	60 *3*
γ_ϵ	343.5 *3*	75 *4*
γ_ϵ[M1+E2]	393.4 *3*	100 *5*

$^{187}_{82}$Pb(15.2 *3* s)

Mode: ϵ, α
Δ: -15060 *300* keV syst
SpA: 3.88×10^{9} Ci/g
Prod: ^{142}Nd(^{48}Ti,3n); ^{107}Ag(^{84}Kr,p3n)

Alpha Particles (^{187}Pb)

α(keV)	α(rel)
5993 *10*	40 *4*
6194 *10*	60 *6*

Photons (^{187}Pb)

γ_{mode}	γ(keV)	γ(rel)
γ_α(E2)	67.5 *3*	
γ_α	187.1 *10*	
γ_α	195.1 *10*	
γ_α(M1)	208.1 *3*	
γ_α(M1)	275.6 *3*	
γ_ϵ(E2+21%M1)	299.33 *24*	100 *5*
γ_ϵ	309.4 *3*	1.33 *7*
γ_ϵ[M1+E2]	448.53 *24*	1.33 *7*
γ_ϵ	493.6 *3*	2.67 *13*
γ_ϵ	617.2 *3*	2.67 *13*
γ_ϵ	645.4 *3*	1.00 *7*
γ_ϵ[E2]	747.87 *24*	1.00 *7*
γ_ϵ	865.8 *3*	<0.7

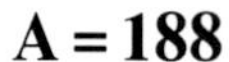

A = 188

NDS 33, 273 (1981)

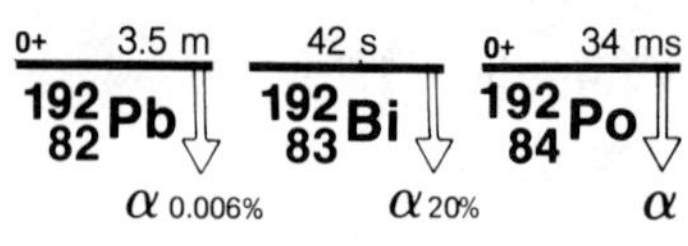

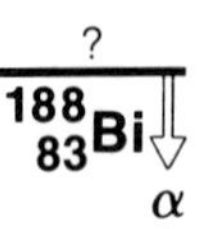

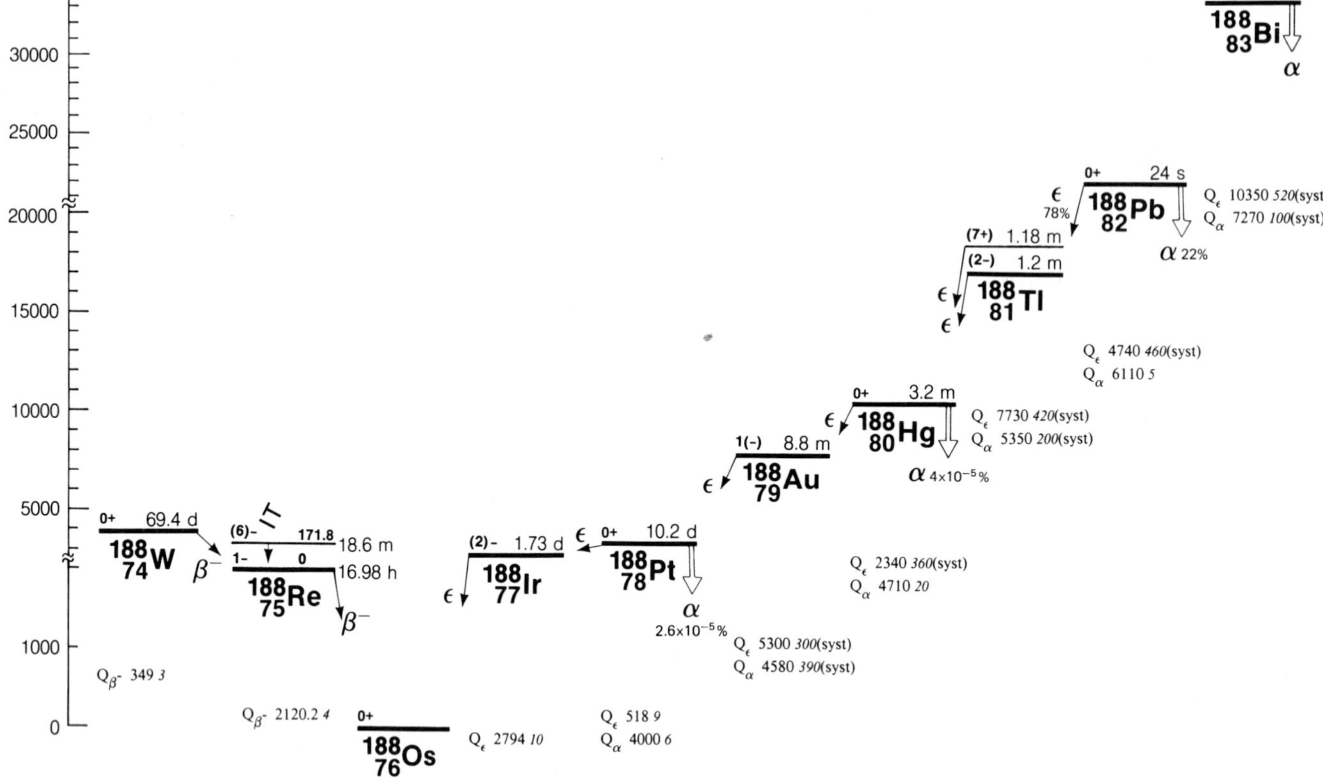

$^{188}_{74}$W (69.4 5 d)

Mode: β-
Δ: -38676 4 keV
SpA: 1.001×10^4 Ci/g
Prod: multiple n-capture from ^{186}W

Photons (^{188}W)

⟨γ⟩=1.89 4 keV

γ_{mode}	γ(keV)	γ(%)†
Re L_ℓ	7.604	0.0020 4
Re L_α	8.646	0.046 7
Re L_η	9.027	0.00067 10
Re L_β	10.089	0.074 13
Re L_γ	11.916	0.017 3
Re $K_{\alpha2}$	59.717	0.051 6
Re $K_{\alpha1}$	61.141	0.088 11
γ M1+0.5%E2	63.580 18	0.109 16
Re $K_{\beta1}$'	69.214	0.030 4
Re $K_{\beta2}$'	71.470	0.0076 9
γ M1,E2	85.31 3	0.0024 8
γ M1(+13%E2)	105.90 6	<0.0012 ?
γ M1(+18%E2)	141.78 3	0.0064 8
γ [E2]	169.48 6	<1.2×10^{-5}
γ E1	207.86 4	0.0080 16
γ M1(+21%E2)	227.093 16	0.221 8
γ M1(+21%E2)	290.672 12	0.402 12

† 4.2% uncert(syst)

Atomic Electrons (^{188}W)

⟨e⟩=0.65 5 keV

e_{bin}(keV)	⟨e⟩(keV)	e(%)
11 - 49	0.032	0.25 3
51	0.135	0.26 4
52	0.0154	0.030 5
53 - 60	0.0057	0.0106 20
61	0.042	0.069 11
62 - 106	0.028	0.041 7
129 - 142	0.0040	0.0030 5
155	0.130	0.083 15
157 - 206	0.000157	8.0 10 ×10^{-5}
207 - 208	8.8 ×10^{-6}	4.2 7 ×10^{-6}
215	0.031	0.014 4
217	0.0023	~0.0011
219	0.17	0.077 14
224 - 227	0.0105	0.0047 9
278	0.031	0.0111 21
279 - 291	0.019	0.0066 12

Continuous Radiation (^{188}W)

⟨β-⟩=99 keV;⟨IB⟩=0.035 keV

E_{bin}(keV)		⟨ ⟩(keV)	(%)
0 - 10	β-	0.342	6.9
	IB	0.0052	
10 - 20	β-	0.98	6.6
	IB	0.0044	0.031
20 - 40	β-	3.66	12.3
	IB	0.0072	0.025

Continuous Radiation (^{188}W) (continued)

E_{bin}(keV)		⟨ ⟩(keV)	(%)
40 - 100	β-	21.2	31.0
	IB	0.0122	0.020
100 - 300	β-	71	42.9
	IB	0.0061	0.0047
300 - 349	β-	1.43	0.457
	IB	4.7×10^{-6}	1.53×10^{-6}

$^{188}_{75}$Re(16.98 2 h)

Mode: β-
Δ: -39025 3 keV
SpA: 9.817×10^5 Ci/g
Prod: ^{187}Re(n,γ)

Photons (^{188}Re)

⟨γ⟩=57.4 9 keV

γ_{mode}	γ(keV)	γ(%)†
Os L_ℓ	7.822	0.052 5
Os L_α	8.904	1.16 7
Os L_η	9.337	0.0223 14
Os L_β	10.453	1.39 9
Os L_γ	12.149	0.242 19

Photons (^{188}Re)
(continued)

γ_{mode}	γ(keV)	γ(%)†
Os $K_{\alpha2}$	61.485	1.34 5
Os $K_{\alpha1}$	63.000	2.31 9
Os $K_{\beta1}$'	71.313	0.80 3
Os $K_{\beta2}$'	73.643	0.202 8
γ E2	155.064 17	14.9 4
γ E2	312.05 3	0.0037 5
γ E2	322.948 24	0.0157 11
γ E2	350.22 5	0.0032 6
γ M1,E2	385.50 5	9.9 14 $\times10^{-5}$
γ E2	453.346 22	0.0705 19
γ E2+0.8%M1+E0	478.013 15	1.04 3
γ	486.11 4	0.078 3
γ [E2]	491.74 5	0.00059 14
γ E2	514.83 4	0.0048 5
γ	538.09 5	8.1 13 $\times10^{-5}$
γ	557.73 12	0.00093 16
γ E1	624.1 4	0.0029 5
γ E2	633.077 21	1.25 10
γ E2+3.9%M1	634.996 25	0.105 18
γ [M1+E2]	652.40 10	0.0008 3
γ [M1+E2]	667.41 4	0.00055 15
γ E1	672.539 21	0.111 3
γ E2	810.53 24	0.00074 23
γ M1,E2	824.40 4	0.0121 21
γ E1(+1.2%M2)	829.522 21	0.408 11
γ E2	845.10 3	0.0071 8
γ E2	931.357 21	0.562 14
γ [E2]	979.46 4	0.00069 17
γ [M2]	984.59 3	0.0051 12
γ M1(+<20%E2)	1017.64 3	0.0147 10
γ E2	1071.3 3	0.00080 16
γ E2	1132.40 3	0.088 4
γ M1,E2	1149.83 4	0.018 3
γ M1,E2	1151.07 5	0.017 4
γ M1(+<20%E2)	1174.62 3	0.0194 13
γ E2	1191.97 8	0.0138 9
γ M1(+25%E2)	1209.90 4	0.0029 3
γ [M1+E2]	1302.41 4	0.0045 7
γ (E2)	1304.89 4	0.0060 7
γ (E1)	1307.533 24	0.0117 24
γ M1(+E2)	1308.05 5	0.067 3
γ E2	1323.11 3	0.0092 11
γ [E2]	1329.68 4	0.003 1
γ (E2)	1457.47 4	0.0201 11
γ [E2]	1549.3 3	0.0026 5
γ [E2]	1610.41 3	0.098 4
γ M1,E2	1652.63 3	0.0042 4
γ [E2]	1669.98 8	0.0106 8
γ M1,E2	1687.91 4	0.00031 4
γ [M1+E2]	1786.06 5	0.0207 11
γ [M1+E2]	1802.22 9	0.0374 16
γ [E2]	1807.69 4	0.0015 3
γ	1842.97 5	6.7 10 $\times10^{-5}$
γ [M1+E2]	1865.5 7	0.0056 5 ?
γ [E2]	1941.12 5	0.00208 21
γ (E2)	1957.28 10	0.0159 10
γ [E2]	2020.6 7	0.00177 21 ?

† 3.8% uncert(syst)

Atomic Electrons (^{188}Re)
⟨e⟩=15.79 22 keV

e_{bin}(keV)	⟨e⟩(keV)	e(%)
11 - 60	0.77	6.1 4
61 - 71	0.0151	0.0238 14
81	3.92	4.82 15
142	0.784	0.551 17
143	4.11	2.88 9
144	3.09	2.14 7
152	1.30	0.85 3
153	0.87	0.568 18
154	0.0497	0.0322 10
155	0.579	0.374 12
238 - 276	0.00275	0.00110 9
299 - 348	0.00177	0.000564 24
350 - 386	0.00579	0.00153 5

Atomic Electrons (^{188}Re)
(continued)

e_{bin}(keV)	⟨e⟩(keV)	e(%)
404 - 453	0.096	0.0235 23
464 - 513	0.0357	0.0076 4
514 - 561	0.078	0.0140 11
579 - 624	0.0212	0.00344 17
630 - 673	0.0064	0.00101 5
737 - 771	0.0098	0.00130 19
798 - 845	0.0021	0.00026 3
857 - 906	0.0226	0.00264 9
911 - 944	0.0080	0.00086 7
966 - 1016	0.00054	5.5 11 $\times10^{-5}$
1017 - 1060	0.00303	0.000286 14
1068 - 1101	0.0032	0.00029 6
1118 - 1164	0.00194	0.000172 10
1172 - 1210	0.000202	1.71 7 $\times10^{-5}$
1229 - 1256	0.0040	0.00033 10
1289 - 1330	0.00086	6.6 15 $\times10^{-5}$
1445 - 1538	0.00273	0.000178 10
1546 - 1642	0.00096	6.0 4 $\times10^{-5}$
1650 - 1734	0.0020	0.000113 21
1769 - 1867	0.00062	3.5 4 $\times10^{-5}$
1883 - 1955	0.00047	2.48 14 $\times10^{-5}$
2008 - 2018	7.7 $\times10^{-6}$	3.8 4 $\times10^{-7}$

Continuous Radiation (^{188}Re)
⟨β-⟩=765 keV;⟨IB⟩=1.46 keV

E_{bin}(keV)		⟨ ⟩(keV)	(%)
0 - 10	β-	0.0256	0.51
	IB	0.030	
10 - 20	β-	0.077	0.51
	IB	0.030	0.21
20 - 40	β-	0.314	1.05
	IB	0.058	0.20
40 - 100	β-	2.31	3.28
	IB	0.161	0.25
100 - 300	β-	25.4	12.5
	IB	0.42	0.24
300 - 600	β-	100	22.1
	IB	0.40	0.095
600 - 1300	β-	425	46.1
	IB	0.34	0.042
1300 - 2120	β-	212	14.0
	IB	0.028	0.0020

$^{188}_{75}$Re(18.6 *1* min)

Mode: IT
Δ: -38853 3 keV
SpA: 5.38$\times10^{7}$ Ci/g
Prod: ^{187}Re(n,γ)

Photons (^{188}Re)
⟨γ⟩=74.8 22 keV

γ_{mode}	γ(keV)	γ(%)†
Re L_ℓ	7.604	0.92 11
Re L_α	8.646	21.0 19
Re L_η	9.027	0.186 21
Re L_β	10.133	21 3
Re L_γ	11.899	4.1 6
Re $K_{\alpha2}$	59.717	18.1 15
Re $K_{\alpha1}$	61.141	31 3
γ M1+0.5%E2	63.580 18	21.6 10
Re $K_{\beta1}$'	69.214	10.5 9
Re $K_{\beta2}$'	71.470	2.69 23
γ M1(+4.1%E2)	92.40 7	5.2 3

Photons (^{188}Re)
(continued)

γ_{mode}	γ(keV)	γ(%)†
γ M1(+13%E2)	105.90 6	10.8 5
γ [E2]	155.98 7	0.62 21
γ [E2]	169.48 6	~0.10

† 0.97% uncert(syst)

Atomic Electrons (^{188}Re)
⟨e⟩=95 3 keV

e_{bin}(keV)	⟨e⟩(keV)	e(%)
2 - 5	2.95	82 4
11	5.6	54 7
12	1.9	16.1 25
13	3.3	26 3
13 - 16	2.24	16.0 9
21	5.8	28.2 21
34	12.7	37 5
47 - 49	0.71	1.47 13
51	26.9	53 3
52	3.07	5.9 4
53 - 60	1.35	2.5 3
61	8.4	13.9 8
62	0.0133	0.0216 21
63	2.52	4.00 23
64 - 69	0.143	0.221 12
80	3.9	4.9 7
82 - 92	1.9	2.2 4
93	5.0	5.4 8
94 - 106	5.8	5.8 19
143 - 169	0.46	0.31 5

$^{188}_{76}$Os(stable)

Δ: -41145 3 keV
%: 13.3 2

$^{188}_{77}$Ir(1.729 *21* d)

Mode: ε
Δ: -38350 11 keV
SpA: 4.02$\times10^{5}$ Ci/g
Prod: alphas on Re; ^{189}Os(p,2n); deuterons on Os

Photons (^{188}Ir)
⟨γ⟩=2099 43 keV

γ_{mode}	γ(keV)	γ(%)†
Os L_ℓ	7.822	0.60 9
Os L_α	8.904	13.4 18
Os L_η	9.337	0.185 24
Os L_β	10.471	13.7 19
Os L_γ	12.203	2.4 4
Os $K_{\alpha2}$	61.485	26 3
Os $K_{\alpha1}$	63.000	44 6
Os $K_{\beta1}$'	71.313	15.2 20
Os $K_{\beta2}$'	73.643	3.9 5
γ E2	155.064 17	29.7 24
γ	272.17 8	0.026 4
γ E2	312.05 3	0.193 18
γ E2	322.948 24	1.62 13
γ E2	332.66 3	0.072 7
γ E2	350.22 5	0.24 4
γ	383.50 8	0.035 6

Photons (^{188}Ir) (continued)

γ_{mode}	γ(keV)	γ(%)†
γ M1,E2	385.50 5	0.235 22
γ	389.98 15	0.019 4
γ	412.00 15	0.012 4
γ	413.74 5	0.035 6
γ M1,E2	424.75 7	0.034 6
γ E2	448.11 7	0.075 15
γ E2	453.346 22	0.033 3
γ E2+0.8%M1+E0	478.013 15	14.8 4
γ E2	487.72 4	0.212 22
γ [E2]	491.74 5	0.050 9
γ E2	514.83 4	0.138 12
γ	522.36 9	0.021 6
γ	534.07 14	0.013 6
γ	538.09 5	0.194 22
γ M1(+<50%E2)	566.64 6	0.212 21
γ M1	586.48 15	0.035 6
γ [E2]	594.16 5	0.096 12
γ	596.55 12	0.026 6
γ	600.96 15	0.028 7
γ E2	623.79 8	0.26 3
γ E2	633.077 21	17.8 22
γ E2+3.9%M1	634.996 25	5.5 8
γ M1(+<45%E2)	641.63 4	0.39 4
γ	646.18 15	0.031 9
γ [M1+E2]	652.40 10	0.021 7
γ	663.46 6	0.063 10
γ [M1+E2]	667.41 4	0.047 10
γ E1	672.539 21	1.42 7
γ	695.47 15	0.035 6
γ	703.49 10	0.025 9
γ	719.63 5	0.022 7
γ	730.59 10	0.076 12
γ M1(+<39%E2)	736.54 5	0.29 3
γ M1(+<26%E2)	747.34 8	0.049 9
γ [E1]	752.12 4	0.075 12
γ M1(+<39%E2)	757.24 5	0.39 4
γ	763.91 7	0.031 7
γ	776.84 25	0.053 18
γ	777.97 20	0.096 18
γ M1,E2	781.94 20	0.17 3
γ	794.21 5	0.037 7
γ E2	810.67 3	0.171 18
γ M1,E2	824.40 4	1.03 9
γ E1(+1.2%M2)	829.522 21	5.2 3
γ E2	845.10 3	0.29 5
γ [M1+E2]	886.22 6	0.253 25
γ	895.30 5	0.140 15
γ	899.93 7	0.109 12
γ [M1+E2]	909.83 5	0.034 9
γ E2	931.357 21	0.262 24
γ	933.99 20	0.15 3
γ	935.3 2	0.17 3
γ M1(+<50%E2)	939.58 7	0.66 5
γ [M1+E2]	947.11 7	0.129 13
γ	972.18 20	0.028 9
γ [E2]	979.46 4	0.059 12
γ [M2]	984.59 3	0.065 12
γ M1(+<39%E2)	987.48 4	0.93 9
γ [E2]	999.31 10	0.071 12
γ M1	1012.68 5	0.121 18
γ M1(+<20%E2)	1017.64 3	1.07 7
γ	1052.29 5	0.034 10
γ M1(+<50%E2)	1096.57 7	1.46 12
γ [M1]	1128.29 5	0.063 7
γ	1132.5 4	0.028 10
γ M1,E2	1142.54 5	0.37 7
γ M1,E2	1149.83 4	0.51 8
γ M1(+<20%E2)	1174.62 3	1.42 11
γ M1(+25%E2)	1209.90 4	7.0 6
γ	1251.63 7	0.026 9
γ [E2]	1286.27 7	0.032 12
γ (M1)	1295.46 8	0.134 13
γ [M1+E2]	1302.41 4	0.39 5
γ (E2)	1304.89 4	0.171 20
γ (E1)	1307.533 24	0.150 21
γ E2	1323.11 3	0.38 5
γ [E2]	1329.68 4	0.22 7
γ E2(+>34%M1)	1331.96 10	0.47 4
γ	1336.42 15	0.140 13
γ	1349.58 15	0.063 9
γ	1414.75 7	0.076 7
γ	1430 1	0.25 6 ?
γ M1,E2	1435.59 8	1.48 12
γ	1445.8 10	0.41 6 ?

Photons (^{188}Ir) (continued)

γ_{mode}	γ(keV)	γ(%)†
γ M1(+<50%E2)	1452.44 8	1.06 9
γ (E2)	1457.47 4	1.71 11
γ [M1+E2]	1461.94 6	0.41 4 ?
γ M1,E2	1465.49 4	1.35 13
γ [E2]	1487.03 10	0.075 10
γ	1530.30 5	0.23 3
γ E1	1558.76 6	0.87 7
γ	1571.74 7	0.115 16
γ M1,E2	1574.58 7	2.63 21
γ M1,E2	1618.92 6	0.49 4
γ M1,E2	1652.63 3	0.304 23
γ M1,E2	1687.91 4	0.73 6
γ (E2)	1704.9 10	1.04 15 ?
γ E1+0.8%M2	1715.74 6	6.1 5
γ	1726.80 8	0.118 12
γ [E2]	1773.99 7	0.060 7
γ	1782.87 10	0.068 7
γ [M1+E2]	1802.22 9	0.98 6
γ [E2]	1807.69 4	0.108 19
γ M1+E2+E0	1809.97 10	0.34 3
γ	1842.97 5	0.160 15
γ	1887.39 22	0.081 12
γ	1904.0 4	0.135 13
γ	1930.45 8	0.29 3
γ (E2)	1944.03 5	3.9 3
γ (E2)	1957.28 10	0.42 3
γ	1971.7 5	0.168 16
γ M1+E2	2011.02 10	0.61 5
γ	2040.80 25	0.49 4
γ M1,E2	2049.74 7	5.0 4
γ M1+4.2%E2	2059.64 4	7.0 5
γ	2068.66 8	0.057 9
γ [M1+E2]	2096.93 6	5.7 7
γ	2099.10 5	4.8 6
γ	2131.25 15	0.27 3
γ	2133.8 5	0.096 12
γ	2144.83 22	0.160 16
γ	2171.12 13	0.072 9
γ [M1+E2]	2192.46 14	0.35 6
γ E1+M2	2193.75 6	2.0 4
γ M1	2214.71 4	18.7 13
γ	2219.1 3	0.191 19
γ	2221.97 15	0.228 22
γ [E2]	2251.99 7	0.35 3
γ	2260.87 10	0.082 10
γ	2286.31 15	0.076 10
γ	2299.89 22	0.013 6
γ	2305.57 18	0.137 18
γ	2326.18 13	0.085 10
γ	2336.0 3	0.046 9
γ (M1)	2347.52 14	0.64 7
γ	2365.40 22	0.073 13
γ	2374.2 3	0.031 7
γ	2385.7 4	0.019 6
γ	2394.51 12	0.165 25
γ	2406.3 3	0.035 7
γ	2426.79 23	0.021 6
γ	2460.63 18	0.23 3
γ	2467.49 21	~0.007
γ	2486.9 3	0.031 9
γ	2504.91 25	0.119 6
γ	2520.46 22	~0.012
γ	2565.6 5	~0.012
γ	2581.85 23	0.041 10
γ	2622.55 21	0.081 12

† 4.1% uncert(syst)

Atomic Electrons (^{188}Ir)

⟨e⟩=46.7 13 keV

e_{bin}(keV)	⟨e⟩(keV)	e(%)
11	3.9	36 6
12	2.4	19 3
13 - 62	2.6	9.4 10
63 - 71	0.118	0.175 14
81	7.8	9.6 8
142	1.56	1.10 9

Atomic Electrons (^{188}Ir) (continued)

e_{bin}(keV)	⟨e⟩(keV)	e(%)
143	8.2	5.7 5
144	6.1	4.3 4
152	2.58	1.70 14
153	1.73	1.13 9
154 - 198	1.26	0.81 6
238 - 276	0.261	0.104 7
299 - 348	0.24	0.075 11
350 - 399	0.036	0.0098 23
400 - 403	0.0015	~0.0004
404	1.16	0.288 9
409 - 453	0.039	0.0091 8
460 - 509	0.529	0.113 5
510 - 559	1.07	0.192 23
561 - 601	0.43	0.076 9
611 - 660	0.46	0.074 4
661 - 709	0.109	0.0160 17
717 - 766	0.22	0.030 6
767 - 817	0.071	0.0088 15
819 - 866	0.107	0.0125 17
873 - 922	0.107	0.0118 10
923 - 972	0.116	0.0123 7
974 - 1023	0.122	0.0121 7
1039 - 1086	0.069	0.0064 14
1094 - 1143	0.53	0.047 7
1147 - 1178	0.0219	0.00188 13
1197 - 1241	0.124	0.0103 12
1249 - 1297	0.051	0.0040 4
1299 - 1348	0.0134	0.00101 14
1349 - 1392	0.25	0.0179 22
1402 - 1451	0.039	0.0027 3
1452 - 1548	0.15	0.0103 21
1556 - 1653	0.180	0.0111 14
1675 - 1772	0.121	0.0070 7
1780 - 1877	0.117	0.0063 5
1883 - 1976	0.20	0.0102 17
1986 - 2085	0.59	0.029 4
2086 - 2183	0.71	0.033 4
2187 - 2284	0.140	0.0063 4
2287 - 2384	0.0115	0.00049 8
2387 - 2484	0.0080	0.00033 10
2492 - 2579	0.0028	0.00011 3
2610 - 2620	0.00025	10 4 $\times 10^{-6}$

Continuous Radiation (^{188}Ir)

⟨β+⟩=2.19 keV;⟨IB⟩=0.58 keV

E_{bin}(keV)		⟨ ⟩(keV)	(%)
0 - 10	β+	5.3×10^{-9}	6.3×10^{-8}
	IB	0.00039	
10 - 20	β+	3.39×10^{-7}	2.01×10^{-6}
	IB	0.00145	0.0128
20 - 40	β+	1.46×10^{-5}	4.39×10^{-5}
	IB	0.0023	0.0069
40 - 100	β+	0.00115	0.00144
	IB	0.35	0.57
100 - 300	β+	0.077	0.0344
	IB	0.035	0.020
300 - 600	β+	0.52	0.114
	IB	0.051	0.0118
600 - 1300	β+	1.47	0.171
	IB	0.097	0.0111
1300 - 2500	β+	0.117	0.0085
	IB	0.036	0.0024
	Σβ+		0.33

$^{188}_{78}$Pt(10.2 *3* d)

Mode: ε, α(2.6 *3* ×10^{-5} %)
Δ: -37832 *7* keV
SpA: 6.81×10^{4} Ci/g
Prod: ^{191}Ir(p,4n)

Alpha Particles (^{188}Pt)

α(keV)
3919 *7*

Photons (^{188}Pt)

⟨γ⟩=206 *5* keV

γ$_{mode}$	γ(keV)	γ(%)†
Ir L$_{\ell}$	8.042	0.85 *10*
Ir L$_{\alpha}$	9.167	18.8 *17*
Ir L$_{\eta}$	9.650	0.246 *22*
Ir L$_{\beta}$	10.818	19.6 *21*
Ir L$_{\gamma}$	12.655	3.7 *5*
γM1+0.3%E2	41.89 *4*	0.52 *6*
γM1+32%E2	54.74 *4*	0.76 *8*
Ir K$_{\alpha2}$	63.286	29.3 *21*
Ir K$_{\alpha1}$	64.896	50 *4*
Ir K$_{\beta1}$'	73.452	17.4 *12*
Ir K$_{\beta2}$'	75.861	4.5 *3*
γM1,E2	92.75 *12*	~0.023
γE2+34%M1	96.64 *4*	0.17 *7*
γM1(+0.5%E2)	98.30 *4*	0.34 *3*
γE2	132.75 *7*	0.25 *5*
γM1(+0.02%E2)	140.20 *5*	2.33 *12*
γM1+0.7%E2	187.50 *7*	19.4 *10*
γM1	194.94 *5*	18.6 *10*
γE1	197.75 *13*	<0.06
γE2+43%M1	280.25 *12*	0.31 *4*
γE1	283.06 *7*	~0.10
γE1	290.50 *9*	0.11 *4*
γE1	381.36 *7*	7.5 *4*
γE1	423.25 *7*	4.36 *23*
γE1	478.00 *7*	1.8 *3*

† uncert(syst) 3.6% for

Atomic Electrons (^{188}Pt)

⟨e⟩=78.7 *18* keV

e$_{bin}$(keV)	⟨e⟩(keV)	e(%)
11	4.6	41 *5*
13	4.4	34 *4*
17 - 41	3.97	12.7 *8*
42	2.7	6.5 *7*
44	2.8	6.4 *7*
49 - 51	0.75	1.49 *11*
52	2.4	4.5 *5*
53 - 63	1.63	2.85 *17*
64	3.06	4.8 *3*
65 - 98	1.06	1.26 *15*
111	19.1	17.2 *9*
119	17.7	14.9 *8*
120 - 140	1.61	1.26 *6*
174	4.42	2.54 *14*
175 - 176	0.478	0.273 *14*
182	4.35	2.40 *13*
184 - 214	3.12	1.65 *5*
267 - 305	0.323	0.108 *6*
347 - 381	0.214	0.0604 *25*
402 - 423	0.088	0.0216 *22*
465 - 478	0.0123	0.0026 *3*

Continuous Radiation (^{188}Pt)

⟨IB⟩=0.33 keV

E$_{bin}$(keV)		⟨ ⟩(keV)	(%)
10 - 20	IB	0.00165	0.0145
20 - 40	IB	0.0018	0.0054
40 - 100	IB	0.32	0.49
100 - 300	IB	0.0097	0.0073
300 - 330	IB	2.7×10^{-6}	8.7×10^{-7}

$^{188}_{79}$Au(8.84 *6* min)

Mode: ε
Δ: -32530 *300* keV syst
SpA: 1.131×10^{8} Ci/g
Prod: protons on Pt; daughter ^{188}Hg

Photons (^{188}Au)

γ$_{mode}$	γ(keV)	γ(rel)
γ	86.9 *3*	0.52 *12*
γ	132.04 *19*	0.51 *7*
γ	182.26 *25*	0.37 *8*
γ [E2]	193.05 *7*	0.17 *4*
γ [M1+E2]	197.43 *7*	0.21 *7*
γ	221.5 *7*	0.04 *1*
γ [E1]	234.70 *6*	0.24 *7*
γ	237.0 *7*	0.20 *2*
γ	238.4 *7*	0.13 *2*
γ	244.8 *7*	0.03 *1*
γ	253.1 *7*	0.13 *2*
γ E2	265.652 *16*	100 *3*
γ	295.17 *10*	1.00 *8*
γ [M1+E2]	313.05 *9*	0.16 *2*
γ E2	316.14 *7*	0.95 *9*
γ M1+41%E2	319.8 *7*	0.13 *2*
γ E2+0.5%M1	330.410 *22*	4.98 *19*
γ E2	340.054 *15*	23.9 *8*
γ	358.9 *7*	~0.15
γ	373.1 *4*	0.25 *8*
γ E2+39%M1	376.21 *6*	0.51 *9*
γ	381.2 *7*	0.08 *2*
γ E2	405.136 *19*	9.1 *3*
γ [E1]	413.48 *4*	0.51 *3* ?
γ M1,E2	414.10 *15*	0.9 *1*
γ (E2)	426.14 *7*	0.33 *9*
γ E2	444.11 *5*	1.08 *9*
γ (M1+E2+E0)	447.5 *7*	0.11 *2*
γ (M1)	452.4 *7*	0.06 *2*
γ M1(+16%E2)	455.4 *7*	0.16 *2*
γ M1,E2	465.2 *7*	0.17 *1*
γ	470.0 *7*	0.06 *1*
γ	470.7 *4*	~0.22
γ E2	479.18 *15*	1.28 *13*
γ	491.8 *7*	0.05 *1*
γ (E0+M1+E2)	497.95 *10*	0.54 *14*
γ	523.5 *4*	0.38 *10*
γ	529.3 *9*	~0.17
γ E2	533.10 *7*	5.86 *22*
γ E1	541.0 *3*	0.73 *11*
γ (M1)	549.8 *7*	0.10 *2*
γ	553.5 *5*	0.26 *8*
γ	557.8 *15*	<0.16
γ	589.1 *7*	~0.04
γ (M1)	591.54 *11*	0.15 *2* ?
γ E2	605.705 *16*	16.3 *6*
γ	618.4 *7*	0.10 *2*
γ (E2)	641.54 *6*	0.65 *11*
γ	667.5 *3*	0.65 *11*
γ [M1+E2]	670.463 *22*	8.0 *3*
γ E1	678.81 *4*	2.01 *12*
γ E2	689.27 *7*	0.39 *11*
γ M1(+20%E2)	695.38 *10*	0.19 *2*
γ E0+M1+E2	706.62 *6*	0.91 *11*
γ (M1)	713.1 *7*	0.12 *2*

Photons (^{188}Au) (continued)

γ$_{mode}$	γ(keV)	γ(rel)
γ	725.9 *7*	0.13 *2*
γ	735.64 *23*	0.43 *19*
γ (E0+M1+E2)	749.1 *4*	0.47 *12*
γ E0	798.76 *7*	<0.10
γ	813.7 *7*	0.08 *2*
γ [E2]	819.23 *15*	0.34 *12*
γ	821.4 *7*	0.08 *3*
γ (E0+M1+E2)	840.3 *7*	<0.050
γ (E0+M1+E2)	849.24 *5*	0.71 *25*
γ [E2]	856.87 *11*	0.11 *2*
γ [M1+E2]	874.16 *9*	0.59 *14*
γ	883.7 *6*	0.38 *14*
γ M1(+E2+E0)	921.95 *11*	0.87 *13*
γ	933.5 *6*	0.40 *13*
γ E2(+<2.7%M1)	948.72 *8*	2.44 *17*
γ (E1)	976.98 *8*	2.06 *17*
γ	1004.0 *7*	0.12 *2*
γ	1006.8 *7*	0.20 *2*
γ M1,E2	1013.2 *4*	0.72 *17*
γ	1017.67 *12*	1.70 *23*
γ [M1+E2]	1019.68 *7*	0.77 *22*
γ E0+M1+E2	1046.67 *5*	1.93 *19*
γ	1067.9 *7*	0.20 *3*
γ	1079.4 *4*	0.36 *15*
γ E1	1083.94 *4*	6.6 *3*
γ E2	1114.89 *5*	4.98 *19*
γ [E2]	1139.49 *9*	0.36 *12*
γ (E1)	1170.03 *6*	2.63 *19*
γ [M1+E2]	1204.57 *9*	1.63 *17*
γ	1214.37 *8*	0.48 *3* ?
γ	1243.7 *13*	0.34 *7*
γ E0(+M1+E2)	1262.00 *11*	1.26 *18*
γ	1284.1 *7*	0.26 *2*
γ	1306.19 *18*	0.81 *17*
γ E2	1312.33 *6*	2.96 *21*
γ	1332.6 *7*	0.40 *16*
γ	1340.9 *13*	<0.2
γ	1348.08 *12*	1.17 *17*
γ	1352.0 *4*	0.82 *17*
γ [M1+E2]	1359.73 *7*	4.13 *22*
γ	1364.8 *8*	0.39 *18*
γ	1377.2 *4*	0.61 *14*
γ	1405.0 *7*	0.23 *3*
γ	1408.56 *21*	0.96 *15*
γ	1451.1 *6*	0.61 *16*
γ	1476.4 *8*	0.32 *15*
γ	1484.25 *16*	0.88 *16*
γ	1496.7 *7*	0.20 *2*
γ (E1)	1510.08 *6*	2.74 *18*
γ [E2]	1527.65 *11*	0.72 *15*
γ [M1+E2]	1544.62 *9*	2.32 *18*
γ	1555.0 *3*	1.13 *16*
γ	1565.4 *3*	0.41 *15*
γ E0+M1+E2	1596.6 *3*	0.68 *15*
γ	1603.2 *7*	0.22 *5*
γ M1	1625.38 *7*	~0.9
γ	1669.3 *5*	0.46 *16*
γ	1691.0 *10*	0.30 *14*
γ	1696.78 *13*	0.56 *15*
γ	1703.0 *7*	0.27 *3*
γ E0(+M1+E2)	1723.4 *7*	<0.16
γ	1760.1 *4*	1.14 *20*
γ	1782.1 *4*	1.09 *24*
γ	1810.6 *5*	0.53 *15*
γ	1846.9 *4*	0.83 *15*
γ	1882.22 *16*	1.48 *19*
γ	1905.5 *3*	0.65 *17*
γ	1917.3 *3*	0.90 *17*
γ	1944.2 *3*	0.85 *19*
γ	1993.7 *4*	0.85 *16*
γ	2009.9 *6*	0.61 *17*
γ	2029.62 *11*	2.54 *23*
γ	2052.8 *11*	~0.26
γ	2161.1 *8*	0.51 *18*
γ	2179.7 *5*	0.92 *19*
γ	2231.52 *12*	2.62 *22*
γ	2258.68 *18*	~0.20
γ	2295.27 *11*	1.28 *19*
γ	2392.6 *6*	0.66 *18*
γ	2428.5 *7*	0.62 *17*
γ	2440.66 *12*	1.08 *20*
γ	2446.51 *22*	1.38 *20*
γ	2509.0 *5*	0.89 *18*

Photons (^{188}Au) (continued)

γ_{mode}	γ(keV)	γ(rel)
γ	2532.2 5	0.75 21
γ	2626.10 16	1.8 4
γ	2780.71 12	2.23 21
γ	2994.31 16	0.80 16
γ	3128.8 10	~1

$^{188}_{80}$Hg(3.25 *15* min)

Mode: ε, α(3.7 *8* $\times 10^{-5}$ %)
Δ: -30200 *200* keV syst
SpA: 3.07×10^{8} Ci/g
Prod: ^{197}Au(p,10n); protons on Pb

Alpha Particles (^{188}Hg)

α(keV)
4610 20

Photons (^{188}Hg)

γ_{mode}	γ(keV)	γ(rel)
γ(E1)	66.4 6	
γ(E1)	82.2 6	
γM1,E2	98.5 5	12 4
γ(M1)	114.3 5	37 4
γ	134.2 7	11.4 24
γ	142.0 7	20 4
γ	144.1 7	9.3 20
γE2	155.4 7	9 3
γ	182.0 7	18.5 23
γM1,E2	185.4 7	13.0 21
γM1,E2	189.4 5	100 7
γ	203.9 5	24 8 ?
γ	253.6 7	10 4
γ	263.2 6	9.6 22
γ	303.7 5	
γ	331.1 7	~5
γ	333.3 7	~6
γ	335.6 7	~10
γ	344.9 7	~10
γ(M1)	452.6 6	~7
γM1,E2	523.4 7	~19
γM1	544.0 7	~4

$^{188}_{81}$Tl(1.18 *17* min)

Mode: ε
Δ: -22470 *370* keV syst
SpA: 8.4×10^{8} Ci/g
Prod: ^{181}Ta(^{16}O,9n)

$^{188}_{81}$Tl(1.183 *17* min)

Mode: ε
Δ: >-22470 keV syst
SpA: 8.4×10^{8} Ci/g
Prod: ^{181}Ta(^{16}O,9n); protons on Pb; ^{165}Ho(^{28}Si,5n); ^{159}Tb(^{32}S,3n); ^{197}Au(^{3}He,12n)

Photons (^{188}Tl)

⟨γ⟩=1752 *27* keV

γ_{mode}	γ(keV)	γ(%)†
Hg L_{ℓ}	8.722	0.57 12
Hg L_{α}	9.980	11.7 24
Hg L_{η}	10.647	0.15 3
Hg L_{β}	11.930	10.9 23
Hg L_{γ}	13.942	2.0 4
Hg $K_{\alpha2}$	68.893	20 4
Hg $K_{\alpha1}$	70.818	34 7
Hg $K_{\beta1}'$	80.124	12.1 24
Hg $K_{\beta2}'$	82.780	3.3 7
γ	146.85 24	<1 ?
γ	153.44 15	0.95 9
γ	167.26 19	0.43 4
γM1(+4.1%E2)	203.07 6	1.12 9
γ	215.51 7	0.52 22
γM1,E2	247.50 10	1.63 9
γ[E2]	269.29 10	1.20 9
γ	281.39 10	0.69 7
γ(E2)	291.61 8	3.4 3
γE2	301.16 6	4.7 3
γE2	326.84 7	9.2 4
γ	358.9 10	0.44 3
γ[E2]	381.26 11	0.43 4
γE2(+<10%M1)	385.67 10	3.2 3
γ	387.39 20	0.26 4
γ	398.09 20	0.52 5
γE2	412.82 7	86
γE2(+<20%M1)	417.79 10	0.95 9
γE1	424.02 8	3.4 3
γE2(+<20%M1)	442.97 10	1.63 17
γ	445.79 10	0.86 9
γM1,E2	450.15 7	0.43 4
γM1,E2	452.60 9	2.41 17
γE2	460.57 10	7.1 4
γE0+M1+E2	468.16 6	4.9 3
γ[E1]	478.99 15	0.69 7
γ	499.4 4	0.95 10
γE2	504.23 7	22.8 14
γ(E2)	519.7 4	0.39 4
γM1,E2	534.97 8	1.12 9
γE2	569.20 7	3.4 3
γE2+18%M1	573.92 7	3.9 3
γE2	591.93 6	59 3
γE1	621.80 17	0.60 6
γE2+39%M1	627.16 8	1.46 9
γE2	645.09 12	2.06 17
γ[M1+E2]	682.42 10	0.172 17
γ(E1)	692.06 9	2.15 17
γE2(+<45%M1)	699.68 11	2.8 3
γ[E1]	701.61 9	0.77 17
γ	710.4 3	0.172 17
γ	714.03 18	0.34 4 ?
γ	745.59 20	0.43 5
γ	764.49 10	0.69 7
γM1(+10%E2)	769.61 8	1.72 17
γE2	772.27 7	11.6 5
γ	789.66 12	0.43 4
γE2	795.00 6	9.7 5
γ	804.34 11	0.52 5 ?
γM1,E2	826.57 7	2.32 17
γE2(+<20%M1)	835.0 4	0.69 7
γE1	837.70 10	1.12 9
γ	841.49 15	1.55 15
γ	873.79 10	0.52 9
γE2	880.98 7	7.4 10
γ(M1)	885.49 10	0.73 8
γE1	904.68 8	10.6 6
γ[E2]	913.13 10	0.258 9
γ	928.32 8	1.38 9

Photons (^{188}Tl) (continued)

γ_{mode}	γ(keV)	γ(%)†
γM1,E2	947.75 11	0.43 3
γ[E2]	1009.26 10	0.172 17
γM1,E2	1042.08 7	3.01 17
γ	1057.69 10	0.95 9
γ	1071.94 18	0.26 3
γ	1170.9 3	2.1 3
γ	1239.39 10	0.34 3
γ	1272.49 10	0.77 9 ?
γ	1305.86 11	0.86 9
γ	1445.49 10	0.95 9
γ[E2]	1477.41 9	0.86 9

† 3.5% uncert(syst)

Atomic Electrons (^{188}Tl)

⟨e⟩=55.2 *20* keV

e_{bin}(keV)	⟨e⟩(keV)	e(%)
12	2.9	24 5
14	1.7	12 3
15 - 64	1.8	5.1 10
65 - 84	1.7	2.3 10
120	1.19	0.99 9
132 - 155	1.7	1.3 5
164	0.9	~0.5
165 - 213	1.5	0.78 13
215 - 235	0.96	0.43 8
244	1.22	0.50 4
245 - 292	1.41	0.51 5
298 - 327	1.79	0.58 5
330	8.5	2.58 12
335 - 384	2.3	0.64 11
385	1.8	0.46 18
386 - 397	0.038	0.0095 20
398	1.45	0.365 16
399	1.53	0.384 17
401 - 420	2.15	0.53 4
421	1.85	0.44 3
422 - 468	1.6	0.35 7
475 - 507	1.56	0.317 22
509	4.15	0.82 4
516 - 562	0.64	0.116 19
566 - 615	1.76	0.303 8
617 - 662	0.54	0.087 10
668 - 714	1.70	0.244 13
721 - 733	0.06	~0.008
741	1.5	~0.21
742 - 791	0.95	0.125 23
792 - 841	1.14	0.142 23
845 - 892	0.32	0.037 10
898 - 948	0.055	0.0060 16
959 - 1008	0.32	0.033 13
1009 - 1058	0.073	0.007 2
1060 - 1088	0.12	~0.011
1156 - 1189	0.08	0.007 3
1223 - 1272	0.06	~0.0045
1291 - 1306	0.009	~0.0007
1362 - 1394	0.07	0.0050 21
1431 - 1477	0.015	0.0011 3

Continuous Radiation (^{188}Tl)

⟨β+⟩=498 keV;⟨IB⟩=33 keV

E_{bin}(keV)		⟨ ⟩(keV)	(%)
0 - 10	β+	1.27×10^{-8}	1.50×10^{-7}
	IB	0.0141	
10 - 20	β+	8.8×10^{-7}	5.2×10^{-6}
	IB	0.0153	0.108
20 - 40	β+	4.15×10^{-5}	0.000124
	IB	0.029	0.099
40 - 100	β+	0.00376	0.00469
	IB	0.44	0.64
100 - 300	β+	0.332	0.145

Continuous Radiation (^{188}Tl) (continued)

E_{bin}(keV)		⟨ ⟩(keV)	(%)
	IB	0.30	0.167
300 - 600	β+	3.43	0.73
	IB	0.54	0.122
600 - 1300	β+	38.0	3.84
	IB	2.6	0.27
1300 - 2500	β+	186	9.8
	IB	9.6	0.51
2500 - 5000	β+	269	8.4
	IB	19	0.55
5000 - 6221	β+	0.111	0.00220
	IB	0.62	0.0119
	Σβ+		23

$^{188}_{82}$Pb(24.5 *15* s)

Mode: ϵ(78 *7* %), α(22 *7* %)
Δ: -17720 *280* keV syst
SpA: 2.42×10^{9} Ci/g
Prod: ^{155}Gd(^{40}Ar,7n); protons on Th; ^{182}W(^{16}O,10n)

Alpha Particles (^{188}Pb)

α(keV)
5980 *10*

Photons (^{188}Pb)

⟨γ⟩=310 *48* keV

γ_{mode}	γ(keV)	γ(%)†
γ_ϵ[E1]	185.0 *5*	49 *10*
γ_ϵ[E1]	758.2 *5*	29 *6*

† uncert(syst): 9.0% for ϵ

$^{188}_{83}$Bi($t_{1/2}$ unknown)

Mode: α
Prod: ^{107}Ag(^{84}Kr,3n)

Alpha Particles (^{188}Bi)

α(keV)	α(rel)
6820 *20*	85 *9*
7005 *25*	15 *9*

A = 189

NDS **34**, 537 (1981)

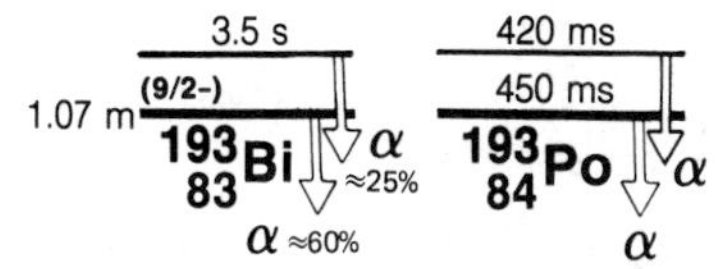

Q$_\alpha$ 6310 *100*(syst)
Q$_\alpha$ 7110 *50*

<1.5 s
189 83 Bi
α
Q$_\epsilon$ 8190 *480*(syst)
Q$_\alpha$ 7267 *21*

51 s
ϵ ≈ 99.6%
189 82 Pb
α ≈0.4%

(9/2-) 1.4 m
(1/2+) 2.3 m
ϵ
ϵ
189 81 Tl
Q$_\epsilon$ 6500 *490*(syst)
Q$_\alpha$ 5840 *50*

13/2+
8.6 m
3/2-
7.6 m
ϵ
ϵ
189 80 Hg
Q$_\epsilon$ 5200 *360*(syst)

11/2- 247.3
IT
4.6 m
1/2+ 0
28.7 m
ϵ
ϵ
189 79 Au
Q$_\epsilon$ 4200 *200*

10.9 h 3/2-
ϵ
189 78 Pt
Q$_\epsilon$ 2700 *200*(syst)

(3/2-)
11.5 m
189 74 W
β⁻
Q$_{\beta^-}$ 2500 *200*

1.01 d (5/2+)
189 75 Re
β⁻
Q$_{\beta^-}$ 1008 *9*

5.8 h 9/2- 30.83
IT
3/2- 0
189 76 Os

3/2+ 13.2 d
ϵ
189 77 Ir
Q$_\epsilon$ 1961 *16*
Q$_\epsilon$ 535 *14*

0
1000
5000
5000
5000
5000
5000

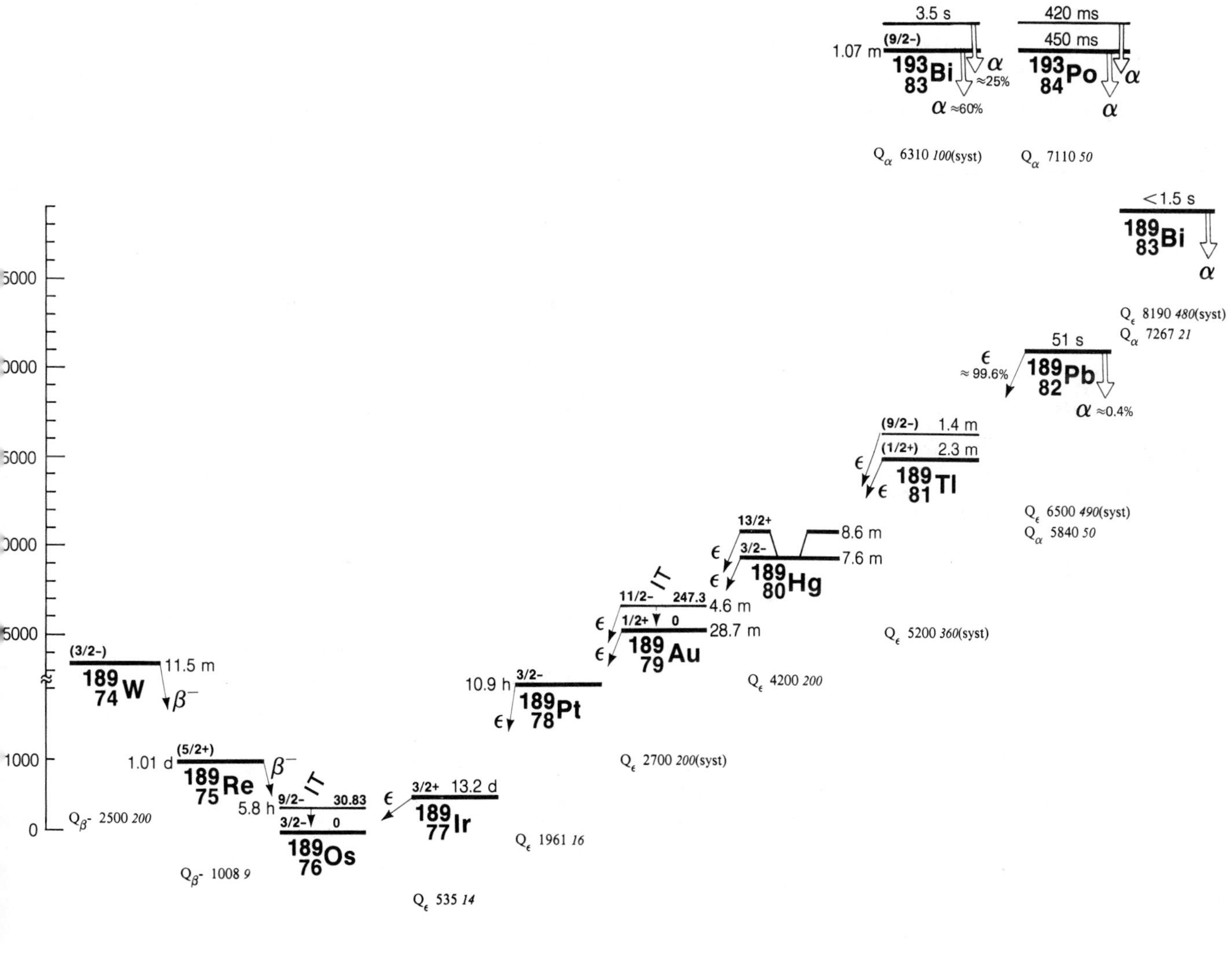

$^{189}_{74}$W (11.5 *3* min)

Mode: β-

Δ: -35490 *200* keV

SpA: 8.65×10^{7} Ci/g

Prod: ^{192}Os(n,α)

Photons (^{189}W)

γ_{mode}	γ(keV)	γ(rel)
γ	94 *5*	3.0 *6*
γ	130 *2*	12.0 *24*
γ	178 *2*	13 *3*
γ	222 *8*	3.0 *6*
γ	258 *3*	100 *20*
γ	360 *8*	10 *2*
γ	417 *4*	96 *19*
γ	550 *10*	28 *6*
γ	855 *15*	20 *4*
γ	955 *20*	17 *3*

$^{189}_{75}$Re(1.013 *17* d)

Mode: β-

Δ: -37987 *10* keV

SpA: 6.82×10^{5} Ci/g

Prod: ^{186}W(α,p); neutrons on Os; ^{192}Os(d,αn)

Photons (^{189}Re)

⟨γ⟩=60.1 *14* keV

γ_{mode}	γ(keV)	γ(%)†
Os L_ℓ	7.822	0.151 *16*
Os L_α	8.904	3.39 *24*
Os L_η	9.337	0.033 *3*
Os L_β	10.491	3.0 *3*
Os L_γ	12.246	0.52 *7*
γ M1+1.2%E2	25.713 *17*	0.0021 *3*
γ E2	33.348 *18*	0.0016 *3*
γ M1+0.2%E2	36.190 *16*	0.13 *4*
γ M1+0.6%E2	56.482 *14*	0.079 *7*
γ M1+0.6%E2	59.061 *15*	0.110 *14*
γ [M1+E2]	59.192 *24*	0.033 *9*
Os $K_{\alpha2}$	61.485	1.52 *4*
Os $K_{\alpha1}$	63.000	2.62 *8*
γ M1+25%E2	69.538 *13*	0.84 *3*
Os $K_{\beta1}$'	71.313	0.90 *3*
Os $K_{\beta2}$'	73.643	0.228 *7*
γ	90.00 *11*	0.018 *7*
γ M1+9.3%E2	95.251 *15*	0.033 *3*
γ M1	100.80 *5*	~0.0037
γ M1	117.25 *4*	0.011 *4*
γ	118.60 *6*	~0.007
γ [E2]	121.440 *23*	0.032 *3*
γ [E2]	124.150 *20*	0.032 *3*
γ [M1]	132.26 *4*	0.0052 *15*
γ M1+27%E2	138.290 *21*	0.0245 *21*
γ M1+50%E2	147.153 *22*	1.35 *4*
γ M1+50%E2	149.862 *18*	0.856 *18*
γ [M1]	152.04 *4*	0.028 *3*
γ M1	160.92 *4*	0.031 *4*
γ M1	164.003 *22*	0.0242 *19*
γ E2	166.93 *4*	0.0085 *18*
γ	175.1 *6*	~0.037
γ (M1)	180.632 *20*	0.0200 *16*
γ (E2)	185.858 *24*	2.06 *4*

Photons (^{189}Re) (continued)

γ_{mode}	γ(keV)	γ(%)†
γ E2+37%M1	188.567 *22*	0.542 *11*
γ E2+23%M1	197.351 *22*	0.117 *8*
γ E2	206.345 *18*	0.052 *4*
γ [M1]	211.23 *4*	0.015 *3*
γ (E2)	216.691 *21*	6.02 *14*
γ [M1]	218.04 *5*	~0.037
γ E2	219.401 *17*	5.0 *1*
γ	223.80 *6*	0.022 *7*
γ E2+34%M1	233.541 *21*	0.095 *11*
γ [E2]	239.693 *21*	0.015 *4*
γ E2	245.050 *23*	3.8
γ [M1]	256.19 *10*	0.0070 *19*
γ [M1]	265.26 *5*	0.020 *3*
γ [E2]	270.55 *4*	0.0104 *11*
γ [M1]	274.02 *5*	0.0092 *19*
γ E2+26%M1	275.883 *18*	0.343 *20*
γ [M1]	296.26 *4*	0.030 *7*
γ [M1]	306.6 *4*	~0.007
γ [M1]	323.71 *3*	0.011 *4*
γ	332.90 *11*	0.018 *7*
γ [M1]	343.42 *5*	0.050 *7*
γ [M1]	351.04 *5*	~0.037
γ [M1]	366.05 *4*	0.044 *4*
γ [E2]	380.19 *3*	0.026 *7*
γ [E2]	382.90 *4*	0.037 *7*
γ [E2]	388.45 *10*	~0.007
γ [M1]	397.08 *4*	0.115 *11*
γ [M1]	402.48 *5*	0.0061 *11*
γ [M1]	403.55 *5*	0.081 *7*
γ [E2]	427.92 *4*	0.067 *11*
γ [M1]	429.26 *5*	0.067 *11*
γ	432.0 *9*	~0.011
γ [M1]	438.67 *5*	0.011 *4*
γ [M1]	438.8 *4*	0.018 *7*
γ [M1]	440.96 *4*	~0.007
γ M1	454.66 *5*	0.096 *11*
γ M1	462.61 *5*	0.155 *4*
γ M1	480.37 *5*	0.033 *11*
γ [M1]	483.30 *4*	~0.007
γ [M1]	497.44 *4*	0.041 *11*
γ M1	498.80 *5*	0.078 *7*
γ M1	504.34 *3*	0.27 *3*
γ M1	530.05 *3*	0.063 *4*
γ M1	549.91 *5*	0.063 *4*
γ M1	563.40 *3*	0.55 *4*
γ M1	599.59 *3*	0.32 *3*

† 22% uncert(syst)

Atomic Electrons (^{189}Re)

⟨e⟩=22.3 *4* keV

e_{bin}(keV)	⟨e⟩(keV)	e(%)
11	0.97	8.9 *10*
12 - 18	0.91	6.6 *5*
20	1.28	6.42 *22*
21 - 27	0.72	3.2 *6*
28	0.674	2.39 *8*
29 - 56	1.44	3.6 *5*
57	1.81	3.17 *24*
58	0.034	0.058 *9*
59	0.97	1.66 *15*
60 - 66	0.251	0.382 *14*
67	0.57	0.85 *5*
68 - 71	0.245	0.354 *20*
73	0.98	1.34 *4*
76	0.614	0.807 *22*
77 - 111	0.29	0.32 *4*
112	0.493	0.440 *21*
113 - 141	1.474	1.130 *16*
143	1.18	0.83 *4*
144 - 145	0.204	0.142 *13*
146	0.99	0.678 *20*
147 - 170	0.264	0.175 *8*
171	0.65	0.38 *6*

Atomic Electrons (^{189}Re) (continued)

e_{bin}(keV)	⟨e⟩(keV)	e(%)
172 - 203	1.182	0.660 *14*
204	0.75	0.370 *17*
205	0.006	~0.0029
206	0.524	0.254 *10*
207 - 256	2.10	0.959 *23*
258 - 307	0.113	0.041 *4*
309 - 358	0.116	0.0349 *21*
363 - 409	0.113	0.0293 *15*
415 - 463	0.137	0.0317 *17*
467 - 504	0.154	0.0314 *17*
517 - 563	0.094	0.0176 *11*
587 - 600	0.0139	0.00235 *17*

Continuous Radiation (^{189}Re)

⟨β-⟩=295 keV;⟨IB⟩=0.29 keV

E_{bin}(keV)		⟨ ⟩(keV)	(%)
0 - 10	β-	0.090	1.80
	IB	0.0141	
10 - 20	β-	0.270	1.80
	IB	0.0134	0.093
20 - 40	β-	1.09	3.63
	IB	0.025	0.087
40 - 100	β-	7.7	11.0
	IB	0.062	0.097
100 - 300	β-	68	34.7
	IB	0.116	0.069
300 - 600	β-	149	34.6
	IB	0.050	0.0128
600 - 1008	β-	69	9.7
	IB	0.0048	0.00074

$^{189}_{76}$Os(stable)

Δ: -38995 *3* keV

%: 16.1 *3*

$^{189}_{76}$Os(5.8 *1* h)

Mode: IT

Δ: -38964 *3* keV

SpA: 2.86×10^{6} Ci/g

Prod: daughter ^{189}Ir

Photons (^{189}Os)

⟨γ⟩=1.87 *10* keV

γ_{mode}	γ(keV)	γ(%)
Os L_ℓ	7.822	0.62 *6*
Os L_α	8.904	13.9 *10*
Os L_η	9.337	0.0127 *9*
Os L_β	10.686	4.9 *5*
Os L_γ	12.376	0.48 *7*
γ M3	30.834 *19*	0.000313 *14*

Atomic Electrons (^{189}Os)

⟨e⟩=27.4 *8* keV

e_{bin}(keV)	⟨e⟩(keV)	e(%)
11	4.6	42 *5*
12	0.052	0.42 *5*
13	0.46	3.5 *4*
18	2.11	11.8 *6*
20	11.8	59 *3*
28	6.2	21.9 *11*
29	0.166	0.57 *3*
30	1.70	5.6 *3*
31	0.345	1.12 *6*

$^{189}_{77}$Ir(13.2 *1* d)

Mode: ϵ

Δ: -38460 *14* keV

SpA: 5.23×10^4 Ci/g

Prod: daughter ^{189}Pt; ^{187}Re(α,2n); ^{190}Os(p,2n)

Photons (^{189}Ir)

⟨γ⟩=78 *6* keV

γ_{mode}	γ(keV)	γ(%)†
Os L_ℓ	7.822	0.74 *13*
Os L_α	8.904	17 *3*
Os L_η	9.337	0.22 *4*
Os L_β	10.464	18 *3*
Os L_γ	12.249	3.4 *6*
γ M1+1.2%E2	25.713 *17*	0.024 *3*
γ E2	33.348 *18*	0.0066 *13*
γ M1+0.2%E2	36.190 *16*	0.67 *6*
γ M1+0.6%E2	56.482 *14*	0.122 *12*
γ M1+0.6%E2	59.061 *15*	1.26 *11*
γ [M1+E2]	59.192 *24*	0.11 *4*
Os $K_{\alpha2}$	61.485	22 *4*
Os $K_{\alpha1}$	63.000	38 *7*
γ M1+25%E2	69.538 *13*	3.5 *4*
Os $K_{\beta1}$'	71.313	13.0 *25*
Os $K_{\beta2}$'	73.643	3.3 *6*
γ M1+9.3%E2	95.251 *15*	0.38 *3*
γ	97.8 *5*	<0.006
γ [E2]	121.440 *23*	0.00274 *22*
γ [E2]	124.150 *20*	0.0033 *3*
γ M1+27%E2	138.290 *21*	0.074 *6*
γ M1+50%E2	147.153 *22*	0.114 *5*
γ M1+50%E2	149.862 *18*	0.087 *3*
γ M1	164.003 *22*	0.073 *3*
γ (M1)	180.632 *20*	0.0310 *24*
γ (E2)	185.858 *24*	0.175 *7*
γ E2+37%M1	188.567 *22*	0.0552 *20*
γ E2+23%M1	197.351 *22*	0.35 *3*
γ E2	206.345 *18*	0.081 *3*
γ (E2)	216.691 *21*	0.509 *18*
γ E2	219.401 *17*	0.510 *17*
γ E2+34%M1	233.541 *21*	0.29 *3*
γ [E2]	239.693 *21*	0.023 *7*
γ E2	245.050 *23*	5.87
γ E2+26%M1	275.883 *18*	0.534 *25*
γ [M1]	343.42 *5*	0.054 *3*
γ	369.1 *5*	<0.006
γ [M1]	402.48 *5*	0.0066 *6*
γ [M1]	438.67 *5*	0.012 *4*

† 3.3% uncert(syst)

Atomic Electrons (^{189}Ir)

⟨e⟩=42.5 *17* keV

e_{bin}(keV)	⟨e⟩(keV)	e(%)
11	4.3	39 *8*
12	2.6	21 *4*
13	1.8	14.2 *22*
15 - 18	0.23	1.36 *20*
20	1.04	5.2 *4*
21 - 22	1.22	5.6 *6*
23	2.27	9.8 *10*
24 - 44	2.87	9.3 *4*
46	2.0	4.4 *6*
47 - 56	3.0	6.0 *15*
57	7.6	13.3 *17*
58	0.36	0.61 *10*
59	4.2	7.1 *10*
60 - 66	1.25	1.93 *15*
67	2.4	3.6 *4*
68 - 115	1.90	2.52 *17*
118 - 168	0.67	0.480 *24*
170 - 219	1.62	0.896 *22*
221 - 270	1.136	0.480 *8*
273 - 295	0.0177	0.0065 *4*
329 - 369	0.0118	0.0034 *4*
390 - 439	0.00139	0.00034 *6*

Continuous Radiation (^{189}Ir)

⟨IB⟩=0.37 keV

E_{bin}(keV)		⟨ ⟩(keV)	(%)
10 - 20	IB	0.00147	0.0132
20 - 40	IB	0.0021	0.0064
40 - 100	IB	0.34	0.55
100 - 300	IB	0.020	0.0129
300 - 535	IB	0.0032	0.00091

$^{189}_{78}$Pt(10.89 *11* h)

Mode: ϵ

Δ: -36499 *12* keV

SpA: 1.523×10^6 Ci/g

Prod: ^{191}Ir(p,3n); descendant ^{189}Hg

Photons (^{189}Pt)

⟨γ⟩=307 *9* keV

γ_{mode}	γ(keV)	γ(%)†
Ir L_ℓ	8.042	0.84 *16*
Ir L_α	9.167	19 *3*
Ir L_η	9.650	0.25 *4*
Ir L_β	10.824	18 *3*
Ir L_γ	12.616	3.2 *6*
γ M1+12%E2	62.711 *25*	0.049 *22*
Ir $K_{\alpha2}$	63.286	32 *6*
Ir $K_{\alpha1}$	64.896	54 *10*
γ M2	71.65 *3*	0.055 *11*
Ir $K_{\beta1}$'	73.452	19 *4*
Ir $K_{\beta2}$'	75.861	4.8 *9*
γ M1+2.8%E2	82.190 *22*	1.72 *17*
γ E2+6.2%M1	94.312 *22*	4.5 *3*
γ M1+27%E2	113.791 *21*	1.71 *12*
γ [M1]	130.50 *8*	0.044 *11*
γ M1+2.1%E2	141.142 *22*	2.49 *13*
γ M1+43%E2	176.501 *22*	0.58 *3*
γ [M1]	181.24 *5*	0.066 *11*
γ M1+24%E2	186.67 *3*	1.32 *13*
γ (M1)	190.86 *5*	0.066 *16*
γ (M1)	203.85 *3*	0.29 *3*

Photons (^{189}Pt)
(continued)

γ_{mode}	γ(keV)	γ(%)†
γ [E2]	212.56 *7*	0.060 *17* ?
γ E2	223.33 *3*	0.88 *6*
γ E2	243.46 *4*	4.1 *6*
γ E2	252.0 *3*	0.025 *8*
γ E3	258.32 *3*	0.187 *17*
γ	263.1 *3*	~0.022
γ (E2)	284.56 *9*	0.093 *16* ?
γ E2	288.30 *8*	0.039 *6*
γ E2	300.46 *3*	2.20 *11*
γ [E1]	307.00 *7*	0.018 *3*
γ M1+26%E2	317.644 *25*	1.92 *10*
γ (M1)	343.2 *2*	0.071 *17*
γ (M1)	351.17 *10*	0.077 *17*
γ	352.6 *3*	0.093 *16*
γ	383.8 *3*	0.049 *11*
γ [M1]	384.8 *3*	0.031 *6*
γ M1	403.75 *3*	0.84 *4*
γ E1	430.96 *7*	0.093 *16*
γ (M1)	459.6 *3*	0.049 *11*
γ	484.6 *4*	0.016 *3*
γ [E1]	493.67 *7*	0.19 *4*
γ (M1)	530.48 *6*	0.138 *11*
γ (E1)	539.83 *9*	0.170 *16* ?
γ M1+3.1%E2	544.89 *3*	3.41 *14*
γ M1+4.5%E2	568.83 *4*	4.17 *17*
γ [M1]	576.94 *8*	0.026 *6*
γ [M1]	584.99 *5*	0.041 *6*
γ (M1)	594.61 *5*	0.063 *7*
γ M1	607.60 *3*	4.80 *19*
γ [M1]	616.18 *4*	0.026 *4*
γ	623.15 *10*	0.15 *2*
γ M1+43%E2	627.08 *3*	1.39 *6*
γ [E1]	640.99 *8*	0.033 *11*
γ (M1)	644.27 *6*	0.37 *3*
γ (E1)	651.63 *5*	0.063 *6*
γ	655.57 *8*	0.028 *4*
γ [M1]	673.56 *18*	0.026 *4* ?
γ [M1]	678.89 *4*	0.026 *4*
γ (M1)	698.37 *4*	0.127 *11*
γ	708.4 *3*	0.021 *4*
γ [E1]	714.35 *6*	0.044 *11*
γ M1+35%E2	721.39 *3*	5.5
γ E1	733.82 *6*	0.21 *3*
γ (M1)	735.75 *5*	0.22 *4*
γ	744.18 *13*	0.025 *4*
γ	751.00 *13*	0.036 *6*
γ [M1]	755.75 *18*	0.041 *6* ?
γ	765.52 *9*	0.021 *4*
γ	772.0 *3*	0.011 *3*
γ	778.1 *5*	0.009 *3*
γ [E1]	782.13 *8*	0.035 *6*
γ	788.58 *9*	~0.044
γ	788.84 *5*	~0.033
γ (M1)	792.68 *4*	0.80 *5*
γ (M1)	798.46 *5*	0.116 *11*
γ (M1)	808.32 *5*	0.032 *7*
γ E1	810.97 *7*	0.077 *11*
γ	820.0 *7*	0.009 *3*
γ E1	828.14 *5*	0.18 *3*
γ E2+46%M1	836.00 *8*	0.049 *5* ?
γ [E1]	855.88 *8*	0.014 *3*
γ (E1,E2)	880.14 *15*	0.049 *8*
γ E2+48%M1	885.59 *6*	0.104 *11*
γ	892.5 *3*	0.023 *3*
γ [M1]	902.63 *5*	0.019 *4*
γ (E2)	912.25 *5*	0.024 *5*
γ	913.93 *10*	0.008 *3*
γ E1	924.76 *7*	0.165 *11*
γ [M1]	929.72 *9*	0.023 *7*
γ	933.9 *4*	0.016 *3*
γ	951.42 *15*	0.012 *3*
γ	955.5 *3*	0.010 *3*
γ M1	986.62 *9*	0.042 *7* ?
γ (M1)	992.43 *9*	0.047 *7*
γ M1+E2+E0	1002.74 *8*	0.030 *6* ?
γ E1	1007.90 *5*	0.057 *8*
γ [E1]	1011.85 *20*	0.029 *4*
γ [M1+E2]	1026.73 *5*	0.204 *11*
γ M1+E2+E0	1033.70 *13*	0.028 *4*
γ (M1)	1044.08 *6*	0.043 *7*
γ (E2)	1070.37 *11*	~0.028 ?
γ [E1]	1070.61 *5*	~0.022
γ (E1,E2)	1080.83 *8*	0.044 *7*
γ [M1]	1089.44 *6*	0.016 *3*

Photons (^{189}Pt) (continued)

γ_{mode}	γ(keV)	γ(%)†
γ (M1)	1106.22 9	0.104 11
γ (E2)	1108.92 6	0.055 6
γ	1123.8 3	0.0115 22
γ	1143.9 3	0.018 6
γ	1157.96 9	0.0115 22
γ	1159.69 15	0.0115 22
γ	1168.03 6	0.022 3
γ (E1)	1184.40 5	0.021 3
γ	1194.76 10	0.008 3
γ [E1]	1198.49 12	0.0105 22
γ (M1)	1230.74 6	0.044 7
γ	1240.55 8	0.012 3
γ M1+43%E2	1254.02 6	0.214 11
γ	1292.56 9	0.0066 11 ?
γ	1300.01 15	0.010 3
γ	1305.22 12	0.0077 22
γ (E1)	1312.28 12	0.037 6
γ E1	1323.65 7	0.159 11
γ (M1)	1344.53 6	0.039 6
γ	1356.4 4	0.015 3
γ (E2)	1362.72 15	0.041 6
γ	1382.20 15	0.035 5
γ	1387.81 16	0.039 6
γ	1395.16 6	0.018 4
γ [E1]	1405.84 7	0.035 5
γ	1408.5 4	0.034 5
γ	1423.06 13	0.024 5
γ	1444.40 8	0.014 5
γ (E1,E2)	1446.36 12	0.035 6
γ (E2)	1457.87 7	0.127 11
γ	1476.51 15	~0.17
γ	1477.35 7	~0.06
γ	1485.2 4	0.0094 16
γ	1496.31 15	0.044 11 ?
γ E1	1501.60 16	0.060 11
γ	1504.0 4	0.020 4
γ	1509.07 12	0.0099 17
γ	1528.55 12	0.037 7
γ	1536.85 13	0.0071 17
γ	1554.8 5	0.007 3
γ	1558.19 8	~0.011
γ	1559.02 15	~0.011
γ (E2)	1571.66 6	0.044 4
γ	1579.9 7	0.0049 22
γ	1593.2 3	0.021 3
γ	1601.7 4	0.0082 16
γ	1610.20 9	0.0127 16 ?
γ	1613.9 5	0.020 3
γ	1622.86 12	0.008 3
γ	1635.4 11	0.012 4
γ [E1]	1638.25 11	0.018 4 ?
γ	1653.36 15	0.031 4
γ	1672.84 15	0.0110 16
γ	1683.7 4	0.0148 22
γ	1688.43 8	0.0099 22
γ [E1]	1720.44 11	0.0132 22 ?
γ	1740.2 5	0.0105 16
γ	1767.15 15	0.033 3
γ	1786.0 5	0.008 3
γ	1792.0 5	0.011 3
γ	1798.0 5	0.016 3
γ	1802.22 8	0.013 4
γ	1811.3 11	~0.008
γ [E1]	1814.75 11	~0.007 ?

† 38% uncert(syst)

Atomic Electrons (^{189}Pt)

⟨e⟩=58.5 17 keV

e_{bin}(keV)	⟨e⟩(keV)	e(%)
6	0.95	15.6 16
11	5.1	46 9
13	4.1	31 6
18	0.92	5.1 8
38	1.84	4.9 4
49 - 54	1.62	3.2 4
58	1.3	2.2 5

Atomic Electrons (^{189}Pt) (continued)

e_{bin}(keV)	⟨e⟩(keV)	e(%)
59 - 64	1.48	2.4 3
65	3.16	4.9 3
68	0.38	0.56 11
69	2.05	3.0 4
70 - 80	0.96	1.28 14
81	7.7	9.4 8
82	0.059	0.072 14
83	6.2	7.5 6
91	2.13	2.33 19
92	1.83	1.99 16
94 - 105	3.16	3.21 15
111	1.47	1.33 15
112 - 147	2.04	1.58 8
163 - 213	1.85	1.06 7
220 - 268	2.27	0.96 5
271 - 318	0.56	0.188 11
328 - 374	0.342	0.104 5
381 - 429	0.100	0.0254 13
430 - 473	0.83	0.177 8
480 - 529	0.93	0.188 8
530 - 577	1.65	0.305 12
579 - 628	0.311	0.0517 18
630 - 679	0.72	0.112 18
680 - 725	0.308	0.043 3
731 - 780	0.039	0.0051 3
781 - 828	0.0258	0.0032 3
833 - 881	0.0116	0.00135 18
882 - 932	0.0146	0.00160 17
933 - 981	0.022	0.0024 6
983 - 1033	0.0203	0.00199 18
1041 - 1089	0.0036	0.00034 9
1092 - 1141	0.0045	0.00041 7
1142 - 1187	0.0161	0.00137 14
1192 - 1241	0.0042	0.00034 4
1243 - 1292	0.0072	0.00057 5
1294 - 1343	0.0062	0.00047 12
1344 - 1393	0.0069	0.00050 7
1394 - 1443	0.012	0.0008 4
1444 - 1494	0.0053	0.00036 9
1496 - 1591	0.0061	0.00040 6
1597 - 1691	0.0031	0.00019 4
1707 - 1804	0.0022	0.00012 3
1808 - 1812	1.0×10^{-5}	$\sim 6 \times 10^{-7}$

Continuous Radiation (^{189}Pt)

⟨β+⟩=2.02 keV;⟨IB⟩=1.36 keV

E_{bin}(keV)		⟨ ⟩(keV)	(%)
0 - 10	β+	2.26×10^{-8}	2.66×10^{-7}
	IB	0.00028	
10 - 20	β+	1.45×10^{-6}	8.6×10^{-6}
	IB	0.00153	0.0133
20 - 40	β+	6.3×10^{-5}	0.000188
	IB	0.0020	0.0062
40 - 100	β+	0.00481	0.0061
	IB	0.39	0.60
100 - 300	β+	0.274	0.125
	IB	0.053	0.029
300 - 600	β+	1.18	0.268
	IB	0.17	0.038
600 - 1300	β+	0.56	0.081
	IB	0.60	0.067
1300 - 1961	IB	0.138	0.0095
	Σβ+		0.48

$^{189}_{79}$Au(28.7 3 min)

Mode: ε

Δ: -33800 200 keV syst

SpA: 3.47×10^{7} Ci/g

Prod: daughter ^{189}Hg(7.6 min); daughter ^{189}Hg(8.6 min); ^{181}Ta(^{12}C,4n); protons on Pt

Photons (^{189}Au)

⟨γ⟩=804 29 keV

γ_{mode}	γ(keV)	γ(%)†
Pt L_{ℓ}	8.266	0.90 20
Pt L_{α}	9.435	19 4
Pt L_{η}	9.975	0.26 6
Pt L_{β}	11.186	19 4
Pt L_{γ}	13.047	3.3 7
γ (E2)	39.45 3	0.029 5
γ M1+13%E2	45.07 5	0.06 2
γ M1+4.6%E2	45.712 25	0.20 3
Pt $K_{\alpha2}$	65.122	33 7
Pt $K_{\alpha1}$	66.831	56 12
Pt $K_{\beta1}'$	75.634	20 4
Pt $K_{\beta2}'$	78.123	5.1 11
γ M1+7.0%E2	82.18 4	~0.7
γ M1+4.5%E2	88.441 25	1.4 5
γ	92.27 10	1.35 21
γ	110.77 10	0.37 8
γ M1+11%E2	126.25 4	0.71 10
γ (M1)	133.52 4	1.36 15
γ [M1]	176.25 4	0.52 25
γ	186.47 10	0.52 25
γ E2	194.92 9	0.75 15
γ M1+38%E2	215.70 3	3.1 5
γ E2	218.7 10	1.9 4
γ E2	221.96 4	5.0 6
γ M1	225.56 9	2.1 5
γ E2	231.2 10	1.2 3
γ	253.67 20	~2
γ	256.07 10	0.94 25
γ E2+33%M1	259.77 4	1.1 3
γ M1	260.6 3	~0.10
γ M1	262 1	0.73 23
γ M1	265.7 10	0.52 22
γ	297.5 10	3.0 5
γ [M1]	302.50 4	~0.6
γ	309.6 10	1.0 4
γ	329.57 20	1.4 4
γ	332.5 10	0.79 25
γ [M1]	341.94 4	1.5 4
γ M1	348.21 4	8.3 7
γ [M1]	359.08 9	0.8 3
γ M1+E2	441.25 8	7.3 10
γ M1	447.52 9	11.0 11
γ	483.82 10	0.83 25
γ	523.27 10	~0.8
γ M1	529.54 10	6.7 11
γ (E1)	631.28 15	2.4 6
γ E1	713.30 9	20.8
γ	802.0 10	1.2 4
γ E1	812.61 12	12.8 17
γ	827.8 4	1.29 25
γ	902.0 20	1.29 25
γ E1	1072.37 11	5.6 10
γ	1085.0 10	~0.6
γ E1	1160.81 11	7.0 11
γ	1176.9 7	3.3 8

† 5.3% uncert(syst)

Atomic Electrons (^{189}Au)

⟨e⟩=68 4 keV

e_{bin}(keV)	⟨e⟩(keV)	e(%)
3 - 10	2.6	54 15
12	5.2	45 10
13	3.2	24 6
14	1.5	~11
26 - 54	7.9	22.4 20
55	2.1	3.8 5
61 - 74	2.1	3.1 6
75	1.5	2.0 7
77 - 78	0.7	~0.9
79	1.6	~2
80 - 126	6.0	6.3 18
130 - 134	0.225	0.172 16
137	2.0	1.4 3
140 - 144	1.29	0.9 1
147	1.8	1.2 3
153 - 202	3.9	2.2 5
204 - 218	2.38	1.13 8
219	1.2	~0.6
220 - 266	3.3	1.34 25
270	4.0	1.47 13
281 - 330	1.3	0.45 11
332 - 359	1.20	0.354 24
363	1.6	~0.4
369	3.7	1.0 1
405 - 448	1.6	0.38 6
451	1.8	0.39 6
470 - 518	0.38	0.075 12
520 - 553	0.163	0.030 4
617 - 635	0.459	0.072 4
699 - 734	0.44	0.061 12
749 - 799	0.16	~0.021
800 - 828	0.13	~0.016
888 - 902	0.021	~0.0023
994 - 1007	0.12	0.012 3
1058 - 1098	0.30	0.027 11
1147 - 1177	0.057	0.0049 17

Continuous Radiation (^{189}Au)

⟨β+⟩=27.6 keV;⟨IB⟩=2.3 keV

E_{bin}(keV)		⟨ ⟩(keV)	(%)
0 - 10	β+	7.9×10^{-8}	9.3×10^{-7}
	IB	0.0013	
10 - 20	β+	5.1×10^{-6}	3.04×10^{-5}
	IB	0.0027	0.021
20 - 40	β+	0.000225	0.00068
	IB	0.0039	0.0127
40 - 100	β+	0.0176	0.0221
	IB	0.41	0.62
100 - 300	β+	1.03	0.465
	IB	0.072	0.040
300 - 600	β+	5.8	1.28
	IB	0.21	0.046
600 - 1300	β+	19.3	2.21
	IB	0.87	0.095
1300 - 2500	β+	1.47	0.106
	IB	0.69	0.043
2500 - 2700	IB	0.00039	1.53×10^{-5}
	Σβ+		4.1

$^{189}_{79}$Au(4.59 *11* min)

Mode: ε, IT

Δ: -33553 200 keV syst

SpA: 2.16×10^{8} Ci/g

Prod: protons on Pt; ^{181}Ta(^{12}C,4n); daughter ^{189}Hg(8.6 min)

Photons (^{189}Au)

γ_{mode}	γ(keV)	γ(rel)
γ_ϵE2	166.7 2	100 5
γ_ϵ	321.1 5	19 7

$^{189}_{80}$Hg(7.6 *1* min)

Mode: ε

Δ: -29600 280 keV syst

SpA: 1.308×10^{8} Ci/g

Prod: protons on Pb; daughter ^{189}Tl(1.4 min)

Photons (^{189}Hg) *

γ_{mode}	γ(keV)	γ(rel)
γ M1	78.17 12	100 12
γ E2	166.61 9	56 7
γ E1	176.30 13	8.3 6
γ [M1]	194.01 14	~2
γ M1+37%E2	200.51 17	4.6 5
γ M1	203.90 13	45 3
γ E2	236.62 11	45 3
γ E2+31%M1	238.70 12	64 4
γ E2	248.59 13	61 4
γ M1	263.87 14	13.5 12
γ M1	279.11 13	9.8 6
γ M1	297.76 13	54 5
γ [E2]	308.56 17	1.35 17
γ [M1]	339.69 15	1.50 22
γ M1	398.74 14	26.3 20
γ M1	445.71 12	9.8 6
γ [E2]	502.57 15	32 3
γ E2	512.40 17	60 15
γ E1	522.61 15	6.0 7
γ [M1]	540.20 20	25 6
γ E2	575.03 14	12.8 15
γ E2	637.44 15	5.3 6
γ [E1]	751.32 13	5.0 8
γ [M1]	757.6 2	5.4 5
γ [E1]	855.06 16	3.4 8
γ [E1]	1014.98 17	8.3 8
γ [E1]	1049.08 15	4.5 9
γ [E1]	1058.97 16	5.1 11
γ	1097.3 2	1.1 3
γ	1291.31 21	1.5 5
γ	1301.2 2	3.6 8
γ [E2]	1312.96 19	1.8 5
γ [E1]	1590.00 16	1.8 4
γ [M1+E2]	1766.30 15	7.5 15

* see also ^{189}Hg(8.6 min)

$^{189}_{80}$Hg(8.6 *1* min)

Mode: ε

Δ: -29600 280 keV syst

SpA: 1.156×10^{8} Ci/g

Prod: ^{197}Au(p,9n); protons on Pb; ^{181}Ta(^{16}O,p7n)

Photons (^{189}Hg)

γ_{mode}	γ(keV)	γ(rel)
γ[M1]	44.69 13	<0.25
γM1	78.17 12	63 8
γ[M1]	151.4 2	2.9 7
γE2	166.61 9	36 4
γE1	176.30 13	5.2 4
γ[M1]	194.01 14	~1.0
γM1	203.90 13	28.6 19
γ(M1)	217.93 18	5.0 5
γE2	229.24 13	5.2 10
γE2	236.62 11	28.6 20
γE2+31%M1	238.70 12	40 3
γE2	248.59 13	38.6 25
γE1	253.05 15	10.0 14
γM1	263.87 14	8.6 8
γ	268.51 20	1.71 24 *
γM1	279.11 13	6.2 4
γM1	291.1 3	1.67 24 *
γ[M1]	293.72 21	3.0 7
γM1	297.76 13	34 3
γ[E2]	308.56 17	0.86 11
γE2	321.09 12	100 12
γM1	326.44 17	2.7 4
γ	330.4 4	2.3 3 *
γ[M1]	339.69 15	0.95 14
γ(M1)	351.57 16	4.8 14
γE2	355.76 22	7.6 10
γE1	360.24 19	5.7 10
γ	363.2 3	2.0 3 *
γM1	375.66 15	2.7 5
γM1	378.12 14	7.6 10
γ(E2)	384.46 14	9.5 10
γM1	387.69 13	36 3
γM1	395.81 20	7.1 5 *
γM1	398.74 14	16.7 13
γE2	409.81 20	2.6 3 *
γM1	418.1 4	1.62 24 *
γM1	419.8 3	5.2 10 *
γM1	429.1 3	2.4 5
γE2	434.56 13	47 4
γM1	445.71 12	6.2 4
γ	452.01 20	4.3 5 *
γE2	459.04 15	10.0 10
γ[E2]	465.23 19	1.3 3
γ	467.7 4	1.8 4 *
γ	474.7 4	1.24 * 24
γE2	483.94 16	11.9 19
γ	487.2 4	2.1 4 *
γM1	499.83 16	10.0 14
γ[E2]	502.57 15	20.5 19
γE2	512.40 17	38 10
γM1	517.0 3	1.4 3 *
γE1	522.61 15	3.8 4
γ	538.5 3	8.1 19 *
γ	540.02 25	16 4
γE2	555.61 20	3.9 4 *
γ	557.8 4	1.00 19 *
γM1	565.52 11	48 3
γ	586.2 3	2.0 4 *
γ(E2)	600.10 15	21.9 24
γM1	603.03 22	5.0 7
γM1	614.74 13	8.1 10
γM1+E2	624.3 3	2.5 3 *
γ	631.0 3	2.8 4 *
γ	633.9 3	3.1 5 *
γE2	637.44 15	3.3 4
γ	641.9 3	1.19 24 *
γ	648.79 25	0.90 24
γE2	651.93 25	6.0 7
γE2	659.11 20	8.8 9
γM1	663.99 20	3.9 5
γ	670.39 14	2.0 5
γE2	675.91 20	8.3 8
γE2+24%M1	686.31 20	4.5 4
γ	694.8 3	1.90 24 *
γ	697.0 3	1.33 24
γ	702.7 4	1.05 24 *
γE2	704.56 15	2.0 4
γ	716.5 3	1.62 24 *
γ	730.6 3	1.8 3 *
γ	734.9 3	2.5 5 *
γE1	736.99 12	17.1 14
γ	746.9 3	1.95 24 *
γ	749.4 3	1.4 3 *

Photons (^{189}Hg)
(continued)

γ_mode	γ(keV)	γ(rel)
γE2	770.8 4	4.0 5
γM1	771.9 3	3.1 5 *
γ	775.7 3	1.43 24 *
γ	781.5 4	1.1 3 *
γ	782.9 4	0.86 24 *
γ	786.8 3	2.9 9 *
γ	788.66 17	1.4 4
γ	800.34 23	4 1
γ	802.0 3	2.9 6 *
γ	809.7 4	2.2 4 *
γ	816.5 4	0.67 19 *
γ	821.79 20	1.5 3
γ	833.72 20	0.57 19
γ	841.6 3	2.4 4 *
γ	847.5 3	2.4 4 *
γ[M1]	850.31 14	1.14 24
γ	853.09 22	1.4 4
γ	857.8 4	1.5 3 *
γ	869.0 3	2.6 5 *
γ	874.25 16	1.00 19
γ	884.4 3	0.76 19 *
γ	888.4 3	0.76 14 *
γ	894.8 4	0.52 14 *
γ	900.6 3	1.9 5 *
γ	903.2 5	0.71 19 *
γ	910.8 4	1.7 3 *
γ	912.6 3	0.95 24
γ	918.7 3	1.9 4 *
γ[E1]	926.49 21	2.6 4
γ	933.2 5	1.05 24 *
γ	936.6 3	2.6 5 *
γ[M1+E2]	941.18 13	6.7 10
γ	951.9 3	3.3 6 *
γ	954.92 17	1.8 3
γ	958.8 3	2.3 3 *
γ	969.0 4	0.71 19 *
γ	972.3 3	2.3 3 *
γ[E2]	978.34 25	2.5 4
γ	981.3 3	1.29 24 *
γ	995.8 4	1.7 4 *
γ	999.8 4	1.7 4 *
γ	1003.8 5	2.1 5 *
γ[M1+E2]	1007.5 4	2.7 5
γ	1026.7 3	1.29 24 *
γ	1035.0 3	3.2 6 *
γ[E1]	1051.9 3	0.48 19
γ[E1]	1057.45 24	1.5 4
γ	1064.9 3	1.19 24 *
γ	1068.88 22	0.71 24
γ	1075.1 4	1.5 4 *
γ	1083.6 3	2.5 5 *
γ	1087.9 3	0.86 24 *
γ[E1]	1092.75 21	1.9 4
γ	1105.9 3	1.3 3 *
γ	1109.7 4	0.38 10 *
γ	1115.5 3	1.14 24 *
γ	1122.54 15	4.8 10
γ	1126.50 17	1.9 4
γ	1129.7 3	0.86 24 *
γ	1135.11 20	5.2 5 *
γ[M1+E2]	1145.25 18	1.8 3
γ	1152.4 4	0.86 24 *
γ	1155.41 25	1.5 4
γ[E1]	1159.87 20	4.8 10
γ	1165.1 3	2.0 4 *
γ	1172.03 16	4.8 5
γ	1179.58 22	1.5 3
γ	1187.91 21	1.4 4
γ	1190.71 25	0.52 14
γ	1197.99 18	1.62 24
γ	1204.2 4	0.57 19 *
γ	1212.81 19	2.0 4
γ	1215.9 4	0.57 14 *
γ	1221.9 3	2.1 4 *
γ	1228.7 3	1.0 3 *
γ	1234.89 20	0.86 19
γ	1238.3 4	0.67 19 *
γ	1253.50 16	3.2 5
γ	1257.3 4	1.1 3 *
γ	1260.1 3	1.4 3 *
γ	1269.7 4	1.0 3 *
γ	1272.9 3	1.1 3 *
γ	1276.8 4	0.62 19 *

Photons (^{189}Hg)
(continued)

γ_mode	γ(keV)	γ(rel)
γ	1279.4 18	14 4 *
γ	1286.5 4	1.1 3 *
γ	1288.0 3	3.7 7 *
γ	1292.76 16	1.5 4
γ	1293.6 4	1.0 3 *
γ	1321.8 4	0.62 19 *
γ	1331.4 3	1.24 24 *
γ	1335.0 4	0.57 19 *
γ	1348.4 4	0.67 24 *
γ	1352.4 3	1.3 4 *
γ	1364.7 3	1.3 3 *
γ	1372.8 4	0.62 19 *
γ	1379.36 16	5.2 10
γ	1395.2 3	0.67 19 *
γ	1398.0 3	0.90 24 *
γ	1408.5 3	0.71 24 *
γ	1411.8 5	0.33 14 *
γ	1421.4 4	0.76 24 *
γ	1428.4 3	2.9 6 *
γ	1432.1 3	1.3 3 *
γ	1435.0 4	0.7 3 *
γ	1440.6 4	0.48 19 *
γ[E1]	1444.97 16	1.3 4
γ	1448.1 3	1.1 3 *
γ[E1]	1452.65 18	1.4 4
γ	1460.01 16	1.3 4
γ	1464.0 5	1.0 3 *
γ	1475.9 3	1.5 4 *
γ	1482.78 16	2.1 4
γ	1487.67 19	0.48 14
γ	1491.3 4	0.62 19 *
γ	1496.67 25	1.0 3
γ	1498.4 4	1.0 3 *
γ	1511.86 16	7.1 14
γ	1519.0 3	0.86 24 *
γ	1522.97 17	1.8 4
γ	1528.9 4	0.67 19 *
γ	1533.1 4	0.48 19 *
γ[E1]	1544.63 15	0.48 19
γ	1556.2 3	1.3 3 *
γ	1559.67 16	2.0 4
γ	1568.29 21	1.1 3
γ	1571.4 4	0.67 19 *
γ	1582.9 4	0.43 14 *
γ	1594.84 18	1.24 24
γ	1599.59 20	0.90 19
γ	1605.2 3	0.86 19 *
γ	1626.27 14	3.4 7
γ	1630.64 18	2.1 4
γ	1634.89 21	1.05 24
γ	1649.04 15	5.2 10
γ	1659.2 4	0.90 19 *
γ	1663.8 3	1.19 24 *
γ	1671.85 16	2.3 5
γ	1675.9 3	1.19 24 *
γ	1692.02 15	6.2 10
γ	1703.6 3	0.76 14 *
γ	1723.6 4	0.67 19 *
γ[E2]	1745.35 17	0.95 24
γ	1782.8 3	1.4 3 *
γ	1786.2 3	2.1 4 *
γ	1795.2 3	2.0 4 *
γ	1803.9 3	0.57 14 *
γ	1826.7 3	0.62 14 *
γ	1845.8 4	1.2 3 *
γ	1863.4 3	1.29 24 *
γ	1895.93 24	1.24 24
γ	1915.7 3	0.90 24 *
γ	1922.23 16	4.8 10
γ	1929.5 3	2.7 5 *
γ	1932.3 3	2.7 5 *
γ	1944.3 3	5.7 10 *
γ	1951.2 4	0.62 14 *
γ	1986.9 3	1.6 3 *
γ	1994.10 16	0.71 19
γ[E1]	2010.49 15	3.4 7
γ	2016.9 4	3.0 8 *
γ	2021.5 4	11 3 *
γ	2025.53 15	16 4
γ	2034.14 19	19 4
γ	2081.6 4	0.71 19 *
γ	2090.9 4	1.4 4 *
γ	2093.7 3	1.9 4 *

Photons (^{189}Hg)
(continued)

γ_mode	γ(keV)	γ(rel)
γ	2102.1 4	1.2 3 *
γ	2106.2 4	0.81 19 *
γ	2182.1 4	0.71 19 *

* with ^{189}Hg(7.6 min) or ^{189}Hg(8.6 min)

$^{189}_{81}$Tl(2.3 *2* min)

Mode: ε
Δ: -24400 *360* keV syst
SpA: 4.3×10^{8} Ci/g
Prod: ^{181}Ta(^{16}O,8n)

Photons (^{189}Tl)

γ_mode	γ(keV)	γ(rel)
γ	333.7 5	100 13
γ	451.0 5	49 9
γ	522.3 5	27 7
γ	942.2 5	69 16

$^{189}_{81}$Tl(1.4 *1* min)

Mode: ε
Δ: >-24400 keV syst
SpA: 7.1×10^{8} Ci/g
Prod: ^{181}Ta(^{16}O,8n); protons on Pb

Photons (^{189}Tl)

γ_mode	γ(keV)	γ(rel)
γ	215.6 5	90 7
γ	228.4 5	50 5
γ	317.5 5	100 10
γ	335.0 5	~63 ?
γ	445.2 5	14 3

$^{189}_{82}$Pb(51 *3* s)

Mode: ε(~99.6 %), α(~0.4 %)
Δ: -17900 *320* keV syst
SpA: 1.16×10^{9} Ci/g
Prod: ^{155}Gd(^{40}Ar,6n); protons on Th
^{182}W(^{16}O,9n)

Alpha Particles (^{189}Pb)

α(keV)
5730 10

$^{189}_{83}$Bi(<1.5 s)

Mode: α
Δ: -9710 *350* keV syst
SpA: >3×10^{10} Ci/g
Prod: ^{159}Tb(^{40}Ar,10n)

Alpha Particles (^{189}Bi)

α(keV)
6670 *10*

A = 190

NDS **35**, 525 (1982)

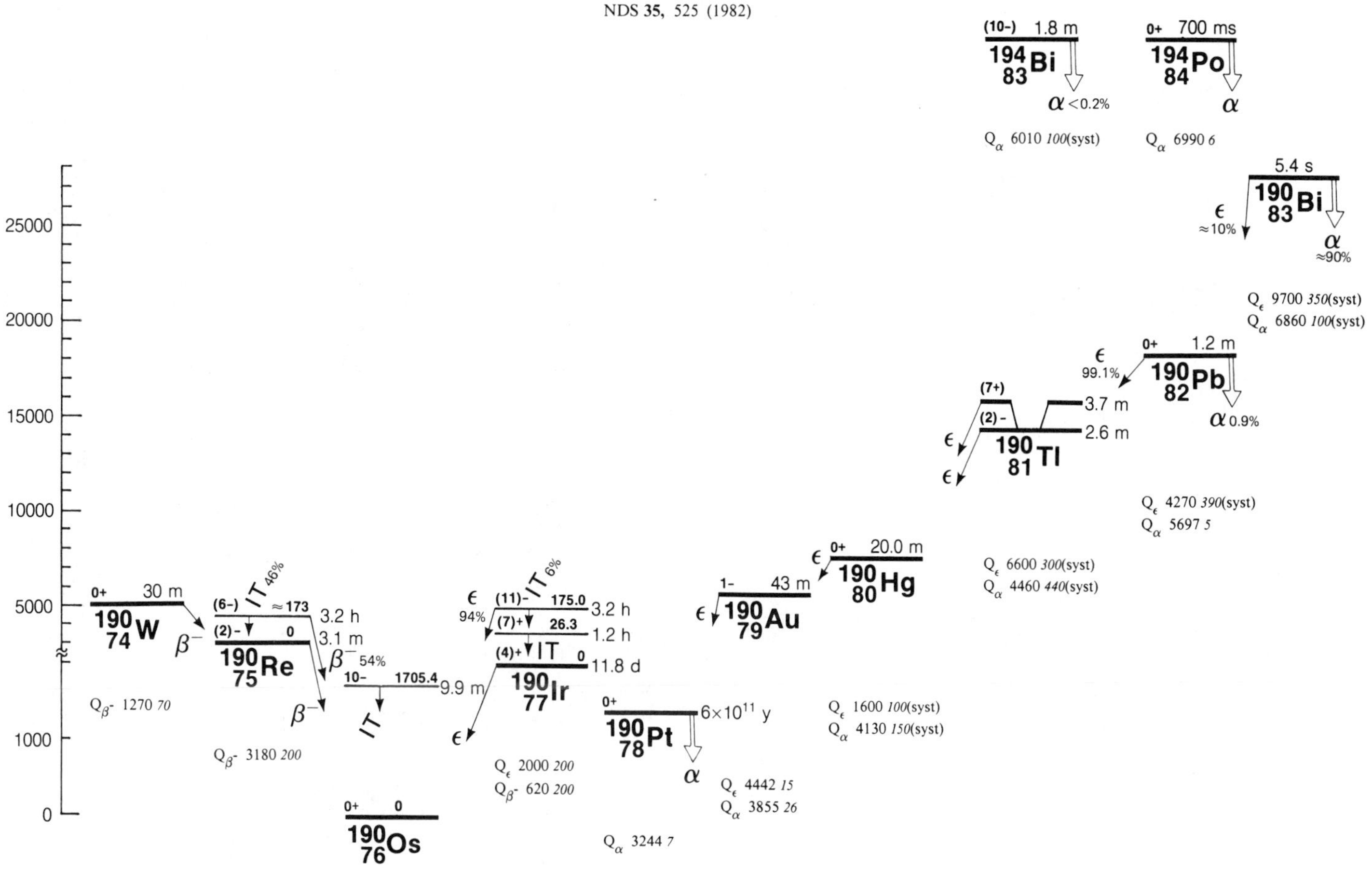

(10-) 1.8 m $^{194}_{83}$Bi α <0.2%
Q_α 6010 *100*(syst)

0+ 700 ms $^{194}_{84}$Po α
Q_α 6990 *6*

5.4 s $^{190}_{83}$Bi ε ≈10% α ≈90%
Q_ϵ 9700 *350*(syst)
Q_α 6860 *100*(syst)

0+ 1.2 m $^{190}_{82}$Pb ε 99.1% α 0.9%
Q_ϵ 4270 *390*(syst)
Q_α 5697 *5*

(7+) 3.7 m; (2)- 2.6 m $^{190}_{81}$Tl ε ε
Q_ϵ 6600 *300*(syst)
Q_α 4460 *440*(syst)

0+ 20.0 m $^{190}_{80}$Hg ε
Q_ϵ 1600 *100*(syst)
Q_α 4130 *150*(syst)

1- 43 m $^{190}_{79}$Au ε
Q_ϵ 4442 *15*
Q_α 3855 *26*

0+ 6×10^{11} y $^{190}_{78}$Pt α
Q_α 3244 *7*

(11)- 175.0 3.2 h IT 6% ε 94%; (7)+ 26.3 1.2 h IT; (4)+ 0 11.8 d $^{190}_{77}$Ir ε
Q_ϵ 2000 *200*
Q_{β^-} 620 *200*

10- 1705.4 9.9 m IT; 0+ 0 $^{190}_{76}$Os

(6-) ≈173 3.2 h IT 46%; (2)- 0 3.1 m β$^-$ 54%; $^{190}_{75}$Re β$^-$
Q_{β^-} 3180 *200*

0+ 30 m $^{190}_{74}$W β$^-$
Q_{β^-} 1270 *70*

Scale: 0, 1000, 5000, 10000, 15000, 20000, 25000

$^{190}_{74}$W (30.0 *15* min)

Mode: β-
Δ: -34270 *210* keV
SpA: 3.30×10^7 Ci/g
Prod: ^{192}Os(n,2pn); ^{192}Os(p,3p)

Photons (^{190}W)

⟨γ⟩=150 *7* keV

γ_{mode}	γ(keV)	γ(%)
Re L_ℓ	7.604	0.64 *8*
Re L_α	8.646	14.8 *13*
Re L_η	9.027	0.189 *17*
Re L_β	10.128	14.9 *16*
Re L_γ	11.810	2.6 *3*
Re $K_{\alpha2}$	59.717	31.3 *24*
Re $K_{\alpha1}$	61.141	54 *4*
Re $K_{\beta1}$'	69.214	18.2 *14*
Re $K_{\beta2}$'	71.470	4.6 *4*
γ(M1)	157.6 *1*	39 *4*
γ(M2)	162.1 *1*	11 *1*

Atomic Electrons (^{190}W)

⟨e⟩=161 *8* keV

e_{bin}(keV)	⟨e⟩(keV)	e(%)
11	4.4	42 *5*
12 - 61	5.9	34 *3*
66 - 69	0.108	0.162 *11*
86	41.0	48 *5*
90	59.4	66 *7*
145	10.1	6.9 *8*
146 - 147	1.01	0.69 *7*
150	23.7	15.8 *16*
152 - 158	6.6	4.3 *3*
159	6.1	3.8 *4*
160 - 162	2.98	1.85 *11*

Continuous Radiation (^{190}W)

⟨β-⟩=314 keV;⟨IB⟩=0.31 keV

E_{bin}(keV)		⟨ ⟩(keV)	(%)
0 - 10	β-	0.084	1.67
	IB	0.0150	
10 - 20	β-	0.252	1.68
	IB	0.0143	0.099
20 - 40	β-	1.02	3.40
	IB	0.027	0.093
40 - 100	β-	7.3	10.4
	IB	0.066	0.103
100 - 300	β-	69	34.7
	IB	0.125	0.074
300 - 600	β-	163	37.6
	IB	0.054	0.0138
600 - 950	β-	73	10.5
	IB	0.0043	0.00067

$^{190}_{75}$Re(3.1 *3* min)

Mode: β-

Δ: -35540 *200* keV

SpA: 3.2×10^{8} Ci/g

Prod: ^{192}Os(d,α); ^{190}Os(n,p); ^{192}Os(γ,np)

Photons (^{190}Re)

⟨γ⟩=1351 *20* keV

γ_{mode}	γ(keV)	γ(%)†
Os L_ℓ	7.822	0.121 *12*
Os L_α	8.904	2.70 *17*
Os L_η	9.337	0.047 *3*
Os L_β	10.458	3.07 *22*
Os L_γ	12.163	0.53 *4*
Os $K_{\alpha2}$	61.485	4.12 *15*
Os $K_{\alpha1}$	63.000	7.1 *3*
Os $K_{\beta1}'$	71.313	2.44 *9*
Os $K_{\beta2}'$	73.643	0.618 *24*
γ E2	186.7134 *20*	48.4 *18*
γ E2+3.4%M1	198.052 *14*	1.69 *15*
γ E2	199.339 *18*	0.31 *6*
γ (E2)	207.817 *22*	0.14 *3*
γ E2(+3.1%M1)	208.185 *14*	0.95 *11*
γ E1	223.790 *7*	25.7 *9*
γ E2	361.131 *6*	14.5 *9*
γ E2+1.4%M1	371.265 *5*	21.4 *11*
γ E2	397.390 *13*	8.1 *3*
γ E2+24%M1	407.155 *19*	9.6 *3*
γ E2+10%M1	407.524 *14*	5.6 *6*
γ E1	431.607 *22*	17.9 *8*
γ E2	557.978 *5*	28.8 *13*
γ E2+1.3%M1	569.316 *13*	24.5 *11*
γ E2	605.206 *20*	16.4 *6*
γ [M1+E2]	615.34 *2*	0.193 *9*
γ E1	630.945 *20*	18.8 *15*
γ E2	768.654 *14*	2.77 *12*
γ E1+<0.4%M2	828.996 *20*	23.4 *14*
γ E1(+M2)	839.129 *21*	7.9 *3*
γ [E2]	976.470 *19*	0.023 *5*
γ (E1)	1200.26 *2*	3.04 *16*
γ	1386.971 *20*	1.13 *7*
γ	1397.08 *21*	0.15 *5*
γ E2	1437.42 *24*	0.66 *13*
γ [E2]	1447.56 *24*	0.23 *7*
γ	1596.42 *21*	0.15 *4*
γ	1794.47 *21*	0.53 *13*
γ [M1+E2]	1808.69 *24*	0.18 *5*
γ	2165.73 *21*	0.06 *3*

† 5.2% uncert(syst)

Atomic Electrons (^{190}Re)

⟨e⟩=55.1 *14* keV

e_{bin}(keV)	⟨e⟩(keV)	e(%)
11 - 60	1.81	13.8 *10*
61 - 71	0.046	0.073 *4*
113	11.1	9.8 *4*
124 - 134	0.69	0.54 *5*
150	1.64	1.09 *6*
174	9.2	5.28 *22*
176	4.98	2.83 *12*
184	3.76	2.04 *9*
185 - 224	2.58	1.34 *4*
287	1.55	0.54 *4*
297	2.29	0.77 *4*
324	0.77	0.238 *10*
333	1.4	0.43 *11*
334 - 371	3.26	0.93 *3*
384 - 432	1.29	0.325 *21*
484	1.88	0.389 *19*
495	1.61	0.325 *16*
531 - 569	2.63	0.482 *12*
592 - 631	0.417	0.0695 *16*
755 - 769	1.7	~0.22
816 - 839	0.39	0.047 *23*
964 - 976	0.00022	2.3 *3* $\times10^{-5}$
1187 - 1200	0.0080	0.00067 *3*
1313 - 1323	0.04	~0.0034
1364 - 1397	0.034	0.0025 *5*
1424 - 1523	0.010	0.00066 *20*
1583 - 1594	0.0008	~5×10^{-5}
1721 - 1806	0.023	0.0013 *5*
2092 - 2163	0.0014	~7×10^{-5}

Continuous Radiation (^{190}Re)

⟨β-⟩=649 keV;⟨IB⟩=1.10 keV

E_{bin}(keV)		⟨ ⟩(keV)	(%)
0 - 10	β-	0.0301	0.60
	IB	0.027	
10 - 20	β-	0.091	0.61
	IB	0.026	0.18
20 - 40	β-	0.374	1.24
	IB	0.051	0.18
40 - 100	β-	2.81	3.98
	IB	0.139	0.22
100 - 300	β-	31.5	15.5
	IB	0.35	0.20
300 - 600	β-	119	26.4
	IB	0.30	0.073
600 - 1300	β-	410	45.6
	IB	0.20	0.025
1300 - 1793	β-	85	6.0
	IB	0.0055	0.00040

$^{190}_{75}$Re(3.2 *2* h)

Mode: β-(54 *3* %), IT(46 *3* %)

Δ: ~- 35367 keV

SpA: 5.2×10^{6} Ci/g

Prod: ^{192}Os(d,α); ^{190}Os(n,p); ^{193}Ir(n,α)

Photons (^{190}Re)

⟨γ⟩=928 *13* keV

γ_{mode}	γ(keV)	γ(%)†
Os L_ℓ	7.822	0.099 *12*
Os L_α	8.904	2.21 *22*
Os L_η	9.337	0.037 *4*
Os L_β	10.461	2.46 *25*
Os L_γ	12.169	0.43 *5*
Re $K_{\alpha2}$	59.717	6.7 *19*
Re $K_{\alpha1}$	61.141	12 *3*
Os $K_{\alpha2}$	61.485	3.5 *3*
Os $K_{\alpha1}$	63.000	6.1 *5*
Re $K_{\beta1}'$	69.214	3.9 *11*
Os $K_{\beta1}'$	71.313	2.10 *18*
Re $K_{\beta2}'$	71.470	1.0 *3*
Os $K_{\beta2}'$	73.643	0.53 *5*
$\gamma_{\beta-}$M1+18%E2	97.83 *6*	0.0185 *24*
$\gamma_{\beta-}$	108.65 *15*	0.31 *3*
$\gamma_{\beta-}$	114.16 *15*	0.19 *3*
γ_{IT} M1(+28%E2)	119.12 *5*	11.4 *10*
$\gamma_{\beta-}$	127.25 *6*	0.40 *4*
$\gamma_{\beta-}$	163.12 *6*	0.53 *3*
$\gamma_{\beta-}$	182.1 *3*	0.79 *8*
$\gamma_{\beta-}$E2	186.7134 *20*	28.1 *12*
$\gamma_{\beta-}$M1	190.45 *7*	0.041 *6*
$\gamma_{\beta-}$E2+30%M1	196.88 *5*	0.38 *3*
$\gamma_{\beta-}$E2+3.4%M1	198.052 *14*	0.96 *9*
$\gamma_{\beta-}$E2	199.339 *18*	0.24 *4*
$\gamma_{\beta-}$	200.0 *3*	1.28 *8*
$\gamma_{\beta-}$(E2)	207.817 *22*	0.13 *3*
$\gamma_{\beta-}$E2(+3.1%M1)	208.185 *14*	0.54 *7*
$\gamma_{\beta-}$E1	223.790 *7*	0.58 *3*
$\gamma_{\beta-}$E1	235.55 *5*	0.083 *8*
$\gamma_{\beta-}$[M1+E2]	242.27 *6*	0.145 *12*
$\gamma_{\beta-}$	252.63 *20*	0.128 *14*
$\gamma_{\beta-}$	255.17 *10*	0.49 *4*
$\gamma_{\beta-}$E2(+<14%M1)	282.94 *4*	2.25 *15*
$\gamma_{\beta-}$	285.05 *12*	0.45 *6*
$\gamma_{\beta-}$E2+17%M1	288.28 *7*	0.48 *3*
$\gamma_{\beta-}$(E2)	294.70 *4*	~0.6
$\gamma_{\beta-}$	315.54 *8*	1.02 *6*
$\gamma_{\beta-}$[M1+E2]	321.80 *8*	0.77 *8* ?
$\gamma_{\beta-}$	344.07 *10*	1.22 *6*
$\gamma_{\beta-}$E2	361.131 *6*	12.2 *10*
$\gamma_{\beta-}$E2+1.4%M1	371.265 *5*	10.5 *5*
$\gamma_{\beta-}$	379.46 *20*	0.73 *6* ?
$\gamma_{\beta-}$E1	379.99 *7*	0.225 *13*
$\gamma_{\beta-}$[E1]	387.15 *8*	0.87 *8*
$\gamma_{\beta-}$[M1+E2]	390.17 *5*	4.9 *3*
$\gamma_{\beta-}$(M1)	394.67 *22*	0.47 *8* ?
$\gamma_{\beta-}$E2	397.390 *13*	6.2 *4*
$\gamma_{\beta-}$E2+24%M1	407.155 *19*	8.6 *4*
$\gamma_{\beta-}$E2+10%M1	407.524 *14*	4.3 *5*
$\gamma_{\beta-}$[E1]	420.67 *5*	0.182 *11*
$\gamma_{\beta-}$[E1]	426.00 *7*	0.0077 *23*
$\gamma_{\beta-}$E1	431.607 *22*	0.404 *21*
$\gamma_{\beta-}$E2	447.83 *5*	2.80 *13*
$\gamma_{\beta-}$[E1]	477.82 *6*	0.36 *5*
$\gamma_{\beta-}$E2	485.16 *7*	0.147 *14*
$\gamma_{\beta-}$E2	490.76 *4*	3.56 *16*
$\gamma_{\beta-}$E2	502.54 *7*	3.42 *10*
$\gamma_{\beta-}$E1(+0.3%M2)	518.49 *4*	6.6 *5*
$\gamma_{\beta-}$	539.19 *25*	0.22 *6*
$\gamma_{\beta-}$E2	557.978 *5*	14.1 *6*
$\gamma_{\beta-}$[E2]	559.13 *10*	5.5 *8* ?
$\gamma_{\beta-}$E2+1.3%M1	569.316 *13*	14.0 *9*
$\gamma_{\beta-}$E2	605.206 *20*	14.8 *6*
$\gamma_{\beta-}$[M1+E2]	615.34 *2*	0.174 *9*
$\gamma_{\beta-}$E2	616.09 *13*	1.04 *8*
$\gamma_{\beta-}$[E1]	628.48 *6*	0.083 *10*
$\gamma_{\beta-}$E1	630.945 *20*	0.42 *4*
$\gamma_{\beta-}$[E1]	631.29 *8*	0.16 *3*
$\gamma_{\beta-}$[M1+E2]	632.44 *7*	~1
$\gamma_{\beta-}$E2+25%M1	656.01 *5*	1.19 *7*
$\gamma_{\beta-}$[E1]	668.27 *8*	0.016 *4*
$\gamma_{\beta-}$[E2]	673.11 *5*	9.5 *5*
$\gamma_{\beta-}$	675.22 *12*	0.53 *20*
$\gamma_{\beta-}$E2	690.10 *4*	1.37 *9*
$\gamma_{\beta-}$[E1]	708.94 *7*	0.022 *3*
$\gamma_{\beta-}$E1(+<3.8%M2)	726.31 *4*	0.73 *7*
$\gamma_{\beta-}$	739.71 *10*	0.77 *4*
$\gamma_{\beta-}$	753.0 *2*	0.041 *11* ?
$\gamma_{\beta-}$E2	768.654 *14*	2.13 *14*

Photons (^{190}Re)
(continued)

γ_{mode}	γ(keV)	γ(%)†
$\gamma_{\beta-}$[E1]	821.75 *9*	0.098 *8*
$\gamma_{\beta-}$[E1]	827.82 *5*	0.063 *12*
$\gamma_{\beta-}$E1+<0.4%M2	828.996 *20*	0.53 *4*
$\gamma_{\beta-}$E1(+M2)	839.129 *21*	0.178 *8*
$\gamma_{\beta-}$[M1+E2]	864.97 *9*	0.63 *4*
$\gamma_{\beta-}$[E2]	880.93 *5*	0.79 *8*
$\gamma_{\beta-}$	889.05 *15*	0.45 *8*
$\gamma_{\beta-}$[M1+E2]	905.64 *8*	1.32 *14*
$\gamma_{\beta-}$[E1]	916.76 *7*	0.038 *4*
$\gamma_{\beta-}$	952.34 *20*	0.164 *19* ?
$\gamma_{\beta-}$	958.16 *12*	1.99 *12*
$\gamma_{\beta-}$	965.1 *4*	0.43 *8*
$\gamma_{\beta-}$[E2]	976.470 *19*	0.021 *4*
$\gamma_{\beta-}$	1010.75 *19*	0.61 *8* ?
$\gamma_{\beta-}$E1	1036.00 *5*	0.270 *17*
$\gamma_{\beta-}$[M1+E2]	1113.46 *8*	0.41 *4*
$\gamma_{\beta-}$	1123.70 *4*	0.0062 *12*
$\gamma_{\beta-}$E1	1133.83 *4*	0.084 *7*
$\gamma_{\beta-}$	1150.39 *20*	0.058 *13* ?
$\gamma_{\beta-}$	1155.6 *5*	0.27 *4*
$\gamma_{\beta-}$	1160.53 *20*	0.047 *13* ?
$\gamma_{\beta-}$	1165.98 *12*	0.41 *5*
$\gamma_{\beta-}$	1194.2 *3*	0.205 *20*
$\gamma_{\beta-}$(E1)	1200.26 *2*	0.069 *4*
$\gamma_{\beta-}$	1250.2 *4*	0.28 *5*
$\gamma_{\beta-}$	1265.9 *5*	0.083 *24*
$\gamma_{\beta-}$[E2]	1312.80 *8*	0.65 *10*
$\gamma_{\beta-}$E1(+M2)	1324.28 *7*	0.160 *19*
$\gamma_{\beta-}$	1382.7 *5*	0.069 *24*
$\gamma_{\beta-}$	1386.971 *20*	0.0257 *17*
$\gamma_{\beta-}$(M2)	1397.13 *5*	0.0166 *12*
$\gamma_{\beta-}$	1494.96 *4*	0.0121 *14*
$\gamma_{\beta-}$	1499.4 *5*	0.051 *16*
$\gamma_{\beta-}$[M1+E2]	1520.98 *8*	0.083 *24*
$\gamma_{\beta-}$	1536.5 *8*	0.049 *14*
$\gamma_{\beta-}$	1564.9 *3*	0.12 *3*
$\gamma_{\beta-}$	1573.7 *3*	0.26 *7*
$\gamma_{\beta-}$	1616.4 *3*	0.14 *4*
$\gamma_{\beta-}$	1725.2 *5*	0.059 *18*
$\gamma_{\beta-}$	1745.4 *5*	0.041 *12*
$\gamma_{\beta-}$	1882.3 *7*	0.03 *1*
$\gamma_{\beta-}$	2023.4 *8*	~0.02

† uncert(syst): 6.6% for IT, 5.3% for β-

Atomic Electrons (^{190}Re)
⟨e⟩=52 *4* keV

e_{bin}(keV)	⟨e⟩(keV)	e(%)
56 - 103	1.2	1.5 *4*
106	0.10	~0.09
107	6.0	~6
108	0.4	~0.4
109	1.7	~2
111 - 112	0.05	~0.05
113	6.4	5.7 *3*
114 - 115	0.21	~0.18
116	1.6	~1
117 - 163	2.7	2.2 *6*
168 - 171	0.27	0.16 *7*
174	5.33	3.06 *15*
176	2.89	1.64 *8*
177 - 182	0.29	~0.16
184	2.24	1.22 *6*
185 - 234	2.43	1.24 *11*
235 - 284	1.3	0.49 *15*
285 - 286	0.0118	0.0041 *7*
287	1.30	0.45 *4*
288 - 295	0.019	0.007 *3*
297	1.12	0.377 *20*
303 - 332	2.3	0.71 *20*
333	1.3	0.39 *11*
334 - 382	2.40	0.68 *4*
384 - 432	1.59	0.395 *22*
435 - 484	1.32	0.279 *12*
485 - 531	2.34	0.461 *19*
536 - 582	1.07	0.193 *18*
592 - 635	0.99	0.164 *9*

Atomic Electrons (^{190}Re)
(continued)

e_{bin}(keV)	⟨e⟩(keV)	e(%)
643 - 690	0.30	0.046 *7*
695 - 742	0.122	0.0174 *15*
748 - 791	0.12	0.015 *4*
807 - 854	0.18	0.021 *6*
862 - 906	0.19	0.021 *10*
914 - 963	0.07	0.007 *3*
964 - 1011	0.007	~0.0007
1023 - 1060	0.025	0.0024 *9*
1077 - 1124	0.049	0.0044 *15*
1126 - 1166	0.008	0.00073 *24*
1176 - 1200	0.017	~0.0014
1237 - 1266	0.035	0.0028 *9*
1300 - 1324	0.013	0.00096 *20*
1370 - 1397	0.0012	8.5 *23* $\times10^{-5}$
1421 - 1463	0.0067	0.00047 *14*
1482 - 1571	0.019	0.0012 *4*
1603 - 1672	0.0034	0.00021 *8*
1712 - 1808	0.0012	7 *3* $\times10^{-5}$
1869 - 1950	0.0006	~3$\times10^{-5}$
2010 - 2021	8.6 $\times10^{-5}$	~4$\times10^{-6}$

Continuous Radiation (^{190}Re)
⟨β-⟩=329 keV;⟨IB⟩=0.55 keV

E_{bin}(keV)		⟨ ⟩(keV)	(%)
0 - 10	β-	0.0200	0.397
	IB	0.0138	
10 - 20	β-	0.060	0.402
	IB	0.0134	0.093
20 - 40	β-	0.247	0.82
	IB	0.026	0.090
40 - 100	β-	1.84	2.61
	IB	0.070	0.109
100 - 300	β-	19.9	9.9
	IB	0.17	0.099
300 - 600	β-	70	15.7
	IB	0.145	0.035
600 - 1300	β-	190	21.7
	IB	0.097	0.0122
1300 - 2500	β-	46.7	3.03
	IB	0.0082	0.00055
2500 - 2805	β-	0.109	0.00422
	IB	2.3$\times10^{-6}$	9.0$\times10^{-8}$

$^{190}_{76}$Os(stable)

Δ: -38717 *3* keV
%: 26.4 *4*

$^{190}_{76}$Os(9.9 *1* min)

Mode: IT
Δ: -37012 *3* keV
SpA: 9.99$\times10^{7}$ Ci/g
Prod: daughter ^{190}Ir(3.2 h); ^{189}Os(n,γ)

Photons (^{190}Os)
⟨γ⟩=1588 *17* keV

γ_{mode}	γ(keV)	γ(%)
Os L_ℓ	7.822	0.59 *6*
Os L_α	8.904	13.2 *10*
Os L_η	9.337	0.147 *11*
Os L_β	10.466	14.4 *15*
Os L_γ	12.290	2.9 *4*
γ M2+0.9%E3	38.90 *10*	0.081 *3*
Os $K_{\alpha2}$	61.485	5.66 *16*
Os $K_{\alpha1}$	63.000	9.7 *3*
Os $K_{\beta1}'$	71.313	3.35 *10*
Os $K_{\beta2}'$	73.643	0.85 *3*
γ E2	186.7134 *20*	70.2 *14*
γ E2	361.131 *6*	94.9 *19*
γ E2	502.54 *7*	97.8 *20*
γ E2	616.09 *13*	98.6 *20*

Atomic Electrons (^{190}Os)
⟨e⟩=113.8 *12* keV

e_{bin}(keV)	⟨e⟩(keV)	e(%)
11	3.1	28 *3*
12 - 13	3.4	26.7 *21*
26	10.7	41.4 *17*
27	2.4	9.2 *11*
28	7.0	25.0 *15*
36	7.2	20.1 *12*
37 - 71	2.94	7.3 *3*
113	16.1	14.2 *4*
174	13.3	7.65 *22*
176	7.23	4.11 *12*
184	5.45	2.96 *8*
185 - 187	1.63	0.875 *21*
287	10.1	3.53 *10*
348 - 361	5.60	1.595 *22*
429	7.15	1.67 *5*
490 - 503	2.74	0.555 *11*
542	5.82	1.07 *3*
603 - 616	1.87	0.309 *5*

$^{190}_{77}$Ir(11.78 *10* d)

Mode: ε
Δ: -36720 *200* keV
SpA: 5.83$\times10^{4}$ Ci/g
Prod: ^{189}Os(d,n); ^{190}Os(p,n); ^{187}Re(α,n); ^{192}Os(p,3n); ^{192}Os(d,4n)

Photons (^{190}Ir)
⟨γ⟩=1490 *18* keV

γ_{mode}	γ(keV)	γ(%)†
Os L_ℓ	7.822	0.61 *8*
Os L_α	8.904	13.6 *13*
Os L_η	9.337	0.188 *18*
Os L_β	10.469	14.2 *15*
Os L_γ	12.212	2.5 *3*
Os $K_{\alpha2}$	61.485	26.0 *22*
Os $K_{\alpha1}$	63.000	45 *4*
Os $K_{\beta1}'$	71.313	15.4 *13*
Os $K_{\beta2}'$	73.643	3.9 *3*
γ M1+18%E2	97.83 *6*	0.097 *12*
γ E2	186.7134 *20*	53.4 *21*
γ M1	190.45 *7*	0.140 *21*
γ E2+30%M1	196.88 *5*	3.5 *3*
γ E2+3.4%M1	198.052 *14*	2.01 *17*

Photons (^{190}Ir)
(continued)

γ_{mode}	γ(keV)	γ(%)†
γ E2	199.339 18	0.25 5
γ (E2)	207.817 22	0.35 7
γ E2(+3.1%M1)	208.185 14	1.13 12
γ E1	223.790 7	3.83 13
γ E1	235.55 5	0.43 4
γ [M1+E2]	242.27 6	0.032 3
γ	248.2 3	0.123 21
γ E2(+<14%M1)	282.94 4	0.50 4
γ E2+17%M1	288.28 7	1.63 10
γ (E2)	294.70 4	~3
γ E2	361.131 6	13.2 4
γ E2+1.4%M1	371.265 5	22.8 12
γ E1	379.99 7	2.06 9
γ (M1)	394.5 4	0.044 16
γ E2	397.390 13	6.62 18
γ E2+24%M1	407.155 19	23.9 8
γ E2+10%M1	407.524 14	4.6 5
γ [E1]	420.67 5	1.67 7
γ [E1]	426.00 7	0.026 8
γ E1	431.607 22	2.65 11
γ E2	447.83 5	2.69 13
γ [E1]	477.82 6	1.88 20
γ E2	485.16 7	0.50 6
γ E2	490.76 4	0.78 3
γ E2	502.54 7	1.28 7
γ E1(+0.3%M2)	518.49 4	34.6 18
γ E2	557.978 5	30.6 9
γ E2+1.3%M1	569.316 13	29.1 9
γ E2	605.206 20	40.9 12
γ [M1+E2]	615.34 2	0.481 21
γ [E1]	628.48 6	0.76 8
γ E1	630.945 20	2.79 22
γ [E1]	631.29 8	0.86 16
γ E2+25%M1	656.01 5	1.14 6
γ [E1]	668.27 8	0.054 12
γ E2	690.10 4	0.302 21
γ [E1]	708.94 7	0.075 10
γ E1(+<3.8%M2)	726.31 4	3.84 21
γ	731.1 4	0.037 12
γ [E1]	740.20 11	0.197 16
γ	753.0 2	0.025 6
γ E2	768.654 14	2.27 8
γ [E1]	821.75 9	0.331 20
γ [E1]	827.82 5	0.57 11
γ E1+<0.4%M2	828.996 20	3.48 20
γ E1(+M2)	839.129 21	1.17 4
γ [E1]	916.76 7	0.128 13
γ [E1]	948.01 11	0.070 12
γ	952.34 20	0.101 13
γ [E2]	976.470 19	0.058 11
γ E1	1036.00 5	2.47 13
γ	1123.70 4	0.032 6
γ E1	1133.83 4	0.44 3
γ [E1]	1147.35 11	0.135 14
γ	1150.39 20	0.036 8
γ	1160.53 20	0.029 8
γ (E1)	1200.26 2	0.452 20
γ E1(+M2)	1324.28 7	0.54 6
γ [E1]	1355.54 11	0.070 7
γ	1386.971 20	0.169 10
γ (M2)	1397.13 5	0.152 10
γ	1494.96 4	0.063 7

† 5.2% uncert(syst)

Atomic Electrons (^{190}Ir)
⟨e⟩=73.5 15 keV

e_{bin}(keV)	⟨e⟩(keV)	e(%)
11	3.9	36 5
12	2.4	19.4 24
13 - 62	2.85	10.8 9
63 - 98	0.258	0.335 21
113	12.2	10.8 5
117 - 162	2.7	2.14 23
168	0.014	~0.008
174	10.2	5.8 3
176	5.49	3.12 14

Atomic Electrons (^{190}Ir)
(continued)

e_{bin}(keV)	⟨e⟩(keV)	e(%)
177 - 180	0.031	0.0176 24
184	4.8	2.59 16
185 - 234	3.9	1.98 16
235 - 284	0.47	0.17 5
285 - 295	1.55	0.538 24
297	2.44	0.82 5
306 - 324	0.721	0.224 7
333	3.6	1.1 3
334 - 382	2.96	0.837 24
384 - 432	2.08	0.52 5
435 - 483	1.13	0.254 21
484	2.00	0.414 15
485 - 492	0.0335	0.0068 3
495	1.91	0.386 14
499 - 518	0.235	0.046 4
531	2.46	0.463 16
541 - 582	1.60	0.288 9
592 - 635	0.857	0.144 3
643 - 690	0.166	0.0255 16
695 - 742	0.140	0.0200 7
748 - 769	0.29	~0.038
809 - 843	0.06	0.008 3
874 - 917	0.010	0.0011 5
935 - 976	0.0390	0.00406 22
1023 - 1060	0.0150	0.00144 11
1073 - 1122	0.0059	0.00054 14
1123 - 1161	0.0074	0.00066 4
1187 - 1200	0.00119	0.000100 4
1250 - 1282	0.04	~0.003
1311 - 1356	0.035	0.0027 6
1374 - 1421	0.0077	0.00055 10
1482 - 1495	0.00041	2.8 14 ×10^{-5}

Continuous Radiation (^{190}Ir)
⟨IB⟩=0.40 keV

E_{bin}(keV)		⟨ ⟩(keV)	(%)
10 - 20	IB	0.00162	0.0146
20 - 40	IB	0.0022	0.0067
40 - 100	IB	0.35	0.56
100 - 300	IB	0.020	0.0127
300 - 600	IB	0.021	0.0049
600 - 1045	IB	0.0046	0.00068

$^{190}_{77}$Ir(1.2 h)

Mode: IT
Δ: -36694 200 keV
SpA: 1.4×10^7 Ci/g
Prod: ^{190}Os(p,n)

Photons (^{190}Ir)
⟨γ⟩=1.99 10 keV

γ_{mode}	γ(keV)	γ(%)
Ir L_ℓ	8.042	0.67 7
Ir L_α	9.167	14.8 10
Ir L_η	9.650	0.0098 7
Ir L_β	11.059	4.8 4
Ir L_γ	12.797	0.37 5
γ M3	26.3 1	0.000101

Atomic Electrons (^{190}Ir)
⟨e⟩=22.8 7 keV

e_{bin}(keV)	⟨e⟩(keV)	e(%)
11	4.8	43 5
13	1.61	12.4 7
15	9.2	61 3
23	0.87	3.78 17
24	4.53	19.1 9
26	1.79	6.9 3

$^{190}_{77}$Ir(3.2 2 h)

Mode: ε(94.4 10 %), IT(5.6 10 %)
Δ: -36545 200 keV
SpA: 5.2×10^6 Ci/g
Prod: ^{187}Re(α,n); deuterons on Os; ^{190}Os(p,n); ^{191}Ir(n,2n)

Photons (^{190}Ir)
⟨γ⟩=1629 17 keV

γ_{mode}	γ(keV)	γ(%)†
Os L_ℓ	7.822	1.02 10
Os L_α	8.904	23.0 15
Os L_η	9.337	0.271 18
Os L_β	10.469	24.2 22
Os L_γ	12.264	4.6 5
γ_ϵM2+0.9%E3	38.90 10	0.081 3 *
Os $K_{\alpha2}$	61.485	25.9 6
Os $K_{\alpha1}$	63.000	44.5 11
Ir $K_{\alpha2}$	63.286	0.329 16
Ir $K_{\alpha1}$	64.896	0.56 3
Os $K_{\beta1}'$	71.313	15.3 4
Ir $K_{\beta1}'$	73.452	0.195 10
Os $K_{\beta2}'$	73.643	3.88 10
Ir $K_{\beta2}'$	75.861	0.0502 25
γ_ϵ	116.7 10	
γ_{IT} M4	148.7 1	0.0113 5
γ_ϵE2	186.7134 20	69.9 14 *
γ_ϵ	206.6 10	
γ_ϵE2	361.131 6	94.4 19 *
γ_ϵE2	502.54 7	97.3 19 *
γ_ϵE2	616.09 13	98.2 20 *

† uncert(syst): 1.1% for ε, 18% for IT
* with ^{190}Os(9.9 min) in equilib

Atomic Electrons (^{190}Ir)
⟨e⟩=104.8 10 keV

e_{bin}(keV)	⟨e⟩(keV)	e(%)
27	2.4	9.2 11
28	7.0	24.8 14
36	7.2	20.0 12
37 - 73	5.35	11.4 3
113	16.0	14.2 4
135 - 136	1.70	1.26 5
137	2.61	1.90 9
146 - 149	1.85	1.26 4
174	13.3	7.61 22
176	7.19	4.09 12
184	5.42	2.95 8
185 - 187	1.62	0.871 21
287	10.1	3.52 10
348 - 361	5.58	1.588 22
429	7.11	1.66 5
490 - 503	2.72	0.553 11
542	5.80	1.07 3
603 - 616	1.87	0.308 5

Continuous Radiation (^{190}Ir)

⟨IB⟩=0.36 keV

E_{bin}(keV)		⟨ ⟩(keV)	(%)
10 - 20	IB	0.00140	0.0126
20 - 40	IB	0.0021	0.0062
40 - 100	IB	0.33	0.54
100 - 300	IB	0.020	0.0126
300 - 470	IB	0.0018	0.00055

$^{190}_{78}$Pt(6.0 *10* ×10^{11} yr)

Mode: α

Δ: -37338 *7* keV

%: <0.02

Prod: natural source

Alpha Particles (^{190}Pt)

α(keV)
3175 *20*

$^{190}_{79}$Au(42.8 *10* min)

Mode: ε

Δ: -32896 *16* keV

SpA: 2.31×10^{7} Ci/g

Prod: daughter ^{190}Hg; ^{191}Ir(α,5n); protons on Pt

Photons (^{190}Au)

⟨γ⟩=1990 *42* keV

γ$_{mode}$	γ(keV)	γ(%)†
γ E2(+<12%M1)	179.63 *5*	0.128 *14*
γ [M1+E2]	192.28 *10*	0.085 *7*
γ	206.2 *3*	0.071 *7*
γ	225.4 *3*	~0.36
γ E2	282.04 *6*	0.77 *6*
γ E2	286.28 *5*	0.383 *21*
γ E2	295.93 *4*	71 *6*
γ E2(+<1.0%M1)	301.95 *3*	25.1 *20*
γ E2(+<15%M1)	319.08 *4*	5.5 *6*
γ E2	323.32 *5*	1.14 *12*
γ	398.1 *3*	~0.09 ?
γ E2	441.40 *5*	4.3 *6*
γ E2+24%M1	460.8 *3*	0.31 *5*
γ E2	465.91 *6*	0.48 *11*
γ M1	478.56 *9*	0.32 *11*
γ E2+34%M1	523.45 *13*	0.43 *8*
γ	530.7 *3*	0.043 *4*
γ	586.8 *3*	0.106 *7*
γ E2	597.87 *4*	10.1
γ M1(+24%E2)	605.36 *5*	1.70 *21*
γ E1	616.41 *13*	1.85 *21*
γ E2+21%M1	621.03 *4*	2.9 *4*
γ E2	625.26 *5*	3.5 *4*
γ E1,E2	634.1 *3*	~0.21
γ [E2]	658.19 *10*	0.121 *14*
γ E0+E2+M1	675.4 *3*	0.227 *21*
γ M1	729.97 *20*	0.92 *16*
γ E1	756.50 *20*	0.14 *4*
γ M1	778.89 *17*	0.26 *7*
γ	797.64 *9*	0.14 *4*
γ E1	816.19 *21*	0.57 *11*
γ M1+E0+E2	864.5 *3*	0.121 *14*
γ M1	869.4 *3*	0.34 *8*
γ	907.31 *5*	2.3 *3*
γ M1	976.9 *3*	~0.14
γ M1+40%E2	987.44 *20*	1.06 *16*
γ M1	1002.6 *4*	0.21 *4*
γ E2(+<37%M1)	1005.10 *24*	0.92 *21*
γ	1054.8 *6*	1.0 *3*
γ E1	1057.80 *14*	4.0 *4*
γ	1099.59 *9*	1.21 *21*
γ E1	1139.51 *21*	1.8 *3*
γ M1	1154.3 *4*	0.32 *8*
γ (E2)	1161.1 *4*	0.72 *16*
γ (E2)	1203.23 *6*	0.50 *14*
γ E1	1279.30 *17*	0.91 *16*
γ (E2)	1303.7 *5*	0.99 *21*
γ M1	1307.04 *24*	0.51 *12*
γ M1	1345.7 *4*	0.77 *10*
γ E2	1395.51 *9*	2.7 *4*
γ E1	1402.1 *5*	0.67 *12*
γ E1	1441.45 *21*	4.0 *4*
γ [E2]	1461.76 *20*	0.59 *6* ?
γ M1	1501.0 *6*	0.40 *12*
γ	1528.2 *8*	0.24 *8*
γ E1	1581.24 *17*	1.07 *21*
γ M1	1602.96 *24*	0.43 *11*
γ M1	1664.2 *5*	0.48 *16*
γ E1	1672.35 *24*	0.71 *8*
γ M1	1739.4 *6*	1.14 *21*
γ M1(+<39%E2)	1760.8 *4*	0.97 *12*
γ M1(+<33%E2)	1785.08 *19*	2.3 *3*
γ E1	1802.8 *5*	0.48 *8*
γ M1	1835.8 *7*	1.0 *3*
γ (E1)	1864.63 *24*	0.99 *21*
γ M1	1880.2 *5*	0.62 *12*
γ M1	1921.1 *5*	1.07 *21*
γ M1,E2	2087.02 *19*	1.14 *21*
γ E1	2097.4 *3*	2.2 *3*
γ E1	2132.9 *3*	0.97 *16*
γ E1	2212.9 *5*	1.49 *21*
γ (E2)	2382.95 *19*	5.1 *4*
γ E1	2416.5 *3*	4.8 *4*
γ	2448.6 *5*	0.54 *12*
γ E1	2452.0 *3*	1.28 *21*
γ E1	2469.99 *24*	1.21 *21*
γ (E2)	2497.6 *5*	1.29 *17*
γ E1+E2	2658.5 *5*	2.4 *3*
γ E1	2680.4 *6*	0.75 *12*
γ E1	2685.5 *4*	1.36 *15*
γ E1	2753.9 *3*	4.3 *4*
γ E1	2771.93 *24*	0.73 *13*
γ M1,E2	2875.4 *5*	0.81 *16*
γ M1,E2	2881.8 *5*	0.78 *16*
γ M1+E2	2942.4 *6*	0.76 *16*
γ	2956.8 *9*	0.9 *3*
γ	2959.9 *9*	1.1 *4*
γ E1	2981.4 *4*	1.14 *21*
γ	3082.1 *5*	0.75 *16*
γ	3095.2 *6*	0.48 *10*
γ	3178.6 *6*	0.64 *14*
γ	3199.8 *6*	0.57 *16*
γ	3355.9 *6*	0.71 *14*

† approximate

$^{190}_{80}$Hg(20.0 *5* min)

Mode: ε

Δ: -31300 *100* keV syst

SpA: 4.95×10^{7} Ci/g

Prod: ^{197}Au(p,8n)

Photons (^{190}Hg)

⟨γ⟩=98 *16* keV

γ$_{mode}$	γ(keV)	γ(%)†
γ M1+0.5%E2	29.11 *18*	0.76 *15*
γ	48.5 *3*	0.22 *4*
γ M1	100.55 *18*	0.32 *7*
γ	125.54 *25*	<0.05 ?
γ	129.64 *18*	<0.43
γ (M1)	129.66 *19*	<1
γ	133.4 *3*	~0.27 ?
γ	135.44 *25*	~0.5 ?
γ	137.3 *3*	~0.43 ?
γ E1	142.55 *18*	54 *11*
γ	146.5 *3*	0.27 *5* ?
γ M1(+<26%E2)	154.83 *18*	2.0 *4*
γ	162.5 *3*	0.054 *11*
γ E2+~34%M1	165.4 *3*	0.76 *15* ?
γ (E1)	171.66 *19*	3.8 *8*
γ	182.3 *3*	0.22 *4* ?
γ	284.8 *3*	0.5 *1*
γ	373.82 *25*	0.22 *4* ?
γ	385.03 *18*	0.16 *3*
γ	637.9 *3*	0.49 *10*

† 22% uncert(syst)

$^{190}_{81}$Tl(2.6 *3* min)

Mode: ε

Δ: -24700 *320* keV syst

SpA: 3.8×10^{8} Ci/g

Prod: daughter ^{190}Pb; protons on Pb

Photons (^{190}Tl) *

⟨γ⟩=595 *15* keV

γ$_{mode}$	γ(keV)	γ(%)†
γ E2	416.38 *21*	88.0
γ E2	625.33 *24*	12.2 *12*
γ M1(+E2)	683.70 *19*	10.2 *9*
γ E2	1100.07 *21*	7.5 *7*

† 1.7% uncert(syst)

* see also ^{190}Tl(3.7 min)

$^{190}_{81}$Tl(3.7 *3* min)

Mode: ε

Δ: -24700 *320* keV syst

SpA: 2.67×10^{8} Ci/g

Prod: ^{181}Ta(^{16}O,7n); protons on Pb

Photons (^{190}Tl)

⟨γ⟩=1893 *63* keV

γ$_{mode}$	γ(keV)	γ(%)†
γ E2	197.11 *21*	4.6 *5*
γ (E2)	240.34 *22*	3.5 *4*
γ [E2]	257.1 *3*	1.55 *16*
γ E1	305.38 *23*	15.5 *16*
γ	313.8 *3*	1.37 *14* *
γ	346.6 *3*	1.28 *13* ?
γ E2	370.31 *24*	4.5 *5*
γ E2	416.38 *21*	91.2
γ [M1+E2]	437.45 *21*	0.91 *9*
γ	445.2 *3*	1.09 *11* *
γ [M1+E2]	458.92 *21*	1.0 *1* *

Photons (^{190}Tl)
(continued)

γ_{mode}	γ(keV)	γ(%)†
γ	478.58 *24*	1.19 *12*
γ E2	544.0 *3*	5.7 *6* ?
γ E2(+~10%M1)	557.15 *22*	6.4 *6*
γ E2+~34%M1	615.52 *24*	4.3 *4*
γ E2	625.33 *24*	82 *8*
γ M1(+E2)	683.70 *19*	6.6 *6*
γ [E2]	692.2 *3*	4.7 *5*
γ E2	731.31 *23*	37 *4*
γ	751.5 *3*	1.19 *12* *
γ E1	839.58 *21*	24.0 *24*
γ	931.1 *3*	1.09 *11* *
γ E2	1100.07 *21*	4.8 *4*
γ	1121.7 *3*	1.28 *13* *
γ [M1+E2]	1142.62 *21*	3.8 *4* *
γ	1194.8 *3*	2.28 *23* *
γ [M1+E2]	1240.85 *20*	1.37 *14*
γ [E1]	1277.03 *22*	1.28 *13*
γ	1323.6 *3*	1.82 *18* *
γ	1348.1 *3*	0.82 *8* *
γ [E2]	1558.99 *21*	1.0 *1* * ?

† 1.4% uncert(syst)
* with ^{190}Tl(3.7 min) or ^{190}Tl(2.6 min)

$^{190}_{82}$Pb(1.2 *1* min)

Mode: ϵ(99.1 *2* %), α(0.9 *2* %)
Δ: -20430 *230* keV syst
SpA: 8.2×10^8 Ci/g
Prod: ^{182}W(^{16}O,8n); ^{181}Ta(^{19}F,10n); ^{155}Gd(^{40}Ar,5n); protons on Th

Alpha Particles (^{190}Pb)

α(keV)
5580 *10*

Photons (^{190}Pb)

$\langle\gamma\rangle$=691 *28* keV

γ_{mode}	γ(keV)	γ(%)†
Tl L_ℓ	8.953	0.91 *17*
Tl L_α	10.259	18 *3*
Tl L_η	10.994	0.25 *5*
Tl L_β	12.310	17 *3*
Tl L_γ	14.404	3.2 *6*
γ_ϵ[M1+E2]	59.36 *15*	
Tl $K_{\alpha2}$	70.832	30 *4*
Tl $K_{\alpha1}$	72.873	50 *7*
γ_ϵ	78.4 *3*	
Tl $K_{\beta1}$'	82.434	18 *3*
Tl $K_{\beta2}$'	85.185	5.0 *7*
γ_ϵ	102.06 *11*	0.65 *9*
γ_ϵ(M1+E2)	118.99 *20*	0.37 *9*
γ_ϵM1+E2	123.02 *12*	0.66 *8*
γ_ϵ[E1]	140.85 *20*	~4
γ_ϵ	142.60 *22*	~7
γ_ϵM1+E2	151.37 *9*	8.92 *18*
γ_ϵM1,E2	158.33 *15*	1.70 *15*
γ_ϵM1(+E2)	162.38 *20*	0.49 *17*
γ_ϵ	193.34 *15*	1.29 *18*
γ_ϵM1	210.73 *13*	3.6 *9*
γ_ϵ	265.62 *22*	~0.17 ?
γ_ϵM1	274.39 *9*	3.0 *5*
γ_ϵ[E1]	362.90 *14*	1.9 *3*
γ_ϵM1	376.45 *8*	7.0 *9*

Photons (^{190}Pb)
(continued)

γ_{mode}	γ(keV)	γ(%)†
γ_ϵ	381.84 *15*	1.82 *18*
γ_ϵE1	565.98 *11*	4.61 *23*
γ_ϵM1(+E2)	598.51 *17*	8.0 *6*
γ_ϵ	739.59 *15*	4.1 *5*
γ_ϵ(E1)	791.06 *11*	2.9 *3*
γ_ϵE1	942.43 *8*	34 *3*
γ_ϵ	1235.68 *15*	4.5 *5*
γ_ϵ	1854.7 *3*	0.69 *18*

† uncert(syst): 8.9% for ϵ

$^{190}_{83}$Bi(5.4 *5* s)

Mode: α(~90 %), ϵ(~10 %)
Δ: -10730 *310* keV syst
SpA: 1.032×10^{10} Ci/g
Prod: ^{159}Tb(^{40}Ar,9n); ^{181}Ta(^{20}Ne,11n)

Alpha Particles (^{190}Bi)

α(keV)
6450 *10*

A = 191

NDS **30**, 653 (1980)

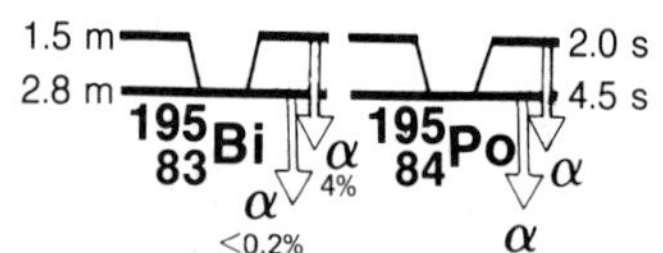

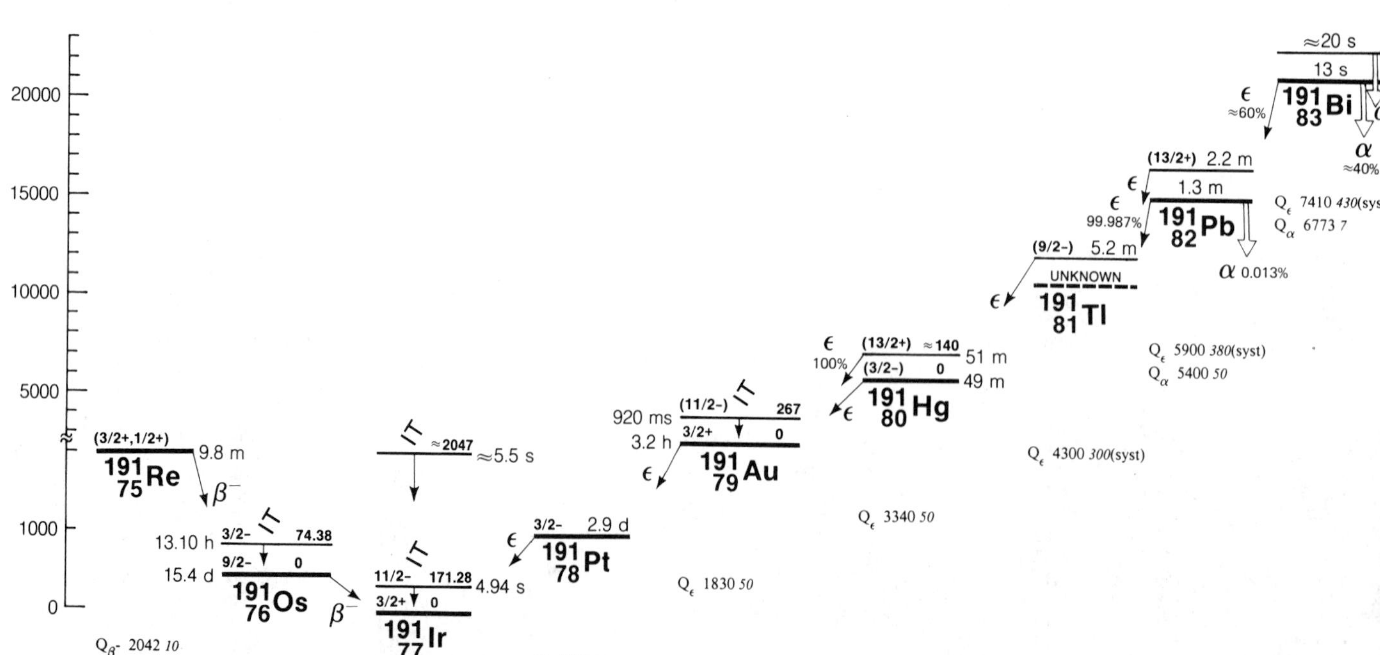

$^{191}_{75}$Re(9.8 *5* min)

Mode: β-
Δ: -34361 *11* keV
SpA: 1.00×10^8 Ci/g
Prod: neutrons on Os; γ rays on Os

$^{191}_{76}$Os(15.4 *1* d)

Mode: β-
Δ: -36403 *4* keV
SpA: 4.44×10^4 Ci/g
Prod: ^{190}Os(n,γ)

Photons (^{191}Os)

⟨γ⟩=74.9 *21* keV

$γ_{mode}$	γ(keV)	γ(%)†
Ir $L_ℓ$	8.042	0.80 *9*
Ir $L_α$	9.167	17.6 *13*
Ir $L_η$	9.650	0.299 *22*
Ir $L_β$	10.814	19.1 *15*
Ir $L_γ$	12.575	3.3 *3*
Ir $K_{α2}$	63.286	16.3 *10*
Ir $K_{α1}$	64.896	28.0 *17*
Ir $K_{β1}$'	73.452	9.6 *6*
Ir $K_{β2}$'	75.861	2.48 *15*
γ M1+43%E2	82.431 *5*	~0.026 *
γ M1+12%E2	129.434 *4*	25.7 *13* *

† 7.8% uncert(syst)
* with ^{191}Ir(4.94 s) in equilib

Atomic Electrons (^{191}Os)

⟨e⟩=93.9 *23* keV

e_{bin}(keV)	⟨e⟩(keV)	e(%)
6	0.009	~0.14
11	5.2	46 *5*
13	4.3	34 *4*
28	0.208	0.73 *4*
29	10.1	35.0 *17*
31	10.9	35.7 *18*
34 - 36	0.078	0.22 *3*
39	8.0	20.4 *10*
40 - 52	4.00	9.5 *3*
53	31.3	59 *3*
54 - 82	0.71	1.15 *6*
116	10.1	8.7 *5*
117	2.6	2.3 *3*
118 - 129	6.2	5.0 *3*

Continuous Radiation (^{191}Os)

⟨β-⟩=37.5 keV;⟨IB⟩=0.0053 keV

E_{bin}(keV)		⟨ ⟩(keV)	(%)
0 - 10	β-	0.89	18.1
	IB	0.0018	
10 - 20	β-	2.37	15.9
	IB	0.00124	0.0088
20 - 40	β-	7.6	25.9
	IB	0.00139	0.0050
40 - 100	β-	23.2	37.0

Continuous Radiation (^{191}Os) (continued)

E_{bin}(keV)		⟨ ⟩(keV)	(%)
	IB	0.00085	0.00155
100 - 141	β-	3.39	3.07
	IB	1.05×10^{-5}	9.9×10^{-6}

$^{191}_{76}$Os(13.10 *5* h)

Mode: IT
Δ: -36329 *4* keV
SpA: 1.252×10^6 Ci/g
Prod: ^{190}Os(n,γ)

Photons (^{191}Os)

⟨γ⟩=7.41 *20* keV

$γ_{mode}$	γ(keV)	γ(%)†
Os $L_ℓ$	7.822	0.53 *6*
Os $L_α$	8.904	11.9 *9*
Os $L_η$	9.337	0.048 *4*
Os $L_β$	10.556	7.1 *7*
Os $L_γ$	12.322	1.13 *15*
Os $K_{α2}$	61.485	2.40 *13*
Os $K_{α1}$	63.000	4.13 *22*
Os $K_{β1}$'	71.313	1.42 *8*
Os $K_{β2}$'	73.643	0.36 *2*
γ M3+0.3%E4	74.38 *1*	0.074

† 14% uncert(syst)

Atomic Electrons (^{191}Os)

⟨e⟩=59.7 *16* keV

e_{bin}(keV)	⟨e⟩(keV)	e(%)
11	3.5	33 *4*
12	0.43	3.5 *4*
13	0.86	6.6 *7*
48	0.0166	0.035 *4*
49	0.027	0.056 *6*
50	0.0169	0.034 *4*
51	0.031	0.062 *7*
52	0.0147	0.028 *3*
58	0.024	0.041 *5*
59	0.0088	0.0149 *17*
60	0.0159	0.026 *3*
61	12.6	20.6 *10*
62	2.20	3.55 *18*
63	0.0028	0.0045 *5*
64	26.1	41.1 *21*
68	0.0031	0.0045 *5*
69	0.0026	0.0038 *4*
70	0.00130	0.00185 *21*
71	4.21	5.9 *3*
72	9.5	13.2 *7*

$^{191}_{77}$Ir(stable)

Δ: -36716 *4* keV
%: 37.3 *5*

$^{191}_{77}$Ir(4.94 *3* s)

Mode: IT
Δ: -36545 *4* keV
SpA: 1.116×10^{10} Ci/g
Prod: daughter ^{191}Os(15.4 d); ^{192}Os(p,2n); ^{191}Ir(n,n')

Photons (^{191}Ir)

⟨γ⟩=74.9 *21* keV

$γ_{mode}$	γ(keV)	γ(%)†
Ir $L_ℓ$	8.042	0.80 *9*
Ir $L_α$	9.167	17.6 *13*
Ir $L_η$	9.650	0.299 *22*
Ir $L_β$	10.814	19.1 *15*
Ir $L_γ$	12.575	3.3 *3*
Ir $K_{α2}$	63.286	16.3 *10*
Ir $K_{α1}$	64.896	28.0 *17*
Ir $K_{β1}$'	73.452	9.6 *6*
Ir $K_{β2}$'	75.861	2.48 *15*
γ M1+43%E2	82.431 *5*	~0.026
γ M1+12%E2	129.434 *4*	25.7 *13*

† 7.8% uncert(syst)

Atomic Electrons (^{191}Ir)

⟨e⟩=93.9 *23* keV

e_{bin}(keV)	⟨e⟩(keV)	e(%)
6	0.009	~0.14
11	5.2	46 *5*
13	4.3	34 *4*
28	0.208	0.73 *4*
29	10.1	35.0 *17*
31	10.9	35.7 *18*
34 - 36	0.078	0.22 *3*
39	8.0	20.4 *10*
40 - 52	4.00	9.5 *3*
53	31.3	59 *3*
54 - 82	0.71	1.15 *6*
116	10.1	8.7 *5*
117	2.6	2.3 *3*
118 - 129	6.2	5.0 *3*

$^{191}_{77}$Ir(5.5 *7* s)

Mode: IT
Δ: ~-34669 keV
SpA: 1.01×10^{10} Ci/g
Prod: ^{192}Os(p,2nγ); ^{192}Os(d,3nγ)

$^{191}_{78}$Pt(2.9 *1* d)

Mode: ϵ
Δ: -35710 *8* keV
SpA: 2.36×10^5 Ci/g
Prod: protons on Ir; ^{191}Ir(d,2n); ^{190}Pt(n,γ)

Photons (^{191}Pt)

⟨γ⟩=296 12 keV

γ_{mode}	γ(keV)	γ(%)†
Ir L_ℓ	8.042	1.00 22
Ir L_α	9.167	22 5
Ir L_η	9.650	0.30 6
Ir L_β	10.825	22 4
Ir L_γ	12.617	3.8 8
Ir $K_{\alpha2}$	63.286	39 8
Ir $K_{\alpha1}$	64.896	68 14
Ir $K_{\beta1}$'	73.452	23 5
Ir $K_{\beta2}$'	75.861	6.0 12
γ M1+43%E2	82.431 5	4.9 5
γ [M1]	85.22 4	0.060 7
γ M1+1.9%E2	96.552 8	3.28 16
γ M1+12%E2	129.434 4	3.2 3
γ [M1]	138.48 5	~0.024
γ M1	140.918 15	0.075 11
γ M1	172.219 15	3.52 16
γ M1+39%E2	178.984 9	1.02 5
γ	186.8 10	~0.040
γ M1+27%E2	187.708 19	0.416 24
γ [E2]	195.55 16	0.0032 8
γ M1	208.95 6	0.136 24
γ M1+12%E2	213.93 16	0.0088 24
γ E2	219.69 4	0.82 4
γ [M1]	221.768 16	0.116 12
γ M1	223.70 5	0.112 12
γ [M1]	244.64 17	0.0032 8
γ E2	267.96 5	0.78 8
γ E2	268.771 16	1.65 16
γ	272 1	
γ [E2]	343.37 16	0.013 4
γ M1	351.202 16	3.36 16
γ M1	359.927 15	6.0 3
γ [M1]	396.66 6	0.010 3
γ [M1]	404.50 17	~0.011
γ [M1]	408.88 6	0.096 16
γ M1	409.476 14	8.0 4
γ [M1]	411.40 5	0.0096 24
γ [M1]	445.14 3	0.054 6
γ (M1)	456.479 15	3.36 16
γ [M1]	458.57 7	0.043 8
γ [E1]	479.95 4	0.057 6
γ (M1+E2)	494.69 3	0.060 6
γ	501.4 10	~0.010
γ M1	538.910 14	13.7 7
γ (M1)	541.69 3	0.37 4
γ M1	568.88 6	0.053 4
γ [E1]	576.50 4	0.118 9
γ [M1]	583.62 5	0.076 6
γ [M1]	588.00 7	0.136 10
γ	604.6 10	
γ [M1]	618.43 6	0.009 3
γ (M1)	624.13 3	1.41 7
γ [M1]	633.17 5	0.0240 24
γ	636 1	~0.006
γ [E1]	658.93 4	0.0152 16
γ	667 2	~0.0048
γ [M1]	680.17 5	0.0069 14
γ	686.6 5	0.0008 3
γ [M1]	747.86 6	0.0042 8
γ	756.6 5	0.0016 5
γ [M1]	762.61 5	0.0120 16
γ	765 2	~0.010
γ	805.99 14	0.0038 7
γ	853.6 4	0.00104 8
γ	935.42 14	0.0120 16

† 25% uncert(syst)

Atomic Electrons (^{191}Pt)

⟨e⟩=71.9 20 keV

e_{bin}(keV)	⟨e⟩(keV)	e(%)
6	1.73	27 3
9	0.046	0.50 6
11	6.1	55 12
13	4.8	37 8
20	3.87	18.9 10
28 - 52	1.84	4.0 4
53	3.9	7.3 8
54 - 65	1.48	2.5 3
69	2.9	4.1 4
70	5.8	8.4 9
71	5.2	7.3 8
72 - 79	0.87	1.11 10
80	3.2	4.0 4
82	1.20	1.47 15
83	2.34	2.81 15
84 - 95	1.15	1.28 5
96	4.04	4.20 21
97 - 146	3.88	3.33 14
148 - 196	2.43	1.45 4
198 - 245	0.278	0.132 4
255 - 275	1.81	0.67 3
284	2.54	0.90 5
321 - 332	0.0086	0.0026 7
333	2.86	0.86 5
335 - 384	2.08	0.575 19
385 - 434	0.73	0.183 7
442 - 459	0.251	0.0562 25
463	3.26	0.70 4
466 - 512	0.150	0.0312 23
525 - 575	1.08	0.202 7
576 - 625	0.068	0.0111 5
630 - 679	0.00136	0.000207 24
680 - 689	0.0027	0.00039 11
730 - 778	0.0011	0.00015 4
793 - 841	7.8×10^{-5}	$\sim 1 \times 10^{-5}$
842 - 859	0.0008	$\sim 9 \times 10^{-5}$
922 - 935	0.00017	$\sim 2 \times 10^{-5}$

Continuous Radiation (^{191}Pt)

⟨IB⟩=0.48 keV

E_{bin}(keV)		⟨ ⟩(keV)	(%)
10 - 20	IB	0.00166	0.0146
20 - 40	IB	0.0020	0.0060
40 - 100	IB	0.39	0.61
100 - 300	IB	0.034	0.020
300 - 600	IB	0.038	0.0091
600 - 1006	IB	0.0097	0.00143

$^{191}_{79}$Au(3.18 8 h)

Mode: ε

Δ: -33880 50 keV

SpA: 5.16×10^6 Ci/g

Prod: protons on Pt; ^{191}Ir(α,4n); ^{192}Pt(d,3n)

Photons (^{191}Au)

⟨γ⟩=559 25 keV

γ_{mode}	γ(keV)	γ(%)†
Pt L_ℓ	8.266	0.9 3
Pt L_α	9.435	19 7
Pt L_η	9.975	0.25 9
Pt L_β	11.184	19 7
Pt L_γ	13.071	3.5 13
γ (M1+2.0%E2)	24.406 10	~0.030 ?
γ M1+0.1%E2	30.410 9	~0.26
γ M2	48.385 10	~0.014
Pt $K_{\alpha2}$	65.122	31 11
Pt $K_{\alpha1}$	66.831	52 19
Pt $K_{\beta1}$'	75.634	18 7
Pt $K_{\beta2}$'	78.123	4.8 17
γ M1+6.2%E2	87.753 16	0.192 16
γ E2	91.124 16	~1
γ M1+39%E2	106.379 19	0.048 16
γ	122.72 10	0.048 16
γ M1+40%E2	126.955 14	0.144 16
γ	132.01 10	0.34 3
γ M1+~5.0%E2	132.913 17	1.15 10
γ	133.71 10	0.26 3
γ M1+15%E2	136.115 13	0.64 6
γ	142.52 10	0.224 16
γ M1	145.909 25	0.096 16
γ M1(+E2)	147.497 22	0.82 5
γ [M1]	156.965 18	0.46 6
γ [E2]	157.319 18	0.64 10
γ M1+21%E2	158.82 3	1.39 10
γ M1+22%E2	166.525 13	3.12 22
γ E2	192.796 19	0.24 3
γ M1+~8.3%E2	194.132 17	2.58 18
γ	202.44 8	0.096 16
γ [M1]	206.400 22	~0.8
γ [M1]	206.408 23	~1
γ M1+8.8%E2	210.11 4	0.56 5
γ [M1]	223.540 23	0.208 16
γ M1+29%E2	244.390 24	0.91 6
γ M1+20%E2	247.487 22	0.70 5
γ E2	253.950 22	2.38 16
γ E2	263.070 15	1.46 10
γ M1+~8.3%E2	268.337 24	1.86 14
γ M1+50%E2	271.652 22	2.37 16
γ M1	277.897 21	6.8 5
γ E2	280.410 22	2.77 18
γ M1+~14%E2	283.920 16	6.3 4
γ M1+35%E2	293.480 14	2.67 18
γ	316.02 15	~0.08
γ [M1]	332.01 5	0.144 16
γ M1(+E2)	340.36 5	0.144 16
γ	347.55 10	0.51 5
γ E2	353.904 23	2.93 19
γ [M1]	359.27 5	0.16 3
γ E2+46%M1	368.85 4	0.42 3
γ [M1]	376.55 3	0.40 3
γ E2+35%M1	386.928 18	3.39 22
γ M1+~14%E2	390.299 19	2.56 18
γ E2	399.859 19	4.5 3
γ M1+E2	408.19 4	0.86 11
γ M1+43%E2	410.08 4	0.42 8
γ [E1]	411.83 12	0.24 5
γ M1+15%E2	413.78 3	3.47 22
γ M1(+E2)	421.45 3	3.25 22
γ	427.34 11	0.208 16
γ	432.43 19	0.144 16
γ M1+26%E2	442.30 3	0.56 5
γ M1	446.70 5	0.34 3
γ	450.70 21	0.14 3
γ [E1]	451.36 15	1.26 10
γ (M1+E2)	451.86 3	1.26 10
γ [E1]	460.85 5	0.22 3
γ M1	467.05 8	0.70 10
γ M1+~39%E2	478.051 16	3.7 3
γ E2	487.612 17	2.61 18
γ M1	495.72 11	0.54 5
γ [E1]	499.36 4	0.43 5
γ M1	525.82 4	0.82 6
γ	532.64 6	0.37 5
γ	535.26 23	0.13 3
γ E1	538.70 5	0.77 13
γ [M1]	544.30 4	0.14 3
γ E2	557.58 4	0.21 3
γ (M1)	561.41 11	0.064 16
γ M1+48%E2	565.15 4	0.46 5
γ	568.30 19	0.16 3
γ [M1]	574.71 4	0.16 3
γ	580.32 12	~0.08
γ E1	586.46 3	16
γ	595.91 19	0.32 3
γ	608.4 4	
γ M1+21%E2	616.27 19	0.35 5
γ M1+40%E2	620.19 3	1.02 10
γ [E1]	625.99 10	0.85 13 ?
γ (M1)	627.71 14	0.19 5
γ	634.6 5	

Photons (^{191}Au)
(continued)

γ_{mode}	γ(keV)	γ(%)†
γ	648.03 *14*	0.10 *3*
γ M1+49%E2	659.72 *10*	0.27 *3*
γ M1	669.6 *3*	0.064 *16*
γ E1	674.22 *3*	6.4 *5*
γ	680.8 *3*	0.064 *16*
γ M1	702.01 *10*	0.51 *5*
γ M1	732.42 *10*	0.096 *16*
γ	734.5 *3*	0.128 *16*
γ	751.6 *5*	
γ [M1]	767.69 *3*	0.19 *3*
γ [E1]	780.59 *3*	0.24 *3*
γ E1	792.86 *3*	0.67 *6*
γ [E1]	820.12 *3*	0.34 *3*
γ	829.9 *4*	0.064 *16*
γ E1	835.7 *1*	0.64 *3*
γ	839.7 *7*	
γ	854.3 *4*	0.064 *16*
γ M1	870.6 *4*	0.096 *16*
γ	878.6 *5*	
γ [M1]	881.15 *11*	0.176 *16*
γ [M1]	896.73 *11*	0.128 *16*
γ M1	920.68 *11*	0.080 *16*
γ	924.1 *6*	
γ	929.2 *6*	
γ	971.3 *7*	
γ	981.7 *7*	
γ	985.4 *7*	
γ	1006.3 *6*	0.064 *16*
γ (M1)	1023.0 *3*	0.096 *16*
γ	1028.0 *6*	0.112 *16*
γ	1036.00 *16*	0.080 *16*
γ [E1]	1064.51 *3*	0.144 *16*
γ [E1]	1074.07 *3*	0.160 *16*
γ	1086.9 *8*	0.064 *16*
γ	1096.8 *6*	0.112 *16*
γ	1101.9 *6*	0.128 *16*
γ [E1]	1113.60 *10*	0.24 *3*
γ	1161.2 *6*	0.18 *3*
γ [M1]	1165.07 *11*	0.19 *3*
γ	1199.3 *3*	0.144 *16*
γ	1259.6 *6*	0.27 *3*
γ	1302.3 *8*	0.14 *3*

† approximate

Atomic Electrons (^{191}Au)
⟨e⟩=77 *4* keV

e_{bin}(keV)	⟨e⟩(keV)	e(%)
9 - 11	0.43	4.2 *14*
12	4.8	42 *15*
13	3.2	24 *9*
14	1.1	8 *3*
17	1.3	~8
19 - 54	5.2	13.8 *24*
55	1.9	3.5 *7*
58 - 77	3.0	4.7 *8*
78	2.8	~4
79	0.73	0.93 *14*
80	4.0	5.0 *17*
84 - 87	0.082	0.096 *11*
88	4.6	5.3 *12*
89 - 115	0.80	0.8 *3*
116	2.44	2.11 *15*
118 - 127	1.5	1.25 *24*
128	2.0	~2
129 - 176	6.5	4.3 *3*
179 - 192	2.38	1.28 *6*
193	1.5	0.77 *17*
194 - 199	0.153	0.078 *7*
200	4.4	2.21 *15*
202 - 204	0.53	0.26 *3*
206	3.61	1.76 *12*
207 - 255	2.59	1.13 *11*
257 - 304	4.53	1.67 *6*
309 - 357	4.7	1.44 *15*
359 - 408	2.91	0.75 *6*
409 - 459	1.03	0.241 *16*

Atomic Electrons (^{191}Au)
(continued)

e_{bin}(keV)	⟨e⟩(keV)	e(%)
460 - 508	1.00	0.204 *17*
512 - 561	0.44	0.082 *9*
562 - 609	0.362	0.061 *4*
612 - 661	0.180	0.0285 *22*
663 - 702	0.066	0.0096 *8*
714 - 757	0.048	0.0066 *6*
764 - 809	0.046	0.0058 *6*
816 - 859	0.033	0.0039 *4*
867 - 909	0.0107	0.00122 *7*
917 - 958	0.026	0.0027 *7*
986 - 1035	0.030	0.0030 *7*
1036 - 1085	0.012	~0.0011
1086 - 1121	0.026	0.0024 *5*
1147 - 1197	0.019	0.0016 *7*
1199 - 1248	0.008	~0.0007
1256 - 1302	0.0019	0.00015 *6*

$^{191}_{79}$Au(920 *110* ms)

Mode: IT
Δ: -33613 *50* keV
SpA: 4.5×10^{10} Ci/g
Prod: daughter ^{191}Hg(51 m); deuterons on Pt

Photons (^{191}Au)
⟨γ⟩=200 *13* keV

γ_{mode}	γ(keV)	γ(%)†
Au L_{ℓ}	8.494	0.69 *9*
Au L_{α}	9.704	14.6 *16*
Au L_{η}	10.309	0.066 *10*
Au L_{β}	11.668	6.9 *9*
Au L_{γ}	13.463	0.77 *13*
Au $K_{\alpha2}$	66.991	5.5 *11*
Au $K_{\alpha1}$	68.806	9.3 *19*
Au $K_{\beta1}'$	77.859	3.3 *7*
Au $K_{\beta2}'$	80.428	0.86 *17*
γ E2	241.14 *18*	13.7 *11*
γ M1+43%E2	252.67 *17*	60 *5*

† 3.3% uncert(syst)

Atomic Electrons (^{191}Au)
⟨e⟩=64 *7* keV

e_{bin}(keV)	⟨e⟩(keV)	e(%)
8 - 9	1.9	~22
11	4.1	37 *9*
12	5.5	46 *6*
13 - 57	2.78	19.1 *16*
63 - 77	0.173	0.26 *3*
160	2.36	1.47 *12*
172	31.3	18 *4*
227	1.60	0.70 *6*
229	0.68	0.296 *24*
238	7.0	2.9 *6*
239	2.4	1.00 *25*
240 - 253	4.6	1.84 *25*

$^{191}_{80}$Hg(49 *10* min)

Mode: ε
Δ: -30540 *70* keV
SpA: 2.0×10^{7} Ci/g
Prod: daughter ^{191}Tl; protons on Pb

Photons (^{191}Hg)

γ_{mode}	γ(keV)	γ(rel)
γM1	196.56 *18*	50 *5*
γE2	224.95 *17*	45 *5*
γE2	241.14 *18*	22.9 *18* *
γM1+43%E2	252.67 *17*	100 *8* *
γ[M1+E2]	331.7 *5*	29 *4* ?
γ	521.7 *6*	6 *3*
γ	778.3 *10*	~5

* with ^{191}Au(920 ms) in equilib

$^{191}_{80}$Hg(50.8 *15* min)

Mode: ε
Δ: ~-30400 keV
SpA: 1.94×10^{7} Ci/g
Prod: ^{197}Au(p,7n); ^{197}Au(d,8n)

Photons (^{191}Hg)
⟨γ⟩=1407 *42* keV

γ_{mode}	γ(keV)	γ(%)†
Au L_{ℓ}	8.494	1.1 *5*
Au L_{α}	9.704	23 *10*
Au L_{η}	10.309	0.22 *9*
Au L_{β}	11.588	17 *7*
Au L_{γ}	13.485	2.7 *11*
Au $K_{\alpha2}$	66.991	28 *11*
Au $K_{\alpha1}$	68.806	47 *19*
Au $K_{\beta1}'$	77.859	17 *7*
Au $K_{\beta2}'$	80.428	4.4 *18*
γM1	196.56 *18*	2.5 *3*
γE2	224.95 *17*	3.3 *4*
γE2	241.14 *18*	12.4 *10* *
γM1+43%E2	252.67 *17*	54 *4* *
γM1	274.2 *6*	13 *3*
γ	301.7 *10*	0.93 *22*
γ	331.7 *9*	4.2 *10*
γM1	343.3 *6*	1.4 *6*
γM1+E2	357.0 *4*	4.7 *9*
γE2	371.1 *4*	6.1 *7*
γM1	410.0 *5*	2.7 *6*
γE2	420.3 *5*	17.9 *18*
γM1+E2	441.1 *10*	1.1 *3*
γ	457.9 *10*	0.55 *17*
γ	487.1 *10*	
γ	511.1 *8*	2.2 *4*
γE2	521.5 *5*	3.8 *5* ?
γ(E2)	536.1 *5*	7.8 *9*
γ	545.7 *5*	1.5 *3*
γ	546.1 *10*	0.50 *16*
γ	549.6 *10*	2.0 *5*
γ[M1+E2]	552.8 *6*	1.5 *3*
γM1	578.7 *4*	17.0 *17*
γ	597.1 *10*	0.77 *17*
γM1+E2	610.7 *5*	4.5 *9* ?
γ	637.2 *7*	1.4 *3*
γ[M1+E2]	645.3 *5*	0.82 *22*
γ[E2]	662.7 *5*	1.1 *3*
γ[E2]	671.1 *7*	2.9 *7*
γ[E2]	678.9 *6*	2.1 *4*
γ	683.3 *10*	1.3 *3*
γM1+E2	689.7 *6*	2.1 *6*
γ	718.3 *6*	3.0 *4*
γ	732.4 *7*	0.50 *11*
γ[E2]	777.6 *5*	1.6 *4*
γ	820.1 *6*	1.2 *3*

Photons (^{191}Hg)
(continued)

γ_{mode}	γ(keV)	γ(%)†
γ	829.1 *10*	0.77 *22*
γ	863.7 *10*	0.77 *22*
γ	887.1 *8*	1.8 *4*
γ	897.1 *10*	0.55 *17*
γ	908.4 *10*	0.71 *22*
γ	938.4 *10*	1.1 *3*
γ	949.5 *10*	0.55 *17*
γ	954.9 *10*	0.60 *17*
γ	970.1 *10*	0.77 *22*
γ	996.5 *8*	2.3 *6*
γ[M1+E2]	1002.5 *5*	1.5 *4* ?
γ	1040.8 *10*	0.50 *16*
γ	1045.7 *10*	0.60 *17*
γ	1057.1 *10*	0.77 *22*
γ	1091.4 *10*	1.2 *3*
γ	1110.0 *7*	1.7 *4* ?
γ	1165.1 *10*	0.82 *22*
γ	1257.2 *10*	0.50 *16*
γ	1284.9 *7*	1.9 *5*
γ	1307.4 *10*	0.93 *22*
γ	1329.8 *6*	2.8 *7*
γ	1338.2 *10*	1.0 *3*
γ	1348.1 *10*	0.55 *17*
γ	1354.8 *10*	0.66 *17*
γ	1364.8 *10*	0.82 *22*
γ	1383.8 *10*	0.55 *17*
γ	1391.9 *10*	1.0 *3*
γ	1434.1 *10*	0.71 *17*
γ	1441.2 *10*	0.77 *17*
γ	1447.9 *10*	1.5 *4*
γ	1473.2 *10*	0.60 *17*
γ	1488.2 *7*	0.71 *17*
γ	1491.4 *10*	0.93 *22*
γ	1503.7 *10*	1.9 *5*
γ	1511.8 *10*	0.77 *17*
γ	1525.1 *10*	0.66 *17*
γ	1532.6 *10*	1.0 *3*
γ	1548.9 *10*	1.9 *5*
γ	1632.9 *10*	0.55 *17*
γ	1672.1 *10*	0.88 *22*
γ	1739.5 *10*	1.4 *3*
γ	1759.7 *10*	0.93 *22*
γ	1863.6 *7*	0.60 *17* ?
γ	1908.4 *6*	1.5 *4* ?

† approximate

* with ^{191}Au(920 ms) in equilib

Atomic Electrons (^{191}Hg)
$\langle e \rangle$=105 *8* keV

e_{bin}(keV)	$\langle e \rangle$(keV)	e(%)
2 - 9	1.4	~18
11	2.9	~26
12	7.3	61 *28*
13	0.8	~6
14	3.5	25 *10*
52 - 77	2.1	3.6 *5*
116	2.6	2.3 *3*
144 - 160	2.73	1.74 *12*
172	28.2	16 *4*
182 - 185	0.69	0.38 *5*
193	9.2	4.7 *10*
194 - 229	3.3	1.49 *14*
238	6.3	2.7 *6*
239 - 274	10.9	4.3 *6*
276	2.2	~0.8
287 - 332	2.5	0.79 *12*
340 - 377	3.3	0.95 *15*
396 - 444	1.9	0.47 *7*
446 - 472	1.4	0.31 *9*
497	0.07	~0.014
498	4.3	0.85 *9*
499 - 547	1.7	0.33 *14*
549 - 598	1.91	0.33 *4*
599 - 645	0.9	0.14 *5*
648 - 697	0.36	0.053 *9*

Atomic Electrons (^{191}Hg)
(continued)

e_{bin}(keV)	$\langle e \rangle$(keV)	e(%)
704 - 748	0.25	0.034 *14*
763 - 808	0.24	~0.030
815 - 864	0.21	0.025 *10*
869 - 916	0.34	0.037 *15*
922 - 970	0.23	0.025 *8*
976 - 1011	0.17	0.018 *7*
1026 - 1057	0.12	~0.012
1077 - 1110	0.08	0.007 *3*
1151 - 1176	0.034	~0.0029
1204 - 1249	0.26	0.021 *10*
1254 - 1303	0.19	0.015 *4*
1304 - 1353	0.13	0.010 *3*
1354 - 1392	0.13	0.010 *4*
1407 - 1452	0.25	0.018 *5*
1459 - 1552	0.15	0.010 *4*
1591 - 1679	0.11	0.0068 *25*
1725 - 1783	0.032	0.0018 *7*
1828 - 1906	0.06	~0.0030

Continuous Radiation (^{191}Hg)
$\langle\beta+\rangle$=52 keV;$\langle$IB$\rangle$=3.3 keV

E_{bin}(keV)		$\langle\ \rangle$(keV)	(%)
0 - 10	$\beta+$	3.79×10^{-8}	4.46×10^{-7}
	IB	0.0021	
10 - 20	$\beta+$	2.54×10^{-6}	1.51×10^{-5}
	IB	0.0035	0.026
20 - 40	$\beta+$	0.000116	0.000349
	IB	0.0053	0.018
40 - 100	$\beta+$	0.0099	0.0124
	IB	0.40	0.58
100 - 300	$\beta+$	0.73	0.325
	IB	0.084	0.047
300 - 600	$\beta+$	5.7	1.23
	IB	0.22	0.048
600 - 1300	$\beta+$	29.0	3.14
	IB	1.06	0.113
1300 - 2500	$\beta+$	17.0	1.10
	IB	1.48	0.086
2500 - 3264	IB	0.096	0.0036
	$\Sigma\beta+$		5.8

$^{191}_{81}$Tl(5.22 *16* min)

Mode: ϵ
Δ: -26240 *310* keV syst
SpA: 1.88×10^{8} Ci/g
Prod: ^{182}W(^{14}N,5n); protons on Hg; protons on Pb; ^{181}Ta(^{16}O,6n)

Photons (^{191}Tl)

γ_{mode}	γ(keV)	γ(rel)
γ	~50.0	
γ	215.71 *12*	100
γ	264.66 *12*	51 *3*
γ	281.0 *6*	7 *3*
γ	284.4 *5*	9 *4* ?
γ	323.1 *5*	~7
γ M1+E2	325.6 *3*	67 *7* ?
γ	335.91 *25*	45 *5*
γ	374.3 *5*	16 *6*
γ	378.1 *3*	27 *5*
γ (M1+E2)	477.5 *3*	11 *2*
γ M1+2.1%E2	535.1 *4*	~10

Photons (^{191}Tl)
(continued)

γ_{mode}	γ(keV)	γ(rel)
γ	563.1 *3*	20 *4*
γ	574.7 *6*	8 *3*
γ	579.7 *4*	35 *8*
γ	606.1 *6*	~6
γ	615.3 *4*	12 *2*
γ	639.0 *4*	16 *3*

$^{191}_{82}$Pb(1.33 *8* min)

Mode: ϵ(99.987 *5* %), α(0.013 *5* %)
Δ: -20340 *220* keV syst
SpA: 7.4×10^{8} Ci/g
Prod: ^{181}Ta(^{19}F,9n); ^{182}W(^{16}O,7n)
^{181}Ta(^{19}F,9n); ^{182}W(^{16}O,7n)

Alpha Particles (^{191}Pb)

α(keV)
5290 *20*

$^{191}_{82}$Pb(2.18 *8* min)

Mode: ϵ
Δ: -20340 *220* keV syst
SpA: 4.50×10^{8} Ci/g
Prod: ^{48}Ti on ^{150}Sm

$^{191}_{83}$Bi(13 *1* s)

Mode: ϵ(~60 %), α(~40 %)
Δ: -12940 *370* keV syst
SpA: 4.4×10^{9} Ci/g
Prod: ^{159}Tb(^{40}Ar,8n); ^{181}Ta(^{20}Ne,10n); ^{203}Tl(^{3}He,15n)

Alpha Particles (^{191}Bi)

α(keV)
6320 *10*

$^{191}_{83}$Bi(20 *15* s)

Mode: α
Δ: -12940 *370* keV syst
SpA: 2.9×10^{9} Ci/g
Prod: ^{181}Ta(^{20}Ne,10n)

Alpha Particles (^{191}Bi)

α(keV)	α(rel)
6630 *20*	30
6860 *20*	70

A = 192

NDS **40**, 425 (1983)

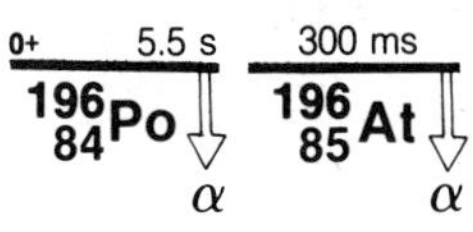

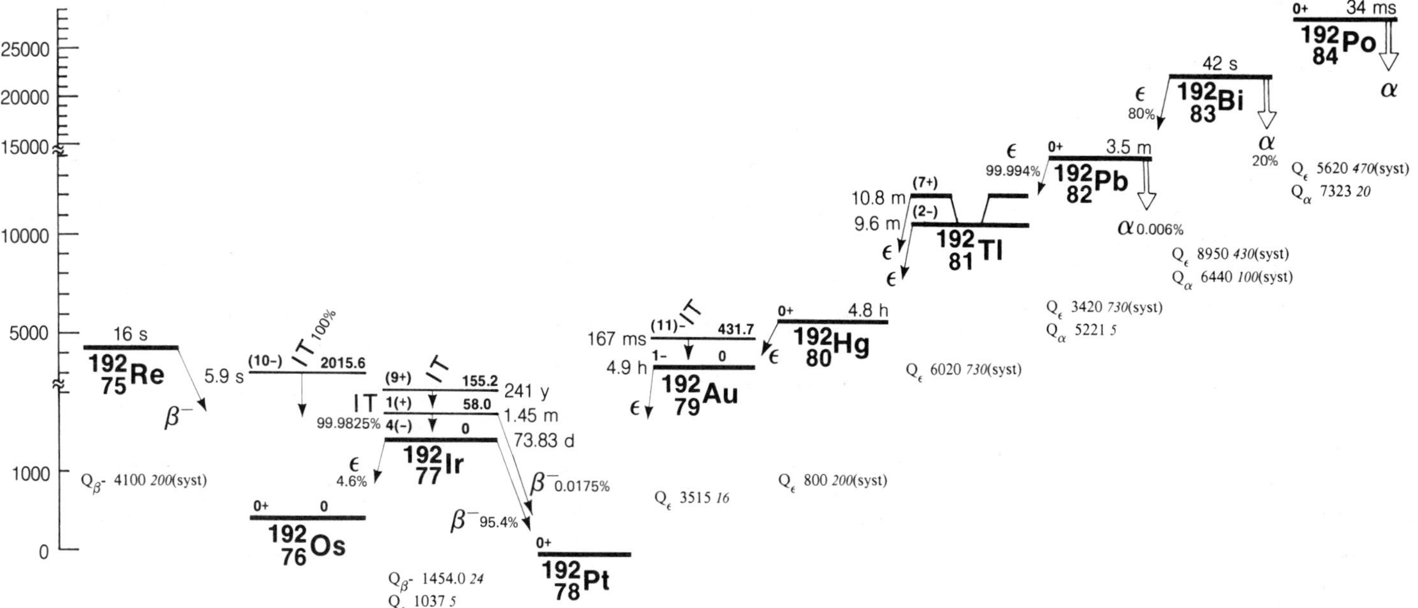

$^{192}_{75}$Re(16 *1* s)

Mode: β-

Δ: -31790 *200* keV syst

SpA: 3.59×10^{9} Ci/g

Prod: ^{192}Os(n,p)

Photons (^{192}Re)

γ_{mode}	γ(keV)
γ E2	205.79581 *6*
γ E2+5.8%M1	283.2671 *6*
γ [E2]	467.3 *4*
γ E2	489.0626 *6*
γ [E2]	750.5 *4*

$^{192}_{76}$Os(stable)

Δ: -35893 *4* keV

%: 41.0 *3*

$^{192}_{76}$Os(5.9 *1* s)

Mode: IT

Δ: -33877 *4* keV

SpA: 9.40×10^{9} Ci/g

Prod: ^{192}Os(n,n')

Photons (^{192}Os)

⟨γ⟩=1890 *53* keV

γ_{mode}	γ(keV)	γ(%)†
Os L_ℓ	7.822	0.32 *5*
Os L_α	8.904	7.2 *8*
Os L_η	9.337	0.136 *15*
Os L_β	10.452	8.7 *10*
Os L_γ	12.162	1.54 *19*
Os $K_{\alpha2}$	61.485	9.8 *3*
Os $K_{\alpha1}$	63.000	16.9 *6*
Os $K_{\beta1}$'	71.313	5.80 *19*
Os $K_{\beta2}$'	73.643	1.47 *5*
γ E2+9.1%M1	201.3805 *6*	7.5 *3*
γ E2	205.79581 *6*	65.9
γ [M1]	218.564 *13*	0.33 *7*
γ [M1]	233.985 *17*	1.8 *4*
γ [M1]	247.73 *13*	0.53 *11*
γ E2+5.8%M1	283.2671 *6*	8.0 *4*
γ [M1]	292.547 *8*	5.1 *4*
γ (E3)	302.55 *6*	53 *3*
γ (M2)	307.09 *9*	7.1 *5*
γ [M1]	322.9 *4*	1.3 *3*
γ E2+17%M1	329.348 *7*	1.64 *24*
γ E2	374.5204 *8*	24 *4*
γ [M1]	379.222 *10*	2.0 *3*
γ [E2]	420.601 *7*	6.5 *5*
γ [M1]	452.549 *13*	~5
γ [E2]	453.205 *15*	59 *3*
γ E2+1.5%M1	484.6473 *4*	51.7 *22*
γ E2	489.0626 *6*	13.7 *6*
γ [M1]	502.87 *20*	0.92 *18*
γ [E2]	508.81 *22*	13 *3*
γ [E2]	555.68 *10*	1.4 *3*
γ [M1]	563.332 *15*	9.8 *7*
γ [E2]	569.43 *9*	70 *7*
γ [E2]	575.7 *4*	1.05 *21*
γ [E2]	580.602 *10*	4.0 *10*
γ [E2]	606.00 *18*	10 *3*
γ [E2]	619.40 *21*	12.5 *13*
γ [M1]	623.95 *21*	1.3 *3*

Photons (^{192}Os)

(continued)

γ_{mode}	γ(keV)	γ(%)†
γ [E2]	671.769 *12*	1.5 *4*
γ [E2]	703.867 *7*	0.59 *13*
γ	1000.1 *4*	2.3 *7*

† 1.1% uncert(syst)

Atomic Electrons (^{192}Os)

⟨e⟩=169 *3* keV

e_{bin}(keV)	⟨e⟩(keV)	e(%)
11 - 60	9.4	50 *5*
61 - 71	0.110	0.174 *10*
128	2.03	1.59 *13*
132	13.7	10.39 *23*
145 - 191	3.6	2.10 *19*
193	9.50	4.91 *11*
195	4.72	2.42 *5*
198 - 223	9.8	4.72 *13*
229	19.7	8.6 *6*
231 - 232	0.069	0.030 *7*
233	12.6	5.4 *5*
234 - 283	2.46	0.93 *6*
289	0.118	0.041 *4*
290	23.4	8.1 *6*
291	4.3 ×10^{-5}	1.49 *13* ×10^{-5}
292	8.1	2.78 *21*
293 - 299	4.5	1.54 *9*
300	7.3	2.45 *18*
301 - 347	7.7	2.53 *16*
362 - 377	1.49	0.41 *4*
379	6.2	1.6 *3*
408 - 410	0.203	0.050 *3*
411	4.07	0.99 *5*
415 - 453	4.6	1.07 *7*

Atomic Electrons (^{192}Os) (continued)

e_{bin}(keV)	⟨e⟩(keV)	e(%)
472 - 492	4.15	0.86 *4*
496	4.7	0.95 *10*
498 - 546	1.81	0.34 *4*
550 - 598	2.58	0.459 *21*
603 - 630	0.366	0.060 *3*
659 - 704	0.034	0.0051 *8*
987 - 1000	0.027	~0.0027

$^{192}_{77}$Ir(73.831 *8* d)

Mode: β-(95.4 *1* %), ϵ(4.6 *1* %)
Δ: -34857 *5* keV
SpA: 9211.3 Ci/g
Prod: ^{191}Ir(n,γ); ^{192}Os(d,2n)

Photons (^{192}Ir)

⟨γ⟩=813 *7* keV

γ_{mode}	γ(keV)	γ(%)†
Os L_ℓ	7.822	0.027 *3*
Pt L_ℓ	8.266	0.076 *7*
Os L_α	8.904	0.60 *4*
Os L_η	9.337	0.0083 *5*
Pt L_α	9.435	1.64 *9*
Pt L_η	9.975	0.0270 *15*
Os L_β	10.469	0.63 *5*
Pt L_β	11.174	1.77 *11*
Os L_γ	12.213	0.113 *11*
Pt L_γ	13.025	0.317 *25*
Os $K_{\alpha2}$	61.485	1.16 *5*
Os $K_{\alpha1}$	63.000	2.00 *8*
Pt $K_{\alpha2}$	65.122	2.66 *7*
Pt $K_{\alpha1}$	66.831	4.56 *12*
Os $K_{\beta1}$'	71.313	0.69 *3*
Os $K_{\beta2}$'	73.643	0.174 *7*
Pt $K_{\beta1}$'	75.634	1.59 *4*
Pt $K_{\beta2}$'	78.123	0.415 *11*
$\gamma_{\beta-}$E2+21%M1	136.34347 *18*	0.181 *8*
$\gamma_{\beta-}$[E1]	177.00 *3*	0.0073 *7*
γ_ϵE2+9.1%M1	201.3805 *6*	0.455 *10*
γ_ϵE2	205.79581 *6*	3.18 *7*
γ_ϵ[M1+E2]	219.221 *7*	~0.0016
γ_ϵE2+5.8%M1	283.2671 *6*	0.252 *14*
$\gamma_{\beta-}$E2+1.9%M1	295.9582 *1*	28.3 *5*
$\gamma_{\beta-}$E2+1.8%M1	308.45689 *11*	29.3 *5*
$\gamma_{\beta-}$E2	316.50789 *11*	83.0 *17*
γ_ϵE2+17%M1	329.348 *7*	0.0160 *22*
γ_ϵE2	374.5204 *8*	0.709 *14*
$\gamma_{\beta-}$M1+E2	416.4714 *5*	0.667 *13*
γ_ϵ[E2]	420.601 *7*	0.064 *6*
$\gamma_{\beta-}$E2	468.07151 *18*	47.7 *10*
γ_ϵE2+1.5%M1	484.6473 *4*	3.13 *6*
$\gamma_{\beta-}$	485.60 *19*	0.0022 *9*
γ_ϵE2	489.0626 *6*	0.432 *9*
$\gamma_{\beta-}$E2	588.5845 *5*	4.47 *9*
$\gamma_{\beta-}$E1	593.48 *3*	0.0438 *14*
$\gamma_{\beta-}$E2+28%M1	604.41463 *12*	8.23 *14*
$\gamma_{\beta-}$E2	612.46561 *13*	5.34 *9*
γ_ϵ[E2]	703.867 *7*	0.0058 *13*
$\gamma_{\beta-}$E2	884.5418 *5*	0.284 *6*
$\gamma_{\beta-}$E1	1061.55 *3*	0.0523 *10*
$\gamma_{\beta-}$E2	1090.01 *19*	0.0011 *4*
$\gamma_{\beta-}$	1378.05 *3*	0.0016 *4*

† uncert(syst): 3.1% for ϵ, 0.47% for β-

Atomic Electrons (^{192}Ir)

⟨e⟩=45.2 *4* keV

e_{bin}(keV)	⟨e⟩(keV)	e(%)
11 - 60	1.44	10.1 *6*
61 - 110	0.099	0.154 *7*
122 - 165	0.998	0.767 *18*
174 - 217	1.102	0.562 *8*
218	4.15	1.91 *5*
219	2.2 $\times10^{-5}$	~1 $\times10^{-5}$
230	4.07	1.77 *5*
236	0.011	<0.009
238	10.6	4.46 *13*
241 - 284	2.49	0.882 *14*
293 - 294	0.637	0.217 *6*
295	1.88	0.639 *17*
296 - 302	0.647	0.218 *5*
303	4.52	1.49 *4*
304	0.0003	<0.0002
305	1.86	0.608 *17*
306 - 312	0.326	0.1063 *20*
313	1.15	0.365 *10*
314 - 362	1.04	0.33 *3*
364 - 375	0.01519	0.00412 *6*
390	3.96	1.02 *3*
398 - 421	0.331	0.081 *5*
454 - 489	1.88	0.411 *6*
510 - 534	1.52	0.290 *8*
575 - 612	0.474	0.0801 *14*
630 - 665	0.00035	5.5 *12* $\times10^{-5}$
690 - 739	0.00013	1.8 *5* $\times10^{-5}$
754 - 768	6.7 $\times10^{-6}$	~9 $\times10^{-7}$
871 - 885	0.00349	0.000400 *8*
977 - 1012	0.00088	8.9 *3* $\times10^{-5}$
1041 - 1090	0.000181	1.72 *4* $\times10^{-5}$
1364 - 1378	1.4 $\times10^{-5}$	~1 $\times10^{-6}$

Continuous Radiation (^{192}Ir)

⟨β-⟩=171 keV;⟨IB⟩=0.123 keV

E_{bin}(keV)		⟨ ⟩(keV)	(%)
0 - 10	β-	0.177	3.54
	IB	0.0086	
10 - 20	β-	0.52	3.48
	IB	0.0079	0.055
20 - 40	β-	2.03	6.8
	IB	0.0141	0.049
40 - 100	β-	13.1	19.0
	IB	0.031	0.049
100 - 300	β-	84	44.5
	IB	0.040	0.025
300 - 600	β-	70	18.0
	IB	0.0053	0.00153
600 - 669	β-	0.74	0.119
	IB	3.5 $\times10^{-6}$	5.7 $\times10^{-7}$

$^{192}_{77}$Ir(1.45 *5* min)

Mode: IT(99.9825 %), β-(0.0175 %)
Δ: -34799 *5* keV
SpA: 6.73 $\times10^{8}$ Ci/g
Prod: ^{191}Ir(n,γ); ^{192}Os(d,2n)

Photons (^{192}Ir)

⟨γ⟩=2.47 *11* keV

γ_{mode}	γ(keV)	γ(%)
Ir L_ℓ	8.042	0.41 *5*
Pt L_ℓ	8.266	8.7 *11* $\times10^{-6}$
Ir L_α	9.167	9.0 *7*
Pt L_α	9.435	0.000188 *19*
Ir L_η	9.650	0.201 *14*
Pt L_η	9.975	3.2 *3* $\times10^{-6}$
Ir L_β	10.797	11.6 *8*
Pt L_β	11.173	0.000206 *22*
Ir L_γ	12.546	2.04 *16*
Pt L_γ	13.023	3.7 *4* $\times10^{-5}$
γ_{IT} E3	58.0 *4*	0.0392 *16*
Pt $K_{\alpha2}$	65.122	0.00030 *3*
Pt $K_{\alpha1}$	66.831	0.00052 *4*
Pt $K_{\beta1}$'	75.634	0.000181 *16*
Pt $K_{\beta2}$'	78.123	4.7 *4* $\times10^{-5}$
$\gamma_{\beta-}$E2+1.9%M1	295.9582 *1*	0.0033 *3*
$\gamma_{\beta-}$E2	316.50789 *11*	0.0161 *16*
$\gamma_{\beta-}$E2	612.46561 *13*	0.00061 *6*

Atomic Electrons (^{192}Ir)

⟨e⟩=54.4 *13* keV

e_{bin}(keV)	⟨e⟩(keV)	e(%)
11	2.7	24 *3*
12	5.2 $\times10^{-5}$	0.00045 *6*
13	2.8	21.4 *23*
14	5.8 $\times10^{-6}$	4.2 *6* $\times10^{-5}$
45	17.3	38.3 *17*
47	16.2	34.6 *15*
51 - 54	1.16 $\times10^{-5}$	2.23 *18* $\times10^{-5}$
55	11.2	20.3 *9*
56 - 57	2.19	3.85 *13*
58	2.06	3.57 *16*
61 - 75	9.6 $\times10^{-6}$	1.51 *9* $\times10^{-5}$
218 - 238	0.00255	0.00109 *9*
282 - 317	0.00193	0.00064 *3*
599 - 612	1.34 $\times10^{-5}$	2.23 *16* $\times10^{-6}$

Continuous Radiation (^{192}Ir)

⟨β-⟩=0.078 keV;⟨IB⟩=0.000101 keV

E_{bin}(keV)		⟨ ⟩(keV)	(%)
0 - 10	β-	9.8 $\times10^{-6}$	0.000196
	IB	3.5 $\times10^{-6}$	
10 - 20	β-	2.96 $\times10^{-5}$	0.000197
	IB	3.4 $\times10^{-6}$	2.3 $\times10^{-5}$
20 - 40	β-	0.000121	0.000402
	IB	6.4 $\times10^{-6}$	2.2 $\times10^{-5}$
40 - 100	β-	0.00088	0.00126
	IB	1.68 $\times10^{-5}$	2.6 $\times10^{-5}$
100 - 300	β-	0.0089	0.00445
	IB	3.7 $\times10^{-5}$	2.2 $\times10^{-5}$
300 - 600	β-	0.0267	0.0060
	IB	2.5 $\times10^{-5}$	6.1 $\times10^{-6}$
600 - 1300	β-	0.0401	0.00493
	IB	8.8 $\times10^{-6}$	1.19 $\times10^{-6}$
1300 - 1512	β-	0.00081	6.0 $\times10^{-5}$
	IB	1.46 $\times10^{-8}$	1.10 $\times10^{-9}$

$^{192}_{77}$Ir(241 *9* yr)

Mode: IT
Δ: -34702 *5* keV
SpA: 7.7 Ci/g
Prod: ^{191}Ir(n,γ)

Photons (^{192}Ir)

⟨γ⟩=3.42 *23* keV

γ_{mode}	γ(keV)	γ(%)
Ir L_ℓ	8.042	0.37 *6*
Ir L_α	9.167	8.2 *11*
Ir L_η	9.650	0.28 *3*
Ir L_β	10.769	15.4 *19*
Ir L_γ	12.550	2.9 *4*
Ir $K_{\alpha2}$	63.286	0.191 *21*
Ir $K_{\alpha1}$	64.896	0.33 *4*
Ir $K_{\beta1}$'	73.452	0.113 *12*
Ir $K_{\beta2}$'	75.861	0.029 *3*
γ (E5)	155.16 *12*	0.097 *4*

Atomic Electrons (^{192}Ir)

⟨e⟩=168 *10* keV

e_{bin}(keV)	⟨e⟩(keV)	e(%)
11	2.3	20 *3*
13	3.9	30 *5*
49	0.00133	0.0027 *4*
50	0.0022	0.0044 *7*
51	0.00134	0.0026 *4*
52	0.0024	0.0047 *7*
54	0.00115	0.0021 *3*
60	0.0020	0.0033 *5*
61	0.00072	0.00119 *17*
62	0.00170	0.0027 *4*
63	0.00055	0.00088 *13*
64	0.00031	0.00049 *7*
65	9.8×10^{-5}	0.000152 *22*
70	0.00025	0.00035 *5*
71	0.00019	0.00027 *4*
72	9.6×10^{-5}	0.000133 *20*
73	0.000126	0.000173 *25*
79	0.55	0.69 *7*
142	78.0	55 *6*
144	40.5	28 *3*
152	24.3	15.9 *17*
153	18.9	12.4 *13*

$^{192}_{78}$Pt(stable)

Δ: -36311 *5* keV
%: 0.79 *5*

$^{192}_{79}$Au(4.94 *9* h)

Mode: ε
Δ: -32796 *17* keV
SpA: 3.30×10^6 Ci/g
Prod: daughter ^{192}Hg; ^{191}Ir(α,3n); protons on Pt

Photons (^{192}Au)

⟨γ⟩=1843 *29* keV

γ_{mode}	γ(keV)	γ(%)†
Pt L_ℓ	8.266	0.55 *14*
Pt L_α	9.435	12 *3*
Pt L_η	9.975	0.15 *4*
Pt L_β	11.189	11 *3*
Pt L_γ	13.055	2.0 *5*
Pt $K_{\alpha2}$	65.122	22 *5*
Pt $K_{\alpha1}$	66.831	38 *9*
Pt $K_{\beta1}$'	75.634	13 *3*
Pt $K_{\beta2}$'	78.123	3.5 *8*
γE2+21%M1	136.34347 *18*	0.0217 *11*
γ[E1]	177.00 *3*	0.130 *12*
γ	185.77 *20*	0.058 *6*
γ	225.95 *9*	0.191 *6*
γ[M1]	244.13 *7*	0.041 *4*
γ(M1)	249.88 *12*	0.278 *17*
γ	275.17 *13*	0.128 *12*
γE2+1.9%M1	295.9582 *1*	22.7 *4*
γE2+1.8%M1	308.45689 *11*	3.50 *6*
γE2	316.50789 *11*	58.0 *12*
γ	361.37 *20*	0.25 *3*
γ[E2]	381.27 *8*	~0.046
γ	393.72 *23*	0.029 *12*
γM1+E2	416.4714 *5*	0.045 *3*
γ	421.3 *10*	0.012 *6*
γ	433.55 *16*	0.058 *6*
γ	441.7 *4*	~0.023
γ[M1]	452.57 *14*	0.029 *12*
γ	455.8 *10*	0.029 *12*
γE2	468.07151 *18*	1.75 *4*
γ(E2+27%M1)	477.13 *13*	1.08 *11*
γ	485.60 *19*	0.070 *6*
γ	496.25 *22*	0.041 *17*
γ[E1]	504.13 *12*	0.070 *6*
γ	517.8 *2*	0.41 *6*
γ[E2]	525.44 *20*	~0.035
γ	541.7 *10*	0.04 *2*
γ[E1]	544.38 *8*	0.046 *17*
γ[M1+E2]	556.6 *3*	0.052 *23*
γ	560.5 *10*	~0.04
γ	569.3 *2*	0.23 *5*
γ(E2)	582.74 *5*	2.68 *5*
γE2	588.5845 *5*	0.304 *17*
γE1	593.48 *3*	0.780 *21*
γ	599.32 *15*	~0.06
γE2+28%M1	604.41463 *12*	0.987 *18*
γE2	612.46561 *13*	4.27 *7*
γ(E2+22%M1)	624.30 *15*	0.064 *6*
γ	633.9 *5*	0.035 *12*
γ	638.2 *5*	0.035 *12*
γ	641.47 *20*	~0.023
γ	647.4 *4*	
γ	649.2 *3*	0.035 *6*
γ(M1+31%E2)	655.55 *6*	0.157 *6*
γ[M1+E2]	660.47 *10*	~0.023
γ	665.63 *11*	0.029 *12*
γ[E1]	668.95 *16*	0.110 *23*
γ	678.4 *9*	0.029 *12*
γ	683.60 *14*	0.19 *6*
γ[E2]	690.03 *11*	0.197 *6*
γ	695.47 *12*	0.017 *6*
γ	705.05 *20*	
γ	710.23 *9*	0.087 *6*
γ	727.41 *20*	
γ[E1]	735.1 *3*	0.035 *12*
γ	745.93 *12*	
γ(M1)	746.93 *12*	0.203 *12*
γ	750.64 *17*	0.046 *17*
γM1	759.14 *9*	1.65 *3*
γ(M1)	764.87 *20*	0.064 *6*
γ	769.8 *1*	0.035 *4*
γ	777.6 *9*	0.04 *2*
γ	780.4 *9*	~0.03
γ	791.4 *5*	0.04 *2*
γ[E1]	795.47 *17*	0.029 *12*
γ[E2]	799.03 *9*	0.093 *6*
γ	815.83 *14*	0.046 *6*
γ	819.4 *10*	~0.06
γ[E1]	822.72 *14*	0.383 *12*
γ(M1)	826.72 *17*	0.064 *6*
γ	830.1 *10*	0.058 *23*
γ	833.4 *10*	0.064 *23*

Photons (^{192}Au) (continued)

γ_{mode}	γ(keV)	γ(%)†
γ	836.99 *18*	0.070 *23*
γ	843.6 *7*	0.064 *23*
γ	856.61 *14*	~0.023
γ(M1)	865.33 *11*	0.104 *6*
γ(E2)	872.72 *13*	0.122 *6*
γ(E2)	878.69 *5*	0.818 *17*
γE2	884.5418 *5*	0.0193 *12*
γ[E1]	890.24 *24*	0.09 *3*
γ(M1)	896.22 *12*	0.215 *6*
γ	901.8 *3*	0.029 *3*
γ[E1]	903.83 *25*	~0.041
γ	906.8 *10*	0.075 *23*
γ	910.1 *10*	~0.029
γ	916.90 *17*	
γ	933.9 *10*	0.53 *17*
γ(E2)	934.45 *10*	0.522 *17*
γ(E2+24%M1)	936.17 *8*	0.30 *2*
γ	948.26 *21*	0.09 *3*
γ(E2+~24%M1)	960.17 *10*	0.157 *12*
γ	961.58 *11*	
γ[M1+E2]	964.01 *6*	0.365 *17*
γ(E2)	969.20 *16*	0.075 *15*
γ	972.71 *19*	0.162 *12*
γ	973.59 *17*	0.157 *17*
γ	985.2 *6*	0.052 *17*
γ(M1)	991.43 *12*	0.046 *6*
γ	996.3 *4*	0.191 *12*
γ	1000.3 *5*	
γ	1002.01 *22*	
γ[E1]	1008.75 *13*	
γ	1016.66 *17*	0.041 *5*
γ[M1+E2]	1036.70 *14*	0.35 *6*
γ(M1)	1052.55 *16*	0.116 *12*
γ(M1)	1055.0 *3*	0.058 *6*
γE1	1061.55 *3*	0.93 *2*
γ(M1)	1066.16 *22*	0.046 *17*
γ	1069.97 *19*	0.070 *23*
γ	1084.27 *17*	<0.06
γE2	1090.01 *19*	0.035 *4*
γ	1093.96 *18*	0.099 *23*
γ[E2]	1095.66 *13*	
γ	1097.4 *7*	0.093 *23*
γ	1102.66 *19*	~0.023
γ(E2)	1108.25 *15*	0.064 *6*
γ	1114.10 *16*	0.064 *12*
γ	1116.68 *17*	0.133 *12*
γ[M1+E2]	1122.82 *6*	1.102 *22*
γ	1127.01 *12*	1.46 *3*
γ	1132.44 *11*	0.041 *9*
γ(M1)	1140.46 *6*	2.61 *6*
γ	1173.6 *3*	0.035 *4*
γ	1181.84 *11*	0.075 *17*
γ	1185.0 *10*	~0.041
γ	1189.23 *13*	~0.046
γ	1197.82 *24*	0.099 *23*
γ	1204.30 *11*	
γ(M1+40%E2)	1207.17 *13*	0.139 *12*
γ	1226.69 *18*	
γ	1228.64 *8*	
γ[E2]	1230.4 *1*	0.029 *12*
γ	1238.6 *10*	~0.029
γ	1241.34 *16*	~0.029
γ(M1+46%E2)	1250.66 *7*	0.215 *12*
γ(E2)	1257.54 *22*	0.099 *6*
γ(M1)	1259.97 *6*	0.032 *6*
γ	1264.17 *20*	0.027 *5*
γ(M1)	1267.78 *13*	0.070 *6*
γ[E2]	1270.62 *13*	0.110 *12*
γ	1278.09 *11*	~0.07
γ(E1)	1282.03 *13*	0.41 *3*
γ(M1)	1288.1 *3*	0.081 *6*
γ	1291.20 *25*	0.087 *6*
γ[E1]	1295.79 *14*	
γ	1302.5 *4*	0.041 *17*
γ	1307.94 *12*	0.09 *3*
γ(E2)	1313.33 *11*	0.429 *12*
γ(M1)	1316.02 *8*	0.220 *23*
γ	1321.9 *3*	0.15 *6*
γ	1325.6 *10*	0.09 *4*
γ	1330.1 *7*	0.09 *3*
γ	1336.49 *10*	0.49 *3*
γ(M1)	1353.20 *14*	0.058 *6*
γ	1364.20 *17*	0.128 *6*
γ	1376.7 *7*	

Photons (^{192}Au) (continued)

γ_{mode}	γ(keV)	γ(%)†
γ	1378.05 3	0.028 8
γ	1383.9 8	0.13 4
γ(E2)	1387.00 7	0.371 17
γ[M1+E2]	1394.68 23	0.09 4
γ	1398.32 11	
γ	1407.7 10	0.08 4
γ	1409.9 10	0.08 4
γ[E2]	1414.73 6	<0.23
γ(M1+~33%E2)	1416.65 15	0.15 6
γ(E1)	1423.07 6	2.98 6
γ(E2)	1428.76 23	0.122 12
γ[E1]	1435.51 7	0.215 12
γ(M1)	1439.33 6	0.058 12
γ	1441.1 4	
γ(E2+49%M1)	1450.00 18	0.162 12
γ[M1+E2]	1461.54 12	~0.09
γ	1467.2 5	~0.07
γ	1469.1 8	0.09 4
γ	1474.3 8	0.10 5
γ(M1)	1477.22 25	0.32 9
γ(M1)	1487.60 15	0.070 23
γ	1492.6 9	0.09 3
γ	1504.91 17	0.162 12
γ	1506.49 19	0.10 4
γ	1511.36 13	
γ[E1]	1511.75 11	
γ	1514.33 24	~0.035
γ	1517.64 18	0.128 12
γ	1521.9 7	0.16 5
γ	1530.1 8	0.07 3
γ	1537.09 8	0.423 17
γ[M1+E2]	1559.12 7	0.638 17
γ[M1+E2]	1563.73 13	0.23 7
γ[E2]	1566.57 13	0.55 9
γ[E2]	1576.47 6	2.26 6
γ[E1]	1577.98 13	
γ[E1]	1579.75 15	0.52 3
γ[M1+E2]	1588.31 14	0.180 12
γ[E1]	1604.24 14	~0.07
γ	1608.3 10	~0.035
γ[M1+E2]	1624.48 8	1.67 3
γ	1629.6 4	
γ[E2]	1632.96 22	0.041 17
γ	1636.2 10	0.07 3
γ	1644.94 10	0.487 17
γ	1649.6 6	0.16 4
γ	1655.2 10	0.052 23
γ	1659.80 19	0.145 12
γ[E1]	1664.30 13	0.052 23
γ	1671.9 8	0.09 4
γ	1674.89 16	~0.06
γ	1678.4 10	0.06 3
γ	1683.47 13	0.128 12
γ	1685.9 10	~0.09
γ[E2]	1688.75 11	~0.09
γ	1693.71 15	0.110 12
γ	1706.77 11	1.92 6
γ[M1+E2]	1723.19 6	3.11 9
γ[E1]	1731.47 7	0.67 4
γ[E1]	1739.57 6	0.209 12
γ	1742.1 9	0.06 3
γ	1757.5 4	
γ[E2]	1763.04 7	0.90 12
γ	1769.5 9	~0.09
γ	1778.9 5	0.16 7
γ	1781.8 9	0.10 4
γ	1787.03 10	1.12 17
γ[M1+E2]	1796.06 15	~0.08
γ	1813.59 18	0.43 9
γ	1822.99 19	0.15 7
γ	1833.05 8	3.2 4
γ	1840.93 18	0.13 6
γ[M1+E2]	1855.07 7	0.17 7
γ	1860.21 21	0.041 17
γ	1868.3 9	0.08 4
γ	1872.5 5	0.26 12
γ	1874.9 8	0.26 12
γ[E2]	1880.24 13	0.11 4
γ[E1]	1894.49 13	0.08 4
γ[E1]	1900.20 14	0.09 4
γ	1915.5 10	0.06 3
γ	1920.82 17	3.2 4
γ	1935.81 12	~0.10
γ	1940.90 10	1.62 20
γ	1947.0 10	0.08 4
γ	1950.23 10	0.51 10
γ	1961.9 9	0.06 3
γ[M1+E2]	1968.88 11	~0.07
γ[E1]	1972.75 13	~0.24
γ[E2]	1972.8 3	~0.18
γ	1979.42 13	0.93 12
γ	1989.4 8	0.09 3
γ	1993.6 8	0.16 5
γ	1999.4 10	~0.15
γ	2002.73 11	0.90 17
γ[M1+E2]	2017.34 23	
γ[M1+E2]	2019.14 6	1.48 17
γ	2024.6 10	~0.05
γ	2035.4 4	0.71 9
γ	2052.13 20	~0.05
γ	2055.4 7	~0.041
γ[E2]	2058.99 7	0.32 6
γ	2068.2 9	~0.05
γ[E2]	2074.00 12	0.16 6
γ	2082.99 10	0.88 12
γ[M1+E2]	2092.01 15	~0.12
γ	2097.2 9	
γ	2106.67 12	0.86 12
γ	2117.74 25	~0.041
γ	2130.10 18	0.35 9
γ	2134.1 6	0.11 5
γ	2136.89 18	0.26 5
γ	2147.30 15	0.08 4
γ	2149.55 8	0.12 3
γ	2156.17 21	0.20 5
γ[E2]	2167.23 24	0.21 7
γ[E2]	2171.58 7	0.86 17
γ	2173.5 5	
γ	2176.2 10	~0.10
γ	2192.8 7	
γ[E2]	2200.77 14	
γ	2206.7 10	0.052 23
γ	2216.77 17	0.53 7
γ[M1+E2]	2221.54 22	
γ	2233.6 10	0.8 3
γ[E2]	2236.94 8	5.6 6
γ	2244.26 12	~0.23
γ	2257.40 10	0.035 7
γ	2260.1 10	0.035 7
γ	2262.2 6	
γ[E1]	2268.71 13	0.12 4
γ	2272.0 10	~0.046
γ	2275.3 10	~0.029
γ[M1+E2]	2277.33 11	0.09 4
γ	2284.1 8	0.08 3
γ	2286.9 4	0.12 4
γ	2296.22 12	0.16 6
γ	2298.12 15	0.16 6
γ[M1+E2]	2313.30 23	0.49 9
γ	2319.23 11	1.59 17
γ[M1]	2335.65 6	1.75 17
γ	2341.9 4	0.08 4
γ	2346.8 7	0.12 4
γ[E1]	2359.19 12	~0.05
γ	2378.3 10	~0.06
γ	2399.49 10	
γ[E2]	2408.52 15	0.16 4
γ	2414.3 5	0.24 5
γ	2423.17 12	1.13 15
γ	2431.8 7	0.046 17
γ[M1+E2]	2440.9 3	0.09 3
γ	2446.3 9	~0.023
γ	2453.39 18	0.104 23
γ	2458.8 8	0.075 23
γ	2464.0 9	0.052 17
γ	2467.6 9	~0.023
γ	2483.7 8	0.041 17
γ	2487.7 8	~0.046
γ	2497.9 9	0.029 12
γ	2503.4 7	0.046 17
γ	2508.7 7	0.058 17
γ	2512.0 10	~0.06
γ[M1+E2]	2517.50 22	0.43 12
γ	2533.1 5	0.20 6
γ	2540.21 12	0.30 9
γ[E1]	2543.75 14	0.24 9
γ	2551.2 12	0.052 23
γ	2560.5 10	0.029 12
γ[M1+E2]	2573.28 11	0.27 6
γ	2592.17 12	~0.05
γ	2599.1 10	~0.029
γ	2601.9 9	0.09 3
γ	2610.4 8	0.14 6
γ	2614.63 15	0.38 6
γ	2624.6 6	0.09 3
γ[E2]	2629.80 23	0.24 8
γ[M1+E2]	2635.29 24	0.59 9
γ	2640.8 10	0.038 17
γ	2646.7 11	0.043 17
γ	2658.4 4	0.10 3
γ	2665.8 15	0.07 3
γ[E1]	2675.70 12	
γ	2692.5 20	0.07 3
γ	2709.7 10	0.075 23
γ	2718.5 5	0.31 6
γ	2747.59 16	0.075 23
γ[E2]	2757.4 3	~0.032
γ	2780.6 6	0.054 23
γ	2795.1 5	0.10 4
γ	2828.91 10	0.14 3
γ[E1]	2839.71 14	0.57 12
γ	2870.2 9	0.064 17
γ	2872.7 9	0.064 17
γ	2878.5 9	0.015 6
γ[E2]	2889.79 11	0.024 9
γ	2903.1 10	0.032 12
γ	2908.68 12	0.020 9
γ	2926.4 20	~0.017
γ[E2]	2951.79 24	0.087 17
γ	2964.9 20	0.030 15
γ	2969.3 20	~0.015
γ	2985.1 20	0.018 9
γ	2998.9 20	0.066 17
γ	3012.3 20	0.035 12
γ	3022.1 20	0.035 12
γ	3043.55 16	0.067 17
γ	3064.1 20	~0.017
γ	3127.1 20	0.028 12
γ	3145.42 10	0.15 4
γ[E1]	3156.21 14	~0.015

† 12% uncert(syst)

Atomic Electrons (^{192}Au)

⟨e⟩=33.0 10 keV

e_{bin}(keV)	⟨e⟩(keV)	e(%)
12	3.2	28 7
13	1.9	14 3
14 - 63	1.9	6.3 10
64 - 107	0.43	0.63 10
122 - 171	0.35	0.22 5
172 - 214	0.10	0.052 23
218	3.32	1.53 4
223 - 237	0.552	0.240 7
238	7.41	3.11 9
241 - 282	0.611	0.218 7
283	0.96	0.34 3
284 - 297	1.418	0.488 7
303	3.17	1.05 3
305	1.04	0.342 10
306 - 308	0.0389	0.01269 24
313	0.800	0.255 7
314 - 363	0.653	0.206 6
367 - 416	0.33	0.085 10
418 - 466	0.18	0.041 11
467 - 516	0.32	0.063 10
517 - 566	0.421	0.079 3
567 - 613	0.273	0.046 3
617 - 666	0.024	0.0037 11
667 - 714	0.332	0.0488 16
717 - 766	0.123	0.0165 19
767 - 816	0.070	0.0088 5
817 - 865	0.14	0.016 5
867 - 913	0.088	0.0099 21
918 - 967	0.074	0.0079 18

Atomic Electrons (^{192}Au) (continued)

e_{bin}(keV)	⟨e⟩(keV)	e(%)
968 - 1017	0.056	0.0057 9
1019 - 1068	0.36	0.034 4
1069 - 1117	0.154	0.0138 22
1119 - 1169	0.073	0.0064 5
1170 - 1219	0.075	0.0063 7
1224 - 1271	0.078	0.0063 13
1274 - 1323	0.048	0.0037 5
1325 - 1374	0.080	0.0059 5
1375 - 1424	0.057	0.0041 7
1425 - 1475	0.106	0.0073 13
1476 - 1575	0.27	0.0179 17
1576 - 1676	0.24	0.015 3
1677 - 1777	0.23	0.014 4
1778 - 1878	0.20	0.011 3
1881 - 1980	0.17	0.0089 15
1981 - 2080	0.110	0.0054 10
2089 - 2182	0.20	0.0093 9
2190 - 2287	0.158	0.0070 9
2293 - 2389	0.062	0.0027 6
2395 - 2492	0.039	0.0016 3
2495 - 2590	0.050	0.00198 24
2596 - 2696	0.017	0.00065 12
2698 - 2794	0.0140	0.00051 8
2800 - 2897	0.0061	0.00021 3
2900 - 2999	0.0045	0.00015 3
3001 - 3078	0.0030	0.00010 4
3113 - 3154	0.00053	1.7 6 $\times 10^{-5}$

Continuous Radiation (^{192}Au)

⟨β+⟩=55 keV;⟨IB⟩=2.7 keV

E_{bin}(keV)		⟨ ⟩(keV)	(%)
0 - 10	β+	2.23×10^{-8}	2.63×10^{-7}
	IB	0.0022	
10 - 20	β+	1.47×10^{-6}	8.7×10^{-6}
	IB	0.0036	0.027
20 - 40	β+	6.6×10^{-5}	0.000199
	IB	0.0056	0.019
40 - 100	β+	0.0056	0.0070
	IB	0.40	0.61
100 - 300	β+	0.427	0.189
	IB	0.080	0.045
300 - 600	β+	3.62	0.78
	IB	0.18	0.040
600 - 1300	β+	24.9	2.61
	IB	0.66	0.072
1300 - 2500	β+	25.7	1.59
	IB	1.19	0.066
2500 - 3515	IB	0.20	0.0072
	Σβ+		5.2

$^{192}_{79}$Au(167 ms)

Mode: IT
Δ: -32364 17 keV
SpA: 8×10^{10} Ci/g
Prod: ^{193}Ir(^{3}He,4n); alphas on Ir

Photons (^{192}Au)

γ_{mode}	γ(keV)
γ	18.0 5
γM1+0.7%E2	31.7 5
γM1+0.4%E2	41.0 4
γE3	59.8 5
γM2	62.8 4
γM1+3%E2	89.5 4

Photons (^{192}Au) (continued)

γ_{mode}	γ(keV)
γE3	103.8 4
γE2	107.5 4
γM1+50%E2	128.9 4
γE2	146.9 4

$^{192}_{80}$Hg(4.85 20 h)

Mode: ϵ
Δ: -32000 200 keV syst
SpA: 3.37×10^{6} Ci/g
Prod: ^{197}Au(p,6n); protons on Pb

Photons (^{192}Hg)

⟨γ⟩=279 8 keV

γ_{mode}	γ(keV)	γ(%)†
Au L_ℓ	8.494	1.14 16
Au L_α	9.704	24 3
Au L_η	10.309	0.30 4
Au L_β	11.546	24 3
Au L_γ	13.539	4.7 7
γM1+0.7%E2	31.61 5	1.26 7
Au $K_{\alpha2}$	66.991	32 4
Au $K_{\alpha1}$	68.806	54 6
Au $K_{\beta1}'$	77.859	19.0 21
Au $K_{\beta2}'$	80.428	5.1 6
γ	95.0 5	0.10 3
γM1+33%E2	99.4 4	0.66 17
γE1	101.9 3	1.15 24
γM1+33%E2	104.5 5	0.70 17
γM1	105.3 4	0.17 4
γM1+35%E2	109.4 5	0.60 8
γM1	114.5 3	0.74 17
γM1+25%E2	116.5 3	0.28 3
γM1	120.1 3	0.77 17
γM1+31%E2	135.9 3	0.87 17
γE1	139.0 3	1.0 4
γM1	142.5 3	0.66 13
γM1+38%E2	146.1 3	0.99 21
γM1	157.3 3	7.0
γE1	186.4 3	3.3 6
γ(M1)	204.6 3	0.83 17
γM1	245.5 3	1.7 3
γM1	262.6 3	0.66 21
γE1	274.87 25	50.4 20
γM1	279.3 4	0.42 6
γ	303.1 5	~0.17
γE1	306.48 25	5.4 6
γM1	436.6 4	0.66 10

† 8.6% uncert(syst)

Atomic Electrons (^{192}Hg)

⟨e⟩=62.0 21 keV

e_{bin}(keV)	⟨e⟩(keV)	e(%)
12	4.9	41 6
14	5.1	37 5
17	6.0	34.9 21
18 - 25	3.7	18.5 22
28	2.77	9.8 6
29 - 76	9.8	22.4 19
77	9.3	12.1 12
81 - 129	8.3	8.2 11
131 - 142	0.9	0.7 3
143	2.8	1.9 3
144 - 157	1.27	0.84 6
165	1.38	0.84 14

Atomic Electrons (^{192}Hg) (continued)

e_{bin}(keV)	⟨e⟩(keV)	e(%)
172 - 193	0.81	0.44 9
194	2.69	1.39 6
199 - 245	1.10	0.50 4
248 - 295	1.09	0.414 18
300 - 306	0.023	0.0075 15
422 - 437	0.063	0.0148 17

Continuous Radiation (^{192}Hg)

⟨IB⟩=0.47 keV

E_{bin}(keV)		⟨ ⟩(keV)	(%)
10 - 20	IB	0.0020	0.0169
20 - 40	IB	0.00165	0.0051
40 - 100	IB	0.43	0.63
100 - 300	IB	0.029	0.019
300 - 600	IB	0.0068	0.0019
600 - 800	IB	1.8×10^{-5}	2.8×10^{-6}

$^{192}_{81}$Tl(9.6 4 min)

Mode: ϵ
Δ: -25970 710 keV syst
SpA: 1.02×10^{8} Ci/g
Prod: protons on Pb; protons on U

Photons (^{192}Tl(9.6 min + 10.8 min))

γ_{mode}	γ(keV)	γ(rel)
γE2	133.25 9	4.34 23
γE1	174.10 9	9.6 5
γ	201.8 3	0.33 5
γ	204.6 3	0.16 5
γE2+23%M1	239.33 20	2.96 18
γE2	246.93 20	1.19 7
γM1+34%E2	312.03 20	1.18 8
γM1+35%E2	323.83 20	2.51 14
γM1	343.20 15	1.9 4
γE2+39%M1	375.33 20	0.87 6
γE1	384.05 15	2.52 14
γE2+44%M1	397.6 3	0.92 9
γE2	422.89 10	100
γ	445.3 3	1.34 16
γ	451.8 3	2.14 21
γ	456.6 3	1.2 4
γ	472.3 3	0.48 6
γM1+25%E2	478.04 18	0.60 7
γ	479.6 3	0.85 8
γ	535.6 3	0.52 6
γ	544.23 20	0.91 8
γM1	559.63 20	1.63 13
γE2	584.23 10	1.92 16
γ	596.03 20	1.09 12
γE2	619.53 13	1.79 12
γE2	634.94 9	76 4
γE2	644.23 20	1.7 3
γM1+35%E2	675.52 9	2.51 23
γE2+25%M1	690.94 9	6.7 4
γ	714.8 3	1.83 20
γ	718.17 18	1.41 16
γ	733.13 20	1.05 16
γE2	745.60 9	26.8 14
γ	774.16 17	1.15 11
γE1	786.45 9	31.7 16
γ	796.8 3	0.66 11
γ	808.83 20	0.58 9
γ	856.13 20	0.36 11
γ	857.33 20	0.58 12

Photons (^{192}Tl(9.6 min + 10.8 min)) (continued)

γ_{mode}	γ(keV)	γ(rel)
γ	867.3 3	0.65 9
γ	919.43 20	0.77 11
γ	999.1 3	0.41 7 ?
γM1	1112.99 17	4.9 3
γ[E2]	1113.83 13	1.62 8
γ	1129.4 4	0.48 9
γ	1171.2 4	0.46 7
γ	1250.6 3	1.11 9
γ	1266.0 3	0.84 9
γ	1284.9 3	0.58 6
γ	1345.2 3	0.91 10
γ	1365.6 3	1.06 10
γ	1375.6 3	0.85 9
γ	1380.1 3	0.65 9
γ	1421.93 20	2.21 19
γ	1485.85 25	0.47 9
γ	1633.59 19	0.66 11
γ	1659.04 19	0.90 13
γ	1687.1 3	0.76 10
γ	1694.6 4	0.25 8
γ	1727.0 5	0.68 12
γ	1759.2 4	0.43 13
γ	1854.2 3	0.22 5
γ	1860.7 5	0.26 5
γ	1908.74 24	0.45 7
γ	1926.8 5	0.28 6
γ	2024.8 4	~0.13
γ	2053.1 6	<0.38
γ	2056.47 21	<0.38 ?
γ	2081.92 21	0.02 1
γ	2116.4 3	0.57 13
γ	2147.5 6	0.34 7
γ	2167.5 5	0.32 7
γ	2262.8 4	0.35 7
γ	2277.0 3	~0.09
γ	2300.1 4	0.42 7

$^{192}_{81}$Tl(10.8 *2* min)

Mode: ϵ
Δ: -25970 *710* keV syst
SpA: 9.06×10^7 Ci/g
Prod: protons on Hg; ^{181}Ta(^{16}O,5n); protons on Pb

see ^{192}Tl(9.6 min) for γ rays

$^{192}_{82}$Pb(3.5 *1* min)

Mode: ϵ(99.994 *1* %), α(0.006 *1* %)
Δ: -22550 *200* keV syst
SpA: 2.79×10^8 Ci/g
Prod: ^{181}Ta(^{19}F,8n); protons on Th

Alpha Particles (^{192}Pb)

α(keV)
5112 5

Photons (^{192}Pb)

γ_{mode}	γ(keV)	γ(rel)
γ	144.5 3	3.0 5
γ	167.5 1	29.0 20
γ	179.2 3	3.0 12
γ	213.1 3	7.7 6
γ	214.9 3	~11
γ	250.7 2	9.5 9
γ	269.5 3	7.4 7
γ	323.7 5	
γ	343.1 3	
γ	371.0 2	~17
γ[M1]	404.5 3	6.5 9
γ	414.1 3	13.0 20
γ E2+27%M1	608.2 1	38 3
γ[E1]	781.6 3	18.0 20
γ	1195.4 2	100 6

$^{192}_{83}$Bi(42 *5* s)

Mode: ϵ(80 %), α(20 %)
Δ: -13600 *380* keV syst
SpA: 1.39×10^9 Ci/g
Prod: ^{181}Ta(^{20}Ne,9n); ^{159}Tb(^{40}Ar,7n); ^{203}Tl(^{3}He,14n)

Alpha Particles (^{192}Bi)

α(keV)
6065 10

$^{192}_{84}$Po(34 *3* ms)

Mode: α
Δ: -7980 *280* keV syst
SpA: 8.5×10^{10} Ci/g
Prod: ^{182}W(^{20}Ne,10n)

Alpha Particles (^{192}Po)

α(keV)
7170 20

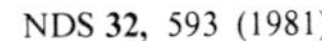

A = 193

NDS **32**, 593 (1981)

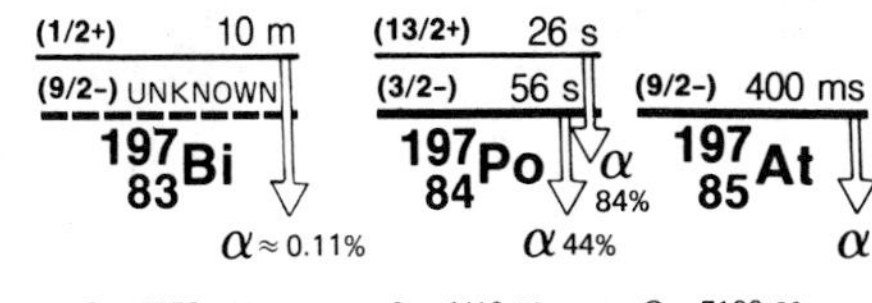

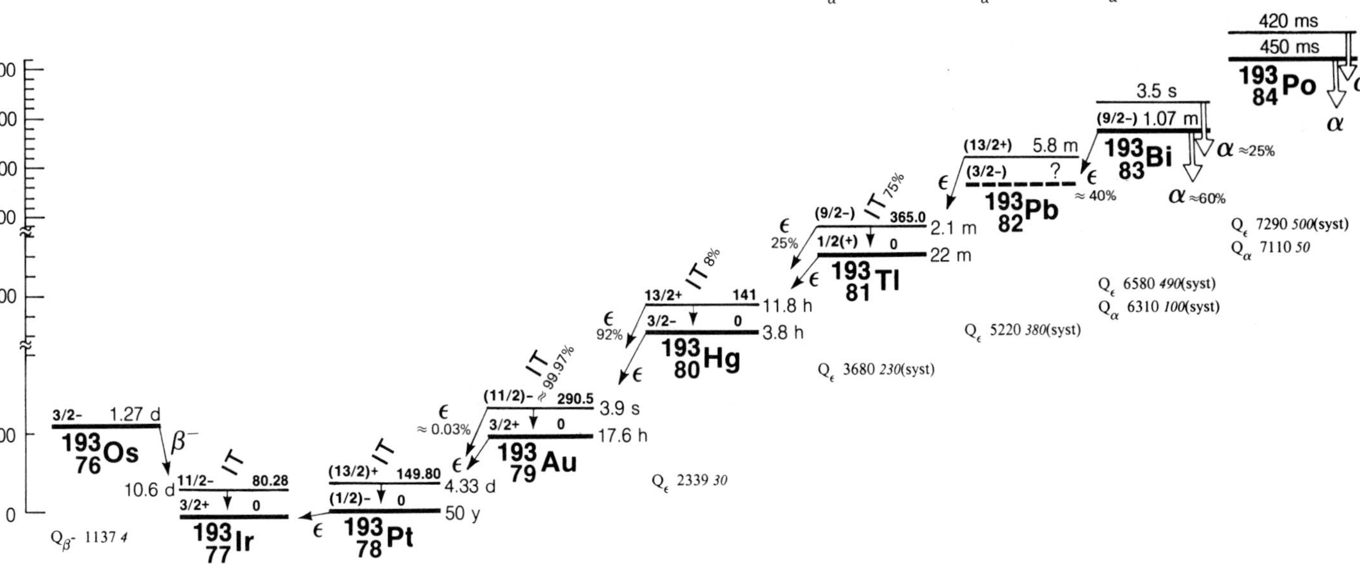

$^{193}_{76}$Os(1.271 *17* d)

Mode: β-
Δ: -33406 *4* keV
SpA: 5.32×10^{5} Ci/g
Prod: ^{192}Os(n,γ)

Photons (^{193}Os)

⟨γ⟩=68.7 *16* keV

γ_{mode}	γ(keV)	γ(%)†
$\gamma_{\beta-}$[M1+E2]	41.136 *12*	0.348 *24*
Ir $K_{\alpha2}$	63.286	3.71 *14*
Ir $K_{\alpha1}$	64.896	6.37 *25*
$\gamma_{\beta-}$M1+28%E2	73.033 *7*	3.2 *5*
Ir $K_{\beta1}$'	73.452	2.20 *9*
Ir $K_{\beta2}$'	75.861	0.566 *23*
γM1+4.0%E2	96.842 *21*	0.099 *8*
γM1	98.646 *25*	0.0166 *24*
γM1+1.9%E2	107.014 *9*	0.64 *3*
γ[M1]	135.86 *6*	0.00043 *12*
γM1+9.3%E2	138.912 *7*	4.27 *20*
γM1	142.149 *8*	0.075 *8*
γM1+7.9%E2	154.773 *23*	0.030 *4*
γM1+34%E2	180.047 *10*	0.182 *20*
γ	181 *1*	<0.00032
γM1+3.5%E2	181.807 *22*	0.194 *20*
γ[M1]	197.38 *4*	0.0047 *16*
γ[M1]	201.51 *12*	~0.0028
γM1+14%E2	218.81 *12*	0.0087 *20*
γE2	219.13 *5*	0.277 *20*
γM1+5.2%E2	234.60 *5*	0.051 *7*
γM1+E2	251.614 *23*	0.217 *16*
γM1+0.5%E2	280.453 *16*	1.24 *6*
γ(E2)	288.821 *22*	0.142 *12*

Photons (^{193}Os) (continued)

γ_{mode}	γ(keV)	γ(%)†
γ	290 *1*	<0.00047
γ(E2)	298.82 *4*	0.186 *16*
γ	316.93 *10*	0.0010 *3*
γM1+2.5%E2	321.589 *15*	1.28 *6*
γ[M1]	333.24 *5*	~0.0028
γ[M1]	337.37 *13*	~0.0012
γ[M1]	350.26 *3*	0.0071 *24*
γ[E2]	357.72 *12*	0.010 *3*
γM1+3.3%E2	361.854 *23*	0.296 *24*
γ[M1]	377.295 *22*	0.071 *8*
γ(E1)	378.52 *7*	0.0016 *4*
γ[M1]	379.18 *4*	0.014 *4*
γM1+4.6%E2	387.467 *15*	1.26 *6*
γ	413.77 *10*	0.0047 *16*
γ[E1]	418.18 *7*	0.0071 *16*
γ[E2]	418.430 *22*	0.055 *6*
γM1+6.8%E2	420.32 *4*	0.166 *12*
γM1+12%E2	440.97 *4*	0.092 *6*
γM1+23%E2	460.500 *15*	3.95 *20*
γ(M1)	484.308 *21*	0.170 *12*
γ[E2]	486.20 *4*	0.011 *6*
γ	512.41 *10*	~0.0016
γ[M1]	515.05 *5*	0.0111 *20*
γ	516.55 *15*	~0.0024
γ(E1)	525.19 *7*	0.0158 *16*
γM1+36%E2	532.067 *24*	0.083 *6*
γ[M1+E2]	556.18 *5*	0.0032 *8*
γ(M1)	557.341 *21*	1.30 *12*
γM1+2.3%E2	559.23 *4*	0.49 *5*
γ[E1]	560.33 *7*	0.0028 *8*
γ[M1]	573.202 *24*	0.0194 *20*
γ(E1)	598.23 *7*	0.0007 *3*
γ[M1]	639.081 *24*	0.0075 *12*
γ	668.3 *10*	0.0008 *4*
γ[M1+E2]	695.09 *5*	0.0028 *6*
γ	709.96 *10*	0.0021 *5*
γ[M1]	712.114 *24*	0.0154 *24*
γ	735.36 *10*	0.0011 *3*
γ	775.84 *10*	~0.00039

Photons (^{193}Os) (continued)

γ_{mode}	γ(keV)	γ(%)†
γ	778.50 *15*	0.0017 *4*
γ	784.25 *12*	0.00067 *16*
γ	801.23 *10*	~0.00032
γ	848.87 *10*	0.0043 *8*
γ	874.27 *10*	0.019 *3*
γ	891.26 *12*	0.0028 *4*

† 3.8% uncert(syst)

Atomic Electrons (^{193}Os)

⟨e⟩=52 *9* keV

e_{bin}(keV)	⟨e⟩(keV)	e(%)
11	2.6	~23
13	2.5	~19
21 - 23	0.136	0.65 *5*
28	5.9	~21
30	6.1	<41
31	0.84	2.73 *15*
38	2.0	~5
39	2.1	<11
40 - 54	1.4	~3
60	6.7	11.2 *17*
61	0.0140	0.0230 *24*
62	3.6	5.8 *9*
63	5.1	8.2 *4*
64 - 66	0.104	0.158 *16*
70	2.9	4.2 *6*
71 - 107	2.02	2.41 *14*
121 - 123	0.0046	0.0038 *12*
125	1.52	1.21 *6*
126 - 176	1.47	1.07 *4*
177 - 223	0.91	0.443 *20*
231 - 280	0.89	0.353 *15*

Atomic Electrons (^{193}Os) (continued)

e_{bin}(keV)	⟨e⟩(keV)	e(%)
282 - 331	0.81	0.265 *10*
333 - 379	0.219	0.0606 *20*
384 - 430	1.08	0.281 *15*
436 - 486	0.69	0.147 *7*
497 - 546	0.081	0.0150 *10*
547 - 596	0.0248	0.00446 *25*
598 - 639	0.0033	0.00052 *7*
655 - 702	0.00081	0.000117 *22*
707 - 735	0.00024	3.4 *7* $\times10^{-5}$
762 - 801	0.0017	~0.00021
815 - 863	0.00048	6 *3* $\times10^{-5}$
871 - 891	0.00011	1.3 *6* $\times10^{-5}$

Continuous Radiation (^{193}Os)

⟨β-⟩=347 keV;⟨IB⟩=0.37 keV

E_{bin}(keV)		⟨ ⟩(keV)	(%)
0 - 10	β-	0.079	1.58
	IB	0.0163	
10 - 20	β-	0.238	1.59
	IB	0.0156	0.108
20 - 40	β-	0.96	3.20
	IB	0.029	0.102
40 - 100	β-	6.9	9.8
	IB	0.074	0.116
100 - 300	β-	63	32.0
	IB	0.148	0.088
300 - 600	β-	155	35.5
	IB	0.078	0.020
600 - 1137	β-	121	16.3
	IB	0.0130	0.0019

$^{193}_{77}$Ir(stable)

Δ: -34543 *4* keV

%: 62.7 *5*

$^{193}_{77}$Ir(10.60 *11* d)

Mode: IT

Δ: -34463 *4* keV

SpA: 6.38$\times10^{4}$ Ci/g

Prod: multiple n-capture from ^{191}Ir; daughter ^{193}Os

Photons (^{193}Ir)

⟨γ⟩=2.30 *10* keV

γ_{mode}	γ(keV)	γ(%)
Ir L_ℓ	8.042	0.62 *6*
Ir L_α	9.167	13.7 *9*
Ir L_η	9.650	0.0229 *16*
Ir L_β	10.995	5.5 *5*
Ir L_γ	12.757	0.60 *8*
Ir $K_{\alpha2}$	63.286	0.131 *6*
Ir $K_{\alpha1}$	64.896	0.225 *11*
Ir $K_{\beta1}'$	73.452	0.077 *4*
Ir $K_{\beta2}'$	75.861	0.0200 *10*
γ M4	80.28 *4*	0.00457 *18*

Atomic Electrons (^{193}Ir)

⟨e⟩=76.8 *19* keV

e_{bin}(keV)	⟨e⟩(keV)	e(%)
4	0.0197	0.473 *21*
11	4.3	38 *4*
13	0.64	4.8 *5*
49	0.00091	0.00185 *20*
50	0.00150	0.0030 *3*
51	0.00091	0.00178 *19*
52	0.00168	0.0032 *4*
54	0.00079	0.00146 *16*
60	0.00136	0.00228 *25*
61	0.00049	0.00081 *9*
62	0.00116	0.00187 *20*
63	0.00038	0.00060 *7*
64	0.000215	0.00033 *4*
65	6.7 $\times10^{-5}$	0.000104 *11*
67	9.9	14.8 *7*
69	37.3	54.0 *24*
70	0.000170	0.00024 *3*
71	0.000130	0.000183 *20*
72	6.6 $\times10^{-5}$	9.1 *10* $\times10^{-5}$
73	8.6 $\times10^{-5}$	0.000118 *13*
77	3.81	4.94 *22*
78	14.9	19.1 *9*
80	5.9	7.4 *3*

$^{193}_{78}$Pt(50 *9* yr)

Mode: ϵ

Δ: -34487 *4* keV

SpA: 37 Ci/g

Prod: ^{192}Pt(n,γ)

Photons (^{193}Pt)

⟨γ⟩=2.1 *3* keV

γ_{mode}	γ(keV)	γ(%)
Ir L_ℓ	8.042	0.31 *7*
Ir L_α	9.167	6.8 *15*
Ir L_η	9.650	0.073 *15*
Ir L_β	10.785	10.7 *25*
Ir L_γ	12.791	2.6 *6*

Continuous Radiation (^{193}Pt)

⟨IB⟩=1.52$\times10^{-5}$ keV

E_{bin}(keV)		⟨ ⟩(keV)	(%)
10 - 20	IB	1.21$\times10^{-5}$	0.000109
20 - 40	IB	1.09$\times10^{-6}$	3.8$\times10^{-6}$
40 - 56	IB	2.0$\times10^{-7}$	4.5$\times10^{-7}$

$^{193}_{78}$Pt(4.33 *3* d)

Mode: IT

Δ: -34337 *4* keV

SpA: 1.562$\times10^{5}$ Ci/g

Prod: ^{193}Ir(d,2n); ^{192}Pt(n,γ)

Photons (^{193}Pt)

⟨γ⟩=12.6 *3* keV

γ_{mode}	γ(keV)	γ(%)
Pt L_ℓ	8.266	0.66 *7*
Pt L_α	9.435	14.2 *10*
Pt L_η	9.975	0.061 *4*
Pt L_β	11.273	7.9 *7*
γ M1+0.02%E2	12.645 *8*	0.67 *3*
Pt L_γ	13.134	1.13 *13*
Pt $K_{\alpha2}$	65.122	4.19 *21*
Pt $K_{\alpha1}$	66.831	7.2 *4*
Pt $K_{\beta1}'$	75.634	2.51 *12*
Pt $K_{\beta2}'$	78.123	0.65 *3*
γ M4	135.50 *3*	0.110 *4*

Atomic Electrons (^{193}Pt)

⟨e⟩=130 *3* keV

e_{bin}(keV)	⟨e⟩(keV)	e(%)
9	5.54	59 *3*
10	0.96	9.9 *16*
11	0.0087	0.083 *12*
12	6.1	52 *5*
13	1.08	8.3 *7*
14	0.62	4.4 *5*
51	0.076	0.149 *16*
52	0.0029	0.0055 *6*
53	0.029	0.055 *6*
54	0.053	0.099 *11*
55	0.025	0.045 *5*
57	8.7	15.2 *7*
61	0.026	0.042 *5*
62	0.031	0.051 *6*
63	0.0027	0.0042 *5*
64	0.041	0.064 *7*
65	0.0084	0.0129 *14*
66	0.0069	0.0104 *11*
67	0.00218	0.0033 *4*
72	0.0056	0.0077 *8*
73	0.0038	0.0052 *6*
74	0.00214	0.0029 *3*
75	0.0032	0.0042 *5*
122	26.5	21.7 *10*
124	47.5	38.3 *17*
132	8.5	6.4 *3*
133	16.7	12.6 *6*
135	7.8	5.8 *3*
135	0.0331	0.0245 *11*

$^{193}_{79}$Au(17.65 *15* h)

Mode: ϵ

Δ: -33490 *100* keV syst

SpA: 9.20$\times10^{5}$ Ci/g

Prod: ^{191}Ir(α,2n); deuterons on Pt; daughter ^{193}Hg(3.8 h); protons on Pt

Photons (^{193}Au)

⟨γ⟩=175 *9* keV

γ_{mode}	γ(keV)	γ(%)†
Pt L_ℓ	8.266	0.79 *17*
Pt L_α	9.435	17 *3*
Pt L_η	9.975	0.22 *4*
Pt L_β	11.190	16 *3*
γ M1+0.02%E2	12.645 *8*	0.22 *3*
Pt L_γ	13.056	2.8 *6*

Photons (^{193}Au)
(continued)

γ_{mode}	γ(keV)	γ(%)†
γ M1+0.1%E2	37.677 *20*	0.0232 *23*
γ M1(+0.03%E2)	44.354 *18*	0.058 *6*
γ M1+15%E2	49.15 *3*	0.012 *3*
γ M1+0.7%E2	52.197 *18*	0.015 *3*
Pt $K_{\alpha2}$	65.122	31 *6*
Pt $K_{\alpha1}$	66.831	53 *10*
γ [M1]	73.655 *16*	0.101 *9*
Pt $K_{\beta1}$'	75.634	19 *4*
Pt $K_{\beta2}$'	78.123	4.8 *9*
γ M1+42%E2	99.879 *10*	0.125 *20*
γ	110.29 *5*	~0.8
γ M1+11%E2	112.524 *7*	~2
γ M1+19%E2	114.178 *7*	0.79 *14*
γ M1(+1.0%E2)	118.010 *14*	0.557 *11*
γ M1+30%E2	119.65 *3*	0.18 *4*
γ M1(+1.0%E2)	155.687 *18*	0.35 *8*
γ M1+11%E2	173.534 *17*	2.9
γ	180.01 *20*	~0.06
γ M1+10%E2	186.179 *16*	10.1 *6*
γ E2+~14%M1	187.833 *16*	0.9 *4*
γ M1+~25%E2	206.890 *24*	0.090 *20*
γ E2	215.41 *10*	0.10 *3*
γ E2	215.91 *10*	
γ E2+27%M1	221.410 *24*	0.078 *17*
γ E2	230.534 *15*	0.54 *6*
γ (E2)	232.188 *15*	0.54 *6*
γ [M1]	251.24 *3*	~0.26
γ M1+16%E2	255.566 *19*	6.7 *6*
γ E2(+>27%M1)	259.087 *22*	0.20 *9*
γ E2+25%M1	268.211 *18*	3.9 *3*
γ E2	269.865 *18*	0.84 *17*
γ M1	281.77 *10*	0.16 *3*
γ M1(+<38%E2)	290.38 *7*	0.09 *4*
γ E2+~24%M1	303.442 *23*	0.27 *8*
γ (M1)	317.77 *4*	0.23 *5*
γ M1	324.900 *22*	0.35 *6*
γ [M1]	334.73 *7*	~0.06
γ	344.1 *9*	0.026 *12*
γ	369.91 *20*	0.061 *15*
γ E2+33%M1	377.097 *19*	0.51 *7*
γ	383.4 *4*	~0.023
γ E2	387.61 *9*	0.38 *5*
γ	401.3 *3*	0.11 *3*
γ M1+~38%E2	408.39 *7*	0.13 *3*
γ	421.3 *4*	~0.05
γ (M1)	424.779 *23*	0.15 *3*
γ	431.4 *3*	0.029 *9*
γ M1+~43%E2	437.424 *22*	0.49 *9*
γ M1	439.078 *22*	1.91 *15*
γ	445 *1*	<0.023
γ (M1)	459.21 *20*	~0.015
γ	464.1 *5*	~0.029
γ (E2)	476.975 *21*	0.46 *9*
γ (E2+~33%M1)	478.41 *15*	0.12 *3*
γ	483 *1*	~0.015
γ M1(+E2)	489.620 *20*	0.23 *5*
γ (E2)	491.275 *20*	0.70 *12*
γ	505.67 *20*	0.096 *17*
γ	508.27 *7*	0.055 *15*
γ (E2)	520.91 *7*	0.078 *17*
γ (E2)	522.56 *7*	0.073 *15*
γ	529.7 *4*	0.038 *9*
γ	577.61 *20*	0.043 *5*
γ	628.56 *25*	0.067 *8*
γ	685 *1*	0.021 *6*
γ	698 *1*	0.064 *14*
γ	730 *1*	0.020 *6*
γ	743 *1*	0.035 *12*
γ	845 *2*	0.070 *23*
γ	1124 *4*	~0.046

† 10% uncert(syst)

Atomic Electrons (^{193}Au)
⟨e⟩=55 *3* keV

e_{bin}(keV)	⟨e⟩(keV)	e(%)
9	1.85	20 *3*
10 - 11	0.32	3.3 *7*
12	5.3	45 *9*
13	2.7	21 *4*
14 - 33	1.8	8.9 *23*
34	2.7	8 *4*
35 - 77	5.2	10.8 *7*
86 - 88	0.209	0.240 *23*
95	3.0	3.2 *6*
96 - 98	0.8	~0.8
99	1.9	~2
100 - 107	1.6	1.6 *3*
108	10.0	9.2 *7*
109 - 156	1.9	1.6 *3*
160 - 171	1.21	0.75 *12*
172	2.33	1.36 *9*
173 - 176	1.10	0.64 *10*
177	4.3	2.40 *24*
178 - 188	1.11	0.60 *4*
190	1.1	0.59 *17*
191 - 240	0.83	0.386 *25*
242 - 290	2.48	0.99 *6*
292 - 341	0.33	0.108 *13*
342 - 390	0.93	0.258 *16*
395 - 443	0.41	0.099 *9*
444 - 492	0.075	0.0159 *18*
494 - 530	0.015	0.0030 *10*
550 - 578	0.009	~0.0016
607 - 652	0.012	0.0020 *9*
665 - 698	0.005	~0.0008
716 - 743	0.0012	0.00017 *8*
831 - 845	0.0012	~0.00015
1110 - 1124	0.0005	$\sim 5\times10^{-5}$

Continuous Radiation (^{193}Au)
⟨IB⟩=0.53 keV

E_{bin}(keV)		⟨ ⟩(keV)	(%)
10 - 20	IB	0.0017	0.0147
20 - 40	IB	0.0017	0.0053
40 - 100	IB	0.41	0.62
100 - 300	IB	0.042	0.024
300 - 600	IB	0.057	0.0134
600 - 1000	IB	0.0145	0.0021

$^{193}_{79}$Au(3.9 *3* s)

Mode: IT(~99.97 %), ϵ(~0.03 %)
Δ: -33200 *100* keV syst
SpA: 1.37×10^{10} Ci/g
Prod: daughter ^{193}Hg(11.8 h); protons on Pt

Photons (^{193}Au)
⟨γ⟩=198 *20* keV

γ_{mode}	γ(keV)	γ(%)†
Au L_ℓ	8.494	0.66 *8*
Au L_α	9.704	14.0 *13*
Au L_η	10.309	0.256 *21*
Au L_β	11.539	15.5 *13*
Au L_γ	13.442	2.8 *3*
γE3	32.36 *10*	0.00105 *4*
γM1+18%E2	38.210 *19*	0.071 *3*
Au $K_{\alpha2}$	66.991	6.8 *9*
Au $K_{\alpha1}$	68.806	11.5 *15*

Photons (^{193}Au)
(continued)

γ_{mode}	γ(keV)	γ(%)†
Au $K_{\beta1}$'	77.859	4.0 *5*
Au $K_{\beta2}$'	80.428	1.07 *14*
γE2	219.91 *4*	3.34 *17*
γM1+21%E2	258.12 *4*	66 *8*

† 2.5% uncert(syst)

Atomic Electrons (^{193}Au)
⟨e⟩=91 *6* keV

e_{bin}(keV)	⟨e⟩(keV)	e(%)
12	4.0	33 *4*
14	3.3	24 *3*
18	0.109	0.60 *3*
19	6.0	32.2 *14*
20	7.1	34.6 *15*
24 - 26	1.39	5.53 *25*
29	2.95	10.1 *5*
30	3.73	12.6 *6*
32 - 77	3.36	9.6 *3*
139	0.62	0.449 *24*
177	42.5	24 *3*
206 - 220	0.99	0.471 *15*
244	10.4	4.3 *6*
246 - 258	4.2	1.64 *18*

$^{193}_{80}$Hg(3.80 *15* h)

Mode: ϵ
Δ: -31150 *100* keV syst
SpA: 4.27×10^{6} Ci/g
Prod: ^{197}Au(p,5n); protons on Pb

Photons (^{193}Hg)

γ_{mode}	γ(keV)
γ M1+18%E2	38.210 *19*
γ M1	186.77 *5*
γ E2	218.2 *3*
γ E2	219.91 *4*
γ [M1+E2]	224.98 *5*
γ M1+21%E2	258.12 *4*
γ	381.8 *10*
γ	574.5 *12*
γ	581.4 *10*
γ	762.0 *10*
γ	855.5 *10*
γ	1040.2 *10*
γ	1078.2 *10*

$^{193}_{80}$Hg(11.8 *2* h)

Mode: ϵ(92 %), IT(8 %)
Δ: -31009 *100* keV syst
SpA: 1.376×10^{6} Ci/g
Prod: ^{197}Au(p,5n)

Photons (^{193}Hg)

$\langle\gamma\rangle$=1192 36 keV

γ_{mode}	γ(keV)	γ(%)†
Au L_ℓ	8.494	1.15 21
Hg L_ℓ	8.722	0.098 10
Au L_α	9.704	24 4
Hg L_α	9.980	2.04 13
Au L_η	10.309	0.38 6
Hg L_η	10.647	0.0110 9
Au L_β	11.547	25 4
Hg L_β	11.952	1.51 15
Au L_γ	13.464	4.4 7
Hg L_γ	14.080	0.28 4
γ_ϵE3	32.36 10	0.00096 10 *
γ_ϵM1+18%E2	38.210 19	0.034 3 *
γ_{IT} M1+0.4%E2	39.49 13	0.316 13
Au $K_{\alpha2}$	66.991	27 5
Au $K_{\alpha1}$	68.806	47 9
Hg $K_{\alpha2}$	68.893	0.061 3
Hg $K_{\alpha1}$	70.818	0.104 5
Au $K_{\beta1}$'	77.859	16 3
Hg $K_{\beta1}$'	80.124	0.0366 18
Au $K_{\beta2}$'	80.428	4.3 8
Hg $K_{\beta2}$'	82.780	0.0100 5
γ_{IT} M4	101.5 10	0.00127 5
γ_ϵ[E2]	165.7 3	~0.35
γ_ϵE2	218.2 3	5.3 8
γ_ϵE2	219.91 4	3.0 3 *
γ_ϵM1+21%E2	258.12 4	60 6 *
γ_ϵ(E2+~28%M1)	281.7 4	0.86 25
γ_ϵE2+31%M1	290.9 3	1.6 3
γ_ϵ(E2+37%M1)	299.9 4	0.40 8
γ_ϵ(E2)	342.0 5	2.17 18
γ_ϵ(E2)	345.5 3	1.21 10
γ_ϵ(E2)	360.6 10	0.32 6
γ_ϵ(E2)	364.5 5	1.66 13
γ_ϵ(M1+~39%E2)	382.7 4	4.3 3
γ_ϵ(E2+40%M1)	394.2 3	3.13 15
γ_ϵE2	407.87 23	25 5
γ_ϵ	429.6 5	0.6 1
γ_ϵ	431.1 10	0.25 5
γ_ϵ	442.2 10	0.18 8
γ_ϵ(M1+21%E2)	462.3 3	1.11 15
γ_ϵ	488.1 5	0.34 5
γ_ϵ(M1)	499.9 4	2.55 20
γ_ϵ(M1+22%E2)	510.4 4	0.7 4
γ_ϵM1+40%E2	535.6 3	2.04 25
γ_ϵ(M1)	537.8 4	1.99 25
γ_ϵ	546.0 10	0.38 8
γ_ϵE2+0.01%M1	573.52 24	14.2 10
γ_ϵM1	600.85 25	2.14 25
γ_ϵ	614.1 5	0.50 13
γ_ϵ	623.7 10	0.33 8
γ_ϵ	626.8 5	0.33 8
γ_ϵ	639.1 4	0.63 10
γ_ϵ	669.2 10	0.43 13 ?
γ_ϵ(M1)	675.4 5	0.71 10
γ_ϵ[E2]	685.1 3	0.78 10
γ_ϵ	693.2 10	0.16 5
γ_ϵ(M1)	701.3 3	0.38 13
γ_ϵ	706.3 5	0.63 10
γ_ϵ[M1+E2]	712.4 3	0.53 8
γ_ϵ	725.6 10	0.40 8
γ_ϵ	728.2 5	0.18 8
γ_ϵ	732.4 5	0.25 6
γ_ϵ	739.2 4	0.43 10
γ_ϵ	739.7 3	0.43 10
γ_ϵ	767.0 3	~0.18
γ_ϵ	817.0 7	0.26 8
γ_ϵ	828.0 10	0.35 10
γ_ϵ	856.5 7	0.20 8
γ_ϵ(M1)	861.2 5	0.71 18
γ_ϵE2+44%M1	870.1 5	1.46 23
γ_ϵE2+43%M1	878.07 25	2.7 4
γ_ϵM1	913.5 4	1.08 23
γ_ϵM1+18%E2	932.6 3	6.7 10
γ_ϵE2+44%M1	995.1 3	1.59 15
γ_ϵ	1004.2 10	0.15 5
γ_ϵ	1036.0 5	0.93 18
γ_ϵ	1049.0 4	0.30 8
γ_ϵ	1052.4 5	0.6 1
γ_ϵ	1076.4 4	0.28 8
γ_ϵ	1094.2 10	0.25 10 ?

Photons (^{193}Hg)

(continued)

γ_{mode}	γ(keV)	γ(%)†
γ_ϵ(E2)	1111.3 4	1.8 6
γ_ϵ	1119.4 10	0.45 10
γ_ϵ	1133.4 4	0.16 5
γ_ϵ	1160.7 4	0.35 8
γ_ϵ(E2+32%M1)	1173.2 5	2.0 6
γ_ϵ	1185.2 10	~0.10
γ_ϵ	1197.2 10	0.11 5
γ_ϵ	1200.2 10	~0.08
γ_ϵ	1232.5 4	0.98 18
γ_ϵ(M1)	1242.0 3	~2
γ_ϵ	1242.2 5	~0.23
γ_ϵ	1250.4 6	0.033 13
γ_ϵ	1255.2 10	0.101 25
γ_ϵ	1263.2 10	0.23 5
γ_ϵ	1268.1 5	0.32 5
γ_ϵ	1277.7 6	0.11 4
γ_ϵ[M1+E2]	1285.94 25	0.62 8
γ_ϵ	1296.9 7	0.20 6
γ_ϵ	1315.4 4	0.68 13
γ_ϵ(M1+43%E2)	1326.4 4	2.14 23
γ_ϵ	1340.4 3	1.51 25
γ_ϵ	1340.5 3	0.63 20
γ_ϵ	1352.6 7	0.23 5
γ_ϵ	1355.0 10	0.25 5
γ_ϵ	1366.1 4	1.44 20
γ_ϵ	1387.8 10	0.081 25
γ_ϵ	1392.1 7	0.156 25
γ_ϵ	1395.5 5	0.81 13
γ_ϵ	1401.2 10	0.13 5
γ_ϵ	1407.8 4	1.03 15
γ_ϵ	1415.9 10	0.08 4
γ_ϵ	1433.7 4	0.55 11
γ_ϵ	1443.4 6	0.26 5
γ_ϵ	1451.2 10	~0.06 ?
γ_ϵ	1455.2 10	~0.05 ?
γ_ϵ	1462.5 7	0.47 6
γ_ϵ	1478.0 10	0.44 9
γ_ϵ	1481.7 10	0.15 4
γ_ϵ(M1)	1487.0 4	1.5 3
γ_ϵ	1500.7 20	~0.07
γ_ϵ	1504.9 5	0.62 15
γ_ϵ	1518.3 7	0.42 8
γ_ϵ	1526.2 5	0.6 1
γ_ϵ	1536.7 20	0.13 5
γ_ϵ	1540.6 20	~0.06
γ_ϵ	1557.8 7	0.14 5
γ_ϵ	1601.8 20	0.15 4
γ_ϵ	1605.7 20	0.17 8
γ_ϵ	1626.0 5	0.45 9
γ_ϵ	1640.4 4	1.6 3
γ_ϵ(M1)	1649.9 3	1.26 25
γ_ϵ	1660.7 20	0.09 4
γ_ϵ	1664.3 20	0.055 25
γ_ϵ	1695.6 20	0.063 25
γ_ϵ	1734.2 4	0.19 5
γ_ϵ	1748.3 4	0.34 10
γ_ϵ	1815.7 5	0.083 25
γ_ϵ	1817.2 20	0.18 6
γ_ϵ	1851.3 6	0.15 5
γ_ϵ	1864.2 20	0.11 4
γ_ϵ	1871.2 20	0.13 5
γ_ϵ	1918.2 20	0.30 6
γ_ϵ	1924.2 20	0.22 6
γ_ϵ	1926.1 7	0.17 5
γ_ϵ	1955.2 20	0.08 4
γ_ϵ	1965.7 7	0.13 4

† 13% uncert(syst)
* with ^{193}Au(3.9 s) in equilib

Atomic Electrons (^{193}Hg)

$\langle e\rangle$=120 5 keV

e_{bin}(keV)	$\langle e\rangle$(keV)	e(%)
12	6.9	58 10
14	5.3	38 7
15 - 18	0.327	2.03 15
19	5.5	29 3
20	6.5	32 3
24 - 27	2.15	8.6 4
29	2.7	9.2 9
30	3.4	11.5 12
32 - 80	5.1	12.6 7
85 - 87	1.31	1.51 9
89	3.54	3.97 18
98 - 139	3.98	3.57 13
151 - 166	0.26	0.17 6
177	38.3	21.6 25
201 - 220	3.35	1.61 10
244	9.4	3.9 5
246 - 291	4.8	1.85 15
296 - 345	4.9	1.53 16
346 - 394	2.18	0.58 6
396 - 442	1.62	0.39 3
448 - 497	2.32	0.49 3
498 - 546	0.93	0.178 18
558 - 604	0.82	0.143 14
609 - 658	0.40	0.063 14
659 - 706	0.16	0.024 6
709 - 755	0.10	0.014 5
764 - 805	0.48	0.061 6
814 - 861	0.95	0.112 14
864 - 913	0.095	0.0109 9
914 - 955	0.36	0.039 5
968 - 1013	0.12	0.012 4
1022 - 1064	0.13	0.013 3
1073 - 1121	0.17	0.016 4
1130 - 1174	0.28	0.024 9
1182 - 1232	0.12	0.0100 23
1235 - 1284	0.30	0.024 6
1285 - 1329	0.21	0.016 5
1335 - 1384	0.089	0.0066 18
1385 - 1433	0.16	0.0117 21
1437 - 1486	0.081	0.0056 14
1487 - 1584	0.17	0.011 3
1587 - 1684	0.051	0.0031 7
1692 - 1790	0.023	0.0013 4
1801 - 1885	0.029	0.0016 5
1904 - 1963	0.0050	0.00026 7

Continuous Radiation (^{193}Hg)

$\langle\beta+\rangle$=5.4 keV;$\langle$IB$\rangle$=1.07 keV

E_{bin}(keV)		$\langle\ \rangle$(keV)	(%)
0 - 10	β+	2.54×10^{-8}	2.99×10^{-7}
	IB	0.00035	
10 - 20	β+	1.69×10^{-6}	1.00×10^{-5}
	IB	0.0019	0.0159
20 - 40	β+	7.6×10^{-5}	0.000228
	IB	0.0019	0.0060
40 - 100	β+	0.0061	0.0077
	IB	0.40	0.58
100 - 300	β+	0.379	0.171
	IB	0.044	0.026
300 - 600	β+	2.11	0.468
	IB	0.099	0.022
600 - 1300	β+	2.87	0.375
	IB	0.36	0.039
1300 - 2190	IB	0.165	0.0109
	$\Sigma\beta$+		1.02

$^{193}_{81}$Tl(21.6 *8* min)

Mode: ϵ
Δ: -27460 *210* keV
SpA: 4.51×10^{7} Ci/g
Prod: ^{184}W(^{14}N,5n); protons on Hg; protons on Pb

Photons (^{193}Tl)

γ_{mode}	γ(keV)	γ(rel)
γ M1+0.4%E2	39.49 *13*	3.20 *16*
γ (M1+12%E2)	50.24 *13*	11 *5*
γ (E2)	207.77 *20*	19.5 *10*
γ (E2+~24%M1)	274.43 *13*	13.5 *13*
γ (M1+23%E2)	284.90 *11*	21.6 *10*
γ (M1+26%E2)	294.08 *15*	4.3 *5*
γ (M1+16%E2)	324.39 *9*	100
γ (M1+16%E2)	335.14 *9*	26.1 *11*
γ (E2+27%M1)	344.03 *10*	41.7 *18*
γ	369.8 *5*	~2
γ (E2)	374.63 *14*	7.6 *9*
γ (M1+~30%E2)	398.6 *4*	6.9 *10*
γ (E2)	493.55 *15*	12.1 *7*
γ (M1+36%E2)	543.3 *7*	3.8 *9*
γ	574.9 *5*	3.8 *6*
γ (M1+31%E2)	636.4 *3*	18 *7*
γ	652.9 *3*	10 *4*
γ	655.0 *5*	~7
γ (M1+22%E2)	676.13 *19*	48 *4*
γ (M1+29%E2)	692.3 *4*	20.9 *16*
γ (E2+~8.9%M1)	713.2 *3*	6.0 *7*
γ	720.0 *5*	1.7 *8*
γ (M1)	752.6 *3*	11.6 *17*
γ (M1+~26%E2)	759.1 *7*	6.5 *15*
γ (M1+46%E2)	770.7 *3*	12.9 *8*
γ	773.9 *6*	1.6 *7*
γ	783.0 *15*	4.0 *16*
γ (E2+40%M1)	821.23 *20*	9.4 *5*
γ	942.1 *5*	1.8 *8*
γ	994.78 *25*	11.0 *11*
γ	1014.4 *3*	8.9 *10*
γ	1044.7 *3*	59 *6*
γ	1064.3 *4*	7.1 *5*
γ	1086.2 *6*	~2
γ	1130.3 *3*	12.3 *13*
γ	1145.8 *4*	4.2 *8*
γ	1152.0 *4*	4.9 *9*
γ	1205.52 *21*	10.2 *12*
γ	1229.2 *6*	2.5 *10*
γ	1236.1 *4*	4.6 *12*
γ	1255.76 *21*	10.3 *19*
γ	1337.6 *4*	5.6 *10*
γ	1360.8 *4*	4.8 *9*
γ	1430.7 *4*	4.5 *9*
γ	1474.7 *7*	2.6 *10*
γ	1483.8 *3*	3.4 *10*
γ	1523.3 *3*	8.0 *19*
γ	1540.66 *22*	8.8 *20*
γ	1580.15 *22*	45 *10*

$^{193}_{81}$Tl(2.11 *15* min)

Mode: IT(75 %), ϵ(25 %)
Δ: -27088 *210* keV
SpA: 4.6×10^{8} Ci/g
Prod: ^{181}Ta(^{16}O,4n); ^{185}Re(^{12}C,4n)

Photons (^{193}Tl)

⟨γ⟩=3.9 *4* keV

γ_{mode}	γ(keV)	γ(%)
Tl L_ℓ	8.953	0.044 *9*
Tl L_α	10.259	0.88 *16*
Tl L_η	10.994	0.0129 *22*
Tl L_β	12.307	0.87 *16*
Tl L_γ	14.398	0.16 *3*
Tl $K_{\alpha2}$	70.832	1.4 *3*
Tl $K_{\alpha1}$	72.873	2.4 *4*
Tl $K_{\beta1}'$	82.434	0.86 *15*
Tl $K_{\beta2}'$	85.185	0.24 *4*
γ_{IT}	365.0 *5*	90.1

$^{193}_{82}$Pb(5.8 *2* min)

Mode: ϵ
Δ: >-22250 keV syst
SpA: 1.68×10^{8} Ci/g
Prod: ^{181}Ta(^{19}F,7n); ^{16}O on W

Photons (^{193}Pb)

γ_{mode}	γ(keV)	γ(rel)
γ [M1+E2]	324.3 *4*	2.3 *5*
γ E2+28%M1	365.0 *5*	*
γ (M1+40%E2)	392.2 *4*	19 *4*
γ	406.5 *5*	2.2 *4*
γ (M1+13%E2)	431.5 *4*	0.72 *14*
γ	466.7 *5*	<1
γ	666.2 *5*	1.8 *4*
γ (E2)	716.5 *4*	6.0 *12*
γ E1+3.5%M2	736.1 *5*	4.6 *9*
γ (E2)	755.8 *4*	2.3 *5*
γ	1113.5 *5*	0.54 *11*

* with ^{193}Tl(2.1 min)

$^{193}_{83}$Bi(1.07 *7* min)

Mode: α(~60 %), ϵ(~40 %)
Δ: -15660 *370* keV syst
SpA: 9.1×10^{8} Ci/g
Prod: ^{181}Ta(^{20}Ne,8n); ^{185}Re(^{16}O,8n); ^{203}Tl(^{3}He,13n)

Alpha Particles (^{193}Bi)

α(keV)
5910 *5*

$^{193}_{83}$Bi(3.5 *2* s)

Mode: α(~25 %)
Δ: >-15660 keV syst
SpA: 1.52×10^{10} Ci/g
Prod: ^{181}Ta(^{20}Ne,8n); ^{185}Re(^{16}O,8n); ^{203}Tl(^{3}He,13n)

Alpha Particles (^{193}Bi)

⟨α⟩=1617.0 keV

α(keV)	α(%)
6180 *20*	1.0
6480 *10*	24

$^{193}_{84}$Po(450 *150* ms)

Mode: α
Δ: -8370 *330* keV syst
SpA: 6.6×10^{10} Ci/g
Prod: ^{185}Re(^{19}F,11n); ^{182}W(^{20}Ne,9n)

Alpha Particles (^{193}Po)

α(keV)
6980 *20*

$^{193}_{84}$Po(420 *100* ms)

Mode: α
Δ: >-8370 keV syst
SpA: 6.8×10^{10} Ci/g
Prod: ^{185}Re(^{19}F,11n); ^{182}W(^{20}Ne,9n)

Alpha Particles (^{193}Po)

α(keV)
6995

A = 194

NDS 22, 433 (1977)

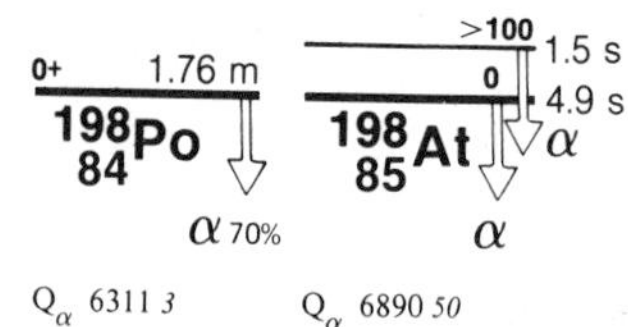

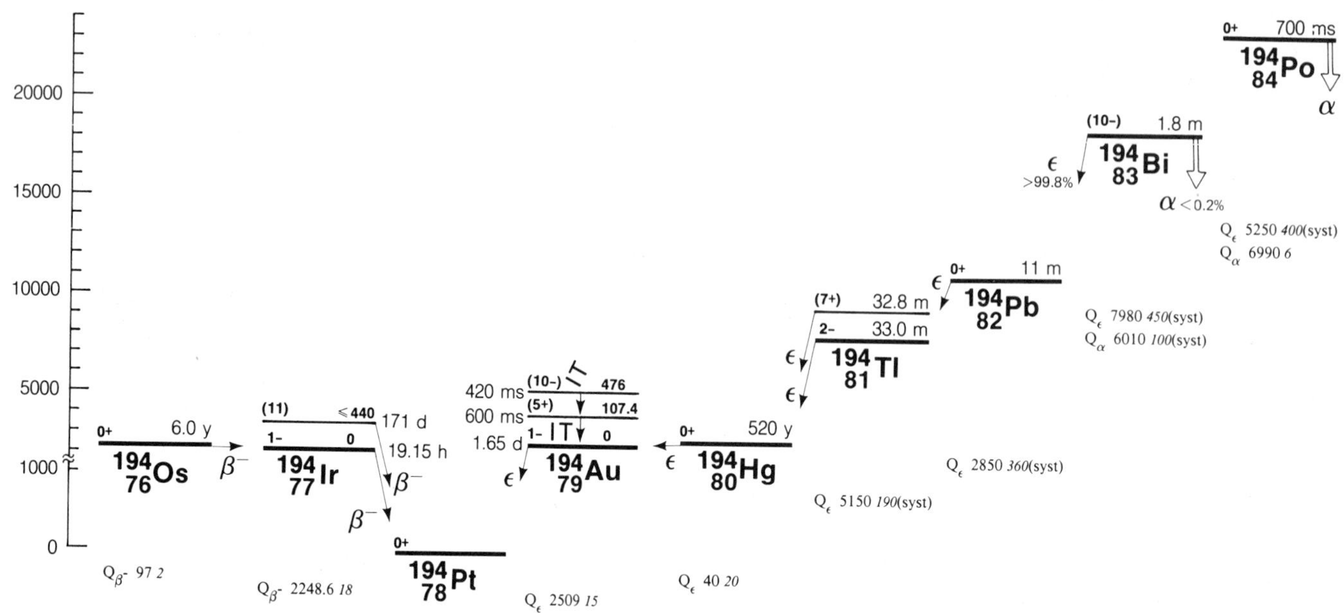

$^{194}_{76}$Os(6.0 2 yr)

Mode: β-
Δ: -32441 5 keV
SpA: 307 Ci/g
Prod: multiple n-capture from ^{192}Os

Photons (^{194}Os)

⟨γ⟩=1.77 23 keV

γ_{mode}	γ(keV)	γ(%)†
Ir L_ℓ	8.042	0.108 25
Ir L_α	9.167	2.4 5
Ir L_η	9.650	0.028 6
Ir L_β	10.783	3.8 9
Ir L_γ	12.782	0.91 22
γ M1	43.1 3	2.3
Ir $K_{\alpha2}$	63.286	0.0108 7
Ir $K_{\alpha1}$	64.896	0.0186 12
Ir $K_{\beta1}$'	73.452	0.0064 4
Ir $K_{\beta2}$'	75.861	0.00165 11
γ M1	82.3 10	0.00421 25

† 22% uncert(syst)

Atomic Electrons (^{194}Os)

⟨e⟩=10.8 15 keV

e_{bin}(keV)	⟨e⟩(keV)	e(%)
6	0.00242	0.0391 25
11	0.020	0.18 4
13	0.94	7.0 16
30	6.9	23 5
32	0.073	0.23 5
40	2.1	5.4 11
41	0.026	0.065 13
42	0.52	1.23 25
43	0.18	0.43 9
49	7.6 ×10^{-5}	0.000154 18
50	0.000124	0.00025 3
51	7.6 ×10^{-5}	0.000147 17
52	0.000139	0.00027 3
54	6.5 ×10^{-5}	0.000121 14
60	0.000113	0.000189 22
61	4.1 ×10^{-5}	6.7 8 ×10^{-5}
62	9.6 ×10^{-5}	0.000155 18
63	3.1 ×10^{-5}	5.0 6 ×10^{-5}
64	1.78 ×10^{-5}	2.8 3 ×10^{-5}
65	5.6 ×10^{-6}	8.6 10 ×10^{-6}
69	0.0044	0.0064 4
70	1.41 ×10^{-5}	2.01 24 ×10^{-5}
71	5.3 ×10^{-5}	7.4 6 ×10^{-5}
72	5.5 ×10^{-6}	7.6 9 ×10^{-6}
73	7.1 ×10^{-6}	9.8 12 ×10^{-6}
79	0.00117	0.00147 9
80	1.30 ×10^{-5}	1.63 10 ×10^{-5}
82	0.000372	0.00045 3

Continuous Radiation (^{194}Os)

⟨β-⟩=21.5 keV;⟨IB⟩=0.0019 keV

E_{bin}(keV)		⟨ ⟩(keV)	(%)
0 - 10	β-	1.55	32.5
	IB	0.00093	
10 - 20	β-	3.53	24.0
	IB	0.00048	0.0035
20 - 40	β-	8.0	28.1
	IB	0.00037	0.00138
40 - 97	β-	8.3	15.4
	IB	8.8×10^{-5}	0.00018

$^{194}_{77}$Ir(19.15 3 h)

Mode: β-
Δ: -32538 4 keV
SpA: 8.435×10^{5} Ci/g
Prod: ^{193}Ir(n,γ); daughter ^{194}Os

Photons (^{194}Ir)

⟨γ⟩=90.4 14 keV

γ_{mode}	γ(keV)	γ(%)†
Pt L_ℓ	8.266	0.0069 7
Pt L_α	9.435	0.149 9
Pt L_η	9.975	0.00247 15
Pt L_β	11.174	0.162 11
Pt L_γ	13.025	0.0289 24
Pt $K_{\alpha2}$	65.122	0.244 9
Pt $K_{\alpha1}$	66.831	0.417 15
Pt $K_{\beta1}$'	75.634	0.146 5
Pt $K_{\beta2}$'	78.123	0.0380 14
γ [M1+E2]	111.449 25	0.0017 5
γ [E2]	189.306 24	~0.0007
γ [E1]	189.65 4	~0.012
γ E1	202.94 5	0.0040 4
γ (E2)	244.792 17	0.0084 10
γ E1	250.14 3	0.00089 18
γ E2+0.5%M1	293.551 9	2.52 9
γ E2	300.755 11	0.350 14
γ E1	318.15 3	0.0092 14
γ E2	328.458 11	13.0 4
γ E2	364.869 13	0.0413 12
γ (E2)	418.26 5	0.00109 14
γ E2	482.856 22	0.0452 18
γ (E1)	530.204 18	0.0158 6
γ (E2)	589.200 13	0.138 4
γ E2	594.305 12	0.0617 19
γ E2	607.56 5	0.0039 4
γ (E1)	621.19 3	0.0089 10
γ E2	622.009 13	0.335 13
γ E2	645.163 12	1.16 3

Photons (^{194}Ir)
(continued)

γ_{mode}	γ(keV)	γ(%)†
γ [M1+E2]	699.39 *3*	~0.0025
γ (E2)	700.65 *3*	0.025 *4*
γ E1	810.498 *19*	0.0024 *3*
γ M1	846.90 *6*	0.00050 *6*
γ M1(+E2)	855.86 *5*	0.00049 *19*
γ [E2]	857.22 *3*	0.0070 *8*
γ [E2]	859.32 *4*	0.0018 *6*
γ (M1+E2)	889.954 *15*	0.0504 *17*
γ E2	925.22 *3*	0.0131 *7*
γ E2	938.713 *13*	0.595 *17*
γ E2+~34%M1	1000.15 *3*	0.0465 *17*
γ E2	1007.65 *7*	0.00070 *21*
γ M1	1048.63 *3*	0.0262 *10*
γ E1	1104.048 *19*	0.0258 *10*
γ [E2]	1119.10 *7*	0.00104 *22*
γ (M1)	1141.05 *6*	~0.00021
γ E2	1150.77 *3*	0.593 *19*
γ M1	1156.61 *5*	0.0020 *3*
γ E1	1175.365 *17*	0.0596 *19*
γ M1+45%E2	1183.504 *15*	0.305 *10*
γ M1,E2	1186.33 *6*	0.0072 *5*
γ E2	1218.77 *3*	0.0551 *20*
γ E2+37%M1+E0	1293.70 *3*	0.046 *3*
γ M1(+E2)	1302.22 *4*	0.00024 *4*
γ E2	1308.41 *7*	0.00127 *12*
γ M1+E2	1342.18 *3*	0.0376 *13*
γ M1	1421.68 *3*	0.00063 *7*
γ (E2)	1431.09 *13*	0.0018 *4*
γ	1432.504 *21*	0.00113 *23*
γ M1+E2	1441.80 *6*	0.00151 *17*
γ M1+E2	1450.16 *5*	0.00156 *10*
γ E2	1463.52 *7*	0.0057 *10*
γ E1	1468.914 *17*	0.189 *7*
γ M1	1487.09 *6*	0.0168 *8*
γ M1	1492.06 *6*	0.00157 *16*
γ [E2]	1511.960 *18*	0.024 *3*
γ [M1]	1512.19 *8*	0.0131 *18*
γ (M1)	1518.77 *14*	0.00165 *23*
γ [E2]	1565.15 *8*	0.0205 *9*
γ M1	1595.77 *4*	0.00161 *15*
γ M1	1601.95 *7*	0.00196 *17*
γ E2	1622.15 *3*	0.063 *3*
γ M1,E2	1670.63 *3*	0.0056 *3*
γ	1675.25 *17*	0.00086 *14*
γ M1+E2	1715.23 *3*	0.00130 *9*
γ	1724.64 *13*	0.00080 *10*
γ M1	1735.35 *6*	0.00245 *22*
γ [E2]	1757.06 *8*	0.00043 *8*
γ (E2)	1778.62 *5*	0.00024 *9*
γ	1780.64 *6*	0.0052 *4*
γ M1	1785.61 *6*	0.00386 *23*
γ E1	1797.370 *19*	0.0173 *8*
γ M1,E2	1805.74 *8*	0.0324 *17*
γ	1812.60 *25*	0.00044 *13*
γ M1	1829.46 *8*	0.00185 *19*
γ M1	1924.23 *4*	0.00177 *13*
γ M1	2043.68 *3*	0.0070 *4*
γ [E2]	2063.80 *6*	8 *3* $\times 10^{-5}$
γ M1	2114.07 *6*	0.00257 *14*

† 15% uncert(syst)

Atomic Electrons (^{194}Ir)
$\langle e \rangle$=4.19 *9* keV

e_{bin}(keV)	$\langle e \rangle$(keV)	e(%)
12 - 55	0.089	0.65 *4*
61 - 109	0.0106	0.0149 *17*
111 - 125	0.0015	0.0013 *6*
166 - 203	0.0022	0.00128 *15*
215	0.357	0.166 *7*
222 - 249	0.0495	0.0222 *9*
250	1.59	0.637 *24*
280	0.165	0.0591 *25*
282 - 307	0.174	0.0605 *11*
315	0.649	0.206 *8*
316	2.5 $\times 10^{-6}$	~8 $\times 10^{-7}$
317	0.195	0.0614 *23*

Atomic Electrons (^{194}Ir)
(continued)

e_{bin}(keV)	$\langle e \rangle$(keV)	e(%)
318	2.3 $\times 10^{-6}$	~7 $\times 10^{-7}$
325	0.164	0.0504 *19*
326 - 365	0.122	0.0372 *10*
404 - 452	0.00413	0.00101 *4*
469 - 517	0.0148	0.00291 *9*
519 - 544	0.0215	0.00395 *17*
567	0.0701	0.0124 *4*
575 - 622	0.0137	0.00227 *6*
631 - 645	0.0232	0.00366 *7*
686 - 732	0.00056	8.1 *9* $\times 10^{-5}$
769 - 812	0.0048	0.00060 *25*
833 - 878	0.0273	0.00318 *11*
887 - 936	0.00973	0.001052 *24*
937 - 970	0.00297	0.000307 *12*
986 - 1035	0.00152	0.000150 *3*
1037 - 1078	0.0222	0.00207 *8*
1090 - 1139	0.0244	0.00220 *10*
1140 - 1186	0.00766	0.000660 *17*
1205 - 1230	0.0027	0.000225 *24*
1264 - 1308	0.0024	0.00019 *6*
1328 - 1372	0.00073	5.4 *9* $\times 10^{-5}$
1385 - 1391	0.00259	0.000186 *7*
1401	0.29	0.021 *5*
1408 - 1456	0.00312	0.000219 *14*
1457 - 1554	0.060	0.0041 *10*
1562 - 1661	0.00091	5.6 *5* $\times 10^{-5}$
1662 - 1754	0.0018	0.000107 *20*
1765 - 1846	0.00043	2.4 *3* $\times 10^{-5}$
1910 - 1985	0.000285	1.45 *7* $\times 10^{-5}$
2030 - 2111	0.000166	8.1 *3* $\times 10^{-6}$

Continuous Radiation (^{194}Ir)
$\langle\beta$-$\rangle$=807 keV;$\langle$IB$\rangle$=1.61 keV

E_{bin}(keV)		$\langle\ \rangle$(keV)	(%)
0 - 10	β-	0.0249	0.497
	IB	0.032	
10 - 20	β-	0.075	0.500
	IB	0.031	0.22
20 - 40	β-	0.305	1.02
	IB	0.060	0.21
40 - 100	β-	2.23	3.18
	IB	0.168	0.26
100 - 300	β-	24.5	12.1
	IB	0.44	0.25
300 - 600	β-	94	20.8
	IB	0.43	0.103
600 - 1300	β-	412	44.4
	IB	0.40	0.049
1300 - 2249	β-	275	17.6
	IB	0.047	0.0032

$^{194}_{77}$Ir(171 *11* d)

Mode: β-
Δ: <-32098 keV
SpA: 3936 Ci/g
Prod: multiple n-capture from ^{191}Ir; ^{196}Pt(d,α)

Photons (^{194}Ir)
$\langle\gamma\rangle$=2143 *43* keV

γ_{mode}	γ(keV)	γ(%)†
Pt L_ℓ	8.266	0.214 *23*
Pt L_α	9.435	4.6 *3*
Pt L_η	9.975	0.090 *6*
Pt L_β	11.165	5.5 *4*
Pt L_γ	13.004	0.98 *8*
Pt $K_{\alpha2}$	65.122	4.66 *18*
Pt $K_{\alpha1}$	66.831	8.0 *3*
Pt $K_{\beta1}$'	75.634	2.79 *11*
Pt $K_{\beta2}$'	78.123	0.73 *3*
γ (E2)	111.7 *5*	8.9 *4*
γ [E2]	189.306 *24*	~2
γ	324.1 *5*	~2
γ E2	328.458 *11*	93 *5*
γ E2	338.9 *5*	55 *3*
γ E1	390.8 *5*	35.1 *18*
γ E2	482.856 *22*	97 *5*
γ E1	562.4 *5*	35.2 *18*
γ E2	600.6 *5*	62 *3*
γ E2	687.9 *5*	59 *3*
γ	1011.9 *5*	3.6 *2*

† 10% uncert(syst)

Atomic Electrons (^{194}Ir)
$\langle e \rangle$=83.7 *14* keV

e_{bin}(keV)	$\langle e \rangle$(keV)	e(%)
12 - 55	4.6	26.4 *16*
61 - 75	0.148	0.231 *11*
98	9.5	9.7 *6*
100	7.3	7.3 *5*
108	0.189	0.174 *11*
109	4.6	4.2 *3*
110 - 112	1.81	1.63 *22*
175 - 189	0.68	0.38 *10*
246	0.6	~0.23
250	11.4	4.55 *24*
261	6.5	2.50 *14*
310 - 313	1.45	0.47 *4*
315	4.64	1.47 *8*
317 - 324	1.44	0.45 *3*
325	2.26	0.70 *4*
326 - 339	4.16	1.26 *4*
377 - 391	0.324	0.085 *3*
404	7.8	1.93 *10*
469 - 484	4.30	0.905 *23*
522	4.03	0.77 *4*
549 - 598	1.54	0.265 *9*
599 - 601	0.080	0.0133 *6*
610	3.36	0.55 *3*
674 - 688	1.06	0.157 *5*
998 - 1012	0.05	~0.005

$^{194}_{78}$Pt(stable)

Δ: -34787 *4* keV
%: 32.9 *5*

$^{194}_{79}$Au(1.646 *21* d)

Mode: ϵ
Δ: -32278 *16* keV
SpA: 4.09×10^5 Ci/g
Prod: deuterons on Pt; ^{193}Ir(α,3n); protons on Pt; ^{194}Pt(d,2n)

Photons (^{194}Au)

$\langle\gamma\rangle$=1051 49 keV

γ_{mode}	γ(keV)	γ(%)†
Pt L_{ℓ}	8.266	0.57 12
Pt L_{α}	9.435	12.3 25
Pt L_{η}	9.975	0.16 3
Pt L_{β}	11.188	11.7 24
Pt L_{γ}	13.059	2.1 5
γ M1	49.72 9	~0.025
γ (M1)	59.5 5	~0.0050
Pt $K_{\alpha2}$	65.122	23 5
Pt $K_{\alpha1}$	66.831	39 8
γ	69.2 3	
Pt $K_{\beta1}$'	75.634	14 3
Pt $K_{\beta2}$'	78.123	3.6 7
γ [M1]	101.41 7	~0.0044
γ (M1)	106.38 7	~0.008
γ [M1+E2]	111.449 25	0.0042 13
γ M1	140.19 10	0.060 6
γ M1	151.67 7	0.063 13
γ M1	162.6 5	0.022 4
γ M1+~5.4%E2	163.96 5	0.126 13
γ M1	171.79 4	0.063 6
γ [M1]	173.10 19	~0.006
γ [E2]	189.306 24	~0.019 ?
γ [E1]	189.65 4	~0.038
γ [M1+E2]	197.76 13	
γ E1	202.94 5	0.33 3
γ	215.6 5	
γ M1,E2	224.0 5	0.033 6
γ M1	239.56 8	0.057 6
γ (E2)	244.792 17	0.0167 22
γ E1	250.14 3	0.032 6
γ [M1+E2]	253.60 5	
γ (E2)	285.07 7	0.055 16
γ [E1]	290.69 6	
γ [M1+E2]	291.25 5	
γ E2+0.5%M1	293.551 9	11.1 6
γ E2	300.755 11	0.87 6
γ E1	318.15 3	0.33 5
γ E2	328.458 11	63
γ E2	364.869 13	1.47 6
γ [M1+E2]	412.27 4	
γ (E2)	418.26 5	0.088 7
γ (E1)	449.37 5	0.170 13
γ E2	482.856 22	1.17 6
γ (M1+E2)	528.82 5	1.7 3
γ (E1)	530.204 18	0.56 3
γ (M1+E2)	544.85 4	0.026 9
γ (E2)	562.7 2	0.081 6
γ (E2)	589.200 13	0.275 13
γ (M1)	593.32 4	0.35 7
γ E2	594.305 12	0.153 11
γ E2	607.56 5	0.315 19
γ (E1)	621.19 3	0.74 9
γ E2	622.009 13	1.48 10
γ E2	645.163 12	2.29 9
γ M1	668.25 4	0.115 7
γ M1	675.96 7	0.0617 19
γ [M1+E2]	699.39 3	~0.008
γ (E2)	700.65 3	0.050 8
γ [M1+E2]	702.49 25	
γ M1	703.52 3	0.44 4
γ M1	736.25 4	0.131 10
γ E1	810.498 19	0.198 22
γ M1	818.89 7	0.032 6
γ [E2]	844.4 3	
γ M1	846.90 6	0.0542 19
γ M1(+E2)	855.86 5	0.11 4
γ [E2]	857.22 3	0.0171 21
γ [E2]	859.32 4	0.060 18
γ (M1+E2)	889.954 15	0.100 6
γ	894.3 3	<0.09
γ E2	925.22 3	0.285 18
γ E2	938.713 13	1.17 5
γ M1	948.31 3	2.34 12
γ E2+~34%M1	1000.15 3	0.150 16
γ E2	1007.65 7	0.088 25
γ M1(+E2)	1030.94 7	0.019 6
γ E1	1038.57 5	0.32 4
γ M1	1048.63 3	0.87 4
γ [E2]	1082.27 13	~0.032
γ E1	1104.048 19	2.13 9
γ [E2]	1119.10 7	0.132 25

Photons (^{194}Au) (continued)

γ_{mode}	γ(keV)	γ(%)†
γ (M1)	1141.05 6	~0.025
γ E2	1150.77 3	1.44 8
γ M1	1156.61 5	0.44 4
γ E1	1175.365 17	2.12 8
γ M1+45%E2	1183.504 15	0.61 3
γ M1,E2	1186.33 6	0.0572 25
γ E1,E2	1195.0 3	0.08 3
γ E2	1218.77 3	1.20 6
γ M1(+E2)	1291.69 25	0.113 25
γ E2+37%M1+E0	1293.70 3	0.149 17
γ M1(+E2)	1302.22 4	0.28 4
γ E2	1308.41 7	0.161 18
γ [E1]	1316.99 11	0.08 3
γ [E1]	1339.32 5	~0.30
γ M1+E2	1342.18 3	1.25 7
γ M1	1421.68 3	0.34 3
γ (E2)	1431.09 13	0.17 3
γ	1432.504 21	0.094 20
γ M1+E2	1441.80 6	0.184 16
γ M1+E2	1450.16 5	0.340 19
γ E2	1463.52 7	0.79 17
γ E1	1468.914 17	6.7 3
γ M1	1487.09 6	0.134 9
γ M1	1492.06 6	0.170 17
γ M1,E2	1500.52 12	0.397 19
γ [E2]	1511.960 18	0.047 6
γ [M1]	1512.19 8	0.074 21
γ (M1)	1518.5 2	0.069 19
γ M1	1562.85 20	0.321 25
γ (M1)	1592.44 25	~1
γ [M1]	1593.47 3	~0.6
γ M1	1595.77 4	1.88 22
γ M1	1601.95 7	0.248 25
γ E1	1617.74 11	0.21 3
γ E2	1622.15 3	0.204 19
γ E1	1632.87 5	0.243 16
γ M1,E2	1670.63 3	0.186 12
γ M1	1676.10 7	0.142 16
γ M1,E2	1689.83 12	0.176 25
γ M1+E2	1715.23 3	0.70 5
γ	1724.64 13	0.073 10 ?
γ M1	1735.35 6	0.298 22
γ	1744.3 5	0.033 9
γ [E2]	1757.06 8	0.059 14
γ (E2)	1778.62 5	0.052 12
γ	1780.64 6	0.041 10
γ M1	1785.61 6	0.42 3
γ E1	1797.370 19	0.61 3
γ M1,E2	1803.1 5	0.19 6
γ M1,E2	1805.74 8	0.18 5
γ	1812.8 6	<0.07
γ	1817.0 3	0.038 13
γ M1	1829.46 8	0.246 19
γ E2	1835.33 7	0.416 25
γ	1856.39 20	0.032 3
γ (M1)	1885.99 25	~2
γ [M1]	1887.02 3	~2
γ E1	1911.29 11	0.132 13
γ M1	1924.23 4	2.08 10
γ M1(+E2)	1958.71 18	0.170 19
γ M1	1969.65 7	0.454 25
γ M1+E2+E0	1983.37 12	0.0378 25
γ M1	2043.68 3	3.78 17
γ [E2]	2063.80 6	0.0101 25
γ	2069.0 4	0.0221 6
γ [M1+E2]	2083.7 4	0.0365 7
γ M1	2114.07 6	0.277 8
γ M1	2215.47 3	0.176 13
γ M1	2298.10 7	0.029 5
γ E2	2311.83 12	0.175 10
γ M1	2365.56 20	0.048 7
γ	2397.5 4	0.0040 8
γ M1	2412.2 4	0.018 3

† 9.5% uncert(syst)

Atomic Electrons (^{194}Au)

$\langle$e$\rangle$=30.3 21 keV

e_{bin}(keV)	$\langle$e$\rangle$(keV)	e(%)
12	3.2	28 6
13	1.9	14 3
14 - 63	2.1	7.2 10
64 - 111	0.76	1.03 9
125 - 173	0.259	0.177 10
175 - 212	0.031	0.0161 23
215	1.57	0.73 4
221 - 249	0.152	0.067 4
250	7.7	3.1 6
271 - 274	0.0052	0.0019 3
280	0.73	0.261 14
282 - 307	0.839	0.292 7
315	3.1	1.00 20
316	8.8 $\times 10^{-5}$	2.8 4 $\times 10^{-5}$
317	0.94	0.30 6
318	8.3 $\times 10^{-5}$	2.6 4 $\times 10^{-5}$
325	0.79	0.24 5
326 - 371	0.69	0.208 25
404 - 452	0.40	0.09 4
466 - 516	0.22	0.043 9
517 - 563	0.156	0.0290 22
567 - 610	0.234	0.0404 12
618 - 666	0.164	0.0259 11
668 - 704	0.0197	0.00284 18
722 - 769	0.0204	0.00274 14
777 - 819	0.027	0.0034 10
833 - 881	0.338	0.0389 17
883 - 929	0.0272	0.00295 17
934 - 970	0.155	0.0162 5
986 - 1036	0.0529	0.00516 19
1037 - 1082	0.103	0.0096 5
1090 - 1139	0.093	0.0084 3
1140 - 1189	0.24	~0.021
1192 - 1239	0.044	0.0036 5
1253 - 1302	0.11	0.009 3
1303 - 1343	0.037	0.0028 3
1353 - 1401	0.5	~0.04
1408 - 1456	0.075	0.0052 5
1457 - 1554	0.57	0.038 15
1560 - 1659	0.110	0.0068 7
1662 - 1757	0.081	0.0047 3
1765 - 1854	0.9	~0.05
1872 - 1970	0.34	0.018 6
1972 - 2070	0.21	~0.010
2072 - 2152	0.14	~0.007
2202 - 2300	0.04	~0.0017
2309 - 2401	0.007	~0.00030
2409 - 2410	2.0 $\times 10^{-5}$	8.2 15 $\times 10^{-7}$

Continuous Radiation (^{194}Au)

$\langle\beta+\rangle$=10.3 keV;$\langle$IB$\rangle$=1.66 keV

E_{bin}(keV)		$\langle\ \rangle$(keV)	(%)
0 - 10	$\beta+$	2.55×10^{-8}	3.00×10^{-7}
	IB	0.00060	
10 - 20	$\beta+$	1.68×10^{-6}	9.9×10^{-6}
	IB	0.0021	0.018
20 - 40	$\beta+$	7.5×10^{-5}	0.000225
	IB	0.0026	0.0082
40 - 100	$\beta+$	0.0061	0.0077
	IB	0.40	0.61
100 - 300	$\beta+$	0.424	0.189
	IB	0.050	0.028
300 - 600	$\beta+$	2.85	0.62
	IB	0.130	0.029
600 - 1300	$\beta+$	6.9	0.83
	IB	0.62	0.066
1300 - 2500	$\beta+$	0.150	0.0111
	IB	0.46	0.029
2500 - 2509	IB	6.8×10^{-11}	1.24×10^{-10}
	$\Sigma\beta+$		1.66

$^{194}_{79}$Au(600 *8* ms)

Mode: IT
Δ: -32171 *16* keV
SpA: 5.75×10^{10} Ci/g
Prod: ^{195}Pt(p,2n); ^{193}Ir(α,3n); ^{194}Pt(p,n); ^{194}Pt(d,2n)

Photons (^{194}Au)

⟨γ⟩=11.5 *11* keV

γ_{mode}	γ(keV)	γ(%)
Au L_ℓ	8.494	1.5 *3*
Au L_α	9.704	32 *5*
Au L_η	10.309	0.31 *7*
Au L_β	11.541	34 *6*
Au L_γ	13.611	7.2 *14*
γ (M2)	26.9 *5*	0.0128 *6*
γ M1+2.1%E2	35.19 *7*	2.4 *5*
γ M1+2.0%E2	45.32 *7*	5.3 *11*

Atomic Electrons (^{194}Au)

⟨e⟩=85 *7* keV

e_{bin}(keV)	⟨e⟩(keV)	e(%)
12	3.9	33 *7*
13	6.2	49 *3*
14	7.1	50 *10*
15	3.79	25.3 *16*
21	13.9	66 *17*
23	6.3	27 *9*
24	1.87	7.8 *5*
25	0.0288	0.117 *8*
26	1.37	5.2 *3*
27	0.296	1.10 *7*
31	15.7	51 *10*
32	10.3	32 *10*
33	3.0	9 *4*
34	0.99	2.9 *6*
35	1.1	3.0 *13*
42	6.3	15 *3*
43	1.0	2.4 *10*
45	2.4	5.3 *13*

$^{194}_{79}$Au(420 *10* ms)

Mode: IT
Δ: -31802 *16* keV
SpA: 6.78×10^{10} Ci/g
Prod: ^{195}Pt(p,2n); ^{193}Ir(α,3n); ^{194}Pt(p,n); ^{194}Pt(d,2n)

Photons (^{194}Au)

γ_{mode}	γ(keV)	γ(%)†
γ (M2)	26.9 *5*	*
γ [M1]	33.63 *10*	0.50 *17*
γ M1+2.1%E2	35.19 *7*	*
γ M1+2.0%E2	45.32 *7*	*
γ E2	128.58 *9*	25.0
γ E2	137.16 *9*	16 *1*
γ E2+29%M1	162.21 *10*	9.3 *8*
γ M1+26%E2	170.78 *9*	25.3 *15*

† 8.0% uncert(syst)
* with ^{194}Au(600 ms)

$^{194}_{80}$Hg(520 *20* yr)

Mode: ε
Δ: -32238 *25* keV
SpA: 3.54 Ci/g
Prod: ^{197}Au(p,4n); ^{194}Pt(α,4n)

Photons (^{194}Hg)

⟨γ⟩=2.18 *14* keV

γ_{mode}	γ(keV)	γ(%)
Au L_ℓ	8.494	0.36 *4*
Au L_α	9.704	7.6 *6*
Au L_η	10.309	0.071 *4*
Au L_β	11.523	9.6 *11*
Au L_γ	13.654	2.2 *3*

$^{194}_{81}$Tl(33.0 *5* min)

Mode: ε
Δ: -27090 *190* keV syst
SpA: 2.94×10^{7} Ci/g
Prod: protons on Hg; daughter ^{194}Pb; protons on Pb

Photons (^{194}Tl)

γ_{mode}	γ(keV)	γ(rel)*
γM1	395.1 *3*	2.1 *4*
γM1	404.0 *4*	2.5 *5*
γE2	428.12 *23*	~100
γE2	636.11 *24*	~23
γM1+E2	644.95 *18*	13 *3*
γ[M1+E2]	1040.1 *3*	5.3 *16*
γ[E2]	1073.1 *3*	4.2 *16*

* see also ^{194}Tl(32.8 min)

$^{194}_{81}$Tl(32.8 *2* min)

Mode: ε
Δ: >-27090 keV syst
SpA: 2.954×10^{7} Ci/g
Prod: protons on Hg; protons on Pb

Photons (^{194}Tl)

⟨γ⟩=2057 *390* keV

γ_{mode}	γ(keV)	γ(%)†
γE2	96.83 *7*	7.7 *15*
γ[M1]	98.82 *10*	0.62 *23*
γ	107.12 *20*	0.77 *15* *
γE1	110.88 *8*	6.4 *15*
γE2	208.84 *17*	6.2 *15*
γM1+E2	218.9 *5*	1.00 *23* *
γM1+6.1%E2	227.90 *8*	6.6 *8*
γE2	233.02 *15*	2.1 *3* *
γM1	238.9 *7*	0.92 *23* *
γM1(+E2)	255.32 *9*	9.2 *15*
γM1	283.83 *15*	1.8 *3*
γM1(+E2)	298.02 *14*	2.08 *23*
γ	299.4 *5*	1.0 *3* *
γE2	319.72 *10*	3.93 *8* *

Photons (^{194}Tl) (continued)

γ_{mode}	γ(keV)	γ(%)†
γM1+E2	352.15 *10*	1.7 *3*
γM1	366.34 *14*	1.8 *3*
γ[E1]	380.39 *15*	1.39 *23*
γE2	428.12 *23*	~96
γM1	446.4 *7*	2.8 *5* *
γM1+E2	450.97 *14*	5.0 *12*
γM1+E2	462.4 *7*	4.6 *15* *
γM1	464.16 *19*	2.3 *8*
γM1	553.34 *13*	4.6 *12*
γ	600.4 *7*	1.7 *5* *
γE2	636.11 *24*	~99
γM1+E2	650.17 *14*	6.9 *15*
γ[E1]	664.22 *15*	1.2 *3*
γE2	734.90 *22*	22 *5*
γE1	748.94 *22*	77 *15*

† 13% uncert(syst)
* with ^{194}Tl(32.8 or 33.0 min)

$^{194}_{82}$Pb(11 *2* min)

Mode: ε
Δ: -24240 *300* keV syst
SpA: 8.8×10^{7} Ci/g
Prod: protons on Tl

Photons (^{194}Pb)

γ_{mode}	γ(keV)
γ(M1)	204 *2*

$^{194}_{83}$Bi(1.75 *25* min)

Mode: ε(>99.8 %), α(<0.2 %)
Δ: -16260 *330* keV syst
SpA: 5.5×10^{8} Ci/g
Prod: ^{181}Ta(^{20}Ne,7n); ^{181}Ta(^{22}Ne,9n); ^{185}Re(^{16}O,7n)

Alpha Particles (^{194}Bi)

α(keV)
5640 *30*

Photons (^{194}Bi)

⟨γ⟩=2014 *200* keV

γ_{mode}	γ(keV)	γ(%)
γ(E2)	166.1 *3*	46 *3*
γ(E1)	174.0 *3*	27.9 *20*
γ(E1)	280.2 *3*	70 *5*
γ(E2)	421 *1*	55 *10*
γ(E2)	575.4 *5*	87 *8*
γ(E2)	965.0 *5*	100

$^{194}_{84}$Po(700 *100* ms)

Mode: α
Δ: -11010 *230* keV syst
SpA: 5.3×10^{10} Ci/g
Prod: ^{185}Re(^{19}F,10n)

Alpha Particles (^{194}Po)

α(keV)
6846 *6*

A = 195

NDS **23**, 607 (1978)

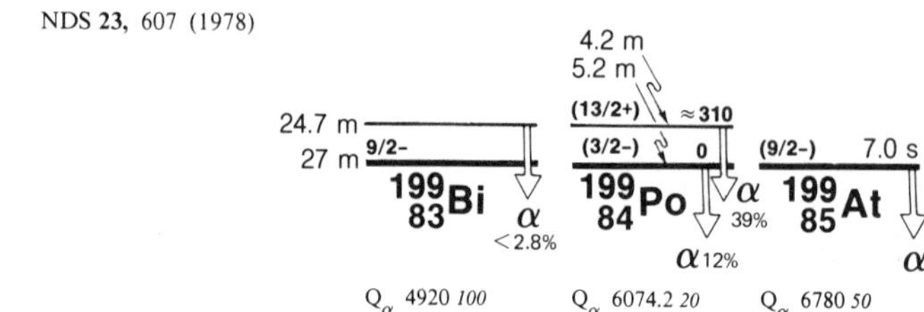

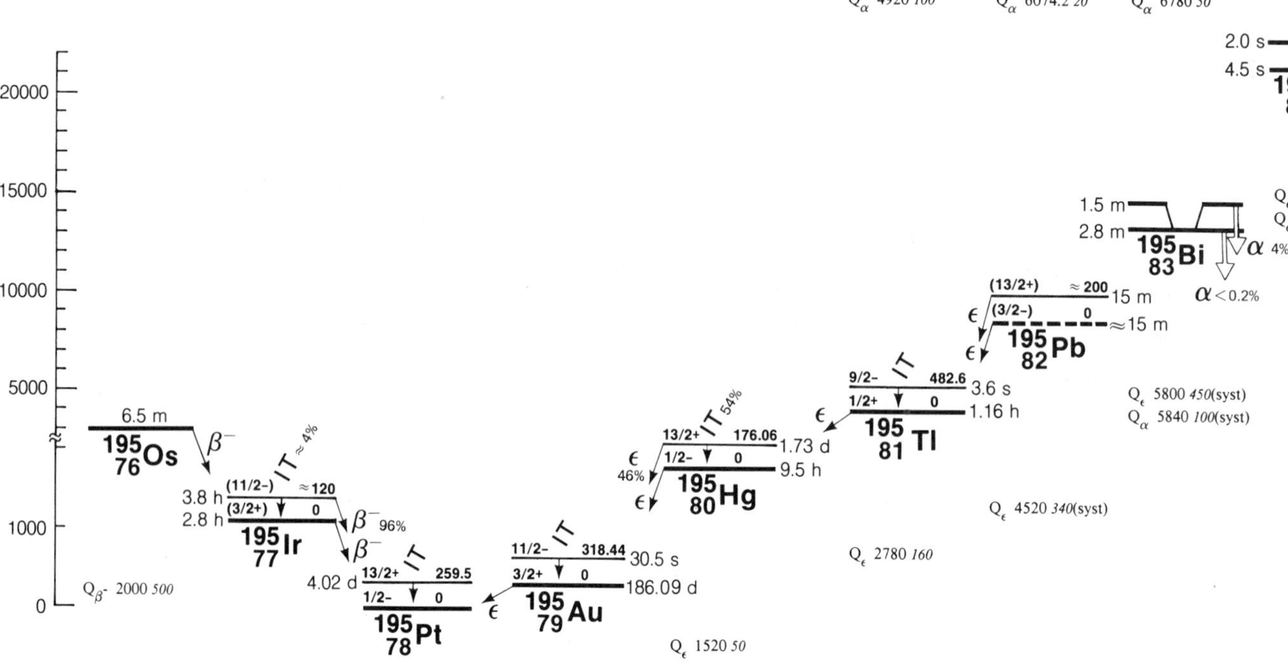

$^{195}_{76}$Os(6.5 min)

Mode: β-
Δ: -29700 *500* keV
SpA: 1.5×10^{8} Ci/g
Prod: ^{198}Pt(n,α)

$^{195}_{77}$Ir(2.8 *1* h)

Mode: β-
Δ: -31702 *13* keV
SpA: 5.74×10^{6} Ci/g
Prod: ^{192}Os(α,p); ^{198}Pt(d,αn); ^{195}Pt(n,p); ^{196}Pt(γ,p); ^{198}Pt(p,α)

Photons (^{195}Ir) *

⟨γ⟩=59 *3* keV

γ_{mode}	γ(keV)	γ(%)†
Pt L_ℓ	8.266	0.62 *8*
Pt L_α	9.435	13.4 *15*
Pt L_η	9.975	0.163 *17*
Pt L_β	11.172	14.7 *19*
Pt L_γ	13.131	3.0 *4*
γ M1+0.05%E2	30.882 *6*	1.33 *13*
Pt $K_{\alpha2}$	65.122	16.3 *16*
Pt $K_{\alpha1}$	66.831	28 *3*
Pt $K_{\beta1}'$	75.634	9.8 *10*
Pt $K_{\beta2}'$	78.123	2.5 *3*
γ M1+1.7%E2	98.883 *10*	9.7 *10*
γ E2	129.765 *10*	1.25 *12*
γ M1+13%E2	211.36 *3*	2.40 *24*

† 26% uncert(syst)
* see also ^{195}Ir(3.8 h)

Atomic Electrons (^{195}Ir)

⟨e⟩=45.3 *17* keV

e_{bin}(keV)	⟨e⟩(keV)	e(%)
12	2.4	21 *3*
13	1.66	12.5 *17*
14	1.58	11.4 *15*
17	6.1	36 *4*
18 - 19	0.79	4.4 *4*
20	11.7	57 *6*
28	2.6	9.4 *10*
29 - 75	2.42	5.7 *3*
85	7.3	8.6 *9*
86	0.94	1.10 *11*
87	0.28	0.33 *4*
96	2.23	2.33 *24*
97 - 130	2.63	2.33 *11*
133	1.99	1.49 *15*
197 - 211	0.71	0.356 *24*

Continuous Radiation (^{195}Ir)

$\langle\beta$-$\rangle$=335 keV;$\langle$IB$\rangle$=0.34 keV

E_{bin}(keV)		$\langle\ \rangle$(keV)	(%)
0 - 10	β-	0.078	1.55
	IB	0.0158	
10 - 20	β-	0.233	1.56
	IB	0.0152	0.105
20 - 40	β-	0.95	3.15
	IB	0.028	0.099
40 - 100	β-	6.8	9.7
	IB	0.071	0.111
100 - 300	β-	65	32.7
	IB	0.139	0.083
300 - 600	β-	164	37.6
	IB	0.067	0.0169
600 - 1118	β-	98	13.7
	IB	0.0075	0.00113

$^{195}_{77}$Ir(3.8 *2* h)

Mode: β-(96 *2* %), IT(4 *2* %)
Δ: ~-31582 keV
SpA: 4.23×10^6 Ci/g
Prod: ^{192}Os(α,p); ^{198}Pt(d,αn); ^{198}Pt(p,α)

Photons (^{195}Ir)

$\langle\gamma\rangle$=418 *17* keV

γ_{mode}	γ(keV)	γ(%)†
$\gamma_{\beta-}$	27.8 *4*	0.23 *5* *
$\gamma_{\beta-}$M1+0.05%E2	30.882 *6*	
$\gamma_{\beta-}$[M1]	60.52 *8*	<1
$\gamma_{\beta-}$M1+1.7%E2	98.883 *10*	
$\gamma_{\beta-}$[M1]	115.10 *13*	0.10 *3*
$\gamma_{\beta-}$(M1)	119.12 *10*	0.38 *10*
$\gamma_{\beta-}$M4	129.72 *15*	#
$\gamma_{\beta-}$E2	129.765 *10*	
$\gamma_{\beta-}$[E1]	130.57 *16*	~0.08
$\gamma_{\beta-}$M1+2.9%E2	140.38 *5*	0.87 *6*
$\gamma_{\beta-}$	147.62 *15*	0.10 *3* *
$\gamma_{\beta-}$[M1]	149.87 *8*	0.17 *4*
$\gamma_{\beta-}$E2+27%M1	172.78 *7*	4.99 *19*
$\gamma_{\beta-}$	178.48 *25*	0.086 *19* *
$\gamma_{\beta-}$	197.46 *25*	0.096 *19* *
$\gamma_{\beta-}$E2+42%M1	199.47 *4*	0.58 *10*
$\gamma_{\beta-}$M1	201.83 *9*	1.44 *10*
$\gamma_{\beta-}$[M1]	210.39 *6*	0.26 *5*
$\gamma_{\beta-}$M1+13%E2	211.36 *3*	~2
$\gamma_{\beta-}$M1(+E2)	215.99 *6*	0.86 *10*
$\gamma_{\beta-}$[M1]	223.59 *10*	~0.010
$\gamma_{\beta-}$[M1]	235.43 *8*	0.26 *5*
$\gamma_{\beta-}$[E2]	238.29 *5*	0.125 *25*
$\gamma_{\beta-}$E2	239.26 *5*	1.66 *8*
$\gamma_{\beta-}$M1	243.89 *5*	0.77 *5*
$\gamma_{\beta-}$M1(+E2)	251.71 *8*	1.82 *19*
$\gamma_{\beta-}$(M1+E2)	255.78 *6*	0.86 *10*
$\gamma_{\beta-}$(M1+E2)	259.37 *7*	0.77 *10*
$\gamma_{\beta-}$	264.5 *5*	0.058 *10* *
$\gamma_{\beta-}$(E1)	267.18 *12*	0.58 *10*
$\gamma_{\beta-}$	270.7 *3*	0.038 *10* *
$\gamma_{\beta-}$	277.0 *4*	0.038 *10* *
$\gamma_{\beta-}$[M1]	282.59 *11*	0.058 *19*
$\gamma_{\beta-}$(M1+E2)	287.88 *12*	1.0 *3*
$\gamma_{\beta-}$M1(+E2)	290.25 *7*	1.92 *10*
$\gamma_{\beta-}$[M1]	306.18 *9*	0.84 *17*
$\gamma_{\beta-}$[E2]	306.48 *7*	1.4 *3*
$\gamma_{\beta-}$E2+32%M1	319.88 *5*	9.6 *5*
$\gamma_{\beta-}$[M1]	325.49 *5*	0.77 *10*
$\gamma_{\beta-}$[M1]	332.47 *12*	0.08 *3*
$\gamma_{\beta-}$[E2]	350.77 *5*	1.02 *14*
$\gamma_{\beta-}$M1	356.37 *5*	1.82 *19*
$\gamma_{\beta-}$E2	359.29 *6*	4.6 *3*
$\gamma_{\beta-}$E2+30%M1	364.90 *5*	9.5 *3*

Photons (^{195}Ir) (continued)

γ_{mode}	γ(keV)	γ(%)†
$\gamma_{\beta-}$M1(+E2)	373.46 *9*	1.1 *3*
$\gamma_{\beta-}$M1	378.30 *6*	1.0 *3*
$\gamma_{\beta-}$	383.3 *3*	0.24 *5*
$\gamma_{\beta-}$	385.22 *20*	0.163 *19* *
$\gamma_{\beta-}$[M1]	387.24 *9*	0.32 *3*
$\gamma_{\beta-}$	389.87 *15*	0.58 *19* *
$\gamma_{\beta-}$	395.7 *3*	<0.10 *
$\gamma_{\beta-}$[E2]	401.36 *9*	0.29 *10*
$\gamma_{\beta-}$E2	409.19 *6*	1.44 *10*
$\gamma_{\beta-}$	413.62 *20*	0.26 *7* *
$\gamma_{\beta-}$	419.71 *15*	0.38 *10* *
$\gamma_{\beta-}$	422.68 *8*	0.134 *19*
$\gamma_{\beta-}$[E2]	425.42 *8*	0.67 *19*
$\gamma_{\beta-}$	427.82 *20*	0.67 *10* *
$\gamma_{\beta-}$E2	433.07 *9*	9.6 *19*
$\gamma_{\beta-}$[E2]	440.06 *10*	0.21 *5*
$\gamma_{\beta-}$	445.66 *9*	0.47 *6*
$\gamma_{\beta-}$M1(+E2)	456.05 *7*	0.78 *5*
$\gamma_{\beta-}$	463.6 *3*	0.058 *19* *
$\gamma_{\beta-}$[E2]	475.49 *7*	0.15 *3*
$\gamma_{\beta-}$M1(+E2)	481.10 *7*	2.69 *19*
$\gamma_{\beta-}$[E2]	495.84 *7*	0.51 *5*
$\gamma_{\beta-}$	498.47 *14*	0.115 *19*
$\gamma_{\beta-}$[E2]	506.18 *8*	0.64 *5*
$\gamma_{\beta-}$[E2]	513.84 *8*	0.029 *10*
$\gamma_{\beta-}$	524.5 *3*	0.067 *10* *
$\gamma_{\beta-}$	526.7 *3*	0.029 *10* *
$\gamma_{\beta-}$	530.1 *3*	0.067 *19* *
$\gamma_{\beta-}$	534.12 *20*	0.35 *5* *
$\gamma_{\beta-}$	537.4 *3*	0.15 *5* *
$\gamma_{\beta-}$	540.4 *3*	~0.019 *
$\gamma_{\beta-}$	544.5 *5*	~0.019 *
$\gamma_{\beta-}$	548.1 *3*	0.029 *10* *
$\gamma_{\beta-}$[M1]	565.54 *6*	0.22 *5*
$\gamma_{\beta-}$M1+E2	575.29 *6*	1.5 *3*
$\gamma_{\beta-}$[E2]	596.43 *6*	0.21 *5*
$\gamma_{\beta-}$	611.1 *3*	0.048 *10* *
$\gamma_{\beta-}$	613.6 *3*	~0.019 *
$\gamma_{\beta-}$	616.5 *3*	0.19 *4* *
$\gamma_{\beta-}$	619.2 *3*	0.048 *10* *
$\gamma_{\beta-}$E2	684.78 *5*	9.6 *5*
$\gamma_{\beta-}$[E2]	691.48 *7*	0.048 *10*
$\gamma_{\beta-}$	715.52 *20*	0.058 *19* *
$\gamma_{\beta-}$	723.72 *20*	~0.19 *
$\gamma_{\beta-}$	750.82 *20*	0.20 *5* *
$\gamma_{\beta-}$	784.1 *3*	0.086 *10* *
$\gamma_{\beta-}$[E2]	800.98 *6*	1.06 *10*

† uncert(syst): 54% for IT, 15% for β-
* with ^{195}Ir(3.8 h) or ^{195}Ir(2.8 h)
\# with ^{195}Pt(4.02 d)

Continuous Radiation (^{195}Ir)

$\langle\beta$-$\rangle$=204 keV;$\langle$IB$\rangle$=0.165 keV

E_{bin}(keV)		$\langle\ \rangle$(keV)	(%)
0 - 10	β-	0.172	3.45
	IB	0.0100	
10 - 20	β-	0.507	3.38
	IB	0.0093	0.065
20 - 40	β-	1.98	6.6
	IB	0.0168	0.059
40 - 100	β-	12.7	18.3
	IB	0.039	0.062
100 - 300	β-	75	40.1
	IB	0.063	0.038
300 - 600	β-	83	19.7
	IB	0.024	0.0061
600 - 979	β-	31.4	4.50
	IB	0.0020	0.00031

$^{195}_{78}$Pt(stable)

Δ: -32821 *4* keV
%: 33.8 *5*

$^{195}_{78}$Pt(4.02 *1* d)

Mode: IT
Δ: -32562 *4* keV
SpA: 1.666×10^5 Ci/g
Prod: ^{194}Pt(n,γ); ^{194}Pt(d,p)

Photons (^{195}Pt)

$\langle\gamma\rangle$=76.3 *17* keV

γ_{mode}	γ(keV)	γ(%)†
Pt L_ℓ	8.266	1.46 *16*
Pt L_α	9.435	31.4 *25*
Pt L_η	9.975	0.277 *19*
Pt L_β	11.195	28 *3*
Pt L_γ	13.144	5.4 *7*
γ M1+0.05%E2	30.882 *6*	2.28 *14*
Pt $K_{\alpha2}$	65.122	22.5 *11*
Pt $K_{\alpha1}$	66.831	38.5 *19*
Pt $K_{\beta1}'$	75.634	13.5 *7*
Pt $K_{\beta2}'$	78.123	3.51 *18*
γ M1+1.7%E2	98.883 *10*	11.4 *6*
γ M4	129.72 *15*	0.085 *5*
γ E2	129.765 *10*	2.80 *17*
γ M1+2.9%E2	140.38 *5*	0.0294 *14*
γ M1+13%E2	211.36 *3*	0.0389 *23*
γ E2	239.26 *5*	0.0558 *25*

† 5.3% uncert(syst)

Atomic Electrons (^{195}Pt)

$\langle e\rangle$=175 *4* keV

e_{bin}(keV)	$\langle e\rangle$(keV)	e(%)
6 - 9	0.011	0.13 *6*
12	6.6	57 *6*
13 - 16	5.8	43 *3*
17	10.5	62 *4*
18 - 19	1.38	7.7 *5*
20	13.7	67 *4*
25 - 27	0.0036	0.014 *6*
28	4.5	16.1 *11*
29 - 31	1.51	5.0 *3*
51	7.8	15.2 *10*
52 - 75	1.34	2.27 *9*
85	8.5	10.1 *5*
86 - 99	4.88	5.24 *17*
116	26.4	22.8 *14*
118	48.2	41 *3*
126	6.6	5.2 *3*
127	17.9	14.1 *9*
128	0.99	0.78 *5*
129	6.8	5.2 *3*
130 - 161	1.26	0.97 *6*
197 - 239	0.0234	0.0109 *3*

$^{195}_{79}$Au(186.09 *4* d)

Mode: ε

Δ: -32591 *4* keV

SpA: 3598.3 Ci/g

Prod: deuterons on Pt; ^{193}Ir(α,2n); ^{195}Pt(p,n)

Photons (^{195}Au)

⟨γ⟩=86 *3* keV

γ_{mode}	γ(keV)	γ(%)†
Pt L_ℓ	8.266	1.02 *12*
Pt L_α	9.435	22.1 *20*
Pt L_η	9.975	0.263 *22*
Pt L_β	11.175	24 *3*
Pt L_γ	13.128	4.7 *6*
γ M1+0.05%E2	30.882 *6*	0.75 *3*
Pt $K_{\alpha2}$	65.122	29 *2*
Pt $K_{\alpha1}$	66.831	50 *4*
Pt $K_{\beta1}$'	75.634	17.4 *12*
Pt $K_{\beta2}$'	78.123	4.5 *3*
γ M1+1.7%E2	98.883 *10*	10.9 *5*
γ E2	129.765 *10*	0.817 *22*
γ E2+42%M1	199.47 *4*	0.0086 *7*
γ M1+13%E2	211.36 *3*	0.0109 *11*

† 5.5% uncert(syst)

Atomic Electrons (^{195}Au)

⟨e⟩=44.8 *11* keV

e_{bin}(keV)	⟨e⟩(keV)	e(%)
12	4.2	36 *4*
13	2.7	20.3 *24*
14	2.5	17.8 *21*
17	3.47	20.4 *10*
18 - 19	0.451	2.52 *13*
20	13.1	64 *3*
28	1.47	5.3 *3*
29 - 75	2.90	5.91 *20*
85	8.2	9.6 *5*
86 - 87	1.37	1.60 *7*
96	2.50	2.61 *14*
97 - 133	2.07	1.87 *4*
186 - 211	0.0063	0.00323 *19*

Continuous Radiation (^{195}Au)

⟨IB⟩=0.25 keV

E_{bin}(keV)		⟨ ⟩(keV)	(%)
10 - 20	IB	0.0046	0.040
20 - 40	IB	0.0024	0.0074
40 - 100	IB	0.24	0.37
100 - 230	IB	0.00076	0.00066

$^{195}_{79}$Au(30.5 *2* s)

Mode: IT

Δ: -32273 *4* keV

SpA: 1.875×10^{9} Ci/g

Prod: daughter ^{195}Hg(1.73 d); protons on Pt; deuterons on Pt

Photons (^{195}Au)

⟨γ⟩=202 *4* keV

γ_{mode}	γ(keV)	γ(%)†
Au L_ℓ	8.494	0.64 *7*
Au L_α	9.704	13.6 *9*
Au L_η	10.309	0.280 *17*
Au L_β	11.531	16.5 *11*
Au L_γ	13.440	3.02 *22*
γ E3	56.61 *3*	0.0296 *6*
γ M1+17%E2	61.432 *24*	0.173 *12*
Au $K_{\alpha2}$	66.991	6.72 *23*
Au $K_{\alpha1}$	68.806	11.4 *4*
Au $K_{\beta1}$'	77.859	4.01 *14*
Au $K_{\beta2}$'	80.428	1.06 *4*
γ E2	200.40 *3*	1.70 *12*
γ M1+21%E2	261.83 *3*	67.9 *14*
γ M4	318.44 *4*	0.041 *5*

† 10% uncert(syst)

Atomic Electrons (^{195}Au)

⟨e⟩=116.4 *15* keV

e_{bin}(keV)	⟨e⟩(keV)	e(%)
12	3.8	32 *3*
14 - 42	4.0	27 *3*
43	16.3	38.0 *11*
45	15.4	34.4 *10*
47 - 52	0.84	1.75 *7*
53	5.99	11.2 *3*
54	5.96	11.1 *3*
55 - 77	4.36	7.69 *17*
120	0.344	0.287 *20*
181	42.9	23.7 *7*
186 - 200	0.670	0.353 *13*
238	0.63	0.26 *3*
247	8.75	3.53 *10*
248 - 262	5.81	2.28 *3*
304 - 318	0.73	0.237 *13*

$^{195}_{80}$Hg(9.5 *5* h)

Mode: ε

Δ: -31070 *50* keV

SpA: 1.69×10^{6} Ci/g

Prod: daughter ^{195}Tl; ^{197}Au(p,3n); ^{194}Pt(^{3}He,2n)

Photons (^{195}Hg)

⟨γ⟩=203 *9* keV

γ_{mode}	γ(keV)	γ(%)†
Au L_ℓ	8.494	0.99 *18*
Au L_α	9.704	21 *3*
Au L_η	10.309	0.31 *5*
Au L_β	11.542	22 *4*
Au L_γ	13.506	4.2 *7*
γ M1+17%E2	61.432 *24*	6.4 *4*
Au $K_{\alpha2}$	66.991	23 *5*
Au $K_{\alpha1}$	68.806	39 *8*
Au $K_{\beta1}$'	77.859	14 *3*
Au $K_{\beta2}$'	80.428	3.7 *7*
γ M1+~2.5%E2	180.11 *3*	1.95 *9*
γ E2	200.40 *3*	0.037 *3*
γ E2	207.05 *3*	1.62 *18*
γ E2+~17%M1	241.54 *3*	0.071 *11*
γ M1+21%E2	261.83 *3*	1.44 *11*
γ	330.5 *3*	0.013 *3*
γ	360.21 *24*	0.013 *3*

Photons (^{195}Hg)

(continued)

γ_{mode}	γ(keV)	γ(%)†
γ [M1+E2]	401.70 *9*	0.0050 *17*
γ M1	439.50 *9*	0.127 *11*
γ	546.41 *19*	0.0013 *6*
γ E2	585.11 *5*	2.04 *8*
γ M1+23%E2	599.66 *3*	1.83 *6*
γ [E1]	671.10 *10*	0.025 *3*
γ	716.77 *18*	0.008 *4*
γ	778.20 *18*	0.0006 *3*
γ M1	779.77 *3*	7.0
γ (M1)	811.43 *11*	0.019 *6*
γ M1(+E2)	821.08 *5*	0.30 *3*
γ M1,E2	841.20 *4*	0.28 *6*
γ M1,E2	841.37 *5*	0.37 *6*
γ	861.0 *4*	0.0049 *14*
γ [E1]	869.06 *7*	0.0036 *13*
γ E2+~34%M1	910.58 *5*	0.064 *12*
γ M1+32%E2	930.87 *5*	0.43 *4*
γ	989.10 *10*	0.011 *4*
γ (M1)	1009.39 *10*	0.028 *6*
γ M1	1021.48 *5*	0.193 *20*
γ [E1]	1049.17 *7*	0.020 *6*
γ (E2)	1082.91 *5*	0.071 *9*
γ [M1+E2]	1091.77 *23*	0.008 *4*
γ M1	1110.98 *5*	1.48 *13*
γ M1	1172.42 *5*	1.28 *11*
γ	1189.50 *10*	0.022 *4*
γ	1250.93 *10*	0.015 *3*
γ	1263.17 *19*	0.0050 *14*
γ [M1+E2]	1292.17 *23*	0.0050 *11*
γ	1324.60 *19*	0.0015 *5*
γ	1339.8 *5*	0.00049 *21*
γ [M1+E2]	1353.60 *23*	0.0101 *22*
γ	1368.3 *4*	0.0013 *4*
γ	1371.6 *3*	0.0018 *5*
γ	1433.0 *3*	0.0008 *3*
γ	1443.12 *24*	0.0010 *3*

† 10% uncert(syst)

Atomic Electrons (^{195}Hg)

⟨e⟩=61.3 *19* keV

e_{bin}(keV)	⟨e⟩(keV)	e(%)
12	4.8	41 *8*
14	4.7	34 *6*
42 - 45	0.68	1.55 *22*
47	10.0	21.3 *15*
48	10.1	21.2 *16*
50	9.3	18.7 *15*
52 - 57	1.34	2.49 *25*
58	6.0	10.3 *7*
59	2.95	5.0 *4*
61	2.79	4.6 *3*
63 - 77	0.73	1.11 *11*
99	2.24	2.25 *11*
120 - 168	0.98	0.66 *4*
177 - 207	1.71	0.92 *5*
227 - 262	0.342	0.137 *7*
279 - 328	0.022	0.0071 *12*
330 - 360	0.048	0.0135 *12*
387 - 437	0.0117	0.00274 *19*
439 - 466	0.0011	0.00024 *8*
504 - 546	0.503	0.098 *4*
571 - 600	0.143	0.0245 *6*
636 - 671	0.0010	~0.00016
697	5.8 ×10^{-5}	~8×10^{-6}
699	1.14	0.163 *17*
702 - 740	0.034	0.0045 *21*
760 - 809	0.33	0.043 *4*
811 - 860	0.066	0.0078 *10*
861 - 910	0.0019	0.00021 *7*
911 - 941	0.0345	0.0037 *3*
968 - 1011	0.0082	0.00082 *7*
1018 - 1049	0.147	0.0142 *13*
1069 - 1111	0.150	0.0137 *10*
1158 - 1189	0.0263	0.00227 *15*
1211 - 1261	0.00058	4.7 *14* ×10^{-5}
1262 - 1311	0.00076	6.0 *20* ×10^{-5}

Atomic Electrons (^{195}Hg) (continued)

e_{bin}(keV)	⟨e⟩(keV)	e(%)
1313 - 1360	0.00018	1.4 *4* $\times 10^{-5}$
1362 - 1371	4.6 $\times 10^{-5}$	~3 $\times 10^{-6}$
1419 - 1443	1.6 $\times 10^{-5}$	1.1 *5* $\times 10^{-6}$

Continuous Radiation (^{195}Hg)

⟨IB⟩=0.91 keV

E_{bin}(keV)		⟨ ⟩(keV)	(%)
10 - 20	IB	0.0018	0.0150
20 - 40	IB	0.00159	0.0048
40 - 100	IB	0.44	0.64
100 - 300	IB	0.052	0.030
300 - 600	IB	0.134	0.030
600 - 1300	IB	0.28	0.034
1300 - 1520	IB	0.00168	0.000126

$^{195}_{80}$Hg(1.73 *3* d)

Mode: IT(54.2 *20* %), ϵ(45.8 *20* %)
Δ: -30894 *50* keV
SpA: 3.86 $\times 10^{5}$ Ci/g
Prod: ^{197}Au(p,3n); ^{194}Pt(^{3}He,2n)

Photons (^{195}Hg)

⟨γ⟩=200 *4* keV

γ_{mode}	γ(keV)	γ(%)†
Au L_ℓ	8.494	0.43 *5*
Hg L_ℓ	8.722	0.86 *9*
Au L_α	9.704	9.0 *8*
Hg L_α	9.980	17.9 *13*
Au L_η	10.309	0.164 *13*
Hg L_η	10.647	0.126 *10*
Au L_β	11.537	10.2 *9*
Hg L_β	11.930	15.8 *17*
Au L_γ	13.455	1.86 *18*
Hg L_γ	14.079	3.2 *4*
γ_{IT} M1+0.06%E2	16.203 *19*	0.155 *20*
γ_{IT} M1+0.1%E2	37.085 *19*	1.85 *6*
γ_{IT} E2	53.288 *20*	0.0092 *7*
γ_ϵE3	56.61 *3*	0.0136 *5* *
γ_ϵM1+17%E2	61.432 *24*	0.091 *15* *
Au $K_{\alpha2}$	66.991	8.0 *7*
Au $K_{\alpha1}$	68.806	13.7 *13*
Hg $K_{\alpha2}$	68.893	1.28 *5*
Hg $K_{\alpha1}$	70.818	2.17 *9*
Au $K_{\beta1}$'	77.859	4.8 *4*
Hg $K_{\beta1}$'	80.124	0.76 *3*
Au $K_{\beta2}$'	80.428	1.27 *12*
Hg $K_{\beta2}$'	82.780	0.208 *9*
γ_ϵ[E1]	90.34 *14*	0.00068 *23*
γ_{IT} M4	122.78 *3*	0.0283 *8*
γ_ϵM1	172.37 *5*	0.057 *8*
γ_ϵE2	200.40 *3*	0.83 *6* *
γ_ϵE2	207.05 *3*	0.39 *8*
γ_ϵM1+21%E2	261.83 *3*	32.4 *11* *
γ_ϵM1	279.06 *7*	0.15 *4*
γ_ϵ[M1]	287.55 *9*	0.0113 *23*
γ_ϵ	308.5 *3*	0.0063 *15*
γ_ϵM4	318.44 *4*	0.0189 *21* *
γ_ϵ	325.10 *20*	0.008 *3*
γ_ϵ	338.19 *10*	0.034 *5*
γ_ϵM1	368.53 *4*	0.355 *14*
γ_ϵM1	386.37 *6*	0.29 *3*
γ_ϵE2	387.91 *4*	2.31 *8*
γ_ϵ[M1+E2]	401.67 *6*	0.016 *4*

Photons (^{195}Hg) (continued)

γ_{mode}	γ(keV)	γ(%)†
γ_ϵE2+~15%M1	419.03 *5*	0.065 *5*
γ_ϵM1	441.54 *15*	0.039 *6*
γ_ϵM1	452.06 *4*	0.219 *14*
γ_ϵ[E1]	461.89 *19*	0.0045 *15*
γ_ϵM1	467.36 *4*	0.303 *17*
γ_ϵ[E1]	518.56 *12*	0.029 *6*
γ_ϵM1	525.77 *4*	0.53 *3*
γ_ϵ	531.88 *16*	0.0010 *4*
γ_ϵ	540.24 *17*	0.0020 *6*
γ_ϵ[M1+E2]	542.45 *11*	0.016 *4*
γ_ϵ(E2)	549.38 *9*	0.053 *5*
γ_ϵ[E2]	556.67 *18*	0.047 *16*
γ_ϵM1	560.28 *3*	7.5
γ_ϵM1+29%E2	575.58 *3*	0.229 *23*
γ_ϵ(M1)	578.13 *17*	0.034 *6*
γ_ϵ	628.29 *15*	0.020 *3*
γ_ϵ	637.8 *3*	0.0060 *23*
γ_ϵ	658.77 *21*	0.0060 *23*
γ_ϵ[E2]	665.44 *5*	0.059 *5*
γ_ϵM1	680.73 *4*	0.233 *23*
γ_ϵ(E2)	693.26 *15*	0.047 *7*
γ_ϵM1	698.15 *6*	0.069 *7*
γ_ϵ	701.1 *6*	0.008 *4*
γ_ϵ	703.4 *6*	0.0038 *15*
γ_ϵ	710.9 *4*	0.0024 *10*
γ_ϵ	720.8 *5*	~0.0010
γ_ϵM1	726.77 *20*	0.051 *5*
γ_ϵM1	749.5 *1*	0.029 *5*
γ_ϵE2	754.90 *6*	0.059 *5*
γ_ϵ	792.02 *20*	0.017 *3*
γ_ϵ	847.25 *13*	0.023 *6*
γ_ϵM1	853.11 *5*	0.28 *3*
γ_ϵ	897.3 *4*	0.011 *3*
γ_ϵ(E2)	899.14 *20*	0.042 *8*
γ_ϵ	946.3 *3*	0.007 *3* ?
γ_ϵM1	961.95 *6*	0.226 *19*
γ_ϵM1	1027.64 *4*	0.122 *14*
γ_ϵ	1040.6 *4*	0.0019 *8*
γ_ϵ	1086.06 *5*	0.051 *8*
γ_ϵM1	1241.01 *5*	0.670 *13*
γ_ϵ	1287.05 *20*	0.0068 *14*

† uncert(syst): 6.6% for ϵ, 6.7% for IT
* with ^{195}Au(30.5 s) in equilib

Atomic Electrons (^{195}Hg)

⟨e⟩=139.6 *16* keV

e_{bin}(keV)	⟨e⟩(keV)	e(%)
4 - 10	0.07	1.8 *7*
12	4.8	40 *5*
13 - 16	6.8	49 *4*
22	7.8	35.3 *13*
23 - 42	7.57	22.7 *5*
43	7.5	17.5 *8*
45	7.1	15.8 *7*
47 - 92	8.61	15.8 *4*
108	9.9	9.2 *3*
109	2.12	1.95 *7*
110	26.2	23.7 *8*
119	4.08	3.42 *12*
120	9.8	8.2 *3*
122	3.72	3.05 *10*
123 - 172	0.81	0.655 *23*
181	20.5	11.3 *5*
186 - 228	0.58	0.300 *18*
238 - 244	0.29	0.122 *14*
247	4.18	1.69 *7*
248 - 297	2.98	1.162 *22*
304 - 338	0.72	0.233 *9*
354 - 402	0.416	0.111 *3*
405 - 454	0.198	0.0446 *21*
455 - 505	2.04	0.425 *18*
507 - 556	0.405	0.075 *3*
557 - 600	0.177	0.0311 *12*
613 - 659	0.0287	0.0046 *3*
662 - 711	0.0250	0.00371 *24*
712 - 755	0.00425	0.00058 *3*

Atomic Electrons (^{195}Hg) (continued)

e_{bin}(keV)	⟨e⟩(keV)	e(%)
767 - 792	0.043	0.0055 *6*
817 - 866	0.0128	0.00153 *14*
881 - 899	0.0280	0.0032 *3*
932 - 962	0.0197	0.00208 *17*
1005 - 1040	0.0060	0.00059 *22*
1072 - 1086	0.0007	~6 $\times 10^{-5}$
1160 - 1206	0.0564	0.00486 *14*
1227 - 1275	0.0121	0.000985 *22*
1284 - 1287	1.6 $\times 10^{-5}$	~1 $\times 10^{-6}$

Continuous Radiation (^{195}Hg)

⟨IB⟩=0.166 keV

E_{bin}(keV)		⟨ ⟩(keV)	(%)
10 - 20	IB	0.00040	0.0033
20 - 40	IB	0.00034	0.00103
40 - 100	IB	0.091	0.134
100 - 300	IB	0.0105	0.0060
300 - 600	IB	0.025	0.0055
600 - 1300	IB	0.039	0.0049
1300 - 1377	IB	1.10 $\times 10^{-5}$	8.3 $\times 10^{-7}$

$^{195}_{81}$Tl(1.16 *5* h)

Mode: ϵ
Δ: -28290 *150* keV
SpA: 1.39 $\times 10^{7}$ Ci/g
Prod: ^{196}Hg(d,3n); protons on Hg; protons on Pb; ^{197}Au(^{3}He,5n)

Photons (^{195}Tl)

⟨γ⟩=1186 *19* keV

γ_{mode}	γ(keV)	γ(%)†
Hg L_ℓ	8.722	1.09 *21*
Hg L_α	9.980	23 *4*
Hg L_η	10.647	0.26 *5*
Hg L_β	11.915	23 *4*
Hg L_γ	14.023	4.7 *9*
γM1+0.1%E2	37.085 *19*	2.7 *3*
Hg $K_{\alpha2}$	68.893	23 *5*
Hg $K_{\alpha1}$	70.818	39 *8*
Hg $K_{\beta1}$'	80.124	14 *3*
Hg $K_{\beta2}$'	82.780	3.8 *8*
γ[M1+E2]	109.74 *4*	~0.06
γ[E2]	131.11 *4*	0.15 *3*
γ[E2]	185.17 *4*	~0.024
γ(E2)	197.12 *10*	0.243 *21*
γM1(+E2)	225.93 *3*	1.28 *9*
γM1(+E2)	242.135 *25*	4.3 *3*
γM1(+E2)	247.30 *3*	1.29 *8*
γ	252.9 *25*	~0.019
γM1(+E2)	263.50 *3*	0.88 *5*
γE2+35%M1	279.22 *3*	3.7 *3*
γ[M1]	294.91 *4*	0.14 *4*
γE2	300.59 *3*	2.40 *16*
γ[M1+E2]	316.28 *4*	~0.09
γ[M1+E2]	320.95 *5*	0.17 *5*
γ[M1+E2]	321.43 *4*	~0.05
γ	325.62 *10*	~0.021
γ[M1+E2]	326.11 *5*	0.15 *4*
γM1	357.04 *3*	0.46 *4*
γ	369.24 *9*	0.21 *3*
γ(E2)	373.24 *3*	0.60 *4*
γ[M1+E2]	396.36 *6*	~0.07
γ[M1+E2]	403.93 *5*	0.12 *3*

Photons (^{195}Tl)
(continued)

γ_{mode}	γ(keV)	γ(%)†
γ(M1)	408.35 *10*	~0.06
γ(M1)	409.09 *5*	0.16 *3*
γ	426 *1*	
γ	456.37 *10*	0.11 *3*
γ[E2]	463.97 *5*	0.11 *3*
γ	471.7 *4*	0.13 *5*
γ	481.46 *9*	0.18 *5*
γ[E2]	482.81 *6*	0.35 *4*
γ(M1+E2)	485.34 *5*	0.43 *4*
γ	501.50 *15*	0.08 *3*
γ[E2]	511.28 *5*	0.48 *16*
γ(M1+E2)	542.21 *3*	1.05 *8*
γ[M1]	544.12 *6*	0.28 *4*
γ	544.86 *10*	~0.018
γM1	547.36 *3*	0.67 *4*
γM1	558.41 *3*	2.55 *14*
γM1	563.57 *3*	10.5 *5*
γ	572.21 *15*	0.075 *21*
γ	582.06 *10*	0.11 *3*
γ	586.21 *6*	~0.038
γM1	592.55 *5*	1.20 *13*
γ[E2]	594.26 *5*	0.22 *3*
γ[M1+E2]	595.50 *4*	0.09 *3*
γ[M1]	600.65 *3*	0.65 *9*
γM1	613.92 *5*	0.71 *8*
γE2	621.02 *4*	0.22 *3*
γ	628.0 *3*	~0.033
γ[M1+E2]	642.39 *4*	0.15 *3*
γ(M1)	655.56 *7*	0.46 *7*
γ	656.91 *7*	0.17 *5*
γ	658.89 *13*	~0.033
γ	663.3 *3*	0.06 *3*
γ	675.59 *11*	0.06 *3*
γ[M1+E2]	704.00 *4*	0.187 *21*
γ[E2]	711.27 *5*	0.80 *4*
γ(M1)	725.36 *4*	0.63 *8*
γ(M1)	727.47 *5*	0.64 *8*
γ[M1]	733.97 *6*	0.64 *5*
γ	738.25 *10*	~0.041
γ	741.9 *2*	0.064 *16*
γ	755.79 *20*	0.101 *21*
γ	761.44 *4*	0.17 *3*
γE0+M1	764.56 *5*	0.41 *4*
γ[M1]	777.65 *4*	1.11 *16*
γ	784.14 *7*	0.08 *3*
γ	788.02 *7*	0.092 *21*
γ	793.08 *14*	0.039 *16*
γ[M1+E2]	800.29 *6*	~0.039
γ[M1+E2]	805.45 *6*	0.18 *4*
γM1	814.73 *4*	1.89 *10*
γ	821.23 *10*	~0.06
γ	821.55 *14*	0.10 *3*
γ	835.04 *17*	0.07 *3*
γ	849.37 *10*	~0.048
γ	849.69 *10*	~0.13
γ[M1+E2]	856.05 *5*	0.30 *3*
γ	861.13 *9*	0.152 *21*
γ[M1+E2]	868.32 *4*	~0.040
γ	871.61 *14*	0.048 *21*
γ	882.79 *12*	~0.09
γM1	884.52 *3*	10.0 *5*
γ(M1)	893.14 *5*	0.89 *6*
γ	899.95 *14*	0.034 *16*
γM1	921.61 *3*	2.25 *13*
γ	928.10 *10*	0.37 *4*
γ	948.05 *6*	~0.08
γ[M1+E2]	951.30 *4*	0.18 *5*
γ	953.20 *6*	~0.05
γM1	967.50 *4*	2.14 *12*
γ	980.32 *13*	0.125 *16*
γ	991.5 *4*	~0.042
γ	999.87 *20*	0.081 *21*
γ[M1+E2]	1004.58 *4*	0.152 *21*
γ	1010.11 *7*	0.33 *4*
γ	1013.95 *7*	~0.046
γ	1052.83 *12*	~0.046
γ	1053.78 *15*	~0.06
γ	1063.45 *9*	0.13 *3*
γ	1067.24 *7*	0.40 *4*
γ	1087.06 *9*	0.057 *16*
γ	1093.09 *7*	0.200 *21*
γ	1100.36 *5*	2.34 *13*
γ	1103.27 *9*	~0.13
γ	1114.03 *14*	0.10 *3*

Photons (^{195}Tl)
(continued)

γ_{mode}	γ(keV)	γ(%)†
γ	1121.55 *8*	0.32 *6*
γM1+E2	1121.72 *5*	2.1 *3*
γ	1140.35 *9*	0.54 *8*
γ	1142.18 *10*	0.047 *21*
γ	1164.27 *12*	0.09 *3*
γ	1193.09 *25*	0.10 *3*
γ	1199.96 *8*	0.14 *3*
γ	1210.83 *15*	0.10 *3*
γ	1216.50 *11*	0.23 *3*
γ	1222.45 *13*	0.059 *21*
γ	1242.68 *11*	0.082 *21*
γ	1248.11 *6*	0.16 *3*
γ	1250.4 *4*	~0.048
γ	1259.54 *13*	0.068 *21*
γM1	1269.48 *6*	2.44 *13*
γ	1279.5 *4*	~0.037
γ	1288.47 *15*	0.101 *21*
γ	1315.5 *4*	0.085 *21*
γ	1326.24 *25*	0.100 *21*
γ	1337.03 *14*	~0.042
γ[E2]	1347.65 *5*	1.17 *6*
γ	1362.13 *12*	~0.15
γ	1363.52 *9*	~0.12
γM1	1363.86 *5*	8.4 *4*
γ	1374.73 *14*	0.095 *16*
γ	1383.44 *11*	0.195 *21*
γ	1390.60 *12*	0.34 *3*
γ	1394.2 *5*	~0.049
γ(E0+M1)	1400.94 *5*	0.066 *16*
γ	1411.91 *12*	0.07 *3*
γ	1415.72 *14*	0.08 *3*
γ	1419.19 *7*	~0.038
γ	1435.46 *14*	0.62 *5*
γ	1443.5 *4*	~0.10
γ	1447.3 *4*	~0.10
γ	1456.76 *11*	0.15 *5*
γ	1461.91 *11*	~0.05
γ	1487.6 *5*	0.08 *4*
γ	1490.32 *11*	0.47 *4*
γ	1511.61 *5*	1.45 *7*
γ	1519.17 *12*	~0.034
γ	1530.81 *15*	0.165 *16*
γ	1548.4 *3*	0.21 *6*
γ	1548.70 *5*	0.13 *4*
γ	1552.17 *15*	0.26 *3*
γ	1560.5 *5*	0.048 *16*
γ	1584.3 *7*	~0.044
γ	1588.4 *3*	0.11 *3*
γ	1591.65 *10*	0.21 *3*
γ	1609.6 *7*	~0.05
γ	1613.94 *24*	0.10 *3*
γ	1620.11 *10*	~0.07
γ	1627.01 *9*	0.58 *5*
γ	1656.7 *7*	0.08 *3*
γ	1660.11 *23*	0.18 *4*
γ	1664.10 *9*	0.59 *6*
γ	1668.2 *7*	~0.06
γ	1674.80 *15*	~0.06
γ	1677.60 *14*	0.09 *3*
γ	1688.24 *12*	0.16 *5*
γ	1689.54 *9*	0.16 *5*
γ	1696.16 *15*	0.10 *4*
γ	1698.52 *10*	~0.06
γ	1705.74 *9*	1.60 *8*
γ	1714.10 *7*	0.59 *8*
γ	1714.68 *14*	0.39 *9*
γ	1726.7 *7*	0.052 *21*
γ	1735.47 *7*	0.108 *21*
γ	1742.83 *9*	0.30 *3*
γ	1756.82 *11*	0.32 *3*
γ	1772.8 *7*	~0.06
γ	1778.10 *15*	0.25 *6*
γ	1778.19 *11*	0.78 *11*
γ	1794.31 *15*	0.41 *3*
γ	1824.81 *14*	0.09 *3*
γ	1833.33 *15*	0.047 *16*
γ	1842.19 *19*	0.44 *3*
γ	1856.07 *24*	0.10 *3*
γ	1859.7 *7*	0.07 *3*
γ	1871.9 *7*	0.043 *16*
γ	1879.27 *19*	0.045 *16*
γ	1893.16 *24*	0.054 *21*
γ	1902.6 *5*	~0.042
γ	1907.60 *19*	~0.036

Photons (^{195}Tl)
(continued)

γ_{mode}	γ(keV)	γ(%)†
γ	1912.60 *9*	0.14 *3*
γ	1917.75 *9*	0.091 *21*
γ	1929.58 *13*	0.077 *16*
γ	1944.4 *5*	0.039 *16*
γ	1950.95 *13*	0.183 *21*
γ	1961.40 *7*	0.107 *21*
γ	1969.1 *5*	~0.033
γ	1977.60 *7*	1.76 *8*
γ	2004.12 *11*	0.17 *4*
γ	2004.51 *12*	0.13 *4*
γ	2014.69 *7*	0.89 *4*
γ	2020.32 *11*	0.10 *3*
γ	2025.82 *11*	0.37 *3*
γ	2032.3 *3*	0.129 *16*
γ	2057.40 *11*	0.32 *3*
γ	2060.4 *7*	0.10 *3*
γ	2083.9 *3*	0.084 *21*
γ	2088.0 *7*	~0.024
γ	2101.4 *4*	0.105 *21*
γ	2105.3 *5*	~0.032
γ	2119.72 *14*	0.093 *21*
γ	2123.9 *3*	~0.025
γ	2141.08 *13*	0.48 *4*
γ	2147.4 *7*	~0.05
γ	2149.60 *15*	0.19 *3*
γ	2172.3 *7*	~0.046
γ	2177.7 *3*	0.20 *3*
γ	2193.08 *13*	0.055 *21*
γ	2202.32 *23*	~0.07
γ	2207.67 *19*	~0.040
γ	2209.8 *7*	0.09 *3*
γ	2212.66 *9*	0.21 *3*
γ	2218.52 *23*	0.087 *21*
γ	2229.03 *19*	0.090 *21*
γ	2234.03 *9*	0.244 *21*
γ	2251.75 *11*	0.11 *3*
γ	2255.60 *23*	0.22 *3*
γ	2267.96 *11*	0.201 *21*
γ	2274.4 *3*	0.15 *3*
γ	2283.73 *11*	<0.6 ?
γ	2285.1 *3*	<0.6
γ	2286.4 *7*	<0.6 ?
γ	2301.3 *3*	0.124 *16*
γ	2322.8 *7*	0.037 *10*
γ	2334.8 *7*	~0.023
γ	2338.4 *3*	0.072 *16*
γ	2363.1 *3*	0.32 *5*
γ	2366.0 *3*	0.54 *5*
γ	2383.21 *14*	0.44 *4*
γ	2388.0 *5*	0.17 *3*
γ	2391.73 *15*	0.26 *3*
γ	2419.8 *3*	<0.15 ?
γ	2420.30 *14*	<0.15
γ	2425.0 *5*	0.058 *21*
γ	2428.82 *15*	~0.018
γ	2452.9 *7*	~0.032
γ	2456.9 *3*	0.109 *21*
γ	2459.96 *9*	~0.041
γ	2471.16 *19*	0.151 *16*
γ	2476.16 *9*	0.176 *16*
γ	2488.1 *5*	0.057 *7*
γ	2505.6 *7*	~0.045
γ	2508.25 *19*	0.09 *3*
γ	2513.24 *9*	0.47 *3*
γ	2520.5 *5*	0.032 *7*
γ	2533.1 *5*	0.043 *7*
γ	2557.1 *5*	0.026 *7*
γ	2565.3 *7*	0.044 *11*
γ	2569.4 *7*	0.052 *11*
γ	2599.4 *5*	~0.015
γ	2606.1 *5*	0.024 *7*
γ	2632.0 *7*	0.020 *7*
γ	2645.5 *7*	0.020 *7*
γ	2686.7 *7*	0.015 *5*
γ	2811.2 *7*	0.014 *5*
γ	2854.7 *7*	0.015 *5*
γ	2959.0 *7*	0.015 *5*

† 9.5% uncert(syst)

Atomic Electrons (^{195}Tl)

⟨e⟩=55 *3* keV

e_{bin}(keV)	⟨e⟩(keV)	e(%)
4 - 9	0.07	1.1 *5*
12	3.5	28 *7*
13	0.29	2.3 *3*
14	2.5	17 *4*
15	2.3	15.6 *25*
16 - 21	0.079	0.45 *14*
22	11.6	52 *6*
23	1.30	5.7 *7*
25 - 27	0.33	1.3 *3*
34	4.6	13.8 *16*
35 - 80	3.44	7.5 *5*
95 - 143	1.2	0.9 *4*
159	2.3	~1
164 - 195	1.2	0.7 *3*
196	1.33	0.68 *9*
197 - 245	2.4	1.07 *21*
246 - 295	1.50	0.56 *5*
297 - 345	0.35	0.109 *12*
353 - 402	0.35	0.091 *20*
403 - 452	0.068	0.016 *4*
453 - 479	1.26	0.27 *3*
480	2.94	0.61 *3*
481 - 530	0.63	0.123 *10*
531 - 580	1.39	0.252 *9*
581 - 630	0.198	0.033 *3*
639 - 681	0.78	0.118 *8*
689 - 738	0.68	0.095 *6*
739 - 788	0.20	0.026 *4*
789 - 800	0.067	0.0084 *13*
801	1.46	0.183 *10*
802 - 850	0.51	0.061 *4*
852 - 900	0.68	0.078 *3*
907 - 955	0.178	0.0192 *22*
964 - 1013	0.085	0.0086 *24*
1014 - 1063	0.33	0.032 *8*
1064 - 1112	0.074	0.0068 *14*
1113 - 1162	0.051	0.0045 *12*
1163 - 1213	0.241	0.0203 *12*
1214 - 1260	0.051	0.0041 *4*
1265 - 1314	0.76	0.059 *3*
1315 - 1364	0.65	0.049 *9*
1369 - 1417	0.13	0.0093 *22*
1418 - 1466	0.09	0.006 *3*
1469 - 1558	0.079	0.0052 *12*
1569 - 1665	0.18	0.011 *3*
1671 - 1770	0.11	0.0065 *16*
1773 - 1869	0.037	0.0020 *4*
1876 - 1975	0.13	0.0067 *19*
1977 - 2076	0.044	0.0022 *5*
2080 - 2179	0.043	0.0020 *4*
2181 - 2280	0.057	0.0026 *7*
2281 - 2380	0.051	0.0022 *5*
2384 - 2482	0.028	0.0012 *3*
2485 - 2567	0.0062	0.00025 *5*
2585 - 2684	0.00067	2.6 *8* $\times 10^{-5}$
2728 - 2808	0.00057	2.1 *8* $\times 10^{-5}$
2840 - 2876	0.00031	1.1 *5* $\times 10^{-5}$
2944 - 2956	5.2 $\times 10^{-5}$	1.8 *8* $\times 10^{-6}$

Continuous Radiation (^{195}Tl)

⟨β+⟩=20.8 keV;⟨IB⟩=1.69 keV

E_{bin}(keV)		⟨ ⟩(keV)	(%)
0 - 10	β+	7.3 $\times 10^{-8}$	8.6 $\times 10^{-7}$
	IB	0.00098	
10 - 20	β+	4.93 $\times 10^{-6}$	2.92 $\times 10^{-5}$
	IB	0.0027	0.021
20 - 40	β+	0.000222	0.00067
	IB	0.0031	0.0101
40 - 100	β+	0.0174	0.0219
	IB	0.46	0.65
100 - 300	β+	0.93	0.426
	IB	0.066	0.038
300 - 600	β+	5.07	1.12
	IB	0.157	0.035
600 - 1300	β+	13.1	1.50
	IB	0.59	0.065
1300 - 2500	β+	1.75	0.124
	IB	0.42	0.026
2500 - 2743	IB	0.00133	5.2 $\times 10^{-5}$
	Σβ+		3.2

$^{195}_{81}$Tl(3.6 *4* s)

Mode: IT

Δ: -27807 *150* keV

SpA: 1.46$\times 10^{10}$ Ci/g

Prod: daughter ^{195}Pb(15 min); ^{187}Re(^{12}C,4n)

Photons (^{195}Tl)

⟨γ⟩=358 *14* keV

γ_{mode}	γ(keV)	γ(%)†
Tl L_ℓ	8.953	0.51 *6*
Tl L_α	10.259	10.4 *8*
Tl L_η	10.994	0.287 *21*
Tl L_β	12.276	15.5 *12*
Tl L_γ	14.354	3.1 *3*
Tl $K_{\alpha2}$	70.832	1.9 *3*
Tl $K_{\alpha1}$	72.873	3.1 *4*
Tl $K_{\beta1}'$	82.434	1.11 *15*
Tl $K_{\beta2}'$	85.185	0.31 *4*
γ E3	98.97 *16*	0.617 *25*
γ E2+24%M1	383.66 *12*	91 *4*

† 20% uncert(syst)

Atomic Electrons (^{195}Tl)

⟨e⟩=121 *4* keV

e_{bin}(keV)	⟨e⟩(keV)	e(%)
13 - 60	6.3	45 *4*
66 - 82	0.059	0.085 *6*
84	36.1	42.9 *19*
86	25.4	29.4 *13*
95	0.416	0.436 *20*
96	19.1	19.9 *9*
97	0.369	0.382 *17*
98	5.09	5.17 *23*
99	1.21	1.22 *6*
298	18.9	6.3 *9*
368	3.5	0.94 *14*
369 - 384	4.66	1.24 *6*

$^{195}_{82}$Pb(~15 min)

Mode: ϵ

Δ: -23770 *300* keV syst

SpA: ~6$\times 10^{7}$ Ci/g

Prod: ^{203}Tl(p,9n); ^{181}Ta(^{19}F,5n)

Photons (^{195}Pb)

γ_{mode}	γ(keV)	γ(rel)
γ(M1)	344.99 *21*	2.64 *21*
γE2+24%M1	383.66 *12*	100
γ[M1+E2]	393.8 *3*	7 *3*
γ[E2]	777.4 *3*	5.8 *18*
γM1	835.3 *3*	2.8 *3*
γM1(+E2)	871.2 *4*	2.6 *4*
γ(E2)	884.1 *5*	3.7 *16*
γ	1304.6 *6*	0.69 *19*

$^{195}_{82}$Pb(15.0 *12* min)

Mode: ϵ

Δ: ~-23570 *300* keV syst

SpA: 6.4$\times 10^{7}$ Ci/g

Prod: ^{203}Tl(p,9n); ^{181}Ta(^{19}F,5n)

Photons (^{195}Pb)

γ_{mode}	γ(keV)	γ(rel)
γE3	98.97 *16*	0.73 *6* *
γM1	236.90 *16*	0.041 *4*
γM1	305.74 *15*	0.85 *9*
γM1+13%E2	313.40 *10*	6.5 *4*
γM1+E2	325.89 *14*	0.62 *7*
γE2+24%M1	383.66 *12*	100 *2* *
γ(M1+E2)	392.7 *4*	0.7 *3*
γM1+15%E2	394.13 *10*	41
γM1	419.81 *22*	0.59 *7*
γM1+11%E2	428.62 *10*	4.24 *25*
γM1	442.62 *15*	0.77 *8*
γ	533.95 *19*	0.49 *25*
γM1	533.95 *13*	~2
γM1	539.5 *2*	1.24 *11*
γM1+30%E2	607.3 *3*	7.8 *9*
γM1(+E2)	630.53 *16*	3.05 *25*
γ	672.6 *9*	0.66 *21*
γM1	691.19 *14*	2.72 *21*
γE2	707.54 *12*	13.1 *8*
γ	717.4 *4*	0.53 *10*
γE2	734.36 *15*	1.5 *4*
γ	739.68 *17*	0.36 *6*
γE2	742.02 *11*	3.95 *25*
γM1	754.50 *16*	0.78 *16*
γM1	801.33 *18*	1.89 *16*
γE2	815.3 *3*	2.3 *6*
γE2	821.3 *4*	~0.6
γ(E1)	847.1 *4*	~0.7
γM1(+E2)	848.53 *18*	2.8 *7*
γ	877.9 *4*	0.9 *3*
γM1(+E2)	878.34 *18*	22.6 *15*
γ	889.9 *9*	<3
γM1	912.7 *3*	1.45 *10*
γ(E2)	928.09 *13*	3.7 *5*
γ(M1+E2)	937.9 *4*	0.79 *12*
γ	979.1 *3*	1.44 *16*
γ[E2]	1001.5 *3*	1.29 *11*
γM1(+E2)	1067.90 *15*	5.9 *4*
γ	1133.82 *16*	0.76 *8*
γ	1242.66 *19*	0.31 *7* ?
γ	1630.5 *7*	0.92 *12*
γ	1929.8 *7*	1.09 *7*

* with ^{195}Tl(3.6 s) in equilib

$^{195}_{83}$Bi(2.8 *3* min)

Mode: α(<0.2 %)
Δ: -17980 *320* keV syst
SpA: 3.4×10^8 Ci/g
Prod: ^{181}Ta(^{20}Ne,6n); ^{185}Re(^{16}O,6n)

Alpha Particles (^{195}Bi)

α(keV)
5450 *20*

$^{195}_{83}$Bi(1.50 *8* min)

Mode: α(4 %)
Δ: -17980 *320* keV syst
SpA: 6.4×10^8 Ci/g
Prod: deuterons on Pb; ^{203}Tl(^{3}He,11n); ^{181}Ta(^{20}Ne,6n); ^{159}Tb(^{40}Ar,4n); ^{185}Re(^{16}O,6n)

Alpha Particles (^{195}Bi)

α(keV)
6110 *10*

$^{195}_{84}$Po(4.5 *5* s)

Mode: α
Δ: -11170 *220* keV syst
SpA: 1.19×10^{10} Ci/g
Prod: ^{185}Re(^{19}F,9n)

Alpha Particles (^{195}Po)

α(keV)
6609 *5*

$^{195}_{84}$Po(2.0 *2* s)

Mode: α
Δ: -11170 *220* keV syst
SpA: 2.44×10^{10} Ci/g
Prod: ^{185}Re(^{19}F,9n)

Alpha Particles (^{195}Po)

α(keV)
6699 *5*

A = 196

NDS **28**, 485 (1979)

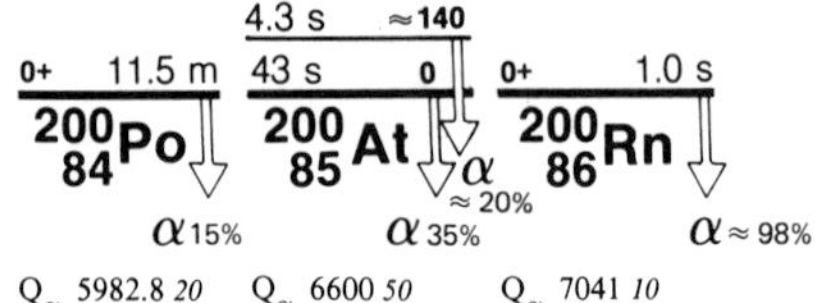

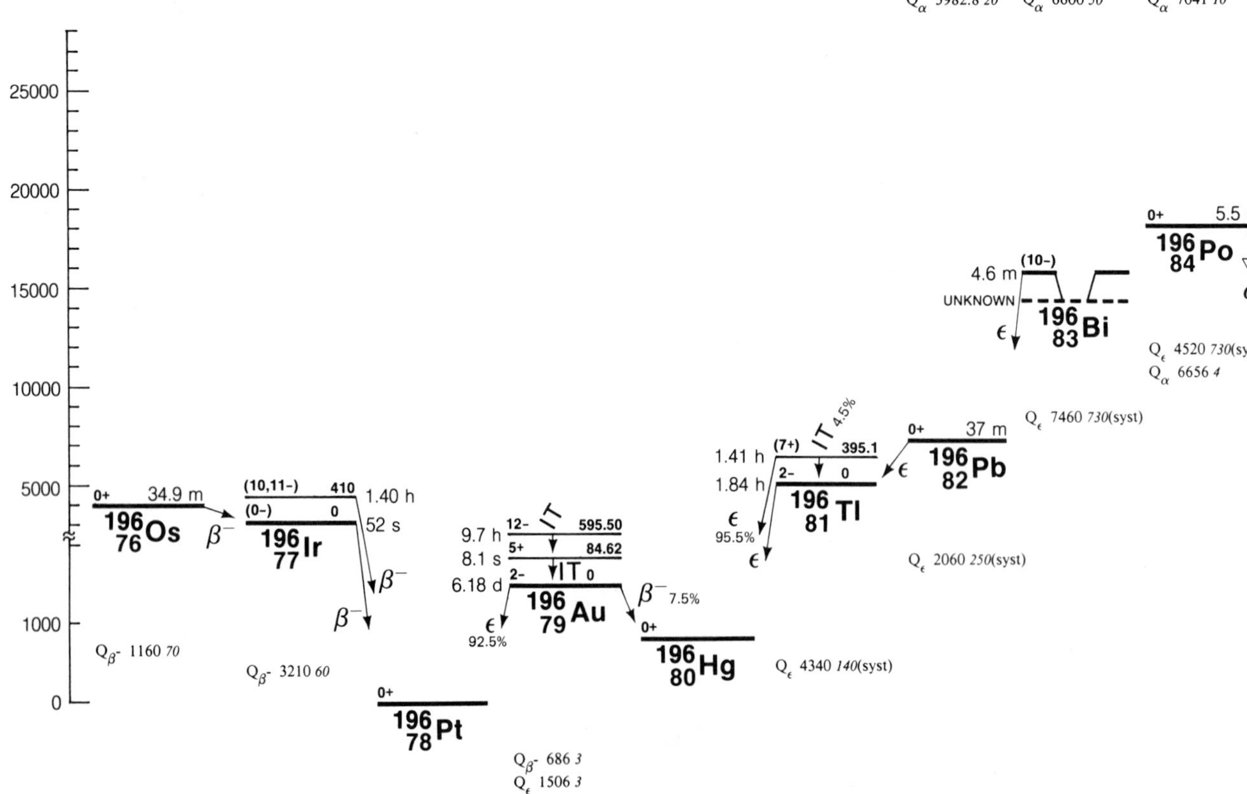

$^{196}_{76}$Os(34.9 *2* min)

Mode: β-

Δ: -28300 *40* keV

SpA: 2.748×10^{7} Ci/g

Prod: ^{198}Pt(n,2pn)

Photons (^{196}Os)

⟨γ⟩=69.5 *15* keV

γ_{mode}	γ(keV)	γ(%)†
γ	126.05 *20*	5.3 *3*
γ[E1]	200.75 *20*	0.56 *5*
γ	206.95 *16*	2.4 *1*
γ	257.65 *20*	2.3 *1*
γ	307.9 *4*	0.43 *8*
γ[E1]	315.20 *17*	2.5 *1*
γ[E1]	407.70 *17*	5.9 *2*
γ[E1]	522.15 *20*	0.78 *10*
γ	586.05 *20*	0.59 *13*
γ	629.0 *4*	1.6 *1*

† 5.0% uncert(syst)

$^{196}_{77}$Ir(52 *2* s)

Mode: β-

Δ: -29460 *60* keV

SpA: 1.10×10^{9} Ci/g

Prod: ^{196}Pt(n,p)

Photons (^{196}Ir)

⟨γ⟩=224 *15* keV

γ_{mode}	γ(keV)	γ(%)†
Pt L_{ℓ}	8.266	0.0091 *15*
Pt L_{α}	9.435	0.20 *3*
Pt L_{η}	9.975	0.0031 *5*
Pt L_{β}	11.177	0.21 *3*
Pt L_{γ}	13.029	0.037 *6*
Pt $K_{\alpha2}$	65.122	0.33 *4*
Pt $K_{\alpha1}$	66.831	0.57 *7*
Pt $K_{\beta1}$'	75.634	0.20 *3*
Pt $K_{\beta2}$'	78.123	0.052 *7*
γ M1+3.7%E2+E0	332.87 *5*	4.35 *19*
γ E2	355.58 *5*	19
γ E2	446.7 *3*	4.5 *4*
γ (E2)	688.45 *7*	0.0012 *3*
γ E2	779.6 *3*	10.4 *4*
γ[E2]	1047.1 *10*	0.95 *19*
γ[E2]	1231 *1*	0.34 *8*
γ[E2]	1468.5 *11*	0.83 *13*
γ[E2]	1563.9 *10*	0.89 *19*

† 6.3% uncert(syst)

Atomic Electrons (^{196}Ir)

⟨e⟩=5.9 *5* keV

e_{bin}(keV)	⟨e⟩(keV)	e(%)
12 - 55	0.114	0.84 *9*
61 - 75	0.0106	0.0166 *12*
254	0.65	0.256 *13*
277	2.1	0.76 *15*
319 - 333	0.388	0.120 *3*
342	0.79	0.23 *5*

Atomic Electrons (^{196}Ir) (continued)

e_{bin}(keV)	⟨e⟩(keV)	e(%)
344	0.21	0.062 *12*
352	0.083	0.024 *5*
353	0.17	0.049 *10*
355 - 356	0.077	0.022 *4*
368	0.40	0.108 *9*
433 - 447	0.187	0.0428 *24*
675 - 688	2.1 ×10^{-5}	3.1 *11* ×10^{-6}
701	0.529	0.075 *3*
766 - 780	0.153	0.0200 *6*
1033 - 1047	0.0093	0.00090 *12*
1217 - 1231	0.0028	0.00023 *3*
1455 - 1552	0.036	0.0024 *4*
1561	0.00101	6.5 *14* ×10^{-5}

Continuous Radiation (^{196}Ir)

⟨β-⟩=1170 keV;⟨IB⟩=3.0 keV

E_{bin}(keV)		⟨ ⟩(keV)	(%)
0 - 10	β-	0.0127	0.252
	IB	0.041	
10 - 20	β-	0.0385	0.256
	IB	0.040	0.28
20 - 40	β-	0.159	0.53
	IB	0.079	0.27
40 - 100	β-	1.21	1.71
	IB	0.22	0.35
100 - 300	β-	14.4	7.0
	IB	0.64	0.36
300 - 600	β-	62	13.6
	IB	0.70	0.166
600 - 1300	β-	343	36.2
	IB	0.90	0.106
1300 - 2500	β-	664	37.3
	IB	0.37	0.023
2500 - 3210	β-	86	3.20
	IB	0.0062	0.00023

$^{196}_{77}$Ir(1.40 *2* h)

Mode: β-

Δ: -29050 *130* keV

SpA: 1.142×10^{7} Ci/g

Prod: ^{198}Pt(d,α)

Photons (^{196}Ir)

⟨γ⟩=2465 *100* keV

γ_{mode}	γ(keV)	γ(%)†
Pt L_{ℓ}	8.266	0.40 *6*
Pt L_{α}	9.435	8.6 *10*
Pt L_{η}	9.975	0.178 *21*
Pt L_{β}	11.162	10.6 *13*
Pt L_{γ}	12.995	1.90 *24*
Pt $K_{\alpha2}$	65.122	5.9 *4*
Pt $K_{\alpha1}$	66.831	10.1 *7*
Pt $K_{\beta1}$'	75.634	3.54 *24*
Pt $K_{\beta2}$'	78.123	0.92 *6*
γ E2	103.17 *19*	16.3 *19*
γ	340.6 *3*	1.54 *19*
γ E2	355.58 *5*	94 *3*
γ E1	393.32 *15*	97.0 *19*
γ	420.62 *25*	2.50 *10*
γ E2	446.96 *18*	94.1 *19*
γ E2	521.14 *13*	96
γ[E2]	552.63 *25*	0.63 *4*

Photons (^{196}Ir) (continued)

γ_{mode}	γ(keV)	γ(%)†
γ	566.2 *5*	0.29 *10*
γ	615.7 *5*	0.42 *5*
γ	633.41 *18*	1.10 *5*
γ E2	647.12 *16*	91 *3*
γ	673.7 *3*	0.17 *4*
γ	693.76 *19*	4.2 *3*
γ	721.9 *3*	0.64 *7*
γ	727.13 *19*	2.59 *10*
γ	760.4 *3*	0.75 *5*
γ	835.39 *17*	6.34 *19*
γ	849.09 *21*	0.51 *5*
γ	867.9 *4*	0.46 *4*
γ	886.8 *7*	0.106 *19*
γ	892.8 *7*	0.192 *19*
γ	904.4 *7*	0.096 *19*
γ	914.46 *18*	0.29 *5*
γ	925.8 *7*	0.067 *19*
γ	1024.23 *25*	0.25 *3*
γ	1067.7 *3*	0.071 *17*
γ	1080.37 *22*	0.115 *19*
γ	1116.5 *11*	0.070 *21*
γ	1281.4 *7*	0.059 *23*
γ	1340.88 *23*	0.28 *3*
γ	1355.4 *3*	0.058 *10*
γ	1393.8 *3*	0.071 *14*
γ	1482.50 *19*	2.30 *19*

† 3.1% uncert(syst)

Atomic Electrons (^{196}Ir)

⟨e⟩=119 *4* keV

e_{bin}(keV)	⟨e⟩(keV)	e(%)
12 - 55	7.9	51 *4*
61 - 89	1.55	1.82 *18*
90	21.4	24 *3*
92	18.3	20.0 *24*
100	6.4	6.4 *8*
101	5.4	5.3 *6*
102 - 103	3.5	3.4 *4*
254 - 262	0.4	~0.17
277	10.5	3.80 *14*
315	3.57	1.13 *3*
319 - 341	0.15	~0.05
342	4.4	1.29 *16*
344 - 356	2.73	0.780 *15*
369	8.21	2.23 *6*
379 - 421	1.04	0.269 *23*
433 - 435	2.91	0.672 *12*
443	7.1	1.6 *3*
444 - 488	1.04	0.232 *8*
507 - 555	3.1	0.60 *6*
563 - 566	0.0024	~0.0004
569	5.49	0.97 *4*
581 - 630	0.5	~0.08
631 - 680	2.20	0.34 *3*
682 - 727	0.17	0.024 *9*
747 - 789	0.6	~0.08
808 - 856	0.18	0.022 *9*
865 - 914	0.014	0.0015 *4*
922 - 946	0.015	~0.0016
989 - 1038	0.019	0.0019 *7*
1054 - 1103	0.0032	0.00030 *10*
1105 - 1116	0.00020	~2×10^{-5}
1262 - 1281	0.015	~0.0011
1315 - 1355	0.0058	0.00044 *18*
1380 - 1404	0.08	~0.006
1469 - 1483	0.018	~0.0012

Continuous Radiation (^{196}Ir)

⟨β-⟩=337 keV;⟨IB⟩=0.37 keV

E_{bin}(keV)		⟨ ⟩(keV)	(%)
0 - 10	β-	0.106	2.12
	IB	0.0157	
10 - 20	β-	0.314	2.10
	IB	0.0150	0.104
20 - 40	β-	1.24	4.14
	IB	0.028	0.098
40 - 100	β-	8.2	11.8
	IB	0.071	0.111
100 - 300	β-	61	31.3
	IB	0.145	0.085
300 - 600	β-	137	31.2
	IB	0.082	0.020
600 - 1152	β-	130	17.3
	IB	0.0161	0.0022

$^{196}_{78}$Pt(stable)

Δ: -32671 *4* keV

%: 25.3 *5*

$^{196}_{79}$Au(6.183 *10* d)

Mode: ε(92.5 *5* %), β-(7.5 *5* %)

Δ: -31165 *5* keV

SpA: 1.0775×10^5 Ci/g

Prod: ^{196}Pt(d,2n); ^{196}Pt(p,n); ^{195}Pt(d,n); ^{193}Ir(α,n); ^{197}Au(n,2n)

Photons (^{196}Au)

⟨γ⟩=471 *62* keV

γ_{mode}	γ(keV)	γ(%)†
Pt L_ℓ	8.266	0.51 *6*
Hg L_ℓ	8.722	0.00169 *19*
Pt L_α	9.435	11.0 *9*
Pt L_η	9.975	0.140 *12*
Hg L_α	9.980	0.035 *3*
Hg L_η	10.647	0.00056 *5*
Pt L_β	11.189	10.4 *10*
Hg L_β	11.918	0.037 *3*
Pt L_γ	13.056	1.84 *20*
Hg L_γ	13.924	0.0068 *7*
Pt $K_{\alpha2}$	65.122	20.5 *13*
Pt $K_{\alpha1}$	66.831	35.0 *22*
Hg $K_{\alpha2}$	68.893	0.056 *4*
Hg $K_{\alpha1}$	70.818	0.095 *6*
Pt $K_{\beta1}$'	75.634	12.3 *8*
Pt $K_{\beta2}$'	78.123	3.19 *21*
Hg $K_{\beta1}$'	80.124	0.0334 *22*
Hg $K_{\beta2}$'	82.780	0.0091 *6*
γ_ϵ[M1]	326.43 *21*	0.050 *11*
γ_ϵM1+3.7%E2+E0	332.87 *5*	22.9 *5*
γ_ϵE2	355.58 *5*	87
γ_ϵE1	393.32 *15*	0.0101 *5*
$\gamma_{\beta-}$E2	425.64 *7*	7.2
γ_ϵ	432.03 *22*	0.0067 *6*
γ_ϵE2	521.14 *13*	0.389 *9*
γ_ϵ(E1+M2)	570.19 *19*	0.0069 *5*
γ_ϵ(M1)	659.3 *2*	0.0037 *3*
γ_ϵ[E2]	672.8 *3*	0.0027 *3*
γ_ϵ(E2)	688.45 *7*	0.0061 *17*
γ_ϵE1	758.46 *16*	0.0443 *17*
γ_ϵ	914.46 *18*	3.0 *5* $\times10^{-5}$
γ_ϵ[M1+E2]	1005.6 *3*	0.0027 *3*
γ_ϵE1	1091.33 *15*	0.149 *6*

Photons (^{196}Au)
(continued)

γ_{mode}	γ(keV)	γ(%)†
γ_ϵE2	1361.2 *3*	0.00043 *17*
γ_ϵ[E3]	1446.90 *16*	0.00078 *17*

† uncert(syst): 1.2% for ε, 6.8% for β-

Atomic Electrons (^{196}Au)

⟨e⟩=31.0 *23* keV

e_{bin}(keV)	⟨e⟩(keV)	e(%)
12	3.1	27 *3*
13	1.79	13.5 *16*
14 - 63	1.82	6.3 *5*
64 - 80	0.38	0.58 *4*
248	0.028	0.011 *3*
254	3.65	1.43 *5*
277	10.4	3.7 *8*
313 - 333	2.18	0.677 *12*
342	3.9	1.13 *23*
343	0.69	0.202 *13*
344	1.05	0.30 *6*
352	0.41	0.116 *23*
353	0.85	0.24 *5*
354 - 393	0.38	0.107 *19*
411 - 443	0.411	0.099 *4*
492 - 521	0.0148	0.0029 *5*
556 - 594	0.0016	0.00028 *8*
610 - 659	0.00058	9.3 *18* $\times10^{-5}$
661 - 688	0.00110	0.000162 *7*
745 - 758	0.000206	2.76 *9* $\times10^{-5}$
901 - 927	0.00021	2.3 *10* $\times10^{-5}$
992 - 1013	0.00251	0.000248 *11*
1077 - 1091	0.000499	4.62 *15* $\times10^{-5}$
1347 - 1369	5.4 $\times10^{-5}$	4.0 *8* $\times10^{-6}$
1433 - 1447	1.35 $\times10^{-5}$	9.4 *13* $\times10^{-7}$

Continuous Radiation (^{196}Au)

⟨β-⟩=5.4 keV;⟨IB⟩=0.59 keV

E_{bin}(keV)		⟨ ⟩(keV)	(%)
0 - 10	β-	0.0350	0.70
	IB	0.00028	
10 - 20	β-	0.100	0.67
	IB	0.00023	0.00160
20 - 40	β-	0.365	1.22
	IB	0.00034	0.00120
40 - 100	β-	1.90	2.80
	IB	0.00046	0.00078
100 - 300	β-	3.00	2.11
	IB	0.000116	9.4×10^{-5}
600 - 1300	IB	0.00153	0.0132
1300 - 1506	IB	0.00160	0.0048

$^{196}_{79}$Au(8.1 *2* s)

Mode: IT

Δ: -31080 *5* keV

SpA: 6.81×10^9 Ci/g

Prod: daughter ^{196}Au(9.7 h); ^{197}Au(n,2n)

Photons (^{196}Au)

⟨γ⟩=3.13 *13* keV

γ_{mode}	γ(keV)	γ(%)†
Au L_ℓ	8.494	0.43 *5*
Au L_α	9.704	9.1 *7*
Au L_η	10.309	0.242 *17*
Au L_β	11.516	13.3 *10*
Au L_γ	13.427	2.52 *20*
Au $K_{\alpha2}$	66.991	0.0232 *11*
Au $K_{\alpha1}$	68.806	0.0395 *19*
Au $K_{\beta1}$'	77.859	0.0139 *7*
Au $K_{\beta2}$'	80.428	0.00368 *19*
γ E3	84.624 *20*	0.300

† 3.3% uncert(syst)

Atomic Electrons (^{196}Au)

⟨e⟩=80.8 *18* keV

e_{bin}(keV)	⟨e⟩(keV)	e(%)
4	0.00327	0.084 *4*
12	2.6	21.6 *24*
14	3.1	22.4 *25*
52	0.000165	0.00032 *4*
53	0.00028	0.00053 *6*
54	0.000159	0.00029 *3*
55	0.00029	0.00052 *6*
57	0.000134	0.00024 *3*
63	0.000146	0.000232 *25*
64	0.000174	0.00027 *3*
65	4.6 $\times10^{-5}$	7.0 *8* $\times10^{-5}$
66	0.000212	0.00032 *4*
67	2.9 $\times10^{-5}$	4.3 *5* $\times10^{-5}$
68	4.3 $\times10^{-5}$	6.3 *7* $\times10^{-5}$
69	6.8 $\times10^{-6}$	9.9 *11* $\times10^{-6}$
70	0.78	1.11 *5*
71	28.8	40.6 *18*
73	23.1	31.8 *14*
74	2.32 $\times10^{-5}$	3.1 *3* $\times10^{-5}$
75	2.9 $\times10^{-5}$	3.8 *4* $\times10^{-5}$
76	3.7 $\times10^{-6}$	4.9 *5* $\times10^{-6}$
77	2.6 $\times10^{-5}$	3.4 *4* $\times10^{-5}$
81	9.4	11.5 *5*
82	7.8	9.6 *4*
84	4.5	5.36 *24*
85	0.79	0.93 *4*

$^{196}_{79}$Au(9.7 *1* h)

Mode: IT

Δ: -30570 *5* keV

SpA: 1.648×10^6 Ci/g

Prod: ^{196}Pt(d,2n); ^{197}Au(n,2n); ^{197}Au(p,pn)

Photons (^{196}Au)

⟨γ⟩=234 *6* keV

γ_{mode}	γ(keV)	γ(%)†
Au L_ℓ	8.494	1.81 *24*
Au L_α	9.704	38 *4*
Au L_η	10.309	0.60 *5*
Au L_β	11.545	40 *4*
Au L_γ	13.469	7.2 *7*
γ (E2)	19.881 *9*	0.0025 *5*
γ M1	30.671 *9*	0.19 *4*
γ (E2)	50.552 *11*	0.0078 *16*
Au $K_{\alpha2}$	66.991	23.0 *21*

Photons (^{196}Au) (continued)

γ_{mode}	γ(keV)	γ(%)†
Au $K_{\alpha1}$	68.806	39 *4*
Au $K_{\beta1}'$	77.859	13.7 *13*
Au $K_{\beta2}'$	80.428	3.6 *3*
γ E3	84.624 *20*	0.298 *15*
γ M1	137.669 *15*	1.2 *4*
γ E2	147.780 *18*	47.2 *24*
γ M1	168.340 *14*	7.0 *4*
γ M4	174.874 *20*	0.42 *8*
γ M1+30%E2	188.221 *15*	34.4 *16*
γ (E2)	285.448 *22*	4.0 *4*
γ E2+<10%M1	316.119 *21*	2.66 *23*

† 10% uncert(syst)

Atomic Electrons (^{196}Au)

$\langle e\rangle$=368 *16* keV

e_{bin}(keV)	$\langle e\rangle$(keV)	e(%)
4 - 8	1.65	23 *3*
12	10.2	85 *11*
14 - 63	14.5	82 *8*
64 - 66	0.43	0.66 *6*
67	11.0	16.4 *9*
68 - 70	0.83	1.18 *6*
71	28.7	40.4 *22*
73	23.0	31.6 *17*
74 - 88	31.2	37.3 *10*
94	24.6	26 *5*
107	29.2	27 *5*
123 - 133	3.34	2.54 *19*
134	20.0	14.9 *8*
135	0.016	0.012 *4*
136	14.5	10.7 *6*
137 - 156	16.1	11.0 *4*
161	37.5	23 *5*
163	47.7	29 *6*
165 - 171	9.9	5.8 *11*
172	17.6	10.2 *21*
173	0.79	0.46 *9*
174	16.6	9.6 *22*
175 - 205	7.1	3.9 *6*
235 - 283	0.98	0.384 *20*
285 - 316	0.322	0.106 *6*

$^{196}_{80}$Hg(stable)

Δ: -31851 *5* keV

%: 0.14 *10*

$^{196}_{81}$Tl(1.84 *3* h)

Mode: ε

Δ: -27520 *140* keV syst

SpA: 8.69×10^6 Ci/g

Prod: daughter ^{196}Pb; protons on Hg; ^{197}Au(α,5n)

Photons (^{196}Tl)

$\langle\gamma\rangle$=1627 *36* keV

γ_{mode}	γ(keV)	γ(%)†
Hg L_ℓ	8.722	0.58 *5*
Hg L_α	9.980	12.1 *6*
Hg L_η	10.647	0.155 *9*
Hg L_β	11.931	11.1 *7*
Hg L_γ	13.944	2.00 *17*
Hg $K_{\alpha2}$	68.893	21.1 *6*
Hg $K_{\alpha1}$	70.818	35.9 *11*
Hg $K_{\beta1}'$	80.124	12.6 *4*
Hg $K_{\beta2}'$	82.780	3.45 *11*
γ	329.6 *4*	0.33 *5*
γ	354.3 *4*	1.25 *19*
γ E2	425.64 *7*	84 *4*
γ	532.4 *5*	0.115 *19*
γ E2+37%M1	610.5 *3*	11.9 *10*
γ E2	635.1 *3*	9.8 *10*
γ	704.9 *10*	1.34 *19*
γ [E2]	714.4 *4*	1.25 *19*
γ [M1+E2]	739.1 *4*	0.37 *6*
γ (E2)	753.8 *3*	1.44 *19*
γ	778.5 *3*	1.15 *19*
γ	808.9 *5*	0.56 *9*
γ	860.6 *5*	0.44 *9*
γ	885.2 *5*	0.17 *3*
γ (E2)	893.2 *8*	0.25 *4*
γ	958.0 *5*	0.23 *4*
γ (E2)	964.8 *4*	3.6 *4*
γ	976.0 *8*	0.42 *6*
γ	1024.9 *15*	0.86 *19*
γ IF(E2)	1036.1 *3*	2.6 *3*
γ	1064.6 *12*	0.31 *5*
γ	1104.9 *11*	0.57 *9*
γ	1136.2 *13*	0.26 *4*
γ	1263.2 *7*	0.80 *12*
γ	1289.0 *15*	1.15 *19*
γ [M1+E2]	1349.5 *4*	1.15 *19*
γ (M1)	1388.9 *3*	2.5 *3*
γ	1418.6 *20*	0.64 *14*
γ	1434.6 *11*	1.44 *19*
γ	1459.2 *11*	0.66 *10*
γ (M1)	1495.7 *4*	8.2 *9*
γ	1553.0 *7*	4.8 *5*
γ	1586.5 *8*	2.3 *3*
γ	1621.3 *20*	4.9 *6*
γ	1696.6 *20*	3.0 *4*
γ [E2]	1775.2 *4*	2.8 *4*
γ	1844.8 *5*	1.92 *19*
γ	1978.6 *7*	0.78 *12*
γ	2012.1 *8*	3.7 *4*
γ	2049.1 *20*	1.15 *19*
γ	2069.7 *11*	1.06 *19*
γ	2102.6 *18*	1.15 *19*
γ	2127.2 *18*	2.8 *4*
γ	2148.9 *20*	0.96 *19*
γ	2211.9 *20*	3.4 *4*
γ	2227.9 *7*	1.25 *19*
γ	2392.6 *20*	1.7 *3*

† 10% uncert(syst)

Atomic Electrons (^{196}Tl)

$\langle e\rangle$=27.1 *9* keV

e_{bin}(keV)	$\langle e\rangle$(keV)	e(%)
12	3.0	24.5 *25*
14	1.75	12.3 *13*
15 - 59	1.36	4.6 *3*
65 - 80	0.67	0.99 *5*
247 - 271	0.5	~0.18
315 - 342	0.13	~0.040
343	8.0	2.34 *13*
351 - 354	0.031	~0.009
411	2.73	0.66 *4*
413	0.58	0.141 *8*
422	0.68	0.162 *9*
423 - 449	0.443	0.104 *5*
518 - 520	0.004	~0.0008
527	1.6	0.30 *6*

Atomic Electrons (^{196}Tl) (continued)

e_{bin}(keV)	$\langle e\rangle$(keV)	e(%)
529 - 552	0.64	0.117 *12*
596 - 635	0.91	0.149 *23*
656 - 705	0.30	0.044 *14*
711 - 754	0.11	~0.015
764 - 810	0.11	0.014 *5*
846 - 893	0.23	0.026 *4*
942 - 982	0.25	0.026 *5*
1010 - 1053	0.10	0.009 *3*
1061 - 1105	0.009	0.0008 *4*
1121 - 1136	0.0034	~0.00031
1180 - 1206	0.10	~0.008
1236 - 1277	0.32	~0.025
1285 - 1335	0.25	0.019 *4*
1336 - 1385	0.17	0.012 *4*
1386 - 1435	0.59	0.041 *5*
1444 - 1493	0.31	0.021 *9*
1495 - 1584	0.33	0.022 *10*
1606 - 1694	0.24	0.015 *5*
1760 - 1842	0.09	0.0051 *22*
1896 - 1987	0.19	0.010 *4*
1997 - 2090	0.17	0.0082 *25*
2099 - 2198	0.15	0.0070 *25*
2200 - 2225	0.010	0.00044 *16*
2310 - 2390	0.047	0.0020 *10*

Continuous Radiation (^{196}Tl)

$\langle\beta+\rangle$=172 keV;$\langle$IB$\rangle$=6.1 keV

E_{bin}(keV)		$\langle\ \rangle$(keV)	(%)
0 - 10	β+	3.85×10^{-8}	4.53×10^{-7}
	IB	0.0062	
10 - 20	β+	2.64×10^{-6}	1.56×10^{-5}
	IB	0.0076	0.054
20 - 40	β+	0.000124	0.000371
	IB	0.013	0.045
40 - 100	β+	0.0110	0.0137
	IB	0.44	0.63
100 - 300	β+	0.89	0.392
	IB	0.153	0.086
300 - 600	β+	7.9	1.69
	IB	0.31	0.070
600 - 1300	β+	58	6.1
	IB	1.45	0.153
1300 - 2500	β+	101	5.8
	IB	2.8	0.157
2500 - 3914	β+	3.89	0.149
	IB	0.90	0.031
	Σβ+		14.1

$^{196}_{81}$Tl(1.41 *2* h)

Mode: ε(95.5 *5* %), IT(4.5 *5* %)

Δ: -27125 *140* keV syst

SpA: 1.134×10^7 Ci/g

Prod: protons on Hg; ^{197}Au(α,5n)

Photons (^{196}Tl)

$\langle\gamma\rangle$=1128 *90* keV

γ_{mode}	γ(keV)	γ(%)†
γ_{IT} [M1]	33.9 *3*	0.018 *8*
γ_ϵE2	83.62 *9*	3.1 *9*
γ_{IT} M4	120.3 *3*	0.00201 *8*
γ_ϵ	222.5 *10*	~2
γ_{IT} M1+<26%E2	240.9 *3*	<0.9
γ_{IT} (M1)	274.8 *4*	2.71 *16*
γ_ϵ	301.1 *12*	~4

Photons (^{196}Tl) (continued)

γ_{mode}	γ(keV)	γ(%)†
γ_ϵE2	425.64 *7*	91 *14*
γ_ϵ	504.8 *5*	~6
γ_ϵ	588.4 *5*	<3
γ_ϵE2+37%M1	610.5 *3*	5.1 *4*
γ_ϵE2	635.1 *3*	51 *8*
γ_ϵE1	695.0 *5*	41 *6*
γ_ϵE2	723.1 *6*	~2
γ_ϵ[E2]	1036.1 *3*	1.12 *18*

† uncert(syst): 0.52% for ϵ, 11% for IT

$^{196}_{82}$Pb(37 *3* min)

Mode: ϵ
Δ: -25450 *200* keV syst
SpA: 2.59×10^{7} Ci/g
Prod: ^{203}Tl(p,8n)

Photons (^{196}Pb)

γ_{mode}	γ(keV)	γ(rel)
γ[M1]	113.5 *5*	~0.17
γ[M1]	126.4 *5*	~0.6
γ[M1]	127.4 *6*	1.1 *6*
γ[M1]	175.0 *5*	~2
γE2	192.3 *4*	23 *4*

Photons (^{196}Pb) (continued)

γ_{mode}	γ(keV)	γ(rel)
γM1+<26%E2	240.9 *3*	27 *5*
γ[M1]	241.0 *6*	~3
γM1	253.8 *4*	74 *7*
γ[M1]	302.4 *6*	~1
γ(M1)	367.3 *5*	38 *12*
γ(M1)	494.7 *6*	~18
γ(M1)	503.6 *20*	100 *10*

$^{196}_{83}$Bi(4.6 *5* min)

Mode: ϵ
Δ: -17990 *700* keV
SpA: 2.08×10^{8} Ci/g
Prod: ^{181}Ta(^{20}Ne,5n); ^{181}Ta(^{22}Ne,7n)

Photons (^{196}Bi)

γ_{mode}	γ(keV)	γ(rel)
γ(E2)	137.6 *3*	10 *2*
γ(E1)	336.8 *3*	16 *2*
γ(E2)	372.0 *6*	46 *5*
γ[E2]	688.0 *5*	62 *5*
γ[E2]	1048.6 *5*	100

$^{196}_{84}$Po(5.5 *5* s)

Mode: α
Δ: -13470 *200* keV syst
SpA: 9.8×10^{9} Ci/g
Prod: ^{209}Bi(p,14n); ^{185}Re(^{19}F,8n)

Alpha Particles (^{196}Po)

α(keV)
6520 *4*

$^{196}_{85}$At(300 *100* ms)

Mode: α
Δ: -3970 *380* keV syst
SpA: 7.4802×10^{10} Ci/g
Prod: ^{185}Re(^{20}Ne,9n)

Alpha Particles (^{196}At)

α(keV)
7055 *7*

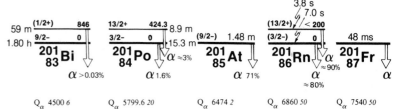

59 m (1/2+) 846
1.80 h 9/2− 0
$^{201}_{83}$Bi α >0.03%
13/2+ 424.3 8.9 m
3/2− 0 15.3 m
$^{201}_{84}$Po α ≈3% α 1.6%
(9/2−) 1.48 m
$^{201}_{85}$At α 71%
3.8 s 7.0 s
(13/2+) < 200
(3/2−) 0
$^{201}_{86}$Rn α ≈90% α ≈80%
48 ms
$^{201}_{87}$Fr α
Q_α 4500 *6* Q_α 5799.6 *20* Q_α 6474 *2* Q_α 6860 *50* Q_α 7540 *50*

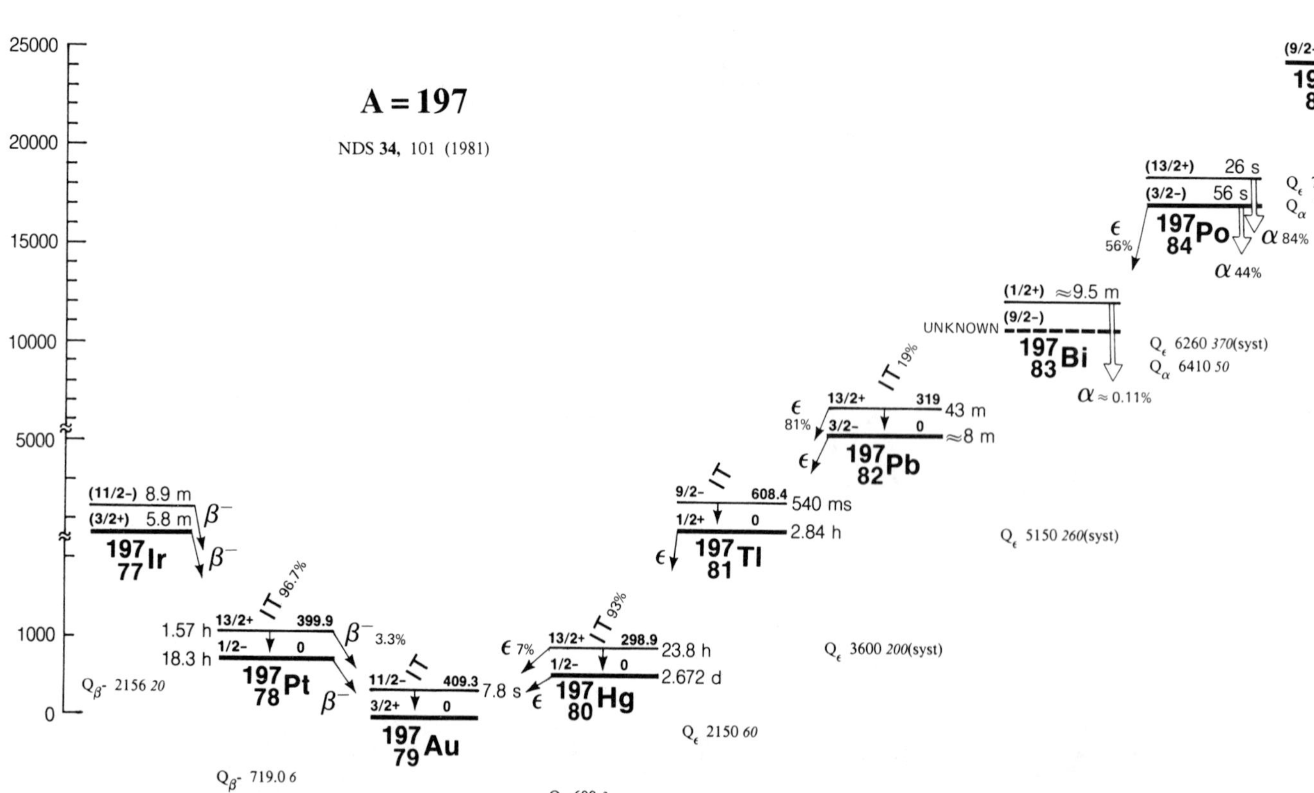

$^{197}_{77}$Ir(5.8 *5* min)

Mode: β-
Δ: -28290 *21* keV
SpA: 1.64×10^8 Ci/g
Prod: ^{198}Pt(n,pn); ^{198}Pt(γ,p); ^{198}Pt(p,2p)

Photons (^{197}Ir(8.9 + 5.8 min))

γ_{mode}	γ(keV)	γ(rel)
γE2	53.104 *20*	8.6 *17*
γ	71.72 *18*	
γ	135.41 *3*	27 *5*
γ	157.53 *7*	3.9 *8*
γ	228.13 *17*	4.3 *9*
γ	229.09 *20*	2.3 *5*
γ[M1+E2]	246.74 *6*	8.2 *16*
γ	269.23 *10*	12.7 *25*
γ	274.1 *3*	2.5 *5*
γ[E2]	299.85 *6*	24 *5*
γ	339.63 *21*	5.4 *11*
γ	340.47 *25*	4.7 *9*
γM4	346.81 *25*	*
γ	378.63 *5*	38 *8*
γ	404.27 *8*	5.8 *12*
γ	406.13 *13*	9.3 *19*
γ	430.87 *7*	61 *12*
γ	457.37 *8*	37 *7*
γ	470.03 *4*	100 *20*
γ	496.75 *6*	36 *7*
γ	509.45 *25*	21 *4*
γ	527.49 *5*	24 *5*
γ	534.18 *13*	5.9 *12*
γ	539.51 *8*	15 *3*
γ	542.33 *9*	10.6 *21*
γ	563.80 *22*	4.5 *9*
γ	644.5 *5*	
γ	708.49 *21*	5 *1*
γ	715.63 *12*	9.1 *18*
γ	739.1 *4*	1.20 *24*
γ	791.99 *22*	5 *1*
γ	809.43 *6*	32 *6*
γ	816.23 *6*	45 *9*
γ	849.80 *17*	6.6 *13*
γ	861.83 *15*	6.0 *12*
γ	866.75 *8*	13 *3*
γ	873.62 *11*	6.4 *13*
γ	888.17 *20*	4.6 *9*
γ	939.72 *8*	21 *4*
γ	987.44 *9*	15 *3*
γ	1008.3 *3*	1.7 *3*
γ	1053.8 *3*	1.8 *4*
γ	1062.5 *3*	1.9 *4*
γ	1343.53 *10*	21 *4*

* with ^{197}Pt(1.57 h)

$^{197}_{77}$Ir(8.9 *3* min)

Mode: β-
Δ: >-28290 keV
SpA: 1.07×10^8 Ci/g
Prod: ^{198}Pt(n,pn); ^{198}Pt(γ,p); ^{198}Pt(p,2p)

see ^{198}Ir(5.8 min) for γ rays

$^{197}_{78}$Pt(18.3 *3* h)

Mode: β-
Δ: -30446 *4* keV
SpA: 8.69×10^5 Ci/g
Prod: ^{196}Pt(n,γ)

Photons (^{197}Pt)

⟨γ⟩=25.4 *15* keV

γ_{mode}	γ(keV)	γ(%)†
Au L_ℓ	8.494	0.34 *7*
Au L_α	9.704	7.2 *13*
Au L_η	10.309	0.106 *22*
Au L_β	11.524	9.1 *18*
Au L_γ	13.573	1.9 *4*
Au $K_{\alpha 2}$	66.991	1.01 *10*
Au $K_{\alpha 1}$	68.806	1.72 *17*
γ M1+8.7%E2	77.343 *9*	17.1 *17*
Au $K_{\beta 1}$'	77.859	0.60 *6*
Au $K_{\beta 2}$'	80.428	0.160 *16*
γ M1+1.9%E2	191.362 *10*	3.7
γ E2(+<8.0%M1)	268.705 *13*	0.231 *20*

† 11% uncert(syst)

Atomic Electrons (^{197}Pt)

⟨e⟩=54 *4* keV

e_{bin}(keV)	⟨e⟩(keV)	e(%)
12	0.85	7.1 *19*
14	2.0	14 *3*
52 - 57	0.045	0.082 *5*
63	19.6	31 *3*
64	7.8	12 *3*
65	5.7	9 *3*
66 - 69	0.0127	0.0191 *20*
74	7.6	10.3 *16*
75	1.7	2.3 *8*
76	0.000162	0.00021 *3*
77	3.0	3.8 *7*
111	4.0	3.6 *4*
177 - 191	1.49	0.83 *6*
254 - 269	0.0384	0.0149 *7*

Continuous Radiation (^{197}Pt)

⟨β-⟩=195 keV;⟨IB⟩=0.129 keV

E_{bin}(keV)		⟨ ⟩(keV)	(%)
0 - 10	β-	0.156	3.12
	IB	0.0098	
10 - 20	β-	0.464	3.09
	IB	0.0091	0.063
20 - 40	β-	1.84	6.1
	IB	0.0163	0.057
40 - 100	β-	12.4	17.8
	IB	0.037	0.058
100 - 300	β-	90	47.0
	IB	0.050	0.031
300 - 600	β-	89	22.7
	IB	0.0077	0.0022
600 - 719	β-	0.96	0.154
	IB	7.5×10^{-6}	1.22×10^{-6}

$^{197}_{78}$Pt(1.573 *13* h)

Mode: IT(96.7 *4* %), β-(3.3 *4* %)
Δ: -30046 *4* keV
SpA: 1.011×10^7 Ci/g
Prod: ^{196}Pt(n,γ); ^{196}Pt(d,p)

Photons (^{197}Pt)

⟨γ⟩=83 *2* keV

γ_{mode}	γ(keV)	γ(%)†
Pt L_ℓ	8.266	0.90 *9*
Au L_ℓ	8.494	0.018 *7*
Pt L_α	9.435	19.5 *13*
Au L_α	9.704	0.39 *15*
Pt L_η	9.975	0.310 *21*
Au L_η	10.309	0.010 *4*
Pt L_β	11.177	20.4 *16*
Au L_β	11.517	0.55 *23*
Pt L_γ	13.023	3.6 *3*
Au L_γ	13.439	0.11 *5*
γ_{IT} E2	53.104 *20*	1.07 *4*
Pt $K_{\alpha 2}$	65.122	13.4 *7*
Pt $K_{\alpha 1}$	66.831	23.0 *11*
Au $K_{\alpha 2}$	66.991	0.23 *3*
Au $K_{\alpha 1}$	68.806	0.40 *6*
Pt $K_{\beta 1}$'	75.634	8.0 *4*
Au $K_{\beta 1}$'	77.859	0.139 *19*
Pt $K_{\beta 2}$'	78.123	2.09 *11*
Au $K_{\beta 2}$'	80.428	0.037 *5*
$\gamma_{\beta -}$E3	130.17 *8*	0.105 *4* *
$\gamma_{\beta -}$E2	201.77 *6*	0.035 *4* *
$\gamma_{\beta -}$M1+14%E2	279.11 *6*	2.3 *
γ_{IT} M4	346.81 *25*	11.1 *4*
$\gamma_{\beta -}$M4	409.28 *8*	0.0035 *12* *

† uncert(syst): 10% for IT, 12% for β-
* with ^{197}Au(7.8 s) in equilib

Atomic Electrons (^{197}Pt)

⟨e⟩=313 *7* keV

e_{bin}(keV)	⟨e⟩(keV)	e(%)
12 - 39	10.4	83 *6*
40	13.9	35.0 *16*
42	14.8	35.6 *16*
49	0.05	~0.11
50	9.1	18.2 *8*
51 - 77	3.96	7.39 *21*
116 - 130	3.7	3.1 *9*
187 - 202	1.45	0.73 *10*
265 - 267	0.34	0.128 *18*
268	130.4	48.6 *22*
276 - 279	0.109	0.039 *5*
329	0.028	~0.009
333	49.9	15.0 *7*
334	12.7	3.82 *17*
335	28.9	8.6 *4*
344	25.2	7.3 *3*
345 - 347	8.2	2.36 *9*
395 - 409	0.022	0.0056 *14*

Continuous Radiation (^{197}Pt)

$\langle\beta-\rangle$=7.3 keV;$\langle$IB$\rangle$=0.0053 keV

E_{bin}(keV)		$\langle\ \rangle$(keV)	(%)
0 - 10	β-	0.00433	0.086
	IB	0.00037	
10 - 20	β-	0.0129	0.086
	IB	0.00034	0.0024
20 - 40	β-	0.052	0.172
	IB	0.00062	0.0022
40 - 100	β-	0.356	0.51
	IB	0.00143	0.0023
100 - 300	β-	2.85	1.47
	IB	0.0021	0.00133
300 - 600	β-	3.88	0.95
	IB	0.00044	0.000121
600 - 710	β-	0.167	0.0266
	IB	1.53×10^{-6}	2.5×10^{-7}

$^{197}_{79}$Au(stable)

Δ: -31165 *4* keV
%: 100

$^{197}_{79}$Au(7.8 *1* s)

Mode: IT
Δ: -30756 *4* keV
SpA: 7.03×10^{9} Ci/g
Prod: daughter ^{197}Hg(23.8 h); daughter ^{197}Pt(1.57 h); ^{197}Au(n,n'); ^{197}Au(γ,γ')

Photons (^{197}Au)

$\langle\gamma\rangle$=232 *31* keV

γ_{mode}	γ(keV)	γ(%)†
Au L_ℓ	8.494	0.55 *8*
Au L_α	9.704	11.8 *13*
Au L_η	10.309	0.29 *3*
Au L_β	11.518	16.6 *17*
Au L_γ	13.441	3.2 *3*
Au $K_{\alpha2}$	66.991	7.3 *10*
Au $K_{\alpha1}$	68.806	12.4 *17*
γ M1+8.7%E2	77.343 *9*	0.30 *5*
Au $K_{\beta1}'$	77.859	4.3 *6*
Au $K_{\beta2}'$	80.428	1.15 *16*
γ E3	130.17 *8*	3.12 *21*
γ E2	201.77 *6*	1.10 *16*
γ M1+14%E2	279.11 *6*	73 *11*
γ M4	409.28 *8*	0.11 *3*

† 0.71% uncert(syst)

Atomic Electrons (^{197}Au)

$\langle e\rangle$=180 *8* keV

e_{bin}(keV)	$\langle e\rangle$(keV)	e(%)
12	3.1	26 *4*
14	3.8	28 *4*
49 - 77	2.92	5.3 *4*
116	48.6	42 *3*
118	30.6	25.8 *18*
121	0.22	0.18 *3*
127	23.3	18.3 *13*
128 - 129	0.71	0.55 *3*

Atomic Electrons (^{197}Au)
(continued)

e_{bin}(keV)	$\langle e\rangle$(keV)	e(%)
130	7.1	5.5 *4*
187 - 190	0.31	0.166 *15*
198	45.4	23 *4*
199 - 202	0.100	0.050 *6*
265	10.4	3.9 *6*
267 - 279	3.8	1.38 *16*

$^{197}_{80}$Hg(2.6725 *21* d)

Mode: ε
Δ: -30565 *5* keV
SpA: 2.4801×10^{5} Ci/g
Prod: ^{197}Au(p,n); ^{197}Au(d,2n)

Photons (^{197}Hg)

$\langle\gamma\rangle$=70 *6* keV

γ_{mode}	γ(keV)	γ(%)
Au L_ℓ	8.494	0.90 *19*
Au L_α	9.704	19 *4*
Au L_η	10.309	0.25 *5*
Au L_β	11.540	20 *4*
Au L_γ	13.544	4.0 *8*
Au $K_{\alpha2}$	66.991	21 *4*
Au $K_{\alpha1}$	68.806	35 *8*
γ M1+8.7%E2	77.343 *9*	18.0 *4*
Au $K_{\beta1}'$	77.859	12 *3*
Au $K_{\beta2}'$	80.428	3.3 *7*
γ M1+1.9%E2	191.362 *10*	0.608 *20*
γ E2(+<8.0%M1)	268.705 *13*	0.0378 *17*

Atomic Electrons (^{197}Hg)

$\langle e\rangle$=57 *3* keV

e_{bin}(keV)	$\langle e\rangle$(keV)	e(%)
12	3.6	30 *7*
14	4.3	31 *6*
52 - 57	0.91	1.68 *19*
63	20.9	33.2 *15*
64	8.2	13 *3*
65	5.9	9 *3*
66 - 69	0.26	0.39 *7*
74	8.0	10.9 *11*
75	1.8	2.4 *8*
76	0.0033	0.0044 *10*
77	3.1	4.1 *6*
177 - 191	0.244	0.135 *4*
254 - 269	0.00629	0.00244 *7*

Continuous Radiation (^{197}Hg)

$\langle$IB$\rangle$=0.47 keV

E_{bin}(keV)		$\langle\ \rangle$(keV)	(%)
10 - 20	IB	0.0020	0.0169
20 - 40	IB	0.00166	0.0051
40 - 100	IB	0.43	0.63
100 - 300	IB	0.029	0.019
300 - 600	IB	0.0050	0.00143

$^{197}_{80}$Hg(23.8 *1* h)

Mode: IT(93.0 *7* %), ε(7.0 *7* %)
Δ: -30266 *5* keV
SpA: 6.68×10^{5} Ci/g
Prod: ^{197}Au(p,n); ^{197}Au(d,2n); alphas on Pt

Photons (^{197}Hg)

$\langle\gamma\rangle$=93 *3* keV

γ_{mode}	γ(keV)	γ(%)†
Au L_ℓ	8.494	0.078 *10*
Hg L_ℓ	8.722	0.92 *9*
Au L_α	9.704	1.65 *17*
Hg L_α	9.980	19.2 *11*
Au L_η	10.309	0.031 *3*
Hg L_η	10.647	0.216 *14*
Au L_β	11.532	1.96 *20*
Hg L_β	11.941	16.1 *12*
Au L_γ	13.469	0.37 *4*
Hg L_γ	13.945	2.80 *25*
Au $K_{\alpha2}$	66.991	1.9 *2*
Au $K_{\alpha1}$	68.806	3.2 *4*
Hg $K_{\alpha2}$	68.893	9.5 *4*
Hg $K_{\alpha1}$	70.818	16.2 *6*
Au $K_{\beta1}'$	77.859	1.13 *12*
Hg $K_{\beta1}'$	80.124	5.69 *21*
Au $K_{\beta2}'$	80.428	0.30 *3*
Hg $K_{\beta2}'$	82.780	1.55 *6*
γ_ϵE3	130.17 *8*	0.222 *9* *
γ_{IT} E2	133.96 *3*	34.1
γ_{IT} M4	164.95 *7*	0.266 *10*
γ_ϵE2	201.77 *6*	0.073 *9* *
γ_ϵM1+14%E2	279.11 *6*	4.9 *
γ_ϵM4	409.28 *8*	0.0074 *25* *

† uncert(syst): 10% for ε, 11% for IT
* with ^{197}Au(7.8 s) in equilib

Atomic Electrons (^{197}Hg)

$\langle e\rangle$=214 *3* keV

e_{bin}(keV)	$\langle e\rangle$(keV)	e(%)
12 - 49	8.9	68 *5*
51	7.2	14.1 *6*
52 - 80	0.87	1.45 *5*
82	16.6	20.2 *9*
116 - 119	7.88	6.70 *18*
120	21.7	18.2 *8*
121	0.0147	0.0122 *14*
122	16.3	13.4 *6*
127 - 130	2.82	2.20 *7*
131	10.7	8.2 *4*
132 - 134	3.59	2.69 *10*
150	26.3	17.5 *8*
151	6.2	4.14 *18*
153	46.1	30.2 *13*
161	8.1	5.05 *22*
162	17.1	10.5 *5*
163	0.83	0.510 *22*
164	6.7	4.10 *18*
165 - 202	4.6	2.50 *22*
265 - 279	0.95	0.35 *4*
395 - 409	0.047	0.0117 *20*

$^{197}_{81}$Tl(2.84 *4* h)

Mode: ϵ

Δ: -28420 *60* keV

SpA: 5.60×10^{6} Ci/g

Prod: ^{197}Au(α,4n); ^{198}Hg(d,3n); protons on Hg; protons on Pb

Photons (^{197}Tl)

⟨γ⟩=442 *11* keV

γ_{mode}	γ(keV)	γ(%)†
Hg L_ℓ	8.722	0.78 *16*
Hg L_α	9.980	16 *3*
Hg L_η	10.647	0.21 *4*
Hg L_β	11.930	15 *3*
Hg L_γ	13.946	2.7 *6*
γ (M1)	18.171 *25*	0.0079 *18*
Hg $K_{\alpha2}$	68.893	28 *5*
Hg $K_{\alpha1}$	70.818	47 *9*
Hg $K_{\beta1}$'	80.124	16 *3*
Hg $K_{\beta2}$'	82.780	4.5 *8*
γ E2	133.96 *3*	2.00 *15*
γ M1	152.13 *3*	7.2 *5*
γ (M1+E2)	156.30 *6*	0.30 *6*
γ M1	173.76 *6*	0.246 *18*
γ	249.27 *11*	0.068 *9*
γ M1	269.53 *7*	0.52 *4*
γ M1	277.60 *6*	0.262 *19*
γ (E2)	307.72 *6*	1.4 *3*
γ (M1)	308.43 *6*	2.2 *4*
γ (M1)	397.47 *7*	0.099 *12*
γ M1	404.82 *8*	0.210 *17*
γ (M1,E2)	423.74 *13*	0.18 *3*
γ M1	425.83 *5*	12.9 *9*
γ M1	433.19 *5*	2.50 *18*
γ (M1+E2)	444.00 *5*	0.57 *4*
γ (M1+E2)	451.36 *5*	1.06 *8*
γ M1(+E2)	483.68 *7*	0.245 *18*
γ M1(+E2)	545.13 *7*	0.134 *12*
γ	547.2 *3*	0.040 *14*
γ	548.49 *17*	0.037 *9*
γ M1	577.96 *5*	4.4 *3*
γ (M1)	584.77 *8*	0.41 *5*
γ (M1)	585.32 *6*	0.49 *5*
γ M1	639.99 *5*	0.86 *7*
γ (M1)	645.51 *7*	0.152 *17*
γ	658.16 *5*	0.103 *13*
γ M1(+E2)	674.35 *8*	1.46 *10*
γ M1	676.73 *24*	0.59 *4*
γ E2	701.56 *7*	1.02 *8*
γ [M1+E2]	740.36 *7*	0.031 *5*
γ [M1+E2]	758.53 *6*	0.115 *12*
γ	771.32 *6*	0.085 *10*
γ M1(+E2)	792.12 *5*	1.75 *14*
γ	830.66 *7*	0.052 *7*
γ (M1)	852.30 *7*	0.028 *9*
γ M1	857.16 *6*	2.05 *14*
γ M1	892.49 *6*	1.14 *8*
γ (M1)	901.60 *6*	0.37 *3*
γ M1	982.78 *7*	1.25 *9*
γ (E0+M1)	1009.29 *6*	0.39 *3*
γ	1108.40 *7*	0.092 *18*
γ	1129.90 *8*	0.062 *8*
γ	1145.23 *17*	0.103 *13*
γ	1255.00 *8*	0.45 *8*
γ	1255.71 *7*	0.32 *7*
γ (M1)	1285.49 *6*	0.83 *7*
γ (M1,E2)	1385.28 *7*	1.25 *9*
γ (M1)	1411.30 *5*	4.5 *4*
γ	1429.47 *5*	0.88 *7*
γ (M1)	1437.62 *6*	0.59 *4*
γ	1541.59 *6*	0.214 *17*
γ	1563.43 *5*	0.076 *8*
γ	1693.71 *6*	0.68 *5*

† 12% uncert(syst)

Atomic Electrons (^{197}Tl)

⟨e⟩=45.3 *14* keV

e_{bin}(keV)	⟨e⟩(keV)	e(%)
3 - 6	0.047	1.4 *3*
12	4.0	32 *7*
14	2.3	17 *3*
15 - 59	2.3	7.5 *10*
65 - 68	0.70	1.05 *13*
69	10.3	14.9 *11*
70 - 119	0.84	1.00 *25*
120	1.27	1.06 *9*
122 - 134	1.83	1.45 *7*
137	3.12	2.27 *17*
138 - 186	2.27	1.48 *9*
194	0.196	0.101 *8*
225	1.6	0.72 *13*
234 - 278	0.195	0.075 *4*
293 - 341	0.77	0.253 *25*
343	5.4	1.57 *11*
350	1.02	0.291 *22*
361 - 410	0.49	0.13 *5*
411 - 451	1.79	0.429 *20*
462 - 484	0.053	0.011 *4*
495	1.20	0.242 *18*
502 - 548	0.25	0.050 *6*
557 - 594	0.93	0.162 *20*
618 - 664	0.188	0.030 *3*
671 - 709	0.26	0.037 *15*
726 - 774	0.330	0.043 *3*
777 - 818	0.268	0.033 *3*
827 - 857	0.073	0.0087 *5*
878 - 926	0.384	0.0422 *24*
968 - 1009	0.073	0.0074 *6*
1025 - 1062	0.015	0.0015 *6*
1094 - 1143	0.0034	0.0003 *1*
1144 - 1173	0.040	~0.0034
1202 - 1251	0.078	0.0065 *6*
1252 - 1285	0.0168	0.00132 *10*
1302 - 1346	0.45	0.034 *4*
1355 - 1399	0.115	0.0084 *5*
1408 - 1438	0.035	0.0025 *3*
1458 - 1551	0.014	0.0009 *4*
1560 - 1611	0.024	~0.0015
1679 - 1691	0.0049	~0.00029

Continuous Radiation (^{197}Tl)

⟨β+⟩=8.8 keV;⟨IB⟩=1.61 keV

E_{bin}(keV)		⟨ ⟩(keV)	(%)
0 - 10	β+	3.28×10^{-8}	3.85×10^{-7}
	IB	0.00049	
10 - 20	β+	2.23×10^{-6}	1.32×10^{-5}
	IB	0.0022	0.018
20 - 40	β+	0.000103	0.000309
	IB	0.0022	0.0069
40 - 100	β+	0.0087	0.0109
	IB	0.46	0.66
100 - 300	β+	0.59	0.266
	IB	0.063	0.036
300 - 600	β+	3.61	0.80
	IB	0.170	0.037
600 - 1300	β+	4.58	0.61
	IB	0.65	0.072
1300 - 2150	IB	0.26	0.017
	Σβ+		1.68

$^{197}_{81}$Tl(540 *10* ms)

Mode: IT

Δ: -27812 *60* keV

SpA: 5.97×10^{10} Ci/g

Prod: daughter ^{197}Pb(43 min); ^{197}Au(α,4n)

Photons (^{197}Tl)

⟨γ⟩=435 *14* keV

γ_{mode}	γ(keV)	γ(%)†
Tl L_ℓ	8.953	0.36 *4*
Tl L_α	10.259	7.3 *6*
Tl L_η	10.994	0.219 *16*
Tl L_β	12.270	11.8 *9*
Tl L_γ	14.363	2.43 *20*
Tl $K_{\alpha2}$	70.832	4.8 *3*
Tl $K_{\alpha1}$	72.873	8.2 *5*
Tl $K_{\beta1}$'	82.434	2.88 *18*
Tl $K_{\beta2}$'	85.185	0.81 *5*
γ E3	222.63 *5*	30.5 *12*
γ E2+27%M1	385.78 *6*	91 *4*

† 6.9% uncert(syst)

Atomic Electrons (^{197}Tl)

⟨e⟩=172 *4* keV

e_{bin}(keV)	⟨e⟩(keV)	e(%)
13 - 60	4.6	32.4 *25*
66 - 82	0.154	0.221 *10*
137	14.7	10.7 *5*
207	6.0	2.89 *13*
208	57.6	27.7 *12*
210	26.7	12.7 *6*
219	17.9	8.2 *4*
220	8.0	3.62 *16*
222	6.8	3.08 *14*
223	1.24	0.558 *25*
300	19.9	6.6 *9*
370 - 386	8.2	2.20 *15*

$^{197}_{82}$Pb(~8 min)

Mode: ϵ

Δ: -24820 *210* keV syst

SpA: ~1.2×10^{8} Ci/g

Prod: daughter ^{197}Pb(43 min); ^{187}Re(^{14}N,4n)

Photons (^{197}Pb)

⟨γ⟩=1651 *110* keV

γ_{mode}	γ(keV)	γ(%)†
Tl L_ℓ	8.953	0.7 *3*
Tl L_α	10.259	14 *7*
Tl L_η	10.994	0.18 *9*
Tl L_β	12.313	13 *6*
Tl L_γ	14.411	2.4 *12*
Tl $K_{\alpha2}$	70.832	24 *11*
Tl $K_{\alpha1}$	72.873	41 *19*
Tl $K_{\beta1}$'	82.434	14 *7*
Tl $K_{\beta2}$'	85.185	4.0 *19*
γ M1	375.38 *6*	14.2 *11*
γ E2+27%M1	385.78 *6*	~56
γ M1	394.81 *10*	2.3 *5*
γ M1	520.65 *7*	1.57 *18*
γ	538.8 *5*	0.27 *13*
γ E2	761.16 *6*	14.8 *12*
γ M1	770.19 *10*	2.8 *4*
γ	815.24 *21*	1.19 *17*
γ (E2)	843.99 *12*	2.9 *3*
γ E1	871.57 *11*	6.8 *5*
γ M1+E2	885.85 *10*	1.0 *2*
γ M1+E2	896.04 *7*	5.6 *14*
γ M1	901.73 *10*	3.5 *4*
γ M1	913.43 *8*	2.4 *3*

Photons (^{197}Pb) (continued)

γ_{mode}	γ(keV)	γ(%)†
γ	1003.79 12	1.8 2
γ	1063.71 8	~0.7
γ (E2)	1088.12 12	2.32 20
γ (E2)	1092.81 9	3.7 3
γ	1140.02 11	2.1 2
γ	1147.62 10	0.78 10
γ M1+E2	1155.97 9	4.2 3
γ (M1)	1208.95 12	2.26 21
γ (M1)	1219.38 12	1.86 19
γ M1+E2	1261.24 9	9.2 7
γ	1277.11 10	1.64 25
γ	1281.82 8	0.44 13
γ M1+E2	1288.81 8	5.4 4
γ	1382.86 20	0.82 13
γ (M1)	1584.36 7	2.1 2
γ	1647.01 9	2.6 2
γ	1662.89 10	0.68 11
γ	1674.59 8	3.3 3
γ	1727.01 20	0.54 8
γ	1853.97 9	6.8 5
γ	1975.71 13	1.34 12
γ	2043.65 9	1.86 16
γ	2112.69 12	1.21 11
γ	2143.63 17	0.72 9
γ	2345.52 8	4.9 4
γ	2429.43 10	0.52 6

† 20% uncert(syst)

Atomic Electrons (^{197}Pb)

⟨e⟩=45 7 keV

e_{bin}(keV)	⟨e⟩(keV)	e(%)
13	3.4	27 13
15	2.5	17 8
55 - 82	1.8	2.9 6
290	7.6	2.63 21
300	12.0	~4
309	1.15	0.37 8
360	1.44	0.40 3
361 - 363	0.141	0.039 3
370	2.2	~0.6
371 - 395	3.6	0.96 20
435 - 453	0.58	0.133 17
505 - 539	0.148	0.029 3
676 - 685	1.40	0.207 16
730 - 770	0.68	0.090 14
786 - 831	1.8	0.22 4
840 - 889	0.27	0.031 7
892 - 918	0.28	0.031 11
978 - 1007	0.34	0.034 5
1048 - 1092	0.57	0.054 16
1123 - 1156	0.52	0.046 4
1176 - 1219	1.2	0.10 3
1246 - 1288	0.24	0.019 4
1368 - 1382	0.009	~0.0006
1499 - 1589	0.42	0.027 7
1632 - 1724	0.076	0.0046 14
1768 - 1851	0.27	~0.015
1890 - 1973	0.10	0.0054 22
2027 - 2110	0.072	0.0035 12
2128 - 2141	0.0041	0.00019 9
2260 - 2344	0.16	0.007 3
2414 - 2426	0.0025	0.00010 5

$^{197}_{82}$Pb(43 1 min)

Mode: ε(81 2 %), IT(19 2 %)

Δ: -24501 210 keV syst

SpA: 2.22×10^{7} Ci/g

Prod: ^{203}Tl(p,7n); protons on U; ^{187}Re(^{14}N,4n)

Photons (^{197}Pb)

⟨γ⟩=1150 140 keV

γ_{mode}	γ(keV)	γ(%)†
Tl L_ℓ	8.953	0.86 25
Pb L_ℓ	9.185	0.217 21
Tl L_α	10.259	17 5
Pb L_α	10.541	4.3 3
Tl L_η	10.994	0.33 7
Pb L_η	11.349	0.0371 23
Tl L_β	12.293	20 5
Pb L_β	12.696	3.9 4
Tl L_γ	14.387	3.9 9
Pb L_γ	14.995	0.79 9
Tl $K_{\alpha2}$	70.832	23 8
Pb $K_{\alpha2}$	72.803	1.89 8
Tl $K_{\alpha1}$	72.873	40 13
Pb $K_{\alpha1}$	74.969	3.17 14
Tl $K_{\beta1}$'	82.434	14 5
Pb $K_{\beta1}$'	84.789	1.12 5
γ_{IT} M1	84.9 2	4.84 11
Tl $K_{\beta2}$'	85.185	3.9 13
Pb $K_{\beta2}$'	87.632	0.323 15
γ_ϵE3	222.63 5	24.6 18 *
γ_{IT} M4	234.4 7	0.293 10
γ_ϵ[M1+E2]	274.50 8	0.36 4 ?
γ_ϵM1	290.67 8	1.13 10
γ_ϵM1+~1.9%E2	298.91 10	0.28 5
γ_ϵ(M1)	301.38 20	0.38 5
γ_ϵM1+2.9%E2	307.94 6	3.0 4
γ_ϵM1+~0.8%E2	321.03 8	0.196 20
γ_ϵE2+27%M1	385.78 6	~74 *
γ_ϵM1+20%E2	387.61 5	25.1 23
γ_ϵM1	393.55 7	0.58 10
γ_ϵ	400.05 20	0.193 25
γ_ϵ(M1)	413.90 9	0.29 8
γ_ϵM1+5.1%E2	416.07 6	2.29 20
γ_ϵM1	446.14 10	0.57 6
γ_ϵ[M1+E2]	457.06 20	
γ_ϵ	457.86 13	
γ_ϵ	485.33 18	0.103 10
γ_ϵM1	488.13 8	1.38 10
γ_ϵE2	492.71 8	0.75 8
γ_ϵM1	499.84 13	0.40 8
γ_ϵE2+50%M1	505.80 7	1.05 10
γ_ϵM1(+E2)	514.96 19	0.38 5
γ_ϵM1	528.52 11	0.43 5
γ_ϵ(M1)	545.14 9	0.20 5
γ_ϵM1	550.64 11	0.27 4
γ_ϵM1+38%E2	557.61 19	3.5 3
γ_ϵM1	575.05 14	0.25 5
γ_ϵ	608.4 15	
γ_ϵE2	611.75 8	0.83 10
γ_ϵM1+E2	616.36 9	1.1 1
γ_ϵ(M1)	628.07 9	0.33 5
γ_ϵ(M1+E2)	649.72 7	0.70 8
γ_ϵM1	650.80 10	0.95 10
γ_ϵE2	695.54 6	9.5 8
γ_ϵM1	706.73 7	1.58 20
γ_ϵE2	714.98 9	1.1 1
γ_ϵE2	724.00 6	3.7 3
γ_ϵE2	737.10 8	0.73 5
γ_ϵM1+32%E2	774.20 7	14.1 10
γ_ϵ	792.40 10	0.93 15
γ_ϵM1	809.62 7	0.50 5
γ_ϵM1	829.81 13	0.30 8
γ_ϵE2	835.66 10	1.03 8
γ_ϵE2+49%M1	893.41 8	2.13 25
γ_ϵ	895.99 16	0.80 18
γ_ϵM1+E2	920.15 10	1.16 10
γ_ϵM1+E2	944.92 11	0.80 8
γ_ϵE2+45%M1	957.66 7	5.9 5
γ_ϵ	984.97 9	0.41 4
γ_ϵE2+43%M1	998.51 7	2.39 18
γ_ϵ[M1+E2]	1014.67 7	0.068 20
γ_ϵ(M1+E2)	1029.36 13	0.65 6
γ_ϵ	1037.40 9	0.40 5
γ_ϵ(E2)	1063.62 8	~0.7
γ_ϵM1+E2	1074.22 9	0.75 8
γ_ϵ(M1)	1080.3 3	0.18 3
γ_ϵE2+41%M1	1117.55 6	3.21 25
γ_ϵ	1194.79 19	0.158 20
γ_ϵ	1278.7 3	0.40 8
γ_ϵM1(+E2)	1345.26 7	0.88 10

Photons (^{197}Pb) (continued)

γ_{mode}	γ(keV)	γ(%)†
γ_ϵ	1371.56 8	0.30 3
γ_ϵ	1376.31 24	0.22 3
γ_ϵ	1425.00 8	0.35 3
γ_ϵ	1455.14 18	0.56 5
γ_ϵ(E2)	1479.53 13	0.98 8
γ_ϵ[E2]	1505.16 7	0.12 3
γ_ϵ	1516.36 8	0.29 3
γ_ϵ	1571.64 24	0.21 3
γ_ϵ	1759.16 8	0.158 23
γ_ϵ	1787.46 12	0.27 3
γ_ϵ	1824.30 8	0.54 5
γ_ϵ	2175.07 13	0.29 3
γ_ϵ	2211.90 8	0.23 3

† uncert(syst): 20% for ε, 11% for IT

* with ^{197}Tl(540 ms) in equilib

Atomic Electrons (^{197}Pb)

⟨e⟩=225 10 keV

e_{bin}(keV)	⟨e⟩(keV)	e(%)
13 - 62	10.8	73 13
66 - 68	0.35	0.52 11
69	7.0	10.1 4
70 - 85	3.83	4.88 13
137	11.9	8.7 6
146	9.9	6.7 3
189 - 207	5.9	2.84 19
208	46.6	22.4 17
210	21.5	10.3 8
213 - 216	0.47	0.219 24
219	24.5	11.2 7
220	6.4	2.93 22
221	8.9	4.01 15
222	7.6	3.4 3
223 - 272	9.03	3.91 11
274 - 299	0.86	0.298 22
300	15.8	~5
301	0.0048	0.00158 22
302	10.8	3.6 5
304 - 331	1.68	0.52 4
361 - 407	10.9	2.9 5
410 - 460	0.80	0.186 22
465 - 514	1.25	0.262 25
515 - 564	0.71	0.131 24
565 - 614	0.92	0.155 10
615 - 652	0.78	0.123 8
680 - 724	2.7	0.39 6
733 - 780	0.64	0.085 9
789 - 836	0.46	0.056 12
859 - 907	0.73	0.083 13
913 - 958	0.44	0.048 7
970 - 1017	0.18	0.018 4
1022 - 1071	0.27	0.026 5
1072 - 1117	0.065	0.0059 10
1179 - 1194	0.024	~0.0020
1260 - 1291	0.08	0.0067 21
1330 - 1376	0.060	0.0044 16
1394 - 1442	0.058	0.0042 7
1451 - 1513	0.021	0.0014 4
1556 - 1569	0.0018	~0.00012
1674 - 1773	0.036	0.0021 8
1775 - 1821	0.0042	0.00023 11
2090 - 2172	0.016	0.0007 3
2197 - 2209	0.0012	~6×10^{-5}

$^{197}_{83}$Bi(9.5 *10* min)

Mode: α(~0.11 %)
Δ: >-19170 keV
SpA: 1.00×10^{8} Ci/g
Prod: protons on Pb; ^{181}Ta(^{20}Ne,4n); ^{203}Tl(^{3}He,9n); ^{187}Re(^{16}O,6n)

Alpha Particles (^{197}Bi)

α(keV)
5770 *10*

$^{197}_{84}$Po(56 *3* s)

Mode: ϵ(56 *7* %), α(44 *7* %)
Δ: -13410 *320* keV syst
SpA: 1.02×10^{9} Ci/g
Prod: ^{19}F on Re; ^{209}Bi(p,13n)

Alpha Particles (^{197}Po)

α(keV)
6282 *4*

$^{197}_{84}$Po(26 *2* s)

Mode: α(84 *9* %)
Δ: >-13206 keV syst
SpA: 2.17×10^{9} Ci/g
Prod: ^{19}F on Re; ^{209}Bi(p,13n); Ne on W

Alpha Particles (^{197}Po)

α(keV)
6385 *3*

$^{197}_{85}$At(400 *100* ms)

Mode: α
Δ: -6140 *380* keV syst
SpA: 6.8×10^{10} Ci/g
Prod: ^{185}Re(^{20}Ne,8n)

Alpha Particles (^{197}At)

α(keV)
6959 *5*

A = 198

NDS **40**, 301 (1983)

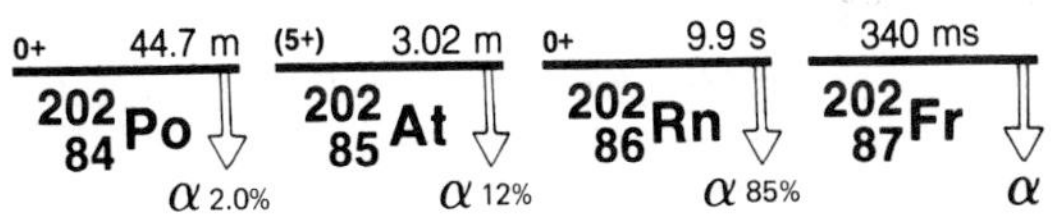

Qα 5701 *2* Qα 6354 *2* Qα 6771 *3* Qα 7590 *110*

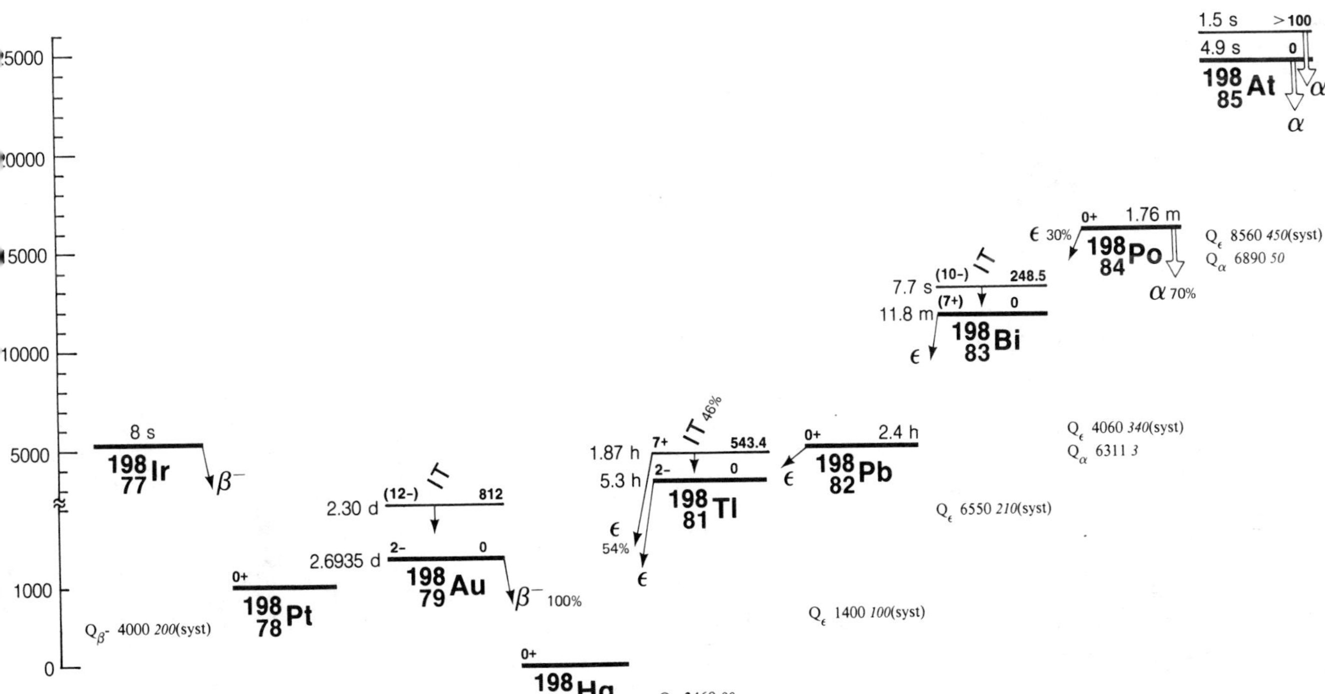

$^{198}_{77}$Ir(8 *1* s)

Mode: β-
Δ: -25930 *200* keV syst
SpA: 6.8×10^{9} Ci/g
Prod: ^{198}Pt(n,p)

Photons (^{198}Ir)

γ_{mode}	γ(keV)	γ(rel)
γ [E2]	407.4 *3*	76
γ [E2]	507.0 *3*	100

$^{198}_{78}$Pt(stable)

Δ: -29930 *6* keV
%: 7.2 *2*

$^{198}_{79}$Au(2.6935 *4* d)

Mode: β-
Δ: -29606 *4* keV
SpA: 2.4484×10^{5} Ci/g
Prod: ^{197}Au(n,γ); ^{198}Pt(p,n)

Photons (^{198}Au)

⟨γ⟩=402.6 *4* keV

γ_{mode}	γ(keV)	γ(%)†
Hg L_{ℓ}	8.722	0.0245 *23*
Hg L_{α}	9.980	0.508 *24*
Hg L_{η}	10.647	0.0082 *5*
Hg L_{β}	11.918	0.53 *3*
Hg L_{γ}	13.923	0.099 *8*
Hg $K_{\alpha 2}$	68.893	0.804 *23*
Hg $K_{\alpha 1}$	70.818	1.37 *4*
Hg $K_{\beta 1}$'	80.124	0.481 *14*
Hg $K_{\beta 2}$'	82.780	0.131 *4*
γ E2	411.8045 *10*	95.50
γ E2+47%M1	675.8875 *16*	0.802 *16*
γ E2	1087.6905 *18*	0.159 *3*

† 0.10% uncert(syst)

Atomic Electrons (^{198}Au)

⟨e⟩=15.24 *20* keV

e_{bin}(keV)	⟨e⟩(keV)	e(%)
12 - 59	0.275	1.89 *13*
65 - 80	0.0256	0.0379 *18*
329	9.47	2.88 *6*
397	1.62	0.408 *8*
398	1.71	0.431 *9*
400	0.747	0.187 *4*
408	0.384	0.0941 *19*
409	0.661	0.162 *3*
410 - 412	0.323	0.0785 *13*
661 - 676	0.0283	0.00427 *22*
1073 - 1088	0.00171	0.000159 *3*

Continuous Radiation (^{198}Au)

⟨β-⟩=406 keV;⟨IB⟩=0.49 keV

E_{bin}(keV)		⟨ ⟩(keV)	(%)
0 - 10	β-	0.061	1.21
	IB	0.019	
10 - 20	β-	0.183	1.22
	IB	0.018	0.125
20 - 40	β-	0.75	2.48
	IB	0.034	0.118
40 - 100	β-	5.4	7.7
	IB	0.088	0.137
100 - 300	β-	54	27.1
	IB	0.19	0.110
300 - 600	β-	159	36.0
	IB	0.115	0.028
600 - 1300	β-	187	24.3
	IB	0.027	0.0037
1300 - 1617	β-	0.0104	0.00075
	IB	3.8×10^{-7}	2.8×10^{-8}

$^{198}_{79}$Au(2.30 *4* d)

Mode: IT
Δ: -28794 *4* keV
SpA: 2.87×10^{5} Ci/g
Prod: ^{200}Hg(d,α); ^{197}Au(d,p); ^{198}Hg(n,p); ^{196}Pt(α,pn); ^{198}Pt(d,2n)

Photons (^{198}Au)

⟨γ⟩=577 *39* keV

γ_{mode}	γ(keV)	γ(%)†
Au L_{ℓ}	8.494	1.45 *18*
Au L_{α}	9.704	31 *3*
Au L_{η}	10.309	0.30 *3*
Au L_{β}	11.579	23.9 *23*
Au L_{γ}	13.495	4.0 *4*
Au $K_{\alpha 2}$	66.991	27.8 *18*
Au $K_{\alpha 1}$	68.806	47 *3*
Au $K_{\beta 1}$'	77.859	16.6 *11*
Au $K_{\beta 2}$'	80.428	4.4 *3*
γE1	97.21 *5*	69 *7*
γ(M4)	115.2 *15*	0.039 *4*
γE2	180.31 *5*	65 *7*
γM1(+1.2%E2)	204.10 *6*	50 *5*
γE2	214.89 *5*	77
γ	333.82 *15*	15 *4*

† 1.3% uncert(syst)

Atomic Electrons (^{198}Au)

⟨e⟩=284 *10* keV

e_{bin}(keV)	⟨e⟩(keV)	e(%)
12	8.4	70 *9*
14 - 63	12.5	67 *5*
64 - 97	6.1	7.3 *4*
100	14.1	14.1 *14*
101	20.0	19.8 *21*
103	46.7	45 *5*
112	23.0	20.5 *22*
113 - 115	8.8	7.7 *6*
123	51.0	41 *5*
134	14.7	10.9 *22*
166	3.0	1.79 *18*
167	14.0	8.4 *9*
168	9.4	5.6 *6*

Atomic Electrons (^{198}Au) (continued)

e_{bin}(keV)	⟨e⟩(keV)	e(%)
177 - 180	9.4	5.3 *3*
190	13.0	6.8 *8*
192	0.16	0.081 *24*
201	15.5	7.7 *14*
202 - 215	13.2	6.3 *8*
319 - 334	1.6	~0.5

$^{198}_{80}$Hg(stable)

Δ: -30979 *4* keV
%: 10.02 *7*

$^{198}_{81}$Tl(5.3 *5* h)

Mode: ϵ
Δ: -27520 *80* keV
SpA: 3.0×10^{6} Ci/g
Prod: daughter ^{198}Pb; ^{197}Au(α,3n); deuterons on Hg

Photons (^{198}Tl)

⟨γ⟩=2001 *43* keV

γ_{mode}	γ(keV)	γ(%)†
Hg L_{ℓ}	8.722	0.64 *6*
Hg L_{α}	9.980	13.3 *6*
Hg L_{η}	10.647	0.17 *1*
Hg L_{β}	11.930	12.3 *8*
Hg L_{γ}	13.946	2.22 *19*
Hg $K_{\alpha 2}$	68.893	23.1 *7*
Hg $K_{\alpha 1}$	70.818	39.2 *12*
Hg $K_{\beta 1}$'	80.124	13.8 *4*
Hg $K_{\beta 2}$'	82.780	3.77 *12*
γ [M1+E2]	234.80 *14*	0.45 *7*
γ	238.29 *16*	0.25 *5*
γ	318.9 *4*	~0.05
γ	325.0 *4*	0.09 *3*
γ [M1+E2]	331.76 *11*	0.59 *8*
γ	336.8 *3*	0.08 *3*
γ	350.42 *18*	0.08 *3*
γ [E2]	370.92 *14*	0.29 *4*
γ	376.67 *20*	0.20 *4*
γ E2	411.8045 *10*	82 *7*
γ	437.36 *20*	0.19 *4*
γ	448.96 *19*	0.12 *4*
γ	480.77 *16*	0.41 *4*
γ	497.92 *23*	0.22 *4*
γ	503.96 *24*	0.09 *3*
γ	511.1 *3*	1.05 *16*
γ [M1+E2]	513.58 *16*	0.26 *5*
γ	525.93 *24*	0.33 *4*
γ	550.36 *24*	0.10 *3*
γ [E2]	563.95 *14*	0.31 *5*
γ E1	587.25 *17*	0.20 *4*
γ	596.82 *14*	1.00 *11*
γ	617.1 *5*	0.10 *3* ?
γ	620.9 *3*	0.08 *3*
γ E2	636.73 *11*	10.1 *7*
γ	664.94 *25*	0.13 *3*
γ E2+47%M1	675.8875 *16*	11.9
γ	704.5 *3*	0.17 *3*
γ	712.0 *3*	0.07 *3*
γ [M1+E2]	744.94 *17*	0.13 *6*
γ	747.24 *25*	~0.08
γ	758.28 *24*	0.39 *10*
γ [M1+E2]	759.59 *13*	1.45 *13*

Photons (^{198}Tl)
(continued)

γ_{mode}	γ(keV)	γ(%)†
γ [M1+E2]	771.19 16	0.153 22
γ	786.40 25	0.28 4
γ	789.85 15	0.49 5
γ [M1+E2]	798.75 14	1.07 8
γ [E2]	810.35 18	0.17 3
γ [E2]	853.03 22	0.14 3
γ	876.79 18	0.27 3
γ	883.32 16	~0.09 ?
γ	898.6 3	~0.07
γ	912.16 24	0.09 3
γ	922.48 19	0.20 3
γ [M1+E2]	941.41 16	0.62 6
γ	951.7 5	0.06 2
γ E2	989.72 23	0.72 7
γ M1(+~0.2%E2)	1007.64 11	2.7 3
γ	1045.08 24	0.21 7
γ [M1+E2]	1046.04 23	0.46 3
γ	1066.71 18	0.22 3
γ	1074.3 6	0.044 16 ?
γ E2	1087.6905 18	2.35 6
γ	1090.04 22	0.65 25
γ	1121.1 4	0.14 3
γ	1131.8 3	0.23 3
γ	1136.73 20	0.37 3
γ	1144.94 18	0.22 3
γ M1(+7.9%E2)	1200.68 11	9.7 10
γ	1208.54 17	0.41 10
γ [E2]	1219.2 3	1.08 10
γ	1232.56 22	0.21 3
γ	1243.91 22	0.32 4
γ [M1+E2]	1273.17 14	0.36 4
γ M1(+0.9%E2)	1312.33 14	4.7 5
γ	1341.5 5	0.065 22
γ	1364.23 16	0.32 4
γ	1368.3 5	0.22 4
γ	1398.46 15	0.078 22
γ [E2]	1416.95 23	0.33 14
γ M1(+2.9%E2)	1420.82 17	8.0 9
γ M1(+3.0%E2)	1435.48 13	3.5 4
γ M1(+5.0%E2)	1447.08 16	4.3 4
γ	1474.9 6	~0.22
γ	1476.70 16	0.24 11
γ	1487.65 21	0.34 15
γ (M1+6.0%E2)	1489.75 21	2.6 3
γ	1514.81 24	0.21 3
γ	1548.53 20	0.109 22
γ	1559.21 16	0.92 10
γ	1593.57 16	2.13 22
γ	1595.8 6	0.35 11
γ [E2]	1612.48 11	0.96 4
γ	1636.9 4	0.26 4
γ	1643.55 25	0.32 4
γ	1659.03 25	1.69 16
γ	1682.7 3	0.050 22
γ	1698.0 5	0.22 3
γ	1702.1 10	0.098 22
γ	1720.96 24	2.8 3
γ [E2]	1734.28 22	0.094 22
γ	1757.64 24	0.45 7
γ	1765.92 22	0.98 10
γ	1797.49 12	0.50 7
γ [E2]	1832.62 17	4.3 4
γ [M1+E2]	1856.0 3	0.48 11
γ [E2]	1858.88 16	0.77 11
γ	1875.49 25	0.66 7
γ	1884.43 17	0.054 22
γ	1899.45 21	2.22 22
γ	1908.45 22	0.142 22
γ	1925.5 3	0.071 16
γ [M1+E2]	1949.05 14	0.120 22
γ	2040.11 16	8.4 9
γ [M1+E2]	2053.68 20	0.174 22
γ	2074.35 15	0.58 7
γ	2109.8 5	0.098 22
γ	2140.7 5	0.094 22
γ	2152.58 16	0.52 6
γ	2169.44 24	0.153 22
γ	2177.73 22	0.054 22
γ	2190.69 24	2.7 3
γ	2209.29 12	0.41 4
γ	2232.5 6	0.065 22
γ	2250.2 8	0.076 22
γ [E2]	2267.8 3	0.028 11
γ	2283.1 6	0.49 11

Photons (^{198}Tl)
(continued)

γ_{mode}	γ(keV)	γ(%)†
γ	2287.29 25	0.44 11
γ	2319.5 5	0.15 2
γ [M1+E2]	2371.0 2	0.53 5
γ	2396.3 6	0.028 9
γ	2404.4 7	0.014 7 ?
γ	2413.8 3	0.34 3
γ	2423.75 23	0.22 3
γ	2433.6 4	0.083 16
γ	2449.9 6	0.051 11
γ	2457.1 6	0.064 11
γ [E2]	2465.48 20	0.62 7
γ	2486.15 15	1.12 11
γ	2529.1 8	0.026 7
γ	2542.9 6	0.032 7
γ	2551.1 8	0.028 7
γ	2564.0 5	0.09 3
γ	2564.38 16	0.12 3
γ	2601.4 3	0.229 22
γ	2612.5 3	0.218 22
γ	2694.9 6	0.040 8
γ	2700.8 5	0.078 10
γ	2710.4 8	0.022 7
γ	2716.2 6	0.037 8
γ	2752.9 6	0.026 5
γ [E2]	2782.8 2	0.44 4
γ	2816.2 7	0.041 8
γ	2825.6 3	0.118 16
γ	2835.55 23	0.024 5
γ	2845.4 4	0.061 9
γ	2861.7 6	0.036 8
γ	2868.9 6	0.026 5
γ	2894.3 6	0.049 7
γ	2954.7 6	0.008 3
γ	2975.8 5	0.032 5
γ	2986.9 8	0.029 6
γ	3022.4 7	0.020 4
γ	3095.8 10	0.028 6
γ	3128.0 6	0.017 4
γ	3138.9 15	0.008 3 ?
γ	3164.7 6	0.026 4

† 9.2% uncert(syst)

Atomic Electrons (^{198}Tl)
$\langle e \rangle$=31.6 10 keV

e_{bin}(keV)	$\langle e \rangle$(keV)	e(%)
12	3.3	27 3
14	1.92	13.5 14
15 - 59	1.5	5.1 4
65 - 80	0.74	1.09 5
152 - 155	0.37	~0.24
220 - 267	0.47	0.20 7
288 - 328	0.18	0.058 19
329	8.1	2.47 20
331 - 377	0.12	0.032 12
397	1.38	0.35 3
398	1.55	0.39 5
400 - 449	2.18	0.53 4
466 - 514	0.30	0.061 24
522 - 552	0.038	0.007 3
554	0.66	0.120 8
560 - 587	0.05	~0.008
593	1.60	0.27 4
594 - 637	0.275	0.044 3
650 - 698	0.68	0.102 16
700 - 747	0.25	0.035 11
755 - 800	0.104	0.013 3
807 - 853	0.030	0.0037 17
858 - 908	0.10	0.012 3
909 - 952	0.35	0.038 4
962 - 1008	0.29	0.029 4
1030 - 1078	0.100	0.0094 22
1084 - 1117	0.0121	0.00111 23
1118	0.87	0.078 9
1119 - 1161	0.101	0.0088 20
1186 - 1233	0.63	0.052 4
1240 - 1285	0.034	0.0026 10

Atomic Electrons (^{198}Tl)
(continued)

e_{bin}(keV)	$\langle e \rangle$(keV)	e(%)
1297 - 1341	0.78	0.059 5
1349 - 1398	0.60	0.044 4
1402 - 1447	0.45	0.0317 22
1460 - 1556	0.23	0.015 4
1560 - 1656	0.23	0.014 5
1668 - 1763	0.22	0.0125 18
1773 - 1872	0.18	0.0101 25
1873 - 1971	0.25	~0.013
1991 - 2086	0.09	0.0044 13
2095 - 2188	0.10	0.0048 19
2194 - 2288	0.049	0.0022 5
2305 - 2403	0.062	0.0026 6
2409 - 2483	0.016	0.00066 14
2514 - 2612	0.013	0.00051 14
2618 - 2713	0.0129	0.00048 5
2733 - 2832	0.0094	0.00034 6
2833 - 2904	0.0017	5.7 14 $\times 10^{-5}$
2939 - 3020	0.0011	3.7 10 $\times 10^{-5}$
3045 - 3136	0.0010	3.3 9 $\times 10^{-5}$
3150 - 3162	8.6 $\times 10^{-5}$	2.7 11 $\times 10^{-6}$

Continuous Radiation (^{198}Tl)
$\langle\beta+\rangle$=6.1 keV;$\langle$IB$\rangle$=1.40 keV

E_{bin}(keV)		$\langle\ \rangle$(keV)	(%)
0 - 10	β+	1.19 $\times 10^{-8}$	1.40 $\times 10^{-7}$
	IB	0.00034	
10 - 20	β+	8.1 $\times 10^{-7}$	4.78 $\times 10^{-6}$
	IB	0.0021	0.017
20 - 40	β+	3.68 $\times 10^{-5}$	0.000111
	IB	0.0019	0.0060
40 - 100	β+	0.00302	0.00379
	IB	0.47	0.67
100 - 300	β+	0.184	0.083
	IB	0.061	0.035
300 - 600	β+	0.98	0.216
	IB	0.149	0.033
600 - 1300	β+	3.20	0.357
	IB	0.43	0.049
1300 - 2500	β+	1.73	0.109
	IB	0.26	0.0153
2500 - 3460	IB	0.025	0.00092
	Σβ+		0.77

$^{198}_{81}$Tl(1.87 3 h)

Mode: ε(54 2 %), IT(46 2 %)
Δ: -26977 80 keV
SpA: 8.46×10^6 Ci/g
Prod: ^{197}Au(α,3n); ^{198}Hg(d,2n)

Photons (^{198}Tl)
$\langle\gamma\rangle$=1205 41 keV

γ_{mode}	γ(keV)	γ(%)†
Hg L_ℓ	8.722	0.61 11
Tl L_ℓ	8.953	0.42 7
Hg L_α	9.980	12.7 20
Tl L_α	10.259	8.6 12
Hg L_η	10.647	0.20 4
Tl L_η	10.994	0.088 12
Hg L_β	11.920	13.1 24
Tl L_β	12.321	7.2 10
Hg L_γ	13.916	2.4 5
Tl L_γ	14.445	1.3 2
γ_{IT} M1	23.11 9	0.028 6

Photons (^{198}Tl) (continued)

γ_{mode}	γ(keV)	γ(%)†
γ_ϵE2	47.79 5	0.25 8
Hg $K_{\alpha2}$	68.893	14.0 8
Hg $K_{\alpha1}$	70.818	23.8 13
Tl $K_{\alpha2}$	70.832	8.8 10
Tl $K_{\alpha1}$	72.873	14.9 17
Hg $K_{\beta1}$'	80.124	8.4 5
Tl $K_{\beta1}$'	82.434	5.3 6
Hg $K_{\beta2}$'	82.780	2.28 13
Tl $K_{\beta2}$'	85.185	1.48 17
γ_ϵ	149.37 23	0.15 5
γ_ϵM1(+E2)	194.7 3	0.77 16
γ_ϵM1(+E2)	215.65 20	1.26 21
γ_ϵM1(+E2)	226.30 16	5.4 8
γ_ϵ	227.6 3	1.4 3
γ_ϵ	249.9 4	0.32 13 ?
γ_{IT} M1	259.52 6	2.8 4
γ_{IT} M4	260.8 3	1.28 22
γ_ϵM1	274.09 16	1.52 21
γ_{IT} M1	282.62 10	28 3
γ_ϵ	292.9 3	0.21 8
γ_ϵ	375.67 24	0.77 16
γ_ϵ	390.64 24	1.68 21
γ_ϵE2	411.8045 10	57 5
γ_ϵ	422.3 4	0.89 16
γ_ϵ	423.46 24	1.08 16
γ_ϵM1	441.95 17	2.2 3
γ_ϵ	489.74 17	4.5 5
γ_ϵM1	519.21 23	3.6 4
γ_ϵ	531.7 5	0.53 11
γ_ϵ	541.1 4	0.79 11
γ_ϵ	567.00 23	0.21 8
γ_ϵE1	587.25 17	52.5
γ_ϵ	606.3 3	0.27 11
γ_ϵE2	636.73 11	57 5
γ_ϵ	698.1 4	0.77 11
γ_ϵ	744.3 5	0.32 13
γ_ϵE2	767.4 3	1.12 16
γ_ϵ	832.6 3	0.45 8
γ_ϵ	898.6 4	0.85 13
γ_ϵ	1050.3 5	0.25 11 ?
γ_ϵ	1281.6 5	0.37 13
γ_ϵ	1392.1 4	0.39 11

† uncert(syst): 3.7% for ϵ, 7.9% for IT

Atomic Electrons (^{198}Tl)

$\langle e \rangle$=189 9 keV

e_{bin}(keV)	$\langle e \rangle$(keV)	e(%)
8 - 33	9.7	74 7
34	5.3	16 5
36	5.7	16 5
44 - 82	6.8	14 3
112 - 149	5.2	3.8 17
167 - 174	2.6	1.52 24
175	33.6	19 3
180 - 195	1.5	0.79 14
197	21.9	11.1 11
201 - 244	3.1	1.4 4
245	21.3	8.7 15
246	5.5	2.2 4
247	0.010	~0.004
248	19.9	8.0 14
249 - 256	0.147	0.058 9
257	7.6	3.0 5
258	6.4	2.5 4
259 - 308	12.2	4.6 3
329	5.7	1.73 16
339 - 388	1.5	0.43 9
390 - 439	5.6	1.37 19
440 - 489	0.50	0.11 4
490 - 539	1.80	0.355 14
540 - 587	4.1	0.74 6
591 - 637	1.44	0.231 16
661 - 698	0.12	0.017 5
729 - 767	0.07	0.009 4
815 - 833	0.08	~0.010

Atomic Electrons (^{198}Tl) (continued)

e_{bin}(keV)	$\langle e \rangle$(keV)	e(%)
884 - 899	0.016	~0.0018
1035 - 1050	0.0038	~0.00036
1267 - 1309	0.022	~0.0017
1377 - 1392	0.0038	~0.00028

$^{198}_{82}$Pb(2.4 *1* h)

Mode: ϵ

Δ: -26120 *130* keV syst

SpA: 6.6×10^6 Ci/g

Prod: ^{203}Tl(p,6n)

Photons (^{198}Pb)

$\langle \gamma \rangle$=363 18 keV

γ_{mode}	γ(keV)	γ(%)†
Tl L_ℓ	8.953	0.86 9
Tl L_α	10.259	17.4 10
Tl L_η	10.994	0.238 18
Tl L_β	12.309	16.6 13
Tl L_γ	14.409	3.1 3
Tl $K_{\alpha2}$	70.832	27.8 11
Tl $K_{\alpha1}$	72.873	47.0 18
Tl $K_{\beta1}$'	82.434	16.6 7
Tl $K_{\beta2}$'	85.185	4.66 19
γ M1	116.93 8	1.17 18
γ E2	173.43 8	18
γ M1	259.52 6	5.8 7
γ (M1+E2)	266.84 8	0.86 17
γ E2+17%M1	290.36 6	36 5
γ M1(+E2)	382.14 7	5.6 7
γ M1+E2	389.46 8	0.54 18
γ M1(+E2)	397.71 9	2.9 5
γ (M1)	467.69 10	0.72 18
γ M1	575.04 7	3.1 4
γ [M1]	605.88 9	0.6 3
γ M1	648.98 7	1.80 18
γ M1	743.1 3	1.5 3
γ M1(+E2)	865.40 7	5.9 5

† 17% uncert(syst)

Atomic Electrons (^{198}Pb)

$\langle e \rangle$=70 4 keV

e_{bin}(keV)	$\langle e \rangle$(keV)	e(%)
13	4.2	33 4
15	3.2	21.8 25
16 - 60	3.0	7.9 9
66 - 82	0.89	1.27 5
88	3.7	4.2 9
92 - 138	1.37	1.30 15
158	0.90	0.57 11
159	5.3	3.3 7
161	3.5	2.2 4
170	2.7	1.6 3
171 - 173	0.86	0.50 10
174	5.1	2.9 4
181 - 204	0.6	~0.32
205	8.9	4.3 10
213 - 261	1.77	0.71 7
263 - 275	1.8	0.66 15
276	1.8	0.64 10
278	0.87	0.31 5
280	8.7	3.1 6
287 - 312	4.4	1.5 5
350 - 398	3.6	1.01 13

Atomic Electrons (^{198}Pb) (continued)

e_{bin}(keV)	$\langle e \rangle$(keV)	e(%)
452 - 490	0.97	0.200 22
520 - 563	0.77	0.138 17
571 - 606	0.091	0.0156 25
634 - 658	0.40	0.062 9
728 - 743	0.07	0.0096 14
850 - 865	0.16	0.019 7

Continuous Radiation (^{198}Pb)

$\langle$IB$\rangle$=0.68 keV

E_{bin}(keV)		$\langle\ \rangle$(keV)	(%)
10 - 20	IB	0.0021	0.017
20 - 40	IB	0.00138	0.0042
40 - 100	IB	0.49	0.68
100 - 300	IB	0.051	0.031
300 - 600	IB	0.076	0.018
600 - 1227	IB	0.057	0.0077

$^{198}_{83}$Bi(11.85 *18* min)

Mode: ϵ

Δ: -19570 *160* keV

SpA: 8.01×10^7 Ci/g

Prod: alphas on Tl; protons on U; protons on Pb

Photons (^{198}Bi)

γ_{mode}	γ(keV)	γ(rel)
γ (E2)	90.0 1	8 3
γ M1(+E2)	138.00 24	2.3 5
γ	157.9 3	<0.9
γ E1	197.69 19	62 7
γ	247.80 24	2.2 4
γ E2	317.98 18	27 4
γ M1(+E2)	434.18 22	7.3 15
γ M1	546.2 4	2.7 5
γ E2	562.40 19	66 9
γ	917.5 6	5.5 11
γ E2	1063.49 24	100 14

$^{198}_{83}$Bi(7.7 *5* s)

Mode: IT

Δ: -19322 *160* keV

SpA: 7.1×10^9 Ci/g

Prod: ^{191}Ir(^{12}C,5n); ^{181}Ta(^{22}Ne,5n)

Photons (^{198}Bi)

$\langle \gamma \rangle$=107 4 keV

γ_{mode}	γ(keV)	γ(%)
Bi L_ℓ	9.419	0.29 3
Bi L_α	10.828	5.6 4
Bi L_η	11.712	0.197 14
Bi L_β	13.063	10.2 8
Bi L_γ	15.330	2.21 18

Photons (^{198}Bi)
(continued)

γ_{mode}	γ(keV)	γ(%)
Bi $K_{\alpha2}$	74.814	2.96 *15*
Bi $K_{\alpha1}$	77.107	4.97 *25*
Bi $K_{\beta1}'$	87.190	1.76 *9*
Bi $K_{\beta2}'$	90.128	0.52 *3*
γ E3	248.5 *5*	39.0 *16*

Atomic Electrons (^{198}Bi)
$\langle e\rangle$=140 *3* keV

e_{bin}(keV)	$\langle e\rangle$(keV)	e(%)
13	1.26	9.4 *10*
16	2.24	14.3 *16*
58	0.057	0.098 *11*
59	0.00185	0.0031 *3*
61	0.052	0.085 *9*
64	0.0153	0.024 *3*
70	0.0199	0.028 *3*
71	0.0139	0.0196 *22*
72	0.0096	0.0134 *15*
73	0.0177	0.024 *3*
74	0.0141	0.0190 *21*
75	0.00198	0.0027 *3*
76	0.0046	0.0060 *7*
77	0.00158	0.00206 *23*
83	0.0042	0.0050 *6*
84	0.0025	0.0029 *3*
85	0.00044	0.00052 *6*
86	0.0025	0.0030 *3*

Atomic Electrons (^{198}Bi)
(continued)

e_{bin}(keV)	$\langle e\rangle$(keV)	e(%)
87	0.00084	0.00098 *11*
158	16.7	10.6 *5*
232	7.1	3.04 *14*
233	56.4	24.2 *11*
235	23.3	9.9 *5*
245	24.7	10.1 *5*
246	0.256	0.104 *5*
248	8.1	3.25 *15*

$^{198}_{84}$Po(1.76 *3* min)

Mode: α(70 *8* %), ϵ(30 *8* %)
Δ: -15500 *300* keV syst
SpA: 5.38×10^8 Ci/g
Prod: ^{209}Bi(p,12n); ^{12}C on Pt; ^{19}F on Re; Ne on W

Alpha Particles (^{198}Po)

α(keV)
6182 *4*

$^{198}_{85}$At(4.9 *5* s)

Mode: α
Δ: -6940 *340* keV syst
SpA: 1.08×10^{10} Ci/g
Prod: ^{185}Re(^{20}Ne,7n)

Alpha Particles (^{198}At)

α(keV)
6748 *5*

$^{198}_{85}$At(1.5 *3* s)

Mode: α
Δ: >-6840 keV syst
SpA: 3.0×10^{10} Ci/g
Prod: ^{185}Re(^{20}Ne,7n)

Alpha Particles (^{198}At)

α(keV)
6847 *5*

A = 199

NDS **24**, 57 (1978)

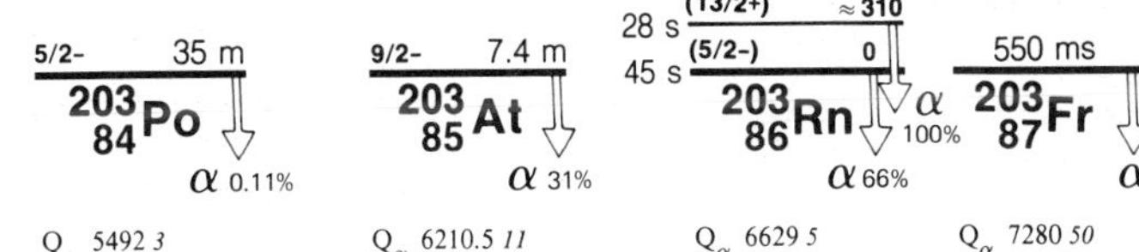

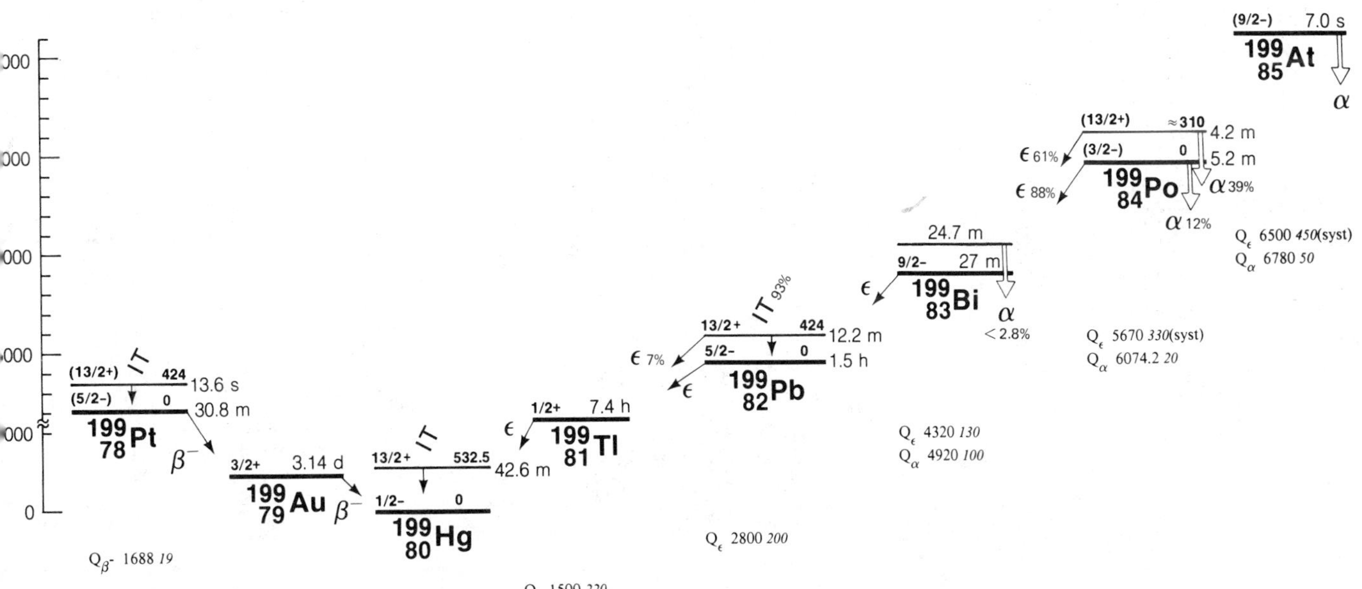

$^{199}_{78}$Pt(30.8 *4* min)

Mode: β-
Δ: -27430 *19* keV
SpA: 3.07×10^{7} Ci/g
Prod: ^{198}Pt(n,γ)

Photons (^{199}Pt)

⟨γ⟩=202 *5* keV

γ_{mode}	γ(keV)	γ(%)†
Au L_ℓ	8.494	0.085 *13*
Au L_α	9.704	1.81 *23*
Au L_η	10.309	0.021 *3*
Au L_β	11.542	1.9 *3*
Au L_γ	13.570	0.38 *6*
γ M2	55.14 *3*	0.016 *3*
Au $K_{\alpha2}$	66.991	1.20 *6*
Au $K_{\alpha1}$	68.806	2.05 *11*
γ M1+5.6%E2	77.184 *21*	1.5 *3*
Au $K_{\beta1}$'	77.859	0.72 *4*
Au $K_{\beta2}$'	80.428	0.191 *10*
γ [E2]	170.11 *3*	~0.022
γ [M1]	176.68 *3*	~0.028
γ E2	185.78 *3*	3.26 *18*
γ E1	191.692 *24*	2.38 *15*
γ M1+21%E2	219.336 *25*	0.39 *3*
γ M1+20%E2	225.902 *25*	0.170 *16*
γ [E2]	239.88 *3*	0.181 *15*
γ [E1]	240.91 *3*	<0.06
γ M1+8.1%E2	246.443 *22*	2.16 *15*
γ [E2]	297.99 *3*	0.049 *18*
γ M1+26%E2	317.061 *24*	4.87 *25*
γ [M1]	323.627 *24*	0.25 *3*
γ [E1]	417.59 *3*	0.39 *3*
γ [M1]	425.33 *4*	0.17 *3*
γ E2	465.779 *25*	0.93 *4*
γ M1+7.9%E2	468.10 *3*	0.99 *13*
γ M1+1.5%E2	474.67 *3*	1.15 *7*
γ E2	493.74 *2*	5.7 *3*
γ [M1]	505.5 *3*	0.075 *13*
γ M1+25%E2	542.963 *22*	14.8 *7*
γ [M1+E2]	610.20 *13*	0.015 *6*
γ	644.67 *4*	0.089 *9*
γ	651.23 *4*	~0.007
γ	665.36 *10*	0.061 *10*
γ M1+<26%E2	714.545 *25*	1.86 *10*
γ [M1+E2]	746.34 *14*	0.038 *7*
γ [M1+E2]	752.91 *14*	0.044 *9*
γ [M1+E2]	780.32 *13*	0.037 *7*
γ [M1+E2]	786.88 *13*	0.034 *7*
γ M1(+<17%E2)	791.729 *24*	1.07 *7*
γ	835.47 *9*	0.021 *4*
γ	842.04 *10*	0.019 *4*
γ	891.11 *4*	0.024 *6*
γ [M1+E2]	902.51 *19*	0.010 *3*
γ [M1]	968.29 *4*	1.10 *9*
γ [M1+E2]	992.78 *14*	0.013 *6*
γ [M1+E2]	1072.62 *19*	0.018 *6*
γ [M1+E2]	1079.18 *19*	0.009 *3*
γ [M1+E2]	1103.94 *13*	0.025 *7*
γ	1144.9 *10*	0.062 *12*
γ	1159.10 *10*	0.009 *3*
γ	1249.4 *3*	0.009 *3*
γ	1292.6 *7*	0.027 *9*

† 14% uncert(syst)

Atomic Electrons (^{199}Pt)

⟨e⟩=21.9 *10* keV

e_{bin}(keV)	⟨e⟩(keV)	e(%)
12 - 14	0.70	5.4 *6*
41	1.08	2.6 *5*
43 - 57	1.09	2.24 *24*
63	2.3	3.7 *8*
64 - 69	0.34	0.52 *17*
74	0.73	0.99 *23*
75 - 96	0.28	0.35 *5*
105	0.70	0.66 *4*
111 - 160	0.65	0.49 *3*
162 - 165	0.009	0.006 *3*
166	1.65	1.00 *10*
167 - 171	0.148	0.086 *5*
172	0.64	0.372 *22*
173 - 222	1.09	0.594 *18*
223 - 235	0.44	0.189 *16*
236	2.26	0.96 *8*
237 - 286	0.28	0.117 *12*
295 - 298	0.0015	0.00051 *15*
303	0.51	0.169 *15*
305 - 345	0.320	0.100 *6*
385 - 425	1.28	0.321 *15*
451 - 461	0.161	0.0352 *22*
462	3.3	0.71 *16*
463 - 505	0.258	0.0537 *13*
529	0.63	0.12 *3*
531 - 571	0.23	0.042 *7*
585 - 634	0.291	0.046 *3*
637 - 672	0.012	0.0018 *6*
700 - 749	0.227	0.0321 *19*
750 - 792	0.040	0.0052 *4*
810 - 842	0.0035	0.00043 *19*
877 - 912	0.132	0.0149 *14*
954 - 998	0.0313	0.00326 *22*
1023 - 1072	0.006	0.00053 *25*
1076 - 1104	0.0009	8 *4* $\times10^{-5}$
1131 - 1169	0.0013	0.00011 *5*
1212 - 1249	0.0013	~0.00011
1278 - 1292	0.00027	~2 $\times10^{-5}$

Continuous Radiation (^{199}Pt)

⟨β-⟩=514 keV;⟨IB⟩=0.76 keV

E_{bin}(keV)		⟨ ⟩(keV)	(%)
0 - 10	β-	0.0487	0.97
	IB	0.022	
10 - 20	β-	0.147	0.98
	IB	0.022	0.151
20 - 40	β-	0.60	1.99
	IB	0.041	0.144
40 - 100	β-	4.37	6.2
	IB	0.111	0.17
100 - 300	β-	44.5	22.2
	IB	0.26	0.150
300 - 600	β-	138	31.1
	IB	0.20	0.048
600 - 1300	β-	296	34.4
	IB	0.103	0.0136
1300 - 1688	β-	31.0	2.21
	IB	0.00137	0.000101

$^{199}_{78}$Pt(13.6 *4* s)

Mode: IT
Δ: -27006 *19* keV
SpA: 4.06×10^{9} Ci/g
Prod: ^{198}Pt(n,γ)

Photons (^{199}Pt)

⟨γ⟩=342 *13* keV

γ_{mode}	γ(keV)	γ(%)†
Pt L_ℓ	8.266	0.47 *14*
Pt L_α	9.435	10 *3*
Pt L_η	9.975	0.12 *3*
Pt L_β	11.151	14 *4*
Pt L_γ	13.189	3.2 *10*
γ M1	32 *2*	2.8 *9*
Pt $K_{\alpha2}$	65.122	1.99 *10*
Pt $K_{\alpha1}$	66.831	3.40 *17*
Pt $K_{\beta1}$'	75.634	1.19 *6*
Pt $K_{\beta2}$'	78.123	0.310 *16*
γ E3	391.93 *14*	85 *3*

† 4.7% uncert(syst)

Atomic Electrons (^{199}Pt)

⟨e⟩=78 *4* keV

e_{bin}(keV)	⟨e⟩(keV)	e(%)
12 - 13	1.34	10.6 *15*
14	2.4	17 *6*
18	12.4	69 *21*
19 - 20	1.4	7.4 *21*
29	5.1	18 *5*
30 - 75	1.9	5.7 *14*
314	22.5	7.2 *3*
378	4.81	1.27 *6*
379	13.6	3.60 *16*
380	4.61	1.21 *5*
389	6.1	1.57 *7*
390 - 392	1.87	0.477 *18*

$^{199}_{79}$Au(3.139 *7* d)

Mode: β-
Δ: -29119 *4* keV
SpA: 2.090×10^{5} Ci/g
Prod: multiple n-capture from ^{197}Au; daughter ^{199}Pt; ^{198}Pt(d,n)

Photons (^{199}Au)

⟨γ⟩=89.4 *13* keV

γ_{mode}	γ(keV)	γ(%)†
Hg L_ℓ	8.722	0.241 *23*
Hg L_α	9.980	5.0 *3*
Hg L_η	10.647	0.096 *6*
Hg L_β	11.906	6.0 *4*
Hg L_γ	13.922	1.14 *9*
γ M1+0.2%E2	49.828 *4*	0.329 *9*
Hg $K_{\alpha2}$	68.893	4.79 *14*
Hg $K_{\alpha1}$	70.818	8.13 *24*
Hg $K_{\beta1}$'	80.124	2.86 *9*
Hg $K_{\beta2}$'	82.780	0.781 *25*
γ E2	158.376 *4*	36.9 *7*
γ M1+11%E2	208.2040 *10*	8.37 *23*

† 1.9% uncert(syst)

Atomic Electrons (^{199}Au)

⟨e⟩=57.5 *7* keV

e_{bin}(keV)	⟨e⟩(keV)	e(%)
12 - 59	4.18	23.1 *13*
65 - 71	0.136	0.204 *11*
75	8.10	10.8 *3*
76 - 80	0.0169	0.0218 *12*
125	8.1	6.50 *22*
144	15.4	10.7 *3*
146	9.4	6.41 *18*
155	4.22	2.72 *8*
156	2.71	1.74 *5*
158	2.11	1.34 *4*
193	1.91	0.99 *3*
194 - 208	1.173	0.582 *13*

Continuous Radiation (^{199}Au)

⟨β-⟩=87 keV;⟨IB⟩=0.029 keV

E_{bin}(keV)		⟨ ⟩(keV)	(%)
0 - 10	β-	0.399	8.0
	IB	0.0045	
10 - 20	β-	1.14	7.6
	IB	0.0038	0.027
20 - 40	β-	4.26	14.3
	IB	0.0060	0.021
40 - 100	β-	23.3	34.2
	IB	0.0096	0.0157
100 - 300	β-	55	35.0
	IB	0.0047	0.0034
300 - 453	β-	3.08	0.91
	IB	6.3×10^{-5}	2.0×10^{-5}

$^{199}_{80}$Hg(stable)

Δ: -29571 *4* keV
%: 16.84 *11*

$^{199}_{80}$Hg(42.6 *5* min)

Mode: IT
Δ: -29039 *4* keV
SpA: 2.22×10^{7} Ci/g
Prod: ^{198}Hg(d,p); ^{196}Pt(α,n); ^{200}Hg(n,2n); ^{199}Hg(n,n')

Photons (^{199}Hg)

⟨γ⟩=186 *5* keV

γ_{mode}	γ(keV)	γ(%)†
Hg L_ℓ	8.722	0.78 *8*
Hg L_α	9.980	16.2 *12*
Hg L_η	10.647	0.239 *17*
Hg L_β	11.922	16.3 *13*
Hg L_γ	13.937	3.0 *3*
Hg $K_{\alpha2}$	68.893	18.1 *12*
Hg $K_{\alpha1}$	70.818	30.7 *20*
Hg $K_{\beta1}$'	80.124	10.8 *7*
Hg $K_{\beta2}$'	82.780	2.95 *19*
γ E2	158.376 *4*	53.0
γ M4	374.1 *1*	13.9 *11*
γ E2	413.87 *4*	0.028 *6*

† 1.9% uncert(syst)

Atomic Electrons (^{199}Hg)

⟨e⟩=350 *13* keV

e_{bin}(keV)	⟨e⟩(keV)	e(%)
12 - 59	8.1	57 *4*
65 - 71	0.51	0.77 *5*
75	11.6	15.5 *4*
76 - 80	0.064	0.082 *5*
144	22.1	15.4 *4*
146	13.5	9.2 *3*
155 - 158	12.98	8.33 *14*
291	144.9	50 *4*
331	0.0027	0.00082 *17*
359	55.6	15.5 *12*
360	14.3	4.0 *3*
362	29.8	8.2 *7*
371	27.7	7.5 *6*
372 - 414	9.1	2.45 *13*

$^{199}_{81}$Tl(7.42 *8* h)

Mode: ϵ
Δ: -28070 *220* keV
SpA: 2.122×10^{6} Ci/g
Prod: ^{197}Au(α,2n); ^{199}Hg(d,2n)

Photons (^{199}Tl)

⟨γ⟩=249 *4* keV

γ_{mode}	γ(keV)	γ(%)†
Hg L_ℓ	8.722	0.80 *7*
Hg L_α	9.980	16.7 *8*
Hg L_η	10.647	0.214 *12*
Hg L_β	11.928	15.7 *11*
Hg L_γ	13.954	2.87 *25*
γ (M1)	36.837 *20*	0.0111 *22*
γ M1+0.2%E2	49.828 *4*	0.479 *11*
γ (M1)	51.95 *4*	0.023 *5*
Hg $K_{\alpha2}$	68.893	27.3 *8*
Hg $K_{\alpha1}$	70.818	46.4 *14*
Hg $K_{\beta1}$'	80.124	16.3 *5*
Hg $K_{\beta2}$'	82.780	4.46 *14*
γ E2	158.376 *4*	4.92 *25*
γ (M1)	195.31 *3*	0.258 *25*
γ [M1]	205.66 *4*	0.010 *4*
γ M1+11%E2	208.2040 *10*	12.2 *4*
γ [M1+E2]	245.13 *3*	<0.037
γ (M1)	247.259 *17*	9.2 *5*
γ [M1]	255.49 *4*	0.012 *4*
γ [M1]	258.11 *4*	0.071 *7*
γ (M1)	284.095 *17*	2.19 *11*
γ [M1]	294.94 *3*	0.052 *5*
γ (E2)	297.086 *17*	0.34 *4*
γ M1+4.6%E2	333.923 *18*	1.75 *9*
γ [M1]	336.54 *5*	0.140 *14*
γ [M1]	346.90 *4*	0.132 *14*
γ M1+9.3%E2	403.51 *3*	1.54 *11*
γ E2	413.87 *4*	0.197 *25*
γ (M1)	455.463 *17*	12.3 *6*
γ [M1]	470.77 *5*	0.038 *7*
γ (M1)	492.299 *17*	1.51 *7*
γ [M1]	542.20 *3*	0.258 *25*
γ [M1]	592.03 *3*	0.10 *4*
γ [M1]	728.87 *4*	0.044 *5*
γ [M1]	750.41 *3*	1.03 *5*
γ [M1]	765.71 *4*	~0.012
γ [M1]	807.30 *5*	0.049 *5*
γ [M1]	817.66 *5*	0.406 *25*
γ [M1+E2]	1012.97 *4*	1.75 *9*
γ [M1+E2]	1062.79 *4*	0.246 *25*
γ [M1+E2]	1221.17 *4*	0.030 *4*

† 33% uncert(syst)

Atomic Electrons (^{199}Tl)

⟨e⟩=56.0 *10* keV

e_{bin}(keV)	⟨e⟩(keV)	e(%)
7 - 10	0.0007	0.010 *4*
12	4.0	32 *3*
14	2.38	16.7 *17*
15 - 34	0.78	5.1 *5*
35	1.34	3.83 *12*
36 - 80	4.06	6.99 *15*
112 - 123	0.31	0.27 *3*
125	11.9	9.5 *4*
144	2.06	1.43 *8*
146 - 162	2.47	1.64 *5*
164	8.0	4.9 *3*
172 - 192	0.172	0.096 *6*
193	2.78	1.44 *6*
194 - 196	0.605	0.311 *11*
201	1.59	0.79 *5*
202 - 231	1.20	0.583 *18*
232	1.73	0.74 *5*
233 - 282	2.41	0.96 *3*
283 - 332	1.02	0.321 *16*
333 - 347	0.0279	0.0083 *4*
372	4.7	1.25 *8*
388 - 414	0.71	0.176 *8*
441 - 490	1.40	0.312 *14*
491 - 540	0.051	0.0099 *19*
541 - 590	0.0069	0.0012 *3*
591 - 592	0.00036	6.0 *19* $\times10^{-5}$
646 - 683	0.203	0.0304 *18*
714 - 763	0.122	0.0166 *7*
765 - 814	0.0164	0.00204 *11*
815 - 818	0.00089	0.000109 *7*
980 - 1013	0.053	0.0053 *14*
1048 - 1063	0.0044	0.00042 *14*
1206 - 1221	0.00044	3.6 *12* $\times10^{-5}$

Continuous Radiation (^{199}Tl)

⟨IB⟩=0.87 keV

E_{bin}(keV)		⟨ ⟩(keV)	(%)
10 - 20	IB	0.0019	0.0157
20 - 40	IB	0.00146	0.0045
40 - 100	IB	0.47	0.67
100 - 300	IB	0.055	0.032
300 - 600	IB	0.126	0.028
600 - 1300	IB	0.22	0.026
1300 - 1500	IB	0.0020	0.000147

$^{199}_{82}$Pb(1.50 *17* h)

Mode: ϵ
Δ: -25270 *80* keV
SpA: 1.05×10^{7} Ci/g
Prod: ^{203}Tl(p,5n); ^{200}Hg(^{3}He,4n)

Photons (^{199}Pb)

⟨γ⟩=1447 *32* keV

γ_{mode}	γ(keV)	γ(%)†
Tl L_ℓ	8.953	0.73 *7*
Tl L_α	10.259	14.8 *7*
Tl L_η	10.994	0.191 *11*
Tl L_β	12.312	13.6 *9*
Tl L_γ	14.412	2.50 *22*
Tl $K_{\alpha2}$	70.832	24.7 *8*
Tl $K_{\alpha1}$	72.873	41.8 *13*
Tl $K_{\beta1}$'	82.434	14.7 *5*
Tl $K_{\beta2}$'	85.185	4.14 *14*

Photons (^{199}Pb)
(continued)

γ_{mode}	γ(keV)	γ(%)†
γ	120.74 6	0.082 25
γ	130.86 20	0.041 16
γ	152.27 20	0.09 3
γ	202.3 3	0.041 16
γ	223.53 15	0.10 4
γ	240.76 7	0.230 16
γ	267.22 8	0.57 3
γ	312.4 7	0.041 16
γ	319.18 9	0.13 4
γ	344.02 11	0.053 21
γ M1+15%E2	353.46 3	14.0 7
γ	361.50 7	0.49 16
γ E2+29%M1	366.98 3	65 3
γ	390.33 9	0.25 4
γ	400.66 4	1.89 16
γ	431.25 11	0.29 7
γ	433.27 8	0.25 7
γ	477.07 9	0.26 7
γ	495.12 8	0.54 8
γ	503.39 11	0.16 5
γ	521.40 5	0.62 12
γ	537.13 20	0.098 25
γ	575.06 14	0.148 25
γ	605.8 6	~0.06
γ	641.98 18	0.090 16
γ	685.33 20	0.139 16
γ [E2]	720.45 3	9.5 4
γ	724.64 16	0.16 3
γ	735.4 3	0.16 3
γ	754.12 4	2.33 12
γ	762.16 6	3.28 16
γ	777.28 9	0.44 4
γ	781.71 6	2.72 16
γ	792.6 4	0.082 25
γ	833.93 7	0.27 4
γ	838.81 10	1.31 10
γ	874.86 5	2.39 12
γ	911.73 8	0.54 8
γ	938.09 5	3.09 16
γ	984.7 3	0.08 3
γ	995.80 10	0.17 3
γ	1005.24 14	1.97 10
γ	1029.38 6	2.37 12
γ	1048.23 8	0.43 7
γ	1052.79 9	0.41 7
γ	1115.62 6	0.93 8
γ	1121.11 4	2.20 12
γ	1135.17 5	11.5 6
γ	1161.44 8	1.26 8
γ	1170.83 8	0.45 4
γ	1177.3 4	0.098 25
γ	1187.39 7	0.68 7
γ	1209.76 10	0.377 25
γ	1215.3 3	0.148 25
γ	1239.23 7	3.10 12
γ	1265.19 8	0.262 25
γ	1291.55 5	0.45 4
γ	1311.37 9	0.57 8
γ	1325.8 3	0.27 4
γ	1328.43 10	0.27 4
γ	1358.70 14	0.51 7
γ	1382.84 6	4.18 16
γ	1401.69 8	1.48 8
γ	1482.60 6	0.20 3
γ	1502.15 5	3.14 12
γ	1506.22 19	0.31 3
γ	1517.20 8	0.68 5
γ	1524.29 8	0.24 6
γ	1531.40 8	0.75 5
γ	1554.37 7	0.123 25
γ	1563.43 15	0.115 25
γ	1577.6 5	0.082 16
γ	1592.69 8	0.39 4
γ	1602.74 9	0.60 5
γ	1610.73 8	0.84 6
γ	1632.17 8	0.131 25
γ	1647.21 13	0.16 3
γ	1658.54 5	8.2
γ	1695.28 10	0.49 4
γ	1725.68 14	0.10 3
γ	1749.82 6	3.39 16
γ	1768.67 8	0.33 8
γ	1793.2 3	0.33 6
γ	1840.1 4	0.115 16

Photons (^{199}Pb)
(continued)

γ_{mode}	γ(keV)	γ(%)†
γ	1859.68 19	0.19 4
γ	1891.27 8	0.59 4
γ	1898.38 9	0.12 4
γ	1930.20 10	0.16 4
γ	1959.67 8	0.098 16
γ	1977.71 8	0.107 16
γ	2000.67 12	0.271 25
γ	2019.73 15	0.139 25
γ	2031.81 9	0.34 7
γ	2042.7 3	0.19 5
γ	2046.9 6	0.19 5
γ	2062.44 9	0.148 16
γ	2067.02 19	0.098 8
γ	2078.53 20	0.148 16
γ	2090.33 20	0.238 25
γ	2100.4 3	0.066 16
γ	2160.2 3	0.066 16
γ	2180.5 3	0.066 16
γ	2207.23 15	0.123 25
γ	2226.66 19	~0.041
γ	2237.64 9	0.90 6
γ	2244.4 3	0.049 16
γ	2303.8 3	0.074 16
γ	2341.7 3	0.221 16
γ	2362.0 3	0.131 25
γ	2367.65 13	0.123 25
γ	2399.3 3	0.115 16
γ	2434.00 19	0.221 16
γ	2547.5 3	0.041 16
γ	2566.9 4	0.041 16
γ	2643.2 3	0.049 16
γ	2752.0 7	~0.025

† 12% uncert(syst)

Atomic Electrons (^{199}Pb)
⟨e⟩=48.8 15 keV

e_{bin}(keV)	⟨e⟩(keV)	e(%)
13	3.5	28 3
15	2.6	17.6 18
35 - 82	2.03	3.4 3
105 - 152	0.36	0.29 10
155 - 202	0.39	~0.22
208 - 255	0.24	0.10 3
258 - 267	0.05	~0.018
268	7.2	2.69 13
276	0.15	~0.05
281	15.5	5.5 3
297 - 331	0.6	~0.19
338	1.38	0.409 21
339 - 351	0.79	0.23 3
352	4.35	1.23 7
353 - 401	2.61	0.72 3
410 - 452	0.31	0.07 3
462 - 509	0.10	0.021 7
518 - 562	0.045	0.009 4
571 - 606	0.023	~0.0039
627 - 673	0.89	0.14 3
677 - 725	0.9	0.13 5
732 - 781	0.39	0.052 17
782 - 831	0.29	~0.037
832 - 875	0.31	~0.037
896 - 944	0.41	0.044 19
963 - 1005	0.10	0.010 4
1014 - 1052	1.0	~0.09
1076 - 1124	0.35	0.031 11
1130 - 1177	0.25	~0.022
1180 - 1227	0.11	0.009 3
1236 - 1279	0.070	0.0055 19
1288 - 1328	0.29	~0.022
1343 - 1389	0.062	0.0045 18
1397 - 1446	0.24	0.017 7
1467 - 1565	0.15	0.0098 23
1573 - 1664	0.55	0.034 15
1680 - 1778	0.068	0.0039 10
1781 - 1879	0.037	0.0020 7
1883 - 1981	0.048	0.0025 6

Atomic Electrons (^{199}Pb)
(continued)

e_{bin}(keV)	⟨e⟩(keV)	e(%)
1985 - 2078	0.025	0.0012 3
2085 - 2178	0.032	0.0015 6
2192 - 2291	0.020	0.00088 21
2300 - 2396	0.011	0.00047 14
2419 - 2481	0.0029	0.00012 4
2532 - 2631	0.0016	6 3 $\times 10^{-5}$
2640 - 2737	0.0006	$\sim 2\times 10^{-5}$
2739 - 2749	2.1×10^{-5}	$\sim 8\times 10^{-7}$

Continuous Radiation (^{199}Pb)
⟨β+⟩=5.7 keV;⟨IB⟩=1.53 keV

E_{bin}(keV)		⟨ ⟩(keV)	(%)
0 - 10	β+	1.16×10^{-8}	1.37×10^{-7}
	IB	0.00033	
10 - 20	β+	8.1×10^{-7}	4.78×10^{-6}
	IB	0.0022	0.018
20 - 40	β+	3.80×10^{-5}	0.000114
	IB	0.0018	0.0057
40 - 100	β+	0.00330	0.00413
	IB	0.49	0.69
100 - 300	β+	0.235	0.105
	IB	0.063	0.037
300 - 600	β+	1.55	0.340
	IB	0.150	0.033
600 - 1300	β+	3.87	0.463
	IB	0.50	0.056
1300 - 2500	β+	0.0299	0.00225
	IB	0.32	0.020
2500 - 2800	IB	0.00018	6.9×10^{-6}
	Σβ+		0.91

$^{199}_{82}$Pb(12.2 3 min)

Mode: IT(93 %), ϵ(7 %)
Δ: -24846 80 keV
SpA: 7.74×10^{7} Ci/g
Prod: ^{203}Tl(p,5n); ^{200}Hg(^{3}He,4n)

Photons (^{199}Pb)

γ_{mode}	γ(keV)	γ(%)
γ_ϵ	145.2 15	
γ_ϵ	323.5 4	
γ_ϵE2+29%M1	366.98 3	
γ_ϵE3	382.4 6	
γ_ϵ	387.18 10	
γ_ϵ	416.6 4	
γ_{IT}M4	424.1 8	17.5
γ_ϵ	896.2 4	
γ_ϵ	947.4 17	
γ_ϵ	1223.3 15	
γ_ϵ	1602.28 20	
γ_ϵ	1891.1 5	
γ_ϵ	2399.3 4	
γ_ϵ	2613.0 4	
γ_ϵ	2752.0 4	

$^{199}_{83}$Bi(27 *1* min)

Mode: ϵ
Δ: -20940 *120* keV
SpA: 3.50×10^{7} Ci/g
Prod: deuterons on Pb; protons on Pb

Photons (^{199}Bi)

γ_{mode}	γ(keV)
γM4	424.1 *8*

$^{199}_{83}$Bi(24.70 *15* min)

Mode: α(<2.8 %)
Δ: >-20516 keV
SpA: 3.824×10^{7} Ci/g
Prod: deuterons on Pb; protons on Pb

Alpha Particles (^{199}Bi)

α(keV)
5484 *5*

$^{199}_{84}$Po(5.2 *1* min)

Mode: ϵ(88 *2* %), α(12 *2* %)
Δ: -15270 *300* keV syst
SpA: 1.82×10^{8} Ci/g
Prod: ^{209}Bi(p,11n); ^{19}F on Re; ^{197}Au(^{10}B,8n)

Alpha Particles (^{199}Po)

α(keV)
5952 *2*

Photons (^{199}Po)

γ_{mode}	γ(keV)	γ(%)†
γ_ϵ(M1)	187.7 *5*	7.5 *14*
γ_ϵ(E3)	229.0 *4*	4.8 *9*
γ_ϵ(M1)	233.5 *5*	5.5 *9*
γ_ϵ(M1)	246.0 *4*	4.2 *14*
γ_ϵ(M1)	260.8 *4*	4.0 *12*
γ_ϵ(E2)	361.6 *4*	22 *4*
γ_ϵ(M1)	397.8 *4*	4.2 *14*
γ_ϵ(M2)	475.0 *4*	7.0 *14*
γ_ϵ(E2)	506.7 *4*	3.7 *12*
γ_ϵ(E2)	998.4 *4*	15.4 *14*
γ_ϵ(M1)	1021.4 *6*	24.2 *23*
γ_ϵ(M1)	1034.4 *4*	46.6 *23*

† 2.3% uncert(syst)

$^{199}_{84}$Po(4.2 *1* min)

Mode: ϵ(61 *4* %), α(39 *4* %)
Δ: ~-14960 keV syst
SpA: 2.25×10^{8} Ci/g
Prod: ^{209}Bi(p,11n); ^{19}F on Re; ^{197}Au(^{10}B,8n)

Alpha Particles (^{199}Po)

α(keV)
6059 *3*

Photons (^{199}Po)

γ_{mode}	γ(keV)	γ(%)†
γ_ϵ(M1)	274.2 *5*	7.4 *22*
γ_ϵ	473.4 *4*	
γ_ϵ(M1)	499.8 *5*	25.3 *25*
γ_ϵ(M1)	1002.0 *5*	60 *3*

† 6.6% uncert(syst)

$^{199}_{85}$At(7.0 *1* s)

Mode: α
Δ: -8770 *330* keV syst
SpA: 7.71×10^{9} Ci/g
Prod: ^{185}Re(^{20}Ne,6n)

Alpha Particles (^{199}At)

α(keV)
6643 *3*

A = 200

NDS **26**, 81 (1979)

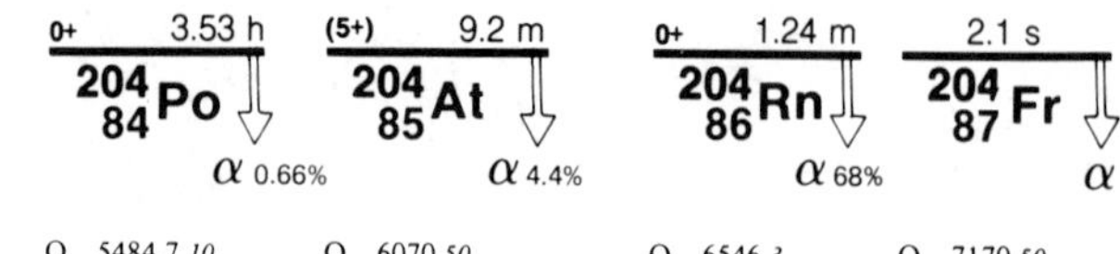

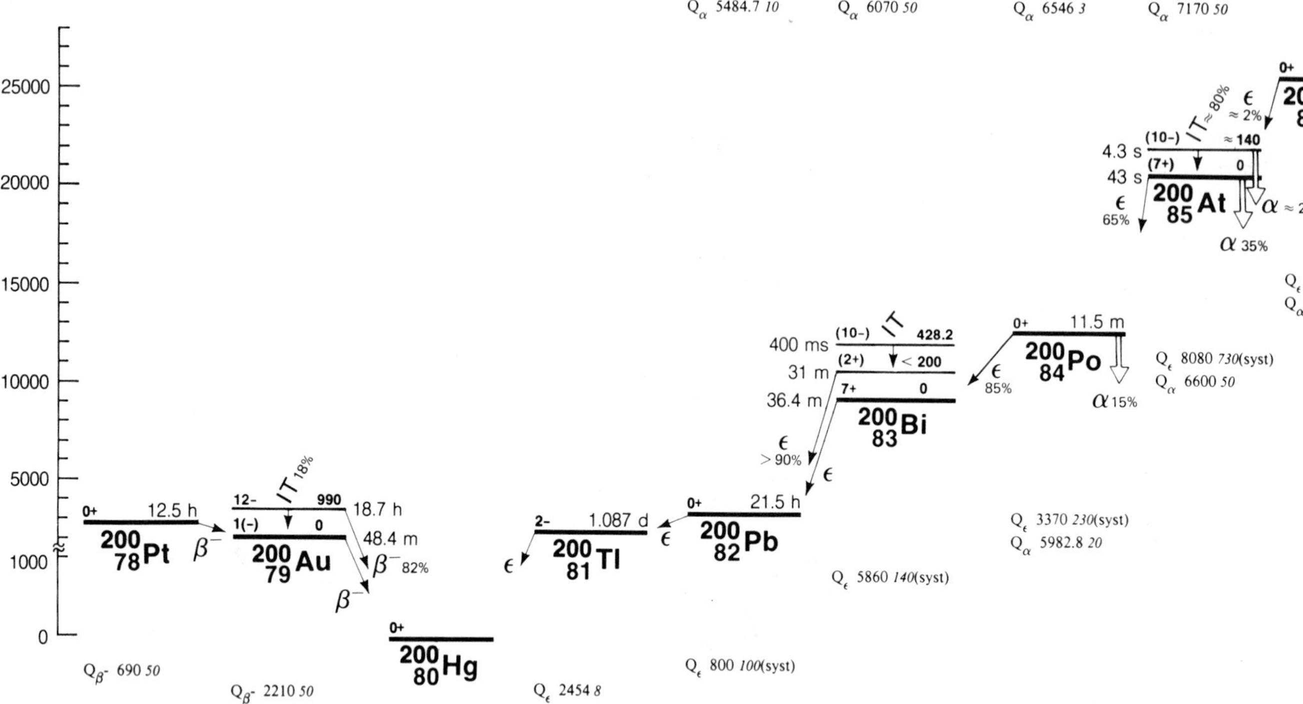

$^{200}_{78}$Pt(12.5 *3* h)

Mode: β-
Δ: -26625 *21* keV
SpA: 1.25×10^6 Ci/g
Prod: multiple n-capt on ^{198}Pt; ^{198}Pt(t,p)

Photons (^{200}Pt)

⟨γ⟩=59.2 *12* keV

γ_{mode}	γ(keV)	γ(%)†
Au L_ℓ	8.494	0.51 *6*
Au L_α	9.704	10.8 *10*
Au L_η	10.309	0.250 *19*
Au L_β	11.502	18.6 *19*
Au L_γ	13.551	4.1 *5*
γ [M1]	25.21 *6*	0.132 *13*
γ [M1]	27.43 *4*	0.037 *12*
γ [M1]	43.67 *3*	0.77 *4*
γ [M1]	59.98 *3*	2.22 *13*
Au $K_{\alpha2}$	66.991	6.16 *24*
Au $K_{\alpha1}$	68.806	10.5 *4*
γ [M1]	76.22 *3*	12.9
Au $K_{\beta1}$'	77.859	3.68 *14*
Au $K_{\beta2}$'	80.428	0.98 *4*
γ [M1]	86.53 *4*	0.028 *12*
γ [M1]	97.51 *5*	0.121 *15*
γ [M1]	103.65 *4*	0.99 *5*
γ [M1]	135.90 *9*	3.12 *18*
γ [M1]	137.68 *13*	0.22 *3*
γ [M1]	139.91 *9*	0.07 *3*
γ [M1]	146.66 *9*	0.47 *4*
γ [M1]	150.67 *9*	0.24 *4*

Photons (^{200}Pt) (continued)

γ_{mode}	γ(keV)	γ(%)†
γ [M1]	165.03 *6*	0.061 *19*
γ [M1]	166.00 *13*	0.49 *5*
γ [M1]	167.34 *9*	0.36 *5*
γ [M1]	179.57 *9*	0.045 *8*
γ [M1]	183.58 *9*	0.058 *8*
γ [M1]	189.06 *5*	0.11 *4*
γ [M1]	200.04 *4*	0.65 *4*
γ [M1]	212.6 *3*	0.017 *7*
γ [M1]	218.17 *21*	0.032 *10*
γ [M1]	227.47 *4*	1.99 *10*
γ [M1]	232.73 *5*	0.086 *7*
γ [M1]	239.54 *9*	0.076 *10*
γ [M1]	243.56 *9*	0.053 *15*
γ [M1]	243.71 *3*	2.40 *15*
γ [M1]	251.46 *22*	0.062 *10*
γ [M1]	286.57 *4*	0.034 *7*
γ [M1]	292.70 *4*	0.264 *17*
γ [M1]	303.68 *3*	0.157 *10*
γ [M1]	314.00 *4*	0.123 *9*
γ [M1]	330.24 *3*	1.06 *7*
γ [M1]	390.21 *3*	0.289 *17*
γ [M1]	408.74 *6*	0.022 *5*
γ [M1]	468.72 *6*	0.250 *14*

† 10% uncert(syst)

Atomic Electrons (^{200}Pt)

⟨e⟩=63.1 *21* keV

e_{bin}(keV)	⟨e⟩(keV)	e(%)
6 - 13	2.01	17.5 *13*
14	2.7	19.3 *22*
16 - 27	2.10	9.4 *5*
29	2.43	8.3 *6*

Atomic Electrons (^{200}Pt) (continued)

e_{bin}(keV)	⟨e⟩(keV)	e(%)
30 - 44	1.41	3.68 *15*
46	4.7	10.3 *7*
48 - 54	0.204	0.393 *20*
55	4.6	8.3 *6*
57	1.72	3.0 *3*
58 - 60	0.54	0.91 *12*
62	18.3	30 *3*
63 - 72	1.34	2.01 *13*
73	5.1	7.0 *7*
74 - 122	6.76	7.18 *25*
123 - 146	1.20	0.91 *5*
147	1.82	1.24 *8*
148 - 162	0.50	0.324 *23*
163	2.07	1.27 *10*
164 - 212	1.03	0.521 *19*
213 - 251	2.01	0.87 *3*
272 - 318	0.351	0.115 *4*
327 - 376	0.072	0.0211 *10*
378 - 409	0.095	0.0245 *15*
454 - 469	0.0211	0.00462 *23*

Continuous Radiation (^{200}Pt)

⟨β-⟩=175 keV;⟨IB⟩=0.108 keV

E_{bin}(keV)		⟨ ⟩(keV)	(%)
0 - 10	β-	0.190	3.80
	IB	0.0089	
10 - 20	β-	0.56	3.74
	IB	0.0081	0.057
20 - 40	β-	2.20	7.3
	IB	0.0144	0.050
40 - 100	β-	14.3	20.5
	IB	0.031	0.050

Continuous Radiation (^{200}Pt)
(continued)

E_{bin}(keV)		⟨ ⟩(keV)	(%)
100 - 300	β-	88	46.9
	IB	0.040	0.025
300 - 600	β-	69	17.7
	IB	0.0055	0.00156
600 - 700	β-	0.77	0.123
	IB	6.0×10^{-6}	9.8×10^{-7}

$^{200}_{79}$Au(48.4 *3* min)

Mode: β-
Δ: -27320 *50* keV
SpA: 1.942×10^{7} Ci/g
Prod: ^{202}Hg(d,α); ^{203}Tl(n,α); neutrons on Hg

Photons (^{200}Au)
⟨γ⟩=273 *12* keV

γ_{mode}	γ(keV)	γ(%)†
Hg L_ℓ	8.722	0.0099 *18*
Hg L_α	9.980	0.21 *3*
Hg L_η	10.647	0.0037 *6*
Hg L_β	11.910	0.24 *4*
Hg L_γ	13.923	0.045 *8*
Hg $K_{\alpha2}$	68.893	0.33 *5*
Hg $K_{\alpha1}$	70.818	0.56 *9*
γ [M1]	76.84 *10*	0.0066 *13*
Hg $K_{\beta1}$'	80.124	0.20 *3*
Hg $K_{\beta2}$'	82.780	0.054 *8*
γ (M1)	115.75 *13*	0.066 *23*
γ [M1]	147.98 *11* ?	~0.0032
γ M1	203.18 *13*	0.00170 *19*
γ (M1)	251.98 *8*	0.0076 *15*
γ M1	289.44 *7*	0.0140 *16*
γ M1	309.21 *8*	0.0071 *12*
γ [M1+E2]	341.55 *13*	~0.0023
γ E2	367.993 *10*	19
γ [M1+E2]	376.81 *8*	~0.005
γ M1	387.40 *8*	0.0080 *19*
γ [M1+E2]	398.78 *14*	0.0027 *7*
γ (M1+E2)	464.24 *9*	0.009 *3*
γ (M1+E2)	476.87 *11*	0.0014 *4*
γ M1	540.99 *8*	0.047 *6*
γ [E2]	544.30 *8*	$<3\times10^{-5}$
γ [E2]	564.08 *6*	0.055 *6*
γ E2	579.32 *5*	0.040 *6*
γ [M1]	601.54 *8*	0.015 *6*
γ (E2)	612.14 *7*	0.0123 *21*
γ (M1+E2)	628.78 *7*	0.027 *3*
γ [E2]	646.21 *7*	0.026 *6*
γ E2	661.45 *4*	0.391 *23*
γ M1	688.97 *8*	0.026 *5*
γ [E2]	694.27 *8*	0.0025 *10*
γ (E2)	701.61 *10*	0.0056 *17*
γ [M1+E2]	718.35 *13*	~0.006
γ [E2]	783.74 *11*	0.0025 *10*
γ E2+28%M1	886.18 *5*	0.142 *12*
γ [E2]	935.65 *7*	~0.0017
γ E2	1147.24 *11*	0.123 *9*
γ M1+22%E2	1202.44 *8*	0.180 *13*
γ M1+6.1%E2	1205.75 *7*	<0.013 ?
γ M1	1225.53 *5*	10.6 *5*
γ E2	1254.17 *5*	0.069 *8*
γ M1+14%E2	1262.99 *7*	3.11 *16*
γ M1+<0.2%E2	1273.58 *7*	0.168 *15*
γ M1+<45%E2	1350.42 *8*	0.033 *4*
γ E2+46%M1+E0	1363.06 *10*	0.015 *4*
γ	1494.8 *10*	<0.0047 ?
γ	1507.1 *10*	<0.0047 ?
γ M1	1514.96 *5*	0.111 *9*
γ M1	1570.43 *8*	0.42 *4*

Photons (^{200}Au)
(continued)

γ_{mode}	γ(keV)	γ(%)†
γ [E2]	1573.74 *7*	$<2\times10^{-5}$
γ [E2]	1593.52 *5*	~0.11
γ E2+<20%M1	1604.53 *12*	0.15 *3*
γ M1,E2	1630.98 *7*	0.31 *3*
γ (M1+<39%E2)	1693.40 *13*	0.076 *15*
γ M1	1718.41 *8*	0.074 *10*

† 14% uncert(syst)

Atomic Electrons (^{200}Au)
⟨e⟩=5.6 *3* keV

e_{bin}(keV)	⟨e⟩(keV)	e(%)
12 - 59	0.21	1.04 *17*
62 - 103	0.080	0.094 *25*
112 - 148	0.021	0.019 *6*
169 - 206	0.017	0.0090 *20*
226 - 275	0.010	0.0040 *11*
277	1.57×10^{-5}	5.7 *7* $\times10^{-6}$
285	2.1	0.74 *10*
286 - 329	0.008	0.0028 *9*
338 - 342	6.9×10^{-5}	$\sim2\times10^{-5}$
353	0.37	0.104 *15*
354	0.47	0.134 *19*
356	0.22	0.061 *9*
362 - 364	0.088	0.024 *3*
365	0.19	0.051 *7*
366 - 399	0.089	0.024 *3*
449 - 496	0.0215	0.0046 *4*
518 - 567	0.016	0.0031 *5*
576 - 625	0.0333	0.0057 *3*
626 - 675	0.0110	0.00169 *11*
677 - 718	0.00080	0.000115 *20*
769 - 803	0.0108	0.00135 *19*
853 - 886	0.0028	0.00032 *3*
921 - 936	2.2×10^{-5}	2.3 *9* $\times10^{-6}$
1064 - 1119	0.0050	0.00047 *4*
1123 - 1135	0.0015	0.00014 *5*
1142	0.97	0.085 *5*
1144 - 1171	0.0029	0.000249 *25*
1180	0.248	0.0210 *18*
1188 - 1206	0.0180	0.00151 *12*
1211	0.163	0.0135 *7*
1213 - 1260	0.106	0.0086 *3*
1261 - 1280	0.0070	0.00055 *4*
1336 - 1363	0.00069	5.1 *5* $\times10^{-5}$
1412 - 1432	0.0077	0.00054 *5*
1480 - 1579	0.058	0.0038 *4*
1581 - 1679	0.0127	0.00078 *8*
1681 - 1716	0.00100	5.9 *6* $\times10^{-5}$

Continuous Radiation (^{200}Au)
⟨β-⟩=735 keV;⟨IB⟩=1.45 keV

E_{bin}(keV)		⟨ ⟩(keV)	(%)
0 - 10	β-	0.0420	0.84
	IB	0.029	
10 - 20	β-	0.126	0.84
	IB	0.028	0.20
20 - 40	β-	0.507	1.69
	IB	0.055	0.19
40 - 100	β-	3.59	5.1
	IB	0.152	0.23
100 - 300	β-	33.2	16.7
	IB	0.40	0.23
300 - 600	β-	92	20.9
	IB	0.38	0.091
600 - 1300	β-	356	38.3
	IB	0.36	0.044
1300 - 2260	β-	249	15.9
	IB	0.043	0.0030

$^{200}_{79}$Au(18.7 *5* h)

Mode: β-(82 *2* %), IT(18 *2* %)
Δ: -26330 *86* keV
SpA: 8.38×10^{5} Ci/g
Prod: ^{198}Pt(α,pn); ^{202}Hg(d,α)

Photons (^{200}Au)
⟨γ⟩=2015 *62* keV

γ_{mode}	γ(keV)	γ(%)†
Au L_ℓ	8.494	0.30 *9*
Hg L_ℓ	8.722	0.26 *3*
Au L_α	9.704	6.3 *18*
Hg L_α	9.980	5.4 *5*
Au L_η	10.309	0.09 *3*
Hg L_η	10.647	0.108 *10*
Au L_β	11.538	6.9 *20*
Hg L_β	11.907	6.5 *6*
Au L_γ	13.535	1.3 *4*
Hg L_γ	13.904	1.22 *13*
γ_{IT} [M1]	59.98 *3*	2.9 *6*
Au $K_{\alpha2}$	66.991	6.5 *21*
Au $K_{\alpha1}$	68.806	11 *4*
Hg $K_{\alpha2}$	68.893	5.9 *3*
Hg $K_{\alpha1}$	70.818	10.1 *6*
Au $K_{\beta1}$'	77.859	3.9 *12*
Hg $K_{\beta1}$'	80.124	3.54 *20*
Au $K_{\beta2}$'	80.428	1.0 *3*
Hg $K_{\beta2}$'	82.780	0.97 *6*
γ_{IT}	84.1 *4*	0.62 *23*
γ_{IT}	101.33 *12*	0.69 *15*
γ_{IT}	105.32 *12*	0.85 *23*
$\gamma_{\beta-}$(E2)	111.14 *10*	1.8 *5*
γ_{IT}	120.18 *12*	1.0 *3*
γ_{IT}	133.13 *12*	2.9 *5*
γ_{IT}	137.2 *3*	1.2 *5*
γ_{IT}	144.5 *3*	1.0 *4*
γ_{IT}	145.97 *20*	3.5 *5*
$\gamma_{\beta-}$E2	181.19 *8*	55 *5*
γ_{IT} [M1]	218.41 *12*	1.6 *3*
$\gamma_{\beta-}$E1	255.88 *7*	71 *5*
γ_{IT}	332.7 *4*	12.1 *23*
$\gamma_{\beta-}$E2	367.993 *10*	77.1
$\gamma_{\beta-}$E2	497.78 *10*	73 *5*
$\gamma_{\beta-}$E2	579.32 *5*	72 *5*
$\gamma_{\beta-}$E2	759.51 *9*	66 *5*
$\gamma_{\beta-}$E1	904.25 *10*	7.7 *8*

† uncert(syst): 11% for IT, 2.7% for β-

Atomic Electrons (^{200}Au)
⟨e⟩=136 *7* keV

e_{bin}(keV)	⟨e⟩(keV)	e(%)
3 - 12	2.8	25 *6*
14	2.8	20 *4*
15 - 39	2.2	8 *4*
46	6.2	14 *3*
48 - 93	15.4	24 *6*
96 - 97	2.4	2.4 *7*
98	11.9	12.1 *12*
99 - 146	15.1	12.6 *25*
166	2.58	1.55 *13*
167	12.9	7.7 *7*
169	8.5	5.0 *4*
173	4.1	2.38 *19*
178	6.6	3.7 *3*
179 - 218	2.56	1.38 *9*
241 - 244	0.97	0.400 *24*
252	3.7	~1
253 - 256	0.143	0.056 *3*
285	8.6	3.01 *10*
318 - 333	2.1	0.61 *8*
353 - 368	5.77	1.62 *3*
415	6.0	1.45 *11*
484 - 495	1.58	0.325 *14*

Atomic Electrons (^{200}Au) (continued)

e_{bin}(keV)	⟨e⟩(keV)	e(%)
496	5.1	1.03 8
497 - 498	0.161	0.0323 21
564 - 579	1.99	0.351 15
676	3.7	0.55 5
747 - 760	0.354	0.0469 24
821 - 889	0.148	0.0180 18
890 - 904	0.0103	0.00115 7

Continuous-Radiation (^{200}Au)

⟨β-⟩=152 keV;⟨IB⟩=0.096 keV

E_{bin}(keV)		⟨ ⟩(keV)	(%)
0 - 10	β-	0.134	2.67
	IB	0.0077	
10 - 20	β-	0.397	2.65
	IB	0.0071	0.050
20 - 40	β-	1.57	5.2
	IB	0.0127	0.044
40 - 100	β-	10.6	15.2
	IB	0.028	0.044
100 - 300	β-	76	39.5
	IB	0.036	0.023
300 - 600	β-	64	16.7
	IB	0.0044	0.00125
600 - 608	β-	0.00290	0.000482
	IB	5.4×10^{-10}	8.9×10^{-11}

$^{200}_{80}$Hg(stable)

Δ: -29529 4 keV

%: 23.13 11

$^{200}_{81}$Tl(1.087 4 d)

Mode: ϵ

Δ: -27075 9 keV

SpA: 6.003×10^{5} Ci/g

Prod: deuterons on Hg; ^{197}Au(α,n); daughter of ^{200}Pb

Photons (^{200}Tl)

⟨γ⟩=1308 24 keV

γ_{mode}	γ(keV)	γ(%)†
Hg L_ℓ	8.722	0.66 8
Hg L_α	9.980	13.6 12
Hg L_η	10.647	0.174 16
Hg L_β	11.930	12.6 13
Hg L_γ	13.949	2.3 3
Hg $K_{\alpha2}$	68.893	23.2 18
Hg $K_{\alpha1}$	70.818	39 3
γ[M1]	76.84 10	0.031 4
Hg $K_{\beta1}'$	80.124	13.9 11
Hg $K_{\beta2}'$	82.780	3.8 3
γ(M1)	115.75 13	0.017 6
γ(M1)	116.55 9	0.11 4
γ[M1]	147.98 11 ?	0.013 4
γE2+~20%M1	182.16 8	0.052 17
γM1	203.18 13	0.0061 17
γ(M1)	251.98 8	0.28 4
γ[M1]	252.34 9	0.096 17
γM1	272.12 10	~0.035
γM1	289.44 7	0.51 4
γM1	309.21 8	0.26 4
γ[M1]	338.75 15	0.026 9
γ[M1+E2]	341.55 13	~0.017
γE2	367.993 10	87.2
γ[M1+E2]	376.81 8	~0.0013
γ[M1+E2]	383.38 17	0.052 17
γM1	387.40 8	0.16 3
γM1	398.3 10	~0.023 ?
γ[M1+E2]	398.78 14	0.021 4
γ(M1+E2)	464.24 9	0.038 13
γ	468.8 3	0.061 17
γ(M1+E2)	476.87 11	0.32 4
γ[M1+E2]	480.31 21	0.08 4
γM1,E2	485.7 10	~0.015
γ(M1+E2)	496.22 14	0.08 4
γ[M1+E2]	520.91 15	0.026 9
γM1+45%E2	521.51 8	0.25 4
γM1	533.68 13	0.044 17
γM1	540.99 8	0.030 6
γ[E2]	544.30 8	0.06 3
γ(M1+E2)	556.77 14	0.10 4
γ[E2]	564.08 6	0.0176 22
γE2	579.32 5	13.8 7
γM1	591.69 8	0.29 5
γ[M1]	601.54 8	0.0039 15
γ(E2)	612.14 7	0.24 3
γ(M1+E2)	628.78 7	0.99 7
γ[E2]	646.21 7	0.0084 18
γE2	661.45 4	2.28 12
γM1	688.97 8	0.116 22
γ[E2]	694.27 8	0.049 20
γ(E2)	701.61 10	1.29 10
γM1	711.82 8	0.27 4
γ[M1+E2]	718.35 13	~0.04
γ[M1+E2]	720.33 20	~0.04
γ[E2]	783.74 11	0.57 17
γM1	787.18 21	1.03 17
γM1+0.7%E2	828.37 7	10.8 6
γ	873.02 13	~0.017
γ	873.8 3	0.052 17
γE2+28%M1	886.18 5	1.99 11
γM1+E2	898.55 8	0.62 4
γ[E2]	935.65 7	~0.06
γM1+~39%E2	975.1 3	0.08 3
γ(M1)	1027.20 19	0.061 17
γE2	1147.24 11	0.12 4
γ(M1)	1167.12 15	0.10 4
γ	1179.89 13	<0.035 ?
γ	1180.6 3	0.11 4
γM1+22%E2	1202.44 8	0.115 20
γM1+6.1%E2	1205.75 7	29.9 17
γM1	1225.53 5	3.37 16
γE2	1254.17 5	0.97 7
γM1+14%E2	1262.99 7	0.79 6
γM1+<0.2%E2	1273.58 7	3.31 19
γ[M1+E2]	1291.14 8	0.60 5
γ[E2]	1341.4 3	0.043 17
γM1+<45%E2	1350.42 8	0.147 12
γE2+46%M1+E0	1363.06 10	3.4 4
γ[M1+E2]	1366.49 21	0.9 3
γM1+16%E2	1407.69 6	1.45 13
γ[M1+E2]	1477.87 8	0.152 13
γM1	1514.96 5	4.05 24
γM1	1570.43 8	0.27 4
γ[E2]	1573.74 7	~0.05
γ[E2]	1593.52 5	~0.036
γE2+>20%M1	1604.53 12	1.17 10
γM1,E2	1630.98 7	0.080 7
γ(M1+<39%E2)	1693.40 13	0.079 7
γM1	1718.41 8	0.331 25
γ(M1)	1746.44 14	0.057 6
γ(M1)	1759.20 12	0.14 4
γ	1759.9 3	~0.044 ?
γ	1783.5 4	0.014 4
γ[M1+E2]	1861.3 3	0.008 4
γ	1870.61 22	0.031 5
γE2	1906.33 17	0.114 10
γ[M1+E2]	1920.7 3	0.065 8
γ[M1+E2]	1928.5 4	~0.006
γ	1963.8 3	0.016 4
γM1	1975.4 6	0.045 6
γE2+>34%M1	2002.1 4	0.043 6
γM1+E2	2020.7 5	0.029 4
γ[M1]	2229.3 3	~0.0035
γ	2274.32 17	0.017 4
γ[E2]	2288.7 3	~0.0035
γM1	2296.5 4	0.033 4

† 0.46% uncert(syst)

Atomic Electrons (^{200}Tl)

⟨e⟩=36.1 6 keV

e_{bin}(keV)	⟨e⟩(keV)	e(%)
12	3.3	27 3
14	1.95	13.7 17
15 - 63	1.78	6.0 5
65 - 114	0.92	1.29 8
115 - 148	0.024	0.019 4
167 - 206	0.74	0.40 4
226 - 275	0.38	0.157 12
277	0.00057	0.000207 17
285	9.70	3.40 7
286 - 329	0.187	0.062 6
335 - 342	0.0015	0.00046 14
353	1.69	0.478 10
354	2.18	0.616 13
356	1.008	0.283 6
362 - 403	1.79	0.489 14
413 - 462	0.12	0.027 6
463 - 495	0.039	0.008 3
496	0.98	0.198 11
506 - 546	0.27	0.051 16
549 - 599	0.552	0.0966 25
600 - 649	0.251	0.040 3
658 - 706	0.27	0.038 5
708 - 720	0.0037	0.00053 6
745	1.73	0.233 14
769 - 791	0.204	0.026 3
803 - 853	0.61	0.075 5
858 - 899	0.062	0.0070 8
921 - 963	0.0116	0.00123 23
972 - 1017	0.0021	0.00021 4
1024 - 1064	0.0053	0.00050 13
1084 - 1119	0.027	0.0025 7
1123	2.71	0.241 15
1132 - 1171	0.350	0.0306 14
1176 - 1226	1.06	0.089 3
1239 - 1289	0.41	0.032 3
1290 - 1339	0.101	0.0077 7
1341 - 1366	0.067	0.0050 5
1393 - 1432	0.36	0.0250 21
1463 - 1561	0.139	0.0092 5
1567 - 1663	0.0364	0.00225 10
1676 - 1771	0.0166	0.00098 14
1778 - 1868	0.0073	0.00040 6
1881 - 1973	0.0067	0.00035 3
1987 - 2018	0.00050	2.5 3 $\times10^{-5}$
2146 - 2226	0.0019	8.5 14 $\times10^{-5}$
2259 - 2294	0.00035	1.52 22 $\times10^{-5}$

Continuous Radiation (^{200}Tl)

⟨β+⟩=1.84 keV;⟨IB⟩=1.01 keV

E_{bin}(keV)		⟨ ⟩(keV)	(%)
0 - 10	β+	7.6×10^{-9}	8.9×10^{-8}
	IB	0.00018	
10 - 20	β+	5.1×10^{-7}	3.04×10^{-6}
	IB	0.0020	0.0167
20 - 40	β+	2.37×10^{-5}	7.1×10^{-5}
	IB	0.00163	0.0051
40 - 100	β+	0.00198	0.00249
	IB	0.46	0.66
100 - 300	β+	0.132	0.059
	IB	0.050	0.030

Continuous Radiation (^{200}Tl)
(continued)

E_{bin}(keV)		⟨ ⟩(keV)	(%)
300 - 600	β+	0.76	0.168
	IB	0.088	0.020
600 - 1300	β+	0.94	0.122
	IB	0.27	0.029
1300 - 2454	β+	0.00484	0.000363
	IB	0.134	0.0088
	Σβ+		0.35

$^{200}_{82}$Pb(21.5 *4* h)

Mode: ε
Δ: -26270 *100* keV syst
SpA: 7.29×10^5 Ci/g
Prod: ^{203}Tl(p,4n); ^{202}Hg(^{3}He,5n)

Photons (^{200}Pb)
⟨γ⟩=209 *3* keV

γ_{mode}	γ(keV)	γ(%)†
Tl L_ℓ	8.953	1.06 *11*
Tl L_α	10.259	21.5 *14*
Tl L_η	10.994	0.326 *23*
Tl L_β	12.303	21.9 *17*
Tl L_γ	14.401	4.1 *4*
γ M1	32.75 *2*	0.027 *4*
Tl $K_{\alpha2}$	70.832	30.4 *17*
Tl $K_{\alpha1}$	72.873	51 *3*
Tl $K_{\beta1}'$	82.434	18.1 *10*
Tl $K_{\beta2}'$	85.185	5.1 *3*
γ M1	109.542 *19*	0.48 *7*
γ M1	142.292 *21*	3.16 *17*
γ E2	147.629 *20*	37.7 *10*
γ [M1]	154.88 *5*	0.047 *17* ?
γ M1	161.36 *4*	0.30 *3*
γ [M1+E2]	193.35 *3*	0.033 *13*
γ M1	235.625 *22*	4.30 *13*
γ E2+45%M1	257.170 *21*	4.46 *13*
γ M1	268.375 *21*	3.96 *17*
γ M1	289.16 *5*	1.1 *3*
γ M1	289.921 *25*	1.7 *3*
γ M1	302.89 *3*	0.17 *3*
γ (M1)	315.48 *4*	0.22 *3*
γ [M1]	348.23 *4*	0.16 *3*
γ [M1]	377.917 *23*	0.027 *10*
γ M1	450.52 *3*	3.33
γ [M1]	457.77 *4*	0.117 *20*
γ [M1]	525.55 *3*	0.42 *3*
γ M1	605.40 *4*	0.56 *4*

† 2.4% uncert(syst)

Atomic Electrons (^{200}Pb)
⟨e⟩=95.3 *13* keV

e_{bin}(keV)	⟨e⟩(keV)	e(%)
13	5.2	41 *5*
15	4.3	29 *3*
17 - 56	1.50	5.0 *4*
57	4.9	8.6 *5*
58 - 60	0.73	1.24 *11*
62	7.8	12.5 *4*
66 - 110	2.10	2.71 *12*
127 - 132	4.14	3.19 *9*
133	18.9	14.2 *5*
135	13.4	9.9 *3*
139 - 142	0.65	0.469 *21*
144	5.88	4.08 *14*

Atomic Electrons (^{200}Pb)
(continued)

e_{bin}(keV)	⟨e⟩(keV)	e(%)
145	3.92	2.71 *9*
146	0.129	0.089 *10*
147	2.59	1.76 *6*
148 - 149	0.467	0.317 *10*
150	4.25	2.83 *10*
151 - 181	2.26	1.32 *10*
183	3.34	1.82 *9*
190 - 232	3.7	1.76 *25*
233 - 277	2.78	1.09 *4*
285 - 334	0.269	0.092 *8*
336 - 378	1.44	0.395 *12*
435 - 458	0.514	0.117 *3*
510 - 526	0.189	0.0364 *22*
590 - 605	0.0377	0.0064 *3*

Continuous Radiation (^{200}Pb)
⟨IB⟩=0.56 keV

E_{bin}(keV)		⟨ ⟩(keV)	(%)
10 - 20	IB	0.0022	0.018
20 - 40	IB	0.00141	0.0044
40 - 100	IB	0.49	0.68
100 - 300	IB	0.041	0.026
300 - 600	IB	0.026	0.0066
600 - 900	IB	0.00080	0.000120

$^{200}_{83}$Bi(36.4 *5* min)

Mode: ε
Δ: -20410 *100* keV
SpA: 2.58×10^7 Ci/g
Prod: protons on Pb; ^{204}Pb(d,6n)

Photons (^{200}Bi)
⟨γ⟩=2289 *27* keV

γ_{mode}	γ(keV)	γ(%)
Pb L_ℓ	9.185	0.9 *3*
Pb L_α	10.541	18 *5*
Pb L_η	11.349	0.24 *7*
Pb L_β	12.699	17 *5*
Pb L_γ	14.883	3.1 *10*
Pb $K_{\alpha2}$	72.803	27 *8*
Pb $K_{\alpha1}$	74.969	46 *14*
Pb $K_{\beta1}'$	84.789	16 *5*
Pb $K_{\beta2}'$	87.632	4.7 *14*
γ M1	97.98 *16*	0.3 *1*
γ M1	103.17 *11*	1.3 *2*
γ M1	114.41 *16*	1.2 *2*
γ M1	201.15 *14*	0.9 *2*
γ E2	245.18 *8*	46 *3*
γ M1	273.43 *19* ?	1.2 *2*
γ M1	294.39 *18*	0.9 *3*
γ M1	303.42 *16*	2.2 *2*
γ	344.36 *19*	~0.50
γ (M1+E2)	348.36 *12*	2.5 *3*
γ	353.6 *10*	0.40 *8*
γ E1	419.73 *8*	91 *3*
γ E2	462.33 *8*	98 *3*
γ M1	480.30 *19*	2.3 *2*
γ (M1)	494.31 *20*	1.20 *24*
γ	539.2 *3*	1.7 *2*
γ M1	545.51 *15*	4.5 *5*
γ [M1]	642.74 *20*	0.8 *2*
γ M1	647.78 *22*	2.6 *2*
γ	781.0 *5*	2.0 *3*

Photons (^{200}Bi)
(continued)

γ_{mode}	γ(keV)	γ(%)
γ [E1]	789.1 *3*	1.0 *2*
γ	811.0 *7*	0.70 *14*
γ	837.0 *4*	~2
γ	902.6 *10*	1.0 *2*
γ	931.7 *3*	2.6 *4*
γ	935.0 *4*	1.4 *3*
γ	979.8 *10*	0.70 *14*
γ	992.9 *10*	2.9 *6*
γ E2	1026.61 *13*	100
γ	1101.4 *10*	1.10 *22*

Atomic Electrons (^{200}Bi)
⟨e⟩=79.8 *25* keV

e_{bin}(keV)	⟨e⟩(keV)	e(%)
10	0.26	2.6 *9*
13	4.2	32 *10*
15	4.1	27 *7*
16	0.56	3.5 *11*
26	1.7	6.6 *11*
56 - 75	5.3	7.9 *16*
80 - 114	6.4	6.8 *6*
157	7.6	4.8 *3*
185 - 206	2.6	1.21 *14*
215 - 229	3.27	1.47 *9*
230	4.8	2.09 *14*
232	2.63	1.14 *8*
241	0.39	0.163 *11*
242	2.05	0.85 *6*
243 - 291	3.0	1.1 *3*
292 - 331	0.19	0.061 *11*
332	3.7	1.11 *8*
333 - 354	0.23	0.068 *20*
374	9.1	2.42 *9*
392 - 420	1.39	0.341 *24*
446 - 451	4.0	0.90 *6*
458	1.92	0.42 *4*
459 - 494	1.24	0.268 *7*
523 - 560	1.39	0.254 *18*
627 - 648	0.224	0.0354 *22*
693 - 723	0.32	~0.05
749 - 798	0.23	~0.03
807 - 847	0.49	0.06 *3*
887 - 935	0.4	0.044 *22*
939	4.80	0.511 *10*
964 - 980	0.06	~0.006
989 - 1026	1.39	0.137 *6*
1086 - 1101	0.018	~0.0016

Continuous Radiation (^{200}Bi)
⟨β+⟩=121 keV;⟨IB⟩=7.8 keV

E_{bin}(keV)		⟨ ⟩(keV)	(%)
0 - 10	β+	2.83×10^{-8}	3.33×10^{-7}
	IB	0.0045	
10 - 20	β+	2.01×10^{-6}	1.19×10^{-5}
	IB	0.0065	0.047
20 - 40	β+	9.8×10^{-5}	0.000292
	IB	0.0098	0.034
40 - 100	β+	0.0090	0.0112
	IB	0.58	0.79
100 - 300	β+	0.74	0.326
	IB	0.146	0.084
300 - 600	β+	6.5	1.40
	IB	0.32	0.072
600 - 1300	β+	47.3	4.95
	IB	1.7	0.18
1300 - 2500	β+	66	3.96
	IB	3.9	0.22
2500 - 3791	β+	0.470	0.0183
	IB	1.08	0.039
	Σβ+		10.7

$^{200}_{83}$Bi(31 *2* min)

Mode: ϵ(>90 %)
Δ: -20210 *100* keV
SpA: 3.03×10^{7} Ci/g
Prod: daughter of ^{200}Po

Photons (^{200}Bi)

$\langle\gamma\rangle$=1213 *45* keV

γ_{mode}	γ(keV)	γ(%)†
γ E2	245.18 *8*	4.4 *3* ?
γ M1	273.43 *19* ?	<0.44
γ E1	419.73 *8*	20.3 *10*
γ E2	462.33 *8*	35.6 *19*
γ	712.65 *9*	1.54 *9*
γ E2	1026.61 *13*	86 *4*
γ	1739.26 *14*	3.68 *17*

† 1.0% uncert(syst)

$^{200}_{83}$Bi(400 *50* ms)

Mode: IT
Δ: -19982 *100* keV
SpA: 6.7×10^{10} Ci/g
Prod: ^{193}Ir(^{12}C,5n)

Photons (^{200}Bi)

$\langle\gamma\rangle$=370 *5* keV

γ_{mode}	γ(keV)	γ(%)
Bi L_ℓ	9.419	0.082 *8*
Bi L_α	10.828	1.61 *9*
Bi L_η	11.712	0.0417 *24*
Bi L_β	13.072	2.34 *15*
Bi L_γ	15.347	0.49 *4*
Bi $K_{\alpha2}$	74.814	1.83 *6*
Bi $K_{\alpha1}$	77.107	3.08 *9*
Bi $K_{\beta1}'$	87.190	1.09 *3*
Bi $K_{\beta2}'$	90.128	0.323 *11*
γ E3	428.2 *10*	85.1 *9*

Atomic Electrons (^{200}Bi)

$\langle e\rangle$=58.0 *7* keV

e_{bin}(keV)	$\langle e\rangle$(keV)	e(%)
13 - 61	0.91	5.9 *4*
64 - 87	0.0676	0.094 *4*
338	22.1	6.56 *15*
412	21.2	5.15 *12*
415	4.38	1.057 *24*
424	1.45	0.341 *8*
425	5.53	1.30 *3*
426 - 428	2.29	0.536 *9*

$^{200}_{84}$Po(11.5 *1* min)

Mode: ϵ(85 *2* %), α(15 *2* %)
Δ: -17040 *200* keV syst
SpA: 8.17×10^{7} Ci/g
Prod: ^{12}C on Pt; ^{204}Pb(^{3}He,7n); ^{209}Bi(p,10n); ^{187}Re(^{19}F,6n)

Alpha Particles (^{200}Po)

α(keV)
5863 *2*

Photons (^{200}Po)

$\langle\gamma\rangle$=902 *15* keV

γ_{mode}	γ(keV)	γ(%)†
γ_ϵ	53.30 *12*	0.95 *10*
γ_ϵ	101.99 *18*	0.17 *3*
γ_ϵ	145.17 *12*	1.26 *7*
γ_ϵ	147.48 *12*	4.45 *24*
γ_ϵ	151.7 *18*	0.24 *7*
γ_ϵ	154.25 *18*	0.34 *3* ?
γ_ϵ	204.98 *10*	1.50 *7*
γ_ϵ	224.96 *13*	0.92 *7*
γ_ϵ	260.2 *4*	0.68 *3*
γ_ϵ	272.38 *11*	0.37 *3*
γ_ϵ	327.92 *9*	2.62 *14*
γ_ϵ	395.32 *11*	0.41 *3*
γ_ϵ	421.8 *4*	1.36 *20*
γ_ϵ	430.07 *12*	4.76 *24*
γ_ϵ	434.28 *12*	9.3 *5*
γ_ϵ	488.3 *4*	0.34 *7*
γ_ϵ	492.7 *7*	
γ_ϵ	551.36 *13*	1.97 *10*
γ_ϵ	575.25 *15*	0.41 *7*
γ_ϵ	581.76 *15*	0.54 *10*
γ_ϵ	590.03 *12*	0.99 *3*
γ_ϵ	599.6 *6*	~0.41
γ_ϵ	617.60 *10*	19.7 *10*
γ_ϵ	662.46 *20*	1.12 *7*
γ_ϵ	670.9 *1*	34.0 *17*
γ_ϵ	692.03 *18*	0.65 *17*
γ_ϵ	695.61 *16*	5.5 *3*
γ_ϵ	720.66 *20*	0.82 *7*
γ_ϵ	730.36 *13*	1.05 *7*
γ_ϵ	755.9 *6*	0.68 *7*
γ_ϵ	777.4 *6*	0.51 *3*
γ_ϵ	796.56 *11*	7.9 *4*
γ_ϵ	818.39 *16*	0.58 *7*
γ_ϵ	849.86 *11*	4.9 *3*
γ_ϵ	875.88 *13*	1.8 *1*
γ_ϵ	895.76 *20*	1.46 *10*
γ_ϵ	914.56 *20*	1.16 *7*
γ_ϵ	918.9 *4*	0.27 *3*
γ_ϵ	931.86 *20*	0.78 *7*
γ_ϵ	945.52 *10*	1.09 *10*
γ_ϵ	1003.2 *4*	0.82 *10*
γ_ϵ	1084.46 *20*	3.81 *20*
γ_ϵ	1106.7 *4*	0.65 *7*
γ_ϵ	1145.6 *6*	0.34 *7*
γ_ϵ	1165.06 *20*	0.65 *3*
γ_ϵ	1172.86 *20*	1.09 *7*
γ_ϵ	1271.20 *13*	0.58 *3*
γ_ϵ	1285.64 *12*	1.16 *7*
γ_ϵ	1387.63 *21*	1.02 *10*
γ_ϵ	1398.4 *12*	0.37 *7*
γ_ϵ	1408.5 *4*	0.54 *3*
γ_ϵ	1438.3 *4*	0.37 *3*
γ_ϵ	1560.5 *4*	0.51 *3*
γ_ϵ	1651.56 *20*	0.65 *3*
γ_ϵ	1750.26 *20*	0.78 *14*
γ_ϵ	1801.96 *20*	1.26 *10*

† 2.3% uncert(syst)

$^{200}_{85}$At(43 *2* s)

Mode: ϵ(65 *8* %), α(35 *8* %)
Δ: -8970 *700* keV
SpA: 1.30×10^{9} Ci/g
Prod: ^{197}Au(^{12}C,9n); ^{185}Re(^{20}Ne,5n)

Alpha Particles (^{200}At)

$\langle\alpha\rangle$=2252 keV

α(keV)	α(%)
6412 *2*	21
6465 *2*	14
6574 *5*	0.21

$^{200}_{85}$At(4.3 *3* s)

Mode: IT(~80 %), α(~20 %)
Δ: -8830 *700* keV
SpA: 1.21×10^{10} Ci/g
Prod: ^{185}Re(^{20}Ne,5n)

Alpha Particles (^{200}At)

α(keV)
6536 *4*

$^{200}_{86}$Rn(1.0 *2* s)

Mode: α(~98 %), ϵ(~2 %)
Δ: -4000 *200* keV syst
SpA: 4.1×10^{10} Ci/g
Prod: protons on Th; ^{16}O on Pt; ^{14}N on Au

Alpha Particles (^{200}Rn)

α(keV)
6909 *8*

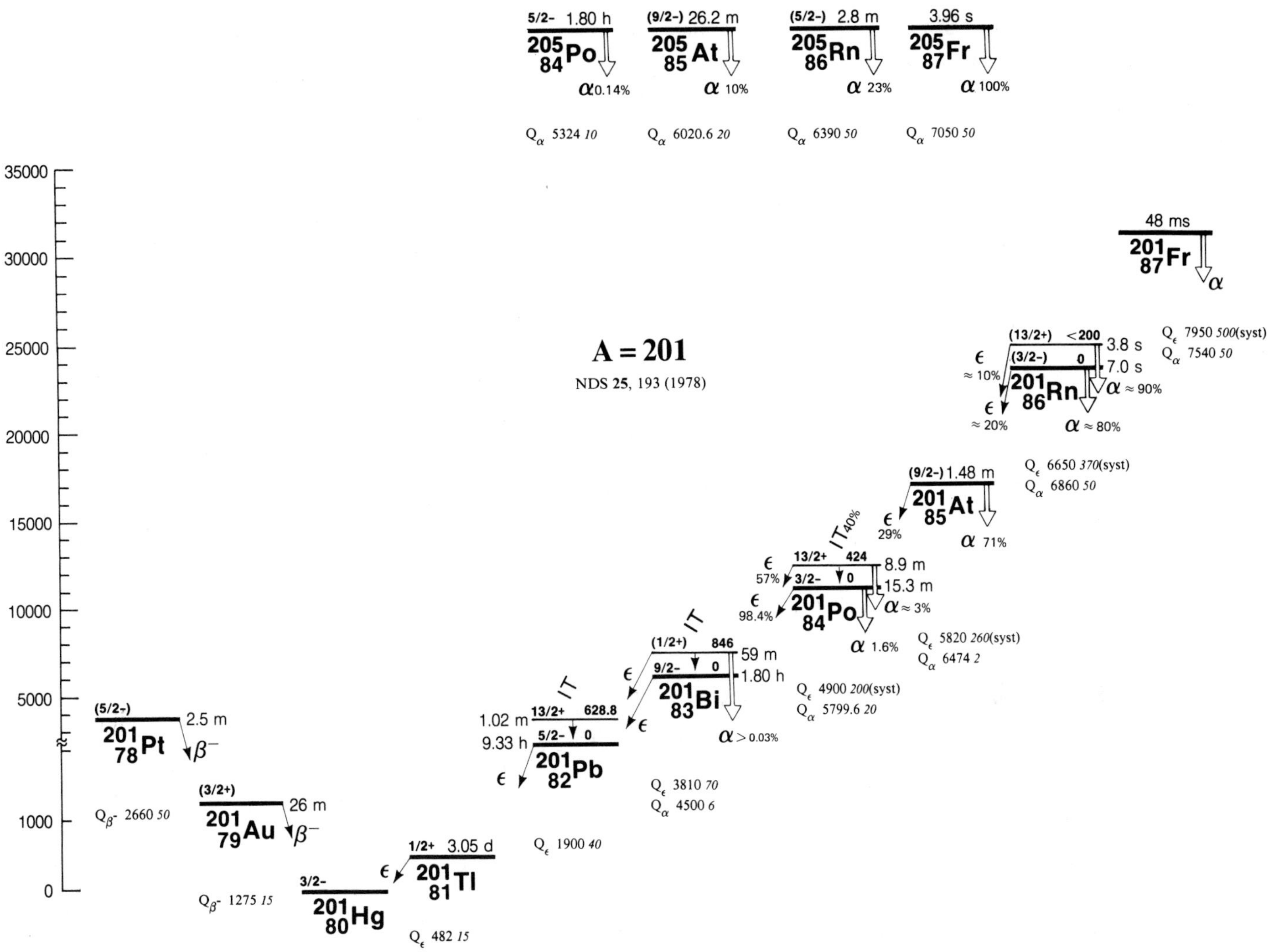

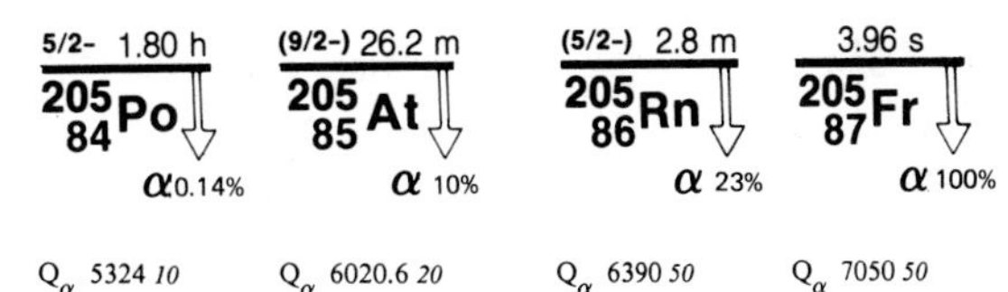

Q_α 5324 *10* Q_α 6020.6 *20* Q_α 6390 *50* Q_α 7050 *50*

A = 201

NDS **25**, 193 (1978)

35000 30000 25000 20000 15000 10000 5000 1000 0

48 ms $^{201}_{87}$Fr α — Q_ϵ 7950 *500*(syst) Q_α 7540 *50*

$^{201}_{86}$Rn: (13/2+) <200 3.8 s; (3/2-) 0 7.0 s; ϵ ≈ 10%; ϵ ≈ 20%; α ≈ 90%; α ≈ 80% — Q_ϵ 6650 *370*(syst) Q_α 6860 *50*

$^{201}_{85}$At: (9/2-) 1.48 m; ϵ 29%; α 71% — Q_ϵ 5820 *260*(syst) Q_α 6474 *2*

$^{201}_{84}$Po: 13/2+ 424 8.9 m; 3/2- 0 15.3 m; IT 40%; ϵ 57%; ϵ 98.4%; α ≈ 3%; α 1.6% — Q_ϵ 4900 *200*(syst) Q_α 5799.6 *20*

$^{201}_{83}$Bi: (1/2+) 846 59 m; 9/2- 0 1.80 h; IT; ϵ; α > 0.03% — Q_ϵ 3810 *70* Q_α 4500 *6*

$^{201}_{82}$Pb: 13/2+ 628.8 1.02 m; 5/2- 0 9.33 h; IT; ϵ — Q_ϵ 1900 *40*

$^{201}_{81}$Tl: 1/2+ 3.05 d; ϵ — Q_ϵ 482 *15*

$^{201}_{80}$Hg: 3/2-

$^{201}_{79}$Au: (3/2+) 26 m; β^- — Q_{β^-} 1275 *15*

$^{201}_{78}$Pt: (5/2-) 2.5 m; β^- — Q_{β^-} 2660 *50*

$^{201}_{78}$Pt(2.5 *1* min)

Mode: β-
Δ: -23750 *50* keV
SpA: 3.73×10^8 Ci/g
Prod: ^{204}Hg(n,α)

Photons (^{201}Pt)

γ_{mode}	γ(keV)
γ[M1+E2]	70 *5*
γ	152 *9*
γ	222 *10*
γ	1760 *20*

$^{201}_{79}$Au(26 *1* min)

Mode: β-
Δ: -26413 *16* keV
SpA: 3.60×10^7 Ci/g
Prod: neutrons on Hg; ^{202}Hg(γ,p)

Photons (^{201}Au)

⟨γ⟩=53 *8* keV

γ_{mode}	γ(keV)	γ(%)†
γ [M1+E2]	1.58 *4*	
γ [M1]	27.0 *10*	
γ M1+0.01%E2	30.54 *3*	
γ M1+0.02%E2	32.12 *3*	
Hg $K_{\alpha2}$	68.893	0.79 *6*
Hg $K_{\alpha1}$	70.818	1.34 *10*
Hg $K_{\beta1}'$	80.124	0.47 *3*
Hg $K_{\beta2}'$	82.780	0.129 *9*
γ M1+0.4%E2	135.28 *3*	0.29 *3*
γ M1+<14%E2	165.82 *4*	0.0174 *18*
γ M1+0.6%E2	167.40 *4*	1.03 *9*
γ	352.26 *20*	0.36 *8*
γ [M1]	385.23 *15*	0.65 *8*
γ	438.2 *3*	0.32 *8*
γ	517.0 *3*	1.32 *13*
γ	520.51 *16*	0.55 *10*
γ	526.86 *20*	0.71 *8*
γ	540.99 *20*	<2
γ	542.57 *20*	<2
γ	552.63 *16*	0.84 *10*
γ [M1+E2]	613.1 *3*	1.22 *11*
γ	645.0 *4*	0.67 *8*
γ [M1+E2]	732.3 *4*	0.44 *8*

† 16% uncert(syst)

Atomic Electrons (^{201}Au)

⟨e⟩=4.6 *4* keV

e_{bin}(keV)	⟨e⟩(keV)	e(%)
52	0.43	0.83 *10*
53 - 83	0.080	0.125 *13*
84	1.35	1.61 *14*
120	0.153	0.127 *15*
121 - 152	0.066	0.050 *5*
153	0.41	0.269 *24*
154 - 167	0.145	0.088 *6*
269	0.11	~0.04
302	0.31	0.103 *13*
337 - 351	0.017	~0.0049
352 - 385	0.15	0.043 *18*
423 - 426	0.016	~0.004
434	0.23	~0.05
435 - 444	0.22	~0.05
458	0.18	<0.08
470	0.13	~0.028
502 - 518	0.08	0.015 *5*
520 - 529	0.05	~0.009
530	0.19	~0.036
537 - 562	0.13	~0.024
598 - 645	0.074	0.012 *4*
649 - 717	0.05	~0.008
718 - 732	0.0050	0.00069 *20*

Continuous Radiation (^{201}Au)

⟨β-⟩=412 keV;⟨IB⟩=0.51 keV

E_{bin}(keV)		⟨ ⟩(keV)	(%)
0 - 10	β-	0.070	1.41
	IB	0.019	
10 - 20	β-	0.205	1.37
	IB	0.018	0.126
20 - 40	β-	0.79	2.65
	IB	0.034	0.119
40 - 100	β-	5.4	7.7
	IB	0.089	0.139
100 - 300	β-	53	26.4
	IB	0.19	0.113
300 - 600	β-	152	34.5
	IB	0.122	0.030
600 - 1270	β-	200	25.5
	IB	0.032	0.0045

$^{201}_{80}$Hg(stable)

Δ: -27687 *4* keV

%: 13.22 *11*

$^{201}_{81}$Tl(3.046 *8* d)

Mode: ϵ

Δ: -27205 *16* keV

SpA: 2.133×10^{5} Ci/g

Prod: daughter ^{201}Pb(9.33 h); deuterons on Hg

Photons (^{201}Tl)

⟨γ⟩=92 *4* keV

γ_{mode}	γ(keV)	γ(%)†
Hg L_{ℓ}	8.722	0.88 *11*
Hg L_{α}	9.980	18.2 *17*
Hg L_{η}	10.647	0.222 *22*
Hg L_{β}	11.925	17.5 *19*
Hg L_{γ}	13.982	3.3 *4*
γ [M1]	27.0 *10*	<0.0030
γ M1+0.01%E2	30.54 *3*	0.22 *2*
γ M1+0.02%E2	32.12 *3*	0.22 *2*
Hg $K_{\alpha2}$	68.893	26.9 *24*
Hg $K_{\alpha1}$	70.818	46 *4*
Hg $K_{\beta1}'$	80.124	16.1 *14*
Hg $K_{\beta2}'$	82.780	4.4 *4*
γ M1+0.4%E2	135.28 *3*	2.67 *13*
γ M1+<14%E2	165.82 *4*	0.16 *1*
γ M1+0.6%E2	167.40 *4*	9.4 *11*

† 5.0% uncert(syst)

Atomic Electrons (^{201}Tl)

⟨e⟩=48 *10* keV

e_{bin}(keV)	⟨e⟩(keV)	e(%)
31	10.0	<64
13	1.1	<17
12	3.8	31 *4*
13	0.0009	~0.007
14	2.3	16.5 *22*
15 - 32	5.03	27.8 *14*
52	3.99	7.6 *4*
53 - 83	2.23	3.62 *15*
84	12.4	14.8 *17*

Atomic Electrons (^{201}Tl) (continued)

e_{bin}(keV)	⟨e⟩(keV)	e(%)
120	1.41	1.17 *6*
121 - 152	0.609	0.455 *17*
153	3.8	2.5 *3*
154 - 167	1.34	0.81 *7*

Continuous Radiation (^{201}Tl)

⟨IB⟩=0.47 keV

E_{bin}(keV)		⟨ ⟩(keV)	(%)
10 - 20	IB	0.0023	0.019
20 - 40	IB	0.00158	0.0048
40 - 100	IB	0.44	0.64
100 - 300	IB	0.022	0.0156
300 - 486	IB	0.00157	0.00042

$^{201}_{82}$Pb(9.33 *3* h)

Mode: ϵ

Δ: -25300 *40* keV

SpA: 1.67×10^{6} Ci/g

Prod: ^{203}Tl(p,3n); ^{203}Tl(d,4n)

Photons (^{201}Pb)

⟨γ⟩=759 *18* keV

γ_{mode}	γ(keV)	γ(%)†
Tl L_{ℓ}	8.953	0.78 *10*
Tl L_{α}	10.259	15.7 *15*
Tl L_{η}	10.994	0.203 *21*
Tl L_{β}	12.312	14.6 *16*
Tl L_{γ}	14.415	2.7 *3*
Tl $K_{\alpha2}$	70.832	25.9 *23*
Tl $K_{\alpha1}$	72.873	44 *4*
Tl $K_{\beta1}'$	82.434	15.4 *14*
Tl $K_{\beta2}'$	85.185	4.3 *4*
γ [M1]	119.62 *5*	0.021 *6*
γ [M1]	124.15 *6*	0.044 *9*
γ [M1]	129.89 *6*	0.111 *16*
γ [M1]	155.27 *6*	0.142 *24*
γ [M1]	202.82 *5*	0.070 *9*
γ [M1]	231.92 *5*	0.100 *14*
γ [M1]	241.12 *5*	0.174 *24*
γ M1	285.16 *5*	0.158 *24*
γ [M1]	302.71 *6*	0.011 *3* ?
γ [M1]	309.11 *6*	0.040 *6*
γ [M1]	322.44 *5*	0.077 *11*
γ E2+36%M1	331.20 *3*	79 *5*
γ [M1]	341.54 *5*	0.103 *14*
γ M1	345.07 *4*	0.285 *24*
γ M1+1.3%E2	361.31 *3*	9.9 *5*
γ M1	381.38 *5*	0.229 *13*
γ (M1)	394.91 *5*	0.165 *13*
γ M1	406.07 *4*	2.01 *11*
γ	450.44 *20*	0.10 *3*
γ [M1]	465.00 *4*	0.356 *16*
γ	482.10 *5*	0.067 *11*
γ	514.53 *5*	0.17 *3*
γ [M1]	541.03 *5*	0.261 *16*
γ M1	546.33 *5*	0.269 *16*
γ	562.86 *10*	0.032 *7*
γ	573.47 *5*	0.11 *4*
γ M1(+5.9%E2)	584.62 *4*	3.56 *16*
γ E2	597.64 *6*	0.324 *24*
γ [M1]	637.99 *4*	0.43 *5*
γ E2	692.52 *3*	4.27 *16*
γ (M1)	708.77 *5*	0.77 *4*
γ [M1+E2]	727.54 *5*	0.142 *16*

Photons (^{201}Pb) (continued)

γ_{mode}	γ(keV)	γ(%)†
γ (M1+E0)	753.41 *5*	0.152 *14*
γ (M1)	767.38 *4*	3.16 *16*
γ [M1]	787.45 *4*	0.59 *7*
γ E2	803.69 *4*	1.51 *11*
γ (M1+13%E2)	826.32 *4*	2.36 *13*
γ M1(+<20%E2)	907.64 *5*	5.7 *3*
γ M1(+25%E2)	945.94 *4*	7.4 *6*
γ	947.10 *5*	~0.47
γ	969.33 *20*	0.087 *16*
γ	979.53 *5*	~0.032
γ	999.30 *4*	0.64 *4*
γ	1010.3 *3*	0.017 *4*
γ	1019.99 *21*	0.025 *7*
γ	1062.87 *7*	0.070 *9*
γ (M1+13%E2)	1070.09 *5*	1.14 *11*
γ [E2]	1088.85 *5*	0.86 *6*
γ E2(+M1)	1098.58 *4*	1.83 *11*
γ M1	1114.72 *5*	0.166 *16*
γ	1124.94 *10*	0.0103 *24*
γ (M1+48%E2)	1148.76 *4*	0.76 *5*
γ	1157.52 *4*	0.134 *16*
γ	1219.44 *14*	0.0245 *24*
γ (M1+42%E2)	1238.84 *5*	1.18 *7*
γ [M1+E2]	1277.14 *4*	1.63 *11*
γ	1286.32 *12*	0.065 *4*
γ (M1)	1308.41 *5*	0.55 *3*
γ	1330.50 *5*	~0.024
γ	1340.84 *5*	0.43 *3*
γ	1377.0 *5*	0.055 *24*
γ	1381.30 *21*	0.019 *4*
γ	1401.29 *5*	0.132 *7*
γ	1420.05 *5*	0.020 *8*
γ	1424.18 *6*	0.092 *6*
γ	1445.92 *6*	0.0364 *24*
γ [E2]	1479.96 *4*	0.164 *9*
γ	1486.25 *10*	0.0190 *24*
γ	1550.64 *14*	0.0047 *7*
γ	1587.6 *5*	0.0026 *9*
γ	1617.52 *12*	0.0245 *24*
γ	1630.9 *6*	0.0025 *7*
γ	1639.61 *5*	0.0035 *8*
γ	1672.04 *5*	0.0261 *24*
γ	1679.01 *13*	0.0042 *6*
γ	1755.38 *6*	0.0113 *10*
γ	1813.1 *3*	0.0045 *9*

† 0.63% uncert(syst)

Atomic Electrons (^{201}Pb)

⟨e⟩=60.9 *19* keV

e_{bin}(keV)	⟨e⟩(keV)	e(%)
13	3.7	29 *4*
15	2.8	18.9 *25*
34 - 60	1.71	3.2 *4*
66 - 116	1.16	1.60 *10*
117 - 156	0.49	0.345 *22*
187 - 232	0.263	0.127 *10*
237 - 241	0.064	0.027 *4*
246	24.6	10.0 *7*
256 - 273	0.263	0.101 *7*
276	5.6	2.02 *11*
281 - 310	0.234	0.078 *3*
316	4.7	1.50 *11*
317	2.23	0.70 *5*
319 - 326	1.92	0.60 *3*
327	1.14	0.348 *25*
328 - 369	3.14	0.93 *3*
378 - 406	0.445	0.115 *5*
429 - 478	0.25	0.056 *7*
479 - 526	1.04	0.209 *13*
528 - 573	0.326	0.058 *3*
581 - 625	0.523	0.086 *3*
634 - 682	0.74	0.109 *6*
689 - 738	0.268	0.0379 *22*
739 - 787	0.337	0.045 *3*
788 - 826	0.82	0.100 *6*
860 - 908	1.11	0.128 *23*

Atomic Electrons (^{201}Pb)
(continued)

e_{bin}(keV)	⟨e⟩(keV)	e(%)
914 - 957	0.27	0.029 5
964 - 1013	0.26	0.026 7
1016 - 1063	0.091	0.0086 8
1066 - 1114	0.060	0.0055 12
1121 - 1157	0.091	0.0079 16
1192 - 1238	0.18	0.015 4
1245 - 1294	0.059	0.0047 14
1296 - 1340	0.020	0.0015 5
1360 - 1409	0.0113	0.00081 12
1412 - 1446	0.00064	4.5 16 $\times10^{-5}$
1465 - 1554	0.0030	0.00020 5
1572 - 1670	0.0021	0.00013 5
1675 - 1752	0.00024	1.4 6 $\times10^{-5}$
1798 - 1810	3.2 $\times10^{-5}$	~2 $\times10^{-6}$

Continuous Radiation (^{201}Pb)
⟨β+⟩=0.097 keV;⟨IB⟩=0.86 keV

E_{bin}(keV)		⟨ ⟩(keV)	(%)
0 - 10	β+	5.9 $\times10^{-9}$	7.0 $\times10^{-8}$
	IB	9.6 $\times10^{-5}$	
10 - 20	β+	4.02 $\times10^{-7}$	2.38 $\times10^{-6}$
	IB	0.0022	0.017
20 - 40	β+	1.82 $\times10^{-5}$	5.5 $\times10^{-5}$
	IB	0.00140	0.0043
40 - 100	β+	0.00138	0.00174
	IB	0.49	0.68
100 - 300	β+	0.054	0.0255
	IB	0.052	0.031
300 - 600	β+	0.0419	0.0117
	IB	0.100	0.022
600 - 1300	IB	0.21	0.025
1300 - 1857	IB	0.0033	0.00025
	Σβ+		0.039

$^{201}_{82}$Pb(1.02 3 min)

Mode: IT
Δ: -24671 40 keV
SpA: 9.1 $\times10^{8}$ Ci/g
Prod: daughter ^{201}Bi(1.80 h); ^{203}Tl(p,3n)

Photons (^{201}Pb)
⟨γ⟩=366 69 keV

γ_{mode}	γ(keV)	γ(%)†
Pb L_ℓ	9.185	0.31 7
Pb L_α	10.541	6.2 13
Pb L_η	11.349	0.079 16
Pb L_β	12.700	5.8 12
Pb L_γ	14.900	1.08 24
Pb $K_{\alpha2}$	72.803	8.7 18
Pb $K_{\alpha1}$	74.969	15 3
Pb $K_{\beta1}$'	84.789	5.1 10
Pb $K_{\beta2}$'	87.632	1.5 3
γ M4	628.8 5	54 11

† 0.92% uncert(syst)

Atomic Electrons (^{201}Pb)
⟨e⟩=263 35 keV

e_{bin}(keV)	⟨e⟩(keV)	e(%)
13	1.4	10.6 24
15	0.83	5.5 12
16	0.25	1.6 4
56	0.063	0.111 25
57	0.103	0.18 4
58	0.0055	0.0096 22
59	0.057	0.096 22
60	0.099	0.17 4
62	0.046	0.074 17
68	0.031	0.045 10
69	0.066	0.096 21
70	0.030	0.044 10
71	0.052	0.073 16
72	0.045	0.063 14
73	0.0022	0.0031 7
74	0.014	0.018 4
75	0.0046	0.0062 14
80	0.0036	0.0045 10
81	0.013	0.017 4
82	0.0039	0.0047 11
83	0.0033	0.0040 9
84	0.0066	0.0078 18
541	167.9	31 6
613	45.8	7.5 15
614	11.0	1.8 4
616	11.0	1.8 4
625	14.7	2.3 5
626	3.3	0.53 11
628	4.7	0.76 15
629	1.09	0.17 4

$^{201}_{83}$Bi(1.80 5 h)

Mode: ϵ
Δ: -21490 60 keV
SpA: 8.66 $\times10^{6}$ Ci/g
Prod: protons on Pb

Photons (^{201}Bi)

γ_{mode}	γ(keV)	γ(rel)
γ M4	628.8 5	100
γ	785.9 4	59 10
γ	901.5 5	40 8
γ	935.7 4	38 8
γ	1013.8 7	27 5
γ	1325.5 10	32 6 ?

$^{201}_{83}$Bi(59.1 6 min)

Mode: ϵ, IT, α(>0.03 %)
Δ: -20644 60 keV
SpA: 1.583 $\times10^{7}$ Ci/g
Prod: protons on Pb; protons on Bi

Alpha Particles (^{201}Bi)

α(keV)
5240 6

Photons (^{201}Bi)

γ_{mode}	γ(keV)	γ(rel)
Bi L_ℓ	9.419	0.20 4
Bi L_α	10.828	3.9 8
Bi L_η	11.712	0.070 15
Bi L_β	13.083	4.5 9
Bi L_γ	15.373	0.9 2
Bi $K_{\alpha2}$	74.814	5.4 11
Bi $K_{\alpha1}$	77.107	9.1 18
Bi $K_{\beta1}$'	87.190	3.2 7
Bi $K_{\beta2}$'	90.128	0.96 19
γ_{IT} E5+47%M4	846 1	100
γ	1215 5 ?	

Atomic Electrons (^{201}Bi)

e_{bin}(keV)	e(rel)
13	6.1 14
16	5.6 13
58	0.18 4
59	0.0057 13
61	0.16 4
64	0.044 10
70	0.052 12
71	0.036 8
72	0.025 6
73	0.044 10
74	0.035 8
75	0.0049 11
76	0.0110 25
77	0.0038 8
83	0.0093 21
84	0.0054 12
85	0.00095 21
86	0.0054 12
87	0.0018 4
755	19 4
830	8.3 17
833	0.48 10
842	2.9 6

$^{201}_{84}$Po(15.3 2 min)

Mode: ϵ(98.4 3 %), α(1.6 3 %)
Δ: -16590 210 keV syst
SpA: 6.11 $\times10^{7}$ Ci/g
Prod: ^{209}Bi(p,9n); ^{197}Au(^{10}B,6n); ^{12}C on Pt; daughter ^{201}At

Alpha Particles (^{201}Po)

α(keV)
5683 2

Photons (^{201}Po)

$\langle\gamma\rangle$=1218 *39* keV

γ_{mode}	γ(keV)	γ(%)†
Bi L_ℓ	9.419	0.84 *23*
Bi L_α	10.828	16 *4*
Bi L_η	11.712	0.28 *7*
Bi L_β	13.088	18 *4*
Bi L_γ	15.362	3.5 *9*
Bi $K_{\alpha2}$	74.814	23 *7*
Bi $K_{\alpha1}$	77.107	38 *11*
Bi $K_{\beta1}$'	87.190	13 *4*
Bi $K_{\beta2}$'	90.128	4.0 *12*
γ_ϵM1+E2	205.6 *3*	8.0 *11* ?
γ_ϵM1	222.9 *4*	5.5 *11* ?
γ_ϵE3	225.0 *4*	12.0 *12* ?
γ_ϵM1(+E2)	239.0 *5*	8.1 *11*
γ_ϵ(E2)	428.5 *4*	8.9 *11*
γ_ϵM1	537.5 *4*	5.5 *11*
γ_ϵM1	552.0 *4*	6.3 *11*
γ_ϵ[E1]	639.0 *4*	5.4 *11* ?
γ_ϵE2	848.3 *5*	13.5 *13* ?
γ_ϵM1	890.4 *4*	54 *3*
γ_ϵM1	904.8 *5*	29.2 *16*
γ_ϵ(E2)	1163.9 *5*	3.7 *11* ?
γ_ϵ(M1)	1206.0 *5*	3.3 *11*

† 11% uncert(syst)

Atomic Electrons (^{201}Po)

$\langle e\rangle$=125 *8* keV

e_{bin}(keV)	$\langle e\rangle$(keV)	e(%)
13	3.7	28 *8*
16 - 61	4.4	23 *6*
64 - 87	0.84	1.16 *13*
115	5.9	~5
132	6.6	5.0 *10*
134	5.5	4.1 *4*
148	5.2	~4
189 - 192	2.2	1.1 *5*
202 - 207	2.8	1.4 *4*
209	28.6	13.7 *14*
210	0.0118	0.0056 *11*
212	11.4	5.4 *6*
219 - 220	0.44	0.20 *4*
221	8.2	3.7 *4*
222	3.6	1.61 *17*
223 - 239	6.6	2.9 *7*
338 - 412	0.90	0.27 *3*
413 - 461	4.8	1.06 *12*
521 - 552	1.28	0.240 *21*
623 - 639	0.035	0.0056 *9*
758	0.78	0.103 *10*
800	9.7	1.22 *7*
814	5.2	0.64 *4*
832 - 848	1.75	0.201 *10*
874 - 905	3.57	0.404 *12*
1073 - 1115	0.56	0.051 *12*
1148 - 1193	0.113	0.0096 *20*
1202 - 1206	0.020	0.0017 *5*

Continuous Radiation (^{201}Po)

$\langle\beta+\rangle$=200 keV;$\langle$IB$\rangle$=8.7 keV

E_{bin}(keV)		$\langle\ \rangle$(keV)	(%)
0 - 10	β+	2.95×10^{-8}	3.47×10^{-7}
	IB	0.0072	
10 - 20	β+	2.14×10^{-6}	1.26×10^{-5}
	IB	0.0087	0.062
20 - 40	β+	0.000106	0.000317
	IB	0.0147	0.051
40 - 100	β+	0.0101	0.0126
	IB	0.48	0.65
100 - 300	β+	0.89	0.391
	IB	0.18	0.101
300 - 600	β+	8.4	1.80
	IB	0.34	0.075
600 - 1300	β+	69	7.1
	IB	1.7	0.18
1300 - 2500	β+	118	6.9
	IB	4.4	0.24
2500 - 5000	β+	4.41	0.167
	IB	1.48	0.052
	$\Sigma\beta$+		16.3

$^{201}_{84}$Po(8.9 *2* min)

Mode: ϵ(57 *15* %), IT(40 *14* %), α(~ 3 %)
Δ: -16166 *210* keV syst
SpA: 1.050×10^{8} Ci/g
Prod: ^{209}Bi(p,9n); ^{197}Au(^{10}B,6n); ^{12}C on Pt

Alpha Particles (^{201}Po)

α(keV)
5786 *2*

Photons (^{201}Po)

$\langle\gamma\rangle$=644 *35* keV

γ_{mode}	γ(keV)	γ(%)†
Bi L_ℓ	9.419	0.36 *12*
Po L_ℓ	9.658	1.2 *3*
Bi L_α	10.828	7.0 *24*
Po L_α	11.119	24 *5*
Bi L_η	11.712	0.09 *3*
Po L_η	12.085	0.29 *6*
Bi L_β	13.099	6.5 *22*
Po L_β	13.504	21 *5*
Bi L_γ	15.379	1.2 *4*
Po L_γ	15.902	4.1 *9*
Bi $K_{\alpha2}$	74.814	11 *4*
Po $K_{\alpha2}$	76.858	26 *5*
Bi $K_{\alpha1}$	77.107	19 *6*
Po $K_{\alpha1}$	79.290	44 *9*
Bi $K_{\beta1}$'	87.190	6.7 *22*
Po $K_{\beta1}$'	89.639	15 *3*
Bi $K_{\beta2}$'	90.128	2.0 *7*
Po $K_{\beta2}$'	92.673	4.7 *10*
γ_ϵM1	272.6 *4*	2.8 *9*
γ_ϵM1	412.3 *5*	15.1 *15*
γ_{IT} M4	417.9 *3*	33 *7*
γ_ϵM1	967.0 *5*	33.6 *17*

† uncert(syst): 26% for ϵ, 35% for IT

Atomic Electrons (^{201}Po)

$\langle e\rangle$=630 *68* keV

e_{bin}(keV)	$\langle e\rangle$(keV)	e(%)
4 - 17	12.2	89 *15*
58 - 89	2.83	4.3 *3*
182 - 256	2.6	1.5 *5*
257 - 273	0.27	0.102 *22*
322	8.2	2.6 *3*
325	302.8	93 *19*
397 - 399	0.156	0.039 *4*
401	121.9	30 *6*
402	32.0	8.0 *16*
404	59.0	15 *3*
408 - 412	0.56	0.136 *10*
414	42.1	10.2 *20*
415	18.6	4.5 *9*
417 - 418	20.4	4.9 *8*
876 - 951	5.4	0.62 *3*
954 - 967	0.314	0.0326 *13*

Continuous Radiation (^{201}Po)

$\langle\beta+\rangle$=218 keV;$\langle$IB$\rangle$=7.4 keV

E_{bin}(keV)		$\langle\ \rangle$(keV)	(%)
0 - 10	β+	1.75×10^{-8}	2.05×10^{-7}
	IB	0.0073	
10 - 20	β+	1.27×10^{-6}	7.5×10^{-6}
	IB	0.0081	0.057
20 - 40	β+	6.3×10^{-5}	0.000189
	IB	0.0147	0.051
40 - 100	β+	0.0061	0.0075
	IB	0.29	0.39
100 - 300	β+	0.54	0.239
	IB	0.155	0.088
300 - 600	β+	5.4	1.15
	IB	0.26	0.059
600 - 1300	β+	50.0	5.1
	IB	1.16	0.122
1300 - 2500	β+	137	7.6
	IB	3.4	0.18
2500 - 4457	β+	24.6	0.90
	IB	2.1	0.072
	$\Sigma\beta$+		15.0

$^{201}_{85}$At(1.48 *5* min)

Mode: α(71 *7* %), ϵ(29 *7* %)
Δ: -10770 *180* keV
SpA: 6.28×10^{8} Ci/g
Prod: ^{197}Au(^{12}C,8n); ^{185}Re(^{20}Ne,4n); ^{187}Re(^{20}Ne,6n)

Alpha Particles (^{201}At)

α(keV)
6344 *3*

$^{201}_{86}$Rn(7.0 *4* s)

Mode: α(~ 80 %), ϵ(~ 20 %)
Δ: -4120 *330* keV syst
SpA: 7.6×10^{9} Ci/g
Prod: ^{197}Au(^{14}N,10n); protons on Th; ^{16}O on Pt

Alpha Particles (^{201}Rn)

α(keV)
6721 *8*

$^{201}_{86}$Rn(3.8 *4* s)

Mode: α(~ 90 %), ε(~ 10 %)
Δ: -3920 *330* keV syst
SpA: 1.35×10^{10} Ci/g
Prod: ^{197}Au(^{14}Ne,10n); ^{16}O on Pt
protons on Th

Alpha Particles (^{201}Rn)

α(keV)
6770 *8*

$^{201}_{87}$Fr(48 *15* ms)

Mode: α
Δ: 3830 *380* keV syst
SpA: 8.1×10^{10} Ci/g
Prod: protons on ^{238}U

Alpha Particles (^{201}Fr)

α(keV)
7388 *15*

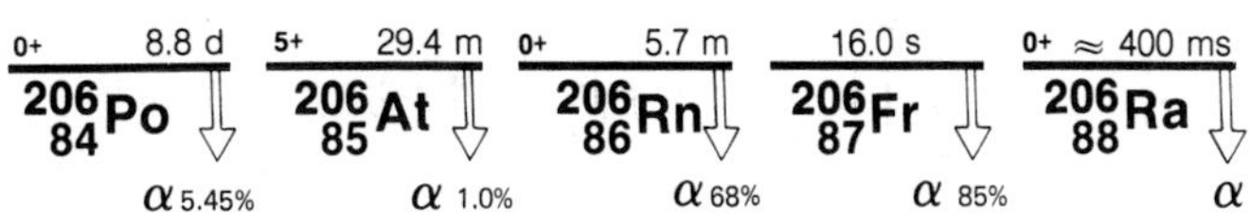

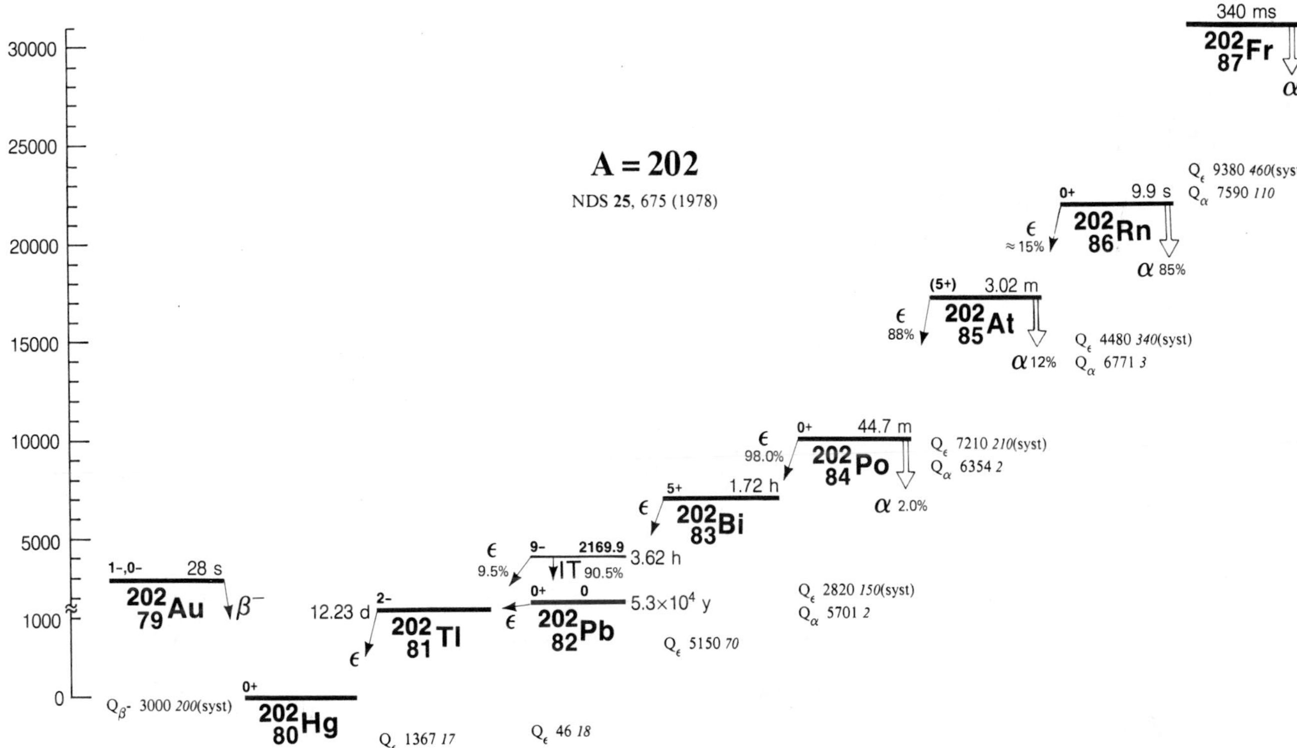

$^{202}_{79}$Au(28 *2* s)

Mode: β-
Δ: -24370 *200* keV syst
SpA: 1.97×10^{9} Ci/g
Prod: ^{202}Hg(n,p); ^{205}Tl(n,α)

Photons (^{202}Au)

⟨γ⟩=152 *14* keV

γ_{mode}	γ(keV)	γ(%)†
Hg L_ℓ	8.722	0.0030 *5*
Hg L_α	9.980	0.062 *10*
Hg L_η	10.647	0.00094 *16*
Hg L_β	11.922	0.062 *11*
Hg L_γ	13.928	0.0114 *21*
Hg $K_{\alpha2}$	68.893	0.102 *16*
Hg $K_{\alpha1}$	70.818	0.17 *3*
Hg $K_{\beta1}$'	80.124	0.061 *10*
Hg $K_{\beta2}$'	82.780	0.017 *3*
γ	388.5 *4*	0.10 *2*
γ E2	439.567 *10*	10 *2*
γ M1+45%E2	520.14 *7*	1.1 *3*
γ	786.4 *5*	0.059 *10*
γ	908.6 *4*	1.8 *3*

Photons (^{202}Au)
(continued)

γ_{mode}	γ(keV)	γ(%)†
γ E2	959.71 *7*	0.14 *5*
γ [E2]	1125.4 *4*	2.5 *5*
γ [E2]	1203.7 *4*	2.1 *5*
γ	1306.5 *5*	2.3 *5*

† approximate

Atomic Electrons (^{202}Au)

⟨e⟩=2.3 3 keV

e_{bin}(keV)	⟨e⟩(keV)	e(%)
12 - 59	0.033	0.22 3
65 - 80	0.0032	0.0048 4
305	0.026	~0.008
356	0.93	0.26 5
374 - 388	0.008	~0.0022
425	0.31	0.072 14
427	0.063	0.015 3
436	0.076	0.018 4
437	0.25	0.057 16
439 - 440	0.029	0.0066 13
505 - 520	0.064	0.0127 23
772 - 786	0.0014	~0.00018
825	0.14	~0.017
877 - 909	0.040	~0.0045
945 - 960	0.0018	0.00019 7
1042	0.102	0.0098 20
1111 - 1113	0.020	0.0018 3
1121	0.082	0.0073 18
1122 - 1125	0.0061	0.00055 9
1189 - 1204	0.020	0.0017 3
1223	0.11	~0.009
1292 - 1306	0.025	~0.0019

Continuous Radiation (^{202}Au)

⟨β-⟩=1230 keV;⟨IB⟩=3.3 keV

E_{bin}(keV)		⟨ ⟩(keV)	(%)
0 - 10	β-	0.0118	0.234
	IB	0.042	
10 - 20	β-	0.0357	0.238
	IB	0.042	0.29
20 - 40	β-	0.147	0.491
	IB	0.082	0.28
40 - 100	β-	1.12	1.59
	IB	0.23	0.36
100 - 300	β-	13.4	6.5
	IB	0.67	0.37
300 - 600	β-	58	12.7
	IB	0.75	0.18
600 - 1300	β-	327	34.4
	IB	0.99	0.115
1300 - 2500	β-	709	39.3
	IB	0.45	0.027
2500 - 3300	β-	122	4.51
	IB	0.0105	0.00040

$^{202}_{80}$Hg(stable)

Δ: -27370 4 keV

%: 29.80 14

$^{202}_{81}$Tl(12.23 2 d)

Mode: ε

Δ: -26003 17 keV

SpA: 5.285×10^4 Ci/g

Prod: ^{202}Hg(d,2n); ^{201}Hg(d,n); ^{203}Tl(d,t)

Photons (^{202}Tl)

⟨γ⟩=467 5 keV

γ_{mode}	γ(keV)	γ(%)†
Hg L_ℓ	8.722	0.64 6
Hg L_α	9.980	13.2 6
Hg L_η	10.647	0.167 10
Hg L_β	11.930	12.2 8
Hg L_γ	13.949	2.20 19
Hg $K_{\alpha2}$	68.893	22.6 7
Hg $K_{\alpha1}$	70.818	38.4 12
Hg $K_{\beta1}$'	80.124	13.5 4
Hg $K_{\beta2}$'	82.780	3.69 12
γ E2	439.567 10	91.4 10
γ M1+45%E2	520.14 7	0.9 3
γ E2	959.71 7	0.12 3

† 1.1% uncert(syst)

Atomic Electrons (^{202}Tl)

⟨e⟩=20.5 5 keV

e_{bin}(keV)	⟨e⟩(keV)	e(%)
12	3.2	26 3
14	1.88	13.2 14
15 - 59	1.50	5.2 4
65 - 80	0.72	1.06 5
356	8.49	2.38 5
425	2.79	0.657 15
427	0.572	0.134 3
436	0.699	0.160 4
437 - 440	0.61	0.140 14
505 - 520	0.053	0.0105 21
877	0.0055	0.00063 15
945 - 960	0.0015	0.00016 3

Continuous Radiation (^{202}Tl)

⟨IB⟩=0.62 keV

E_{bin}(keV)		⟨ ⟩(keV)	(%)
10 - 20	IB	0.0019	0.0161
20 - 40	IB	0.00148	0.0045
40 - 100	IB	0.46	0.67
100 - 300	IB	0.049	0.029
300 - 600	IB	0.073	0.017
600 - 1300	IB	0.027	0.0037
1300 - 1368	IB	2.2×10^{-6}	1.65×10^{-7}

$^{202}_{82}$Pb(5.3 3 $\times10^4$ yr)

Mode: ε

Δ: -25957 11 keV

SpA: 0.0337 Ci/g

Prod: ^{203}Tl(d,3n)

Photons (^{202}Pb)

⟨γ⟩=2.2 3 keV

γ_{mode}	γ(keV)	γ(%)
Tl L_ℓ	8.953	0.76 17
Tl L_α	10.259	16 3
Tl L_η	10.994	0.0028 6
Tl L_β	12.549	4.3 9
Tl L_γ	14.570	0.085 21

Continuous Radiation (^{202}Pb)

⟨IB⟩=4.3×10^{-6} keV

E_{bin}(keV)		⟨ ⟩(keV)	(%)
10 - 20	IB	4.1×10^{-6}	3.3×10^{-5}
20 - 40	IB	3.4×10^{-8}	1.41×10^{-7}
40 - 46	IB	2.9×10^{-10}	2.0×10^{-9}

$^{202}_{82}$Pb(3.62 3 h)

Mode: IT(90.5 5 %), ε(9.5 5 %)

Δ: -23787 11 keV

SpA: 4.29×10^6 Ci/g

Prod: ^{203}Tl(d,3n); ^{203}Tl(p,2n)

Photons (^{202}Pb)

⟨γ⟩=1991 78 keV

γ_{mode}	γ(keV)	γ(%)†
Tl L_ℓ	8.953	0.091 21
Pb L_ℓ	9.185	0.18 4
Tl L_α	10.259	1.9 4
Pb L_α	10.541	3.6 8
Tl L_η	10.994	0.025 5
Pb L_η	11.349	0.09 3
Tl L_β	12.309	1.8 4
Pb L_β	12.672	5.1 14
Tl L_γ	14.411	0.33 7
Pb L_γ	14.859	1.1 3
Tl $K_{\alpha2}$	70.832	3.0 6
Pb $K_{\alpha2}$	72.803	2.29 9
Tl $K_{\alpha1}$	72.873	5.1 11
Pb $K_{\alpha1}$	74.969	3.84 16
Tl $K_{\beta1}$'	82.434	1.8 4
Pb $K_{\beta1}$'	84.789	1.3[illegible]
Tl $K_{\beta2}$'	85.185	0.51 11
Pb $K_{\beta2}$'	87.632	0.392 17
γ_{IT} E1	125.21 4	0.62 9
γ_{IT} E4	129.47 6	0.040 16
γ_ϵM1	148.85 10	0.22 8
γ_ϵM1	212.00 6	0.75 15
γ_{IT} M1+<14%E2	240.25 3	0.183 25
γ_ϵE2	241.17 9	0.84 15
γ_{IT} (M1+<50%E2)	292.07 5	0.032 6
γ_ϵM1	335.63 10	0.22 5
γ_ϵM1	390.02 6	6.2 5
γ_{IT} [E1]	417.28 4	0.27 5
γ_{IT} E2	422.19 3	86 5
γ_ϵE3	459.79 7	8.6 5
γ_ϵE2	490.55 7	9.1 5
γ_{IT} (M1)	532.32 4	0.054 10
γ_{IT} E5	546.75 7	0.12 4
γ_ϵ(E2)	602.02 6	0.60 5
γ_{IT} E1	657.53 3	32.4 15
γ_{IT} (E2)	662.44 4	0.052 9
γ_{IT} E5	787.00 6	50
γ_{IT} E2	954.51 4	0.98 14
γ_{IT} E2	960.71 5	92 8

† uncert(syst): 5.3% for ε, 1.9% for IT

Atomic Electrons (^{202}Pb)

⟨e⟩=124 6 keV

e_{bin}(keV)	⟨e⟩(keV)	e(%)
13 - 62	3.0	20 4
63 - 112	0.53	0.80 18
114	10.7	9 4
116	5.9	5.1 20
121 - 125	0.0090	0.0074 6
126	6.1	4.9 18
127 - 156	3.0	2.3 6
197 - 241	0.59	0.276 25
250 - 292	0.15	0.058 12
304 - 333	3.2	1.04 9
334	8.6	2.58 14
335 - 377	2.63	0.70 4
386 - 422	6.31	1.54 4
444 - 491	2.89	0.639 20
516 - 534	0.30	0.056 12
570 - 602	0.87	0.153 7
642 - 662	0.199	0.0308 11
699	32.0	4.58 24
771	19.2	2.49 13
772	7.7	1.00 5
774	3.85	0.50 3
867	0.050	0.0057 8
873	4.6	0.53 4
939 - 961	1.33	0.140 7
1295	4.7×10^{-5}	3.6 6 $\times 10^{-6}$
1367 - 1382	1.73×10^{-5}	1.26 12 $\times 10^{-6}$

Continuous Radiation (^{202}Pb)

⟨IB⟩=0.057 keV

E_{bin}(keV)		⟨ ⟩(keV)	(%)
10 - 20	IB	0.00020	0.00163
20 - 40	IB	0.000132	0.00040
40 - 100	IB	0.047	0.065
100 - 300	IB	0.0046	0.0028
300 - 600	IB	0.0048	0.00114
600 - 876	IB	0.00075	0.000114

$^{202}_{83}$Bi(1.72 5 h)

Mode: ϵ

Δ: -20810 70 keV

SpA: 9.0×10^6 Ci/g

Prod: daughter ^{202}Po; protons on Pb

Photons (^{202}Bi)

γ_{mode}	γ(keV)	γ(rel)
γ M1	80.81 5	0.75 11
γ M1	97.56 5	0.24 4
γ E1	125.21 4	•
γ	127.21 14	0.099 15
γ E4	129.47 6	•
γ M1	158.20 15	0.35 5
γ E2	168.15 4	4.8 3
γ M1	195.67 10	0.29 4
γ M1	198.13 15	0.089 13
γ	204.79 15	0.27 4
γ M1	216.04 10	0.23 3
γ M1	222.80 3	0.69 10
γ M1	232.09 4	0.34 5
γ M1+<14%E2	240.25 3	•
γ M1	248.96 4	3.07 18
γ M1	285.62 12	0.22 3
γ (M1+<50%E2)	292.07 5	•
γ	316.2 4	0.10 3
γ	318.20 5	0.10 3
γ M1	320.18 5	3.12 18
γ M1(+E2)	342.10 5	0.43 6
γ M1	346.51 3	4.6 3
γ (M1)	348.73 5	0.60 20
γ E2(+39%M1)	358.03 4	0.30 5
γ M1	369.30 4	0.50 7
γ (M1)	387.06 5	0.139 20
γ M1(+E2)	412.31 18	0.29 4
γ [E1]	417.28 4	•
γ E2	422.19 3	•
γ M1	438.26 5	1.55 23
γ (M1)	504.27 22	0.28 4
γ M1	514.46 9	1.62 24
γ M1	529.72 5	0.41 6
γ (M1)	532.32 4	•
γ	535.03 6	0.169 25
γ E5	546.75 7	•
γ E2+35%M1	569.31 3	4.8 3
γ M1	578.60 4	7.3 4
γ M1(+E2)	582.35 4	0.96 14
γ	591.64 8	0.139 20
γ (M1)	599.34 10	0.53 8
γ	632.04 17	0.18 3
γ E1,E2	644.50 4	0.67 10
γ E1	657.53 3	•
γ (E2)	662.44 4	•
γ M1+E2	666.64 11	0.84 13
γ M1+E2	671.05 12	0.45 7
γ M1	676.22 3	1.9 3
γ [E2]	690.33 5	0.19 3
γ (M1)	701.88 7	1.0 3
γ	705.6 5	0.22 7
γ	714.64 9	0.27 4
γ	717.1 3	0.169 25
γ (M1)	763.89 14	0.55 8
γ M1	768.58 6	0.68 10
γ (M1)	783.50 9	0.33 5
γ E5	787.00 6	•
γ	788.4 5	0.74 11
γ	802.32 8	0.42 6 ?
γ	825.68 5	0.22 8
γ E2	852.61 7	2.28 14
γ [M1+E2]	858.48 4	1.64 25
γ	871.3 3	0.139 21
γ (E2)	876.23 4	1.06 16
γ	899.02 4	0.34 5
γ	904.33 6	0.30 5
γ	915.2 3	0.149 22
γ M1	927.34 4	7.1 4
γ	942.13 7	1.19 20
γ E2	954.51 4	•
γ E2	960.71 5	•
γ (E1)	983.69 5	0.88 13
γ	997.9 4	0.30 5
γ M1(+E2)	1004.54 4	0.85 13
γ	1035.22 10	0.50 20
γ	1052.90 13	0.31 5
γ	1062.88 18	0.139 21
γ	1072.66 8	0.83 13
γ	1103.67 13	0.38 6
γ	1108.7 3	0.19 6
γ	1111.86 20	0.24 8
γ	1117.44 20	0.169 25
γ [E2]	1127.48 11	0.32 5
γ	1134.48 5	0.21 3
γ	1144.31 20	0.50 7
γ	1150.75 9	0.40 6
γ	1163.5 4	0.18 3
γ	1165.02 9	0.159 20
γ	1173.66 17	0.15 5
γ	1192.9 3	0.099 15
γ	1197.57 16	0.20 3
γ (M1)	1206.29 6	0.58 9
γ	1211.25 6	0.22 3
γ	1224.50 8	1.56 23 ?
γ	1226.84 4	0.45 15
γ E2	1236.12 10	0.59 9
γ (M1)	1245.53 3	2.79 17
γ	1291.2 3	0.099 15
γ	1295.46 8	0.20 3
γ	1313.63 15	0.18 3
γ	1336.73 8	0.26 4
γ	1350.89 8	0.40 6
γ	1358.59 16	0.40 6
γ	1363.06 9	0.20 3
γ [E2]	1367.73 11	0.50 20
γ	1375.47 16	0.159 24
γ	1420.76 10	0.60 9
γ	1433.48 15	0.29 4
γ	1439.21 21	0.159 24
γ	1487.14 13	0.22 3
γ	1495.12 11	0.159 24
γ	1512.51 10	0.25 8 ?
γ [E1]	1516.01 5	0.72 11
γ	1523.72 20	0.27 4
γ	1526.9 3	0.17 6
γ	1556.67 5	1.9 3
γ	1563.39 17	0.18 3
γ	1584.87 5	0.69 20
γ	1585.48 9	0.69 20
γ	1615.29 15	0.159 24
γ	1619.69 15	0.23 3
γ	1623.38 15	0.18 3
γ	1635.77 9	0.169 25
γ	1695.0 4	0.139 21
γ	1715.98 20	0.079 12
γ	1730.9 3	0.30 5
γ	1754.1 4	0.18 6
γ	1757.5 3	0.37 6
γ	1780.56 6	0.68 10
γ	1790.59 18	0.26 4
γ	1807.99 10	0.40 6
γ	1813.74 20	0.15 5
γ	1833.29 13	0.28 4
γ	1839.6 3	0.069 10
γ	1848.77 14	0.20 3
γ	1858.84 15	0.149 22
γ	1882.26 20	0.28 4
γ	1957.01 16	0.36 5
γ	1989.79 20	0.21 3
γ	1998.4 2	0.109 16
γ	2003.17 20	0.079 12
γ	2016.5 3	0.079 12
γ	2059.25 9	0.36 5
γ	2100.51 8	0.20 3
γ	2153.25 14	0.20 3
γ	2197.83 7	0.099 15
γ	2277.32 15	0.159 24
γ	2286.4 3	0.069 10
γ	2322.59 13	0.22 3
γ	2340.76 7	0.20 7
γ	2365.9 4	0.060 9
γ	2435.3 5	0.069 10
γ	2559.6 5	0.109 16
γ	2640.8 4	0.040 6
γ	2660.90 13	0.119 18
γ	2685.1 6	0.030 5
γ	2734.6 3	0.040 6
γ	2779.6 3	0.040 6
γ	2784.44 20	0.060 9
γ	2868.64 20	0.040 6
γ	2945.0 3	0.060 9
γ	2966.94 20	0.119 18
γ	3058.7 4	0.069 10
γ	3138.9 4	0.040 6
γ	3210.81 15	0.119 18
γ	3217.2 4	0.024 4
γ	3236.7 4	0.040 6
γ	3244.34 20	0.079 12
γ	3259.4 4	0.030 5
γ	3266.7 5	0.079 12
γ	3316.58 16	0.129 19
γ	3322.24 20	0.079 12
γ	3359.8 5	0.023 3
γ	3406.6 4	0.024 4
γ	3498.6 5	0.099 15
γ	3520.7 3	0.020 3

• with ^{202}Pb(3.62 h)

$^{202}_{84}$Po(44.7 *5* min)

Mode: ϵ(98.0 *2* %), α(2.0 *2* %)
Δ: -17990 *130* keV syst
SpA: 2.082×10^{7} Ci/g
Prod: ^{209}Bi(p,8n); ^{12}C on Pt; daughter ^{202}At

Alpha Particles (^{202}Po)

α(keV)
5588 *2*

Photons (^{202}Po)

⟨γ⟩=810 *43* keV

γ_{mode}	γ(keV)	γ(%)†
γ_ϵ(M1)	40.99 *16*	3.01 *20*
γ_ϵM1,E2	65.02 *20*	1.99 *10*
γ_ϵM1	165.56 *15*	8.7 *5*
γ_ϵM1	213.48 *16*	3.36 *25*
γ_ϵM1	215.75 *17*	1.63 *10*
γ_ϵM1,E2	251.58 *16*	1.53 *10*
γ_ϵM1(+E2)	315.82 *20*	14.3 *8*
γ_ϵM1+E2	336.52 *20*	1.99 *15*
γ_ϵ(M1+E2)	427.54 *18*	1.63 *10*
γ_ϵ(M1)	458.14 *16*	3.77 *25*
γ_ϵM1	506.07 *16*	4.4 *3*
γ_ϵ(M1)	551.25 *17*	1.78 *10*
γ_ϵE1	597.72 *20*	2.60 *15*
γ_ϵM1	624.8 *4*	1.73 *15*
γ_ϵM1	643.30 *22*	3.57 *25*
γ_ϵM2	688.4 *5*	51
γ_ϵE1	712.63 *20*	4.6 *3*
γ_ϵE1	716.81 *21*	6.1 *4*
γ_ϵE1	785.3 *5*	1.99 *15*
γ_ϵM1	790.3 *5*	7.2 *4*
γ_ϵM1	808.3 *5*	1.58 *10*
γ_ϵ(E2)	828.4 *4*	1.89 *15*
γ_ϵM1+E2	973.56 *25*	4.9 *4*
γ_ϵM1	1168.4 *5*	1.99 *15*
γ_ϵM1	1214.8 *5*	1.73 *15*

† 12% uncert(syst)

$^{202}_{85}$At(3.02 *5* min)

Mode: ϵ(88.0 *8* %), α(12.0 *8* %)
Δ: -10790 *160* keV
SpA: 3.08×10^{8} Ci/g
Prod: ^{197}Au(^{12}C,7n); ^{187}Re(^{20}Ne,5n)

Alpha Particles (^{202}At)

⟨α⟩=740.2 *4* keV

α(keV)	α(%)
6135 *2*	7.68 *24*
6228 *2*	4.32 *24*

Photons (^{202}At)

⟨γ⟩=1223 *63* keV

γ_{mode}	γ(keV)	γ(%)†
γ_ϵ(E2)	441.3 *3*	41 *13*
γ_ϵ(E2)	569.7 *4*	81 *4*
γ_ϵ(E2)	675.3 *5*	86.6

† 0.91% uncert(syst)

$^{202}_{86}$Rn(9.85 *20* s)

Mode: α(85 *15* %), ϵ(∼ 15 %)
Δ: -6310 *300* keV syst
SpA: 5.48×10^{9} Ci/g
Prod: ^{197}Au(^{14}N,9n); ^{16}O on Pt; protons on Th

Alpha Particles (^{202}Rn)

α(keV)
6636 *3*

$^{202}_{87}$Fr(340 *40* ms)

Mode: α
Δ: 3080 *350* keV syst
SpA: 7.0×10^{10} Ci/g

Alpha Particles (^{202}Fr)

α(keV)
7250 *20*

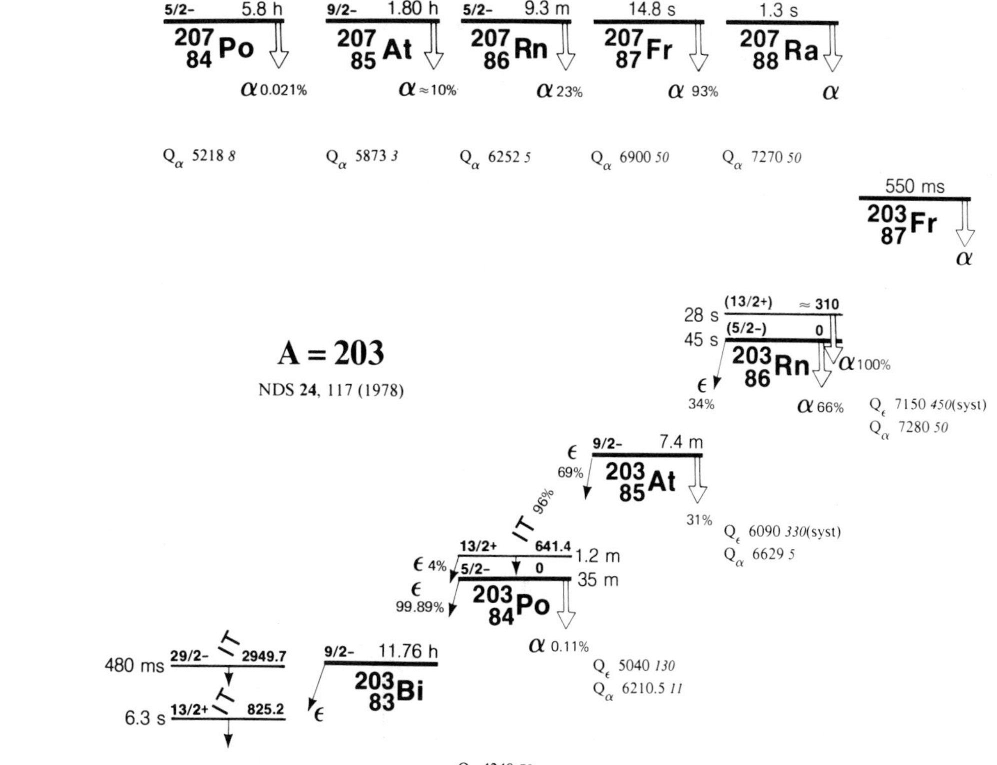

$^{203}_{79}$Au(53 *2* s)

Mode: β-
Δ: -23153 *16* keV
SpA: 1.04×10^9 Ci/g
Prod: ^{204}Hg(γ,p); ^{204}Hg(n,d)

Photons (^{203}Au)

γ_{mode}	γ(keV)	γ(%)
γ	690.0 *25*	~10 ?

$^{203}_{80}$Hg(46.60 *2* d)

Mode: β-
Δ: -25292 *5* keV
SpA: 1.3803×10^4 Ci/g
Prod: ^{202}Hg(n,γ)

Photons (^{203}Hg)

⟨γ⟩=237.7 *22* keV

γ_{mode}	γ(keV)	γ(%)†
Tl L_ℓ	8.953	0.116 *10*
Tl L_α	10.259	2.36 *10*
Tl L_η	10.994	0.0354 *19*
Tl L_β	12.305	2.36 *14*
Tl L_γ	14.396	0.44 *3*
Tl $K_{\alpha2}$	70.832	3.72 *8*
Tl $K_{\alpha1}$	72.873	6.30 *14*
Tl $K_{\beta1}$'	82.434	2.22 *5*
Tl $K_{\beta2}$'	85.185	0.623 *16*
γ E2+43%M1	279.188 *3*	81.5

† 0.98% uncert(syst)

Atomic Electrons (^{203}Hg)

⟨e⟩=40.3 *3* keV

e_{bin}(keV)	⟨e⟩(keV)	e(%)
13	0.58	4.6 *5*
15	0.47	3.2 *3*
55	0.070	0.127 *13*
56	0.00242	0.0043 *4*
58	0.069	0.119 *12*
60	0.0204	0.034 *3*
66	0.0131	0.0198 *20*
67	0.028	0.042 *4*
68	0.0132	0.0195 *20*

Atomic Electrons (^{203}Hg) (continued)

e_{bin}(keV)	⟨e⟩(keV)	e(%)
69	0.0209	0.030 *3*
70	0.0214	0.031 *3*
71	0.00093	0.00131 *13*
72	0.0067	0.0093 *9*
73	0.00115	0.00158 *16*
78	0.0039	0.0049 *5*
79	0.0034	0.0043 *4*
80	0.00168	0.00211 *21*
81	0.00228	0.0028 *3*
82	0.00193	0.00236 *24*
194	25.89	13.37 *13*
264	8.80	3.33 *3*
267	1.673	0.628 *6*
275	2.65	0.962 *9*

Continuous Radiation (^{203}Hg)

⟨β-⟩=58 keV;⟨IB⟩=0.0124 keV

E_{bin}(keV)		⟨ ⟩(keV)	(%)
0 - 10	β-	0.58	11.8
	IB	0.0029	
10 - 20	β-	1.63	10.9
	IB	0.0023	0.0161
20 - 40	β-	5.8	19.5
	IB	0.0032	0.0113
40 - 100	β-	26.6	39.9

Continuous Radiation (^{203}Hg)
(continued)

E_{bin}(keV)		⟨ ⟩(keV)	(%)
	IB	0.0035	0.0061
100 - 212	β-	23.1	17.9
	IB	0.00046	0.00040

$^{203}_{81}$Tl(stable)

Δ: -25784 *4* keV

%: 29.524 *9*

$^{203}_{82}$Pb(2.169 *4* d)

Mode: ε

Δ: -24811 *9* keV

SpA: 2.966×10^5 Ci/g

Prod: ^{203}Tl(d,2n); daughter ^{203}Bi

Photons (^{203}Pb)

⟨γ⟩=310.4 *25* keV

γ_{mode}	γ(keV)	γ(%)†
Tl L_ℓ	8.953	0.76 *7*
Tl L_α	10.259	15.3 *7*
Tl L_η	10.994	0.200 *11*
Tl L_β	12.311	14.3 *9*
Tl L_γ	14.416	2.65 *22*
Tl $K_{\alpha2}$	70.832	25.0 *5*
Tl $K_{\alpha1}$	72.873	42.3 *9*
Tl $K_{\beta1}$'	82.434	14.9 *3*
Tl $K_{\beta2}$'	85.185	4.19 *10*
γ E2+43%M1	279.188 *3*	80.1
γ M1+0.1%E2	401.314 *10*	3.44 *16*
γ E2	680.501 *10*	0.70 *8*

† 1.0% uncert(syst)

Atomic Electrons (^{203}Pb)

⟨e⟩=48.9 *6* keV

e_{bin}(keV)	⟨e⟩(keV)	e(%)
13	3.6	29 *3*
15	2.8	18.8 *19*
55 - 82	1.90	3.08 *13*
194	25.45	13.14 *13*
264	8.65	3.28 *3*
267	1.644	0.617 *6*
275	2.60	0.945 *10*
316	1.68	0.53 *3*
386 - 401	0.453	0.116 *4*
595	0.045	0.0075 *9*
665 - 681	0.0156	0.00233 *15*

Continuous Radiation (^{203}Pb)

⟨IB⟩=0.56 keV

E_{bin}(keV)		⟨ ⟩(keV)	(%)
10 - 20	IB	0.0022	0.018
20 - 40	IB	0.00141	0.0043
40 - 100	IB	0.49	0.68
100 - 300	IB	0.042	0.027
300 - 600	IB	0.024	0.0062
600 - 695	IB	0.000148	2.4×10^{-5}

$^{203}_{82}$Pb(6.3 *2* s)

Mode: IT

Δ: -23986 *9* keV

SpA: 8.4×10^9 Ci/g

Prod: daughter ^{203}Bi; ^{204}Pb(n,2n)

Photons (^{203}Pb)

⟨γ⟩=656 *11* keV

γ_{mode}	γ(keV)	γ(%)†
Pb L_ℓ	9.185	0.151 *14*
Pb L_α	10.541	3.01 *15*
Pb L_η	11.349	0.0394 *23*
Pb L_β	12.700	2.83 *19*
Pb L_γ	14.896	0.53 *5*
Pb $K_{\alpha2}$	72.803	4.49 *14*
Pb $K_{\alpha1}$	74.969	7.56 *24*
Pb $K_{\beta1}$'	84.789	2.67 *9*
Pb $K_{\beta2}$'	87.632	0.77 *3*
γ [E2]	633.80 *22*	<0.30 ?
γ M1+E2	820.32 *19*	6.4 *9*
γ M4	825.24 *10*	71.5

† 1.4% uncert(syst)

Atomic Electrons (^{203}Pb)

⟨e⟩=171 *3* keV

e_{bin}(keV)	⟨e⟩(keV)	e(%)
13 - 62	1.41	9.0 *6*
68 - 84	0.143	0.200 *8*
546	0.010	~0.0019
618 - 634	0.0040	0.00065 *17*
732	0.8	~0.11
737	117.7	16.0 *4*
804 - 807	0.15	0.019 *9*
809	27.7	3.43 *8*
810	6.08	0.751 *18*
812 - 820	4.13	0.508 *12*
821	6.90	0.840 *20*
822 - 825	6.02	0.732 *11*

$^{203}_{82}$Pb(480 *20* ms)

Mode: IT

Δ: -21861 *9* keV

SpA: 6.1×10^{10} Ci/g

Prod: ^{204}Hg(α,5n)

Photons (^{203}Pb)

⟨γ⟩=1910 *48* keV

γ_{mode}	γ(keV)	γ(%)†
γ (E3)	153.42 *17*	5.4 *3*
γ [M1]	173.83 *18*	1.3 *2*
γ [M1]	217.65 *19*	0.9 *2*
γ	231.9 *3*	1.5 *2*
γ	238.5 *3*	4.9 *12*
γ (M1)	239.48 *18*	10 *3*
γ E2	258.49 *9*	83 *3*
γ [M1]	280.33 *15*	4.0 *5*
γ [M1]	454.15 *17*	1.0 *2*
γ M1	634.41 *16*	20.5 *10*
γ [E2]	678.23 *16*	3.7 *4*
γ M1+E2	820.32 *19*	*
γ M4	825.24 *10*	*
γ E2	838.65 *10*	100
γ [E2]	852.05 *15*	4.3 *4*
γ E2	873.89 *9*	51 *2*
γ M4	1027.31 *18*	14.7 *6*

† 5.0% uncert(syst)

* with ^{203}Pb(6.3 s)

$^{203}_{83}$Bi(11.76 *5* h)

Mode: ε

Δ: -21590 *40* keV

SpA: 1.313×10^6 Ci/g

Prod: ^{206}Pb(p,4n)

Photons (^{203}Bi)

⟨γ⟩=2360 *46* keV

γ_{mode}	γ(keV)	γ(%)†
Pb L_ℓ	9.185	0.83 *19*
Pb L_α	10.541	16 *4*
Pb L_η	11.349	0.21 *5*
Pb L_β	12.699	15 *3*
Pb L_γ	14.903	2.9 *7*
γM1	60.02 *3*	0.33 *3*
Pb $K_{\alpha2}$	72.803	25 *5*
Pb $K_{\alpha1}$	74.969	42 *9*
Pb $K_{\beta1}$'	84.789	15 *3*
Pb $K_{\beta2}$'	87.632	4.3 *9*
γ[M1]	100.5 *4*	0.056 *11*
γ[E1]	119.8 *5*	0.056 *11*
γE2	126.51 *24*	1.20 *24*
γE2+30%M1	137.0 *4*	0.25 *5*
γ	157.2 *5*	0.059 *12*
γ	166.3 *10*	0.086 *17*
γM1	186.53 *23*	3.11 *15*
γ	195.8 *6*	0.086 *17*
γE2+37%M1	202.7 *5*	0.074 *15*
γ	212.6 *6*	0.27 *5*
γ	220.4 *10*	0.030 *6*
γE2+37%M1	252.3 *5*	0.104 *21*
γM1(+21%E2)	264.2 *3*	5.2 *3*
γE2+30%M1	271.5 *6*	0.14 *3*
γ	296.0 *4*	0.083 *17*
γ	299.4 *10*	0.115 *23*
γ[M1+E2]	306.2 *4*	0.030 *6*
γ	311.1 *10*	0.065 *13*
γ	321.9 *6*	0.15 *3*
γ	325.5 *4*	0.065 *13*
γ	331.8 *5*	0.21 *4*
γ	337.7 *5*	0.17 *3*
γ	339.2 *5*	0.16 *3*
γ	349.1 *4*	0.127 *25*
γE2+37%M1	375.4 *5*	0.36 *7*
γM1	378.3 *4*	0.29 *6*
γM1	381.8 *4*	1.3 *3*
γM1	392.7 *6*	0.33 *7*
γM1	406.3 *5*	0.37 *7*

Photons (^{203}Bi)
(continued)

γ_{mode}	γ(keV)	γ(%)†
γ	416.2 10	0.092 18
γE2	422.3 5	0.38 8
γ	429.1 10	0.024 5
γ	433.2 5	0.13 3
γ	449.3 5	0.038 8
γ	452.8 10	0.098 20
γ	459.5 10	0.062 12
γM1	462.2 10	0.17 4
γM1	465.8 10	0.083 17
γM1	468.6 5	0.22 5
γ	476.8 6	0.086 17
γ	484.1 5	0.27 5
γ	486.5 4	0.115 23
γ[M1+E2]	489.9 5	0.110 22
γM1	498.7 5	0.66 13
γ	501.0 5	0.19 4
γ	508.2 10	0.16 3
γ(E1)	513.8 4	<1.0
γ[M1+E2]	531.3 5	0.092 18
γ	542.5 5	0.22 5
γ	547 1	0.17 3
γ	559.0 5	0.16 3
γM1	569.4 4	1.22 24
γM1	590.9 10	0.127 25
γ(E2)	595.1 4	0.47 9
γ	618.8 6	0.36 7
γ	621.1 10	0.41 8
γ	624.0 10	0.17 3
γM1+45%E2	626.8 10	0.36 7
γE2+30%M1	633.7 4	1.3 3
γ[E2]	633.80 22	<1 ?
γ	647.1 10	0.15 3
γ[E1]	650.8 4	0.044 9
γ	657.3 6	0.22 4
γ	664.7 4	0.107 21
γ[M1+E2]	674.7 5	0.101 20
γ	697.4 10	0.16 3
γ	704.9 5	0.14 3
γM1+32%E2	719.5 4	0.40 8
γE2	722.4 3	4.77 24
γ(E2)	740.1 5	0.38 8
γ	746.5 5	<1
γ	759 1	0.27 5
γ[E1]	768.6 4	0.41 8
γ	772.6 5	0.21 4
γ	779.9 10	0.115 23
γ	788.2 10	0.092 18
γE2	816.2 5	4.0 8
γM1+E2	820.32 19	29.6
γM4	825.24 10	14.6 7
γE2	847.3 3	8.5 17
γ	861.2 6	0.13 3
γE2	866.6 5	1.5 3
γE1,E2	869.1 5	0.49 10
γ	871.1 10	0.23 5
γ	879.8 6	0.089 18
γE2	896.9 3	13 3
γ	904.2 10	0.21 4
γ	906.7 10	0.25 5
γ	911.7 10	0.22 4
γ	924.6 5	0.20 4
γ	927.8 6	0.20 4
γM1	933.4 4	1.4 3
γE1,E2	935.9 5	0.74 15
γ[M1+E2]	951.2 4	0.23 5
γ	974.6 4	0.083 17
γ	982.8 5	0.18 4
γ	985.1 5	0.31 6
γ	995.2 10	0.15 3
γE1	1000.3 4	0.98 20
γ	1006.9 10	0.110 22
γ	1024.3 10	0.121 24
γM1	1033.9 3	8.8 18
γ	1044.0 10	0.24 5
γ	1058.8 10	0.104 21
γ	1068.1 5	0.60 12
γ	1069.8 5	0.70 14
γ	1074.9 5	0.29 6
γ	1087.6 5	0.39 8
γ	1092.2 6	0.17 3
γ	1096.7 10	0.115 23
γM1	1112.2 5	0.72 14
γM1,E2	1120.2 4	0.72 14
γM1	1124.0 10	0.30 6

Photons (^{203}Bi)
(continued)

γ_{mode}	γ(keV)	γ(%)†
γ	1143.8 10	0.104 21
γ	1151.8 5	0.14 3
γ	1152.8 6	0.20 4
γ	1167.5 5	0.16 3
γ[E1]	1177.0 4	0.110 22
γM1,E2	1184.8 5	0.49 10
γ	1187.9 6	0.127 25
γE2	1198.6 4	2.0 4
γM1	1203.1 4	1.5 3
γ[M1+E2]	1206.0 4	0.15 3
γ	1214.4 5	0.22 4
γE2,E1	1223.5 4	0.73 15
γ	1228.4 4	0.22 4
γM1	1246.5 5	0.53 11
γM1	1253.7 5	1.23 25
γ	1261.9 10	0.115 23
γ	1273.7 5	0.095 19
γ	1303.3 10	0.49 10
γ	1307.6 10	0.17 3
γ	1311.2 5	0.124 25
γ	1338.0 4	0.37 8
γ	1343.4 10	0.17 3
γ[E2]	1349.9 4	0.110 22
γ	1358.2 10	0.059 12
γ[E1]	1365.7 5	0.124 25
γ	1370.3 5	0.36 7
γ	1374.2 10	0.107 21
γ	1381.3 10	0.28 6
γ	1385.6 10	0.38 8
γ	1395.6 10	0.31 6
γ	1407.8 5	0.94 19
γ	1409.9 10	0.71 14
γ	1417.0 6	0.19 4
γ	1421.5 6	0.25 5
γ	1431 1	0.121 24
γ(E2)	1438.5 5	0.64 13
γ[E1]	1464.7 4	0.61 12
γ	1469.2 4	0.43 9
γ	1496.2 5	0.54 11
γE1	1506.7 3	3.67 18
γ[E1]	1510.3 4	0.35 7
γM1	1536.5 4	7.5 4
γ[E1]	1550.5 4	0.77 15
γ	1552.3 4	1.5 3
γ	1562.7 4	0.14 3
γ	1576.3 6	0.14 3
γ	1578.5 10	0.038 8
γ	1582 1	0.030 6
γ	1589.4 10	0.20 4
γM1,E2	1592.9 5	1.09 22
γ	1608.4 10	0.121 24
γ[E1]	1634.0 3	0.65 13
γ	1647.0 5	0.104 21
γE1	1679.6 3	8.8 4
γ	1716.4 10	0.55 11
γE1	1719.8 3	3.40 15
γ	1739.0 10	0.25 5
γ[E2]	1743.6 4	0.25 5
γ(E1)	1748.5 4	1.9 4
γ[E1]	1770.9 3	0.51 10
γ[M2]	1780.1 4	0.041 8
γ	1787.3 5	0.19 4
γ	1799.8 6	0.92 18
γ[E1]	1802.5 4	0.92 18
γ	1812.4 10	0.080 16
γ[E1]	1816.5 3	0.40 8
γ	1841.5 6	0.49 10
γE1	1847.5 3	11.4 6
γ[E1]	1856.7 4	0.29 6
γE2	1888.2 3	1.9 4
γE1,E2	1893.1 3	8.2 4
γ	1908.2 6	0.34 7
γ(E2)	1928.2 10	1.12 22
γ	1930.9 10	0.16 3
γ	1939.3 10	0.127 25
γ	1951.8 10	0.033 7
γ	1968 1	0.030 6
γE2	1983.1 5	0.88 18
γM1,E2	1991.1 10	0.118 24
γE1	2000.8 5	0.83 17
γE1	2011.4 5	1.8 4
γ	2075.2 10	0.036 7
γ	2078.3 10	0.46 9
γ	2083.6 5	0.059 12

Photons (^{203}Bi)
(continued)

γ_{mode}	γ(keV)	γ(%)†
γ	2113.2 10	0.115 23
γ[E3]	2118.3 5	0.17 4
γ	2144.2 10	0.23 5
γ[E3]	2158.5 5	0.038 8
γ	2181.5 6	0.14 3
γ	2196.3 10	0.024 5
γ	2224.9 5	0.17 3
γ	2270.3 10	0.033 7
γ	2331.6 10	0.33 7
γ	2362.2 10	0.041 8
γ	2371.8 4	0.041 8
γ	2428.9 10	0.26 5
γ[E3]	2526.9 3	0.030 6
γ[E3]	2567.5 10	0.030 6
γ	2584.3 10	0.065 13
γ	2651.0 10	0.024 5
γ[M2]	2667.8 3	0.059 12
γ	2682.9 10	0.0089 18
γ[M2]	2713.4 3	0.024 5
γ	2716.7 10	0.0030 6
γ	2884.4 10	0.0059 12
γ	2945.4 10	0.021 4

† 5.1% uncert(syst)

Atomic Electrons (^{203}Bi)

⟨e⟩=75.5 20 keV

e_{bin}(keV)	⟨e⟩(keV)	e(%)
4 - 5	0.017	0.40 15
13	3.7	28 7
15	2.2	15 3
16 - 62	3.4	9.7 10
68 - 98	0.99	1.35 17
99	4.21	4.27 22
100 - 144	3.5	3.0 3
150 - 173	1.38	0.81 4
176	4.0	2.3 4
180 - 223	0.78	0.41 5
234 - 283	1.84	0.73 10
284 - 332	1.52	0.51 6
333 - 382	0.68	0.187 17
389 - 438	0.59	0.145 23
440 - 489	0.67	0.142 21
490 - 539	0.34	0.065 16
540 - 588	0.44	0.078 22
590 - 638	0.55	0.087 7
641 - 690	0.19	0.029 13
692 - 731	0.40	0.055 7
732	1.69	0.231 13
733 - 736	0.0014	0.00019 7
737	24.0	3.26 18
738 - 786	0.64	0.083 13
787 - 807	0.463	0.0576 24
809	6.4	0.79 6
810 - 859	5.38	0.657 17
860 - 909	0.32	0.036 4
911 - 959	1.33	0.14 3
962 - 1009	0.19	0.020 6
1011 - 1060	0.49	0.048 5
1062 - 1111	0.22	0.020 3
1112 - 1161	0.30	0.026 4
1162 - 1211	0.22	0.0186 24
1212 - 1261	0.12	0.0098 24
1262 - 1311	0.09	0.0073 20
1320 - 1368	0.15	0.011 4
1369 - 1419	0.14	0.0102 18
1421 - 1469	0.67	0.046 4
1475 - 1574	0.26	0.0171 19
1575 - 1667	0.268	0.0166 12
1676 - 1774	0.27	0.0157 16
1776 - 1875	0.35	0.019 4
1877 - 1976	0.121	0.0064 8
1978 - 2075	0.050	0.0025 5
2080 - 2178	0.015	0.00072 18
2180 - 2274	0.012	0.00053 23
2284 - 2369	0.010	0.00042 17
2413 - 2512	0.0052	0.00021 5

Atomic Electrons (^{203}Bi) (continued)

e_{bin}(keV)	⟨e⟩(keV)	e(%)
2514 - 2595	0.0060	0.00023 *4*
2625 - 2714	0.0035	0.000134 *16*
2796 - 2881	0.00055	1.9 *8* $\times10^{-5}$
2930 - 2942	8.5 $\times10^{-5}$	2.9 *14* $\times10^{-6}$

Continuous Radiation (^{203}Bi)

⟨β+⟩=1.18 keV;⟨IB⟩=0.90 keV

E_{bin}(keV)		⟨ ⟩(keV)	(%)
0 - 10	β+	3.89 $\times10^{-9}$	4.58 $\times10^{-8}$
	IB	0.000139	
10 - 20	β+	2.73 $\times10^{-7}$	1.62 $\times10^{-6}$
	IB	0.0025	0.020
20 - 40	β+	1.29 $\times10^{-5}$	3.88 $\times10^{-5}$
	IB	0.00141	0.0044
40 - 100	β+	0.00111	0.00139
	IB	0.49	0.67
100 - 300	β+	0.070	0.0317
	IB	0.043	0.027
300 - 600	β+	0.369	0.082
	IB	0.058	0.0129
600 - 1300	β+	0.74	0.090
	IB	0.20	0.022
1300 - 2370	β+	0.000447	3.41 $\times10^{-5}$
	IB	0.098	0.0063
	Σβ+		0.21

$^{203}_{84}$Po(34.8 *14* min)

Mode: ϵ(99.89 *2* %), α(0.11 *2* %)
Δ: -17350 *80* keV
SpA: 2.66$\times10^{7}$ Ci/g
Prod: ^{209}Bi(p,7n); daughter ^{203}At

Alpha Particles (^{203}Po)

α(keV)
5384 *3*

Photons (^{203}Po)

⟨γ⟩=1498 *36* keV

γ_{mode}	γ(keV)	γ(%)†
γ	140.2 *2*	0.28 *11*
γM1	175.16 *6*	3.0 *3*
γM1	182.31 *7*	~0.11
γE3	189.51 *8*	3.9 *3*
γM1	197.45 *8*	0.56 *17*
γM2(+10%E3)	204.65 *6*	0.50 *17*
γE2(+7.0%M1)	214.77 *6*	14.5 *11*
γ[M1]	261.88 *8*	1.2 *4*
γE2	389.93 *8*	1.12 *11*
γE1	419.42 *6*	2.46 *22*
γ(M1)	443.4 *3*	0.28 *11*
γE2	486.1 *1*	2.13 *17*
γM1	647.76 *7*	2.07 *17*
γ(M1)	743.01 *7*	0.62 *11*
γ	798.92 *8*	0.62 *11*
γ(M1)	822.91 *7*	2.41 *17*
γM1+E2	883.5 *10*	2.0 *8*
γE2	893.50 *8*	19.0 *11*
γM1(+20%E2)	908.63 *7*	56
γ	918.17 *7*	0.84 *11*
γ	955.3 *4*	0.34 *6*
γ	974.08 *9*	0.50 *11*
γ	1037.69 *8*	0.28 *11*
γM1	1090.95 *7*	19.6 *11*
γ	1096.0 *5*	0.78 *11*
γ	1123.9 *1*	1.62 *17*
γ	1132.94 *9*	0.56 *11*
γ	1138.1 *4*	0.17 *6*
γ	1150.1 *4*	~0.22
γ	1178.2 *2*	0.34 *11*
γ	1188.85 *10*	~0.22
γ	1201.30 *12*	~0.22
γE1	1242.33 *7*	4.7 *3*
γ	1264.0 *1*	0.95 *11*
γ	1277.1 *2*	0.62 *11*
γ	1307.2 *4*	~0.28
γ	1314.5 *3*	0.50 *17*
γ	1337.58 *9*	2.96 *22*
γ	1352.82 *8*	1.40 *17*
γ	1416.7 *8*	~0.39
γ	1419.5 *8*	~0.17
γ	1466.0 *5*	0.39 *17*
γ	1475.71 *13*	0.62 *17*
γ	1490.3 *4*	~0.45
γ	1511.4 *3*	0.45 *11*
γ	1552.2 *4*	0.45 *17*
γ	1568.5 *4*	0.56 *17*
γ	1598.42 *10*	0.50 *11*
γ	1601.7 *5*	0.22 *6*
γ	1615.3 *6*	0.22 *6*
γ	1622.2 *3*	0.50 *11*
γ	1658.02 *13*	0.50 *11*
γ	1666.3 *5*	0.22 *6*
γ	1673.16 *14*	0.50 *11*
γ	1716.2 *6*	0.34 *11*
γ	1758.3 *4*	<0.22
γ	1780.73 *9*	0.67 *11*
γ	1795.87 *10*	0.56 *11*
γ	1804.9 *4*	0.22 *6*
γ	1817.55 *15*	1.06 *11*
γ	1830.1 *7*	0.28 *6*
γ	1909.8 *4*	<0.11
γ	1914.2 *3*	~0.11
γ	1930.8 *5*	0.90 *22*
γ	1936.2 *6*	0.17 *6*
γ	1960.4 *5*	~0.11
γ	1970.7 *4*	0.17 *6*
γ	1991.0 *3*	~0.11
γ	2029.5 *3*	0.56 *11*
γ	2032.32 *15*	0.39 *11*
γ	2086.8 *3*	~0.22
γ	2189.4 *7*	~0.11
γ	2197.7 *3*	0.22 *6*
γ	2236.96 *15*	0.56 *11*
γ	2373.7 *3*	~0.22
γ	2477.7 *6*	~0.11
γ	2529.5 *4*	0.17 *6*
γ	2665.6 *6*	<0.11
γ	2728.8 *4*	~0.11
γ	2916.4 *4*	0.22 *6*
γ	2952.2 *4*	0.17 *6*

† 5.4% uncert(syst)

Continuous Radiation (^{203}Po)

⟨β+⟩=67 keV;⟨IB⟩=4.9 keV

E_{bin}(keV)		⟨ ⟩(keV)	(%)
0 - 10	β+	2.40 $\times10^{-8}$	2.82 $\times10^{-7}$
	IB	0.0026	
10 - 20	β+	1.73 $\times10^{-6}$	1.02 $\times10^{-5}$
	IB	0.0045	0.033
20 - 40	β+	8.5 $\times10^{-5}$	0.000256
	IB	0.0060	0.020
40 - 100	β+	0.0080	0.0100
	IB	0.52	0.69
100 - 300	β+	0.66	0.293
	IB	0.109	0.063
300 - 600	β+	5.6	1.21
	IB	0.25	0.055
600 - 1300	β+	34.9	3.71
	IB	1.33	0.141
1300 - 2500	β+	26.1	1.67
	IB	2.4	0.136
2500 - 3347	IB	0.25	0.0094
	Σβ+		6.9

$^{203}_{84}$Po(1.2 *2* min)

Mode: IT(95.5 *10* %), ϵ(4.5 *10* %)
Δ: -16709 *80* keV
SpA: 7.7$\times10^{8}$ Ci/g
Prod: ^{197}Au(^{11}B,5n); ^{190}Os(^{18}O,5n)

Photons (^{203}Po)

⟨γ⟩=408 *17* keV

γ_{mode}	γ(keV)	γ(%)†
Bi L_{ℓ}	9.419	0.037 *15*
Po L_{ℓ}	9.658	0.33 *4*
Bi L_{α}	10.828	0.7 *3*
Po L_{α}	11.119	6.4 *5*
Bi L_{η}	11.712	0.010 *4*
Po L_{η}	12.085	0.086 *7*
Bi L_{β}	13.099	0.7 *3*
Po L_{β}	13.501	6.1 *5*
Bi L_{γ}	15.380	0.13 *5*
Po L_{γ}	15.892	1.18 *12*
Bi $K_{\alpha2}$	74.814	1.2 *5*
Po $K_{\alpha2}$	76.858	8.4 *5*
Bi $K_{\alpha1}$	77.107	2.0 *8*
Po $K_{\alpha1}$	79.290	14.1 *8*
Bi $K_{\beta1}$'	87.190	0.7 *3*
Po $K_{\beta1}$'	89.639	5.0 *3*
Bi $K_{\beta2}$'	90.128	0.20 *8*
Po $K_{\beta2}$'	92.673	1.52 *9*
γ_{ϵ}(M1)	261.5 *4*	0.53 *18*
γ_{ϵ}(M1)	577.0 *5*	2.4 *3*
γ_{IT} M4	641.4 *1*	50.5 *25*
γ_{ϵ}(M1)	904.9 *5*	4.4 *4*

† uncert(syst): 22% for ϵ, 2.5% for IT

Atomic Electrons (^{203}Po)

⟨e⟩=265 *9* keV

e_{bin}(keV)	⟨e⟩(keV)	e(%)
13 - 62	3.02	19.0 *15*
63 - 89	0.455	0.64 *3*
171	0.53	0.31 *10*
245 - 261	0.17	0.070 *17*
486	0.81	0.167 *20*
548	164.7	30.0 *16*
561 - 577	0.208	0.037 *3*
624	46.9	7.5 *4*
625	11.5	1.84 *10*
628	11.2	1.78 *10*
637	12.1	1.89 *10*
638	6.5	1.02 *6*
639 - 641	6.32	0.99 *4*
814	0.78	0.096 *10*
889 - 905	0.187	0.0210 *18*

Continuous Radiation (^{203}Po)

$\langle\beta+\rangle$=8.5 keV;$\langle$IB$\rangle$=0.39 keV

E_{bin}(keV)		$\langle\ \rangle$(keV)	(%)
0 - 10	β+	1.35×10^{-9}	1.59×10^{-8}
	IB	0.00031	
10 - 20	β+	9.8×10^{-8}	5.8×10^{-7}
	IB	0.00038	0.0027
20 - 40	β+	4.87×10^{-6}	1.46×10^{-5}
	IB	0.00063	0.0022
40 - 100	β+	0.000463	0.00058
	IB	0.023	0.031
100 - 300	β+	0.0406	0.0178
	IB	0.0079	0.0045
300 - 600	β+	0.381	0.081
	IB	0.0153	0.0034
600 - 1300	β+	3.02	0.314
	IB	0.080	0.0084
1300 - 2500	β+	4.88	0.285
	IB	0.20	0.0111
2500 - 3976	β+	0.204	0.0078
	IB	0.064	0.0022
	$\Sigma\beta$+		0.71

$^{203}_{85}$At(7.37 *20* min)

Mode: ϵ(69 *3* %), α(31 *3* %)
Δ: -12310 *120* keV
SpA: 1.26×10^{8} Ci/g
Prod: ^{197}Au(^{12}C,6n); protons on Th

Alpha Particles (^{203}At)

α(keV)
6088 *1*

Photons (^{203}At)

γ_{mode}	γ(keV)	γ(rel)
γ_ϵ	145.8 *1*	14.1 *9*
γ_ϵ	152.1 *1*	5.0 *7*
γ_ϵ	154.7 *1*	4.1 *4*
γ_ϵ	204.4 *2*	5 *2*
γ_ϵ	206.6 *1*	9.0 *8*
γ_ϵ	245.9 *2*	47.6 *24*
γ_ϵ	361.6 *3*	23.1 *14*
γ_ϵ	390.3 *2*	8.1 *8*
γ_ϵ	416.9 *1*	14.3 *11*
γ_ϵ	487.3 *1*	6.0 *18*
γ_ϵ	531.9 *1*	18.2 *18*
γ_ϵ	608.8 *1*	20.4 *18*
γ_ϵ	639.3 *1*	97 *5*
γ_ϵ	641.4 *1*	53 *3*
γ_ϵ	656.2 *1*	29.6 *18*
γ_ϵ	737.9 *1*	41.6 *21*
γ_ϵ	845.8 *1*	30.0 *21*
γ_ϵ	880.4 *1*	40.7 *24*
γ_ϵ	1002.0 *1*	86 *4*
γ_ϵ	1034.0 *1*	100 *5*

$^{203}_{86}$Rn(45 *3* s)

Mode: α(66 *9* %), ϵ(34 *9* %)
Δ: -6220 *300* keV syst
SpA: 1.23×10^{9} Ci/g
Prod: ^{197}Au(^{14}N,8n); protons on Th

Alpha Particles (^{203}Rn)

α(keV)
6498 *5*

$^{203}_{86}$Rn(28 *2* s)

Mode: α
Δ: -5910 *340* keV syst
SpA: 1.96×10^{9} Ci/g
Prod: ^{197}Au(^{14}N,8n); protons on Th

Alpha Particles (^{203}Rn)

α(keV)
6548 *3*

$^{203}_{87}$Fr(550 *20* ms)

Mode: α
Δ: 930 *330* keV syst
SpA: 5.74×10^{10} Ci/g
Prod: ^{197}Au(^{16}O,10n)

Alpha Particles (^{203}Fr)

α(keV)
7132 *5*

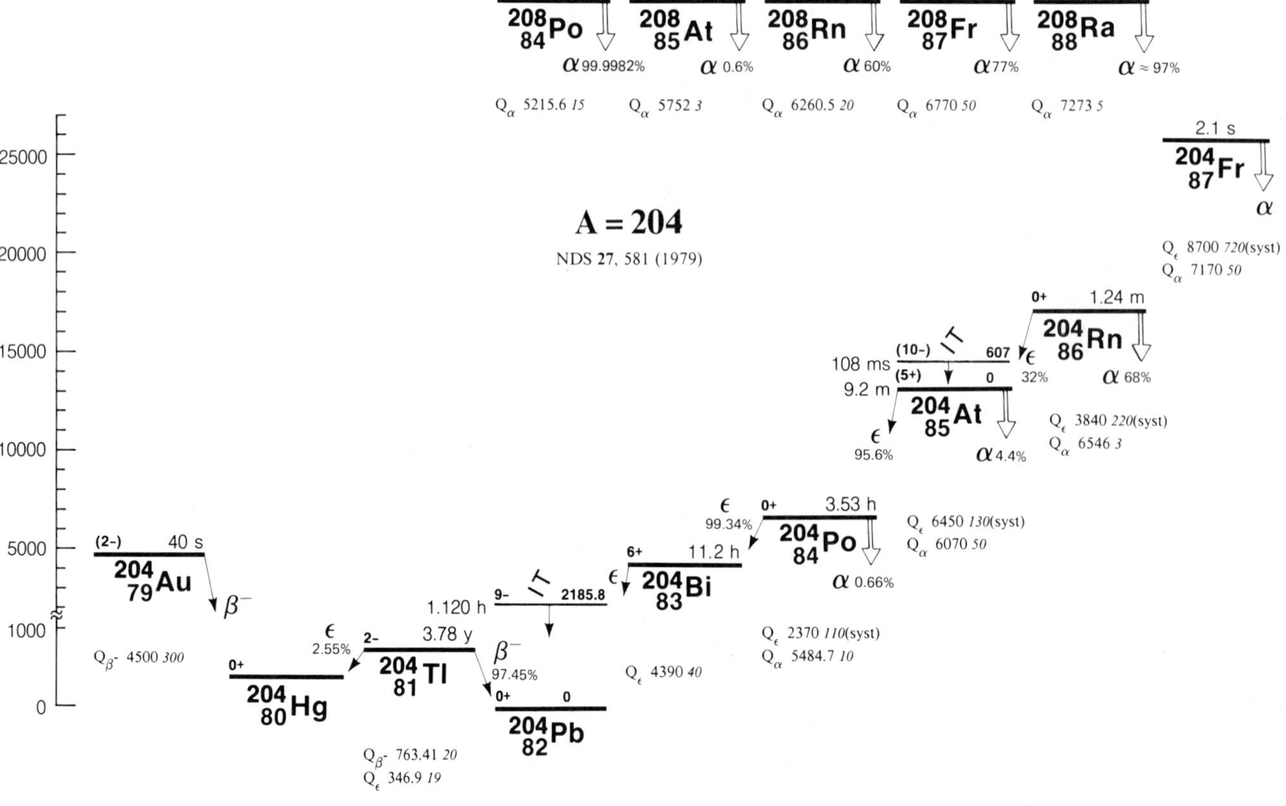

$^{204}_{79}$Au(40 *3* s)

Mode: β-
Δ: -20220 *300* keV
SpA: 1.37×10^{9} Ci/g
Prod: ^{204}Hg(n,p)

Photons (^{204}Au)

⟨γ⟩=1946 *130* keV

γ_{mode}	γ(keV)	γ(%)
γ (E2)	436.47 *19*	91
γ	691.8 *3*	23 *3*
γ	722.9 *3*	22 *3*
γ	1391.9 *4*	24 *3*
γ	1404.8 *7*	~4 ?
γ	1414.7 *3*	8 *3*
γ	1511.2 *4*	28 *4*
γ	1553.0 *4*	8 *3*
γ	1704.1 *6*	4.8 *14*
γ	1828.4 *4*	2.8 *9*
γ	1841.3 *7*	2.7 *9*

$^{204}_{80}$Hg(stable)

Δ: -24716 *4* keV
%: 6.85 *5*

$^{204}_{81}$Tl(3.78 *2* yr)

Mode: β-(97.45 *5* %), ϵ(2.55 *5* %)
Δ: -24369 *4* keV
SpA: 463.6 Ci/g
Prod: ^{203}Tl(n,γ)

Photons (^{204}Tl)

⟨γ⟩=1.11 *12* keV

γ_{mode}	γ(keV)	γ(%)†
Hg L_{ℓ}	8.722	0.015 *3*
Hg L_{α}	9.980	0.32 *7*
Hg L_{η}	10.647	0.0035 *7*
Hg L_{β}	11.932	0.29 *6*
Hg L_{γ}	13.989	0.053 *12*
Hg $K_{\alpha2}$	68.893	0.41 *8*
Hg $K_{\alpha1}$	70.818	0.70 *14*
Hg $K_{\beta1}'$	80.124	0.25 *5*
Hg $K_{\beta2}'$	82.780	0.068 *14*

† 2.0% uncert(syst)

Atomic Electrons (^{204}Tl)

⟨e⟩=0.155 *18* keV

e_{bin}(keV)	⟨e⟩(keV)	e(%)
12	0.069	0.56 *13*
14	0.036	0.25 *6*
15	0.019	0.13 *3*
53	0.0030	0.0056 *12*
54	0.0048	0.0088 *20*
55	0.00027	0.00050 *11*
56	0.0028	0.0050 *11*
57	0.0050	0.0088 *20*

Atomic Electrons (^{204}Tl) (continued)

e_{bin}(keV)	⟨e⟩(keV)	e(%)
59	0.0023	0.0040 *9*
65	0.0044	0.0067 *15*
66	0.0014	0.0021 *5*
67	0.0012	0.0018 *4*
68	0.0036	0.0053 *12*
69	0.00033	0.00048 *11*
70	0.00076	0.00108 *24*
71	0.00012	0.00018 *4*
76	0.00042	0.00056 *12*
77	0.00051	0.00066 *15*
78	6.5 ×10^{-5}	8.4 *19* ×10^{-5}
79	0.00034	0.00043 *10*
80	0.00012	0.00015 *3*

Continuous Radiation (^{204}Tl)

⟨β-⟩=238 keV;⟨IB⟩=0.20 keV

E_{bin}(keV)		⟨ ⟩(keV)	(%)
0 - 10	β-	0.133	2.67
	IB	0.0117	
10 - 20	β-	0.39	2.60
	IB	0.0110	0.076
20 - 40	β-	1.53	5.09
	IB	0.020	0.070
40 - 100	β-	10.0	14.4
	IB	0.048	0.075
100 - 300	β-	75	38.5
	IB	0.079	0.048
300 - 600	β-	133	31.3
	IB	0.023	0.0062
600 - 763	β-	18.5	2.87
	IB	0.00034	5.4×10^{-5}

$^{204}_{82}$Pb(stable)

Δ: -25132 *4* keV

%: 1.4 *1*

$^{204}_{82}$Pb(1.120 *5* h)

Mode: IT

Δ: -22946 *4* keV

SpA: 1.371×10^{7} Ci/g

Prod: daughter ^{204}Bi; ^{203}Tl(d,n)

Photons (^{204}Pb)

⟨γ⟩=2102 *210* keV

γ_{mode}	γ(keV)	γ(%)†
Pb L_ℓ	9.185	0.095 *14*
Pb L_α	10.541	1.89 *23*
Pb L_η	11.349	0.039 *5*
Pb L_β	12.680	2.3 *3*
Pb L_γ	14.864	0.47 *7*
Pb $K_{\alpha2}$	72.803	2.7 *3*
Pb $K_{\alpha1}$	74.969	4.5 *5*
Pb $K_{\beta1}'$	84.789	1.59 *18*
Pb $K_{\beta2}'$	87.632	0.46 *5*
γ [M4]	120.44 *9* ?	<0.6 ?
γ [E2]	209.13 *16* ?	
γ M1	289.26 *7*	0.079 *9*
γ [E5]	368.21 *8* ?	<2
γ E2	374.81 *6*	89 *16*
γ E5	622.48 *8*	~1
γ	664.07 *9* ?	
γ E2	899.22 *7*	99 *16*
γ E5	911.75 *7*	94 *15*
γ (E4)	1274.03 *9*	0.0119 *22*

† 5.1% uncert(syst)

Atomic Electrons (^{204}Pb)

⟨e⟩=109 *8* keV

e_{bin}(keV)	⟨e⟩(keV)	e(%)
13 - 62	1.04	6.7 *7*
68 - 84	0.085	0.119 *7*
201	0.065	0.032 *4*
273 - 286	0.0187	0.0068 *6*
287	10.0	3.5 *6*
288 - 289	0.00120	0.00042 *4*
359 - 375	7.4	2.03 *18*
534	1.1	~0.21
607 - 619	2.0	0.33 *13*
811	5.3	0.65 *11*
824	42.6	5.2 *8*
883 - 895	1.37	0.155 *17*
896	8.5	0.95 *15*
897	16.9	1.9 *3*
898	0.069	0.0077 *13*
899	3.0	0.33 *5*
908	9.4	1.04 *17*
1186	0.0018	0.00015 *3*
1258 - 1274	0.00073	5.8 *6* $\times10^{-5}$

$^{204}_{83}$Bi(11.22 *10* h)

Mode: ε

Δ: -20740 *40* keV

SpA: 1.369×10^{6} Ci/g

Prod: ^{206}Pb(p,3n); ^{203}Tl(α,3n); ^{204}Pb(d,2n)

Photons (^{204}Bi)

⟨γ⟩=3177 *81* keV

γ_{mode}	γ(keV)	γ(%)†
Pb $K_{\alpha2}$	72.803	26 *5*
Pb $K_{\alpha1}$	74.969	44 *8*
γ E2	78.61 *6*	0.33 *5*
γ M1	80.32 *7*	0.76 *12*
Pb $K_{\beta1}'$	84.789	16 *3*
Pb $K_{\beta2}'$	87.632	4.5 *9*
γ	91.00 *10*	<0.024
γ	92.30 *10*	<0.024
γ	96.64 *15*	~0.032
γ (M1)	100.41 *6*	0.167 *24*
γ [M1]	109.64 *8*	~0.12
γ [M4]	120.44 *9* ?	<0.08 •
γ	139.7 *3*	0.13 *4*
γ M1+45%E2	140.89 *7*	0.94 *10*
γ [M1]	141.06 *9*	0.048 *16*
γ	145.5 *3*	0.032 *8*
γ [E2]	147.16 *7*	0.13 *3*
γ	149.7 *3*	0.071 *24*
γ	165.02 *15*	0.13 *3*
γ (M1)	168.5 *3*	0.13 *4*
γ [M1]	168.77 *8*	0.048 *8* ?
γ M1+21%E2	169.90 *7*	0.28 *4*
γ M1	176.17 *6*	1.11 *12*
γ [E2]	209.13 *16* ?	•
γ	211.3 *3*	
γ (M1)	212.68 *15*	0.29 *6*
γ M1	216.08 *7*	1.43 *16*
γ [M1]	216.18 *10*	0.36 *3*
γ E2(+6.9%M1)	219.51 *8*	2.30 *24*
γ M1	222.34 *6*	0.94 *10*
γ	224.9 *3*	<0.12
γ	227.56 *15*	<0.21
γ M1	240.62 *9*	0.31 *4*
γ M1	249.09 *6*	2.06 *24*
γ M1	251.0 *3*	0.151 *16*
γ [M1+E2]	257.36 *10*	0.079 *16*
γ M1	289.26 *7*	2.8 *4* •
γ E2+30%M1	291.36 *9*	0.95 *16*
γ	304.64 *10*	0.127 *24*
γ [E1]	320.97 *12*	0.135 *24*
γ M1	330.85 *15*	0.37 *8*
γ [E2]	332.73 *10*	0.151 *16*
γ	336.41 *10*	0.058 *12*
γ (M1)	340.69 *15*	0.119 *24*
γ [E5]	368.21 *8* ?	<0.32 •
γ [E2]	368.5 *10*	0.159 *16*
γ [E1]	368.96 *9*	0.49 *10*
γ E2	374.81 *6*	81 *6* •
γ	386.8 *10*	0.064 *16*
γ (M1)	405.86 *9*	0.24 *3*
γ M1	412.52 *8*	0.29 *4*
γ M1	421.76 *7*	1.08 *11*
γ [M1+E2]	432.61 *9*	0.048 *16*
γ M1	438.52 *9*	0.79 *16* ?
γ E1	440.56 *5*	2.5 *4*
γ	447.30 *9*	0.48 *6*
γ	456.00 *11*	0.127 *24*
γ	461.81 *14*	0.119 *24*
γ (M1)	468.63 *10*	0.51 *6*
γ (M1)	473.60 *15*	~0.10
γ (M1)	477.86 *9*	0.17 *3*
γ (M1+~50%E2)	502.04 *8*	0.82 *10*
γ [E2]	510.87 *9*	0.43 *6*
γ [E2]	514.45 *9*	0.294 *24*
γ M1	522.16 *7*	0.64 *8*
γ [E2]	522.48 *10*	0.151 *16*
γ M1	532.84 *7*	1.35 *16*
γ E2+30%M1	543.54 *6*	<0.27
γ (M1)	548.91 *7*	0.44 *7*
γ (M1)	585.16 *7*	0.32 *5*

Photons (^{204}Bi) (continued)

γ_{mode}	γ(keV)	γ(%)†
γ [M1+E2]	595.08 *7*	0.37 *6*
γ (M1)	598.16 *8*	0.41 *6*
γ	604.83 *15*	0.21 *3*
γ [M1+E2]	611.91 *8*	0.25 *4*
γ	618.37 *11*	0.29 *3*
γ E5	622.48 *8*	0.19 *9* •
γ	631.98 *15*	0.079 *16*
γ (M1+E0)	654.95 *8*	0.127 *24*
γ M1	661.61 *7*	2.6 *4*
γ M1+E2	663.38 *9*	<0.8
γ	664.07 *9* ?	<0.048 •
γ M1	670.85 *6*	10.6 *10*
γ (M1)	683.47 *11*	0.22 *3*
γ (M1)	690.93 *8*	0.95 *10*
γ [M1+E2]	705.66 *16*	0.198 *24*
γ (M1)	709.25 *7*	1.43 *24*
γ (M1)	710.45 *9*	1.43 *24*
γ M1	718.48 *8*	0.88 *10*
γ M1+40%E2	725.19 *8*	0.91 *6*
γ M1	736.15 *9*	0.68 *8*
γ (M1)	745.23 *10*	0.73 *11*
γ	753.93 *8*	1.06 *11*
γ (M1)	765.53 *9*	0.50 *6*
γ (E2)	771.25 *6*	0.40 *6*
γ M1+21%E2	791.30 *7*	3.2 *3*
γ	821.34 *9*	0.60 *9*
γ [E1]	823.01 *13*	0.52 *8*
γ [E2]	827.98 *8*	0.50 *8*
γ M1	832.00 *9*	0.95 *16*
γ (M1)	834.25 *6*	1.03 *16*
γ	840.99 *8*	0.25 *4*
γ M1(+20%E2)	847.25 *7*	0.79 *16*
γ E2	899.22 *7*	98 *8* •
γ E5	911.75 *7*	13.6 *16* •
γ (M1)	912.31 *8*	11.1 *16*
γ E2	918.34 *8*	10.8 *8*
γ	924.19 *11*	0.071 *16*
γ	934.28 *14*	0.29 *4*
γ	940.54 *15*	0.10 *5*
γ	950.43 *15*	0.159 *24*
γ	958.87 *15*	0.19 *3*
γ	964.42 *15*	0.52 *6*
γ	971.3 *10*	0.262 *24*
γ [M1+E2]	974.28 *8*	0.44 *3*
γ E1	984.09 *5*	58.0 *14*
γ [E3]	990.36 *6*	1.11 *16*
γ	1013.74 *12*	0.079 *16*
γ (M1)	1022.03 *25*	0.19 *3*
γ [E1]	1027.04 *9*	0.071 *16*
γ	1037.41 *9*	0.39 *6*
γ M1	1043.68 *8*	1.27 *16*
γ (M1)	1049.14 *25*	0.20 *3*
γ	1054.7 *4*	0.087 *8*
γ	1056.65 *25*	0.17 *3*
γ	1060.27 *25*	0.52 *8*
γ E1	1064.41 *8*	0.95 *16*
γ	1091.80 *11*	0.095 *16*
γ [E1]	1095.51 *9*	0.21 *3*
γ [E1]	1102.17 *8*	0.51 *11*
γ [E1]	1105.1 *1*	0.198 *24*
γ E1	1111.40 *6*	1.43 *16*
γ	1120.6 *10*	0.036 *8*
γ	1127.60 *25*	0.27 *4*
γ (M1)	1133.21 *10*	0.84 *10*
γ M1(+E2)	1139.47 *9*	0.60 *8*
γ (M1)	1146.64 *25*	0.17 *3*
γ (M1)	1157.75 *8*	0.53 *7*
γ	1165.66 *12*	0.056 *10*
γ (M1)	1167.11 *25*	0.183 *24*
γ	1181.55 *12*	0.071 *16*
γ	1199.08 *25*	0.24 *4*
γ M1	1203.92 *8*	2.1 *3*
γ E1	1211.81 *6*	3.1 *4*
γ (M1)	1232.93 *9*	0.41 *6*
γ	1240.4 *10*	~0.13
γ	1259.20 *21*	0.44 *7*
γ	1261.67 *8*	0.135 *24*
γ (E4)	1274.03 *9*	0.0108 *21* •
γ [E1]	1274.81 *7*	2.2 *5*
γ (M1)	1290.71 *25*	0.18 *3*
γ (M1)	1299.13 *13*	0.127 *24*
γ (M1)	1328.14 *13*	0.39 *6*
γ (M1)	1334.60 *9*	0.31 *5*
γ (M1)	1348.02 *11*	0.214 *24* ?

Photons (^{204}Bi)
(continued)

γ_{mode}	γ(keV)	γ(%)†
γ[E1]	1352.87 9	0.48 4
γ[E2]	1354.16 18	0.87 16
γ(M1)	1373.83 7	0.43 7
γ(E2)	1380.09 6	0.21 3
γ(M1)	1383.67 24	0.159 24
γE1	1414.83 7	0.97 11
γ	1446.22 25	0.087 16
γ	1447.62 25	0.21 3
γ	1453.4 10	0.043 10
γ(M1)	1466.46 25	0.36 6
γ(M1)	1469.45 16	<0.40
γ	1475.30 13	0.087 16
γ(M1)	1487.53 12	0.21 3
γ(M1+E2)	1517.63 25	0.36 6
γ(M1)	1524.19 9	0.91 10
γ(M1+~50%E2)	1536.64 25	0.41 6
γ(M1)	1569.23 22	0.135 24
γE2,E1	1572.88 9	0.23 4
γ(M1)	1589.50 21	0.35 5
γ(M1+E2)	1607.14 10	0.22 3 ?
γ	1612.18 22	0.20 8
γ	1614.16 12	0.143 16
γ(E2)	1639.48 25	0.37 7
γE2,E1	1645.70 7	0.68 11
γ(M1)	1652.10 11	0.56 9
γ(M1+E2)	1654.94 7	0.56 9 ?
γ[M1+E2]	1658.88 15	0.079 8
γ	1665.06 22	0.061 6
γ[E1]	1669.11 9	0.079 24
γ[E3]	1675.21 16	0.064 16
γ	1679.9 10	0.041 8
γ[M1+E2]	1685.63 15	0.151 24
γ(M1)	1689.34 12	0.59 7
γ	1696.92 17	0.058 9
γ	1700.24 25	0.25 4
γE1	1703.55 9	1.98 24
γ(M1)	1709.88 17	0.095 16
γ(M1)	1715.32 25	0.135 24
γ(M1)	1731.80 15	0.60 6
γ(E2)	1749.88 11	0.29 5
γE1	1755.34 6	1.23 15
γ(M1)	1760.81 15	0.21 3
γ	1778.89 11	0.28 5
γ(M1)	1779.79 19	0.33 6
γ	1785.97 25	0.040 8
γ	1791.3 10	0.037 8
γ	1794.5 5	0.053 8
γ	1797.01 25	0.095 24
γ	1804.05 25	0.056 8
γ[E1]	1818.34 7	0.54 6
γ	1826.65 25	0.60 6
γ	1836.50 11	0.071 16
γ[M1+E2]	1850.78 17	0.075 8
γ(M1)	1857.04 17	0.27 5
γ	1881.86 25	0.10 3
γ	1891.2 5	0.029 6
γ	1891.5 10	0.073 7
γE1	1896.40 9	1.35 16
γ(M1)	1907.97 15	0.17 3
γ	1916.53 25	0.048 8
γ	1926.05 9	0.59 6
γ	1930.9 4	0.016 4
γ(M1)	1941.36 9	0.79 10
γ	1952.4 8	0.087 24
γ	1956.84 25	0.34 7
γ[E1]	1958.37 8	0.40 6
γ[E1]	1964.75 10	0.37 5
γ	1988.10 25	0.032 8
γ	2009.22 12	0.079 16
γ	2014.2 5	0.048 8
γ(E2)	2028.02 22	0.143 24
γ	2046.13 12	0.071 16
γ	2064.43 13	0.032 8 ?
γ	2084.3 5	~0.024
γ	2092.7 5	0.032 8
γ[E1]	2100.76 16	0.040 8
γ	2137.7 5	0.032 8
γ	2170.13 14	0.040 8
γ	2172.3 5	0.032 8
γ	2177.0 5	0.048 8
γ	2184.87 13	0.032 8
γ	2250.8 5	0.016 4
γ	2252.2 10	0.024 4
γ	2263.52 8	<0.30
γ	2279.5 5	0.040 8
γ	2313.27 23	0.071 16
γ	2324.9 10	0.015 3
γ	2326.1 5	0.040 8
γ	2433.0 4	0.032 8
γ	2447.7 10	0.016 3
γ	2450.8 5	0.048 8
γ	2471.5 3	0.025 4
γ	2472.7 5	0.040 8
γ	2475.9 4	0.032 8
γ	2494.85 8	~0.016 ?
γ	2517.79 8	0.198 24
γ	2566.22 10	0.095 16
γ	2655.2 5	0.040 16
γ	2668.3 5	~0.016
γ	2680.82 18	0.37 5
γ	2686.91 10	0.27 4
γ	2722.2 4	0.032 8
γ	2758.9 5	0.095 16
γ	2765.2 4	~0.016
γ	2794.4 5	~0.016
γ[E1]	2802.40 22	~0.016
γ	2837.42 10	0.21 3
γ	2855:47 12	~0.016
γ	2864.7 5	<0.016
γ	2898.23 22	<0.016
γ	2948.1 5	<0.016
γ	2955.6 5	~0.016
γ	2976.17 11	~0.016
γ	3012.10 22	0.024 8

† 2.5% uncert(syst)
* with ^{204}Pb(1.12 h) in equilib

Atomic Electrons (^{204}Bi)
⟨e⟩=90.7 20 keV

e_{bin}(keV)	⟨e⟩(keV)	e(%)
3 - 12	0.18	1.6 3
13	4.0	31 6
15	2.6	17 3
16 - 65	6.1	13.9 14
66 - 110	6.8	8.8 5
124 - 127	1.1	0.90 21
128	2.3	1.8 3
130 - 160	3.5	2.46 21
161	2.1	1.30 16
162 - 200	1.05	0.57 4
201	2.3	1.16 17
203 - 253	3.12	1.43 9
254 - 286	0.83	0.30 3
287	9.1	3.18 23
288 - 337	1.05	0.326 22
338 - 356	0.49	0.140 22
359	1.78	0.50 6
360	2.31	0.64 5
362 - 410	3.31	0.89 3
412 - 461	1.41	0.323 19
462 - 511	0.46	0.093 10
512 - 549	0.48	0.092 19
567 - 582	0.91	0.159 22
583	2.68	0.46 4
584 - 632	1.53	0.248 23
637 - 683	1.64	0.249 21
687 - 736	0.95	0.135 16
738 - 787	0.74	0.099 9
788 - 810	0.024	0.0031 11
811	5.2	0.64 5
812 - 823	0.078	0.0095 10
824	8.0	0.97 12
825 - 871	0.69	0.083 6
876 - 895	1.47	0.166 11
896	2.74	0.31 3
897	2.5	0.28 3
898 - 947	2.32	0.257 20
948 - 990	0.62	0.064 5
998 - 1047	0.240	0.0232 21
1048 - 1096	0.17	0.0163 24
1098 - 1146	0.39	0.035 4
1150 - 1199	0.140	0.0119 20
1200 - 1249	0.127	0.0104 8
1255 - 1299	0.139	0.0109 8
1312 - 1361	0.065	0.0049 7
1364 - 1412	0.078	0.0056 13
1414 - 1463	0.127	0.0088 9
1464 - 1560	0.128	0.0084 9
1564 - 1662	0.264	0.0164 11
1664 - 1763	0.144	0.0085 10
1764 - 1854	0.132	0.0072 8
1864 - 1962	0.060	0.0032 5
1972 - 2071	0.0068	0.00034 7
2077 - 2176	0.011	0.00053 22
2181 - 2276	0.0059	0.00026 8
2297 - 2388	0.0054	0.00023 5
2407 - 2505	0.009	0.00038 12
2514 - 2599	0.016	0.00062 21
2634 - 2719	0.0069	0.00026 7
2743 - 2842	0.0065	0.00023 9
2849 - 2945	0.0014	4.9 15 $\times10^{-5}$
2952 - 3009	0.00017	5.8 19 $\times10^{-6}$

Continuous Radiation (^{204}Bi)
⟨β+⟩=1.27 keV;⟨IB⟩=1.04 keV

E_{bin}(keV)		⟨ ⟩(keV)	(%)
0 - 10	β+	3.59×10^{-9}	4.22×10^{-8}
	IB	0.000140	
10 - 20	β+	2.53×10^{-7}	1.50×10^{-6}
	IB	0.0025	0.019
20 - 40	β+	1.21×10^{-5}	3.62×10^{-5}
	IB	0.00141	0.0044
40 - 100	β+	0.00106	0.00133
	IB	0.51	0.69
100 - 300	β+	0.073	0.0327
	IB	0.057	0.035
300 - 600	β+	0.420	0.093
	IB	0.114	0.025
600 - 1300	β+	0.64	0.079
	IB	0.27	0.031
1300 - 2500	β+	0.129	0.0088
	IB	0.083	0.0054
2500 - 2946	IB	0.00066	2.6×10^{-5}
	Σβ+		0.21

$^{204}_{84}$Po(3.53 2 h)

Mode: ϵ(99.34 1 %), α(0.66 1 %)
Δ: -18360 100 keV syst
SpA: 4.352×10^{6} Ci/g
Prod: ^{209}Bi(p,6n); daughter ^{204}At(9.2 m); ^{196}Pt(^{12}C,4n); ^{204}Pb(^{3}He,3n)

Alpha Particles (^{204}Po)

α(keV)
5377 1

Photons (^{204}Po)

$\langle\gamma\rangle$=1316 *28* keV

γ_{mode}	γ(keV)	γ(%)†
Bi L_ℓ	9.419	1.7 *4*
Bi L_α	10.828	32 *7*
Bi L_η	11.712	0.39 *7*
Bi L_β	13.089	32 *8*
Bi L_γ	15.432	6.4 *17*
γ_ϵ(M1)	63.12 *12*	~9
Bi $K_{\alpha2}$	74.814	39 *4*
Bi $K_{\alpha1}$	77.107	65 *7*
Bi $K_{\beta1}'$	87.190	23.0 *23*
Bi $K_{\beta2}'$	90.128	6.8 *7*
γ_ϵM1+E2	116.6 *3*	0.80 *12* ?
γ_ϵM1	122.6 *3*	1.5 *6*
γ_ϵM1	131.0 *3*	1.05 *18*
γ_ϵM1	137.0 *3*	12.0 *6*
γ_ϵ[E1]	149.4 *4*	0.92 *18* ?
γ_ϵ[M1]	151.6 *9*	0.68 *18* ?
γ_ϵ	170.4 *10*	0.89 *25* ?
γ_ϵE1	203.68 *7*	3.4 *3*
γ_ϵM1	230.03 *7*	2.99 *6*
γ_ϵM1	270.24 *14*	31
γ_ϵM1	304.9 *3*	3.4 *3*
γ_ϵM1	316.72 *12*	4.9 *6*
γ_ϵM1	427.0 *8*	2.2 *3*
γ_ϵM1	451.94 *7*	2.7 *6*
γ_ϵ	460.1 *10*	1.54 *25*
γ_ϵ	486.1 *10*	0.43 *12*
γ_ϵ	492.1 *10*	0.34 *15*
γ_ϵM1	534.9 *3*	12.9 *6*
γ_ϵM1	540.1 *4*	1.76 *25*
γ_ϵ	566.8 *10*	0.52 *15*
γ_ϵ	587.3 *10*	0.49 *15*
γ_ϵ	590.1 *10*	0.28 *9*
γ_ϵ	609.1 *10*	0.25 *6*
γ_ϵ	619.1 *10*	~0.12
γ_ϵ	680.77 *13*	9.2 *9*
γ_ϵ	695.1 *4*	2.5 *3* ?
γ_ϵ	705.6 *10*	0.43 *12*
γ_ϵ	712.3 *10*	0.46 *22*
γ_ϵ	751.4 *10*	1.23 *22*
γ_ϵE1	762.65 *7*	11.4 *6*
γ_ϵ	791.6 *10*	0.68 *9*
γ_ϵ	817.7 *3*	0.95 *15*
γ_ϵ	854.4 *10*	0.28 *12*
γ_ϵ	866.1 *10*	0.18 *6*
γ_ϵE1	884.45 *12*	33.9 *15*
γ_ϵ	905.1 *10*	1.5 *3*
γ_ϵ	954.1 *10*	~0.22
γ_ϵ	964.7 *10*	0.22 *9*
γ_ϵ	968.4 *10*	0.34 *12*
γ_ϵ	974.1 *10*	~0.22
γ_ϵE1	1016.25 *10*	24.6 *15*
γ_ϵM1	1040.0 *3*	10.8 *6*
γ_ϵ	1046.8 *10*	0.34 *15*
γ_ϵ	1060.1 *10*	0.34 *15*
γ_ϵ	1127.6 *5*	0.43 *12*
γ_ϵ[E2]	1176.9 *4*	0.22 *9*
γ_ϵ	1179.9 *10*	0.62 *12*
γ_ϵ	1236.6 *10*	~0.06
γ_ϵ	1319.1 *10*	~0.031
γ_ϵ	1333.1 *10*	~0.08
γ_ϵ	1350.1 *10*	~0.12
γ_ϵ	1363.5 *10*	0.40 *12*
γ_ϵ	1401.1 *10*	~0.06
γ_ϵ	1419.4 *3*	~0.09 ?
γ_ϵ	1454.1 *10*	~0.06
γ_ϵ	1460.7 *10*	~0.15
γ_ϵ	1589.1 *10*	~0.12
γ_ϵ	1611.1 *10*	0.15 *6*
γ_ϵ	1640.1 *10*	~0.09
γ_ϵ	1953.1 *10*	~0.06

†13% uncert(syst)

Atomic Electrons (^{204}Po)

$\langle e\rangle$=164 *16* keV

e_{bin}(keV)	$\langle e\rangle$(keV)	e(%)
13	5.6	42 *6*
16	5.9	37 *10*
26 - 40	4.8	14 *4*
46	20.0	43.0 *24*
47	23.6	~50
50 - 58	0.94	1.7 *3*
59	7.1	~12
60 - 109	9.2	12.5 *23*
113 - 120	2.0	1.7 *4*
121	9.0	7.4 *4*
122 - 170	8.4	6.09 *25*
180	29.5	16 *3*
187 - 217	3.8	1.78 *13*
226	4.1	1.81 *22*
227 - 230	0.0777	0.0339 *8*
254	6.5	2.6 *5*
255 - 304	4.8	1.73 *14*
305 - 336	1.42	0.43 *5*
361 - 411	2.1	0.57 *13*
414 - 439	0.34	0.079 *14*
444	5.0	1.12 *7*
447 - 492	0.93	0.20 *3*
497 - 540	1.65	0.317 *25*
550 - 596	1.4	~0.24
603 - 622	0.5	~0.08
661 - 710	1.0	0.15 *4*
711 - 760	0.21	0.029 *13*
762 - 804	0.83	0.105 *7*
814 - 863	0.17	~0.020
864 - 905	0.29	0.033 *5*
926 - 974	2.13	0.225 *11*
1000 - 1047	0.51	0.050 *3*
1056 - 1089	0.05	~0.005
1111 - 1161	0.013	0.0012 *5*
1163 - 1179	0.011	~0.0009
1220 - 1260	0.014	~0.0012
1273 - 1320	0.027	~0.0021
1329 - 1370	0.023	0.0017 *7*
1385 - 1419	0.0019	0.00013 *6*
1438 - 1460	0.0025	~0.00017
1499 - 1598	0.019	0.0013 *5*
1607 - 1637	0.0012	~7×10^{-5}
1863 - 1950	0.0026	~0.00014

$^{204}_{85}$At(9.2 *2* min)

Mode: ϵ(95.6 *2* %), α(4.4 *2* %)
Δ: -11920 *90* keV
SpA: 1.001×10^8 Ci/g
Prod: ^{197}Au(^{12}C,5n); ^{209}Bi(α,9n)

Alpha Particles (^{204}At)

α(keV)
5951 *2*

Photons (^{204}At)

$\langle\gamma\rangle$=1772 *42* keV

γ_{mode}	γ(keV)	γ(%)†
Po L_ℓ	9.658	~0.6
Po L_α	11.119	~11
Po L_η	12.085	~0.15
Po L_β	13.501	~11
Po L_γ	15.892	~2
Po $K_{\alpha2}$	76.858	~16
Po $K_{\alpha1}$	79.290	~27
Po $K_{\beta1}'$	89.639	~10
Po $K_{\beta2}'$	92.673	~3
γ_ϵ	154.4 *10* ?	
γ_ϵ	311.4 *10* ?	
γ_ϵ[M1]	327.1 *7*	4.7 *9*
γ_ϵ	336.7 *7*	5.6 *9*
γ_ϵ[E2]	425.4 *3*	66 *5*
γ_ϵ	490.4 *10*	4.7 *9*
γ_ϵ[E2]	515.6 *3*	90 *5*
γ_ϵ[E1]	588.8 *10*	8.5 *9*
γ_ϵ	608.4 *6*	19.8 *19*
γ_ϵ[E2]	683.7 *5*	94.07
γ_ϵ	762.1 *7*	4.7 *9*
γ_ϵ	842.7 *7*	8.5 *19*

†0.23% uncert(syst)

Atomic Electrons (^{204}At)

$\langle e\rangle$=55 *5* keV

e_{bin}(keV)	$\langle e\rangle$(keV)	e(%)
14	2.4	~17
16 - 65	2.6	13 *5*
72 - 89	0.53	0.70 *15*
234	3.8	1.6 *3*
244	2.3	~1.0
310 - 327	1.8	0.58 *19*
332	6.9	2.06 *17*
333 - 337	0.23	~0.07
403	2.5	0.6 *3*
409	1.56	0.38 *3*
412 - 421	0.92	0.222 *14*
422	8.6	2.04 *14*
423 - 425	0.299	0.070 *4*
473 - 496	0.62	0.13 *4*
499	2.68	0.54 *4*
502 - 513	1.24	0.244 *9*
515	3.9	~0.8
516	0.0057	0.00110 *7*
572 - 589	0.064	0.0110 *9*
591	7.3	1.24 *14*
592 - 608	0.37	0.06 *3*
667	1.75	0.262 *10*
669 - 684	1.5	0.22 *8*
745 - 762	1.2	~0.16
826 - 843	0.25	~0.03

Continuous Radiation (^{204}At)

$\langle\beta+\rangle$=460 keV;$\langle$IB$\rangle$=14.4 keV

E_{bin}(keV)		$\langle\ \rangle$(keV)	(%)
0 - 10	$\beta+$	2.44×10^{-8}	2.86×10^{-7}
	IB	0.0148	
10 - 20	$\beta+$	1.80×10^{-6}	1.06×10^{-5}
	IB	0.0161	0.112
20 - 40	$\beta+$	9.1×10^{-5}	0.000273
	IB	0.030	0.102
40 - 100	$\beta+$	0.0090	0.0112
	IB	0.47	0.58
100 - 300	$\beta+$	0.83	0.364
	IB	0.30	0.17
300 - 600	$\beta+$	8.4	1.79
	IB	0.48	0.110
600 - 1300	$\beta+$	82	8.4
	IB	2.0	0.21
1300 - 2500	$\beta+$	270	14.6
	IB	6.0	0.32
2500 - 5000	$\beta+$	99	3.48
	IB	5.1	0.167
5000 - 5082	IB	8.4×10^{-6}	1.68×10^{-7}
	$\Sigma\beta+$		29

$^{204}_{85}$At(108 *10* ms)

Mode: IT
Δ: -11313 *92* keV
SpA: 8.0×10^{10} Ci/g
Prod: ^{193}Ir(^{16}O,5n); ^{197}Au(^{12}C,5n); ^{196}Pt(^{15}N,7n)

Photons (^{204}At)

⟨γ⟩=550 *6* keV

γ_{mode}	γ(keV)	γ(%)†
At L_ℓ	9.897	0.044 *5* •
At L_α	11.414	0.83 *5* •
At L_η	12.466	0.0186 *13* •
At L_β	13.907	1.08 *8* •
At L_γ	16.371	0.229 *20* •
At $K_{\alpha2}$	78.947	1.06 *5* •
At $K_{\alpha1}$	81.517	1.76 *8* •
At $K_{\beta1}'$	92.136	0.63 *3* •
At $K_{\beta2}'$	95.265	0.194 *9* •
γ (E3)	587.3 *2*	93 *1*

† 1.1% uncert(syst)
• from 587 keV transition

Atomic Electrons (^{204}At)

⟨e⟩=36.1 *9* keV

e_{bin}(keV)	⟨e⟩(keV)	e(%)
14	0.174	1.23 *13*
17	0.209	1.24 *13*
61	0.0215	0.035 *4*
62	0.00065	0.00105 *11*
64	0.0068	0.0106 *12*
65	0.0115	0.0178 *19*
67	0.0053	0.0079 *9*
74	0.0075	0.0101 *11*
75	0.0053	0.0071 *8*
76	0.0033	0.0043 *5*
77	0.0038	0.0050 *5*
78	0.0072	0.0093 *10*
79	0.00114	0.00144 *16*
80	0.00035	0.00043 *5*
81	0.00186	0.00230 *25*
87	0.00121	0.00139 *15*
88	0.00094	0.00106 *12*
89	0.00038	0.00043 *5*
90	0.00032	0.00036 *4*
91	0.00087	0.00095 *10*
92	0.000154	0.000168 *18*
492	18.5	3.77 *16*
570	4.26	0.75 *3*
571	6.9	1.20 *5*
573	1.45	0.253 *10*
583	2.96	0.507 *21*
584	0.417	0.071 *3*
585	0.00349	0.000597 *25*
586	0.79	0.134 *6*
587	0.333	0.0566 *23*

$^{204}_{86}$Rn(1.24 *3* min)

Mode: α(68 *4* %), ε(32 *4* %)
Δ: -8070 *200* keV syst
SpA: 7.40×10^{8} Ci/g
Prod: ^{197}Au(^{14}N,7n); ^{16}O on Pt; ^{12}C on Hg; protons on Th

Alpha Particles (^{204}Rn)

α(keV)
6417 *3*

$^{204}_{87}$Fr(2.1 *2* s)

Mode: α
Δ: 630 *690* keV
SpA: 2.24×10^{10} Ci/g
Prod: ^{197}Au(^{16}O,9n)

Alpha Particles (^{204}Fr)

α(keV)	α(rel)
6967 *5*	30 *6*
7027 *5*	70 *6*

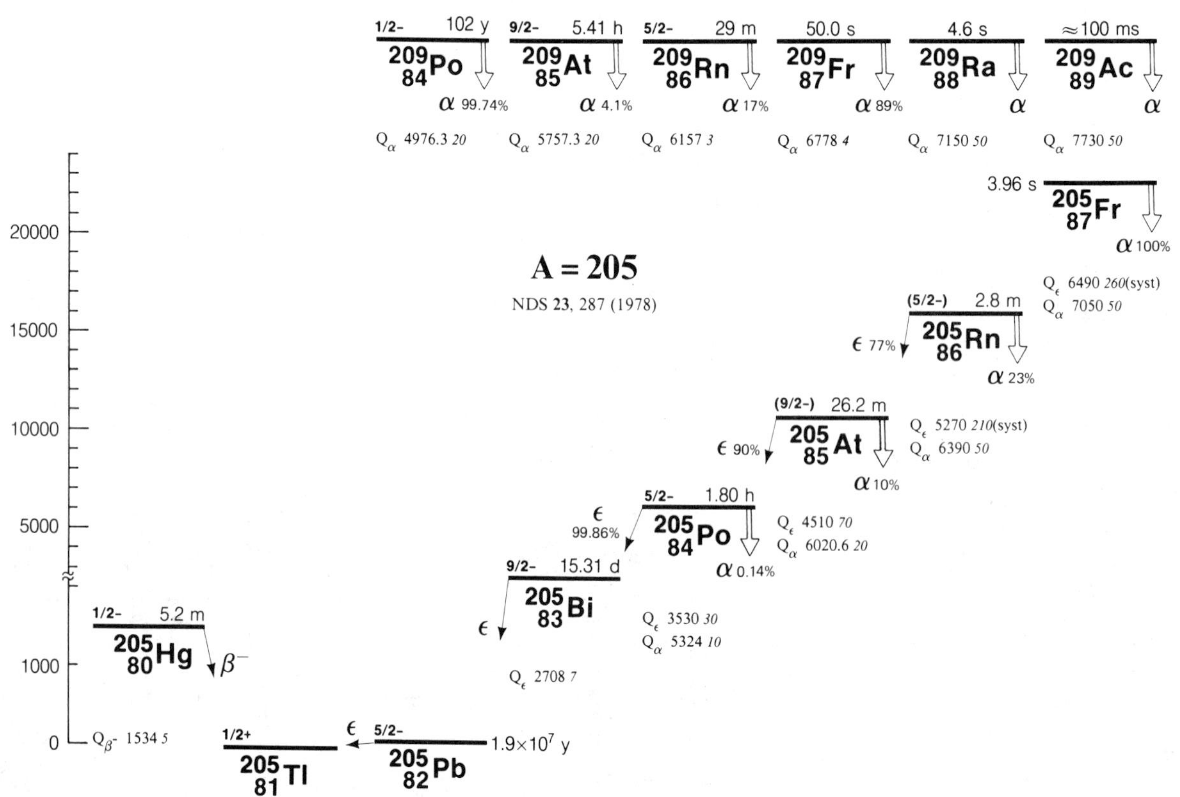

$^{205}_{80}$Hg(5.2 *1* min)

Mode: β-

Δ: -22312 *7* keV

SpA: 1.76×10^{8} Ci/g

Prod: ^{204}Hg(n,γ); ^{204}Hg(d,p)

Photons (^{205}Hg)

⟨γ⟩=4.8 *9* keV

γ_{mode}	γ(keV)	γ(%)†
γ E2+29%M1	203.58 *3*	2.2
γ M1+0.5%E2	415.4 *3*	0.0130 *18*
γ E2	619.0 *3*	0.0020 *4*
γ [M1+E2]	721.1 *5*	0.0011 *3*
γ E2	937.1 *6*	0.0020 *4*
γ [M1+E2]	1014.9 *4*	0.00068 *22*
γ [M1+E2]	1136.5 *4*	0.0046 *11*
γ M1(+10%E2)	1140.7 *6*	0.0010 *4*
γ [M1]	1218.5 *4*	0.0062 *11*
γ [M1+E2]	1230.3 *5*	0.00051 *22*
γ [M1+E2]	1340.0 *4*	0.00033 *11*
γ [M1]	1433.8 *5*	0.0044 *11*

† 45% uncert(syst)

$^{205}_{81}$Tl(stable)

Δ: -23846 *4* keV

%: 70.476 *9*

$^{205}_{82}$Pb($1.9\ 3 \times 10^{7}$ yr)

Mode: ϵ

Δ: -23792 *4* keV

SpA: 9.1×10^{-5} Ci/g

Prod: ^{204}Pb(n,γ)

Photons (^{205}Pb)

⟨γ⟩=2.35 *7* keV

γ_{mode}	γ(keV)	γ(%)
Tl L_ℓ	8.953	0.79 *7*
Tl L_α	10.259	16.1 *6*
Tl L_η	10.994	0.0063 *3*
Tl L_β	12.524	4.84 *25*
Tl L_γ	14.571	0.194 *25*

Atomic Electrons (^{205}Pb)

⟨e⟩=5.0 *5* keV

e_{bin}(keV)	⟨e⟩(keV)	e(%)
13	4.8	38 *4*
15	0.169	1.11 *11*

Continuous Radiation (^{205}Pb)

⟨IB⟩=1.12×10^{-5} keV

E_{bin}(keV)		⟨ ⟩(keV)	(%)
10 - 20	IB	9.95×10^{-6}	8.1×10^{-5}
20 - 40	IB	5.8×10^{-7}	2.1×10^{-6}
40 - 60	IB	1.39×10^{-7}	3.0×10^{-7}

$^{205}_{83}$Bi(15.31 *4* d)

Mode: ϵ

Δ: -21085 *8* keV

SpA: 4.160×10^{4} Ci/g

Prod: ^{206}Pb(d,3n); daughter ^{205}Po

Photons (^{205}Bi)

⟨γ⟩=933 *14* keV

γ_{mode}	γ(keV)	γ(%)†
Pb L_ℓ	9.185	0.70 *8*
Pb L_α	10.541	13.9 *11*
Pb L_η	11.349	0.172 *15*
Pb L_β	12.700	13.0 *13*
Pb L_γ	14.912	2.5 *3*
γ M2	26.234 *10*	0.000137 *12*
Pb $K_{\alpha2}$	72.803	20.6 *15*
Pb $K_{\alpha1}$	74.969	34.7 *25*
Pb $K_{\beta1}'$	84.789	12.3 *9*
Pb $K_{\beta2}'$	87.632	3.5 *3*
γ E1	90.06 *4*	0.0109 *22*
γ	112.72 *10*	0.00087 *6*
γ M1+3.8%E2	115.19 *4*	0.0075 *3*
γ	122.6 *10*	0.00068 *6*
γ [E2]	127.22 *4*	0.00031 *6*
γ [M1]	129.62 *5*	0.00062 *6*
γ	148.82 *20*	0.00053 *16*
γ [M1]	165.07 *4*	0.00156 *9*
γ [E1]	170.74 *4*	0.00040 *9*
γ M1	185.21 *4*	0.0095 *6*
γ M1	205.86 *4*	0.0025 *3*
γ M1	221.03 *3*	0.00311 *19*
γ M1	235.80 *5*	0.0057 *3*
γ	248.48 *5*	~0.00019
γ [E1]	259.52 *5*	~0.0050
γ M1	260.49 *3*	0.109 *3*
γ M1	262.83 *3*	0.0364 *12*
γ [E2]	277.14 *4*	0.0016 *3*
γ M1	282.30 *4*	0.0426 *9*
γ M1	284.17 *3*	0.169 *3*
γ [M1]	284.23 *5*	0.0031 *9*
γ E3	310.41 *3*	0.0104 *3*
γ (M1)	312.84 *5*	0.0025 *9*
γ (M1)	313.41 *4*	0.0034 *9*
γ	339.27 *20*	0.00109 *16*
γ M1(+23%E2)	349.58 *3*	0.0563 *11*
γ	354.58 *5*	0.00171 *19*
γ [M1+E2]	361.27 *5*	0.0031 *9*
γ	361.82 *6*	0.0030 *9*
γ [M1+E2]	445.24 *5*	0.0014 *6*
γ	476.52 *6*	0.0023 *3*
γ	487.99 *5*	0.0039 *5*
γ M1	493.67 *4*	0.0373 *8*
γ (M1)	498.62 *4*	0.0093 *16*
γ	498.89 *5*	~0.0040
γ (M1)	499.58 *5*	0.0062 *16*
γ [M1+E2]	503.33 *4*	~0.0008
γ M1+14%E2	511.53 *3*	0.0855 *17*
γ E1	549.86 *3*	0.295 *6*
γ [M1+E2]	561.31 *3*	0.0053 *5*
γ M1	570.60 *3*	0.434 *9*
γ (M1+21%E2)	573.90 *3*	0.0622 *12*
γ [M1]	576.24 *3*	0.0188 *6*
γ E2	579.74 *3*	0.54 *11*

Photons (^{205}Bi)
(continued)

γ_{mode}	γ(keV)	γ(%)†
γ [E1]	605.98 *3*	0.0025 *4*
γ M1(+27%E2)	626.72 *3*	0.0585 *12*
γ	646.05 *5*	0.0065 *3*
γ	661.27 *4*	0.0028 *5*
γ	668.6 *6*	~0.0019
γ	669.8 *12*	<0.0022
γ	683.5 *3*	0.0026 *3*
γ (M1+43%E2)	688.54 *4*	0.0227 *9*
γ	701.20 *8*	0.016 *6*
γ E2+2.9%M1	703.47 *3*	31.1
γ	703.48 *22*	
γ [E1]	704.75 *5*	0.038 *9*
γ (M1+45%E2)	717.38 *3*	0.0311 *6*
γ (M1)	720.60 *4*	0.0143 *9*
γ	723.09 *5*	0.0028 *12*
γ	723.79 *5*	0.0152 *12*
γ	729.47 *4*	0.0065 *4*
γ E2	744.81 *4*	0.0697 *16*
γ [M1+E2]	757.14 *13*	0.012 *5*
γ E2	759.10 *4*	0.104 *5*
γ E2	761.45 *4*	0.068 *3*
γ	764.83 *5*	0.0009 *4*
γ [E1]	771.04 *4*	0.0047 *4* ?
γ	777.87 *15*	0.0073 *9*
γ [E2]	780.92 *3*	0.0572 *11*
γ [M1+E2]	788.19 *5*	0.0100 *16*
γ [E1]	789.35 *5*	~0.0019
γ [M1+E2]	795.70 *3*	0.0140 *6*
γ [E1]	800.83 *4*	0.0190 *6*
γ [M1+E2]	806.50 *4*	0.0159 *12*
γ E2	813.96 *5*	0.0470 *12*
γ	828.23 *4*	0.0289 *12*
γ [E1]	831.09 *5*	0.0040 *9*
γ [E1]	842.57 *5*	0.0022 *6*
γ [M1+E2]	848.25 *5*	0.0026 *4*
γ [M1+E2]	852.91 *3*	0.0072 *5*
γ	860.16 *3*	0.0435 *9*
γ	871.94 *4*	0.0417 *9*
γ E1	890.15 *3*	0.0678 *14*
γ (M1+<50%E2)	894.53 *4*	0.0622 *12*
γ	901.90 *4*	0.0129 *5*
γ M1	910.89 *3*	0.164 *3*
γ	922.22 *8*	0.0053 *3*
γ	931.52 *15*	0.0039 *5*
γ [E1]	950.83 *3*	0.0389 *9*
γ [M1+E2]	971.57 *3*	0.0280 *6*
γ	978.34 *8*	0.0040 *6*
γ	987.57 *5*	0.009 *3*
γ E2	987.64 *3*	1.61 *3*
γ	988.98 *5*	<0.0031
γ	989.78 *5*	0.0075 *25*
γ [E1]	992.57 *4*	0.009 *3*
γ	1001.55 *3*	0.026 *4*
γ [M1+E2]	1001.95 *4*	0.027 *4*
γ [M1+E2]	1002.91 *5*	0.007 *3*
γ (M1)	1013.31 *4*	0.0082 *19*
γ M4	1013.87 *3*	0.0058 *12*
γ M1	1014.38 *4*	0.0914 *19*
γ	1031.52 *5*	0.0034 *11*
γ [E2]	1038.11 *4*	0.0114 *9*
γ M1	1043.75 *3*	0.751 *15*
γ [M1+E2]	1060.89 *4*	0.0044 *5*
γ	1063.92 *15*	0.0025 *5*
γ [E1]	1066.02 *4*	0.0109 *5*
γ (M1)	1072.36 *4*	0.0302 *6*
γ	1075.12 *10*	0.0011 *5*
γ [E1]	1107.76 *4*	0.0099 *9*
γ M1	1190.04 *4*	0.226 *6*
γ [E2]	1199.59 *5*	0.0190 *12*
γ	1208.59 *5*	0.0512 *10*
γ [E1]	1216.28 *4*	0.0101 *5*
γ [M1+E2]	1256.71 *13*	~0.0022
γ [E2]	1262.43 *3*	0.0062 *6*
γ	1264.71 *5*	0.0050 *22*
γ [M1+E2]	1264.78 *3*	0.0124 *22*
γ	1265.9 *3*	0.0047 *12*
γ	1277.22 *20*	0.0038 *4*
γ E2	1351.52 *3*	0.106 *3*
γ	1438.72 *20*	0.0117 *6*
γ [E2]	1499.16 *4*	0.0171 *14*
γ (M1)	1501.52 *4*	0.0227 *14*
γ [E2]	1513.00 *5*	0.0070 *12*

Photons (^{205}Bi) (continued)

γ_{mode}	γ(keV)	γ(%)†
γ [E1]	1521.43 3	0.0199 12
γ	1548.88 5	0.0280 16
γ E2	1551.31 3	0.0970 25
γ [E1]	1563.17 4	0.0165 9
γ [E1]	1577.54 3	0.0166 9
γ [E2]	1593.05 4	0.0115 8
γ M1	1614.35 3	0.228 5
γ [E1]	1619.28 4	0.0367 16
γ	1676.4 3	0.0033 6
γ	1756.4 3	0.0218 12
γ [M1+E2]	1760.04 13	0.012 3
γ M1	1764.35 4	32.5 7
γ M1	1775.83 4	0.399 8
γ	1815.66 18	0.0014 5
γ	1818.01 18	
γ [M1+E2]	1818.02 13	0.0047 4
γ E1	1861.71 3	0.617 12
γ E1	1903.45 4	0.247 5
γ	1965.97 8	0.00081 16
γ	2003.3 5	0.00037 16
γ (M2)	2565.17 3	0.00423 22
γ (M2)	2606.91 4	0.00187 19

† 3.2% uncert(syst)

Atomic Electrons (^{205}Bi)

⟨e⟩=14.0 5 keV

e_{bin}(keV)	⟨e⟩(keV)	e(%)
10 - 11	0.082	0.79 7
13	3.0	23 3
15	1.77	11.7 14
16	0.66	4.1 5
22 - 71	1.43	2.63 10
72 - 120	0.263	0.338 20
122 - 171	0.0155	0.0105 4
172 - 221	0.322	0.174 3
222 - 271	0.125	0.0488 20
273 - 313	0.0296	0.0103 5
323 - 362	0.0107	0.0032 4
389 - 432	0.0518	0.0124 6
441 - 491	0.177	0.0367 10
492 - 539	0.064	0.0128 17
545 - 593	0.0609	0.0109 4
596 - 614	0.0088	0.0015 4
615	2.16	0.350 15
617 - 666	0.0177	0.0028 4
667 - 686	0.0138	0.00205 20
688	0.508	0.074 3
689 - 732	0.246	0.0351 11
740 - 789	0.019	0.0025 7
791 - 840	0.0392	0.00480 17
842 - 891	0.0085	0.00097 21
892 - 938	0.114	0.0125 5
944 - 993	0.131	0.0136 3
997 - 1046	0.0292	0.00285 7
1048 - 1095	0.00104	9.9 5 $\times10^{-5}$
1102 - 1128	0.0292	0.00264 22
1169 - 1216	0.0087	0.00074 7
1241 - 1277	0.00461	0.000365 16
1336 - 1351	0.0016	0.00012 3
1411 - 1438	0.00308	0.000217 12
1461 - 1560	0.0239	0.00158 7
1562 - 1661	0.00377	0.000236 7
1663 - 1673	0.0014	9 4 $\times10^{-5}$
1676	2.02	0.121 3
1688 - 1747	0.0251	0.00149 4
1748	0.318	0.0182 5
1749 - 1847	0.1213	0.00689 13
1849 - 1915	0.000976	5.20 13 $\times10^{-5}$
1950 - 2000	8.0 $\times10^{-6}$	4.1 16 $\times10^{-7}$
2477 - 2562	0.00059	2.36 12 $\times10^{-5}$
2591 - 2604	3.4 $\times10^{-5}$	1.32 11 $\times10^{-6}$

Continuous Radiation (^{205}Bi)

⟨β+⟩=0.506 keV;⟨IB⟩=0.89 keV

E_{bin}(keV)		⟨ ⟩(keV)	(%)
0 - 10	β+	2.59 $\times10^{-9}$	3.04 $\times10^{-8}$
	IB	0.000124	
10 - 20	β+	1.82 $\times10^{-7}$	1.08 $\times10^{-6}$
	IB	0.0028	0.022
20 - 40	β+	8.7 $\times10^{-6}$	2.60 $\times10^{-5}$
	IB	0.00142	0.0044
40 - 100	β+	0.00076	0.00095
	IB	0.50	0.67
100 - 300	β+	0.0506	0.0227
	IB	0.052	0.032
300 - 600	β+	0.268	0.060
	IB	0.093	0.021
600 - 1300	β+	0.187	0.0264
	IB	0.19	0.022
1300 - 2500	IB	0.054	0.0037
2500 - 2707	IB	9.5 $\times10^{-6}$	3.7 $\times10^{-7}$
	Σβ+		0.110

$^{205}_{84}$Po(1.80 4 h)

Mode: ε(99.86 3 %), α(0.14 3 %)
Δ: -17560 30 keV
SpA: 8.49$\times10^{6}$ Ci/g
Prod: ^{209}Bi(p,5n); ^{204}Pb(α,3n)

Alpha Particles (^{205}Po)

α(keV)
5220 10

Photons (^{205}Po)

⟨γ⟩=1564 30 keV

γ_{mode}	γ(keV)	γ(%)†
Bi L_ℓ	9.419	0.89 13
Bi L_α	10.828	17.5 22
Bi L_η	11.712	0.23 3
Bi L_β	13.098	16.3 22
Bi L_γ	15.387	3.1 5
γ(M1)	22.58 8	~0.024
γ(E1)	24.96 10	0.26 6
Bi $K_{\alpha2}$	74.814	27 3
Bi $K_{\alpha1}$	77.107	45 5
Bi $K_{\beta1}'$	87.190	15.8 17
Bi $K_{\beta2}'$	90.128	4.7 5
γ	122.71 10	~0.026
γM1	128.83 6	1.11 7
γM1	150.39 8	0.33 7
γM1	151.41 7	1.33 7
γM1	158.41 10	0.155 18
γE2+16%M1	212.02 7	3.59 18
γ	222.54 7	0.181 22
γ	225.50 13	0.126 15
γ(M1+49%E2)	248.18 8	0.14 3
γM1	261.08 7	4.03 22
γ[M1]	335.03 8	0.18 4
γM1	358.81 10	0.18 4
γ(M1+45%E2)	381.81 10	0.15 4
γM1	454.09 8	0.18 4
γE2	473.10 7	0.89 7
γ[E3]	495.95 8	~0.15 ?
γ	559.2 3	~0.07
γE2	599.83 9	2.63 15
γM1	614.26 7	1.59 7
γM2	624.78 6	1.04 7
γ	679.91 20	0.26 11
γ	713.37 15	0.59 7

Photons (^{205}Po) (continued)

γ_{mode}	γ(keV)	γ(%)†
γ[M1+E2]	715.16 10	0.33 4
γM1	783.01 20	0.22 7
γM1+46%E2	795.88 8	0.78 7
γE1	836.81 6	19.2 11
γM1+30%E2	849.83 7	25.5 15
γ(M1)	859.51 13	0.19 6
γE2	872.41 7	37.0
γ	959.91 20	0.22 7
γM1(+20%E2)	1001.24 7	28.8 15
γ	1026.8 3	0.11 4
γE2	1044.05 8	0.63 7
γ	1060.5 4	~0.11
γ	1103.2 3	~0.15
γ[E1]	1120.46 9	~0.11
γ(M1)	1183.51 10	1.22 11
γ	1187.2 6	0.11 4
γ	1194.54 13	~0.22
γ	1211.4 5	~0.05
γ	1221.2 3	~0.11
γ	1224.6 4	~0.07
γ(M1+44%E2)	1239.13 10	4.6 3
γ	1241.97 16	0.37 15
γ	1267.3 4	0.22 7
γ	1276.6 7	0.11 4
γ	1301.9 3	0.22 7
γ	1309.7 4	0.11 4
γ	1323.38 13	0.48 4
γ	1336.27 9	1.33 11
γ	1392.7 3	0.55 7
γ	1418.8 3	~0.07
γ	1422.3 5	~0.07
γ	1454.4 6	~0.05
γ	1470.9 4	0.15 4
γ	1486.5 4	0.15 4
γ	1487.1 5	0.15 4
γ[M1+E2]	1513.71 20	2.11 18
γ	1520.1 9	~0.06
γ	1529.6 3	0.22 7
γ	1546.1 4	0.26 4
γ[E1]	1551.97 10	2.92 18
γ	1570.8 3	0.15 4
γ	1575.21 20	0.78 7
γ	1578.27 13	0.63 7
γ	1610.91 20	0.26 7
γ	1622.8 3	0.15 4
γ	1638.4 7	~0.07
γ	1674.11 20	0.26 7
γ	1701.53 15	0.22 4
γ	1707.10 14	0.48 7
γ	1712.05 17	0.18 7
γ	1715.7 3	0.30 7
γ	1724.10 15	0.33 4
γ	1729.68 14	1.55 11
γ	1800.21 20	0.15 4
γ	1808.1 5	0.11 4
γ	1811.25 15	1.18 11
γ	1836.8 3	~0.11
γ	1949.94 15	0.18 4
γ	1957.51 20	0.26 4
γ	2020.31 11	0.15 4
γ	2036.6 9	~0.07
γ	2101.35 15	0.18 4
γ	2126.21 20	0.22 4
γ	2168.91 20	0.37 4
γ	2174.7 3	0.22 4
γ	2190.1 6	0.078 18
γ	2223.8 3	0.26 4
γ	2256.51 20	0.037 15
γ	2265.1 8	0.09 3
γ	2268.2 6	0.15 4
γ	2298.6 4	~0.07
γ	2320.6 15	~0.07
γ	2338.3 12	0.11 4
γ	2342.6 3	<0.07
γ	2432.4 4	0.11 4
γ	2573.92 15	0.18 4
γ	2694.7 6	~0.033
γ	2769.5 4	0.074 22

† 5.4% uncert(syst)

Atomic Electrons (^{205}Po)

$\langle e \rangle$=43.9 *11* keV

e_{bin}(keV)	$\langle e \rangle$(keV)	e(%)
6 - 12	0.45	6.1 *24*
13	3.9	29 *4*
16	3.0	19 *3*
19 - 32	0.26	1.3 *4*
38	1.81	4.7 *3*
58 - 60	1.08	1.83 *24*
61	2.65	4.4 *3*
64 - 113	2.20	2.59 *10*
115 - 120	0.018	0.015 *7*
121	1.3	1.0 *3*
122 - 158	2.15	1.58 *13*
171	4.04	2.37 *14*
196	0.88	0.45 *7*
199 - 244	0.91	0.44 *4*
245	1.13	0.46 *4*
246 - 291	0.52	0.196 *13*
319 - 366	0.178	0.051 *6*
368 - 405	0.121	0.031 *5*
438 - 483	0.112	0.024 *4*
492 - 534	1.50	0.285 *12*
543 - 589	0.11	0.018 *6*
596 - 625	0.53	0.086 *12*
664 - 713	0.20	0.029 *4*
714 - 746	0.53	0.071 *8*
759	3.93	0.52 *3*
767 - 780	0.069	0.0089 *15*
782	2.09	0.268 *15*
783 - 823	0.101	0.0124 *14*
833 - 872	1.68	0.199 *6*
911	3.8	0.42 *4*
936 - 986	0.75	0.076 *6*
988 - 1031	0.239	0.0239 *19*
1040 - 1088	0.0062	0.00058 *23*
1090 - 1134	0.186	0.0170 *21*
1149 - 1198	0.46	0.040 *4*
1205 - 1254	0.23	0.018 *5*
1260 - 1309	0.042	~0.0033
1310 - 1336	0.027	0.0021 *9*
1364 - 1409	0.034	0.0024 *8*
1415 - 1461	0.20	0.014 *4*
1467 - 1565	0.15	0.0102 *24*
1567 - 1661	0.15	0.009 *3*
1670 - 1746	0.09	0.0051 *19*
1784 - 1867	0.028	0.0016 *5*
1930 - 2023	0.018	0.0009 *3*
2033 - 2123	0.030	0.0015 *5*
2133 - 2230	0.026	0.0012 *3*
2240 - 2339	0.008	0.00034 *12*
2342 - 2429	0.0036	~0.00015
2483 - 2571	0.006	0.00022 *11*
2604 - 2692	0.0026	10 *5* $\times10^{-5}$
2753 - 2766	0.00035	1.3 *6* $\times10^{-5}$

Continuous Radiation (^{205}Po)

$\langle\beta+\rangle$=14.3 keV;$\langle$IB$\rangle$=2.7 keV

E_{bin}(keV)		$\langle\ \rangle$(keV)	(%)
0 - 10	$\beta+$	1.75×10^{-8}	2.06×10^{-7}
	IB	0.00068	
10 - 20	$\beta+$	1.26×10^{-6}	7.4×10^{-6}
	IB	0.0029	0.022
20 - 40	$\beta+$	6.2×10^{-5}	0.000185
	IB	0.0023	0.0077
40 - 100	$\beta+$	0.0057	0.0071
	IB	0.57	0.76
100 - 300	$\beta+$	0.434	0.193
	IB	0.084	0.050
300 - 600	$\beta+$	3.14	0.68
	IB	0.21	0.046
600 - 1300	$\beta+$	10.3	1.18
	IB	0.98	0.105
1300 - 2500	$\beta+$	0.510	0.0371
	IB	0.84	0.052
2500 - 3494	IB	0.00025	9.7×10^{-6}
	$\Sigma\beta+$		2.1

$^{205}_{85}$At(26.2 *5* min)

Mode: ϵ(90 *2* %), α(10 *2* %)
Δ: -13050 *60* keV
SpA: 3.50×10^{7} Ci/g
Prod: ^{197}Au(^{12}C,4n); ^{209}Bi(α,8n); daughter ^{205}Rn

Alpha Particles (^{205}At)

α(keV)
5902 *2*

Photons (^{205}At)

$\langle\gamma\rangle$=681 *34* keV

γ_{mode}	γ(keV)	γ(%)†
Po $K_{\alpha2}$	76.858	28 *3* *
Po $K_{\alpha1}$	79.290	47 *5* *
Po $K_{\beta1}$'	89.639	16.2 *16* *
Po $K_{\beta2}$'	92.673	5.1 *5* *
γ_ϵ	123.50 *20*	0.08 *3*
γ_ϵ(E2)	143.30 *10*	0.78 *8*
γ_ϵM1	154.29 *15*	2.38 *16*
γ_ϵM2	161.03 *13*	1.24 *11*
γ_ϵ	165.90 *10*	0.24 *5*
γ_ϵ	178.80 *10*	0.11 *3*
γ_ϵ	202.70 *10*	0.32 *8*
γ_ϵ	230.30 *10*	0.22 *5*
γ_ϵ	275.80 *20*	0.13 *5*
γ_ϵM1	311.42 *13*	3.43 *24*
γ_ϵ	312.90 *20*	0.40 *8*
γ_ϵ	316.50 *10*	0.35 *8*
γ_ϵ	337.10 *20*	0.19 *5*
γ_ϵM1	361.26 *13*	0.76 *8*
γ_ϵ	365.10 *10*	0.51 *11*
γ_ϵ	384.93 *12*	0.81 *8*
γ_ϵ	395.8 *3*	~0.38
γ_ϵ(M1+E2)	448.77 *12*	1.43 *14*
γ_ϵ	462.7 *1*	0.49 *11*
γ_ϵ	488.0 *1*	0.40 *8*
γ_ϵ	506.4 *3*	~0.13
γ_ϵM1	516.6 *1*	0.73 *14*
γ_ϵE2	520.71 *14*	3.54 *7*
γ_ϵ	529.2 *2*	0.35 *11*
γ_ϵ	554.2 *2*	0.43 *11*
γ_ϵ	577.3 *1*	0.46 *8*
γ_ϵ	583.9 *2*	0.24 *11*
γ_ϵ	587.4 *1*	0.35 *11*
γ_ϵ	595.7 *2*	~0.27
γ_ϵE1	617.89 *14*	1.86 *16*
γ_ϵ(M1+E2)	628.94 *18*	4.6 *5*
γ_ϵ	659.6 *1*	2.00 *16*
γ_ϵE2	669.56 *13*	8.1 *5*
γ_ϵ(E2)	672.9 *1*	2.97 *6*
γ_ϵ	691.5 *3*	0.30 *11*
γ_ϵE2	719.39 *11*	27
γ_ϵ	757.0 *2*	0.43 *11*
γ_ϵ[M1+E2]	783.23 *11*	1.59 *16*
γ_ϵ	789.1 *1*	1.11 *22*
γ_ϵ	792.4 *2*	0.51 *16*
γ_ϵ	806.8 *2*	0.46 *11*
γ_ϵ	872.4 *5*	~2
γ_ϵ	929.2 *3*	0.32 *8*
γ_ϵ	942.3 *3*	0.24 *8*
γ_ϵ	976.0 *3*	0.68 *13*
γ_ϵ	1026.4 *3*	0.81 *16*
γ_ϵ(E2)	1031.6 *3*	2.05 *19*
γ_ϵ	1082.5 *3*	0.46 *8*
γ_ϵ	1091.3 *3*	0.22 *5*
γ_ϵ	1171.1 *3*	0.57 *11*
γ_ϵ	1246.1 *3*	0.30 *8*
γ_ϵ	1252.2 *3*	0.46 *11*
γ_ϵ	1307.6 *3*	0.84 *13*
γ_ϵ	1325.4 *3*	1.13 *14*
γ_ϵ	1398.8 *3*	0.59 *13*
γ_ϵ	1414.1 *3*	0.27 *8*
γ_ϵ	1442.5 *4*	0.27 *11*

Photons (^{205}At)
(continued)

γ_{mode}	γ(keV)	γ(%)†
γ_ϵ	1475.8 *3*	0.59 *13*
γ_ϵ	1479.2 *3*	0.59 *13*
γ_ϵ	1761.5 *4*	0.40 *11*
γ_ϵ	2031.8 *4*	0.27 *8*
γ_ϵ	2050.8 *4*	0.62 *13*

† 17% uncert(syst)
* exp values

$^{205}_{86}$Rn(2.83 *12* min)

Mode: ϵ(77 *2* %), α(23 *2* %)
Δ: -7780 *220* keV syst
SpA: 3×10^{8} Ci/g
Prod: ^{197}Au(^{14}N,6n); ^{16}O on Pt; ^{12}C on Hg

Alpha Particles (^{205}Rn)

$\langle\alpha\rangle$=1441.7 keV

α(keV)	α(%)
6123 *3*	0.023
6262 *3*	23

Photons (^{205}Rn)

γ_{mode}	γ(keV)	γ(rel)
γ_ϵ(E2)	265.2 *7*	100
γ_ϵ	355.3 *8*	3.7 *7*
γ_ϵ	464.8 *8*	25 *5*
γ_ϵ	620.5 *8*	25 *5*
γ_ϵ	675.3 *10*	20 *4*
γ_ϵ	730.0 *8*	20 *4*

$^{205}_{87}$Fr(3.96 *4* s)

Mode: α
Δ: -1290 *170* keV
SpA: 1.275×10^{10} Ci/g
Prod: ^{197}Au(^{16}O,8n); protons on Th

Alpha Particles (^{205}Fr)

α(keV)
6914 *5*

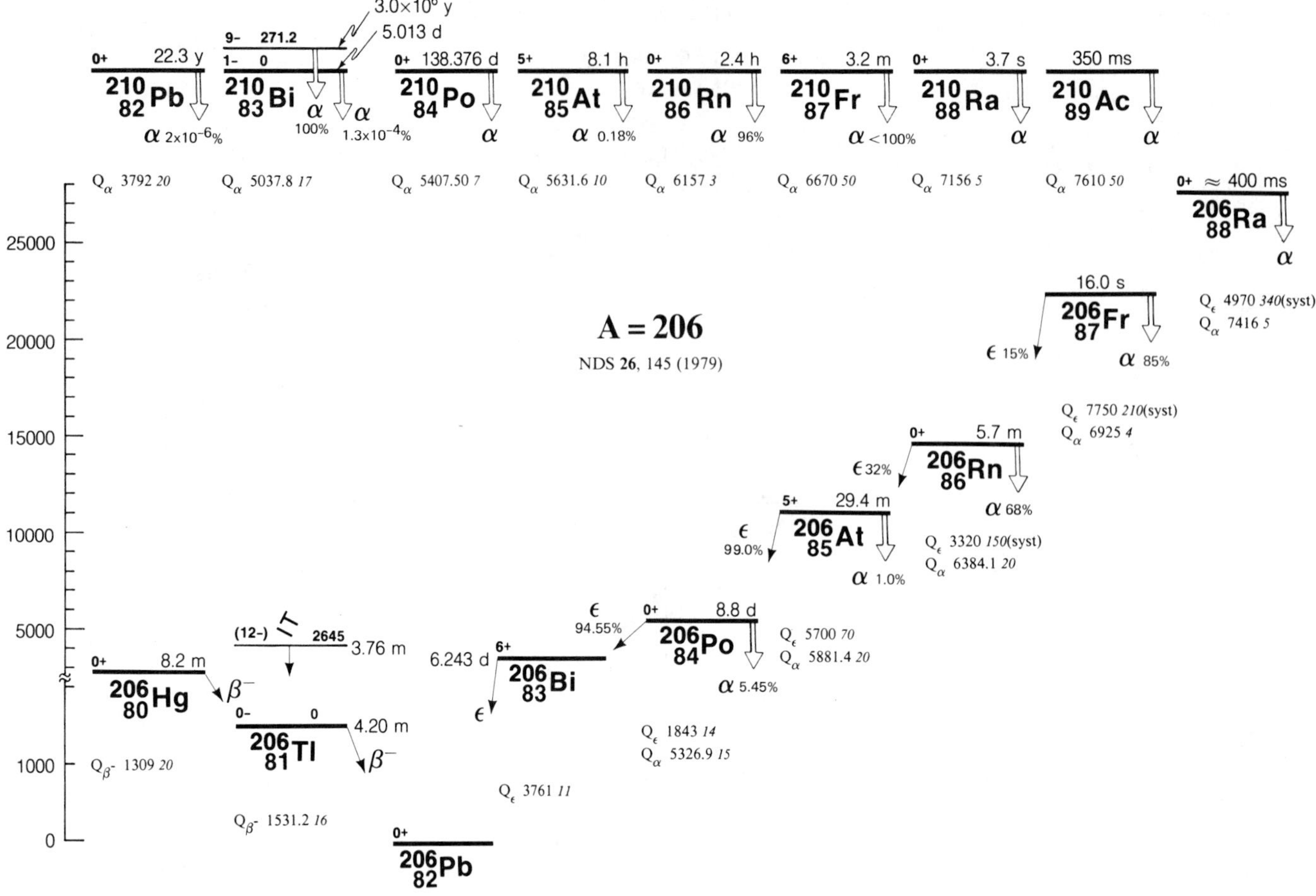

$^{206}_{80}$Hg(8.15 *10* min)

Mode: β-
Δ: -20969 *21* keV
SpA: 1.119×10^{8} Ci/g
Prod: daughter ^{210}Pb; ^{208}Pb(p,3p)

Photons (^{206}Hg)

⟨γ⟩=107 *17* keV

γ_{mode}	γ(keV)	γ(%)†
Tl L_ℓ	8.953	0.072 *15*
Tl L_α	10.259	1.5 *3*
Tl L_η	10.994	0.019 *4*
Tl L_β	12.313	1.3 *3*
Tl L_γ	14.407	0.24 *5*
Tl $K_{\alpha2}$	70.832	2.5 *5*
Tl $K_{\alpha1}$	72.873	4.2 *8*
Tl $K_{\beta1}'$	82.434	1.5 *3*
Tl $K_{\beta2}'$	85.185	0.42 *8*
γ M1	305.25 *16*	27
γ M1	344.96 *19*	0.52 *10*
γ [M1]	384.06 *23*	0.0038 *10*
γ M1	650.21 *17*	2.5 *5*

† 19% uncert(syst)

Atomic Electrons (^{206}Hg)

⟨e⟩=27 *4* keV

e_{bin}(keV)	⟨e⟩(keV)	e(%)
13 - 60	0.72	4.7 *7*
66 - 82	0.080	0.114 *10*
220	19.2	8.7 *18*
259	0.32	0.122 *22*
290	3.9	1.3 *3*
291 - 299	0.39	0.134 *25*
302	1.04	0.34 *7*
303 - 345	0.42	0.136 *18*
369 - 384	0.00053	0.00014 *6*
565	0.60	0.107 *20*
635 - 650	0.147	0.023 *3*

Continuous Radiation (^{206}Hg)

⟨β-⟩=401 keV;⟨IB⟩=0.48 keV

E_{bin}(keV)		⟨ ⟩(keV)	(%)
0 - 10	β-	0.064	1.28
	IB	0.018	
10 - 20	β-	0.193	1.29
	IB	0.018	0.123
20 - 40	β-	0.78	2.61
	IB	0.033	0.116
40 - 100	β-	5.7	8.1
	IB	0.087	0.135
100 - 300	β-	56	28.0
	IB	0.18	0.108
300 - 600	β-	156	35.4
	IB	0.113	0.028

Continuous Radiation (^{206}Hg) (continued)

E_{bin}(keV)		⟨ ⟩(keV)	(%)
600 - 1300	β-	182	23.3
	IB	0.030	0.0042
1300 - 1313	β-	0.00247	0.000189
	IB	6.5×10^{-10}	5.0×10^{-11}

$^{206}_{81}$Tl(4.20 *2* min)

Mode: β-
Δ: -22278 *4* keV
SpA: 2.170×10^{8} Ci/g
Prod: ^{205}Tl(n,γ); ^{209}Bi(n,α); daughter ^{210}Bi(3.0×10^{6} y)

Photons (^{206}Tl)

⟨γ⟩=0.099 *9* keV

γ_{mode}	γ(keV)	γ(%)
Pb L_ℓ	9.185	0.00050 *13*
Pb L_α	10.541	0.0100 *25*
Pb L_η	11.349	0.00014 *3*
Pb L_β	12.710	0.0088 *22*
Pb L_γ	14.836	0.0015 *4*

Photons (^{206}Tl)
(continued)

γ_{mode}	γ(keV)	γ(%)
Pb $K_{\alpha2}$	72.803	0.020 *5*
Pb $K_{\alpha1}$	74.969	0.033 *8*
Pb $K_{\beta1}$'	84.789	0.012 *3*
Pb $K_{\beta2}$'	87.632	0.0034 *8*
γ E2	803.13 *5*	0.0055 *5*

Atomic Electrons (^{206}Tl)
$\langle e \rangle$=0.77 *18* keV

e_{bin}(keV)	$\langle e \rangle$(keV)	e(%)
13	0.0028	0.022 *6*
15 - 62	0.0025	0.012 *3*
68 - 84	0.00062	0.00087 *9*
715	0.00032	4.5 *4* $\times10^{-5}$
787 - 803	0.000101	1.28 *7* $\times10^{-5}$
1077	0.76	0.070 *17*
1149 - 1165	0.0013	~0.00012

Continuous Radiation (^{206}Tl)
$\langle\beta\text{-}\rangle$=536 keV;$\langle$IB$\rangle$=0.79 keV

E_{bin}(keV)		$\langle\ \rangle$(keV)	(%)
0 - 10	β-	0.0407	0.81
	IB	0.023	
10 - 20	β-	0.123	0.82
	IB	0.023	0.157
20 - 40	β-	0.503	1.67
	IB	0.043	0.151
40 - 100	β-	3.73	5.3
	IB	0.116	0.18
100 - 300	β-	40.1	19.8
	IB	0.28	0.159
300 - 600	β-	139	31.0
	IB	0.21	0.051
600 - 1300	β-	339	39.6
	IB	0.096	0.0129
1300 - 1526	β-	13.2	0.97
	IB	0.00026	2.0$\times10^{-5}$

$^{206}_{81}$Tl(3.76 *4* min)

Mode: IT
Δ: -19633 *4* keV
SpA: 2.42$\times10^{8}$ Ci/g
Prod: ^{204}Hg(α,pn); ^{204}Hg(^{7}Li,αn)

Photons (^{206}Tl)
$\langle\gamma\rangle$=2628 *110* keV

γ_{mode}	γ(keV)	γ(%)†
Tl L_{ℓ}	8.953	0.44 *6*
Tl L_{α}	10.259	9.0 *9*
Tl L_{η}	10.994	0.155 *14*
Tl L_{β}	12.297	9.8 *10*
Tl L_{γ}	14.389	1.87 *20*
Tl $K_{\alpha2}$	70.832	12 *1*
Tl $K_{\alpha1}$	72.873	20.4 *18*
Tl $K_{\beta1}$'	82.434	7.2 *6*
Tl $K_{\beta2}$'	85.185	2.02 *18*
γ (E2)	216.6 *10*	89 *6*
γ (M1)	247.7 *10*	9.7 *19*
γ E2	266.15 *17*	86 *7*
γ (E1)	453.4 *8*	94 *7*
γ (M1)	457.6 *8*	22 *4*
γ (M4)	564.4 *8*	13 *4*
γ E2	686.6 *6*	100 *9*
γ (E5)	1021.9 *8*	76 *7*
γ (E3)	1140.0 *8*	7.5 *15*

† 8.0% uncert(syst)

Atomic Electrons (^{206}Tl)
$\langle e \rangle$=248 *17* keV

e_{bin}(keV)	$\langle e \rangle$(keV)	e(%)
13 - 60	4.7	32 *3*
66 - 82	0.384	0.55 *3*
131	16.4	12.5 *10*
162	9.0	5.6 *11*
181	13.3	7.4 *6*
201	3.4	1.70 *13*
202	12.7	6.3 *5*
204	7.5	3.7 *3*
213 - 248	11.4	5.2 *3*
251	8.9	3.5 *3*
253 - 266	7.7	2.99 *14*
368	3.3	0.89 *8*
372	8.9	2.4 *5*
438 - 458	3.2	0.71 *7*
479	49.8	10 *3*
549	14.1	2.6 *8*
550 - 564	15.0	2.7 *5*
601	6.4	1.06 *10*
671 - 687	2.21	0.327 *17*
936	28.5	3.0 *3*
1007	13.9	1.39 *13*
1009 - 1054	6.7	0.66 *5*
1125 - 1140	0.22	0.020 *3*

$^{206}_{82}$Pb(stable)

Δ: -23809 *4* keV
%: 24.1 *1*

$^{206}_{83}$Bi(6.243 *3* d)

Mode: ϵ
Δ: -20048 *11* keV
SpA: 1.0153$\times10^{5}$ Ci/g
Prod: ^{206}Pb(d,2n); ^{207}Pb(p,2n); daughter ^{206}Po

Photons (^{206}Bi)
$\langle\gamma\rangle$=3279 *18* keV

γ_{mode}	γ(keV)	γ(%)†
Pb L_{ℓ}	9.185	1.04 *10*
Pb L_{α}	10.541	20.7 *10*
Pb L_{η}	11.349	0.267 *16*
Pb L_{β}	12.700	19.4 *13*
Pb L_{γ}	14.899	3.6 *3*
Pb $K_{\alpha2}$	72.803	32.4 *10*
Pb $K_{\alpha1}$	74.969	54.4 *16*
Pb $K_{\beta1}$'	84.789	19.2 *6*
Pb $K_{\beta2}$'	87.632	5.55 *18*
γ M1(+<1.7%E2)	123.53 *3*	0.0227 *20*
γ M1(+<3.8%E2)	157.25 *6*	0.036 *4*
γ M1(+<1.9%E2)	158.52 *4*	0.082 *8*
γ M1(+<0.8%E2)	184.026 *23*	15.8 *3*
γ E3	202.51 *4*	0.044 *4*
γ M1(+<3.5%E2)	234.241 *24*	0.241 *12*
γ M1(+<1.2%E2)	262.76 *3*	3.02 *6*
γ M1(+E2)	313.68 *3*	0.359 *10*
γ M1+0.07%E2	343.527 *25*	23.4 *5*
γ M1(+<11%E2)	386.29 *3*	0.516 *10*
γ M1(+0.1%E2)	398.035 *22*	10.74 *21*
γ	435.04 *7*	0.0227 *20*
γ	442.179 *25*	0.038 *4*
γ	452.86 *3*	0.156 *8*
γ [M1+E2]	463.28 *6*	0.053 *5*
γ M1	480.44 *5*	0.089 *9*
γ M1(+0.8%E2)	497.00 *3*	15.3 *3*
γ E3	516.19 *3*	40.7 *8*
γ M1+0.1%E2	537.48 *3*	30.4 *6*
γ [M1+E2]	555.29 *6*	0.038 *4*
γ	576.38 *3*	0.112 *10*
γ E2	582.06 *3*	0.485 *25*
γ M1(+<5.9%E2)	620.53 *3*	5.76 *12*
γ M1(+<7.8%E2)	632.28 *3*	4.47 *9*
γ M1(+<16%E2)	657.21 *3*	1.91 *4*
γ M1	664.46 *5*	0.098 *5*
γ (M1)+E2	739.31 *6*	0.157 *8*
γ M1(+E2)	754.94 *6*	0.527 *11*
γ E1	784.57 *3*	0.536 *11*
γ E2	803.13 *5*	98.90
γ (M1)+E2	841.29 *4*	0.186 *9*
γ E2	881.00 *3*	66.2 *13*
γ M1(+0.09%E2)	895.03 *3*	15.7 *3*
γ [E2]	915.15 *6*	0.031 *3*
γ [E1]	963.84 *6*	0.037 *4*
γ M1(+0.03%E2)	1018.56 *3*	7.60 *15*
γ	1025.31 *4*	0.043 *4*
γ	1047.58 *10*	0.056 *6*
γ	1093.34 *10*	0.070 *7*
γ E1	1098.25 *3*	13.5 *3*
γ E1	1142.40 *3*	0.111 *5*
γ	1180.65 *5*	0.066 *7*
γ M1(+E2)	1194.69 *4*	0.277 *15*
γ E2	1202.59 *3*	0.105 *6*
γ	1208.79 *10*	0.049 *5*
γ E1	1246.58 *4*	0.084 *8*
γ E1	1281.57 *4*	0.065 *7*
γ E1	1332.49 *3*	0.282 *15*
γ E1	1405.10 *4*	1.43 *3*
γ	1420.25 *10*	0.043 *4*
γ	1459.93 *10*	0.080 *8*
γ	1496.21 *8*	0.176 *10*
γ [E1]	1560.26 *3*	0.378 *20*
γ E1	1565.30 *5*	0.304 *15*
γ (M1)+E2	1588.35 *7*	0.041 *4*
γ E1	1595.25 *3*	5.01 *10*
γ E1	1718.78 *3*	31.8 *6*
γ E1	1844.84 *7*	0.569 *25*
γ E1	1878.98 *5*	2.01 *4*
γ E1	1903.78 *4*	0.349 *15*
γ	1963.2 *3*	0.0109 *20*
γ	2023.40 *4*	0.0129 *20*
γ	2439.0 *4*	0.0049 *20*
γ	2476.25 *4*	0.0148 *20*
γ (M1)	2599.77 *4*	0.130 *10*
γ	2759.98 *5*	0.0138 *20*

† 0.10% uncert(syst)

Atomic Electrons (^{206}Bi)
$\langle e \rangle$=133.1 *14* keV

e_{bin}(keV)	$\langle e \rangle$(keV)	e(%)
13	4.6	36 *4*
15 - 62	5.1	26.2 *20*
68 - 84	1.20	1.69 *7*
96	21.5	22.4 *10*
108 - 157	0.378	0.269 *12*
158	0.0037	0.00235 *24*
168	5.9	3.49 *16*

Atomic Electrons (^{206}Bi) (continued)

e_{bin}(keV)	⟨e⟩(keV)	e(%)
169 - 203	5.91	3.34 8
218 - 250	0.91	0.38 5
256	15.3	5.98 17
259 - 301	0.53	0.188 14
310	5.7	1.85 9
311 - 354	4.42	1.34 3
365 - 398	1.74	0.452 16
409	5.9	1.45 7
419 - 427	0.0034	~0.0008
428	8.48	1.98 6
429 - 448	0.013	0.0030 15
449	10.6	2.36 7
450 - 497	1.65	0.340 12
500	1.87	0.374 11
501	3.52	0.703 20
503 - 552	8.53	1.630 22
553 - 582	0.51	0.090 5
605 - 654	0.815	0.132 3
655 - 697	0.09	0.014 6
715	5.74	0.802 16
723 - 772	0.049	0.0066 19
781 - 790	1.378	0.1750 25
793	3.57	0.450 13
799 - 841	3.07	0.381 15
865 - 914	1.68	0.193 3
915 - 964	1.06	0.114 5
1003 - 1047	0.492	0.0489 12
1054 - 1098	0.059	0.0054 3
1107 - 1142	0.029	0.0026 9
1159 - 1208	0.0102	0.00086 15
1231 - 1279	0.00498	0.00040 2
1281 - 1330	0.0230	0.00174 5
1357	0.006	~0.00045
1389 - 1420	0.013	0.0009 4
1444 - 1493	0.0124	0.00084 8
1494 - 1592	0.0897	0.00590 15
1631 - 1716	0.516	0.0314 8
1757 - 1842	0.0392	0.00219 5
1863 - 1960	0.0066	0.000355 20
2008 - 2020	8.5×10^{-5}	$\sim4\times10^{-6}$
2351 - 2436	0.00051	$2.1\ 9\times10^{-5}$
2460 - 2512	0.0046	0.000185 16
2584 - 2672	0.00123	$4.7\ 6\times10^{-5}$
2744 - 2757	6.2×10^{-5}	$2.2\ 10\times10^{-6}$

Continuous Radiation (^{206}Bi)

⟨IB⟩=0.55 keV

E_{bin}(keV)		⟨ ⟩(keV)	(%)
10 - 20	IB	0.0027	0.021
20 - 40	IB	0.00141	0.0044
40 - 100	IB	0.51	0.69
100 - 300	IB	0.029	0.021
300 - 600	IB	0.0064	0.00151
600 - 1300	IB	0.0040	0.00053
1300 - 1378	IB	7.7×10^{-7}	5.8×10^{-8}

$^{206}_{84}$Po(8.8 *1* d)

Mode: ϵ(94.55 *5* %), α(5.45 *5* %)
Δ: -18206 *11* keV
SpA: 7.20×10^{4} Ci/g
Prod: ^{209}Bi(p,4n); ^{204}Pb(α,2n); ^{206}Pb(α,4n)

Alpha Particles (^{206}Po)

α(keV)
5223.4 15

Photons (^{206}Po)

⟨γ⟩=1186 *10* keV

γ_{mode}	γ(keV)	γ(%)†
Bi L_ℓ	9.419	1.45 20
Bi L_α	10.828	28 3
Bi L_η	11.712	0.48 6
Bi L_β	13.090	30 4
Bi L_γ	15.357	5.9 8
γ_ϵE2	59.911 16	1.23 12
Bi $K_{\alpha2}$	74.814	27 3
Bi $K_{\alpha1}$	77.107	46 5
γ_ϵM1+<0.2%E2	82.833 19	0.091 12
Bi $K_{\beta1}$'	87.190	16.3 18
Bi $K_{\beta2}$'	90.128	4.8 5
γ_ϵM1(+<0.2%E2)	117.577 21	0.132 20
γ_ϵM1(+<1.0%E2)	129.648 15	0.036 6
γ_ϵM1(+<0.9%E2)	140.499 16	0.14 3
γ_ϵM1	144.187 19	0.054 10
γ_ϵM1(+<0.7%E2)	146.196 20	0.11 3
γ_ϵM1(+<1.0%E2)	170.529 16	0.32 3
γ_ϵM1(+<2.6%E2)	171.32 3	0.101 9
γ_ϵM1(+<3.5%E2)	178.212 19	0.044 8
γ_ϵM1(+<3.1%E2)	180.802 17	0.100 12
γ_ϵM1	210.703 20	0.041 14
γ_ϵ[M1+E2]	224.903 21	~0.026
γ_ϵM1+<2.5%E2	281.957 15	0.84 3
γ_ϵM1(+<0.6%E2)	286.437 20	23.8 5
γ_ϵM1	292.809 17	0.043 3
γ_ϵM1(+<2.8%E2)	311.559 20	4.24 9
γ_ϵM1	322.838 18	0.123 7
γ_ϵM1	324.746 23	0.098 6
γ_ϵM1(+<1.1%E2)	338.444 22	19.2 4
γ_ϵM1	343.968 22	0.059 18
γ_ϵM1	354.89 2	0.395 16
γ_ϵM1(+<10%E2)	369.090 21	0.174 10
γ_ϵM1(+<7.7%E2)	381.232 21	0.176 11
γ_ϵM1(+<16%E2)	452.486 17	0.322 14
γ_ϵM1	457.755 24	0.154 9
γ_ϵM1(+<6.2%E2)	463.337 19	1.79 7
γ_ϵM1(+<16%E2)	468.932 23	0.259 11
γ_ϵM1(+<2.3%E2)	511.340 22	24.1 5
γ_ϵM1(+<2.3%E2)	522.52 3	15.7 3
γ_ϵM1	533.541 22	0.085 7
γ_ϵM1	544.40 3	0.036 14
γ_ϵM1(+<10%E2)	554.671 22	1.56 6
γ_ϵM1(+<13%E2)	579.793 23	1.06 4
γ_ϵ[M1+E2]	591.98 6	0.044 5
γ_ϵM1	645.56 6	0.352 14
γ_ϵ	664.09 8	0.022 8
γ_ϵM1(+<15%E2)	668.713 24	0.86 3
γ_ϵE2	677.728 22	1.47 6
γ_ϵM1	693.835 25	0.205 12
γ_ϵE2	722.09 3	0.067 12
γ_ϵM1(+<4.2%E2)	807.375 21	22.7 5
γ_ϵE2	818.227 22	1.04 4
γ_ϵ[M1+E2]	826.355 25	<0.15
γ_ϵM1	837.21 3	0.099 7
γ_ϵE2	860.960 25	3.54 14
γ_ϵE2	866.230 24	0.036 3
γ_ϵ[E2]	877.508 24	0.021 3
γ_ϵE2	902.631 24	0.247 9
γ_ϵM1(+<12%E2)	980.27 3	7.08 14
γ_ϵM1(+<17%E2)	1007.156 22	3.06 11
γ_ϵE2	1018.007 24	0.36 10
γ_ϵM1(+<5.0%E2)	1032.278 23	32.9 7
γ_ϵE2	1043.129 24	0.288 11
γ_ϵM1	1114.49 6	0.295 18
γ_ϵE2	1191.02 3	0.472 18
γ_ϵ	1193.90 16	0.041 8
γ_ϵE2	1318.713 24	0.649 25
γ_ϵE2	1452.93 6	0.058 3
γ_ϵE2	1496.92 3	0.252 9

† 13% uncert(syst)

Atomic Electrons (^{206}Po)

⟨e⟩=150 *3* keV

e_{bin}(keV)	⟨e⟩(keV)	e(%)
13	6.7	50 7
16	5.8	37 5
27 - 39	0.26	0.88 12
44	15.7	36 4
46	15.0	32 3
50 - 54	0.33	0.64 11
56	5.3	9.5 10
57	5.0	8.8 9
58 - 104	6.6	10.3 5
113 - 163	0.68	0.51 3
164 - 195	0.90	0.482 16
196	21.1	10.8 5
197 - 234	3.53	1.60 7
248	13.6	5.49 25
253 - 269	0.487	0.184 7
270	4.55	1.68 8
271 - 320	3.47	1.214 23
321 - 369	4.39	1.34 4
373 - 381	0.90	0.241 12
421	9.5	2.27 10
432	6.0	1.40 6
436 - 469	0.87	0.189 7
489 - 538	4.55	0.905 20
539 - 588	0.548	0.0960 25
589 - 632	0.077	0.0127 6
642 - 691	0.1112	0.0168 4
693 - 709	0.00197	0.00028 3
717	4.61	0.64 3
718 - 747	0.093	0.0127 15
770 - 818	1.36	0.172 5
821 - 866	0.0694	0.00820 24
874 - 917	1.45	0.162 6
927	0.018	0.0020 6
942	4.66	0.495 22
953 - 1002	0.340	0.0349 9
1003 - 1043	1.16	0.113 4
1098 - 1114	0.0329	0.00299 21
1175 - 1193	0.0061	0.00052 4
1228	0.0270	0.00220 10
1302 - 1318	0.00670	0.000513 15
1398	0.0118	0.00084 3
1437 - 1484	0.00229	0.000155 5
1493 - 1496	0.000524	$3.51\ 11\times10^{-5}$

Continuous Radiation (^{206}Po)

⟨IB⟩=0.56 keV

E_{bin}(keV)		⟨ ⟩(keV)	(%)
10 - 20	IB	0.0025	0.020
20 - 40	IB	0.00120	0.0037
40 - 100	IB	0.50	0.67
100 - 300	IB	0.037	0.025
300 - 600	IB	0.0139	0.0033
600 - 911	IB	0.0029	0.00044

$^{206}_{85}$At(29.4 *3* min)

Mode: ϵ(99.04 *9* %), α(0.96 *9* %)
Δ: -12500 *70* keV
SpA: 3.10×10^{7} Ci/g
Prod: ^{197}Au(^{12}C,3n); ^{209}Bi(α,7n); ^{209}Bi(^{3}He,6n); daughter ^{206}Rn; ^{14}N on Pt

Alpha Particles (^{206}At)

α(keV)
5703 *2*

Photons (^{206}At)

$\langle\gamma\rangle$=2294 *45* keV

γ_{mode}	γ(keV)	γ(%)†
Bi L_ℓ	9.419	~0.025
Po L_ℓ	9.658	0.76 *7*
Bi L_α	10.828	~0.5
Po L_α	11.119	14.5 *8*
Bi L_η	11.712	~0.011
Po L_η	12.085	0.198 *13*
Bi L_β	13.078	~0.6
Po L_β	13.501	13.9 *10*
Bi L_γ	15.348	~0.13
Po L_γ	15.891	2.7 *3*
γ_αM1,E2	65.02 *20*	0.15 *3*
Po $K_{\alpha2}$	76.858	21.1 *9*
Po $K_{\alpha1}$	79.290	35.1 *14*
Po $K_{\beta1}$'	89.639	12.4 *5*
Po $K_{\beta2}$'	92.673	3.79 *16*
γ_ϵ	154.5 *3*	0.49 *10*
γ_ϵM1(+E2)	198.01 *6*	1.56 *20*
γ_ϵE1	201.86 *6*	5.4 *6*
γ_ϵ(E2)	233.54 *6*	3.1 *3*
γ_ϵM1	256.58 *4*	4.4 *4*
γ_ϵE2	268.39 *9*	1.27 *10*
γ_ϵE2	275.64 *11*	2.05 *20*
γ_ϵM1(+E2)	278.99 *4*	2.6 *3*
γ_ϵM1	317.35 *16*	0.49 *10*
γ_ϵE2	342.50 *7*	1.46 *20*
γ_ϵ	373.46 *9*	0.39 *10*
γ_ϵM1	380.89 *11*	0.78 *10*
γ_ϵE2	386.78 *6*	2.6 *3*
γ_ϵE2	395.61 *4*	48 *3*
γ_ϵ	399.87 *8*	0.68 *10*
γ_ϵ	416.44 *7*	1.27 *10*
γ_ϵM1	444.94 *6*	1.27 *10*
γ_ϵE2	477.16 *3*	86 *4*
γ_ϵ	498.6 *4*	0.59 *10*
γ_ϵE1	527.42 *6*	2.9 *3*
γ_ϵ	565.62 *6*	3.2 *3*
γ_ϵ	599.38 *14*	0.39 *10*
γ_ϵM1	614.45 *4*	6.1 *6*
γ_ϵE2	700.71 *3*	98
γ_ϵ	704.64 *6*	6.0 *6*
γ_ϵ	709.37 *23*	0.59 *10*
γ_ϵ	729.19 *9*	~0.20
γ_ϵ	729.28 *5*	0.98 *10*
γ_ϵE2	733.75 *4*	10.1 *7*
γ_ϵ	738.11 *6*	1.17 *10*
γ_ϵ	747.57 *5*	<0.20
γ_ϵ	796.65 *11*	1.17 *10*
γ_ϵ	802.55 *15*	~0.20
γ_ϵ	824.24 *8*	1.27 *10*
γ_ϵ	868.31 *4*	7.6 *8*
γ_ϵ	912.01 *9*	0.59 *10*
γ_ϵ	923.03 *6*	5.6 *6*
γ_ϵ	927.29 *7*	0.98 *10*
γ_ϵ	939.30 *7*	1.95 *20*
γ_ϵ	955.17 *7*	1.46 *20*
γ_ϵ	961.23 *6*	1.37 *10*
γ_ϵ	976.37 *10*	1.37 *10*
γ_ϵ	1008.27 *6*	1.76 *20*
γ_ϵ	1013.94 *7*	2.9 *3*
γ_ϵ	1048.11 *7*	2.24 *20*
γ_ϵ	1059.38 *4*	3.4 *4*
γ_ϵ	1071.83 *19*	~0.20
γ_ϵ	1087.81 *15*	0.68 *10*
γ_ϵ	1094.91 *7*	0.68 *10*
γ_ϵ	1124.89 *4*	1.85 *20*
γ_ϵ	1196.91 *11*	1.46 *20*
γ_ϵ	1257.58 *10*	1.17 *10*
γ_ϵ	1290.63 *7*	0.68 *10*
γ_ϵ	1292.92 *6*	0.68 *10*
γ_ϵ	1294.94 *12*	0.68 *10*
γ_ϵ	1349.57 *14*	0.68 *10*
γ_ϵ	1446.16 *7*	1.27 *10*

Photons (^{206}At) (continued)

γ_{mode}	γ(keV)	γ(%)†
γ_ϵ	1493.05 *8*	~0.20
γ_ϵ	1637.46 *9*	1.17 *10*
γ_ϵ	1736.27 *7*	0.88 *10*
γ_ϵ	1745.61 *20*	0.68 *10*
γ_ϵ	1855.9 *7*	0.39 *10*
γ_ϵ	1899.93 *8*	0.49 *10*
γ_ϵ	1909.48 *9*	0.59 *10*
γ_ϵ	1928.22 *19*	0.68 *10*
γ_ϵ	1938.13 *7*	1.27 *10*
γ_ϵ	2075.6 *5*	0.39 *10*
γ_ϵ	2116.24 *15*	0.49 *10*
γ_ϵ	2218.80 *11*	0.49 *10*
γ_ϵ	2271.15 *8*	0.29 *10*
γ_ϵ	2298.89 *8*	0.78 *10*
γ_ϵ	2319.01 *11*	0.49 *10*
γ_ϵ	2495.22 *21*	~0.20
γ_ϵ	2558.89 *15*	0.39 *10*
γ_ϵ	2566.7 *14*	~0.20
γ_ϵ	2592.5 *10*	<0.20

† uncert(syst): 9.4% for α, 5.1% for ϵ

Atomic Electrons (^{206}At)

$\langle e\rangle$=74 *3* keV

e_{bin}(keV)	$\langle e\rangle$(keV)	e(%)
13	0.11	<2
14	3.1	22.3 *24*
16	2.1	12.7 *21*
17 - 65	4.0	9 *3*
72 - 109	2.3	2.5 *10*
138 - 155	1.0	0.74 *22*
163	4.8	2.9 *3*
175 - 185	1.3	0.70 *19*
186	1.5	~0.8
188 - 234	1.84	0.85 *8*
240 - 288	3.9	1.53 *13*
294 - 301	0.39	0.133 *13*
303	5.3	1.76 *11*
304 - 352	1.5	0.45 *13*
357 - 378	0.36	0.099 *11*
379	2.42	0.64 *4*
380 - 383	0.67	0.175 *15*
384	8.2	2.13 *12*
386 - 431	1.50	0.38 *4*
434	1.7	~0.4
441 - 485	5.5	1.17 *13*
494 - 514	0.5	~0.10
521	2.06	0.40 *4*
523 - 566	0.33	0.06 *3*
582 - 601	0.42	0.071 *7*
608	6.8	1.12 *6*
610 - 654	2.1	0.34 *13*
684	1.74	0.255 *14*
687 - 735	1.74	0.25 *4*
737 - 786	0.9	~0.12
789 - 834	0.8	~0.10
846 - 895	0.8	0.10 *3*
896 - 945	0.7	0.08 *3*
947 - 995	0.6	0.07 *3*
997 - 1046	0.37	0.037 *13*
1047 - 1094	0.049	0.0046 *15*
1104 - 1124	0.14	~0.013
1164 - 1202	0.24	0.020 *7*
1241 - 1290	0.09	0.007 *3*
1291 - 1336	0.010	~0.0008
1345 - 1353	0.07	~0.005
1400 - 1446	0.027	0.0019 *9*
1476 - 1544	0.06	~0.0038
1621 - 1720	0.09	0.0053 *21*
1722 - 1816	0.066	0.0037 *13*
1835 - 1934	0.10	0.0055 *20*
1935 - 2023	0.030	0.0015 *7*
2059 - 2126	0.022	0.0010 *5*
2178 - 2268	0.053	0.0024 *8*

Atomic Electrons (^{206}At) (continued)

e_{bin}(keV)	$\langle e\rangle$(keV)	e(%)
2282 - 2316	0.008	0.00036 *13*
2402 - 2499	0.025	0.0010 *4*
2542 - 2589	0.0039	0.00015 *5*

Continuous Radiation (^{206}At)

$\langle\beta+\rangle$=245 keV;$\langle$IB$\rangle$=9.2 keV

E_{bin}(keV)		$\langle\ \rangle$(keV)	(%)
0 - 10	β+	2.27×10^{-8}	2.66×10^{-7}
	IB	0.0083	
10 - 20	β+	1.67×10^{-6}	9.9×10^{-6}
	IB	0.0099	0.070
20 - 40	β+	8.4×10^{-5}	0.000252
	IB	0.0169	0.058
40 - 100	β+	0.0082	0.0102
	IB	0.48	0.60
100 - 300	β+	0.73	0.322
	IB	0.20	0.115
300 - 600	β+	7.0	1.49
	IB	0.35	0.079
600 - 1300	β+	60	6.2
	IB	1.60	0.169
1300 - 2500	β+	152	8.4
	IB	4.2	0.23
2500 - 4509	β+	25.5	0.94
	IB	2.4	0.081
	$\Sigma\beta$+		17

$^{206}_{86}$Rn(5.67 *17* min)

Mode: α(68 *3* %), ϵ(32 *3* %)

Δ: -9180 *130* keV syst

SpA: 1.61×10^{8} Ci/g

Prod: ^{197}Au(^{14}N,5n); ^{12}C on Hg; protons on Th; ^{16}O on Pt

Alpha Particles (^{206}Rn)

α(keV)
6258 *3*

Photons (^{206}Rn)

γ_{mode}	γ(keV)	γ(rel)
γ_ϵ	61.70 *9*	14.3 *15*
γ_ϵ	96.85 *10*	5.2 *6*
γ_ϵ	100.91 *22*	3.6 *4*
γ_ϵ	133.72 *14*	5.0 *6*
γ_ϵ	186.2 *3*	6.8 *8*
γ_ϵ	195.43 *14*	11.7 *15*
γ_ϵ	208.03 *15*	26 *3*
γ_ϵ	213.1 *4*	11.0 *17*
γ_ϵ	215.1 *4*	4.0 *8*
γ_ϵ	290.6 *3*	7.5 *10*
γ_ϵ	301.91 *18*	53 *7*
γ_ϵ	324.46 *13*	100
γ_ϵ	350.4 *3*	12.7 *20*
γ_ϵ	371.07 *22*	52 *6*
γ_ϵ	386.16 *13*	63 *6*
γ_ϵ	435.64 *15*	5.5 *16*
γ_ϵ	443.9 *3*	28 *3*
γ_ϵ	458.2 *6*	5 *1*
γ_ϵ	465.45 *25*	3.7 *6*

Photons (^{206}Rn) (continued)

γ_{mode}	γ(keV)	γ(rel)
γ_ϵ	482.25 *22*	59 *7*
γ_ϵ	485.3 *3*	31 *4*
γ_ϵ	497.34 *14*	104 *15*
γ_ϵ	527.15 *24*	25 *3*
γ_ϵ	536.3 *3*	17 *3*
γ_ϵ	631.8 *3*	15.0 *23*
γ_ϵ	642.9 *7*	6.7 *10*
γ_ϵ	716.6 *7*	6.8 *10*
γ_ϵ	738.2 *6*	15.5 *20*
γ_ϵ	756.8 *6*	10.7 *15*
γ_ϵ	772.8 *3*	60 *7*
γ_ϵ	794.8 *4*	10.0 *17*

$^{206}_{87}$Fr(16.0 *1* s)

Mode: α(85 *3* %), ϵ(15 *3* %)
Δ: -1440 *160* keV
SpA: 3.350×10^{9} Ci/g
Prod: ^{197}Au(^{16}O,7n); ^{203}Tl(^{12}C,9n); protons on Th

Alpha Particles (^{206}Fr)

α(keV)
6789 *5*

$^{206}_{88}$Ra(400 *200* ms)

Mode: α
Δ: 3540 *300* keV syst
SpA: 7×10^{10} Ci/g
Prod: ^{197}Au(^{19}F,10n); ^{22}Ne on Pt

Alpha Particles (^{206}Ra)

α(keV)
7272 *5*

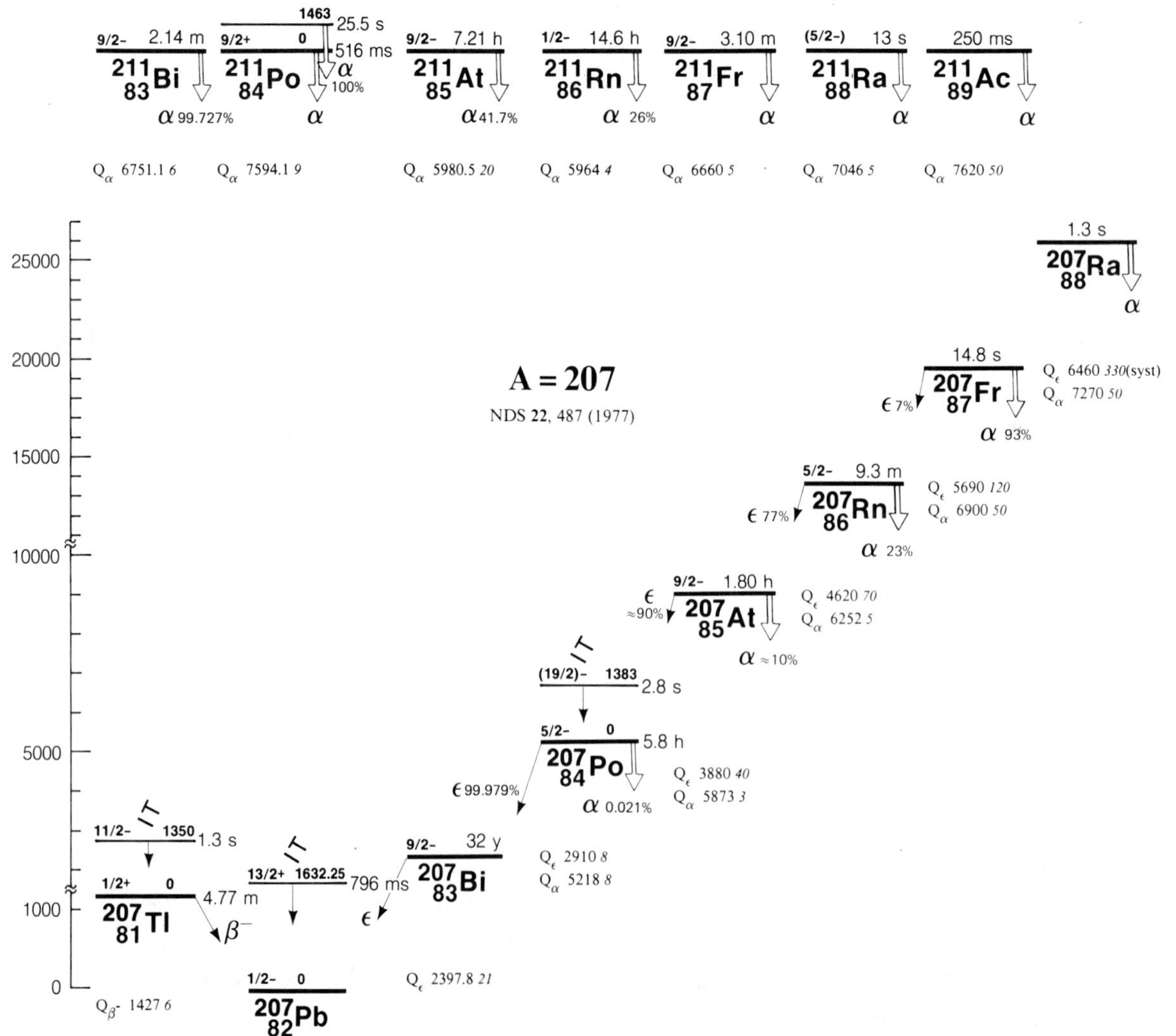

$^{207}_{81}$Tl(4.77 *2* min)

Mode: β-
Δ: -21048 *6* keV
SpA: 1.902×10^{8} Ci/g
Prod: descendant ^{227}Ac

Photons (^{207}Tl)

⟨γ⟩=2.2 *4* keV

γ_{mode}	γ(keV)	γ(%)
Pb L_{ℓ}	9.185	4.5 *8* ×10^{-5}
Pb L_{α}	10.541	0.00090 *15*
Pb L_{η}	11.349	1.17 *20* ×10^{-5}
Pb L_{β}	12.703	0.00082 *14*
Pb L_{γ}	14.886	0.00015 *3*
Pb $K_{\alpha2}$	72.803	0.00150 *24*
Pb $K_{\alpha1}$	74.969	0.0025 *4*
Pb $K_{\beta1}$'	84.789	0.00089 *14*
Pb $K_{\beta2}$'	87.632	0.00026 *4*
γ [M1]	328.08 *7*	0.0015 *5*
γ E2	569.150 *19*	<0.0010
γ M1+0.9%E2	897.23 *7*	0.24 *4*

Atomic Electrons (^{207}Tl)

⟨e⟩=0.051 *7* keV

e_{bin}(keV)	⟨e⟩(keV)	e(%)
13 - 62	0.00043	0.0028 *4*
68 - 84	4.8 ×10^{-5}	6.7 *5* ×10^{-5}
240	0.0010	~0.00043
312 - 328	0.00030	10 *3* ×10^{-5}
481	0.00015	~3×10^{-5}
553 - 569	6.3 ×10^{-5}	1.1 *3* ×10^{-5}
809	0.040	0.0049 *8*
881	0.0066	0.00075 *13*
882 - 884	0.00056	6.4 *10* ×10^{-5}
893	0.0015	0.00017 *3*
894 - 897	0.00070	7.8 *9* ×10^{-5}

Continuous Radiation (^{207}Tl)

⟨β-⟩=493 keV;⟨IB⟩=0.68 keV

E_{bin}(keV)		⟨ ⟩(keV)	(%)
0 - 10	β-	0.0460	0.92
	IB	0.022	
10 - 20	β-	0.139	0.92
	IB	0.021	0.147
20 - 40	β-	0.57	1.89
	IB	0.040	0.140
40 - 100	β-	4.18	5.9
	IB	0.107	0.166
100 - 300	β-	44.2	21.9
	IB	0.25	0.143
300 - 600	β-	147	32.8
	IB	0.18	0.043
600 - 1300	β-	295	35.4
	IB	0.068	0.0090
1300 - 1422	β-	2.58	0.194
	IB	2.0×10^{-5}	1.50×10^{-6}

$^{207}_{81}$Tl(1.33 *11* s)

Mode: IT
Δ: -19698 *7* keV
SpA: 3.2×10^{10} Ci/g
Prod: ^{208}Pb(t,α)

Photons (^{207}Tl)

⟨γ⟩=1168 *48* keV

γ_{mode}	γ(keV)	γ(%)†
Tl L_{ℓ}	8.953	0.219 *21*
Tl L_{α}	10.259	4.4 *3*
Tl L_{η}	10.994	0.058 *4*
Tl L_{β}	12.312	4.1 *3*
Tl L_{γ}	14.410	0.75 *7*
Tl $K_{\alpha2}$	70.832	7.4 *3*
Tl $K_{\alpha1}$	72.873	12.5 *6*
Tl $K_{\beta1}$'	82.434	4.40 *20*
Tl $K_{\beta2}$'	85.185	1.24 *6*
γ M1+6.8%E2	350.1 *24*	79 *4*
γ (M4)	1000.0 *25*	87 *4*

† 0.26% uncert(syst)

Atomic Electrons (^{207}Tl)

⟨e⟩=181 *6* keV

e_{bin}(keV)	⟨e⟩(keV)	e(%)
13 - 60	2.18	14.3 *11*
66 - 82	0.235	0.338 *14*
265	44.4	16.8 *9*
335	9.5	2.84 *15*
337 - 350	3.23	0.93 *3*
914	89.0	9.7 *6*
985	22.7	2.30 *15*
987	1.91	0.193 *12*
996	4.6	0.46 *3*
997 - 1000	3.54	0.355 *15*

$^{207}_{82}$Pb(stable)

Δ: -22476 *4* keV
%: 22.1 *1*

$^{207}_{82}$Pb(796 *2* ms)

Mode: IT
Δ: -20844 *4* keV
SpA: 4.571×10^{10} Ci/g
Prod: daughter ^{207}Bi; ^{207}Pb(n,n'); ^{208}Pb(γ,n); ^{208}Pb(n,2n)

Photons (^{207}Pb)

⟨γ⟩=1506 *55* keV

γ_{mode}	γ(keV)	γ(%)†
Pb L_{ℓ}	9.185	0.093 *9*
Pb L_{α}	10.541	1.84 *12*
Pb L_{η}	11.349	0.0249 *18*
Pb L_{β}	12.699	1.75 *14*
Pb L_{γ}	14.890	0.33 *3*

Photons (^{207}Pb)
(continued)

γ_{mode}	γ(keV)	γ(%)†
Pb $K_{\alpha2}$	72.803	2.86 *14*
Pb $K_{\alpha1}$	74.969	4.80 *24*
Pb $K_{\beta1}$'	84.789	1.70 *9*
Pb $K_{\beta2}$'	87.632	0.49 *3*
γ E2	569.150 *19*	98 *5*
γ M4+0.1%E5	1063.10 *2*	89 *4*

† 0.23% uncert(syst)

Atomic Electrons (^{207}Pb)

⟨e⟩=126 *5* keV

e_{bin}(keV)	⟨e⟩(keV)	e(%)
13 - 62	0.88	5.6 *4*
68 - 84	0.091	0.127 *6*
481	7.5	1.57 *8*
553 - 569	3.23	0.580 *16*
975	84.4	8.7 *5*
1047	17.9	1.71 *9*
1048 - 1050	5.16	0.492 *20*
1059	4.37	0.413 *22*
1060 - 1063	2.95	0.278 *11*

$^{207}_{83}$Bi(32.2 *13* yr)

Mode: ϵ
Δ: -20078 *5* keV
SpA: 53.6 Ci/g
Prod: deuterons on Pb; daughter ^{211}At

Photons (^{207}Bi)

⟨γ⟩=1539 *17* keV

γ_{mode}	γ(keV)	γ(%)†
Pb L_{ℓ}	9.185	0.75 *7*
Pb L_{α}	10.541	14.9 *8*
Pb L_{η}	11.349	0.184 *11*
Pb L_{β}	12.699	14.0 *10*
Pb L_{γ}	14.914	2.65 *25*
Pb $K_{\alpha2}$	72.803	21.7 *7*
Pb $K_{\alpha1}$	74.969	36.5 *12*
Pb $K_{\beta1}$'	84.789	12.9 *4*
Pb $K_{\beta2}$'	87.632	3.72 *13*
γ [M1]	328.08 *7*	0.0009 *3*
γ E2	569.150 *19*	97.8
γ M1+0.9%E2	897.23 *7*	0.147 *10*
γ M4+0.1%E5	1063.10 *2*	74.9 *15*
γ E2	1441.63 *8*	0.147 *20*
γ M1+0.8%E2	1769.71 *4*	6.85 *20*

† 0.51% uncert(syst)

Atomic Electrons (^{207}Bi)

⟨e⟩=115.9 *21* keV

e_{bin}(keV)	⟨e⟩(keV)	e(%)
13 - 62	6.7	43 *3*
68 - 84	0.69	0.96 *4*
240	0.0006	~0.00026
312 - 328	0.00018	5.8 *21* ×10^{-5}

Atomic Electrons (^{207}Bi) (continued)

e_{bin}(keV)	⟨e⟩(keV)	e(%)
481	7.53	1.57 3
553 - 569	3.23	0.580 6
809	0.0243	0.00300 21
881 - 897	0.0057	0.00065 3
975	71.5	7.33 21
1047	15.1	1.45 4
1048 - 1050	4.37	0.417 9
1059	3.7	0.349 10
1060 - 1063	2.50	0.235 5
1354	0.0055	0.00040 5
1426 - 1441	0.00129	9.0 9 $\times10^{-5}$
1682 - 1767	0.511	0.0301 9

Continuous Radiation (^{207}Bi)

⟨β+⟩=0.0464 keV;⟨IB⟩=0.64 keV

E_{bin}(keV)		⟨ ⟩(keV)	(%)
0 - 10	β+	4.60×10^{-10}	5.4×10^{-9}
	IB	7.7×10^{-5}	
10 - 20	β+	3.23×10^{-8}	1.91×10^{-7}
	IB	0.0021	0.0166
20 - 40	β+	1.53×10^{-6}	4.57×10^{-6}
	IB	0.00120	0.0037
40 - 100	β+	0.000130	0.000163
	IB	0.48	0.66
100 - 300	β+	0.0079	0.00357
	IB	0.043	0.028
300 - 600	β+	0.0312	0.0072
	IB	0.040	0.0093
600 - 1300	β+	0.0072	0.00109
	IB	0.061	0.0069
1300 - 1835	IB	0.0108	0.00076
	Σβ+		0.0120

$^{207}_{84}$Po(5.83 7 h)

Mode: ϵ(99.979 %), α(0.021 2 %)
Δ: -17168 9 keV
SpA: 2.60×10^{6} Ci/g
Prod: ^{209}Bi(p,3n); ^{206}Pb(α,3n)

Alpha Particles (^{207}Po)

α(keV)
5115.0 25

Photons (^{207}Po)

⟨γ⟩=1327 37 keV

γ_{mode}	γ(keV)	γ(%)†
Bi L_ℓ	9.419	0.80 16
Bi L_α	10.828	16 3
Bi L_η	11.712	0.20 4
Bi L_β	13.099	14 3
Bi L_γ	15.382	2.7 6
Bi $K_{\alpha2}$	74.814	24 5
Bi $K_{\alpha1}$	77.107	41 8
Bi $K_{\beta1}$'	87.190	15 3
Bi $K_{\beta2}$'	90.128	4.3 8
γM1+8.3%E2	99.93 11	0.13 3
γM1(+<8.3%E2)	149.69 11	0.090 13
γM1	156.08 7	~0.09
γE2+5.4%M1	158.02 8	0.56 4
γ	177.7 10	<0.018
γ	214.4 10	<0.036
γ(E1)	222.11 8	~1
γ(M1)	222.76 12	~0.27
γ(M1)	224.11 8	0.20 4
γM1	249.62 7	1.62 9
γE1	297.24 9	1.01 5
γM1	307.52 8	0.65 5
γM1	330.13 14	0.23 4 ?
γM1	345.21 16	2.00 11
γ(E1)	369.53 8	1.93 13
γM1	405.70 6	10.1 5
γE2	503.22 18	0.18 7
γE2+~41%M1	531.63 10	0.6 3
γM1	629.81 7	1.48 14
γM1+E2	669.56 7	0.50 7
γE1	687.64 7	2.03 16
γ(M1)	698.31 9	<0.11 ?
γM1	742.63 6	29.2 14
γ(M1)	770.60 12	0.72 5
γE2+~20%M1	892.32 11	0.43 9
γE1	911.76 7	18.0 9
γ(M1)	947.93 8	1.08 9 ?
γM1	992.25 7	60 3
γ[M1+E2]	1020.22 11	0.18 4
γE2	1148.33 6	6.1 5
γM1(+E0)	1211.3 5	0.131 22 ?
γ[M2]	1317.45 7	0.061 20
γ(M1)	1360.43 14	0.63 5 ?
γE2	1372.44 8	1.39 11
γ[E2]	1377.0 5	0.16 4
γ	1586 1	0.108 18
γ	1613 2	~0.036
γE1	1662.66 17	0.40 7
γ	1746 1	0.054 18
γ(E2)	1762.85 11	0.25 5
γ	1846.82 20	0.34 4
γ	1926 1	0.061 11
γ	1953 1	0.081 11
γ[E3]	2060.08 8	1.44 11

† 5.6% uncert(syst)

Atomic Electrons (^{207}Po)

⟨e⟩=45.7 12 keV

e_{bin}(keV)	⟨e⟩(keV)	e(%)
9	0.099	1.05 24
13	3.5	26 6
16	2.6	17 4
58 - 100	2.53	3.8 3
124 - 158	1.46	1.05 16
159	1.71	1.08 6
161 - 210	0.24	0.118 25
211 - 250	1.34	0.59 3
255	1.39	0.55 3
279 - 314	0.301	0.103 5
315	5.7	1.80 10
317 - 366	0.444	0.133 6
367 - 413	1.60	0.407 17
441 - 490	0.13	0.029 13
499 - 539	0.48	0.090 8
579 - 627	0.26	0.043 10
629	0.0066	0.00106 9
652	6.9	1.06 6
653 - 698	0.201	0.0297 25
726	1.18	0.162 9
727 - 771	0.552	0.0748 24
820	0.418	0.051 3
857 - 898	0.253	0.0292 20
902	9.4	1.04 6
908 - 948	0.081	0.0087 11
976	1.55	0.159 9
977 - 1020	0.666	0.0675 23
1058	0.28	0.0267 24
1121 - 1148	0.089	0.0079 4
1195 - 1227	0.018	0.0015 4
1270 - 1317	0.128	0.0100 6
1344 - 1377	0.0292	0.00216 10
1495 - 1583	0.013	0.0009 3
1597 - 1672	0.0121	0.00073 15
1730 - 1760	0.015	~0.0009
1830 - 1923	0.008	0.00045 15
1937 - 1970	0.082	0.0042 4
2044 - 2057	0.0222	0.00109 7

Continuous Radiation (^{207}Po)

⟨β+⟩=2.24 keV;⟨IB⟩=1.58 keV

E_{bin}(keV)		⟨ ⟩(keV)	(%)
0 - 10	β+	1.15×10^{-8}	1.35×10^{-7}
	IB	0.00017	
10 - 20	β+	8.3×10^{-7}	4.88×10^{-6}
	IB	0.0024	0.018
20 - 40	β+	4.00×10^{-5}	0.000120
	IB	0.00137	0.0043
40 - 100	β+	0.00354	0.00443
	IB	0.55	0.73
100 - 300	β+	0.236	0.106
	IB	0.071	0.043
300 - 600	β+	1.19	0.267
	IB	0.161	0.036
600 - 1300	β+	0.81	0.113
	IB	0.61	0.067
1300 - 2163	IB	0.18	0.0122
	Σβ+		0.49

$^{207}_{84}$Po(2.8 2 s)

Mode: IT
Δ: -15785 9 keV
SpA: 1.72×10^{10} Ci/g
Prod: ^{209}Bi(p,3n); ^{204}Pb(α,n)

Photons (^{207}Po)

⟨γ⟩=1077 82 keV

γ_{mode}	γ(keV)	γ(%)†
Po L_ℓ	9.658	0.77 8
Po L_α	11.119	14.7 10
Po L_η	12.085	0.313 23
Po L_β	13.488	18.4 14
Po L_γ	15.859	3.8 4
Po $K_{\alpha2}$	76.858	16.9 9
Po $K_{\alpha1}$	79.290	28.2 14
Po $K_{\beta1}$'	89.639	10.0 5
Po $K_{\beta2}$'	92.673	3.05 16
γ E3	268.2 10	45 4
γ M2	300.74 6	33 3
γ E2	814.48 6	99 10

†<0.1% uncert(syst)

$^{207}_{85}$At(1.80 4 h)

Mode: ϵ(~ 90 %), α(~ 10 %)
Δ: -13290 40 keV
SpA: 8.41×10^{6} Ci/g
Prod: ^{209}Bi(α,6n); daughter ^{207}Rn; ^{197}Au(^{12}C,2n); protons on Th

Alpha Particles (^{207}At)

α(keV)
5758 3

Photons (^{207}At)

⟨γ⟩=1326 36 keV

γmode	γ(keV)	γ(%)†
Po K$_{\alpha2}$	76.858	28 8 ‡
Po K$_{\alpha1}$	79.290	46 14 ‡
Po K$_{\beta1}$'	89.639	16 5 ‡
Po K$_{\beta2}$'	92.673	5.0 15 ‡
γ$_ϵ$M1+50%E2	168.01 7	1.10 9
γ$_ϵ$M1+50%E2	191.29 6	0.43 7
γ$_ϵ$M1+50%E2	236.57 6	0.73 6
γ$_ϵ$	286.83 10	0.23 3
γ$_ϵ$	292.83 10	0.27 3
γ$_ϵ$M2	300.74 6	9.7 7
γ$_ϵ$	316.03 10	0.13 3
γ$_ϵ$[M1+E2]	324.49 7	0.50 3
γ$_ϵ$	338.63 10	0.10 3
γ$_ϵ$M1(+14%E2)	357.33 7	1.83 10
γ$_ϵ$	365.53 10	0.20 3
γ$_ϵ$[M1+E2]	393.04 7	0.47 7
γ$_ϵ$	411.33 8	0.43 3
γ$_ϵ$	422.08 12	1.37 10
γ$_ϵ$M1	456.81 7	1.30 10
γ$_ϵ$	459.66 8	1.10 7
γ$_ϵ$E2	467.20 5	5.26 11
γ$_ϵ$	498.43 10	0.27 3
γ$_ϵ$	503.33 10	0.20 3
γ$_ϵ$[M3]	519.86 8	0.27 7 ?
γ$_ϵ$	520.83 10	0.57 7
γ$_ϵ$M1	529.91 6	2.53 17
γ$_ϵ$M1	583.40 7	1.53 10
γ$_ϵ$E2+10%M1	588.42 6	14.7 10
γ$_ϵ$	617.23 10	1.13 10
γ$_ϵ$	626.80 7	1.67 13
γ$_ϵ$M1	637.40 7	1.73 13
γ$_ϵ$	641.28 8	0.43 7
γ$_ϵ$M1	648.09 6	3.30 20
γ$_ϵ$E2+41%M1	658.48 6	5.0 3
γ$_ϵ$E2(+E1)	670.65 7	2.80 20
γ$_ϵ$E1	675.21 7	4.9 3
γ$_ϵ$M1	693.64 11	1.60 13
γ$_ϵ$M1	721.19 5	4.9 3
γ$_ϵ$	755.13 7	0.47 7
γ$_ϵ$	764.93 20	0.33 3
γ$_ϵ$[E1]	767.94 7	0.17 3
γ$_ϵ$	789.73 10	0.30 7
γ$_ϵ$	798.83 20	0.20 7
γ$_ϵ$E2	814.48 6	33
γ$_ϵ$	852.77 9	0.20 3 ?
γ$_ϵ$	862.34 8	0.43 7
γ$_ϵ$	881.10 8	0.67 7 ?
γ$_ϵ$M1	907.21 7	4.0 7
γ$_ϵ$	932.23 10	0.23 3
γ$_ϵ$	954.83 20	~0.13
γ$_ϵ$	960.59 9	1.67 13
γ$_ϵ$[E1]	994.01 6	1.67 7
γ$_ϵ$	1015.73 20	0.23 3
γ$_ϵ$(E1)	1021.56 9	0.63 7
γ$_ϵ$	1042.33 20	0.17 3
γ$_ϵ$M1	1054.22 8	0.80 7
γ$_ϵ$E1(+E2)	1077.80 7	1.37 10
γ$_ϵ$	1086.93 10	0.27 7
γ$_ϵ$E2	1115.29 6	3.3 3

Photons (^{207}At) (continued)

γmode	γ(keV)	γ(%)†
γ$_ϵ$	1131.80 7	0.30 3
γ$_ϵ$	1171.82 7	0.83 7
γ$_ϵ$	1174.53 20	0.33 3
γ$_ϵ$	1188.39 5	1.23 10
γ$_ϵ$	1193.73 20	0.33 7
γ$_ϵ$	1225.82 7	1.07 10
γ$_ϵ$	1242.83 10	0.63 7
γ$_ϵ$	1245.50 9	0.43 7
γ$_ϵ$	1264.03 10	0.37 7
γ$_ϵ$	1283.33 10	0.83 7
γ$_ϵ$	1396.40 6	1.03 10
γ$_ϵ$	1410.03 20	0.87 7
γ$_ϵ$	1413.23 20	0.73 7
γ$_ϵ$	1493.43 20	0.13 3
γ$_ϵ$	1511.1 3	0.40 7
γ$_ϵ$	1548.49 7	0.80 7
γ$_ϵ$	1552.7 3	0.33 7
γ$_ϵ$	1557.03 20	0.23 3
γ$_ϵ$	1642.10 8	0.93 13
γ$_ϵ$(M1)	1676.82 7	1.96 17
γ$_ϵ$	1690.83 20	0.23 3
γ$_ϵ$	1712.70 9	1.03 10
γ$_ϵ$	1716.53 20	0.80 10
γ$_ϵ$	1731.03 10	2.8 3
γ$_ϵ$	1772.82 20	0.50 7
γ$_ϵ$	1781.93 20	0.43 7
γ$_ϵ$	1786.83 10	0.70 7
γ$_ϵ$	1805.4 3	0.63 7
γ$_ϵ$	1826.03 20	0.10 3
γ$_ϵ$	1858.4 4	0.053 17
γ$_ϵ$	1876.1 3	0.23 7
γ$_ϵ$	1928.6 3	0.083 20
γ$_ϵ$	2016.7 3	0.50 7
γ$_ϵ$	2029.83 20	0.20 3
γ$_ϵ$	2046.6 4	0.10 3
γ$_ϵ$	2053.0 3	0.17 3
γ$_ϵ$	2075.2 3	0.47 10
γ$_ϵ$	2155.5 4	0.057 20
γ$_ϵ$	2202.8 5	0.08 3
γ$_ϵ$	2343.3 4	0.70 7
γ$_ϵ$	2393.8 3	0.13 3 ?
γ$_ϵ$	2486.6 5	0.097 20
γ$_ϵ$	2496.2 5	0.063 20
γ$_ϵ$	2558.5 5	0.27 7
γ$_ϵ$	2566.4 5	0.27 7
γ$_ϵ$	2712.23 20	0.93 10
γ$_ϵ$	2862.2 5	0.050 17

† 11% uncert(syst)
‡ exp values

$^{207}_{86}$Rn(9.3 2 min)

Mode: ϵ(77 2 %), α(23 2 %)
Δ: -8670 80 keV
SpA: 9.76×10^{7} Ci/g
Prod: ^{197}Au(^{14}N,4n); protons on Th; ^{16}O on Pt; ^{11}B on Tl

Alpha Particles (^{207}Rn)

⟨α⟩=1408.9 13 keV

α(keV)	α(%)
5995 4	0.023 7
6068 3	0.152 5
6126 3	22.825 9

Photons (^{207}Rn)

⟨γ⟩=813 39 keV

γmode	γ(keV)	γ(%)†
At L$_{\ell}$	9.897	0.45 7
At L$_{\alpha}$	11.414	8.4 11
At L$_{\eta}$	12.466	0.118 16
At L$_{\beta}$	13.918	8.0 12
At L$_{\gamma}$	16.392	1.57 24
At K$_{\alpha2}$	78.947	12.6 16
At K$_{\alpha1}$	81.517	21 3
At K$_{\beta1}$'	92.136	7.4 10
At K$_{\beta2}$'	95.265	2.3 3
γ$_ϵ$	188.24 17	0.25 5
γ$_ϵ$	233.83 20	0.68 14
γ$_ϵ$	243.07 25	0.15 5
γ$_ϵ$	245.93 24	0.15 4
γ$_ϵ$	295.5 3	0.40 7
γ$_ϵ$	308.03 17	0.17 4
γ$_ϵ$(M1)	329.47 4	3.0 3
γ$_ϵ$	337.6 4	~0.13
γ$_ϵ$E2(+~8.0%M1)	344.55 4	45
γ$_ϵ$	350.17 21	0.55 14
γ$_ϵ$	360.97 24	0.10 5
γ$_ϵ$M1	367.67 8	2.5 3
γ$_ϵ$	377.98 14	0.68 14
γ$_ϵ$	380.3 6	0.26 9
γ$_ϵ$M1	402.67 4	11.8 12
γ$_ϵ$	417.73 20	0.45 9
γ$_ϵ$	436.3 3	0.30 8
γ$_ϵ$	443.5 4	0.19 6
γ$_ϵ$	445.96 12	0.50 9
γ$_ϵ$	472.1 3	0.20 6
γ$_ϵ$	475.63 20	0.68 18
γ$_ϵ$	477.57 16	0.35 8
γ$_ϵ$	485.2 3	0.24 7
γ$_ϵ$	486.9 5	0.29 8
γ$_ϵ$	520.91 14	~0.14
γ$_ϵ$	523.95 9	~0.23
γ$_ϵ$	535.00 20	0.35 9
γ$_ϵ$	536.94 15	0.28 8
γ$_ϵ$	547.03 20	0.33 9
γ$_ϵ$	553.23 10	1.18 23
γ$_ϵ$	559.06 23	0.21 5
γ$_ϵ$	561.92 23	0.35 7
γ$_ϵ$	566.22 13	0.29 9
γ$_ϵ$	573.4 4	0.17 6
γ$_ϵ$	580.30 19	0.35 9
γ$_ϵ$	598.6 3	0.24 8
γ$_ϵ$	604.04 15	0.20 5
γ$_ϵ$	610.14 15	0.38 8
γ$_ϵ$	616.50 22	0.22 6
γ$_ϵ$	620.73 20	0.32 9
γ$_ϵ$	628.73 7	1.09 23
γ$_ϵ$	631.59 9	2.9 3
γ$_ϵ$	636.0 4	~0.14
γ$_ϵ$	638.1 4	~0.14
γ$_ϵ$	643.31 17	1.23 23
γ$_ϵ$	647.23 10	1.8 4
γ$_ϵ$	655.6 4	0.19 7
γ$_ϵ$	660.43 20	0.86 18
γ$_ϵ$	672.0 3	0.64 18
γ$_ϵ$E2	674.02 4	~8
γ$_ϵ$E2	674.03 5	~4
γ$_ϵ$	685.83 10	1.23 23
γ$_ϵ$	687.48 18	0.64 18
γ$_ϵ$	691.3 3	0.12 5
γ$_ϵ$	697.14 9	2.4 4
γ$_ϵ$	700.53 10	0.44 9
γ$_ϵ$	712.83 20	0.59 18
γ$_ϵ$	739.8 5	0.23 8
γ$_ϵ$M1+E2	747.23 5	14.1 14
γ$_ϵ$	751.6 4	0.45 9
γ$_ϵ$	754.57 25	0.27 6
γ$_ϵ$	763.64 24	0.09 3
γ$_ϵ$	768.66 17	0.28 9
γ$_ϵ$	775.44 12	2.0 3
γ$_ϵ$	780.92 19	~0.07
γ$_ϵ$	787.98 22	0.21 6
γ$_ϵ$	792.27 14	0.17 5
γ$_ϵ$	799.60 20	0.22 6
γ$_ϵ$	804.8 3	0.20 9
γ$_ϵ$	806.36 20	0.28 10
γ$_ϵ$	820.7 4	0.23 9
γ$_ϵ$	823.37 20	0.20 7

Photons (^{207}Rn) (continued)

γ_{mode}	γ(keV)	γ(%)†
γ_ϵ	847.5 *3*	0.31 *12*
γ_ϵ	853.43 *9*	2.3 *5*
γ_ϵ	861.18 *22*	0.15 *6*
γ_ϵ	865.47 *14*	0.23 *10*
γ_ϵ	873.5 *7*	0.24 *11*
γ_ϵ	880.24 *16*	0.18 *7*
γ_ϵ	884.5 *3*	0.30 *11*
γ_ϵ	884.8 *3*	0.17 *6* ?
γ_ϵ	892.7 *7*	1.00 *18*
γ_ϵ	908.63 *10*	~1
γ_ϵ	919.8 *3*	0.29 *12*
γ_ϵ	923.2 *6*	~0.13
γ_ϵ	939.62 *15*	0.35 *11*
γ_ϵ	948.05 *17*	0.30 *9*
γ_ϵ	951.70 *18*	0.37 *14*
γ_ϵ	973.28 *8*	2.5 *5*
γ_ϵ	983.0 *5*	0.24 *9*
γ_ϵ	985.8 *3*	0.38 *14*
γ_ϵ	990.49 *17*	0.34 *11*
γ_ϵ	993.36 *17*	0.50 *14*
γ_ϵ	999.23 *20*	1.18 *18*
γ_ϵ	1083.0 *7*	~0.27
γ_ϵ	1121.1 *5*	0.23 *9*
γ_ϵ	1129.7 *5*	0.20 *9*
γ_ϵ	1172.0 *4*	
γ_ϵ	1176.40 *17*	
γ_ϵ	1190.65 *22*	0.23 *9*
γ_ϵ	1224.79 *16*	0.59 *14*
γ_ϵ	1254.63 *20*	0.336 *14*
γ_ϵ	1326.6 *7*	
γ_ϵ	1474.3 *7*	
γ_ϵ	1478.8 *3*	
γ_ϵ	1507.5 *6*	0.45 *18*
γ_ϵ	1522.8 *4*	0.68 *18*
γ_ϵ	1539.49 *14*	0.59 *23*
γ_ϵ	1799.51 *19*	
γ_ϵ	1805.12 *17*	
γ_ϵ	2576.6 *3*	0.30 *9*

† 34% uncert(syst)

$^{207}_{87}$Fr(14.8 *1* s)

Mode: α(93 *3* %), ϵ(7 *3* %)
Δ: -2980 *110* keV
SpA: 3.598×10^9 Ci/g
Prod: ^{197}Au(^{16}O,6n); ^{203}Tl(^{12}C,8n); protons on Th

Alpha Particles (^{207}Fr)

α(keV)
6766 *5*

$^{207}_{88}$Ra(1.3 *2* s)

Mode: α
Δ: 3480 *310* keV syst
SpA: 3.2×10^{10} Ci/g
Prod: ^{197}Au(^{19}F,9n); ^{192}Pt(^{22}Ne,7n)

Alpha Particles (^{207}Ra)

α(keV)
7133 *5*

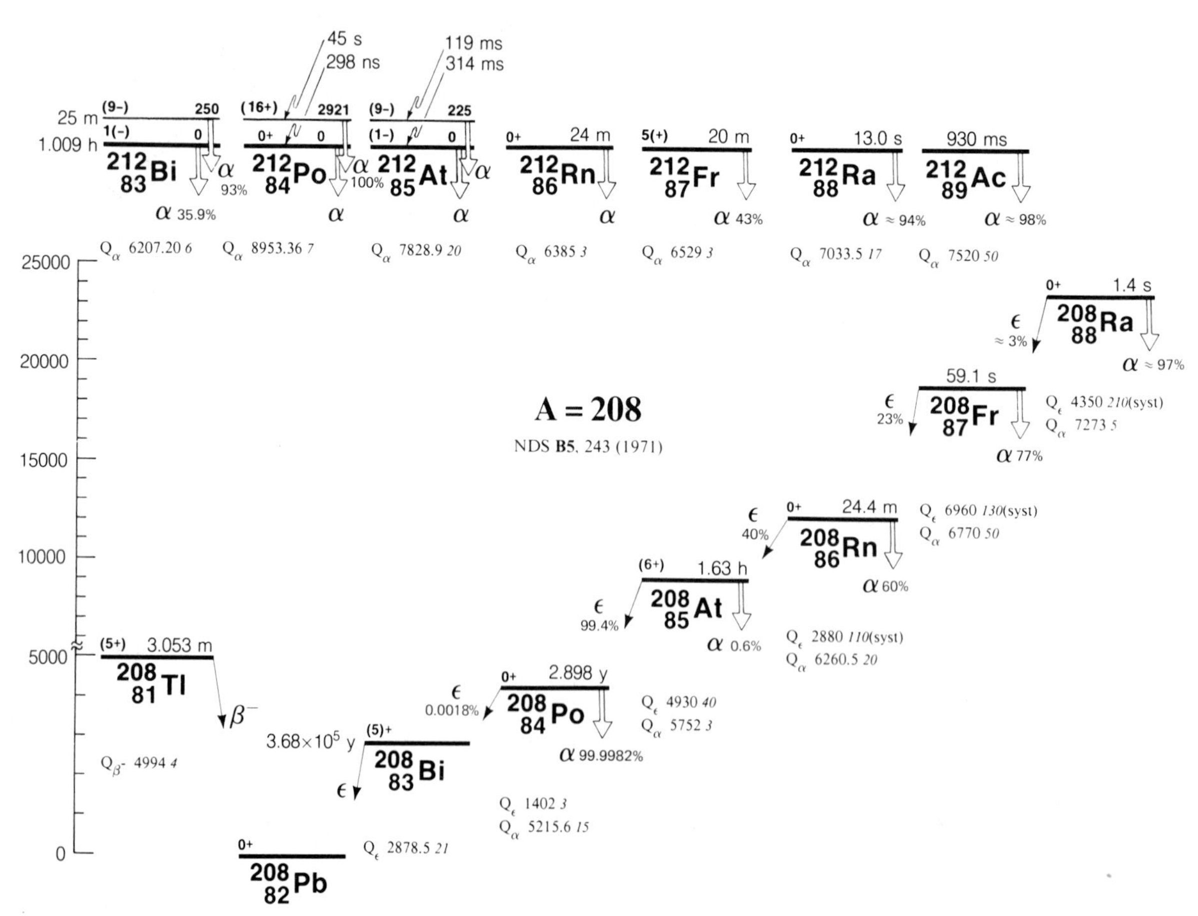

$^{208}_{81}$Tl(3.053 *3* min)

Mode: β-
Δ: -16778 *5* keV
SpA: 2.956×10^{8} Ci/g
Prod: descendant ^{228}Th

Photons (^{208}Tl)

⟨γ⟩=3374 *18* keV

γ_{mode}	γ(keV)	γ(%)†
Pb L_ℓ	9.185	0.065 *6*
Pb L_α	10.541	1.30 *6*
Pb L_η	11.349	0.0174 *10*
Pb L_β	12.701	1.22 *8*
Pb L_γ	14.884	0.224 *19*
Pb $K_{\alpha2}$	72.803	2.14 *7*
Pb $K_{\alpha1}$	74.969	3.60 *11*
Pb $K_{\beta1}$'	84.789	1.27 *4*
Pb $K_{\beta2}$'	87.632	0.367 *13*
γ M1	211.31 *9*	0.170 *20*
γ M1	233.32 *6*	0.31 *3*
γ M1	252.45 *6*	0.80 *5*
γ M1	277.28 *6*	6.8 *3*
γ (M1)	485.78 *8*	0.050 *5*
γ M1+2.8%E2	510.606 *19*	21.6 *9*
γ E2	583.022 *22*	86 *3*
γ [M1+E2]	587.82 *20*	~0.040
γ	650.14 *17*	0.036 *5*
γ [M1+E2]	705.24 *17*	0.022 *4*
γ M1+8.8%E2	721.91 *9*	0.203 *14*
γ	748.58 *20*	0.043 *4*
γ M1(+E2)	763.06 *6*	1.64 *9*
γ [M1+E2]	821.14 *19*	0.040 *4*
γ M1+0.06%E2	860.30 *6*	12.0 *4*
γ [M1+E2]	883.27 *16*	0.031 *3*
γ	927.42 *17*	0.125 *11*
γ [M1+E2]	982.52 *17*	0.197 *15*
γ (E2)	1093.63 *3*	0.37 *4*
γ	1125.6 *4*	0.0050 *20*
γ [M1+E2]	1160.55 *17*	0.011 *3*
γ [M1+E2]	1185.1 *3*	0.017 *5*
γ [M1+E2]	1282.7 *3*	0.052 *5*
γ	1381.0 *5*	0.007 *3*
γ	1647.4 *7*	~0.0020
γ [M1+E2]	1743.57 *17*	~0.0020
γ E3	2614.35 *10*	99.79

†<0.1% uncert(syst)

Atomic Electrons (^{208}Tl)

⟨e⟩=37.9 *6* keV

e_{bin}(keV)	⟨e⟩(keV)	e(%)
13 - 62	0.63	4.0 *3*
68 - 84	0.068	0.095 *4*
123 - 164	1.31	0.86 *4*
189	5.9	3.10 *15*
195 - 239	0.377	0.169 *7*
249 - 252	0.062	0.0251 *13*
261	1.25	0.480 *23*
262 - 277	0.579	0.213 *6*
398	0.0201	0.0051 *5*
423	7.9	1.87 *9*
470 - 486	0.0053	0.00111 *9*
495	8.0	1.62 *7*
498 - 511	0.516	0.102 *4*
562	0.005	~0.0009
567	1.09	0.193 *8*
568 - 617	1.62	0.283 *6*
634 - 675	0.27	0.041 *18*
689 - 736	0.017	0.0024 *5*
745 - 763	0.058	0.008 *3*

Atomic Electrons (^{208}Tl) (continued)

e_{bin}(keV)	⟨e⟩(keV)	e(%)
772	2.12	0.275 *11*
795 - 844	0.368	0.0436 *19*
845 - 895	0.17	0.0198 *12*
912 - 927	0.0027	~0.00029
967 - 1006	0.022	0.0022 *3*
1038 - 1081	0.0046	0.00043 *6*
1090 - 1125	0.0024	0.00022 *7*
1145 - 1185	0.00051	4.4 *12* $\times10^{-5}$
1195	0.0036	0.00030 *13*
1267 - 1293	0.0012	9 *3* $\times10^{-5}$
1365 - 1381	8.1 $\times10^{-5}$	~6 $\times10^{-6}$
1559 - 1656	0.00020	1.2 *6* $\times10^{-5}$
1728 - 1741	2.0 $\times10^{-5}$	1.2 *6* $\times10^{-6}$
2526	4.36	0.173 *4*
2598 - 2611	1.149	0.0442 *6*

Continuous Radiation (^{208}Tl)

⟨β-⟩=560 keV;⟨IB⟩=0.87 keV

E_{bin}(keV)		⟨ ⟩(keV)	(%)
0 - 10	β-	0.0403	0.80
	IB	0.024	
10 - 20	β-	0.122	0.81
	IB	0.023	0.162
20 - 40	β-	0.498	1.66
	IB	0.045	0.156
40 - 100	β-	3.68	5.2
	IB	0.121	0.19
100 - 300	β-	39.2	19.4
	IB	0.29	0.167
300 - 600	β-	134	29.9
	IB	0.23	0.056
600 - 1300	β-	337	38.8
	IB	0.130	0.0169
1300 - 2377	β-	45.4	3.18
	IB	0.0028	0.00021

$^{208}_{82}$Pb(stable)

Δ: -21772 *4* keV
%: 52.4 *1*

$^{208}_{83}$Bi(3.68 *4* $\times10^{5}$ yr)

Mode: ε
Δ: -18894 *5* keV
SpA: 0.00467 Ci/g
Prod: ^{209}Bi(n,2n)

Photons (^{208}Bi)

⟨γ⟩=2668 *6* keV

γ_{mode}	γ(keV)	γ(%)†
Pb L_ℓ	9.185	0.67 *15*
Pb L_α	10.541	13 *3*
Pb L_η	11.349	0.16 *3*
Pb L_β	12.698	13 *3*
Pb L_γ	14.921	2.4 *5*
Pb $K_{\alpha2}$	72.803	19 *4*
Pb $K_{\alpha1}$	74.969	32 *6*
Pb $K_{\beta1}$'	84.789	11.3 *23*

Photons (^{208}Bi) (continued)

γ_{mode}	γ(keV)	γ(%)†
Pb $K_{\beta2}$'	87.632	3.3 *7*
γ E3	2614.35 *10*	99.79

†<0.1% uncert(syst)

Atomic Electrons (^{208}Bi)

⟨e⟩=12.0 *7* keV

e_{bin}(keV)	⟨e⟩(keV)	e(%)
13	2.7	21 *5*
15	1.7	10.9 *24*
16	0.72	4.5 *10*
56	0.14	0.24 *5*
57	0.23	0.40 *9*
58	0.012	0.021 *5*
59	0.12	0.21 *5*
60	0.22	0.36 *8*
62	0.101	0.16 *4*
68	0.067	0.098 *22*
69	0.14	0.21 *5*
70	0.066	0.095 *21*
71	0.114	0.16 *4*
72	0.099	0.14 *3*
73	0.0049	0.0067 *15*
74	0.030	0.040 *9*
75	0.0101	0.014 *3*
80	0.0079	0.0098 *22*
81	0.029	0.036 *8*
82	0.0084	0.0103 *23*
83	0.0072	0.0087 *19*
84	0.014	0.017 *4*
2526	4.37	0.173 *4*
2598	0.695	0.0267 *5*
2599	0.203	0.00782 *16*
2601	0.0288	0.001108 *22*
2611	0.224	0.00858 *17*

Continuous Radiation (^{208}Bi)

⟨IB⟩=0.49 keV

E_{bin}(keV)		⟨ ⟩(keV)	(%)
10 - 20	IB	0.0032	0.025
20 - 40	IB	0.00149	0.0046
40 - 100	IB	0.47	0.64
100 - 266	IB	0.0138	0.0139

$^{208}_{84}$Po(2.898 *2* yr)

Mode: α(99.9982 *2* %), ε(0.00181 *18* %)
Δ: -17492 *5* keV
SpA: 593.1 Ci/g
Prod: ^{209}Bi(d,3n); ^{209}Bi(p,2n)

Alpha Particles (^{208}Po)

⟨α⟩=5115.7 *20* keV

α(keV)	α(%)
4233.8 *20*	0.00024 *7*
5115.8 *20*	100

Photons (^{208}Po)

$\langle\gamma\rangle$=0.0164 *12* keV

γ_{mode}	γ(keV)	γ(%)†
Pb L_ℓ	9.185	1.3 *4* $\times10^{-8}$
Bi L_ℓ	9.419	2.9 *7* $\times10^{-5}$
Pb L_α	10.541	2.7 *8* $\times10^{-7}$
Bi L_α	10.828	0.00057 *13*
Pb L_η	11.349	3.8 *11* $\times10^{-9}$
Bi L_η	11.712	6.4 *14* $\times10^{-6}$
Pb L_β	12.700	2.6 *8* $\times10^{-7}$
Bi L_β	13.083	0.00057 *14*
Pb L_γ	14.880	4.8 *14* $\times10^{-8}$
Bi L_γ	15.463	0.00012 *3*
γ_ϵ[M1]	31.45 *13*	1.38 *20* $\times10^{-5}$
γ_ϵ[M1]	63.18 *13*	~0.00014
Pb $K_{\alpha2}$	72.803	4.4 *13* $\times10^{-7}$
Bi $K_{\alpha2}$	74.814	0.00052 *9*
Pb $K_{\alpha1}$	74.969	7.4 *22* $\times10^{-7}$
Bi $K_{\alpha1}$	77.107	0.00088 *15*
Pb $K_{\beta1}'$	84.789	2.6 *8* $\times10^{-7}$
Bi $K_{\beta1}'$	87.190	0.00031 *5*
Pb $K_{\beta2}'$	87.632	7.5 *22* $\times10^{-8}$
Bi $K_{\beta2}'$	90.128	9.2 *16* $\times10^{-5}$
γ_ϵ(M1)	291.3 *5*	0.00099
γ_ϵ(M1)	538.9 *4*	0.00021 *4*
γ_ϵ(M1)	570.4 *4*	0.00057 *11*
γ_ϵ(M1)	602.1 *4*	0.00044 *9*
γ_ϵ[E2]	861.7 *5*	0.00031 *6*
γ_αE2	899.22 *7*	0.00024

† uncert(syst): 9.9% for ϵ; 29% for α

Atomic Electrons (^{208}Po)

$\langle e\rangle$=0.00260 *23* keV

e_{bin}(keV)	$\langle e\rangle$(keV)	e(%)
13	7.6 $\times10^{-5}$	0.00057 *11*
15	8.6 $\times10^{-5}$	0.00057 *9*
16	0.00012	0.00074 *18*
18 - 31	5.7 $\times10^{-5}$	0.000204 *21*
47	0.00038	~0.0008
50 - 58	1.3 $\times10^{-5}$	2.4 *5* $\times10^{-5}$
59	0.00011	~0.00019
60 - 87	6.9 $\times10^{-5}$	0.00011 *3*
201	0.00086	0.00043 *5*
275	0.000184	6.7 *7* $\times10^{-5}$
276 - 291	8.5 $\times10^{-5}$	2.99 *19* $\times10^{-5}$
448	7.8 $\times10^{-5}$	1.7 *4* $\times10^{-5}$
480	0.00020	4.1 *8* $\times10^{-5}$
512	0.00014	2.8 *6* $\times10^{-5}$
523 - 570	7.1 $\times10^{-5}$	1.30 *15* $\times10^{-5}$
586 - 602	3.6 $\times10^{-5}$	6.1 *10* $\times10^{-6}$
788	3.0 $\times10^{-5}$	3.8 *7* $\times10^{-6}$
845 - 886	8.3 $\times10^{-6}$	9.6 *10* $\times10^{-7}$
895 - 899	9.1 $\times10^{-7}$	1.02 *18* $\times10^{-7}$

Continuous Radiation (^{208}Po)

$\langle IB\rangle$=1.04 $\times10^{-5}$ keV

E_{bin}(keV)		$\langle\ \rangle$(keV)	(%)
10 - 20	IB	4.9 $\times10^{-8}$	3.8 $\times10^{-7}$
20 - 40	IB	2.3 $\times10^{-8}$	7.2 $\times10^{-8}$
40 - 100	IB	9.7 $\times10^{-6}$	1.28 $\times10^{-5}$
100 - 300	IB	6.5 $\times10^{-7}$	4.6 $\times10^{-7}$
300 - 478	IB	3.9 $\times10^{-8}$	1.29 $\times10^{-8}$

$^{208}_{85}$At(1.63 *3* h)

Mode: ϵ(99.45 *7* %), α(0.55 *7* %)

Δ: -12560 *40* keV

SpA: 9.24$\times10^{6}$ Ci/g

Prod: ^{209}Bi(α,5n); daughter ^{212}Fr

Alpha Particles (^{208}At)

$\langle\alpha\rangle$=31.05 keV

α(keV)	α(%)
5507 *10*	0.0011
5586 *3*	0.005
5626 *10*	0.012
5641 *3*	0.5

Photons (^{208}At)

$\langle\gamma\rangle$=2700 *160* keV

γ_{mode}	γ(keV)	γ(%)†
Po L_ℓ	9.658	0.92 *21*
Po L_α	11.119	18 *4*
Po L_η	12.085	0.28 *6*
Po L_β	13.497	18 *4*
Po L_γ	15.875	3.7 *8*
Po $K_{\alpha2}$	76.858	23 *5*
Po $K_{\alpha1}$	79.290	38 *8*
Po $K_{\beta1}'$	89.639	13 *3*
Po $K_{\beta2}'$	92.673	4.1 *9*
γ	148.0 *8*	<0.49
γE2	177.0 *6*	46 *9*
γM1+E2	206.0 *8*	5.3 *11*
γ	252.1 *21*	<0.49
γ	332.1 *21*	<0.49
γM1+E2	517.0 *7*	7.0 *14*
γE2	631.1 *8*	4.3 *9*
γE2	660.1 *10*	90 *18*
γE2	685.2 *10*	97.89
γM1	808.1 *8*	8.4 *17*
γ(E1)	845.0 *7*	21 *4*
γ	887.1 *21*	2.1 *4*
γ(E2)	896.1 *10*	6.0 *12*
γ	986.1 *21*	~9
γ(M1)	993.1 *7*	~14
γ	1013.1 *21*	2.9 *6*
γM1	1028.1 *10*	27 *6*
γ	1113.1 *21*	
γ	1184.1 *21*	1.4 *3*
γ	1199.1 *8*	1.27 *25*
γ	1231.1 *21*	3.3 *7*
γ	1280.1 *8*	3.8 *8*
γ	1362.1 *7*	0.98 *20*
γ	1439.1 *21*	1.17 *23*
γ	1457.1 *8*	1.27 *25*
γ	1539.1 *7*	1.6 *3*
γ	1581.1 *21*	0.88 *18*
γ	1801.1 *21*	0.78 *16*
γ	1869.1 *21*	0.49 *10*
γ	2028.1 *21*	1.5 *3*
γ	2199.1 *21*	0.49 *10*
γ	2486.1 *21*	0.88 *18*
γ	2636.1 *21*	2.3 *5*

†0.12% uncert(syst)

Atomic Electrons (^{208}At)

$\langle e\rangle$=104 *6* keV

e_{bin}(keV)	$\langle e\rangle$(keV)	e(%)
5 - 9	2.11	34 *4*
14	3.8	28 *6*
16	2.8	17 *4*
17 - 65	1.9	5.8 *10*
72 - 79	0.66	0.89 *9*
84	8.4	10.0 *20*
85 - 89	0.083	0.096 *11*
113	4.1	~4
131 - 160	3.0	1.9 *4*
161	16.4	10.2 *20*
163	10.2	6.2 *13*
173	5.3	3.0 *6*
174	3.0	1.7 *4*
176 - 206	5.5	3.0 *7*
235 - 252	0.21	~0.09
315 - 332	0.05	~0.015
424	1.8	~0.4
500 - 538	0.9	0.17 *5*
567	6.6	1.16 *23*
592	6.92	1.169 *23*
614 - 660	2.8	0.43 *4*
668 - 715	4.6	0.66 *5*
752 - 794	1.07	0.14 *3*
803 - 845	0.57	0.070 *9*
870 - 900	3.3	0.37 *15*
920	0.27	~0.029
935	4.4	0.47 *9*
969 - 1014	1.6	0.16 *3*
1024 - 1028	0.24	0.023 *4*
1091 - 1138	0.43	0.038 *19*
1167 - 1215	0.34	~0.029
1217 - 1266	0.06	~0.005
1269 - 1280	0.07	~0.006
1345 - 1364	0.15	0.011 *5*
1422 - 1457	0.11	~0.008
1488 - 1578	0.07	0.0047 *23*
1708 - 1798	0.06	0.0034 *15*
1852 - 1935	0.06	~0.0029
2011 - 2106	0.027	0.0013 *6*
2182 - 2196	0.0034	~0.00015
2393 - 2483	0.029	~0.0012
2543 - 2633	0.07	0.0028 *14*

Continuous Radiation (^{208}At)

$\langle\beta+\rangle$=39.4 keV;$\langle IB\rangle$=7.6 keV

E_{bin}(keV)		$\langle\ \rangle$(keV)	(%)
0 - 10	β+	5.7 $\times10^{-9}$	6.7 $\times10^{-8}$
	IB	0.00144	
10 - 20	β+	4.19 $\times10^{-7}$	2.47 $\times10^{-6}$
	IB	0.0034	0.024
20 - 40	β+	2.12 $\times10^{-5}$	6.3 $\times10^{-5}$
	IB	0.0037	0.0123
40 - 100	β+	0.00205	0.00256
	IB	0.53	0.62
100 - 300	β+	0.183	0.080
	IB	0.099	0.059
300 - 600	β+	1.71	0.366
	IB	0.26	0.058
600 - 1300	β+	13.4	1.39
	IB	1.7	0.18
1300 - 2500	β+	22.2	1.28
	IB	4.0	0.22
2500 - 4227	β+	1.99	0.074
	IB	1.01	0.035
	$\Sigma\beta$+		3.2

$^{208}_{86}$Rn(24.35 *13* min)

Mode: α(60 *7* %), ϵ(40 *7* %)
Δ: -9680 *100* keV syst
SpA: 3.712×10^7 Ci/g
Prod: ^{197}Au(^{14}N,3n); protons on ^{232}Th

Alpha Particles (^{208}Rn)

⟨α⟩=3684.1 *20* keV

α(keV)	α(%)
5469.7 *24*	0.00282 *24*
6140.2 *24*	59.99718 *24*

$^{208}_{87}$Fr(59.1 *3* s)

Mode: α(77 *3* %), ϵ(23 *3* %)
Δ: -2720 *70* keV
SpA: 9.12×10^8 Ci/g
Prod: ^{197}Au(^{16}O,5n); ^{203}Tl(^{12}C,7n); ^{205}Tl(^{12}C,9n)

Alpha Particles (^{208}Fr)

α(keV)
6636 *5*

$^{208}_{88}$Ra(1.4 *4* s)

Mode: α(∼ 97 %), ϵ(∼ 3 %)
Δ: 1630 *200* keV syst
SpA: 3.1×10^{10} Ci/g
Prod: ^{197}Au(^{19}F,8n); ^{206}Pb(^{12}C,10n); ^{194}Pt(^{22}Ne,8n); ^{192}Pt(^{22}Ne,6n)

Alpha Particles (^{208}Ra)

α(keV)
7133 *5*

A = 209

NDS 22, 545 (1977)

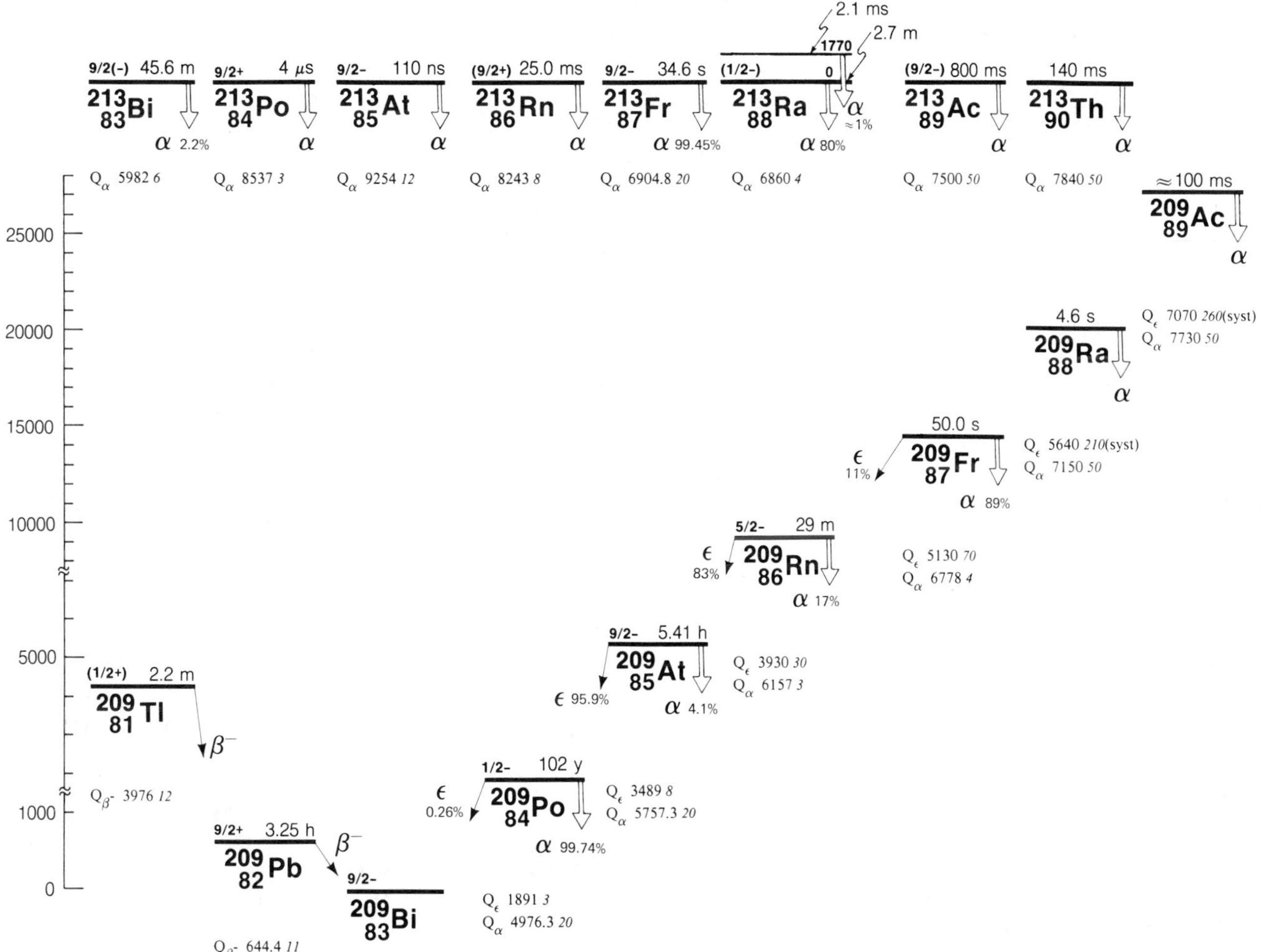

$^{209}_{81}$Tl(2.20 *7* min)

Mode: β-
Δ: -13662 *12* keV
SpA: 4.08×10^{8} Ci/g
Prod: descendant ^{233}U; descendant ^{225}Ac

Photons (^{209}Tl)

⟨γ⟩=2032 *190* keV

γ_{mode}	γ(keV)	γ(%)†
Pb L_ℓ	9.185	0.19 *3*
Pb L_α	10.541	3.7 *5*
Pb L_η	11.349	0.051 *7*
Pb L_β	12.703	3.5 *5*
Pb L_γ	14.871	0.63 *10*
Pb $K_{\alpha2}$	72.803	6.0 *8*
Pb $K_{\alpha1}$	74.969	10.1 *13*
Pb $K_{\beta1}$'	84.789	3.6 *5*
Pb $K_{\beta2}$'	87.632	1.03 *14*
γ E1	117 *1*	81 *12*
γ [E2]	467 *2*	81 *12*
γ [E2]	1566 *4*	98 *12*

† 6.0% uncert(syst)

Atomic Electrons (^{209}Tl)

⟨e⟩=28.4 *15* keV

e_{bin}(keV)	⟨e⟩(keV)	e(%)
13	0.93	7.1 *12*
15 - 16	0.65	4.3 *6*
29	5.6	19 *3*
56 - 84	0.448	0.71 *4*
101	2.2	2.2 *3*
102	0.82	0.81 *12*
104 - 117	2.14	1.94 *16*
379	7.4	2.0 *3*
451	1.30	0.29 *4*
452	1.22	0.27 *4*
454 - 467	1.49	0.32 *3*
1478	3.4	0.23 *3*
1550 - 1563	0.77	0.049 *4*

Continuous Radiation (^{209}Tl)

⟨β-⟩=659 keV;⟨IB⟩=1.13 keV

E_{bin}(keV)		⟨ ⟩(keV)	(%)
0 - 10	β-	0.0297	0.59
	IB	0.027	
10 - 20	β-	0.090	0.60
	IB	0.027	0.19
20 - 40	β-	0.369	1.23
	IB	0.052	0.18
40 - 100	β-	2.77	3.93
	IB	0.141	0.22
100 - 300	β-	31.0	15.3
	IB	0.35	0.20
300 - 600	β-	117	26.0
	IB	0.31	0.074
600 - 1300	β-	412	45.8
	IB	0.21	0.027
1300 - 1824	β-	95	6.6
	IB	0.0069	0.00050

$^{209}_{82}$Pb(3.253 *14* h)

Mode: β-
Δ: -17638 *4* keV
SpA: 4.609×10^{6} Ci/g
Prod: ^{208}Pb(d,p); descendant ^{233}U; descendant ^{225}Ac; ^{208}Pb(n,γ)

Continuous Radiation (^{209}Pb)

⟨β-⟩=198 keV;⟨IB⟩=0.131 keV

E_{bin}(keV)		⟨ ⟩(keV)	(%)
0 - 10	β-	0.152	3.03
	IB	0.0100	
10 - 20	β-	0.452	3.01
	IB	0.0092	0.064
20 - 40	β-	1.80	6.0
	IB	0.0165	0.058
40 - 100	β-	12.2	17.5
	IB	0.037	0.059
100 - 300	β-	90	47.0
	IB	0.051	0.032
300 - 600	β-	92	23.5
	IB	0.0076	0.0021
600 - 645	β-	0.440	0.072
	IB	1.02×10^{-6}	1.69×10^{-7}

$^{209}_{83}$Bi(>1.0×10^{19} yr)

Δ: -18282 *4* keV
%: 100

$^{209}_{84}$Po(102 *5* yr)

Mode: α(99.74 *3* %), ϵ(0.26 *3* %)
Δ: -16391 *5* keV
SpA: 16.8 Ci/g
Prod: ^{209}Bi(d,2n); ^{209}Bi(p,n)

Alpha Particles (^{209}Po)

⟨α⟩=4866 *3* keV

α(keV)	α(%)
4135.3 *21* ?	0.00056 *4*
4316.9 *21*	0.00015 *4*
4624.3 *21*	0.565 *14*
4879.8 *21*	99.171 *20*
4882.1 *21*	

Photons (^{209}Po)

⟨γ⟩=3.1 *3* keV

γ_{mode}	γ(keV)	γ(%)†
Pb L_ℓ	9.185	0.0011 *3*
Bi L_ℓ	9.419	0.0019 *3*
Pb L_α	10.541	0.021 *6*
Bi L_α	10.828	0.037 *5*
Pb L_η	11.349	0.00027 *8*
Bi L_η	11.712	0.00046 *6*
Pb L_β	12.703	0.019 *6*
Bi L_β	13.098	0.034 *5*
Pb L_γ	14.886	0.0036 *10*
Bi L_γ	15.400	0.0065 *10*
Pb $K_{\alpha2}$	72.803	0.035 *10*
Bi $K_{\alpha2}$	74.814	0.053 *6*
Pb $K_{\alpha1}$	74.969	0.059 *17*
Bi $K_{\alpha1}$	77.107	0.088 *11*
Pb $K_{\beta1}$'	84.789	0.021 *6*
Bi $K_{\beta1}$'	87.190	0.031 *4*
Pb $K_{\beta2}$'	87.632	0.0060 *17*
Bi $K_{\beta2}$'	90.128	0.0093 *11*
γ_αM1	260.49 *3*	0.17 *6*
γ_αM1	262.83 *3*	0.058 *20*
γ_ϵM1+33%E2	896.4 *2*	0.25

† uncert(syst): <0.1% for α, 12% for ϵ

Atomic Electrons (^{209}Po)

⟨e⟩=0.36 *6* keV

e_{bin}(keV)	⟨e⟩(keV)	e(%)
13	0.013	0.097 *21*
15 - 64	0.0136	0.069 *9*
68 - 87	0.00278	0.0038 *3*
172	0.16	0.09 *3*
175	0.054	0.031 *11*
245	0.039	0.016 *6*
247	0.012	0.0049 *17*
248 - 263	0.018	0.0071 *15*
806	0.035	0.0044 *6*
880 - 896	0.0088	0.00099 *9*

Continuous Radiation (^{209}Po)

⟨IB⟩=0.0018 keV

E_{bin}(keV)		⟨ ⟩(keV)	(%)
10 - 20	IB	6.1×10^{-6}	4.7×10^{-5}
20 - 40	IB	3.1×10^{-6}	9.6×10^{-6}
40 - 100	IB	0.00143	0.0019
100 - 300	IB	0.000142	9.1×10^{-5}
300 - 600	IB	0.000142	3.3×10^{-5}
600 - 995	IB	6.5×10^{-5}	9.4×10^{-6}

$^{209}_{85}$At(5.41 *5* h)

Mode: ϵ(95.9 *5* %), α(4.1 *5* %)
Δ: -12902 *8* keV
SpA: 2.77×10^{6} Ci/g
Prod: ^{209}Bi(α,4n)

Alpha Particles (^{209}At)

⟨α⟩=231.74 keV

α(keV)	α(%)
5116 *2* ?	~0.0041
5647 *2*	4.1

Photons (^{209}At)

⟨γ⟩=2277 *24* keV

γ_{mode}	γ(keV)	γ(%)†
Po L_ℓ	9.658	1.35 *20*
Po L_α	11.119	26 *3*
Po L_η	12.085	0.39 *5*
Po L_β	13.499	26 *3*
Po L_γ	15.880	5.1 *7*
Po $K_{\alpha2}$	76.858	34 *4*
Po $K_{\alpha1}$	79.290	56 *7*
Po $K_{\beta1}$'	89.639	19.8 *23*
γ_ϵE2	90.82 *6*	1.84 *20*
Po $K_{\beta2}$'	92.673	6.1 *7*
γ_ϵM1+38%E2	104.22 *7*	2.4 *4*
γ_ϵE2	112.95 *7*	0.118 *18*
γ_ϵ[E2]	151.52 *7*	0.082 *11*
γ_ϵ	191.05 *20*	0.41 *5*
γ_ϵM1+16%E2	195.05 *6*	22.6 *10*
γ_ϵM1+12%E2	233.62 *7*	0.96 *6*
γ_ϵE1	239.16 *6*	12.4 *5*
γ_ϵE1	321.08 *7*	0.63 *3*
γ_ϵM1+14%E2	388.87 *6*	0.49 *3*
γ_ϵ	415.99 *24*	0.055 *18*
γ_ϵE2	545.03 *7*	91.0
γ_ϵ[E2]	551.03 *5*	4.91 *18*
γ_ϵ(M1+10%E2)	552.48 *7*	1.55 *18*
γ_ϵ[E2]	554.61 *15*	0.59 *7*
γ_ϵE2+19%M1	596.44 *7*	0.66 *4*
γ_ϵE2	630.37 *6*	0.68 *3*
γ_ϵE2	666.24 *6*	1.87 *6*
γ_ϵE2	781.89 *6*	83.5 *22*
γ_ϵE1	790.20 *6*	63.5 *17*
γ_ϵ(M1)	815.69 *7*	0.23 *3*
γ_ϵ(M1+E2)	817.65 *20*	0.16 *3*
γ_ϵ	826.9 *3*	0.045 *9*
γ_ϵM1(+20%E2)	854.41 *15*	0.58 *4*
γ_ϵM1(+6.4%E2)	863.99 *6*	2.07 *8*
γ_ϵE1	903.15 *6*	3.65 *10*
γ_ϵ	910.8 *5*	0.07 *1*
γ_ϵ	922.25 *20*	0.070 *9*
γ_ϵ	939.64 *22*	0.045 *9*
γ_ϵE1	985.24 *5*	0.85 *12*
γ_ϵ	999.65 *20*	0.155 *9*
γ_ϵ	1008.5 *4*	0.035 *8*
γ_ϵ	1037.9 *4*	0.027 *6*
γ_ϵ	1043.5 *2*	0.109 *18*
γ_ϵ	1074.69 *9*	0.200 *18*
γ_ϵ	1092.9 *4*	0.045 *6*
γ_ϵ	1096.05 *20*	0.14 *3*
γ_ϵE2+33%M1	1103.51 *7*	5.40 *16*
γ_ϵ(M2)	1136.77 *7*	0.068 *9*
γ_ϵ(M1+E2)	1141.36 *7*	0.33 *3*
γ_ϵE2	1147.47 *7*	1.36 *9*
γ_ϵ[E1]	1148.85 *8*	0.78 *9*
γ_ϵE2	1170.76 *6*	3.09 *9*
γ_ϵE2	1175.39 *7*	1.91 *9*
γ_ϵ	1183.15 *20*	0.136 *18*
γ_ϵ	1192.81 *8*	0.155 *18*
γ_ϵ	1213.8 *11*	0.43 *4* ?
γ_ϵ(M1+E2)	1217.28 *6*	1.11 *6*
γ_ϵE1	1262.61 *6*	1.89 *6*
γ_ϵ	1289.8 *7*	0.27 *7*
γ_ϵ	1311.67 *10*	0.051 *6*
γ_ϵ	1333.5 *3*	0.14 *6*
γ_ϵ	1343.0 *3*	0.065 *6*
γ_ϵ	1357.01 *8*	0.164 *9*
γ_ϵ	1409.1 *6*	0.017 *7*
γ_ϵ	1411.2 *4*	0.053 *7*
γ_ϵ	1426.80 *12*	0.028 *6*
γ_ϵ(E2+~37%M1)	1446.05 *6*	0.54 *3*
γ_ϵ[E1]	1456.44 *7*	0.118 *9*
γ_ϵ	1479.07 *19*	0.040 *4*
γ_ϵ[E1]	1484.74 *9*	0.091 *9*
γ_ϵ	1490.86 *7*	0.273 *18*
γ_ϵ	1510.1 *10*	0.055 *18*
γ_ϵ	1533.15 *20*	0.17 *3*
γ_ϵ[E1]	1537.72 *7*	0.49 *3*
γ_ϵE1	1575.56 *7*	0.86 *6*
γ_ϵ(E1)	1581.68 *6*	1.79 *6*
γ_ϵ	1622.45 *10*	0.173 *9*
γ_ϵ[E1]	1651.48 *7*	0.041 *4*
γ_ϵ(E2)	1687.35 *10*	0.373 *18*
γ_ϵ	1729.84 *23*	0.0118 *18*
γ_ϵ	1745.87 *9*	0.084 *5*

Photons (^{209}At) (continued)

γ_{mode}	γ(keV)	γ(%)†
γ_ϵM2	1767.13 *5*	0.51 *4*
γ_ϵ	1786.55 *10*	0.118 *9*
γ_ϵ	1804.15 *10*	0.056 *7*
γ_ϵ	1810.05 *20*	0.043 *5*
γ_ϵ	1861.5 *5*	0.0073 *18*
γ_ϵ	1947.8 *4*	0.0136 *18*
γ_ϵ	2105.1 *10*	0.036 *9*
γ_ϵ	2109.43 *19*	0.041 *3*
γ_ϵ	2204.1 *10*	0.036 *9*
γ_ϵ	2245.9 *6*	0.0064 *9*
γ_ϵ	2292.4 *5*	0.015 *4*
γ_ϵ[E3]	2319.60 *8*	0.0073 *18*
γ_ϵ	2343.0 *4*	0.019 *5*
γ_ϵ[E3]	2357.45 *8*	0.0055 *18*
γ_ϵ	2363.56 *8*	0.0136 *18*
γ_ϵ	2368.4 *4*	0.0109 *18*
γ_ϵ[E3]	2433.37 *7*	0.0136 *18*
γ_ϵ	2448.1 *10*	~0.018
γ_ϵ	2527.76 *10*	0.0027 *9*
γ_ϵ	2555.5 *4*	~0.0018
γ_ϵ	2589.0 *4*	0.019 *3*
γ_ϵ	2645.7 *3*	0.012 *3*
γ_ϵ	2654.46 *19*	0.0027 *9*

† 0.53% uncert(syst)

Atomic Electrons (^{209}At)

⟨e⟩=113 *3* keV

e_{bin}(keV)	⟨e⟩(keV)	e(%)
11	1.5	14 *4*
14	5.6	41 *6*
16	3.8	24 *4*
17 - 65	2.4	8.1 *9*
72 - 74	0.75	1.03 *8*
75	6.5	8.7 *10*
76	0.163	0.21 *3*
77	5.3	6.8 *8*
78 - 86	0.139	0.170 *13*
87	3.9	4.5 *9*
88	3.8	4.3 *13*
89 - 101	5.1	5.4 *11*
102	29.5	29.0 *20*
103 - 152	2.72	2.08 *17*
174 - 177	0.19	~0.11
178	8.2	4.6 *3*
179 - 190	2.5	1.40 *18*
191	2.55	1.34 *12*
192 - 239	1.87	0.91 *5*
296 - 323	0.32	0.107 *12*
372 - 416	0.093	0.0248 *20*
452	8.04	1.78 *4*
458 - 503	1.16	0.250 *19*
528 - 573	4.52	0.846 *9*
580 - 628	0.052	0.0087 *8*
629 - 666	0.0570	0.00873 *18*
689	5.57	0.81 *3*
697 - 734	1.70	0.243 *8*
761 - 811	2.92	0.379 *6*
812 - 861	0.150	0.0178 *15*
863 - 911	0.062	0.0070 *14*
915 - 950	0.017	0.0019 *9*
968 - 1010	0.50	0.049 *5*
1021 - 1061	0.144	0.0137 *16*
1071 - 1120	0.393	0.0362 *15*
1121 - 1170	0.26	0.023 *5*
1171 - 1219	0.062	0.0052 *17*
1240 - 1289	0.035	0.0028 *9*
1295 - 1343	0.012	0.0009 *3*
1353 - 1398	0.057	0.0042 *8*
1405 - 1454	0.030	0.0021 *5*
1455 - 1534	0.060	0.0040 *5*
1558 - 1653	0.030	0.00186 *18*
1670 - 1770	0.122	0.0072 *5*
1773 - 1858	0.0018	10 *3* $\times10^{-5}$
1931 - 2016	0.0029	0.00014 *6*
2088 - 2187	0.0022	0.00010 *4*
2188 - 2279	0.0027	0.00012 *3*

Atomic Electrons (^{209}At) (continued)

e_{bin}(keV)	⟨e⟩(keV)	e(%)
2288 - 2365	0.0018	7.6 *18* $\times10^{-5}$
2416 - 2514	0.0010	4.0 *12* $\times10^{-5}$
2524 - 2586	0.00051	2.0 *8* $\times10^{-5}$
2629 - 2651	8.2 $\times10^{-5}$	3.1 *14* $\times10^{-6}$

Continuous Radiation (^{209}At)

⟨IB⟩=0.83 keV

E_{bin}(keV)		⟨ ⟩(keV)	(%)
10 - 20	IB	0.0023	0.017
20 - 40	IB	0.00104	0.0032
40 - 100	IB	0.48	0.63
100 - 300	IB	0.063	0.040
300 - 600	IB	0.100	0.023
600 - 1300	IB	0.153	0.019
1300 - 2159	IB	0.030	0.0020

$^{209}_{86}$Rn(28.5 *10* min)

Mode: ε(83 *2* %), α(17 *2* %)
Δ: -8970 *30* keV
SpA: 3.16×10^{7} Ci/g
Prod: daughter ^{213}Ra(2.7 m); protons on Th

Alpha Particles (^{209}Rn)

⟨α⟩=1026.5 *6* keV

α(keV)	α(%)
5660 *3*	0.0041 *3*
5887 *3*	0.037 *3*
5898 *3*	0.024 *3*
6038.7 *21*	16.935 *3*

Photons (^{209}Rn)

⟨γ⟩=1054 *13* keV

γ_{mode}	γ(keV)	γ(%)†
At L_ℓ	9.897	0.77 *15*
At L_α	11.414	14 *3*
At L_η	12.466	0.20 *4*
At L_β	13.918	14 *3*
At L_γ	16.396	2.7 *5*
At $K_{\alpha2}$	78.947	22 *4*
At $K_{\alpha1}$	81.517	36 *7*
At $K_{\beta1}$'	92.136	12.7 *25*
At $K_{\beta2}$'	95.265	3.9 *8*
γ_ϵ[M1+E2]	182.21 *5*	0.25 *6*
γ_ϵ	188.5 *3*	0.12 *3*
γ_ϵ[M1+E2]	202.4 *3*	<0.11
γ_ϵ	276.02 *25*	0.36 *8*
γ_ϵM1(+5.5%E2)	279.33 *7*	1.12 *12*
γ_ϵ[M1]	286.54 *6*	0.31 *10*
γ_ϵE2+27%M1	296.68 *9*	0.33 *4*
γ_ϵM1(+22%E2)	303.03 *5*	0.57 *14*
γ_ϵM1	337.53 *3*	14.7 *4*
γ_ϵ	357.50 *15*	0.33 *10*
γ_ϵ(M1)	380.94 *13*	0.57 *17*
γ_ϵM1(+8.5%E2)	386.39 *5*	
γ_ϵM1(+6.9%E2)	386.63 *19*	2.09 *13* •
γ_ϵE2	408.41 *3*	51.0
γ_ϵM1+18%E2	461.54 *6*	1.46 *8*
γ_ϵ(M1+E2)	526.9 *5*	0.21 *7*

Photons (^{209}Rn) (continued)

γ_{mode}	γ(keV)	γ(%)†
γ_ϵM1+48%E2	577.23 8	0.99 7
γ_ϵM1(+16%E2)	599.71 9	0.59 6
γ_ϵ(M1)	605.5 5	0.173 17
γ_ϵE2(+<4.2%M1)	672.93 4	3.32 11
γ_ϵ(E0+M1+E2)	685.02 10	1.18 14
γ_ϵM1(+8.0%E2)	689.42 4	9.8 3
γ_ϵ(M1+E2)	696.0 3	0.240 24
γ_ϵE1,E2	705.62 20	0.27 3
γ_ϵM1(+7.4%E2)	722.6 3	0.419 24
γ_ϵ	731.1 3	0.260 15
γ_ϵM1(+18%E2)	745.94 3	23.1 7
γ_ϵE0+M1	761.71 8	0.57 4
γ_ϵ(E2)	794.81 5	3.41 23
γ_ϵE1,E2	855.90 4	4.9 3
γ_ϵ	868.4 3	~0.34
γ_ϵ	872.39 6	0.71 25
γ_ϵ	948.8 10	
γ_ϵ	951.6 8	0.25 5 ‡
γ_ϵE2(+<22%M1)	986.10 8	0.55 6
γ_ϵ(E2+32%M1)	1021.6 5	0.19 3
γ_ϵ	1027.67 20	0.19 4
γ_ϵE1	1038.11 5	4.22 22
γ_ϵE1	1054.60 5	1.66 9
γ_ϵE2+<18%M1	1059.57 15	0.57 4
γ_ϵ(M1+37%E2)	1065.67 7	1.72 9
γ_ϵ	1081.34 5	~0.11
γ_ϵ	1085.1 10	0.15 6
γ_ϵ	1097.83 5	0.240 22
γ_ϵ	1110.5 3	0.13 6
γ_ϵ	1119.1 10	0.051 22
γ_ϵ	1129.41 18	0.173 23
γ_ϵ	1135.3 3	0.21 3
γ_ϵ	1158.93 6	0.85 6
γ_ϵ	1187.1 3	0.42 3
γ_ϵ	1207.79 6	0.246 23
γ_ϵ	1278.9 5	0.091 12
γ_ϵ	1291.3 7	0.047 12
γ_ϵ	1298.3 4	0.220 22
γ_ϵ	1317.44 7	<0.11
γ_ϵ	1323.1 6	0.041 20
γ_ϵ	1338.7 3	0.157 21
γ_ϵ	1341.14 6	0.50 4
γ_ϵ	1377.1 6	0.096 9
γ_ϵE2(+<41%M1)	1394.52 8	0.99 4
γ_ϵ	1415.6 3	<0.11
γ_ϵ	1429.14 21	<0.11
γ_ϵ	1471.6 3	0.145 17
γ_ϵ	1497.9 8	0.096 15
γ_ϵ	1500.3 3	0.087 11
γ_ϵ	1543.17 10	0.82 6
γ_ϵ	1592.28 23	0.191 12
γ_ϵ	1597.8 3	0.099 12
γ_ϵ	1603.1 10	0.051 13
γ_ϵ	1608.77 24	0.065 9
γ_ϵ	1615.34 9	0.063 13
γ_ϵ	1631.83 9	0.056 16
γ_ϵ	1635.1 10	0.055 16
γ_ϵ	1665.6 10	0.071 14
γ_ϵ	1669.33 8	0.120 16
γ_ϵ	1692.9 6	0.049 13
γ_ϵ	1700.3 6	0.045 12
γ_ϵ	1709.07 15	0.68 6
γ_ϵ	1722.08 14	0.142 16
γ_ϵ	1727.77 19	<0.11
γ_ϵ	1740.8 3	0.036 13
γ_ϵ	1746.32 20	0.148 12
γ_ϵ	1770.94 13	0.194 12
γ_ϵ	1774.6 3	0.099 11
γ_ϵ	1778.0 3	~0.08
γ_ϵ	1786.8 3	<0.11
γ_ϵ	1796.6 3	0.167 11
γ_ϵ	1835.7 3	<0.11
γ_ϵ	1875.1 10	0.047 19
γ_ϵ	1887.1 10	0.065 25
γ_ϵ	1912.1 7	0.062 25
γ_ϵ	1925.8 3	0.275 17
γ_ϵ	1950.1 10	0.050 11
γ_ϵ	1953.73 6	0.132 13
γ_ϵ	2043.01 20	<0.11
γ_ϵ	2074.69 25	0.095 11
γ_ϵ	2114.16 19	0.335 12
γ_ϵ	2121.3 5	0.076 15
γ_ϵ	2130.6 10	0.067 13
γ_ϵ	2133.9 10	0.057 9

Photons (^{209}Rn) (continued)

γ_{mode}	γ(keV)	γ(%)†
γ_ϵ	2145.1 10	0.061 19
γ_ϵ	2150.17 18	0.154 15
γ_ϵ	2161.0 3	0.087 11
γ_ϵ	2176.6 4	0.073 12
γ_ϵ	2195.6 4	0.111 11
γ_ϵ	2205.3 10	0.063 12
γ_ϵ	2232.8 3	<0.11
γ_ϵ	2281.69 23	0.112 11
γ_ϵ	2290.89 21	0.074 12
γ_ϵ	2307.38 21	~0.027
γ_ϵ	2317.3 4	0.085 12
γ_ϵ	2346.03 20	0.118 12
γ_ϵ	2394.89 20	0.086 10
γ_ϵ	2413.7 3	0.069 8
γ_ϵ	2426.58 25	~0.033
γ_ϵ	2446.85 17	0.067 11
γ_ϵ	2453.7 3	0.061 12
γ_ϵ	2463.34 17	0.101 12
γ_ϵ	2475.6 4	0.101 15
γ_ϵ	2485.8 4	0.097 25
γ_ϵ	2536.8 7	0.041 11
γ_ϵ	2555.8 3	0.116 9
γ_ϵ	2638.9 6	0.072 14
γ_ϵ	2642.78 21	0.32 3
γ_ϵ	2646.5 8	0.082 16
γ_ϵ	2656.2 3	0.061 13
γ_ϵ	2667.1 10	0.031 11
γ_ϵ	2695.0 6	0.046 11
γ_ϵ	2749.87 17	0.050 11
γ_ϵ	2762.6 7	0.060 10
γ_ϵ	2798.73 17	0.068 8
γ_ϵ	2824.4 10	~0.027
γ_ϵ	2832.5 3	~0.027
γ_ϵ	2881.4 3	0.046 11
γ_ϵ	2937.1 10	0.031 11
γ_ϵ	2942.1 3	0.133 21
γ_ϵ	2980.30 21	0.020 7
γ_ϵ	3008.0 3	0.056 7
γ_ϵ	3088.6 10	0.041 13
γ_ϵ	3118.1 15	~0.016
γ_ϵ	3123.6 15	~0.016
γ_ϵ	3136.26 17	0.131 13
γ_ϵ	3143.1 3	0.064 8
γ_ϵ	3169.6 15	0.020 6
γ_ϵ	3183.1 15	0.022 6
γ_ϵ	3218.9 3	0.025 6

† 2.7% uncert(syst)
* 386γ + 387γ
‡ 949γ + 952γ

Atomic Electrons (^{209}Rn)

$\langle e\rangle$=56.1 22 keV

e_{bin}(keV)	$\langle e\rangle$(keV)	e(%)
14	3.1	22 10
17	2.4	14 6
61 - 107	2.0	2.8 5
165 - 207	2.5	1.36 17
242	12.3	5.07 18
259 - 289	1.30	0.48 8
291	2.1	0.7 3
292 - 303	0.060	0.020 3
313	5.59	1.79 6
320	2.58	0.81 3
321 - 370	2.49	0.72 5
372 - 391	1.28	0.328 16
392	1.49	0.382 14
394 - 431	1.75	0.436 13
444 - 482	0.44	0.095 11
504 - 527	0.27	0.054 6
560 - 591	1.06	0.18 3
594	2.84	0.48 4
595 - 635	0.23	0.037 7
650	5.5	0.84 11
655 - 703	1.40	0.207 9

Atomic Electrons (^{209}Rn) (continued)

e_{bin}(keV)	$\langle e\rangle$(keV)	e(%)
705 - 748	1.49	0.204 17
757 - 795	0.43	0.056 17
838 - 872	0.12	0.015 4
890 - 937	0.086	0.0094 19
942 - 989	0.41	0.043 4
1002 - 1051	0.15	0.015 3
1052 - 1097	0.13	0.012 6
1102 - 1145	0.041	0.0037 17
1155 - 1204	0.041	0.0035 12
1205 - 1245	0.05	~0.0041
1261 - 1309	0.083	0.0064 6
1313 - 1360	0.019	0.0014 6
1363 - 1412	0.038	0.0027 6
1413 - 1457	0.05	~0.0032
1467 - 1540	0.043	0.0028 8
1570 - 1666	0.073	0.0045 15
1675 - 1775	0.046	0.0027 7
1779 - 1873	0.029	0.0016 5
1883 - 1979	0.010	0.00054 18
2018 - 2118	0.044	0.0021 5
2120 - 2219	0.014	0.00063 16
2222 - 2318	0.013	0.00059 15
2329 - 2423	0.016	0.00066 15
2429 - 2523	0.0074	0.00030 9
2533 - 2630	0.019	0.00075 21
2632 - 2729	0.0065	0.00024 6
2732 - 2829	0.0030	0.00011 3
2841 - 2939	0.0066	0.00023 7
2963 - 3047	0.0062	0.00020 6
3071 - 3169	0.0028	9.1 18 $\times10^{-5}$
3179 - 3215	0.00013	4.2 16 $\times10^{-6}$

Continuous Radiation (^{209}Rn)

$\langle\beta+\rangle$=39.0 keV;$\langle$IB$\rangle$=4.7 keV

E_{bin}(keV)		$\langle\ \rangle$(keV)	(%)
0 - 10	β+	1.07 $\times10^{-8}$	1.26 $\times10^{-7}$
	IB	0.00154	
10 - 20	β+	8.0 $\times10^{-7}$	4.74 $\times10^{-6}$
	IB	0.0036	0.026
20 - 40	β+	4.12 $\times10^{-5}$	0.000123
	IB	0.0037	0.0127
40 - 100	β+	0.00405	0.00504
	IB	0.52	0.66
100 - 300	β+	0.355	0.156
	IB	0.096	0.058
300 - 600	β+	3.16	0.68
	IB	0.21	0.047
600 - 1300	β+	20.5	2.17
	IB	1.20	0.126
1300 - 2500	β+	15.0	0.96
	IB	2.4	0.135
2500 - 3486	IB	0.27	0.0099
	$\Sigma\beta$+		4.0

$^{209}_{87}$Fr(50.0 *3* s)

Mode: α(89 *3* %), ϵ(11 *3* %)
Δ: -3840 *60* keV
SpA: 1.072×10^{9} Ci/g
Prod: ^{197}Au(^{16}O,4n); ^{203}Tl(^{12}C,6n) ^{205}Tl(^{12}C,8n); protons on ^{232}Th

Alpha Particles (^{209}Fr)

α(keV)
6646 *5*

$^{209}_{88}$Ra(4.6 *2* s)

Mode: α
Δ: 1790 *220* keV syst
SpA: 1.09×10^{10} Ci/g
Prod: ^{197}Au(^{19}F,7n); ^{194}Pt(^{22}Ne,7n); ^{192}Pt(^{22}Ne,5n)

Alpha Particles (^{209}Ra)

α(keV)
7008 *5*

$^{209}_{89}$Ac(100 *50* ms)

Mode: α
Δ: 8870 *180* keV
SpA: 8×10^{10} Ci/g
Prod: ^{197}Au(^{20}Ne,8n)

Alpha Particles (^{209}Ac)

α(keV)
7585 *15*

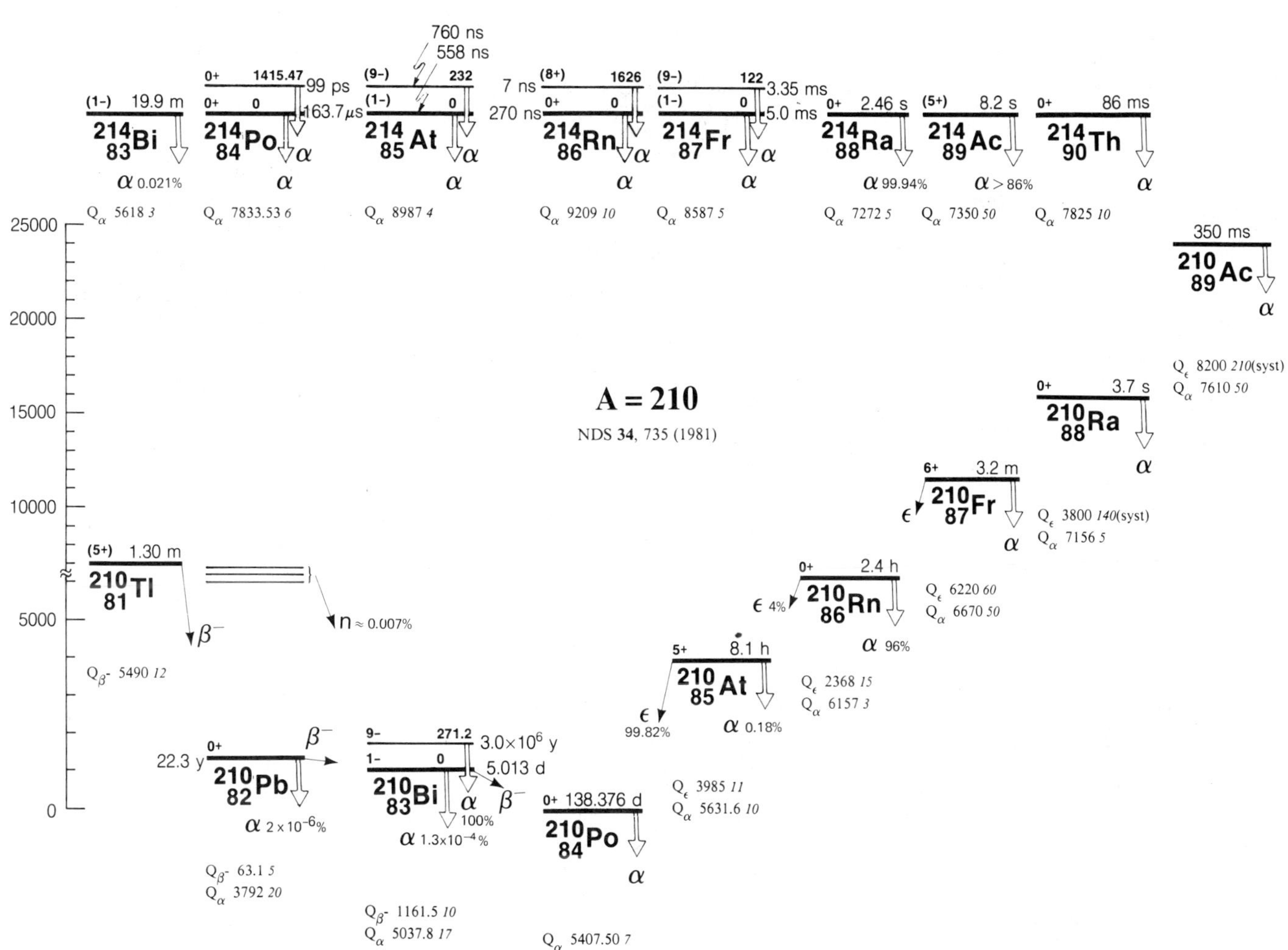

$^{210}_{81}$Tl(1.30 *3* min)

Mode: β-, β-n(~ 0.007 %)
Δ: -9263 *13* keV
SpA: 6.86×10^{8} Ci/g
Prod: descendant ^{226}Ra

Photons (^{210}Tl)

⟨γ⟩=2732 *170* keV

γ_{mode}	γ(keV)	γ(%)†
Pb L_ℓ	9.185	0.41 *10*
Pb L_α	10.541	8.3 *19*
Pb L_η	11.349	0.20 *5*
Pb L_β	12.678	11 *3*
Pb L_γ	14.836	2.2 *5*
Pb $K_{\alpha2}$	72.803	2.8 *4*
Pb $K_{\alpha1}$	74.969	4.6 *6*
γ (E2)	81 *3*	2.0 *4*
Pb $K_{\beta1}$'	84.789	1.64 *22*

Photons (^{210}Tl)
(continued)

γ_{mode}	γ(keV)	γ(%)†
Pb $K_{\beta2}$'	87.632	0.47 *7*
γ (E2)	95 *3*	~4
γ E2	298.1 *10*	79 *10*
γ (M1)	354 *10*	~4
γ	380 *10*	~3
γ	478 *20*	~2
γ	668 *20*	~2
γ E2	797.88 *10*	98.96
γ [M1+E2]	860 *16*	6.9 *20*
γ	908 *30*	~3
γ (E1)	1068 *10*	12 *5*
γ [E2]	1110 *13*	6.9 *20*

Photons (^{210}Tl)
(continued)

γ_{mode}	γ(keV)	γ(%)†
γ	1208 *17*	17 *4*
γ	1314 *12*	21 *5*
γ [M1+E2]	1408 *13*	4.9 *20*
γ	1538 *30*	~2
γ [E1]	1588 *22*	~2
γ	1648 *30*	~2
γ	2008 *30*	6.9 *20*
γ	2088 *30*	4.9 *20*
γ [M1+E2]	2268 *12*	~3
γ [M1+E2]	2358 *22*	8 *3*
γ	2428 *19*	9 *3*

† < 0.1% uncert(syst)

Atomic Electrons (^{210}Tl)
⟨e⟩=94 *8* keV

e_{bin}(keV)	⟨e⟩(keV)	e(%)
7 - 56	4.6	34 *6*
57 - 65	0.36	0.57 *8*
66	8.2	12.4 *25*
68	7.2	10.6 *22*
69 - 77	0.161	0.22 *3*
78	4.7	6.0 *12*
79	0.46	0.6 *3*
80	10.2	13 *6*
81	0.86	1.06 *22*
82	7.9	~10
83 - 91	0.14	0.15 *7*
92	5.2	~6
93 - 95	1.8	1.9 *6*
210	11.0	5.2 *7*
266	2.5	~0.9
282	2.1	0.76 *10*
283	4.5	1.59 *20*
285 - 298	6.3	2.2 *4*
338 - 380	1.0	0.30 *9*
390	0.4	~0.11
462 - 478	0.13	~0.028
580	0.3	~0.05
652 - 668	0.07	~0.011
710	5.77	0.813 *16*
772 - 820	2.9	0.37 *7*
844 - 893	0.25	0.030 *11*
895 - 908	0.018	~0.0019
1004	0.53	0.053 *13*
1052 - 1097	0.109	0.0101 *20*
1106 - 1120	1.1	~0.10
1193 - 1226	1.4	~0.11
1298 - 1320	0.57	0.043 *17*
1392 - 1407	0.07	0.0050 *22*
1450 - 1535	0.14	~0.009
1560 - 1645	0.10	~0.007
1920 - 2005	0.42	0.021 *10*
2072 - 2085	0.031	~0.0015
2180 - 2270	0.39	0.017 *6*
2340 - 2424	0.33	0.014 *6*

Continuous Radiation (^{210}Tl)
⟨β-⟩=1199 keV;⟨IB⟩=3.3 keV

E_{bin}(keV)		⟨ ⟩(keV)	(%)
0 - 10	β-	0.0170	0.338
	IB	0.040	
10 - 20	β-	0.051	0.343
	IB	0.040	0.28
20 - 40	β-	0.212	0.70
	IB	0.078	0.27
40 - 100	β-	1.60	2.27
	IB	0.22	0.34
100 - 300	β-	18.2	9.0
	IB	0.63	0.35
300 - 600	β-	73	16.0

Continuous Radiation (^{210}Tl)
(continued)

E_{bin}(keV)		⟨ ⟩(keV)	(%)
	IB	0.70	0.165
600 - 1300	β-	320	34.5
	IB	0.95	0.111
1300 - 2500	β-	479	26.5
	IB	0.59	0.035
2500 - 4389	β-	307	10.3
	IB	0.090	0.0032

$^{210}_{82}$Pb(22.3 *2* yr)

Mode: β-, α(2.2 *7* $\times 10^{-6}$ %)
Δ: -14752 *4* keV
SpA: 76.3 Ci/g
Prod: descendant ^{226}Ra

Alpha Particles (^{210}Pb)

α(keV)
3720 *20*

Photons (^{210}Pb)
⟨γ⟩=4.67 *17* keV

γ_{mode}	γ(keV)	γ(%)†
Bi L_ℓ	9.419	0.47 *5*
Bi L_α	10.828	9.2 *6*
Bi L_η	11.712	0.079 *5*
Bi L_β	13.066	10.4 *11*
Bi L_γ	15.537	2.5 *3*
γM1	46.52 *2*	4.05

† 5.0% uncert(syst)

Atomic Electrons (^{210}Pb)
⟨e⟩=27.7 *5* keV

e_{bin}(keV)	⟨e⟩(keV)	e(%)
13	0.038	0.29 *3*
16	2.05	12.6 *13*
30	15.9	52.9 *15*
31	1.62	5.26 *15*
33	0.151	0.457 *13*
43	5.88	13.8 *4*
44	0.00745	0.0170 *5*
46	2.02	4.41 *12*
47	0.0490	0.105 *3*

Continuous Radiation (^{210}Pb)
⟨β-⟩=6.5 keV;⟨IB⟩=0.00025 keV

E_{bin}(keV)		⟨ ⟩(keV)	(%)
0 - 10	β-	3.11	83
	IB	0.00018	
10 - 20	β-	1.38	10.5
	IB	5.0 $\times 10^{-5}$	0.00037
20 - 40	β-	1.56	5.6
	IB	2.2 $\times 10^{-5}$	8.8 $\times 10^{-5}$
40 - 63	β-	0.483	1.06
	IB	8.5 $\times 10^{-7}$	2.0 $\times 10^{-6}$

$^{210}_{83}$Bi(5.013 *5* d)

Mode: β-, α(0.000132 *10* %)
Δ: -14815 *4* keV
SpA: 1.2403 $\times 10^{5}$ Ci/g
Prod: ^{209}Bi(n,γ); descendant ^{226}Ra

Alpha Particles (^{210}Bi)

α(keV)	α(rel)
4648.3 *14*	60
4686.7 *14*	40

Photons (^{210}Bi)
⟨γ⟩=0.00029 *12* keV

γ_{mode}	γ(keV)	γ(%)†
γ_αE2	266.15 *17*	~4 $\times 10^{-5}$
γ_αM1	305.25 *16*	~6 $\times 10^{-5}$

† 7.6% uncert(syst)

Continuous Radiation (^{210}Bi)
⟨β-⟩=389 keV;⟨IB⟩=0.45 keV

E_{bin}(keV)		⟨ ⟩(keV)	(%)
0 - 10	β-	0.064	1.28
	IB	0.018	
10 - 20	β-	0.193	1.29
	IB	0.017	0.120
20 - 40	β-	0.78	2.61
	IB	0.033	0.114
40 - 100	β-	5.7	8.1
	IB	0.084	0.131
100 - 300	β-	57	28.3
	IB	0.18	0.104
300 - 600	β-	162	36.7
	IB	0.102	0.025
600 - 1161	β-	164	21.7
	IB	0.021	0.0029

$^{210}_{83}$Bi(3.00 20 $\times 10^6$ yr)

Mode: α

Δ: -14544 4 keV

SpA: 0.00057 Ci/g

Prod: ^{209}Bi(n,γ)

Alpha Particles (^{210}Bi)

$\langle\alpha\rangle$=4905 2 keV

α(keV)	α(%)
4107.4 20	0.002
4225.7 20	0.0008
4272.5 9	0.006 1
4420.1 6	0.29 3
4569.3 6	3.9 1
4583.6 6	1.4 1
4907.6 6	38.8 7
4946.0 6	55.5 10

Photons (^{210}Bi)

$\langle\gamma\rangle$=260 26 keV

γ_{mode}	γ(keV)	γ(%)†
Tl L_ℓ	8.953	0.123 20
Tl L_α	10.259	2.5 4
Tl L_η	10.994	0.038 6
Tl L_β	12.304	2.5 4
Tl L_γ	14.394	0.47 8
Tl $K_{\alpha2}$	70.832	3.9 5
Tl $K_{\alpha1}$	72.873	6.6 9
Tl $K_{\beta1}$'	82.434	2.3 3
Tl $K_{\beta2}$'	85.185	0.65 9
γ E2	266.15 17	50
γ M1	305.25 16	28 6
γ M1	330.30 23	0.73 15
γ M1	344.96 19	0.76 13
γ M1	369.40 23	0.63 13
γ [M1]	384.06 23	0.0055 11
γ	457.6 20	
γ M1	536.1 3	0.25 5
γ [E2]	635.55 24	0.010 2
γ M1	650.21 17	3.6 6
γ E2	686.6 6	0.0060 12
γ [M1+E2]	734.3 19	
γ	1121.1 19	

† 14% uncert(syst)

Atomic Electrons (^{210}Bi)

$\langle$e$\rangle$=47 4 keV

e_{bin}(keV)	$\langle$e$\rangle$(keV)	e(%)
13 - 60	1.29	8.6 10
66 - 82	0.123	0.177 12
181	7.7	4.3 6
220	19.7	9.0 18
245	0.47	0.19 4
251	5.2	2.1 3
253	2.0	0.78 11
259 - 262	0.83	0.32 4
263	1.55	0.59 8
264 - 284	0.95	0.35 3
290	4.0	1.4 3
291 - 332	2.04	0.67 8
341 - 384	0.127	0.036 4
451	0.083	0.018 4
521 - 565	0.90	0.16 3
601 - 650	0.21	0.034 4
671 - 687	0.000132	1.96 22 $\times 10^{-5}$

$^{210}_{84}$Po(138.376 2 d)

Mode: α

Δ: -15977 4 keV

SpA: 4493.44 Ci/g

Prod: daughter ^{210}Bi(5.013 d)

Alpha Particles (^{210}Po)

$\langle\alpha\rangle$=5304.43 12 keV

α(keV)	α(%)
4516.57 13	0.00107 2
5304.38 12	100

Photons (^{210}Po)

γ_{mode}	γ(keV)	γ(%)
γ E2	803.13 5	0.00106 2

$^{210}_{85}$At(8.1 4 h)

Mode: ϵ(99.825 20 %), α(0.175 20 %)

Δ: -11992 11 keV

SpA: 1.84$\times 10^6$ Ci/g

Prod: ^{209}Bi(α,3n)

Alpha Particles (^{210}At)

$\langle\alpha\rangle$= 9.557 4 keV

α(keV)	α(%)
5131 2	0.00066 23
5177.8 6	0.00037 10
5242 3	0.00158 18
5360.9 8	0.049 4
5386.2 8	0.0081 5
5442.5 6	0.050 3
5454.3 6	0.0007 1
5465.0 6	0.0126 5
5523.8 6	0.0534 16

Photons (^{210}At)

$\langle\gamma\rangle$=2959 48 keV

γ_{mode}	γ(keV)	γ(%)†
Bi L_ℓ	9.419	0.00065 12
Po L_ℓ	9.658	1.13 15
Bi L_α	10.828	0.0128 22
Po L_α	11.119	21.7 23
Bi L_η	11.712	0.000146 25
Po L_η	12.085	0.35 4
Bi L_β	13.084	0.0129 24
Po L_β	13.497	23 3
Bi L_γ	15.458	0.0027 5
Po L_γ	15.875	4.5 6
γ_ϵE2	46.56 4	0.13 3
γ_αE2	59.911 16	
Bi $K_{\alpha2}$	74.814	0.0122 19
Po $K_{\alpha2}$	76.858	23.9 18
Bi $K_{\alpha1}$	77.107	0.021 3
γ_ϵM1	77.21 12	0.027 13
Po $K_{\alpha1}$	79.290	40 3
γ_αM1+<0.2%E2	82.833 19	0.0123 25
γ_ϵE2	83.60 8	0.0308 20

Photons (^{210}At) (continued)

γ_{mode}	γ(keV)	γ(%)†
Bi $K_{\beta1}$'	87.190	0.0073 11
Po $K_{\beta1}$'	89.639	14.1 10
Bi $K_{\beta2}$'	90.128	0.0022 3
γ_ϵ(E2)	92.26 17	0.0011 3
Po $K_{\beta2}$'	92.673	4.3 3
γ_α(M1)	106.1 7	0.0044 9
γ_ϵ(M1)	112.19 14	~0.030
γ_ϵM1	116.22 6	0.65 6
γ_αM1	140.2 8	0.0016 3
γ_α(M1)	166.0 7	0.0028 6
γ_ϵM1	201.94 12	0.149 20
γ_ϵE2	245.35 8	79 4
γ_ϵM1	250.46 15	0.21 4
γ_αM1+<2.5%E2	281.957 15	
γ_αM1	292.809 17	
γ_ϵM1	298.90 10	0.109 20
γ_ϵM1	316.86 15	0.169 10
γ_ϵ	334.35 20	0.050 10
γ_ϵM1	402.12 9	0.775 20
γ_ϵM1	499.08 11	0.149 10
γ_ϵE1	506.85 9	0.685 20
γ_ϵM1	518.34 9	0.149 10
γ_ϵE1	527.58 7	1.14 4
γ_ϵE1	584.06 11	0.338 20
γ_ϵM1	602.54 15	0.119 20
γ_ϵM1	615.30 11	0.358 20
γ_ϵE1	623.07 9	0.427 20
γ_ϵM1	630.86 15	0.308 20
γ_ϵE2	639.48 11	0.258 20
γ_ϵE1	643.80 8	0.457 20
γ_ϵM1	701.01 9	0.467 20
γ_ϵ[E1]	721.54 16	0.10 4 ?
γ_ϵM1	724.74 15	0.21 3
γ_ϵ[E1]	798.75 17	0.060 20 ?
γ_ϵM1	817.23 9	1.71 5
γ_ϵM1	852.73 11	1.38 5
γ_ϵ(M1+E2)	869.48 13	0.129 20
γ_ϵM1(+E2)	881.32 11	0.219 20
γ_ϵ[E2]	909.21 7	0.09 3
γ_ϵM1	929.94 9	0.75 3
γ_ϵM1	955.77 7	1.80 6
γ_ϵ[E1]	960.08 11	<0.040 ?
γ_ϵ[M1+E2]	964.92 10	0.16 4
γ_ϵM1	976.50 9	0.80 4
γ_ϵ[E2]	1041.60 11	0.30 4
γ_ϵ[E1]	1045.91 10	0.16 3
γ_ϵ[E1]	1087.16 14	0.22 3
γ_ϵE2	1181.43 9	99.3 25
γ_ϵ(E2)	1201.12 9	0.159 20
γ_ϵ(E1)	1205.43 12	0.79 3
γ_ϵ[E1]	1289.10 12	0.516 20
γ_ϵ[E1]	1324.08 10	0.467 20
γ_ϵE1	1436.78 6	29.0 13
γ_ϵE1	1483.35 5	46.5 20
γ_ϵ[E1]	1543.6 3	0.030 10 ?
γ_ϵ[E1]	1553.00 7	0.169 10
γ_ϵE1	1599.56 6	13.4 6
γ_ϵ	1648.45 20	0.072 8 ?
γ_ϵ[E1]	1684.82 15	0.026 4
γ_ϵ[E1]	1955.12 9	0.407 20
γ_ϵ[E1]	2001.68 8	0.109 10
γ_ϵ[E1]	2052.08 11	0.071 3
γ_ϵ	2226.12 21	0.046 3
γ_ϵ	2247.03 11	0.026 4 ?
γ_ϵ[E1]	2254.01 9	1.52 5
γ_ϵ	2266.9 3	0.029 5
γ_ϵ	2272.68 21	0.348 10
γ_ϵ[E2]	2290.28 19	0.012 3 ?
γ_ϵ	2306.26 13	0.0368 20
γ_ϵ	2352.82 13	0.139 10
γ_ϵ[E3]	2386.85 14	0.0079 20

† uncert(syst): 11% for α, 2.5% for ϵ

Atomic Electrons (^{210}At)

⟨e⟩=74.9 19 keV

e_{bin}(keV)	⟨e⟩(keV)	e(%)
13	0.0018	0.0133 25
14	4.7	34 5
16	3.5	21 3
17 - 23	1.75	8.8 7
30	4.1	13 3
33	4.2	13 3
42	0.039	0.091 18
43	2.9	6.8 14
44	0.038	0.087 18
46	1.00	2.2 4
47 - 96	2.36	3.50 14
98 - 140	1.19	1.14 7
150	0.0015	0.0010 2
152	12.8	8.4 5
153 - 202	0.32	0.19 3
216	0.241	0.112 10
228	2.91	1.28 7
229	9.9	4.33 23
232	5.2	2.26 12
234 - 241	0.83	0.343 25
242	4.22	1.75 9
243 - 285	1.70	0.69 3
295 - 334	0.533	0.173 5
385 - 434	0.327	0.0807 18
482 - 530	0.234	0.0457 19
538 - 586	0.142	0.0262 13
589 - 637	0.259	0.0422 17
638 - 687	0.0259	0.00380 16
697 - 725	0.403	0.0558 19
760 - 803	0.404	0.053 3
813 - 856	0.240	0.0288 8
863 - 909	0.484	0.0557 21
913 - 962	0.158	0.0168 4
963 - 994	0.0127	0.00130 7
1025 - 1073	0.0057	0.00056 5
1083 - 1087	0.000212	1.95 22 $\times10^{-5}$
1088	4.69	0.431 14
1108 - 1112	0.0220	0.00198 11
1164 - 1205	1.255	0.1075 21
1231 - 1275	0.0095	0.00077 4
1285 - 1324	0.00213	0.000163 6
1372	1.22	0.089 3
1420 - 1467	0.218	0.0150 4
1470 - 1555	0.250	0.0166 7
1583 - 1682	0.0417	0.00263 10
1862 - 1959	0.0090	0.000479 21
1985 - 2049	0.000469	2.34 12 $\times10^{-5}$
2133 - 2231	0.035	0.0016 3
2233 - 2303	0.011	0.00049 12
2336 - 2384	0.0010	4.2 18 $\times10^{-5}$

Continuous Radiation (^{210}At)

⟨IB⟩=0.67 keV

E_{bin}(keV)		⟨ ⟩(keV)	(%)
10 - 20	IB	0.0024	0.018
20 - 40	IB	0.00105	0.0033
40 - 100	IB	0.47	0.62
100 - 300	IB	0.061	0.038
300 - 600	IB	0.085	0.019
600 - 1300	IB	0.053	0.0073
1300 - 1660	IB	0.00039	2.9 $\times10^{-5}$

$^{210}_{86}$Rn(2.4 *1* h)

Mode: α(96 *1* %), ε(4 *1* %)

Δ: -9623 *11* keV

SpA: 6.2×10^6 Ci/g

Prod: protons on ^{232}Th

Alpha Particles (^{210}Rn)

⟨α⟩=5797.9 26 keV

α(keV)	α(%)
5352.1 17	0.0054 3
6039.5 17	95.9942 10

Photons (^{210}Rn)

⟨γ⟩=59.1 22 keV

γ_{mode}	γ(keV)	γ(%)†
At L_ℓ	9.897	0.078 13
At L_α	11.414	1.47 22
At L_η	12.466	0.017 3
At L_β	13.903	1.48 24
At L_γ	16.471	0.32 6
γ_ϵM1	72.65 8	0.62 5
At $K_{\alpha2}$	78.947	1.37 22
At $K_{\alpha1}$	81.517	2.3 4
At $K_{\beta1}'$	92.136	0.81 13
At $K_{\beta2}'$	95.265	0.25 4
γ_ϵ	147.19 20	0.0151 16
γ_ϵM1	190.44 5	0.139 7
γ_ϵM1	196.25 7	0.325 20
γ_ϵM1	225.13 7	0.049 5
γ_ϵM1	233.24 6	0.500 15
γ_ϵM1	238.11 6	0.144 8
γ_ϵ[M1+E2]	239.55 7	~0.016 ?
γ_ϵ	248.0 10	~0.008 ?
γ_ϵM1	255.61 8	0.090 7
γ_ϵM1	283.82 6	0.049 5
γ_ϵM1	307.44 8	0.074 5
γ_ϵM1	314.09 10	0.353 16
γ_ϵ[M1+E2]	331.49 7	0.0164 25
γ_ϵM1	360.07 7	0.046 3
γ_ϵM1	396.54 7	0.044 3
γ_ϵM1	423.55 6	0.148 5
γ_ϵM1	437.89 10	0.033 3
γ_ϵM1+28%E2	458.24 6	1.6
γ_ϵM1	472.79 6	0.121 5
γ_ϵ	484.5 10	~0.008 ?
γ_ϵ(M1+E2)	488.72 8	0.0361 16
γ_ϵM1	496.20 8	0.141 5
γ_ϵM1	521.93 7	0.118 8
γ_ϵ(M1+E2)	540.55 9	0.041 4
γ_ϵM1	571.04 6	0.823 25
γ_ϵ	591.89 10	0.049 5
γ_ϵ(M1+E2)	598.18 7	0.066 5
γ_ϵM1	648.68 6	0.820 25
γ_ϵ	673.0 10	~0.033 ?
γ_ϵ	689.0 10	~0.016 ?
γ_ϵE2	696.35 6	0.284 10
γ_ϵM1+E2	721.33 8	0.066 3
γ_ϵM1	756.61 6	0.162 7
γ_ϵM1	761.48 6	0.525 16
γ_ϵE2	767.29 7	0.323 10
γ_ϵ[E2]	796.17 7	~0.013 ?
γ_ϵE2	804.28 6	0.139 13
γ_ϵ	828.0 10	~0.006 ?
γ_ϵ	837.98 10	~0.015 ?
γ_ϵ(E2)	879.59 10	0.048 5
γ_ϵ[M1+E2]	912.28 9	0.061 8
γ_ϵ(E1)	914.23 8	0.289 13
γ_ϵE2	957.73 7	0.279 15
γ_ϵM1(+E2)	964.11 8	0.133 11
γ_ϵE2	980.17 6	0.271 13
γ_ϵE2	994.71 5	0.287 15
γ_ϵ	1164.5 10	~0.010 ?
γ_ϵE1	1198.05 8	0.225 13
γ_ϵM1	1202.49 10	0.044 5
γ_ϵ	1743.0 10	

† 28% uncert(syst)

Atomic Electrons (^{210}Rn)

⟨e⟩=8.5 3 keV

e_{bin}(keV)	⟨e⟩(keV)	e(%)
14	0.20	1.4 3
17	0.27	1.6 3
51	0.014	~0.027
55	1.45	2.63 22
56 - 67	0.219	0.380 24
68	0.42	0.62 5
69 - 92	0.262	0.364 20
95	0.221	0.233 12
101	0.50	0.50 3
129 - 133	0.081	0.063 10
138	0.654	0.475 17
142 - 190	0.619	0.382 15
192 - 216	0.323	0.153 4
218	0.323	0.148 8
219 - 267	0.239	0.101 4
270 - 317	0.161	0.0539 17
327 - 360	0.121	0.0365 12
363	0.69	0.19 6
377 - 426	0.239	0.0597 24
434 - 473	0.24	0.055 11
475	0.331	0.0697 25
479 - 526	0.062	0.013 3
536 - 541	0.0008	0.00015 7
553	0.275	0.0496 18
554 - 601	0.124	0.0219 12
626 - 675	0.290	0.0442 15
679 - 721	0.0215	0.00308 23
732 - 779	0.0569	0.0076 3
782 - 828	0.021	0.0026 6
834 - 880	0.033	0.0039 10
884 - 914	0.0349	0.00391 16
940 - 981	0.0174	0.00181 18
990 - 995	0.00119	0.000120 4
1069 - 1107	0.0111	0.00101 10
1147 - 1195	0.00216	0.000183 15
1197 - 1202	0.00037	3.12 23 $\times10^{-5}$

Continuous Radiation (^{210}Rn)

⟨IB⟩=0.026 keV

E_{bin}(keV)		⟨ ⟩(keV)	(%)
10 - 20	IB	9.9×10^{-5}	0.00072
20 - 40	IB	3.8×10^{-5}	0.000119
40 - 100	IB	0.021	0.027
100 - 300	IB	0.0024	0.00159
300 - 600	IB	0.0020	0.00048
600 - 880	IB	0.00028	4.2×10^{-5}

$^{210}_{87}$Fr(3.18 *6* min)

Mode: α, ε

Δ: -3410 *50* keV

SpA: 2.81×10^8 Ci/g

Prod: ^{203}Tl(^{12}C,5n); ^{205}Tl(^{12}C,7n); ^{197}Au(^{16}O,3n)

Alpha Particles (^{210}Fr)

α(keV)
6543 5

Photons (^{210}Fr)

γ_{mode}	γ(keV)	γ(rel)
γ_ϵ(E2)	119.8 *10*	
γ_ϵE2	203.0 *8*	35 *2*
γ_ϵ	256.2 *10*	11 *2*
γ_ϵ	425.2 *10*	10 *3*
γ_ϵ	461 *1*	11 *3*
γ_ϵE2	643.8 *8*	100
γ_ϵ	733 *1*	10 *3*
γ_ϵE2	817.5 *8*	60 *6*
γ_ϵ(E2)	900.8 *7*	30 *3*

$^{210}_{88}$Ra(3.7 *2* s)

Mode: α
Δ: 400 *130* keV syst
SpA: 1.32×10^{10} Ci/g
Prod: ^{197}Au(^{19}F,6n); ^{206}Pb(^{12}C,8n); ^{194}Pt(^{22}Ne,6n); ^{192}Pt(^{22}Ne,4n)

Alpha Particles (^{210}Ra)

α(keV)
7020 *5*

$^{210}_{89}$Ac(350 *50* ms)

Mode: α
Δ: 8590 *170* keV
SpA: 6.68×10^{10} Ci/g
Prod: ^{197}Au(^{20}Ne,7n); ^{203}Tl(^{16}O,9n)

Alpha Particles (^{210}Ac)

α(keV)
7462 *8*

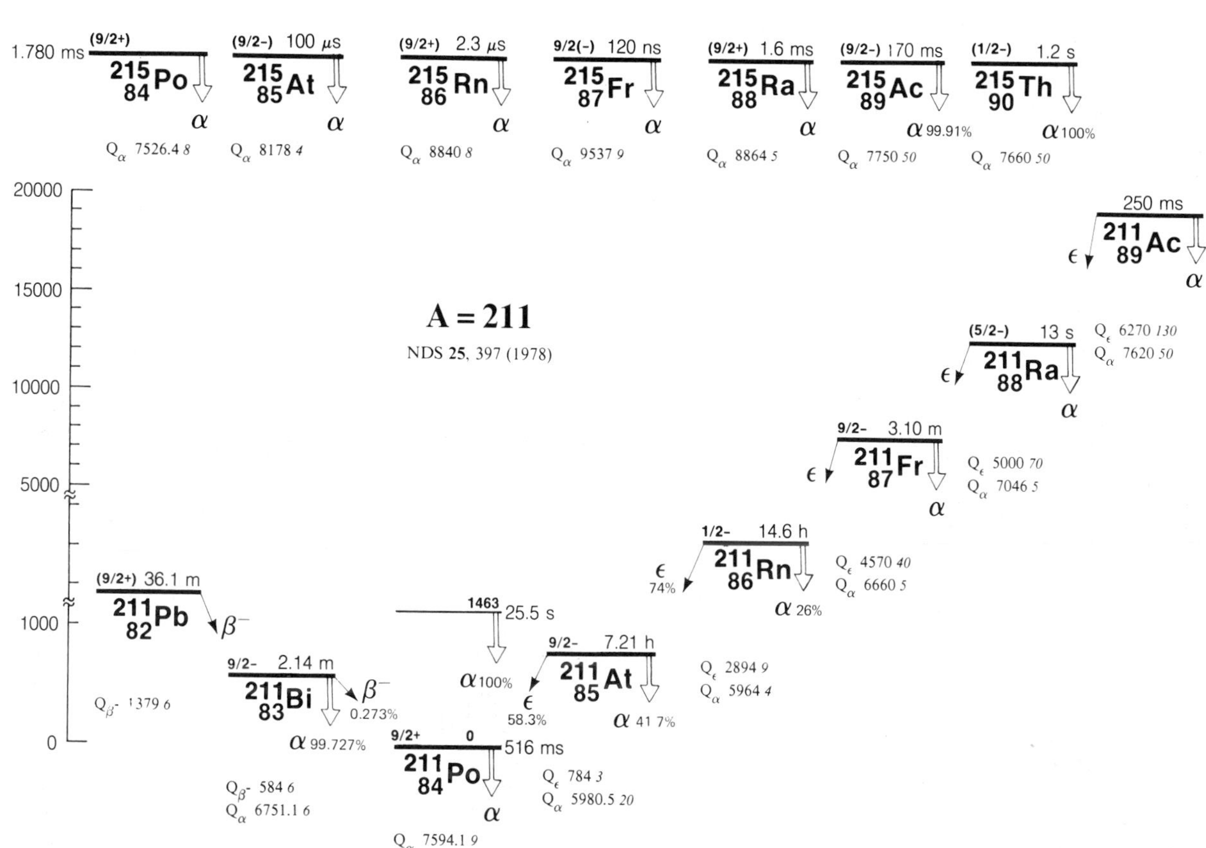

$^{211}_{82}$Pb(36.1 *2* min)

Mode: β-
Δ: -10493 *4* keV
SpA: 2.468×10^{7} Ci/g
Prod: descendant ^{227}Ac

Photons (^{211}Pb)
⟨γ⟩=67.8 *12* keV

γ_{mode}	γ(keV)	γ(%)
Bi L_ℓ	9.419	0.019 *5*
Bi L_α	10.828	0.37 *9*
Bi L_η	11.712	0.006 *2*
Bi L_β	13.082	0.41 *12*
Bi L_γ	15.397	0.09 *3*
γ M1	65.516 *8*	0.077 *5*
Bi $K_{\alpha2}$	74.814	0.256 *25*
Bi $K_{\alpha1}$	77.107	0.43 *4*
γ	81.0 *4*	0.045 *11*
γ	83.81 *17*	0.057 *9*
Bi $K_{\beta1}$'	87.190	0.152 *15*
γ	88.2 *4*	0.017 *4*

Photons (^{211}Pb)
(continued)

γ_{mode}	γ(keV)	γ(%)
Bi $K_{\beta2}$'	90.128	0.045 *4*
γ	94.3 *5*	0.011 *3*
γ E2+24%M1	94.89 *13*	0.018 *3*
γ	97.3 *4*	0.0115 *18*
γ [M1+50%E2]	313.85 *9*	0.031 *4*
γ [M1+50%E2]	343.02 *8*	0.034 *5*
γ	361.49 *3* ?	
γ E2+34%M1	404.86 *3*	3.83 *11*
γ M1+0.8%E2	427.00 *3*	1.72 *8*
γ	430.26 *20*	0.0064 *19*
γ	478.0 *7*	0.013 *3*
γ	481.1 *7*	0.026 *5*
γ	492.0 *7*	0.014 *4*
γ	500.4 *9*	0.011 *3*
γ	504.33 *17*	0.015 *3*

Photons (^{211}Pb) (continued)

γ_{mode}	γ(keV)	γ(%)
γ	546.2 *10* ?	
γ	609.62 *12*	0.023 *4*
γ	675.33 *9*	0.0064 *17*
γ M1+32%E2	704.51 *8*	0.481 *24*
γ M1(+21%E2)	766.35 *3*	0.71 *4*
γ M1	831.86 *3*	3.81 *11*
γ	865.81 *16*	0.0064 *10*
γ	951 *1*	~0.022
γ	1014.48 *12*	0.0179 *14*
γ	1080.19 *9*	0.0153 *15*
γ	1090.5 *9*	0.0026 *6* ?
γ	1103.4 *4*	0.0051 *8*
γ [M1+E2]	1109.37 *8*	0.147 *10*
γ	1196.61 *20*	0.0128 *13*
γ	1234.3 *4*	0.0013 *3*
γ	1270.67 *16*	0.0089 *9*

Atomic Electrons (^{211}Pb)

⟨e⟩=5.3 *3* keV

e_{bin}(keV)	⟨e⟩(keV)	e(%)
4 - 16	0.16	1.19 *25*
49	0.180	0.366 *25*
50 - 64	0.089	0.153 *7*
65	0.16	~0.25
66 - 67	0.05	~0.07
68	0.2	<0.6
70 - 97	0.59	0.76 *21*
237	0.028	0.0120 *12*
297 - 313	0.0058	0.00192 *14*
314	1.00	0.320 *19*
327 - 330	0.0041	0.00126 *18*
336	0.89	0.265 *13*
339 - 387	0.006	0.0017 *8*
388	0.191	0.049 *3*
389 - 410	0.213	0.054 *3*
411	0.185	0.0450 *22*
414 - 462	0.066	0.0155 *10*
465 - 504	0.0049	0.0010 *4*
519	0.004	~0.0008
585 - 614	0.096	0.016 *3*
659 - 675	0.00025	~4×10^{-5}
676	0.136	0.020 *3*
688 - 705	0.024	0.0035 *4*
741	0.76	0.103 *4*
750 - 775	0.035	0.0046 *5*
815 - 865	0.186	0.0228 *6*
866	2.0×10^{-6}	~2×10^{-7}
924 - 951	0.0020	~0.00022
990 - 1019	0.015	0.0015 *6*
1064 - 1109	0.0044	0.00041 *12*
1144 - 1193	0.0008	~7×10^{-5}
1194 - 1234	2.9×10^{-5}	2.4 *11* $\times10^{-6}$
1254 - 1270	0.00013	~1×10^{-5}

Continuous Radiation (^{211}Pb)

⟨β-⟩=447 keV;⟨IB⟩=0.59 keV

E_{bin}(keV)		⟨ ⟩(keV)	(%)
0 - 10	β-	0.063	1.26
	IB	0.020	
10 - 20	β-	0.189	1.26
	IB	0.019	0.134
20 - 40	β-	0.76	2.53
	IB	0.037	0.128
40 - 100	β-	5.4	7.6
	IB	0.097	0.150
100 - 300	β-	49.7	25.0
	IB	0.22	0.126
300 - 600	β-	143	32.2
	IB	0.149	0.037
600 - 1300	β-	248	30.2

Continuous Radiation (^{211}Pb) (continued)

E_{bin}(keV)		⟨ ⟩(keV)	(%)
	IB	0.050	0.0071
1300 - 1373	β-	0.56	0.0422
	IB	2.0×10^{-6}	1.50×10^{-7}

$^{211}_{83}$Bi(2.14 *2* min)

Mode: α(99.727 *4* %), β-(0.273 *3* %)

Δ: -11872 *6* keV

SpA: 4.15×10^{8} Ci/g

Prod: descendant ^{227}Ac

Alpha Particles (^{211}Bi)

⟨α⟩=6550 *5* keV

α(keV)	α(%)
6279 *4*	15.96 *20*
6623 *4*	83.77 *20*

Photons (^{211}Bi)

⟨γ⟩=46.7 *12* keV

γ_{mode}	γ(keV)	γ(%)†
Tl L_ℓ	8.953	0.0216 *20*
Tl L_α	10.259	0.438 *22*
Tl L_η	10.994	0.0057 *4*
Tl L_β	12.313	0.40 *3*
Tl L_γ	14.407	0.073 *6*
Tl $K_{\alpha2}$	70.832	0.75 *3*
Tl $K_{\alpha1}$	72.873	1.27 *4*
Tl $K_{\beta1}'$	82.434	0.447 *16*
Tl $K_{\beta2}'$	85.185	0.126 *5*
γM1+6.8%E2	350.1 *24*	12.8 *3*

†1.3% uncert(syst)

Atomic Electrons (^{211}Bi)

⟨e⟩=9.42 *21* keV

e_{bin}(keV)	⟨e⟩(keV)	e(%)
13	0.107	0.85 *9*
15	0.077	0.52 *5*
55	0.0142	0.026 *3*
56	0.00049	0.00087 *9*
58	0.0139	0.0239 *25*
60	0.0041	0.0068 *7*
66	0.0027	0.0040 *4*
67	0.0056	0.0084 *9*
68	0.0027	0.0039 *4*
69	0.0042	0.0061 *6*
70	0.0043	0.0062 *6*
71	0.000187	0.00026 *3*
72	0.00136	0.00188 *20*
73	0.000231	0.00032 *3*
78	0.00078	0.00100 *10*
79	0.00068	0.00087 *9*
80	0.00034	0.00043 *4*
81	0.00046	0.00057 *6*
82	0.00039	0.00048 *5*
265	7.13	2.70 *8*
335	1.53	0.456 *13*
337	0.0228	0.00676 *19*
346	0.329	0.095 *3*

Atomic Electrons (^{211}Bi) (continued)

e_{bin}(keV)	⟨e⟩(keV)	e(%)
347	0.0468	0.0135 *4*
348	0.000162	4.65 *12* $\times10^{-5}$
349	0.098	0.0280 *8*
350	0.0228	0.00653 *18*

Continuous Radiation (^{211}Bi)

⟨β-⟩=0.480 keV;⟨IB⟩=0.00029 keV

E_{bin}(keV)		⟨ ⟩(keV)	(%)
0 - 10	β-	0.000484	0.0097
	IB	2.4×10^{-5}	
10 - 20	β-	0.00143	0.0096
	IB	2.2×10^{-5}	0.000156
20 - 40	β-	0.0057	0.0189
	IB	4.0×10^{-5}	0.000139
40 - 100	β-	0.0377	0.054
	IB	8.6×10^{-5}	0.000136
100 - 300	β-	0.256	0.135
	IB	0.000104	6.7×10^{-5}
300 - 579	β-	0.179	0.0478
	IB	1.01×10^{-5}	3.0×10^{-6}

$^{211}_{84}$Po(516 *3* ms)

Mode: α

Δ: -12457 *4* keV

SpA: 5.70×10^{10} Ci/g

Prod: ^{208}Pb(α,n); descendant ^{227}Ac

Alpha Particles (^{211}Po)

⟨α⟩=7443.1 *15* keV

α(keV)	α(%)
6570.4 *11*	0.537 *19*
6892.3 *11*	0.546 *19*
7450.6 *11*	98.92 *3*

Photons (^{211}Po)

⟨γ⟩=7.7 *3* keV

γ_{mode}	γ(keV)	γ(%)†
γ [M1]	328.08 *7*	0.0032 *11*
γ E2	569.150 *19*	0.534
γ M1+0.9%E2	897.23 *7*	0.52 *3*

† 3.6% uncert(syst)

$^{211}_{84}$Po(25.5 *3* s)

Mode: α

Δ: -10994 *7* keV

SpA: 2.069×10^{9} Ci/g

Prod: ^{208}Pb(α,n); ^{209}Bi(α,pn); ^{207}Pb(^{12}C,2α); ^{208}Pb(^{12}C,2αn)

Alpha Particles (^{211}Po)

$\langle\alpha\rangle$=7401 8 keV

α(keV)	α(%)
7273 8	91.05 15
7994 8	1.66 3
8316 8	0.25 2
8875 8	7.04 14

Photons (^{211}Po)

$\langle\gamma\rangle$=1431 7 keV

γ_{mode}	γ(keV)	γ(%)†
Pb L_ℓ	9.185	0.087 8
Pb L_α	10.541	1.74 8
Pb L_η	11.349	0.0235 13
Pb L_β	12.699	1.66 11
Pb L_γ	14.890	0.31 3
Pb $K_{\alpha2}$	72.803	2.70 7
Pb $K_{\alpha1}$	74.969	4.54 12
Pb $K_{\beta1}$'	84.789	1.60 4
Pb $K_{\beta2}$'	87.632	0.463 14
γ [M1]	328.08 7	0.010 3
γ E2	569.150 19	92.0 2 •
γ M1+0.9%E2	897.23 7	1.65 11
γ M4	1063.10 2	83.20 •

† 0.17% uncert(syst)
• with ^{207}Pb(796 ms) in equilib

Atomic Electrons (^{211}Po)

$\langle e\rangle$=119.3 16 keV

e_{bin}(keV)	$\langle e\rangle$(keV)	e(%)
13 - 62	0.83	5.3 4
68 - 84	0.086	0.120 5
240	0.007	~0.0029
312 - 328	0.0020	0.00064 23
481	7.09	1.47 3
553 - 569	3.04	0.546 6
809	0.273	0.0338 24
881 - 897	0.064	0.0073 4
975	79.4	8.14 16
1047	16.8	1.61 3
1048 - 1050	4.85	0.463 7
1059	4.11	0.388 8
1060 - 1063	2.77	0.261 4

$^{211}_{85}$At(7.214 7 h)

Mode: ϵ(58.3 2 %), α(41.7 2 %)
Δ: -11672 5 keV
SpA: 2.0588×10^6 Ci/g
Prod: ^{209}Bi(α,2n)

Alpha Particles (^{211}At)

$\langle\alpha\rangle$=2447.1 keV

α(keV)	α(%)
4992.3 10	0.0004
5139.2 10	0.0009 3
5210.9 10	0.0036 7
5867.7 10	41.7

Photons (^{211}At)

$\langle\gamma\rangle$=36.3 4 keV

γ_{mode}	γ(keV)	γ(%)†
Po L_ℓ	9.658	0.43 4
Po L_α	11.119	8.3 4
Po L_η	12.085	0.107 6
Po L_β	13.500	7.9 5
Po L_γ	15.905	1.55 14
Po $K_{\alpha2}$	76.858	11.58 24
Po $K_{\alpha1}$	79.290	19.3 4
Po $K_{\beta1}$'	89.639	6.84 14
Po $K_{\beta2}$'	92.673	2.09 5
γ_αM1+E2	669.56 7	0.0034 6
γ_ϵM1(+28%E2)	687.0 1	0.246 13
γ_αM1	742.63 6	0.0009 3

† uncert(syst): 3.5% for α, 3.6% for ϵ

Atomic Electrons (^{211}At)

$\langle e\rangle$=4.03 20 keV

e_{bin}(keV)	$\langle e\rangle$(keV)	e(%)
13	5.5 ×10^{-6}	4.1 20 ×10^{-5}
14	1.66	12.0 12
16	1.01	6.2 6
17	0.41	2.40 24
58 - 59	0.086	0.146 15
60	0.146	0.244 24
61 - 62	0.083	0.133 12
63	0.130	0.206 21
64 - 89	0.436	0.590 21
594	0.056	0.0094 6
652 - 687	0.0150	0.00222 10
726 - 743	5.4 ×10^{-5}	7.4 18 ×10^{-6}

Continuous Radiation (^{211}At)

$\langle$IB$\rangle$=0.33 keV

E_{bin}(keV)		$\langle\ \rangle$(keV)	(%)
10 - 20	IB	0.00140	0.0104
20 - 40	IB	0.00062	0.0019
40 - 100	IB	0.27	0.36
100 - 300	IB	0.032	0.021
300 - 600	IB	0.025	0.0060
600 - 791	IB	0.00150	0.00023

$^{211}_{86}$Rn(14.6 2 h)

Mode: ϵ(74 1 %), α(26 1 %)
Δ: -8779 10 keV
SpA: 1.017×10^6 Ci/g
Prod: protons on Th; ^{209}Bi(^{16}O,5p9n); ^{208}Pb(^{16}O,4p9n); C on Tl

Alpha Particles (^{211}Rn)

$\langle\alpha\rangle$=1504.5 3 keV

α(keV)	α(%)
5052 1	0.00016 5
5181.7 10	0.00068 5
5273.8 10	0.0039 3
5465.5 10	0.0036 3
5619.0 10	0.70 5
5783.8 10	16.4 3
5851.1 10	8.8 3

Photons (^{211}Rn)

$\langle\gamma\rangle$=1907 41 keV

γ_{mode}	γ(keV)	γ(%)†
Po L_ℓ	9.658	0.110 14
At L_ℓ	9.897	0.70 8
Po L_α	11.119	2.11 22
At L_α	11.414	13.2 10
Po L_η	12.085	0.055 6
At L_η	12.466	0.186 15
Po L_β	13.489	2.9 3
At L_β	13.915	12.9 12
Po L_γ	15.821	0.59 6
At L_γ	16.406	2.6 3
γ_αE2	68.55 6	0.44 4
Po $K_{\alpha2}$	76.858	0.042 5
At $K_{\alpha2}$	78.947	17.8 11
Po $K_{\alpha1}$	79.290	0.070 8
At $K_{\alpha1}$	81.517	29.6 19
Po $K_{\beta1}$'	89.639	0.025 3
At $K_{\beta1}$'	92.136	10.5 7
Po $K_{\beta2}$'	92.673	0.0075 8
At $K_{\beta2}$'	95.265	3.26 22
γ_ϵM1(+21%E2)	116.13 10	0.092 9
γ_αM1+50%E2	168.01 7	0.097 13
γ_ϵE2+12%M1	168.77 6	6.8 4
γ_ϵ	175.97 7	0.064 14
γ_ϵM1(+27%E2)	191.88 6	0.92 5
γ_αM1+50%E2	236.57 6	0.065 8
γ_ϵE2	250.22 6	6.1 3
γ_ϵM1(+4.6%E2)	262.07 6	0.225 23
γ_ϵM2	350.47 7	0.40 3
γ_ϵE1	370.49 8	1.38 9
γ_ϵE1	416.32 6	3.54 18
γ_ϵE2	442.09 6	23.4 14
γ_ϵ	592.29 7	0.267 23
γ_ϵM1+18%E2	674.12 7	46.0
γ_ϵ(E1+5.0%M2)	678.39 6	29.4 14
γ_ϵ	684.59 5	0.60 5
γ_ϵM1+40%E2	853.37 6	4.69 23
γ_ϵE2	866.00 7	8.0 4
γ_ϵE2	934.81 6	3.72 18
γ_ϵM1+50%E2	946.66 6	5.1 14
γ_ϵM1(+<45%E2)	947.44 7	16.5 18
γ_ϵ	992.49 9	~1 ?
γ_ϵ	1012.51 6	0.216 18
γ_ϵ	1045.13 11	0.060 18
γ_ϵ	1115.43 7	~0.09 ?
γ_ϵE2	1126.68 6	22.5 14
γ_ϵ	1181.28 7	1.47 9
γ_ϵ	1242.70 9	0.069 14
γ_ϵ	1318.45 10	0.129 14
γ_ϵE1	1362.98 5	33.1 18
γ_ϵ	1434.58 9	0.069 9
γ_ϵ	1482.90 20	0.041 9
γ_ϵ	1531.75 7	~0.046
γ_ϵ	1538.95 8	4.8 5
γ_ϵ	1654.50 20	0.028 5
γ_ϵ	1686.0 4	~0.013
γ_ϵ	1747.0 4	0.0051 18
γ_ϵ	1805.07 7	0.119 23
γ_ϵ	1947.9 4	0.011 3
γ_ϵ	1953.4 4	0.0074 23
γ_ϵ	1992.57 10	0.51 3
γ_ϵ	2113.4 4	0.0055 18
γ_ϵ	2128.72 8	0.0046 18 ?
γ_ϵ	2219.9 3	0.0101 18
γ_ϵ	2405.1 5	0.0041 14
γ_ϵ	2486.0 4	0.0041 9
γ_ϵ	2696.3 5	0.0017 6

† uncert(syst): 19% for α, 2.9% for ϵ

Atomic Electrons (^{211}Rn)

⟨e⟩=65.9 *16* keV

e_{bin}(keV)	⟨e⟩(keV)	e(%)
14	3.2	23 *3*
16	0.57	3.5 *5*
17	2.3	13.7 *16*
20	0.103	0.5 *1*
52	3.9	7.4 *6*
55	3.4	6.3 *5*
59 - 64	0.53	0.85 *7*
65	2.50	3.8 *3*
66 - 72	0.90	1.33 *10*
73	2.4	3.3 *5*
74 - 116	2.02	2.3 *3*
151	0.89	0.59 *8*
152	2.81	1.85 *12*
154	0.98	0.63 *4*
155	1.72	1.11 *7*
158 - 192	2.99	1.78 *8*
220 - 262	3.02	1.24 *4*
305	0.209	0.069 *3*
333 - 336	0.316	0.095 *6*
346	2.49	0.72 *5*
347 - 370	0.0511	0.0145 *5*
399 - 442	1.68	0.391 *16*
497	0.06	~0.011
575	0.011	~0.0020
578	12.5	2.16 *17*
583	1.9	0.33 *11*
588 - 592	0.10	~0.018
657	2.48	0.38 *3*
660 - 685	1.41	0.211 *22*
763	1.24	0.163 *15*
836 - 851	1.15	0.136 *21*
852	2.2	0.26 *3*
853 - 897	0.18	~0.021
917 - 949	0.83	0.089 *7*
975 - 1020	0.048	~0.0048
1028 - 1045	1.15	0.112 *7*
1086 - 1126	0.44	0.040 *8*
1147 - 1181	0.033	0.0028 *14*
1223 - 1267	0.59	0.047 *3*
1301 - 1349	0.102	0.0076 *5*
1359 - 1387	0.0307	0.00226 *16*
1417 - 1466	0.27	~0.019
1469 - 1559	0.06	~0.0039
1590 - 1683	0.0013	8 *4* $\times10^{-5}$
1709 - 1802	0.007	~0.00038
1852 - 1950	0.021	~0.0011
1975 - 2033	0.0047	0.00024 *11*
2096 - 2125	0.00042	~2 $\times10^{-5}$
2202 - 2216	7.4 $\times10^{-5}$	~3 $\times10^{-6}$
2309 - 2402	0.00027	1.2 *5* $\times10^{-5}$
2469 - 2483	2.6 $\times10^{-5}$	~1 $\times10^{-6}$
2601 - 2693	5.3 $\times10^{-5}$	~2 $\times10^{-6}$

Continuous Radiation (^{211}Rn)

⟨IB⟩=0.46 keV

E_{bin}(keV)		⟨ ⟩(keV)	(%)
10 - 20	IB	0.0026	0.019
20 - 40	IB	0.00088	0.0028
40 - 100	IB	0.43	0.54
100 - 300	IB	0.028	0.021
300 - 413	IB	0.00036	0.000138

$^{211}_{87}$Fr(3.10 *2* min)

Mode: α, ϵ
Δ: -4200 *40* keV
SpA: 2.869×10^{8} Ci/g
Prod: ^{203}Tl(^{12}C,4n); ^{205}Tl(^{12}C,6n); ^{197}Au(^{18}O,4n); ^{197}Au(^{16}O,2n)

Alpha Particles (^{211}Fr)

α(keV)
6534 *5*

Photons (^{211}Fr)

γ_{mode}	γ(keV)	γ(rel)
γ_ϵ	220.0 *8*	9 *2*
γ_ϵ[M1+E2]	279.9 *10*	34 *3*
γ_ϵ[E1]	438.9 *10*	20 *3*
γ_ϵ[E2]	538.9 *10*	100
γ_ϵ	762.0 *8*	5 *1*
γ_ϵ[E2]	916.9 *10*	55 *5*
γ_ϵ[E1]	981.9 *8*	20 *3*

$^{211}_{88}$Ra(13 *2* s)

Mode: α, ϵ
Δ: 800 *80* keV
SpA: 4.0×10^{9} Ci/g
Prod: ^{197}Au(^{19}F,5n); ^{206}Pb(^{12}C,7n); ^{194}Pt(^{22}Ne,5n); ^{192}Pt(^{22}Ne,3n)

Alpha Particles (^{211}Ra)

α(keV)
6912 *5*

$^{211}_{89}$Ac(250 *50* ms)

Mode: α, ϵ
Δ: 7070 *120* keV
SpA: 7.2×10^{10} Ci/g
Prod: ^{197}Au(^{20}Ne,6n); ^{203}Tl(^{16}O,8n)

Alpha Particles (^{211}Ac)

α(keV)
7480 *8*

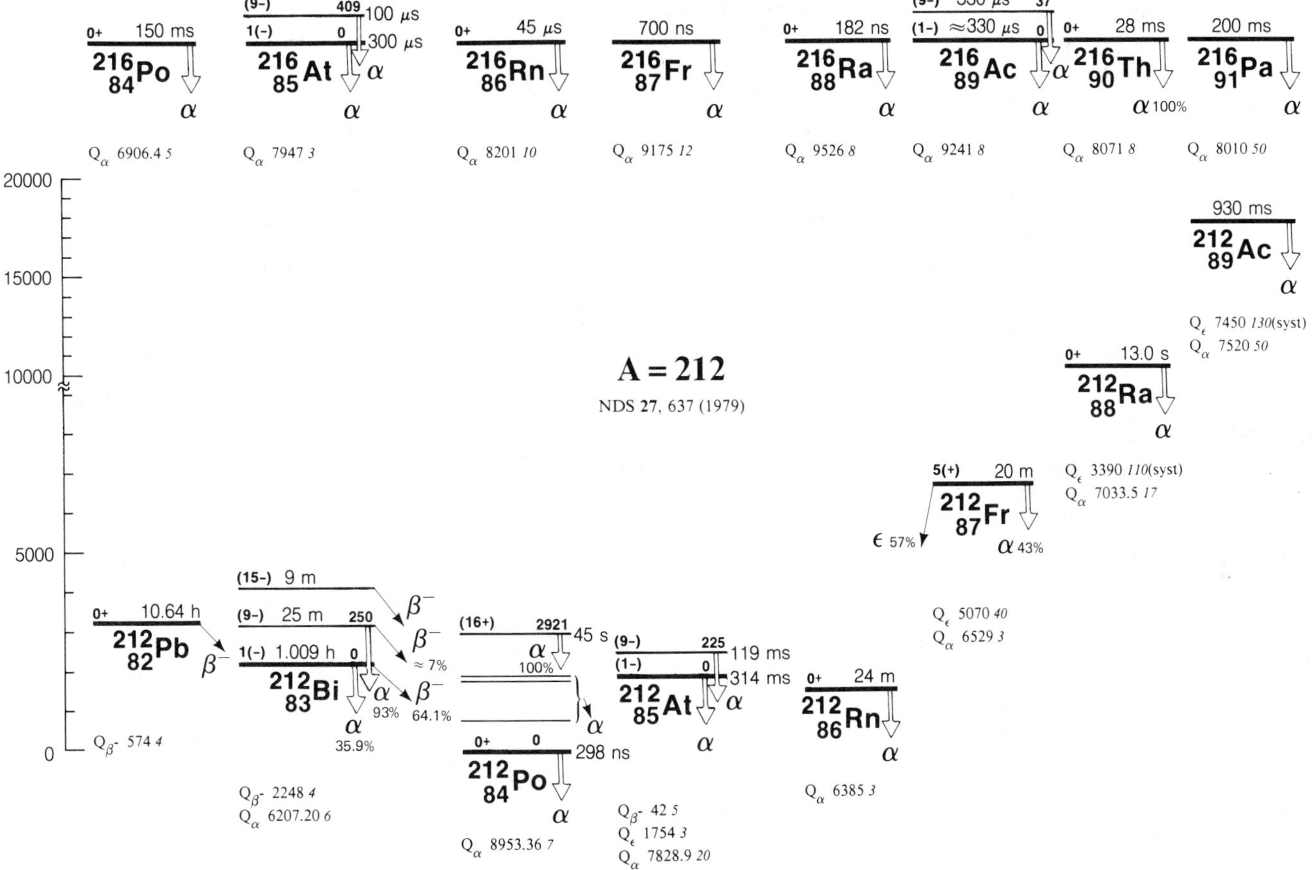

$^{212}_{82}$Pb(10.64 *1* h)

Mode: β-
Δ: -7572 *6* keV
SpA: 1.3893×10^{6} Ci/g
Prod: descendant ^{228}Th

Photons (^{212}Pb)

⟨γ⟩=145 *3* keV

γ_{mode}	γ(keV)	γ(%)†
Bi L_ℓ	9.419	0.34 *3*
Bi L_α	10.828	6.6 *3*
Bi L_η	11.712	0.087 *5*
Bi L_β	13.099	6.0 *4*
Bi L_γ	15.376	1.13 *10*
Bi $K_{\alpha2}$	74.814	10.5 *4*
Bi $K_{\alpha1}$	77.107	17.7 *6*
Bi $K_{\beta1}$'	87.190	6.27 *22*
Bi $K_{\beta2}$'	90.128	1.86 *7*
γ M1(+0.06%E2)	115.122 *7*	0.591 *18*
γ [E2]	123.456 *8* ?	
γ E2	164.15 *10*	~0.005
γ M1	176.577 *12*	0.051 *6*
γ M1	238.578 *4*	43.6 *11*
γ M1(+5.2%E2)	300.034 *10*	3.34 *10*
γ [M1]	415.156 *12*	0.028 *7*

† 1.4% uncert(syst)

Atomic Electrons (^{212}Pb)

⟨e⟩=74.2 *16* keV

e_{bin}(keV)	⟨e⟩(keV)	e(%)
13	1.52	11.3 *12*
16 - 64	2.41	11.2 *8*
70 - 115	1.23	1.34 *4*
148	48.7	32.9 *10*
151 - 177	0.036	0.0219 *22*
210	2.71	1.29 *9*
222	11.4	5.14 *16*
223 - 225	1.20	0.536 *16*
235	3.14	1.34 *4*
236 - 284	1.68	0.662 *21*
287 - 325	0.236	0.079 *5*
399 - 415	0.0042	0.00105 *21*

Continuous Radiation (^{212}Pb)

⟨β-⟩=101 keV;⟨IB⟩=0.039 keV

E_{bin}(keV)		⟨ ⟩(keV)	(%)
0 - 10	β-	0.353	7.1
	IB	0.0052	
10 - 20	β-	1.02	6.8
	IB	0.0045	0.032
20 - 40	β-	3.79	12.7
	IB	0.0073	0.026
40 - 100	β-	21.3	31.2
	IB	0.0129	0.021
100 - 300	β-	67	40.2
	IB	0.0087	0.0061
300 - 573	β-	7.9	2.13
	IB	0.00040	0.000119

$^{212}_{83}$Bi(1.0092 *10* h)

Mode: β-(64.06 *6* %), α(35.94 *6* %), β-α(0.023 %)
Δ: -8146 *5* keV
SpA: 1.4646×10^{7} Ci/g
Prod: descendant ^{228}Th

Alpha Particles (^{212}Bi)

⟨α⟩=2174.0 *4* keV

α(keV)	α(%)
5302 *3*	4.0 *4* ×10^{-5}
5344 *14*	0.0004
5481.5 *4*	0.005
5606.78 *12*	0.4
5625.68 *17*	0.06
5768.36 *8*	0.600 *7*
6051.00 *3*	25.23 *7*
6090.09 *3*	9.63 *7*
9496.9 *7*	4×10^{-5} *
10427 *4*	1.6×10^{-5} *
10548.7 *7*	0.00017 *

* Long-range α (from levels in ^{212}Po)

Photons (^{212}Bi)

$\langle\gamma\rangle$=106.1 *15* keV

γ_{mode}	γ(keV)	γ(%)†
Tl L_ℓ	8.953	0.150 *16*
Po L_ℓ	9.658	0.0021 *7*
Tl L_α	10.259	3.04 *21*
Tl L_η	10.994	0.0283 *18*
Po L_α	11.119	0.041 *13*
Po L_η	12.085	0.00062 *22*
Tl L_β	12.282	3.5 *4*
Po L_β	13.501	0.041 *14*
Tl L_γ	14.563	0.82 *11*
Po L_γ	15.869	0.008 *3*
γ_αM1	39.846 *5*	1.100 *25*
Tl $K_{\alpha2}$	70.832	0.075 *12*
Tl $K_{\alpha1}$	72.873	0.127 *21*
Po $K_{\alpha2}$	76.858	0.060 *16*
Po $K_{\alpha1}$	79.290	0.10 *3*
Tl $K_{\beta1}$'	82.434	0.045 *7*
Tl $K_{\beta2}$'	85.185	0.0125 *21*
Po $K_{\beta1}$'	89.639	0.035 *10*
Po $K_{\beta2}$'	92.673	0.011 *3*
γ_α	124.1 *10*	<0.032
$\gamma_{\beta-}$	130 *10*	<0.036 ?
γ_α	143.99 *20*	0.010 *4*
γ_α(E2)	164.0 *6*	~0.005
$\gamma_{\beta-}$	190 *10*	<0.050
γ_αM1+0.5%E2	288.08 *8*	0.341 *14*
γ_α	295.09 *20*	0.024 *6*
γ_αM1+0.7%E2	327.93 *8*	0.137 *4*
γ_α(M1)	433.51 *17*	0.014 *4*
γ_αM1	452.77 *12*	0.363 *11*
γ_αE2	473.36 *17*	0.050 *4*
γ_α(M1)	492.62 *12*	0.0061 *12* ?
γ_α[E2]	580.4 *4*	0.00079 *16* ?
$\gamma_{\beta-}$	600 *10*	
γ_α[M1+E2]	620.3 *4*	0.0036 *7* ?
$\gamma_{\beta-}$E2	727.25 *5*	6.65 *14*
$\gamma_{\beta-}$M1+0.2%E2	785.51 *5*	1.107 *25*
$\gamma_{\beta-}$M1	893.42 *6*	0.367 *11*
$\gamma_{\beta-}$	952.17 *10*	0.176 *9*
$\gamma_{\beta-}$[E2]	1074.1 *5*	~0.016
$\gamma_{\beta-}$M1+E2	1078.69 *10*	0.535 *18*
$\gamma_{\beta-}$	1300 *10*	<0.014
$\gamma_{\beta-}$[E2]	1512.75 *6*	0.313 *25*
$\gamma_{\beta-}$M1	1620.66 *6*	1.51 *5*
$\gamma_{\beta-}$	1679.42 *11*	0.068 *11*
$\gamma_{\beta-}$	1805.93 *11*	0.111 *22*
$\gamma_{\beta-}$	2200 *10*	<0.036

† uncert(syst): 0.17% for α, <0.1% for β-

Atomic Electrons (^{212}Bi)

$\langle e\rangle$=10.50 *18* keV

e_{bin}(keV)	$\langle e\rangle$(keV)	e(%)
13 - 14	0.035	0.27 *4*
15	0.71	4.7 *5*
16 - 17	0.0074	0.046 *14*
24	4.58	18.7 *6*
25	0.462	1.84 *6*
27	0.0481	0.177 *5*
36	1.74	4.81 *15*
37	0.037	~0.10
39	0.50	1.29 *8*
40 - 89	0.135	0.323 *17*
97 - 144	0.10	0.09 *3*
149 - 190	0.022	0.013 *5*
203	0.261	0.129 *6*
210 - 242	0.097	0.041 *4*
273 - 315	0.101	0.0359 *14*
324 - 367	0.162	0.0445 *15*
388 - 434	0.0084	0.00209 *16*
437 - 480	0.0418	0.00946 *23*
489 - 535	0.0008	0.00015 *7*
565 - 608	0.00015	2.5 *11* $\times10^{-5}$
617 - 620	3.9 $\times10^{-5}$	6 *3* $\times10^{-6}$
634	0.449	0.0708 *21*
692 - 727	0.424	0.0605 *12*

Atomic Electrons (^{212}Bi) (continued)

e_{bin}(keV)	$\langle e\rangle$(keV)	e(%)
769 - 800	0.136	0.0173 *4*
859 - 893	0.035	0.0040 *16*
935 - 981	0.0051	0.0005 *3*
986	0.05	0.005 *3*
1057 - 1078	0.013	0.0012 *4*
1207	0.0005	<9 $\times10^{-5}$
1283 - 1300	0.00012	~9 $\times10^{-6}$
1420	0.0123	0.00087 *8*
1496 - 1586	0.129	0.0084 *3*
1604 - 1676	0.0271	0.00168 *6*
1708 - 1803	0.16	0.0091 *15*
2107 - 2197	0.0008	~4 $\times10^{-5}$

Continuous Radiation (^{212}Bi)

$\langle\beta-\rangle$=492 keV;$\langle$IB$\rangle$=0.97 keV

E_{bin}(keV)		$\langle\ \rangle$(keV)	(%)
0 - 10	β-	0.0204	0.405
	IB	0.019	
10 - 20	β-	0.061	0.408
	IB	0.019	0.132
20 - 40	β-	0.248	0.83
	IB	0.037	0.128
40 - 100	β-	1.80	2.57
	IB	0.102	0.158
100 - 300	β-	18.1	9.0
	IB	0.27	0.152
300 - 600	β-	61	13.5
	IB	0.26	0.061
600 - 1300	β-	249	26.9
	IB	0.24	0.029
1300 - 2246	β-	162	10.4
	IB	0.028	0.0019

$^{212}_{83}$Bi(25 *1* min)

Mode: α(93 *4* %), β-(7 *4* %)
Δ: -7896 *5* keV
SpA: 3.55×10^7 Ci/g
Prod: ^{48}Ca on ^{238}U; ^{48}Ca on ^{248}Cm; ^{40}Ar on ^{238}U

Alpha Particles (^{212}Bi)

α(keV)	α(rel)
6300 *7*	~40
6340 *7*	~53

$^{212}_{83}$Bi(9 *1* min)

Mode: β-
Δ: >-7896 keV
SpA: 9.8×10^7 Ci/g
Prod: ^{48}Ca on ^{238}U; ^{48}Ca on ^{248}Cm; ^{40}Ar on ^{238}U

$^{212}_{84}$Po(298 *3* ns)

Mode: α
Δ: -10394 *4* keV
SpA: 7.68×10^{10} Ci/g
Prod: descendant ^{228}Th

Alpha Particles (^{212}Po)

α(keV)
8784.37 *7*

$^{212}_{84}$Po(45.1 *6* s)

Mode: α(100 %)
Δ: -7473 *16* keV
SpA: 1.171×10^9 Ci/g
Prod: ^{209}Bi(α,p); ^{208}Pb(^{11}B,^{7}Li)

Alpha Particles (^{212}Po)

$\langle\alpha\rangle$=11436 keV

α(keV)	α(%)
6842 ?	<5
7127 ?	<3
7971 ?	<0.15
8514 *9*	2.05 *9*
9086 *9*	1.00 *4*
11650 *20*	97

Photons (^{212}Po)

$\langle\gamma\rangle$=80 *10* keV

γ_{mode}	γ(keV)	γ(%)
γ E2	583.022 *22*	~2
γ E3	2614.35 *10*	2.6 *3*

$^{212}_{85}$At(314 *2* ms)

Mode: α
Δ: -8640 *5* keV
SpA: 6.83×10^{10} Ci/g
Prod: ^{209}Bi(α,n)

Alpha Particles (^{212}At)

$\langle\alpha\rangle$=7662 keV

α(keV)	α(%)
6629	0.14
6763	
6765	0.05
6805	0.07
7058	0.4
7088	0.6
7177	0.15
7618 *2*	15
7681 *2*	84

$^{212}_{85}$At(119 *3* ms)

Mode: α
Δ: -8415 *5* keV
SpA: 7.65×10^{10} Ci/g
Prod: ^{209}Bi(α,n)

Alpha Particles (^{212}At)

$\langle\alpha\rangle$= 7848 keV

α(keV)	α(%)
6821	0.3
6954	0.05
7032	0.13
7261	0.3
7282	0.4
7309	0.07
7397	0.3
7837	65
7897	33

$^{212}_{86}$Rn(24 *2* min)

Mode: α

Δ: -8682 *6* keV

SpA: 3.7×10^{7} Ci/g

Prod: daughter ^{212}Fr; protons on Th

Alpha Particles (^{212}Rn)

$\langle\alpha\rangle$=6259 *5* keV

α(keV)	α(%)
5587 *4*	0.050 *5*
6260 *4*	99.950 *5*

$^{212}_{87}$Fr(20.0 *6* min)

Mode: ε(57 *2* %), α(43 *2* %)

Δ: -3610 *40* keV

SpA: 4.43×10^{7} Ci/g

Prod: protons on Th; ^{205}Tl(^{12}C,5n)

Alpha Particles (^{212}Fr)

$\langle\alpha\rangle$=2754 *1* keV

α(keV)	α(%)
5828 *8*	~0.022
5983 *4*	~0.043
6076 *3*	~0.17
6127 *3*	0.43 *9*
6173 *4*	0.5
6183 *3*	0.6
6260.7 *14*	16.3 *9*
6335.3 *15*	4
6343.2 *14*	1.3
6382.6 *14*	10.3 *9*
6405.6 *14*	9.5 *9*

Photons (^{212}Fr)

$\langle\gamma\rangle$=1084 *60* keV

γ_{mode}	γ(keV)	γ(%)†
At L_ℓ	9.897	0.38 *12*
Rn L_ℓ	10.137	0.7 *3*
At L_α	11.414	7.2 *23*
Rn L_α	11.713	14 *6*
At L_η	12.466	0.073 *21*
Rn L_η	12.855	0.26 *10*

Photons (^{212}Fr)
(continued)

γ_{mode}	γ(keV)	γ(%)†
At L_β	13.890	7.9 *25*
Rn L_β	14.337	16 *7*
At L_γ	16.522	1.9 *6*
Rn L_γ	16.904	3.3 *14*
γ_αM1(+0.04%E2)	23.5 *4*	~0.16
γ_αM1(+0.03%E2)	40.1 *4*	0.25 *3*
γ_αM1(+1.9%E2)	71.7 *5*	0.55 *4*
At $K_{\alpha2}$	78.947	2.4 *6*
Rn $K_{\alpha2}$	81.067	14 *6*
At $K_{\alpha1}$	81.517	3.9 *9*
Rn $K_{\alpha1}$	83.787	24 *10*
γ_α(M1)	84.1 *4*	0.63 *13*
At $K_{\beta1}$'	92.136	1.4 *3*
Rn $K_{\beta1}$'	94.677	8 *4*
At $K_{\beta2}$'	95.265	0.43 *10*
γ_ϵ	97.5 *5*	2.7 *6*
Rn $K_{\beta2}$'	97.907	2.7 *11*
γ_αM1(+20%E2)	124.2 *4*	1.77 *17*
γ_ϵE2	138.3 *1*	7.7 *4*
γ_α(E2+~16%M1)	147.7 *4*	0.21 *4*
γ_ϵE2	227.72 *10*	42.2 *21*
γ_ϵ	309.1 *2*	1.01 *8*
γ_ϵ	311.5 *2*	1.35 *13*
γ_ϵ	322.5 *2*	0.42 *8*
γ_ϵ	358.2 *5*	0.21 *4*
γ_ϵ	422.0 *5*	0.80 *8*
γ_ϵ	532.0 *5*	2.8 *3*
γ_ϵ	564.4 *5*	0.97 *8*
γ_ϵ	620.1 *5*	0.80 *8*
γ_ϵ[M1+E2]	801.9 *15*	3.5 *4*
γ_ϵ	824.0 *15*	0.76 *8*
γ_ϵ	902.2 *15*	0.67 *8*
γ_ϵ[E1]	1047.3 *14*	7.2 *7*
γ_ϵ	1178.4 *20*	1.35 *17*
γ_ϵ[E1]	1185.6 *14*	13.9 *17*
γ_ϵE2	1274.8 *20*	46 *4*

† uncert(syst): 8.5% for α, 10% for ε

Atomic Electrons (^{212}Fr)

$\langle e\rangle$=91 *7* keV

e_{bin}(keV)	⟨e⟩(keV)	e(%)
6 - 14	1.7	24 *10*
15	2.9	20 *9*
17	3.9	23 *8*
18 - 26	3.4	16 *3*
28	2.3	8.2 *20*
36 - 78	6.4	12.1 *7*
79	2.3	~3
80	4.7	~6
81	0.025	0.030 *11*
83	3.4	~4
84 - 120	7.5	7 *3*
121	7.4	6.1 *4*
123	0.17	0.14 *7*
124	5.0	4.01 *23*
129	6.8	5.3 *3*
130 - 133	0.25	0.187 *24*
134	2.30	1.72 *9*
135 - 148	2.88	2.11 *7*
210	9.8	4.66 *25*
211	0.5	~0.25
213	4.9	2.3 *4*
223	0.48	0.217 *12*
224	3.7	1.63 *17*
225 - 260	1.44	0.63 *4*
291 - 340	0.9	0.30 *11*
341 - 358	0.016	0.0047 *23*
404 - 434	0.8	~0.19
466 - 515	0.39	~0.08
517 - 564	0.30	~0.06
602 - 620	0.05	~0.008
707	0.7	~0.10
784 - 824	0.27	0.035 *12*
884 - 902	0.021	~0.0024

Atomic Electrons (^{212}Fr)
(continued)

e_{bin}(keV)	⟨e⟩(keV)	e(%)
949	0.155	0.0163 *18*
1029 - 1047	0.034	0.00334 *25*
1085	0.39	0.036 *9*
1160 - 1185	2.32	0.197 *18*
1257 - 1274	0.60	0.047 *3*

Continuous Radiation (^{212}Fr)

$\langle\beta+\rangle$=39.5 keV;⟨IB⟩=3.1 keV

E_{bin}(keV)		⟨ ⟩(keV)	(%)
0 - 10	β+	9.7×10^{-9}	1.13×10^{-7}
	IB	0.00151	
10 - 20	β+	7.4×10^{-7}	4.34×10^{-6}
	IB	0.0028	0.020
20 - 40	β+	3.83×10^{-5}	0.000114
	IB	0.0034	0.0116
40 - 100	β+	0.00383	0.00477
	IB	0.36	0.44
100 - 300	β+	0.339	0.149
	IB	0.076	0.047
300 - 600	β+	2.96	0.64
	IB	0.152	0.034
600 - 1300	β+	18.4	1.96
	IB	0.83	0.088
1300 - 2500	β+	17.7	1.10
	IB	1.49	0.085
2500 - 3478	IB	0.22	0.0081
	Σβ+		3.8

$^{212}_{88}$Ra(13.0 *2* s)

Mode: α

Δ: -220 *100* keV syst

SpA: 3.99×10^{9} Ci/g

Prod: ^{197}Au(^{19}F,4n); ^{206}Pb(^{12}C,6n); ^{194}Pt(^{22}Ne,4n); ^{192}Pt(^{22}Ne,2n)

Alpha Particles (^{212}Ra)

α(keV)
6901 *2*

$^{212}_{89}$Ac(930 *50* ms)

Mode: α

Δ: 7230 *90* keV

SpA: 4.03×10^{10} Ci/g

Prod: ^{203}Tl(^{16}O,7n); ^{197}Au(^{20}Ne,5n)

Alpha Particles (^{212}Ac)

α(keV)
7379 *8*

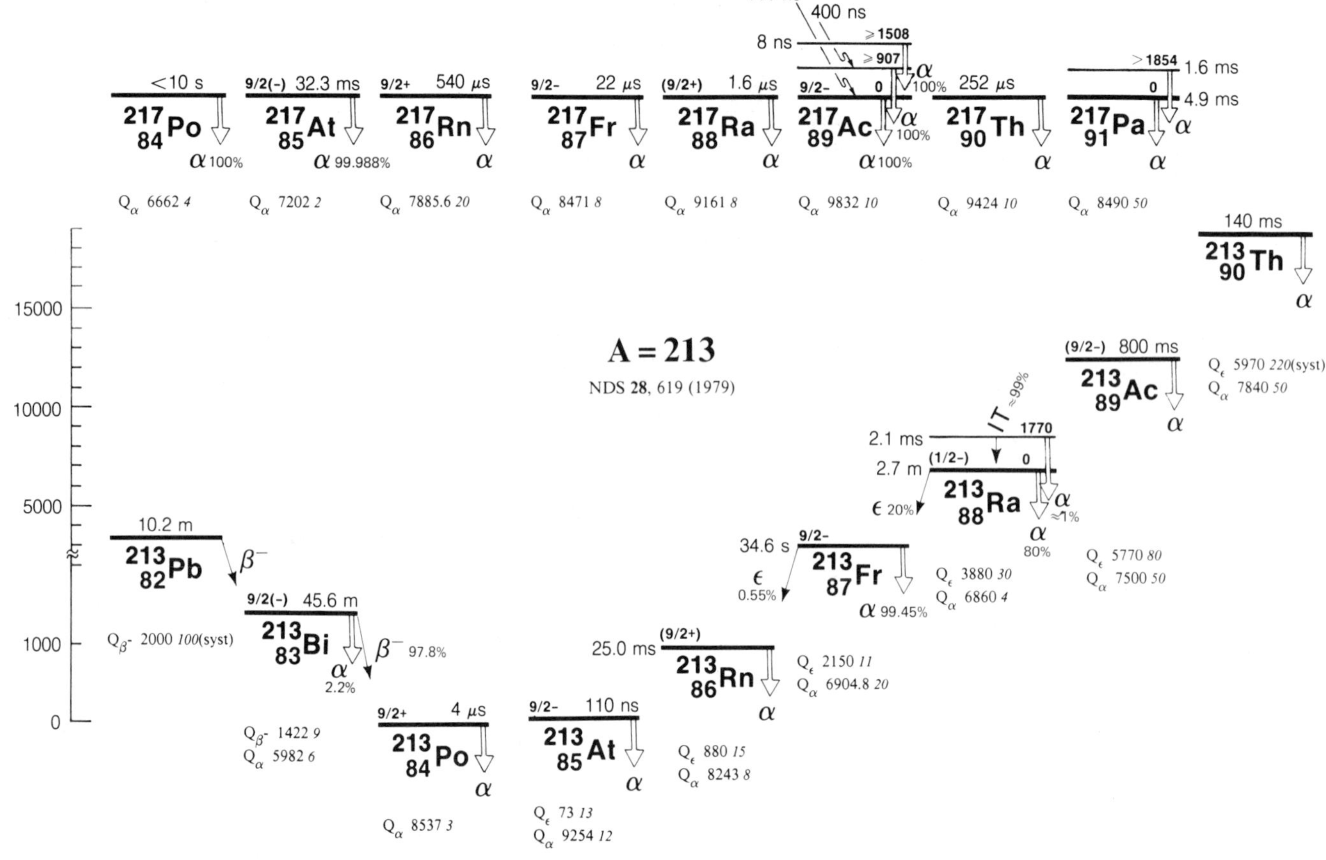

$^{213}_{82}$Pb(10.2 *3* min)

Mode: β-

Δ: -3250 *100* keV syst

SpA: 8.65×10^7 Ci/g

Prod: descendant ^{221}Rn

$^{213}_{83}$Bi(45.59 *6* min)

Mode: β-(97.84 *13* %), α(2.16 *13* %)

Δ: -5254 *11* keV

SpA: 1.9361×10^7 Ci/g

Prod: descendant ^{229}Th

Alpha Particles (^{213}Bi)

⟨α⟩=126.8 *3* keV

α(keV)	α(%)
5549 *10*	0.16 *3*
5869 *10*	2.01 *11*

Photons (^{213}Bi)

⟨γ⟩=82.5 *17* keV

γ_{mode}	γ(keV)	γ(%)†
Tl L_ℓ	8.953	~0.00023
Po L_ℓ	9.658	0.0258 *24*
Tl L_α	10.259	~0.005
Tl L_η	10.994	$\sim 6 \times 10^{-5}$
Po L_α	11.119	0.495 *25*
Po L_η	12.085	0.0067 *4*
Tl L_β	12.311	~0.004
Po L_β	13.504	0.46 *3*
Tl L_γ	14.402	~0.0008
Po L_γ	15.878	0.088 *8*
Tl $K_{\alpha 2}$	70.832	~0.008
Tl $K_{\alpha 1}$	72.873	~0.013
Po $K_{\alpha 2}$	76.858	0.78 *3*
Po $K_{\alpha 1}$	79.290	1.29 *5*
Tl $K_{\beta 1}'$	82.434	~0.005
Tl $K_{\beta 2}'$	85.185	~0.0013
Po $K_{\beta 1}'$	89.639	0.459 *17*
Po $K_{\beta 2}'$	92.673	0.140 *6*
$\gamma_{\beta-}$(M1)	292.79 *6*	0.44 *6*
γ_α	323.81 *5*	0.185 *13*
$\gamma_{\beta-}$M1	440.34 *2*	16.5 *4*
$\gamma_{\beta-}$	659.72 *5*	0.087 *8*
$\gamma_{\beta-}$	807.27 *4*	0.261 *8*
$\gamma_{\beta-}$	1100.06 *5*	0.282 *12*

† uncert(syst): 7.3% for α, 6.6% for β-

Atomic Electrons (^{213}Bi)

⟨e⟩=12.3 *3* keV

e_{bin}(keV)	⟨e⟩(keV)	e(%)
13 - 61	0.212	1.35 *10*
62 - 89	0.0433	0.0619 *22*
200	0.41	0.20 *3*
238 - 279	0.16	0.063 *24*
289 - 324	0.057	0.019 *4*
347	8.9	2.55 *8*
423	1.69	0.400 *12*
424 - 427	0.169	0.0399 *11*
436	0.449	0.103 *3*
437 - 440	0.154	0.0349 *9*
567	0.014	~0.0025
643 - 660	0.0039	~0.0006
714	0.03	~0.005
790 - 807	0.008	~0.0011
1007	0.023	~0.0023
1083 - 1100	0.005	~0.0005

Continuous Radiation (^{213}Bi)

⟨β-⟩=444 keV;⟨IB⟩=0.59 keV

E_{bin}(keV)		⟨ ⟩(keV)	(%)
0 - 10	β-	0.055	1.09
	IB	0.020	
10 - 20	β-	0.165	1.10
	IB	0.019	0.133
20 - 40	β-	0.67	2.23
	IB	0.037	0.127
40 - 100	β-	4.84	6.9
	IB	0.096	0.149
100 - 300	β-	48.2	24.1

Continuous Radiation (^{213}Bi) (continued)

E_{bin}(keV)		⟨ ⟩(keV)	(%)
	IB	0.22	0.125
300 - 600	β-	146	32.9
	IB	0.148	0.036
600 - 1300	β-	242	29.5
	IB	0.053	0.0070
1300 - 1420	β-	1.88	0.141
	IB	1.40×10^{-5}	1.06×10^{-6}

$^{213}_{84}$Po(4.2 *8* μs)

Mode: α
Δ: -6676 *5* keV
SpA: 7.6×10^{10} Ci/g
Prod: daughter ^{213}Bi; descendant ^{225}Ac

Alpha Particles (^{213}Po)

⟨α⟩=8375 *7* keV

α(keV)	α(%)
7614 *10*	0.003 *1*
8375 *5*	100

$^{213}_{85}$At(110 *20* ns)

Mode: α
Δ: -6603 *13* keV
SpA: 7.6×10^{10} Ci/g
Prod: descendant ^{229}Np; descendant ^{225}Pa

Alpha Particles (^{213}At)

α(keV)
9080 *12*

$^{213}_{86}$Rn(25.0 *2* ms)

Mode: α
Δ: -5723 *9* keV
SpA: 7.64×10^{10} Ci/g
Prod: daughter ^{217}Ra; daughter ^{213}Fr

Alpha Particles (^{213}Rn)

⟨α⟩=8082 keV

α(keV)	α(%)
7552 *8*	1.0
8087 *8*	99

$^{213}_{87}$Fr(34.6 *3* s)

Mode: α(99.45 *3* %), ε(0.55 *3* %)
Δ: -3573 *9* keV
SpA: 1.516×10^{9} Ci/g
Prod: ^{205}Tl(^{12}C,4n); ^{208}Pb(^{11}B,6n); protons on Th

Alpha Particles (^{213}Fr)

α(keV)
6775 *2*

$^{213}_{88}$Ra(2.74 *6* min)

Mode: α(80 *5* %), ε(20 *5* %)
Δ: 310 *30* keV
SpA: 3.21×10^{8} Ci/g
Prod: ^{206}Pb(^{12}C,5n); ^{194}Pt(^{22}Ne,3n); protons on Th; ^{209}Bi(^{11}B,7n)

Alpha Particles (^{213}Ra)

⟨α⟩=5352 *3* keV

α(keV)	α(%)
6409 *5*	~0.32
6521 *3*	4.8 *8*
6622 *3*	39.2 *16*
6730 *3*	36.0 *16*

Photons (^{213}Ra)

⟨γ⟩=9.6 *19* keV

γ_{mode}	γ(keV)	γ(%)†
γ_α[M1+E2]	102.40 *15*	0.31 *10*
γ_α[E2]	110.10 *9*	6.4 *16*
γ_α[M1+E2]	212.50 *15*	1.1 *3*

† 31% uncert(syst)

$^{213}_{88}$Ra(2.1 *1* ms)

Mode: IT(~ 99 %), α(~ 1 %)
Δ: 2080 *31* keV
SpA: 7.6×10^{10} Ci/g
Prod: ^{209}Bi(^{10}B,6n); ^{204}Pb(^{12}C,3n); ^{206}Pb(^{12}C,5n)

Alpha Particles (^{213}Ra)

α(keV)	α(rel)
8259 *4*	~3
8359 *4*	28 *6*
8467 *4*	69 *7*

Photons (^{213}Ra)

⟨γ⟩=1635 *34* keV

γ_{mode}	γ(keV)	γ(%)†
Rn L_ℓ	10.137	~0.00045
Ra L_ℓ	10.622	0.42 *5*
Rn L_α	11.713	~0.008
Ra L_α	12.325	7.5 *6*
Rn L_η	12.855	0.00018 *9*
Ra L_η	13.662	0.222 *17*
Rn L_β	14.339	0.010 *5*
Ra L_β	15.228	11.4 *9*
Rn L_γ	16.882	0.0022 *11*
Ra L_γ	17.962	2.59 *23*
Rn $K_{\alpha2}$	81.067	~0.006
Rn $K_{\alpha1}$	83.787	~0.011
Ra $K_{\alpha2}$	85.429	3.70 *19*
Ra $K_{\alpha1}$	88.471	6.1 *3*
Rn $K_{\beta1}$'	94.677	~0.004
Rn $K_{\beta2}$'	97.907	~0.0012
Ra $K_{\beta1}$'	99.915	2.18 *11*
γ_α[M1+E2]	102.40 *15*	0.0026 *11*
Ra $K_{\beta2}$'	103.341	0.71 *4*
γ_α[E2]	110.10 *9*	0.005 *2*
γ_{IT} (E2)	160.87 *5*	43.7 *19*
γ_α[M1+E2]	212.50 *15*	0.0088 *22*
γ_{IT} (E2)	546.35 *5*	99 *3*
γ_{IT} (E2)	1062.5 *2*	95 *3*

Atomic Electrons (^{213}Ra)

⟨e⟩=104.2 *24* keV

e_{bin}(keV)	⟨e⟩(keV)	e(%)
4 - 17	1.63	10.6 *12*
18	2.03	11.0 *13*
19	0.104	0.54 *6*
57	5.9	10.4 *6*
62 - 110	0.328	0.433 *18*
114	0.007	~0.007
142	33.1	23.3 *14*
145	18.2	12.5 *7*
156	9.6	6.2 *4*
157	5.5	3.48 *21*
158	0.119	0.076 *5*
160	4.16	2.60 *15*
161 - 210	1.11	0.69 *4*
212	0.00035	~0.00017
442	9.4	2.13 *10*
527 - 546	5.48	1.030 *25*
959	5.8	0.60 *3*
1043 - 1062	1.77	0.169 *5*

$^{213}_{89}$Ac(800 *50* ms)

Mode: α
Δ: 6090 *80* keV
SpA: 4.4×10^{10} Ci/g
Prod: ^{197}Au(^{20}Ne,4n); ^{203}Tl(^{16}O,6n); ^{209}Bi(^{12}C,8n); ^{205}Tl(^{16}O,8n)

Alpha Particles (^{213}Ac)

α(keV)
7364 *8*

$^{213}_{90}$Th(140 *40* ms)

Mode: α
Δ: 12060 *230* keV syst
SpA: 7.6×10^{10} Ci/g
Prod: ^{206}Pb(^{16}O,9n)

Alpha Particles (^{213}Th)

α(keV)
7692 *10* ?

A = 214

NDS **21**, 437 (1977)

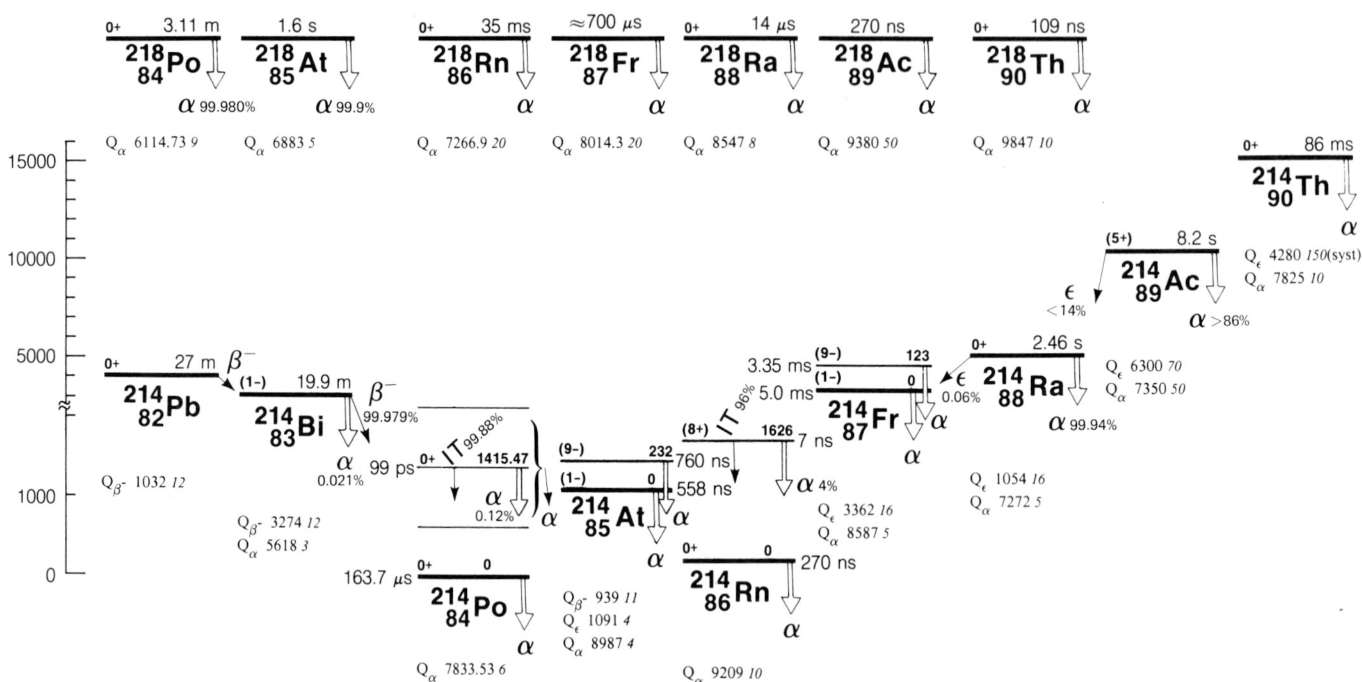

$^{214}_{82}$Pb(26.8 *9* min)

Mode: β-
Δ: -188 *3* keV
SpA: 3.28×10^{7} Ci/g
Prod: descendant ^{226}Ra

Photons (^{214}Pb)

⟨γ⟩=250 *3* keV

γ_{mode}	γ(keV)	γ(%)†
Bi L_ℓ	9.419	0.29 *3*
Bi L_α	10.828	5.8 *3*
Bi L_η	11.712	0.069 *4*
Bi L_β	13.088	5.7 *5*
Bi L_γ	15.439	1.16 *12*
γ M1(+E2)	53.172 *18*	1.10 *5*
Bi $K_{\alpha2}$	74.814	6.52 *22*
Bi $K_{\alpha1}$	77.107	11.0 *4*
Bi $K_{\beta1}$'	87.190	3.88 *13*
Bi $K_{\beta2}$'	90.128	1.15 *4*
γ	137.4 *4*	0.059 *20*
γ	141.3 *8*	~0.039
γ	196.3 *6*	0.049 *20*
γ [M1+E2]	205.77 *6*	<0.015 ?
γ	238.40 *7*	<0.015 ?
γ M1(+E2)	241.92 *3*	7.46 *15*
γ [M1]	258.94 *6*	0.55 *2*
γ [M1+E2]	274.56 *6*	0.32 *5*
γ M1+7.7%E2	295.091 *24*	19.2 *4*
γ	298.69 *5*	<0.020 ?
γ	305.50 *6*	~0.023
γ	314.2 *5*	0.079 *20*
γ	324.3 *6*	~0.020
γ M1(+E2)	351.87 *4*	37.1 *7*
γ	462.05 *25*	0.17 *3*
γ	470.8 *6*	~0.010 ?
γ	480.32 *6*	0.338 *18*
γ	487.13 *5*	0.439 *14*
γ	511.0 *5*	0.029 *10*
γ	533.50 *6*	0.190 *14*
γ	538.7 *5*	~0.0049
γ [E1]	543.91 *3*	0.023 *8*
γ [E1]	580.06 *5*	0.364 *15*
γ	765.9 *6*	0.079 *20* ?
γ [E1]	785.827 *20*	1.09 *3*
γ [E1]	838.999 *18*	0.587 *22*

† 1.7% uncert(syst)

Atomic Electrons (^{214}Pb)

⟨e⟩=73.9 *15* keV

e_{bin}(keV)	⟨e⟩(keV)	e(%)
13 - 16	2.01	13.7 *10*
37	3.93	10.7 *6*
40 - 87	2.29	4.37 *20*
106 - 148	0.18	0.15 *5*
151	8.2	5.41 *24*
168 - 203	0.76	0.44 *7*
205	15.4	7.5 *3*
206 - 225	0.05	~0.023
226	2.09	0.93 *4*
229 - 259	0.96	0.398 *17*
261	25.2	9.6 *4*
271 - 275	0.021	~0.008
279	3.68	1.32 *7*
282 - 324	1.32	0.454 *19*
335	5.03	1.50 *7*
336 - 380	2.35	0.679 *21*
390 - 420	0.19	~0.05
443 - 490	0.12	0.026 *9*
495 - 544	0.014	0.0026 *13*
564 - 580	0.00265	0.000467 *19*
690	0.036	0.0052 *12*
748 - 786	0.0216	0.00287 *24*
823 - 839	0.00298	0.000361 *15*

Continuous Radiation (^{214}Pb)

$\langle\beta-\rangle$=220 keV;$\langle$IB$\rangle$=0.163 keV

E_{bin}(keV)		$\langle\ \rangle$(keV)	(%)
0 - 10	β-	0.147	2.95
	IB	0.0109	
10 - 20	β-	0.435	2.91
	IB	0.0102	0.071
20 - 40	β-	1.71	5.7
	IB	0.018	0.065
40 - 100	β-	11.3	16.3
	IB	0.043	0.067
100 - 300	β-	84	43.4
	IB	0.065	0.040
300 - 600	β-	112	27.4
	IB	0.0152	0.0041
600 - 1024	β-	10.3	1.51
	IB	0.00054	8.4×10^{-5}

$^{214}_{83}$Bi(19.9 *4* min)

Mode: β-(99.979 *1* %), α(0.021 *1* %), β-α(0.003 %)

Δ: -1219 *12* keV

SpA: 4.41×10^{7} Ci/g

Prod: descendant ^{226}Ra

Alpha Particles (^{214}Bi)

$\langle\alpha\rangle$=1.428 keV

α(keV)	α(%)
4941 *2*	5.3 *11* $\times10^{-5}$
5023 *2*	4.4 *8* $\times10^{-5}$
5184 *1*	0.000128 *13*
5268 *3*	0.001218 *21*
5450 *2*	0.01132 *6*
5513 *3*	0.00823 *6*
8286 *6*	0.00012 •
8429 *6*	6×10^{-5} •
8949 *8*	2.0×10^{-5} •
9080 *6*	0.0022 •
9319 *6*	5×10^{-5} •
9377 *8*	2.0×10^{-5} •
9499 *6*	0.00010 •
9669 *8*	4×10^{-5} •
9801 *6*	0.00012 •
9906 *6*	7×10^{-5} •
10081 *6*	0.00014 •
10149 *8*	2.0×10^{-5} •
10331 *6*	8×10^{-5} •
10504 *10*	2.0×10^{-5} •

• Long-range α (from levels in ^{214}Po)

Photons (^{214}Bi)

$\langle\gamma\rangle$=1508 *9* keV

γ_{mode}	γ(keV)	γ(%)†
Po L_ℓ	9.658	0.0200 *20*
Po L_α	11.119	0.383 *23*
Po L_η	12.085	0.0055 *4*
Po L_β	13.502	0.37 *3*
Po L_γ	15.874	0.071 *7*
γ_α	62.5	
Po $K_{\alpha2}$	76.858	0.58 *3*
Po $K_{\alpha1}$	79.290	0.97 *5*

Photons (^{214}Bi)
(continued)

γ_{mode}	γ(keV)	γ(%)†
Po $K_{\beta1}$'	89.639	0.345 *16*
Po $K_{\beta2}$'	92.673	0.105 *5*
γ_α	191.1	
$\gamma_{\beta-}$	273.7 *4*	0.18 *3*
$\gamma_{\beta-}$	280.93 *12*	0.081 *14*
$\gamma_{\beta-}$	286.9 *6*	0.032 *7*
$\gamma_{\beta-}$	304.42 *12*	0.034 *12*
$\gamma_{\beta-}$	333.60 *12*	0.095 *18*
$\gamma_{\beta-}$	334.9 *5*	0.057 *11*
$\gamma_{\beta-}$	338.5 *6*	0.039 *8*
$\gamma_{\beta-}$	347.1 *8*	~0.06
$\gamma_{\beta-}$	364.2 *6*	0.0063 *13*
$\gamma_{\beta-}$	376.6 *6*	0.0049 *25*
$\gamma_{\beta-}$[M1+E2]	386.834 *24*	0.36 *6*
$\gamma_{\beta-}$[M1]	388.95 *3*	0.41 *5*
$\gamma_{\beta-}$	394.0 *10*	0.009 *4*
$\gamma_{\beta-}$	396.00 *12*	0.030 *6*
$\gamma_{\beta-}$	405.73 *3*	0.167 *10*
$\gamma_{\beta-}$	426.5 *5*	0.11 *3*
$\gamma_{\beta-}$	440.4 *6*	0.029 *6*
$\gamma_{\beta-}$	454.832 *23*	0.318 *14*
$\gamma_{\beta-}$	469.76 *3*	0.133 *8*
$\gamma_{\beta-}$[M1+E2]	474.51 *11*	0.118 *12*
$\gamma_{\beta-}$	494.6 *10*	~0.009
$\gamma_{\beta-}$	502.2 *6*	0.018 *4*
$\gamma_{\beta-}$	520.4 *6*	0.0057 *24*
$\gamma_{\beta-}$	525.0 *6*	0.016 *6*
$\gamma_{\beta-}$	536.93 *20*	0.071 *9*
$\gamma_{\beta-}$[M1+E2]	542.84 *11*	0.085 *9*
$\gamma_{\beta-}$	547.1 *3*	0.032 *7*
$\gamma_{\beta-}$	572.67 *3*	0.082 *6*
$\gamma_{\beta-}$	596.0 *8*	~0.012
$\gamma_{\beta-}$E2	609.311 *13*	46.1 •
$\gamma_{\beta-}$	615.77 *6*	0.07 *3*
$\gamma_{\beta-}$	617.1 *3*	0.034 *12*
$\gamma_{\beta-}$	626.4 *6*	0.0049 *25*
$\gamma_{\beta-}$	631.2 *4*	0.017 *6*
$\gamma_{\beta-}$	633.14 *4*	0.060 *6*
$\gamma_{\beta-}$	639.36 *20*	0.031 *5*
$\gamma_{\beta-}$	649.18 *4*	0.059 *7*
$\gamma_{\beta-}$	660.75 *11*	0.043 *17*
$\gamma_{\beta-}$E1	665.442 *17*	1.56 *5*
$\gamma_{\beta-}$	683.21 *6*	0.079 *9*
$\gamma_{\beta-}$	687.7 *6*	~0.006
$\gamma_{\beta-}$	693.3 *5*	~0.006
$\gamma_{\beta-}$	697.89 *25*	0.037 *7*
$\gamma_{\beta-}$[M1]	703.07 *3*	0.472 *20*
$\gamma_{\beta-}$	710.84 *10*	0.0747 *20*
$\gamma_{\beta-}$E2	719.856 *23*	0.403 *20*
$\gamma_{\beta-}$	723.32 *20*	~0.045
$\gamma_{\beta-}$	727.8 *8*	0.0157 *20*
$\gamma_{\beta-}$	733.64 *10*	0.047 *6*
$\gamma_{\beta-}$[M1+E2]	740.87 *3*	~0.039
$\gamma_{\beta-}$	752.843 *22*	0.133 *10*
$\gamma_{\beta-}$E2+8.6%M1	768.350 *15*	4.88 *10*
$\gamma_{\beta-}$[E1]	786.42 *4*	0.31 *10*
$\gamma_{\beta-}$	799.75 *15*	0.041 *6*
$\gamma_{\beta-}$E2	806.155 *17*	1.23 *3* •
$\gamma_{\beta-}$	814.87 *4*	0.040 *7*
$\gamma_{\beta-}$M1	821.166 *23*	0.150 *16*
$\gamma_{\beta-}$[M1]	826.44 *11*	0.092 *7*
$\gamma_{\beta-}$	832.34 *20*	0.023 *5*
$\gamma_{\beta-}$	847.2 *4*	0.017 *6*
$\gamma_{\beta-}$	904.33 *4*	0.105 *14*
$\gamma_{\beta-}$	915.8 *4*	0.023 *6*
$\gamma_{\beta-}$M1+19%E2	934.039 *18*	3.16 *6*
$\gamma_{\beta-}$	943.3 *4*	0.017 *6*
$\gamma_{\beta-}$	964.07 *3*	0.383 *20*
$\gamma_{\beta-}$	976.2 *10*	~0.023
$\gamma_{\beta-}$	989.2 *6*	~0.012
$\gamma_{\beta-}$	1013.4 *10*	~0.010
$\gamma_{\beta-}$	1020.5 *6*	~0.012
$\gamma_{\beta-}$[E1]	1032.22 *4*	0.096 *18*
$\gamma_{\beta-}$	1038.0 *6*	0.017 *8*
$\gamma_{\beta-}$	1045.4 *6*	0.029 *7*
$\gamma_{\beta-}$[M1+E2]	1051.950 *22*	0.315 *14*
$\gamma_{\beta-}$	1067.30 *4*	0.029 *12*
$\gamma_{\beta-}$[E1]	1070.02 *4*	0.285 *20*
$\gamma_{\beta-}$	1103.7 *3*	~0.10
$\gamma_{\beta-}$	1104.766 *25*	0.080 *4*
$\gamma_{\beta-}$M1+10%E2	1120.273 *18*	15.0 *3*
$\gamma_{\beta-}$	1130.6 *3*	0.045 *12*
$\gamma_{\beta-}$	1133.65 *3*	0.255 *17*

Photons (^{214}Bi)
(continued)

γ_{mode}	γ(keV)	γ(%)†
$\gamma_{\beta-}$M1+24%E2	1155.183 *19*	1.69 *5*
$\gamma_{\beta-}$	1172.93 *4*	0.058 *15*
$\gamma_{\beta-}$	1173.04 *10*	0.058 *15*
$\gamma_{\beta-}$	1207.674 *21*	0.460 *18*
$\gamma_{\beta-}$	1226.8 *6*	0.027 *9*
$\gamma_{\beta-}$	1230.84 *17*	~0.022
$\gamma_{\beta-}$M1	1238.107 *25*	5.92 *12*
$\gamma_{\beta-}$M1	1280.952 *20*	1.47 *5*
$\gamma_{\beta-}$	1303.76 *7*	0.121 *11*
$\gamma_{\beta-}$	1317.02 *12*	0.086 *10*
$\gamma_{\beta-}$	1330.0 *6*	~0.011
$\gamma_{\beta-}$	1341.5 *3*	~0.023
$\gamma_{\beta-}$	1353.0 *8*	0.0045 *12*
$\gamma_{\beta-}$E2	1377.659 *18*	4.02 *9*
$\gamma_{\beta-}$[E1]	1385.295 *23*	0.78 *3*
$\gamma_{\beta-}$	1392.5 *4*	0.019 *9*
$\gamma_{\beta-}$	1401.48 *4*	1.39 *4*
$\gamma_{\beta-}$[E2]	1407.97 *4*	2.48 *5*
$\gamma_{\beta-}$	1419.7 *6*	0.0051 *13*
$\gamma_{\beta-}$	1471.1 *6*	~0.012
$\gamma_{\beta-}$	1479.19 *10*	0.069 *8*
$\gamma_{\beta-}$[M1+E2]	1509.22 *3*	2.19 *6*
$\gamma_{\beta-}$	1538.49 *6*	0.41 *6*
$\gamma_{\beta-}$	1543.347 *22*	0.35 *5*
$\gamma_{\beta-}$M1	1583.22 *3*	0.72 *3*
$\gamma_{\beta-}$[M1+E2]	1594.78 *11*	0.265 *20*
$\gamma_{\beta-}$	1599.30 *6*	0.334 *20*
$\gamma_{\beta-}$	1636.6 *4*	0.019 *6*
$\gamma_{\beta-}$	1657.36 *20*	0.07 *3*
$\gamma_{\beta-}$E2	1661.258 *25*	1.15 *4*
$\gamma_{\beta-}$	1683.99 *4*	0.236 *20*
$\gamma_{\beta-}$E2	1729.580 *21*	3.05 *6*
$\gamma_{\beta-}$M1	1764.490 *22*	15.9 *3*
$\gamma_{\beta-}$	1782.1 *10*	0.016 *6*
$\gamma_{\beta-}$	1814.01 *24*	0.012 *5*
$\gamma_{\beta-}$[E1]	1838.37 *4*	0.383 *20*
$\gamma_{\beta-}$	1847.41 *3*	2.12 *7*
$\gamma_{\beta-}$	1873.112 *20*	0.226 *20*
$\gamma_{\beta-}$	1890.259 *24*	0.089 *11*
$\gamma_{\beta-}$	1896.28 *16*	0.177 *20*
$\gamma_{\beta-}$	1898.9 *3*	0.063 *24*
$\gamma_{\beta-}$	1935.8 *4*	0.051 *23*
$\gamma_{\beta-}$	1994.7 *15*	~0.005
$\gamma_{\beta-}$	2004.5 *10*	~0.0029
$\gamma_{\beta-}$	2010.79 *4*	0.049 *5*
$\gamma_{\beta-}$	2021.7 *3*	0.019 *3*
$\gamma_{\beta-}$	2052.93 *15*	0.070 *7*
$\gamma_{\beta-}$	2085.0 *5*	0.010 *3*
$\gamma_{\beta-}$	2089.55 *14*	0.056 *6*
$\gamma_{\beta-}$	2109.91 *7*	0.087 *8*
$\gamma_{\beta-}$M1	2118.53 *3*	1.21 *3*
$\gamma_{\beta-}$	2147.80 *6*	0.016 *3*
$\gamma_{\beta-}$	2176.8 *6*	0.0039 *20*
$\gamma_{\beta-}$	2192.52 *4*	0.061 *11*
$\gamma_{\beta-}$M1	2204.09 *11*	4.99 *10*
$\gamma_{\beta-}$	2251.2 *4*	~0.007
$\gamma_{\beta-}$	2259.7 *4*	0.009 *4*
$\gamma_{\beta-}$	2266.67 *20*	0.018 *3*
$\gamma_{\beta-}$	2270.0 *25*	~0.0029
$\gamma_{\beta-}$	2284.4 *7*	0.0051 *3*
$\gamma_{\beta-}$	2293.29 *4*	0.324 *20*
$\gamma_{\beta-}$	2312.2 *4*	0.012 *3*
$\gamma_{\beta-}$	2324.8 *10*	~0.0019
$\gamma_{\beta-}$	2331.2 *3*	0.022 *3*
$\gamma_{\beta-}$	2360.9 *6*	0.0019 *6*
$\gamma_{\beta-}$	2369.3 *6*	~0.0029
$\gamma_{\beta-}$	2376.99 *20*	0.0118 *20*
$\gamma_{\beta-}$	2390.9 *6*	0.0020 *6*
$\gamma_{\beta-}$	2423.31 *24*	0.0059 *10*
$\gamma_{\beta-}$E1	2447.68 *4*	1.55 *3*
$\gamma_{\beta-}$	2482.417 *23*	0.0021 *9*
$\gamma_{\beta-}$	2505.58 *16*	0.0059 *10*
$\gamma_{\beta-}$	2551.0 *10*	~0.00039
$\gamma_{\beta-}$	2604.5 *10*	0.00045 *12*
$\gamma_{\beta-}$	2631.0 *3*	~0.0009
$\gamma_{\beta-}$	2662.23 *15*	0.00029 *15*
$\gamma_{\beta-}$	2694.67 *12*	0.0324 *20*
$\gamma_{\beta-}$	2698.86 *14*	0.0028 *9*
$\gamma_{\beta-}$	2719.21 *7*	0.0018 *5*
$\gamma_{\beta-}$	2769.99 *20*	0.0256 *20*
$\gamma_{\beta-}$	2786.09 *20*	0.0059 *10*
$\gamma_{\beta-}$	2827.0 *3*	0.0025 *4*
$\gamma_{\beta-}$	2860.9 *8*	0.00034 *17*
$\gamma_{\beta-}$	2880.4 *3*	0.0092 *12*

Photons (^{214}Bi) (continued)

γ_{mode}	γ(keV)	γ(%)†
$\gamma_{\beta-}$	2893.59 *20*	0.0064 *11*
$\gamma_{\beta-}$	2922.09 *20*	0.0157 *10*
$\gamma_{\beta-}$	2928.7 *8*	0.0012 *6*
$\gamma_{\beta-}$	2934.9 *8*	0.00057 *24*
$\gamma_{\beta-}$	2940.5 *3*	0.0017 *6*
$\gamma_{\beta-}$	2978.79 *20*	0.0147 *10*
$\gamma_{\beta-}$	2988.7 *10*	0.0011 *4*
$\gamma_{\beta-}$	2999.99 *20*	0.0088 *20*
$\gamma_{\beta-}$	3053.89 *20*	0.0226 *20*
$\gamma_{\beta-}$	3081.7 *3*	0.0043 *8*
$\gamma_{\beta-}$	3093.9 *8*	0.00051 *17*
$\gamma_{\beta-}$	3136.3 *10*	0.00034 *12*
$\gamma_{\beta-}$	3142.6 *4*	0.0016 *6*
$\gamma_{\beta-}$	3160.5 *7*	0.00051 *23*
$\gamma_{\beta-}$	3183.6 *4*	0.0015 *5*
$\gamma_{\beta-}$	3233.3 *25*	~0.00020
$\gamma_{\beta-}$	3269.7 *25*	~10×10^{-5}

†1.7% uncert(syst)

• with ^{214}Po(99 ps) in equilib

Atomic Electrons (^{214}Bi)

⟨e⟩=21.4 *6* keV

e_{bin}(keV)	⟨e⟩(keV)	e(%)
14 - 63	0.176	1.07 *8*
65 - 89	0.0220	0.0298 *12*
181 - 211	0.17	~0.09
241 - 288	0.18	0.072 *24*
291 - 338	0.54	0.18 *4*
343 - 392	0.31	0.084 *24*
393 - 441	0.051	0.012 *4*
444 - 492	0.09	0.020 *6*
494 - 511	0.0037	~0.0007
516	3.57	0.692 *18*
517 - 559	0.065	0.012 *4*
568 - 590	0.064	0.0112 *21*
592	0.622	0.105 *3*
593 - 641	1.127	0.187 *4*
644 - 692	0.46	0.068 *5*
693 - 741	0.182	0.0253 *13*
749 - 798	0.153	0.0201 *7*
799 - 833	0.035	0.0043 *12*
841	0.50	0.060 *3*
843 - 891	0.04	~0.005
896 - 945	0.132	0.0143 *6*
947 - 996	0.053	0.0055 *19*
997 - 1024	0.015	~0.0015
1027	1.97	0.192 *10*
1028 - 1070	0.23	0.021 *3*
1080 - 1128	0.51	0.046 *3*
1129 - 1141	0.040	0.0035 *5*
1145	0.716	0.0626 *18*
1151 - 1194	0.188	0.0159 *7*
1204 - 1248	0.184	0.0150 *7*
1260 - 1308	0.31	0.024 *5*
1313 - 1317	0.103	0.0079 *4*
1322	4.9	0.37 *4*
1325 - 1374	0.043	0.00315 *8*
1375 - 1398	0.044	0.0032 *8*
1399	1.02	0.073 *15*
1400 - 1445	0.17	0.012 *4*
1450 - 1543	0.153	0.0103 *14*
1564 - 1658	0.196	0.0122 *6*
1667 - 1670	0.0019	~0.00011
1671	1.14	0.0683 *19*
1680 - 1779	0.36	0.021 *3*
1780 - 1879	0.047	0.0026 *7*
1880 - 1978	0.085	0.0044 *8*
1981 - 2076	0.089	0.00441 *25*
2081 - 2179	0.275	0.0130 *4*
2187 - 2284	0.068	0.0031 *3*
2289 - 2388	0.0196	0.00083 *3*
2389 - 2489	0.00391	0.000161 *6*
2492 - 2591	5.3 ×10^{-5}	2.1 *8* ×10^{-6}
2600 - 2696	0.0019	7.0 *21* ×10^{-5}
2702 - 2800	0.00059	2.1 *5* ×10^{-5}

Atomic Electrons (^{214}Bi) (continued)

e_{bin}(keV)	⟨e⟩(keV)	e(%)
2810 - 2908	0.0011	3.9 *9* ×10^{-5}
2912 - 3001	0.00072	2.4 *8* ×10^{-5}
3037 - 3133	0.00021	6.9 *17* ×10^{-6}
3138 - 3230	1.6 ×10^{-5}	5.2 *15* ×10^{-7}
3253 - 3266	4.1 ×10^{-7}	~1×10^{-8}

Continuous Radiation (^{214}Bi)

⟨β-⟩=641 keV;⟨IB⟩=1.21 keV

E_{bin}(keV)		⟨ ⟩(keV)	(%)
0 - 10	β-	0.0439	0.87
	IB	0.026	
10 - 20	β-	0.132	0.88
	IB	0.025	0.17
20 - 40	β-	0.54	1.79
	IB	0.049	0.169
40 - 100	β-	3.92	5.6
	IB	0.132	0.20
100 - 300	β-	39.8	19.8
	IB	0.33	0.19
300 - 600	β-	125	28.1
	IB	0.30	0.071
600 - 1300	β-	293	33.7
	IB	0.26	0.032
1300 - 2500	β-	158	9.0
	IB	0.085	0.0053
2500 - 3270	β-	21.4	0.79
	IB	0.0017	6.6 ×10^{-5}

$^{214}_{84}$Po(163.69 *13* μs)

Mode: α
Δ: -4494 *4* keV
SpA: 7.606×10^{10} Ci/g
Prod: descendant ^{226}Ra; descendant ^{230}U

Alpha Particles (^{214}Po)

⟨α⟩=7686.79 *11* keV

α(keV)	α(%)
6611.5 *10*	6 *2*×10^{-5}
6903.96 *12*	0.010
7686.90 *6*	99.9895 *6*

Photons (^{214}Po)

⟨γ⟩=0.083 *5* keV

γ_{mode}	γ(keV)	γ(%)
γ E2	298.1 *10*	5 *2*×10^{-5}
γ E2	797.88 *10*	0.0104 *6*

$^{214}_{84}$Po(99 *3* ps)

Mode: IT(99.875 *7* %), α(0.125 *7* %)
Δ: -3079 *4* keV
SpA: 7.61×10^{10} Ci/g
Prod: descendant ^{226}Ra

Alpha Particles (^{214}Po)

α(keV)
9080 *6*

Photons (^{214}Po)

⟨γ⟩=1033 *16* keV

γ_{mode}	γ(keV)	γ(%)†
γ_{IT} E2	609.311 *13*	72.6 *15*
γ_{IT} E2	806.155 *17*	73.3 *15*

†<0.1% uncert(syst)

$^{214}_{85}$At(558 *8* ns)

Mode: α
Δ: -3403 *6* keV
SpA: 7.61×10^{10} Ci/g
Prod: descendant ^{226}Pa

Alpha Particles (^{214}At)

⟨α⟩=8819 keV

α(keV)	α(%)
8272 *15*	<0.1
8482 *15*	<0.2
8819 *4*	100

$^{214}_{85}$At(760 *15* ns)

Mode: α
Δ: -3171 *6* keV
SpA: 7.61×10^{10} Ci/g
Prod: daughter ^{218}Fr

Alpha Particles (^{214}At)

α(keV)
8762 *5*

$^{214}_{86}$Rn(270 *20* ns)

Mode: α
Δ: -4342 *11* keV
SpA: 7.6×10^{10} Ci/g
Prod: descendant ^{222}Th

Alpha Particles (^{214}Rn)

α(keV)
9037 *10*

$^{214}_{86}$Rn(7.3 *15* ns)

Mode: IT(95.9 *7* %), α(4.1 *7* %)
Δ: -2716 *11* keV
SpA: 7.6×10^{10} Ci/g
Prod: ^{208}Pb(^{12}C,α2n)

Alpha Particles (^{214}Rn)

α(keV)
10630 *30*

$^{214}_{87}$Fr(5.0 *2* ms)

Mode: α
Δ: -980 *12* keV
SpA: 7.6×10^{10} Ci/g
Prod: daughter ^{218}Ac; ^{208}Pb(^{11}B,5n); daughter ^{214}Ra

Alpha Particles (^{214}Fr)

⟨α⟩=8407 keV

α(keV)	α(%)
7409 *3*	0.3
7605 *8*	1.0
7940 *3*	1.0
8355 *3*	4.7
8427 *3*	93

$^{214}_{87}$Fr(3.35 *5* ms)

Mode: α
Δ: -857 *12* keV
SpA: 7.61×10^{10} Ci/g
Prod: ^{208}Pb(^{11}B,5n)

Alpha Particles (^{214}Fr)

⟨α⟩=8500.5 keV

α(keV)	α(%)
7341 *8*	0.05
7594 *5*	0.5
7708 *5*	1.1
7963 *5*	0.7
8046 *5*	0.9
8476 *4*	51
8547 *4*	46

$^{214}_{88}$Ra(2.46 *3* s)

Mode: α(99.94 *1* %), ϵ(0.06 *1* %)
Δ: 74 *12* keV
SpA: 1.868×10^{10} Ci/g
Prod: ^{209}Bi(^{11}B,6n); ^{206}Pb(^{12}C,4n); ^{194}Pt(^{22}Ne,2n); protons on Th

Alpha Particles (^{214}Ra)

α(keV)
7136 *4*

$^{214}_{89}$Ac(8.2 *2* s)

Mode: α(>86 %), ϵ(<14 %)
Δ: 6370 *70* keV
SpA: 6.16×10^{9} Ci/g
Prod: ^{203}Tl(^{16}O,5n); ^{209}Bi(^{12}C,7n); ^{205}Tl(^{16}O,7n); ^{197}Au(^{20}Ne,3n)

Alpha Particles (^{214}Ac)

α(keV)	α(rel)
7002 *15*	3.4 *9*
7082 *5*	37.8 *17*
7214 *5*	44.7 *17*

$^{214}_{90}$Th(86 *16* ms)

Mode: α
Δ: 10650 *130* keV syst
SpA: 7.6×10^{10} Ci/g
Prod: ^{206}Pb(^{16}O,8n)

Alpha Particles (^{214}Th)

α(keV)
7677 *10*

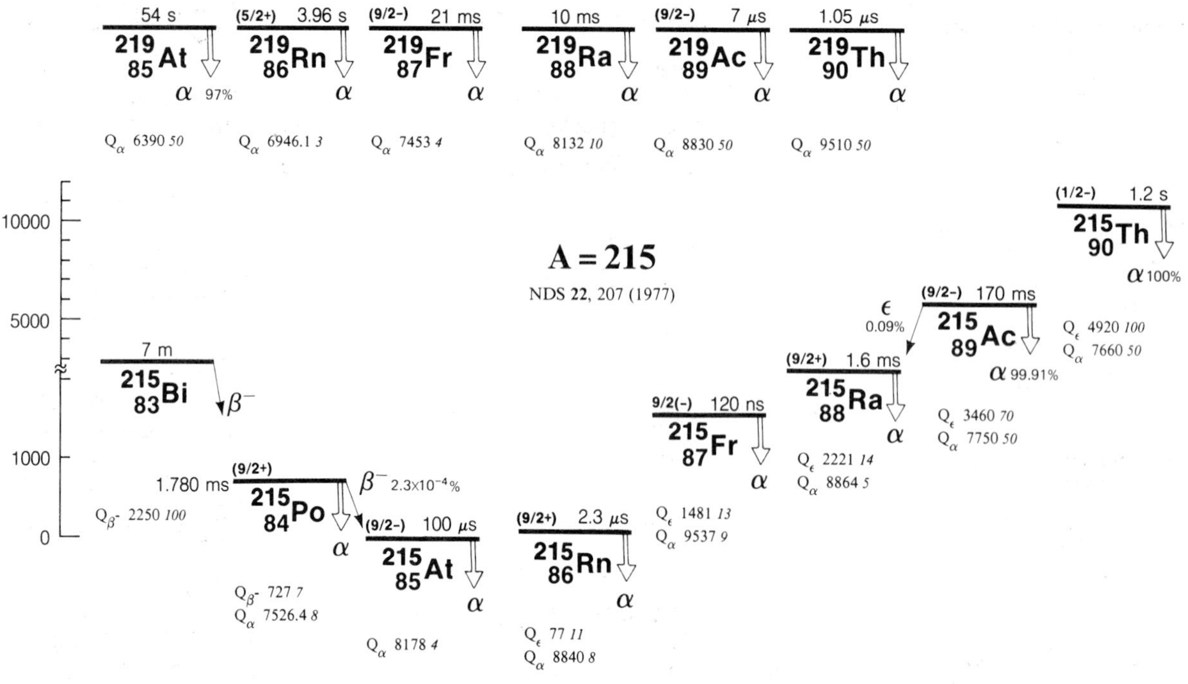

$^{215}_{83}$Bi(7.4 *6* min)

Mode: β-
Δ: 1710 *100* keV
SpA: 1.181×10^{8} Ci/g
Prod: descendant ^{227}Ac; natural source

$^{215}_{84}$Po(1.780 *4* ms)

Mode: α, β-(0.00023 *2* %)
Δ: -542 *4* keV
SpA: 7.570×10^{10} Ci/g
Prod: descendant ^{227}Ac

Alpha Particles (^{215}Po)

⟨α⟩=7386.4 *8* keV

α(keV)	α(%)
6950	~0.02
6957	~0.03
7386.4 *8*	100

Photons (^{215}Po)

γ_{mode}	γ(keV)	γ(%)
γ	438.8	~0.04

$^{215}_{85}$At(100 *20* μs)

Mode: α
Δ: -1269 *7* keV
SpA: 7.6×10^{10} Ci/g
Prod: descendant ^{227}Pa

Alpha Particles (^{215}At)

⟨α⟩=8023 *9* keV

α(keV)	α(%)
7626 *6*	0.05 *2*
8023 *6*	99.95

Photons (^{215}At)

γ_{mode}	γ(keV)	γ(%)
γ E2+34%M1	404.86 *3*	0.045 *18*

$^{215}_{86}$Rn(2.3 *1* μs)

Mode: α
Δ: -1192 *9* keV
SpA: 7.6×10^{10} Ci/g
Prod: descendant ^{227}U; descendant ^{223}Th

Alpha Particles (^{215}Rn)

α(keV)
8674 *8*

$^{215}_{87}$Fr(120 *20* ns)

Mode: α
Δ: 289 *10* keV
SpA: 7.6×10^{10} Ci/g
Prod: ^{208}Pb(^{11}B,4n); ^{209}Bi(^{12}C,α2n); descendant ^{223}Pa

Alpha Particles (^{215}Fr)

α(keV)
9360 *8*

$^{215}_{88}$Ra(1.59 *9* ms)

Mode: α
Δ: 2510 *11* keV
SpA: 7.6×10^{10} Ci/g
Prod: ^{209}Bi(^{11}B,5n); daughter ^{219}Th

Alpha Particles (^{215}Ra)

⟨α⟩=8675 *6* keV

α(keV)	α(%)
7883 *6*	2.8 *4*
8171 *3*	1.4 *4*
8700 *3*	95.9 *10*

$^{215}_{89}$Ac(170 *10* ms)

Mode: α(99.91 *2* %), ε(0.09 *2* %)
Δ: 5970 *60* keV
SpA: 7.4×10^{10} Ci/g
Prod: ^{203}Tl(^{16}O,4n); ^{205}Tl(^{16}O,6n); ^{209}Bi(^{12}C,6n)

Alpha Particles (^{215}Ac)

α(keV)
7604 *5*

$^{215}_{90}$Th(1.2 *2* s)

Mode: α
Δ: 10890 *100* keV
SpA: 3.3×10^{10} Ci/g
Prod: ^{206}Pb(^{16}O,7n)

Alpha Particles (^{215}Th)

⟨α⟩=7442 *11* keV

α(keV)	α(%)
7333 *10*	8 *3*
7395 *8*	52 *3*
7524 *8*	40 *3*

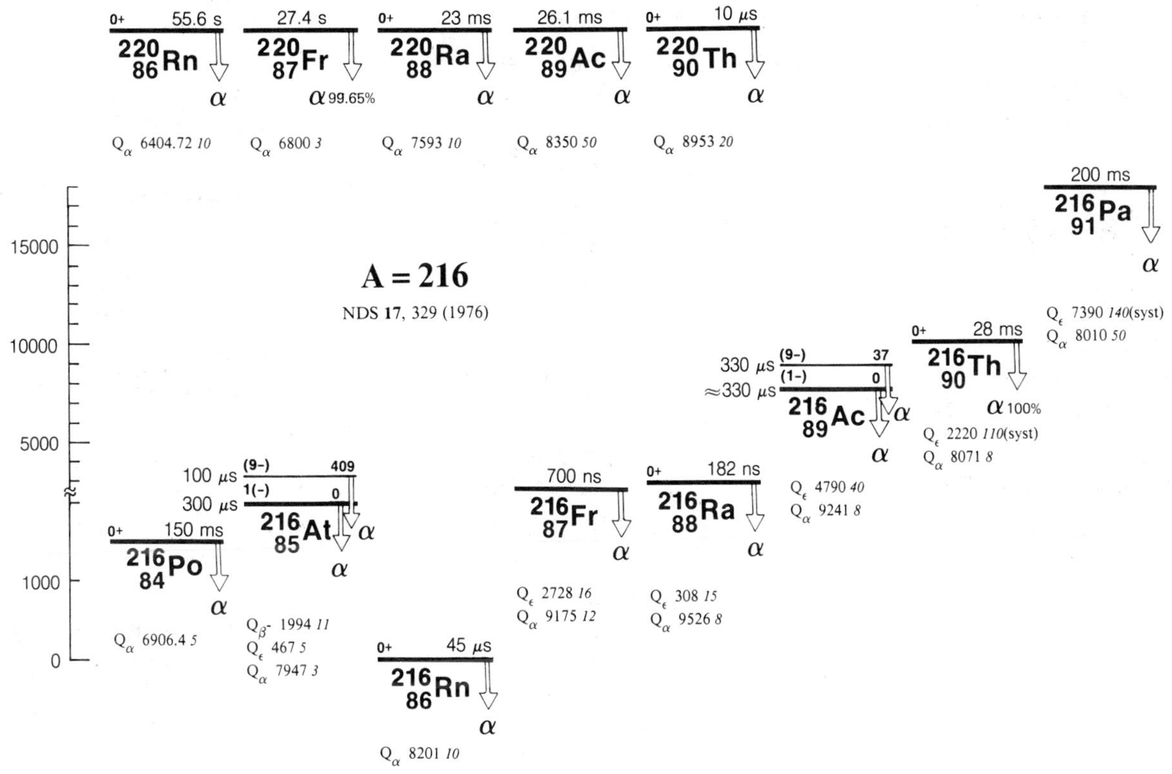

0+ 55.6 s $^{220}_{86}$Rn α Q$_α$ 6404.72 *10*
27.4 s $^{220}_{87}$Fr α 99.65% Q$_α$ 6800 *3*
0+ 23 ms $^{220}_{88}$Ra α Q$_α$ 7593 *10*
26.1 ms $^{220}_{89}$Ac α Q$_α$ 8350 *50*
0+ 10 μs $^{220}_{90}$Th α Q$_α$ 8953 *20*

A = 216

NDS **17**, 329 (1976)

15000
10000
5000
1000
0

200 ms $^{216}_{91}$Pa α Q$_ε$ 7390 *140*(syst) Q$_α$ 8010 *50*
0+ 28 ms $^{216}_{90}$Th α 100% Q$_ε$ 2220 *110*(syst) Q$_α$ 8071 *8*
330 μs (9−) 37; ≈330 μs (1−) 0 $^{216}_{89}$Ac α α Q$_ε$ 4790 *40* Q$_α$ 9241 *8*
0+ 182 ns $^{216}_{88}$Ra α Q$_ε$ 308 *15* Q$_α$ 9526 *8*
700 ns $^{216}_{87}$Fr α Q$_ε$ 2728 *16* Q$_α$ 9175 *12*
0+ 45 μs $^{216}_{86}$Rn α Q$_α$ 8201 *10*
100 μs (9−) 409; 300 μs 1(−) 0 $^{216}_{85}$At α α Q$_{β^-}$ 1994 *11* Q$_ε$ 467 *5* Q$_α$ 7947 *3*
0+ 150 ms $^{216}_{84}$Po α Q$_α$ 6906.4 *5*

$^{216}_{84}$Po(150 ms)

Mode: α
Δ: 1759 *6* keV
SpA: 7×10^{10} Ci/g
Prod: descendant ^{228}Th

Alpha Particles (^{216}Po)

⟨α⟩=6778.5 *7* keV

α(keV)	α(%)
5985	0.0021 *4*
6778.5 *5*	99.9979

Photons (^{216}Po)

$γ_{mode}$	γ(keV)	γ(%)
γ	804.9	0.0018 *4*

$^{216}_{85}$At(300 *30* μs)

Mode: α
Δ: 2226 *6* keV
SpA: 7.5×10^{10} Ci/g
Prod: descendant ^{228}Pa

Alpha Particles (^{216}At)

⟨α⟩=7793 keV

α(keV)	α(%)
7238	0.06
7315	0.085
7393	0.23
7482	<0.1

Alpha Particles (^{216}At)
(continued)

α(keV)	α(%)
7565	
7595	0.20•
7697	2.1
7800 *3*	97

• 7565α + 7595α

$^{216}_{85}$At(100 μs syst)

Mode: α
Δ: 2635 *50* keV
SpA: 8×10^{10} syst Ci/g
Prod: descendant ^{224}Ac

Alpha Particles (^{216}At)

α(keV)
7960 ?

$^{216}_{86}$Rn(45 *5* μs)

Mode: α
Δ: 232 *11* keV
SpA: 7.5×10^{10} Ci/g
Prod: descendant ^{228}U; descendant ^{224}Th

Alpha Particles (^{216}Rn)

α(keV)
8050 *10*

$^{216}_{87}$Fr(700 *20* ns)

Mode: α
Δ: 2960 *13* keV
SpA: 7.54×10^{10} Ci/g
Prod: daughter ^{220}Ac

Alpha Particles (^{216}Fr)

α(keV)
9005 *10*

$^{216}_{88}$Ra(182 *10* ns)

Mode: α
Δ: 3269 *10* keV
SpA: 7.5×10^{10} Ci/g
Prod: ^{208}Pb(^{12}C,4n); ^{209}Bi(^{11}B,4n); descendant ^{224}U

Alpha Particles (^{216}Ra)

α(keV)
9349 *8*

$^{216}_{89}$Ac(~330 μs)

Mode: α
Δ: 8060 *40* keV
SpA: 7.5×10^{10} Ci/g
Prod: ^{209}Bi(^{12}C,5n)

Alpha Particles (^{216}Ac)
⟨α⟩=9062 keV

α(keV)	α(%)
8990 *20*	10
9070 *8*	90

$^{216}_{89}$Ac(330 *20* μs)

Mode: α
Δ: 8097 *41* keV
SpA: 7.5×10^{10} Ci/g
Prod: ^{209}Bi(^{12}C,5n); ^{205}Tl(^{16}O,5n)

Alpha Particles (^{216}Ac)
⟨α⟩=8995 keV

α(keV)	α(%)
8198 *8*	1.7
8283 *8*	2.5
9028 *5*	49.2
9106 *5*	46.2

$^{216}_{90}$Th(28 *2* ms)

Mode: α
Δ: 10270 *100* keV syst
SpA: 7.5×10^{10} Ci/g
Prod: ^{206}Pb(^{16}O,6n)

Alpha Particles (^{216}Th)

α(keV)
7921 *8*

$^{216}_{91}$Pa(200 *40* ms)

Mode: α
Δ: 17660 *100* keV
SpA: 7.3×10^{10} Ci/g
Prod: ^{189}Os(^{31}P,4n); ^{190}Os(^{31}P,5n); ^{197}Au(^{24}Mg,5n)

Alpha Particles (^{216}Pa)

α(keV)
7720
7820
7920

A = 217

NDS 28, 639 (1979)

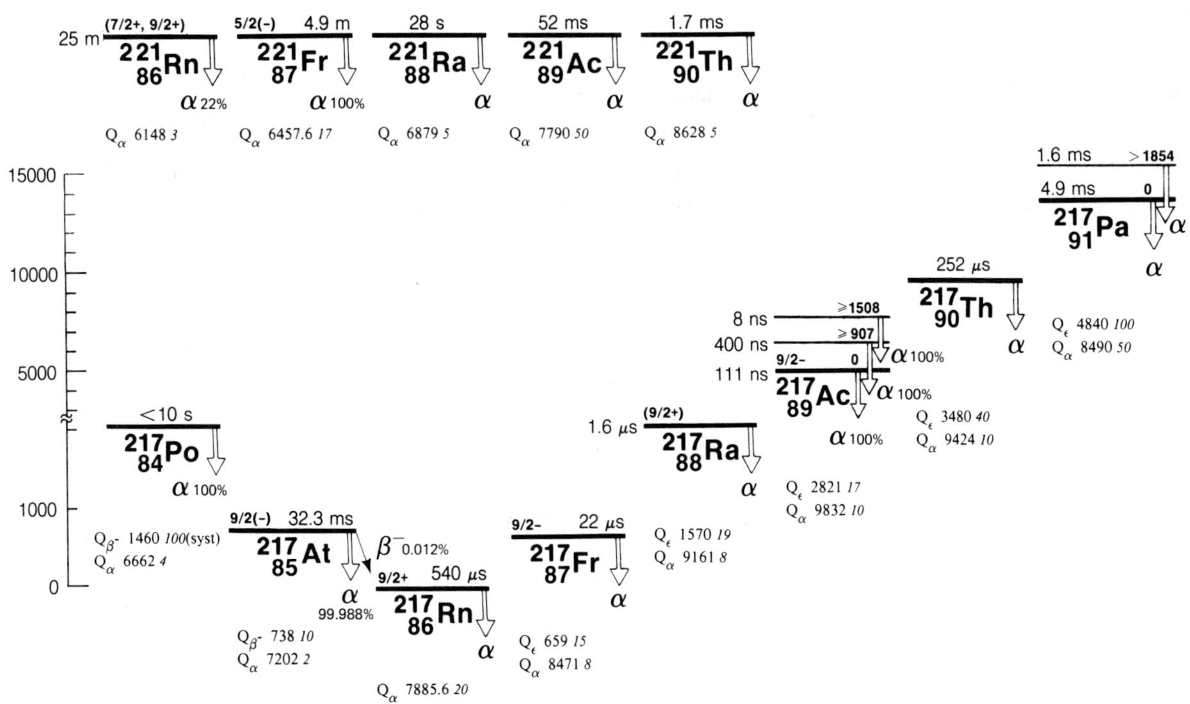

$^{217}_{84}$Po(<10 s)

Mode: α
Δ: 5830 *100* keV syst
SpA: >5×10^9 Ci/g
Prod: daughter ^{221}Ra

Alpha Particles (^{217}Po)

α(keV)
6539 *4*

$^{217}_{85}$At(32.3 *4* ms)

Mode: α(99.988 *4* %), β-(0.012 *4* %)
Δ: 4373 *11* keV
SpA: 7.50×10^{10} Ci/g
Prod: descendant ^{225}Ac

Alpha Particles (^{217}At)

⟨α⟩=7065 *4* keV

α(keV)	α(%)
6483 *5*	~0.02
6609 *7*	~0.01
6812 *3*	0.060 *20*
7067 *3*	99.89 *10*

Photons (^{217}At)

γ_{mode}	γ(keV)
γ	140 *1* ?
γ	166 *1* ?
γ	218 *1*
γ	259.5 *8*
γ	334.5 *8*
γ	375 *1* ?
γ	455 *1* ?
γ	594.0 *8*

$^{217}_{86}$Rn(540 *50* μs)

Mode: α
Δ: 3634 *6* keV
SpA: 7.5×10^{10} Ci/g
Prod: descendant ^{229}U

Alpha Particles (^{217}Rn)

⟨α⟩=7742 keV

α(keV)	α(%)
7500 ?	0.10
7742 *4*	100

$^{217}_{87}$Fr(22 *5* μs)

Mode: α
Δ: 4293 *15* keV
SpA: 7.5×10^{10} Ci/g
Prod: descendant ^{225}Pa; descendant ^{229}Np; descendant ^{221}Ac

Alpha Particles (^{217}Fr)

α(keV)
8315 *8*

$^{217}_{88}$Ra(1.6 *2* μs)

Mode: α
Δ: 5863 *12* keV
SpA: 7.5×10^{10} Ci/g
Prod: descendant ^{221}Th; ^{209}Bi(^{11}B,3n)

Alpha Particles (^{217}Ra)

α(keV)
8992 *8*

$^{217}_{89}$Ac(111 *7* ns)

Mode: α
Δ: 8684 *13* keV
SpA: 7.5×10^{10} Ci/g
Prod: ^{208}Pb(^{14}N,5n)

Alpha Particles (^{217}Ac)

α(keV)	α(%)
9650 *10*	100

$^{217}_{89}$Ac(8 *2* ns)

Mode: α
Δ: >10192 keV
SpA: 7.5×10^{10} Ci/g
Prod: ^{208}Pb(^{14}N,5n)

Alpha Particles (^{217}Ac)

α(keV)	α(%)
11130	100

$^{217}_{91}$Pa(4.9 *5* ms)

Mode: α
Δ: 17000 *90* keV
SpA: 7.5×10^{10} Ci/g
Prod: ^{203}Tl(^{20}Ne,6n); ^{206}Pb(^{20}Ne,p8n)

Alpha Particles (^{217}Pa)

α(keV)
8340 *10*

$^{217}_{89}$Ac(400 *100* ns)

Mode: α
Δ: >9591 keV
SpA: 7.5×10^{10} Ci/g
Prod: ^{208}Pb(^{14}N,5n)

Alpha Particles (^{217}Ac)

α(keV)	α(%)
10540	100

$^{217}_{90}$Th(252 *7* μs)

Mode: α
Δ: 12160 *40* keV
SpA: 7.50×10^{10} Ci/g
Prod: ^{206}Pb(^{16}O,5n)

Alpha Particles (^{217}Th)

α(keV)
9250 *10*

$^{217}_{91}$Pa(1.6 *8* ms)

Mode: α
Δ: >18854 keV
SpA: 8×10^{10} Ci/g
Prod: ^{181}Ta(^{40}Ar,4n)

Alpha Particles (^{217}Pa)

α(keV)
10160 *20*

A = 218

NDS **21**, 467 (1977)

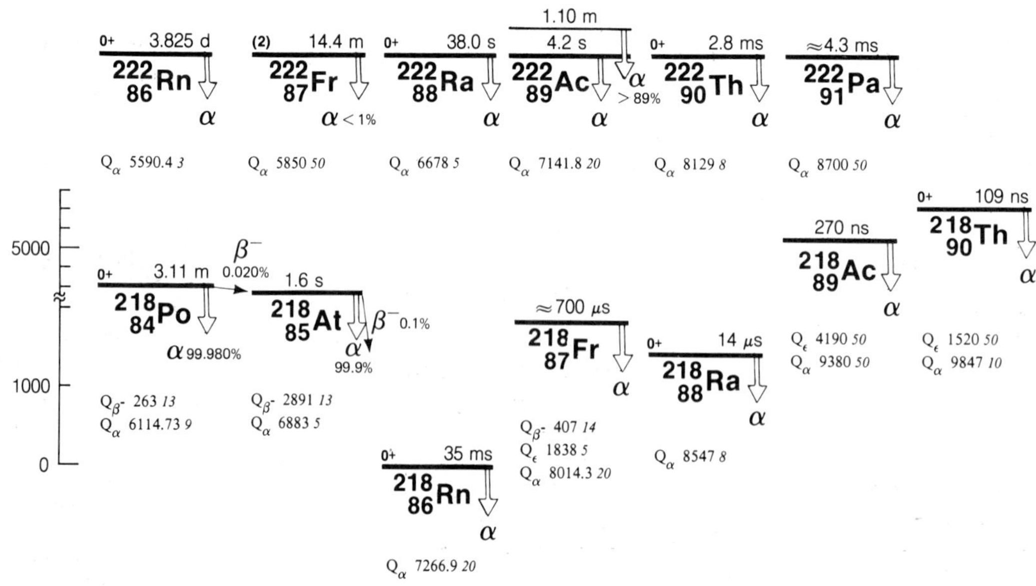

$^{218}_{84}$Po(3.11 *2* min)

Mode: α(99.980 *2* %), β-(0.020 *2* %)
Δ: 8352 *3* keV
SpA: 2.768×10^{8} Ci/g
Prod: descendant ^{226}Ra;

Alpha Particles (^{218}Po)

⟨α⟩=6001.34 *9* keV

α(keV)	α(%)
5181 *2*	0.0011
6002.55 *9*	100

$^{218}_{85}$At(1.6 *4* s)

Mode: α(99.9 %), β-(0.1 %)
Δ: 8089 *13* keV
SpA: 2.6×10^{10} Ci/g
Prod: daughter ^{218}Po

Alpha Particles (^{218}At)

⟨α⟩=6688 keV

α(keV)	α(%)
6654 *6*	6
6695 *4*	90
6748 *4*	4

$^{218}_{86}$Rn(35 *6* ms)

Mode: α
Δ: 5198 *5* keV
SpA: 7.5×10^{10} Ci/g
Prod: descendant ^{230}U

Alpha Particles (^{218}Rn)

⟨α⟩=7129 *2* keV

α(keV)	α(%)
6534.9 *14*	0.16 *4*
7133.1 *14*	99.8 *1*

Photons (^{218}Rn)

γ_{mode}	γ(keV)	γ(%)
γ	609.31 *6*	0.124 *5*

$^{218}_{87}$Fr(700 *600* μs)

Mode: α
Δ: 7036 *6* keV
SpA: 7×10^{10} Ci/g
Prod: descendant ^{226}Pa

Alpha Particles (^{218}Fr)

⟨α⟩=7845.9 keV

α(keV)	α(%)
7384 *10*	0.5
7542 *15*	1.0
7572 *10*	5
7732 *10*	0.5
7867 *2*	93

$^{218}_{88}$Ra(14 *2* μs)

Mode: α
Δ: 6630 *14* keV
SpA: 7.5×10^{10} Ci/g
Prod: daughter ^{222}Th

Alpha Particles (^{218}Ra)

α(keV)
8390 *8*

$^{218}_{89}$Ac(270 *40* ns)

Mode: α
Δ: 10820 *50* keV
SpA: 7.5×10^{10} Ci/g
Prod: daughter ^{222}Pa

Alpha Particles (^{218}Ac)

α(keV)
9205 *15*

$^{218}_{90}$Th(109 *13* ns)

Mode: α
Δ: 12346 *16* keV
SpA: 7.5×10^{10} Ci/g
Prod: ^{206}Pb(^{16}O,4n); ^{209}Bi(^{14}N,5n)

Alpha Particles (^{218}Th)

α(keV)
9665 *10*

A = 219

NDS 22, 223 (1977)

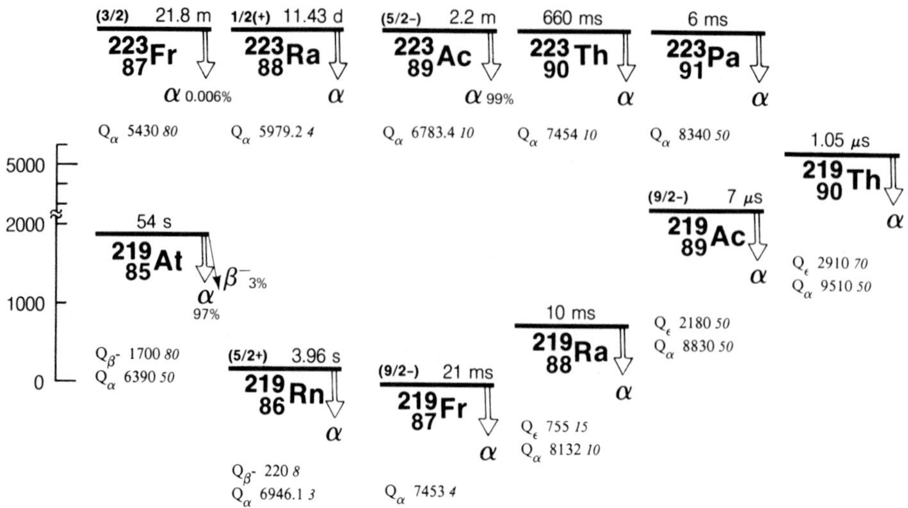

$^{219}_{85}$At(54 *6* s)

Mode: α(97 *1* %), β-(3 *1* %)
Δ: 10520 *80* keV
SpA: 9.5×10^{8} Ci/g
Prod: descendant ^{227}Ac; natural source

Alpha Particles (^{219}At)

α(keV)
6275 *50*

$^{219}_{86}$Rn(3.96 *5* s)

Mode: α
Δ: 8829 *4* keV
SpA: 1.193×10^{10} Ci/g
Prod: descendant ^{227}Th

Alpha Particles (^{219}Rn)

⟨α⟩=6812 *2* keV

α(keV)	α(%)
5782.8 *8*	~0.0010
5947.8 *11*	0.004
6000.3 *13*	0.0044 *5*
6101.0 *7*	0.0030 *3*
6148 *25*	~0.0026
6155.3 *9*	0.0174 *18*
6222.0 *8*	0.0026 *3*
6313.0 *5*	0.054 *6*
6425.0 *3*	7.5 *5*
6530.9 *4*	0.12 *1*
6553.1 *3*	12.2 *7*

Alpha Particles (^{219}Rn) (continued)

α(keV)	α(%)
6819.3 *3*	80.9 *10*

Photons (^{219}Rn)

⟨γ⟩=56.0 *16* keV

γ_{mode}	γ(keV)	γ(%)†
Po L_ℓ	9.658	0.0211 *23*
Po L_α	11.119	0.41 *3*
Po L_η	12.085	0.0075 *6*
Po L_β	13.495	0.45 *4*
Po L_γ	15.856	0.091 *9*
Po $K_{\alpha2}$	76.858	0.52 *3*
Po $K_{\alpha1}$	79.290	0.87 *5*
Po $K_{\beta1}'$	89.639	0.310 *19*
Po $K_{\beta2}'$	92.673	0.094 *6*
γ	115.3 *5*	0.0033 *15* ?
γ M1+25%E2	130.57 *6*	0.126 *8*
γ	221.9 *4*	0.030 *4* ?
γ E2+6.3%M1	271.13 *5*	9.9
γ	293.8 *3*	0.065 *5*
γ	324.1 *7*	<0.006 ?
γ	337.3 *7*	<0.008 ?
γ	372.0 *11*	<0.010 ?
γ	376.9 *6*	~0.007 ?
γ	379.4 *8*	~0.00030 ?
γ	387.9 *6*	~0.0004 ?
γ E2	401.70 *6*	6.64 *21*
γ	437.9 *6*	<0.028 ?
γ	515.7 *4*	~0.04
γ	540.1 *8*	~0.006
γ	563.1 *13*	<0.0030
γ	608.5 *7*	~0.004
γ	666 *4*	<0.008 •
γ	676.4 *8*	~0.006
γ	834.2 *13*	~0.0010
γ	887.7 *11*	0.0015 *7*
γ	1055.8 *7*	~0.0006

† 5.0% uncert(syst)
• possible doublet

Atomic Electrons (^{219}Rn)

⟨e⟩=6.36 *20* keV

e_{bin}(keV)	⟨e⟩(keV)	e(%)
14 - 22	0.182	1.21 *10*
37	0.163	0.43 *4*
59 - 101	0.046	0.066 *4*
111 - 131	0.214	0.181 *20*
178	2.04	1.15 *10*
201 - 244	0.05	~0.023
254	0.45	0.176 *16*
255	0.86	0.338 *19*
257	0.423	0.165 *9*
267	0.348	0.130 *9*
268 - 308	0.295	0.109 *5*
309	0.73	0.235 *9*
310 - 358	0.006	~0.0018
360 - 384	0.0012	0.00032 *16*
385	0.324	0.084 *3*
387 - 435	0.225	0.0568 *25*
437 - 470	0.0018	~0.0004
499 - 547	0.0040	~0.0008
549 - 595	0.0019	~0.00033
604 - 652	0.00020	~3×10^{-5}
659 - 676	0.00031	~5×10^{-5}
741	0.00012	~2×10^{-5}
795 - 834	0.00019	~2×10^{-5}
871 - 888	4.1 ×10^{-5}	~5×10^{-6}
963	5.2 ×10^{-5}	~5×10^{-6}
1039 - 1055	1.2 ×10^{-5}	~1×10^{-6}

$^{219}_{87}$Fr(21 *1* ms)

Mode: α
Δ: 8609 *8* keV
SpA: 7.4×10^{10} Ci/g
Prod: descendant ^{227}Pa

Alpha Particles (^{219}Fr)

$\langle\alpha\rangle$=7293 keV

α(keV)	α(%)
6802.7 *20*	0.25
6845.6 *25*	0.05
6957 *3*	~0.02
6967.2 *20*	0.6
7146.1 *20*	0.25 *7*
7313.6 *19*	99

Photons (^{219}Fr)

γ_{mode}	γ(keV)
γ(M1)	171 *3*
γ	182 *3*
γ	353 *3*
γ	477 *3*
γ	520 *3*

$^{219}_{88}$Ra(10 *3* ms)

Mode: α
Δ: 9365 *14* keV
SpA: 7.4×10^{10} Ci/g
Prod: descendant ^{227}U; ^{208}Pb(^{16}O,αn); ^{208}Pb(^{14}N,p2n)

Alpha Particles (^{219}Ra)

$\langle\alpha\rangle$=7786 *12* keV

α(keV)	α(%)
7680 *10*	65 *5*
7982 *9*	35 *2*

$^{219}_{89}$Ac(7 *2* μs)

Mode: α
Δ: 11540 *50* keV
SpA: 7.4×10^{10} Ci/g
Prod: daughter ^{223}Pa

Alpha Particles (^{219}Ac)

α(keV)
8664 *10*

$^{219}_{90}$Th(1.05 *3* μs)

Mode: α
Δ: 14450 *50* keV
SpA: 7.43×10^{10} Ci/g
Prod: ^{206}Pb(^{16}O,3n)

Alpha Particles (^{219}Th)

α(keV)
9340 *20*

A = 220

NDS **17**, 341 (1976)

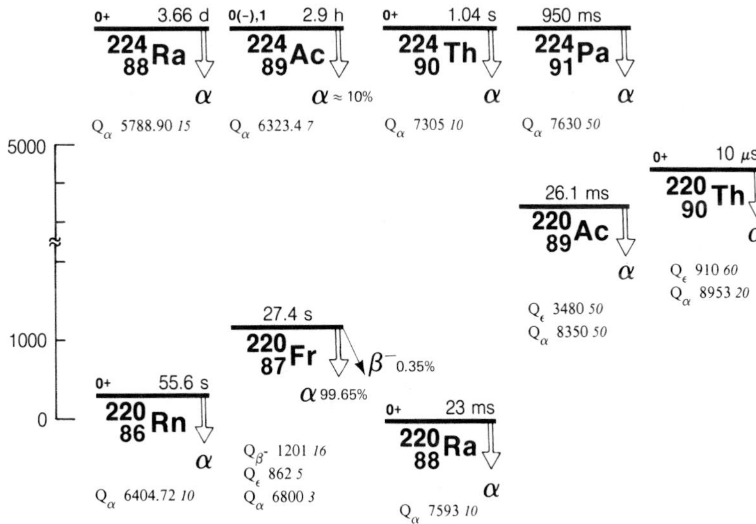

$^{220}_{86}$Rn(55.6 *1* s)

Mode: α
Δ: 10589 *6* keV
SpA: 9.166×10^{8} Ci/g
Prod: descendant ^{228}Th

Alpha Particles (^{220}Rn)

⟨α⟩=6287.9 *2* keV

α(keV)	α(%)
5748.6 *5*	0.07 *2*
6288.29 *10*	99.93 *2*

Photons (^{220}Rn)

γ$_{mode}$	γ(keV)	γ(%)
γ	549.7 *5*	0.070 *14*

$^{220}_{87}$Fr(27.4 *3* s)

Mode: α(99.65 %), β-(0.35 %)
Δ: 11451 *7* keV
SpA: 1.848×10^{9} Ci/g
Prod: descendant ^{228}Pa

Alpha Particles (^{220}Fr)

⟨α⟩=6603 keV

α(keV)	α(%)
6314 *3*	~0.015
6374 *3*	~0.010
6389.5 *9*	0.3
6412.6 *9*	1.2
6438 *2*	0.24
6455 *3*	~0.010
6482.6 *9*	1.3
6490.5 *21*	0.6
6519.5 *9*	~0.6
6527.2 *8*	3
6534.8 *9*	2.5
6581.8 *8*	10
6630.5 *21*	6
6641.7 *8*	12
6685.9 *8*	61

Photons (^{220}Fr)

γ$_{mode}$	γ(keV)	γ(rel)
γ	45.0 *3*	2.3 *5*
γ	61.0 *4*	0.43 *9*
γ	99.5 *5*	0.090 *18*
γ	106.0 *4*	1.7 *3*
γ	116.6 *4*	
γ	118.5 *5*	<0.17 ‡
γ	124.4 *4*	0.17 *3* ?
γ	132.4 *4*	0.19 *4* ?
γ	140.2 *4*	
γ	142.5 *5*	<0.26 ? §
γ	153.9 *4*	1.00 *20*
γ	161.7 *4*	1.5 *3*
γ	207.1 *5*	0.050 *10* ?

‡ 116.6γ + 118.5γ
§ 140.2γ + 142.5γ

$^{220}_{88}$Ra(23 *5* ms)

Mode: α
Δ: 10250 *15* keV
SpA: 7.4×10^{10} Ci/g
Prod: daughter ^{224}Th; descendant ^{228}U

Alpha Particles (^{220}Ra)

⟨α⟩=7415 keV

α(keV)	α(%)
6998 *7*	1.0
7455 *7*	99

Photons (^{220}Ra)

γ$_{mode}$	γ(keV)	γ(%)
γ	465 *4*	1.0

$^{220}_{89}$Ac(26.1 *5* ms)

Mode: α
Δ: 13730 *50* keV
SpA: 7.40×10^{10} Ci/g
Prod: daughter ^{224}Pa; ^{208}Pb(^{15}N,3n)

Alpha Particles (^{220}Ac)

⟨α⟩=7709 *30* keV

α(keV)	α(%)
7610 *20*	23 *5*
7680 *20*	21 *5*
7790 *10*	13 *2*
7850 *10*	24 *2*
7985 *10*	~4
8005 *10*	~5
8060 *10*	6 *1*
8195 *10*	3 *1*

$^{220}_{90}$Th(9.7 *6* μs)

Mode: α
Δ: 14646 *23* keV
SpA: 7.4×10^{10} Ci/g
Prod: ^{208}Pb(^{16}O,4n); ^{207}Pb(^{16}O,3n); daughter ^{224}U

Alpha Particles (^{220}Th)

α(keV)
8790 *20*

A = 221

NDS **27**, 681 (1979)

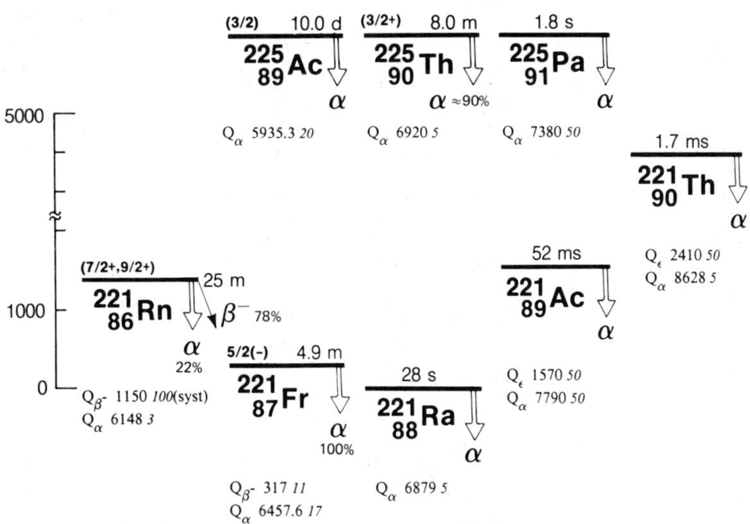

$^{221}_{86}$Rn(25 *2* min)

Mode: β-(78 *1* %), α(22 *1* %)
Δ: 14410 *100* keV syst
SpA: 3.4×10^{7} Ci/g
Prod: protons on Th

Alpha Particles (^{221}Rn)

⟨α⟩=1318.1 keV

α(keV)	α(%)
5778 *3*	1.8
5788 *3*	2.2
6037 *3*	18

Photons (^{221}Rn)

⟨γ⟩=120 *5* keV

γ_{mode}	γ(keV)	γ(%)†
Po L_ℓ	9.658	~0.011
Fr L_ℓ	10.381	~0.8
Po L_α	11.119	~0.22
Fr L_α	12.017	~14
Po L_η	12.085	~0.0032
Fr L_η	13.255	~0.27
Po L_β	13.502	~0.21
Fr L_β	14.775	~16
Po L_γ	15.871	~0.04
Fr L_γ	17.439	~3
$\gamma_{\beta-}$M1(+E2)	36.65 *4*	>0.008
$\gamma_{\beta-}$M1(+E2)	38.51 *3*	>0.010
$\gamma_{\beta-}$[E1]	49.20 *4*	0.071 *10*
$\gamma_{\beta-}$[E1]	57.77 *4*	0.023 *7*
$\gamma_{\beta-}$M1	62.93 *4*	0.143 *17*
$\gamma_{\beta-}$M1+17%E2	64.23 *4*	0.267 *20*
$\gamma_{\beta-}$E2	69.85 *4*	0.041 *9*
$\gamma_{\beta-}$E2	71.71 *4*	0.119 *12*
$\gamma_{\beta-}$E2(+<41%M1)	73.60 *6*	~0.0044
$\gamma_{\beta-}$E1	73.84 *4*	0.53 *3*

Photons (^{221}Rn) (continued)

γ_{mode}	γ(keV)	γ(%)†
$\gamma_{\beta-}$(M1+~20%E2)	74.90 *6*	0.17 *7*
Po $K_{\alpha2}$	76.858	~0.32
Po $K_{\alpha1}$	79.290	~0.5
Fr $K_{\alpha2}$	83.229	8.2 *8*
Fr $K_{\alpha1}$	86.105	13.6 *12*
$\gamma_{\beta-}$M1	87.39 *4*	0.21 *4*
Po $K_{\beta1}$'	89.639	~0.19
Po $K_{\beta2}$'	92.673	~0.06
$\gamma_{\beta-}$M1(+E2)	94.87 *4*	0.12 *5*
$\gamma_{\beta-}$M1+37%E2	96.17 *5*	~0.022
Fr $K_{\beta1}$'	97.272	4.9 *5*
$\gamma_{\beta-}$M1+3.3%E2	99.58 *4*	0.16 *3*
$\gamma_{\beta-}$E1	99.82 *6*	2.8 *6*
Fr $K_{\beta2}$'	100.599	1.56 *15*
$\gamma_{\beta-}$(M1+~26%E2)	100.88 *3*	0.29 *7*
$\gamma_{\beta-}$M1+E2	103.44 *4*	0.050 *13*
$\gamma_{\beta-}$M1+22%E2	108.36 *3*	2.19 *13*
$\gamma_{\beta-}$[E1]	111.56 *3*	2.15 *12*
$\gamma_{\beta-}$	119.87 *10*	0.29 *4*
$\gamma_{\beta-}$[E1]	123.76 *5*	~0.08
$\gamma_{\beta-}$M1+E2	124.81 *6*	~0.022
$\gamma_{\beta-}$(E1)	126.23 *7*	0.22 *3*
$\gamma_{\beta-}$	129.19 *4*	0.22 *3*
$\gamma_{\beta-}$	133.70 *10*	0.54 *10* ?
$\gamma_{\beta-}$(E1)	135.01 *7*	0.65 *6*
$\gamma_{\beta-}$(M1+~39%E2)	144.67 *5*	0.55 *20*
$\gamma_{\beta-}$(E1)	145.15 *4*	0.69 *14*
$\gamma_{\beta-}$E1	150.08 *3*	4.4 *3*
$\gamma_{\beta-}$[E1]	152.64 *4*	~0.2 ?
$\gamma_{\beta-}$[E1]	153.93 *4*	0.83 *6*
$\gamma_{\beta-}$M1+7.5%E2	157.24 *3*	0.23 *4*
$\gamma_{\beta-}$	168.88 *13*	0.27 *3*
$\gamma_{\beta-}$	170.91 *4*	0.47 *4*
$\gamma_{\beta-}$E1	178.39 *3*	0.86 *6*
$\gamma_{\beta-}$[E1]	186.12 *5*	~0.009
$\gamma_{\beta-}$E1	186.39 *4*	20.4 *12*
$\gamma_{\beta-}$E1	187.99 *5*	~0.2
$\gamma_{\beta-}$M1+37%E2	195.75 *3*	0.106 *23*
$\gamma_{\beta-}$	197.79 *12*	0.74 *9*
$\gamma_{\beta-}$(E1)	216.86 *4*	2.3 *3*
$\gamma_{\beta-}$[E1]	224.64 *4*	~0.036
$\gamma_{\beta-}$	240.76 *4*	0.70 *7*
$\gamma_{\beta-}$	253.29 *6*	
$\gamma_{\beta-}$[E1]	253.51 *4*	0.55 *9*
γ_α	254.2 *3*	~2
$\gamma_{\beta-}$	256.24 *5*	0.33 *8*

Photons (^{221}Rn) (continued)

γ_{mode}	γ(keV)	γ(%)†
γ_α(M1+E2)	264.68 *4*	1.14 *8*
$\gamma_{\beta-}$	273.5 *25*	0.46 *20*
$\gamma_{\beta-}$(E1)	279.27 *3*	1.85 *12*

† uncert(syst): 9.8% for α, 8.5% for β-

Atomic Electrons (^{221}Rn)

⟨e⟩=56 *6* keV

e_{bin}(keV)	⟨e⟩(keV)	e(%)
7	1.23	17.0 *13*
10 - 14	0.11	1.0 *3*
15	2.8	~19
16 - 17	0.039	~0.24
18	2.6	~15
19	2.2	~12
20	0.23	<2
21	1.8	<18
22	1.8	<17
23	2.0	~9
24 - 32	1.0	~3
33	1.3	~4
34 - 43	3.0	8 *4*
44	1.2	2.7 *7*
45	0.055	0.122 *15*
46	1.5	3.2 *6*
48 - 53	1.39	2.8 *5*
54	1.09	2.03 *21*
55 - 56	1.22	2.2 *5*
57	1.27	2.2 *4*
58 - 84	6.8	9.8 *10*
85	1.58	1.85 *17*
86 - 89	0.37	0.43 *15*
90	4.0	4.4 *5*
91 - 103	3.0	3.1 *8*
104	1.12	1.08 *11*
105 - 154	5.7	4.6 *8*
155 - 160	0.34	~0.22
161	1.3	<2
163 - 213	2.7	1.6 *5*
214 - 262	1.5	0.64 *20*
264 - 279	0.07	0.027 *12*

Continuous Radiation (^{221}Rn)

⟨β-⟩=236 keV;⟨IB⟩=0.23 keV

E_{bin}(keV)		⟨ ⟩(keV)	(%)
0 - 10	β-	0.072	1.43
	IB	0.0113	
10 - 20	β-	0.215	1.44
	IB	0.0108	0.075
20 - 40	β-	0.87	2.89
	IB	0.020	0.070
40 - 100	β-	6.1	8.8
	IB	0.050	0.077
100 - 300	β-	55	28.0
	IB	0.092	0.055
300 - 600	β-	120	27.9
	IB	0.040	0.0101
600 - 1120	β-	54	7.6
	IB	0.0041	0.00062

$^{221}_{87}$Fr(4.9 *2* min)

Mode: α
Δ: 13255 *11* keV
SpA: 1.73×10^8 Ci/g
Prod: descendant ^{229}Th

Alpha Particles (^{221}Fr)

⟨α⟩=6357 *1* keV

α(keV)	α(%)
5689 *3*	~0.002
5697 *4* ?	~0.0010
5774.4 *7*	0.06 *1*
5783 *4*	0.005 *2*
5813 *3*	~0.004
5925.1 *7*	0.03 *1*
5939.3 *7*	0.17 *3*
5965.9 *7*	0.08 *1*
5979.7 *7*	0.49 *3*
6037 *3*	~0.003
6075.1 *7*	0.15 *3*
6127.0 *7*	15.1 *2*
6243.3 *7*	1.34 *10*
6341.0 *7*	83.4 *8*

Photons (^{221}Fr)

⟨γ⟩=27.7 *8* keV

γ_{mode}	γ(keV)	γ(%)†
At L_ℓ	9.897	0.042 *6*
At L_α	11.414	0.79 *10*
At L_η	12.466	0.0167 *17*
At L_β	13.911	0.96 *11*
At L_γ	16.361	0.199 *24*
At $K_{\alpha 2}$	78.947	0.85 *11*
At $K_{\alpha 1}$	81.517	1.41 *18*
At $K_{\beta 1}$'	92.136	0.50 *7*
At $K_{\beta 2}$'	95.265	0.155 *20*
γ	97.2 *3*	~0.019
γ M1(+14%E2)	99.51 *13*	0.10 *4*
γ (M1)	118.47 *13*	~0.035
γ (M1+E2)	149.99 *20*	0.07 *3*
γ	171.29 *20*	0.07 *3*
γ E2	217.98 *4*	10.9 *4*
γ	282.54 *15*	~0.009
γ	324.10 *20*	~0.017
γ	359.09 *20*	~0.035
γ	382.05 *15*	~0.035
γ	409.1 *2*	0.13 *4*

† 6.9% uncert(syst)

Atomic Electrons (^{221}Fr)

⟨e⟩=8.44 *19* keV

e_{bin}(keV)	⟨e⟩(keV)	e(%)
4 - 14	0.21	2.2 *4*
17	0.183	1.08 *15*
23 - 67	0.15	0.41 *16*
74 - 118	0.53	0.62 *13*
122	1.83	1.50 *6*
133 - 171	0.16	0.11 *4*
187	0.005	<0.005
200	0.477	0.238 *9*
201	2.16	1.07 *4*
204	1.18	0.581 *22*
214	0.73	0.340 *13*
215	0.350	0.163 *6*
217	0.286	0.132 *5*
218 - 266	0.092	0.041 *7*
268 - 313	0.06	~0.020
320 - 368	0.010	~0.0028
378 - 409	0.017	~0.004

$^{221}_{88}$Ra(28 *2* s)

Mode: α
Δ: 12938 *8* keV
SpA: 1.80×10^9 Ci/g
Prod: descendant ^{229}U

Alpha Particles (^{221}Ra)

⟨α⟩=6640 *5* keV

α(keV)	α(%)
6160 *25* ?	~0.3
6254 *10*	0.7 *3*
6400 *25* ?	~0.3
6460 *25* ?	~0.4
6578 *5*	3 *1*
6585 *3*	8 *1*
6608 *3*	35 *2*
6669 *3*	21 *2*
6758 *3*	31 *2*

Photons (^{221}Ra)

⟨γ⟩=43 *4* keV

γ_{mode}	γ(keV)	γ(%)
γ	90.2 *19*	15 *2*
γ	152.3 *19*	13 *2*
γ	176.1 *19*	2.0 *5*
γ	220 *10*	~0.10
γ	294 *10*	~0.6
γ	321 *10*	~0.7
γ	416 *10*	~0.50

$^{221}_{89}$Ac(52 *2* ms)

Mode: α
Δ: 14500 *50* keV
SpA: 7.4×10^{10} Ci/g
Prod: descendant ^{225}Pa; ^{208}Pb(^{19}F,α2n); ^{205}Tl(^{22}Ne,α2n); ^{208}Pb(^{16}O,p2n); ^{208}Pb(^{18}O,p4n)

Alpha Particles (^{221}Ac)

⟨α⟩=7720 *19* keV

α(keV)	α(%)
7170 *10*	~2
7375 *10*	~10
7440 *15*	20 *5*
7645 *10*	70 *10*

$^{221}_{90}$Th(1.68 *6* ms)

Mode: α
Δ: 16916 *13* keV
SpA: 7.4×10^{10} Ci/g
Prod: ^{208}Pb(^{16}O,3n); ^{208}Pb(^{22}Ne,α5n); ^{208}Pb(^{20}Ne,α3n); ^{209}Bi(^{19}F,α3n)

Alpha Particles (^{221}Th)

⟨α⟩=8330 *10* keV

α(keV)	α(%)
7733 *8*	6 *1*
8146 *5*	56 *3*
8472 *5*	39 *2*

A = 222

NDS **21**, 479 (1977)

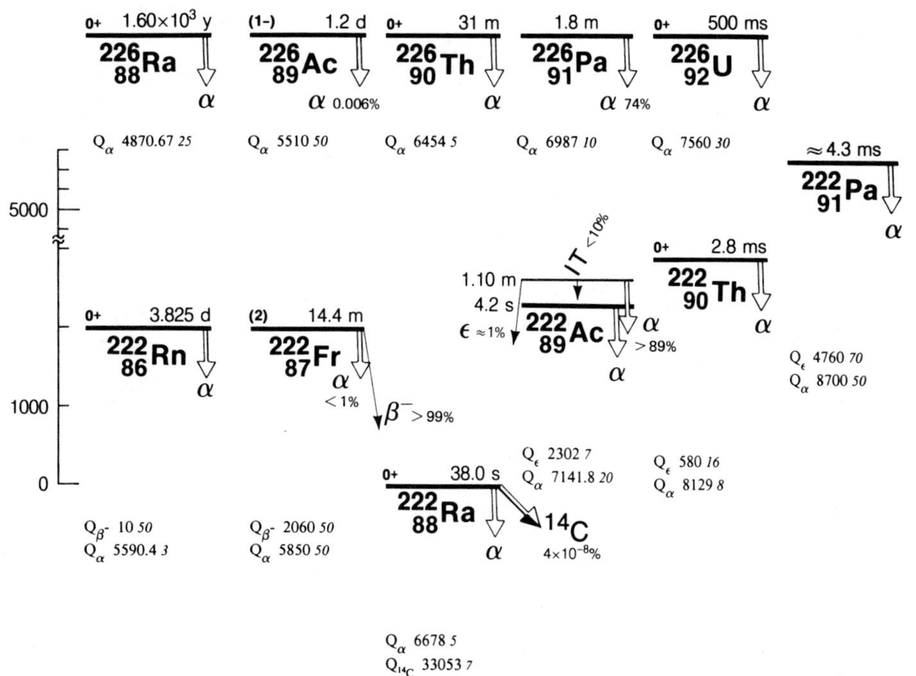

$^{222}_{86}$Rn(3.825 *4* d)

Mode: α
Δ: 16367 *3* keV
SpA: 1.5377×10^{5} Ci/g
Prod: natural source

Alpha Particles (^{222}Rn)

⟨α⟩=5489.2 *4* keV

α(keV)	α(%)
4827 *4* ?	~0.0005
4987 *1*	0.08
5489.7 *3*	99.92 *1*

Photons (^{222}Rn)

γ$_{mode}$	γ(keV)	γ(%)
γ	510 *2*	0.070 *2*

$^{222}_{87}$Fr(14.4 *4* min)

Mode: β-(>99 %), α(<1 %)
Δ: 16360 *50* keV
SpA: 5.88×10^{7} Ci/g
Prod: protons on Th

Continuous Radiation (^{222}Fr)

⟨β-⟩=625 keV;⟨IB⟩=1.04 keV

E$_{bin}$(keV)		⟨ ⟩(keV)	(%)
0 - 10	β-	0.0323	0.64
	IB	0.026	
10 - 20	β-	0.098	0.65
	IB	0.026	0.18
20 - 40	β-	0.400	1.33
	IB	0.049	0.17
40 - 100	β-	2.98	4.23
	IB	0.134	0.21
100 - 300	β-	32.7	16.1
	IB	0.33	0.19
300 - 600	β-	120	26.7
	IB	0.29	0.069
600 - 1300	β-	394	44.0
	IB	0.18	0.024
1300 - 1784	β-	75	5.3
	IB	0.0047	0.00034

$^{222}_{88}$Ra(38.0 *5* s)

Mode: α, ^{14}C(3.7 *6* ×10^{-8} %)
Δ: 14301 *6* keV
SpA: 1.325×10^{9} Ci/g
Prod: descendant ^{230}U; protons on Th

Alpha Particles (^{222}Ra)

⟨α⟩=6543 *3* keV

α(keV)	α(%)
5730.5 *19*	0.0040 *3*
5772.8 *19*	0.0041 *5*
5913.9 *19*	0.0043 *4*
6237.0 *19*	3.05 *5*
6555.7 *19*	96.9 *1*

Photons (^{222}Ra)

⟨γ⟩=9.2 *3* keV

γ$_{mode}$	γ(keV)	γ(%)
Rn L$_{\ell}$	10.137	0.00216 *21*
Rn L$_{\alpha}$	11.713	0.0400 *23*
Rn L$_{\eta}$	12.855	0.00088 *6*
Rn L$_{\beta}$	14.337	0.050 *4*
Rn L$_{\gamma}$	16.890	0.0106 *9*
Rn K$_{\alpha2}$	81.067	0.0460 *19*
Rn K$_{\alpha1}$	83.787	0.076 *3*
Rn K$_{\beta1}$'	94.677	0.0270 *11*
Rn K$_{\beta2}$'	97.907	0.0086 *4*
γ	143.71 *20*	<0.00033
γ E2	324.49 *5*	2.77 *8*
γ	329.07 *19*	0.0043 *4*
γ [E1]	472.78 *10*	0.0040 *3*
γ [E2]	515.88 *9*	0.0015 *1*
γ	840.37 *10*	0.0025 *2*

Atomic Electrons (^{222}Ra)

$\langle e \rangle$=0.844 *15* keV

e_{bin}(keV)	$\langle e \rangle$(keV)	e(%)
15 - 64	0.0193	0.116 *8*
66 - 94	0.00256	0.00347 *15*
126 - 144	0.00028	~0.00021
226	0.366	0.162 *6*
231	0.0021	~0.0009
306	0.080	0.0261 *9*
307	0.175	0.0571 *20*
310	0.076	0.0245 *9*
311 - 314	0.0006	~0.00021
320	0.0684	0.0214 *8*
321 - 323	0.0276	0.00859 *25*
324	0.0248	0.0077 *3*
325 - 329	0.00022	~7×10^{-5}
394	0.000293	7.5 *4* $\times10^{-5}$
455 - 501	0.000101	2.11 *8* $\times10^{-5}$
511 - 516	2.08 $\times10^{-5}$	4.06 *18* $\times10^{-6}$
742	0.0003	~5×10^{-5}
822 - 840	8.9 $\times10^{-5}$	~1×10^{-5}

$^{222}_{89}$Ac(4.2 *5* s)

Mode: α
Δ: 16603 *7* keV
SpA: 1.12×10^{10} Ci/g
Prod: daughter ^{226}Pa; ^{226}Ra(p,5n); ^{208}Pb(^{18}O,p3n)

Alpha Particles (^{222}Ac)

$\langle \alpha \rangle$=7010 *3* keV

α(keV)	α(%)
6967 *10*	6 *1*
7013 *2*	94 *1*

$^{222}_{89}$Ac(1.10 *5* min)

Mode: α(>89 %), IT(<10 %), ϵ(~ 1 %)
Δ: 16603 *7* keV
SpA: 7.7×10^{8} Ci/g
Prod: ^{208}Pb(^{18}O,p3n); ^{209}Bi(^{18}O,αn); protons on Th

Alpha Particles (^{222}Ac)

$\langle \alpha \rangle$=6090 *19* keV

α(keV)	α(%)
6460 *20*	~2
6710 *20*	~7
6750 *20*	13 *4*
6810 *20*	24 *9*
6840 *20*	~9
6890 *20*	13 *4*
6970 *20*	7 *3*
7000 *20*	13 *4*

$^{222}_{90}$Th(2.8 *3* ms)

Mode: α
Δ: 17183 *16* keV
SpA: 7.3×10^{10} Ci/g
Prod: ^{208}Pb(^{16}O,2n); ^{209}Bi(^{19}F,α2n); ^{208}Pb(^{20}Ne,α2n)

Alpha Particles (^{222}Th)

α(keV)
7982 *8*

$^{222}_{91}$Pa(~4.3 ms)

Mode: α
Δ: 21940 *70* keV
SpA: 7.3×10^{10} Ci/g
Prod: ^{209}Bi(^{16}O,3n); ^{206}Pb(^{19}F,3n)

Alpha Particles (^{222}Pa)

$\langle \alpha \rangle$~ 8318 keV

α(keV)	α(%)
~8180	~50
~8330	~20
~8540	~30

A = 223

NDS **22**, 243 (1977)

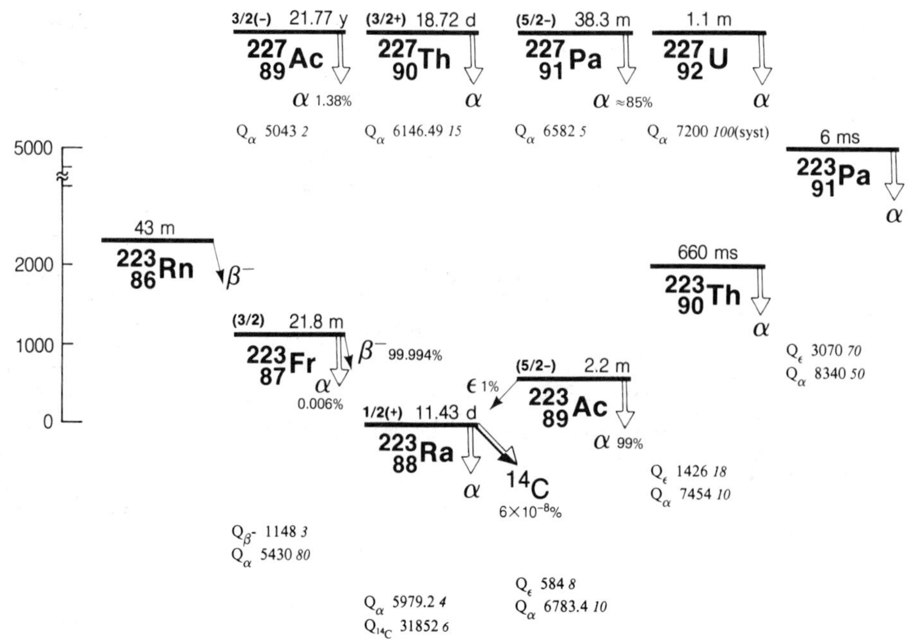

$^{223}_{86}$Rn(43 5 min)

Mode: β-
SpA: 1.96×10^{7} Ci/g
Prod: protons on Th

$^{223}_{87}$Fr(21.8 4 min)

Mode: β-(99.994 1 %), α(0.006 1 %)
Δ: 18381 4 keV
SpA: 3.87×10^{7} Ci/g
Prod: natural source

Alpha Particles (^{223}Fr)

α(keV)
5340 80

Photons (^{223}Fr)

⟨γ⟩=63 4 keV

γ_{mode}	γ(keV)	γ(%)†
γ[E1]	6.38 5	<0.07
Ra L_ℓ	10.622	0.91 23
Ra L_α	12.325	16 4
Ra L_η	13.662	0.30 8
γ	15 3	
Ra L_β	15.213	19 5
Ra L_γ	18.016	4.3 11
γ[E1]	20.27 4	0.8 3
γ	25.22 10 ?	
γ[M1]	29.58 4	0.025 6
γM1+14%E2	29.869 23	0.03
γM1+6.8%E2	31.566 10	<0.27
γ(E1)	43.73 4	0.0033 11
γE2	48.27 5	0.00017 3
γE1	49.85 3	~0.8
γE1	50.14 4	33
γ(M1)	51.29 6	$2.3\ 8\times10^{-5}$
γ(M1)	54.18 8	~0.00014
γE2	61.435 24	<0.26
γM1+17%E2	68.75 4	<0.47
γE1	79.72 3	8.9 13
Ra $K_{\alpha2}$	85.429	2.4 5
Ra $K_{\alpha1}$	88.471	4.0 8
γ[M1]	99.56 6	~0.00022
Ra $K_{\beta1}$'	99.915	1.4 3
γE2	100.31 4	<1.0
Ra $K_{\beta2}$'	103.341	0.47 9
γ	134.50 7	0.56 5
γE2+20%M1	173.38 5	0.128 12
γ[E1]	184.68 6	0.29 3
γE2	204.20 5	0.0039 8
γM1	204.95 5	1.13 19
γE2	206.04 5	0.0033 10
γ[E1]	206.39 4	$<9\times10^{-5}$
γE1	210.58 5	0.019 3
γ[M1+E2]	224.68 3	0.00011 4
γM1	234.82 5	3.7 5
γE1	235.97 4	0.09 3
γ	245.52 8	
γM1+E2	250.12 5	0.0053 20
γE1	254.66 5	0.014 3
γE2	256.24 3	0.051 16
γ(E1)	262.72 5	0.0061 19
γM1(+E2)	272.95 4	0.0083 15
γM1	279.70 5	0.0010 4
γ[E1]	284.24 5	0.0008 4
γM1+20%E2	286.11 3	0.0121 5

Photons (^{223}Fr) (continued)

γ_{mode}	γ(keV)	γ(%)†
γ	289.60 8	0.25 6
γ[E1]	292.30 6	0.0036 12
γ(E1)	299.97 4	0.030 9
γM1+5.9%E2	304.51 4	0.018 3
γ	307.89 7	0.030 9 ?
γM1	312.57 5	0.029 8
γ	319.18 7	0.54 4
γ(E1)	329.84 4	0.039 12
γM1+25%E2	334.38 4	0.0168 24
γ	339.45 7	0.066 13 ?
γE2+41%M1	342.44 5	0.023 7
γ	369.32 8	0.107 9
γ	448.00 20	~0.0009
γ	452.52 16	~0.0020
γ	457.07 16	~0.0014
γ	480.25 13	0.0012 5
γ	482.14 14	0.0018 6
γ	493.15 12	0.0022 7
γ	507.64 12	0.0031 7
γ	515.96 14	0.0023 7
γ	524.02 14	0.0019 5
γ	536.88 12	0.0044 15
γ	552.09 16	0.0046 16
γ	555.91 12	0.0017 4
γ	568.76 16	0.037 10
γ	575.80 22	0.029 9
γ	578.20 15	0.0017 5
γ	588.17 13	0.00024 11
γ	596.44 21	~0.0022
γ	607.20 13	0.0014 3
γ	621.92 15	0.00037 14
γ	643.93 24	~0.0027
γ	648.43 13	$8\ 3\times10^{-5}$
γ	691.46 22	0.009 3
γ	704.19 24	~0.0044
γ	707.18 16	0.0008 3
γ	718.22 12	0.00024 8
γ	722.60 25	0.025 12
γ	723.86 16	0.049 10
γ	733.9 3	~0.007
γ	734.60 14	0.0020 7
γ	736.76 16	0.0014 6
γ	746.24 10	0.024 5
γ	748.56 24	~0.016
γ	753.44 16	0.0062 17 ?
γ	754.17 25	0.0066 25 ?
γ	756.79 19	<0.013 ?
γ	757.03 16	0.016 4 ?
γ	762.30 12	0.0020 4
γ	764.52 10	0.079 17
γ	772.85 12	0.00051 25
γ	775.82 10	0.40 4
γ	780.59 12	0.0025 5
γ	784.04 25	0.0066 25 ?
γ	786.90 16	0.0021 5 ?
γ	788.36 19	<0.0027
γ	792.6 6	<0.0046 ?
γ	793.12 12	0.00012 5
γ	796.96 14	0.0101 18
γ	803.58 15	0.058 8
γ	808.26 14	~0.0006
γ	812.15 12	0.0207 20
γ	817.31 24	~0.005
γ	818.23 19	<0.0020
γ	822.99 12	0.010 3
γ	825.96 10	0.050 14 ?
γ	825.97 10	0.0013 3
γ	828.31 15	0.0012 4
γ	828.53 14	0.00013 6
γ	837.36 24	0.0068 19
γ	842.02 12	0.0049 9
γ	846.56 21	0.035 8 ?
γ	848.88 24	~0.0033 ?
γ	857.89 15	0.00037 14
γ	858.40 14	0.0031 7
γ	863.2 5	0.0020 8 ?
γ	867.23 24	0.0012 4 ?
γ	876.14 21	0.041 7 ?
γ	878.16 15	0.00068 12 ?
γ	891.0 7	~0.0020 ?
γ	892.8 5	0.0013 5 ?
γ	896.41 21	0.019 3

Photons (^{223}Fr) (continued)

γ_{mode}	γ(keV)	γ(%)†
γ	908.03 15	0.0130 19
γ	926.28 21	0.0016 5

† approximate

Atomic Electrons (^{223}Fr)

⟨e⟩=53 5 keV

e_{bin}(keV)	⟨e⟩(keV)	e(%)
5 - 14	4.1	35 8
15	2.9	19 5
16	1.5	~9
17	0.05	~0.31
18	2.6	14 4
19 - 30	4.2	17 3
31	3.0	10 4
32	1.7	5.5 11
33 - 34	0.06	0.18 8
35	2.1	6.0 12
36 - 42	0.07	~0.16
43	2.2	<10
44 - 45	0.70	1.5 3
46	3.1	~7
47 - 49	0.71	1.47 24
50	1.4	2.7 13
51 - 81	4.6	7.5 21
82	1.6	~2
83 - 84	0.018	0.022 4
85	1.2	~1
87 - 100	1.2	~1
101	2.0	2.0 3
102 - 130	1.1	~0.9
131	5.9	4.5 6
132 - 181	0.29	0.19 5
182 - 215	1.5	0.76 20
216	1.74	0.81 10
217 - 266	0.76	0.32 3
267 - 316	0.25	0.08 3
318 - 366	0.052	0.015 5
368 - 412	0.0048	0.0012 5
420 - 467	0.014	~0.0032
472 - 520	0.011	~0.0022
521 - 570	0.006	0.0011 5
571 - 620	0.018	~0.0030
621 - 669	0.031	0.0047 24
672 - 721	0.10	~0.014
722 - 771	0.044	0.0059 21
772 - 821	0.020	0.0025 8
822 - 871	0.0041	0.00048 16
872 - 911	0.0015	0.00017 6
921 - 926	1.4×10^{-5}	$\sim2\times10^{-6}$

Continuous Radiation (^{223}Fr)

⟨β-⟩=342 keV;⟨IB⟩=0.36 keV

E_{bin}(keV)		⟨ ⟩(keV)	(%)
0 - 10	β-	0.080	1.60
	IB	0.0161	
10 - 20	β-	0.241	1.61
	IB	0.0154	0.107
20 - 40	β-	0.97	3.23
	IB	0.029	0.101
40 - 100	β-	6.9	9.8
	IB	0.073	0.114
100 - 300	β-	63	31.9
	IB	0.145	0.086
300 - 600	β-	159	36.3
	IB	0.074	0.019
600 - 1097	β-	112	15.4
	IB	0.0108	0.00156

$^{223}_{88}$Ra(11.43 *2* d)

Mode: α, ^{14}C(6.1 *10* ×10^{-8} %)
Δ: 17234 *4* keV
SpA: 5.123×10^{4} Ci/g
Prod: daughter ^{227}Th; protons on Th

Alpha Particles (^{223}Ra)

⟨α⟩=5697 *2* keV

α(keV)	α(%)
5014.4 *20*	~0.00045
5025.5 *20*	~0.0006
5036 *2*	~0.00040
5056 *2*	~0.00020
5086.2 *20*	~0.00030
5112.4 *20*	~0.0006
5134.8 *20*	~0.0017
5151.8 *20*	0.021
5172.8 *20*	0.026
5211.6 *20*	0.005
5236.3 *20*	0.04
5258.8 *20*	0.04
5282.8 *20*	0.09
5287.3 *10*	0.15
5338.7 *10*	~0.13
5365.6 *10*	~0.13
5433.6 *5*	2.27 *20*
5481	~0.008
5501.6 *10*	1.00 *15*
5540 *1*	9.2 *3*
5606.9 *3*	24.2 *4*
5716.4 *3*	52.5 *8*
5747.2 *4*	9.5 *6*
5833 ?	0.05
5857.5 *10*	0.32 *4*
5865 ?	<0.02
5871.6 *10*	0.85 *4*

Photons (^{223}Ra)

⟨γ⟩=134.8 *12* keV

γ_{mode}	γ(keV)	γ(%)
Rn L_ℓ	10.137	0.55 *5*
Rn L_α	11.713	10.2 *5*
Rn L_η	12.855	0.145 *8*
Rn L_β	14.341	9.8 *6*
Rn L_γ	16.918	1.94 *16*
γ (E2)	31.87 *6*	0.00010 *2*
Rn $K_{\alpha2}$	81.067	15.2 *4*
Rn $K_{\alpha1}$	83.787	25.2 *6*
Rn $K_{\beta1}'$	94.677	8.90 *22*
Rn $K_{\beta2}'$	97.907	2.84 *8*
γ [E1]	103.85 *9*	0.017 *5*
γ (M1)	106.66 *6*	0.022 *3*
γ (E2)	110.80 *4*	0.048 *4*
γ	114.5 *5*	~0.009
γ M1+1.5%E2	122.31 *6*	1.190 *24*
γ	131.1 *3*	~0.005
γ M1+1.5%E2	144.18 *3*	3.26 *7*
γ M1	154.18 *3*	5.59 *11*
γ M1+3.8%E2	158.59 *3*	0.688 *14*
γ	165.50 *19*	~0.005
γ	175.55 *5*	0.017 *4*
γ	176.8 *3*	~0.004
γ [E1]	177.35 *9*	0.047 *8*
γ M1+25%E2	179.69 *4*	0.153 *12*
γ	199.43 *25*	~0.003
γ	220.38 *21*	0.033 *4*
γ	245.19 *21*	0.009 *3*
γ	249.49 *9*	0.037 *9*
γ	251.12 *25*	0.035 *15*
γ	251.58 *20*	0.031 *10*
γ	254.98 *4*	0.032 *10*
γ	255.74 *25*	~0.005
γ	260.40 *17*	~0.006
γ M1+2.5%E2	269.39 *3*	13.6 *3*

Photons (^{223}Ra) (continued)

γ_{mode}	γ(keV)	γ(%)
γ (E1)	288.15 *9*	0.154 *6*
γ M1+2.8%E2	323.88 *3*	3.90 *9*
γ	328.49 *5*	0.198 *10*
γ (E2)	333.87 *4*	0.099 *6*
γ M1	338.28 *4*	2.78 *7*
γ	342.90 *5*	0.20 *3*
γ	361.80 *8*	0.043 *4*
γ	368.78 *20*	~0.008
γ	369.48 *10*	~0.02
γ M1	371.80 *7*	0.490 *12*
γ	373.3 *3*	<0.0050 ?
γ	375.98 *10*	<0.0050 ?
γ	383.12 *17*	~0.004
γ	388.3 *3*	0.014 *5*
γ	391.0 *8*	~0.003
γ	393.48 *10*	<0.0050 ?
γ	430.53 *6*	0.019 *5*
γ	432.34 *9*	0.033 *8*
γ (M1)	444.94 *5*	1.27 *6*
γ	481.58 *10*	<0.0050 ?
γ	487.59 *24*	0.010 *2*
γ	527.30 *17*	0.071 *5*
γ	598.69 *17*	0.090 *6*
γ	609.08 *18*	0.063 *8*
γ	623.38 *10*	~0.008
γ	631.98 *10*	<0.0050 ?
γ	711.2 *7*	0.0035 *10*

Atomic Electrons (^{223}Ra)

⟨e⟩=73.1 *16* keV

e_{bin}(keV)	⟨e⟩(keV)	e(%)
4 - 14	3.5	~53
15	2.23	15.3 *16*
16 - 18	1.68	9.6 *8*
24	1.80	7.51 *21*
27 - 33	0.020	0.068 *17*
46	5.89	12.9 *4*
56	10.3	18.5 *5*
60 - 108	4.38	5.69 *10*
110 - 122	0.59	0.498 *24*
126	2.64	2.09 *6*
127 - 131	0.357	0.281 *14*
136	4.09	3.00 *9*
137 - 165	4.07	2.79 *5*
171	15.8	9.2 *3*
172 - 220	0.066	0.035 *4*
225	3.60	1.60 *5*
227 - 238	0.15	~0.06
240	2.49	1.04 *3*
241 - 250	0.10	~0.04
251	3.74	1.49 *5*
252 - 295	2.30	0.871 *16*
306 - 355	2.94	0.907 *20*
356 - 393	0.041	0.0110 *9*
412 - 445	0.250	0.058 *5*
464 - 513	0.039	~0.008
523 - 534	0.0036	~0.0007
581 - 629	0.011	0.0018 *8*
631	1.0 ×10^{-5}	~2×10^{-6}
693 - 711	0.03210450	3.194000 *15*

$^{223}_{89}$Ac(2.2 *1* min)

Mode: α(99 %), ϵ(1 %)
Δ: 17818 *8* keV
SpA: 3.82×10^{8} Ci/g
Prod: daughter ^{227}Pa

Alpha Particles (^{223}Ac)

⟨α⟩=6563 *2* keV

α(keV)	α(%)
5967 *3*	0.03
6023 *3*	0.010
6082.7 *20*	0.03
6134.9 *20*	0.12 *3*
6140 *3*	~0.030
6163.0 *25*	0.05
6177.7 *20*	0.94 *15*
6205.9 *25*	~0.030
6223 *4*	~0.006
6235.8 *20*	0.09
6281 ?	0.05
6292.8 *15*	0.47 *6*
6325.7 *15*	0.30 *5*
6332.5 *20*	0.14 *6*
6341.8 *20*	0.049 *10*
6360.5 *15*	0.22 *3*
6396.8 *15*	0.129 *20*
6448.6 *15*	0.20 *3*
6455 *3*	~0.06
6473.0 *15*	3.1 *3*
6523.2 *20*	~0.6
6528.4 *15*	3.1 *3*
6563 *1*	13.6 *10*
6582 *3*	~0.30
6646 *1*	44 *4*
6660.9 *10*	31 *3*

Photons (^{223}Ac)

⟨γ⟩=3.9 keV

γ_{mode}	γ(keV)	γ(%)
γ	72.5 *9*	~0.20
γ(M1+E2)	83.9 *7*	~0.17
γ	92.7 *7*	~0.17
γ(M1+E2)	99.0 *6*	~0.20
γ(M1)	119.7 *13*	~0.030
γ(M1)	125.3 *15*	~0.040
γ	176.6 *8*	~0.049
γ	191.7 *7*	~0.25
γ	206.7 *11*	~0.06
γ	215.8 *9*	~0.15
γ	268.4 *13*	~0.049
γ	305.7 *11*	~0.08
γ	358.8 *11*	~0.10
γ	373.9 *11*	~0.049
γ	432.7 *14*	~0.07
γ	434.7 *9*	~0.07 ?
γ	476.8 *15*	~0.14

$^{223}_{90}$Th(660 ms)

Mode: α
Δ: 19243 *17* keV
SpA: 5×10^{10} Ci/g
Prod: daughter ^{227}U; ^{208}Pb(^{18}O,3n)

Alpha Particles (^{223}Th)

⟨α⟩=7299 *19* keV

α(keV)	α(%)
7287 *10*	60 *10*
7317 *10*	40 *10*

$^{223}_{91}$Pa(6.5 *10* ms)

Mode: α
Δ: 22310 *70* keV
SpA: 7.3×10^{10} Ci/g
Prod: ^{208}Pb(^{19}F,4n); ^{205}Tl(^{22}Ne,4n); ^{209}Bi(^{20}Ne,α2n)

Alpha Particles (^{223}Pa)

⟨α⟩=8092 *13* keV

α(keV)	α(%)
8006 *10*	55 *5*
8196 *10*	45 *5*

A = 224

NDS **17**, 351 (1976)

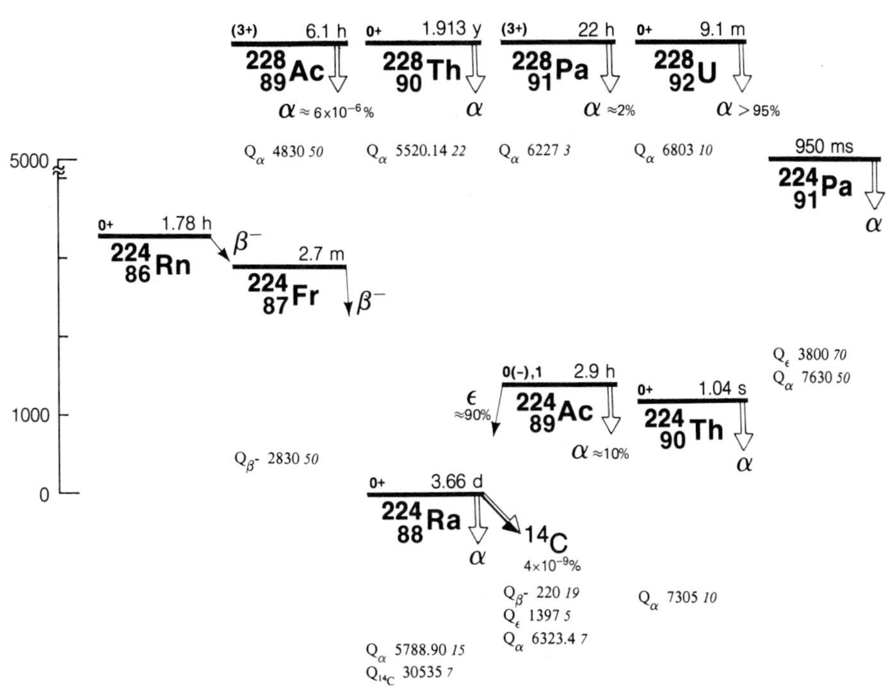

$^{224}_{86}$Rn(1.78 *5* h)

Mode: β-
SpA: 7.84×10^{6} Ci/g
Prod: protons on Th

Photons (^{224}Rn)

γmode	γ(keV)	γ(rel)
γ	108.5 *5*	3.3 *6*
γ	113.2 *3*	3.3 *6*
γ	156.4 *2*	1.8 *4*
γ	168.67 *15*	1.9 *4*
γ	200.7 *3*	~0.9
γ	202.6 *3*	4.4 *4*
γ	209.90 *15*	2.3 *5*
γ	256.2 *3*	3.0 *6*
γ	260.1 *1*	23 *1*
γ	265.50 *15*	21 *1*
γ	273 *1*	~0.4 ? *
γ	302.0 *3*	0.8 *2* ? *
γ	306.2 *3*	~0.50 ? *
γ	318.8 *6*	0.71 *15* ? *
γ	342.7 *3*	~0.6 ? *
γ	371.9 *3*	3.5 *2* ? *

Photons (^{224}Rn) (continued)

γmode	γ(keV)	γ(rel)
γ	374.2 *3*	1.8 *2* ? *
γ	380.2 *5*	~0.4 ? *
γ	387.2 *3*	1.7 *2* ? *
γ	398 *1*	5.6 *8* ? *
γ	402 *1*	5.7 *8* ? *
γ	408.7 *10*	1.4 *5* ? *

* with ^{224}Rn or ^{224}Fr

$^{224}_{87}$Fr(2.67 *20* min)

Mode: β-
Δ: 21630 *50* keV
SpA: 3.14×10^{8} Ci/g
Prod: daughter ^{224}Rn; protons on Th

Photons (^{224}Fr)

γmode	γ(keV)	γ(rel)
γ E2	84.26 *5*	
γ E1	131.50 *6*	83 *9*
γ (E2)	166.43 *16*	2.5 *5*
γ [E1]	205.75 *16*	13.7 *10*
γ E1	215.75 *6*	179 *8*
γ	325.2 *3*	4.5 *3*
γ	327.8 *3*	3.5 *7*
γ	334.91 *19*	5.5 *5*
γ	383.1 *3*	3.2 *2*
γ	414.37 *25*	4.4 *3*
γ	417.3 *3*	~0.6 ?
γ [E1]	762.51 *22*	12.4 *10*
γ [E2]	801.8 *3*	5.9 *5*
γ	831.62 *23*	1.7 *2*
γ [E1]	836.76 *21*	58 *2*
γ	873.9 *3*	3.5 *4*
γ	881.01 *24*	5.0 *5*
γ [M1+E2]	968.26 *21*	5.3 *5*
γ [E2]	1052.51 *21*	2.7 *3*
γ	1161.96 *20*	5.2 *6*
γ	1173.7 *6*	0.9 *3*
γ	1185.8 *6*	1.30 *25*
γ	1207.1 *10*	
γ	1219.90 *24*	~1
γ	1298.26 *25*	4.8 *6*
γ	1304.5 *10*	

Photons (^{224}Fr)
(continued)

γ_{mode}	γ(keV)	γ(rel)
γ	1340.2 3	25 3
γ	1351.40 23	5.7 10
γ	1377.72 20	17.2 20
γ	1435.65 23	10.0 15
γ	1567.7 3	3.3 6
γ	1573.5 4	2.5 5
γ	1579.7 10	
γ	1621.3 3	1.8 4
γ	1652.1 5	6 1
γ	1658.5 6	1.3 3
γ	1670.4 6	1.3 3
γ	1705.5 3	1.6 4
γ	1712.62 22	1.6 4
γ	1735.5 10	
γ	1780.9 10	
γ	1826.5 10	
γ	1992.5 10	
γ	2031.7 8	
γ	2163.2 8	

$^{224}_{88}$Ra(3.66 4 d)

Mode: α, ^{14}C(4.3 12 $\times 10^{-9}$ %)
Δ: 18803 6 keV
SpA: 1.593×10^{5} Ci/g
Prod: daughter ^{228}Th; protons on Th

Alpha Particles (^{224}Ra)
$\langle\alpha\rangle$=5675 1 keV

α(keV)	α(%)
5034 10	0.003
5047.1 7	0.007
5164 5	0.007
5449.10 22	4.9 4
5685.56 20	95.1 4

Photons (^{224}Ra)
$\langle\gamma\rangle$=10 3 keV

γ_{mode}	γ(keV)	γ(%)
Rn L_ℓ	10.137	0.0073 22
Rn L_α	11.713	0.13 4
Rn L_η	12.855	0.0034 10
Rn L_β	14.336	0.18 5
Rn L_γ	16.881	0.039 12
Rn $K_{\alpha2}$	81.067	0.12 4
Rn $K_{\alpha1}$	83.787	0.20 6
Rn $K_{\beta1}'$	94.677	0.072 20
Rn $K_{\beta2}'$	97.907	0.023 7
γ E2	240.76 10	3.9 11
γ	290 5	~0.009
γ [E1]	409.3 7	~0.004
γ [E1]	650.1 7	~0.007

Atomic Electrons (^{224}Ra)
$\langle e\rangle$=2.2 3 keV

e_{bin}(keV)	$\langle e\rangle$(keV)	e(%)
15 - 64	0.067	0.41 8
66 - 94	0.0069	0.0093 12
142	0.62	0.43 12
192	0.005	<0.005
223	0.78	0.35 10
226	0.32	0.14 4
236	0.042	0.018 5
237	0.26	0.11 3
238	0.0018	0.00077 22
240	0.082	0.034 10
241 - 290	0.023	0.0093 24
311	0.00017	~6×10^{-5}
391 - 409	4.9×10^{-5}	1.2 4 $\times10^{-5}$
552	0.00021	~4×10^{-5}
632 - 650	5.2×10^{-5}	8 3 $\times10^{-6}$

$^{224}_{89}$Ac(2.9 2 h)

Mode: ϵ(~ 90 %), α(~ 10 %)
Δ: 20200 7 keV
SpA: 4.8×10^{6} Ci/g
Prod: daughter ^{228}Pa

Alpha Particles (^{224}Ac)

α(keV)	α(rel)
5640.5 25	0.06
5709 2	0.12
5718 2	0.12
5739.0 25	~0.05
5768 2	0.24
5775.8 20	0.06
5802.9 25	~0.02
5836.8 15	0.26
5840.7 3	0.55
5850.8 25	~0.07
5852.9 6	0.25
5860.1 5	0.75
5867.2 10	0.1
5875.1 4	1.7
5901.6 14	0.14
5908.1 7	0.15
5915.0 4	0.96
5941.1 4	4.4
5957.7 20	~0.03
5967.7 14	0.18
6000.1 4	6.7
6013.7 4	1.4
6056.1 3	21.9
6138.3 3	25.6
6154.7 4	1.03
6203.6 3	11.9
6210.3 3	20.4

Photons (^{224}Ac)
$\langle\gamma\rangle$=202 22 keV

γ_{mode}	γ(keV)	γ(%)†
Fr L_ℓ	10.381	~0.3
Ra L_ℓ	10.622	1.0 5
Fr L_α	12.017	~6
Ra L_α	12.325	18 9
Fr L_η	13.255	~0.13
Ra L_η	13.662	0.31 15
Fr L_β	14.778	~7
Ra L_β	15.217	19 9
Fr L_γ	17.418	~2
Ra L_γ	18.000	4.1 20
γ_α	24.8 3	0.0090 18
γ_α(M1)	36.32 24	0.0060 12
γ_α(M1)	48.54 23	0.067 13
γ_α	53.6 5	~0.0050
γ_α	56.2 5	0.0030 6 ?
γ_α	56.6 4	~0.002
γ_α	61.1 3	0.034 7
γ_α	66.5 3	0.005 1
γ_α	67.2 4	0.05 1
γ_α	73.0 4	~0.010
γ_α	73.3 3	~0.002
γ_α	78.4 5	0.031 6 ?
Fr $K_{\alpha2}$	83.229	0.17 3
γ_α	83.7 3	0.17 3
γ_ϵE2	84.26 5	1.10 22
Ra $K_{\alpha2}$	85.429	22 10
Fr $K_{\alpha1}$	86.105	0.28 4
Ra $K_{\alpha1}$	88.471	35 16
γ_α	92.5 3	0.05 1
Fr $K_{\beta1}'$	97.272	0.101 15
Ra $K_{\beta1}'$	99.915	13 6
Fr $K_{\beta2}'$	100.599	0.032 5
Ra $K_{\beta2}'$	103.341	4.1 19
γ_α	103.5 3	0.010 2
γ_αE1	108.5 3	0.075 15
γ_αE1	128.8 3	0.080 16
γ_ϵE1	131.50 6	20 3
γ_αE1	141.0 3	0.32 6
γ_αE1	144.8 3	0.15 3
γ_αE1	150.2 3	0.084 17
γ_αE1	157.1 3	0.53 11
γ_α	164.5 4	0.0010 2
γ_α	165.5 4	0.037 7
γ_α	170.0 3	~0.002
γ_α	187.1 5	0.0040 8
γ_α	193.4 3	0.0040 8
γ_α	200.2 3	~0.002
γ_αM1	200.3 5	0.021 4
γ_α	207.2 4	0.030 6
γ_ϵE1	215.75 6	44 5
γ_αM1	225.6 3	0.032 6
γ_α	252.2 3	0.021 4
γ_αM1	261.9 3	0.18 4
γ_α	271.4 5	0.0120 24 ?
γ_α(M1)	274.1 3	0.010 2
γ_α	288.5 3	0.005 1
γ_α	300.7 3	0.016 3
γ_α	327.8 3	0.0060 12
γ_α(M1)	364.2 3	0.022 4
γ_α	376.4 3	0.0060 12

† approximate

Atomic Electrons (^{224}Ac)
$\langle e\rangle$=45 7 keV

e_{bin}(keV)	$\langle e\rangle$(keV)	e(%)
6 - 7	0.9	~13
10	1.3	<26
15	5.2	~34
18	4.0	~22
19 - 65	7.6	25 6
66	7.2	~11
67 - 68	0.036	~0.05
69	5.7	~8
70 - 79	1.0	1.3 4
80	4.0	~5
81 - 82	0.23	0.28 10
83	1.2	~1
84 - 111	0.9	1.0 3
112	3.5	3.1 4
113 - 162	1.17	0.91 6
163 - 213	1.37	0.69 6
215 - 263	0.219	0.094 7
267 - 313	0.012	0.0041 13
323 - 372	0.0073	0.0021 3
373 - 376	8.1×10^{-5}	~2×10^{-5}

$^{224}_{90}$Th(1.04 *5* s)

Mode: α
Δ: 19980 *18* keV
SpA: 3.53×10^{10} Ci/g
Prod: daughter ^{228}U; ^{208}Pb(^{22}Ne,α2n)

Alpha Particles (^{224}Th)

⟨α⟩=7102 *8* keV

α(keV)	α(%)
6706 *6*	~0.4
6768 *5*	1.2 *4*
6997 *5*	19 *3*
7170 *5*	79 *3*

Photons (^{224}Th)

⟨γ⟩=23 *4* keV

γ_{mode}	γ(keV)	γ(%)
Ra L_ℓ	10.622	0.058 *14*
Ra L_α	12.325	1.03 *24*
Ra L_η	13.662	0.031 *7*
Ra L_β	15.228	1.6 *4*
Ra L_γ	17.962	0.36 *8*
Ra $K_{\alpha2}$	85.429	0.54 *12*

Photons (^{224}Th) (continued)

γ_{mode}	γ(keV)	γ(%)
Ra $K_{\alpha1}$	88.471	0.88 *19*
Ra $K_{\beta1}$'	99.915	0.32 *7*
Ra $K_{\beta2}$'	103.341	0.103 *22*
γ E2	175.9 *18*	9 *2*
γ [E1]	233.5 *23*	~0.4
γ [E2]	297 *3*	0.3 *1*
γ [E1]	409.4 *23*	0.8 *3*

Atomic Electrons (^{224}Th)

⟨e⟩=13.1 *14* keV

e_{bin}(keV)	⟨e⟩(keV)	e(%)
15	0.22	1.4 *4*
18	0.28	1.5 *4*
19 - 67	0.027	0.095 *19*
70	0.0091	0.0131 *23*
72	1.3	1.8 *4*
73 - 99	0.0208	0.0253 *22*
130	0.026	~0.02
157	5.2	3.3 *7*
160	2.7	1.7 *4*
171	1.5	0.87 *20*
172	0.80	0.47 *10*
173	0.017	0.0100 *22*
175	0.63	0.36 *8*
176 - 218	0.22	0.121 *23*
229 - 278	0.043	0.016 *4*
281 - 305	0.069	0.023 *5*
390 - 409	0.0107	0.0027 *6*

$^{224}_{91}$Pa(950 *150* ms)

Mode: α
Δ: 23780 *70* keV
SpA: 3.8×10^{10} Ci/g
Prod: ^{232}Th(p,9n); ^{205}Tl(^{22}Ne,3n); ^{208}Pb(^{19}F,3n)

Alpha Particles (^{224}Pa)

α(keV)	α(%)
7490 *10*	~100
7880 *20* ?	
7960 *20* ?	

A = 225

NDS 27, 701 (1979)

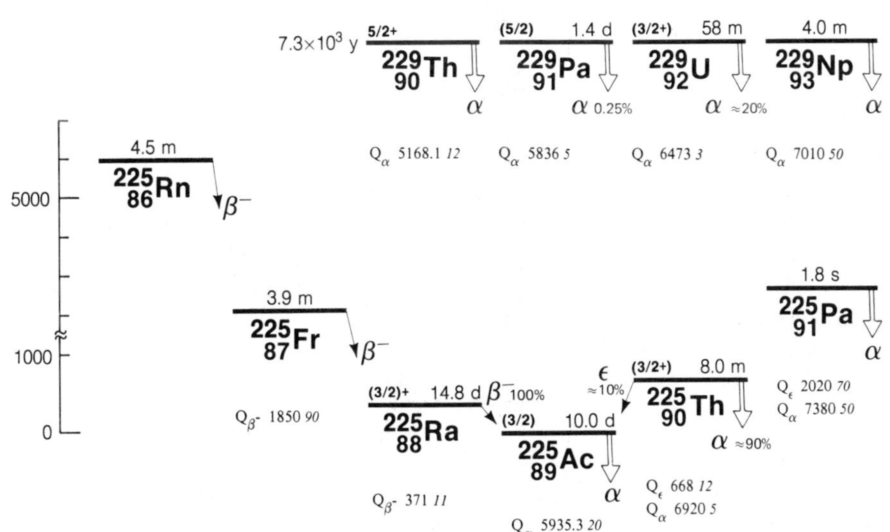

$^{225}_{86}$Rn(4.5 *3* min)

Mode: β-
SpA: 1.85×10^{8} Ci/g
Prod: protons on Th

$^{225}_{87}$Fr(3.9 *2* min)

Mode: β-
Δ: 23840 *90* keV
SpA: 2.14×10^{8} Ci/g
Prod: daughter ^{225}Rn

$^{225}_{88}$Ra(14.8 *2* d)

Mode: β-
Δ: 21987 *3* keV
SpA: 3.92×10^{4} Ci/g
Prod: descendant ^{233}U; ^{226}Ra(n,2n); descendant ^{229}Th

Photons (^{225}Ra)

⟨γ⟩=13.7 *24* keV

γ_{mode}	γ(keV)	γ(%)†
Ac L_ℓ	10.871	0.32 *7*
Ac L_α	12.636	5.5 *12*
Ac L_η	14.082	0.091 *19*
Ac L_β	15.662	6.1 *13*
Ac L_γ	18.593	1.4 *3*
γ E1	40.34 *17*	29 *6*

† 14% uncert(syst)

Atomic Electrons (^{225}Ra)

⟨e⟩=11.7 *9* keV

e_{bin}(keV)	⟨e⟩(keV)	e(%)
16	0.94	5.9 *13*
19	0.75	3.9 *9*
20	2.0	9.8 *20*
21	2.0	9.2 *19*
24	2.6	10.7 *22*
35	0.72	2.0 *4*
36	1.5	4.1 *8*
37	0.33	0.89 *18*
39	0.63	1.6 *3*
40	0.29	0.74 *15*

Continuous Radiation (^{225}Ra)

⟨β-⟩=94 keV;⟨IB⟩=0.032 keV

E_{bin}(keV)		⟨ ⟩(keV)	(%)
0 - 10	β-	0.354	7.1
	IB	0.0049	
10 - 20	β-	1.02	6.8
	IB	0.0042	0.029
20 - 40	β-	3.86	12.9
	IB	0.0067	0.024
40 - 100	β-	22.2	32.3
	IB	0.0111	0.018
100 - 300	β-	66	40.6
	IB	0.0053	0.0040
300 - 362	β-	0.93	0.294
	IB	3.9×10^{-6}	1.27×10^{-6}

$^{225}_{89}$Ac(10.0 *1* d)

Mode: α
Δ: 21615 *11* keV
SpA: 5.80×10^{4} Ci/g
Prod: descendant ^{233}U; ^{226}Ra(d,3n); descendant ^{229}Th; daughter ^{229}Pa

Alpha Particles (^{225}Ac)

⟨α⟩=5750 *160* keV

α(keV)	α(%)
4901 *5*	0.0020 *5*
5020 *25* ?	<0.0010
5030 *25* ?	<0.0010
5066 *5*	0.003 *1*
5091 *4*	0.006 *1*
5130 *5*	0.0020 *8*
5160 *5*	0.0020 *8*
5192 *25* ?	<0.002
5201 *5*	0.0020 *5*
5211 *3*	0.030 *3*
5238 *4*	0.0030 *8*
5271 *4*	0.009 *2*
5286.6 *8*	0.23 *1*
5322 *3*	0.068 *8*
5342 *25* ?	<0.0010
5355 *25* ?	<0.0010
5377 *25* ?	<0.0010
5391 *4*	~0.0010
5411 *4*	0.0020 *5*
5427 *4*	0.008 *3*
5437 *4*	0.07 *2*
5444 *3*	0.13 *1*
5468 *25* ?	<0.0010
5489 *4*	0.0020 *7*
5497 *4*	0.003 *1*
5519 *25* ?	<0.02
5526 *5*	0.010 *2*
5539.3 *8*	0.015
5544 *4*	0.03
5554.5 *8*	0.10
5562 *4*	0.03
5598.4 *8*	0.04
5608.1 *8*	1.1 *1*
5636.5 *8*	4.5 *3*
5681.4 *8*	1.4 *2*
5722.3 *8*	2.9 *5*
5731.0 *8*	10.0 *1*
5790.9 *8*	8.6 *9*
5792.8 *8*	18.1 *20*
5803.3 *8*	0.3
5828.8 *8*	51.6 *15*

Photons (^{225}Ac)

⟨γ⟩=17.6 *6* keV

γ_{mode}	γ(keV)	γ(%)
Fr L_ℓ	10.381	0.43 *10*
γ	10.67 *6*	
Fr L_α	12.017	7.8 *16*
γ	12.53 *6*	
Fr L_η	13.255	0.16 *4*
Fr L_β	14.772	9.6 *21*
Fr L_γ	17.449	2.1 *5*
γ [E2]	25.98 *5*	~0.0015
γ M1(+E2)	36.65 *4*	~0.015
γ M1(+E2)	38.51 *3*	~0.009
γ [E1]	49.20 *4*	0.0108 *17*
γ	53.82 *10*	0.019 *4*
γ [E1]	57.77 *4*	0.0042 *13*
γ M1	62.93 *4*	0.55 *5*
γ M1+17%E2	64.23 *4*	0.056 *6*
γ E2	69.85 *4*	0.0051 *12*
γ	70.92 *10*	~0.010
γ E2	71.71 *4*	0.0147 *21*
γ E2(+<41%M1)	73.60 *6*	~0.017
γ E1	73.84 *4*	0.324 *25*
γ (M1+~20%E2)	74.90 *6*	0.036 *16*
γ	82.9 *10*	0.15 *4*
Fr $K_{\alpha2}$	83.229	1.26 *13*
Fr $K_{\alpha1}$	86.105	2.08 *22*
γ M1	87.39 *4*	0.29 *3*
γ M1(+E2)	94.87 *4*	0.16 *6*
γ M1+37%E2	96.17 *5*	~0.030
Fr $K_{\beta1}$'	97.272	0.75 *8*
γ M1+3.3%E2	99.58 *4*	0.63 *14*
γ E1	99.82 *6*	1.68 *20*
Fr $K_{\beta2}$'	100.599	0.240 *25*
γ (M1+~26%E2)	100.88 *3*	0.06 *3*
γ M1+E2	103.44 *4*	0.009 *3*
γ M1+22%E2	108.36 *3*	0.27 *3*
γ [E1]	111.56 *3*	0.327 *24*
γ	119.92 *10*	0.06 *1*
γ [E1]	123.76 *5*	0.19 *2*
γ M1+E2	124.81 *6*	0.05 *1*
γ (E1)	126.23 *7*	0.014 *4*
γ	129.19 *4*	0.0043 *16*
γ (E1)	135.01 *7*	0.04 *1*
γ	138.2 *10*	~0.02
γ (E1)	145.15 *4*	0.126 *25*
γ E1	150.08 *3*	0.67 *5*
γ [E1]	152.64 *4*	~0.037
γ [E1]	153.93 *4*	0.153 *23*
γ M1+7.5%E2	157.24 *3*	0.31 *3*
γ	170.91 *4*	0.009 *3*
γ E1	178.39 *3*	0.016 *6*
γ [E1]	186.12 *5*	~0.02
γ E1	187.99 *5*	0.46 *5*
γ M1+37%E2	195.75 *3*	0.145 *20*
γ	198.72 *7*	~0.02
γ	216.2 *10*	0.34 *10*
γ (E1)	216.86 *4*	0.42 *8* ?
γ [E1]	224.64 *4*	0.08 *1*
γ	240.76 *4*	0.013 *5*
γ	248.7 *10*	~0.02
γ [E1]	253.51 *4*	0.100 *10*
γ (E1)	279.27 *3*	0.035 *11*
γ	285.5 *10*	~0.010
γ	452.43 *10*	0.11 *1*
γ	480.5 *10*	0.03 *1*
γ	513.50 *11*	~0.010
γ	526.03 *11*	~0.010

Atomic Electrons (^{225}Ac)

⟨e⟩=25.7 *18* keV

e_{bin}(keV)	⟨e⟩(keV)	e(%)
7 - 11	0.81	9.0 *24*
15	1.5	10 *3*
18	1.5	8.2 *21*
19	1.5	8 *4*
20	0.017	~0.09
21	0.6	~3
22	1.7	~8
23	0.7	3.1 *15*
24 - 32	0.7	2.4 *9*
33	0.5	~2
34 - 35	0.63	1.8 *7*
36	0.6	~2
37 - 43	0.49	1.3 *6*
44	1.94	4.4 *4*
45 - 55	1.25	2.6 *4*
56	0.83	1.5 *4*
57	0.18	0.32 *8*
58	0.60	1.03 *9*
59 - 68	1.9	2.9 *9*
69	0.73	1.06 *14*
70 - 80	1.3	1.7 *6*
81	1.12	1.4 *3*
82 - 130	3.6	3.8 *4*
131 - 181	0.58	0.39 *4*
182 - 231	0.27	0.13 *7*
234 - 283	0.011	0.0043 *11*
284 - 286	0.00020	~7×10^{-5}
357	0.05	~0.013
412 - 452	0.019	0.0043 *19*
462 - 511	0.0046	0.0010 *4*
512 - 526	0.00028	~5×10^{-5}

$^{225}_{90}$Th(8.0 *5* min)

Mode: α(~ 90 %), ϵ(~ 10 %)
Δ: 22283 *9* keV
SpA: 1.04×10^{8} Ci/g
Prod: daughter ^{229}U; ^{231}Pa(p,α3n); ^{84}Kr on Th

Alpha Particles (^{225}Th)

α(keV)	α(rel)
6312 *5*	~2
6345 *5*	~2
6440.6 *24*	15 *1*
6479.0 *22*	43 *1*
6501 *3*	14 *1*
6627 *3*	3 *1*
6650 *5*	3 *1*
6700 *5*	~2
6743 *3*	7 *1*
6795.8 *23*	9 *1*

Photons (^{225}Th)

⟨γ⟩=127 *10* keV

γ_{mode}	γ(keV)	γ(%)†
γ_α	54 *4*	
γ_α	135 *2*	0.36 *7*
γ_α	151.0 *19*	0.90 *18* ?
γ_α	246.1 *19*	4.5 *9*
γ_α	305 *2*	0.90 *18*
γ_α	322.5 *18*	27 *3*
γ_α	361.6 *24*	4.5 *9*
γ_α	450 *2*	0.90 *18*
γ_α	490 *2*	0.90 *18*

† approximate

$^{225}_{91}$Pa(1.8 *3* s)

Mode: α
Δ: 24310 *70* keV
SpA: 2.3×10^{10} Ci/g
Prod: ^{232}Th(p,8n); ^{232}Th(d,9n); daughter ^{229}Np; ^{209}Bi(^{22}Ne,α2n)

Alpha Particles (^{225}Pa)

⟨α⟩=7230 *13* keV

α(keV)	α(%)
7195 *10* ?	30 *10*
7245 *10*	70 *10*

A = 226

NDS **20**, 119 (1977)

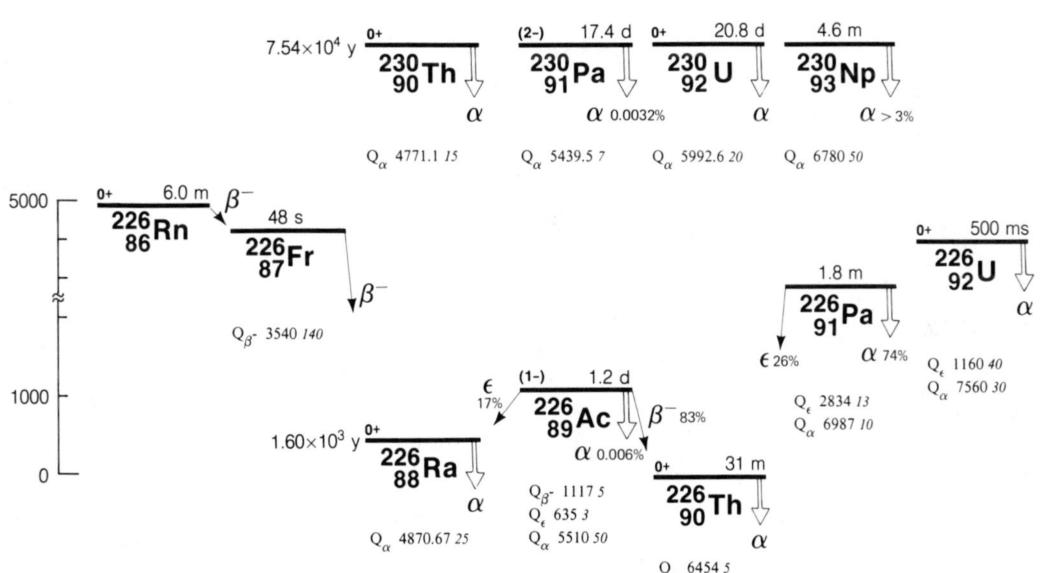

$^{226}_{86}$Rn(6.0 *5* min)

Mode: β-
SpA: 1.39×10^{8} Ci/g
Prod: protons on Th

$^{226}_{87}$Fr(48 *1* s)

Mode: β-
Δ: 27200 *140* keV
SpA: 1.033×10^{9} Ci/g
Prod: protons on Th

Photons (^{226}Fr)

γ_{mode}	γ(keV)	γ(rel)
γ	67.6758 *20*	
γ	186.057 *4*	66 *4*
γ	253.732 *4*	83 *5*
γ	553.9 *20*	1.3 *6*
γ	571.9 *20*	2.2 *8*
γ	587.9 *20*	1.9 *7*
γ	621.9 *20*	2.0 *8*
γ	793.9 *20*	2.4 *10*
γ	835.9 *20*	~1
γ	942.9 *20*	3.9 *12*
γ	981.9 *20*	5.0 *14*
γ	1006.9 *20*	~15
γ	1048.9 *20*	4.8 *14*
γ	1321.9 *20*	8.5 *20*
γ	1388.9 *20*	4.4 *14*

$^{226}_{88}$Ra(1600 *7* yr)

Mode: α
Δ: 23663 *3* keV
SpA: 0.989 Ci/g
Prod: natural source

Alpha Particles (^{226}Ra)

⟨α⟩=4774 *1* keV

α(keV)	α(%)
4159.9 *7*	0.00027 *5*
4194.0 *6*	0.0010 *1*
4343.7 *6*	0.0065 *3*
4601.4 *6*	5.55 *5*
4784.2 *6*	94.45 *5*

Photons (^{226}Ra)

⟨γ⟩=6.74 *13* keV

γ_{mode}	γ(keV)	γ(%)
Rn L_ℓ	10.137	0.0143 *14*
Rn L_α	11.713	0.265 *15*
Rn L_η	12.855	0.0073 *4*
Rn L_β	14.336	0.383 *24*
Rn L_γ	16.874	0.084 *6*
Rn $K_{\alpha2}$	81.067	0.176 *6*

Photons (^{226}Ra) (continued)

γ_{mode}	γ(keV)	γ(%)
Rn $K_{\alpha1}$	83.787	0.292 *10*
Rn $K_{\beta1}'$	94.677	0.103 *4*
Rn $K_{\beta2}'$	97.907	0.0329 *13*
γ E2	186.11 *11*	3.28 *7*
γ [E2]	262.41 *16*	0.0054 *3*
γ [E1]	414.72 *12*	0.00039 *8*
γ	449.5 *3*	0.00027 *5*
γ [E1]	600.83 *12*	0.00061 *12*

Atomic Electrons (^{226}Ra)

⟨e⟩=3.53 *5* keV

e_{bin}(keV)	⟨e⟩(keV)	e(%)
15	0.060	0.41 *4*
17	0.070	0.41 *4*
18 - 66	0.0109	0.0329 *24*
69 - 84	0.00608	0.0079 *3*
88	0.548	0.625 *18*
89 - 94	0.00066	0.00073 *4*
164	0.00082	0.00050 *3*
168	0.181	0.107 *3*
169	1.18	0.698 *20*
172	0.689	0.402 *11*
182	0.384	0.211 *6*
183	0.207	0.113 *3*
185	0.157	0.0850 *24*
186	0.0402	0.0216 *6*
244 - 262	0.00161	0.000645 *22*
345	0.00010	~3×10^{-5}
397 - 446	3.2 ×10^{-5}	~7×10^{-6}
447 - 450	1.8 ×10^{-6}	~4×10^{-7}
502	1.9 ×10^{-5}	3.9 *8* ×10^{-6}
583 - 601	4.9 ×10^{-6}	8.4 *11* ×10^{-7}

$^{226}_{89}$Ac(1.2 d)

Mode: β-(82.8 *16* %), ϵ(17.2 *16* %)
α(0.006 *2* %)
Δ: 24298 *3* keV
SpA: 5×10^{5} Ci/g
Prod: ^{226}Ra(d,2n); ^{232}Th(p,α3n)

Alpha Particles (^{226}Ac)

α(keV)
5399 *5*

Photons (^{226}Ac)

⟨γ⟩=137 *5* keV

γ_{mode}	γ(keV)	γ(%)†
Ra L_ℓ	10.622	0.20 *5*
Th L_ℓ	11.118	0.27 *6*
Ra L_α	12.325	3.6 *8*
Th L_α	12.952	4.6 *10*
Ra L_η	13.662	0.061 *14*
Th L_η	14.511	0.13 *3*
Ra L_β	15.216	3.9 *9*
Th L_β	16.159	6.4 *15*
Ra L_γ	18.006	0.84 *20*
Th L_γ	19.100	1.4 *4*
γ_ϵE2	67.6758 *20*	0.11 *3*

Photons (^{226}Ac) (continued)

γ_{mode}	γ(keV)	γ(%)†
$\gamma_{\beta-}$E2	72.18 *3*	0.56 *13*
Ra $K_{\alpha2}$	85.429	3.6 *8*
Ra $K_{\alpha1}$	88.471	6.0 *13*
Th $K_{\alpha2}$	89.955	1.10 *4*
Th $K_{\alpha1}$	93.350	1.78 *7*
Ra $K_{\beta1}'$	99.915	2.1 *5*
Ra $K_{\beta2}'$	103.341	0.70 *15*
Th $K_{\beta1}'$	105.362	0.64 *3*
Th $K_{\beta2}'$	108.990	0.214 *9*
$\gamma_{\beta-}$E1	158.16 *3*	17.3 *5*
γ_ϵE1	186.057 *4*	4.57 *24*
$\gamma_{\beta-}$E1	230.34 *4*	29.6 *18*
γ_ϵE1	253.732 *4*	5.8 *3*

† uncert(syst): 10% for β-, 13% for ϵ

Atomic Electrons (^{226}Ac)

⟨e⟩=30.7 *24* keV

e_{bin}(keV)	⟨e⟩(keV)	e(%)
15	0.72	4.6 *11*
16	1.00	6.1 *14*
18 - 19	0.67	3.6 *7*
20	1.1	5.7 *14*
48	0.040	0.082 *21*
49	2.4	4.9 *7*
52	7.7	15 *4*
56	5.3	9.5 *23*
63 - 66	0.93	1.45 *24*
67	2.5	3.7 *9*
68	1.8	2.7 *7*
69 - 70	0.132	0.19 *3*
71	1.2	1.6 *4*
72 - 105	0.86	1.10 *11*
121	1.94	1.61 *11*
138 - 186	1.43	0.969 *20*
210 - 254	1.04	0.478 *15*

Continuous Radiation (^{226}Ac)

⟨β-⟩=259 keV;⟨IB⟩=0.39 keV

E_{bin}(keV)		⟨ ⟩(keV)	(%)
0 - 10	β-	0.074	1.46
	IB	0.0124	
10 - 20	β-	0.220	1.47
	IB	0.0124	0.086
20 - 40	β-	0.89	2.96
	IB	0.022	0.077
40 - 100	β-	6.3	9.0
	IB	0.169	0.22
100 - 300	β-	57	28.9
	IB	0.117	0.073
300 - 600	β-	129	29.9
	IB	0.048	0.0120
600 - 1117	β-	66	9.2
	IB	0.0053	0.00080

$^{226}_{90}$Th(31 min)

Mode: α
Δ: 23180 *6* keV
SpA: 3×10^{7} Ci/g
Prod: daughter ^{230}U

Alpha Particles (^{226}Th)

$\langle\alpha\rangle$=6306 *1* keV

α(keV)	α(%)
5330.7 *5*	0.00015 *5*
5439.7 *4*	0.00034 *5*
5872.2 *4*	0.00020 *2*
6025.8 *4*	0.205 *8*
6041.4 *4*	0.187 *6*
6099.7 *4*	1.27 *5*
6228.3 *4*	22.8 *2*
6337.5 *4*	75.5 *3*

Photons (^{226}Th)

$\langle\gamma\rangle$=8.70 *25* keV

γ_{mode}	γ(keV)	γ(%)
Ra L_ℓ	10.622	0.136 *16*
Ra L_α	12.325	2.41 *20*
Ra L_η	13.662	0.075 *6*
Ra L_β	15.230	3.8 *3*
Ra L_γ	17.954	0.86 *8*
Ra $K_{\alpha2}$	85.429	0.303 *18*
Ra $K_{\alpha1}$	88.471	0.50 *3*
Ra $K_{\beta1}$'	99.915	0.179 *11*
Ra $K_{\beta2}$'	103.341	0.058 *4*
γ E2	111.10 *3*	3.29 *20*
γ [E1]	131.00 *4*	0.278 *13*
γ	172.3 *3*	0.00020 *2*
γ E2	190.28 *5*	0.109 *6*
γ E1	206.21 *5*	0.189 *8*
γ E1	242.10 *4*	0.87 *4*
γ [E1]	671.83 *25*	0.00028 *3*
γ [E2]	707.6 *3*	6 *2* $\times10^{-5}$
γ [E1]	782.8 *3*	9 *3* $\times10^{-5}$
γ [E2]	802.83 *25*	6 *2* $\times10^{-5}$

Atomic Electrons (^{226}Th)

$\langle e\rangle$=20.8 *7* keV

e_{bin}(keV)	⟨e⟩(keV)	e(%)
7 - 15	0.61	4.5 *4*
18	0.69	3.7 *4*
19 - 68	0.039	0.154 *11*
69 - 92	0.44	0.49 *3*
93	7.6	8.2 *5*
94 - 95	0.00049	0.00052 *5*
96	5.5	5.7 *4*
97 - 106	0.154	0.146 *9*
107	4.0	3.76 *24*
108	0.0394	0.0365 *23*
110	1.16	1.05 *7*
111 - 157	0.379	0.331 *18*
167 - 206	0.109	0.0617 *18*
223 - 242	0.0228	0.0100 *3*
580	1.35 $\times10^{-5}$	2.3 *3* $\times10^{-6}$
653 - 699	1.07 $\times10^{-5}$	1.56 *25* $\times10^{-6}$
703 - 708	5.3 $\times10^{-7}$	7.5 *18* $\times10^{-8}$
764 - 803	2.3 $\times10^{-6}$	2.9 *6* $\times10^{-7}$

$^{226}_{91}$Pa(1.8 *2* min)

Mode: α(74 *5* %), ϵ(26 *5* %)
Δ: 26015 *12* keV
SpA: 4.6$\times10^8$ Ci/g
Prod: ^{232}Th(p,7n)

Alpha Particles (^{226}Pa)

$\langle\alpha\rangle$=5064 keV

α(keV)	α(%)
6728 *10*	0.7
6823 *10*	35
6863 *10*	39

$^{226}_{92}$U (500 *200* ms)

Mode: α
Δ: 27170 *30* keV
SpA: 5.4$\times10^{10}$ Ci/g
Prod: ^{232}Th(α,10n)

Alpha Particles (^{226}U)

α(keV)
7430 *30*

A = 227

NDS **22**, 275 (1977)

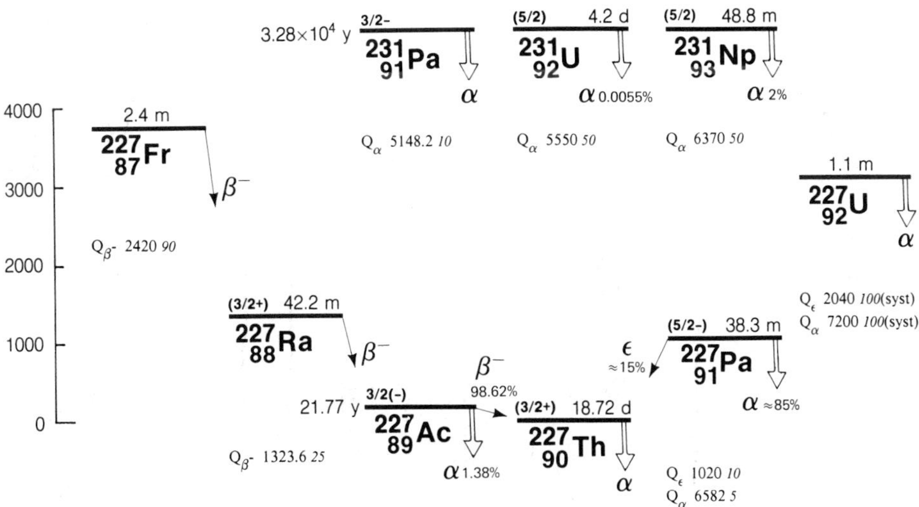

$^{227}_{87}$Fr(2.4 2 min)

Mode: β-
Δ: 29590 90 keV
SpA: 3.4×10^8 Ci/g
Prod: protons on U; protons on Th

Photons (^{227}Fr)

γmode	γ(keV)	γ(rel)
γ	37.9 1	0.82 19
γ	64.085 10	<3
γ	64.267 2	18.3 19
γ	76.1 1	0.19 3
γ	90.035 2	49 5
γ	100.1 2	3.4 7
γ	101.894 2	4.5 7
γ	107.306 4	0.75 7
γ	113.03 5	0.11 4
γ	120.709 8	1.76 19
γ	123.17 5	0.17 2
γ	131.0 1	0.14 2
γ	135.280 8	0.9 3
γ	135.525 5	1.3 5
γ	139.65 10	0.09 2
γ	145.139 16	0.09 1
γ	149.3 1	0.19 2
γ	151.553 25	0.16 2
γ	153.272 15	0.14 2
γ	159.37 5	0.28 3
γ	161.052 13	0.54 5
γ	163.563 7	1.98 19
γ	173.9 3	0.09 2
γ	175.228 14	~0.6
γ	175.867 4	1.3 3
γ	178.47 10	0.28 4
γ	182.394 15	0.34 4
γ	187.27 4	0.09 3
γ	194.255 10	0.19 2
γ	200.85 5	0.75 7
γ	202.394 4	0.10 2
γ	204.30 1	3.7 3
γ	206.539 5	0.67 7
γ	225.47 4	0.09 2
γ	237.2 2	0.07 2
γ	249.3 1	0.08 2
γ	251.9 1	0.08 2
γ	263.7 1	0.06 2
γ	270.6 1	0.06 1
γ	284.31 3	0.19 4
γ	291.55 5	0.93 7
γ	294.52 11	0.30 4
γ	296.7 1	~0.04
γ	306.2 2	~0.04
γ	321.763 25	0.12 2
γ	347.251 18	0.26 2
γ	348.803 25	0.10 1
γ	350.85 5	0.25 2
γ	362.28 5	0.28 2
γ	369.669 8	4.00 22
γ	373.12 6	0.09 2
γ	379.5 1	0.12 2
γ	381.556 15	2.54 19
γ	384.348 8	2.28 19
γ	391.57 2	3.29 22
γ	403.19 10	0.42 3
γ	413.029 11	0.39 3
γ	422.0 2	0.09 2
γ	433.824 9	6.4 4
γ	437.07 3	0.13 2
γ	438.768 18	0.19 3
γ	445.76 10	0.13 3
γ	449.26 3	0.46 4
γ	469.8 3	0.13 3
γ	473.33 6	0.17 4
γ	475.016 23	0.62 6
γ	497.0 2	0.10 2

Photons (^{227}Fr) (continued)

γmode	γ(keV)	γ(rel)
γ	498.4 3	0.09 2
γ	514.8 2	2.4 4
γ	523.75 10	0.15 2
γ	555.15 10	3.8 3
γ	573.84 10	0.58 4
γ	585.80 5	37.4 19
γ	596.76 21	0.09 2
γ	600.4 5	0.07 2
γ	629.75 7	0.14 2
γ	641.77 11	0.09 2
γ	736.9 5	0.17 2
γ	743.4 5	0.08 2
γ	789.2 5	0.15 2
γ	795.3 5	0.09 2
γ	810.4 5	0.09 2
γ	846.8 5	0.13 2
γ	889.3 5	0.18 3
γ	908.5 5	0.07 2
γ	937.0 5	0.12 2
γ	944.6 5	0.10 3
γ	981.3 5	0.04 1
γ	983.5 5	0.05 1
γ	993.0 5	0.20 2
γ	996.1 5	0.07 2
γ	1005.0 5	0.09 2
γ	1013.4 5	0.09 2
γ	1048.4 5	0.12 2
γ	1147.4 5	0.06 2
γ	1178.2 5	0.09 2
γ	1190.3 5	0.08 2
γ	1217.0 5	0.15 2
γ	1247.4 5	0.04 1
γ	1272.4 5	0.03 1
γ	1307.1 5	0.56 6
γ	1313.7 5	0.11 2
γ	1342.4 5	0.31 3
γ	1347.4 5	0.67 7
γ	1354.3 8	0.04 1
γ	1365.1 5	0.12 2
γ	1378.1 5	0.49 5
γ	1384.9 5	0.24 3
γ	1432.3 5	0.07 1
γ	1455.3 5	0.25 3
γ	1468.1 5	0.35 4
γ	1474.9 5	0.04 1

$^{227}_{88}$Ra(42.2 5 min)

Mode: β-
Δ: 27173 3 keV
SpA: 1.963×10^7 Ci/g
Prod: ^{226}Ra(n,γ)

Photons (^{227}Ra)

⟨γ⟩=163 5 keV

γmode	γ(keV)	γ(%)
Ac L_ℓ	10.871	1.3 4
Ac L_α	12.636	23 7
Ac L_η	14.082	0.30 6
Ac L_β	15.644	25 7
γ [E1]	16.397 16	~0.46
Ac L_γ	18.652	5.9 18
γ [M1]	18.998 16	~0.17
γ [M1]	24.56 5	0.30 15 ?
γ E1	27.396 9	17 4
γ M1+~3.8%E2	29.996 14	0.036 7
γ [E1]	46.393 15	0.19 4
γ	56.90 4	0.024 5 ?
γ (E2)	74.209 17	

Photons (^{227}Ra) (continued)

γmode	γ(keV)	γ(%)
Ac $K_{\alpha2}$	87.673	2.6 5
Ac $K_{\alpha1}$	90.886	4.3 8
Ac $K_{\beta1}'$	102.613	1.5 3
Ac $K_{\beta2}'$	106.137	0.50 9
γ	146.70 10	<0.30 ?
γ	198.96 4	0.065 25
γ	209.60 10	0.110 22
γ	218.21 10	0.21 4
γ	220.20 7	0.21 4
γ [E1]	226.77 4	0.030 6
γ	228.16 10	0.42 8
γ	230.62 7	~1.4
γ	232.45 13	0.30 6
γ	242.22 10	0.030 6
γ E2+37%M1	243.17 4	0.55 6
γ [M2]	245.438 21	0.016 3
γ [E1]	245.77 4	0.081 18
γ E2	255.858 21	0.222 24
γ	258.43 6	2.0 4
γ	259.72 20	0.030 6
γ M1+34%E2	273.17 4	0.67 7
γ	277.43 6	2.9 6
γ E1	283.673 21	3.5 4
γ E2+46%M1	300.070 19	5.3 5
γ E1	302.671 19	3.8 5
γ (E1)	327.23 5	0.29 5
γ M1+50%E2	330.066 18	2.9 3
γ	341.17 4	0.22 4 ?
γ [E1]	351.59 4	~0.030 ?
γ E2+10%M1	354.63 5	0.78 13
γ M1+27%E2	379.40 3	0.47 7
γ	388.98 5	0.078 16 ?
γ [E1]	395.80 4	0.027 6
γ	398.38 7	<0.18 ?
γ	398.40 4	0.095 16
γ M1	407.98 4	2.5 3
γ	428.38 8	0.093 19 ?
γ	435.38 4	0.24 4
γ	468.03 9	0.27 5
γ	471.33 10	0.27 5
γ	478.22 13	0.090 18
γ	487.03 9	2.5 5
γ	490.89 13	0.15 3
γ	501.33 10	1.05 21
γ	509.88 13	
γ	516.45 13	1.5 3
γ	535.45 13	0.66 13
γ	543 25	0.27 5
γ	611.72 21	1.3 3
γ	639.12 21	0.24 5 ?
γ	652.22 14	0.24 5 ?
γ	671.22 14	0.16 3 ?
γ	760.17 17	0.13 3 ?
γ	790.17 17	0.16 3 ?
γ	828.33 24	0.030 6 ?
γ	836.27 21	0.10 2 ?
γ	847.33 24	<0.06 ?
γ	863.67 21	0.16 3 ?
γ	874.72 24	0.090 18 ?

Atomic Electrons (^{227}Ra)

⟨e⟩=54 4 keV

e_{bin}(keV)	⟨e⟩(keV)	e(%)
5	2.4	~51
8	2.1	26 5
9 - 11	0.58	5.6 8
12	2.5	21 5
13	0.05	~0.42
14	2.4	17 8
15	0.049	~0.32
16	2.6	16 3
18	0.7	~4
19	2.4	13 3
20	4.6	23 11
21 - 22	0.61	2.7 5

Atomic Electrons (^{227}Ra)
(continued)

e_{bin}(keV)	⟨e⟩(keV)	e(%)
23	2.4	10 3
24 - 72	3.7	12.5 16
75 - 121	1.0	1.0 4
126 - 149	1.1	0.85 24
152	1.6	~1
153 - 166	0.69	0.42 4
171	2.1	~1
177 - 191	0.30	0.17 3
193	3.6	1.84 19
194 - 222	0.87	0.43 10
223	2.01	0.90 11
224 - 273	3.4	1.3 4
274 - 300	2.19	0.77 6
301	2.2	0.71 10
302 - 351	1.20	0.38 4
352 - 379	0.41	0.11 4
380	0.9	~0.24
383 - 431	1.9	0.47 15
432 - 481	0.48	0.10 4
482 - 531	0.7	0.14 7
532 - 564	0.17	0.031 15
592 - 639	0.13	0.022 11
647 - 683	0.07	0.011 5
722 - 770	0.08	0.011 4
771 - 820	0.007	0.0009 4
823 - 872	0.014	0.0016 6
873 - 875	0.00025	~3×10^{-5}

$^{227}_{89}$Ac(21.77 *3* yr)

Mode: β-(98.62 *1* %), α(1.38 *1* %)
Δ: 25849 *3* keV
SpA: 72.34 Ci/g
Prod: daughter ^{227}Ra; natural source

Alpha Particles (^{227}Ac)
⟨α⟩=67.33 *2* keV

α(keV)	α(%)
4362 4	~4×10^{-5}
4420 5	8×10^{-5}
4442.1 10	0.0007
4456 7	~7×10^{-5}
4509 5	~4×10^{-5}
4578 7	~4×10^{-5}
4586 4	0.00014
4591 4	0.00028
4711 4	0.0040 18
4735 4	0.0011 3
4766 3	0.023 4
4782.3 11	0.0011
4793.5 10	0.012 3
4819.3 11	0.0011 3
4852.8 10	0.059 14
4869.5 10	0.087 7
4897 3	0.0018 4
4938.1 10	0.52 3
4950.7 10	0.65 3

Photons (^{227}Ac)
⟨γ⟩=0.168 *12* keV

γ_{mode}	γ(keV)	γ(%)†
$\gamma_{\beta\text{-}}$(E2)	9.30 9	0.00013 3
Th L_ℓ	11.118	0.0073 17
Th L_α	12.952	0.12 3
Th L_η	14.511	0.00112 25
$\gamma_{\beta\text{-}}$(M1)	15.20 9	0.035 8
Th L_β	16.074	0.15 4
Th L_γ	19.308	0.04 1
$\gamma_{\beta\text{-}}$M1	24.50 12	0.0035 7
γ_α	46.13 10	0.0014 3 ?
γ_α	69.13 10	0.0055 11 ?
γ_α(E1)	69.83 10	0.017 3
γ_α(E1)	99.7 5	0.032 6
γ_α	106.1 15	0.0014 3 ?
γ_α	120.9 6	0.0055 11 ?
γ_α	133.8 5	0.0028 6 ?
γ_α	147.2 4	0.0083 17
γ_α	160.0 4	0.019 4
γ_α	171.4 5	0.0041 8

† uncert(syst): 0.72% for α, <0.1% for β-

Atomic Electrons (^{227}Ac)*
⟨e⟩=3.0 *3* keV

e_{bin}(keV)	⟨e⟩(keV)	e(%)
5	0.93	18 4
6	0.029	0.49 10
8	0.72	8.7 18
9	0.23	2.6 5
10	0.67	6.7 15
11	0.0050	0.045 10
12	0.0013	0.0111 25
14	0.25	1.8 4
15	0.084	0.56 13
16	0.00036	0.0022 5
19	0.028	0.14 3
20	0.031	0.15 3
21	4.7 ×10^{-5}	0.00022 5
23	0.0101	0.044 9
24	0.0034	0.014 3

* with β-

Continuous Radiation (^{227}Ac)
⟨β-⟩=9.5 keV;⟨IB⟩=0.00037 keV

E_{bin}(keV)		⟨ ⟩(keV)	(%)
0 - 10	β-	2.65	60
	IB	0.00030	
10 - 20	β-	3.89	27.3
	IB	6.1×10^{-5}	0.00047
20 - 40	β-	2.91	11.5
	IB	8.4×10^{-6}	3.7×10^{-5}
40 - 44	β-	0.0135	0.0330
	IB	6.7×10^{-10}	1.58×10^{-9}

$^{227}_{90}$Th(18.718 *20* d)

Mode: α
Δ: 25805 *4* keV
SpA: 3.073×10^4 Ci/g
Prod: daughter ^{227}Ac

Alpha Particles (^{227}Th)
⟨α⟩=5902 *2* keV

α(keV)	α(%)
5031.5 7	0.00031 2
5056.06 24	0.00023 2
5084.05 25	0.000150 15
5111.9 5	0.00028 2
5128.26 22	0.00062 5
5146.19 17	0.00410 8
5169.7 5	0.00170 17
5174.96 25	0.00120 24
5194.95 15	0.0038 3
5211.03 14	0.0070 3
5229.73 14	0.0098 3
5248.80 17	0.00320 16
5265.18 17	0.0026 2
5322 4	0.00024 10
5337.8 4	~0.0002
5365.0 25	0.00066 3
5409 3	0.00044 7
5458.6 7	0.00270 5
5480.4 22	0.0012 1
5510.8 3	0.0166 3
5533.61 13	0.021 2
5586.28 15	0.176 6
5601.09 15	0.170 17
5613.54 10	0.216 8
5621.70 19	0.0070 4
5640.54 17	0.0179 15
5668.58 9	2.06 12
5675.39 10	0.057 4
5693.90 11	1.5 1
5701.50 8	3.63 20
5709.72 8	8.2 3
5714.18 8	4.89 20
5728.9 3	0.0342 25
5757.14 7	20.3 10
5762.94 10	0.228 10
5795.21 10	0.311 5
5807.52 9	1.27 2
5866.72 9	2.42 10
5910.31 8	0.174 8
5916.58 8	0.78 3
5959.88 8	3.00 15
5977.85 7	23.4 10
5988.94 8	<0.005
6008.85 7	2.90 15
6038.20 7	24.5 10

Photons (^{227}Th)
⟨γ⟩=111.3 *24* keV

γ_{mode}	γ(keV)	γ(%)†
Ra L_ℓ	10.622	0.98 20
Ra L_α	12.325	17 3
Ra L_η	13.662	0.36 7
Ra L_β	15.218	21 4
Ra L_γ	17.997	4.8 10
γ [E1]	20.27 4	0.20 6
γ	20.64 9	
γ [M1]	29.58 4	0.0060 12
γ M1+14%E2	29.869 23	0.10 3
γ M1+6.8%E2	31.566 10	0.08 2
γ [E1]	33.52 6	~0.014 ?
γ [E1]	40.16 18	0.020 4
γ [E1]	41.88 6	0.040 8
γ	43.53 10	0.05 2 ?
γ (E1)	43.73 4	0.23 3
γ M1+20%E2	44.08 4	0.007 3
γ M1	44.37 5	0.013 6
γ [E1]	44.44 6	
γ E2	48.27 5	0.010 1
γ E1	49.85 3	~0.20
γ E1	50.14 4	8.5 3
γ (M1+15%E2)	50.82 6	0.016 7
γ (M1)	51.29 6	0.0030 6
γ (M1)	54.18 8	~0.008
γ M1	56.03 8	~0.0050 ?
γ M1+15%E2	56.63 5	~0.007 ?

Photons (^{227}Th)
(continued)

γ_{mode}	γ(keV)	γ(%)†
γ	59.5 6	0.009 3
γ E2	61.435 24	0.08 2
γ [E1]	62.05 16	0.0020 4 ?
γ [E1]	62.36 4	0.24 3
γ M1+5.0%E2	62.54 8	0.009 4
γ [E1]	64.37 8	0.028 10
γ	66.1 6	~0.006
γ	66.3 6	0.007 3
γ [E1]	68.70 14	0.0057 11 ?
γ M1+17%E2	68.75 4	0.04 1
γ	69.65 11	0.009 3 ?
γ [M1+E2]	72.80 7	0.028 6 ?
γ E2+15%M1	73.66 5	0.019 3
γ	75.10 10	~0.009 ?
γ E1	79.72 3	2.1 1
Ra $K_{\alpha2}$	85.429	1.87 12
Ra $K_{\alpha1}$	88.471	3.08 19
γ [E1]	89.85 9	0.0034 7
γ E1	93.93 4	1.40 12
γ (E2)	94.90 6	0.012 2
γ (E1)	96.06 7	0.06 1
γ [M1]	99.56 6	~0.013
γ	99.64 17	~0.002
Ra $K_{\beta1}$'	99.915	1.10 7
γ E2	100.31 4	0.086 10
γ [M1+E2]	102.45 8	0.00120 24
Ra $K_{\beta2}$'	103.341	0.359 23
γ	104.9 11	~0.04 ?
γ (M1+E2)	107.92 6	0.007 2
γ	109.1 3	~0.006 ?
γ (E2)	110.58 13	0.005 1 ?
γ	112.54 17	~0.008 ?
γ E2	113.12 5	0.15 5
γ E1	113.19 6	0.56 11
γ E1	117.17 6	0.17 1
γ [E1]	117.30 14	0.012 3
γ [E2]	123.55 7	0.008 2
γ [E2]	124.48 7	~0.0050 ?
γ [E1]	124.60 16	~0.003
γ	134.50 7	0.0252 25
γ	140.33 25	0.032 6
γ (E1)	141.44 7	0.13 1
γ	150.02 8	~0.010
γ	162.18 16	0.007 2
γ [E2]	168.19 6	0.014 2
γ [E1]	169.98 16	~0.003
γ [E2]	171.21 12	~0.0010 ?
γ E2+20%M1	173.38 5	0.0153 18
γ	175.23 12	0.018 4 ?
γ [E2]	179.28 11	~0.002 ?
γ [E1]	184.68 6	0.035 4
γ [E1]	197.61 14	0.012 3
γ [E1]	200.48 8	0.020 4 ?
γ [M1]	201.64 7	0.020 4
γ	202.4 6	0.006 2 ?
γ E2	204.20 5	0.23 3
γ M1	204.95 5	0.134 22
γ E2	206.04 5	0.23 2
γ [E1]	206.39 4	<0.012 ?
γ E1	210.58 5	1.13 8
γ M1+10%E2	212.56 5	0.070 15
γ	212.69 14	0.017 5
γ	218.77 7	0.06 1
γ [E1]	218.94 5	0.04 1
γ [M1+E2]	224.68 3	0.015 3
γ [M1+E2]	230.17 16	~0.0007 ?
γ M1	234.82 5	0.44 4
γ E1	235.97 4	11.2 6
γ [E1]	246.07 6	0.011 2
γ	249.36 18	0.007 2 ?
γ M1+E2	250.12 5	0.37 9
γ M1	250.33 7	0.13 2
γ M1(+E2)	252.46 6	0.11 2
γ E1	254.66 5	0.8 1
γ E2	256.24 3	6.7 5
γ (E1)	262.72 5	0.10 1
γ [E1]	266.30 11	0.0025 5
γ (M1+E2)	267.82 17	~0.010
γ (E1)	270.34 14	0.0080 16
γ [M1]	270.76 8	0.032 12
γ M1(+E2)	272.95 4	0.49 4
γ	275.2 7	~0.0015 ?
γ M1	279.70 5	0.07 2
γ [M1]	281.00 15	0.0070 14
γ M1+20%E2	281.31 5	0.16 1
γ [E1]	284.24 5	0.05 2
γ	285.48 14	0.055 10
γ M1+20%E2	286.11 3	1.59 5
γ	289.5 3	~0.009 ?
γ	289.60 8	0.011 3 ?
γ [E1]	292.30 6	0.06 1
γ M1(+E2)	296.54 5	0.43 3
γ (E1)	299.97 4	2.10 20
γ (M1+E2)	300.34 9	0.20 8
γ M1+5.9%E2	304.51 4	1.09 13
γ	307.89 7	0.0013 4
γ (M1+E2)	308.49 8	0.015 2
γ M1	312.57 5	0.47 5
γ (M1)	314.78 13	0.03 1
γ (E1)	314.82 5	0.46 6
γ	318.58 16	0.006 2
γ	319.18 7	0.024 3
γ (M1)	325.08 15	0.006 1
γ [M1+E2]	326.12 6	0.005 2 ?
γ (E1)	329.84 4	2.73 16
γ M1+25%E2	334.38 4	0.99 9
γ [E1]	339.10 11	~0.0014 ?
γ	339.45 7	0.0030 7
γ E2+41%M1	342.44 5	0.38 5
γ [E1]	346.39 5	0.0075 15
γ	348.4 6	0.006 2 ?
γ (E1)	350.48 8	0.11 1
γ (M1)	352.57 8	0.010 2
γ	362.55 18	0.005 1
γ	369.32 8	0.0048 6
γ [E1]	370.85 8	0.0070 14 ?
γ [E1]	374.93 15	0.0014 5 ?
γ [M2]	376.26 5	0.006 2 ?
γ [M1]	382.15 8	0.006 1
γ (M1)	383.47 10	0.048 6
γ	392.24 17	0.009 2
γ	398.60 14	~0.009 ?
γ [E1]	402.42 8	0.012 4 ?
γ [E2]	415.09 13	0.0017 4 ?
γ [E1]	432.29 8	0.005 1
γ	448.00 20	~0.00015 ?
γ	452.52 16	~0.00010 ?
γ	457.07 16	~7 $\times10^{-5}$?
γ	461.9 10	~5 $\times10^{-5}$?
γ	466.2 7	~5 $\times10^{-5}$?
γ	480.25 13	0.0003 1 ?
γ	482.14 14	0.00014 4 ?
γ	493.15 12	0.00055 8
γ	507.64 12	0.00040 8
γ	515.96 14	0.00018 4
γ	524.02 14	0.00015 3
γ	534.9 12	0.00010 3 ?
γ	536.88 12	0.0011 2
γ	552.09 16	0.00023 5
γ	555.91 12	0.00022 4
γ	568.76 16	0.0006 1 ?
γ	575.80 22	0.00013 3
γ	578.20 15	0.00013 3
γ	588.17 13	6 2 $\times10^{-5}$?
γ	596.44 21	~1 $\times10^{-5}$
γ	607.20 13	0.00018 4
γ	621.92 15	6 2 $\times10^{-5}$
γ	623.52 23	0.00016 4
γ	633.2 4	0.00013 3
γ	641.45 24	2.0 6 $\times10^{-5}$
γ	643.93 24	5 2 $\times10^{-5}$
γ	648.43 13	2.0 6 $\times10^{-5}$?
γ	662.8 4	6 2 $\times10^{-5}$
γ	691.46 22	4 1 $\times10^{-5}$
γ	704.19 24	8 2 $\times10^{-5}$?
γ	707.18 16	4 1 $\times10^{-5}$
γ	718.22 12	3 1 $\times10^{-5}$?
γ	722.60 25	0.00038 11 ?
γ	723.86 16	0.0008 2
γ	733.9 3	0.00010 4
γ	734.60 14	0.00016 5
γ	736.76 16	7 2 $\times10^{-5}$?
γ	746.24 10	9.1 20 $\times10^{-5}$
γ	748.56 24	0.00030 6 ?
γ	753.44 16	0.00010 3
γ	754.17 25	0.00010 3 ?
γ	756.79 19	0.00020 5 ?
γ	757.03 16	0.00078 19
γ	762.30 12	0.00026 5
γ	764.52 10	0.00030 6
γ	772.85 12	0.00013 5
γ	775.82 10	0.00153 17
γ	780.59 12	0.00032 6
γ	784.04 25	0.00010 3
γ	786.90 16	0.000105 23 ?
γ	788.36 19	4 1 $\times10^{-5}$?
γ	792.6 6	4 1 $\times10^{-5}$?
γ	793.12 12	3 1 $\times10^{-5}$?
γ	796.96 14	0.00079 9
γ	803.58 15	0.00095 9
γ	808.26 14	~5 $\times10^{-5}$?
γ	812.15 12	0.0026 3
γ	817.31 24	0.00010 3
γ	818.23 19	3 1 $\times10^{-5}$?
γ	822.99 12	0.0025 3
γ	825.96 10	0.00019 5
γ	828.31 15	0.00020 5 ?
γ	828.53 14	1.0 4 $\times10^{-5}$?
γ	837.36 24	0.00040 4
γ	842.02 12	0.00062 6
γ	846.56 21	0.00016 3
γ	848.88 24	6 2 $\times10^{-5}$?
γ	854.2 5	7.0 14 $\times10^{-5}$?
γ	857.89 15	6 2 $\times10^{-5}$
γ	858.40 14	0.00024 4
γ	863.2 5	2.0 8 $\times10^{-5}$?
γ	867.23 24	7.1 13 $\times10^{-5}$
γ	876.14 21	0.00018 4
γ	878.16 15	0.000111 23
γ	891.0 7	2.0 6 $\times10^{-5}$?
γ	892.8 5	1.3 4 $\times10^{-5}$
γ	896.41 21	8.3 15 $\times10^{-5}$
γ	908.03 15	0.0021 3
γ	909.85 24	1.5 6 $\times10^{-5}$
γ	920.06 23	1.2 3 $\times10^{-5}$
γ	926.28 21	7 2 $\times10^{-6}$
γ	938.34 22	1.0 4 $\times10^{-5}$
γ	941.41 24	7.2 11 $\times10^{-5}$
γ	958.6 3	6.2 12 $\times10^{-5}$
γ	969.91 22	3.0 6 $\times10^{-5}$
γ	971.28 24	1.0 5 $\times10^{-5}$
γ	990.2 3	3.5 9 $\times10^{-5}$
γ	994.9 7	~7 $\times10^{-6}$
γ	999.78 23	3 1 $\times10^{-5}$
γ	1015.1 7	1.5 4 $\times10^{-5}$?
γ	1020.0 3	2.0 6 $\times10^{-5}$
γ	1024.8 7	1.5 5 $\times10^{-5}$

† 20% uncert(syst)

Atomic Electrons (^{227}Th)

$\langle e \rangle$=54 *3* keV

e_{bin}(keV)	$\langle e \rangle$(keV)	e(%)
4 - 10	0.103	1.37 *25*
11	2.7	25 *7*
12 - 13	1.36	10.7 *19*
14	2.4	17 *5*
15	3.3	22 *5*
16 - 17	0.86	5.3 *13*
18	3.2	17 *4*
19 - 24	0.85	4.2 *9*
25	3.0	~12
26	1.3	4.9 *15*
27	0.75	2.8 *7*
28	1.8	~6
29 - 42	5.8	17.5 *25*
43	1.6	3.8 *14*
44 - 45	0.57	1.3 *3*
46	1.6	3.4 *8*
47 - 97	5.7	9.0 *9*
98 - 147	3.4	2.8 *3*
149 - 151	0.15	~0.10
152	1.01	0.67 *5*
153 - 181	0.8	0.44 *18*
182	1.71	0.94 *5*
183 - 200	1.1	0.55 *15*
201	1.22	0.61 *7*
202 - 237	2.81	1.26 *6*
238	1.05	0.44 *3*
239 - 287	3.08	1.19 *6*
288 - 338	0.85	0.274 *13*
339 - 388	0.036	0.0101 *8*
389 - 439	0.0016	0.00038 *11*
442 - 490	0.00053	0.00011 *4*
492 - 540	0.00034	6.5 *21* $\times10^{-5}$
545 - 593	0.00012	2.2 *8* $\times10^{-5}$
595 - 645	0.00043	7 *3* $\times10^{-5}$
647 - 693	0.0009	0.00013 *5*
699 - 748	0.0015	0.00020 *8*
749 - 798	0.00043	5.5 *14* $\times10^{-5}$
799 - 848	0.0006	~7 $\times10^{-5}$
849 - 896	0.00010	1.2 *5* $\times10^{-5}$
901 - 943	3.1 $\times10^{-5}$	3.4 *13* $\times10^{-6}$
951 - 1000	4.5 $\times10^{-6}$	4.7 *13* $\times10^{-7}$
1001 - 1024	1.2 $\times10^{-6}$	1.2 *5* $\times10^{-7}$

$^{227}_{91}$Pa(38.3 *3* min)

Mode: α(~ 85 %), ϵ(~ 15 %)
Δ: 26825 *10* keV
SpA: 2.162×10^{7} Ci/g
Prod: ^{232}Th(d,7n); ^{232}Th(p,6n)

Alpha Particles (^{227}Pa)

α(keV)	α(rel)
6299 *10*	0.7
6326 *10*	0.3
6335 *4*	0.6
6357 *4*	7
6376 *10*	2.2
6401 *4*	8
6416 *4*	13
6423 *10*	10
6465 *4*	43

Photons (^{227}Pa)

$\langle \gamma \rangle$=6.4 *10* keV

γ_{mode}	γ(keV)	γ(%)
γ_α[M1+E2]	43 *11*	
γ_α	50.0 *24*	<2
γ_α[E1]	64.9 *10*	5.3 *11*
γ_α	66.9 *10*	1.02 *20*
γ_α[E1]	110.0 *10*	1.7 *3*

$^{227}_{92}$U (1.1 *3* min)

Mode: α
Δ: 28870 *100* keV syst
SpA: 7.5×10^{8} Ci/g
Prod: ^{232}Th(α,9n); ^{231}Pa(p,5n); ^{208}Pb(^{22}Ne,3n)

Alpha Particles (^{227}U)

α(keV)
6870 *20*

$^{227}_{93}$Np(1.0 min)

see ^{228}Np(1.0 min)

A = 228

NDS **17**, 367 (1976)

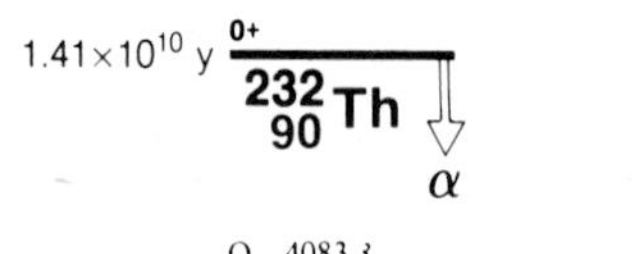

0+ 70 y
232
92 U
α
Q$_α$ 5413.59 15

0+ 34 m
232
94 Pu
α ≤20%
Q$_α$ 6716 10

55 s
232
95 Am
α <100%

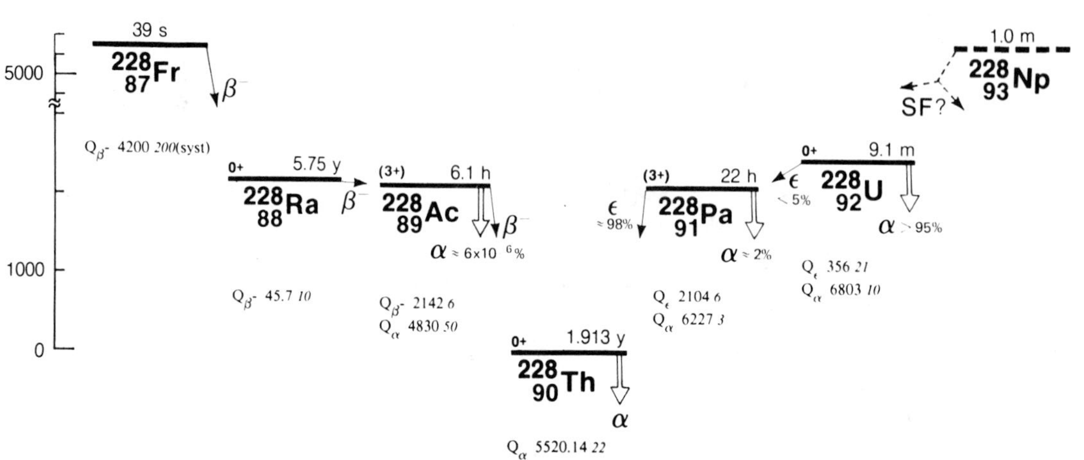

$^{228}_{87}$Fr(39 *1* s)

Mode: β-
Δ: 33140 *200* keV syst
SpA: 1.26×10^9 Ci/g
Prod: protons on Th; protons on U

$^{228}_{88}$Ra(5.75 *3* yr)

Mode: β-
Δ: 28936 *4* keV
SpA: 272.7 Ci/g
Prod: natural source

Photons (^{228}Ra)

$γ_{mode}$	γ(keV)	γ(rel)
γ	6.3 *10*	
γ	6.7 *10*	
γ	12.76 *3*	19 *4*
γ	13.5 *10*	100
γ	15.15 *2*	10 *2*
γ	16.18 *3*	45 *5*
γ	18.8 *4*	13 *3*
γ	19.4 *4*	~0.9 ?
γ	26.4 *10*	
γ	30.5 *10*	

$^{228}_{89}$Ac(6.13 *9* h)

Mode: β-, α(5.5 *22* ×10^{-6} %)
Δ: 28890 *4* keV
SpA: 2.24×10^6 Ci/g
Prod: natural source

Alpha Particles (^{228}Ac)

α(keV)
4270 *40*

Photons (^{228}Ac)

⟨γ⟩=992 *33* keV

$γ_{mode}$	γ(keV)	γ(%)†
Th $L_ℓ$	11.118	0.89 *12*
Th $L_α$	12.952	15.1 *17*
Th $L_η$	14.511	0.39 *5*
Th $L_β$	16.154	20 *3*
Th $L_γ$	19.113	4.6 *7*
γE2	57.81 *5*	0.525 *14*
Th $K_{α2}$	89.955	3.4 *8*
Th $K_{α1}$	93.350	5.6 *13*
γM1	99.55 *8*	1.3 *5*
Th $K_{β1}$'	105.362	2.0 *5*
Th $K_{β2}$'	108.990	0.67 *15*
γE2	129.03 *7*	2.9 *9*
γ[E1]	135.68 *17*	~0.017
γE1	141.19 *19*	0.049 *20*
γ(E1)	146.06 *10*	0.29 *3*
γE1	153.89 *10*	0.84 *11*
γ	174.18 *23*	0.032 *12* ?
γE0+M1+E2	184.72 *14*	0.14 *5*
γE2	191.29 *17*	0.12 *4*

Photons (^{228}Ac) (continued)

$γ_{mode}$	γ(keV)	γ(%)†
γ[E1]	199.54 *8*	0.28 *3*
γ(E1)	204.37 *11*	0.157 *20*
γE1	209.39 *7*	4.1 *8*
γ	210.76 *19*	~0.23 ?
γ	216.24 *9*	~0.8
γ	220.49 *14*	~0.014
γM1+25%E2	223.72 *8*	0.066 *17*
γ	232.3 *3*	~0.06
γ	257.29 *18*	0.032 *9*
γ	263.57 *10*	0.058 *17*
γE1	270.26 *8*	3.8 *11*
γ(M1)	279.3 *3*	0.23 *7*
γM1+35%E2	282.02 *8*	0.089 *23*
γ[E2]	321.9 *4*	0.25 *5*
γE2	327.67 *9*	~0.13
γE1	328.07 *9*	3.5 *12*
γE1(+M2)	332.48 *9*	0.47 *9*
γE1	338.42 *6*	12.4 *23*
γM1+E2	340.94 *16*	0.52 *5*
γ(M1)	356.83 *20*	0.020 *8*
γ	372.3 *10*	~0.009
γ[E2]	377.94 *17*	<0.13
γ	388.9 *12*	~0.015
γ	396.9 *10*	0.032 *13*
γ	399.2 *5*	0.037 *15*
γE2	409.62 *8*	2.20 *20*
γ	416.2 *10*	~0.017
γ	419.23 *14*	~0.026
γ(M1)	440.49 *17*	0.15 *3*
γ	449.57 *19*	0.065 *17*
γ	460.84 *15*	~0.05
γE2	463.10 *7*	4.6 *3*
γ	471.4 *10*	0.035 *15*
γ	474.32 *19*	0.029 *12*
γ(E1)	478.20 *24*	0.24 *5*
γ	481.8 *3*	~0.009
γ	492.36 *18*	~0.02
γ	498.26 *12*	~0.043
γE1	503.7 *4*	0.21 *4*
γ(M1+E2)	509.17 *11*	0.49 *12*
γ	515.2 *3*	0.043 *12*
γ	519.97 *20*	0.078 *23*

Photons (^{228}Ac)
(continued)

γ_{mode}	γ(keV)	γ(%)†
γ	523.18 15	0.12 3
γ	540.5 10	0.029 9
γ(E1)	546.36 23	0.22 4
γ	555.3 10	0.049 15
γM1+E2	562.65 10	1.01 14
γM1	570.2 6	0.19 6
γM1	572.5 3	0.19 5
γ	581.52 19	<3
γ	583.28 19	0.15 4 ?
γ[E1]	615.9 5	0.09 3
γM1+E2	619.88 23	0.11 3
γ	623.59 14	<0.16
γ	629.93 25	0.050 10
γM1	640.7 3	0.061 19
γ[E2]	649.19 12	~0.043
γ	651.44 21	0.101 3 ?
γ	660.4 3	<0.026
γ	666.4 10	0.046 15
γ	673.86 18	0.10 3
γE2	677.07 12	~0.9
γ[E2]	687.58 24	<0.08 ?
γ	688.05 24	<0.08
γE0(+M1+E2)	692.50 23	0.0034 17
γM1	701.80 11	0.19 4
γM1+E2	707.49 13	0.15 4
γ[M1+E2]	726.63 10	0.87 22
γ	737.7 4	0.041 12
γM1	739.23 19	<0.006
γM1	744.32 22	<0.010
γM1	755.28 8	1.32 22
γM1+E2	772.28 7	1.09 7
γ[E2]	774.0 4	~0.09
γ[E2]	782.12 7	0.59 8
γM1	791.05 16	0.021 9
γ	791.2 4	~0.029
γM1+E2	794.79 11	4.6 3
γ[M1+E2]	816.62 24	0.035 12
γ	825.3 5	0.058 12
γE2	830.59 10	0.63 6
γE2	835.60 9	1.71 15
γE2	840.44 10	0.94 6
γM1	853.60 18	0.014 6
γM1	870.47 15	0.066 13
γ[E2]	874.43 24	0.084 17
γ	877.65 19	~0.02
γ	884.0 3	0.10 5
γ	887.46 19	0.020 6
γE2	904.29 15	0.89 6
γE2	911.16 3	29 3
γ	919.19 22	0.028 8
γM1	922.28 14	0.021 8
γM1	923.95 14	<0.048
γ	931.1 10	~0.015
γE1	940.65 18	<0.08
γ[E2]	943.9 5	0.107 20
γ	948.4 5	0.122 23
γ	958.30 23	0.31 6
γE2	964.64 8	5.8 5
γE2	968.97 5	17.4 17
γM1+E2	975.76 12	<0.27
γ	979.7 10	0.023 9
γ	987.87 16	~0.34
γ	987.94 16	0.19 3
γ	1016.12 23	0.025 7
γ[E2]	1019.7 5	0.025 7
γ[E2]	1033.32 15	0.155 19
γ[E1]	1039.97 10	0.055 8
γ	1041.35 17	~0.032
γ	1054.30 9	0.032 7
γ[E1]	1065.05 10	0.171 16
γ[M1+E2]	1095.87 14	0.132 19
γ	1104.01 11	0.017 6 ?
γE1	1110.70 8	0.387 21
γ	1116.97 17	0.061 17
γ[E1]	1135.39 12	~0.012
γ[M1+E2]	1142.7 5	~0.010
γ[E2]	1153.69 14	0.159 23
γ	1163.6 3	0.075 12
γ	1174.79 18	0.029 9
γ	1216.4 4	0.026 9
γ	1245.23 9	~0.09
γE2	1246.60 17	0.57 6
γ	1249.81 12	~0.06
γ	1277.5 5	0.020 6
γ	1286.6 3	0.119 17
γM1+E2	1309.6 10	0.021 6
γ	1314.76 20	0.021 6
γ	1347.6 4	~0.013
γ	1357.6 4	0.029 9
γ	1374.26 8	~0.020
γ	1415.7 4	0.026 9
γE2	1431.0 10	0.032 9
γ	1451.2 3	0.018 6
γE2	1459.19 12	1.06 10
γ	1468.8 5	0.020 6
γ[E1]	1481.2 5	0.020 6
γ	1496.0 3	1.05 10
γ	1501.44 25	0.58 6
γE1	1503.87 18	0.029 13
γ	1528.9 6	0.067 14
γE2	1537.40 10	0.049 11
γ	1548.50 13	0.043 9
γ	1557.0 4	0.20 4
γE1,E2	1572.03 19	0.050 24
γ	1573.3 4	0.055 12
γ	1580.3 3	0.71 10
γ(E2)	1588.23 12	3.6 3
γ	1610.0 5	<0.05
γ	1625.1 3	0.32 6
γ	1630.47 25	1.95 4
γ	1638.1 3	0.54 11
γM1	1666.43 9	0.20 3
γ	1677.9 4	0.067 14
γ	1686.0 4	0.104 20
γ	1702.3 4	0.067 14
γM1+E2	1706.07 15	0.014 3
γ	1713.25 18	0.0047 21
γ	1724.24 10	<0.10
γ	1739.1 5	0.0194 6
γ	1741.2 10	0.015 4
γ	1750.9 10	~0.009
γM1+E2	1757.88 12	0.0414 24
γ	1784.5 10	0.0096 3
γ	1823.47 17	0.051 8
γM1+E2	1835.10 14	0.041 8
γM1+E2	1842.29 18	0.048 4
γ(E2)	1871.2 8	0.026 5
γE2	1886.92 12	0.118 7
γ	1900.10 18	0.0046 13
γ	1907.5 3	0.047 3
γ	1930.3 8	0.025 5
γ	1952.50 17	0.072 9
γ	1965.3 3	0.024 3

† 14% uncert(syst)

Atomic Electrons (^{228}Ac)
⟨e⟩=88 4 keV

e_{bin}(keV)	⟨e⟩(keV)	e(%)
16	3.1	19.1 25
19	0.15	0.77 23
20	3.6	18.1 24
26 - 37	0.411	1.10 4
38	12.1	31.8 11
42	11.2	27.0 9
44	0.052	0.117 16
53	4.64	8.8 3
54	4.13	7.7 3
56	0.0530	0.094 3
57	2.47	4.35 15
58 - 74	0.89	1.48 8
75	5.0	7 3
77	0.016	0.021 5
79	3.0	3.7 14
80 - 107	3.1	3.2 9
109	5.2	4.8 14
111	0.014	<0.025
113	3.2	2.8 8
114 - 161	4.1	3.2 5
164	2.3	1.4 6
165 - 213	1.8	0.96 18
215 - 264	2.4	1.1 4
265 - 313	0.8	0.28 9
316 - 365	1.29	0.38 4
367 - 416	0.63	0.16 4
418 - 467	1.1	0.24 6
468 - 517	0.7	~0.14
518 - 567	0.51	0.092 24
568 - 617	0.42	0.07 3
618 - 668	0.86	0.13 3
669 - 718	1.2	~0.18
719 - 768	0.54	0.073 6
769 - 817	2.58	0.32 3
819 - 868	1.80	0.210 16
869 - 918	0.89	0.10 1
919 - 968	0.63	0.067 3
969 - 1017	0.057	0.0058 14
1018 - 1065	0.028	0.0027 8
1075 - 1123	0.013	0.0012 3
1126 - 1174	0.057	0.0050 10
1177 - 1226	0.025	0.0021 8
1227 - 1276	0.015	0.0012 3
1277 - 1321	0.0059	0.00045 16
1327 - 1374	0.062	0.0046 4
1386 - 1435	0.14	~0.010
1439 - 1488	0.30	0.020 3
1491 - 1590	0.31	0.020 7
1593 - 1690	0.073	0.0045 13
1693 - 1790	0.019	0.00111 18
1798 - 1896	0.014	0.00076 20
1902 - 1961	0.0017	9 3 $\times 10^{-5}$

Continuous Radiation (^{228}Ac)
⟨β-⟩=391 keV;⟨IB⟩=0.52 keV

E_{bin}(keV)		⟨ ⟩(keV)	(%)
0 - 10	β-	0.099	1.97
	IB	0.017	
10 - 20	β-	0.292	1.95
	IB	0.0168	0.116
20 - 40	β-	1.15	3.84
	IB	0.032	0.110
40 - 100	β-	7.7	11.1
	IB	0.082	0.127
100 - 300	β-	60	30.7
	IB	0.18	0.104
300 - 600	β-	121	27.9
	IB	0.126	0.031
600 - 1300	β-	171	20.6
	IB	0.066	0.0085
1300 - 2079	β-	29.4	1.99
	IB	0.0031	0.00022

$^{228}_{90}$Th(1.913 *2* yr)

Mode: α

Δ: 26748 6 keV

SpA: 819.6 Ci/g

Prod: natural source; daughter ^{232}U; daughter ^{228}Ac

Alpha Particles (^{228}Th)

$\langle\alpha\rangle$=5399 24 keV

α(keV)	α(%)
5138.41 20	~0.05
5177.04 20	0.18
5211.36 14	0.4
5340.54 12	26.7 2
5423.32 13	72.7 4

Photons (^{228}Th)

$\langle\gamma\rangle$=3.4 3 keV

γ_{mode}	γ(keV)	γ(%)†
Ra L_ℓ	10.622	0.18 4
Ra L_α	12.325	3.1 7
Ra L_η	13.662	0.094 20
Ra L_β	15.230	4.7 10
Ra L_γ	17.952	1.06 23
γ E2	84.26 5	1.21
Ra $K_{\alpha2}$	85.429	0.0175 11
Ra $K_{\alpha1}$	88.471	0.0288 19
Ra $K_{\beta1}$'	99.915	0.0103 7
Ra $K_{\beta2}$'	103.341	0.00336 23
γ E1	131.50 6	0.128 14
γ (E2)	166.43 16	0.082 11
γ [E1]	205.75 16	0.028 7
γ E1	215.75 6	0.277 7

† 5% uncert(syst)

Atomic Electrons (^{228}Th)

$\langle e\rangle$=20.1 21 keV

e_{bin}(keV)	$\langle e\rangle$(keV)	e(%)
15	0.71	4.6 11
18	0.85	4.6 11
19 - 65	0.29	0.51 8
66	6.9	10.5 22
67	1.06 $\times10^{-5}$	1.58 19 $\times10^{-5}$
69	5.6	8.1 17
70 - 79	0.098	0.12 3
80	4.0	5 1
81 - 82	0.044	0.054 11
83	1.14	1.4 3
84 - 131	0.35	0.40 8
147 - 197	0.124	0.080 6
200 - 216	0.00335	0.00161 3

$^{228}_{91}$Pa(22 1 h)

Mode: ϵ(~ 98 %), α(~ 2 %)
Δ: 28852 7 keV
SpA: 6.2$\times10^5$ Ci/g
Prod: ^{232}Th(p,5n); ^{232}Th(d,6n); ^{230}Th(d,4n)

Alpha Particles (^{228}Pa)

α(keV)	α(rel)
5711	0.020
5756	0.05
5760	0.028
5765	0.04
5779	0.028
5799	0.23
5805	0.15
5843	0.008
5858	0.006
5874	0.028
5907	0.022
5922	0.016
5941	0.010
5947	0.012
5975	0.06
5982	0.06
5989	0.022
5998	0.006
6011	0.016
6028	0.18
6041	0.05
6066	0.020
6078	0.4
6091	0.05
6105	0.25
6118	0.22

Photons (^{228}Pa)

$\langle\gamma\rangle$=1091 24 keV*

γ_{mode}	γ(keV)	γ(%)
Th L_ℓ	11.118	1.7 4
Th L_α	12.952	30 7
Th L_η	14.511	0.61 13
Th L_β	16.141	35 8
Th L_γ	19.136	7.8 18
γ_ϵE2	57.81 5	0.52 8
Th $K_{\alpha2}$	89.955	22 6
Th $K_{\alpha1}$	93.350	35 9
γ_ϵM1	99.55 8	0.8 4
Th $K_{\beta1}$'	105.362	13 3
Th $K_{\beta2}$'	108.990	4.3 11
γ_ϵE2	129.03 7	2.85 15
γ_α	130 1	27 ‡
γ_ϵ	132.1 6	0.46 7
γ_ϵ[E1]	135.68 17	~0.05
γ_ϵ	138.39 20	0.40 6
γ_ϵE1	141.19 19	0.16 7
γ_ϵ(E1)	146.06 10	0.32 3
γ_α	150 1	34 ‡
γ_ϵE1	153.89 10	0.37 5
γ_α	170 1	11 ‡
γ_ϵ	174.18 23	~0.008
γ_ϵM1	177.96 18	<0.08 ?
γ_ϵE0+M1+E2	184.72 14	0.034 13
γ_ϵE2	191.29 17	0.27 4
γ_ϵ[E1]	199.54 8	0.31 4
γ_α	200 1	14 ‡
γ_ϵ(E1)	204.37 11	0.48 4
γ_ϵE1	209.39 7	1.68 15
γ_ϵ	210.76 19	~0.37
γ_ϵ	216.24 9	0.87 9
γ_α	220 1	10 ‡
γ_ϵ	220.49 14	~0.18
γ_ϵM1+25%E2	223.72 8	0.92 8
γ_ϵ	223.80 24	
γ_α	240 1	55 ‡
γ_ϵ	240.3 8	~0.10
γ_ϵ	255.1 10	~0.30
γ_ϵ	257.29 18	0.09 3
γ_ϵ	263.57 10	0.17 5
γ_ϵE1	270.26 8	2.1 1
γ_ϵ(M1)	278.1 10	<0.12
γ_ϵ(M1)	279.3 3	0.058 21
γ_α	280 1	49 ‡
γ_ϵM1+35%E2	282.02 8	1.22 7
γ_α	310 1	100 ‡
γ_ϵ[E2]	321.9 4	0.063 17
γ_ϵE2	327.67 9	~2
γ_ϵE1	328.07 9	1.9 4
γ_ϵE1(+M2)	332.48 9	1.57 14
γ_ϵE1	338.42 6	5.1 3
γ_ϵM1+E2	340.94 16	1.52 11
γ_α	345 1	21 ‡
γ_ϵ(M1)	356.83 20	0.012 6
γ_ϵ[E2]	377.94 17	<0.08
γ_ϵE2	409.62 8	6.4 7
γ_ϵ	419.23 14	~0.018
γ_ϵ(M1)	440.49 17	0.09 3
γ_ϵ	449.57 19	0.22 8
γ_ϵ	460.84 15	~0.8
γ_ϵE2	463.10 7	13.2 6
γ_ϵ	474.32 19	0.046 19
γ_ϵ	481.8 3	~0.12
γ_ϵ	492.36 18	~0.014
γ_ϵ	498.26 12	~0.6
γ_ϵ(M1+E2)	509.17 11	0.30 10
γ_ϵ	519.97 20	0.12 4
γ_ϵ	523.18 15	0.08 3
γ_ϵ	525.1 6	0.18 6
γ_ϵ	547.6 6	0.12 3
γ_ϵ	556.04 11	0.18 4
γ_ϵM1+E2	562.65 10	0.61 14
γ_ϵ	571.19 20	0.57 6
γ_ϵM1	572.5 3	0.40 10
γ_ϵ	581.52 19	1.02 24 ?
γ_ϵM1+E2	589.3 8	<0.08
γ_ϵM1	602.6 8	<0.06
γ_ϵ	615.0 4	0.096 24
γ_ϵM1+E2	619.88 23	0.25 7
γ_ϵ	624.20 13	0.0078 12
γ_ϵ	629.93 25	<0.044
γ_ϵM1	640.7 3	0.13 6
γ_ϵ[E2]	649.19 12	~0.048
γ_ϵ	650.6 4	0.25 4
γ_ϵM1+E2	663.4 6	0.36 5
γ_ϵE2	668.29 22	0.39 6
γ_ϵ	673.86 18	0.16 5
γ_ϵE2	677.07 12	0.6 3
γ_ϵ	688.05 24	<0.19
γ_ϵE0(+M1+E2)	692.50 23	0.054 24
γ_ϵM1	701.80 11	0.18 5
γ_ϵM1+E2	707.49 13	0.47 14
γ_ϵ	718.19 20	0.240 12
γ_ϵ[M1+E2]	726.63 10	0.38 10
γ_ϵM1	737.77 22	<0.09
γ_ϵM1	739.23 19	<0.09
γ_ϵM1	744.32 22	<0.12
γ_ϵ	750.52 22	0.21 4
γ_ϵM1	755.28 8	1.24 8
γ_ϵM1+E2	772.28 7	1.21 7
γ_ϵE2	776.59 20	0.42 5
γ_ϵ[E2]	782.12 7	0.33 5
γ_ϵM1	791.05 16	0.27 3
γ_ϵM1+E2	794.79 11	2.03 9
γ_ϵ	796.65 23	<0.24 ?
γ_ϵM1	802.1 5	<0.09
γ_ϵE2	818.1 8	~0.6
γ_ϵ	823.6 10	~0.24
γ_ϵE2	830.59 10	1.94 9
γ_ϵE2	835.60 9	2.77 13
γ_ϵE2	840.44 10	1.04 5
γ_ϵM1	853.60 18	~0.19
γ_ϵM1	870.47 15	1.05 6
γ_ϵ	877.65 19	~0.07
γ_ϵ	884.0 3	0.34 7
γ_ϵ	887.46 19	0.014 4
γ_ϵE1	888.7 5	0.78 18
γ_ϵ	894.46 12	2.6 9
γ_ϵE2	904.29 15	2.83 21
γ_ϵE2	911.16 3	16.0 7
γ_ϵ	919.19 22	0.019 6
γ_ϵM1	922.28 14	~0.27
γ_ϵM1	923.95 14	~0.36
γ_ϵE1	940.65 18	~0.6
γ_ϵ	945.7 8	1.8 6
γ_ϵ	958.30 23	0.7 3
γ_ϵE2	964.64 8	9.4 9
γ_ϵE2	968.97 5	9.7 13
γ_ϵM1+E2	975.76 12	1.56 9
γ_ϵ	987.87 16	0.240 12
γ_ϵ	987.94 16	0.240 12
γ_ϵ	1016.12 23	<0.08
γ_ϵ	1018.5 3	<0.51

Photons (^{228}Pa) (continued)

γ_{mode}	γ(keV)	γ(%)
γ_ϵ	1018.7 3	0.21 3
γ_ϵ[E2]	1033.32 15	0.49 3
γ_ϵ[E1]	1039.97 10	0.170 22
γ_ϵ	1041.35 17	~0.022
γ_ϵ	1046.2 8	0.036 12
γ_ϵ	1054.30 9	0.44 14
γ_ϵ[E1]	1065.05 10	0.076 5
γ_ϵ	1070.3 5	0.114 24
γ_ϵ[M1+E2]	1095.87 14	0.033 6
γ_ϵ	1104.0 8	0.018 6
γ_ϵ	1104.01 11	0.051 17
γ_ϵE1	1110.70 8	0.428 17
γ_ϵ	1116.97 17	0.076 22
γ_ϵ	1118.7 6	0.042 6
γ_ϵ[E1]	1135.39 12	~0.007
γ_ϵ[E2]	1153.69 14	0.040 8
γ_ϵM1+E2	1164.5 6	0.072 6
γ_ϵ	1174.79 18	0.036 11
γ_ϵM1,E2	1184.2 3	~0.024
γ_ϵM1,E2	1194.8 10	0.018 6
γ_ϵ	1237.8 6	0.084 12
γ_ϵ	1245.23 9	~0.25
γ_ϵE2	1246.60 17	0.90 4
γ_ϵ	1249.81 12	~0.039
γ_ϵM1	1252.4 3	0.018 6
γ_ϵ	1273.1 6	0.078 12
γ_ϵM1+E2	1288.1 4	0.120 6
γ_ϵE1+M2	1298.1 4	0.114 12
γ_ϵM1+E2	1311.1 6	0.054 18
γ_ϵ	1314.76 20	0.034 10
γ_ϵ	1373.59 17	<0.26
γ_ϵ	1374.26 8	<0.048
γ_ϵE2	1420.7 6	0.096 6
γ_ϵE2	1431.8 6	0.138 12
γ_ϵE1	1453.9 6	0.120 6
γ_ϵE2	1459.19 12	0.72 3
γ_ϵ	1464.29 20	~0.06
γ_ϵE2	1481.5 6	0.090 6
γ_ϵ	1487.6 6	~0.05
γ_ϵE1,E2	1496.0 4	0.168 12
γ_ϵ	1501.44 25	0.030 5
γ_ϵE1	1503.87 18	0.096 6
γ_ϵ	1522.7 3	0.048 6
γ_ϵE1,E2	1529.07 18	0.174 12
γ_ϵE2	1537.40 10	0.046 13
γ_ϵ	1548.50 13	0.097 6
γ_ϵM1+E2	1557.3 3	0.276 12
γ_ϵE1,E2	1572.03 19	0.17 3
γ_ϵE1	1580.5 3	0.38 4
γ_ϵ(E2)	1588.23 12	2.44 11
γ_ϵM1	1610.3 4	0.072 6
γ_ϵM1+E2	1618.48 17	0.138 18
γ_ϵM1	1621.3 4	0.258 24
γ_ϵ	1630.47 25	0.102 12
γ_ϵE2	1638.4 3	0.096 12
γ_ϵM1	1666.43 9	0.187 12
γ_ϵ	1676.30 18	0.030 6
γ_ϵE2	1685.9 4	0.144 12
γ_ϵ	1701.1 6	0.060 6
γ_ϵM1+E2	1706.07 15	0.216 12
γ_ϵ	1713.25 18	0.016 6
γ_ϵ	1724.24 10	<0.10
γ_ϵ	1725.3 6	0.024 6
γ_ϵ	1733.8 8	0.042 12
γ_ϵM1+E2	1738.46 17	0.64 3
γ_ϵ	1751.34 12	0.030 6
γ_ϵM1+E2	1757.88 12	0.54 3
γ_ϵ	1772.8 6	0.030 6
γ_ϵM1+E2	1784.9 3	0.084 6
γ_ϵ	1794.4 3	0.090 6
γ_ϵM1+E2	1807.2 4	0.0324 24
γ_ϵ	1823.47 17	0.036 5
γ_ϵ	1829.7 4	0.054 12
γ_ϵM1+E2	1835.10 14	0.64 4
γ_ϵM1+E2	1842.29 18	0.163 12
γ_ϵM1+E2	1842.7 3	0.162 12
γ_ϵ	1866.0 6	0.048 6
γ_ϵ(E2)	1871.1 6	0.090 12
γ_ϵ	1880.37 10	0.138 12
γ_ϵE2	1886.92 12	1.55 9
γ_ϵ	1900.10 18	0.016 3
γ_ϵ	1907.5 3	0.059 3
γ_ϵ	1918.2 6	0.0168 18

Photons (^{228}Pa) (continued)

γ_{mode}	γ(keV)	γ(%)
γ_ϵ	1924.5 6	0.0084 18
γ_ϵ	1936.2 4	0.0144 18
γ_ϵ	1952.50 17	0.051 3
γ_ϵ	1958.7 4	0.022 3
γ_ϵ	1965.3 3	0.029 4

* with ϵ
‡ rel int

Atomic Electrons (^{228}Pa)*

$\langle e \rangle$=102 5 keV

e_{bin}(keV)	$\langle e \rangle$(keV)	e(%)
16	5.8	36 8
19	0.145	0.75 4
20	5.9	30 7
22 - 37	1.2	~4
38	12.1	32 5
42	11.1	27 4
44	0.023	0.052 6
53	4.6	8.7 14
54	4.1	7.6 12
56	0.053	0.093 15
57	2.5	4.3 7
58 - 77	3.0	4.4 8
79	1.8	2.3 10
80 - 107	3.2	3.3 9
109	5.1	4.6 3
111 - 112	0.9	~0.8
113	3.10	2.75 15
114 - 162	7.1	5.7 5
164 - 213	4.0	2.2 3
215 - 222	0.65	0.30 8
223	2.8	~1
224 - 273	2.1	0.9 4
274 - 323	2.8	0.9 3
324 - 373	2.7	0.78 14
374 - 421	1.3	0.33 7
424 - 473	2.4	0.54 8
474 - 523	0.31	0.063 20
524 - 573	0.67	0.122 25
576 - 625	0.47	0.078 21
626 - 674	1.06	0.16 3
676 - 725	1.0	0.14 4
726 - 775	1.14	0.152 16
776 - 825	2.3	0.28 5
826 - 875	2.3	0.27 4
876 - 925	0.82	0.092 11
926 - 975	0.70	0.074 8
976 - 1025	0.071	0.0071 17
1026 - 1075	0.044	0.0042 12
1076 - 1119	0.011	0.0010 3
1128 - 1175	0.11	0.0097 24
1178 - 1227	0.066	0.0055 20
1229 - 1278	0.032	0.0026 10
1282 - 1322	0.0177	0.00135 13
1344 - 1392	0.067	0.0050 6
1394 - 1443	0.034	0.0024 5
1444 - 1493	0.171	0.0116 10
1495 - 1594	0.149	0.0097 7
1596 - 1694	0.16	0.0099 18
1696 - 1791	0.195	0.0112 14
1798 - 1896	0.052	0.0028 3
1898 - 1961	0.0020	0.000103 25

* with ϵ

Continuous Radiation (^{228}Pa)

$\langle\beta+\rangle$=0.72 keV; $\langle$IB$\rangle$=1.27 keV

E_{bin}(keV)		$\langle\ \rangle$(keV)	(%)
0 - 10	β+	1.68×10^{-9}	1.97×10^{-8}
	IB	8.4×10^{-5}	
10 - 20	β+	1.36×10^{-7}	8.0×10^{-7}
	IB	0.0041	0.026
20 - 40	β+	7.5×10^{-6}	2.23×10^{-5}
	IB	0.00100	0.0033
40 - 100	β+	0.00078	0.00096
	IB	0.56	0.64
100 - 300	β+	0.061	0.0273
	IB	0.28	0.24
300 - 600	β+	0.356	0.079
	IB	0.085	0.019
600 - 1300	β+	0.305	0.0423
	IB	0.25	0.028
1300 - 2050	IB	0.083	0.0057
	$\Sigma\beta$+		0.150

$^{228}_{92}$U (9.1 2 min)

Mode: α(>95 %), ϵ(<5 %)
Δ: 29208 21 keV
SpA: 9.06×10^{7} Ci/g
Prod: ^{232}Th(α,8n)

Alpha Particles (^{228}U)

α(keV)	α(rel)
6404 6	0.6
6440 5	0.7
6589 5	29
6681 6	70

Photons (^{228}U)

$\langle\gamma\rangle$=4.8 6 keV

γ_{mode}	γ(keV)	γ(%)†
Th L_ℓ	11.118	~0.18
Th L_α	12.952	~3
Th L_η	14.511	~0.10
Th L_β	16.165	~5
Th L_γ	19.096	~1
Th $K_{\alpha2}$	89.955	0.027 5
Th $K_{\alpha1}$	93.350	0.044 9
γ_α[E2]	95 4	1.63 23
Th $K_{\beta1}'$	105.362	0.016 3
Th $K_{\beta2}'$	108.990	0.0053 11
γ_α[E1]	152 3	0.19 5
γ_α[E2]	187 3	0.29 10
γ_α[E1]	246 3	0.38 10

†13% uncert(syst)

Atomic Electrons (^{228}U)*

$\langle e \rangle$=22 5 keV

e_{bin}(keV)	$\langle e \rangle$(keV)	e(%)
16	0.6	~4
20	0.9	~4
42 - 74	0.35	0.48 23
75	7.6	~10
77	0.00013	0.00016 4
78	5.7	~7
84 - 89	0.13	0.14 7

Atomic Electrons (^{228}U)*
(continued)

e_{bin}(keV)	⟨e⟩(keV)	e(%)
90	4.3	~5
91 - 92	0.044	~0.048
93	0.7	~0.8
94	0.9	~0.9
95 - 137	0.032	0.024 *5*
147 - 187	0.34	0.19 *4*
226 - 246	0.0105	0.0046 *7*

* with α

$^{228}_{93}$Np(1.00 *8* min)

activity attributed to ^{228}Np or ^{227}Np
Mode: SF
SpA: 8×10^{8} Ci/g
Prod: ^{22}Ne on ^{209}Bi

A = 229

NDS **24**, 263 (1978)

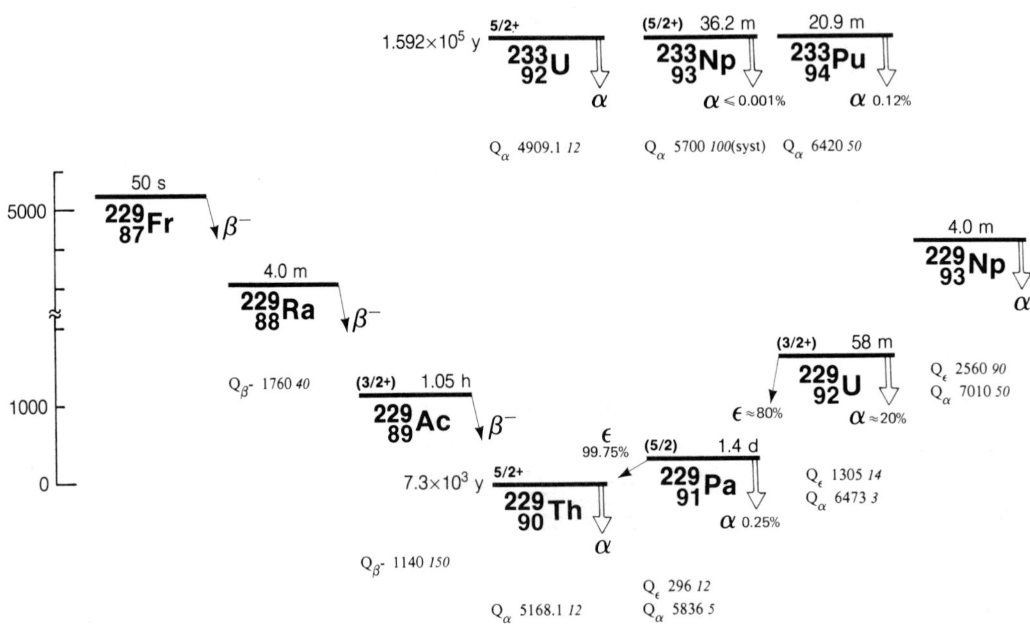

$^{229}_{87}$Fr(50 *20* s)

Mode: β-
SpA: 1×10^{9} Ci/g
Prod: protons on Th

$^{229}_{88}$Ra(4.0 *2* min)

Mode: β-
Δ: 32480 *160* keV
SpA: 2.05×10^{8} Ci/g
Prod: ^{228}Ra(n,γ); protons on Th

$^{229}_{89}$Ac(1.045 *8* h)

Mode: β-
Δ: 30720 *150* keV
SpA: 1.309×10^{7} Ci/g
Prod: daughter ^{229}Ra; ^{232}Th(γ,p2n)

Photons (^{229}Ac)

$γ_{mode}$	γ(keV)	γ(rel)
γ M1	29.191 *12*	~2
γ [M1+E2]	42.656 *14*	
γ E2	71.847 *16*	~3
γ	74.500 *20*	8.1 *16*
γ [E1]	100.008 *12*	0.14 *4*
γ [E1]	103.375 *12*	0.21 *6*
γ	117.155 *16*	15.5 *23*
γ [E1]	135.328 *14*	34 *4*
γ	146.347 *17*	35 *4*
γ [E1]	155.989 *15*	0.12 *4*
γ [E1]	164.512 *10*	100 *10*
γ	168.98 *3*	0.17 *4*
γ	170.794 *18*	0.36 *9*
γ	172.35 *3*	0.072 *21*
γ	174.162 *16*	0.48 *14*
γ [E2]	223.357 *22*	0.067 *20*
γ	239.25 *20*	4.1 *12*
γ (M1)	245.294 *14*	8.9 *13*
γ (M1)	248.661 *15*	9.2 *20*
γ	252.01 *8*	24 *3*
γ [E2]	261.88 *8*	39 *5*
γ [M1+E2]	274.692 *16*	1.20 *25*
γ (M1)	278.059 *14*	2.5 *6*
γ	284.95 *20*	4.3 *10*
γ	287.84 *12*	~6
γ [M1+E2]	287.949 *11*	2.7 *6*
γ (M1)	291.317 *9*	10.6 *19*
γ (M1)	317.133 *9*	22 *3*

Photons (^{229}Ac)
(continued)

$γ_{mode}$	γ(keV)	γ(rel)
γ (M1)	320.508 *9*	6.4 *11*
γ	322.85 *20*	3.5 *8*
γ	332.03 *9*	2.0 *6*
γ	365.65 *20*	2.7 *6*
γ	404.64 *8*	8.7 *17*
γ	406.53 *9*	5.9 *14*
γ	422.81 *8*	6.8 *12*
γ	435.93 *9*	6.3 *12*
γ	449.18 *9*	15.9 *24*
γ	478.37 *9*	17 *3*
γ	508.55 *20*	3 *1*
γ	526.71 *8*	6.3 *12*
γ	539.96 *8*	20 *3*
γ	562.53 *12*	6.4 *12*
γ	569.16 *8*	91 *12*
γ	575.79 *12*	5.2 *10*
γ	604.97 *12*	23 *3*

$^{229}_{90}$Th(7340 *160* yr)

Mode: α
Δ: 29580 *3* keV
SpA: 0.213 Ci/g
Prod: daughter ^{233}U

Alpha Particles (^{229}Th)

⟨α⟩=4862 keV

α(keV)	α(%)
4478 *3*	~0.005
4484 *3*	~0.009
4598 *3*	~0.007
4667 *25* ?	~0.0010
4689.5 *5*	0.15
4692 *3*	~0.08
4737 *25* ?	~0.010
4748 *25* ?	~0.005
4754 *25*	~0.05
4761.8 *5*	0.63
4797.9 *5*	1.27
4809 *25*	~0.22
4814.0 *5*	9.30 *8*
4833 *25* ?	~0.29
4838.1 *5*	4.8
4845.2 *5*	56.2 *2*
4852 *25* ?	~0.030
4861 *25*	0.18
4865 *25* ?	~0.030
4878 *25* ?	~0.030
4900.8 *5*	10.20 *8*
4930.1 *5*	0.11
4967.8 *5*	5.97 *6*
4978.6 *5*	3.17 *4*
5035.4 *5*	0.24
5050 *25*	5.2
5052.5 *5*	1.6
5077.4 *5*	0.010

Photons (^{229}Th)

⟨γ⟩=34 *3* keV

γ_{mode}	γ(keV)	γ(%)
γ (M1)	17.349 *21*	~0.17
γ (M1+~23%E2)	25.373 *18*	~0.036
γ	30.28 *10*	
γ E1	31.37 *14*	4.1 *8*
γ	37.78 *10*	
γ (M1+~8.3%E2)	42.722 *22*	~0.16
γ [M1+E2]	53.18 *10*	
γ (M1)	56.58 *3*	~0.33
γ (M1+~8.3%E2)	68.14 *4*	~0.10
γ (E2)	68.88 *3*	~0.11
γ (E2)	75.17 *5*	~0.5
γ	75.28 *10*	
γ (M1+~33%E2)	86.23 *3*	~0.38
γ [M1]	86.44 *4*	~3
γ M1(+E2)	107.16 *4*	~0.8
γ [M1]	124.51 *4*	~1
γ [M1]	124.72 *4*	~0.6
γ M1(+E2)	131.92 *4*	~0.33
γ	132.58 *10*	
γ	134.78 *10*	
γ	135.69 *7*	
γ M1	137.02 *3*	1.6 *3*
γ	140.28 *20*	
γ M1(+E2)	142.98 *6*	~0.43
γ	147.78 *10*	
γ [E1]	148.37 *14*	~1
γ	150.2 *3*	
γ	151.29 *9*	
γ [M1]	154.37 *3*	~0.7
γ [M1]	156.46 *4*	~1
γ	158.50 *7*	
γ	161.6 *3*	
γ	165.7 *3*	

Photons (^{229}Th) (continued)

γ_{mode}	γ(keV)	γ(%)
γ [M1]	172.90 *8*	~0.22
γ [E2]	179.74 *4*	~0.51 ?
γ (M1+~45%E2)	183.96 *8*	~0.23
γ	190.18 *20*	
γ M1	193.59 *3*	4.6 *9*
γ	204.9 *3*	
γ M1	210.94 *3*	3.3 *7*
γ [M1]	218.15 *5*	~0.14
γ [M1]	236.32 *4*	~0.036
γ [M1+E2]	242.69 *5*	
γ	243.53 *5*	
γ	261.0 *5*	
γ	290.0 *5*	

$^{229}_{91}$Pa(1.4 *4* d)

Mode: ϵ(99.75 %), α(0.25 %)
Δ: 29876 *12* keV
SpA: 4.1×10^{5} Ci/g
Prod: ^{230}Th(d,3n); ^{232}Th(p,4n); daughter ^{229}U

Alpha Particles (^{229}Pa)

⟨α⟩=14 keV

α(keV)	α(%)
5319 *5* ?	0.00012
5412 *5*	0.00025
5421 *25* ?	0.00018
5479 *4*	0.004
5500 *5*	0.0018
5516 *5*	0.0015
5535.8 *18*	0.021
5564.6 *18*	0.010
5579.5 *18*	0.09
5589 *4*	0.012
5614.0 *18*	0.03
5629.0 *18*	0.024
5668.9 *18*	0.05
5693.3 *18*	0.0005
5732.9 *18*	<0.0012

Photons (^{229}Pa)

γ_{mode}	γ(keV)
γ_α	24.77 *20*
γ_α	30.33 *13*
γ_α	34.78 *14*
γ_αE1	40.34 *17*
γ_α	40.64 *19*
γ_α	41 *4*
γ_ϵM1+14%E2	42.441 *15*
γ_α	44.48 *20*
γ_α	50.43 *23*
γ_α	54 *4*
γ_α	65.11 *14*
γ_α	68.13 *16*
γ_α	75.41 *16*
γ_α	78 *4*
γ_α	79.58 *16*
γ_α	80.74 *22*
γ_α	91.07 *20*
γ_α	94.91 *16*
γ_α	112.03 *16*
γ_α	115.84 *17*
γ_α(M1+E2)	121.08 *16*
γ_α	125.84 *17*
γ_α	135.55 *14*

Photons (^{229}Pa) (continued)

γ_{mode}	γ(keV)
γ_α	140.94 *20*
γ_α	156.18 *19*
γ_α	170.32 *13*
γ_α	180.15 *16*

$^{229}_{92}$U (58 *3* min)

Mode: ϵ(~ 80 %), α(~ 20 %)
Δ: 31181 *10* keV
SpA: 1.42×10^{7} Ci/g
Prod: ^{232}Th(α,7n)

Alpha Particles (^{229}U)

α(keV)	α(rel)
6185 ?	~1
6223 *3*	3 *1*
6260 ?	~1
6297 *3*	11 *1*
6332 *3*	20 *2*
6360 *3*	64 *2*

$^{229}_{93}$Np(4.0 *2* min)

Mode: α
Δ: 33740 *90* keV
SpA: 2.05×10^{8} Ci/g
Prod: ^{233}U(p,5n)

Alpha Particles (^{229}Np)

α(keV)
6890 *20*

A = 230

NDS **20**, 139 (1977)

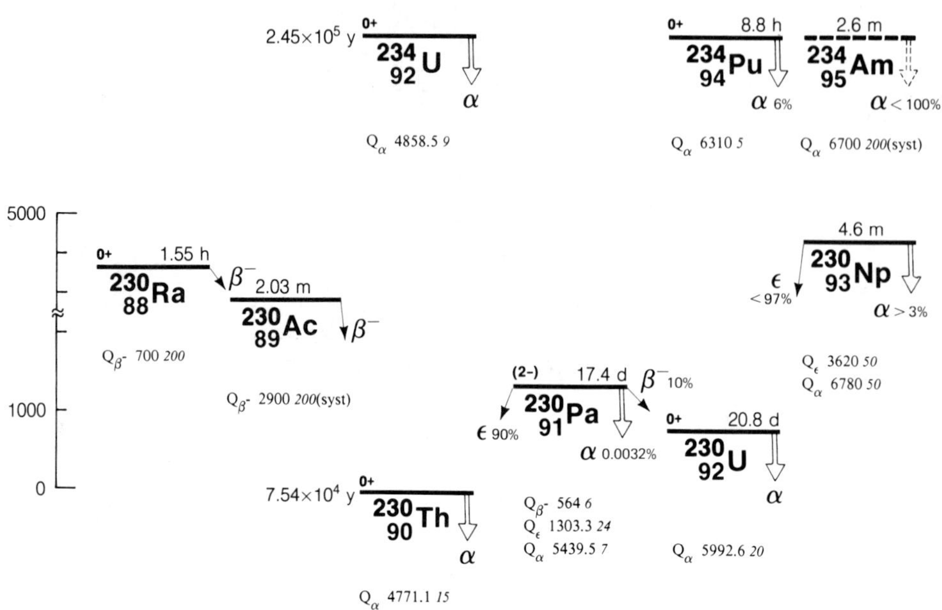

$^{230}_{88}$Ra(1.55 *5* h)

Mode: β-
Δ: 34460 *280* keV syst
SpA: 8.8×10^{6} Ci/g
Prod: ^{232}Th(d,3pn); ^{232}Th(n,2pn)

Photons (^{230}Ra)

γ_{mode}	γ(keV)	γ(rel)
γ	49.24 *9*	2.6 *5*
γ	63.06 *7*	40 *2*
γ	72.05 *7*	113 *6*
γ	92.11 *6*	21 *5*
γ	101.09 *6*	16 *3*
γ	110.74 *7*	3.1 *3*
γ	117.3 *3*	<0.30
γ	125.3 *3*	<0.30
γ	132.59 *10*	<0.30
γ	134.33 *8*	4.5 *5*
γ	138.75 *20*	0.6 *2*
γ	141.42 *11*	0.7 *2*
γ	147.95 *7*	5.6 *3*
γ	151.56 *9*	2.0 *2*
γ	177.58 *14*	0.6 *2*
γ	184.15 *7*	11.5 *3*
γ	189.23 *7*	16.7 *5*
γ	192.71 *8*	1.4 *3*
γ	197.19 *11*	1.3 *2*
γ	198.22 *7*	3.4 *4*
γ	202.85 *6*	30.8 *10*
γ	211.84 *6*	11.3 *3*
γ	236.3 *3*	<0.30

Photons (^{230}Ra) (continued)

γ_{mode}	γ(keV)	γ(rel)
γ	247.4 *3*	<0.30
γ	251.58 *9*	9.9 *5*
γ	255.77 *9*	1.6 *4*
γ	259.75 *9*	3.4 *5*
γ	274.70 *13*	1.0 *3*
γ	285.25 *7*	18.2 *7*
γ	288.31 *12*	1.0 *3*
γ	292.97 *9*	4.3 *6*
γ	296.03 *11*	1.2 *2*
γ	297.65 *20*	0.5 *2*
γ	316.52 *8*	1.0 *2*
γ	363.9 *3*	0.6 *2*
γ	375.79 *13*	0.3 *1*
γ	412.95 *10*	0.9 *2*
γ	437.64 *18*	0.3 *1*
γ	440.08 *13*	0.3 *1*
γ	448.98 *7*	15.0 *5*
γ	457.97 *7*	18.5 *6*
γ	469.77 *7*	29.3 *10*
γ	473.5 *3*	<0.30
γ	478.75 *7*	24.2 *10*
γ	484.30 *17*	1.9 *4*
γ	490.55 *10*	4.3 *8*
γ	509.23 *9*	6.5 *6*
γ	536.83 *11*	1.2 *2*

$^{230}_{89}$Ac(2.03 *5* min)

Mode: β-
Δ: 33760 *200* keV syst
SpA: 4.009×10^{8} Ci/g
Prod: daughter ^{230}Ra; ^{232}Th(γ,pn)

Photons (^{230}Ac)

⟨γ⟩=546 *5* keV

γ_{mode}	γ(keV)	γ(%)†
Th L_ℓ	11.118	0.38 *10*
Th L_α	12.952	6.5 *16*
Th L_η	14.511	0.18 *5*
Th L_β	16.160	9.0 *23*
Th L_γ	19.097	2.0 *5*
γ E2	53.23 *3*	0.20 *5*
Th $K_{\alpha2}$	89.955	0.51 *8*
Th $K_{\alpha1}$	93.350	0.82 *13*
Th $K_{\beta1}'$	105.362	0.30 *5*
Th $K_{\beta2}'$	108.990	0.099 *15*
γ E2	120.912 *20*	0.33 *3*
γ [E1]	170.56 *5*	
γ [M1+E2]	183.92 *9*	
γ [E1]	253.54 *9*	0.00108 *24*
γ [E1]	274.33 *6*	0.0028 *8*
γ [E1]	294.13 *7*	0.0013 *6*
γ [E1]	297.92 *7*	~0.0048
γ [E1]	316.99 *8*	0.0047 *9*
γ [M1,E2]	332.07 *7*	0.013 *4*
γ	364.00 *12*	0.041 *16*
γ (E2)	374.73 *9*	0.008 *3*
γ E2	380.16 *5*	0.0087 *17*
γ	388.32 *9*	0.098 *8*

Photons (^{230}Ac) (continued)

γ_{mode}	γ(keV)	γ(%)†
γ E1	397.69 5	0.40 3
γ M1+~26%E2	399.96 6	0.0220 21
γ [E1]	401.69 8	0.0026 7
γ	423.36 6	0.090 16
γ M1	440.56 17	0.042 13
γ M1+30%E2	443.79 3	0.160 10
γ [E1]	448.96 5	0.123 25
γ E1	454.97 3	8.9 4
γ M1+~14%E2	463.59 5	0.029 3
γ [E2]	503.53 5	0.19 3
γ [E2]	504.19 16	0.027 5
γ [M1+E2]	507.52 7	~0.024
γ E1	508.20 3	5.14 14
γ E1	518.60 5	0.42 3
γ E2	571.16 6	0.118 12
γ E2	581.78 8	0.53 4
γ	600.73 20	0.082 25
γ [E2]	607.29 4	0.021 10
γ E0+E2+M1	624.44 5	0.147 15
γ	628.85 9	0.238 25
γ	635.6 3	0.08 3
γ [M1+E2]	651.68 7	0.040 10
γ	671.43 20	0.18 3
γ [E2]	677.67 5	0.177 15
γ E2	728.20 4	0.47 4
γ	735.20 9	0.098 25
γ	750.9 1	0.148 25
γ E2	772.59 7	0.24 4
γ E2	781.43 4	0.37 3
γ	788.99 7	0.533 25
γ	798.03 10	0.115 16
γ	816.73 10	0.320 16
γ [E2]	835.60 5	0.020 7
γ [E1]	838.25 16	0.010 4
γ [M1+E2]	839.94 9	0.180 16
γ	867.15 6	0.467 25
γ [E1]	877.89 10	0.082 8
γ [E1]	892.75 5	0.69 3
γ E1	898.76 3	0.177 17
γ	913.84 9	0.098 25
γ E1	918.56 5	0.287 25
γ	939.23 20	0.08 3
γ	946.33 20	0.08 3
γ E1	951.99 4	0.833 25
γ E2	956.51 5	0.429 23
γ E1	959.16 16	0.189 16
γ	963.03 20	0.11 4
γ	968.06 17	0.10 4
γ	973.53 20	0.11 4
γ	977.48 8	0.057 25
γ	982.0 5	<0.025
γ	987.0 5	<0.025
γ	991.21 8	0.098 25 ?
γ	999.10 9	0.22 4
γ E2	1009.74 5	0.256 17
γ E1	1026.13 6	0.160 14
γ	1043.44 10	0.082 25
γ	1045.3 3	~0.049
γ	1052.99 13	0.15 3
γ	1065.5 3	0.041 16
γ	1068.72 12	0.066 25
γ	1093.8 4	0.15 4
γ	1106.2 3	0.057 16
γ	1147.78 10	0.394 25 ?
γ	1187.53 9	0.107 25
γ	1198.97 13	0.15 3
γ	1212.03 20	0.15 3
γ [E2]	1226.81 5	0.96 5
γ	1243.96 7	3.50 8
γ	1252.64 12	0.066 16
γ	1267.05 7	0.41 4
γ	1268.2 3	0.21 4
γ	1302.58 6	0.541 25
γ	1306.2 4	0.11 3
γ	1311.28 22	0.15 3
γ	1322.12 6	0.71 4
γ [M1+E2]	1347.72 5	1.57 4
γ	1375.35 6	1.21 4
γ	1394.53 9	0.16 4
γ [E2]	1400.95 5	0.33 3
γ	1432.45 8	0.197 16
γ	1455.59 11	0.15 4
γ	1524.6 3	0.08 3
γ [E2]	1536.6 3	0.066 25

Photons (^{230}Ac) (continued)

γ_{mode}	γ(keV)	γ(%)†
γ	1573.53 20	0.189 16
γ	1585.35 10	0.172 16
γ	1597.23 20	0.131 25
γ	1625.05 11	0.115 25 ?
γ	1636.73 14	0.16 3
γ	1642.50 9	0.090 16
γ [E2]	1675.54 9	0.082 16
γ	1691.70 8	0.60 3
γ	1695.73 9	0.238 25
γ	1717.53 10	0.64 3
γ	1722.02 7	0.66 3
γ	1728 1	<0.025
γ	1732 1	<0.025
γ	1757.55 6	0.869 25
γ	1770.76 10	0.098 25
γ	1775.25 7	1.12 4
γ	1787.1 5	0.025 8
γ	1789.4 5	0.025 8
γ	1797.17 17	0.082 16
γ	1800.4 3	0.066 16
γ	1805.0 5	0.049 16
γ	1810.78 6	0.172 16
γ	1817.7 3	0.057 16
γ	1839.63 20	0.090 16
γ	1853.8 5	0.049 16
γ	1869.0 3	0.164 25
γ	1896.66 7	0.525 25
γ	1902.73 9	0.746 25
γ	1913.80 10	0.558 25
γ	1920.37 10	0.426 16
γ	1949.89 7	1.255 25
γ	1956.92 9	0.435 25
γ	1967.03 10	0.082 16
γ	1971.55 13	0.041 16
γ	1973.60 10	0.041 16
γ	2000.94 8	0.36 3
γ	2010.15 9	0.074 16
γ	2024.78 13	0.049 16
γ	2069.57 9	0.295 25
γ	2084.93 20	0.254 25
γ	2098.61 10	0.525 25
γ	2122.80 9	0.582 25
γ	2150 1	0.033 8
γ	2151.84 10	0.033 8
γ	2187.7 3	0.066 16
γ	2203.0 5	0.049 16
γ	2229.84 22	0.090 16
γ	2233.0 5	0.049 16
γ	2245.4 3	~0.033
γ	2263 1	~0.016
γ	2277.0 5	0.041 16
γ	2283.07 22	0.115 8
γ	2298.6 3	0.057 8
γ	2314.39 15	0.041 8
γ	2330.5 5	0.057 8
γ	2356.8 5	~0.016
γ	2517 1	<0.016

† 27% uncert(syst)

Atomic Electrons (^{230}Ac)

⟨e⟩=35 3 keV

e_{bin}(keV)	⟨e⟩(keV)	e(%)
11	0.0094	0.083 9
16	1.4	8.9 23
20	1.6	8.0 20
33	0.19	0.57 14
34	5.9	18 4
37	5.6	15 4
48	2.3	4.8 12
49	2.1	4.2 11
50	0.053	0.11 3
52	1.3	2.5 6
53 - 102	1.14	1.48 19
103 - 144	1.01	0.91 7
165 - 212	0.029	~0.014
222 - 271	0.07	0.029 14

Atomic Electrons (^{230}Ac) (continued)

e_{bin}(keV)	⟨e⟩(keV)	e(%)
273 - 317	0.14	0.048 21
327 - 376	0.58	0.169 14
377 - 424	0.322	0.080 4
427 - 472	0.199	0.0445 15
483 - 514	0.110	0.022 5
515	3.0	0.58 8
517 - 519	0.07	~0.013
525	2.6	~0.50
526 - 571	0.12	0.022 9
577 - 603	0.016	0.0028 8
604	0.72	0.12 3
605 - 652	0.8	0.14 6
655 - 688	0.19	0.028 14
707 - 756	0.14	0.019 9
757 - 804	0.19	0.024 10
809 - 858	0.16	0.018 4
862 - 911	0.11	0.013 4
913 - 962	0.093	0.010 3
963 - 1010	0.048	0.0049 19
1021 - 1068	0.06	~0.006
1073 - 1117	0.109	0.0099 21
1127 - 1171	0.5	~0.04
1178 - 1228	0.25	0.020 7
1232 - 1268	0.33	0.026 10
1282 - 1331	0.12	0.0093 21
1343 - 1391	0.058	0.0043 15
1393 - 1439	0.018	0.0013 5
1450 - 1533	0.066	0.0044 13
1553 - 1648	0.22	0.014 4
1655 - 1754	0.16	0.010 3
1755 - 1853	0.26	0.014 4
1857 - 1955	0.089	0.0047 11
1957 - 2053	0.10	0.0048 15
2064 - 2153	0.031	0.0015 4
2167 - 2267	0.019	0.00087 22
2272 - 2353	0.0021	9 3 $\times10^{-5}$
2407 - 2501	0.0004	~2 $\times10^{-5}$
2512	1.6 $\times10^{-5}$	~6 $\times10^{-7}$

Continuous Radiation (^{230}Ac)

⟨β-⟩=876 keV;⟨IB⟩=2.0 keV

E_{bin}(keV)		⟨ ⟩(keV)	(%)
0 - 10	β-	0.0302	0.60
	IB	0.033	
10 - 20	β-	0.091	0.61
	IB	0.032	0.22
20 - 40	β-	0.370	1.23
	IB	0.063	0.22
40 - 100	β-	2.70	3.84
	IB	0.18	0.27
100 - 300	β-	27.7	13.8
	IB	0.47	0.27
300 - 600	β-	91	20.3
	IB	0.48	0.114
600 - 1300	β-	328	35.5
	IB	0.54	0.064
1300 - 2500	β-	414	24.1
	IB	0.157	0.0098
2500 - 2900	β-	12.6	0.484
	IB	0.00036	1.40 $\times10^{-5}$

$^{230}_{90}$Th(7.54 3 $\times10^{4}$ yr)

Mode: α

Δ: 30858.6 24 keV

SpA: 0.02062 Ci/g

Prod: natural source

Alpha Particles (^{230}Th)

$\langle\alpha\rangle$=4665 1 keV

α(keV)	α(%)
3829.4 17	~1×10^{-6}
3877.8 16	~3×10^{-6}
4248.9 6	8.9 20×10^{-6}
4278.2 7	8 2×10^{-6}
4371.7 6	0.00097 13
4438.3 6	~0.030
4479.8 6	~0.12
4621.1 6	23.4 1
4687.6 6	76.3 3

Photons (^{230}Th)

$\langle\gamma\rangle$=0.371 15 keV

γ_{mode}	γ(keV)	γ(%)†
γ E2	67.6758 20	0.376 21
γ	109.97 7	5.9 5×10^{-5}
γ E2	143.876 4	0.0486 22
γ E1	186.057 4	0.0088 4
γ	205.1 5	5 1×10^{-6}
γ	235.01 10	8.1 20×10^{-6}
γ E1	253.732 4	0.0111 5
γ	253.84 7	0.00085 9
γ	551.8 8	1.01 20×10^{-6}
γ	570.5 5	3 1×10^{-6}
γ	620.0 8	1.01 20×10^{-6}

† 6.0% uncert(syst)

$^{230}_{91}$Pa(17.4 5 d)

Mode: ε(90.5 6 %), β-(9.5 6 %)
α(0.0032 1 %)
Δ: 32162 3 keV
SpA: 3.26×10^{4} Ci/g
Prod: ^{232}Th(p,3n); ^{232}Th(d,4n); ^{230}Th(d,2n)

Alpha Particles (^{230}Pa)

$\langle\alpha\rangle$=0.170 1 keV

α(keV)	α(%)
4766 2	~6×10^{-6}
4798 5	<16×10^{-7}
4934 3	~1×10^{-5}
4973 2	2.2 6×10^{-5}
5060 3	~1×10^{-5}
5084 2	2.2 6×10^{-5}
5119 3	1.9 6×10^{-5}
5153 2	1.3 3×10^{-5}
5183 3	1.6 6×10^{-5}
5216.8 15	1.6 3×10^{-5}
5268.6 7	0.000112 16
5275.8 7	9.6 16×10^{-5}
5287.6 15	10 3×10^{-5}
5300.7 7	0.00054 10
5312.2 7	0.00042 10
5326.4 7	0.00058 10
5339.9 10	0.00048 16
5344.9 7	0.00074 16

Photons (^{230}Pa)

$\langle\gamma\rangle$=656 9 keV

γ_{mode}	γ(keV)	γ(%)†
Th L_ℓ	11.118	1.31 19
U L_ℓ	11.620	0.082 13
Th L_α	12.952	22 3
U L_α	13.600	1.35 18
Th L_η	14.511	0.42 6
U L_η	15.400	0.034 5
Th L_β	16.136	25 4
U L_β	17.130	1.72 25
Th L_γ	19.145	5.6 9
U L_γ	20.293	0.38 6
$\gamma_{\beta-}$E2	51.77 5	0.030 3
γ_ϵE2	53.23 3	0.235 21
γ_ϵ[E2]	63.63 5	
Th $K_{\alpha2}$	89.955	18.3 23
Th $K_{\alpha1}$	93.350	30 4
U $K_{\alpha2}$	94.651	0.00189 17
U $K_{\alpha1}$	98.434	0.0030 3
Th $K_{\beta1}$'	105.362	10.7 13
Th $K_{\beta2}$'	108.990	3.6 5
U $K_{\beta1}$'	111.025	0.00111 10
U $K_{\beta2}$'	114.866	0.00037 3
$\gamma_{\beta-}$[E2]	116.0 6	<0.023 ?
γ_ϵE2	120.912 20	0.34 5
γ_ϵ[E1]	170.56 5	
γ_ϵ[E2]	175.91 6	
γ_ϵ[M1+E2]	183.92 9	
γ_ϵ	194.6 12 ?	
γ_ϵ	197.1 12 ?	
γ_ϵE0+E2+M1	228.30 6	<0.005
γ_ϵ[E1]	253.54 9	0.0096 19
$\gamma_{\beta-}$[E1]	253.6 3	0.015 3 ?
γ_ϵ	266.0 6	0.0069 14
γ_ϵ[E1]	274.33 6	0.10 3
γ_ϵ[E1]	294.13 7	0.037 16
γ_ϵ[E1]	297.92 7	~0.043
γ_ϵ[E1]	302.09 9	0.0117 23
$\gamma_{\beta-}$[E1]	314.81 10	0.106 13
γ_ϵ[E1]	316.99 8	0.16 3
γ_ϵ[M1,E2]	332.07 7	0.053 16
γ_ϵ[E1]	346.47 7	<0.027
$\gamma_{\beta-}$[E1]	366.58 10	0.086 13
$\gamma_{\beta-}$[E1]	369.6 5	0.033 7
γ_ϵ(E2)	374.73 9	0.032 11
γ_ϵ[M1+E2]	375.0 7	~0.011 ?
γ_ϵE2	380.16 5	0.29 5
γ_ϵE1	397.69 5	1.82 10
γ_ϵM1+~26%E2	399.96 6	0.61 3
γ_ϵ[E1]	401.69 8	0.023 5
γ_ϵM1	440.56 17	0.11 3
γ_ϵM1+30%E2	443.79 3	5.4 3
γ_ϵ[E1]	450.24 8	~0.011 ?
γ_ϵE1	454.97 3	6.08 12
γ_ϵM1+~14%E2	463.59 5	0.80 5
γ_ϵ[E2]	503.53 5	0.059 8
γ_ϵ[E2]	504.19 16	0.069 11
γ_ϵ[M1+E2]	507.52 7	~0.21
γ_ϵE1	508.20 3	3.49 12
γ_ϵE1	518.60 5	1.91 10
γ_ϵ	533.0 12 ?	
γ_ϵ	536.0 12 ?	
γ_ϵM1+50%E2	556.07 6	0.192 21
γ_ϵE2	571.16 6	1.05 5
γ_ϵE2	581.78 8	0.128 11
γ_ϵ[E2]	607.29 4	0.08 4
γ_ϵE2	619.70 6	0.160 21
γ_ϵE0+E2+M1	624.44 5	0.044 6
γ_ϵ[M1+E2]	651.68 7	0.016 3
γ_ϵ[E2]	677.67 5	0.054 7
γ_ϵE2	728.20 4	1.84 8
γ_ϵE2	772.59 7	0.10 3
γ_ϵE2	781.43 4	1.44 5
γ_ϵ[E2]	835.60 5	0.08 3
γ_ϵ[E1]	838.25 16	0.026 11
γ_ϵE1	898.76 3	6.0 5
γ_ϵE1	918.56 5	8.0 4
γ_ϵE1	951.99 4	28.2 5
γ_ϵ[E1]	953.76 6	0.16 4
γ_ϵE2	956.51 5	1.75 14
γ_ϵE1	959.16 16	0.48 11
γ_ϵ	970.0 12	0.013 4

Photons (^{230}Pa)
(continued)

γ_{mode}	γ(keV)	γ(%)†
γ_ϵ[M1+E2]	999.5 7	0.012 4
γ_ϵE2	1009.74 5	1.04 5
γ_ϵE1	1026.13 6	1.43 6
γ_ϵE1	1074.67 6	0.732 21

† uncert(syst): 6.8% for ε, 13% for β-

Atomic Electrons (^{230}Pa)

$\langle e\rangle$=48.6 18 keV

e_{bin}(keV)	⟨e⟩(keV)	e(%)
11	0.0096	0.085 14
16	4.3	26 4
17	0.27	1.58 24
20	4.2	21 3
21 - 30	0.32	1.48 20
31	1.11	3.6 4
33	0.224	0.68 6
34	7.0	21 2
35	1.06	3.1 4
37	6.7	18.2 17
46 - 47	0.88	1.86 21
48	2.8	5.8 5
49	2.49	5.1 5
50 - 51	0.33	0.65 5
52	1.60	3.1 3
53 - 102	2.68	3.49 18
103 - 144	1.03	0.93 8
156 - 199	0.023	0.013 4
207 - 255	0.06	~0.026
258 - 302	0.598	0.208 9
309 - 332	0.11	0.033 9
334	3.4	1.0 4
341 - 385	1.02	0.286 11
393 - 441	1.5	0.37 6
442 - 492	0.52	0.115 6
498 - 542	1.62	0.31 4
551 - 600	0.088	0.0158 5
602 - 652	0.55	0.091 12
657 - 678	0.127	0.0189 8
708 - 756	0.079	0.0111 4
761 - 809	0.431	0.0541 23
815 - 860	0.893	0.106 3
878 - 919	0.203	0.0225 6
932 - 980	0.248	0.0264 5
983 - 1025	0.0326	0.00326 8
1054 - 1069	0.00420	0.000641 10

Continuous Radiation (^{230}Pa)

⟨β-⟩=14.1 keV;⟨IB⟩=0.82 keV

E_{bin}(keV)		⟨ ⟩(keV)	(%)
0 - 10	β-	0.0214	0.428
	IB	0.00078	
10 - 20	β-	0.063	0.418
	IB	0.0048	0.031
20 - 40	β-	0.242	0.81
	IB	0.0021	0.0070
40 - 100	β-	1.53	2.21
	IB	0.51	0.59
100 - 300	β-	8.7	4.68
	IB	0.23	0.21
300 - 600	β-	3.59	1.01
	IB	0.032	0.0074
600 - 1250	IB	0.033	0.0044

$^{230}_{92}$U (20.8 d)

Mode: α
Δ: 31598 *6* keV
SpA: 3×10^4 Ci/g
Prod: daughter ^{230}Pa

Alpha Particles (^{230}U)

⟨α⟩=5867.7 *4* keV

α(keV)	α(%)
5056.5 *7*	~7 ×10^{-5}
5097.6 *4*	~0.00030
5446.3 *4*	~0.00025
5449.1 *5*	~6 ×10^{-5}
5533 *2*	~0.00010
5544 *1*	0.00054 *5*
5586.6 *3*	0.0115 *10*
5662.4 *3*	0.26 *3*
5666.3 *3*	0.38 *4*
5817.8 *3*	32.0 *2*
5888.7 *3*	67.4 *4*

Photons (^{230}U)

⟨γ⟩=2.90 *13* keV

γ_{mode}	γ(keV)	γ(%)†
Th L_ℓ	11.118	0.24 *3*
Th L_α	12.952	4.1 *4*
Th L_η	14.511	0.124 *15*
Th L_β	16.163	6.0 *7*
Th L_γ	19.095	1.38 *17*
γ E2	72.18 *3*	0.60 *4*
γ [E1]	81.08 *10*	0.00044 *10*
Th $K_{\alpha2}$	89.955	0.0130 *6*
Th $K_{\alpha1}$	93.350	0.0211 *10*
Th $K_{\beta1}$'	105.362	0.0076 *4*
Th $K_{\beta2}$'	108.990	0.00254 *13*
γ [E2]	154.21 *3*	0.126 *6*
γ E1	158.16 *3*	0.071 *3*
γ [E2]	221.0 *5*	4.6 *9* ×10^{-5}
γ [E1]	223.9 *3*	0.00024 *6*
γ E1	230.34 *4*	0.121 *5*
γ [E1]	235.29 *10*	0.0106 *7*
γ [E1]	539.5 *7*	3.2 *14* ×10^{-5}
γ [E1]	574.8 *3*	0.00030 *4*
γ [E1]	616.6 *7*	~4 ×10^{-5}

† 8.8% uncert(syst)

Atomic Electrons (^{230}U)

⟨e⟩=21.0 *6* keV

e_{bin}(keV)	⟨e⟩(keV)	e(%)
16	0.90	5.5 *7*
20	1.07	5.5 *7*
45 - 49	0.0176	0.0386 *20*
52	7.0	13.4 *9*
56	5.7	10.2 *7*
61 - 65	4.2 ×10^{-5}	6.8 *12* ×10^{-5}
67	2.40	3.57 *23*
68	1.94	2.85 *18*
69 - 70	0.048	0.070 *5*
71	1.24	1.75 *11*
72	0.357	0.50 *3*
73 - 121	0.0087	0.0075 *3*
126 - 158	0.284	0.203 *7*
201 - 235	0.00403	0.00188 *5*
461	1.23 ×10^{-5}	2.7 *3* ×10^{-6}
507 - 555	3.7 ×10^{-6}	7.0 *13* ×10^{-7}
559 - 600	1.18 ×10^{-6}	2.06 *22* ×10^{-7}
611 - 617	8.5 ×10^{-8}	1.4 *4* ×10^{-8}

$^{230}_{93}$Np(4.6 *3* min)

Mode: ε(<97 %), α(>3 %)
Δ: 35220 *50* keV
SpA: 1.77×10^8 Ci/g
Prod: ^{233}U(p,4n)

Alpha Particles (^{230}Np)

α(keV)
6660 *20*

A = 231

NDS **21**, 91 (1977)

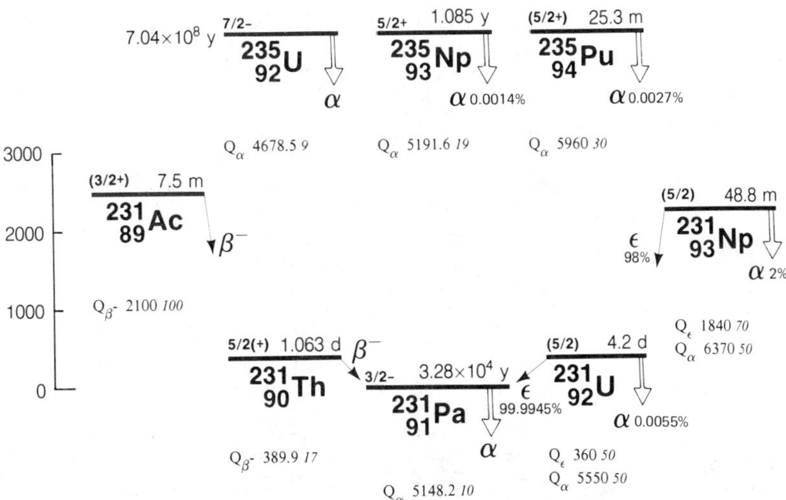

$^{231}_{89}$Ac(7.5 *1* min)

Mode: β-
Δ: 35910 *100* keV
SpA: 1.084×10^8 Ci/g
Prod: ^{232}Th(γ,p); ^{232}Th(n,pn)

Photons (^{231}Ac)

γ_{mode}	γ(keV)	γ(rel)
γ [M1+E2]	19.594 *10*	
γ [M1+E2]	26.14 *15*	
γ [E2+~20%M1]	41.954 *12*	
γ [M1+E2]	50.69 *15*	
γ	68.97 *20*	~5
γ [E1]	143.786 *12*	9.0 *8*
γ (E1)	185.739 *5*	45 *5*
γ [M1]	198.933 *19*	6.6 *8*
γ [M1]	221.406 *20*	52 *5*
γ [M1]	240.886 *22*	11.3 *20*
γ [E2]	247.55 *15*	1.1 *2*
γ [M1+E2]	272.10 *15*	7 *1*
γ [E1]	282.50 *15*	100 *10*
γ [E1]	307.05 *15*	80 *6*
γ	350.87 *20*	2.4 *4*
γ [E2]	368.86 *10*	38 *4*
γ [E1]	372.29 *12*	4.4 *7*
γ	375.87 *20*	4.1 *6*
γ [E2]	388.36 *12*	1.1 *2*
γ	397.07 *20*	1.9 *3*
γ	400.37 *20*	2.6 *4*
γ [M1+E2]	407.96 *12*	8.5 *12*
γ	503.67 *20*	1.1 *2*
γ [E3]	512.65 *10*	1.2 *2*
γ	528.27 *20*	2.3 *4*
γ [M2]	554.6 *1*	3.9 *6*

$^{231}_{90}$Th(1.0633 *13* d)

Mode: β-
Δ: 33812.1 *24* keV
SpA: 5.316×10^5 Ci/g
Prod: ^{230}Th(n,γ)

Photons (^{231}Th)

⟨γ⟩=29 *5* keV

γ_{mode}	γ(keV)	γ(%)†
Pa L_ℓ	11.372	~3
Pa L_α	13.274	~49
Pa L_η	14.953	0.40 *4*
Pa L_β	16.559	37 *14*
γ (M1)	17.187 *12*	0.22 *7*
γ M1+3.7%E2	18.051 *12*	<0.33
Pa L_γ	19.811	7.4 *11*
γ E1	25.642 *11*	14.6 *10*
γ [E1]	42.828 *14*	0.058 *4*
γ [M1+E2]	44.060 *15*	0.0007 *2*
γ E2	58.562 *11*	0.48 *2*
γ M1+19%E2	63.834 *12*	0.023 *2*
γ E2	68.50 *3*	0.0057 *14*
γ [E1]	72.767 *12*	0.251 *15*
γ [M1+E2]	77.680 *19*	
γ (M1+7.3%E2)	81.229 *12*	0.90 *4*
γ (M1+3.6%E2)	82.093 *12*	0.49 *9*
γ E1	84.203 *9*	6.6 *3*
γ (E1)	89.954 *8*	0.94 *6*
Pa $K_{\alpha2}$	92.279	0.39 *3*
γ (E1)	93.07 *3*	0.049 *5*

Photons (^{231}Th) (continued)

γ_{mode}	γ(keV)	γ(%)†
Pa $K_{\alpha1}$	95.863	0.63 *5*
γ M1+20%E2	99.28 *1*	0.120 *7*
γ (E1)	102.255 *12*	0.41 *3*
γ [E1]	105.803 *18*	0.0071 *7*
γ [E1]	106.581 *20*	0.017 *1*
Pa $K_{\beta1}$'	108.166	0.228 *19*
Pa $K_{\beta2}$'	111.897	0.076 *6*
γ [M1+E2]	115.595 *12*	0.0010 *2*
γ (E1)	116.827 *14*	0.0207 *13*
γ E1	124.922 *12*	0.056 *3*
γ E1	134.014 *14*	0.024 *1*
γ M1+12%E2	135.681 *17*	0.078 *5*
γ [E1]	136.705 *19*	0.0042 *2*
γ [M1+E2]	140.537 *20*	0.00071 *7*
γ [M1+E2]	145.063 *14*	0.0058 *4*
γ M1+39%E2	145.927 *13*	0.032 *2*
γ M1+6.9%E2	163.114 *11*	0.155 *9*
γ [E2]	164.97 *3*	0.0039 *4*
γ [E1]	169.637 *18*	0.0012 *1*
γ [M1+E2]	174.157 *9*	0.0183 *11*
γ E1	183.483 *11*	0.0329 *13*
γ [E1]	188.756 *12*	0.0032 *2*
γ E1	217.934 *19*	0.040 *3*
γ [E1]	235.985 *18*	0.0092 *6*
γ [E1]	240.25 *4*	0.00028 *3*
γ [M1+E2]	242.508 *22*	0.00084 *8*
γ [E1]	249.56 *4*	0.00078 *8*
γ [E1]	250.43 *4*	0.00065 *7*
γ [E1]	267.62 *4*	0.00116 *13*
γ [M1+E2]	274.14 *4*	3 *1* $\times 10^{-5}$
γ [E1]	308.75 *4*	0.00039 *4*
γ M1+40%E2	311.00 *3*	0.0029 *2*
γ [E1]	317.93 *4*	8 *1* $\times 10^{-5}$
γ [M1+E2]	320.188 *19*	0.00011 *1*
γ [M1+E2]	351.82 *4*	7 *1* $\times 10^{-5}$

† 5.0% uncert(syst)

Atomic Electrons (^{231}Th)

⟨e⟩=94 *15* keV

e_{bin}(keV)	⟨e⟩(keV)	e(%)
3 - 10	2.28	50 *4*
12	3.8	32 *11*
13	6.0	~47
14	4.1	~30
15 - 16	1.4	9 *3*
17	12.4	<148
18	1.1	<12
20	5.5	27.0 *24*
21 - 37	3.0	13.2 *13*
38	11.4	29.9 *14*
39 - 41	0.022	~0.05
42	10.4	24.8 *11*
43 - 53	0.93	1.89 *21*
54	8.0	14.9 *7*
55 - 59	3.26	5.65 *18*
60	2.83	4.7 *3*
61 - 62	2.8	4.5 *6*
63	4.50	7.1 *4*
64	3.26	5.1 *3*
65 - 78	2.59	3.5 *3*
79	3.17	4.02 *22*
80 - 129	1.28	1.46 *9*
130 - 179	0.372	0.257 *17*
180 - 229	0.0049	0.0024 *8*
230 - 274	0.00034	0.00014 *3*
288 - 335	0.0013	0.00044 *14*
346 - 352	5.7×10^{-6}	1.6 *7* $\times 10^{-6}$

Continuous Radiation (^{231}Th)

⟨β-⟩=79 keV;⟨IB⟩=0.023 keV

E_{bin}(keV)		⟨ ⟩(keV)	(%)
0 - 10	β-	0.460	9.2
	IB	0.0041	
10 - 20	β-	1.31	8.7
	IB	0.0034	0.024
20 - 40	β-	4.76	16.0
	IB	0.0051	0.018
40 - 100	β-	24.6	36.4
	IB	0.0076	0.0126
100 - 300	β-	47.5	31.6
	IB	0.0026	0.0020
300 - 390	β-	0.00397	0.00128
	IB	1.36×10^{-8}	4.3×10^{-9}

$^{231}_{91}$Pa(3.276 *11* $\times 10^4$ yr)

Mode: α
Δ: 33422 *3* keV
SpA: 0.04724 Ci/g
Prod: natural source

Alpha Particles (^{231}Pa)

⟨α⟩=4923 keV

α(keV)	α(%)
4413.1 *8* ?	0.0018
4505.6 *6*	0.003
4566.0 *5*	0.008
4597.2 *5*	0.015
4630.9 *5*	0.10
4640.3 *5*	0.10
4678.1 *5*	1.5
4710.2 *5*	1.0
4734.3 *5*	8.4
4791.9 *5*	0.04
4851.3 *5*	1.4
4899 *2*	0.0020
4933.9 *5*	3
4950.5 *5*	22.8
4975.5 *5*	0.4
4985.8 *5*	1.4
5013.1 *5*	25.4
5029.2 *5*	20
5031.8 *5*	~2.5
5058.7 *5*	11

Photons (^{231}Pa)

⟨γ⟩=39.9 *14* keV

γ_{mode}	γ(keV)	γ(%)†
Ac L_ℓ	10.871	1.2 *3*
γ	10.9 *5*	
γ	12.4 *5*	
Ac L_α	12.636	22 *4*
Ac L_η	14.082	0.37 *7*
γ	14.14 *10*	
γ	15.5 *5*	
Ac L_β	15.659	25 *5*
γ [E1]	16.397 *16*	0.38 *5* ?
γ	18.2 *5*	
Ac L_γ	18.607	5.9 *12*
γ [M1]	18.998 *16*	0.33 *2*
γ	19.6 *10*	
γ	22.7 *10*	
γ [M1+E2]	23.55 *5*	~0.00017
γ [M1]	24.56 *5*	~0.030 ?

Photons (^{231}Pa)
(continued)

γ_{mode}	γ(keV)	γ(%)†
γ (M1)	25.464 *23*	~0.09
γ	27 *1* ?	
γ E1	27.396 *9*	9.3
γ M1+~3.8%E2	29.996 *14*	0.092 *9*
γ	31.04 *5*	~0.009
γ	31.58 *5*	0.007 *2*
γ	34 *1*	
γ [E1]	35.884 *20*	0.016 *2*
γ M1+~1.4%E2	38.235 *15*	0.149 *15*
γ	39.61 *4*	~0.0014
γ	40.006 *20*	0.012 *2*
γ	42.52 *5*	0.006 *1*
γ	43.09 *5*	0.007 *2*
γ [M1]	44.213 *15*	0.060 *7*
γ [E1]	46.393 *15*	0.208 *4*
γ [M1]	50.97 *4*	0.0014 *6*
γ [M1]	52.763 *17*	0.085 *9*
γ [E1]	54.632 *15*	0.081 *8*
γ	56.80 *4*	0.006 *1*
γ	56.90 *4*	0.011 *3*
γ E2	57.200 *22*	0.015 *2*
γ E2	57.233 *16*	0.025 *3*
γ [E1]	60.526 *24*	0.007 *1*
γ E2	63.700 *21*	0.050 *5*
γ	70.54 *5*	0.007 *1*
γ [M1]	71.92 *6*	~0.002 ?
γ [M1]	72.57 *8*	~0.004 ?
γ (E2)	74.209 *17*	0.025 *3*
γ (M1)	77.405 *24*	0.068 *7*
γ	82 *3* ?	
Ac $K_{\alpha2}$	87.673	0.476 *18*
Ac $K_{\alpha1}$	90.886	0.78 *3*
γ	96 *1* ?	
γ E2	96.976 *21*	0.088 *9*
γ (E2)	100.96 *4*	0.032 *5*
Ac $K_{\beta1}$'	102.613	0.280 *11*
γ [E2]	102.87 *3*	<0.013 ?
γ	106.0 *5* ?	
Ac $K_{\beta2}$'	106.137	0.092 *4*
γ [E2]	124.69 *6*	0.005 *2* ?
γ [E2]	144.50 *8*	~0.004
γ	146.70 *10*	<18 ×10^{-5}
γ	198.96 *4*	0.0056 *21*
γ	220.20 *7*	0.00024 *8*
γ [E1]	226.77 *4*	0.0026 *6*
γ	228.16 *10*	0.00024 *9*
γ	230.62 *7*	<0.0032
γ	232.45 *13*	<0.0012
γ	242 *25*	0.008 *1*
γ E2+37%M1	243.17 *4*	0.047 *4*
γ [M2]	245.438 *21*	0.007 *1* ?
γ [E1]	245.77 *4*	0.0069 *14*
γ E2	255.858 *21*	0.100 *7*
γ	258.43 *6*	0.0023 *6*
γ E2+31%M1	260.29 *3*	0.182 *9*
γ M1+34%E2	273.17 *4*	0.057 *4*
γ E1+13%M2	277.17 *3*	0.062 *4*
γ	277.43 *6*	0.0033 *11*
γ E1	283.673 *21*	1.59 *10*
γ	286.70 *10*	
γ E2+46%M1	300.070 *19*	2.38 *12*
γ E1	302.638 *24*	0.64 *8*
γ E1	302.671 *19*	1.71 *16*
γ	310.19 *10*	0.0014 *5*
γ E2+29%M1	313.058 *24*	0.113 *9*
γ	318.1 *7*	~0.002
γ (E1)	327.23 *5*	0.0301 *19*
γ M1+50%E2	330.066 *18*	1.32 *10*
γ E1+3.8%M2	340.873 *24*	0.174 *21*
γ	341.17 *4*	0.022 *5*
γ [E1]	351.59 *4*	
γ E2+10%M1	354.63 *5*	0.080 *8*
γ M1+13%E2	357.270 *23*	0.154 *13*
γ	359.51 *7*	0.0090 *7*
γ [M1+E2]	364.03 *4*	0.0074 *7*
γ	375.17 *10*	
γ M1+27%E2	379.40 *3*	0.046 *4*
γ	384.97 *7*	0.0041 *4*
γ [E2]	387.267 *23*	0.0005 *2* ?
γ	388.98 *5*	0.00107 *23*
γ [E1]	391.85 *4*	0.0068 *6*
γ [E1]	395.80 *4*	0.0026 *3*
γ	398.40 *4*	0.0093 *8*

Photons (^{231}Pa)
(continued)

γ_{mode}	γ(keV)	γ(%)†
γ M1	407.98 *4*	0.034 *3*
γ [E1]	410.84 *4*	0.0019 *2* ?
γ	427.12 *10*	<0.12
γ	435.38 *4*	0.0033 *4*
γ [M1+E2]	438.24 *4*	0.0041 *4*
γ	439 *1*	
γ	471.33 *10*	0.00015 *6*
γ	478.22 *13*	7.7 *23* ×10^{-5}
γ	490.89 *13*	<0.0006
γ	501.33 *10*	0.0006 *2*
γ	509.88 *13*	
γ	516.45 *13*	0.0013 *3*
γ	535.45 *13*	0.00056 *18*
γ	543.14 *10*	
γ	546.9 *5*	0.0005 *2* ?
γ	572.3 *5*	0.0005 *2* ?
γ	582.7 *5* ?	
γ	610.6 *5* ?	

† 20% uncert(syst)

Atomic Electrons (^{231}Pa)
⟨e⟩=48 *3* keV

e_{bin}(keV)	⟨e⟩(keV)	e(%)
4 - 6	1.1	19 *8*
8	1.14	14 *3*
9 - 11	1.43	13.7 *12*
12	2.4	~21
13	0.045	0.34 *5*
14	4.7	33.4 *22*
15	0.8	~6
16	3.7	24 *8*
17	0.00118	0.0070 *9*
18	2.53	14.0 *11*
19	3.7	19 *4*
20	2.4	12 *4*
21 - 22	1.0	4.8 *24*
23	1.2	5.0 *19*
24	1.6	~7
25	0.93	3.7 *5*
26	1.1	4.0 *19*
27	1.3	~5
28 - 37	3.2	10.0 *19*
38	1.3	3.5 *13*
39 - 88	7.3	14.2 *12*
90 - 139	0.49	0.49 *3*
140 - 183	0.320	0.195 *6*
193	1.60	0.83 *4*
194 - 243	1.21	0.55 *3*
244 - 293	1.01	0.37 *1*
294 - 341	0.77	0.250 *11*
344 - 393	0.0408	0.0112 *6*
394 - 440	0.016	0.0038 *17*
451 - 501	0.00039	8 *4* ×10^{-5}
511 - 556	0.00018	3.3 *12* ×10^{-5}
567 - 572	1.1 ×10^{-5}	~2×10^{-6}

$^{231}_{92}$U (4.2 *1* d)

Mode: ε(99.9945 %), α(0.0055 %)
Δ: 33780 *50* keV
SpA: 1.35×10^{5} Ci/g
Prod: ^{230}Th(α,3n); ^{231}Pa(d,2n); ^{232}Th(α,5n)

Alpha Particles (^{231}U)

α(keV)
5455

Photons (^{231}U)
⟨γ⟩=81 *16* keV

γ_{mode}	γ(keV)	γ(%)
Pa L_ℓ	11.372	2.2 *7*
Pa L_α	13.274	37 *12*
Pa L_η	14.953	0.55 *17*
Pa L_β	16.588	41 *13*
Pa L_γ	19.803	10 *3*
γE1	25.642 *11*	13.4 *15*
γ[E1]	42.828 *14*	
γE2	58.562 *11*	0.44 *9*
γE2	68.50 *3*	0.0055 *11*
γ[E1]	72.767 *12*	
γ[M1+E2]	77.680 *19*	
γ(M1+7.3%E2)	81.229 *12*	0.013 *3*
γ(M1+3.6%E2)	82.093 *12*	0.0074 *19*
γE1	84.203 *9*	6.0 *5*
γ(E1)	89.954 *8*	
Pa $K_{\alpha2}$	92.279	17 *8*
γ(E1)	93.07 *3*	
Pa $K_{\alpha1}$	95.863	27 *14*
γM1+20%E2	99.28 *1*	0.00179 *23*
γ(E1)	102.255 *12*	
γ[E1]	105.803 *18*	0.000106 *16*
Pa $K_{\beta1}$'	108.166	10 *5*
Pa $K_{\beta2}$'	111.897	3.3 *16*
γ[M1+E2]	115.595 *12*	
γE1	124.922 *12*	0.00083 *10*
γ[E1]	136.705 *19*	~0.08
γ[E2]	164.97 *3*	
γ[M1+E2]	174.157 *9*	
γE1	183.483 *11*	0.00049 *6*
γE1	217.934 *19*	~0.8
γ[E1]	235.985 *18*	~0.18
γ[M1+E2]	242.508 *22*	~0.017
γM1+40%E2	311.00 *3*	~0.06
γ[M1+E2]	320.188 *19*	~0.0022

Continuous Radiation (^{231}U)
⟨IB⟩=0.76 keV

E_{bin}(keV)		⟨ ⟩(keV)	(%)
10 - 20	IB	0.0061	0.038
20 - 40	IB	0.00114	0.0038
40 - 100	IB	0.43	0.50
100 - 282	IB	0.32	0.30

$^{231}_{93}$Np(48.8 *2* min)

Mode: ε(98 *1* %), α(2 *1* %)
Δ: 35620 *50* keV
SpA: 1.668×10^{7} Ci/g
Prod: ^{233}U(d,4n); ^{235}U(d,6n)

Alpha Particles (^{231}Np)

α(keV)
6280

Photons (^{231}Np)

γ_{mode}	γ(keV)	γ(rel)
γ_ϵ	45.0 *3*	
γ_ϵ	262.9 *3*	2.84 *10*
γ_ϵ	347.5 *3*	3.63 *20*
γ_ϵ	370.30 *22*	9.8
γ_ϵ	375.1 *3*	0.64 *3*
γ_ϵ	415.25 *25*	0.28 *6*
γ_ϵ	420.1 *3*	1.05 *11*
γ_ϵ	435.8 *3*	0.28 *6*
γ_ϵ	480.7 *3*	0.61 *12*
γ_ϵ	483.8 *5*	1.6 *3*
γ_ϵ	714.6 *4*	0.24 *3*
γ_ϵ	736.91 *24*	1.23 *7*
γ_ϵ	785.6 *3*	0.186 *10*
γ_ϵ	836.4 *4*	0.40 *6*
γ_ϵ	851.1 *3*	0.70 *3*
γ_ϵ	1107.21 *24*	0.54 *5*

A = 232

NDS **36**, 367 (1982)

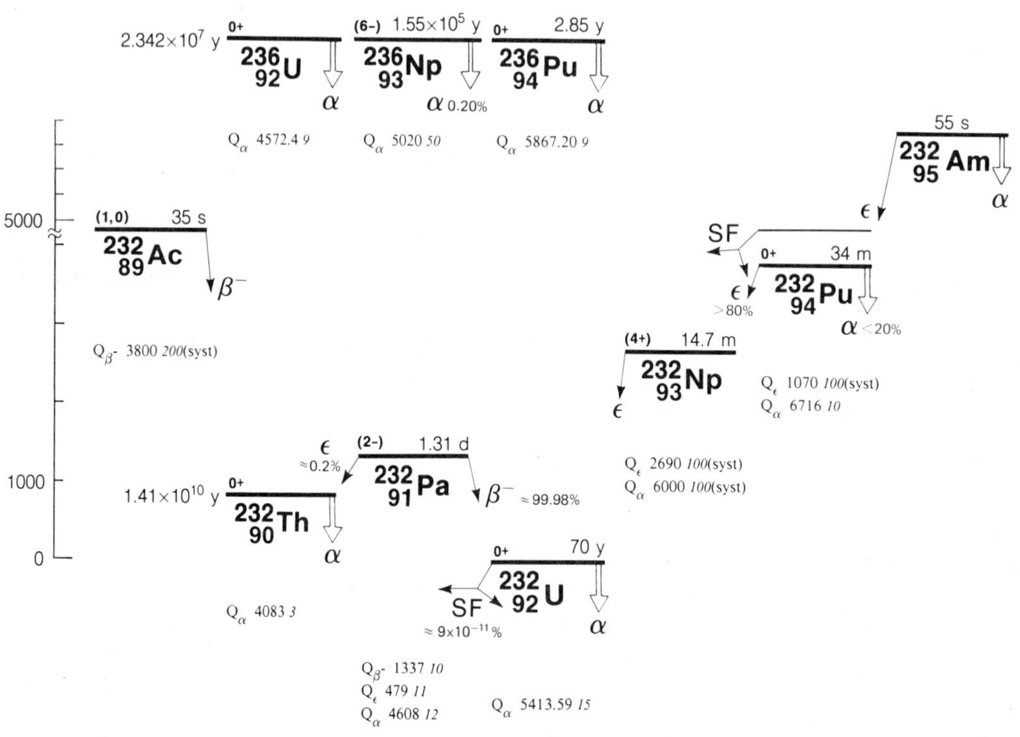

$^{232}_{89}$Ac(35 5 s)

Mode: β-
Δ: 39240 200 keV syst
SpA: 1.38×10^{9} Ci/g
Prod: ^{232}Th(n,p)

$^{232}_{90}$Th(1.405 6 $\times10^{10}$ yr)

Mode: α
Δ: 35444.4 21 keV
SpA: Ci/g
Prod: natural source
%: 100

Alpha Particles (^{232}Th)

$\langle\alpha\rangle$=4005 6 keV

α(keV)	α(%)
3830 10	0.20 8
3952 5	23 3
4010 5	77 3

Photons (^{232}Th)

$\langle\gamma\rangle$=0.17 4 keV

γ_{mode}	γ(keV)	γ(%)
γ E2	59.0 10	0.190 25
γ (E2)	124 11	~0.043

$^{232}_{91}$Pa(1.31 2 d)

Mode: β-(~ 99.98 %), ϵ(~ 0.2 %)
Δ: 35923 11 keV
SpA: 4.30×10^{5} Ci/g
Prod: ^{231}Pa(n,γ); ^{232}Th(d,2n); ^{232}Th(p,n)

Photons (^{232}Pa)

$\langle\gamma\rangle$=941 20 keV

γ_{mode}	γ(keV)	γ(%)†
U L_{ℓ}	11.620	1.15 21
U L_{α}	13.600	19 3
U L_{η}	15.400	0.48 8
U L_{β}	17.130	24 4
U L_{γ}	20.295	5.4 10
$\gamma_{\beta-}$E2	47.579 9	0.21 4
$\gamma_{\beta-}$[E1]	80.24 7	0.15 3
U $K_{\alpha2}$	94.651	1.10 4
U $K_{\alpha1}$	98.434	1.76 7
$\gamma_{\beta-}$[E1]	105.48 3	1.65 19
$\gamma_{\beta-}$E2	109.001 8	2.8 3
U $K_{\beta1}$'	111.025	0.644 24
U $K_{\beta2}$'	114.866	0.217 9
$\gamma_{\beta-}$M1,E2	132.24 7	0.013 5
$\gamma_{\beta-}$[E1]	139.53 3	0.58 5

Photons (^{232}Pa) (continued)

γ_{mode}	γ(keV)	γ(%)†
$\gamma_{\beta-}$E1	150.096 3	10.8 5
$\gamma_{\beta-}$E2	164.7 5	0.029 5
$\gamma_{\beta-}$[E2]	175.57 24	0.0097 19
$\gamma_{\beta-}$M1,E2	176.85 7	~0.0039
$\gamma_{\beta-}$[E1]	184.142 6	1.3 3
$\gamma_{\beta-}$[E1]	282.33 7	~0.010
$\gamma_{\beta-}$E2+4.7%M1	387.919 4	7.0 3
$\gamma_{\beta-}$E2+21%M1	421.965 5	2.52 19
$\gamma_{\beta-}$E2+12%M1	453.693 5	8.61 19
$\gamma_{\beta-}$E1	472.426 5	4.16 19
$\gamma_{\beta-}$E1	515.653 6	5.53 19
$\gamma_{\beta-}$E1	563.231 7	3.66 19
$\gamma_{\beta-}$E1	581.427 6	6.0 3
$\gamma_{\beta-}$(E2)	643.68 24	<0.019 ?
$\gamma_{\beta-}$E2	710.249 7	0.222 10
$\gamma_{\beta-}$[E2]	734.59 7	0.029 6
$\gamma_{\beta-}$E2	754.86 3	0.49 3
$\gamma_{\beta-}$[M1+E2]	814.15 7	0.17 4
$\gamma_{\beta-}$E2	819.250 6	7.48 14
$\gamma_{\beta-}$E2	863.86 3	2.17 14
$\gamma_{\beta-}$E2	866.829 8	5.77 18
$\gamma_{\beta-}$E1	894.390 7	19.8 4
$\gamma_{\beta-}$[M3]	911.44 3	0.0116 10 ?
$\gamma_{\beta-}$[E2]	923.15 7	0.0403 25
$\gamma_{\beta-}$E1	969.345 6	41.6 19
$\gamma_{\beta-}$E1	1003.391 7	0.158 8
$\gamma_{\beta-}$M2	1016.924 8	0.0135 19
$\gamma_{\beta-}$	1050.969 10	0.0165 19
$\gamma_{\beta-}$E1	1055.2 3	0.068 4
$\gamma_{\beta-}$E1,E2	1085.28 9	0.0229 19
$\gamma_{\beta-}$E1	1125.53 16	0.213 10
$\gamma_{\beta-}$E1,E2	1132.86 9	0.020 3
$\gamma_{\beta-}$E1	1164.2 3	0.015 3
$\gamma_{\beta-}$[M2]	1173.11 16	<0.0048

† 0.52% uncert(syst)

Atomic Electrons (^{232}Pa)

$\langle e\rangle$=79 3 keV

e_{bin}(keV)	$\langle e\rangle$(keV)	e(%)
17	3.7	22 4
21	3.9	19 3
22 - 26	0.47	1.89 25
27	10.2	38 7
30	10.2	33 6
34	0.57	1.64 8
42	4.5	10.7 20
43	4.1	9.4 17
44 - 87	4.2	8.3 9
88	9.6	10.9 11
89 - 91	0.0141	0.0158 13
92	6.6	7.2 8
93 - 103	0.241	0.235 20
104	3.0	2.9 3
105	2.18	2.08 22
106 - 155	2.75	2.42 16
156 - 184	0.087	0.053 5
261 - 306	2.02	0.71 4
338 - 387	3.24	0.92 5
388 - 437	1.56	0.372 12
448 - 495	0.791	0.173 5
498 - 546	0.0532	0.0101 3
558 - 595	0.45	0.079 7
619 - 666	0.111	0.0170 21
670 - 717	0.72	0.102 5
729 - 755	0.674	0.090 3
779 - 819	0.889	0.1130 23
842 - 891	1.58	0.185 7
893 - 940	0.0260	0.00288 15
948 - 996	0.294	0.0309 11
998 - 1047	0.0090	0.00088 6
1049 - 1085	0.0022	0.00021 6
1104 - 1152	0.00195	0.000175 15
1156 - 1172	0.00012	10 3 $\times10^{-6}$

Continuous Radiation (^{232}Pa)

$\langle\beta-\rangle$=92 keV;$\langle$IB$\rangle$=0.034 keV

E_{bin}(keV)		$\langle\ \rangle$(keV)	(%)
0 - 10	β-	0.385	7.7
	IB	0.0047	
10 - 20	β-	1.11	7.4
	IB	0.0040	0.028
20 - 40	β-	4.13	13.8
	IB	0.0064	0.023
40 - 100	β-	23.0	33.7
	IB	0.0106	0.017
100 - 300	β-	58	36.9
	IB	0.0065	0.0045
300 - 600	β-	2.29	0.53
	IB	0.0018	0.00043
600 - 1289	β-	2.88	0.367
	IB	0.00046	6.4×10^{-5}

$^{232}_{92}$U (68.9 10 yr)

Mode: α, SF(9 7 $\times10^{-11}$ %)
Δ: 34586 6 keV
SpA: 22.4 Ci/g
Prod: daughter ^{232}Pa; ^{232}Th(α,4n)

Alpha Particles (^{232}U)

$\langle\alpha\rangle$=5306.5 2 keV

α(keV)	α(%)
4502.9 5	2.4 7 $\times10^{-5}$
4930.95 10	0.00021 3
4948.73 20	0.00017 3
4997.94 11	0.0029 2
5136.72 10	0.28 2
5263.53 8	31.2 4
5320.34 8	68.6 4

Photons (^{232}U)

$\langle\gamma\rangle$=0.24 3 keV

γ_{mode}	γ(keV)	γ(%)
γ E2	57.81 5	0.21 4
γ E2	129.03 7	0.075 15
γ E1	141.19 19	5.4 23 $\times10^{-5}$
γ E2	191.29 17	3.4 3 $\times10^{-5}$
γ E1	209.39 7	1.3 3 $\times10^{-5}$
γ E1	270.26 8	0.0038 5
γ E1	328.07 9	0.0034 7
γ E1(+M2)	332.48 9	0.00051 3
γ E1	338.42 6	4.05 14 $\times10^{-5}$
γ (E1)	478.20 24	1.6 6 $\times10^{-6}$
γ E1	503.7 4	2.0 6 $\times10^{-5}$
γ (E1)	546.36 23	~1 $\times10^{-6}$
γ [E2]	774.0 4	~8 $\times10^{-6}$
γ [M1+E2]	816.62 24	~2 $\times10^{-7}$
γ	840 1	
γ [E2]	874.43 24	5.5 11 $\times10^{-7}$

$^{232}_{93}$Np(14.7 3 min)

Mode: ϵ
Δ: 37280 100 keV syst
SpA: 5.51×10^{7} Ci/g
Prod: ^{235}U(d,5n); ^{238}U(d,8n); ^{233}U(d,3n)

Photons (^{232}Np)

⟨γ⟩=1191 38 keV

γ_{mode}	γ(keV)	γ(%)†
U L_ℓ	11.620	2.0 5
U L_α	13.600	32 7
U L_η	15.400	0.61 17
U L_β	17.101	35 9
U L_γ	20.329	7.7 21
γ E2	47.579 9	~0.15
γ [E1]	80.24 7	0.0065 15
U $K_{\alpha2}$	94.651	23.7 20
U $K_{\alpha1}$	98.434	38 3
γ [E1]	105.48 3	0.0123 25
γ E2	109.001 8	~2
U $K_{\beta1}$'	111.025	13.9 12
U $K_{\beta2}$'	114.866	4.7 4
γ M1,E2	132.24 7	0.055 21
γ [E1]	139.53 3	0.024 4
γ [E1]	142.71 17	0.42 10
γ E1	150.096 3	0.081 14
γ E2	164.7 5	0.31 10
γ [E2]	175.57 24	0.042 9
γ M1,E2	176.85 7	~0.035
γ [E1]	184.142 6	0.053 14
γ [E2]	222.95 18	2.24 16
γ [E2]	282.24 17	19.8 10
γ [E1]	282.33 7	~7×10^{-5}
γ [E2]	326.85 17	52
γ	376.6 3	1.25 10
γ E2+4.7%M1	387.919 4	0.052 9
γ E2+21%M1	421.965 5	0.106 15
γ E2+12%M1	453.693 5	0.065 11
γ E2	710.249 7	0.97 6
γ E2	754.86 3	4.49 25
γ [M1+E2]	814.15 7	4.1 3
γ E2	819.250 6	32.6 12
γ E2	863.86 3	19.7 9
γ E2	866.829 8	25.1 11
γ E1	894.390 7	0.83 10
γ [M3]	911.44 3	0.105 11
γ [E2]	923.15 7	0.96 9
γ [E1]	941.2 4	1.6 3
γ E1	969.345 6	0.31 5
γ E1	1003.391 7	0.0066 9
γ	1016.4 4	0.57 5
γ M2	1016.924 8	0.000102 22
γ [M1+E2]	1037.10 17	3.28 21
γ	1050.969 10	0.00069 12
γ E1,E2	1085.28 9	0.99 5
γ E1	1125.53 16	1.46 16
γ E1,E2	1132.86 9	0.88 10
γ	1146.3 5	0.36 5 ?
γ [M2]	1173.11 16	<0.033
γ	1193.68 17	0.36 10
γ	1935.6 5	0.42 5 ?

† 9.6% uncert(syst)

Atomic Electrons (^{232}Np)

⟨e⟩=99 7 keV

e_{bin}(keV)	⟨e⟩(keV)	e(%)
17	6.0	35 9
21	4.8	23 7
22 - 26	0.99	4.4 7
27	7.4	~28
30	7.3	~24
34	0.0042	~0.012
42	3.3	~8
43	2.9	~7
44 - 87	3.9	7.2 18
88	5.3	6 3
89 - 91	0.30	0.34 3
92	3.6	3.9 20
93 - 142	4.9	4.7 11
143 - 165	0.75	0.50 8
167	2.79	1.68 11
170 - 206	1.30	0.65 4
211	7.1	3.4 7
217 - 260	1.42	0.583 25
261	4.1	1.6 4
265 - 305	5.6	1.96 14
306	5.6	1.8 4
310	2.1	0.67 14
321 - 367	4.0	1.24 13
371 - 421	0.11	~0.029
422 - 454	0.0091	0.0021 5
595 - 639	0.490	0.078 4
688 - 699	1.1	~0.15
704	2.77	0.393 17
705 - 754	3.88	0.518 17
755 - 802	1.28	0.161 19
808 - 854	1.63	0.196 6
858 - 908	0.57	0.065 9
910 - 952	0.6	~0.07
964 - 1013	0.11	0.011 3
1015 - 1064	0.27	0.027 8
1068 - 1116	0.33	0.030 8
1120 - 1169	0.019	0.0017 7
1170 - 1193	0.110	0.0093 18
1820 - 1918	0.032	~0.0018
1930 - 1931	0.0013	~7×10^{-5}

Continuous Radiation (^{232}Np)

⟨IB⟩=1.60 keV

E_{bin}(keV)		⟨ ⟩(keV)	(%)
10 - 20	IB	0.0038	0.023
20 - 40	IB	0.00087	0.0029
40 - 100	IB	0.49	0.56
100 - 300	IB	0.54	0.47
300 - 600	IB	0.161	0.036
600 - 1300	IB	0.36	0.042
1300 - 2500	IB	0.048	0.0029
2500 - 2690	IB	2.6×10^{-5}	1.02×10^{-6}

$^{232}_{94}$Pu(34.1 7 min)

Mode: ϵ(>80 %), α(<20 %)
Δ: 38349 23 keV
SpA: 2.38×10^{7} Ci/g
Prod: ^{233}U(α,5n); ^{235}U(α,7n)

Alpha Particles (^{232}Pu)

α(keV)	α(rel)
6542 10	38
6600 10	62

$^{232}_{95}$Am(55 7 s)

Mode: ϵ, α, ϵSF
SpA: 8.8×10^{8} Ci/g
Prod: ^{230}Th(^{10}B,8n)

A = 233

NDS **24**, 289 (1978)

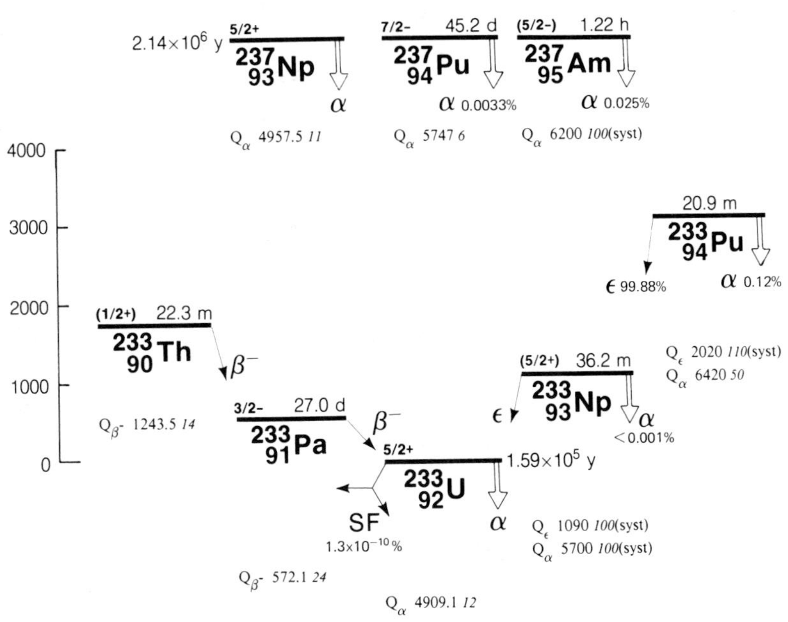

$^{233}_{90}$Th(22.3 *2* min)

Mode: β-
Δ: 38729.4 *21* keV
SpA: 3.62×10^{7} Ci/g
Prod: ^{232}Th(n,γ)

Photons (^{233}Th)

⟨γ⟩=36.4 *18* keV

γ_{mode}	γ(keV)	γ(%)
Pa L_ℓ	11.372	0.21 *4*
Pa L_α	13.274	3.6 *5*
Pa L_η	14.953	0.085 *13*
Pa L_β	16.629	4.6 *7*
Pa L_γ	19.725	1.07 *18*
γ E1	29.378 *9*	2.6 *4*
γ [E1]	46.57 *3*	
γ E2	57.149 *15*	0.054 *11*
γ (E2)	63.95 *3*	0.0008 *3*
γ [M1+E2]	70.62 *3*	0.00080 *16*
γ [M1+E2]	74.46 *4*	0.052 *10*
γ E1	86.528 *14*	2.6 *4*
γ [E1]	88.05 *4*	0.21 *5*
Pa $K_{\alpha2}$	92.279	0.51 *4*
γ E1	94.723 *21*	0.90 *10*
Pa $K_{\alpha1}$	95.863	0.83 *6*
γ	105.22 *10*	0.043 *9*
Pa $K_{\beta1}$'	108.166	0.301 *23*
γ M1+<4.6%E2	108.69 *3*	0.00070 *14*
Pa $K_{\beta2}$'	111.897	0.100 *8*
γ [M1+E2]	115.19 *3*	0.0020 *4*
γ M1+10%E2	117.689 *22*	0.0015 *3*
γ E1	131.09 *3*	0.065 *8*
γ [M1+E2]	134.23 *3*	0.0024 *5*
γ M1+14%E2	143.227 *21*	0.0144 *23*

Photons (^{233}Th) (continued)

γ_{mode}	γ(keV)	γ(%)
γ M1+12%E2	151.423 *24*	0.0088 *14*
γ [E1]	153.58 *8*	0.066 *13*
γ E1	155.263 *21*	0.00090 *18*
γ [E1]	162.51 *4*	0.172 *23*
γ [E1]	162.58 *8*	0.15 *3*
γ [E1]	169.18 *4*	0.34 *5*
γ [E1]	170.78 *8*	0.13 *3*
γ (M1+E2)	179.03 *12*	0.038 *8*
γ [E1]	180.801 *25*	0.00080 *16*
γ (M1+E2)	186.68 *8*	0.034 *7*
γ M1	190.54 *7*	0.13 *3*
γ E1	195.04 *3*	0.154 *20*
γ E1	201.72 *3*	0.032 *4*
γ (M1+E2)	210.47 *12*	0.035 *7*
γ	211.32 *20*	0.019 *4*
γ E1	212.412 *19*	0.0014 *3*
γ (M1+E2)	216.67 *11*	0.015 *3*
γ M1(+E2)	226.05 *13*	0.023 *5*
γ [E1]	237.950 *25*	0.0023 *5*
γ [E1]	246.12 *9*	
γ [E2]	250.63 *8*	0.0047 *9*
γ [M1+E2]	252.90 *11*	0.0120 *24*
γ [M1+E2]	257.30 *8*	0.068 *14*
γ	278.7 *4*	0.0078 *16*
γ (M1+E2)	285.43 *11*	0.021 *4*
γ [M1+E2]	347.52 *11*	0.0120 *24*
γ M1	359.89 *11*	0.120 *24*
γ [E1]	361.31 *9*	0.038 *8*
γ [M1+E2]	368.09 *11*	0.0047 *9*
γ [M1+E2]	377.21 *9*	0.038 *8*
γ [M1+E2]	399.32 *20*	0.014 *3*
γ	408.8 *5*	0.0038 *8*
γ (M1)	412.6 *3*	0.013 *3*
γ	418.4 *5*	0.0120 *24*
γ (M1)	430.76 *19*	0.023 *5*
γ	433.2 *3*	0.015 *3*
γ	435.0 *5*	
γ (M1+8.3%E2)	441.17 *9*	0.23 *5*
γ (M1+6.8%E2)	447.84 *9*	0.15 *3*
γ	454.2 *5*	0.040 *8*
γ M1	459.31 *11*	1.4 *3*

Photons (^{233}Th) (continued)

γ_{mode}	γ(keV)	γ(%)
γ	467.51 *11*	0.018 *4*
γ	473.9 *5*	0.0035 *7*
γ M1	490.75 *11*	0.17 *3*
γ	497.1 *4*	0.021 *4*
γ M1(+E2)	498.94 *11*	0.21 *4*
γ	505.5 *6*	0.0049 *10*
γ	513.4 *4*	0.020 *4*
γ	517.0 *4*	0.0068 *14*
γ	526.55 *9*	0.0063 *13*
γ	531.8 *4*	0.0042 *8*
γ	552.09 *9*	0.024 *5*
γ	554.9 *5*	0.0035 *7*
γ	562.79 *9*	0.070 *14*
γ [M1+E2]	573.57 *10*	0.042 *8*
γ M1(+E2)	595.32 *9*	0.16 *3*
γ [M1+E2]	599.1 *1*	0.047 *9*
γ [M1+E2]	609.8 *1*	0.085 *17*
γ [M1+E2]	642.33 *10*	0.028 *6*
γ [M1+E2]	663.2 *3*	0.0024 *5*
γ (M1)	669.78 *8*	0.68 *14*
γ	677.98 *8*	0.087 *17*
γ	681.2 *6*	0.016 *3*
γ	698.5 *6*	0.0120 *24*
γ	703.7 *6*	0.0110 *22*
γ [E2]	707.80 *10*	0.0120 *24*
γ [M1+E2]	716.79 *10*	0.056 *11*
γ (M1)	724.99 *10*	0.087 *17*
γ [E1]	740.89 *10*	0.031 *6*
γ	744.9 *5*	0.0068 *14*
γ	751.6 *6*	0.0024 *5*
γ	757.83 *9*	0.042 *8*
γ	764.50 *8*	0.120 *24*
γ	774.0 *4*	0.014 *3*
γ [M1+E2]	783.08 *18*	0.0061 *12*
γ	784.2 *5*	0.0049 *10*
γ [E1]	804.84 *10*	0.031 *6*
γ	806.29 *25*	0.013 *3*
γ [E1]	811.52 *10*	0.0078 *16*
γ [M1+E2]	815.61 *18*	0.028 *6*
γ	816.99 *25*	0.016 *3*
γ	832.0 *3*	0.0081 *16*

Photons (^{233}Th)
(continued)

γ_{mode}	γ(keV)	γ(%)
γ	846.8 7	0.0014 3
γ	849.52 25	0.0047 9
γ	870.7 7	0.0021 4
γ	874.0 5	0.0062 12
γ [E2]	881.07 18	0.0078 16
γ [M1+E2]	890.07 18	0.14 3
γ [M1+E2]	898.27 18	0.0033 7
γ	918.9 5	<0.007
γ	935.2 7	0.049 10
γ	941.9 8	0.0078 16
γ	948.08 25	0.0075 15
γ	955 1	0.0054 11
γ	960.8 8	0.0068 14
γ	962.8 9	0.0014 3
γ	968.2 9	0.0110 22
γ [E1]	978.12 18	0.0075 15
γ [E1]	984.79 18	0.0014 3
γ	994 1	0.00094 19
γ	1001 1	0.00120 24
γ	1007 1	0.0028 6
γ	1011 1	0.0040 8
γ	1026.5 10	0.0081 16
γ	1092.5 10	<0.007
γ	1144 1	0.0029 6
γ	1201 1	<0.007

Atomic Electrons (^{233}Th)
⟨e⟩=17.5 8 keV

e_{bin}(keV)	⟨e⟩(keV)	e(%)
5	0.75	14 3
6 - 13	1.2	15 5
17	0.64	3.8 6
18	0.0025	0.0133 17
20	0.69	3.4 5
21 - 22	0.111	0.53 8
24	0.38	1.6 3
25 - 36	0.40	1.52 19
37	1.4	3.8 8
39	0.016	0.042 8
40	1.3	3.2 6
41 - 50	0.059	0.12 3
52	0.54	1.04 21
53	0.58	1.1 4
54 - 64	1.0	1.7 7
65	0.74	1.13 20
66	1.28	1.9 3
67 - 116	1.90	2.4 3
117 - 167	0.25	0.17 4
168 - 217	0.31	0.171 18
218 - 265	0.27	0.111 14
269 - 318	0.062	0.020 3
321 - 346	0.39	0.117 15
347	1.21	0.35 7
348 - 396	0.35	0.093 12
397 - 446	0.43	0.100 13
447 - 496	0.37	0.079 8
497 - 546	0.079	0.0155 24
547 - 555	0.0079	0.00144 23
557	0.35	0.062 13
558 - 606	0.113	0.019 4
608 - 658	0.16	0.026 5
659 - 708	0.069	0.0101 11
711 - 760	0.019	0.0026 8
761 - 811	0.057	0.0074 13
812 - 861	0.016	0.0020 9
864 - 902	0.0150	0.00172 24
914 - 963	0.0053	0.0006 3
964 - 1010	0.0012	~0.00012
1021 - 1031	0.0004	~4×10^{-5}
1071 - 1092	0.0006	~6×10^{-5}
1123 - 1143	9.6×10^{-5}	~9×10^{-6}
1180 - 1200	0.00012	~1×10^{-5}

Continuous Radiation (^{233}Th)
⟨β-⟩=394 keV;⟨IB⟩=0.47 keV

E_{bin}(keV)		⟨ ⟩(keV)	(%)
0 - 10	β-	0.069	1.38
	IB	0.018	
10 - 20	β-	0.208	1.39
	IB	0.017	0.121
20 - 40	β-	0.84	2.79
	IB	0.033	0.114
40 - 100	β-	6.0	8.5
	IB	0.085	0.132
100 - 300	β-	56	28.2
	IB	0.18	0.106
300 - 600	β-	153	34.7
	IB	0.109	0.027
600 - 1245	β-	178	23.0
	IB	0.026	0.0037

$^{233}_{91}$Pa(27.0 1 d)

Mode: β-
Δ: 37485.9 24 keV
SpA: 2.076×10^4 Ci/g
Prod: daughter ^{233}Th; ^{232}Th(d,n)

Photons (^{233}Pa)
⟨γ⟩=204 23 keV

γ_{mode}	γ(keV)	γ(%)†
U L_ℓ	11.620	1.07 20
U L_α	13.600	18 3
U L_η	15.400	0.23 5
U L_β	17.058	19 4
γ M1+1.7%E2	17.269 18	~0.0036
U L_γ	20.420	4.5 9
γ M1+2.4%E2	28.578 17	0.065 7
γ M1+27%E2	40.415 10	0.036 7
γ [E1]	41.718 20	0.013 4
γ [M1+E2]	51.92 13	<0.0007
γ (E2)	58.074 19	<0.0036
γ M1	75.343 10	1.17 7
γ M1	86.652 10	1.76 22
γ [E2]	92.34 13	<0.0036
U $K_{\alpha2}$	94.651	10.2 16
U $K_{\alpha1}$	98.434	16 3
γ M1+6.9%E2	103.921 16	0.69 7
U $K_{\beta1}'$	111.025	6.0 10
U $K_{\beta2}'$	114.866	2.0 3
γ [E2]	248.25 13	0.054 22
γ [E1]	258.46 3	0.0054 5
γ E2	271.597 20	0.28 3
γ [E1]	298.87 3	0.0253 22
γ M1	300.175 18	6.2 3
γ M1	312.012 19	36
γ M1+18%E2	340.590 18	4.2 3
γ E2	375.518 18	0.58 11
γ E2	398.663 21	1.19 14
γ E2+20%M1	415.932 18	1.51 14

† 5.6% uncert(syst)

Atomic Electrons (^{233}Pa)
⟨e⟩=130 12 keV

e_{bin}(keV)	⟨e⟩(keV)	e(%)
2 - 51	11.1	62 6
52 - 53	0.019	~0.04
54	5.4	10.0 7
55 - 58	0.043	0.074 20
65	5.8	9.0 11

Atomic Electrons (^{233}Pa)
(continued)

e_{bin}(keV)	⟨e⟩(keV)	e(%)
66 - 110	9.6	12.1 6
133 - 156	0.048	0.032 4
183	0.00148	0.00081 8
185	9.9	5.4 3
196	55.0	28 6
225	4.8	2.2 4
226 - 272	0.230	0.091 7
277 - 283	3.00	1.08 5
290	13.9	4.8 10
291 - 300	3.1	1.04 11
306	3.5	1.15 23
307 - 355	3.9	1.21 11
358 - 399	0.429	0.111 5
410 - 416	0.074	0.0179 11

Continuous Radiation (^{233}Pa)
⟨β-⟩=64 keV;⟨IB⟩=0.017 keV

E_{bin}(keV)		⟨ ⟩(keV)	(%)
0 - 10	β-	0.59	11.9
	IB	0.0033	
10 - 20	β-	1.64	11.0
	IB	0.0026	0.018
20 - 40	β-	5.7	19.3
	IB	0.0038	0.0134
40 - 100	β-	25.0	37.8
	IB	0.0052	0.0087
100 - 300	β-	27.9	19.6
	IB	0.0025	0.0017
300 - 573	β-	3.04	0.82
	IB	0.000163	4.8×10^{-5}

$^{233}_{92}$U (1.592 20 $\times10^5$ yr)

Mode: α, SF(1.3 3 $\times10^{-10}$ %)
Δ: 36914 3 keV
SpA: 0.00964 Ci/g
Prod: daughter ^{233}Pa

Alpha Particles (^{233}U)
⟨α⟩=4814 1 keV

α(keV)	α(%)
4307.2 8	0.0009
4404 10	0.0003
4406.2 8	0.0004
4457 10	0.0028
4465.2 8	0.003
4483 10	0.0014
4503.2 8	0.0010
4509.7 8	0.012
4513.0 8	0.018
4541.7 8	0.004
4567.3 8	0.0028
4573.2 8	0.0023
4591.4 8	0.007
4611.3 8	0.006
4616.0 8	0.004
4626 10 ?	<0.004
4632.3 8	0.010
4641 10	0.003
4654.0 8	~0.005
4664.2 8	0.04
4680.9 8	0.010
4687 10	0.0028
4701.4 8	0.06
4729.2 8	1.6

Alpha Particles (^{233}U) (continued)

α(keV)	α(%)
4751 10	0.010
4754.1 8	0.16
4758 10	0.016
4783.0 8	13.2 2
4796.0 8	0.28
4804 10	0.05
4824.7 8	84.4 5

Photons (^{233}U)

⟨γ⟩=1.29 24 keV

γ_{mode}	γ(keV)	γ(%)†
Th L_ℓ	11.118	0.15 6
Th L_α	12.952	2.6 10
Th L_η	14.511	0.057 23
Th L_β	16.141	3.3 12
Th L_γ	19.148	0.8 3
γ[M1+E2]	25.304 24	0.00121 24
γM1	29.191 12	0.0068 14
γ[M1]	31.24 4	0.00027 5 ?
γ	32.27 20	0.00099 20
γ[M1+E2]	37.85 4	0.00036 7
γM1+14%E2	42.441 15	0.060 12
γ	50.59 6	
γ[M1+E2]	52.613 15	0.00025 8
γ[M1+E2]	53.559 20	0.0043 3
γM1+17%E2	54.702 22	0.014 3
γ	63.69 4	3.3 7 $\times10^{-5}$
γ	65.51 10	
γ(M1+24%E2)	66.11 4	0.00084 17
γ[E2]	67.960 21	0.00032 6
γ	68.97 3	0.000108 21
γ[M1+E2]	70.35 4	0.00060 12
γE2	71.847 16	0.0027 5
γ[E2]	72.821 22	0.00059 12
γ	74.500 20	0.0015 3
γ	76.32 3	0.00039 8
γ	77.10 4	0.00072 14 ?
γ[M1+E2]	78.41 9	6.0 12 $\times10^{-5}$
γ[M1+E2]	82.957 16	0.00018 4
γ	84.27 20	7.5 25 $\times10^{-5}$
γ[M1+E2]	85.402 20	0.00019 4
γ	86.74 15	0.00013 3
γ	87.24 11	0.00019 4
γ	88.43 8	0.00044 9
Th $K_{\alpha2}$	89.955	0.0104 8
γ	90.994 24	0.00033 7
Th $K_{\alpha1}$	93.350	0.0169 14
γ[E2]	96.215 18	0.0014 3
γE2	97.143 21	0.022 4
γ[E1]	100.008 12	5.5 11 $\times10^{-5}$
γ	101.80 4	9.0 18 $\times10^{-5}$
γ[E1]	103.375 12	0.00010 2
Th $K_{\beta1}$'	105.362	0.0061 5
Th $K_{\beta2}$'	108.990	0.00203 17
γ	109.47 10	0.00031 3
γ[E1]	111.935 25	0.00047 5
γ	114.28 7	0.00025 8
γ[E2]	116.26 9	0.00021 4
γ[E1]	117.155 16	0.0028 4
γ[E1]	118.97 3	0.0032 6
γE2	120.81 4	0.0021 4
γ[E2]	123.91 4	0.00065 13
γ[E2]	125.398 18	6.5 13 $\times10^{-5}$
γ	129.2 1	7.0 14 $\times10^{-5}$
γ	131.20 4	3.3 7 $\times10^{-5}$
γ[E1]	135.328 14	0.0022 4
γ	138.5 10	~2 $\times10^{-5}$
γ	139.76 5	0.000105 21
γ	141.58 6 ?	
γ[E2]	144.52 9	0.00030 7
γ[E1]	145.286 15	0.0016 3
γ[E1]	146.347 17	0.0063 10
γ	148.15 3	0.00036 7
γ	149.80 12	0.000127 25

Photons (^{233}U) (continued)

γ_{mode}	γ(keV)	γ(%)†
γ[E1]	152.621 13 ?	
γ[E2]	153.31 4	5.5 11 $\times10^{-5}$
γ	154.69 10	0.00015 3
γ[E1]	155.989 15	5.8 11 $\times10^{-5}$
γ	162.52 6	7.5 23 $\times10^{-5}$
γ[E1]	164.512 10	0.0066 10
γ[E1]	165.493 23	0.00039 8
γ	168.98 3	6.7 13 $\times10^{-5}$
γ	170.794 18	0.00014 3
γ	172.35 3	3.5 7 $\times10^{-5}$
γ	174.162 16	0.00023 5
γ	176.10 7	4.3 9 $\times10^{-5}$
γ	177.78 6	2.0 4 $\times10^{-5}$
γ[E2]	184.17 9	2.5 5 $\times10^{-5}$
γ	185.782 20	4.0 8 $\times10^{-5}$
γ[E1]	187.942 9	0.0020 4
γ	192.12 3	4.0 8 $\times10^{-5}$
γ	200.67 10	1.0 2 $\times10^{-6}$
γ	205.99 9	6.5 13 $\times10^{-5}$
γ[E1]	208.149 18	0.0025
γ	212.29 8	0.00014 3
γ	216.08 6	0.00066 13
γ[E1]	217.133 14	0.0035 7
γ	217.61 3	5 1 $\times10^{-5}$
γ	219.424 19	0.00015 3
γ[E2]	223.357 22	3.3 7 $\times10^{-5}$
γ	225.0 3	1.0 2 $\times10^{-5}$
γ	226.74 4	1.0 2 $\times10^{-5}$
γ	228.07 10	2.0 4 $\times10^{-5}$
γ[M1+E2]	230.086 19	6.7 13 $\times10^{-5}$
γ	236.39 21	5 1 $\times10^{-5}$
γ[M1+E2]	240.364 19	0.00038 8
γ(M1)	245.294 14	0.0035 5
γ(M1)	248.661 15	0.0045 12
γ	252.50 11	3.8 8 $\times10^{-5}$
γ	255.92 4	4.3 9 $\times10^{-5}$
γ	259.30 4	0.00018 4
γ	260.40 10	0.000108 21
γ[E2]	261.88 8	0.00031 6
γ[M1+E2]	268.619 20	0.00025 5
γ	272.27 8	6.3 13 $\times10^{-5}$
γ[M1+E2]	274.692 16	0.00047 9
γ(M1)	278.059 14	0.00122 22
γ	284.22 15	1.0 2 $\times10^{-5}$
γ[M1+E2]	287.949 11	0.00108 20
γ(M1)	291.317 9	0.0052 8
γ[M1+E2]	293.923 16	0.00014 3
γ	295.2 5	2.3 5 $\times10^{-5}$
γ	302.84 10	7.0 14 $\times10^{-5}$
γ	309.37 9	7.3 15 $\times10^{-5}$
γ	311.38 8	2.8 10 $\times10^{-5}$?
γ(M1)	317.133 9	0.0088 12
γ(M1)	320.508 9	0.0031 5
γ[M1+E2]	323.321 16	0.00084 17
γ[M1+E2]	328.69 4	6.5 13 $\times10^{-5}$
γ[M1+E2]	336.579 12	0.00059 12
γ	339.64 8	7.5 25 $\times10^{-6}$
γ	351.782 10	4.5 9 $\times10^{-5}$
γ[M1+E2]	354.00 4	5.8 11 $\times10^{-5}$
γ(M1)	365.762 9	0.00082 16
γ[M1+E2]	383.40 4	9.5 19 $\times10^{-5}$
γ	384.0 5	~2 $\times10^{-5}$
γ	394.34 8	7.5 25 $\times10^{-6}$
γ[E2]	396.65 4	8.3 17 $\times10^{-6}$
γ	402.37 20	8.3 17 $\times10^{-6}$?
γ	406.7 3	5.3 11 $\times10^{-6}$?
γ	416.4 12	1.00 25 $\times10^{-5}$
γ	436.78 8	~5 $\times10^{-6}$
γ	449.6 15	1.00 25 $\times10^{-5}$
γ	459.77 20	8.3 17 $\times10^{-6}$?
γ	471.3 13	1.5 3 $\times10^{-5}$
γ	479 3	1.5 3 $\times10^{-5}$
γ	484.05 9	4.3 9 $\times10^{-6}$?
γ	514.0 5 ?	
γ	537.6 5 ?	
γ	540.27 20	5.3 11 $\times10^{-6}$?

Photons (^{233}U) (continued)

γ_{mode}	γ(keV)	γ(%)†
γ	545.1 3	2.4 5 $\times10^{-6}$?
γ	562.8 5 ?	
γ	569.37 20	3.8 8 $\times10^{-6}$?
γ	578.47 20	5 1 $\times10^{-6}$?
γ	620.87 20	2.3 5 $\times10^{-6}$?
γ	656.97 20	3.0 6 $\times10^{-6}$?
γ	707.47 20	2.7 6 $\times10^{-6}$?
γ	710.8 5 ?	
γ	826.3 5 ?	
γ	867.9 4	2.1 4 $\times10^{-6}$?
γ	920.0 10 ?	
γ	1003.0 10	8.3 17 $\times10^{-6}$?
γ	1055.0 10 ?	
γ	1119.0 10	8.3 17 $\times10^{-6}$?

† 20% uncert(syst)

Atomic Electrons (^{233}U)

⟨e⟩=5.5 7 keV

e_{bin}(keV)	⟨e⟩(keV)	e(%)
5 - 8	0.12	~2
9	0.25	~3
11 - 15	0.07	~0.6
16	0.5	~3
17 - 18	0.017	~0.09
20	0.7	3.4 16
21	0.12	<1
22	0.42	1.9 5
23	0.55	2.4 10
24 - 25	0.14	~0.6
26	0.54	2.1 10
27 - 33	0.15	0.52 22
34	0.14	~0.40
35 - 36	0.078	0.22 5
37	0.23	0.6 3
38	0.52	1.4 6
39 - 40	0.006	0.016 8
41	0.18	0.45 17
42 - 91	0.62	1.06 12
92 - 141	0.113	0.113 10
142 - 191	0.0132	0.0076 9
192 - 241	0.0225	0.0106 10
242 - 291	0.0066	0.00250 19
292 - 340	0.0060	0.00199 17
345 - 394	0.00035	0.000101 13
395 - 445	8.3 $\times10^{-6}$	2.0 6 $\times10^{-6}$
446 - 484	7.9 $\times10^{-6}$	1.7 6 $\times10^{-6}$
511 - 559	2.9 $\times10^{-6}$	5.4 20 $\times10^{-7}$
562 - 605	1.0 $\times10^{-6}$	~2 $\times10^{-7}$
616 - 657	2.8 $\times10^{-7}$	~4 $\times10^{-8}$
687 - 707	1.9 $\times10^{-7}$	~3 $\times10^{-8}$
847 - 893	1.3 $\times10^{-6}$	~1 $\times10^{-7}$
983 - 1009	1.3 $\times10^{-6}$	~1 $\times10^{-7}$
1099 - 1118	2.6 $\times10^{-7}$	~2 $\times10^{-8}$

$^{233}_{93}$Np(36.2 *1* min)

Mode: ε, α(<0.001 %)

Δ: 38000 100 keV syst

SpA: 2.229$\times10^{7}$ Ci/g

Prod: ^{233}U(d,2n); ^{235}U(d,4n)

Alpha Particles (^{233}Np)

α(keV)
5530 ?

Photons (^{233}Np)

⟨γ⟩=90 *21* keV

γ_{mode}	γ(keV)	γ(%)†
U L_ℓ	11.620	~1
U L_α	13.600	18 *9*
U L_η	15.400	~0.24
U L_β	17.068	~17
U L_γ	20.369	~4
γ[M1+E2]	21.94 *11*	
γM1+2.4%E2	28.578 *17*	0.00086 *11*
γM1+27%E2	40.415 *10*	
γ[E1]	41.718 *20*	0.00017 *5*
γ[M1+E2]	51.92 *13*	
γ[E2]	92.34 *13*	
U $K_{\alpha2}$	94.651	22 *11*
U $K_{\alpha1}$	98.434	35 *17*
U $K_{\beta1}$'	111.025	13 *6*
U $K_{\beta2}$'	114.866	4.3 *21*
γ	205.2 *4*	0.0231 *21*
γ[E1]	225.67 *20*	0.036 *3*
γ[E1]	228.47 *14*	0.046 *3*
γ[M1+E2]	234.47 *18*	0.154 *7*
γ	242.4 *4*	0.084 *7*
γ[E1]	247.61 *18*	0.040 *5*
γ[E2]	248.25 *13*	0.0007 *3*
γ[M1+E2]	256.69 *24*	0.042 *3*
γ[E1]	258.46 *3*	0.101 *6*
γE2	271.597 *20*	0.0055 *12*
γ[E1]	280.39 *11*	0.127 *5*
γ[E1]	298.87 *3*	0.48 *3*
γM1	300.175 *18*	0.082 *6*
γM1	312.012 *19*	0.70
γ[E1]	320.81 *11*	0.074 *3*
γM1+18%E2	340.590 *18*	0.055 *3*
γ	392.9 *4*	0.0175 *14*
γ	425.5 *4*	0.058 *4*
γ[M1+E2]	504.94 *25*	0.041 *6*
γ[M1+E2]	506.06 *18*	0.154 *21*
γ[M1+E2]	546.48 *18*	0.280 *14*
γ[M1+E2]	556.87 *24*	0.028 *3*
γ[M1+E2]	597.28 *24*	0.0266 *21*
γ	644.3 *5*	0.0077 *7*
γ	665.8 *4*	0.0168 *14*

† approximate

Continuous Radiation (^{233}Np)

⟨IB⟩=1.18 keV

E_{bin}(keV)		⟨ ⟩(keV)	(%)
10 - 20	IB	0.0039	0.024
20 - 40	IB	0.00089	0.0030
40 - 100	IB	0.48	0.55
100 - 300	IB	0.53	0.46
300 - 600	IB	0.100	0.023
600 - 1100	IB	0.058	0.0082

$^{233}_{94}$Pu(20.9 *4* min)

Mode: ϵ(99.88 *5* %), α(0.12 *5* %)

Δ: 40020 *50* keV

SpA: 3.86×10^{7} Ci/g

Prod: ^{233}U(α,4n)

Alpha Particles (^{233}Pu)

α(keV)
6300 *20*

Photons (^{233}Pu)

γ_{mode}	γ(keV)	γ(rel)
γ[M1+E2]	34.40 *19*	
γ	150.3 *3*	15.2 *16*
γ	180.4 *3*	12.0 *25*
γ	191.12 *25*	13.0 *18*
γ	207.65 *25*	23.8 *20*
γ	221.8 *3*	11.8 *12*
γ	235.30 *23*	100
γ	247.4 *3*	7.2 *19*
γ	457.10 *21*	10.2 *21*
γ	472.9 *3*	7.2 *15*
γ	478.07 *23*	13.8 *17*
γ	500.25 *21*	39 *4*
γ	503.81 *24*	20.7 *24*
γ	512.47 *23*	13 *3*
γ	524.27 *21*	13.0 *21*
γ	534.65 *18*	90 *4*
γ	558.67 *22*	26.9 *25*
γ	583.3 *3*	8.6 *20*
γ	688.0 *3*	33 *3*
γ	725.8 *3*	9.0 *14*
γ	830.8 *3*	11.1 *13*
γ	977.92 *23*	13.4 *23*
γ	991.75 *22*	23.0 *23*
γ	1000.79 *23*	18 *4*
γ	1003.9 *3*	31 *3*
γ	1012.32 *23*	28 *3*
γ	1028.1 *3*	6.6 *17*
γ	1035.2 *3*	5.7 *15*

A = 234

NDS **21**, 493 (1977)

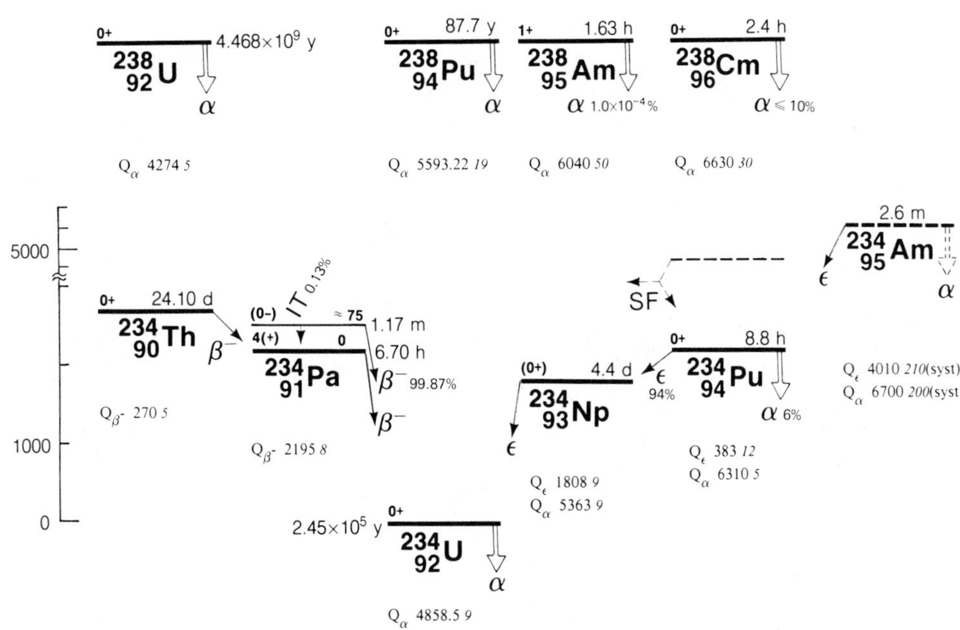

$^{234}_{90}$Th(24.10 3 d)

Mode: β-
Δ: 40607 6 keV
SpA: 2.315×10⁴ Ci/g
Prod: natural source

Photons (^{234}Th)

⟨γ⟩=9.4 3 keV

γ_{mode}	γ(keV)	γ(%)†
Pa L_ℓ	11.372	0.23 3
Pa L_α	13.274	3.9 4
Pa L_η	14.953	0.049 6
Pa L_β	16.572	4.4 6
Pa L_γ	19.839	1.08 17
γ M1+0.5%E2	20.019 20	0.010 3
γ E2	29.490 20	0.00146 10
γ	57.75 10	~0.005
γ M1+10%E2	62.862 18	0.019 3
γ E1	63.288 18	3.8
γ (M1+1.3%E2)	73.90 4	0.0156 16 *
γ (E1)	74.0 1	0.0426 21
γ [E1]	83.31 3	0.070 3
γ (M1)	87.02 6	0.0073 10
Pa $K_{\alpha2}$	92.279	~0.008
γ M1	92.35 3	2.72 5
γ E1	92.78 3	2.69 5
Pa $K_{\alpha1}$	95.863	~0.013
γ (M1)	103.35 10	~0.0035
γ (M1)	103.71 6	~0.006
γ	108.00 5	0.0059 5
Pa $K_{\beta1}$'	108.166	~0.005
Pa $K_{\beta2}$'	111.897	~0.0016
γ [E1]	112.80 3	0.242 5
γ	184.8 10	~0.012

† 7.7% uncert(syst)
* with ^{234}Pa(1.17 min) in equilib

Atomic Electrons (^{234}Th)

⟨e⟩=15.8 3 keV

e_{bin}(keV)	⟨e⟩(keV)	e(%)
8 - 20	1.47	10.2 5
21	0.42	2.01 21
22 - 70	1.80	4.6 3
71	7.44	10.4 3
72	0.98	1.36 6
73 - 86	0.150	0.194 16
87	2.47	2.84 9
88 - 90	0.062	0.071 12
91	0.71	0.78 3
92 - 113	0.275	0.296 11
164 - 185	0.015	~0.009

$^{234}_{91}$Pa(6.70 5 h)

Mode: β-
Δ: 40337 8 keV
SpA: 1.999×10⁶ Ci/g
Prod: natural source

Photons (^{234}Pa)

⟨γ⟩=1903 56 keV

γ_{mode}	γ(keV)	γ(%)†
U L_ℓ	11.620	2.6 4
U L_α	13.600	43 6
U L_η	15.400	0.99 15
U L_β	17.120	52 8
U L_γ	20.309	11.8 19
γ E2	34.37 4	0.0036 10
γ E2	43.470 9	0.12 3
γ E2	45.28 6	0.011 2
γ (E2)	58.31 18	0.009 3
γ E1	62.78 4	3.2 2
γ E2+45%M1	67.22 22	0.058 20
γ	69.9 10	~0.23
γ E2	79.65 6	~0.27
U $K_{\alpha2}$	94.651	15.7 10
U $K_{\alpha1}$	98.434	25.3 16
γ E2	99.852 10	4.8 10
γ [E2]	103.7 3	0.12 4
U $K_{\beta1}$'	111.025	9.2 6
U $K_{\beta2}$'	114.866	3.11 21
γ E2	125.53 20	1.0 3
γ E1	131.29 17	20
γ (M1)	134.3 3	0.21 7
γ	137.7 5	0.15 3 ?
γ E2+39%M1	140.15 9	0.96 9
γ [M1+E2]	143.9 4	0.35 5
γ [E2]	150.0 4	~0.2
γ E2	152.69 3	6.7 5
γ [E1]	157.96 16	0.7 2
γ	165.8 5	0.09 2
γ M1	170.75 17	0.5 1
γ [M1+E2]	174.52 9	0.20 5
γ M1	185.92 16	2.0 3
γ (M1)	193.55 25	0.6 1
γ E0+E2+M1	196.43 18	0.070 14
γ [M1+E2]	199.46 20	0.48 16
γ E2	200.96 5	1.1 3
γ E2+31%M1	203.20 3	1.12 17
γ	210.7 5	0.05 1
γ (M1)	219.80 10	~0.2
γ E2+45%M1	226.63 22	5.9 12
γ M1	227.17 16	5.5 11
γ M1	245.16 18	0.9 2
γ E1	248.94 18	2.8 3
γ	267.1 8	0.17 3
γ (M1)	272.15 17	1.0 2
γ [E2]	274.45 21	~0.30
γ [E1]	276.83 22	0.26 7
γ [E2]	278.10 19	0.070 14
γ [E2]	286.40 25	0.14 3
γ [E2]	289.25 19	0.110 22 ?
γ E2+45%M1	293.85 19	3.9 3
γ [E1]	299.08 10	
γ	309.6 8	0.10 2
γ [M1+E2]	313.83 24	~0.30
γ [E2]	316.29 20	0.120 24
γ	320.7 8	0.120 24
γ [E1]	330.0 4	~0.30

Photons (^{234}Pa)
(continued)

γ_{mode}	γ(keV)	γ(%)†
γ (M1+E2)	330.62 23	~0.6
γ E2	352.16 15	0.6 1
γ M1	370.00 23	2.9 3
γ M1+21%E2	371.86 23	1.3 2
γ	409.8 4	~0.4
γ [E2]	416.0 4	0.10 2 ?
γ	426.8 4	~0.6
γ	432.8 5	0.011 3
γ [M1+E2]	446.1 3	0.120 24
γ E2+37%M1	458.65 24	1.5 1
γ [M1+E2]	461.3 4	0.16 3 ?
γ	467.5 10	~0.4
γ [E1]	472.32 17	0.24 5
γ	473.5 10	0.18 4
γ [E1]	478.7 6	0.30 6
γ	480.4 8	0.4 1
γ	482.5 7	0.3 1
γ [M1+E2]	498.8 3	0.10 2
γ [E1]	506.69 17	1.6 3
γ [M1+E2]	513.6 3	1.3 2
γ	520.2 5	0.60 12
γ [E1]	521.09 23	0.90 18
γ [M1]	526.20 24	
γ (M1)	527.04 25	0.6 2
γ	533.2 10	0.20 4
γ	537.1 10	0.16 3
γ [M1+E2]	557.78 24	0.26 5
γ (M1+33%E2)	565.2 4	1.4 3
γ (M1)	568.29 25	3.0 6
γ (M1)	569.47 17	10.7 21
γ [E1]	574.8 3	~2 ?
γ	585.8 8	0.15 3
γ	596.5 3	<0.7 ?
γ [E1]	602.1 4	0.9 4 ?
γ (M1)	611.5 4	~0.8
γ [E2]	617.0 3	0.20 4
γ (M1+E2)	624.31 21	0.8 2
γ	627.5 5	0.8 2
γ (M1)	632.0 4	0.40 8
γ [M1+E2]	634.4 4	0.30 6
γ	639.7 10	0.20 4
γ	643.2 10	0.20 4
γ [E1]	646.84 19	0.30 6
γ (M1)	653.57 25	0.9 4
γ	655.0 8	0.6 2
γ	658.0 5	0.9 4
γ	660.6 10	0.30 6
γ	664.8 10	1.3 4
γ [E1]	666.64 25	1.6 4
γ [E1]	669.7 3	1.4 4
γ	683.3 8	0.24 5
γ [E1]	688.08 20	0.28 6
γ (M1)	692.7 4	1.5 5
γ M1(+E2)	698.85 25	4.6 3
γ [E1]	706.02 9	3.0 4
γ [E2]	708.40 7	0.024
γ	711.2 8	0.20 4 ?
γ	713.8 8	0.16 5 ?
γ [M1+E2]	721.33 25	
γ M1(+E2)	733.22 25	8.6 8
γ [M1+E2]	738.0 4	1.0 4
γ E1	742.817 20	2.4 3
γ	746.5 8	0.13 3
γ [E2]	755.2 5	~1
γ [E2]	760.1 3	0.16 3 ?
γ [M1+E2]	766.369 20	0.30 6
γ (E2)	766.412 14	0.080 16
γ [M1+E2]	768.5 4	0.56 11
γ	777.9 10	0.20 4
γ [E1]	780.4 4	1.1 4
γ [E2]	783.38 4	0.5 1
γ (E1)	786.287 20	1.43 20
γ	793.6 10	1.5 3
γ [M1+E2]	795.3 3	3.8 5
γ E0+E2	804.52 13	0.40 8
γ [E1]	805.87 9	3.4 4
γ E0+E2	808.25 7	0.07
γ	812.5 15	0.5 1
γ [E1]	819.33 25	2.6 5
γ	824.0 8	<2
γ [M1+E2]	825.81 24	4.0 8
γ [E1]	831.43 18	5.5 7
γ	842.4 7	0.14 3
γ	844.4 10	0.5 2

Photons (^{234}Pa)
(continued)

γ_{mode}	γ(keV)	γ(%)†
γ [E2]	851.72 7	0.12
γ [M1+E2]	873.4 3	0.120 24
γ E2	876.34 22	~4
γ [M1+E2]	880.47 25	4.0 8
γ [E1]	880.53 4	9.0 18
γ E2	883.24 4	12.0 16
γ [E1]	898.65 20	4.1 8
γ [E2]	904.38 13	0.5 2
γ	920 1	~0.4
γ E2	925.66 24	11 2
γ [E1]	925.81 6	2.9 6
γ E2	926.71 4	9.0 11
γ [E2]	942.05 10	
γ E1	946.02 3	8.1 9
γ [M1+E2]	948.0 3	8.0 16
γ	960 1	0.10 2
γ [M1+E2]	967.33 5	0.60 12 ?
γ [M1+E2]	978.8 10	~1
γ [E2]	980.32 25	~2
γ [E1]	980.39 4	~3
γ [E2]	982.85 20	
γ [E1]	984.12 18	1.9 6
γ [M1+E2]	1022.23 23	~0.6
γ [E2]	1029.03 22	0.8 3
γ [M1+E2]	1041.90 10	
γ	1044.9 10	0.5 1
γ	1074.6 10	0.25 8
γ [M1+E2]	1082.7 2	0.75 9
γ [E2]	1085.37 10	
γ	1108.5 8	0.3 1
γ [M1+E2]	1122.09 23	0.5 1
γ [E2]	1126.17 20	0.80 9
γ [E1]	1153.6 6	0.3 1
γ [E2]	1171.3 6	0.24 9
γ	1208 1	~0.30
γ [M1+E2]	1216.4 6	0.37 6
γ	1229 1	0.30 6 ?
γ (E2)	1241.40 18	~0.5
γ	1251 1	0.30 6 ?
γ	1277.4 8	0.20 7
γ M1	1293.0 3	0.6 2
γ M1	1352.85 17	1.7 5
γ	1358.5 10	0.120 24
γ M1	1394.09 18	3.0 9
γ	1399.7 10	0.23 5
γ [E2]	1427.33 17	0.20 4
γ [M1+E2]	1445.7 3	0.4 1
γ [M1+E2]	1452.70 17	1.0 2
γ	1460 1	0.30 6 ?
γ [E2]	1493.94 18	0.20 6
γ	1516 1	0.40 8
γ M1+E2	1527.20 24	
γ	1549.4 10	0.10 4
γ M1	1558.7 3	
γ M1	1570.67 24	
γ [M1+E2]	1580.02 17	0.17 8
γ [E2]	1585.3 5	0.25 10
γ (E2)	1594.3 4	0.6 2
γ (M1)	1602.2 3	
γ	1627.9 10	0.15 3
γ (E2)	1638 1	~0.4
γ	1656 3	~0.15
γ (M1)	1668.2 7	1.2 2
γ	1686.2 10	0.5 2
γ [M1+E2]	1694.2 4	1.2 5
γ	1699.8 10	0.15 5
γ	1719.5 10	~0.02
γ [M1+E2]	1738.0 5	~0.10
γ	1741.7 10	~0.10
γ	1750.1 15	0.05 2
γ	1756 1	0.25 5
γ	1768.0 7	~0.06
γ	1772.3 15	~0.10
γ	1797.3 10	0.3 1
γ	1828 1	0.030 6
γ [E2]	1837.9 5	0.06 2
γ	1850 1	0.05 2
γ	1872.8 10	0.07 3
γ	1890.1 10	0.19 6
γ	1897.1 10	0.15 4
γ	1905 1	0.28 6
γ	1926.0 6	0.5 2
γ	1937.7 10	~0.05
γ	1959 1	~0.0010

Photons (^{234}Pa)
(continued)

γ_{mode}	γ(keV)	γ(%)†
γ	1988 1	~0.002
γ	1998 1	~0.0010

† 30% uncert(syst)

Atomic Electrons (^{234}Pa)

⟨e⟩=265 9 keV

e_{bin}(keV)	⟨e⟩(keV)	e(%)
10 - 16	1.09	7.5 15
17	8.7	51 8
19	0.32	1.7 6
21	8.0	38 6
22	1.0	4.4 12
23	7.5	33 8
24 - 25	1.35	5.5 6
26	7.8	30 7
28 - 37	2.4	7.8 16
38	3.5	9.2 23
39 - 69	21.2	43 6
70	4.7	6.7 12
72 - 78	5.0	6.5 11
79	22.1	28 6
80 - 82	0.20	0.25 11
83	16.1	19 4
84 - 94	2.8	3.2 7
95	7.1	7.5 16
96	5.2	5.5 12
97 - 98	0.32	0.33 8
99	3.6	3.6 8
100 - 110	6.1	5.8 8
111	6.0	5.4 11
112	11.5	10.3 21
113 - 131	8.5	6.9 6
132	7.7	5.8 5
133 - 135	0.88	0.66 13
136	4.6	3.4 3
137 - 147	0.98	0.69 12
148	3.7	2.49 19
149 - 176	7.7	4.9 4
178	3.4	1.91 25
179 - 203	4.4	2.3 3
205	5.7	2.8 6
206 - 253	7.2	3.3 3
254	3.6	1.42 15
255 - 304	4.0	1.48 12
305 - 355	3.1	0.90 14
357 - 406	2.1	0.54 15
408 - 453	3.9	0.89 13
454	7.4	1.6 3
455 - 504	1.2	0.24 6
505 - 554	5.2	0.97 14
556 - 606	3.6	0.62 18
607 - 655	3.9	0.6 3
656 - 706	2.6	0.39 11
707 - 756	2.5	0.35 11
757 - 806	3.6	0.46 10
807 - 856	4.0	0.48 13
859 - 907	1.9	0.22 4
908 - 958	0.96	0.103 24
959 - 1008	0.45	0.046 12
1011 - 1057	0.122	0.0119 18
1061 - 1109	0.17	0.016 5
1113 - 1162	0.15	0.013 5
1166 - 1215	0.16	0.013 4
1216 - 1260	0.37	0.030 8
1271 - 1312	0.62	0.049 13
1330 - 1379	0.41	0.030 6
1383 - 1432	0.11	0.0077 23
1434 - 1479	0.094	0.0065 13
1488 - 1584	0.41	0.026 5
1589 - 1683	0.16	0.0098 18
1689 - 1782	0.059	0.0034 9
1789 - 1888	0.07	0.0037 17
1892 - 1984	0.010	~0.0005
1992 - 1994	2.9 $\times 10^{-6}$	~1 $\times 10^{-7}$

Continuous Radiation (^{234}Pa)

⟨β-⟩=215 keV;⟨IB⟩=0.18 keV

E_{bin}(keV)		⟨ ⟩(keV)	(%)
0 - 10	β-	0.182	3.66
	IB	0.0105	
10 - 20	β-	0.53	3.55
	IB	0.0098	0.068
20 - 40	β-	2.04	6.8
	IB	0.018	0.062
40 - 100	β-	13.0	18.8
	IB	0.041	0.065
100 - 300	β-	81	43.2
	IB	0.067	0.041
300 - 600	β-	76	18.5
	IB	0.028	0.0071
600 - 1259	β-	42.0	5.6
	IB	0.0054	0.00077

$^{234}_{91}$Pa(1.17 *1* min)

Mode: β-(99.87 *2* %), IT(0.13 *2* %)
Δ: ~40412 keV
SpA: 6.83×10^{8} Ci/g
Prod: natural source

Photons (^{234}Pa)

⟨γ⟩=12.1 *9* keV

γ_{mode}	γ(keV)	γ(%)†
Pa L_ℓ	11.372	0.00108 *17*
U L_ℓ	11.620	0.0061 *15*
Pa L_α	13.275	0.0181 *23*
U L_α	13.600	0.099 *23*
Pa L_η	14.953	0.00017 *3*
U L_η	15.400	0.0013 *3*
Pa L_β	16.547	0.021 *3*
U L_β	17.069	0.092 *22*
Pa L_γ	19.891	0.0056 *10*
U L_γ	20.366	0.020 *5*
$\gamma_{\beta-}$E1	62.78 *4*	0.0028 *3*
γ_{IT} (M1+1.3%E2)	73.90 *4*	0.0109 *5*
U $K_{\alpha2}$	94.651	0.120 *25*
U $K_{\alpha1}$	98.434	0.19 *4*
$\gamma_{\beta-}$E2	99.852 *10*	0.00037 *3*
U $K_{\beta1}$'	111.025	0.070 *15*
U $K_{\beta2}$'	114.866	0.024 *5*
$\gamma_{\beta-}$E2+39%M1	140.15 *9*	0.00084 *7*
$\gamma_{\beta-}$	184.6 *4*	0.00120 *10*
$\gamma_{\beta-}$[E1]	192.74 *4*	5.4 *24* $\times10^{-5}$
$\gamma_{\beta-}$	193.4 *8*	~0.00044
$\gamma_{\beta-}$[M1+E2]	199.46 *20*	0.00040 *8*
$\gamma_{\beta-}$E2+31%M1	203.20 *3*	0.00098 *20*
$\gamma_{\beta-}$[M1]	210.6 *4*	0.00092 *10*
$\gamma_{\beta-}$E0	234.59 *3*	<0.0010
$\gamma_{\beta-}$[E1]	236.03 *12*	~3 $\times10^{-5}$
$\gamma_{\beta-}$	243.5 *8*	0.00035 *7*
$\gamma_{\beta-}$	247.7 *8*	0.00067 *15*
$\gamma_{\beta-}$[E1]	258.18 *3*	0.0567 *23*
$\gamma_{\beta-}$[E2]	274.45 *21*	~0.00025
$\gamma_{\beta-}$[E1]	276.83 *22*	0.00022 *4*
$\gamma_{\beta-}$[E1]	299.08 *10*	0.00044 *9*
$\gamma_{\beta-}$	311.0 *10*	0.00036 *7*
$\gamma_{\beta-}$[E2]	316.29 *20*	0.000100 *25*
$\gamma_{\beta-}$	338.4 *6*	0.00079 *16*
$\gamma_{\beta-}$	357.9 *3*	0.00056 *12*
$\gamma_{\beta-}$	360.8 *7*	0.00048 *10*
$\gamma_{\beta-}$	387.8 *4*	0.00070 *12*
$\gamma_{\beta-}$[E2]	387.87 *9*	0.00030 *4*
$\gamma_{\beta-}$M1+33%E2	450.92 *4*	0.00211 *17*
$\gamma_{\beta-}$	453.6 *4*	0.00170 *20*
$\gamma_{\beta-}$	456.5 *5*	0.00050 *10*
$\gamma_{\beta-}$	468.21 *22*	0.00165 *16*
$\gamma_{\beta-}$[M1+E2]	476.0 *3*	0.00200 *20*

Photons (^{234}Pa)
(continued)

γ_{mode}	γ(keV)	γ(%)†
$\gamma_{\beta-}$	507.3 *4*	0.00110 *10*
$\gamma_{\beta-}$	509.11 *20*	0.00150 *20*
$\gamma_{\beta-}$(M1)	516.8 *3*	1.2 *3* $\times10^{-5}$
$\gamma_{\beta-}$[M1]	526.20 *24*	<3 $\times10^{-5}$
$\gamma_{\beta-}$[E1]	544.1 *3*	0.0026 *3*
$\gamma_{\beta-}$	557.3 *10*	0.00050 *11*
$\gamma_{\beta-}$(M1)	557.7 *3*	1.4 *4* $\times10^{-5}$
$\gamma_{\beta-}$(M1)	557.7 *3*	1.4 *3* $\times10^{-5}$
$\gamma_{\beta-}$	572.0 *3*	0.00061 *12*
$\gamma_{\beta-}$[E1]	625.5 *3*	0.00100 *10*
$\gamma_{\beta-}$	647.7 *8*	0.00110 *10*
$\gamma_{\beta-}$	649.1 *3*	0.00075 *15*
$\gamma_{\beta-}$[E2]	655.1 *3*	0.00097 *10*
$\gamma_{\beta-}$	671.5 *4*	0.00026 *6*
$\gamma_{\beta-}$[M1+E2]	674.24 *18*	0.00045 *9*
$\gamma_{\beta-}$	683.0 *3*	0.00040 *8*
$\gamma_{\beta-}$	691.0 *3*	0.0055 *5*
$\gamma_{\beta-}$[E2]	695.9 *3*	0.00110 *10*
$\gamma_{\beta-}$	700.4 *4*	0.00056 *11*
$\gamma_{\beta-}$	701.85 *20*	0.0054 *5*
$\gamma_{\beta-}$[E1]	706.02 *9*	0.00271 *24*
$\gamma_{\beta-}$[E2]	708.40 *7*	0.00078 *10*
$\gamma_{\beta-}$[M1+E2]	718.94 *25*	2.4 *5* $\times10^{-5}$
$\gamma_{\beta-}$[M1+E2]	721.33 *25*	2.3 *11* $\times10^{-5}$
$\gamma_{\beta-}$[M1+E2]	732.7 *5*	0.00091 *10*
$\gamma_{\beta-}$	740.9 *4*	0.0071 *7*
$\gamma_{\beta-}$E1	742.817 *20*	0.0565 *21*
$\gamma_{\beta-}$(M1)	750.5 *3*	1.4 *6* $\times10^{-5}$
$\gamma_{\beta-}$	760.3 *10*	0.00110 *10*
$\gamma_{\beta-}$(E2)	766.412 *14*	0.207 *8*
$\gamma_{\beta-}$	782.8 *4*	0.0053 *5*
$\gamma_{\beta-}$[E2]	783.38 *4*	4.9 *11* $\times10^{-5}$
$\gamma_{\beta-}$(E1)	786.287 *20*	0.0342 *12*
$\gamma_{\beta-}$[M1]	792.3 *3*	~6 $\times10^{-6}$
$\gamma_{\beta-}$[E1]	805.87 *9*	0.00309 *25*
$\gamma_{\beta-}$E0+E2	808.25 *7*	0.00219 *17*
$\gamma_{\beta-}$[E2]	818.0 *3*	0.00070 *20*
$\gamma_{\beta-}$[E1]	826.08 *20*	0.00099 *20*
$\gamma_{\beta-}$	831.5 *20*	~0.0025 ?
$\gamma_{\beta-}$	844.4 *5*	0.00076 *16*
$\gamma_{\beta-}$[E2]	851.72 *7*	0.0042 *3*
$\gamma_{\beta-}$[E1]	866.98 *18*	0.00075 *15*
$\gamma_{\beta-}$[M1+E2]	881.1 *3*	0.0027 *3*
$\gamma_{\beta-}$	882.5 *3*	0.0013 *3*
$\gamma_{\beta-}$E2	883.24 *4*	0.00117 *13*
$\gamma_{\beta-}$	887.5 *5*	0.0052 *5*
$\gamma_{\beta-}$[M1+E2]	921.96 *18*	0.0083 *8*
$\gamma_{\beta-}$E2	926.71 *4*	0.00088 *8*
$\gamma_{\beta-}$	936.8 *6*	0.0013 *3*
$\gamma_{\beta-}$[E2]	942.05 *10*	0.00217 *18*
$\gamma_{\beta-}$E1	946.02 *3*	0.0071 *6*
$\gamma_{\beta-}$	959.8 *3*	0.00060 *20*
$\gamma_{\beta-}$[E2]	982.85 *20*	
$\gamma_{\beta-}$[E1]	995.0 *3*	0.0029 *5*
$\gamma_{\beta-}$E2	1001.00 *3*	0.65
$\gamma_{\beta-}$[M1+E2]	1041.90 *10*	0.00097 *8*
$\gamma_{\beta-}$[E1]	1059.72 *19*	0.00077 *15*
$\gamma_{\beta-}$[E2]	1062.11 *19*	0.00140 *10*
$\gamma_{\beta-}$[M1+E2]	1082.7 *2*	0.00063 *9*
$\gamma_{\beta-}$[E2]	1085.37 *10*	0.00035 *5*
$\gamma_{\beta-}$[E2]	1120.5 *5*	0.00120 *10*
$\gamma_{\beta-}$[M1+E2]	1125.16 *18*	0.0021 *4*
$\gamma_{\beta-}$[E2]	1126.17 *20*	0.00067 *10*
$\gamma_{\beta-}$	1174.1 *4*	0.00134 *13*
$\gamma_{\beta-}$E1	1193.74 *3*	0.0090 *5*
$\gamma_{\beta-}$	1220.0 *15*	0.00070 *20*
$\gamma_{\beta-}$E1	1237.21 *3*	0.0036 *3*
$\gamma_{\beta-}$	1353.0 *15*	0.00044 *9*
$\gamma_{\beta-}$E1	1392.0 *3*	0.0018 *4*
$\gamma_{\beta-}$	1414.3 *4*	0.00150 *10*
$\gamma_{\beta-}$E1	1435.4 *3*	0.0057 *5*
$\gamma_{\beta-}$	1457.4 *3*	0.0013 *3*
$\gamma_{\beta-}$	1500.9 *3*	0.00090 *18* ?
$\gamma_{\beta-}$	1510.11 *20*	0.0091 *6*
$\gamma_{\beta-}$M1+E2	1527.20 *24*	0.00155 *20*
$\gamma_{\beta-}$	1549.2 *4*	0.00130 *10*
$\gamma_{\beta-}$	1553.58 *20*	0.0063 *5*
$\gamma_{\beta-}$M1	1558.7 *3*	0.00054 *11*
$\gamma_{\beta-}$M1	1570.67 *24*	0.00076 *11*
$\gamma_{\beta-}$	1592.6 *4*	0.0027 *3*
$\gamma_{\beta-}$(M1)	1602.2 *3*	0.00029 *7*
$\gamma_{\beta-}$[E1]	1667.3 *3*	0.00058 *12*

Photons (^{234}Pa)
(continued)

γ_{mode}	γ(keV)	γ(%)†
$\gamma_{\beta-}$	1693.7 *5*	0.00032 *6*
$\gamma_{\beta-}$	1720.5 *15*	0.00023 *10*
$\gamma_{\beta-}$	1732.2 *15*	0.00130 *20*
$\gamma_{\beta-}$[E2]	1737.8 *3*	0.0142 *6*
$\gamma_{\beta-}$	1759.1 *15*	0.00160 *20*
$\gamma_{\beta-}$	1765.7 *3*	0.0061 *6*
$\gamma_{\beta-}$	1796.2 *6*	0.00022 *4*
$\gamma_{\beta-}$	1809.2 *3*	0.0030 *3*
$\gamma_{\beta-}$	1820.0 *6*	0.00083 *18*
$\gamma_{\beta-}$	1831.7 *4*	0.0112 *4*
$\gamma_{\beta-}$	1863.5 *6*	0.00085 *17*
$\gamma_{\beta-}$[E1]	1867.97 *18*	0.0053 *5*
$\gamma_{\beta-}$	1875.2 *4*	0.0055 *5*
$\gamma_{\beta-}$	1894.1 *4*	0.00150 *10*
$\gamma_{\beta-}$[E1]	1911.44 *18*	0.0037 *4*
$\gamma_{\beta-}$[E1]	1926.4 *5*	0.00031 *6*
$\gamma_{\beta-}$	1937.6 *4*	0.00210 *20*
$\gamma_{\beta-}$[E1]	1969.9 *5*	0.00039 *8*

† uncert(syst): 17% for IT, 14% for β-

Atomic Electrons (^{234}Pa)

⟨e⟩=3.7 *6* keV

e_{bin}(keV)	⟨e⟩(keV)	e(%)
17 - 63	0.086	0.28 *3*
69 - 110	0.042	0.055 *4*
118 - 167	0.0084	0.0061 *7*
171 - 219	0.015	0.0081 *18*
222 - 270	0.0036	0.0015 *4*
271 - 317	0.0008	~0.00029
321 - 371	0.0049	0.0014 *4*
382 - 430	0.0017	0.00042 *18*
432 - 476	0.0016	0.00037 *9*
486 - 534	0.0010	0.00020 *8*
536 - 585	0.0024	~0.00042
586 - 635	0.0058	0.0009 *3*
638 - 687	0.0220	0.00336 *24*
689 - 693	0.00033	4.8 *5* $\times10^{-5}$
694	2.7	0.39 *9*
695 - 743	0.0025	0.00034 *9*
745 - 787	0.0112	0.00148 *14*
788	0.65	0.082 *23*
789 - 835	0.044	0.0055 *19*
839 - 884	0.00080	9 *3* $\times10^{-5}$
885	0.049	0.0055 *8*
886 - 935	0.00089	10 *3* $\times10^{-5}$
936 - 984	0.0136	0.00138 *14*
989 - 1039	0.0048	0.00048 *4*
1040 - 1085	0.00048	4.5 *14* $\times10^{-5}$
1099 - 1125	0.00031	2.8 *8* $\times10^{-5}$
1152 - 1199	0.00012	10 *3* $\times10^{-6}$
1203 - 1237	7.4 $\times10^{-5}$	~6 $\times10^{-6}$
1276 - 1320	0.00032	2.5 *9* $\times10^{-5}$
1331 - 1375	0.00015	~1 $\times10^{-5}$
1385 - 1435	0.0013	~9 $\times10^{-5}$
1436 - 1484	0.0010	7 *3* $\times10^{-5}$
1487 - 1585	0.00059	3.9 *10* $\times10^{-5}$
1587 - 1681	0.0015	9.0 *22* $\times10^{-5}$
1688 - 1787	0.0021	0.00012 *4*
1788 - 1877	0.00058	3.2 *9* $\times10^{-5}$
1889 - 1966	5.5 $\times10^{-5}$	2.9 *10* $\times10^{-6}$

Continuous Radiation (^{234}Pa)

$\langle\beta-\rangle$=819 keV;$\langle$IB$\rangle$=1.66 keV

E_{bin}(keV)		$\langle\ \rangle$(keV)	(%)
0 - 10	β-	0.0229	0.456
	IB	0.032	
10 - 20	β-	0.069	0.462
	IB	0.031	0.22
20 - 40	β-	0.285	0.95
	IB	0.061	0.21
40 - 100	β-	2.13	3.03
	IB	0.170	0.26
100 - 300	β-	23.9	11.8
	IB	0.45	0.26
300 - 600	β-	94	20.7
	IB	0.44	0.105
600 - 1300	β-	410	44.3
	IB	0.42	0.051
1300 - 2281	β-	289	18.4
	IB	0.055	0.0036

$^{234}_{92}$U (2.454 *6* $\times10^5$ yr)

Mode: α
Δ: 38142.0 *22* keV
SpA: 0.006225 Ci/g
Prod: daughter ^{238}Pu; descendant ^{234}Th
%: 0.0055 *5*

Alpha Particles (^{234}U)

$\langle\alpha\rangle$=4773 *2* keV

α(keV)	α(%)
4110.0 *11*	7×10^{-6}
4151.9 *11*	2.6×10^{-5}
4276.6 *11*	4 *1* $\times10^{-5}$
4604.9 *11*	0.24 *3*
4723.8 *11*	27.5 *15*
4776.1 *11*	72.5 *20*

Photons (^{234}U)

$\langle\gamma\rangle$=0.113 *7* keV

γ_{mode}	γ(keV)	γ(%)
γ E2	53.23 *3*	0.119 *10*
γ E2	120.912 *20*	0.041 *4*
γ E1	454.97 *3*	2.6 *3* $\times10^{-5}$
γ [E2]	503.53 *5*	~1×10^{-6}
γ E1	508.20 *3*	1.47 *17* $\times10^{-5}$
γ E2	581.78 *8*	1.2 *5* $\times10^{-5}$
γ E0+E2+M1	624.44 *5*	~8×10^{-7}
γ [E2]	677.67 *5*	~1×10^{-6}
γ	740 *1*	<3×10^{-6} ?

$^{234}_{93}$Np(4.4 *1* d)

Mode: ε
Δ: 39950 *9* keV
SpA: 1.27×10^5 Ci/g
Prod: ^{233}U(d,n); ^{235}U(d,3n); ^{235}U(p,2n); ^{233}U(α,p2n)

Photons (^{234}Np)

$\langle\gamma\rangle$=1104 *37* keV

γ_{mode}	γ(keV)	γ(%)†
U L_ℓ	11.620	1.07 *19*
U L_α	13.600	17 *3*
U L_η	15.400	0.22 *4*
U L_β	17.059	17 *3*
U L_γ	20.401	3.8 *8*
γ E2	43.470 *9*	
U $K_{\alpha2}$	94.651	18 *3*
U $K_{\alpha1}$	98.434	29 *5*
γ E2	99.852 *10*	
U $K_{\beta1}$'	111.025	10.7 *17*
U $K_{\beta2}$'	114.866	3.6 *6*
γ [E1]	192.74 *4*	0.033 *15*
γ	233.56 *20*	
γ [E1]	236.03 *12*	~0.0028
γ	238.6 *4*	
γ	247.5 *4*	0.11 *4*
γ [E1]	258.18 *3*	0.126 *15*
γ	265.8 *5*	<0.038 ?
γ	297.6 *5*	~0.019 ?
γ [E1]	299.08 *10*	~0.019
γ	311.4 *10*	~0.038
γ [E2]	387.87 *9*	0.186 *25*
γ M1+33%E2	450.92 *4*	1.30 *12*
γ	484.4 *8*	0.09 *4*
γ	485.1 *10* ?	
γ (M1)	516.8 *3*	0.38 *6*
γ [M1]	526.20 *24*	<0.19
γ (M1)	557.7 *3*	0.47 *6*
γ [E1]	625.5 *3*	1.10 *16*
γ [E1]	706.02 *9*	0.19 *4*
γ [E2]	708.40 *7*	0.046 *16*
γ [M1+E2]	721.33 *25*	0.17 *8*
γ E1	742.817 *20*	5.0 *4*
γ (M1)	750.5 *3*	0.47 *19*
γ (E2)	766.412 *14*	0.56 *9*
γ (E1)	786.287 *20*	3.00 *21*
γ [M1]	792.3 *3*	~0.19
γ [E1]	805.87 *9*	0.22 *4*
γ E0+E2	808.25 *7*	0.13 *4*
γ [E2]	851.72 *7*	0.24 *7*
γ [E2]	942.05 *10*	~0.09
γ	945.7 *10*	0.56 *19*
γ E2	1001.00 *3*	1.44 *18*
γ [M1+E2]	1041.90 *10*	~0.09
γ [E2]	1085.37 *10*	~0.032
γ	1105.0 *20* ?	
γ E1	1193.74 *3*	5.5 *4*
γ E1	1237.21 *3*	2.23 *21*
γ E1	1392.0 *3*	2.0 *4*
γ E1	1435.4 *3*	6.3 *5*
γ M1+E2	1527.20 *24*	11.5 *10*
γ M1	1558.7 *3*	18.4
γ M1	1570.67 *24*	5.6 *5*
γ (M1)	1602.2 *3*	9.7 *11*

† 8.5% uncert(syst)

Continuous Radiation (^{234}Np)

$\langle\beta+\rangle$=0.175 keV;$\langle$IB$\rangle$=1.08 keV

E_{bin}(keV)		$\langle\ \rangle$(keV)	(%)
0 - 10	β+	8.8×10^{-10}	1.03×10^{-8}
	IB	7.1×10^{-5}	
10 - 20	β+	7.4×10^{-8}	4.33×10^{-7}
	IB	0.0060	0.037
20 - 40	β+	4.15×10^{-6}	1.24×10^{-5}
	IB	0.00112	0.0038
40 - 100	β+	0.000434	0.00054
	IB	0.42	0.48
100 - 300	β+	0.0310	0.0140
	IB	0.42	0.35
300 - 600	β+	0.122	0.0282
	IB	0.055	0.0123
600 - 1300	β+	0.0217	0.00334
	IB	0.160	0.018
1300 - 1808	IB	0.023	0.00160
	Σβ+		0.046

$^{234}_{94}$Pu(8.8 *1* h)

Mode: ε(94 %), α(6 %)
Δ: 40333 *8* keV
SpA: 1.522×10^6 Ci/g
Prod: ^{233}U(α,3n); ^{235}U(α,5n)

Alpha Particles (^{234}Pu)

$\langle\alpha\rangle$=372 keV

α(keV)	α(%)
6035 *3*	0.024
6149 *3*	1.9
6200 *3*	4

$^{234}_{95}$Am(2.6 *2* min)

Mode: ε, α
Δ: 44340 *210* keV syst
SpA: 3.08×10^8 Ci/g
Prod: ^{230}Th(^{10}B,6n); ^{230}Th(^{11}B,7n); ^{10}B on ^{233}U; ^{11}B on ^{233}U

Alpha Particles (^{234}Am)

α(keV)
6460?

A = 235

NDS **21**, 117 (1977)

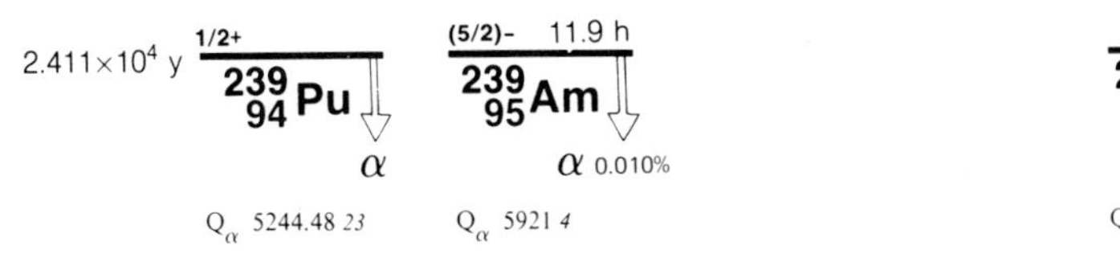

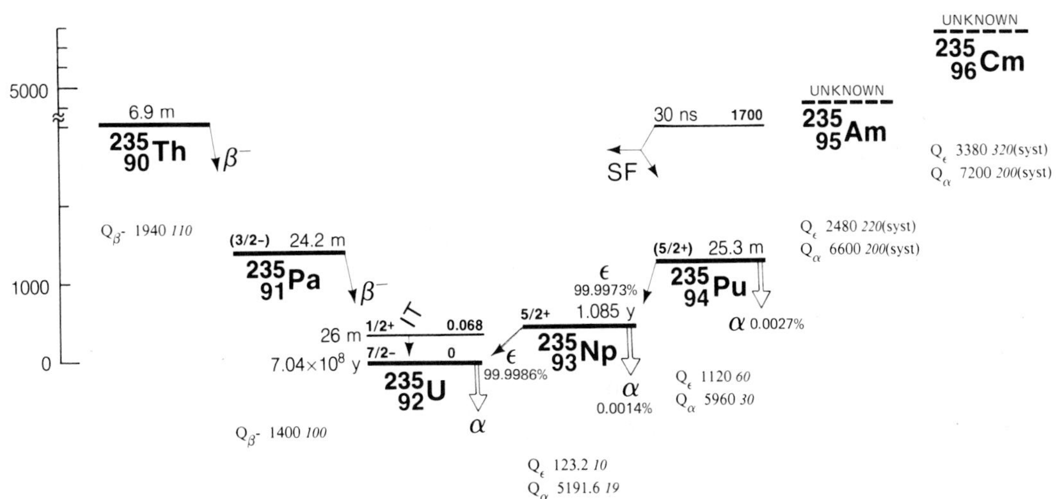

$^{235}_{90}$Th(6.9 *2* min)

Mode: β-
Δ: 44250 *50* keV
SpA: 1.16×10^{8} Ci/g
Prod: ^{238}U(n,α)

Photons (^{235}Th)

γ$_{mode}$	γ(keV)
γ	416.2 *10*
γ	659.4 *10*
γ	727.2 *10*
γ	747 *1*
γ	931.8 *10*

$^{235}_{91}$Pa(24.2 *3* min)

Mode: β-
Δ: 42320 *100* keV
SpA: 3.31×10^{7} Ci/g
Prod: daughter ^{235}Th; ^{238}U(γ,p2n); ^{235}U(n,p); ^{236}U(n,pn)

Photons (^{235}Pa)

γ$_{mode}$	γ(keV)
γ [M1]	30.08 *3*
γ M1+46%E2	38.674 *19*
γ E2	51.622 *20*
γ (E2)	68.75 *3*
γ E1	129.283 *20*
γ	131.8 *10*
γ M1(+E2)	344.94 *3*
γ M1+50%E2	375.018 *21*
γ M1(+E2)	380.17 *3*
γ M1(+E2)	393.12 *4*
γ M1+49%E2	413.691 *21*
γ (E2)	637.96 *8*
γ [E1]	645.98 *7*
γ [E1]	652.18 *7*
γ [E1]	658.93 *7*

Continuous Radiation (^{235}Pa)

⟨β-⟩=470 keV;⟨IB⟩=0.63 keV

E$_{bin}$(keV)		⟨ ⟩(keV)	(%)
0 - 10	β-	0.052	1.03
	IB	0.021	
10 - 20	β-	0.156	1.04
	IB	0.020	0.141
20 - 40	β-	0.63	2.11
	IB	0.039	0.134
40 - 100	β-	4.63	6.6
	IB	0.102	0.158
100 - 300	β-	47.2	23.5
	IB	0.23	0.134
300 - 600	β-	147	33.1
	IB	0.162	0.040
600 - 1300	β-	268	32.5
	IB	0.059	0.0079
1300 - 1407	β-	1.59	0.120
	IB	1.0×10^{-5}	7.6×10^{-7}

$^{235}_{92}$U (7.037 *11* ×10^{8} yr)

Mode: α
Δ: 40915.5 *22* keV
SpA: 1.922×10^{-6} Ci/g
Prod: natural source
%: 0.7200 *12*

Alpha Particles (^{235}U)

⟨α⟩=4378 *5* keV

α(keV)	α(%)
4152.5 *9*	0.9 *2*
4215.7 *9*	5.7 *6*
4225.8 *9* ?	~0.9
4271 *5*	~0.40
4285.3 *9* ?	
4323.7 *9*	4.6 *5*
4344 ?	~2
4364.1 *9*	~11
4370 *4* ?	~6
4395.2 *9*	55 *3*
4414.4 *9*	2.1 *2*
4437.8 *9*	~0.7
4502.5 *9*	1.7 *2*
4555.8 *9*	4.2 *3*
4597.0 *9*	5.0 *5*

Photons (^{235}U)

⟨γ⟩=156 *12* keV

γ$_{mode}$	γ(keV)	γ(%)†
Th L$_{\ell}$	11.118	~1
Th L$_{\alpha}$	12.952	~22
Th L$_{\eta}$	14.511	0.22 *6*
Th L$_{\beta}$	16.119	15 *6*
Th L$_{\gamma}$	19.118	2.6 *7*

Photons (^{235}U)
(continued)

γ_{mode}	γ(keV)	γ(%)†
γ [M1+~20%E2]	31.585 14	0.016 5
γ [M1]	41.13 18	~0.03
γ [E2+~20%M1]	41.954 12	~0.04
γ [E2]	51.179 14	~0.020
γ [M1+E2]	54.201 18	<0.030
γ [M1+E2]	64.348 21	~0.020
γ [E2]	72.71 18	~0.11
γ [E1]	74.98 3	0.06 1
Th $K_{\alpha2}$	89.955	3.36 21
Th $K_{\alpha1}$	93.350	5.5 3
γ [M1+E2]	95.72 8	
γ [E2]	96.154 16	0.086 11
Th $K_{\beta1}$'	105.362	1.98 12
Th $K_{\beta2}$'	108.990	0.66 4
γ [E1]	109.178 15	1.5 2
γ [E1]	116.11 18	~0.07
γ [E2]	119.98 3	~0.026
γ [M1+E2]	136.72 6	~0.012
γ [E1]	140.763 19	0.22 3
γ [E1]	143.786 12	10.5 8
γ [E1]	146.988 23	
γ [M1]	150.957 14	0.076 10
γ [E1]	163.379 12	4.7 4
γ [E1]	172.27 5	~0.010
γ [E1]	181.89 18 ?	
γ [M1]	182.542 15	0.40 5
γ (E1)	185.739 5	53 7
γ [E1]	194.964 9	0.59 6
γ [M1]	198.933 19	0.040 5
γ [M1]	202.135 13	1.0 1
γ [E1]	205.333 9	4.7 4
γ [M1]	215.305 19	0.027 3
γ [M1]	221.406 20	0.10 1
γ [M1]	228.77 4	0.008 3
γ [M1]	233.52 3	0.04 1
γ [M1]	240.886 22	0.068 13
γ [M1]	246.890 20	0.06 2
γ [E2]	266.483 20	0.006 2
γ [M1]	275.47 3	0.049 5
γ [M1]	281.45 5	~0.006
γ [M1]	282.98 4	0.005 2
γ [M1]	289.57 4	~0.007
γ [E1]	290.29 4	
γ [E1]	291.720 19	~0.0030
γ	301.72 10	~0.0050
γ	310.71 6	~0.004
γ [M1]	317.12 8	~0.0010
γ [E2]	343.78 4	~0.0030
γ [E1]	345.921 14	0.038 5
γ [E1]	356.068 23	~0.0050
γ [E1]	387.874 13	0.038 5
γ	390.32 20	0.04 1 ?
γ [E1]	410.269 19	~0.0030
γ	448.42 6	~0.0010
γ	455.12 10	~0.008
γ	517.2 10	~0.0004
γ	742.52 20	~0.0004 ?
γ	794.72 10	~0.0006 ?

† 10% uncert(syst)

Atomic Electrons (^{235}U)
$\langle e \rangle$=42 6 keV

e_{bin}(keV)	$\langle e \rangle$(keV)	e(%)
6 - 14	0.62	5.2 12
15	1.8	~12
16	6.6	~41
18 - 19	1.2	~6
20	2.1	10 3
21	0.38	~2
22	2.0	~9
25	0.0018	~0.007
26	2.2	8 4
27 - 37	4.1	12 3
38	1.1	~3
39 - 52	2.2	5.0 12
53	1.3	~2
54 - 55	0.33	0.61 9
56	1.0	~2
59 - 75	2.2	3.2 5
76	4.0	5.3 7
77 - 91	0.71	0.84 6
92	2.11	2.28 25
93 - 141	1.89	1.65 8
142 - 163	0.66	0.43 3
165	0.98	0.59 8
166 - 215	2.79	1.55 7
216 - 265	0.157	0.068 6
266 - 314	0.039	~0.014
316 - 356	0.005	~0.0016
367 - 410	0.009	~0.0024
428 - 455	0.0014	~0.00032
497 - 517	4.8 $\times10^{-5}$	~10 $\times10^{-6}$
685 - 726	0.00014	~2 $\times10^{-5}$
737 - 778	3.1 $\times10^{-5}$	~4 $\times10^{-6}$
790 - 795	8.4 $\times10^{-6}$	~1 $\times10^{-6}$

$^{235}_{92}$U (26 2 min)

Half-life dependent on chemical environment
Mode: IT
Δ: 40915.6 22 keV
SpA: 3.08$\times10^{7}$ Ci/g
Prod: daughter ^{239}Pu

$^{235}_{93}$Np(1.085 3 yr)

Mode: ϵ(99.9986 2 %), α(0.0014 2 %)
Δ: 41038.7 24 keV
SpA: 1402 Ci/g
Prod: ^{235}U(d,2n); daughter ^{235}Pu; ^{233}U(α,pn); ^{235}U(α,p3n)

Alpha Particles (^{235}Np)
$\langle \alpha \rangle$=0.070 1 keV

α(keV)	α(%)
4806 7	~1 $\times10^{-6}$
4861.1 9	9.8 14 $\times10^{-6}$
4923.8 9	0.000161 7
4937.7 10	~8 $\times10^{-6}$
4994.5 9	~8 $\times10^{-5}$
5003.7 9 ?	<7 $\times10^{-6}$
5004.5 9	0.00034 11
5021.4 9	0.00074 11
5046.6 9	2.5 4 $\times10^{-5}$
5095.2 9	~3 $\times10^{-6}$
5104.2 9	2.1 3 $\times10^{-5}$

Photons (^{235}Np)
$\langle \gamma \rangle$=7.6 4 keV

γ_{mode}	γ(keV)	γ(%)†
Pa L_ℓ	11.372	2.7 10 $\times10^{-5}$
U L_ℓ	11.620	0.83 10
Pa L_α	13.274	0.00044 16
U L_α	13.600	13.5 12
Pa L_η	14.953	6.8 20 $\times10^{-6}$
U L_η	15.400	0.113 12
Pa L_β	16.589	0.00051 17
U L_β	17.017	15.9 22
Pa L_γ	19.805	0.00012 4
U L_γ	20.509	4.4 7
γ_αE1	25.642 11	0.000214 19
γ_α[E1]	42.828 14	
γ_αE2	58.562 11	8 3 $\times10^{-6}$
γ_αM1+19%E2	63.834 12	~1 $\times10^{-7}$
γ_αE2	68.50 3	
γ_α[E1]	72.767 12	~2 $\times10^{-7}$
γ_α(M1+7.3%E2)	81.229 12	2.07 13 $\times10^{-5}$
γ_α(M1+3.6%E2)	82.093 12	1.14 23 $\times10^{-5}$
γ_αE1	84.203 9	9.7 6 $\times10^{-5}$
γ_α(E1)	89.954 8	~6 $\times10^{-7}$
Pa $K_{\alpha2}$	92.279	1.8 6 $\times10^{-6}$
γ_α(E1)	93.07 3	5.2 12 $\times10^{-7}$?
U $K_{\alpha2}$	94.651	0.596 12
Pa $K_{\alpha1}$	95.863	2.9 10 $\times10^{-6}$
U $K_{\alpha1}$	98.434	0.957 19
γ_αM1+20%E2	99.28 1	2.8 3 $\times10^{-6}$
γ_α(E1)	102.255 12	4.3 9 $\times10^{-6}$
γ_α[E1]	105.803 18	1.64 21 $\times10^{-7}$
Pa $K_{\beta1}$'	108.166	1.0 4 $\times10^{-6}$
γ_α(E2)	110.8 4	1.3 4 $\times10^{-6}$
U $K_{\beta1}$'	111.025	0.350 7
Pa $K_{\beta2}$'	111.897	3.5 12 $\times10^{-7}$
U $K_{\beta2}$'	114.866	0.118 3
γ_α[M1+E2]	115.595 12	<12 $\times10^{-10}$
γ_αE1	124.922 12	1.29 11 $\times10^{-6}$
γ_αM1+12%E2	135.681 17	~4 $\times10^{-7}$
γ_α[M1+E2]	145.063 14	~3 $\times10^{-8}$
γ_αM1+39%E2	145.927 13	~1 $\times10^{-7}$
γ_αM1+6.9%E2	163.114 11	~7 $\times10^{-7}$
γ_α[E2]	164.97 3	~2 $\times10^{-9}$
γ_α[E1]	169.637 18	~5 $\times10^{-9}$
γ_α[M1+E2]	174.157 9	~1 $\times10^{-8}$
γ_αE1	183.483 11	7.6 6 $\times10^{-7}$
γ_α[E1]	188.756 12	~1 $\times10^{-8}$

† 14% uncert(syst)

Atomic Electrons (^{235}Np)
$\langle e \rangle$=2.94 23 keV

e_{bin}(keV)	$\langle e \rangle$(keV)	e(%)
2 - 16	0.00013	0.0015 5
17	0.085	0.50 5
18 - 20	9.0 $\times10^{-5}$	0.00044 16
21	0.55	2.6 3
22	2.26	10.4 10
23 - 72	0.0059	0.0087 7
73 - 121	0.0430	0.0517 20
124 - 173	1.9 $\times10^{-6}$	1.3 3 $\times10^{-6}$
174 - 189	9.1 $\times10^{-9}$	5.1 3 $\times10^{-9}$

Continuous Radiation (^{235}Np)

⟨IB⟩=0.143 keV

E_{bin}(keV)		⟨ ⟩(keV)	(%)
10 - 20	IB	0.0140	0.086
20 - 40	IB	0.00165	0.0058
40 - 100	IB	0.052	0.066
100 - 124	IB	0.075	0.055

$^{235}_{94}$Pu(25.3 *5* min)

Mode: ε(99.9973 *10* %), α(0.0027 *5* %)
Δ: 42160 *60* keV
SpA: 3.16×10^7 Ci/g
Prod: ^{235}U(α,4n); ^{233}U(α,2n)

Alpha Particles (^{235}Pu)

α(keV)
5850 *20*

Photons (^{235}Pu)

⟨γ⟩=94 *9* keV

γ_{mode}	γ(keV)	γ(%)†
Np L_ℓ	11.871	1.5 *3*
Np L_α	13.927	24 *5*
Np L_η	15.861	0.30 *7*
Np L_β	17.549	24 *6*
Np L_γ	21.027	5.4 *13*
γ M1+~1.4%E2	34.36 *9*	0.23 *4*
γ [M1+E2]	44.8 *4*	
γ E1	49.26 *10*	2.36 *6*
γ [E2]	79.2 *4*	
Np $K_{\alpha2}$	97.066	21 *5*
Np $K_{\alpha1}$	101.059	34 *7*
Np $K_{\beta1}'$	113.944	13 *3*
Np $K_{\beta2}'$	117.891	4.2 *9*
γ [M2]	722.2 *3*	0.0048 *12*
γ [E1]	740.1 *3*	0.0120 *24*
γ [E1]	745.33 *22*	0.090 *8*
γ [E1]	756.6 *3*	0.479 *17*
γ [E1]	779.69 *22*	0.037 *4*
γ [E1]	784.9 *3*	0.0072 *24*
γ [E1]	819.3 *3*	<0.0024 ?
γ [M2]	857.9 *4*	0.011 *4*
γ	902.7 *3*	0.0180 *24*
γ	910.38 *22*	0.164 *5*
γ	937.1 *3*	0.011 *4*
γ	940.9 *3*	0.114 *5*
γ	944.74 *22*	0.114 *5*

† 17% uncert(syst)

Atomic Electrons (^{235}Pu)

⟨e⟩=15.7 *13* keV

e_{bin}(keV)	⟨e⟩(keV)	e(%)
12	2.3	20 *4*
13	0.65	5.1 *10*
17	0.49	2.9 *6*
18	3.3	19 *4*
22	3.6	16 *4*
27 - 28	0.260	0.954 *22*
29	1.7	6.1 *11*
30 - 32	0.42	1.37 *16*
33	0.62	1.9 *4*
34 - 83	1.38	2.34 *20*
91 - 113	0.74	0.78 *8*
604 - 638	0.0253	0.0040 *3*
661 - 705	0.0031	0.00045 *6*
717 - 763	0.016	0.0021 *5*
767 - 816	0.04	~0.005
818 - 858	0.05	~0.006
880 - 927	0.021	0.0023 *9*
931 - 945	0.0033	0.00035 *17*

Continuous Radiation (^{235}Pu)

⟨IB⟩=1.23 keV

E_{bin}(keV)		⟨ ⟩(keV)	(%)
10 - 20	IB	0.0041	0.025
20 - 40	IB	0.00090	0.0031
40 - 100	IB	0.47	0.53
100 - 300	IB	0.59	0.51
300 - 600	IB	0.103	0.024
600 - 1130	IB	0.064	0.0089

A = 236

NDS **36**, 402 (1982)

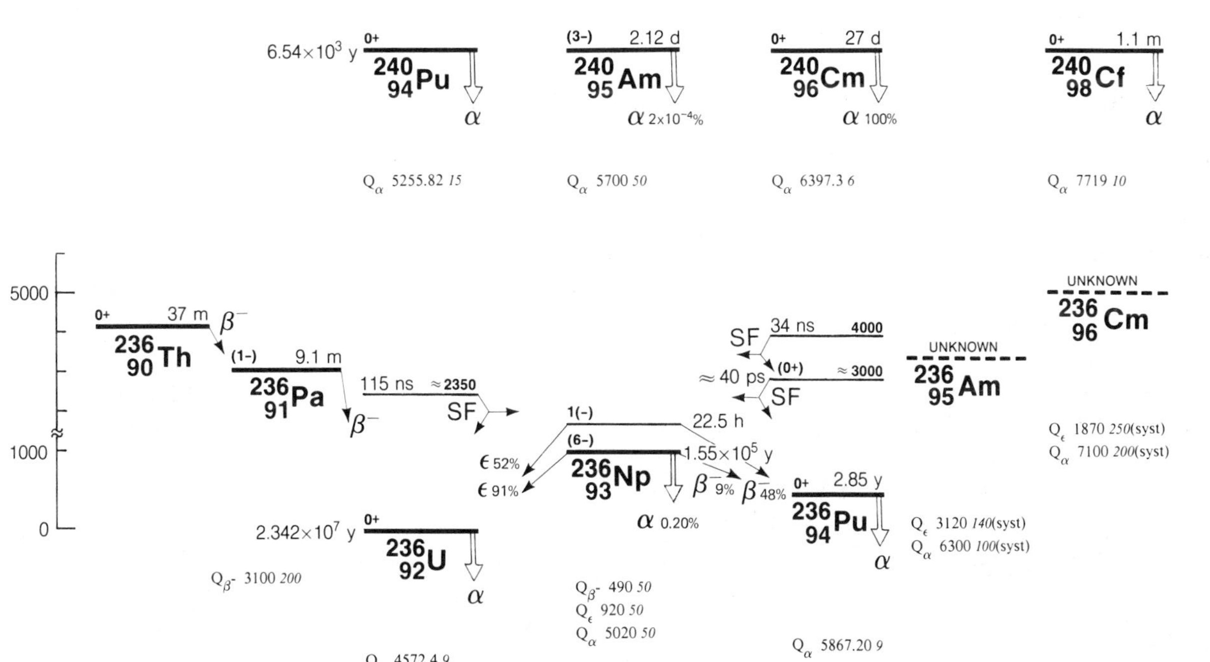

$^{236}_{90}$Th(37.1 *15* min)

Mode: β-
SpA: 2.15×10^{7} Ci/g
Prod: ^{238}U(γ,2p); ^{238}U(p,3p)

Photons (^{236}Th)

⟨γ⟩=21 *4* keV

γ_{mode}	γ(keV)	γ(%)†
Pa L_ℓ	11.372	0.28 *10*
Pa L_α	13.274	4.6 *16*
Pa L_η	14.953	0.051 *20*
Pa L_β	16.567	4.7 *17*
Pa L_γ	19.837	1.1 *4*
Pa $K_{\alpha2}$	92.279	3.4 *16*
Pa $K_{\alpha1}$	95.863	5.6 *25*
Pa $K_{\beta1}$'	108.166	2.0 *9*
γ [M1]	110.7 *5*	4.0 *11*
Pa $K_{\beta2}$'	111.897	0.7 *3*
γ [M1]	112.7 *5*	1.0 *5*
γ [E1]	131.6 *10*	~0.8
γ [E1]	229.6 *10*	~0.8

† 34% uncert(syst)

Atomic Electrons (^{236}Th)

⟨e⟩=17 *3* keV

e_{bin}(keV)	⟨e⟩(keV)	e(%)
2	0.118865	5
17	0.50	3.0 *14*
19 - 21	0.81	3.9 *9*
70 - 89	0.21	0.27 *5*
90	9.0	10 *3*
91	0.020	0.022 *10*
92	2.2	2.4 *11*
94 - 104	0.064	0.067 *15*
105	2.3	2.2 *6*
106	0.28	0.26 *7*
107	0.6	0.6 *3*
109	0.72	0.66 *18*
110 - 132	0.34	0.30 *6*
208 - 230	0.024	0.011 *3*

$^{236}_{91}$Pa(9.1 *2* min)

Mode: β-
Δ: 45540 *200* keV
SpA: 8.75×10^{7} Ci/g
Prod: ^{238}U(d,α); ^{236}U(n,p)

Photons (^{236}Pa)

⟨γ⟩=505 *44* keV

γ_{mode}	γ(keV)	γ(%)†
U L_ℓ	11.620	0.087 *14*
U L_α	13.600	1.42 *20*
U L_η	15.400	0.018 *3*
U L_β	17.060	1.30 *21*
U L_γ	20.384	0.28 *5*
γ E2	45.244 *6*	
γ (E2)	56.5 *5*	
γ	69.1 *10*	
U $K_{\alpha2}$	94.651	1.64 *21*
U $K_{\alpha1}$	98.434	2.6 *3*
γ E2	104.235 *5*	
U $K_{\beta1}$'	111.025	0.97 *12*
U $K_{\beta2}$'	114.866	0.33 *4*
γ E2+30%M1	244.0 *5*	0.15 *3*
γ [M1+E2]	279.42 *24*	0.45 *8*
γ (E2)	300.4 *4*	0.079 *17*
γ [M1]	366.7 *7*	~0.45
γ [M1]	423.1 *6*	0.51 *10*
γ	526.8 *10*	
γ E3	538.11 *5*	0.34 *5*
γ	550.7 *10*	
γ [E1]	594.6 *5*	~0.15
γ E1+M2	642.35 *5*	29
γ E1	687.59 *5*	7.8 *9*
γ [E1]	698.8 *5*	
γ	740.6 *10*	
γ	861 *1*	
γ	870.7 *10*	
γ [E2]	873.94 *12*	~0.30
γ	917.1 *10*	
γ E1	921.77 *24*	0.26 *5*
γ E1	942.8 *4*	0.38 *7*
γ E1	967.01 *24*	0.57 *10*
γ	991.2 *10*	
γ	1023.1 *10*	
γ [E1]	1065.5 *6*	0.24 *5*
γ	1155.9 *10*	
γ	1177.9 *10*	
γ	1226.2 *10*	
γ	1235.2 *10*	
γ	1284.2 *10*	
γ	1291.8 *10*	
γ	1518.1 *10*	
γ	1560.1 *10*	1.9 *4*
γ	1587.2 *10*	
γ	1617.4 *7*	0.87 *17*
γ	1662.6 *7*	0.51 *10*
γ	1749.1 *10*	
γ	1763.0 *7*	5.4 *11*
γ	1773.9 *10*	
γ	1808.2 *7*	2.0 *4*
γ	1865.5 *10*	
γ	1907.4 *10*	
γ	1927 *1*	
γ	1934.6 *10*	
γ	1948.6 *10*	
γ	1973.2 *10*	
γ	1981.2 *10*	
γ	2041.6 *7*	1.6 *3*
γ [E2]	2078.1 *7*	0.090 *18*
γ	2086.8 *7*	0.81 *16*
γ	2182.3 *7*	0.045 *9*

† 33% uncert(syst)

Atomic Electrons (^{236}Pa)

⟨e⟩=43 *4* keV

e_{bin}(keV)	⟨e⟩(keV)	e(%)
17 - 22	0.43	2.30 *25*
72 - 110	0.132	0.162 *9*
128 - 164	0.5	~0.34
185 - 227	0.085	0.039 *9*
238 - 283	0.9	0.34 *12*
295 - 308	0.54	0.18 *4*
345 - 367	0.20	0.057 *20*
401 - 423	0.27	0.065 *7*
479 - 521	0.112	0.0218 *21*
527	17.0	3.2 *6*
533 - 538	0.041	0.0077 *8*
572	9.8	1.72 *19*
573 - 595	0.0016	0.00028 *9*
621	5.5	0.89 *18*
637	1.9	0.29 *6*
666	3.6	0.53 *6*
682	1.20	0.176 *19*
758 - 806	0.032	0.0042 *16*
827 - 874	0.037	0.0044 *5*
900 - 946	0.0074	0.00080 *6*
950 - 967	0.0075	0.00079 *14*
1044 - 1065	0.00153	0.000146 *21*
1445 - 1543	0.27	~0.018
1547 - 1645	0.07	0.0047 *22*
1647 - 1746	0.6	~0.037
1757 - 1804	0.053	0.0030 *14*
1926 - 2024	0.16	0.008 *4*
2036 - 2083	0.019	0.0009 *4*
2161 - 2178	0.0006	~3×10^{-5}

Continuous Radiation (^{236}Pa)

⟨β-⟩=591 keV;⟨IB⟩=1.25 keV

E_{bin}(keV)		⟨ ⟩(keV)	(%)
0 - 10	β-	0.0188	0.373
	IB	0.023	
10 - 20	β-	0.057	0.378
	IB	0.022	0.155
20 - 40	β-	0.232	0.77
	IB	0.043	0.150
40 - 100	β-	1.72	2.45
	IB	0.121	0.19
100 - 300	β-	18.6	9.2
	IB	0.32	0.18
300 - 600	β-	68	15.0
	IB	0.32	0.076
600 - 1300	β-	259	28.1
	IB	0.33	0.039
1300 - 2500	β-	238	14.4
	IB	0.072	0.0046
2500 - 3100	β-	6.2	0.233
	IB	0.00034	1.31×10^{-5}

$^{236}_{92}$U (2.342 *3* ×10^{7} yr)

Mode: α
Δ: 42441.7 *21* keV
SpA: 6.508×10^{-5} Ci/g
Prod: ^{235}U(n,γ)

Alpha Particles (^{236}U)

⟨α⟩=4479 *3* keV

α(keV)	α(%)
4332 *8*	0.26 *1*
4445 *5*	26 *4*
4494 *3*	74 *4*

Photons (^{236}U)

⟨γ⟩=1.5 *2* keV

γ_{mode}	γ(keV)	γ(%)†
Th L_ℓ	11.118	0.20 *5*
Th L_α	12.952	3.4 *7*
Th L_η	14.511	0.095 *21*
Th L_β	16.161	4.7 *11*
Th L_γ	19.094	1.05 *24*

Photons (^{236}U) (continued)

γ_{mode}	γ(keV)	γ(%)†
γ E2	49.369 *9*	0.078
γ E2	112.750 *15*	0.019 *2*

† 17% uncert(syst)

Atomic Electrons (^{236}U)

$\langle e \rangle$=10.8 *10* keV

e_{bin}(keV)	$\langle e \rangle$(keV)	e(%)
16	0.77	4.7 *11*
20	0.82	4.2 *9*
29	0.093	0.32 *7*
30	3.0	10.0 *20*
33	2.9	8.8 *18*
44	0.042	0.096 *19*
45	2.3	5.1 *10*
46	0.029	0.062 *12*
48	0.68	1.4 *3*
49	0.20	0.41 *8*

$^{236}_{93}$Np(1.550 *10* $\times 10^5$ yr)

Mode: ϵ(91 *2* %), β-(8.9 *20* %), α(0.20 *5* %)
Δ: 43370 *50* keV
SpA: 0.00977 Ci/g
Prod: ^{235}U(d,n)

Photons (^{236}Np)

$\langle \gamma \rangle$=141 *12* keV

γ_{mode}	γ(keV)	γ(%)†
U L_ℓ	11.620	2.8 *5*
U L_α	13.600	46 *7*
U L_η	15.400	1.07 *17*
U L_β	17.118	57 *9*
U L_γ	20.316	12.9 *22*
γ_ϵE2	45.244 *6*	0.150 *9*
U $K_{\alpha2}$	94.651	21 *4*
U $K_{\alpha1}$	98.434	33 *7*
γ_ϵE2	104.235 *5*	7.5 *15*
U $K_{\beta1}$'	111.025	12.1 *24*
U $K_{\beta2}$'	114.866	4.1 *8*
γ_ϵE2	160.312 *8*	28 *6*

† 2.2% uncert(syst)

Atomic Electrons (^{236}Np)

$\langle e \rangle$=188 *11* keV

e_{bin}(keV)	$\langle e \rangle$(keV)	e(%)
17	8.6	50 *8*
21	8.6	41 *6*
22 - 23	1.13	5.1 *8*
24	8.4	34.4 *22*
28	8.5	30.3 *19*
40 - 82	15.2	33.4 *16*
83	29.4	35 *7*
87	20.6	24 *5*
88 - 98	0.64	0.70 *6*
99	10.0	10.1 *20*
100	6.7	6.7 *14*
101	0.16	0.16 *3*
103	4.8	4.6 *9*
104 - 110	1.5	1.4 *3*
139	29.5	21 *4*
143	15.2	10.6 *21*
155	8.9	5.7 *12*
156 - 160	9.8	6.2 *8*

Continuous Radiation (^{236}Np)

$\langle \beta$-$\rangle$=8.7 keV;$\langle$IB$\rangle$=0.90 keV

E_{bin}(keV)		$\langle\ \rangle$(keV)	(%)
0 - 10	β-	0.0310	0.62
	IB	0.00050	
10 - 20	β-	0.090	0.60
	IB	0.0045	0.027
20 - 40	β-	0.339	1.13
	IB	0.00148	0.0051
40 - 100	β-	1.96	2.86
	IB	0.43	0.49
100 - 300	β-	6.2	3.76
	IB	0.45	0.40
300 - 600	β-	0.097	0.0311
	IB	0.0147	0.0039
600 - 771	IB	0.00017	2.7×10^{-5}

$^{236}_{93}$Np(22.5 *4* h)

Mode: ϵ(52 *1* %), β-(48 *1* %)
Δ: >43370 keV
SpA: 5.90×10^5 Ci/g
Prod: ^{235}U(d,n); ^{235}U(α,p2n)

Photons (^{236}Np)

$\langle \gamma \rangle$=49.7 *14* keV

γ_{mode}	γ(keV)	γ(%)†
U L_ℓ	11.620	0.64 *10*
Pu L_ℓ	12.124	0.081 *19*
U L_α	13.600	10.4 *14*
Pu L_α	14.262	1.3 *3*
U L_η	15.400	0.15 *3*
Pu L_η	16.333	0.030 *7*
U L_β	17.077	10.1 *18*
Pu L_β	18.136	1.5 *4*
U L_γ	20.359	2.2 *4*
Pu L_γ	21.554	0.34 *8*
γ_{β}-E2	44.59 *10*	0.0110 *22*
γ_ϵE2	45.244 *6*	~0.014
U $K_{\alpha2}$	94.651	11.1 *6*
U $K_{\alpha1}$	98.434	17.8 *10*
U $K_{\beta1}$'	111.025	6.5 *4*
U $K_{\beta2}$'	114.866	2.20 *13*
γ_ϵE3	538.11 *5*	0.0109 *9*
γ_ϵE1+M2	642.35 *5*	0.92
γ_ϵE1	687.59 *5*	0.250 *5*

† uncert(syst): 6.0% for ϵ, 2.1% for β-

Atomic Electrons (^{236}Np)

$\langle e \rangle$=11.0 *9* keV

e_{bin}(keV)	$\langle e \rangle$(keV)	e(%)
17	1.8	10.7 *18*
18	0.24	1.3 *3*
21	1.21	5.8 *12*
22	1.32	6.0 *11*
23	0.031	0.13 *7*
24	0.8	~3
27	0.73	2.8 *6*
28	0.8	~3
39	0.35	0.90 *18*
40	0.7	1.7 *7*
41	0.33	~0.8
42 - 91	1.23	2.1 *3*
92 - 110	0.201	0.208 *10*
516 - 521	0.00339	0.00066 *4*
527	0.54	0.103 *11*
533 - 572	0.314	0.0550 *11*
621 - 666	0.349	0.055 *3*
682	0.0382	0.00561 *11*

Continuous Radiation (^{236}Np)

$\langle \beta$-$\rangle$=67 keV;$\langle$IB$\rangle$=0.59 keV

E_{bin}(keV)		$\langle\ \rangle$(keV)	(%)
0 - 10	β-	0.110	2.20
	IB	0.0035	
10 - 20	β-	0.324	2.16
	IB	0.0053	0.035
20 - 40	β-	1.26	4.21
	IB	0.0058	0.020
40 - 100	β-	8.0	11.6
	IB	0.26	0.30
100 - 300	β-	43.8	23.9
	IB	0.28	0.24
300 - 600	β-	13.8	3.96
	IB	0.034	0.0082
600 - 920	IB	0.0066	0.00102

$^{236}_{94}$Pu(2.851 *8* yr)

Mode: α
Δ: 42879 *6* keV
SpA: 531.3 Ci/g
Prod: daughter ^{236}Np; ^{235}U(α,3n)

Alpha Particles (^{236}Pu)

$\langle \alpha \rangle$=5752 *1* keV

α(keV)	α(%)
5045.6 *7* ?	$<2\times10^{-5}$
5088.2 *7*	0.0006
5149.4 *7* ?	$<15\times10^{-6}$
5214.1 *7*	0.00027 *3*
5451.9 *8*	0.0020
5613.8 *7*	0.18
5721.0 *7*	31.7 *9*
5767.7 *7*	68.1 *8*

Photons (^{236}Pu)

$\langle\gamma\rangle$=2.0 *3* keV

γ_{mode}	γ(keV)	γ(%)†
U L_ℓ	11.620	0.29 *7*
U L_α	13.600	4.7 *10*
U L_η	15.400	0.12 *3*
U L_β	17.129	5.9 *13*
U L_γ	20.292	1.3 *3*
γ E2	47.579 *9*	0.069 *14*
γ E2	109.001 *8*	0.0120 *24*
γ E2	164.7 *5*	0.00066 *13*
γ E1	515.653 *6*	0.000159 *20*
γ E1	563.231 *7*	0.000106 *13*
γ (E2)	643.68 *24*	0.00024 *5*

† 30% uncert(syst)

Atomic Electrons (^{236}Pu)

$\langle e\rangle$=12.6 *11* keV

e_{bin}(keV)	$\langle e\rangle$(keV)	e(%)
17	0.95	5.5 *12*
21	0.96	4.6 *10*
22 - 26	0.127	0.51 *9*
27	3.3	12.5 *25*
30	3.3	10.8 *22*
42	1.5	3.5 *7*
43	1.3	3.0 *6*
44	0.034	0.077 *16*
46	0.44	0.95 *19*
47	0.60	1.3 *3*
48 - 95	0.107	0.154 *19*
97 - 144	0.032	0.030 *3*
147 - 165	0.00074	0.00048 *5*
400 - 448	1.08 $\times10^{-5}$	2.59 *24* $\times10^{-6}$
494 - 542	2.6 $\times10^{-5}$	5.0 *9* $\times10^{-6}$
546 - 576	0.0011	~0.00019
622 - 670	0.00029	~4$\times10^{-5}$

A = 237

NDS **23**, 71 (1978)

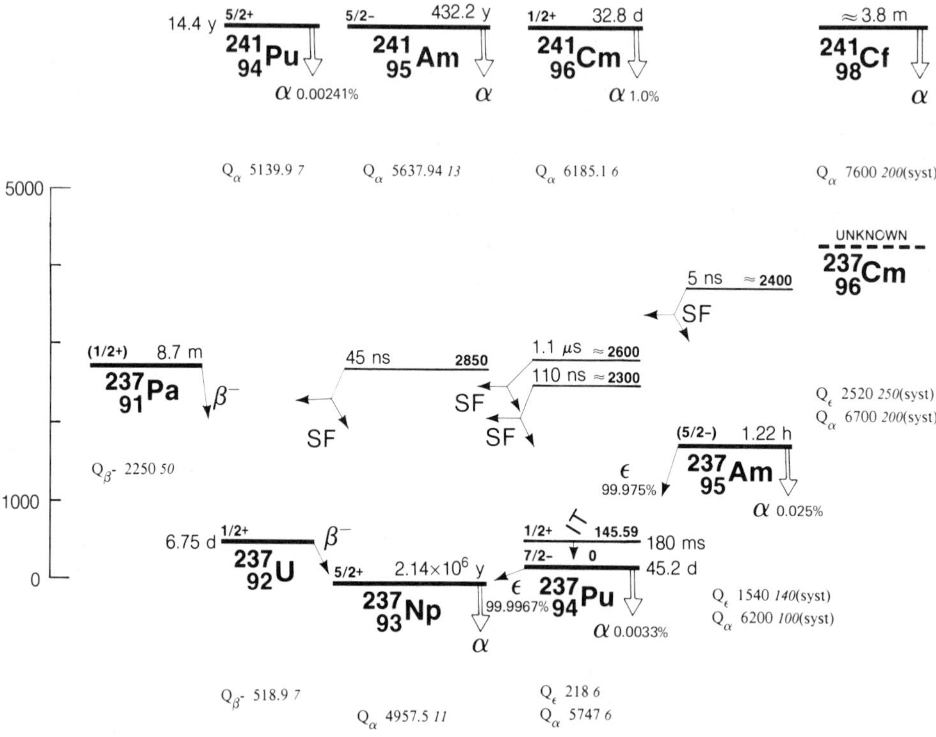

$^{237}_{91}$Pa(8.7 *2* min)

Mode: β-
Δ: 47640 *50* keV
SpA: 9.11×10^{7} Ci/g
Prod: ^{238}U(d,2pn); ^{238}U(γ,p); ^{238}U(n,pn)

Photons (^{237}Pa)

⟨γ⟩=604 *59* keV

γ_{mode}	γ(keV)	γ(%)
γ [M1+E2]	44.887 *11*	
γ	179.05 *14*	0.17 *3*
γ [M1+E2]	310.09 *14*	1.73 *24*
γ [E1]	498.64 *11*	2.4 *3*
γ [E1]	529.32 *14*	14.8 *15*
γ [E1]	540.71 *14*	9.3 *9*
γ [E1]	543.52 *11*	0.24 *10*
γ [E1]	554.91 *11*	1.53 *17*
γ	701.0 *5*	~0.14
γ	722.57 *12*	0.82 *14*
γ	733.96 *12*	0.65 *14*
γ [E1]	847.1 *5*	0.51 *17*
γ [E1]	853.61 *12*	34
γ [E1]	865.00 *12*	15.5 *3*
γ	1333.3 *4*	~0.17
γ	1344.7 *4*	~0.10
γ	1396.0 *4*	~0.17
γ	1407.4 *4*	~0.10

Continuous Radiation (^{237}Pa)

⟨β-⟩=572 keV;⟨IB⟩=0.91 keV

E_{bin}(keV)		⟨ ⟩(keV)	(%)
0 - 10	β-	0.0424	0.84
	IB	0.024	
10 - 20	β-	0.128	0.85
	IB	0.024	0.164
20 - 40	β-	0.52	1.74
	IB	0.045	0.158
40 - 100	β-	3.83	5.5
	IB	0.122	0.19
100 - 300	β-	40.2	19.9
	IB	0.30	0.170
300 - 600	β-	134	29.9
	IB	0.24	0.058
600 - 1300	β-	325	37.7
	IB	0.149	0.019
1300 - 2250	β-	68	4.45
	IB	0.0103	0.00069

$^{237}_{92}$U (6.75 *1* d)

Mode: β-
Δ: 45387.2 *22* keV
SpA: 8.162×10^{4} Ci/g
Prod: ^{236}U(n,γ); ^{238}U(n,2n)

Photons (^{237}U)

⟨γ⟩=144 *9* keV

γ_{mode}	γ(keV)	γ(%)†
Np L_ℓ	11.871	1.5 *3*
γ M1+0.1%E2	13.804 *16*	0.101 *4*
Np L_α	13.927	25 *4*
Np L_η	15.861	0.48 *7*
Np L_β	17.592	30 *5*
Np L_γ	20.990	7.2 *12*
γ E1	26.3445 *10*	2.29 *13*
γ M1+1.8%E2	33.1920 *14*	0.11 *5*
γ (M1+~43%E2)	42.64 *3*	
γ M1+14%E2	43.415 *10*	0.033 *3*
γ E1	51.013 *18*	0.20 *9*
γ E1	59.5364 *10*	32.8 *18*
γ E1	64.817 *13*	1.16 *12*
γ [E1]	69.760 *10*	
γ (E2)	75.83 *3*	
Np $K_{\alpha2}$	97.066	16 *3*
Np $K_{\alpha1}$	101.059	26 *4*
γ E1	102.952 *10*	0.0087 *9*
Np $K_{\beta1}$'	113.944	9.6 *15*
γ	114.08 *5*	
Np $K_{\beta2}$'	117.891	3.3 *5*
γ E2	164.593 *11*	1.83 *6*
γ M1+2.4%E2	208.008 *9*	22
γ E2	221.812 *17*	0.0204 *13*
γ M2	234.352 *9*	0.0194 *13*
γ E1+19%M2	267.544 *9*	0.712 *14*
γ [E2]	292.76 *3*	0.0027 *4*
γ E2	332.361 *13*	1.20 *5*
γ M1+18%E2	335.401 *14*	0.097 *5*
γ (E2)	337.727 *19*	0.0086 *7*
γ M1(+<8.8%E2)	368.592 *14*	0.0455 *24*
γ M1+12%E2	370.919 *19*	0.108 *6*

† 11% uncert(syst)

Atomic Electrons (^{237}U)

⟨e⟩=121 *9* keV

e_{bin}(keV)	⟨e⟩(keV)	e(%)
5 - 17	7.7	83 *5*
18	3.4	19 *3*
21	0.78	3.7 *4*
22	5.5	25 *4*
23 - 37	4.7	14.4 *13*
38	6.0	15.8 *9*
39 - 51	2.30	5.4 *3*
54	4.08	7.6 *4*
55 - 85	1.49	2.32 *12*
89	50.0	56 *9*
91 - 116	0.70	0.71 *5*
142 - 174	5.02	3.37 *6*
186	20.4	11.0 *18*
190 - 200	0.21	0.112 *18*
202	4.7	2.3 *4*
203 - 250	3.7	1.73 *14*
252 - 293	0.297	0.115 *6*
310 - 353	0.425	0.133 *3*
363 - 371	0.0178	0.00486 *23*

Continuous Radiation (^{237}U)

⟨β-⟩=66 keV;⟨IB⟩=0.016 keV

E_{bin}(keV)		⟨ ⟩(keV)	(%)
0 - 10	β-	0.51	10.3
	IB	0.0034	
10 - 20	β-	1.45	9.7
	IB	0.0027	0.019
20 - 40	β-	5.2	17.5
	IB	0.0039	0.0140
40 - 100	β-	25.7	38.2
	IB	0.0050	0.0086
100 - 252	β-	32.9	23.9
	IB	0.00105	0.00087

$^{237}_{93}$Np(2.140 *10* $\times 10^{6}$ yr)

Mode: α
Δ: 44868.3 *21* keV
SpA: 0.000705 Ci/g
Prod: daughter ^{237}U

Alpha Particles (^{237}Np)

⟨α⟩=4760 *4* keV

α(keV)	α(%)
4386 *25*	0.020
4513.5 *5*	~0.04
4574.7 *5*	0.05
4577.9 *5*	0.40 *4*
4595 *2*	0.08
4598.4 *5*	0.34 *4*
4639.5 *5*	6.18 *12*
4659.2 *20*	0.6
4664.6 *5*	3.32 *10*
4697.1 *7*	0.48 *20*
4707.1 *5*	1.0
4712.9 *5*	0.13
4741.4 *20*	0.019
4766.1 *5*	8 *3*
4771.5 *5*	25 *6*
4788.4 *5*	47 *9*
4804.0 *5*	1.6
4817.3 *5*	2.5 *4*
4862.9 *20*	0.24
4866.9 *5*	~0.3
4873.4 *5*	2.6 *2*

Photons (^{237}Np)

⟨γ⟩=32.7 *20* keV

γ_{mode}	γ(keV)	γ(%)†
Pa L_ℓ	11.372	1.15 *15*
Pa L_α	13.274	19.2 *19*
Pa L_η	14.953	0.47 *6*
Pa L_β	16.632	25 *3*
Pa L_γ	19.718	5.8 *8*
γ E1	29.378 *9*	12.9 *17*
γ [E1]	46.57 *3*	0.133 *19*
γ E2	57.149 *15*	0.39 *4*
γ [M1+E2]	62.66 *4*	~0.012
γ (E2)	63.95 *3*	0.016 *4*
γ [M1+E2]	70.62 *3*	0.016 *4* ?
γ [M1+E2]	74.46 *4*	0.0111 *24*
γ E1	86.528 *14*	12.6
γ [E1]	88.05 *4*	0.18 *3*
Pa $K_{\alpha2}$	92.279	1.59 *13*
γ E1	94.723 *21*	0.76 *8*
Pa $K_{\alpha1}$	95.863	2.58 *21*
γ [E2]	106.13 *5*	0.056 *6*
Pa $K_{\beta1}$'	108.166	0.94 *8*
γ M1+<4.6%E2	108.69 *3*	0.073 *13*
Pa $K_{\beta2}$'	111.897	0.31 *3*
γ [M1+E2]	115.19 *3*	0.0025 *7* ?
γ M1+10%E2	117.689 *22*	0.161 *19*
γ E1	131.09 *3*	0.085 *8*
γ [M1+E2]	134.23 *3*	0.067 *7*
γ [E1]	140.61 *7*	0.018 *5* ?
γ M1+14%E2	143.227 *21*	0.39 *3*
γ M1+12%E2	151.423 *24*	0.24 *3*
γ [E2]	153.72 *10*	0.0069 *16*
γ E1	155.263 *21*	0.092 *8*
γ [E1]	162.51 *4*	0.037 *3*
γ [E1]	169.18 *4*	0.072 *7*
γ [M1+E2]	170.67 *5*	0.019 *3*
γ	172.56 *20*	0.0068 *18*
γ [M1+E2]	176.09 *4*	0.020 *4*
γ [E1]	180.801 *25*	0.023 *3*
γ [E1]	186.7 *5*	0.0067 *13*
γ [M1+E2]	191.46 *5*	0.027 *4*

Photons (^{237}Np)
(continued)

γ_{mode}	γ(keV)	γ(%)†
γ [M1+E2]	193.29 3	0.051 6
γ [M1+E2]	194.74 8	0.049 14 ?
γ E1	195.04 3	0.201 18
γ [M1+E2]	196.89 4	0.025 4
γ [M1+E2]	200.17 7	0.0035 13
γ E1	201.72 3	0.042 4
γ [E1]	202.85 10	~0.0036
γ [E1]	209.19 4	0.0155 24
γ E1	212.412 19	0.152 19
γ [M1+E2]	214.08 4	0.042 5
γ [E1]	222.66 3	<0.0033
γ [E1]	229.98 5	0.0131 24
γ [E1]	237.950 25	0.064 7
γ	248.91 10	0.0048 13
γ [M1+E2]	256.98 10	0.0067 17
γ [M1+E2]	262.41 10	0.0067 14
γ [E2]	279.60 10	<0.0033

† 10% uncert(syst)

Atomic Electrons (^{237}Np)
⟨e⟩=64.0 21 keV

e_{bin}(keV)	⟨e⟩(keV)	e(%)
5 - 13	2.96	28.7 22
17	3.5	21 3
18	0.0032	0.0174 16
20	3.9	19.0 23
21 - 22	0.54	2.6 4
24	1.88	7.8 10
25 - 36	2.66	9.6 7
37	10.1	27 3
39	0.44	1.14 16
40	9.3	23.0 21
41 - 50	0.96	2.1 5
52	4.0	7.6 7
53	3.4	6.5 6
54	0.24	~0.4
56	2.17	3.9 4
57 - 64	1.12	1.9 3
65	3.6	5.6 9
66	6.0	9.0 16
67 - 116	5.4	6.4 6
117 - 166	1.53	1.17 7
167 - 214	0.25	0.14 3
217 - 263	0.013	0.0056 12
274 - 280	0.00017	~6 $\times 10^{-5}$

$^{237}_{94}$Pu(45.17 *6* d)

Mode: ϵ(99.9967 3 %), α(0.0033 3 %)
Δ: 45086 6 keV
SpA: 1.2197$\times 10^{4}$ Ci/g
Prod: ^{235}U(α,2n); ^{237}Np(d,2n)

Alpha Particles (^{237}Pu)
⟨α⟩=0.18 keV

α(keV)	α(%)
5155 4	0.00018
5259 4	2.0 $\times 10^{-5}$
5302 4	0.0004
5334 4	0.0015
5356 4	0.0006
5559 4 ?	
5610 4	
5650 4	0.0007

Photons (^{237}Pu)
⟨γ⟩=53.6 23 keV

γ_{mode}	γ(keV)	γ(%)†
Np L_ℓ	11.871	1.21 16
Np L_α	13.927	19.5 22
Np L_η	15.861	0.23 3
Np L_β	17.541	21 3
Np L_γ	21.060	5.1 8
γ_ϵE1	26.3445 10	0.230 7
γ_ϵM1+1.8%E2	33.1920 14	0.083 3
γ_ϵ(M1+~43%E2)	42.64 3	
γ_ϵM1+14%E2	43.415 10	
γ_ϵM1+17%E2	55.528 23	
γ_ϵE1	59.5364 10	3.30 8
γ_ϵ[E1]	69.760 10	
γ_ϵ(E2)	75.83 3	
Np $K_{\alpha2}$	97.066	12.6 11
γ_ϵE2	98.944 21	
Np $K_{\alpha1}$	101.059	20.1 18
γ_ϵE1	102.952 10	
Np $K_{\beta1}$'	113.944	7.4 7
Np $K_{\beta2}$'	117.891	2.50 22
γ_ϵ[E1]	125.288 21	
γ_α[E1]	198.60 20	5.2 7 $\times 10^{-5}$
γ_α[M1+E2]	204.81 11	2.3 6 $\times 10^{-5}$
γ_α[E1]	228.47 14	0.000258 11
γ_α[E1]	258.46 3	0.000110 6
γ_α[E1]	261.53 16	0.000129 8
γ_α[E1]	280.39 11	0.000717 14
γ_α[E1]	298.87 3	0.000517 13
γ_α[E1]	305.39 20	2.1 6 $\times 10^{-5}$
γ_α[E1]	313.45 16	0.000198 10
γ_α[E1]	320.81 11	0.000419 12
γ_α[E1]	411.34 14	1.2 4 $\times 10^{-5}$
γ_α[E1]	463.26 11	2.4 7 $\times 10^{-5}$
γ_α[E1]	503.68 11	4.9 9 $\times 10^{-5}$

† uncert(syst): 9.1% for α, 4.2% for ϵ

Atomic Electrons (^{237}Pu)
⟨e⟩=10.5 5 keV

e_{bin}(keV)	⟨e⟩(keV)	e(%)
5 - 9	0.0722	1.24 3
11	0.83	7.7 3
12	0.28	2.43 25
16 - 17	0.24	1.55 23
18	2.0	11.3 15
21	0.0493	0.237 7
22	3.4	15.2 19
23	0.00729	0.0322 10
27	0.512	1.87 7
28 - 37	0.92	2.88 11
38	0.575	1.52 4
42	0.166	0.396 9
54	0.410	0.758 18

Atomic Electrons (^{237}Pu)
(continued)

e_{bin}(keV)	⟨e⟩(keV)	e(%)
55 - 101	1.01	1.27 6
105 - 146	0.051	0.046 3
165 - 211	0.000143	7.8 7 $\times 10^{-5}$
223 - 263	2.21 $\times 10^{-5}$	8.83 22 $\times 10^{-6}$
275 - 321	3.08 $\times 10^{-5}$	1.065 18 $\times 10^{-5}$
348 - 394	3.3 $\times 10^{-6}$	8.8 14 $\times 10^{-7}$
406 - 446	3.0 $\times 10^{-7}$	6.8 16 $\times 10^{-8}$
458 - 504	7.1 $\times 10^{-7}$	1.46 16 $\times 10^{-7}$

Continuous Radiation (^{237}Pu)
⟨IB⟩=0.66 keV

E_{bin}(keV)		⟨ ⟩(keV)	(%)
10 - 20	IB	0.0090	0.054
20 - 40	IB	0.00142	0.0050
40 - 100	IB	0.32	0.37
100 - 218	IB	0.33	0.30

$^{237}_{94}$Pu(180 *20* ms)

Mode: IT
Δ: 45232 6 keV
SpA: 6.7$\times 10^{10}$ Ci/g
Prod: daughter ^{241}Cm

Photons (^{237}Pu)
⟨γ⟩=9.6 4 keV

γ_{mode}	γ(keV)	γ(%)†
Pu L_ℓ	12.124	0.73 9
Pu L_α	14.262	11.6 10
Pu L_η	16.333	0.43 5
Pu L_β	18.181	20.0 21
Pu L_γ	21.559	4.9 6
Pu $K_{\alpha2}$	99.522	0.110 5
Pu $K_{\alpha1}$	103.734	0.175 7
Pu $K_{\beta1}$'	116.930	0.063 3
Pu $K_{\beta2}$'	120.974	0.0219 10
γ E3	145.586 7	1.85 6

† 3.0% uncert(syst)

Atomic Electrons (^{237}Pu)
⟨e⟩=131.1 24 keV

e_{bin}(keV)	⟨e⟩(keV)	e(%)
18 - 24	5.4	26.2 19
76 - 116	0.0089	0.0103 5
122	3.62	2.95 11
123	55.9	45.3 16
128	25.8	20.2 7
140	20.1	14.3 5
141	8.5	6.06 22
142	0.72	0.508 18
144	8.3	5.76 21
145 - 146	2.74	1.88 6

$^{237}_{95}$Am(1.217 *17* h)

Mode: ϵ(99.975 *3* %), α(0.025 *3* %)
Δ: 46630 *140* keV syst
SpA: 1.087×10^7 Ci/g
Prod: ^{239}Pu(p,3n); ^{239}Pu(d,4n); ^{237}Np(α,4n); ^{237}Np(^{3}He,3n)

Alpha Particles (^{237}Am)

α(keV)
6042 *5*

Photons (^{237}Am)

⟨γ⟩=367 *9* keV

γ_{mode}	γ(keV)	γ(%)†
Pu L_ℓ	12.124	1.7 *3*
Pu L_α	14.262	28 *4*
Pu L_η	16.333	0.44 *7*
Pu L_β	18.077	29 *4*
Pu L_γ	21.623	6.5 *11*
γM1+3.7%E2	40.747 *6*	0.027 *5*
γM1+18%E2	45.723 *7*	0.0088 *22*
γM1+5.9%E2	47.71 *3*	0.0052 *12*
γ(E2)	55.632 *9*	0.019 *3*
γ(E2)	68.79 *5*	0.032 *6*
γ(M1)	79.045 *14*	0.20 *3*
Pu $K_{\alpha2}$	99.522	25 *3*
Pu $K_{\alpha1}$	103.734	40 *5*
Pu $K_{\beta1}$'	116.930	14.6 *19*
Pu $K_{\beta2}$'	120.974	5.0 *7*
γ[M1+E2]	124.00 *4*	~0.040
γ(M1)	124.769 *14*	0.28 *5*
γ[M1+E2]	127.61 *6*	0.110 *20*
γE3	145.586 *7*	0.48 *4* §
γ[M1+E2]	158.18 *7*	0.070 *20*
γ(M1+33%E2)	179.976 *20*	0.24 *5*
γM1+33%E2	183.59 *7*	0.19 *5*
γ[M1+E2]	193.29 *7*	0.09 *3*
γM1+20%E2	203.04 *4*	0.42 *5*
γM1+13%E2	206.66 *6*	0.33 *4*
γ(M1)	214.98 *4*	0.24 *5*
γ(M1)	224.89 *4*	0.24 *5*
γ[M1+E2]	228.98 *18*	0.15 *5*
γ(M1+30%E2)	248.77 *4*	0.59 *6*
γ[M1+E2]	252.04 *18*	0.15 *5*
γM1+~33%E2	252.38 *6*	0.27 *7*
γ[E1]	273.30 *4*	0.76 *5*
γE1	280.263 *14*	47.3 *20*
γE1	321.010 *15*	1.40 *10*
γE1	390.74 *7*	0.55 *4*
γ(E1)	407.87 *6*	0.63 *5*
γE1	425.85 *7*	1.94 *12*
γM1	435.40 *13*	0.25 *4*
γE1	438.45 *7*	8.3 *4*
γ[E1]	453.26 *18*	0.100 *20*
γM1	455.69 *19*	0.090 *20*
γE1	473.55 *7*	4.3 *3*
γM1	501.08 *13*	0.28 *4*
γM1	504.69 *12*	0.19 *4*
γM1(+E2)	648.53 *21*	0.26 *4*
γM1(+E2)	655.34 *20*	1.30 *13*
γM1	696.23 *21*	0.20 *4*
γ	720.4 *3*	0.24 *5*
γ	743.5 *5*	0.27 *5*
γ	792.0 *5*	0.16 *4*
γ[E1]	861.25 *12*	0.37 *4*
γ[E1]	908.95 *12*	2.60 *15*
γ	1000.7 *3*	0.19 *5*

† 9.0% uncert(syst)
§ with ^{237}Pu(180 ms) in equilib

Atomic Electrons (^{237}Am)

⟨e⟩=70.6 *20* keV

e_{bin}(keV)	⟨e⟩(keV)	e(%)
5 - 10	0.72	10 *3*
18	4.8	26 *4*
22	3.4	15.3 *22*
23	1.4	6.2 *11*
24 - 73	7.7	17.7 *11*
74 - 122	7.5	8.0 *6*
123	14.6	11.8 *10*
124 - 127	1.0	0.8 *3*
128	6.7	5.3 *5*
130 - 136	0.68	0.52 *16*
140	5.2	3.7 *3*
141	2.22	1.57 *13*
142	0.187	0.132 *11*
144	2.16	1.49 *13*
145 - 157	1.00	0.68 *8*
158	2.96	1.87 *12*
160 - 209	2.41	1.30 *10*
210 - 258	1.91	0.79 *8*
262 - 304	0.640	0.230 *5*
314 - 352	1.00	0.309 *18*
368 - 416	0.66	0.170 *15*
417 - 456	0.159	0.0364 *19*
468 - 505	0.165	0.034 *3*
527 - 574	0.7	~0.13
599 - 647	0.32	0.051 *20*
648 - 697	0.14	0.021 *7*
698 - 744	0.044	0.0061 *22*
769 - 792	0.094	0.0120 *13*
838 - 887	0.05	~0.006
891 - 909	0.0060	0.00067 *3*
978 - 1000	0.010	~0.0010

Continuous Radiation (^{237}Am)

⟨IB⟩=1.38 keV

E_{bin}(keV)		⟨ ⟩(keV)	(%)
10 - 20	IB	0.0043	0.025
20 - 40	IB	0.00094	0.0033
40 - 100	IB	0.46	0.51
100 - 300	IB	0.67	0.58
300 - 600	IB	0.117	0.027
600 - 1300	IB	0.116	0.0153
1300 - 1550	IB	8.0×10^{-5}	6.0×10^{-6}

A = 238

NDS **38**, 277 (1983)

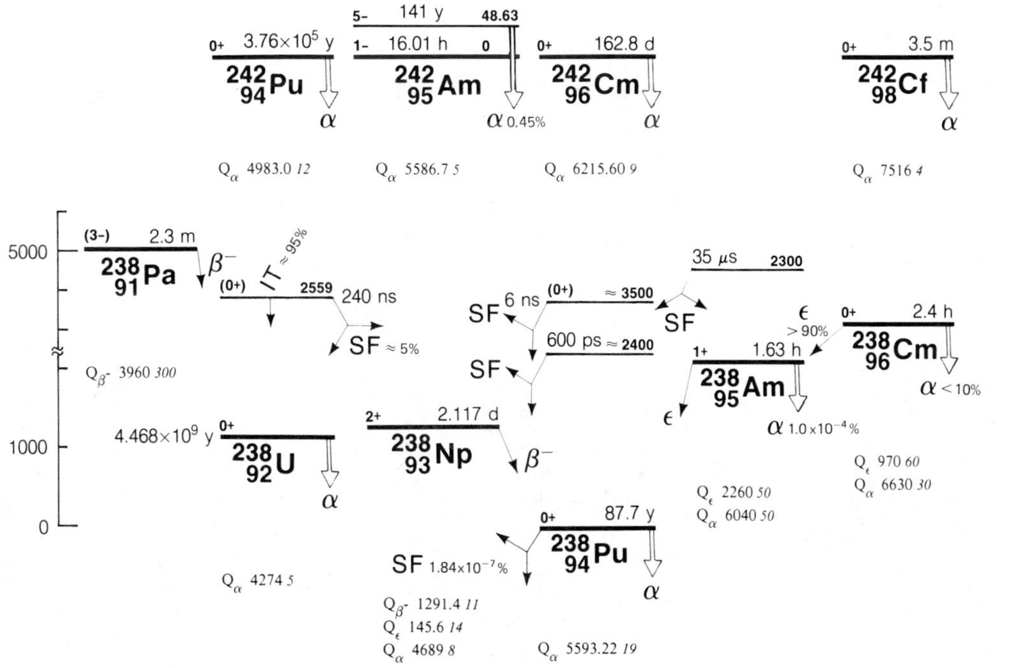

$^{238}_{91}$Pa(2.3 *1* min)

Mode: β-
Δ: 51270 *300* keV
SpA: 3.43×10^{8} Ci/g
Prod: ^{238}U(n,p)

Photons (^{238}Pa)

γmode	γ(keV)	γ(rel)
γ[M1+E2]	40.5 *4*	
γ[E2]	44.915 *13*	
γ[E1]	68.1 *5*	
γ[E1]	68.8 *4*	7.0 *14*
γ[E2]	103.50 *4*	12.0 *24*
γ[E1]	109.3 *4*	
γ[M1+E2]	130.8 *5*	
γ	142.6 *10*	
γ	154.3 *10*	3.0 *6*
γ[E2]	158.80 *8*	4.0 *8*
γ	164.9 *10*	2.0 *4*
γ[M1+E2]	171.3 *5* ?	3.0 *6*
γ[M1+E2]	178.5 *5*	11.0 *22*
γ	189.4 *10*	
γ	193.3 *10*	2.0 *4*
γ[M1+E2]	197.7 *4*	9.0 *18*
γ	212.9 *10*	
γ[M1+E2]	217.9 *5*	14 *3*
γ	221.9 *10*	4.0 *8*
γ	228.8 *10*	
γ[E2]	238.3 *4*	
γ[M1+E2]	250.6 *5*	7.0 *14*
γ[M1+E2]	258.7 *4*	8.0 *16*
γ[M1+E2]	269.8 *5*	12.0 *24*
γ	276.0 *10*	
γ	289.2 *5*	4.0 *8*
γ	293.0 *10*	12.0 *24*
γ	301.8 *10*	2.0 *4*
γ	316.9 *5*	7.0 *14*
γ	322.0 *10*	
γ	329.5 *10*	
γ	347.1 *6*	
γ	353.3 *10*	
γ	372.7 *5*	6 †
γ	374.8 *6*	
γ	377.0 *10*	
γ[M1+E2]	396.4 *4*	18 *4*
γ	407.5 *10*	9.0 *18*
γ	422.2 *10*	6.0 *12*
γ[M1+E2]	432.6 *6* ?	
γ[M1+E2]	436.9 *4*	16 *3*
γ	442.9 *10*	
γ[M1+E2]	448.4 *4*	76 *15*
γ	455.9 *6* ?	
γ	459.6 *10*	
γ	465.6 *10*	2.0 *4*
γ	476.1 *5*	19 *4*
γ[E2]	488.9 *4*	20 *4*
γ	501.9 *5*	26 *5*
γ[E1]	508.0 *6* ?	
γ[M1+E2]	510.9 *6*	
γ[E1]	519.2 *8*	
γ[M1+E2]	547.1 *4*	40 *8*
γ	557.7 *5*	
γ[E1]	569.9 *6*	6 *
γ	572.1 *10*	
γ[E1]	583.5 *4*	41 *8*
γ[M1+E2]	605.7 *5*	10 *2*
γ[E1]	615.2 *5*	8.0 *16*
γ	623.6 *10*	19 *4*
γ[E1]	635.0 *4*	88 *18*
γ[M1+E2]	646.2 *5*	9.0 *18*
γ	659.8 *10*	
γ	667.5 *6* ?	
γ[E1]	678.0 *8* ?	
γ[E1]	680.0 *4*	73 *15*
γ[E1]	687.0 *4*	54 *11*
γ[E2]	744.8 *5*	
γ	749.2 *6*	

Photons (^{238}Pa) (continued)

γmode	γ(keV)	γ(rel)
γ[M1+E2]	765.3 *4*	4.0 *8*
γ	769.0 *10*	
γ	797.5 *10*	
γ[M1+E2]	805.8 *4*	44 *9*
γ	818.1 *10*	
γ	823.2 *5*	9.0 *18*
γ	836.7 *10*	
γ	839.6 *10*	
γ[E1]	849.1 *5*	14 *3*
γ	863.7 *5*	54 *11*
γ[E1]	874.6 *5*	9.0 *18*
γ[E1]	885.7 *4*	45 *9*
γ[E1]	904.9 *5*	23 *5*
γ[M1+E2]	911.1 *4*	19 *4*
γ[E2]	911.8 *5* ?	
γ[E1]	930.6 *4*	6 *
γ	932.5 *5*	
γ[M1+E2]	943.5 *4*	7.0 *14*
γ[E1]	952.6 *5*	21 *4*
γ[M1+E2]	957.1 *5*	18 *4*
γ	961.0 *10*	
γ	967.0 *10*	4 *
γ	969.0 *10*	
γ	979.6 *10*	
γ[M1+E2]	984.3 *5*	7.0 *14*
γ	991.1 *10*	
γ[E2]	995.5 *5*	10 *2*
γ[E2]	1003.5 *5*	
γ[M1+E2]	1014.6 *4*	100 *
γ[M1+E2]	1015.3 *5*	
γ[M1+E2]	1019.0 *6*	10 *
γ[E1]	1020.4 *3*	
γ	1032.9 *10*	
γ	1036.1 *10*	
γ[M1+E2]	1042.6 *5*	8.0 *16*
γ[E2]	1060.2 *5*	45 *
γ[M1+E2]	1060.6 *5*	
γ	1071.0 *10*	
γ	1074.0 *10*	
γ[E1]	1083.4 *3*	50 *10*
γ	1090.2 *10*	
γ[E2]	1094.6 *5*	5 *1*
γ	1112.0 *10*	2.0 *4*
γ	1113.0 *10*	4.0 *8*
γ[E2]	1122.5 *6*	5 *
γ[E1]	1123.9 *3*	
γ	1138.4 *6*	2.0 *4* ?
γ	1159.5 *10*	5 *
γ	1161.5 *10*	
γ	1178.9 *6*	6.0 *12*
γ	1214.8 *10*	6.0 *12*
γ	1224.0 *10*	6.0 *12*
γ	1233.5 *10*	
γ	1306.4 *10*	
γ	1311.7 *10*	
γ	1325.2 *10*	
γ[M1+E2]	1332.0 *6*	5 *1*
γ	1336.7 *10*	
γ	1359.3 *10*	
γ	1364.0 *10*	
γ	1368.8 *10*	5 *1*
γ	1376.7 *6*	4.0 *8*
γ[M1+E2]	1383.9 *6*	7.0 *14*
γ	1394.0 *10*	
γ	1410.0 *10*	3 *
γ	1413.0 *4*	
γ	1420.0 *10*	
γ	1496.5 *5*	8.0 *16*
γ	1507.1 *10*	
γ	1516.5 *4*	
γ[E1]	1527.0 *3*	4.0 *8*
γ	1600.0 *5*	
γ	1611.0 *10*	3.0 *6*
γ	1620.0 *10*	
γ[E1]	1626.1 *4*	
γ[E1]	1630.5 *3*	
γ	1647.5 *10*	
γ[E1]	1729.6 *4*	
γ	1737.0 *10*	
γ	1752.0 *10*	
γ[E1]	1785.7 *4*	
γ	1804.0 *10*	

Photons (^{238}Pa) (continued)

γmode	γ(keV)	γ(rel)
γ	1841.0 *10*	
γ	1872.5 *10*	
γ[E1]	1889.2 *4*	17 *3*
γ	1907.0 *10*	
γ	1976.0 *10*	
γ	1985.5 *10*	
γ	1996.7 *7*	4.0 *8*
γ	2013.0 *10*	3.0 *6*
γ[E1]	2018.9 *5*	7.0 *14*
γ	2048.0 *10*	
γ	2081.0 *10*	
γ	2089.0 *10*	
γ	2126.0 *10*	
γ	2529.0 *10*	2.0 *4*

* combined intensity for doublet
† 373γ + 375γ + 377γ

$^{238}_{92}$U (4.468 *5* ×10^{9} yr)

Mode: α
Δ: 47306.0 *21* keV
SpA: Ci/g
Prod: natural source
%: 99.2745 *15*

Alpha Particles (^{238}U)

⟨α⟩=4194 *5* keV

α(keV)	α(%)
4039 *5*	0.23 *7*
4147 *5*	23 *4*
4196 *5*	77 *4*

Photons (^{238}U)

⟨γ⟩=1.30 *15* keV

γmode	γ(keV)	γ(%)
Th L_{ℓ}	11.118	0.18 *4*
Th L_{α}	12.952	3.0 *6*
Th L_{η}	14.511	0.083 *17*
Th L_{β}	16.161	4.1 *8*
Th L_{γ}	19.094	0.93 *19*
γ E2	49.55 *6*	0.070 *12*
γ[E2]	110.5	0.024 *8*

Atomic Electrons (^{238}U)

⟨e⟩= 9.5 *7* keV

e_{bin}(keV)	⟨e⟩(keV)	e(%)
16	0.68	4.2 *8*
20	0.72	3.7 *7*
29	0.082	0.28 *5*
30	2.6	8.8 *15*
33	2.6	7.7 *13*
44	0.037	0.084 *14*
45	1.04	2.3 *4*
46	1.00	2.2 *4*
48	0.31	0.65 *11*
49	0.43	0.89 *15*
50	0.026	0.052 *9*

$^{238}_{93}$Np(2.117 *2* d)

Mode: β-
Δ: 47451.6 *21* keV
SpA: 2.5916×10^{5} Ci/g
Prod: ^{237}Np(n,γ); ^{238}U(d,2n); ^{238}U(p,n); ^{238}U(^{3}He,p2n)

Photons (^{238}Np)

⟨γ⟩=647 *13* keV

γ_{mode}	γ(keV)	γ(%)†
Pu L_ℓ	12.124	0.82 *10*
Pu L_α	14.262	12.9 *12*
Pu L_η	16.333	0.31 *4*
Pu L_β	18.136	15.6 *17*
Pu L_γ	21.556	3.4 *4*
γ E2	44.077 *18*	0.090 *10*
Pu $K_{\alpha2}$	99.522	0.214 *9*
γ E2	101.889 *17*	0.27 *1*
Pu $K_{\alpha1}$	103.734	0.341 *14*
γ [E2]	114.4 *4*	0.006 *1*
Pu $K_{\beta1}$'	116.930	0.123 *5*
γ (M1)	119.90 *7*	0.108 *6*
Pu $K_{\beta2}$'	120.974	0.0426 *19*
γ [E1]	132.51 *7*	0.0028 *3*
γ [E2]	157.42 *6*	~0.0010
γ [E1]	173.92 *8*	0.026 *1*
γ (M2)	220.88 *11*	0.0034 *4*
γ E2	301.39 *5*	0.012 *1*
γ M1+E2	319.30 *7*	0.009 *1*
γ [E1]	321.67 *6*	0.0013 *6*
γ M1+E2	324.05 *5*	0.016 *1*
γ [E1]	336.35 *8*	0.00025 *3*
γ E2+15%M1	357.67 *3*	0.053 *3*
γ [E1]	377.96 *6*	0.0033 *6*
γ [M1]	380.34 *5*	0.012 *1*
γ [M1]	421.18 *6*	0.023 *1*
γ [E1]	459.88 *8*	~0.0030
γ E1	515.43 *4*	0.043 *2*
γ E1	561.03 *3*	0.1140 *23*
γ E1	605.11 *3*	0.081 *4*
γ [E1]	617.30 *6*	~0.009
γ E1	617.32 *4*	0.066 *5*
γ [E2]	837.10 *5*	0.028 *2*
γ E2	882.581 *22*	0.88 *3*
γ (E2)	897.38 *7*	0.008 *1*
γ E1	918.707 *23*	0.59 *2*
γ E2	923.989 *16*	2.90 *9*
γ [E1]	924.1 *4*	0.05 *1*
γ [E1]	936.60 *5*	0.40 *1*
γ E0+E2	938.99 *5*	0.026 *3*
γ [E1]	941.37 *4*	0.54 *2*
γ E1	962.784 *22*	0.70 *2*
γ	968.9 *4*	0.017 *6*
γ E2	984.470 *16*	27.8
γ E2	1025.878 *16*	9.6 *6*
γ E2	1028.546 *17*	20.3 *8*

† 3.0% uncert(syst)

Atomic Electrons (^{238}Np)

⟨e⟩=39.3 *9* keV

e_{bin}(keV)	⟨e⟩(keV)	e(%)
11	6.1×10^{-5}	0.00057 *5*
18	2.4	13.3 *15*
21	0.229	1.09 *6*
22	9.0	41 *3*
23	0.070	0.30 *3*
26	6.8	26.0 *15*
36 - 38	0.125	0.329 *19*
39	3.13	8.1 *5*
40	2.98	7.5 *4*
43	1.90	4.4 *3*
44 - 79	0.64	1.38 *7*
80	1.35	1.69 *7*
81	0.0033	0.0041 *4*
84	0.93	1.11 *5*
86 - 135	1.56	1.58 *4*
139 - 180	0.0046	0.0029 *3*
198 - 236	0.067	0.032 *8*
256 - 304	0.058	0.0205 *16*
306 - 355	0.0179	0.0054 *6*
356 - 403	0.0163	0.00429 *16*
415 - 460	0.00729	0.00169 *4*
483 - 515	0.0067	0.00136 *8*
538 - 587	0.00210	0.000378 *7*
594 - 617	0.00107	0.000179 *12*
715 - 761	0.079	0.0104 *4*
776 - 820	1.02	0.125 *22*
831 - 860	0.048	0.0056 *4*
863	2.31	0.268 *10*
865 - 906	0.88	0.098 *6*
907	1.65	0.182 *8*
913 - 962	0.86	0.090 *6*
963 - 1010	0.951	0.0954 *19*
1020 - 1028	0.223	0.0218 *6*

Continuous Radiation (^{238}Np)

⟨β-⟩=190 keV;⟨IB⟩=0.19 keV

E_{bin}(keV)		⟨ ⟩(keV)	(%)
0 - 10	β-	0.340	6.8
	IB	0.0089	
10 - 20	β-	0.96	6.5
	IB	0.0083	0.058
20 - 40	β-	3.51	11.8
	IB	0.0149	0.052
40 - 100	β-	18.2	26.9
	IB	0.035	0.055
100 - 300	β-	43.5	26.8
	IB	0.069	0.040
300 - 600	β-	55	12.4
	IB	0.042	0.0104
600 - 1248	β-	69	8.8
	IB	0.0108	0.00148

$^{238}_{94}$Pu(87.74 *9* yr)

Mode: α, SF(1.84 *5* $\times10^{-7}$ %)
Δ: 46160.2 *22* keV
SpA: 17.119 Ci/g
Prod: daughter ^{238}Np; daughter ^{242}Cm

Alpha Particles (^{238}Pu)

⟨α⟩=5487.1 *2* keV

α(keV)	α(%)
4432.10 *14*	$\sim1\times10^{-6}$
4472.31 *12*	$\sim2\times10^{-6}$
4492.58 *12*	5×10^{-7}
4526.37 *12*	3×10^{-7}
4567.31 *16*	1.0×10^{-7}
4579 ?	2.0×10^{-5}
4588.09 *12*	1.2×10^{-5}
4661.81 *14*	$<2\times10^{-5}$
4664.16 *14*	3×10^{-7}
4702.95 *12*	5×10^{-5}
4726.15 *12*	2.2×10^{-5}
5010.60 *13*	$\sim4\times10^{-6}$
5208.18 *12*	~0.003
5358.30 *12*	0.10 *3*
5456.47 *12*	28.3 *6*
5499.21 *12*	71.6 *6*

Photons (^{238}Pu)

⟨γ⟩=1.76 *10* keV

γ_{mode}	γ(keV)	γ(%)†
U L_ℓ	11.620	0.25 *3*
U L_α	13.600	4.2 *3*
U L_η	15.400	0.102 *11*
U L_β	17.128	5.2 *5*
U L_γ	20.292	1.15 *13*
γE2	43.470 *9*	0.0390 *8*
γE1	62.78 *4*	$\sim4\times10^{-8}$
γE2	99.852 *10*	0.00724 *20*
γE2+39%M1	140.15 *9*	$\sim10\times10^{-9}$
γE2	152.69 *3*	0.00101 *20*
γ[M1+E2]	174.52 *9*	5 *1* $\times10^{-9}$
γ[E2]	192.74 *8*	$<4\times10^{-8}$
γE2	200.96 *5*	4.1 *10* $\times10^{-6}$
γE2+31%M1	203.20 *3*	$\sim1\times10^{-8}$
γ[E1]	236.03 *12*	10 *5* $\times10^{-9}$?
γ[E1]	258.18 *3*	1.19 *12* $\times10^{-7}$
γ[E1]	299.08 *10*	1.3 *3* $\times10^{-7}$
γ[E1]	706.02 *9*	1.51 *15* $\times10^{-7}$
γ[E2]	708.40 *7*	3.8 *4* $\times10^{-7}$
γE1	742.817 *20*	7.8 *5* $\times10^{-6}$
γ(E2)	766.412 *14*	3.3 *3* $\times10^{-5}$
γ[E2]	783.38 *4*	4.6 *10* $\times10^{-8}$
γ(E1)	786.287 *20*	4.7 *3* $\times10^{-6}$
γ[E1]	805.87 *9*	1.72 *15* $\times10^{-7}$
γE0+E2	808.25 *7*	1.06 *8* $\times10^{-6}$
γ[E2]	851.72 *7*	2.03 *17* $\times10^{-6}$
γ[E1]	880.53 *4*	2.3 *3* $\times10^{-7}$
γE2	883.24 *4*	1.11 *9* $\times10^{-6}$
γ[E2]	904.38 *13*	1.0 *2* $\times10^{-7}$
γE2	926.71 *4*	8.2 *7* $\times10^{-7}$
γ[E2]	942.05 *10*	6.4 *5* $\times10^{-7}$
γE1	946.02 *3*	1.30 *11* $\times10^{-7}$
γ[E1]	980.39 *4*	$\sim8\times10^{-8}$
γE2	1001.00 *3*	1.37 *15* $\times10^{-6}$
γ[M1+E2]	1041.90 *10*	2.89 *25* $\times10^{-7}$
γ[E2]	1085.37 *10*	1.05 *14* $\times10^{-7}$

† 2.0% uncert(syst)

Atomic Electrons (^{238}Pu)

⟨e⟩= 9.92 17 keV

e_{bin}(keV)	⟨e⟩(keV)	e(%)
13	1.1 $\times 10^{-8}$	~8 $\times 10^{-8}$
17	0.85	4.9 5
21	0.83	4.0 4
22	0.094	0.433 17
23	2.44	10.8 3
25	6.8 $\times 10^{-9}$	~3 $\times 10^{-8}$
26	2.53	9.6 3
29 - 37	8.1 $\times 10^{-5}$	0.00022 4
38	1.15	2.99 9
39	1.05	2.69 8
40 - 41	0.0275	0.0690 20
42	0.658	1.55 4
43	0.202	0.467 13
46 - 95	0.0692	0.0840 18
96 - 143	0.0168	0.0168 4
147 - 196	0.00081	0.00055 8
197 - 241	7.3 $\times 10^{-7}$	3.7 5 $\times 10^{-7}$
253 - 299	3.9 $\times 10^{-9}$	1.40 13 $\times 10^{-9}$
590 - 627	2.97 $\times 10^{-7}$	4.8 3 $\times 10^{-8}$
651 - 694	0.00042	6.0 13 $\times 10^{-5}$
700 - 749	1.23 $\times 10^{-6}$	1.66 12 $\times 10^{-7}$
761 - 808	0.00010	1.3 4 $\times 10^{-5}$
811 - 860	1.90 $\times 10^{-7}$	2.31 10 $\times 10^{-8}$
861 - 910	1.63 $\times 10^{-7}$	1.85 13 $\times 10^{-8}$
920 - 970	8.7 $\times 10^{-8}$	9 3 $\times 10^{-9}$
975 - 1021	4.8 $\times 10^{-8}$	4.8 6 $\times 10^{-9}$
1025 - 1068	5.6 $\times 10^{-9}$	5.4 14 $\times 10^{-10}$
1080 - 1085	6.2 $\times 10^{-10}$	5.7 6 $\times 10^{-11}$

$^{238}_{95}$Am(1.63 3 h)

Mode: ε, α(0.00010 4 %)
Δ: 48420 50 keV
SpA: 8.06 $\times 10^{6}$ Ci/g
Prod: ^{239}Pu(p,2n); ^{239}Pu(d,3n); ^{237}Np(α,3n)

Alpha Particles (^{238}Am)

α(keV)
5940

Photons (^{238}Am)

⟨γ⟩=898 57 keV

γ_{mode}	γ(keV)	γ(%)†
Pu L_ℓ	12.124	1.7 4
Pu L_α	14.262	27 6
Pu L_η	16.333	0.43 9
Pu L_β	18.080	27 6
Pu L_γ	21.607	5.9 14
γ_ϵE2	44.077 18	0.064 9
Pu $K_{\alpha2}$	99.522	22 5
γ_ϵE2	101.889 17	~0.24
Pu $K_{\alpha1}$	103.734	36 8
Pu $K_{\beta1}$'	116.930	13 3
Pu $K_{\beta2}$'	120.974	4.4 10
γ_ϵE2	301.39 5	0.50 4
γ_ϵ[E1]	321.67 6	~0.00017
γ_ϵM1+E2	324.05 5	0.062 8
γ_ϵ[E1]	336.35 8	~0.018
γ_ϵE2+15%M1	357.67 3	2.10 14
γ_ϵ[E1]	377.96 6	~0.00043
γ_ϵ[M1]	380.34 5	0.045 14
γ_ϵE1	515.43 4	0.39 4

Photons (^{238}Am) (continued)

γ_{mode}	γ(keV)	γ(%)†
γ_ϵE1	561.03 3	10.9 7
γ_ϵ	565.8 3	0.154 20
γ_ϵ(M1+E2)	574.37 14	0.11 3
γ_ϵ[M1+E2]	597.04 14	0.146 17
γ_ϵE1	605.11 3	7.6 5
γ_ϵE1	617.32 4	0.73 6
γ_ϵ	633.0 5	~0.06
γ_ϵ[E1]	653.30 14	~0.06
γ_ϵE0+E2+M1	658.47 12	0.179 20
γ_ϵE1	679.79 13	0.255 25
γ_ϵE1	821.48 24	0.31 3
γ_ϵ[E2]	837.10 5	~0.0037
γ_ϵE2	882.581 22	~0.0036
γ_ϵ	884.3 3	0.132 17
γ_ϵ(E2)	897.38 7	0.56 6
γ_ϵ(M1)	908.8 2	0.218 20
γ_ϵE1	918.707 23	23.0 14
γ_ϵ	935.2 3	~0.08
γ_ϵE0+E2	938.99 5	0.0034 7
γ_ϵ[E1]	941.37 4	2.24 14
γ_ϵ[M1+E2]	954.71 14	~0.08
γ_ϵE1	962.784 22	28
γ_ϵ[E2]	983.07 5	~0.022
γ_ϵE2	984.470 16	0.115 16
γ_ϵE0+E2+M1	1016.14 12	0.28 3
γ_ϵ(E2)	1097.3 3	0.31 3
γ_ϵ[E2]	1118.25 15	0.176 19
γ_ϵ	1130.3 4	0.050 8
γ_ϵ	1174.4 4	0.042 8
γ_ϵE2	1184.58 17	0.53 4
γ_ϵE0+E2+M1	1220.14 15	0.140 17
γ_ϵ	1226.4 3	0.045 8
γ_ϵ	1231.3 3	0.042 8
γ_ϵM1	1236.94 21	0.249 22
γ_ϵM1	1266.2 3	1.68 14
γ_ϵM1	1293.22 21	0.31 3
γ_ϵ	1368.8 5	~0.06
γ_ϵE1	1403.13 19	0.67 6
γ_ϵE0+E2+M1	1414.20 21	~0.028
γ_ϵE1	1447.21 19	0.42 3
γ_ϵ[E2]	1450.37 23	~0.06 ?
γ_ϵ[E2]	1458.28 21	0.123 14
γ_ϵ[E1]	1515.74 14	0.115 14
γ_ϵ[M1+E2]	1552.26 23	0.073 11
γ_ϵ[E1]	1559.82 14	0.095 14
γ_ϵE1	1577.17 12	2.88 22
γ_ϵ[E1]	1592.28 13	0.48 5
γ_ϵ[E2]	1596.33 23	~0.022
γ_ϵ	1607.1 3	0.095 14
γ_ϵ[E1]	1621.25 12	~0.017
γ_ϵE1	1636.36 13	1.26 11
γ_ϵ	1651.2 3	0.017 6
γ_ϵ	1682.24 21	0.48 5
γ_ϵ	1726.32 21	0.28 3
γ_ϵ	1739.4 3	0.028 6
γ_ϵ	1761.5 4	0.090 11
γ_ϵ	1783.5 3	0.059 14
γ_ϵ	1789.0 5	~0.011
γ_ϵ	1835.1 5	0.034 8

† 11% uncert(syst)

Atomic Electrons (^{238}Am)

⟨e⟩=40.3 19 keV

e_{bin}(keV)	⟨e⟩(keV)	e(%)
18	4.7	26 6
21	0.149	0.71 10
22	7.7	35 6
23	0.83	3.6 9
26	4.4	16.9 23
38	0.082	0.21 3
39	2.0	5.3 7
40	1.9	4.9 7
43	1.24	2.9 4
44 - 93	2.16	3.05 25
94 - 116	0.86	0.88 6
180 - 202	0.14	0.07 3
236 - 283	0.91	0.37 7

Atomic Electrons (^{238}Am) (continued)

e_{bin}(keV)	⟨e⟩(keV)	e(%)
295 - 340	0.53	0.162 17
352 - 394	0.205	0.057 4
439 - 483	0.93	0.203 20
492 - 532	0.057	0.011 4
537	1.02	0.19 3
538 - 587	0.44	0.080 8
591 - 640	0.32	0.051 9
647 - 680	0.057	0.0087 20
700 - 720	0.17	0.024 5
763 - 803	0.88	0.111 7
813 - 819	0.15	0.019 4
820	2.1	0.26 6
821 - 866	0.88	0.105 20
874 - 893	0.0408	0.0046 3
894	1.16	0.129 18
896 - 945	1.31	0.143 18
949 - 998	0.39	0.039 6
1005 - 1053	0.11	0.011 3
1063 - 1110	0.78	0.071 10
1112 - 1161	0.50	0.044 3
1162 - 1211	0.26	0.022 3
1213 - 1262	0.134	0.0108 8
1265 - 1305	0.38	0.029 5
1325 - 1368	0.0235	0.00176 19
1380 - 1429	0.091	0.0065 14
1430 - 1475	0.092	0.0063 5
1485 - 1584	0.10	0.0068 22
1585 - 1678	0.059	0.0036 13
1703 - 1784	0.013	0.0007 3
1812 - 1831	0.0007	~4 $\times 10^{-5}$

Continuous Radiation (^{238}Am)

⟨β+⟩=0.74 keV;⟨IB⟩=1.7 keV

E_{bin}(keV)		⟨ ⟩(keV)	(%)
0 - 10	β+	6.8 $\times 10^{-10}$	7.9 $\times 10^{-9}$
	IB	7.0 $\times 10^{-5}$	
10 - 20	β+	5.9 $\times 10^{-8}$	3.45 $\times 10^{-7}$
	IB	0.0043	0.025
20 - 40	β+	3.47 $\times 10^{-6}$	1.03 $\times 10^{-5}$
	IB	0.00100	0.0035
40 - 100	β+	0.000398	0.000493
	IB	0.46	0.51
100 - 300	β+	0.0362	0.0160
	IB	0.68	0.58
300 - 600	β+	0.257	0.056
	IB	0.141	0.032
600 - 1300	β+	0.449	0.057
	IB	0.32	0.037
1300 - 2256	IB	0.096	0.0063
	Σβ+		0.13

$^{238}_{96}$Cm(2.4 1 h)

Mode: ε(>90 %), α(<10 %)
Δ: 49390 30 keV
SpA: 5.49 $\times 10^{6}$ Ci/g
Prod: ^{239}Pu(α,5n); ^{238}Pu(α,4n)

Alpha Particles (^{238}Cm)

α(keV)
6520 50

A = 239

NDS **21**, 153 (1977)

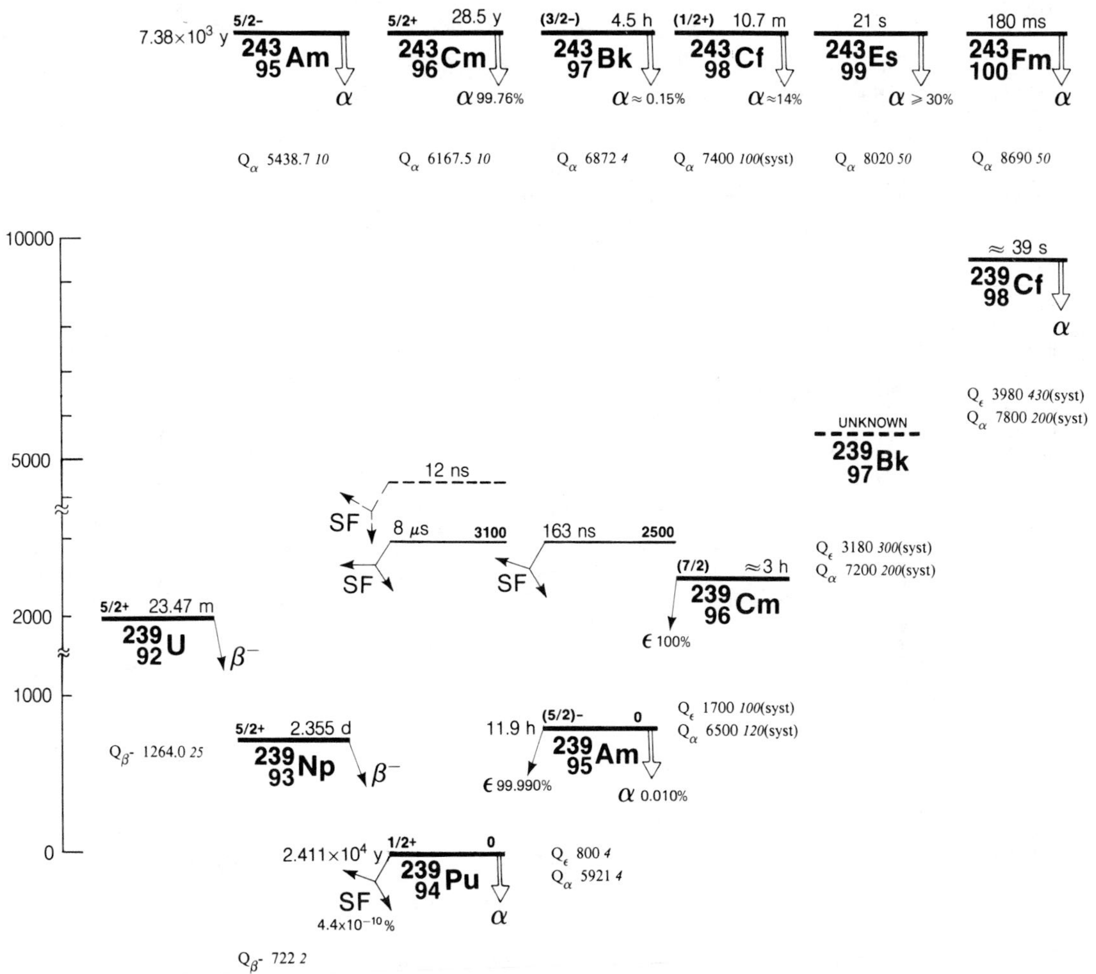

$^{239}_{92}$U (23.47 *5* min)

Mode: β-
Δ: 50570.9 *21* keV
SpA: 3.351×10^7 Ci/g
Prod: ^{238}U(n,γ)

Photons (^{239}U)

⟨γ⟩=51.6 *18* keV

γ_{mode}	γ(keV)	γ(%)†
γ E1	43.537 *3*	4.44 *14*
γ [E1]	50.8 *18*	
γ M1+~39%E2	55.40 *4*	
γ [E2]	71.21 *3*	
γ E1	74.672 *3*	52.2 *24*
γ E1	86.57 *3*	0.068 *9*
γ (E2)	98.43 *4*	
γ	111.16 *4*	0.020 *4*
γ E1	117.70 *3*	0.125 *17*
γ E1	141.97 *4*	
γ [E1]	170.14 *5*	
γ	186.136 *16*	0.033 *7*
γ	187.383 *21*	0.0068 *13*
γ	191.938 *24*	0.0028 *6*
γ	196.85 *11*	0.0022 *4*
γ	201.19 *5*	0.0021 *4*
γ	231.70 *11*	0.0031 *6*
γ	255.36 *4*	0.0028 *6*
γ	258.45 *4*	0.0026 *5*
γ	260.809 *17*	0.0023 *5*
γ	296.88 *4*	0.0016 *
γ	296.95 *3*	
γ	301.98 *3*	0.0015 *3*
γ	304.34 *6*	0.0018 *4*
γ	312.05 *3*	0.0055 *11*
γ	321.71 *16*	0.00120 *24*
γ	343.74 *11*	0.0019 *4*
γ	345.14 *3*	0.0031 *6*
γ	363.10 *21*	0.00085 *17*
γ [M1+E2]	373.519 *17*	0.021 *4*
γ	378.074 *19*	0.0110 *22*
γ	381.53 *3*	0.0015 *3*
γ	395.30 *4*	0.0015 *3*
γ	399.59 *9*	0.0005 *1*
γ [E2]	407.70 *5*	0.0035 *7*
γ	434.71 *3*	0.0044 *9*
γ [E1]	448.191 *17*	0.0085 *17*
γ	455.63 *5*	0.0032 *6*
γ	474.26 *9*	0.0035 *7*
γ [E1]	486.868 *20*	0.060 *12*
γ	492.69 *3*	0.0047 *9*
γ	499.17 *5*	0.0015 *3*
γ [E2]	504.76 *4*	0.0044 *9*
γ	513.84 *6*	0.00070 *14*
γ [E1]	518.004 *20*	0.0030 *6*
γ	522.12 *4*	0.0023 *5*
γ	530.30 *5*	0.00120 *24*
γ	532.77 *3*	0.0021 *4*
γ	535.01 *15*	0.0013 *3*
γ	539.9 *18*	
γ	544.55 *3*	0.0035 *7*
γ [E1]	548.30 *4*	0.0021 *4*
γ	563.90 *3*	0.00125 *25*
γ	566.11 *9*	0.0021 *4*
γ	577.52 *3*	0.0018 ‡
γ	577.91 *5*	
γ	587.579 *18*	
γ	587.77 *4*	0.024 *5* §
γ	602.68 *4*	0.0049 *10*
γ	624.01 *3*	0.0070 *14*
γ	631.116 *18*	0.069 *12*
γ	646.16 *4*	0.0023 *5*
γ	658.13 *6*	0.00060 *12*
γ	662.251 *18*	0.20 *3*
γ	664.092 *23*	0.0095 *19*
γ	695.227 *23*	0.0044 *9*
γ	700.93 *9*	0.0021 *4*
γ	703.42 *3*	0.0028 *6*
γ	707.25 *4*	0.0029 *6*

Photons (^{239}U) (continued)

γ_{mode}	γ(keV)	γ(%)†
γ	710.24 4	0.00125 25
γ	714.09 7	0.0040 8
γ	722.87 4	0.028 6
γ	730.92 4	0.0125 25
γ	745.74 6	0.0039 8
γ	748.05 3	0.105 21
γ	752.85 11	0.0015 3
γ	772.88 4	0.0035 7
γ	774.73 4	0.017 3
γ	779.56 4	0.00120 24
γ	788.13 3	0.0056 11
γ	791.3 11	0.0090 18
γ	793.49 6	0.0030 6
γ	812.96 3	0.080 16
γ	819.27 3	0.15 3
γ	831.86 4	0.0035 7
γ	840.3 3	0.0047 9
γ	844.10 3	0.17 3
γ	846.53 3	0.039 8
γ	848.88 6	0.0017 3 ?
γ	863.44 5	0.00075 15
γ	867.27 5	0.0016 3
γ [M1+E2]	874.41 4	0.0039 8
γ [M1+E2]	876.14 4	0.0021 4
γ	884.52 3	0.0125 25
γ	889.558 22	0.024 5
γ	895.37 5	0.0019 4
γ [M1+E2]	917.44 4	0.0032 6
γ [E1]	920.90 4	0.0029 6
γ	922.67 4	0.00120 24
γ	928.06 3	0.0060 12 ?
γ [M1+E2]	931.54 4	0.0049 10
γ	933.095 22	0.037 7
γ	939.0 3	0.00040 8
γ	959.20 3	0.0070 14
γ [E1]	960.97 4	0.018 4
γ	964.230 23	0.090 18
γ	965.70 3	0.0022 4
γ [E2]	974.57 3	0.0040 8
γ [E1]	992.11 4	0.0032 6
γ [E1]	1018.11 3	<0.0011
γ	1040.37 3	0.00110 22
γ	1065.83 3	0.00070 14
γ	1078.88 16	0.0016 3
γ	1096.96 3	0.0027 5
γ	1122.8 3	0.00070 14
γ	1161.40 21	0.0010 2
γ	1196.90 16	0.00095 19
γ	1204.90 21	0.0016 3

† 10% uncert(syst)
* 296.88γ + 296.95γ
‡ 577.5γ + 577.9γ
§ 587.8γ + 587.6γ

Continuous Radiation (^{239}U)

$\langle\beta$-$\rangle$=395 keV;$\langle$IB$\rangle$=0.47 keV

E_{bin}(keV)		$\langle\ \rangle$(keV)	(%)
0 - 10	β-	0.066	1.32
	IB	0.018	
10 - 20	β-	0.199	1.33
	IB	0.017	0.121
20 - 40	β-	0.81	2.69
	IB	0.033	0.115
40 - 100	β-	5.8	8.3
	IB	0.085	0.133
100 - 300	β-	56	28.0
	IB	0.18	0.106
300 - 600	β-	157	35.5
	IB	0.108	0.027
600 - 1264	β-	175	22.8
	IB	0.025	0.0035

$^{239}_{93}$Np(2.355 4 d)

Mode: β-
Δ: 49307 3 keV
SpA: 2.320$\times10^5$ Ci/g
Prod: daughter ^{239}U; daughter ^{243}Am

Photons (^{239}Np)

$\langle\gamma\rangle$=174.0 25 keV

γ_{mode}	γ(keV)	γ(%)†
Pu L_ℓ	12.124	1.48 17
Pu L_α	14.262	23.6 19
Pu L_η	16.333	0.38 5
Pu L_β	18.080	25 3
Pu L_γ	21.622	5.6 8
γ M1+3.6%E2	44.665 3	0.104 11
γ M1+20%E2	49.4152 19	0.105 14
γ E2	57.2759 18	0.148 16
γ M1+E2	57.278 4	~0.006
γ E1	61.462 4	0.96 14
γ E2	67.8462 22	0.090 23
γ M1+20%E2	88.057 18	0.0059 20
Pu $K_{\alpha2}$	99.522	15.0 5
γ E2	101.943 5	~0.0007
Pu $K_{\alpha1}$	103.734	23.9 8
γ E1+0.3%M2	106.1272 23	22.7 11
γ E2	106.488 18	0.048 9
Pu $K_{\beta1}'$	116.930	8.6 3
Pu $K_{\beta2}'$	120.974	2.98 11
γ M1(+<6.3%E2)	124.439 8	0.0104 14
γ M1	166.364 18	0.017 7
γ M1+1.7%E2	181.716 7	0.111 10
γ M1+0.08%E2	209.7554 17	3.32 13
γ M1+23%E2	226.382 7	0.34 3
γ M1+0.6%E2	228.1865 15	10.7 4
γ M1+1.7%E2	254.421 4	0.099 9
γ M1	272.852 4	0.075 7
γ M1+2.6%E2	277.6016 16	14.2 6
γ E2	285.4624 15	0.74 3
γ (M1+E2)	311.698 5	~0.0014
γ E1+2.3%M2	315.8825 23	1.59 9
γ [E2]	322.267 4	0.0057 19
γ E1+0.1%M2	334.3136 23	2.05 13
γ [E1]	392.3 5	0.0016 3
γ [E1]	429.87 18	0.0037 6
γ [E1]	434.9 3	0.0123 25
γ [M1]	436.137 8	0.00083 17
γ [E1]	448.30 18	0.00025 5
γ [M1]	454.568 8	0.00124 23
γ [E1]	461.9 4	0.0016 3
γ [E1]	469.7 4	0.00106 21
γ [E1]	484.3 3	0.00103 21
γ [E1]	492.2 3	0.0059 12
γ [E1]	497.71 18	0.0032 5
γ [E1]	498.8 5	0.0031 6
γ [E2]	503.983 8	0.0014 3

† 2.8% uncert(syst)

Atomic Electrons (^{239}Np)

$\langle$e$\rangle$=134.1 22 keV

e_{bin}(keV)	$\langle$e$\rangle$(keV)	e(%)
1 - 17	0.24	3.1 4
18	3.7	20.5 24
22	4.3	19.2 24
23 - 34	4.3	15.9 12
35	4.6	13.2 15
38	0.047	0.121 18
39	4.8	12.2 14
40 - 87	16.7	29.2 14
88	8.3	9.4 4
93 - 105	2.80	2.78 11
106	25.0	23.5 10
110 - 151	0.460	0.341 18

Atomic Electrons (^{239}Np) (continued)

e_{bin}(keV)	$\langle$e$\rangle$(keV)	e(%)
156	27.6	17.7 8
159 - 204	4.98	2.64 8
205	8.5	4.15 17
206 - 253	5.21	2.38 5
254	7.9	3.11 14
255 - 304	4.62	1.72 4
308 - 348	0.073	0.0233 16
362 - 412	0.00094	0.00025 3
413 - 462	0.00095	0.00022 5
464 - 504	0.00033	6.9 8 $\times10^{-5}$

Continuous Radiation (^{239}Np)

$\langle\beta$-$\rangle$=118 keV;$\langle$IB$\rangle$=0.052 keV

E_{bin}(keV)		$\langle\ \rangle$(keV)	(%)
0 - 10	β-	0.292	5.9
	IB	0.0061	
10 - 20	β-	0.85	5.7
	IB	0.0054	0.038
20 - 40	β-	3.24	10.9
	IB	0.0090	0.032
40 - 100	β-	19.4	28.2
	IB	0.0170	0.027
100 - 300	β-	79	45.2
	IB	0.0138	0.0092
300 - 600	β-	15.4	4.27
	IB	0.00083	0.00024
600 - 714	β-	0.228	0.0363
	IB	2.2$\times10^{-6}$	3.5$\times10^{-7}$

$^{239}_{94}$Pu(2.411 3 $\times10^4$ yr)

Mode: α, SF(4.4$\times10^{-10}$ %)
Δ: 48584.9 22 keV
SpA: 0.06204 Ci/g
Prod: daughter ^{239}Np

Alpha Particles (^{239}Pu)

$\langle\alpha\rangle$=5101 1 keV

α(keV)	α(%)
4058.2 4	<2$\times10^{-8}$
4115.9 4	7 2$\times10^{-8}$
4179.6 4	5 1$\times10^{-8}$
4291.2 4	5 1$\times10^{-8}$
4325.8 4	~3$\times10^{-7}$
4348.1 4	5$\times10^{-8}$
4363.0 4	~5$\times10^{-8}$
4389.1 4	1.2 1$\times10^{-6}$
4390.9 4	2.0$\times10^{-7}$
4399.0 4	2.5 8$\times10^{-5}$
4447.1 4	2.6$\times10^{-6}$
4463.5 4	1.08 3$\times10^{-5}$
4466.2 4	3.8 3$\times10^{-6}$
4495.8 4	~4$\times10^{-7}$
4502.1 4	2.4 1$\times10^{-6}$
4507.6 4	2.7 1$\times10^{-5}$
4526.0 4	<8$\times10^{-6}$
4528.3 4	3.0 1$\times10^{-6}$
4533.0 4	2.5 7$\times10^{-6}$
4557.7 4	~0.00020
4613.5 4	~2$\times10^{-5}$

Alpha Particles (^{239}Pu)
(continued)

α(keV)	α(%)
4631.3 4	0.0007 2
4671.8 4	<9 $\times 10^{-6}$
4689.2 4	0.00046 3
4717.3 4	0.0004 1
4736.0 4	0.0053 3
4747.8 4	0.00079 4
4769.0 4	0.0012 2
4794.7 4	0.0010 2
4804.3 4 ?	0.00036 15
4822.4 4 ?	<10 $\times 10^{-5}$
4828.3 4	0.0037 2
4865.8 4	0.0011 1
4869.3 4	0.0007 3
4910.6 4	~0.002
4933.9 4	0.004 1
4961.7 4	0.0054 2
4987.1 4	0.007 2
4987.7 4	<0.010
5007.6 4	0.017 2
5028.4 4	0.008 3
5054.2 4	0.023 7
5075.2 4	0.056 6
5104.7 4	10.6 13
5110.1 4	<0.030
5142.8 4	15.1 2
5155.5 4	73.2 7

Photons (^{239}Pu)
$\langle\gamma\rangle$=0.0660 6 keV

γ_{mode}	γ(keV)	γ(%)†
γ [M1]	30.08 3	0.000105 11
γ M1+46%E2	38.674 19	0.00586 12
γ M1+E2	42.09 3	~0.00010 ?
γ (M1+1.9%E2)	46.22 3	0.00051 2
γ [M1]	47.52 3	2.9 7 $\times 10^{-5}$
γ E2	51.622 20	0.0208 4
γ [M1]	54.02 3	0.000169 7
γ M1(+3.6%E2)	56.82 3	0.000930 19
γ (M1+E2)	65.70 4	3.3 5 $\times 10^{-5}$
γ M1+E2	67.69 4	0.000142 7
γ (M1+E2)	68.73 4	~0.00012
γ (E2)	68.75 3	<0.0006
γ (M1+E2)	74.89 4	3.8 6 $\times 10^{-5}$
γ (M1+~20%E2)	77.593 22	0.000427 9
γ [M1]	78.37 6	0.000169 8
γ (M1+E2)	89.60 3	1.3 $\times 10^{-5}$ §
γ [M1]	89.68 10	
U $K_{\alpha2}$	94.651	0.00367 4 *
U $K_{\alpha1}$	98.434	0.00590 6 *
γ [E2]	98.81 4	0.0013 1
γ E2	103.05 4	0.000179 9
U $K_{\beta1}$'	111.025	0.00225 2 *
U $K_{\beta2}$'	114.866	0.000559 6*
γ E2	115.36 4	0.000676 14
γ (M1+~50%E2)	116.266 24	0.000596 12
γ (M1+E2)	119.68 3	3.5 $\times 10^{-5}$ §
γ [E1]	119.72 4	
γ [E1]	122.34 4	~3 $\times 10^{-6}$?
γ [M1]	123.64 5	1.68 17 $\times 10^{-5}$
γ [E2]	124.51 4	6.2 2 $\times 10^{-5}$
γ [E1]	125.15 3	5.8 3 $\times 10^{-5}$
γ E1	129.283 20	0.00620 12
γ [M1]	141.65 4	3.11 15 $\times 10^{-5}$
γ [M1+E2]	143.62 4	5.7 12 $\times 10^{-6}$
γ E2	144.18 8	0.000286 10
γ [E2]	146.05 6	0.000113 4
γ [E2]	158.35 3	8 3 $\times 10^{-6}$
γ [E2]	160.07 13	5 2 $\times 10^{-6}$
γ [M1]	161.44 3	0.000130 4
γ [E2]	168.05 9	3.8 10 $\times 10^{-6}$
γ [E1]	171.37 3	0.0001090 22
γ [E2]	173.70 3	4.0 9 $\times 10^{-6}$
γ [E1]	179.17 3	6.39 13 $\times 10^{-5}$
γ [M1]	184.22 19	1.6 6 $\times 10^{-6}$
γ [E1]	188.04 5	9.6 10 $\times 10^{-6}$
γ [M1]	189.34 4	7.76 16 $\times 10^{-5}$
γ [M1]	195.66 4	0.0001070 21
γ	197.98 10	5.0 8 $\times 10^{-6}$
γ M1(+E2)	203.52 3	0.000560 11
γ [M1+E2]	218.45 19	~1 $\times 10^{-6}$
γ [E1]	225.39 3	1.63 7 $\times 10^{-5}$
γ [M1]	237.75 4	1.5 2 $\times 10^{-5}$
γ M1+E2	242.12 4	8.3 12 $\times 10^{-6}$
γ [M1]	243.36 4	2.32 12 $\times 10^{-5}$
γ [E1]	244.86 4	5.2 8 $\times 10^{-6}$
γ [M1]	248.88 4	7.5 5 $\times 10^{-6}$
γ [M1]	255.34 3	8.03 16 $\times 10^{-5}$
γ [M1]	263.90 4	2.55 13 $\times 10^{-5}$
γ [E1]	265.81 7	3.0 7 $\times 10^{-6}$
γ [M1+E2]	281.12 3	2.1 7 $\times 10^{-6}$
γ [M1+E2]	285.27 4	1.5 6 $\times 10^{-6}$
γ [M1]	297.43 3	5.0 1 $\times 10^{-5}$
γ [M1]	302.89 3	5.7 6 $\times 10^{-6}$
γ [M1]	307.82 3	6.2 6 $\times 10^{-6}$
γ [E1]	311.68 4	2.74 6 $\times 10^{-5}$
γ [M1+E2]	313.39 10	2.0 8 $\times 10^{-6}$
γ [M1]	316.40 6	1.41 7 $\times 10^{-5}$
γ [M1+E2]	319.79 3	5.7 $\times 10^{-5}$ §
γ [E1]	320.81 4	
γ (M1)	323.77 4	5.39 11 $\times 10^{-5}$
γ E1	332.81 3	0.000505 10
γ M1	336.08 5	0.0001130 23
γ M1	341.50 4	6.63 13 $\times 10^{-5}$
γ M1(+E2)	344.94 3	0.00057 §
γ (M1)	344.98 3	
γ [E2]	354.02 4	8 3 $\times 10^{-7}$
γ [M1]	361.84 3	1.17 4 $\times 10^{-5}$
γ [E1]	367.03 4	8.72 17 $\times 10^{-5}$
γ [E1]	368.51 4	8.96 18 $\times 10^{-5}$
γ M1+50%E2	375.018 21	0.00158 3
γ M1(+E2)	380.17 3	0.000307 6
γ (M1)	382.68 6	0.00026 §
γ (M1)	382.71 4	
γ (M1)	392.50 4	0.000116 11
γ M1(+E2)	393.12 4	0.000444 15
γ [E1]	399.46 6	6.1 3 $\times 10^{-6}$
γ (M1)	410.94 7	~8 $\times 10^{-6}$
γ	412.83 20	~2 $\times 10^{-8}$
γ M1+49%E2	413.691 21	0.00151 3
γ M1+~50%E2	422.57 3	0.000119 5
γ E1+E2	426.64 3	2.29 7 $\times 10^{-5}$
γ [E1]	430.16 4	4.9 4 $\times 10^{-6}$
γ [E1]	445.68 6	9.5 $\times 10^{-6}$ §
γ [E1]	445.78 10	
γ M1+~39%E2	451.44 3	0.000192 4
γ [M1]	457.57 6	1.54 11 $\times 10^{-6}$
γ [E2]	461.25 3	2.0 2 $\times 10^{-6}$
γ [E1]	463.80 20	1.6 5 $\times 10^{-7}$
γ [E1]	474.26 3	9 3 $\times 10^{-8}$
γ [E2]	481.52 3	4.77 10 $\times 10^{-6}$
γ [E1]	486.99 4	2.5 4 $\times 10^{-7}$
γ [E1]	493.25 8	8.8 5 $\times 10^{-7}$
γ	538.90 20	3.0 6 $\times 10^{-7}$
γ [E1]	550.60 9	4.0 4 $\times 10^{-7}$
γ	557.7 3	6 2 $\times 10^{-8}$
γ [E1]	586.27 8	1.4 3 $\times 10^{-7}$
γ [E2]	598.05 9	1.7 $\times 10^{-6}$ §
γ	598.07 5	
γ (M1)	606.9 3	1.5 3 $\times 10^{-7}$
γ [E1]	612.93 8	8.1 3 $\times 10^{-7}$
γ M1+E2	617.43 9	2.04 8 $\times 10^{-6}$
γ	618.94 6	2.44 10 $\times 10^{-6}$?
γ M1	624.80 20	4.0 2 $\times 10^{-7}$ §
γ [E1]	624.94 8	
γ M1	633.19 10	2.32 7 $\times 10^{-6}$
γ (E2)	637.96 8	2.50 8 $\times 10^{-6}$
γ [M1]	640.15 10	7.9 2 $\times 10^{-6}$ §
γ	640.15 5	
γ [E1]	645.98 7	1.45 3 $\times 10^{-5}$
γ [E1]	649.41 9	8 2 $\times 10^{-7}$
γ [E1]	652.18 7	6.4 2 $\times 10^{-6}$
γ (E2)	654.88 9	2.1 2 $\times 10^{-6}$
γ [E1]	658.93 7	9.5 2 $\times 10^{-6}$
γ M1(+E2)	664.62 8	1.54 6 $\times 10^{-6}$
γ (M1)	674.26 9	5.4 $\times 10^{-7}$ §
γ [M1+E2]	674.6 3	
γ	686.16 10	8.9 4 $\times 10^{-7}$
γ [E1]	690.85 7	5.5 4 $\times 10^{-7}$
γ	701.00 20	6.0 $\times 10^{-7}$ §
γ [M1+E2]	701.10 9	
γ [E1]	703.80 7	3.88 8 $\times 10^{-6}$
γ	714.58 20	9.0 9 $\times 10^{-8}$
γ	717.74 5	2.67 8 $\times 10^{-6}$
γ [E1]	727.84 14	1.09 9 $\times 10^{-6}$
γ	756.42 5	3.37 7 $\times 10^{-6}$
γ	769.37 5	1.100 22 $\times 10^{-5}$
γ [M1]	779.46 14	1.33 7 $\times 10^{-7}$
γ	787.30 20	8.4 6 $\times 10^{-8}$
γ	793.00 20	2.5 5 $\times 10^{-8}$
γ	796.50 20	3.2 5 $\times 10^{-8}$
γ	803.30 20	4.4 7 $\times 10^{-8}$
γ	808.20 20	1.47 12 $\times 10^{-7}$
γ	813.90 20	6.2 7 $\times 10^{-8}$
γ M1	821.10 20	5.1 6 $\times 10^{-8}$
γ	828.80 20	1.38 9 $\times 10^{-7}$
γ	831.93 15	2.5 5 $\times 10^{-8}$
γ M1	839.0 10	~4 $\times 10^{-8}$
γ	843.80 20	1.60 15 $\times 10^{-7}$
γ [E1]	879.00 20	5.0 9 $\times 10^{-8}$
γ	891.10 20	7.7 12 $\times 10^{-8}$
γ	940.10 20	4.2 7 $\times 10^{-8}$
γ	956.40 20	5.6 10 $\times 10^{-8}$
γ	979.52 14	2.3 4 $\times 10^{-8}$
γ [M1+E2]	986.7 3	1.2 5 $\times 10^{-8}$
γ	992.47 14	2.3 5 $\times 10^{-8}$
γ	1005.63 15	1.2 4 $\times 10^{-8}$
γ	1057.32 15	4.5 7 $\times 10^{-8}$?

† 3.0% uncert(syst)
§ combined intensity for doublet
* exp value

$^{239}_{95}$Am(11.9 1 h)

Mode: ϵ(99.990 1 %), α(0.010 1 %)
Δ: 49385 5 keV
SpA: 1.102 $\times 10^{6}$ Ci/g
Prod: ^{239}Pu(p,n); ^{239}Pu(d,2n); ^{237}Np(α,2n)

Alpha Particles (^{239}Am)
$\langle\alpha\rangle$=0.5755 2 keV

α(keV)	α(%)
5680 2	0.000198 3
5734 2	0.001375 7
5776.1 18	0.00837 4
5824.5 18	3.3 2 $\times 10^{-5}$

Photons (^{239}Am)

$\langle\gamma\rangle$=242 9 keV

γ_{mode}	γ(keV)	γ(%)†
Pu L_ℓ	12.124	2.9 4
Pu L_α	14.262	46 6
Pu L_η	16.333	0.67 9
Pu L_β	18.068	45 7
Pu L_γ	21.627	10.0 16
γ_ϵM1+3.6%E2	44.665 3	0.089 8
γ_αE1	49.26 10	0.005 1
γ_ϵM1+20%E2	49.4152 19	0.109 9
γ_ϵE2	57.2759 18	0.153 16
γ_ϵM1+E2	57.278 4	0.023 4
γ_ϵE1	61.462 4	0.0020 4
γ_ϵE2	67.8462 22	0.130 13
γ_ϵM1+20%E2	88.057 18	0.00062 24
Pu $K_{\alpha2}$	99.522	38 4
γ_ϵE2	101.943 5	~0.008
Pu $K_{\alpha1}$	103.734	61 7
γ_ϵE1+0.3%M2	106.1272 23	0.047 5
γ_ϵE2	106.488 18	0.0050 10
Pu $K_{\beta1}$'	116.930	22.2 24
Pu $K_{\beta2}$'	120.974	7.7 9
γ_ϵM1(+<6.3%E2)	124.439 8	0.100 10
γ_ϵM1	166.364 18	0.014 6
γ_ϵM1+1.7%E2	181.716 7	1.08 6
γ_ϵM1+0.08%E2	209.7554 17	3.52 12
γ_ϵM1+23%E2	226.382 7	3.30 19
γ_ϵM1+0.6%E2	228.1865 15	11.4 4
γ_ϵM1+1.7%E2	254.421 4	0.084 5
γ_ϵM1	272.852 4	0.064 4
γ_ϵM1+2.6%E2	277.6016 16	15.0 5
γ_ϵE2	285.4624 15	0.79 3
γ_ϵ(M1+E2)	311.698 5	0.0164 15
γ_ϵE1+2.3%M2	315.8825 23	0.0033 3
γ_ϵ[E2]	322.267 4	0.0049 17
γ_ϵE1+0.1%M2	334.3136 23	0.0042 4
γ_ϵ[E1]	429.87 18	0.0017 3
γ_ϵ[M1]	436.137 8	0.0080 10
γ_ϵ[E1]	448.30 18	0.00012 3
γ_ϵ[M1]	454.568 8	0.0120 12
γ_ϵ[E1]	497.71 18	0.00147 24
γ_ϵ[E2]	503.983 8	0.0140 14

† uncert(syst): 10% for α, 10% for ϵ

Atomic Electrons (^{239}Am)

$\langle e\rangle$=161 3 keV

e_{bin}(keV)	$\langle e\rangle$(keV)	e(%)
1 - 17	1.9	24 6
18	7.4	41 6
22	6.3	28 4
23 - 34	5.3	20.3 15
35	5.2	14.7 25
38	9.6×10^{-5}	~0.00025
39	5.1	13.1 22
40 - 87	18.9	35.0 13
88	8.7	9.8 4
93 - 104	1.50	1.55 8
105	6.1	5.8 6
106	26.5	25.0 9
110 - 151	0.594	0.464 17
156	29.2	18.8 7
159 - 182	2.06	1.26 6
187	3.60	1.93 7
190 - 204	3.62	1.78 11
205	9.0	4.40 16
206 - 253	6.41	2.92 6
254	8.4	3.29 13
255 - 304	4.85	1.80 4
306 - 334	0.0257	0.0080 7
376 - 425	0.0041	0.00103 8
426 - 475	0.0053	0.00122 8
480 - 504	0.00148	0.000304 16

Continuous Radiation (^{239}Am)

$\langle IB\rangle$=1.02 keV

E_{bin}(keV)		$\langle\ \rangle$(keV)	(%)
10 - 20	IB	0.0055	0.032
20 - 40	IB	0.00109	0.0039
40 - 100	IB	0.43	0.47
100 - 300	IB	0.58	0.51
300 - 600	IB	0.0046	0.00141
600 - 743	IB	2.5×10^{-5}	3.8×10^{-6}

$^{239}_{96}$Cm(~ 3 h)

Mode: ϵ

Δ: 51090 100 keV syst

SpA: ~ 4×10^6 Ci/g

Prod: ^{239}Pu(α,4n)

Photons (^{239}Cm)

γ_{mode}	γ(keV)
γ [M1+E2]	40.7 5
γ (E1)	146.6 5
γ (E1)	187.4 4

$^{239}_{98}$Cf(~39 s)

Mode: α

Δ: 58240 320 keV syst

SpA: 1.2×10^9 Ci/g

Prod: daughter ^{243}Fm

Alpha Particles (^{239}Cf)

α(keV)
7630 25

A = 240

NDS **20**, 218 (1977)

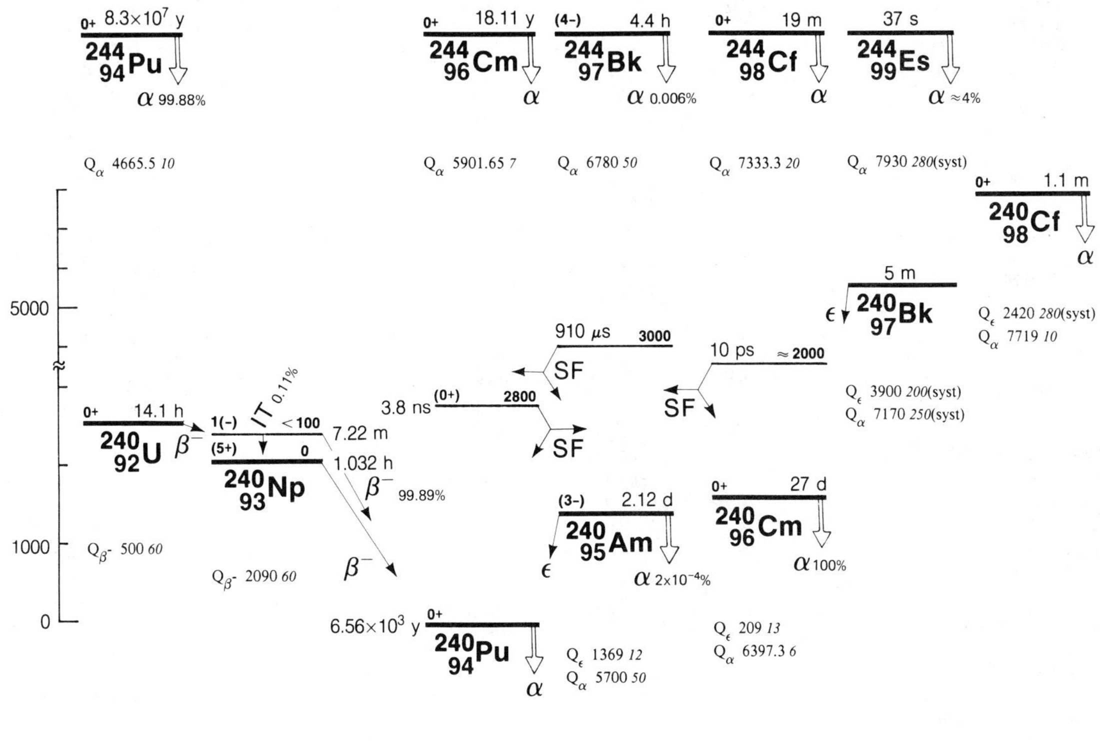

$^{240}_{92}$U (14.1 *2* h)

Mode: β-
Δ: 52711 *5* keV
SpA: 9.26×10^5 Ci/g
Prod: multiple n-capture from ^{238}U

Photons (^{240}U)

⟨γ⟩=7.3 *7* keV

γ_{mode}	γ(keV)	γ(%)
Np L_ℓ	11.871	0.97 *16*
Np L_α	13.927	15.5 *23*
Np L_η	15.861	0.116 *18*
Np L_β	17.500	18 *3*
Np L_γ	21.139	5.1 *10*
γ M1	44.10 *7*	1.69 *20*

Atomic Electrons (^{240}U)

⟨e⟩=30 *3* keV

e_{bin}(keV)	⟨e⟩(keV)	e(%)
18	0.034	0.19 *3*
22	19.8	91 *11*
26	0.103	0.39 *5*
38	6.4	16.6 *20*
39	0.76	1.96 *24*
40	0.048	0.121 *14*
43	2.2	5.1 *6*
44	0.72	1.64 *20*

$^{240}_{93}$Np(1.032 *3* h)

Mode: β-
Δ: 52210 *60* keV
SpA: 1.265×10^7 Ci/g
Prod: ^{238}U(α,pn); ^{238}U(^{3}He,p)

Photons (^{240}Np)

⟨γ⟩=1193 *72* keV

γ_{mode}	γ(keV)	$\gamma(\%)^{\dagger}$
γ E2	42.817 *6*	
γ E2	98.855 *11*	
γ	134.6 *10*	0.40 *8*
γ (M1+E2)	147.1 *3*	1.5 *3*
γ (E2)	152.622 *20*	9.0 *18*
γ	175 *1*	6.5 *13*
γ	182.6 *10*	1.0 *2*
γ (M1+E2)	193.06 *22*	7.3 *15*
γ (M1+E2)	270.8 *3*	9.0 *18*
γ	280.14 *14*	<0.7
γ (M1)	295.6 *3*	0.70 *14*
γ (M1+E2)	306.9 *3*	1.5 *3*
γ [E2]	309.97 *6*	<0.15
γ [E2]	361.51 *6*	<0.12
γ (E1)	448.16 *20*	18 *3*
γ	462.2 *10*	1.5 *3*
γ [E2]	466.89 *18*	2.2 *4*
γ [E1]	507.22 *4*	1.9 *3*
γ (M1+E2)	566.39 *17*	29 *6*
γ (E1)	600.78 *20*	22 *4*
γ [E1]	606.07 *4*	1.78 *22*
γ	847 *1*	5 *1*
γ [E1]	867.44 *20*	9.0 *18*
γ	884.9 *10*	4.0 *8*
γ E2	888.78 *4*	1.37 *11*
γ [E1]	896.3 *3*	14 *3*
γ [E1]	916.04 *5*	1.5 *3*
γ [M2]	958.86 *6*	<0.017
γ [E1]	959.14 *20*	2.5 *5*
γ [E1]	974.11 *17*	23 *5*
γ E2	987.64 *4*	4.0 *3*
γ	1074.4 *10*	1.0 *2*
γ	1088.49 *16*	0.55 *18*
γ	1131.31 *16*	0.70 *14*
γ	1137.63 *14*	<0.25
γ	1163 *1*	0.70 *14*
γ [E1]	1167.17 *20*	5 *1*
γ [M1+E2]	1180.25 *16*	0.70 *6*
γ [E2]	1223.06 *16*	0.24 *3*

† 20% uncert(syst)

Continuous Radiation (^{240}Np)

⟨β-⟩=273 keV;⟨IB⟩=0.24 keV

E_{bin}(keV)		⟨ ⟩(keV)	(%)
0 - 10	β-	0.104	2.08
	IB	0.0133	
10 - 20	β-	0.312	2.08
	IB	0.0126	0.087
20 - 40	β-	1.25	4.17
	IB	0.023	0.081
40 - 100	β-	8.8	12.5
	IB	0.056	0.088
100 - 300	β-	76	38.5
	IB	0.098	0.059
300 - 600	β-	148	34.8
	IB	0.035	0.0089
600 - 872	β-	39.1	5.8
	IB	0.00154	0.00024

$^{240}_{93}$Np(7.22 *2* min)

Mode: β-(99.89 *3* %), IT(0.11 *3* %)
Δ: <52310 keV
SpA: 1.084×10^8 Ci/g
Prod: daughter ^{240}U

Photons (^{240}Np)

⟨γ⟩=328 *7* keV

γ_{mode}	γ(keV)	γ(%)†
$\gamma_{\beta-}$E2	42.817 *6*	0.08 *3*
$\gamma_{\beta-}$	66.49 *10*	0.27 *3*
$\gamma_{\beta-}$E2	98.855 *11*	0.17 *3*
$\gamma_{\beta-}$(E2)	152.622 *20*	
$\gamma_{\beta-}$[E2]	189.45 *9*	0.250 *20*
$\gamma_{\beta-}$[E1]	251.43 *5*	0.93 *6*
$\gamma_{\beta-}$(E1)	263.33 *6*	1.09 *6*
$\gamma_{\beta-}$	280.14 *14*	0.014 *3*
$\gamma_{\beta-}$[E2]	289.17 *7*	0.017 *4*
$\gamma_{\beta-}$	296.3 *3*	0.0050 *20*
$\gamma_{\beta-}$[E1]	302.96 *4*	1.18 *5*
$\gamma_{\beta-}$[E2]	309.97 *6*	0.052 *10*
$\gamma_{\beta-}$[M1]	340.70 *7*	0.074 *10*
$\gamma_{\beta-}$[E2]	361.51 *6*	0.041 *10*
$\gamma_{\beta-}$	475.0 *3*	0.009 *4*
$\gamma_{\beta-}$	496.8 *3*	0.011 *5*
$\gamma_{\beta-}$[E1]	507.22 *4*	0.79 *4*
$\gamma_{\beta-}$	518.4 *3*	0.0050 *20*
$\gamma_{\beta-}$E1	554.54 *3*	22.3 *10*
$\gamma_{\beta-}$E1	597.35 *3*	12.6 *6*
$\gamma_{\beta-}$[E1]	606.07 *4*	0.74 *4*
$\gamma_{\beta-}$[M1+E2]	658.63 *7*	<0.018 ?
$\gamma_{\beta-}$E2	758.64 *4*	1.17 *5*
$\gamma_{\beta-}$[E1]	789.57 *7*	0.210 *20*
$\gamma_{\beta-}$(E1)	813.42 *14*	0.211 *25*
$\gamma_{\beta-}$E2	817.87 *6*	1.28 *6*
$\gamma_{\beta-}$	837.49 *20*	0.0050 *20*
$\gamma_{\beta-}$[E1]	841.11 *7*	0.166 *12*
$\gamma_{\beta-}$[M1+E2]	857.50 *4*	0.47 *3*
$\gamma_{\beta-}$[E2]	890.78 *7*	0.015 *4*
$\gamma_{\beta-}$	895.24 *7*	0.051 *10*
$\gamma_{\beta-}$[E2]	900.31 *4*	0.130 *20*
$\gamma_{\beta-}$[E1]	910.05 *7*	0.170 *20*
$\gamma_{\beta-}$[E1]	916.04 *5*	1.04 *6*
$\gamma_{\beta-}$	928.58 *10*	0.170 *20*
$\gamma_{\beta-}$	938.05 *7*	1.29 *5*
$\gamma_{\beta-}$[M1+E2]	942.31 *6*	0.110 *20*
$\gamma_{\beta-}$[E1]	959.14 *20*	~0.0047 ?
$\gamma_{\beta-}$[E1]	961.58 *7*	0.144 *10*
$\gamma_{\beta-}$	986.3 *7*	~0.020
$\gamma_{\beta-}$	989.19 *20*	0.074 *10*
$\gamma_{\beta-}$	1029.0 *3*	0.0045 *22*
$\gamma_{\beta-}$[E2]	1046.95 *10*	0.100 *10*
$\gamma_{\beta-}$	1061.72 *24*	0.036 *7*
$\gamma_{\beta-}$	1072.2 *3*	0.025 *8*
$\gamma_{\beta-}$	1088.49 *16*	0.039 *10*

Photons (^{240}Np)
(continued)

γ_{mode}	γ(keV)	γ(%)†
$\gamma_{\beta-}$	1093.78 *23*	
$\gamma_{\beta-}$[M1+E2]	1094.66 *15*	~0.023 §
$\gamma_{\beta-}$	1113.25 *24*	0.015 *4*
$\gamma_{\beta-}$	1131.31 *16*	0.050 *10*
$\gamma_{\beta-}$[E2]	1137.48 *15*	~0.010 §
$\gamma_{\beta-}$	1137.63 *14*	
$\gamma_{\beta-}$[M1+E2]	1180.25 *16*	~0.024 §
$\gamma_{\beta-}$	1180.45 *14*	
$\gamma_{\beta-}$	1198.0 *3*	~0.006
$\gamma_{\beta-}$[E2]	1223.06 *16*	0.0166 *19*
$\gamma_{\beta-}$	1305.57 *23*	0.029 *7*
$\gamma_{\beta-}$	1321.2 *3*	0.029 *10*
$\gamma_{\beta-}$	1357.11 *23*	~0.019
$\gamma_{\beta-}$[E2]	1417.27 *7*	~0.014
$\gamma_{\beta-}$	1428.29 *20*	0.029 *10*
$\gamma_{\beta-}$	1435.6 *3*	~0.004
$\gamma_{\beta-}$[E1]	1445.33 *7*	0.36 *3*
$\gamma_{\beta-}$	1483.11 *10*	0.035 *9*
$\gamma_{\beta-}$[E1]	1488.15 *7*	0.210 *20*
$\gamma_{\beta-}$[E1]	1496.85 *6*	1.31 *7*
$\gamma_{\beta-}$[M1+E2]	1516.12 *7*	0.010 *3*
$\gamma_{\beta-}$[E1]	1539.67 *6*	0.79 *10*
$\gamma_{\beta-}$[E2]	1558.94 *7*	0.0040 *10*
$\gamma_{\beta-}$	1568.49 *20*	0.0040 *10*
$\gamma_{\beta-}$	1584.5 *4*	0.019 *8*
$\gamma_{\beta-}$	1590.45 *9*	0.091 *12*
$\gamma_{\beta-}$	1604.4 *3*	0.032 *10*
$\gamma_{\beta-}$	1607.39 *20*	0.045 *10*
$\gamma_{\beta-}$	1633.26 *9*	0.144 *15*
$\gamma_{\beta-}$	1667.78 *24*	0.0160 *20*
$\gamma_{\beta-}$	1752.49 *20*	0.0048 *12*
$\gamma_{\beta-}$	1765.19 *20*	0.0067 *11*
$\gamma_{\beta-}$	1874.69 *20*	0.0080 *20*
$\gamma_{\beta-}$	1911.64 *23*	0.014 *3*
$\gamma_{\beta-}$	1954.46 *23*	0.0027 *12*

† 5.0% uncert(syst)
§ combined intensity for doublet

Continuous Radiation (^{240}Np)

⟨β-⟩=630 keV;⟨IB⟩=1.10 keV

E_{bin}(keV)		⟨ ⟩(keV)	(%)
0 - 10	β-	0.0401	0.80
	IB	0.026	
10 - 20	β-	0.121	0.80
	IB	0.025	0.18
20 - 40	β-	0.489	1.63
	IB	0.049	0.170
40 - 100	β-	3.55	5.06
	IB	0.133	0.21
100 - 300	β-	35.9	17.9
	IB	0.33	0.19
300 - 600	β-	116	26.0
	IB	0.29	0.070
600 - 1300	β-	349	39.1
	IB	0.22	0.028
1300 - 2180	β-	124	8.2
	IB	0.017	0.00120

$^{240}_{94}$Pu(6563 *7* yr)

Mode: α
Δ: 50122.4 *21* keV
SpA: 0.22696 Ci/g
Prod: multiple n-capture from ^{238}U; ^{239}Pu(n,γ)

Alpha Particles (^{240}Pu)

⟨α⟩=5154.9 *1* keV

α(keV)	α(%)
4217.28 *24*	<2×10^{-7}
4264.31 *12*	~6×10^{-7}
4492.03 *6*	2.0 *2*×10^{-5}
4654.62 *5*	4.7 *5*×10^{-5}
4863.53 *5*	0.00113 *2*
5021.17 *5*	0.071 *1*
5123.66 *5*	26.39 *21*
5168.15 *5*	73.5 *4*

Photons (^{240}Pu)

⟨γ⟩=0.0286 *5* keV

γ_{mode}	γ(keV)	γ(%)†
γ E2	45.244 *6*	0.0450 *9*
U $K_{\alpha2}$	94.651	6.1 *6*×10^{-5} *
U $K_{\alpha1}$	98.434	1.00 *5*×10^{-5} *
γ E2	104.235 *5*	0.00700 *14*
U $K_{\beta1}$'	111.025	3.6 *4*×10^{-5} *
U $K_{\beta2}$'	114.866	~1×10^{-5}
γ E2	160.312 *8*	0.00042 *8*
γ E2	212.46 *4*	3.0 *6*×10^{-5} *
γ E3	538.11 *5*	1.2 *3*×10^{-7}
γ E1+M2	642.35 *5*	1.0 *2*×10^{-5}
γ E1	687.59 *5*	2.7 *6*×10^{-6}
γ [E1]	698.8 *5*	<3×10^{-8}
γ [E2]	873.94 *12*	5.8 *6*×10^{-7}
γ E1	967.01 *24*	<5×10^{-8}

† 1.0% uncert(syst)
* exp value

$^{240}_{95}$Am(2.117 *13* d)

Mode: ϵ, α(0.00019 *7* %)
Δ: 51491 *12* keV
SpA: 2.570×10^5 Ci/g
Prod: ^{239}Pu(d,n); ^{239}Pu(α,p2n); ^{237}Np(α,n)

Alpha Particles (^{240}Am)

⟨α⟩=0.010210 *3* keV

α(keV)	α(%)
5286 *3*	2.34 *19*×10^{-6}
5337 *2*	2.28 *8*×10^{-5}
5378 *1*	0.0001649 *19*

Photons (^{240}Am)

⟨γ⟩=1031 *15* keV

γ_{mode}	γ(keV)	γ(%)†
Pu L_ℓ	12.124	2.2 *3*
Pu L_α	14.262	35 *3*
Pu L_η	16.333	0.62 *8*
Pu L_β	18.092	38 *5*
Pu L_γ	21.612	8.7 *12*
γ_ϵE2	42.817 *6*	0.09 *1*
γ_ϵE2	98.855 *11*	1.5 *2*

Photons (^{240}Am) (continued)

γ_{mode}	γ(keV)	γ(%)†
Pu $K_{\alpha2}$	99.522	17.6 *8*
Pu $K_{\alpha1}$	103.734	28.1 *13*
Pu $K_{\beta1}'$	116.930	10.2 *5*
Pu $K_{\beta2}'$	120.974	3.51 *18*
γ_ϵ(E2)	152.622 *20*	0.012 *3*
γ_ϵ[E1]	249.8 *7*	0.020 *3*
γ_ϵ[E1]	251.43 *5*	0.0064 *8*
γ_ϵ[E1]	302.96 *4*	0.0081 *10*
γ_ϵ[E2]	309.97 *6*	<0.009
γ_ϵ[E1]	343.4 *7*	0.049 *5*
γ_ϵ[E2]	361.51 *6*	<0.007
γ_ϵ	382 *1*	0.053 *5*
γ_ϵ(E1)	448.16 *20*	0.013 *4*
γ_ϵ[E1]	507.22 *4*	0.073 *5*
γ_ϵE1	554.54 *3*	0.0100 *8*
γ_ϵE1	597.35 *3*	0.0056 *5*
γ_ϵ(E1)	600.78 *20*	0.015 *5*
γ_ϵ[E1]	606.07 *4*	0.068 *5*
γ_ϵ[E2]	698.0 *7*	0.035 *8* ?
γ_ϵE2	758.64 *4*	0.0080 *23*
γ_ϵ[M1+E2]	857.50 *4*	0.0032 *10*
γ_ϵE2	888.78 *4*	25.1 *5*
γ_ϵ[E2]	900.31 *4*	0.0009 *3*
γ_ϵ[E1]	916.04 *5*	0.090 *5*
γ_ϵ[M1+E2]	934.6 *3*	0.025 *3*
γ_ϵ[E2]	937.9 *5*	0.007 *3*
γ_ϵ[M2]	958.86 *6*	<0.0010
γ_ϵ[E1]	959.14 *20*	0.040 *4*
γ_ϵE2	987.64 *4*	73.4 *15*
γ_ϵ[E2]	1033.4 *3*	0.010 *1*
γ_ϵ[M1+E2]	1036.11 *21*	0.016 *2*
γ_ϵ[M1+E2]	1090.5 *5*	0.0031 *6*
γ_ϵ[M1+E2]	1094.66 *15*	0.016 *1*
γ_ϵ[M1+E2]	1120.27 *24*	0.011 *1*
γ_ϵ[M1+E2]	1134.96 *21*	0.049 *3*
γ_ϵ[E2]	1137.48 *15*	0.0073 *19*
γ_ϵ[M1+E2]	1180.25 *16*	0.0102 *8*
γ_ϵ[E2]	1189.4 *5*	~0.0005
γ_ϵ	1195.21 *24*	0.0026 *5*
γ_ϵ[M1+E2]	1219.12 *24*	0.035 *2*
γ_ϵ[E2]	1223.06 *16*	0.0071 *10*
γ_ϵ	1294.06 *24*	0.009 *1*

† 5.0% uncert(syst)

Atomic Electrons (^{240}Am)

$\langle e\rangle$=68.6 *21* keV

e_{bin}(keV)	$\langle e\rangle$(keV)	e(%)
18	5.5	31 *4*
20	0.226	1.14 *13*
21	6.5	32 *4*
22	4.8	22 *3*
23	1.33	5.8 *6*
25	6.8	28 *3*
31	0.00072	0.0023 *6*
37	3.3	8.9 *10*
38	3.0	7.8 *9*
39 - 76	3.47	7.4 *4*
77	8.3	10.8 *15*
81	6.0	7.5 *10*
86	0.074	0.087 *9*
93	3.0	3.3 *4*
94 - 135	4.3	4.5 *3*
147 - 188	0.0126	0.0083 *12*
222 - 260	0.04	~0.016
280 - 326	0.0031	0.00103 *15*
337 - 385	0.021	0.006 *3*
425 - 448	0.00061	0.000142 *11*
476 - 507	0.0045	0.00093 *7*
531 - 579	0.0038	0.00066 *14*
583 - 606	0.00086	0.000146 *7*
637 - 680	0.0022	0.00033 *7*
692 - 741	0.0017	~0.00023
753 - 759	0.00010	1.4 *5* $\times10^{-5}$
767	2.21	0.288 *8*
779 - 816	0.009	0.0012 *5*
834 - 857	0.0018	0.00022 *4*

Atomic Electrons (^{240}Am) (continued)

e_{bin}(keV)	$\langle e\rangle$(keV)	e(%)
866	6.51	0.752 *21*
867 - 916	0.519	0.0593 *10*
917 - 965	1.66	0.172 *5*
969 - 1018	0.721	0.0734 *14*
1027 - 1077	0.0034	0.00032 *11*
1085 - 1134	0.010	0.0009 *4*
1136 - 1185	0.0018	~0.00016
1186 - 1222	0.0019	0.00016 *6*
1271 - 1293	0.00032	~3$\times10^{-5}$

Continuous Radiation (^{240}Am)

$\langle$IB$\rangle$=0.94 keV

E_{bin}(keV)		$\langle\ \rangle$(keV)	(%)
10 - 20	IB	0.0063	0.037
20 - 40	IB	0.00119	0.0043
40 - 100	IB	0.41	0.45
100 - 300	IB	0.52	0.47
300 - 600	IB	5.1$\times10^{-5}$	1.45$\times10^{-5}$
600 - 772	IB	3.6$\times10^{-7}$	5.7$\times10^{-8}$

$^{240}_{96}$Cm(27 *1* d)

Mode: α
Δ: 51701 *6* keV
SpA: 2.02$\times10^4$ Ci/g
Prod: ^{239}Pu(α,3n)

Alpha Particles (^{240}Cm)

$\langle\alpha\rangle$=6245 *1* keV

α(keV)	α(%)
5989	0.014
6147	0.05
6247.8 *6*	28.8 *6*
6290.6 *6*	70.6 *6*

Photons (^{240}Cm)

$\langle\gamma\rangle$=1.93 *11* keV

γ_{mode}	γ(keV)	γ(%)
Pu L_ℓ	12.124	0.28 *3*
Pu L_α	14.262	4.5 *3*
Pu L_η	16.333	0.105 *11*
Pu L_β	18.136	5.3 *5*
Pu L_γ	21.554	1.18 *14*
γ E2	44.59 *10*	0.0380 *11*

Atomic Electrons (^{240}Cm)

$\langle e\rangle$=10.01 *23* keV

e_{bin}(keV)	$\langle e\rangle$(keV)	e(%)
18	0.84	4.6 *5*
21	0.086	0.402 *15*
22	3.31	14.8 *8*
23	0.0177	0.076 *8*
27	2.53	9.5 *3*

Atomic Electrons (^{240}Cm) (continued)

e_{bin}(keV)	$\langle e\rangle$(keV)	e(%)
39	1.21	3.11 *11*
40	1.08	2.70 *10*
41	0.0280	0.0688 *25*
43	0.696	1.61 *6*
44	0.188	0.425 *15*
45	0.0291	0.0653 *24*

$^{240}_{97}$Bk(5 *2* min)

Mode: ϵ
Δ: 55600 *200* keV syst
SpA: 1.6$\times10^8$ Ci/g
Prod: ^{232}Th(^{14}N,6n)

$^{240}_{98}$Cf(1.06 *15* min)

Mode: α
Δ: 58020 *200* keV syst
SpA: 7.4$\times10^8$ Ci/g
Prod: ^{233}U(^{12}C,5n)

Alpha Particles (^{240}Cf)

α(keV)
7590 *10*

A = 241

NDS **23**, 123 (1978)

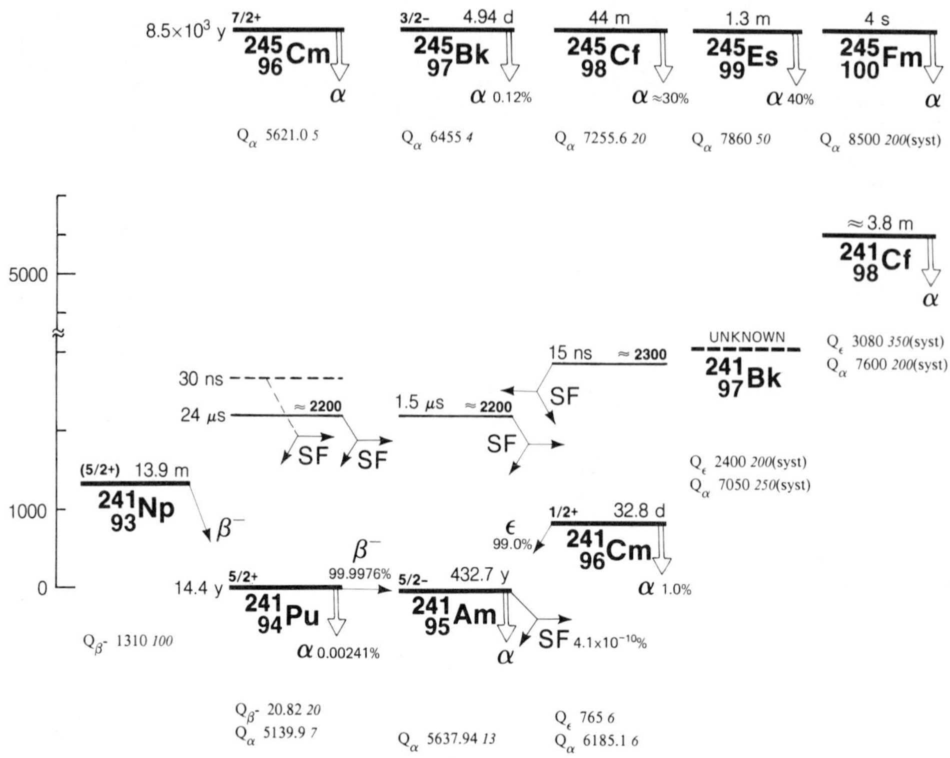

$^{241}_{93}$Np(13.9 *2* min)

Mode: β-
Δ: 54260 *100* keV
SpA: 5.61×10^{7} Ci/g
Prod: ^{238}U(α,p)

Photons (^{241}Np)

γmode	γ(keV)	γ(rel)
γ[M1+E2]	132.99 *3*	29.2 *15*
γ[M1+E2]	174.94 *4*	100
γ	280 *1*	

$^{241}_{94}$Pu(14.4 *2* yr)

Mode: β-(99.99759 *4* %), α(0.00241 *4* %)
Δ: 52952.0 *21* keV
SpA: 103.0 Ci/g
Prod: multiple n-capture from ^{238}U; multiple n-capture from ^{239}Pu

Alpha Particles (^{241}Pu)

⟨α⟩=0.1178 *1* keV

α(keV)	α(%)
4693 *6*	~7×10^{-7}
4733 *25*	~7×10^{-7}
4743 *5*	~2×10^{-6}
4784.5 *12*	~5×10^{-6}
4797.5 *11*	2.89 *24*×10^{-5}
4853.2 *7*	0.000292 *5*
4896.6 *7*	0.002005 *12*
4969 ?	
4972.4 *7*	3.13 *24*×10^{-5}
4998.5 *7*	9.9 *12*×10^{-6}
4999 ?	
5042.7 *7*	2.5×10^{-5}
5053.9 *7*	8×10^{-6}

Photons (^{241}Pu)

⟨γ⟩=0.001354 *17* keV

γmode	γ(keV)	γ(%)†
γ	38.57 *10*	9.6 *19*×10^{-6}
γ[M1+E2]	44.16 *9*	4.10 *14*×10^{-6}
γ[M1+E2]	44.887 *11*	8.2 *10*×10^{-7}
γ[E2]	56.275 *20*	2.46 *14*×10^{-6}
γ[M1+E2]	56.6 *10*	9.6 *19*×10^{-7}
γ[E2]	71.49 *8*	2.8 *3*×10^{-6}
γ[M1+E2]	77.04 *8*	2.17 *7*×10^{-5}
U K$_{\alpha2}$	94.651	0.000280 *7* *

Photons (^{241}Pu)
(continued)

γmode	γ(keV)	γ(%)†
U K$_{\alpha1}$	98.434	0.000448 *9* *
γ[M1+E2]	103.653 *5*	0.0000991 *20*
U K$_{\beta1}$'	111.025	0.00016 *1* *
γE1	114.0 *10*	6.0 *12*×10^{-6}
U K$_{\beta2}$'	114.866	4.46 *12*×10^{-5} *
γ[M1+E2]	121.21 *11*	6.7 *7*×10^{-7}
γ[M1+E2]	148.540 *10*	0.000183 *4*
γ[E2]	159.928 *19*	6.60 *14*×10^{-6}

† 1.7% uncert(syst)
* exp value

Continuous Radiation (^{241}Pu)

⟨β-⟩=5.2 keV;⟨IB⟩=0.000106 keV

E$_{bin}$(keV)		⟨ ⟩(keV)	(%)
0 - 10	β-	3.43	86
	IB	0.000104	
10 - 20	β-	1.81	14.2
	IB	2.5×10^{-6}	2.1×10^{-5}
20 - 21	β-	0.00127	0.0063
	IB	4.7×10^{-12}	2.4×10^{-11}

$^{241}_{95}$Am(432.7 5 yr)

Mode: α, SF(4.1 1 $\times10^{-10}$ %)
Δ: 52931.2 21 keV
SpA: 3.428 Ci/g
Prod: daughter ^{241}Pu

Alpha Particles (^{241}Am)

⟨α⟩=5480 1 keV

α(keV)	α(%)
4757.56 16	
4800.92 9	9 $\times10^{-5}$
4834.28 9	0.0007
4883 6 ?	
4959 6 ?	
5004 3	0.00010
5055.58 10	
5066 3	0.00014
5092.31 9	~0.00040
5099.36 13	~0.00040
5107 5 ?	
5117.30 10	0.0004
5133 4 ?	
5155.32 9	0.0007
5179.49 9	0.0003
5181.78 8	0.0009
5190.59 13	0.0006
5217.41 8	
5225.24 9	0.0013
5236 5 ?	
5244.29 9	0.0024
5281.15 8	0.0005
5322.06 9	0.015 5
5388.40 9	1.4 2
5416.54 13	~0.010
5443.01 8	12.8 2
5469.68 9 ?	<0.04
5485.70 8	85.2 8
5511.61 8	0.20 5
5544.25 8	0.34 5

Photons (^{241}Am)

⟨γ⟩=28.7 6 keV

γ_{mode}	γ(keV)	γ(%)
Np L_ℓ	11.871	0.81 10
γ M1+0.1%E2	13.804 16	
Np L_α	13.927	13.0 12
Np L_η	15.861	0.33 4
Np L_β	17.611	20.2 24
Np L_γ	20.997	5.2 7
γ E1	26.3445 10	2.4 1
γ	32.2 10	0.0174 4
γ M1+1.8%E2	33.1920 14	0.12 1
γ (M1+~43%E2)	42.64 3	~0.0010
γ M1+14%E2	43.415 10	0.073 10
γ E1	51.013 18	2.5 5 $\times10^{-5}$
γ [M1+E2]	54.04 11	~0.0006
γ M1+17%E2	55.528 23	0.020 5
γ [E2]	56.86 3	
γ E1	59.5364 10	35.7
γ E1	64.817 13	0.00014 3
γ (M1+18%E2)	67.46 3	~0.00044
γ [E1]	69.760 10	~0.024
γ (E2)	75.83 3	~0.00011
γ [M1+E2]	79.09 4	
γ [M1+E2]	92.09 10	
γ [E2]	96.67 11	4.7 16 $\times10^{-5}$?

Photons (^{241}Am) (continued)

γ_{mode}	γ(keV)	γ(%)
Np $K_{\alpha2}$	97.066	0.00126 11
γ E2	98.944 21	0.020 2
Np $K_{\alpha1}$	101.059	0.00201 17
γ [M1+E2]	101.43 6	
γ E1	102.952 10	0.0195 10
γ [E2]	109.76 4	~2 $\times10^{-5}$?
Np $K_{\beta1}'$	113.944	0.00074 6
Np $K_{\beta2}'$	117.891	0.000249 22
γ E2	122.989 25	0.0010 1
γ [E1]	125.288 21	0.00400 10
γ	139.40 20	4.5 10 $\times10^{-6}$
γ E2	146.549 20	0.00046 2
γ [E1]	150.110 25	8.0 6 $\times10^{-5}$
γ	154.4 3	8.0 16 $\times10^{-6}$?
γ	156.4 3	1.20 24 $\times10^{-5}$?
γ [M2]	158.480 21	1.4 5 $\times10^{-6}$?
γ	161.7 3	9.0 18 $\times10^{-6}$?
γ E2	164.593 11	6.6 6 $\times10^{-5}$
γ [M1+E2]	165.92 3	2.4 2 $\times10^{-5}$
γ E2	169.56 3	0.000168 18
γ	175.09 10	1.8 1 $\times10^{-5}$
γ	190.40 10	2.2 5 $\times10^{-6}$
γ [E2]	191.90 4	2.16 18 $\times10^{-7}$
γ	197.0 4	1.3 3 $\times10^{-5}$?
γ	201.70 14	<20 $\times10^{-7}$?
γ	203.9 10	2.9 6 $\times10^{-6}$
γ M1+2.4%E2	208.008 9	0.00079 13
γ [M1+E2]	221.45 3	4.2 2 $\times10^{-5}$
γ E2	221.812 17	
γ [M1+E2]	232.86 7	4.6 6 $\times10^{-6}$
γ M2	234.352 9	7.0 8 $\times10^{-7}$
γ	242.4 4	3.0 6 $\times10^{-5}$?
γ	245.0 3	3.0 6 $\times10^{-5}$?
γ [M1+E2]	246.66 6	2.4 6 $\times10^{-6}$
γ	249.10 16	4 2 $\times10^{-6}$?
γ	260.90 10	~10 $\times10^{-7}$
γ [M1+E2]	264.87 3	9 1 $\times10^{-6}$
γ E1+19%M2	267.544 9	2.56 24 $\times10^{-5}$
γ	271.58 10	6.42 18 $\times10^{-7}$
γ [M1+E2]	275.68 5	6.6 10 $\times10^{-6}$
γ [E1]	291.21 3	3.1 4 $\times10^{-6}$
γ [E2]	292.76 3	1.38 17 $\times10^{-5}$
γ	294.9 5	5.5 11 $\times10^{-5}$?
γ [M1+E2]	300.10 10	<20 $\times10^{-7}$?
γ	304.2 10	1.0 4 $\times10^{-6}$
γ	310.3 6	1.5 3 $\times10^{-5}$?
γ	317 5	
γ (M1+~26%E2)	322.54 3	0.000152 3
γ [E2]	329.71 11	<2 $\times10^{-5}$?
γ E2	332.361 13	0.00015 3
γ M1+18%E2	335.401 14	0.000495 10
γ (E2)	337.727 19	4.2 3 $\times10^{-6}$
γ [E1]	349.46 10	1.2 5 $\times10^{-6}$?
γ	358.36 10	1.2 3 $\times10^{-6}$
γ M1(+<8.8%E2)	368.592 14	0.000231 14
γ M1+12%E2	370.919 19	5.26 19 $\times10^{-5}$
γ (M1)	376.58 10	0.000138 5
γ [M1+E2]	383.75 4	2.8 2 $\times10^{-5}$
γ	390.6 4	5.9 10 $\times10^{-6}$
γ	406.4 10	1.4 6 $\times10^{-6}$?
γ [M1+E2]	419.22 10	2.8 1 $\times10^{-5}$
γ [M1+E2]	426.39 5	2.46 10 $\times10^{-5}$
γ	429.8 10	~1 $\times10^{-6}$
γ	435.3 10	~2 $\times10^{-6}$
γ	442.75 20	3.5 10 $\times10^{-6}$
γ [E2]	452.41 10	2.4 3 $\times10^{-6}$?

Photons (^{241}Am) (continued)

γ_{mode}	γ(keV)	γ(%)
γ [M1+E2]	454.67 6	9.7 8 $\times10^{-6}$
γ [M1+E2]	459.58 5	3.6 6 $\times10^{-6}$
γ	468.0 10	2.9 6 $\times10^{-6}$
γ	486.3 10	1.0 5 $\times10^{-6}$
γ	512.00 20	1.15 25 $\times10^{-6}$?
γ [E1]	514.20 6	2.6 10 $\times10^{-6}$
γ	522.0 10	9 3 $\times10^{-7}$
γ	525.7 10	8.0 16 $\times10^{-5}$?
γ [M1+E2]	574.05 14	~1 $\times10^{-6}$
γ	582.6 10	~2 $\times10^{-7}$?
γ	586.5 10	1.3 4 $\times10^{-6}$
γ	590.3 10	2.8 3 $\times10^{-6}$
γ [M1+E2]	597.41 5	7.4 20 $\times10^{-6}$
γ [M1+E2]	619.011 24	5.90 20 $\times10^{-5}$
γ	627.2 10	5.6 20 $\times10^{-7}$
γ	633 1	1.3 3 $\times10^{-6}$
γ [M1+E2]	641.51 14	7 3 $\times10^{-6}$
γ [M1+E2]	652.94 4	3.8 3 $\times10^{-5}$
γ (E0+M1+E2)	662.426 23	0.00036 2
γ	669.9 10	3.8 12 $\times10^{-7}$
γ	675.8 10	6.4 20 $\times10^{-7}$
γ [E1]	680.06 5	3.1 4 $\times10^{-6}$
γ [E1]	688.770 23	3 1 $\times10^{-5}$
γ	693.5 10	3.7 2 $\times10^{-6}$
γ [M1+E2]	696.36 4	5.0 5 $\times10^{-6}$ §
γ [M1+E2]	697.03 14	
γ	710.95 16	6.4 6 $\times10^{-6}$
γ [E1]	721.962 23	6.0 12 $\times10^{-5}$
γ [E1]	722.70 4	0.00013 3
γ	729.5 10	1.3 2 $\times10^{-6}$
γ	731.5 10	4.7 15 $\times10^{-7}$
γ	737.30 16	8.0 8 $\times10^{-6}$
γ [E1]	755.89 4	7.6 8 $\times10^{-6}$
γ	759.5 10	1.70 20 $\times10^{-6}$
γ	763.4 10	1.98 6 $\times10^{-7}$
γ [E1]	766.79 14	5.0 6 $\times10^{-6}$
γ	770.49 16	5.0 5 $\times10^{-6}$
γ	772.1 10	2.66 20 $\times10^{-6}$
γ	777.2 10	~6 $\times10^{-8}$
γ	780.5 10	2.5 10 $\times10^{-7}$
γ	788.8 10	3.9 10 $\times10^{-7}$
γ	801.9 2	1.3 2 $\times10^{-6}$
γ	811.8 10	6 1 $\times10^{-7}$
γ	819.3 10	4.0 7 $\times10^{-7}$
γ	822.6 10	2.2 6 $\times10^{-7}$
γ	828.5 10	2.4 6 $\times10^{-7}$
γ	851.5 10	3.8 7 $\times10^{-7}$
γ	854.7 10	2.0 4 $\times10^{-7}$
γ	859.2 10	8.4 24 $\times10^{-8}$
γ	862.6 10	5.3 10 $\times10^{-7}$
γ	872.7 20	7 2 $\times10^{-7}$
γ	887.5 10	2.2 5 $\times10^{-7}$
γ	898.4 10	7 3 $\times10^{-8}$
γ	902.5 10	3.0 5 $\times10^{-7}$
γ	912.4 10	2.5 5 $\times10^{-7}$
γ	922.2 10	1.9 4 $\times10^{-7}$
γ	928.8 10	~5 $\times10^{-8}$
γ	945.7 10	~5 $\times10^{-8}$
γ	955.7 10	5.8 7 $\times10^{-7}$

†1.4% uncert(syst)
§ 696.4γ + 697.0γ

Atomic Electrons (^{241}Am)

$\langle e \rangle$=30.4 *6* keV

e_{bin}(keV)	$\langle e \rangle$(keV)	e(%)
5 - 9	0.754	12.9 *4*
11	1.21	11.2 *10*
12 - 17	0.76	5.8 *6*
18	1.18	6.7 *7*
20	0.006	~0.031
21	1.13	5.4 *6*
22	4.5	20.5 *23*
23 - 26	0.83	3.3 *5*
27	0.74	2.71 *23*
28 - 36	1.21	3.9 *3*
37	2.93	7.90 *13*
38	6.96	18.3 *6*
39 - 41	0.32	0.83 *13*
42	2.08	4.96 *17*
43 - 53	0.24	0.51 *7*
54	4.48	8.29 *14*
55 - 104	1.033	1.71 *3*
105 - 154	0.00581	0.00485 *19*
155 - 204	0.00143	0.00077 *8*
207 - 256	0.00116	0.00051 *4*
257 - 306	0.00033	0.000122 *11*
307 - 355	0.00049	0.000151 *8*
357 - 406	7.3×10^{-5}	1.96 *17* $\times 10^{-5}$
407 - 456	4.5×10^{-5}	$\sim 1 \times 10^{-5}$
458 - 507	3.7×10^{-5}	7 *4* $\times 10^{-6}$
508 - 557	0.00015	$\sim 3 \times 10^{-5}$
560 - 610	2.0×10^{-5}	3.4 *8* $\times 10^{-6}$
611 - 659	5.3×10^{-5}	8 *3* $\times 10^{-6}$
661 - 710	7.0×10^{-6}	1.03 *24* $\times 10^{-6}$
711 - 760	2.3×10^{-6}	3.1 *7* $\times 10^{-7}$
761 - 810	6.0×10^{-7}	7.7 *19* $\times 10^{-8}$
811 - 859	2.5×10^{-7}	2.9 *13* $\times 10^{-8}$
861 - 909	7.3×10^{-8}	8.3 *23* $\times 10^{-9}$
911 - 956	3.9×10^{-8}	4.2 *21* $\times 10^{-9}$

$^{241}_{96}$Cm(32.8 *2* d)

Mode: ϵ(99.0 *1* %), α(1.0 *1* %)
Δ: 53696 *6* keV
SpA: 1.652×10^4 Ci/g
Prod: ^{239}Pu(α,2n)

Alpha Particles (^{241}Cm)

$\langle \alpha \rangle$=59.29 *1* keV

α(keV)	α(%)
5684.5 *4*	0.0022 *5*
5717.7 *4*	~0.0008
5785 *3*	~0.0007
5861.5 *4*	0.0014 *5*
5884.2 *4*	0.118 *4*
5914 *4*	0.0012 *5*
5929.1 *4*	0.181 *5*
5938.9 *4*	0.689 *10*
5978 *3*	0.0028 *7*
6035.1 *4*	0.0012 *4*
6082.0 *4*	0.0015 *5*

Photons (^{241}Cm)

$\langle \gamma \rangle$=481 *14* keV

γ_{mode}	γ(keV)	γ(%)†
γ_ϵ[M1+E2]	15.2282 *20*	0.055 *3*
γ_ϵ[E1]	28.99 *3*	0.030 *6*
γ_ϵ[M1+E2]	32.639 *3*	0.21 *3*
γ_ϵ[M1+E2]	41.176 *3*	0.017 *4*
Am $K_{\alpha2}$	102.026	22.3 *10*
Am $K_{\alpha1}$	105.472	35.2 *16*
Am $K_{\beta1}$'	119.960	12.9 *6*
Am $K_{\beta2}$'	124.123	4.44 *22*
γ_ϵ[M1+E2]	132.413 *5*	3.86 *20*
γ_αE3	145.586 *7*	*
γ_ϵ[E2]	147.641 *5*	0.020 *8*
γ_ϵ[E1]	151.29 *3*	~0.020
γ_ϵ[E1]	164.707 *10*	0.48 *8*
γ_ϵ[M1+E2]	165.052 *5*	2.97 *20*
γ_ϵ[M1+E2]	180.280 *5*	0.48 *4*
γ_ϵ[E1]	205.883 *10*	2.66 *14*
γ_ϵ[E1]	265.927 *9*	0.40 *4*
γ_ϵ[E1]	298.565 *10*	0.079 *20*
γ_ϵ[E2]	410.81 *10*	0.086 *9*
γ_ϵ[E2]	417.22 *3*	0.65 *4*
γ_ϵ[E2]	430.633 *9*	4.06 *20*
γ_ϵ[E1]	430.978 *10*	~0.040
γ_ϵ[M1+E2]	447.35 *4*	0.119 *15*
γ_ϵ[M1+E2]	463.272 *9*	1.23 *8*
γ_ϵ[M1+E2]	464.35 *7*	0.085 *14*
γ_ϵ[M1+E2]	471.809 *9*	71 *3*
γ_ϵ[M1+E2]	504.448 *9*	0.59 *4*
γ_ϵ[E2]	595.684 *10*	0.015 *3*
γ_ϵ[M2]	623.10 *3*	0.012 *3*
γ_ϵ[M1+E2]	636.860 *10*	1.53 *11*
γ_ϵ[E2]	652.088 *10*	0.040 *10*
γ_ϵ[E1]	653.23 *4*	0.149 *10*
γ_ϵ[E1]	670.24 *8*	0.57 *4*

† uncert(syst): 10% for α, 4.0% for ϵ
* with ^{237}Pu(180 ms)

Continuous Radiation (^{241}Cm)

$\langle$IB$\rangle$=3.6 keV

E_{bin}(keV)		$\langle\ \rangle$(keV)	(%)
10 - 20	IB	0.68	3.9
20 - 40	IB	0.095	0.37
40 - 100	IB	1.48	1.9
100 - 293	IB	1.3	2.5

$^{241}_{98}$Cf(3.8 *7* min)

Mode: α
Δ: 59170 *290* keV syst
SpA: 2.1×10^8 Ci/g
Prod: ^{233}U(^{12}C,4n); ^{235}U(^{12}C,6n); ^{234}U(^{12}C,5n)

Alpha Particles (^{241}Cf)

α(keV)
7335 *5*

A = 242

NDS **21**, 615 (1977)

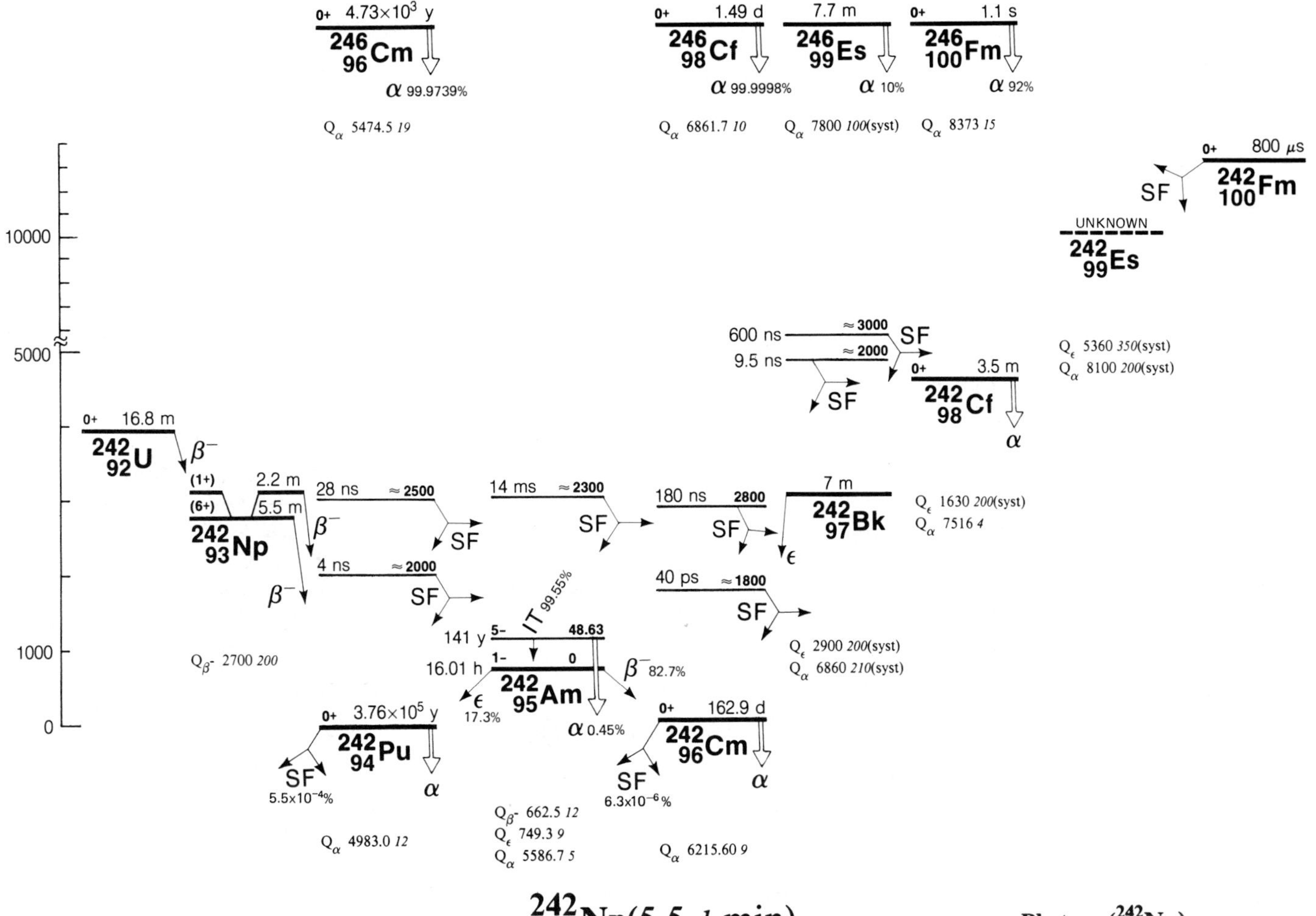

$^{242}_{92}$U (16.8 *5* min)

Mode: β-

SpA: 4.62×10^{7} Ci/g

Prod: ^{244}Pu(n,2pn)

Photons (^{242}U)

⟨γ⟩=38.5 *11* keV

γ_{mode}	γ(keV)	γ(%)†
γ	55.58 *5*	3.75 *15*
γ	67.60 *5*	9.20 *25*
γ	160.4 *1*	0.75 *20*
γ	182.0 *1*	0.70 *5*
γ	220.4 *3*	0.15 *5*
γ	226.23 *7*	~0.10
γ	238.25 *8*	~0.2
γ	274.14 *10*	0.11 *4*
γ	304.43 *15*	0.34 *8*
γ	320.6 *1*	0.20 *5*
γ	329.72 *9*	0.75 *5*
γ	530.67 *15*	~0.2
γ	572.94 *8*	1.8 *1*
γ	584.96 *8*	1.85 *10*

† 8.0% uncert(syst)

$^{242}_{93}$Np(5.5 *1* min)

Mode: β-

Δ: 57410 *200* keV

SpA: 1.41×10^{8} Ci/g

Prod: ^{244}Pu(n,p2n); ^{242}Pu(n,p)

Photons (^{242}Np)

γ_{mode}	γ(keV)	γ(rel)
γ E2	44.533 *9*	
γ [E2]	102.76 *16*	
γ [E2]	159.10 *8*	32 *2*
γ [E1]	265.1 *1*	24 *2*
γ [M1+E2]	785.70 *8*	100
γ [E2]	944.80 *8*	63 *3*
γ	1104 *1*	0.6 *2*

$^{242}_{93}$Np(2.2 *2* min)

Mode: β-

Δ: 57410 *200* keV

SpA: 3.5×10^{8} Ci/g

Prod: daughter ^{242}U

Photons (^{242}Np)

⟨γ⟩=250 *8* keV

γ_{mode}	γ(keV)	γ(%)†
γ E2	44.533 *9*	
γ [E2]	102.76 *16*	
γ	620.90 *10*	0.9 *1*
γ [M1+E2]	647.69 *24*	0.275 *25*
γ	681.7 *4*	0.15 *5*
γ [E1]	685.28 *10*	0.35 *5*
γ [E1]	736.20 *4*	5 *1*
γ [E1]	780.74 *4*	2.65 *5*
γ [E1]	788.04 *18* ?	
γ [E1]	813.95 *8*	1.2 *1*
γ [M1+E2]	948.30 *20*	0.085 *25*
γ [E1]	1007.56 *16*	0.150 *25*
γ	1034.60 *16*	0.28 *5*
γ	1039.33 *23*	0.110 *25*
γ	1093.82 *10*	1.15 *10*
γ [E1]	1110.33 *16*	0.35 *5*
γ	1123.36 *17*	0.25 *5*
γ [E1]	1137.37 *9*	1.25 *5*
γ	1172.3 *3*	0.15 *3*
γ	1181.90 *9*	0.150 *25*
γ	1240.2 *4*	0.13 *5*
γ [E1]	1383.90 *24*	0.13 *5*
γ [E1]	1473.40 *6*	2.25 *5*
γ [E1]	1517.94 *6*	1.20 *5*
γ [E1]	1551.14 *8*	0.35 *5*
γ	1814.00 *20*	0.175 *25*
γ	1827.36 *23*	0.115 *25*
γ	1859.57 *17*	0.550 *25*
γ	1874.56 *10*	0.25 *5*
γ	1905.53 *14*	0.275 *25*
γ	1925.68 *14*	0.225 *25*
γ	1950.06 *14*	0.740 *25*
γ	1970.21 *14*	0.525 *25*

Photons (^{242}Np)
(continued)

γ_{mode}	γ(keV)	γ(%)†
γ	1984.8 5	0.05 1
γ	1992.4 3	0.20 1
γ	2042.7 7	0.04 1
γ	2061.4 10	0.03 1
γ	2077.1 5	0.065 15
γ	2201.8 4	0.060 15
γ	2246.4 4	0.045 15
γ	2358.2 5	~0.050
γ	2370.8 5	~0.050

† 8.0% uncert(syst)

Continuous Radiation (^{242}Np)
$\langle\beta-\rangle$=894 keV;$\langle$IB$\rangle$=2.0 keV

E_{bin}(keV)		$\langle\ \rangle$(keV)	(%)
0 - 10	β-	0.0288	0.57
	IB	0.034	
10 - 20	β-	0.086	0.58
	IB	0.033	0.23
20 - 40	β-	0.348	1.16
	IB	0.064	0.22
40 - 100	β-	2.49	3.54
	IB	0.18	0.28
100 - 300	β-	24.2	12.1
	IB	0.49	0.28
300 - 600	β-	82	18.3
	IB	0.50	0.119
600 - 1300	β-	353	37.9
	IB	0.55	0.066
1300 - 2500	β-	428	25.4
	IB	0.138	0.0089
2500 - 2700	β-	2.90	0.114
	IB	3.1×10^{-5}	1.23×10^{-6}

$^{242}_{94}$Pu(3.763 *20* $\times10^5$ yr)

Mode: α, SF(0.000550 *6* %)
Δ: 54713.9 *21* keV
SpA: 0.003926 Ci/g
Prod: multiple n-capture from ^{238}U; multiple n-capture from ^{239}Pu; daughter ^{242}Am(16.01 h)

Alpha Particles (^{242}Pu)
$\langle\alpha\rangle$=4890 *1* keV

α(keV)	α(%)
4598.4 7	0.0013 5
4754.6 7	0.098 17
4856.4 7	22.4 20
4900.6 7	78 3

Photons (^{242}Pu)
$\langle\gamma\rangle$=1.39 *19* keV

γ_{mode}	γ(keV)	γ(%)†
U L_ℓ	11.620	0.20 5
U L_α	13.600	3.3 7
U L_η	15.400	0.081 18
U L_β	17.128	4.1 9
U L_γ	20.292	0.91 21
γ [E2]	44.915 13	0.036
γ [E2]	103.50 4	0.0078 8
γ [E2]	158.80 8	0.00045 15

† 14% uncert(syst)

Atomic Electrons (^{242}Pu)
$\langle$e$\rangle$=8.1 *7* keV

e_{bin}(keV)	$\langle$e$\rangle$(keV)	e(%)
17	0.66	3.9 9
21	0.66	3.1 7
22	0.0110	0.051 11
23	0.068	0.29 6
24	2.1	8.6 17
28	2.1	7.5 15
39	0.034	0.087 18
40	0.91	2.3 5
41	0.88	2.2 4
43	0.0111	0.025 5
44	0.53	1.21 25
45	0.16	0.35 7

$^{242}_{95}$Am(16.01 *2* h)

Mode: β-(82.7 *3* %), ϵ(17.3 *3* %)
Δ: 55463.2 *22* keV
SpA: 8.088×10^5 Ci/g
Prod: ^{241}Am(n,γ); multiple n-capture from ^{238}U; multiple n-capture from ^{239}Pu

Photons (^{242}Am)
$\langle\gamma\rangle$=18.0 *16* keV

γ_{mode}	γ(keV)	γ(%)†
Pu L_ℓ	12.124	0.30 6
Cm L_ℓ	12.633	0.48 11
Pu L_α	14.262	4.8 8
Cm L_α	14.939	7.3 16
Pu L_η	16.333	0.078 13
Cm L_η	17.314	0.17 4
Pu L_β	18.082	4.9 9
Cm L_β	19.191	8.6 19
Pu L_γ	21.608	1.07 20
Cm L_γ	22.878	1.9 4
$\gamma_{\beta-}$E2	42.2 1	0.039
γ_ϵE2	44.533 9	0.0137 5
Pu $K_{\alpha2}$	99.522	3.6 7
Pu $K_{\alpha1}$	103.734	5.8 12
Pu $K_{\beta1}'$	116.930	2.1 4
Pu $K_{\beta2}'$	120.974	0.72 15

† uncert(syst): 1.7% for ϵ, 0.36% for β-

Atomic Electrons (^{242}Am)
$\langle$e$\rangle$=19.1 *13* keV

e_{bin}(keV)	$\langle$e$\rangle$(keV)	e(%)
18	0.93	5.2 10
19	4.5	24 5
21	0.0313	0.146 6
22	1.50	6.8 6
23	3.6	16 3
24	1.3	5.5 12
25	0.030	0.12 3
26	0.92	3.46 13
36	1.8	4.9 10
37	1.6	4.2 8
38	0.040	0.106 21
39	0.439	1.13 4
40	0.391	0.98 4
41	1.08	2.7 5
42 - 86	0.83	1.76 16
93 - 116	0.128	0.131 11

Continuous Radiation (^{242}Am)
$\langle\beta-\rangle$=159 keV;$\langle$IB$\rangle$=0.30 keV

E_{bin}(keV)		$\langle\ \rangle$(keV)	(%)
0 - 10	β-	0.133	2.66
	IB	0.0080	
10 - 20	β-	0.395	2.64
	IB	0.0082	0.056
20 - 40	β-	1.56	5.2
	IB	0.0134	0.047
40 - 100	β-	10.5	15.0
	IB	0.108	0.133
100 - 300	β-	74	39.0
	IB	0.150	0.120
300 - 600	β-	71	18.2
	IB	0.0106	0.0029
600 - 749	β-	0.401	0.065
	IB	7.4×10^{-5}	1.21×10^{-5}

$^{242}_{95}$Am(141 *2* yr)

Mode: IT(99.55 *2* %), α(0.45 *2* %)
Δ: 55511.8 *22* keV
SpA: 10.48 Ci/g
Prod: ^{241}Am(n,γ)

Alpha Particles (^{242}Am)
$\langle\alpha\rangle$=23.2 keV

α(keV)	α(%)
5064.2 5	0.0010
5088.5 3	0.0009
5141.3 4	0.026
5207.04 21	0.4
5314.5 4	0.003
5367.37 21	0.005
5409.76 21	0.005

Photons (^{242}Am)

$\langle\gamma\rangle$=4.9 *3* keV

γ_{mode}	γ(keV)	γ(%)†
Am L_ℓ	13.377	0.69 *8*
Am L_α	15.599	10.6 *8*
Am L_η	16.819	0.24 *3*
Am L_β	18.928	12.2 *12*
Am L_γ	22.207	2.7 *3*
γ_α[M1+E2]	41.7 *3*	
γ_{IT} E4	48.63 *5*	0.000134 *4*
γ_αE1	49.163 *4*	0.18 *4*
γ_α(E1)	66.694 *20*	0.021 *4*
γ_α[M1+E2]	66.9 *4*	
γ_α[M1+E2]	67.6 *4*	0.0072 *14*
γ_α[E1]	73.1 *5*	0.0058 *12* ?
γ_αM1	86.48 *3*	0.036 *7*
γ_α[E1]	92.27 *6*	0.0040 *8*
γ_α[E1]	109.44 *9*	0.024 *5*
γ_α[E1]	111.0 *4*	0.0027 *5*
γ_α[E2]	121.3 *3*	0.0058 *12*
γ_α[E1]	134.97 *6*	0.0104 *21*
γ_α[E1]	135.64 *3*	0.0095 *19*
γ_α[E1]	152.54 *6*	0.0013 *3*
γ_α	153.63 *6*	0.0045 *9*
γ_α[M1+E2]	163.04 *4*	0.023 *5*
γ_α[E1]	194.43 *5*	
γ_α[E2]	206.14 *5*	

†uncert(syst): <0.1% for IT, 4.4% for α

Atomic Electrons (^{242}Am)

$\langle e\rangle$=40.3 *6* keV

e_{bin}(keV)	$\langle e\rangle$(keV)	e(%)
20	2.10	10.7 *11*
23	1.84	8.0 *9*
24	0.0201	0.085 *9*
25	0.112	0.452 *16*
26	6.33	24.6 *9*
29	6.60	22.7 *8*
42	0.126	0.296 *11*
43	5.5	12.8 *5*
44	3.58	8.2 *3*
45	6.93	15.5 *6*
47	3.16	6.67 *24*
48	3.29	6.86 *25*
49	0.681	1.40 *5*

$^{242}_{96}$Cm(162.94 *6* d)

Mode: α, SF(6.3 *2* $\times10^{-6}$ %)
Δ: 54800.7 *22* keV
SpA: 3311.4 Ci/g
Prod: multiple n-capture from ^{238}U; multiple n-capture from ^{239}Pu; daughter ^{242}Am(16.01 h)

Alpha Particles (^{242}Cm)

$\langle\alpha\rangle$=6043.4 *2* keV

α(keV)	α(%)
5146.14 *8*	1.6 $\times10^{-6}$
5187.06 *10*	5.2 *15* $\times10^{-5}$
5517.84 *7*	0.00024 *5*
5608.0 *10*	2.0 $\times10^{-5}$
5814.57 *9*	0.0046 *5*
5969.39 *7*	0.035 *2*
6069.59 *7*	25.0 *5*
6112.94 *7*	74.0 *5*

Photons (^{242}Cm)

$\langle\gamma\rangle$=1.75 *10* keV

γ_{mode}	γ(keV)	γ(%)†
Pu L_ℓ	12.124	0.25 *3*
Pu L_α	14.262	4.0 *3*
Pu L_η	16.333	0.095 *10*
Pu L_β	18.136	4.8 *5*
Pu L_γ	21.554	1.07 *12*
γE2	44.077 *18*	0.0325 *10*
γE2	101.889 *17*	0.0025 *3*
γ[E2]	157.42 *6*	0.00140 *15*
γ[E2]	210.0 *10*	2.0 *4* $\times10^{-5}$
γE2	301.39 *5*	1.3 *4* $\times10^{-7}$
γ[E1]	336.35 *8*	6.7 *20* $\times10^{-7}$
γE2+15%M1	357.67 *3*	5.9 *18* $\times10^{-7}$
γ[E1]	459.88 *8*	5.7 *24* $\times10^{-8}$
γE1	515.43 *4*	4.5 *3* $\times10^{-6}$
γE1	561.03 *3*	0.00015 *3*
γE1	605.11 *3*	0.00011 *2*
γE1	617.32 *4*	1.0 *2* $\times10^{-5}$
γ[E2]	837.10 *5*	1.8 *3* $\times10^{-7}$
γE2	882.581 *22*	6.2 *12* $\times10^{-8}$
γ(E2)	897.38 *7*	3 *1* $\times10^{-5}$
γE1	918.707 *23*	5.4 *5* $\times10^{-7}$
γE0+E2	938.99 *5*	1.8 *3* $\times10^{-7}$
γE1	962.784 *22*	5.3 *5* $\times10^{-7}$
γ[M1+E2]	979.85 *17*	2.5 *5* $\times10^{-7}$
γ[E2]	983.07 *5*	5.0 *9* $\times10^{-7}$
γE2	984.470 *16*	2.0 *3* $\times10^{-6}$
γE2	1028.546 *17*	1.58 *16* $\times10^{-6}$
γ[E2]	1081.74 *17*	4.9 *16* $\times10^{-8}$
γ[E2]	1118.25 *15*	1.7 *8* $\times10^{-7}$
γE2	1184.58 *17*	4.9 *6* $\times10^{-7}$
γE0+E2+M1	1220.14 *15*	2.9 *5* $\times10^{-7}$

† 8.8% uncert(syst)

Atomic Electrons (^{242}Cm)

$\langle e\rangle$=8.95 *20* keV

e_{bin}(keV)	$\langle e\rangle$(keV)	e(%)
18	0.76	4.2 *5*
21	0.076	0.362 *13*
22	2.94	13.4 *7*
23	0.0159	0.069 *7*
26	2.25	8.6 *3*
38	0.0417	0.109 *4*
39	1.04	2.70 *10*
40	0.99	2.50 *9*
43	0.630	1.47 *5*
44	0.186	0.425 *15*
79	0.00066	0.00084 *11*
80	0.0125	0.0157 *20*
84	0.0086	0.0102 *13*
96	0.0043	0.0045 *6*
97	0.0029	0.0029 *4*
98	6.8 $\times10^{-5}$	7.0 *9* $\times10^{-5}$
100	7.9 $\times10^{-5}$	7.9 *10* $\times10^{-5}$
101	0.0020	0.00201 *25*
102	0.00062	0.00061 *8*

$^{242}_{97}$Bk(7.0 *13* min)

Mode: ϵ
Δ: 57700 *200* keV syst
SpA: 1.11$\times10^{8}$ Ci/g
Prod: ^{235}U(^{11}B,4n); ^{232}Th(^{15}N,5n)

$^{242}_{98}$Cf(3.5 *2* min)

Mode: α
Δ: 59330 *30* keV
SpA: 2.22$\times10^{8}$ Ci/g
Prod: ^{233}U(^{12}C,3n); ^{236}U(^{12}C,6n); ^{235}U(^{12}C,5n); ^{234}U(^{12}C,4n)

Alpha Particles (^{242}Cf)

$\langle\alpha\rangle$~7378 keV

α(keV)	α(%)
7351 *6*	~20
7385 *4*	~80

$^{242}_{100}$Fm(800 *200* μs)

Mode: SF
SpA: 6.7$\times10^{10}$ Ci/g
Prod: ^{204}Pb(^{40}Ar,2n)

A = 243

NDS **33**, 79 (1981)

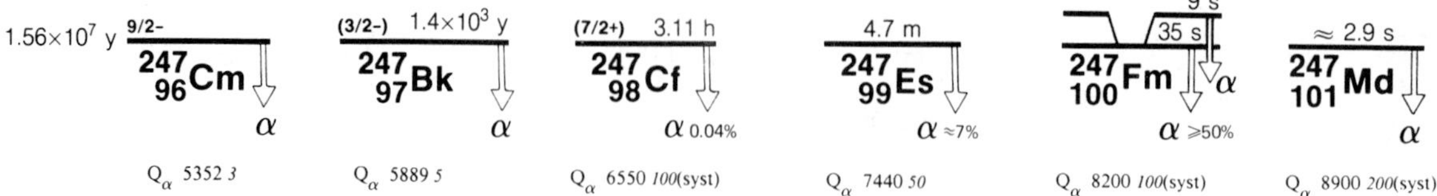

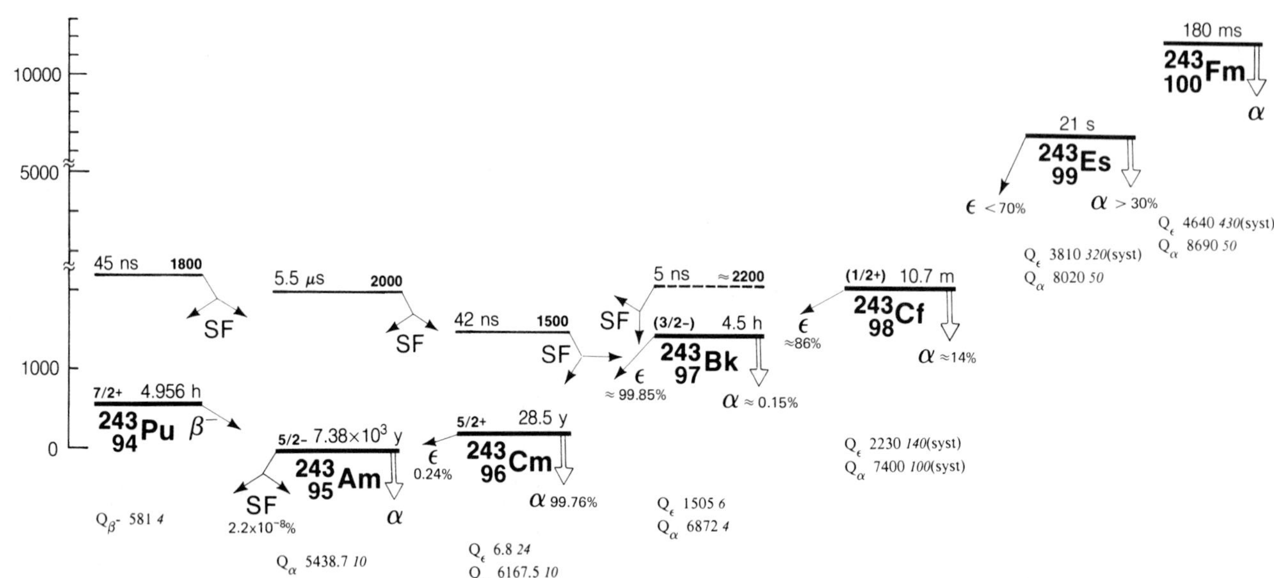

$^{243}_{94}$Pu(4.956 *5* h)

Mode: β-
Δ: 57751 *3* keV
SpA: 2.602×10^{6} Ci/g
Prod: ^{242}Pu(n,γ)

Photons (^{243}Pu)

⟨γ⟩=26 *4* keV

γ_{mode}	γ(keV)	γ(%)†
Am L_ℓ	13.377	0.37 *15*
Am L_α	15.599	5.7 *23*
Am L_η	16.819	0.09 *4*
Am L_β	18.884	6.2 *25*
Am L_γ	22.290	1.5 *6*
γ [E1]	41.75 *18*	0.76 *7*
γ M1+~7.8%E2	42.20 *22*	~0.08
γ [M1+E2]	54.1 *4*	<0.023
γ [E1]	67.06 *25*	~0.23
γ E1	83.95 *16*	23 *5*
γ (E2)	96.3 *4*	0.0138 *23*
γ [E1]	101.3 *3*	<0.037
Am $K_{\alpha2}$	102.026	0.113 *15*
Am $K_{\alpha1}$	105.472	0.178 *24*
γ [E1]	109.27 *17*	0.161 *16*
Am $K_{\beta1}'$	119.960	0.066 *9*
Am $K_{\beta2}'$	124.123	0.022 *3*
γ [M1+E2]	322.11 *25*	0.0276 *23*
γ [M1+E2]	343.2 *5*	~0.0014
γ (M1+E2)	356.37 *21*	0.131 *12*
γ M1	381.68 *21*	0.55 *5*
γ [M1+E2]	388.91 *25*	0.0046 *7*
γ	407.0 *5*	~0.0009
γ [M1+E2]	423.17 *23*	0.0122 *14*
γ [E2]	448.5 *3*	~0.00023
γ [E1]	465.64 *21*	<0.00023

† 8.7% uncert(syst)

Atomic Electrons (^{243}Pu)

⟨e⟩=12.5 *12* keV

e_{bin}(keV)	⟨e⟩(keV)	e(%)
6 - 15	0.6	~6
18	0.8	4.3 *21*
19	0.6	3.2 *15*
20	1.0	~5
21 - 22	0.24	~1
23	1.2	5.5 *24*
24	0.40	~2
25 - 35	0.9	3.0 *14*
36	0.7	1.9 *9*
37 - 40	0.28	0.7 *3*
41	0.28	0.7 *3*
42 - 54	0.46	1.0 *4*
60	0.94	1.6 *3*
61	0.72	1.18 *24*
62 - 63	0.0039	0.0062 *25*
64	0.69	1.08 *22*
65 - 77	0.164	0.22 *3*
78	0.49	0.63 *13*
79 - 119	0.61	0.74 *7*
197 - 231	0.14	~0.06
257	0.83	0.32 *3*
264 - 303	0.024	~0.008
316 - 365	0.29	0.082 *10*
366 - 407	0.084	0.0222 *14*
417 - 466	0.0011	0.00025 *12*

Continuous Radiation (^{243}Pu)

⟨β-⟩=161 keV;⟨IB⟩=0.091 keV

E_{bin}(keV)		⟨ ⟩(keV)	(%)
0 - 10	β-	0.211	4.22
	IB	0.0082	
10 - 20	β-	0.62	4.12
	IB	0.0075	0.052
20 - 40	β-	2.38	8.0
	IB	0.0131	0.046
40 - 100	β-	15.0	21.7
	IB	0.028	0.044
100 - 300	β-	92	49.0
	IB	0.031	0.020
300 - 581	β-	51	13.9
	IB	0.0026	0.00076

$^{243}_{95}$Am(7380 *40* yr)

Mode: α, SF(2.2 *2* ×10^{-8} %)
Δ: 57171 *3* keV
SpA: 0.1993 Ci/g
Prod: multiple n-capture from ^{238}U; multiple n-capture from ^{239}Pu

Alpha Particles (^{243}Am)

⟨α⟩=5265.6 keV

α(keV)	α(%)
4698.7 *5*	0.0017 *5*
4918.4 *5*	9×10^{-5}
4930 *3*	0.00018
4946 *3*	0.0003
4997 *3*	0.0016 †
5008 *3*	

Alpha Particles (^{243}Am)
(continued)

α(keV)	α(%)
5029 *3*	0.0022 §
5037.6 *19*	
5088 *3*	0.004
5112.7 *5*	0.005
5179.8 *5*	1.1
5234.3 *5*	11
5276.6 *5*	88
5319.4 *5*	0.12
5350.0 *5*	0.16

† 4997α + 5008α
§ 5029α + 5038α

Photons (^{243}Am)

⟨γ⟩=48.1 *9* keV

γ_{mode}	γ(keV)	γ(%)†
γ M1+0.1%E2	31.136 *4*	0.066 *6*
γ M1+13%E2	43.03 *3*	0.058 *12*
γ E1	43.537 *3*	5.10 *22*
γ [E1]	50.8 *18*	0.0026 *5*
γ M1+~39%E2	55.40 *4*	0.0092 *18*
γ E1	74.672 *3*	60.0 *12*
γ E1	86.57 *3*	0.30 *3*
γ (E2)	98.43 *4*	~0.008
γ E1	117.70 *3*	0.55 *7*
γ E1	141.97 *4*	0.114 *12*
γ [E1]	170.14 *5*	0.00120 *24*
γ [E1]	195.3 *11*	0.00084 *17*
γ	220.0 *10*	
γ	544.55 *3*	1.7 *4* ×10^{-5}
γ	587.579 *18*	
γ	631.116 *18*	0.00033 *5*
γ	662.251 *18*	0.00095 *15*

† 10% uncert(syst)

$^{243}_{96}$Cm(28.5 *2* yr)

Mode: α(99.76 *4* %), ε(0.24 *4* %)
Δ: 57177.3 *24* keV
SpA: 51.6 Ci/g
Prod: multiple n-capture from ^{238}U; multiple n-capture from ^{239}Pu

Alpha Particles (^{243}Cm)

⟨α⟩=5838 keV

α(keV)	α(%)
5226 *15*	0.0004
5267 *3*	0.0015
5316 *3*	0.0010
5323 *3*	0.003
5332 *3*	0.003
5519.7 *7*	0.0020
5532 *3* ?	0.006
5537 *3*	0.0020
5569.4 *5*	0.007
5575 *3* ?	0.007
5582.6 *6*	~0.009
5587 *3*	~0.020
5593 *3* ?	0.010
5604.6 *6*	<0.010
5612 *3*	~0.030
5622 *10* ?	0.06
5639 *3*	0.14
5646 *3* ?	0.03
5681.5 *5*	0.20
5685.6 *5*	1.6

Alpha Particles (^{243}Cm)
(continued)

α(keV)	α(%)
5713 *5* ?	<0.040
5742.0 *5*	10.57 *20*
5785.9 *5*	73.3 *10*
5876 *3*	0.6
5905.6 *5*	0.10
5992.2 *5*	6.48 *20*
6010.3 *5*	1.0
6058.9 *5*	5
6066.6 *5*	1.5

Photons (^{243}Cm)

⟨γ⟩=131.6 *15* keV

γ_{mode}	γ(keV)	γ(%)†
Pu L_ℓ	12.124	1.16 *14*
Pu L_α	14.262	18.5 *17*
Pu L_η	16.333	0.28 *4*
Pu L_β	18.074	18.7 *24*
Pu L_γ	21.624	4.2 *6*
γ_αM1+3.6%E2	44.665 *3*	0.116 *12*
γ_αM1+20%E2	49.4152 *19*	0.064 *7*
γ_αE2	57.2759 *18*	0.090 *6*
γ_αM1+E2	57.278 *4*	0.024 *8*
γ_αE1	61.462 *4*	0.0111 *18*
γ_αE2	67.8462 *22*	
γ_αM1+20%E2	88.057 *18*	
Pu $K_{\alpha2}$	99.522	14.4 *4*
γ_αE2	101.943 *5*	~0.009
Pu $K_{\alpha1}$	103.734	23.0 *6*
γ_αE1+0.3%M2	106.1272 *23*	0.262 *23*
γ_αE2	106.488 *18*	
Pu $K_{\beta1}$'	116.930	8.32 *24*
Pu $K_{\beta2}$'	120.974	2.87 *9*
γ_αM1	166.364 *18*	0.019 *8*
γ_αM1+0.08%E2	209.7554 *17*	3.27 *9*
γ_αM1+0.6%E2	228.1865 *15*	10.56 *25*
γ_αM1+1.7%E2	254.421 *4*	0.110 *8*
γ_αM1	272.852 *4*	0.083 *7*
γ_αM1+2.6%E2	277.6016 *16*	14.0 *4*
γ_αE2	285.4624 *15*	0.733 *18*
γ_α(M1+E2)	311.698 *5*	0.0175 *20*
γ_αE1+2.3%M2	315.8825 *23*	0.0183 *13*
γ_α[E2]	322.267 *4*	0.0063 *21*
γ_αE1+0.1%M2	334.3136 *23*	0.0237 *16*
γ_α[E1]	392.3 *5*	
γ_α[E1]	429.87 *18*	
γ_α[E1]	434.9 *3*	
γ_α[E1]	448.30 *18*	
γ_α[E1]	461.9 *4*	
γ_α[E1]	469.7 *4*	
γ_α[E1]	484.3 *3*	
γ_α[E1]	492.2 *3*	
γ_α[E1]	497.71 *18*	
γ_α[E1]	498.8 *5*	
γ_α	640.0 *25*	
γ_α	680.0 *25*	
γ_α	720.0 *25*	
γ_α	740.0 *25*	
γ_α	760.0 *25*	

†<0.1% uncert(syst)

Atomic Electrons (^{243}Cm)

⟨e⟩=112.9 *18* keV

e_{bin}(keV)	⟨e⟩(keV)	e(%)
18	3.0	16.4 *20*
22	3.6	16.5 *22*
23 - 38	6.2	20 *3*
39	3.5	9.0 *16*
40 - 86	5.7	10.8 *9*
88	8.1	9.2 *3*
93 - 105	0.51	0.53 *3*

Atomic Electrons (^{243}Cm)
(continued)

e_{bin}(keV)	⟨e⟩(keV)	e(%)
106	24.7	23.2 *7*
110 - 151	0.491	0.361 *18*
156	27.2	17.5 *6*
160 - 166	0.113	0.069 *3*
187	3.35	1.79 *6*
190 - 204	0.93	0.455 *16*
205	8.4	4.09 *13*
206 - 253	4.92	2.24 *4*
254	7.8	3.07 *10*
255 - 304	4.52	1.68 *3*
306 - 334	0.0041	0.0013 *4*

$^{243}_{97}$Bk(4.5 *2* h)

Mode: ε(~ 99.85 %), α(~ 0.15 %)
Δ: 58682 *6* keV
SpA: 2.87×10^{6} Ci/g
Prod: ^{241}Am(α,2n); ^{242}Cm(d,n); ^{243}Am(α,4n)

Alpha Particles (^{243}Bk)

⟨α⟩= 9.8 *3* keV

α(keV)	α(%)
6182 *4*	0.0058 *8*
6210 *3*	0.0204 *13*
6394 *25*	~0.00030
6446 *5*	0.0010 *3*
6502 *4*	0.0104 *10*
6542 *4*	0.0291 *19*
6573.8 *22*	0.0384 *24*
6605 *5*	~0.0010
6666 *4*	~0.0018
6718.0 *22*	0.0188 *13*
6758.1 *22*	0.0231 *15*

Photons (^{243}Bk)

⟨γ⟩=176 *41* keV

γ_{mode}	γ(keV)	γ(%)
γ_α[M1+E2]	40.7 *5*	~0.006
γ_ϵE2	87.4 *1*	
γ_α(E1)	146.6 *5*	0.012 *5*
γ_α(E1)	187.4 *4*	0.060 *15*
γ_α[M1]	557 *4*	0.015 *3*
γ_ϵ	755 *2*	10.0 *20*
γ_ϵ	840 *40*	3.0 *6*
γ_ϵ	946 *2*	~8

$^{243}_{98}$Cf(10.7 *5* min)

Mode: ε(~ 86 %), α(~ 14 %)
Δ: 60910 *140* keV syst
SpA: 7.2×10^{7} Ci/g
Prod: ^{235}U(^{12}C,4n); ^{236}U(^{12}C,5n); ^{238}U(^{12}C,7n); ^{242}Cm(^{3}He,2n)

Alpha Particles (^{243}Cf)

$\langle\alpha\rangle$~993 keV

α(keV)	α(%)
7060 *10*	~10
7170 *10* ?	~4

$^{243}_{99}$Es(21 *2* s)

Mode: ε(<70 %), α(>30 %)
Δ: 64720 *290* keV syst
SpA: 2.17×10^{9} Ci/g
Prod: ^{233}U(^{15}N,5n)

Alpha Particles (^{243}Es)

α(keV)
7890 *20*

$^{243}_{100}$Fm(180 *80* ms)

Mode: α
Δ: 69360 *320* keV syst
SpA: 7×10^{10} Ci/g
Prod: ^{206}Pb(^{40}Ar,3n)

Alpha Particles (^{243}Fm)

α(keV)
8546 *25*

A = 244

NDS **34**, 619 (1981)

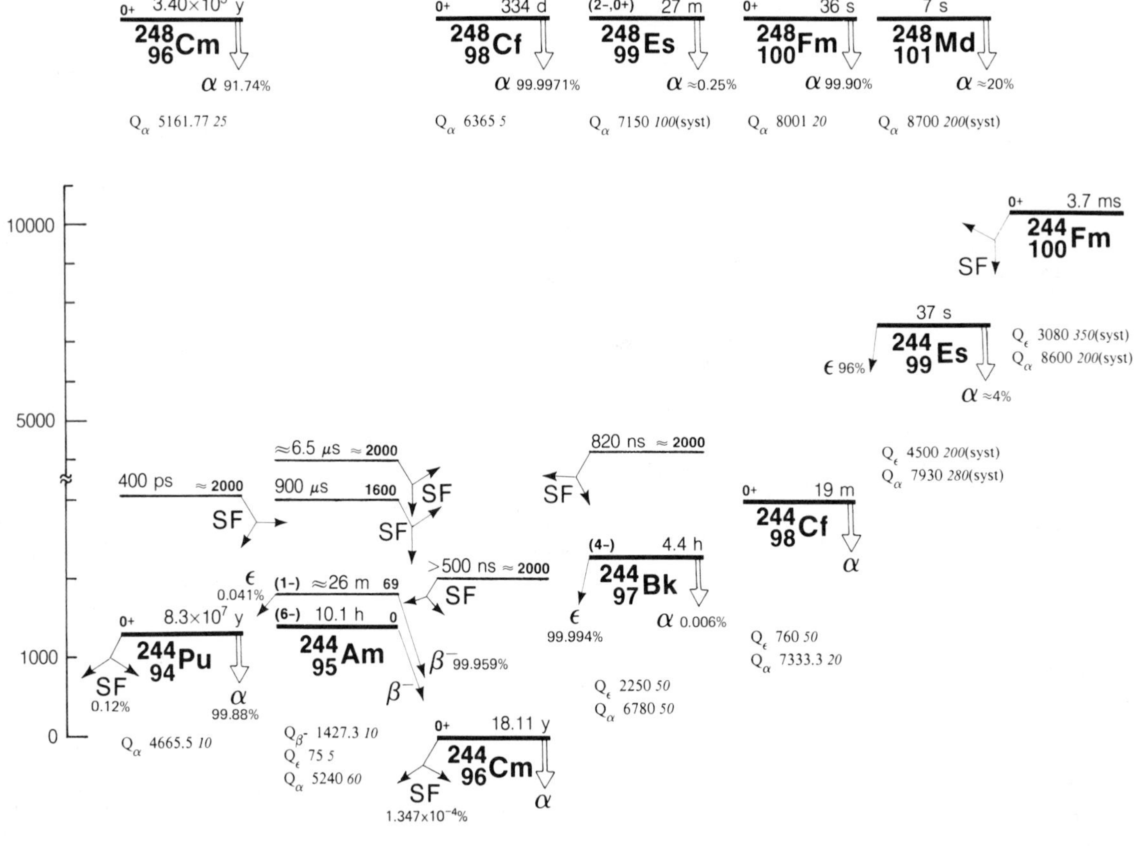

$^{244}_{94}$Pu(8.26 *9* ×10^{7} yr)

Mode: α(99.875 *6* %), SF(0.125 *6* %)
Δ: 59801 *5* keV
SpA: 1.759×10^{-5} Ci/g
Prod: multiple n-capture from ^{238}U; multiple n-capture from ^{239}Pu; multiple n-capture from ^{242}Pu

Alpha Particles (^{244}Pu)

$\langle\alpha\rangle$=4575 *1* keV

α(keV)	α(%)
4546 *1*	19.4 *8*
4589 *1*	80.5 *8*

Photons (^{244}Pu)

$\langle\gamma\rangle$=0.118 *7* keV

γ_{mode}	γ(keV)	γ(%)†
U L_ℓ	11.620	0.0172 *19*
U L_α	13.600	0.281 *22*
U L_η	15.400	0.0069 *8*
U L_β	17.128	0.35 *4*
U L_γ	20.292	0.077 *9*
γ_α[E2]	43.9 *8*	0.00277 *6*

† 7.2% uncert(syst)

Atomic Electrons (^{244}Pu)

$\langle e \rangle$=0.673 *15* keV

e_{bin}(keV)	$\langle e \rangle$(keV)	e(%)
17	0.057	0.33 *4*
21	0.056	0.27 *3*
22	0.0065	0.0293 *16*
23	0.168	0.73 *3*
27	0.174	0.65 *3*
38	0.00284	0.0074 *3*
39	0.076	0.196 *9*
40	0.074	0.187 *8*
42	0.00092	0.00216 *10*
43	0.0447	0.104 *5*
44	0.0132	0.0302 *14*

$^{244}_{95}$Am(10.1 *1* h)

Mode: β-

Δ: 59876.3 *23* keV

SpA: 1.272×10^{6} Ci/g

Prod: ^{243}Am(n,γ)

Photons (^{244}Am)

$\langle \gamma \rangle$=806 keV

γ_{mode}	γ(keV)	γ(%)
Cm L_ℓ	12.633	2.5
Cm L_α	14.939	38
Cm L_η	17.314	1.0
Cm L_β	19.212	50
Cm L_γ	22.883	12
γ E2	42.90 *10*	0.09
γ E2	99.4 *1*	5
Cm $K_{\alpha2}$	104.586	2.3 *8*
Cm $K_{\alpha1}$	109.271	3.6 *12*
Cm $K_{\beta1}'$	123.059	1.3 *5*
Cm $K_{\beta2}'$	127.344	0.46 *16*
γ E2	154.0 *8*	18
γ [E2]	206.0 *19*	0.26
γ [E2]	540.0 *18*	0.4
γ M1+46%E2	746.0 *8*	67
γ E2	900.0 *8*	28

Atomic Electrons (^{244}Am)

$\langle e \rangle$=225 keV

e_{bin}(keV)	$\langle e \rangle$(keV)	e(%)
18	0.27	1
19	13.8	72
24	15.7	66
25 - 43	11.9	32
75	1.7	2
76	30.3	40
78 - 79	0.048	0.061
80	20.7	26
81 - 90	0.044	0.051
93	10.9	12
95	7.1	8
97 - 129	9.7	9
130	27.9	21
135	15.2	11
148	9.5	6
149 - 187	10.7	7.1
200 - 206	0.11	0.054
515 - 540	0.043	0.0082
618	26.6	4
721	5.5	0.8
722 - 772	6.1	0.81
875 - 900	1.2	0.13

Continuous Radiation (^{244}Am)

$\langle \beta\text{-} \rangle$=109 keV;$\langle IB \rangle$=0.043 keV

E_{bin}(keV)		$\langle\ \rangle$(keV)	(%)
0 - 10	β-	0.303	6.1
	IB	0.0057	
10 - 20	β-	0.88	5.9
	IB	0.0050	0.035
20 - 40	β-	3.36	11.2
	IB	0.0081	0.029
40 - 100	β-	20.1	29.3
	IB	0.0146	0.024
100 - 300	β-	79	45.9
	IB	0.0091	0.0066
300 - 385	β-	5.4	1.67
	IB	4.1×10^{-5}	1.35×10^{-5}

$^{244}_{95}$Am($\sim$ 26 min)

Mode: β-(99.959 *3* %), ϵ(0.041 *3* %)

Δ: 59945 *10* keV

SpA: $\sim 3\times10^{7}$ Ci/g

Prod: ^{243}Am(n,γ)

Photons (^{244}Am)

γ_{mode}	γ(keV)	γ(%)
$\gamma_{\beta\text{-}}$E2	42.90 *10*	$\sim$0.018
$\gamma_{\beta\text{-}}$	994 *14* ?	
$\gamma_{\beta\text{-}}$	1036 *14* ?	

Continuous Radiation (^{244}Am)

$\langle \beta\text{-} \rangle$=501 keV;$\langle IB \rangle$=0.71 keV

E_{bin}(keV)		$\langle\ \rangle$(keV)	(%)
0 - 10	β-	0.0473	0.94
	IB	0.022	
10 - 20	β-	0.143	0.95
	IB	0.021	0.148
20 - 40	β-	0.58	1.93
	IB	0.041	0.142
40 - 100	β-	4.25	6.0
	IB	0.109	0.169
100 - 300	β-	43.9	21.8
	IB	0.25	0.146
300 - 600	β-	142	31.9
	IB	0.19	0.045
600 - 1300	β-	303	35.8
	IB	0.078	0.0105
1300 - 1496	β-	7.3	0.54
	IB	0.000112	8.4×10^{-6}

$^{244}_{96}$Cm(18.11 *2* yr)

Mode: α, SF(0.0001347 *2* %)

Δ: 58449.0 *21* keV

SpA: 80.90 Ci/g

Prod: multiple n-capture from ^{238}U; multiple n-capture from ^{239}Pu; multiple n-capture from ^{243}Am

Alpha Particles (^{244}Cm)

$\langle \alpha \rangle$=5796.5 *1* keV

α(keV)	α(%)
4919.42 *5*	9 *4* $\times10^{-5}$
4958.40 *7*	$\sim$0.0002
5217.41 *4*	0.00012 *5*
5313 *3*	$\sim4\times10^{-5}$
5515.49 *3*	0.0035 *1*
5665.61 *3*	0.022 *1*
5762.84 *3*	23.6 *2*
5804.96 *3*	76.4 *2*

Photons (^{244}Cm)

$\langle \gamma \rangle$=1.6 *5* keV

γ_{mode}	γ(keV)	γ(%)†
Pu L_ℓ	12.124	$\sim$0.23
Pu L_α	14.262	$\sim$4
Pu L_η	16.333	$\sim$0.09
Pu L_β	18.135	$\sim$4
Pu L_γ	21.554	$\sim$1.0
γ E2	42.817 *6*	$\sim$0.025
γ E2	98.855 *11*	$\sim$0.0015
γ (E2)	152.622 *20*	0.00098 *10*
γ [E1]	251.43 *5*	1.21 *10* $\times10^{-5}$
γ (E1)	263.33 *6*	5.7 *3* $\times10^{-5}$
γ [E2]	289.17 *7*	$<8\times10^{-9}$
γ [E1]	302.96 *4*	1.54 *8* $\times10^{-5}$
γ [M1]	340.70 *7*	$<3\times10^{-8}$
γ [E1]	507.22 *4*	7.6 *6* $\times10^{-6}$
γ E1	554.54 *3*	7.9 *5* $\times10^{-5}$
γ E1	597.35 *3*	4.5 *4* $\times10^{-5}$
γ [E1]	606.07 *4*	7.2 *4* $\times10^{-6}$
γ E2	758.64 *4*	1.53 *9* $\times10^{-5}$
γ E2	817.87 *6*	6.7 *4* $\times10^{-5}$
γ [M1+E2]	857.50 *4*	6.1 *5* $\times10^{-6}$
γ	895.24 *7*	$<2\times10^{-8}$
γ [E2]	900.31 *4*	1.7 *3* $\times10^{-6}$
γ [E1]	916.04 *5*	$<3\times10^{-7}$?
γ	938.05 *7*	$<6\times10^{-7}$?

† 10% uncert(syst)

$^{244}_{97}$Bk(4.35 *15* h)

Mode: ϵ(99.994 *2* %), α(0.006 *2* %)

Δ: 60690 *50* keV

SpA: 2.95×10^{6} Ci/g

Prod: ^{243}Am(α,3n); ^{244}Cm(d,2n); ^{244}Cm(p,n); ^{241}Am(α,n)

Alpha Particles (^{244}Bk)

$\langle\alpha\rangle\sim$0.40 keV

α(keV)	α(%)
6625 *4*	~0.0030
6667 *4*	~0.0030

Photons (^{244}Bk)

γ_{mode}	γ(keV)	γ(rel)
γ	144.5 *3*	7.4 *7*
γ E2	154.0 *8*	3.7 *4*
γ	177.0 *3*	4.2 *5*
γ	187.6 *3*	16.5 *15*
γ	217.6 *3*	100
γ	233.8 *4*	2.9 *4*
γ	333.5 *5*	10.0 *15*
γ	490.5 *5*	18.0 *20*
γ M1+46%E2	746.0 *8*	8.0 *10*
γ	870 *1*	7.0 *10*
γ	891.5 *10*	114 *12*
γ	910 *1*	3.0 *5*
γ	921.5 *10*	22 *3*
γ	988 *1*	5.0 *10*
γ	1042 *2*	~3
γ	1136 *2*	~1
γ	1141 *2*	~2
γ	1153 *1*	9.5 *14*
γ	1173 *2*	~0.7
γ	1178 *1*	5.0 *8*
γ	1205 *2*	~1.0
γ		
γ	1211 *2*	~1
γ	1233 *1*	4.0 *8*
γ	1252 *1*	3.0 *6*
γ	1333 *1*	1.2 *4*
γ	1505 *5*	~3

$^{244}_{98}$Cf(19.4 *6* min)

Mode: α
Δ: 61459 *6* keV
SpA: 3.97×10^{7} Ci/g
Prod: ^{244}Cm(α,4n); ^{242}Cm(α,2n); ^{238}U(^{12}C,6n); ^{236}U(^{12}C,4n)

Alpha Particles (^{244}Cf)

$\langle\alpha\rangle$=7200 *6* keV

α(keV)	α(%)
7168 *5*	25 *3*
7210 *5*	75 *3*

$^{244}_{99}$Es(37 *4* s)

Mode: ε(96 *3* %), α(4 *3* %)
Δ: 65960 *200* keV syst
SpA: 1.24×10^{9} Ci/g
Prod: ^{233}U(^{15}N,4n)

Alpha Particles (^{244}Es)

α(keV)
7570 *20*

$^{244}_{100}$Fm(3.7 *4* ms)

Mode: SF
Δ: 69040 *280* keV syst
SpA: 6.7×10^{10} Ci/g
Prod: ^{233}U(^{16}O,5n); ^{206}Pb(^{40}Ar,2n)

A = 245

NDS **33**, 119 (1981)

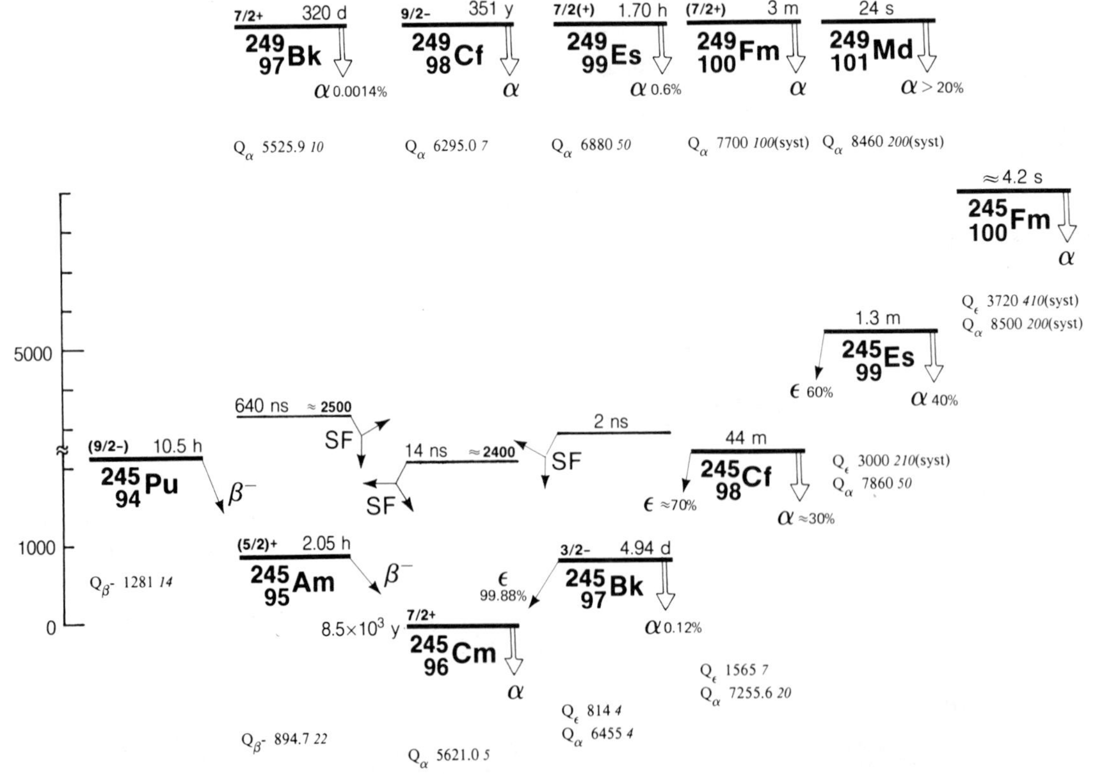

$^{245}_{94}$Pu(10.5 1 h)

Mode: β-
Δ: 63174 14 keV
SpA: 1.218×10^{6} Ci/g
Prod: ^{244}Pu(n,γ); multiple n-capture from ^{238}U; multiple n-capture from ^{239}Pu

Photons (^{245}Pu)

⟨γ⟩=397 11 keV

γ_{mode}	γ(keV)	γ(%)†
Am L_ℓ	13.377	0.46 11
Am L_α	15.599	7.2 16
Am L_η	16.819	0.096 22
Am L_β	18.907	6.6 15
Am L_γ	22.282	1.4 4
γ[E1]	28.05 16	~0.7
Am $K_{\alpha2}$	102.026	7.8 16
Am $K_{\alpha1}$	105.472	12.3 25
Am $K_{\beta1}'$	119.960	4.5 9
Am $K_{\beta2}'$	124.123	1.6 3
γ	277.1 5	~0.017
γ(M1+35%E2)	280.430 15	1.28 14
γ	293.3 5	~0.017
γ[E1]	299.48 16	~0.017
γ M1+22%E2	308.320 8	4.9 5
γ M1+25%E2	327.525 8	25.5 23
γ	333.2 3	~0.034
γ[M1]	341.09 14	0.101 17
γ[M1]	348.884 13	0.96 10
γ[M1+E2]	358.04 18	0.063 17
γ(M1)	376.773 3	3.2 3
γ[M1]	387.97 3	0.29 7
γ	392.8 4	~0.07
γ[E2]	395.978 12	0.10 3
γ[E2]	412.032 22	0.49 5
γ	423.3 3	<0.034
γ[M1]	428.540 19	0.52 5
γ	439.1 10	~0.034
γ[M1]	445.55 8	0.30 5
γ	450.1 10	~0.034
γ[M1]	475.48 19	0.059 25
γ	479.9 10	~0.02
γ[M1]	482.09 3	~0.014
γ	486.4 6	~0.034
γ(E2)	491.689 9	2.7 3
γ	512.07 5	~0.034
γ	514.70 20	0.17 3
γ	518.3 5	0.051 17
γ[M1]	525.21 8	0.27 3
γ	530.7 3	~0.034
γ	548.74 14	~0.034
γ(E2)	560.142 12	5.4 5
γ	591.73 4	0.17 3
γ[M1+E2]	593.66 8	~0.034
γ	598.9 3	0.12 3
γ	624.5 4	0.22 3
γ M1(+E2)	630.199 14	2.7 3
γ	642.1 5	<0.034
γ	657.3 7	~0.14
γ	660.18 4	0.85 12
γ	662.3 7	0.08 3
γ	669.46 11	0.34 5
γ	687.7 8	~0.034
γ	691.1 5	<0.034
γ	696.85 14	0.08 3

Photons (^{245}Pu)
(continued)

γ_{mode}	γ(keV)	γ(%)†
γ	701.8 3	~0.07
γ	708.08 20	0.27 5
γ	712.1 5	<0.034
γ[E1]	730.28 13	0.19 3
γ	733.6 4	0.08 3
γ	737.92 11	0.22 5
γ	740.3 7	0.14 5
γ	743.80 20	0.15 3
γ	750.2 10	<0.034
γ	758.3 8	~0.034
γ[E1]	762.98 12	0.71 7
γ[E1]	766.81 12	0.35 5
γ	776.76 20	0.20 3
γ	781.6 3	~0.07
γ	786.64 14	0.37 5
γ[E1]	796.50 13	0.25 7
γ[E2]	800.00 4	1.57 17
γ[E1]	817.14 14	0.85 9
γ	822.0 7	0.08 3
γ[E2]	823.18 14	<0.034 ?
γ[E1]	833.04 12	0.26 5
γ[M1+E2]	840.572 15	1.28 14
γ[E1]	859.62 16	0.51 5
γ[M1+E2]	868.461 9	0.12 3
γ[M1+E2]	870.06 4	~0.07
γ	874.09 8	0.14 3
γ	879.7 4	0.051 17
γ[E1]	887.20 14	0.71 9
γ	899.69 18	~0.034
γ	901.98 8	~0.051
γ[M1+E2]	910.629 20	1.39 14
γ	917.18 14	0.08 3
γ	923.1 6	0.051 17
γ	925.5 10	~0.017
γ	930.90 17	~0.051
γ[M1+E2]	938.518 16	1.01 17
γ	940.61 4	~0.25
γ	945.3 5	0.051 17
γ	953.85 19	~0.017
γ[E2]	957.724 16	0.98 10
γ	964.1 7	0.042 17
γ	968.50 4	~0.034
γ	972.7 5	0.08 3
γ	975.1 10	~0.25
γ	977.28 14	0.39 17
γ	982.5 7	0.08 3
γ	987.70 4	1.32 14
γ	996.33 18	0.20 3
γ	1001.1 10	~0.025
γ	1005.17 14	0.27 10
γ	1007.41 20	0.41 10
γ	1013.3 3	0.10 3
γ	1018.35 11	1.03 14
γ	1023.53 18	0.54 10
γ	1028.3 10	0.017 7
γ	1036.3 8	0.008 3
γ	1040.68 23	~0.007
γ	1042.5 8	0.015 5
γ	1051.1 3	0.0051 17
γ	1079.2 10	0.0051 17
γ	1083.16 24	0.034 7
γ	1093.8 7	0.014 5
γ	1098.0 3	0.017 5
γ	1111.20 19	0.054 7
γ	1138.6 3	0.042 7
γ	1166.5 3	0.051 7

† 10% uncert(syst)

Atomic Electrons (^{245}Pu)

⟨e⟩=89 12 keV

e_{bin}(keV)	⟨e⟩(keV)	e(%)
9 - 28	2.5	12.3 19
77 - 119	0.62	0.70 5
152 - 174	1.9	1.2 3
183	7.4	4.0 11
203	35.3	17 6
208 - 233	1.87	0.84 9
252	4.9	1.94 22
253 - 280	1.53	0.58 8
285	2.5	0.86 24
287 - 303	0.80	0.27 5
304	10.1	3.3 10
305 - 318	2.6	0.85 22
321	2.8	0.87 25
322 - 371	4.8	1.40 12
372 - 420	0.90	0.23 3
422 - 471	1.18	0.266 23
472 - 521	1.6	~0.32
524 - 572	1.1	0.20 6
574 - 623	0.69	0.11 4
624 - 673	0.54	0.085 22
674 - 721	0.8	0.11 4
724 - 774	0.22	0.029 7
775 - 823	1.0	0.12 4
827 - 877	0.7	0.08 3
878 - 927	0.70	0.08 3
929 - 979	0.20	0.021 6
981 - 1031	0.18	0.018 5
1032 - 1079	0.012	~0.0011
1082 - 1119	0.0047	0.00043 19
1132 - 1166	0.0028	~0.00025

Continuous Radiation (^{245}Pu)

⟨β-⟩=252 keV;⟨IB⟩=0.23 keV

E_{bin}(keV)		⟨ ⟩(keV)	(%)
0 - 10	β-	0.172	3.44
	IB	0.0121	
10 - 20	β-	0.501	3.34
	IB	0.0114	0.079
20 - 40	β-	1.92	6.4
	IB	0.021	0.073
40 - 100	β-	11.9	17.1
	IB	0.051	0.079
100 - 300	β-	68	36.3
	IB	0.092	0.055
300 - 600	β-	110	25.5
	IB	0.042	0.0106
600 - 1234	β-	60	8.3
	IB	0.0053	0.00078

$^{245}_{95}$Am(2.05 1 h)

Mode: β-
Δ: 61893 3 keV
SpA: 6.24×10^{6} Ci/g
Prod: daughter ^{245}Pu

Photons (^{245}Am)

$\langle\gamma\rangle$=32.2 *17* keV

γ_{mode}	γ(keV)	γ(%)†
Cm L_ℓ	12.633	0.26 *4*
Cm L_α	14.939	4.0 *5*
Cm L_η	17.314	0.046 *7*
Cm L_β	19.064	3.9 *6*
Cm L_γ	23.002	0.94 *16*
γ [M1]	42.872 *18*	0.057 *9*
γ E2+39%M1	54.81 *3*	<0.0006 ?
γ	78.3 *5*	
Cm $K_{\alpha2}$	104.586	3.6 *4*
Cm $K_{\alpha1}$	109.271	5.7 *6*
Cm $K_{\beta1}$'	123.059	2.1 *2*
Cm $K_{\beta2}$'	127.344	0.73 *7*
γ	140 *5*	
γ	153 *5*	
γ E2	198.19 *3*	0.033 *4*
γ M1(+<33%E2)	241.06 *3*	0.33 *5*
γ M1+2.6%E2	252.998 *25*	6.1
γ M1+8.1%E2	295.870 *24*	0.23 *4*

† 9.8% uncert(syst)

Atomic Electrons (^{245}Am)

$\langle e\rangle$=26.6 *16* keV

e_{bin}(keV)	$\langle e\rangle$(keV)	e(%)
18	0.63	3.5 *6*
19 - 55	1.71	7.0 *5*
70 - 119	0.88	0.85 *8*
120 - 122	0.0052	0.0043 *4*
125	14.7	11.8 *12*
168 - 217	0.80	0.44 *5*
222	0.018	0.0082 *13*
228	4.8	2.12 *21*
229 - 241	0.76	0.33 *3*
247	1.43	0.58 *6*
248 - 296	0.79	0.306 *18*

Continuous Radiation (^{245}Am)

$\langle\beta$-$\rangle$=259 keV;$\langle$IB$\rangle$=0.22 keV

E_{bin}(keV)		$\langle\ \rangle$(keV)	(%)
0 - 10	β-	0.116	2.31
	IB	0.0126	
10 - 20	β-	0.346	2.31
	IB	0.0119	0.083
20 - 40	β-	1.38	4.60
	IB	0.022	0.077
40 - 100	β-	9.5	13.7
	IB	0.053	0.083
100 - 300	β-	78	40
	IB	0.090	0.055
300 - 600	β-	134	31.9
	IB	0.031	0.0079
600 - 895	β-	35.3	5.2
	IB	0.00155	0.00025

$^{245}_{96}$Cm(8500 *100* yr)

Mode: α
Δ: 60998.0 *22* keV
SpA: 0.1717 Ci/g
Prod: multiple n-capture from ^{238}U; multiple n-capture from ^{239}Pu; multiple n-capture from ^{243}Am; daughter ^{245}Bk

Alpha Particles (^{245}Cm)

$\langle\alpha\rangle$=5363 *1* keV

α(keV)	α(%)
5151 *25*	<0.0050
5159 *25*	<0.004
5235 *10*	0.3
5273 ?	0.07
5303.8 *10*	5.0 *1*
5362.0 *7*	93.2 *5*
5370 *25* ?	
5436 *10*	0.04
5492.7 *11*	0.8
5533.1 *11*	0.6

Photons (^{245}Cm)

$\langle\gamma\rangle$=117 *8* keV

γ_{mode}	γ(keV)	γ(%)†
Pu L_ℓ	12.124	~3
Pu L_α	14.262	~51
Pu L_η	16.333	~1.0
Pu L_β	18.103	~56
Pu L_γ	21.591	~12
γ [M1+E2]	41.95 *3*	0.350 *17*
γ [M1+E2]	53.74 *6*	0.067 *4*
γ [M1+E2]	56.81 *6*	0.0361 *19*
γ [M1+E2]	65.36 *6*	0.011 *4*
γ [M1+E2]	69.17 *6*	0.007 *3*
γ [M1+E2]	79.25 *6*	0.150 *9*
γ	89.58 *6*	0.022 *3*
γ	93.82 *6*	0.036 *4*
Pu $K_{\alpha2}$	99.522	21.1 *12*
Pu $K_{\alpha1}$	103.734	33.6 *20*
Pu $K_{\beta1}$'	116.930	12.2 *7*
Pu $K_{\beta2}$'	120.974	4.2 *3*
γ [M1]	132.99 *3*	2.77 *14*
γ [M1+E2]	136.06 *6*	0.112 *7*
γ [M1+E2]	139.81 *6*	0.0057 *19*
γ [M1+E2]	161.6 *1*	0.009 *4*
γ	165.3 *1*	0.009 *4*
γ [M1]	174.94 *4*	9.5
γ	185.8 *1*	0.010 *4*
γ [M1+E2]	189.82 *6*	0.193 *12*
γ	210.6 *2*	0.0066 *19*
γ	232.7 *2*	0.015 *4*

† 7.0% uncert(syst)

Atomic Electrons (^{245}Cm)

$\langle e\rangle$=134 *24* keV

e_{bin}(keV)	$\langle e\rangle$(keV)	e(%)
11 - 14	3.23	28.7 *19*
18	8.8	~49
19	2.2	~12
20	13.6	~69
22	7.6	~34
23	1.3	5.4 *14*
24	14.2	<119
31 - 35	2.3	~7
36	9.0	~25
37	6.3	<34
38 - 40	1.0	~3
41	4.0	~10
42 - 52	3.5	8 *4*
53	23.7	45 *4*
54 - 103	5.8	8.5 *21*
104	0.0030	0.0029 *3*
110	5.5	5.0 *3*
111 - 147	3.58	2.90 *19*
152	11.9	7.8 *6*
153 - 193	6.7	4.02 *20*
205 - 233	0.016	~0.007

$^{245}_{97}$Bk(4.94 *3* d)

Mode: ϵ(99.88 *1* %), α(0.12 *1* %)
Δ: 61812 *5* keV
SpA: 1.079×10^5 Ci/g
Prod: ^{243}Am(α,2n); ^{244}Cm(d,n); ^{242}Cm(α,p); ^{244}Cm(α,p2n)

Alpha Particles (^{245}Bk)

$\langle\alpha\rangle$=7.51 *2* keV

α(keV)	α(%)
5796 *12*	<0.0016
5814 *12* ?	
5853.0 *5*	<0.0044
5861.0 *12*	0.0032 *5*
5885.1 *5*	0.026
5897 *12* ?	
5979 *6*	9.6 *24* $\times10^{-5}$
6035.0 *12*	0.00066 *10*
6079 *12*	
6083.0 *12*	0.0080 *6*
6113.0 *12*	0.0062 *10*
6117.6 *9*	0.0120 *12*
6146.7 *5*	0.0248 *24*
6193.0 *12*	0.00144 *12*
6257.1 *5*	0.00180 *12*
6308.7 *5*	0.0146 *24*
6349.2 *5*	0.0181 *6*

Photons (^{245}Bk)

$\langle\gamma\rangle$=235 *6* keV

γ_{mode}	γ(keV)	γ(%)†
Cm L_ℓ	12.633	2.23 *24*
Cm L_α	14.939	34 *3*
Cm L_η	17.314	0.44 *5*
Cm L_β	19.081	32 *4*
Cm L_γ	22.967	7.3 *10*
γ_ϵ[M1]	42.872 *18*	
Am $K_{\alpha2}$	102.026	0.0026 *12*
γ_ϵE2	103.25 *9*	0.40 *8*
Cm $K_{\alpha2}$	104.586	35.7 *17*
Am $K_{\alpha1}$	105.472	0.0042 *18*
Cm $K_{\alpha1}$	109.271	56 *3*
Am $K_{\beta1}$'	119.960	0.0015 *7*
Cm $K_{\beta1}$'	123.059	20.6 *10*
Am $K_{\beta2}$'	124.123	0.00053 *23*
Cm $K_{\beta2}$'	127.344	7.2 *4*
γ_α[E1]	164.707 *10*	0.0076 *13*
γ_α[E1]	194.3 *9*	~0.0012
γ_ϵE2	198.19 *3*	0.158 *19*
γ_α[E1]	205.883 *10*	0.042 *6*
γ_ϵM1+2.6%E2	252.998 *25*	29.1 *19*
γ_α[E1]	265.927 *9*	0.00015 *3*
γ_ϵ[E1]	272.4 *3*	0.012 *3*
γ_ϵM1	350.65 *10*	0.076 *7*
γ_ϵE1	365.97 *7*	0.36 *3*
γ_ϵE1	380.95 *10*	2.40 *17*
γ_ϵM1	385.15 *9*	0.57 *4*
γ_ϵ(M1)	407.95 *20*	0.028 *6*
γ_ϵ[E1]	408.84 *7*	0.19 *3*
γ_α[E2]	430.633 *9*	0.0015 *3*
γ_α[M1+E2]	471.809 *9*	0.026 *5*
γ_ϵ[E2]	488.39 *12*	0.014 *3*

† uncert(syst): 8.3% for α, 7.5% for ϵ

Atomic Electrons (^{245}Bk)

$\langle e\rangle$=127 *5* keV

e_{bin}(keV)	$\langle e\rangle$(keV)	e(%)
19	5.1	27 *3*
20 - 23	0.00065	0.0031 *10*
24	3.3	13.8 *16*
25 - 70	1.56	6.3 *7*
77 - 122	8.5	9.8 *8*
125	70.3	56 *4*
141 - 190	0.181	0.103 *8*
192 - 222	0.212	0.100 *6*
228	23.1	10.1 *7*
229 - 246	3.12	1.36 *9*
247	6.8	2.76 *19*
248 - 281	3.82	1.51 *6*
306 - 351	0.077	0.023 *4*
356 - 405	0.412	0.113 *5*
406 - 452	0.007	0.0015 *7*
464 - 488	0.0037	0.00078 *20*

Continuous Radiation (^{245}Bk)

$\langle$IB$\rangle$=1.14 keV

E_{bin}(keV)		$\langle\ \rangle$(keV)	(%)
10 - 20	IB	0.0056	0.031
20 - 40	IB	0.00132	0.0050
40 - 100	IB	0.31	0.34
100 - 300	IB	0.81	0.70
300 - 561	IB	0.0083	0.0022

$^{245}_{98}$Cf(43.6 *8* min)

Mode: ϵ(~ 70 %), α(~ 30 %)
Δ: 63377 *6* keV
SpA: 1.76×10^7 Ci/g
Prod: ^{244}Cm(α,3n); ^{242}Cm(α,n); ^{238}U(^{12}C,5n)

Alpha Particles (^{245}Cf)

α(keV)
6886
6983
7036
7084
7137 *2*

$^{245}_{99}$Es(1.33 *15* min)

Mode: ϵ(60 *10* %), α(40 *10* %)
Δ: 66380 *210* keV syst
SpA: 5.7×10^8 Ci/g
Prod: ^{235}U(^{14}N,4n); ^{238}U(^{14}N,7n); ^{237}Np(^{12}C,4n); ^{240}Pu(^{10}B,5n); ^{241}Am(^{12}C,α4n)

Alpha Particles (^{245}Es)

α(keV)
7730 *20*

$^{245}_{100}$Fm(4.2 *13* s)

Mode: α
Δ: 70100 *350* keV syst
SpA: 1.0×10^{10} Ci/g
Prod: ^{233}U(^{16}O,4n)

Alpha Particles (^{245}Fm)

α(keV)
8150 *20*

A = 246

NDS **32**, 92 (1981)

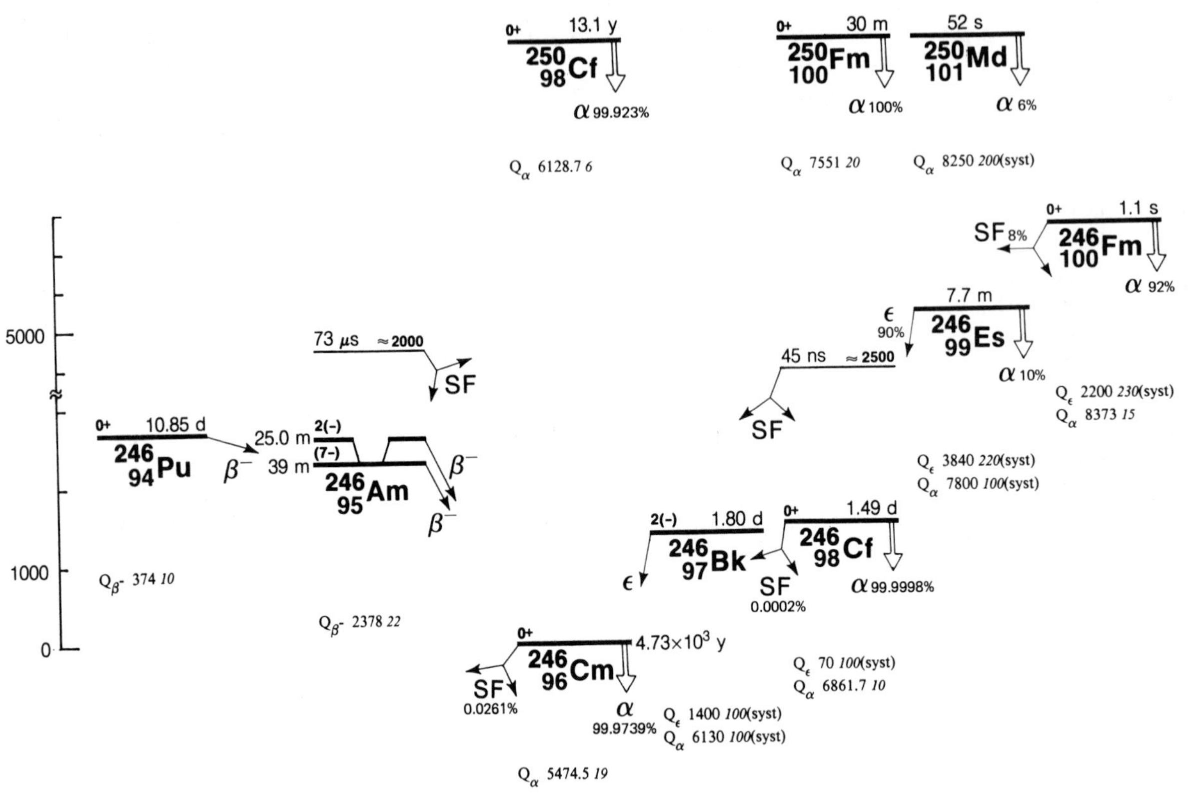

$^{246}_{94}$Pu(10.85 *2* d)

Mode: β-
Δ: 65365 *19* keV
SpA: 4.892×10^{4} Ci/g
Prod: multiple n-capture from ^{238}U

Photons (^{246}Pu)

⟨γ⟩=143 *5* keV

γ_{mode}	γ(keV)	γ(%)†
Am L_ℓ	13.377	1.5 *3*
Am L_α	15.599	24 *4*
Am L_η	16.819	0.35 *9*
Am L_β	18.901	23 *5*
Am L_γ	22.280	5.2 *12*
γ [E1]	27.561 *18*	3.5 *4*
γ [E1]	43.792 *15*	25.0 *13*
γ	66.581 *16*	0.255 *18*
γ	75.608 *15*	0.180 *25*
Am $K_{\alpha2}$	102.026	13.9 *9*
Am $K_{\alpha1}$	105.472	21.9 *14*
Am $K_{\beta1}$'	119.960	8.1 *5*
Am $K_{\beta2}$'	124.123	2.76 *18*
γ	149.384 *23*	~0.06
γ	158.411 *23*	0.035 *8*
γ (M1)	179.925 *14*	9.7 *5*
γ	188.952 *17*	0.047 *8*
γ	216.513 *23*	0.112 *18*
γ (E1)	223.717 *15*	23.5 *18*

Photons (^{246}Pu) (continued)

γ_{mode}	γ(keV)	γ(%)†
γ	232.744 *18*	0.080 *13*
γ	255.533 *16*	0.230 *18*
γ	299.324 *18*	0.030 *8*

† 12% uncert(syst)

Atomic Electrons (^{246}Pu)

⟨e⟩=82 *5* keV

e_{bin}(keV)	⟨e⟩(keV)	e(%)
8 - 16	2.7	24 *10*
20	5.2	27 *4*
21	1.73	8.3 *6*
22	0.146	0.67 *8*
23	2.9	13 *3*
24	3.3	13.5 *15*
26 - 43	4.0	10.4 *20*
44	2.8	~6
47	2.4	~5
52 - 53	1.8	~3
55	25.1	46 *3*
56 - 105	8.1	11 *3*
108 - 155	0.7	0.56 *24*
156	12.8	8.2 *5*
157 - 169	1.57	1.00 *7*
174	3.9	2.24 *16*
175 - 224	2.58	1.36 *6*
227 - 276	0.22	0.09 *4*
280 - 299	0.006	0.0019 *9*

Continuous Radiation (^{246}Pu)

⟨β-⟩=53 keV;⟨IB⟩=0.0125 keV

E_{bin}(keV)		⟨ ⟩(keV)	(%)
0 - 10	β-	0.72	14.7
	IB	0.0027	
10 - 20	β-	1.96	13.1
	IB	0.0021	0.0145
20 - 40	β-	6.5	21.9
	IB	0.0029	0.0102
40 - 100	β-	23.5	36.3
	IB	0.0036	0.0061
100 - 300	β-	20.6	13.8
	IB	0.00129	0.00100
300 - 330	β-	0.097	0.0315
	IB	1.43×10^{-7}	4.7×10^{-8}

${}^{246}_{95}$Am(39 *3* min)

Mode: β-
Δ: 64991 *22* keV
SpA: 1.96×10^{7} Ci/g
Prod: ^{244}Pu(α,d); ^{244}Pu(^{3}He,p)

Photons (^{246}Am)

⟨γ⟩=695 *38* keV

γ_{mode}	γ(keV)	γ(%)†
Cm L_ℓ	12.633	3.1 *4*
Cm L_α	14.939	46 *5*
Cm L_η	17.314	1.28 *17*
Cm L_β	19.218	63 *8*
Cm L_γ	22.883	14.8 *20*
γ E2	42.850 *3*	0.092 *3*
γ [M1+E2]	49.7 *13*	
γ [M1+E2]	77.7 *13*	
γ [E2]	81.66 *4*	~0.0031
γ E2	99.166 *20*	4.8 *11*
Cm $K_{\alpha2}$	104.586	2.91 *22*
Cm $K_{\alpha1}$	109.271	4.6 *4*
Cm $K_{\beta1}$'	123.059	1.68 *13*
Cm $K_{\beta2}$'	127.344	0.59 *5*
γ [E2]	127.4 *5*	~2
γ [E2]	147.6 *19*	
γ (E2)	152.92 *15*	25 *3*
γ (E2)	204.6 *8*	36 *4*
γ [E1]	628.9 *10*	2.7 *5*
γ [E1]	678.6 *8*	53
γ [E1]	685.9 *14*	~2 ?
γ [E1]	755.8 *4*	13.3 *11*
γ [E1]	781.31 *3*	4.0 *4*
γ [E1]	838.9 *14*	~2 ?

† 9.4% uncert(syst)

Atomic Electrons (^{246}Am)

⟨e⟩=289 *12* keV

e_{bin}(keV)	⟨e⟩(keV)	e(%)
18	0.269	1.47 *5*
19	14.9	78 *6*
24	17.6	74 *7*
25 - 63	12.3	33.9 *9*
75	1.7	2.2 *5*
76	34.1	45 *10*
77 - 79	0.035	0.045 *8*
80	20.7	26 *6*
81 - 90	0.059	0.068 *5*
93	10.8	12 *3*
94 - 128	28.9	28 *3*
129	40.4	31 *4*
134	22.0	16.4 *18*
147	13.7	9.3 *10*
148	7.0	4.7 *5*
149 - 180	11.5	7.3 *4*
181	22.9	12.7 *14*
186	10.6	5.7 *6*
198 - 205	15.5	7.8 *5*
550 - 558	2.16	0.39 *4*
604 - 653	0.66	0.105 *7*
654 - 686	0.62	0.095 *6*
711 - 758	0.24	0.033 *5*
762 - 781	0.0118	0.00152 *11*
814 - 839	0.020	0.0025 *8*

Continuous Radiation (^{246}Am)

⟨β-⟩=392 keV;⟨IB⟩=0.46 keV

E_{bin}(keV)		⟨ ⟩(keV)	(%)
0 - 10	β-	0.066	1.32
	IB	0.018	
10 - 20	β-	0.199	1.33
	IB	0.017	0.121
20 - 40	β-	0.80	2.68
	IB	0.033	0.114
40 - 100	β-	5.8	8.3
	IB	0.085	0.132
100 - 300	β-	56	28.3
	IB	0.18	0.105
300 - 600	β-	158	35.8
	IB	0.106	0.026
600 - 1199	β-	171	22.3
	IB	0.023	0.0033

${}^{246}_{95}$Am(25.0 *2* min)

Mode: β-
Δ: 64991 *22* keV
SpA: 3.057×10^{7} Ci/g
Prod: daughter ^{246}Pu

Photons (^{246}Am)

⟨γ⟩=973 *14* keV

γ_{mode}	γ(keV)	γ(%)†
Cm L_ℓ	12.633	0.7 *3*
Cm L_α	14.939	11 *5*
Cm L_η	17.314	0.25 *11*
Cm L_β	19.185	13 *6*
Cm L_γ	22.885	2.9 *12*
γ E2	42.850 *3*	~0.049
γ [E2]	81.66 *4*	0.0027 *6*
γ E2	99.166 *20*	0.165 *12*
Cm $K_{\alpha2}$	104.586	1.17 *23*
Cm $K_{\alpha1}$	109.271	1.9 *4*
Cm $K_{\beta1}$'	123.059	0.68 *14*
Cm $K_{\beta2}$'	127.344	0.24 *5*
γ	150.81 *14*	0.0079 *15*
γ (E2)	152.92 *15*	0.0044 *15*
γ [M1+E2]	170.926 *25*	0.049 *20*
γ [M1+E2]	227.07 *3*	0.015 *5* ?
γ [E1]	228.65 *4*	0.037 *7*
γ [M1+E2]	237.189 *21*	0.143 *7*
γ [M1+E2]	238.605 *22*	0.146 *7*
γ [M1+E2]	244.006 *24*	0.679 *25*
γ [M1+E2]	244.82 *3*	~0.006
γ [M1+E2]	251.583 *24*	0.0027 *5*
γ [M1+E2]	261.759 *23*	0.156 *5*
γ [M1+E2]	263.201 *21*	0.0333 *22*
γ [M1+E2]	266.91 *3*	~0.0049
γ M1+21%E2	270.02 *3*	1.02 *3*
γ	270.96 *8*	~0.0049
γ [M1]	277.17 *7*	0.0020 *7*
γ [M1+E2]	287.771 *21*	0.128 *4*
γ [E1]	289.04 *3*	0.0047 *12*
γ [M1+E2]	293.25 *4*	0.0044 *12*
γ	302.96 *5*	0.0069 *7*
γ [E1]	306.00 *8*	~0.0012
γ [M1+E2]	321.04 *3*	0.0185 *12*
γ	325.60 *6*	0.0059 *10*
γ [M1+E2]	327.63 *4*	0.0030 *10*
γ	329.87 *14*	0.0032 *10*
γ [M1+E2]	343.88 *4*	0.0257 *10*
γ	347.26 *4*	0.0240 *12*
γ [M1+E2]	354.39 *3*	0.0064 *10*
γ [E1]	360.46 *3*	0.0566 *22*
γ	361.85 *7*	0.0121 *15*
γ	370.84 *6*	0.0042 *10*
γ [E2]	373.343 *23*	0.0207 *12*
γ [M1+E2]	377.12 *4*	0.0027 *10*

Photons (^{246}Am)
(continued)

γ_{mode}	γ(keV)	γ(%)†
γ [M1+E2]	381.29 *8*	0.0015 *5*
γ M1	383.76 *3*	0.0185 *20*
γ [M1+E2]	397.91 *3*	0.0082 *12*
γ E1	401.654 *22*	0.264 *7*
γ [M1+E2]	408.115 *24*	0.0101 *10*
γ [E1]	414.18 *3*	0.0104 *12*
γ [M1+E2]	421.065 *25*	0.0220 *17*
γ [E1]	422.96 *6*	0.0040 *17*
γ	434.85 *4*	0.009 *3*
γ [E2]	443.30 *4*	0.0035 *10*
γ [E2]	447.08 *3*	~0.0012
γ [E1]	451.07 *8*	0.0025 *10*
γ [E1]	456.01 *3*	0.0138 *17*
γ [E2]	460.88 *9*	0.0032 *12*
γ [M1+E2]	465.67 *3*	0.0254 *20*
γ [E1]	469.42 *3*	0.0101 *12*
γ [E2]	472.44 *3*	0.0363 *17*
γ M1	476.95 *3*	0.0212 *15*
γ [M1+E2]	486.92 *7*	0.0091 *20*
γ M1	488.828 *23*	0.091 *4*
γ M1	493.466 *24*	0.107 *4*
γ [M1+E2]	505.51 *3*	0.0121 *22*
γ E1,E2	507.21 *3*	0.067 *3*
γ M1	514.840 *24*	0.086 *4*
γ [M1+E2]	516.62 *3*	0.0099 *25*
γ M1	522.53 *5*	0.0457 *20*
γ M1	524.960 *21*	0.073 *3*
γ M1	528.67 *3*	0.0148 *15*
γ M1(+<50%E2)	542.99 *3*	0.040 *5*
γ M1(+<50%E2)	554.32 *8*	0.0220 *17*
γ M1(+<50%E2)	554.68 *3*	0.0146 *15*
γ M1	566.14 *3*	0.0425 *25*
γ	577.73 *6*	0.0084 *22*
γ	580.58 *7*	0.0084 *22*
γ E2+10%M1	602.61 *4*	0.232 *12*
γ E1	610.24 *4*	0.044 *7*
γ [E1]	636.77 *4*	0.012 *3*
γ E2+19%M1	649.494 *23*	0.366 *12*
γ M1+E0	656.53 *3*	0.012 *3*
γ [E2]	670.37 *4*	0.008 *3*
γ [E1]	677.90 *4*	0.045 *4*
γ E2+39%M1	684.266 *23*	0.583 *20*
γ M1	698.16 *4*	0.116 *7*
γ M1	717.257 *23*	0.252 *10*
γ E1	724.79 *3*	0.212 *7*
γ [M1+E2]	732.33 *4*	0.015 *4*
γ E1	734.420 *21*	1.16 *3*
γ M1+E0	745.049 *23*	0.235 *7*
γ [M1+E2]	747.68 *4*	0.025 *5*
γ	751.24 *11*	0.035 *12*
γ M1+E0	752.029 *22*	0.82 *3*
γ E1	759.56 *3*	0.640 *20*
γ	776.3 *3*	0.0040 *12*
γ M1	779.821 *24*	0.067 *10*
γ [E1]	781.31 *3*	0.168 *12*
γ [M1+E2]	791.87 *3*	0.064 *12*
γ E1	798.814 *17*	24.7
γ M1+E0	810.2 *3*	0.006 *3*
γ [E1]	819.99 *3*	~0.0037 ?
γ [M1+E2]	829.34 *3*	0.018 *4*
γ E1	833.585 *17*	1.79 *5*
γ E1,E2	904.37 *3*	0.0571 *22*
γ [E2]	924.99 *16*	0.009 *3*
γ E1,E2	939.14 *3*	0.077 *5*
γ	960.28 *6*	0.0054 *22*
γ [M2]	962.848 *25*	<0.0010
γ [E2]	982.26 *3*	0.0137 *14*
γ	986.002 *23*	0.96 *3*
γ [M1]	1023.45 *3*	0.039 *4*
γ E1	1036.002 *18*	12.7 *4*
γ [E1]	1045.09 *3*	0.018 *3*
γ E1	1062.014 *19*	17.1 *4*
γ E1	1078.852 *18*	27.6 *9*
γ E2	1081.425 *20*	0.35 *3*
γ E1	1085.168 *21*	1.52 *5*
γ	1102.50 *20*	0.0035 *10*
γ [M2]	1104.864 *19*	<0.0007
γ	1113.60 *20*	0.0067 *12*
γ E2	1122.62 *3*	0.099 *5*
γ E2	1124.275 *20*	0.259 *10*
γ [E2]	1131.88 *7*	0.0109 *12*
γ	1148.62 *6*	0.0183 *15*
γ [E1]	1158.42 *3*	0.0124 *10*
γ	1167.72 *5*	0.0249 *15*

Photons (^{246}Am) (continued)

γ_{mode}	γ(keV)	γ(%)†
γ [E2]	1177.08 8	0.0037 10
γ	1198.18 5	0.0306 15
γ	1203.20 20	0.0049 17
γ E1	1206.927 19	0.148 5
γ [E2]	1210.57 5	0.0111 17
γ [M1+E2]	1237.21 7	0.0072 10
γ E1	1249.777 19	0.148 5
γ [E1]	1257.59 3	0.0388 25
γ M1(+E0)	1274.72 4	0.267 7
γ	1297.34 5	0.0104 12
γ	1303.20 11	0.0086 10
γ [E1]	1306.02 3	0.0062 10
γ [E1]	1323.772 20	0.025 5
γ [E2]	1336.37 7	0.0183 12
γ E1	1348.87 3	0.120 4
γ [M1+E2]	1367.34 4	0.016 3
γ	1379.3 4	~0.0017
γ [E1]	1383.91 3	0.0054 10
γ [M1+E2]	1409.06 3	0.0333 17
γ [M1+E2]	1435.58 4	0.0257 25
γ M1	1451.91 3	0.0452 20
γ [M1+E2]	1459.20 3	0.0094 10
γ [M1+E2]	1466.50 4	0.0074 10
γ E1	1479.466 23	0.227 7
γ [E1]	1483.077 21	0.0205 17
γ	1486.90 7	~0.0020
γ	1497.0 4	~0.0006
γ	1509.35 4	~0.0007
γ [E1]	1528.99 3	0.222 10
γ [M1+E2]	1530.88 5	0.025 5
γ [E2]	1538.79 5	0.0014 5
γ	1540.60 20	~0.0007
γ	1545.0 5	0.0022 10
γ E1,E2	1550.00 20	0.052 25
γ E1	1550.840 21	0.272 25
γ E1,E2	1552.22 16	0.052 25
γ [M1+E2]	1558.37 3	0.0168 17
γ E1,E2	1561.31 3	0.095 4
γ [M1+E2]	1570.35 5	0.0153 12
γ M1	1573.73 5	0.0482 20
γ E1,E2	1578.631 21	0.077 3
γ	1586.07 7	~0.0049
γ E1	1590.68 3	0.52 4
γ	1601.22 3	0.0027 12 ?
γ E1,E2	1604.16 3	0.102 4
γ [E1]	1616.34 8	0.0030 7
γ M1	1618.80 3	0.115 4
γ E1,E2	1628.15 3	0.055 3
γ M1	1637.95 5	0.161 20
γ [E1]	1659.18 8	0.0126 10
γ M1(+E2)	1661.65 3	0.225 7
γ [M1+E2]	1669.52 4	0.0156 10
γ [E2]	1680.80 5	0.00106 20
γ	1690.15 16	0.0012 5
γ [M1+E2]	1714.59 4	0.00215 22
γ M1	1737.95 3	0.111 7
γ [E2]	1756.07 9	0.00141 22
γ	1759.31 5	0.0212 17
γ	1764.10 13	0.00089 20
γ	1769.47 7	0.0020 4
γ	1778.90 5	0.0222 12
γ [E2]	1780.80 3	0.0040 10
γ	1793.86 6	0.00037 12
γ	1801.53 6	0.0094 10
γ	1805.06 6	0.00091 22
γ [M1+E2]	1813.76 4	0.00272 25
γ	1821.75 5	0.0015 3
γ	1827.38 5	0.0190 15
γ	1832.66 10	0.00047 22
γ	1836.71 6	0.0047 5
γ [M1+E2]	1843.90 3	0.0089 7
γ [M1+E2]	1855.24 9	0.0015 5
γ	1858.48 5	0.00077 12
γ	1863.26 13	0.00094 15
γ	1866.46 5	0.0049 7
γ	1870.23 5	0.00099 20
γ	1875.51 10	0.00084 20
γ	1881.70 4	0.0074 7
γ [M1+E2]	1886.75 3	0.0124 7
γ [E2]	1898.09 9	0.00044 10
γ	1904.23 6	0.00121 15
γ	1909.31 5	0.00141 15
γ	1924.55 4	0.0082 7
γ	1940.48 7	0.00054 10
γ	1944.79 15	~0.00035
γ	1953.6 5	~10×10⁻⁵
γ	1974.2 3	0.00030 10
γ	1983.33 7	0.00030 10
γ	1989.64 6	0.00104 20
γ	2000.3 5	~0.00012
γ	2029.40 6	0.00116 12
γ	2032.49 6	0.0010 4
γ	2058.18 6	0.00143 10
γ	2065.00 20	0.00042 10
γ	2068.69 8	0.00151 10
γ	2083.10 20	0.00032 7
γ	2091.4 3	0.00020 5
γ	2103.19 5	0.00156 15
γ	2123.66 7	0.00259 17
γ	2128.56 6	0.00128 12
γ	2140.2 3	0.00022 5
γ	2146.04 5	0.00304 17
γ	2149.50 20	0.00047 7
γ	2156.05 17	0.00035 7
γ	2168.33 7	0.00109 10
γ	2184.79 15	0.00027 5
γ	2203.4 5	7.4 25 $\times 10^{-5}$
γ	2234.4 3	0.00015 5
γ	2259.2 4	~10 $\times 10^{-5}$
γ	2287.0 6	~5 $\times 10^{-5}$

† 4.1% uncert(syst)

Atomic Electrons (^{246}Am)

⟨e⟩=37 5 keV

e_{bin}(keV)	⟨e⟩(keV)	e(%)
10 - 18	0.7	~5
19	5.9	31 15
22 - 23	0.016	~0.07
24	6.1	26 12
25 - 35	0.55	1.9 8
37	2.1	~6
38	1.9	~5
39 - 75	1.9	4.5 13
76	1.06	1.40 11
77 - 110	2.10	2.31 25
116	0.9	~0.8
117 - 139	0.29	~0.22
142	1.9	1.36 20
143 - 178	0.31	0.20 9
193 - 241	1.5	0.66 17
242 - 292	1.33	0.53 5
293 - 342	0.19	0.060 11
343 - 392	0.417	0.114 4
393 - 442	0.253	0.062 3
443 - 492	0.183	0.0387 25
493 - 542	0.212	0.0408 24
543 - 591	0.492	0.086 4
596 - 645	1.08	0.173 6
646 - 670	0.131	0.020 3
671	0.88	0.132 6
672 - 721	0.245	0.0351 14
723 - 772	0.259	0.0353 19
773 - 820	0.277	0.0355 11
823 - 858	0.07	0.008 3
880 - 925	0.412	0.0454 15
933 - 944	0.522	0.0559 16
951	0.84	0.088 3
953 - 1000	0.134	0.0139 9
1004 - 1049	0.59	0.057 3
1054 - 1104	0.286	0.0270 7
1105 - 1154	0.186	0.0163 12
1156 - 1205	0.027	0.0023 5
1206 - 1254	0.0253	0.00204 14
1256 - 1305	0.017	0.0013 3
1307 - 1356	0.026	0.00192 19
1359 - 1408	0.0118	0.00084 16
1411 - 1460	0.041	0.0029 3
1461 - 1560	0.118	0.0079 10
1562 - 1657	0.053	0.00331 25
1662 - 1760	0.0140	0.00081 11
1762 - 1860	0.0028	0.00015 4
1861 - 1960	0.00123	6.5 13 $\times 10^{-5}$
1963 - 2062	0.00096	4.8 13 $\times 10^{-5}$
2063 - 2162	0.00023	1.1 3 $\times 10^{-5}$
2163 - 2262	7.3 $\times 10^{-6}$	3.3 10 $\times 10^{-7}$
2263 - 2282	2.4 $\times 10^{-7}$	~1 $\times 10^{-8}$

Continuous Radiation (^{246}Am)

⟨β-⟩=454 keV;⟨IB⟩=0.63 keV

E_{bin}(keV)		⟨ ⟩(keV)	(%)
0 - 10	β-	0.061	1.21
	IB	0.020	
10 - 20	β-	0.183	1.22
	IB	0.019	0.135
20 - 40	β-	0.74	2.46
	IB	0.037	0.129
40 - 100	β-	5.3	7.6
	IB	0.098	0.151
100 - 300	β-	51	25.7
	IB	0.22	0.128
300 - 600	β-	144	32.6
	IB	0.156	0.038
600 - 1300	β-	227	27.6
	IB	0.073	0.0095
1300 - 2335	β-	25.8	1.66
	IB	0.0047	0.00031

$^{246}_{96}$Cm(4730 *10* yr)

Mode: α(99.97386 5 %), SF(0.02614 5 %)

Δ: 62613 3 keV

SpA: 0.3072 Ci/g

Prod: multiple n-capture from ^{238}U; multiple n-capture from ^{239}Pu; multiple n-capture from ^{244}Cm; daughter ^{250}Cf

Alpha Particles (^{246}Cm)

⟨α⟩=5376 4 keV

α(keV)	α(%)
5343 3	21.0 10
5386 3	79.0 10

Photons (^{246}Cm)

⟨γ⟩=1.39 9 keV

γ_{mode}	γ(keV)	γ(%)
Pu L_ℓ	12.124	0.203 25
Pu L_α	14.262	3.2 3
Pu L_η	16.333	0.076 9
Pu L_β	18.136	3.9 4
Pu L_γ	21.554	0.85 11
γ E2	44.533 9	0.0273 15

Atomic Electrons (^{246}Cm)

⟨e⟩=7.23 23 keV

e_{bin}(keV)	⟨e⟩(keV)	e(%)
18	0.60	3.3 4
21	0.062	0.291 17
22	2.39	10.7 8
23	0.0128	0.055 6
26	1.82	6.9 4
39	0.88	2.25 13
40	0.78	1.95 11
41	0.0202	0.050 3
43	0.50	1.16 7
44	0.136	0.308 18
45	0.0204	0.046 3

$^{246}_{97}$Bk(1.80 2 d)

Mode: ϵ

Δ: 64010 100 keV syst

SpA: 2.95×10^5 Ci/g

Prod: ^{244}Cm(α,pn); ^{243}Am(α,n)

Photons (^{246}Bk)

⟨γ⟩=852 33 keV

γ_{mode}	γ(keV)	γ(%)†
Cm L_ℓ	12.633	2.4 4
Cm L_α	14.939	36 6
Cm L_η	17.314	0.63 12
Cm L_β	19.134	38 7
Cm L_γ	22.932	8.8 18
γ E2	42.850 3	~0.08
γ E2	99.166 20	0.220 15
Cm $K_{\alpha2}$	104.586	18.8 20
Cm $K_{\alpha1}$	109.271	30 3
Cm $K_{\beta1}$'	123.059	10.9 12
Cm $K_{\beta2}$'	127.344	3.8 4
γ [M1+E2]	237.189 21	0.0195 15
γ [M1+E2]	251.583 24	0.00070 14
γ [M1+E2]	263.201 21	0.0057 5
γ [E1]	289.04 3	0.018 5
γ E1	734.420 21	3.17 15
γ E1	798.814 17	61 4
γ E1	833.585 17	4.90 22
γ [M2]	962.848 25	$<17\times10^{-5}$
γ [E2]	982.26 3	0.226 20
γ [E1+E2]	986.002 23	0.248 21
γ [M1]	1023.45 3	0.151 18
γ E1	1036.002 18	1.73 10
γ E1	1062.014 19	2.9 2
γ E1	1078.852 18	3.75 22
γ E2	1081.425 20	5.8 4
γ E1	1085.168 21	0.39 3
γ [M2]	1104.864 19	$<13\times10^{-5}$
γ E2	1122.62 3	0.38 5
γ E2	1124.275 20	4.3 3

† 5.0% uncert(syst)

Atomic Electrons (^{246}Bk)

⟨e⟩=48 4 keV

e_{bin}(keV)	⟨e⟩(keV)	e(%)
10 - 18	1.3	~9
19	12.4	65 13
24	11.9	50 10
25 - 35	2.1	8.0 17
37	3.6	9.9 20
38	3.2	8.4 17
39 - 75	3.0	7.2 8
76	1.37	1.82 13
79 - 123	3.63	4.09 17
135 - 161	0.008	~0.006
213 - 262	0.028	0.013 5
263 - 289	0.00060	0.00022 7
606	0.119	0.0196 10
671	2.18	0.325 22
705 - 734	0.205	0.0289 12
774 - 815	0.64	0.082 3
827 - 858	0.048	0.0056 9
895 - 944	0.203	0.0222 12
951 - 1000	1.04	0.108 5
1004 - 1043	0.0350	0.00341 12
1054 - 1104	0.330	0.0307 9
1105 - 1123	0.0397	0.00355 18

Continuous Radiation (^{246}Bk)

⟨IB⟩=1.13 keV

E_{bin}(keV)		⟨ ⟩(keV)	(%)
10 - 20	IB	0.0062	0.035
20 - 40	IB	0.00144	0.0055
40 - 100	IB	0.31	0.34
100 - 300	IB	0.78	0.67
300 - 600	IB	0.017	0.0040
600 - 1300	IB	0.0162	0.0021
1300 - 1307	IB	6.7×10^{-12}	4.2×10^{-11}

$^{246}_{98}$Cf(1.487 21 d)

Mode: α(99.99980 2 %), SF(0.00020 2 %)

Δ: 64087.3 25 keV

SpA: 3.57×10^5 Ci/g

Prod: ^{244}Cm(α,2n); ^{238}U(^{12}C,4n)

Alpha Particles (^{246}Cf)

⟨α⟩=6740 1 keV

α(keV)	α(%)
6471 4	~0.016
6615.6 10	0.180 20
6708.6 7	21.80 20
6750.1 7	78.00 20

Photons (^{246}Cf)

⟨γ⟩=1.62 13 keV

γ_{mode}	γ(keV)	γ(%)
Cm L_ℓ	12.633	0.24 3
Cm L_α	14.939	3.6 4
Cm L_η	17.314	0.083 12
Cm L_β	19.191	4.2 6
Cm L_γ	22.878	0.94 14
γ E2	42.2 1	0.0188 17

Photons (^{246}Cf)
(continued)

γ_{mode}	γ(keV)	γ(%)
γ [E2]	94.5 12	0.0076 8
γ [E2]	147 4	~0.0036

Atomic Electrons (^{246}Cf)

⟨e⟩=7.0 3 keV

e_{bin}(keV)	⟨e⟩(keV)	e(%)
18	0.057	0.32 3
19	2.18	11.7 13
23	1.68	7.2 7
24	0.64	2.7 4
25	0.0148	0.060 8
36	0.86	2.37 23
37	0.76	2.03 19
38	0.0195	0.051 5
41	0.52	1.27 12
42	0.161	0.38 4
70	0.0030	0.0043 5
71	0.056	0.079 9
76	0.039	0.052 6
88	0.00120	0.00136 16
89	0.0192	0.0216 25
90	0.0137	0.0153 18
93	0.0102	0.0109 13
94	0.0032	0.0034 4
95	3.4×10^{-6}	$3.6\ 4\times10^{-6}$

$^{246}_{99}$Es(7.7 5 min)

Mode: ϵ(90.1 18 %), α(9.9 18 %)

Δ: 67930 220 keV syst

SpA: 9.9×10^7 Ci/g

Prod: ^{238}U(^{14}N,6n); ^{241}Am(^{12}C,α3n)

Alpha Particles (^{246}Es)

α(keV)
7350 20

$^{246}_{100}$Fm(1.1 2 s)

Mode: α(92 3 %), SF(8 3 %)

Δ: 70130 40 keV

SpA: 3.1×10^{10} Ci/g

Prod: ^{235}U(^{16}O,5n); ^{239}Pu(^{12}C,5n); ^{233}U(^{18}O,5n); ^{208}Pb(^{40}Ar,2n)

Alpha Particles (^{246}Fm)

α(keV)
8240 20

A = 247

NDS **33**, 161 (1981)

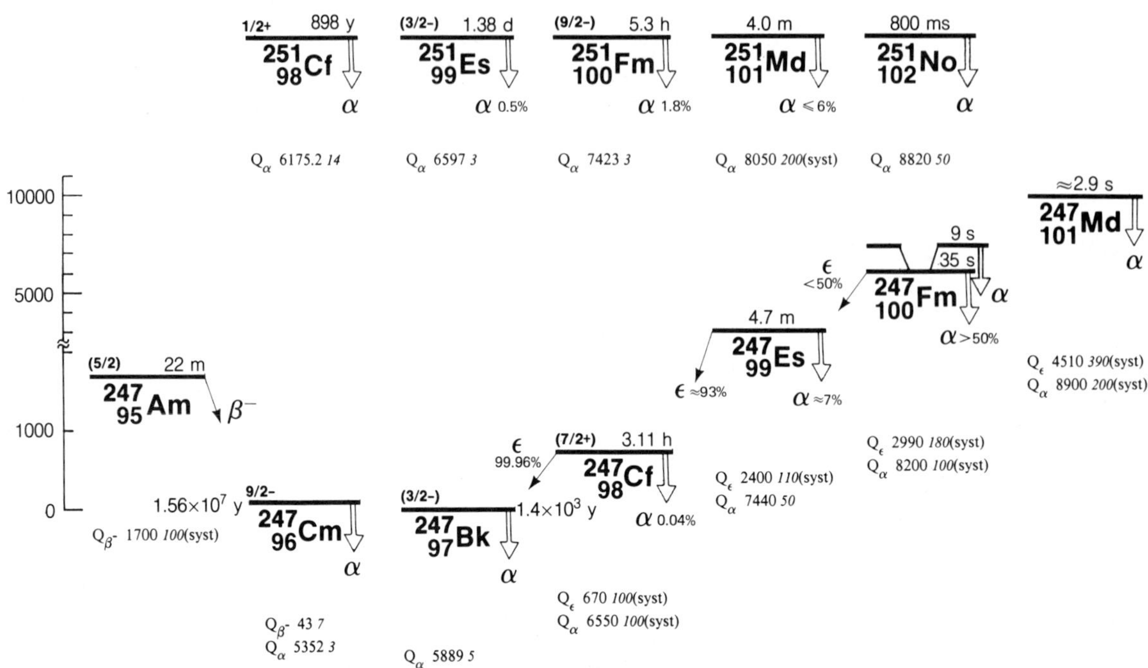

$^{247}_{95}$Am(22 *3* min)

Mode: β-
Δ: 67230 *100* keV syst
SpA: 3.5×10⁷ Ci/g
Prod: ^{244}Pu(α,p)

Photons (^{247}Am)

⟨γ⟩=135 *17* keV

γ_{mode}	γ(keV)	γ(%)†
Cm L_ℓ	12.633	0.9 *3*
Cm L_α	14.939	14 *4*
Cm L_η	17.314	0.16 *5*
Cm L_β	19.066	13 *4*
Cm L_γ	22.979	2.9 *9*
Cm $K_{\alpha2}$	104.586	14 *4*
Cm $K_{\alpha1}$	109.271	21 *6*
Cm $K_{\beta1}$'	123.059	7.9 *22*
Cm $K_{\beta2}$'	127.344	2.7 *8*
γ M2	226.7 *7*	5.8 *16*
γ [E1]	285.3 *2*	23

† 22% uncert(syst)

Atomic Electrons (^{247}Am)

⟨e⟩=103 *15* keV

e_{bin}(keV)	⟨e⟩(keV)	e(%)
19 - 25	3.9	19 *4*
79 - 97	0.68	0.81 *11*
98	45.7	46 *13*
99 - 122	0.31	0.30 *3*
157	1.39	0.88 *4*
202	28.5	14 *4*
203	4.1	2.0 *6*
208	3.5	1.7 *5*
220	8.0	3.6 *10*
221 - 266	7.0	3.1 *5*
279 - 285	0.171	0.0611 *17*

Continuous Radiation (^{247}Am)

⟨β-⟩=490 keV;⟨IB⟩=0.68 keV

E_{bin}(keV)		⟨ ⟩(keV)	(%)
0 - 10	β-	0.0488	0.97
	IB	0.022	
10 - 20	β-	0.147	0.98
	IB	0.021	0.146
20 - 40	β-	0.60	1.99
	IB	0.040	0.139
40 - 100	β-	4.38	6.2
	IB	0.107	0.165
100 - 300	β-	45.0	22.4
	IB	0.24	0.142
300 - 600	β-	144	32.4
	IB	0.18	0.043
600 - 1300	β-	291	34.7
	IB	0.069	0.0094
1300 - 1473	β-	4.85	0.362
	IB	6.1×10⁻⁵	4.6×10⁻⁶

$^{247}_{96}$Cm(1.56 *5* ×10⁷ yr)

Mode: α
Δ: 65528 *5* keV
SpA: 9.2×10⁻⁵ Ci/g
Prod: multiple n-capture from ^{238}U; multiple n-capture from ^{239}Pu; multiple n-capture from ^{244}Cm; ^{246}Cm(n,γ)

Alpha Particles (^{247}Cm)

⟨α⟩=4947.5 *25* keV

α(keV)	α(%)
4818 *4*	4.7 *3*
4869.0 *18*	71 *1*
4941 *4*	1.6 *2*
4982.0 *19*	2.0 *2*
5143.6 *19*	1.2 *2*
5210.4 *19*	5.7 *5*
5265.9 *18*	13.8 *7*

Photons (^{247}Cm)

⟨γ⟩=315 *25* keV

γ_{mode}	γ(keV)	γ(%)
Pu $K_{\alpha2}$	99.522	1.30 *15* ‡
Pu $K_{\alpha1}$	103.734	2.1 *2* ‡
Pu $K_{\beta1}$'	116.930	0.80 *13* ‡
Pu $K_{\beta2}$'	120.974	~0.30 ‡
γ [E1]	279.2 *8*	3.4 *7*
γ M1	288.6 *7*	2.0 *3*
γ [E1]	347.1 *8*	~1
γ E1	403.5 *5*	72 *6*

‡ exp value

$^{247}_{97}$Bk(1380 *250* yr)

Mode: α
Δ: 65485 *6* keV
SpA: 1.05 Ci/g
Prod: daughter ^{247}Cf; ^{244}Cm(α,p); ^{245}Cm(α,pn); ^{246}Cm(α,p2n)

Alpha Particles (^{247}Bk)

⟨α⟩=5566 *5* keV

α(keV)	α(%)
5456 *5*	1.5 *2*
5501 *5*	7 *1*
5532 *5*	45 *2*
5608.5 *21*	~0.40
5653.5 *20*	5.5 *6*
5687.2 *20*	13 *1*
5712.1 *20*	17 *1*
5753.2 *20*	4.3 *4*
5794.7 *20*	5.5 *5*

Photons (^{247}Bk)

⟨γ⟩=114 *45* keV

γ_{mode}	γ(keV)	γ(%)
γ [E1]	41.75 *18*	~1
γ E1	83.95 *16*	~40
γ (M1+E2)	268 *5*	~30

$^{247}_{98}$Cf(3.11 *3* h)

Mode: ϵ(99.965 *7* %), α(0.035 *7* %)
Δ: 66150 *100* keV syst
SpA: 4.08×10^6 Ci/g
Prod: ^{244}Cm(α,n); ^{238}U(^{14}N,p4n); ^{245}Cm(α,2n); ^{246}Cm(α,3n)

Alpha Particles (^{247}Cf)

α(keV)
6301 *5*

Photons (^{247}Cf)

⟨γ⟩=87 *9* keV

γ_{mode}	γ(keV)	γ(%)†
γ [M1+1.5%E2]	29.88 *11*	~0.22
γ [M2]	40.81 *11*	0.0017 *4*
γ [M1+E2]	42.0 *2*	~0.15
Bk $K_{\alpha2}$	107.181	21 *4*
Bk $K_{\alpha1}$	112.121	32 *6*
Bk $K_{\beta1}'$	126.216	12.0 *23*
Bk $K_{\beta2}'$	130.611	4.2 *8*
γ M1	294.12 *9*	0.98 *7*
γ [E1]	305.05 *13*	0.0089 *14*
γ [E1]	334.93 *12*	0.028 *3*
γ [M1]	337.3 *5*	~0.006 ?
γ [M1+E2]	363.91 *20*	0.0110 *20*
γ [M1+E2]	376.21 *10*	0.071 *7*
γ (E1)	407.00 *9*	0.190 *20*
γ M1+31%E2	417.92 *9*	0.34 *3*

Photons (^{247}Cf) (continued)

γ_{mode}	γ(keV)	γ(%)†
γ M1+49%E2	447.81 *8*	0.55 *4*
γ [M1+E2]	459.5 *3*	0.0110 *16*

† 13% uncert(syst)

Atomic Electrons (^{247}Cf)

⟨e⟩=4.64 *22* keV

e_{bin}(keV)	⟨e⟩(keV)	e(%)
163	2.27	1.40 *10*
173 - 206	0.014	0.007 *3*
232 - 245	0.08	~0.032
269	0.68	0.253 *19*
270 - 281	0.093	0.0342 *22*
286	0.38	0.13 *3*
288	0.200	0.069 *5*
289 - 315	0.081	0.0276 *15*
316	0.44	0.138 *25*
318 - 364	0.041	0.012 *5*
370 - 418	0.172	0.043 *5*
423	0.136	0.032 *5*
428 - 460	0.060	0.0136 *12*

Continuous Radiation (^{247}Cf)

⟨IB⟩=1.25 keV

E_{bin}(keV)		⟨ ⟩(keV)	(%)
10 - 20	IB	0.0054	0.030
20 - 40	IB	0.00163	0.0065
40 - 100	IB	0.26	0.28
100 - 300	IB	0.97	0.83
300 - 600	IB	0.0135	0.0038
600 - 629	IB	5.1×10^{-6}	5.5×10^{-7}

$^{247}_{99}$Es(4.7 *3* min)

Mode: ϵ(~ 93 %), α(~ 7 %)
Δ: 68550 *50* keV
SpA: 1.62×10^8 Ci/g
Prod: ^{238}U(^{14}N,5n); ^{241}Am(^{12}C,α2n)

Alpha Particles (^{247}Es)

α(keV)
7320 *30*

$^{247}_{100}$Fm(35 *4* s)

Mode: α(>50 %), ϵ(<50 %)
Δ: 71540 *170* keV syst
SpA: 1.29×10^9 Ci/g
Prod: ^{239}Pu(^{12}C,4n)

Alpha Particles (^{247}Fm)

α(keV)	α(rel)
7870 *50*	~70
7930 *50*	~30

$^{247}_{100}$Fm(9.2 *23* s)

Mode: α
Δ: 71540 *170* keV syst
SpA: 4.8×10^9 Ci/g
Prod: ^{239}Pu(^{12}C,4n)

Alpha Particles (^{247}Fm)

α(keV)
8180 *30*

$^{247}_{101}$Md(2.9 *15* s)

Mode: α
Δ: 76040 *350* keV syst
SpA: 1.4×10^{10} Ci/g
Prod: ^{209}Bi(^{40}Ar,2n)

Alpha Particles (^{247}Md)

α(keV)
8428 *25*

A = 248

NDS **32**, 119 (1981)

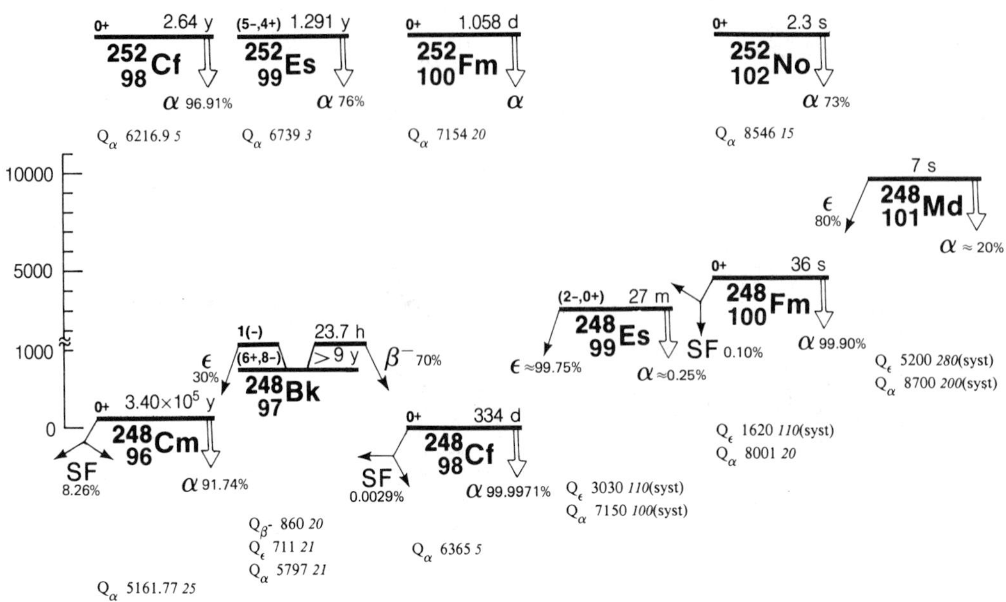

$^{248}_{96}$Cm(3.40 *3* $\times 10^5$ yr)

Mode: α(91.74 *3* %), SF(8.26 *3* %)
Δ: 67388 *5* keV
SpA: 0.00424 Ci/g
Prod: daughter ^{252}Cf
multiple n-capture from ^{238}U;
multiple n-capture from ^{239}Pu;
multiple n-capture from ^{244}Cm

Alpha Particles (^{248}Cm)

⟨α⟩=4652.4 *3* keV

α(keV)	α(%)
4776.0 *15*	<0.009
4931.1 *5*	0.070 *11*
5034.93 *25*	16.54 *17*
5078.45 *25*	75.1 *4*

$^{248}_{97}$Bk(>9 yr)

decay not observed
Δ: 68099 *21* keV
SpA: <160 Ci/g
Prod: ^{246}Cm(α,pn)

$^{248}_{97}$Bk(23.7 *2* h)

Mode: β-(70 *5* %), ε(30 *5* %)
Δ: 68099 *21* keV
SpA: 5.33$\times 10^5$ Ci/g
Prod: ^{247}Bk(n,γ); ^{245}Cm(α,p)

Photons (^{248}Bk)

⟨γ⟩=55 *6* keV

γ_{mode}	γ(keV)	γ(%)†
Cm L_ℓ	12.633	0.40 *10*
Cf L_ℓ	13.146	~0.28
Cm L_α	14.939	6.0 *14*
Cf L_α	15.636	~4
Cm L_η	17.314	0.079 *20*
Cf L_η	18.347	~0.10
Cm L_β	19.083	5.7 *14*
Cf L_β	20.303	~5
Cm L_γ	22.966	1.3 *3*
Cf L_γ	24.273	~1
$\gamma_{\beta-}$[E2]	41.3 *10*	~0.016
γ_ϵE2	43.399 *25*	~0.002
Cm $K_{\alpha2}$	104.586	6.2 *12*
Cm $K_{\alpha1}$	109.271	9.8 *20*
Cf $K_{\alpha2}$	109.826	0.0160 *6*
Cf $K_{\alpha1}$	115.032	0.0249 *9*
Cm $K_{\beta1}$'	123.059	3.6 *7*
Cm $K_{\beta2}$'	127.344	1.25 *25*
Cf $K_{\beta1}$'	129.436	0.0093 *3*
Cf $K_{\beta2}$'	133.949	0.00328 *13*
$\gamma_{\beta-}$E1	550.7 *1*	5.0 *10*

† uncert(syst): 17% for ε, 7.1% for β-

Atomic Electrons (^{248}Bk)

⟨e⟩=10.5 *15* keV

e_{bin}(keV)	⟨e⟩(keV)	e(%)
15	0.06	~0.37
16	1.5	~10
19	0.93	4.9 *12*
20	0.9	~4
21	1.7	~8
24	0.76	3.2 *10*
25	1.0	4.1 *18*
26	0.018	~0.07
35	1.0	~3
36	0.8	~2
37 - 39	0.20	0.52 *18*
40	0.6	~1
41 - 90	0.53	0.94 *23*
95 - 129	0.222	0.217 *17*
416	0.229	0.0552 *16*
525 - 551	0.0750	0.01415 *24*

Continuous Radiation (^{248}Bk)

⟨β-⟩=174 keV;⟨IB⟩=0.52 keV

E_{bin}(keV)		⟨ ⟩(keV)	(%)
0 - 10	β-	0.095	1.90
	IB	0.0085	
10 - 20	β-	0.281	1.88
	IB	0.0095	0.064
20 - 40	β-	1.10	3.68
	IB	0.0151	0.053
40 - 100	β-	7.2	10.4
	IB	0.134	0.162
100 - 300	β-	53	27.2
	IB	0.32	0.26
300 - 600	β-	93	21.9
	IB	0.027	0.0071
600 - 860	β-	20.4	3.06
	IB	0.00076	0.000120

$^{248}_{98}$Cf(334 *3* d)

Mode: α(99.9971 *3* %), SF(0.0029 *3* %)
Δ: 67239 *6* keV
SpA: 1579 Ci/g
Prod: ^{245}Cm(α,n); ^{246}Cm(α,2n); ^{247}Cm(α,3n); ^{248}Cm(α,4n); ^{238}U(^{14}N,p3n); daughter ^{248}Bk(23.7 h); daughter ^{248}Es; daughter ^{252}Fm

Alpha Particles (^{248}Cf)

⟨α⟩=6255 *6* keV

α(keV)	α(%)
6220 *5*	17.0 *5*
6262 *5*	83.0 *5*

$^{248}_{99}$Es(27 *3* min)

Mode: ϵ(~ 99.75 %), α(~ 0.25 %)
Δ: 70270 *110* keV syst
SpA: 2.8×10^7 Ci/g
Prod: ^{249}Cf(d,3n); ^{249}Bk(α,5n); ^{249}Bk(^{3}He,4n)

Alpha Particles (^{248}Es)

α(keV)
6870 *10*

$^{248}_{100}$Fm(36 *3* s)

Mode: α(99.90 *5* %), SF(0.10 *5* %)
Δ: 71885 *21* keV
SpA: 1.25×10^9 Ci/g
Prod: ^{240}Pu(^{12}C,4n); ^{238}U(^{16}O,6n)

Alpha Particles (^{248}Fm)

⟨α⟩=7854 keV

α(keV)	α(%)
7830 *20*	20
7870 *20*	80

$^{248}_{101}$Md(7 *3* s)

Mode: ϵ(80 *10* %), α(~ 20 %)
Δ: 77080 *280* keV syst
SpA: 6×10^9 Ci/g
Prod: ^{241}Am(^{12}C,5n)

Alpha Particles (^{248}Md)

⟨α⟩~1666 keV

α(keV)	α(%)
8320 *20*	~15
8360 *30*	~5

A = 249

NDS **34**, 8 (1981)

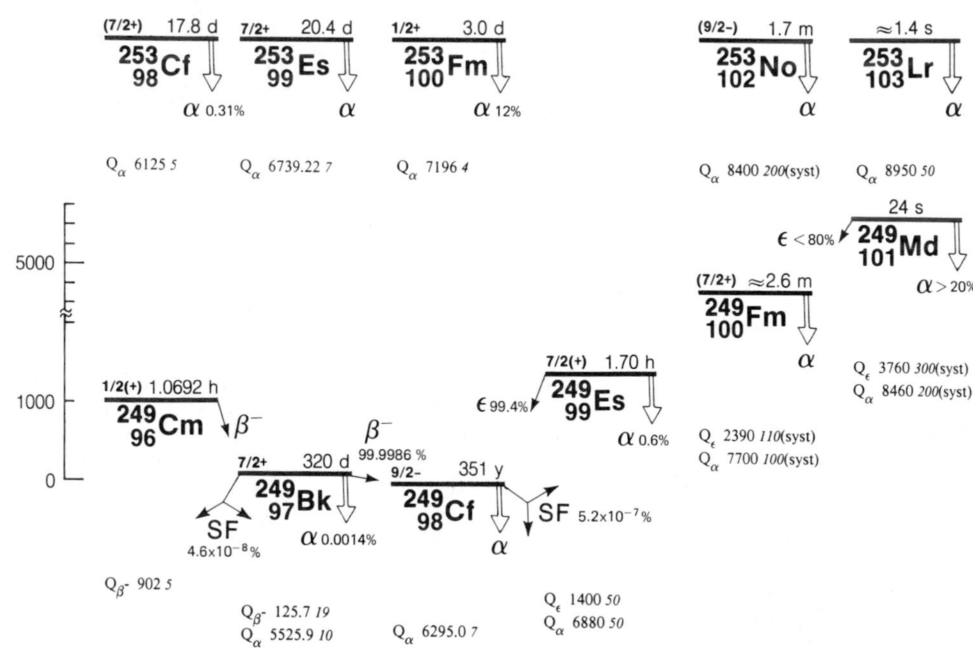

$^{249}_{96}$Cm(1.0692 *5* h)

Mode: β-
Δ: 70746 *5* keV
SpA: 1.1770×10^{7} Ci/g
Prod: ^{248}Cm(n,γ); multiple n-capture from ^{238}U; multiple n-capture from ^{239}Pu; multiple n-capture from ^{244}Cm

Photons (^{249}Cm)

$\langle\gamma\rangle$=18.5 *7* keV

γ_{mode}	γ(keV)	γ(%)
γ (M1+E2)	84.98 *6*	0.0054 *5*
γ (E1)	136.88 *6*	0.039 *3*
γ (E1)	158.63 *9*	0.0029 *4*
γ (E1)	168.91 *8*	0.0022 *2*
γ (E1)	180.54 *4*	0.020 *2*
γ (E1)	191.60 *5*	0.0100 *9*
γ [E1]	306.57 *9*	6.4 *11* $\times10^{-6}$
γ [E2]	347.38 *8*	4.3 *7* $\times10^{-5}$
γ [E1]	349.55 *9*	3.3 *6* $\times10^{-5}$
γ E1+18%M2	368.79 *4*	0.35 *2*
γ [E1]	380.42 *8*	9 *3* $\times10^{-5}$
γ M1+E2	389.18 *7*	0.0063 *8*
γ [M1]	421.21 *8*	0.0092 *10*
γ (E2)	475.48 *4*	0.0072 *12*
γ (M1+E2)	518.46 *4*	0.089 *6*
γ (E2)	529.52 *5*	0.0070 *8*
γ (M1+E2)	549.33 *4*	0.029 *4*
γ M1+35%E2	560.39 *5*	0.84 *6*
γ (E2)	603.44 *6*	0.0064 *9*
γ E2+>15%M1	621.91 *5*	0.182 *13*
γ E2+30%M1	634.31 *5*	1.5 *1*
γ E2+37%M1	652.77 *5*	0.143 *10*

Continuous Radiation (^{249}Cm)

$\langle\beta$-$\rangle$=271 keV;$\langle$IB$\rangle$=0.24 keV

E_{bin}(keV)		$\langle\ \rangle$(keV)	(%)
0 - 10	β-	0.114	2.27
	IB	0.0132	
10 - 20	β-	0.338	2.25
	IB	0.0124	0.087
20 - 40	β-	1.34	4.47
	IB	0.023	0.080
40 - 100	β-	9.1	13.1
	IB	0.056	0.088
100 - 300	β-	74	37.7
	IB	0.098	0.060
300 - 600	β-	144	33.7
	IB	0.036	0.0092
600 - 893	β-	43.2	6.4
	IB	0.0019	0.00030

$^{249}_{97}$Bk(320 *6* d)

Mode: β-(99.99855 *8* %), α(0.00145 *8* %), SF(4.6 *2* $\times10^{-8}$ %)
Δ: 69844 *3* keV
SpA: 1639 Ci/g
Prod: multiple n-capture from ^{238}U; multiple n-capture from ^{239}Pu; multiple n-capture from ^{244}Cm

Alpha Particles (^{249}Bk)

$\langle\alpha\rangle$=0.0778 *5* keV

α(keV)	α(%)
5046.6 *6*	1.0 *4* $\times10^{-6}$
5114.0 *6*	~3 $\times10^{-5}$
5151 ?	6 $\times10^{-7}$
5248.4 *6* ?	~1 $\times10^{-6}$
5313.6 *6*	~1 $\times10^{-6}$
5322 *10* ?	7 $\times10^{-7}$
5350.0 *6*	~2 $\times10^{-5}$
5389.9 *6*	0.00023 *4*
5417.4 *6*	0.00108 *7*
5436.3 *6*	7 *3* $\times10^{-5}$

Photons (^{249}Bk)

$\langle\gamma\rangle$=0.000085 *9* keV

γ_{mode}	γ(keV)	γ(%)†
Am $K_{\alpha2}$	102.026	4.4 *11* $\times10^{-6}$
Am $K_{\alpha1}$	105.472	6.9 *18* $\times10^{-6}$
Am $K_{\beta1}$'	119.960	2.6 *7* $\times10^{-6}$
Am $K_{\beta2}$'	124.123	8.8 *23* $\times10^{-7}$
γ_α(M1+35%E2)	280.430 *15*	8.7 *17* $\times10^{-7}$
γ_α[E1]	299.48 *16*	~1 $\times10^{-8}$
γ_αM1+22%E2	308.320 *8*	3.3 *5* $\times10^{-6}$
γ_αM1+25%E2	327.525 *8*	1.72 *24* $\times10^{-5}$

† 5.5% uncert(syst)

Atomic Electrons (^{249}Bk)

$\langle$e$\rangle$=0.000043 *9* keV

e_{bin}(keV)	$\langle$e$\rangle$(keV)	e(%)
155 - 174	1.2 $\times10^{-6}$	~8 $\times10^{-7}$
183	5.0 $\times10^{-6}$	2.7 *8* $\times10^{-6}$
203	2.4 $\times10^{-5}$	1.2 *4* $\times10^{-5}$
257 - 280	6.8 $\times10^{-7}$	2.6 *10* $\times10^{-7}$
285	1.7 $\times10^{-6}$	5.8 *18* $\times10^{-7}$
289 - 303	4.8 $\times10^{-7}$	1.6 *4* $\times10^{-7}$
304	6.4 $\times10^{-6}$	2.1 *7* $\times10^{-6}$
305	1.3 $\times10^{-6}$	4.3 *14* $\times10^{-7}$
307 - 308	4.1 $\times10^{-7}$	~1 $\times10^{-7}$
321	1.6 $\times10^{-6}$	5.1 *16* $\times10^{-7}$
322 - 328	1.22 $\times10^{-6}$	3.7 *8* $\times10^{-7}$

Continuous Radiation (^{249}Bk)

$\langle\beta$-$\rangle$=32.9 keV;$\langle$IB$\rangle$=0.0041 keV

E_{bin}(keV)		$\langle\ \rangle$(keV)	(%)
0 - 10	β-	1.01	20.6
	IB	0.00156	
10 - 20	β-	2.64	17.8
	IB	0.00101	0.0072
20 - 40	β-	8.1	27.6
	IB	0.00104	0.0038
40 - 100	β-	20	33.0
	IB	0.00049	0.00096
100 - 126	β-	1.08	1.02
	IB	1.42 $\times10^{-6}$	1.36 $\times10^{-6}$

$^{249}_{98}$Cf(350.6 *21* yr)

Mode: α, SF(5.2 *1* $\times10^{-7}$ %)
Δ: 69717.9 *23* keV
SpA: 4.095 Ci/g
Prod: daughter ^{249}Bk; multiple n-capture from ^{238}U; multiple n-capture from ^{239}Pu; multiple n-capture from ^{244}Cm

Alpha Particles (^{249}Cf)

$\langle\alpha\rangle$=5834 *4* keV

α(keV)	α(%)
5201 *25*	<5 $\times10^{-5}$
5238 *25*	<0.0002
5273 *25*	<1.2 $\times10^{-4}$
5307 *25*	<1.3 $\times10^{-4}$
5341 *25*	~0.00040
5355 *25*	
5359 *25*	
5370 *25*	~0.00013
5422 *25*	<0.00029
5431 *25*	~0.0026
5433 *3*	~0.0008
5471 *3* ?	~0.00020
5483 *3*	~0.00032
5503.4 *3*	<0.038
5532 *3*	~0.00021
5560.53 *25*	~0.05
5566 *25* ?	
5604 *3*	~0.0010
5616 *3*	~0.007
5645 *3*	~0.0015
5658 *3*	~0.00010
5693.1 *3*	~0.2
5704 *2*	~0.030
5758.20 *24*	3.7 *10*
5783.98 *24*	~0.26
5811.97 *24*	84.4 *20*
5848.86 *24*	1.0 *3*
5902.93 *24*	2.8 *4*
5945.11 *24*	4.0 *6*
5999.8 *9*	~0.04
6074.49 *24*	~0.24
6140.11 *24*	1.11 *20*
6194.04 *24*	2.17 *20*

Photons (^{249}Cf)

$\langle\gamma\rangle$=326 *8* keV

γ_{mode}	γ(keV)	γ(%)†
Cm L_{ℓ}	12.633	0.59 *14*
Cm L_{α}	14.939	8.9 *19*
Cm L_{η}	17.314	0.18 *5*
Cm L_{β}	19.158	10.4 *25*
Cm L_{γ}	22.929	2.5 *6*
γ[E1]	37.501 *20*	0.0195 *11*
γ[M1]	42.872 *18*	0.033 *2*
γM1+31%E2	54.65 *6*	0.026 *7*
γE2+39%M1	54.81 *3*	0.15 *4* ?
γ[M1+E2]	54.95 *3*	<0.026
γ[M1]	65.94 *4*	0.0178 *20*
γM1+41%E2	66.69 *4*	0.026 *3*
γ[E1]	92.45 *3*	0.195 *4*
Cm $K_{\alpha2}$	104.586	2.08 *6*
Cm $K_{\alpha1}$	109.271	3.28 *10*
γ[E2]	120.85 *10*	<0.044
γ[E2]	121.50 *4*	0.045 *3*
Cm $K_{\beta1}$'	123.059	1.20 *4*
Cm $K_{\beta2}$'	127.344	0.419 *14*
γE2	198.19 *3*	0.0134 *16*
γ[M1]	229.32 *4*	0.0396 *20*

Photons (^{249}Cf)
(continued)

γ_{mode}	γ(keV)	γ(%)†
γM1(+<33%E2)	241.06 3	0.192 4
γM1+2.6%E2	252.998 25	2.47 5
γ[M1]	255.54 9	0.043 7
γ[E1]	266.82 3	0.36 11 §
γ	267.36 10	
γM1+8.1%E2	295.870 24	0.136 6
γ[E1]	321.47 5	0.066 3
γE1	333.51 3	14.4 3
γE1	388.32 3	66
γ[E1]	390.87 9	0.016 3
γ[E1]	406.06 10	0.0108 8
γ[E1]	589.06 8	0.0024 4
γ[E1]	643.86 9	0.013 5
γ	670.1 5	~0.0007 ?
γ	680.1 5	0.0046 9 ?
γ[E1]	701.93 10	0.0059 12 ?
γ	718.6 3	0.0084 8 ?
γ	740.1 13	0.0066 13 ?
γ	760.1 13	0.020 4 ?
γ	770.1 13	~0.026 ?
γ	990.1 13	~0.0006 ?

† 3.0% uncert(syst)
§ 266.8γ + 267.4γ

Atomic Electrons (^{249}Cf)
⟨e⟩=37 3 keV

e_{bin}(keV)	⟨e⟩(keV)	e(%)
13 - 18	0.37	2.02 14
19	1.4	7.3 18
24	1.4	5.7 15
25	0.32	1.30 20
30	0.9	3.0 11
31	4.5	15 5
32 - 33	0.00167	0.0052 3
36	4.1	11 4
37 - 48	1.33	3.1 5
49	1.9	4.0 15
50	1.6	3.2 12
51 - 100	2.4	4.1 7
101 - 122	0.79	0.72 6
125	5.96	4.78 14
127 - 175	0.41	0.267 21
179 - 225	1.06	0.510 13
228	1.97	0.860 25
229 - 256	1.197	0.494 9
260	3.51	1.35 5
261 - 310	0.350	0.118 3
315 - 364	0.804	0.225 7
365 - 406	0.617	0.164 3
516 - 565	0.0026	~0.0005
570 - 619	0.006	~0.0009
620 - 669	0.016	~0.0025
670 - 719	0.0017	0.00025 11
721 - 770	0.0050	~0.0007
966 - 990	4.0 $\times 10^{-5}$	~4×10^{-6}

$^{249}_{99}$Es(1.703 *10* h)

Mode: ε(99.43 8 %), α(0.57 8 %)
Δ: 71110 50 keV
SpA: 7.39×10^{6} Ci/g
Prod: ^{249}Cf(d,2n); ^{249}Bk(α,4n); ^{249}Cf(α,p3n)

Alpha Particles (^{249}Es)

α(keV)
6770 5

Photons (^{249}Es)
⟨γ⟩=408 15 keV

γ_{mode}	γ(keV)	γ(%)†
Cf L_{ℓ}	13.146	1.8 4
Cf L_{α}	15.636	27 5
Cf L_{η}	18.347	0.35 8
Cf L_{β}	20.154	27 6
Cf L_{γ}	24.372	6.4 15
γ M1	43.00 4	0.033 7
γ M1	55.15 4	0.032 7
γ M1+22%E2	58.02 4	0.044 7
γ M1+7.8%E2	62.48 4	0.126 10
γ M1(+E2)	63.44 5	<0.07 ?
γ M1(+<3.1%E2)	63.46 4	0.058 7
Cf $K_{\alpha2}$	109.826	24 5
Cf $K_{\alpha1}$	115.032	37 7
Cf $K_{\beta1}$'	129.436	14 3
Cf $K_{\beta2}$'	133.949	4.9 10
γ [E2]	136.36 8	<0.08 ?
γ [M1]	136.41 5	0.065 15
γ M2+13%E3	144.97 5	0.179 20
γ M1+21%E2	191.56 5	0.40 3
γ	191.64 10	<0.40 ?
γ E2+2.0%M1	234.56 5	0.26 3
γ M1	255.02 6	0.11 3
γ M1+15%E2	298.02 5	0.56 4
γ [E1]	370.23 8	0.14 4
γ E1	375.07 6	3.3 3
γ E1	379.53 5	40.4 25
γ [E1]	433.69 8	0.062 10
γ E1	437.56 5	0.75 5
γ [E1]	442.99 6	0.031 9
γ [M1+E2]	506.95 9	0.040 10
γ M1	564.96 8	0.209 17
γ [M2]	570.10 9	<0.07 ?
γ M1	570.39 7	0.054 14
γ [E1]	609.07 9	~0.030
γ [E1]	625.25 8	<0.26
γ M1(+36%E2)	628.41 7	0.209 17
γ [E1]	664.22 9	~0.010
γ [E1]	668.25 8	~0.24
γ [E1]	707.22 8	~0.030
γ [E2]	766.19 16	0.099 10
γ E2	789.71 7	1.14 9
γ E2	813.22 7	9.1 6
γ [M1+E2]	819.98 8	~0.010
γ	820.05 11	<0.010 ?
γ [M1+E2]	840.07 15	0.096 9
γ E2	852.19 7	0.87 7
γ [E2]	862.97 7	0.018 6
γ [M1+E2]	902.56 15	0.018 6
γ E1	945.46 6	0.239 20
γ	1000.5 5	0.020 8
γ E1	1007.94 6	0.73 6
γ	1021.4 5	~0.010
γ	1093.08 17	0.031 6
γ	1137.10 11	~0.010
γ	1199.58 11	0.016 6
γ [M1+E2]	1205.2 4	0.030 6
γ M1(+37%E2)	1218.5 1	1.50 10
γ	1238.05 17	0.023 3
γ [M1+E2]	1267.7 4	0.0060 20
γ	1304.3 3	0.038 4

† 11% uncert(syst)

Atomic Electrons (^{249}Es)
⟨e⟩=39.8 19 keV

e_{bin}(keV)	⟨e⟩(keV)	e(%)
10 - 18	1.04	8.5 9
20	3.6	18 4
23	0.0024	0.0105 22
25	2.6	10.3 24
26	1.6	6.0 11
29 - 35	1.12	3.6 4
36	1.26	3.5 4
37	1.1	3.0 9
38	1.1	~3
39 - 56	2.3	4.8 12
57	2.0	3.5 15
58 - 107	2.5	3.2 5
108 - 117	0.65	0.58 13
119	3.1	2.6 3
120	1.7	1.4 3
121 - 123	0.058	0.047 6
125	1.05	0.84 12
126 - 136	0.20	0.15 5
138	0.97	0.70 8
139 - 145	1.57	1.11 11
163	1.17	0.72 9
166 - 215	1.7	0.94 19
228 - 240	0.40	0.169 16
245	2.21	0.90 6
248 - 298	0.64	0.231 18
299 - 345	0.045	0.0148 9
349 - 380	0.99	0.275 13
408 - 443	0.39	0.091 23
474 - 507	0.16	0.032 10
529 - 572	0.15	0.028 6
583 - 631	0.064	0.0104 19
638 - 668	0.130	0.0198 16
678	1.00	0.147 10
681 - 728	0.14	0.019 4
740 - 790	0.453	0.0578 24
793 - 840	0.229	0.0282 10
843 - 886	0.047	0.0054 7
896 - 945	0.0028	0.00030 3
958 - 1007	0.018	0.0018 7
1015 - 1020	0.00018	~2×10^{-5}
1065 - 1112	0.41	0.038 7
1117 - 1136	0.0014	~0.00013
1169 - 1218	0.128	0.0107 14
1231 - 1279	0.0021	~0.00017
1284 - 1303	0.0005	~4×10^{-5}

Continuous Radiation (^{249}Es)
⟨IB⟩=1.53 keV

E_{bin}(keV)		⟨ ⟩(keV)	(%)
10 - 20	IB	0.0044	0.024
20 - 40	IB	0.0022	0.0093
40 - 100	IB	0.19	0.21
100 - 300	IB	1.16	0.97
300 - 600	IB	0.092	0.021
600 - 1300	IB	0.080	0.0104
1300 - 1400	IB	3.4×10^{-5}	2.5×10^{-6}

$^{249}_{100}$Fm(2.6 *7* min)

Mode: α
Δ: 73500 *100* keV syst
SpA: 2.9×10^{8} Ci/g
Prod: ^{238}U(^{16}O,5n)

Alpha Particles (^{249}Fm)

α(keV)
7530 *20*

$^{249}_{101}$Md(24 *4* s)

Mode: ϵ(<80 %), α(>20 %)
Δ: 77260 *290* keV syst
SpA: 1.9×10^{9} Ci/g
Prod: ^{241}Am(^{12}C,4n); ^{243}Am(^{12}C,6n)

Alpha Particles (^{249}Md)

α(keV)
8030 *20*

A = 250

NDS **32**, 134 (1981)

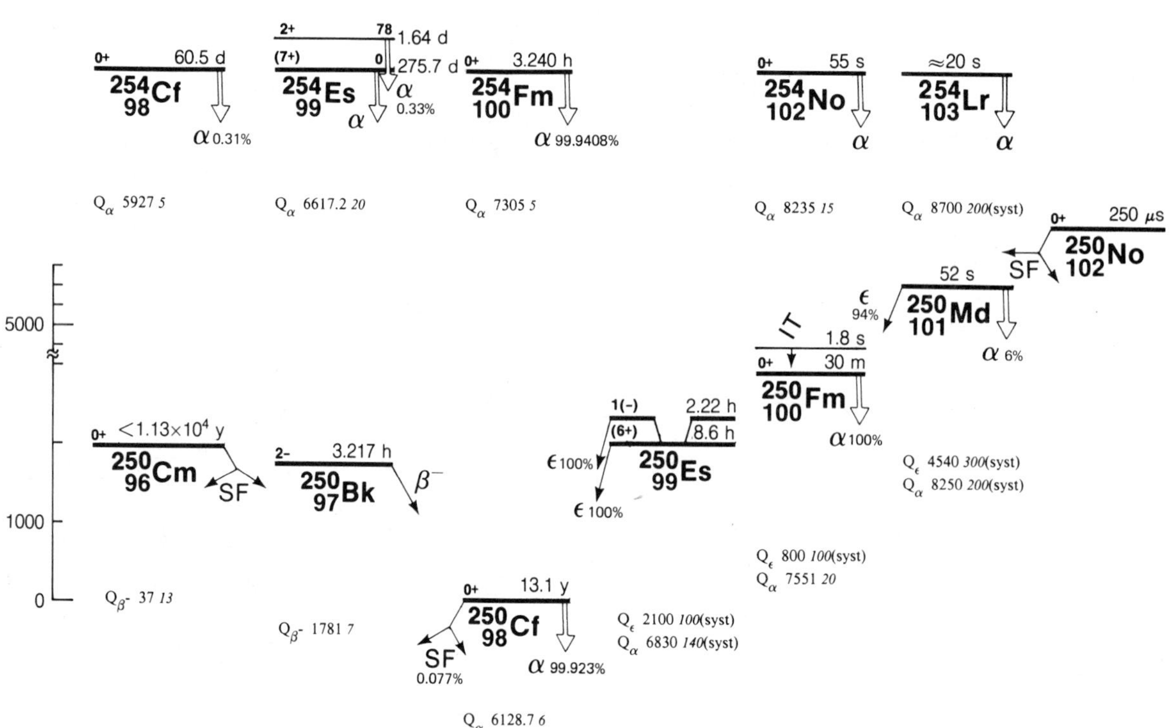

$^{250}_{96}$Cm(<1.13 ×10^4 yr)

Mode: SF
Δ: 72985 *11* keV
SpA: >0.127 Ci/g
Prod: multiple n-capture from ^{238}U

$^{250}_{97}$Bk(3.217 *4* h)

Mode: β-
Δ: 72948 *7* keV
SpA: 3.896×10^6 Ci/g
Prod: daughter ^{254}Es(275.7 d); ^{249}Bk(n,γ)

Photons (^{250}Bk)

⟨γ⟩=898 *35* keV

γ_{mode}	γ(keV)	γ(%)†
Cf L_ℓ	13.146	0.59 *8*
Cf L_α	15.636	8.7 *9*
Cf L_η	18.347	0.21 *3*
Cf L_β	20.304	10.5 *12*
Cf L_γ	24.275	2.4 *3*
γ E2	42.721 *5*	0.038 *3*
γ E2	99.157 *6*	0.128 *7*
Cf $K_{\alpha2}$	109.826	0.306 *17*
Cf $K_{\alpha1}$	115.032	0.47 *3*
γ [E1]	119.383 *23*	0.00067 *22*
γ [E1]	125.968 *17*	0.0063 *5*
Cf $K_{\beta1}$'	129.436	0.177 *10*
Cf $K_{\beta2}$'	133.949	0.063 *4*
γ [E1]	160.291 *17*	0.0284 *18*
γ [E1]	165.475 *23*	0.00135 *18*
γ [E1]	199.798 *23*	0.00108 *13*
γ (M1+E2)	303.95 *3*	0.00230 *23*
γ [M1+E2]	555.21 *10*	0.0063 *5* ?
γ (M1+25%E2)	586.63 *3*	0.0065 *5*
γ (M1+10%E2)	626.137 *23*	0.0242 *13*
γ [E1]	786.43 *3*	0.0049 *9*
γ E1	828.830 *17*	0.117 *6*
γ [E2]	889.964 *12*	1.53 *3*
γ (M1+E2)	929.470 *16*	1.238 *25*
γ E2	989.121 *11*	45 *3*
γ (E2)	1028.627 *16*	4.88 *13*
γ E2	1031.842 *12*	35.6 *7*
γ [E2]	1047.51 *3*	0.00235 *19*
γ [M2]	1068.09 *3*	0.00058 *9* ?
γ	1098.35 *16*	0.00054 *9* ?
γ [E2]	1102.61 *3*	0.00035 *9*
γ	1103.5 *25*	0.00054 *23*
γ (E2)	1111.50 *9*	0.00108 *9*
γ [E1]	1132.782 *21*	0.0190 *9*
γ E0+E2	1146.66 *3*	0.0126 *6*
γ [E2]	1154.75 *3*	0.0072 *4*
γ E1	1167.25 *3*	0.0275 *13*
γ	1175.5 *25*	0.0068 *13*
γ E1	1175.503 *21*	0.0366 *19*
γ [M1+E2]	1201.77 *3*	0.00472 *24*
γ (E2)	1223.90 *4*	0.00279 *18*
γ [E2]	1244.49 *3*	0.00131 *8*
γ E0+E2	1253.91 *3*	0.00167 *13*
γ	1279.20 *23*	0.00081 *9* ?
γ [E2]	1296.63 *3*	0.00067 *9*
γ	1302.89 *22*	0.00045 *9*
γ	1312.94 *6*	0.00148 *9*
γ	1342.73 *5*	0.00189 *14*
γ	1368.61 *5*	0.00315 *23*
γ	1385.45 *5*	0.00202 *14*
γ	1411.33 *5*	0.00059 *14* ?
γ [E2]	1516.098 *23*	0.00121 *9*
γ [M1+E2]	1553.27 *9*	0.00054 *13*
γ E2	1615.255 *22*	0.0456 *21*
γ	1633.17 *24*	0.00054 *9* ?
γ [M1+E2]	1652.42 *9*	0.00099 *9*
γ E2	1657.976 *22*	0.0271 *12*

† 1.8% uncert(syst)

Atomic Electrons (^{250}Bk)

⟨e⟩=30.8 *7* keV

e_{bin}(keV)	⟨e⟩(keV)	e(%)
17	0.122	0.73 *5*
18	3.31	18.8 *14*
20	1.48	7.4 *9*
23	3.6	15.6 *12*
25	1.61	6.4 *8*
26 - 31	0.049	0.187 *21*
36	1.99	5.5 *4*
38	1.75	4.6 *3*
41	1.17	2.84 *21*
42 - 73	0.44	0.97 *6*
74	0.96	1.29 *7*
79 - 129	1.47	1.68 *5*
134 - 180	0.0048	~0.0031
193 - 200	1.41 ×10^{-5}	7.3 *6* ×10^{-6}
278 - 304	0.0019	0.0007 *3*
420 - 452	0.009	0.0022 *9*
491 - 535	0.023	0.0047 *6*
548 - 587	0.0024	0.00042 *7*
600 - 626	0.0074	0.00122 *10*
651 - 694	0.00454	0.00066 *4*
755 - 804	0.6	0.07 *4*
809 - 829	0.000363	4.41 *16* ×10^{-5}
854	4.5	0.52 *4*
864 - 894	0.552	0.062 *3*
897	3.45	0.385 *11*
903 - 933	0.13	0.015 *7*
963	0.91	0.095 *8*
964 - 1012	2.30	0.232 *5*
1019 - 1067	0.420	0.0409 *8*
1072 - 1122	0.00127	0.000114 *6*
1126 - 1175	0.048	0.0043 *21*
1176 - 1225	0.0010	9 *3* ×10^{-5}
1228 - 1277	0.013	0.0011 *5*
1278 - 1323	0.00016	1.2 *5* ×10^{-5}
1336 - 1385	0.00038	2.8 *9* ×10^{-5}
1386 - 1418	9.6 ×10^{-5}	~7×10^{-6}
1480 - 1548	0.00564	0.000377 *14*
1589 - 1653	0.00170	0.000106 *3*

Continuous Radiation (^{250}Bk)

⟨β-⟩=262 keV;⟨IB⟩=0.25 keV

E_{bin}(keV)		⟨ ⟩(keV)	(%)
0 - 10	β-	0.124	2.48
	IB	0.0126	
10 - 20	β-	0.370	2.47
	IB	0.0119	0.083
20 - 40	β-	1.47	4.90
	IB	0.022	0.076
40 - 100	β-	10.0	14.4
	IB	0.053	0.083
100 - 300	β-	79	40.9
	IB	0.092	0.056
300 - 600	β-	121	29.1
	IB	0.040	0.0100
600 - 1300	β-	43.3	5.2
	IB	0.0168	0.0021
1300 - 1781	β-	6.7	0.467
	IB	0.00040	2.9×10^{-5}

$^{250}_{98}$Cf(13.08 *9* yr)

Mode: α(99.923 *3* %), SF(0.077 *3* %)
Δ: 71167 *3* keV
SpA: 109.3 Ci/g
Prod: multiple n-capture from ^{238}U; multiple n-capture from ^{239}Pu; multiple n-capture from ^{244}Cm; daughter ^{250}Bk; daughter ^{254}Fm

Alpha Particles (^{250}Cf)

⟨α⟩=6020 *1* keV

α(keV)	α(%)
5740.8 *5*	~0.010
5891.3 *4*	0.3
5988.9 *4*	15.1 *12*
6031.0 *4*	84.5 *12*

Photons (^{250}Cf)

⟨γ⟩=1.12 *9* keV

γ_{mode}	γ(keV)	γ(%)
Cm L_ℓ	12.633	0.163 *22*
Cm L_α	14.939	2.5 *3*
Cm L_η	17.314	0.058 *8*
Cm L_β	19.192	2.9 *4*
Cm L_γ	22.878	0.65 *9*
γ E2	42.850 *3*	0.0140 *14*

Atomic Electrons (^{250}Cf)

⟨e⟩=4.91 *23* keV

e_{bin}(keV)	⟨e⟩(keV)	e(%)
18	0.041	0.226 *19*
19	1.56	8.2 *8*
24	1.65	6.9 *7*
25	0.0103	0.042 *5*
37	0.61	1.66 *14*
38	0.54	1.41 *12*
39	0.0138	0.036 *3*
41	0.199	0.48 *4*
42	0.174	0.42 *4*
43	0.112	0.262 *22*

$^{250}_{99}$Es(8.6 *1* h)

Mode: ϵ
Δ: 73270 *100* keV syst
SpA: 1.458×10^6 Ci/g
Prod: ^{249}Bk(α,3n); ^{249}Cf(d,n); ^{249}Cf(α,t); ^{249}Bk(^{3}He,2n)

Photons (^{250}Es)

$\langle\gamma\rangle$=1241 *35* keV

γ_{mode}	γ(keV)	γ(%)
Cf L_ℓ	13.146	7.2 *9*
Cf L_α	15.636	105 *10*
Cf L_η	18.347	1.65 *19*
Cf L_β	20.195	118 *15*
Cf L_γ	24.370	30 *4*
γ M1+15%E2	34.323 *5*	~0.06
γ M1(+2.1%E2)	41.771 *5*	0.29 *3*
γ E2	42.721 *5*	0.09 *1*
γ M1+14%E2	46.092 *4*	0.19 *2*
γ M1+26%E2	55.603 *5*	0.20 *2*
γ M1+14%E2	56.529 *8*	0.09 *1*
γ M1+3.9%E2	61.667 *5*	0.85 *7*
γ M1(+13%E2)	66.759 *10*	0.05 *2*
γ M1(+25%E2)	79.996 *11*	0.11 *3*
γ E2	80.415 *6*	0.29 *3*
γ M1(+0.2%E2)	82.283 *6*	2.6 *2*
γ M1(+3.6%E2)	85.088 *6*	1.07 *9*
γ E2	99.157 *6*	0.80 *7*
γ E2	102.621 *8*	0.20 *2*
γ M1(+7.4%E2)	103.438 *6*	0.71 *6*
Cf $K_{\alpha2}$	109.826	46 *4*
Cf $K_{\alpha1}$	115.032	71 *6*
γ [E1]	119.383 *23*	6.5 *22* $\times10^{-5}$
Cf $K_{\beta1}$'	129.436	26.7 *22*
Cf $K_{\beta2}$'	133.949	9.4 *8*
γ M1(+<1.0%E2)	140.691 *6*	4.7 *3*
γ M1(+13%E2)	146.755 *7*	0.22 *6*
γ E2	154.34 *5*	0.31 *7*
γ [E1]	165.475 *23*	0.000130 *21*
γ [E1]	184.01 *5*	0.47 *7*
γ [E1]	199.798 *23*	0.000104 *16*
γ M1+15%E2	222.974 *8*	1.85 *13*
γ M1+50%E2	246.86 *5*	3.8 *2*
γ	299.60 *20*	1.00 *9*
γ M1+46%E2	303.39 *5*	22.3 *11*
γ E2+5.0%M1	349.48 *5*	20.4 *9*
γ E2	383.81 *5*	14.0 *7*
γ [E1]	712.28 *5*	1.34 *9*
γ E1	763.996 *17*	4.0 *2*
γ E1	810.088 *18*	9.1 *5*
γ E1	828.830 *17*	74 *4*
γ E1	863.153 *17*	5.1 *3*
γ [E1]	866.617 *19*	1.3 *1*
γ (M1+E2)	929.470 *16*	0.119 *10*
γ (E2)	1028.627 *16*	0.47 *4*

Atomic Electrons (^{250}Es)

$\langle e\rangle$=303 *5* keV

e_{bin}(keV)	$\langle e\rangle$(keV)	e(%)
6 - 17	11.3	119 *7*
18	6.7	38.3 *22*
19	0.0089	0.046 *10*
20	14.5	73 *8*
21 - 22	2.4	11.2 *24*
23	7.3	31.9 *18*
25	12.3	49 *6*
26	8.9	34 *4*
28 - 35	10.6	34 *3*
36	13.3	37 *3*
37 - 54	19.7	47 *3*
55	7.4	13.4 *21*
56	16.0	28.4 *20*
57	2.3	4.0 *5*
59	6.1	10.3 *9*
60 - 73	6.6	10.7 *18*
74	7.7	10.4 *9*
75 - 114	45.8	53.1 *16*
115	12.3	10.8 *8*
116 - 165	11.0	8.2 *8*
168	30.4	18.0 *15*
174 - 223	11.3	5.32 *23*
227 - 274	4.1	1.67 *18*
277	9.5	3.4 *3*
278 - 324	14.4	4.79 *15*
330 - 379	6.79	1.93 *4*

Atomic Electrons (^{250}Es) (continued)

e_{bin}(keV)	$\langle e\rangle$(keV)	e(%)
380 - 384	0.309	0.081 *3*
629 - 675	0.498	0.075 *3*
686 - 732	3.00	0.431 *23*
738 - 785	0.114	0.0148 *5*
790 - 838	0.88	0.109 *5*
841 - 867	0.0285	0.00334 *11*
894 - 929	0.059	0.0065 *8*
1003 - 1028	0.0192	0.0019 *1*

Continuous Radiation (^{250}Es)

$\langle$IB$\rangle$=1.32 keV

E_{bin}(keV)		$\langle\ \rangle$(keV)	(%)
10 - 20	IB	0.0049	0.026
20 - 40	IB	0.0024	0.0102
40 - 100	IB	0.19	0.21
100 - 300	IB	1.11	0.93
300 - 600	IB	0.0135	0.0037
600 - 642	IB	5.7×10^{-7}	3.2×10^{-7}

$^{250}_{99}$Es(2.22 *5* h)

Mode: ε

Δ: 73270 *100* keV syst

SpA: 5.65×10^{6} Ci/g

Prod: ^{249}Bk(α,3n); ^{249}Bk(^{3}He,2n)

Photons (^{250}Es)

$\langle\gamma\rangle$=550 *14* keV

γ_{mode}	γ(keV)	γ(%)†
Cf L_ℓ	13.146	1.78 *22*
Cf L_α	15.636	26.0 *25*
Cf L_η	18.347	0.40 *5*
Cf L_β	20.192	26 *3*
Cf L_γ	24.334	5.9 *9*
γ E2	42.721 *5*	0.0277 *19*
γ E2	99.157 *6*	0.034 *4*
Cf $K_{\alpha2}$	109.826	21.7 *17*
Cf $K_{\alpha1}$	115.032	34 *3*
γ [E1]	119.383 *23*	3.5 *14* $\times10^{-5}$
γ [E1]	125.968 *17*	0.00188 *19*
Cf $K_{\beta1}$'	129.436	12.6 *10*
Cf $K_{\beta2}$'	133.949	4.5 *4*
γ [E1]	160.291 *17*	0.0085 *7*
γ [E1]	165.475 *23*	7.1 *19* $\times10^{-5}$
γ [E1]	199.798 *23*	5.7 *15* $\times10^{-5}$
γ (M1+E2)	303.95 *3*	0.096 *11*
γ (M1+25%E2)	586.63 *3*	0.259 *24*
γ (M1+10%E2)	626.137 *23*	0.97 *8*
γ (M1)	659.68 *20*	0.48 *9*
γ [E1]	786.43 *3*	0.20 *4*
γ (M1+E2)	802.88 *20*	0.44 *9*
γ E1	828.830 *17*	5.5 *9*
γ [E2]	889.964 *12*	0.457 *24*
γ (M1+E2)	929.470 *16*	0.065 *15*
γ E2	989.121 *11*	13.5 *9*
γ (E2)	1028.627 *16*	0.26 *7*
γ E2	1031.842 *12*	10.6 *5*
γ [E2]	1047.51 *3*	0.038 *7*
γ	1068.2 *5*	<0.10 ?
γ [E2]	1102.61 *3*	0.092 *23*
γ (E2)	1111.50 *9*	0.27 *4*
γ [E1]	1132.782 *21*	0.80 *5*
γ E0+E2	1146.66 *3*	0.20 *3*
γ [E2]	1154.75 *3*	0.099 *20*
γ E1	1167.25 *3*	2.97 *20*
γ E1	1175.503 *21*	1.54 *8*
γ [M1+E2]	1201.77 *3*	1.24 *8*
γ (E2)	1223.90 *4*	0.33 *3*

Photons (^{250}Es) (continued)

γ_{mode}	γ(keV)	γ(%)†
γ [E2]	1244.49 *3*	0.345 *23*
γ E0+E2	1253.91 *3*	~0.049
γ [E2]	1296.63 *3*	0.0093 *22*
γ [E2]	1516.098 *23*	0.049 *5*
γ E2	1615.255 *22*	1.83 *12*
γ E2	1657.976 *22*	1.08 *7*

† 3.0% uncert(syst)

Atomic Electrons (^{250}Es)

$\langle e\rangle$=34.3 *11* keV

e_{bin}(keV)	$\langle e\rangle$(keV)	e(%)
17	0.091	0.54 *3*
18	2.46	14.0 *9*
20	4.0	20.2 *25*
23	2.65	11.6 *7*
25	3.0	12.1 *15*
26	0.93	3.6 *5*
31	2.8×10^{-6}	$\sim9\times10^{-6}$
36	1.48	4.07 *25*
38	1.30	3.45 *21*
41 - 90	2.50	4.41 *25*
92 - 140	1.08	1.04 *7*
146 - 195	0.12	~0.07
196 - 200	2.1×10^{-7}	1.0 *4* $\times10^{-7}$
278 - 304	0.08	~0.028
452 - 491	1.09	0.23 *3*
525 - 567	0.49	0.092 *16*
580 - 626	0.32	0.052 *5*
634 - 668	0.32	0.049 *20*
755 - 804	0.17	0.022 *4*
809 - 829	0.0172	0.00209 *21*
854	1.33	0.156 *10*
864 - 894	0.048	0.0055 *8*
897	1.03	0.115 *6*
903 - 933	0.024	~0.0026
963 - 1012	1.01	0.103 *3*
1019 - 1067	1.36	0.132 *24*
1077 - 1126	0.102	0.0092 *8*
1127 - 1131	0.22	0.019 *5*
1132	5.2	0.46 *6*
1134 - 1182	0.11	0.009 *3*
1195 - 1240	0.046	0.0038 *8*
1241	1.3	0.101 *22*
1242 - 1290	0.12	0.009 *3*
1292 - 1296	1.7×10^{-5}	1.3 *6* $\times10^{-6}$
1480 - 1523	0.216	0.0145 *7*
1589 - 1653	0.0658	0.00409 *14*

Continuous Radiation (^{250}Es)

$\langle\beta+\rangle$=0.76 keV;$\langle$IB$\rangle$=2.3 keV

E_{bin}(keV)		$\langle\ \rangle$(keV)	(%)
0 - 10	β+	8.0×10^{-10}	9.3×10^{-9}
	IB	6.6×10^{-5}	
10 - 20	β+	7.4×10^{-8}	4.34×10^{-7}
	IB	0.0041	0.023
20 - 40	β+	4.66×10^{-6}	1.38×10^{-5}
	IB	0.0021	0.0090
40 - 100	β+	0.00057	0.00071
	IB	0.20	0.21
100 - 300	β+	0.053	0.0235
	IB	1.20	1.00
300 - 600	β+	0.343	0.076
	IB	0.157	0.035
600 - 1300	β+	0.362	0.0492
	IB	0.51	0.056
1300 - 2100	IB	0.19	0.0128
	Σβ+		0.149

$^{250}_{100}$Fm(30 *3* min)

Mode: α
Δ: 74063 *20* keV
SpA: 2.51×10^{7} Ci/g
Prod: ^{249}Cf(α,3n); ^{238}U(^{16}O,4n)

Alpha Particles (^{250}Fm)

α(keV)
7430 *30*

$^{250}_{100}$Fm(1.8 *1* s)

Mode: IT
Δ: 74063 *20* keV
SpA: 2.08×10^{10} Ci/g
Prod: ^{249}Cf(α,3n)

$^{250}_{101}$Md(52 *6* s)

Mode: ε(94 *3* %), α(6 *3* %)
Δ: 78600 *300* keV syst
SpA: 8.62×10^{8} Ci/g
Prod: ^{243}Am(^{13}C,6n); ^{243}Am(^{12}C,5n); ^{241}Am(^{16}O,α3n); ^{240}Pu(^{15}N,5n)

Alpha Particles (^{250}Md)

α(keV)	α(rel)
7750 *20*	~4
7820 *30*	~2

$^{250}_{102}$No(250 *50* μs)

Mode: SF
SpA: 6.5×10^{10} Ci/g
Prod: ^{233}U(^{22}Ne,5n)

A = 251

NDS **34**, 35 (1981)

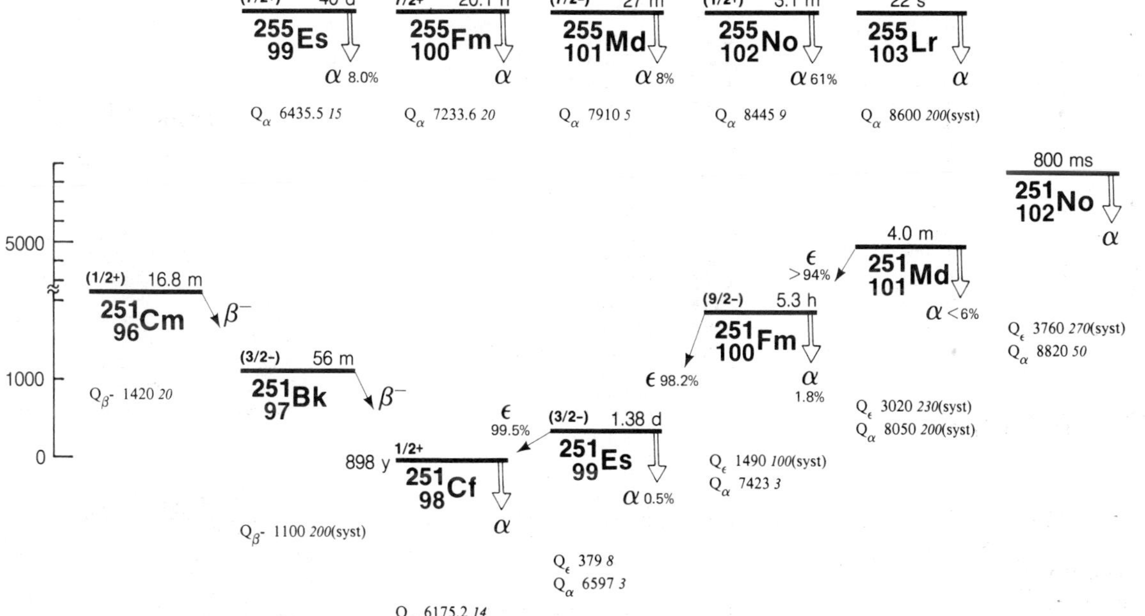

$^{251}_{96}$Cm(16.8 *2* min)

Mode: β-
Δ: 76650 *200* keV syst
SpA: 4.46×10^{7} Ci/g
Prod: ^{250}Cm(n,γ)

Photons (^{251}Cm)

⟨γ⟩=102 *3* keV

γ_{mode}	γ(keV)	γ(%)†
γ[M1]	233.5 *3*	0.45 *4*
γ[E1]	311.6 *3*	0.52 *6*
γ[M1]	389.65 *24*	1.28 *9*
γ[M1]	415.81 *21*	0.16 *7*
γ[M1]	422.13 *24*	0.72 *9*
γ[M1]	435.65 *23*	0.15 *5*
γ[M1]	438.1 *3*	1.24 *8*
γ[M1]	529.91 *22*	1.62 *12*
γ[M1]	542.55 *23*	10.9 *3*
γ[M1]	562.39 *21*	1.0 *3*
γ	945.71 *22*	~0.15
γ[M1]	978.19 *20*	1.00 *13*

† 20% uncert(syst)

Continuous Radiation (^{251}Cm)

⟨β-⟩=421 keV;⟨IB⟩=0.54 keV

E_{bin}(keV)		⟨ ⟩(keV)	(%)
0 - 10	β-	0.067	1.33
	IB	0.019	
10 - 20	β-	0.201	1.34
	IB	0.018	0.127
20 - 40	β-	0.81	2.70
	IB	0.035	0.121
40 - 100	β-	5.8	8.2
	IB	0.091	0.141
100 - 300	β-	54	27.2
	IB	0.20	0.116
300 - 600	β-	145	33.0
	IB	0.132	0.032
600 - 1300	β-	214	26.1
	IB	0.045	0.0061
1300 - 1420	β-	1.56	0.117
	IB	1.16×10^{-5}	8.8×10^{-7}

$^{251}_{97}$Bk(56 *2* min)

Mode: β-
Δ: 75230 *200* keV syst
SpA: 1.34×10^{7} Ci/g
Prod: daughter ^{255}Es;
daughter ^{251}Cm

Photons (^{251}Bk)

γ_{mode}	γ(keV)	γ(rel)
γ M1+6.9%E2	24.812 *12*	
γ(M1)	152.82 *10*	39 *4*
γ(M1)	177.63 *10*	100 *12*

$^{251}_{98}$Cf(898 *44* yr)

Mode: α
Δ: 74128 *5* keV
SpA: 1.59 Ci/g
Prod: multiple n-capture from ^{238}U;
multiple n-capture from ^{239}Pu;
multiple n-capture from ^{244}Cm;
daughter ^{255}Fm

Alpha Particles (^{251}Cf)

⟨α⟩=5664 *7* keV

α(keV)	α(%)
5501 *5*	0.3 *1*
5564.8 *7*	1.5 *2*
5603 *7*	~0.22
5632 *1*	4.5 *10*
5648 *1*	3.5 *13*
5677.3 *6*	35 *1*
5738 *7*	1.0 *3*
5762 *3*	3.8 *4*
5793.7 *7*	2.0 *3*
5812.4 *8*	4.2 *4*
5851.4 *6*	27 *1*
5941.4 *12*	0.6 *1*
6014.0 *7*	11.6 *5*
6074.4 *7*	2.7 *3*

Photons (^{251}Cf)

⟨γ⟩=128 *7* keV

γ_{mode}	γ(keV)	γ(%)
Cm L_{ℓ}	12.633	1.8 *4*
Cm L_{α}	14.939	27 *6*
Cm L_{η}	17.314	0.44 *11*
Cm L_{β}	19.117	30 *7*
Cm L_{γ}	22.967	7.3 *19*
γ[M1]	61.4 *3*	0.56 *22*
γ	68.3 *10*	~0.2 ?
γ	73.3 *10*	~0.30 ?
γ	83.3 *10*	~0.10 ?
Cm L_{γ}	104.586	16 *3*
Cm L_{γ}	109.271	26 *4*
Cm $K_{\beta1}$'	123.059	9.4 *15*
Cm $K_{\beta2}$'	127.344	3.3 *5*
γ[E2]	135.2 *10*	~0.10 ?
γ	144.3 *10*	~0.10 ?
γ	154.3 *10*	~0.2 ?
γ E2	176.87 *10*	17.7 *15*
γ	214.3 *10*	~0.2 ?
γ M2	226.7 *7*	6.3 *11*
γ	255.3 *10*	~0.2 ?
γ	262.3 *10*	~0.2
γ[E1]	266.3 *3*	0.5 *2*
γ	270.3 *10*	~0.2 ?
γ[E1]	285.3 *2*	1.4 *3*
γ	291.2 *3*	~0.4

Atomic Electrons (^{251}Cf)

⟨e⟩=187 *12* keV

e_{bin}(keV)	⟨e⟩(keV)	e(%)
7 - 55	27.2	81 *13*
57 - 97	9.1	13 *4*
98	50.0	51 *9*
99 - 148	2.7	2.1 *5*
149 - 152	2.04	1.34 *12*
153	17.8	11.6 *10*
154 - 157	0.097	0.062 *14*

Atomic Electrons (^{251}Cf)
(continued)

e_{bin}(keV)	⟨e⟩(keV)	e(%)
158	8.9	5.6 *5*
163	0.5	<0.6
171	6.0	3.5 *3*
172 - 195	6.5	3.71 *19*
202	31.3	15 *3*
203	4.5	2.2 *4*
208 - 214	3.9	1.9 *4*
220	8.8	4.0 *7*
221 - 270	7.9	3.5 *3*
272 - 291	0.10	~0.036

$^{251}_{99}$Es(1.38 *4* d)

Mode: ε(99.51 *12* %), α(0.49 *12* %)
Δ: 74507 *7* keV
SpA: 3.78×10^{5} Ci/g
Prod: ^{249}Bk(α,2n)

Alpha Particles (^{251}Es)

⟨α⟩=31.77 *1* keV

α(keV)	α(%)
6410.4 *14*	0.016 *3*
6421.4 *13*	0.015 *3*
6451.7 *13*	0.016 *3*
6462.5 *13*	0.046 *5*
6491.9 *13*	0.397 *8*

Photons (^{251}Es)

⟨γ⟩=6.6 *7* keV

γ_{mode}	γ(keV)	γ(%)
γ M1+6.9%E2	24.812 *12*	
γ E2	47.806 *12*	
γ[M1]	129.83 *10*	~0.6
γ(M1)	152.82 *10*	0.91 *9*
γ[M1]	163.85 *13*	~0.10
γ(M1)	177.63 *10*	2.4 *3*
γ[M1]	186.84 *13*	~0.0050

$^{251}_{100}$Fm(5.30 *8* h)

Mode: ε(98.20 *13* %), α(1.80 *13* %)
Δ: 76000 *100* keV syst
SpA: 2.36×10^{6} Ci/g
Prod: ^{249}Cf(α,2n); ^{238}U(^{18}O,5n)

Alpha Particles (^{251}Fm)

$\langle\alpha\rangle$=122.89 *3* keV

α(keV)	α(%)
6578.9 *10*	0.0047 *7*
6638.0 *9*	0.0101 *11*
6681 *4*	0.0013 *5*
6720 *3*	0.0079 *7*
6762.9 *12*	0.0068 *11*
6781.7 *8*	0.086 *4*
6832.4 *8*	1.566 *16*
6884.8 *8*	0.0306 *18*
6928.1 *8*	0.0324 *18*
7107.4 *8*	~0.0009
7185.0 *8*	0.0052 *5*
7251.0 *8*	0.0167 *14*
7305.1 *8*	0.0270 *18*

Photons (^{251}Fm)

$\langle\gamma\rangle$=156 *4* keV

γ_{mode}	γ(keV)	γ(%)†
Es L_ℓ	13.403	1.55 *21*
Es L_α	15.991	22.2 *24*
Es L_η	18.884	0.28 *4*
Es L_β	20.704	22 *3*
Es L_γ	25.093	5.2 *9*
γ_ϵM1+3.5%E2	47.52 *5*	0.118 *10*
γ_αM1+21%E2	55.00 *10*	0.0104 *14*
γ_αM1+17%E2	67.09 *10*	0.0050 *9*
Es $K_{\alpha2}$	112.526	21.4 *15*
Es $K_{\alpha1}$	118.012	32.9 *23*
γ_αE2	122.09 *11*	0.0050 *9*
Es $K_{\beta1}$'	132.727	12.4 *9*
Es $K_{\beta2}$'	137.358	4.4 *3*
γ_ϵM1(+18%E2)	281.45 *7*	0.071 *5*
γ_ϵM1+23%E2	307.44 *7*	0.035 *3*
γ_α[E1]	331.0 *3*	0.0063 *13*
γ_ϵM1(+4.1%E2)	349.87 *6*	0.82 *4*
γ_αE1	358.30 *8*	0.315 *20*
γ_α[M1]	372.2 *3*	0.0045 *9*
γ_ϵM1	375.86 *7*	0.363 *20*
γ_αM1	383.2 *3*	0.0196 *20*
γ_ϵM1+21%E2	385.33 *20*	0.030 *3*
γ_ϵ(E1)	405.62 *7*	0.99 *5*
γ_α[E1]	409.89 *19*	0.0090 *13*
γ_αE1	425.39 *8*	0.95 *5*
γ_ϵM1+15%E2	429.73 *10*	0.082 *7*
γ_ϵ(E1)	453.14 *7*	1.45 *8*
γ_ϵM1(+23%E2)	461.03 *7*	0.090 *6*
γ_ϵ[E2]	461.3 *4*	<0.017 ?
γ_α[E1]	476.98 *19*	0.0097 *14*
γ_αE1	480.39 *8*	0.392 *20*
γ_ϵ[M1]	487.03 *9*	0.0108 *20*
γ_α[M1]	496.1 *10*	~0.0014
γ_α[E1]	615.9 *7*	~0.0009
γ_α[E1]	623.0 *6*	0.0013 *4*
γ_ϵM1,E2	663.88 *9*	0.017 *3*
γ_α[E1]	678.0 *6*	0.0047 *11*
γ_α[E1]	683.0 *7*	~0.0007
γ_ϵ[M1+E2]	694.5 *3*	0.015 *3*
γ_ϵM1(+25%E2)	706.54 *10*	0.025 *3*
γ_ϵM1+27%E2	722.11 *7*	0.081 *7*
γ_ϵM1(+13%E2)	769.63 *7*	0.393 *20*
γ_ϵM1,E2	774.96 *9*	~0.041
γ_ϵ[M1+E2]	775.04 *7*	~0.09
γ_ϵ	786.3 *5*	0.0079 *20*
γ_ϵ	796.6 *5*	0.0118 *20*
γ_ϵ	826.8 *5*	~0.0049
γ_ϵE2+25%M1	833.27 *6*	0.63 *4*
γ_ϵM1+38%E2	843.46 *7*	0.118 *10*
γ_ϵ	858.5 *5*	~0.0049
γ_ϵE2	880.79 *6*	2.19 *11*
γ_ϵ(E2)	901.69 *6*	0.128 *10*
γ_ϵ[M1]	1056.41 *10*	0.0059 *10*
γ_ϵ[M1]	1124.91 *8*	0.027 *3*
γ_ϵ[M1]	1150.91 *8*	0.0147 *20*
γ_ϵ[M1]	1183.14 *7*	0.073 *5*
γ_ϵ	1193.03 *14*	0.0088 *10*
γ_ϵ[M1]	1209.14 *7*	0.034 *3*

Photons (^{251}Fm)
(continued)

γ_{mode}	γ(keV)	γ(%)†
γ_ϵ[E2]	1230.66 *7*	0.137 *10*
γ_ϵ	1242.92 *20*	0.0051 *5*
γ_ϵ	1245.55 *24*	0.0061 *6*
γ_ϵ	1251.26 *13*	0.0187 *16*
γ_ϵ[E2]	1256.65 *7*	0.064 *5*
γ_ϵ	1279.0 *4*	0.0035 *4* ?
γ_ϵ	1293.07 *24*	0.037 *3*
γ_ϵ	1298.77 *13*	0.041 *3*
γ_ϵ	1301.15 *19*	0.0098 *10*
γ_ϵ	1348.66 *19*	0.0032 *3*

† uncert(syst): 7.2% for α, 0.13% for ϵ

Atomic Electrons (^{251}Fm)

$\langle e\rangle$=20.5 *10* keV

e_{bin}(keV)	$\langle e\rangle$(keV)	e(%)
6	1.0	~15
7	0.7	~10
8	0.7	~8
11 - 19	0.017	~0.13
20	3.0	14.9 *21*
21	1.44	6.9 *6*
22	0.46	2.1 *9*
24 - 25	0.014	~0.06
26	2.1	8.1 *12*
27	1.6	5.8 *13*
30 - 40	0.09	~0.26
41	0.94	2.3 *4*
42 - 91	1.42	2.4 *3*
92 - 132	1.04	0.99 *4*
143 - 169	0.23	0.15 *3*
211	1.78	0.84 *6*
237	0.75	0.314 *18*
247 - 291	0.35	0.128 *11*
300 - 315	0.085	0.0271 *17*
323	0.62	0.193 *22*
324 - 371	0.598	0.172 *6*
374 - 423	0.097	0.0247 *15*
424 - 467	0.084	0.0191 *20*
480 - 525	0.011	~0.0021
556 - 584	0.079	0.0136 *25*
631 - 681	0.35	0.056 *10*
686 - 722	0.24	0.035 *8*
742 - 791	0.368	0.049 *3*
792 - 839	0.082	0.0101 *13*
842 - 881	0.130	0.0151 *4*
895 - 918	0.0048	0.00052 *6*
987 - 1031	0.0193	0.00194 *17*
1036 - 1071	0.046	0.0044 *3*
1092 - 1141	0.030	0.0028 *3*
1144 - 1192	0.031	0.0027 *8*
1202 - 1251	0.0102	0.00084 *8*
1252 - 1300	0.0056	0.00044 *16*
1322 - 1348	0.00017	~1×10^{-5}

Continuous Radiation (^{251}Fm)

$\langle$IB$\rangle$=1.8 keV

E_{bin}(keV)		$\langle\ \rangle$(keV)	(%)
10 - 20	IB	0.0034	0.018
20 - 40	IB	0.0031	0.0137
40 - 100	IB	0.123	0.136
100 - 300	IB	1.32	1.08
300 - 600	IB	0.146	0.033
600 - 1300	IB	0.25	0.031
1300 - 1482	IB	0.00101	7.5×10^{-5}

$^{251}_{101}$Md(4.0 *5* min)

Mode: ϵ(>94 %), α(<6 %)
Δ: 79020 *210* keV syst
SpA: 1.87×10^{8} Ci/g
Prod: ^{243}Am(^{13}C,5n); ^{243}Am(^{12}C,4n); ^{15}N on ^{240}Pu

Alpha Particles (^{251}Md)

α(keV)
7550 *20*

$^{251}_{102}$No(800 *300* ms)

Mode: α
Δ: 82780 *180* keV syst
SpA: 3.8×10^{10} Ci/g
Prod: ^{244}Cm(^{12}C,5n)

Alpha Particles (^{251}No)

α(keV)	α(rel)
8600 *20*	80
8680 *20*	20

A = 252

NDS 32, 158 (1981)

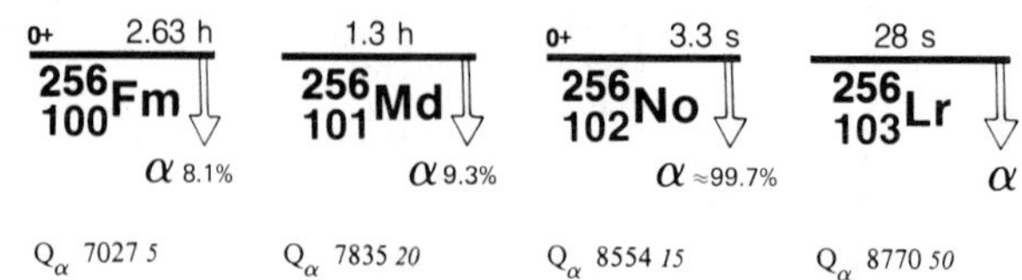

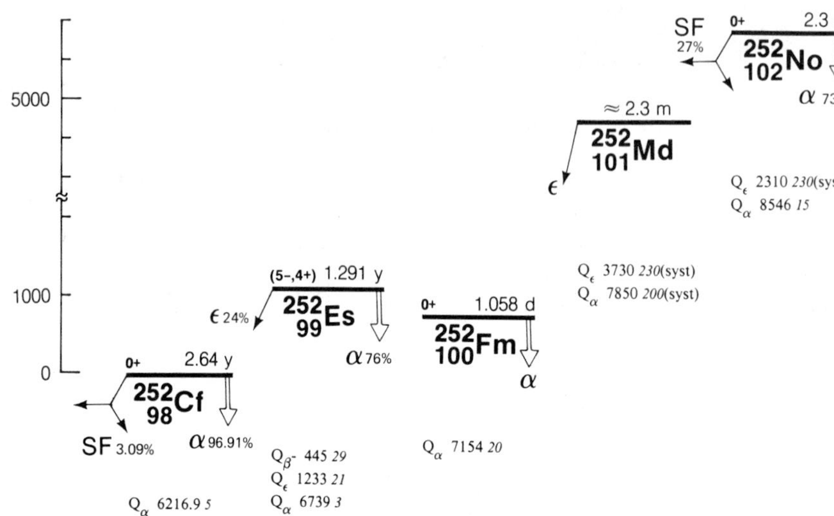

$^{252}_{98}$Cf(2.645 *8* yr)

Mode: α(96.908 *8* %), SF(3.092 *8* %)
Δ: 76030 *5* keV
SpA: 536.3 Ci/g
Prod: multiple n-capture from ^{238}U; multiple n-capture from ^{239}Pu; multiple n-capture from ^{244}Cm

Alpha Particles (^{252}Cf)

⟨α⟩=5930.8 *5* keV

α(keV)	α(%)
5616 *10*	~6×10^{-5}
5824 *8*	~0.0019
5977.1 *11*	0.23 *4*
6075.6 *4*	15.2 *3*
6118.4 *4*	81.6 *3*

Photons (^{252}Cf)

⟨γ⟩=1.14 *8* keV

γ_{mode}	γ(keV)	γ(%)†
Cm L_ℓ	12.633	0.163 *20*
Cm L_α	14.939	2.47 *23*
Cm L_η	17.314	0.058 *7*
Cm L_β	19.193	2.9 *3*
Cm L_γ	22.878	0.66 *8*
γ E2	43.399 *25*	0.0148 *9*
γ (E2)	100.2 *10*	~0.013
γ (E2)	155 *8*	0.0019 *4*

† 4.0% uncert(syst)

Atomic Electrons (^{252}Cf)

⟨e⟩=5.14 *17* keV

e_{bin}(keV)	⟨e⟩(keV)	e(%)
19	0.48	2.5 *3*
20	1.14	5.8 *4*
24	1.66	6.9 *5*
25	0.0105	0.043 *5*
37	0.61	1.64 *10*
39	0.55	1.43 *9*
42	0.368	0.88 *5*
43	0.116	0.269 *17*
76	0.0043	~0.006
77	0.08	~0.10
81	0.05	~0.06
94	0.028	~0.029
95	0.018	~0.018
96	0.00042	~0.00044
98	0.0005	~0.0006
99	0.013	~0.013
100	0.0042	~0.0042

$^{252}_{99}$Es(1.291 *5* yr)

Mode: α(76 *4* %), ε(24 *2* %)
Δ: 77263 *21* keV
SpA: 1098 Ci/g
Prod: ^{249}Bk(α,n); ^{252}Cf(d,2n)

Alpha Particles (^{252}Es)

⟨α⟩=5018 *1* keV

α(keV)	α(%)
5944 *4*	0.030 *11*
5986 *4*	0.038 *11*
6018 *4*	0.091 *23*
6051.0 *12*	0.78 *7*
6110.9 *13*	0.091 *23*
6156 *5*	~0.030
6181 *5*	0.061 *23*
6215 *5*	0.076 *23*
6238.3 *12*	0.43 *4*
6266 *3*	0.57 *5*
6298 *5*	~0.030
6375 *5*	0.053 *23*
6422.4 *12*	0.34 *4*
6462.8 *12*	0.19 *3*
6482.7 *12*	1.66 *7*
6498.7 *12*	0.24 *3*
6562.1 *12*	10.34 *23*
6631.6 *12*	61.0 *7*

Photons (^{252}Es)

⟨γ⟩=257 *11* keV

γ_{mode}	γ(keV)	γ(%)†
Bk L_ℓ	12.890	~2
Cf L_ℓ	13.146	0.67 *9*
Bk L_α	15.285	~24
Cf L_α	15.636	9.7 *11*
Bk L_η	17.826	~0.6
Cf L_η	18.347	0.151 *20*
Bk L_β	19.734	~30
Cf L_β	20.194	10.2 *14*
Bk L_γ	23.590	~7
Cf L_γ	24.351	2.5 *4*
γ_ϵE2	45.69 *5*	0.026 *3*
γ_α	52.28 *5*	0.55 *5*
γ_α(E1)	64.37 *5*	0.274 *23* ?

Photons (^{252}Es) (continued)

γ_{mode}	γ(keV)	γ(%)†
γ_α(E2)	70.60 *5*	0.122 *15*
γ_α(E2)	80.65 *8*	0.032 *5*
γ_ϵE1	102.30 *4*	1.88 *13*
γ_ϵE2	105.99 *4*	0.139 *13*
Bk $K_{\alpha2}$	107.181	0.23 *3*
Cf $K_{\alpha2}$	109.826	2.49 *20*
Bk $K_{\alpha1}$	112.121	0.36 *5*
Cf $K_{\alpha1}$	115.032	3.9 *3*
Bk $K_{\beta1}$'	126.216	0.132 *18*
Cf $K_{\beta1}$'	129.436	1.44 *12*
Bk $K_{\beta2}$'	130.611	0.046 *7*
Cf $K_{\beta2}$'	133.949	0.51 *4*
γ_ϵE1	139.00 *5*	13.9 *10*
γ_α	149.05 *20*	0.020 *3*
γ_α[E2]	151.25 *8*	0.073 *7*
γ_ϵM1	164.99 *7*	0.144 *16*
γ_αM1	193.45 *10*	0.052 *6*
γ_αM1	227.9 *4*	0.027 *5*
γ_α	230.8 *4*	0.024 *5*
γ_α	325.9 *4*	0.024 *5*
γ_αM1	377.4 *3*	0.122 *15*
γ_αM1	399.6 *3*	0.122 *15*
γ_αM1	418.4 *3*	0.220 *23*
γ_α	428.2 *5*	~0.009
γ_α	452.3 *5*	0.030 *8*
γ_α	522.9 *10*	~0.012
γ_α	529.1 *7*	0.053 *8*
γ_α	547.9 *25*	~0.008
γ_α	589.9 *7*	0.084 *8*
γ_ϵ[M1+E2]	693.98 *7*	0.46 *3*
γ_ϵ[E1]	715.77 *6*	0.86 *5*
γ_ϵ[M1+E2]	748.54 *24*	0.076 *13*
γ_ϵ[M1+E2]	759.08 *6*	0.51 *4*
γ_ϵ[E1]	765.27 *10*	0.183 *16*
γ_ϵ[E1]	785.07 *6*	18.3 *10*
γ_ϵ[M1+E2]	799.97 *7*	1.49 *10*
γ_ϵ[E2]	804.78 *7*	0.39 *3*
γ_ϵ[M1+E2]	818.08 *5*	0.75 *6*
γ_ϵ[E1]	821.77 *5*	0.35 *3*
γ_ϵ[E2]	854.53 *24*	0.034 *8*
γ_ϵ[M1+E2]	924.07 *5*	2.41 *16*

† uncert(syst): 5.3% for α, 12% for ϵ

Atomic Electrons (^{252}Es)

$\langle e \rangle$=84 *22* keV

e_{bin}(keV)	$\langle e \rangle$(keV)	e(%)
17	0.017	~0.10
19	3.8	~19
20	1.44	7.2 *10*
21	1.95	9.5 *10*
24	4.3	~18
25	1.6	6.5 *25*
26	2.5	9.6 *10*
27	2.8	~10
28	13.6	<98
30	0.34	1.12 *12*
33	12.6	<77
39 - 45	2.97	7.2 *4*
46	9.8	~21
47	5.3	<22
48	0.13	~0.3
51	5.8	~11
52 - 101	7.3	10.8 *23*
102 - 151	2.08	1.70 *5*
158 - 206	0.32	0.183 *18*
209 - 246	0.24	0.098 *11*
268 - 307	0.56	0.200 *16*
319 - 358	0.088	0.026 *7*
371 - 418	0.27	0.070 *9*
422 - 458	0.05	~0.012
498 - 547	0.016	0.0032 *15*
548 - 590	0.27	~0.05
614 - 650	0.96	0.15 *3*
665 - 714	1.0	0.15 *7*
715 - 764	0.24	0.032 *5*
765 - 814	1.2	0.15 *7*
815 - 855	0.010	0.0012 *4*
898 - 924	0.26	0.029 *13*

Continuous Radiation (^{252}Es)

$\langle$IB$\rangle$=0.24 keV

E_{bin}(keV)		$\langle\ \rangle$(keV)	(%)
10 - 20	IB	0.0018	0.0100
20 - 40	IB	0.00084	0.0037
40 - 100	IB	0.039	0.042
100 - 300	IB	0.19	0.166
300 - 387	IB	7.4×10^{-6}	2.3×10^{-6}

$^{252}_{100}$Fm(1.0579 *21* d)

Mode: α

Δ: 76817 *21* keV

SpA: 4.8978×10^{5} Ci/g

Prod: ^{252}Cf(α,4n); ^{250}Cf(α,2n); ^{249}Cf(α,n); ^{238}U(^{18}O,4n)

Alpha Particles (^{252}Fm)

$\langle\alpha\rangle$~7034 keV

α(keV)	α(%)
6999 *9*	~15
7040 *9*	~85

$^{252}_{101}$Md(2.3 *8* min)

Mode: ϵ

Δ: 80540 *230* keV syst

SpA: 3.2×10^{8} Ci/g

Prod: ^{238}U(^{19}F,5n); ^{243}Am(^{13}C,4n)

$^{252}_{102}$No(2.30 *22* s)

Mode: α(73.1 *19* %), SF(26.9 *19* %)

Δ: 82856 *26* keV

SpA: 1.68×10^{10} Ci/g

Prod: ^{244}Cm(^{12}C,4n); ^{244}Cm(^{13}C,5n); ^{241}Am(^{15}N,4n); ^{239}Pu(^{18}O,5n); ^{235}U(^{22}Ne,5n); ^{48}Ca on Pb

Alpha Particles (^{252}No)

$\langle\alpha\rangle$~6144 keV

α(keV)	α(%)
8372 *8*	~18
8415 *6*	~55

A = 253

NDS 34, 58 (1981)

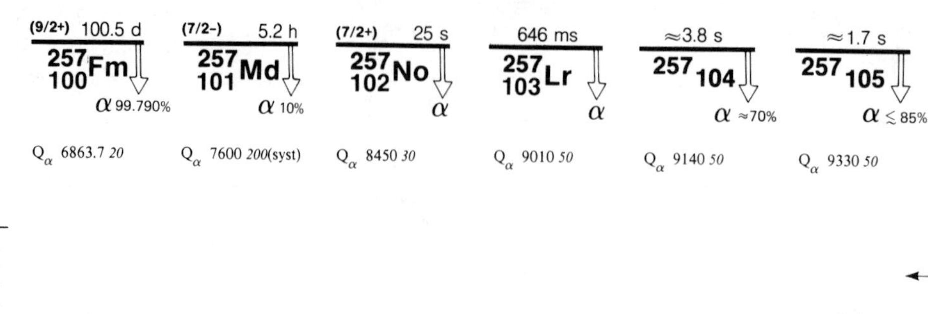

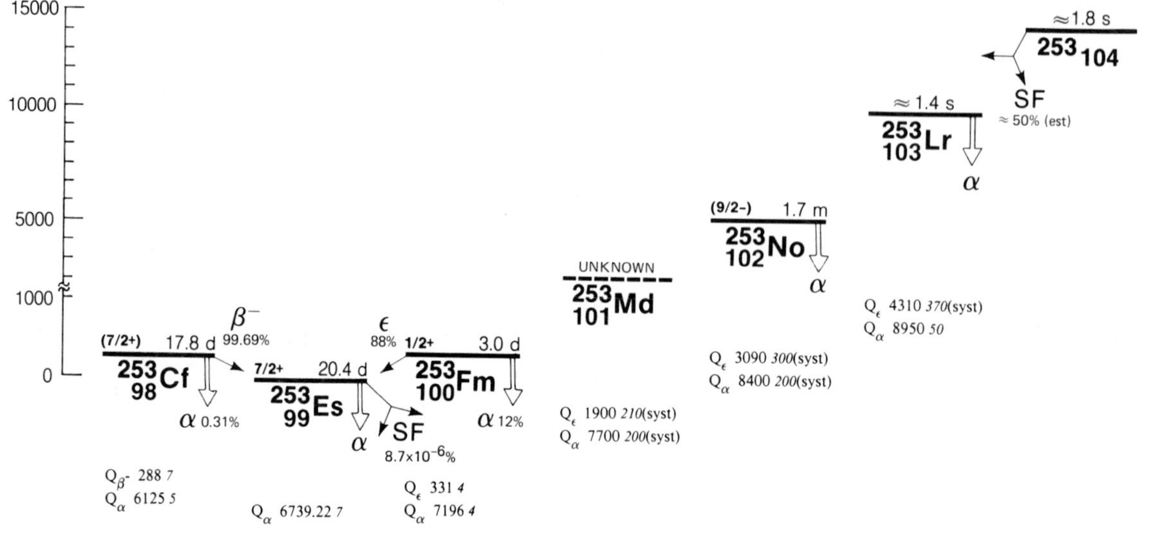

$^{253}_{98}$Cf(17.81 *8* d)

Mode: β-(99.69 *4* %), α(0.31 *4* %)
Δ: 79296 *7* keV
SpA: 2.898×10^4 Ci/g
Prod: multiple n-capture from ^{238}U; multiple n-capture from ^{239}Pu; multiple n-capture from ^{244}Cm

Alpha Particles (^{253}Cf)

⟨α⟩=18.52 *3* keV

α(keV)	α(%)
5921 *5*	0.016 *6*
5979 *5*	0.294 *6*

Continuous Radiation (^{253}Cf)

⟨β-⟩=78 keV;⟨IB⟩=0.022 keV

E_{bin}(keV)		⟨ ⟩(keV)	(%)
0 - 10	β-	0.440	8.8
	IB	0.0040	
10 - 20	β-	1.26	8.4
	IB	0.0034	0.023
20 - 40	β-	4.63	15.5
	IB	0.0051	0.018
40 - 100	β-	24.5	36.2
	IB	0.0074	0.0124
100 - 288	β-	47.2	31.8
	IB	0.0023	0.0019

$^{253}_{99}$Es(20.4 *1* d)

Mode: α, SF(8.7 *3* $\times10^{-6}$ %)
Δ: 79008 *3* keV
SpA: 2.530×10^4 Ci/g
Prod: daughter ^{253}Cf

Alpha Particles (^{253}Es)

⟨α⟩=6626.9 *1* keV

α(keV)	α(%)
5715.3 *10*	8 *3* $\times10^{-5}$
5876.7 *10* ?	2.7 *8* $\times10^{-5}$
5935 *4*	~4 $\times10^{-5}$
5944 *3*	0.00015 *5*
5974 ?	~6 $\times10^{-5}$
6017.7 *7*	0.00018 *5*
6035.8 *6*	0.00029 *7*
6044.6 *3*	0.00040 *9*
6072.60 *10* ?	<5 $\times10^{-5}$
6083.48 *9*	0.00025 *5*
6099.34 *22*	0.0034 *2*
6122.0 *7*	0.00078 *8*
6165.36 *21*	0.015 *1*
6210.59 *11*	0.039 *2*
6218.20 *9*	~0.0015
6228.72 *13* ?	0.00012 *4*
6249.72 *9*	0.045 *2*
6265.8 *5*	0.00080 *8*
6325 *3* ?	0.0004 *1*
6354.05 *10*	0.0082 *4*
6407.09 *7*	0.013 *1*
6431.39 *10*	0.061 *3*
6479.35 *7*	0.085 *3*
6497.19 *9*	0.26 *1*
6540.46 *6*	0.85 *2*
6551.43 *9*	0.71 *2*
6591.60 *6*	6.6 *1*
6593.74 *9*	~0.7

Alpha Particles (^{253}Es) (continued)

α(keV)	α(%)
6624.12 *9*	~0.8
6632.74 *5*	89.8 *2*

Photons (^{253}Es)

⟨γ⟩=0.292 *14* keV

γ_{mode}	γ(keV)	γ(%)†
γ M1,E2	30.87 *3*	~0.007
γ M1+~1.4%E2	41.80 *4*	~0.050
γ M1(+~0.6%E2)	42.985 *25*	~0.009
γ M1+~2.2%E2	51.958 *25*	~0.0048
γ M1+~1.9%E2	55.12 *3*	~0.0019
γ M1,E2	62.09 *3*	~0.0010
γ [M1+E2]	66.86 *4*	~0.0010
γ [M1]	73.42 *4*	<0.0008
γ E2	73.85 *3*	~0.0008
γ [M1]	78.58 *6*	~0.00028
γ E2	93.76 *4*	~0.0007
γ E2	98.10 *4*	0.00097 *19*
γ E2	114.05 *4*	~0.0006
γ E2	121.97 *4*	~0.0007
γ [E2]	135.52 *4*	~0.00015
γ (E1)	136.88 *6*	4.6 *7* $\times10^{-5}$
γ E2	145.44 *5*	0.00028 *3*
γ (E1)	158.63 *9*	1.4 *3* $\times10^{-7}$
γ [E1]	162.79 *9*	~1 $\times10^{-5}$
γ [E2]	168.3 *5*	~5 $\times10^{-5}$
γ (E1)	168.91 *8*	2.6 *4* $\times10^{-6}$
γ (E1)	180.54 *4*	2.3 *4* $\times10^{-5}$
γ (E1)	191.60 *5*	4.8 *10* $\times10^{-7}$
γ	283.4 *10*	6.9 *14* $\times10^{-6}$
γ [E1]	291.21 *12*	3.5 *4* $\times10^{-5}$
γ [E1]	306.57 *9*	2.7 *3* $\times10^{-5}$

Photons (^{253}Es) (continued)

γ_{mode}	γ(keV)	γ(%)†
γ [E2]	312.70 22	~4 $\times 10^{-6}$
γ [E2]	319.05 21	3.7 4 $\times 10^{-5}$
γ [E2]	335.18 10	0.000149 15
γ [M1]	337.17 21	5.4 5 $\times 10^{-5}$
γ [E1]	346.33 12	0.000172 17
γ [E2]	347.38 8	0.00018 2
γ [E1]	349.55 9	0.000140 15
γ M1+E2	368.3 3	~0.00015
γ E1+18%M2	368.79 4	~0.00017
γ [E1]	380.42 8	0.00038 12
γ M1	381.14 21	0.0056 5
γ M1	386.12 22	0.0013 5
γ M1	387.14 10	0.0181 18
γ M1+E2	389.18 7	0.0264 5
γ [M1]	392.28 21	~0.00013 ?
γ [E1]	404.25 23	5.6 6 $\times 10^{-6}$
γ [M1]	421.21 8	0.000106 21
γ [E2]	425.2 7	0.00024 5
γ M1	428.94 9	0.0060 6
γ [M1]	433.10 21	0.0029 3
γ M1	441.7 3	8.5 8 $\times 10^{-5}$
γ M1+23%E2	448.22 22	0.00069 7
γ [M1+E2]	468.8 6	8.7 17 $\times 10^{-5}$
γ M1(+<14%E2)	474.89 21	0.00035 4
γ (E2)	475.48 4	8.4 18 $\times 10^{-6}$
γ [M1+E2]	477.1 7	0.000120 24
γ M1	500.18 22	0.000168 20
γ [M1+E2]	503.8 3	2.0 2 $\times 10^{-5}$
γ	514.4 10	7.0 14 $\times 10^{-6}$
γ (M1+E2)	518.46 4	0.000104 13
γ [M1+E2]	523.9 6	5.7 11 $\times 10^{-5}$
γ (E2)	529.52 5	3.3 8 $\times 10^{-7}$
γ [E2]	541.97 22	2.4 3 $\times 10^{-5}$
γ (M1+E2)	549.33 4	3.4 5 $\times 10^{-5}$
γ M1+35%E2	560.39 5	4.0 8 $\times 10^{-5}$
γ [M1+E2]	566.9 6	5.1 10 $\times 10^{-5}$
γ [M1+E2]	585.3 7	3.7 7 $\times 10^{-5}$
γ [M1+E2]	616.1 7	4.2 8 $\times 10^{-6}$
γ	663.8 10	3.5 7 $\times 10^{-6}$
γ	768.2 10	2.9 6 $\times 10^{-5}$
γ	900.2 10	8.3 17 $\times 10^{-6}$
γ	932.2 10	4.6 9 $\times 10^{-5}$
γ	946.3 10	7.6 15 $\times 10^{-6}$
γ	1040.4 10	2.8 6 $\times 10^{-6}$
γ	1075 1	5.2 10 $\times 10^{-6}$
γ	1106.4 10	1.4 3 $\times 10^{-6}$

† 10% uncert(syst)

$^{253}_{100}$Fm(3.00 *12* d)

Mode: ϵ(88 *1* %), α(12 *1* %)
Δ: 79339 *5* keV
SpA: 1.72$\times 10^{5}$ Ci/g
Prod: ^{252}Cf(α,3n)

Alpha Particles (^{253}Fm)

⟨α⟩=820.7 *6* keV

α(keV)	α(%)
6490 25	~0.036
6544.3 19	0.18 5
6633 4	0.31 6
6653 4	0.29 5
6675.7 14	2.78 11
6846.7 13	1.01 6
6868 25	~0.11
6901.0 13	1.18 6
6943.3 13	5.12 13

Alpha Particles (^{253}Fm) (continued)

α(keV)	α(%)
7024.5 13	0.80 5
7086.0 13	0.156 24

Photons (^{253}Fm)

⟨γ⟩=59.6 *16* keV

γ_{mode}	γ(keV)	γ(%)†
Es $K_{\alpha 2}$	112.526	13.0 3
Es $K_{\alpha 1}$	118.012	20.0 5
Es $K_{\beta 1}$'	132.727	7.51 17
Es $K_{\beta 2}$'	137.358	2.66 7
γ_αM2+13%E3	144.97 5	0.192 24
γ_αE2	271.9 4	2.6 5
γ_α[E2]	405.4 15	~0.08

† 17% uncert(syst)

Atomic Electrons (^{253}Fm)*

⟨e⟩=5.6 *3* keV

e_{bin}(keV)	⟨e⟩(keV)	e(%)
20	1.77	8.7 9
26	1.34	5.2 5
27	1.37	5.1 5
85	0.118	0.139 14
86	0.243	0.28 3
87	0.0068	0.0079 8
91	0.074	0.081 8
92	0.109	0.118 12
98	0.048	0.049 5
105	0.126	0.120 12
106	0.077	0.073 7
107	0.031	0.029 3
108	0.0062	0.0057 6
110	0.042	0.038 4
111	0.075	0.068 7
112	0.0072	0.0064 7
113	0.029	0.026 3
114	0.0036	0.0032 3
116	0.0110	0.0095 10
117	0.0093	0.0079 8
118	0.0046	0.0039 4
124	0.0074	0.0060 6
125	0.0144	0.0115 12
126	0.0043	0.0034 3
127	0.0060	0.0048 5
128	0.0032	0.00252 25
129	0.00155	0.00120 12
130	0.0131	0.0101 10
131	0.0045	0.0034 4
132	0.00180	0.00136 14

* with ϵ

Continuous Radiation (^{253}Fm)

⟨IB⟩=0.93 keV

E_{bin}(keV)		⟨ ⟩(keV)	(%)
10 - 20	IB	0.0055	0.030
20 - 40	IB	0.0047	0.021
40 - 100	IB	0.096	0.108
100 - 281	IB	0.82	0.69

$^{253}_{102}$No(1.7 *3* min)

Mode: α
Δ: 84330 *220* keV syst
SpA: 4.4$\times 10^{8}$ Ci/g
Prod: ^{244}Cm(^{13}C,4n); ^{246}Cm(^{12}C,5n); ^{242}Pu(^{16}O,5n); ^{239}Pu(^{18}O,4n)

Alpha Particles (^{253}No)

α(keV)
8010 20

$^{253}_{103}$Lr(1.4 *8* s)

Mode: α
Δ: 88640 *290* keV syst
SpA: 2.5$\times 10^{10}$ Ci/g
Prod: daughter ^{257}105

Alpha Particles (^{253}Lr)

α(keV)
8721 20
8805 20

$^{253}_{104}$104(~1.8 s)

Mode: SF(50 % est)
SpA: ~ 2.1$\times 10^{10}$ Ci/g
Prod: ^{206}Pb(^{50}Ti,3n)

A = 254

NDS 32, 171 (1981)

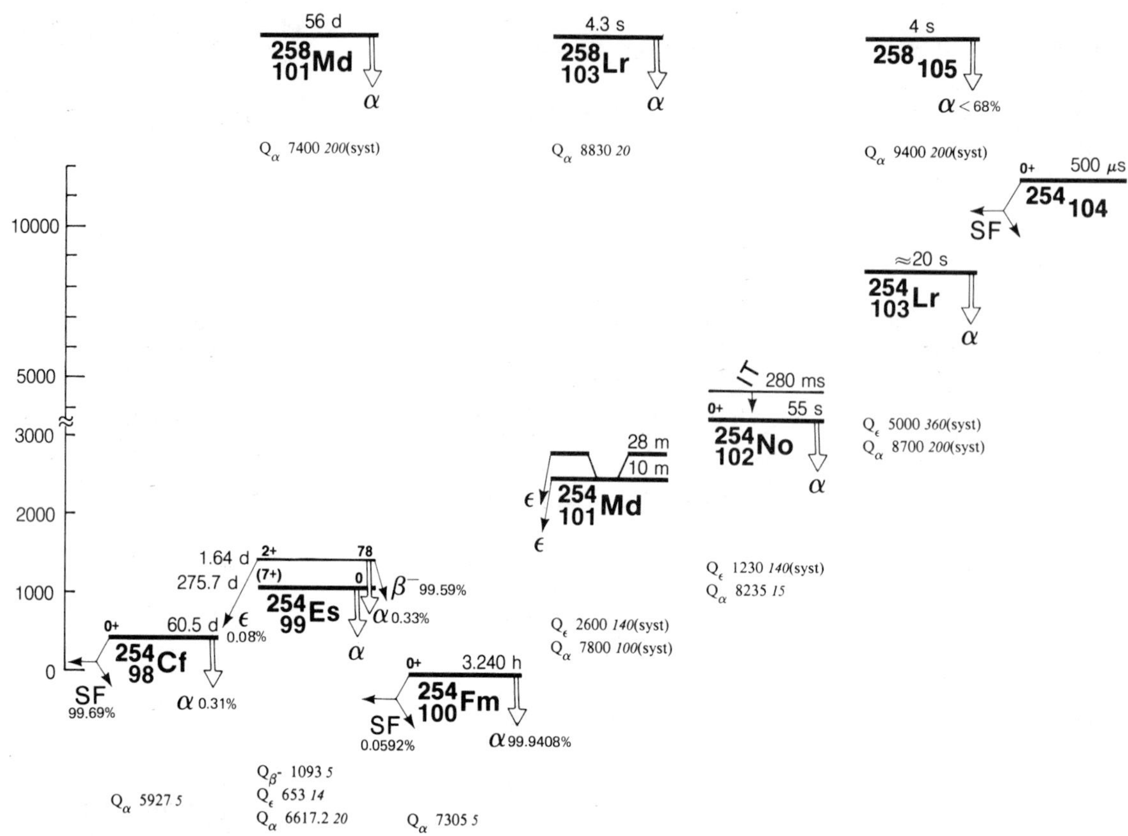

$^{254}_{98}$Cf(60.5 *2* d)

Mode: SF(99.690 *16* %), α(0.310 *16* %)
Δ: 81337 *12* keV
SpA: 8497 Ci/g
Prod: multiple n-capture from ^{238}U; multiple n-capture from ^{239}Pu; multiple n-capture from ^{244}Cm; multiple n-capture from ^{252}Cf; daughter ^{254}Es(1.64 d); ^{253}Cf(n,γ)

Alpha Particles (^{254}Cf)

⟨α⟩=18.06 *2* keV

α(keV)	α(%)
5694	0.0015
5792 *5*	0.053 *6*
5834 *5*	0.256 *6*

$^{254}_{99}$Es(275.7 *5* d)

Mode: α
Δ: 81990 *8* keV
SpA: 1865 Ci/g
Prod: multiple n-capture from ^{238}U; multiple n-capture from ^{239}Pu; multiple n-capture from ^{241}Am; multiple n-capture from ^{242}Pu; multiple n-capture from ^{244}Cm; multiple n-capture from ^{252}Cf; ^{253}Es(n,γ)

Alpha Particles (^{254}Es)

⟨α⟩=6405 *2* keV

α(keV)	α(%)
5993.6 *22*	~0.030
6046.6 *13*	0.16
6104.4 *14*	0.34 *2*
6177	~0.020
6184	~0.06
6194.2 *13*	~0.04
6258	<0.02
6268.0 *10*	0.22 *4*
6273.5 *12*	0.14 *2*
6324 *2*	0.04 *1*
6347.5 *10*	0.75 *5*
6357.3 *12*	2.6 *3*
6379 *3*	<0.010
6385.1 *13*	<0.10
6416.1 *10*	1.8 *1*
6426.6 *12*	93.1 *10*
6438.1 *13*	~0.030

Alpha Particles (^{254}Es) (continued)

α(keV)	α(%)
6480.1 *13*	0.23 *4*
6481.1 *13*	<0.050
6515.0 *13*	~0.0046

Photons (^{254}Es)

⟨γ⟩=2.79 *20* keV

γ_{mode}	γ(keV)	γ(%)†
γM1,E2	34.45 *8*	~0.017
γ(M2)	35.47 *8*	~0.005
γM1,E2	42.60 *10*	0.15
γE1	65.0 *12*	2.0 *2*
γM1(+E2)	69.71 *10*	~0.014
γM1,E2	70.40 *10*	~0.044
γM1,E2	80.81 *10*	<0.027
γM1,E2	85.1 *1*	<0.034
γ[E2]	150.52 *14*	0.020 *3*
γ	236.1 *14*	0.0080 *16* ?
γ	247.8 *14*	0.025 *5*
γ	264.9 *20*	0.05 *1*
γ	278.8 *20*	0.030 *6*
γ	285.2 *15*	0.010 *2*
γ	305.8 *11*	0.070 *14*
γ	316.7 *13*	0.15 *3*
γ	344.0 *13*	0.0090 *18*
γ	348.9 *20*	0.0070 *14*
γ	375.5 *11*	0.015 *3*
γ	386.1 *14*	0.05 *1*

† 10% uncert(syst)

$^{254}_{99}$Es(1.638 *8* d)

Mode: β-(99.590 *12* %), α(0.33 *1* %), ε(0.078 *6* %)
Δ: 82068 *8* keV
SpA: 3.139×10^5 Ci/g
Prod: multiple n-capture from ^{238}U; multiple n-capture from ^{239}Pu; multiple n-capture from ^{244}Cm; multiple n-capture from ^{252}Cf; ^{253}Es(n,γ)

Alpha Particles (^{254}Es)

⟨α⟩=21.11 *5* keV

α(keV)	α(%)
6280 *3*	0.00053 *10*
6297 *2*	0.00158 *20*
6325 *2*	0.0073 *7*
6357 *2*	0.0274 *16*
6382.1 *9*	0.248 *3*
6417.8 *9*	0.0059 *7*
6455 *3*	0.00040 *13*
6460.9 *9*	0.00205 *23*
6467.2 *9*	~0.00026
6513.9 *9*	<0.009
6555.8 *9*	0.0191 *13*
6590.7 *9*	0.0132 *16*

Photons (^{254}Es)

⟨γ⟩=474 *19* keV

γ_{mode}	γ(keV)	γ(%)†
Fm L_ℓ	13.660	0.75 *12*
Fm L_α	16.351	10.6 *14*
Fm L_η	19.433	0.27 *4*
Fm L_β	21.479	13.2 *18*
Fm L_γ	25.747	3.1 *5*
$\gamma_{\beta-}$[M1+E2]	39.872 *10*	
$\gamma_{\beta-}$E2	44.979 *10*	0.049 *5*
γ_α(E1)	50.24 *5*	0.0092 *10*
γ_α(E1)	71.47 *5*	0.043 *4*
γ_α(M1)	80.05 *7*	0.0036 *4*
γ_α(M1)	91.04 *17*	0.00066 *20*
γ_α(M1)	96.44 *8*	0.0056 *6*
γ_α(M1)	104.25 *15*	0.0102 *10*
$\gamma_{\beta-}$E2	104.351 *12*	0.179 *17*
Fm $K_{\alpha2}$	115.279	0.502 *23*
Fm $K_{\alpha1}$	121.058	0.77 *4*
γ_α(M1)	125.49 *16*	0.00049 *20*
Fm $K_{\beta1}'$	136.088	0.292 *13*
Fm $K_{\beta2}'$	140.840	0.104 *5*
γ_α(E1)	175.72 *15*	0.0028 *6*
γ_α(E1)	177.50 *8*	0.056 *6*
γ_αE1	211.95 *7*	0.096 *10*
$\gamma_{\beta-}$E2	544.444 *22*	0.90 *8*
$\gamma_{\beta-}$E2	584.316 *22*	2.89 *20*
$\gamma_{\beta-}$E2	648.795 *21*	28.9 *20*
$\gamma_{\beta-}$E2	688.667 *22*	12.4 *9*
$\gamma_{\beta-}$E2	693.774 *22*	24.7 *17*

† uncert(syst): 3.2% for α, 20% for β-

Atomic Electrons (^{254}Es)

⟨e⟩=38.5 *9* keV

e_{bin}(keV)	⟨e⟩(keV)	e(%)
17	0.161	0.93 *10*
18	4.1	22.6 *23*
21	1.74	8.4 *12*
24	4.3	18.0 *19*
27	2.0	7.6 *10*
28	0.071	0.26 *3*
38	2.6	6.7 *7*
40	2.15	5.4 *6*
43 - 77	2.07	4.7 *3*
78	1.33	1.71 *17*
83 - 132	2.01	2.18 *11*
133 - 135	0.00074	0.00055 *4*
403 - 442	0.51	0.118 *7*
507	3.8	0.74 *5*
517 - 544	0.132	0.0253 *11*
547	1.58	0.290 *22*
552	3.13	0.57 *4*
557 - 584	0.368	0.0653 *23*
621	0.96	0.154 *11*
622	1.02	0.164 *12*
628 - 673	3.49	0.531 *15*
681 - 694	0.94	0.137 *5*

Continuous Radiation (^{254}Es)

⟨β-⟩=200 keV;⟨IB⟩=0.163 keV

E_{bin}(keV)		⟨ ⟩(keV)	(%)
0 - 10	β-	0.194	3.87
	IB	0.0098	
10 - 20	β-	0.57	3.80
	IB	0.0091	0.063
20 - 40	β-	2.21	7.4
	IB	0.0163	0.057
40 - 100	β-	14.1	20.4
	IB	0.037	0.059
100 - 300	β-	81	43.7
	IB	0.059	0.036
300 - 600	β-	61	15.0
	IB	0.026	0.0065
600 - 1171	β-	40.9	5.4
	IB	0.0048	0.00071

$^{254}_{100}$Fm(3.240 *2* h)

Mode: α(99.9408 *2* %), SF(0.0592 *2* %)
Δ: 80897 *6* keV
SpA: 3.8078×10^6 Ci/g
Prod: daughter ^{254}Es(1.64 d)

Alpha Particles (^{254}Fm)

⟨α⟩=7171 *5* keV

α(keV)	α(%)
6898 *4*	0.004
7050 *4*	0.90 *10*
7147 *4*	14.0 *10*
7189 *4*	84.9 *10*

Photons (^{254}Fm)

⟨γ⟩=1.31 *17* keV

γ_{mode}	γ(keV)	γ(%)
Cf L_ℓ	13.146	0.18 *4*
Cf L_α	15.636	2.6 *5*
Cf L_η	18.347	0.064 *14*
Cf L_β	20.307	3.2 *7*
Cf L_γ	24.273	0.73 *16*
γE2	42.721 *5*	0.0118
γE2	99.157 *6*	0.037 *7*
γE2	154.34 *5*	0.0010 *3*

Atomic Electrons (^{254}Fm)

⟨e⟩=5.5 *4* keV

e_{bin}(keV)	⟨e⟩(keV)	e(%)
17	0.038	0.23 *5*
18	1.03	5.9 *12*
20	0.45	2.2 *5*
23	1.11	4.9 *10*
25	0.49	2.0 *4*
26	0.0119	0.046 *10*
36	0.62	1.7 *3*
38	0.55	1.4 *3*
41	0.36	0.88 *18*
42	0.102	0.24 *5*
43	0.017	0.040 *8*
73	0.016	0.022 *4*
74	0.28	0.37 *8*
79	0.18	0.23 *5*
92	0.0066	0.0072 *14*
93	0.096	0.103 *21*
94	0.064	0.068 *14*
95	0.0015	0.0016 *3*
97	0.0021	0.0022 *4*
98	0.048	0.049 *10*
99	0.015	0.015 *3*

$^{254}_{101}$Md(10 *3* min)

Mode: ε
Δ: 83490 *140* keV syst
SpA: 7.4×10^7 Ci/g
Prod: ^{253}Es(α,3n)

$^{254}_{101}$Md(28 *8* min)

Mode: ε
Δ: 83490 *140* keV syst
SpA: 2.6×10^7 Ci/g
Prod: ^{253}Es(α,3n)

$^{254}_{102}$No(55 *5* s)

Mode: α
Δ: 84723 *26* keV
SpA: 8.0×10^{8} Ci/g
Prod: ^{246}Cm(^{12}C,4n); ^{246}Cm(^{13}C,5n); ^{244}Cm(^{13}C,3n); ^{242}Pu(^{16}O,4n); ^{243}Am(^{15}N,4n); ^{238}U(^{22}Ne,6n)

Alpha Particles (^{254}No)

α(keV)
8100 *20*

$^{254}_{102}$No(280 *40* ms)

Mode: IT
Δ: 84723 *26* keV
SpA: 5.9×10^{10} Ci/g
Prod: ^{249}Cf(^{12}C,α3n); ^{246}Cm(^{12}C,4n)

$^{254}_{103}$Lr(20 *10* s)

Mode: α
Δ: 89730 *360* keV syst
SpA: 2.2×10^{9} Ci/g
Prod: daughter ^{258}105

Alpha Particles (^{254}Lr)

α(keV)
8455 *20*

$^{254}_{104}$104(500 *200* μs)

Mode: SF
SpA: 6×10^{10} Ci/g
Prod: ^{206}Pb(^{50}Ti,2n)

A = 255

NDS **34**, 70 (1981)

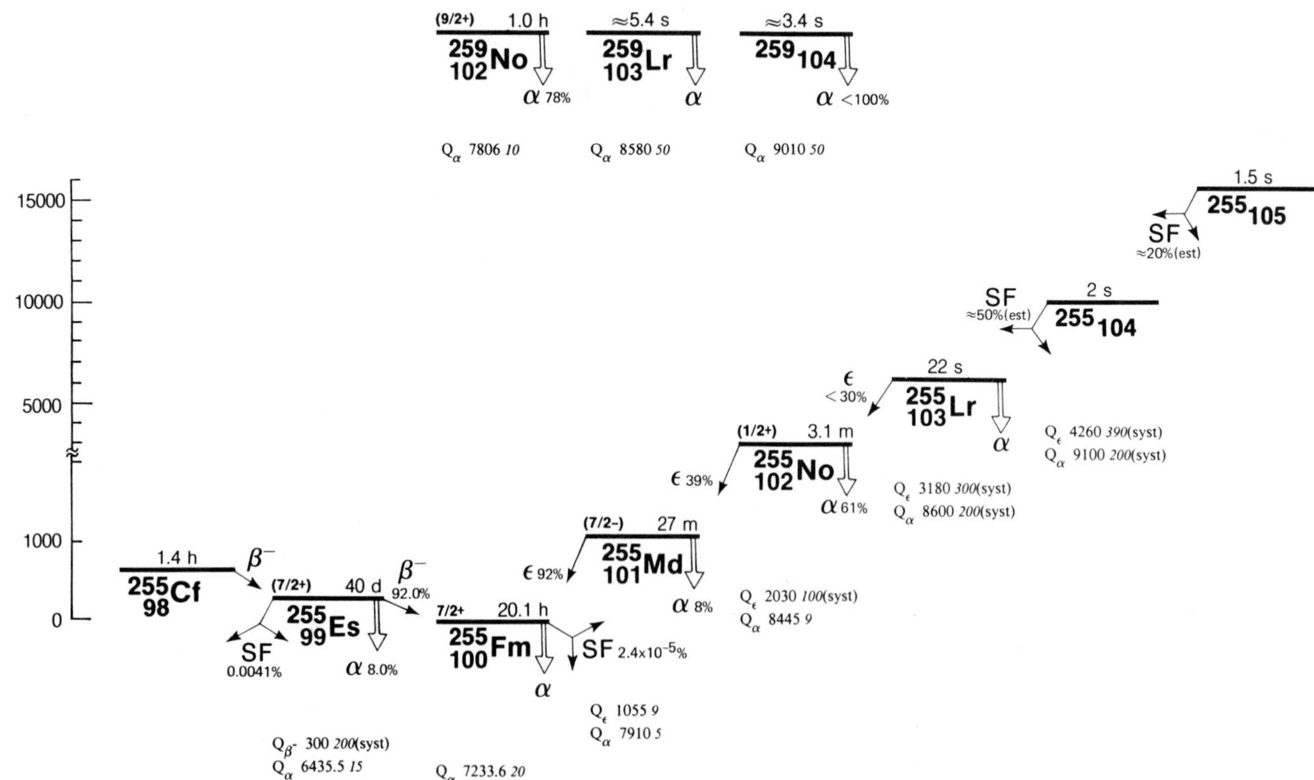

$^{255}_{98}$Cf(1.4 *3* h)

Mode: β-
SpA: 8.7×10^6 Ci/g
Prod: ^{254}Cf(n,γ)

$^{255}_{99}$Es(39.8 *12* d)

Mode: β-(92.0 *4* %), α(8.0 *4* %), SF(0.0041 *2* %)
Δ: 84090 *200* keV syst
SpA: 1.29×10^4 Ci/g
Prod: multiple n-capture from ^{238}U; multiple n-capture from ^{239}Pu; multiple n-capture from ^{244}Cm; multiple n-capture from ^{252}Cf; multiple n-capture from ^{253}Es

Alpha Particles (^{255}Es)

$\langle\alpha\rangle$=504 keV

α(keV)	α(%)
6069.7 *15*	0.008
6213 *10*	0.20
6260 *10*	0.8
6299.5 *15*	7

Photons (^{255}Es)

γ_{mode}	γ(keV)
γ_α	32.6 ?

$^{255}_{100}$Fm(20.07 *7* h)

Mode: α, SF(2.4 *10* $\times10^{-5}$ %)
Δ: 83787 *5* keV
SpA: 6.123×10^5 Ci/g
Prod: daughter ^{255}Es

Alpha Particles (^{255}Fm)

$\langle\alpha\rangle$=7091.0 *5* keV

α(keV)	α(%)
6488.2 *5*	0.0030 *5*
6546.3 *5*	0.014 *2*
6591.7 *5*	0.017 *2*
6621 *3*	0.0022 *5*
6692 *3*	0.005 *2*
6699.7 *6*	0.036 *2*
6709.8 *5*	0.013 *1*
6741 *3*	0.0012 *4*
6762.6 *5*	0.016 *2*
6807.0 *5*	0.110 *6*
6812.9 *5*	0.0020 *5*
6836.2 *5*	0.008 *1*
6872.8 *5*	0.008 *1*
6891.6 *5*	0.62 *1*
6893.1 *5*	~0.010 ?
6918.8 *5*	0.017 *2*
6952.3 *5*	0.022 *4*
6963.5 *5*	5.04 *6*
6983.0 *5*	0.13 *1*
7022.5 *5*	93.4 *3*
7080.0 *5*	0.40 *3*
7102.7 *5*	0.090 *9*
7127.1 *5*	0.070 *7*

Photons (^{255}Fm)

$\langle\gamma\rangle$=14.9 *16* keV

γ_{mode}	γ(keV)	γ(%)†
Cf L_ℓ	13.146	2.3 *5*
Cf L_α	15.636	34 *7*
Cf L_η	18.347	0.37 *6*
Cf L_β	20.110	29 *6*
γ M1	22.994 *11*	0.13 *3*
Cf L_γ	24.360	6.3 *12*
γ M1+6.9%E2	24.812 *12*	0.09 *2*
γ E2	47.806 *12*	0.019 *2*
γ M1	57.878 *14*	0.110 *22*
γ M1	58.453 *13*	0.67 *13*
γ M1	59.978 *14*	0.12 *2*
γ M1(+6.9%E2)	63.9 *3*	0.0008 *3*
γ M1	73.02 *3*	0.029 *2*
γ E2	80.872 *18*	0.27 *5*
γ E2	81.447 *14*	0.81 *16*
γ [M1]	85.98 *9*	0.0066 *6*
γ [E2]	98.61 *16*	0.0019 *4*
γ [M1]	98.73 *15*	~0.0010
Cf $K_{\alpha2}$	109.826	0.0266 *16*
Cf $K_{\alpha1}$	115.032	0.0413 *24*
Cf $K_{\beta1}'$	129.436	0.0154 *9*
γ [M1]	129.83 *10*	~0.0013
γ (E1)	131.04 *11*	0.028 *3*
γ [E2]	132.00 *13*	0.0021 *4*
γ [E2]	133.00 *3*	0.0085 *9*
Cf $K_{\beta2}'$	133.949	0.0054 *3*
γ [E2]	149.18 *20*	0.00065 *6*
γ [M1]	152.69 *12*	~0.0002
γ (M1)	152.82 *10*	0.00187 *24*
γ [E2]	159.00 *9*	0.0036 *4*
γ [M1]	163.85 *13*	0.0020 *2*
γ [M1]	172.80 *16*	0.00030 *3*
γ (M1)	177.63 *10*	0.0048 *4*
γ [M1]	184.71 *15*	0.00080 *8*
γ [M1]	186.84 *13*	0.00010 *2*
γ E1	204.06 *11*	0.024 *2*
γ [M1]	210.57 *12*	0.00030 *4*
γ [M1]	213.53 *18*	~0.00010
γ [E2]	233.56 *12*	0.00026 *5*
γ	264.04 *11*	0.0010 *1*
γ [E1]	268.0 *3*	0.00018 *4*
γ [M1]	270.87 *19*	0.00032 *4*
γ [M1]	285.60 *16*	0.00039 *4*
γ [E1]	327.9 *3*	0.00023 *4*
γ [M1]	329.9 *3*	0.00044 *8*
γ [M1]	331.71 *17*	<0.0033
γ (M1)	366.34 *13*	0.0050 *5*
γ (M1)	378.43 *14*	0.0023 *3*
γ [M1]	390.77 *25*	0.00040 *8*
γ [M1]	409.89 *23*	0.00012 *4*
γ [M1]	423.85 *12*	0.00060 *6*
γ [M1]	437.72 *12*	0.00135 *14*
γ [M1]	483.83 *12*	0.00039 *4*
γ [M1]	496.17 *12*	0.00017 *4*
γ [M1]	502.73 *25*	~4×10^{-5}
γ [M1]	519.16 *12*	0.00015 *4*
γ [M1+E2]	543.46 *23*	0.00024 *6*

† 10% uncert(syst)

Atomic Electrons (^{255}Fm)

$\langle e\rangle$=83 *5* keV

e_{bin}(keV)	$\langle e\rangle$(keV)	e(%)
1 - 17	2.9	33 *3*
18	3.7	20 *6*
19	0.0033	0.018 *4*
20	7.6	38 *10*
21 - 22	1.10	5.2 *11*
23	2.9	12.9 *25*
24	1.1	4.7 *15*
25	3.1	12.3 *21*
26 - 30	2.6	9.7 *10*
32	7.5	23 *5*
33 - 51	4.6	11.8 *8*
52	2.9	5.6 *11*
53 - 55	1.26	2.3 *3*
56	15.2	27 *5*
57 - 60	1.47	2.6 *4*
61	2.8	4.6 *9*
62	8.2	13 *3*
63 - 74	0.225	0.32 *3*
75	5.8	7.8 *16*
76	4.0	5.3 *11*
77 - 79	0.56	0.71 *12*
80	2.6	3.2 *6*
81 - 130	1.03	1.24 *23*
131 - 180	0.0484	0.0332 *10*
182 - 229	0.010	~0.005
231 - 280	0.0170	0.0071 *6*
281 - 331	0.0067	0.0022 *7*
332 - 378	0.0069	0.00196 *10*
384 - 433	0.00163	0.00040 *4*
436 - 484	0.000329	7.1 *5* $\times10^{-5}$
489 - 538	0.00015	3.0 *7* $\times10^{-5}$
539 - 543	4.7 $\times10^{-6}$	~9×10^{-7}

$^{255}_{101}$Md(27 *2* min)

Mode: ε(92 *2* %), α(8 *2* %)
Δ: 84842 *8* keV
SpA: 2.73×10^7 Ci/g
Prod: ^{253}Es(α,2n); ^{254}Es(α,3n); ^{11}B on ^{252}Cf; ^{12}C on ^{252}Cf; ^{13}C on ^{252}Cf

Alpha Particles (^{255}Md)

α(keV)
7326 *5*

Photons (^{255}Md)

$\langle\gamma\rangle$=32.9 *12* keV

γ_{mode}	γ(keV)	γ(%)†
γ_αM1+21%E2	385.33 *20*	0.090 *9*
γ_α(E1)	405.62 *7*	2.93 *15*
γ_αM1+15%E2	429.73 *10*	0.244 *22*
γ_α(E1)	453.14 *7*	4.30 *24*
γ_α[E2]	461.3 *4*	<0.05 ?

† 25% uncert(syst)

$^{255}_{102}$No(3.1 *2* min)

Mode: α(61.4 *25* %), ϵ(38.6 *25* %)
Δ: 86870 *100* keV syst
SpA: 2.37×10^{8} Ci/g
Prod: ^{246}Cm(^{13}C,4n); ^{248}Cm(^{12}C,5n); ^{242}Pu(^{18}O,5n); ^{238}U(^{22}Ne,5n); ^{249}Cf(^{12}C,α2n)

Alpha Particles (^{255}No)

⟨α⟩=4971 *7* keV

α(keV)	α(%)
7620 *10*	1.72 *25*
7717 *11*	1.47 *25*
7771 *7*	5.5 *4*
7879 *11*	2.58 *25*
7927 *7*	7.31 *12*
8007 *11*	3.9 *7*
8077 *9*	7.3 *4*
8121 *6*	27.9 *8*
8266 *8*	3.1 *6*
8312 *9*	1.17 *6*

$^{255}_{103}$Lr(22 *4* s)

Mode: α, ϵ(<30 %)
Δ: 90050 *290* keV syst
SpA: 2.0×10^{9} Ci/g
Prod: ^{243}Am(^{16}O,4n); ^{249}Cf(^{10}B,4n); ^{249}Cf(^{11}B,5n)

Alpha Particles (^{255}Lr)

⟨α⟩=8394 *26* keV

α(keV)	α(%)
8370 *13*	60 *10*
8429 *20*	40 *10*

$^{255}_{104}$104(2 s)

Mode: SF(50 % est)
Δ: 94310 *270* keV syst
SpA: 1.9×10^{10} Ci/g
Prod: ^{207}Pb(^{50}Ti,2n)

$^{255}_{105}$105(1.5 s)

Mode: SF(20 % est)
SpA: 2.4×10^{10} Ci/g
Prod: ^{207}Pb(^{51}V,3n); ^{206}Pb(^{51}V,2n)

A = 256

NDS **32**, 184 (1981)

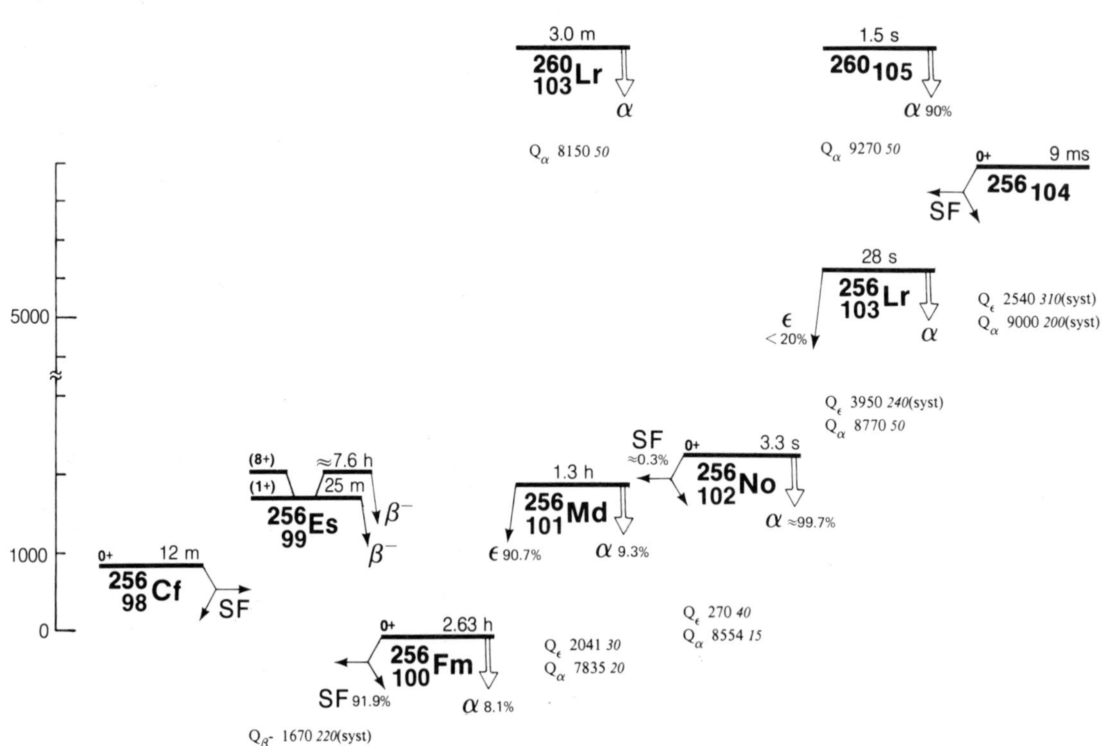

$^{256}_{98}$Cf(12.3 *12* min)

Mode: SF
SpA: 6.0×10^{7} Ci/g
Prod: ^{254}Cf(t,p)

$^{256}_{99}$Es(25.0 *24* min)

Mode: β-
Δ: 87160 *220* keV syst
SpA: 2.9×10^{7} Ci/g
Prod: ^{255}Es(n,γ)

$^{256}_{99}$Es(~7.6 h)

Mode: β-
Δ: 87160 *220* keV syst
SpA: ~1.6×10^{6} Ci/g
Prod: ^{254}Es(t,p)

Photons (^{256}Es)

γ_{mode}	γ(keV)
γ (E2)	111.6 *5*
γ (E2)	172.8 *5*
γ	190.3 *5*
γ	199.6 *5*
γ	218.5 *5*
γ (E2)	231.5 *4*
γ	417.9 *5*
γ	634.3 *5*
γ	862.2 *4*
γ	1093.7 *4*

$^{256}_{100}$Fm(2.627 *22* h)

Mode: SF(91.9 *3* %), α(8.1 *3* %)
Δ: 85482 *7* keV
SpA: 4.66×10^{6} Ci/g
Prod: daughter ^{256}Es(25 min); daughter ^{256}Md; ^{253}Es(α,p); ^{255}Fm(n,γ)

Alpha Particles (^{256}Fm)

α(keV)
6915 *5*

$^{256}_{101}$Md(1.27 *7* h)

Mode: ϵ(90.7 *5* %), α(9.3 *5* %)
Δ: 87522 *29* keV
SpA: 9.7×10^{6} Ci/g
Prod: ^{253}Es(α,n); ^{11}B on ^{252}Cf; ^{12}C on ^{252}Cf; ^{13}C on ^{252}Cf

Alpha Particles (^{256}Md)

⟨α⟩=675 *3* keV

α(keV)	α(%)
7140 *7*	1.49 *19*
7210 *7*	5.9 *4*
7330 *30*	0.37 *9*
7460 *30*	0.46 *9*
7490 *20*	0.56 *9*
7670 *30* ?	~0.19
7720 *20*	0.37 *9*

Photons (^{256}Md)

γ_{mode}	γ(keV)
γ_{α}(M1)	400 *20*

$^{256}_{102}$No(3.3 *2* s)

Mode: α(~99.7 %), SF(~0.3 %)
Δ: 87796 *26* keV
SpA: 1.20×10^{10} Ci/g
Prod: ^{248}Cm(^{12}C,4n); ^{248}Cm(^{13}C,5n); ^{246}Cm(^{13}C,3n); ^{242}Pu(^{18}O,4n); ^{238}U(^{22}Ne,4n)

Alpha Particles (^{256}No)

α(keV)
8430 *20*

$^{256}_{103}$Lr(28 *3* s)

Mode: α, ϵ(<20 %)
Δ: 91740 *240* keV syst
SpA: 1.55×10^{9} Ci/g
Prod: ^{243}Am(^{18}O,5n); ^{15}N on ^{246}Cm; ^{12}C on ^{249}Bk; ^{10}B on ^{249}Cf; ^{11}B on ^{249}Cf

Alpha Particles (^{256}Lr)

⟨α⟩=8446 *23* keV

α(keV)	α(%)
8319 *15*	6.6 *15*
8390 *15*	20 *2*
8430 *15*	37 *3*
8475 *15*	13.3 *15*
8520 *15*	19.1 *15*
8635 *20*	4 *1*

$^{256}_{104}$104(9 *2* ms)

Mode: SF
Δ: 94280 *200* keV syst
SpA: 6.4×10^{10} Ci/g
Prod: ^{208}Pb(^{50}Ti,2n)

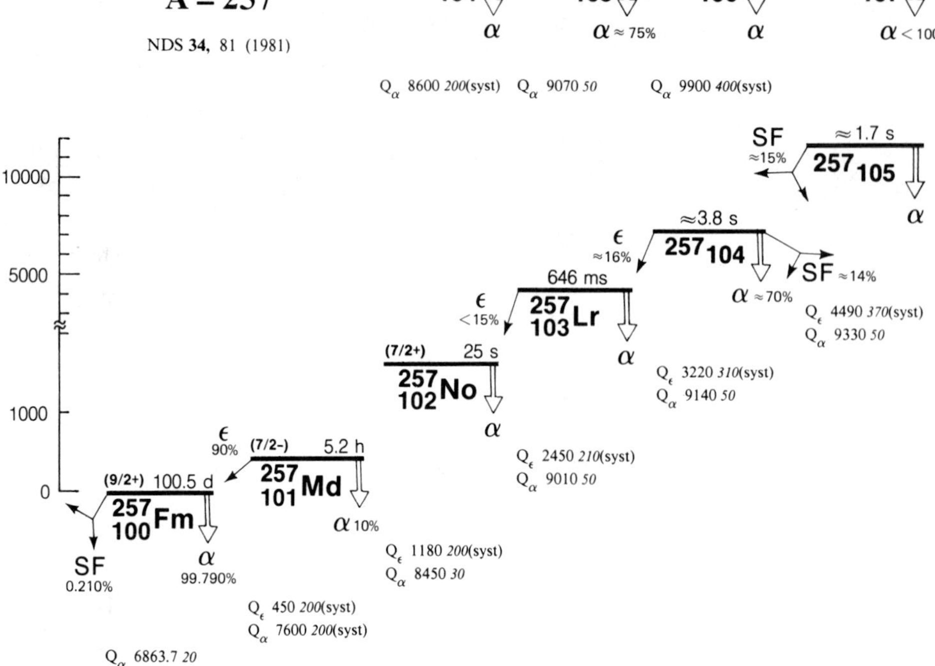

$^{257}_{100}$Fm(100.5 *2* d)

Mode: α(99.790 *5* %), SF(0.210 *5* %)
Δ: 88585 *7* keV
SpA: 5055 Ci/g
Prod: multiple n-capture from ^{242}Pu; multiple n-capture from ^{243}Am; multiple n-capture from ^{244}Cm; ^{11}B on ^{252}Cf; ^{12}C on ^{252}Cf; ^{13}C on ^{252}Cf

Alpha Particles (^{257}Fm)

⟨α⟩=6511.0 *15* keV

α(keV)	α(%)
6346 *5*	0.30 *10*
6440.5 *15*	2.00 *20*
6519.4 *13*	93.5 *7*
6622.3 *13* ?	<1
6696.1 *13*	3.39 *20*
6756.7 *13*	0.60 *6*

Photons (^{257}Fm)

⟨γ⟩=111 *3* keV

γ_{mode}	γ(keV)	γ(%)†
Cf L_ℓ	13.146	1.8 *3*
Cf L_α	15.636	26 *3*
Cf L_η	18.347	0.30 *5*
Cf L_β	20.139	29 *5*
Cf L_γ	24.419	7.8 *14*
γ M1	61.63 *8*	1.50 *15*
γ [M1]	74.96 *8*	0.26 *3*
γ M1(+E2)	80.2 *10*	0.100 *20*
γ [M1+E2]	96 *5*	0.30 *6*
γ [M1]	104.46 *8*	0.70 *7*

Photons (^{257}Fm)
(continued)

γ_{mode}	γ(keV)	γ(%)†
Cf L_γ	109.826	14.4 *9*
Cf L_γ	115.032	22.3 *14*
Cf $K_{\beta1}$'	129.436	8.3 *5*
Cf $K_{\beta2}$'	133.949	2.95 *20*
γ [E2]	136.59 *10*	
γ M1(+50%E2)	179.42 *7*	8.7 *7*
γ M1(+20%E2)	241.05 *8*	10.3 *7*

† 10% uncert(syst)

Atomic Electrons (^{257}Fm)

⟨e⟩=121 *4* keV

e_{bin}(keV)	⟨e⟩(keV)	e(%)
20 - 26	7.2	31 *3*
36	13.5	38 *4*
37 - 42	1.63	4.4 *4*
44	12.1	27.2 *24*
49 - 54	2.4	4.8 *7*
55	6.5	11.9 *24*
57 - 76	7.5	11 *3*
78	3.0	3.8 *4*
79 - 105	4.4	4.7 *8*
106	23.7	22.3 *18*
107 - 129	0.259	0.231 *10*
153	8.2	5.4 *5*
154	5.9	3.8 *3*
160	2.40	1.51 *13*
173	4.1	2.38 *21*
174 - 179	2.67	1.51 *8*
215	9.0	4.2 *3*
216 - 241	6.7	2.94 *11*

$^{257}_{101}$Md(5.2 *5* h)

Mode: ϵ(90 *4* %), α(10 *4* %)
Δ: 89030 *200* keV syst
SpA: 2.34×10^{6} Ci/g
Prod: ^{11}B on ^{252}Cf; ^{12}C on ^{252}Cf; ^{13}C on ^{252}Cf; ^{255}Es(α,2n); ^{254}Es(α,n)

Alpha Particles (^{257}Md)

α(keV)
7064 *5*
7250 ?

$^{257}_{102}$No(25 *2* s)

Mode: α
Δ: 90220 *30* keV
SpA: 1.73×10^{9} Ci/g
Prod: ^{248}Cm(^{13}C,4n); ^{248}Cm(^{12}C,3n); ^{246}Cm(^{12}C,n); ^{246}Cm(^{13}C,2n); B on Cf

Alpha Particles (^{257}No)

α(keV)	α(rel)
8220 *20*	55 *3*
8270 *20* ?	26 *2*
8320 *20* ?	19 *2*

$^{257}_{103}$Lr(646 *25* ms)

Mode: α, ε(<15 %)
Δ: 92670 *210* keV syst
SpA: 4.17×10^{10} Ci/g
Prod: ^{249}Cf(^{11}B,3n); ^{249}Cf(^{15}N,α3n); ^{249}Cf(^{14}N,α2n)

Alpha Particles (^{257}Lr)

⟨α⟩=8852 *17* keV

α(keV)	α(%)
8800 *13*	18 *2*
8864 *12*	82 *2*

$^{257}_{104}$104(3.8 *8* s)

Mode: α(~70 %), ε(~16 %), SF(~14 %)
Δ: 95890 *230* keV syst
SpA: 1.06×10^{10} Ci/g
Prod: ^{249}Cf(^{12}C,4n)

Alpha Particles (^{257}104)

⟨α⟩= 7182 keV

α(keV)	α(%)
8553	4
8615	12
8663	9
8722	10
8774	18
8951	13
9013	17

Photons (^{257}104)

γ_{mode}	γ(keV)
γ_α	117

$^{257}_{105}$105(1.7 *6* s)

Mode: α, SF(15 *11* %)
Δ: 100390 *300* keV syst
SpA: 2.12×10^{10} Ci/g
Prod: ^{209}Bi(^{50}Ti,2n); ^{208}Pb(^{51}V,2n); ^{205}Tl(^{54}Cr,2n)

A = 258

NDS **32**, 194 (1981)

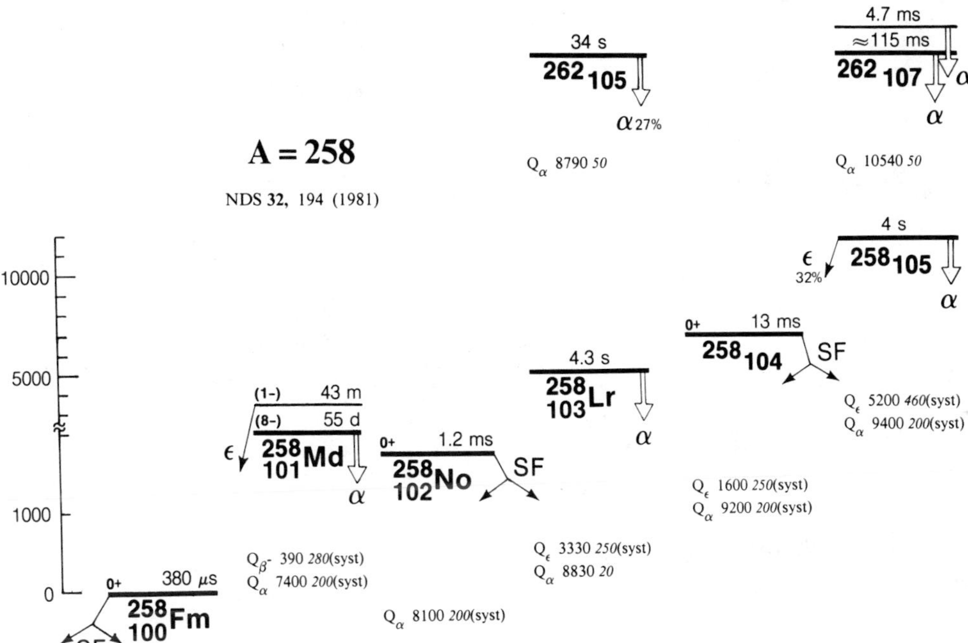

$^{258}_{100}$Fm(380 *60* μs)

Mode: SF
SpA: 6.31×10^{10} Ci/g
Prod: ^{257}Fm(d,p)

$^{258}_{101}$Md(55 *4* d)

Mode: α
Δ: 91820 *200* keV syst
SpA: 9202 Ci/g
Prod: ^{255}Es(α,n)

Alpha Particles (^{258}Md)

⟨α⟩=6737 *20* keV

α(keV)	α(%)
6716 *5*	72 *10*
6790 *10*	28 *6*

$^{258}_{101}$Md(43 *4* min)

Mode: ε
Δ: 91820 *200* keV syst
SpA: 1.69×10^{7} Ci/g
Prod: ^{255}Es(α,n)

$^{258}_{102}$No(1.2 ms)

Mode: SF
Δ: 91420 *200* keV syst
SpA: 6×10^{10} Ci/g
Prod: ^{248}Cm(^{13}C,3n)

$^{258}_{103}$Lr(4.3 *5* s)

Mode: α
Δ: 94750 *140* keV syst
SpA: 9.4×10^{9} Ci/g
Prod: ^{246}Cm(^{15}N,3n); ^{248}Cm(^{15}N,5n); ^{249}Cf(^{15}N,α2n); ^{244}Pu(^{19}F,5n); ^{249}Bk(^{12}C,3n); B on Cf

Alpha Particles (^{258}Lr)

⟨α⟩=8601 *27* keV

α(keV)	α(%)
8565 *25*	~20
8595 *10*	46 *3*
8621 *10*	25 *10*
8654 *10*	9 *2*

$^{258}_{104}$104(13 *3* ms)

Mode: SF
Δ: 96350 *200* keV syst
SpA: 6.3×10^{10} Ci/g
Prod: ^{249}Cf(^{12}C,3n); ^{249}Cf(^{13}C,4n); ^{246}Cm(^{16}O,4n)

$^{258}_{105}$105(4.0 *11* s)

Mode: α, ϵ(32 *15* %)
Δ: 101550 *410* keV syst
SpA: 1.0×10^{10} Ci/g
Prod: ^{209}Bi(^{50}Ti,n)

Alpha Particles (^{258}105)

α(keV)
9018 *25*
9087 *20*
9180 *20*

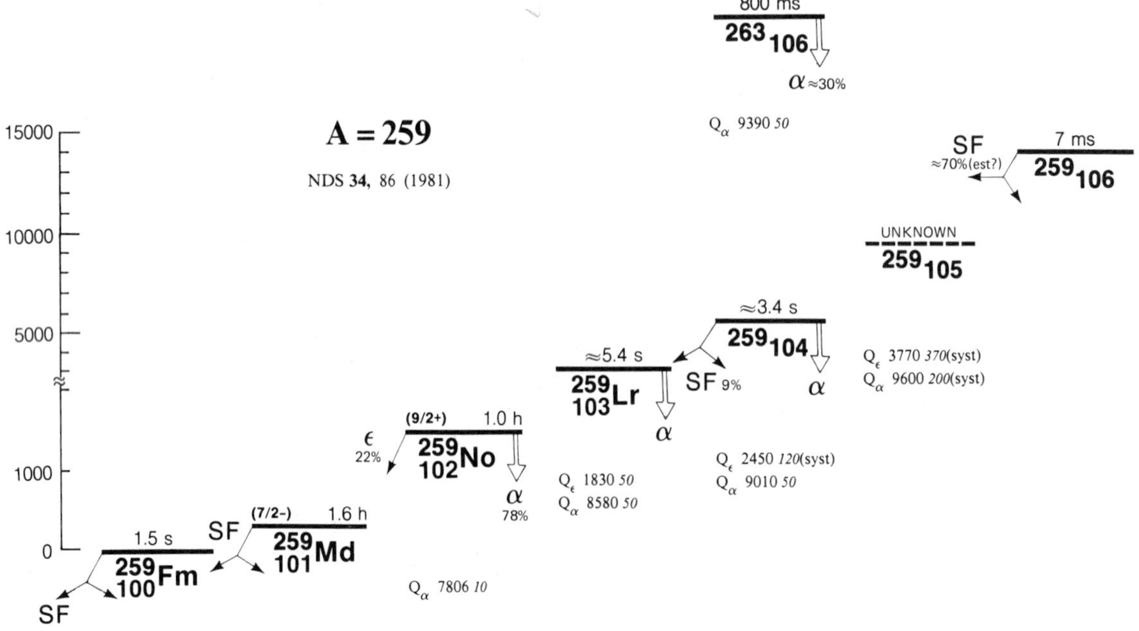

$^{259}_{100}$Fm(1.5 *3* s)

Mode: SF
SpA: 2.3×10^{10} Ci/g
Prod: ^{257}Fm(t,p)

$^{259}_{101}$Md(1.6 *4* h)

Mode: SF
SpA: 7.6×10^{6} Ci/g
Prod: daughter ^{259}No

$^{259}_{102}$No(1.00 *8* h)

Mode: α(78 %), ϵ(22 %)
Δ: 94018 *11* keV
SpA: 1.21×10^{7} Ci/g
Prod: ^{248}Cm(^{18}O,α3n)

Alpha Particles (^{259}No)

⟨α⟩=5879 *17* keV

α(keV)	α(%)
7455 *10*	10 *3*
7500 *10*	30 *8*
7533 *10*	18 *6*
7605 *10*	11 *4*
7685 *10*	9 *3*

$^{259}_{103}$Lr(5.4 *8* s)

Mode: α
Δ: 95850 *50* keV
SpA: 7.6×10^{9} Ci/g
Prod: ^{248}Cm(^{15}N,4n); ^{250}Cf(^{15}N,α2n)

Alpha Particles (^{259}Lr)

α(keV)
8450 *20*

$^{259}_{104}$104(3.4 *17* s)

Mode: α, SF(9 *3* %)
Δ: 98300 *110* keV syst
SpA: 1.2×10^{10} Ci/g
Prod: ^{249}Cf(^{13}C,3n); ^{248}Cm(^{16}O,5n); ^{246}Cm(^{18}O,5n); ^{245}Cm(^{18}O,4n); ^{242}Pu(^{22}Ne,5n)

Alpha Particles (^{259}104)

α(keV)	α(rel)
8770 *15*	~60
8865 *15*	~40

$^{259}_{106}$106(7 *3* ms)

Mode: SF(70 syst %)
SpA: 6×10^{10} Ci/g
Prod: ^{208}Pb(^{54}Cr,3n); ^{207}Pb(^{54}Cr,2n)

A = 260

NDS **32**, 199 (1981)

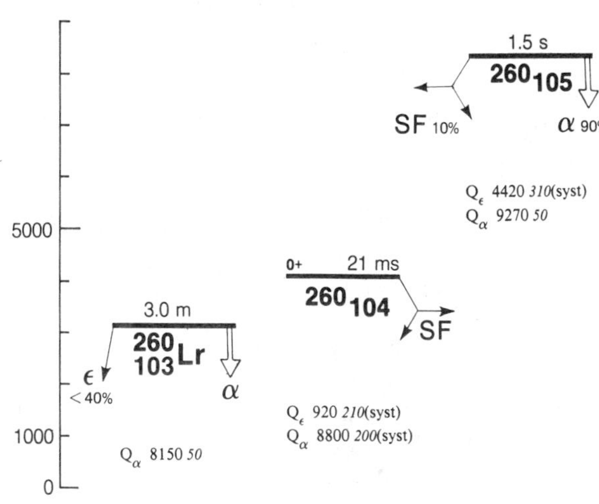

$^{260}_{103}$Lr(3.0 *5* min)

Mode: α, ϵ(<40 %)
Δ: 98100 *60* keV
SpA: 2.4×10^{8} Ci/g
Prod: ^{248}Cm(^{15}N,3n); ^{18}O on ^{249}Bk

Alpha Particles (^{260}Lr)

α(keV)
8030 *20*

$^{260}_{104}$104(21 *1* ms)

Mode: SF
Δ: 99020 *200* keV syst
SpA: 6.3×10^{10} Ci/g
Prod: ^{249}Bk(^{15}N,4n); ^{248}Cm(^{16}O,4n); ^{249}Cf(^{18}O,α3n)

$^{260}_{105}$105(1.52 *13* s)

Mode: α(90.4 *6* %), SF(9.6 *6* %)
Δ: 103440 *240* keV syst
SpA: 2.29×10^{10} Ci/g
Prod: ^{249}Cf(^{15}N,4n); ^{243}Am(^{22}Ne,5n)

Alpha Particles (^{260}105)

⟨α⟩=8201 *21* keV

α(keV)	α(%)
9047 *14*	48 *5*
9082 *14*	25 *3*
9128 *17*	17 *3*

A = 261

NDS **34**, 91 (1981)

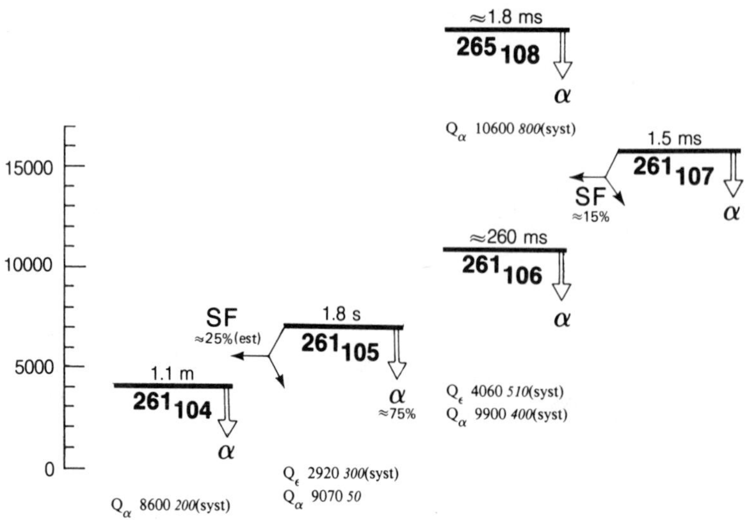

$^{261}_{104}$104(1.08 *17* min)

Mode: α
Δ: 101240 *200* keV syst
SpA: 6.6×10^{8} Ci/g
Prod: ^{248}Cm(^{18}O,5n)

Alpha Particles (^{261}104)

α(keV)
8280 *20*

$^{261}_{105}$105(1.8 *4* s)

Mode: α(~75 %), SF(~25 %)
Δ: 104160 *220* keV syst
SpA: 2.0×10^{10} Ci/g
Prod: ^{250}Cf(^{15}N,4n); ^{249}Bk(^{16}O,4n); ^{243}Am(^{22}Ne,4n)

Alpha Particles (^{261}105)

α(keV)
8930

$^{261}_{106}$106(~260 ms)

Mode: α
Δ: 108220 *460* keV syst
SpA: ~ 5.8×10^{10} Ci/g
Prod: ^{208}Pb(^{54}Cr,n)

Alpha Particles (^{261}106)

α(keV)
9560 *30*

$^{261}_{107}$107(1.5 *5* ms)

Mode: α, SF(~15 %)
SpA: 6.2×10^{10} Ci/g
Prod: ^{209}Bi(^{54}Cr,2n); ^{208}Pb(^{55}Mn,2n); ^{205}Tl(^{58}Fe,2n)

A = 262

NDS **32**, 202 (1981)

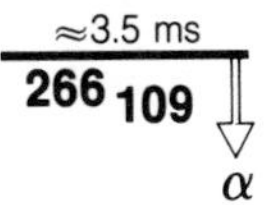

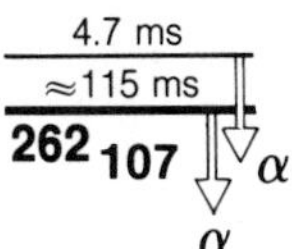

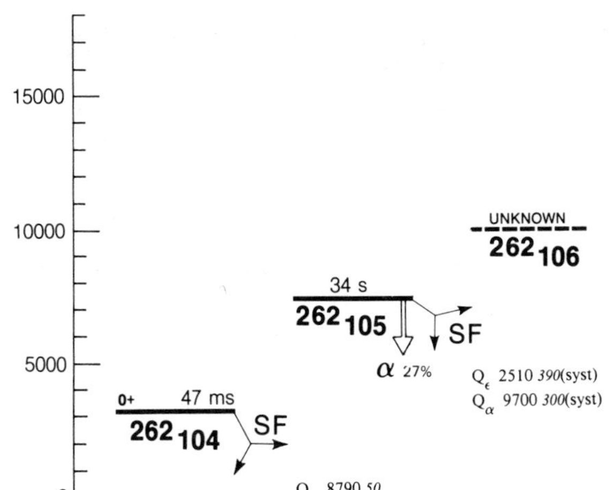

$^{262}_{104}$104(47 *5* ms)

Mode: SF
SpA: 6.2×10^{10} Ci/g
Prod: ^{248}Cm(^{18}O,4n); ^{244}Pu(^{22}Ne,4n)

$^{262}_{105}$105(34 *4* s)

Mode: SF, α(27 *5* %)
Δ: 105970 *150* keV syst
SpA: 1.25×10^{9} Ci/g
Prod: ^{249}Bk(^{18}O,5n)

Alpha Particles (^{262}105)

⟨α⟩=2290 keV

α(keV)	α(%)
8450 *20*	20
8530 *20*	4
8670 *20*	2.4

$^{262}_{107}$107(~115 ms)

Mode: α
Δ: 114510 *420* keV syst
SpA: ~6×10^{10} Ci/g
Prod: ^{209}Bi(^{54}Cr,n)

Alpha Particles (^{262}107)

α(keV)
9704 *50*

$^{262}_{107}$107(4.7 *20* ms)

Mode: α
Δ: 114510 *420* keV syst
SpA: 6×10^{10} Ci/g
Prod: ^{209}Bi(^{54}Cr,n)

Alpha Particles (^{262}107)

α(keV)
10376 *35*

A = 263

NDS **34**, 91 (1981)

$^{263}_{106}$106(800 *200* ms)

Mode: SF(~70 %), α(~30 %)
Δ: 110120 *120* keV syst
SpA: 3.6×10^{10} Ci/g
Prod: ^{249}Cf(^{18}O,4n)

Alpha Particles (^{263}106)

⟨α⟩~2724 keV

α(keV)	α(%)
9060 *40*	~27
9250 *40*	~3

A = 265

ZP **A317**, 235 (1984)

$^{265}_{108}108$(~1.8 ms)

Mode: α
Δ: 121240 *920* keV syst
SpA: ~6×10^{10} Ci/g
Prod: ^{208}Pb(^{58}Fe,n)

Alpha Particles (^{265}108)

α(keV)
10360 *30*

A = 266

ZP **A315**, 145 (1984)

$^{266}_{109}109$ (~3.5 ms)

Mode: α
Δ: 128210 *420* keV syst
SpA: ~6×10^{10} Ci/g
Prod: ^{209}Bi(^{58}Fe,n)

Alpha Particles (^{266}109)

α(keV)
11100 *40*

APPENDIX A. PHYSICAL CONSTANTS*

Quantity	Symbol, equation	Value	Uncert. (ppm)
speed of light	c	$2.997\ 924\ 58(1.2) \times 10^{10}$ cm s^{-1} (see note**)	0.004
Planck constant	h	$6.626\ 176(36) \times 10^{-27}$ erg s	5.4
Planck constant, reduced	$\hbar = h/2\pi$	$1.054\ 588\ 7(57) \times 10^{-27}$ erg s	5.4
		$= 6.582\ 173(17) \times 10^{-22}$ MeV s	2.6
electron charge magnitude	e	$4.803\ 242(14) \times 10^{-10}$ esu	2.9
		$= 1.602\ 189\ 2(46) \times 10^{-19}$ coulomb	2.9
conversion constant	$\hbar c$	197.328 58(51) MeV fm	2.6
conversion constant	$(\hbar c)^2$	0.389 385 7(20) GeV^2 mbarn	5.2
electron mass	m_e	0.511 003 4(14) MeV/c^2 = $9.109\ 534(47) \times 10^{-28}$	2.8, 5.1
proton mass	m_p	938.279 6(27) MeV/c^2 = $1.672\ 648\ 5(86) \times 10^{-24}$	2.8, 5.1
		= 1.007 276 470(11) amu = 1836.151 52(70) m_e	0.011, 0.38
neutron mass	m_n	939.573 1(27) MeV/c^2 = $1.674\ 954\ 3(86) \times 10^{-24}$	2.8, 5.1
		= 1.008 665 012(37) amu	0.037
deuteron mass	m_d	1875.628 0(53) MeV/c^2	2.8
atomic mass unit (amu)	(mass ^{12}C atom)/12 = (1 g)/N_A	931.501 6(26) MeV/c^2 = $1.660\ 565\ 5(86) \times 10^{-24}$	2.8, 5.1
electron charge to mass ratio	e/m_e	$5.272\ 764(15) \times 10^{17}$ esu g^{-1}	2.8
		$= 1.758\ 804\ 7(49) \times 10^{8}$ coulomb g^{-1}	2.8
quantum of magnetic flux	hc/e	$4.135\ 701(11) \times 10^{-15}$ joule s coulomb^{-1}	2.6
	h/e	$1.379\ 521\ 5(36) \times 10^{-17}$ erg s esu^{-1}	2.6
Josephson frequency-voltage ratio	$2e/h$	$4.835\ 939(13) \times 10^{14}$ cycles s^{-1} v^{-1}	2.6
Faraday constant	F	$9.648\ 456(27) \times 10^{4}$ coulomb mol^{-1}	2.8
fine structure constant	$\alpha = e^2/\hbar c$	1/137.036 04(11)	0.82
classical electron radius	$r_e = e^2/m_e c^2$	2.817 938 0(70) fm	2.5
electron Compton wavelength	$\lambda_e = \hbar/m_e c = r_e \alpha^{-1}$	$3.861\ 590\ 5(64) \times 10^{-11}$ cm	1.6
proton Compton wavelength	$\lambda_p = \hbar/m_p c$	$2.103\ 089\ 2(36) \times 10^{-14}$ cm	1.7
neutron Compton wavelength	$\lambda_n = \hbar/m_n c$	$2.100\ 194\ 1(35) \times 10^{-14}$ cm	1.7
Bohr radius ($m_{nucleus} = \infty$)	$a_\infty = \hbar^2/m_e e^2 = r_e \alpha^{-2}$	$0.529\ 177\ 06(44) \times 10^{-8}$ cm	0.82
Rydberg energy	$hcR_\infty = m_e e^4/2\hbar^2 = m_e c^2 \alpha^2/2$	13.605 804(36) eV	2.6
Thomson cross section	$\sigma_T = 8\pi r_e^2/3$	0.665 244 8(33) barn	4.9
Bohr magneton	$\mu_B = e\hbar/2m_e c$	$5.788\ 378\ 5(95) \times 10^{-15}$ MeV gauss^{-1}	1.6
nuclear magneton	$\mu_N = e\hbar/2m_p c$	$3.152\ 451\ 5(53) \times 10^{-18}$ MeV gauss^{-1}	1.7
electron cyclotron frequency/field	$\omega^e_{cycl}/B = e/m_e c$	$1.758\ 804\ 7(49) \times 10^{7}$ radian s^{-1} gauss^{-1}	2.8
proton cyclotron frequency/field	$\omega^p_{cycl}/B = e/m_p c$	$9.578\ 756(28) \times 10^{3}$ radian s^{-1} gauss^{-1}	2.8
gravitational constant	G_N	$6.672\ 0(41) \times 10^{-8}$ cm^3 g^{-1} s^{-2}	615
grav. acceleration, sea level, 45° lat.	g	980.62 cm s^{-2}	—
Fermi coupling constant	$G_F/(\hbar c)^3$	$1.166\ 37(2) \times 10^{-5}$ GeV^{-2}	17
Avogadro number	N_A	$6.022\ 045(31) \times 10^{23}$ mol^{-1}	5.1
molar gas constant, ideal gas at STP	R	$8.314\ 41(26) \times 10^{7}$ erg mol^{-1} K^{-1}	31
Boltzmann constant	k	$1.380\ 662(44) \times 10^{-16}$ erg K^{-1}	32
		$= 8.617\ 35(28) \times 10^{-5}$ eV K^{-1}	32
molar volume, ideal gas at STP	$N_A k$(273.15 K)/(1 atmosphere)	22 413.83(70) cm^3 mol^{-1}	31
Stefan-Boltzmann constant	$\sigma = \pi^2 k^4/60\hbar^3 c^2$	$5.670\ 32(71) \times 10^{-5}$ erg s^{-1} cm^{-2} K^{-4}	125
first radiation constant	$2\pi hc^2$	$3.741\ 832(20) \times 10^{-5}$ erg cm^2 s^{-1}	5.4
second radiation constant	hc/k	1.438 786(45) cm K	31

Abbreviations for units

amu	atomic mass unit
cm	centimeter
esu	electrostatic unit
eV	electron volt
fm	Fermi
g	gram
K	degree Kelvin
mol	mole
s	second
v	volt

Useful constants and conversion factors	
π= 3.141 592 653 589 793 238	1 coulomb = 2.997 924 58$\times 10^9$ esu
e = 2.718 281 828 459 045 235	1 tesla = 10^4 gauss
γ = 0.577 215 664 901 532 861	1 atm. = 1.013 25$\times 10^6$ dyne/ cm^2
1 in = 2.54 cm	0° C = 273.15 K
1 Å = 10^{-8} cm	1 sidereal year = 3.155 815 0$\times 10^7$ s
1 fm = 10^{-13} cm	1 tropical year $\simeq$ 3.155 69$\times 10^7$ s
1 barn = 10^{-24} cm^2	1 light year = 9.460 528$\times 10^{17}$ cm
1 newton = 10^5 dyne	1 parsec = 3.261 633 light year
1 joule = 10^7 erg	1 astro. unit = 1.495 979$\times 10^{13}$ cm
1 eV = 1.602 189 2$\times 10^{-12}$ erg	1 curie = 3.7$\times 10^{10}$ disintegration/ s
1 eV/ c^2 = 1.782 676$\times 10^{-33}$ g	1 rad = 100 erg/ g of tissue
1 cal = 4.184 joule	1 roentgen = 1 esu/0.001293 g of air

*) Revised by Barry N. Taylor,[1] based mainly on the "1973 Least-Squares Adjustment of the Fundamental Constants."[2] The figures in parentheses give the 1-standard-deviation uncertainties in the last digits of the main numbers; the uncertainties in parts per million (ppm) are given in the last column. The uncertainties of the output values of a least-squares adjustment are in general correlated, and the laws of error propagation must be used in calculating additional quantities.

The set of constants resulting from the 1973 adjustment of Cohen and Taylor[2] has been recommended for international use by CODATA (Committee on Data for Science and Technology), and is the most up-to-date, generally accepted set currently available. Since the publication of the 1973 adjustment, new experiments have yielded better values for some of the constants: N_A = 6.022 097 8(63)$\times 10^{23}$ mol^{-1} (1.04 ppm); α^{-1} = 137.035 963(15) (0.11 ppm); and m_p/m_e = 1836.152 470(76) (0.041 ppm). However, since a change in the measured value of one constant usually leads to changes in the adjusted values of others, one must be cautious in using together the values from the 1973 adjustment and the results of more recent experiments.

**) In October 1983, the Conférence Générale des Poids et Mesures adopted a new definition of the meter. The meter is the length of the path traveled by light in vacuum during a time interval of 1/299 792 458 s. Thus the speed of light is *defined* to be 299 792 458 m/s. A discussion of this change is given in a commentary by B.W. Petley.[3]

This table was adapted, with permission, from one that appeared in the 1984 *Review of Particle Properties*.[4]

---

[1]B.N. Taylor, private communication (1984); private communication (1985).

[2]E.R. Cohen and B.N. Taylor, *J. Phys. Chem. Ref. Data* **2**, 663 (1973).

[3]B.W. Petley, *Nature* **303**, 373 (1983).

[4]*Review of Particle Properties*, Particle Data Group, *Rev. Mod. Phys.* **56**, No. 2, Part II, S31 (1984).

APPENDIX B. NUCLEAR SPECTROSCOPY STANDARDS

1. Gamma-ray Absolute Intensity Standards

Table 1 lists absolute γ-ray intensity standards for several sources, evaluated by Vaninbroukx.[1] One group of sources (designated primary) includes isotopes which have accurately determined absolute γ-ray intensities and well characterized decay schemes, and which are available with high isotopic purity. The other group of sources (designated secondary) includes isotopes which may be utilized for detector calibrations, but which do not fulfil all of the criteria mentioned for the primary. Columns 1, 2, and 3 show the isotope names (with (p) for primary or (s) for secondary appended), half-lives, and γ-ray energies, respectively. Column 4 lists the absolute intensities per 100 disintegrations, with their corresponding uncertainties (in italics) in the last significant digits.

[1] R. Vaninbroukx, *Emission Probabilities of Selected Gamma Rays for Radionuclides Used as Detector-Calibration Standards*, report presented at the Advisory Group Meeting of the International Atomic Energy Agency (IAEA), Vienna (1985).

Table 1. Gamma-ray Absolute Intensities for Some Standard Sources

Source	Half-life	E_γ (keV)	I_γ (%)
^{7}Be (s)	53.3 d	477.6	10.45 *10*
^{22}Na (p)	2.602 y	1274.5	99.94 *2*
^{24}Na (p)	14.659 h	1368.6	99.994 *3*
		2754.0	99.881 *8*
^{46}Sc (p)	83.83 d	889.3	99.984 *1*
		1120.5	99.987 *1*
^{51}Cr (p)	27.704 d	320.1	9.85 *9*
^{54}Mn (p)	312.2 d	834.8	99.975 *1*
^{56}Mn (s)	2.578 h	846.8	98.87 *3*
		1810.7	27.2 *8*
		2113.0	14.3 *4*
^{56}Co (s)	77.7 d	846.8	99.925 *6*
		1037.8	14.11 *5*
		1238.3	66.3 *5*
		1360.2	4.26 *2*
		1771.3	15.48 *4*
		2034.8	7.76 *4*
		2598.5	16.96 *4*
		3523.4	7.70 *12*
^{57}Co (p)	271.77 d	14.4	9.3 *2*
		122.1	85.63 *15*
		136.5	10.58 *8*
^{58}Co (p)	70.92 d	810.8	99.44 *2*
^{60}Co (p)	5.271 y	1173.2	99.89 *2*
		1332.5	99.983 *1*
^{65}Zn (p)	244.1 d	1115.5	50.65 *20*
^{85}Sr (p)	64.84 d	514.0	98.8 *5*
^{88}Y (p)	106.61 d	898.0	94.2 *4*
		1836.0	99.30 *5*
^{94}Nb (p)	2.0×10^4 y	702.6	99.82 *1*
		871.1	99.89 *1*
^{95}Zr (s)	64.02 d	724.2	44.15 *20*
		756.7	54.50 *25*
^{95}Nb (p)	34.97 d	765.8	99.80 *2*
^{99}Tc (s)	6.006 h	140.5	89.0 *2*
^{109}Cd (p)	1.267 y	88.0	3.68 *5*
^{110}Ag (s)	249.76 d	446.8	3.72 *3*
		657.8	94.4 *1*
^{110}Ag (s) (continued)		677.6	10.40 *8*
		687.0	6.44 *3*
		706.7	16.6 *1*
		744.3	4.70 *4*
		763.9	22.39 *8*
		818.0	7.32 *4*
		884.7	72.7 *3*
		937.5	34.31 *12*
		1384.3	24.25 *8*
		1475.8	3.99 *2*
		1505.0	13.04 *4*
^{111}In (s)	2.807 d	171.3	90.2 *3*
		245.3	94.0 *2*
^{113}In (p)	1.658 h	391.7	64.9 *2*
^{115}In (p)	4.486 h	336.2	45.9 *2*
^{124}Sb (s)	60.20 d	602.7	98.0 *1*
		645.9	7.3 *1*
		722.8	11.3 *2*
		1691.0	48.5 *3*
		2090.9	5.66 *9*
^{125}I (p)	60.1 d	35.5	6.6 *1*
^{133}Ba (s)	10.54 y	53.2	2.19 *3*
		79.6	2.62 *7*
		81.0	34.1 *5*
		276.4	7.16 *4*
		302.9	18.31 *7*
		356.0	62.00 *14*
		383.9	8.92 *5*
^{134}Cs (p)	2.062 y	604.7	97.63 *4*
		795.8	85.52 *4*
^{137}Cs (p)	30.0 y	661.7	85.2 *1*
^{139}Ce (p)	137.7 d	165.9	79.9 *1*
^{141}Ce (p)	32.50 d	145.4	48.6 *4*
^{152}Eu (s)	13.33 y	121.8	28.40 *23*
		244.7	7.51 *7*
		344.3	26.58 *19*
		411.1	2.23 *2*
		444.0	3.12 *3*
^{152}Eu (s) (continued)		778.9	12.96 *7*
		964.1	14.62 *6*
		1085.9	10.14 *6*
		1112.1	13.54 *6*
		1408.0	20.85 *8*
^{182}Ta (s)	115.0 d	100.1	14.23 *25*
		152.4	7.02 *8*
		222.1	7.57 *8*
		1121.3	35.3 *2*
		1189.0	16.42 *10*
		1221.4	27.20 *22*
		1231.0	11.57 *8*
^{192}Ir (s)	73.83 d	296.0	28.7 *1*
		308.5	29.8 *1*
		316.5	83.0 *3*
		468.1	47.7 *2*
		588.6	4.49 *2*
		604.4	8.11 *4*
		612.5	5.28 *3*
^{198}Au (p)	2.6935 d	411.8	95.56 *7*
^{203}Hg (p)	46.60 d	279.2	81.50 *8*
^{207}Bi (s)	32 y	569.7	97.9 *1*
		1063.7	74.1 *3*
		1770.2	6.87 *4*
^{232}U (s)*	70 y	238.6	43.5 *2*
		241.0	4.04 *5*
		510.8	8.25 *14*
		583.1	30.60 *17*
		727.2	6.62 *6*
		860.4	4.50 *5*
		2614.6	35.86 *6*
^{241}Am (p)	432.7 y	26.3	2.41 *5*
		59.5	35.9 *3*
^{243}Am (s)&	7.38×10^3 y	43.5	6.05 *13*
		74.7	68.6 *15*
		106.1	27.2 *2*
		228.3	11.27 *18*
		277.6	14.38 *21*

* With decay daughter isotopes in equilibrium

& With ^{239}Np in equilibrium

2. Gamma-ray Energy Standards

Table 2 lists some γ-ray energy standards from the evaluation of Helmer, et al.[1] for calibration of γ-ray measurements. Most of the isotopes given here have half-lives of more than 30 days, and are commercially available. γ-ray energies are based on the *gold standard*, the 411.8044 *11*[2] keV transition in ^{198}Au decay. Uncertainties are intended to represent one standard deviation, and include the 2.6 ppm uncertainty in the definition of the electron volt relative to wavelength. γ-ray energies from the decays of ^{51}Cr, ^{57}Co, ^{60}Co, ^{137}Cs, ^{152}Eu, ^{182}Ta, ^{192}Ir, ^{198}Au, and ^{203}Hg are from absolute wavelength or curved-crystal spectrometer measurements, and are thus tied directly to the 411.8 keV γ ray of ^{198}Au. The remaining energies are from the measurements of small γ-ray energy differences with Ge detectors. Columns 1 and 2 show the isotope names and half-lives, respectively. Column 3 lists the γ-ray energies, with their corresponding uncertainties (in italics) in the last significant digits.

[1]R.G. Helmer, P.H.M. Van Assche, and C. Van Der Leun, *At. Data and Nucl. Data Tables* **24**, *39 (1979).*

[2]E.G. Kessler, R.D. Deslattes, A. Henins, and W.C. Sauder, *Phys.Rev.Lett.***40**, *171 (1978).*

Table 2. Gamma-ray Energies for Some Standard Sources

Source	Half-life	E_γ (keV)
^{7}Be	53.3 d	477.605 *3*
^{22}Na	2.602 y	1274.542 *7*
^{24}Na	14.659 h	1368.633 *6*
		2754.030 *14*
^{46}Sc	83.83 d	889.277 *3*
		1120.545 *4*
^{51}Cr	27.704 d	320.0842 *9*
^{54}Mn	312.2 d	834.843 *6*
^{59}Fe	44.50 d	1099.251 *4*
		1291.596 *7*
^{56}Co	77.7 d	846.764 *6*
		1037.844 *4*
		1175.099 *8*
		1238.287 *6*
		1360.206 *6*
		1771.350 *15*
		1810.722 *17*
		1963.714 *12*
		2015.179 *11*
		2034.759 *11*
		2113.107 *12*
		2212.921 *10*
		2598.460 *10*
		3009.596 *17*
		3201.954 *14*
		3253.417 *14*
		3272.998 *14*
		3451.154 *13*
^{57}Co	271.77 d	122.06135 *30*
		136.4743 *5*
^{60}Co	5.271 y	1173.238 *4*
		1332.502 *5*
^{65}Zn	244.1 d	1115.546 *4*
^{88}Y	106.6 d	898.042 *4*
		1836.063 *13*
^{95}Zr	64.02 d	724.199 *5*
^{94}Nb	2.0×10^{4} y	702.645 *6*
		871.119 *4*
^{108}Ag	127.0 y	433.936 *4*
		614.281 *4*
		722.929 *4*
^{110}Ag	249.76 d	446.811 *3*
		620.360 *3*
		657.762 *2*
		677.623 *2*
		687.015 *3*
		706.682 *3*
		744.277 *3*
		763.944 *3*
		818.031 *4*
		884.685 *3*
		937.493 *4*
		1384.300 *4*
		1475.788 *6*
		1505.040 *5*
		1562.032 *5*
^{124}Sb	60.20 d	602.730 *3*
		645.855 *2*
		713.781 *5*
		722.786 *4*
		790.712 *7*
		968.201 *4*
		1045.131 *4*
		1325.512 *6*
		1368.164 *7*
		1436.563 *7*
		1690.980 *6*
		2090.942 *8*
^{137}Cs	30.0 y	661.660 *3*
^{144}Ce	284.9 d	696.510 *3*
		1489.160 *5*
		2185.662 *7*
^{152}Eu	13.33 y	121.7824 *4*
		244.6989 *10*
		344.2811 *19*
^{153}Gd	241.6 d	69.6734 *2*
		97.4316 *3*
		103.1807 *3*
^{170}Tm	128.6 d	84.2551 *3*
^{182}Ta	115.0 d	67.75001 *20*
		84.6808 *3*
		100.10653 *30*
		113.6723 *4*
^{182}Ta (continued)		116.4186 *7*
		152.46308 *5*
		156.3874 *5*
		179.3948 *5*
		198.3530 *6*
		222.1099 *6*
		229.3220 *9*
		264.0755 *8*
		1121.301 *5*
		1189.050 *5*
		1221.408 *5*
		1231.016 *5*
		1257.418 *5*
		1273.730 *5*
		1289.156 *5*
		1373.836 *5*
		1387.402 *5*
^{192}Ir	73.83 d	136.3434 *5*
		205.7955 *5*
		295.95821 *80*
		308.45685 *80*
		316.5080 *8*
		416.4719 *12*
		468.0715 *12*
		484.5779 *13*
		588.5851 *16*
		604.4145 *16*
		612.4657 *16*
		884.5423 *20*
^{198}Au	2.6935 d	411.8044 *11*
		675.8875 *19*
		1087.6905 *30*
^{203}Hg	46.60 d	279.1967 *12*
^{207}Bi	32 y	569.702 *2*
		1063.662 *4*
		1770.237 *10*
^{228}Th	1.913 y	238.632 *2*
		583.191 *2*
		860.564 *5*
		893.408 *5*
		1620.735 *10*
		2614.533 *13*

3. Alpha-particle Energy Standards

Table 3 lists some α-particle energy standards from the evaluation of Rytz[1] for calibration of α-particle measurements. These recommended energies were determined, as described by Rytz,[1] from an adjustment of experimental values to several absolute energy standards. The alpha sources selected for this table all have convenient half-lives, and are presented in order of increasing α-particle energy. Columns 1, 2, and 4 show the source names, half-lives, and absolute (per 100 disintegrations) α-particle intensities, respectively, taken from the alpha particle tables in the *Table of Radioactive Isotopes.* Column 3 lists the recommended α-particle energies, with their corresponding uncertainties (in italics) in the last significant digits.

[1]A. Rytz, *At. Data and Nucl. Data Tables* **23**, 39 (1979).

Table 3. Alpha-particle Energies for Some Standard Sources

Source	Half-life	E_α (keV)	I_α (%)
^{147}**Sm**	1.06×10^{11} y	2234 *3*	
^{154}**Dy**	$\sim3\times10^{6}$ y	2870 *5*	
^{148}**Gd**	75 y	3182.708 *24*	
^{232}**Th**	1.41×10^{10} y	4013 *3*	77
		3954 *8*	23
^{238}**U**	4.468×10^{9} y	4197 *5*	77
		4150 *5*	23
^{236}**U**	2.342×10^{7} y	4494 *3*	74
		4445 *5*	26
^{235}**U**	7.04×10^{8} y	4599 *2*	5.0
		4400 *2*	55
		4374 *4*	6
		4368 *3*	11
		4218 *2*	5.7
^{230}**Th**	7.54×10^{4} y	4687.7 *15*	76.3
		4621.2 *15*	23.4
^{234}**U**	2.45×10^{5} y	4774.8 *9*	72
		4722.6 *9*	28
^{231}**Pa**	3.28×10^{4} y	5059 *1*	11
		5028 *1*	20
		5014 *1*	25.4
		4952 *1*	22.8
		4736 *1*	8.4
^{239}**Pu**	2.411×10^{4} y	5156.6 *4*	73.2
		5143.8 *8*	15.1
		5105.0 *8*	11
^{240}**Pu**	6.54×10^{3} y	5168.17 *15*	73.5
		5123.68 *23*	26.4
^{243}**Am**	7.38×10^{3} y	5275.3 *5*	88
		5233.0 *5*	11
^{210}**Po**	138.376 d	5304.38 *7*	
^{241}**Am**	432.2 y	5485.60 *12*	85.2
		5442.90 *13*	12.8
^{238}**Pu**	87.7 y	5499.07 *20*	71.6
		5456.3 *2*	28.3
^{244}**Cm**	18.11 y	5804.82 *5*	76.4
		5762.70 *5*	23.6
^{243}**Cm**	28.5 y	6066 *1*	1.5
		6058 *1*	5
		5992 *1*	6.5
		5785.0 *5*	73
		5741.1 *5*	10.6
^{242}**Cm**	162.8 d	6112.77 *8*	74.0
		6069.42 *12*	25.0
^{254}**Es**	275.7 d	6428.6 *10*	93
^{253}**Es**	20.4 d	6632.57 *5*	89.8
		6591.4 *5*	6.6

APPENDIX C. ATOMIC DATA

1. Theoretical Internal Conversion Coefficients

The following graphs provide selected theoretical conversion coefficients for *M1, M2, M3, M4, E1, E2, E3*, and *E4* transitions, to an accuracy of 3% to 5%. For atomic numbers $Z=10$ and 20, the graphs show *K*-shell conversion coefficients from Band, et al.[1] For $Z=30$ through $Z=100$, they show total conversion coefficients, at intervals of 5 atomic numbers, from calculations by Rösel, et al.[2]

Discontinuities in the plots of total conversion coefficients occur at the binding energies of the *K* atomic shells, and the graphs at these energies indicate only the change in the conversion coefficient due to the presence of the *K*-shell edge. Between the *K* binding energy and that energy plus 1 *keV*, the plots are straight lines connecting the two extreme values. Consequently, conversion coefficients should not be interpolated within the regions of the graphs spanning these 1-*keV* gaps.

The *K* binding energies used by Rösel, et al.[2] for calculating conversion coefficients are from Bearden and Burr.[3] The newer and generally more precise *K* binding energies of Porter and Freedman[4] are somewhat different; and for some elements with $Z \geq 84$,[5] differ by more than 2 *keV*. One should be aware that these differences may significantly affect conversion coefficients near the *K* binding energy.

Except for the *K*-shell plots, which were taken from the *Table of Isotopes*, 7th edition,[6] the plots are copies of selected graphs from a new compilation by Cole, et al.[7] These were provided by the authors and with the permission of the publisher for use in the *Table of Radioactive Isotopes.*

---

[1]I.M. Band, M.B. Trzhaskovskaya, and M.A. Listengarten, *At. Data and Nucl. Data Tables* **18**, 433 (1976).

[2]F. Rösel, H.M. Fries, K. Alder, and H.C. Pauli, *At. Data and Nucl. Data Tables* **21**, 91 (1978) [for Z=30-67]; **21**, 291 (1978) [for Z=68-104].

[3]J.A. Bearden and A.F. Burr, *Rev. Mod. Phys.* **39**, 125 (1967).

[4]F.T. Porter and M.S. Freedman, *J. Phys. Chem. Ref. Data.* **7**, 1267 (1978).

[5]M.R. Schmorak, private communication (1982).

[6]*Table of Isotopes*, 7th edition, C.M. Lederer and V.S. Shirley, editors; E. Browne, J.M. Dairiki, and R.E. Doebler, principal authors; A.A. Shihab-Eldin, L.J. Jardine, J.K. Tuli, and A.B. Buyrn, authors; John Wiley and Sons, Inc., New York (1978).

[7]*Graphical Representation of K-Shell and Total Internal Conversion Coefficients from Z=30-104*, J.D. Cole, W. Lourens, J.H. Hamilton, and B. van Nooijen, Delft University Press, Delft, The Netherlands (1984).

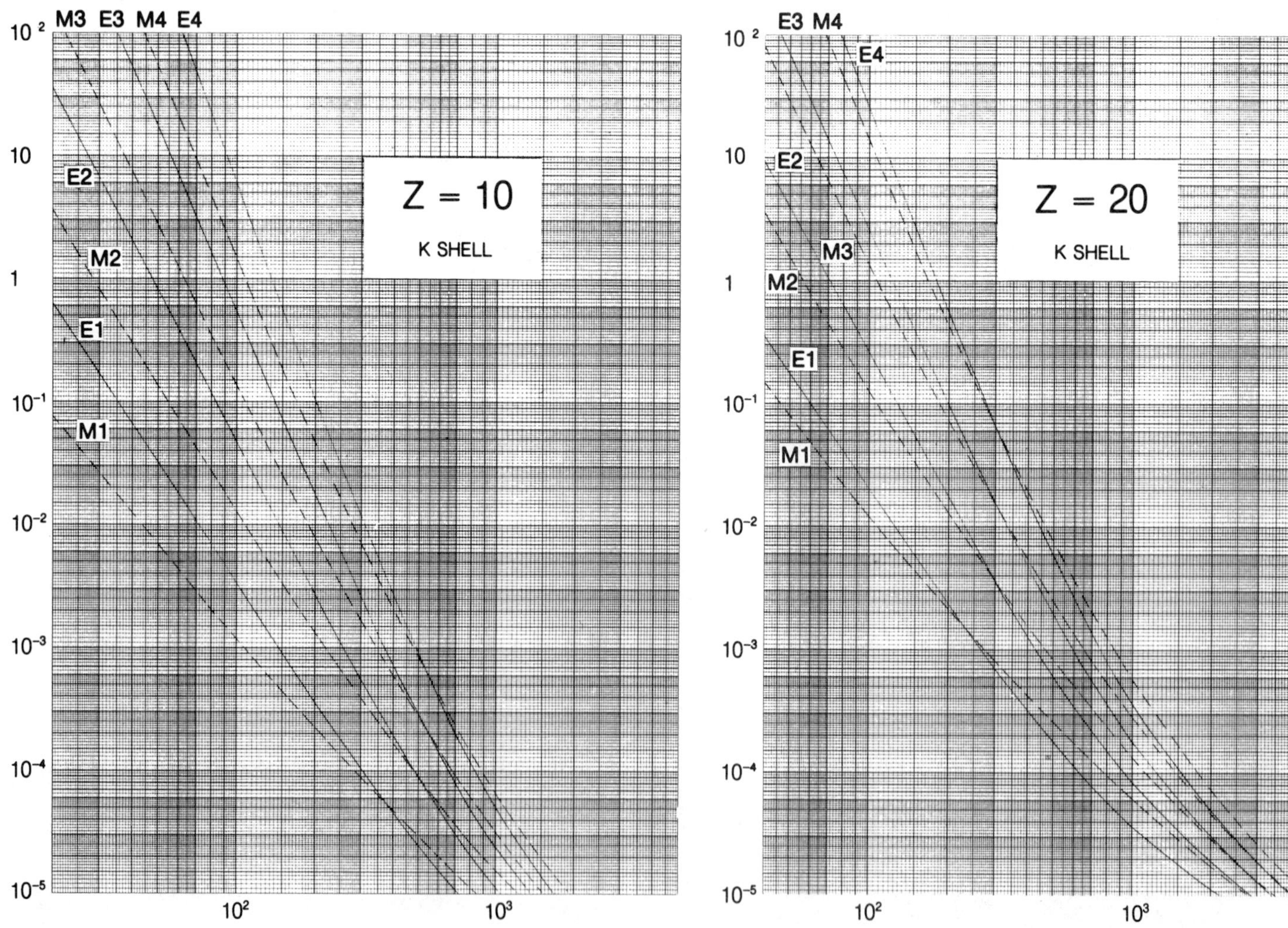

Theoretical Internal Conversion Coefficients

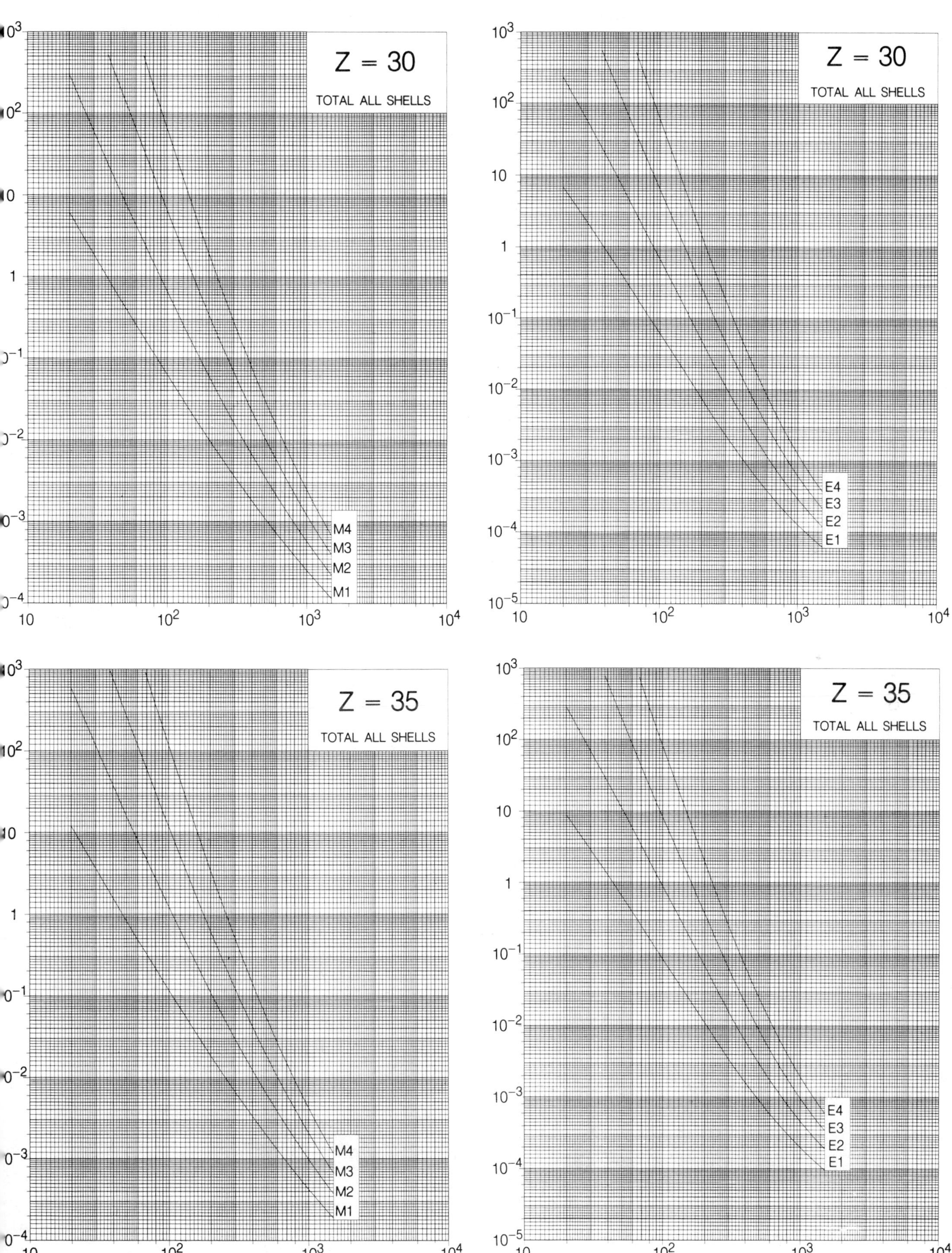

Theoretical Internal Conversion Coefficients

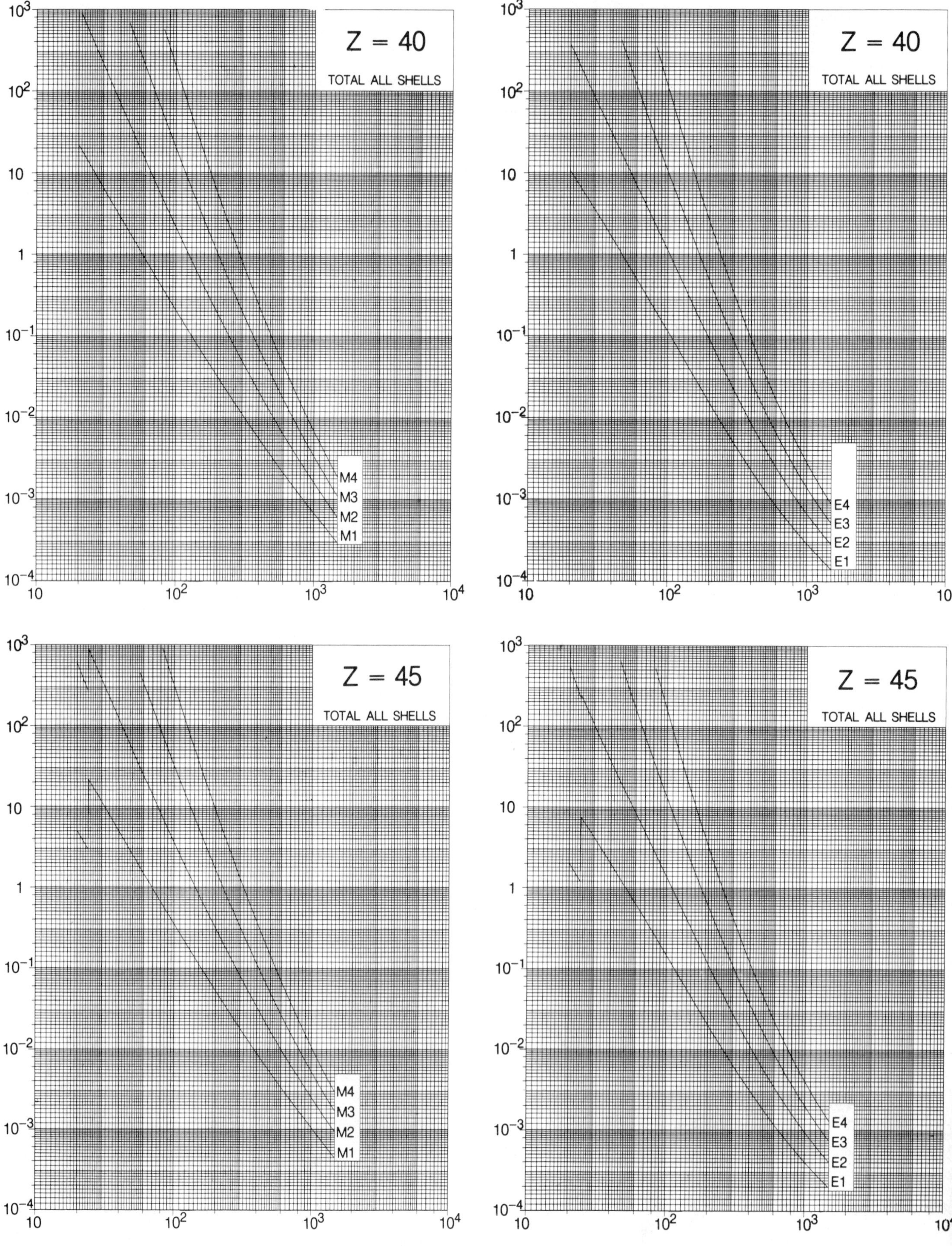

Theoretical Internal Conversion Coefficients

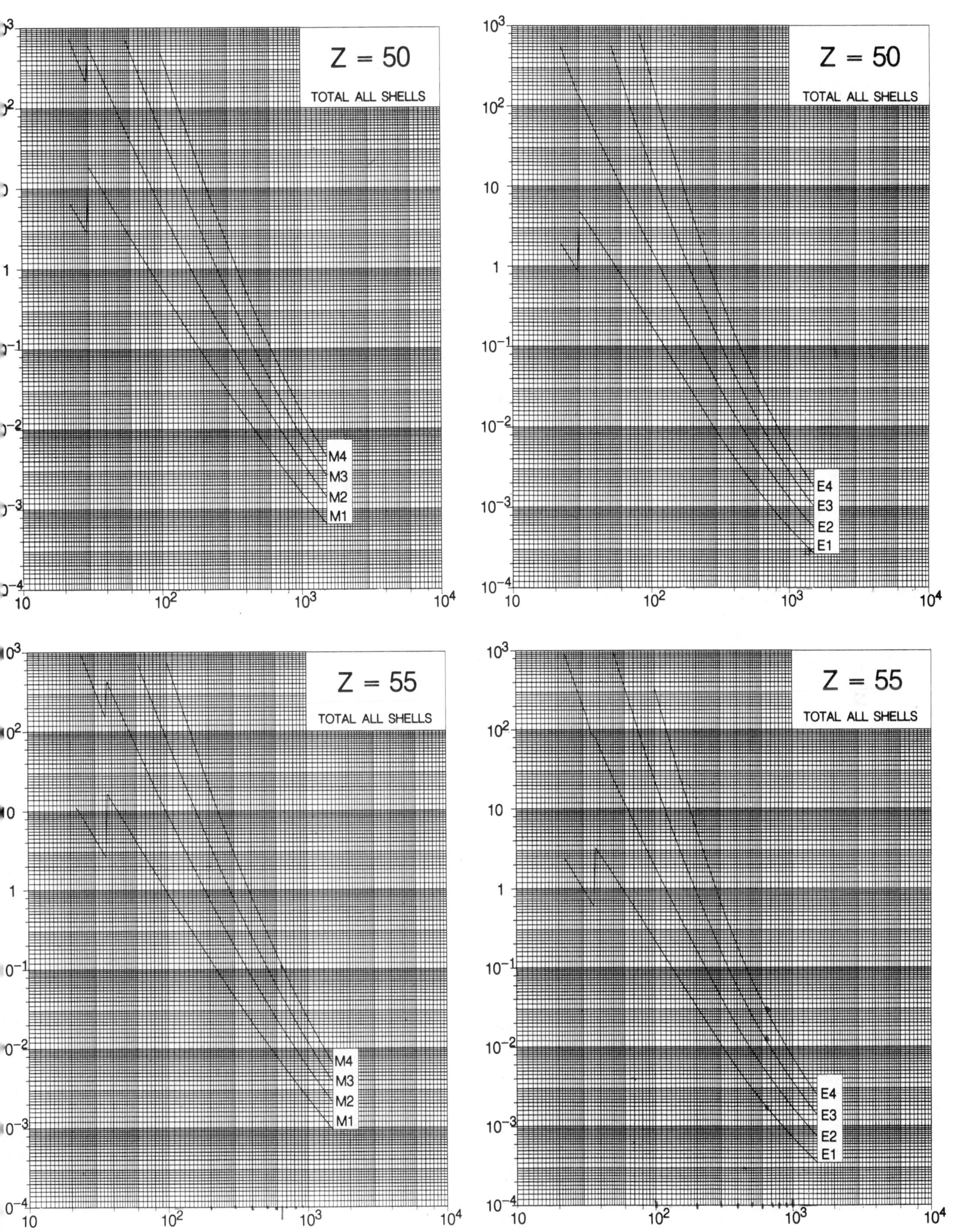

Theoretical Internal Conversion Coefficients

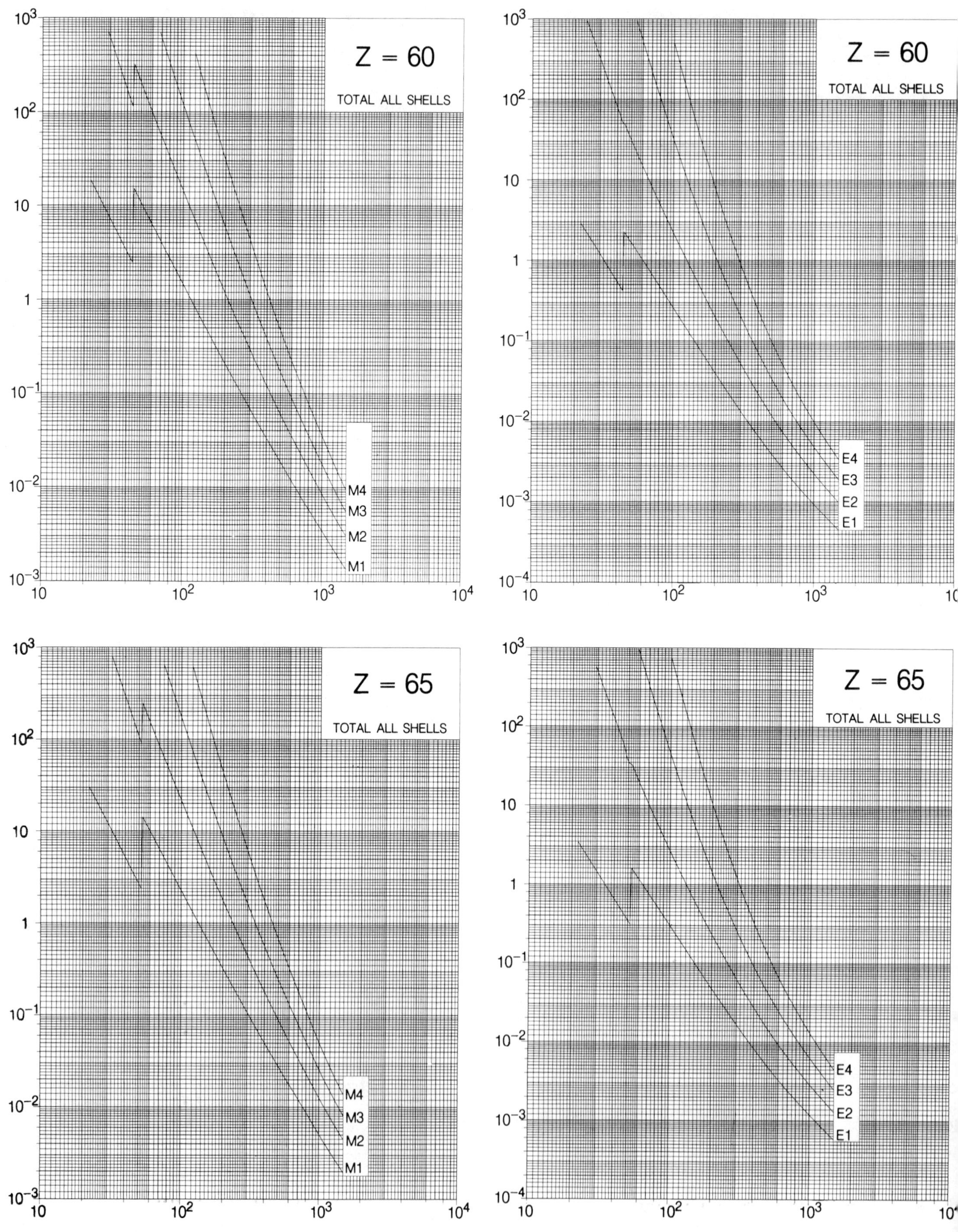

Theoretical Internal Conversion Coefficients

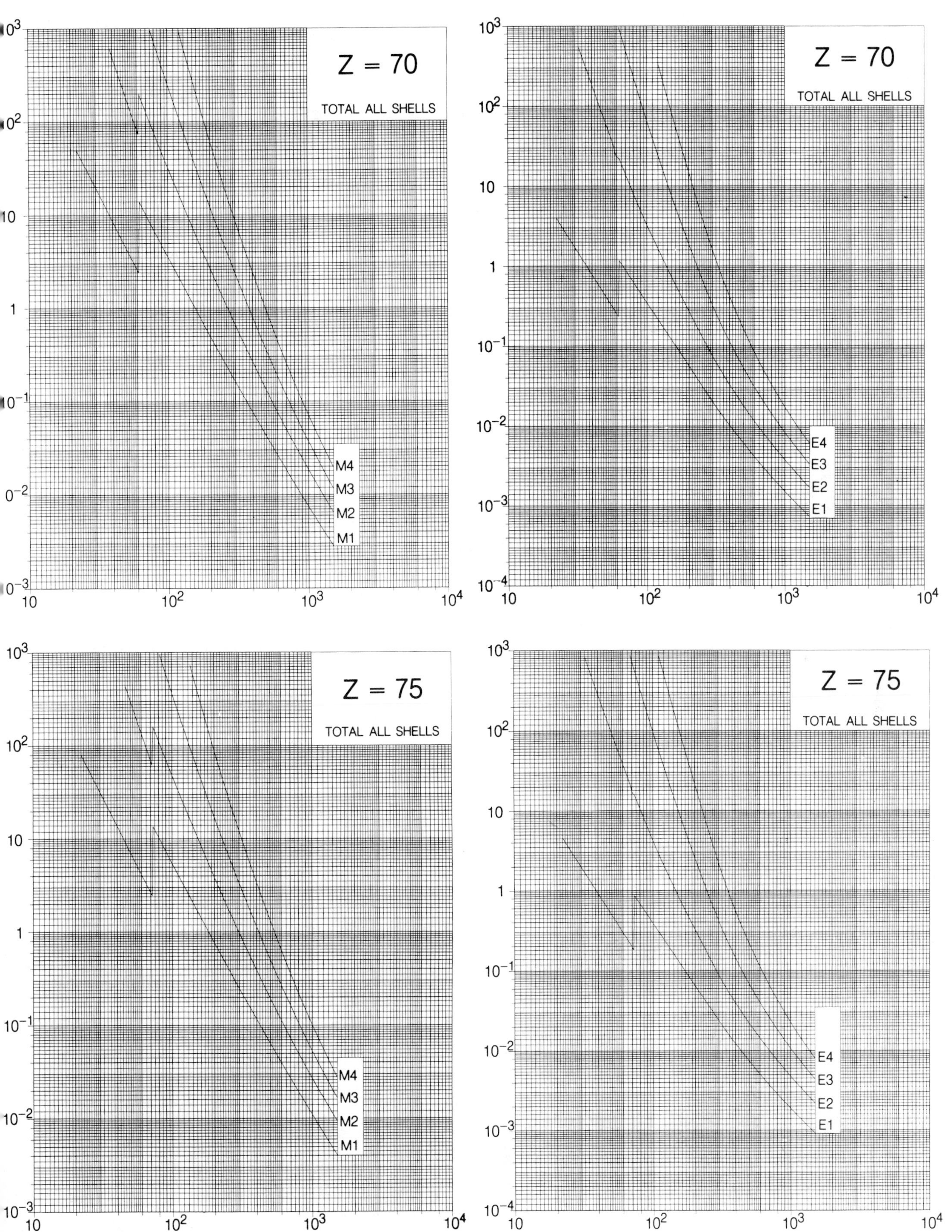

Theoretical Internal Conversion Coefficients

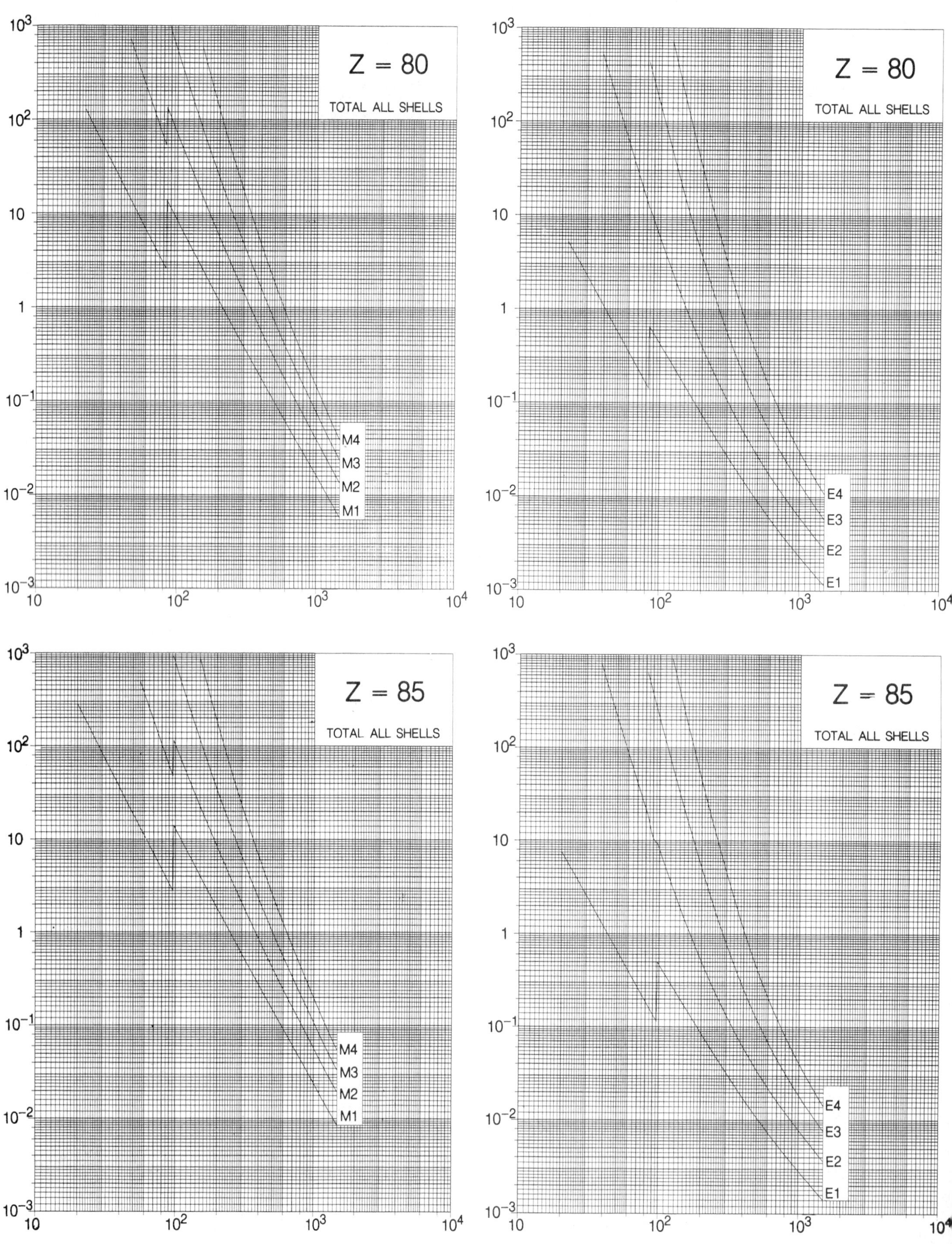

Theoretical Internal Conversion Coefficients

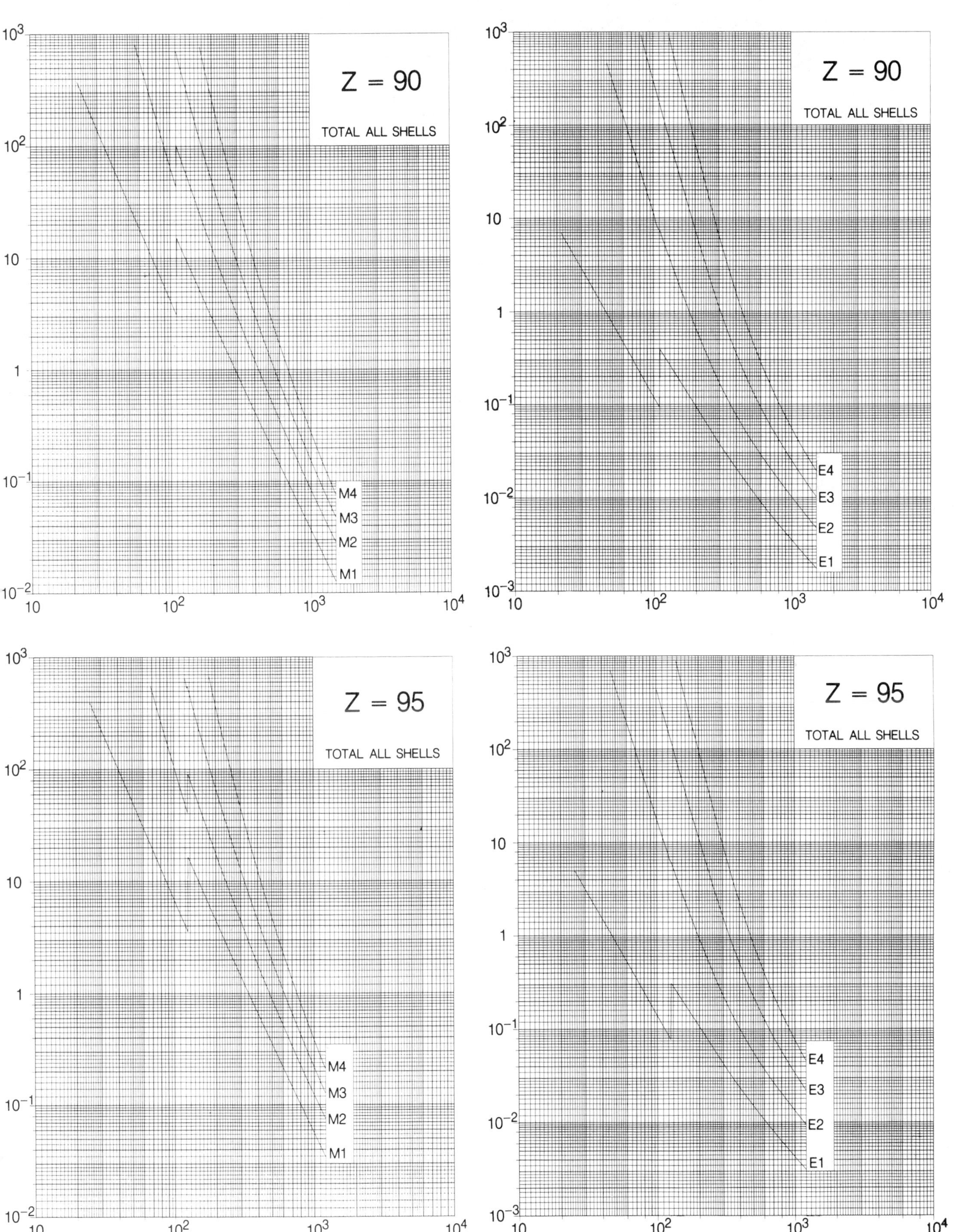

Theoretical Internal Conversion Coefficients

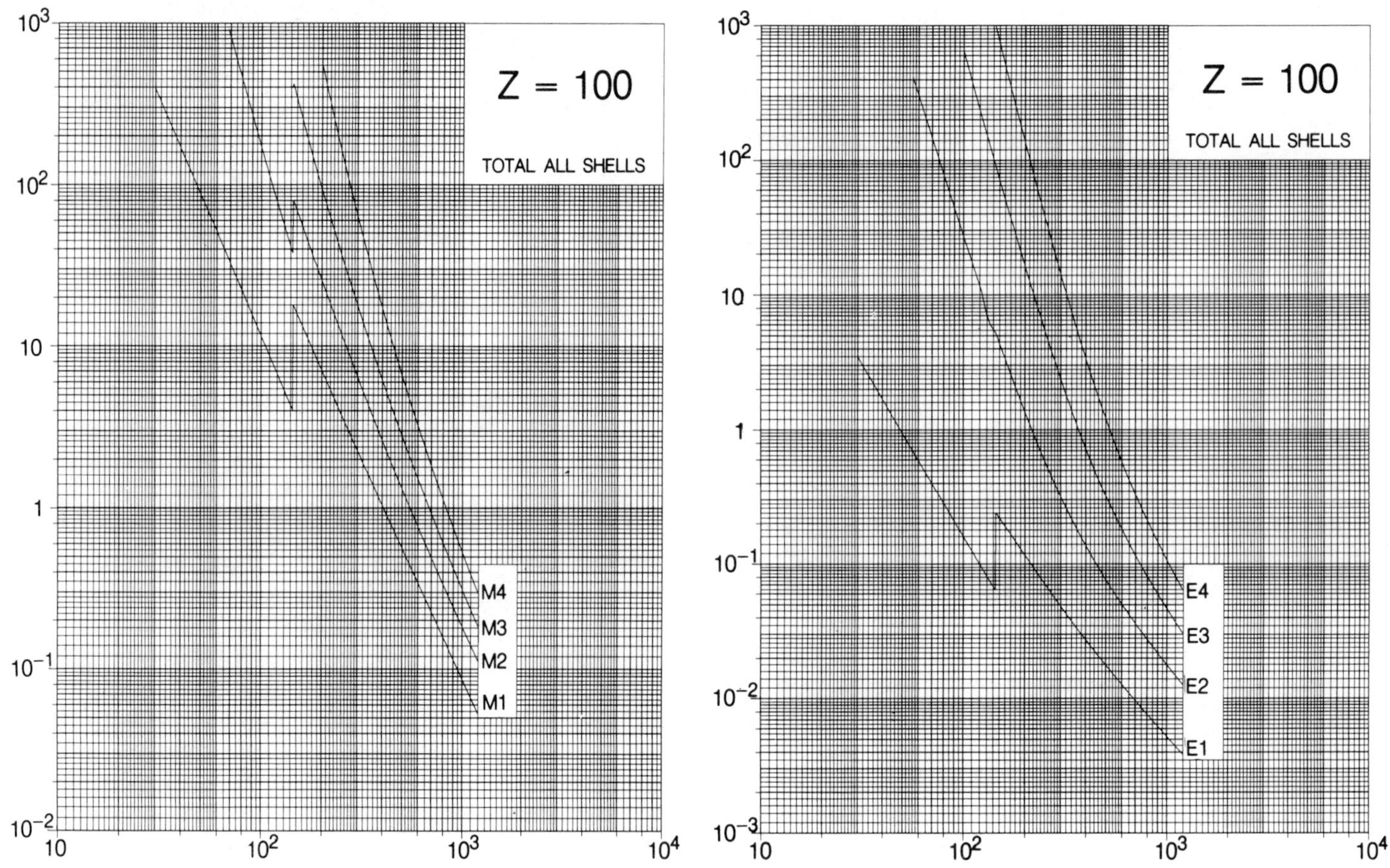

2. Electron-Capture Subshell Ratios

Electron-capture subshell ratios can be calculated as described below,[1] using decay-scheme information and the squared amplitudes of the bound-state electron radial wavefunctions given in Table 1.

The electron-capture transition probability per unit time from all atomic shells is

$$\lambda=\frac{G_\beta^2}{2\pi^3}\sum_x n_x C_x F_x \ , \tag{1}$$

where G_β is the fundamental weak coupling constant, n_x is the relative occupation number for partially filled shells ($n_x=1$ for closed shells), C_x contains the nuclear matrix elements, and the summation extends over all atomic subshells x from which electrons can be captured. The function F_x, which corresponds to the integrated Fermi function of β decay, is given by

$$F_x=\frac{\pi}{2}q_x^2\beta_x^2 B_x \ . \tag{2}$$

Here $q_x(=W_0+W_x)$ is the neutrino energy (neglecting the atomic recoil energy), W_0 (= $Q_\epsilon - E_\ell$) is the electron-capture transition energy, Q_ϵ is the electron-capture decay energy (the difference between the atomic masses of the parent and daughter nuclei), E_ℓ is the energy of the level populated in the daughter nucleus, W_x (= $1-|E_x'|$) is the energy of the bound electron in the parent atom, E_x' is the electron binding energy in the parent atom, β_x is the Coulomb amplitude of the bound-state electron radial wavefunction, and B_x is the associated electron exchange and overlap correction. W_0, W_x, and E_x' are in units of the electron rest-mass energy $m_e c^2$.

After a power series expansion of the wavefunctions, the (L - 1)-forbidden-unique electron-capture transition probability λ from all atomic shells becomes

$$\lambda=M_L\ \frac{(2L-2)!!}{(2L-1)!!}\sum_x \frac{n_x p_x^{2(k_x-1)} q_x^{2(L-k_x+1)}\ \beta_x^2 B_x}{(2k_x-1)![2(L-k_x)+1]\,!}\ , \tag{3}$$

where L is the electron-capture transition angular momentum, p_x $(=(1-W_x^2)^{1/2})$ is the bound electron linear momentum, M_L contains the nuclear matrix elements, and

$$k_x=\begin{cases}1 \text{ for capture from the } K,\ L_1,\ L_2,\ M_1,\ M_2,\ \cdots \text{ atomic shells}\\ 2 \text{ for capture from the } L_3,\ M_3,\ M_4,\ \cdots \text{ atomic shells}\\ 3 \text{ for capture from the } M_5,\ N_5,\ N_6,\ \cdots \text{ atomic shells}\end{cases}$$

Table 1 gives the squared amplitudes $\beta_x^2 B_x p_x^{2(k_x-1)}$ of the bound-state electron radial wavefunctions for Z=1-102, derived from calculations by Bambynek et al.[1] These are used to calculate electron-capture subshell ratios as follows.

For allowed decay, the terms for $k_x=1$ in equation (3) predominate, and the transition probability becomes

$$\lambda^{(k_x=1)}=M_1^2\ (n_K q_K^2\beta_K^2 B_K+n_{L1}q_{L1}^2\beta_{L1}^2 B_{L1}+n_{L2}q_{L2}^2\beta_{L2}^2 B_{L2}+...)\ . \tag{4}$$

The electron-capture subshell ratios are then easy to derive, e.g.,

$$\frac{\lambda_{L1}}{\lambda_K}=\frac{n_{L1}q_{L1}^2\beta_{L1}^2 B_{L1}}{n_K q_K^2\beta_K^2 B_K}\ . \tag{5}$$

For first-forbidden unique transitions (L=2), the contribution of subshells with k_x=1,2 leads to

$$\lambda=\lambda^{(k_x=1)}+\lambda^{(k_x=2)}\ , \tag{6}$$

where

$$\lambda^{(k_x=1)}=M_2^2\ (n_K q_K^4\beta_K^2+n_{L1}q_{L1}^4\beta_{L1}^2 B_{L1}+...)\ , \tag{7}$$

[1] W. Bambynek, H. Behrens, M.H. Chen, B. Crasemann, M.L. Fitzpatrick, K.W.D. Ledingham, H. Genz, M. Mutterer, and R.L. Intemann, *Rev. Mod. Phys.* **49**, 77 (1977).

and

$$\lambda^{(k_x=2)} = M_2^2 \; (n_{L3}q_{L3}^2p_{L3}^2\beta_{L3}^2B_{L3} + n_{M3}q_{M3}^2p_{M3}^2\beta_{M3}^2B_{M3} + \ldots) \; . \tag{8}$$

The L_1/K capture ratio is then

$$\frac{\lambda_{L1}}{\lambda_K} = \frac{n_{L1}q_{L1}^4\beta_{L1}^2B_{L1}}{n_Kq_K^4\beta_K^2B_K} \; . \tag{9}$$

Expressions for the L_2/K, M_1/K, L_2/L_1, and M_1/L_1 capture ratios may be derived in an analogous manner. The L_3/L_1 capture ratio, on the other hand, is given by

$$\frac{\lambda_{L3}}{\lambda_{L1}} = \frac{n_{L3}p_{L3}^2q_{L3}^2\beta_{L3}^2B_{L3}}{n_{L1}q_{L1}^4\beta_{L1}^2B_{L1}} \; , \tag{10}$$

as are the other $k_x = 2$ to $k_x = 1$ electron-capture subshell ratios.

Table 1. Squared Amplitudes of the Bound-State Electron Radial Wavefunctions ($\beta_x^2 B_x p_x^{2(k_x-1)}$)

	$_1$H	$_2$He	$_3$Li	$_4$Be	$_5$B	$_6$C	$_7$N	$_8$O	$_9$F	$_{10}$Ne	$_{11}$Na	$_{12}$Mg
K	1.023×10^{-6}	6.975×10^{-6}	2.877×10^{-5}	7.579×10^{-5}	1.576×10^{-4}	2.844×10^{-4}	4.670×10^{-4}	7.147×10^{-4}	0.001038	0.001448	0.001955	0.002574
L_1			3.778×10^{-6}	8.519×10^{-6}	1.692×10^{-5}	2.783×10^{-5}	4.185×10^{-5}	6.072×10^{-5}	8.630×10^{-5}	1.198×10^{-4}	1.688×10^{-4}	2.227×10^{-4}
L_2							4.508×10^{-9}	1.016×10^{-8}	2.045×10^{-8}	3.785×10^{-8}	7.431×10^{-8}	1.343×10^{-7}
L_3											2.881×10^{-7}	5.180×10^{-7}

	$_{13}$Al	$_{14}$Si	$_{15}$P	$_{16}$S	$_{17}$Cl	$_{18}$Ar	$_{19}$K	$_{20}$Ca	$_{21}$Sc	$_{22}$Ti	$_{23}$V	$_{24}$Cr
K	0.003319	0.004205	0.005246	0.006461	0.007865	0.009475	0.01132	0.01342	0.01578	0.01843	0.02142	0.02477
L_1	2.952×10^{-4}	3.830×10^{-4}	4.882×10^{-4}	6.127×10^{-4}	7.587×10^{-4}	9.273×10^{-4}	0.001121	0.001339	0.001595	0.001885	0.002214	0.002585
L_2	2.287×10^{-7}	3.416×10^{-7}	5.343×10^{-7}	8.061×10^{-7}	1.180×10^{-6}	1.684×10^{-6}	2.353×10^{-6}	3.229×10^{-6}	4.357×10^{-6}	5.796×10^{-6}	7.609×10^{-6}	9.888×10^{-6}
L_3	8.747×10^{-7}	1.293×10^{-6}	2.013×10^{-6}	3.022×10^{-6}	4.402×10^{-6}	6.249×10^{-6}	8.667×10^{-6}	1.179×10^{-5}	1.580×10^{-5}	2.085×10^{-5}	2.716×10^{-5}	3.496×10^{-5}
M_1	2.425×10^{-5}	3.723×10^{-5}	5.279×10^{-5}	7.161×10^{-5}	9.398×10^{-5}	1.199×10^{-4}	1.567×10^{-4}	1.973×10^{-4}	2.430×10^{-4}	2.937×10^{-4}	3.505×10^{-4}	4.080×10^{-4}
M_2			3.110×10^{-8}	5.450×10^{-8}	8.930×10^{-8}	1.392×10^{-7}	2.309×10^{-7}	3.609×10^{-7}	5.178×10^{-7}	7.217×10^{-7}	9.842×10^{-7}	1.287×10^{-6}
M_3									1.874×10^{-6}	2.578×10^{-6}	3.463×10^{-6}	4.491×10^{-6}
M_4												5.799×10^{-10}

	$_{25}$Mn	$_{26}$Fe	$_{27}$Co	$_{28}$Ni	$_{29}$Cu	$_{30}$Zn	$_{31}$Ga	$_{32}$Ge	$_{33}$As	$_{34}$Se	$_{35}$Br	$_{36}$Kr
K	0.02848	0.03263	0.03723	0.04230	0.04791	0.05407	0.06086	0.06827	0.07642	0.08538	0.09516	0.1058
L_1	0.003001	0.003466	0.003983	0.004563	0.005202	0.005906	0.006697	0.007573	0.008533	0.009586	0.01075	0.01202
L_2	1.270×10^{-5}	1.618×10^{-5}	2.042×10^{-5}	2.559×10^{-5}	3.184×10^{-5}	3.927×10^{-5}	4.812×10^{-5}	5.866×10^{-5}	7.115×10^{-5}	8.585×10^{-5}	1.031×10^{-4}	1.233×10^{-4}
L_3	4.447×10^{-5}	5.603×10^{-5}	6.994×10^{-5}	8.663×10^{-5}	1.064×10^{-4}	1.298×10^{-4}	1.571×10^{-4}	1.892×10^{-4}	2.266×10^{-4}	2.699×10^{-4}	3.198×10^{-4}	3.776×10^{-4}
M_1	4.851×10^{-4}	5.649×10^{-4}	6.534×10^{-4}	7.527×10^{-4}	8.525×10^{-4}	9.833×10^{-4}	0.001126	0.001289	0.001473	0.001680	0.001913	0.002172
M_2	1.732×10^{-6}	2.245×10^{-6}	2.875×10^{-6}	3.646×10^{-6}	4.508×10^{-6}	5.705×10^{-6}	7.195×10^{-6}	9.039×10^{-6}	1.128×10^{-5}	1.398×10^{-5}	1.724×10^{-5}	2.113×10^{-5}
M_3	5.951×10^{-6}	7.654×10^{-6}	9.728×10^{-6}	1.224×10^{-5}	1.505×10^{-5}	1.884×10^{-5}	2.348×10^{-5}	2.910×10^{-5}	3.589×10^{-5}	4.398×10^{-5}	5.361×10^{-5}	6.498×10^{-5}
M_4	1.061×10^{-9}	1.602×10^{-9}	2.355×10^{-9}	3.387×10^{-9}	4.358×10^{-9}	6.621×10^{-9}	9.770×10^{-9}	1.407×10^{-8}	1.989×10^{-8}	2.760×10^{-8}	3.770×10^{-8}	5.077×10^{-8}
M_5							3.604×10^{-8}	5.180×10^{-8}	7.279×10^{-8}	1.004×10^{-7}	1.365×10^{-7}	1.828×10^{-7}
N_1									1.378×10^{-4}	1.748×10^{-4}	2.167×10^{-4}	2.638×10^{-4}

	$_{37}$Rb	$_{38}$Sr	$_{39}$Y	$_{40}$Zr	$_{41}$Nb	$_{42}$Mo	$_{43}$Tc	$_{44}$Ru	$_{45}$Rh	$_{46}$Pd	$_{47}$Ag	$_{48}$Cd
K	0.1174	0.1301	0.1441	0.1592	0.1756	0.1934	0.2129	0.2339	0.2568	0.2817	0.3087	0.3376
L_1	0.01342	0.01497	0.01665	0.01853	0.02057	0.02279	0.02522	0.02791	0.03085	0.03396	0.03752	0.04131
L_2	1.469×10^{-4}	1.745×10^{-4}	2.065×10^{-4}	2.437×10^{-4}	2.866×10^{-4}	3.363×10^{-4}	3.937×10^{-4}	4.595×10^{-4}	5.350×10^{-4}	6.219×10^{-4}	7.212×10^{-4}	8.341×10^{-4}
L_3	4.435×10^{-4}	5.194×10^{-4}	6.056×10^{-4}	7.042×10^{-4}	8.155×10^{-4}	9.423×10^{-4}	0.001086	0.001246	0.001427	0.001631	0.001858	0.002111
M_1	0.002463	0.002783	0.003138	0.003525	0.003960	0.004438	0.004965	0.005547	0.006188	0.006886	0.007666	0.008505
M_2	2.579×10^{-5}	3.131×10^{-5}	3.786×10^{-5}	4.557×10^{-5}	5.463×10^{-5}	6.519×10^{-5}	7.753×10^{-5}	9.189×10^{-5}	1.086×10^{-4}	1.278×10^{-4}	1.502×10^{-4}	1.760×10^{-4}
M_3	7.827×10^{-5}	9.380×10^{-5}	1.119×10^{-4}	1.327×10^{-4}	1.568×10^{-4}	1.844×10^{-4}	2.160×10^{-4}	2.522×10^{-4}	2.933×10^{-4}	3.400×10^{-4}	3.930×10^{-4}	4.528×10^{-4}
M_4	6.761×10^{-8}	8.903×10^{-8}	1.163×10^{-7}	1.505×10^{-7}	1.934×10^{-7}	2.458×10^{-7}	3.106×10^{-7}	3.892×10^{-7}	4.850×10^{-7}	6.003×10^{-7}	7.401×10^{-7}	9.075×10^{-7}
M_5	2.422×10^{-7}	3.175×10^{-7}	4.120×10^{-7}	5.294×10^{-7}	6.745×10^{-7}	8.532×10^{-7}	1.071×10^{-6}	1.336×10^{-6}	1.657×10^{-6}	2.043×10^{-6}	2.503×10^{-6}	3.050×10^{-6}
N_1	3.333×10^{-4}	4.145×10^{-4}	4.982×10^{-4}	5.895×10^{-4}	6.800×10^{-4}	7.881×10^{-4}	9.186×10^{-4}	7.025×10^{-4}	0.001184	0.001333	0.001520	0.001731
N_2	2.306×10^{-6}	3.229×10^{-6}	4.253×10^{-6}	5.467×10^{-6}	6.768×10^{-6}	8.525×10^{-6}	1.076×10^{-5}	1.310×10^{-5}	1.602×10^{-5}	1.928×10^{-5}	2.349×10^{-5}	2.851×10^{-5}
N_3			1.263×10^{-5}	1.611×10^{-5}	1.988×10^{-5}	2.448×10^{-5}	3.033×10^{-5}	3.604×10^{-5}	4.323×10^{-5}	5.090×10^{-5}	6.104×10^{-5}	7.296×10^{-5}
N_4						1.658×10^{-8}	2.596×10^{-8}	3.246×10^{-8}	4.380×10^{-8}	5.370×10^{-8}	7.585×10^{-8}	1.042×10^{-7}

Table 1. Squared Amplitudes of the Bound-State Electron Radial Wavefunctions ($\beta_x^2 B_x p_x^{2(k_x-1)}$) (continued)

	$_{49}$In	$_{50}$Sn	$_{51}$Sb	$_{52}$Te	$_{53}$I	$_{54}$Xe	$_{55}$Cs	$_{56}$Ba	$_{57}$La	$_{58}$Ce	$_{59}$Pr	$_{60}$Nd
K	0.3691	0.4030	0.4399	0.4793	0.5231	0.5696	0.6204	0.6747	0.7341	0.7986	0.8685	0.9441
L_1	0.04547	0.04994	0.05488	0.06014	0.06609	0.07238	0.07937	0.08691	0.09531	0.1044	0.1144	0.1252
L_2	9.645×10^{-4}	0.001112	0.001280	0.001471	0.001691	0.001937	0.002219	0.002539	0.002902	0.003315	0.003783	0.004310
L_3	0.002397	0.002713	0.003065	0.003458	0.003894	0.004375	0.004910	0.005504	0.006158	0.006882	0.007683	0.008560
M_1	0.009439	0.01045	0.01156	0.01277	0.01411	0.01556	0.01718	0.01893	0.02087	0.02300	0.02535	0.02790
M_2	2.058×10^{-4}	2.400×10^{-4}	2.793×10^{-4}	3.242×10^{-4}	3.764×10^{-4}	4.355×10^{-4}	5.036×10^{-4}	5.809×10^{-4}	6.698×10^{-4}	7.714×10^{-4}	8.876×10^{-4}	0.001019
M_3	5.204×10^{-4}	5.964×10^{-4}	6.821×10^{-4}	7.780×10^{-4}	8.864×10^{-4}	0.001007	0.001142	0.001292	0.001460	0.001648	0.001858	0.002090
M_4	1.108×10^{-6}	1.347×10^{-6}	1.630×10^{-6}	1.964×10^{-6}	2.361×10^{-6}	2.826×10^{-6}	3.373×10^{-6}	4.012×10^{-6}	4.760×10^{-6}	5.634×10^{-6}	6.651×10^{-6}	7.826×10^{-6}
M_5	3.698×10^{-6}	4.464×10^{-6}	5.367×10^{-6}	6.424×10^{-6}	7.665×10^{-6}	9.108×10^{-6}	1.079×10^{-5}	1.274×10^{-5}	1.499×10^{-5}	1.762×10^{-5}	2.067×10^{-5}	2.414×10^{-5}
N_1	0.001966	0.002230	0.002524	0.002848	0.003216	0.003614	0.004066	0.004565	0.005116	0.005682	0.006263	0.006933
N_2	3.459×10^{-5}	4.183×10^{-5}	5.031×10^{-5}	6.023×10^{-5}	7.197×10^{-5}	8.556×10^{-5}	1.015×10^{-4}	1.200×10^{-4}	1.416×10^{-4}	1.649×10^{-4}	1.901×10^{-4}	2.203×10^{-4}
N_3	8.699×10^{-5}	1.033×10^{-4}	1.223×10^{-4}	1.441×10^{-4}	1.693×10^{-4}	1.982×10^{-4}	2.309×10^{-4}	2.680×10^{-4}	3.103×10^{-4}	3.540×10^{-4}	3.994×10^{-4}	4.529×10^{-4}
N_4	1.402×10^{-7}	1.854×10^{-7}	2.422×10^{-7}	3.123×10^{-7}	3.986×10^{-7}	5.034×10^{-7}	6.312×10^{-7}	7.856×10^{-7}	9.730×10^{-7}	1.175×10^{-6}	1.390×10^{-6}	1.663×10^{-6}
N_5	4.644×10^{-7}	6.131×10^{-7}	7.951×10^{-7}	1.017×10^{-6}	1.289×10^{-6}	1.616×10^{-6}	2.012×10^{-6}	2.488×10^{-6}	3.057×10^{-6}	3.658×10^{-6}	4.291×10^{-6}	5.084×10^{-6}
N_6			1.297×10^{-10}	1.613×10^{-10}	1.972×10^{-10}	2.373×10^{-10}	2.977×10^{-10}	3.678×10^{-10}	4.398×10^{-10}	5.203×10^{-10}	6.170×10^{-10}	8.309×10^{-10}
N_7							1.941×10^{-14}	2.604×10^{-14}	3.293×10^{-14}	1.274×10^{-12}	1.330×10^{-12}	1.468×10^{-12}

	$_{61}$Pm	$_{62}$Sm	$_{63}$Eu	$_{64}$Gd	$_{65}$Tb	$_{66}$Dy	$_{67}$Ho	$_{68}$Er	$_{69}$Tm	$_{70}$Yb	$_{71}$Lu	$_{72}$Hf
K	1.025	1.113	1.208	1.309	1.420	1.539	1.668	1.808	1.961	2.122	2.300	2.490
L_1	0.1369	0.1498	0.1638	0.1790	0.1956	0.2138	0.2335	0.2553	0.2789	0.3045	0.3329	0.3632
L_2	0.004906	0.005579	0.006345	0.007199	0.008184	0.009281	0.01052	0.01193	0.01352	0.01530	0.01732	0.01959
L_3	0.009527	0.01059	0.01176	0.01303	0.01446	0.01600	0.01770	0.01955	0.02158	0.02379	0.02622	0.02886
M_1	0.03067	0.03373	0.03705	0.04066	0.04464	0.04899	0.05377	0.05901	0.06475	0.07096	0.07781	0.08525
M_2	0.001168	0.001338	0.001532	0.001750	0.002001	0.002283	0.002605	0.002970	0.003385	0.003851	0.004382	0.004979
M_3	0.002346	0.002630	0.002946	0.003291	0.003676	0.004098	0.004565	0.005080	0.005647	0.006268	0.006953	0.007703
M_4	9.186×10^{-6}	1.075×10^{-5}	1.258×10^{-5}	1.465×10^{-5}	1.707×10^{-5}	1.983×10^{-5}	2.299×10^{-5}	2.661×10^{-5}	3.075×10^{-5}	3.545×10^{-5}	4.079×10^{-5}	4.685×10^{-5}
M_5	2.813×10^{-5}	3.269×10^{-5}	3.789×10^{-5}	4.377×10^{-5}	5.051×10^{-5}	5.812×10^{-5}	6.676×10^{-5}	7.652×10^{-5}	8.756×10^{-5}	9.997×10^{-5}	1.140×10^{-4}	1.296×10^{-4}
N_1	0.007660	0.008457	0.009331	0.01032	0.01132	0.01247	0.01373	0.01511	0.01662	0.01825	0.02012	0.02216
N_2	2.547×10^{-4}	2.938×10^{-4}	3.389×10^{-4}	3.916×10^{-4}	4.479×10^{-4}	5.138×10^{-4}	5.891×10^{-4}	6.748×10^{-4}	7.723×10^{-4}	8.822×10^{-4}	0.001011	0.001158
N_3	5.123×10^{-4}	5.781×10^{-4}	6.516×10^{-4}	7.364×10^{-4}	8.227×10^{-4}	9.221×10^{-4}	0.001032	0.001154	0.001288	0.001435	0.001604	0.001791
N_4	1.982×10^{-6}	2.352×10^{-6}	2.783×10^{-6}	3.317×10^{-6}	3.855×10^{-6}	4.516×10^{-6}	5.277×10^{-6}	6.152×10^{-6}	7.157×10^{-6}	8.303×10^{-6}	9.685×10^{-6}	1.129×10^{-5}
N_5	5.997×10^{-6}	7.045×10^{-6}	8.263×10^{-6}	9.739×10^{-6}	1.126×10^{-5}	1.308×10^{-5}	1.516×10^{-5}	1.753×10^{-5}	2.022×10^{-5}	2.327×10^{-5}	2.689×10^{-5}	3.103×10^{-5}
N_6	1.099×10^{-9}	1.433×10^{-9}	1.852×10^{-9}	2.553×10^{-9}	2.999×10^{-9}	3.769×10^{-9}	4.703×10^{-9}	5.833×10^{-9}	7.194×10^{-9}	8.824×10^{-9}	1.133×10^{-8}	1.442×10^{-8}
N_7	1.616×10^{-12}	1.777×10^{-12}	6.049×10^{-9}	8.475×10^{-9}	9.791×10^{-9}	1.229×10^{-8}	1.531×10^{-8}	1.895×10^{-8}	2.331×10^{-8}	2.851×10^{-8}	3.675×10^{-8}	4.678×10^{-8}
O_1							0.001708	0.001873	0.002054	0.002249	0.002573	0.002935
O_2											1.252×10^{-4}	1.492×10^{-4}

	$_{73}$Ta	$_{74}$W	$_{75}$Re	$_{76}$Os	$_{77}$Ir	$_{78}$Pt	$_{79}$Au	$_{80}$Hg	$_{81}$Tl	$_{82}$Pb	$_{83}$Bi	$_{84}$Po
K	2.697	2.920	3.163	3.422	3.707	4.012	4.346	4.704	5.090	5.511	5.970	6.472
L_1	0.3969	0.4335	0.4733	0.5167	0.5648	0.6165	0.6740	0.7355	0.8033	0.8781	0.9605	1.051
L_2	0.02216	0.02504	0.02831	0.03198	0.03615	0.04083	0.04614	0.05214	0.05885	0.06647	0.07510	0.08490
L_3	0.03175	0.03490	0.03833	0.04206	0.04615	0.05058	0.05543	0.06073	0.06644	0.07268	0.07943	0.08682
M_1	0.09343	0.1023	0.1122	0.1228	0.1345	0.1474	0.1617	0.1770	0.1938	0.2125	0.2329	0.2557
M_2	0.005659	0.006426	0.007298	0.008277	0.009394	0.01065	0.01209	0.01370	0.01552	0.01759	0.01995	0.02263
M_3	0.008527	0.009431	0.01042	0.01151	0.01270	0.01400	0.01543	0.01699	0.01869	0.02055	0.02260	0.02483
M_4	5.374×10^{-5}	6.155×10^{-5}	7.039×10^{-5}	8.037×10^{-5}	9.168×10^{-5}	1.044×10^{-4}	1.189×10^{-4}	1.350×10^{-4}	1.533×10^{-4}	1.738×10^{-4}	1.970×10^{-4}	2.231×10^{-4}
M_5	1.472×10^{-4}	1.670×10^{-4}	1.891×10^{-4}	2.138×10^{-4}	2.414×10^{-4}	2.722×10^{-4}	3.067×10^{-4}	3.448×10^{-4}	3.873×10^{-4}	4.346×10^{-4}	4.871×10^{-4}	5.455×10^{-4}
N_1	0.02442	0.02691	0.02964	0.03263	0.03597	0.03962	0.04371	0.04818	0.05306	0.05849	0.06448	0.07117
N_2	0.001327	0.001519	0.001739	0.001989	0.002276	0.002601	0.002975	0.003397	0.003879	0.004431	0.005062	0.005784
N_3	0.002000	0.002231	0.002488	0.002771	0.003086	0.003435	0.003819	0.004242	0.004707	0.005221	0.005789	0.006416
N_4	1.314×10^{-5}	1.529×10^{-5}	1.775×10^{-5}	2.056×10^{-5}	2.380×10^{-5}	2.749×10^{-5}	3.172×10^{-5}	3.653×10^{-5}	4.202×10^{-5}	4.828×10^{-5}	5.541×10^{-5}	6.353×10^{-5}
N_5	3.577×10^{-5}	4.118×10^{-5}	4.736×10^{-5}	5.437×10^{-5}	6.234×10^{-5}	7.140×10^{-5}	8.165×10^{-5}	9.312×10^{-5}	1.060×10^{-4}	1.206×10^{-4}	1.370×10^{-4}	1.554×10^{-4}
N_6	1.819×10^{-8}	2.279×10^{-8}	2.842×10^{-8}	3.520×10^{-8}	4.336×10^{-8}	5.314×10^{-8}	6.479×10^{-8}	7.862×10^{-8}	9.501×10^{-8}	1.144×10^{-7}	1.374×10^{-7}	1.644×10^{-7}
N_7	5.895×10^{-8}	7.365×10^{-8}	9.123×10^{-8}	1.123×10^{-7}	1.375×10^{-7}	1.673×10^{-7}	2.028×10^{-7}	2.448×10^{-7}	2.943×10^{-7}	3.526×10^{-7}	4.210×10^{-7}	5.011×10^{-7}
O_1	0.003345	0.003803	0.004320	0.004893	0.005538	0.006397	0.006993	0.007935	0.008978	0.01016	0.01149	0.01299
O_2	1.773×10^{-4}	2.099×10^{-4}	2.505×10^{-4}	2.976×10^{-4}	3.528×10^{-4}	4.140×10^{-4}	4.877×10^{-4}	5.762×10^{-4}	6.811×10^{-4}	8.052×10^{-4}	9.491×10^{-4}	0.001117
O_3			3.490×10^{-4}	4.020×10^{-4}	4.616×10^{-4}	5.217×10^{-4}	5.956×10^{-4}	6.860×10^{-4}	7.902×10^{-4}	9.087×10^{-4}	0.001044	0.001197
O_4									4.614×10^{-6}	5.745×10^{-6}	7.128×10^{-6}	8.757×10^{-6}
O_5											1.723×10^{-5}	2.097×10^{-5}

Table 1. Squared Amplitudes of the Bound-State Electron Radial Wavefunctions ($\beta_x^2 B_x p_x^{2(k_x-1)}$) (continued)

	$_{85}$At	$_{86}$Rn	$_{87}$Fr	$_{88}$Ra	$_{89}$Ac	$_{90}$Th	$_{91}$Pa	$_{92}$U	$_{93}$Np	$_{94}$Pu	$_{95}$Am	$_{96}$Cm
K	5.068	7.560	8.197	8.879	9.650	10.419	11.321	12.234	13.298	14.427	15.633	16.941
L_1	1.152	1.253	1.372	1.499	1.646	1.796	1.971	2.153	2.365	2.591	2.838	3.109
L_2	0.09605	0.1079	0.1220	0.1378	0.1562	0.1759	0.1994	0.2248	0.2549	0.2886	0.3263	0.3691
L_3	0.09487	0.1033	0.1127	0.1229	0.1341	0.1459	0.1591	0.1730	0.1885	0.2051	0.2229	0.2423
M_1	0.2807	0.3058	0.3359	0.3681	0.4053	0.4429	0.4875	0.5332	0.5872	0.6449	0.7080	0.7767
M_2	0.02569	0.02895	0.03284	0.03721	0.04231	0.04779	0.05432	0.06141	0.06983	0.07925	0.08984	0.1019
M_3	0.02729	0.02987	0.03278	0.03593	0.03943	0.04311	0.04726	0.05166	0.05657	0.06188	0.06762	0.07386
M_4	2.524×10^{-4}	2.844×10^{-4}	3.212×10^{-4}	3.622×10^{-4}	4.087×10^{-4}	4.595×10^{-4}	5.177×10^{-4}	5.813×10^{-4}	6.538×10^{-4}	7.342×10^{-4}	8.236×10^{-4}	9.231×10^{-4}
M_5	6.104×10^{-4}	6.807×10^{-4}	7.599×10^{-4}	8.473×10^{-4}	9.446×10^{-4}	0.001050	0.001169	0.001298	0.001442	0.001600	0.001772	0.001962
N_1	0.07861	0.08615	0.09510	0.1048	0.1159	0.1273	0.1407	0.1547	0.1710	0.1887	0.2079	0.2291
N_2	0.006613	0.007504	0.008571	0.009777	0.01119	0.01272	0.01454	0.01654	0.01891	0.02157	0.02458	0.02802
N_3	0.007109	0.007845	0.008677	0.009587	0.01060	0.01168	0.01290	0.01419	0.01565	0.01723	0.01896	0.02084
N_4	7.277×10^{-5}	8.298×10^{-5}	9.481×10^{-5}	1.081×10^{-4}	1.234×10^{-4}	1.402×10^{-4}	1.596×10^{-4}	1.811×10^{-4}	2.057×10^{-4}	2.332×10^{-4}	2.640×10^{-4}	2.986×10^{-4}
N_5	1.761×10^{-4}	1.989×10^{-4}	2.247×10^{-4}	2.535×10^{-4}	2.859×10^{-4}	3.216×10^{-4}	3.617×10^{-4}	4.058×10^{-4}	4.554×10^{-4}	5.103×10^{-4}	5.710×10^{-4}	6.382×10^{-4}
N_6	1.962×10^{-7}	2.329×10^{-7}	2.763×10^{-7}	3.269×10^{-7}	3.860×10^{-7}	4.539×10^{-7}	5.340×10^{-7}	6.255×10^{-7}	7.323×10^{-7}	8.558×10^{-7}	9.961×10^{-7}	1.157×10^{-6}
N_7	5.946×10^{-7}	7.023×10^{-7}	8.284×10^{-7}	9.742×10^{-7}	1.144×10^{-6}	1.337×10^{-6}	1.561×10^{-6}	1.816×10^{-6}	2.111×10^{-6}	2.446×10^{-6}	2.830×10^{-6}	3.266×10^{-6}
O_1	0.01469	0.01646	0.01857	0.02090	0.02358	0.02639	0.02958	0.03300	0.03698	0.04126	0.04602	0.05136
O_2	0.001314	0.001531	0.001794	0.002096	0.002454	0.002851	0.003303	0.003813	0.004421	0.005096	0.005886	0.006801
O_3	0.001371	0.001561	0.001777	0.002017	0.002288	0.002584	0.002895	0.003241	0.003629	0.004040	0.004500	0.005021
O_4	1.067×10^{-5}	1.286×10^{-5}	1.546×10^{-5}	1.848×10^{-5}	2.205×10^{-5}	2.613×10^{-5}	3.011×10^{-5}	3.492×10^{-5}	4.046×10^{-5}	4.626×10^{-5}	5.349×10^{-5}	6.201×10^{-5}
O_5	2.530×10^{-5}	3.021×10^{-5}	3.601×10^{-5}	4.268×10^{-5}	5.040×10^{-5}	5.910×10^{-5}	6.787×10^{-5}	7.810×10^{-5}	8.967×10^{-5}	1.019×10^{-4}	1.160×10^{-4}	1.327×10^{-4}
O_6			1.939×10^{-9}	2.367×10^{-9}	2.831×10^{-9}	3.324×10^{-9}	4.439×10^{-8}	5.772×10^{-8}	7.344×10^{-8}	8.496×10^{-8}	1.062×10^{-7}	1.385×10^{-7}
O_7					6.591×10^{-13}	8.052×10^{-13}	9.564×10^{-12}	1.076×10^{-11}	1.214×10^{-11}	1.326×10^{-11}	2.825×10^{-7}	3.795×10^{-7}
O_8							9.295×10^{-13}	1.090×10^{-12}	1.279×10^{-12}	1.456×10^{-12}	1.489×10^{-11}	1.709×10^{-11}
O_9												8.657×10^{-15}

	$_{97}$Bk	$_{98}$Cf	$_{99}$Es	$_{100}$Fm	$_{101}$Md	$_{102}$No
K	18.391	19.970	21.673	23.601	25.626	27.879
L_1	3.413	3.747	4.112	4.529	4.974	5.469
L_2	0.4183	0.4742	0.5374	0.6112	0.6933	0.7882
L_3	0.2634	0.2862	0.3109	0.3380	0.3669	0.3984
M_1	0.8544	0.9392	1.033	1.140	1.253	1.382
M_2	0.1157	0.1315	0.1494	0.1702	0.1935	0.2204
M_3	0.08071	0.08816	0.09624	0.1052	0.1147	0.1252
M_4	0.001035	0.001159	0.001298	0.001453	0.001625	0.001817
M_5	0.002172	0.002402	0.002654	0.002933	0.003237	0.003571
N_1	0.2529	0.2792	0.3080	0.3409	0.3762	0.4160
N_2	0.03198	0.03650	0.04165	0.04767	0.05441	0.06223
N_3	0.02292	0.02519	0.02766	0.03040	0.03335	0.03660
N_4	3.375×10^{-4}	3.813×10^{-4}	4.302×10^{-4}	4.858×10^{-4}	5.473×10^{-4}	6.166×10^{-4}
N_5	7.128×10^{-4}	7.955×10^{-4}	8.867×10^{-4}	9.883×10^{-4}	0.001100	0.001223
N_6	1.342×10^{-6}	1.553×10^{-6}	1.795×10^{-6}	2.072×10^{-6}	2.386×10^{-6}	2.745×10^{-6}
N_7	3.768×10^{-6}	4.336×10^{-6}	4.980×10^{-6}	5.714×10^{-6}	6.542×10^{-6}	7.484×10^{-6}
O_1	0.05724	0.06384	0.07113	0.07948	0.08852	0.09872
O_2	0.007850	0.009065	0.01046	0.01209	0.01394	0.01609
O_3	0.005563	0.006176	0.006847	0.007593	0.008401	0.009293
O_4	7.100×10^{-5}	8.155×10^{-5}	9.345×10^{-5}	1.070×10^{-4}	1.222×10^{-4}	1.395×10^{-4}
O_5	1.494×10^{-4}	1.690×10^{-4}	1.908×10^{-4}	2.152×10^{-4}	2.422×10^{-4}	2.723×10^{-4}
O_6	1.602×10^{-7}	1.942×10^{-7}	2.340×10^{-7}	2.804×10^{-7}	3.342×10^{-7}	3.967×10^{-7}
O_7	4.272×10^{-7}	5.171×10^{-7}	6.207×10^{-7}	7.408×10^{-7}	8.782×10^{-7}	1.033×10^{-6}
O_8	1.868×10^{-11}	2.090×10^{-11}	2.336×10^{-11}	2.617×10^{-11}	2.922×10^{-11}	3.267×10^{-11}

3. Atomic-Electron Binding Energies

The binding energies given in Table 2 are those reported by Larkins,[1] mainly from the compilations of Sevier[2] for $Z \leqslant 83$, and of Porter and Freedman,[3] for $Z \geqslant 84$. All binding energies listed are for solid systems referenced to the Fermi level, excepting those for Ne, Cl, Ar, Br, Kr, Xe, and Rn. These latter binding energies are for vapor-phase systems referenced to the vacuum level.

The binding energies are accurate to within at least 1 to 2 *eV* for most of the subshells in the lighter elements, and for the outer orbitals in the heavier elements. Uncertainties may be as large as 10 or 20 *eV* for the inner orbitals in the high-*Z* elements, and changes in chemical state can lead to substantial shifts in the binding energies of non-valence shells.[4] Bearden and Burr[5] reevaluated existing data on x-ray emission wavelengths, and discussed binding energies determined from atomic energy-level differences.

---

[1]F.B. Larkins, *At. Data and Nucl. Data Tables* **20**, 313 (1977).

[2]K.D. Sevier, *Low Energy Electron Spectrometry*, John Wiley and Sons, Inc., New York (1972).

[3]F.T. Porter and M.S. Freedman, *J. Phys. Chem. Ref. Data* **7**, 1267 (1978).

[4]D.A. Shirley, R.L. Martin, S.P. Kowalczyk, F.R. McFeely, and L. Ley, *Phys. Rev.* **B15**, 544 (1977).

[5]J.A. Bearden and A.F. Burr, *Rev. Mod. Phys.* **39**, 125 (1967).

Table 2. Atomic-Electron Binding Energies

El	K	L_1	L_2	L_3	M_1	M_2	M_3	M_4	M_5	N_1	N_2	N_3	N_4	N_5
1 H	0.0136													
2 He	0.0246													
3 Li	0.0548	0.0053												
4 Be	0.1121	0.0080												
5 B	0.1880	0.0126	0.0047	0.0047										
6 C	0.2838	0.0180	0.0064	0.0064										
7 N	0.4016	0.0244	0.0092	0.0092										
8 O	0.5320	0.0285	0.0071	0.0071										
9 F	0.6854	0.0340	0.0086	0.0086										
10 Ne	0.8701	0.0485	0.0217	0.0216										
11 Na	1.0721	0.0633	0.0311	0.0311	0.0007									
12 Mg	1.3050	0.0894	0.0514	0.0514	0.0021									
13 Al	1.5596	0.1177	0.0732	0.0727	0.0007	0.0055	0.0055							
14 Si	1.8389	0.1487	0.0995	0.0989	0.0076	0.0030	0.0030							
15 P	2.1455	0.1893	0.1362	0.1353	0.0162	0.0099	0.0099							
16 S	2.4720	0.2292	0.1654	0.1642	0.0158	0.0080	0.0080							
17 Cl	2.8224	0.2702	0.2016	0.2000	0.0175	0.0068	0.0068							
18 Ar	3.2060	0.3263	0.2507	0.2486	0.0292	0.0159	0.0158							
19 K	3.6074	0.3771	0.2963	0.2936	0.0339	0.0178	0.0178							
20 Ca	4.0381	0.4378	0.3500	0.3464	0.0437	0.0254	0.0254							
21 Sc	4.4928	0.5004	0.4067	0.4022	0.0538	0.0323	0.0323	0.0066	0.0066					
22 Ti	4.9664	0.5637	0.4615	0.4555	0.0603	0.0346	0.0346	0.0037	0.0037					
23 V	5.4651	0.6282	0.5205	0.5129	0.0665	0.0378	0.0378	0.0022	0.0022					
24 Cr	5.9892	0.6946	0.5837	0.5745	0.0741	0.0425	0.0425	0.0023	0.0023					
25 Mn	6.5390	0.7690	0.6514	0.6403	0.0839	0.0486	0.0486	0.0033	0.0033					
26 Fe	7.1120	0.8461	0.7211	0.7081	0.0929	0.0540	0.0540	0.0036	0.0036					
27 Co	7.7089	0.9256	0.7936	0.7786	0.1007	0.0595	0.0595	0.0029	0.0029					
28 Ni	8.3328	1.0081	0.8719	0.8547	0.1118	0.0681	0.0681	0.0036	0.0036					
29 Cu	8.9789	1.0961	0.9510	0.9311	0.1198	0.0736	0.0736	0.0016	0.0016					
30 Zn	9.6586	1.1936	1.0428	1.0197	0.1359	0.0866	0.0866	0.0081	0.0081					
31 Ga	10.3671	1.2977	1.1423	1.1154	0.1581	0.1068	0.1029	0.0174	0.0174	0.0015	0.0008	0.0008		
32 Ge	11.1031	1.4143	1.2478	1.2167	0.1800	0.1279	0.1208	0.0287	0.0287	0.0050	0.0023	0.0023		
33 As	11.8667	1.5265	1.3586	1.3231	0.2035	0.1464	0.1405	0.0412	0.0412	0.0085	0.0025	0.0025		
34 Se	12.6578	1.6539	1.4762	1.4358	0.2315	0.1682	0.1619	0.0567	0.0567	0.0120	0.0056	0.0056		
35 Br	13.4737	1.7820	1.5960	1.5499	0.2565	0.1893	0.1815	0.0701	0.0690	0.0273	0.0052	0.0046		
36 Kr	14.3256	1.9210	1.7272	1.6749	0.2921	0.2218	0.2145	0.0950	0.0938	0.0275	0.0147	0.0140		
37 Rb	15.1997	2.0651	1.8639	1.8044	0.3221	0.2474	0.2385	0.1118	0.1103	0.0293	0.0148	0.0140		
38 Sr	16.1046	2.2163	2.0068	1.9396	0.3575	0.2798	0.2691	0.1350	0.1331	0.0377	0.0199	0.0199		
39 Y	17.0384	2.3725	2.1555	2.0800	0.3936	0.3124	0.3003	0.1596	0.1574	0.0454	0.0256	0.0256	0.0024	0.0024
40 Zr	17.9976	2.5316	2.3067	2.2223	0.4303	0.3442	0.3305	0.1824	0.1800	0.0513	0.0287	0.0287	0.0030	0.0030
41 Nb	18.9856	2.6977	2.4647	2.3705	0.4684	0.3784	0.3630	0.2074	0.2046	0.0581	0.0339	0.0339	0.0032	0.0032
42 Mo	19.9995	2.8655	2.6251	2.5202	0.5046	0.4097	0.3923	0.2303	0.2270	0.0618	0.0348	0.0348	0.0018	0.0018
43 Tc	21.0440	3.0425	2.7932	2.6769	0.5440	0.4449	0.4250	0.2564	0.2529	0.0680	0.0389	0.0389	0.0020	0.0020
44 Ru	22.1172	3.2240	2.9669	2.8379	0.5850	0.4828	0.4606	0.2836	0.2794	0.0749	0.0431	0.0431	0.0020	0.0020
45 Rh	23.2199	3.4119	3.1461	3.0038	0.6271	0.5210	0.4962	0.3117	0.3070	0.0810	0.0479	0.0479	0.0025	0.0025
El	K	L_1	L_2	L_3	M_1	M_2	M_3	M_4	M_5	N_1	N_2	N_3	N_4	N_5

Table 2. Atomic-Electron Binding Energies (continued)

El	K	L_1	L_2	L_3	M_1	M_2	M_3	M_4	M_5	N_1	N_2	N_3	N_4	N_5
46 Pd	24.3503	3.6043	3.3303	3.1733	0.6699	0.5591	0.5315	0.3400	0.3347	0.0864	0.0511	0.0511	0.0015	0.0015
47 Ag	25.5140	3.8058	3.5237	3.3511	0.7175	0.6024	0.5714	0.3728	0.3667	0.0952	0.0626	0.0559	0.0033	0.0033
48 Cd	26.7112	4.0180	3.7270	3.5375	0.7702	0.6507	0.6165	0.4105	0.4037	0.1076	0.0669	0.0669	0.0093	0.0093
49 In	27.9399	4.2375	3.9380	3.7301	0.8256	0.7022	0.6643	0.4508	0.4431	0.1219	0.0774	0.0774	0.0162	0.0162
50 Sn	29.2001	4.4647	4.1561	3.9288	0.8838	0.7564	0.7144	0.4933	0.4848	0.1365	0.0886	0.0886	0.0239	0.0239
51 Sb	30.4912	4.6983	4.3804	4.1322	0.9437	0.8119	0.7656	0.5369	0.5275	0.1520	0.0984	0.0984	0.0314	0.0314
52 Te	31.8138	4.9392	4.6120	4.3414	1.0060	0.8697	0.8187	0.5825	0.5721	0.1683	0.1102	0.1102	0.0398	0.0398
53 I	33.1694	5.1881	4.8521	4.5571	1.0721	0.9305	0.8746	0.6313	0.6194	0.1864	0.1227	0.1227	0.0496	0.0496
54 Xe	34.5644	5.4528	5.1037	4.7822	1.1487	1.0021	0.9406	0.6894	0.6767	0.2133	0.1455	0.1455	0.0695	0.0675
55 Cs	35.9846	5.7143	5.3594	5.0119	1.2171	1.0650	0.9976	0.7395	0.7255	0.2308	0.1723	0.1616	0.0788	0.0765
56 Ba	37.4406	5.9888	5.6236	5.2470	1.2928	1.1367	1.0622	0.7961	0.7807	0.2530	0.1918	0.1797	0.0925	0.0899
57 La	38.9246	6.2663	5.8906	5.4827	1.3613	1.2044	1.1234	0.8485	0.8317	0.2704	0.2058	0.1914	0.0989	0.0989
58 Ce	40.4430	6.5488	6.1642	5.7234	1.4346	1.2728	1.1854	0.9013	0.8833	0.2896	0.2233	0.2072	0.1100	0.1100
59 Pr	41.9906	6.8348	6.4404	5.9643	1.5110	1.3374	1.2422	0.9511	0.9310	0.3045	0.2363	0.2176	0.1132	0.1132
60 Nd	43.5689	7.1260	6.7215	6.2079	1.5753	1.4028	1.2974	0.9999	0.9777	0.3152	0.2433	0.2246	0.1175	0.1175
61 Pm	45.1840	7.4279	7.0128	6.4593	1.6500	1.4714	1.3569	1.0515	1.0269	0.3310	0.2420	0.2420	0.1204	0.1204
62 Sm	46.8342	7.7368	7.3118	6.7162	1.7228	1.5407	1.4198	1.1060	1.0802	0.3457	0.2656	0.2474	0.1290	0.1290
63 Eu	48.5190	8.0520	7.6171	6.9769	1.8000	1.6139	1.4806	1.1606	1.1309	0.3602	0.2839	0.2566	0.1332	0.1332
64 Gd	50.2391	8.3756	7.9303	7.2428	1.8808	1.6883	1.5440	1.2172	1.1852	0.3758	0.2885	0.2709	0.1405	0.1405
65 Tb	51.9957	8.7080	8.2516	7.5140	1.9675	1.7677	1.6113	1.2750	1.2412	0.3979	0.3102	0.2850	0.1470	0.1470
66 Dy	53.7885	9.0458	8.5806	7.7901	2.0468	1.8418	1.6756	1.3325	1.2949	0.4163	0.3318	0.2929	0.1542	0.1542
67 Ho	55.6177	9.3942	8.9178	8.0711	2.1283	1.9228	1.7412	1.3915	1.3514	0.4357	0.3435	0.3066	0.1610	0.1610
68 Er	57.4855	9.7513	9.2643	8.3579	2.2065	2.0058	1.8118	1.4533	1.4093	0.4491	0.3662	0.3200	0.1767	0.1676
69 Tm	59.3896	10.1157	9.6169	8.6480	2.3068	2.0898	1.8845	1.5146	1.4677	0.4717	0.3859	0.3366	0.1796	0.1796
70 Yb	61.3323	10.4864	9.9782	8.9436	2.3981	2.1730	1.9498	1.5763	1.5278	0.4872	0.3967	0.3435	0.1981	0.1849
71 Lu	63.3138	10.8704	10.3486	9.2441	2.4912	2.2635	2.0236	1.6394	1.5885	0.5062	0.4101	0.3593	0.2048	0.1950
72 Hf	65.3508	11.2707	10.7394	9.5607	2.6009	2.3654	2.1076	1.7164	1.6617	0.5381	0.4370	0.3804	0.2238	0.2137
73 Ta	67.4164	11.6815	11.1361	9.8811	2.7080	2.4687	2.1940	1.7932	1.7351	0.5655	0.4648	0.4045	0.2413	0.2293
74 W	69.5250	12.0998	11.5440	10.2068	2.8196	2.5749	2.2810	1.8716	1.8092	0.5950	0.4916	0.4253	0.2588	0.2454
75 Re	71.6764	12.5267	11.9587	10.5353	2.9317	2.6816	2.3673	1.9489	1.8829	0.6250	0.5179	0.4444	0.2737	0.2602
76 Os	73.8708	12.9680	12.3850	10.8709	3.0485	2.7922	2.4572	2.0308	1.9601	0.6543	0.5465	0.4682	0.2894	0.2728
77 Ir	76.1110	13.4185	12.8241	11.2152	3.1737	2.9087	2.5507	2.1161	2.0404	0.6901	0.5771	0.4943	0.3114	0.2949
78 Pt	78.3948	13.8805	13.2726	11.5638	3.2976	3.0270	2.6453	2.2015	2.1211	0.7240	0.6076	0.5191	0.3307	0.3138
79 Au	80.7249	14.3528	13.7336	11.9187	3.4249	3.1478	2.7430	2.2911	2.2057	0.7588	0.6437	0.5454	0.3520	0.3339
80 Hg	83.1023	14.8393	14.2087	12.2839	3.5616	3.2785	2.8471	2.3849	2.2949	0.8030	0.6810	0.5769	0.3785	0.3593
81 Tl	85.5304	15.3467	14.6979	12.6575	3.7041	3.4157	2.9566	2.4851	2.3893	0.8455	0.7213	0.6090	0.4066	0.3862
82 Pb	88.0045	15.8608	15.2000	13.0352	3.8507	3.5542	3.0664	2.5856	2.4840	0.8936	0.7639	0.6445	0.4352	0.4129
83 Bi	90.5259	16.3875	15.7111	13.4186	3.9991	3.6963	3.1769	2.6876	2.5796	0.9382	0.8053	0.6789	0.4636	0.4400
84 Po	93.1000	16.9280	16.2370	13.8100	4.1520	3.8440	3.2930	2.7940	2.6800	0.9870	0.8510	0.7150	0.4950	0.4690
85 At	95.7240	17.4820	16.7760	14.2070	4.3100	3.9940	3.4090	2.9010	2.7810	1.0380	0.8970	0.7510	0.5270	0.4990
86 Rn	98.3970	18.0480	17.3280	14.6100	4.4730	4.1500	3.5290	3.0120	2.8840	1.0900	0.9440	0.7900	0.5580	0.5300
87 Fr	101.1300	18.6340	17.8990	15.0250	4.6440	4.3150	3.6560	3.1290	2.9940	1.1480	0.9990	0.8340	0.5970	0.5670
88 Ra	103.9150	19.2320	18.4840	15.4440	4.8220	4.4830	3.7850	3.2480	3.1050	1.2080	1.0550	0.8790	0.6360	0.6030
89 Ac	106.7560	19.8460	19.0810	15.8700	4.9990	4.6550	3.9150	3.3700	3.2190	1.2690	1.1120	0.9240	0.6760	0.6400
90 Th	109.6500	20.4720	19.6930	16.3000	5.1820	4.8310	4.0460	3.4910	3.3320	1.3300	1.1680	0.9670	0.7130	0.6770
91 Pa	112.5960	21.1050	20.3140	16.7330	5.3610	5.0010	4.1740	3.6060	3.4420	1.3830	1.2170	1.0040	0.7430	0.7080
92 U	115.6020	21.7580	20.9480	17.1680	5.5480	5.1810	4.3040	3.7260	3.5500	1.4410	1.2710	1.0430	0.7790	0.7370
93 Np	118.6690	22.4270	21.6000	17.6100	5.7390	5.3660	4.4350	3.8490	3.6640	1.5010	1.3280	1.0850	0.8160	0.7710
94 Pu	121.7910	23.1040	22.2660	18.0570	5.9330	5.5470	4.5630	3.9700	3.7750	1.5590	1.3800	1.1230	0.8460	0.7980
95 Am	124.9820	23.8080	22.9520	18.5100	6.1330	5.7390	4.6980	4.0960	3.8900	1.6200	1.4380	1.1650	0.8800	0.8290
96 Cm	128.2410	24.5260	23.6510	18.9700	6.3370	5.9370	4.8380	4.2240	4.0090	1.6840	1.4980	1.2070	0.9160	0.8620
97 Bk	131.5560	25.2560	24.3710	19.4350	6.5450	6.1380	4.9760	4.3530	4.1270	1.7480	1.5580	1.2490	0.9550	0.8980
98 Cf	134.9390	26.0100	25.1080	19.9070	6.7610	6.3450	5.1160	4.4840	4.2470	1.8130	1.6200	1.2920	0.9910	0.9300
99 Es	138.3960	26.7820	25.8650	20.3840	6.9810	6.5580	5.2590	4.6170	4.3680	1.8830	1.6830	1.3360	1.0290	0.9650
100 Fm	141.9260	27.5740	26.6410	20.8680	7.2080	6.7760	5.4050	4.7520	4.4910	1.9520	1.7490	1.3790	1.0670	1.0000
101 Md	146.5260	28.3870	27.4380	21.3560	7.4400	7.0010	5.5520	4.8890	4.6150	2.0240	1.8160	1.4240	1.1050	1.0340
102 No	149.2080	29.2210	28.2550	21.8510	7.6780	7.2310	5.7020	5.0280	4.7410	2.0970	1.8850	1.4690	1.1450	1.0700
103 Lr	152.9700	30.0830	29.1030	22.3590	7.9300	7.4740	5.8600	5.1760	4.8760	2.1800	1.9630	1.5230	1.1920	1.1120
104	156.2880	30.8810	29.9860	22.9070	8.1610	7.7380	6.0090	5.3360	5.0140	2.2370	2.0350	1.5540	1.2330	1.1490
El	K	L_1	L_2	L_3	M_1	M_2	M_3	M_4	M_5	N_1	N_2	N_3	N_4	N_5

Table 2. Atomic-Electron Binding Energies (continued)

El	N_6	N_7	O_1	O_2	O_3	O_4	O_5	O_6	O_7	P_1	P_2	P_3	P_4	P_5
48 Cd			0.0022	0.0022	0.0022									
49 In			0.0001	0.0008	0.0008									
50 Sn			0.0009	0.0011	0.0011									
51 Sb			0.0067	0.0021	0.0021									
52 Te			0.0116	0.0023	0.0023									
53 I			0.0136	0.0033	0.0033									
54 Xe			0.0234	0.0134	0.0121									
55 Cs			0.0227	0.0131	0.0114									
56 Ba			0.0291	0.0166	0.0146									
57 La			0.0323	0.0144	0.0144									
58 Ce	0.0001	0.0001	0.0378	0.0198	0.0198									
59 Pr	0.0020	0.0020	0.0374	0.0223	0.0223									
60 Nd	0.0015	0.0015	0.0375	0.0211	0.0211									
61 Pm	0.0040	0.0040	0.0380	0.0220	0.0220									
62 Sm	0.0055	0.0055	0.0374	0.0213	0.0213									
63 Eu			0.0318	0.0220	0.0220									
64 Gd	0.0001	0.0001	0.0361	0.0203	0.0203									
65 Tb	0.0026	0.0026	0.0390	0.0254	0.0254									
66 Dy	0.0042	0.0042	0.0629	0.0263	0.0263									
67 Ho	0.0037	0.0037	0.0512	0.0203	0.0203									
68 Er	0.0043	0.0043	0.0598	0.0294	0.0294									
69 Tm	0.0053	0.0053	0.0532	0.0323	0.0323									
70 Yb	0.0063	0.0063	0.0541	0.0234	0.0234									
71 Lu	0.0069	0.0069	0.0568	0.0280	0.0280	0.0046	0.0046							
72 Hf	0.0171	0.0171	0.0649	0.0381	0.0306	0.0066	0.0066							
73 Ta	0.0275	0.0256	0.0711	0.0449	0.0364	0.0057	0.0057							
74 W	0.0379	0.0358	0.0771	0.0468	0.0356	0.0061	0.0061							
75 Re	0.0481	0.0457	0.0828	0.0456	0.0346	0.0035	0.0035							
76 Os	0.0538	0.0510	0.0837	0.0580	0.0454									
77 Ir	0.0640	0.0610	0.0952	0.0630	0.0505	0.0038	0.0038							
78 Pt	0.0745	0.0711	0.1017	0.0653	0.0510	0.0021	0.0021							
79 Au	0.0878	0.0841	0.1078	0.0717	0.0587	0.0025	0.0025							
80 Hg	0.1040	0.0999	0.1203	0.0840	0.0650	0.0098	0.0078							
81 Tl	0.1231	0.1188	0.1363	0.0996	0.0730	0.0153	0.0131							
82 Pb	0.1412	0.1363	0.1473	0.1048	0.0830	0.0218	0.0192			0.0031	0.0007	0.0007		
83 Bi	0.1624	0.1571	0.1593	0.1168	0.0930	0.0265	0.0244			0.0080	0.0027	0.0027		
84 Po	0.1840	0.1780	0.1760	0.1320	0.1020	0.0340	0.0300			0.0090	0.0040	0.0010		
85 At	0.2060	0.1990	0.1920	0.1440	0.1130	0.0410	0.0370			0.0130	0.0060	0.0010		
86 Rn	0.2290	0.2220	0.2080	0.1580	0.1230	0.0480	0.0430			0.0160	0.0080	0.0020		
87 Fr	0.2580	0.2490	0.2290	0.1780	0.1380	0.0600	0.0550			0.0240	0.0140	0.0070		0.0038
88 Ra	0.2870	0.2790	0.2510	0.1970	0.1530	0.0720	0.0660			0.0310	0.0200	0.0120		0.0047
89 Ac	0.3160	0.3070	0.2720	0.2170	0.1680	0.0840	0.0760			0.0370	0.0240	0.0150	0.0044	0.0054
90 Th	0.3440	0.3350	0.2900	0.2360	0.1800	0.0940	0.0870			0.0410	0.0240	0.0170	0.0055	0.0059
91 Pa	0.3660	0.3550	0.3050	0.2450	0.1880	0.0970	0.0900	0.0073		0.0430	0.0270	0.0170	0.0046	0.0056
92 U	0.3890	0.3790	0.3240	0.2570	0.1940	0.1040	0.0950	0.0085		0.0440	0.0270	0.0170	0.0046	0.0057
93 Np	0.4140	0.4030	0.3380	0.2740	0.2060	0.1090	0.1010	0.0097		0.0470	0.0290	0.0180	0.0046	0.0058
94 Pu	0.4360	0.4240	0.3500	0.2830	0.2130	0.1130	0.1020	0.0070		0.0460	0.0290	0.0160		0.0054
95 Am	0.4610	0.4460	0.3650	0.2980	0.2190	0.1160	0.1060	0.0079	0.0066	0.0480	0.0290	0.0160		0.0055
96 Cm	0.4840	0.4700	0.3830	0.3130	0.2290	0.1240	0.1100	0.0129	0.0113	0.0500	0.0300	0.0160	0.0045	0.0061
97 Bk	0.5110	0.4950	0.3990	0.3260	0.2370	0.1300	0.1170	0.0140	0.0122	0.0520	0.0320	0.0160	0.0044	0.0062
98 Cf	0.5380	0.5200	0.4160	0.3410	0.2450	0.1370	0.1220	0.0105	0.0087	0.0540	0.0330	0.0170		0.0057
99 Es	0.5640	0.5460	0.4340	0.3570	0.2550	0.1420	0.1270	0.0113	0.0094	0.0570	0.0350	0.0170		0.0058
100 Fm	0.5910	0.5720	0.4520	0.3730	0.2620	0.1490	0.1330	0.0170	0.0147	0.0590	0.0360	0.0170	0.0042	0.0065
101 Md	0.6180	0.5970	0.4710	0.3890	0.2720	0.1540	0.1370	0.0129	0.0105	0.0610	0.0370	0.0170		0.0059
102 No	0.6450	0.6240	0.4900	0.4060	0.2800	0.1610	0.1420	0.0136	0.0111	0.0630	0.0380	0.0180		0.0060
103 Lr	0.6800	0.6580	0.5160	0.4290	0.2960	0.1740	0.1540	0.0199	0.0170	0.0710	0.0440	0.0210	0.0039	0.0069
104	0.7250	0.7010	0.5350	0.4480	0.3190	0.1900	0.1710	0.0260	0.0228	0.0820	0.0550	0.0330	0.0050	0.0075
El	N_6	N_7	O_1	O_2	O_3	O_4	O_5	O_6	O_7	P_1	P_2	P_3	P_4	P_5

4. Fluorescence and Coster-Kronig Yields

Table 3 lists atomic yields for the K and L shells of elements with $Z=5-110$ from the evaluation of Krause.[1] The yields are for singly ionized atoms, and do not include corrections for solid state, chemical, or multiple ionization effects. These corrections are expected to be small[2] for all except the lighter elements, and small throughout the spectra, excepting at the onsets and cutoffs of Coster-Kronig transitions.

Fluorescence yields (ω_K, ω_{L1},...) represent the probabilities for the filling of vacancies in the corresponding atomic shells by radiative processes, and are used for calculating x-ray and Auger-electron intensities. The intrashell radiative yields f'_{12} and f'_{13} represent the probabilities for x-ray emission per vacancy in the L_1 subshell, resulting in subsequent vacancies in the L_2 and L_3 subshells, respectively. These intrashell radiative yields are included in ω_{L1}; however, because $f'_{12} << f'_{13}$, the yield f'_{12} is not listed separately. The Coster-Kronig yields f_{12}, f_{13}, and f_{23} represent the probabilities for electron emission per vacancy in lower L subshells, resulting in vacancies in higher L subshells. Finally, Auger yields (a) represent the probabilities for electron emission per vacancy in given atomic shells, resulting in vacancies in higher atomic shells. Auger yields are not given explicitly in Table 3, but can be calculated for any shell i from the equation

$$a_i=1-\omega_i-f_i \ ,$$

with $f_1=f_{12}+f_{13}$ and $f_2=f_{23}$.

The atomic yields in Table 3 are based on both experimental and theoretical information. Estimates of their percentage uncertainties, listed for various ranges of atomic numbers, are given in Table 4. These are based on the presumed and stated reliabilities of the input data or calculations, the number or lack of measurements, and the degrees of compatibility of the different relevant data.

---

[1]M.O. Krause, *J. Phys. Chem. Ref. Data* **8**, 307 (1979).

[2]S.T. Manson, J.L. Dehmer, and M. Inokuti, *Bull. Am. Phys. Soc.* **22**, 1332 (1977).

Table 3. Fluorescence and Coster-Kronig Yields

El	ω_K	ω_{L_1}	ω_{L_2}	ω_{L_3}	f_{12}	f_{13}	f_{13}'	f_{23}
5 B	0.0017							
6 C	0.0028							
7 N	0.0052							
8 O	0.0083							
9 F	0.013							
10 Ne	0.018							
11 Na	0.023							
12 Mg	0.030	0.000029	0.0012	0.0012	0.32	0.64	0.000020	
13 Al	0.039	0.000026	0.00075	0.00075	0.32	0.64	0.000016	
14 Si	0.050	0.000030	0.00037	0.00038	0.32	0.64	0.000014	
15 P	0.063	0.000039	0.00031	0.00031	0.32	0.63	0.000012	
16 S	0.078	0.000074	0.00026	0.00026	0.32	0.62	0.000014	
17 Cl	0.097	0.00012	0.00024	0.00024	0.32	0.62	0.000014	
18 Ar	0.118	0.00018	0.00022	0.00022	0.31	0.62	0.000013	
19 K	0.140	0.00024	0.00027	0.00027	0.31	0.62	0.000012	
20 Ca	0.163	0.00031	0.00033	0.00033	0.31	0.61	0.000014	
21 Sc	0.188	0.00039	0.00084	0.00084	0.31	0.60	0.000014	
22 Ti	0.214	0.00047	0.0015	0.0015	0.31	0.59	0.000015	
23 V	0.243	0.00058	0.0026	0.0026	0.31	0.58	0.000016	
24 Cr	0.275	0.00071	0.0037	0.0037	0.31	0.57	0.000018	
25 Mn	0.308	0.00084	0.0050	0.0050	0.30	0.58	0.000019	
26 Fe	0.340	0.0010	0.0063	0.0063	0.30	0.57	0.000021	
27 Co	0.373	0.0012	0.0077	0.0077	0.30	0.56	0.000023	
28 Ni	0.406	0.0014	0.0086	0.0093	0.30	0.55	0.000024	0.028
29 Cu	0.440	0.0016	0.0100	0.011	0.30	0.54	0.000026	0.028
30 Zn	0.474	0.0018	0.011	0.012	0.29	0.54	0.000028	0.026
31 Ga	0.507	0.0021	0.012	0.013	0.29	0.53	0.000030	0.032
32 Ge	0.535	0.0024	0.013	0.015	0.28	0.53	0.000032	0.050
33 As	0.562	0.0028	0.014	0.016	0.28	0.53	0.000034	0.063
34 Se	0.589	0.0032	0.016	0.018	0.28	0.52	0.000036	0.076
35 Br	0.618	0.0036	0.018	0.020	0.28	0.52	0.000038	0.088
36 Kr	0.643	0.0041	0.020	0.022	0.27	0.52	0.000041	0.100
37 Rb	0.667	0.0046	0.022	0.024	0.27	0.52	0.000044	0.109
38 Sr	0.690	0.0051	0.024	0.026	0.27	0.52	0.000047	0.117
39 Y	0.710	0.0059	0.026	0.028	0.26	0.52	0.000052	0.126
40 Zr	0.730	0.0068	0.028	0.031	0.26	0.52	0.000058	0.132
41 Nb	0.747	0.0094	0.031	0.034	0.10	0.61	0.000078	0.137
42 Mo	0.765	0.0100	0.034	0.037	0.10	0.61	0.000081	0.141
43 Tc	0.780	0.011	0.037	0.040	0.10	0.61	0.000088	0.144
44 Ru	0.794	0.012	0.040	0.043	0.10	0.61	0.000096	0.148
45 Rh	0.808	0.013	0.043	0.046	0.10	0.60	0.000100	0.150
46 Pd	0.820	0.014	0.047	0.049	0.10	0.60	0.00011	0.151
47 Ag	0.831	0.016	0.051	0.052	0.10	0.59	0.00012	0.153
48 Cd	0.843	0.018	0.056	0.056	0.10	0.59	0.00014	0.155
49 In	0.853	0.020	0.061	0.060	0.10	0.59	0.00016	0.157
50 Sn	0.862	0.037	0.065	0.064	0.17	0.27	0.00030	0.157
51 Sb	0.870	0.039	0.069	0.069	0.17	0.28	0.00032	0.156
52 Te	0.877	0.041	0.074	0.074	0.18	0.28	0.00034	0.155
53 I	0.884	0.044	0.079	0.079	0.18	0.28	0.00037	0.154
54 Xe	0.891	0.046	0.083	0.085	0.19	0.28	0.00040	0.154
55 Cs	0.897	0.049	0.090	0.091	0.19	0.28	0.00043	0.154
56 Ba	0.902	0.052	0.096	0.097	0.19	0.28	0.00047	0.153
57 La	0.907	0.055	0.103	0.104	0.19	0.29	0.00051	0.153
58 Ce	0.912	0.058	0.110	0.111	0.19	0.29	0.00055	0.153
59 Pr	0.917	0.061	0.117	0.118	0.19	0.29	0.00060	0.153
60 Nd	0.921	0.064	0.124	0.125	0.19	0.30	0.00066	0.152
61 Pm	0.925	0.066	0.132	0.132	0.19	0.30	0.00072	0.151
62 Sm	0.929	0.071	0.140	0.139	0.19	0.30	0.00079	0.150
63 Eu	0.932	0.075	0.149	0.147	0.19	0.30	0.00087	0.149
64 Gd	0.935	0.079	0.158	0.155	0.19	0.30	0.00096	0.147
65 Tb	0.938	0.083	0.167	0.164	0.19	0.30	0.0011	0.145
66 Dy	0.941	0.089	0.178	0.174	0.19	0.30	0.0012	0.143
67 Ho	0.944	0.094	0.189	0.182	0.19	0.30	0.0013	0.142
68 Er	0.947	0.100	0.200	0.192	0.19	0.30	0.0014	0.140
69 Tm	0.949	0.106	0.211	0.201	0.19	0.29	0.0016	0.139
70 Yb	0.951	0.112	0.222	0.210	0.19	0.29	0.0018	0.138
71 Lu	0.953	0.120	0.234	0.220	0.19	0.28	0.0020	0.136
72 Hf	0.955	0.128	0.246	0.231	0.18	0.28	0.0023	0.135
73 Ta	0.957	0.137	0.258	0.243	0.18	0.28	0.0026	0.134
74 W	0.958	0.147	0.270	0.255	0.17	0.28	0.0028	0.133
75 Re	0.959	0.144	0.283	0.268	0.16	0.33	0.0030	0.130
76 Os	0.961	0.130	0.295	0.281	0.16	0.39	0.0029	0.128
77 Ir	0.962	0.120	0.308	0.294	0.15	0.45	0.0028	0.126
78 Pt	0.963	0.114	0.321	0.306	0.14	0.50	0.0028	0.124
79 Au	0.964	0.107	0.334	0.320	0.14	0.53	0.0028	0.122
80 Hg	0.965	0.107	0.347	0.333	0.13	0.56	0.0030	0.120
81 Tl	0.966	0.107	0.360	0.347	0.13	0.57	0.0032	0.118
82 Pb	0.967	0.112	0.373	0.360	0.12	0.58	0.0035	0.116
83 Bi	0.968	0.117	0.387	0.373	0.11	0.58	0.0038	0.113
84 Po	0.968	0.122	0.401	0.386	0.11	0.58	0.0042	0.111
85 At	0.969	0.128	0.415	0.399	0.10	0.59	0.0047	0.111
86 Rn	0.969	0.134	0.429	0.411	0.10	0.58	0.0052	0.110
87 Fr	0.970	0.139	0.443	0.424	0.10	0.58	0.0058	0.109
88 Ra	0.970	0.146	0.456	0.437	0.09	0.58	0.0064	0.108
89 Ac	0.971	0.153	0.468	0.450	0.09	0.58	0.0071	0.108
90 Th	0.971	0.161	0.479	0.463	0.09	0.57	0.0078	0.108
91 Pa	0.972	0.162	0.472	0.476	0.08	0.58	0.0084	0.139
92 U	0.972	0.176	0.467	0.489	0.08	0.57	0.0097	0.167
93 Np	0.973	0.187	0.466	0.502	0.07	0.57	0.011	0.192
94 Pu	0.973	0.205	0.464	0.514	0.05	0.56	0.013	0.198
95 Am	0.974	0.218	0.471	0.526	0.05	0.55	0.014	0.203
96 Cm	0.974	0.228	0.479	0.539	0.04	0.55	0.016	0.200
97 Bk	0.975	0.236	0.485	0.550	0.04	0.54	0.017	0.198
98 Cf	0.975	0.244	0.490	0.560	0.03	0.54	0.019	0.197
99 Es	0.975	0.253	0.497	0.570	0.03	0.54	0.021	0.196
100 Fm	0.976	0.263	0.506	0.579	0.03	0.53	0.023	0.194
101 Md	0.976	0.272	0.515	0.588	0.02	0.53	0.026	0.191
102 No	0.976	0.280	0.524	0.596	0.02	0.52	0.028	0.189
103 Lr	0.977	0.282	0.533	0.604	0.01	0.53	0.030	0.185
104	0.977	0.291	0.544	0.611	0.01	0.52	0.033	0.181
105	0.977	0.300	0.553	0.618	0.01	0.51	0.035	0.178
106	0.978	0.310	0.562	0.624		0.51	0.038	0.174
107	0.978	0.320	0.573	0.630		0.50	0.042	0.171
108	0.978	0.331	0.584	0.635		0.50	0.046	0.165
109	0.978	0.343	0.590	0.640		0.49	0.050	0.163
110	0.979	0.354	0.598	0.644		0.48	0.054	0.158

Table 4. Estimated Percentage Uncertainties for Fluorescence and Coster-Kronig Yields

Z(range)	ω_K	ω_{L_1}	ω_{L_2}	ω_{L_3}	f_{12}	f_{13}	f_{23}
5-10	40-10[a]						
10-20	10-5	>30[a]	>25[a]	>25[a]	10[a]	5[a]	
20-30	5-3	30[a]	25[a,b]	25	15[a]	10[a]	40[a]
30-40	3	30[b]	25	20	15	10	30-20
40-50	2	30-20[b]	25-10	20-10	20[b]	10[b]	20
50-60	2-1	20-15	10	10-5	20	15	20
60-70	1	15	10-5	5	15	10	20-15[a]
70-80	1	15[b]	5	5-3	20[b]	10-5[b]	15
80-90	<1	15	5	3	10	5	15
90-100	<1	15-20	10[b]	3-5	10-50	5-10	15[b]
100-110	1	20	10	5	50-100	15	20

[a] In these regions, yields for molecules and solids may differ from those for atoms by more than the values quoted.

[b] Near breaks in the yield curves, uncertainties may exceed those listed.

5. X-Ray Energies and Intensities

Tables 7a, 7b, 7c, and 7d list energies and intensities for x-rays with intensities greater than 0.001 per 100 primary vacancies in the K, L_1, L_2, and L_3 atomic shells, respectively. The first column shows the Siegbahn notations for the x-ray transitions (the associations with initial and final atomic-shell vacancies are given in Table 6). The following columns give, for each element, the x-ray energies in *keV* (boldface) rounded to the nearest *eV*, and their corresponding intensities directly below. Intensities for the L x-rays are totals from both primary and secondary atomic-shell vacancies.

X-ray energies have been determined from the differences between the corresponding atomic-shell binding energies reported by Larkins.[1] Energies of complex x-ray transitions, e.g., $L_{\beta 2,15}$, are unweighted averages of those for the single-line components.

X-ray intensities have been determined from the experimental relative emission probabilities of Salem, et al.,[2] and the atomic yields of Krause.[3] The theoretical emission probabilities of Scofield[4] were occasionally used whenever experimental values were not available.

The relative intensities of x-rays from the same initial atomic shells are independent of the processes creating the shell vacancies. Tables 7a-7d may therefore be used to separate experimentally unresolved or complex x-ray intensities from the photon tables of the *Table of Radioactive Isotopes.* Table 5 shows the initial atomic shells and their associated x-rays, and the procedure below illustrates the separation of an x-ray peak.

Table 5

Atomic shell	Associated x-rays
K	$K_{\alpha 1}$, $K_{\alpha 2}$, $K_{\alpha 3}$, $K_{\beta 1}$, $K_{\beta 2}$, $K_{\beta 3}$, $K_{\beta 4}$, $K_{\beta 5}$, $KO_{2,3}$, $KP_{2,3}$
L_1	$L_{\beta 3}$, $L_{\beta 4}$, $L_{\gamma 2}$, $L_{\gamma 3}$
L_2	$L_{\beta 1}$, L_{η}, $L_{\gamma 1}$, $L_{\gamma 6}$
L_3	$L_{\alpha 1}$, $L_{\alpha 2}$, $L_{\beta 2,15}$, $L_{\beta 5}$, $L_{\beta 6}$, L_{ℓ}

The single-line x-ray intensity of a specific transition i from an initial atomic shell j is

$$I_{ji} = \frac{I}{I^0} I_{ji}^0 , \qquad (1)$$

where I is the measured (or photon-table) intensity value of a single or complex x-ray transition from atomic-shell j, I^0 is the intensity of the same x-ray transition from Tables 7a-7d, and I_{ji}^0 is the intensity of the specific i x-ray transition from atomic-shell j, also from Tables 7a-7d. As an example, the uranium $K_{\beta 1}$ intensity per 100 disintegrations of ^{235}Np is

$$I_{K\beta 1} = \frac{I_{K_{\alpha 1}}}{I_{K_{\alpha 1}}^0} I_{K_{\beta 1}}^0 = \frac{0.957}{45.1} 10.70 = 0.227\% . \qquad (2)$$

$I_{K_{\alpha 1}}$ is from the photons table for ^{235}Np on page 235-2, and $I_{K_{\alpha 1}}^0$ and $I_{K_{\beta 1}}^0$ are from Table 7a. Calculations for the L_1 atomic shell may be more complex, because none of the x-ray transitions in the photon tables are associated exclusively with this shell.

---

[1]F.B. Larkins, *At. Data and Nucl. Data Tables* **20**, 313, (1977).

[2]S.I. Salem, S.L. Panossian, and R.A. Krause, At. Data and Nucl. Data Tables **14**, *91 (1974).*

[3]M.O. Krause, *J. Phys. Chem. Ref. Data* **8**, 307 (1979).

[4]J.H. Scofield, *At. Data and Nucl. Data Tables* **14**, 121 (1974).

Table 6. Notations for X-ray Transitions

Classical designation (Siegbahn notation)	Associated initial - final shell vacancies
K_{α_1}	K - L_3
K_{α_2}	K - L_2
K_{α_3}	K - L_1
K_{β_1}	K - M_3
K_{β_2}	K - N_2N_3
K_{β_3}	K - M_2
K_{β_4}	K - N_4N_5
K_{β_5}	K - M_4M_5
$KO_{2,3}$	K - O_2O_3
$KP_{2,3}$	K - P_2P_3
L_{α_1}	L_3 - M_5
L_{α_2}	L_3 - M_4
L_{β_1}	L_2 - M_4
$L_{\beta_{2,15}}$	L_3 - N_4N_5
L_{β_3}	L_1 - M_3
L_{β_4}	L_1 - M_2
L_{β_5}	L_3 - O_4O_5
L_{β_6}	L_3 - N_1
L_{γ_1}	L_2 - N_4
L_{γ_2}	L_1 - N_2
L_{γ_3}	L_1 - N_3
L_{γ_6}	L_2 - O_4
L_{η}	L_2 - M_1
L_{ℓ}	L_3 - M_1
Group designation	**Associated transitions**
K_{β_1}'	$K_{\beta_1} + K_{\beta_3} + K_{\beta_5}$
K_{β_2}'	$K_{\beta_2} + K_{\beta_4} + \ldots$
L_{α}	$L_{\alpha_1} + L_{\alpha_2}$
L_{β}	$L_{\beta_1} + L_{\beta_{2,15}} + L_{\beta_3} + L_{\beta_4} + L_{\beta_5} + L_{\beta_6}$
L_{γ}	$L_{\gamma_1} + L_{\gamma_2} + L_{\gamma_3} + L_{\gamma_6}$

Table 7a. X-ray Energies and Intensities (per 100 K-Shell Vacancies)

	$_{5}$B	$_{6}$C	$_{7}$N	$_{8}$O	$_{9}$F	$_{10}$Ne	$_{11}$Na	$_{12}$Mg	$_{13}$Al	$_{14}$Si	$_{15}$P	$_{16}$S	$_{17}$Cl	$_{18}$Ar	$_{19}$K
$K_{\alpha1}$	**0.183** 0.11 *5*	**0.277** 0.19 *8*	**0.392** 0.35 *14*	**0.525** 0.55 *22*	**0.677** 0.9 *4*	**0.849** 1.20 *12*	**1.041** 1.53 *16*	**1.254** 2.0 *2*	**1.487** 2.6 *3*	**1.740** 3.3 *3*	**2.010** 4.1 *4*	**2.308** 5.0 *5*	**2.622** 6.1 *6*	**2.957** 7.3 *7*	**3.314** 8.5 *9*
$K_{\alpha2}$	**0.183** 0.056 *23*	**0.277** 0.09 *4*	**0.392** 0.17 *7*	**0.525** 0.28 *11*	**0.677** 0.43 *17*	**0.848** 0.60 *6*	**1.041** 0.77 *8*	**1.254** 1.00 *10*	**1.486** 1.29 *13*	**1.739** 1.64 *17*	**2.009** 2.04 *21*	**2.307** 2.49 *25*	**2.621** 3.0 *3*	**2.955** 3.6 *4*	**3.311** 4.3 *4*
$K_{\beta1}$									**1.554** 0.0155 *16*	**1.836** 0.056 *6*	**2.136** 0.122 *12*	**2.464** 0.229 *23*	**2.816** 0.38 *4*	**3.190** 0.58 *6*	**3.590** 0.79 *8*
$K_{\beta3}$									**1.554** 0.0079 *8*	**1.836** 0.028 *3*	**2.136** 0.062 *6*	**2.464** 0.116 *12*	**2.816** 0.192 *20*	**3.190** 0.30 *3*	**3.590** 0.40 *4*
$L_{\beta1}$														**0.251** 0.011 *3*	**0.296** 0.013 *4*
$L_{\beta3}$														**0.310** 0.0038 *13*	**0.359** 0.0050 *17*
$L_{\beta4}$														**0.310** 0.0024 *9*	**0.359** 0.0010 *5*

	$_{20}$Ca	$_{21}$Sc	$_{22}$Ti	$_{23}$V	$_{24}$Cr	$_{25}$Mn	$_{26}$Fe	$_{27}$Co	$_{28}$Ni	$_{29}$Cu	$_{30}$Zn	$_{31}$Ga	$_{32}$Ge	$_{33}$As	$_{34}$Se
$K_{\alpha1}$	**3.692** 9.8 *4*	**4.091** 11.3 *5*	**4.511** 12.8 *6*	**4.952** 14.5 *7*	**5.415** 16.4 *7*	**5.899** 18.3 *8*	**6.404** 20.2 *9*	**6.930** 22.1 *10*	**7.478** 24.0 *11*	**8.048** 26.0 *12*	**8.639** 28.0 *10*	**9.252** 29.8 *11*	**9.886** 31.3 *11*	**10.544** 32.7 *12*	**11.222** 34.1 *12*
$K_{\alpha2}$	**3.688** 4.93 *22*	**4.086** 5.68 *25*	**4.505** 6.4 *3*	**4.945** 7.3 *3*	**5.405** 8.3 *4*	**5.888** 9.3 *4*	**6.391** 10.2 *5*	**6.915** 11.2 *5*	**7.461** 12.2 *6*	**8.028** 13.3 *6*	**8.616** 14.3 *5*	**9.225** 15.2 *6*	**9.855** 16.1 *6*	**10.508** 16.8 *6*	**11.182** 17.6 *6*
$K_{\beta1}$	**4.013** 1.02 *5*	**4.461** 1.22 *6*	**4.932** 1.42 *6*	**5.427** 1.64 *7*	**5.947** 1.84 *8*	**6.490** 2.14 *10*	**7.058** 2.40 *11*	**7.649** 2.65 *12*	**8.265** 2.88 *13*	**8.905** 3.10 *14*	**9.572** 3.39 *12*	**10.264** 3.70 *13*	**10.982** 3.98 *14*	**11.726** 4.25 *15*	**12.496** 4.54 *16*
$K_{\beta2}$												**10.366** 0.0314 *11*	**11.101** 0.097 *4*	**11.864** 0.194 *7*	**12.652** 0.323 *12*
$K_{\beta3}$	**4.013** 0.519 *23*	**4.461** 0.62 *3*	**4.932** 0.72 *3*	**5.427** 0.84 *4*	**5.947** 0.94 *4*	**6.490** 1.09 *5*	**7.058** 1.23 *6*	**7.649** 1.36 *6*	**8.265** 1.48 *7*	**8.905** 1.59 *7*	**9.572** 1.74 *6*	**10.260** 1.90 *7*	**10.975** 2.05 *7*	**11.720** 2.19 *8*	**12.490** 2.34 *8*
$K_{\beta5}$							**7.108** 0.00127 *7*	**7.706** 0.00188 *11*	**8.329** 0.00264 *15*	**8.977** 0.00365 *21*	**9.651** 0.00504 *25*	**10.350** 0.0063 *3*	**11.074** 0.0078 *4*	**11.826** 0.0095 *5*	**12.601** 0.0116 *6*
$L_{\alpha1}$		**0.396** 0.026 *7*	**0.452** 0.063 *16*	**0.511** 0.12 *3*	**0.572** 0.19 *5*	**0.637** 0.26 *7*	**0.704** 0.33 *8*	**0.776** 0.41 *10*	**0.851** 0.50 *13*	**0.929** 0.60 *15*	**1.012** 0.65 *13*	**1.098** 0.70 *14*	**1.188** 0.81 *16*	**1.282** 0.87 *17*	**1.379** 0.98 *20*
$L_{\alpha2}$		**0.396** 0.0028 *7*	**0.452** 0.0070 *18*	**0.511** 0.013 *3*	**0.572** 0.021 *5*	**0.637** 0.029 *7*	**0.704** 0.037 *9*	**0.776** 0.045 *11*	**0.851** 0.056 *14*	**0.929** 0.066 *17*	**1.012** 0.072 *15*	**1.098** 0.077 *16*	**1.188** 0.090 *18*	**1.282** 0.096 *19*	**1.379** 0.108 *22*
$L_{\beta1}$	**0.350** 0.016 *4*	**0.400** 0.020 *5*	**0.458** 0.050 *12*	**0.518** 0.096 *24*	**0.581** 0.15 *4*	**0.648** 0.20 *5*	**0.717** 0.25 *6*	**0.791** 0.31 *8*	**0.868** 0.34 *9*	**0.949** 0.39 *10*	**1.035** 0.42 *11*	**1.125** 0.46 *12*	**1.219** 0.49 *12*	**1.317** 0.52 *13*	**1.420** 0.58 *15*
$L_{\beta3}$	**0.412** 0.0062 *19*	**0.468** 0.0075 *23*	**0.529** 0.009 *3*	**0.590** 0.010 *3*	**0.652** 0.012 *4*	**0.720** 0.014 *4*	**0.792** 0.016 *5*	**0.866** 0.018 *5*	**0.940** 0.020 *6*	**1.022** 0.021 *6*	**1.107** 0.023 *7*	**1.195** 0.024 *7*	**1.294** 0.025 *7*	**1.386** 0.027 *8*	**1.492** 0.029 *9*
$L_{\beta4}$	**0.412** 0.0039 *12*	**0.468** 0.0048 *15*	**0.529** 0.0056 *17*	**0.590** 0.0067 *20*	**0.652** 0.0079 *24*	**0.720** 0.009 *3*	**0.792** 0.010 *3*	**0.866** 0.012 *4*	**0.940** 0.013 *4*	**1.022** 0.014 *4*	**1.107** 0.015 *5*	**1.191** 0.016 *5*	**1.286** 0.016 *5*	**1.380** 0.018 *5*	**1.486** 0.019 *6*
$L_{\beta6}$		**0.402** 0.0017 *4*	**0.456** 0.0018 *5*	**0.513** 0.0022 *6*		**0.640** 0.0023 *6*	**0.708** 0.0022 *6*	**0.779** 0.0022 *6*	**0.855** 0.0022 *6*		**1.020** 0.0021 *4*	**1.114** 0.0027 *5*	**1.212** 0.0033 *7*	**1.315** 0.0038 *8*	**1.424** 0.0045 *9*
$L_{\gamma3}$												**1.297** 0.0012 *4*	**1.412** 0.0042 *13*	**1.524** 0.0047 *15*	**1.648** 0.0051 *16*
L_{η}		**0.353** 0.020 *5*	**0.401** 0.022 *6*	**0.454** 0.026 *7*	**0.510** 0.025 *6*	**0.568** 0.026 *7*	**0.628** 0.028 *7*	**0.693** 0.028 *7*	**0.760** 0.026 *7*	**0.831** 0.028 *7*	**0.907** 0.029 *7*	**0.984** 0.030 *8*	**1.068** 0.031 *8*	**1.155** 0.031 *8*	**1.245** 0.034 *9*
L_{ℓ}		**0.348** 0.026 *7*	**0.395** 0.029 *8*	**0.446** 0.034 *9*	**0.500** 0.033 *9*	**0.556** 0.038 *10*	**0.615** 0.040 *11*	**0.678** 0.043 *11*	**0.743** 0.045 *12*	**0.811** 0.048 *13*	**0.884** 0.047 *10*	**0.957** 0.048 *10*	**1.037** 0.052 *11*	**1.120** 0.053 *11*	**1.204** 0.056 *12*

	$_{35}$Br	$_{36}$Kr	$_{37}$Rb	$_{38}$Sr	$_{39}$Y	$_{40}$Zr	$_{41}$Nb	$_{42}$Mo	$_{43}$Tc	$_{44}$Ru	$_{45}$Rh	$_{46}$Pd	$_{47}$Ag	$_{48}$Cd	$_{49}$In
$K_{\alpha1}$	**11.924** 35.6 *13*	**12.651** 36.8 *13*	**13.395** 38.0 *14*	**14.165** 39.1 *14*	**14.958** 40.1 *14*	**15.775** 41.0 *12*	**16.615** 41.8 *12*	**17.479** 42.6 *12*	**18.367** 43.3 *12*	**19.279** 44.0 *12*	**20.216** 44.6 *13*	**21.177** 45.1 *13*	**22.163** 45.6 *13*	**23.174** 46.1 *13*	**24.210** 45.3 *13*
$K_{\alpha2}$	**11.878** 18.4 *7*	**12.598** 19.0 *7*	**13.336** 19.7 *7*	**14.098** 20.3 *7*	**14.883** 20.9 *8*	**15.691** 21.4 *6*	**16.521** 21.9 *6*	**17.374** 22.4 *6*	**18.251** 22.8 *6*	**19.150** 23.2 *7*	**20.074** 23.5 *7*	**21.020** 23.9 *7*	**21.990** 24.2 *7*	**22.984** 24.5 *7*	**24.002** 24.5 *7*
$K_{\alpha3}$													**21.708** 0.00100 *4*	**22.693** 0.00115 *4*	**23.702** 0.00135 *5*
$K_{\beta1}$	**13.292** 4.84 *17*	**14.111** 5.12 *19*	**14.961** 5.39 *19*	**15.836** 5.63 *20*	**16.738** 5.89 *21*	**17.667** 6.15 *17*	**18.623** 6.35 *18*	**19.607** 6.61 *19*	**20.619** 6.80 *19*	**21.657** 6.99 *20*	**22.724** 7.18 *20*	**23.819** 7.35 *21*	**24.943** 7.52 *21*	**26.095** 7.69 *22*	**27.276** 7.85 *22*
$K_{\beta2}$	**13.469** 0.484 *18*	**14.311** 0.676 *24*	**15.185** 0.85 *3*	**16.085** 1.00 *4*	**17.013** 1.13 *4*	**17.969** 1.25 *4*	**18.952** 1.33 *4*	**19.965** 1.45 *4*	**21.005** 1.54 *4*	**22.074** 1.64 *5*	**23.172** 1.72 *5*	**24.299** 1.79 *5*	**25.455** 1.88 *5*	**26.644** 1.98 *6*	**27.863** 2.09 *6*
$K_{\beta3}$	**13.284** 2.50 *9*	**14.104** 2.64 *10*	**14.952** 2.78 *10*	**15.825** 2.91 *10*	**16.726** 3.04 *11*	**17.653** 3.17 *9*	**18.607** 3.28 *9*	**19.590** 3.41 *10*	**20.599** 3.51 *10*	**21.634** 3.61 *10*	**22.699** 3.71 *10*	**23.791** 3.81 *11*	**24.912** 3.90 *11*	**26.060** 3.99 *11*	**27.238** 4.07 *12*
$K_{\beta4}$							**18.982** 0.0010 *5*	**19.998** 0.0015 *7*	**21.042** 0.0023 *11*	**22.115** 0.0032 *16*	**23.217** 0.0043 *21*	**24.349** 0.006 *3*	**25.511** 0.007 *3*	**26.702** 0.008 *4*	**27.924** 0.010 *5*
$K_{\beta5}$	**13.404** 0.0139 *7*	**14.231** 0.0162 *8*	**15.089** 0.0186 *9*	**15.971** 0.0215 *11*	**16.880** 0.0244 *12*	**17.816** 0.0275 *12*	**18.780** 0.0305 *14*	**19.771** 0.0341 *15*	**20.789** 0.0377 *17*	**21.836** 0.0418 *19*	**22.911** 0.0446 *20*	**24.013** 0.0496 *22*	**25.144** 0.0547 *25*	**26.304** 0.060 *3*	**27.493** 0.065 *3*
$KO_{2,3}$															**27.939** 0.0170 *18*
$L_{\alpha1}$	**1.481** 1.09 *22*	**1.581** 1.20 *24*	**1.694** 1.3 *3*	**1.806** 1.4 *3*	**1.923** 1.5 *3*	**2.042** 1.66 *25*	**2.166** 1.8 *3*	**2.293** 1.9 *3*	**2.424** 2.0 *3*	**2.558** 2.1 *3*	**2.697** 2.3 *3*	**2.839** 2.4 *4*	**2.984** 2.5 *4*	**3.134** 2.6 *4*	**3.287** 2.8 *4*

Table 7a. X-ray Energies and Intensities (per 100 K-Shell Vacancies) (continued)

	$_{35}$Br	$_{36}$Kr	$_{37}$Rb	$_{38}$Sr	$_{39}$Y	$_{40}$Zr	$_{41}$Nb	$_{42}$Mo	$_{43}$Tc	$_{44}$Ru	$_{45}$Rh	$_{46}$Pd	$_{47}$Ag	$_{48}$Cd	$_{49}$In
continued															
$L_{\alpha2}$	**1.480** 0.121 *24*	**1.580** 0.13 *3*	**1.693** 0.15 *3*	**1.805** 0.16 *3*	**1.920** 0.17 *3*	**2.040** 0.19 *3*	**2.163** 0.20 *3*	**2.290** 0.21 *3*	**2.420** 0.23 *3*	**2.554** 0.24 *4*	**2.692** 0.25 *4*	**2.833** 0.26 *4*	**2.978** 0.28 *4*	**3.127** 0.29 *5*	**3.279** 0.31 *5*
$L_{\beta1}$	**1.526** 0.64 *16*	**1.632** 0.69 *17*	**1.752** 0.75 *19*	**1.872** 0.81 *20*	**1.996** 0.86 *21*	**2.124** 0.90 *14*	**2.257** 0.93 *14*	**2.395** 1.00 *15*	**2.537** 1.07 *16*	**2.683** 1.14 *17*	**2.834** 1.20 *18*	**2.990** 1.28 *19*	**3.151** 1.39 *21*	**3.317** 1.51 *23*	**3.487** 1.63 *25*
$L_{\beta2,15}$					**2.078** 0.0044 *9*	**2.219** 0.0116 *18*	**2.367** 0.056 *9*	**2.518** 0.100 *15*	**2.675** 0.150 *23*	**2.836** 0.20 *3*	**3.001** 0.24 *4*	**3.172** 0.28 *4*	**3.348** 0.32 *5*	**3.528** 0.38 *6*	**3.714** 0.43 *7*
$L_{\beta3}$	**1.601** 0.029 *9*	**1.707** 0.030 *9*	**1.827** 0.031 *9*	**1.947** 0.032 *10*	**2.072** 0.035 *11*	**2.201** 0.038 *10*	**2.335** 0.049 *12*	**2.473** 0.048 *12*	**2.617** 0.050 *13*	**2.763** 0.052 *13*	**2.916** 0.053 *13*	**3.073** 0.054 *14*	**3.234** 0.059 *15*	**3.402** 0.062 *16*	**3.573** 0.065 *16*
$L_{\beta4}$	**1.593** 0.020 *6*	**1.699** 0.020 *6*	**1.818** 0.022 *7*	**1.936** 0.022 *7*	**2.060** 0.025 *7*	**2.187** 0.026 *7*	**2.319** 0.034 *9*	**2.456** 0.034 *9*	**2.598** 0.035 *9*	**2.741** 0.035 *9*	**2.891** 0.035 *9*	**3.045** 0.036 *9*	**3.203** 0.038 *10*	**3.367** 0.039 *10*	**3.535** 0.041 *10*
$L_{\beta6}$	**1.523** 0.0052 *11*	**1.647** 0.0060 *12*	**1.775** 0.0069 *14*	**1.902** 0.0079 *16*	**2.035** 0.0088 *18*	**2.171** 0.0100 *15*	**2.312** 0.0110 *17*	**2.458** 0.0119 *18*	**2.609** 0.0128 *20*	**2.763** 0.0139 *21*	**2.923** 0.0149 *23*	**3.087** 0.0157 *24*	**3.256** 0.0166 *25*	**3.430** 0.018 *3*	**3.608** 0.019 *3*
$L_{\gamma1}$					**2.153** 0.012 *3*	**2.304** 0.030 *5*	**2.462** 0.041 *6*	**2.623** 0.055 *8*	**2.791** 0.068 *10*	**2.965** 0.083 *13*	**3.144** 0.109 *17*	**3.329** 0.137 *21*	**3.520** 0.147 *23*	**3.718** 0.161 *25*	**3.922** 0.18 *3*
$L_{\gamma2}$	**1.777** 0.0013 *4*	**1.906** 0.0018 *6*	**2.050** 0.0022 *7*	**2.196** 0.0027 *8*	**2.347** 0.0031 *9*	**2.503** 0.0036 *9*	**2.664** 0.0050 *13*	**2.831** 0.0051 *13*	**3.004** 0.0055 *14*	**3.181** 0.0059 *15*	**3.364** 0.0062 *16*	**3.553** 0.0065 *17*	**3.743** 0.0073 *19*	**3.951** 0.0080 *20*	**4.160** 0.0087 *22*
$L_{\gamma3}$	**1.777** 0.0052 *16*	**1.907** 0.0054 *17*	**2.051** 0.0058 *18*	**2.196** 0.0061 *19*	**2.347** 0.0066 *20*	**2.503** 0.0072 *19*	**2.664** 0.0095 *24*	**2.831** 0.0095 *24*	**3.004** 0.010 *3*	**3.181** 0.011 *3*	**3.364** 0.011 *3*	**3.553** 0.011 *3*	**3.750** 0.012 *3*	**3.951** 0.013 *3*	**4.160** 0.014 *4*
L_{η}	**1.339** 0.036 *9*	**1.435** 0.037 *9*	**1.542** 0.039 *10*	**1.649** 0.04 *1*	**1.762** 0.041 *10*	**1.876** 0.041 *6*	**1.996** 0.041 *6*	**2.120** 0.043 *7*	**2.249** 0.044 *7*	**2.382** 0.046 *7*	**2.519** 0.046 *7*	**2.660** 0.048 *7*	**2.806** 0.051 *8*	**2.957** 0.054 *8*	**3.112** 0.056 *9*
L_{ℓ}	**1.293** 0.060 *13*	**1.383** 0.063 *14*	**1.482** 0.067 *14*	**1.582** 0.069 *15*	**1.686** 0.073 *16*	**1.792** 0.078 *13*	**1.902** 0.082 *14*	**2.016** 0.085 *15*	**2.133** 0.089 *15*	**2.253** 0.092 *16*	**2.377** 0.095 *16*	**2.503** 0.097 *17*	**2.634** 0.101 *17*	**2.767** 0.107 *18*	**2.905** 0.112 *19*

	$_{50}$Sn	$_{51}$Sb	$_{52}$Te	$_{53}$I	$_{54}$Xe	$_{55}$Cs	$_{56}$Ba	$_{57}$La	$_{58}$Ce	$_{59}$Pr	$_{60}$Nd	$_{61}$Pm	$_{62}$Sm	$_{63}$Eu	$_{64}$Gd
$K_{\alpha1}$	**25.271** 45.7 *10*	**26.359** 46.0 *10*	**27.472** 46.2 *11*	**28.612** 46.4 *11*	**29.782** 46.6 *11*	**30.973** 46.7 *11*	**32.194** 46.7 *11*	**33.442** 46.8 *11*	**34.720** 47.0 *11*	**36.026** 47.1 *11*	**37.361** 47.2 *10*	**38.725** 47.3 *10*	**40.118** 47.5 *10*	**41.542** 47.6 *10*	**42.996** 47.5 *10*
$K_{\alpha2}$	**25.044** 24.7 *6*	**26.111** 24.9 *6*	**27.202** 25.0 *6*	**28.317** 25.2 *6*	**29.461** 25.3 *6*	**30.625** 25.5 *6*	**31.817** 25.6 *6*	**33.034** 25.7 *6*	**34.279** 25.9 *6*	**35.550** 26.1 *6*	**36.847** 26.2 *6*	**38.171** 26.3 *6*	**39.522** 26.4 *6*	**40.902** 26.6 *6*	**42.309** 26.6 *6*
$K_{\alpha3}$	**24.735** 0.00154 *5*	**25.793** 0.00179 *6*	**26.875** 0.00203 *6*	**27.981** 0.00227 *7*	**29.112** 0.00262 *8*	**30.270** 0.00296 *9*	**31.452** 0.00334 *11*	**32.658** 0.00373 *12*	**33.894** 0.00422 *13*	**35.156** 0.00472 *15*	**36.443** 0.00531 *16*	**37.756** 0.00580 *18*	**39.097** 0.00678 *21*	**40.467** 0.00727 *22*	**41.864** 0.00824 *25*
$K_{\beta1}$	**28.486** 7.99 *18*	**29.726** 8.09 *18*	**30.995** 8.21 *18*	**32.295** 8.34 *19*	**33.624** 8.42 *19*	**34.987** 8.53 *19*	**36.378** 8.63 *19*	**37.801** 8.70 *19*	**39.258** 8.83 *20*	**40.748** 8.9 *2*	**42.272** 8.97 *19*	**43.827** 9.08 *19*	**45.414** 9.15 *19*	**47.038** 9.21 *19*	**48.695** 9.30 *19*
$K_{\beta2}$	**29.111** 2.19 *5*	**30.393** 2.28 *5*	**31.704** 2.37 *5*	**33.047** 2.47 *6*	**34.419** 2.55 *6*	**35.818** 2.64 *6*	**37.255** 2.73 *6*	**38.726** 2.81 *6*	**40.228** 2.84 *6*	**41.764** 2.88 *7*	**43.335** 2.93 *6*	**44.942** 2.98 *6*	**46.578** 3.02 *6*	**48.249** 3.05 *6*	**49.959** 3.11 *6*
$K_{\beta3}$	**28.444** 4.15 *9*	**29.679** 4.20 *9*	**30.944** 4.26 *10*	**32.239** 4.32 *10*	**33.562** 4.36 *10*	**34.920** 4.42 *10*	**36.304** 4.47 *10*	**37.720** 4.51 *10*	**39.170** 4.57 *10*	**40.653** 4.61 *10*	**42.166** 4.65 *10*	**43.713** 4.69 *10*	**45.293** 4.73 *10*	**46.905** 4.76 *10*	**48.551** 4.81 *10*
$K_{\beta4}$	**29.176** 0.012 *6*	**30.460** 0.013 *6*	**31.774** 0.015 *7*	**33.120** 0.017 *8*	**34.496** 0.019 *9*	**35.907** 0.021 *10*	**37.349** 0.023 *11*	**38.826** 0.025 *12*	**40.333** 0.027 *13*	**41.877** 0.028 *14*	**43.451** 0.030 *15*	**45.064** 0.032 *16*	**46.705** 0.034 *17*	**48.386** 0.036 *18*	**50.099** 0.038 *19*
$K_{\beta5}$	**28.711** 0.070 *3*	**29.959** 0.071 *3*	**31.236** 0.075 *3*	**32.544** 0.081 *3*	**33.881** 0.086 *4*	**35.252** 0.091 *4*	**36.652** 0.100 *4*	**38.085** 0.105 *4*	**39.551** 0.110 *5*	**41.050** 0.116 *5*	**42.580** 0.121 *5*	**44.145** 0.130 *5*	**45.741** 0.136 *6*	**47.373** 0.141 *6*	**49.038** 0.146 *6*
$KO_{2,3}$	**29.199** 0.049 *5*	**30.489** 0.092 *10*	**31.812** 0.147 *15*	**33.166** 0.212 *22*	**34.552** 0.29 *3*	**35.972** 0.35 *4*	**37.425** 0.40 *4*	**38.910** 0.45 *5*	**40.423** 0.42 *4*	**41.968** 0.42 *4*	**43.548** 0.42 *4*	**45.162** 0.42 *4*	**46.813** 0.42 *4*	**48.497** 0.42 *4*	**50.219** 0.45 *5*
$L_{\alpha1}$	**3.444** 2.9 *3*	**3.605** 3.0 *3*	**3.769** 3.2 *3*	**3.938** 3.4 *4*	**4.106** 3.6 *4*	**4.286** 3.8 *4*	**4.466** 4.1 *4*	**4.651** 4.3 *5*	**4.840** 4.6 *5*	**5.033** 4.9 *5*	**5.230** 5.2 *3*	**5.432** 5.4 *3*	**5.636** 5.7 *3*	**5.846** 6.0 *3*	**6.058** 6.3 *4*
$L_{\alpha2}$	**3.435** 0.32 *3*	**3.595** 0.34 *4*	**3.759** 0.36 *4*	**3.926** 0.38 *4*	**4.093** 0.40 *4*	**4.272** 0.43 *4*	**4.451** 0.45 *5*	**4.634** 0.48 *5*	**4.822** 0.51 *5*	**5.013** 0.54 *6*	**5.208** 0.57 *3*	**5.408** 0.60 *3*	**5.610** 0.63 *4*	**5.816** 0.67 *4*	**6.026** 0.70 *4*
$L_{\beta1}$	**3.663** 1.75 *18*	**3.843** 1.84 *19*	**4.029** 1.96 *20*	**4.221** 2.07 *21*	**4.414** 2.16 *22*	**4.620** 2.32 *24*	**4.828** 2.47 *25*	**5.042** 2.6 *3*	**5.263** 2.8 *3*	**5.489** 3.0 *3*	**5.722** 3.12 *23*	**5.961** 3.31 *24*	**6.206** 3.5 *3*	**6.457** 3.7 *3*	**6.713** 3.9 *3*
$L_{\beta2,15}$	**3.905** 0.46 *5*	**4.101** 0.52 *6*	**4.302** 0.58 *6*	**4.508** 0.64 *7*	**4.714** 0.71 *7*	**4.934** 0.77 *8*	**5.156** 0.84 *9*	**5.384** 0.91 *10*	**5.613** 0.97 *10*	**5.851** 1.03 *11*	**6.090** 1.10 *7*	**6.339** 1.15 *7*	**6.587** 1.21 *7*	**6.844** 1.27 *8*	**7.102** 1.32 *8*
$L_{\beta3}$	**3.750** 0.113 *23*	**3.933** 0.114 *23*	**4.121** 0.114 *23*	**4.314** 0.116 *23*	**4.512** 0.115 *23*	**4.717** 0.116 *23*	**4.927** 0.118 *24*	**5.143** 0.120 *24*	**5.363** 0.120 *24*	**5.593** 0.120 *24*	**5.829** 0.121 *18*	**6.071** 0.119 *18*	**6.317** 0.122 *18*	**6.571** 0.125 *19*	**6.832** 0.126 *19*
$L_{\beta4}$	**3.708** 0.071 *14*	**3.886** 0.070 *14*	**4.070** 0.069 *14*	**4.258** 0.070 *14*	**4.451** 0.069 *14*	**4.649** 0.069 *14*	**4.852** 0.070 *14*	**5.062** 0.071 *14*	**5.276** 0.071 *14*	**5.497** 0.071 *14*	**5.723** 0.072 *11*	**5.956** 0.071 *11*	**6.196** 0.073 *11*	**6.438** 0.075 *11*	**6.687** 0.077 *12*
$L_{\beta5}$								**5.483** 0.0091 *9*							**7.243** 0.0108 *6*
$L_{\beta6}$	**3.792** 0.0205 *21*	**3.980** 0.0225 *23*	**4.173** 0.0245 *25*	**4.371** 0.026 *3*	**4.569** 0.029 *3*	**4.781** 0.031 *3*	**4.994** 0.034 *4*	**5.212** 0.036 *4*	**5.434** 0.039 *4*	**5.660** 0.042 *4*	**5.893** 0.0454 *25*	**6.128** 0.049 *3*	**6.370** 0.053 *3*	**6.617** 0.058 *3*	**6.867** 0.063 *4*
$L_{\gamma1}$	**4.132** 0.206 *22*	**4.349** 0.226 *24*	**4.572** 0.25 *3*	**4.802** 0.28 *3*	**5.034** 0.30 *3*	**5.281** 0.33 *4*	**5.531** 0.36 *4*	**5.792** 0.39 *4*	**6.054** 0.43 *5*	**6.327** 0.46 *5*	**6.604** 0.50 *4*	**6.892** 0.54 *4*	**7.183** 0.58 *4*	**7.484** 0.62 *5*	**7.790** 0.67 *5*
$L_{\gamma2}$	**4.376** 0.016 *3*	**4.600** 0.016 *3*	**4.829** 0.017 *4*	**5.065** 0.018 *4*	**5.307** 0.018 *4*	**5.542** 0.019 *4*	**5.797** 0.020 *4*	**6.060** 0.020 *4*	**6.326** 0.021 *4*	**6.599** 0.021 *4*	**6.883** 0.022 *4*	**7.186** 0.022 *4*	**7.471** 0.023 *4*	**7.768** 0.023 *4*	**8.087** 0.024 *4*
$L_{\gamma3}$	**4.376** 0.025 *5*	**4.600** 0.025 *5*	**4.829** 0.026 *5*	**5.065** 0.027 *6*	**5.307** 0.027 *6*	**5.553** 0.028 *6*	**5.809** 0.028 *6*	**6.075** 0.029 *6*	**6.342** 0.030 *6*	**6.617** 0.030 *6*	**6.901** 0.031 *5*	**7.186** 0.031 *5*	**7.489** 0.032 *5*	**7.795** 0.033 *5*	**8.105** 0.034 *6*
$L_{\gamma6}$								**5.891** 0.0042 *5*							**7.930** 0.0055 *6*
L_{η}	**3.272** 0.059 *6*	**3.437** 0.060 *6*	**3.606** 0.063 *6*	**3.780** 0.064 *7*	**3.955** 0.065 *7*	**4.142** 0.068 *7*	**4.331** 0.070 *7*	**4.529** 0.073 *7*	**4.730** 0.075 *8*	**4.929** 0.078 *8*	**5.146** 0.081 *6*	**5.363** 0.083 *6*	**5.589** 0.086 *6*	**5.817** 0.089 *7*	**6.049** 0.092 *7*
L_{ℓ}	**3.045** 0.114 *15*	**3.189** 0.122 *16*	**3.335** 0.129 *17*	**3.485** 0.136 *18*	**3.634** 0.145 *19*	**3.795** 0.155 *20*	**3.954** 0.163 *21*	**4.121** 0.176 *23*	**4.289** 0.189 *24*	**4.453** 0.20 *3*	**4.633** 0.213 *20*	**4.809** 0.225 *22*	**4.993** 0.237 *23*	**5.177** 0.252 *24*	**5.362** 0.266 *25*

Table 7a. X-ray Energies and Intensities (per 100 K-Shell Vacancies) (continued)

	$_{65}$Tb	$_{66}$Dy	$_{67}$Ho	$_{68}$Er	$_{69}$Tm	$_{70}$Yb	$_{71}$Lu	$_{72}$Hf	$_{73}$Ta	$_{74}$W	$_{75}$Re	$_{76}$Os	$_{77}$Ir	$_{78}$Pt	$_{79}$Au
$K_{\alpha1}$	**44.482** 47.5 10	**45.998** 47.5 10	**47.547** 47.5 10	**49.128** 47.5 10	**50.742** 47.4 10	**52.389** 47.4 10	**54.070** 47.3 10	**55.790** 47.1 10	**57.535** 47.2 10	**59.318** 47.0 10	**61.141** 46.9 10	**63.000** 46.7 10	**64.896** 46.7 10	**66.831** 46.5 10	**68.806** 46.4 10
$K_{\alpha2}$	**43.744** 26.7 6	**45.208** 26.8 6	**46.700** 26.9 6	**48.221** 27.0 6	**49.773** 27.2 6	**51.354** 27.2 6	**52.965** 27.3 6	**54.611** 27.3 6	**56.280** 27.3 6	**57.981** 27.4 6	**59.718** 27.4 6	**61.486** 27.4 6	**63.287** 27.4 6	**65.122** 27.4 6	**66.991** 27.5 6
$K_{\alpha3}$	**43.288** 0.0092 3	**44.743** 0.0102 3	**46.224** 0.0111 3	**47.734** 0.0126 4	**49.274** 0.0135 4	**50.846** 0.0145 4	**52.443** 0.0159 5	**54.080** 0.0173 5	**55.735** 0.0192 6	**57.425** 0.0206 6	**59.150** 0.0224 7	**60.903** 0.0242 7	**62.693** 0.0261 8	**64.514** 0.0298 9	**66.372** 0.0326 10
$K_{\beta1}$	**50.384** 9.44 19	**52.113** 9.58 20	**53.877** 9.68 20	**55.674** 9.77 20	**57.505** 9.86 20	**59.383** 9.99 20	**61.290** 10.1 2	**63.243** 10.20 21	**65.222** 10.30 21	**67.244** 10.30 21	**69.309** 10.40 21	**71.414** 10.60 21	**73.560** 10.60 22	**75.749** 10.70 22	**77.982** 10.70 22
$K_{\beta2}$	**51.698** 3.15 7	**53.476** 3.20 7	**55.293** 3.24 7	**57.142** 3.28 7	**59.028** 3.32 7	**60.962** 3.38 7	**62.929** 3.42 7	**64.942** 3.48 7	**66.982** 3.53 7	**69.067** 3.58 7	**71.195** 3.63 7	**73.363** 3.71 8	**75.575** 3.75 8	**77.831** 3.81 8	**80.130** 3.84 8
$K_{\beta3}$	**50.228** 4.88 10	**51.947** 4.95 10	**53.695** 5.0 1	**55.480** 5.06 10	**57.300** 5.11 10	**59.159** 5.18 10	**61.050** 5.21 10	**62.985** 5.28 11	**64.948** 5.32 11	**66.950** 5.35 11	**68.995** 5.42 11	**71.079** 5.48 11	**73.202** 5.52 11	**75.368** 5.56 11	**77.577** 5.57 11
$K_{\beta4}$	**51.849** 0.040 20	**53.634** 0.042 21	**55.457** 0.045 22	**57.313** 0.047 23	**59.210** 0.049 24	**61.141** 0.051 25	**63.114** 0.05 3	**65.132** 0.06 3	**67.181** 0.06 3	**69.273** 0.06 3	**71.409** 0.07 3	**73.590** 0.07 3	**75.808** 0.07 3	**78.073** 0.08 4	**80.382** 0.08 4
$K_{\beta5}$	**50.738** 0.156 6	**52.475** 0.166 7	**54.246** 0.176 7	**56.054** 0.186 8	**57.898** 0.195 8	**59.780** 0.204 8	**61.700** 0.213 9	**63.662** 0.222 9	**65.652** 0.232 9	**67.685** 0.241 10	**69.760** 0.25 1	**71.875** 0.259 10	**74.033** 0.268 11	**76.233** 0.276 11	**78.476** 0.285 11
$KO_{2,3}$	**51.970** 0.42 4	**53.762** 0.42 4	**55.597** 0.42 4	**57.456** 0.42 4	**59.357** 0.42 4	**61.309** 0.41 4	**63.286** 0.44 5	**65.316** 0.46 5	**67.376** 0.49 5	**69.484** 0.51 5	**71.636** 0.54 6	**73.819** 0.56 6	**76.054** 0.57 6	**78.337** 0.60 6	**80.660** 0.62 6
$L_{\alpha1}$	**6.273** 6.7 4	**6.495** 7.1 4	**6.720** 7.4 4	**6.949** 7.7 4	**7.180** 8.1 5	**7.416** 8.3 4	**7.656** 8.6 4	**7.899** 8.9 4	**8.146** 9.3 4	**8.398** 9.7 4	**8.652** 10.1 5	**8.911** 10.4 5	**9.175** 10.9 5	**9.443** 11.2 5	**9.713** 11.6 5
$L_{\alpha2}$	**6.239** 0.74 4	**6.458** 0.78 4	**6.680** 0.82 5	**6.905** 0.86 5	**7.133** 0.90 5	**7.367** 0.93 4	**7.605** 0.97 5	**7.844** 1.00 5	**8.088** 1.04 5	**8.335** 1.08 5	**8.586** 1.13 5	**8.840** 1.17 5	**9.099** 1.22 6	**9.362** 1.26 6	**9.628** 1.30 6
$L_{\beta1}$	**6.977** 4.1 3	**7.248** 4.4 3	**7.526** 4.7 3	**7.811** 4.9 4	**8.102** 5.2 4	**8.402** 5.4 3	**8.709** 5.7 3	**9.023** 6.0 3	**9.343** 6.2 3	**9.672** 6.5 4	**10.010** 6.7 4	**10.354** 7.0 4	**10.708** 7.2 4	**11.071** 7.5 4	**11.443** 7.8 4
$L_{\beta2,15}$	**7.367** 1.38 8	**7.636** 1.45 9	**7.910** 1.49 9	**8.186** 1.55 9	**8.468** 1.59 10	**8.752** 1.62 8	**9.044** 1.75 9	**9.342** 1.90 10	**9.646** 2.05 10	**9.955** 2.19 11	**10.268** 2.31 12	**10.590** 2.44 13	**10.912** 2.57 13	**11.242** 2.69 14	**11.576** 2.82 14
$L_{\beta3}$	**7.097** 0.127 19	**7.370** 0.131 20	**7.653** 0.132 20	**7.940** 0.133 20	**8.231** 0.137 21	**8.537** 0.139 21	**8.847** 0.144 22	**9.163** 0.147 22	**9.488** 0.151 23	**9.819** 0.159 24	**10.159** 0.152 23	**10.511** 0.131 20	**10.868** 0.118 18	**11.235** 0.109 16	**11.610** 0.099 15
$L_{\beta4}$	**6.940** 0.078 12	**7.204** 0.081 12	**7.471** 0.083 12	**7.746** 0.085 13	**8.026** 0.088 13	**8.313** 0.091 14	**8.607** 0.096 15	**8.905** 0.100 15	**9.213** 0.104 16	**9.525** 0.112 17	**9.845** 0.109 16	**10.176** 0.096 15	**10.510** 0.088 13	**10.854** 0.083 13	**11.205** 0.078 12
$L_{\beta5}$							**9.240** 0.0103 5	**9.554** 0.0268 12	**9.875** 0.0372 17	**10.201** 0.0483 22	**10.532** 0.091 4	**10.871** 0.138 6	**11.211** 0.179 8	**11.562** 0.222 10	**11.916** 0.268 12
$L_{\beta6}$	**7.116** 0.068 4	**7.374** 0.074 4	**7.635** 0.079 4	**7.909** 0.087 5	**8.176** 0.092 5	**8.456** 0.098 5	**8.738** 0.103 5	**9.023** 0.108 5	**9.316** 0.114 5	**9.612** 0.121 6	**9.910** 0.132 6	**10.217** 0.143 7	**10.525** 0.152 7	**10.840** 0.160 7	**11.160** 0.170 8
$L_{\gamma1}$	**8.105** 0.71 5	**8.426** 0.76 6	**8.757** 0.82 6	**9.088** 0.88 7	**9.437** 0.93 7	**9.780** 0.99 6	**10.144** 1.04 6	**10.516** 1.10 6	**10.895** 1.16 7	**11.285** 1.22 7	**11.685** 1.29 8	**12.096** 1.35 8	**12.513** 1.41 8	**12.942** 1.47 9	**13.382** 1.56 9
$L_{\gamma2}$	**8.398** 0.025 4	**8.714** 0.025 4	**9.051** 0.026 4	**9.385** 0.026 4	**9.730** 0.028 5	**10.090** 0.029 5	**10.460** 0.030 5	**10.834** 0.031 5	**11.217** 0.033 5	**11.608** 0.035 6	**12.009** 0.034 6	**12.421** 0.030 5	**12.841** 0.028 5	**13.273** 0.027 4	**13.709** 0.025 4
$L_{\gamma3}$	**8.423** 0.035 6	**8.753** 0.037 6	**9.088** 0.038 6	**9.431** 0.039 6	**9.779** 0.040 7	**10.143** 0.041 7	**10.511** 0.044 7	**10.890** 0.045 7	**11.277** 0.047 8	**11.675** 0.050 8	**12.082** 0.049 8	**12.500** 0.043 7	**12.924** 0.039 6	**13.361** 0.037 6	**13.807** 0.034 6
$L_{\gamma6}$							**10.344** 0.0063 6	**10.733** 0.0167 16	**11.130** 0.030 3	**11.538** 0.047 4	**11.955** 0.081 8	**12.385** 0.112 11	**12.820** 0.145 14	**13.270** 0.180 17	**13.731** 0.209 20
L_{η}	**6.284** 0.095 7	**6.534** 0.099 7	**6.789** 0.103 8	**7.058** 0.106 8	**7.310** 0.110 8	**7.580** 0.114 6	**7.857** 0.119 6	**8.139** 0.124 7	**8.428** 0.130 7	**8.724** 0.136 7	**9.027** 0.142 8	**9.337** 0.148 8	**9.650** 0.155 8	**9.975** 0.163 9	**10.309** 0.172 9
L_{ℓ}	**5.546** 0.28 3	**5.743** 0.30 3	**5.943** 0.32 3	**6.151** 0.34 3	**6.341** 0.36 3	**6.545** 0.37 3	**6.753** 0.39 4	**6.960** 0.41 4	**7.173** 0.43 4	**7.387** 0.46 4	**7.604** 0.49 4	**7.822** 0.52 5	**8.042** 0.55 5	**8.266** 0.58 5	**8.494** 0.61 6

	$_{80}$Hg	$_{81}$Tl	$_{82}$Pb	$_{83}$Bi	$_{84}$Po	$_{85}$At	$_{86}$Rn	$_{87}$Fr	$_{88}$Ra	$_{89}$Ac	$_{90}$Th	$_{91}$Pa	$_{92}$U	$_{93}$Np	$_{94}$Pu
$K_{\alpha1}$	**70.818** 46.3 9	**72.873** 46.3 9	**74.969** 46.2 9	**77.107** 46.2 9	**79.290** 46.1 9	**81.517** 46.1 9	**83.787** 46.0 9	**86.105** 45.8 9	**88.471** 45.7 9	**90.886** 45.5 9	**93.350** 45.4 9	**95.863** 45.3 9	**98.434** 45.1 9	**101.059** 45.1 9	**103.734** 45.1 9
$K_{\alpha2}$	**68.894** 27.5 6	**70.832** 27.6 6	**72.805** 27.7 6	**74.815** 27.7 6	**76.863** 27.7 6	**78.948** 27.9 6	**81.069** 27.9 6	**83.231** 27.9 6	**85.431** 28.0 6	**87.675** 28.1 6	**89.957** 28.1 6	**92.282** 28.1 6	**94.654** 28.2 6	**97.069** 28.3 6	**99.525** 28.4 6
$K_{\alpha3}$	**68.263** 0.0358 11	**70.184** 0.0395 12	**72.144** 0.0428 13	**74.138** 0.0474 14	**76.172** 0.196 6	**78.242** 0.0571 17	**80.349** 0.0616 19	**82.496** 0.0675 20	**84.683** 0.0732 22	**86.910** 0.0791 24	**89.178** 0.085 3	**91.491** 0.091 3	**93.844** 0.099 3	**96.242** 0.105 3	**98.687** 0.114 3
$K_{\beta1}$	**80.255** 10.70 22	**82.574** 10.70 22	**84.938** 10.70 22	**87.349** 10.70 21	**89.807** 10.70 21	**92.315** 10.70 21	**94.868** 10.60 21	**97.474** 10.70 21	**100.130** 10.70 21	**102.841** 10.70 21	**105.604** 10.70 21	**108.422** 10.70 22	**111.298** 10.70 22	**114.234** 10.70 22	**117.228** 10.70 22
$K_{\beta2}$	**82.473** 3.87 8	**84.865** 3.90 8	**87.300** 3.91 8	**89.784** 3.93 8	**92.317** 3.95 8	**94.900** 3.97 8	**97.530** 3.98 8	**100.214** 4.01 8	**102.948** 4.04 8	**105.738** 4.07 8	**108.582** 4.10 8	**111.486** 4.13 8	**114.445** 4.15 8	**117.463** 4.17 8	**120.540** 4.18 8
$K_{\beta3}$	**79.824** 5.59 11	**82.115** 5.59 11	**84.450** 5.58 11	**86.830** 5.59 11	**89.256** 5.57 11	**91.730** 5.58 11	**94.247** 5.56 11	**96.815** 5.58 11	**99.432** 5.59 11	**102.101** 5.61 11	**104.819** 5.61 11	**107.595** 5.64 11	**110.421** 5.65 11	**113.303** 5.65 11	**116.244** 5.44 11
$K_{\beta4}$	**82.733** 0.08 4	**85.134** 0.09 4	**87.580** 0.09 4	**90.074** 0.09 4	**92.618** 0.09 4	**95.211** 0.10 5	**97.853** 0.10 5	**100.548** 0.10 5	**103.295** 0.11 5	**106.098** 0.11 5	**108.955** 0.11 5	**111.870** 0.12 6	**114.844** 0.12 6	**117.876** 0.12 6	**120.969** 0.13 6
$K_{\beta5}$	**80.762** 0.294 12	**83.093** 0.303 12	**85.470** 0.312 12	**87.892** 0.321 13	**90.363** 0.330 13	**92.883** 0.339 14	**95.449** 0.349 14	**98.069** 0.358 14	**100.738** 0.362 15	**103.462** 0.371 15	**106.239** 0.380 15	**109.072** 0.389 16	**111.964** 0.397 16	**114.912** 0.405 16	**117.918** 0.413 16
$KO_{2,3}$	**83.028** 0.64 7	**85.444** 0.67 7	**87.911** 0.70 7	**90.421** 0.73 8	**92.983** 0.76 8	**95.595** 0.78 8	**98.257** 0.81 8	**100.972** 0.84 9	**103.740** 0.86 9	**106.563** 0.89 9	**109.442** 0.90 9	**112.380** 0.93 10	**115.377** 0.95 10	**118.429** 0.97 10	**121.543** 0.99 10
$KP_{2,3}$		**85.530** 0.0059 6	**88.003** 0.0165 17	**90.522** 0.031 3	**93.095** 0.049 5	**95.717** 0.070 7	**98.389** 0.094 10	**101.118** 0.114 12	**103.899** 0.132 13	**106.738** 0.146 15	**109.630** 0.160 16	**112.575** 0.156 16	**115.580** 0.159 16	**118.646** 0.162 17	**121.768** 0.157 16
$L_{\alpha1}$	**9.989** 12.0 5	**10.268** 12.4 5	**10.551** 12.8 5	**10.839** 13.2 5	**11.130** 13.6 5	**11.426** 13.9 5	**11.726** 14.2 5	**12.031** 14.6 5	**12.339** 14.9 6	**12.651** 15.3 6	**12.968** 15.6 7	**13.291** 16.2 8	**13.618** 16.8 8	**13.946** 17.3 8	**14.282** 17.9 9
$L_{\alpha2}$	**9.899** 1.35 5	**10.172** 1.39 5	**10.450** 1.44 5	**10.731** 1.48 6	**11.016** 1.52 6	**11.306** 1.56 6	**11.598** 1.59 6	**11.896** 1.63 6	**12.196** 1.67 6	**12.500** 1.71 6	**12.809** 1.75 8	**13.127** 1.82 8	**13.442** 1.89 9	**13.761** 1.94 9	**14.087** 2.00 10
$L_{\beta1}$	**11.824** 8.0 4	**12.213** 8.3 4	**12.614** 8.5 5	**13.024** 8.8 5	**13.443** 9.1 5	**13.875** 9.4 5	**14.316** 9.7 5	**14.770** 9.9 5	**15.236** 10.2 6	**15.711** 10.4 6	**16.202** 10.7 11	**16.708** 10.5 11	**17.222** 10.3 11	**17.751** 10.3 10	**18.296** 10.3 10

Table 7a. X-ray Energies and Intensities (per 100 K-Shell Vacancies) (continued)

	$_{80}$Hg	$_{81}$Tl	$_{82}$Pb	$_{83}$Bi	$_{84}$Po	$_{85}$At	$_{86}$Rn	$_{87}$Fr	$_{88}$Ra	$_{89}$Ac	$_{90}$Th	$_{91}$Pa	$_{92}$U	$_{93}$Np	$_{94}$Pu
continued															
$L_{\beta2,15}$	**11.915** 2.94 *13*	**12.261** 3.06 *13*	**12.611** 3.18 *14*	**12.967** 3.28 *14*	**13.328** 3.40 *15*	**13.694** 3.52 *15*	**14.066** 3.65 *16*	**14.443** 3.74 *16*	**14.825** 3.86 *17*	**15.212** 3.97 *17*	**15.605** 4.09 *21*	**16.008** 4.27 *22*	**16.410** 4.45 *23*	**16.817** 4.59 *24*	**17.235** 4.77 *25*
$L_{\beta3}$	**11.992** 0.097 *15*	**12.390** 0.094 *14*	**12.794** 0.095 *14*	**13.211** 0.096 *15*	**13.635** 0.107 *16*	**14.073** 0.101 *15*	**14.519** 0.106 *16*	**14.978** 0.106 *16*	**15.447** 0.110 *17*	**15.931** 0.111 *17*	**16.426** 0.116 *20*	**16.931** 0.113 *19*	**17.454** 0.122 *21*	**17.992** 0.124 *21*	**18.541** 0.135 *23*
$L_{\beta4}$	**11.561** 0.077 *12*	**11.931** 0.077 *12*	**12.307** 0.080 *12*	**12.691** 0.083 *13*	**13.084** 0.095 *14*	**13.488** 0.092 *14*	**13.898** 0.099 *15*	**14.319** 0.102 *15*	**14.749** 0.109 *16*	**15.191** 0.113 *17*	**15.641** 0.122 *21*	**16.104** 0.121 *21*	**16.577** 0.134 *23*	**17.061** 0.140 *24*	**17.557** 0.16 *3*
$L_{\beta5}$	**12.275** 0.315 *12*	**12.643** 0.362 *13*	**13.015** 0.411 *15*	**13.393** 0.458 *17*	**13.778** 0.506 *19*	**14.168** 0.556 *21*	**14.565** 0.605 *23*	**14.967** 0.683 *25*	**15.375** 0.71 *3*	**15.790** 0.76 *3*	**16.209** 0.81 *4*	**16.639** 0.87 *4*	**17.069** 0.94 *4*	**17.505** 1.17 *6*	**17.950** 1.06 *5*
$L_{\beta6}$	**11.481** 0.180 *7*	**11.812** 0.190 *7*	**12.142** 0.200 *8*	**12.480** 0.210 *8*	**12.823** 0.220 *8*	**13.169** 0.230 *9*	**13.520** 0.239 *9*	**13.877** 0.251 *9*	**14.236** 0.263 *10*	**14.601** 0.273 *10*	**14.970** 0.284 *13*	**15.350** 0.300 *14*	**15.727** 0.318 *15*	**16.109** 0.333 *16*	**16.498** 0.349 *17*
$L_{\gamma1}$	**13.830** 1.63 *10*	**14.291** 1.71 *10*	**14.765** 1.78 *10*	**15.248** 1.87 *11*	**15.742** 1.95 *11*	**16.249** 2.05 *12*	**16.770** 2.15 *13*	**17.302** 2.25 *13*	**17.848** 2.34 *14*	**18.405** 2.42 *14*	**18.980** 2.5 *3*	**19.571** 2.5 *3*	**20.169** 2.5 *3*	**20.784** 2.5 *3*	**21.420** 2.5 *3*
$L_{\gamma2}$	**14.158** 0.025 *4*	**14.625** 0.026 *4*	**15.097** 0.027 *4*	**15.582** 0.029 *5*	**16.077** 0.033 *5*	**16.585** 0.033 *5*	**17.104** 0.036 *6*	**17.635** 0.038 *6*	**18.177** 0.041 *7*	**18.734** 0.044 *7*	**19.304** 0.048 *9*	**19.888** 0.048 *9*	**20.487** 0.055 *10*	**21.099** 0.059 *11*	**21.724** 0.067 *12*
$L_{\gamma3}$	**14.262** 0.034 *6*	**14.738** 0.033 *5*	**15.216** 0.034 *6*	**15.709** 0.035 *6*	**16.213** 0.040 *6*	**16.731** 0.038 *6*	**17.258** 0.040 *7*	**17.800** 0.041 *7*	**18.353** 0.044 *7*	**18.922** 0.045 *7*	**19.505** 0.048 *9*	**20.101** 0.047 *9*	**20.715** 0.052 *9*	**21.342** 0.054 *10*	**21.981** 0.059 *11*
$L_{\gamma6}$	**14.199** 0.248 *23*	**14.683** 0.28 *3*	**15.178** 0.31 *3*	**15.685** 0.34 *3*	**16.203** 0.38 *4*	**16.735** 0.41 *4*	**17.280** 0.44 *4*	**17.839** 0.47 *4*	**18.412** 0.50 *5*	**18.997** 0.52 *5*	**19.599** 0.54 *7*	**20.217** 0.53 *7*	**20.844** 0.53 *7*	**21.491** 0.53 *7*	**22.153** 0.53 *7*
L_{η}	**10.647** 0.180 *10*	**10.994** 0.188 *10*	**11.349** 0.196 *11*	**11.712** 0.207 *11*	**12.085** 0.218 *12*	**12.466** 0.228 *12*	**12.855** 0.238 *13*	**13.255** 0.247 *13*	**13.662** 0.255 *14*	**14.082** 0.266 *14*	**14.511** 0.28 *3*	**14.953** 0.28 *3*	**15.400** 0.27 *3*	**15.861** 0.28 *3*	**16.333** 0.28 *3*
L_{ℓ}	**8.722** 0.65 *6*	**8.953** 0.68 *6*	**9.184** 0.72 *6*	**9.420** 0.75 *7*	**9.658** 0.79 *7*	**9.897** 0.82 *7*	**10.137** 0.86 *7*	**10.381** 0.89 *8*	**10.622** 0.93 *8*	**10.871** 0.98 *8*	**11.118** 1.02 *9*	**11.372** 1.08 *10*	**11.620** 1.14 *10*	**11.871** 1.20 *11*	**12.124** 1.25 *11*

	$_{95}$Am	$_{96}$Cm	$_{97}$Bk	$_{98}$Cf	$_{99}$Es	$_{100}$Fm	$_{101}$Md	$_{102}$No	$_{103}$Lr	104
$K_{\alpha1}$	**106.472** 44.9 *9*	**109.271** 44.8 *9*	**112.121** 44.6 *9*	**115.032** 44.4 *9*	**118.012** 44.3 *9*	**121.058** 44.2 *9*	**125.170** 44.1 *9*	**127.357** 44.0 *9*	**130.611** 43.8 *9*	**133.381** 43.6 *9*
$K_{\alpha2}$	**102.030** 28.5 *6*	**104.590** 28.5 *6*	**107.185** 28.7 *6*	**109.831** 28.7 *6*	**112.531** 28.8 *6*	**115.285** 28.9 *6*	**119.088** 29.0 *6*	**120.953** 29.0 *6*	**123.867** 29.1 *6*	**126.302** 29.2 *6*
$K_{\alpha3}$	**101.174** 0.123 *4*	**103.715** 0.132 *4*	**106.300** 0.145 *4*	**108.929** 0.158 *5*	**111.614** 0.171 *5*	**114.352** 0.184 *6*	**118.139** 0.201 *6*	**119.987** 0.218 *7*	**122.887** 0.235 *7*	**125.407** 0.252 *8*
$K_{\beta1}$	**120.284** 10.70 *21*	**123.403** 10.60 *21*	**126.580** 10.70 *21*	**129.823** 10.70 *22*	**133.137** 10.70 *22*	**136.521** 10.80 *22*	**140.974** 10.70 *22*	**143.506** 10.70 *22*	**147.110** 10.80 *22*	**150.279** 10.80 *22*
$K_{\beta2}$	**123.680** 4.19 *8*	**126.889** 4.19 *8*	**130.152** 4.22 *9*	**133.483** 4.26 *9*	**136.887** 4.27 *9*	**140.362** 4.31 *9*	**144.906** 4.30 *9*	**147.531** 4.32 *9*	**151.227** 4.35 *9*	**154.494** 4.36 *9*
$K_{\beta3}$	**119.243** 5.64 *11*	**122.304** 5.63 *11*	**125.418** 5.67 *11*	**128.594** 5.70 *11*	**131.838** 5.71 *11*	**135.150** 5.74 *12*	**139.525** 5.73 *12*	**141.977** 5.75 *12*	**145.496** 5.78 *12*	**148.550** 5.79 *12*
$K_{\beta4}$	**124.127** 0.13 *6*	**127.352** 0.13 *6*	**130.630** 0.14 *7*	**133.979** 0.14 *7*	**137.399** 0.14 *7*	**140.892** 0.15 *7*	**145.456** 0.15 *7*	**148.100** 0.15 *7*	**151.818** 0.16 *8*	**155.097** 0.16 *8*
$K_{\beta5}$	**120.989** 0.421 *17*	**124.124** 0.429 *17*	**127.316** 0.437 *18*	**130.573** 0.449 *18*	**133.904** 0.454 *18*	**137.304** 0.457 *18*	**141.774** 0.465 *19*	**144.323** 0.472 *19*	**147.944** 0.479 *19*	**151.113** 0.486 *19*
$KO_{2,3}$	**124.723** 1.0 *1*	**127.970** 1.02 *10*	**131.274** 1.04 *11*	**134.646** 1.05 *11*	**138.090** 1.06 *11*	**141.608** 1.08 *11*	**146.195** 1.09 *11*	**148.865** 1.10 *11*	**152.607** 1.12 *11*	**155.904** 1.13 *12*
$KP_{2,3}$	**124.955** 0.158 *16*	**128.210** 0.169 *17*	**131.524** 0.170 *17*	**134.908** 0.162 *17*	**138.363** 0.163 *17*	**141.889** 0.163 *17*	**146.490** 0.164 *17*	**149.171** 0.165 *17*	**152.926** 0.174 *18*	**156.236** 0.183 *19*
$L_{\alpha1}$	**14.620** 18.2 *9*	**14.961** 18.5 *9*	**15.308** 18.8 *9*	**15.660** 19.0 *9*	**16.016** 19.2 *9*	**16.377** 19.4 *11*	**16.741** 19.6 *11*	**17.110** 19.7 *11*	**17.483** 19.8 *11*	**17.893** 19.8 *11*
$L_{\alpha2}$	**14.414** 2.04 *10*	**14.746** 2.08 *10*	**15.082** 2.1 *1*	**15.423** 2.12 *10*	**15.767** 2.15 *10*	**16.116** 2.17 *13*	**16.467** 2.19 *13*	**16.823** 2.20 *13*	**17.183** 2.22 *13*	**17.571** 2.22 *13*
$L_{\beta1}$	**18.856** 10.4 *11*	**19.427** 10.6 *11*	**20.018** 10.7 *11*	**20.624** 10.8 *11*	**21.248** 11.0 *11*	**21.889** 11.2 *11*	**22.549** 11.4 *12*	**23.227** 11.5 *12*	**23.927** 11.7 *12*	**24.650** 12.0 *12*
$L_{\beta2,15}$	**17.655** 4.9 *3*	**18.081** 5.0 *3*	**18.509** 5.1 *3*	**18.946** 5.2 *3*	**19.387** 5.3 *3*	**19.834** 5.3 *3*	**20.286** 5.4 *3*	**20.744** 5.5 *3*	**21.207** 5.5 *4*	**21.716** 5.6 *4*
$L_{\beta3}$	**19.110** 0.137 *24*	**19.688** 0.142 *24*	**20.280** 0.142 *24*	**20.894** 0.145 *25*	**21.523** 0.15 *3*	**22.169** 0.15 *3*	**22.835** 0.15 *3*	**23.519** 0.15 *3*	**24.223** 0.15 *3*	**24.872** 0.15 *3*
$L_{\beta4}$	**18.069** 0.16 *3*	**18.589** 0.17 *3*	**19.118** 0.18 *3*	**19.665** 0.19 *3*	**20.224** 0.21 *4*	**20.798** 0.21 *4*	**21.386** 0.23 *5*	**21.990** 0.24 *5*	**22.609** 0.24 *5*	**23.143** 0.26 *5*
$L_{\beta5}$	**18.399** 1.11 *5*	**18.853** 1.16 *6*	**19.312** 1.19 *6*	**19.777** 1.23 *6*	**20.249** 1.26 *6*	**20.727** 1.29 *8*	**21.210** 1.32 *8*	**21.700** 1.35 *8*	**22.195** 1.38 *8*	**22.727** 1.41 *8*
$L_{\beta6}$	**16.890** 0.361 *17*	**17.286** 0.373 *18*	**17.687** 0.385 *18*	**18.094** 0.396 *19*	**18.501** 0.407 *19*	**18.916** 0.419 *24*	**19.332** 0.430 *25*	**19.754** 0.44 *3*	**20.179** 0.45 *3*	**20.670** 0.46 *3*
$L_{\gamma1}$	**22.072** 2.6 *3*	**22.735** 2.7 *3*	**23.416** 2.7 *3*	**24.117** 2.8 *3*	**24.836** 2.8 *3*	**25.574** 2.9 *3*	**26.333** 3.0 *3*	**27.110** 3.1 *3*	**27.911** 3.1 *3*	**28.753** 3.2 *3*
$L_{\gamma2}$	**22.370** 0.073 *13*	**23.028** 0.079 *14*	**23.698** 0.082 *15*	**24.390** 0.087 *16*	**25.099** 0.093 *17*	**25.825** 0.096 *20*	**26.571** 0.102 *21*	**27.336** 0.109 *23*	**28.120** 0.109 *23*	**28.846** 0.116 *24*
$L_{\gamma3}$	**22.643** 0.062 *11*	**23.319** 0.065 *12*	**24.007** 0.065 *12*	**24.718** 0.068 *12*	**25.446** 0.071 *13*	**26.195** 0.071 *15*	**26.963** 0.073 *15*	**27.752** 0.075 *16*	**28.560** 0.074 *15*	**29.327** 0.076 *16*
$L_{\gamma6}$	**22.836** 0.54 *7*	**23.527** 0.55 *7*	**24.241** 0.57 *7*	**24.971** 0.59 *8*	**25.723** 0.60 *8*	**26.492** 0.63 *8*	**27.284** 0.65 *8*	**28.094** 0.68 *9*	**28.929** 0.70 *9*	**29.796** 0.73 *9*
L_{η}	**16.819** 0.28 *3*	**17.314** 0.29 *3*	**17.826** 0.30 *3*	**18.347** 0.30 *3*	**18.884** 0.31 *3*	**19.433** 0.31 *3*	**19.998** 0.32 *3*	**20.577** 0.33 *3*	**21.173** 0.34 *3*	**21.825** 0.35 *4*
L_{ℓ}	**12.377** 1.31 *12*	**12.633** 1.36 *12*	**12.890** 1.40 *13*	**13.146** 1.45 *13*	**13.403** 1.49 *14*	**13.660** 1.53 *15*	**13.916** 1.57 *15*	**14.173** 1.61 *16*	**14.429** 1.64 *16*	**14.746** 1.68 *16*

Table 7b. X-ray Energies and Intensities (per 100 L_1-Shell Vacancies)

	$_{13}$Al	$_{14}$Si	$_{15}$P	$_{16}$S	$_{17}$Cl	$_{18}$Ar	$_{19}$K	$_{20}$Ca	$_{21}$Sc	$_{22}$Ti	$_{23}$V	$_{24}$Cr	$_{25}$Mn	$_{26}$Fe
$L_{\alpha1}$									**0.396** 0.023 *6*	**0.452** 0.056 *15*	**0.511** 0.11 *3*	**0.572** 0.16 *4*	**0.637** 0.23 *6*	**0.704** 0.29 *8*
$L_{\alpha2}$									**0.396** 0.0025 *7*	**0.452** 0.0061 *16*	**0.511** 0.012 *3*	**0.572** 0.018 *5*	**0.637** 0.025 *7*	**0.704** 0.032 *9*
$L_{\beta1}$	**0.073** 0.024 *7*	**0.099** 0.012 *4*	**0.136** 0.010 *3*	**0.165** 0.0083 *25*	**0.202** 0.0077 *23*	**0.251** 0.0068 *20*	**0.296** 0.0084 *25*	**0.350** 0.010 *3*	**0.400** 0.013 *3*	**0.458** 0.032 *8*	**0.518** 0.063 *16*	**0.581** 0.098 *25*	**0.648** 0.13 *3*	**0.717** 0.17 *4*
$L_{\beta3}$	**0.112** 0.0016 *6*	**0.146** 0.0018 *6*	**0.179** 0.0024 *8*	**0.221** 0.0045 *16*	**0.263** 0.007 *3*	**0.310** 0.011 *4*	**0.359** 0.015 *5*	**0.412** 0.019 *6*	**0.468** 0.024 *7*	**0.529** 0.029 *9*	**0.590** 0.035 *11*	**0.652** 0.043 *13*	**0.720** 0.051 *15*	**0.792** 0.061 *18*
$L_{\beta4}$	**0.112** 0.0010 *4*	**0.146** 0.0012 *4*	**0.179** 0.0015 *5*	**0.221** 0.0029 *10*	**0.263** 0.0046 *16*	**0.310** 0.0070 *24*	**0.359** 0.009 *3*	**0.412** 0.012 *4*	**0.468** 0.015 *5*	**0.529** 0.018 *6*	**0.590** 0.023 *7*	**0.652** 0.028 *8*	**0.720** 0.033 *10*	**0.792** 0.039 *12*
$L_{\beta6}$									**0.402** 0.0015 *4*	**0.456** 0.0016 *4*	**0.513** 0.0019 *5*		**0.640** 0.0020 *6*	**0.708** 0.0019 *5*
L_{η}									**0.353** 0.013 *3*	**0.401** 0.014 *4*	**0.454** 0.017 *4*	**0.510** 0.017 *4*	**0.568** 0.017 *4*	**0.628** 0.019 *5*
L_{ℓ}									**0.348** 0.023 *7*	**0.395** 0.025 *7*	**0.446** 0.030 *8*	**0.500** 0.029 *8*	**0.556** 0.033 *9*	**0.615** 0.035 *10*

	$_{27}$Co	$_{28}$Ni	$_{29}$Cu	$_{30}$Zn	$_{31}$Ga	$_{32}$Ge	$_{33}$As	$_{34}$Se	$_{35}$Br	$_{36}$Kr	$_{37}$Rb	$_{38}$Sr	$_{39}$Y	$_{40}$Zr
$L_{\alpha1}$	**0.776** 0.35 *10*	**0.851** 0.43 *12*	**0.929** 0.51 *14*	**1.012** 0.55 *12*	**1.098** 0.59 *13*	**1.188** 0.69 *15*	**1.282** 0.75 *17*	**1.379** 0.83 *19*	**1.481** 0.93 *21*	**1.581** 1.03 *23*	**1.694** 1.13 *25*	**1.806** 1.2 *3*	**1.923** 1.3 *3*	**2.042** 1.5 *3*
$L_{\alpha2}$	**0.776** 0.039 *11*	**0.851** 0.048 *13*	**0.929** 0.056 *15*	**1.012** 0.061 *14*	**1.098** 0.066 *15*	**1.188** 0.077 *17*	**1.282** 0.083 *18*	**1.379** 0.092 *20*	**1.480** 0.103 *23*	**1.580** 0.114 *25*	**1.693** 0.13 *3*	**1.805** 0.14 *3*	**1.920** 0.15 *3*	**2.040** 0.16 *3*
$L_{\beta1}$	**0.791** 0.21 *5*	**0.868** 0.24 *6*	**0.949** 0.28 *7*	**1.035** 0.30 *8*	**1.125** 0.33 *8*	**1.219** 0.34 *9*	**1.317** 0.37 *9*	**1.420** 0.42 *11*	**1.526** 0.48 *12*	**1.632** 0.51 *13*	**1.752** 0.57 *14*	**1.872** 0.62 *15*	**1.996** 0.64 *16*	**2.124** 0.68 *10*
$L_{\beta2,15}$													**2.078** 0.0038 *9*	**2.219** 0.0103 *18*
$L_{\beta3}$	**0.866** 0.073 *22*	**0.940** 0.084 *25*	**1.022** 0.10 *3*	**1.107** 0.11 *3*	**1.195** 0.12 *4*	**1.294** 0.13 *4*	**1.386** 0.15 *5*	**1.492** 0.17 *5*	**1.601** 0.19 *6*	**1.707** 0.21 *6*	**1.827** 0.24 *7*	**1.947** 0.26 *8*	**2.072** 0.30 *9*	**2.201** 0.34 *9*
$L_{\beta4}$	**0.866** 0.047 *14*	**0.940** 0.056 *17*	**1.022** 0.064 *19*	**1.107** 0.072 *22*	**1.191** 0.082 *25*	**1.286** 0.09 *3*	**1.380** 0.10 *3*	**1.486** 0.11 *3*	**1.593** 0.13 *4*	**1.699** 0.15 *4*	**1.818** 0.16 *5*	**1.936** 0.18 *5*	**2.060** 0.21 *6*	**2.187** 0.24 *6*
$L_{\beta6}$	**0.779** 0.0019 *5*	**0.855** 0.0019 *5*		**1.020** 0.0018 *4*	**1.114** 0.0022 *5*	**1.212** 0.0028 *6*	**1.315** 0.0033 *7*	**1.424** 0.0038 *9*	**1.523** 0.0045 *10*	**1.647** 0.0051 *11*	**1.775** 0.0060 *13*	**1.902** 0.0069 *15*	**2.035** 0.0077 *17*	**2.171** 0.0088 *16*
$L_{\gamma1}$													**2.153** 0.0089 *22*	**2.304** 0.022 *3*
$L_{\gamma2}$						**1.412** 0.0014 *4*	**1.524** 0.0032 *10*	**1.648** 0.0054 *17*	**1.777** 0.009 *3*	**1.906** 0.013 *4*	**2.050** 0.017 *5*	**2.196** 0.022 *7*	**2.347** 0.026 *8*	**2.503** 0.033 *9*
$L_{\gamma3}$					**1.297** 0.0061 *19*	**1.412** 0.022 *7*	**1.524** 0.026 *8*	**1.648** 0.031 *9*	**1.777** 0.034 *11*	**1.907** 0.039 *12*	**2.051** 0.044 *13*	**2.196** 0.049 *15*	**2.347** 0.056 *17*	**2.503** 0.065 *17*
L_{η}	**0.693** 0.019 *5*	**0.760** 0.018 *5*	**0.831** 0.020 *5*	**0.907** 0.020 *5*	**0.984** 0.021 *5*	**1.068** 0.022 *5*	**1.155** 0.022 *6*	**1.245** 0.025 *6*	**1.339** 0.027 *7*	**1.435** 0.027 *7*	**1.542** 0.029 *7*	**1.649** 0.030 *8*	**1.762** 0.030 *8*	**1.876** 0.031 *5*
L_{ℓ}	**0.678** 0.037 *11*	**0.743** 0.039 *11*	**0.811** 0.041 *12*	**0.884** 0.041 *10*	**0.957** 0.041 *10*	**1.037** 0.045 *11*	**1.120** 0.045 *11*	**1.204** 0.048 *11*	**1.293** 0.051 *12*	**1.383** 0.054 *13*	**1.482** 0.058 *14*	**1.582** 0.060 *14*	**1.686** 0.063 *15*	**1.792** 0.069 *13*

	$_{41}$Nb	$_{42}$Mo	$_{43}$Tc	$_{44}$Ru	$_{45}$Rh	$_{46}$Pd	$_{47}$Ag	$_{48}$Cd	$_{49}$In	$_{50}$Sn	$_{51}$Sb	$_{52}$Te	$_{53}$I	$_{54}$Xe
$L_{\alpha1}$	**2.166** 1.8 *3*	**2.293** 1.9 *3*	**2.424** 2.0 *4*	**2.558** 2.1 *4*	**2.697** 2.2 *4*	**2.839** 2.4 *4*	**2.984** 2.4 *4*	**3.134** 2.6 *5*	**3.287** 2.8 *5*	**3.444** 1.44 *25*	**3.605** 1.6 *3*	**3.769** 1.7 *3*	**3.938** 1.8 *3*	**4.106** 2.0 *3*
$L_{\alpha2}$	**2.163** 0.20 *4*	**2.290** 0.21 *4*	**2.420** 0.22 *4*	**2.554** 0.24 *4*	**2.692** 0.25 *5*	**2.833** 0.26 *5*	**2.978** 0.27 *5*	**3.127** 0.29 *5*	**3.279** 0.31 *6*	**3.435** 0.16 *3*	**3.595** 0.18 *3*	**3.759** 0.19 *3*	**3.926** 0.20 *3*	**4.093** 0.22 *4*
$L_{\beta1}$	**2.257** 0.28 *4*	**2.395** 0.31 *5*	**2.537** 0.33 *5*	**2.683** 0.36 *6*	**2.834** 0.38 *6*	**2.990** 0.41 *6*	**3.151** 0.45 *7*	**3.317** 0.49 *8*	**3.487** 0.53 *8*	**3.663** 0.96 *10*	**3.843** 1.01 *11*	**4.029** 1.15 *12*	**4.221** 1.22 *13*	**4.414** 1.35 *14*
$L_{\beta2,15}$	**2.367** 0.055 *10*	**2.518** 0.099 *18*	**2.675** 0.15 *3*	**2.836** 0.20 *4*	**3.001** 0.23 *4*	**3.172** 0.28 *5*	**3.348** 0.32 *6*	**3.528** 0.37 *7*	**3.714** 0.43 *8*	**3.905** 0.23 *4*	**4.101** 0.27 *5*	**4.302** 0.31 *5*	**4.508** 0.34 *6*	**4.714** 0.38 *7*
$L_{\beta3}$	**2.335** 0.47 *12*	**2.473** 0.50 *13*	**2.617** 0.55 *14*	**2.763** 0.60 *15*	**2.916** 0.65 *16*	**3.073** 0.71 *18*	**3.234** 0.81 *20*	**3.402** 0.91 *23*	**3.573** 1.01 *25*	**3.750** 1.9 *4*	**3.933** 2.0 *4*	**4.121** 2.1 *4*	**4.314** 2.2 *4*	**4.512** 2.3 *5*
$L_{\beta4}$	**2.319** 0.33 *8*	**2.456** 0.35 *9*	**2.598** 0.38 *10*	**2.741** 0.41 *10*	**2.891** 0.44 *11*	**3.045** 0.46 *12*	**3.203** 0.52 *13*	**3.367** 0.58 *15*	**3.535** 0.63 *16*	**3.708** 1.16 *23*	**3.886** 1.21 *24*	**4.070** 1.26 *25*	**4.258** 1.3 *3*	**4.451** 1.4 *3*
$L_{\beta6}$	**2.312** 0.0108 *20*	**2.458** 0.0118 *21*	**2.609** 0.0127 *23*	**2.763** 0.0139 *25*	**2.923** 0.015 *3*	**3.087** 0.016 *3*	**3.256** 0.016 *3*	**3.430** 0.018 *3*	**3.608** 0.019 *4*	**3.792** 0.0104 *18*	**3.980** 0.0118 *20*	**4.173** 0.0130 *22*	**4.371** 0.0139 *24*	**4.569** 0.015 *3*
$L_{\gamma1}$	**2.462** 0.0125 *19*	**2.623** 0.017 *3*	**2.791** 0.021 *3*	**2.965** 0.026 *4*	**3.144** 0.035 *5*	**3.329** 0.044 *7*	**3.520** 0.047 *7*	**3.718** 0.052 *8*	**3.922** 0.060 *9*	**4.132** 0.113 *12*	**4.349** 0.125 *13*	**4.572** 0.146 *15*	**4.802** 0.164 *17*	**5.034** 0.189 *20*
$L_{\gamma2}$	**2.664** 0.048 *12*	**2.831** 0.052 *13*	**3.004** 0.060 *15*	**3.181** 0.068 *18*	**3.364** 0.076 *19*	**3.553** 0.084 *22*	**3.743** 0.10 *3*	**3.951** 0.12 *3*	**4.160** 0.13 *4*	**4.376** 0.26 *5*	**4.600** 0.28 *6*	**4.829** 0.31 *6*	**5.065** 0.34 *7*	**5.307** 0.36 *8*
$L_{\gamma3}$	**2.664** 0.091 *23*	**2.831** 0.098 *25*	**3.004** 0.11 *3*	**3.181** 0.12 *3*	**3.364** 0.13 *3*	**3.553** 0.15 *4*	**3.750** 0.17 *4*	**3.951** 0.19 *5*	**4.160** 0.22 *6*	**4.376** 0.41 *9*	**4.600** 0.44 *9*	**4.829** 0.47 *10*	**5.065** 0.51 *11*	**5.307** 0.54 *11*
L_{η}	**1.996** 0.0127 *19*	**2.120** 0.0133 *20*	**2.249** 0.0139 *21*	**2.382** 0.0144 *22*	**2.519** 0.0147 *22*	**2.660** 0.0154 *23*	**2.806** 0.0163 *25*	**2.957** 0.017 *3*	**3.112** 0.018 *3*	**3.272** 0.032 *3*	**3.437** 0.033 *3*	**3.606** 0.037 *4*	**3.780** 0.038 *4*	**3.955** 0.040 *4*
L_{ℓ}	**1.902** 0.081 *16*	**2.016** 0.085 *17*	**2.133** 0.088 *17*	**2.253** 0.092 *18*	**2.377** 0.094 *18*	**2.503** 0.097 *19*	**2.634** 0.100 *20*	**2.767** 0.106 *21*	**2.905** 0.112 *22*	**3.045** 0.058 *11*	**3.189** 0.064 *12*	**3.335** 0.068 *13*	**3.485** 0.072 *13*	**3.634** 0.078 *15*

Table 7b. X-ray Energies and Intensities (per 100 L_1-Shell Vacancies) (continued)

	$_{55}$Cs	$_{56}$Ba	$_{57}$La	$_{58}$Ce	$_{59}$Pr	$_{60}$Nd	$_{61}$Pm	$_{62}$Sm	$_{63}$Eu	$_{64}$Gd	$_{65}$Tb	$_{66}$Dy	$_{67}$Ho	$_{68}$Er
$L_{\alpha1}$	**4.286** 2.1 4	**4.466** 2.2 4	**4.651** 2.4 4	**4.840** 2.6 4	**5.033** 2.8 5	**5.230** 3.0 3	**5.432** 3.2 3	**5.636** 3.3 4	**5.846** 3.5 4	**6.058** 3.7 4	**6.273** 3.9 4	**6.495** 4.2 4	**6.720** 4.4 5	**6.949** 4.6 5
$L_{\alpha2}$	**4.272** 0.23 4	**4.451** 0.24 4	**4.634** 0.27 5	**4.822** 0.29 5	**5.013** 0.31 5	**5.208** 0.33 4	**5.408** 0.35 4	**5.610** 0.37 4	**5.816** 0.39 4	**6.026** 0.41 4	**6.239** 0.44 5	**6.458** 0.46 5	**6.680** 0.49 5	**6.905** 0.51 5
$L_{\beta1}$	**4.620** 1.46 15	**4.828** 1.55 16	**5.042** 1.66 18	**5.263** 1.77 19	**5.489** 1.88 20	**5.722** 1.99 15	**5.961** 2.11 16	**6.206** 2.24 17	**6.457** 2.37 18	**6.713** 2.51 19	**6.977** 2.66 21	**7.248** 2.83 22	**7.526** 3.00 23	**7.811** 3.17 25
$L_{\beta2,15}$	**4.934** 0.42 7	**5.156** 0.46 8	**5.384** 0.51 9	**5.613** 0.54 9	**5.851** 0.58 10	**6.090** 0.64 7	**6.339** 0.67 7	**6.587** 0.70 8	**6.844** 0.74 8	**7.102** 0.77 8	**7.367** 0.81 9	**7.636** 0.86 9	**7.910** 0.88 10	**8.186** 0.92 10
$L_{\beta3}$	**4.717** 2.5 5	**4.927** 2.6 5	**5.143** 2.7 6	**5.363** 2.9 6	**5.593** 3.0 6	**5.829** 3.2 5	**6.071** 3.2 5	**6.317** 3.5 5	**6.571** 3.6 6	**6.832** 3.8 6	**7.097** 4.0 6	**7.370** 4.3 6	**7.653** 4.5 7	**7.940** 4.7 7
$L_{\beta4}$	**4.649** 1.5 3	**4.852** 1.5 3	**5.062** 1.6 3	**5.276** 1.7 3	**5.497** 1.8 4	**5.723** 1.9 3	**5.956** 1.9 3	**6.196** 2.1 3	**6.438** 2.2 3	**6.687** 2.3 4	**6.940** 2.4 4	**7.204** 2.6 4	**7.471** 2.8 4	**7.746** 3.0 5
$L_{\beta5}$			**5.483** 0.0051 9							**7.243** 0.0063 7				
$L_{\beta6}$	**4.781** 0.017 3	**4.994** 0.018 3	**5.212** 0.020 4	**5.434** 0.022 4	**5.660** 0.024 4	**5.893** 0.026 3	**6.128** 0.029 3	**6.370** 0.031 3	**6.617** 0.034 4	**6.867** 0.037 4	**7.116** 0.040 4	**7.374** 0.044 5	**7.635** 0.047 5	**7.909** 0.052 6
$L_{\gamma1}$	**5.281** 0.209 22	**5.531** 0.225 24	**5.792** 0.25 3	**6.054** 0.27 3	**6.327** 0.29 3	**6.604** 0.318 24	**6.892** 0.34 3	**7.183** 0.37 3	**7.484** 0.40 3	**7.790** 0.43 3	**8.105** 0.46 4	**8.426** 0.49 4	**8.757** 0.53 4	**9.088** 0.56 4
$L_{\gamma2}$	**5.542** 0.40 8	**5.797** 0.44 9	**6.060** 0.47 10	**6.326** 0.5 1	**6.599** 0.54 11	**6.883** 0.57 9	**7.186** 0.59 10	**7.471** 0.64 10	**7.768** 0.68 11	**8.087** 0.73 12	**8.398** 0.77 13	**8.714** 0.83 13	**9.051** 0.88 14	**9.385** 0.93 15
$L_{\gamma3}$	**5.553** 0.58 12	**5.809** 0.62 13	**6.075** 0.67 14	**6.342** 0.71 15	**6.617** 0.75 16	**6.901** 0.80 13	**7.186** 0.84 14	**7.489** 0.91 15	**7.795** 0.97 16	**8.105** 1.03 17	**8.423** 1.10 18	**8.753** 1.19 19	**9.088** 1.27 21	**9.431** 1.37 22
$L_{\gamma6}$			**5.891** 0.0027 3							**7.930** 0.0035 4				
L_{η}	**4.142** 0.043 4	**4.331** 0.044 5	**4.529** 0.046 5	**4.730** 0.048 5	**4.929** 0.050 5	**5.146** 0.052 4	**5.363** 0.053 4	**5.589** 0.055 4	**5.817** 0.057 4	**6.049** 0.059 4	**6.284** 0.061 4	**6.534** 0.064 5	**6.789** 0.066 5	**7.058** 0.068 5
L_{ℓ}	**3.795** 0.083 16	**3.954** 0.088 16	**4.121** 0.098 18	**4.289** 0.106 20	**4.453** 0.113 21	**4.633** 0.124 16	**4.809** 0.131 17	**4.993** 0.139 18	**5.177** 0.147 19	**5.362** 0.156 20	**5.546** 0.166 22	**5.743** 0.178 23	**5.943** 0.187 25	**6.151** 0.20 3

	$_{69}$Tm	$_{70}$Yb	$_{71}$Lu	$_{72}$Hf	$_{73}$Ta	$_{74}$W	$_{75}$Re	$_{76}$Os	$_{77}$Ir	$_{78}$Pt	$_{79}$Au	$_{80}$Hg	$_{81}$Tl	$_{82}$Pb
$L_{\alpha1}$	**7.180** 4.7 5	**7.416** 4.8 4	**7.656** 4.9 4	**7.899** 5.1 4	**8.146** 5.3 4	**8.398** 5.5 4	**8.652** 6.7 5	**8.911** 8.1 7	**9.175** 9.6 8	**9.443** 11.0 9	**9.713** 12.1 10	**9.989** 13.2 8	**10.268** 13.9 8	**10.551** 14.6 9
$L_{\alpha2}$	**7.133** 0.52 6	**7.367** 0.54 4	**7.605** 0.55 4	**7.844** 0.57 4	**8.088** 0.59 5	**8.335** 0.62 5	**8.586** 0.75 6	**8.840** 0.91 7	**9.099** 1.08 9	**9.362** 1.23 10	**9.628** 1.35 11	**9.899** 1.48 9	**10.172** 1.56 9	**10.450** 1.64 10
$L_{\beta1}$	**8.102** 3.3 3	**8.402** 3.51 20	**8.709** 3.69 21	**9.023** 3.67 21	**9.343** 3.83 22	**9.672** 3.77 22	**10.010** 3.70 22	**10.354** 3.84 24	**10.708** 3.74 25	**11.071** 3.6 3	**11.443** 3.7 3	**11.824** 3.59 24	**12.213** 3.70 25	**12.614** 3.53 25
$L_{\beta2,15}$	**8.468** 0.92 10	**8.752** 0.94 8	**9.044** 0.99 8	**9.342** 1.08 9	**9.646** 1.16 10	**9.955** 1.25 10	**10.268** 1.53 13	**10.590** 1.90 16	**10.912** 2.28 19	**11.242** 2.64 22	**11.576** 2.93 25	**11.915** 3.23 21	**12.261** 3.42 22	**12.611** 3.62 23
$L_{\beta3}$	**8.231** 4.9 8	**8.537** 5.2 8	**8.847** 5.5 8	**9.163** 5.8 9	**9.488** 6.2 9	**9.819** 6.6 10	**10.159** 6.4 10	**10.511** 5.7 9	**10.868** 5.2 8	**11.235** 4.9 7	**11.610** 4.5 7	**11.992** 4.4 7	**12.390** 4.4 7	**12.794** 4.5 7
$L_{\beta4}$	**8.026** 3.2 5	**8.313** 3.4 5	**8.607** 3.7 6	**8.905** 4.0 6	**9.213** 4.3 7	**9.525** 4.6 7	**9.845** 4.6 7	**10.176** 4.2 6	**10.510** 3.9 6	**10.854** 3.7 6	**11.205** 3.5 5	**11.561** 3.6 5	**11.931** 3.6 5	**12.307** 3.8 6
$L_{\beta5}$			**9.240** 0.0059 5	**9.554** 0.0152 12	**9.875** 0.0212 16	**10.201** 0.0275 22	**10.532** 0.061 5	**10.871** 0.107 9	**11.211** 0.159 13	**11.562** 0.218 18	**11.916** 0.278 23	**12.275** 0.345 21	**12.643** 0.405 25	**13.015** 0.47 3
$L_{\beta6}$	**8.176** 0.053 6	**8.456** 0.057 4	**8.738** 0.058 5	**9.023** 0.061 5	**9.316** 0.065 5	**9.612** 0.069 5	**9.910** 0.087 7	**10.217** 0.111 9	**10.525** 0.135 11	**10.840** 0.157 13	**11.160** 0.177 14	**11.481** 0.198 12	**11.812** 0.213 13	**12.142** 0.228 14
$L_{\gamma1}$	**9.437** 0.60 5	**9.780** 0.64 4	**10.144** 0.68 4	**10.516** 0.68 4	**10.895** 0.71 4	**11.285** 0.71 4	**11.685** 0.71 4	**12.096** 0.74 5	**12.513** 0.73 4	**12.942** 0.71 4	**13.382** 0.75 5	**13.830** 0.73 5	**14.291** 0.77 5	**14.765** 0.74 5
$L_{\gamma2}$	**9.730** 1.00 16	**10.090** 1.07 17	**10.460** 1.15 19	**10.834** 1.24 20	**11.217** 1.33 21	**11.608** 1.43 23	**12.009** 1.42 23	**12.421** 1.31 21	**12.841** 1.23 20	**13.273** 1.19 19	**13.709** 1.14 18	**14.158** 1.16 19	**14.625** 1.20 19	**15.097** 1.29 21
$L_{\gamma3}$	**9.779** 1.46 23	**10.143** 1.55 25	**10.511** 1.7 3	**10.890** 1.8 3	**11.277** 1.9 3	**11.675** 2.1 3	**12.082** 2.0 3	**12.500** 1.9 3	**12.924** 1.7 3	**13.361** 1.6 3	**13.807** 1.55 25	**14.262** 1.55 25	**14.738** 1.55 25	**15.216** 1.6 3
$L_{\gamma6}$			**10.344** 0.0041 4	**10.733** 0.0103 10	**11.130** 0.0184 17	**11.538** 0.027 3	**11.955** 0.044 4	**12.385** 0.061 6	**12.820** 0.075 7	**13.270** 0.087 8	**13.731** 0.101 10	**14.199** 0.111 11	**14.683** 0.126 12	**15.178** 0.127 12
L_{η}	**7.310** 0.071 5	**7.580** 0.074 4	**7.857** 0.077 4	**8.139** 0.076 4	**8.428** 0.080 4	**8.724** 0.079 4	**9.027** 0.078 4	**9.337** 0.081 4	**9.650** 0.080 4	**9.975** 0.079 4	**10.309** 0.083 5	**10.647** 0.081 4	**10.994** 0.084 5	**11.349** 0.081 4
L_{ℓ}	**6.341** 0.21 3	**6.545** 0.217 24	**6.753** 0.221 24	**6.960** 0.23 3	**7.173** 0.25 3	**7.387** 0.26 3	**7.604** 0.32 4	**7.822** 0.40 5	**8.042** 0.49 5	**8.266** 0.57 6	**8.494** 0.63 7	**8.722** 0.71 7	**8.953** 0.76 8	**9.184** 0.81 8

Table 7b. X-ray Energies and Intensities (per 100 L_1-Shell Vacancies) (continued)

	$_{83}$Bi	$_{84}$Po	$_{85}$At	$_{86}$Rn	$_{87}$Fr	$_{88}$Ra	$_{89}$Ac	$_{90}$Th	$_{91}$Pa	$_{92}$U	$_{93}$Np	$_{94}$Pu	$_{95}$Am	$_{96}$Cm
$L_{\alpha1}$	**10.839** 15.1 *9*	**11.130** 15.5 *9*	**11.426** 16.2 *10*	**11.726** 16.4 *10*	**12.031** 16.8 *10*	**12.339** 17.3 *10*	**12.651** 17.8 *11*	**12.968** 17.9 *16*	**13.291** 18.8 *17*	**13.618** 19.0 *17*	**13.946** 19.4 *17*	**14.282** 19.6 *17*	**14.620** 19.7 *17*	**14.961** 20.1 *18*
$L_{\alpha2}$	**10.731** 1.69 *10*	**11.016** 1.74 *10*	**11.306** 1.82 *11*	**11.598** 1.84 *11*	**11.896** 1.89 *11*	**12.196** 1.94 *12*	**12.500** 1.99 *12*	**12.809** 2.01 *18*	**13.127** 2.10 *19*	**13.442** 2.13 *19*	**13.761** 2.17 *19*	**14.087** 2.19 *19*	**14.414** 2.20 *19*	**14.746** 2.25 *20*
$L_{\beta1}$	**13.024** 3.34 *24*	**13.443** 3.44 *25*	**13.875** 3.23 *24*	**14.316** 3.32 *25*	**14.770** 3.4 *3*	**15.236** 3.15 *25*	**15.711** 3.2 *3*	**16.202** 3.3 *5*	**16.708** 2.9 *4*	**17.222** 2.8 *4*	**17.751** 2.5 *4*	**18.296** 1.8 *3*	**18.856** 1.8 *3*	**19.427** 1.44 *24*
$L_{\beta2,15}$	**12.967** 3.75 *24*	**13.328** 3.90 *25*	**13.694** 4.1 *3*	**14.066** 4.2 *3*	**14.443** 4.3 *3*	**14.825** 4.5 *3*	**15.212** 4.6 *3*	**15.605** 4.7 *4*	**16.008** 4.9 *5*	**16.410** 5.0 *5*	**16.817** 5.1 *5*	**17.235** 5.2 *5*	**17.655** 5.3 *5*	**18.081** 5.4 *5*
$L_{\beta3}$	**13.211** 4.6 *7*	**13.635** 4.8 *7*	**14.073** 4.9 *7*	**14.519** 5.0 *8*	**14.978** 5.1 *8*	**15.447** 5.3 *8*	**15.931** 5.4 *8*	**16.426** 5.6 *10*	**16.931** 5.5 *10*	**17.454** 5.9 *10*	**17.992** 6.2 *10*	**18.541** 6.6 *11*	**19.110** 6.9 *12*	**19.688** 7.0 *12*
$L_{\beta4}$	**12.691** 4.0 *6*	**13.084** 4.2 *6*	**13.488** 4.4 *7*	**13.898** 4.7 *7*	**14.319** 4.9 *8*	**14.749** 5.2 *8*	**15.191** 5.5 *8*	**15.641** 5.9 *10*	**16.104** 5.9 *10*	**16.577** 6.5 *11*	**17.061** 7.0 *12*	**17.557** 7.7 *13*	**18.069** 8.2 *14*	**18.589** 8.6 *15*
$L_{\beta5}$	**13.393** 0.52 *3*	**13.778** 0.58 *4*	**14.168** 0.65 *4*	**14.565** 0.70 *4*	**14.967** 0.79 *5*	**15.375** 0.82 *5*	**15.790** 0.88 *5*	**16.209** 0.93 *8*	**16.639** 1.01 *9*	**17.069** 1.06 *10*	**17.505** 1.31 *12*	**17.950** 1.16 *10*	**18.399** 1.20 *11*	**18.853** 1.26 *11*
$L_{\beta6}$	**12.480** 0.239 *15*	**12.823** 0.251 *15*	**13.169** 0.268 *16*	**13.520** 0.275 *17*	**13.877** 0.290 *18*	**14.236** 0.305 *18*	**14.601** 0.318 *19*	**14.970** 0.33 *3*	**15.350** 0.35 *3*	**15.727** 0.36 *3*	**16.109** 0.37 *3*	**16.498** 0.38 *3*	**16.890** 0.39 *3*	**17.286** 0.40 *4*
$L_{\gamma1}$	**15.248** 0.71 *4*	**15.742** 0.74 *5*	**16.249** 0.71 *5*	**16.770** 0.74 *5*	**17.302** 0.77 *5*	**17.848** 0.72 *5*	**18.405** 0.75 *5*	**18.980** 0.77 *9*	**19.571** 0.68 *8*	**20.169** 0.68 *8*	**20.784** 0.60 *7*	**21.420** 0.43 *5*	**22.072** 0.44 *5*	**22.735** 0.36 *5*
$L_{\gamma2}$	**15.582** 1.38 *22*	**16.077** 1.49 *24*	**16.585** 1.6 *3*	**17.104** 1.7 *3*	**17.635** 1.8 *3*	**18.177** 2.0 *3*	**18.734** 2.1 *4*	**19.304** 2.3 *4*	**19.888** 2.4 *4*	**20.487** 2.7 *5*	**21.099** 2.9 *5*	**21.724** 3.3 *6*	**22.370** 3.7 *7*	**23.028** 3.9 *7*
$L_{\gamma3}$	**15.709** 1.7 *3*	**16.213** 1.8 *3*	**16.731** 1.9 *3*	**17.258** 1.9 *3*	**17.800** 2.0 *3*	**18.353** 2.1 *3*	**18.922** 2.2 *4*	**19.505** 2.3 *4*	**20.101** 2.3 *4*	**20.715** 2.5 *5*	**21.342** 2.7 *5*	**21.981** 2.9 *5*	**22.643** 3.1 *6*	**23.319** 3.2 *6*
$L_{\gamma6}$	**15.685** 0.130 *12*	**16.203** 0.143 *14*	**16.735** 0.140 *13*	**17.280** 0.151 *14*	**17.839** 0.161 *15*	**18.412** 0.153 *15*	**18.997** 0.160 *15*	**19.599** 0.165 *21*	**20.217** 0.146 *19*	**20.844** 0.145 *19*	**21.491** 0.127 *16*	**22.153** 0.091 *12*	**22.836** 0.092 *12*	**23.527** 0.075 *10*
L_{η}	**11.712** 0.079 *4*	**12.085** 0.083 *5*	**12.466** 0.078 *4*	**12.855** 0.082 *4*	**13.255** 0.085 *5*	**13.662** 0.079 *4*	**14.082** 0.082 *5*	**14.511** 0.086 *9*	**14.953** 0.075 *8*	**15.400** 0.075 *8*	**15.861** 0.066 *7*	**16.333** 0.047 *5*	**16.819** 0.048 *5*	**17.314** 0.040 *4*
L_{ℓ}	**9.420** 0.86 *8*	**9.658** 0.90 *9*	**9.897** 0.96 *9*	**10.137** 0.98 *10*	**10.381** 1.03 *10*	**10.622** 1.08 *11*	**10.871** 1.14 *11*	**11.118** 1.17 *14*	**11.372** 1.25 *15*	**11.620** 1.29 *15*	**11.871** 1.34 *16*	**12.124** 1.37 *16*	**12.377** 1.41 *17*	**12.633** 1.47 *17*

	$_{97}$Bk	$_{98}$Cf	$_{99}$Es	$_{100}$Fm	$_{101}$Md	$_{102}$No	$_{103}$Lr	104
$L_{\alpha1}$	**15.308** 20.1 *18*	**15.660** 20.4 *18*	**16.016** 20.8 *18*	**16.377** 21 *3*	**16.741** 21 *3*	**17.110** 21 *3*	**17.483** 22 *3*	**17.893** 22 *3*
$L_{\alpha2}$	**15.082** 2.25 *20*	**15.423** 2.29 *20*	**15.767** 2.33 *21*	**16.116** 2.3 *4*	**16.467** 2.4 *4*	**16.823** 2.4 *4*	**17.183** 2.4 *4*	**17.571** 2.4 *4*
$L_{\beta1}$	**20.018** 1.45 *24*	**20.624** 1.1 *2*	**21.248** 1.11 *20*	**21.889** 1.1 *3*	**22.549** 0.8 *3*	**23.227** 0.8 *3*	**23.927** 0.39 *20*	**24.650** 0.4 *2*
$L_{\beta2,15}$	**18.509** 5.4 *5*	**18.946** 5.6 *5*	**19.387** 5.7 *5*	**19.834** 5.7 *9*	**20.286** 5.8 *9*	**20.744** 5.9 *9*	**21.207** 6.1 *9*	**21.716** 6.1 *9*
$L_{\beta3}$	**20.280** 7.1 *12*	**20.894** 7.2 *12*	**21.523** 7.3 *13*	**22.169** 7.4 *15*	**22.835** 7.5 *15*	**23.519** 7.5 *15*	**24.223** 7.4 *15*	**24.872** 7.4 *15*
$L_{\beta4}$	**19.118** 9.1 *16*	**19.665** 9.5 *16*	**20.224** 10.0 *17*	**20.798** 10.5 *21*	**21.386** 11.1 *22*	**21.990** 11.6 *23*	**22.609** 11.8 *24*	**23.143** 12.4 *25*
$L_{\beta5}$	**19.312** 1.28 *11*	**19.777** 1.32 *12*	**20.249** 1.37 *12*	**20.727** 1.39 *21*	**21.210** 1.43 *22*	**21.700** 1.44 *22*	**22.195** 1.51 *23*	**22.727** 1.54 *23*
$L_{\beta6}$	**17.687** 0.41 *4*	**18.094** 0.43 *4*	**18.501** 0.44 *4*	**18.916** 0.45 *7*	**19.332** 0.46 *7*	**19.754** 0.47 *7*	**20.179** 0.49 *8*	**20.670** 0.50 *8*
$L_{\gamma1}$	**23.416** 0.37 *5*	**24.117** 0.28 *4*	**24.836** 0.29 *4*	**25.574** 0.29 *5*	**26.333** 0.20 *4*	**27.110** 0.21 *4*	**27.911** 0.10 *3*	**28.753** 0.11 *3*
$L_{\gamma2}$	**23.698** 4.1 *7*	**24.390** 4.3 *8*	**25.099** 4.5 *8*	**25.825** 4.8 *10*	**26.571** 5 *1*	**27.336** 5.3 *11*	**28.120** 5.4 *11*	**28.846** 5.6 *12*
$L_{\gamma3}$	**24.007** 3.3 *6*	**24.718** 3.4 *6*	**25.446** 3.4 *6*	**26.195** 3.5 *7*	**26.963** 3.6 *8*	**27.752** 3.7 *8*	**28.560** 3.6 *8*	**29.327** 3.7 *8*
$L_{\gamma6}$	**24.241** 0.077 *10*	**24.971** 0.059 *8*	**25.723** 0.060 *8*	**26.492** 0.063 *9*	**27.284** 0.044 *7*	**28.094** 0.046 *7*	**28.929** 0.024 *4*	**29.796** 0.025 *4*
L_{η}	**17.826** 0.040 *4*	**18.347** 0.031 *3*	**18.884** 0.031 *3*	**19.433** 0.032 *4*	**19.998** 0.0216 *25*	**20.577** 0.022 *3*	**21.173** 0.0113 *15*	**21.825** 0.0116 *15*
L_{ℓ}	**12.890** 1.51 *18*	**13.146** 1.56 *18*	**13.403** 1.62 *19*	**13.660** 1.6 *3*	**13.916** 1.7 *3*	**14.173** 1.7 *3*	**14.429** 1.8 *3*	**14.746** 1.8 *3*

Table 7c. X-ray Energies and Intensities (per 100 L_2-Shell Vacancies)

	$_{13}$Al	$_{14}$Si	$_{15}$P	$_{16}$S	$_{17}$Cl	$_{18}$Ar	$_{19}$K	$_{20}$Ca	$_{21}$Sc	$_{22}$Ti	$_{23}$V	$_{24}$Cr	$_{25}$Mn	$_{26}$Fe
$L_{\beta1}$	**0.073** 0.075 *23*	**0.099** 0.037 *11*	**0.136** 0.031 *9*	**0.165** 0.026 *8*	**0.202** 0.024 *7*	**0.251** 0.022 *7*	**0.296** 0.027 *8*	**0.350** 0.033 *8*	**0.400** 0.042 *11*	**0.458** 0.10 *3*	**0.518** 0.21 *5*	**0.581** 0.32 *8*	**0.648** 0.44 *11*	**0.717** 0.57 *14*
L_{η}									**0.353** 0.042 *10*	**0.401** 0.046 *12*	**0.454** 0.055 *14*	**0.510** 0.054 *13*	**0.568** 0.058 *14*	**0.628** 0.062 *16*

	$_{27}$Co	$_{28}$Ni	$_{29}$Cu	$_{30}$Zn	$_{31}$Ga	$_{32}$Ge	$_{33}$As	$_{34}$Se	$_{35}$Br	$_{36}$Kr	$_{37}$Rb	$_{38}$Sr	$_{39}$Y	$_{40}$Zr
$L_{\alpha1}$		**0.851** 0.022 *12*	**0.929** 0.026 *14*	**1.012** 0.026 *8*	**1.098** 0.035 *11*	**1.188** 0.064 *20*	**1.282** 0.09 *3*	**1.379** 0.12 *4*	**1.481** 0.15 *5*	**1.581** 0.19 *6*	**1.694** 0.22 *7*	**1.806** 0.26 *8*	**1.923** 0.30 *10*	**2.042** 0.35 *9*
$L_{\alpha2}$		**0.851** 0.0024 *13*	**0.929** 0.0029 *16*	**1.012** 0.0029 *9*	**1.098** 0.0039 *12*	**1.188** 0.0071 *23*	**1.282** 0.010 *3*	**1.379** 0.013 *4*	**1.480** 0.017 *5*	**1.580** 0.021 *7*	**1.693** 0.025 *8*	**1.805** 0.029 *9*	**1.920** 0.033 *11*	**2.040** 0.039 *10*
$L_{\beta1}$	**0.791** 0.71 *18*	**0.868** 0.80 *20*	**0.949** 0.93 *23*	**1.035** 1.0 *3*	**1.125** 1.1 *3*	**1.219** 1.2 *3*	**1.317** 1.3 *3*	**1.420** 1.5 *4*	**1.526** 1.7 *4*	**1.632** 1.9 *5*	**1.752** 2.1 *5*	**1.872** 2.3 *6*	**1.996** 2.5 *6*	**2.124** 2.6 *4*
$L_{\beta2,15}$														**2.219** 0.0025 *6*
$L_{\beta6}$											**1.775** 0.0012 *4*	**1.902** 0.0015 *5*	**2.035** 0.0018 *6*	**2.171** 0.0021 *5*
$L_{\gamma1}$													**2.153** 0.034 *9*	**2.304** 0.086 *13*
L_{η}	**0.693** 0.064 *16*	**0.760** 0.061 *15*	**0.831** 0.067 *17*	**0.907** 0.070 *18*	**0.984** 0.074 *18*	**1.068** 0.077 *19*	**1.155** 0.080 *20*	**1.245** 0.088 *22*	**1.339** 0.095 *24*	**1.435** 0.102 *25*	**1.542** 0.11 *3*	**1.649** 0.11 *3*	**1.762** 0.12 *3*	**1.876** 0.119 *18*
L_{ℓ}		**0.743** 0.0019 *11*	**0.811** 0.0021 *12*	**0.884** 0.0019 *6*	**0.957** 0.0024 *8*	**1.037** 0.0041 *13*	**1.120** 0.0052 *17*	**1.204** 0.0067 *22*	**1.293** 0.008 *3*	**1.383** 0.010 *3*	**1.482** 0.011 *4*	**1.582** 0.013 *4*	**1.686** 0.015 *5*	**1.792** 0.016 *4*

	$_{41}$Nb	$_{42}$Mo	$_{43}$Tc	$_{44}$Ru	$_{45}$Rh	$_{46}$Pd	$_{47}$Ag	$_{48}$Cd	$_{49}$In	$_{50}$Sn	$_{51}$Sb	$_{52}$Te	$_{53}$I	$_{54}$Xe
$L_{\alpha1}$	**2.166** 0.39 *10*	**2.293** 0.43 *11*	**2.424** 0.47 *12*	**2.558** 0.51 *13*	**2.697** 0.55 *14*	**2.839** 0.58 *15*	**2.984** 0.62 *15*	**3.134** 0.67 *17*	**3.287** 0.72 *18*	**3.444** 0.76 *17*	**3.605** 0.81 *18*	**3.769** 0.86 *19*	**3.938** 0.9 *2*	**4.106** 0.97 *22*
$L_{\alpha2}$	**2.163** 0.043 *11*	**2.290** 0.048 *12*	**2.420** 0.052 *13*	**2.554** 0.056 *14*	**2.692** 0.061 *15*	**2.833** 0.064 *16*	**2.978** 0.069 *17*	**3.127** 0.074 *19*	**3.279** 0.080 *20*	**3.435** 0.085 *19*	**3.595** 0.090 *20*	**3.759** 0.095 *21*	**3.926** 0.100 *23*	**4.093** 0.107 *24*
$L_{\beta1}$	**2.257** 2.9 *4*	**2.395** 3.1 *5*	**2.537** 3.4 *5*	**2.683** 3.6 *5*	**2.834** 3.8 *6*	**2.990** 4.1 *6*	**3.151** 4.5 *7*	**3.317** 4.9 *7*	**3.487** 5.3 *8*	**3.663** 5.6 *6*	**3.843** 6.0 *6*	**4.029** 6.4 *7*	**4.221** 6.8 *7*	**4.414** 7.1 *7*
$L_{\beta2,15}$	**2.367** 0.012 *3*	**2.518** 0.022 *6*	**2.675** 0.034 *9*	**2.836** 0.047 *12*	**3.001** 0.057 *14*	**3.172** 0.068 *17*	**3.348** 0.08 *2*	**3.528** 0.095 *24*	**3.714** 0.11 *3*	**3.905** 0.12 *3*	**4.101** 0.14 *3*	**4.302** 0.15 *4*	**4.508** 0.17 *4*	**4.714** 0.19 *4*
$L_{\beta6}$	**2.312** 0.0024 *6*	**2.458** 0.0027 *7*	**2.609** 0.0029 *7*	**2.763** 0.0033 *8*	**2.923** 0.0036 *9*	**3.087** 0.0038 *10*	**3.256** 0.0041 *10*	**3.430** 0.0046 *12*	**3.608** 0.0050 *13*	**3.792** 0.0055 *12*	**3.980** 0.0060 *13*	**4.173** 0.0065 *15*	**4.371** 0.0070 *16*	**4.569** 0.0076 *17*
$L_{\gamma1}$	**2.462** 0.125 *19*	**2.623** 0.17 *3*	**2.791** 0.21 *3*	**2.965** 0.26 *4*	**3.144** 0.35 *5*	**3.329** 0.44 *7*	**3.520** 0.47 *7*	**3.718** 0.52 *8*	**3.922** 0.60 *9*	**4.132** 0.67 *7*	**4.349** 0.73 *8*	**4.572** 0.81 *9*	**4.802** 0.91 *10*	**5.034** 0.99 *10*
L_{η}	**1.996** 0.127 *19*	**2.120** 0.133 *20*	**2.249** 0.139 *21*	**2.382** 0.144 *22*	**2.519** 0.147 *22*	**2.660** 0.154 *23*	**2.806** 0.163 *25*	**2.957** 0.17 *3*	**3.112** 0.18 *3*	**3.272** 0.189 *19*	**3.437** 0.195 *20*	**3.606** 0.204 *21*	**3.780** 0.210 *21*	**3.955** 0.213 *22*
L_{ℓ}	**1.902** 0.018 *5*	**2.016** 0.019 *5*	**2.133** 0.020 *5*	**2.253** 0.022 *6*	**2.377** 0.023 *6*	**2.503** 0.024 *6*	**2.634** 0.025 *7*	**2.767** 0.027 *7*	**2.905** 0.029 *8*	**3.045** 0.031 *7*	**3.189** 0.032 *8*	**3.335** 0.034 *8*	**3.485** 0.036 *9*	**3.634** 0.039 *9*

	$_{55}$Cs	$_{56}$Ba	$_{57}$La	$_{58}$Ce	$_{59}$Pr	$_{60}$Nd	$_{61}$Pm	$_{62}$Sm	$_{63}$Eu	$_{64}$Gd	$_{65}$Tb	$_{66}$Dy	$_{67}$Ho	$_{68}$Er
$L_{\alpha1}$	**4.286** 1.03 *23*	**4.466** 1.09 *24*	**4.651** 1.2 *3*	**4.840** 1.2 *3*	**5.033** 1.3 *3*	**5.230** 1.38 *25*	**5.432** 1.5 *3*	**5.636** 1.5 *3*	**5.846** 1.6 *3*	**6.058** 1.7 *3*	**6.273** 1.7 *3*	**6.495** 1.8 *3*	**6.720** 1.9 *3*	**6.949** 2.0 *4*
$L_{\alpha2}$	**4.272** 0.11 *3*	**4.451** 0.12 *3*	**4.634** 0.13 *3*	**4.822** 0.14 *3*	**5.013** 0.15 *3*	**5.208** 0.15 *3*	**5.408** 0.16 *3*	**5.610** 0.17 *3*	**5.816** 0.18 *3*	**6.026** 0.18 *3*	**6.239** 0.19 *3*	**6.458** 0.20 *4*	**6.680** 0.21 *4*	**6.905** 0.22 *4*
$L_{\beta1}$	**4.620** 7.7 *8*	**4.828** 8.2 *8*	**5.042** 8.7 *9*	**5.263** 9.3 *10*	**5.489** 9.9 *10*	**5.722** 10.5 *8*	**5.961** 11.1 *8*	**6.206** 11.8 *9*	**6.457** 12.5 *9*	**6.713** 13.2 *10*	**6.977** 14 *1*	**7.248** 14.9 *11*	**7.526** 15.8 *11*	**7.811** 16.7 *12*
$L_{\beta2,15}$	**4.934** 0.21 *5*	**5.156** 0.22 *5*	**5.384** 0.24 *6*	**5.613** 0.26 *6*	**5.851** 0.28 *6*	**6.090** 0.29 *5*	**6.339** 0.31 *6*	**6.587** 0.32 *6*	**6.844** 0.33 *6*	**7.102** 0.35 *6*	**7.367** 0.36 *6*	**7.636** 0.37 *7*	**7.910** 0.38 *7*	**8.186** 0.39 *7*
$L_{\beta5}$			**5.483** 0.0024 *6*							**7.243** 0.0028 *5*				
$L_{\beta6}$	**4.781** 0.0084 *19*	**4.994** 0.009 *2*	**5.212** 0.0097 *22*	**5.434** 0.0105 *24*	**5.660** 0.0113 *25*	**5.893** 0.0122 *22*	**6.128** 0.0131 *23*	**6.370** 0.0141 *25*	**6.617** 0.015 *3*	**6.867** 0.016 *3*	**7.116** 0.018 *3*	**7.374** 0.019 *3*	**7.635** 0.020 *4*	**7.909** 0.022 *4*
$L_{\gamma1}$	**5.281** 1.10 *12*	**5.531** 1.19 *12*	**5.792** 1.30 *14*	**6.054** 1.43 *15*	**6.327** 1.55 *16*	**6.604** 1.67 *13*	**6.892** 1.81 *14*	**7.183** 1.94 *15*	**7.484** 2.10 *16*	**7.790** 2.25 *17*	**8.105** 2.40 *18*	**8.426** 2.59 *20*	**8.757** 2.78 *21*	**9.088** 2.97 *23*
$L_{\gamma6}$			**5.891** 0.0140 *18*							**7.930** 0.0185 *20*				
L_{η}	**4.142** 0.224 *23*	**4.331** 0.233 *24*	**4.529** 0.242 *25*	**4.730** 0.25 *3*	**4.929** 0.26 *3*	**5.146** 0.272 *20*	**5.363** 0.28 *2*	**5.589** 0.288 *21*	**5.817** 0.300 *22*	**6.049** 0.311 *23*	**6.284** 0.321 *23*	**6.534** 0.335 *24*	**6.789** 0.347 *25*	**7.058** 0.36 *3*
L_{ℓ}	**3.795** 0.041 *10*	**3.954** 0.044 *10*	**4.121** 0.047 *11*	**4.289** 0.051 *12*	**4.453** 0.054 *13*	**4.633** 0.057 *11*	**4.809** 0.060 *12*	**4.993** 0.063 *12*	**5.177** 0.067 *13*	**5.362** 0.070 *14*	**5.546** 0.073 *14*	**5.743** 0.077 *15*	**5.943** 0.081 *16*	**6.151** 0.085 *17*

	$_{69}$Tm	$_{70}$Yb	$_{71}$Lu	$_{72}$Hf	$_{73}$Ta	$_{74}$W	$_{75}$Re	$_{76}$Os	$_{77}$Ir	$_{78}$Pt	$_{79}$Au	$_{80}$Hg	$_{81}$Tl	$_{82}$Pb
$L_{\alpha1}$	**7.180** 2.0 *4*	**7.416** 2.1 *3*	**7.656** 2.2 *3*	**7.899** 2.2 *4*	**8.146** 2.3 *4*	**8.398** 2.4 *4*	**8.652** 2.5 *4*	**8.911** 2.5 *4*	**9.175** 2.6 *4*	**9.443** 2.6 *4*	**9.713** 2.7 *4*	**9.989** 2.7 *4*	**10.268** 2.8 *4*	**10.551** 2.8 *4*
$L_{\alpha2}$	**7.133** 0.23 *4*	**7.367** 0.24 *4*	**7.605** 0.24 *4*	**7.844** 0.25 *4*	**8.088** 0.26 *4*	**8.335** 0.27 *4*	**8.586** 0.27 *4*	**8.840** 0.28 *4*	**9.099** 0.29 *5*	**9.362** 0.29 *5*	**9.628** 0.30 *5*	**9.899** 0.31 *5*	**10.172** 0.31 *5*	**10.450** 0.32 *5*
$L_{\beta1}$	**8.102** 17.6 *13*	**8.402** 18.5 *10*	**8.709** 19.4 *10*	**9.023** 20.4 *11*	**9.343** 21.3 *11*	**9.672** 22.2 *12*	**10.010** 23.1 *12*	**10.354** 24.0 *13*	**10.708** 24.9 *13*	**11.071** 25.8 *14*	**11.443** 26.7 *14*	**11.824** 27.6 *15*	**12.213** 28.5 *15*	**12.614** 29.4 *16*
$L_{\beta2,15}$	**8.468** 0.40 *7*	**8.752** 0.41 *7*	**9.044** 0.44 *7*	**9.342** 0.47 *8*	**9.646** 0.51 *8*	**9.955** 0.54 *9*	**10.268** 0.56 *9*	**10.590** 0.59 *9*	**10.912** 0.61 *10*	**11.242** 0.63 *10*	**11.576** 0.65 *10*	**11.915** 0.67 *10*	**12.261** 0.69 *11*	**12.611** 0.70 *11*

Table 7c. X-ray Energies and Intensities (per 100 L_2-Shell Vacancies) (continued)

	$_{69}$Tm	$_{70}$Yb	$_{71}$Lu	$_{72}$Hf	$_{73}$Ta	$_{74}$W	$_{75}$Re	$_{76}$Os	$_{77}$Ir	$_{78}$Pt	$_{79}$Au	$_{80}$Hg	$_{81}$Tl	$_{82}$Pb
continued														
$L_{\beta5}$			**9.240** 0.0026 *4*	**9.554** 0.0067 *11*	**9.875** 0.0092 *15*	**10.201** 0.0120 *19*	**10.532** 0.022 *4*	**10.871** 0.033 *5*	**11.211** 0.042 *7*	**11.562** 0.052 *8*	**11.916** 0.062 *10*	**12.275** 0.072 *11*	**12.643** 0.081 *12*	**13.015** 0.091 *14*
$L_{\beta6}$	**8.176** 0.023 *4*	**8.456** 0.025 *4*	**8.738** 0.026 *4*	**9.023** 0.027 *4*	**9.316** 0.029 *4*	**9.612** 0.030 *5*	**9.910** 0.032 *5*	**10.217** 0.034 *5*	**10.525** 0.036 *6*	**10.840** 0.038 *6*	**11.160** 0.039 *6*	**11.481** 0.041 *6*	**11.812** 0.043 *7*	**12.142** 0.044 *7*
$L_{\gamma1}$	**9.437** 3.16 *24*	**9.780** 3.36 *20*	**10.144** 3.55 *21*	**10.516** 3.75 *22*	**10.895** 3.96 *23*	**11.285** 4.17 *24*	**11.685** 4.4 *3*	**12.096** 4.6 *3*	**12.513** 4.9 *3*	**12.942** 5.1 *3*	**13.382** 5.4 *3*	**13.830** 5.6 *3*	**14.291** 5.9 *3*	**14.765** 6.2 *4*
$L_{\gamma6}$			**10.344** 0.0214 *20*	**10.733** 0.057 *5*	**11.130** 0.102 *10*	**11.538** 0.160 *15*	**11.955** 0.28 *3*	**12.385** 0.38 *4*	**12.820** 0.50 *5*	**13.270** 0.62 *6*	**13.731** 0.72 *7*	**14.199** 0.86 *8*	**14.683** 0.97 *9*	**15.178** 1.06 *10*
L_η	**7.310** 0.37 *3*	**7.580** 0.388 *21*	**7.857** 0.406 *22*	**8.139** 0.424 *23*	**8.428** 0.445 *24*	**8.724** 0.466 *25*	**9.027** 0.49 *3*	**9.337** 0.51 *3*	**9.650** 0.54 *3*	**9.975** 0.56 *3*	**10.309** 0.59 *3*	**10.647** 0.62 *3*	**10.994** 0.65 *4*	**11.349** 0.68 *4*
L_ℓ	**6.341** 0.090 *18*	**6.545** 0.094 *16*	**6.753** 0.098 *17*	**6.960** 0.102 *18*	**7.173** 0.108 *19*	**7.387** 0.114 *20*	**7.604** 0.119 *21*	**7.822** 0.124 *22*	**8.042** 0.129 *23*	**8.266** 0.135 *24*	**8.494** 0.141 *25*	**8.722** 0.147 *25*	**8.953** 0.15 *3*	**9.184** 0.16 *3*

	$_{83}$Bi	$_{84}$Po	$_{85}$At	$_{86}$Rn	$_{87}$Fr	$_{88}$Ra	$_{89}$Ac	$_{90}$Th	$_{91}$Pa	$_{92}$U	$_{93}$Np	$_{94}$Pu	$_{95}$Am	$_{96}$Cm
$L_{\alpha1}$	**10.839** 2.9 *4*	**11.130** 2.9 *5*	**11.426** 3.0 *5*	**11.726** 3.0 *5*	**12.031** 3.1 *5*	**12.339** 3.1 *5*	**12.651** 3.2 *5*	**12.968** 3.3 *5*	**13.291** 4.3 *7*	**13.618** 5.4 *8*	**13.946** 6.3 *10*	**14.282** 6.6 *10*	**14.620** 6.9 *11*	**14.961** 7.0 *11*
$L_{\alpha2}$	**10.731** 0.32 *5*	**11.016** 0.32 *5*	**11.306** 0.33 *5*	**11.598** 0.34 *5*	**11.896** 0.35 *5*	**12.196** 0.35 *5*	**12.500** 0.36 *6*	**12.809** 0.37 *6*	**13.127** 0.49 *8*	**13.442** 0.60 *9*	**13.761** 0.70 *11*	**14.087** 0.74 *12*	**14.414** 0.78 *12*	**14.746** 0.78 *12*
$L_{\beta1}$	**13.024** 30.4 *16*	**13.443** 31.3 *17*	**13.875** 32.3 *17*	**14.316** 33.2 *18*	**14.770** 34.1 *18*	**15.236** 35.0 *19*	**15.711** 35.8 *19*	**16.202** 37 *4*	**16.708** 36 *4*	**17.222** 35 *4*	**17.751** 35 *4*	**18.296** 35 *4*	**18.856** 36 *4*	**19.427** 36 *4*
$L_{\beta2,15}$	**12.967** 0.71 *11*	**13.328** 0.73 *11*	**13.694** 0.75 *12*	**14.066** 0.77 *12*	**14.443** 0.79 *12*	**14.825** 0.81 *13*	**15.212** 0.84 *13*	**15.605** 0.86 *14*	**16.008** 1.15 *18*	**16.410** 1.41 *22*	**16.817** 1.7 *3*	**17.235** 1.8 *3*	**17.655** 1.9 *3*	**18.081** 1.9 *3*
$L_{\beta5}$	**13.393** 0.099 *15*	**13.778** 0.108 *17*	**14.168** 0.119 *18*	**14.565** 0.129 *20*	**14.967** 0.144 *22*	**15.375** 0.148 *23*	**15.790** 0.159 *25*	**16.209** 0.17 *3*	**16.639** 0.23 *4*	**17.069** 0.30 *5*	**17.505** 0.42 *7*	**17.950** 0.39 *6*	**18.399** 0.42 *7*	**18.853** 0.44 *7*
$L_{\beta6}$	**12.480** 0.045 *7*	**12.823** 0.047 *7*	**13.169** 0.049 *8*	**13.520** 0.051 *8*	**13.877** 0.053 *8*	**14.236** 0.055 *9*	**14.601** 0.058 *9*	**14.970** 0.060 *9*	**15.350** 0.081 *13*	**15.727** 0.101 *16*	**16.109** 0.120 *19*	**16.498** 0.13 *2*	**16.890** 0.138 *22*	**17.286** 0.141 *22*
$L_{\gamma1}$	**15.248** 6.4 *4*	**15.742** 6.7 *4*	**16.249** 7.1 *4*	**16.770** 7.4 *4*	**17.302** 7.7 *5*	**17.848** 8.0 *5*	**18.405** 8.3 *5*	**18.980** 8.6 *9*	**19.571** 8.5 *9*	**20.169** 8.5 *9*	**20.784** 8.6 *9*	**21.420** 8.6 *9*	**22.072** 8.8 *9*	**22.735** 9.0 *9*
$L_{\gamma6}$	**15.685** 1.18 *11*	**16.203** 1.30 *12*	**16.735** 1.40 *13*	**17.280** 1.51 *14*	**17.839** 1.61 *15*	**18.412** 1.70 *16*	**18.997** 1.77 *17*	**19.599** 1.84 *23*	**20.217** 1.82 *23*	**20.844** 1.81 *23*	**21.491** 1.81 *23*	**22.153** 1.81 *23*	**22.836** 1.84 *23*	**23.527** 1.87 *24*
L_η	**11.712** 0.71 *4*	**12.085** 0.75 *4*	**12.466** 0.78 *4*	**12.855** 0.82 *4*	**13.255** 0.85 *5*	**13.662** 0.88 *5*	**14.082** 0.91 *5*	**14.511** 0.95 *10*	**14.953** 0.94 *10*	**15.400** 0.94 *10*	**15.861** 0.95 *10*	**16.333** 0.95 *10*	**16.819** 0.97 *10*	**17.314** 0.99 *10*
L_ℓ	**9.420** 0.16 *3*	**9.658** 0.17 *3*	**9.897** 0.18 *3*	**10.137** 0.18 *3*	**10.381** 0.19 *3*	**10.622** 0.20 *3*	**10.871** 0.21 *4*	**11.118** 0.22 *4*	**11.372** 0.29 *5*	**11.620** 0.36 *6*	**11.871** 0.43 *8*	**12.124** 0.47 *8*	**12.377** 0.50 *9*	**12.633** 0.51 *9*

	$_{97}$Bk	$_{98}$Cf	$_{99}$Es	$_{100}$Fm	$_{101}$Md	$_{102}$No	$_{103}$Lr	104
$L_{\alpha1}$	**15.308** 7.1 *11*	**15.660** 7.1 *11*	**16.016** 7.2 *11*	**16.377** 7.2 *15*	**16.741** 7.2 *15*	**17.110** 7.2 *15*	**17.483** 7.1 *15*	**17.893** 7.0 *15*
$L_{\alpha2}$	**15.082** 0.79 *12*	**15.423** 0.80 *13*	**15.767** 0.81 *13*	**16.116** 0.81 *17*	**16.467** 0.81 *17*	**16.823** 0.81 *17*	**17.183** 0.80 *17*	**17.571** 0.79 *16*
$L_{\beta1}$	**20.018** 36 *4*	**20.624** 37 *4*	**21.248** 37 *4*	**21.889** 38 *4*	**22.549** 38 *4*	**23.227** 39 *4*	**23.927** 39 *4*	**24.650** 40 *4*
$L_{\beta2,15}$	**18.509** 1.9 *3*	**18.946** 1.9 *3*	**19.387** 2.0 *3*	**19.834** 2.0 *4*	**20.286** 2.0 *4*	**20.744** 2.0 *4*	**21.207** 2.0 *4*	**21.716** 2.0 *4*
$L_{\beta5}$	**19.312** 0.45 *7*	**19.777** 0.46 *7*	**20.249** 0.47 *7*	**20.727** 0.48 *10*	**21.210** 0.49 *10*	**21.700** 0.49 *10*	**22.195** 0.50 *10*	**22.727** 0.5 *1*
$L_{\beta6}$	**17.687** 0.145 *23*	**18.094** 0.149 *23*	**18.501** 0.153 *24*	**18.916** 0.16 *3*	**19.332** 0.16 *3*	**19.754** 0.16 *3*	**20.179** 0.16 *3*	**20.670** 0.16 *3*
$L_{\gamma1}$	**23.416** 9.2 *10*	**24.117** 9.4 *10*	**24.836** 9.5 *10*	**25.574** 9.8 *10*	**26.333** 10 *1*	**27.110** 10.2 *11*	**27.911** 10.5 *11*	**28.753** 10.7 *11*
$L_{\gamma6}$	**24.241** 1.93 *25*	**24.971** 1.98 *25*	**25.723** 2.0 *3*	**26.492** 2.1 *3*	**27.284** 2.2 *3*	**28.094** 2.3 *3*	**28.929** 2.4 *3*	**29.796** 2.5 *3*
L_η	**17.826** 1.01 *10*	**18.347** 1.02 *10*	**18.884** 1.04 *11*	**19.433** 1.06 *11*	**19.998** 1.08 *11*	**20.577** 1.11 *11*	**21.173** 1.13 *12*	**21.825** 1.16 *12*
L_ℓ	**12.890** 0.53 *9*	**13.146** 0.54 *10*	**13.403** 0.56 *10*	**13.660** 0.57 *13*	**13.916** 0.58 *13*	**14.173** 0.59 *13*	**14.429** 0.59 *13*	**14.746** 0.59 *13*

Table 7d. X-ray Energies and Intensities (per 100 L_3-Shell Vacancies)

	$_{21}$Sc	$_{22}$Ti	$_{23}$V	$_{24}$Cr	$_{25}$Mn	$_{26}$Fe	$_{27}$Co	$_{28}$Ni	$_{29}$Cu	$_{30}$Zn	$_{31}$Ga	$_{32}$Ge	$_{33}$As	$_{34}$Se
$L_{\alpha1}$	**0.396** 0.038 *10*	**0.452** 0.094 *24*	**0.511** 0.19 *5*	**0.572** 0.29 *7*	**0.637** 0.39 *10*	**0.704** 0.51 *13*	**0.776** 0.63 *16*	**0.851** 0.77 *19*	**0.929** 0.92 *23*	**1.012** 1.01 *20*	**1.098** 1.10 *22*	**1.188** 1.3 *3*	**1.282** 1.4 *3*	**1.379** 1.5 *3*
$L_{\alpha2}$	**0.396** 0.0042 *11*	**0.452** 0.010 *3*	**0.511** 0.020 *5*	**0.572** 0.032 *8*	**0.637** 0.044 *11*	**0.704** 0.056 *14*	**0.776** 0.070 *18*	**0.851** 0.086 *22*	**0.929** 0.10 *3*	**1.012** 0.112 *23*	**1.098** 0.122 *25*	**1.188** 0.14 *3*	**1.282** 0.15 *3*	**1.379** 0.17 *3*
$L_{\beta6}$	**0.402** 0.0026 *6*	**0.456** 0.0027 *7*	**0.513** 0.0033 *8*	**0.574** 0.0012 *3*	**0.640** 0.0035 *9*	**0.708** 0.0034 *9*	**0.779** 0.0034 *9*	**0.855** 0.0034 *9*	**0.931** 0.0013 *3*	**1.020** 0.0032 *7*	**1.114** 0.0042 *8*	**1.212** 0.0052 *11*	**1.315** 0.0060 *12*	**1.424** 0.0071 *14*
L_ℓ	**0.348** 0.039 *10*	**0.395** 0.043 *11*	**0.446** 0.051 *13*	**0.500** 0.051 *13*	**0.556** 0.057 *15*	**0.615** 0.062 *16*	**0.678** 0.067 *18*	**0.743** 0.069 *18*	**0.811** 0.075 *20*	**0.884** 0.074 *16*	**0.957** 0.076 *16*	**1.037** 0.082 *18*	**1.120** 0.083 *18*	**1.204** 0.088 *19*

Table 7d. X-ray Energies and Intensities (per 100 L_3-Shell Vacancies) (continued)

	$_{35}$Br	$_{36}$Kr	$_{37}$Rb	$_{38}$Sr	$_{39}$Y	$_{40}$Zr	$_{41}$Nb	$_{42}$Mo	$_{43}$Tc	$_{44}$Ru	$_{45}$Rh	$_{46}$Pd	$_{47}$Ag	$_{48}$Cd
$L_{\alpha1}$	**1.481** 1.7 3	**1.581** 1.9 4	**1.694** 2.1 4	**1.806** 2.2 5	**1.923** 2.4 5	**2.042** 2.7 4	**2.166** 2.9 4	**2.293** 3.1 5	**2.424** 3.2 5	**2.558** 3.4 5	**2.697** 3.6 6	**2.839** 3.8 6	**2.984** 4.0 6	**3.134** 4.3 7
$L_{\alpha2}$	**1.480** 0.19 4	**1.580** 0.21 4	**1.693** 0.23 5	**1.805** 0.25 5	**1.920** 0.27 5	**2.040** 0.29 5	**2.163** 0.32 5	**2.290** 0.34 5	**2.420** 0.36 5	**2.554** 0.38 6	**2.692** 0.40 6	**2.833** 0.43 7	**2.978** 0.45 7	**3.127** 0.48 7
$L_{\beta2,15}$					**2.078** 0.0070 14	**2.219** 0.018 3	**2.367** 0.088 13	**2.518** 0.159 24	**2.675** 0.24 4	**2.836** 0.32 5	**3.001** 0.38 6	**3.172** 0.45 7	**3.348** 0.52 8	**3.528** 0.61 9
$L_{\beta6}$	**1.523** 0.0082 16	**1.647** 0.0094 19	**1.775** 0.0109 22	**1.902** 0.0125 25	**2.035** 0.014 3	**2.171** 0.0159 24	**2.312** 0.017 3	**2.458** 0.019 3	**2.609** 0.020 3	**2.763** 0.022 3	**2.923** 0.024 4	**3.087** 0.025 4	**3.256** 0.027 4	**3.430** 0.030 5
L_{ℓ}	**1.293** 0.094 20	**1.383** 0.099 21	**1.482** 0.105 23	**1.582** 0.110 24	**1.686** 0.115 25	**1.792** 0.124 21	**1.902** 0.130 22	**2.016** 0.135 23	**2.133** 0.141 24	**2.253** 0.147 25	**2.377** 0.15 3	**2.503** 0.16 3	**2.634** 0.17 3	**2.767** 0.17 3

	$_{49}$In	$_{50}$Sn	$_{51}$Sb	$_{52}$Te	$_{53}$I	$_{54}$Xe	$_{55}$Cs	$_{56}$Ba	$_{57}$La	$_{58}$Ce	$_{59}$Pr	$_{60}$Nd	$_{61}$Pm	$_{62}$Sm
$L_{\alpha1}$	**3.287** 4.6 7	**3.444** 4.9 5	**3.605** 5.2 5	**3.769** 5.5 6	**3.938** 5.9 6	**4.106** 6.3 6	**4.286** 6.7 7	**4.466** 7.1 7	**4.651** 7.6 8	**4.840** 8.1 8	**5.033** 8.6 9	**5.230** 9.1 5	**5.432** 9.6 5	**5.636** 10.1 5
$L_{\alpha2}$	**3.279** 0.51 8	**3.435** 0.54 6	**3.595** 0.58 6	**3.759** 0.61 6	**3.926** 0.65 7	**4.093** 0.70 7	**4.272** 0.74 8	**4.451** 0.79 8	**4.634** 0.84 9	**4.822** 0.90 9	**5.013** 0.95 10	**5.208** 1.01 5	**5.408** 1.07 6	**5.610** 1.12 6
$L_{\beta2,15}$	**3.714** 0.71 11	**3.905** 0.78 8	**4.101** 0.88 9	**4.302** 1.00 10	**4.508** 1.10 12	**4.714** 1.22 13	**4.934** 1.34 14	**5.156** 1.47 15	**5.384** 1.59 17	**5.613** 1.70 18	**5.851** 1.82 19	**6.090** 1.94 11	**6.339** 2.04 12	**6.587** 2.14 13
$L_{\beta5}$									**5.483** 0.0159 16					
$L_{\beta6}$	**3.608** 0.032 5	**3.792** 0.035 4	**3.980** 0.038 4	**4.173** 0.042 4	**4.371** 0.045 5	**4.569** 0.050 5	**4.781** 0.054 6	**4.994** 0.059 6	**5.212** 0.064 7	**5.434** 0.069 7	**5.660** 0.074 8	**5.893** 0.080 4	**6.128** 0.087 5	**6.370** 0.094 5
L_{ℓ}	**2.905** 0.18 3	**3.045** 0.194 25	**3.189** 0.21 3	**3.335** 0.22 3	**3.485** 0.23 3	**3.634** 0.25 3	**3.795** 0.27 3	**3.954** 0.28 4	**4.121** 0.31 4	**4.289** 0.33 4	**4.453** 0.35 5	**4.633** 0.38 4	**4.809** 0.40 4	**4.993** 0.42 4

	$_{63}$Eu	$_{64}$Gd	$_{65}$Tb	$_{66}$Dy	$_{67}$Ho	$_{68}$Er	$_{69}$Tm	$_{70}$Yb	$_{71}$Lu	$_{72}$Hf	$_{73}$Ta	$_{74}$W	$_{75}$Re	$_{76}$Os
$L_{\alpha1}$	**5.846** 10.7 6	**6.058** 11.3 6	**6.273** 12.0 7	**6.495** 12.7 7	**6.720** 13.3 7	**6.949** 14.1 8	**7.180** 14.7 8	**7.416** 15.3 7	**7.656** 15.9 7	**7.899** 16.5 7	**8.146** 17.3 8	**8.398** 18.0 8	**8.652** 18.8 8	**8.911** 19.6 9
$L_{\alpha2}$	**5.816** 1.19 6	**6.026** 1.25 7	**6.239** 1.33 7	**6.458** 1.41 8	**6.680** 1.48 8	**6.905** 1.56 8	**7.133** 1.64 9	**7.367** 1.71 8	**7.605** 1.78 8	**7.844** 1.85 8	**8.088** 1.93 9	**8.335** 2.02 9	**8.586** 2.11 9	**8.840** 2.19 10
$L_{\beta2,15}$	**6.844** 2.25 13	**7.102** 2.35 14	**7.367** 2.47 14	**7.636** 2.61 15	**7.910** 2.69 16	**8.186** 2.81 16	**8.468** 2.90 17	**8.752** 2.96 15	**9.044** 3.22 16	**9.342** 3.52 18	**9.646** 3.80 19	**9.955** 4.09 20	**10.268** 4.33 22	**10.590** 4.59 23
$L_{\beta5}$		**7.243** 0.0192 10							**9.240** 0.0190 9	**9.554** 0.0495 22	**9.875** 0.069 3	**10.201** 0.090 4	**10.532** 0.171 8	**10.871** 0.259 12
$L_{\beta6}$	**6.617** 0.103 6	**6.867** 0.112 6	**7.116** 0.122 7	**7.374** 0.133 7	**7.635** 0.144 8	**7.909** 0.157 9	**8.176** 0.168 9	**8.456** 0.179 8	**8.738** 0.189 8	**9.023** 0.200 9	**9.316** 0.212 10	**9.612** 0.225 10	**9.910** 0.246 11	**10.217** 0.268 12
L_{ℓ}	**5.177** 0.45 4	**5.362** 0.47 5	**5.546** 0.51 5	**5.743** 0.54 5	**5.943** 0.57 5	**6.151** 0.61 6	**6.341** 0.65 6	**6.545** 0.68 6	**6.753** 0.72 6	**6.960** 0.76 7	**7.173** 0.81 7	**7.387** 0.86 8	**7.604** 0.91 8	**7.822** 0.97 9

	$_{77}$Ir	$_{78}$Pt	$_{79}$Au	$_{80}$Hg	$_{81}$Tl	$_{82}$Pb	$_{83}$Bi	$_{84}$Po	$_{85}$At	$_{86}$Rn	$_{87}$Fr	$_{88}$Ra	$_{89}$Ac	$_{90}$Th
$L_{\alpha1}$	**9.175** 20.4 9	**9.443** 21.1 10	**9.713** 22.0 10	**9.989** 22.8 8	**10.268** 23.7 9	**10.551** 24.4 9	**10.839** 25.2 9	**11.130** 26.0 9	**11.426** 26.8 10	**11.726** 27.5 10	**12.031** 28.2 10	**12.339** 29 1	**12.651** 29.8 11	**12.968** 30.6 14
$L_{\alpha2}$	**9.099** 2.28 10	**9.362** 2.37 11	**9.628** 2.46 11	**9.899** 2.55 9	**10.172** 2.65 10	**10.450** 2.74 10	**10.731** 2.83 10	**11.016** 2.91 10	**11.306** 3.00 11	**11.598** 3.08 11	**11.896** 3.16 11	**12.196** 3.25 12	**12.500** 3.34 12	**12.809** 3.42 15
$L_{\beta2,15}$	**10.912** 4.83 24	**11.242** 5.07 25	**11.576** 5.3 3	**11.915** 5.58 24	**12.261** 5.82 25	**12.611** 6.1 3	**12.967** 6.3 3	**13.328** 6.5 3	**13.694** 6.8 3	**14.066** 7.0 3	**14.443** 7.3 3	**14.825** 7.5 3	**15.212** 7.8 3	**15.605** 8.0 4
$L_{\beta5}$	**11.211** 0.337 15	**11.562** 0.418 19	**11.916** 0.506 23	**12.275** 0.597 22	**12.643** 0.688 25	**13.015** 0.79 3	**13.393** 0.88 3	**13.778** 0.97 4	**14.168** 1.07 4	**14.565** 1.17 4	**14.967** 1.32 5	**15.375** 1.37 5	**15.790** 1.48 5	**16.209** 1.58 7
$L_{\beta6}$	**10.525** 0.286 13	**10.840** 0.302 13	**11.160** 0.321 14	**11.481** 0.342 12	**11.812** 0.362 13	**12.142** 0.381 14	**12.480** 0.401 15	**12.823** 0.422 15	**13.169** 0.442 16	**13.520** 0.462 17	**13.877** 0.486 18	**14.236** 0.511 18	**14.601** 0.533 19	**14.970** 0.556 25
L_{ℓ}	**8.042** 1.03 9	**8.266** 1.09 10	**8.494** 1.15 10	**8.722** 1.22 10	**8.953** 1.29 11	**9.184** 1.36 12	**9.420** 1.44 12	**9.658** 1.51 13	**9.897** 1.58 13	**10.137** 1.65 14	**10.381** 1.73 15	**10.622** 1.82 15	**10.871** 1.91 16	**11.118** 2.00 18

	$_{91}$Pa	$_{92}$U	$_{93}$Np	$_{94}$Pu	$_{95}$Am	$_{96}$Cm	$_{97}$Bk	$_{98}$Cf	$_{99}$Es	$_{100}$Fm	$_{101}$Md	$_{102}$No	$_{103}$Lr	104
$L_{\alpha1}$	**13.291** 31.3 14	**13.618** 32.1 14	**13.946** 32.6 15	**14.282** 33.5 15	**14.620** 34.2 15	**14.961** 35.0 16	**15.308** 35.6 16	**15.660** 36.2 16	**16.016** 36.7 16	**16.377** 37.2 20	**16.741** 37.7 20	**17.110** 38.1 20	**17.483** 38.5 21	**17.893** 38.8 21
$L_{\alpha2}$	**13.127** 3.51 16	**13.442** 3.59 16	**13.761** 3.66 16	**14.087** 3.76 17	**14.414** 3.83 17	**14.746** 3.92 17	**15.082** 3.99 18	**15.423** 4.05 18	**15.767** 4.11 18	**16.116** 4.17 22	**16.467** 4.22 23	**16.823** 4.27 23	**17.183** 4.31 23	**17.571** 4.35 23
$L_{\beta2,15}$	**16.008** 8.2 4	**16.410** 8.5 4	**16.817** 8.7 4	**17.235** 9.0 5	**17.655** 9.2 5	**18.081** 9.4 5	**18.509** 9.6 5	**18.946** 9.8 5	**19.387** 10.0 5	**19.834** 10.2 6	**20.286** 10.4 6	**20.744** 10.6 6	**21.207** 10.8 6	**21.716** 11.0 6
$L_{\beta5}$	**16.639** 1.69 8	**17.069** 1.79 8	**17.505** 2.20 10	**17.950** 1.99 9	**18.399** 2.08 9	**18.853** 2.19 10	**19.312** 2.27 10	**19.777** 2.34 10	**20.249** 2.41 11	**20.727** 2.48 13	**21.210** 2.55 14	**21.700** 2.62 14	**22.195** 2.70 15	**22.727** 2.77 15
$L_{\beta6}$	**15.350** 0.58 3	**15.727** 0.61 3	**16.109** 0.63 3	**16.498** 0.65 3	**16.890** 0.68 3	**17.286** 0.70 3	**17.687** 0.73 3	**18.094** 0.76 3	**18.501** 0.78 4	**18.916** 0.80 4	**19.332** 0.83 5	**19.754** 0.85 5	**20.179** 0.87 5	**20.670** 0.90 5
L_{ℓ}	**11.372** 2.09 19	**11.620** 2.18 19	**11.871** 2.25 20	**12.124** 2.35 21	**12.377** 2.46 22	**12.633** 2.57 23	**12.890** 2.67 24	**13.146** 2.76 25	**13.403** 2.9 3	**13.660** 2.9 3	**13.916** 3.0 3	**14.173** 3.1 3	**14.429** 3.2 3	**14.746** 3.3 3

6. Auger-Electron Intensities

Table 8 lists intensities for K-Auger electrons whose intensities are greater than 0.001 per 100 vacancies in the K atomic shell. The first column identifies the Auger transitions, using the notation $K-XY$, where X and Y represent the inner atomic shells involved in the K-Auger-electron emission process. The following columns give, for each element, the K-Auger-electron intensities (I_{K-XY}). These have been derived from theoretical emission probabilities through the relationship

$$I_{K-XY} = 100\,(1-\omega_K)\frac{P_{K-XY}}{\sum P_{K-XY}}\,, \tag{1}$$

where P_{K-XY} is the theoretical emission probability of a $K-XY$ Auger electron, from Chen, et al.,[1] ω_K is the K fluorescence yield, from Krause,[2] and the summation is over all Auger electrons which are energetically possible.

The tables of atomic electrons in the *Table of Radioactive Isotopes* do not list single-line Auger-electron intensities. These may be calculated on an absolute scale per 100 disintegrations, however, by normalizing the Auger-electron intensities (per 100 vacancies in the K atomic shell) through the K-x-ray absolute intensities from the photons tables. The $K-XY$-Auger-electron intensity is then given by

$$I_{K-XY} = \frac{I_{Ki}}{I^0_{Ki}}\, I^0_{K-XY}\,, \tag{2}$$

where I_{Ki} is the absolute intensity of the specific K-x-ray transition i from the photons tables, I^0_{Ki} is the intensity of the same K-x-ray transition from Table 7a, and I^0_{K-XY} is the corresponding Auger-electron intensity from Table 8. As an example, the uranium I_{K-L1L1} absolute Auger-electron intensity per 100 disintegrations of ^{235}Np is

$$I_{K-L1L1} = \frac{I_{K\alpha 1}}{I^0_{K\alpha 1}} I^0_{K-L1L1} = \frac{0.957}{45.1} 0.33 = 0.0070\,.$$

$I_{K\alpha 1}$ is from the photons table for ^{235}Np on page 235-2, and $I^0_{K\alpha 1}$ and I^0_{K-L1L1} are from Table 7a and Table 8, respectively.

Approximate Auger-electron energies can be calculated with the empirical equations of Dillman.[3] The average energy for a $K-L_iX$ Auger transition is given by

$$\bar{E}(K-L_iX) = E_K - E_{Li} - E_{M3} - 0.75(E_{M3+} - E_{M3})\,, \tag{3}$$

and for higher atomic shells, by

$$\bar{E}(K-XY) = E_K - 2E_{M3} - 0.75(E_{M3+} - E_{M3})\,. \tag{4}$$

E_{Li} is the binding energy of the L_i atomic shell for the element, E_K and E_{M3} are the corresponding binding energies of the K and M_3 atomic shells, E_{M3+} is the binding energy of the M_3 atomic subshell for the next higher element, and X and Y are designations for the higher atomic subshells. For more precise Auger-electron energies, one is referred to the publication of Larkins.[4]

---

[1]M.H. Chen, B. Crasemann, and H. Mark, *At. Data and Nucl. Data Tables* **24**, 13 (1979).

[2]M.O. Krause, *J. Phys. Chem. Ref. Data* **8**, 307 (1979).

[3]*EDISTR - A Computer Program to Obtain a Nuclear Decay Data Base for Radiation Dosimetry,* L.T. Dillman, Oak Ridge National Lab., Report **ORNL/TM-6689** (1980).

[4]F.B. Larkins, *At. Data and Nucl. Data Tables,* **20**, 313 (1977).

Table 8. Auger-Electron Intensities per 100 K-Shell Vacancies

	$_{18}$Ar	$_{19}$K	$_{20}$Ca	$_{21}$Sc	$_{22}$Ti	$_{23}$V	$_{24}$Cr	$_{25}$Mn	$_{26}$Fe	$_{27}$Co	$_{28}$Ni	$_{29}$Cu	$_{30}$Zn	$_{31}$Ga	$_{32}$Ge
K-L_1L_1	6.1	5.8	5.6	5.4	5.2	5.0	4.8	4.6	4.4	4.2	4.0	3.8	3.6	3.4	3.3
K-L_1L_2	6.7	6.4	6.2	6.0	5.8	5.6	5.4	5.1	4.9	4.7	4.5	4.3	4.0	3.8	3.7
K-L_1L_3	12.8	12.2	11.7	11.3	10.9	10.4	10.0	9.4	9.0	8.5	8.0	7.5	7.1	6.7	6.3
K-L_1M_1	1.35	1.39	1.40	1.38	1.36	1.32	1.28	1.23	1.17	1.12	1.07	1.02	0.97	0.94	0.91
K-L_1M_2	0.63	0.69	0.72	0.71	0.70	0.69	0.68	0.65	0.62	0.60	0.57	0.55	0.53	0.52	0.51
K-L_1M_3	1.20	1.30	1.35	1.34	1.31	1.28	1.24	1.19	1.13	1.08	1.02	0.97	0.92	0.89	0.86
K-L_1M_4								0.030	0.029	0.029	0.029	0.029	0.029	0.032	0.033
K-L_1M_5								0.0076	0.0176	0.026	0.032	0.037	0.040	0.042	0.044
K-L_1N_1			0.159	0.150	0.141	0.132	0.123	0.113	0.097	0.085	0.078	0.075	0.074	0.083	0.089
K-L_2L_2	1.23	1.18	1.13	1.10	1.07	1.03	0.99	0.94	0.90	0.85	0.81	0.76	0.72	0.68	0.64
K-L_2L_3	30	29	28	27	26	24.9	23.8	22.5	21.4	20.3	19.1	18.0	16.8	15.8	14.9
K-L_2M_1	0.65	0.67	0.67	0.66	0.65	0.63	0.61	0.59	0.56	0.54	0.51	0.49	0.46	0.45	0.43
K-L_2M_2	0.221	0.239	0.248	0.247	0.244	0.239	0.233	0.224	0.214	0.204	0.195	0.185	0.176	0.170	0.165
K-L_2M_3	2.53	2.7	2.8	2.8	2.7	2.7	2.6	2.46	2.33	2.21	2.10	1.98	1.88	1.81	1.74
K-L_2M_4								0.043	0.042	0.041	0.041	0.041	0.041	0.043	0.044
K-L_2M_5								0.028	0.065	0.096	0.119	0.137	0.148	0.156	0.162
K-L_2N_1			0.074	0.070	0.066	0.062	0.057	0.053	0.045	0.040	0.037	0.035	0.034	0.038	0.041
K-L_3L_3	17.2	16.5	15.7	15.2	14.6	14.0	13.3	12.6	12.0	11.3	10.6	9.9	9.3	8.6	8.0
K-L_3M_1	1.25	1.28	1.28	1.25	1.22	1.18	1.13	1.08	1.02	0.97	0.92	0.86	0.81	0.72	0.66
K-L_3M_2	2.53	2.7	2.8	2.8	2.7	2.7	2.6	2.46	2.34	2.22	2.10	1.99	1.88	1.66	1.54
K-L_3M_3	2.9	3.1	3.2	3.2	3.1	3.0	2.9	2.8	2.6	2.50	2.37	2.24	2.11	1.88	1.74
K-L_3M_4								0.194	0.188	0.184	0.181	0.180	0.179	0.144	0.132
K-L_3M_5			0.142	0.083	0.046	0.028	0.0241	0.033	0.080	0.117	0.146	0.167	0.181	0.145	0.133
K-L_3N_1								0.098	0.083	0.073	0.066	0.062	0.060	0.037	0.030
K-M_1M_1	0.075	0.084	0.089	0.088	0.087	0.086	0.083	0.081	0.077	0.074	0.071	0.068	0.066	0.052	0.047
K-M_1M_2	0.063	0.072	0.078	0.079	0.080	0.079	0.078	0.075	0.071	0.068	0.065	0.063	0.060	0.046	0.040
K-M_1M_3	0.117	0.137	0.149	0.149	0.147	0.145	0.141	0.136	0.129	0.123	0.117	0.112	0.107	0.087	0.078
K-M_1N_1			0.0213	0.0199	0.0187	0.0175	0.0163	0.0151	0.0130	0.0115	0.0107	0.0103	0.0104	0.0032	
K-M_2M_3	0.217	0.26	0.29	0.30	0.29	0.29	0.28	0.27	0.26	0.245	0.234	0.223	0.214	0.177	0.160
K-M_2N_1			0.0106	0.0100	0.0094	0.0088	0.0082	0.0076	0.0065	0.0058	0.0054	0.0052	0.0052	0.00214	0.00120
K-M_3M_3	0.125	0.153	0.170	0.172	0.171	0.169	0.165	0.158	0.149	0.140	0.133	0.126	0.121	0.10	0.091
K-M_3M_4								0.0176	0.0171	0.0169	0.0169	0.0170	0.0173	0.0123	0.0108
K-M_3M_5								0.00252	0.0070	0.0106	0.0135	0.0157	0.0173	0.0123	0.0108
K-M_3N_1			0.0177	0.0166	0.0156	0.0146	0.0136	0.0126	0.0108	0.0096	0.0089	0.0086	0.0086	0.0043	0.0030

	$_{33}$As	$_{34}$Se	$_{35}$Br	$_{36}$Kr	$_{37}$Rb	$_{38}$Sr	$_{39}$Y	$_{40}$Zr	$_{41}$Nb	$_{42}$Mo	$_{43}$Tc	$_{44}$Ru	$_{45}$Rh	$_{46}$Pd	$_{47}$Ag
K-L_1L_1	3.1	2.9	2.6	2.40	2.27	2.14	2.02	1.90	1.79	1.67	1.58	1.49	1.40	1.32	1.25
K-L_1L_2	3.5	3.2	3.0	2.7	2.55	2.41	2.28	2.15	2.03	1.90	1.80	1.71	1.61	1.53	1.45
K-L_1L_3	5.9	5.5	4.9	4.4	4.1	3.9	3.6	3.3	3.1	2.9	2.7	2.49	2.31	2.15	2.01
K-L_1M_1	0.88	0.83	0.77	0.71	0.68	0.65	0.63	0.60	0.57	0.54	0.51	0.49	0.47	0.44	0.42
K-L_1M_2	0.49	0.47	0.44	0.41	0.39	0.38	0.36	0.35	0.33	0.32	0.31	0.29	0.28	0.27	0.26
K-L_1M_3	0.83	0.78	0.72	0.66	0.63	0.60	0.57	0.54	0.51	0.48	0.45	0.43	0.40	0.38	0.36
K-L_1M_4	0.034	0.034	0.033	0.031	0.031	0.030	0.030	0.029	0.028	0.027	0.026	0.0255	0.0245	0.0237	0.0229
K-L_1M_5	0.045	0.044	0.041	0.039	0.039	0.038	0.037	0.036	0.034	0.033	0.032	0.030	0.029	0.027	0.026
K-L_1N_1	0.094	0.096	0.094	0.092	0.096	0.098	0.099	0.098	0.096	0.094	0.091	0.089	0.086	0.084	0.081
K-L_1N_2			0.040	0.040	0.044	0.047	0.049	0.050	0.049	0.048	0.048	0.047	0.047	0.046	0.045
K-L_1N_3			0.049	0.065	0.071	0.074	0.076	0.076	0.075	0.073	0.071	0.068	0.066	0.064	0.062
K-L_1O_1													0.0042	0.0033	0.0036
K-L_2L_2	0.60	0.56	0.50	0.46	0.43	0.40	0.37	0.34	0.32	0.30	0.28	0.26	0.239	0.223	0.208
K-L_2L_3	13.9	12.9	11.5	10.4	9.7	9.0	8.3	7.7	7.1	6.6	6.1	5.7	5.2	4.8	4.5
K-L_2M_1	0.42	0.39	0.36	0.34	0.32	0.31	0.30	0.28	0.27	0.26	0.246	0.235	0.223	0.214	0.206
K-L_2M_2	0.158	0.149	0.137	0.126	0.120	0.114	0.108	0.102	0.097	0.091	0.085	0.080	0.075	0.071	0.067
K-L_2M_3	1.66	1.57	1.44	1.32	1.25	1.18	1.12	1.05	0.99	0.93	0.87	0.82	0.76	0.72	0.68
K-L_2M_4	0.045	0.044	0.043	0.041	0.040	0.039	0.039	0.037	0.037	0.035	0.034	0.032	0.031	0.029	0.028
K-L_2M_5	0.164	0.162	0.155	0.147	0.145	0.141	0.137	0.132	0.128	0.122	0.117	0.111	0.106	0.101	0.096
K-L_2N_1	0.043	0.044	0.044	0.043	0.045	0.046	0.046	0.045	0.044	0.043	0.042	0.041	0.040	0.039	0.038
K-L_2N_2			0.0123	0.0130	0.0140	0.0144	0.0146	0.0144	0.0139	0.0133	0.0130	0.0126	0.0122	0.0119	0.0116
K-L_2N_3			0.098	0.128	0.138	0.144	0.147	0.145	0.141	0.136	0.132	0.128	0.122	0.117	0.112
K-L_2N_5													0.0075	0.0095	0.0108
K-L_2O_1													0.00188	0.00145	0.00160
K-L_3L_3	7.5	6.9	6.3	5.7	5.3	4.9	4.5	4.1	3.8	3.5	3.2	3.0	2.7	2.53	2.34
K-L_3M_1	0.63	0.62	0.61	0.61	0.56	0.51	0.47	0.44	0.41	0.39	0.36	0.34	0.32	0.30	0.28
K-L_3M_2	1.47	1.45	1.44	1.46	1.33	1.22	1.13	1.05	0.99	0.92	0.86	0.81	0.76	0.71	0.67
K-L_3M_3	1.66	1.63	1.59	1.60	1.46	1.34	1.24	1.15	1.08	1.0	0.94	0.88	0.82	0.77	0.72

Table 8. Auger-Electron Intensities per 100 K-Shell Vacancies (continued)

	$_{33}$As	$_{34}$Se	$_{35}$Br	$_{36}$Kr	$_{37}$Rb	$_{38}$Sr	$_{39}$Y	$_{40}$Zr	$_{41}$Nb	$_{42}$Mo	$_{43}$Tc	$_{44}$Ru	$_{45}$Rh	$_{46}$Pd	$_{47}$Ag
continued															
K-L_3M_4	0.138	0.158	0.183	0.214	0.194	0.177	0.164	0.154	0.147	0.140	0.133	0.127	0.120	0.114	0.108
K-L_3M_5	0.140	0.160	0.185	0.217	0.196	0.179	0.167	0.156	0.150	0.142	0.135	0.129	0.122	0.116	0.110
K-L_3N_1	0.036	0.052	0.073	0.098	0.088	0.081	0.075	0.070	0.068	0.065	0.062	0.060	0.057	0.054	0.052
K-L_3N_2			0.130	0.202	0.181	0.165	0.154	0.145	0.141	0.135	0.132	0.128	0.122	0.117	0.112
K-L_3N_3			0.108	0.222	0.20	0.182	0.170	0.160	0.154	0.148	0.143	0.138	0.132	0.125	0.120
K-L_3N_4				0.0090	0.0058	0.0045	0.0049	0.0065	0.010	0.0121	0.0125	0.0127	0.0127	0.0124	0.0124
K-L_3N_5										0.00181	0.0041	0.0064	0.0085	0.0109	0.0124
K-L_3O_1				0.0120	0.0055	0.00170			0.00132	0.0036	0.0033	0.0031	0.0028	0.00229	0.00241
K-M_1M_1	0.046	0.050	0.056	0.064	0.058	0.052	0.049	0.046	0.045	0.043	0.041	0.040	0.038	0.037	0.036
K-M_1M_2	0.041	0.046	0.054	0.064	0.058	0.052	0.049	0.046	0.045	0.043	0.042	0.040	0.039	0.038	0.037
K-M_1M_3	0.077	0.081	0.089	0.099	0.090	0.082	0.076	0.071	0.068	0.065	0.062	0.059	0.056	0.053	0.050
K-M_1N_1	0.00249	0.0071	0.0135	0.0210	0.0185	0.0167	0.0157	0.0151	0.0153	0.0151	0.0148	0.0145	0.0141	0.0139	0.0136
K-M_1N_2			0.0045	0.0090	0.0078	0.0071	0.0067	0.0065	0.0066	0.0066	0.0067	0.0067	0.0066	0.0065	0.0064
K-M_1N_3			0.0067	0.0140	0.0125	0.0113	0.0106	0.0101	0.0099	0.0097	0.0094	0.0091	0.0089	0.0089	0.0088
K-M_2M_3	0.158	0.168	0.182	0.202	0.183	0.167	0.155	0.145	0.139	0.132	0.126	0.119	0.113	0.107	0.102
K-M_2N_1	0.00193	0.0040	0.0067	0.010	0.0088	0.0079	0.0075	0.0072	0.0073	0.0073	0.0072	0.0072	0.0071	0.0069	0.0068
K-M_2N_3			0.0123	0.028	0.0249	0.0226	0.0212	0.0201	0.0199	0.0194	0.0189	0.0185	0.0179	0.0173	0.0168
K-M_3M_3	0.089	0.095	0.102	0.113	0.102	0.094	0.087	0.081	0.077	0.073	0.069	0.066	0.062	0.059	0.056
K-M_3M_4	0.0119	0.0150	0.0191	0.0240	0.0215	0.0195	0.0182	0.0173	0.0169	0.0163	0.0159	0.0153	0.0146	0.0139	0.0132
K-M_3M_5	0.0119	0.0150	0.0191	0.0240	0.0215	0.0195	0.0182	0.0173	0.0169	0.0163	0.0159	0.0153	0.0146	0.0139	0.0132
K-M_3N_1	0.0042	0.0072	0.0112	0.0160	0.0143	0.0130	0.0121	0.0115	0.0113	0.0109	0.0105	0.0102	0.0099	0.0098	0.0096
K-M_3N_2			0.0168	0.028	0.0249	0.0226	0.0212	0.0201	0.0198	0.0194	0.0192	0.0189	0.0183	0.0178	0.0172
K-M_3N_3			0.0135	0.031	0.028	0.0252	0.0235	0.0223	0.0218	0.0212	0.0209	0.0204	0.0198	0.0190	0.0184

	$_{48}$Cd	$_{49}$In	$_{50}$Sn	$_{51}$Sb	$_{52}$Te	$_{53}$I	$_{54}$Xe	$_{55}$Cs	$_{56}$Ba	$_{57}$La	$_{58}$Ce	$_{59}$Pr	$_{60}$Nd	$_{61}$Pm	$_{62}$Sm
K-L_1L_1	1.17	1.11	1.05	1.0	0.95	0.91	0.86	0.82	0.79	0.76	0.73	0.69	0.67	0.64	0.61
K-L_1L_2	1.37	1.30	1.24	1.18	1.14	1.09	1.04	1.0	0.97	0.94	0.90	0.87	0.85	0.82	0.79
K-L_1L_3	1.85	1.73	1.61	1.51	1.42	1.33	1.24	1.17	1.10	1.04	0.98	0.92	0.87	0.82	0.78
K-L_1M_1	0.40	0.38	0.36	0.35	0.34	0.32	0.31	0.30	0.29	0.28	0.27	0.26	0.248	0.240	0.231
K-L_1M_2	0.248	0.238	0.229	0.221	0.214	0.207	0.20	0.193	0.189	0.184	0.179	0.173	0.169	0.165	0.161
K-L_1M_3	0.34	0.32	0.30	0.28	0.27	0.26	0.241	0.228	0.218	0.208	0.197	0.187	0.179	0.170	0.162
K-L_1M_4	0.0217	0.0208	0.0199	0.0192	0.0185	0.0178	0.0170	0.0165	0.0160	0.0155	0.0149	0.0143	0.0138	0.0133	0.0128
K-L_1M_5	0.0245	0.0233	0.0221	0.0211	0.0202	0.0190	0.0180	0.0171	0.0164	0.0157	0.0149	0.0141	0.0135	0.0128	0.0121
K-L_1N_1	0.079	0.077	0.075	0.073	0.071	0.069	0.067	0.066	0.065	0.063	0.061	0.059	0.057	0.056	0.054
K-L_1N_2	0.044	0.044	0.043	0.043	0.042	0.042	0.041	0.041	0.041	0.040	0.039	0.038	0.038	0.037	0.036
K-L_1N_3	0.059	0.058	0.056	0.054	0.053	0.051	0.049	0.048	0.047	0.045	0.043	0.041	0.039	0.038	0.036
K-L_1O_1	0.0061	0.0079	0.0090	0.0095	0.0099	0.0101	0.0104	0.0112	0.0116	0.0112	0.0108	0.0103	0.0100	0.0096	0.0092
K-L_1O_2			0.0037	0.0042	0.0046	0.0047	0.0051	0.0058	0.0062	0.0061	0.0060	0.0058	0.0056	0.0053	0.0051
K-L_1O_3					0.0027	0.0045	0.0058	0.0066	0.0070	0.0066	0.0063	0.0059	0.0056	0.0053	0.0050
K-L_2L_2	0.192	0.179	0.167	0.156	0.147	0.138	0.128	0.121	0.114	0.108	0.101	0.095	0.090	0.085	0.080
K-L_2L_3	4.1	3.8	3.6	3.3	3.1	2.9	2.7	2.50	2.35	2.21	2.06	1.92	1.81	1.70	1.59
K-L_2M_1	0.195	0.187	0.179	0.173	0.167	0.161	0.155	0.150	0.146	0.142	0.138	0.133	0.130	0.127	0.124
K-L_2M_2	0.063	0.059	0.056	0.052	0.049	0.047	0.044	0.041	0.039	0.037	0.035	0.033	0.032	0.030	0.029
K-L_2M_3	0.63	0.59	0.55	0.52	0.49	0.46	0.43	0.41	0.39	0.36	0.34	0.32	0.31	0.29	0.27
K-L_2M_4	0.027	0.0255	0.0243	0.0233	0.0223	0.0213	0.0203	0.0192	0.0184	0.0177	0.0170	0.0162	0.0156	0.0150	0.0143
K-L_2M_5	0.091	0.086	0.081	0.077	0.074	0.070	0.066	0.063	0.060	0.057	0.054	0.051	0.049	0.046	0.044
K-L_2N_1	0.037	0.036	0.035	0.034	0.034	0.033	0.033	0.032	0.032	0.031	0.030	0.029	0.029	0.028	0.028
K-L_2N_2	0.0111	0.0107	0.0103	0.0100	0.0097	0.0093	0.0090	0.0087	0.0084	0.0081	0.0077	0.0073	0.0070	0.0066	0.0062
K-L_2N_3	0.107	0.103	0.099	0.096	0.093	0.089	0.086	0.083	0.080	0.076	0.072	0.069	0.065	0.062	0.058
K-L_2N_5	0.0113	0.0116	0.0118	0.0120	0.0121	0.0122	0.0122	0.0121	0.0120	0.0116	0.0112	0.0107	0.0103	0.0098	0.0093
K-L_2O_1	0.0030	0.0038	0.0044	0.0044	0.0046	0.0046	0.0048	0.0053	0.0056	0.0056	0.0054	0.0053	0.0051	0.0049	0.0047
K-L_2O_3					0.0048	0.0079	0.0099	0.0112	0.0118	0.0111	0.0105	0.0098	0.0092	0.0087	0.0081
K-L_3L_3	2.14	1.97	1.82	1.69	1.57	1.46	1.35	1.25	1.17	1.10	1.02	0.95	0.89	0.83	0.77
K-L_3M_1	0.26	0.247	0.232	0.219	0.207	0.195	0.183	0.173	0.165	0.156	0.148	0.140	0.133	0.126	0.120
K-L_3M_2	0.62	0.58	0.54	0.51	0.48	0.45	0.42	0.40	0.38	0.36	0.34	0.32	0.30	0.28	0.27
K-L_3M_3	0.67	0.62	0.58	0.54	0.51	0.48	0.45	0.42	0.40	0.37	0.35	0.33	0.31	0.29	0.27
K-L_3M_4	0.101	0.095	0.090	0.085	0.081	0.076	0.072	0.067	0.064	0.061	0.057	0.054	0.051	0.048	0.046
K-L_3M_5	0.103	0.097	0.092	0.087	0.082	0.078	0.073	0.069	0.065	0.062	0.058	0.055	0.052	0.049	0.047
K-L_3N_1	0.050	0.047	0.046	0.044	0.042	0.040	0.038	0.037	0.036	0.034	0.032	0.031	0.029	0.028	0.026
K-L_3N_2	0.107	0.103	0.099	0.095	0.092	0.088	0.084	0.081	0.078	0.075	0.071	0.067	0.064	0.061	0.057
K-L_3N_3	0.114	0.109	0.105	0.101	0.097	0.094	0.089	0.086	0.082	0.078	0.074	0.070	0.067	0.063	0.059
K-L_3N_4	0.0126	0.0128	0.0131	0.0135	0.0137	0.0136	0.0134	0.0131	0.0128	0.0123	0.0118	0.0112	0.0108	0.0103	0.0098
K-L_3N_5	0.0126	0.0128	0.0131	0.0135	0.0137	0.0137	0.0136	0.0134	0.0132	0.0126	0.0120	0.0114	0.0109	0.0104	0.0099

Table 8. Auger-Electron Intensities per 100 K-Shell Vacancies (continued)

	$_{48}$Cd	$_{49}$In	$_{50}$Sn	$_{51}$Sb	$_{52}$Te	$_{53}$I	$_{54}$Xe	$_{55}$Cs	$_{56}$Ba	$_{57}$La	$_{58}$Ce	$_{59}$Pr	$_{60}$Nd	$_{61}$Pm	$_{62}$Sm
continued															
K-L_3O_1	0.0037	0.0047	0.0053	0.0057	0.0059	0.0059	0.0060	0.0063	0.0064	0.0061	0.0058	0.0054	0.0051	0.0049	0.0046
K-L_3O_2			0.0084	0.0091	0.0097	0.0097	0.0101	0.0113	0.0118	0.0111	0.0104	0.0098	0.0092	0.0087	0.0082
K-L_3O_3					0.0051	0.0082	0.0104	0.0117	0.0124	0.0116	0.0108	0.010	0.0094	0.0088	0.0083
K-M_1M_1	0.034	0.033	0.032	0.030	0.029	0.028	0.027	0.026	0.026	0.0250	0.0243	0.0235	0.0229	0.0222	0.0214
K-M_1M_2	0.036	0.034	0.033	0.032	0.032	0.031	0.030	0.029	0.029	0.028	0.028	0.027	0.027	0.026	0.026
K-M_1M_3	0.048	0.045	0.043	0.041	0.039	0.037	0.035	0.034	0.032	0.031	0.030	0.028	0.027	0.026	0.0246
K-M_1N_1	0.0133	0.0130	0.0128	0.0125	0.0124	0.0122	0.0120	0.0118	0.0116	0.0113	0.0110	0.0107	0.0104	0.0102	0.0099
K-M_1N_2	0.0063	0.0063	0.0062	0.0062	0.0062	0.0062	0.0062	0.0062	0.0062	0.0061	0.0061	0.0060	0.0059	0.0058	0.0057
K-M_1N_3	0.0085	0.0083	0.0081	0.0079	0.0078	0.0076	0.0074	0.0072	0.0070	0.0067	0.0064	0.0061	0.0059	0.0057	0.0055
K-M_2M_3	0.096	0.091	0.086	0.081	0.077	0.073	0.069	0.066	0.063	0.060	0.057	0.054	0.051	0.049	0.046
K-M_2N_1	0.0067	0.0066	0.0065	0.0065	0.0065	0.0065	0.0065	0.0063	0.0062	0.0061	0.0061	0.0060	0.0059	0.0058	0.0058
K-M_2N_3	0.0164	0.0160	0.0156	0.0152	0.0148	0.0143	0.0138	0.0134	0.0130	0.0125	0.0119	0.0114	0.0109	0.0104	0.0099
K-M_3M_3	0.052	0.049	0.047	0.044	0.042	0.040	0.038	0.036	0.034	0.032	0.031	0.029	0.028	0.026	0.0247
K-M_3M_4	0.0126	0.0121	0.0115	0.0110	0.0105	0.0100	0.0094	0.0090	0.0086	0.0082	0.0078	0.0074	0.0071	0.0068	0.0065
K-M_3M_5	0.0126	0.0121	0.0115	0.0110	0.0105	0.0100	0.0094	0.0090	0.0086	0.0082	0.0078	0.0074	0.0071	0.0068	0.0065
K-M_3N_1	0.0091	0.0087	0.0084	0.0082	0.0081	0.0078	0.0076	0.0074	0.0072	0.0069	0.0066	0.0063	0.0061	0.0058	0.0056
K-M_3N_2	0.0165	0.0160	0.0156	0.0153	0.0150	0.0146	0.0140	0.0136	0.0132	0.0127	0.0122	0.0117	0.0112	0.0107	0.0102
K-M_3N_3	0.0178	0.0173	0.0168	0.0165	0.0161	0.0156	0.0150	0.0145	0.0140	0.0135	0.0129	0.0123	0.0118	0.0113	0.0107

	$_{63}$Eu	$_{64}$Gd	$_{65}$Tb	$_{66}$Dy	$_{67}$Ho	$_{68}$Er	$_{69}$Tm	$_{70}$Yb	$_{71}$Lu	$_{72}$Hf	$_{73}$Ta	$_{74}$W	$_{75}$Re	$_{76}$Os	$_{77}$Ir
K-L_1L_1	0.60	0.58	0.56	0.54	0.51	0.49	0.48	0.47	0.45	0.44	0.42	0.42	0.41	0.40	0.39
K-L_1L_2	0.78	0.76	0.74	0.72	0.70	0.68	0.67	0.66	0.65	0.63	0.62	0.62	0.62	0.60	0.60
K-L_1L_3	0.74	0.70	0.67	0.63	0.60	0.56	0.54	0.51	0.49	0.46	0.44	0.43	0.41	0.39	0.38
K-L_1M_1	0.225	0.219	0.212	0.206	0.198	0.191	0.187	0.182	0.178	0.173	0.167	0.166	0.164	0.158	0.157
K-L_1M_2	0.158	0.155	0.152	0.149	0.145	0.141	0.140	0.138	0.136	0.134	0.131	0.131	0.130	0.126	0.126
K-L_1M_3	0.155	0.149	0.142	0.135	0.129	0.122	0.117	0.113	0.108	0.103	0.099	0.096	0.094	0.089	0.086
K-L_1M_4	0.0124	0.0120	0.0116	0.0112	0.0108	0.0103	0.010	0.0097	0.0094	0.0091	0.0087	0.0086	0.0084	0.0080	0.0079
K-L_1M_5	0.0115	0.0110	0.0105	0.0100	0.0095	0.0090	0.0086	0.0082	0.0078	0.0074	0.0070	0.0068	0.0066	0.0062	0.0060
K-L_1N_1	0.053	0.051	0.050	0.048	0.047	0.045	0.044	0.043	0.042	0.041	0.040	0.040	0.040	0.039	0.039
K-L_1N_2	0.036	0.035	0.034	0.034	0.033	0.032	0.032	0.032	0.031	0.031	0.031	0.031	0.031	0.031	0.031
K-L_1N_3	0.035	0.033	0.032	0.031	0.029	0.028	0.027	0.026	0.0248	0.0238	0.0229	0.0224	0.0220	0.0210	0.0206
K-L_1O_1	0.0089	0.0086	0.0083	0.0080	0.0077	0.0073	0.0071	0.0070	0.0071	0.0071	0.0071	0.0073	0.0074	0.0074	0.0075
K-L_1O_2	0.0050	0.0049	0.0047	0.0046	0.0045	0.0042	0.0042	0.0042	0.0043	0.0044	0.0045	0.0048	0.0050	0.0050	0.0052
K-L_1O_3	0.0047	0.0045	0.0043	0.0040	0.0038	0.0035	0.0034	0.0032	0.0032	0.0032	0.0032	0.0033	0.0033	0.0033	0.0033
K-L_2L_2	0.076	0.072	0.068	0.065	0.061	0.057	0.055	0.052	0.050	0.047	0.045	0.043	0.042	0.039	0.038
K-L_2L_3	1.50	1.42	1.34	1.25	1.17	1.10	1.04	0.99	0.93	0.88	0.82	0.79	0.76	0.71	0.68
K-L_2M_1	0.122	0.119	0.117	0.114	0.111	0.108	0.107	0.106	0.104	0.103	0.101	0.101	0.101	0.099	0.099
K-L_2M_2	0.027	0.026	0.0248	0.0235	0.0222	0.0210	0.0201	0.0192	0.0183	0.0174	0.0165	0.0160	0.0155	0.0147	0.0142
K-L_2M_3	0.26	0.248	0.235	0.222	0.210	0.197	0.188	0.179	0.170	0.161	0.153	0.148	0.143	0.134	0.129
K-L_2M_4	0.0138	0.0132	0.0127	0.0121	0.0115	0.0110	0.0106	0.0102	0.0098	0.0094	0.0090	0.0088	0.0086	0.0082	0.0079
K-L_2M_5	0.042	0.040	0.038	0.036	0.034	0.032	0.031	0.029	0.028	0.026	0.0250	0.0241	0.0233	0.0219	0.0210
K-L_2N_1	0.027	0.027	0.026	0.026	0.0253	0.0247	0.0245	0.0243	0.0240	0.0237	0.0234	0.0236	0.0238	0.0233	0.0235
K-L_2N_2	0.0060	0.0057	0.0055	0.0052	0.0050	0.0047	0.0045	0.0043	0.0041	0.0039	0.0038	0.0037	0.0036	0.0034	0.0033
K-L_2N_3	0.056	0.053	0.051	0.048	0.046	0.043	0.041	0.039	0.037	0.036	0.034	0.033	0.032	0.031	0.030
K-L_2N_5	0.0089	0.0086	0.0082	0.0078	0.0074	0.0071	0.0068	0.0066	0.0063	0.0060	0.0058	0.0056	0.0055	0.0053	0.0051
K-L_2O_1	0.0046	0.0045	0.0044	0.0043	0.0042	0.0040	0.0040	0.0039	0.0040	0.0041	0.0041	0.0043	0.0044	0.0044	0.0045
K-L_2O_3	0.0077	0.0073	0.0068	0.0064	0.0059	0.0053	0.0050	0.0047	0.0047	0.0047	0.0047	0.0048	0.0048	0.0047	0.0048
K-L_3L_3	0.73	0.68	0.64	0.60	0.56	0.52	0.49	0.46	0.44	0.41	0.38	0.36	0.35	0.33	0.31
K-L_3M_1	0.115	0.109	0.104	0.099	0.094	0.089	0.085	0.081	0.078	0.074	0.071	0.069	0.067	0.063	0.061
K-L_3M_2	0.253	0.241	0.228	0.215	0.203	0.190	0.181	0.173	0.164	0.155	0.146	0.141	0.136	0.128	0.123
K-L_3M_3	0.26	0.247	0.233	0.219	0.206	0.193	0.183	0.174	0.165	0.156	0.147	0.141	0.136	0.127	0.122
K-L_3M_4	0.043	0.041	0.039	0.037	0.034	0.032	0.031	0.029	0.028	0.026	0.0246	0.0237	0.0227	0.0213	0.0203
K-L_3M_5	0.044	0.042	0.040	0.038	0.035	0.033	0.031	0.030	0.028	0.027	0.0250	0.0240	0.0231	0.0215	0.0206
K-L_3N_1	0.0254	0.0244	0.0233	0.0222	0.0211	0.0199	0.0192	0.0184	0.0177	0.0169	0.0162	0.0158	0.0154	0.0146	0.0142
K-L_3N_2	0.055	0.052	0.050	0.047	0.044	0.042	0.040	0.038	0.036	0.035	0.033	0.032	0.031	0.029	0.028
K-L_3N_3	0.056	0.054	0.051	0.048	0.045	0.042	0.040	0.038	0.037	0.035	0.033	0.032	0.031	0.029	0.028
K-L_3N_4	0.0094	0.0090	0.0085	0.0081	0.0076	0.0072	0.0069	0.0066	0.0063	0.0060	0.0057	0.0056	0.0054	0.0051	0.0050
K-L_3N_5	0.0096	0.0092	0.0087	0.0083	0.0078	0.0073	0.0070	0.0066	0.0064	0.0061	0.0058	0.0056	0.0055	0.0052	0.0050
K-L_3O_1	0.0043	0.0041	0.0039	0.0036	0.0034	0.0032	0.0031	0.0030	0.0030	0.0029	0.0029	0.0029	0.0029	0.0028	0.0028
K-L_3O_2	0.0078	0.0074	0.0070	0.0066	0.0061	0.0055	0.0051	0.0049	0.0049	0.0049	0.0049	0.0050	0.0049	0.0047	0.0047
K-L_3O_3	0.0078	0.0073	0.0069	0.0064	0.0059	0.0053	0.0050	0.0047	0.0048	0.0048	0.0047	0.0048	0.0047	0.0045	0.0045
K-M_1M_1	0.0210	0.0205	0.020	0.0195	0.0189	0.0183	0.0179	0.0176	0.0171	0.0167	0.0163	0.0162	0.0160	0.0155	0.0154
K-M_1M_2	0.0254	0.0251	0.0247	0.0243	0.0239	0.0234	0.0232	0.0231	0.0228	0.0226	0.0222	0.0224	0.0225	0.0221	0.0222
K-M_1M_3	0.0237	0.0228	0.0219	0.0209	0.0199	0.0190	0.0183	0.0176	0.0169	0.0162	0.0155	0.0152	0.0148	0.0141	0.0137

Table 8. Auger-Electron Intensities per 100 K-Shell Vacancies (continued)

	$_{63}$Eu	$_{64}$Gd	$_{65}$Tb	$_{66}$Dy	$_{67}$Ho	$_{68}$Er	$_{69}$Tm	$_{70}$Yb	$_{71}$Lu	$_{72}$Hf	$_{73}$Ta	$_{74}$W	$_{75}$Re	$_{76}$Os	$_{77}$Ir
continued															
K-M_1N_1	0.0098	0.0097	0.0095	0.0092	0.0090	0.0086	0.0085	0.0083	0.0082	0.0080	0.0078	0.0078	0.0078	0.0076	0.0076
K-M_1N_2	0.0057	0.0057	0.0056	0.0055	0.0054	0.0053	0.0053	0.0053	0.0053	0.0052	0.0052	0.0053	0.0053	0.0053	0.0053
K-M_1N_3	0.0053	0.0051	0.0049	0.0047	0.0045	0.0043	0.0042	0.0040	0.0039	0.0038	0.0036	0.0036	0.0035	0.0034	0.0033
K-M_2M_3	0.045	0.043	0.041	0.039	0.037	0.034	0.033	0.032	0.030	0.029	0.027	0.026	0.026	0.0242	0.0233
K-M_2N_1	0.0057	0.0056	0.0055	0.0054	0.0053	0.0052	0.0052	0.0051	0.0051	0.0051	0.0050	0.0051	0.0051	0.0050	0.0051
K-M_2N_3	0.0096	0.0091	0.0087	0.0083	0.0079	0.0075	0.0072	0.0070	0.0067	0.0064	0.0061	0.0059	0.0058	0.0055	0.0054
K-M_3M_3	0.0236	0.0225	0.0215	0.0204	0.0193	0.0182	0.0174	0.0166	0.0158	0.0151	0.0143	0.0138	0.0134	0.0126	0.0122
K-M_3M_4	0.0062	0.0059	0.0056	0.0054	0.0051	0.0048	0.0046	0.0044	0.0042	0.0039	0.0038	0.0036	0.0035	0.0033	0.0032
K-M_3M_5	0.0062	0.0059	0.0056	0.0054	0.0051	0.0048	0.0046	0.0044	0.0041	0.0039	0.0037	0.0036	0.0035	0.0033	0.0032
K-M_3N_1	0.0055	0.0053	0.0051	0.0049	0.0047	0.0044	0.0043	0.0041	0.0040	0.0038	0.0037	0.0036	0.0036	0.0034	0.0034
K-M_3N_2	0.0098	0.0094	0.0090	0.0086	0.0082	0.0078	0.0075	0.0072	0.0069	0.0066	0.0063	0.0062	0.0061	0.0058	0.0056
K-M_3N_3	0.0103	0.0098	0.0093	0.0089	0.0084	0.0080	0.0077	0.0073	0.0070	0.0067	0.0064	0.0063	0.0061	0.0058	0.0056

	$_{78}$Pt	$_{79}$Au	$_{80}$Hg	$_{81}$Tl	$_{82}$Pb	$_{83}$Bi	$_{84}$Po	$_{85}$At	$_{86}$Rn	$_{87}$Fr	$_{88}$Ra	$_{89}$Ac	$_{90}$Th	$_{91}$Pa	$_{92}$U
K-L_1L_1	0.39	0.38	0.37	0.37	0.36	0.35	0.35	0.35	0.35	0.34	0.35	0.34	0.34	0.33	0.33
K-L_1L_2	0.60	0.60	0.59	0.59	0.58	0.58	0.59	0.59	0.60	0.59	0.61	0.60	0.61	0.61	0.62
K-L_1L_3	0.36	0.35	0.34	0.32	0.31	0.30	0.29	0.28	0.28	0.27	0.26	0.250	0.246	0.234	0.231
K-L_1M_1	0.155	0.152	0.150	0.148	0.145	0.142	0.144	0.141	0.143	0.140	0.141	0.138	0.140	0.136	0.137
K-L_1M_2	0.125	0.125	0.125	0.125	0.126	0.125	0.128	0.127	0.130	0.129	0.132	0.131	0.133	0.132	0.135
K-L_1M_3	0.084	0.081	0.078	0.076	0.073	0.070	0.070	0.067	0.067	0.064	0.063	0.061	0.060	0.057	0.057
K-L_1M_4	0.0077	0.0075	0.0073	0.0071	0.0069	0.0067	0.0067	0.0064	0.0064	0.0062	0.0062	0.0059	0.0059	0.0056	0.0056
K-L_1M_5	0.0057	0.0055	0.0053	0.0050	0.0048	0.0046	0.0045	0.0043	0.0042	0.0040	0.0039	0.0037	0.0036	0.0034	0.0033
K-L_1N_1	0.038	0.038	0.038	0.037	0.037	0.036	0.037	0.036	0.037	0.036	0.037	0.036	0.037	0.036	0.037
K-L_1N_2	0.031	0.031	0.031	0.031	0.031	0.031	0.032	0.032	0.033	0.033	0.034	0.034	0.034	0.034	0.035
K-L_1N_3	0.0201	0.0196	0.0191	0.0186	0.0180	0.0175	0.0175	0.0170	0.0169	0.0164	0.0163	0.0158	0.0157	0.0151	0.0150
K-L_1O_1	0.0076	0.0077	0.0078	0.0079	0.0080	0.0081	0.0083	0.0084	0.0086	0.0086	0.0089	0.0089	0.0092	0.0091	0.0094
K-L_1O_2	0.0054	0.0056	0.0058	0.0060	0.0062	0.0063	0.0066	0.0067	0.0071	0.0072	0.0076	0.0078	0.0081	0.0082	0.0085
K-L_1O_3	0.0034	0.0034	0.0034	0.0035	0.0035	0.0035	0.0035	0.0035	0.0035	0.0035	0.0036	0.0036	0.0036	0.0036	0.0036
K-L_2L_2	0.036	0.035	0.034	0.032	0.031	0.030	0.029	0.028	0.028	0.026	0.026	0.0246	0.0242	0.0230	0.0226
K-L_2L_3	0.65	0.62	0.59	0.56	0.54	0.51	0.50	0.47	0.46	0.44	0.43	0.40	0.39	0.37	0.36
K-L_2M_1	0.099	0.099	0.099	0.098	0.098	0.097	0.100	0.099	0.102	0.101	0.103	0.102	0.105	0.103	0.106
K-L_2M_2	0.0137	0.0132	0.0127	0.0122	0.0117	0.0112	0.0111	0.0106	0.0105	0.010	0.0099	0.0094	0.0093	0.0088	0.0087
K-L_2M_3	0.124	0.119	0.114	0.109	0.105	0.10	0.098	0.094	0.092	0.088	0.086	0.082	0.080	0.076	0.075
K-L_2M_4	0.0077	0.0075	0.0072	0.0070	0.0068	0.0065	0.0065	0.0063	0.0062	0.0060	0.0060	0.0057	0.0057	0.0054	0.0054
K-L_2M_5	0.0202	0.0193	0.0185	0.0177	0.0168	0.0160	0.0157	0.0149	0.0146	0.0139	0.0136	0.0128	0.0125	0.0118	0.0115
K-L_2N_1	0.0236	0.0237	0.0238	0.0239	0.0239	0.0239	0.0246	0.0246	0.0253	0.0253	0.026	0.026	0.027	0.026	0.027
K-L_2N_2	0.0032	0.0031	0.0030	0.0029	0.0028	0.0027	0.0027	0.0026	0.0026	0.00249	0.00247	0.00236	0.00234	0.00224	0.00221
K-L_2N_3	0.029	0.028	0.027	0.026	0.0248	0.0238	0.0236	0.0227	0.0225	0.0215	0.0213	0.0203	0.0201	0.0192	0.0189
K-L_2N_5	0.0050	0.0048	0.0047	0.0045	0.0043	0.0042	0.0041	0.0040	0.0039	0.0038	0.0037	0.0036	0.0035	0.0033	0.0033
K-L_2O_1	0.0047	0.0048	0.0049	0.0051	0.0052	0.0053	0.0055	0.0056	0.0058	0.0059	0.0062	0.0062	0.0065	0.0066	0.0070
K-L_2O_3	0.0048	0.0048	0.0048	0.0047	0.0047	0.0046	0.0047	0.0047	0.0047	0.0047	0.0047	0.0047	0.0047	0.0045	0.0045
K-L_3L_3	0.29	0.28	0.27	0.253	0.240	0.227	0.221	0.209	0.203	0.192	0.187	0.175	0.170	0.160	0.156
K-L_3M_1	0.059	0.057	0.055	0.053	0.051	0.049	0.049	0.047	0.046	0.044	0.044	0.042	0.041	0.039	0.039
K-L_3M_2	0.118	0.113	0.108	0.104	0.099	0.094	0.093	0.088	0.086	0.082	0.080	0.076	0.075	0.070	0.069
K-L_3M_3	0.117	0.112	0.107	0.102	0.097	0.092	0.091	0.086	0.084	0.080	0.078	0.074	0.073	0.068	0.067
K-L_3M_4	0.0194	0.0185	0.0177	0.0168	0.0160	0.0151	0.0148	0.0140	0.0137	0.0129	0.0126	0.0118	0.0115	0.0108	0.0105
K-L_3M_5	0.0196	0.0187	0.0178	0.0169	0.0160	0.0151	0.0148	0.0140	0.0136	0.0128	0.0124	0.0117	0.0113	0.0106	0.0103
K-L_3N_1	0.0138	0.0134	0.0130	0.0127	0.0122	0.0118	0.0118	0.0114	0.0113	0.0109	0.0108	0.0104	0.0103	0.0099	0.0098
K-L_3N_2	0.027	0.026	0.0254	0.0244	0.0234	0.0224	0.0222	0.0212	0.0209	0.020	0.0197	0.0187	0.0184	0.0175	0.0171
K-L_3N_3	0.027	0.026	0.0252	0.0242	0.0232	0.0223	0.0220	0.0210	0.0208	0.0198	0.0195	0.0186	0.0183	0.0174	0.0171
K-L_3N_4	0.0048	0.0046	0.0044	0.0043	0.0041	0.0039	0.0039	0.0037	0.0037	0.0035	0.0035	0.0033	0.0032	0.0031	0.0030
K-L_3N_5	0.0049	0.0047	0.0045	0.0043	0.0041	0.0039	0.0039	0.0037	0.0037	0.0035	0.0034	0.0033	0.0032	0.0030	0.0029
K-L_3O_1	0.0028	0.0027	0.0027	0.0027	0.0027	0.0026	0.0026	0.0026	0.0026	0.0026	0.0026	0.00255	0.0026	0.00248	0.00248
K-L_3O_2	0.0047	0.0047	0.0046	0.0046	0.0046	0.0046	0.0045	0.0044	0.0044	0.0043	0.0043	0.0042	0.0042	0.0041	0.0041
K-L_3O_3	0.0045	0.0045	0.0044	0.0045	0.0044	0.0044	0.0044	0.0043	0.0043	0.0042	0.0042	0.0042	0.0042	0.0041	0.0041
K-M_1M_1	0.0152	0.0151	0.0149	0.0147	0.0145	0.0142	0.0145	0.0142	0.0144	0.0141	0.0143	0.0140	0.0141	0.0138	0.0140
K-M_1M_2	0.0223	0.0223	0.0223	0.0223	0.0223	0.0222	0.0228	0.0227	0.0234	0.0232	0.0239	0.0237	0.0243	0.0241	0.0247
K-M_1M_3	0.0133	0.0129	0.0125	0.0121	0.0117	0.0113	0.0113	0.0109	0.0108	0.0104	0.0103	0.0099	0.0098	0.0094	0.0092
K-M_1N_1	0.0075	0.0075	0.0074	0.0073	0.0073	0.0072	0.0074	0.0073	0.0074	0.0073	0.0074	0.0073	0.0074	0.0073	0.0074
K-M_1N_2	0.0054	0.0054	0.0054	0.0055	0.0055	0.0055	0.0057	0.0057	0.0059	0.0059	0.0061	0.0061	0.0063	0.0063	0.0065
K-M_1N_3	0.0032	0.0031	0.0031	0.0030	0.0029	0.0028	0.0028	0.0028	0.0028	0.0027	0.0027	0.0026	0.00255	0.00246	0.00245
K-M_2M_3	0.0225	0.0217	0.0208	0.020	0.0192	0.0183	0.0181	0.0173	0.0170	0.0162	0.0160	0.0152	0.0149	0.0141	0.0139
K-M_2N_1	0.0051	0.0051	0.0052	0.0052	0.0052	0.0052	0.0054	0.0054	0.0055	0.0055	0.0057	0.0057	0.0059	0.0058	0.0060
K-M_2N_3	0.0052	0.0051	0.0049	0.0047	0.0046	0.0044	0.0044	0.0043	0.0043	0.0041	0.0040	0.0037	0.0036	0.0035	0.0035

Table 8. Auger-Electron Intensities per 100 K-Shell Vacancies (continued)

	$_{78}$Pt	$_{79}$Au	$_{80}$Hg	$_{81}$Tl	$_{82}$Pb	$_{83}$Bi	$_{84}$Po	$_{85}$At	$_{86}$Rn	$_{87}$Fr	$_{88}$Ra	$_{89}$Ac	$_{90}$Th	$_{91}$Pa	$_{92}$U
continued															
K-M_3M_3	0.0117	0.0113	0.0108	0.0104	0.010	0.0096	0.0095	0.0090	0.0089	0.0085	0.0084	0.0080	0.0078	0.0074	0.0073
K-M_3M_4	0.0031	0.0030	0.0028	0.0027	0.0026	0.00249	0.00246	0.00234	0.00230	0.00219	0.00215	0.00203	0.00199	0.00188	0.00184
K-M_3M_5	0.0030	0.0029	0.0028	0.0027	0.00254	0.00241	0.00238	0.00226	0.00222	0.00210	0.00205	0.00192	0.00187	0.00177	0.00173
K-M_3N_1	0.0033	0.0032	0.0031	0.0030	0.0029	0.0028	0.0028	0.0028	0.0028	0.0027	0.0027	0.0026	0.00255	0.00244	0.00243
K-M_3N_2	0.0055	0.0053	0.0051	0.0050	0.0048	0.0046	0.0046	0.0044	0.0044	0.0042	0.0042	0.0040	0.0039	0.0038	0.0037
K-M_3N_3	0.0055	0.0053	0.0051	0.0050	0.0048	0.0046	0.0046	0.0044	0.0044	0.0042	0.0042	0.0040	0.0039	0.0038	0.0037

	$_{93}$Np	$_{94}$Pu	$_{95}$Am	$_{96}$Cm	$_{97}$Bk	$_{98}$Cf	$_{99}$Es	$_{100}$Fm	$_{101}$Md	$_{102}$No	$_{103}$Lr	104
K-L_1L_1	0.32	0.33	0.32	0.32	0.31	0.31	0.31	0.30	0.30	0.31	0.29	0.30
K-L_1L_2	0.61	0.62	0.61	0.62	0.61	0.62	0.64	0.62	0.63	0.65	0.63	0.64
K-L_1L_3	0.219	0.216	0.204	0.201	0.190	0.186	0.183	0.173	0.170	0.166	0.157	0.154
K-L_1M_1	0.134	0.135	0.131	0.132	0.128	0.130	0.131	0.126	0.127	0.128	0.124	0.124
K-L_1M_2	0.133	0.135	0.133	0.136	0.133	0.136	0.139	0.136	0.138	0.141	0.137	0.139
K-L_1M_3	0.054	0.053	0.051	0.050	0.048	0.047	0.046	0.044	0.043	0.043	0.040	0.040
K-L_1M_4	0.0053	0.0053	0.0050	0.0050	0.0047	0.0047	0.0046	0.0044	0.0043	0.0042	0.0040	0.0039
K-L_1M_5	0.0031	0.0030	0.0028	0.0027	0.0026	0.00249	0.00242	0.00225	0.00219	0.00213	0.00198	0.00193
K-L_1N_1	0.036	0.036	0.035	0.036	0.035	0.035	0.036	0.034	0.035	0.035	0.034	0.034
K-L_1N_2	0.035	0.036	0.035	0.036	0.036	0.036	0.037	0.037	0.037	0.038	0.037	0.038
K-L_1N_3	0.0144	0.0143	0.0137	0.0136	0.0129	0.0128	0.0127	0.0121	0.0120	0.0119	0.0112	0.0111
K-L_1O_1	0.0093	0.0095	0.0094	0.0096	0.0094	0.0096	0.0099	0.0096	0.0098	0.010	0.0098	0.0099
K-L_1O_2	0.0086	0.0089	0.0089	0.0092	0.0091	0.0094	0.0097	0.0096	0.0099	0.0101	0.0100	0.0102
K-L_1O_3	0.0035	0.0035	0.0034	0.0034	0.0033	0.0033	0.0033	0.0032	0.0032	0.0032	0.0030	0.0030
K-L_2L_2	0.0214	0.0210	0.0199	0.0195	0.0184	0.0181	0.0177	0.0167	0.0164	0.0161	0.0151	0.0148
K-L_2L_3	0.34	0.33	0.31	0.30	0.28	0.27	0.26	0.247	0.240	0.233	0.216	0.210
K-L_2M_1	0.104	0.107	0.105	0.108	0.106	0.108	0.110	0.108	0.110	0.112	0.109	0.111
K-L_2M_2	0.0082	0.0081	0.0077	0.0076	0.0072	0.0070	0.0069	0.0065	0.0064	0.0063	0.0060	0.0059
K-L_2M_3	0.071	0.069	0.065	0.064	0.060	0.058	0.057	0.054	0.052	0.051	0.048	0.047
K-L_2M_4	0.0051	0.0051	0.0049	0.0048	0.0046	0.0045	0.0045	0.0043	0.0042	0.0042	0.0040	0.0039
K-L_2M_5	0.0108	0.0105	0.0098	0.0096	0.0089	0.0087	0.0084	0.0078	0.0076	0.0073	0.0068	0.0066
K-L_2N_1	0.027	0.028	0.027	0.028	0.028	0.029	0.029	0.029	0.029	0.030	0.029	0.030
K-L_2N_2	0.00211	0.00209	0.00199	0.00197	0.00187	0.00185	0.00184	0.00175	0.00173	0.00171	0.00163	0.00161
K-L_2N_3	0.0180	0.0177	0.0168	0.0165	0.0156	0.0153	0.0150	0.0142	0.0139	0.0136	0.0128	0.0125
K-L_2N_5	0.0031	0.0031	0.0029	0.0028	0.0027	0.0026	0.00255	0.00240	0.00235	0.00229	0.00215	0.00210
K-L_2O_1	0.0070	0.0073	0.0073	0.0075	0.0075	0.0077	0.0079	0.0078	0.0079	0.0081	0.0079	0.0080
K-L_2O_3	0.0043	0.0043	0.0042	0.0042	0.0040	0.0040	0.0040	0.0039	0.0039	0.0039	0.0037	0.0038
K-L_3L_3	0.146	0.141	0.132	0.128	0.119	0.115	0.111	0.103	0.100	0.096	0.089	0.086
K-L_3M_1	0.037	0.036	0.035	0.034	0.032	0.032	0.031	0.030	0.029	0.029	0.027	0.026
K-L_3M_2	0.065	0.063	0.059	0.058	0.054	0.053	0.052	0.048	0.047	0.046	0.043	0.042
K-L_3M_3	0.063	0.061	0.058	0.056	0.053	0.051	0.050	0.047	0.045	0.044	0.041	0.040
K-L_3M_4	0.0099	0.0096	0.0090	0.0087	0.0081	0.0078	0.0076	0.0070	0.0068	0.0066	0.0061	0.0059
K-L_3M_5	0.0096	0.0093	0.0086	0.0083	0.0077	0.0074	0.0071	0.0066	0.0063	0.0061	0.0056	0.0053
K-L_3N_1	0.0093	0.0092	0.0088	0.0087	0.0082	0.0081	0.0080	0.0076	0.0075	0.0073	0.0069	0.0068
K-L_3N_2	0.0162	0.0159	0.0150	0.0147	0.0138	0.0136	0.0133	0.0125	0.0122	0.0119	0.0112	0.0110
K-L_3N_3	0.0162	0.0159	0.0150	0.0147	0.0139	0.0136	0.0133	0.0125	0.0122	0.0120	0.0112	0.0110
K-L_3N_4	0.0028	0.0028	0.0026	0.0026	0.00243	0.00237	0.00231	0.00216	0.00210	0.00205	0.00191	0.00185
K-L_3N_5	0.0028	0.0027	0.00253	0.00247	0.00232	0.00227	0.00222	0.00208	0.00204	0.0020	0.00187	0.00184
K-L_3O_1	0.00239	0.00238	0.00228	0.00228	0.00218	0.00217	0.00216	0.00207	0.00206	0.00205	0.00195	0.00194
K-L_3O_2	0.0040	0.0039	0.0037	0.0037	0.0035	0.0034	0.0033	0.0031	0.0029	0.0028	0.0026	0.00248
K-L_3O_3	0.0039	0.0039	0.0037	0.0037	0.0035	0.0034	0.0033	0.0031	0.0030	0.0030	0.0028	0.0027
K-M_1M_1	0.0137	0.0138	0.0134	0.0136	0.0132	0.0133	0.0134	0.0130	0.0130	0.0131	0.0127	0.0127
K-M_1M_2	0.0244	0.0250	0.0246	0.0252	0.0248	0.0254	0.026	0.0254	0.026	0.026	0.026	0.026
K-M_1M_3	0.0088	0.0087	0.0083	0.0082	0.0078	0.0077	0.0076	0.0073	0.0072	0.0071	0.0067	0.0067
K-M_1N_1	0.0072	0.0073	0.0072	0.0073	0.0071	0.0072	0.0073	0.0071	0.0072	0.0073	0.0070	0.0071
K-M_1N_2	0.0064	0.0066	0.0065	0.0067	0.0066	0.0068	0.0069	0.0068	0.0070	0.0071	0.0069	0.0071
K-M_1N_3	0.00236	0.00235	0.00225	0.00223	0.00213	0.00211	0.00209	0.00199	0.00197	0.00195	0.00184	0.00182
K-M_2M_3	0.0131	0.0128	0.0121	0.0119	0.0112	0.0109	0.0107	0.0101	0.0099	0.0096	0.0090	0.0089
K-M_2N_1	0.0059	0.0061	0.0060	0.0062	0.0061	0.0062	0.0064	0.0063	0.0064	0.0065	0.0064	0.0065
K-M_2N_3	0.0034	0.0034	0.0032	0.0031	0.0029	0.0028	0.0027	0.00242	0.00228	0.00213	0.00189	0.00173
K-M_3M_3	0.0069	0.0068	0.0064	0.0063	0.0059	0.0058	0.0057	0.0053	0.0052	0.0051	0.0047	0.0046
K-M_3M_4	0.00174	0.00170	0.00161	0.00158	0.00149	0.00146	0.00144	0.00136	0.00133	0.00131	0.00124	0.00122
K-M_3M_5	0.00163	0.00159	0.00149	0.00144	0.00135	0.00130	0.00126	0.00117	0.00112	0.00108		
K-M_3N_1	0.00232	0.00231	0.00220	0.00219	0.00209	0.00207	0.00206	0.00196	0.00194	0.00193	0.00183	0.00182
K-M_3N_2	0.0036	0.0035	0.0033	0.0033	0.0031	0.0031	0.0030	0.0028	0.0028	0.0027	0.0026	0.00254
K-M_3N_3	0.0036	0.0035	0.0034	0.0033	0.0031	0.0031	0.0031	0.0029	0.0029	0.0028	0.0027	0.0026

APPENDIX D. ABSORPTION OF RADIATION IN MATTER

1. Absorption of Photons in Matter

The decrease in intensity of a parallel beam of photons traversing an absorber of thickness d is given by[1]

$$I = I_0 \, 2^{\frac{d}{d_{1/2}}} , \tag{1}$$

where I_0 and I are the beam intensity before and after passing through the absorber, and $d_{1/2}$ is the half-thickness of the absorber. The quantity $d_{1/2}$ can be expressed in terms of the photon energy and the properties of the absorbing material[1]:

$$d_{1/2} = \frac{ln 2}{N_\rho} \sum_i \frac{A_i}{p_i \sigma_i(E)} , \tag{2}$$

where N is Avogadro's number, ρ is the density of the material, and A_i, p_i, and $\sigma_i(E)$ are, respectively, the (average) atomic masses, the fractions by mass, and the "atomic cross sections" of the elements of which the absorber is composed. The atomic cross section includes contributions from coherent (Rayleigh) and incoherent (Compton) scattering, as well as absorption by the photoeffect and by positron-electron pair production.

The following graphs (Figs. 1a -1c) give values of the half-thickness expressed in terms of mass per unit area ($\rho \times d_{1/2}$) for some commonly used elemental absorbers and for water. These curves are based on tables derived from evaluated experimental data and theoretical calculations.[2] The accuracy of the half-thickness is better than 5% below 5 *keV*, and 2% between 5 *keV* and 10 *MeV*.[3]

Two points should be noted: (1) the cross sections do not include photonuclear effects, which can be as large as 5 to 10% of the atomic cross section between 10 and 30 *MeV*,[3] and (2) the attenuation calculated by the above formula gives only the decrease in intensity of the original beam. The total radiation downstream of the absorber is larger, due to the presence of scattered photons, and to the creation of secondary photons by a variety of processes, including fluorescence (x-rays), and positron annihilation radiation.

---

[1]A.H. Wapstra, G.J. Nijgh, and R. van Lieshout, *Nuclear Spectroscopy Tables*, North-Holland Publ. Co., Amsterdam (1959).

[2]E.F. Plechaty, D.E. Cullen, and R.J. Howerton, *Univ. of California Radiation Lab.*, Report **UCRL-50400**, Vol. 6, Rev. 1 (1975).

[3]J.H. Hubbell, Radiat. Res. **70**, 58 (1977).

Fig. 1a. Absorption of photons of energy up to 100 Mev

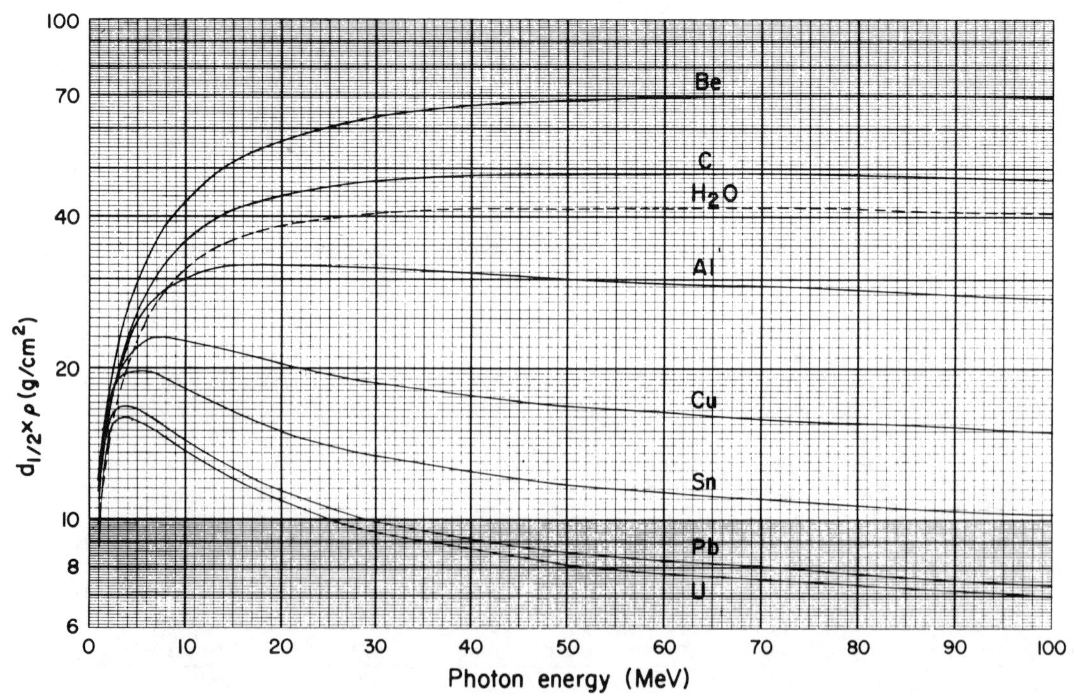

Fig. 1b. Absorption of photons in beryllium, water, aluminum, and lead (1 keV to 5 Mev)

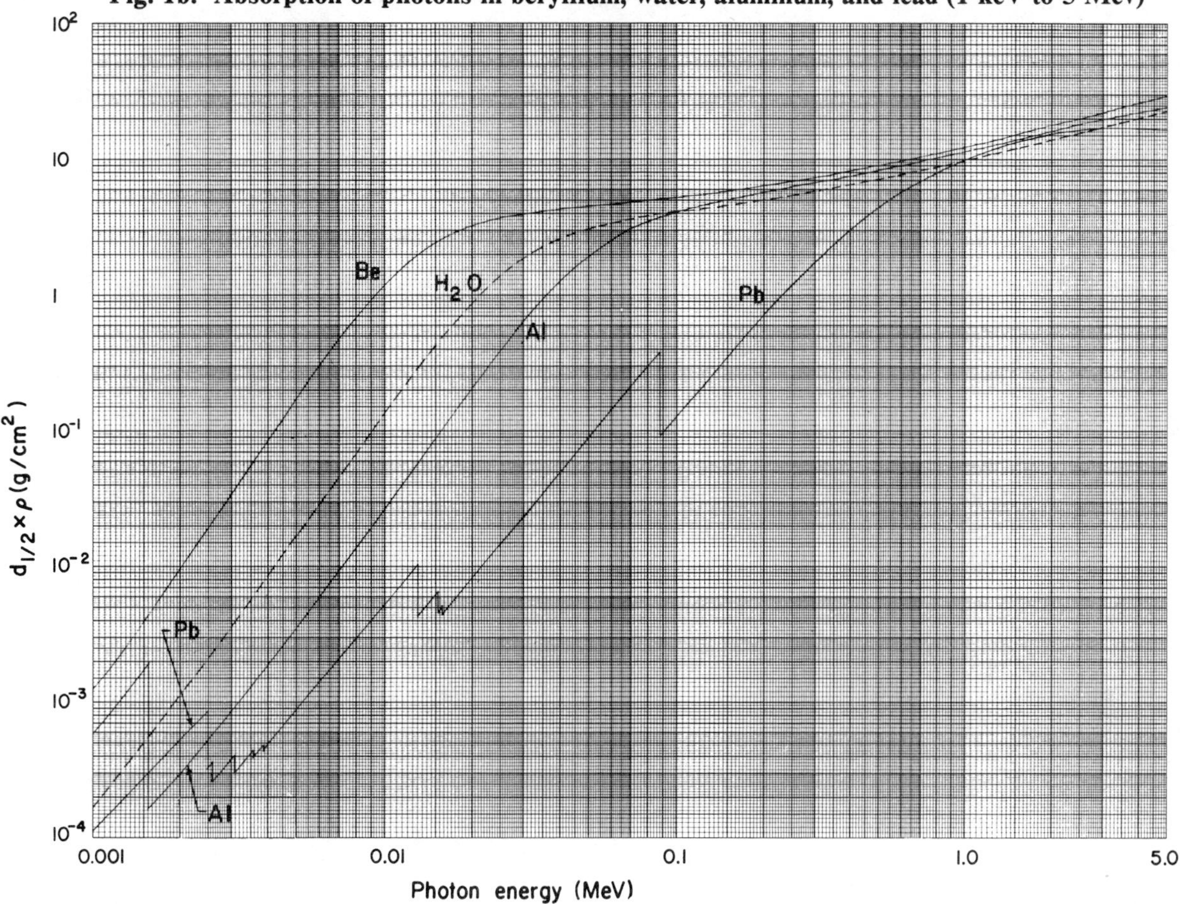

Fig. 1c. Absorption of photons in carbon, copper, tin, and uranium (1 keV to 5 Mev)

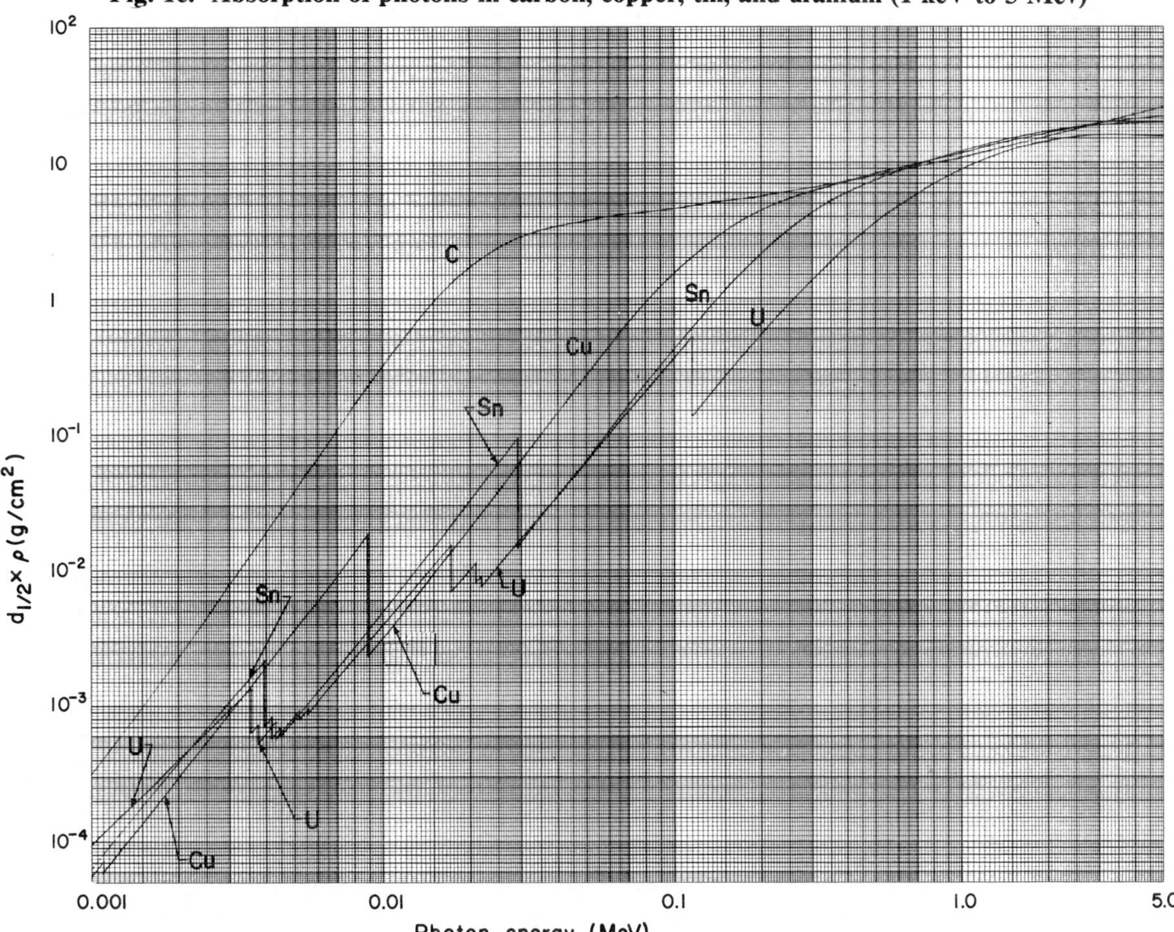

2. Range and Stopping Power for Charged Particles

Figure 2a gives the range and stopping power for electrons in several media. These curves are based on the tables of Pages, et al.[1] Corrections for the density effect and the bremsstrahlung contribution to the stopping power have been included. As in the case of heavy charged particles, the ranges are *path lengths.* Because the lighter electrons scatter more readily, range straggling is greater, and the average range is considerably shorter.

Figures 2b and 2c give the range (R) and the stopping power (S), respectively, for some charged particles in several different stopping media. The curves were calculated by Bichsel,[2] using a computer program based on Bethe's stopping-power formula[3] taking into account the so-called shell-correction terms.[4] Parameters in the formula have been adjusted to fit experimental data. It should be noted that *the abscissa is in units of MeV per atomic mass unit (amu)* of the particle, and that *curves for successively heavier incident particles are displaced vertically by a decade.* Separate ordinate scales are given for each particle.

Since the underlying equations have been verified to about 1%, the graphs are generally reliable to within reading accuracy, with the following caveats: (1) at energies below 1 *MeV per amu,* the assumptions underlying the calculated stopping power are not valid; the curves are less accurate at such energies, particularly below 0.5 *MeV per amu;* (2) at the highets energies ($\geqslant$1000 *MeV per amu*), there is a relativistic correction for polarization of the medium (the "density effect") that has not been included; its effect is to decrease the stopping power (increase the range); (3) the stopping power includes only energy loss by ionization; other effects (nuclear collisions, bremsstrahlung, etc.) are not included; the calculated range is the integral of S^{-1} or *path length*; the average range will be slightly shorter, mainly due to small-angle scattering; and (4) the stopping-power formula is valid for condensed media; it should not be used for gaseous stopping media.

---

[1]L. Pages, E. Bertel, H. Joffre, and L. Sklavenetis, *Atomic Data* **A4**, 1 (1972).

[2]H. Bichsel, *Univ. of California Radiation Laboratory,* Report **UCRL-17538** (1967).

[3]H.A. Bethe, *Quantenmechanik der Ein- und Zwei-Electronenprobleme,* in *Handbuch der Physik,* H. Geiger and K. Scheel, eds., **Vol. 24-I**, 2nd ed., Springer-Verlag, Berlin (1933); U. Fano, *Ann. Rev. Nucl. Sci.* **13**, 1 (1963).

[4]M.C. Walske, *Phys. Rev.* **88**, 1283 (1952).

Fig. 2a. Range and stopping power for electrons in liquid hydrogen, water, beryllium, and aluminum

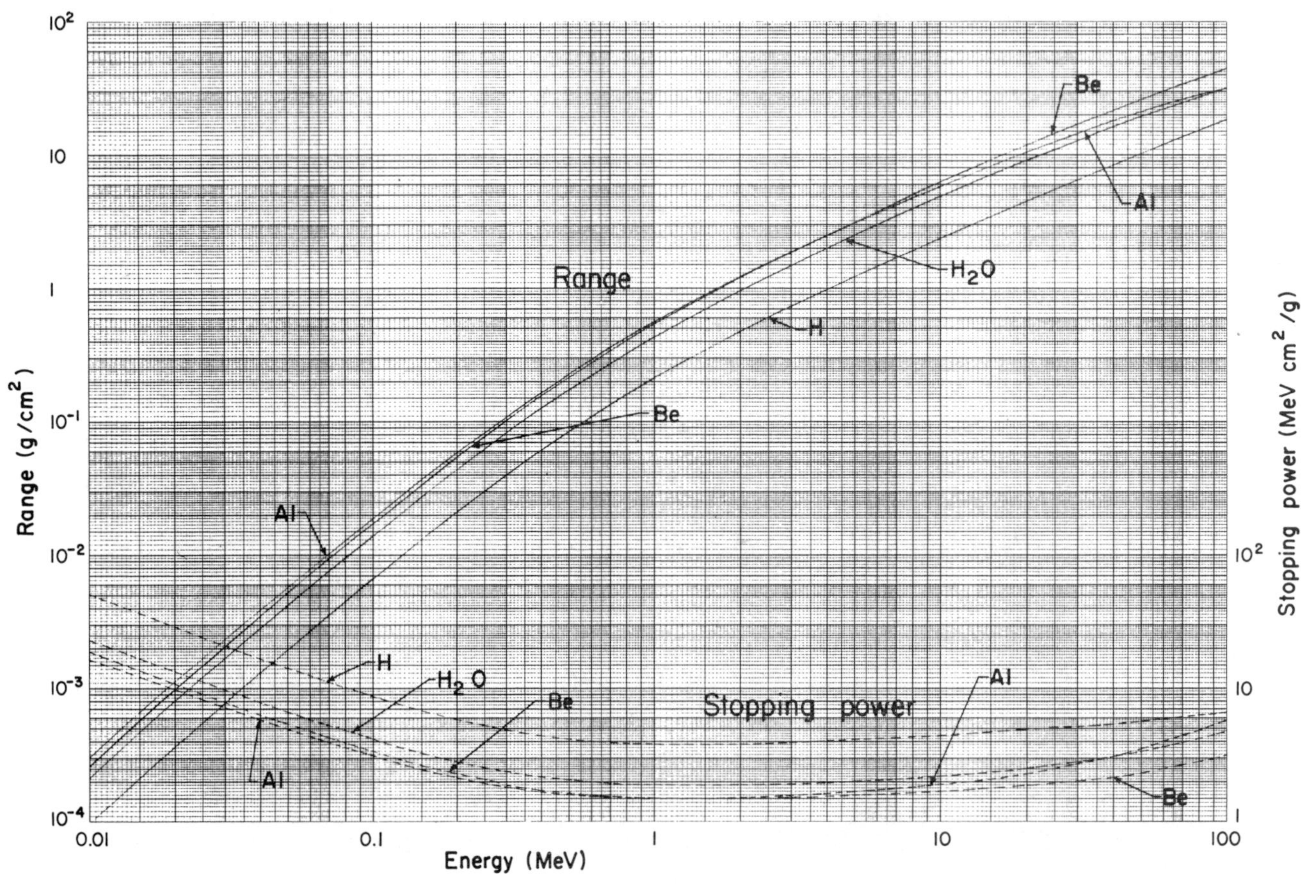

Fig. 2b. Range of charged particles in liquid hydrogen, carbon, water, aluminum, iron, and lead

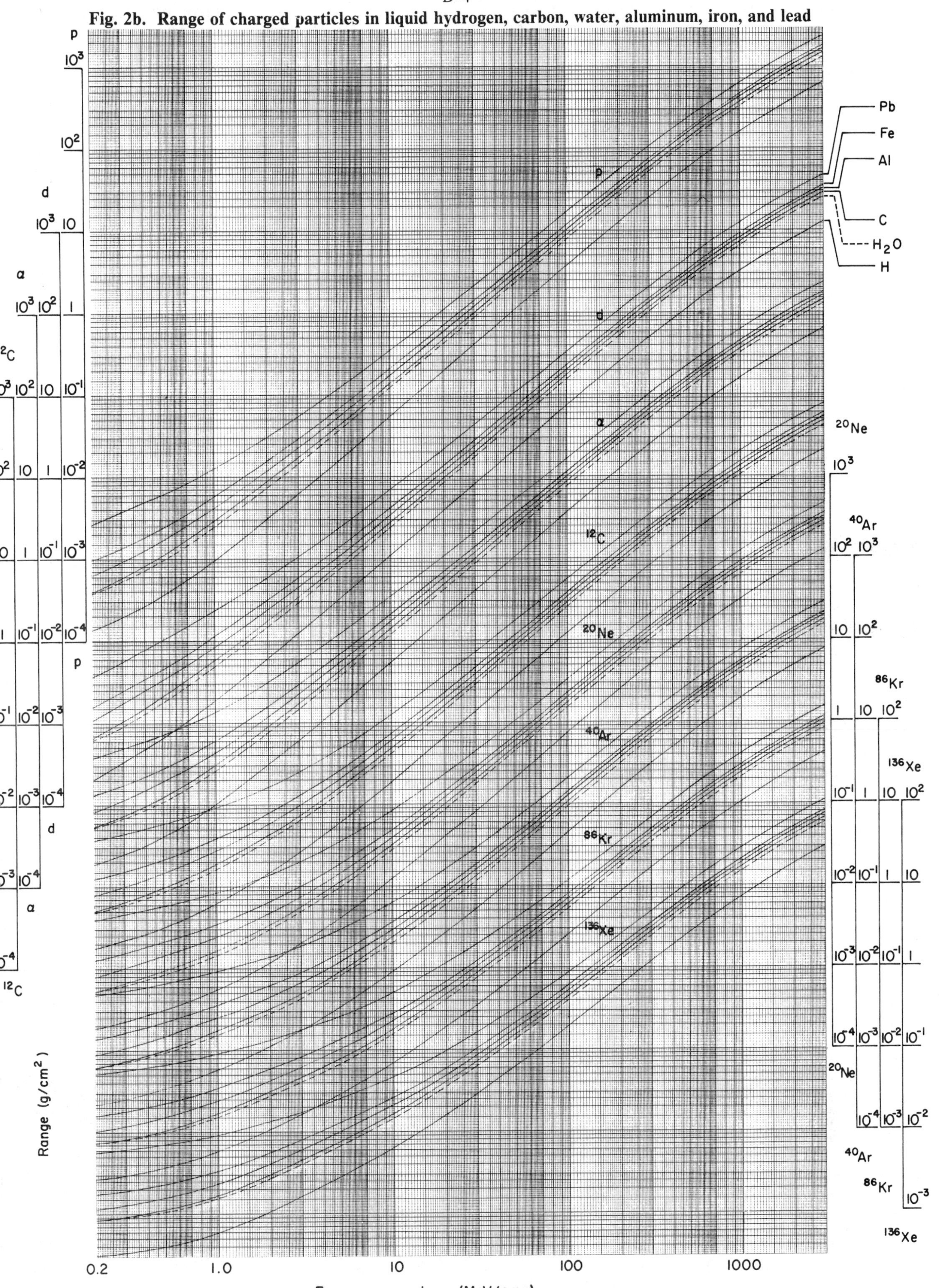

Fig. 2c. Stopping power for charged particles in liquid hydrogen, carbon, water, aluminum, iron, and lead

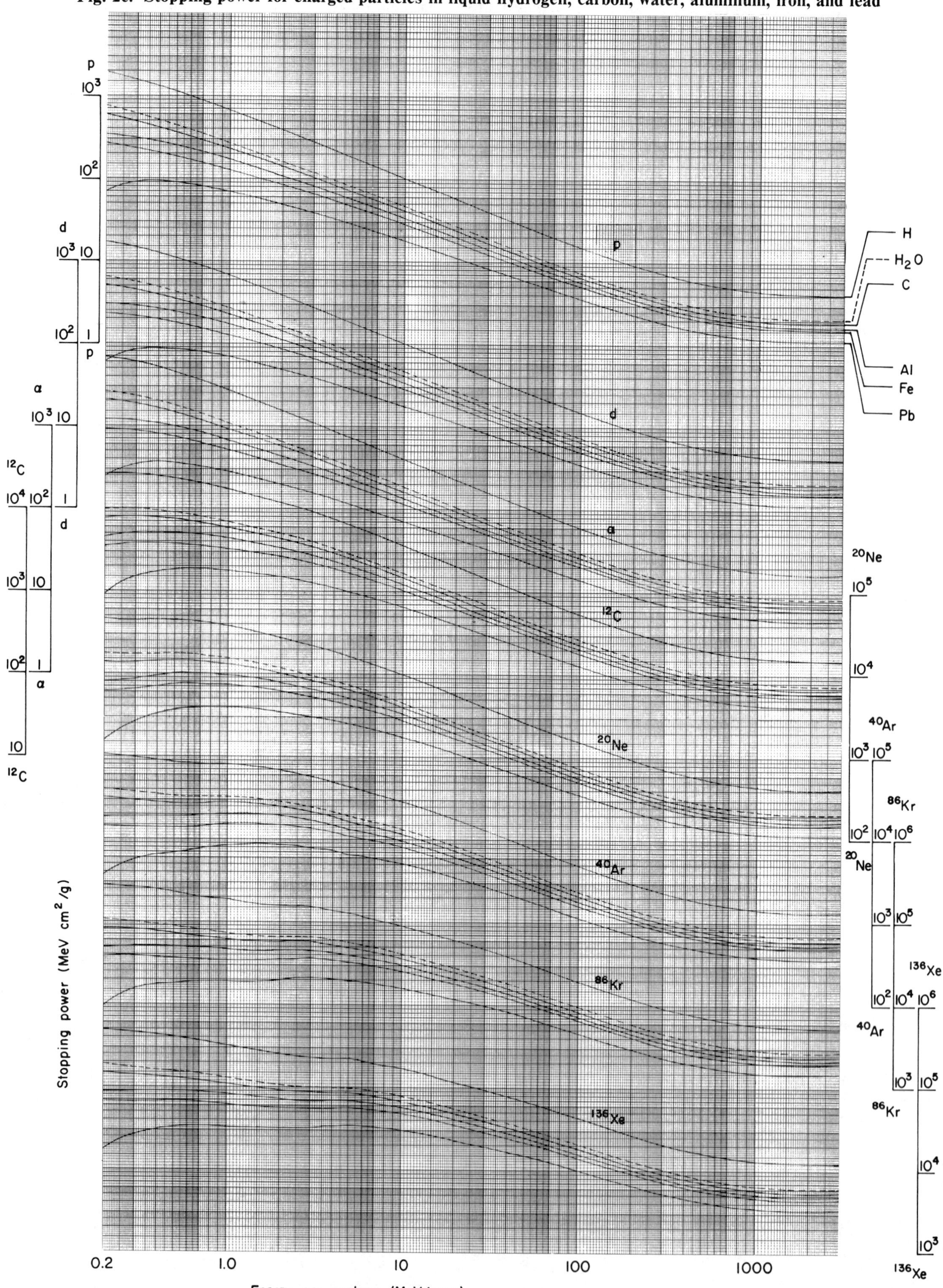

3. Positron Annihilation-in-Flight

The absolute intensity of 511-keV photons per 100 disintegrations ($\gamma^{\pm}(\%)$) from positrons annihilating at thermal energies in an absorber is

$$\gamma^{\pm}(\%)=2(\beta^{+}(\%) - \beta_f^{+}(\%)) , \tag{1}$$

where $\beta^{+}(\%)$ and $\beta_f^{+}(\%)$ are the absolute intensities of the emitted positrons and of those which have annihilated-in-flight, respectively.

There is significant probability for annihilation-in-flight, resulting in either *one quantum annihilation* (*OQA*) or *two quanta annihilation* (*TQA*), with continuous photon energy distributions. The maximum photon energy is E_0+1, where E_0 is the maximum positron kinetic energy (endpoint) in units of the electron rest-mass energy m_ec^2. The *OQA* probability for annihilation-in-flight of a positron with energy E, by collision with an atomic electron, is given by Bethe[1] as

$$^{OQA}\Phi(E,Z)=\frac{2\pi\alpha^4Z^5r_0^2\left\{E^2+\frac{2}{3}E+\frac{4}{3}-(E+2)\ln[E+(E^2-1)^{1/2}]\,(E^2-1)^{-1/2}\right\}}{(E+1)^2(E^2-1)^{1/2}} , \tag{2}$$

and the *TQA* probability, as

$$^{TQA}\Phi(E)=\frac{\pi r_0^2\left\{(E^2+4E+1)\ln[E+(E^2-1)^{1/2}] - (E+3)(E^2-1)^{1/2}\right\}}{(E^2-1)(E+1)} . \tag{3}$$

Here, r_0 (= 2.82×10^{-13} cm) is the classical electron radius, α ($\simeq 1/137$) is the fine structure constant, and Z is the atomic number of the absorber. Positron energies are given in units of the electron rest-mass energy (m_ec^2).

The *OQA* probability given in equation (2) is generally small, but for high-Z absorbing materials, it can be as much as approximately 16% that of the *TQA* probability. Equation (2) takes into account collisions between positrons and electrons from the K atomic shell only, but the *TQA* probability given in equation (3) includes collisions with electrons from all atomic shells. The total probability for annihilation-in-flight by positrons with energy E is given by[2]

$$P(E,Z)=\frac{N\rho}{A}\int_0^E[Z\,^{TQA}\Phi(E)+2\,^{OQA}\Phi(E,Z)]\,(-dE/dx)^{-1}dE . \tag{4}$$

In this equation, N is Avogadro's number, and A is the atomic weight, ρ is the density, and dE/dx is the stopping power of the absorber.

Figure 3 shows the total probability for annihilation-in-flight of fully absorbed positrons in Be, Al, Ag, W, Pb and U, as calculated from equation (4). The integrations were done numerically, and the probabilities were corrected for the reappearance of annihilated positrons at lower energies. The stopping powers (dE/dx) were calculated for both collision- and bremsstrahlung-energy losses (for collisions, as described by Nelms,[3] with mean excitation energies and density-effect corrections of Berger and Seltzer[4]; for bremsstrahlung, as described by Koch and Motz[5]).

---

[1]H.A. Bethe, *Proc. Roy. Soc. (London)* **A150**, 129 (1935).

[2]G. Azuelos and J.E. Kitching, *At. Data and Nucl. Data Tables* **17**, 103 (1976).

[3]*Energy Loss and Range of Electrons and Positrons,* A.T. Nelms, National Bureau of Standards, Circular **577** (1956).

[4]*Tables of Energy Losses and Ranges of Electrons and Positrons,* M.J. Berger and S.M. Seltzer, National Aeronautics and Space Administration, Report **NASA SP-3012** (1964).

[5]H.W. Koch and J.W. Motz, *Rev. Mod. Phys.* **31**, 920 (1959).

The probabilities given in Fig. 3 do not include positron energy distributions; and therefore can be used, along with the positron data from the *Table of Radioactive Isotopes*, for calculating absolute intensities of 511-*keV* radiation ($\gamma^{\pm}(\%)$) from positrons annihilating in absorbers. As an example, the absolute intensity of this radiation from the pure positron emitter ^{8}B, in a lead absorber, can be determined as follows, by combining information from the table of continuous radiation for ^{8}B and from Fig. 3. The two quantities required for solving equation (1) are $\beta^{+}(\%)$ (in this case equal to 100), and $\beta_{f}^{+}(\%)$. The latter is the sum of the positron-in-flight contributions (β_{fi}^{+}) for each energy interval (bin), where $\beta_{fi}^{+}=P(E_{av})\times\beta^{+}(\%)\times10^{-2}$. The first three columns in the table below show data necessary for the calculations, taken from the continuous radiation table for ^{8}B. The fourth column shows the average positron bin energy (E_{av}), calculated from columns 2 and 3 ($E_{av}=10^{2}\langle\beta^{+}\rangle/\beta^{+}(\%)$). The fifth column shows the annihilation-in-flight probabilities ($P(E_{av})$) read from Fig. 3 for each E_{av}, and the last column shows the resulting β_{fi}^{+} values. When these are summed to give $\beta_{f}^{+}(\%)=12.6$, the absolute intensity of 511-*keV* radiation in equation (1) becomes

$$\gamma^{\pm}(\%)=2(100-12.6)=174.8\% .$$

$^{8}_{5}$**B** (770 *3* ms)

E_{bin} (*keV*)	$\langle\beta^{+}\rangle$ (*keV*)	$\beta^{+}(\%)$ %	E_{av} (*keV*)	$P(E_{av})$ %	β_{fi}^{+} %
0 - 10	0.00483	0.073	6.6	-	-
10 - 20	0.027	0.176	15.3	0.1	-
20 - 40	0.16	0.52	31.0	0.3	0.002
40 - 100	1.51	2.13	54.0	0.4	0.009
100 - 300	7.0	4.14	175.0	1.0	0.041
300 - 600	0.84	0.185	454.0	2.5	0.0050
600 - 1300	9.5	0.95	1000.0	5.2	0.049
1300 - 2500	79.0	3.98	1975.0	8.3	0.33
2500 - 5000	736.0	18.9	3894.0	11.5	2.17
5000 - 7500	1757.0	28.0	6275.0	13.6	3.81
7500 - 15500	3957.0	41.0	9651.0	15.2	6.23

$\beta_{f}^{+}(\%)=$ 12.6

Fig. 3. Probability for Annihilation-in-Flight of fully absorbed positrons in Be, Al, Ag, W, Pb, and U

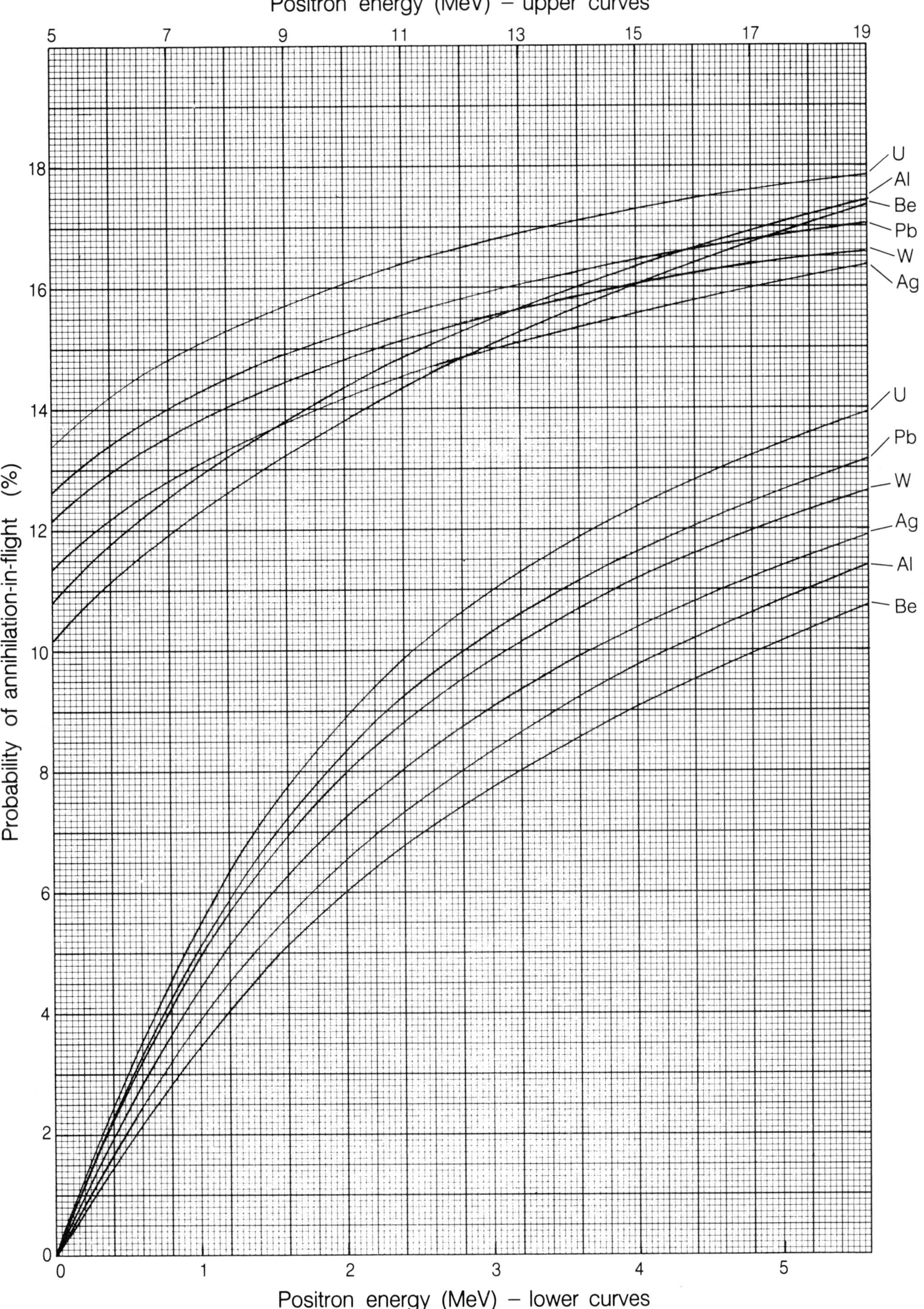

4. Experimental Average Radiation Energies per Disintegration

Average radiation energies are commonly used for calculating the absorbed dose from internally distributed radioactive isotopes[1] and the heat generated by fission products and actinides in fuel elements after reactor shutdown.[2]

Values for the experimental average radiation energies per disintegration are ordered first by atomic number (Z), then by mass number (A), and are then distinguished by half-life. They are given in MeV in the fourth column of Table I, rounded from text values to report 3 significant figures, and coded to show the following:

1. Average electromagnetic radiation energy per disintegration (γ), including γ rays, x-rays, internal bremsstrahlung, and (for β+ emitters) annihilation radiation. This latter is calculated from the positron percentage branching (β+(%)) times 0.01022.
2. Average α-particle energy per disintegration (α), including all prompt α particles.
3. Average atomic electrons energy per disintegration (e-), including β-, conversion, and Auger electrons.
4. Average positron energy per disintegration (β+). This is followed by the positron percentage branching (%(β+)).

For isotopes with large decay energies (Q-values), the experimental average energies per disintegration are systematically low for γ rays and high for β particles. The observation of this effect for fission products with Q-values >5 Mev[2] emphasizes the importance of being cautious when using average energies for isotopes with these large Q-values.

Average energies from Table I may be used to estimate absorbed dose under equilibrium conditions.[1] For a radioactive isotope uniformly distributed in an infinite homogeneous absorbing material, the absorbed dose under equilibrium conditions is [1]

$$D_{eq} = \tilde{C}\sum_i \Delta_i = \tilde{C}k\sum_i <E_i> = \tilde{C}k<E> .$$

Here

$\tilde{C} = \int Cdt$ is the time integral of the activity per unit mass,

$\Delta_i = k<E_i>$, where $<E_i>$ is the average energy per disintegration for the i-th radiation, and the numerical factor k depends on the choice of units, and

$<E> = \sum_i <E_i>$ is the *total* average energy per disintegration given in Table I.

The recommended unit system is the *Système International d'Unités (SI)*. The factor k is 1 and Δ_i is $1.602\times10^{-13}<E>$ in this system, where $<E>$ is in MeV and Δ_i in *joules*. Alternatively, $k = 2.13$ and $\Delta_i = 2.13<E>$, where $<E>$ is in MeV and Δ_i in $g\ rad/\mu Ci\ h$.

For decay heat calculations in reactor fuel elements, see the comprehensive work of Toshida and Nakasima[2] and references therein.

---

[1] *Methods of Assessment of Absorbed Dose in Clinical Use of Radionuclides*, International Commission on Radiation Units and Measurements (ICRU), Report **32** (1979).

[2] T. Yoshida and R. Nakasima, *J. Sci. Technol. (Tokyo)* **18**, 393 (1981).

Table I. Average Energies per Disintegration

Z	El	A	Half-life	⟨E⟩ (MeV), or %β+
1	**H**	3	12.3 yr	1.12×10^{-7} γ 5.70×10^{-3} e-
2	**He**	6	807 ms	4.60×10^{-3} γ 1.57 e-
		8	119 ms	8.70×10^{-1} γ 4.34 e-
3	**Li**	8	838 ms	3.70×10^{-2} γ 6.24 e-
		9	178.3 ms	3.40×10^{-2} γ 5.73 e-
		11	8.7 ms	1.52 γ 8.58 e-
4	**Be**	7	53.3 d	4.98×10^{-2} γ
		10	1.6×10^{6} yr	1.26×10^{-4} γ 2.03×10^{-1} e-
		11	13.8 s	1.48 γ 4.65 e-
		12	24 ms	3.20×10^{-2} γ 5.62 e-
5	**B**	8	770 ms	1.06 γ 15.8 α 6.55 β+ 1.00×10^{2} %(β+)
		12	20.20 ms	9.50×10^{-2} γ 6.36 e-
		13	17.4 ms	3.18×10^{-1} γ 6.32 e-
		14	16 ms	6.25 γ 7.12 e-
6	**C**	10	19.26 s	1.75 γ 8.09×10^{-1} β+ 1.00×10^{2} %(β+)
		11	20.39 min	1.02 γ 3.85×10^{-1} β+ 99.8%(β+)
		14	5.73×10^{3} yr	8.40×10^{-6} γ 4.95×10^{-2} e-
		15	2.449 s	3.62 γ 2.87 e-
		16	747 ms	7.10×10^{-3} γ 2.05 e-
7	**N**	12	11.00 ms	1.17 γ 7.73 β+ 1.00×10^{2} %(β+)
		13	9.965 min	1.02 γ 4.91×10^{-1} β+ 99.8%(β+)
		16	7.13 s	4.60 γ 2.69 e-
		16	7.6 us	1.20×10^{-1} γ 1.66×10^{-5} e-
		17	4.173 s	4.18×10^{-2} γ 1.71 e-
		18	630 ms	5.35 γ 3.89 e-
8	**O**	13	8.9 ms	1.10 γ 7.89 β+ 1.00×10^{2} %(β+)
		14	1.1768 min	3.32 γ 7.76×10^{-1} β+ 99.9%(β+)
		15	2.037 min	1.02 γ 7.35×10^{-1} β+ 99.9%(β+)
		19	26.9 s	9.45×10^{-1} γ 1.74 e-
		20	13.6 s	1.06 γ 1.20 e-
8	**O**	21	3.4 s	2.96 γ 2.32 e-
9	**F**	17	1.075 min	1.02 γ 7.39×10^{-1} β+ 99.8%(β+)
		18	1.830 h	9.90×10^{-1} γ 2.42×10^{-1} β+ 96.9%(β+)
		20	11.0 s	1.64 γ 2.48 e-
		21	4.32 s	5.56×10^{-1} γ 2.35 e-
		22	4.23 s	5.40 γ 2.36 e-
10	**Ne**	17	109 ms	1.08 γ 3.63 β+ 1.00×10^{2} %(β+)
		18	1.672 s	1.11 γ 1.50 β+ 1.00×10^{2} %(β+)
		19	17.22 s	1.02 γ 9.63×10^{-1} β+ 99.9%(β+)
		23	37.2 s	1.71×10^{-1} γ 1.90 e-
		24	3.38 min	5.43×10^{-1} γ 8.03×10^{-1} e-
		25	602 ms	3.91×10^{-1} γ
11	**Na**	20	446 ms	2.39 γ 4.76 β+ 1.00×10^{2} %(β+)
		21	22.48 s	1.04 γ 1.10 β+ 99.9%(β+)
		22	2.602 yr	2.19 γ 1.95×10^{-1} β+ 89.4%(β+)
		24	14.659 h	4.12 γ 5.54×10^{-1} e-
		24	20.2 ms	4.72×10^{-1} γ 8.30×10^{-4} e-
		25	1.0 min	4.39×10^{-1} γ 1.50 e-
		26	1.07 s	2.18 γ 3.34 e-
		27	304 ms	1.15 γ 3.63 e-
		28	30.5 ms	1.17 γ 6.10 e-
		29	45 ms	1.85 γ 4.86 e-
		30	50 ms	2.86 γ 5.69 e-
		31	17.0 ms	8.43×10^{-1} γ
		32	14 ms	2.14 γ
		33	8.2 ms	3.30×10^{-1} γ
12	**Mg**	21	122 ms	1.48 γ 4.72 β+ 1.00×10^{2} %(β+)
		22	3.86 s	1.73 γ 1.37 β+ 1.00×10^{2} %(β+)
		23	11.32 s	1.06 γ 1.34 β+ 99.9%(β+)
		27	9.46 min	9.14×10^{-1} γ 7.02×10^{-1} e-
		28	20.90 h	1.37 γ 1.52×10^{-1} e-
		29	1.1 s	1.87 γ 2.54 e-
12	**Mg**	30	325 ms	5.83×10^{-1} γ
		31	250 ms	1.75 γ
		32	120 ms	8.69×10^{-1} γ
13	**Al**	24	2.07 s	9.50 γ 2.03 β+ 99.9%(β+)
		24	130 ms	8.77×10^{-1} γ 9.56×10^{-1} β+ 18.0%(β+)
		25	7.18 s	1.04 γ 1.45 β+ 1.00×10^{2} %(β+)
		26	$7.2.\times10^{5}$ yr	2.68 γ 4.00×10^{-1} β+ 82.0%(β+)
		26	6.345 s	1.02 γ 1.44 β+ 99.9%(β+)
		28	2.2406 min	1.78 γ 1.24 e-
		29	6.6 min	1.38 γ 9.77×10^{-1} e-
		30	3.68 s	3.50 γ 2.30 e-
		32	35 ms	5.37×10^{-1} γ 5.85 e-
14	**Si**	24	100 ms	1.40×10^{-2} γ 3.18 e-
		26	2.21 s	1.26 γ 1.62 β+ 99.9%(β+)
		27	4.16 s	1.03 γ 1.72 β+ 99.9%(β+)
		31	2.622 h	1.81×10^{-3} γ 5.95×10^{-1} e-
		32	104 yr	1.66×10^{-5} γ 6.90×10^{-2} e-
15	**P**	28	270.3 ms	4.74 γ 4.59 β+ 1.00×10^{2} %(β+)
		29	4.14 s	1.05 γ 1.77 β+ 99.9%(β+)
		30	2.498 min	1.03 γ 1.43 β+ 99.9%(β+)
		32	14.282 d	1.18×10^{-3} γ 6.95×10^{-1} e-
		33	25.3 d	2.00×10^{-5} γ 7.60×10^{-2} e-
		34	12.4 s	3.43×10^{-1} γ 2.31 e-
		35	47 s	1.57 γ 9.88×10^{-1} e-
		36	5.9 s	4.61 γ
16	**S**	29	187.0 ms	2.23 γ 4.06 β+ 99.0%(β+)
		30	1.18 s	1.61 γ 2.08 β+ 99.8%(β+)
		31	2.584 s	1.04 γ 2.00 β+ 99.9%(β+)
		35	87.5 d	8.60×10^{-6} γ 4.86×10^{-2} e-
		37	5.05 min	2.93 γ 8.00×10^{-1} e-
		38	2.84 h	1.73 γ 4.80×10^{-1} e-

Table I. Average Energies per Disintegration (continued)

Z	El	A	Half-life	⟨E⟩ (MeV), or %β+
16	**S**	39	11.5 s	1.79 γ
				2.28 e-
17	**Cl**	32	298.0 ms	4.79 γ
				3.77 β+
				1.00×10^{2} %(β+)
		33	2.511 s	1.05 γ
				2.08 β+
				99.7%(β+)
		34	1.5262 s	1.03 γ
				2.05 β+
				99.9%(β+)
		34	32.23 min	2.11 γ
				5.55×10^{-3} e-
				4.55×10^{-1} β+
				54.0%(β+)
		36	3.01×10^{5} yr	3.55×10^{-4} γ
				2.46×10^{-1} e-
				7.50×10^{-6} β+
				1.49×10^{-2} %(β+)
		38	37.24 min	1.43 γ
				1.54 e-
		38	715 ms	6.71×10^{-1} γ
		39	55.6 min	1.45 γ
				8.21×10^{-1} e-
		40	1.35 min	4.04 γ
				1.57 e-
18	**Ar**	33	173.0 ms	1.45 γ
				3.96 β+
				99.9%(β+)
		34	844.0 ms	1.87 γ
				2.29 β+
				99.9%(β+)
		35	1.775 s	1.06 γ
				2.27 β+
				99.9%(β+)
		37	35.04 d	3.69×10^{-4} γ
		39	269 yr	1.47×10^{-4} γ
				2.18×10^{-1} e-
		41	1.83 h	1.28 γ
				4.64×10^{-1} e-
		42	33 yr	1.66×10^{-4} γ
				2.33×10^{-1} e-
		43	5.4 min	1.54 γ
				1.36 e-
		44	11.87 min	1.83 γ
				6.60×10^{-1} e-
		45	21.5 s	2.98 γ
				1.58 e-
19	**K**	35	190 ms	4.31 γ
				3.57 β+
				1.00×10^{2} %(β+)
		36	342 ms	5.50 γ
				3.52 β+
				1.00×10^{2} %(β+)
		37	1.23 s	1.07 γ
				2.35 β+
				99.9%(β+)
		38	7.64 min	3.19 γ
				1.20 β+
				99.5%(β+)
		38	925 ms	1.03 γ
				2.32 β+
				99.9%(β+)
		40	1.28×10^{9} yr	1.57×10^{-1} γ
				4.55×10^{-1} e-
		42	12.360 h	2.94×10^{-1} γ
				1.42 e-
		43	22.3 h	9.64×10^{-1} γ
				3.14×10^{-1} e-
		44	22.1 min	2.38 γ
				1.41 e-
		45	17.8 min	1.86 γ
				9.93×10^{-1} e-
		46	1.9 min	3.00 γ
				2.32 e-
19	**K**	47	17.5 s	2.60 γ
				1.80 e-
		48	6.8 s	6.79 γ
				2.63 e-
20	**Ca**	37	173 ms	1.15 γ
				3.27 β+
				1.00×10^{2} %(β+)
		38	435 ms	1.38 γ
				2.43 β+
				99.9%(β+)
		39	860 ms	1.03 γ
				2.56 β+
				99.9%(β+)
		41	1.03×10^{5} yr	4.31×10^{-4} γ
		45	164 d	2.10×10^{-5} γ
				7.70×10^{-2} e-
		47	4.536 d	1.06 γ
				3.45×10^{-1} e-
		49	8.72 min	3.17 γ
				8.70×10^{-1} e-
		50	14 s	1.79 γ
				1.35 e-
		51	10 s	2.75×10^{3} γ
				2.11×10^{3} e-
21	**Sc**	40	182 ms	7.12 γ
				3.39 β+
				1.00×10^{2} %(β+)
		41	596 ms	1.03 γ
				2.54 β+
				99.8%(β+)
		42	681 ms	1.03 γ
				2.51 β+
				99.9%(β+)
		42	1.03 min	4.21 γ
				1.25 β+
				99.3%(β+)
		43	3.89 h	9.83×10^{-1} γ
				4.50×10^{-4} e-
				4.19×10^{-1} β+
				88.0%(β+)
		44	3.93 h	2.13 γ
				9.40×10^{-4} e-
				5.97×10^{-1} β+
				94.0%(β+)
		44	2.442 d	2.81×10^{-1} γ
				3.26×10^{-2} e-
		45	316 ms	3.18×10^{-4} γ
				5.20×10^{-3} e-
		46	83.83 d	2.01 γ
				1.12×10^{-1} e-
		46	18.70 s	8.86×10^{-2} γ
				5.23×10^{-2} e-
		47	3.341 d	1.08×10^{-1} γ
				1.62×10^{-1} e-
		48	1.821 d	3.35 γ
				2.21×10^{-1} e-
		49	57.4 min	2.64×10^{-3} γ
				8.22×10^{-1} e-
		50	1.71 min	3.20 γ
				1.63 e-
		50	350 ms	2.46×10^{-1} γ
				8.80×10^{-3} e-
		51	12.4 s	2.36 γ
				1.84 e-
22	**Ti**	41	80 ms	1.11 γ
				3.43 β+
				1.00×10^{2} %(β+)
		42	199 ms	1.40 γ
				2.61 β+
				1.00×10^{2} %(β+)
		43	490 ms	1.03 γ
				2.72 β+
				99.9%(β+)
		44	47 yr	1.38×10^{-1} γ
				1.09×10^{-2} e-
22	**Ti**	45	3.08 h	8.73×10^{-1} γ
				5.30×10^{-4} e-
				3.73×10^{-1} β+
				85.0%(β+)
		51	5.76 min	3.71×10^{-1} γ
				8.68×10^{-1} e-
		52	1.7 min	1.28×10^{-1} γ
				7.65×10^{-1} e-
		53	33 s	1.97 γ
				1.34 e-
23	**V**	46	422.3 ms	1.03 γ
				2.81 β+
				99.9%(β+)
		47	32.6 min	1.00 γ
				1.28×10^{-4} e-
				8.03×10^{-1} β+
				97.0%(β+)
		48	15.976 d	2.91 γ
				2.01×10^{-3} e-
				1.44×10^{-1} β+
				50.0%(β+)
		49	330 d	9.23×10^{-4} γ
				3.60×10^{-3} e-
		52	3.75 min	1.45 γ
				1.07 e-
		53	1.61 min	1.04 γ
				1.00 e-
		54	49.8 s	4.11 γ
				1.31 e-
		55	6.5 s	6.98×10^{-1} γ
				2.39 e-
24	**Cr**	46	260 ms	1.03 γ
				3.09 β+
				1.00×10^{2} %(β+)
		47	460 ms	1.03 γ
				3.01 β+
				99.9%(β+)
		48	21.56 h	4.36×10^{-1} γ
				1.80×10^{-2} e-
				1.33×10^{-3} β+
				1.47%(β+)
		49	42.1 min	1.06 γ
				5.95×10^{-1} β+
				93.0%(β+)
		51	27.704 d	3.26×10^{-2} γ
				3.08×10^{-3} e-
		55	3.497 min	3.27×10^{-3} γ
				1.10 e-
		56	5.9 min	7.09×10^{-2} γ
				6.30×10^{-1} e-
		57	21 s	4.45×10^{-1} γ
				1.94 e-
25	**Mn**	49	384 ms	1.05 γ
				3.13 β+
				99.9%(β+)
		50	283.0 ms	1.03 γ
				3.10 β+
				99.9%(β+)
		50	1.75 min	4.37 γ
				4.98×10^{-4} e-
				1.52 β+
				99.0%(β+)
		51	46.2 min	9.99×10^{-1} γ
				9.30×10^{-5} e-
				9.35×10^{-1} β+
				97.0%(β+)
		52	5.591 d	3.45 γ
				2.67×10^{-3} e-
				7.10×10^{-2} β+
				29.0%(β+)
		52	21.1 min	2.42 γ
				4.01×10^{-4} e-
				1.13 β+
				97.0%(β+)
		53	3.74×10^{6} yr	1.39×10^{-3} γ

Table I. Average Energies per Disintegration (continued)

Z	El	A	Half-life	⟨E⟩ (MeV), or %β+
25	**Mn**	53	3.74×10^{6} yr	3.20×10^{-3} e-
		54	312.2 d	8.36×10^{-1} γ 3.40×10^{-3} e-
		56	2.578 h	1.69 γ 8.30×10^{-1} e-
		57	1.45 min	7.81×10^{-2} γ 1.11 e-
		58	1.09 min	2.37 γ 1.75 e-
		60	1.8 s	2.70 γ 2.65 e-
26	**Fe**	51	245 ms	1.04 γ 3.29 β+ 99.9%(β+)
		52	8.28 h	7.47×10^{-1} γ 7.50×10^{-3} e- 1.89×10^{-1} β+ 56.0%(β+)
		52	46 s	5.01 γ 1.98 β+ 99.7%(β+)
		53	8.51 min	1.18 γ 3.54×10^{-4} e- 1.11 β+ 97.0%(β+)
		53	2.6 min	3.01 γ
		55	2.73 yr	1.63×10^{-3} γ 3.80×10^{-3} e-
		59	44.50 d	1.19 γ 1.18×10^{-1} e-
		60	1×10^{5} yr	1.18×10^{-3} γ 5.10×10^{-2} e-
		61	6.0 min	1.39 γ 1.09 e-
27	**Co**	53	262 ms	1.04 γ 3.43 β+ 99.9%(β+)
		53	247 ms	1.02 γ 3.45 β+ 98.4%(β+)
		54	193.2 ms	1.04 γ 3.40 β+ 99.9%(β+)
		54	1.46 min	3.97 γ 2.05 β+ 99.5%(β+)
		55	17.53 h	2.00 γ 1.60×10^{-3} e- 4.30×10^{-1} β+ 76.0%(β+)
		56	77.7 d	3.58 γ 3.60×10^{-3} e- 1.20×10^{-1} β+ 20.0%(β+)
		57	271.77 d	1.25×10^{-1} γ 1.76×10^{-2} e-
		58	70.92 d	9.77×10^{-1} γ 3.60×10^{-3} e- 3.00×10^{-2} β+ 14.9%(β+)
		58	9.2 h	1.83×10^{-3} γ 2.08×10^{-2} e-
		60	5.271 yr	2.50 γ 9.60×10^{-2} e-
		60	10.47 min	6.60×10^{-3} γ 5.36×10^{-2} e-
		61	1.650 h	9.84×10^{-2} γ 7.39×10^{-3} e-
		62	1.50 min	1.60 γ 1.64 e-
		62	13.91 min	2.69 γ 1.06 e-
		63	27.4 s	1.23×10^{-1} γ 1.57 e-
		64	300 ms	1.96×10^{-1} γ 3.29 e-
28	**Ni**	55	189 ms	1.04 γ 3.62 β+ 99.9%(β+)
		56	6.10 d	1.72 γ 6.90×10^{-3} e-
		57	1.503 d	1.92 γ 4.40×10^{-3} e- 1.43×10^{-1} β+ 40.0%(β+)
		59	7.5×10^{4} yr	2.62×10^{-3} γ 4.10×10^{-3} e- 3.73×10^{-9} β+ 1.50×10^{-5} %(β+)
		63	100 yr	1.12×10^{-6} γ 1.71×10^{-2} e-
		65	2.520 h	5.50×10^{-1} γ 6.32×10^{-1} e-
		66	2.28 d	1.55×10^{-5} γ 6.50×10^{-2} e-
29	**Cu**	58	3.20 s	1.56 γ 1.19×10^{-3} e- 3.30 β+ 99.8%(β+)
		59	1.36 min	1.45 γ 2.54×10^{-4} e- 1.49 β+ 98.0%(β+)
		60	23.2 min	3.90 γ 7.30×10^{-4} e- 8.94×10^{-1} β+ 92.0%(β+)
		61	3.41 h	8.23×10^{-1} γ 2.20×10^{-3} e- 3.06×10^{-1} β+ 61.0%(β+)
		62	9.74 min	1.00 γ 9.40×10^{-5} e- 1.28 β+ 97.0%(β+)
		64	12.701 h	1.92×10^{-1} γ 7.29×10^{-2} e- 4.98×10^{-2} β+ 18.0%(β+)
		66	5.10 min	8.05×10^{-2} γ 1.08 e-
		67	2.580 d	1.15×10^{-1} γ 1.56×10^{-1} e-
		68	31 s	9.90×10^{-1} γ 1.49 e-
		68	3.75 min	1.01 γ 1.88×10^{-1} e-
		69	3.0 min	2.24×10^{-1} γ 1.03 e-
		70	4.5 s	1.14×10^{-2} γ 2.78 e-
		70	46 s	2.84 γ
30	**Zn**	57	40 ms	8.10×10^{-1} γ
		59	184 ms	1.08 γ 3.79 β+ 99.9%(β+)
		60	2.38 min	1.40 γ 1.86×10^{-2} e- 1.20 β+ 98.0%(β+)
		61	1.485 min	1.53 γ 2.70×10^{-4} e- 1.86 β+ 98.0%(β+)
		62	9.26 h	4.39×10^{-1} γ 1.04×10^{-2} e- 2.17×10^{-2} β+ 8.40%(β+)
		63	38.1 min	1.11 γ 3.66×10^{-4} e- 9.19×10^{-1} β+ 93.0%(β+)
30	**Zn**	65	244.1 d	5.85×10^{-1} γ 4.40×10^{-3} e- 2.08×10^{-3} β+ 1.46%(β+)
		69	56 min	3.16×10^{-4} γ 3.21×10^{-1} e-
		69	13.76 h	4.16×10^{-1} γ 2.26×10^{-2} e-
		71	2.4 min	3.08×10^{-1} γ 1.05 e-
		71	3.94 h	1.58 γ 5.41×10^{-1} e-
		72	1.938 d	1.52×10^{-1} γ 1.01×10^{-1} e-
		74	1.58 min	2.00×10^{-1} γ 9.18×10^{-1} e-
		78	1.5 s	1.53 γ 1.82 e-
31	**Ga**	62	116.1 ms	1.04 γ 3.86 β+ 99.9%(β+)
		63	32.4 s	1.37 γ 4.30×10^{-4} e- 1.89 β+ 99.0%(β+)
		64	2.63 min	3.41 γ 4.34×10^{-4} e- 1.77 β+ 98.0%(β+)
		65	15.2 min	1.17 γ 2.32×10^{-2} e- 7.91×10^{-1} β+ 89.0%(β+)
		66	9.5 h	2.46 γ 2.13×10^{-3} e- 9.86×10^{-1} β+ 57.0%(β+)
		67	3.261 d	1.55×10^{-1} γ 3.32×10^{-2} e-
		68	1.135 h	9.50×10^{-1} γ 5.00×10^{-4} e- 7.40×10^{-1} β+ 89.0%(β+)
		70	21.15 min	8.66×10^{-3} γ 6.44×10^{-1} e-
		72	14.10 h	2.71 γ 5.02×10^{-1} e-
		73	4.87 h	3.53×10^{-1} γ 4.98×10^{-1} e-
		74	8.1 min	3.07 γ 9.79×10^{-1} e-
		74	9.5 s	4.31×10^{-2} γ 1.32×10^{-2} e-
		76	33 s	2.81 γ 1.79 e-
		78	5.09 s	2.55 γ 2.59 e-
		79	3.0 s	2.09 γ 2.15 e-
		80	1.66 s	2.79 γ 3.54 e-
		81	1.23 s	2.08 γ
32	**Ge**	64	1.06 min	3.65×10^{-1} γ
		65	31 s	1.84 γ 5.10×10^{-3} e- 2.05 β+ 99.0%(β+)
		66	2.26 h	6.86×10^{-1} γ 1.86×10^{-2} e- 8.20×10^{-2} β+ 24.0%(β+)
		67	18.7 min	1.43 γ 2.84×10^{-3} e- 1.18 β+ 93.0%(β+)
		68	270.8 d	4.12×10^{-3} γ

Table I. Average Energies per Disintegration (continued)

Z	El	A	Half-life	⟨E⟩ (MeV), or %β+
32	**Ge**	68	270.8 d	4.40×10^{-3} e-
		69	1.627 d	9.56×10^{-1} γ
				3.69×10^{-3} e-
				1.16×10^{-1} β+
				24.0%(β+)
		71	11.2 d	4.17×10^{-3} γ
				4.50×10^{-3} e-
		73	499 ms	1.11×10^{-2} γ
				5.48×10^{-2} e-
		75	1.380 h	3.55×10^{-2} γ
				4.21×10^{-1} e-
		75	48 s	5.68×10^{-2} γ
				8.08×10^{-2} e-
		77	11.30 h	1.02 γ
				6.72×10^{-1} e-
		77	53 s	6.73×10^{-2} γ
				9.48×10^{-1} e-
		78	1.47 h	2.77×10^{-1} γ
				2.27×10^{-1} e-
		79	19.1 s	3.14×10^{-1} γ
				1.69 e-
		79	39 s	1.78 γ
				1.26 e-
		80	29.5 s	6.02×10^{-1} γ
				1.05 e-
		82	4.6 s	1.08 γ
33	**As**	67	42 s	1.48 γ
				2.00×10^{-2} e-
				2.08 β+
				1.00×10^{2} %(β+)
		68	2.53 min	3.70 γ
				8.41×10^{-4} e-
				2.01 β+
				99.0%(β+)
		69	15.2 min	1.15 γ
				4.61×10^{-3} e-
				1.22 β+
				94.0%(β+)
		70	52.6 min	4.10 γ
				1.99×10^{-3} e-
				8.63×10^{-1} β+
				90.0%(β+)
		71	2.70 d	5.84×10^{-1} γ
				1.69×10^{-2} e-
				1.04×10^{-1} β+
				30.0%(β+)
		72	1.083 d	1.78 γ
				1.15×10^{-2} e-
				1.02 β+
				88.0%(β+)
		73	80.3 d	1.58×10^{-2} γ
				5.94×10^{-2} e-
		74	17.78 d	7.64×10^{-1} γ
				1.39×10^{-1} e-
				1.28×10^{-1} β+
				29.0%(β+)
		76	1.097 d	4.22×10^{-1} γ
				1.07 e-
		77	1.618 d	8.06×10^{-3} γ
				2.26×10^{-1} e-
		78	1.512 h	1.33 γ
				1.23 e-
		79	9.0 min	5.97×10^{-2} γ
				1.02 e-
		80	15.2 s	6.45×10^{-1} γ
				2.25 e-
		81	33 s	1.43×10^{-1} γ
				1.62 e-
		82	19.1 s	6.24×10^{-1} γ
				3.28 β+
		82	14.0 s	2.73 γ
				1.97 β+
		84	5.5 s	1.59 γ
				4.04 e-
		85	2.03 s	3.58×10^{-1} γ
34	**Se**	69	27.4 s	1.61 γ
34	**Se**	69	27.4 s	2.44 β+
				1.00×10^{2} %(β+)
		70	41.1 min	9.94×10^{-1} γ
				4.14×10^{-2} e-
				4.47×10^{-1} β+
				70.0%(β+)
		71	4.74 min	1.61 γ
				3.19×10^{-3} e-
				1.39 β+
				96.0%(β+)
		72	8.4 d	3.42×10^{-2} γ
				2.20×10^{-2} e-
		73	7.2 h	1.10 γ
				1.80×10^{-2} e-
				3.68×10^{-1} β+
				66.0%(β+)
		73	40 min	2.54×10^{-1} γ
				1.86×10^{-2} e-
				1.55×10^{-1} β+
				21.0%(β+)
		75	119.77 d	3.92×10^{-1} γ
				1.42×10^{-2} e-
		77	17.4 s	8.74×10^{-2} γ
				7.24×10^{-2} e-
		79	$65.\times10^{3}$ yr	9.40×10^{-6} γ
				5.30×10^{-2} e-
		79	3.91 min	1.37×10^{-2} γ
				8.15×10^{-2} e-
		81	18.5 min	1.09×10^{-2} γ
				6.11×10^{-1} e-
		81	57.28 min	1.50×10^{-2} γ
				8.81×10^{-2} e-
		83	22.5 min	2.56 γ
				4.78×10^{-1} e-
		83	1.173 min	9.58×10^{-1} γ
				1.26 e-
		84	3.2 min	4.21×10^{-1} γ
				5.39×10^{-1} e-
		85	32 s	2.22 γ
				1.75 e-
35	**Br**	71	21 s	1.56×10^{-1} γ
		72	1.31 min	2.91 γ
				5.50×10^{-2} e-
				2.76 β+
				99.9%(β+)
		73	3.4 min	5.07×10^{-1} γ
		74	25.3 min	4.56 γ
				3.14×10^{-3} e-
				1.01 β+
				89.0%(β+)
		74	42 min	4.09 γ
				2.38×10^{-3} e-
				1.41 β+
				91.0%(β+)
		75	1.62 h	1.20 γ
				6.80×10^{-3} e-
				4.99×10^{-1} β+
				71.0%(β+)
		76	16.2 h	2.78 γ
				3.37×10^{-3} e-
				6.42×10^{-1} β+
				54.0%(β+)
		76	1.31 s	4.00×10^{-2} γ
				6.52×10^{-2} e-
		77	2.3765 d	3.31×10^{-1} γ
				7.80×10^{-3} e-
				1.12×10^{-3} β+
				7.40×10^{-1} %(β+)
		77	4.3 min	1.93×10^{-2} γ
				8.58×10^{-2} e-
		78	6.46 min	1.03 γ
				4.80×10^{-4} e-
				1.02 β+
				92.0%(β+)
		79	4.86 s	1.59×10^{-1} γ
				4.80×10^{-2} e-
35	**Br**	80	17.68 min	7.69×10^{-2} γ
				7.16×10^{-1} e-
				8.00×10^{-3} β+
				2.20%(β+)
		80	4.42 h	2.41×10^{-2} γ
				6.08×10^{-2} e-
		82	1.471 d	2.64 γ
				1.39×10^{-1} e-
		82	6.1 min	7.39×10^{-3} γ
				6.98×10^{-2} e-
		83	2.39 h	7.76×10^{-3} γ
				3.26×10^{-1} e-
		84	31.8 min	1.72 γ
				1.25 e-
		84	6.0 min	2.83 γ
				8.97×10^{-1} e-
		85	2.87 min	6.84×10^{-2} γ
				1.04 e-
		86	55.7 s	3.26 γ
				1.93 e-
		87	55.7 s	4.08 γ
				1.87 e-
		88	16.7 s	3.07 γ
				2.78 e-
		89	4.37 s	1.67 γ
		90	1.92 s	1.54 γ
36	**Kr**	72	17.2 s	3.59×10^{-1} γ
				5.00×10^{-3} e-
		73	27 s	5.61×10^{-1} γ
		74	11.5 min	1.18 γ
				7.13×10^{-1} β+
				87.0%(β+)
		75	4.3 min	1.26 γ
				1.21×10^{-2} e-
				1.55 β+
				93.0%(β+)
		76	14.8 h	4.26×10^{-1} γ
				1.67×10^{-2} e-
		77	1.24 h	1.11 γ
				2.01×10^{-2} e-
				6.81×10^{-1} β+
				87.0%(β+)
		79	1.460 d	2.58×10^{-1} γ
				5.30×10^{-3} e-
				1.85×10^{-2} β+
				7.10%(β+)
		79	50 s	4.00×10^{-2} γ
				8.91×10^{-2} e-
		81	2.1×10^{5} yr	1.65×10^{-2} γ
				4.80×10^{-3} e-
		81	13 s	1.30×10^{-1} γ
				5.95×10^{-2} e-
		83	1.86 h	2.57×10^{-3} γ
				3.82×10^{-2} e-
		85	10.72 yr	2.40×10^{-3} γ
				2.51×10^{-1} e-
		85	4.48 h	1.56×10^{-1} γ
				2.55×10^{-1} e-
		87	1.27 h	7.96×10^{-1} γ
				1.33 e-
		88	2.84 h	1.96 γ
				3.65×10^{-1} e-
		89	3.16 min	1.83 γ
				1.38 e-
		90	32.3 s	1.24 γ
				1.31 e-
		91	8.57 s	1.75 γ
				2.08 e-
		92	1.84 s	1.46 γ
				2.09 e-
		93	1.29 s	2.29 γ
				2.82 e-
37	**Rb**	74	64.9 ms	1.04 γ
				4.46 β+
				1.00×10^{2} %(β+)
		77	3.7 min	1.81 γ

Table I. Average Energies per Disintegration (continued)

Z	El	A	Half-life	⟨E⟩ (MeV), or %β+
37	**Rb**	77	3.7 min	1.66×10^{-2} e- 1.65 β+ 98.0%(β+)
		78	17.7 min	4.15 γ 3.10×10^{-3} e- 1.30 β+ 90.0%(β+)
		78	5.7 min	3.23 γ 2.50×10^{-3} e- 1.51 β+ 88.0%(β+)
		79	22.9 min	1.44 γ 4.39×10^{-2} e- 7.43×10^{-1} β+ 82.0%(β+)
		80	34 s	1.19 γ 3.90×10^{-4} e- 2.04 β+ 98.0%(β+)
		81	4.58 h	6.43×10^{-1} γ 6.06×10^{-2} e- 1.32×10^{-1} β+ 31.0%(β+)
		81	30.6 min	1.22×10^{-2} γ
		82	1.273 min	1.09 γ 3.19×10^{-4} e- 1.41 β+ 95.0%(β+)
		82	6.47 h	2.77 γ 6.20×10^{-3} e- 9.10×10^{-2} β+ 23.0%(β+)
		83	86.2 d	5.07×10^{-1} γ 3.65×10^{-2} e-
		84	32.9 d	8.89×10^{-1} γ 1.71×10^{-2} e- 1.44×10^{-1} β+ 26.0%(β+)
		84	20.26 min	3.76×10^{-1} γ
		86	18.66 d	9.57×10^{-2} γ 6.67×10^{-1} e-
		86	1.017 min	5.46×10^{-1} γ 9.90×10^{-3} e-
		87	4.80×10^{10} yr	8.2×10^{-2} e-
		88	17.8 min	6.37×10^{-1} γ 2.07 e-
		89	15.2 min	2.07 γ 1.02 e-
		90	2.55 min	1.85 γ 1.98 e-
		90	4.3 min	3.30 γ 1.40 e-
		91	58.4 s	2.34 γ 1.58 e-
		92	4.50 s	5.37×10^{-1} γ 3.53 e-
		93	5.85 s	1.42 γ 2.71 e-
		94	2.73 s	2.74 γ 3.24 e-
		95	384 ms	6.12×10^{-1} γ
		96	199 ms	2.03 γ 4.08 e-
		99	59 ms	3.50×10^{-2} γ
38	**Sr**	79	2.2 min	1.76×10^{-1} γ
		80	1.77 h	4.12×10^{-1} γ 6.60×10^{-3} e- 3.01×10^{-2} β+ 7.20%(β+)
		81	22.2 min	1.39 γ 8.10×10^{-3} e- 9.67×10^{-1} β+ 86.0%(β+)
		82	25.6 d	7.86×10^{-3} γ 4.76×10^{-3} e-
		83	1.350 d	7.86×10^{-1} γ
38	**Rb**	83	1.350 d	2.61×10^{-2} e- 1.23×10^{-1} β+ 24.0%(β+)
		83	5.0 s	2.28×10^{-1} γ 3.12×10^{-2} e-
		85	64.84 d	5.18×10^{-1} γ 8.40×10^{-3} e-
		85	1.126 h	2.16×10^{-1} γ 1.40×10^{-2} e-
		87	2.80 h	3.21×10^{-1} γ 6.74×10^{-2} e-
		89	50.6 d	9.86×10^{-4} γ 5.83×10^{-1} e-
		90	28.5 yr	1.24×10^{-4} γ 1.96×10^{-1} e-
		91	9.5 h	1.05 γ 6.67×10^{-1} e-
		92	2.71 h	1.34 γ 1.77×10^{-1} e-
		93	7.4 min	2.22 γ 7.87×10^{-1} e-
		94	1.235 min	1.43 γ 8.35×10^{-1} e-
		95	25.1 s	1.02 γ 2.30 e-
		96	1.06 s	9.32×10^{-1} γ 1.99 e-
		97	441 ms	1.86 γ 2.66 e-
		98	650 ms	1.44×10^{-1} γ 2.56 e-
39	**Y**	81	1.20 min	3.70×10^{-3} γ
		82	9.5 s	1.85×10^{-1} γ
		83	7.1 min	1.38 γ 1.38 β+ 95.0%(β+)
		83	2.85 min	1.40 γ 3.31×10^{-2} e- 1.39 β+ 94.0%(β+)
		84	40 min	3.79 γ 2.43×10^{-3} e- 1.14 β+ 93.0%(β+)
		84	4.6 s	1.30 γ 3.30×10^{-4} e- 2.34 β+ 99.0%(β+)
		85	2.68 h	1.08 γ 4.83×10^{-1} β+ 66.0%(β+)
		85	4.9 h	1.35 γ 3.70×10^{-3} e- 5.64×10^{-1} β+ 57.0%(β+)
		86	14.74 h	3.59 γ 5.50×10^{-3} e- 2.19×10^{-1} β+ 33.0%(β+)
		86	48 min	2.20×10^{-1} γ 2.04×10^{-2} e- 2.88×10^{-3} β+ 4.40×10^{-1} %(β+)
		87	3.35 d	7.89×10^{-1} γ 7.55×10^{-2} e- 4.20×10^{-4} β+ 2.10×10^{-1} %(β+)
		87	12.9 h	3.07×10^{-1} γ 7.48×10^{-2} e- 4.00×10^{-3} β+ 7.50×10^{-1} %(β+)
		88	106.61 d	2.69 γ 5.40×10^{-3} e- 7.90×10^{-4} β+ 2.20×10^{-1} %(β+)
		89	16.06 s	9.01×10^{-1} γ
39	**Y**	89	16.06 s	7.70×10^{-3} e-
		90	2.671 d	2.00×10^{-3} γ 9.34×10^{-1} e-
		90	3.19 h	6.34×10^{-1} γ 4.75×10^{-2} e-
		91	58.5 d	4.56×10^{-3} γ 6.03×10^{-1} e-
		91	49.71 min	5.28×10^{-1} γ 2.79×10^{-2} e-
		92	3.54 h	2.57×10^{-1} γ 1.43 e-
		93	10.25 h	9.19×10^{-2} γ 1.17 e-
		93	820 ms	8.60×10^{-1} γ 7.90×10^{-2} e-
		94	18.7 min	7.78×10^{-1} γ 1.81 e-
		95	10.3 min	1.29 γ 1.35 e-
		96	6.2 s	9.80×10^{-4} γ 1.49×10^{-1} e-
		96	9.6 s	3.97 γ 4.92×10^{-3} e-
		97	3.7 s	1.83 γ 2.16 e-
		97	1.21 s	1.80 γ 2.41 e-
		98	640 ms	9.07×10^{-1} γ 3.74 e-
		98	2.0 s	3.04 γ 2.63 e-
		99	1.5 s	8.18×10^{-1} γ 3.12 e-
		100	940 ms	6.07×10^{-1} γ
40	**Zr**	83	44 s	1.41×10^{-1} γ
		85	7.86 min	1.48 γ 2.10×10^{-3} e- 1.33 β+ 92.0%(β+)
		86	16.5 h	2.94×10^{-1} γ 2.99×10^{-2} e-
		87	14.0 s	2.38×10^{-1} γ 9.80×10^{-2} e-
		88	83.4 d	3.92×10^{-1} γ 1.53×10^{-2} e- 1.80×10^{-2} %(β+)
		89	3.268 d	1.16 γ 1.15×10^{-2} e- 9.00×10^{-2} β+ 23.0%(β+)
		89	4.18 min	6.33×10^{-1} γ 2.47×10^{-2} e- 7.10×10^{-3} β+ 1.51%(β+)
		90	809 ms	2.02 γ 1.67×10^{-2} e-
		93	1.5×10^{6} yr	1.84×10^{-3} γ 4.71×10^{-2} e-
		95	64.02 d	7.33×10^{-1} γ 1.20×10^{-1} e-
		97	16.90 h	8.70×10^{-1} γ 7.11×10^{-1} e-
		98	30.7 s	1.90×10^{-3} γ 9.07×10^{-1} e-
		99	2.1 s	8.60×10^{-1} γ 1.55 e-
41	**Nb**	84	12 s	7.06×10^{-1} γ
		87	3.8 min	1.24 γ 1.04×10^{-1} e- 1.66 β+ 96.0%(β+)
		88	14.3 min	3.20 γ 5.11×10^{-2} e- 1.18 β+
		88	7.8 min	3.79 γ 3.70×10^{-3} e-

Table I. Average Energies per Disintegration (continued)

Z	El	A	Half-life	⟨E⟩ (MeV), or %β+
41	**Nb**	88	7.8 min	1.43 β+ 94.0%(β+)
		89	1.10 h	1.89 γ 2.84×10^{-2} e- 7.89×10^{-1} β+ 82.0%(β+)
		89	2.0 h	1.40 γ 1.71×10^{-3} e- 1.09 β+ 76.0%(β+)
		90	14.60 h	4.20 γ 4.55×10^{-2} e- 3.50×10^{-1} β+ 53.0%(β+)
		90	18.8 s	8.24×10^{-2} γ 3.94×10^{-2} e-
		91	680 yr	1.24×10^{-2} γ 4.90×10^{-3} e- 1.78×10^{-4} β+ 1.64×10^{-1} %(β+)
		91	62 d	5.10×10^{-2} γ 9.00×10^{-2} e-
		92	3.6×10^{7} yr	1.51 γ 7.30×10^{-3} e-
		92	10.15 d	9.69×10^{-1} γ 5.60×10^{-3} e- 5.08×10^{-5} β+ 5.80×10^{-2} %(β+)
		93	13.6 yr	1.84×10^{-3} γ 2.81×10^{-2} e-
		94	2.0×10^{4} yr	1.57 γ 1.46×10^{-1} e-
		94	6.26 min	1.17×10^{-2} γ 3.49×10^{-2} e-
		95	34.97 d	7.64×10^{-1} γ 4.35×10^{-2} e-
		95	3.61 d	7.12×10^{-2} γ 1.71×10^{-1} e-
		96	23.4 h	2.46 γ 2.54×10^{-1} e-
		97	1.20 h	6.67×10^{-1} γ 4.67×10^{-1} e-
		97	1.0 min	7.28×10^{-1} γ 1.56×10^{-2} e-
		98	2.9 s	9.07×10^{-2} γ 1.96 e-
		98	51.3 min	2.71 γ 7.89×10^{-1} e-
		99	15.0 s	1.68×10^{-1} γ
		99	2.6 min	7.55×10^{-1} γ 1.47 e-
		106	1.02 s	1.00 γ
42	**Mo**	90	5.67 h	8.38×10^{-1} γ 7.90×10^{-2} e- 1.22×10^{-1} β+ 26.0%(β+)
		91	15.49 min	9.86×10^{-1} γ 3.16×10^{-4} e- 1.45 β+ 94.0%(β+)
		91	1.09 min	1.39 γ 1.22×10^{-2} e- 5.41×10^{-1} β+ 44.0%(β+)
		93	3.5×10^{3} yr	1.25×10^{-2} γ 3.16×10^{-2} e-
		93	6.8 h	2.31 γ 1.03×10^{-1} e-
		99	2.7477 d	2.73×10^{-1} γ 4.08×10^{-1} e-
		101	14.6 min	1.51 γ 5.12×10^{-1} e-
		102	11.3 min	1.89×10^{-2} γ 3.51×10^{-1} e-
		104	1.00 min	1.47×10^{-1} γ 6.85×10^{-1} e-
43	**Tc**	90	8.3 s	7.40×10^{-1} γ 6.70×10^{-4} e-
		90	49.2 s	3.10 γ
		91	3.14 min	2.48 γ 1.11×10^{-3} e- 1.63 β+ 91.0%(β+)
		91	3.3 min	1.94 γ 1.82×10^{-2} e- 1.97 β+ 96.0%(β+)
		92	4.4 min	3.92 γ 3.89×10^{-2} e- 1.70 β+ 93.0%(β+)
		93	2.75 h	1.58 γ 5.00×10^{-3} e- 3.96×10^{-2} β+ 11.4%(β+)
		93	44 min	7.74×10^{-1} γ 7.60×10^{-2} e-
		94	4.88 h	2.67 γ 8.20×10^{-3} e- 3.95×10^{-2} β+ 11.0%(β+)
		94	52 min	1.95 γ 2.44×10^{-3} e- 7.53×10^{-1} β+ 70.0%(β+)
		95	20.0 h	7.98×10^{-1} γ 6.10×10^{-3} e-
		95	61 d	7.19×10^{-1} γ 1.39×10^{-2} e- 8.90×10^{-4} β+ 3.10×10^{-1} %(β+)
		96	4.3 d	2.51 γ 8.20×10^{-3} e-
		96	52 min	4.80×10^{-2} γ 2.62×10^{-2} e-
		97	2.6×10^{6} yr	1.18×10^{-2} γ 4.90×10^{-3} e-
		97	90 d	9.40×10^{-3} γ 8.50×10^{-2} e-
		98	4.2×10^{6} yr	1.39 γ 1.23×10^{-1} e-
		99	2.13×10^{5} yr	2.60×10^{-5} γ 8.50×10^{-2} e-
		99	6.006 h	1.24×10^{-1} γ 1.42×10^{-2} e-
		100	15.8 s	8.65×10^{-2} γ
		101	14.2 min	3.38×10^{-1} γ 4.77×10^{-1} e-
		102	5.3 s	8.76×10^{-2} γ 1.95 e-
		102	4.4 min	2.49 γ 7.80×10^{-1} e-
		103	54 s	2.38×10^{-1} γ 9.83×10^{-1} e-
		104	18.3 min	2.00 γ 1.61 e-
		105	7.7 min	5.39×10^{-1} γ 1.43 e-
		107	21.2 s	5.21×10^{-1} γ 1.86 e-
		108	5.2 s	9.01×10^{-1} γ 3.12 e-
44	**Ru**	92	3.65 min	1.97 γ 9.70×10^{-2} e- 6.18×10^{-1} β+ 53.0%(β+)
		93	1.0 min	1.59×10^{-1} γ
		93	10.8 s	1.13 γ
		94	52 min	5.20×10^{-1} γ 8.30×10^{-3} e-
		95	1.64 h	1.24 γ 8.19×10^{-3} e-
44	**Ru**	95	1.64 h	7.20×10^{-2} β+ 13.7%(β+)
		97	2.88 d	2.40×10^{-1} γ 1.26×10^{-2} e-
		103	39.25 d	4.85×10^{-1} γ 1.10×10^{-1} e-
		105	4.44 h	7.38×10^{-1} γ 4.41×10^{-1} e-
		106	1.020 yr	3.90×10^{-7} γ 1.00×10^{-2} e-
		107	3.75 min	2.06×10^{-1} γ 1.23 e-
		108	4.55 min	6.10×10^{-2} γ
45	**Rh**	94	1.18 min	2.71 γ
		94	25.8 s	2.87 γ
		95	5.0 min	2.47 γ 2.62×10^{-3} e- 8.99×10^{-1} β+ 72.0%(β+)
		95	2.0 min	8.82×10^{-1} γ 4.29×10^{-2} e- 1.41×10^{-1} β+ 8.80%(β+)
		96	9.6 min	3.98 γ 6.08×10^{-3} e- 8.38×10^{-1} β+ 75.0%(β+)
		96	1.51 min	1.21 γ 2.58×10^{-2} e- 5.72×10^{-1} β+ 36.0%(β+)
		97	31 min	7.44×10^{-1} γ
		97	46 min	1.90 γ
		98	8.7 min	1.76 γ 2.22×10^{-3} e- 1.32 β+ 90.0%(β+)
		98	3.5 min	2.31 γ 9.28×10^{-1} β+ 82.0%(β+)
		99	16 d	5.72×10^{-1} γ 4.30×10^{-2} e- 1.45×10^{-2} β+ 4.20%(β+)
		99	4.7 h	6.27×10^{-1} γ 1.05×10^{-2} e- 2.84×10^{-2} β+ 8.10%(β+)
		100	20.8 h	2.76 γ 7.70×10^{-3} e- 5.40×10^{-2} β+ 4.70%(β+)
		100	4.7 min	2.43×10^{-1} γ
		101	3.3 yr	3.00×10^{-1} γ 2.67×10^{-2} e-
		101	4.34 d	3.04×10^{-1} γ 1.98×10^{-2} e-
		102	2.9 yr	2.16 γ 1.15×10^{-2} e-
		102	207.0 d	5.02×10^{-1} γ 8.46×10^{-2} e- 7.70×10^{-2} β+ 14.3%(β+)
		103	56.12 min	1.65×10^{-3} γ 3.75×10^{-2} e-
		104	42.3 s	1.52×10^{-2} γ 9.83×10^{-1} e-
		104	4.34 min	4.40×10^{-2} γ 8.38×10^{-2} e-
		105	1.473 d	7.71×10^{-2} γ 1.53×10^{-1} e-
		105	45 s	3.45×10^{-2} γ 9.48×10^{-2} e-
		106	29.8 s	2.10×10^{-1} γ 1.41 e-
		106	2.17 h	2.88 γ

Table I. Average Energies per Disintegration (continued)

Z	El	A	Half-life	⟨E⟩ (MeV), or %$\beta+$
45	**Rh**	106	2.17 h	3.14×10^{-1} e-
		107	21.7 min	3.13×10^{-1} γ 4.37×10^{-1} e-
		108	6.0 min	2.26 γ 7.90×10^{-3} e-
		108	16.8 s	3.25×10^{-1} γ 1.78 e-
		109	1.33 min	3.14×10^{-1} γ 9.31×10^{-1} e-
		110	3.2 s	2.27×10^{-1} γ
		110	28.5 s	2.59 γ 1.09 e-
46	**Pd**	97	3.1 min	1.55 γ
		98	17.7 min	4.52×10^{-1} γ 1.29×10^{-2} $\beta+$ 4.00%($\beta+$)
		99	21.4 min	1.34 γ 1.75×10^{-2} e- 4.30×10^{-1} $\beta+$ 49.0%($\beta+$)
		101	8.47 h	3.53×10^{-1} γ 1.41×10^{-2} e- 1.74×10^{-2} $\beta+$ 5.10%($\beta+$)
		103	16.97 d	1.63×10^{-2} γ 4.25×10^{-2} e-
		107	6.5×10^{6} yr	3.30×10^{-7} γ 9.30×10^{-3} e-
		107	21.3 s	1.52×10^{-1} γ 6.28×10^{-2} e-
		109	13.7 h	1.21×10^{-2} γ 4.38×10^{-1} e-
		109	4.69 min	1.11×10^{-1} γ 7.74×10^{-2} e-
		111	23.4 min	4.67×10^{-2} γ 8.32×10^{-1} e-
		111	5.5 h	3.59×10^{-1} γ 2.04×10^{-1} e-
		114	2.4 min	1.49×10^{-2} γ 5.31×10^{-1} e-
		116	12.72 s	1.52×10^{-1} γ
47	**Ag**	98	46.7 s	2.38 γ
		99	2.07 min	1.40 γ
		100	2.0 min	2.50 γ
		100	2.3 min	1.77 γ
		101	11.1 min	1.53 γ 8.70×10^{-3} e- 8.04×10^{-1} $\beta+$ 69.0%($\beta+$)
		101	3.1 s	1.56×10^{-1} γ 1.16×10^{-1} e-
		102	12.9 min	3.41 γ 5.18×10^{-3} e- 9.57×10^{-1} $\beta+$ 78.0%($\beta+$)
		102	7.7 min	1.65 γ 2.00×10^{-3} e- 4.62×10^{-1} $\beta+$ 39.0%($\beta+$)
		103	1.095 h	8.51×10^{-1} γ 1.88×10^{-2} e- 1.81×10^{-1} $\beta+$ 28.0%($\beta+$)
		103	5.7 s	3.76×10^{-2} γ 9.60×10^{-2} e-
		104	1.15 h	2.72 γ 9.70×10^{-3} e- 8.60×10^{-2} $\beta+$ 15.7%($\beta+$)
		104	33.5 min	1.25 γ 5.00×10^{-3} e- 5.08×10^{-1} $\beta+$ 43.0%($\beta+$)
		105	41.29 d	5.09×10^{-1} γ
47	**Ag**	105	41.29 d	1.87×10^{-2} e- 1.31×10^{-6} $\beta+$ 8.00×10^{-4} %($\beta+$)
		105	7.23 min	5.73×10^{-3} γ 2.47×10^{-2} e- 3.69×10^{-6} $\beta+$ 2.30×10^{-3} %($\beta+$)
		106	24.0 min	7.06×10^{-1} γ 2.56×10^{-3} e- 4.92×10^{-1} $\beta+$ 59.0%($\beta+$)
		106	8.46 d	2.87 γ 1.24×10^{-2} e-
		107	44.3 s	1.26×10^{-2} γ 8.00×10^{-2} e-
		108	2.37 min	1.95×10^{-2} γ 6.08×10^{-1} e- 1.12×10^{-3} $\beta+$ 2.80×10^{-1} %($\beta+$)
		108	127.0 yr	1.68 γ 1.51×10^{-2} e-
		109	39.6 s	1.10×10^{-2} γ 7.65×10^{-2} e-
		110	24.6 s	3.40×10^{-2} γ 1.18 e-
		110	249.76 d	2.74 γ 7.55×10^{-2} e-
		111	7.45 d	2.70×10^{-2} γ 3.55×10^{-1} e-
		111	1.08 min	7.48×10^{-3} γ 5.64×10^{-2} e-
		112	3.14 h	6.95×10^{-1} γ 1.38 e-
		113	5.37 h	7.34×10^{-2} γ 7.61×10^{-1} e-
		113	1.145 min	1.23×10^{-1} γ 1.41×10^{-1} e-
		114	4.6 s	1.07×10^{-1} γ 2.17 e-
		115	20.0 min	4.86×10^{-1} γ 1.11 e-
		116	2.68 min	2.15 γ 1.68 e-
		116	10.4 s	1.81 γ 1.81 e-
		117	1.21 min	1.31 γ 1.31 e-
		117	5.34 s	7.53×10^{-1} γ 1.83 e-
		118	4.0 s	2.06 γ 2.83 e-
		118	2.8 s	1.24 γ 1.38 e-
		119	2.1 s	1.34 γ 1.87 e-
		122	480.0 ms	1.12 γ
48	**Cd**	101	1.2 min	1.70 γ
		102	5.5 min	9.54×10^{-1} γ 1.25×10^{-2} e- 9.20×10^{-2} $\beta+$ 27.0%($\beta+$)
		103	7.7 min	2.10 γ 2.18×10^{-2} e- 3.48×10^{-1} $\beta+$ 32.0%($\beta+$)
		104	57.7 min	2.59×10^{-1} γ 2.96×10^{-2} e- 1.34×10^{-4} $\beta+$ 7.50×10^{-1} %($\beta+$)
		105	55.5 min	1.25 γ 3.10×10^{-2} e- 2.04×10^{-1} $\beta+$ 27.0%($\beta+$)
		107	6.5 h	3.41×10^{-2} γ 8.50×10^{-2} e- 2.84×10^{-4} $\beta+$
48	**Cd**	107	6.5 h	2.00×10^{-1} %($\beta+$)
		109	1.267 yr	2.60×10^{-2} γ 8.13×10^{-2} e-
		111	48.6 min	2.85×10^{-1} γ 1.10×10^{-1} e-
		113	9.3 15 yr	3.00×10^{-5} γ 9.10×10^{-2} e-
		113	13.7 yr	1.83×10^{-4} γ 1.83×10^{-1} e-
		115	2.228 d	3.70×10^{-1} γ 4.93×10^{-1} e-
		115	44.6 d	3.39×10^{-2} γ 6.02×10^{-1} e-
		117	2.49 h	1.09 γ 4.44×10^{-1} e-
		117	3.36 h	2.04 γ 2.38×10^{-1} e-
		118	50.3 min	1.90×10^{-4} γ 2.44×10^{-1} e-
		119	2.69 min	1.78 γ 1.03×10^{-2} e-
		119	2.2 min	2.39 γ
		120	50.8 s	1.26×10^{-3} γ 7.08×10^{-1} e-
		124	900.0 ms	1.36×10^{-1} γ 1.77×10^{-2} e-
49	**In**	104	1.7 min	3.17 γ 5.22×10^{-3} e- 1.90 $\beta+$ 88.0%($\beta+$)
		105	43 s	6.36×10^{-1} γ 3.80×10^{-2} e-
		106	6.2 min	2.67 γ
		106	5.2 min	2.94 γ 3.25×10^{-3} e- 1.61 $\beta+$ 86.0%($\beta+$)
		107	32.4 min	1.55 γ 1.18×10^{-2} e- 3.16×10^{-1} $\beta+$ 35.0%($\beta+$)
		107	50 s	6.41×10^{-1} γ 3.73×10^{-2} e-
		108	40 min	2.76 γ 4.86×10^{-3} e- 6.94×10^{-1} $\beta+$ 53.0%($\beta+$)
		108	58 min	3.22 γ 1.35×10^{-2} e- 1.61×10^{-1} $\beta+$ 27.0%($\beta+$)
		109	4.2 h	6.72×10^{-1} γ 1.59×10^{-2} e- 2.85×10^{-2} $\beta+$ 7.90%($\beta+$)
		109	1.3 min	6.10×10^{-1} γ 4.16×10^{-2} e-
		109	210 ms	2.09 γ
		110	1.15 h	1.57 γ 4.08×10^{-3} e- 6.23×10^{-1} $\beta+$ 62.0%($\beta+$)
		110	4.9 h	3.10 γ 1.25×10^{-2} e- 7.80×10^{-5} $\beta+$ 3.30×10^{-2} %($\beta+$)
		111	2.807 d	4.05×10^{-1} γ 3.39×10^{-2} e-
		111	7.7 min	4.69×10^{-1} γ 6.74×10^{-2} e-
		112	14.4 min	2.57×10^{-1} γ 9.59×10^{-2} e- 1.51×10^{-1} $\beta+$ 22.0%($\beta+$)
		112	20.9 min	3.42×10^{-2} γ 1.21×10^{-1} e-
		113	1.658 h	2.55×10^{-1} γ

Table I. Average Energies per Disintegration (continued)

Z	El	A	Half-life	⟨E⟩ (MeV), or %β+
49	**In**	113	1.658 h	1.34×10^{-1} e-
		114	1.198 min	3.50×10^{-3} γ 7.73×10^{-1} e-
		114	49.51 d	9.40×10^{-2} γ 1.43×10^{-1} e-
		115	4.4×10^{14} yr	8.00×10^{-5} γ 1.53×10^{-1} e-
		115	4.486 h	1.63×10^{-1} γ 1.74×10^{-1} e-
		116	14.10 s	2.34×10^{-2} γ 1.36 e-
		116	54.2 min	2.47 γ 3.15×10^{-1} e-
		116	2.18 s	6.26×10^{-2} γ 9.40×10^{-2} e-
		117	44 min	6.92×10^{-1} γ 2.67×10^{-1} e-
		117	1.94 h	9.13×10^{-2} γ 2.70×10^{-1} e-
		118	5.0 s	8.34×10^{-2} γ 1.71 e-
		118	4.40 min	2.72 γ 5.49×10^{-1} e-
		118	8.5 s	3.99×10^{-2} γ 1.06×10^{-1} e-
		119	2.4 min	7.70×10^{-1} γ 6.32×10^{-1} e-
		119	18.0 min	9.40×10^{-3} γ 1.04 e-
		120	2.9 s	3.55×10^{-1} γ 2.21 e-
		120	44 s	2.98 γ 9.51×10^{-1} e-
		121	23 s	9.29×10^{-1} γ 9.85×10^{-1} e-
		121	3.9 min	6.84×10^{-2} γ 1.52 e-
		122	10.8 s	3.61 γ 4.57×10^{-2} e-
		122	1.5 s	6.29×10^{-1} γ
		122	10 s	3.13 γ 3.47×10^{-3} e-
		123	6.0 s	1.11 γ 1.37 e-
		123	47.8 s	7.29×10^{-2} γ 2.02 e-
		124	3.17 s	2.66 γ 1.97 e-
		124	2.4 s	3.54 γ 1.72 e-
		125	2.33 s	1.29 γ
		126	1.4 s	4.32 γ 2.52 e-
		126	1.5 s	2.81 γ 1.97 e-
		127	1.12 s	1.77 γ 2.15 e-
		127	3.76 s	5.12×10^{-1} γ 2.67 e-
		128	900 ms	3.08 γ 2.85 e-
		128	900 ms	1.58 γ 2.62 e-
		129	590.0 ms	2.18 γ 2.48 e-
50	**Sn**	106	2.1 min	1.19 γ 2.38×10^{-2} e- 8.40×10^{-2} β+ 20.0%(β+)
		108	10.3 min	6.42×10^{-1} γ 3.10×10^{-2} e-
		109	18.0 min	2.32 γ 1.97×10^{-2} e- 6.90×10^{-2} β+ 9.50%(β+)
		110	4.1 h	2.93×10^{-1} γ
50	**Sn**	110	4.1 h	1.31×10^{-2} e-
		111	35 min	5.05×10^{-1} γ 3.99×10^{-3} e- 1.95×10^{-1} β+ 31.0%(β+)
		113	115.09 d	2.80×10^{-1} γ 1.39×10^{-1} e-
		113	21.4 min	1.35×10^{-2} γ 5.82×10^{-2} e-
		117	13.61 d	1.58×10^{-1} γ 1.61×10^{-1} e-
		119	293 d	1.14×10^{-2} γ 7.83×10^{-2} e-
		121	1.128 d	4.70×10^{-5} γ 1.15×10^{-1} e-
		121	55 yr	5.03×10^{-3} γ 3.51×10^{-2} e-
		123	129.2 d	7.66×10^{-3} γ 5.23×10^{-1} e-
		123	40.1 min	1.42×10^{-1} γ 4.79×10^{-1} e-
		125	9.64 d	3.14×10^{-1} γ 8.11×10^{-1} e-
		125	9.52 min	3.48×10^{-1} γ 8.06×10^{-1} e-
		126	1×10^{5} yr	5.70×10^{-2} γ 1.25×10^{-1} e-
		127	2.10 h	1.86 γ 5.19×10^{-1} e-
		128	59.1 min	6.02×10^{-1} γ 2.52×10^{-1} e-
		128	6.5 s	2.01 γ 8.22×10^{-2} e-
		130	3.7 min	9.65×10^{-1} γ 4.75×10^{-1} e-
		132	40 s	1.28 γ 7.32×10^{-1} e-
51	**Sb**	109	17 s	2.88 γ
		110	23.0 s	3.77 γ 2.61×10^{-3} e- 1.97 β+ 94.0%(β+)
		111	1.25 min	1.58 γ 2.39×10^{-2} e- 1.36 β+ 87.0%(β+)
		112	51.4 s	2.83 γ 2.03×10^{-3} e- 1.77 β+ 90.0%(β+)
		113	6.7 min	1.27 γ 2.14×10^{-2} e- 7.01×10^{-1} β+ 66.0%(β+)
		114	3.49 min	2.68 γ 3.20×10^{-3} e- 1.17 β+ 79.0%(β+)
		115	32.1 min	8.87×10^{-1} γ 7.80×10^{-3} e- 2.24×10^{-1} β+ 33.0%(β+)
		116	16 min	2.23 γ 4.07×10^{-3} e- 4.91×10^{-1} β+ 50.0%(β+)
		116	1.00 h	3.21 γ 6.10×10^{-2} e- 1.24×10^{-1} β+ 22.0%(β+)
		117	2.80 h	1.86×10^{-1} γ 2.43×10^{-2} e- 4.45×10^{-3} β+ 1.70%(β+)
		118	3.6 min	8.12×10^{-1} γ 1.46×10^{-3} e-
51	**Sb**	118	3.6 min	8.82×10^{-1} β+ 74.0%(β+)
		118	5.00 h	2.58 γ 3.10×10^{-2} e- 2.60×10^{-4} β+
		119	1.59 d	2.33×10^{-2} γ 2.40×10^{-2} e-
		120	15.89 min	4.92×10^{-1} γ 3.11×10^{-3} e- 3.26×10^{-1} β+ 44.0%(β+)
		120	5.76 d	2.47 γ 4.44×10^{-2} e-
		122	2.70 d	4.34×10^{-1} γ 5.66×10^{-1} e-
		122	4.21 min	7.04×10^{-2} γ 9.20×10^{-2} e-
		124	60.20 d	1.85 γ 3.90×10^{-1} e-
		124	1.6 min	4.40×10^{-1} γ 1.12×10^{-1} e-
		124	20.2 min	2.16×10^{-4} γ 2.54×10^{-2} e-
		125	2.73 yr	4.43×10^{-1} γ 1.26×10^{-1} e-
		126	12.4 d	2.75 γ 3.53×10^{-1} e-
		126	19.0 min	1.55 γ 6.32×10^{-1} e-
		126	11 s	2.49×10^{-4} γ 2.14×10^{-2} e-
		127	3.85 d	6.64×10^{-1} γ 3.14×10^{-1} e-
		128	9.01 h	3.11 γ 5.06×10^{-1} e-
		128	10.4 min	1.91 γ 9.58×10^{-1} e-
		129	4.40 h	1.36 γ 3.94×10^{-1} e-
		130	38 min	3.27 γ 7.03×10^{-1} e-
		130	6.3 min	2.65 γ 1.01 e-
		131	23.03 min	1.81 γ 5.82×10^{-1} e-
		132	4.15 min	2.59 γ 1.34 e-
		132	2.8 min	2.61 γ 1.24 e-
		134	850 ms	1.80×10^{-2} γ 3.79 e-
		134	10.4 s	2.05 γ 2.87 e-
52	**Te**	113	1.7 min	1.32 γ 2.44×10^{-3} e- 1.65 β+
		115	5.8 min	2.07 γ 5.15×10^{-3} e- 6.15×10^{-1} β+ 53.0%(β+)
		115	6.7 min	2.49 γ 5.60×10^{-3} e- 5.50×10^{-1} β+ 46.0%(β+)
		116	2.49 h	8.31×10^{-2} γ 5.74×10^{-2} e- 1.18×10^{-3} β+ 6.00×10^{-1} %(β+)
		117	1.03 h	1.53 γ 5.95×10^{-3} e- 1.98×10^{-1} β+ 25.0%(β+)
		117	103 ms	2.63×10^{-1} γ
		118	6.00 d	1.99×10^{-2} γ 5.50×10^{-3} e-
		119	16.05 h	7.72×10^{-1} γ

Table I. Average Energies per Disintegration (continued)

Z	El	A	Half-life	⟨E⟩ (MeV), or %β+
52	**Te**	119	16.05 h	7.80×10^{-3} e-
				5.80×10^{-3} β+
				2.10%(β+)
		119	4.69 d	1.51 γ
				1.53×10^{-2} e-
				2.47×10^{-3} β+
		121	16.8 d	5.77×10^{-1} γ
				9.10×10^{-3} e-
		121	154 d	2.17×10^{-1} γ
				7.69×10^{-2} e-
				3.09×10^{-6} β+
				2.10×10^{-3} %(β+)
		123	1.3×10^{13} yr	2.56×10^{-4} γ
				2.60×10^{-3} e-
		123	119.7 d	1.48×10^{-1} γ
				1.02×10^{-1} e-
		125	58 d	3.60×10^{-2} γ
				1.11×10^{-1} e-
		127	9.4 h	4.96×10^{-3} γ
				2.24×10^{-1} e-
		127	109 d	1.11×10^{-2} γ
				8.21×10^{-2} e-
		129	1.160 h	6.31×10^{-2} γ
				5.42×10^{-1} e-
		129	33.6 d	3.73×10^{-2} γ
				2.66×10^{-1} e-
		131	25.0 min	4.22×10^{-1} γ
				7.21×10^{-1} e-
		131	1.2 d	1.42 γ
				5.23×10^{-2} e-
		132	3.26 d	2.34×10^{-1} γ
				1.02×10^{-1} e-
		133	12.4 min	9.54×10^{-1} γ
				7.97×10^{-1} e-
		133	55.4 min	1.70 γ
				6.00×10^{-1} e-
		134	42 min	8.58×10^{-1} γ
				2.43×10^{-1} e-
		136	17.5 s	2.17 γ
				1.22 e-
53	**I**	116	2.9 s	1.07 γ
				3.70×10^{-4} e-
				3.02 β+
				97.0%(β+)
		118	14.3 min	9.23×10^{-1} γ
		119	19.1 min	8.58×10^{-1} γ
				1.54×10^{-2} e-
				4.95×10^{-1} β+
				53.0%(β+)
		120	1.35 h	1.74 γ
				4.29×10^{-3} e-
		120	53 min	4.47 γ
				1.10×10^{-2} e-
		121	2.12 h	4.22×10^{-1} γ
				1.97×10^{-2} e-
				6.10×10^{-2} β+
				13.2%(β+)
		122	3.6 min	9.53×10^{-1} γ
				1.96×10^{-3} e-
				1.09 β+
				77.0%(β+)
		123	13.2 h	1.73×10^{-1} γ
				2.76×10^{-2} e-
		124	4.18 d	1.09 γ
				6.50×10^{-3} e-
				1.88×10^{-1} β+
				23.0%(β+)
		125	60.1 d	4.24×10^{-2} γ
				1.79×10^{-2} e-
		126	13.0 d	4.33×10^{-1} γ
				1.40×10^{-1} e-
				5.50×10^{-3} β+
				1.15%(β+)
		128	24.99 min	9.15×10^{-2} γ
				7.38×10^{-1} e-
				3.13×10^{-6} β+
53	**I**	128	24.99 min	2.80×10^{-3} %(β+)
		129	1.57×10^{7} yr	2.48×10^{-2} γ
				5.56×10^{-2} e-
		130	12.36 h	2.14 γ
				2.96×10^{-1} e-
		130	9.0 min	1.21×10^{-1} γ
				1.91×10^{-1} e-
		131	8.040 d	3.82×10^{-1} γ
				1.92×10^{-1} e-
		132	2.284 h	2.29 γ
				4.94×10^{-1} e-
		132	1.39 h	3.19×10^{-1} γ
				1.61×10^{-1} e-
		133	20.8 h	6.08×10^{-1} γ
				4.10×10^{-1} e-
		133	9 s	1.58 γ
				5.51×10^{-2} e-
		134	52.6 min	2.61 γ
				6.18×10^{-1} e-
		134	3.50 min	2.86×10^{-1} γ
				9.01×10^{-2} e-
		135	6.55 h	1.65 γ
				3.77×10^{-1} e-
		136	1.40 min	2.46 γ
				1.97 e-
		136	45 s	2.14 γ
				2.19 e-
		137	24.5 s	1.14 γ
				1.96 e-
		138	6.4 s	1.89 γ
				2.59 e-
54	**Xe**	120	40 min	4.33×10^{-1} γ
				4.48×10^{-2} e-
				1.04×10^{-2} β+
				2.70%(β+)
		121	40 min	1.82 γ
				3.32×10^{-2} e-
				4.61×10^{-1} β+
				44.0%(β+)
		122	20.1 h	6.81×10^{-2} γ
				9.70×10^{-3} e-
		123	2.08 h	6.49×10^{-1} γ
				3.74×10^{-2} e-
				1.51×10^{-1} β+
				23.0%(β+)
		125	16.9 h	2.69×10^{-1} γ
				3.23×10^{-2} e-
				1.34×10^{-3} β+
				6.90×10^{-1} %(β+)
		125	57 s	1.16×10^{-1} γ
				1.37×10^{-1} e-
		127	36.41 d	2.71×10^{-1} γ
				3.07×10^{-2} e-
		127	1.15 min	1.68×10^{-1} γ
				1.29×10^{-1} e-
		129	8.89 d	5.13×10^{-2} γ
				1.83×10^{-1} e-
		131	11.9 d	2.00×10^{-2} γ
				1.43×10^{-1} e-
		133	5.24 d	4.63×10^{-2} γ
				1.36×10^{-1} e-
		133	2.19 d	4.14×10^{-2} γ
				1.92×10^{-1} e-
		134	290 ms	1.84 γ
				6.95×10^{-2} e-
		135	9.10 h	2.49×10^{-1} γ
				3.19×10^{-1} e-
		135	15.6 min	4.32×10^{-1} γ
				9.84×10^{-2} e-
		137	3.82 min	1.96×10^{-1} γ
				1.70 e-
		138	14.1 min	1.13 γ
				6.39×10^{-1} e-
		139	39.7 s	8.94×10^{-1} γ
				1.78 e-
		140	13.6 s	1.15 γ
54	**Xe**	140	13.6 s	1.22 e-
55	**Cs**	116	3.8 s	2.86 γ
				3.77 β+
				99.0%(β+)
		122	4.5 min	2.60 γ
		122	21 s	7.26×10^{-1} γ
		123	5.87 min	9.38×10^{-1} γ
				1.77×10^{-2} e-
				1.00 β+
				73.0%(β+)
		124	30.8 s	1.26 γ
				4.30×10^{-3} e-
				1.95 β+
				91.0%(β+)
		124	6.3 s	3.05×10^{-1} γ
				1.14×10^{-1} e-
		125	45 min	7.41×10^{-1} γ
				9.70×10^{-3} e-
				3.27×10^{-1} β+
				38.0%(β+)
		126	1.64 min	1.14 γ
				4.37×10^{-3} e-
				1.32 β+
				81.0%(β+)
		127	6.2 h	4.00×10^{-1} γ
				1.76×10^{-2} e-
				1.42×10^{-2} β+
				3.50%(β+)
		128	3.62 min	8.99×10^{-1} γ
				3.39×10^{-3} e-
				8.69×10^{-1} β+
				69.0%(β+)
		129	1.336 d	2.81×10^{-1} γ
				1.66×10^{-2} e-
		130	29.21 min	5.13×10^{-1} γ
				5.38×10^{-3} e-
				3.94×10^{-1} β+
				44.0%(β+)
		131	9.69 d	2.30×10^{-2} γ
				5.70×10^{-3} e-
		132	6.48 d	7.28×10^{-1} γ
				1.29×10^{-2} e-
				3.09×10^{-3} β+
				1.50%(β+)
		134	2.062 yr	1.55 γ
				1.64×10^{-1} e-
		134	2.91 h	2.68×10^{-2} γ
				1.09×10^{-1} e-
		135	3.0×10^{6} yr	1.18×10^{-5} γ
				5.60×10^{-2} e-
		135	53 min	1.59 γ
				3.66×10^{-2} e-
		136	13.16 d	2.17 γ
				1.35×10^{-1} e-
		137	30.0 yr	5.66×10^{-1} γ
				2.50×10^{-1} e-
		138	32.2 min	2.36 γ
				1.24 e-
		138	2.9 min	4.20×10^{-1} γ
				3.19×10^{-1} e-
		139	9.27 min	3.29×10^{-1} γ
				1.65 e-
		140	1.062 min	2.26 γ
				1.75 e-
		141	24.9 s	1.01 γ
				1.88 e-
		143	1.78 s	7.01×10^{-1} γ
				2.76 e-
		144	1.02 s	1.08 γ
				1.66×10^{-2} e-
		146	343 ms	8.17×10^{-1} γ
				2.29×10^{-2} e-
56	**Ba**	124	12 min	3.48×10^{-1} γ
		126	1.67 h	5.72×10^{-1} γ
				1.55×10^{-2} e-
				2.96×10^{-3} β+

Table I. Average Energies per Disintegration (continued)

Z	El	A	Half-life	⟨E⟩ (MeV), or %β+
56	**Ba**	127	13 min	7.30×10^{-1} γ 2.45×10^{-2} e- 5.75×10^{-1} β+ 54.0%(β+)
		128	2.43 d	6.74×10^{-2} γ 8.30×10^{-3} e-
		131	11.8 d	4.58×10^{-1} γ 4.46×10^{-2} e- 7.10×10^{-7} β+ 7.20×10^{-4} %(β+)
		131	14.6 min	7.66×10^{-2} γ 1.09×10^{-1} e-
		133	10.54 yr	4.04×10^{-1} γ 5.47×10^{-2} e-
		133	1.621 d	6.68×10^{-2} γ 2.19×10^{-1} e-
		135	1.20 d	6.00×10^{-2} γ 2.07×10^{-1} e-
		136	306 ms	1.92 γ 1.07×10^{-1} e-
		137	2.552 min	5.99×10^{-1} γ 6.52×10^{-2} e-
		139	1.41 h	4.54×10^{-2} γ 8.99×10^{-1} e-
		140	12.75 d	1.83×10^{-1} γ 3.10×10^{-1} e-
		141	18.27 min	8.47×10^{-1} γ 9.67×10^{-1} e-
		142	10.6 min	1.04 γ 3.78×10^{-1} e-
		143	14.5 s	2.31×10^{-1} γ 1.72 e-
		145	4.0 s	3.10×10^{-1} γ 2.02 e-
		148	607 ms	1.91×10^{-1} γ
57	**La**	128	5.0 min	2.91 γ 2.05×10^{-2} e- 1.20 β+ 76.0%(β+)
		129	11.6 min	1.00 γ 2.19×10^{-2} e- 6.75×10^{-1} β+ 62.0%(β+)
		129	560 ms	5.07×10^{-2} γ 1.25×10^{-1} e-
		130	8.7 min	2.18 γ 1.26×10^{-2} e- 1.19 β+ 76.0%(β+)
		131	59 min	6.80×10^{-1} γ 2.99×10^{-2} e- 1.79×10^{-1} β+ 25.0%(β+)
		132	4.8 h	2.17 γ 1.05×10^{-2} e- 5.43×10^{-1} β+ 42.0%(β+)
		132	24.3 min	4.91×10^{-1} γ
		133	3.91 h	8.16×10^{-2} γ 6.50×10^{-3} e-
		134	6.4 min	7.24×10^{-1} γ 2.41×10^{-3} e- 7.56×10^{-1} β+ 63.0%(β+)
		135	19.5 h	3.62×10^{-2} γ 6.00×10^{-3} e- 1.05×10^{-5} β+ 1.17×10^{-2} %(β+)
		136	9.87 min	4.18×10^{-1} γ 3.84×10^{-3} e- 2.99×10^{-1} β+ 36.0%(β+)
		137	6×10^{4} yr	2.54×10^{-2} γ 5.90×10^{-3} e-
		138	1.06×10^{11} yr	1.24 γ 2.84×10^{-2} e-
57	**La**	140	1.6780 d	2.32 γ 5.34×10^{-1} e-
		142	1.54 h	2.49 γ 8.42×10^{-1} e-
		143	14.1 min	9.66×10^{-2} γ 1.31 e-
		144	40.9 s	2.17 γ 1.61 e-
		145	25 s	6.43×10^{-1} γ 1.48 e-
		146	6.3 s	1.49 γ 2.16 e-
		146	10.0 s	1.31 γ
		148	1.05 s	1.25 γ 3.69×10^{-2} e-
58	**Ce**	131	10 min	7.88×10^{-1} γ
		132	3.5 h	2.73×10^{-1} γ 1.69×10^{-2} e-
		134	3.16 d	2.79×10^{-2} γ 6.30×10^{-3} e-
		135	17.8 h	8.28×10^{-1} γ 2.56×10^{-2} e- 3.18×10^{-3} β+ 1.19%(β+)
		135	20 s	5.47×10^{-2} γ 1.09×10^{-1} e-
		137	9.0 h	4.14×10^{-2} γ 1.38×10^{-2} e- 1.30×10^{-5} β+ 1.36×10^{-2} %(β+)
		137	1.43 d	5.50×10^{-2} γ 2.02×10^{-1} e-
		139	137.7 d	1.60×10^{-1} γ 3.31×10^{-2} e-
		139	56.4 s	7.00×10^{-1} γ 5.47×10^{-2} e-
		141	32.50 d	7.71×10^{-2} γ 1.71×10^{-1} e-
		143	1.38 d	2.75×10^{-1} γ 4.39×10^{-1} e-
		144	284.9 d	1.92×10^{-2} γ 9.18×10^{-2} e-
		145	3.0 min	7.74×10^{-1} γ 6.91×10^{-1} e-
		146	13.5 min	2.89×10^{-1} γ 4.31×10^{-2} e-
		148	56 s	3.18×10^{-1} γ 6.33×10^{-1} e-
59	**Pr**	136	13.1 min	2.10 γ 9.60×10^{-3} e- 7.40×10^{-1} β+ 57.0%(β+)
		137	1.28 h	3.72×10^{-1} γ 5.40×10^{-3} e- 1.90×10^{-1} β+ 25.0%(β+)
		138	1.45 min	8.24×10^{-1} γ 1.90×10^{-3} e- 1.16 β+ 75.0%(β+)
		138	2.1 h	2.49 γ 5.18×10^{-2} e- 1.76×10^{-1} β+ 24.0%(β+)
		139	4.41 h	1.24×10^{-1} γ 5.60×10^{-3} e- 3.96×10^{-2} β+ 7.90%(β+)
		140	3.39 min	5.50×10^{-1} γ 4.07×10^{-3} e- 5.44×10^{-1} β+ 51.0%(β+)
		142	19.13 h	6.00×10^{-2} γ 8.09×10^{-1} e-
		143	13.58 d	3.10×10^{-4} γ 3.15×10^{-1} e-
59	**Pr**	144	17.28 min	3.20×10^{-2} γ 1.21 e-
		144	7.2 min	1.21×10^{-2} γ 4.60×10^{-2} e-
		145	5.98 h	1.59×10^{-2} γ 6.77×10^{-1} e-
		146	24.2 min	1.02 γ 1.30 e-
		147	13.4 min	8.64×10^{-1} γ 7.81×10^{-1} e-
		148	2.27 min	8.87×10^{-1} γ 1.79 e-
		148	2.0 min	9.43×10^{-1} γ 1.73 e-
		149	2.3 min	1.80×10^{-1} γ
60	**Nd**	134	8.5 min	5.06×10^{-1} γ 2.39×10^{-2} e- 1.51×10^{-1} β+ 17.0%(β+)
		135	12 min	1.27 γ 8.50×10^{-2} e- 9.10×10^{-1} β+ 65.0%(β+)
		136	50.6 min	2.95×10^{-1} γ 6.70×10^{-2} e- 2.44×10^{-2} β+ 5.20%(β+)
		137	38 min	1.13 γ 5.40×10^{-2} e- 2.92×10^{-1} β+ 28.0%(β+)
		137	1.6 s	3.70×10^{-1} γ 1.44×10^{-1} e-
		138	5.0 h	4.39×10^{-2} γ 7.30×10^{-3} e-
		139	29.7 min	4.18×10^{-1} γ 6.40×10^{-3} e- 2.01×10^{-1} β+ 25.0%(β+)
		139	5.5 h	1.38 γ 7.80×10^{-2} e- 3.00×10^{-2} β+ 4.00%(β+)
		140	3.37 d	2.87×10^{-2} γ 6.00×10^{-3} e-
		141	2.49 h	7.68×10^{-2} γ 6.10×10^{-3} e- 9.10×10^{-3} β+ 2.50%(β+)
		141	1.04 min	6.95×10^{-1} γ 6.20×10^{-2} e-
		147	10.98 d	1.41×10^{-1} γ 2.69×10^{-1} e-
		149	1.73 h	3.85×10^{-1} γ 5.04×10^{-1} e-
		151	12.4 min	9.17×10^{-1} γ 6.20×10^{-1} e-
61	**Pm**	136	1.8 min	2.73 γ 1.89×10^{-2} e- 2.06 β+ 89.0%(β+)
		137	2.4 min	1.59 γ 1.18×10^{-1} e- 7.77×10^{-1} β+ 55.0%(β+)
		139	4.15 min	9.14×10^{-1} γ 5.80×10^{-3} e- 1.05 β+ 68.0%(β+)
		139	180 ms	8.51×10^{-2} γ 1.03×10^{-1} e-
		140	9.2 s	9.06×10^{-2} γ 1.42×10^{-1} α 1.30×10^{-2} e- 2.04 β+
		140	5.95 min	3.05 γ

Table I. Average Energies per Disintegration (continued)

Z	El	A	Half-life	⟨E⟩ (MeV), or %β+
61	Pm	140	5.95 min	2.94×10^{-2} e-
				9.82×10^{-1} β+
				68.0%(β+)
		141	20.90 min	7.56×10^{-1} γ
				4.00×10^{-3} e-
				6.30×10^{-1} β+
				52.0%(β+)
		142	40.5 s	8.79×10^{-1} γ
				1.45×10^{-3} e-
				1.36 β+
				78.0%(β+)
		143	265 d	3.16×10^{-1} γ
				7.30×10^{-3} e-
		144	363 d	1.56 γ
				1.61×10^{-2} e-
		145	17.7 yr	3.32×10^{-2} γ
				1.36×10^{-2} e-
		146	5.53 yr	7.54×10^{-1} γ
				9.28×10^{-2} e-
		147	2.6234 yr	1.86×10^{-5} γ
				6.20×10^{-2} e-
		148	5.37 d	5.75×10^{-1} γ
				7.24×10^{-1} e-
		148	41.3 d	1.99 γ
				1.69×10^{-1} e-
		149	2.212 d	1.11×10^{-2} γ
				3.67×10^{-1} e-
		150	2.68 h	1.49 γ
				7.92×10^{-1} e-
		151	1.183 d	3.21×10^{-1} γ
				3.05×10^{-1} e-
		152	4.1 min	1.55×10^{-1} γ
				1.39 e-
		152	7.5 min	1.51 γ
				8.85×10^{-1} e-
		154	1.7 min	1.91 γ
				8.90×10^{-1} e-
		154	2.7 min	2.00 γ
				9.15×10^{-1} e-
62	Sm	139	9.5 s	2.83×10^{-1} γ
				1.64×10^{-1} e-
		141	10.2 min	1.41 γ
				1.15×10^{-2} e-
				6.90×10^{-1} β+
				53.0%(β+)
		141	22.6 min	1.99 γ
				5.49×10^{-2} e-
				3.76×10^{-1} β+
				34.0%(β+)
		142	1.208 h	9.65×10^{-2} γ
				5.90×10^{-3} e-
				2.77×10^{-2} β+
				5.70%(β+)
		143	8.83 min	5.39×10^{-1} γ
				3.58×10^{-3} e-
				4.97×10^{-1} β+
				46.0%(β+)
		143	1.10 min	6.85×10^{-1} γ
				7.12×10^{-2} e-
				9.60×10^{-4} β+
				9.60×10^{-2} %(β+)
		145	340 d	6.52×10^{-2} γ
				2.93×10^{-2} e-
		151	90 yr	6.71×10^{-5} γ
				1.25×10^{-1} e-
		153	1.946 d	6.28×10^{-2} γ
				2.70×10^{-1} e-
		155	22.1 min	1.04×10^{-1} γ
				5.68×10^{-1} e-
63	Eu	141	40 s	1.20 γ
				4.80×10^{-3} e-
				1.82 β+
				85.0%(β+)
		141	3.3 s	5.80×10^{-2} γ
		142	2.4 s	1.17 γ
				2.72 β+
63	Eu	142	2.4 s	94.0%(β+)
		142	1.22 min	3.44 γ
				3.22×10^{-2} e-
				1.45 β+
				83.0%(β+)
		143	2.63 min	1.11 γ
				5.00×10^{-3} e-
				1.29 β+
				72.0%(β+)
		144	10.2 s	1.10 γ
				1.02×10^{-3} e-
				2.06 β+
				87.0%(β+)
		146	4.59 d	2.22 γ
		147	24 d	4.97×10^{-1} γ
				4.01×10^{-2} e-
				1.01×10^{-3} β+
				3.60×10^{-1} %(β+)
		148	54.5 d	2.17 γ
				2.70×10^{-2} e-
				9.30×10^{-4} β+
				2.30×10^{-1} %(β+)
		149	93.1 d	6.38×10^{-2} γ
				8.40×10^{-3} e-
		150	12.6 h	4.34×10^{-2} γ
				3.00×10^{-1} e-
				2.16×10^{-3} β+
				3.90×10^{-1} %(β+)
		150	36 yr	1.50 γ
				2.63×10^{-2} e-
		152	13.33 yr	1.16 γ
				1.27×10^{-1} e-
				8.70×10^{-5} β+
				2.70×10^{-2} %(β+)
		152	9.32 h	3.07×10^{-1} γ
				5.07×10^{-1} e-
				3.11×10^{-5} β+
				7.70×10^{-3} %(β+)
		152	1.60 h	7.61×10^{-2} γ
				6.64×10^{-2} e-
		154	8.8 yr	1.25 γ
				2.79×10^{-1} e-
		154	46.0 min	8.40×10^{-2} γ
		155	4.96 yr	6.30×10^{-2} γ
				6.50×10^{-2} e-
		156	15.2 d	1.33 γ
				4.25×10^{-1} e-
		158	46.0 min	1.08 γ
				9.68×10^{-1} e-
64	Gd	143	39 s	3.65×10^{-1} γ
		143	1.83 min	2.11 γ
				8.70×10^{-2} e-
				1.10 β+
				67.0%(β+)
		145	23.4 min	1.79 γ
		145	1.42 min	6.72×10^{-1} γ
				1.13×10^{-1} e-
				6.50×10^{-2} β+
				3.50%(β+)
		146	48.3 d	2.56×10^{-1} γ
				1.26×10^{-1} e-
		147	1.588 d	1.33 γ
				5.80×10^{-2} e-
				8.00×10^{-4} β+
				1.57×10^{-1} %(β+)
		149	9.2 d	4.18×10^{-1} γ
				6.64×10^{-2} e-
		151	120 d	6.44×10^{-2} γ
				3.09×10^{-2} e-
		153	241.6 d	1.02×10^{-1} γ
				3.99×10^{-2} e-
		159	18.6 h	5.30×10^{-2} γ
				3.11×10^{-1} e-
		161	3.7 min	3.92×10^{-1} γ
				5.84×10^{-1} e-
		162	8.6 min	4.22×10^{-1} γ
64	Gd	162	8.6 min	3.38×10^{-1} e-
65	Tb	145	30 s	1.56 γ
		146	8 s	2.37×10^{-1} γ
		146	23 s	2.86 γ
		147	1.6 h	1.60 γ
				1.56×10^{-2} e-
				5.18×10^{-1} β+
				42.0%(β+)
		147	1.8 min	1.45 γ
		148	1.00 h	1.85 γ
		148	2.20 min	2.63 γ
		149	4.15 h	1.28 γ
				6.63×10^{-1} α
				3.60×10^{-2} e-
		149	4.16 min	7.44×10^{-1} γ
		150	3.3 h	2.11 γ
				1.33×10^{-2} e-
				5.24×10^{-1} β+
				41.0%(β+)
		150	6.0 min	2.38 γ
				3.60×10^{-2} e-
		151	17.6 h	8.94×10^{-1} γ
				3.24×10^{-4} α
				7.30×10^{-2} e-
				5.60×10^{-3} β+
				1.26%(β+)
		151	25 s	8.03×10^{-2} γ
		152	17.5 h	1.31 γ
				2.00×10^{-2} e-
				2.06×10^{-1} β+
				18.0%(β+)
		152	4.3 min	7.54×10^{-1} γ
				1.42×10^{-1} e-
				2.93×10^{-3} β+
				6.60×10^{-1} %(β+)
		153	2.34 d	3.10×10^{-1} γ
				4.22×10^{-2} e-
				1.95×10^{-4} β+
				9.00×10^{-2} %(β+)
		154	21.4 h	2.38 γ
				4.40×10^{-2} e-
				9.60×10^{-3} β+
				1.15%(β+)
		154	23 h	2.29 γ
				1.12×10^{-1} e-
		155	5.3 d	1.84×10^{-1} γ
				7.30×10^{-2} e-
		156	5.3 d	1.82 γ
				8.47×10^{-2} e-
		156	5.0 h	4.38×10^{-3} γ
				8.32×10^{-2} e-
		157	150 yr	9.61×10^{-3} γ
				3.70×10^{-3} e-
		158	150 yr	7.87×10^{-1} γ
				1.13×10^{-1} e-
		158	10.5 s	2.37×10^{-2} γ
				8.42×10^{-2} e-
		160	72.3 d	1.13 γ
				2.54×10^{-1} e-
		161	6.91 d	3.39×10^{-2} γ
				1.97×10^{-1} e-
		162	7.7 min	1.10 γ
				5.44×10^{-1} e-
		163	19.5 min	7.85×10^{-1} γ
				3.20×10^{-1} e-
		164	3.0 min	2.18 γ
				7.72×10^{-1} e-
66	Dy	145	18 s	7.70×10^{-2} γ
		146	29 s	1.65×10^{-1} γ
		146	150 ms	2.92 γ
				9.10×10^{-2} e-
		147	59 s	2.42×10^{-1} γ
		150	7.17 min	2.98×10^{-1} γ
				6.60×10^{-3} e-
		151	16.9 min	2.30×10^{-1} γ

Table I. Average Energies per Disintegration (continued)

Z	El	A	Half-life	⟨E⟩ (MeV), or %β+
66	**Dy**	152	2.38 h	2.87×10^{-1} γ 1.17×10^{-2} e-
		153	6.4 h	6.97×10^{-1} γ 3.26×10^{-4} α 6.62×10^{-2} e- 1.04×10^{-2} β+ 2.20%(β+)
		155	10.0 h	7.00×10^{-1} γ 1.96×10^{-2} e- 5.60^{-3} β+ 1.42%(β+)
		157	8.1 h	3.41×10^{-1} γ 1.24×10^{-2} e-
		159	144.4 d	4.58×10^{-2} γ 1.16×10^{-2} e-
		165	2.33 h	2.66×10^{-2} γ 4.49×10^{-1} e-
		165	1.26 min	1.92×10^{-2} γ 1.05×10^{-1} e-
		166	3.400 d	4.00×10^{-2} γ 1.59×10^{-1} e-
		167	6.2 min	5.35×10^{-1} γ 7.38×10^{-1} e-
67	**Ho**	146	3.9 s	2.25 γ
		147	5.8 s	8.39×10^{-1} γ
		150	28 s	2.36 γ
		152	2.4 min	7.60×10^{-1} γ
		152	52.3 s	1.99 γ
		154	3.2 min	1.74 γ
		155	48 min	4.15×10^{-1} γ 2.50×10^{-2} e- 2.01×10^{-1} β+ 22.0%(β+)
		157	12.6 min	5.22×10^{-1} γ 5.60×10^{-2} e- 2.87×10^{-2} β+ 5.10%(β+)
		159	33 min	3.30×10^{-1} γ 5.22×10^{-2} e-
		159	8.3 s	1.00×10^{-1} γ 1.06×10^{-1} e-
		160	25.6 min	1.78 γ
		161	2.48 h	5.90×10^{-2} γ 3.20×10^{-2} e-
		161	6.7 s	1.04×10^{-1} γ 1.06×10^{-1} e-
		162	15 min	1.70×10^{-1} γ 3.87×10^{-2} e- 2.07×10^{-2} β+ 4.20%(β+)
		162	1.13 h	5.32×10^{-1} γ
		163	1.09 s	2.37×10^{-1} γ 6.11×10^{-2} e-
		164	29.0 min	2.95×10^{-2} γ 2.25×10^{-2} e-
		164	38 min	4.77×10^{-2} γ 8.70×10^{-2} e-
		166	1.117 d	3.03×10^{-2} γ 7.40×10^{-1} e-
		166	1.2×10^{3} yr	1.75 γ 1.30×10^{-1} e-
		167	3.1 h	3.65×10^{-1} γ 2.30×10^{-1} e-
		168	3.0 min	8.71×10^{-1} γ 7.25×10^{-1} e-
		169	4.7 min	6.70×10^{-1} γ
		170	2.8 min	2.00 γ 8.30×10^{-1} e-
68	**Er**	158	2.25 h	8.10×10^{-2} γ
		159	36.0 min	8.96×10^{-1} γ 4.20×10^{-2} e- 4.42×10^{-2} β+ 7.50%(β+)
		160	1.191 d	3.70×10^{-2} γ
68	**Er**	160	1.191 d	6.50×10^{-3} e-
		161	3.24 h	9.98×10^{-1} γ 4.56×10^{-2} e- 4.91×10^{-4} β+ 1.44×10^{-1} %(β+)
		163	1.25 h	4.09×10^{-2} γ 6.70×10^{-3} e- 3.93×10^{-6} β+ 4.00×10^{-3} %(β+)
		165	10.36 h	3.81×10^{-2} γ 6.60×10^{-3} e-
		167	2.28 s	9.70×10^{-2} γ 1.11×10^{-1} e-
		169	9.4 d	5.00×10^{-5} γ 9.96×10^{-2} e-
		171	7.52 h	3.73×10^{-1} γ 4.17×10^{-1} e-
		172	2.05417 d	5.13×10^{-1} γ 1.29×10^{-1} e-
		173	1.4 min	8.33×10^{-1} γ 6.69×10^{-1} e-
69	**Tm**	158	4.02 min	1.86 γ 4.07×10^{-2} e- 1.54 β+ 74.0%(β+)
		159	9.0 min	6.38×10^{-1} γ 6.47×10^{-2} e- 2.96×10^{-1} β+ 24.0%(β+)
		161	38.0 min	1.08 γ 9.81×10^{-2} e-
		162	21.7 min	1.48 γ
		162	24.3 s	2.92×10^{-1} γ 1.02×10^{-1} e- 2.63×10^{-2} β+ 2.60%(β+)
		163	1.81 h	1.24 γ 6.33×10^{-2} e- 1.12×10^{-2} β+ 1.90%(β+)
		164	2.0 min	7.34×10^{-1} γ 2.60×10^{-2} e- 5.23×10^{-1} β+ 40.0%(β+)
		164	5.1 min	3.84×10^{-1} γ 2.06×10^{-3} β+ 4.70×10^{-1} %(β+)
		165	1.2525 d	5.96×10^{-1} γ 4.80×10^{-2} e-
		166	7.7 h	1.91 γ 9.51×10^{-2} e- 1.41×10^{-2} β+ 1.66%(β+)
		167	9.24 d	1.46×10^{-1} γ 1.26×10^{-1} e-
		168	93.1 d	1.15 γ 8.01×10^{-2} e-
		170	128.6 d	5.73×10^{-3} γ 3.30×10^{-1} e-
		171	1.92 yr	6.22×10^{-4} γ 2.55×10^{-2} e-
		172	2.65 d	4.71×10^{-1} γ 5.32×10^{-1} e-
		173	8.24 h	3.87×10^{-1} γ 3.08×10^{-1} e-
		174	5.4 min	1.78 γ 5.13×10^{-1} e-
		175	15.2 min	1.08 γ 7.41×10^{-1} e-
		176	1.9 min	1.95 γ 8.50×10^{-1} e-
70	**Yb**	161	4.2 min	8.12×10^{-1} γ 4.30×10^{-2} e-
		162	18.9 min	1.36×10^{-1} γ 2.61×10^{-2} e-
		163	11.05 min	8.11×10^{-1} γ
70	**Yb**	163	11.05 min	2.83×10^{-1} β+ 28.0%(β+)
		165	9.9 min	3.38×10^{-1} γ 7.33×10^{-2} e- 7.00×10^{-2} β+ 9.80%(β+)
		166	2.3625 d	8.67×10^{-2} γ 3.89×10^{-2} e-
		167	17.5 min	2.63×10^{-1} γ 7.83×10^{-2} e- 1.51×10^{-3} β+ 5.00×10^{-1} %(β+)
		169	32.022 d	3.12×10^{-1} γ 1.12×10^{-1} e-
		169	46.0 s	1.30×10^{-3} γ 2.18×10^{-2} e-
		175	4.19 d	4.01×10^{-2} γ 1.31×10^{-1} e-
		176	11.4 s	8.97×10^{-1} γ 1.53×10^{-1} e-
		177	1.9 h	1.92×10^{-1} γ 4.38×10^{-1} e-
		177	6.41 s	1.49×10^{-1} γ 1.80×10^{-1} e-
71	**Lu**	165	11.8 min	8.58×10^{-1} γ
		166	2.8 min	1.86 γ
		166	1.4 min	1.58 γ 9.10×10^{-2} e- 1.98×10^{-1} β+ 18.0%(β+)
		166	2.1 min	1.74 γ 3.30×10^{-2} e-
		167	52 min	9.80×10^{-1} γ 5.78×10^{-2} e-
		168	5.5 min	7.76×10^{-2} γ 5.00×10^{-2} β+ 7.30%(β+)
		168	6.7 min	1.26×10^{-1} γ 1.08×10^{-1} β+ 12.0%(β+)
		169	1.419 d	1.31 γ 3.12×10^{-2} e- 1.11×10^{-3} β+ 3.00×10^{-1} %(β+)
		169	2.7 min	1.43×10^{-3} γ 2.64×10^{-2} e-
		170	2.00 d	2.68 γ 5.24×10^{-2} e- 2.42×10^{-3} β+ 2.40×10^{-2} %(β+)
		170	670 ms	3.60×10^{-3} γ 8.70×10^{-2} e-
		171	8.24 d	6.56×10^{-1} γ 8.93×10^{-2} e- 1.45×10^{-5} β+ 8.10×10^{-3} %(β+)
		171	1.32 min	1.77×10^{-3} γ 6.63×10^{-2} e-
		172	6.70 d	1.93 γ 1.18×10^{-1} e-
		172	3.7 min	1.31×10^{-3} γ 3.91×10^{-2} e-
		173	1.37 yr	1.03×10^{-1} γ 1.63×10^{-2} e-
		174	3.31 yr	1.33×10^{-1} γ 4.30×10^{-2} e- 4.05×10^{-5} β+ 2.40×10^{-2} %(β+)
		174	142 d	6.01×10^{-2} γ 1.17×10^{-1} e-
		176	3.635 h	1.44×10^{-2} γ 3.89×10^{-2} e-
		176	3.59×10^{10} yr	4.90×10^{-1} γ 1.15×10^{-1} e-
		177	6.71 d	3.52×10^{-2} γ 1.47×10^{-1} e-

Table I. Average Energies per Disintegration (continued)

Z	El	A	Half-life	⟨E⟩ (MeV), or %β+
71	Lu	177	160.9 d	9.98×10^{-1} γ 2.68×10^{-1} e-
		178	28.4 min	1.52×10^{-1} γ 7.54×10^{-1} e-
		178	23.1 min	1.06 γ 5.68×10^{-1} e-
		179	4.6 h	3.22×10^{-2} γ 4.66×10^{-1} e-
		180	5.7 min	1.57 γ
		181	3.5 min	5.42×10^{-1} γ
72	Hf	166	6.8 min	1.88×10^{-1} γ 2.16×10^{-2} e-
		167	2.05 min	7.44×10^{-1} γ 1.23×10^{-2} e- 5.23×10^{-1} β+ 43.0%(β+)
		169	3.24 min	5.53×10^{-1} γ 2.40×10^{-2} e- 1.31×10^{-1} β+ 4.80%(β+)
		170	16.0 h	5.79×10^{-1} γ 9.20×10^{-2} e-
		172	1.87 yr	1.05×10^{-1} γ 1.47×10^{-1} e-
		173	23.6 h	4.01×10^{-1} γ 4.67×10^{-2} e- 4.86×10^{-4} β+ 3.70×10^{-1} %(β+)
		175	70 d	3.64×10^{-1} γ 4.39×10^{-2} e-
		177	1.1 s	1.07 γ 2.40×10^{-1} e-
		177	51.4 min	1.16 γ 2.51×10^{-1} e-
		178	4.0 s	1.01 γ 1.43×10^{-1} e-
		178	31 yr	2.20 γ 2.12×10^{-1} e-
		179	18.7 s	2.43×10^{-1} γ 1.35×10^{-1} e-
		179	25.1 d	8.91×10^{-1} γ 1.54×10^{-1} e-
		180	5.519 h	9.98×10^{-1} γ 1.48×10^{-1} e-
		181	42.4 d	5.18×10^{-1} γ 1.99×10^{-1} e-
		182	9×10^{6} yr	2.39×10^{-1} γ 8.25×10^{-2} e-
		182	1.02 h	9.86×10^{-1} γ 2.28×10^{-1} e-
		183	1.07 h	7.52×10^{-1} γ 4.41×10^{-1} e-
		184	4.12 h	2.51×10^{-1} γ 4.61×10^{-1} e-
73	Ta	168	2.4 min	7.22×10^{-1} γ 7.25×10^{-2} e-
		170	6.8 min	9.99×10^{-1} γ 1.35 β+ 66.0%(β+)
		172	36.8 min	1.69 γ 1.36×10^{-1} e- 2.88×10^{-1} β+ 25.0%(β+)
		173	3.65 h	6.70×10^{-1} γ 7.60×10^{-2} e- 1.62×10^{-1} β+ 23.0%(β+)
		174	1.18 h	9.15×10^{-1} γ 9.90×10^{-2} e- 2.95×10^{-1} β+ 24.0%(β+)
		175	10.5 h	1.06 γ 6.00×10^{-2} e- 2.14×10^{-3} β+ 6.30×10^{-1} %(β+)
		176	8.1 h	2.14 γ
73	Ta	176	8.1 h	7.20×10^{-2} e- 7.20×10^{-3} β+ 8.60×10^{-1} %(β+)
		177	2.358 d	6.78×10^{-2} γ 2.22×10^{-2} e-
		178	9.31 min	1.22×10^{-1} γ 3.22×10^{-2} e- 4.47×10^{-3} β+ 1.11%(β+)
		178	2.45 h	1.16 γ 1.63×10^{-1} e-
		179	1.8 yr	2.85×10^{-2} γ 5.60×10^{-3} e-
		180	8.15 h	4.83×10^{-2} γ 5.38×10^{-2} e-
		182	115.0 d	1.30 γ 2.07×10^{-1} e-
		182	283 ms	1.69×10^{-3} γ 1.22×10^{-2} e-
		182	15.8 min	2.54×10^{-1} γ 5.06×10^{-1} e-
		183	5.1 d	2.95×10^{-1} γ 3.42×10^{-1} e-
		184	8.7 h	1.61 γ 5.36×10^{-1} e-
		185	49 min	1.45×10^{-1} γ 7.37×10^{-1} e-
		186	10.5 min	1.57 γ 9.67×10^{-1} e-
74	W	177	2.21 h	9.11×10^{-1} γ 9.20×10^{-2} e- 7.10×10^{-4} β+ 2.10×10^{-1} %(β+)
		178	21.5 d	1.67×10^{-2} γ 4.60×10^{-3} e-
		179	37.5 min	4.90×10^{-4} γ
		179	6.4 min	5.50×10^{-2} γ 1.64×10^{-1} e-
		181	121.2 d	4.03×10^{-2} γ 8.20×10^{-3} e-
		183	5.15 s	1.28×10^{-1} γ 1.73×10^{-1} e-
		185	75.1 d	1.07×10^{-4} γ 1.27×10^{-1} e-
		185	1.67 min	2.75×10^{-2} γ 1.65×10^{-1} e-
		187	23.9 h	4.30×10^{-1} γ 3.11×10^{-1} e-
		188	69.4 d	1.92×10^{-3} γ 9.97×10^{-2} e-
		190	30 min	1.50×10^{-1} γ 4.75×10^{-1} e-
75	Re	178	13.2 min	1.39 γ 1.06×10^{-1} e- 1.55×10^{-1} β+ 10.7%(β+)
		179	19.7 min	1.09 γ 1.47×10^{-1} e- 8.30×10^{-3} β+ 1.53%(β+)
		180	2.46 min	1.19 γ 8.00×10^{-2} e- 6.60×10^{-2} β+ 8.20%(β+)
		181	20 h	8.14×10^{-1} γ 1.35×10^{-1} e-
		182	2.67 d	1.92 γ 2.13×10^{-1} e-
		182	12.7 h	1.18 γ 7.36×10^{-2} e- 1.42×10^{-3} β+ 1.80×10^{-1} %(β+)
		183	70 d	1.50×10^{-1} γ 1.05×10^{-1} e-
		184	38.0 d	8.93×10^{-1} γ 5.41×10^{-2} e-
75	Re	184	165 d	3.90×10^{-1} γ 1.40×10^{-1} e-
		186	3.777 d	1.97×10^{-2} γ 3.37×10^{-1} e-
		186	2×10^{5} yr	2.10×10^{-2} γ 1.34×10^{-1} e-
		188	16.98 h	5.89×10^{-2} γ 7.81×10^{-1} e-
		188	18.6 min	7.48×10^{-2} γ 9.50×10^{-2} e-
		189	1.01 d	6.04×10^{-2} γ 3.17×10^{-1} e-
		190	3.1 min	1.35 γ 7.04×10^{-1} e-
		190	3.2 h	9.29×10^{-1} γ 3.81×10^{-1} e-
76	Os	169	3.3 s	6.11×10^{-1} α
		181	1.75 h	1.39 γ 7.70×10^{-2} e- 1.96×10^{-2} β+ 2.90%(β+)
		182	21.6 h	4.35×10^{-1} γ 5.32×10^{-2} e-
		183	13.0 h	5.33×10^{-1} γ 7.30×10^{-2} e-
		183	9.9 h	1.03 γ 3.09×10^{-2} e-
		185	93.6 d	7.13×10^{-1} γ 1.66×10^{-2} e-
		189	5.8 h	1.87×10^{-3} γ 2.74×10^{-2} e-
		190	9.9 min	1.59 γ 1.14×10^{-1} e-
		191	15.4 d	7.49×10^{-2} γ 1.31×10^{-1} e-
		191	13.10 h	7.41×10^{-3} γ 5.97×10^{-2} e-
		192	5.9 s	1.89 γ 1.69×10^{-1} e-
		193	1.27 d	6.91×10^{-2} γ 3.99×10^{-1} e-
		194	6.0 yr	1.77×10^{-3} γ 3.23×10^{-2} e-
		196	34.9 min	6.95×10^{-2} γ
77	Ir	180	1.5 min	6.87×10^{-1} γ
		182	15 min	1.31 γ 9.70×10^{-2} e- 7.65×10^{-1} β+ 44.0%(β+)
		184	3.0 h	1.90 γ 1.12×10^{-1} e- 1.55×10^{-1} β+ 12.2%(β+)
		186	15.8 h	1.60 γ 9.58×10^{-2} e- 1.44×10^{-2} β+ 1.64%(β+)
		188	1.73 d	2.10 γ 4.67×10^{-2} e- 2.19×10^{-3} β+ 3.30×10^{-1} %(β+)
		189	13.2 d	7.84×10^{-2} γ 4.25×10^{-2} e-
		190	11.8 d	1.49 γ 7.35×10^{-2} e-
		190	1.2 h	1.99×10^{-3} γ 2.28×10^{-2} e-
		190	3.2 h	1.63 γ 1.05×10^{-1} e-
		191	4.94 s	7.49×10^{-2} γ 9.39×10^{-2} e-
		192	73.83 d	8.13×10^{-1} γ 2.16×10^{-1} e-
		192	1.45 min	2.47×10^{-3} γ 5.45×10^{-2} e-
		192	241 yr	3.42×10^{-3} γ

Table I. Average Energies per Disintegration (continued)

Z	El	A	Half-life	⟨E⟩ (MeV), or %β+
77	**Ir**	192	241 yr	1.68×10^{-1} e-
		193	10.6 d	2.30×10^{-3} γ
				7.68×10^{-2} e-
		194	19.15 h	9.20×10^{-2} γ
				8.11×10^{-1} e-
		194	171 d	2.14 γ
				8.37×10^{-2} e-
		195	2.8 h	5.93×10^{-2} γ
				3.80×10^{-1} e-
		195	3.8 h	4.18×10^{-1} γ
				2.04×10^{-1} e-
		196	52 s	2.27×10^{-1} γ
				1.18 e-
		196	1.40 h	2.46 γ
				4.56×10^{-1} e-
78	**Pt**	175	2.5 s	3.80 α
		176	6.3 s	2.41 α
		177	11 s	4.96×10^{-1} α
		178	21 s	4.02×10^{-1} α
		188	10.2 d	2.06×10^{-1} γ
				7.87×10^{-2} e-
		189	10.9 h	3.13×10^{-1} γ
				5.85×10^{-2} e-
				2.02×10^{-3} β+
				4.80×10^{-1} %(β+)
		191	2.9 d	2.96×10^{-1} γ
				7.19×10^{-2} e-
		193	50 yr	2.10×10^{-3} γ
		193	4.33 d	1.26×10^{-2} γ
				1.30×10^{-1} e-
		195	4.02 d	7.63×10^{-2} γ
				1.75×10^{-1} e-
		197	18.3 h	2.55×10^{-2} γ
				2.49×10^{-1} e-
		197	1.57 h	8.30×10^{-2} γ
				3.20×10^{-1} e-
		199	30.8 min	2.03×10^{-1} γ
				5.36×10^{-1} e-
		199	13.6 s	3.42×10^{-1} γ
				7.80×10^{-2} e-
		200	12.5 h	5.93×10^{-2} γ
				2.38×10^{-1} e-
79	**Au**	181	11.4 s	6.12×10^{-2} α
		184	53 s	1.13×10^{-3} α
		189	28.7 min	8.48×10^{-1} γ
				6.80×10^{-2} e-
				2.76×10^{-2} β+
				4.10%(β+)
		190	43 min	1.99 γ
		191	3.2 h	5.59×10^{-1} γ
				7.70×10^{-2} e-
		191	920 ms	2.00×10^{-1} γ
				6.40×10^{-2} e-
		192	4.9 h	1.90 γ
				3.30×10^{-2} e-
				5.50×10^{-2} β+
				5.20%(β+)
		193	17.6 h	1.76×10^{-1} γ
				5.50×10^{-2} e-
		193	3.9 s	1.98×10^{-1} γ
				9.10×10^{-2} e-
		194	1.65 d	1.07 γ
				3.03×10^{-2} e-
				1.03×10^{-2} β+
				1.66%(β+)
		194	600 ms	1.15×10^{-2} γ
				8.50×10^{-2} e-
		195	186.09 d	8.62×10^{-2} γ
				4.48×10^{-2} e-
		195	30.5 s	2.02×10^{-1} γ
				1.16×10^{-1} e-
		196	6.18 d	4.72×10^{-1} γ
				3.64×10^{-2} e-
		196	8.1 s	3.13×10^{-3} γ
				8.08×10^{-2} e-
		196	9.7 h	2.34×10^{-1} γ
79	**Au**	196	9.7 h	3.68×10^{-1} e-
		197	7.8 s	2.32×10^{-1} γ
				1.80×10^{-1} e-
		198	2.6935 d	4.03×10^{-1} γ
				4.21×10^{-1} e-
		198	2.30 d	5.77×10^{-1} γ
				2.84×10^{-1} e-
		199	3.14 d	8.94×10^{-2} γ
				1.45×10^{-1} e-
		200	48.4 min	2.75×10^{-1} γ
				7.41×10^{-1} e-
		200	18.7 h	2.02 γ
				2.88×10^{-1} e-
		201	26 min	5.35×10^{-2} γ
				4.17×10^{-1} e-
		202	28 s	1.55×10^{-1} γ
				1.23 e-
		204	40 s	1.95 γ
80	**Hg**	182	11 s	5.27×10^{-1} α
		183	8.8 s	7.08×10^{-1} α
		184	30.6 s	6.95×10^{-2} α
		185	50 s	3.11×10^{-1} α
		190	20.0 min	9.80×10^{-2} γ
		191	51 min	1.47 γ
				1.05×10^{-1} e-
				5.20×10^{-2} β+
				5.80%(β+)
		192	4.8 h	2.80×10^{-1} γ
				6.20×10^{-2} e-
		193	11.8 h	1.20 γ
				1.20×10^{-1} e-
				5.40×10^{-3} β+
				1.02%(β+)
		194	520 yr	2.18×10^{-3} γ
		195	9.5 h	2.04×10^{-1} γ
				6.13×10^{-2} e-
		195	1.73 d	2.00×10^{-1} γ
				1.40×10^{-1} e-
		197	2.672 d	7.05×10^{-2} γ
				5.70×10^{-2} e-
		197	23.8 h	9.30×10^{-2} γ
				2.14×10^{-1} e-
		199	42.6 min	1.86×10^{-1} γ
				3.50×10^{-1} e-
		203	46.60 d	2.38×10^{-1} γ
				9.83×10^{-2} e-
		205	5.2 min	4.80×10^{-3} γ
		206	8.2 min	1.08×10^{-1} γ
				4.28×10^{-1} e-
81	**Tl**	186	4 s	3.04×10^{-1} γ
		188	1.2 min	2.02 γ
				5.52×10^{-2} e-
				4.98×10^{-1} β+
				23.0%(β+)
		190	2.6 min	5.95×10^{-1} γ
		190	3.7 min	1.89 γ
		193	2.1 min	3.90×10^{-3} γ
		194	32.8 min	2.06 γ
		195	1.16 h	1.22 γ
				5.50×10^{-2} e-
				2.08×10^{-2} β+
				3.20%(β+)
		195	3.6 s	3.58×10^{-1} γ
				1.21×10^{-1} e-
		196	1.84 h	1.78 γ
				2.71×10^{-2} e-
				1.72×10^{-1} β+
				14.1%(β+)
		196	1.41 h	1.13 γ
		197	2.84 h	4.61×10^{-1} γ
				4.53×10^{-2} e-
				8.80×10^{-3} β+
				1.68%(β+)
		197	540 ms	4.35×10^{-1} γ
				1.72×10^{-1} e-
		198	5.3 h	2.01 γ
81	**Tl**	198	5.3 h	3.16×10^{-2} e-
				6.10×10^{-3} β+
				7.70×10^{-1} %(β+)
		198	1.87 h	1.21 γ
				1.89×10^{-1} e-
		199	7.4 h	2.50×10^{-1} γ
				5.60×10^{-2} e-
		200	1.087 d	1.31 γ
				3.61×10^{-2} e-
				1.84×10^{-3} β+
				3.50×10^{-1} %(β+)
		201	3.05 d	9.25×10^{-2} γ
				4.80×10^{-2} e-
		202	12.23 d	4.68×10^{-1} γ
				2.05×10^{-2} e-
		204	3.78 yr	1.31×10^{-3} γ
				2.38×10^{-1} e-
		206	4.20 min	8.89×10^{-4} γ
				5.37×10^{-1} e-
		206	3.76 min	2.63 γ
				2.48×10^{-1} e-
		207	4.77 min	2.88×10^{-3} γ
				4.93×10^{-1} e-
		207	1.3 s	1.17 γ
				1.81×10^{-1} e-
		208	3.053 min	3.38 γ
				5.98×10^{-1} e-
		209	2.2 min	2.03 γ
				6.87×10^{-1} e-
		210	1.30 min	2.73 γ
				1.29 e-
82	**Pb**	188	24 s	3.10×10^{-1} γ
		190	1.2 min	6.91×10^{-1} γ
		197	8 min	1.65 γ
				4.50×10^{-2} e-
		197	43 min	1.15 γ
				2.25×10^{-1} e-
		198	2.4 h	3.64×10^{-1} γ
				7.00×10^{-2} e-
		199	1.5 h	1.46 γ
				4.88×10^{-2} e-
				5.70×10^{-3} β+
				9.10×10^{-1} %(β+)
		200	21.5 h	2.10×10^{-1} γ
				9.53×10^{-2} e-
		201	9.33 h	7.60×10^{-1} γ
				6.09×10^{-2} e-
				9.70×10^{-5} β+
				3.90×10^{-2} %(β+)
		201	1.02 min	3.66×10^{-1} γ
				2.63×10^{-1} e-
		202	5.3×10^{4} yr	2.20×10^{-3} γ
		202	3.62 h	1.99 γ
				1.24×10^{-1} e-
		203	2.169 d	3.11×10^{-1} γ
				4.89×10^{-2} e-
		203	6.3 s	6.56×10^{-1} γ
				1.71×10^{-1} e-
		203	480 ms	1.91 γ
		204	1.120 h	2.10 γ
				1.09×10^{-1} e-
		205	1.9×10^{7} yr	2.35×10^{-3} γ
				5.00×10^{-3} e-
		207	796 ms	1.51 γ
				1.26×10^{-1} e-
		209	3.25 h	1.31×10^{-4} γ
				1.98×10^{-1} e-
		210	22.3 yr	4.67×10^{-3} γ
				3.42×10^{-2} e-
		211	36.1 min	6.84×10^{-2} γ
				4.52×10^{-1} e-
		212	10.64 h	1.45×10^{-1} γ
				1.75×10^{-1} e-
		214	27 min	2.50×10^{-1} γ
				2.94×10^{-1} e-
83	**Bi**	193	3.5 s	1.62 α

Table I. Average Energies per Disintegration (continued)

Z	El	A	Half-life	⟨E⟩ (MeV), or %β+
83	Bi	194	1.8 min	5.64 γ 2.01 e-
		198	7.7 s	1.07×10^{-1} γ 1.40×10^{-1} e-
		200	36.4 min	2.41 γ 7.98×10^{-2} e- 1.21×10^{-1} β+ 10.7%(β+)
		200	31 min	1.21 γ
		203	11.76 h	2.36 γ 7.55×10^{-2} e- 1.18×10^{-3} β+ 2.10×10^{-1} %(β+)
		204	11.2 h	3.18 γ 9.07×10^{-2} e- 1.27×10^{-3} β+ 2.10×10^{-1} %(β+)
		205	15.31 d	9.35×10^{-1} γ 1.40×10^{-2} e- 5.06×10^{-4} β+ 1.10×10^{-1} %(β+)
		206	6.243 d	3.28 γ 1.33×10^{-1} e-
		207	32 yr	1.54 γ 1.16×10^{-1} e- 4.64×10^{-5} β+ 1.20×10^{-2} %(β+)
		208	3.68×10^{5} yr	2.67 γ 1.20×10^{-2} e-
		210	5.013 d	4.50×10^{-4} γ 3.89×10^{-1} e-
		210	3.0×10^{6} yr	2.60×10^{-1} γ 4.91 α 4.70×10^{-2} e-
		211	2.14 min	4.67×10^{-2} γ 6.55 α 9.90×10^{-3} e-
		212	1.009 h	1.07×10^{-1} γ 2.17 α 5.02×10^{-1} e-
		212	25 min	5.88 α
		213	45.6 min	8.31×10^{-2} γ 1.27×10^{-1} α 4.56×10^{-1} e-
		214	19.9 min	1.51 γ 1.43×10^{-3} α 6.62×10^{-1} e-
84	Po	194	700 ms	6.85 γ
		200	11.5 min	9.02×10^{-1} γ
		201	15.3 min	1.39 γ 1.25×10^{-1} e- 2.00×10^{-1} β+ 16.3%(β+)
		201	8.9 min	8.05×10^{-1} γ 6.30×10^{-1} e- 2.18×10^{-1} β+ 15.0%(β+)
		202	44.7 min	8.10×10^{-1} γ
		203	35 min	1.57 γ 6.70×10^{-2} β+ 6.90%(β+)
		203	1.2 min	4.16×10^{-1} γ 2.65×10^{-1} e- 8.50×10^{-3} β+ 7.10×10^{-1} %(β+)
		204	3.53 h	1.32 γ 1.64×10^{-1} e-
		205	1.80 h	1.59 γ 4.39×10^{-2} e- 1.43×10^{-2} β+ 2.10%(β+)
		206	8.8 d	1.19 γ 1.50×10^{-1} e-
		207	5.8 h	1.33 γ 4.57×10^{-2} e- 2.24×10^{-3} β+
84	Po	207	5.8 h	4.90×10^{-1} %(β+)
		207	2.8 s	1.08 γ
		208	2.898 yr	1.64×10^{-5} γ 5.12 α 2.60×10^{-6} e-
		209	102 yr	3.10×10^{-3} γ 4.87 α 3.60×10^{-4} e-
		210	138.376 d	5.30 α
		211	516 ms	7.70×10^{-3} γ 7.44 α
		211	25.5 s	1.43 γ 7.40 α 1.19×10^{-1} e-
		212	45 s	11.4 α
		213	4 us	8.38 α
		214	163.7 us	8.30×10^{-5} γ 7.69 α
		214	99 ps	1.03 γ
		215	1.780 ms	7.39 α
		216	150 ms	6.78 α
		218	3.11 min	6.00 α
85	At	200	43 s	2.25 γ
		202	3.02 min	1.22 γ 7.40×10^{-1} α
		204	9.2 min	2.08 γ 5.50×10^{-2} e- 4.60×10^{-1} β+ 29.0%(β+)
		204	108 ms	5.50×10^{-1} γ 3.61×10^{-2} e-
		205	26.2 min	6.81×10^{-1} γ
		206	29.4 min	2.48 γ 7.40×10^{-2} e- 2.45×10^{-1} β+ 17.0%(β+)
		207	1.80 h	1.33 γ
		208	1.63 h	2.74 γ 3.11×10^{-2} α 1.04×10^{-1} e- 3.94×10^{-2} β+ 3.20%(β+)
		209	5.41 h	2.28 γ 2.32×10^{-1} α 1.13×10^{-1} e-
		210	8.1 h	2.96 γ 9.60×10^{-3} α 7.49×10^{-2} e-
		211	7.21 h	3.66×10^{-2} γ 2.45 α 4.03×10^{-3} e-
		212	314 ms	7.66 α
		212	119 ms	7.85 α
		214	558 ns	8.82 α
		215	100 us	8.02 α
		216	300 us	7.79 α
		217	32.3 ms	7.07 α
		218	1.6 s	6.69 α
86	Rn	205	2.8 min	1.44 α
		207	9.3 min	8.13×10^{-1} γ 1.41 α
		208	24.4 min	3.68 α
		209	29 min	1.10 γ 1.03 α 5.61×10^{-2} e- 3.90×10^{-2} β+ 4.00%(β+)
		210	2.4 h	5.91×10^{-2} γ 5.80 α 8.50×10^{-3} e-
		211	14.6 h	1.91 γ 1.50 α 6.59×10^{-2} e-
		212	24 min	6.26 α
		213	25.0 ms	8.08 α
		217	540 us	7.74 α
86	Rn	218	35 ms	7.13 α
		219	3.96 s	5.60×10^{-2} γ 6.81 α 6.36×10^{-3} e-
		220	55.6 s	6.29 α
		221	25 min	1.20×10^{-1} γ 1.32 α 2.92×10^{-1} e-
		222	3.825 d	5.49 α
87	Fr	212	20 min	1.13 γ 2.75 α 9.10×10^{-2} e- 3.95×10^{-2} β+ 3.80%(β+)
		214	5.0 ms	8.41 α
		214	3.35 ms	8.50 α
		218	700 us	7.85 α
		219	21 ms	7.29 α
		220	27.4 s	6.60 α
		221	4.9 min	2.77×10^{-2} γ 6.36 α 8.44×10^{-3} e-
		222	14.4 min	1.04×10^{-3} γ 6.25×10^{-1} e-
		223	21.8 min	6.34×10^{-2} γ 3.95×10^{-1} e-
88	Ra	213	2.7 min	9.60×10^{-3} γ 5.35 α
		213	2.1 ms	1.63 γ 8.40×10^{-2} α 1.04×10^{-1} e-
		215	1.6 ms	8.68 α
		219	10 ms	7.79×10^{3} α
		220	23 ms	7.41 α
		221	28 s	4.30×10^{-2} γ 10.0 α
		222	38.0 s	9.20×10^{-3} γ 6.54 α 8.44×10^{-4} e-
		223	11.43 d	1.35×10^{-1} γ 5.70 α 7.31×10^{-2} e-
		224	3.66 d	1.00×10^{-2} γ 5.68 α 2.20×10^{-3} e-
		225	14.8 d	1.37×10^{-2} γ 1.06×10^{-1} e-
		226	1.60×10^{3} yr	6.74×10^{-3} γ 4.77 α 3.53×10^{-3} e-
		227	42.2 min	1.63×10^{-1} γ 5.40×10^{-2} e-
89	Ac	216	330 us	9.06 α
		216	330 us	8.99 α
		217	111 ns	9.65 α
		217	400 ns	10.5 α
		217	8 ns	11.1 α
		220	26.1 ms	7.71 α
		221	52 ms	7.72 α
		222	4.2 s	7.01 α
		222	1.10 min	6.09 α
		223	2.2 min	3.90×10^{-3} γ 6.56 α
		224	2.9 h	2.02×10^{-1} γ 4.50×10^{-2} e-
		225	10.0 d	1.76×10^{-2} γ 2.57×10^{-2} e- 5.75×10^{3} α
		226	1.2 d	1.37×10^{-1} γ 2.89×10^{-1} e-
		227	21.77 yr	1.68×10^{-4} γ 6.73×10^{-2} α 1.25×10^{-2} e-
		228	6.1 h	9.93×10^{-1} γ 4.79×10^{-1} e-

Table I. Average Energies per Disintegration (continued)

Z	El	A	Half-life	⟨E⟩ (MeV), or %β+
89	**Ac**	230	2.03 min	5.48×10^{-1} γ
				9.11×10^{-1} e-
90	**Th**	215	1.2 s	7.44 α
		221	1.7 ms	8.33 α
		223	660 ms	7.30 α
		224	1.04 s	2.30×10^{-2} γ
				7.10 α
				1.31×10^{-2} e-
		225	8.0 min	1.27×10^{-1} γ
		226	31 min	8.70×10^{-3} γ
				6.31 α
				2.08×10^{-2} e-
		227	18.72 d	1.11×10^{-1} γ
				5.90 α
				5.40×10^{-2} e-
		228	1.913 yr	3.40×10^{-3} γ
				5.40 α
				2.01×10^{-2} e-
		229	7.3×10^{3} yr	3.40×10^{-2} γ
				4.86 α
		230	7.54×10^{4} yr	3.71×10^{-4} γ
				4.66 α
		231	1.063 d	2.90×10^{-2} γ
				1.73×10^{-1} e-
		232	1.41×10^{10} yr	1.7×10^{-4} γ
				4.00×10^{3} α
		233	22.3 min	3.69×10^{-2} γ
				4.12×10^{-1} e-
		234	24.10 d	9.40×10^{-3} γ
				1.58×10^{-2} e-
		236	37 min	2.10×10^{-2} γ
				1.70×10^{-2} e-
91	**Pa**	222	4.3 ms	8.32 α
		223	6 ms	8.09 α
		225	1.8 s	7.23 α
		226	1.8 min	5.06 α
		227	38.3 min	6.40×10^{-3} γ
		228	22 h	1.09 γ
				1.02×10^{-1} e-
				7.20×10^{-4} β+
				1.50×10^{-1} %(β+)
		229	1.4 d	1.40×10^{-2} α
		230	17.4 d	6.57×10^{-1} γ
				1.70×10^{-4} α
				6.25×10^{-2} e-
		231	3.28×10^{4} yr	3.99×10^{-2} γ
				4.92 α
				4.80×10^{-2} e-
		232	1.31 d	9.41×10^{-1} γ
				1.71×10^{-1} e-
		233	27.0 d	2.04×10^{-1} γ
				1.94×10^{-1} e-
		234	6.70 h	1.90 γ
				4.80×10^{-1} e-
		234	1.17 min	1.38×10^{-2} γ
				8.23×10^{-1} e-
		235	24.2 min	6.30×10^{-4} γ
				4.70×10^{-1} e-
		236	9.1 min	5.06×10^{-1} γ
				6.34×10^{-1} e-
		237	8.7 min	6.05×10^{-1} γ
				5.72×10^{-1} e-
92	**U**	228	9.1 min	4.80×10^{-3} γ
				2.20×10^{-2} e-
		230	20.8 d	2.90×10^{-3} γ
				5.87 α
				2.10×10^{-2} e-
		231	4.2 d	8.18×10^{-2} γ
		232	70 yr	2.40×10^{-4} γ
				5.31 α
		233	1.59×10^{5} yr	1.29×10^{-3} γ
				4.81 α
				5.50×10^{-3} e-
		234	2.45×10^{5} yr	1.13×10^{-4} γ
				4.77 α
92	**U**	235	7.04×10^{8} yr	1.56×10^{-1} γ
				4.38 α
				4.20×10^{-2} e-
		236	2.342×10^{7} yr	1.50×10^{-3} γ
				4.48 α
				1.08×10^{-2} e-
		237	6.75 d	1.44×10^{-1} γ
				1.87×10^{-1} e-
		238	4.468×10^{9} yr	1.30×10^{-3} γ
				4.19 α
				9.50×10^{-3} e-
		239	23.47 min	5.21×10^{-2} γ
				3.95×10^{-1} e-
		240	14.1 h	7.30×10^{-3} γ
				3.00×10^{-2} e-
		242	16.8 min	3.85×10^{-2} γ
93	**Np**	232	14.7 min	1.19 γ
				9.90×10^{-2} e-
		233	36.2 min	9.12×10^{-2} γ
		234	4.4 d	1.11 γ
				1.17×10^{-3} β+
				4.60×10^{-2} %(β+)
		235	1.085 yr	7.74×10^{-3} γ
				7.00×10^{-5} α
				2.94×10^{-3} e-
		236	1.55×10^{5} yr	1.42×10^{-1} γ
				1.97×10^{-1} e-
		236	22.5 h	5.03×10^{-2} γ
				7.80×10^{-2} e-
		237	2.14×10^{6} yr	3.27×10^{-2} γ
				4.76 α
				6.40×10^{-2} e-
		238	2.117 d	6.47×10^{-1} γ
				2.29×10^{-1} e-
		239	2.355 d	1.74×10^{-1} γ
				2.52×10^{-1} e-
		240	1.032 h	1.19 γ
				2.73×10^{-1} e-
		240	7.22 min	3.29×10^{-1} γ
				6.30×10^{-1} e-
		242	2.2 min	2.52×10^{-1} γ
				8.94×10^{-1} e-
94	**Pu**	234	8.8 h	3.72×10^{-1} α
		235	25.3 min	9.52×10^{-2} γ
				1.57×10^{-2} e-
		236	2.85 yr	2.00×10^{-3} γ
				5.75 α
				1.26×10^{-2} e-
		237	45.2 d	5.43×10^{-2} γ
				1.80×10^{-4} α
				1.05×10^{-2} e-
		237	180 ms	9.60×10^{-3} γ
				1.31×10^{-1} e-
		238	87.7 yr	1.76×10^{-3} γ
				5.49 α
				9.92×10^{-3} e-
		239	2.411×10^{4} yr	6.60×10^{-5} γ
				5.10 α
		240	6.54×10^{3} yr	2.86×10^{-5} γ
				5.16 α
		241	14.4 yr	1.46×10^{-6} γ
				1.18×10^{-4} α
				5.20×10^{-3} e-
		242	3.76×10^{5} yr	1.39×10^{-3} γ
				4.89 α
				8.10×10^{-3} e-
		243	4.956 h	2.61×10^{-2} γ
				1.74×10^{-1} e-
		244	8.3×10^{7} yr	1.18×10^{-4} γ
				4.57 α
				6.73×10^{-4} e-
		245	10.5 h	3.97×10^{-1} γ
				3.41×10^{-1} e-
		246	10.85 d	1.43×10^{-1} γ
				1.35×10^{-1} e-
95	**Am**	237	1.22 h	3.68×10^{-1} γ
				7.06×10^{-2} e-
		238	1.63 h	9.01×10^{-1} γ
				4.03×10^{-2} e-
				7.40×10^{-4} β+
				1.30×10^{-1} %(β+)
		239	11.9 h	2.43×10^{-1} γ
				5.76×10^{-4} α
				1.61×10^{-1} e-
		240	2.12 d	1.03 γ
				1.02×10^{-5} α
				6.86×10^{-2} e-
		241	432.7 yr	2.87×10^{-2} γ
				5.48 α
				3.04×10^{-2} e-
		242	16.01 h	1.83×10^{-2} γ
				1.77×10^{-1} e-
		242	141 yr	4.90×10^{-3} γ
				2.32×10^{-2} α
				4.03×10^{-2} e-
		243	7.38×10^{3} yr	4.81×10^{-2} γ
				5.27 α
		244	10.1 h	8.06×10^{-1} γ
				3.34×10^{-1} e-
		244	26 min	7.10×10^{-4} γ
				5.01×10^{-1} e-
		245	2.05 h	3.24×10^{-2} γ
				2.86×10^{-1} e-
		246	39 min	6.96×10^{-1} γ
				6.81×10^{-1} e-
		246	25.0 min	9.74×10^{-1} γ
				4.91×10^{-1} e-
		247	22 min	1.36×10^{-1} γ
				5.93×10^{-1} e-
96	**Cm**	240	27 d	1.93×10^{-3} γ
				6.24 α
				1.00×10^{-2} e-
		241	32.8 d	4.85×10^{-1} γ
				5.93×10^{-2} α
		242	162.9 d	1.75×10^{-3} γ
				6.04 α
				8.95×10^{-3} e-
		243	28.5 yr	1.32×10^{-1} γ
				5.84 α
				1.13×10^{-1} e-
		244	18.11 yr	1.60×10^{-3} γ
				5.80 α
		245	8.5×10^{3} yr	1.17×10^{-1} γ
				5.36 α
				1.34×10^{-1} e-
		246	4.73×10^{3} yr	1.39×10^{-3} γ
				5.38 α
				7.23×10^{-3} e-
		247	1.56×10^{7} yr	4.95 γ
				3.15×10^{-1} e-
		248	3.4×10^{5} yr	4.65 α
		249	1.0692 h	1.92×10^{-2} γ
				2.73×10^{-1} e-
		251	16.8 min	1.02×10^{-1} γ
				4.21×10^{-1} e-
97	**Bk**	243	4.5 h	1.76×10^{-1} γ
				9.80×10^{-3} α
		244	4.4 h	4.00×10^{-4} α
		245	4.94 d	2.36×10^{-1} γ
				7.51×10^{-3} α
				1.27×10^{-1} e-
		246	1.80 d	8.53×10^{-1} γ
				4.80×10^{-2} e-
		247	1.4×10^{3} yr	5.57 γ
				1.14×10^{-1} e-
		248	23.7 h	5.55×10^{-2} γ
				1.84×10^{-1} e-
		249	320 d	4.18×10^{-6} γ
				7.80×10^{-5} α
				3.29×10^{-2} e-
		250	3.217 h	8.98×10^{-1} γ

Table I. Average Energies per Disintegration (continued)

Z	El	A	Half-life	⟨E⟩ (MeV), or %β+
97	**Bk**	250		2.93×10^{-1} e-
98	**Cf**	242	3.5 min	7.38 α
		243	10.7 min	9.93×10^{-1} α
		244	19 min	7.20 α
		246	1.49 d	6.75 γ
				1.62×10^{-3} e-
		247	3.11 h	8.82×10^{-2} γ
				4.64×10^{-3} e-
		248	333.5 d	6.26 α
		249	351 yr	3.26×10^{-1} γ
				5.83 α
				3.70×10^{-2} e-
		250	13.1 yr	1.12×10^{-3} γ
				6.02 α
				4.91×10^{-3} e-
		251	898 yr	1.28×10^{-1} γ
				5.66 α
				1.87×10^{-1} e-
		252	2.64 yr	1.14×10^{-3} γ
				5.93 α
				5.14×10^{-3} e-
		253	17.8 d	2.20×10^{-5} γ
				1.85×10^{-2} α
				7.80×10^{-2} e-
		254	60.5 d	1.81×10^{-2} α
99	**Es**	249	1.70 h	4.10×10^{-1} γ
				3.98×10^{-2} e-
		250	8.6 h	1.24 γ
				3.03×10^{-1} e-
		250	2.22 h	5.52×10^{-1} γ
				3.43×10^{-2} e-
				7.60×10^{-4} β+
		251	1.38 d	6.60×10^{-3} γ
				3.18×10^{-2} α
		252	1.291 yr	2.57×10^{-1} γ
				5.02 α
				8.40×10^{-2} e-
		253	20.4 d	2.92×10^{-4} γ
				6.63 α
		254	275.7 d	2.79×10^{-3} γ
				6.41 α
		254	1.64 d	4.74×10^{-1} γ
				2.11×10^{-2} α
				2.38×10^{-1} e-
		255	40 d	5.04×10^{-1} α
100	**Fm**	248	36 s	7.85 α
		251	5.3 h	1.58×10^{-1} γ
				1.23×10^{-1} α
				2.05×10^{-2} e-
		252	1.058 d	7.03 α
		253	3.0 d	6.05×10^{-2} γ
				8.21×10^{-1} α
100	**Fm**	253	3.0 d	5.60×10^{-3} e-
		254	3.240 h	1.31×10^{-3} γ
				7.17 α
				5.50×10^{-3} e-
		255	20.1 h	1.49×10^{-2} γ
				7.09 α
				8.30×10^{-2} e-
		257	100.5 d	1.11×10^{-1} γ
				6.55 α
				1.21×10^{-1} e-
101	**Md**	248	7 s	1.67 α
		255	27 min	3.29×10^{-2} γ
		256	1.3 h	6.75×10^{-1} α
		258	55 d	6.74 α
102	**No**	252	2.3 s	6.14 α
		255	3.1 min	4.97 α
		259	1.0 h	5.88 α
103	**Lr**	255	22 s	8.39 α
		256	28 s	8.45 α
		257	646 ms	8.85 α
		258	4.3 s	8.60 α
104		257	3.8 s	7.18 α
105		260	1.5 s	8.20 α
		262	34 s	2.29 α

APPENDIX E. PROPERTIES OF THE ELEMENTS

This table lists the atomic weights, densities, melting and boiling points, first ionization potentials, and specific heats of the elements. Data were taken mostly from the 65th edition of the *CRC Handbook of Chemistry and Physics.*[1] Atomic weights apply to elements as they exist naturally on earth or, in the cases of radium, actinium, thorium, protactinium, and neptunium, to the isotopes which have the longest half-lives. Values in parentheses are the mass numbers for the longest-lived isotopes of some of the radioactive elements. Specific heats are given for the elements at 25°C. Densities for solids and liquids are given at 20°C, unless otherwise indicated by a superscript temperature (in °C); densities for the gaseous elements are for the liquids at their boiling points.

This table was adapted, with permission, from one that appeared in the October 1985 edition of the *X-ray Data Booklet.*[2]

[1]C.R. Hammond, in *CRC Handbook of Chemistry and Physics*, 65th Edition (R.C. Weast, editor, CRC Press, Inc., Boca Raton, Florida (1984)).

[2]D. Vaughan, editor, *X-ray Data Booklet* (published by Center for X-ray Optics, Lawrence Berkeley Laboratory, Berkeley, California (1985)).

Z	Element	Atomic weight	Density (g/cm^3)	Melting point (°C)	Boiling point (°C)	Ionization potential (eV)	Specific heat (cal/g·K)
1	Hydrogen	1.00794	0.0708	−259.14	−252.87	13.598	3.41
2	Helium	4.00260	0.122	−272.2	−268.934	24.587	1.24
3	Lithium	6.941	0.533	180.54	1342	5.392	0.834
4	Beryllium	9.01218	1.845	1278	2970	9.322	0.436
5	Boron	10.81	2.34	2079	2550^{c}	8.298	0.245
6	Carbon	12.011	2.26	3550	3367^{c}	11.260	0.170
7	Nitrogen	14.0067	0.81	−209.86	−195.8	14.534	0.249
8	Oxygen	15.9994	1.14	−218.4	−182.962	13.618	0.219
9	Fluorine	18.998403	1.108	−219.62	−188.14	17.422	0.197
10	Neon	20.179	1.207	−248.67	−246.048	21.564	0.246
11	Sodium	22.98977	0.969	97.81	882.9	5.139	0.292
12	Magnesium	24.305	1.735	648.8	1090	7.646	0.245
13	Aluminum	26.98154	2.6941	660.37	2467	5.986	0.215
14	Silicon	28.0855	2.32^{25}	1410	2355	8.151	0.168
15	Phosphorus	30.97376	1.82	44.1	280	10.486	0.181
16	Sulfur	32.06	2.07	112.8	444.674	10.360	0.175
17	Chlorine	35.453	1.56	−100.98	−34.6	12.967	0.114
18	Argon	39.948	1.40	−189.2	−185.7	15.759	0.124
19	Potassium	39.0983	0.860	63.25	760	4.341	0.180
20	Calcium	40.08	1.55	839	1484	6.113	0.155
21	Scandium	44.9559	2.980^{25}	1541	2831	6.54	0.1173
22	Titanium	47.88	4.53	1660	3287	6.82	0.1248
23	Vanadium	50.9415	$6.10^{18.7}$	1890	3380	6.74	0.116
24	Chromium	51.996	7.18	1857	2672	6.766	0.107
25	Manganese	54.9380	7.43	1244	1962	7.435	0.114
26	Iron	55.847	7.860	1535	2750	7.870	0.1075
27	Cobalt	58.9332	8.9	1495	2870	7.86	0.107
28	Nickel	58.69	8.876^{25}	1453	2732	7.635	0.1061
29	Copper	63.546	8.94	1083.4	2567	7.726	0.0924
30	Zinc	65.38	7.112^{25}	419.58	907	9.394	0.0922
31	Gallium	69.72	$5.877^{29.6}$	29.78	2403	5.999	0.088
32	Germanium	72.59	5.307^{25}	937.4	2830	7.899	0.077
33	Arsenic	74.9216	5.72	$817^{28\ atm}$	613^{c}	9.81	0.0785
34	Selenium	78.96	4.78	217	684.9	9.752	0.0767
35	Bromine	79.904	3.11	−7.2	58.78	11.814	0.0537
36	Krypton	83.80	2.6	−156.6	−152.30	13.999	0.059
37	Rubidium	85.4678	1.529	38.89	686	4.177	0.0860
38	Strontium	87.62	2.54	769	1384	5.695	0.0719
39	Yttrium	88.9059	4.456^{25}	1522	3338	6.38	0.0713
40	Zirconium	91.22	6.494	1852	4377	6.84	0.0660
41	Niobium	92.9064	8.55	2468	4742	6.88	0.0663
42	Molybdenum	95.94	10.20	2617	4612	7.099	0.0597

Z	Element	Atomic weight	Density (g/cm^3)	Melting point (°C)	Boiling point (°C)	Ionization potential (eV)	Specific heat (cal/g·K)
43	Technetium	(98)	11.48[a]	2172	4877	7.28	0.058
44	Ruthenium	101.07	12.39	2310	3900	7.37	0.0569
45	Rhodium	102.9055	12.39	1966	3727	7.46	0.0580
46	Palladium	106.42	12.00	1554	2970	8.34	0.0583
47	Silver	107.8682	10.48	961.93	2212	7.576	0.0562
48	Cadmium	112.41	8.63	320.9	765	8.993	0.0552
49	Indium	114.82	7.30	156.61	2080	5.786	0.0556
50	Tin	118.69	7.30	231.9681	2270	7.344	0.0519
51	Antimony	121.75	6.679	630.74	1950	8.641	0.0495
52	Tellurium	127.60	6.23	449.5	989.8	9.009	0.0481
53	Iodine	126.9045	4.92	113.5	184.35	10.451	0.102
54	Xenon	131.29	3.52	−111.9	−107.1	12.130	0.0378
55	Cesium	132.9054	1.870	28.40	669.3	3.894	0.0575
56	Barium	137.33	3.5	725	1640	5.212	0.0362
57	Lanthanum	138.9055	6.127^{25}	921	3457	5.577	0.0479
58	Cerium	140.12	6.637^{25}	799	3426	5.47	0.0459
59	Praseodymium	140.9077	6.761	931	3512	5.42	0.0467
60	Neodymium	144.24	6.994	1021	3068	5.49	0.0453
61	Promethium	(145)	7.20^{25}	1168	2460	5.55	0.0442
62	Samarium	150.36	7.51	1077	1791	5.63	0.0469
63	Europium	151.96	5.228^{25}	822	1597	5.67	0.0326
64	Gadolinium	157.25	7.8772^{25}	1313	3266	6.14	0.056
65	Terbium	158.9254	8.214	1356	3123	5.85	0.0435
66	Dysprosium	162.50	8.525^{25}	1412	2562	5.93	0.0414
67	Holmium	164.9304	8.769^{25}	1474	2695	6.02	0.0394
68	Erbium	167.26	9.039^{25}	159	2863	6.10	0.0401
69	Thulium	168.9342	9.294^{25}	1545	1947	6.18	0.0382
70	Ytterbium	173.04	6.953	819	1194	6.254	0.0287
71	Lutetium	174.967	9.811^{25}	1663	3395	5.426	0.0285
72	Hafnium	178.49	13.29	2227	4602	7.0	0.028
73	Tantalum	180.9479	16.624	2996	5425	7.89	0.0334
74	Tungsten	183.85	19.3	3410	5660	7.98	0.0322
75	Rhenium	186.207	20.98	3180	5627[b]	7.88	0.0330
76	Osmium	190.2	22.53	3045	5027	8.7	0.0310
77	Iridium	192.22	22.39^{17}	2410	4130	9.1	0.0312
78	Platinum	195.08	21.41	1772	3827	9.0	0.0317
79	Gold	196.9665	18.85	1064.43	3080	9.225	0.0308
80	Mercury	200.59	13.522	−38.842	356.58	10.437	0.0333
81	Thallium	204.383	11.83	303.5	1457	6.108	0.0307
82	Lead	207.2	11.33	327.502	1740	7.416	0.0305
83	Bismuth	208.9804	9.730	271.3	1560	7.289	0.0238
84	Polonium	(209)	9.30	254	962	8.42	0.030
85	Astatine	(210)	—	302	337[b]	—	—
86	Radon	(222)	4.4	−71	−61.8	10.748	0.0224
87	Francium	(223)	—	27	677	—	—
88	Radium	226.0254	5	700	1140	5.279	0.0288
89	Actinium	227.0278	10.05[a]	1050	3200[b]	6.9	—
90	Thorium	232.0381	11.70	1750	4790	—	0.0281
91	Protactinium	231.0359	15.34[a]	<1600	—	—	0.029
92	Uranium	238.0289	18.92	1132.3	3818	—	0.0278
93	Neptunium	237.0482	20.21	640	3902[b]	—	—
94	Plutonium	(244)	19.80	641	3232	5.8	—
95	Americium	(243)	13.64	994	2607	6.0	
96	Curium	(247)	13.49[a]	1340	—	—	—
97	Berkelium	(247)	14[b]	—	—	—	—
98	Californium	(251)	—	—	—	—	—
99	Einsteinium	(252)	—	—	—	—	—
100	Fermium	(257)	—	—	—	—	—
101	Mendelevium	(258)	—	—	—	—	—
102	Nobelium	(259)	—	—	—	—	—
103	Lawrencium	(260)	—	—	—	—	—

[a] Calculated [b] Estimated [c] Sublimes